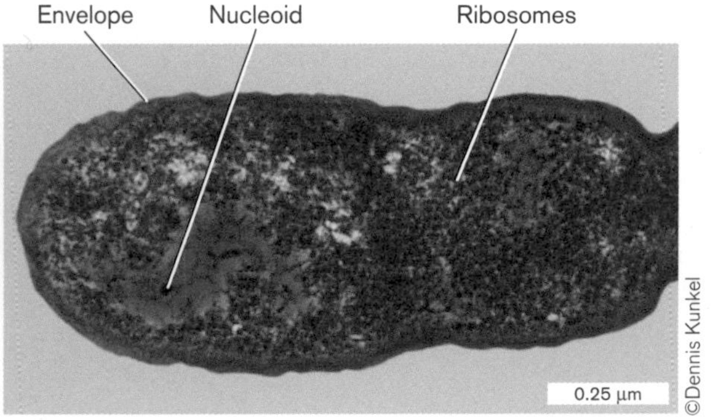

Envelope Nucleoid Ribosomes

0.25 μm

©Dennis Kunkel

Model of a bacterial cell (*Escherichia coli*). **Envelope:** The cell membrane contains embedded proteins for structure and transport. The cell membrane is supported by the cell wall. In this gram-negative cell, the cell wall is coated by the outer membrane, whose sugar chain extensions protect the cell from attack by the immune system or by predators. Plugged into the membranes is the rotary motor of a flagellum. **Cytoplasm:** Molecules of nascent messenger RNA (mRNA) extend out of the nucleoid to the region of the cytoplasm rich in ribosomes. Ribosomes translate the mRNA to make proteins, which are folded by chaperones. **Nucleoid:** The chromosomal DNA is wrapped around binding proteins. Replication by DNA polymerase and transcription by RNA polymerase occur at the same time within the nucleoid. (PDB codes: ribosome, 1GIX,1GIY; DNA-binding protein, 1P78; RNA polymerase, 1MSW)

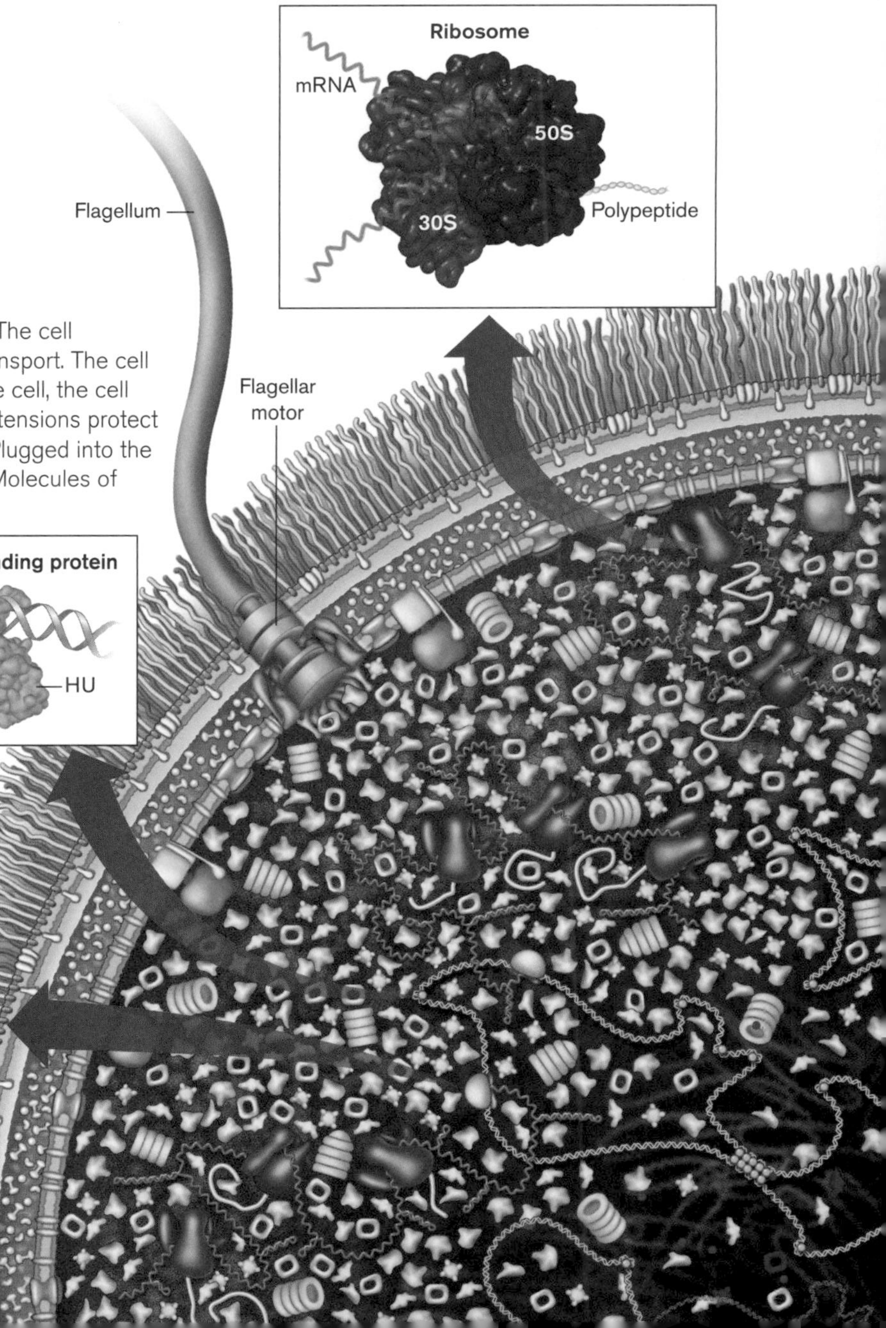

Ribosome

mRNA

50S

30S

Polypeptide

Flagellum

Flagellar motor

DNA-binding protein

HU

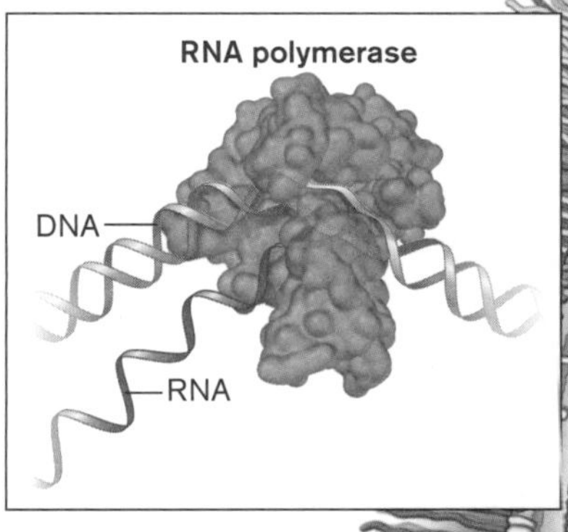

RNA polymerase

DNA

RNA

A Key to the Icons in *Microbiology: An Evolving Science*

WWW | Weblink icons indicate that there is an author-recommended website related to the topic at hand, which can be found at <u>microbiology2.com/links</u>.

▶Ⅱ Animation icons in a figure's caption indicate that there is a process animation to further illustrate that particular figure. The animations can be found at <u>microbiology2.com/animations</u>.

 Visit Norton StudySpace (at <u>microbiology2.com</u>) to access these resources and other review material.

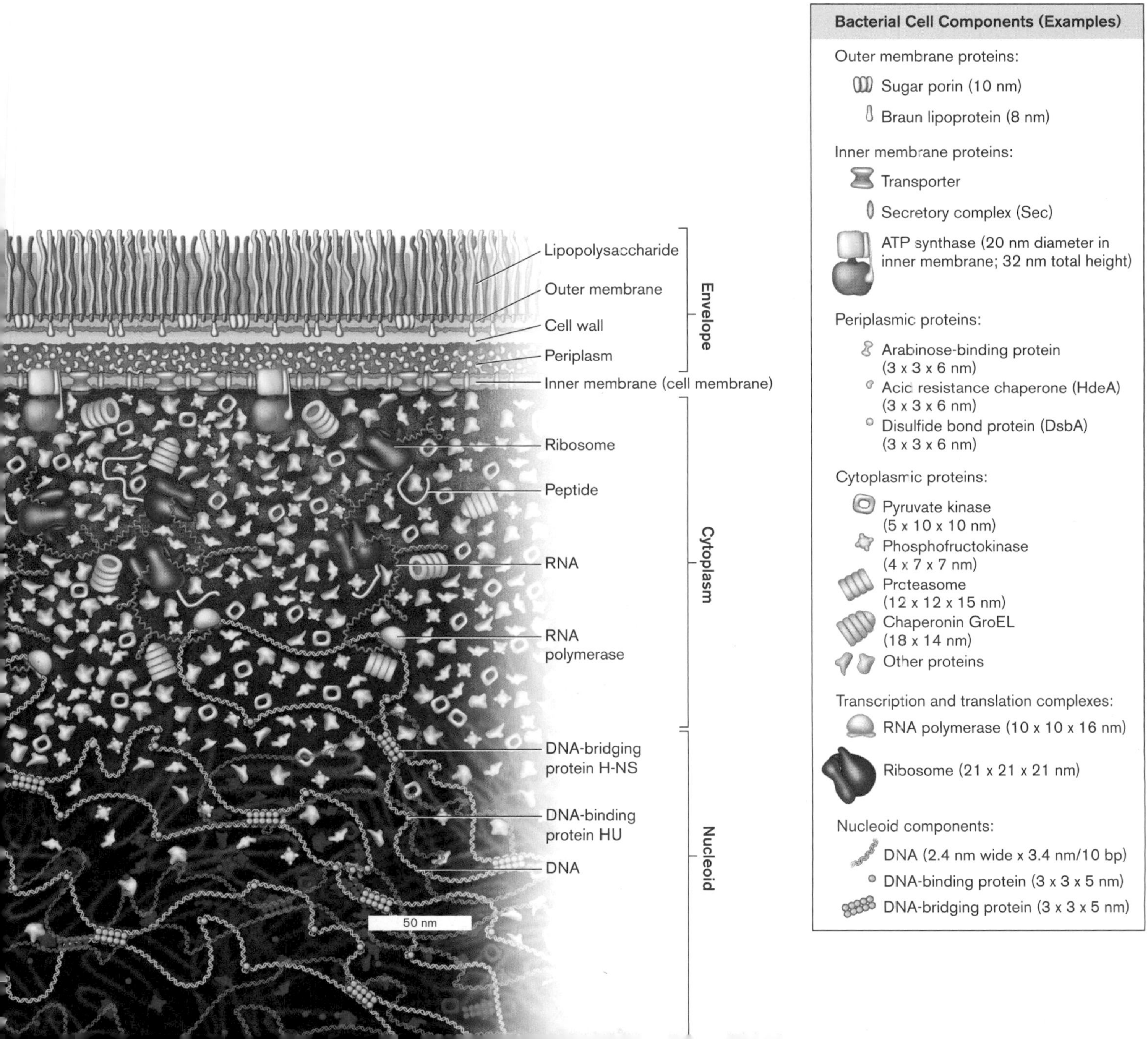

Lipopolysaccharide

Outer membrane

Cell wall

Periplasm — Envelope

Inner membrane (cell membrane)

Ribosome

Peptide

RNA — Cytoplasm

RNA polymerase

DNA-bridging protein H-NS

DNA-binding protein HU — Nucleoid

DNA

50 nm

Bacterial Cell Components (Examples)

Outer membrane proteins:

　Sugar porin (10 nm)

　Braun lipoprotein (8 nm)

Inner membrane proteins:

　Transporter

　Secretory complex (Sec)

　ATP synthase (20 nm diameter in inner membrane; 32 nm total height)

Periplasmic proteins:

　Arabinose-binding protein (3 x 3 x 6 nm)
　Acid resistance chaperone (HdeA) (3 x 3 x 6 nm)
　Disulfide bond protein (DsbA) (3 x 3 x 6 nm)

Cytoplasmic proteins:

　Pyruvate kinase (5 x 10 x 10 nm)
　Phosphofructokinase (4 x 7 x 7 nm)
　Proteasome (12 x 12 x 15 nm)
　Chaperonin GroEL (18 x 14 nm)
　Other proteins

Transcription and translation complexes:

　RNA polymerase (10 x 10 x 16 nm)

　Ribosome (21 x 21 x 21 nm)

Nucleoid components:

　DNA (2.4 nm wide x 3.4 nm/10 bp)
　DNA-binding protein (3 x 3 x 5 nm)
　DNA-bridging protein (3 x 3 x 5 nm)

Microbiology
An Evolving Science
Second Edition

Microbiology
An Evolving Science

Second Edition

Joan L. Slonczewski
Kenyon College

John W. Foster
University of South Alabama

Appendices and Glossary by
Kathy M. Gillen
Kenyon College

W·W·NORTON

NEW YORK · LONDON

W. W. Norton & Company has been independent since its founding in 1923, when William Warder Norton and Mary D. Herter Norton first published lectures delivered at the People's Institute, the adult education division of New York City's Cooper Union. The firm soon expanded its program beyond the Institute, publishing books by celebrated academics from America and abroad. By mid-century, the two major pillars of Norton's publishing program—trade books and college texts—were firmly established. In the 1950s, the Norton family transferred control of the company to its employees, and today—with a staff of four hundred and a comparable number of trade, college, and professional titles published each year—W. W. Norton & Company stands as the largest and oldest publishing house owned wholly by its employees.

Composition by Precision Graphics
Manufacturing by Transcontinental Interglobe
Illustrations by Precision Graphics

Editor: Betsy Twitchell
Developmental editors: Carol Pritchard-Martinez and Philippa Solomon
Senior project editor: Thomas Foley
Copy editor: Stephanie Hiebert
Production manager: Christopher Granville
Art director: Rubina Yeh
Designer: Sandra Watanabe
Managing editor, college: Marian Johnson
Science media editor: Robert Bellinger
Associate editor, emedia: Matthew A. Freeman
Marketing manager: Natasha Zabohonski
Photo editor: Trish Marx
Editorial assistants: Callinda Taylor and Cait Callahan
Photo researcher: Dena Digilio Betz

Library of Congress Cataloging-in-Publication Data
Slonczewski, Joan.
  Microbiology: an evolving science / Joan L. Slonczewski, John W. Foster; appendices and glossary by Kathy M. Gillen. — 2nd ed.
      p. ; cm.
  Includes bibliographical references and index.
  ISBN 978-0-393-93447-2 (hardcover)
  1. Microbiology—Textbooks. I. Foster, John Watkins. II Title.
  [DNLM: 1. Microbiological Phenomena. 2. Anti-Infective Agents—therapeutic use. 3. Bacterial Infections—drug therapy. 4. Genetics, Microbial. QW 4]
  QR41.2.S585 2011
  579—dc22
                                                      2010039775

W. W. Norton & Company, Inc., 500 Fifth Avenue, New York, N.Y. 10110
www.wwnorton.com

W. W. Norton & Company Ltd., Castle House, 75/76 Wells Street, London W1T 3QT

1 2 3 4 5 6 7 8 9

DEDICATION

We rededicate this second edition to the memory of our doctoral research mentors. Joan's doctoral mentor, Bob Macnab, offered an unfailingly rigorous pursuit of bacterial chemotaxis and physiology, and lasting friendship. John was mentored by Al Moat, a gifted microbial physiologist and humorist who instilled in his neophyte students an appreciation for critical thinking and a love for the science of microbiology.

Brief Contents

Contents

PART 1

The Microbial Cell   2

**AN INTERVIEW WITH RITA COLWELL:
THE GLOBAL IMPACT OF MICROBIOLOGY**

CHAPTER 1

Microbial Life: Origin and Discovery   5

CHAPTER 2

Observing the Microbial Cell   39

PART 2
Genes and Genomes   218

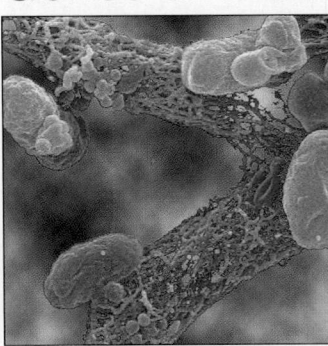

AN INTERVIEW WITH CHRISTINE JACOBS-WAGNER:
THE THRILL OF DISCOVERY IN MOLECULAR MICROBIOLOGY

PART 3
Metabolism and Biochemistry · 454

AN INTERVIEW WITH DAN WOZNIAK:
POLYMER BIOSYNTHESIS MAKES A PATHOGENIC BIOFILM

CHAPTER 13
Energetics and Catabolism · 457

PART 4
Microbial Diversity and Ecology   620

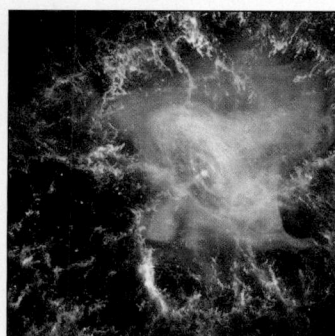

AN INTERVIEW WITH KARL STETTER: ADVENTURES IN MICROBIAL DIVERSITY LEAD TO PRODUCTS IN INDUSTRY

PART 5
Medicine and Immunology   858

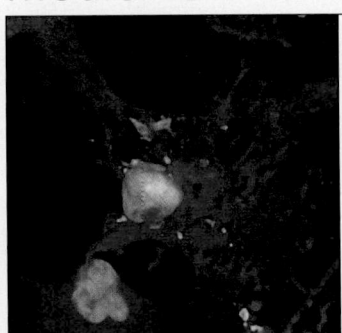

AN INTERVIEW WITH FERRIC FANG:
MOLECULAR MICROBIOLOGY DISSECTS A PATHOGEN

CHAPTER 27

Antimicrobial Chemotherapy 1027

CHAPTER 28

Clinical Microbiology and Epidemiology 1063

APPENDIX 1

Biological Molecules A-1

Preface

In the First Edition of *Microbiology: An Evolving Science*, we set out to write the defining core text of our generation—the book that would inspire undergraduate science majors to embrace the microbial world. The First Edition's emphasis on genetics and ecology, its use of case histories in the medical section, and the balanced depiction of women and minority scientists, including young researchers, drew enthusiastic responses from our more than one hundred adopters.

In this Second Edition, we have maintained that balance between cutting-edge ecology and medicine, while adding new research topics and emerging pathogens, from cold-seep archaeal-bacterial consortia to emerging *E. coli* pathogens and viral gene therapy. We listened to our adopters and have added topics that they requested, such as the molecular biology of chemotaxis (Chapter 10, Molecular Regulation) and an upgraded treatment of immunology (Chapter 24, The Adaptive Immune Response). We have also made a point to relate topics and discussions to current research and current events to keep students interested and informed on the role of microbiology in the world today. One example is connecting the discussion of catabolism to microbe remediation in relation to the Deepwater Horizon spill of 2010 (Chapter 13, Energetics and Catabolism, and Chapter 22, Microbes and the Global Environment).

We have found that writing this textbook is an ongoing, never-ending process requiring the daily harvesting of new, exciting papers from the literature. For microbiology, as the subtitle of this book states, truly is a rapidly evolving science and we have thoroughly embraced this evolving nature in our writing in order to make this Second Edition even more relevant to today's student.

While we have expanded and developed new topics, we also recognized the need to keep the length and "core" of the book to a size reasonable enough for the undergraduate student. So, we committed ourselves to maintaining the same overall page length as the First Edition. But to contain length while adding new material required an innovative solution, so certain topics, considered less central by adopters, have been transferred online as "eTopics." The eTopics (available on the StudySpace website at microbiology2.com) are called out in the text, hyperlinked to the ebook, and their key terms are fully indexed in the printed book. Therefore, returning adopters can be confident of still having access to all of the material they taught from the First Edition, but now they also have new sections on infectious biofilms and horizontal transfer of virulence genes (Chapter 25, Microbial Pathogenesis), new concepts of defining bacterial species, such as the pangenome (Chapter 17, Origins and Evolution), and much more to introduce their students to. A complete list of the eTopics can be found at the end of this preface.

In the Second Edition, we continue to present the story of molecular microbiology and microbial ecology in the tradition of its classical history, from Koch and Pasteur to Winogradsky and Beijerinck. But we have also drawn on all of our experience as researchers and educators (and on the input of hundreds of colleagues) to create a microbiology text for the twenty-first century. This Second Edition includes many refinements recommended by colleagues from around the world, at institutions such as Washington University, University of California-Davis, University of Wisconsin-Madison, Florida State University, University of Toronto, University of Edinburgh, University of Antwerp, Seoul National University, Chinese University of Hong Kong, and many more. We are grateful to you all.

Major Features

Our book targets the science major in biology, microbiology, or biochemistry. Several important features make our book the best text available for undergraduates today:

- **Current research examples and tools throughout the text enrich students' understanding of foundational topics.** Every chapter presents numerous current research examples within the up-to-date framework of molecular biology. Examples of current research include measuring the movement of a single translating ribosome; transplanting a whole genome; determining the "pangenome," the overall set of genes available to a species; and the spectroscopic measurement of carbon flux from microbial communities.

- **Genetics and genomics are presented as the foundation of microbiology.** Molecular genetics and genomics are thoroughly integrated with core topics throughout the book. This approach gives students an understanding of how genomes reveal potential metabolic pathways in diverse organisms, and how genomics and metagenomics reveal the character of microbial communities.

- **Microbial ecology and medical microbiology receive equal emphasis**, with particular attention paid to the merging of these fields. Throughout the book, phenomena are presented with examples from both ecology and medicine; for example, when discussing horizontal transfer of "genomic islands" we present symbiosis islands associated with nitrogen fixation, as well as pathogenicity islands associated with disease (Chapter 9).

- **Unlike most microbiology textbooks, our text provides size scale information for nearly every micrograph.**

- **Viruses are presented in molecular detail and in ecological perspective.** For example, in marine ecosystems, viruses play key roles in limiting algal populations while selecting for species diversity (Chapter 6). Similarly, a constellation of bacteriophages influences enteric flora.

- **Microbial diversity that students can grasp.** We present microbial diversity in a manageable framework that enables students to grasp the essentials of the most commonly presented taxa, the continual discovery of organisms ranging from anammox bacteria to emerging pathogenic *Escherichia* strains.

- **Medical microbiology is presented using the physician-scientist's approach to microbial diseases.** Case histories present how a physician-scientist approaches the interplay between the human immune response and microbial diseases. For example, one case shows how a routine tuberculin test can lead to complications in a sensitized patient.

- **Scientists pursuing research today are presented alongside the traditional icons.** For example, Chapter 1 introduces historical figures such as Koch and Pasteur alongside genome sequencer Claire Fraser-Liggett and postdoctoral researcher Kazem Kashefi growing a hyperthermophile in an autoclave, and undergraduate students conducting transcriptomics in *E. coli*.

■ **Appendices for students in need of review.** Our book assumes a sophomore-level understanding of introductory biology and chemistry. For those in need of review, two appendices summarize the fundamental structure and function of biological molecules and cells.

Organization

The topics in this book are arranged so that students can progressively develop an understanding of microbiology from key concepts and research tools. The chapters of Part 1 present key foundational topics: history, visualization, the bacterial cell, microbial growth and control, and virology.

The six chapters in Part 1 present many topics that are then developed in further detail throughout Parts 2 through 5. Part 2 presents modern genetics and genomics. Part 3 presents cell metabolism and biochemistry, although the chapters in Part 2 are written in such a way that they can be presented before the genetics material if so desired. Part 4 explores microbial ecology and diversity and discusses the roles of microbial communities in local ecosystems and global cycling. And the chapters of Part 5 (Chapters 23–28) present medical and disease microbiology from an investigative perspective, founded on the principles of genetics, metabolism, and microbial ecology.

What's New in the Second Edition?

The Second Edition of *Microbiology: An Evolving Science* has been thoroughly revised and updated. We have added more up-to-the-minute research and current events as well as incorporating much of the generous feedback that we have received from our many reviewers and adopters. The following list highlights some of the more important content changes for the Second Edition:

Three new part-opening interviews highlight contemporary scientists and their research.

Part 2: Christine Jacobs-Wagner, a Howard Hughes Medical Institute investigator, reveals the fascinating life cycle of *Caulobacter crescentus*, a stalked/motile organism of aquatic environments, important for nutrient recycling.

Part 3: Dan Wozniak, professor of internal medicine at The Ohio State University, investigates biofilms of *Pseudomonas aeruginosa*, life-threatening lung pathogen of cystic fibrosis patients.

Part 5: Ferric Fang, editor in chief of the journal *Infection and Immunity*, discovers a molecular "immune system" in *Salmonella enterica*, a pathogen recently in the news due to a half-billion eggs being recalled.

Content changes throughout the text include:

Chapter 1. A new chapter opener features *Prochlorococcus*, the ocean's most abundant prototroph, and the history of smallpox prevention includes new coverage of African contributions.

Chapter 2. A new animation illustrates the fundamentals of microscopy. There is also new coverage of 3D cryo–electron tomography.

Chapter 3. Special Topic 3.1 features research on ampicillin penetrating the outer membrane through porin OmpF.

Chapter 4. The chapter opener highlights an unusual, unculturable organism, and we have included new information on the relationship between biofilms and disease.

Chapter 5. The chapter opener highlights a shape-shifting bacterium found in fermenting foods. There is also new information about the relationship between bacteria and weather, proton circulations, and bacterial resistance to disinfectants.

Chapter 6. The swine flu outbreak is presented, and there is expanded coverage of viral ecology.

Chapter 7. The Chapter opener highlights altered morphology of cell division mutants. There is new coverage of plasmid segregation and metagenomic analysis of environmental populations, as well as a new special topic that discusses next generation DNA sequencing.

Chapter 8. The chapter opener illustrates nucleoid transcription. A new special topic explores how scientists can now track the movement of a single ribosome on mRNA, and there are updates on transcription, translation, and protein secretion.

Chapter 9. Connections between biofilms, conjugation, and pili are highlighted in the new chapter opener. A refined model of conjugation has been included, and a new special topic illustrates how bacterial DNA was found embedded in eukaryotic genomes.

Chapter 10. Chemotaxis is presented as an integrated regulatory system, supplemented by a new animation. There is updated coverage of glucose repression, small RNA regulation and virulence, and toxin-antitoxin systems regulating bacterial physiology.

Chapter 11. The quasispecies concept of viral evolution within a host is presented, and a new section covers viral vectors for gene therapy.

Chapter 12. The new chapter opener highlights the new field of whole genome transplants. New coverage of ChIP-on-Chip analysis and protein interaction networks is also provided.

Chapter 13. Oil spill remediation through bacterial aromatic metabolism is discussed in relation to the Deep Horizon oil spill of 2010.

Chapter 14. A special topic presents *Shewanella* fuel cells, and the explanations of electron transfer reactions are enhanced and updated.

Chapter 15. A new chapter opener features phenazines of colored soil bacteria, molecules that act as antibiotics and enhance biofilm formation. A new figure presents an overview of biosynthesis from carbon fixation to quorum signals.

Chapter 16. There is updated coverage of food safety, including the 2008 Salmonella outbreak in peanut products.

Chapter 17. There is updated coverage of microfossils and biosignatures as evidence of early life, and a new section presents the "pangenome," the sum total genes available to a microbial species.

Chapter 18. The chapter presents emerging forms of metabolism, including a special topic on carbon monoxide oxidation by *Roseobacter* and related marine bacteria.

Chapter 19. Archaea coverage is expanded, including unique turreted viruses of thermophiles, newly discovered deep-branching groups such as ammonia-oxidizing Thaumarchaeota, and cold seeps of exotic sea-floor symbionts.

Chapter 20. Parasite coverage is expanded, including a new special topic on research of the trypanosome cell cycle.

Chapter 21. Current research is presented on community genomics, including a transcriptomic analysis of a Hawaiian marine microbial community. New mutualisms are presented, including benthic clam methanotrophs and thiotrophs, crop plant endophytes, and gut fermenters of termites and vertebrates.

Chapter 22. There is an updated presentation of the controversies over iron fertilization and marine dead zones. Up-to-date coverage is provided of the Deepwater Horizon oil spill and its impact on carbon cycling. There is also a new section on anammox bacteria in nitrogen cycling.

Chapter 23. Toll-like receptors and their induction by infection is highlighted in the new chapter opener. New coverage of pathogen associated molecular patterns (PAMPs) is included, as well as autophagy and its role in clearing intracellular infections.

Chapter 24. We have included enhanced descriptions and discussion of the genetics of antibody production and diversity, the effects of human microbiota on immune system development, and biophotonic imaging of infections in whole animals.

Chapter 25. The new chapter opener shows *Rickettsia typhi*, the cause of Rocky Mountain Spotted fever, making an actin tail to propel itself through the host cytoplasm. We also include a new discussion of the immunopathogenesis of microbial disease.

Chapter 26. Ebola pathogenesis is highlighted in the new chapter opener and in the text. There is also new coverage of tuberculosis as a re-emerging disease, an update on the Columbus New World Theory of Syphilis, and updated Vaccine Schedules.

Chapter 27. The chapter opener highlights emerging community-acquired methicillin-resistant *Staphylococcus Aureus* (MRSA). There is also new coverage of antibiotic effects on normal microbiota, new pyronin antibiotics, the relationship between biofilms and antibiotic persistence, a new special topic highlighting the design of antibiotics that interfere with quorum sensing, and an update on the pathogenesis of the 1918 influenza virus.

Chapter 28. Patient Zero of the recent H1N1 influenza epidemic is highlighted in the chapter opener. A new special topic discusses bioterrorism and quick pathogen detection, and new sections are included that describe the relationship between ecology and infectious disease epidemiology and the evolution of community-acquired MRSA.

Special Features

Throughout our book, special features aid student understanding and stimulate inquiry.

CURRENT RESEARCH

Examples of current research are seamlessly integrated within the up-to-date framework of molecular biology, facilitating the incorporation of the latest research into foundational topics of genetics, physiology, ecology, evolution, and immunology.

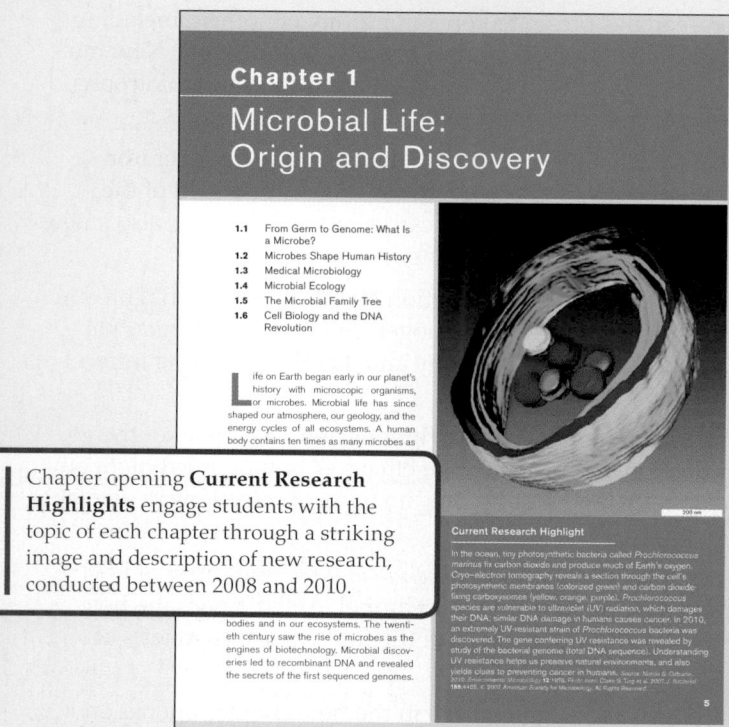

Chapter opening **Current Research Highlights** engage students with the topic of each chapter through a striking image and description of new research, conducted between 2008 and 2010.

Special Topic boxes in every chapter focus on cutting-edge research with an emphasis on experimental detail and put a human face on the dynamic science of microbiology.

Current research examples are then integrated into every chapter on topics such as biofilms, chemotaxis, the human microbiome, and the 2010 Deepwater Horizon oil spill.

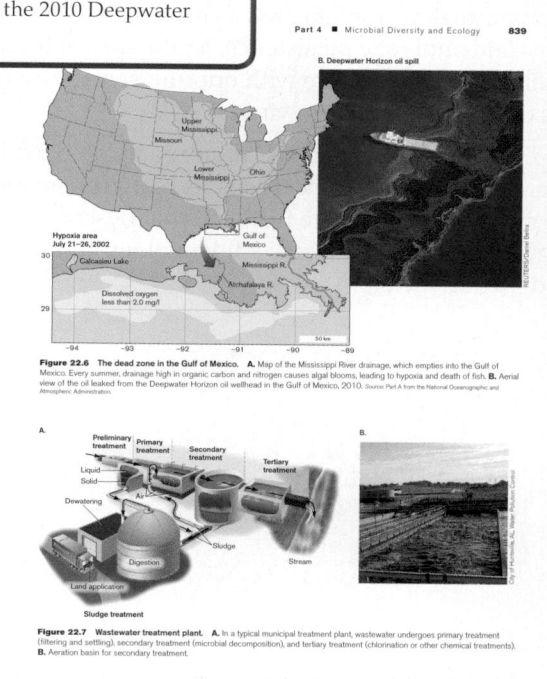

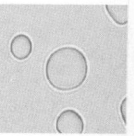
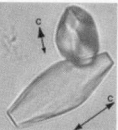

ART PROGRAM

Successful microbiology students must learn to visualize microbial processes and structures that by their very nature occur at an unseen level. *Microbiology*'s extensive and consistently executed art program helps students visualize key microbial processes and showcases the latest structural discoveries.

> Large, engaging figures are accompanied by distinctive bubble captions and numbered labels that help students interpret and analyze microbial processes and structures.

> To convey microbiology as a diverse and dynamic field, photos of iconic and contemporary microbiologists are paired with figures of their most important contributions. Many figures are also accompanied by photos of researchers, graduate students, and even undergraduates doing research—putting a human face to the science.

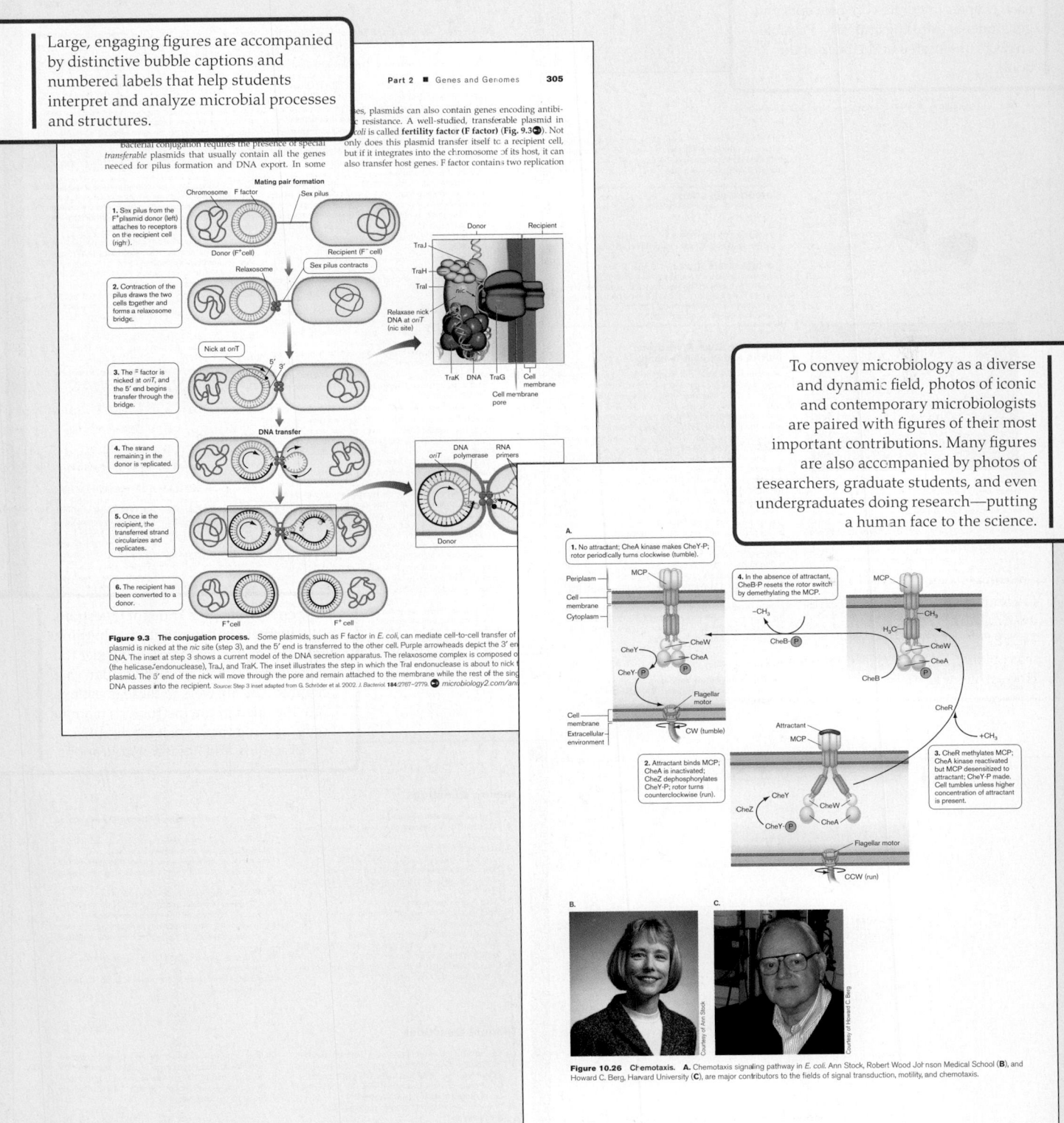

IN TEXT TOOLS

A readable narrative and helpful pedagogical features in every chapter challenge students to review what they've learned, to identify key themes, and to think critically about important questions.

> Ample Thought Questions throughout each chapter integrate core concepts and get students thinking critically. Possible answers are located at the back of the book.

282   **Chapter 8** ■ Transcription, Translation, and Bioinformatics

A.

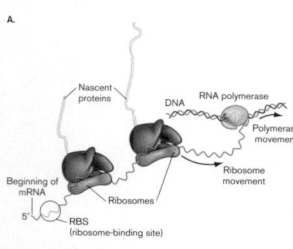

B.

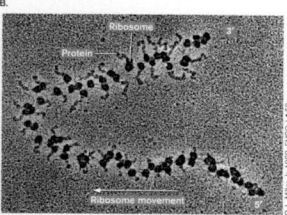

Figure 8.29 Coupled transcription and translation in prokaryotes. A. Ribosomes attach at mRNA ribosome-binding sites and start synthesizing protein before transcription of the gene is complete. **B.** Electron micrograph of a polysome (ribosome is 21 nm across). Several ribosomes may translate a single mRNA molecule at the same time. The beginning (5') end of the mRNA is to the right (at the arrow), and the 3' end is to the left. Note that the synthesized protein molecule grows longer and longer the closer the ribosome gets to the 3' end of the mRNA. The protein molecule is seen most clearly at the end of the mRNA (to the upper right).

machines and destroy the message. The cell handles this problem by modulating transcriptional speed. The rate of RNA synthesis averages about 45 nt per second, which roughly equals the average rate of translation (16 amino acids per second). However, these speeds are not constant. For example, RNA polymerase pauses momentarily at sites rich in GC content (GC base pairs, with three hydrogen bonds, are harder to melt than AT pairs with only two hydrogen bonds). Once sigma factor exits the transcription complex, proteins called NusA and NusG enter the complex and can actually lengthen the pause

in RNA polymerase activity to allow time for the trailing protein-synthesizing ribosome to catch up to the polymerase. Pausing allows the ribosome to follow RNA polymerase closely and protect the RNA.

> **THOUGHT QUESTION 3.7** How might one gene code for two proteins with different amino acid sequences?

> **THOUGHT QUESTION 8.8** Why involve RNA in protein synthesis? Why not translate directly from DNA?

> **THOUGHT QUESTION 8.9** Codon 45 of a 90-codon gene was changed into a translation stop codon, producing a shortened, truncated protein. What kind of mutant cell could produce a full-length protein from the gene *without* removing the stop codon? *Hint:* What molecule recognizes a codon?

Unsticking Stuck Ribosomes: tmRNA and Protein Tagging

Sometimes an RNase will shear off the 3' end of a message, removing the stop codon before the translation is complete. What happens after a ribosome has finished translating this damaged mRNA molecule? Without a stop codon, there is nothing to trigger ribosome release when the ribosome reaches the end of the message. So the ribosome is stuck at the end of the mRNA with a peptidyl-tRNA in the ribosome P site.

The answer to this problem is a translation rescue molecule, the previously mentioned tmRNA, which has the properties of both tRNA and mRNA. One end of the tmRNA molecule has an attached amino acid and a folded structure that resembles a tRNA. This aminoacyl end acts like tRNA, entering the unoccupied ribosome A site, where a peptide bond then forms between the stalled polypeptide (in the P site) and the emergency amino acid on the tmRNA. Another section of the tmRNA is then read as mRNA, which adds a tag of 12 or so amino acids to the now dislodged peptide (SsrA; **Fig. 8.30A**). These amino acids, called a proteolysis tag, predestine the protein for destruction. A stop codon present in tmRNA triggers peptide release and ribosome disassembly. A helper protein (SspB) recognizes the proteolysis tag and brings the useless aberrant polypeptide to the ClpXP protease (discussed shortly) for degradation. **Figure 8.30B** shows this process for tmRNA from *E. coli.*

TO SUMMARIZE:

- **Triplet nucleotide codons** in mRNA encode specific amino acics. **Transfer RNA molecules** interpret the genetic code and bring specific amino acids to the A (acceptor) site in the ribosome.

AT-rich region was horizontally inherited from illicit DNA exchange with another species. (See Chapter 9.)

8.5 Synthesis of ribosomal RNAs and ribosomal proteins causes a major energy drain on the cell. How might the cell regulate synthesis of these molecules when growth rate slows as a result of amino acid limitation? *Hint:* Predict what happens to any translating ribosome when an amino acid is limiting. Look up RelA protein on the Internet.

ANSWER: As energy levels fall, fewer amino acids are made, which means that fewer charged tRNA molecules are assembled, causing ribosomes to pause during translation until finding the right charged tRNA. This pause causes synthesis of a signal molecule called guanosine tetraphosphate (ppGpp), which interacts with RNA polymerase and selectively stops transcription of the rRNA genes. The decrease in rRNA also stops new ribosome synthesis. As cells divide, the ribosomes will dilute to a lower steady-state level. Since less rRNA is made, fewer partners for ribosomal proteins are available and unassembled ribosomal proteins accumulate. These proteins bind sequences in their own mRNA molecules that are similar to the target sequences on rRNA. By binding to their mRNA, these ribosomal proteins can prevent their own translation. (See Chapter 10.)

8.6 While working as a member of a pharmaceutical company's drug discovery team, you find that a soil microbe snatched from the jungles of South America produces an antibiotic that will kill even the most deadly, drug-resistant form of *Enterococcus faecalis*, which is an important cause of heart valve vegetations in bacterial endocarditis. Your experiments indicate that the compound stops protein synthesis. How could you more precisely determine the antibiotic's mode of action? *Hint:* Can you use mutants resistant to the antibiotic?

ANSWER: One way is to take a culture of sensitive bacteria and isolate resistant mutants (bacteria that are not killed by the antibiotic), purify their ribosomes, and separate the 30S and 50S ribosomal subunits. Cross-mix subunits from sensitive and resistant cells (for example, mix 30S subunits from sensitive cells with 50S subunits from resistant cells). Then, measure protein synthesis by the hybrid ribosomes with and without the drug. If resistance is due to an altered ribosomal protein or RNA, the subunit mix containing the altered component will make protein regardless of whether the drug is present. Once identified, the responsible ribosomal subunits from resistant and sensitive cells can be broken down further into their component parts, reconstituted in hybrid form, and again tested for an ability to make protein in the presence of drug. This reductive approach will likely, but not always, uncover the target ribosomal protein or RNA.

8.7 How might one gene code [for] different amino acid sequences?

ANSWER: By having two different translation start sites in different reading frames. While this is not a common occurrence, it happens.

8.8 Why involve RNA in protein synthesis? Why not translate directly from DNA?

ANSWER: Because transcription enables the cell to amplify the gene sequence information into multiple copies of RNA. Amplification means that more ribosomes can be engaged in translating the same protein, causing the concentration of the protein to rise more quickly than if only a single gene were used. The transcriptional process also presents an opportunity to regulate production of a protein.

8.9 Codon 45 of a 90-codon gene was changed into a translation stop codon, producing a shortened, truncated protein. What kind of mutant cell could produce a full-length protein from the gene *without* removing the stop codon? *Hint:* What molecule recognizes a codon?

ANSWER: A mutant in which a tRNA gene has been altered by mutation so that the anticodon of the tRNA "sees" the stop codon as an amino acid codon. This mutated tRNA molecule will transfer its amino acid to the peptide chain. The attached amino acid can be used to bridge the gap caused by the stop codon, and a full-length protein is made. These modified tRNAs are called suppressor tRNAs because they suppress the mutant phenotype.

8.10 An ORF 1,200 bp in length could encode a protein of what size and molecular weight? *Hint:* Find the molecular weight of an average amino acid.

ANSWER: There are three bases per codon, so the ORF can encode 400 amino acids. The average molecular weight of an amino acid is 110 Da. Therefore, the hypothetical protein will be approximately 44,000 Da (44 kDa).

Chapter 9

9.1 Transfer of an F factor from an F⁺ cell to an F⁻ cell converts the recipient to F⁺. Why does transfer of an Hfr not do the same?

ANSWER: The last piece of an Hfr to transfer is the F factor and *oriT*. Only rarely will an entire chromosome transfer from one cell to another, so most Hfr transfers do not result in transfer of *oriT* and thus cannot initiate conjugation.

9.2 Would it be easier to demonstrate generalized transduction using a temperate phage (which generates a lysogen) or a lytic (virulent) phage?

ANSWER: Both can be used, but more care must be taken with a virulent phage. The reason is that a lytic phage can kill

---

> New end-of-chapter Thought Questions further challenge students to think critically by asking them to consider the big-picture concepts introduced in that chapter. The answers to these questions are included in the instructor's manual only, making them great for discussion, or as quiz and homework questions.

CHAPTER REVIEW

Review Questions

1. What are some characteristics of an ORF?
2. What is a DNA sequence alignment, and what can it tell you?
3. Describe the differences between a pair of orthologous genes and a pair of paralogous genes.
4. How can bioinformatics predict a metabolic pathway for an organism that cannot be grown in the laboratory?
5. What defines a promoter?
6. What are sigma factors, and what role do they play in gene expression?
7. Describe the three stages of transcription.
8. Explain degeneracy of the genetic code. What is the wobble in codon-anticodon recognition?
9. Describe the stages of protein synthesis. Why is the ribosome called a ribozyme?
10. Discuss some antibiotics that affect transcription or translation.
11. What is meant by coupled transcription and translation? Does this occur in eukaryotic cells?
12. How do bacterial cells release ribosomes that are stuck onto damaged mRNA molecules lacking termination codons?
13. What can happen to misfolded proteins?
14. Why are only certain proteins secreted from the bacterial cell? What are some secretion mechanisms?
15. In what major way do proteins transported by the twin arginine translocase (TAT) differ from other exported proteins?
16. Compare protein degradation in eukaryotes and prokaryotes.
17. What is annotation? How does it apply to bioinformatics?

Thought Questions

1. The process of transcription will generate positive supercoils in front of the polymerase as it moves along a DNA template. Why doesn't the DNA in front of the polymerase become so knotted that the polymerase can no longer separate the DNA strands?
2. Why do cells secrete some proteins into their environments?
3. Type I protein secretion systems transport certain proteins from the cytoplasm of Gram-negative bacteria directly to the outside of the cell, across two membranes. How might the system "know" which proteins to transport?

Part 1

The Microbial Cell

Rita Colwell, former director of the National Science Foundation.

AN INTERVIEW WITH

RITA COLWELL: THE GLOBAL IMPACT OF MICROBIOLOGY

Rita Colwell is Distinguished Professor at the University of Maryland and Johns Hopkins University and served as director of the U.S. National Science Foundation from 1998 to 2004. Colwell's decades of research on *Vibrio cholerae*, the causative agent of cholera, have revealed its natural ecology, its genome sequence, and ways to control it. Colwell originated the concept of viable but nonculturable microorganisms, microbial cells that metabolize but cannot be cultured in the laboratory. She is now chairman of the board of Canon U.S. Life Sciences, Inc., and she represents the American Society for Microbiology at the United Nations Educational, Scientific and Cultural Organization (UNESCO).

Why did you decide to make a career in microbiology?
I was first inspired by the report of my college roommate at Purdue University about a wonderful bacteriology professor, Dr. Dorothy Powelson, probably one of only two women at Purdue who were full professors at the time. I enrolled in Powelson's course and was truly inspired by this remarkable woman who was so interested in microbiology and made it fascinating for her students.

How did you choose to study *Vibrio cholerae*? What makes this organism interesting?
I chose to study *Vibrio cholerae* as a result of my having become an "expert" on vibrios through my graduate dissertation on marine microorganisms. Vibrios were the most readily culturable of the marine bacteria and were therefore considered the most dominant. Of course, new information indicates that although vibrios are the dominant bacteria in many estuarine areas, there are other organisms that are very difficult to culture that are important as well.

When I took my first faculty position at Georgetown University, a friend of mine at NIH, Dr. John Feeley, suggested that I study *Vibrio cholerae*. What makes *V. cholerae* interesting is that it is a human pathogen of extremely great importance, yet resides naturally in estuaries and coastal areas of the world.

What is it like to study this organism?
Vibrio cholerae is naturally occurring (in the environment outside humans) and therefore can never be eradicated; it carries out important functions in the environment, and significant among these is its ability to digest chitin, the structural component of shellfish and many zooplankton. It is at once a "recycling agent" and a public health threat in the form of the massive epidemics of cholera that it causes.

You led an international collaboration in Bangladesh training women to avoid cholera by filtering water through sari cloth. How did the sari cloth filtration project come about?

It came about through collaboration with the International Centre for Diarrhoeal Disease Research, Bangladesh, located in Dhaka, Bangladesh, and the Matlab Field Laboratory, which is located in the village area of Matlab, Bangladesh. Our work had shown that *Vibrio cholerae* is associated with environmental zooplankton—namely, the copepod. The notion that the copepods are large and could be filtered out and therefore lead to reduced incidence of cholera was a result of my work on the vibrios and the relationships described by my students, notably Dr. Anwar Huq, who did his thesis on *Vibrio cholerae* attachment to copepods. Anwar Huq is now an associate professor at the University of Maryland.

An important collaborator was Nell Roberts, an outstanding public health microbiologist at Lake Charles, Louisiana. Nell, Professor Xu (a colleague from Qingdao, China), and I did the critical experiment showing the presence of *Vibrio cholerae* in water from which blue crabs had been harvested—the cause of an outbreak of cholera in Louisiana back in 1982. We were able to use fluorescent antibody to show the presence of the vibrio on copepods in the water.

From there, the idea of sari cloth came about in searching for a very inexpensive filter for use by village

2

---

acid resistance. In a process analogous to the heat-shock response, bacterial physiology undergoes a major molecular reprogramming in response to hydrogen ion stress. The levels of a large number of proteins increase, while the levels of others decrease. Many of the genes and proteins involved in the acid stress response overlap with other stress response systems, including the heat-shock response. These physiological responses include modifications in membrane lipid composition, enhanced pH homeostasis, and numerous other changes with unclear purpose. Some pathogens, such as *Salmonella*, sense a change in external pH as part of the signal indicating that the bacterium has entered a host cell environment (see **eTopic 5.2**).

TO SUMMARIZE:

- **Hydrogen ion concentration** affects protein structure and function. Thus, enzymes have pH optima, minima, and maxima.
- **Microbes use pH homeostasis mechanisms** to keep their internal pH near neutral when in acidic or alkaline media.
- **Adding weak acids** to certain foods undermines bacterial pH homeostasis mechanisms, thereby preventing food spoilage and killing potential pathogens.
- **Neutralophiles, acidophiles, and alkaliphiles** prefer growth under neutral, low, and high pH conditions, respectively.
- **Acid and alkaline stress responses** result when a given species is placed under pH conditions that slow its growth. The cell increases the levels of proteins designed to mediate pH homeostasis and protect cell constituents.

5.6 Oxygen and Other Electron Acceptors

Many microorganisms can grow in the presence of molecular oxygen (O_2). Some even use oxygen as a terminal electron acceptor in the **electron transport chain**, a group of membrane proteins (cytochromes) that convert energy trapped in nutrients to a biologically useful form. The use of O_2 as the terminal electron acceptor is called **aerobic respiration** (see Chapter 14).

Oxygen Has Benefits and Risks

Electrons pulled from various energy sources (for example, glucose) possess intrinsic energy, an energy that cytochrome proteins can harness. They do this by extracting the energy in incremental stages and using it to move protons out of the cell. This unequal distribution of H^+ across the membrane produces a transmembrane electrochemical gradient, a sort of biobattery called the proton motive force (discussed in Section 14.2). Once the cell has drained as much energy as possible from an electron, that electron must be passed to a diffusible, final electron acceptor molecule floating in the medium. This clears the way for another electron to be passed down the electron transport chain (**Fig. 5.18**).

One such terminal electron acceptor is dissolved oxygen. However, oxygen and its breakdown products are dangerously reactive—a serious problem for all cells. As a result, different species have evolved to either tolerate or avoid oxygen altogether. **Table 5.2** gives examples of microbes that grow at different levels of oxygen.

Figure 5.18 Role of oxygen as a terminal electron acceptor in respiration. This pumping of protons out of the cell by electron transport chains produces more positive charges outside the cell than inside, resulting in an electrochemical gradient (also called proton motive force). An electron from a high-energy source is passed from one member of the chain to the next. With each transfer of an electron to the next member of the chain, more energy is extracted and converted to proton motive force. At the end of the chain, the electron must be passed to a final or terminal electron acceptor, clearing the path for the next electron. This net process is called respiration.

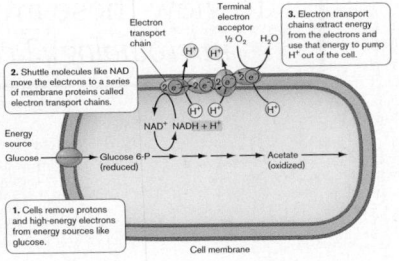

1. Cells remove protons and high-energy electrons from energy sources like glucose.

2. Shuttle molecules like NAD move the electrons to a series of membrane proteins called electron transport chains.

3. Electron transport chains extract energy from the electrons and use that energy to pump H^+ out of the cell.

Electron transport chain

Terminal electron acceptor

$\frac{1}{2} O_2$ H_2O

H^+ H^+ H^+ H^+

Energy source

Glucose → Glucose 6-P (reduced) → Acetate (oxidized)

NAD^+ $NADH + H^+$

Cell membrane

INTERVIEWS WITH PROMINENT SCIENTISTS

Each Part of the book opens with an interview of a prominent microbiologist working today. In each interview, the authors ask the featured scientist questions about everything from how they first became interested in microbiology to how their thought processes and experiments allowed them to make important discoveries. New interviewees include Christine Jacobs-Wagner (Yale University), Ferric Fang (University of Washington) and Daniel Wozniak (The Ohio State University).

TO SUMMARIZE

This feature ensures that students understand key concepts of each section before they continue with the reading.

ENGAGING PROCESS ANIMATIONS MAKE COMPLEX CONCEPTS CLEAR

Each animation topic was chosen by instructors and developed specifically for *Microbiology: An Evolving Science,* in close coordination with the authors. Concepts are presented accurately and with just the right level of detail. Animations are available to students on the free and open StudySpace and are included on the Instructor's Resource Disc for instructors to use in lecture.

Process Animation Topics

Microscopy

Replisome Movement in a Dividing Cell

Chemotaxis: Molecular Events

Phosphotransferase System (PTS) Transport

Endospore Formation

Dilution Streaking Technique

Biofilm Formation

Twitching Motility

Lysis and Lysogeny

DNA Replication

DNA Sequencing

PCR

Supercoiling and Topoisomerases

Rolling Circle Mechanism of Plasmid Replication

Protein Synthesis

Protein Export

SecA-Dependent General Secretion Pathway

ABC Transporters

Recombination

Transposition

DNA Repair Mechanisms: Methyl Mismatch Repair

DNA Repair Mechanisms: Nucleotide Excision Repair

DNA Repair Mechanisms: Base Excision Repair

Bacterial Conjugation

The *lac* Operon

Transcriptional Attenuation

Quorum Sensing

Influenza Virus Entry into a Cell

Influenza Virus Replication

HIV Replication

Herpes Virus Replication

Tagging Proteins for Easy Purification

Real-Time PCR

DNA Shuffling

Construction of a Gene Therapy Vector

A Bacterial Electron Transport System

ATP Synthase Mechanism

Oxygenic Photosynthesis

Agrobacterium: A Plant Gene Transfer Vector

Phylogenetic Trees

Listeria Infection

Light-Driven Ion Pumps and Sensors

Malaria: A Cycle of Transmission between Mosquito and Human

The Basic Inflammatory Response

Phagocytosis

The Activation of the Humoral and Cell-Mediated Pathways

Cholera Toxin Mode of Action

Process of Type III Secretion

Retrograde Movement of Tetanus Toxin to an Inhibitory Neuron

To view these animations, visit
microbiology2.com/animations

ANIMATION: QUORUM SENSING

This animation illustrates quorum sensing, the molecular process by which bacteria chemically converse with each other. Quorum sensing in different forms instructs bacteria to form biofilms, activate stress response, and induce virulence genes.

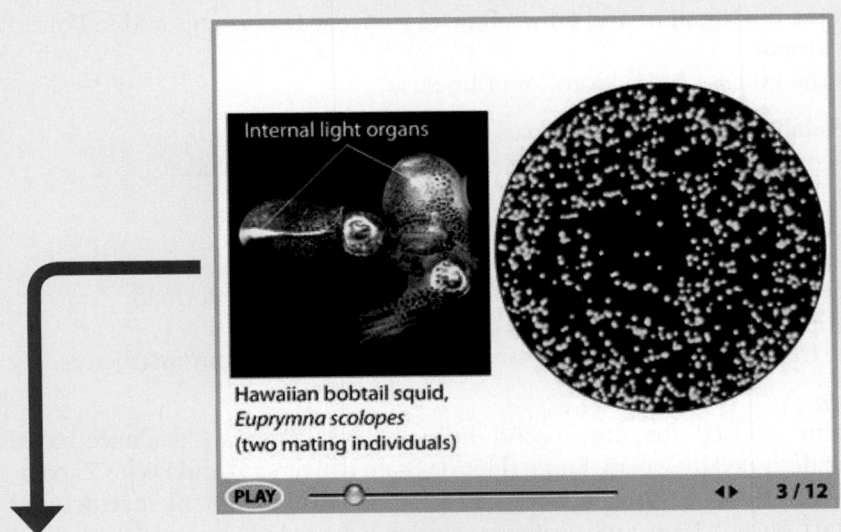

Internal light organs

Hawaiian bobtail squid,
Euprymna scolopes
(two mating individuals)

PLAY ◀▶ 3 / 12

The bacteria glow only under certain conditions. The genes needed to make light are off when bacterial cell densities are low (such as in sea water), but turn on under crowded conditions (such as in the light organ of the squid), when enough cells are together to make a visual impact. How do cells *know* they have achieved an adequate number of nearby individuals—that is, a quorum—before turning on their luminescence genes?

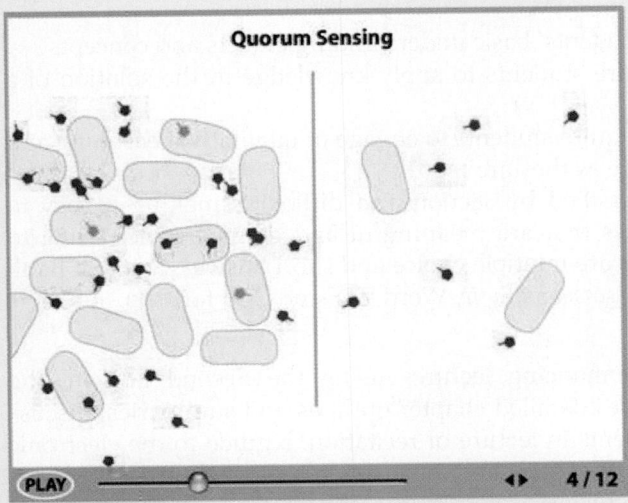

Quorum Sensing

PLAY ◀▶ 4 / 12

It turns out that this phenomenon, called quorum sensing, is only loosely associated with cell numbers. Induction of a quorum-sensing gene system requires the accumulation of a molecule called an autoinducer. After a cell produces an autoinducer, the molecule rapidly diffuses out of the cell. The more cells in a given space, the faster the autoinducer builds up and the more likely it will reenter cells and trigger the luminesence response.

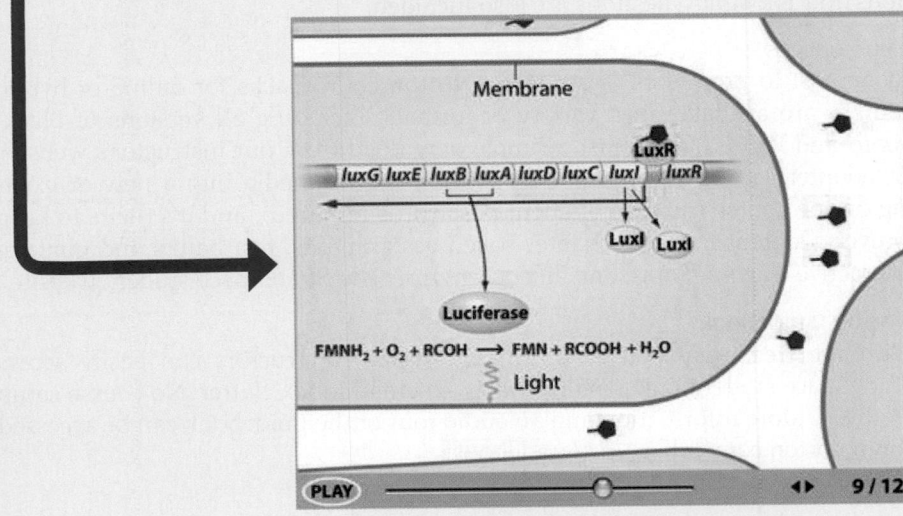

Membrane

LuxR

luxG luxE luxB luxA luxD luxC luxI luxR

LuxI LuxI

Luciferase

$FMNH_2 + O_2 + RCOH \longrightarrow FMN + RCOOH + H_2O$

Light

PLAY ◀▶ 9 / 12

Luciferase catalyzes a redox reaction that produces oxidized and reduced chemical products as well as blue-green light. Because the Lux proteins, like other proteins, require energy to produce, the cells turn this system on only when appropriate—such as when they are crowded together in the light organ of the squid.

Instructor Resources

- The Instructor's Resource Disc includes:
  - PowerPoint Lecture slides now with additional clicker questions and new summary slides, revised for the Second Edition.
  - All of the photographs and drawn figures from the text as jpgs and in Power-Point format.
  - All of the Process Animations for offline use.

- Downloadable Instructor's Resources
  Feature content for use in both the classroom and online, includes:
  - PowerPoint lecture slides
  - Book art in zipped jpeg and PowerPoint formats
  - Free, customizable Blackboard and WebCT coursepacks
  - Test Bank in *ExamView,* WebCT, Blackboard and Word RTF formats
  - StudySpace quizzes in Blackboard and WebCT formats
  For more information and to view samples, visit wwnorton.com/instructors.

- Test Bank
  Thoroughly revised for the Second Edition and using an evidence-based approach designed by Valerie Shute (Florida State University) and Diego Zapata-Rivera (Educational Testing Service), each chapter of the Test Bank is structured around a Concept Map and evaluates students knowledge on three distinct levels:
  - Factual questions test students' basic understanding of facts and concepts
  - Applied questions require students to apply knowledge in the solution of a problem
  - Conceptual questions require students to engage in qualitative reasoning and to explain why things are as they are
  Questions are further classified by section and difficulty, making it easy to construct tests and quizzes that are meaningful and diagnostic according to instructor need. Questions are multiple choice and short answer. The Test Bank is available in *ExamView Assessment Suite,* Word RTF and PDF formats.

- Instructor's Manual
  Full of suggestions for enhancing lectures using the Second Edition, the Instructor's Manual includes detailed chapter outlines and summaries, discussion topics to engage students in lecture or recitation, a guide to the electronic media—including animations, and additional readings to help make lectures more dynamic and interactive. Answers to the end-of-chapter Review Questions and Thought Questions are also included.

- Coursepacks
  At no cost to professors or students, Norton coursepacks for online or hybrid courses are available in a variety of formats, including all versions of Blackboard and WebCT. With just a simple download from our instructor's website, an adopter can bring high-quality Norton digital media into a new or existing online course (no extra student passwords required), and it's theirs to keep forever. Content includes chapter-based assignments, test banks and quizzes, interactive learning tools, and all content from the StudySpace student website.

- Norton Gradebook
  With the free, easy-to-use Norton gradebook, instructors can easily access StudySpace student quiz results and avoid email inbox clutter. No course setup required. More information and an audio tour of the gradebook can be accessed at wwnorton.com/college/nrl/gradebook.

Student Resources

- StudySpace
  Students rely on effective and well-designed online resources to help them succeed in their courses. StudySpace is a free and open website that shows students what they need to know, what they still need to review, and provides them with an organized study plan for mastering the material. Resources for students include:
  - Process animations, including 22 new animations for the Second Edition, are based on textbook art and have been developed with direct input from the authors.
  - New Quiz+ Diagnostic Quizzes provides a customized study plan based on right and wrong quiz answers that offers specific page references and links to the ebook and other online learning tools.
  - New Visual Quizzes based on textbook art ask students to identify regions of a figure to demonstrate understanding of key processes.
  - Detailed Study Plans guide students through mastering the core concepts for each chapter and help them utilize all the online resources for the chapter.
  - Summaries from the textbook
  - Flashcard for vocabulary terms
  - Weblinks throughout the text refer students to internet sites that contain interesting multi-media supplemental information that enhances the textbook content.
  - New Study Microbes page at MicrobeWiki (Kenyon College) offers an evolving list of practice questions and microbial species to know.

- New eTopics on StudySpace supplement textbook content by introducing students to topics that, despite their importance, were removed from the print textbook to make room for new, breaking science. Most chapters in the text include references to specific eTopics and describe how the eTopic relates to the textbook content. There is a complete list of eTopics on the last page of this preface.

- Ebook
  Same great content, one-third the price. An affordable and convenient alternative to the print book, Norton ebooks retain the content and design of the print book and allow students to highlight and take notes with ease, print chapters as needed, and search the text. Norton ebooks are available online and as downloadable pdfs. They can be purchased directly from our website, or with a registration folder that can be sold in the bookstore.

  Chapter Select ebook chapters are available for just $3 each!
  ($30.00 minimum purchase)
  Visit nortonebooks.com for pricing and purchasing details as well as instructions about how to create a custom ebook using Chapter Select.

Acknowledgments

We are very grateful for the help of many people in developing and completing the book, including Norton editors John Byram, Vanessa Drake-Johnson, Mike Wright, and especially Betsy Twitchell, whose heroic efforts assured completion of the Second Edition. Our developmental editors, Philippa Solomon and Carol Pritchard-Martinez, contributed greatly to the clarity of presentation. Philippa's strength in chemistry was invaluable in improving our presentation of metabolism. Trish Marx and photo researcher Dena Digilio Betz did an amazing job of tracking down all kinds of images from sources all over the world. Our colleague Kathy Gillen provided exceptional expertise on review topics for the appendices and wrote outstanding review questions for the student website. Rob Bellinger's coordination of electronic media development has resulted in a superb suite of resources for students and instructors alike. We thank Judith Kandel for authoring an instructor's manual that demonstrates a clear understanding of our goals for the book, and Karen Sullivan, Miguel Cervantes, Kari Murad, and Diuto Esiobu for doing the same with the test bank. We thank Matthew Freeman for editing both. Without Thom Foley's incredible attention to detail, the innumerable moving parts of this project would never have become a finished book. Marian Johnson, Norton's managing editor in the college department, helped coordinate the complex process involved in shaping the manuscript over the years. Chris Granville ably and calmly managed the production and manufacturing of this book. Editorial assistants Callinda Taylor and Cait Callahan coordinated the transfer of many drafts among many people. Steve Dunn and Natasha Zabohonski have been effective advocates for the book in the marketplace. Finally, we thank Roby Harrington, Drake McFeely, and Julia Reidhead for their support of this book over its many years of development.

For the quality of our illustrations we thank the many artists at Precision Graphics, who developed attractive and accurate representations and showed immense patience in getting the details right. We especially thank Terri Hamer and Rebecca Reid for project management; Karen Hawk for the layout of every page in the book; Becky Oles and Kitty Auble for developing the art style and leading the art team; and Jeff Griffin for the rendering of the molecular models based on PDB files, including some near-impossible structures that we requested.

We thank the numerous colleagues over the years who encouraged us in our project, especially the many attendees at the Microbial Stress Gordon Conferences. We greatly appreciate the insightful reviews and discussions of the manuscript provided by our colleagues, and the many researchers who contributed their micrographs and personal photos. We especially thank the American Society for Microbiology journals for providing many valuable resources. Reviewers Bob Bender, Bob Kadner, and Caroline Harwood offered particularly insightful comments on the metabolism and genetics sections, and James Brown offered invaluable assistance in improving the coverage of microbial evolution. Peter Rich was especially thoughtful in providing materials from the archive of Peter Mitchell. We also thank the following reviewers:

Second Edition Reviewers:

Michael Allen, University of North Texas
Gladys Alexandre, University of Tennessee Knoxville
Hazel Barton, Northern Kentucky University
Suzanne S. Barth, University of Texas at Austin
Barry Beutler, The College of Eastern Utah
Michael J. Bidochka, Brock University
Dwayne Boucaud, Quinnipiac University
Derrick Brazill, Hunter College
Graciela Brelles-Mariño, California State Polytechnic University, Pomona
Jay Brewster, Pepperdine University
Linda Bruslind, Oregon State University
Marion Brodhagen, Western Washington University
Alison Buchan, University of Tennessee Knoxville
Jeffrey Byrd, St. Mary's College of Maryland
Silvia T. Cardona, University of Manitoba
Andrea Castillo, Eastern Washington University
Miguel Cervantes-Cervantes, Rutgers University
Tin-Chun Chu, Seton Hall University

Paul Cobine, Auburn University
Tyrrell Conway, University of Oklahoma
Scott Dawson, University of California Davis
Jose de Ondarza, SUNY Plattsburgh
Donald W. Deters, Bowling Green State University
Clarissa Dirks, The Evergreen State College
William T. Doerrler, Louisiana State University
Janet R. Donaldson, Mississippi State University
Xin Fan, West Chester University
Babu Z. Fathepure, Oklahoma State University
Clifton Franklund, Ferris State University
Gregory D. Frederick, University of Mary Hardin-Baylor
Christopher French, University of Edinburgh
Jason M. Fritzler, Stephen F. Austin State University
Katrina Forest, University of Wisconsin–Madison
Kimberley Gilbride, Ryerson University
Stjepko Golubic, Boston University
Enid T. Gonzalez, California State University, Sacramento
John E. Gustafson, New Mexico State University

Lynn E. Hancock, Kansas State University
Martina Hausner, Ryerson University
J. D. Hendrix, Kennesaw State University
Michael C. Hudson, University of North Carolina Charlotte
Jane E. Huffman, East Stroudsburg University
Michael Ibba, Ohio State University
Gilbert H. John, Oklahoma State University
John A. Johnson, University of New Brunswick St. John
Mark C. Johnson, Georgetown College
Carol Ann Jones, University Of California Riverside
Ece Karatan, Appalachian State University
Daniel B. Kearns, Indiana University Bloomington
Robert J. Kearns, University of Dayton
Susan Koval, University of Western Ontario
Deborah Kuzmanovic, University of Michigan
Peter Kennedy, Lewis & Clark College
Greg Kleinheinz, University of Wisconsin Oshkos
Jesse J. Kwiek, Ohio State University
Andrew Lang, Memorial University of Newfoundland
Margaret Liu, University of Michigan
Thomas W. De Lany, Kilgore College
Maia Larios-Sanz, University of St. Thomas
Beth Lazazzera, University of California, Los Angeles
Dr. Lee H. Lee, Montclair State University
Mark Liles, Auburn University
Jun Liu, University of Toronto
Manuel Llano, University of Texas at El Paso
Zhongjing Lu, Kennesaw State University
Aaron Lynne, Sam Houston State University
John C. Makemson, Florida International University
Donna L. Marykwas, California State University Long Beach
Ann G. Matthysse, University of North Carolina at Chapel Hill
Ghislaine Mayer, Virginia Commonwealth University
Robert Maxwell, Georgia State University
William R. McCleary, Brigham Young University

Nancy L. McQueen, California State University, Los Angeles
Scott A. Minnich, University of Idaho
Philip F. Mixter, Washington State University
Christian D. Mohr, University of Minnesota
Craig Moyer, Western Washington University
Scott Mulrooney, Michigan State University
Kari Murad, The College of Saint Rose
William Wiley Navarre, University of Toronto
Ivan J. Oresnik, University of Manitoba
Cleber Costa Ouverney, San Jose State University
Deborah Polayes, George Mason University
Pablo J. Pomposiello, University of Massachusetts Amherst
Joan Press, Brandeis University
Todd P. Primm, Sam Houston State University
Sharon R. Roberts, Auburn University
Michelle Rondon, University of Wisconsin–Madison
Silvia Rossbach, Western Michigan University
Ben Rowley, University of Central Arkansas
Chad R. Sethman, Waynesburg University
Matthew O. Schrenk, East Carolina State University
Anthony Siame, Trinity Western University
Lyle Simmons, University of Michigan
Daniel R. Smith, Seattle University
Garriet W. Smith, University of South Carolina Aiken
Geoffrey B. Smith, New Mexico State University
Ruth Sporer, Rutgers University Camden
Anand Sukhan, Northeastern State University
Karen Sullivan, Louisiana State University
Virginia Stroeher, Bishop's University
Dorothea K. Thompson, Purdue University
Wendy C. Trzyna, Marshall University
Bernard Turcotte, McGill University
Dave Westenberg, Missouri University of Science and Technology
Ann Williams, University of Tampa
Charles F. Wimpee, University of Wisconsin–Milwaukee
Jianping Xu, McMaster University

First Edition Reviewers:

Laurie A. Achenbach, Southern Illinois University, Carbondale
Stephen B. Aley, University of Texas, El Paso
Mary E. Allen, Hartwick College
Shivanthi Anandan, Drexel University
Brandi Baros, Allegheny College
Gail Begley, Northeastern University
Robert A. Bender, University of Michigan
Michael J. Benedik, Texas A&M University
George Bennett, Rice University
Kathleen Bobbitt, Wagner College
James Botsford, New Mexico State University
Nancy Boury, Iowa State University of Science and Technology

Jay Brewster, Pepperdine University
James W. Brown, North Carolina State University
Whitney Brown, Kenyon College undergraduate
Alyssa Bumbaugh, Pennsylvania State University, Altoona
Kathleen Campbell, Emory University
Alana Synhoff Canupp, Paxon School for Advanced Studies, Jacksonville, FL
Jeffrey Cardon, Cornell College
Tyrrell Conway, University of Oklahoma
Vaughn Cooper, University of New Hampshire
Marcia L. Cordts, University of Iowa
James B. Courtright, Marquette University
James F. Curran, Wake Forest University
Paul Dunlap, University of Michigan

David Faguy, University of New Mexico

Bentley A. Fane, University of Arizona

Bruce B. Farnham, Metropolitan State College of Denver

Noah Fierer, University of Colorado, Boulder

Linda E. Fisher, late of the University of Michigan, Dearborn

Robert Gennis, University of Illinois, Urbana-Champaign

Charles Hagedorn, Virginia Polytechnic Institute and State University

Caroline Harwood, University of Washington

Chris Heffelfinger, Yale University graduate student

Joan M. Henson, Montana State University

Michael Ibba, Ohio State University

Nicholas J. Jacobs, Dartmouth College

Douglas I. Johnson, University of Vermont

Robert J. Kadner, late of the University of Virginia

Judith Kandel, California State University, Fullerton

Robert J. Kearns, University of Dayton

Madhukar Khetmalas, University of Central Oklahoma

Dennis J. Kitz, Southern Illinois University, Edwardsville

Janice E. Knepper, Villanova University

Jill Kreiling, Brown University

Donald LeBlanc, Pfizer Global Research and Development (retired)

Robert Lausch, University of South Alabama

Petra Levin, Washington University in St. Louis

Elizabeth A. Machunis-Masuoka, University of Virginia

Stanley Maloy, San Diego State University

John Makemson, Florida International University

Scott B. Mulrooney, Michigan State University

Spencer Nyholm, Harvard University

John E. Oakes, University of South Alabama

Oladele Ogunseitan, University of California, Irvine

Anna R. Oller, University of Central Missouri

Rob U. Onyenwoke, Kenyon College

Michael A. Pfaller, University of Iowa

Joseph Pogliano, University of California, San Diego

Martin Polz, Massachusetts Institute of Technology

Robert K. Poole, University of Sheffield

Edith Porter, California State University, Los Angeles

S. N. Rajagopal, University of Wisconsin, La Crosse

James W. Rohrer, University of South Alabama

Michelle Rondon, University of Wisconsin–Madison

Donna Russo, Drexel University

Pratibha Saxena, University of Texas, Austin

Herb E. Schellhorn, McMaster University

Kurt Schesser, University of Miami

Dennis Schneider, University of Texas, Austin

Margaret Ann Scuderi, Kenyon College

Ann C. Smith Stein, University of Maryland, College Park

John F. Stolz, Duquesne University

Marc E. Tischler, University of Arizona

Monica Tischler, Benedictine University

Beth Traxler, University of Washington

Luc Van Kaer, Vanderbilt University

Lorraine Grace Van Waasbergen, The University of Texas, Arlington

Costantino Vetriani, Rutgers University

Amy Cheng Vollmer, Swarthmore College

Andre Walther, Cedar Crest College

Robert Weldon, University of Nebraska, Lincoln

Christine White-Ziegler, Smith College

Jianping Xu, McMaster University

Finally, we offer special thanks to our families for their support. Joan's husband Michael Barich offered unfailing support and her son Daniel Barich contributed photo research, as well as filling the indispensable role of technical director for the *Microbial Biorealm* website. John's wife Zarrintaj ("Zari") Aliabadi contributed to the text development, especially the sections on medical microbiology and public health.

To the Reader: Thanks!

We greatly appreciate your selection of this book as your introduction to the science of microbiology. Although this is a second edition, we recognize that as the text evolves it can benefit greatly from the input of readers. We welcome your comments, especially if you find text or figures that are in error or unclear. Feel free to contact us at the addresses listed below.

Joan L. Slonczewski
slonczewski@kenyon.edu

John W. Foster
jwfoster@jaguar1.usouthal.edu

eTopic Contents

All eTopics are available on StudySpace at <u>Microbiology2.com</u>

About the **Authors**

JOAN L. SLONCZEWSKI received her B.A. from Bryn Mawr College and her Ph.D. in Molecular Biophysics and Biochemistry from Yale University, where she studied bacterial motility with Robert M. Macnab. After postdoctoral work at the University of Pennsylvania, she has since taught undergraduate microbiology in the Department of Biology at Kenyon College, where she earned a Silver Medal in the National Professor of the Year program of the Council for the Advancement and Support of Education. She has published numerous research articles with undergraduate coauthors on bacterial pH regulation, and has published five science fiction novels including *A Door into Ocean*, which earned the John W. Campbell Memorial Award. She served as At-large Member representing Divisions on the Council Policy Committee of the American Society for Microbiology, and is a member of the Editorial Board of the journal *Applied and Environmental Microbiology*.

JOHN W. FOSTER received his B.S. from the Philadelphia College of Pharmacy and Science (now the University of the Sciences in Philadelphia), and his Ph.D. from Hahnemann University (now Drexel University School of Medicine), also in Philadelphia, where he worked with Albert G. Moat. After postdoctoral work at Georgetown University, he joined the Marshall University School of Medicine in West Virginia; he is currently teaching in the Department of Microbiology and Immunology at the University of South Alabama College of Medicine in Mobile, Alabama. Dr. Foster has coauthored three editions of the textbook *Microbial Physiology* and has published over 100 journal articles describing the physiology and genetics of microbial stress responses. He has served as Chair of the Microbial Physiology and Metabolism division of the American Society for Microbiology and as a member of the editorial advisory board of the journal *Molecular Microbiology*.

Microbiology
An Evolving Science
Second Edition

Part 1

The Microbial Cell

AN INTERVIEW WITH

RITA COLWELL: THE GLOBAL IMPACT OF MICROBIOLOGY

Rita Colwell is Distinguished Professor at the University of Maryland and Johns Hopkins University and served as director of the U.S. National Science Foundation from 1998 to 2004. Colwell's decades of research on *Vibrio cholerae*, the causative agent of cholera, have revealed its natural ecology, its genome sequence, and ways to control it. Colwell originated the concept of viable but nonculturable microorganisms, microbial cells that metabolize but cannot be cultured in the laboratory. She is now chairman of the board of Canon U.S. Life Sciences, Inc., and she represents the American Society for Microbiology at the United Nations Educational, Scientific and Cultural Organization (UNESCO).

Rita Colwell, former director of the National Science Foundation.

Why did you decide to make a career in microbiology?
I was first inspired by the report of my college roommate at Purdue University about a wonderful bacteriology professor, Dr. Dorothy Powelson, probably one of only two women at Purdue who were full professors at the time. I enrolled in Powelson's course and was truly inspired by this remarkable woman who was so interested in microbiology and made it fascinating for her students.

How did you choose to study *Vibrio cholerae*? What makes this organism interesting?
I chose to study *Vibrio cholerae* as a result of my having become an "expert" on vibrios through my graduate dissertation on marine microorganisms. Vibrios were the most readily culturable of the marine bacteria and were therefore considered the most dominant. Of course, new information indicates that although vibrios are the dominant bacteria in many estuarine areas, there are other organisms that are very difficult to culture that are important as well.

When I took my first faculty position at Georgetown University, a friend of mine at NIH, Dr. John Feeley, suggested that I study *Vibrio cholerae*. What makes *V. cholerae* interesting is that it is a human pathogen of extremely great importance, yet resides naturally in estuaries and coastal areas of the world.

What is it like to study this organism?
Vibrio cholerae is naturally occurring (in the environment outside humans) and therefore can never be eradicated; it carries out important functions in the environment, and significant among these is its ability to digest chitin, the structural component of shellfish and many zooplankton. It is at once a "recycling agent" and a public health threat in the form of the massive epidemics of cholera that it causes.

You led an international collaboration in Bangladesh training women to avoid cholera by filtering water through sari cloth. How did the sari cloth filtration project come about?

It came about through collaboration with the International Centre for Diarrhoeal Disease Research, Bangladesh, located in Dhaka, Bangladesh, and the Matlab Field Laboratory, which is located in the village area of Matlab, Bangladesh. Our work had shown that *Vibrio cholerae* is associated with environmental zooplankton—namely, the copepod. The notion that the copepods are large and could be filtered out and therefore lead to reduced incidence of cholera was a result of my work on the vibrios and the relationships described by my students, notably Dr. Anwar Huq, who did his thesis on *Vibrio cholerae* attachment to copepods. Anwar Huq is now an associate professor at the University of Maryland.

An important collaborator was Nell Roberts, an outstanding public health microbiologist at Lake Charles, Louisiana. Nell, Professor Xu (a colleague from Qingdao, China), and I did the critical experiment showing the presence of *Vibrio cholerae* in water from which blue crabs had been harvested—the cause of an outbreak of cholera in Louisiana back in 1982. We were able to use fluorescent antibody to show the presence of the vibrio on copepods in the water.

From there, the idea of sari cloth came about in searching for a very inexpensive filter for use by village

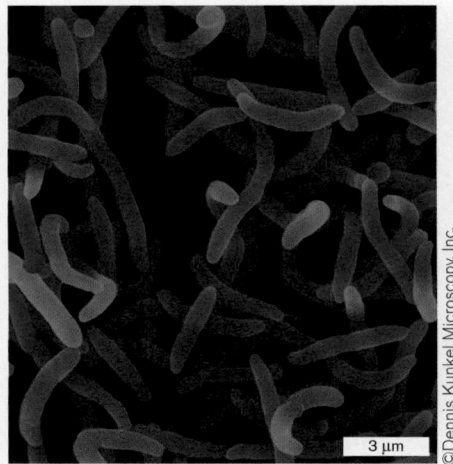
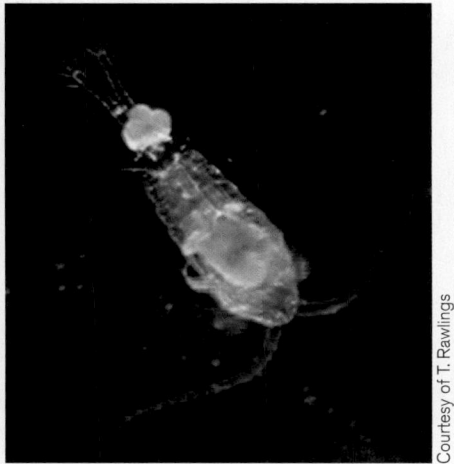

Vibrio cholerae bacteria (left) colonize copepods such as this one (right).

women in Bangladesh. We were able to show that folded sari cloth yielded a 20-micrometer (μm) mesh net. Because plankton are 200 μm or more in size, we could filter them out. The hypothesis that I came up with was that by removing the copepods and associated particulates, we could reduce cholera, which proved to be the case.

What are "viable but nonculturable" organisms?

Viable but nonculturable is a state into which Gram-negative microorganisms transform under adverse conditions in the environment. In this state, the bacteria are unable to be cultured, even though they remain viable and potentially pathogenic. Hence, they pose a public health risk, since routine tests done

A Bangladeshi woman filters water through sari cloth. Colwell's graduate student Anwar Huq compares the filtered and unfiltered water.

in a bacteriology laboratory would be negative for their presence.

What are the challenges of marine microbiology today? How does marine microbiology impact human health?

The challenges of marine microbiology today are to understand and catalog the extraordinary diversity of marine microorganisms. The world's oceans function in large part as a result of the activities of marine organisms. Marine microbiology impacts human health because of the many pathogens naturally occurring in the environment. But more than that, marine microorganisms may well be the cycling agent that keeps the blue planet inhabitable for humans. Marine microorganisms actively cycle carbon, nitrogen, phosphorus, and other elements in our oceans and even play a role in the weather by producing dimethyl sulfoxide (DMSO), which is involved in cloud formation and moisture condensation.

Why did you move to the National Science Foundation? What difference did you make as a microbiologist heading NSF?

I was asked by the president of the United States to serve as director of the National Science Foundation (NSF). It is a position appointed by the president and confirmed by the U.S. Senate. As a microbiologist, I was able to bring a molecular understanding of biology to the NSF, while as an interdisciplinary researcher, I was attuned to the needs of all aspects of science, from astronomy to physics. In the biological sciences, my major impact was in launching the Biocomplexity Initiative, which has been enormously productive and continues to yield new information on biological systems, including those of microbiology.

What do you think are the most exciting areas for students entering microbiology today?

Microbial diversity and microbial population studies are two emerging areas of huge interest that will lead to a better understanding of microbial evolution and development.

What advice do you have for today's students?

Develop an expertise as an undergraduate in some area of science, whether it be biology, chemistry, mathematics, physics, or some other area of science or engineering, and be creative and curious about other disciplines. The world of the future will be interdisciplinary and multidisciplinary.

How does your family relate to your work?

I have been happily married ever since I graduated from college! We have two daughters. One is a medical doctor (pediatrician). She was named an outstanding physician scholar and voted the best physician in her fellowship class by her colleagues. She also worked in Africa on delivery of health care to women in Tanzania for her PhD. We are equally proud of our other daughter, who earned a PhD in evolutionary biology and now works for the U.S. Geological Survey, cataloging rare plants in Yosemite National Park and Forest.

Chapter 1

Microbial Life: Origin and Discovery

1.1 From Germ to Genome: What Is a Microbe?

1.2 Microbes Shape Human History

1.3 Medical Microbiology

1.4 Microbial Ecology

1.5 The Microbial Family Tree

1.6 Cell Biology and the DNA Revolution

Life on Earth began early in our planet's history with microscopic organisms, or microbes. Microbial life has since shaped our atmosphere, our geology, and the energy cycles of all ecosystems. A human body contains ten times as many microbes as it does human cells, including numerous tiny bacteria on the skin and in the digestive tract. Throughout history, humans have had a hidden partnership with microbes ranging from food production and preservation to mining for precious minerals.

Yet throughout most of our history, humans were unaware that microbes even existed. To study these unseen organisms required a microscope, first developed in the 1600s. In the nineteenth century—the "golden age" of microbiology—microscopes revealed the tiny organisms at work in our bodies and in our ecosystems. The twentieth century saw the rise of microbes as the engines of biotechnology. Microbial discoveries led to recombinant DNA and revealed the secrets of the first sequenced genomes.

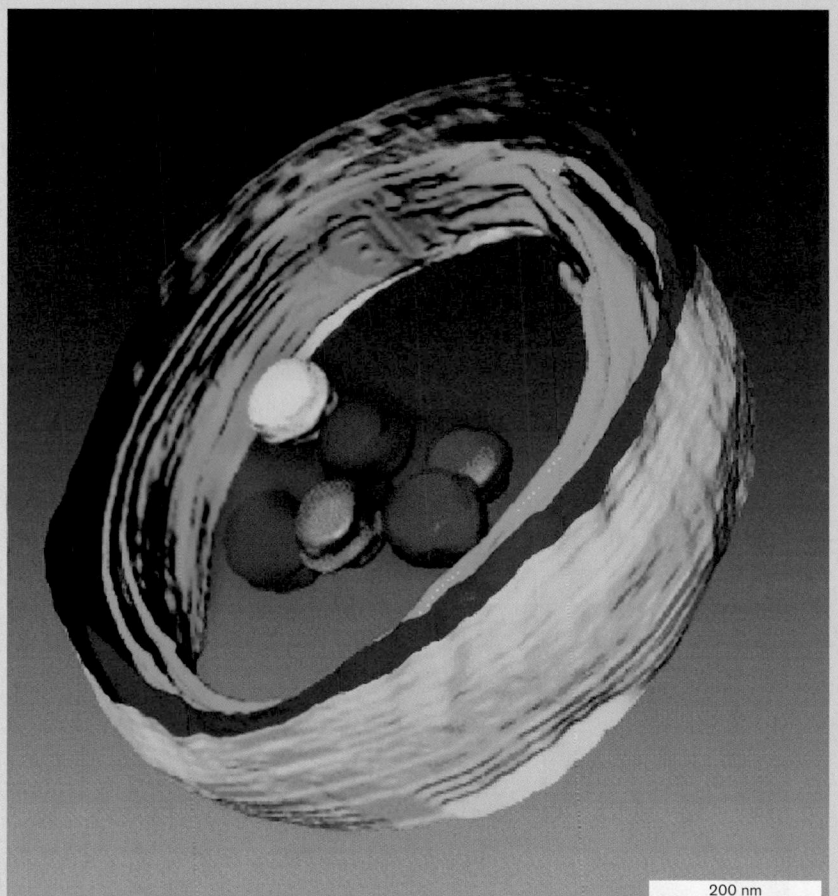

200 nm

Current Research Highlight

In the ocean, tiny photosynthetic bacteria called *Prochlorococcus marinus* fix carbon dioxide and produce much of Earth's oxygen. Cryo–electron tomography reveals a section through the cell's photosynthetic membranes (colorized green) and carbon dioxide-fixing carboxysomes (yellow, orange, purple). *Prochlorococcus* species are vulnerable to ultraviolet (UV) radiation, which damages their DNA; similar DNA damage in humans causes cancer. In 2010, an extremely UV-resistant strain of *Prochlorococcus* bacteria was discovered. The gene conferring UV resistance was revealed by study of the bacterial genome (total DNA sequence). Understanding UV resistance helps us preserve natural environments, and also yields clues to preventing cancer in humans. *Source:* Marcia S. Osburne. 2010. *Environmental Microbiology* **12**:1978. Photo from: Claire S. Ting et al. 2007. *J. Bacteriol.* **189**:4485. © 2007, American Society for Microbiology. All Rights Reserved.

In 2008, the *Phoenix* Mars lander arrived at the north pole of the planet Mars (**Fig. 1.1**). The lander carried scientific instruments to study the history of water in Martian soil and search for evidence of microbial life. Its robotic instruments tested the soil for life-supporting elements such as carbon, nitrogen, phosphorus, and hydrogen. The discovery of surface water in the form of frost supported the possible existence of living microbes.

Why do we care whether microbes exist on Mars? The discovery of life beyond Earth would fundamentally change how we see our place in the universe. The observation of Martian life could yield clues as to the origin of our own biosphere and expand our knowledge of the capabilities of living cells on our own planet. As of this writing, the existence of microbial life on Mars remains unknown, but here on Earth, many terrestrial microbes remain as mysterious as Mars. Barely 0.1% of the microbes in our biosphere can be cultured in the laboratory; even the digestive tract of a newborn infant contains species of bacteria unknown to science. Our "exploration rovers" for microbiology include, for example, new tools of microscopy and the sequencing of microbial DNA.

On Earth, the microscope reveals microbes throughout our biosphere, from the superheated black smoker vents at the ocean floor to the subzero ice fields of Antarctica. Bacteria such as *Escherichia coli* live in our intestinal tract, while algae and cyanobacteria turn ponds green (**Fig. 1.2**). Protists are the predators of the microscopic world. And viruses such as influenza virus cause disease, as do many bacteria and protists.

Yet before microscopes were developed in the seventeenth century, we humans were unaware of the unseen living organisms that surround us, that float in the air we breathe and the water we drink, and that inhabit our own bodies. Microbes generate the very air we breathe, including nitrogen gas and much of the oxygen and carbon dioxide. They fix nitrogen for plants, and they make vitamins, such as vitamin B_{12}. In the ocean, microbes produce biomass for the food web that feeds the fish we eat; and microbes consume toxic wastes such as oil from the Deepwater Horizon spill in 2010. At the same time, virulent pathogens take our lives. Despite all the advances of modern medicine and public health, microbial disease remains the number one cause of human mortality.

In the twentieth century, the science of microbiology exploded with discoveries, creating entire new fields such as genetic engineering. The promise—and pitfalls—were dramatized by Michael Crichton's best-selling science fiction novel and film *The Andromeda Strain* (1969; filmed in 1971). In *The Andromeda Strain*, scientists at a top-secret laboratory race to identify a deadly pathogen from outer space—or perhaps from a biowarfare lab (**Fig. 1.3A**). The film prophetically depicts the computerization of medical research, as well as the emergence of pathogens, such as the human immunodeficiency virus (HIV), that can yet defeat the efforts of advanced science.

Today, we discover surprising new kinds of microbes deep underground and in places previously thought uninhabitable, such as the hot springs of Yellowstone National Park (**Fig. 1.3B**). These microbes shape our biosphere and provide new tools that impact human society. For example, the use of heat-stable bacterial DNA polymerase (a DNA-replicating enzyme) in a technique called the polymerase chain reaction (PCR) allows us to detect minute amounts of DNA in traces of blood or fossil bone. Microbial technologies led us from the discovery of the double helix to the sequence of the human genome, the total genetic information that defines our species.

In Chapter 1, we introduce the concept of a microbe and the question of how microbial life originated. We then survey the history of human discovery of the role microbes play in disease and in our ecosystems. Finally, we address the exciting century of molecular microbiology, in which microbial genetics and genomics have transformed the face of modern biology and medicine.

NASA/JPL-Caltech/Univ. of Arizona/Texas A&M Univ.

Figure 1.1   Is there microbial life on Mars?   On July 14, 2008, the *Phoenix* lander photographed the north pole of the planet Mars. In the soil, the lander observed atmospheric water condensing to frost, supporting the possibility that microbial life exists on Mars.

1.1   From Germ to Genome: What Is a Microbe?

From early childhood, we hear that we are surrounded by microscopic organisms, or "germs," that we cannot see. What are microbes? Our modern concept of

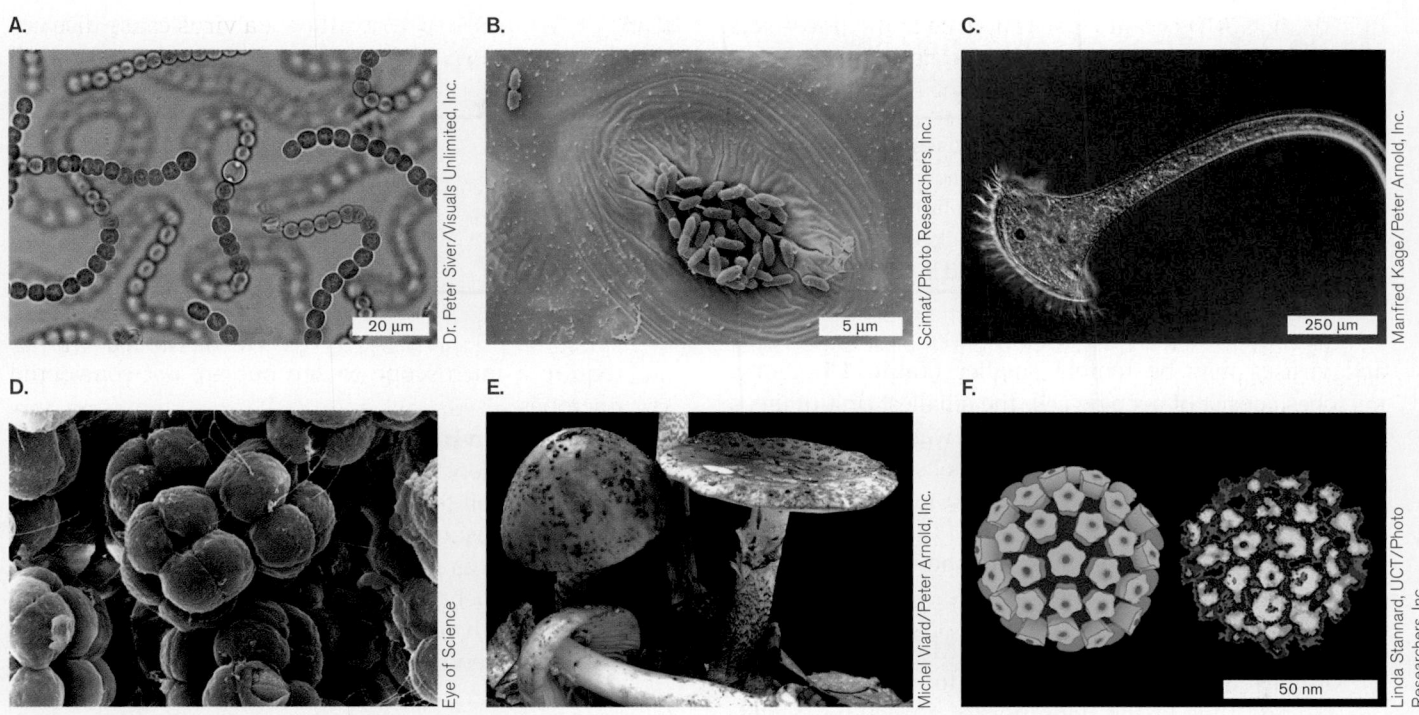

Figure 1.2 Representative microbes. **A.** Filamentous cyanobacteria produce oxygen for planet Earth (dark-field light micrograph). **B.** *Escherichia coli* bacteria colonize the stomata of a lettuce leaf cell (scanning electron microscopy). **C.** *Stentor* is a protist, a eukaryotic microbe. Cilia beat food into its mouth. **D.** Halophilic archaea, a form of life distinct from bacteria and eukaryotes, grow at extremely high salt concentrations. **E.** Mushrooms are multicellular fungi (eukaryotes). They serve the ecosystem as decomposers. **F.** Papillomavirus causes cancer and genital warts, an infectious disease commonly acquired by young adults (model based on electron microscopy).

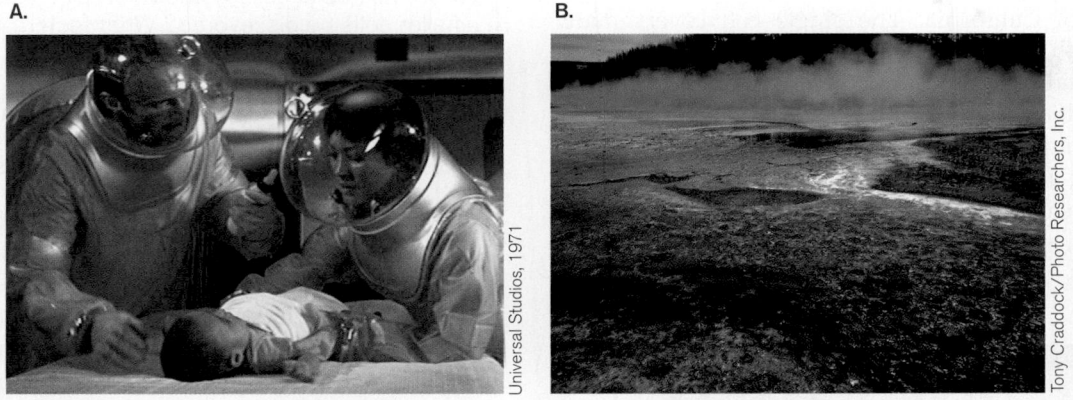

Figure 1.3 Microbial discovery: science fiction and science fact. **A.** In *The Andromeda Strain*, medical scientists try to feed a baby who was infected by a deadly pathogen from outer space. Though the details of the pathogen are imaginary, the film's approach to identifying the mystery organism captures the spirit of actual investigations of emerging diseases. **B.** Yellowstone National Park hot springs are surrounded by mats of colorful microbes that grow above 80°C in waters containing sulfuric acid. Bacteria discovered at Yellowstone produce enzymes used in the polymerase chain reaction (PCR), a technique of DNA amplification for making many copies of DNA.

a microbe has deepened through two major research tools: advanced microscopy and the sequencing of genomic DNA. Microscopy is covered in Chapter 2, and microbial genetics and genomics are presented in Chapters 7–12.

A Microbe Is a Microscopic Organism

A **microbe** is commonly defined as a living organism that requires a microscope to be seen. Microbial cells range in size from millimeters (mm) down to 0.2 micrometer (μm),

Table 1.1   Sizes of some microbes.

Microbe	Description	Approximate size
Varicella-zoster virus 1	Virus that causes chickenpox and shingles	100 nanometers (nm) = 10^{-7} meter (m)
Prochlorococcus	Photosynthetic marine bacteria	500 nm = 5×10^{-7} m
Rhizobium	Bacteria that fix N_2 in symbiosis with leguminous plants	1 micrometer (μm) = 10^{-6} m
Spirogyra	Filamentous algae found in aquatic habitat	40 μm = 4×10^{-5} m (cell width)
Pelomyxa (an ameba)	Protists found in solid or aquatic habitat	5 millimeters (mm)

and viruses may be tenfold smaller (**Table 1.1**). Some microbes consist of a single cell, the smallest unit of life, a membrane-enclosed compartment of water solution containing molecules that carry out metabolism. Each microbe contains in its genome the capacity to reproduce its own kind.

Our simple definition of a microbe, however, leaves us with contradictions.

- **Super-size microbial cells.** Most single-celled organisms require a microscope to render them visible and thus fit the definition of a microbe. Nevertheless, some species of protists and algae, and even some bacterial cells, are large enough to see with the naked eye. The marine sulfur bacterium *Thiomargarita namibiensis*, called the sulfur pearl of Namibia, grows as large as the head of a fruit fly (**Fig. 1.4**). Even more surprising, a single-celled plant, the "killer alga" *Caulerpa taxifolia*, spreads through the coastal waters of California. The single cell covers many acres with its leaflike cell parts.
- **Microbial communities.** Many microbes form complex multicellular assemblages, such as mushrooms, kelps, and biofilms. In these structures, cells are differentiated into distinct types that complement each other's function, as in multicellular organisms. And yet, some multicellular worms and arthropods

require a microscope to see but are *not* considered microbes.

- **Viruses.** A **virus** consists of a noncellular particle containing genetic material that takes over the metabolism of a cell to generate more virus particles. Some viruses consist of only a few molecular parts, whereas others, such as the mimivirus infecting amebas (also spelled "amoebae"), show the size and complexity of a cell. Although viruses are not fully functional cells, the mimivirus genome shows that it evolved from a cell.

NOTE: Each section contains questions to think about. These thought questions may have various answers. Possible responses are posted at the back of the book.

THOUGHT QUESTION 1.1   The minimum size of known microbial cells is about 0.2 μm. Could even smaller cells be discovered? What factors may determine the minimum size of a cell?

THOUGHT QUESTION 1.2   If viruses are not functional cells, are they truly "alive"?

In practice, our definition of a microbe derives from tradition as well as genetic considerations. In this book, we consider microbes to include **prokaryotes** (cells lacking a nucleus, including bacteria and archaea) as well as certain classes of **eukaryotes** (cells with a nucleus) that include simple multicellular forms: algae, fungi, and protists (**Fig. 1.5**). The bacteria, archaea, and eukaryotes—known as the three domains—diverged from a common ancestral cell. We also discuss viruses and related infectious particles (Chapters 6 and 11).

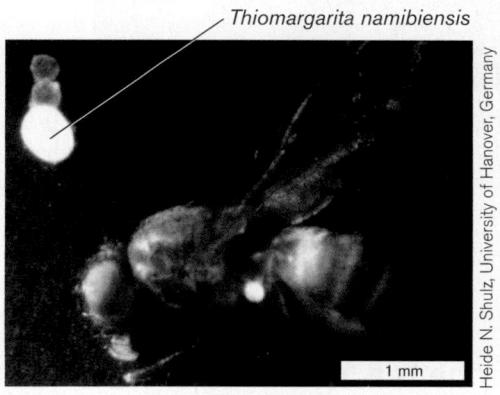

Thiomargarita namibiensis

1 mm

Heide N. Shulz, University of Hanover, Germany

Figure 1.4   Giant microbial cells.   The largest known bacterium, *Thiomargarita namibiensis*, a marine sulfur metabolizer, nearly the size of the head of a fruit fly. *Source:* Reprinted with permission from H. N. Shulz et al. 1999. *Science* **284**:493. © 2005 AAAS.

NOTE: The formal names of the three domains are **Bacteria**, **Archaea**, and **Eukarya**. Members of these domains are called **bacteria** (singular, **bacterium**), **archaea** (singular, **archaeon**), and **eukaryotes** (singular, **eukaryote**), respectively. The microbiology literature includes alternative spellings for some of these terms, such as "archaean" and "eucaryote."

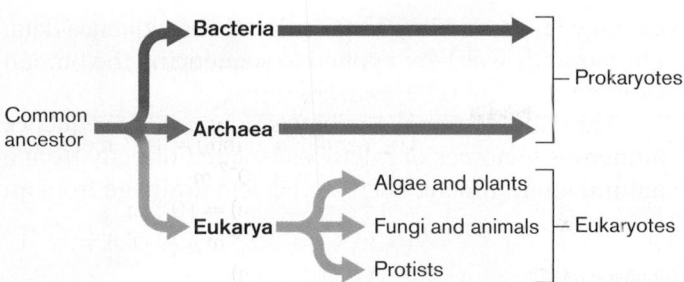

Figure 1.5 Three domains of life. Analysis of DNA sequences reveals the ancient divergence of three domains of living organisms: Bacteria and Archaea (both prokaryotes) and Eukarya (eukaryotes). The color code shown here is used throughout this book to indicate the three domains.

Microbial Genomes Are Sequenced

Our understanding of microbes has grown tremendously through the study of their genomes. A **genome** is the total genetic information contained in an organism's chromosomal DNA (**Fig. 1.6**). By determining the sequence of genes in a microbe's genome, we learn a lot about how that microbe grows and associates with other species. For example, if a microbe's genome includes genes for nitrogenase, a nitrogen-fixing enzyme, that microbe probably can fix nitrogen from the atmosphere into compounds that plants can assimilate into protein. And by comparing DNA sequences, we can measure the degree of relatedness between different species based on the time since they diverged from a common ancestor.

Historically, the first genomes to be sequenced were those of viruses. The first genome whose complete DNA sequence was determined was that of a bacteriophage (a virus that infects bacteria), bacteriophage φX174. The DNA sequence of φX174 was determined in 1977 by Fred Sanger (**Fig. 1.7A**), who shared the 1980 Nobel Prize in Chemistry with Walter Gilbert and Paul Berg for developing the method of DNA sequence analysis. The genome of bacteriophage φX174 includes over

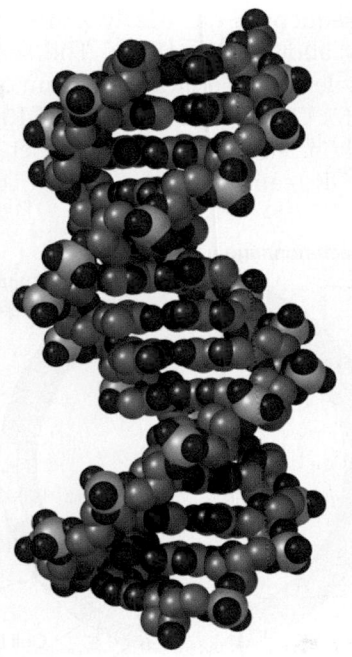

Figure 1.6 DNA. The sequence of base pairs in DNA encodes all the genetic information of an organism.

5,000 base pairs, with just ten genes encoding proteins (**Fig. 1.8A**). Its genome is so compact that several gene sequences actually overlap, sharing the same segment of nucleotides (for example, genes C and K).

WWW   The Nobel Prize website presents the lectures and autobiographies of all Nobel Prize winners, including many who were awarded prizes for advances in microbiology.
microbiology2.com/links

Nearly two decades passed before scientists completed the first genome sequence of a cellular microbe, *Haemophilus influenzae*, a bacterium that causes ear infections and meningitis in children (**Fig. 1.8B**). The strain of

Figure 1.7 Microbial genome sequencers. A. Fred Sanger, who shared the 1980 Nobel Prize in Chemistry for devising the method of DNA sequence analysis that is the basis of genome sequencing. He is reading sequence data from bands of DNA separated by electrophoresis. **B.** Claire Fraser-Liggett, past president of The Institute for Genomic Research (TIGR), which completed the sequences of *H. influenzae* and many other microbial genomes.

A.

B.

H. influenzae sequenced has nearly 2 million base pairs, which specify about 1,700 genes. The sequence of *H. influenzae* was determined by a large team of scientists at The Institute for Genomic Research (TIGR) led by Craig Venter, Hamilton Smith, and Claire Fraser-Liggett (**Fig. 1.7B**). The TIGR team devised a special computational strategy for assembling large amounts of sequence data. This strategy was later applied to sequencing the human genome.

The computational strategy is now used to sequence numerous genomes of microbes isolated directly from a natural environment, such as the acid drainage from an

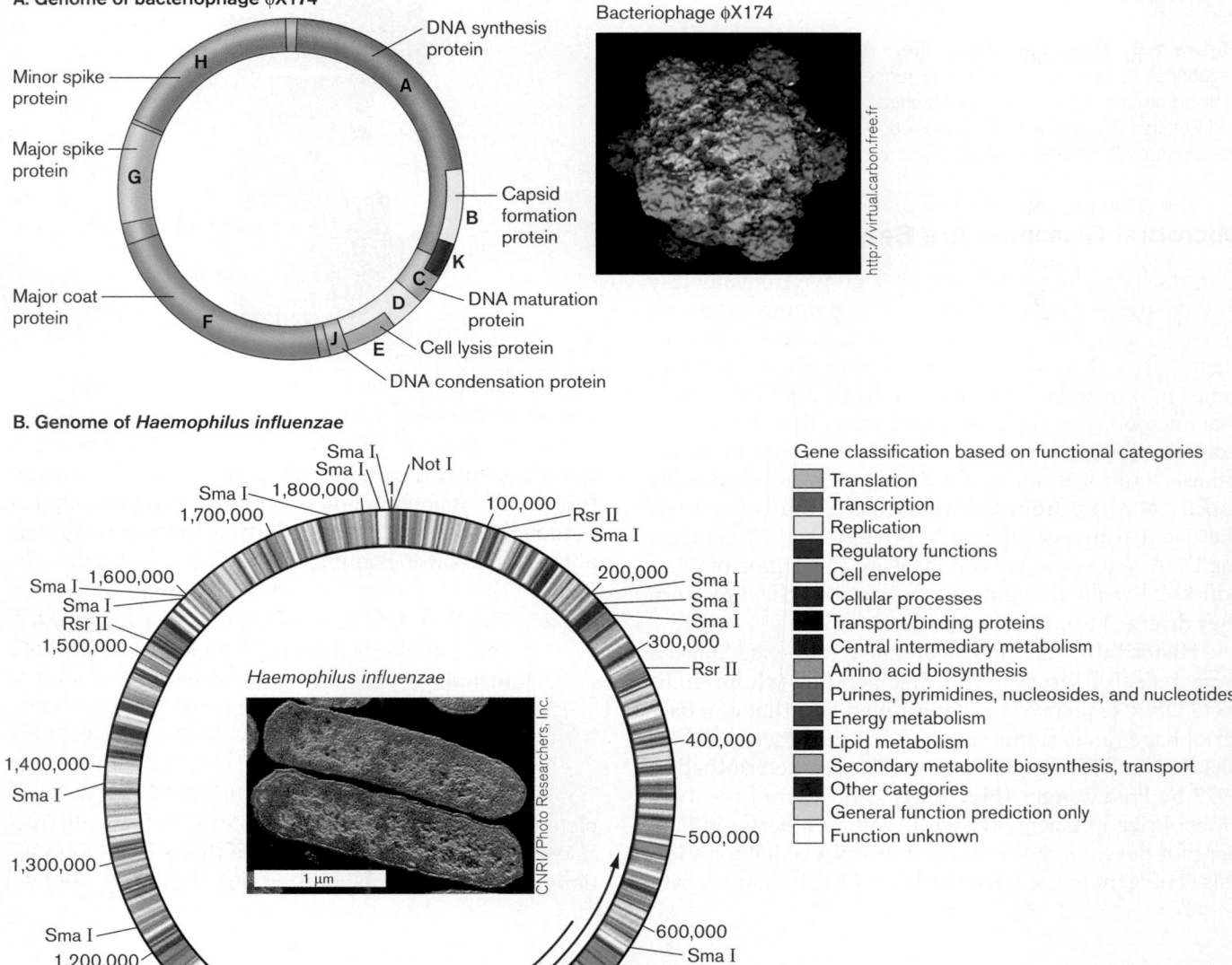

A. Genome of bacteriophage φX174

B. Genome of *Haemophilus influenzae*

Figure 1.8 The first sequenced genomes.   A. The first organism whose genome sequence was determined was bacteriophage φX174, a virus that grows in *Escherichia coli* (virus diameter, 27 nm). The entire DNA sequence of φX174 contains 5,386 base pairs specifying only ten genes (here labeled A–H and J–K), nine of whose functions have since been determined. Note the highly compact genome, with several overlapping genes. **B.** The genome of *Haemophilus influenzae*, a bacterium that causes ear infections and meningitis, was the first DNA sequence completed for a cellular organism (inset, colorized electron micrograph). The genome of *H. influenzae* contains nearly 2 million base pairs specifying approximately 1,743 genes, which are expressed to make protein and RNA products. The annotated sequence of the genome appears on the website of the National Center for Biotechnology Information. Colored bars indicate gene sequences throughout.

iron mine. The environmental collection of sequences is called a **metagenome**. The first metagenome of an acid mine was sequenced by Jill Banfield and co-workers at the University of California at Berkeley in 2004. Since then, metagenomes have been sequenced for microbial communities of medical interest, such as that of the human colon. Metagenome sequences reveal thousands of unknown species (discussed further in Chapter 21).

www | The National Center for Biotechnology Information (NCBI) provides free access to all published genome sequences. *microbiology2.com/links*

The growing availability of sequenced genomes at universities and in industry has generated the new field of comparative genomics, which involves the systematic comparison of all genomic sequences of living species, ranging from microbes to *Homo sapiens*. Comparative genomics reveals a set of core genes shared by all organisms, further evidence that all life on Earth shares a common ancestry.

1.2 Microbes Shape Human History

Throughout most of human history, we were unaware of the microbial world. Microorganisms have shaped human culture since our earliest civilizations. Yeasts and bacteria have made foods such as bread and cheese (**Fig. 1.9A**), as well as alcoholic beverages (discussed in Chapter 16). "Rock-eating" bacteria known as lithotrophs leached copper and other metals from ores exposed by mining, enabling ancient human miners to obtain these metals. The lithotrophic oxidation of minerals for energy generates strong acid, which accelerates breakdown of the ore. Today, about 20% of the world's copper, as well as some uranium and zinc, is produced by bacterial

leaching. Unfortunately, microbial acidification also consumes the stone of ancient monuments (**Fig. 1.9B**)—a process intensified by airborne acidic pollution. Management of microbial corrosion is an important field of applied microbiology.

As humans became aware of microbes, our relationship with the microbial world changed in important ways (**Table 1.2**, pages 14–15). Early microscopists in the seventeenth and eighteenth centuries formulated key concepts of microbial existence, including their means of reproduction and death. In the nineteenth century, the "golden age" of microbiology, key principles of disease pathology and microbial ecology were established that scientists still use today. This period laid the foundation for modern science, in which genetics and molecular biology provide powerful tools for scientists to manipulate microorganisms for medicine and industry.

Microbial Disease Devastates Human Populations

Throughout history, microbial diseases such as tuberculosis and leprosy have profoundly affected human demographics and cultural practices (**Fig. 1.10**). The bubonic plague, which wiped out a third of Europe's population in the fourteenth century, was caused by *Yersinia pestis*, a bacterium spread by rat fleas. Ironically, the plague-induced population decline enabled the social transformation that led to the Renaissance, a period of unprecedented cultural advancement. In the nineteenth century, the bacterium *Mycobacterium tuberculosis* stalked overcrowded cities, and tuberculosis became so common that the pallid appearance of tubercular patients became a symbol of tragic youth in European literature. Today, societies throughout the world have been profoundly shaped by the epidemic of acquired immunodeficiency syndrome (AIDS), caused by the human immunodeficiency virus (HIV). More than 36 million people are living with HIV infection today, and each year 2 million die of AIDS.

Figure 1.9 Production and destruction by microbes.
A. Roquefort cheeses ripening in France. **B.** Statue decaying from the action of lithotrophic microbes. The process is accelerated by acid rain. (Cathedral of Cologne, Germany.)

A.

Bettmann/Corbis

B.

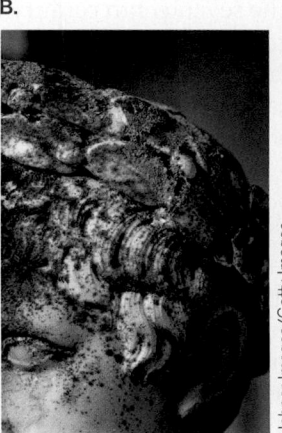

Johner Images/Getty Images

Historians traditionally emphasize the role of warfare in shaping human destiny; and the brilliance of leaders or the advantage of new technology, in determining which civilizations rise or fall. Yet the fate of human societies has often been determined by microbes. For example, much of the native population of North America was exterminated by smallpox introduced by European invaders. Throughout history, more soldiers have died of microbial infections than of wounds in battle. The significance of disease in warfare was first recognized by the British nurse and statistician Florence Nightingale (1820–1910) (**Fig. 1.11A**).

Better known as the founder of professional nursing, Nightingale also founded the science of medical statistics. She used methods invented by French statisticians to demonstrate the high mortality rate due to disease among British soldiers during the Crimean War. To show the deaths of soldiers due to various causes, she devised the "polar area chart" (**Fig. 1.11B**). Blue wedges represent deaths due to infectious disease, red wedges represent deaths due to wounds, and black wedges represent all other causes of death. Infectious disease accounts for more than half of all mortality.

Before Nightingale's study, no one understood the impact of disease on armies, or on other crowded populations, such as cities. Nightingale's statistics convinced the British government to improve army living conditions and to upgrade the standards of army hospitals. In modern epidemiology, statistical analysis continues to serve as a crucial tool in determining the causes of disease.

www | The CDC (Centers for Disease Control and Prevention), in Atlanta, is the U.S. agency for medical information and epidemiology. *microbiology2.com/links*

Figure 1.10   Microbial disease in history and culture. **A.** Medieval church procession to ward off the Black Death (bubonic plague). **B.** The AIDS Memorial Quilt spread before the Washington Monument. Each panel of the quilt memorializes an individual who died of AIDS.

Microscopes Reveal the Microbial World

The seventeenth century was a time of growing inquiry and excitement about the "natural magic" of science and patterns of our world, such as the laws of gravitation and motion formulated by Isaac Newton (1642–1727). Robert Boyle (1627–1691) performed the first controlled experiments on the chemical conversion of matter. Physicians attempted new treatments for disease involving the application of "stone and minerals" (that is, the application of chemicals), what today we would call chemotherapy. Minds were open to

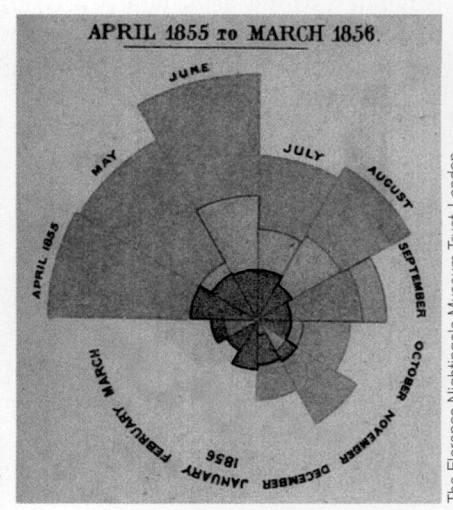

Figure 1.11   Florence Nightingale, founder of medical statistics.   A. Florence Nightingale was the first to use medical statistics to demonstrate the significance of mortality due to disease. **B.** Nightingale's polar area chart of mortality data during the Crimean War.

consider the astounding possibility that our surroundings, indeed our very bodies, were inhabited by tiny living beings.

Robert Hooke observes the microscopic world. The first microscopist to publish a systematic study of the world as seen under a microscope was Robert Hooke (1635–1703). As curator of experiments for the Royal Society of London, Hooke built the first compound microscope—a magnifying instrument containing two or more lenses that multiply their magnification in series. With his microscope, Hooke observed biological materials such as nematode "vinegar eels," mites, and mold filaments, illustrations of which he published in *Micrographia* (1665), the first publication that illustrated objects observed under a microscope (**Fig. 1.12**).

Hooke was the first to observe distinct units of living material, which he called "cells." Hooke first named the units cells because the shape of hollow cell walls in a slice of cork reminded him of the shape of monks' cells in a monastery. But his crude lenses achieved at best 30-fold power (30×), so he never observed single-celled organisms.

Antonie van Leeuwenhoek observes bacteria with a single lens. Hooke's *Micrographia* inspired other microscopists, including Antonie van Leeuwenhoek (1632–1723), who became the first individual to observe single-celled microbes (**Fig. 1.13A**). As a young man, Leeuwenhoek lived in the Dutch city of Delft, where he worked as a cloth draper, a profession that introduced him to magnifying glasses. (The magnifying glasses were used to inspect the quality of the cloth, enabling the worker to count the number of threads.) Later in life, he

Figure 1.12 Hooke's *Micrographia*. An illustration of mold sporangia, drawn by Hooke in 1665, from his observations of objects using a compound microscope.

took up the hobby of grinding ever stronger lenses to see into the world of the unseen.

Leeuwenhoek ground lenses stronger than Hooke's, which he used to build single-lens magnifiers, complete with sample holder and focus adjustment (**Fig. 1.13B**). First he observed insects, including lice and fleas; then the relatively large single cells of protists and algae; then ultimately bacteria. One day he applied his microscope to observe matter extracted from between his teeth. He wrote, "To my great surprise [I] perceived that the aforesaid matter contained very many small living Animals, which moved themselves very extravagantly."

Over the rest of his life, Leeuwenhoek recorded page after page on the movement of microbes, reporting their size and shape so accurately that in many cases we can determine the species he observed (**Fig. 1.13C**).

A.　　　　　　　　　　　**B.**　　　　　　　　　　　**C.**

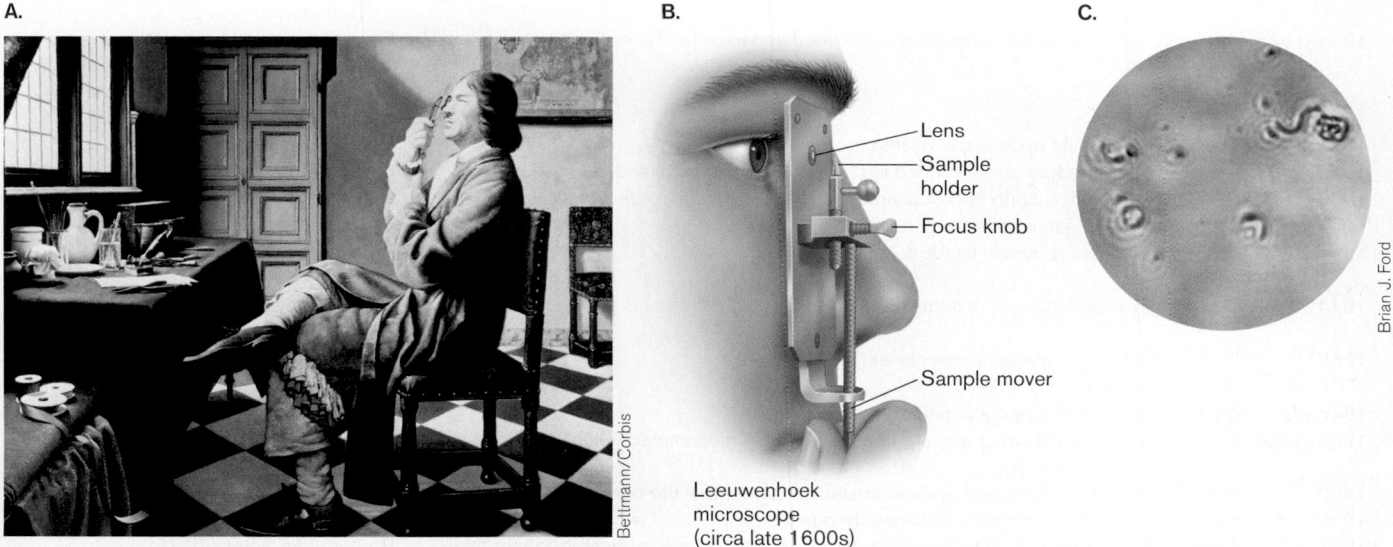

Leeuwenhoek
microscope
(circa late 1600s)

Figure 1.13 Antonie van Leeuwenhoek. A. A portrait of Leeuwenhoek, the first person to observe individual microbes.
B. "Microscope" (magnifying glass) used by Leeuwenhoek. **C.** Spiral bacteria viewed through a replica of Leeuwenhoek's instrument.

Table 1.2 Microbes and human history.

Date	Microbial discovery	Discoverer(s)
Microbes impact human culture without detection		
10,000 BCE	Food and drink are produced by microbial fermentation.	Egyptians, Chinese, and others
1500 BCE	Tuberculosis, polio, leprosy, and smallpox are evident in mummies and tomb art.	Egyptians
50 BCE	Copper is recovered from mine water acidified by sulfur-oxidizing bacteria.	Roman metal workers under Julius Caesar
1546 AD	Syphilis and other diseases are observed to be contagious.	Girolamo Fracastoro (Padua)
Early microscopy and the origin of microbes		
1676	Microbes are observed under a microscope.	Antonie van Leeuwenhoek (Netherlands)
1688	Spontaneous generation is disproved for maggots.	Francesco Redi (Italian)
1717	Smallpox is prevented by inoculation of pox material, a rudimentary form of immunization.	Turkish women taught Lady Mary Montagu, who brought the practice to England
1765	Microbe growth in organic material is prevented by boiling in a sealed flask.	Lazzaro Spallanzani (Padua)
1798	Cowpox vaccination prevents smallpox.	Edward Jenner (England)
1835	Fungus causes disease in silkworms (first pathogen to be demonstrated in animals).	Agostino Bassi de Lodi (Italy)
1847	Chlorine as antiseptic wash for doctor's hands decreases pathogens.	Ignaz Semmelweis (Hungary)
1881	Bacterial spores survive boiling but are killed by cyclic boiling and cooling.	John Tyndall (Ireland)
"Golden age" of microbiology: principles and methods established		
1855	Statistical correlation is shown between sanitation and mortality (Crimean War).	Florence Nightingale (England)
1857	Microbial fermentation produces lactic acid or alcohol.	Louis Pasteur (France)
1864	Microbes fail to appear spontaneously, even in the presence of oxygen.	Louis Pasteur (France)
1866	Microbes are defined as a class distinct from animals and plants.	Ernst Haeckel (Germany)
1867	Antisepsis during surgery prevents patient death.	Joseph Lister (England)
1877	Bacteria are a causative agent in developing anthrax.	Robert Koch (Germany)
1881	The first artificial vaccine is developed (against anthrax).	Louis Pasteur (France)
1882	First pure culture of colonies is grown on solid medium, *Mycobacterium tuberculosis*.	Robert Koch (Germany)
1884	Koch's postulates are published, based on anthrax and tuberculosis.	Robert Koch (Germany)
1884	Gram stain is devised to distinguish bacteria from human cells.	Hans Christian Gram (Netherlands)
1886	Intestinal bacteria include *Escherichia coli*, the future model organism.	Theodor Escherich (Austria)
1889	Bacteria oxidize iron and sulfur (lithotrophy).	Sergei Winogradsky (Russia)
1889	Bacteria isolated from root nodules are proposed to fix nitrogen.	Martinus Beijerinck (Netherlands)
1899	The concept of a virus is proposed to explain tobacco mosaic disease.	Martinus Beijerinck (Netherlands)
Cell biology, biochemistry, and genetics		
1908	Antibiotic chemicals are synthesized and identified (chemotherapy).	Paul Ehrlich (USA)
1911	Cancer in chickens can be caused by a virus.	Peyton Rous (USA)
1917	Bacteriophages are recognized as viruses that infect bacteria.	Frederick Twort (England) and Félix d'Herelle (France)
1924	The ultracentrifuge is invented and used to measure the size of proteins.	Theodor Svedberg (Sweden)
1928	*Streptococcus pneumoniae* bacteria are transformed by a genetic material from dead cells.	Frederick Griffith (England)
1929	Penicillin, the first widely successful antibiotic, is made by a fungus. The molecule is isolated in 1941.	Alexander Fleming (Scotland), Howard Florey (Australia), and Ernst Chain (Germany)
1933–1945	The transmission electron microscope is invented and used to observe cells.	Ernst Ruska and Max Knoll, inventors (Germany); first cells observed by Albert Claude (Belgium), Christian de Duve (Belgium), and George Palade (USA)
1937	The tricarboxylic acid cycle is discovered.	Hans Krebs (England)
1938	The microbial "kingdom" is subdivided into prokaryotes (Monera) and eukaryotes.	Herbert Copeland (USA)
1938	*Bacillus thuringiensis* spray is produced as the first bacterial insecticide.	Insecticide manufacturers (France)
1941	One gene encodes one enzyme in *Neurospora*.	George Beadle and Edward Tatum (USA)
1941	Poliovirus is grown in human tissue culture.	John Enders, Thomas Weller, and Frederick Robbins (USA)
1944	DNA is the genetic material that transforms *S. pneumoniae*.	Oswald Avery, Colin MacLeod, and Maclyn McCarty (USA)
1945	The bacteriophage replication mechanism is elucidated.	Salvador Luria (Italy) and Max Delbrück (Germany), working in the USA
1946	Bacteria transfer DNA by conjugation.	Edward Tatum and Joshua Lederberg (USA)
1946–1956	X-ray diffraction crystal structures are obtained for the first complex biological molecules, penicillin and vitamin B_{12}.	Dorothy Hodgkin, John Bernal, and co-workers (England)
1950	Anaerobic culture technique is devised to study anaerobes of the bovine rumen.	Robert Hungate (USA)
1950	Bacteria can carry latent bacteriophages (lysogeny).	André Lwoff (France)
1951	Transposable elements are discovered in maize and later shown in bacteria, where they play key roles in evolution.	Barbara McClintock (USA)
1952	DNA is injected into a cell by a bacteriophage.	Martha Chase and Alfred Hershey (USA)

Table 1.2 Microbes and human history (*continued*)

Date	Microbial discovery	Discoverer(s)
Molecular biology and recombinant DNA		
1953	The overall structure of DNA is a double helix, based on X-ray diffraction analysis.	Rosalind Franklin and Maurice Wilkins (England)
1953	Double-helical DNA consists of antiparallel chains connected by the hydrogen bonding of AT and GC base pairs.	James Watson (USA) and Francis Crick (England)
1959	Expression of the messenger RNA for the *E. coli lac* operon is regulated by a repressor protein.	Arthur Pardee (England) and François Jacob and Jacques Monod (France)
1960	Radioimmunoassay for detection of biomolecules is developed.	Rosalyn Yalow and Solomon Bernson (USA)
1961	The chemiosmotic theory, which states that biochemical energy is stored in a transmembrane proton gradient, is proposed and tested.	Peter Mitchell and Jennifer Moyle (England)
1966	The genetic code by which DNA information specifies protein sequence is deciphered.	Marshall Nirenberg, H. Gobind Khorana, and others (USA)
1967	Bacteria can grow at temperatures above 80°C in hot springs at Yellowstone National Park.	Thomas Brock (USA)
1968	Serial endosymbiosis is proposed to explain the evolution of mitochondria and chloroplasts.	Lynn Margulis (USA)
1969	Retroviruses contain reverse transcriptase, which copies RNA to make DNA.	Howard Temin, David Baltimore, and Renato Dulbecco (USA)
1972	Inner and outer membranes of Gram-negative bacteria (*Salmonella*) are separated by ultracentrifugation.	Mary Osborn (USA)
1973	A recombinant DNA molecule is made in vitro (in a test tube).	Stanley Cohen, Annie Chang, Robert Helling, and Herbert Boyer (USA)
1974	The bacterial flagellum is driven by a rotary motor.	Howard Berg, Michael Silverman, and Melvin Simon (USA)
1975	mRNA-rRNA base pairing initiates protein synthesis in *E. coli*.	Joan Steitz and Karen Jakes (USA) and Lynn Dalgarno and John Shine (Australia)
1975	The dangers of recombinant DNA are assessed at the Asilomar Conference.	Paul Berg, Maxine Singer, and others (USA)
1975	Monoclonal antibodies are produced indefinitely in tissue culture by hybridomas, antibody-producing cells fused to cancer cells.	George Kohler and Cesar Milstein (USA)
1977	A DNA sequencing method is invented and used to sequence the first genome of a virus.	Fred Sanger, Walter Gilbert, and Allan Maxam (USA)
1977	Archaea are a third domain of life, the others being eukaryotes and bacteria.	Carl Woese (USA)
1978	The first protein catalog is compiled for *E. coli* based on 2-D gels.	Fred Neidhardt, Peter O'Farrell, and colleagues (USA)
1978	Biofilms are a major form of existence of microbes.	William Costerton and others (Canada)
1979	Smallpox is declared eliminated—the culmination of worldwide efforts of immunology, molecular biology, and public health.	The World Health Organization
Genomics, structural biology, and molecular ecology		
1981	Invention of the polymerase chain reaction (PCR) makes available large quantities of DNA.	Kary Mullis (USA)
1981–1986	Self-splicing RNA is discovered in the protist *Tetrahymena*.	Thomas Cech and Sidney Altman (USA)
1982	Archaea are discovered with optimal growth above 100°C.	Karl Stetter (Germany)
1982	Viable but nonculturable bacteria contribute to ecology and pathology.	Rita Colwell, Norman Pace, and others (USA)
1982	Prions, infectious agents consisting solely of protein, are characterized.	Stanley Prusiner (USA)
1983	Human immunodeficiency virus (HIV) is implicated in the development of AIDS.	Luc Montagnier, Robert Gallo, and others (France and USA)
1983	Genes are introduced into plants by use of *Agrobacterium tumefaciens* plasmid vectors.	Eugene Nester, Mary-Dell Chilton, and colleagues (USA)
1984	Acid-resistant *Helicobacter pylori* grow in the stomach, where they cause gastritis.	Barry Marshall and Robin Warren (Australia)
1987	*Geobacter* bacteria that can generate electricity are discovered.	Derek Lovley and colleagues (USA)
1988	Earth's most abundant marine phototroph is *Prochlorococcus*.	Sallie Chisholm and colleagues (USA)
1993	A giant bacterium (*Epulopiscium*)—large enough to see—is identified.	Esther Angert and Norman Pace (USA)
1995	The first genome is sequenced for a cellular organism, *Haemophilus influenzae*.	Craig Venter, Hamilton Smith, Claire Fraser-Liggett, and others (USA)
2001	The ribosome structure is obtained at near-atomic level by X-ray diffraction.	Marat Yusupov, Harry Noller, and colleagues (USA)
2004	Mimivirus genome shows that large DNA viruses evolved from cells.	Didier Raoult and colleagues (France)
2006	First metagenomes are sequenced, from Iron Mountain acid mine drainage and from the Sargasso Sea.	Jill Banfield, Craig Venter, and others (USA)
2006	Vaccine prevents genital human papillomavirus (HPV), the most common sexually transmitted infection.	Patented by Georgetown University and other institutions (USA and Australia)
2009	Self-sustained replication of an RNA enzyme shows how life could have originated as an "RNA world."	Gerald Joyce and Tracey Lincoln (USA)

He performed experiments, comparing, for example, the appearance of "small animals" from his teeth before and after drinking hot coffee. The disappearance of microbes from his teeth after drinking a hot beverage suggested that heat killed microbes—a profoundly important principle for the study and control of microbes ever since.

Ironically, Leeuwenhoek is believed to have died of a disease contracted from sheep whose bacteria he observed. Historians have often wondered why it took so many centuries for Leeuwenhoek and his successors to determine the link between microbes and disease. Although observers such as Agostino Bassi de Lodi (1773–1856) noted isolated cases of microbes associated with pathology (see **Table 1.2**), the very ubiquity of microbes—most of them actually harmless—may have obscured their more deadly roles. In addition, it was hard to distinguish between microbes and the single-celled components of the human body, such as blood cells and sperm. It was not until the nineteenth century that human tissues could be distinguished from microbial cells by the application of differential chemical stains (discussed in Chapter 2).

> **THOUGHT QUESTION 1.3** Why do you think it took so long for humans to connect microbes with infectious disease?

Spontaneous Generation: Do Microbes Have Parents?

The observation of microscopic organisms led priests and philosophers to wonder where they came from. In the eighteenth century, scientists and church leaders intensely debated the question of **spontaneous generation**, the theory that living creatures such as maggots could arise spontaneously, without parental organisms. Chemists of the day tended to support spontaneous generation, as it appeared similar to the changes in matter that could occur when chemicals were mixed. Christian church leaders, however, supported the biblical view that all organisms have "parents" going back to the first week of creation.

The Italian priest Francesco Redi (1626–1697) showed that maggots in decaying meat were the offspring of flies. Meat kept in a sealed container, excluding flies, did not produce maggots. Thus, Redi's experiment argued against spontaneous generation for macroscopic organisms. The meat still putrefied, however, producing microbes that seemed to arise "without parents."

To disprove spontaneous generation of microbes, another Italian priest, Lazzaro Spallanzani (1729–1799),

showed that a sealed flask of meat broth sterilized by boiling failed to grow microbes. Spallanzani also noticed that microbes often appeared in pairs. Were these two parental microbes coupling to produce offspring, or did one microbe become two? By long and tenacious observation, Spallanzani watched a single microbe grow in size until it split in two. Thus, he demonstrated cell fission, the process by which cells arise by the splitting of preexisting cells.

Even Spallanzani's experiments, however, did not put the matter to rest. Proponents of spontaneous generation argued that the microbes in the priest's flask lacked access to oxygen and therefore could not grow. The pursuit of this question was left to future microbiologists, including the famous French microbiologist Louis Pasteur (1822–1895) (**Fig. 1.14A**). In addressing spontaneous generation and related questions, Pasteur and his contemporaries laid the foundations for modern microbiology.

Louis Pasteur reveals the biochemical basis of microbial growth. Pasteur began his scientific career as a chemist and wrote his doctoral thesis on the

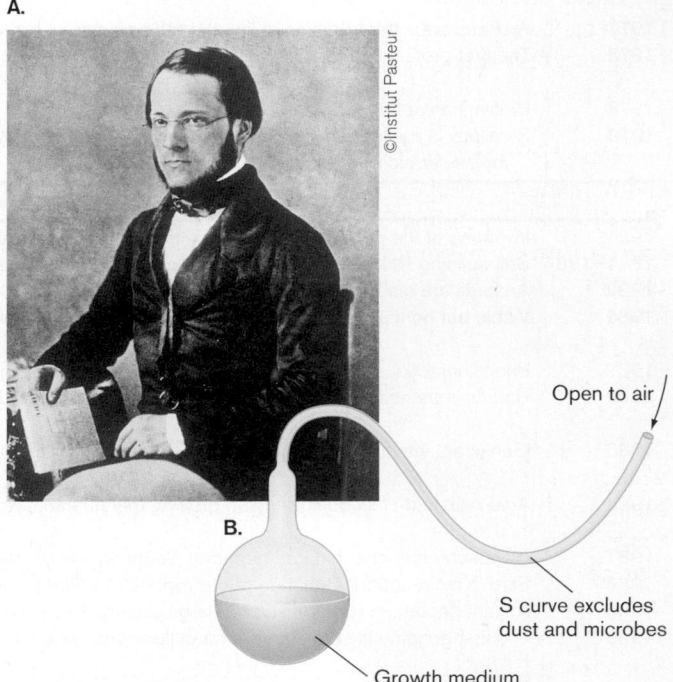

A.

©Institut Pasteur

B.

Open to air

S curve excludes dust and microbes

Growth medium

Figure 1.14   Louis Pasteur, founder of medical microbiology and immunology.   A. Pasteur's contributions to the science of microbiology and immunology earned him lasting fame. **B.** Swan-necked flask. Pasteur showed that in such a flask, after boiling, the contents remain free of microbial growth, despite access to air.

structure of organic crystals. He discovered the fundamental chemical property of chirality, the fact that some organic molecules exist in two forms that differ only by mirror symmetry. In other words, the two structures are mirror images of one another, like the right and left hands. Pasteur found that when microbes were cultured on a nutrient substance containing both mirror forms, only one mirror form was consumed. He concluded that the metabolic preference for one mirror form was a fundamental property of life. Subsequent research has confirmed that most biological molecules, such as DNA and proteins, occur in only one of their mirror forms.

As a chemist, Pasteur was asked to help with a widespread problem encountered by French manufacturers of wine and beer. The production of alcoholic beverages is now known to occur by **fermentation**, a process by which microbes gain energy by converting sugars into alcohol. In the time of Pasteur, however, the conversion of grapes or grain to alcohol was believed to be a spontaneous chemical process. No one could explain why some fermentation mixtures produced vinegar (acetic acid) instead of alcohol. Pasteur discovered that fermentation is actually caused by living yeast, a single-celled fungus. In the absence of oxygen, yeast produces alcohol as a terminal waste product. But when the yeast culture is contaminated with bacteria, the bacteria outgrow the yeast and produce acetic acid instead of alcohol. (Fermentative metabolism is discussed further in Chapter 13.)

Pasteur's work on fermentation led him to test a key claim made by proponents of spontaneous generation. The proponents claimed that Spallanzani's failure to find spontaneous appearance of microbes was due to lack of oxygen. From his studies of yeast fermentation, Pasteur knew that some microbial species do not require oxygen for growth. So he devised an unsealed flask with a long, bent "swan neck" that admitted air but kept the boiled contents free of microbes (**Fig. 1.14B**). The famous swan-necked flasks remained free of microbial growth for many years; but when a flask was tilted to enable contact of broth with microbe-containing dust, growth occurred immediately. Thus, Pasteur disproved that lack of oxygen was the reason for the failure of spontaneous generation in Spallanzani's flasks.

But even Pasteur's work did not prove that microbial growth requires preexisting microbes. The Irish scientist John Tyndall (1820–1893) attempted the same experiment as Pasteur, but sometimes found the opposite result. Tyndall found that the broth sometimes gave rise to microbes, no matter how long it was sterilized by boiling. The microbes appear because some kinds of organic matter, particularly hay infusion, are contaminated with a heat-resistant form of bacteria called endospores (or spores). The spore form can be eliminated only by repeated cycles of boiling and resting, in which the spores germinate to the growing, vegetative form that is killed at 100°C.

It was later discovered that endospores could be killed by boiling under pressure, as in a pressure cooker, which generates higher temperatures than can be obtained at atmospheric pressure. The steam pressure device called the **autoclave** became a standard method for the sterilization of materials required for the controlled study of microbes. (Microbial control and antisepsis are discussed further in Chapter 5.)

Although spontaneous generation has been discredited as a continual source of microbes, at some point in the past the first living organisms must have originated from nonliving materials. The origin of life is explored in **Special Topic 1.1**, and discussed further in Chapter 17.

TO SUMMARIZE:

- **Microbes affected human civilization** for centuries before humans guessed at their existence through their contributions to our environment, food and drink production, and infectious diseases.
- **Robert Hooke and Antonie van Leeuwenhoek** were the first to record observations of microbes through simple microscopes.
- **Spontaneous generation** is the theory that microbes arise spontaneously, without parental organisms. **Lazzaro Spallanzani** showed that microbes arise from preexisting microbes and demonstrated that heat sterilization can prevent microbial growth.
- **Louis Pasteur** discovered the microbial basis of fermentation. He also showed that providing oxygen does not enable spontaneous generation.
- **John Tyndall** showed that repeated cycles of heat were necessary to eliminate spores formed by certain kinds of bacteria.
- **Florence Nightingale** statistically quantified the impact of infectious disease on human populations.

1.3 Medical Microbiology

Over the centuries, thoughtful observers, such as Fracastoro and Bassi (see **Table 1.2**), noted a connection between microbes and disease. Ultimately, researchers developed the **germ theory of disease**, the theory that many diseases are caused by microbes.

The first to establish a scientific basis for determining that a specific microbe causes a specific disease was

Special Topic 1.1   How Did Life Originate?

If all life on Earth shares descent from a microbial ancestor, how did the first microbe arise? The earliest fossil evidence of cells in the geological record appears in sedimentary rock that formed as early as 3.8 billion years ago. Although the nature of the earliest reported fossils remains controversial, it is generally accepted that "microfossils" from over 2 billion years ago were formed by living cells. Moreover, the living cells that formed these microfossils looked remarkably similar to bacterial cells today, forming chains of simple rods or spheres (**Fig. 1**).

The exact composition of the first environment for life is controversial. The components of the first living cells may have formed from spontaneous reactions sparked by ultraviolet absorption or electrical discharge. American chemists Stanley Miller (1930–2007) and Harold C. Urey (1893–1981) argued that the environment of early Earth contained mainly reduced compounds—compounds that have a strong tendency to donate electrons, such as ferrous iron, methane, and ammonia. More recent evidence has modified this view, but it is agreed that the strong electron acceptor oxygen gas (O_2) was absent until the first photosynthetic microbes produced it. Today, all our cells are composed of highly reduced molecules that are readily oxidized (accept electrons from O_2). This seem-ingly hazardous composition may reflect our cellular origin in the chemically reduced environment of early Earth.

In 1953, Miller attempted to simulate the highly reduced conditions of early Earth to test whether ultraviolet absorption or electrical discharge could cause reactions producing the fundamental components of life (**Fig. 2A**). Miller boiled a solution of water containing hydrogen gas, methane, and ammonia and applied an electrical discharge (comparable to a lightning strike). The electrical discharge excites electrons in the molecules and causes them to react. Astonishingly, the reaction produced a number of amino acids, including glycine, alanine, and aspartic acid. A similar experiment in 1961 by Spanish-American researcher Juan Oró (1923–2004) (**Fig. 2B**) combined hydrogen cyanide and ammonia under electrical discharge to obtain adenine, a fundamental component of DNA and of the energy carrier adenosine triphosphate (ATP).

How could early cells have survived the heat and chemically toxic environment of early Earth? Clues may be found in the survival of archaea that thrive under habitat conditions that we consider extreme, such as solutions of boiling sulfuric acid. The specially adapted structures of such microbes may resemble those of the earliest life-forms.

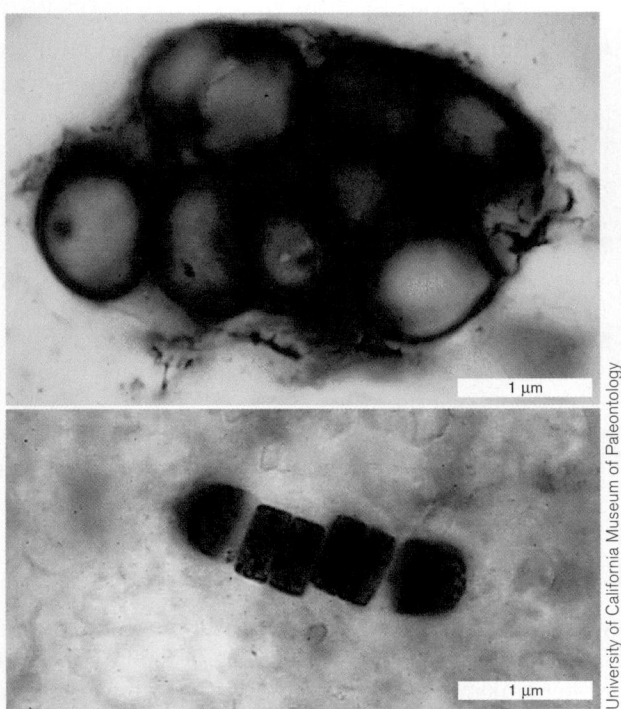

University of California Museum of Paleontology

1 μm

1 μm

Figure 1   Evidence of ancient microbial life.
Microfossils of ancient cyanobacteria from the Bitter Springs Formation, Australia, about 850 million years old.

A.

Bettmann/Corbis

B.

University of Houston

Figure 2   Simulating early Earth's chemistry.
A. Stanley Miller with the apparatus of his early-Earth simulation experiment. **B.** Biochemist Juan Oró demonstrated the formation of adenine and other biochemicals from reaction conditions found in comets.

Research since Miller's day has generated as many questions as answers concerning the origin of life—for example:

- Were the reduced forms of early carbon and nitrogen, such as methane (CH_4) and ammonia (NH_3), supplemented by oxidized forms such as CO_2, spewed out by volcanoes? If oxidized carbon and nitrogen were available, different kinds of early-life chemistry may have occurred.
- How did the origin of informational molecules, such as RNA and DNA, coincide with the origin of metabolism? One possibility is that early metabolism was catalyzed by molecules of RNA instead of protein. RNA molecules capable of catalysis, called ribozymes, were discovered in 1982 by Thomas Cech and Sidney Altman, who earned the Nobel Prize in Chemistry in 1989 (**Fig. 3**). In 2009, researchers made the first self-replicating ribozyme, an RNA that can catalyze reactions and copy itself indefinitely. This achievement supports the theory that early organisms were composed primarily of RNA—the so-called "RNA world."
- Geochemical evidence suggests that cells may have originated on Earth as early as 3.8 billion years ago, when Earth was just barely cool enough to allow the existence

of cells. How could cells have formed so quickly? Could the first cells, in fact, have come from somewhere else?

Most of the molecules that spontaneously formed in Miller's experiments are also found in meteorites and comets. This observation led Oró to propose that the first chemicals of life could have come from outer space, perhaps carried by comets. But could life itself have an extraterrestrial origin? This controversial concept was proposed by British physicists Fred Hoyle (1915–2001) and Chandra Wickramasinghe (1939–). Hoyle and Wickramasinghe argued that features of the infrared spectroscopy of interstellar matter might be explained by the existence of microbes in outer space—microbes that could be brought to Earth by comets or meteors. Alternatively, some scientists propose that the first microbes originated on Mars. Because Mars orbits farther out from the Sun than Earth does, its surface would have cooled before Earth did; therefore, life might have formed on Mars and traveled to Earth on a meteorite. But the "Mars first" explanation says nothing about how life would have arisen on Mars.

Current evidence for the origin and evolution of microbes is discussed in Chapter 17.

A.

Geoffery Wheeler for Howard Hughes Medical Institute

B.

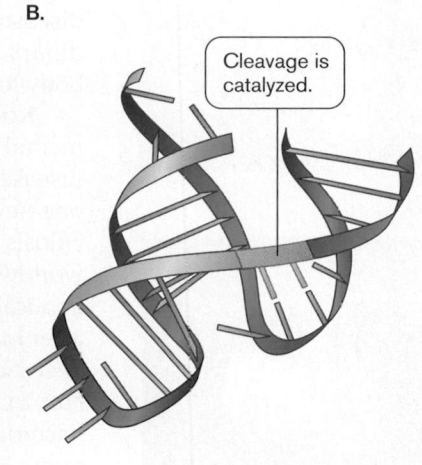

Cleavage is catalyzed.

Figure 3   Tom Cech, discoverer of catalytic RNA.   A. Tom Cech (University of Colorado, Boulder) holding a flask containing *Oxytricha nova*, microbes that make catalytic RNA, the kind of molecule that in early cells may have served both genetic and catalytic functions. **B.** Diagram of a catalytic RNA, where horizontal bars represent bases. The RNA catalyzes cleavage of itself.

the German physician Robert Koch (1843–1910) (**Fig. 1.15**). As a college student, Koch conducted biochemical experiments on his own digestive system. Koch's curiosity about the natural world led him to develop principles and techniques crucial to microbial investigation, including the pure-culture technique and the famous Koch's postulates for identifying the causative agent of a disease. He applied his methods to numerous lethal diseases around the world, including anthrax and tuberculosis in Europe, malaria in Africa and the East Indies, and bubonic plague in India.

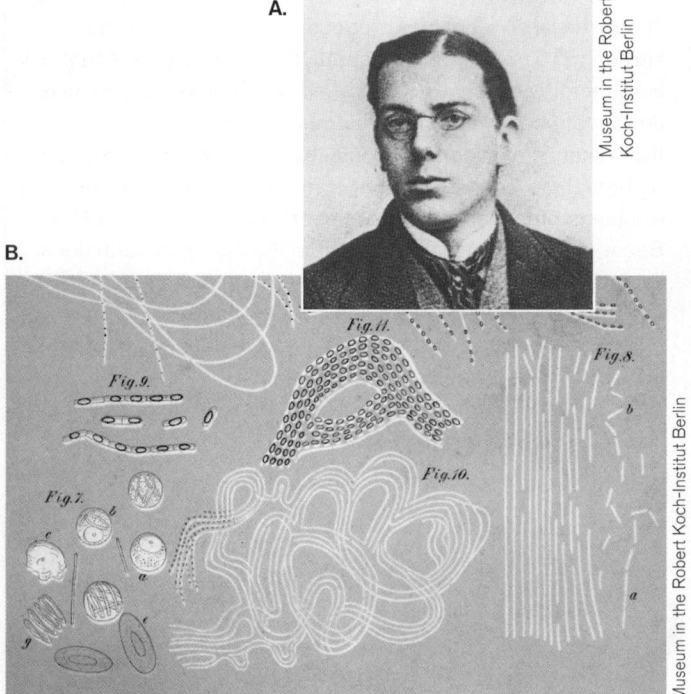

A.

B.

Museum in the Robert Koch-Institut Berlin

Museum in the Robert Koch-Institut Berlin

C.

Museum in the Robert Koch-Institut Berlin

Figure 1.15 Robert Koch, founder of the scientific method of microbiology. A. Robert Koch as a university student. **B.** Koch's sketch of anthrax bacilli in mouse blood. **C.** Koch (second from left) during his visit to New Guinea to investigate malaria.

Growth of Microbes in Pure Culture

Unlike Pasteur, who was a university professor, Koch took up a medical practice in a small Polish-German town. To make space in his home for a laboratory to study anthrax and other deadly diseases, his wife curtained off part of his patients' examining room.

Anthrax interested Koch because its epidemics in sheep and cattle caused economic hardship among local farmers. Today, anthrax is no longer a major problem for agriculture, as its transmission is prevented by effective environmental controls and vaccination. It has, however, gained notoriety as a bioterror agent because anthrax bacteria can survive for long periods in the dormant, desiccated form of an endospore. In 2001, anthrax spores sent through the mail contaminated post offices throughout the northeastern United States, as well as an office building of the United States Senate, causing several deaths.

To investigate whether anthrax was a transmissible disease, Koch used blood from an anthrax-infected carcass to inoculate a rabbit. When the rabbit died, he used the rabbit's blood to inoculate a second rabbit, which then died in turn. The blood of the unfortunate animal had turned black with long, rod-shaped bacilli. Upon introduction of these bacilli into healthy animals, the animals became ill with anthrax. Thus, Koch demonstrated an important principle of epidemiology: the **chain of infection**, or transmission of a disease. In retrospect, his choice of anthrax was fortunate, for the microbes generate disease very quickly, multiply in the blood to an extraordinary concentration, and remain infective outside the body for long periods.

Koch and his colleagues then applied their experimental logic and culture methods to a more challenging disease: tuberculosis. In Koch's day, tuberculosis caused one-seventh of reported deaths in Europe; today, tuberculosis bacteria continue to infect millions of people worldwide. Koch's approach to anthrax, however, was less applicable to tuberculosis, a disease that develops slowly after many years of dormancy. Furthermore, the causative bacteria, *Mycobacterium tuberculosis*, are small and difficult to distinguish from human tissue or from different bacteria of similar appearance associated with the human body. How could Koch prove that a particular bacterium caused a particular disease?

What was needed was to isolate a **pure culture** of microorganisms, a culture grown from a single "parental" cell. This had been done by previous researchers using the laborious process of serial dilution of suspended bacteria until a culture tube contained only a single cell. Alternatively, inoculation of a solid surface such as a sliced potato could produce isolated **colonies**, distinct populations of bacteria, each grown from a single cell. For *M. tuberculosis*, Koch inoculated serum, which then formed a solid gel after heating. Later he refined the solid-substrate technique by adding gelatin to a defined liquid medium, which could then be chilled to form a solid medium in a glass dish. A covered version called the **petri dish** (also called a petri plate) was invented by a colleague, Julius Richard Petri (1852–1921). The petri dish consists of a round dish with vertical walls covered by an inverted dish of slightly larger diameter. Today, the petri dish, generally made of disposable plastic, remains an indispensable part of the microbiological laboratory.

Another improvement in solid-substrate culture was the replacement of gelatin with materials that remain solid at higher temperatures, such as the gelling agent **agar** (a polymer of the sugar galactose). The use of agar was recommended by Angelina Hesse (1850–1934), a microscopist and illustrator, to her husband, Walther Hesse (1846–1911), a young medical colleague of Koch (**Fig. 1.16**). Agar comes from red algae (seaweed), which is used by East Indian birds to build nests; it is the main ingredient in the delicacy "bird's nest soup." Dutch colonists used agar to make jellies and preserves, and a Dutch colonist from Java introduced it to Angelina Hesse. The Hesses used agar to develop the first effective growth medium for tuberculosis bacteria. (Pure culture is discussed further in Chapter 4.)

Note that some kinds of microbes cannot be grown in pure culture without other organisms. For example, viruses can be cultured only in the presence of their host cells. The discovery of viruses is explored at the end of this section.

Figure 1.16 Angelina and Walther Hesse. A. Portrait of the Hesses, who first used agar to make solid plate media for bacterial growth. **B.** Colonies from a streaked agar plate.

Koch's Postulates

For his successful determination of the bacterium that causes tuberculosis, *Mycobacterium tuberculosis*, Koch was awarded the Nobel Prize in Physiology or Medicine in 1905. Koch formulated his famous set of criteria for establishing a causative link between an infectious agent and a disease (**Fig. 1.17**). These four criteria are known as **Koch's postulates**:

1. The microbe is found in all cases of the disease but is absent from healthy individuals.
2. The microbe is isolated from the diseased host and grown in pure culture.
3. When the microbe is introduced into a healthy, susceptible host (or animal model), the same disease occurs.
4. The same strain of microbe is obtained from the newly diseased host. When cultured, the strain shows the same characteristics as before.

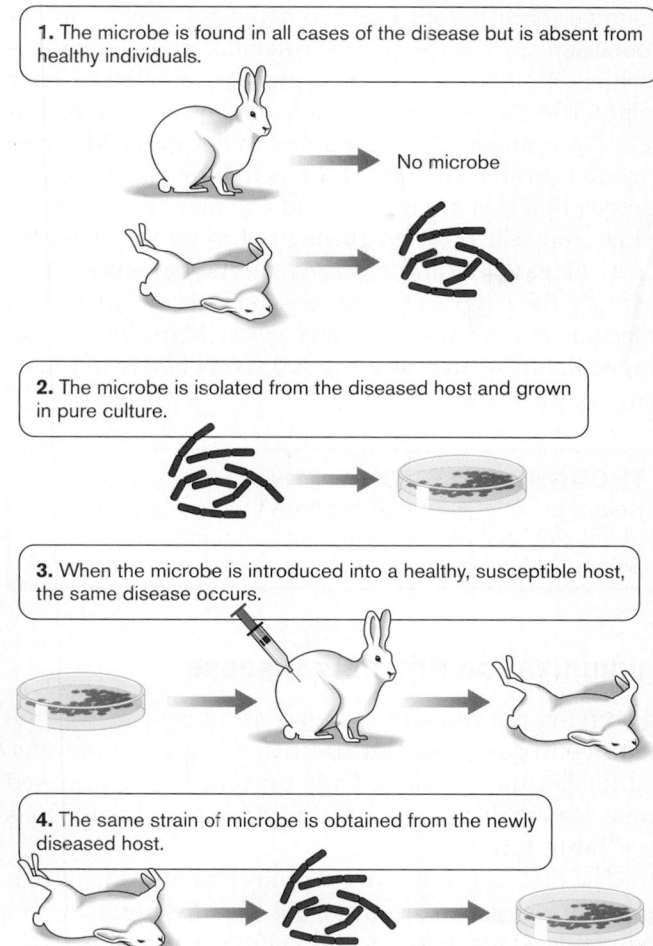

Figure 1.17 Koch's postulates defining the causative agent of a disease.

Koch's postulates continue to be used to determine whether a given strain of microbe causes a disease. Modern examples include Lyme disease, a tick-borne infection that has become widespread in New England and the Mid-Atlantic states; and hantaviral pneumonia, an emerging disease particularly prevalent among Native Americans in the Southwest. Nevertheless, the postulates remain only a guide; individual diseases and pathogens may confound one or more of the criteria. For example, tuberculosis bacteria are now known to cause symptoms in only 10% of the people infected. If Koch had been able to detect these silent bacilli, they would not have fulfilled his first criterion. In the case of AIDS, the concentration of HIV virus is so low that initially no virus could be detected in patients with fully active symptoms. It took the invention of the polymerase chain reaction (PCR), a method of producing any number of copies of DNA or RNA sequences, to detect the presence of HIV.

Another difficulty with AIDS and many other human diseases is the absence of an animal host that exhibits the same disease. In the case of AIDS, even chimpanzees, our closest relatives, are not susceptible, although they exhibit a similar disease from a related pathogen, simian immunodeficiency virus (SIV). Experimentation on humans is prohibited, although in rare instances researchers have voluntarily exposed themselves to a proposed pathogen. For example, Australian researcher Barry Marshall ingested *Helicobacter pylori* to convince skeptical colleagues that this organism could colonize the extremely acidic stomach. *H. pylori* turned out to be the causative agent of gastritis and stomach ulcers, conditions that had long been thought to be caused by stress rather than infection. For the discovery of *H. pylori*, Marshall and colleague Robin Warren won the 2005 Nobel Prize in Physiology or Medicine.

> **THOUGHT QUESTION 1.4** How could you use Koch's postulates to demonstrate the causative agent of influenza? What problems not encountered with anthrax would you need to overcome?

Immunization Prevents Disease

Identifying the cause of a disease is, of course, only the first step to developing an effective therapy and preventing further transmission. Early microbiologists achieved some remarkable insights on how to control pathogens (see **Table 1.2**).

The first clue as to how to protect an individual from a deadly disease came from the dreaded smallpox. In the eighteenth century, smallpox infected a large fraction of the European population, killing or disfiguring many people. In countries of Asia and Africa, however, the incidence of smallpox was decreased by the practice of deliberately inoculating children with material from smallpox pustules. Inoculated children usually developed a mild case of the disease and were protected from smallpox thereafter. The practice of smallpox inoculation was introduced from Turkey to Europe in 1717 by Lady Mary Montagu, a smallpox survivor (**Fig. 1.18A**). While traveling in Turkey, Lady Montagu learned that many elderly women there had perfected the art of inoculation: "The old woman comes with a nut-shell full of the matter of the best sort of small-pox, and asks what vein you please to have opened." During a period outside the host, the virus becomes "attenuated"—that is, loses some of its molecular structure required for infection. The attenuated virus stimulates the immune system with much lower mortality than the fully virulent virus. Lady Montagu arranged for the procedure on her own son, then brought the practice back to England. A similar practice of smallpox inoculation was introduced to the American colonies by a slave, Onesimus, from the Coromantee people of Africa. Onesimus convinced his master, Reverend Dr. Cotton Mather, to promote smallpox inoculation as a defense against an epidemic that was devastating Boston.

Preventive inoculation with smallpox was dangerous, however, as some infected individuals still contracted serious disease and were contagious. Thus, doctors continued to seek a better method of prevention. In England, milkmaids claimed that they were protected from smallpox after they contracted cowpox, a related but much milder disease. This claim was confirmed by English physician Edward Jenner (1749–1823), who deliberately infected patients with matter from cowpox lesions (**Fig. 1.18B**). The practice of cowpox inoculation was called **vaccination**, after the Latin word *vacca*, meaning "cow." To this day, cowpox, or vaccinia virus, remains the basis of the modern smallpox vaccine.

Pasteur was aware of vaccination as he studied the course of various diseases in experimental animals. In the spring of 1879, he was studying fowl cholera, a transmissible disease of chickens with a high death rate. He had isolated and cultured the bacteria responsible, but he left his work during the summer for a long vacation. No refrigeration was available to preserve cultures, and when he returned to work, the aged bacteria failed to cause disease in his chickens. Pasteur then obtained fresh bacteria from an outbreak of disease elsewhere, as well as some new chickens. But the fresh bacteria failed to make the original chickens sick (those that had been exposed to the aged bacteria). All of the new chickens, exposed only to the fresh bacteria, contracted the disease. Grasping the clue from his mistake, Pasteur had the insight to recognize that an attenuated (or "weakened") strain of microbe, altered somehow to eliminate its potency to cause disease, could still confer immunity to the virulent disease-causing form.

A.

B.

C.

Figure 1.18 **Smallpox vaccination.** **A.** Lady Mary Wortley Montagu, shown in Turkish dress. The artist avoided showing Montagu's facial disfigurement from smallpox. **B.** Dr. Edward Jenner, depicted vaccinating 8-year-old James Phipps with cowpox matter from the hand of milkmaid Sarah Nelmes, who had caught the disease from a cow. **C.** Newspaper cartoon depicting public reaction to cowpox vaccination.

Pasteur was the first to recognize the significance of attenuation and extend the principle to other pathogens. We now know that the molecular components of pathogens generate **immunity**, the resistance to a specific disease, by stimulating the **immune system**, an organism's exceedingly complex cellular mechanisms of defense (see Chapter 23). Understanding the immune system awaited the techniques of molecular biology a century later, but nineteenth-century physicians developed several effective examples of **immunization**, the stimulation of an immune response by deliberate inoculation with an attenuated pathogen.

The way to attenuate a strain varies greatly among pathogens. Heat treatment or aging for various periods often turned out to be the most effective approach. The original success of prophylactic smallpox inoculation was due to natural attenuation of the virus during the time between acquisition of smallpox matter from a diseased individual and inoculation of the healthy patient. A far more elaborate treatment was required for the most famous disease for which Pasteur devised a vaccine: rabies.

The rabid dog loomed large in folklore, and the disease was dreaded for its particularly horrible and inevitable course of death. Pasteur's vaccine for rabies required a highly complex series of heat treatments and repeated inoculations. Its success led to his instant fame (**Fig. 1.19**). Grateful survivors of rabies founded the Pasteur Institute for medical research, one of the world's greatest medical research institutions, whose scientists in the twentieth century discovered the virus HIV, which causes AIDS.

Antiseptics and Antibiotics

Before the work of Koch and Pasteur, many patients died of infections transmitted unwittingly by their own doctors. In 1847, Hungarian physician Ignaz Semmelweis (1818–1865) noticed that the death rate of women in childbirth due to puerperal fever was much higher in his own hospital than in a birthing center run by midwives. He guessed that the doctors in his hospital were transmitting pathogens from cadavers that they had dissected. So he ordered the doctors to wash their hands in chlorine, an **antiseptic** agent (a chemical that kills

Figure 1.19   Pasteur cures rabies.   This cartoon in a French newspaper depicts Pasteur protecting children from rabid dogs.

Figure 1.20   Alexander Fleming, discoverer of penicillin.
A. Alexander Fleming in his laboratory. **B.** Fleming's original plate of bacteria with *Penicillium* mold inhibiting the growth of bacterial colonies.

microbes). The mortality rate fell; but this revelation displeased other doctors, who refused to accept Semmelweis's findings.

In 1865, the British surgeon Joseph Lister (1827–1912) noted that half his amputee patients died of sepsis. Lister knew from Pasteur that microbial contamination might be the cause. So he began experiments to develop the use of antiseptic agents, most successfully carbolic acid, to treat wounds and surgical instruments. After initial resistance, Lister's work, with the support of Pasteur and Koch, drew widespread recognition. In the twentieth century, surgeons developed fully **aseptic** environments for surgery—that is, environments completely free of microbes.

Antibiotics. The problem with most antiseptic chemicals that killed microbes was that if taken internally, they also tended to kill the patients. Researchers sought a "magic bullet," an **antibiotic** molecule that killed microbes alone, leaving their host unharmed.

An important step in the search for antibiotics was the realization that microbes themselves produce antibiotic compounds. This conclusion followed from the famous accidental discovery of penicillin by the English medical researcher Alexander Fleming (1881–1955) (**Fig. 1.20A**). In 1929, Fleming was culturing *Staphylococcus*, which infects wounds. He found that one of his plates of *Staphylococcus* was contaminated with a mold, *Penicillium notatum*, which he noticed was surrounded by a clear region, free of *Staphylococcus* colonies (**Fig. 1.20B**). Following up on this observation, Fleming showed that the mold produced a substance that killed bacteria. We now know this substance as penicillin.

In 1941, British biochemists Howard Florey (1898–1968) and Ernst Chain (1906–1979) purified the antibiotic molecule, which we now know inhibits formation of the bacterial cell wall. Penicillin saved the lives of many Allied troops during World War II, the first war in which an antibiotic became available to soldiers.

The second half of the twentieth century saw the discovery of many new and powerful antibiotics. Most of the new antibiotics, however, were produced by obscure strains of bacteria and fungi from endangered ecosystems—a circumstance that focused attention on wilderness preservation worldwide. Furthermore, the widespread and often indiscriminate use of antibiotics has selected for pathogens that are antibiotic resistant. As a result, antibiotics have lost their effectiveness against certain strains of major pathogens. For example, multidrug-resistant *Mycobacterium tuberculosis* is now a serious threat to public health.

Fortunately, biotechnology provides new approaches to antibiotic development, including genetic engineering of microbial producers and artificial design of antimicrobial chemicals. This industry has become ever more critical because the indiscriminate use of antibiotics has led to a "molecular arms race" in which our only hope is to succeed faster than the pathogens develop resistance. (Microbial biosynthesis of antibiotics is discussed in Chapter 15, and the medical use of antibiotics is discussed in Chapter 27.)

THOUGHT QUESTION 1.5   Why do you think some pathogens generate immunity readily, whereas others evade the immune system?

THOUGHT QUESTION 1.6   How do you think microbes protect themselves from the antibiotics they produce?

The Discovery of Viruses

Some kinds of microbes cannot be grown in pure culture without other organisms. For example, viruses can be cultured only in the presence of their host cells. Most viruses are also much smaller than cells. From the nineteenth century on, researchers were puzzled to find contagious diseases whose agent of transmission could pass through a filter of a pore size that blocked known microbial cells: 0.1 μm. One of these researchers was the Dutch plant microbiologist Martinus Beijerinck (1851–1931), who studied tobacco mosaic disease, a condition in which the leaves become mottled and the crop yield is decreased or destroyed altogether. Beijerinck concluded that because the agent of disease passed through a filter that retained bacteria, it could not be a bacterial cell.

The filterable agent was ultimately purified by the American scientist Wendell Stanley (1904–1971), who processed 4,000 kilograms (kg) of infected tobacco leaves and crystallized the infective particle. What he had crystallized was the tobacco mosaic virus, the causative agent of tobacco mosaic disease (**Fig. 1.21A**). The crystallization of a virus particle earned Stanley the 1946 Nobel Prize in Chemistry. The fact that an entity capable of biological reproduction could be inert enough to be crystallized amazed scientists and ultimately led to a new, more mechanical view of living organisms.

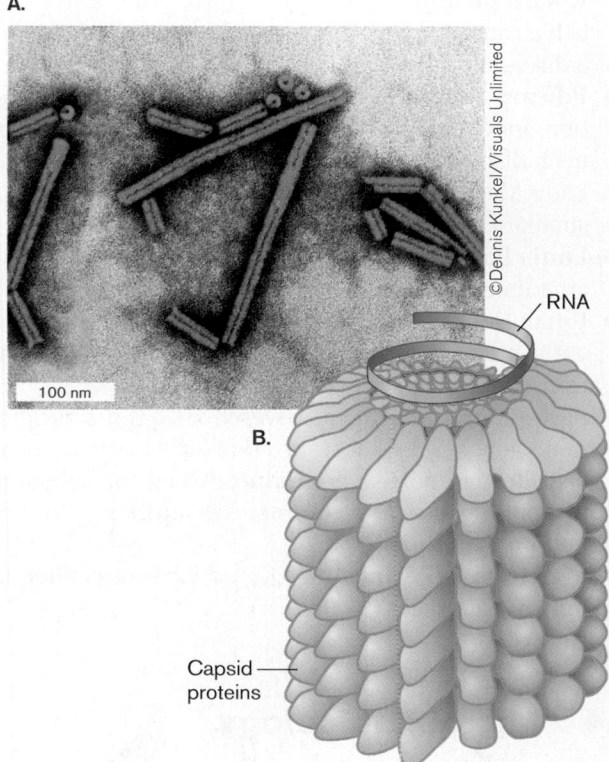

Figure 1.21   Tobacco mosaic virus (TMV).   A. Particles of tobacco mosaic virus (colorized transmission EM). **B.** In TMV, a protein capsid surrounds an RNA chromosome.

The individual particle of tobacco mosaic virus consists of a helical tube of protein subunits containing its genetic material coiled within (**Fig. 1.21B**). Stanley thought that the virus was a catalytic protein, but colleagues later determined that it contained RNA as its genetic material. The structure of the coiled RNA was solved through X-ray crystallography by the British scientist Rosalind Franklin (1920–1958). We now know that all kinds of animals, plants, and microbial cells can be infected by viruses.

Toward the end of the twentieth century, even smaller infective particles were discovered consisting of a single molecule of RNA (viroids) or of protein (prions). Prions are suspected as a factor in the development of Alzheimer's disease. (The infectious processes of viruses, viroids, and prions are discussed in Chapters 6, 11, and 26.)

TO SUMMARIZE:

■ **Robert Koch** devised techniques of pure culture to study a single species of microbe in isolation. A key technique is culture on solid medium using agar, as developed by Angelina and Walther Hesse, in a double-dish container devised by Julius Petri.

- **Koch's postulates** provide a set of criteria to establish a causative link between an infectious agent and a disease.

- **Edward Jenner** established the practice of vaccination, inoculation of cowpox to prevent smallpox. Jenner's discovery was based on earlier observations by Lady Mary Montagu and others that a mild case of smallpox could prevent future cases.

- **Louis Pasteur** developed the first vaccines based on attenuated strains, such as the rabies vaccine.

- **Ignaz Semmelweis** and **Joseph Lister** showed that antiseptics could prevent transmission of pathogens from doctor to patient.

- **Alexander Fleming** discovered that the *Penicillium* mold generated a substance that kills bacteria. **Howard Florey** and **Ernst Chain** purified the substance penicillin, the first commercial antibiotic to save human lives.

- **Martinus Beijerinck** discovered viruses as filterable infective particles.

1.4   Microbial Ecology

Koch's growth of microbes in pure culture was a major advance in technology, enabling the systematic study of microbial physiology and biochemistry. In hindsight, this discovery eclipsed the equally important study of microbial ecology. Microbes cycle the many minerals essential for all life, including all global N_2 and much of the O_2. Yet barely 0.1% of all microbial species can be cultured in the laboratory—and the remainder make up the majority of Earth's entire biosphere. Only the outer skin of Earth supports complex multicellular organisms. The depths of Earth's crust, to at least 2 miles down, as well as the atmosphere 10 miles out into the stratosphere, remain the domain of microbes. So, to a first approximation, Earth's ecology *is* microbial ecology.

Microbes Support Natural Ecosystems

The first microbiologists to culture microbes in the laboratory selected the kinds of nutrients that feed humans, such as beef broth or potatoes. Some of Koch's contemporaries, however, suspected that other kinds of microbes living in soil or wetlands existed on more exotic fare. Soil samples were known to oxidize hydrogen gas, and this activity was eliminated by treatment with heat or acid, suggesting microbial origin. Ammonia in sewage was oxidized by donating electrons to oxygen, forming nitrate. Nitrate formation was eliminated by antibacterial treatment. These findings suggested the existence of microbes that "ate" hydrogen gas or ammonia instead of beef or potatoes, but no one could isolate these microbes in culture.

Among the first to study microbes in natural habitats was the Russian scientist Sergei Winogradsky (1856–1953). Winogradsky waded through marshes to discover microbes with metabolism quite alien from human digestion. For example, Winogradsky discovered that species of the bacterium *Beggiatoa* oxidize hydrogen sulfide (H_2S) to sulfuric acid (H_2SO_4). *Beggiatoa* fixes carbon dioxide into biomass without consuming any organic food. Organisms that feed solely on inorganic minerals are known as **chemolithotrophs**, or **lithotrophs**, discussed further in Chapters 4 and 14.

The lithotrophs studied by Winogradsky could not be grown on Koch's plate media containing agar or gelatin. The bacteria that Winogradsky isolated could grow only on inorganic minerals; in fact, some species are actually poisoned by organic food. For example, nitrifiers convert ammonia to nitrate, forming a crucial part of the nitrogen cycle in natural ecosystems. Winogradsky cultured nitrifiers on a totally inorganic solution containing ammonia and silica gel, which supported no other kind of organism. This experiment was an early example of **enrichment culture**, the use of selective growth media that support certain classes of microbial metabolism while excluding others.

Instead of isolating pure colonies, Winogradsky built a model wetland ecosystem containing regions of enrichment for microbes of diverse metabolism. This model is called the **Winogradsky column** (**Fig. 1.22**). The model consists of a glass tube containing mud (a source of wetland bacteria) mixed with shredded newsprint (an organic carbon source) and calcium salts of sulfate and carbonate (an inorganic carbon source for autotrophs). After exposure to light for several weeks, several zones of color develop, full of mineral-metabolizing bacteria. At the top, cyanobacteria conduct **photosynthesis**, using light

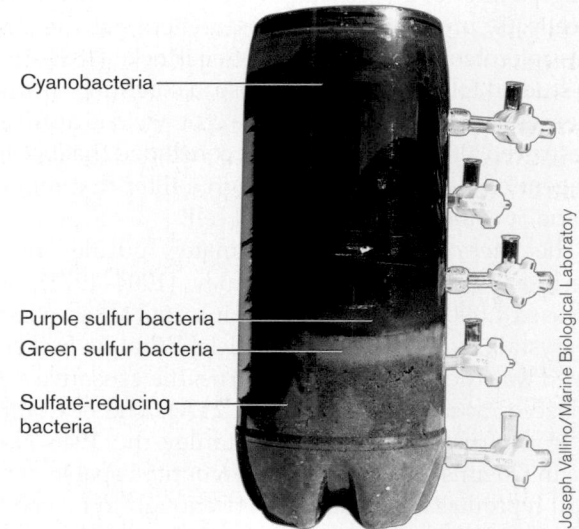

Cyanobacteria

Purple sulfur bacteria

Green sulfur bacteria

Sulfate-reducing bacteria

Joseph Vallino/Marine Biological Laboratory

Figure 1.22   Winogradsky column.   A wetland model ecosystem designed by Sergei Winogradsky.

energy to split water and produce molecular oxygen. Below, purple sulfur bacteria use photosynthesis to split hydrogen sulfide, producing sulfur. At the bottom, with O_2 exhausted, bacteria reduce (donate electrons to) alternative electron acceptors such as sulfate. Sulfate-reducing bacteria produce hydrogen sulfide and precipitate iron.

The gradient from oxygen-rich conditions at the surface to highly reduced conditions below generates a voltage potential, like a battery cell. We now know that the entire Earth's surface acts as a battery—for humans, a potential source of renewable energy.

Winogradsky and later microbial ecologists showed that bacteria perform unique roles in **geochemical cycling**, the global interconversion of inorganic and organic forms of nitrogen, sulfur, phosphorus, and other minerals. Without these essential conversions (nutrient cycles), no plants or animals could live. Bacteria and archaea *fix nitrogen* (N_2) by reducing it to ammonia (NH_3), the form of nitrogen assimilated by plants. This process is called **nitrogen fixation (Fig. 1.23)**. Other bacterial species oxidize ammonium ions (NH_4^+) in several stages back to nitrogen gas.

Bacterial Endosymbiosis with Plants and Animals

Within plant cells, certain bacteria fix nitrogen as **endosymbionts**, organisms living symbiotically inside a larger organism. Endosymbiotic bacteria known as rhizobia induce the roots of legumes to form special nodules to facilitate bacterial nitrogen fixation. Rhizobial endosymbiosis was first observed by Martinus Beijerinck.

> **THOUGHT QUESTION 1.7** Why don't all living organisms fix their own nitrogen?

Microbial endosymbiosis, in a variety of diverse forms, is widespread in all ecosystems. Many interesting cases involve animal or human hosts.

Endosymbiotic microbes make essential nutritional contributions to host animals. Ruminant animals, such as cattle, as well as insects such as termites, require digestive bacteria to break down cellulose and other plant polymers. Even humans obtain about 10% of their nutrition

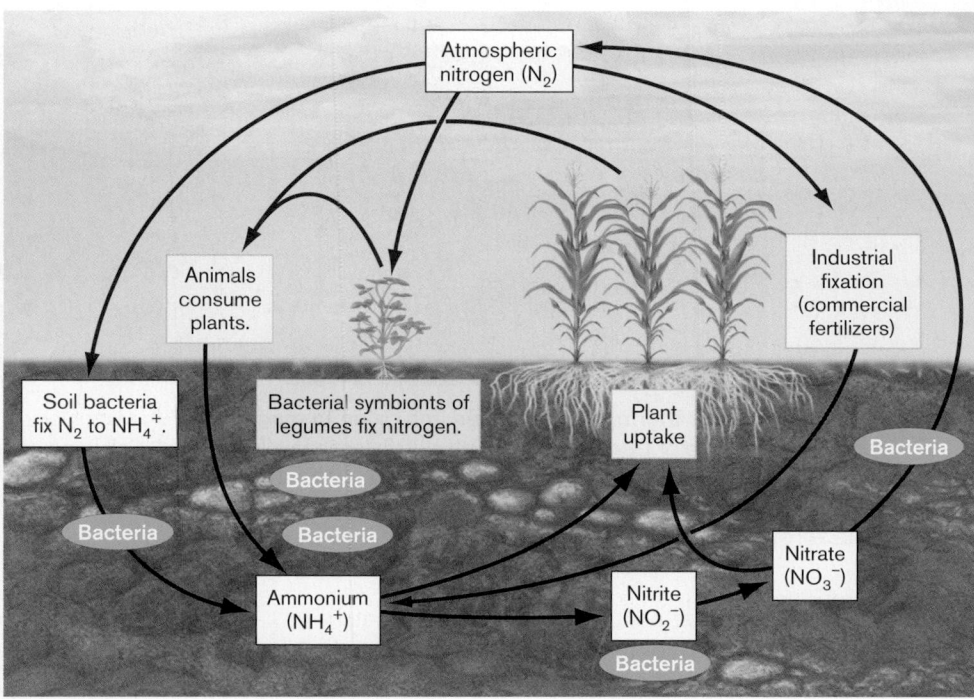

Figure 1.23 The global nitrogen cycle. All life depends on these oxidative and reductive conversions of nitrogen—most of which are performed only by microbes.

from colonic bacteria. Colonic bacteria such as *E. coli* and *Bacteroides* species grow as **biofilms**, organized multispecies communities upon a surface. The biofilm shown in **Figure 1.24** is attached to the surface of a digested food particle. Biofilms play major roles in all ecosystems and within parts of the human body (**Fig. 1.24**).

Many invertebrates, such as hydras and corals, harbor endosymbiotic phototrophs that provide products of photosynthesis in return for protection and nutrients. Other kinds of endosymbiosis involve more than nutrition. In the light organs of squid, luminescent bacteria such as *Vibrio*

Figure 1.24 Intestinal biofilm. Bacteria growing on the surface of a residual food particle aid human digestion. *Source:* Sandra Macfarlane and George T. Macfarlane. 2006. *Appl. Environ. Microbiol.* **72**:6204.

A.

B.

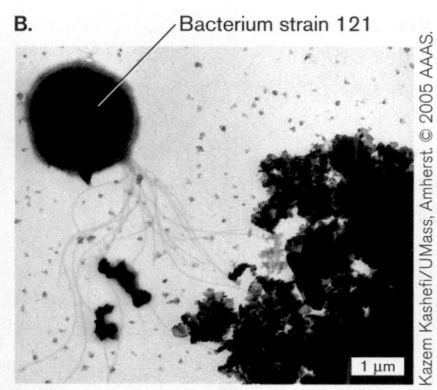

Figure 1.25 **An extreme thermophile reduces iron oxide to magnetite.** **A.** Microbiologist Kazem Kashefi (Michigan State University) pulls a live culture of *Geobacter* out of an autoclave generally used to kill all living organisms at 121°C (250°F). The magnet shows that strain 121 is converting nonmagnetic iron oxide (rust; Fe_2O_3) to the magnetic mineral magnetite (Fe_3O_4). **B.** Strain 121 is a round bacterium with a tuft of rotary flagella (transmission electron micrograph). The black material at right is iron oxide, which feeds the bacteria. *Source:* Part B reprinted with permission from Kashefi et al. 2003. *Science* **301**(5635):934. © 2005 AAAS.

fischeri help the host evade nocturnal predators by matching illumination from the moon. Certain sponges maintain endosymbiotic bacteria that produce antibiotics, preventing infection by pathogens. Similarly, some bacteria that normally inhabit the human skin are believed to protect us from infection. Microbial endosymbiosis has even inspired fictional creations, such as the "breathmicrobes" of Joan Slonczewski's novel *A Door into Ocean*, which acquire oxygen for human hosts swimming underwater. (Microbial endosymbiosis is discussed further in Section 21.2.)

Microbes with unusual properties, such as the ability to digest toxic wastes, have valuable applications in industry and bioremediation. For this reason, microbial ecology is a priority for funding by the National Science Foundation (NSF). The NSF program Life in Extreme Environments supports research aimed at documenting microbes from environments with extreme heat, salinity, acidity, or other factors. One such microbe is a species of *Geobacter* that reduces rust (iron oxide, Fe_2O_3) to the magnetic mineral magnetite (Fe_3O_4) while growing at 121°C, a temperature high enough to kill all other known organisms (**Fig. 1.25**).

TO SUMMARIZE:

■ **Sergei Winogradsky** first developed a system of enrichment culture, the Winogradsky column, to grow microbes from natural environments.
■ **Chemolithotrophs (or lithotrophs)** metabolize inorganic minerals, such as ammonia, instead of the organic nutrients used by the microbes isolated by Koch.
■ **Geochemical cycling** depends on bacteria and archaea that cycle nitrogen, phosphorus, and other minerals throughout the biosphere.
■ **Endosymbionts** are microbes that live within multicellular organisms and provide essential functions

for their hosts, such as nitrogen fixation for legume plants.
■ **Martinus Beijerinck** first demonstrated that nitrogen-fixing rhizobia grow as endosymbionts within leguminous plants.

1.5 The Microbial Family Tree

The bewildering diversity of microbial life-forms presented nineteenth-century microbiologists with a seemingly impossible task of classification. So little was known about life under the lens that natural scientists despaired of ever learning how to distinguish microbial species. The famous classifier of species, Swedish botanist Carl von Linné (Carolus Linnaeus, 1707–1778), called the microbial world "chaos."

Microbes Are a Challenge to Classify

Two challenges faced would-be taxonomists with respect to microbes. The first was that the resolution of the light microscope allows little more than visualizing the outward shape of microbial cells, and vastly different kinds of microbes look more or less alike (visualization of cells and molecules is discussed in Chapter 2). This challenge was overcome as advances in biochemistry and microscopy made it possible to distinguish microbes by metabolism and cell ultrastructure and ultimately by DNA sequence.

The second challenge was that microbes do not readily fit the classic definition of a species—that is, a group of organisms that interbreed. Unlike multicellular eukaryotes, microbes generally reproduce asexually, with only occasional sexual exchange. When they do exchange genes, they may do so with distantly related species (discussed in Chapter 9). Nevertheless, microbiologists devise working definitions of microbial species that enable us to usefully describe populations while being flexible enough to accommodate continual revision and change (discussed in Chapter 17). The most useful definitions are based on genetic similarity. For example, two distinct species generally share no more than 95% similarity of DNA sequence.

NOTE: The actual names of microbial species are often changed to reflect new understanding of genetic relationships. For example, the causative agent of bubonic plague was formerly called *Bacterium pestis* (1896), *Bacillus pestis* (1900), and *Pasteurella pestis* (1923), but it is now called *Yersinia pestis* (1944). The older names, however, still appear in the literature—a point to remember during research. Current and historical nomenclature is compiled at the List of Prokaryotic Names with Standing in Nomenclature (LPSN).

www | List of Prokaryotic Names with Standing in Nomenclature *microbiology2.com/links*

Microbes Include Eukaryotes and Prokaryotes

In the nineteenth century, taxonomists had no access to DNA information. As they tried to incorporate microbes into the tree of life, they faced a conceptual dilemma in that microbes could not be categorized as either animals or plants, which since ancient times had been considered the two "kingdoms" or major categories of life. Taxonomists attempted to apply these categories to microbes—for example, by including algae and fungi with plants. But German naturalist Ernst Haeckel (1834–1919) recognized that microbes differed from both plants and animals in fundamental aspects of their lifestyle, cellular structure, and biochemistry. Haeckel proposed that microscopic organisms constituted a third kind of life—neither animal nor plant—which he called Monera.

In the twentieth century, biochemical studies revealed profound distinctions even within the microbial category. In particular, microbes such as protists and algae contain a nucleus bounded by a nuclear membrane, whereas bacteria

do not. Herbert Copeland (1902–1968) proposed a system of classification that divided Monera into two groups: the eukaryotic protists (protozoa and algae) and the prokaryotic bacteria. Copeland's four-kingdom classification (plants, animals, eukaryotic protists, and prokaryotic bacteria) was later modified by Robert Whittaker (1920–1980) to include fungi as a fifth kingdom of eukaryotic microbes. Whittaker's system thus generated five kingdoms: bacteria, protists, fungi, and the multicellular plants and animals.

Eukaryotes Evolved through Endosymbiosis

The five-kingdom system was modified dramatically by Lynn Margulis, at the University of Massachusetts (**Fig. 1.26**). Margulis tried to explain how it is that eukaryotic cells contain mitochondria and chloroplasts, membranous organelles that possess their own chromosomes. She proposed that eukaryotes evolved by merging with bacteria to form composite cells by intracellular endosymbiosis, in which one cell internalizes

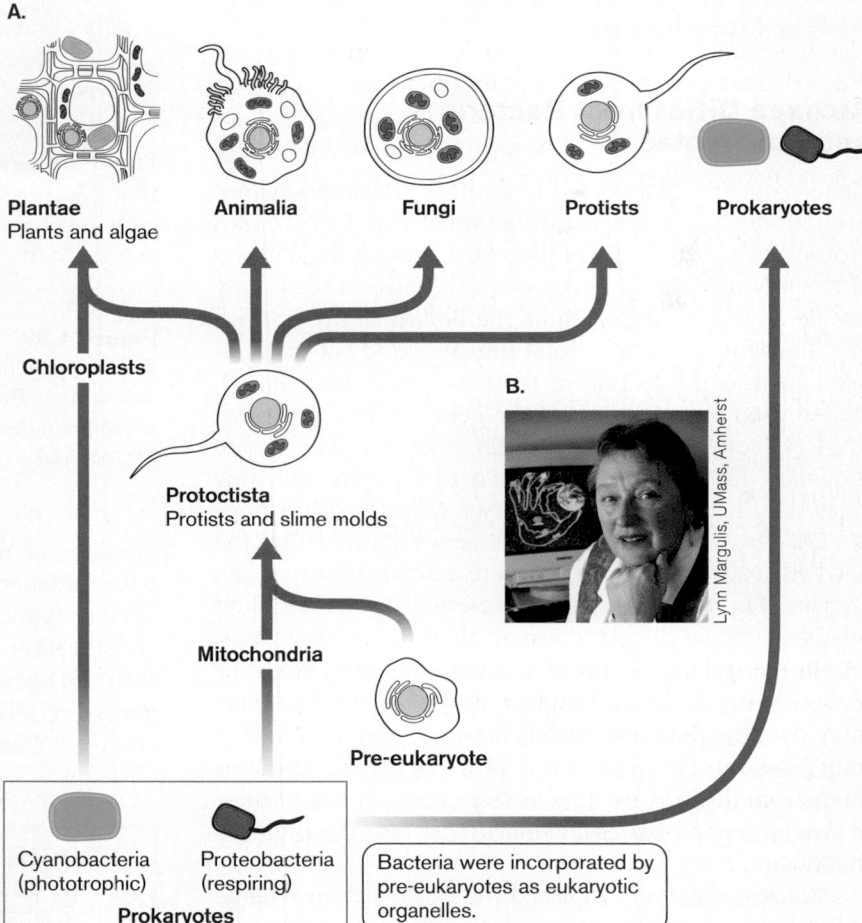

Figure 1.26 Lynn Margulis and the serial endosymbiosis theory. A. Five-kingdom scheme, modified by the endosymbiosis theory. **B.** Lynn Margulis (University of Massachusetts, Amherst) proposed that organelles evolve through endosymbiosis.

another that grows within it. The endosymbiosis may ultimately generate a single organism whose formerly independent members are now incapable of independent existence.

Margulis proposed that early in the history of life, respiring bacteria similar to *E. coli* were engulfed by pre-eukaryotic cells, where they evolved into mitochondria, the eukaryote's respiratory organelles. Similarly, she proposed that a phototroph related to cyanobacteria was taken up by a eukaryote, giving rise to the chloroplasts of phototrophic algae and plants.

The endosymbiosis theory was highly controversial because it implied a **polyphyletic**, or multiple, ancestry of living species, inconsistent with the long-held assumption that species evolve only by divergence from a common ancestor (**monophyletic** ancestry). Ultimately, DNA sequence analysis produced compelling evidence of the bacterial origin of mitochondria and chloroplasts. Both these classes of organelles contain circular molecules of DNA, whose sequences show unmistakable homology (similarity) to those of bacteria. DNA sequences and other evidence established the common ancestry between mitochondria and respiring bacteria and between chloroplasts and cyanobacteria.

Archaea Differ from Bacteria and Eukaryotes

Gene sequence analysis led to another startling advance in our understanding of the evolution of cells. In 1977, Carl Woese, at the University of Illinois, was studying a group of recently discovered prokaryotes that live in seemingly hostile environments, such as the boiling sulfur springs of Yellowstone, or that exhibit unusual kinds of metabolism, such as production of methane (methanogenesis). Woese used the sequence of the gene for 16S ribosomal RNA (16S rRNA) as a "molecular clock," a gene whose sequence differences can be used to measure the time since the divergence of two species (discussed in Chapter 17). The divergence of rRNA genes showed that the newly discovered prokaryotes were a distinct form of life, archaea (**Fig. 1.27**). The archaea resemble bacteria in their relatively simple cell structure, in their lack of a nucleus, and in their ability to grow in a wide range of environments. Many archaea, however, grow in environments more extreme than any bacterium does, such as 110°C at high pressure. The genetic sequences of archaea differ as much from those of bacteria as from those of eukaryotes; in fact, their gene expression machinery is more similar to that of eukaryotes.

Woese's discovery replaced the classification scheme of five kingdoms with three equally distinct groups, now called the three domains: Bacteria, Archaea, and Eukarya (**Fig. 1.28**). In the three-domain model, the bacterial

A.

B.

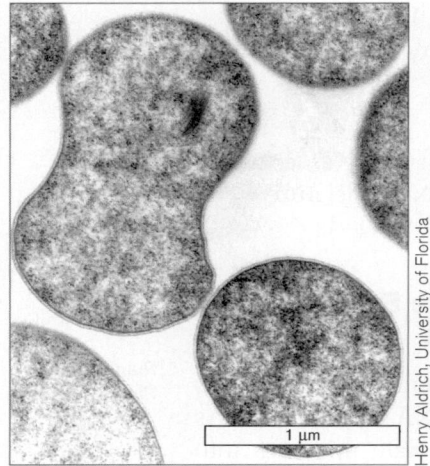

1 µm

Figure 1.27   Archaea, newly discovered life-forms.
A. Finding archaea in the hot-spring Obsidian Pool at Yellowstone. **B.** *Pyrococcus furiosus*, an organism that lives at temperatures above 100°C (transmission electron micrograph).

ancestor of mitochondria derives from ancient proteobacteria (shaded pink), whereas chloroplasts derive from ancient cyanobacteria (shaded green). The three-domain classification is largely supported by the sequences of microbial genomes, although some horizontal transfer of genes occurs both within and between the domains (discussed in Chapter 17).

THOUGHT QUESTION 1.8   What arguments support the classification of Archaea as a third domain of life? What arguments support the classification of archaea and bacteria together, as prokaryotes, distinct from eukaryotes?

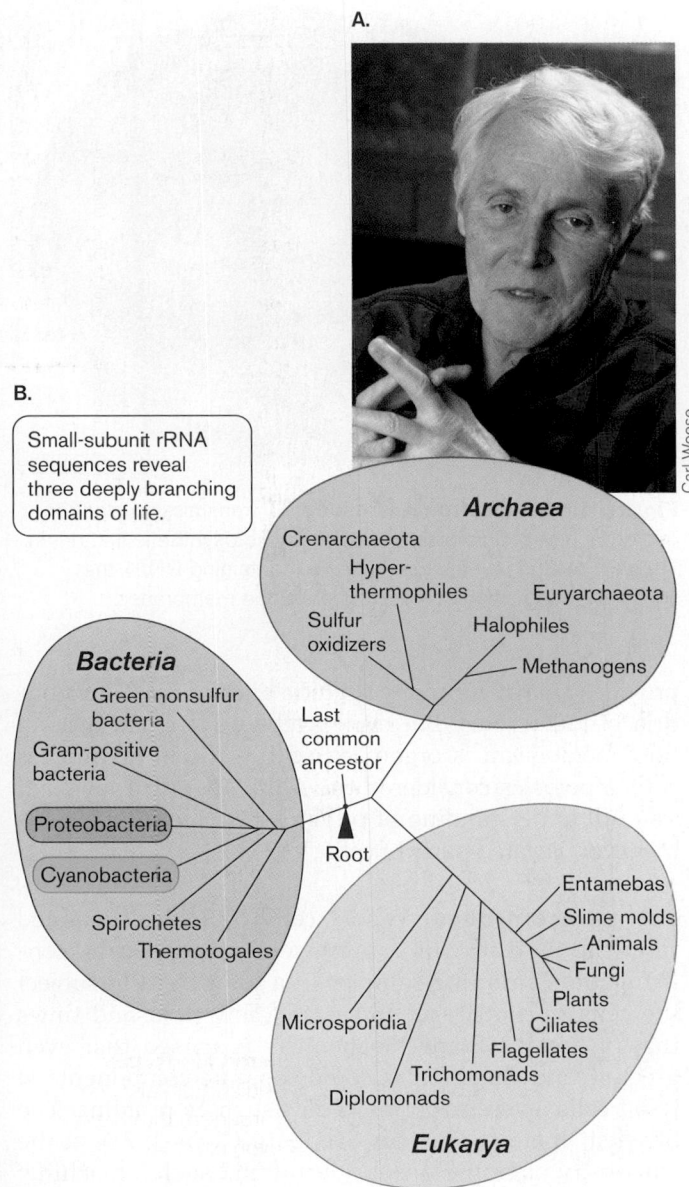

A.

Carl Woese

B.

Small-subunit rRNA sequences reveal three deeply branching domains of life.

Archaea

Crenarchaeota
Hyper-
thermophiles
Euryarchaeota
Sulfur
oxidizers
Halophiles
Methanogens

Bacteria

Green nonsulfur
bacteria
Last
common
ancestor
Gram-positive
bacteria
Proteobacteria
Root
Cyanobacteria
Spirochetes
Thermotogales

Entamebas
Slime molds
Animals
Fungi
Plants
Microsporidia
Ciliates
Flagellates
Trichomonads
Diplomonads

Eukarya

**Figure 1.28 Carl Woese and the three domains of life.
A.** Carl Woese (University of Illinois at Urbana-Champaign) proposed that archaea constitute a third domain of life. **B.** Three domains form a monophyletic tree based on small-subunit rRNA sequences. Examples of branching groups of organisms are shown. The length of each branch approximates the time of divergence from the last common ancestor. For a more detailed tree see Chapter 17.

TO SUMMARIZE:

- **Classifying microbes** was a challenge historically because of the difficulties in observing distinguishing characteristics of different categories.
- **Ernst Haeckel** recognized that microbes constitute a form of life distinct from animals and plants.

- **Herbert Copeland and Robert Whittaker** classified prokaryotes as a form of microbial life distinct from eukaryotic microbes such as protists.
- **Lynn Margulis** proposed that eukaryotic organelles such as mitochondria and chloroplasts evolved by endosymbiosis from prokaryotic cells engulfed by proto-eukaryotes.
- **Carl Woese** discovered a domain of prokaryotes, Archaea, whose genetic sequences diverge equally from those of bacteria and those of eukaryotes. Many, though not all, archaea grow in extreme environments.

1.6 Cell Biology and the DNA Revolution

During the twentieth century, amid world wars and societal transformations, the field of microbiology exploded with new knowledge (see **Table 1.2**). More than 99% of what we know about microbes today was discovered after 1900 by scientists too numerous to cite in this book. Advances in biochemistry and microscopy revealed the fundamental structure and function of cell membranes and proteins. The revelation of the structures of DNA and RNA led to the discovery of the genetic programs of model bacteria such as *E. coli* and the bacteriophage lambda. Beyond microbiology, these advances produced the technology of "recombinant DNA," or genetic engineering, the construction of molecules that combine DNA sequences from unrelated species. These microbial tools offered unprecedented applications to human medicine and industry.

Cell Membranes and Macromolecules

In 1900, the study of cell structure was still limited by the resolution of the light microscope and by the absence of tools that could take apart cells to isolate their components. Both these limitations were overcome by the invention of powerful instruments. Just as society was being transformed by machines ranging from jet airplanes to vacuum cleaners, the study of microbiology was also being transformed by machines. Two instruments had exceptional impact: The **electron microscope** revealed the internal structure of cells (Chapter 2), and the **ultracentrifuge** enabled isolation of subcellular parts (Chapter 3).

The electron microscope. In the 1920s, at the Technical College in Berlin, student Ernst Ruska (1906–1988) was invited to develop an instrument for focusing rays of electrons. Ruska recalled as a child how his father's microscope could magnify fascinating specimens of plants and animals, but that its resolution was limited by the wavelength of light. He was eager to devise lenses that

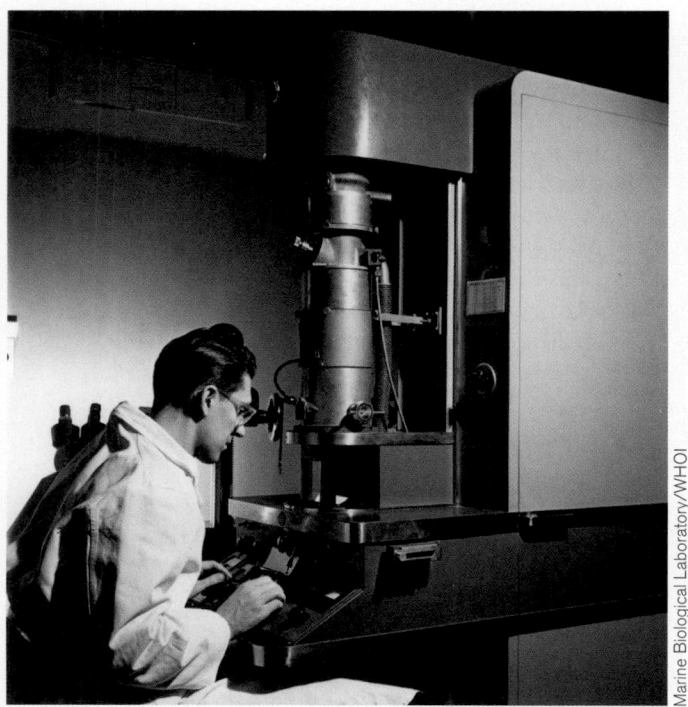

Marine Biological Laboratory/WHOI

Figure 1.29 Electron microscopy. An early transmission electron microscope. The tall column contains a series of magnetic lenses.

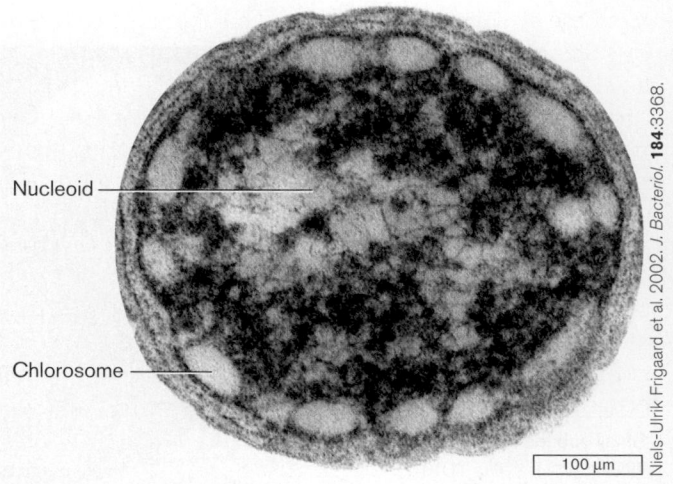

Niels-Ulrik Frigaard et al. 2002. *J. Bacteriol.* **184**:3368.

Figure 1.30 Electron microscopy. Transmission electron micrograph of a *Chlorobium* species, a photosynthetic bacterium. The thin section reveals the nucleoid (containing DNA), the light-harvesting chlorosomes, and envelope membranes.

could focus beams of electrons, with wavelengths far smaller than that of light, to reveal living details never seen before. Ultimately, Ruska built lenses to focus electrons using specially designed electromagnets. Magnetic lenses were used to complete the first electron microscope in 1933 (**Fig. 1.29**). Early transmission electron microscopes achieved about tenfold greater magnification than the light microscope, revealing details such as the ridged shell of a diatom. Further development steadily increased magnification, to as high as a millionfold.

For the first time, cells were seen to be composed of a cytoplasm containing macromolecules and bounded by a phospholipid membrane. By the 1960s, thin sections revealed the entire inner architecture of eukaryotic cells, including intracellular membranes, ribosomes, and organelles such as mitochondria and chloroplasts. Within the smaller cells of bacteria, the DNA-containing nucleoid was revealed, as well as specialized structures, such as photosynthetic organelles called chlorosomes (**Fig. 1.30**).

Subcellular structures, however, raised many questions about cell function that visualization alone could not answer. Biochemists showed that cell function involves numerous chemical transformations mediated by enzymes. A milestone in the study of metabolism was the elucidation by German biochemist Hans Krebs (1900–1981) of the tricarboxylic acid cycle (TCA cycle, or Krebs cycle), by which the products of sugar digestion are converted to carbon dioxide. The tricarboxylic acid cycle

provides energy for many bacteria and for the mitochondria of eukaryotes. But even Krebs understood little of how metabolism is organized within a cell; he and his contemporaries considered the cell a "bag of enzymes." The full understanding of cell structure required experiments on isolated parts of cells.

The ultracentrifuge. Whole cells could be separated from the fluid in which they were suspended by centrifugation, spinning samples in a rotor so as to subject the cells to centrifugal forces of a few thousand times that of gravity. Some biochemists proposed that even greater centrifugal forces could separate components of lysed cells, even macromolecules such as proteins. The Swedish chemist Theodor Svedberg (1884–1971), at the University of Uppsala, set out to build such a machine: the ultracentrifuge.

Svedberg studied the properties of macromolecules such as proteins and polysaccharides. A major challenge was to measure the size of such particles. Svedberg calculated that particle size could be determined from the rate of the movement of particles in a tube subjected to forces hundreds of thousands times that of gravity. So he designed and built rotors spinning at ever-greater rates—rates so high that they required a vacuum to avoid burning up like a space reentry vehicle. Such an advanced centrifuge, used to separate components of cells, is called an ultracentrifuge.

Experiments based on electron microscopy and ultracentrifugation revealed how membranes govern energy transduction within bacteria and within organelles such as mitochondria and chloroplasts. In the 1960s, English biochemists Peter Mitchell (1920–1992) and Jennifer Moyle (1921–) proposed and tested a revolutionary idea called the **chemiosmotic theory**. The chemiosmotic theory states that the reduction-oxidation (redox) reactions of the electron transport system store energy in the form of a gradient of protons (hydrogen ions) across a membrane, such as the bacterial cell membrane or the inner membrane of mitochondria. The energy stored in the proton gradient, in turn, drives the synthesis of ATP.

The chemiosmotic theory earned Mitchell the Nobel Prize in Chemistry in 1978. Nevertheless, the theory took many years to win acceptance by biologists committed to the "bag of enzymes" model of the cell. Ultimately, the role of ion gradients across membranes was recognized as fundamental to all living cells. (Ion gradients in membranes are discussed further in Chapters 3 and 4, and their role in energy transduction is described in Chapter 14.)

Microbial Genetics Leads the DNA Revolution

As the form and function of living cells emerged in the early twentieth century, a largely separate line of research revealed patterns of heredity of cell traits. In eukaryotes, the Mendelian rules of inheritance were rediscovered and connected to the behavior of subcellular structures called chromosomes. Frederick Griffith (1879–1941) showed in 1928 that an unknown substance out of dead bacteria could carry genetic information into living cells, transforming harmless bacteria into a strain capable of killing mice—a process called **transformation**. Some kind of "genetic material" must be inherited to direct the expression of inherited traits, but no one knew what that material was or how its information was expressed.

Then, in 1944, Oswald Avery (1877–1955) and colleagues showed that the genetic material for transformation is deoxyribonucleic acid, or DNA. An obscure acidic polymer, DNA had been previously thought too uniform in structure to carry information; its precise structure was unknown. As World War II raged among nations, among scientists an epic struggle began: the quest for the structure of DNA.

The discovery of the double helix. The tool of choice for macromolecular structure was X-ray crystallography, a method developed by British physicists in the early 1900s. The field of X-ray analysis included an unusual number of women, including Dorothy Hodgkin (1910–1994), who won a Nobel Prize for the structures of penicillin and vitamin B_{12}. In 1953, crystallographer Rosalind Franklin joined a laboratory at King's College to study the structure of DNA (**Fig. 1.31A**). As a woman and as a Jew who supported relief work in Palestine, Franklin felt socially isolated at the male-dominated Protestant university. Nevertheless, the exceptional quality of her X-ray micrographs impressed her colleagues. The X-shaped pattern in her micrograph (**Fig. 1.31B**) showed for the first time that the standard B-form DNA was a double helix. Without Franklin's knowledge, her colleague Maurice Wilkins showed her data to a competitor, James Watson (1928–). The pattern led Watson and Francis Crick (1916–2004) to guess that the four bases of the DNA "alphabet" were paired in the interior of Franklin's double helix (**Fig. 1.31C**). They published their model in the journal *Nature*, while denying that they had used Franklin's data.

The discovery of the double helix earned Watson, Crick, and Wilkins the 1962 Nobel Prize in Physiology or Medicine. Franklin died of ovarian cancer before the Nobel Prize was awarded. Before her death, however, she turned her efforts to the structure of ribonucleic acid (RNA). She determined the form of the RNA chromosome within tobacco mosaic virus, the first viral RNA to be characterized.

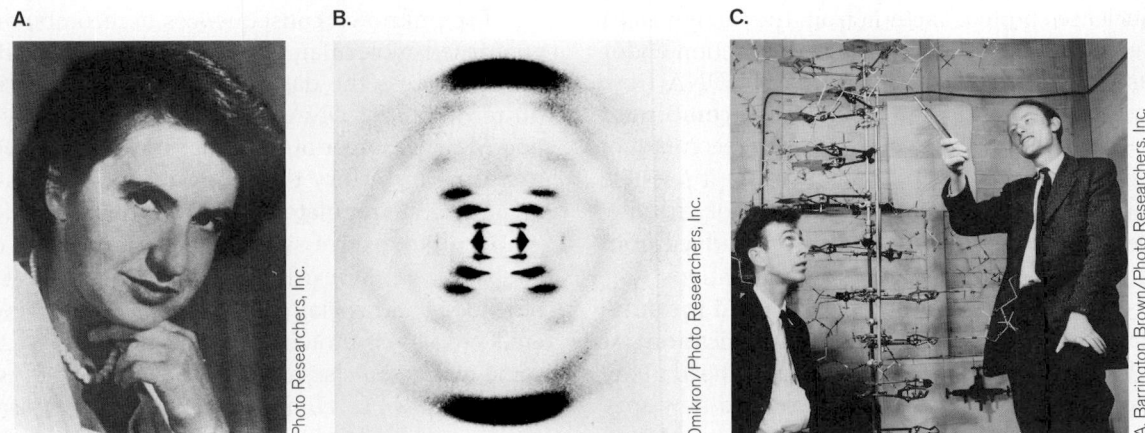

A.   **B.**   **C.**

Figure 1.31   The DNA double helix.   A. Rosalind Franklin discovered that DNA forms a double helix. **B.** X-ray diffraction pattern of DNA, obtained by Rosalind Franklin. **C.** James Watson and Francis Crick discovered the complementary pairing between bases of DNA and the antiparallel form of the double helix.

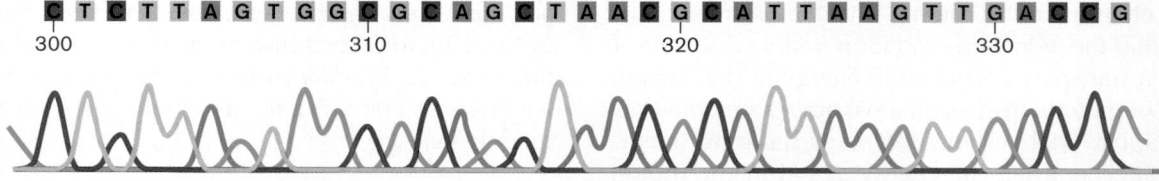

Figure 1.32 A DNA sequence fluorogram. Sequence obtained by an undergraduate at Kenyon College using an ABI DNA sequencer, from genomic DNA of a microbe isolated from a discarded beverage container. Each colored trace represents the intensity of fluorescence of one of the four bases terminating a chain of DNA. Units represent DNA bases.

The structure of DNA base pairs led to the development of techniques for **DNA sequencing**, the reading of the sequence of DNA base pairs. **Figure 1.32** shows an example of DNA sequence data, where each color represents one of the four bases and each peak represents a DNA fragment terminating in that particular base. The order of fragment lengths yields the sequence of bases in one strand. Reading the sequence enabled microbiologists to determine the beginning and endpoint of genes and ultimately entire genomes.

The DNA revolution began with bacteria. What amazed the world about DNA was that such a simple substance, composed of only four types of subunits, is the genetic material that determines all the different organisms on Earth. The promise of this insight was first fulfilled in bacteria and bacteriophages, whose small genomes and short generation times made key experiments possible. Furthermore, naturally occurring mechanisms of gene exchange in bacteria and viruses provided scientists with the key tools for transferring genes, including those of animals and plants. Consider these examples:

- **Restriction endonucleases led to recombinant DNA.** Bacteria make **restriction endonucleases**, enzymes that cut DNA at positions determined by specific short base sequences. In nature, restriction endonucleases protect bacteria from the foreign DNA of viruses. In the test tube, purified restriction endonucleases were used to "cut and paste" DNA from two unrelated organisms, generating **recombinant DNA**. The construction of artificially recombinant DNA, by "gene cloning," ultimately made it possible to transfer genes between the genomes of virtually all types of organisms, using processes derived from natural phenomena of bacterial transformation.
- **A heat-stable DNA polymerase was used for polymerase chain reaction (PCR) amplification of DNA.** While molecular biologists forged ahead in the laboratory, microbial ecologists discovered new species of bacteria and archaea in increasingly extreme habitats. A hot spring in Yellowstone National Park yielded the bacterium *Thermus aquaticus*, whose DNA polymerase could survive many rounds of cycling to near-boiling temperature. The Taq polymerase formed the basis of a multibillion-dollar industry of PCR amplification of DNA, with applications ranging from genome sequencing to forensic identification.
- **Gene regulation discovered in bacteria provided models for animals and plants.** The information carried by DNA was shown to be expressed by transcription to RNA and then translation of RNA to make proteins. Most of the key discoveries of gene expression were made in bacteria and bacteriophages. The regulation of expression of a bacterial gene was first demonstrated for the *lac* operon, a set of contiguous genes in *E. coli*. Transcription of the *lac* operon was shown to be regulated by a repressor protein binding to DNA. DNA-binding proteins were subsequently demonstrated in all classes of living organisms.

Public response. As early as the 1970s, when the DNA revolution was largely limited to bacteria, its implications drew public concern. The use of recombinant DNA to make hybrid organisms—organisms combining DNA from more than one species—seemed "unnatural," although we now know that interspecies gene transfer occurs ubiquitously in nature. Furthermore, recombinant DNA technology raised the specter of placing deadly genes that produce toxins such as botulin into innocuous human flora such as *E. coli*.

The unknown consequences of recombinant DNA so concerned molecular biologists that they held a conference to assess the dangers and restrict experimentation on recombinant DNA. The conference, led by Paul Berg and Maxine Singer at Asilomar in 1975, was possibly the first time in history that a group of scientists organized and agreed to regulate and restrict their own field.

On the positive side, the emerging world of molecular biology drew excitement from students at a time of new ideas and social change. In 1971, the newly discovered process of protein translation by the bacterial ribosome inspired a classic cult film, *Protein Synthesis: An Epic on the Cellular Level*, directed by Gabriel Weiss and choreographed by Jackie Bennington, America's 1969 Junior Miss (**Fig. 1.33**). Introduced by Nobel laureate Paul Berg, the film depicts dancers on the Stanford University football field forming the shape of the ribosome while "messenger

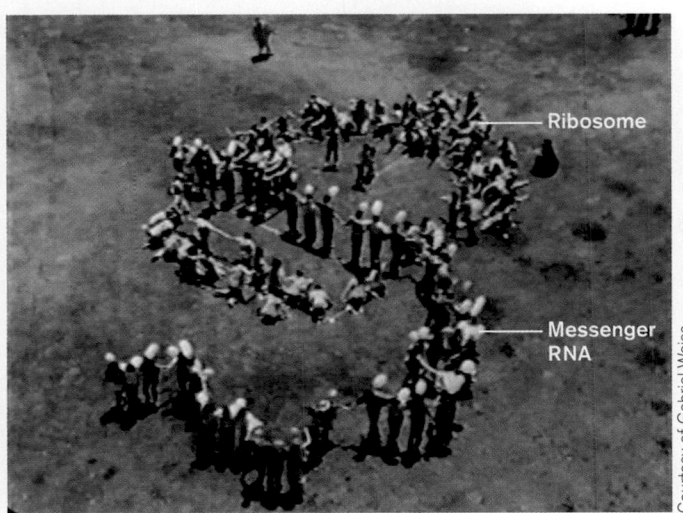

Courtesy of Gabriel Weiss

Figure 1.33 "Protein Jive Sutra." At Stanford University, in 1971, rock dancers represented the process of protein synthesis in the ribosome. The depiction was immortalized in a film produced by chemist Kent Wilson and directed by Gabriel Weiss, with an introduction by Nobel laureate Paul Berg. *biology.kenyon.edu/protein.mp4*

RNA" and "transfer RNAs" meet to assemble a "protein." The dancers sway to "Protein Jive Sutra," performed by a Haight-Ashbury-inspired rock band. Their excitement echoed that of scientists exploring the extraordinary potential of microbial discovery.

THOUGHT QUESTION 1.9 Do you think engineered strains of bacteria should be patentable? What about sequenced genes or genomes?

Microbial Discoveries Transform Medicine and Industry

Twentieth-century microbiology transformed the practice of medicine and generated entire new industries of biotechnology and bioremediation. Following the discovery of penicillin, Americans poured millions of dollars of private and public funds into medical research. The March of Dimes campaign for private donations to prevent polio led to the successful development of a vaccine that has nearly eliminated the disease. With the end of World War II, research on microbes and other aspects of biology drew increasing financial support from U.S. government agencies, such as the National Institutes of Health and the National Science Foundation, as well as from governments of other countries, particularly the European nations and Japan. Further support has come from private foundations, such as the Pasteur Institute, the Wellcome Trust, and the Howard Hughes Medical Institute.

Research in microbiology finds applications in diverse fields (**Table 1.3**), all of which recruit microbiologists (**Fig. 1.34**). Cloned genes in bacteria produce valuable therapeutic proteins, such as insulin for diabetics. Recombinant viruses make safer vaccines. Bacteria and viral recombinant genomes are used to construct transgenic animals and plants. In environmental science, newly discovered microbes provide new ways to bioremediate wastes and control insect pests. On a global level, the management of our planet's biosphere, with the challenges of pollution and global warming, increasingly depends on our understanding of microbial ecology.

WWW | American Society for Microbiology
microbiology2.com/links

TO SUMMARIZE:

- **Genetics of bacteria, bacteriophages, and fungi** in the early twentieth century revealed fundamental insights about gene transmission that apply to all organisms.
- **Structure and function of the genetic material, DNA**, were revealed by a series of experiments in the twentieth century.
- **Molecular microbiology generated key advances**, such as the cloning of the first recombinant molecules and the invention of DNA sequencing technology.

Table 1.3	Fields of research in microbiology.
Field	**Subject of study**
Experimental microbiology	Fundamental questions about microbial form and function, genetics, and ecology
Medical microbiology	The mechanism, diagnosis, and treatment of microbial disease
Epidemiology	Distribution and causes of disease in humans, animals, and plants
Immunology	The immune system and other host defenses against infectious disease
Food and industrial microbiology	Fermented foods, food preservation, and industrial microbial products
Environmental microbiology	Microbial diversity and microbial processes in natural and artificial environments
Bioremediation	The use of microbial metabolism to remediate human wastes and industrial pollutants
Forensic microbiology	Analysis of microbial strains as evidence in criminal investigations
Astrobiology	The origin of life in the universe and the possibility of life outside Earth

Figure 1.34 Microbiologists at work. Students at Kenyon College conduct research on bacterial gene expression.

Joan Slonczewski, Kenyon College

- **Genome sequence determination and bioinformatic analysis** became the tools that shape the study of biology in the twenty-first century.
- **Microbial discoveries transformed medicine and industry.** Biotechnology enables the production of new kinds of pharmaceuticals and industrial products.

Concluding Thoughts

The advances in microbial science raise important questions for society. How can medical research control emerging diseases? How do microbes contribute to global cycles of carbon and nitrogen? How can microbial metabolism clean up polluted environments, such as the vast oil spill from the explosion of the Deepwater Horizon wellhead in 2010?

This book explores the current explosion of knowledge about microbial cells, genetics, and ecology. We introduce the applications of microbial science to human affairs, from medical microbiology to environmental science. Most important, we discuss research methods—how scientists make the discoveries that will shape tomorrow's view of microbiology. In the rest of Part 1, Chapter 2 presents the visualization tools that make possible our increasingly detailed view of the structures of cells (Chapter 3) and viruses (Chapter 6). Chapters 4 and 5 introduce microbial nutrition and growth in diverse habitats, including new information derived from the sequencing of microbial genomes. Throughout, we invite readers to share with us the excitement of discovery in microbiology.

CHAPTER REVIEW

Review Questions

1. Explain the apparent contradictions in defining microbiology as the study of microscopic organisms or the study of single-celled organisms.
2. What is the genome of an organism? How do genomes of viruses differ from those of cellular microbes?
3. Under what conditions might microbial life have originated? What evidence supports current views of microbial origin?
4. List the ways in which microbes have affected human life throughout history.
5. Summarize the key experiments and insights that shaped the controversy over spontaneous generation. What key questions were raised, and how were they answered?
6. Explain how microbes are cultured on liquid and solid media. Compare and contrast the culture methods of Koch and Winogradsky. How did their different approaches to microbial culture address different questions in microbiology?
7. Explain how a series of observations of disease transmission led to the development of immunization to prevent disease.
8. Summarize key historical developments in our view of microbial taxonomy. What attributes of microbes have made them challenging to classify?
9. Explain how various discoveries in "natural" bacterial genetics were used to develop recombinant DNA technology.

Thought Questions

1. How do the Earth's microbes contribute to human health? Include examples of environmental microbes outside the human body, as well as microbes associated with the human body.
2. When oxygen is used up, certain microbes ferment their carbon sources to produce ethanol or acetate. If oxygen is replaced, what do you think happens to the ethanol or acetate?
3. Why do you think so many environmental microbes cannot be cultured in laboratory broth or agar media?
4. Outline the different contributions to medical microbiology and immunology of Louis Pasteur, Robert Koch, and Florence Nightingale. What methods and assumptions did they have in common, and how did they differ?
5. Outline the different contributions to environmental microbiology of Sergei Winogradsky and Martinus Beijerinck. Why did it take longer for the significance of environmental microbiology to be recognized, as compared with pure-culture microbiology?
6. The Part 1 interview of Rita Colwell describes how she and Anwar Huq devised an inexpensive way for Bangladeshi villagers to prevent cholera. What historical discoveries in microbiology, both medical and environmental, laid the foundation for their approach?

Key Terms

agar (21)
antibiotic (24)
antiseptic (23)
Archaea (8)
archaeon (8)
aseptic (24)
autoclave (17)
Bacteria (8)
bacterium (8)
biofilm (27)
chain of infection (20)
chemiosmotic theory (33)
chemolithotroph (26)
colony (20)
DNA sequencing (34)

electron microscope (31)
endosymbiont (27)
enrichment culture (26)
Eukarya (8)
eukaryote (8)
fermentation (17)
genome (9)
geochemical cycling (27)
germ theory of disease (17)
immune system (23)
immunity (immunization) (23)
Koch's postulates (21)
lithotroph (26)
metagenome (11)
microbe (7)

monophyletic (30)
nitrogen fixation (27)
petri dish (21)
photosynthesis (26)
polyphyletic (30)
prokaryote (8)
pure culture (20)
recombinant DNA (34)
restriction endonuclease (34)
spontaneous generation (16)
transformation (33)
ultracentrifuge (31)
vaccination (22)
virus (8)
Winogradsky column (26)

Recommended Reading

Angert, Esther, Kendall D. Clements, and Norman R. Pace. 1993. The largest bacterium. *Nature* **362**:239–241.

Brock, Thomas D. 1999. *Milestones in Microbiology, 1546 to 1940.* ASM Press, Washington, DC.

Brock, Thomas D. 1999. *Robert Koch: A Life in Medicine and Bacteriology.* ASM Press, Washington, DC.

Dinc, Gulten, and Yesim I. Ulman. 2007. The introduction of variolation 'A La Turca' to the West by Lady Mary Montagu and Turkey's contribution to this. *Vaccine* **25**:4261–4265.

Doudna, Jennifer A., and Thomas R. Cech. 2004. The chemical repertoire of natural ribozymes. *Nature* **418**:222–228.

Dubos, Rene. 1998. *Pasteur and Modern Science.* Translated by Thomas Brock. ASM Press, Washington, DC.

Fleishmann, Robert D., Mark D. Adams, Owen White, Rebecca A. Clayton, Ewen F. Kirkness, et al. 1995. Whole-genome random sequencing and assembly of *Haemophilus influenzae* Rd. *Science* **269**:496–512.

Hesse, Wolfgang. 1992. Walther and Angelina Hesse—early contributors to bacteriology. *ASM News* **58**:425–428.

Huq, Anwar, Mohammed Yunus, Syed S. Sohelb, Abbas Bhuiyab, Michael Emche, et al. 2010. Simple sari cloth filtration of water is sustainable and continues to protect villagers from cholera in Matlab, Bangladesh. *mbio* **1**:e00034.

Joklik, Wolfgang K., Lars G. Ljungdahl, Alison D. O'Brien, Alexander von Graevenitz, and Charles Yanofsky (eds.). 1999. *Microbiology: A Centenary Perspective.* ASM Press, Washington, DC.

Macfarlane, Sandra, and George T. Macfarlane. 2006. Composition and metabolic activities of bacterial biofilms colonizing food residues in the human gut. *Applied and Environmental Microbiology* **72**:6204–6211.

Margulis, Lynn. 1968. Evolutionary criteria in Thallophytes: A radical alternative. *Science* **161**:1020–1022.

Raoult, Didier. 2005. The journey from *Rickettsia* to Mimivirus. *ASM News* **71**:278–284.

Sherman, Irwin W. 2006. *The Power of Plagues.* ASM Press, Washington, DC.

Thomas, Gavin. 2005. Microbes in the air: John Tyndall and the spontaneous generation debate. *Microbiology Today* (Nov. 5): 164–167.

Ward, Naomi, and Claire Fraser. 2005. How genomics has affected the concept of microbiology. *Current Opinion in Microbiology* **8**:564–571.

Westall, Frances. 2005. Life on the early Earth: A sedimentary view. *Science* **308**:366–367.

Woese, Carl R., and George E. Fox. 1977. Phylogenetic structure of the prokaryotic domain: The primary kingdoms. *Proceedings of the National Academy of Sciences USA* **74**:5088–5090.

Worden, Alexandra Z., Marie L. Cuvelier, and Douglas H. Bartlett. (2006). In-depth analyses of marine microbial community genomics. *Trends in Microbiology* **14**:331–336.

Zuniga, Elina I., Bumsuk Hahm, Kurt H. Edelmann, and Michael B. A. Oldstone. 2005. Immunosuppressive viruses and dendritic cells: A multifront war. *ASM News* **71**:285–290.

 For further study and review, please visit StudySpace at **microbiology2.com**.

Chapter 2

Observing the Microbial Cell

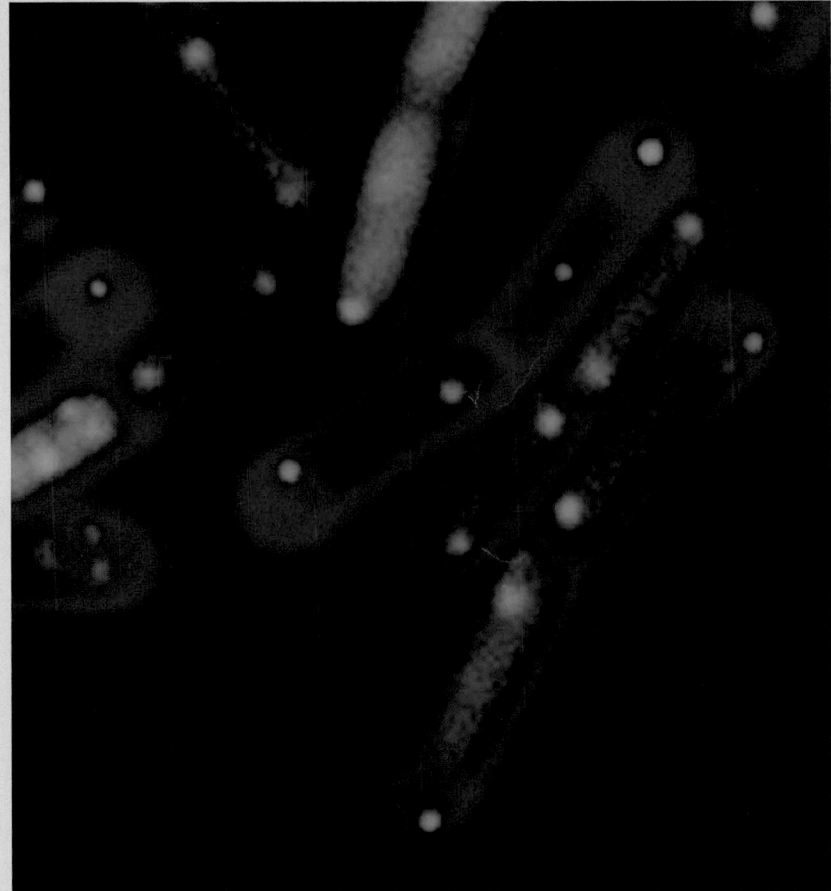

Microscopy reveals the vast realm of bacteria and protists invisible to the unaided eye. The microbial world spans a wide range of size—over several orders of magnitude. For different size ranges, we use different instruments, from the simple bright-field microscope to the electron microscope. The microscope enables us to count the number of microbes in the human bloodstream or in dilute natural environments such as the ocean. It shows how microbes swim and respond to signals such as a new food source.

Advanced forms of microscopy reveal microbes in remarkable detail. Fluorescence microscopy shows how parts function within a living cell. Electron microscopy explores the cell's interior, showing how all the parts of the cell fit together. The electron microscope is used to build models of viruses, and to probe the subcellular structures of membranes, ribosomes, and flagellar motors. Scanning-probe microscopy images live microbes with an unprecedented degree of realism.

Current Research Highlight

Bacillus subtilis bacteria form endospores, observed by fluorescence microscopy. Endospores can survive thousands of years in a dormant state, then germinate to grow and multiply. The cells shown contain a gene expressing a form of green fluorescent protein (GFP) that binds to the DNA origin of replication, seen as a green dot at each pole of the cell. The cell membrane and forespore (red bulge pinching off one pole) are stained with a dye that fluoresces red. This experiment in 2009 showed how DNA replication is organized in the cell to develop an endospore. Cell length = 2–3 μm. *Source:* Lilah Rahn-Lee et al. 2009. *J. Bacteriol.* **191**:3796.

When the microscope of Antonie van Leeuwenhoek first revealed the tiny life-forms on his teeth, scientists in England refused to believe it until their own instruments showed the same. Leeuwenhoek's superior lenses were key to his success. Since the time of Leeuwenhoek, microscopists have devised ever more powerful instruments to search for microbes in unexpected habitats.

An example of such a habitat is the human stomach, long believed too acidic to harbor life. In the 1980s, however, Australian scientist Barry Marshall reported finding a new species of bacterium in the stomach. The bacterium, *Helicobacter pylori*, proved difficult to isolate and culture, and for over a decade medical researchers refused to believe Marshall's report. Ultimately, electron microscopy (EM) confirmed the existence of *H. pylori* in the stomach and helped document its role in gastritis and stomach ulcers. The scanning electron micrograph in **Figure 2.1** shows *H. pylori* colonizing the gastric crypt cells. *H. pylori* bacteria are helical rods (colorized green). The actual size of the bacteria can be inferred from the scale bar at lower right in the figure. The cell contours are well resolved by the scanning beam of electrons—an achievement far beyond that possible with a light microscope.

Today, various kinds of microscopes are used in many settings, from hospitals and veterinary clinics to industrial plants and wastewater treatment facilities. Chapter 2 covers two areas:

- **Light microscopy.** The theory and practice of light microscopy are essential for every student and professional observing microorganisms.
- **Advanced tools for research.** Fluorescence microscopy, electron microscopy, and X-ray crystallography probe ever farther the frontiers of the unseen.

2.1 Observing Microbes

Most microbes are too small to be seen; that is, they are microscopic, requiring the use of a **microscope** to be seen. But why can't we see microbes without magnification? The answer is surprisingly complex. In fact, our definition of "microscopic" is based not on inherent properties of the organism under study, but on the properties of our eyes. What is microscopic actually lies in the eye of the beholder.

Resolution of Objects by Our Eyes

The size at which objects become visible depends on the resolution of the observer's eye. **Resolution** is the smallest distance by which two objects can be separated and still be distinguished. In the eyes of humans and other animals, resolution is achieved by focusing an image on a retina packed with light-absorbing photoreceptor cells (**Fig. 2.2**●). A group of photoreceptors with their linked neurons forms one unit of detection, comparable to a pixel on

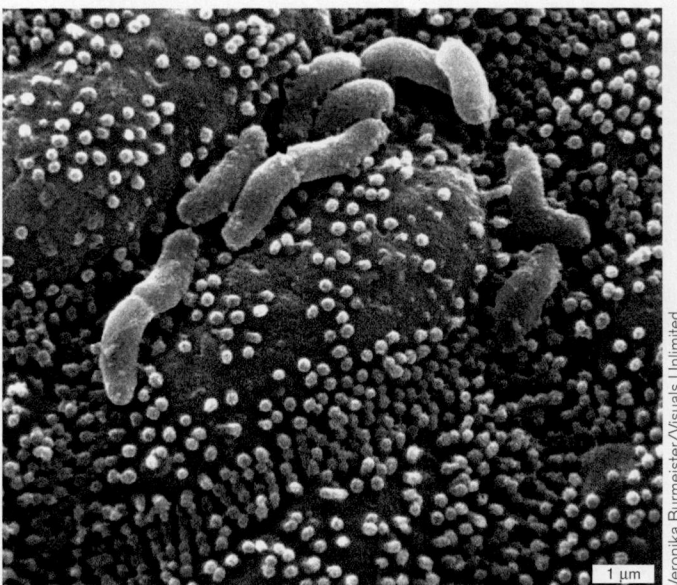

Veronika Burmeister/Visuals Unlimited

Figure 2.1 *Helicobacter pylori* **within the crypt cells of the stomach lining.** Microscopy demonstrated the presence of *H. pylori*, the cause of stomach ulcers, growing on the lining of the human stomach, a location previously believed too acidic to permit microbial growth. Scanning electron micrograph, colorized to indicate bacteria (green).

a computer screen. The distance between two retinal "pixels" limits resolution.

The resolution of the human retina (that is, the length of the smallest object most human eyes can see) is about 150 μm, or one-seventh of a millimeter. In the retinas of eagles, photoreceptors are more closely packed, so an eagle can resolve objects eight times as small or eight times as far away as a human can; hence, the phrase "eagle-eyed" means "sharp-sighted." On the other hand, insect compound eyes have poorer resolution (they can resolve only objects 100-fold larger than those we can resolve) because their receptor cells are farther apart than ours. If a science-fictional giant ameba had eyes with a retinal resolution of 2 meters, it would perceive humans as we do microbes.

NOTE: In this book, we use standard metric units for size:

1 millimeter (mm) = one-thousandth of a meter (m)
$$= 10^{-3} \text{ m}$$
1 micrometer (μm) = one-thousandth of a millimeter
$$= 10^{-6} \text{ m}$$
1 nanometer (nm) = one-thousandth of a micrometer
$$= 10^{-9} \text{ m}$$
1 picometer (pm) = one-thousandth of a nanometer
$$= 10^{-12} \text{ m}$$

Some authors still use the traditional unit angstrom (Å), which equals a tenth of a nanometer, or 10^{-10} m.

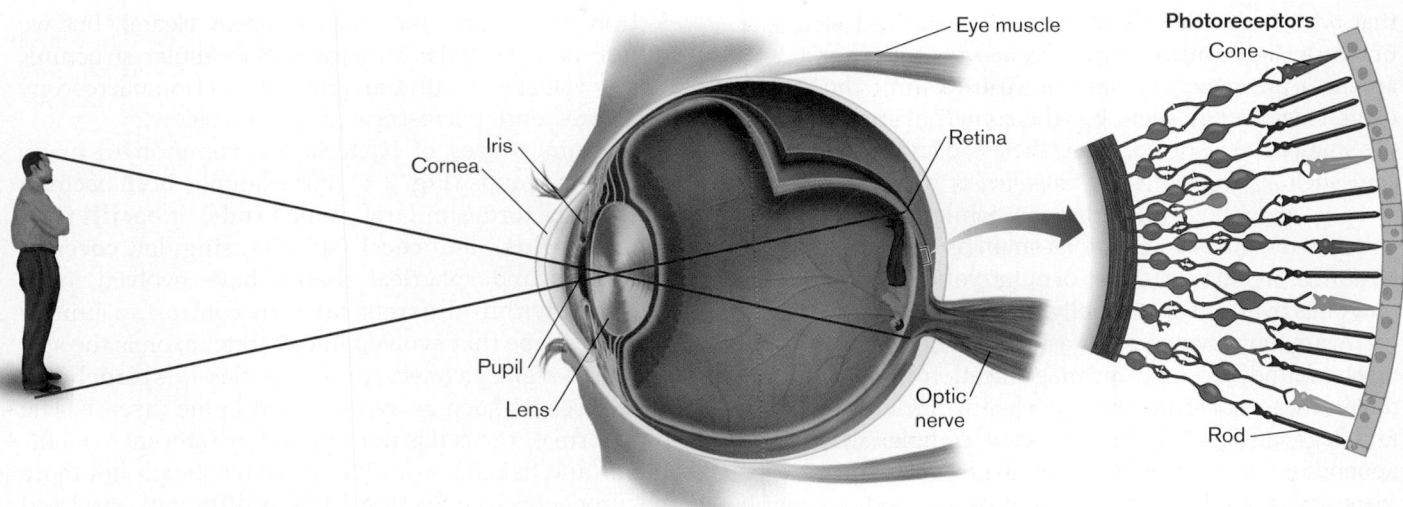

Figure 2.2   Defining the microscopic.   We define what is visible and what is microscopic in terms of the human eye. Within the human eye, the lens focuses an image on the retina. Adjacent light receptors are sufficiently close to resolve the image of a human being, even at a distance where its apparent height shrinks to a millimeter. ▶‖ *microbiology2.com/animations*

Resolution Differs from Detection

Objects of sizes below the resolution limit may yet be *detected*; that is, we can observe their presence as a group. For example, our eyes can detect a large population of microbes, such as a spot of mold on a piece of bread (about a million cells) or a cloudy tube of bacteria in liquid culture (a million cells per milliliter; **Fig. 2.3A**). **Detection**, the ability to determine the presence of an object, differs from resolution. When the unaided eye detects the presence of mold or bacteria, it cannot resolve individual cells. To resolve bacterial cells, such as those of the wetland phototroph *Rhodospirillum rubrum* (**Fig. 2.3B**), requires magnification. In microscopy, **magnification** means to increase the apparent size of an image so as to resolve smaller separations between objects and thus increase the information obtained by our eyes.

Microbial Size and Shape

Different kinds of microbes differ in size, over a range of several orders of magnitude, or powers of ten (**Fig. 2.4**). Eukaryotic microbes are often sufficiently large (10–100 μm)

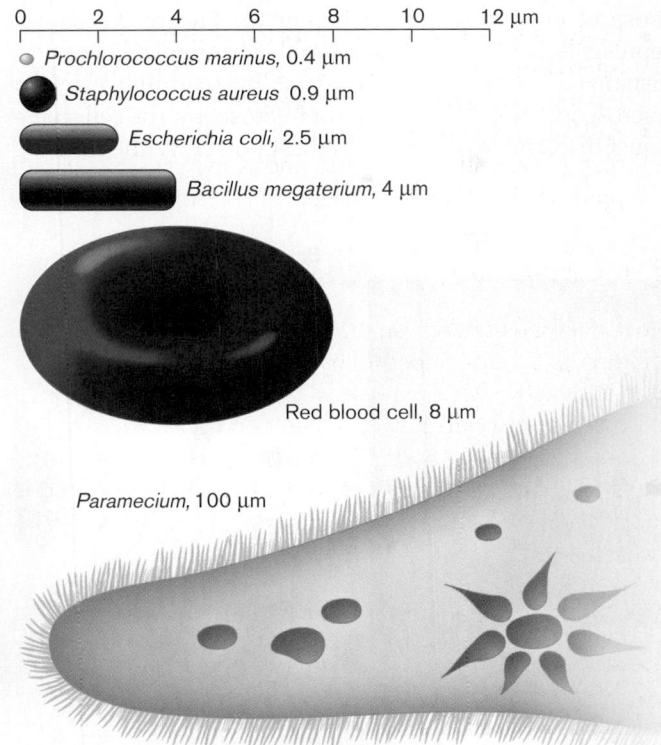

Figure 2.4   Relative sizes of different cells.   Microbial cells come in various sizes, most of which are below the threshold of resolution by the unaided human eye (150 μm). Note that most bacteria are much smaller than the human red blood corpuscle, a small eukaryotic cell.

A.   **B.**

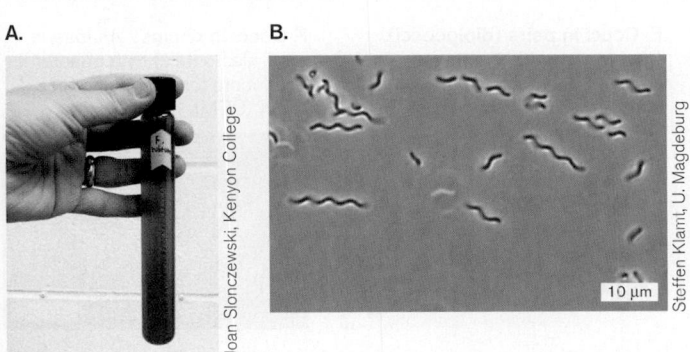

Joan Slonczewski, Kenyon College

Steffen Klamt, U. Magdeburg

10 μm

Figure 2.3   Detecting and resolving bacteria.   A. A tube of bacterial culture, *Rhodospirillum rubrum*. Bacterial cells are detected by turbidity, though not resolved. **B.** Bright-field image. Individual cells of *R. rubrum* are resolved.

that we can resolve their compartmentalized structure under a light microscope. Prokaryotes (bacteria and archaea) are generally smaller (0.4–10 μm); thus, their overall size can be seen, but their internal structures are too small to be resolved. Nevertheless, a few bacterial species, such as *Thiomargarita namibiensis*, are large enough to be seen without a microscope, while many unculturable marine eukaryotes are as small as the smallest bacteria. The actual size ranges of eukaryotic and prokaryotic microbes overlap substantially.

Many eukaryotes, such as protists and algae, can be resolved under low-power magnification to reveal internal and external structures such as the nucleus, vacuoles, and flagella (**Fig. 2.5**). Protists show complex shapes and appendages. For example, an ameba from an aquatic ecosystem shows a large nucleus and pseudopods to engulf prey (**Fig. 2.5A**). Pseudopods can be seen moving by the streaming of their cytoplasm. Another protist readily observed by light microscopy is *Trypanosoma brucei*, an insect-borne blood parasite that causes African sleeping sickness (**Fig. 2.5B**). In the trypanosome, we observe a nucleus and a flagellum. Eukaryotic flagella propel the cell by a whiplike action. Similarly, we can observe the complex internal patterns of chloroplasts in algae and the exceptional variety of crystalline forms in diatoms. For more on microbial eukaryotes, see Chapter 20.

Prokaryotic cell structures are generally simpler than those of eukaryotes (see Chapter 3). **Figure 2.6** shows representative members of several common cell types, as visualized by light microscopy or by scanning electron microscopy. Note that with light microscopy, the cell shape is just discernible under the highest power. With scanning electron microscopy, the shapes appear clearer, but we still see no subcellular structures. Subcellular structures are best visualized with transmission electron microscopy and fluorescence microscopy (discussed below).

Certain shapes of bacteria are common to many taxonomic groups (**Fig. 2.6**). For example, both bacteria and archaea form similarly shaped **rods**, or **bacilli** (singular, **bacillus**), and **cocci** (spheres; singular, **coccus**). Thus, rods and spherical shapes have evolved independently within different taxa. In contrast, a unique bacterial shape that evolved in only one taxon is the spirochete, a tightly coiled spiral. Species of **spirochetes** cause diseases such as syphilis and Lyme disease. The spiral form of the cell is maintained by internal axial filaments and flagella, as well as an outer sheath (for more on spirochetes, see Section 18.7). A different, unrelated spiral form is the "spirillum," a wide, rigid spiral cell that is similar to a rod-shaped bacillus.

A. Filamentous rods (bacilli). *Lactobacillus lactis*, gram-positive bacteria (LM).

B. Rods (bacilli). *Lactobacillus acidophilus*, gram-positive bacteria (SEM).

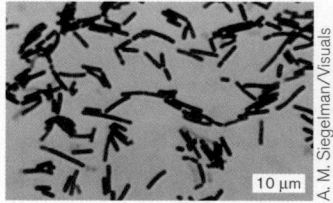

C. Spirochetes. *Borrelia burgdorferi*, cause of Lyme disease, among human blood cells (LM).

D. Spirochetes. *Leptospira interrogans*, cause of leptospirosis in animals and humans (SEM).

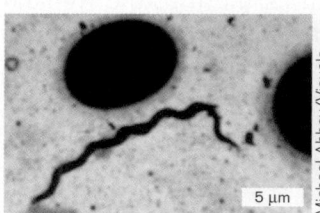
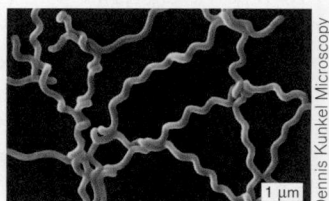

E. Cocci in pairs (diplococci). *Streptococcus pneumoniae*, a cause of pneumonia. Methylene blue stain (LM).

F. Cocci in chains. *Anabaena* spp., filaments of cyanobacteria. Producers for the marine food chain (SEM).

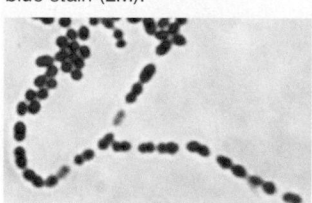
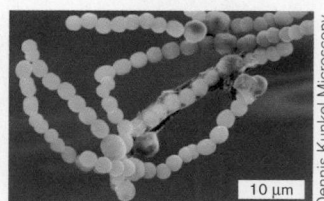

Figure 2.6   Common shapes of bacteria.   The shape of most bacterial cells can be discerned with light microscopy (LM) (**A, C, E**), but their subcellular structures and surface details cannot be seen. Surface detail is revealed by scanning electron microscopy (SEM) (**B, D, F**). These SEM images are colorized to enhance clarity.

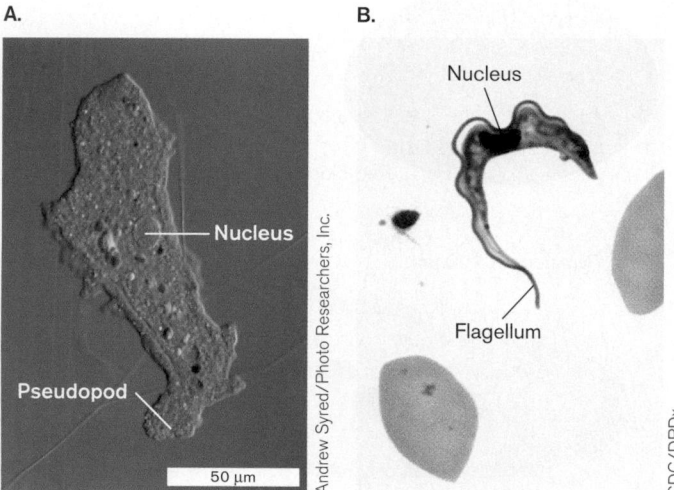

Figure 2.5   Eukaryotic microbial cells.   Eukaryotic microbes are large enough that details of internal and external organelles can be seen under a light microscope. **A.** *Amoeba proteus*. **B.** *Trypanosoma brucei*, cause of sleeping sickness. Length = 14–33 μm.

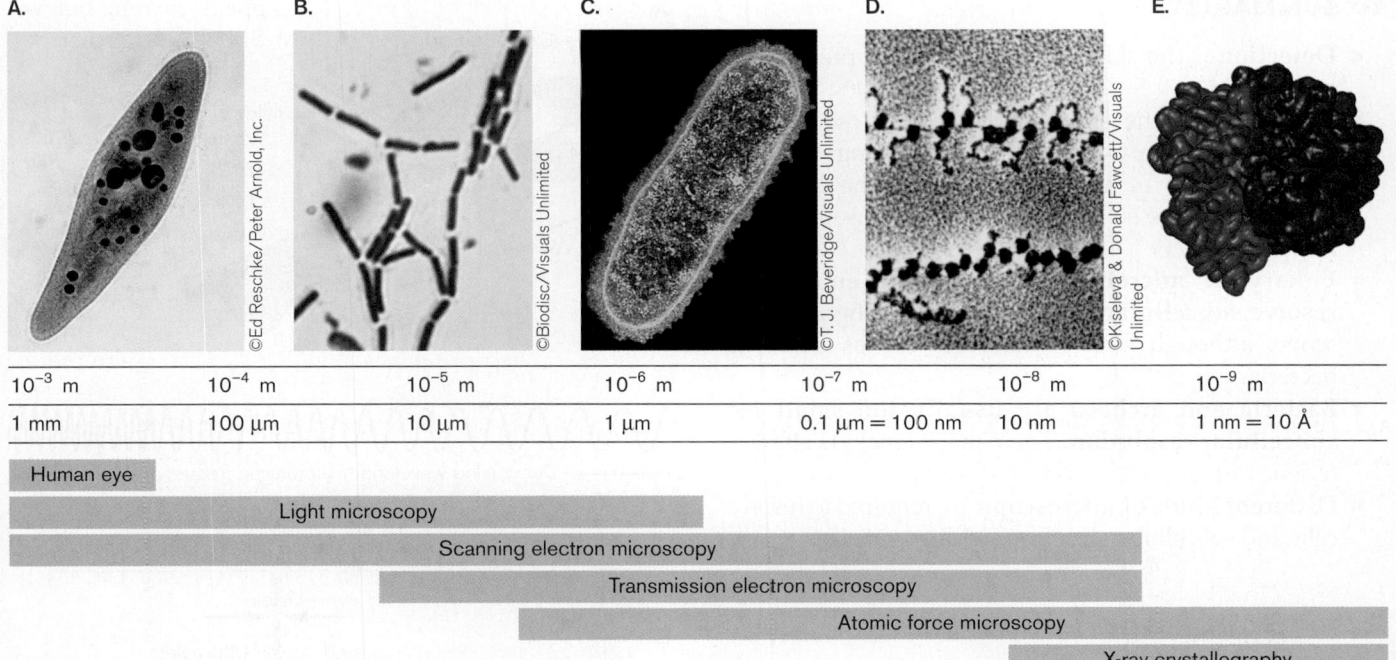

Figure 2.7 **Microscopy and X-ray crystallography, range of resolution.** **A.** Light microscopy reveals internal structures of a paramecium (a eukaryote, 100 μm). LM, magnification 100× at 35 mm size. **B.** *Pseudomonas* sp., rod-shaped bacteria (1–2 μm), are resolved. LM, magnification 1,200×. **C.** Internal structures of the bacterium *Escherichia coli* (1–2 μm) are revealed by transmission electron microscopy (TEM). Magnification 32,000×. **D.** Individual ribosomes are attached to the messenger RNA molecules that are translated to make peptides. TEM, magnification 150,000×. **E.** A ribosome (diameter 21 nm) can be modeled with X-ray crystallography.

NOTE: The genus name *Bacillus* refers to a specific taxonomic group of bacteria, but the term "bacillus" (plural, bacilli) refers to any rod-shaped bacterium or archaeon.

Microscopy for Different Size Scales

To resolve microbes and microbial structures of different sizes requires different kinds of microscopes. **Figure 2.7** shows the different techniques used to resolve microbes and structures of various sizes. For example, a paramecium can be resolved under a light microscope, but an individual ribosome requires electron microscopy.

■ **Light microscopy** resolves images of individual bacteria by their absorption of light. The specimen is commonly viewed as a dark object against a light-filled field, or background; this is called **bright-field microscopy** (seen in **Fig. 2.7A** and **B**). Advanced techniques, based on special properties of light, include dark-field, phase-contrast, and fluorescence microscopy.
■ **Electron microscopy (EM)** uses beams of electrons to resolve details several orders of magnitude smaller than those seen under light microscopy.

In **scanning electron microscopy (SEM)**, the electron beam is scattered from the metal-coated surface of an object, generating an appearance of three-dimensional depth. In **transmission electron microscopy (TEM)** (**Fig. 2.7C** and **D**), the electron beam travels through the object, where the electrons are absorbed by an electron-dense metal stain.

■ **Atomic force microscopy (AFM)** uses intermolecular forces between a probe and an object to map the three-dimensional topography of a cell.
■ **X-ray crystallography** detects the interference pattern of X-rays entering the crystal lattice of a molecule. From the interference pattern, researchers build a computational model of the structure of the individual molecule, such as a protein or a nucleic acid, or even a molecular complex such as a ribosome (**Fig. 2.7E**).

THOUGHT QUESTION 2.1 You have discovered a new kind of microbe, never observed before. What kinds of questions about this microbe might be answered by light microscopy? What questions would be better addressed by electron microscopy?

TO SUMMARIZE:

- **Detection** is the ability to determine the presence of an object.
- **Resolution** is the smallest distance by which two objects can be separated and still be distinguished.
- **Magnification** means an increase in the apparent size of an image so as to resolve smaller separations between objects.
- **Eukaryotic microbes may be large enough to resolve subcellular structures** under a light microscope, although some eukaryotes are as small as bacteria.
- **Bacteria and archaea are usually too small for subcellular resolution.** Their shapes include characteristic forms, such as rods and cocci.
- **Different kinds of microscopy** are required to resolve cells and subcellular structures of different sizes.

2.2 Optics and Properties of Light

Light microscopy directly extends the lens system of our own eyes. Light is part of the spectrum of **electromagnetic radiation (Fig. 2.8)**, a form of energy that is propagated as waves associated with electrical and magnetic fields. Regions of the electromagnetic spectrum are defined by wavelength, which for visible light is about 400–750 nm. Radiation of longer wavelengths includes infrared and radio waves, whereas shorter wavelengths include ultraviolet rays and X-rays.

WWW | Molecular Expressions: Exploring the World of Optics and Microscopy (www.microscopy.fsu.edu) *microbiology2.com/links*

Light Carries Information

All forms of electromagnetic radiation carry information from the objects with which they interact. The information carried by radiation can be used to detect objects; for example, radar (using radio waves) detects a speeding car. All electromagnetic radiation travels through a vacuum at the same speed: about 3×10^8 meters per second (m/s), the speed of light. The speed of light (c) is equal to the wavelength (λ) of the radiation multiplied by its frequency (ν), the number of wave cycles per unit time:

$$c = \lambda\nu$$

Because c is constant, the longer the wavelength λ is, the lower the frequency ν is. Frequency is usually measured in hertz (Hz), reciprocal seconds.

For electromagnetic radiation to resolve an object from neighboring objects or from its surrounding medium, certain conditions must exist:

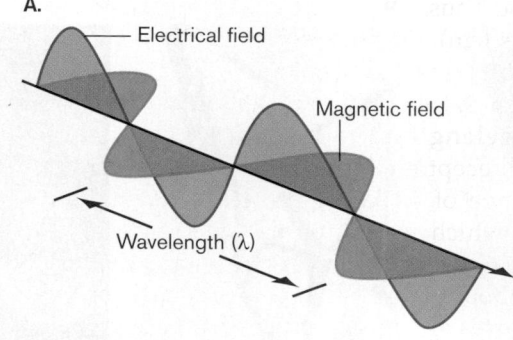

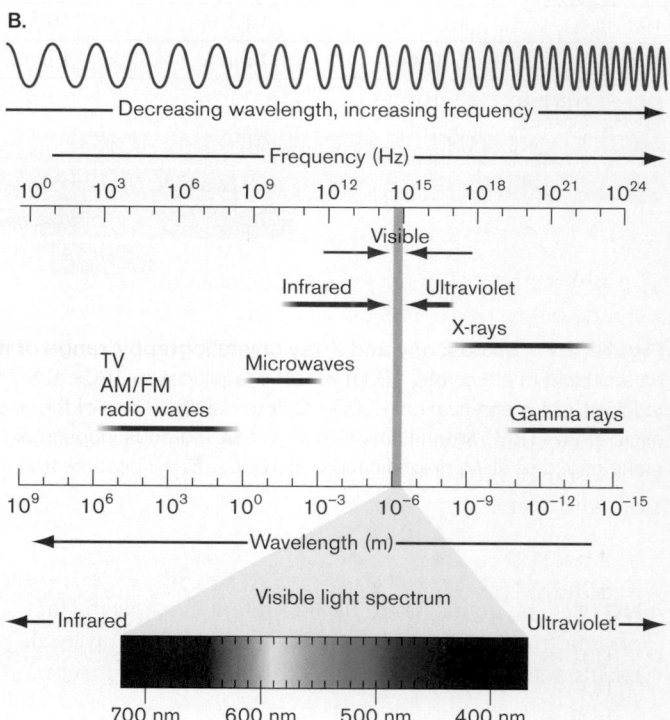

Figure 2.8   Electromagnetic energy.   A. Electromagnetic radiation is composed of electrical and magnetic waves perpendicular to each other. **B.** The electromagnetic spectrum includes the visual range, used in light microscopy.

- **Contrast between the object and its surroundings. Contrast** is the difference in light and dark. If an object and its surroundings absorb or reflect radiation equally, then the object will be undetectable. It is hard to observe a cell of transparent cytoplasm floating in water, because the aqueous cytoplasm and the extracellular water tend to transmit light similarly, producing little contrast.
- **Wavelength smaller than the object.** For an object to be resolved, the wavelength of the radiation must be equal to or smaller than the size of the object. If the wavelength of the radiation is larger than the object, then most of the wave's energy will simply pass through it, like an ocean wave passing around

a dock post. Thus, radar, with a wavelength of 1–100 centimeters (cm), cannot resolve microbes, though it easily resolves cars and people.

- **A detector with sufficient resolution for the given wavelength.** The human eye has a retina with photoreceptors that absorb radiation within a narrow range of wavelengths, 400–750 nm (0.40–0.75 μm), which we define as "visible light." But the distance between our retinal photoreceptors is 150 μm, about 500 times the wavelength of light. Thus, our eyes are unable to access all the information contained in the light that enters. In order for us to exploit the full information capacity of light, the light rays from point sources of light must be spread apart sufficiently to be resolved by our retinal photoreceptors. The spreading of the light rays magnifies the image.

Light Interacts with an Object

The physical behavior of light resembles in some ways a beam of particles and in other ways a waveform. The particles of light are called photons. Each photon has an associated wavelength that determines how the photon will interact with a given object. The combined properties of particle and wave enable light to interact with an object in several different ways: absorption, reflection, refraction, and scattering.

- **Absorption** means that the photon's energy is acquired by the absorbing object (**Fig. 2.9A**). The energy is converted to a different form, usually heat (infrared radiation), a form of electromagnetic radiation of longer wavelength than light. When a microbial specimen absorbs light, it can be observed as a dark spot against a bright field, as in bright-field microscopy. Some molecules that absorb light of a specific wavelength reemit energy not as heat, but as light with a longer wavelength; this is called **fluorescence**. Fluorescence is also used in microscopy (as discussed in Section 2.5).
- **Reflection** means that the wave front redirects from the surface of an object at an angle equal to its incident angle (**Fig. 2.9B**). Reflection of light waves is analogous to the reflection of water waves. Reflection from a silvered mirror or a glass surface is used in the optics of microscopy.
- **Refraction** is the bending of light as it enters a substance that slows its speed (**Fig. 2.9C**). Such a substance is said to be "refractive" and, by definition, has a higher **refractive index** than air. Refraction is the key property that enables a lens to magnify an image.
- **Scattering** occurs when a portion of the wave front is converted to a spherical wave originating from the object (**Fig. 2.9D**). If a large number of particles

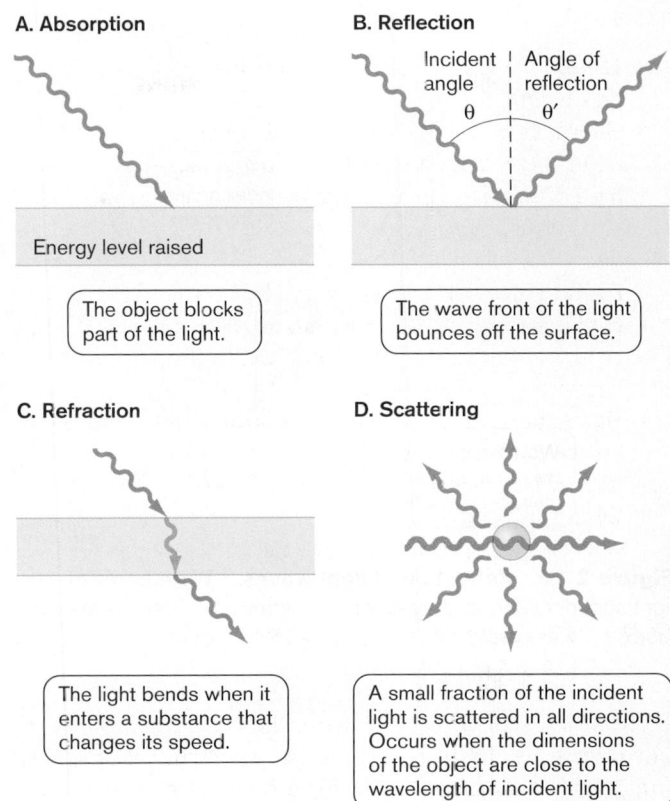

Figure 2.9   Interaction of light with matter.

simultaneously scatter light, a haze is observed—for example, the haze of bacteria suspended in a culture tube. Special optical arrangements (such as dark-field microscopy, discussed in Section 2.4) can use scattered light to detect (but not resolve) microbial shapes smaller than the wavelength of light.

Refraction Enables Magnification

Magnification requires the refraction of light through a medium of high refractive index, such as glass. As a wave front of light enters a refractive material, the region of the wave that first reaches the material is slowed, while the rest of the wave continues at its original speed until it also passes into the refractive material (**Fig. 2.10**). As the entire wave front continues through the refractive material, its path continues at an angle from its original direction. At the opposite face of the refractive material, the wave front resumes its original speed.

> **THOUGHT QUESTION 2.2** Explain what happens to the refracted light wave as it emerges from a piece of glass of even thickness. How do its new speed and direction compare with its original (incident) speed and direction?

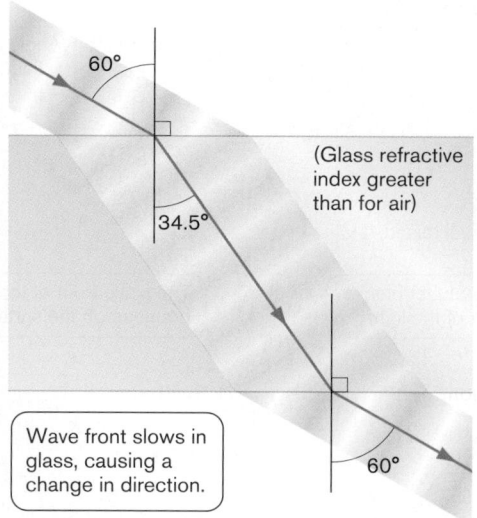

Figure 2.10 **Refraction of light waves.** Wave fronts of light shift direction as they enter a substance of higher refractive index, such as glass.

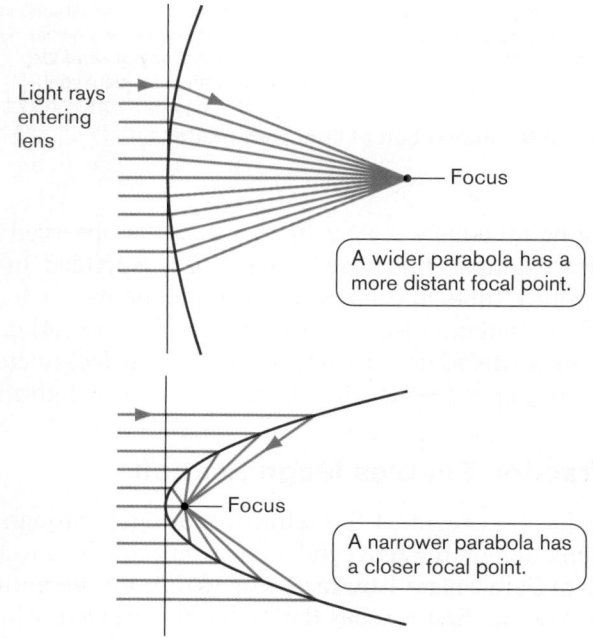

Figure 2.11 **The focus of a parabola.** The wider the parabola, the more distant the focal point.

Refraction magnifies an image when light passes through a refractive material shaped so as to spread its rays, widening the wave front. The shape that accomplishes this widening is that of a parabola. Light rays entering a parabolic surface are focused; that is, they are each bent at an angle such that all rays meet at a certain point, called the focus of the parabola (**Fig. 2.11**).

In a lens, a parabolic surface focuses impinging light rays. Parallel light rays entering the lens emerge at angles so as to intersect each other at the **focal point** (**Fig. 2.12**), equivalent to the focus of the parabola. From the focal point, the light rays continue onward in an expanding wave front. This expansion magnifies the image carried by the wave. The distance from the lens to the focal point (called the focal distance) is determined by the degree of parabolic curvature of the lens and by the refractive index of its material.

In **Figure 2.12**, the object under observation is placed just outside the **focal plane**, a plane containing the focal point of the lens. All light rays from the object are bent by the lens, converging at the opposite focal point. At the focal point, the light rays continue until they converge with nonparallel light rays refracted by the lens. The plane of convergence generates a reversed image of the object. The image is expanded, or magnified, by the spreading out of refracted rays. The distances between parts of the image are enlarged, enabling our eyes to resolve finer details.

> **THOUGHT QUESTION 2.3** Parabolic lenses are generally "biconvex"—that is, curving outward on both sides. What will happen to parallel light rays that pass through a lens that is concave on both sides? Or a lens that is convex on one side and concave with equal curvature on the other?

Magnification and Resolution

The spreading of light rays does not necessarily increase resolution. For example, an image composed of dots does not gain detail when enlarged on a photocopier, nor does an image composed of pixels gain detail when enlarged on a computer screen. In these cases, resolution fails to increase because the individual details of the image expand in proportion to the expansion of the overall image. Magnification without increase in resolution is called **empty magnification**.

The resolution of detail in microscopy is limited by the degree to which details "expand" with magnification. This expansion of detail is limited by the interference of light rays converging at the focal point. In **interference**, two wave fronts interact by addition (amplitudes in phase) or subtraction (amplitudes out of phase) (**Fig. 2.13A**). The result of interference between two waves is a pattern of alternating zones of constructive and destructive interference (brightness and darkness) (**Fig. 2.13B**).

In theory, a perfect lens that focuses all the light from an object should have no interference as its rays converge toward the focal point. In fact, however, the

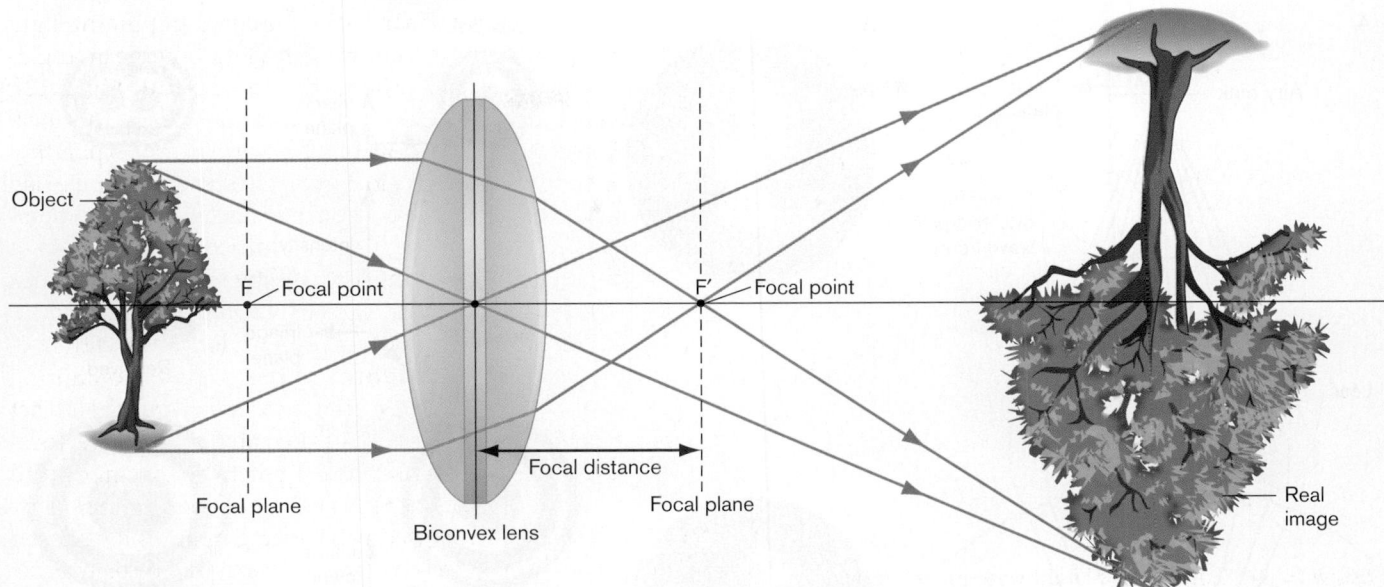

Figure 2.12   Generating an image with a lens.   The object is placed near the focal plane of the lens. All light rays from the object are bent by the lens, converging at the opposite focal point. The light rays continue through the focal point, generating an image of the object that is reversed and magnified.

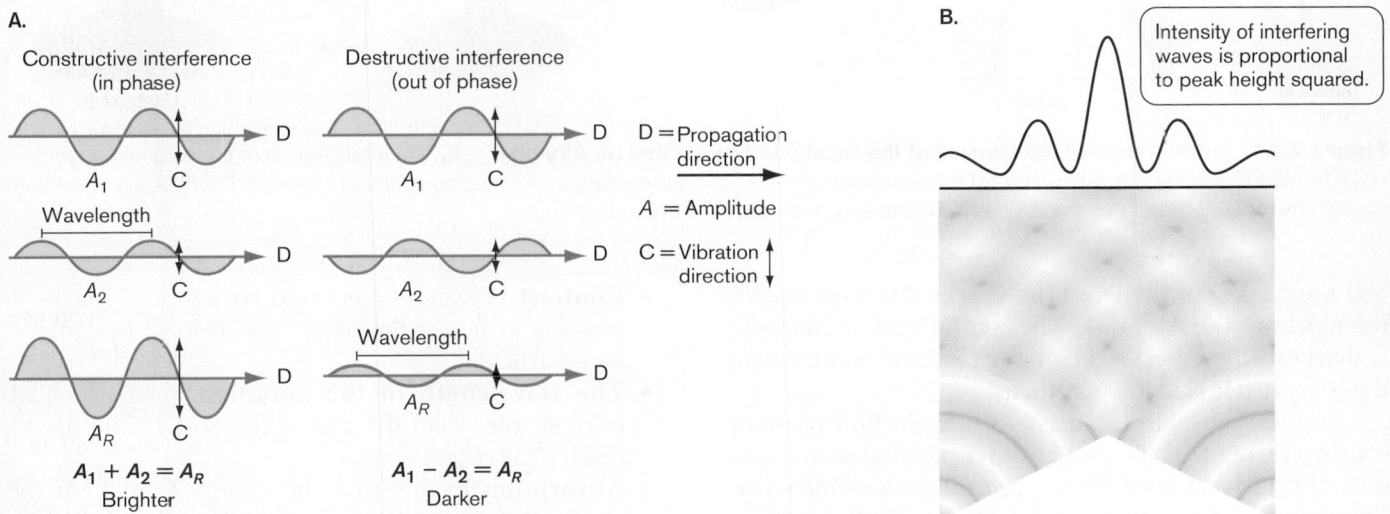

Figure 2.13   Constructive (additive) and destructive (subtractive) interference of light waves.   A. In constructive interference, the peaks of the two wave trains rise together; their amplitudes are additive, forming a wave of greater amplitude. In destructive interference, the peaks of the waves are opposite one another, so their amplitudes cancel, forming a wave of lesser amplitude. **B.** Two wave fronts approaching at an angle generate an interference pattern in which intensity alternately increases and decreases.

outer edges of the lens introduce imperfection. The converging edges of the wave interfere with each other to form alternating regions of light and dark. At the focal point, the interference forms a central sphere of intensity surrounded by a series of concentric spheres of alternating light and dark. The viewing plane intersects the concentric spheres to form an **Airy disk**, a disk containing a bright central peak surrounded by rings of light and dark (**Fig. 2.14A**). The Airy disk was originally

discovered by the astronomer George Airy (1801–1892), who showed that the stars viewed through a telescope could never appear as true point sources, but only as tiny disks of light surrounded by rings. Similarly, under a microscope, a well-focused bright object appears as a bright disk surrounded by faint rings arising from wave interference.

The width of the Airy disk—and hence the limit to resolution of detail—depends on the wavelength of light

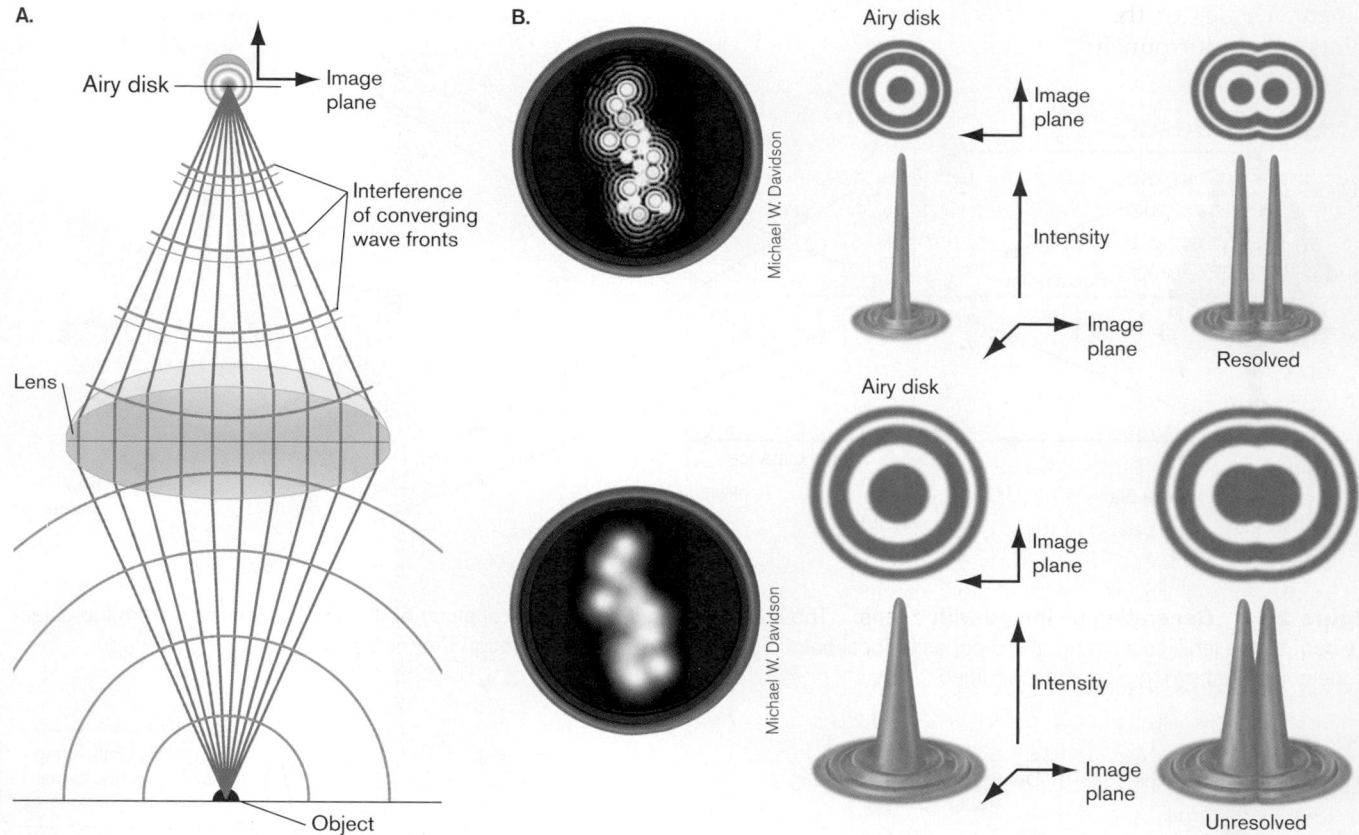

Figure 2.14   Interference of light waves at the focal point generates an Airy disk.   A. The interference of converging spherical wave fronts forms a sphere with a central zone of maximum intensity, surrounded by alternating zones of bright and dark. A plane section through the sphere generates the Airy disk. **B.** The Airy peak and its surrounding rings.

and the quality of the lens. The shorter the wavelength, the narrower the Airy disk. The central peak of intensity is sharpest when the specimen is at the focal point, where the specimen is said to be in **focus**.

Suppose an object consists of a collection of point sources of light. Each point source generates an Airy disk of rings surrounding one central peak of intensity. The width of the central peak increases with distance away from the focal point. This width will define the resolution, or separation distance, between any two points of the object (**Fig. 2.14B**). Full resolution is achieved for two points when there is complete separation of the central peaks of the two Airy disks. This resolution distance determines the degree of detail that can be observed. In practice, any object, such as a stained microbe against a bright field, can be considered a large collection of points of light that act as partially resolved Airy disks.

TO SUMMARIZE:

■ **Electromagnetic radiation** interacts with an object and acquires information we can use to detect the object.

■ **Contrast** between object and background makes it possible to detect the object and resolve its component parts.

■ **The wavelength of the radiation** must be equal to or smaller than the size of the object if we are to resolve the object's shape.

■ **Absorption** means that the energy from light (or other electromagnetic radiation) is acquired by the object.

■ **Reflection** means that the wave front bounces off the surface of a particle at an angle equal to its incident angle.

■ **Refraction** is the bending of light as it enters a substance that slows its speed.

■ **Scattering** occurs when a wave front interacts with an object of smaller dimension than the wavelength. Light scattering enables the detection of objects whose detail cannot be resolved.

2.3   Bright-Field Microscopy

In bright-field microscopy, an object such as a bacterial cell is perceived as a dark silhouette blocking the passage of light. Details of the dark object are defined by the points of light surrounding its edge.

Increasing Resolution

The optics of a modern bright-field microscope are designed to maximize detail under magnification by a lens. In maximizing the resolution, the following factors need to be considered:

- **Wavelength.** The theoretical maximum magnification can be calculated as the resolution distance of our retina (0.15 mm, or 150,000 nm) divided by the average wavelength of light (550 nm, for green). This gives an approximately 300-fold magnification. Additional optical factors can bring this magnification closer to 1,000×, although full resolution is rarely reached in practice. Any greater magnification expands only the width of the interference patterns; the image becomes larger but with no greater resolution (empty magnification).

- **Light and contrast.** For any given lens system, an optimal amount of light yields the highest contrast between the dark specimen and the light background. High contrast is needed to achieve the maximum resolution at high magnification.

- **Lens quality.** All lenses possess inherent **aberrations** that detract from perfect parabolic curvature. Optical properties limit the degree of perfection of a single lens, but manufacturers construct microscopes with a series of lenses that multiply each other's magnification and correct for aberrations.

Let us first consider magnification of an image by a single lens. The principles described would hold for any lens. **Figure 2.15** shows an **objective lens**, a lens situated directly above an object or specimen that we wish to observe at high resolution. How can we maximize the resolution of details in the image?

An object at the focal point of a lens sits at the tip of an inverted light cone formed by rays of light from the lens converging at the object. The angle of the light cone is determined by the curvature and refractive index of the lens. The lens fills an aperture, or hole, for the passage of light; and for a given lens the light cone is defined by an angle θ (theta) projecting from the midline, known as the **angle of aperture**. As θ increases and the horizontal width of the light cone (sin θ) increases, a wider cone of light passes through the specimen. The wider cone of light rays, the smaller the edge interference effects between wave fronts. Thus, the Airy disk interference pattern gets narrower and resolution improves. In other words, as sin θ increases, we can resolve points that are closer together. So the greater the angle of aperture of the lens, the better the resolution.

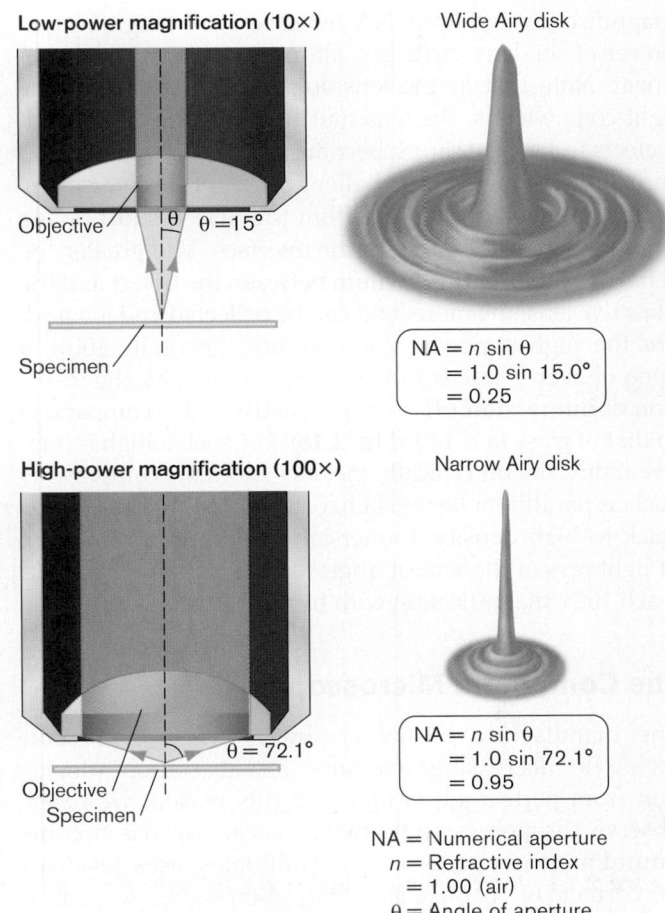

Figure 2.15 Resolution and numerical aperture. The resolution depends on numerical aperture (NA), which equals the refractive index (n) of the medium containing the light cone multiplied by the sine of the angle of the light cone (θ). Higher NA gives greater resolution.

> **THOUGHT QUESTION 2.4** In theory, what angle θ would produce the highest resolution? What practical problem would you have in designing a lens to generate this light cone?

Resolution also depends on the refractive index of the medium containing the light cone, which is usually air. The refractive index is the ratio of the speed of light in a vacuum to its speed in another medium. For air, the refractive index (n) is taken as 1, whereas lens material has a refractive index greater than 1. The lens bends the light spreading at an angle (θ). The product of the refractive index (n) of the medium multiplied by sin θ is the **numerical aperture** (NA):

$$\text{NA} = n \sin \theta$$

In **Figure 2.15**, we see the calculation of NA for an objective lens of magnification 10×, and for a lens of magnification 100×. As NA increases, the magnification power of the lens increases, although the increase is not linear. Note that as the lens strength increases and the light cone widens, the lens must come nearer the object. Defects in lens curvature become more of a problem, and focusing becomes more challenging. As θ becomes very wide, too much of the light from the object is lost owing to refraction at the glass-to-air interface. The greater the refractive index of the medium between the object and the objective lens, the more light can be collected and focused. For the highest-power objective lens, generally 100×, a zone of even refractive index is maintained by the insertion of **immersion oil** with a refractive index comparable to that of glass ($n = 1.5$) (**Fig. 2.16**). For such a high refractive index, the oil typically includes aromatic components such as paraffin or benzyl benzoate, whose ring structures stack to high density. Immersion oil minimizes the loss of light rays at the widest angles and makes it possible to reach 100× magnification with minimal distortion.

The Compound Microscope

The manufacture of higher-power lenses is difficult owing to decreasing tolerance for aberration (deviation from perfect curvature). For this reason, we rarely observe through a single lens. Instead we use a **compound microscope**, a system of multiple lenses designed to correct or compensate for aberration. A typical arrangement of a compound microscope is shown in **Figure 2.17**. In this arrangement, the light source is placed at the bottom, shining upward through a series of lenses, including the condenser, objective, and ocular lenses.

Between the light source and the condenser sits a **diaphragm**, a device to cut the diameter of the light column. Lower-power lenses require operation at lower light levels because the excess light makes it impossible to observe the darkening effect of absorbance by the specimen (contrast). Higher-power lenses require more light and thus an open diaphragm.

Above the diaphragm, the **condenser** consists of one or more lenses that collect a beam of rays from the light source onto a small area of the slide. The condensed light rays generate a narrower Airy interference pattern and thus improve the resolution of the objective lens. The **objective lens** forms a magnified image (*I*) of the object (**Fig. 2.17A**). As the image forms, each light ray traces a path toward a position opposite its point of origin;

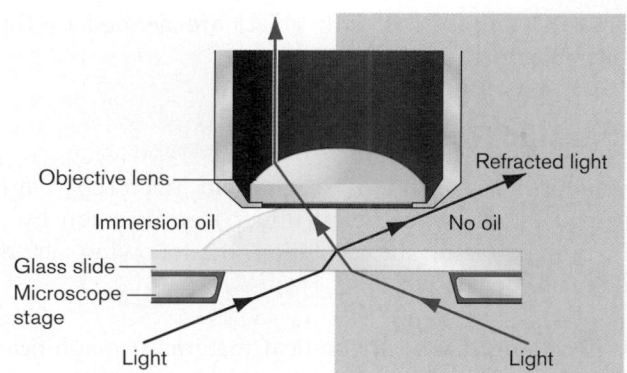

Figure 2.16   Use of immersion oil in microscopy. Immersion oil with a refractive index comparable to that of glass ($n = 1.5$) prevents light rays from bending away from the objective lens. Thus, more light is collected, NA increases, and resolution improves.

thus, the image is mirror-reversed. This mirror reversal must be kept in mind when exploring a field of cells.

The first image (*I*) is then amplified by a secondary magnification step through the **ocular lens**. The final virtual image (*I′*) includes the total magnification of the object.

The nosepiece of a compound microscope typically holds three or four objective lenses of different magnifying power, such as 4×, 10×, 40×, and 100× (requiring immersion oil). These lenses are arranged so as to rotate in turn into the optical column. A high-quality instrument will have the lenses set at different heights from the slide so as to be **parfocal**. In a parfocal system, when

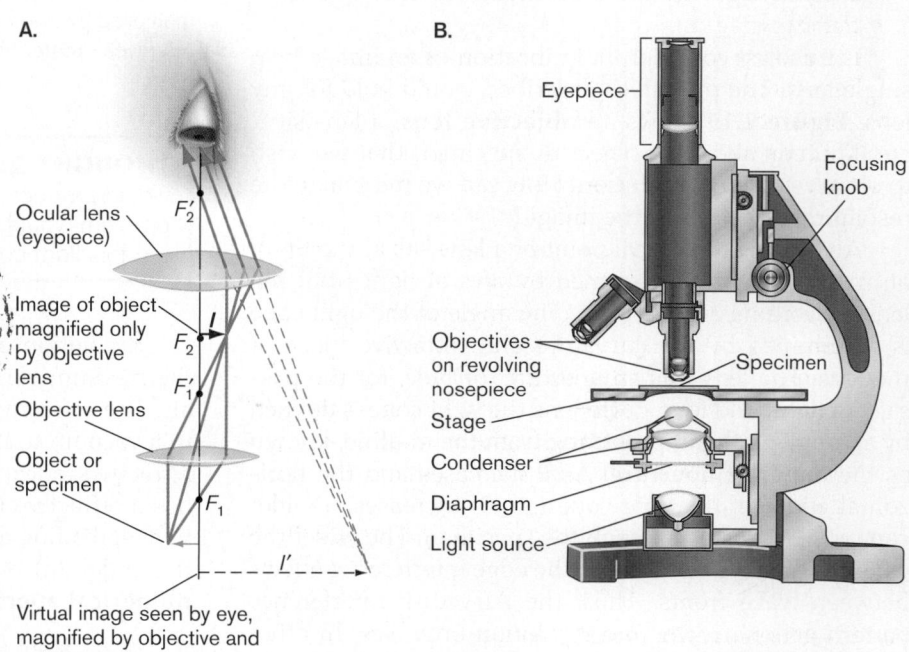

Figure 2.17   Anatomy of a compound microscope.   A. Light path through a compound microscope. **B.** Cutaway view of a compound microscope.

an object is focused with one lens, it remains in focus, or nearly so, when another lens is rotated to replace the first.

NOTE: Objective lenses can be obtained in several different grades of quality, manufactured with different kinds of correction for aberrations. Lenses should feature at minimum the following corrections: "plan" correction for field curvature, to generate a field that appears flat; and "apochromat" correction for spherical and chromatic aberrations.

The magnification factor of the ocular lens is multiplied by the magnification factor of the objective lens to generate the **total magnification** (power). For example, a 10× ocular multiplied by a 100× objective generates 1,000× total magnification.

Observing a specimen under a compound microscope requires several steps:

- **Position the specimen centrally in the optical column.** Only a small area of a slide can be visualized within the field of view of a given lens. The higher the magnification, the smaller the field of view that will be seen.
- **Optimize the amount of light.** At lower power, too much light will wash out the light absorption of the specimen without contributing to magnification. At higher power, more light needs to be collected by the condenser. To optimize light, the condenser must be set at the correct vertical position to focus on the specimen, and the diaphragm must be adjusted to transmit the amount of light that produces the best contrast.
- **Focus the objective lens.** The focusing knob permits adjustment of the focal distance between the objective lens and the specimen on the slide so as to bring the specimen into the focal plane. Typically, we focus first using a low-power objective, which generates a greater **depth of field**—that is, a region along the optical column over which the object appears in reasonable focus. After focusing under low power, we can rotate a higher-power lens into view and then perform a smaller adjustment of focus.

Is the Object in Focus?

An object appears in focus (that is, is situated within the focal plane of the lens) when its edge appears sharp and distinct from the background. At higher power, however,

recognition of the focal plane is a challenge because of Airy-like interference. The shape of the dark object is actually defined by the points of light surrounding its edge. The partial resolution of these points of light generates extra rings of light surrounding an object whose dimensions are close to the resolution limit.

In **Figure 2.18**, we see microscopic observation of *Rhodospirillum rubrum*, photosynthetic bacteria that contribute to wetland productivity. As cells of *R. rubrum* swim in and out of the focal plane, their appearance changes through optical effects. When a bacterium swims out of the focal plane too close to the lens, resolution declines and the image blurs (**Fig. 2.18A**). When the bacterium swims within the focal plane, its image appears sharp, with a bright line along its edge. A helical cell like *R. rubrum* shows well-focused segments alternating with hazier segments that are too close to the lens (**Fig. 2.18B**). When the cell swims too far past the focal plane, the bright interference lines collapse into the object's silhouette, which now appears bright or "hollow" (**Fig. 2.18C**). In fact, the bacterium is not hollow at all; only its image has changed.

When the cell extends across several focal planes, different portions appear out of focus (either too near or too far from the lens). In addition, when the end of a

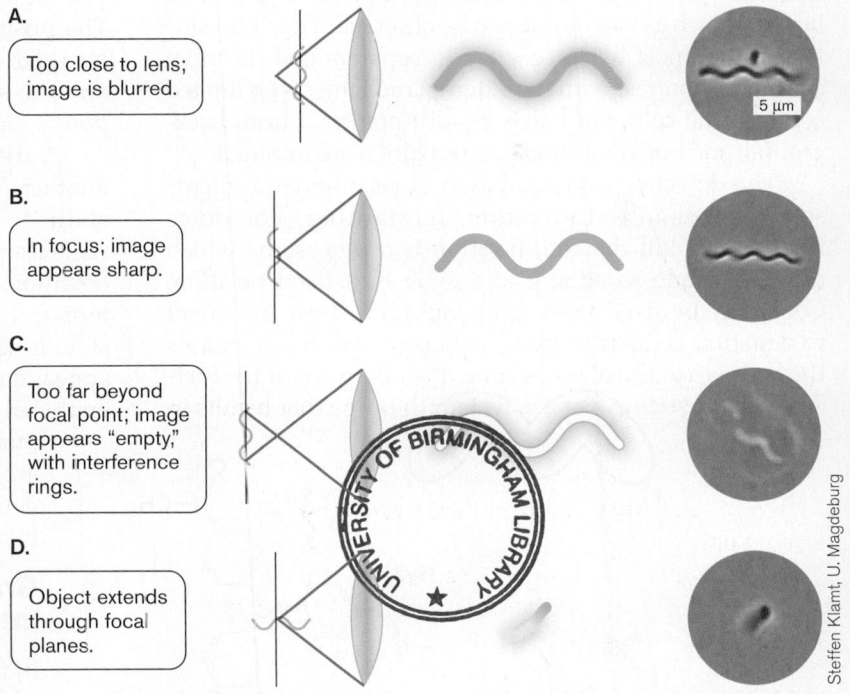

A. Too close to lens; image is blurred.

5 µm

B. In focus; image appears sharp.

C. Too far beyond focal point; image appears "empty," with interference rings.

D. Object extends through focal planes.

Figure 2.18 *Rhodospirillum rubrum* **observed at different levels of focus. A.** When a bacterium swims too close to the lens, its image blurs. **B.** When the bacterium lies within the focal plane, its image appears sharp. If the width of the cell crosses several focal planes, some parts appear sharp, whereas other parts appear blurred. **C.** When the bacterium lies too far from the lens, its image appears "empty" or "hollow," surrounded by rings of Airy-like interference. **D.** When the spirillum extends through the focal plane, different parts show different focal effects (in focus, too near, or too far).

cell points toward the observer, light travels through the length of the cell before reaching the observer, so the cell absorbs more light and appears dark (**Fig. 2.18D**). The observation and identification of motile bacteria swimming in and out of the focal plane present a challenge even to experienced microscopists. The higher the magnification, the narrower the depth of the focal plane; thus, observing swimming organisms requires a trade-off between magnification and depth of field.

> **THOUGHT QUESTION 2.5**   Under starvation conditions, bacteria such as *Bacillus thuringiensis*, the biological insecticide, package their cytoplasm into spores, leaving behind an empty cell wall. Suppose, under a microscope, you observe what appears to be a hollow cell. How can you tell if the cell is indeed hollow or if it is simply out of focus?

Fixation and Staining Improve Resolution and Contrast

The simplest way to observe microbes is to place them in a drop of water on a slide with a coverslip. This is called a **wet mount** preparation. The advantage of the wet mount is that the organism is viewed in as natural a state as possible, without artifacts resulting from chemical treatment; and live behavior such as swimming can be observed. The disadvantage is that most living cells are transparent and therefore show little contrast with the external medium. With limited contrast, the cells can barely be distinguished from background, and both detection and resolution are minimal.

The detection and resolution of cells under a microscope are enhanced by **fixation** and **staining**, procedures that usually kill the cell. Fixation is a process by which cells are made to adhere to a slide in a fixed position. Cells may be fixed with methanol or by heat treatment to denature cellular proteins, whose exposed side chains then adhere to the glass. A stain absorbs much of the incident light, usually over a wavelength range that results in

a distinctive color. The use of chemical stains was developed in the nineteenth century, when German chemists used organic synthesis to invent new coloring agents for women's clothing. Clothing was made of natural fibers such as cotton or wool, so a substance that dyed clothing would be likely to react with biological specimens.

How do stains work? Most stain molecules contain conjugated double bonds or aromatic rings that absorb visible light (**Fig. 2.19**) and one or more positive charges that bind to negative charges such as phosphoryl groups on membrane phospholipids (discussed in Section 3.4). Different stains vary with respect to the strength of their binding and the degree of binding to different parts of the cell.

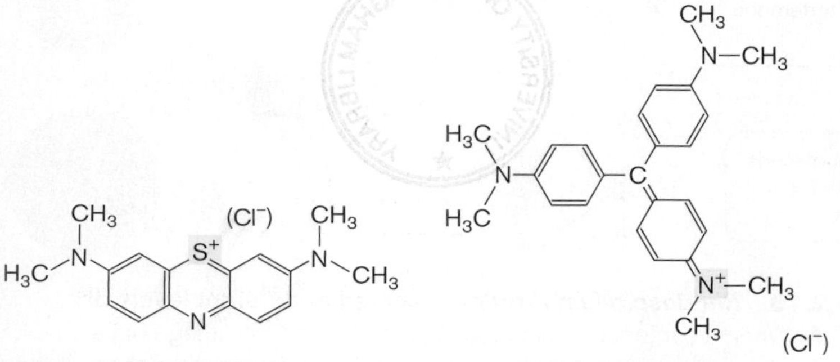

Different Kinds of Stains

A **simple stain** adds dark color specifically to cells, but not to the external medium or surrounding tissue (in the case of pathological samples). The most commonly used simple stain is methylene blue, originally used by Robert Koch to stain bacteria. A typical procedure for fixation and staining is shown in **Figure 2.20**. First, a drop of culture is fixed on a slide by treatment with methanol or by heating on a slide warmer. Either treatment denatures cell proteins, exposing side chains that bind to the glass. The slide is then flooded with methylene blue solution. The positively charged molecule binds to the negatively charged cell envelope. After excess stain is washed off and the slide has been dried, it is observed under high-power magnification using immersion oil.

A **differential stain** stains one kind of cell but not another. The most famous differential stain is the **Gram stain**, devised in 1884 by the Dutch physician Hans Christian Gram (1853–1938). Gram first used the Gram stain to distinguish pneumococcus (*Streptococcus pneumoniae*) bacteria from human lung tissue. A similar use of the Gram stain is seen in **Figure 2.21A**, where *Enterococcus* bacteria appear dark purple against the pink background of human epithelial cells. Other species of bacteria fail to retain the purple stain (**Fig. 2.21B**). Different bacterial species are classified as Gram-positive or Gram-negative, depending on whether they retain the purple stain.

Gram Staining Separates Bacteria into Two Classes

In the Gram stain procedure (**Fig. 2.22**), a dye such as crystal violet binds to the bacteria; it also binds to the surface of human cells, but less strongly. After the excess stain is washed off, a **mordant**, or binding agent, is applied. The mordant used is iodine solution, which contains iodide ions (I^-). The iodide complexes with the positively charged crystal violet molecules trapped inside the cells. The crystal violet–iodide complex is now held more strongly

Figure 2.19   Chemical structure of stains.   Methylene blue and crystal violet are cationic (positively charged) dyes. The positively charged groups react with the bacterial cell envelope, which carries mainly negative charge. Chloride (Cl^-) is the counter-ion.

Methylene blue

Crystal violet

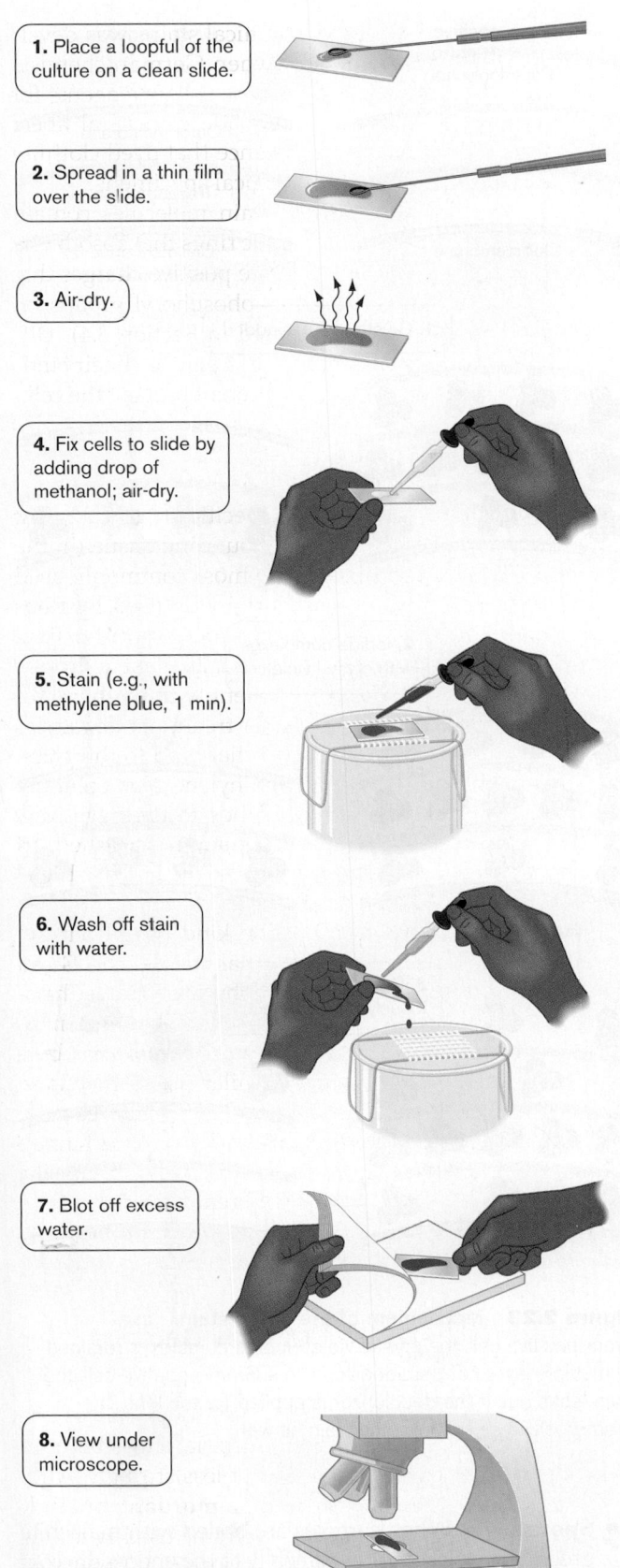

1. Place a loopful of the culture on a clean slide.

2. Spread in a thin film over the slide.

3. Air-dry.

4. Fix cells to slide by adding drop of methanol; air-dry.

5. Stain (e.g., with methylene blue, 1 min).

6. Wash off stain with water.

7. Blot off excess water.

8. View under microscope.

Figure 2.20 **Procedure for simple staining with methylene blue.**

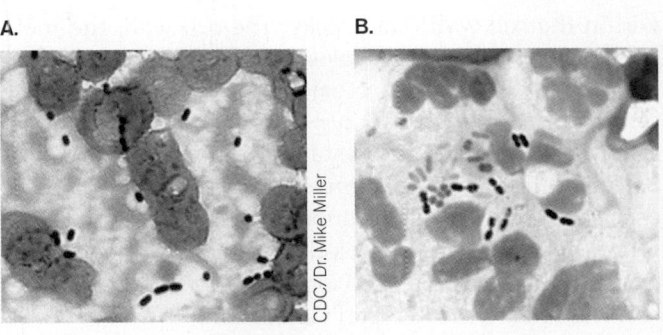

A. CDC/Dr. Mike Miller

B. Dr. Sandra Richter

Figure 2.21 **Gram staining of bacteria (a type of differential stain).** **A.** Gram stain of a sputum specimen from a patient with pneumonia, containing Gram-positive *Enterococcus sp.* (purple diplococci). Cell length, 0.5–1.0 μm. **B.** Gram stain of a gingival specimen containing both Gram-positive and Gram-negative bacteria (purple and pink rods). Cell length, 0.5–1.0 μm. Light microscopy.

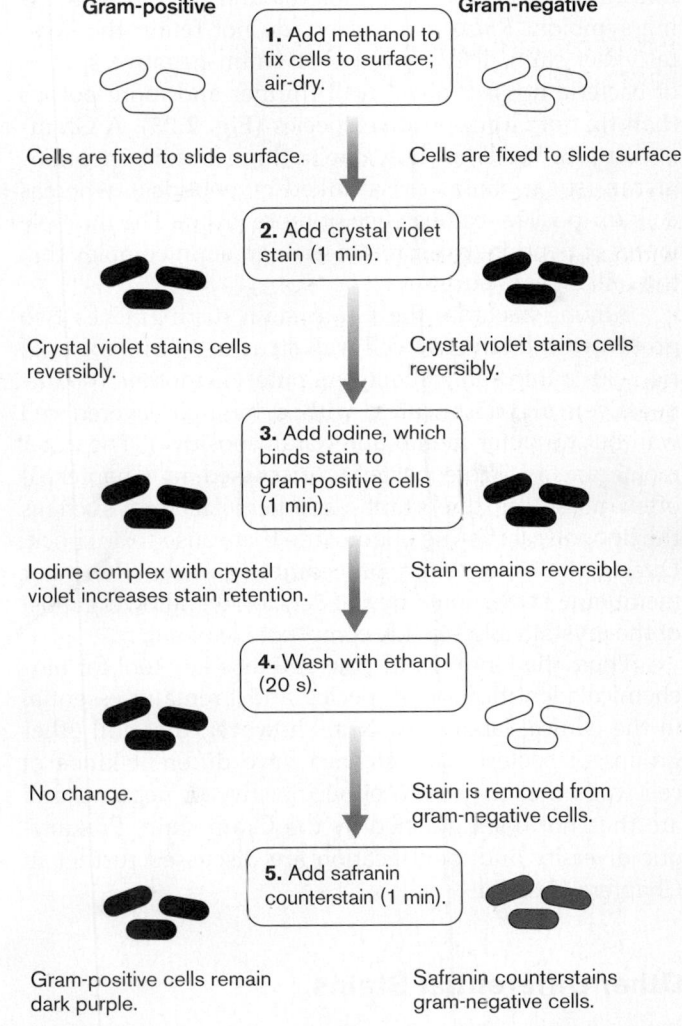

Gram-positive **Gram-negative**

1. Add methanol to fix cells to surface; air-dry.

Cells are fixed to slide surface. Cells are fixed to slide surface.

2. Add crystal violet stain (1 min).

Crystal violet stains cells reversibly. Crystal violet stains cells reversibly.

3. Add iodine, which binds stain to gram-positive cells (1 min).

Iodine complex with crystal violet increases stain retention. Stain remains reversible.

4. Wash with ethanol (20 s).

No change. Stain is removed from gram-negative cells.

5. Add safranin counterstain (1 min).

Gram-positive cells remain dark purple. Safranin counterstains gram-negative cells.

Figure 2.22 **The Gram stain procedure.** The Gram stain distinguishes between Gram-positive cells, with thick cell walls, which retain the crystal violet stain, and Gram-negative cells, with thinner cell walls, which lose the crystal violet stain but are counterstained by safranin.

within the cell wall. The thicker the cell wall, the more crystal violet–iodide molecules are held.

Next, a decolorizer, ethanol, is added for a precise time interval (typically, 20 seconds). The decolorizer removes loosely bound crystal violet–iodide, but Gram-positive cells retain the stain tightly. The **Gram-positive** cells that retain the stain appear dark purple, while the **Gram-negative** cells are colorless. The decolorizer step is critical because if it lasts too long, the Gram-positive cells, too, will release their crystal violet stain.

In the final step, a **counterstain**, safranin, is applied. This process allows the visualization of Gram-negative material, which is stained pale pink by the safranin.

The Gram stain procedure was originally devised to distinguish bacteria (Gram-positive) from human cells (Gram-negative). Microscopists soon discovered, however, that many important species of bacteria, such as the intestinal bacterium *Escherichia coli* and the nitrogen-fixing symbiont *Rhizobium meliloti*, do not retain the crystal violet stain. It turns out that Gram-negative species of bacteria possess a cell wall thinner and more porous than that of Gram-positive species (**Fig. 2.23**). A Gram-negative cell wall has only one to three layers of peptidoglycan (sugar chains cross-linked by peptides), whereas a Gram-positive cell has five or more layers. The multiple layers of peptidoglycan retain enough stain complex that the cell appears purple.

Among bacteria, the Gram stain distinguishes two groups with distinctive cell wall structures: Proteobacteria, with a thin cell wall plus an outer membrane (Gram-negative); and Firmicutes, with a multiple-layered cell wall but no outer membrane (Gram-positive). The outer membrane of Proteobacteria (discussed in Chapter 3) often possesses important pathogenic factors, such as the lipopolysaccharide endotoxins that cause toxic shock. During the Gram stain procedure, however, the outer membrane is disrupted by the decolorizer, allowing most of the crystal violet–iodide complex to leak out.

Thus, the Gram stain emerged as a key tool for biochemical identification of species, and it remains essential in the clinical laboratory. Note, however, that still other groups of bacteria and archaea have different kinds of cell walls that may stain either positive or negative and are thus not distinguished by the Gram stain. Prokaryotic diversity and identification are discussed further in Chapters 18 and 19.

Other Differential Stains

Other differential stains applied to various types of prokaryotes are illustrated in **Figure 2.24**. These include:

- **Acid-fast stain** (Ziehl-Neelsen). Carbolfuchsin specifically stains mycolic acids of *Mycobacterium tuberculosis* and *M. leprae*, the causative agents of tuberculosis and leprosy, respectively (**Fig. 2.24A**).

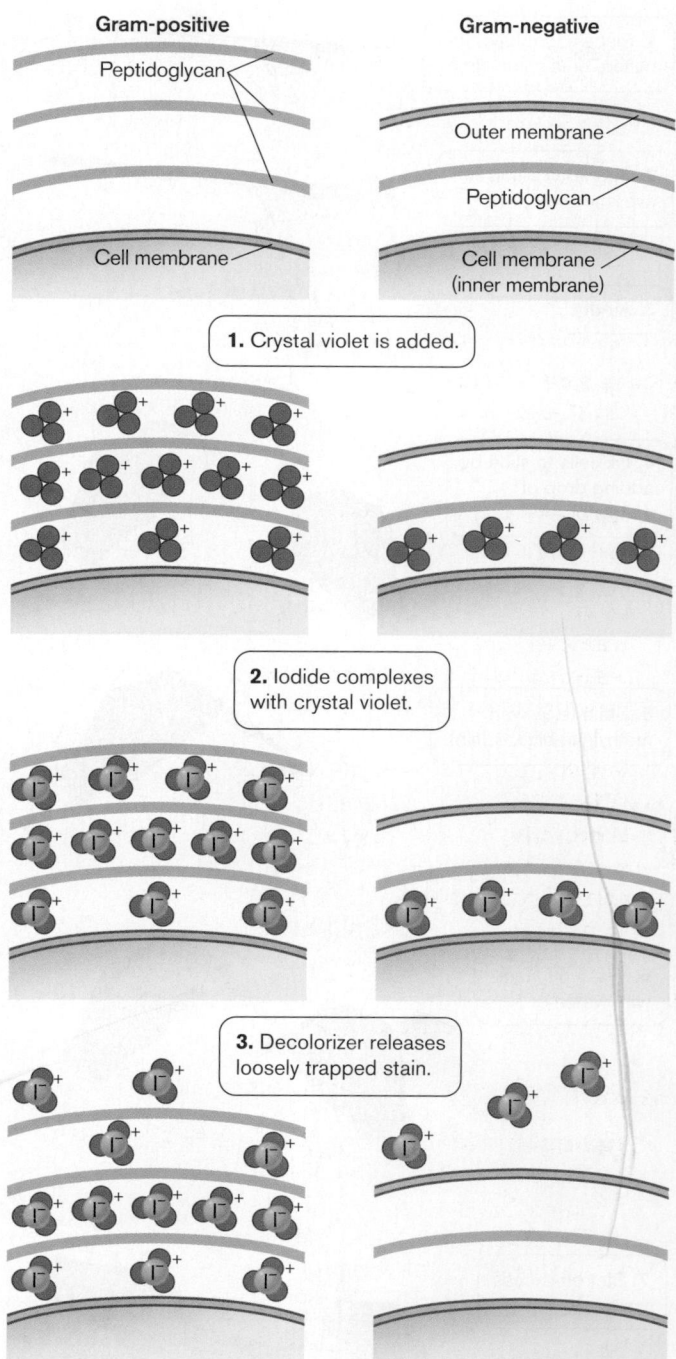

Figure 2.23 Mechanism of the Gram stain. In a Gram-positive cell, the crystal violet–iodide complex is retained by multiple layers of peptidoglycan. In a Gram-negative cell, the stain leaks out. If the decolorizer is applied for too long, the Gram-positive cell will lose its stain as well.

- **Spore stain.** When samples are boiled with malachite green, the stain binds specifically to the endospore coat (**Fig. 2.24B**). It detects spores of *Bacillus* species such as the insecticide *B. thuringiensis* and *B. anthracis*, the cause of anthrax, as well as spores of *Clostridium botulinum*, which produces botulinum toxin.

A.
B.
C.

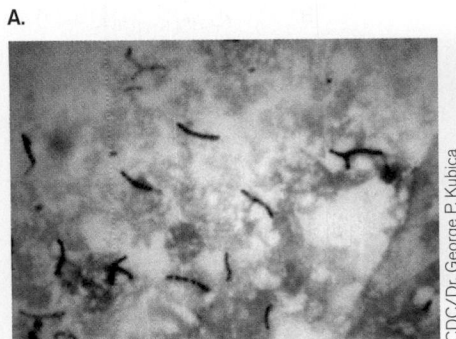
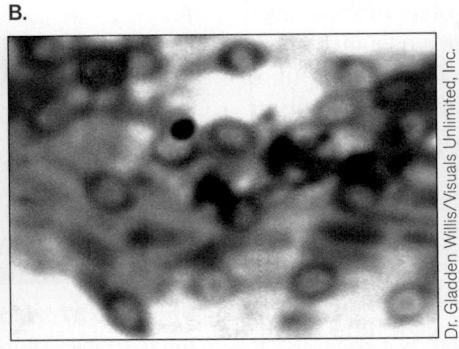
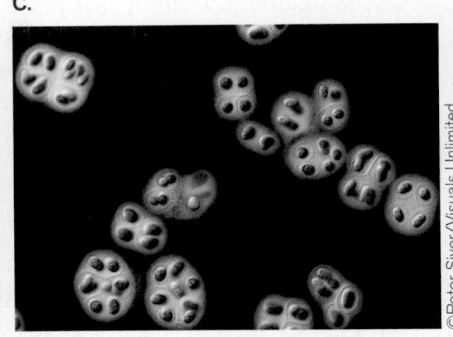

CDC/Dr. George P. Kubica

Dr. Gladden Willis/Visuals Unlimited, Inc.

©Peter Siver/Visuals Unlimited

Figure 2.24 Differential stains. A. *Mycobacterium tuberculosis*, acid-fast stain (stained cells are red, 1–2 μm long). LM, 600×.
B. *Clostridium tetani*, endospore stain (stained endospores are blue-green, 2 μm long). LM, 400×. **C.** *Gloeocapsa*, colonial cyanobacteria, whose individual cells are encased in a thick sheath. Staining with India ink makes the sheath clearly visible. Individual cell size 3–5 μm.

- **Negative stain.** Some bacteria synthesize a capsule of extracellular polysaccharide filaments, which protects the cell from predation or from engulfment by white blood cells (discussed in Section 3.4). The capsule is transparent and invisible in suspended cells. However, a suspension of opaque particles such as India ink can be added to darken the surrounding medium. The particles are excluded by the thick polysaccharide capsule, which thus appears clear against the dark background (**Fig. 2.24C**). This is an example of a negative stain.
- **Antibody stains.** Specialized stains utilize antibodies to identify precise strains of bacteria or even specific molecular components of cells. The antibody (which binds a specific cell protein) is linked to a reactive enzyme for detection or to a fluorophore (fluorescent molecule) for immunofluorescence microscopy (discussed in Section 2.5).

TO SUMMARIZE:

- **In bright-field microscopy**, resolution depends on
  - The wavelength of light, which limits resolution to about 200 nm.
  - The magnifying power of a lens, which depends on its numerical aperture ($n \sin \theta$).
  - The position of the focal plane, the location where the specimen is "in focus" (that is, where the sharpest image is obtained).
- **A compound microscope** achieves magnification and resolution through the objective and ocular lenses.
- **A wet mount** specimen is the only way to observe living microbes.
- **Fixation and staining** of a specimen kills it but improves contrast and resolution.
- **Differential stains** distinguish between different kinds of bacteria with different structural features.
- **The Gram stain** differentiates between two major bacterial taxa: Proteobacteria (Gram-negative) and Firmicutes (Gram-positive). Other bacteria and archaea vary in their Gram stain appearance. Eukaryotes stain negative.

2.4 Dark-Field, Phase-Contrast, and Interference Microscopy

Advanced optical techniques enable us to visualize structures that are difficult or impossible to detect under a bright-field microscope, either because their size is below the limit of resolution of light or because their cytoplasm is transparent. These techniques take advantage of special properties of light waves, including scattering and interference patterns.

One application of **dark-field microscopy** is the detection of pathogenic spirochetes, such as *Treponema pallidum*, the causative organism of syphilis. *T. pallidum* cells are so narrow (0.1 μm) that their shape cannot be fully resolved by light microscopy. Nevertheless, the spiral form of *T. pallidum* can be detected by dark-field microscopy (**Fig. 2.25**).

Dark-Field Microscopy Detects Unresolved Objects

Dark-field optics enables microbes to be visualized as halos of bright light against the darkness, just as stars are

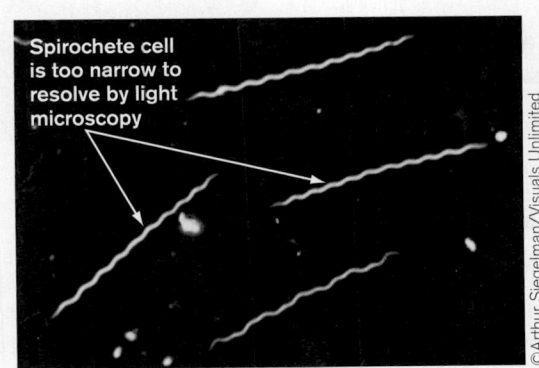

Spirochete cell is too narrow to resolve by light microscopy

©Arthur Siegelman/Visuals Unlimited

Figure 2.25 Dark-field observation of bacteria. *Treponema pallidum* specimen from a patient with syphilis. Note the detection of dust particles. Dark-field light microscopy. Cell length about 10 μm; actual cell width 0.2 μm.

observed against the night sky. A tiny object whose size is well below the wavelength of light, such as a virus particle, can be detected by light scattering. The wave front of scattered light is spherical, like a wave emitted by a point source (see **Fig. 2.9D**).

Light scattering. The scattered wave has a much smaller amplitude than the incident (incoming) wave has. Therefore, with ordinary bright-field optics, scattered light is washed out. Detection of scattered light requires a modified condenser arrangement that excludes all light transmitted directly (**Fig. 2.26**). The condenser contains a "spider light stop," an opaque disk held by three "spider legs" across an open ring. The ring permits only a hollow cone of light to focus on the object. The incident hollow cone converges at the object and then generates an inverted hollow cone radiating outward.

The objective lens is positioned in the central region, where it completely misses the directly transmitted light. For this reason, the field appears dark. However, light scattered by the object radiates outward in a spherical wave. A sector of this spherical wave enters the objective lens and is detected as a halo of light.

An intriguing application of dark-field optics is the study of bacterial motility. Motility is important in bacterial diseases such as urethritis, in which the organism needs to swim up through the urethra. The bacterial swimming apparatus consists of helical filaments called

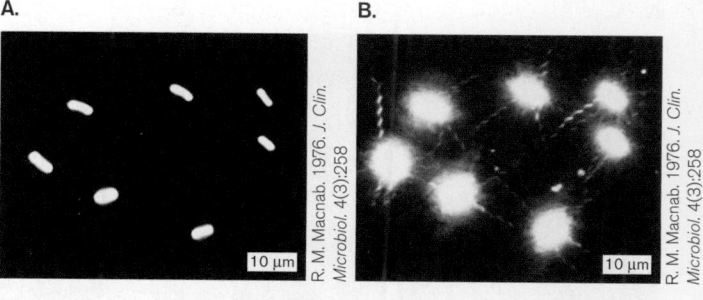

Figure 2.27   Motile bacteria observed under dark-field microscopy.   A. Flagellated *E. coli* observed with low light, which limits scattering. Only cell bodies are detected; no flagella. **B.** The light intensity is increased. Flagella are detected, although their fine structure is not resolved, because their width is below the threshold of resolution by light.

flagella (singular, **flagellum**), which are rotated by a motor device embedded in the bacterial cell wall (for flagellar structure, see Section 3.7). The "swimming strokes" of bacteria were first elucidated by Howard Berg (1934–) and Robert Macnab (1940–2003) using dark-field optics to view the helical flagella. The length of flagella is great enough to be resolved with light waves—but their width is not. Thus, dark-field optics are needed to detect and observe the helical flagella (**Fig. 2.27**). Note, however, that the bacterial cell itself appears "overexposed"; its shape is unresolved owing to the high light intensity.

WWW | Videos of swimming bacteria
microbiology2.com/links

> **THOUGHT QUESTION 2.6**   Some early observers claimed that the rotary motions observed in bacterial flagella could not be distinguished from whiplike patterns, comparable to the motion of eukaryotic flagella. Can you imagine an experiment to distinguish the two and prove that the flagella rotate? *Hint*: Bacterial flagella can get "stuck" to the microscope slide or coverslip.

Limitations of dark-field microscopy. A disadvantage of dark-field microscopy is that any tiny particle, including specks of dust, can scatter light and interfere with visualization of the specimen. Unless the medium is extremely clear, it can be difficult to distinguish microbes of interest from particulates. Other methods of contrast enhancement, such as phase contrast and fluorescence, avoid this difficulty.

Phase-Contrast Microscopy

Phase-contrast microscopy exploits differences in refractive index between the cytoplasm and the surrounding medium or between different organelles. This technique is particularly useful for eukaryotic cells such as amebas,

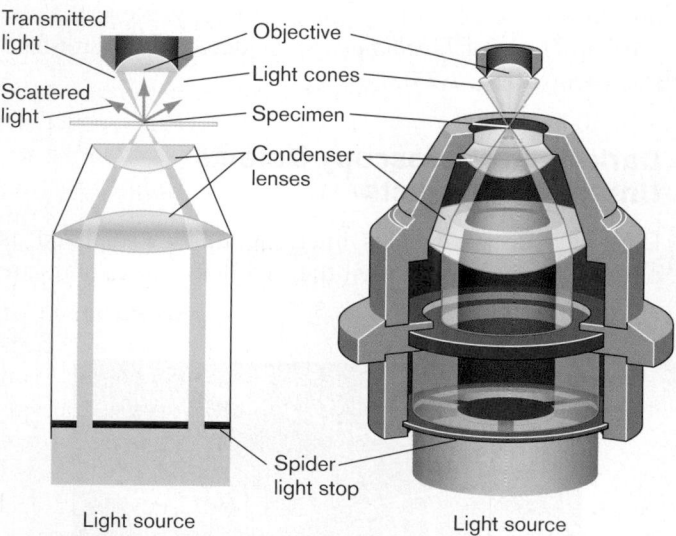

Figure 2.26   A dark-field condenser system, with a spider light stop.   Below the condenser, the spider light stop excludes all but an annular ring of light from the light source. The annular ring converges as a hollow cone of light focused on the specimen. Objects in the specimen scatter light in all directions. The scattered light is collected by the objective lens, but the transmitted light shines outside the range of the objective; thus, in the absence of scattering objects, the field appears dark. Only light scattered by the specimen enters the objective lens.

which contain many intracellular compartments. For example, **Figure 2.28A** shows an image of *Entamoeba histolytica*, a human parasite, obtained by phase-contrast microscopy. In some cells, the nucleus is clearly visible.

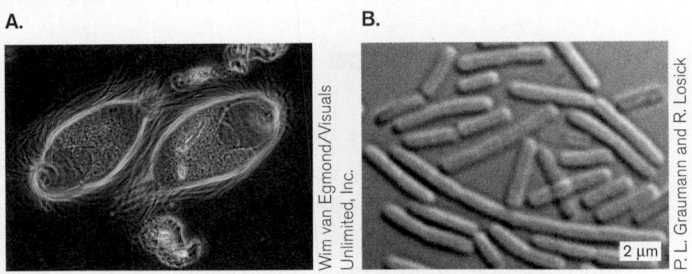

A.   B.

Wim van Egmond/Visuals Unlimited, Inc.

2 µm   P. L. Graumann and R. Losick

Figure 2.28 Phase-contrast and interference effects in microscopy. A. Phase-contrast micrograph of the parasitic protist *Entamoeba histolytica* (15–60 µm). **B.** Nomarski interference micrograph of *Bacillus subtilis*. The apparent three-dimensional effect is illusory. Note that one bacterium appears as if crossed through another. *Source*: Part B reprinted by permission from P. L. Graumann and R. Losick. 2001. *J. Bacteriol.* **183**(13): 4052–4060.

The optical system for phase contrast was invented in the 1930s by the Dutch microscopist Frits Zernike (1888–1966), for which he earned the Nobel Prize in Physics. In this system, slight differences in the refractive index of the various cell components are transformed into differences in the intensity of transmitted light. Zernike's scheme makes use of the fact that living cells have relatively high contrast owing to their high concentration of solutes. Given the size and refractive index of commonly observed cells, light is retarded by approximately one-quarter of a wavelength when it passes through the cell. In other words, after having passed through a cell, light exits the cell about one-quarter of a wavelength behind the phase of light transmitted directly through the medium.

The Zernike optical system is designed to retard the refracted light by an additional one-quarter of a wavelength, so that the light refracted through the cell is slowed by a total of half a wavelength compared with the light transmitted through the medium. When two waves are out of phase by half a wavelength, they produce destructive interference, canceling each other's amplitude (see **Fig. 2.13A**). The result is a region of darkness in the image of the specimen.

As in dark-field microscopy, the light transmitted through the medium in phase-contrast microscopy needs to be separated from the light interacting with the object—in this case, light waves slowed by refraction. This separation is performed by a ring-shaped slit, called an "annular ring," similar in function to the spider light stop. The annular ring stops light from passing directly through the center of the lens system, where the specimen is located, and generates a hollow cone of light, which is focused through the specimen and generates an inverted cone above it (**Fig. 2.29**). Light passing through the specimen, however, is not only retarded; it is also refracted and thus bent into the central region within the inverted cone.

Both the refracted light from the specimen and the outer cone of transmitted light enter the phase plate. The phase plate consists of refractive material that is thinner in the region met by the outer (transmitted) light cone. The refracted light passing

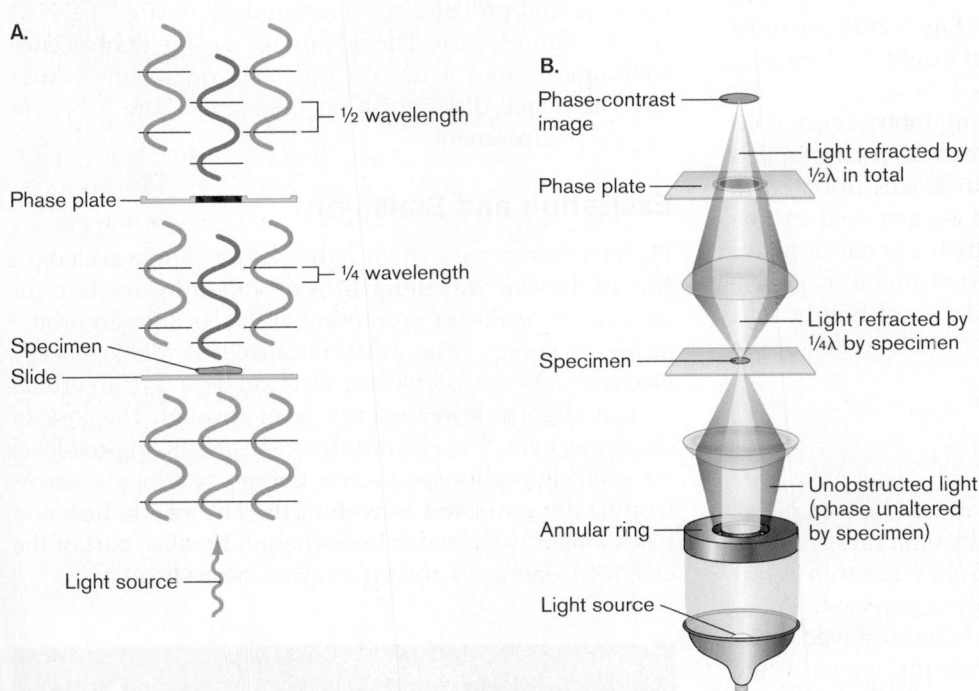

Figure 2.29 Phase-contrast optics. A. The specimen retards light by approximately one-quarter of a wavelength. The phase plate contains a central disk of refractive material that retards light from the specimen by another quarter wavelength, increasing the phase difference to half a wavelength. The light from the specimen and the transmitted light are now fully out of phase; and if they coincide their waveforms cancel, making the specimen appear dark. **B.** In the phase-contrast microscope, the annular ring forms a hollow cone of light. As the light cone passes through the refractive material of the specimen, it is delayed by about one-quarter of a wavelength, and its path bends inward to the central region. This refracted light is surrounded by the hollow cone of unrefracted light. The light refracted by the specimen enters the lens through the dense central disk of the phase plate, which retards the wave by another quarter wavelength. When the transmitted and refracted light cones re-join at the focal point, they are out of phase; their amplitudes cancel each other, and that region of the image appears dark against a bright background.

through the center of the phase plate is retarded by an additional one-quarter wavelength compared with the transmitted light passing through the thinner region on the outside. The overall difference approximates half a wavelength, so that the two waves are out of phase, thus canceling each other's amplitude. When the light from the inner and outer regions focuses at the ocular lens, the amplitudes of the wave trains cancel and produce a region of darkness. In this system, small differences in refractive index can produce dramatic differences in contrast between the offset phases of light.

Interference Microscopy

Other kinds of optical systems have been devised using light interference to enhance cytoplasmic contrast. **Interference microscopy** enhances contrast by superimposing the image of the specimen on a second beam of light that generates interference fringes. The interference pattern produces an illusion of shadowing across the specimen. For example, the shapes of the *Bacillus subtilis* cells illuminated by interference contrast (**Fig. 2.28B**) are more clearly defined than in conventional bright-field microscopy (for comparison, see **Fig. 2.21**).

One optical system for producing interference contrast is based on the Wollaston-Nomarski prism. In this system, light is polarized to obtain waves oriented in one direction. The polarized light is then split by the prism into two separate beams, which are out of phase with each other. The two beams recombine to generate interference patterns whose edges are highly sensitive to slight differences in the refractive index of the specimen.

TO SUMMARIZE:

- **Dark-field microscopy** uses scattered light to detect objects too small to be resolved by light rays.
  Advantage: Extremely small microbes and thin extracellular structures can be detected.
  Limitation: The shape of objects is not resolved. Dust particles easily obscure the image of the specimen.
- **Phase-contrast microscopy** superimposes refracted light and transmitted light shifted out of phase so as to reveal differences in refractive index as patterns of light and dark.
  Advantage: Live cells with transparent cytoplasm, and the organelles of eukaryotes, can be observed with high contrast.
  Limitation: Phase contrast is less effective for organisms whose cytoplasm has a low refractive index.
- **Interference microscopy** superimposes interference bands on an image, accentuating small differences in refractive index.

Advantage: The shape of cells can be defined most clearly.
Limitations: Interference microscopy requires complex optical adjustment and is less effective for organisms with low refractive index.

2.5   Fluorescence Microscopy

In fluorescence microscopy, incident light is absorbed by the specimen and reemitted at a lower energy, thus longer wavelength. Fluorescence microscopy offers a powerful way to detect microbes and subcellular structures while avoiding artifacts caused by dust and other nonspecific materials.

One use of fluorescence microscopy is to assess microbial populations in highly dilute natural environments (**Fig. 2.30**). This technique was used by the Oceanic Microbial Observatory at the Bermuda Biological Station for Research to conduct a long-term study assessing the effects on microbial populations of changing water chemistry due to climate changes such as global warming. Populations of microbes, including viruses, bacteria, and protists, are measured by means of DNA-specific fluorescence. The advantage of this fluorescence technique is that it detects only live organisms whose DNA is intact, distinguishing them from fine debris in natural environments.

Excitation and Emission

Fluorescence occurs when light of a specific wavelength (the **excitation wavelength**) is absorbed by an atom (or molecule) capable of promoting an electron to an orbital of higher energy (**Fig. 2.31**). Because this higher-energy electron state is unstable, the electron decays to an orbital with a slightly lower energy level through the loss of energy as heat. The electron then falls to its original level by emitting a photon of lower energy and longer wavelength, the **emission wavelength**. The emitted photon has a longer wavelength (less energy) because part of the electron's energy of absorption was lost as heat.

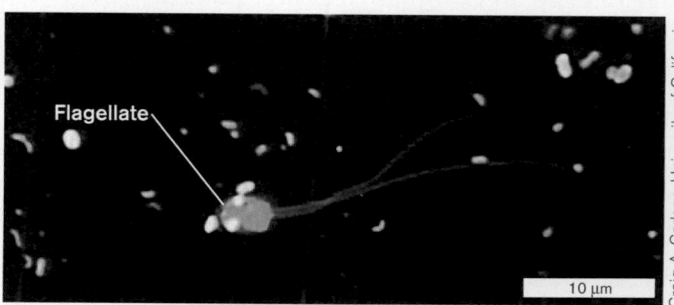

Figure 2.30   Cell counting using fluorescence microscopy. Bacterioplankton and flagellated protists from the surface waters of the Sargasso Sea near Bermuda are enumerated to investigate their contribution to the oceanic carbon cycle. Microbes can be distinguished from inert particles by fluorescence microscopy using the DNA-specific fluorophore DAPI.

A.

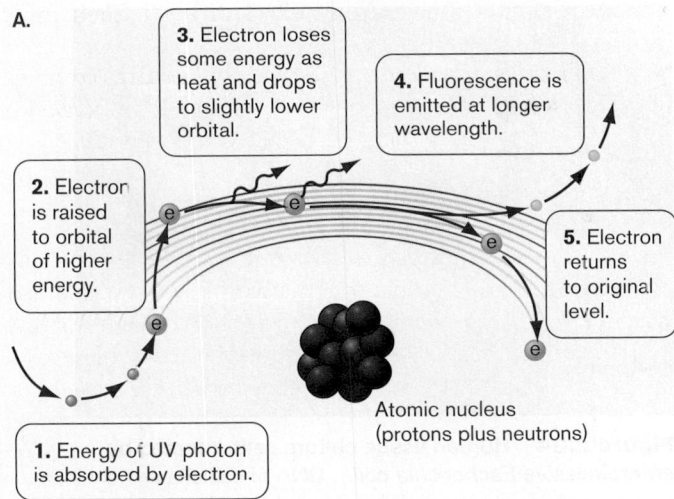

3. Electron loses some energy as heat and drops to slightly lower orbital.

4. Fluorescence is emitted at longer wavelength.

2. Electron is raised to orbital of higher energy.

5. Electron returns to original level.

1. Energy of UV photon is absorbed by electron.

Atomic nucleus (protons plus neutrons)

B.

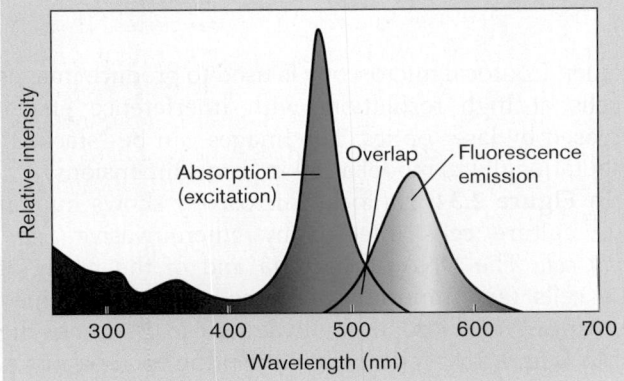

Absorption (excitation)

Overlap

Fluorescence emission

Figure 2.31   Fluorescence. Energy gained from UV absorption is released as heat and as a photon of longer wavelength in the visible region. **A.** Fluorescence on the molecular level. **B.** Comparison of absorption and emission spectra for a fluorophore.

The wavelengths of excitation and emission are determined by the **fluorophore**, the fluorescent molecule used to stain the specimen. For example, the slides for cell counting in the Bermuda study (see **Fig. 2.30**) used the DNA-specific stain 4′,6-diamidino-2-phenylindole (DAPI). The

aromatic rings of DAPI mimic a base pair, enabling intercalation between base pairs of DNA. Another commonly used DNA-specific fluorophore is acridine orange (**Fig. 2.32**). These fluorophores provide an extremely sensitive means of detecting diverse microorganisms in the environment, including those whose dimensions are too small to be resolved.

The optical system for fluorescence microscopy utilizes filters to limit incident light to the wavelength of excitation and emitted light to the wavelength of emission. The wavelengths of excitation and emission are determined by the specific fluorophore used.

Fluorophores for Labeling

Fluorescence can be used to label specific parts of cells. The specificity of the fluorophore can be arranged in several ways:

- **Chemical affinity.** Certain fluorophores have chemical affinity for certain classes of biological molecules; for example, the fluorophore acridine specifically binds DNA and RNA.
- **Labeled antibodies.** Antibodies that specifically bind a cell component are chemically linked to a fluorophore molecule. The use of antibodies linked to fluorophores is known as immunofluorescence.
- **Gene fusion.** Genetic recombination can be used to make a hybrid gene expressing a protein that generates fluorescence, such as green fluorescent protein (GFP).
- **DNA hybridization.** A short sequence of DNA attached to a fluorophore will hybridize to a specific sequence in the genome, thus labeling one position in the chromosome or nucleoid.

Fluorophore labeling is used to study endospore formation in *Bacillus* species (**Fig. 2.33**). The cell envelope and DNA origin of replication are labeled by different fluorophores to track movements of DNA during sporulation. Richard Losick and colleagues applied this technique to show how the DNA origin moves toward one pole of the cell, followed by formation of a septum, a new portion of cell envelope, just behind the developing endospore.

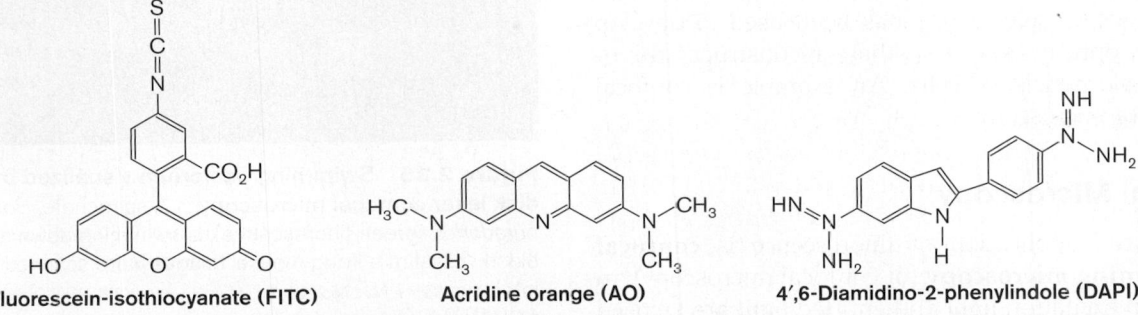

Fluorescein-isothiocyanate (FITC)   Acridine orange (AO)   4′,6-Diamidino-2-phenylindole (DAPI)

Figure 2.32   Fluorescent molecules (fluorophores) commonly used in microscopy. The abundance of conjugated double bonds in these molecules provides closely spaced molecular orbitals that give rise to fluorescence.

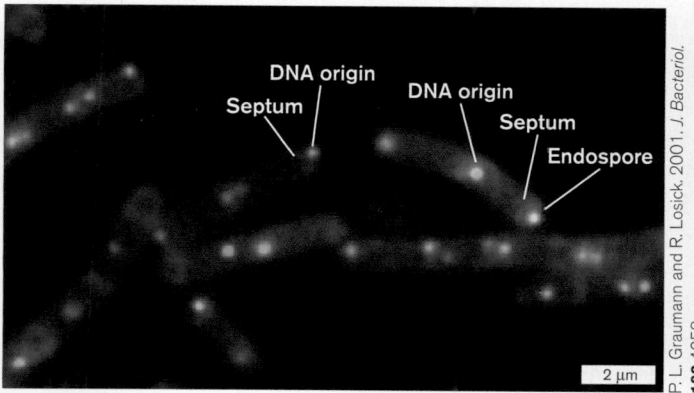

Figure 2.33   Fluorescence micrograph of *Bacillus subtilis* cells during sporulation.   Red fluorescence arises from membrane stained with the dye FM-464. Green fluorescence arises from green fluorescent protein (GFP) bound to the DNA origin of replication. The yellow color occurs where green fluorescence overlaps red.

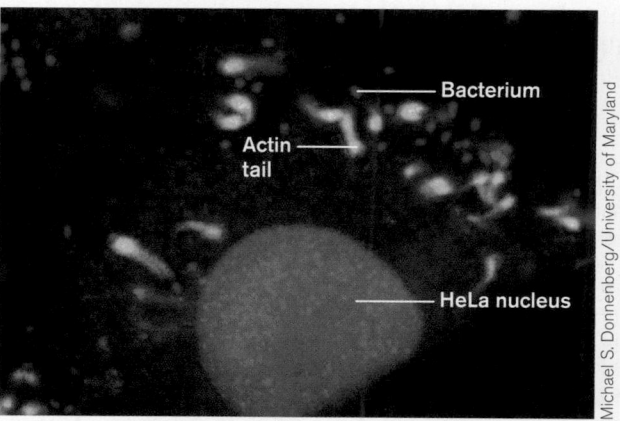

Figure 2.34   Human tissue culture cells infected by enteroinvasive *Escherichia coli*.   DNA of bacteria and of HeLa nuclei are stained blue with DAPI fluorophore. Invading bacteria (blue-stained rods, 1–2 μm long) cause the host to form actin "tails," stained green with the actin-binding FITC fluorophore. Confocal microscopy.

In **Figure 2.33**, the cell envelope is stained red by the fluorescent dye FM-464. The dye FM-464 specifically binds phospholipid membranes. The DNA origin of replication is stained by green fluorescent protein (GFP) fused to a protein that specifically binds the DNA origin of replication. The fluorescence micrographs were taken at two different wavelengths for both excitation and emission to record the two different fluorescent labels. The images were superimposed with two different colors marking envelope and DNA. The result shows how, in sporulation, DNA replication originates not at the midpoint of the cell, as in the vegetative growth of most bacteria, but at the poles where the endospore forms.

> **THOUGHT QUESTION 2.7**   What experiment could you devise to determine the actual order of events in DNA movement toward the pole during formation of an endospore?
>
> **THOUGHT QUESTION 2.8**   Compare and contrast fluorescence microscopy with dark-field microscopy. What similar advantage do they provide, and how do they differ?

Fluorescence microscopy has been used to develop advanced optical systems that reconstruct three-dimensional models of cells. An example is confocal fluorescence microscopy.

Confocal Microscopy

An advanced application of fluorescence is **confocal laser scanning microscopy** (or confocal microscopy), in which both excitation light and emitted light are focused

together. Confocal microscopy is used to produce images of cells at high resolution, with interference effects decreased by laser optics. The images can be "stacked" computationally to model a cell in three dimensions.

In **Figure 2.34**, confocal microscopy shows human tissue culture cells infected by enteroinvasive *Escherichia coli*. The DNA of bacteria and of the nuclei of HeLa cells (an immortal cancer cell line) are stained blue with DAPI fluorophore. Invading bacteria cause the host to form actin "tails," which help the bacteria move. Tails are stained green by the actin-binding fluorophore fluorescein-isothiocyanate (FITC).

Confocal microscopy enables rapid observation of living microbes in real time. **Figure 2.35** shows motility of a spirochete bacterium, *Borrelia burgdorferi*, which causes

Figure 2.35   Swimming bacterium visualized by spinning-disk laser confocal microscopy.   A spirochete, *Borrelia burgdorferi* (green fluorescence), is swimming down a capillary blood vessel of a living mouse. Images were scanned at 1-second intervals.   *Source*: Modified from Tara Moriarty et al. 2008. *PLoS Pathogens* **4**:e1000090, fig. 2.

Lyme disease. The spirochete is swimming through a capillary blood vessel of a living mouse. The spirochete is labeled green by a gene expressing a protein fused to GFP (green fluorescent protein). The vessel walls are labeled red by a fluorescent antibody.

In confocal microscopy, a laser beam is focused onto the specimen and scanned across it in two dimensions—that is, in two planes at right angles to each other (**Fig. 2.36**). The laser beam excites the fluorophore, causing it to emit light at a longer wavelength. The emitted light passes in reverse direction through the objective, where it encounters a dichroic mirror that allows light transmission at the excitation wavelength but reflects light at the wavelength of emission. The reflected emission rays are then focused and pass through a pinhole, which eliminates all unfocused light. A narrow, focused beam, representing one "pixel" of image, enters the photomultiplier tube. The laser beam scans across the specimen to give a two-dimensional pattern of pixels that forms an image. The scanned images can be stacked through a series of focal planes to generate a three-dimensional model.

The scanning feature of confocal microscopy is also used to acquire data for high-throughput experiments in which a large number of chemical reactions are arranged in microscopic quantities in an array. An example of an array is a DNA microarray containing probes for all the genes of a genome. Short segments of DNA are arrayed on a microscope slide such that an entire genome of potential protein-encoding genes may be present on a single slide. DNA microarrays are discussed further in Chapter 12.

TO SUMMARIZE:

■ **Fluorescence microscopy** involves detection of specific cells or cell parts based on fluorescence by a fluorophore.
■ **Cell parts can be labeled** by a fluorophore attached to an antibody stain.

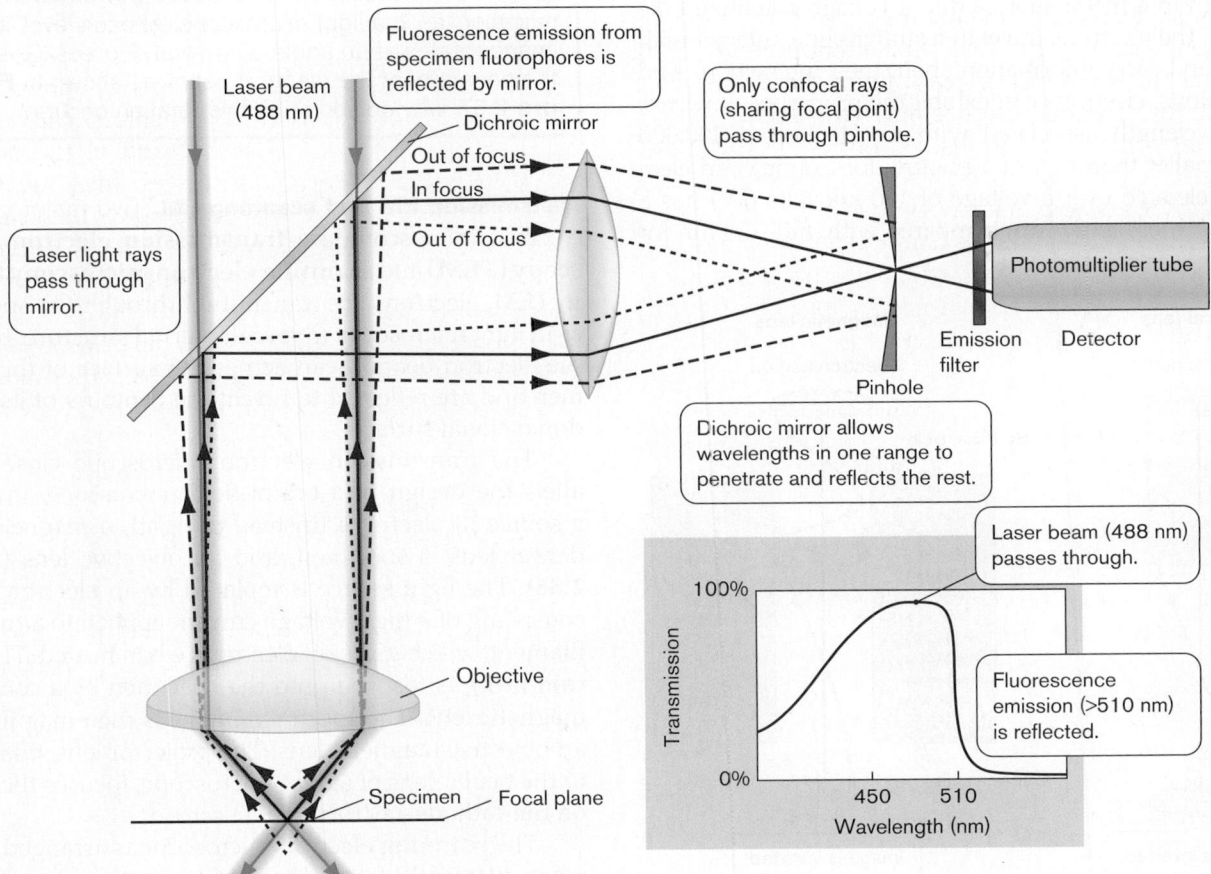

Figure 2.36 Confocal microscopy. In confocal optics, the incident laser beam (blue) passes through a dichroic mirror and reaches the specimen, where its absorption leads to fluorescent emission at a longer wavelength. The fluorescent emission travels back to the dichroic mirror, where it is reflected toward the photomultiplier. Only the confocal rays (those emitted from the focal point) pass through the pinhole and reach the photomultiplier.

■ **Confocal laser scanning microscopy** visualizes cells in three dimensions.

2.6 Electron Microscopy

All cells are built of macromolecular structures. The tool of choice for observing the shape of these macro-molecular structures is **electron microscopy (EM)**. In electron microscopy, beams of electrons are focused to generate images of cell membranes, chromosomes, and ribosomes with a resolution a thousand times that of light microscopy. The modern electron microscope was developed in the 1950s and was popularly known as the centerpiece of any biological research program. In Michael Crichton's film *The Andromeda Strain*, for example, an electron microscope is used to analyze a fictional pathogen from outer space.

Beams of Electrons

How does an electron microscope work? Electrons are ejected from a metal subjected to a voltage potential. Like photons, the electrons travel in a straight line, interact with matter, and carry information about their interaction. And like photons, electrons can exhibit the properties of waves. The wavelength associated with an electron is 100,000 times smaller than that of a photon; for example, an electron accelerated over a voltage of 100 kilovolts (kV) has a wavelength of 0.0037 nm compared with 400–750 nm for

visible light. However, the actual resolution of electrons in microscopy is limited not by the wavelength, but by the aberrations of the lensing systems used to focus electrons. The magnetic lenses used to focus electrons never achieve the precision required to utilize the full potential resolution of the electron beam.

Electrons are focused by means of a magnetic field directed along the line of travel of the beam (**Fig. 2.37**). As a beam of electrons enters the field, it spirals around the magnetic field lines. The shape of the magnet can be designed to generate field lines that will focus the beam of electrons in a manner analogous to the focusing of photons by a refractive lens. The electron beam, however, forms a spiral because electrons travel around magnetic field lines. Because magnetic lenses generate aberrations, a series of corrective magnetic lenses is generally required to obtain a resolution of about 0.2 nm. This represents a resolution a thousand times greater than the 200 nm resolution of light microscopy.

> **THOUGHT QUESTION 2.9** An electron microscope can be focused at successive powers of magnification, as in a light microscope. At each level, the image rotates at an angle of several degrees. Given the geometry of the electron beam, as shown in **Figure 2.37**, why do you think this rotation occurs?

Transmission EM and scanning EM. Two major types of electron microscopy are **transmission electron microscopy (TEM)** and **scanning electron microscopy (SEM)**. In TEM, electrons are transmitted through the specimen as in light microscopy to reveal internal structure. In SEM, the electron beams scan across the surface of the specimen and are reflected to reveal the contours of its three-dimensional surface.

The transmission electron microscope closely parallels the design of a bright-field microscope, including a source of electrons (instead of light), a magnetic condenser lens, a specimen, and an objective lens (**Figure 2.38**). The light source is replaced by an electron source consisting of a high-voltage current applied to a tungsten filament, which gives off electrons when heated. The electron beam is focused onto the specimen by a condenser magnetic lens. The specimen image is then magnified by an objective magnetic lens. The projector lens, analogous to the ocular lens of a light microscope, focuses the image on the cathode-ray tube (CRT) screen.

The scanning electron microscope is arranged somewhat differently from the TEM in that a series of condenser lenses focuses the electron beam onto the surface of the specimen. Reflected electrons are then picked up by a detector (**Fig. 2.39**). In either case, the overall apparatus required for electron microscopy is complex, resembling the bridge of a *Star Trek* spaceship.

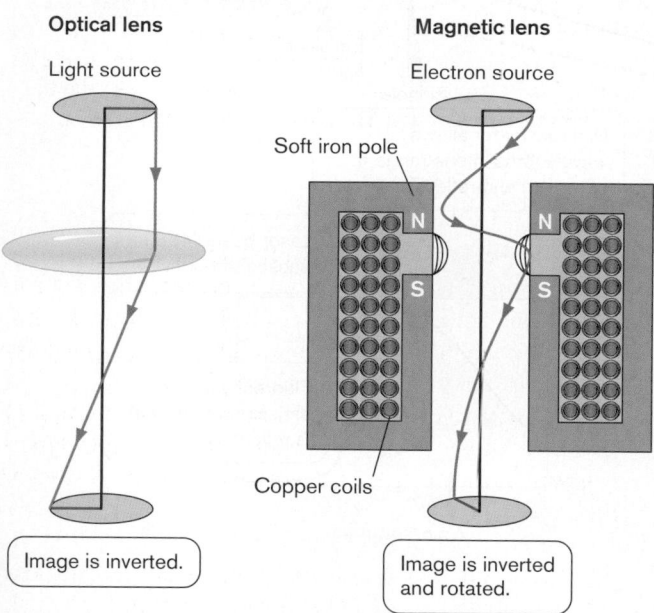

Figure 2.37 A magnetic lens. The beam of electrons spirals around the magnetic field lines. The U-shaped magnet acts as a lens, focusing the spiraling electrons much as a refractive lens focuses light rays.

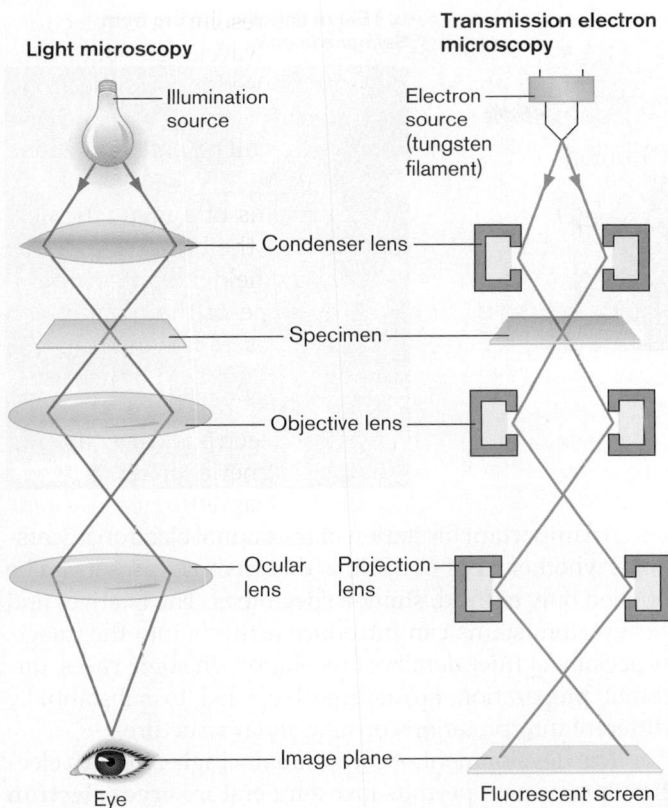

Light microscopy

Illumination source

Transmission electron microscopy

Electron source (tungsten filament)

Condenser lens

Specimen

Objective lens

Ocular lens

Projection lens

Image plane

Eye

Fluorescent screen

Figure 2.38 Transmission electron microscopy. The light source is replaced by an electron source consisting of a high-voltage current applied to a tungsten filament, which gives off electrons when heated. Because of the problem of aberrations in focusing electrons, each magnetic lens shown (condenser, objective, projection) actually represents a series of lenses.

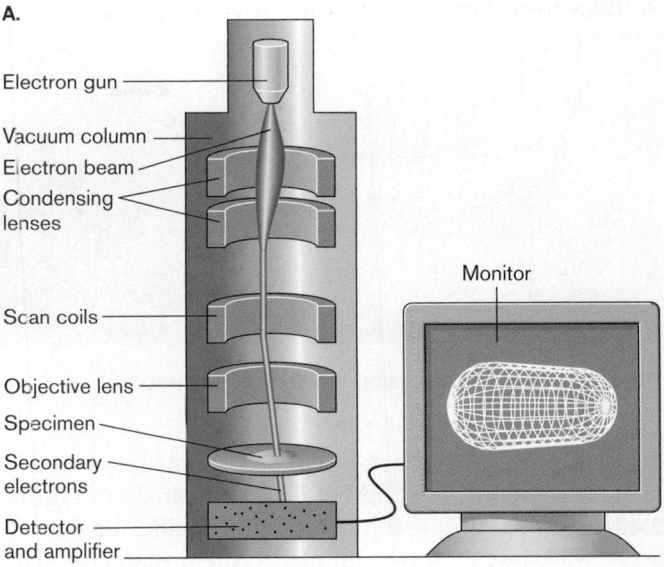

A.

Electron gun

Vacuum column

Electron beam

Condensing lenses

Scan coils

Objective lens

Specimen

Secondary electrons

Detector and amplifier

Monitor

©Museum of Science/Boston

B.

Figure 2.39 Scanning electron microscopy.   A. In the scanning electron microscope (SEM), the electron beam is scanned across a specimen coated in gold, which acts as a source of secondary electrons. The incident electron beam ejects secondary electrons toward a detector, generating an image of the surface of the specimen. **B.** Loading a specimen into the vacuum column.

Sample Preparation

Electron microscopy poses special problems for biological specimens. With the exception of cryo-EM (discussed shortly), the entire optical column must be maintained under vacuum to prevent the electrons from colliding with the gas molecules in air. The requirement for a vacuum precludes the viewing of live specimens, which in any case would be quickly destroyed by the electron beam. Moreover, the structure of most specimens lacks sufficient electron density (ability to scatter electrons) to provide contrast. Thus, the specimen usually requires an electron-dense stain using salts of heavy metals such as gold or uranium. The heavy atoms collect at the surface of cell structures such as membranes, where their electron scatter reveals the outline of the structure.

The specimen can be prepared in one of three ways:

■ **Embedded in a polymer for thin sections.** A special knife called a microtome cuts slices through the specimen, each slice a fraction of a micrometer thick.

■ **Sprayed onto a copper grid.** Except for "electron-dense" structures that scatter electrons, the electron beam penetrates the object as if it were transparent. This method is effective for imaging virus particles and isolated macromolecular complexes.

■ **Flash-frozen (for cryo-EM).** Samples frozen rapidly in refrigerant provide sufficient contrast for detection by a high-intensity electron beam, a recent innovation.

For sectioned samples or for samples sprayed onto a grid, the specimen is treated with a heavy-metal salt such as uranyl acetate. The metal salt is deposited around the biological structures, acting as a negative stain. (For comparison, a negative stain used in light microscopy is the India ink stain, illustrated in **Fig. 2.24C**.)

A. TEM of *Bacillus anthracis* showing envelope and cytoplasm

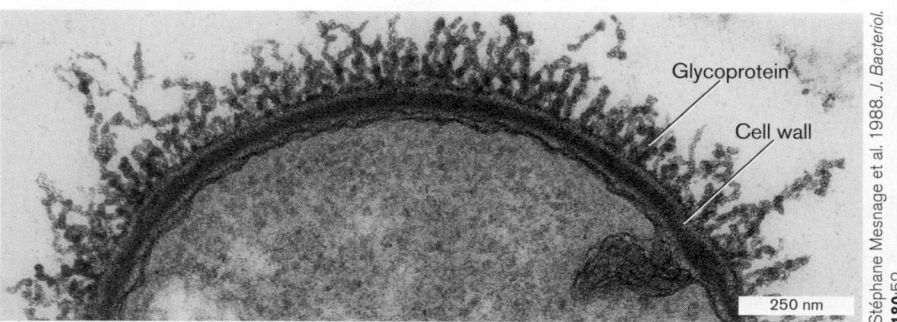

Glycoprotein

Cell wall

250 nm

Stéphane Mesnage et al. 1988. *J. Bacteriol.* **180**:52

B. TEM of flagellar motors from *Salmonella enterica*

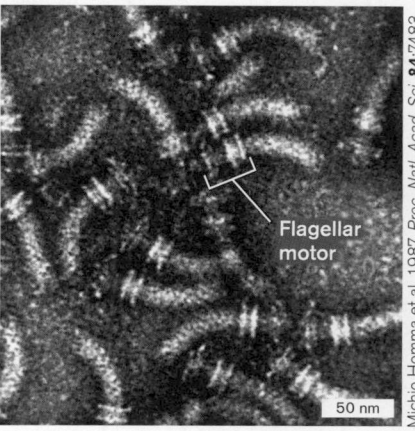

Flagellar motor

50 nm

Michio Homma et al. 1987. *Proc. Natl. Acad. Sci.* **84**:7483.

Figure 2.40   Transmission electron micrographs.

In **Figure 2.40** we see examples of transmission electron micrographs. The TEM of *Bacillus anthracis* in **Figure 2.40A** shows a thin section through a bacillus, including a cell wall, membranes, and glycoprotein filaments. The image includes electron density throughout the depth of the section. In **Figure 2.40B**, flagellar motors have been isolated, or "unplugged," from a bacterial cell envelope and spread on a grid for TEM. The micrograph reveals details of the motor, including the axle and individual rings.

Scanning electron microscopy can show whole cells in three-dimensional view, with much greater resolution than light microscopy; for comparison of the two, see **Figure 2.6**. SEM is particularly effective for visualizing cells within complex communities such as a biofilm. **Figure 2.41** shows *Alcaligenes xylosoxidans* bacilli growing on the interior surface of a catheter inserted in the blood vessel of an immunocompromised patient. The SEM reveals the irregular contours of the bacteria, as well as the fibrin-like filaments deposited among them.

> **THOUGHT QUESTION 2.10**   What kinds of research questions could you investigate using SEM? What questions could you answer using TEM?

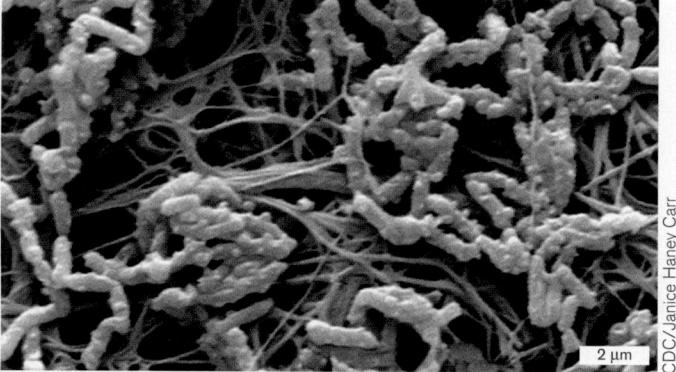

2 µm

CDC/Janice Haney Carr

Figure 2.41   Scanning electron micrograph of a catheter biofilm. *Alcaligenes xylosoxidans* bacilli growing amid fibrin-like filaments on the interior surface of a catheter.

An important limitation of traditional electron microscopy, whether TEM or SEM, is that in most cases it can be applied only to fixed, stained specimens. The fixatives and heavy-atom stains can introduce artifacts into the image, especially at finer details of resolution. In some cases, different preparation procedures have led to substantially different interpretations of subcellular structure.

The development of exceptionally high-strength electron beams now permits low-temperature **cryo–electron microscopy (cryo-EM)**, also known as **electron cryomicroscopy**. In cryo-EM, the specimen is flash-frozen—that is, suspended in water and frozen rapidly in a refrigerant of high heat capacity (ability to absorb heat). The rapid freezing avoids ice crystallization, leaving the water solvent in a glass-like amorphous phase. The specimen retains water content and thus closely resembles its viable form, although it is still ultimately destroyed by electron bombardment. The sample does not require staining, because the high-intensity electron beams can detect smaller signals than in earlier instruments.

Another innovation in cryo-EM is **tomography**, the acquisition of projected images from different angles of a transparent specimen. **Cryo–electron tomography**, or **electron cryotomography**, avoids the need to physically slice the sample for thin-section TEM. The images from tomography are combined digitally to visualize the entire object.

One use of cryo–electron tomography is to generate high-resolution models of virus particles. For these small particles, multiple images can be averaged together by computational analysis. The digitally combined images can achieve high resolution, nearly comparable to that of X-ray crystallography. Wah Chiu at Baylor College of Medicine has pioneered the use of cryo-EM to visualize virus particles at high resolution (**Fig. 2.42A**). An example is the visualization of rice dwarf virus (**Fig. 2.42B**), one of the world's most economically damaging agricultural pathogens.

Cryo-EM is especially important for particles that cannot be crystallized for X-ray diffraction analysis, the

A.

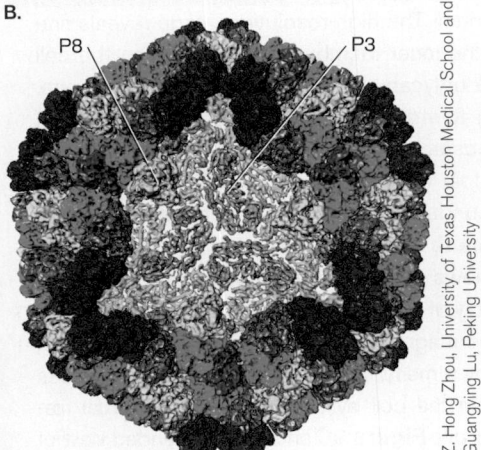

B.

P8   P3

Figure 2.42   Cryo–electron tomography reveals virus structure.   A. Wah Chiu (standing) and Joanita Jakana, at Baylor College of Medicine, using a JEOL 300-kilo-electron-volt (keV) electron cryomicroscope. Chiu's laboratory images virus structures using cryo–electron tomography. **B.** Rice dwarf virus model, at resolution 0.68 nm. The outer shell is composed of 396 subunits (bright colors). A cutaway from the outer shell reveals the inner shell (pale colors) composed of 120 subunits.

most common means of molecular visualization. Because the frozen sample remains hydrated, the biological molecules retain the same conformation as in solution. The sample is imaged without heavy-metal stains, and thus higher resolution is obtained. This technique avoids introducing stain artifacts but generates very low contrast. Repeated scans can be summed computationally to obtain an image at higher resolution.

Cryo–electron tomography is now used to obtain remarkable three-dimensional images of entire cells and subcellular structures (**Special Topic 2.1**).

Interpreting Microscopy Results

The images produced by high-level microscopy can be difficult to interpret. For example, an oval that appears

A.   B.

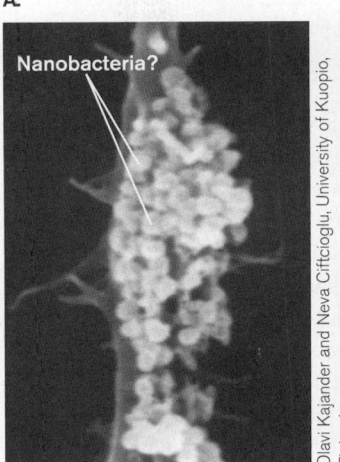   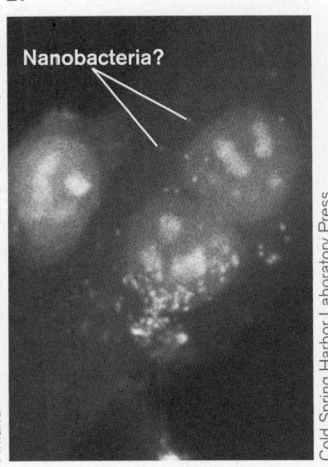

Nanobacteria?   Nanobacteria?

Figure 2.43   Artifacts of microscopy.   A. Objects attached to the surface of a fibroblast were observed by SEM. The objects (200–300 nm in diameter) were identified as "nanobacteria," exceptionally small bacteria inhabiting human blood plasma, but they are now believed to be mineral deposits. **B.** Cultured human cells observed by immunofluorescence microscopy revealed small particles identified as nanobacteria. This observation, however, has not been reproduced by other researchers.

hollow might be interpreted as a cell when in fact it represents a deposit of staining material. A microscopic structure that is interpreted incorrectly is termed an **artifact**. Avoiding artifacts is an important concern in microscopy.

Sometimes, published interpretations generate controversy. For example, in one report electron microscopy and immunofluorescence microscopy were used to test for contamination of cultured white blood cells that grew poorly (**Fig. 2.43**). The SEM images were interpreted to show the presence of tiny prokaryotic cells, 200–300 nm in diameter, attached to cultured fibroblast cells (**Fig. 2.43A**). These proposed cells were termed "nanobacteria." Unfortunately, other researchers were unable to confirm these results and suggested that the objects observed might actually be mineral deposits. Another technique, immunofluorescence microscopy, reportedly showed the invasion of cultured cells by nanobacteria (**Fig. 2.43B**). The technique required use of fluorescent-tagged antibodies against nanobacteria, but the actual specificity of the antibodies was unclear. The existence of small infective particles in the blood remains a question of interest to medical researchers.

THOUGHT QUESTION 2.11   What kinds of experiments could prove or disprove the interpretations of the images of "nanobacteria" in blood plasma?

Special Topic 2.1	Cryo–Electron Tomography Images an Entire Cell in Three Dimensions

Imagine the ability to "see" an entire cell in three dimensions, including organelles as small as ribosomes. The technique to do this, **cryo–electron tomography**, or **electron cryotomography**, was pioneered by Grant Jensen and colleagues at the California Institute of Technology. Cryo–electron tomography enables us to visualize the cell's parts as part of the whole, as they are organized within the living cell.

In cryo–electron tomography, the object is imaged without heavy-metal stains that obscure the surface details. Thus, higher resolution and more accurate details are obtained. To obtain an image in three dimensions, repeated scans are taken, either at different angles or within different focal planes

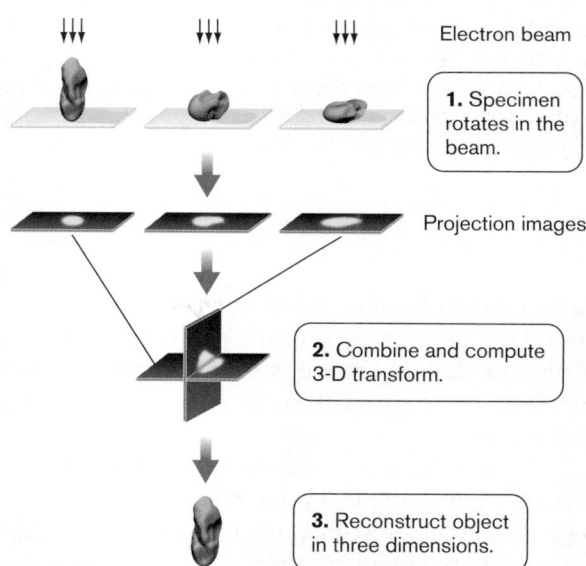

Figure 1.   Three-dimensional image construction in cryo–electron tomography.   Cryo-EM images are obtained in multiple focal planes throughout an object. The images are combined through a mathematical transformation to model the entire object in three dimensions.

(**Figure 1**). Each of the different scans images slightly different parts of the object. The scans are then summed computationally to generate a model in three dimensions.

Jensen used cryo–electron tomography to visualize a magnetotactic bacterium, *Magnetospirillum magneticum*. Magnetotactic bacteria can swim along magnetic field lines because the cell contains a string of magnetic particles composed of the mineral magnetite (iron oxide, Fe_3O_4). But how are these particles of magnetite "grown" and organized within a cell?

In **Figure 2A**, a cryo-EM image shows the section through the bacterium. The high-resolution image reveals fine details, including the inner membrane (equivalent to the cell membrane), peptidoglycan cell wall, and outer membrane, an outer covering found in Gram-negative bacteria. Within the cytoplasm, ribosomes can be resolved. The image also reveals a chain of four dark magnetosomes (particles of magnetite), each surrounded by a vesicle of membrane. How do these vesicles relate to the cell as a whole?

Figure 2B shows a model of the magnetosomes, reconstructed by assembly of multiple cryo-EM scans across the volume of the cell. The magnetosomes are digitally colorized red, each surrounded by a membrane vesicle (green). The vesicles are organized within the cell by a series of protein axial filaments, colorized yellow. **Figure 2C** shows an expanded view of the magnetosomes viewed from the cell interior. In this expansion, the magnetosome vesicles are seen to consist of invaginations from the cell membrane. Thus, the three-dimensional model shows how the magnetite particles are fixed in position by invaginated membranes and held in a line by axial filaments.

Cryo–electron tomography is now used to visualize many kinds of cells and viruses. Jensen's group has used the method to visualize the flagellar motors of spirochete bacteria, and the carboxysomes (carbon dioxide–fixing structures) of marine cyanobacteria. Wah Chiu's group at Baylor has used cryo–electron tomography to model viruses such as herpes virus and human immunodeficiency virus (HIV).

Emerging Methods of Microscopy

New methods of microscopy are emerging that enable nanoscale observation of cell surfaces, in some cases of living cells suspended in water. A general term for these methods is scanning probe microscopy (SPM). SPM methods differ from light and electron microscopy, in which the sample interacts with a beam of light or electrons. Instead, SPM methods measure a physical interaction, such as the "atomic force" between the sample and a sharp tip. **Atomic force microscopy (AFM)** measures the van der Waals forces between the electron shells of adjacent atoms of the cell surface and the sharp tip. AFM is particularly useful to study the surfaces of live bacteria.

A.

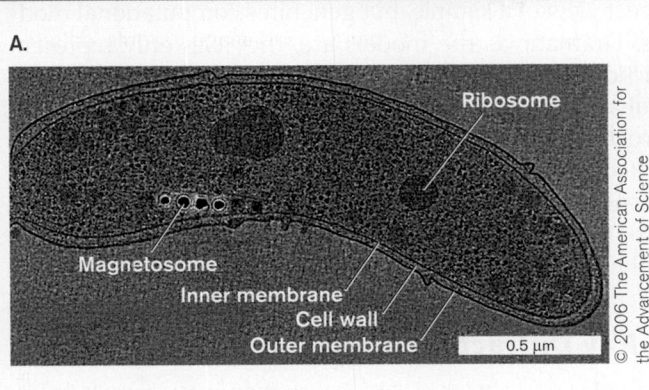

© 2006 The American Association for the Advancement of Science

Ribosome

Magnetosome

Inner membrane

Cell wall

Outer membrane

0.5 µm

B.

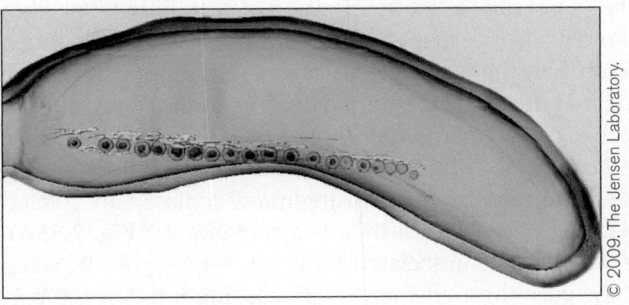

© 2009. The Jensen Laboratory.

C.

© 2009. The Jensen Laboratory.

Figure 2 **The magnetotactic cell visualized by cryo–electron tomography.** **A.** A single cryo-EM scan lengthwise through *Magnetospirillum magneticum*. **B.** Three-dimensional model of *M. magneticum* based on multiple scans. **C.** Expanded view of the cell interior. *Source:* Parts A–C from Arash Komeili et al. 2006. *Science* **311**:242–245.

A.

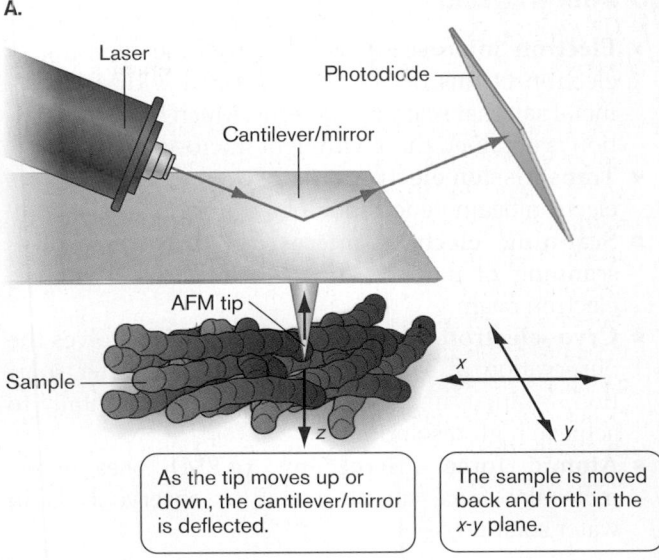

Laser

Photodiode

Cantilever/mirror

AFM tip

Sample

x

z

y

As the tip moves up or down, the cantilever/mirror is deflected.

The sample is moved back and forth in the *x-y* plane.

B.

Nicole Hansmeier et al. 2006. *Microbiology* **152**:923–935.

Figure 2.44 **Atomic force microscopy enables visualization of untreated cells.** **A.** The atomic force microscope (AFM) has a fine-pointed tip attached to a cantilever that moves over a sample. The tip interacts with the sample surface through atomic force. As the tip is pushed away, or pulled into a depression, the cantilever is deflected. The deflection is measured by a laser light beam focused onto the cantilever and reflected into a photodiode detector. **B.** AFM image showing cells of *Corynebacterium glutamicum*, a major industrial producer of amino acids and vitamins. Each cell is 1.0–1.5 µm in length, and about 0.7 µm in diameter.

In AFM, an instrument probes the surface of a sample with a sharp tip a couple of micrometers long and often less than 10 nm in diameter (**Fig. 2.44A**). The tip is located at the free end of a lever that is 100–200 µm long. The lever is deflected by the force between the tip and the sample surface. Deflection of the lever is measured by a laser beam reflected off a cantilever attached to the tip as the sample scans across. The measured deflections allow a computer to map the topography of cells in liquid medium with a resolution below 1 nm.

In the example in **Figure 2.44B**, AFM reveals cells of *Corynebacterium glutamicum*, a bacterium that is used for commercial production of vitamins and amino acids. The cells were observed in water suspension, without stain; and their contours appear without the flattening on a grid required for EM.

TO SUMMARIZE:

- **Electron microscopy** is based on the focusing of electron beams on an object stained with a heavy-metal salt that scatters electrons. Much higher resolution is obtained than with light microscopy.
- **Transmission electron microscopy (TEM)** involves electron beam penetration of a thin sample.
- **Scanning electron microscopy (SEM)** involves scanning of a three-dimensional surface with an electron beam.
- **Cryo–electron microscopy (cryo-EM)** involves the observation of samples flash-frozen in water solution. Multiple images may be combined digitally to achieve high resolution.
- **Atomic force microscopy (AFM)** uses intermolecular force measurement to observe cells in water solution.

2.7 Visualizing Molecules

To understand the structure of cells, ultimately we need to isolate the cell's molecules to observe their individual structure and function. The major tool used at present for molecular visualization is **X-ray diffraction analysis**, or **X-ray crystallography**. In some cases, cryo-EM modeling based on thousands of samples has also reached near-atomic resolution. Another emerging alternative to X-ray diffraction for analysis of small molecules and proteins is nuclear magnetic resonance (NMR). The advantage of NMR is that it presents a dynamic view of molecules in solution.

Unlike microscopy, X-ray diffraction does not present a direct view of a sample, but generates computational models. Dramatic as the models are, they can only represent particular aspects of electron clouds and electron density that are fundamentally "unseeable." That is why molecular structures are represented in different ways that depend on the context—by electron density maps, as models defined by van der Waals radii, or as stick models, for example. Proteins are frequently presented in a cartoon form that shows alpha helix and beta sheet secondary structures.

X-Ray Diffraction Analysis

For substances that can be crystallized, X-ray diffraction makes it possible to fix the position of each individual atom in a molecule, because the wavelengths of X-rays are much shorter than the wavelengths of visible light and are of the same magnitude as the dimensions of atoms. X-ray diffraction, like phase microscopy, is based on the principle of wave interference (see **Fig. 2.13**). The interference pattern is generated when a crystal containing many copies of an isolated molecule is bombarded by a beam of X-rays (**Fig. 2.45A**). The wave fronts associated with the X-rays are diffracted as they pass through the crystal, causing interference patterns. In the crystal, the diffraction pattern is generated by a symmetrical array of many sample molecules (**Fig. 2.45B**). The larger the number of copies of the molecule in the array, the narrower the interference pattern and the greater the

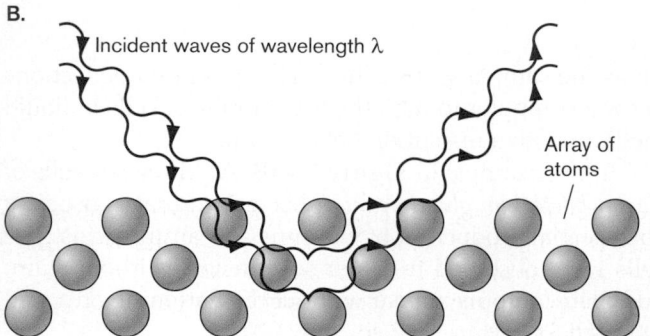

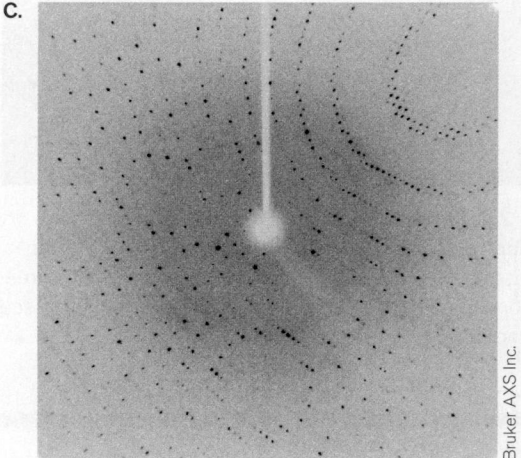

Figure 2.45 **Visualizing molecules by X-ray crystallography.** **A.** Modern apparatus for X-ray crystallography. The X-ray beam is focused onto a crystal, which is rotated over all angles to obtain diffraction patterns. The intensity of the diffracted X-rays is recorded on film or with an electronic detector. **B.** X-rays are diffracted by rows of identical molecules in a crystal. The diffraction pattern is analyzed to generate a model of the individual molecules. **C.** Diffraction pattern from a crystal.

resolution of atoms within the molecule. Diffraction patterns obtained from the passage of X-rays through a crystal (**Fig. 2.45C**) can be analyzed by computation to develop a precise structural model for the molecule, detailing the position of every atom in the structure.

The application of X-ray crystallography to complex biological molecules was pioneered by the Irish crystallographer John Bernal (1901–1971) (**Fig. 2.46A**). Bernal was particularly supportive of women students and colleagues, including Rosalind Franklin (1920–1958), who made important discoveries about DNA and RNA, and Nobel laureate Dorothy Crowfoot Hodgkin (1910–1994). Hodgkin (**Fig. 2.46B**) won the 1964 Nobel Prize in Chemistry for solving the crystal structures of penicillin and vitamin B_{12} (**Fig. 2.46C**). She later solved one of the first protein structures, that of the hormone insulin.

Today, X-ray data undergo digital analysis to generate sophisticated molecular models, such as the one seen in **Figure 2.47** of anthrax lethal factor, a toxin produced by *Bacillus anthracis* that kills the infected host cells. The model for anthrax lethal factor was encoded in a Protein Data Bank (PDB) text file that specifies coordinates for all atoms of the structure. The Protein Data Bank is a growing world database of solved X-ray structures, freely available on the Internet. Visualization software is used to present the structure as a "ribbon" of amino acid residues, color-coded for secondary structure. In **Figure 2.47**, the red-colored coils represent alpha helix structures, whereas the blue arrows represent beta sheets (for a review of these secondary structures, see Appendix 1).

WWW | Protein Data Bank: Research Collaboratory for Structural Bioinformatics, protein and nucleic acid databases

WWW | Biomolecules at Kenyon College: molecular tutorials by undergraduate students on anthrax lethal factor and other proteins, as well as instructions to write your own tutorials. *microbiology2.com/links*

NOTE: Molecular and cellular biology increasingly rely on visualization in three dimensions. Many of the molecules illustrated in our book are based on structural models deposited in the Protein Data Bank, as indicated by the PDB file code. You may view these structures in 3-D by downloading the PDB file and viewing with a free plug-in such as Jmol.

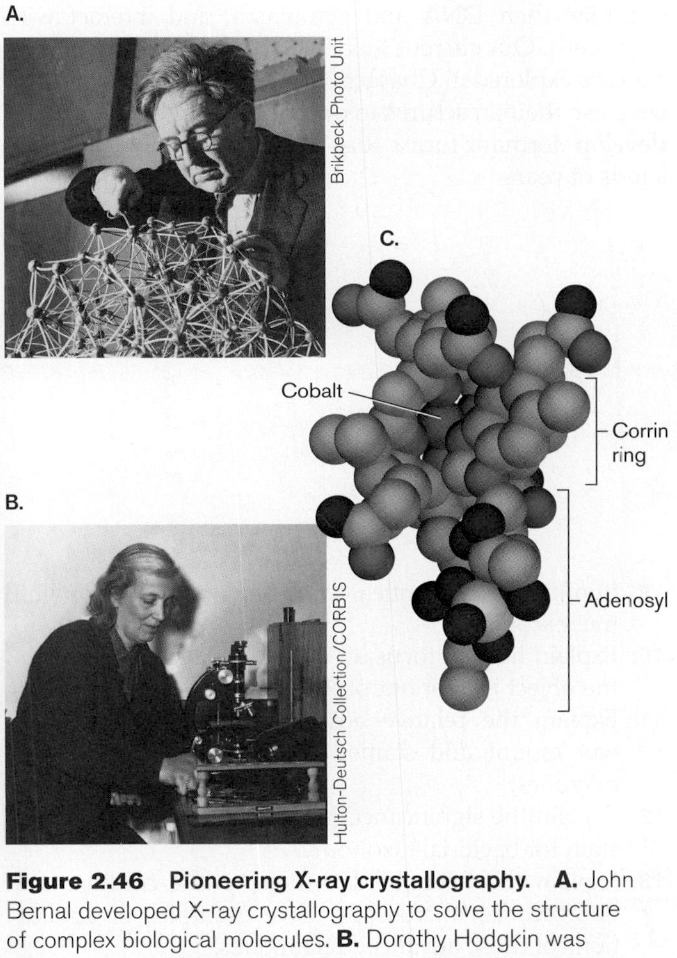

Figure 2.46   Pioneering X-ray crystallography.   A. John Bernal developed X-ray crystallography to solve the structure of complex biological molecules. **B.** Dorothy Hodgkin was awarded the 1964 Nobel Prize in Chemistry for her work in X-ray crystallography. **C.** Vitamin B_{12}, whose structure was originally solved by Dorothy Hodgkin. The corrin ring structure is built around an atom of cobalt (pink). Carbon atoms are gray; oxygen, red; nitrogen, blue; phosphorus, yellow. Hydrogen atoms are omitted for clarity.

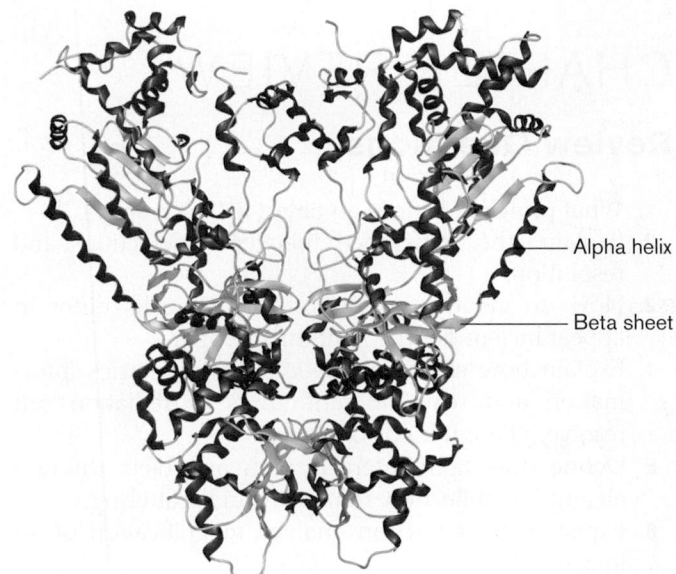

**Figure 2.47   X-ray crystallography of a protein complex, anthrax lethal factor.   The toxin consists of a butterfly-shaped dimer of two peptide chains. This cartoon model is based on X-ray crystallographic data, showing alpha helix (red coils) and beta sheet (blue arrows). (PDB code: 1J7N)

A limitation of X-ray analysis is the unavoidable deterioration of the specimen under bombardment by X-rays. The earliest X-ray diffraction models of molecular complexes such as the ribosome relied heavily on components from thermophilic bacteria and archaea that grow at high temperatures. Because thermophiles have evolved to grow under higher thermal stress, their macromolecular complexes form more stable crystals than do their homologs in organisms growing at moderate temperatures.

X-ray diffraction analysis of crystals from a wide range of sources was made possible by cryocrystallography. In cryocrystallography, as in cryo-EM, crystals are frozen rapidly to liquid nitrogen temperature. The frozen crystals have greatly decreased thermal vibrations and diffusion, thus lessening the radiation damage to the molecules. Models based on cryocrystallography can present multisubunit structures such as the bacterial ribosome complexed with transfer RNAs and messenger RNA. Much of our knowledge of microbial genetics (Chapters 7–12) and metabolism (Chapters 13–16) is based on crystal structures of key macromolecules.

TO SUMMARIZE:

- **X-ray diffraction analysis**, or **X-ray crystallography**, uses X-ray diffraction (interference patterns) from crystallized macromolecules to determine structure at atomic resolution.
- **Cryocrystallography** uses frozen crystals with greatly decreased thermal vibrations and diffusion, enabling the determination of structures of large macromolecular complexes, such as the ribosome.
- **Molecular visualization** by crystallography can only model the "appearance" of a molecule at atomic resolution. Different models emphasize different structural features and levels of resolution.

Concluding Thoughts

The tools of microscopy and molecular visualization described in this chapter have shaped our current understanding of microbial cells—how they grow and divide, organize their DNA and cytoplasm, and interact with other cells. Our current models of cell structure and function are explored in Chapter 3. In Chapter 4, we learn how cells use their structures to obtain energy, reproduce, and develop dormant forms that can remain viable for thousands of years.

CHAPTER REVIEW

Review Questions

1. What principle defines an object as "microscopic"?
2. Explain the difference between detection and resolution.
3. How do eukaryotic and prokaryotic cells differ in appearance under the light microscope?
4. Explain how electromagnetic radiation carries information and why different kinds of radiation can resolve different kinds of objects.
5. Define how light interacts with an object through absorption, reflection, refraction, and scattering.
6. Explain how refraction enables magnification of an image.
7. Explain how magnification increases resolution and why "empty magnification" fails to increase resolution.
8. Explain how angle of aperture and resolution change with increasing lens magnification.
9. Summarize the optical arrangement of a compound microscope.
10. Explain how to focus an object and how to tell when the object is in or out of focus.
11. Explain the relative advantages and limitations of wet mount and stained preparations for observing microbes.
12. Explain the significance (and limitations) of the Gram stain for bacterial taxonomy.
13. Explain the basis of dark-field, phase-contrast, and fluorescence microscopy. Give examples of applications of these advanced techniques.
14. Explain the difference between transmission and scanning electron microscopy and the different applications of each.

Thought Questions

1. Explain which features of bacteria you can study by (a) light microscopy; (b) fluorescence microscopy; (c) scanning EM; (d) transmission EM.
2. Explain how resolution is increased by magnification. Why can't the details be resolved by your unaided eye? Explain why magnification reaches a limit; why can it not go on resolving greater detail?
3. Explain why artifacts appear, even with the best lenses. Explain how you can tell the difference between an optical artifact and an actual feature of an image.
4. How can "detection without resolution" be useful in microscopy? Explain with specific examples.

Key Terms

aberration (49)
absorption (45)
acid-fast stain (54)
Airy disk (47)
angle of aperture (49)
antibody stain (55)
artifact (65)
atomic force microscopy (AFM) (43, 66)
bacillus (42)
bright-field microscopy (43)
coccus (42)
compound microscope (50)
condenser (50)
confocal laser scanning microscopy (60)
contrast (44)
counterstain (54)
cryo–electron microscopy (cryo-EM) (electron cryomicroscopy) (64)
cryo–electron tomography (electron cryotomography) (64, 66)
dark-field microscopy (55)
depth of field (51)
detection (41)
diaphragm (50)
differential stain (52)
electromagnetic radiation (44)

electron microscopy (EM) (43, 62)
emission wavelength (58)
empty magnification (46)
excitation wavelength (58)
fixation (52)
flagellum (56)
fluorescence (45)
fluorophore (59)
focal plane (46)
focal point (46)
focus (48)
Gram-negative (54)
Gram-positive (54)
Gram stain (52)
immersion oil (50)
interference (46)
interference microscopy (58)
light microscopy (43)
magnification (41)
micrometer (μm) (40)
microscope (40)
millimeter (mm) (40)
mordant (52)
nanometer (nm) (40)
negative stain (55)

numerical aperture (49)
objective lens (49, 50)
ocular lens (50)
parfocal (50)
phase-contrast microscopy (56)
picometer (pm) (40)
reflection (45)
refraction (45)
refractive index (45)
resolution (40)
rod (42)
scanning electron microscopy (SEM) (43, 62)
scattering (45)
simple stain (52)
spirochete (42)
spore stain (54)
staining (52)
tomography (64)
total magnification (51)
transmission electron microscopy (TEM) (43, 62)
wet mount (52)
X-ray crystallography (43, 68)
X-ray diffraction analysis (68)

Recommended Reading

Chiu, W., M. L. Baker, W. Jiang, and Z. H. Zhou. 2002. Deriving folds of macromolecular complexes through electron cryomicroscopy and bioinformatics approaches. *Current Opinion in Structural Biology* **12**:263–269.

Graumann, Peter L., and Richard Losick. 2001. Coupling of asymmetric division to polar placement of replication origin regions in *Bacillus subtilis. Journal of Bacteriology* **183**:4052–4060.

Jiang, W., J. Chang, J. Jakana, P. Weigele, J. King, et al. 2006. Structure of epsilon15 bacteriophage reveals genome organization and DNA packaging-injection apparatus. *Nature* **439**:612–616.

Komeili, A., Z. Li, D. K. Newman, and G. J. Jensen. 2006. Magnetosomes are cell membrane invaginations organized by the actin-like protein MamK. *Science* **311**:242–245.

Lucic, Vladan, Friedrich Förster, and Wolfgang Baumeister. 2005. Structural studies by electron tomography: From cells to molecules. *Annual Review of Biochemistry* **74**:833–865.

Matias, Valério R. F., Ashruf Al-Amoudi, Jacques Dubochet, and Terry J. Beveridge. 2003. Cryo-transmission electron microscopy of frozen-hydrated sections of *Escherichia coli* and *Pseudomonas aeruginosa*. *Journal of Bacteriology* **185**:6112–6118.

Murphy, Douglas B. 2001. *Fundamentals of Light Microscopy and Electronic Imaging.* Wiley-Liss, Hoboken, NJ.
Popescu, Aurel, and R. J. Doyle. 1996. The Gram stain after more than a century. *Biotechniques in Histochemistry* **71**:145–151.

For further study and review, please visit StudySpace at **<u>microbiology2.com</u>**.

Chapter 3

Cell Structure and Function

Microbial cells face environmental stress, enduring extreme changes in temperature and salinity; and pathogens face the chemical defenses of their hosts. To meet these challenges, microbes build complex structures, such as a cell envelope with tensile strength comparable to that of steel. Within the cytoplasm, molecular devices such as the ribosome build and expand the cell.

With just a few thousand genes in its genome, how does a bacterial cell grow and reproduce? Bacteria coordinate their DNA replication through the DNA replisome and the cell fission ring. Other devices, such as flagellar propellers, enable microbial cells to compete, to communicate, and even to cooperate in building biofilm communities.

Discoveries of cell form and function have exciting applications for medicine and biotechnology. The structures of ribosomes and cell envelope materials provide targets for new antibiotics. And devices such as the rotary ATP synthase inspire "nanotechnology," the design of molecular machines.

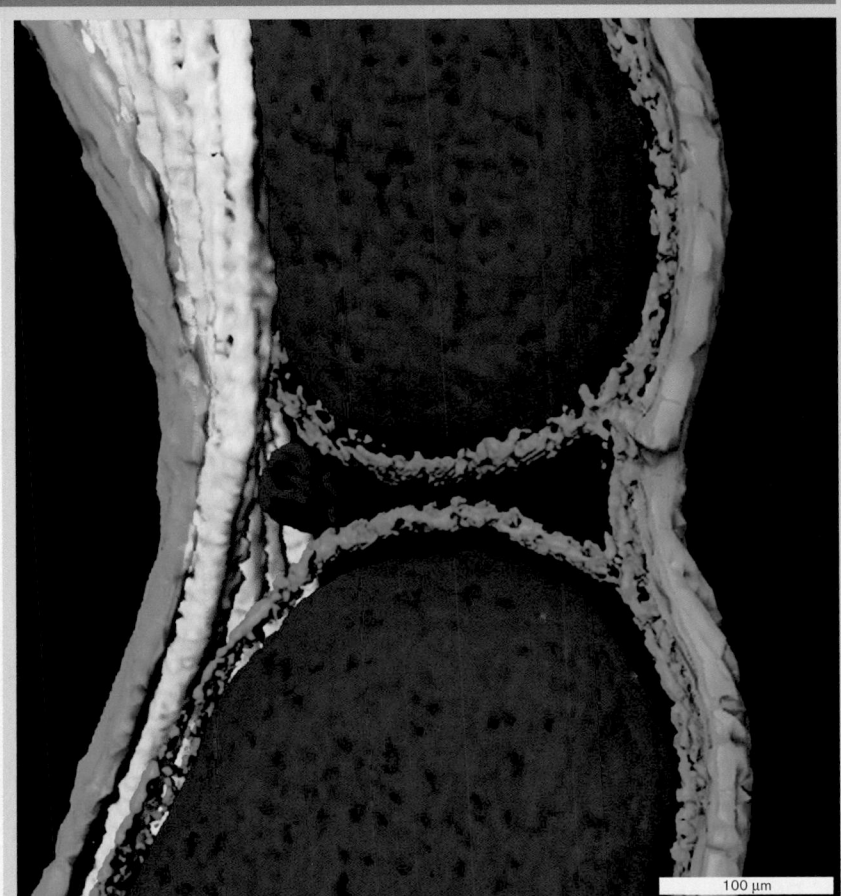

100 µm

Current Research Highlight

A dividing cell of the bacterium *Borrelia afzelii*, which causes Lyme disease in Europe. The cell membrane (colorized purple), cell wall (green), internal flagellar filaments (yellow), and outer sheath (cyan) are visualized by cryo–electron tomography. The internal flagella rotate, causing the entire cell to twist in a helical motion. The cell twisting enables it to penetrate viscous secretions of the host.
Source: Mikhail Kudryashev et al. 2009. *Mol. Microbiol.* **71**:1415. Journal compilation © 2009 Blackwell Publishing.

The microbial cell has formidable tasks: to obtain nutrients faster than its competitors, to protect itself from toxins and predators, and to reproduce itself. These tasks are accomplished by the cell's molecular parts.

To study the cell requires microscopy, as well as isolating cell parts. Chapter 3 presents an overview of the bacterial cell and then explains how such views derive from cell fractionation and genetic analysis. We explore the structures common to most microbial cells, as well as more specialized devices such as light-harvesting complexes and magnetosomes. We dissect the cell's complex outer layers and see how they interact with the replicating

chromosome to accomplish cell fission. Many of the cell's structures offer targets for antibiotic design, as well as opportunities for biotechnology. Overall, this chapter focuses on prokaryotes, with reference to eukaryotes for comparison (eukaryotic microbes are covered in Chapter 20). Our discussion assumes an elementary knowledge of cell biology (reviewed in Appendices 1 and 2).

www| Microbe Wiki provides a quick, readable reference for microbial species, including those mentioned in this book.
microbiology2.com/links

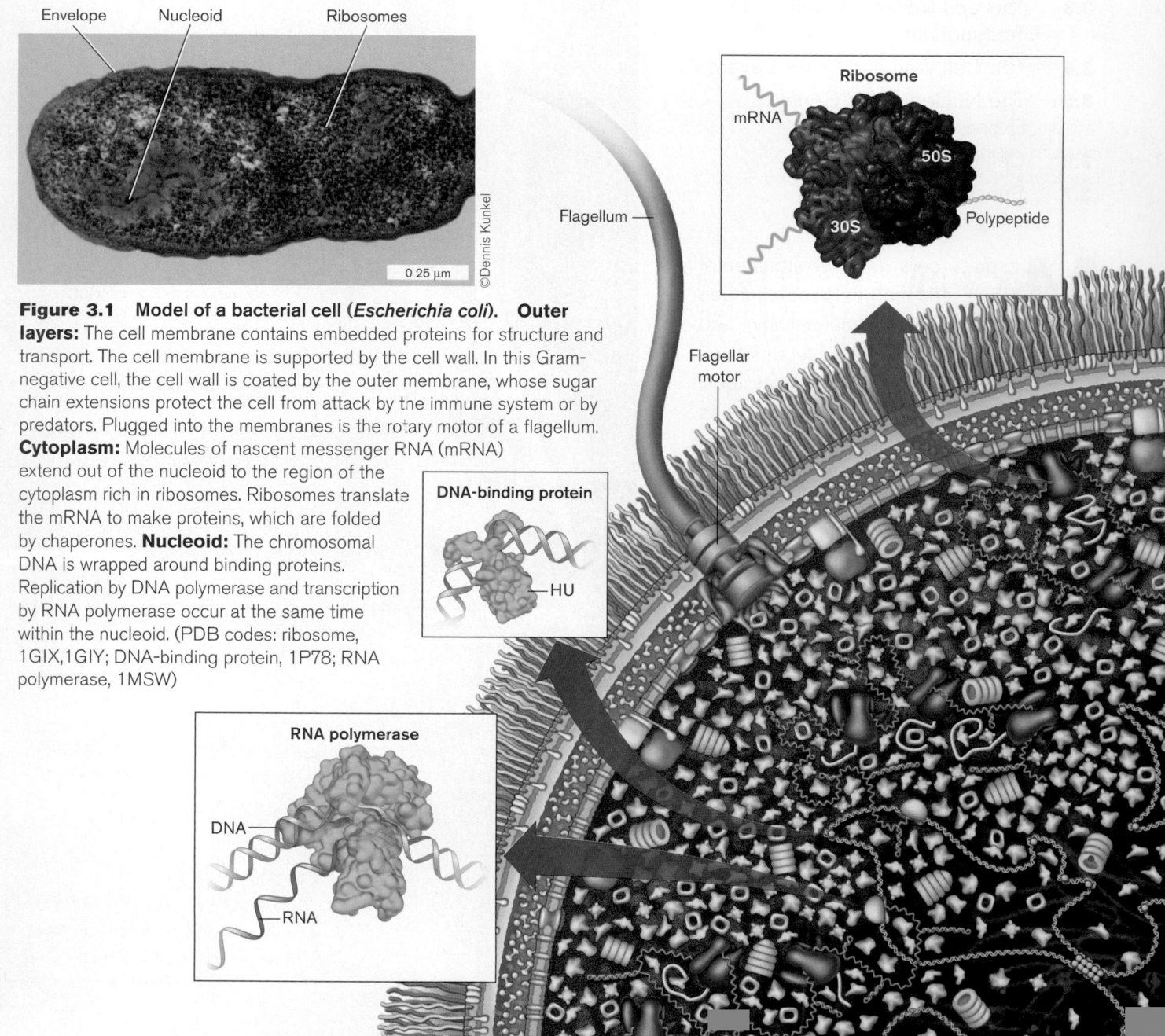

Figure 3.1 **Model of a bacterial cell (*Escherichia coli*). Outer layers:** The cell membrane contains embedded proteins for structure and transport. The cell membrane is supported by the cell wall. In this Gram-negative cell, the cell wall is coated by the outer membrane, whose sugar chain extensions protect the cell from attack by the immune system or by predators. Plugged into the membranes is the rotary motor of a flagellum. **Cytoplasm:** Molecules of nascent messenger RNA (mRNA) extend out of the nucleoid to the region of the cytoplasm rich in ribosomes. Ribosomes translate the mRNA to make proteins, which are folded by chaperones. **Nucleoid:** The chromosomal DNA is wrapped around binding proteins. Replication by DNA polymerase and transcription by RNA polymerase occur at the same time within the nucleoid. (PDB codes: ribosome, 1GIX,1GIY; DNA-binding protein, 1P78; RNA polymerase, 1MSW)

3.1 The Bacterial Cell: An Overview

Prokaryotic cells show a wide range of form, whose diversity is explored in Chapters 18 and 19. At the same time, most prokaryotes share fundamental traits:

- **Thick, complex outer envelope.** The envelope protects the cell from environmental stress and predators. It also mediates exchange with the environment and communication with other organisms.
- **Compact genome.** Prokaryotic genomes are compact, with relatively little noncoding DNA. Small genomes maximize the production of cells from limited resources.
- **Tightly coordinated cell functions.** The cell's subcellular parts work together in a highly coordinate mechanism. Coordinate action enables a high reproduction rate.

In the early twentieth century, the cell was envisioned as a bag of "soup" full of floating ribosomes and enzymes. Modern research shows that, in fact, the cell's parts fit together in a structure that is ordered, though flexible.

A Model of the Bacterial Cell

Here we present a model of the bacterial cell (**Fig. 3.1**). This model offers an interpretation of how the major components of one bacterial cell fit together. The model represents *Escherichia coli*, but its general features apply to many kinds of bacteria. Remember that we cannot literally "see" the molecules within a cell, but microscopy and subcellular analysis generate a remarkably detailed view.

Within the bacterial cell, the cytoplasm consists of a gel-like network composed of proteins and other macromolecules. The cytoplasm is contained by a **cell membrane**. For Gram-negative bacteria, the cell membrane is called the inner membrane, in order to distinguish it from

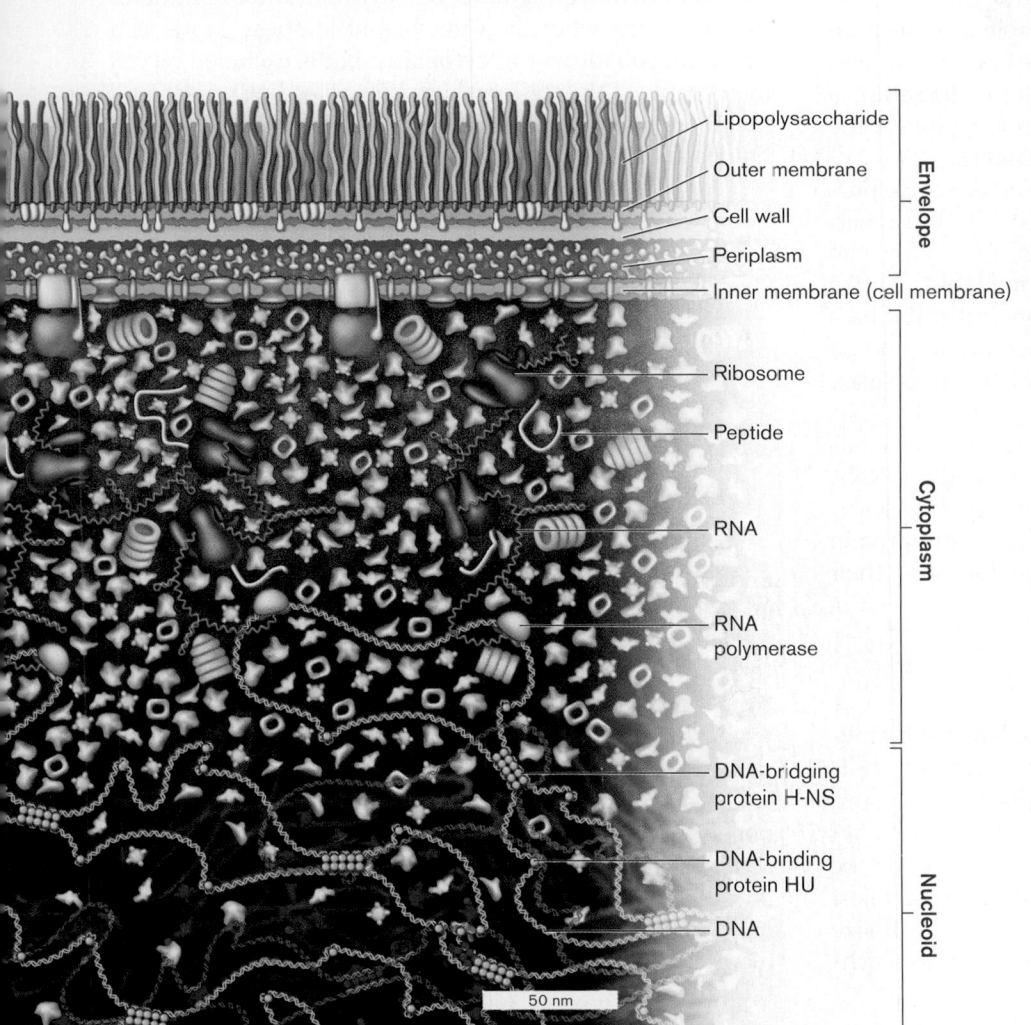

Lipopolysaccharide
Outer membrane
Cell wall — Envelope
Periplasm
Inner membrane (cell membrane)

Ribosome
Peptide
RNA — Cytoplasm
RNA polymerase

DNA-bridging protein H-NS
DNA-binding protein HU — Nucleoid
DNA

50 nm

Bacterial Cell Components (Examples)

Outer membrane proteins:

- Sugar porin (10 nm)
- Braun lipoprotein (8 nm)

Inner membrane proteins:

- Transporter
- Secretory complex (Sec)
- ATP synthase (20 nm diameter in inner membrane; 32 nm total height)

Periplasmic proteins:

- Arabinose-binding protein (3 x 3 x 6 nm)
- Acid resistance chaperone (HdeA) (3 x 3 x 6 nm)
- Disulfide bond protein (DsbA) (3 x 3 x 6 nm)

Cytoplasmic proteins:

- Pyruvate kinase (5 x 10 x 10 nm)
- Phosphofructokinase (4 x 7 x 7 nm)
- Proteasome (12 x 12 x 15 nm)
- Chaperonin GroEL (18 x 14 nm)
- Other proteins

Transcription and translation complexes:

- RNA polymerase (10 x 10 x 16 nm)
- Ribosome (21 x 21 x 21 nm)

Nucleoid components:

- DNA (2.4 nm wide x 3.4 nm/10 bp)
- DNA-binding protein (3 x 3 x 5 nm)
- DNA-bridging protein (3 x 3 x 5 nm)

the additional outer membrane. The membrane is composed of phospholipids, transporter proteins, and other molecules. The cell membrane prevents cytoplasmic proteins from leaking out and maintains gradients of ions and nutrients through transporters. It is covered by a **cell wall**, a rigid structure composed of polysaccharides linked covalently by peptides (peptidoglycan). The cell wall forms a single cage-like molecule that surrounds the cell.

In Gram-negative bacteria such as *E. coli*, the cell wall consists of one or two layers of peptidoglycan, a polymer of sugars and peptides. The cell wall extends within the periplasm, an aqueous layer containing proteins such as sugar-binding proteins. Outside the cell wall lies an **outer membrane** of phospholipids and **lipopolysaccharides (LPS)**, a class of lipids attached to long polysaccharides (sugar chains). The LPS layer may be surrounded by a thick capsule. The capsule polysaccharides form a slippery mucus layer that inhibits phagocytosis by macrophages (see Section 26.3). The cell wall and outer membrane constitute the **envelope**.

The envelope includes cell surface proteins that enable the bacteria to interact with specific host organisms. For example, *E. coli* cell surface proteins enable colonization of the human intestinal epithelium, whereas *Sinorhizobium* cell surface proteins enable colonization of legume plants for nitrogen fixation. Another common external structure is the **flagellum** (plural, **flagella**), a helical protein filament whose rotary motor propels the cell in search of a more favorable environment.

Within the cell, the cell membrane and envelope provide an attachment point for one or more chromosomes. The chromosome is organized within the cytoplasm as a system of looped coils called the **nucleoid**. Unlike the round, compact nucleus of eukaryotic cells, the bacterial nucleoid is not bounded by a membrane, and so the coils of DNA can extend throughout the cytoplasm. Loops of DNA from the nucleoid are transcribed by RNA polymerase to form messenger RNA (mRNA), as well as functional small RNA (sRNA) molecules. As the mRNA transcripts grow, they bind ribosomes to start synthesizing polypeptide chains. As the polypeptides grow, protein complexes called chaperones help them fold into their functional conformations.

Microbial Eukaryotes

How do eukaryotic microbes, such as euglenas and amebas, compare with prokaryotes? Most eukaryotic cells that have been studied are much larger than prokaryotes; some amebas are actually visible to the unaided eye. Yet DNA-based surveys of natural environments, such as the ocean and soil, reveal eukaryotes as small as the smallest bacteria. Thus, eukaryotic cells extend over the full size range encompassed by prokaryotes, in addition to reaching much larger sizes.

Like bacteria, microbial eukaryotes typically possess a thick outer covering. Fungi possess cell walls of polysaccharide, whereas protists have a "pellicle," consisting of membranous layers reinforced by protein microtubules. Other kinds of eukaryotic cell surfaces are reinforced by inorganic materials, such as the silicate shells of diatoms.

Eukaryotic microbes contain subcellular organelles composed of membranes that are more extensive and specialized than those of prokaryotes. Examples include the nuclear membrane, Golgi, mitochondria, and chloroplasts (reviewed in Appendix 2). These organelles perform some of the physiological functions that are provided by organ systems in multicellular organisms, such as circulation, digestion, and excretion. Mitochondria and chloroplasts are the products of ancestral engulfment of prokaryotic cells, followed by evolution of endosymbiosis, as explained in Chapter 17. Eukaryotic microbial diversity is explored in Chapter 20.

Biochemical Composition of Bacteria

The bacterial cell model indicates shape and size but tells little about chemical composition. Chemistry explains, for example, why wiping a surface with ethanol eliminates microbial life, whereas water has little effect. Water is a universal constituent of cytoplasm but is excluded by cell membranes. Ethanol, however, dissolves both polar and nonpolar substances; thus, ethanol disintegrates membranes and destroys the secondary structure of proteins.

All cells share common chemical components:

- **Water**, the fundamental solvent of life
- **Essential ions**, such as potassium, magnesium, and chloride ions
- **Small organic molecules**, such as lipids and sugars, that are incorporated into cell structures and that provide nutrition by catabolism
- **Macromolecules**, such as nucleic acids and proteins, that contain information, catalyze reactions, and mediate transport, among many other functions

Cell composition varies with species, growth phase, and environmental conditions. **Table 3.1** summarizes the chemical components of a cell for the model bacterium *Escherichia coli* during exponential growth.

Small molecules. The *E. coli* cell consists of about 70% water, the essential solvent required to carry out fundamental metabolic reactions and to stabilize proteins. The water solution contains inorganic ions, predominantly potassium, magnesium, and phosphate. Inorganic ions store energy in the form of transmembrane gradients, and they serve essential roles in enzymes. For example, a magnesium ion is required at the active site of RNA polymerase to help catalyze the incorporation of ribonucleotides into RNA.

Table 3.1 **Molecular composition of a bacterial cell, *Escherichia coli*, during balanced exponential growth.[a]**

Component	Percentage of total weight[b]	Approximate number of molecules/cell	Number of different kinds
Water	70	20,000,000,000	1
Proteins	16	2,400,000	2,000[c]
RNA: rRNA, tRNA, and other small RNA (sRNA) molecules,	6	250,000	200
mRNA	0.7	4,000	2,000[c]
Lipids: phospholipids (membrane)	3	25,000,000	50
lipopolysaccharides (outer membrane)	1	1,400,000	1
DNA	1	2^{d}	1
Metabolites and biosynthetic precursors	1.3	50,000,000	1,000
Peptidoglycan (murein sacculus)	0.8	1	1
Inorganic ions	0.1	250,000,000	20
Polyamines (mainly putrescine and spermine)	0.1	6,700,000	2

[a]Values shown are for a hypothetical "average" cell cultured with aeration in glucose medium with minimal salts at 37°C.

[b]The total weight of the cell (including water) is about 10^{-12} gram (g), or 1 picogram (pg).

[c]The number of kinds of mRNA and of proteins is difficult to estimate because some genes are transcribed at extremely low levels and because RNA and proteins include kinds that are rapidly degraded.

[d]In rapidly growing cells, cell fission typically lags approximately one generation behind DNA replication—hence, two identical DNA copies per cell.

Source: Modified from Neidhardt, F., and H. E. Umbarger. 1996. Chemical composition of *Escherichia coli*, p. 14. In F. C. Neidhardt (ed.), *Escherichia coli* and *Salmonella: Cellular and Molecular Biology*, 2nd ed. ASM Press, Washington, DC.

The cell also contains many kinds of small charged organic molecules, such as phospholipids and enzyme cofactors. A major class of organic cations is the **polyamines**, molecules with multiple amine groups that are positively charged when the pH is near neutral. Polyamines balance the negative charges of the cell's DNA, and they stabilize ribosomes during translation.

Macromolecules. Many cells show similar content of water and small molecules, but their specific character is defined by their macromolecules, especially their proteins. Proteins vary among different species; and a given species makes very different proteins, depending on environmental conditions such as temperature, nutrient levels, and entry into a host organism. Individual proteins, encoded by specific genes, are found in very different amounts, from 10 per cell to 10,000 per cell. The total proteins encoded by a genome, capable of expression in the cell, are known collectively as the **proteome**. Early attempts to define the proteome of a cell were conducted by Fred Neidhardt (**Fig. 3.2**) and colleagues at the University of Michigan, who compiled the first protein catalog of *E. coli* using **two-dimensional polyacrylamide gel electrophoresis** (**2-D PAGE** or **2-D gels**) (**Fig. 3.3**).

To obtain 2-D gels, cells are lysed by sonication or detergent to release and solubilize as many proteins as possible. The proteins are then subjected to two forms of separation: isoelectric focusing followed by electrophoresis. In **isoelectric focusing**, proteins are placed in a gel that has a pH gradient, and they migrate along the

Figure 3.2 Proteins of *E. coli*. Fred Neidhardt and colleagues at the University of Michigan, Ann Arbor, used 2-D gel electrophoresis to compile the first protein catalog of *E. coli*.

gradient until they reach the point where the number of positively charged residues equals the number of negatively charged residues. At this point, their net charge is zero (that is, the pH equals the protein's isoelectric point), and the protein stops moving. In **electrophoresis**, proteins deposited in an electrical field migrate toward a positive electrode. The proteins are coated with a reagent (SDS, or sodium dodecyl sulfate) to confer a uniform negative charge so that their distance of migration depends on molecular weight (roughly a logarithmic function). The proteins are then stained for detection, and the protein spots can be identified, either by N-terminal protein sequencing or by mass spectrometry.

In a 2-D gel of *E. coli* (**Fig. 3.3**), about 500 different proteins can be distinguished. The most highly expressed proteins include ribosomal proteins and translation factors such as elongation factor Tu (EF-Tu). Outer membrane proteins such as OmpX and ProX also show high concentration. Conversely, many proteins of the inner (cell) membrane are too insoluble in water to appear, and some important regulators are present at levels too low to be seen. Nevertheless, 2-D gels are useful for identifying changes in protein expression that occur under different environmental conditions or when bacteria are invading host cells.

Protein synthesis is directed by DNA and RNA. The content of nucleic acids in *E. coli* is nearly 8% by weight, much higher than in multicellular eukaryotes. For microbes, the high nucleic acid content is advantageous, allowing the cell to maximize reproduction of its chromosome while minimizing resources for protein-rich cytoplasm. The high level of nucleic acids is actually toxic to human consumers, who lack the enzymes to digest the uric acid waste product of digested nucleotides. That is why most kinds of bacteria cannot be eaten as a major part of the diet.

Other kinds of macromolecules are found in the cell wall and outer membrane. The bacterial cell wall consists of **peptidoglycan**, an organic polymer that constitutes nearly 1% of the cell mass, approximately the same mass as that of DNA. This investment of biomass in the cell wall shows the importance (for most species) of maintaining turgor pressure in dilute environments, where water would otherwise enter by osmosis, causing osmotic shock.

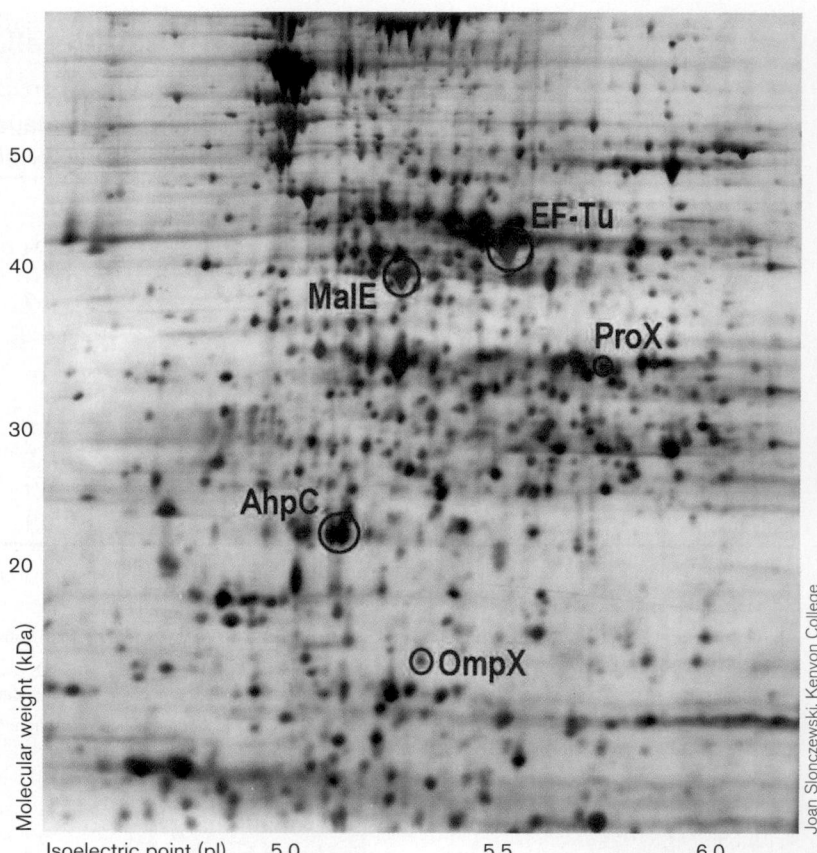

Figure 3.3   Proteins of *E. coli*.   2-D gel of proteins of *Escherichia coli* grown aerobically in a casein-yeast extract medium. Proteins were identified by N-terminal sequence and by mass spectroscopy. EF-Tu is an elongation factor for translation; MalE is an outer membrane maltose-binding protein; ProX and OmpX are periplasmic transporters; and AhpC is an antioxidant stress protein.

■ **The biochemical composition of bacteria** includes relatively high nucleic acid content, as well as proteins, phospholipids, and other organic and inorganic constituents.

■ **Proteins in the cell** vary, depending on the species and environmental conditions.

THOUGHT QUESTION 3.1   Which molecules occur in the greatest number in a bacterial cell? The smallest number? Why does a cell contain 100 times as many lipid molecules as strands of RNA?

TO SUMMARIZE:

■ **Bacterial cells** are protected by a thick cell envelope.

■ **Compact genomes** maximize reproductive potential with minimal resources.

■ **The model bacterial cell** contains a highly ordered cytoplasm in which DNA replication, RNA transcription, and protein synthesis occur coordinately.

3.2 How We Study Cell Parts

How can we determine how all the macromolecules listed in **Table 3.1** interact and work together? Electron microscopy largely defines how we "see" the cell's interior as a whole. But smaller parts, such as the ribosomes, appear only as small, densely packed particles. Furthermore, even higher-resolution images from electron microscopy cannot tell us the chemical composition of ribosomes or how they function in the living cell.

To study ribosome function requires isolation and analysis of subcellular parts:

- **Subcellular fractionation** enables isolation of cell parts such as ribosomes and membranes so that we can study their form and function.
- **Structural analysis** by X-ray crystallography and related methods reveals the form of cell components.
- **Genetic analysis** dissects the function of cell components by constructing mutant cells with altered function.

In this section, we present these methods as applied to a key example of cellular function: that of the bacterial ribosome. Each method has its particular strengths and limitations to reveal form and function.

Isolating Parts of Cells

Cellular components, such as ribosomes and flagellar motors, can be isolated readily from cells. The cells must be broken open by techniques that allow subcellular components to remain intact. Here are examples of such techniques:

- **Mild detergent lysis.** Cells can be lysed with a detergent capable of dissolving membranes but not denaturing proteins.
- **Sonication.** Cells can be lysed by intense ultrasonic vibrations above the range of human hearing.
- **Enzymes.** Enzymes such as lysozyme can break the cell wall, allowing the cell to be lysed by mild osmotic shock.
- **Mechanical disruption.** Cells can be broken open by the application of high pressure (with a French press) or through beating with microscopic beads (with a bead beater).

The ultracentrifuge. Different parts of the cell can be isolated by **subcellular fractionation**. A key tool of subcellular fractionation is the **ultracentrifuge**, a device in which solutions containing cell components are rotated in tubes at high speed. The high rotation rate generates centrifugal forces strong enough to separate subcellular particles (**Fig. 3.4A**). The ultracentrifuge was invented by the Swedish physicist Theodor Svedberg (1884–1971), who won the 1926 Nobel

Figure 3.4 Subcellular fractionation by ultracentrifugation. **A.** An ultracentrifuge is used to fractionate cell components. **B.** *Salmonella* ribosomes (TEM) isolated from the cytoplasm by ultracentrifugation through a linear sucrose density gradient. Polysomes consist of two or more ribosomes attached to mRNA. **C.** High-speed rotation generates high centrifugal forces, measured in units of gravity (*g*). The Svedberg unit (S) offers a measure of particle size based on its rate of travel in a tube subjected to high *g* force. The Svedberg coefficient (number of S units; for example, 30S) is defined in terms of the velocity of the particle in the tube (*v*), the radius of the rotor (*r*), and the rotational velocity (ω). The coefficient of S for a given particle depends on its mass (*m*) and its shape. After centrifugation, fractions are collected from the base of the centrifuge tube. The fractions contain radiolabeled ribosomes. The largest particles (whole ribosomes) sediment near the bottom of the tube, and the smaller particles (separate 50S and 30S subunits) appear in upper fractions.

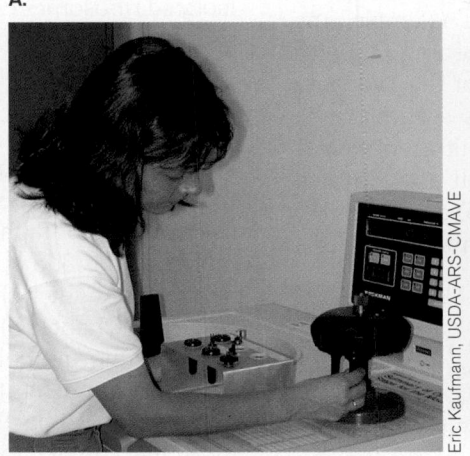

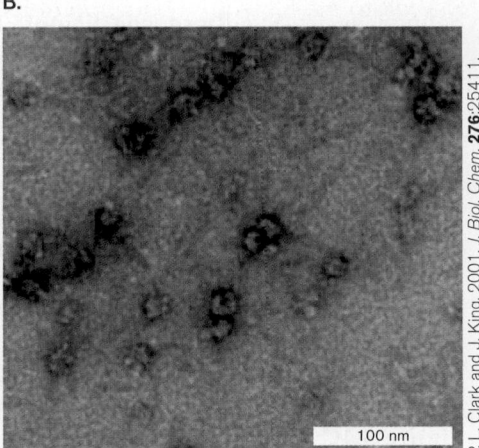

Eric Kaufmann, USDA-ARS-CMAVE

P. L. Clark and J. King. 2001. *J. Biol. Chem.* **276**:25411.

100 nm

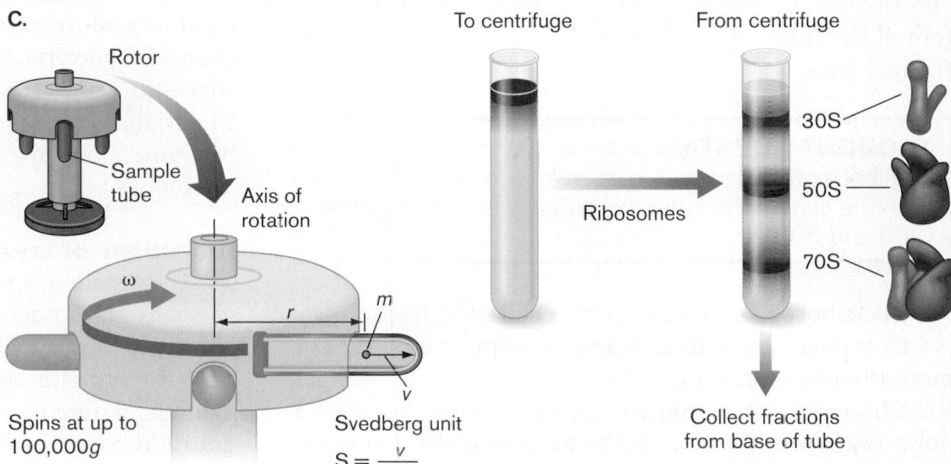

$$S = \frac{v}{\omega^2 r}$$

Prize in Chemistry for the use of ultracentrifugation to separate proteins.

Modern ultracentrifuges have titanium rotors that spin in a vacuum to avoid frictional heating, generating forces up to 100,000 times gravity (100,000*g*). For fractionation, cells are first lysed by one of the methods previously described to obtain a cell **lysate**, a general term for the contents of a broken cell. The cell lysate is placed in tubes containing a high-density solution, such as sucrose or cesium chloride solution, in which suspended particles sediment slowly. Under a high *g* force, however, particles sediment at different rates, depending on their size and density.

The **sedimentation rate** is the rate at which particles of a given size and shape travel to the bottom of the tube under centrifugal force. The sedimentation rate is measured by the Svedberg unit, S, which is given by the particle's rate of sedimentation (*v*), the radius at which the tube rotates (*r*), and the rotational velocity (ω):

$$S = v/(\omega^2 r)$$

For a given type of particle in suspension, the sedimentation rate also depends on the particle's mass and shape. The contribution of particle mass and shape is defined as its **Svedberg coefficient**; for example, the coefficient of the small subunit of the ribosome is 30, for a sedimentation value of 30S. The value of the Svedberg coefficient increases with the average cross-sectional area of the particle. For bacterial ribosomes, ultracentrifugation yields intact ribosomes (70S) as well as separated ribosomal subunits: the large subunit (50S) and the small subunit (30S). Within cells, ribosomes normally exist as a mixture of joined and separate subunits.

To isolate the ribosomes, a cell lysate is layered onto a tube of sucrose solution, whose density decreases the sedimentation rate and increases the separation of particles of different size (**Fig. 3.4C**). The fractions shown in **Figure 3.4C** were drained sequentially from the base of the tube after centrifugation. The heaviest particle, the 70S ribosome, appears in the fractions nearest the bottom of the tube because it travels fastest under the centrifugal force.

THOUGHT QUESTION 3.2 Why does the Svedberg coefficient of the intact ribosome (70S) differ from the sum of the individual subunits of the ribosome (30S and 50S)?

The ribosomes can be seen in collected fractions by electron microscopy (**Fig. 3.4B**). In some cases, two or more ribosomes are connected by a strand of messenger RNA (mRNA). This multiple-ribosome structure, called a polysome, gives our first clue as to the intracellular organization of the translation apparatus within a bacterial cell (see **Fig. 3.1**), where a number of ribosomes translate each mRNA at the same time.

Furthermore, ribosomes isolated by centrifugation can translate messenger RNA in cell-free systems. Experiments in cell-free systems provide the basis of much of our knowledge of protein synthesis (see Chapter 8).

Limitations of subcellular fractionation. Subcellular fractionation yields clues about internal structure but provides little information about processes that require overall integrity of the cell. For example, the role of the transmembrane electrochemical potential, or proton potential, in ATP synthesis was obscured for many years because biochemists were unable to isolate a cytoplasmic complex that generates it. Transmembrane ion gradients were, in fact, observed within membrane vesicles—spheres of membrane isolated by cell disintegration and centrifugation. But it was harder to demonstrate that the entire cell membrane of an intact cell supports a proton potential.

Crystallographic Analysis Reveals Structure

Isolated ribosomes can be crystallized for structure analysis by X-ray diffraction crystallography. A crystallographic model of the 30S ribosomal subunit is shown in **Figure 3.5**. The 30S subunit consists of ribosomal RNA (rRNA) and 27 different proteins, each of which is encoded by a particular gene. The sequences of all the ribosomal genes, with their predicted protein and RNA products, were fitted with X-ray crystallography data to build a three-dimensional model of the 30S subunit. This model shows the channel through which the mRNA passes for translation (discussed in Chapter 8).

A similar model was obtained of the 30S ribosomal subunit bound to streptomycin, an antibiotic produced by *Streptomyces* bacteria. In this model, the streptomycin molecule contacts a particular ribosomal protein, S12. The contact with S12 distorts the structure of the mRNA channel and thus halts protein synthesis. Only the ribosomes of bacteria, not of eukaryotes, have the precise shape to bind streptomycin; thus, this antibiotic kills bacterial pathogens without harming the human patient. Studying ribosome structure enables us to design new antibiotics.

Limitations of crystallography. As discussed in Chapter 2, crystallographic analysis applies only to isolated particles and under conditions in which its full function cannot be observed. The technique can solve structures only for proteins and nucleic acids capable of crystallization. It remains impractical for proteins of flexible, nonrigid structure and for many membrane-soluble proteins.

The ribosome: x-ray diffraction model

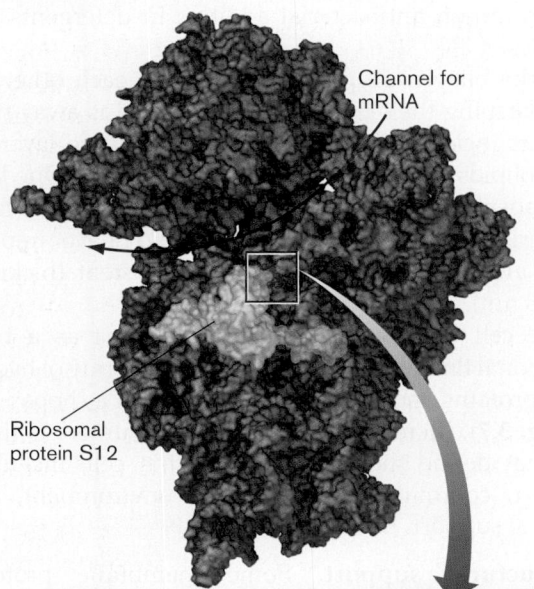

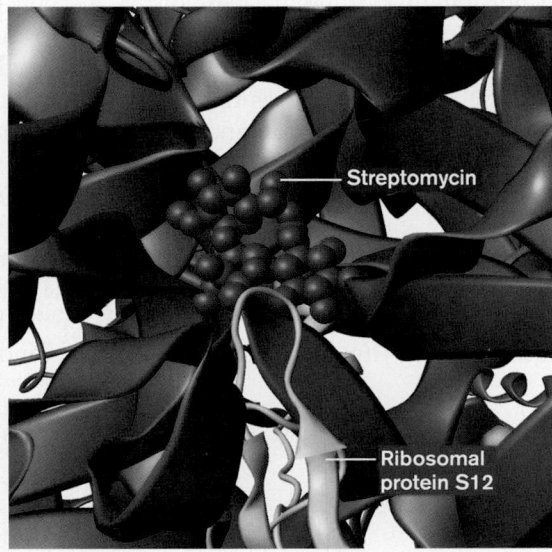

Figure 3.5   X-ray diffraction models relate structure to function.   This X-ray diffraction model of the 30S ribosomal subunit shows the channel through which the mRNA travels. When the ribosome is crystallized with streptomycin, the antibiotic molecule binds to protein S12, near the A site. Streptomycin interferes with the tRNA binding to the A site. (PDB code: 1FJG)

Genetic Analysis Reveals Function

The crystal structure of a macromolecule only yields clues as to function; it does not show the structure in action. The function of cell components can be dissected by **genetic analysis**. In genetic analysis, mutant strains are selected for loss of a given function, or a strain can be intentionally mutated so as to lose or alter a gene.

An example of genetic analysis is the use of ribosome-encoding genes to show how ribosome func-

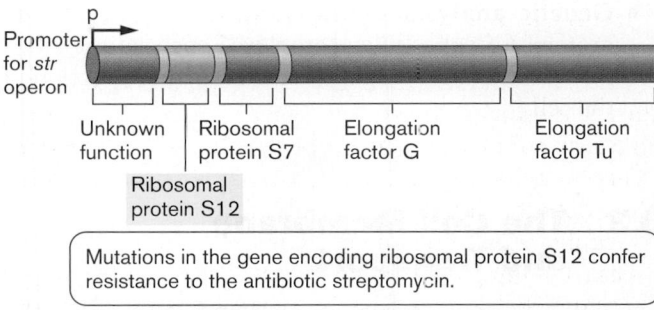

Figure 3.6   Genetic analysis of the streptomycin-resistant ribosome.   The *str* operon has genes encoding components of the ribosome. A mutation in the gene encoding ribosomal protein S12 alters the protein's interaction with streptomycin. The mutant gene makes the bacteria resistant to the antibiotic.

tion is blocked by the antibiotic streptomycin. The genes specifying each ribosomal protein and RNA (rRNA) are organized in operons of genes under one promoter, such as the *str* operon (**Fig. 3.6**). The *str* operon was named for streptomycin resistance. Mutations conferring streptomycin resistance were known for decades before the structural basis was discovered. These mutations were mapped to the gene encoding the protein S12, whose position in the 30S subunit was shown by X-ray crystallography (see **Fig. 3.5**). The S12 protein forms part of the ribosome channel through which mRNA is translated to make a peptide. Mutation in the gene encoding S12 generates a protein of altered shape, permitting function in the presence of streptomycin. Thus, genetic analysis and crystallography were combined to demonstrate the precise mode of action of an important antibiotic.

An exciting extension of genetic analysis is the construction of strains with "reporter genes" fused to genes encoding structures of interest. An example of a reporter gene is green fluorescent protein, which fluoresces at the site of the fused protein. Fluorescent reporter genes enable us to observe the function of proteins within a live cell. In Section 3.6 we discuss the use of fluorescent reporter genes to observe chromosome replication and cell division. The methods of gene manipulation are discussed in Chapters 7 and 12.

> **THOUGHT QUESTION 3.3**   What are the advantages and limitations of biochemical and genetic approaches to deciphering cell structure and function?

TO SUMMARIZE:

- **Subcellular fractionation** isolates cell parts for structural, biochemical, and genetic analysis.
- **X-ray crystallography** shows the three-dimensional form of cell components at the atomic level.

■ **Genetic analysis** shows which genes express the proteins of subcellular complexes such as the ribosome. Mutation of a gene leads to altered function of the cell.

3.3 The Cell Membrane and Transport

The structure that defines the existence of a cell is the cell membrane (**Fig. 3.7**). The cell membrane consists of a phospholipid bilayer containing lipid-soluble proteins. Overall, the membrane serves two kinds of functions: It contains the cytoplasm within the external medium, mediating transport between the two; and it carries many proteins with specific functions, such as biosynthetic enzymes and environmental signal receptors.

Membrane Constituents

Cell membranes are composed of approximately equal parts of phospholipids and proteins. Each phospholipid possesses a charged phosphate-containing "head" that contacts the water interface, as well as a hydrophobic "tail" packed within the bilayer. A phospholipid consists of glycerol with ester links to two fatty acids and a phosphoryl polar head group (for review, see Appendix 2). The polar head group may have a side chain, such as ethanolamine in **phosphatidylethanolamine**, a major phospholipid of *E. coli* (**Fig. 3.8**). Lipid biosynthesis is a key process of cells; for example, a bacterial enzyme for fatty

acid biosynthesis, enoyl reductase, is the target of triclosan, a common antibacterial additive in detergents and cosmetics.

In the bilayer, all phospholipids face each other tail to tail, keeping their hydrophobic side chains away from the water inside and outside the cell. The two layers of phospholipids in the bilayer are called **leaflets**. One leaflet of phospholipids faces the cell interior; the other faces the exterior. As a whole, the phospholipid bilayer imparts fluidity and gives the membrane a consistent thickness (about 8 nm).

The cell membrane can be thought of as a two-dimensional fluid within which float a vast array of hydrophobic proteins and smaller molecules such as hopanoids (see **Fig. 3.7**). Membrane proteins serve numerous functions that define the capabilities of the cell, including transport, communication with the environment, and structural support. Examples include:

■ **Structural support.** Some membrane proteins anchor together different layers of the cell envelope (see Section 3.4). Other proteins attached to membranes form part of the cytoskeleton. Still others form the base of structures that extend out from the cell, such as pili, filaments, and flagella; these enable adherence and motility.

■ **Detection of environmental signals.** In *Vibrio cholerae*, the causative agent of cholera, ToxR is a transmembrane protein whose amino-terminal part reaches into the cytoplasm. When ToxR detects acidity and elevated temperature—signs of the host digestive tract—its amino-terminal domain binds to

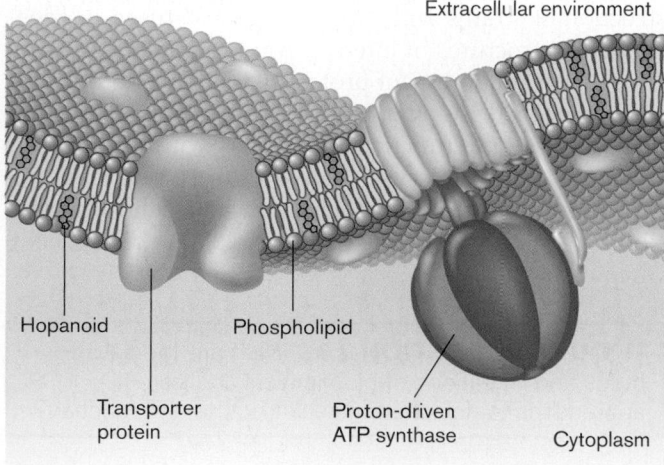

Figure 3.7 Bacterial cell membrane. The cell membrane consists of a phospholipid bilayer, with hydrophobic fatty acid chains directed inward, away from water. The bilayer contains stiffening agents such as hopanoids, which serve the same function as cholesterol in eukaryotic membranes. Half the volume of the membrane consists of proteins.

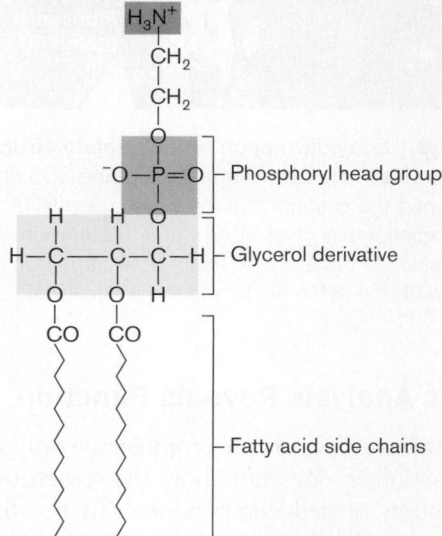

Figure 3.8 Phosphatidylethanolamine. This major bacterial phospholipid consists of glycerol with ester links to two fatty acids and a phosphoethanolamine.

DNA at a sequence that activates the expression of cholera toxin and other virulence factors.

- **Secretion of virulence factors and communication signals.** Membrane proteins form secretion complexes to export toxins and cell signals across the envelope. For example, symbiotic nitrogen-fixing rhizobia require membrane proteins NodI and NodJ to transport nodulation signals out to the host plant roots, where they induce the formation of root nodules containing the bacteria.
- **Ion transport and energy storage.** Transport proteins manage ion flux between the cell and exterior, and store energy in ion gradients. Ion gradients and energy storage are discussed in Chapter 4.

Proteins embedded in a membrane must have a hydrophobic portion that is soluble within the membrane. Typically, several hydrophobic alpha helices thread back and forth through the membrane. Other peptide regions extend outside the membrane, containing charged and polar amino acids that interact favorably with water. The combination of hydrophobic and hydrophilic regions effectively locks the protein into the membrane.

Transport across the Cell Membrane

Overall, the cell membrane acts as a barrier to sequester water-soluble proteins of the cytoplasm. The membrane's specific components, particularly proteins, determine which substances cross the membrane between the cytoplasm and the outside. Selective transport is essential for cell survival; it means the ability to acquire scarce nutrients while excluding toxins.

> **NOTE:** Diffusion and osmosis are presented in detail in Appendix 2. Protein transporters are presented in Chapter 4.

Passive diffusion. Small uncharged molecules, such as O_2, CO_2, and water, easily permeate the membrane. Some molecules, such as ethanol, also disrupt the membrane—an action that can make such molecules toxic to cells. By contrast, strongly polar molecules, such as sugars, generally cannot penetrate the hydrophobic interior of the membrane and require transport mediated by specific proteins. Water molecules permeate the membrane, but their rate of passage is increased by protein channels called aquaporins.

Osmosis. Most cells maintain a concentration of total **solutes** (molecules in solution) that is higher inside the cell than outside. As a result, the internal concentration of water is lower than the concentration outside the cell. Because water can cross the membrane but its charged solutes cannot, water tends to diffuse across the membrane into the cell, causing the expansion of cell volume,

in a process called **osmosis**. The resulting pressure on the cell membrane is called **osmotic pressure** (reviewed in Appendix 2). Osmotic pressure will cause a cell to burst, or lyse, in the absence of a countering pressure such as that provided by the cell wall.

Various solutes cross the membrane by different means. Nonspecific transport, or passive diffusion, requires dissolving in the phospholipid bilayer. Specific transport—for example, of nutrients and toxins—requires specific transport proteins. The presence of specific transporters depends on the microbial species and environmental conditions.

Membrane-permeant weak acids and bases. A special case of movement across cell membranes is that of **membrane-permeant weak acids** and **weak bases**, which exist in equilibrium between charged and uncharged forms:

$$\text{Weak acid:} \quad HA \rightleftharpoons H^+ + A^-$$
$$\text{Weak base:} \quad B + H_2O \rightleftharpoons BH^+ + OH^-$$

Weak acids and weak bases cross the membrane in their uncharged form: HA (weak acid) or B (weak base). On the other side, upon reentering aqueous solution, they dissociate (HA to A^- and H^+) or reassociate with H^+ (B to BH^+). In effect, they conduct acid (H^+) or base (OH^-) across the membrane, causing acid or alkali stress. A high proton concentration outside the cell will increase the amount of uncharged weak acid that can freely enter the cell. Thus, if the H^+ concentration (acidity) outside the cell is higher than inside, it will drive weak acids into the cell.

Many key substances in cellular metabolism are membrane-permeant weak acids and bases, such as acetic acid. Most pharmaceutical drugs—therapeutic agents delivered to our tissues via the bloodstream—are weak acids or bases whose uncharged forms exist at sufficiently low concentration to cross the membrane without disrupting it. Examples of weak acids that deprotonate (acquiring negative charge) at neutral pH include aspirin (acetylsalicylic acid) and penicillin (**Fig. 3.9**). Examples of weak bases that protonate (acquiring positive charge) at neutral pH include Prozac (fluoxetine) and tetracycline.

> **THOUGHT QUESTION 3.4** Amino acids have acidic and basic groups that can dissociate. Why are they *not* membrane-permeant weak acids or weak bases? Why do they fail to cross the phospholipid bilayer?

Transmembrane ion gradients. Molecules that carry a fixed charge, such as hydrogen and sodium ions (H^+ and Na^+), cannot cross the phospholipid bilayer. Such ions usually exist in very different concentrations inside and outside the cell. An **ion gradient** (ratio of concentrations) across the cell membrane can store energy for nutrition, or

The protonated form of a weak acid (RCOOH) crosses the membrane, whereas the deprotonated form (RCOO⁻) does not.

Membrane-soluble

Water-soluble

$$C-OH \rightleftharpoons C-O^- + H^+$$

Aspirin

Penicillin

The protonated form of a weak base (RNH₃⁺) does not cross the membrane, whereas the deprotonated form (RNH₂) does.

Membrane-soluble

Prozac (fluoxetine)

Water-soluble

Tetracycline

Figure 3.9   Common drugs are membrane-permeant weak acids and bases.   In its charged form, each drug is soluble in the bloodstream. The uncharged form is hydrophobic and penetrates the cell membrane.

to drive the transport of other molecules. **The role of ion gradients for storing energy is discussed in Section 4.1.**

Inorganic ions require transport through specific **transport proteins**, or **transporters**. So, too, do organic molecules that carry charge at cytoplasmic pH, such as amino acids and vitamins. For example, the vitamin B_{12} transporter in *E. coli* includes two proteins (**Fig. 3.10**). The proteins spend ATP for energy to drive vitamin B_{12} across the inner cell membrane into the cytoplasm.

Transporters are highly specific to different ions and organic substances under different environmental conditions. Much of the character and ecological niche of a microbe—the nutrients it can consume, the drugs it can resist—depend on the set of transporters encoded by its genome. Transporters evolve in families common to many species; thus, analysis of known genomes predicts the transport capabilities in a newly sequenced genome (discussed in Chapter 8). Organisms that live in complex, changing environments express numerous transporters. For example, the genome of *Streptomyces coelicolor*, a soil-dwelling actinomycete bacterium that produces several antibiotics, shows over 400 different transporters. These include exporters for toxic ions such as arsenate and chromate, exporters for oligosaccharides and peptides, and multiple-drug pumps.

Transport proteins act by various mechanisms, such as a structural channel, or pore, that allows a nutrient molecule to enter the cell or a toxin to be exported. Transport may be passive or active. In **passive transport**, molecules accumulate or dissipate along their concentration gradient. **Active transport**—that is, transport from lower

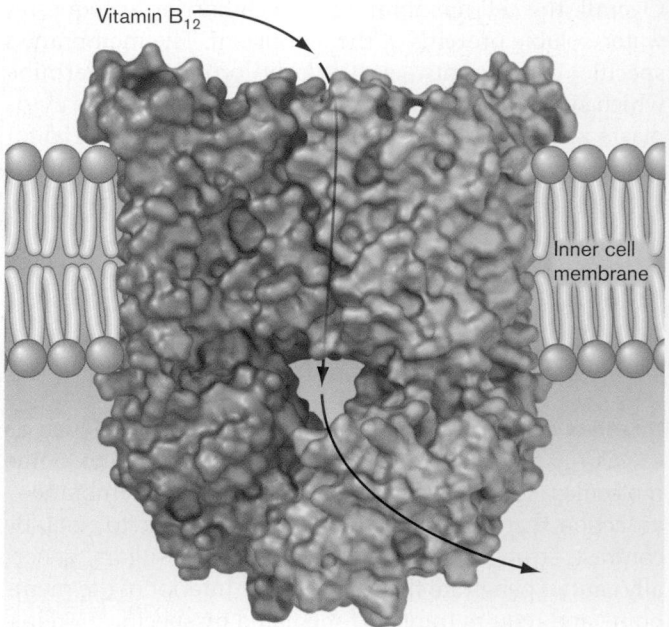

Vitamin B_{12}

Inner cell membrane

Figure 3.10   A membrane-embedded transport protein: the vitamin B_{12} transporter.   The protein complex hydrolyzes ATP to drive transport of vitamin B_{12} across the membrane. PDB ID: 1L7V

to higher concentration—requires the expenditure of energy. The energy for active transport may be obtained by cotransport of another substance down its gradient from higher to lower concentration. For example, many transporters couple uptake of amino acids to uptake of sodium ions. Alternatively, transport may be powered by a coupled chemical reaction that spends energy, such as ATP hydrolysis (**Fig. 3.10**; discussed in Chapter 4).

A medically important example of active transport is that of drug efflux proteins powered by the hydrogen ion gradient. Efflux proteins pump antibiotics such as tetracycline out of the bacterial cell, enabling harmful bacteria to grow in the presence of antibiotics. Pathogens and cancer cells evolve multidrug transporters that enable them to survive chemotherapy.

Membrane Lipids

Membranes require a certain uniformity to maintain structural integrity and function. Yet individual membrane lipids differ remarkably in structure. Membrane lipids help determine whether an organism grows in a Yellowstone hot spring or in the human lungs, where it may cause pneumonia. Phospholipids vary with respect to their phosphoryl head groups and with respect to their hydrocarbon side chains. Some membrane lipids lack phosphate altogether, substituting other polar groups; and some replace mobile side chains with fused rings.

Phosphoryl head groups. At the pH of most biological systems, the phosphoryl head group carries a net negative charge called a **phosphatide**. The negatively charged phosphatide can contain various organic groups, such as glycerol to form **phosphatidylglycerol** (**Fig. 3.11A**). A more complex phosphatide is **cardiolipin**, or **diphosphatidylglycerol**, which is actually a double phospholipid linked by a glycerol (**Fig. 3.11B**). Cardiolipin increases in concentration in bacteria grown to starvation or stationary phase, possibly because its extended structure stabilizes the membrane.

Other phospholipids have a positively charged head group, such as phosphatidylethanolamine (**Fig. 3.11C**). Phospholipids with positive charge or with mixed charges are concentrated in portions of the membrane that interact with DNA, which has negative charge. In addition, some transporter proteins require interaction with positively charged phospholipids.

The fatty acid component of phospholipids also varies greatly among bacteria. Certain soil bacteria and pathogens such as *Mycobacterium tuberculosis* have several hundred different kinds of fatty acids. Fatty acid structures that stiffen the membrane are increased under stress conditions, such as starvation or acidity.

Variant fatty acid structures in phospholipids enhance survival under environmental stress. The most common bacterial fatty acids are hydrogenated chains of varying length, typically between 6 and 22 carbons. These chains pack neatly to form a smooth layer. Some fatty acid chains, however, are partly unsaturated (possess one or more carbon-carbon double bonds). If the unsaturated bond is *cis*, meaning that both alkyl chains are on

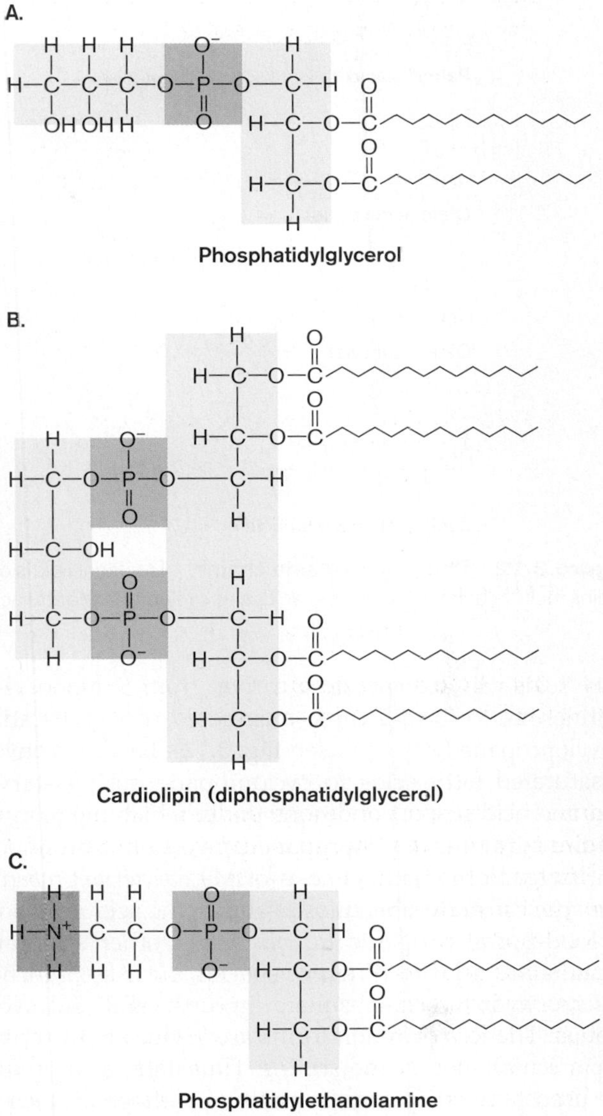

Phosphatidylglycerol

Cardiolipin (diphosphatidylglycerol)

Phosphatidylethanolamine

Figure 3.11 Phospholipid head groups. Bacterial membranes contain phospholipids with several kinds of polar head groups. **A.** Phosphatidylglycerol (negatively charged, a phosphatide). **B.** Cardiolipin (a double phospholipid joined by a third glycerol). **C.** Phosphatidylethanolamine (with a positively charged amine).

the same side of the bond, the unsaturated chain has a "kink," as in oleic acid (**Fig. 3.12**). Because the kinked chains do not pack as closely as the straight hydrocarbon chains do, the membrane is more "fluid." This is why, at room temperature, unsaturated vegetable oils are fluid, whereas highly saturated butterfat is solid. The enhanced fluidity of a kinked phospholipid improves the function of the membrane at low temperature; hence, bacteria can respond to cold and heat by increasing or decreasing their synthesis of unsaturated phospholipids.

Another interesting structural variation is cyclization of part of the chain to form a stiff planar ring with decreased fluidity. The double bond of unsaturated

Palmitic acid

Oleic acid (*trans*)

Oleic acid (*cis*)

Cyclopropane fatty acid

Figure 3.12   Phospholipid side chains.   Bacterial lipid side chains include palmitic acid, oleic acid, and cyclopropane fatty acid.

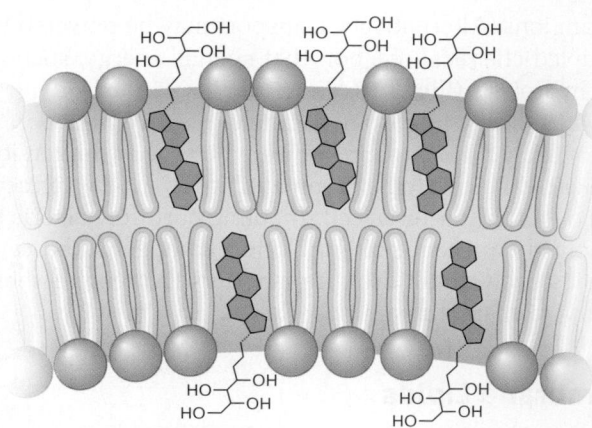

Figure 3.13   Hopanoids add strength to membranes. Hopanoids fit between the fatty acid side chains of membranes and limit their motion, thus stiffening the membrane. A hopanoid has five fused rings and an extended hydroxylated tail.

fatty acids can incorporate a carbon from S-adenosyl-L-methionine to form a three-membered ring, generating a cyclopropane fatty acid (see **Fig. 3.12**). Bacteria convert unsaturated fatty acids to cyclopropane during starvation and acid stress, conditions under which membranes require stiffening. Cyclopropane conversion is an important factor in the pathogenesis of *Mycobacterium tuberculosis* (the cause of tuberculosis) and in the acid resistance of food-borne toxigenic *E. coli*. Many other structural variants are seen in different species, such as branched chains, hydroxyl and sulfate groups, and polycyclic groups. The less common forms are highly characteristic of particular species and strains. Thus, fatty acid profiles are used to identify certain kinds of pathogens, such as *Bacillus anthracis*, the cause of anthrax.

Terpene derivatives stiffen membranes. In addition to phospholipids, membranes include planar molecules that fill gaps between hydrocarbon chains (**Fig. 3.13**). These stiff, planar molecules reinforce the membrane, much as steel rods reinforce concrete. In eukaryotic membranes, the reinforcing agents are sterols, such as **cholesterol**. In bacteria, the same function is filled by pentacyclic (five-ring) hydrocarbon derivatives called **hopanoids**, or **hopanes**. Hopanoids appear in geological sediments, where they indicate ancient bacterial decomposition; they provide useful data for petroleum exploration.

Variations in phospholipid side-chain structures reach their extreme in archaea, many of which inhabit the planet's most extreme environments. All archaeal phospholipids replace the ester link between glycerol and fatty acid with an ether link, C—O—C (**Fig. 3.14**). Ethers are much more stable than esters, which hydrolyze easily in water. Another modification is that archaeal hydrocarbon chains

are branched **terpenoids**, polymeric structures derived from isoprene, in which every fourth carbon extends a methyl branch. The branches strengthen the membrane by limiting movement of the hydrocarbon chains.

The most extreme hyperthermophiles, which live beneath the ocean at 110°C, have terpenoid chains linked at the tails, forming a tetraether monolayer. In some species, the terpenoids cyclize to form cyclopentane rings. These planar rings stiffen the membrane under stress to an even greater extent than the cyclopropyl chains of bacteria. For more on archaeal cells, see Chapter 19.

Both cholesterol and hopanoids are synthesized from the same precursor molecules as the unique lipids of archaea (**Fig. 3.15**). It may be that cholesterol and hopanoids persist in bacteria and eukaryotes as derivatives of lipids once possessed by a common ancestor of bacteria, eukaryotes, and archaea.

TO SUMMARIZE:

■ **The cell membrane** consists of a phospholipid bilayer containing hydrophobic membrane proteins. Membrane proteins serve diverse functions, including transport, cell defense, and cell communication.

■ **Small uncharged molecules**, such as oxygen, can penetrate the cell membrane by diffusion.

■ **Weak acids and weak bases** exist partly in an uncharged form that can diffuse across the membrane and increase or decrease, respectively, the H^+ concentration within the cell.

■ **Polar molecules and charged molecules** require membrane **proteins to mediate** transport. Such facilitated transport can be active or passive.

■ **Active transport** requires input of energy from a chemical reaction or from an ion gradient across the

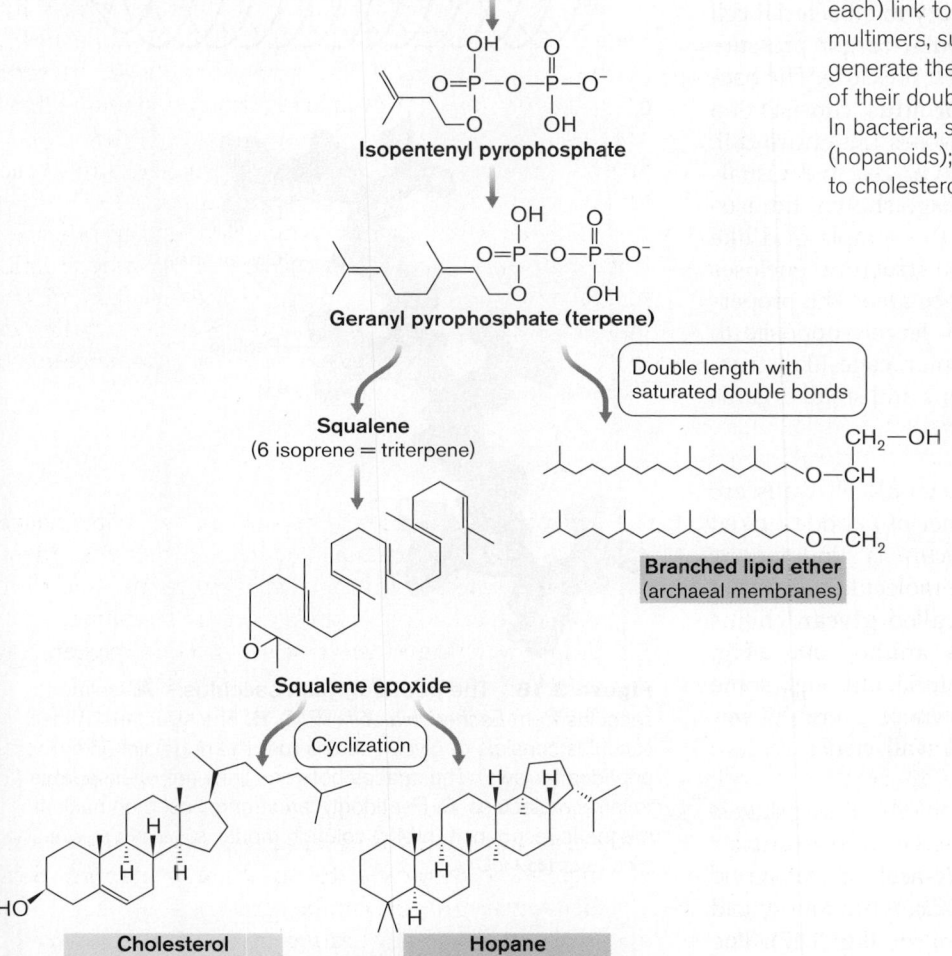

Figure 3.14 Terpene-derived lipids of archaea.   In archaea, the hydrocarbon chains are ether-linked to glycerol, and every fourth carbon has a methyl branch. In some archaea, the tails of the two facing lipids of the bilayer are fused, forming tetraethers; thus, the entire membrane consists of a single layer of molecules (a monolayer).

Glycerol diether

Diethers condensed here

Diglycerol tetraether

Isoprene cyclized to cyclopentane

Cyclopentane rings

Figure 3.15 Synthesis of terpene-derived lipids.   Archaeal lipids are synthesized from isoprene chains. Two isoprene units (five carbons in each) link to form a terpene (ten carbons). Terpenoid multimers, such as the triterpene derivative squalene, generate the lipid chains of archaea, in which most of their double bonds have been hydrogenated. In bacteria, squalene cyclizes to form hopanes (hopanoids); and in eukaryotes, squalene is converted to cholesterol.

Isoprene

Isopentenyl pyrophosphate

Geranyl pyrophosphate (terpene)

Double length with saturated double bonds

Squalene
(6 isoprene = triterpene)

Branched lipid ether
(archaeal membranes)

Squalene epoxide

Cyclization

Cholesterol
(eukaryotic membranes)

Hopane
(bacterial membranes)

membrane. **Ion gradients** generated by membrane pumps store energy for cell functions.

- **Diverse fatty acids** are found in different microbial species and in microbes grown under different environmental conditions.

- **Archaeal membranes have ether-linked terpenoids**, which confer increased stability at high temperature and acidity. Some species have diglycerol tetraethers, which generate a lipid monolayer.

3.4 The Cell Wall and Outer Layers

A few prokaryotes, such as the mycoplasmas, have a cell membrane with no outer layers. In most bacteria and archaea, however, the cell envelope includes at least one structural supporting layer outside the cell membrane. The most common structural support is the cell wall. Many species possess additional coverings, such as an outer membrane and capsule.

The Cell Wall Is a Single Molecule

The cell wall confers shape and rigidity to a bacterial cell and helps it withstand the intracellular turgor pressure that can build up as a result of osmotic pressure. The bacterial cell wall, also known as the **sacculus**, consists of a single interlinked molecule that encloses the entire cell. The sacculus has been isolated from *E. coli* and visualized by TEM (**Fig. 3.16A**). In the image shown, the isolated sacculus appears flattened on the sample grid like a deflated balloon. Its geometrical structure encloses maximal volume with minimal surface area. The properties of the cell wall, or sacculus, are largely opposite to those of the membrane: a unimolecular, cage-like structure (**Fig. 3.16B**), highly porous to ions and small organic molecules.

Peptidoglycan structure. Most bacterial cell walls are composed of peptidoglycan, a polymer of peptide-linked chains of amino sugars. "Peptidoglycan" is synonymous with **murein** ("wall molecule"). The molecule consists of parallel polymers of disaccharides called **glycan** chains cross-linked with peptides of four amino acids (**Fig. 3.17**). Peptidoglycan is unique to bacteria, although some archaea build analogous structures whose overall physical nature is similar. (Archaeal cell wall structures are presented in Chapter 19.)

Glycan chains are linked by peptide cross-bridges. The long chains of peptidoglycan consist of repeating units of the disaccharide composed of *N*-acetylglucosamine (an amino sugar derivative) and *N*-acetylmuramic acid (glucosamine plus a lactic acid group; see **Fig. 3.17**). The

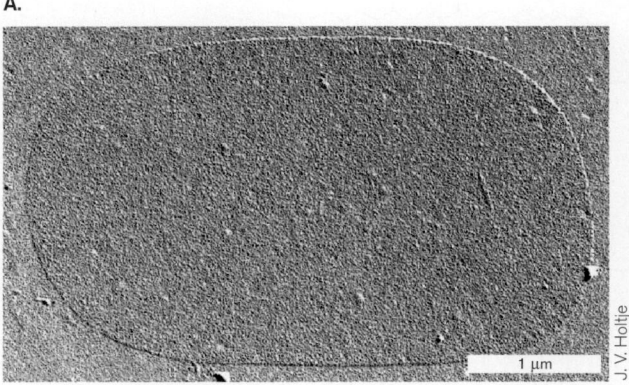

A.

B.   Glycan chain   Peptide cross-bridge

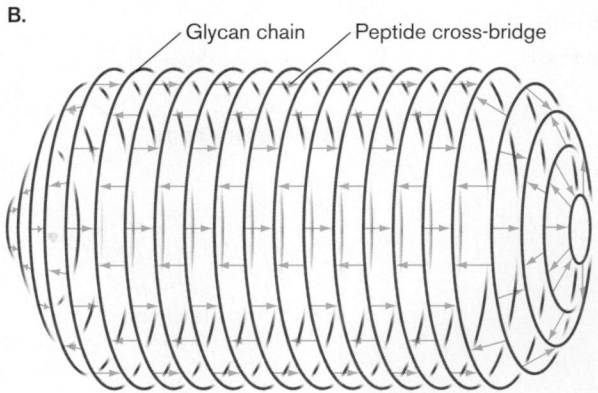

C.

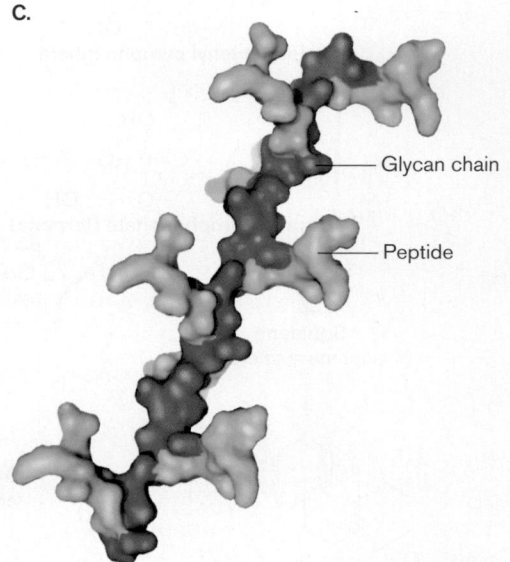

Glycan chain

Peptide

Figure 3.16   The peptidoglycan sacculus.   A. Isolated sacculus from *Escherichia coli* (TEM). **B.** The structure of the sacculus consists of glycan chains (parallel rings) linked by peptides (arrows). The spaces between links are open, porous to large molecules. **C.** Peptidoglycan strand, based on nuclear magnetic resonance (NMR) solution model. *Source:* S. O. Meroueh et al. 2006. *PNAS* **103**:4404.

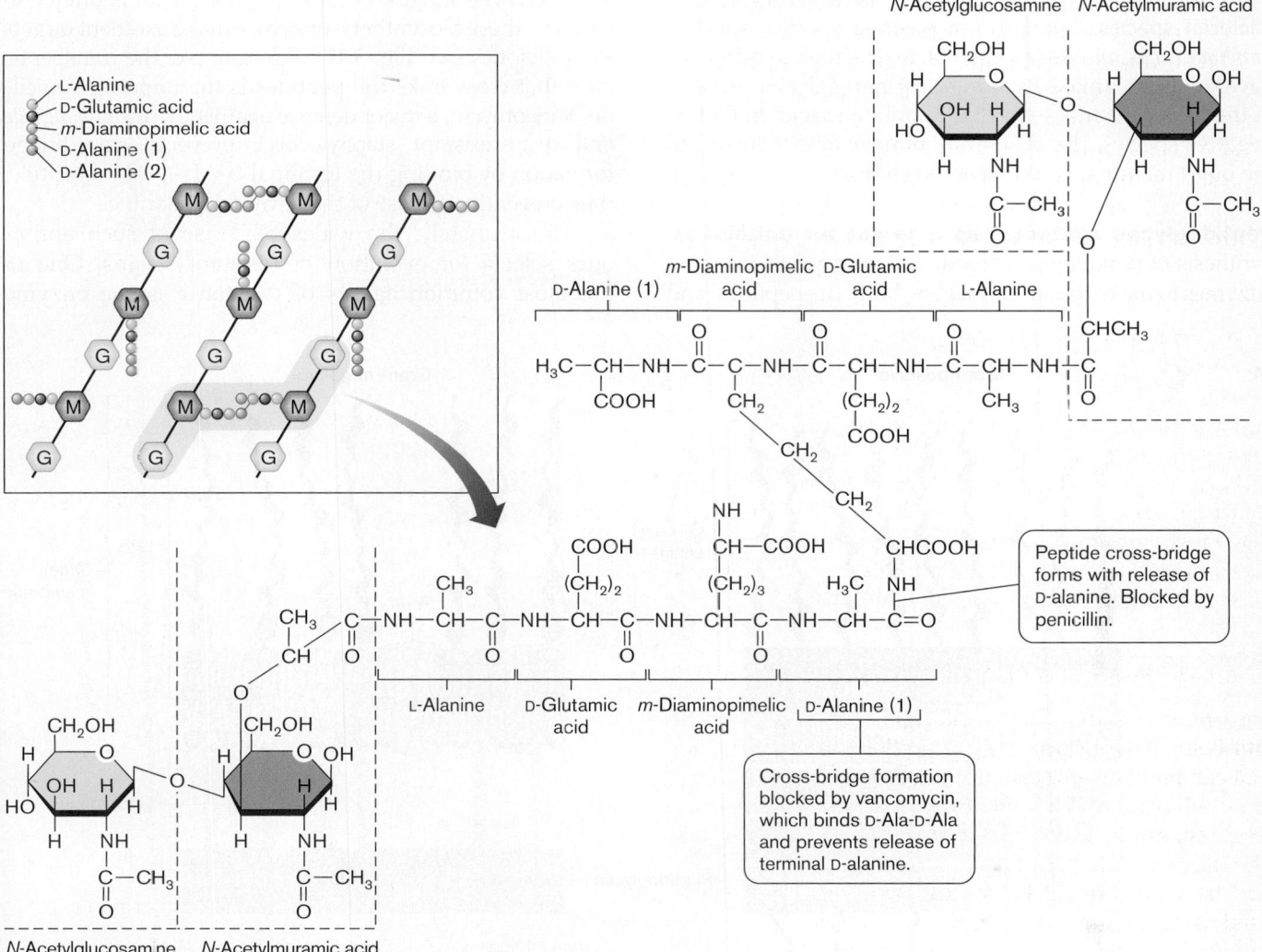

Figure 3.17 Peptidoglycan cross-bridge formation. A disaccharide unit of glycan has an attached peptide of four to six amino acids. The amino terminus of the peptide forms an amide bond with the lactate group of muramic acid. On the peptide, the extra amino group of *m*-diaminopimelic acid can cross-link to the carboxyl terminus of a neighboring peptide. The connection of D-alanine (1) to the peptide is blocked by vancomycin, which binds D-ala (1)–D-ala (2). The cross-bridge formation by transpeptidase is blocked by penicillin.

lactate group of muramic acid forms an amide link with the amino terminus of a short peptide containing four to six amino acid residues. The peptide extension can form **cross-bridges** connecting parallel strands of glycan.

The peptide contains two amino acids in the unusual D mirror form: D-glutamate and D-alanine. A glycan chain with peptide extensions is shown in **Figure 3.16C**, which is based on an NMR solution model. Surprisingly, the actual geometry of the cross-bridge connections and overall sacculus remains unclear, and it is highly controversial in the literature. The third amino acid, *m*-diaminopimelic acid, has an extra amine group, which forms an amide link to a cross-bridged peptide. The amide link

forms with the fourth amino acid of the adjacent peptide, D-alanine (1). The cross-bridge forms by removal of a second D-alanine (2) at the end of the chain. The cross-linked peptides of neighboring glycan strands form the cage of the sacculus.

NOTE: Amino acids occur in two forms that are "mirror opposites," D and L, of which only the L form is incorporated by ribosomes into protein. The D-form amino acids, however, are used by microbes for many nonprotein structural molecules.

The details of peptidoglycan structure vary among bacterial species. Some Gram-positive species, such as *Staphylococcus aureus* (a cause of toxic shock syndrome), have peptides linked by bridges of pentaglycine instead of the D-alanine link to *m*-diaminopimelic acid. In Gram-negative species, the *m*-diaminopimelic acid is linked to the outer membrane, as discussed shortly.

Peptidoglycan synthesis as a target for antibiotics. Synthesis of peptidoglycan requires many genes encoding enzymes to make the special sugars, build the peptides, and

seal the cross-bridges. Because peptidoglycan is unique to bacteria, these biosynthetic enzymes make excellent targets for antibiotics (see **Fig. 3.17**). For example, the transpeptidase that cross-links the peptides is the target of penicillin. Vancomycin, a major defense against *Clostridium difficile* and drug-resistant staphylococci, prevents cross-bridge formation by binding the terminal D-Ala-D-Ala dipeptide, thus preventing release of the terminal D-alanine.

Unfortunately, the widespread use of such antibiotics selects for evolution of resistant strains. One of the most common agents of resistance is the enzyme

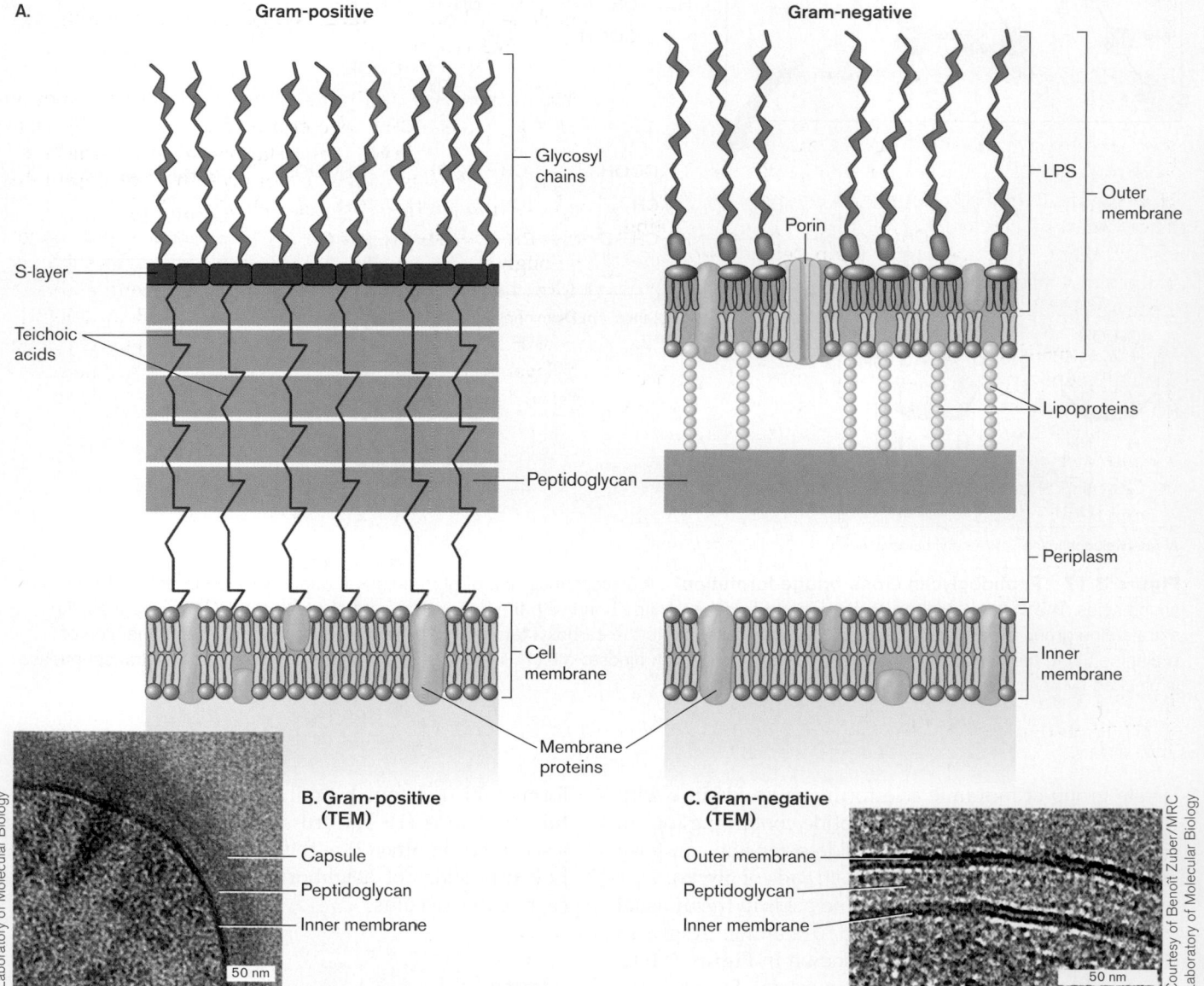

Figure 3.18   Cell envelope: Gram-positive and Gram-negative.   A. The Gram-positive cell has a thick cell wall with multiple layers of peptidoglycan, threaded by teichoic acids. The cell wall may be covered by an S-layer. In some Gram-positive species, carbohydrate filaments form a capsule. The Gram-negative cell has a single layer of peptidoglycan covered by an outer membrane. Some Gram-negative species include an S-layer or a capsule (not shown). The cell membrane of Gram-negative species is called the inner membrane. **B.** Gram-positive envelope of *Bacillus subtilis* (TEM), showing cell membrane, cell wall, and capsule. **C.** Gram-negative envelope of *Pseudomonas aeruginosa* (TEM), showing inner membrane, thin cell wall in the periplasm, and outer membrane.

beta-lactamase, which cleaves the lactam ring of penicillin, rendering it ineffective as an inhibitor of transpeptidase. Strains resistant to vancomycin, on the other hand, contain an altered enzyme that adds lactic acid to the end of the branch peptides in place of the terminal D-alanine. The altered enzyme is no longer blocked by vancomycin. As new forms of drug resistance emerge, researchers continue to seek new antibiotics that target cell wall formation (discussed in Chapter 27).

> **THOUGHT QUESTION 3.5** If the sacculus (cell wall) consists of a single molecule, how do you think it expands as the cell grows?

Gram-Positive and Gram-Negative Bacteria

Most bacteria have additional envelope layers that provide structural support and protection from predators and host defenses (**Fig. 3.18**). Additional molecules are attached to the cell wall and cell membrane, in some cases threading through them. Envelope composition defines two major categories of bacteria distinguished by the Gram stain (discussed in Chapter 2):

- **Gram-positive bacteria** have a thick cell wall with multiple layers of peptidoglycan, interpenetrated by teichoic acids. The phylum Firmicutes consists of Gram-positive species such as *Bacillus thuringiensis* and *Streptococcus pyogenes*, the cause of "strep throat."
- **Gram-negative bacteria** have a thin cell wall with one to three layers of peptidoglycan, enclosed by an outer membrane. The phylum Proteobacteria consists of Gram-negative species such as *Escherichia coli* and nitrogen-fixing rhizobia.

Outside these two groups, however, many bacterial species cannot be classified according to the Gram stain models. For example, archaeal cell envelopes (see Chapter 19) are highly diverse and cannot be distinguished by Gram stain.

> **THOUGHT QUESTION 3.6** The actual thickness of the cell wall is difficult to determine based solely on electron microscopic observation of the envelope layers. Devise a biochemical experiment that can show the number of layers of peptidoglycan in cells of a given species.

The Gram-Positive Cell Envelope

A section of a Gram-positive cell envelope is shown in **Figure 3.18**. The Gram-positive cell wall consists of multiple layers of peptidoglycan, up to 40 in some species. No space appears between the peptidoglycan layers, although the outer band of envelope appears more dense. The peptidoglycan is reinforced by **teichoic acids** threaded through its multiple layers (**Fig. 3.19**). Teichoic acids are chains of phosphodiester-linked glycerol or ribitol, with sugars or amino acids linked to the middle OH groups. The negatively charged cross-threads of teichoic acids, as well as the overall thickness of the Gram-positive cell wall, help retain the Gram stain.

Outside the cell wall, Gram-positive cells are often encased in a slippery **capsule** consisting of loosely bound polysaccharides. The capsule can be visualized under the light microscope with India ink, a negative stain consisting of soot-like carbon particles, which the capsule excludes (shown in **Fig. 2.24C**). Gram-negative cells may also have a capsule.

The S-layer. An additional protective layer commonly found in free-living bacteria and archaea is the surface layer, or **S-layer**. The S-layer is a crystalline layer of thick subunits consisting of protein or glycoprotein (proteins with attached sugars) (**Fig. 3.20**).

Each subunit of the S-layer contains a pore large enough to admit a wide range of molecules. The subunits form a smooth layer on the cell wall or outer membrane (see **Fig. 3.20**). The proteins are arranged in a highly ordered array, either hexagonally or tetragonally. The S-layer is rigid, but it also flexes and allows substances to pass through it in either direction.

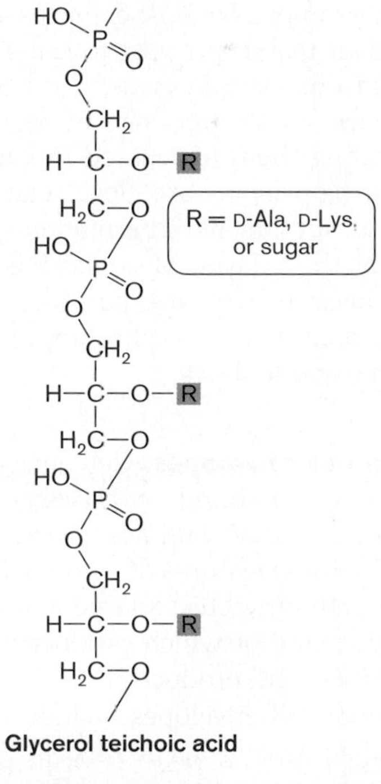

Glycerol teichoic acid

Figure 3.19   Teichoic acids. Teichoic acids in the Gram-positive cell wall consist of glycerol or ribitol phosphodiester chains. The middle hydroxyl group of each glycerol or ribitol is typically linked to D-alanine, to D-lysine, or to a sugar such as galactose or *N*-acetylglucosamine.

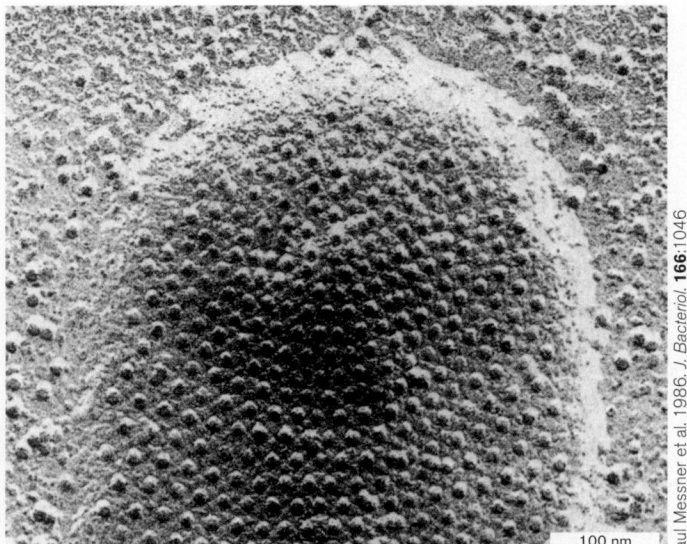

Paul Messner et al. 1986. *J. Bacteriol.* **166**:1046

100 nm

Figure 3.20   The S-layer.   The archaeon *Thermoproteus tenax* has a single tetraether membrane encased by an S-layer (SEM). Note the regular pattern of tiled proteins.

The functions of the S-layer are uncertain, in part because the S-layer is often lost by bacteria after repeated subculturing in the laboratory. The protein subunits serve as attachment platforms for polysaccharide filaments. In natural environments, the tough proteins may deter predators such as amebas or host defenses such as phagocytes. The S-layer may contribute to cell shape and help protect the cell from osmotic stress.

Traits are lost in the absence of selective pressure for genes encoding them (discussed in Chapter 17). For example, the mycoplasmas are close relatives of Grampositive bacteria, yet they have permanently lost their cell walls, as well as the S-layer. Mycoplasmas have no need for cell walls because they are parasites living in host environments, such as the human lung, where they are protected from osmotic shock.

Mycobacterial cell envelopes.   Cell envelopes of exceptional complexity are found in mycobacteria, such as *Mycobacterium tuberculosis* and *Mycobacterium leprae*, the cause of leprosy. Some features of mycobacterial cell walls also appear in actinomycetes, a large and diverse family of soil bacteria, many of which produce antibiotics and other industrially useful products.

Mycobacterial cell envelopes include unusual membrane lipids called mycolic acids. Mycolic acids contain a hydroxy acid backbone with two hydrocarbon chains—one comparable in length to typical membrane lipids (about 20 carbons), the other about threefold longer. The long chain includes ketones, methoxyl groups, and cyclopropane rings. Hundreds of different forms are known.

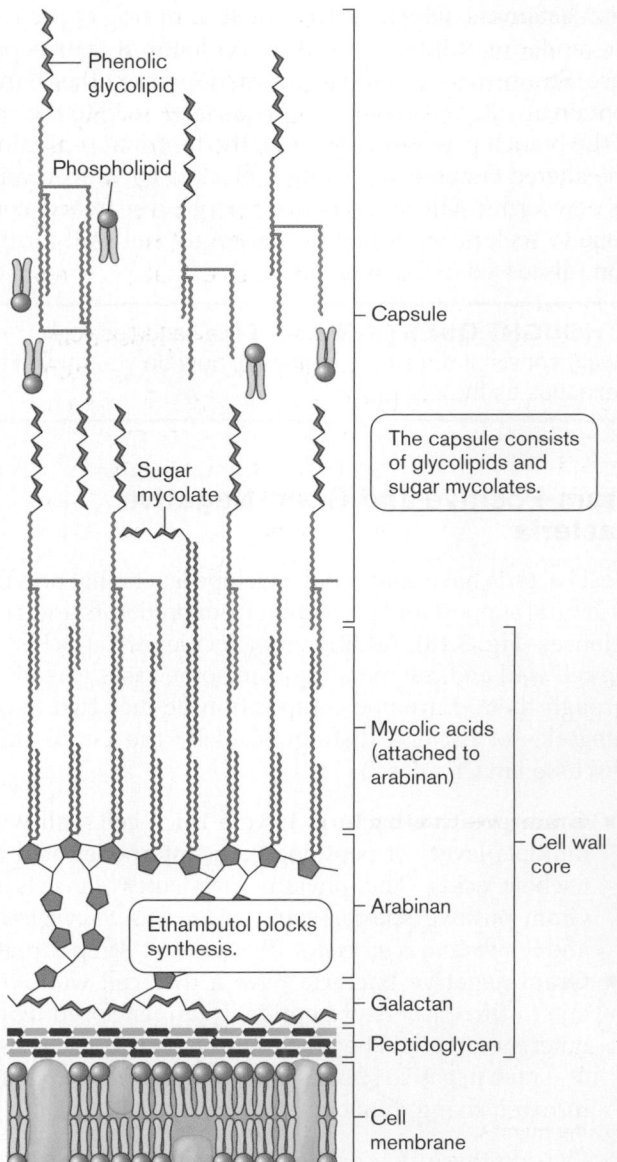

Phenolic glycolipid

Phospholipid

Capsule

The capsule consists of glycolipids and sugar mycolates.

Sugar mycolate

Mycolic acids (attached to arabinan)

Cell wall core

Ethambutol blocks synthesis.

Arabinan

Galactan

Peptidoglycan

Cell membrane

Figure 3.21   Mycobacterial envelope structure.   A complex cell wall includes a peptidoglycan layer linked to a chain of galactose polymer (galactan) and arabinose polymer (arabinan). Arabinan forms ester links to mycolic acids, which form an outer bilayer with phenolic glycolipids. Outside the outer bilayer is a capsule of loosely associated phospholipids and phenolic glycolipids.

The phenolic glycolipids include a phenol group, also linked to sugar chains.

The mycobacterial cell wall includes peptidoglycan linked to chains of galactose, called galactans (**Fig. 3.21**). The galactans are attached to arabinans, polymers of the five-carbon sugar arabinose. The arabinan-galactan polymers are known as arabinogalactans. Arabinogalactan biosynthesis can be inhibited by ethambutol, a major drug against tuberculosis.

The ends of the arabinan chains form ester links to mycolic acids. Mycolic acids provide the basis for acid-fast staining, in which cells retain the dye carbolfuchsin, an important diagnostic test for mycobacteria and actinomycetes (described in Chapter 2). In *M. tuberculosis* and *M. leprae*, the mycolic acids form a kind of bilayer interleaved with phenolic glycolipids. The extreme hydrophobicity of the phenol derivatives generates a waxy surface that prevents phagocytosis by macrophages.

Overall, the thick, waxy envelope excludes many antibiotics and offers exceptional protection from host defenses, enabling the pathogens of tuberculosis and leprosy to colonize their hosts over long periods. However, the mycobacterial envelope also retards uptake of nutrients. As a result, *M. tuberculosis* and *M. leprae* grow extremely slowly and are a challenge to culture in the laboratory.

> **THOUGHT QUESTION 3.7** Why would laboratory culture conditions select for evolution of cells lacking an S-layer?

The Gram-Negative Outer Membrane

A Gram-negative cell envelope is seen in **Fig. 3.18C**. The thin layer of peptidoglycan consists of one or two sheets, based on calculations of molecular density. The peptidoglycan is covered by an outer membrane. The Gram-negative outer membrane confers defensive abilities and toxigenic properties on many pathogens, such as *Salmonella* species and enterohemorrhagic *E. coli* (strains that cause hemorrhaging of the colon). Between the outer and inner (cell) membranes is the periplasm.

Lipoprotein and lipopolysaccharide (LPS). The inward-facing leaflet of the outer membrane has a phospholipid composition similar to that of the cell membrane (in Gram-negative species, it is called the **inner membrane** or inner cell membrane). The outer membrane's inward-facing leaflet includes lipoproteins that connect the outer membrane to the peptide bridges of the cell wall. The major lipoprotein is called **murein lipoprotein**, also known as Braun lipoprotein (**Fig. 3.22A**). Murein lipoprotein consists of a protein with an N-terminal cysteine attached to three fatty acid side chains. The side

A. Gram-negative envelope

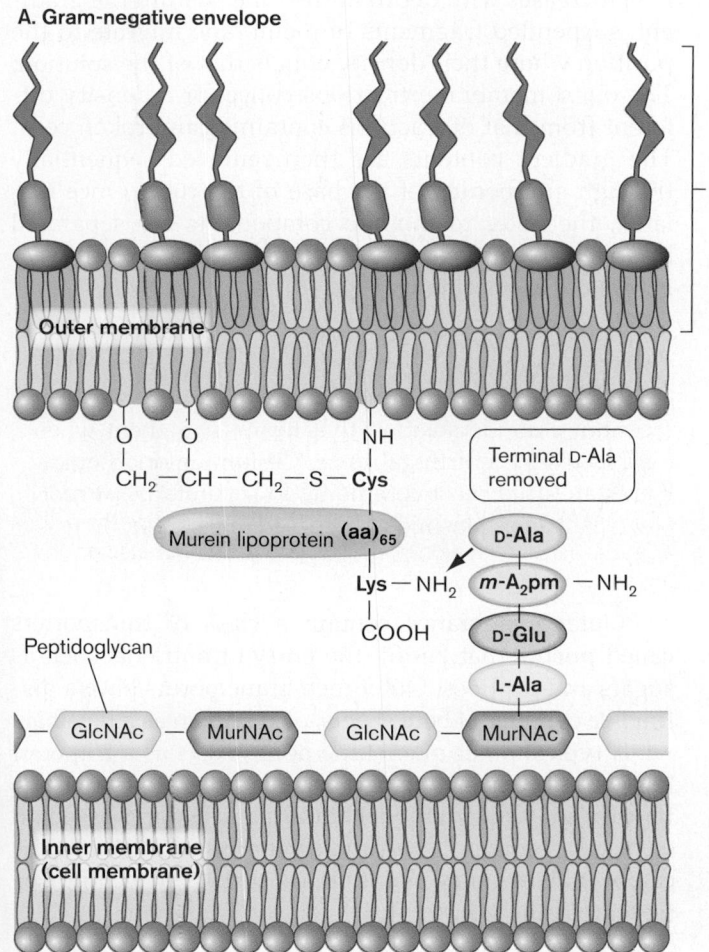

B. Lipopolysaccharide (LPS)

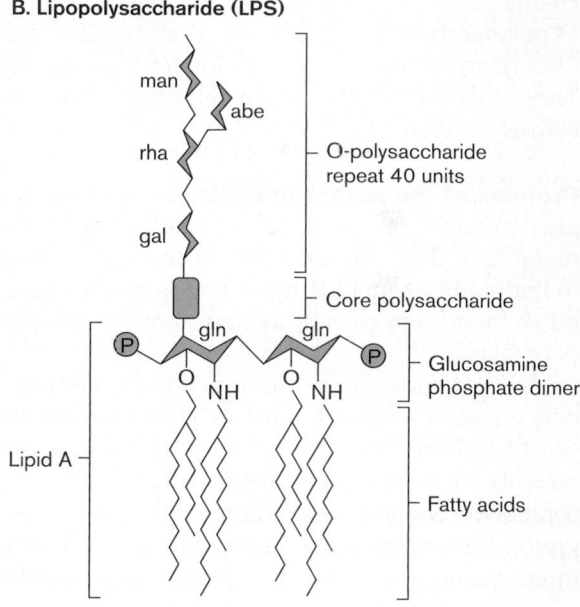

Figure 3.22 Lipoprotein and lipopolysaccharide. A Murein lipoprotein has an N-terminal cysteine triglyceride inserted in the inward-facing leaflet of the outer membrane. The C-terminal lysine forms a peptide bond with the *m*-diaminopimelic acid of the peptidoglycan (murein) cell wall. **B.** Lipopolysaccharide (LPS) consists of branched and unbranched short-chain fatty acids linked to a dimer of phosphoglucosamine. One glucosamine is linked to a core polysaccharide extending out from the cell, which is attached to about 40 repeating units of a polysaccharide known as O antigen.

chains are inserted in the inward-facing leaflet of the outer membrane. The C-terminal lysine forms a peptide bond with the *m*-diaminopimelic acid of peptidoglycan (murein).

The outward-facing leaflet, however, consists of a special kind of phospholipid called **lipopolysaccharide** (**Fig. 3.22B**). The LPS lipids have shorter fatty acid chains than those of the inner cell membrane, and some are branched. LPS is of crucial medical importance because it acts as an **endotoxin**. An endotoxin is a cell component that is harmless as long as the pathogen remains intact; but when released by a lysed cell, endotoxin overstimulates host defenses, inducing potentially lethal endotoxic shock. Thus, antibiotic treatment of an LPS-containing pathogen can kill the cells but can also lead to death of the patient.

In LPS the fatty acids are esterified to **glucosamine**, an amino sugar found also in peptidoglycan. Two glucosamine dimers each condense with two fatty acid chains, making four fatty acid chains in all. Like the glycerol of phospholipid, each glucosamine of LPS connects to a phosphate, whose negative charge interacts with water. One of the glucosamines is attached to a **core polysaccharide**, a sugar chain that extends outside the cell. The core polysaccharide consists of about five sugars with side chains such as phosphoethanolamine. It extends to an **O polysaccharide**, a chain of as many as 200 sugars. The O polysaccharide may extend longer than the cell itself. These chains form a layer that helps bacteria resist phagocytosis by white blood cells.

Proteins of the outer membrane. The outer membrane also contains unique proteins not found in the inner membrane. The various layers of the cell envelope need to maintain distinct identities and properties that distinguish them from each other and from the cytoplasm and periplasm.

The composition of the inner and outer membranes and aqueous compartments is characterized by ultracentrifugation. Fractions containing various cell components show that each protein is confined to a specific component of the cell (**Table 3.2**). For example, the proton-translocating ATP synthase is found only in the inner membrane fractions, whereas sugar-binding proteins are only in the outer membrane.

> **THOUGHT QUESTION 3.8**   Why would the proteins listed in **Table 3.2** be confined to specific cell fractions? Why could a protein not function everywhere in the cell?

Outer membrane proteins of pathogens are important targets for vaccine development. For example, the outer membrane proteins were characterized for

Table 3.2	**Subcellular location of proteins in Gram-negative bacteria.**
Cell fraction	**Proteins (examples)**
Cytoplasm	Glycolytic enzymes
	Biosynthesis of amino acids
	Cytoplasmic chaperones
Inner membrane (cell membrane)	Proton-translocating ATP synthase
	Electron transport chain
	Transporters for ions, sugars and amino acids
Periplasm (periplasmic space)	Sugar taxis receptors
	Periplasmic chaperones
Outer membrane	Porins for sugars and peptides
	Sugar-binding proteins

Borrelia burgdorferi, the cause of Lyme disease. In the experiment shown in **Figure 3.23**, fractions of *Borrelia* outer membrane are separated from whole cells by density gradient centrifugation. The sample is loaded on a tube of sucrose solution whose concentration (and density) increases with depth in the tube. Within the gradient, suspended fragments of membrane migrate to the position where their density equals that of the solution. The outer membrane fractions collect at a density different from that of fractions containing unbroken cells. The gradient contents are then removed sequentially through an opening at the base of the tube. Once isolated, the outer membrane components are separated by gel electrophoresis to reveal distinctive proteins that may serve as vaccine targets.

> **NOTE:** Another type of gradient centrifugation, called equilibrium density gradient centrifugation, uses a cesium chloride solution that forms a gradient when subjected to centrifugal force. Cesium chloride gradients are used most commonly to separate DNA molecules such as plasmids (discussed in Chapter 7).

Outer membranes contain a class of transporters called **porins** that permit the entry of nutrients such as sugars and peptides. Outer membrane porins have a distinctive cylinder of beta sheets, also known as a beta barrel. A typical outer membrane porin exists as a trimer of beta barrels, each of which acts as a pore for nutrients. **Figure 3.24** shows a model of the sucrose porin based on a crystal structure. Unfortunately for the cell, some porins can also allow entry of antibiotics such as penicillin (see **Special Topic 3.1**).

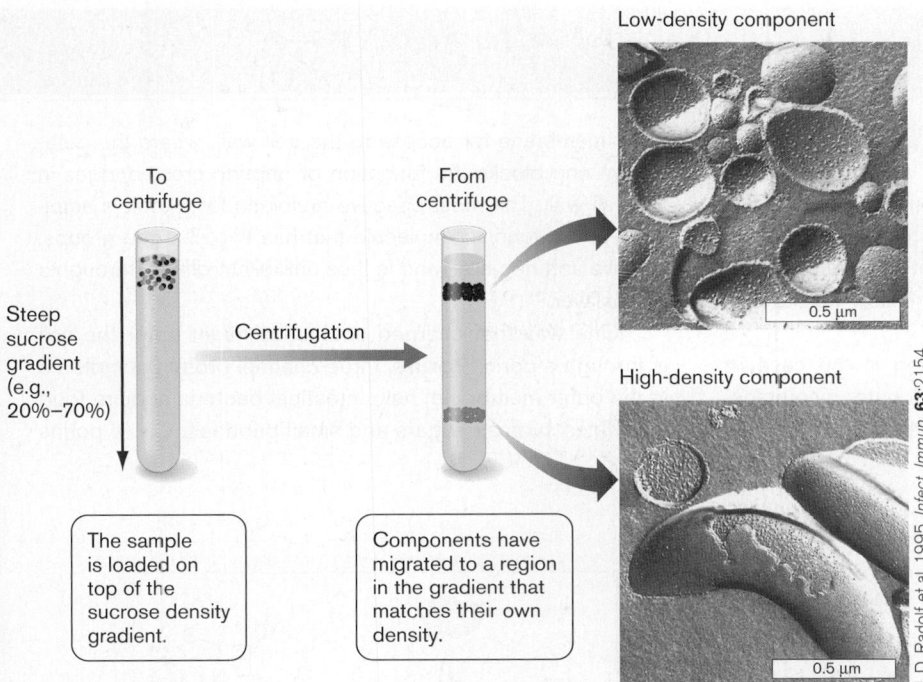

Low-density component

High-density component

0.5 μm

0.5 μm

Figure 3.23 **Outer membrane analysis by centrifugation.** *Borrelia burgdorferi* outer membrane vesicles are separated from the cell by equilibrium density gradient ultracentrifugation. The lighter fractions contain outer membrane vesicles; the denser fractions contain unbroken whole cells (inset images, freeze-fracture TEM).

J. D. Radolf et al. 1995. *Infect. Immun.* **63**:2154

To centrifuge

From centrifuge

Centrifugation

Steep sucrose gradient (e.g., 20%–70%)

The sample is loaded on top of the sucrose density gradient.

Components have migrated to a region in the gradient that matches their own density.

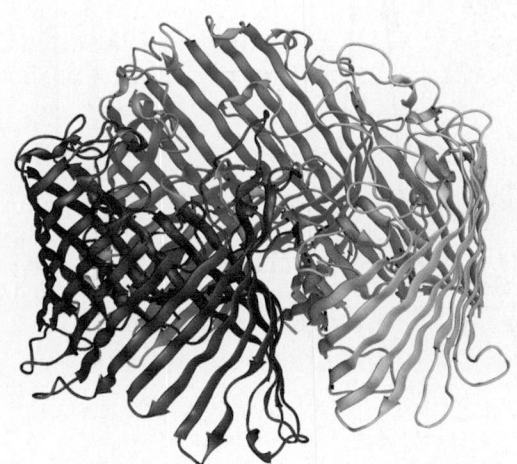

Figure 3.24 **Sucrose porin.** Outer membrane porin (OMP) for sucrose transport in *Salmonella typhimurium* based on X-ray crystallography. The porin comprises three beta barrel channels (red, yellow, and blue). Each transports a molecule of sucrose through its channel. (PDB code: 1AOS)

Cells express different outer membrane porins under different environmental conditions. In a dilute environment, cells express porins of large pore size, maximizing the uptake of nutrients. In a rich environment—for example, within a host—cells down-regulate the expression of large porins and express porins of smaller pore size, selecting only smaller nutrients and avoiding the uptake of toxins. For example, the porin regulation system of Gram-negative bacteria enables them to grow in the intestinal region containing bile salts—a hostile environment for Gram-positive bacteria.

Periplasm. The outer membrane is porous to most ions and many small organic molecules, but it prevents the passage of proteins and other macromolecules. Thus, the region between the inner and outer membranes of Gram-negative cells, including the cell wall, defines a separate membrane-enclosed compartment of the cell known as the periplasm (see **Fig. 3.18**). The periplasm contains specific enzymes and nutrient transporters not found within the cytoplasm, such as periplasmic transporters for sugars, amino acids, or other nutrients. Periplasmic proteins are subjected to fluctuations in pH and salt concentration, because the outer membrane is porous to ions. Some periplasmic proteins help refold proteins unfolded by oxidizing agents or by acidification.

Eukaryotic Microbes: Protection from Osmotic Shock

The problems confronting bacteria, such as adjusting to environmental change, also challenge eukaryotes (Chapter 2). A key example is osmotic shock. Eukaryotic microbes possess their own structures to avoid osmotic shock. Algae form cell walls of cellulose fibers, and fungi form walls of chitin. Diatoms form intricate exoskeletons of silicate.

Special Topic 3.1 | How Antibiotics Cross the Outer Membrane

Ever since Alexander Fleming discovered penicillin produced by a mold, researchers have sought new antibiotics to control disease-causing bacteria. But antibiotics affect different bacterial species very differently. And bacteria quickly evolve resistance to any new drug. An important question is, How do antibiotics get into a bacterial cell? And how do bacteria respond to avoid antibiotics?

The question is particularly interesting in the case of Gram-negative bacteria, which possess an outer membrane. The penicillin class of antibiotics needs to get through the outer membrane for access to the cell wall, where the beta-lactam ring blocks the formation of peptide cross-bridges in the cell wall. The most effective antibiotic for E. coli is ampicillin, a "zwitterionic" molecule that has two charged groups, positive and negative, and is thus unlikely to diffuse through a lipid bilayer.

One way that charged antibiotics might enter the cell is through a porin. **Porins**, three-channel protein complexes in the outer membrane, help intestinal bacteria acquire food molecules such as sugars and small peptides. Could porins

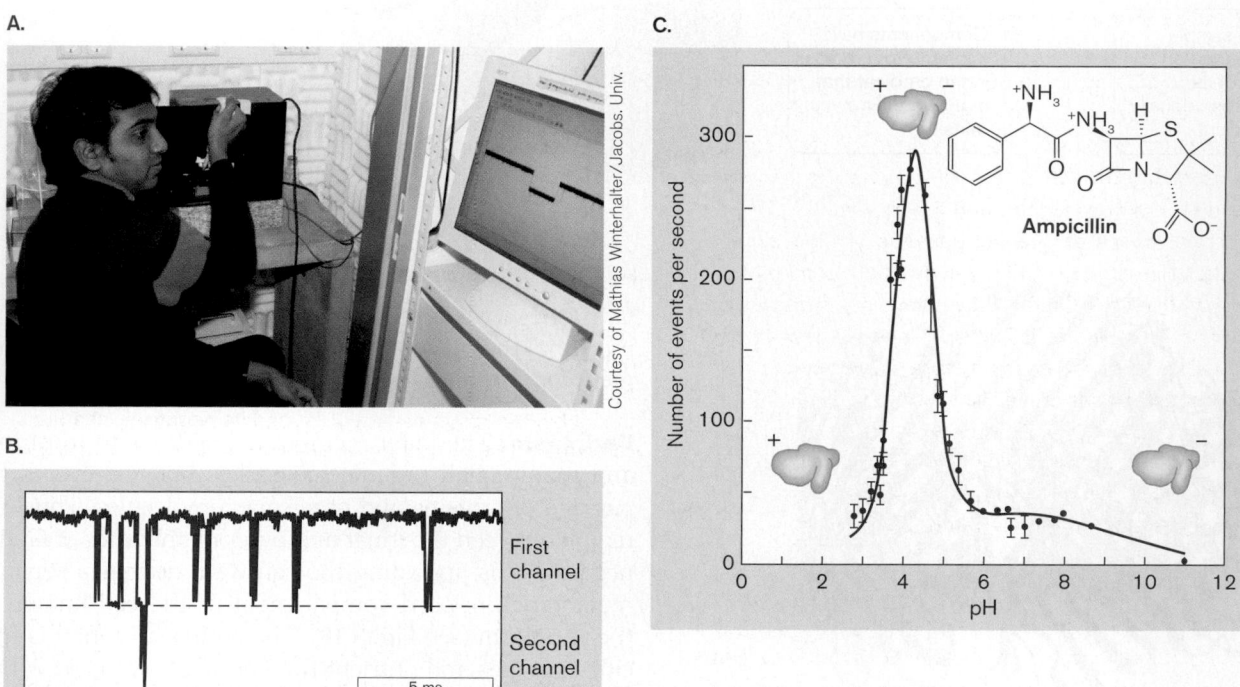

Figure 1 Ampicillin passage rate and pH dependence. A. PhD student Kozhinjampara R. Mahendran measures voltage across a membrane during ampicillin passage. **B.** Passage events are observed as dips in transmembrane current, caused by ampicillin briefly blocking the OmpF channel. **C.** At pH 4.5, where the dual-charged form predominates, ampicillin shows the highest rate of passage through OmpF. *Source*: Parts B and C modified from Ekaterina M. Nestorovich et al. 2002. *PNAS* **99**:9789–9794.

Other eukaryotes, however, such as amebas and ciliated protists, possess a flexible outer coating, the pellicle. The pellicle allows great flexibility of shape; it enables the uptake of large particles by endocytosis and of larger objects, even entire cells, by phagocytosis. Thus, eukaryotic

cells have the nutritional option of engulfing prey, an option unavailable to prokaryotes and to eukaryotes with cell walls. Eukaryotic microbes that lack a cell wall possess a **contractile vacuole** to pump water out of the cell, avoiding osmotic shock (**Fig. 3.25**). The vacuole takes up water from

also allow the entry of more deadly molecules such as ampicillin? Mathias Winterhalter and co-worker Kozhinjampara R. Mahendran (**Fig. 1A**) at the Institute of Pharmacology and Structural Biology in Toulouse, France, tested a hypothesis that ampicillin penetrates the outer membrane through a porin called OmpF.

To observe ampicillin passage through OmpF, a single OmpF trimer (with three protein channels) was solubilized within an artificial membrane bilayer. The bilayer was formed across a chamber of solution containing electrodes. A voltage was applied, and the current was measured. Dips in the current indicated transient blockage of a channel by ampicillin (**Fig. 1B**). When two of the three channels of the trimer were blocked simultaneously, the dip was twice as large. Overall, the rate of ampicillin passage peaked at close to pH 4.5, the pH at which most of the ampicillin has both a positive and a negative charge (**Fig. 1C**). This pH dependence suggests that the dual-charged zwitterionic form of ampicillin is the form that most effectively passes through the channel.

The observations of channel passage yielded clues as to how ampicillin interacts with the channel. Computational modeling based on the OmpF X-ray crystal structure showed that the ampicillin molecule is just small enough to fit through the narrowest part of the channel (**Fig. 2**). Furthermore, the optimized model places ampicillin in a position such that its positively charged amine group binds the carboxylate of glutamate-117, and its negatively charged carboxylate binds the positively charged arginine residues. Thus, the zwitterionic form of ampicillin is a good fit for the channel.

How, then, do Gram-negative bacteria avoid ampicillin and other antibiotics in their surroundings? The most effective resistance mechanism is to acquire a gene encoding beta-lactamase, an enzyme that cleaves ampicillin's beta-lactam ring. But until genetic resistance is established, bacteria have short-term environmental responses that can decrease their exposure. First, bacteria produce positively charged molecules called polyamines, which block the porin channels. Polyamine production is enhanced at low pH,

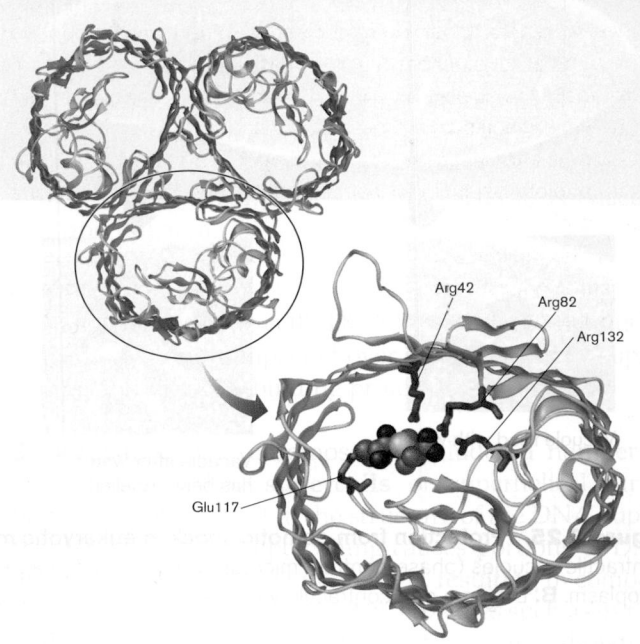

Figure 2   Computational model of OmpF blocked by ampicillin.   The OmpF trimeric channel was modeled after X-ray crystallography. (PDB code: 2OMF) Inset shows an optimized model of amino acid residues contacting ampicillin. Ampicillin's positively charged amine group is attracted to the carboxylate of glutamate-117, and its negatively charged carboxylate is attracted to the arginine residues. *Source:* Ekaterina M. Nestorovich et al. 2002. *PNAS* **99**:9789–9794.

where ampicillin penetrates the channel most effectively. In addition, bacteria decrease expression of the gene that encodes OmpF, and instead up-regulate genes for other porins, with smaller channels that exclude ampicillin. Understanding these molecular mechanisms will enable us to design better antibiotic treatments that overcome bacterial defenses.

the cytoplasm through an elaborate network of intracellular channels, and then expels the water through a pore.

THOUGHT QUESTION 3.9   What do you think are the advantages and disadvantages of a contractile vacuole, compared with a cell wall?

TO SUMMARIZE:

■ **The cell wall maintains turgor pressure.** The cell wall is porous, but its rigid network of covalent bonds protects the cell from osmotic shock.

■ **The Gram-positive cell envelope** has multiple layers of peptidoglycan, threaded by teichoic acids.

Within the cell, replication proceeds outward in both directions around the genome. Thus, bidirectional replication requires two replisomes, one for each replicating fork. A long-standing question has been, Do the two replisomes move oppositely around the DNA, or do they stay in the middle while the DNA helices slide through them?

To answer this question, fluorescence microscopy is used to observe the process of DNA replication within a growing cell of *Bacillus subtilis* (**Fig. 3.31**). The DNA origin of replication (*ori*) and the pair of replisomes are labeled by fluorescence. The origin of replication is labeled blue by a hybrid protein fused to a gene encoding cyan fluorescent protein (CFP). CFP binds to a promoter sequence cloned in *B. subtilis* near its origin site. The replisomes are labeled yellow by a hybrid protein expressed from a gene encoding a DNA polymerase subunit fused to a gene encoding yellow fluorescent protein (YFP). The replisomes usually locate together near the center of the cell, but sometimes they separate, visible as two yellow spots.

The fluorescence data are consistent with a model in which two replisomes are located at the midpoint of the growing cell (**Fig. 3.32**●). Each of the two replisomes forms a replicating fork that directs two daughter strands of DNA toward opposite poles. The two copies

of the DNA origin of replication (green in the figure), attached to the cell envelope, move apart as the cell expands. The termination site (red) remains in the middle of the cell.

The two replisomes continue replication at both forks in the middle of the cell. Finally, as the termination site replicates, the two replisomes separate from the DNA. At

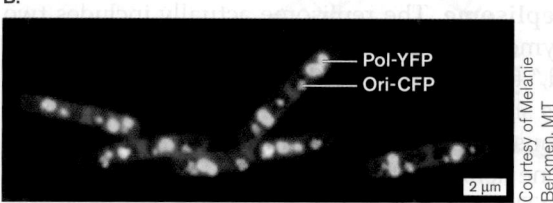

A.

B.

Figure 3.31 The replisome and the DNA origin.
A. Melanie Berkmen, working in the laboratory of Alan Grossman, obtains the fluorescence micrographs shown. **B.** Fluorescence microscopy reveals the DNA origin, labeled blue by a protein fused to cyan fluorescent protein, binding at a sequence near the origin (Ori-CFP). Replisomes are labeled yellow by fusion of a DNA polymerase subunit to yellow fluorescent protein (Pol-YFP) in dividing cells of *Bacillus subtilis*. The cell envelope is labeled red with the membrane stain FM4-64.

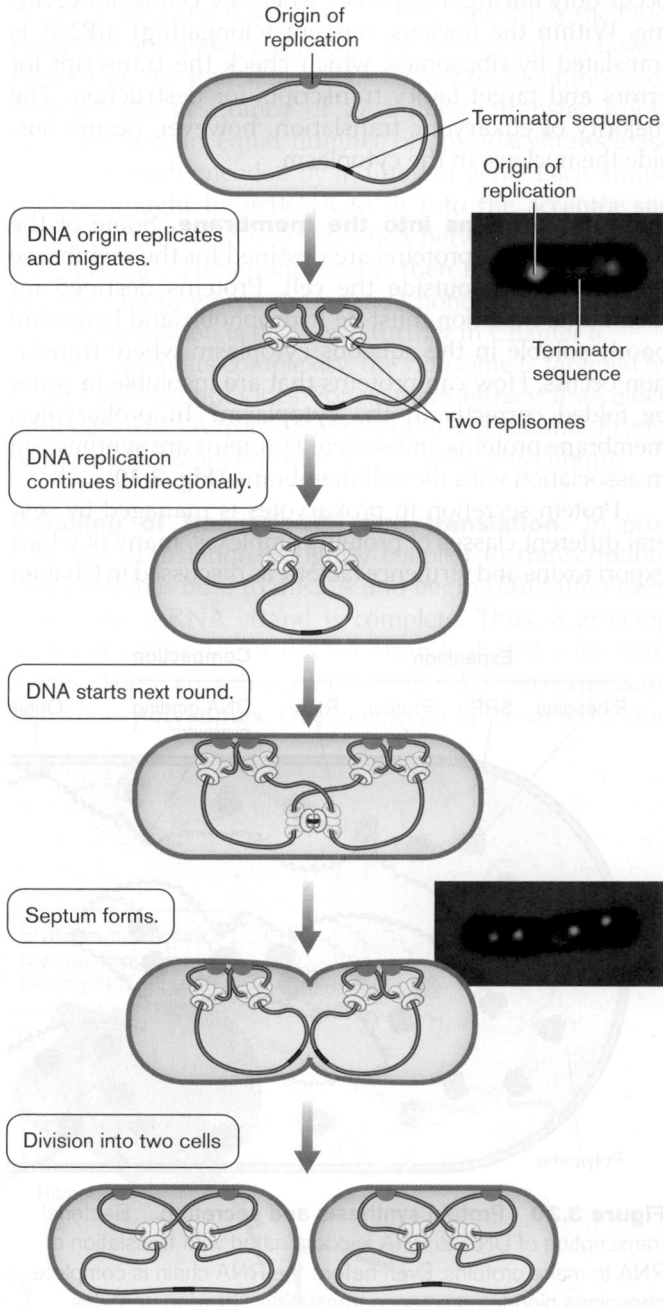

Figure 3.32 Replisome movement within a dividing cell.
The DNA origin-of-replication sites (green) move apart in the expanding cell as the pair of replisomes (yellow) stay near the middle, where they replicate around the entire chromosome, completing the terminator sequence last (red). *Source:* Ivy Lau et al. 2003. *Mol. Microbiol.* **49**:731. ●Ⅱ *microbiology2.com/animations*

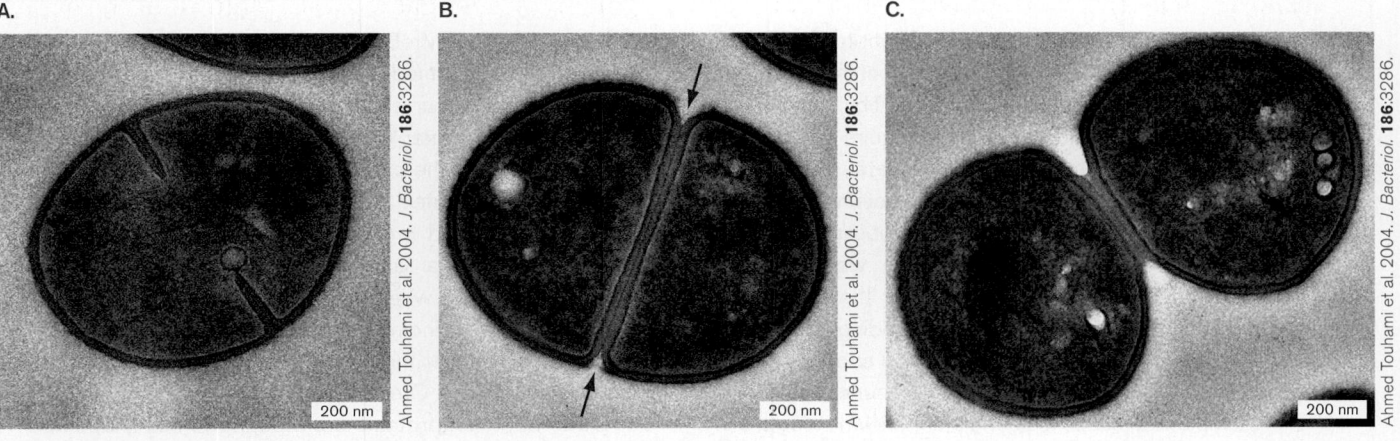

A. **B.** **C.**

Ahmed Touhami et al. 2004. *J. Bacteriol.* **186**:3286.

Figure 3.33 **Septation in *Staphylococcus aureus*.** **A.** Furrows appear in the cell envelope, all around the cell equator, as new cell wall grows inward (TEM). **B.** Two new envelope partitions are complete. **C.** The two daughter cells peel apart. The facing halves of each cell contain entirely new cell wall.

each new *ori* site, however, two pairs of new replisomes have formed. Replication of the new *ori* sites begins, sometimes even before termination of the previous round of replication.

Note that the contents of the cytoplasm must expand coordinately with DNA replication for the cell to generate progeny equivalent to the parent. In a rod-shaped cell, the cell envelope and cell wall must elongate as well, to maintain progeny of even girth and length.

Septation Completes Cell Division

For the cell to divide, DNA replication must be complete. Replication of the DNA termination site triggers growth of the dividing partition of the envelope, called the **septum** (plural, **septa**) The septum grows inward from the sides of the cell, at last constricting and sealing off the two daughter cells. This process is called **septation**. Septation and envelope extension require rapid biosynthesis of all envelope components, including membranes and cell wall. The biosynthetic enzymes required are all of great interest as antibiotic targets. Cell wall biosynthesis poses an interesting theoretical problem: How is it possible to expand the covalent network of the sacculus without breaking links to insert new material, thus weakening the wall? The answer remains unclear.

Septation of spherical cells. In spherical cells (cocci), such as *Staphylococcus aureus*, the process of septation generates most of the new cell envelope to enclose the expanding cytoplasm (**Fig. 3.33**). Furrows form in the cell envelope, in a ring all around the cell equator, as new cell wall grows inward. The wall material must compose two separable partitions. When the partitions are complete, the two progeny cells peel apart. The facing halves of each cell consist of entirely new cell wall.

A.

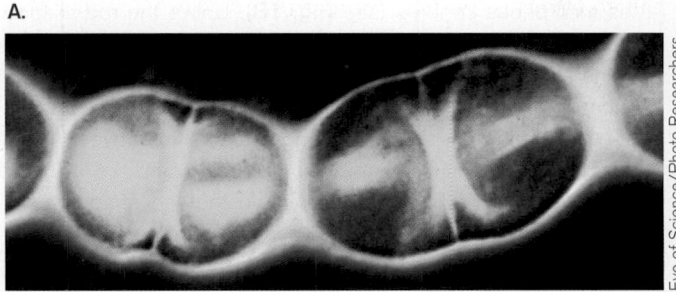

Eye of Science/Photo Researchers, Inc.

B.

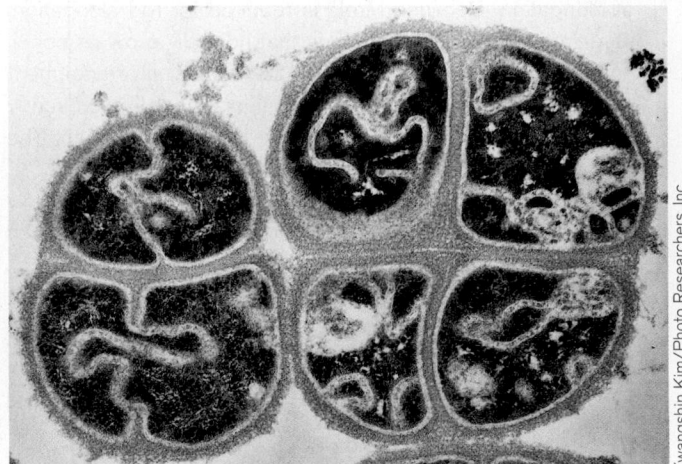

Kwangshin Kim/Photo Researchers, Inc.

Figure 3.34 **Septation orientation determines the arrangement of progeny cells.** **A.** A chain of cocci results from septation in the same plane (*Streptococcus*, 0.5–1.0 µm, colorized TEM). **B.** Septation in two planes generates a tetrad (*Micrococcus tetragenus*, 0.5–1.0 µm, negative stain).

The spatial orientation of septation has a key role in determining the shape and arrangement of cocci. When septation always occurs in parallel planes, such as in *Streptococcus* species, cells form chains (**Fig. 3.34A**).

A major distinction of prokaryotes was thought to be the lack of a cytoskeleton, a structural feature prominent in eukaryotes. Bacterial shape was thought to be maintained by the cell wall. In fact, bacterial shape and cell division require cytoskeletal proteins homologous to eukaryotic cytoskeletal components such as tubulin and actin. The first such protein recognized was the Z ring subunit FtsZ, an analog of tubulin. Additional components of the bacterial cytoskeleton were revealed by several means: gene defects that confer the loss of cell shape, fluorescent labeling of the corresponding gene products in wild-type cells, and genomic analysis showing bacterial homologs of eukaryotic cytoskeletal components.

An example of a "shape loss" mutant is shown in **Figure 1**. The bacterium *Shigella* is an enteric pathogen closely related to *E. coli* that normally grows as a rod, or bacillus (see **Fig. 1A**). The opposite poles of the rod bind specific pole proteins, which are marked in the micrographs by fluorescent-labeled antibodies. A mutant strain of *Shigella* was isolated that grows as amorphous spheres (see **Fig. 1B**). Unlike the rod-shaped cells, the spherical mutants show multiple pole-marker proteins at undefined locations. The mutation eliminates expression of *mreB*, which encodes the MreB cytoskeletal protein. Sequence analysis shows a strong relationship between MreB and the eukaryotic filament-forming protein actin.

How do MreB proteins maintain the rod shape of the bacterium? Like actin filaments in eukaryotes, MreB proteins form elongated structures beneath the cell membrane. **Figure 2A** shows how MreB proteins polymerize in a helical arrangement around the cell, thus determining the axis of elongation. Because MreB is required for rod elongation, its gene is absent from bacteria that normally grow as cocci, such as *Streptococcus*. In **Figure 2B**, the helical arrangement of MreB is observed by confocal fluorescence microscopy of *Caulobacter crescentus*, a pond-dwelling bacterium. The MreB proteins are labeled by fluorescent antibodies.

Crescent-shaped cells such as *C. crescentus* usually possess a third kind of shape-determining protein, CreS, in addition to FtsZ and MreB (see **Fig. 2A**). CreS is a homolog of the eukaryotic intermediate filament protein. CreS polymerizes along the inner curve of a crescent cell, as seen in **Figure 2C**. In the micrograph shown, CreS proteins are labeled by recombination of the *creS* gene with a gene encoding the green fluorescent protein (GFP). The result is a gene fusion expressing the hybrid protein CreS-GFP, which fluoresces green. In the confocal fluorescence micrograph, the CreS-GFP appears green, whereas the cell membrane is labeled red by a membrane-binding fluorophore. The membrane-binding fluorophore reveals the overall crescent outline of the cell and the localization of CreS to the cell's inner curve.

Other bacterial proteins contributing to cell shape and division include the Min proteins, required for equal division of cells, and the Par proteins, which segregate replicated plasmids during cell division. In one group of bacteria, the phylum Verrucomicrobia, a tubulin homolog has been discovered that forms tubules in vitro. Such tubules may contribute to the unusual extended shapes of Verrucomicrobia.

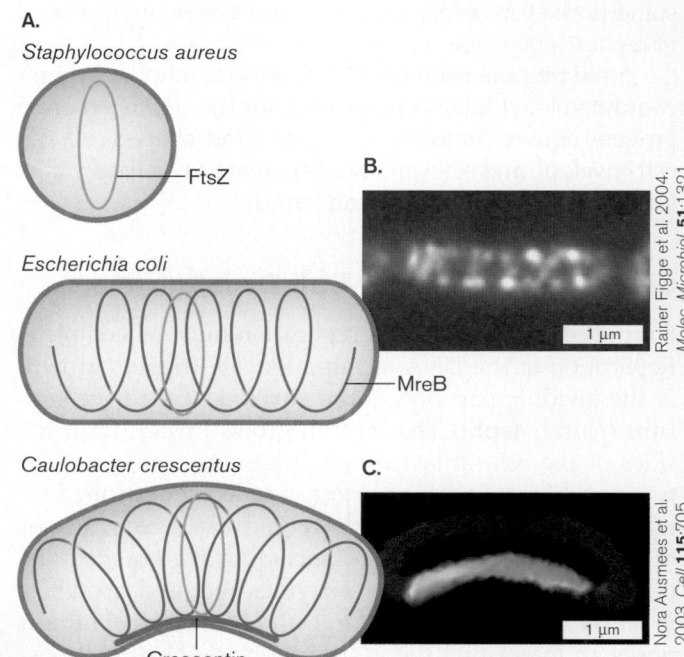

A.
Staphylococcus aureus
— FtsZ

Escherichia coli
— MreB

Caulobacter crescentus
Crescentin

B.
1 μm

Rainer Figge et al. 2004. *Molec. Microbiol.* **51**:1321.

C.
1 μm

Nora Ausmees et al. 2003. *Cell* **115**:705.

Figure 2 Shape-determining proteins. A. Spherical cell size and shape are maintained by FtsZ polymerization to form the Z ring. Elongation of a rod-shaped cell is accomplished by helical polymerization of Mre proteins. Crescent-shaped cells possess a third shape-determining protein, CreS (crescentin), which polymerizes along the inner curve of the crescent. **B.** MreB localizes in a helical coil within *Caulobacter crescentus* cells. Images of MreB immunostaining were obtained by confocal fluorescence microscopy. **C.** Crescentin protein fused to green fluorescent protein (CreS-GFP) localizes to the inner curve of *C. crescentus*. Membrane-specific stain FM4-64 (red fluorescence) localizes to membrane around cell.

A. Parental strain   **B. *mreB* mutant**

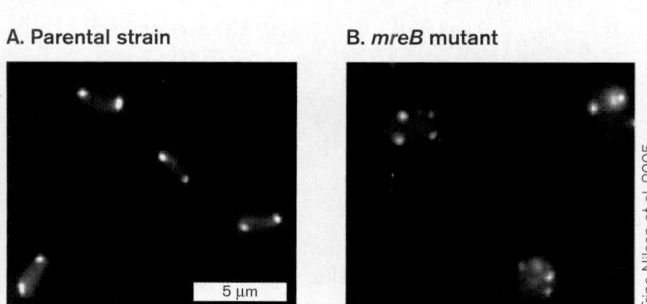

5 μm

Trine Nilsen et al. 2005. *J. Bacteriol.* **187**:6187.

Figure 1 Spherical *mreB* mutants of rod-shaped bacteria. A. Rod-shaped *Shigella* cells show a pole-associated protein (IcsA-GFP) localized to opposite poles. **B.** Mutant MreB cells form bloated spherical cells with multiple sites localizing IcsA-GFP.

If septation occurs in random orientations, or if cells reassociate loosely after septation, they form compact hexagonal arrays similar to the grape clusters portrayed in classical paintings—hence the Greek-derived term **staphylococci**. Such clusters are found in colonies of *Staphylococcus aureus*. If subsequent septation occurs at right angles to the previous division, the cells may form tetrads and even cubical octads called sarcinae (singular, sarcina). Tetrads are formed by *Micrococcus tetragenus*, a cause of pulmonary infections (**Fig. 3.34B**).

Septation of rod-shaped cells. In rod-shaped cells, unlike cocci, cell division requires the envelope to elongate before septation, followed by the formation of a new polar envelope for each progeny cell. The process of septation involves an intricate series of molecular signals that is at the frontier of current research.

Certain mutant strains of *E. coli* form long filaments instead of dividing normally. This filamentation results from a failure to form a septum between cells. The mutant genes causing this behavior were called *fts* for "filamentation temperature sensitive" because the cells divide normally at the **permissive temperature** but fail to septate at the **restrictive temperature**, forming long filaments.

Some of the *fts* genes encode proteins directly involved in formation of the septum. The most dramatic example is the protein FtsZ, which assembles to form the "Z ring," a constriction ring around the equator (**Fig. 3.35**). FtsZ is universally found in bacteria and archaea as a key septation protein. FtsZ is also an ancient homolog of tubulin, which is the major component of the mitotic apparatus in eukaryotes. The discovery of FtsZ is interesting because it implies that the processes of mitosis in eukaryotes and cell division in prokaryotes might have evolved from a common process in an ancestral cell. Other bacterial homologs of eukaryotic structural proteins suggest that bacteria possess a kind of cytoskeleton beneath the cell wall (**Special Topic 3.2**).

What signals the cell as to when and where to form the septum? What keeps the cell from septating at positions other than the equator? In fact, defective gene products can lead cells to divide incorrectly with a septum near the pole of the cell. This incorrect septation generates a "minicell," a small cellular compartment with no DNA. Genes whose defects lead to minicells are called *min* genes. In *E. coli*, the protein encoded by *minD* oscillates in rings from pole to pole (**Fig. 3.36**). The Min proteins somehow regulate formation of the septum, ensuring symmetrical cell division.

Note, however, that normal division of bacteria is not always symmetrical. For example, *Bacillus* species undergo an asymmetrical cell division to form an endospore. In this process, one daughter nucleoid forms an inert endospore capable of remaining dormant but viable for many years (discussed in Chapter 4). Other bacteria expand their cells by polar extension; an example is *Corynebacterium diphtheriae*, the cause of diphtheria. Polar extension occurs at variable rates and direction, and thus generates irregularly shaped rods.

TO SUMMARIZE:

- **The nucleoid** region contains loops of DNA, supercoiled and bound to DNA-binding proteins.
- **DNA is transcribed** in the cytoplasm, often at the same time that it is being replicated.

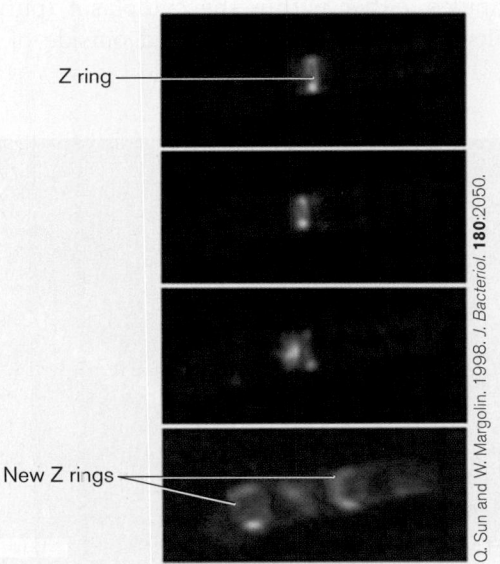

Z ring

New Z rings

O. Sun and W. Margolin. 1998. *J. Bacteriol.* **180**:2050.

Figure 3.35   Septation and the Z ring. Fluorescence microscopy of *E. coli* (1–2 μm) based on FtsZ-GFP, a genetic fusion of FtsZ with green fluorescent protein (GFP). The Z ring of FtsZ subunits forms around the equator of the constricting cell, as the septum grows inward.

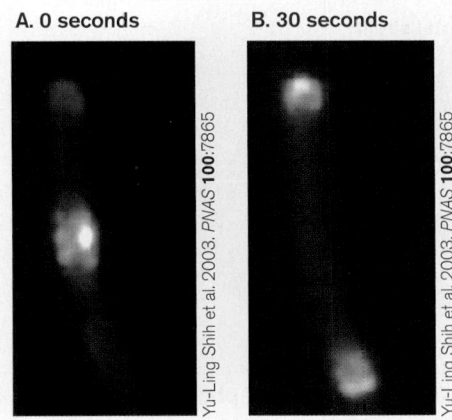

A. 0 seconds

B. 30 seconds

Yu-Ling Shih et al. 2003. *PNAS* **100**:7865

Yu-Ling Shih et al. 2003. *PNAS* **100**:7865

Figure 3.36   MinD protein rings regulate cell division. Fluorescence shows MinD-YFP protein rings in an *E. coli* cell completing division. The rings oscillate from pole to pole.

- **The ribosome translates RNA to make proteins**, which are folded by chaperones and in some cases secreted at the cell membrane.
- **DNA is replicated** bidirectionally by the replisome.
- **Cell expansion and septation** are coordinated with DNA replication.
- **Cell shape** is determined by the FtsZ "Z ring" and other cytoskeletal proteins.

3.7 Specialized Structures

We have introduced the major structures required to contain the cell, organize its contents, maintain its DNA, and express its genes. The cell envelope, the nucleoid, and the gene expression and protein translocation complexes are essential for all living prokaryotes. In addition to these fundamental structures, different species have evolved different kinds of specialized devices adapted to diverse metabolic strategies and environments.

Thylakoids and Carboxysomes

Photosynthetic bacteria, also called phototrophs, need to collect as much light as possible to drive photosynthesis. Light is harvested by protein complexes containing chlorophylls (see Section 14.6). Some of the energy obtained is stored in the form of storage granules.

To maximize their photosynthetic membranes, phototrophs have evolved specialized systems of extensively folded intracellular membrane called **thylakoids** (**Fig. 3.37A**). Thylakoids consist of layers of folded sheets (lamellae) or tubes of membranes packed with photosynthetic proteins and electron carriers. Cyanobacteria

containing thylakoids structurally resemble eukaryotic chloroplasts, which are believed to have evolved from a common ancestor of modern cyanobacteria.

The thylakoids conduct only the "light reaction" of photon absorption and energy storage. The energy obtained is rapidly spent to fix carbon dioxide—a process that occurs within **carboxysomes**. Carboxysomes are polyhedral, protein-covered bodies packed with the enzyme Rubisco for CO_2 fixation.

Aquatic and marine phototrophs, as well as archaea, often possess **gas vesicles** to increase buoyancy and keep themselves high in the water column, near the sunlight (**Fig. 3.37B**). Gas vesicles are specialized vacuoles composed of specific proteins. The vesicles trap and collect gases such as hydrogen or carbon dioxide produced by the cell's metabolism.

Storage Granules

During times of starvation, phototrophs may digest their phycobilisomes for energy and as a source of nitrogen. Alternatively, energy is stored in storage granules composed of glycogen or other polymers, such as polyhydroxybutyrate (PHB) and poly-3-hydroxyalkanoate (PHA). PHB and PHA polymers are of interest as biodegradable plastics, which bacteria are engineered to produce industrially. Similar storage granules are also produced by nonphototrophic soil bacteria.

Another type of storage device is sulfur—granules of elemental sulfur produced by purple and green phototrophs through photolysis of hydrogen sulfide (H_2S). Instead of disposing of the sulfur, the bacteria store it in granules, either within the cytoplasm (purple phototrophs) or as "globules" attached outside of the cell.

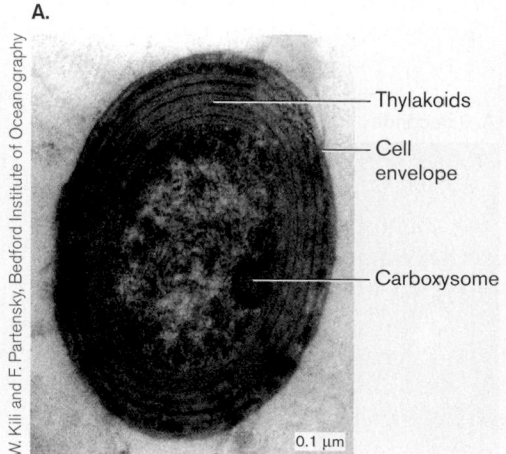
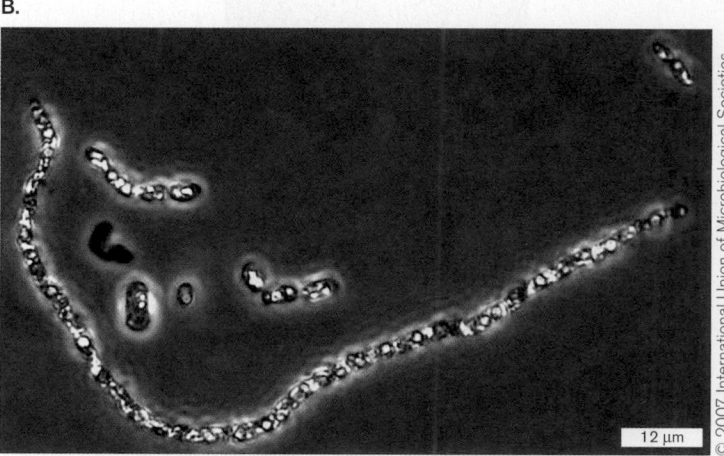

Figure 3.37 Organelles of phototrophs. A. The marine phototroph *Prochlorococcus marinus* (TEM). Beneath the envelope lie the photosynthetic double membranes called thylakoids. Carboxysomes are polyhedral, protein-covered bodies packed with the Rubisco enzyme for CO_2 fixation. **B.** Filaments (chains of cells) of the Gram-negative phototroph *Halochromatium roseum*. Gas vesicles provide buoyancy, enabling the phototroph to remain at the surface of the water, exposed to light. *Source* for B: P. Anil Kumar et al. 2007. *Int. J. Syst. Evol. Microbial.* **57**:2110–2113.

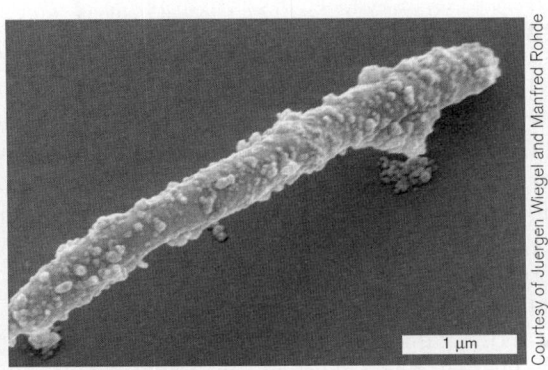

Courtesy of Juergen Wiegel and Manfred Rohde

1 μm

Figure 3.38   External sulfur particles.   Sulfur globules dot the surface of *Thermoanaerobacter sulfurigignens*, an anaerobic thermophilic bacterium that gains energy by reducing thiosulfate ($S_2O_3{}^{2-}$) to elemental sulfur (S^0).

Sulfur-reducing bacteria also make extracellular sulfur globules (**Fig. 3.38**). The sulfur may be usable as an oxidant when reduced substrates become available. Alternatively, the presence of potentially toxic sulfur granules may help cells avoid predation.

Magnetosomes

An unusual structure possessed by magnetotactic species of bacteria (bacteria showing magnetically directed motility) is the magnetosome. **Magnetosomes** are microscopic membrane-embedded crystals of the magnetic mineral magnetite, Fe_3O_4 (**Fig. 3.39**). They are found in anaerobic pond-dwelling organisms such as *Magnetospirillum gryphiswaldense*. The crystals generate a magnetic dipole moment along the length of a bacterium, constraining it to swim along a magnetic field. This magnetic orientation of swimming is called **magnetotaxis**, the ability to sense

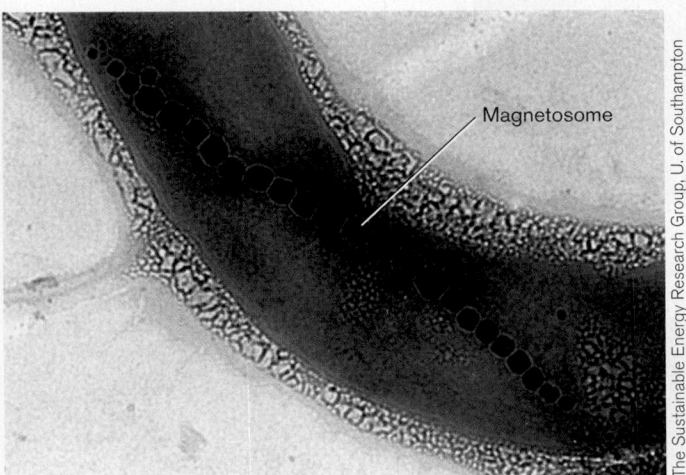

Magnetosome

The Sustainable Energy Research Group, U. of Southampton

Figure 3.39   Magnetotactic bacterium.   A magnetospirillum is a spiral-shaped magnetotactic bacterium. The spiral cell contains a chain of magnetosomes, of 20–120 nm in size (TEM).

and respond to magnetism. Magnetotactic bacteria can be collected by placing a magnet in a jar of pond water; bacteria orienting by the field lines collect nearby.

The natural function of magnetosomes appears to be to orient bacterial swimming toward the bottom of the pond. Magnetotactic organisms are anaerobes, which prefer the lower part of the water column, where oxygen concentration is lowest. Because Earth's magnetic field lines point downward in the northern latitudes, bacteria that are magnetotactic swim "downward" toward magnetic north.

> **THOUGHT QUESTION 3.10**   How would a magnetotactic species have to behave if it were in the Southern Hemisphere instead of the Northern Hemisphere?

Magnetotactic bacteria are being studied for their potential applications in wastewater treatment. Through their anaerobic metabolism, some magnetotactic bacteria accumulate high concentrations of toxic metals from the water. They and the toxic metals they scavenge can then be removed by application of a magnetic field to attract and concentrate the bacteria.

Pili and Stalks

In a favorable habitat, such as a running stream full of fresh nutrients or the epithelial surface of a host, it is advantageous for a cell to adhere to a substrate. **Adherence**, the ability to attach to a substrate, requires specific adherence structures, such as pili (protein filaments). As bacteria grow and proliferate, however, they face the question of whether to stay where they are or leave their present habitat, where nutrients may be depleted and waste products increased. In rapidly changing environments, cell survival requires **motility**, the ability to move and relocate. Motility requires structures such as rotary flagella.

For attachment, the most common bacterial structures are **pili** (singular, **pilus**), or **fimbriae** (singular, **fimbria**), straight filaments of protein monomers called **pilin**. For example, fimbriae attach the oral pathogen *Porphyromonas gingivalis* to gum epithelium, where it causes periodontal disease (**Fig. 3.40**). A different kind of pili, the **sex pili**, attach a "male" donor cell to a "female" recipient cell for transfer of DNA. This process of DNA transfer is called conjugation. The genetic consequences of conjugation are discussed in Chapter 9.

Another kind of attachment organelle is a membrane-embedded extension of the cytoplasm called a **stalk**. The tip of the stalk secretes adhesion factors called **holdfasts**, which firmly attach the bacterium in an environment that has proved favorable. The mechanism of

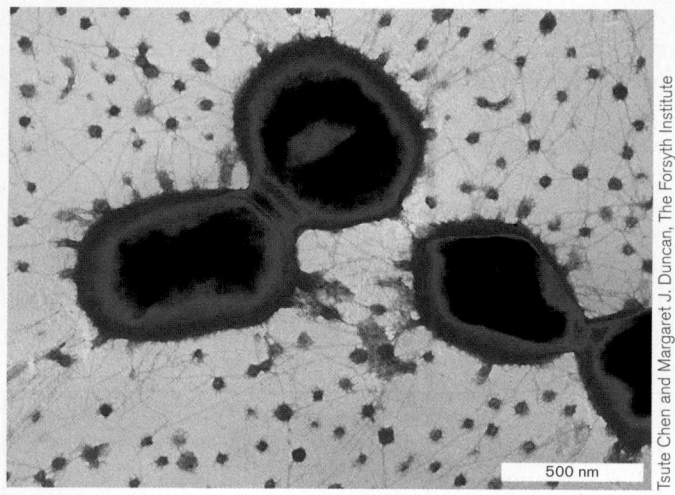

Tsute Chen and Margaret J. Duncan, The Forsyth Institute

500 nm

Figure 3.40   Pili: protein filaments for attachment.
Porphyromonas gingivalis, a causative agent of gum disease or
gingivitis. The *P. gingivalis* cells show fimbriae along with vesicles
budding from the cell's outer membrane. (TEM)

stalk and holdfast attachment has been extensively studied in iron-oxidizing bacteria that interfere with mining
operations by producing massive biofilms. An example is
Gallionella ferruginea, an iron-oxidizing species that grows
a long stalk (**Fig. 3.41**). The long, twisted stalks of adherent *Gallionella* cells become coated by iron hydroxides,
and *Gallionella* contributes a major part of the process of
biomineralization (biological crystallization of minerals)
in iron mines.

Rotary Flagella

Bacteria and archaea that are motile generally swim by
means of rotary **flagella** (singular, **flagellum**). Flagella
are helical propellers that drive the cell forward like the
motor of a boat. Shown originally by Howard Berg at the
California Institute of Technology, the bacterial flagellar
motor was the first rotary device to be discovered in a living organism.

Different bacterial species have different numbers and arrangements of flagella. Peritrichous cells,
such as *E. coli* and *Salmonella* species, have flagella randomly distributed around the cell (**Fig. 3.42A**). The
flagella rotate together in a bundle behind the swimming
cell (**Fig. 3.42B**). Lophotrichous cells, such as *Rhodospirillum rubrum*, have flagella attached at one or both ends. In
monotrichous (polar) species, such as *Pseudomonas aeruginosa*, the cell has a single flagellum at one end.

Each flagellum is a spiral filament of protein monomers called flagellin. The filament is actually rotated by
means of a motor driven by the cell's transmembrane
proton current, the same proton potential that drives the
membrane-embedded ATP synthase. (Alternatively, the
motor is driven by a sodium ion potential, particularly
in marine bacteria such as *Vibrio cholerae*.) The flagellar
motor is embedded in the layers of the cell envelope (**Fig.
3.43**). The motor is observed by electron microscopy, and
its parts are defined by genetic analysis of mutants with
aberrant motility. The motor actually possesses an axle
and rotary parts, all composed of specific proteins. Much
of the structure and function of the motor was elucidated

A.  **B.**

Eleanora I. Robbins, U.S. Geological
Survey

Eleanora I. Robbins, U.S. Geological
Survey

**Figure 3.41   *Gallionella
ferruginea*: iron-oxidizing, stalked
bacteria.   A.** The oval cell of
Gallionella ferruginea generates a
long, twisted stalk about 2 μm wide,
encrusted with iron oxides. **B.** Karen L.
Prestegaard, of the University of
Maryland, studies iron-oxidizing
bacteria attached to iron surfaces,
such as this rod in a stream, where the
bacteria promote rusting, coating the
surface orange.

A. 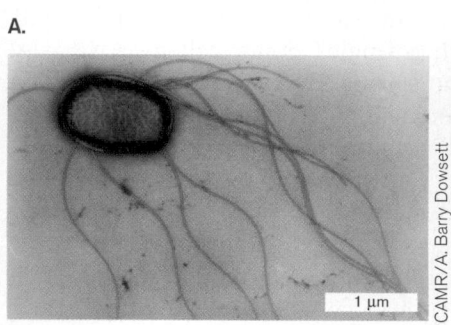 **B.**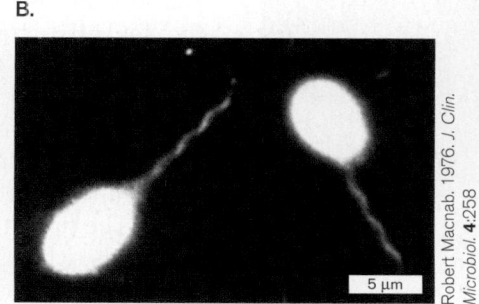

CAMR/A. Barry Dowsett

Robert Macnab. 1976. *J. Clin.
Microbiol.* **4**:258

1 μm

5 μm

**Figure 3.42   Flagellated *Salmonella*
bacteria.   A.** *Salmonella enterica*
has multiple flagella (colorized TEM).
B. The flagella collect in a bundle
behind a swimming cell. Under dark-
field microscopy, the cell body appears
overexposed, about five times
as large as the actual cell.

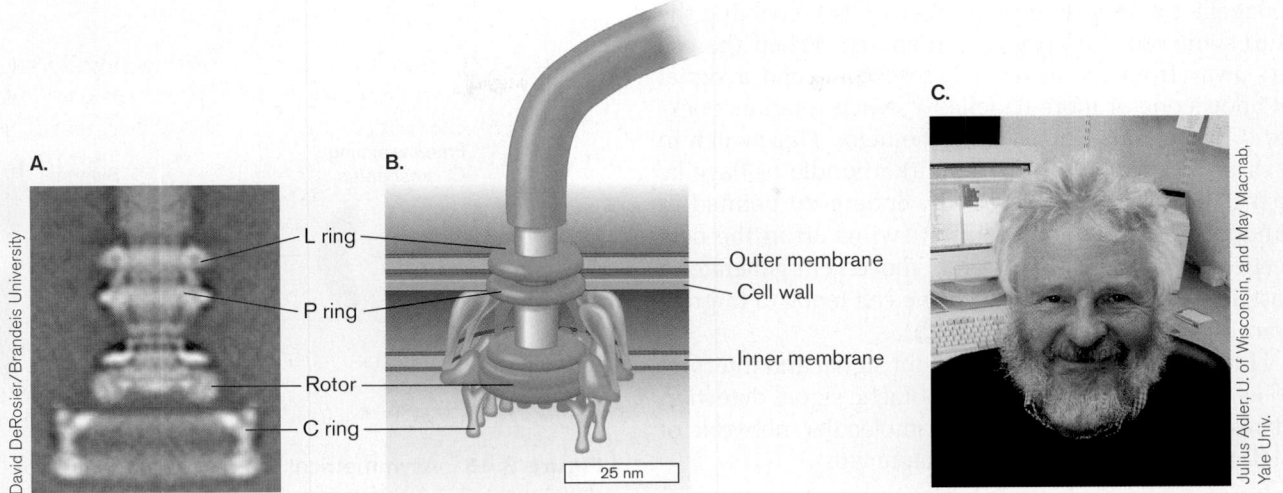

Figure 3.43   The flagellar motor.   A. The basal body, or motor, of the bacterial flagellum (TEM). The image is based on digital reconstruction, in which electron micrographs of purified hook-basal bodies were rotationally averaged. The rings (L, P, and C) correspond to those labeled in the diagram. **B.** Diagram of the flagellar motor including major protein components. **C.** Robert Macnab (Yale University) identified many components of the motor and chemotaxis signaling.

by Scottish microbiologist Robert Macnab (1940–2003) at Yale University.

Flagellar motility benefits the cell by causing the dispersal of progeny, decreasing competition. In addition, most flagellated cells have an elaborate sensory system that enables them to swim toward favorable environments (attractant signals, such as nutrients) and away from inferior environments (repellent signals, such as waste products). This sensory system is known as **chemotaxis**.

Chemotaxis requires a way for the cell to move toward attractants and away from repellents. This movement is accomplished by flagellar rotation either clockwise or counterclockwise relative to the cell (**Fig. 3.44A**⬤). When a cell is swimming toward an attractant chemical,

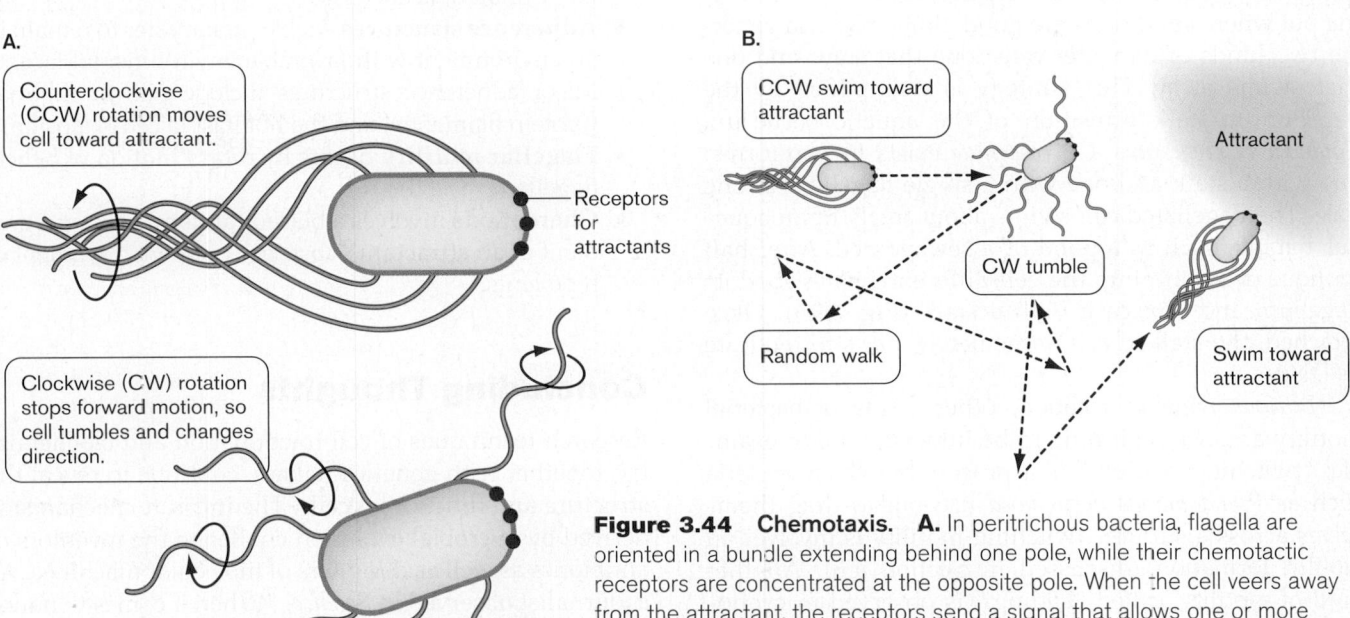

Figure 3.44   Chemotaxis.   A. In peritrichous bacteria, flagella are oriented in a bundle extending behind one pole, while their chemotactic receptors are concentrated at the opposite pole. When the cell veers away from the attractant, the receptors send a signal that allows one or more flagella to switch rotation from counterclockwise (CCW) to clockwise (CW). This switched rotation disrupts the bundle of flagella, causing the cell to tumble briefly before it swims off in a new direction. **B.** The pattern of movement resulting from alternating swimming and tumbling is a "biased random walk" in which the cell sometimes moves randomly but overall tends to migrate toward the attractant. ⬤ *microbiology2.com/animations*

the flagella rotate counterclockwise (CCW), enabling the cell to swim smoothly for a long stretch. When the cell veers away from the attractant, receptors send a signal that allows one or more flagella to switch rotation clockwise (CW), against the twist of the helix. This switch in the direction of rotation disrupts the bundle of flagella, causing the cell to tumble briefly, ending up pointed in a random direction. The cell then swims off in the new direction. The resulting pattern of movement generates a "biased random walk" in which the cell tends to migrate toward the attractant (**Fig. 3.44B⬤**).

How do cells detect an attractant signal and interpret it so as to swim or tumble? Chemotactic signal detection and response require a complex molecular network of proteins, which is discussed in Chapter 10.

> **THOUGHT QUESTION 3.11**   Most laboratory strains of *E. coli* and *Salmonella* commonly used for genetic research lack flagella. Why do you think this is the case? How can a researcher maintain a motile strain?

Note that bacterial flagella differ completely from the whiplike flagella and cilia of eukaryotes. Eukaryotic flagella are much larger structures containing multiple microtubules enclosed by a membrane. They move with a whiplike motion that forms a flat sine wave, propagated by ATP hydrolysis all along the flagellum.

In natural environments, swimming helps a microbe seek better conditions, but there are advantages to staying put where conditions are good. Some bacteria generate two kinds of daughter cells: one that stays and one that swims away. This strategy is exemplified by the flagellum-to-stalk transition of the aquatic bacterium *Caulobacter crescentus*. *C. crescentus* exists in two forms: one with a stalk and one with a single flagellum at one pole. The flagellated cell swims about freely in an aqueous habitat, such as a pond or a sewage bed. After half an hour of swimming, the cell differentiates to shed its flagellum and replace it with a stalk (**Fig. 3.45**). Once attached, the stalked cell immediately starts to replicate its DNA.

Besides flagellar rotation, other forms of bacterial motility are just beginning to be understood. For example, "twitching motility" is a process by which bacteria such as *Pseudomonas aeruginosa* use pili to drag themselves across a surface. Twitching motility is involved in biofilm formation (discussed in Section 4.6). Another kind of motility, called "gliding," is observed in cyanobacteria and in myxobacteria. The structure and mechanism of gliding are unknown.

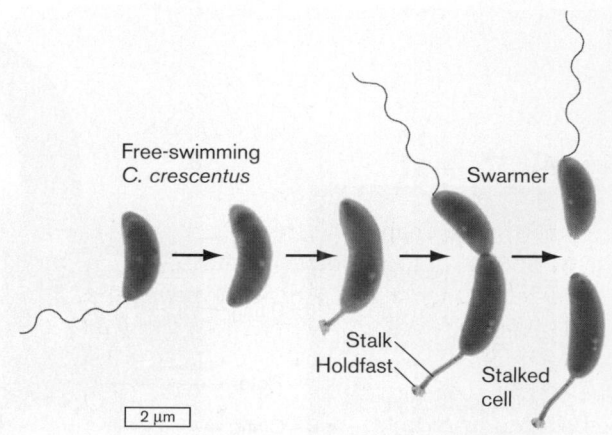

Figure 3.45   Asymmetrical cell division: a model for development.   A swarmer cell of *Caulobacter crescentus* loses its flagellum and grows a stalk. The stalked cell divides to produce a swarmer cell. *Source:* Photo courtesy of Yves Brun, Indiana University.

TO SUMMARIZE:

- **Phototrophs** possess thylakoid membrane organelles packed with photosynthetic apparatus and carboxysomes for carbon dioxide fixation. Other subcellular structures may include sulfur granules from H_2S photolysis and gas vesicles for buoyancy in the water column.
- **Storage granules** store polymers for energy.
- **Magnetosomes** orient the swimming of magnetotactic anaerobic bacteria.
- **Adherence structures** enable prokaryotes to remain in an environment with favorable environmental factors. Major adherence structures include pili, or fimbriae (protein filaments), and the holdfast (a cell extension).
- **Flagellar motility** occurs by rotary motion of helical flagella.
- **Chemotaxis** involves a biased random walk up a gradient of an attractant substance or down a gradient of repellents.

Concluding Thoughts

Research techniques of cell fractionation and biochemistry, together with genetic analysis, continue to reveal the structure and function of cells. The intricate mechanisms derived by microbial evolution challenge the inventors of antibiotics as well as designers of molecular machines. As a journalist observed in *Science*, "When it comes to nanotechnology, physicists, chemists, and materials scientists can't hold a candle to the simplest bacteria."

CHAPTER REVIEW

Review Questions

1. What are the major features of a bacterial cell, and how do they fit together for cell function as a whole?
2. What fundamental traits do most prokaryotes have in common with eukaryotic microbes? What traits are different?
3. Give examples of how our views of ribosome structure and function have emerged from microscopy, cell fractionation, X-ray crystallography, and genetic analysis. Explain the advantages and limitations of each technique.
4. Outline the structure of the peptidoglycan sacculus, and explain how it expands during growth. Cite two different kinds of experimental data that support our current views of the sacculus.
5. Compare and contrast the structure of Gram-positive and Gram-negative cell envelopes. Explain the strengths and weaknesses of each kind of envelope.
6. Outline the process of DNA replication, and explain how it is coordinated with cell wall septation.
7. Explain how DNA transcription to RNA is integrated with translation and protein processing and secretion.
8. What kinds of subcellular structures are found in certain cells with different functions, such as magnetotaxis or photosynthesis?
9. Compare and contrast bacterial structures for attachment and motility. Explain the molecular basis of chemotaxis.

Thought Questions

1. The aquatic bacterium *Caulobacter crescentus* alternates between two cell forms: a cell with a flagellum that swims, and a stalked cell that adheres to particulate matter. The flagellar cell can discard its flagellum to grow a stalk and adhere, and then the stalked cell divides to give one stalked cell and one flagellated cell. What would be the adaptive advantage of this alternating morphology?
2. Suppose that one cell out of a million has a mutant gene blocking S-layer synthesis, and suppose that the mutant strain can grow twice as fast as the S-layered parent. How many generations would it take for the mutant strain to constitute 90% of the population?
3. Explain two different ways that an aquatic phototroph might use its subcellular structures to maximize its access to light. Explain how an aerobe (an organism requiring molecular oxygen for growth) might remain close to the surface, with access to air.
4. How do pathogenic bacteria avoid engulfment by phagocytes of the human bloodstream? How do you think various aspects of the cell structure can prevent phagocytosis?

Key Terms

active transport (84)
adherence (107)
capsule (91)
carboxysome (106)
cardiolipin (85)
cell membrane (75)
cell wall (76)
chemotaxis (109)
cholesterol (86)
contractile vacuole (96)
core polysaccharide (94)
cross-bridge (89)
diphosphatidylglycerol (85)

electrophoresis (77)
endotoxin (94)
envelope (76)
fimbria (107)
flagellum (76, 108)
gas vesicle (106)
genetic analysis (79, 81)
glucosamine (94)
glycan (88)
holdfast (107)
hopanoid (hopane) (86)
inner membrane (93)
ion gradient (83)

isoelectric focusing (77)
leaflet (82)
lipopolysaccharide (LPS) (76, 94)
lysate (80)
magnetosome (107)
magnetotaxis (107)
membrane-permeant weak acid (83)
membrane-permeant weak base (83)
motility (107)
murein (88)
murein lipoprotein (93)
nucleoid (76)
O polysaccharide (94)

Recommended Reading

Ausmees, Nora, Jeffrey R. Kuhn, and Christine Jacobs-Wagner. 2003. The bacterial cytoskeleton: An intermediate filament-like function in cell shape. *Cell* **115**:705–713.

Begic, Sanela, and Elizabeth A. Worobec. 2006. Regulation of *Serratia marcescens ompF* and *ompC* porin genes in response to osmotic stress, salicylate, temperature and pH. *Microbiology* **152**:485–491.

Feucht, Andrea, and Jeff Errington. 2005. *ftsZ* mutations affecting cell division frequency, placement and morphology in *Bacillus subtilis*. *Microbiology* **151**:2053–2064.

Gitai, Zemer, Natalie Dye, and Lucy Shapiro. 2004. An actin-like gene can determine cell polarity in bacteria. *Proceedings of the National Academy of Sciences USA* **101**:8643–8648.

Komeili, Arash, Zhuo Li, Dianne K. Newman, and Grant J. Jensen. 2006. Magnetosomes are cell membrane invaginations organized by the actin-like protein MamK. *Science* **311**:242–245.

Lemon, Katherine P., and Alan D. Grossman. 1998. Localization of bacterial DNA: Evidence for a factory model of replication. *Science* **282**:1516–1519.

Margolin, William. 2008. What does it take to divide a bacterial cell? *Microbe* **3**:329–336.

Matias, Valério R. F., and Terry J. Beveridge. 2007. Cryo–electron microscopy of cell division in *Staphylococcus aureus* reveals a mid-zone between nascent cross walls. *Molecular Microbiology* **64**:195–206.

Nestorovich, Ekaterina M., Christophe Danelon, Mathias Winterhalter, and Sergey M. Bezrukov. 2002. Designed to penetrate: Time-resolved interaction of single antibiotic molecules with bacterial pores. *Proceedings of the National Academy of Sciences USA* **99**:9789–9794.

Nilsen, Trine, Arthur W. Yan, Gregory Gale, and Marcia B. Goldberg. 2005. Presence of multiple sites containing polar material in spherical *Escherichia coli* cells that lack MreB. *Journal of Bacteriology* **187**:6187–6196.

Noji, Hiroyuki, Ryohei Yasuda, Masasuke Yoshida, and Kazuhiko Kinoshita, Jr. 1997. F_1F_0 Direct observation of the rotation of F_1-ATPase. *Nature* **386**:299–302.

Pagès, Jean-Marie M., Chloë E. James, and Mathias Winterhalter. 2008. The porin and the permeating antibiotic: A selective diffusion barrier in Gram-negative bacteria. *Nature Reviews. Microbiology* **6**:893–903.

Parkinson, John S., Peter Ames, and Claudia A. Studdert. 2005. Collaborative signaling by bacterial chemoreceptors. *Current Opinion in Microbiology* **8**:116–121.

Ruiz, Natividad, Daniel Kahne, and Thomas J. Silhavy. 2006. Advances in understanding bacterial outer-membrane biogenesis. *Nature Reviews. Microbiology* **4**:57–66.

Saier, Milton H., Jr. 2008. Structure and evolution of prokaryotic cell envelopes. *Microbe* **3**:323–328.

Schäffer, Christina, and Paul Messner. 2005. The structure of secondary cell wall polymers: How Gram-positive bacteria stick their cell walls together. *Microbiology* **151**:643–651.

Shih, Yu-Ling, Trung Le, and Lawrence Rothfield. 2003. Division site selection in *Escherichia coli* involves dynamic redistribution of Min proteins within coiled structures that extend between the two cell poles. *Proceedings of the National Academy of Sciences USA* **100**:7865–7870.

Sutcliffe, Joyce A. 2005. Improving on nature: Antibiotics that target the ribosome. *Current Opinion in Microbiology* **8**:534–542.

 For further study and review, please visit StudySpace at **microbiology2.com**.

Chapter 4

Bacterial Culture, Growth, and Development

The adage "To eat well is to live well" is as true for microbes as it is for humans. Microorganisms constantly struggle to survive in natural habitats because they compete for food. Yet a single microbial pathogen can multiply within the course of a day to cause deadly illness, and a few algae can abruptly bloom and cover the entire surface of a lake. In each case, the population explosion results from the sudden availability of food. Over eons, bacteria have evolved ingenious strategies to find, acquire, and metabolize a wide assortment of potential food sources, ranging from glucose to mothballs. This metabolic diversity arose from the need to find new sources of food ignored by competitors.

The remarkable plasticity of microbial genomes enables adaptation to new food sources. During the course of DNA replication, gene systems involved in the use of one food naturally undergo duplications and mutations, some of which sculpt new biochemical pathways capable of metabolizing novel substrates. This genomic flexibility raises hopes that we can engineer microbial biochemistry to remediate pollution and produce tomorrow's wonder drugs.

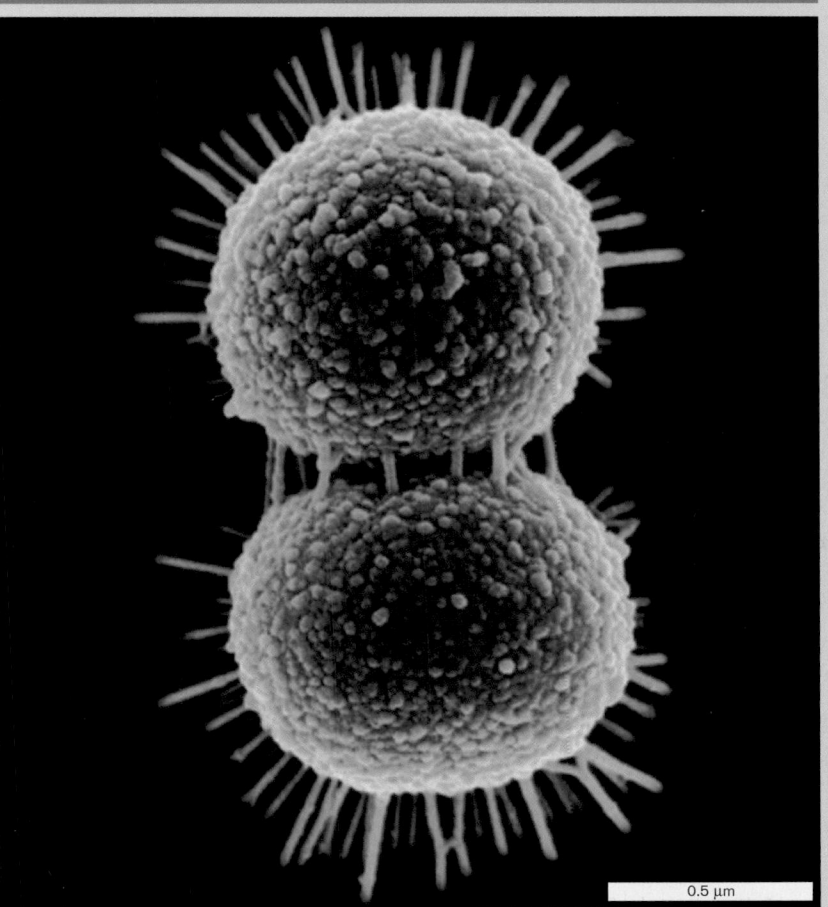

0.5 μm

Current Research Highlight

Culturing the "unculturable." MSC33 is a typical marine "unculturable" microorganism that refuses to grow on standard laboratory media (SEM). It was recently discovered that adding a short peptide will stimulate growth of this organism in the laboratory. The peptide, which is produced by a companion microbe found in the natural environment, does not fulfill a nutritional need of MSC33 but has an apparent signaling function that somehow induces cell division. William Fowle, Northeastern University Source: D. Nichols et al. 2008. *Appl. Environ. Microbiol.* **74**:4889–4897.

Microorganisms need sources of carbon in order to grow. Understanding how they use food to increase their cell mass and, ultimately, cell number enables us to control their growth and manipulate them to make useful products. Yeast, for example, consume glucose and break it down to ethanol and carbon dioxide gas. These end products are merely waste to the yeast but are extremely important to humans who enjoy beer. Brewers have learned to control the amount of sugar supplied to yeast growing in fermentation vats to produce just the right amount of alcohol and CO_2, which causes beer's bubbling carbonation. Many a home brewer has learned the hard way that providing too much sugar (in malt) will cause yeast to make too much CO_2 gas, turning a homemade fermentor into an unpredictable explosive device.

Learning how bacteria grow has also provided a window through which to view the core processes of life. As a result of studying bacterial growth and nutrition, we now know how DNA replicates, how RNA is made, and how proteins are assembled. We have also learned that the availability of nutrients has influenced the evolution of sophisticated microbial processes designed to avoid or survive starvation. For example, bacteria communicate with each other to build elaborate multicellular reproductive structures, such as fruiting bodies, or complex multispecies biofilms, such as those that erode our teeth and rust the surface of ocean liners.

In Chapter 4, we consider the microbial biosphere and the basics of bacterial growth. We'll discuss a variety of questions: How diverse is this biosphere? Is it true that we have identified only about 0.1% of the bacterial species that inhabit Earth? Where do bacteria get their energy? You might be surprised to learn that some bacteria gain energy from light (photosynthesis), whereas others gain energy from oxidizing sulfur. With so many microbes consuming nutrients, how has nature arranged biosystems to avoid depleting Earth of key compounds?

To understand the biochemistry behind metabolic diversity, we need to grow microbes. How do we ensure their growth and measure it? Some bacteria make elaborate, multicellular structures, whereas others form environmentally resistant fortresses called spores. In this chapter, we discuss how microbes obtain energy and nutrients for growth, and how cell populations develop. A more detailed treatment of mechanisms of energy gain and biosynthesis is presented in Chapters 13–15.

4.1 Microbial Nutrition

Bacterial cells, for all their apparent simplicity, are remarkably complex and efficient replication machines. One cell of *Escherichia coli*, for example, can divide into two cells every 20–30 minutes. At a rate of 30 minutes per division, one cell could potentially multiply to over 1×10^{14} cells in 24 hours—that is, 100 trillion organisms! Although 100 trillion cells would weigh only about 1 gram, after another 24 hours (a total of two days) of replicating every 30 minutes, the mass of cells would explode to 10^{14} grams, or 10^7 tons. So why are we not buried under mountains of *E. coli*?

Nutrient Supplies Limit Microbial Growth

One factor limiting growth is limited supplies of nutrients. Microbes commonly encounter environments where essential nutrients are limiting. **Essential nutrients** are those compounds a microbe cannot make itself but must gather from its environment if the cell is to grow and divide. Microbial cells obtain all the essential nutrients for growth from their immediate environment. Consequently, when an environment becomes depleted of one or more essential nutrients, the microorganism will stop growing. How organisms cope with these periods of starvation until nutrients are restored is a rapidly developing field of microbiology.

Table 4.1 **Growth factors and natural habitats of organisms associated with disease.**

Organism	Diseases	Natural habitats	Growth factors
Shigella	Bloody diarrhea	Humans	Nicotinamide (NAD)[a]
Haemophilus	Meningitis, chancroid	Humans and other animal species, upper respiratory tract	Hemin, NAD
Staphylococcus	Boils, osteomyelitis	Widespread	Complex requirement
Abiotrophia	Osteomyelitis	Humans and other animal species	Vitamin K, cysteine
Legionella	Legionnaires' disease	Soil, refrigeration cooling towers	Cysteine
Bordetella	Whooping cough	Humans and other animal species	Glutamate, proline, cystine
Francisella	Tularemia	Wild deer, rabbits	Complex, cysteine
Mycobacterium	Tuberculosis, leprosy	Humans	Nicotinic acid (NAD),[a] alanine
Streptococcus pyogenes	Pharyngitis, rheumatic fever	Humans	Glutamate, alanine

[a]Both nicotinamide and nicotinic acid are derived from NAD, nicotinamide adenine dinucleotide.

All microorganisms require a minimum set of **macronutrients**, nutrients needed in large quantities. Six of these—carbon, nitrogen, phosphorus, hydrogen, oxygen, and sulfur—make up the carbohydrates, lipids, nucleic acids, and proteins of the cell. Four other macronutrients are cations whose roles range from serving as enzyme cofactors (Mg^{2+}, Fe^{2+}, and K^+) to acting as regulatory signal molecules (Ca^{2+}). In addition to macronutrients, all cells require very small amounts of certain trace elements, called **micronutrients**. These include cobalt, copper, manganese, molybdenum, nickel, and zinc, which are ubiquitous contaminants on glassware and in water. As a result, these trace elements are not added to laboratory media unless heroic measures have been taken to first remove the elements from the medium.

Micronutrients are required by cells because these elements are frequently essential components of enzymes or can themselves be components of cofactors. **Cofactors** are small molecules that fit into specific enzymes and aid in the catalytic process. Cobalt, for example, is part of the cofactor vitamin B_{12}. Some organisms, such as the laboratory "workhorse" bacterium *E. coli*, make all their proteins, nucleic acids, and cell wall and membrane

components from this very simple blend of chemical elements and compounds. For many other microbes, this basic set of nutrients is not enough. One example is *Borrelia burgdorferi*, the causative agent of Lyme disease, which requires an extensive mixture of complex organic supplements to grow.

Microbes Evolved to Grow in Different Environments

Based in part on the ecological niche it inhabits, an organism may have evolved to require additional **growth factors** (**Table 4.1**), specific nutrients that are not required by all cells. For example, why should an organism like *Streptococcus pyogenes* make glutamic acid or alanine if those amino acids are readily available in its normal environment (such as the human oral cavity)? Because it never needs to make these compounds, *S. pyogenes*, as a matter of efficiency, has lost the genes whose protein products synthesize glutamic acid and alanine. A **defined minimal medium** contains only those compounds needed for an organism to grow (**Table 4.2**). In the case of *S. pyogenes*, this medium would include glutamic acid

Table 4.2	**Composition of commonly used media.**		
Medium	**Ingredients per liter**		**Organisms cultured**
Luria Bertani (complex)	Bacto tryptone[a]	10 g	Many Gram-negative and Gram-positive organisms
	Bacto yeast extract	5 g	
	NaCl	10 g	
	pH 7		
M9 medium (defined)	Glucose	2.0 g	Gram-negative organisms such as *E. coli*
	Na_2HPO_4	6.0 g (42 mM)	
	KH_2PO_4	3.0 g (22 mM)	
	NH_4Cl	1.0 g (19 mM)	
	NaCl	0.5 g (9 mM)	
	$MgCl_2$	2.0 mM	
	$CaCl_2$	0.1 mM	
	pH 7		
Azotobacter medium (defined)	Mannitol	2.0 g	*Azotobacter*
	K_2HPO_4	0.5 g	
	$MgSO_4 \cdot 7H_2O$	0.2 g	
	$FeSO_4 \cdot 7H_2O$	0.1 g	
Sulfur oxidizers (defined)	NH_4Cl	0.52 g	*Thiobacillus thiooxidans*
	KH_2PO_4	0.28 g	
	$MgSO_4 \cdot 7H_2O$	0.25 g	
	$CaCl_2$	0.07 g	
	Elemental sulfur	1.56 g	
	CO_2	5%	
	pH 3		

[a]Bacto tryptone is a pancreatic digest of casein (bovine milk protein).

A.

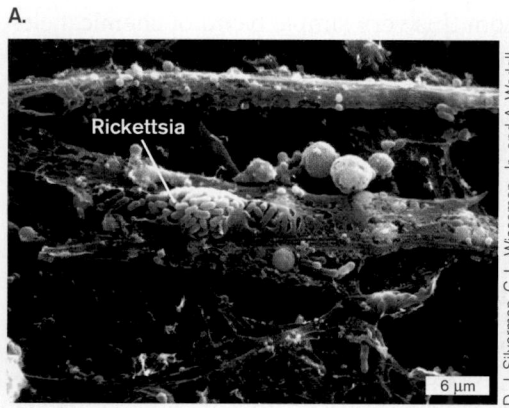

Rickettsia

6 µm

D. J. Silverman, C. L. Wisseman, Jr., and A. Wadell

B.

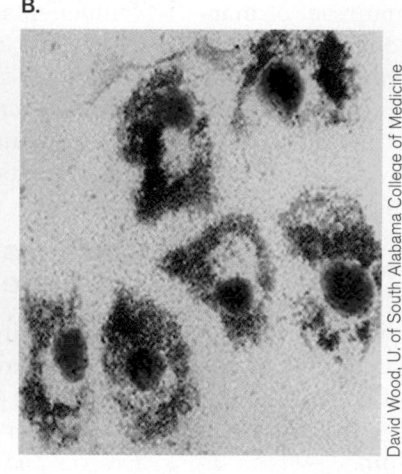

David Wood, U. of South Alabama College of Medicine

Figure 4.1 *Rickettsia prowazekii* **growing within eukaryotic cells.** **A.** *R. prowazekii* growing within the cytoplasm of a chicken embryo fibroblast (SEM). **B.** Giemsa stain of *R. prowazekii* growing within cultured mouse cells. The rickettsia are red; the host cell nucleus is blue.

and alanine in addition to the macro- and micronutrients mentioned earlier. Other organisms have adapted so well to their natural habitat that we still do not know how to grow them in the laboratory. For example, *Rickettsia prowazekii*, the causative agent of epidemic typhus fever, grows only within the cytoplasm of eukaryotic cells (**Fig. 4.1**). This obligate intracellular bacterium lost key pathways needed for independent growth because the host cell supplies them. Despite extensive efforts to grow them outside a host cell (**axenic growth**), cells of *R. prowazekii* have proved uncooperative in this endeavor.

Recently, another supposed obligate intracellular bacterium (*Coxiella burnetii*, the cause of Q fever) was coaxed into dividing axenically. This pathogen normally grows only within acidified phagosome vacuoles. Robert Heinzen and colleagues at Rocky Mountain Laboratories designed an acidified (pH 5) medium containing numerous supplements that would support growth. Finding ways to grow these pathogens axenically will aid studies to identify virulence factors and develop treatments.

Although it will not grow in a defined medium, we can at least grow *Rickettsia* in the laboratory using eggs or animal cell tissue culture. But it is estimated that 99.9% of all the microorganisms on Earth cannot be grown in the laboratory at all. They are considered unculturable. Either these organisms need nutrients we have yet to discover, or they require cooperation with other species in their normal environment (see the chapter opener figure). How do we even know that these microbes exist if we cannot grow them? Evidence for their existence has been revealed only recently, by newly developed tools of molecular biology.

All known microorganisms have a set of genes, associated with ribosomes, whose DNA sequences are highly conserved across the phylogenetic tree. A DNA-amplifying procedure called the polymerase chain reaction (PCR, described in Section 7.6) can be used to screen for the presence of these genes in soil and water samples. Comparing the DNA sequences of the PCR products

with the DNA sequences of similar genes from known organisms reveals that nature harbors many microbes hitherto undiscovered because we cannot grow them in the laboratory. Even though the growth and nutritional requirements of these phantom microbes are unknown, modern genomic techniques can expose their existence. We can gain remarkable insight into the physiology of these "invisible," nonculturable organisms by comparing their gene sequences, mined by PCR, with known gene sequences from culturable organisms that have well-characterized physiologies (discussed in Chapter 8).

Microbes Build Biomass through Autotrophy or Heterotrophy

Maintaining life on this planet is an amazing process. All of Earth's life-forms are based on carbon, which they acquire by delicately choreographed processes that recycle key nutrients. The carbon cycle, a critical part of this process, involves two counterbalancing metabolic groups of organisms: heterotrophs and autotrophs. Their pathways of metabolism are discussed in Chapters 13 and 14.

Heterotrophs (such as *E. coli*) rely on other organisms to form the organic compounds, such as glucose, that they use as carbon sources. During heterotrophic metabolism, organic carbon sources are disassembled to generate energy and then reassembled to make cell constituents such as proteins and carbohydrates. This process converts a large amount of the organic carbon source to CO_2, which is then released to the atmosphere. Thus, left on their own, heterotrophs would deplete the world of organic carbon sources (converting them to unusable CO_2) and starve to death. For life to continue, CO_2 must be recycled.

Autotrophs assimilate CO_2 as a carbon source, reducing it (adding hydrogen atoms; see Chapter 15) to make complex cell constituents made up of C, H, and O (for example, carbohydrates, which have the general formula CH_2O). These organic compounds can later be used as

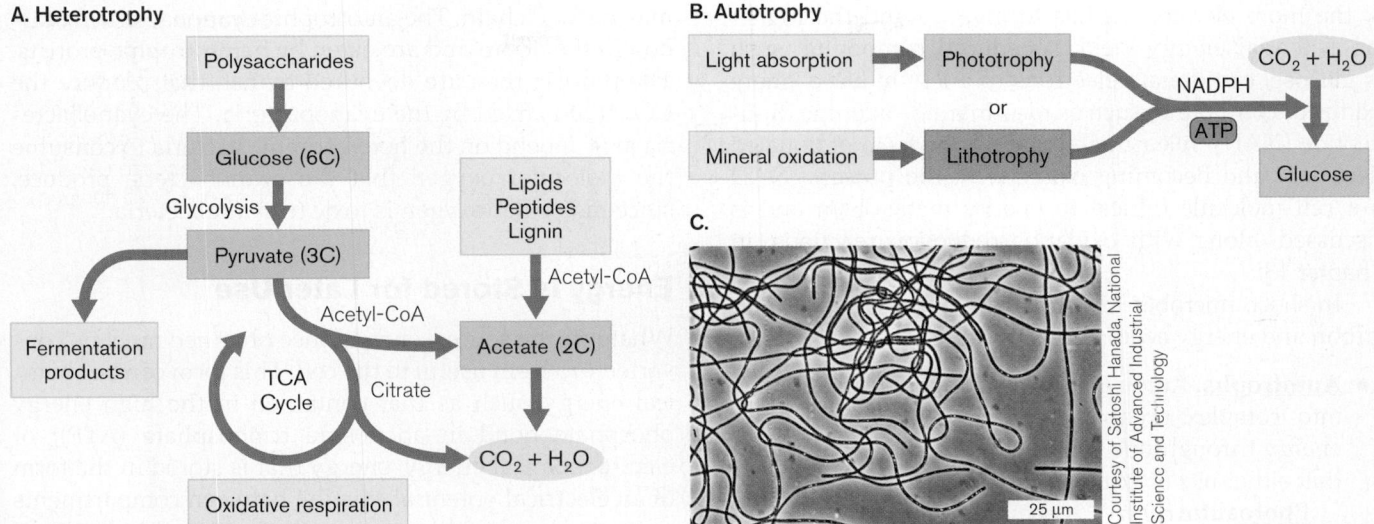

Figure 4.2 The carbon cycle. The carbon cycle requires both autotrophs and heterotrophs. **A.** Heterotrophs gain energy from degrading complex organic compounds, such as polysaccharides, to smaller compounds, such as glucose and pyruvate. The carbon from pyruvate moves through the tricarboxylic acid (TCA) cycle, and CO_2 is released. In the absence of a TCA cycle, the carbon can end up as fermentation products, such as ethanol or acetic acid. **B.** Autotrophs use light energy or energy derived from the oxidation of minerals to capture CO_2 and convert it to complex organic molecules. **C.** *Chloroflexus aggregans*, originally isolated from hot springs in Japan, possesses extraordinary metabolic versatility. It grows anaerobically (without oxygen) as a photoheterotroph and aerobically (with oxygen) as a chemoheterotroph.

carbon sources by heterotrophs. As will be discussed, autotrophs are subclassified by how they obtain energy—into **photoautotrophs**, organisms that use light for photosynthesis; and **chemoautotrophs**, also known as **chemolithotrophs** or **lithotrophs** (literally, "rock-eaters"), organisms that gain energy by oxidizing inorganic substances such as iron or ammonia. Both photoautotrophs and chemoautotrophs carry out the autotrophic process of CO_2 fixation, which forms part of the carbon cycle between autotrophs and heterotrophs (**Fig. 4.2**). You should also realize that some microorganisms can use both heterotrophy and autotrophy to gain energy.

> **THOUGHT QUESTION 4.1** In a mixed ecosystem of autotrophs and heterotrophs, what happens when a heterotroph allows the autotroph to grow and begin to make excess organic carbon?

Microbes Obtain Energy through Phototrophy or Chemotrophy

Although the macronutrients mentioned earlier provide the essential building blocks (for instance, carbon) to make proteins and other cell structures, those synthetic processes require an energy source. Depending on the organism, energy can be obtained from chemical reactions triggered by the absorption of light (**phototrophy**, or photosynthesis) or from oxidation-reduction reactions that transfer electrons from high-energy compounds to

make products of lower energy (**chemotrophy**). Chemotrophic organisms fall into two classes that use different sources of electron donors: chemoautotrophs and chemoheterotrophs. Chemoautotrophs oxidize inorganic chemicals H_2, H_2S, NH_4^+, NO_2^-, and Fe^{2+} for energy. **Chemoheterotrophs** oxidize organic compounds such as sugars to obtain energy.

> **NOTE:** The following prefixes for "-trophy" terms help distinguish different forms of biomass-building and energy-yielding metabolism.
>
> Carbon source for biomass
>
> "Auto-" CO_2 is fixed and assembled into organic molecules.
>
> "Hetero-" Preformed organic molecules are acquired from outside and assembled.
>
> Energy source
>
> "Photo-" Light absorption excites electron to high-energy state.
>
> "Chemo-" Chemical electron donors are oxidized.
>
> Electron source
>
> "Litho-" Inorganic molecules donate electrons.
>
> "Organo-" Organic molecules donate electrons.

In chemotrophy, the amount of energy harvested from oxidizing a compound is directly related to the compound's reduction state. The more reduced the compound is, the more electrons it has to give up and the higher the potential energy yield. A reduced compound, such as glucose, can donate electrons to a less reduced (more oxidized) compound, such as nicotinamide adenine dinucleotide (NAD), releasing energy (in the form of donated electrons) and becoming oxidized in the process. NAD is a cell molecule critical to energy metabolism and is discussed, along with oxidation-reduction reactions, in Chapter 13.

In short, microbes are classified on the basis of their carbon and energy acquisition as follows:

- **Autotrophs.** Autotrophs build biomass by fixing CO_2 into complex organic molecules. Autotrophs gain energy through one of two general metabolic routes that either use or ignore light:

    Photoautotrophy generates energy through light absorption by the photolysis (light-activated breakdown) of H_2O or H_2S. The energy is used to fix CO_2 into biomass.

    Chemoautotrophy (or lithotrophy) produces energy from oxidizing inorganic molecules such as iron, sulfur, or nitrogen. This energy is also used to fix CO_2 into biomass.

- **Heterotrophs.** Heterotrophs break down organic compounds from other organisms to gain energy and to harvest carbon for building their own biomass. Heterotrophic metabolism can be divided into two classes, also based on whether light is involved.

    Photoheterotrophy produces energy through photolysis of organic compounds. Organic compounds are broken down and used to build biomass.

    Chemoheterotrophy (or organotrophy) yields energy and carbon for biomass solely from organic compounds. Chemoorganotrophy is commonly called heterotrophy.

Note that many species, particularly free-living soil and aquatic bacteria, can utilize more than one of these strategies, depending on environmental conditions. They do this by having multiple gene systems that are expressed under different conditions and yield products that carry out different functions. For example, *Rhodospirillum rubrum* grows by photoheterotrophy when light is available and oxygen is absent; but when oxygen is available, the organism grows by respiration, without absorbing light.

The survival and metabolism of any one group of organisms depend on the survival and metabolism of other groups of organisms. Even metazoa (multicellular organisms) rely on microbial metabolism to survive. For example, the cyanobacteria, a type of photosynthetic microorganism that originated 2.5–3.5 billion years ago, produce most of the oxygen we and other metazoa breathe. In fact, cyanobacteria also form the base of the marine food chain. The autotrophic cyanobacteria fix carbon in the ocean and are eaten by heterotrophic protists. The protists then are devoured by fish that produce the CO_2 that is fixed by the cyanobacteria. The cyanobacteria also depend on the heterotrophic bacteria to consume the molecular oxygen that the cyanobacteria produce, since molecular oxygen is toxic to cyanobacteria.

Energy Is Stored for Later Use

Whatever the source, energy once obtained must be converted to a form useful to the cell. This form can be chemical energy, such as that contained in the high-energy phosphate bond in adenosine triphosphate (ATP); or electrochemical energy, energy that is stored in the form of an electrical potential existing between compartments separated by a membrane (see Chapter 14). Energy stored by an electrical potential across the membrane is known as the **membrane potential**. A membrane potential is generated when chemical energy is used to pump protons (and in some cases Na^+) outside of the cell, so that the proton concentration is greater outside the cell than inside. For example, membrane proteins such as cytochrome oxidases use energy from respiration to pump hydrogen ions across the cell membrane, generating a hydrogen ion gradient. This ion movement produces an electrical gradient across the cell membrane, making the inside of the cell more negatively charged than the outside. The hydrogen ion gradient plus the charge difference (voltage potential) across the membrane forms an **electrochemical potential**. Because this electrochemical potential includes the hydrogen ion gradient, it is called the **proton potential**, or **proton motive force**. The energy stored in the proton motive force can be used directly to move nutrients into the cell via specific transport proteins (see Section 3.3), to drive motors that rotate flagella, and to drive synthesis of ATP by a membrane-embedded **ATP synthase** (**Fig. 4.3**).

The membrane-embedded ATP synthase, also called F_1F_o ATP synthase, provides most of the ATP for aerobic respiring cells such as *E. coli*; and essentially the same complex mediates ATP generation in our own mitochondria. ATP synthase is composed of many different proteins. The enzyme includes a channel (F_o) that allows H^+ to move across the membrane and drive rotation of the ATP synthase complex (F_1). Rotation of F_1 mediates the formation of ATP. The role of the proton potential in metabolism is discussed in detail in Chapters 13 and 14.

The idea that a living organism could contain rotating parts was highly controversial when such parts were first discovered in bacterial flagella (discussed in Section 3.7). Before this discovery, scientists had believed that living body parts could not rotate and that only humans had "invented the wheel." The discovery of rotary biomolecules has inspired advances in nanotechnology, the

A.

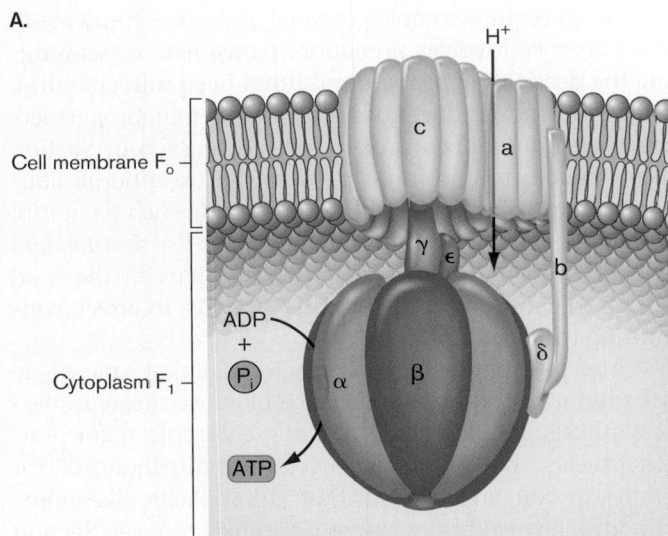

B.

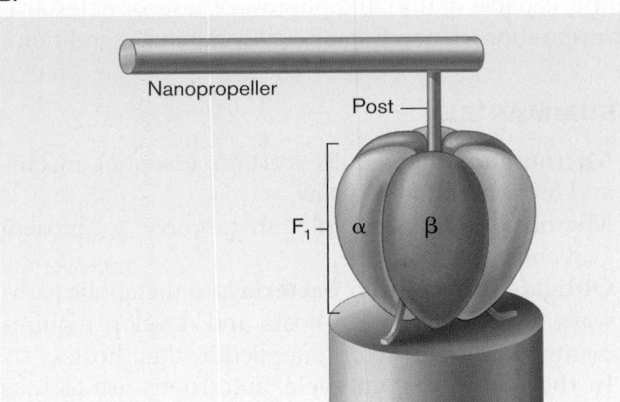

Figure 4.3 Bacterial membrane ATP synthase. A. The F_1F_o complex of ATP synthase is embedded in the cell membrane. The F_o channel/rotor admits hydrogen ions; the H^+ current flow across the membrane is driven by the concentration and charge differences between inside and outside the cell. The flow of H^+ causes the rotation of F_1. Rotation of F_1 drives the conversion of $ADP + P_i$ to ATP. **B.** An artificial "biomolecular motor" was built from an ATP synthase F_1 unit attached to a nickel post and a nanopropeller.

engineering of microscopic devices. For example, a "biomolecular motor" was devised using an ATP synthase F_1 complex to drive a metal submicroscopic propeller (**Fig. 4.3B**). In the future, such biomolecular design may be used to build microscopic robots that enter the bloodstream to perform microsurgery.

> **NOTE:** The name F_1F_o derives from early biochemical studies. F_1 comes from fraction 1 obtained during a purification scheme. F_o comes from this fraction's ability to bind oligomycin.

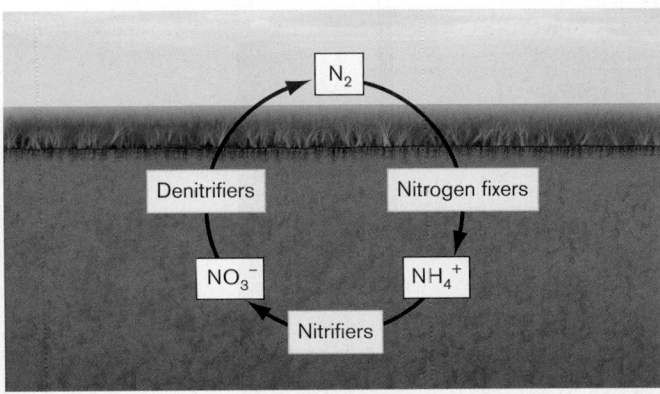

Figure 4.4 The nitrogen cycle. Dinitrogen gas (N_2) is fixed by species of bacteria (nitrogen fixers) that possess the enzyme nitrogenase. Other bacteria (nitrifiers) oxidize ammonia (NH_4^+) to generate energy. Still others (denitrifiers) use oxidized forms of nitrogen, such as nitrate (NO_3^-), as an alternative electron acceptor in place of O_2.

The Nitrogen Cycle Depends on Bacteria and Archaea

Nitrogen is an essential component of proteins, nucleic acids, and other cellular constituents, and as such is required in large amounts by living organisms. Nitrogen gas (N_2) makes up nearly 79% of Earth's atmosphere, but nitrogen gas is unavailable for use by most organisms because the triple bond between the two nitrogen atoms is highly stable and requires considerable energy to be broken. For nitrogen to be used for growth, it must first be "fixed," or converted to ammonium ions (NH_4^+). As with the carbon cycle, various groups of organisms collaborate to interconvert nitrogen gas, ammonium ions, and nitrate ions (NO_3^-) in what is called the nitrogen cycle. One group "fixes" atmospheric nitrogen while other bacteria do the opposite, transforming ammonia to nitrate (**nitrification**) and then converting nitrate to N_2 (**denitrification**) (**Fig. 4.4**). For the environmental significance of nitrogen metabolism, see Chapter 22.

Nitrogen-fixing bacteria may be free-living in soil or water, or they may form symbiotic associations with plants or other organisms. A **symbiont** is an organism that lives in intimate association with a second organism. *Rhizobium, Sinorhizobium,* and *Bradyrhizobium* species, for example, are nitrogen-fixing symbionts in leguminous plants such as soybeans, chickpeas, and clover (**Fig. 4.5**). These plant symbionts take atmospheric nitrogen and convert it to the ammonia that the plant needs to form proteins and other essential compounds. This type of beneficial symbiosis is called mutualism. Although symbionts are the most widely known nitrogen-fixing bacteria, the majority of nitrogen in soil and marine environments is fixed by free-living bacteria and archaea.

Once fixed, how does nitrogen get back into the atmosphere? The nitrifying bacteria, such as *Pseudomonas,*

A.

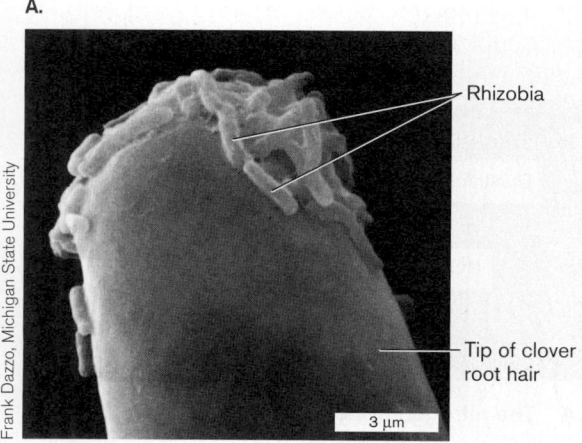

Frank Dazzo, Michigan State University

Rhizobia

Tip of clover
root hair

3 µm

B.

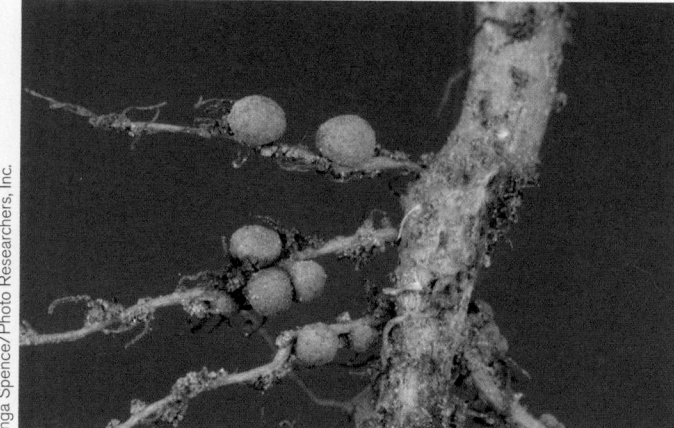

Inga Spence/Photo Researchers, Inc.

Figure 4.5 *Rhizobium* **and a legume. A.** Symbiotic *Rhizobium* cells clustered on a clover root tip (SEM). The rhizobia shown here are clustered on the surface of the root. Soon they will start to invade the root and begin a symbiotic partnership that will benefit both organisms. **B.** Root nodules. After the rhizobia (approx. 0.9 µm × 3 µm) invade the plant root, symbiosis between plant and microbe produces nodules.

Alcaligenes, and *Bacillus*, gain energy by oxidizing ammonia to produce nitrate (a form of chemoautotrophy). Denitrifying bacteria are heterotrophs that use the nitrate provided by the nitrifiers to produce a series of nitrogen compounds, ending with gaseous N_2. The activity of denitrifying bacteria results in a substantial loss of nitrogen into the atmosphere that roughly balances the amount of nitrogen fixation occurring each year. As in the carbon cycle, this, once again, illustrates how nature manages to replenish planet Earth.

Eukaryotic Microbes Include Consumers and Producers

Eukaryotic microbes such as protists and fungi are heterotrophic consumers (discussed in Chapter 21). Protists

and fungi require complex organic molecules for growth; their lifestyle involves predation, parasitism, or scavenging the dead. Most protists and fungi need mitochondria, which are the products of an ancient symbiotic partnering with internalized prokaryotes (discussed in Section 17.6). Heterotrophic fungi possess the exceptional ability to digest complex organic compounds such as lignin, a major component of wood and bark. In marine and aquatic systems, protists form a huge part of the food chain, so eating trout or swordfish means, in effect, consuming trillions of protists.

Just like photosynthetic bacteria, eukaryotic algae are photoautotrophs that produce biomass through photosynthesis. In addition to needing chloroplasts for photosynthesis, however, algae require mitochondria for energy production. (Recall that chloroplasts, like mitochondria, evolved from internalized bacteria; see Section 1.5.) Protist algae such as single-celled *Euglena* are **mixotrophic**, capable of utilizing photosynthesis or heterotrophic respiration, depending on environmental conditions.

TO SUMMARIZE:

- **Microorganisms** require certain essential macro- and micronutrients to grow.
- **Microbial genomes** evolve in response to nutrient availability.
- **Obligate intracellular bacteria** lose metabolic pathways provided by their hosts and develop requirements for growth factors supplied by their hosts.
- **In the global carbon cycle**, autotrophs use CO_2 as a carbon source, either through photosynthesis or through lithotrophy.
- **Autotrophs** make complex organic compounds that are consumed by heterotrophs.
- **Nitrogen fixers**, **nitrifiers**, **and denitrifiers** contribute to the nitrogen cycle.
- **Bacteria or archaea** carry out all of the carbon, nitrogen, and energy reactions just described, whereas eukaryotes carry out only a limited range of heterotrophic and photosynthetic reactions.

4.2 Nutrient Uptake

Whether a microbe is propelled by flagella toward a favorable habitat or, lacking motility, drifts through its environment courtesy of Brownian movement, it must be able to find nutrients and move them across the membrane into the cytoplasm. The membrane, however, presents a daunting obstacle. Membranes are designed to separate what is outside the cell from what is inside. So, for a cell to gain sustenance from the environment, the membrane must be *selectively* permeable to nutrients the cell can use. A few compounds, such as oxygen and carbon dioxide, can passively diffuse across the membrane, but most cannot. Selective permeability is achieved in three ways:

- By the use of substrate-specific carrier proteins (**permeases**) in the membrane
- With the aid of dedicated nutrient-binding proteins that patrol the periplasmic space
- Through the action of membrane-spanning protein channels, or pores, that discriminate between substrates

Microbes must also overcome the problem of low nutrient concentrations in the natural environment (for example, lakes or streams). If the intracellular concentration of nutrients were no greater than the extracellular concentration, the cell would remain starved of most nutrients. To solve this dilemma, most organisms have evolved efficient transport systems that concentrate nutrients inside the cell relative to outside. However, moving molecules against a concentration gradient requires some form of energy.

In contrast to environments where nutrients are available but exist at low concentrations (for example, aqueous environments), certain habitats have plenty of nutrients but those nutrients are locked in a form that cannot be transported into the cell. Starch, a large, complex carbohydrate, is but one example. Many microbes unlock these nutrient vaults by secreting digestive proteins that break down complex carbohydrates or other molecules into smaller compounds that are easier to transport. The amazing techniques that cells use to extrude these large digestive proteins through the membrane and into their surrounding environment are discussed in Chapter 8.

Facilitated Diffusion

Although most transport systems use cellular energy to bring compounds into the cell, a few do not. **Facilitated diffusion** is a type of system that simply uses the concentration gradient of a compound to move that compound across the membrane from a compartment of higher concentration to a compartment of lower concentration. The best these systems can do is to equalize the internal and external concentrations of a solute; facilitated transport cannot move a molecule *against* its gradient.

The most important facilitated diffusion transporters are those of the aquaporin family that transport water and small polar molecules such as glycerol. Glycerol transport is performed by an integral membrane protein

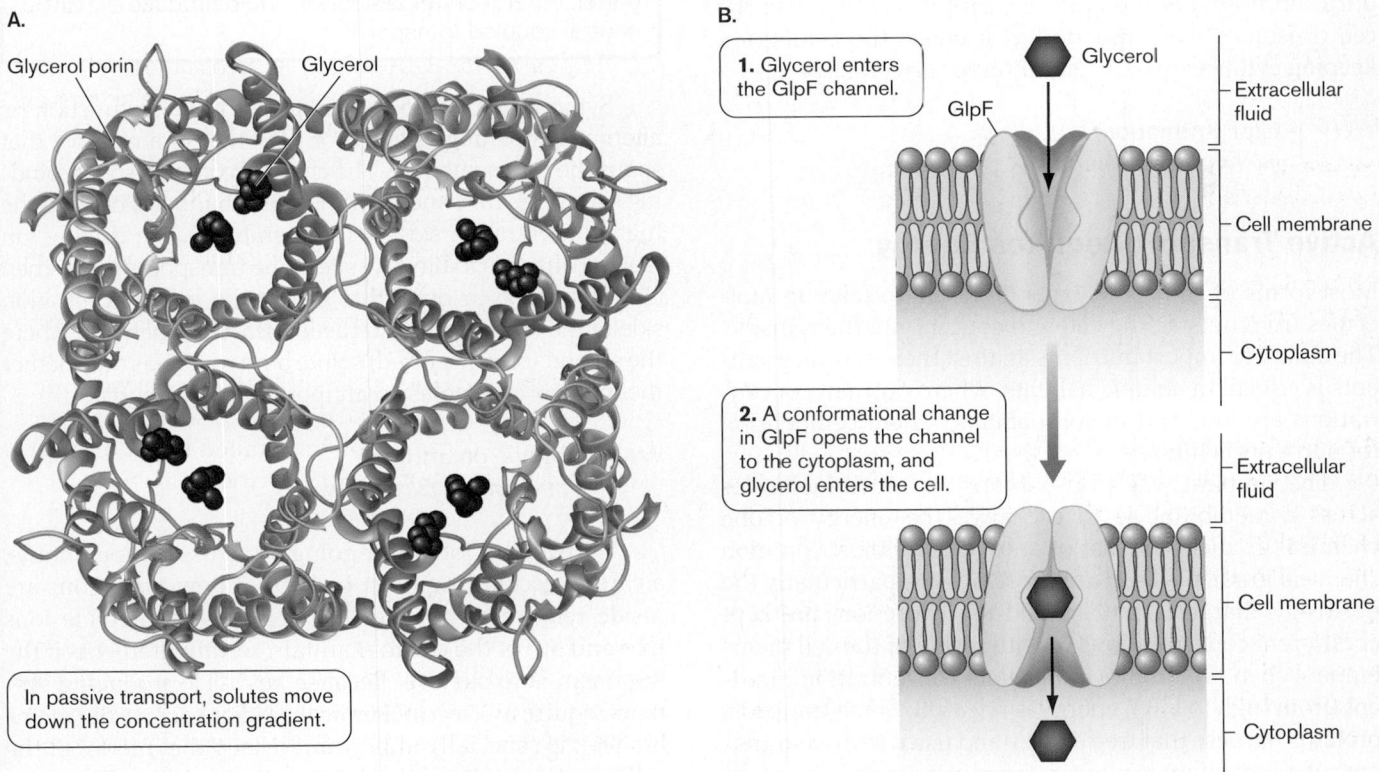

Figure 4.6 Facilitated diffusion. **A.** The glycerol transporter of *E. coli*, viewed from the external side of the membrane, consists of a tetramer of four dimer channels formed by hydrophobic alpha helices alternating back and forth across the cell membrane. Each channel (blue and yellow) contains a glycerol (magenta). (PDB code: 1FX8) **B.** Facilitated diffusion of glycerol through GlpF. The protein facilitates the movement of the compound from outside the cell, where glycerol is at high concentration, to inside the cell, where the concentration of glycerol is low.

in *E. coli* called GlpF. The structure of a glycerol channel is shown in **Figure 4.6A**. This protein channel allows passive uptake of glycerol, a polar molecule useful to cells for energy and for building phospholipids. The glycerol channel complex is viewed from the outer face of the membrane. The complex is a tetramer of four channels. Each channel transports a molecule of glycerol.

In the membrane, GlpF reversibly (and randomly) assumes two conformations. One form exposes the glycerol-binding site to the external environment, whereas the second form exposes this site to the cytoplasm. When the concentration of glycerol is greater outside than inside the cell, the form with the binding site exposed to the exterior is more likely to find and bind glycerol. After binding glycerol, GlpF changes shape, closing itself to the exterior and opening to the interior (**Fig. 4.6B**). Bound glycerol is then released (diffuses) into the cytoplasm (influx). Of course, this form of GlpF could bind a glycerol molecule in the cytoplasm and release it outside the cell. Thus, once the cytoplasmic concentration of glycerol equals the exterior concentration, bound glycerol can be released to either compartment. If the internal concentration exceeds the external concentration, a situation that can arise if a cell is moved from a high- to a low-glycerol environment, glycerol will move out of the cell (efflux). However, facilitated diffusion normally promotes glycerol influx, because the cell consumes the compound as it enters the cytoplasm, keeping cytoplasmic concentrations of glycerol low.

WWW | GlpF animation
microbiology2.com/links

Active Transport Requires Energy

Most forms of transport expend energy to take up molecules from outside the cell and concentrate them inside. The ability to import nutrients against their natural gradients is critical in aquatic habitats, where nutrient concentrations are low, and in soil habitats, where competition for nutrients is high.

The simplest way to use energy to move molecules across a membrane is to exchange the energy of one chemical gradient for that of another. The most common chemical gradients used are those of ions, particularly the positively charged ions Na^+ and K^+. These ions are kept at different concentrations on either side of the cell membrane. When an ion moves *down* its concentration gradient (from high to low), energy is released. Some transport proteins harness that free energy and use it to drive transport of a second molecule *up*, or against, its concentration gradient in the process called **coupled transport**.

The two types of coupled transport systems are **symport**, in which the two molecules travel in the same direction and **antiport**, in which the actively transported molecule moves in the direction *opposite* to the driving ion.

An example of a symporter is the lactose permease LacY of *E. coli*, one of the first transport proteins whose function was elucidated. This work was carried out by the pioneering membrane biochemist H. Ronald Kaback. LacY moves lactose inward, powered by a proton that is also moving inward (symport). LacY proton-driven transport is said to be **electrogenic** because an unequal distribution of charge results (for example, symport of a neutral lactose molecule with H^+ results in net movement of positive charge).

An example of **electroneutral** coupled transport, in which there is no net transfer of charge, is that of the Na^+/H^+ antiporter. The Na^+/H^+ antiporter couples the export of Na^+ with the import of a proton (antiport). Because molecules of like charge are exchanged, there is no net movement of charge. The mechanism by which Na^+/H^+ antiporters function was elucidated by the Israeli biochemist Etana Padan. Sodium exchange is important for all organisms and particularly critical for organisms living in a high-salt habitat (see Section 5.4).

THOUGHT QUESTION 4.2 In what situation would antiport and symport be passive rather than active transport?

THOUGHT QUESTION 4.3 What kind of transporter, other than an antiporter, could produce electroneutral coupled transport?

Symport and antiport transporter proteins function by alternately opening one end or the other of a channel that spans the cell membrane. The channel contains solute-binding sites (**Fig. 4.7A** and **B**). When the channel is open to the high-concentration side of the membrane, the driving ion (solute) attaches to binding sites. The transport protein then changes shape to open that site to the low-concentration side of the membrane, and the ion leaves. When and where the second (cotransported) solute binds depends on whether the transport protein is an antiporter or a symporter.

WWW | Movie on antiport
microbiology2.com/links

With all this ion traffic going on across the membrane, a careful accounting must be kept of how many ions are inside relative to outside. The cell must recirculate ions into and out of the cell to maintain certain gradients if the organism is to survive. Because key ATP-producing systems require an electrochemical gradient across the membrane, it is especially important to keep the interior of the cell negatively charged relative to the exterior. However, because the movement of many compounds is coupled to the import of positive ions, the electrochemical gradient will eventually dissipate, or depolarize, unless positive ions are also exported. Depolarization must be avoided because a depolarized cell loses membrane integrity and

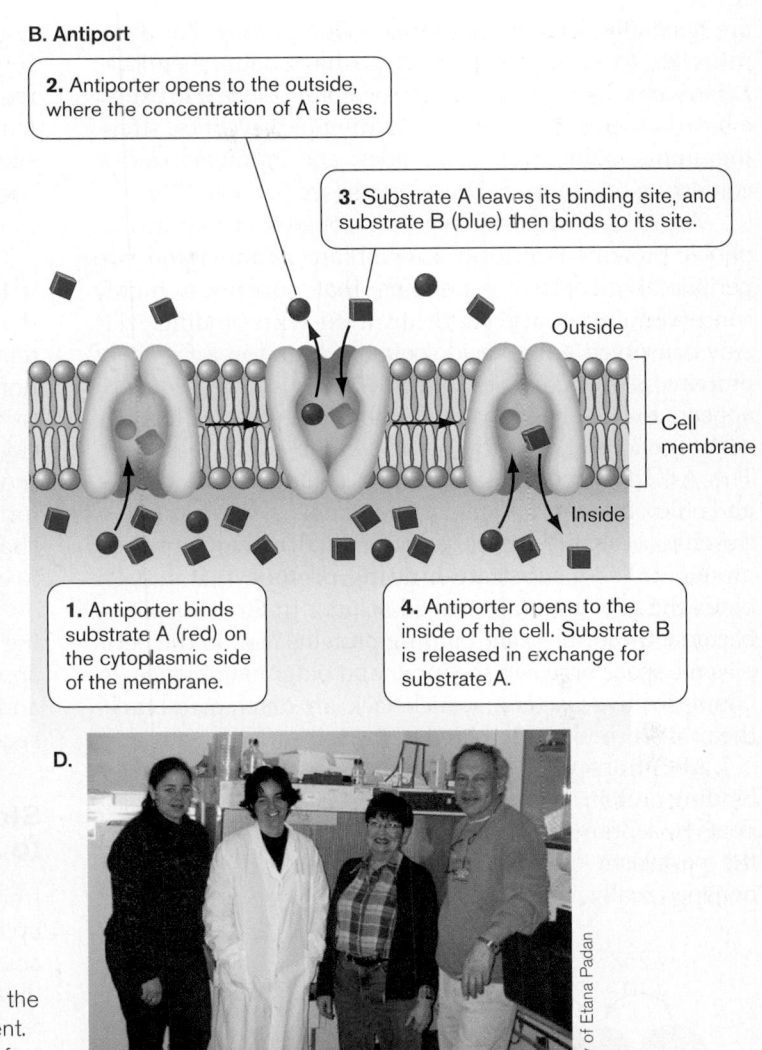

A. Symport

1. Energy is released as one substrate (red) moves down its concentration gradient.

2. This energy moves a second substrate (blue) against its gradient and into the cell.

Outside

Inside

B. Antiport

2. Antiporter opens to the outside, where the concentration of A is less.

3. Substrate A leaves its binding site, and substrate B (blue) then binds to its site.

Outside

Cell membrane

Inside

1. Antiporter binds substrate A (red) on the cytoplasmic side of the membrane.

4. Antiporter opens to the inside of the cell. Substrate B is released in exchange for substrate A.

C.

Courtesy of Ronald Kaback

D.

Courtesy of Etana Padan

Figure 4.7 Coupled transport. A. Symport. **B.** Antiport. Substrate B (blue) is taken up against its gradient because of the energy released by substrate A (red) traveling down its gradient. **C.** Ronald Kaback (left) of UCLA elucidated the mechanism of transport by the proton-driven lactose symporter LacY. **D.** Etana Padan (second from right) of the Hebrew University of Jerusalem dissected the molecular mechanism of the Na^+/H^+ antiporter NhaA.

cannot carry out the simple transport functions needed to sustain life. A healthy cell maintains a proper charge balance by using the electron transport chain to move protons out of the cell and by exchanging negatively and positively charged ions as needed.

In addition to the direct linking of ion transport described for symport and antiport systems, ion circuitry is such that movement of one ion can be linked *indirectly* to movement of another molecule. For example, proton concentrations are typically greater outside than inside the cell. The inwardly directed proton gradient is called proton motive force. Proton motive force can impel the exit of Na^+ through the Na^+/H^+ antiporter. The resulting Na^+ gradient can drive the symport of amino acids into the cell. In this case, Na^+ moves back into the cell down its gradient, and the energy released is tied to the import of an amino acid against its gradient.

ABC Transporters Are Powered by ATP

As we pointed out in Section 3.3, a major function of proton transport is to form the proton motive force that powers ATP synthesis (for details on the proton motive force, see Chapter 14). The energy stored in ATP, whether generated by the proton current or by cytoplasmic means such as fermentation, can drive membrane transport of nutrients.

The largest family of energy-driven transport systems is the <u>A</u>TP-<u>b</u>inding <u>c</u>assette superfamily, also known as **ABC transporters**. These transporters are found in bacteria, archaea, and eukaryotes. It is impressive that nearly 5% of the *E. coli* genome is dedicated to producing the components of 70 different varieties of uptake and efflux ABC transporters. The uptake ABC transporters are critical for transporting nutrients such as maltose, histidine, arabinose, and galactose. The efflux ABC transporters are generally used as multidrug efflux pumps that allow microbes to survive exposures to hazardous chemicals. *Lactococcus*, for example, can use one pump, LmrP, to export a broad range of antibiotics, including tetracyclines, streptogramins, quinolones, macrolides, and aminoglycosides, conferring resistance to those drugs (see Section 27.6).

An ABC transporter typically consists of two hydrophobic proteins that form a membrane channel and two peripheral cytoplasmic proteins that contain a highly conserved amino acid motif involved with binding ATP. Any conserved amino acid sequence found in a family of proteins is viewed as a "cassette" because the sequence appears to have been inserted into those proteins for specific functions (hence the name "ATP-binding cassette"). The ABC transporter superfamily contains both uptake and efflux transport systems. The uptake systems (but not the efflux systems) possess an additional, extracytoplasmic protein, called a **substrate-binding protein**, that initially binds the substrate (also called solute). In Gram-negative bacteria, these substrate-binding proteins float in the periplasmic space between the inner and outer membranes. In Gram-positive bacteria, which lack an outer membrane, the proteins must be tethered to the cell surface.

ABC transport (**Fig. 4.8**) starts with the substrate-binding protein snagging the appropriate solute, either as it floats by a Gram-positive microbe or as the molecule enters the periplasm of a Gram-negative cell. Most substrates nonspecifically enter the periplasm of Gram-negative

organisms through the outer membrane pores, although some high-molecular-weight substrates, like vitamin B_{12}, require the assistance of a specific outer membrane protein to move the substrate into the periplasm. Because substrate-binding proteins have a high affinity for their cognate (matched) solutes, their use increases the efficiency of transport when concentrations of solute are low.

Once united with its solute, the binding protein binds to the periplasmic face of the channel protein and releases the solute, which now moves to a site on the channel protein. This interaction triggers a structural (or conformational) change in the channel protein that is telegraphed to the nucleotide-binding proteins on the cytoplasmic side. On receiving this signal, the nucleotide-binding proteins start hydrolyzing ATP and send a return conformational change through the channel, signaling the channel to open its cytoplasmic side and allow the solute to enter the cell.

There are many different ABC transporters mediating transport of a wide variety of substrates. All of them appear to have arisen from a common ancestral porter and, thus, share a considerable amount of amino acid sequence homology.

Siderophores Are Secreted to Scavenge Iron

Iron, an essential nutrient of most cells, is mostly locked up in nature as $Fe(OH)_3$, which is insoluble and unavailable for transport. Many bacteria and fungi have solved this transport dilemma by synthesizing and secreting specialized molecules called **siderophores** (Greek for "iron bearer") that have a very high affinity for whatever soluble ferric iron is available in the environment. These iron scavenger molecules are produced and sent forth by cells when the intracellular iron concentration is low (**Fig. 4.9**). In most Gram-negative organisms, the siderophore binds iron in the environment, and the siderophore-iron complex then attaches to specific receptors in the outer membrane. At this point, either the iron is released directly and is passed to other transport proteins or the complex is transported across the cytoplasmic membrane by a dedicated ABC transporter. The iron is released intracellularly and reduced to Fe^{2+} for biosynthetic use. Other Gram-negative microorganisms, such as *Neisseria gonorrhoeae* (the causative agent of gonorrhea), do not use siderophores at all but employ receptors on their surface that bind human iron complexes (for example, transferrin or lactoferrin) and wrest the iron from them.

Group Translocation

The ABC transporters we have just considered increase concentrations of solute inside the cell relative to the outside concentration. They move nutrients "uphill" against

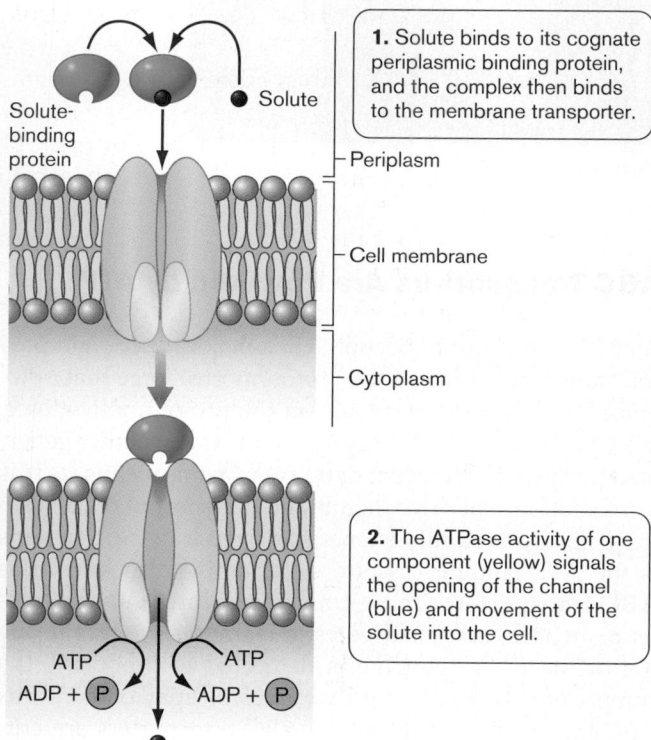

1. Solute binds to its cognate periplasmic binding protein, and the complex then binds to the membrane transporter.

Solute-binding protein

Solute

Periplasm

Cell membrane

Cytoplasm

2. The ATPase activity of one component (yellow) signals the opening of the channel (blue) and movement of the solute into the cell.

ATP   ATP
ADP + Ⓟ   ADP + Ⓟ

Figure 4.8 ABC transporters.

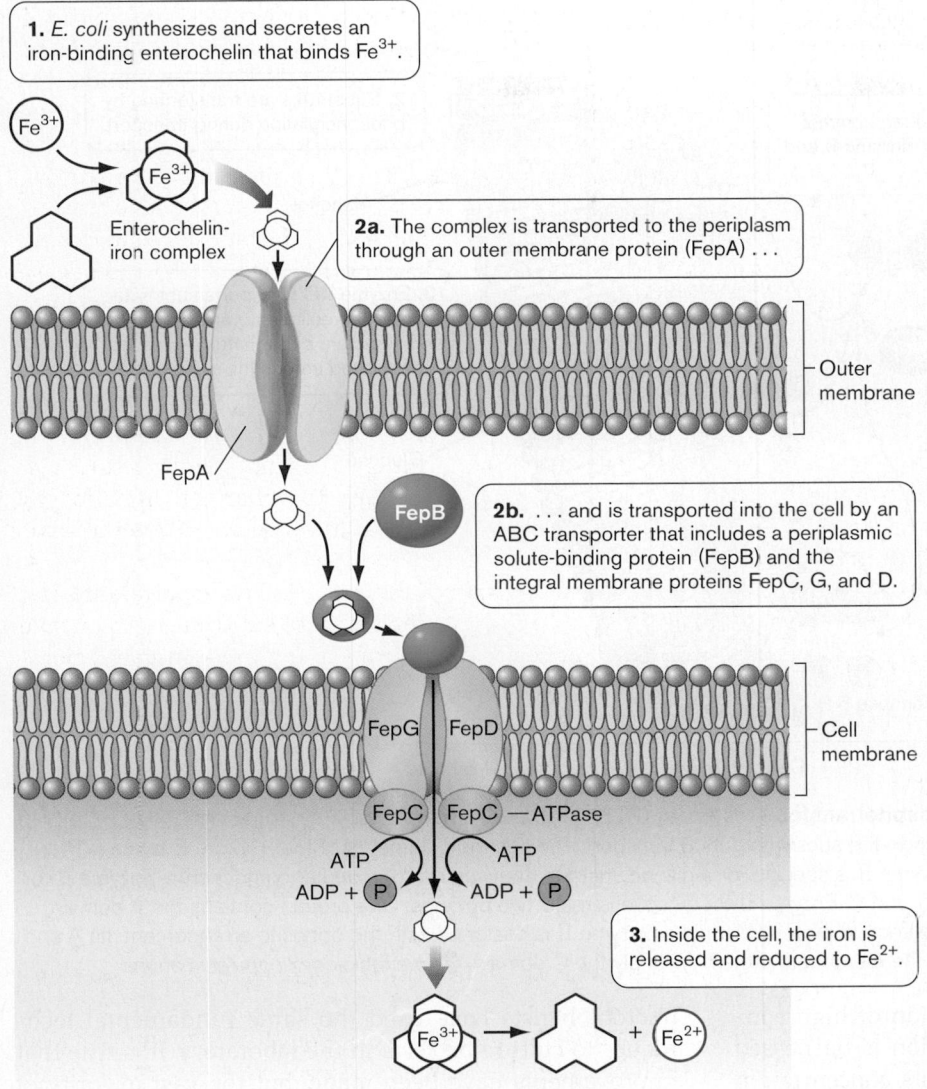

1. *E. coli* synthesizes and secretes an iron-binding enterochelin that binds Fe^{3+}.

Enterochelin-iron complex

2a. The complex is transported to the periplasm through an outer membrane protein (FepA) . . .

Outer membrane

FepA

FepB

2b. . . . and is transported into the cell by an ABC transporter that includes a periplasmic solute-binding protein (FepB) and the integral membrane proteins FepC, G, and D.

FepG | FepD

Cell membrane

FepC | FepC —ATPase

ATP | ATP
ADP + P | ADP + P

3. Inside the cell, the iron is released and reduced to Fe^{2+}.

Fe^{3+} → + Fe^{2+}

Figure 4.9 Siderophores and iron transport.

a concentration gradient. An entirely different system, known as **group translocation**, cleverly accomplishes the same result but without really moving a substance "uphill." Group translocation alters the substrate during transport by attaching a new group (for example, phosphate) on it. Because the modified nutrient inside the cell is chemically different from the related compound outside, the parent solute entering the cell is always moving *down* its concentration gradient, regardless of how much solute has already been transported. Note that this process uses energy to chemically alter the solute. ABC transporters and group translocation systems both involve active transport, but group translocation systems are *not* ABC transporters.

The **phosphotransferase system (PTS)** is a well-characterized group translocation system present in many bacteria. It uses energy from phosphoenolpyruvate

(PEP), an intermediate in glycolysis, to attach a phosphate to specific sugars during their transport into the cell. Glucose, for example, is converted during transport to glucose 6-phosphate. The system has a modular design that accommodates different substrates. Some elements are used by all sugars transported by the PTS; other elements are unique to a given carbohydrate (**Fig. 4.10**). Common elements located in the cytoplasm include Enzyme I (PtsI) and a histidine-rich protein called HPr (PtsH). Enzyme I strips the high-energy phosphate from PEP and passes it to HPr, which in turn distributes the phosphate to various substrate-specific transport proteins called Enzyme II. A typical Enzyme II comprises three domains (A, B, and C) that may be fused together as a single polypeptide or assembled in a variety of combinations. Regardless of the configuration, the phosphorylated HPr (HPr-P) transfers its phosphate to the Enzyme IIA domains/proteins, which relay the phosphate to their cognate Enzyme IIB domains/proteins. Enzyme IIB finally delivers the phosphate to the specific sugar that has been transported into the cell by the Enzyme IIC domain embedded in the cytoplasmic membrane. For example, glucose is transported by Enzyme IIC and converted to glucose 6-phosphate by Enzyme IIB. In Chapter 9, we will see how this physiological system impacts the genetic control of many other systems.

Like prokaryotic cells, eukaryotic cells possess antiporters and symporters and use ABC transport systems as multidrug efflux pumps, but they also employ another process, called endocytosis, which often precedes nutrient transport across membranes. Endocytosis is discussed in **eTopic 4.1** and Appendix 2, Section A2.1.

TO SUMMARIZE:

- **Transport systems** move nutrients across semipermeable membranes.
- **Facilitated diffusion** helps solutes move across a membrane from a region of high concentration to one of lower concentration.
- **Antiporters and symporters** are coupled transport systems in which energy released by moving a

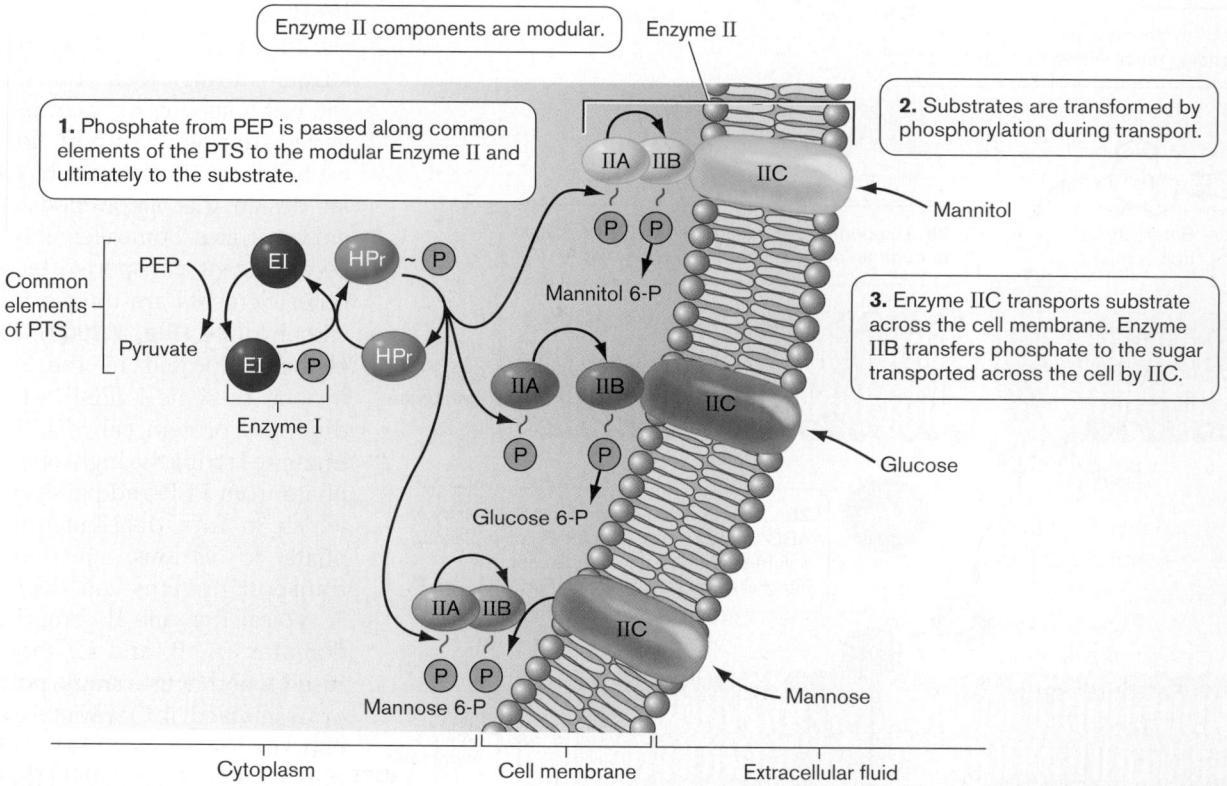

Figure 4.10   Group translocation: the phosphotransferase system (PTS) of *E. coli*.   The phosphate group from phosphoenolpyruvate (PEP) is ultimately passed to the substrate during transport. The common elements of the PTS are Enzyme I (PtsI) and HPr (histidine-rich protein; PtsH). Each Enzyme II is specific for a given substrate and consists of modular components. Enzyme II for mannitol is one protein with three domains: A, B, and C. Enzyme II for glucose is really two proteins: One protein contains the A domain, and the B and C domains are joined to form the second protein. Mannose Enzyme II is designed with the opposite arrangement: Its A and B domains are fused into one protein, whereas the membrane protein is simply the C domain. ◖❚❚ *microbiology2.com/animations*

driving ion (H$^+$ or Na$^+$) from a region of high concentration to one of low concentration is harnessed and used to move a solute against its concentration gradient.

- **ABC transporters** use the energy from ATP hydrolysis to move solutes "uphill," against their concentration gradients.
- **Siderophores** are secreted to bind ferric iron and transport it into the cell, where it is reduced to the more useful ferrous form. Siderophore-iron complexes enter cells with the help of ABC transporters.
- **Group translocation systems** chemically modify the solute during transport.

4.3   Culturing Bacteria

Microbes in nature usually exist in complex, multispecies communities, but for detailed studies they must be grown separately in pure culture. It is nothing short of amazing—and humbling—that after 120 years of trying to grow microbes in the laboratory, we have succeeded in culturing only 0.1% of the microorganisms around us. Since the time of Koch in the late nineteenth century,

microbiologists have used the same fundamental techniques to culture bacteria in the laboratory. It is true that improvements have been made, but the vast majority of the microbial world has yet to be tamed.

Bacteria Are Grown in Culture Media

For those organisms that can be cultured, we have access to a variety of culturing techniques that can be used for different purposes. Bacterial culture media may be either liquid or solid. A liquid, or broth, medium, in which organisms can move about freely, is useful for studying the growth characteristics of a single strain of a single species (that is, a **pure culture**). Liquid media are also convenient for examining growth kinetics and microbial biochemistry at different phases of growth. Solid media, usually gelled with agar, are useful for trying to separate mixtures of different organisms as they are found in the natural environment or in clinical specimens.

Dilution Streaking and Spread Plates

Solid media are basically liquid media to which a solidifying agent has been added. The most versatile and widely used solidifying agent is agar (for the development of agar medium, see Section 1.3). Derived from seaweed, agar forms an unusual gel that liquefies at 100°C but does not solidify until cooled to about 40°C. Liquefied agar medium poured into shallow, covered petri dishes cools and hardens to provide a large, flat surface on which a mixture of microorganisms can be streaked to separate individual cells. Each cell will divide and grow to form a distinct, visible colony of cells (**Fig. 4.11**). As shown in **Figure 4.12**, a drop of liquid culture is collected with an inoculating loop and streaked across the agar plate surface in a pattern called **dilution streaking**. Organisms fall off the loop as it moves along the agar surface. Toward the end of the streak, few bacteria remain on the loop, so at that point individual cells will land and stick to different places on the agar surface. If the medium, whether artificial (for example, laboratory medium) or natural (for example, crab carapace) contains the proper nutrients and growth factors, a single cell will multiply into many millions of offspring, forming a **microcolony**. At first visible only under a microscope, the microcolony grows into a visible droplet called a **colony** (**Figs. 4.12B** and **4.13**). A pure culture of the species or one strain of a species can be obtained by touching a single colony with a sterile inoculating loop and inserting that loop into fresh liquid medium.

It is important to note that the "one cell equals one colony" paradigm does not hold for all bacteria. Organisms such as *Streptococcus* or *Staphylococcus* usually do not exist as single cells, but occur as chains or clusters of several cells. Thus, a cluster of ten *Staphyloccocus* cells will form only one colony on an agar medium and is called a colony-forming unit (CFU).

Another way to isolate pure colonies is the **spread plate** technique. Starting from a liquid culture of bacteria, a series of tenfold dilutions is made, and a small amount of each dilution is placed directly on the surface of separate agar plates (**Fig. 4.14**). The sample is spread over the surface of the plate with a heat-sterilized, bent glass rod. The early dilutions, those containing the most bacteria, will produce **confluent** growth that covers the entire agar surface. Later dilutions, containing fewer and fewer organisms, yield individual colonies. As we shall see later, spread plates not only enable us to isolate pure cultures but also can be used to enumerate the number of **viable** bacteria in the original growth tube. A viable organism is one that successfully replicates to form a colony. **Thus, each colony on an agar plate represents one viable organism present in the original liquid culture.**

There is some disagreement about use of the term "viable cell." Is an organism that does not replicate but continues to metabolize viable? Consider, for instance, that human beings who

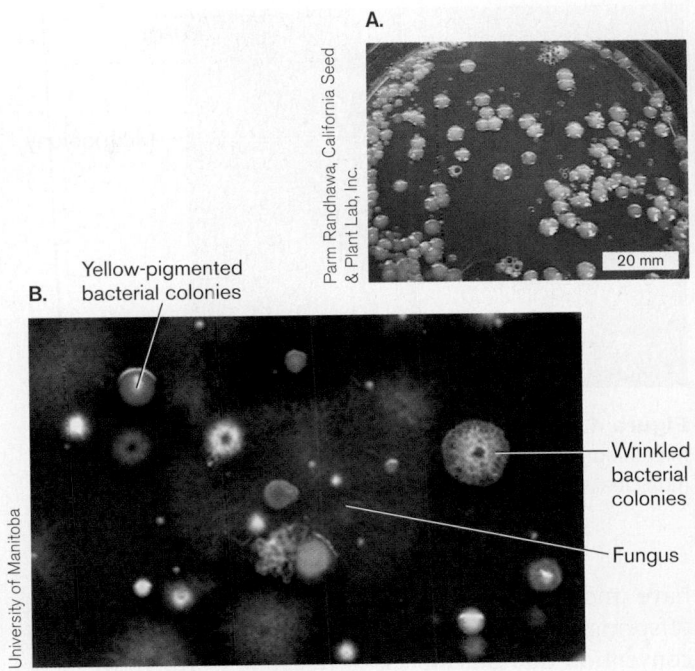

Yellow-pigmented bacterial colonies

Wrinkled bacterial colonies

Fungus

Parm Randhawa, California Seed & Plant Lab, Inc.

University of Manitoba

Figure 4.11 Separation and growth of microbes on an agar surface. A. Colonies (diameter 1–5 mm) of *Acidovorax aveane* separated on an agar plate. This organism is a plant pathogen that causes watermelon fruit blotch. **B.** Mixture of yellow-pigmented bacterial colonies, wrinkled bacterial colonies, and fungus separated by dilution on an agar plate.

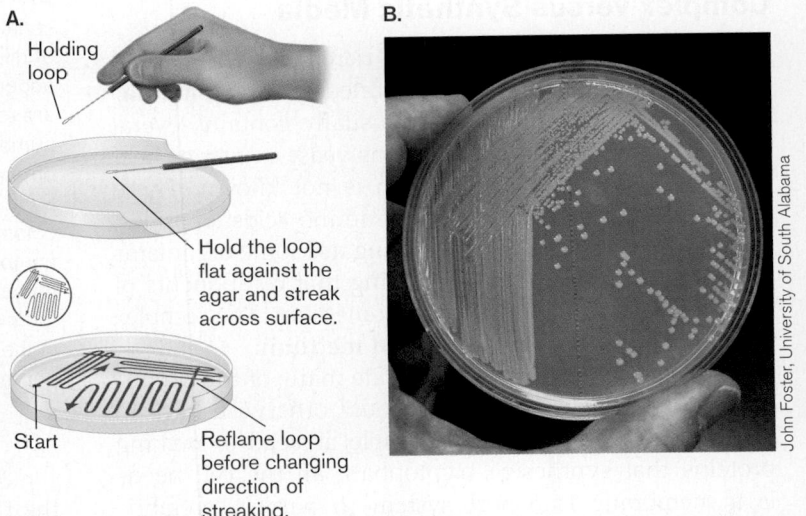

A.

Holding loop

Hold the loop flat against the agar and streak across surface.

Reflame loop before changing direction of streaking.

Start

B.

John Foster, University of South Alabama

Figure 4.12 Dilution streaking technique.   A. A liquid culture is sampled with an inoculating loop and streaked across the plate in three to four areas, with the loop flamed between areas to kill bacteria still clinging to the loop. The result is that dragging the loop across the agar diminishes the number of organisms clinging to the loop until only single cells are deposited at a given location. **B.** *Salmonella enterica* culture obtained by dilution streaking. ▶ *microbiology2.com/animations*

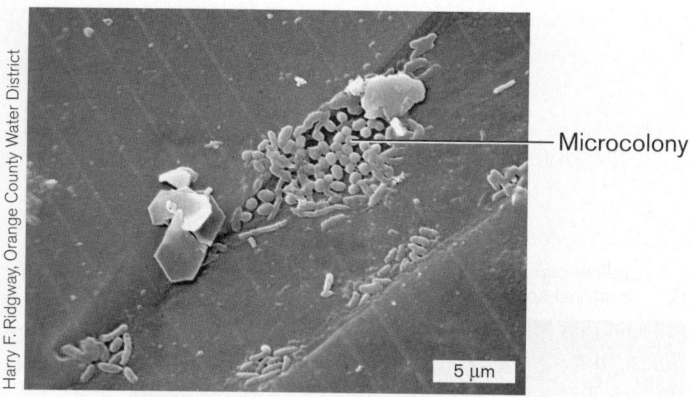

Figure 4.13 Bacterial colonies. A three-day-old microcolony (biofilm) on the surface of a membrane used to treat municipal wastewater in Fountain Valley, California (SEM).

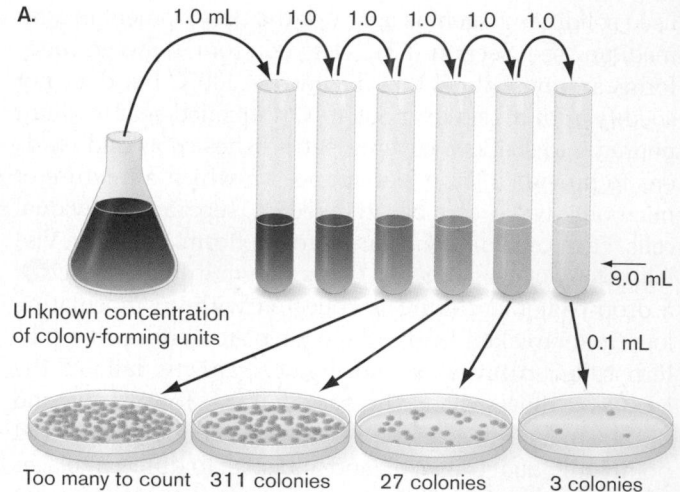

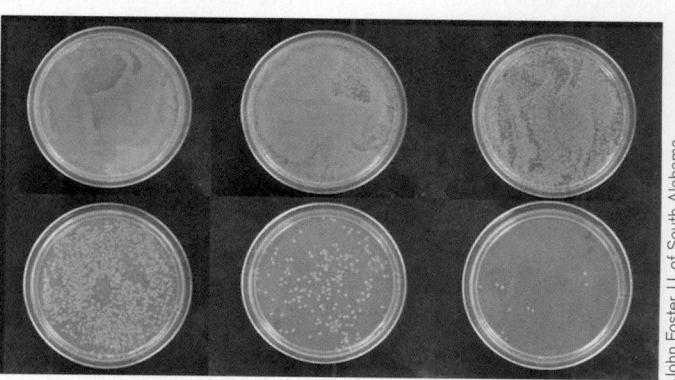

Figure 4.14 Tenfold dilutions, plating, and viable counts. **A.** A culture containing an unknown concentration of cells is serially diluted. One milliliter (ml) of culture is added to 9.0 ml of diluent broth and mixed, and then 1 ml of this $\frac{1}{10}$ dilution is added to another 9.0 ml of diluent (10^{-2} dilution). These steps are repeated for further dilution, each of which lowers the cell number tenfold. After dilution, 0.1 ml of each dilution is spread onto an agar plate. **B.** Plates prepared as in (A) are incubated at 37°C to yield colonies. By multiplying the number of countable colonies (11 colonies on the 10^{-6} plate) by 10, you get the number of cells in 1.0 ml of the 10^{-6} dilution. Multiplying that number by the reciprocal of the dilution factor, you can calculate the number of cells (colony-forming units; CFUs) per milliliter in the original broth tube ($11 \times 10^{1} \times 10^{6} = 1.1 \times 10^{8}$ CFUs per ml).

have undergone sterilization are incapable of producing offspring but they are certainly alive. Nevertheless, by a convention that dates back to the time of Koch, the scientific community considers a cell "viable" if it reproduces to form a colony on a plate. Organisms that appear to metabolize but for some reason cannot replicate have been referred to as dormant or "**viable but nonculturable (VBNC).**" Increasingly, it seems probable that all viable but nonculturable organisms simply appear that way to us because we have not yet discovered the culture conditions necessary for them to reproduce.

Complex versus Synthetic Media

Bacteria can be grown in nutrient-rich but poorly defined **complex media** or in precisely defined **synthetic media**.

Recipes for complex media usually contain several poorly defined ingredients, such as yeast extract or beef extract, whose exact composition is not known. These additives include a rich variety of amino acids, peptides, nucleosides, vitamins, and some sugars. Some organisms are particularly fastidious, requiring that components of blood be added to a basic complex medium. The complex medium is now called an **enriched medium**.

Complex, or rich, media provide many of the chemical building blocks that a cell would otherwise have to synthesize on its own. For example, instead of making proteins that synthesize tryptophan, all the cell needs is a membrane transport system to harvest prefabricated tryptophan from the medium. Likewise, fastidious organisms that require blood in their media may reclaim the heme released from red blood cells as their own, using it as an "enzyme prosthetic group," a group critical to enzyme function (for example, the heme group in cytochromes). All of this saves the scavenging cell a tremendous amount of energy, and as a result, bacteria tend to grow fastest in complex media.

An argument can be made that complex media mimic the rich environments that pathogens encounter in animal hosts. However, the metabolism of a microbe growing in a complex medium is hard to characterize. How would you know whether *E. coli* possesses the ability to make tryptophan if the bacterium grew only in complex media? Questions like this can be investigated by studies of organisms that can be grown in fully defined synthetic media. In preparing a synthetic medium, we start with water and then add various salts, carbon, nitrogen, and energy sources in precise amounts. For self-reliant organisms like *E. coli* or *Bacillus subtilis*, that is all that is needed. Other organisms, such as *Shigella* species or mutant strains of *E. coli* or *B. subtilis*, require additional ingredients to satisfy requirements imposed by the absence of specific metabolic pathways.

> **THOUGHT QUESTION 4.4** What would be the phenotype (growth characteristic) of a cell that lacks the *trp* genes (genes required for the synthesis of tryptophan)? What would be the phenotype of a cell missing the *lac* genes (genes whose products catabolize the carbohydrate lactose)?

Selective and Differential Media Reveal Differences in Metabolism

Microorganisms are remarkably diverse with respect to their metabolic capabilities and resistance to certain toxic agents. These differences are exploited in **selective media**, which favor the growth of one organism over another; and in **differential media**, which expose biochemical differences between two species that grow equally well. For example, Gram-negative bacteria, with their outer membrane, are much more resistant than Gram-positive microbes to detergents like bile salts and certain dyes, such as crystal violet. A solid medium containing bile salts and crystal violet is considered selective because it favors the growth of Gram-negative organisms over Gram-positive ones. On the other hand, a differential medium is needed to distinguish between organisms that differ not in the ability to grow, but in some biochemical aspect. For example, *E. coli* and *Salmonella enterica* are both Gram-negative, but only *E. coli* can ferment lactose. Both can be grown on solid media containing lactose, the dye neutral red, and a nonfermentable carbon source like peptone. *E. coli*, however, ferments the lactose and produces acidic end products. The lower pH surrounding an *E. coli* colony causes the dye to enter the cells, turning the colony red. In stark contrast, *S. enterica*, a major cause of diarrhea, will grow on nonfermentable peptides but cannot ferment lactose. Consequently, acidic end products are not produced by *S. enterica*, so the colonies remain white, their natural color. In this example, growth in differential media easily distinguishes colonies of lactose fermenters from nonfermenters.

Several media used in clinical microbiology are both selective and differential. **MacConkey medium**, for example, selects for growth of Gram-negative bacteria because it contains bile salts and crystal violet, which prevent the growth of Gram-positives. The medium also includes lactose, neutral red, and peptones to differentiate lactose fermenters from nonfermenters. Thus, a culture grown in MacConkey medium will consist of Gram-negative organisms that can be identified as lactose

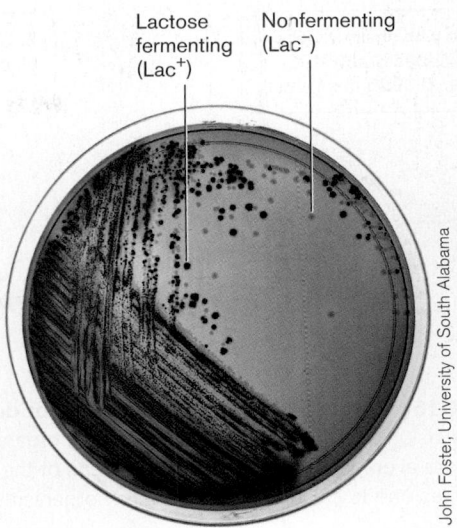

Lactose fermenting (Lac⁺) Nonfermenting (Lac⁻)

John Foster, University of South Alabama

Figure 4.15 MacConkey medium, a culture medium both selective and differential. Dilution streak of a mixture of Lac⁺ and Lac⁻ bacteria. Only Gram-negative bacteria grow on lactose MacConkey (selective). Only a species capable of fermenting lactose produces pink colonies (differential).

fermenters (red colonies) or nonfermenters (uncolored colonies; **Fig. 4.15**). This medium is of particular benefit in diagnosing the etiology (cause) of diarrheal disease because most normal flora are lactose fermenters, whereas two important pathogens, *Salmonella* and *Shigella*, are lactose nonfermenters.

> **THOUGHT QUESTION 4.5** The addition of sheep blood to agar produces a very rich medium called blood agar. Do you think blood agar can be considered a selective medium? A differential medium? *Hint*: Some bacteria can lyse red blood cells.

TO SUMMARIZE:

- **Microbes in nature** usually exist in complex, multispecies communities, but for detailed studies they must be grown separately in pure culture.
- **Bacteria can be cultured** on solid or liquid media.
- **Minimal defined media** contain only those nutrients essential for growth.
- **Complex, or rich, media** contain many nutrients. Other media exploit specific differences between organisms and can be defined as *selective* or *differential*.

4.4 Counting Bacteria

How do we determine if a lake is contaminated with fecal bacteria or whether our peanut butter contains salmonella? We have to identify the organism present, of

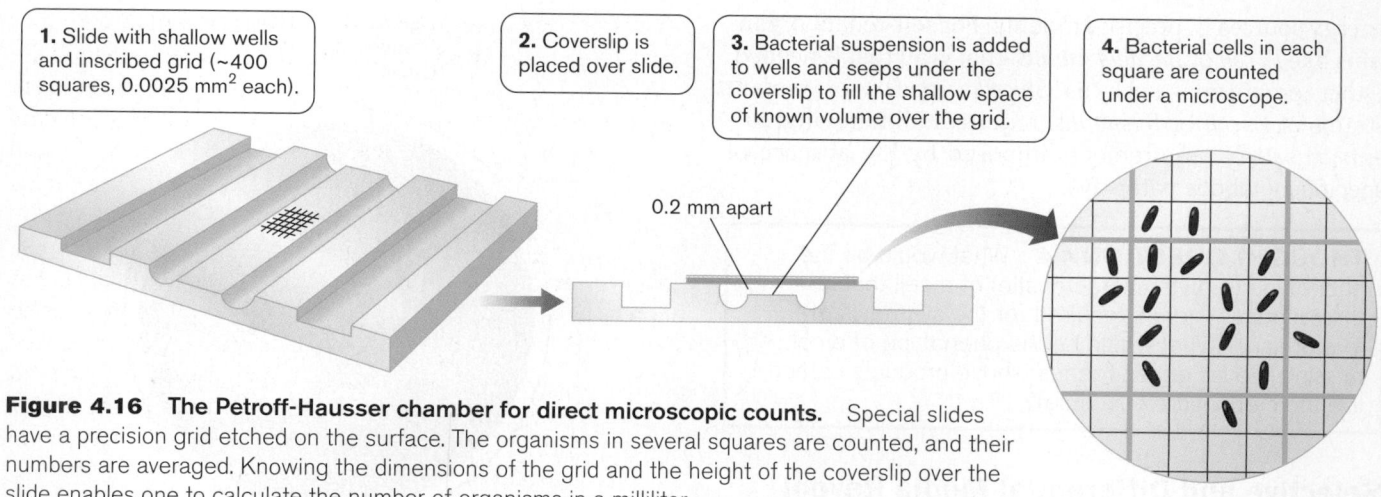

1. Slide with shallow wells and inscribed grid (~400 squares, 0.0025 mm² each).

2. Coverslip is placed over slide.

3. Bacterial suspension is added to wells and seeps under the coverslip to fill the shallow space of known volume over the grid.

4. Bacterial cells in each square are counted under a microscope.

0.2 mm apart

Figure 4.16   The Petroff-Hausser chamber for direct microscopic counts.   Special slides have a precision grid etched on the surface. The organisms in several squares are counted, and their numbers are averaged. Knowing the dimensions of the grid and the height of the coverslip over the slide enables one to calculate the number of organisms in a milliliter.

course, but then we have to count how many there are to understand the extent of contamination. In fact, there are many other reasons why it is important to know the number of bacteria in a sample. These will become obvious as you read further.

Counting or quantifying organisms invisible to the naked eye is surprisingly difficult because each of the available techniques measures a different physical or biochemical aspect of growth. Thus, a cell density value (given as cells per milliliter) derived from one technique will not necessarily agree with the value obtained by a different method.

Direct Counting of Living and Dead Cells

Microorganisms can be counted directly using a microscope. A dilution of a bacterial culture is placed on a special microscope slide called a "hemocytometer" (or, more specifically for bacteria, a "Petroff-Hausser counting chamber;" **Fig. 4.16**). Etched on the surface of the slide is a grid of precise dimensions, and placing a coverslip over the grid forms a space of precise volume. The number of organisms counted within that volume is used to calculate the concentration of cells in the original culture.

However, "seeing" an organism under the microscope does not mean that the organism is alive. Living and dead cells are indistinguishable by this basic approach. Living cells may be distinguished from dead cells by fluorescence microscopy using fluorescent chemical dyes, as discussed in Chapter 2. For example, propidium iodide, a red dye, intercalates between DNA bases but cannot freely penetrate the energized membranes of living cells. Thus, only *dead* cells stain red under a fluorescence scope. Another dye, Syto-9, enters both living and dead cells, staining them both green. By combining Syto-9 with propidium iodide, living and dead cells can be distinguished: Living cells will stain green, whereas dead cells appear orange or

yellow because both dyes enter and Syto-9 (green) plus propidium (red) appears yellow (**Fig. 4.17**).

Direct counting without microscopy can be accomplished by a Coulter counter. In the Coulter counter, a microbial culture is forced through a small orifice, through which flows an electric current. Electrodes placed on both sides measure resistance. Every time a cell passes through the orifice, electrical resistance increases and the cell is counted. The Coulter counter, however, works best with

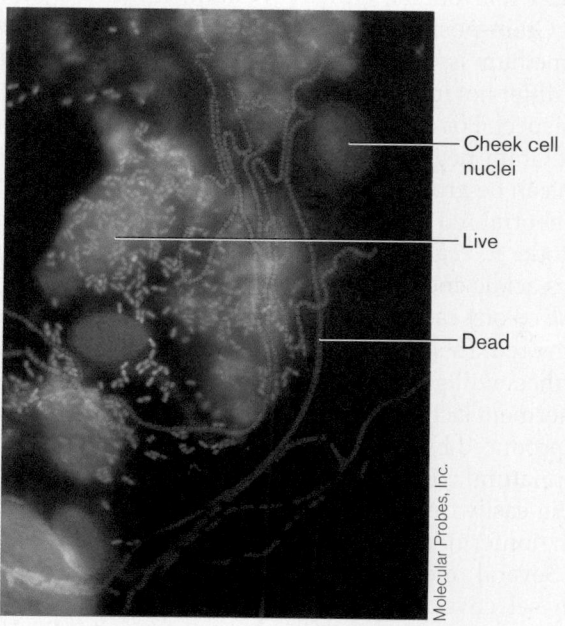

Cheek cell nuclei

Live

Dead

Molecular Probes, Inc.

Figure 4.17   Live-dead stain.   Live and dead bacteria visualized on freshly isolated human cheek epithelial cells using LIVE/DEAD *Bac*Light Bacterial Viability Kit. Dead bacterial cells fluoresce orange or yellow because propidium (red) can enter the cells and intercalate the base pairs of DNA. Live cells fluoresce green because Syto-9 (green) enters the cell. The faint green smears are the outlines of cheek cells.

larger eukaryotic cells, such as red blood cells; the instrument is not generally sensitive enough to detect individual bacteria.

An electronic technique more suited to counting and separating bacterial cells requires an instrument called a **fluorescence-activated cell sorter (FACS)**. In the FACS technique, bacterial cells that synthesize a fluorescent protein (such as cyan fluorescent protein; see Section 3.6) or that have been labeled with a fluorescent antibody or chemical are passed through a small orifice, as in the Coulter counter, and then past a laser (**Fig. 4.18A**).

Detectors measure light scatter in the forward direction—a measure of particle size; and to the side, which indicates shape or granularity. In addition, the laser activates the fluorophore in the fluorescent antibody, and a detector measures fluorescence intensity.

Within a single culture, the FACS technique enables us to use cell size and level of fluorescence (since one subpopulation may fluoresce more or less than another) to identify and count different populations of cells. For example, by placing the green fluorescent protein gene (*gfp*) under the control of the regulatory DNA sequences of a bacterial gene (making a gene fusion), researchers can count the cells expressing that gene by using FACS analysis; this analysis, in turn, makes it possible to determine what conditions allow expression of the gene and whether all cells in the population express that gene at the same time and to the same extent (**Fig. 4.18B**).

Viable Counts

Viable cells, as noted previously, are those that can replicate and form colonies on a plate. To obtain a viable cell count, dilutions of a liquid culture can be plated directly

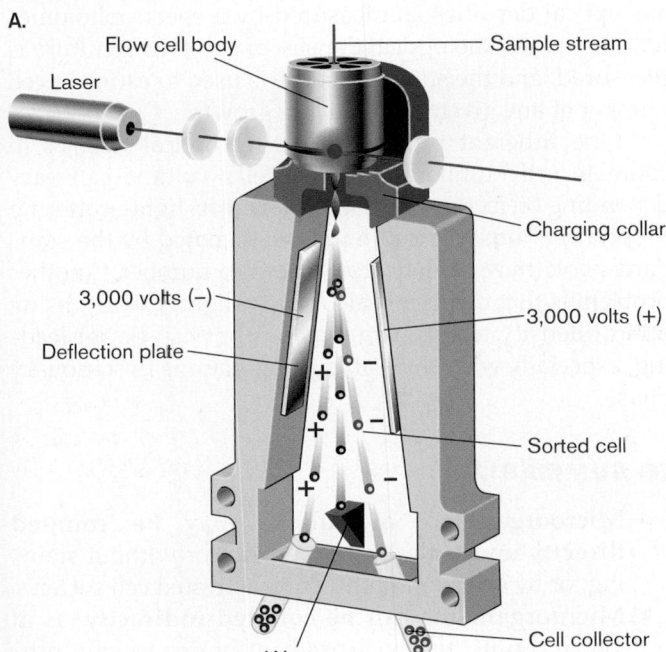

A.

Figure 4.18   Fluorescence-activated cell sorter.
A. Schematic diagram of a FACS apparatus (bidirectional sorting). **B.** Separation of GFP-producing *E. coli* from non-GFP-producing *E. coli*. The low-level fluorescence in the cells on the left is baseline fluorescence (autofluorescence). The scatterplot displays the same FACS data, showing the size distribution of cells (*x*-axis) with respect to level of fluorescence (*y*-axis). The larger cells may be ones that are about to divide.

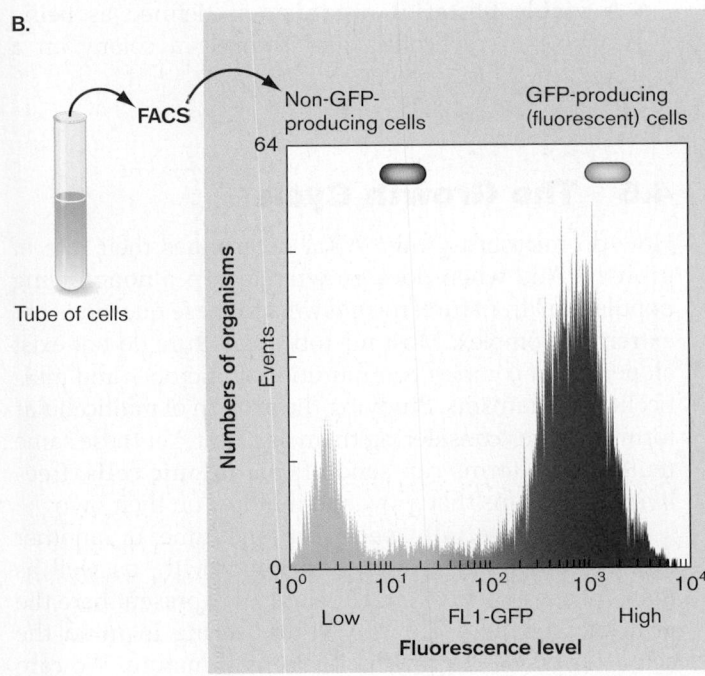

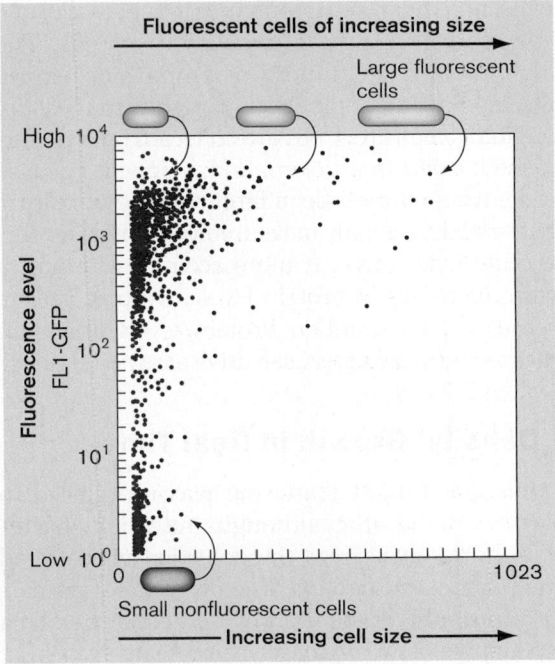

on an agar surface or added to liquid agar cooled to about 42°C–45°C. The agar is subsequently poured into an empty petri plate (**pour plate**), where the agar cools further and solidifies. Because many bacteria resist short exposures to that temperature, individual cells retain the ability to form colonies on, and in, the pour plate. After colonies form, they are counted, and the original cell number is calculated. For example, if 100 colonies are observed in a pour plate made with 100 microliters (μl) of a 10^{-3} dilution of a culture, there are 10^6 organisms per milliliter of the original culture.

Although viable counts are widely used in research, there are problems using this method to measure cell number and determine cultural characteristics. One issue is that colony counting does not reflect cell size or growth stage. Even more problematic, colony counts usually underestimate the number of living cells in a culture. As noted earlier, metabolically active cells that do not form colonies on agar plates will typically not be counted as alive. Cells damaged for one reason or another, while still alive, may be too compromised to divide. Comparing a viable count with a direct count obtained from a live-dead stain can expose the presence of damaged cells. Organisms that grow in chains, such as *Streptococcus*, pose another problem. If individual cells in a chain are not separated prior to plating, each colony seen will have formed from a group of cells, which will cause actual cell numbers to be underestimated. The results of this counting technique are thus reported as colony-forming units (CFUs) rather than cells.

Biochemical Assays

In contrast to methods that visualize individual cells, assays of cell mass, protein content, or metabolic rate measure the overall size of a population of cells. The most straightforward but time-consuming biochemical approach to monitoring population growth is to measure the dry weight of a culture. Cells are collected by centrifugation, washed, dried in an oven, and weighed. Because bacterial cells weigh very little, a large volume of culture must be harvested to obtain measurements, making this technique quite insensitive. A more accurate alternative is to measure increases in protein levels, which correlate with increases in cell number. Protein levels are more easily measured with relatively sensitive assays.

Optical Density: Growth in Real Time

The phenomenon of light scattering was introduced in Chapter 2. We noted that although individual bacteria could not be resolved from the detection of scattered light by the unaided human eye, the presence of numerous bacteria in a tube of medium could be detected as a cloudy appearance. The decrease in intensity of a light beam due to the scattering of light by a suspension of particles is measured as **optical density**. Note that only particles in suspension, such as cells and macromolecules, can scatter light. True solutions, such as a solution of salt or of glucose, are clear even at high concentration.

The optical density of light scattered by bacteria is a very useful tool for estimating population size. The method is quick and easy, but because light scattering is a complex function of cell number and volume, optical density provides only an approximate result that must be corrected using a standard curve. Typically, a standard curve is obtained that plots viable counts for cultures versus optical densities as measured by a spectrophotometer. Thereafter, the optical density of a growing culture is measured, and the standard curve is used to estimate cell number at any given point during growth.

One inherent problem in using optical density to estimate cell numbers is that a cell's volume can vary depending on its growth stage, altering its light-scattering properties. Thus, the cell number estimated by the standard curve may deviate from the true number. Another problem is that dead cells also scatter light. Clearly, using optical density to estimate viable count can be misleading, especially when measuring populations in stationary phase.

TO SUMMARIZE:

- **Microorganisms in culture may be counted directly** under a microscope, with or without staining, or by use of a fluorescence-activated cell sorter.
- **Microorganisms can be counted indirectly**, as in viable counts and measurement of dry weight, protein levels, or optical density.
- **A viable bacterial organism** is defined as being capable of replicating and forming a colony on a solid-medium surface.

4.5 The Growth Cycle

How do microbes grow? What determines their rate of growth? And when does growth start in a nongrowing population? In nature, the answers to these questions are extremely complex. Most microbes in nature do not exist alone, but in complex communities of microbes and multicellular organisms. Studying the growth of multicellular forms requires considering them as a unit. Yet these same multicellular forms can send off **planktonic cells**, free-living organisms that grow and multiply on their own.

Nevertheless, all species at one time or another exhibit both rapid growth and nongrowth, as well as many phases in between. For clarity, we present here the principles of rapid growth, while bearing in mind the actual diversity of growth situations in nature. We care about growth and how fast microbes grow for many reasons. How rapidly a microbe grows influences how fast a pathogen causes disease, or how quickly contaminated food spoils. A bacterial species that consumes oil is useful for cleaning an oil spill only if the organism can grow rapidly.

Survival of the species ultimately depends on its ability to make more of its own kind. A typical bacterium (at least the typical bacterium that can be cultured in the laboratory) grows by increasing in length and mass, which facilitates expansion of its nucleoid as its DNA replicates (see Section 3.6). As DNA replication nears completion, the cell, in response to complex genetic signals, begins to synthesize an equatorial septum that ultimately separates the two daughter cells. In this overall process, called **binary fission**, one parental cell splits into two equal daughter cells (**Fig. 4.19A**). Note, however, that although a majority of culturable bacteria divide symmetrically in two equal halves, some species divide asymmetrically. For example, the bacterium *Caulobacter* forms a stalked cell that remains fixed to a solid surface but reproduces by budding from one end to produce small, unstalked motile cells. The marine organism *Hyphomicrobium* also replicates asymmetrically by budding, releasing a smaller cell from a stalked parent (**Fig. 4.19B**).

Eukaryotic microbes divide by a special form of cell fission involving mitosis, the segregation of pairs of chromosomes within the nucleus (see Section 20.2 and Appendix 2). Some eukaryotes also undergo more complex life cycles involving budding and diverse morphological forms. Nevertheless, a large population of eukaryotic microbes will exhibit the same mathematical functions

of growth seen in bacteria; indeed, large populations of multicellular organisms—we ourselves among them—exhibit these same growth patterns.

Exponential Growth

The process of reproduction has implications for growth not only of the individual, but also of populations. If we assume that growth occurs without limits, what happens to the population? The unlimited growth of any population obeys a simple law: The **growth rate**, or rate of increase in population numbers or biomass, is *proportional* to the population size at a given time. Such a growth rate is called exponential because it generates an exponential curve, a curve whose slope increases continually.

How does binary fission of cells generate an exponential curve? If each cell produces two cells per generation, then the population size at any given time is proportional to 2^n, where the exponent n represents the number of generations (replacement of parents by offspring) that have taken place between two time points. Thus, cell number rises exponentially. Many microbes, however, have replication cycles based on numbers other than 2. For example, some cyanobacteria form cell aggregates that divide by multiple fission, releasing dozens of daughter cells. The cyanobacterium enlarges without dividing, and then suddenly divides many times without separating. The cell mass breaks open to release hundreds of progeny cells.

Note that simple binary fission is not the only kind of reproduction that would generate an exponential curve. Any generation size will yield exponential growth. For

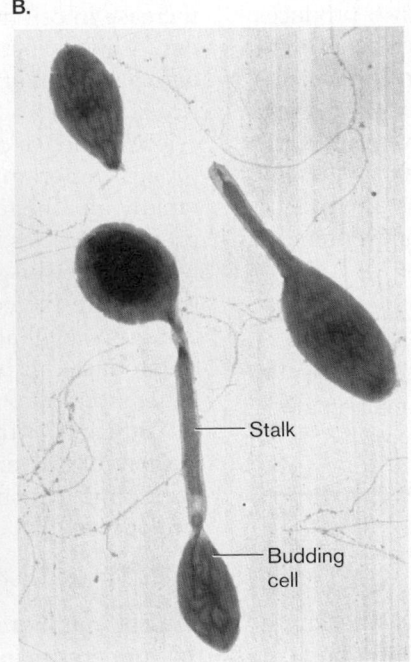

B.

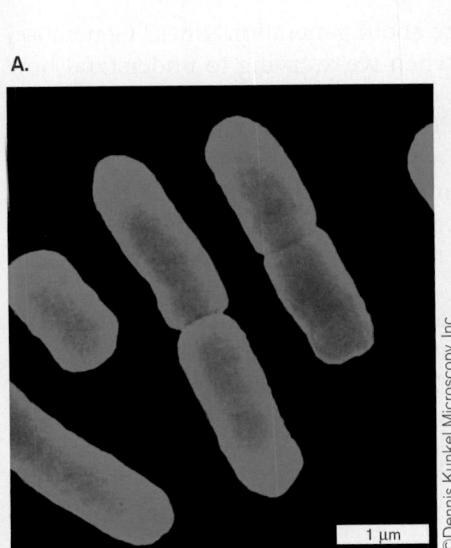

A.

Stalk

Budding cell

1 μm

©Dennis Kunkel Microscopy, Inc.

Ellen Quardokus

Figure 4.19 Symmetrical and asymmetrical cell division. A. Symmetrical cell division, or binary fission, in *Lactobacillus* sp. (SEM). **B.** Budding of the marine bacterium *Hyphomicrobium* (approx. 4 μm long).

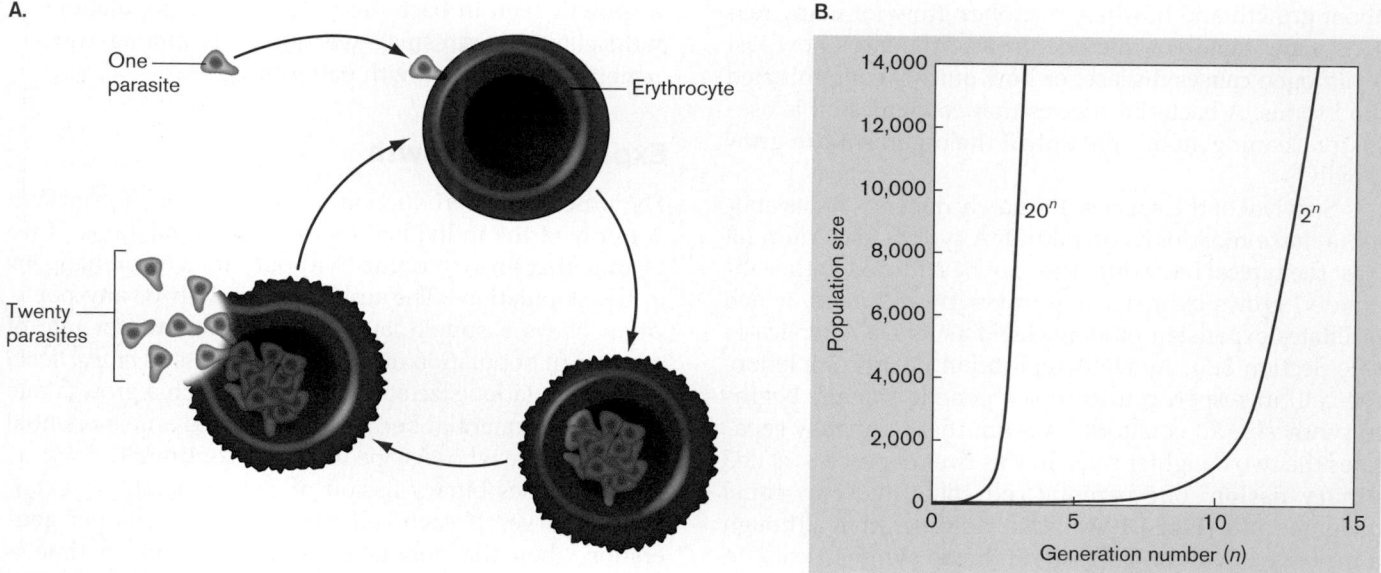

Figure 4.20   Reproduction: from one to many.   A. Exponential growth of *Plasmodium falciparum* produces 20 progeny per generation. **B.** Growth curves for a population that increases 20-fold per generation and for a population that increases 2-fold per generation.

example, the malaria parasite *Plasmodium falciparum* invades a red blood cell, releasing approximately 20 progeny per generation. The rate of increase of the parasite population in the blood is proportional to 20^n, where n is the number of parasite generations (**Fig. 4.20A**). In comparison with bacterial reproduction, which produces only 2 cells per generation, we might expect the growth rate of the parasite to be dramatically faster (assuming the same generation time). Nevertheless, growth of both populations follows the same kind of exponential function, which can be defined by a doubling time (**Fig. 4.20A**). In comparison with bacterial reproduction, which produces only 2 cells per generation, we might expect the growth *rate* of *Plasmodium* to be dramatically faster. In fact, this is the case for the first few generations. But after a few more generations, the bacterial growth curve eventually achieves the same steep slope (rate) that the *Plasmodium* curve has (**Fig. 4.20B**); the two curves are, in fact, graphical representations of the same kind of exponential function.

> **THOUGHT QUESTION 4.6**   A virus such as influenza virus might produce 800 progeny virus particles from one infected host cell. How would you mathematically represent the exponential growth of the virus? What practical factors might limit such growth?

Generation Time

A variable not accounted for in **Figure 4.20B** is the length of time from one generation to the next. In an environment

with unlimited resources, bacteria divide at a constant interval called the **generation time**. The length of that interval varies with respect to many parameters, including the bacterial species, type of medium, temperature, and pH. The generation time for cells in culture is also known as the **doubling time**, because the population of cells doubles over one generation. For example, one cell of *E. coli* placed into a complex medium will divide every 20 minutes. After 1 hour of growth (three generations), that one cell will have become eight (1 to 2, 2 to 4, 4 to 8). Because cell number (*N*) *doubles* with each division, the increase in cell number over time is exponential, not linear. A linear increase would occur if cell number rose by a *fixed* amount after every generation (for example, 1 to 2, 2 to 3, 3 to 4).

Why do we care about generation times? Generation time is important when we're trying to understand how rapidly a pathogen can cause disease symptoms. A fast-growing species will cause disease rapidly, within days, while a slower-growing organism may take weeks or months. In biotechnology, how fast a producing organism grows will affect how quickly a commercially useful by-product can be made. Let's review how we calculate generation time.

Starting with any number of organisms (N_0), the number of organisms after *n* generations will be $N_0 \times 2^n$. For example, a single cell after three generations ($n = 3$) will produce

$$1 \text{ cell} \times 2^3 = 8 \text{ cells}$$

The number of generations that an exponential culture undergoes in a given time period can be calculated if

A.

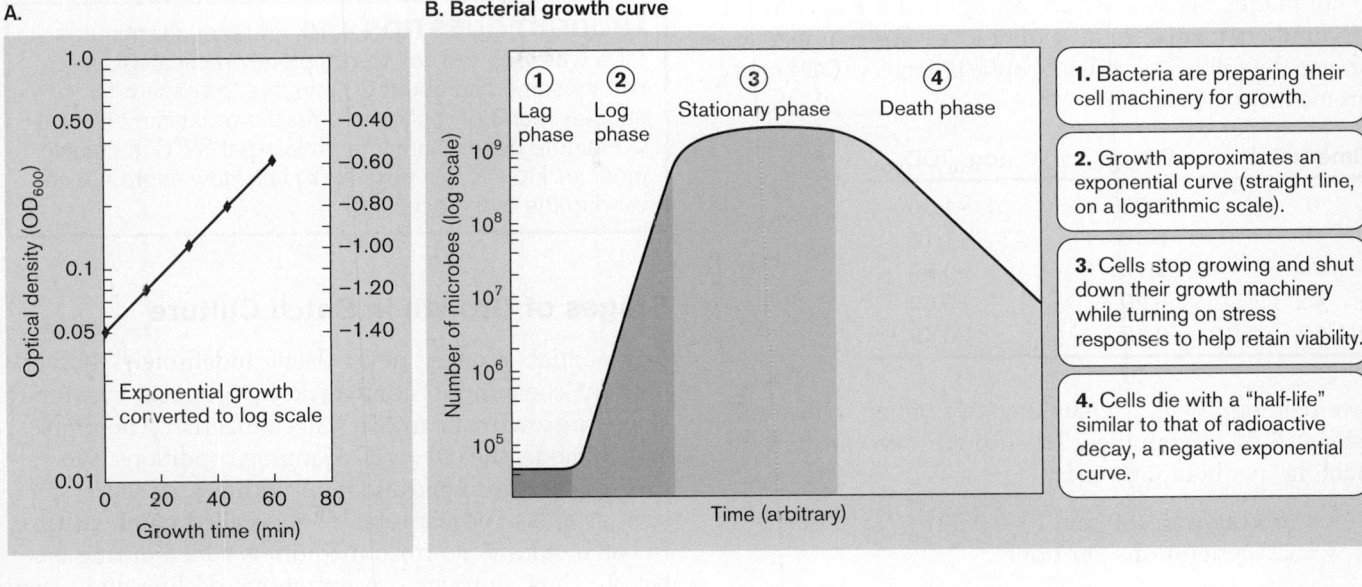

B. Bacterial growth curve

① Lag phase ② Log phase ③ Stationary phase ④ Death phase

Time (arbitrary)

1. Bacteria are preparing their cell machinery for growth.

2. Growth approximates an exponential curve (straight line, on a logarithmic scale).

3. Cells stop growing and shut down their growth machinery while turning on stress responses to help retain viability.

4. Cells die with a "half-life" similar to that of radioactive decay, a negative exponential curve.

C.

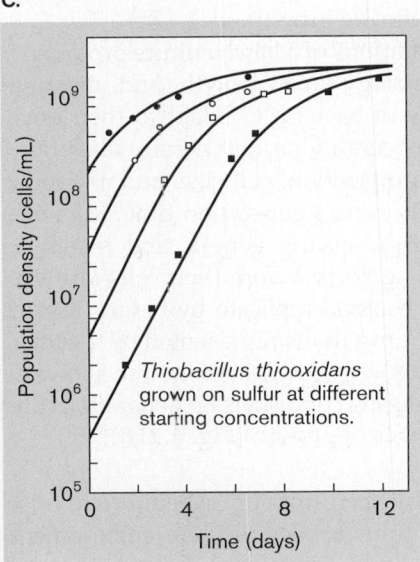

Thiobacillus thiooxidans grown on sulfur at different starting concentrations.

Figure 4.21 Bacterial growth curves. A. Theoretical growth curve of a bacterial suspension measured by optical density (OD) at a wavelength of 600 nm. **B.** Phases of bacterial growth in a typical batch culture. **C.** Published growth curves of *Thiobacillus thiooxidans*, an acidophile that oxidizes sulfur to sulfuric acid. Whatever the starting cell density, the culture grows exponentially until it runs out of sulfur; then it enters the stationary phase. Source: C. Yasuhiro Konishi et al. 1995. *Appl. Environ. Biol.* **61**:3617.

the number of cells at the start of the period (N_0) and the number of cells at the end of the period (N_t) are known. Methods such as viable counts are used to make those determinations. To simplify calculations, instead of using base 2 logarithms, the $\log_2$ expressions are converted to base 10 through division by a factor of $\log_{10} 2$, which is approximately 0.301. Thus:

$$N_t = N_0 \times 2^n$$

becomes

$$\log_{10} N_t = \log_{10} N_0 + n \log_{10} 2$$
$$n \log_{10} 2 = \log_{10} N_t - \log_{10} N_0$$
$$n = (\log_{10} N_t - \log_{10} N_0) \div \log_{10} 2$$
$$n = \log_{10} (N_t/N_0) \div \log_{10} 2$$
$$n = \log_{10} (N_t/N_0) \div 0.301$$

In practice, exponential growth occurs only for a short period when all nutrients are in full supply and the concentration of waste products has not become a limiting factor.

The rate of exponential growth can be expressed as the mean **growth rate constant** (k), which is the number of generations (n) per unit time (usually generations per hour). Even if we do not know the generation time, if we know the number of organisms at time zero (N_0) and the number of organisms after incubation time t (N_t), we can calculate k as follows:

$$k = n/t = (\log_{10} N_t - \log_{10} N_0) \div 0.301t$$
$$k = \log_{10}(N_t/N_0)/0.301t$$

Thus, a culture with a generation time of 20 minutes (0.33 hour) will have a mean growth rate constant of 3 generations per hour ($k = 3$ gen/hr). For a culture doubling every 120 minutes (2 hours), k is 0.5 gen/hr. Note from these examples that the **mean generation time** (g) in hours is the reciprocal of the mean growth rate constant:

$$g = 1/k$$

The growth rate constant can also be calculated from the slope of $\log_{10} N$ over time, where N is a relative measure of culture density, such as the optical density measured in a spectrophotometer (**Fig. 4.21A**). The units of N

do not matter, because we are always looking at ratios of cell numbers relative to an earlier level. For example, we can use the following series of optical density (OD) measurements to measure N:

Time (min)	OD_{600}	$\log_{10} OD_{600}$
0	0.05	−1.30
15	0.08	−1.10
30	0.13	−0.89
45	0.20	−0.69
60	0.33	−0.48

If we plot $\log_{10} OD_{600}$ versus time, we obtain a line with a slope of 0.0136/minutes. The growth rate constant, in doublings per hour, becomes

$$k = (0.0136/\text{min})(60 \text{ min/h}) \div 0.301$$
$$= 2.7 \text{ generations per hour}$$

The steeper the slope, the faster the organisms are dividing.

THOUGHT QUESTION 4.7 Suppose you ingest 20 cells of *Salmonella enterica* in a peanut butter cookie. They all survive the stomach and enter the intestine. Suppose further that the sum total of subsequent bacterial replication and death (caused by the host) produces an average generation time of 2 hours and you will feel sick when there are 1,000,000 bacteria. How much time will elapse before you feel sick?

THOUGHT QUESTION 4.8 Suppose one cell of the nitrogen fixer *Sinorhizobium meliloti* colonizes a plant root. After 5 days (120 hours), there are 10,000 bacteria fixing N_2 within the plant cells. What is the bacterial doubling time?

THOUGHT QUESTION 4.9 **Figure 4.21C** shows growth curves for different population densities of *Thiobacillus thiooxidans* when the concentration of sulfur in the medium is constant. Draw the growth curves you would expect to see if the initial population density were constant but the concentration of sulfur varied.

The mathematics of exponential growth is relatively straightforward, but remember that microbes grow differently in pure culture (very rare in nature) than they do in mixed communities, where neighboring cells produce all kinds of substances that may feed or poison other microbes. In mixed communities, the microbes may grow planktonically (floating in liquid), as in the open ocean, or as a biofilm on solid matter suspended in that ocean. In each instance, the mathematics of exponential growth applies at least until the community reaches a density at which different species begin to compete.

THOUGHT QUESTION 4.10 It takes 40 minutes for a typical *E. coli* cell to completely replicate its chromosome and about 20 minutes to prepare for another round of replication. Yet the organism enjoys a 20-minute generation time growing at 37°C in complex medium. How is this possible? *Hint*: How might the cell overlap the two processes?

Stages of Growth in Batch Culture

Exponential growth never lasts indefinitely, because nutrient consumption and toxic by-products eventually slow the growth rate until it halts altogether. The simplest way to model the effects of changing conditions is to culture bacteria in liquid medium within a closed system, such as a flask or test tube. This is called **batch culture**. In batch culture, no fresh medium is added during incubation; thus, nutrient concentrations decline and waste products accumulate during growth.

The changing conditions of a batch culture profoundly affect bacterial physiology and growth and illustrate the remarkable ability of bacteria to adapt to their environment. As medium conditions deteriorate, alterations occur in membrane composition, cell size, and metabolic pathways, all of which impact generation time. Microbes possess intricate, self-preserving genetic and metabolic mechanisms that slow growth before their cells lose viability. Because many bacteria replicate by binary fission, the plotting of culture growth (as represented by the logarithm of the cell number) versus incubation time allows us to see the effect of changing conditions on generation time and reveals several stages of growth (**Fig. 4.21B**).

Lag phase. Cells transferred from an old culture to fresh growth media need time to detect their environment, express specific genes, and synthesize components needed to institute rapid growth. As a result, bacteria inoculated into fresh media typically experience a lag period, or **lag phase**, where cells do not divide. Several factors influence lag phases. Cells taken from an aged culture may be damaged and require time for repair. Carbon, nitrogen, or energy sources different from those originally used by the seed culture must be sensed, and the appropriate enzyme systems must be synthesized. The length of lag phase varies, depending on the age of the culture, changes in temperature, and the differences between the new and old media (for example, changes in nutrient levels, pH, and salt concentrations). For example, transferring cells from a complex medium to a fresh complex medium results in a very short lag phase, whereas cells grown in a complex medium and then plunged into a minimal defined medium experience a protracted lag phase, during which time they readjust to synthesize all the amino acids, nucleotides, and other metabolites originally supplied by the complex medium.

Early log, or exponential, phase. Once cells have retooled their physiology to accommodate the new environment, they begin to grow exponentially and enter what is called **exponential**, or **logarithmic (log), phase**. Exponential growth is balanced growth, where all cell components are synthesized at constant rates relative to each other. At this stage, cells are growing and dividing at the maximum rate possible based on the medium and growth conditions provided (such as temperature, pH, and osmolarity). Cells are largest at this stage of growth. Early log phase is represented by the linear part of the growth curve. If cell division were synchronized and all cells divided at the same time, the growth curve during this period would appear as a series of steps with cell numbers doubling instantly after every generation time. But batch cultures are not synchronous. Every cell has an equal generation time, but each cell divides at a slightly different moment, making the cell number rise smoothly.

Cells enjoying balanced, exponential growth are temporarily thrown into metabolic chaos (unbalanced growth) when their medium is abruptly changed. Nutritional downshift (moving cells from a good carbon source such as glucose to a poorer carbon source such as succinate) or nutritional upshift (moving cells to a better carbon source) casts cells into unbalanced growth. Downshift to a carbon source with a lower energy yield means not only that a different set of enzymes must be made and employed to use the carbon source, but also that the previous high rate of macromolecular synthesis (such as ribosome synthesis) used to support a fast generation time is now too rapid relative to the lower energy yield. Failure to adjust will lead to increased mistakes in RNA, protein, and DNA synthesis, depletion of key energy stores, and ultimately death.

Microbes, however, possess molecular fail-safe systems that immediately dampen rates of macromolecular synthesis until they come into balance with the rest of metabolism. In contrast, nutritional upshift means that the cell will start making more energy (ATP) than it can use at its current rate of macromolecular synthesis. The same fail-safe mechanisms used during downshift will, during upshift, kick macromolecular synthesis into a higher gear, once again establishing balanced growth. Nutritional upshift causes cells to reenter log phase, but with a shorter generation time.

Late log phase. As cell density (number of cells per milliliter) rises during log phase, the rate of doubling eventually slows, and a new set of growth phase–dependent genes is expressed. At this point, some species can also begin to detect the presence of others by sending and receiving chemical signals in a process known as quorum sensing (discussed in Chapter 10).

Stationary phase. Eventually, cell numbers stop rising, owing to lack of a key nutrient or buildup of waste products. These conditions occur for bacteria grown in a complex medium when cell density rises above 10^9 cells per milliliter, but they can occur at lower cell densities if nutrients are limiting. At this point, the growth curve levels off and the culture enters what is called **stationary phase**. In contrast to bacteria, eukaryotic microorganisms, such as protozoa, enter stationary phase at much lower cell numbers, usually around 10^6 organisms per milliliter. Eukaryotic microbes reach stationary phase sooner than bacteria simply because they are bigger. Bigger cells use more nutrient and run out of it sooner.

If they did not change their physiology, microbes would be very vulnerable once entering stationary phase. Because cells in stationary phase are not as metabolically nimble as cells in exponential phase, damage from oxygen radicals and toxic by-products of metabolism would readily kill them. As an avoidance strategy, some bacteria differentiate into very resistant spores in response to nutrient depletion (see Section 4.7), while other bacteria undergo less dramatic but very effective molecular reprogramming. The microbial model organism *E. coli*, for example, adjusts to stationary phase by decreasing its size, minimizing the volume of its cytoplasm compared with the volume of its nucleoid. Less nutrient is required to sustain the smaller cell. New stress resistance enzymes are also synthesized to handle oxygen radicals, protect DNA and proteins, and increase cell wall strength through increased peptidoglycan cross-linking. As a result, *E. coli* cells in stationary phase become more resistant to heat, osmotic pressure, pH changes, and other stresses that they might encounter while waiting for a new supply of nutrients.

Death phase. Without reprieve in the form of new nutrients, cells in stationary phase will eventually succumb to toxic chemicals present in the environment. Like the growth rate, the **death rate**, the rate at which cells die, is logarithmic. The death rate, however, is a negative exponential function. Recall that the increase in cell number during exponential phase is a positive exponential function of time. In **death phase**, the number of cells that die in a given time period is proportional to the number that existed at the beginning of the time period. Determining microbial death rates is critical to the study of food preservation and to antibiotic development (further discussed in Chapters 5, 16, and 27). Although death curves are basically logarithmic, exact death rates are difficult to define because mutations arise that promote survival, and some cells grow by cannibalizing others. Consequently, the death phase is extremely prolonged. A portion of the cells will often survive for months, years, or even decades.

In **Figure 4.21C**, we see an actual bacterial growth curve from a study of *Thiobacillus thiooxidans*, an organism that oxidizes sulfur to sulfuric acid and grows below pH 1. Growth curves are shown for several different starting concentrations of bacteria. In each case, the bacteria grow exponentially for several days, until they have exhausted the sulfur in their growth medium. Their growth then slows until they enter stationary phase.

THOUGHT QUESTION 4.11 What can happen to the growth curve when a culture medium contains two carbon sources, one a preferred carbon source of growth-limiting concentration and the second, a non-preferred source?

THOUGHT QUESTION 4.12 How would you modify the equations describing microbial growth rate to describe the rate of death?

THOUGHT QUESTION 4.13 Why are cells in log phase larger than cells in stationary phase?

Continuous Culture

In the classic growth curve that develops in closed systems, the exponential phase spans only a few generations. In open systems, however, where fresh medium is continually added to a culture and an equal amount of culture is constantly siphoned off, bacterial populations can be maintained in exponential phase at a constant cell mass for extended periods of time. In this type of growth pattern, known as **continuous culture**, all cells in a population achieve a steady state, which permits detailed analysis of microbial physiology at different growth rates. The **chemostat** is a continuous culture system in which the diluting medium contains a limiting amount of an essential nutrient (**Fig. 4.22**). Increasing the flow rate increases the

amount of nutrient available to the microbe. The more nutrient available, the faster a cell's mass will increase. Because cell division is triggered at a defined cell mass, it follows that the growth rate in a chemostat is directly related to the dilution rate, or flow rate (milliliters per hour divided by the vessel volume). The more nutrient a culture receives as a result of increasing flow rate, the faster those cells can replicate (that is, the shorter the generation time).

The complex relationships among dilution rate, cell mass, and generation time in a chemostat are illustrated in **Figure 4.23**. The curves in this figure represent a typical experimental result, which can vary with organism, limiting nutrient, temperature, and so on. Depending on the experimental conditions, the shapes of the curves may change, but the general relationships will remain similar. Note that at moderate dilution rates (defined differently for each species), an increased flow rate of medium through the system increases division rate (thus, generation time decreases as the cells divide faster). A constant cell mass, or density, is maintained over a range of flow rates because the amount of culture (and cells) removed from the vessel exactly compensates for the increased rate of cell division.

At faster and faster flow rates, cells are eventually removed more quickly than they can be replenished by division, so cell density (cell mass) *decreases* in the vessel—a phenomenon called "washout." Notice, in contrast, that at very *low* dilution (flow) rates, an increase in rate will actually *increase* cell density. The reason is that the nutrient is so limiting at extremely low dilution rates that cell mass cannot increase to the point necessary for division. If more nutrient becomes available, the system can maintain a higher cell mass and cell density because the cells are able to divide faster than they are removed.

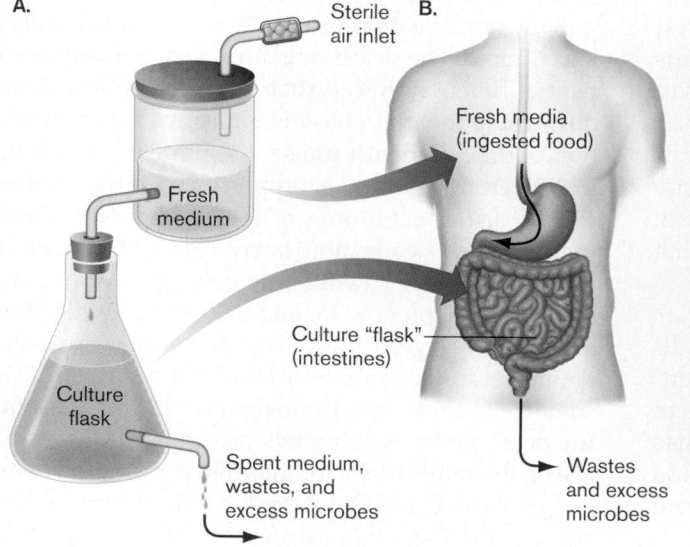

Figure 4.22 Chemostats and continuous culture.   A. The basic chemostat ensures logarithmic growth by constantly adding and removing equal amounts of culture media. **B.** The human gastrointestinal tract is engineered much like a chemostat, in that new nutrients are always arriving from the throat while equal amounts of bacterial culture exit in fecal waste. **C.** A modern chemostat.

With slow flow rates, the removal of fluid is not a significant factor in determining cell density.

The turbidostat is essentially a chemostat in which a photoelectric cell constantly monitors the optical density (turbidity) of the culture. The flow of medium through the vessel is then regulated to maintain a constant turbidity and thus cell density. The turbidostat is most suited to high dilution rates, where the numbers of cells can change quickly and overwhelm a less responsive system; the chemostat, which must be adjusted manually, is better suited to low dilution rates.

Continuous cultures are used to study large numbers of cells at constant growth rate and cell mass for both research and industrial applications. Most bacteria in nature grow at very slow rates because of low nutrient levels—a situation that can be mimicked in a chemostat. The physiology of these cells is quite different from what is typically observed with batch culture.

TO SUMMARIZE:

- **The growth cycle** of organisms grown in liquid batch culture consists of lag phase, log phase, stationary phase, and death phase.
- **The physiology of a bacterial population** changes with growth phase.
- **Continuous culture** can be used to sustain a population of bacteria at a specified growth rate and cell density.

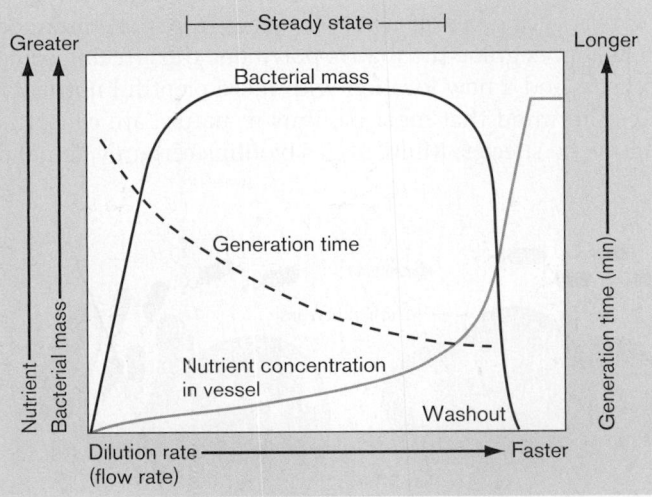

Figure 4.23   Relationships between dilution rate, cell mass, and generation time.   As the dilution rate increases in a chemostat (meaning more nutrient is fed to the culture), the generation time decreases (the cells divide more quickly) and the cell mass of the culture increases. This pattern continues until the rate of dilution exceeds the division rate, at which point cells are washed from the vessel faster than they can be replaced by division and the cell mass decreases. The *y*-axis varies depending on the curve, as labeled in figure.

4.6   Biofilms

Bacteria are typically thought of as unicellular; but in nature, many, if not most, bacteria form specialized, surface-attached communities called **biofilms**. Indeed, within aquatic environments bacteria are found mainly associated with surfaces—a fact that underscores the importance of biofilms in nature. Biofilms also play critical roles in microbial pathogenesis and environmental quality, and cost the nation billions of dollars each year in equipment damage, product contamination, and medical infections. For example, pseudomonad or staphylococcal biofilms can damage ventilators used to assist respiration and can act as direct sources of infection.

Biofilms: Multicellular Microbes?

Biofilms can be constructed by a single species or by multiple, collaborating species and can form on a range of organic or inorganic surfaces (**Fig. 4.24**⬤▯). The Gram-negative bacterium *Pseudomonas aeruginosa*, for example, can form a single-species biofilm on the lungs of patients

A.

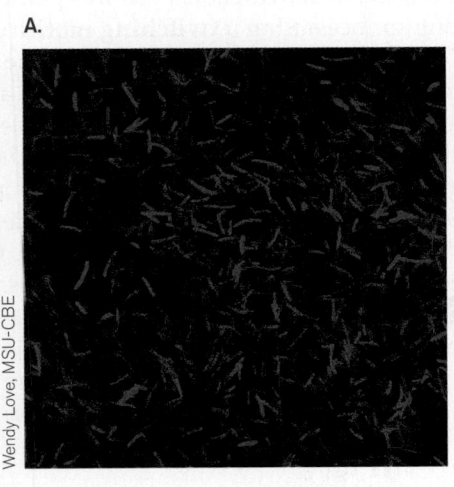

Wendy Love, MSU-CBE

B.

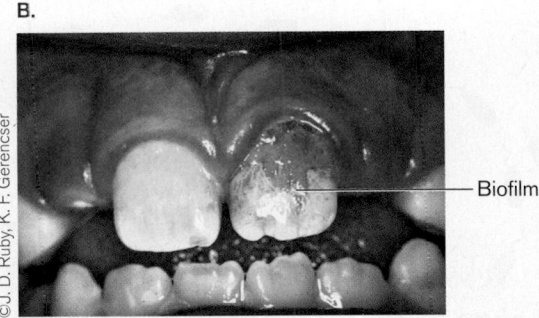

©J. D. Ruby, K. F. Gerencser

— Biofilm

Figure 4.24   Biofilms.   A. A thermophilic microbial mat found attached to a rock in Yellowstone National Park (laser scanning confocal microscope image). The autofluorescence of these cyanobacterial cells of the genus *Synechococcus* was generated with a 568-nm krypton laser and a 590LP filter. **B.** Biofilm on a tooth. The biofilm that forms on teeth is called plaque. ⬤▯
microbiology2.com/animations

with cystic fibrosis or on medical implants. Distinct stages in biofilm development include initiation, maturation, maintenance, and dissolution. Bacterial biofilms form when nutrients are plentiful. The goal of biofilms in nature is to stay where food is plentiful. Why should a microbe travel off to hunt for food when it is already available? Once nutrients become scarce, however, individuals detach from the community to forage for new sources of nutrient.

Biofilms in nature can take many different forms and serve different functions for different species. In addition, formation of biofilms can be cued by different environmental signals in different species. Such signals include pH, iron concentration, temperature, oxygen availability, and the presence of certain amino acids. Nevertheless, a common pattern emerges in the formation of many kinds of biofilms (**Fig. 4.25**).

First, the specific environmental signal induces a genetic program in planktonic cells. The planktonic cells then start to attach to nearby inanimate surfaces by means of flagella, pili, lipopolysaccharides, or other cell surface appendages, and they begin to coat that surface with an organic monolayer of polysaccharides or glycoproteins to which more planktonic cells can attach. At this point, cells may move along surfaces using a **twitching motility** that involves the extension and retraction of a specific type of pilus. Ultimately, they stop moving and firmly attach to the surface. As more and more cells bind to the surface, they can begin to communicate with each other by sending and receiving chemical signals in a process called **quorum sensing**. These chemical signal molecules are continually made and secreted by individual cells. Once the population reaches a certain number (analogous to an organizational "quorum"), the chemical signal reaches a specific concentration that the cells can sense. This concentration triggers genetically regulated changes that cause cells to bind tenaciously to the substrate and to each

other. Quorum sensing serves the biofilm in many ways. Among other functions, quorum sensing has recently been implicated in the increased resistance of biofilms to antibiotics—a subject we explore in **Special Topic 4.1** (see also Special Topic 27.1 and **eTopic 4.2**).

Once a microcolony has been established (see **Fig. 4.25**), the cells form a thick extracellular matrix of polysaccharide polymers and entrapped organic and inorganic materials. These **exopolysaccharides (EPSs)**, such as alginate produced by *P. aeruginosa* and colanic acid produced by *E. coli*, increase the antibiotic resistance of residents within the biofilm. As the biofilm matures, the amalgam of adherent bacteria and matrix takes on complex three-dimensional forms such as columns and streamers, forming channels through which nutrients flow. Sessile (nonmoving) cells in a biofilm chemically "talk" to each other in order to build microcolonies and keep water channels open. Little is known about how a biofilm dissolves, although the process is thought to be triggered by starvation. *P. aeruginosa* produces an alginate lyase that can strip away the EPSs, but the regulatory pathways involved in releasing cells from biofilms are not clear.

Recent evidence suggests that another cell-cell communication molecule may be the actual signal for dispersal. Scientists from the State University of New York have found that an unsaturated fatty acid, *cis*-2-decanoic acid, produced by *P. aeruginosa* during growth can trigger single cell dispersal from biofilms. These researchers propose that as a biofilm forms and grows larger, the fatty acid signal is unable to diffuse away and becomes trapped in biofilm nooks and crannies. At some point it reaches a concentration triggering the release of enzymes from the enmeshed cells that degrade the matrix polymers, thus freeing some cells to find a new location with more plentiful nutrients. Keep in mind that most biofilms in nature are consortia of several species. Multispecies biofilms certainly demand

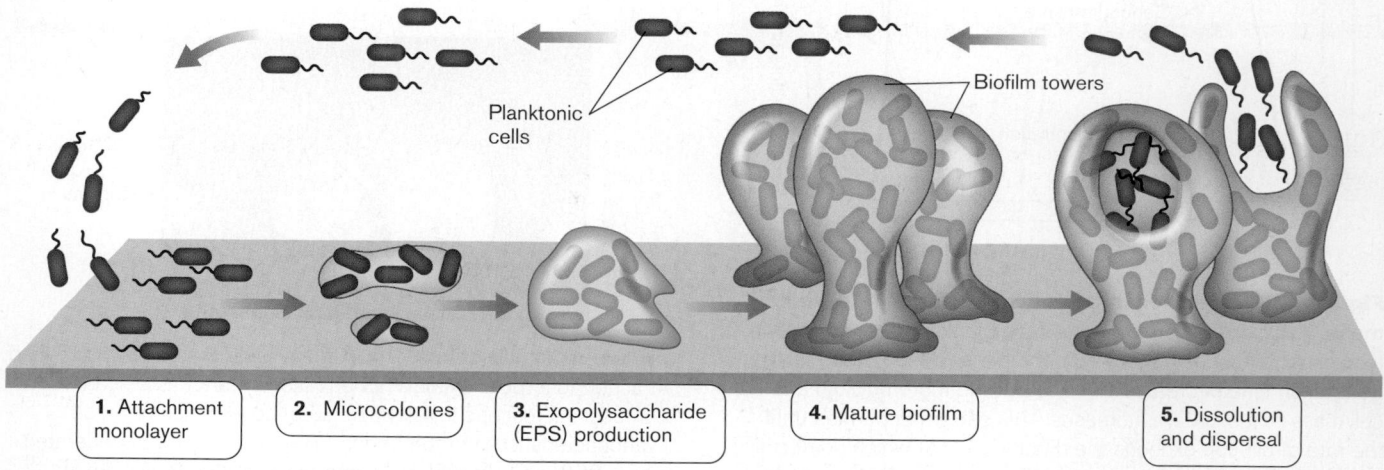

Figure 4.25   Biofilm development.   Biofilm development in *Pseudomonas*. ⏵ *microbiology2.com/animations*

Special Topic 4.1   Biofilms, Disease, and Garlic

When causing diseases of plants, animals, or humans, bacteria preferentially exist in surface-attached biofilms. Attachment to host tissues and multicellular growth are important in many situations, from simple wound infections caused by *Staphylococcus aureus* to colonization of the lungs of cystic fibrosis patients by *Pseudomonas aeruginosa*. One characteristic of bacterial biofilms is a marked increase in antibiotic tolerance (see **eTopic 4.2**). Explanations for increased tolerance of biofilms to antibiotics include reduced penetration of drug into the biofilm and an altered stress-resistant physiology by the cells in the biofilm. Studies using *P. aeruginosa* biofilms clearly demonstrated the importance of quorum sensing for the development of the drug-tolerant state. A chemical signal molecule called an acyl homoserine lactone (AHL) released by cells in the biofilm accumulates and triggers expression of genes that increase antibiotic tolerance. Inhibitors of quorum sensing have been discovered that will make cells in biofilms more sensitive to antibiotics.

More recently, scientists have learned that signal molecules secreted by *P. aeruginosa* in biofilms also make the community more resistant to being killed by polymorphonuclear leukocytes (PMNs). PMNs are white blood cells that become activated and attack bacteria in a typical infection. Surprisingly, a group of scientists from Denmark discovered that garlic extract contains a natural inhibitor of this quorum-sensing response. As shown in **Figure 1**, cells within a typical biofilm are quite resistant to "grazing" by PMNs added after the biofilm has formed. However, PMNs voraciously grazed on biofilms treated with a 2% garlic extract. (A control was performed to show that PMNs treated with garlic extract were unaffected.) It seems that quorum sensing by the biofilm decreased the expression of proteins that would otherwise activate the PMNs. Hence, shutdown of the biofilm's quorum-sensing system by garlic made the bacteria visible to the PMNs. The identity of the quorum-sensing inhibitor is not known, but the study points out that the administration of anti–quorum sensing drugs to patients could lead not only to the development of less persistent biofilms, but also to inhibition of the expression of bacterial virulence determinants that counter host defense systems.

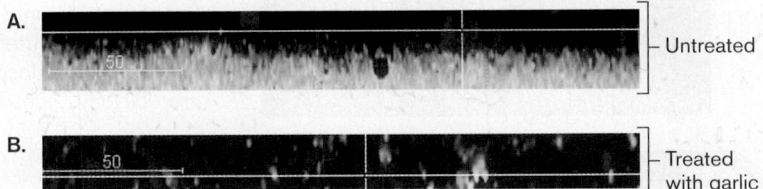

A. — Untreated

B. — Treated with garlic

Figure 1   **Garlic-dependent sensitivity of *Pseudomonas aeruginosa* biofilms toward PMNs.**   Two sets of biofilms of *P. aeruginosa* (green) containing green fluorescent protein were grown in the presence or absence of 2% garlic extract. On day 3 the biofilms were exposed to PMNs (stained red with fluorescent nucleus-staining dye) for 2.5 hours. **A.** Cross section of the untreated biofilm shows that PMNs were unable to penetrate and remained on top. **B.** In striking contrast, 2% garlic–treated biofilms were fully penetrated by PMNs. Size bar = 50 µm. *Source:* Thomas Bjarnsholt et al. 2005. *Microbiology* **151**:3873–3880.

interspecies communication, and individual species may perform specialized tasks in the community.

Organisms adapted to life in extreme environments also form biofilms. Members of Archaea form biofilms in acid mine drainage (pH 0), where they contribute to the recycling of sulfur; and cyanobacterial biofilms are common in thermal springs. Suspended particles called "marine snow" are found in ocean environments and appear to be floating biofilms comprising many organisms that have not yet been identified. The particles appear capable of methanogenesis, nitrogen fixation, and sulfide production, indicating that biofilm architecture can allow anaerobic metabolism to occur in an otherwise aerobic environment.

WWW | Biofilms and twitching motility
      microbiology2.com/links

TO SUMMARIZE:

- **Biofilms** are complex, multicellular, surface-attached microbial communities.
- **Chemical signals** enable bacteria to communicate (quorum sensing) and in some cases to form biofilms.

■ **Biofilm development** involves adherence of cells to a substrate, formation of microcolonies, and, ultimately, formation of complex channeled communities that generate new planktonic cells.

4.7 Cell Differentiation

Many bacteria faced with environmental stress undergo complex molecular reprogramming that includes changes in cell structure. Some species, like *E. coli*, experience relatively simple changes in cell structure, such as the formation of smaller cells or thicker cell surfaces. However, select species undergo elaborate cell differentiation processes. An example is *Caulobacter crescentus*, whose cells convert from the swimming form to the holdfast form before cell division. Each cell cycle then produces one sessile cell attached to its substrate by a holdfast, while its sister cell swims off in search of another habitat.

Other species undergo far more elaborate transformations. The endospore formers generate heat-resistant capsules (spores) that can remain in suspended animation for thousands of years. Yet another group, the actinomycetes, form complex multicellular structures analogous to those of eukaryotes. In this case, cell structure can change radically; and individual, freewheeling members of a species can relinquish their independence and band together, forming multicellular "organisms." The actinomycete that produces streptomycin is one example. These differentiation programs illustrate the distinction between the presence of genes in a bacterial genome and their activation. Genes for differentiation are expressed only when the cell needs to survive stress. We will discuss gene activation in more detail in Chapter 10.

Eukaryotic microbes also undergo highly complex life cycles. For example, *Dictyostelium discoideum* is a seemingly unremarkable ameba that grows as separate,

A.

John Foster, U. of South Alabama

B.

Courtesy of Peter Setlow

Figure 4.26 Endospore formation.
A. Photomicrograph of *Bacillus subtilis* spores. The cells (approx. 2 μm long) are stained with crystal violet. Cells are blue; spore is white.
B. Peter Setlow (right) of the University of Connecticut figured out how proteins regulate the process of differentiation of endospores.
C. Stages of endospore formation. ▶❙❙
microbiology2.com/animations

C.

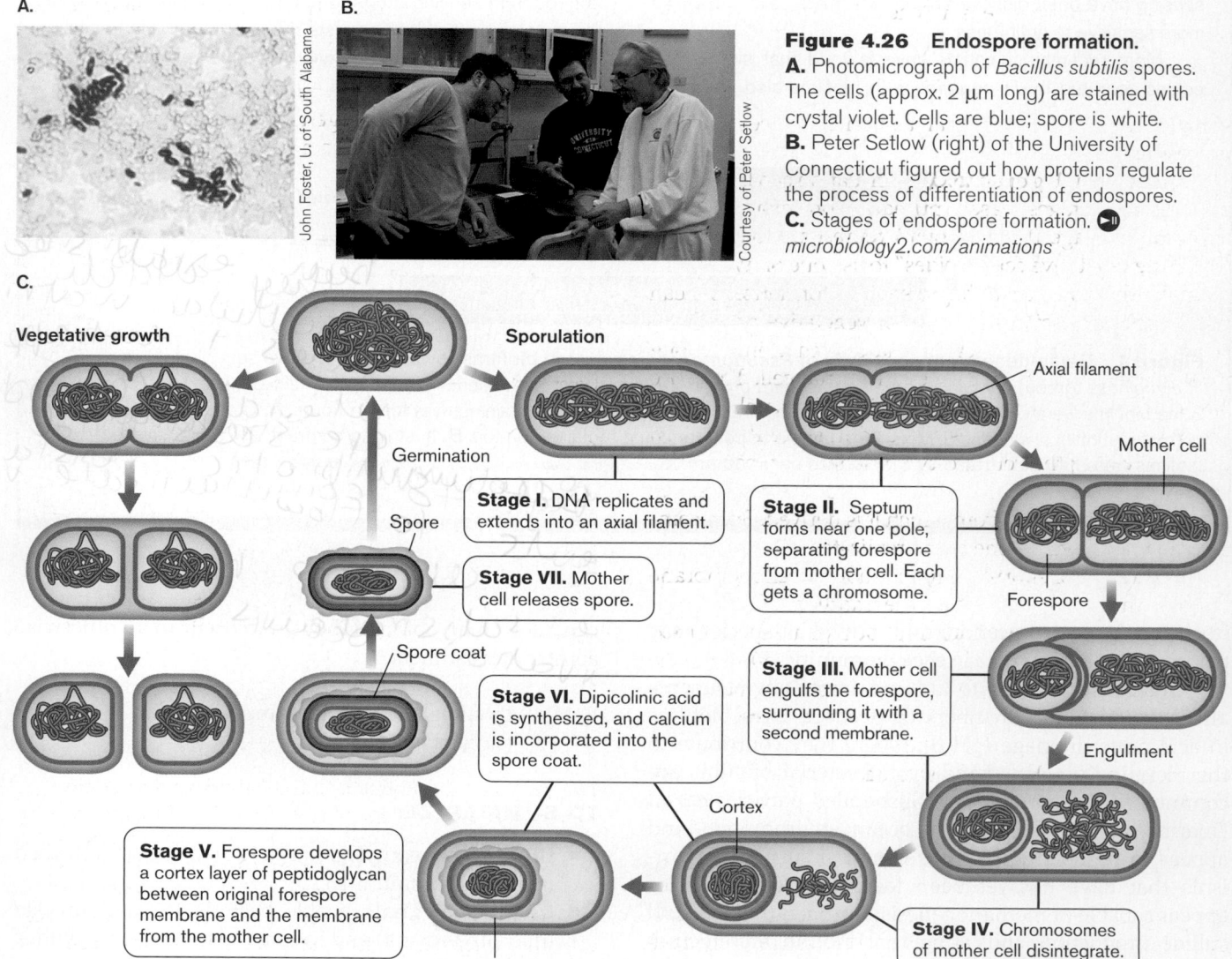

Vegetative growth

Sporulation

Germination

Stage I. DNA replicates and extends into an axial filament.

Spore

Stage VII. Mother cell releases spore.

Spore coat

Stage VI. Dipicolinic acid is synthesized, and calcium is incorporated into the spore coat.

Stage V. Forespore develops a cortex layer of peptidoglycan between original forespore membrane and the membrane from the mother cell.

Exosporangium

Axial filament

Mother cell

Stage II. Septum forms near one pole, separating forespore from mother cell. Each gets a chromosome.

Forespore

Stage III. Mother cell engulfs the forespore, surrounding it with a second membrane.

Engulfment

Cortex

Stage IV. Chromosomes of mother cell disintegrate.

independent cells. However, when challenged by adverse conditions such as starvation, the microbe secretes chemical signal molecules that choreograph a massive interaction of individuals to form complex multicellular structures. The developmental cycle of this organism may be compared to that of other eukaryotic microbes that are human parasites (discussed in Chapter 20).

Endospores Are Bacteria in Suspended Animation

Certain Gram-positive genera, including important pathogens such as *Clostridium tetani* (tetanus), *Clostridium botulinum* (botulism), and *Bacillus anthracis* (anthrax) have the remarkable ability to develop dormant spores that are heat- and desiccation-resistant. Desiccation and heat resistance are properties that make *B. anthracis* spores a potential bioweapon. Note that spores do not grow and do not need nutrients until they germinate.

Most of our knowledge of bacterial sporulation comes from the Gram-positive soil bacterium *Bacillus subtilis*. When growing in rich media, this microbe undergoes normal vegetative growth and can replicate every 30–60 minutes. However, starvation initiates an elaborate 8-hour genetic program that directs an asymmetrical cell division process and ultimately yields a spore (see Section 10.4).

As shown in **Figure 4.26**, sporulation can be divided into discrete stages based primarily on morphological appearance. Stage 0 (not shown) represents the point at which the vegetative cell "decides" to use one of two potential polar division sites to begin septum formation instead of the central division site used for vegetative growth. In stage I, the DNA is replicated and stretched into a long axial filament that spans the length of the cell. There are two chromosome copies at this point. Ultimately, one of the polar division sites wins out, and in stage II septation occurs, dividing the cell into two unequal compartments: the **forespore**, which will ultimately become the spore; and the larger **mother cell**, from which it is derived. Each compartment will contain one of the replicated chromosomes.

In stage III of sporulation, the mother cell membrane engulfs the forespore. Next, the mother cell chromosome is destroyed and a thick peptidoglycan layer (cortex) is placed between the two membranes surrounding the forespore protoplast (stage IV). Layers of coat proteins are then deposited on the outer membrane in stage V. Stage VI completes the development of spore resistance to heat and chemical insults. This last process includes the synthesis of dipicolinic acid and the uptake of calcium into the core of the spore. Finally, the mother cell, now called a sporangium, releases the mature spore (stage VII).

Spores are resistant to many environmental stresses that would kill vegetative cells. The nature of this resistance is due, in part, to desiccation of the spore (they have only 10%–30% of a vegetative cell's water content). But,

as discovered by Peter Setlow and colleagues, spores are also packed with small acid-soluble proteins (SASPs) that bind to and protect DNA. The SASP coat protects the spore's DNA from damage by ultraviolet light and various toxic chemicals.

A fully mature spore can exist in soil for at least 50–100 years, and spores have been known to last thousands of years. Once proper nutrient conditions arise, another genetic program, called **germination**, is triggered to wake the dormant cell, dissolve the spore coat, and release a viable vegetative cell.

Cyanobacteria Differentiate into Nitrogen-Fixing Heterocysts

Some of the autotrophic cyanobacteria, such as *Anabaena*, not only make oxygen through photosynthesis but "fix" atmospheric nitrogen to make ammonia. This is surprising because nitrogenase, the enzyme required to fix nitrogen, is very sensitive to oxygen, so one might expect that photosynthesis and nitrogen fixation would be two mutually exclusive physiological activities. *Anabaena* have solved this dilemma by developing specialized cells, called heterocysts, that function in nitrogen fixation (**Fig. 4.27**). A tightly regulated genetic program converts every tenth photosynthetic cell to a heterocyst, which loses the

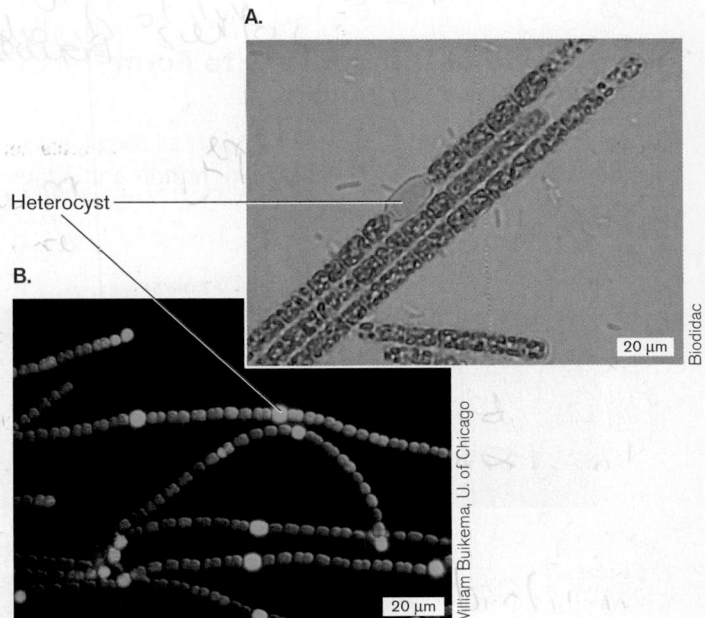

A.

B.

Heterocyst

20 μm

20 μm

William Buikema, U. of Chicago

Biodidac

Figure 4.27   Cyanobacteria and heterocyst formation.
A. Light-microscope image of cyanobacterium *Phanizomenon*.
B. The cyanobacterium *Anabaena*. The expression of genes in heterocysts is different from their expression in other cells. All cells in the figure contain a cyanobacterial gene to which the gene for green fluorescent protein (GFP) has been spliced. Only cells that have formed heterocysts are expressing the fused gene, which makes the cell fluoresce bright green.

capacity to fix CO_2 and forms a specialized envelope to limit O_2 access and allow N_2 fixation. How this organism produces such a precise spacing of heterocysts is currently the subject of intensive research.

Starvation Induces Differentiation into Fruiting Bodies

Certain species of bacteria, in the microbial equivalent of barn raising, produce architectural marvels called fruiting bodies. The Gram-negative species *Myxococcus xanthus* uses a **gliding motility** (involving a type of pilus, not a flagellum) to travel on surfaces as individuals or to move together as a mob (**Fig. 4.28**). Starvation triggers a developmental cycle in which 100,000 or more individuals aggregate, rising into a mound called a fruiting body. At this point, the system resembles a stage in *P. aeruginosa* biofilm formation. However, myxococci within the interior of the fruiting body differentiate into thick-walled, spherical spores that are released into the surroundings. The random dispersal of spores is an attempt to find new sources of nutrients. The changes involved in this differentiation process require many cell-cell interactions and a complex genetic program that we do not fully understand.

WWW | *Myxococcus* fruiting body
microbiology2.com/links

Some Bacteria Differentiate to Form Eukaryotic-like Structures

The actinomycetes (see Chapter 18) such as *Streptomyces* are bacteria that form mycelia and sporangia analogous to the filamentous structures of eukaryotic fungi (**Fig. 4.29**). Several developmental programs tied to nutrient availability are at work in this process (**Fig. 4.30**). Under favorable nutrient conditions, a germ tube emerges from a germinating spore, grows from its tip (tip extension), and forms branches that grow along the surface of its food source. This type of growth produces an intertwined network of long multinucleate filaments (**hyphae**) collectively called substrate **mycelia**. After a few days, new genes are activated that cause the hyphae to grow upward, rising above the surface to form aerial mycelia. Compartments at the tips of these aerial hyphae contain 20–30 copies of the genome. Aerial hyphae stop growing as nutrients decline, triggering a developmental program that synthesizes antibiotics and produces spores (arthrospores) that are fundamentally different from endospores. This program lays down multiple septa that subdivide the compartment into single-genome prespores. The shape of the prespore then changes, its cell wall thickens, and deposits are made in the spore that increase resistance to desiccation. These organisms are of tremendous interest, both for their ability to make antibiotics and for their fascinating developmental programs.

TO SUMMARIZE:

- **Microbial development** involves complex changes in cell forms.
- **Endospore development** in *Bacillus* and *Clostridium* involves production of dormant, stress-resistant endospores.
- **Heterocyst development** enables cyanobacteria to fix nitrogen anaerobically while maintaining oxygenic photosynthesis.

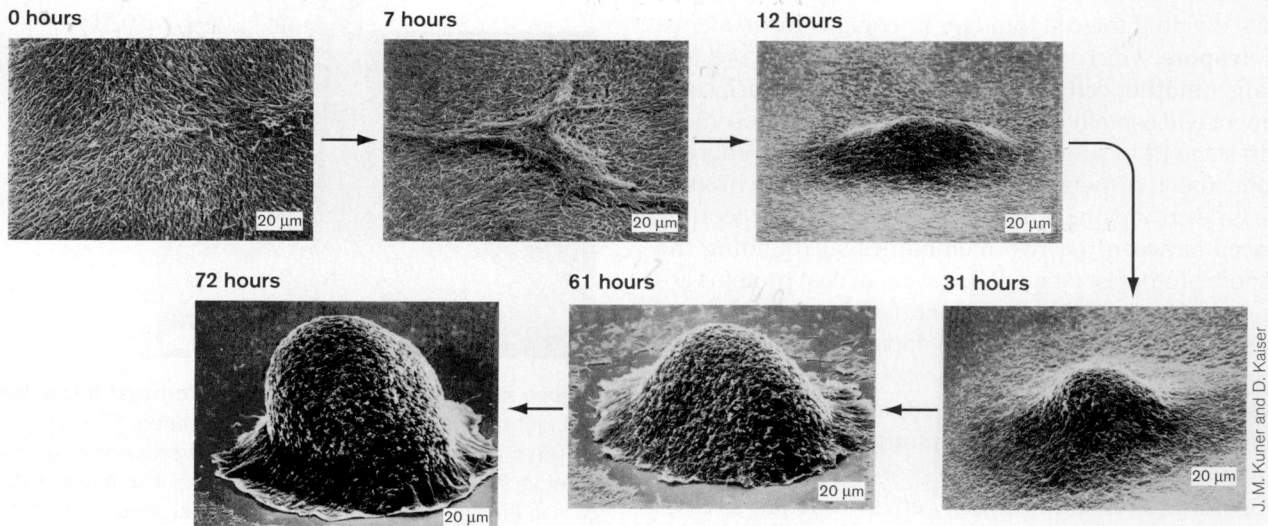

Figure 4.28 *Myxococcus* **swarm erecting a fruiting body.** Approximately 100,000 cells begin to aggregate, and over the course of 72 hours they erect a fruiting body.

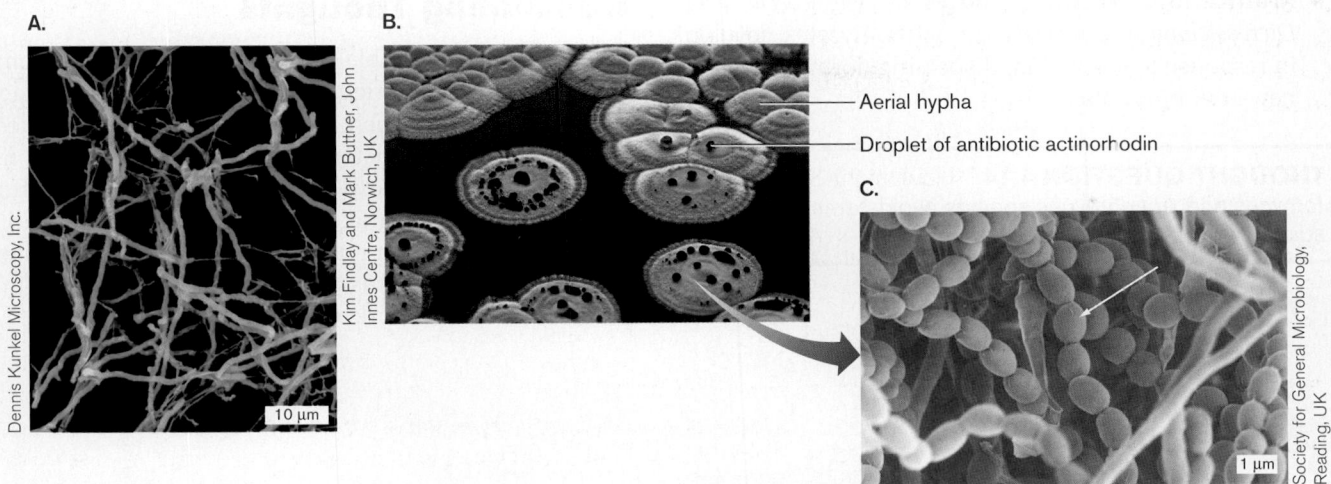

A.

Dennis Kunkel Microscopy, Inc.

10 µm

B.

Kim Findlay and Mark Buttner, John Innes Centre, Norwich, UK

Aerial hypha

Droplet of antibiotic actinorhodin

C.

Society for General Microbiology, Reading, UK

1 µm

Figure 4.29   Mycelia.   A. *Streptomyces lavendulae* substrate mycelia. **B.** Filamentous colonies of *Streptomyces coelicolor*, an actinomycete known for producing antibiotics (blue pigment in water droplets). **C.** *Streptomyces* aerial hyphae. Arrow points to a hyphal spore (approx. 1 µm each).

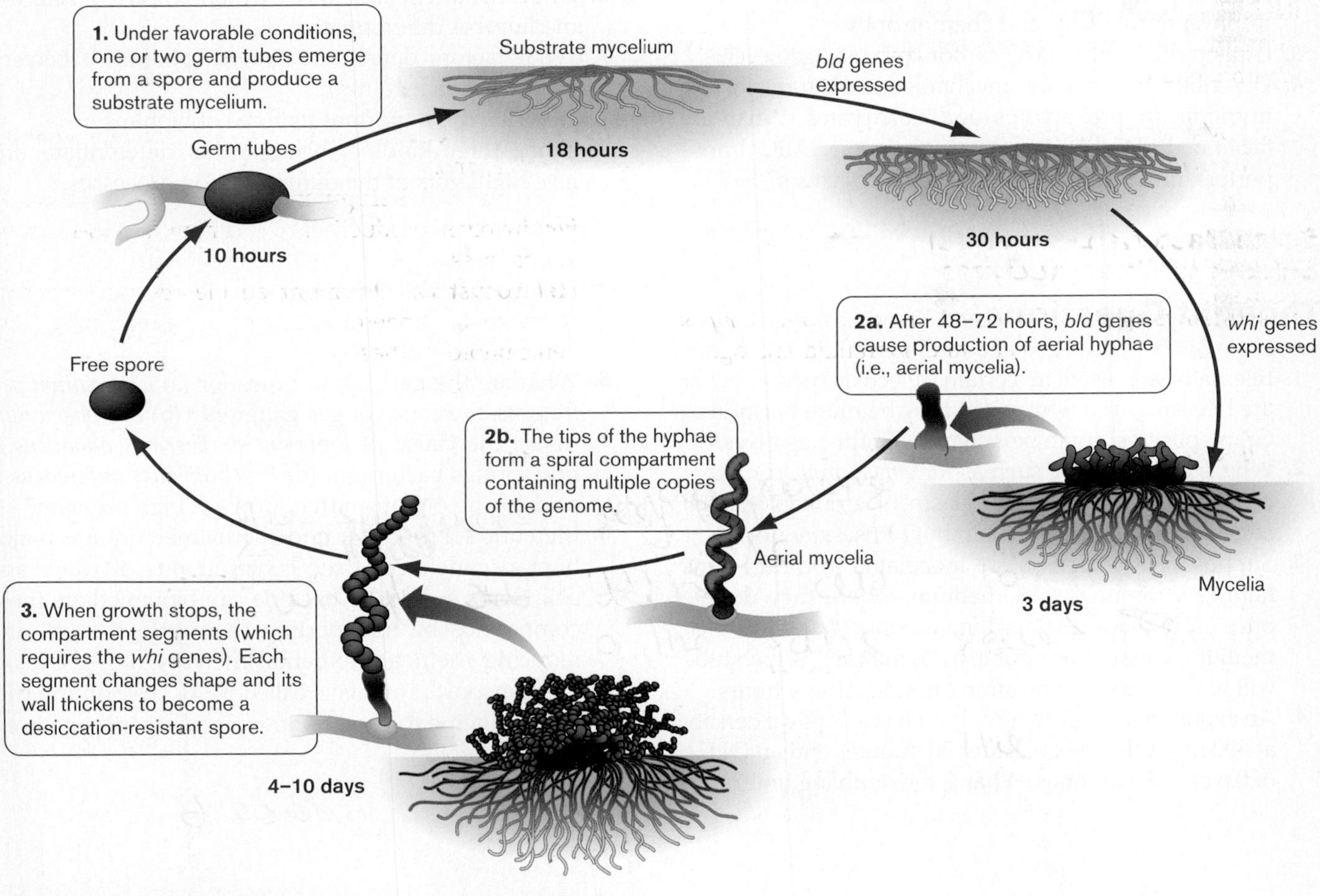

1. Under favorable conditions, one or two germ tubes emerge from a spore and produce a substrate mycelium.

Germ tubes

10 hours

Free spore

Substrate mycelium

18 hours

bld genes expressed

30 hours

whi genes expressed

2a. After 48–72 hours, *bld* genes cause production of aerial hyphae (i.e., aerial mycelia).

2b. The tips of the hyphae form a spiral compartment containing multiple copies of the genome.

Aerial mycelia

Mycelia

3 days

3. When growth stops, the compartment segments (which requires the *whi* genes). Each segment changes shape and its wall thickens to become a desiccation-resistant spore.

4–10 days

Figure 4.30   Developmental cycle of *Streptomyces coelicolor.*

■ **Multicellular fruiting bodies** in *Myxococcus* and *Dictyostelium* and **mycelia** in actinomycetes develop in response to starvation, dispersing dormant cells to new environments.

> **THOUGHT QUESTION 4.14** How might *Streptomyces* and *Actinomyces* species avoid committing suicide when they make their antibiotics?

Concluding Thoughts

All systems of microbial development share a common theme. They are all triggered in response to a change in their environment, such as depletion of resources, desiccation, or changes in temperature. Microbial responses to the environment have major implications for the function of ecosystems and for the microbial communities that inhabit the human body for good or ill. In Chapter 5, we survey the mechanisms of some of the major environmental responses of microbes.

CHAPTER REVIEW

Review Questions

1. What nutrients do microbes need to grow?
2. Explain the differences among autotrophy, heterotrophy, phototrophy, and chemotrophy.
3. Explain the basics of the carbon and nitrogen cycles.
4. Describe the various mechanisms of transporting nutrients in prokaryotes and eukaryotes. What are facilitated diffusion, coupled transport, ABC transporters, group translocation, and endocytosis?

5. Why is it important to grow bacteria in pure culture?
6. Under what circumstances would you use a selective medium? A differential medium?
7. What factors define the growth phases of bacteria grown in batch culture?
8. Describe the important features of biofilms.
9. Name three kinds of bacteria that differentiate, and give highlights of the differentiation processes.

Thought Questions

1. Bile salts are used in certain selective media. What are bile salts, and why might they be more harmful to Gram-positive organisms than to Gram-negatives?
2. Why is an organism, such as *Rickettsia prowazekii*, that can grow only in the cytoplasm of a eukaryotic cell considered a living organism but viruses are not?
3. Suppose 1,000 bacteria are inoculated in a tube containing a minimal salts medium, where they double once an hour; and 10 bacteria are inoculated into rich medium, where they double in 20 minutes. Which tube will have more bacteria after 2 hours? After 4 hours?
4. An exponentially growing culture has an optical density at 600 nm (OD_{600}) of 0.2 after 30 minutes and an OD_{600} of 0.8 after 80 minutes. What is the doubling time?

5. What are the generation times for (a) *Clostridium perfringens*, the cause of gas gangrene; (b) *Mycobacterium leprae*, the cause of leprosy; (c) *Thermus aquaticus*, a hot-springs bacterium; (d) *Psychromonas antarcticus*, a cold-loving organism that grows at high pressure?
6. Mercuric ions (Hg^{2+}) and methylmercury are major, human-generated, toxic contaminants of water and soil. Some species of bacteria can bioremediate these compounds by transporting them into the cell and reducing them to elemental mercury (Hg^0). One of the transport proteins is called MerC. Use an Internet search engine to explore how many organisms have a MerC homolog.

Key Terms

ABC transporter (123)
antiport (122)
ATP synthase (118)
autotroph (116, 118)
axenic growth (116)
batch culture (136)
binary fission (133)
biofilm (139)
chemoautotroph, chemoautotrophy (117, 118)
chemoheterotroph, chemoheterotrophy (117, 118)
chemolithotroph (117)
chemostat (138)
chemotrophy (117)
cofactor (115)
colony (127)
complex medium (128)
confluent (127)
continuous culture (138)
coupled transport (122)
death phase (137)
death rate (137)
defined minimal medium (115)
denitrification (119)
differential medium (129)
dilution streaking (127)
doubling time (134)
electrochemical potential (118)

electrogenic (122)
electroneutral (122)
enriched medium (128)
essential nutrient (114)
exopolysaccharide (EPS) (140)
exponential phase (137)
facilitated diffusion (121)
fluorescence-activated cell sorter (FACS) (131)
forespore (143)
generation time (134)
germination (143)
gliding motility (144)
group translocation (125)
growth factor (115)
growth rate (133)
growth rate constant (135)
heterotroph (116, 118)
hypha (144)
lag phase (136)
lithotroph (117)
logarithmic (log) phase (137)
MacConkey medium (129)
macronutrient (115)
mean generation time (135)
membrane potential (118)
microcolony (127)
micronutrient (115)
mixotrophic (120)

mother cell (143)
mycelium (144)
nitrification (119)
nitrogen-fixing bacterium (119)
optical density (132)
permease (121)
phosphotransferase system (PTS) (125)
photoautotroph, photoautotrophy (117, 118)
photoheterotrophy (118)
phototrophy (117)
planktonic cell (132)
pour plate (132)
proton potential (proton motive force) (118)
pure culture (126)
quorum sensing (140)
selective medium (129)
siderophore (124)
spread plate (127)
stationary phase (137)
substrate-binding protein (124)
symbiont (119)
symport (122)
synthetic medium (128)
twitching motility (140)
viable (127)
viable but nonculturable (VBNC) (128)

Recommended Reading

Angert, Esther R. 2005. Alternatives to binary fission in bacteria. *Nature Reviews. Microbiology* **3**:214–224.

Bassler, Bonnie L., and Richard Losick. 2006. Bacterially speaking. *Cell* **125**:237–246.

Bjarnsholt, T., P. O. Jensen, T. B. Rasmussen, L. Christophersen, H. Calum, et al. 2005. Garlic blocks quorum sensing and promotes rapid clearing of pulmonary *Pseudomonas aeruginosa* infections. *Microbiology* **151**:3873–3880.

Branda, Steven S., Ashlid Vik, Lisa Friedman, and Robert Kolter. 2005. Biofilms: The matrix revisited. *Trends in Microbiology* **13**:20–26.

Davidson, Amy L., and Jue Chen. 2004. ATP-binding cassette transporters in bacteria. *Annual Reviews in Biochemistry* **73**:241–268.

Davies, David G., and Claudia N. H. Marques. 2009. A fatty acid messenger is responsible for inducing dispersion in microbial biofilms. *Journal of Bacteriology* **191**:1393–1403.

Diggle, Stephen P., Ashleigh S. Griffin, Genevieve S. Campbell, and Stuart A. West. 2007. Cooperation and conflict in quorum sensing bacterial populations. *Nature* **450**:411–414.

Errington, John. 2010. From spores to antibiotics. *Microbiology* **156**:1–13.

Finkel, Steven E., and Robert Kolter. 1999. Evolution of microbial diversity during prolonged starvation. *Proceedings of National Academy of Sciences USA* **96**:4023–4027.

Higgins, Christopher F. 2001. ABC transporters: Physiology, structure and mechanism—an overview. *Research in Microbiology* **152**:205–210.

Nichols, D., K. Lewis, J. Orjala, S. Mo, R. Ortenberg, et al. 2008. Short peptide induces an "uncultivable" microorganism to grow in vitro. *Applied and Environmental Microbiology* **74**:4889–4897.

Nystrom, Thomas. 2004. Stationary-phase physiology. *Annual Review of Microbiology* **58**:161–181.

Omsland, Anders, Diane C. Cockrell, Dale Howe, Elizabeth R. Fischer, Kimmo Virtaneva, et al. 2009. Host cell-free growth of the Q fever bacterium *Coxiella burnetii*. *Proceedings of the National Academy of Sciences USA* **106**:4430–4434.

Piggot, Patrick J., and David W. Hilbert. 2004. Sporulation of *Bacillus subtilis*. *Current Opinion in Microbiology* **7**:579–586.

Saier, Milton H., Jr., Bin Wang, Wei Hao Zheng, Eric I. Sun, and Ming Ren Yen. 2010. Mosaic energy-coupled transporters. *Microbe.* **5**:105–109.

Skerker, Jeffrey M., and Michael T. Laub. 2004. Cell-cycle progression and the generation of asymmetry in *Caulobacter crescentus*. *Nature Reviews. Microbiology* **2**:325–337.

For further study and review, please visit StudySpace at **<u>microbiology2.com</u>**.

Chapter 5

Environmental Influences and Control of Microbial Growth

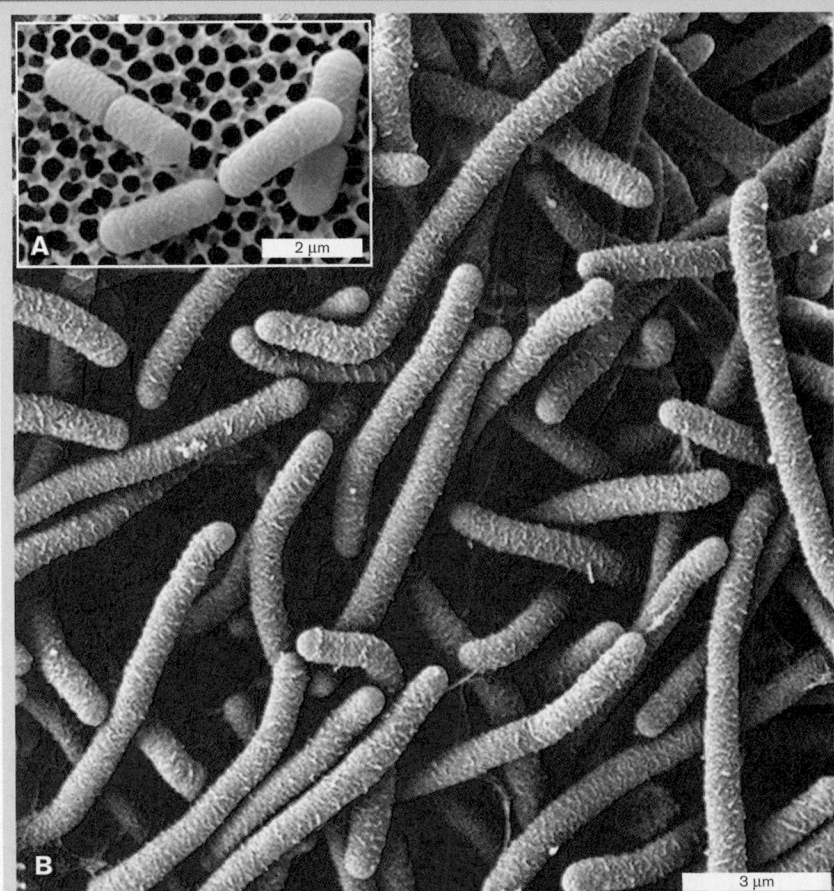

Microbes have both the fastest and the slowest growth rates of any known organism. Some hot-springs bacteria can double in as little as 10 minutes, whereas deep-sea sediment microbes may take as long as 100 years. What determines these differences in growth rate? Nutrition is one factor, but niche-specific physical parameters like temperature, pH, and osmolarity are equally important.

A microbe's physiology is geared to work only within a narrow range of physical parameters. But in nature, the environment can quickly change. Many marine microbes, for instance, can move from deep-sea cold to the searing heat of a thermal vent. How do these organisms survive? Stopgap measures called stress survival responses help, but some species evolve to thrive, not just survive, in extreme environments. How do these so-called extremophiles grow under conditions that seem uninhabitable? In Chapter 5, we explore the limits of microbial growth and show how this knowledge helps us control the microbial world.

Current Research Highlight

New evidence reveals that some bacteria can change shape to cope with harsh environments. *Lactobacillus plantarum* is a Gram-positive species found in fermented foods that can grow under acidic to neutral pH conditions. The SEM shows shape-shifting morphological changes that occur when the organism transitions from growth at pH 6.5 (**A**) to pH 3 (**B**). At pH 3 the organisms become filamentous and develop a rough cell surface. This population is acid resistant and reverts to the less resistant, shorter, smoother cell when returned to pH 6.5. Why the elongated cells are more acid resistant is unknown. *Source*: C. J. Ingham et al. 2008. *Appl. Environ. Microbiol.* **74**:7750–7758. © 2008, American Society for Microbiology. All Rights Reserved.

In 1998, apple and pear growers in Washington and northern Oregon lost crops worth an estimated $68 million owing to outbreaks of fire blight, a devastating bacterial disease. The causative agent, *Erwinia amylovora*, destroys apple and pear trees, making them appear as if they were torched by fire. How could growth of these bacteria stop? Cold might do it. Fire blight only progresses when temperatures rise above 18°C (65°F). How temperature and other environmental conditions affect bacterial growth will be discussed in this chapter. But cold will not kill these bacteria. Fortunately, during active disease, the pathogens can be killed by antibiotics, although only two, streptomycin and oxytetracycline, are approved for use on plants. How antibiotics are used to control bacterial growth will also be introduced in this chapter. But, resistance to streptomycin has already been demonstrated in *Erwinia amylovora*, underscoring why all aspects of bacterial growth must be studied in an effort to control pathogens of plants, animals, and humans.

We begin Chapter 5 by discussing how physical and chemical changes in the environment modify the growth of different groups of microbes. We will also explore how microorganisms adapt to different environments in ways both transient (involving temporary expression of inactive genes) and permanent (modifications of the gene pool). The permanent genetic changes have led to biological diversity. Finally, we will examine the different ways humans try to limit the growth of microorganisms to protect plants, animals, and ourselves.

As you proceed through this chapter, you will encounter two recurring themes: that different groups of microbes live in vastly different environments and that microbes can respond in diverse ways when confronted by conditions outside their niche, or comfort zone.

5.1 Environmental Limits on Growth

With our human frame of reference, we tend to think that "normal" growth conditions are those found at sea level with a temperature between 20°C and 40°C, a near-neutral pH, a salt concentration of 0.9%, and ample nutrients. Any ecological niche outside this window is called "extreme" and the organisms inhabiting them **extremophiles** (R. MacElroy first used the term "extremophile" in 1974). Extremophiles are prokaryotes (bacteria and archaea) that are able to grow in extreme environments. For example, one group of organisms can grow at temperatures above boiling, while another group requires a pH 2 acidic environment to grow. By our definition of "normal" conditions, conditions on Earth when life began were certainly extreme. Consequently, the earliest microbes likely grew in these extreme environments.

Organisms that grow under conditions that seem normal to humans likely evolved from an ancient extremophile that gradually adapted as the environment evolved to that of our present-day Earth.

Be aware that multiple extremes in the environment can be encountered simultaneously. For instance, in Yellowstone National Park, one can find an extreme acid pool next to an extreme alkali pool, both at extremely high temperatures. Thus, extremophiles typically evolve to survive multiple extreme environments.

Extremophiles may provide insight into the workings of extraterrestrial microbes we may one day encounter, since outer space certainly qualifies as an extreme environment. Our experiences with extremophiles should alert us to the dangers of underestimating the precautions necessary in handling extraterrestrial samples. For example, we should not assume that irradiation would be sufficient to sterilize samples from future planetary or interstellar missions. Such treatments do not even kill the extremophile *Deinococcus radiodurans* found on Earth.

WWW Extremophiles
microbiology2.com/links

How do we even begin to study organisms that grow in boiling water or sulfuric acid solutions or organisms that we cannot even culture in the laboratory? How do we dissect the molecular response of organisms to changes in an environment such as acid rain or to changes in the human body? Genome sequences present new opportunities for investigating these questions.

Bioinformatic analysis allows us to study the biology of organisms that we cannot culture. First, genome sequences from novel organisms are amplified by the polymerase chain reaction, or PCR (a technique that multiplies a small sample of DNA; see Section 7.6). PCR amplification can be done directly from the natural environment in which the organisms are found, and the sequences compared with those of model systems, such as *E. coli*, whose biochemistry is well known. Genomic comparison quickly reveals whether an organism under study may possess specific metabolic pathways and regulatory responses.

Techniques that examine the expression of all the genes in a genome or all the proteins in the cytoplasm allow us to study the response of an organism to changing environments. These so-called global approaches can reveal in a single experiment the response of all an organism's genes to a single environmental change, such as change in temperature or pH (**Fig. 5.1**). **DNA microarrays** (see Section 10.9) consist of slides with a grid containing DNA probes for every gene in an organism's genome. They are used to assess which genes are

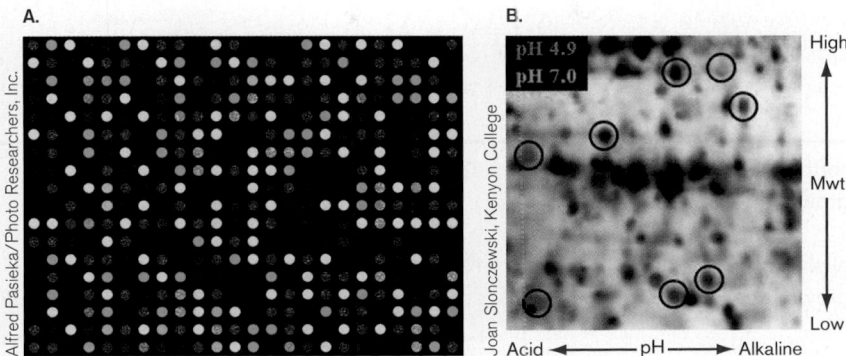

Figure 5.1   Response to environmental stress: global analysis of genes and proteins.   A. This microarray contains DNA probes (small pieces of DNA sequence) representing each gene in the genome. The DNA probes hybridize to preparations of fluorescent-labeled cDNA (complementary DNA) made from whole-cell RNA of cultures grown under different conditions. Each colored disk represents one gene. Red fluorescence indicates RNA expressed under one condition; green indicates expression under the second condition; yellow indicates expression under both conditions. **B.** Two-dimensional gels separate the proteins expressed in the cell. Proteins are separated in one dimension by their charge (isoelectric pH point) and in the second dimension by their molecular weight (Mwt). Gel images are obtained from cultures grown under two different conditions (in this case, *Escherichia coli* K-12 grown at pH 4.9 versus pH 7.0). Pink spots denote proteins expressed only at pH 4.9; green indicates expression only at pH 7.0. Circled spots were chosen for additional studies.

expressed to make RNA in a given organism at a given time or under a given condition. Two-dimensional protein gels (see Section 3.1) achieve two-dimensional separation of proteins based on differences in each protein's isoelectric point (first dimension) and molecular weight (second dimension). They show which proteins the cell produces when cultured in different environments. Knowing what genes and proteins are expressed in a given situation helps elucidate the molecular strategies that microbes use to grow under different conditions and to defend themselves against environmental stresses. These techniques of molecular analysis are discussed further in Chapter 10.

We have already mentioned the fundamental physical conditions (temperature, pH, osmolarity) that define an environment and select for (favor) the growth of specific groups of organisms. And within a given microbial community, each species is further localized to a specific niche defined by a narrower range of environmental factors. Species are thus localized because every protein and macromolecular structure within a cell is affected profoundly by changes in environmental conditions and through reaction with the by-products of oxygen consumption. For example, a single enzyme works best under a unique set of temperature, pH, and salt conditions because those conditions allow it to fold into its optimum shape, or conformation. Deviations from these optimal conditions cause the protein to fold a little differently and become less active. While not all enzymes within a cell boast the same physical optima, these optima must at least be similar and matched to the organism's environment for the organism to function effectively.

As may be surmised from the preceding discussion, microbes are commonly classified by their environmental niche. **Table 5.1** summarizes these environmental classes.

TO SUMMARIZE:

- **Extremophiles** inhabit fringe environments with conditions that do not support human life.
- **The environmental habitat** (such as high salt or acidic pH) inhabited by a particular species is defined by the tolerance of that organism's proteins and other macromolecular structures to the physical conditions within that niche.
- **Global approaches** used to study gene expression allow us to view how organisms respond to changes in their environment.

5.2   Adaptation to Temperature

Unlike humans (and mammals in general), microbes cannot control their temperature; thus, bacterial cell temperature matches that of the immediate environment. Because temperature affects the average rate of molecular motion, changes in temperature impact every aspect of microbial physiology, including membrane fluidity, nutrient transport, DNA stability, RNA stability, and enzyme structure and function. Every organism has an "optimum" temperature at which it grows most quickly, as well as minimum and maximum temperatures that define the limits of growth. These limits are imposed, in part, by the thousands of proteins in a

Table 5.1   **Basic environmental classification of microorganisms.**

Environmental parameter	Classification			
Temperature	Hyperthermophile* (growth above 80°C)	Thermophile* (growth between 50°C and 80°C)	Mesophile (growth between 15°C and 45°C)	Psychrophile* (growth below 15°C)
pH	Alkaliphile* (growth above pH 9)	Neutralophile (growth between pH 5 and pH 8)	Acidophile* (growth below pH 3)	
Osmolarity	Halophile* (growth in high salt, > 2-M NaCl)			
Oxygen	Aerobe (growth only in oxygen)	Facultative (growth with or without oxygen)	Microaerophile (growth only in small amounts of oxygen)	Anaerobe (growth only without oxygen)
Pressure	Barophile* (growth at high pressure, greater than 380 atm)		Barotolerant (growth between 10 and 495 atm)	

*Considered extremophiles.

cell, all of which must function within the same temperature range. The fastest growth rate for a species occurs at temperatures where all of the cell's proteins work most efficiently as a group to produce energy and synthesize cell components. Growth stops when rising temperatures cause critical enzymes or cell structures (such as the cell membrane) to fail. At cold temperatures, growth ceases because enzymatic processes become too sluggish and the cell membrane less fluid. The membrane needs to remain fluid so that it can expand as cells grow larger and so that proteins needed for solute transport can be inserted into the membrane.

Growth Rate and Temperature

In general, microbes that grow at higher temperatures can achieve higher rates of growth. Remarkably, the relationship between the maximum growth temperature and the growth rate constant k (number of generations per hour; see Section 4.5) obeys the Arrhenius equation for simple chemical reactions:

$$\log k = C/T \qquad (5.1)$$

where T is the absolute temperature in kelvins (K), and C is a second constant that combines the gas constant and the average activation energy of cellular reactions. Over a defined temperature range, which differs for each species, growth rate increases (that is, cells divide faster) as temperature increases. If we plot the logarithm of the growth rate versus temperature, we get a

straight line whose slope (C) we can obtain by substituting in equation 5.1:

$$\log (k_2/k_1) = C/(T_1 - T_2)$$

The general result of the Arrhenius equation is that growth rate roughly doubles for every 10°C rise in temperature (**Fig. 5.2A**). This is the same relationship observed for any chemical reaction.

At the upper and lower limits of the growth range, however, the Arrhenius effect breaks down. At high temperatures, critical proteins denature. Lower temperatures decrease membrane fluidity and limit the conformational mobility of enzymes, thereby lowering their activities. As a result, growth fails to occur at temperature extremes. The typical growth temperature range spans about 30°C–40°C, but some organisms have a much narrower range. Within a species, we can identify mutants that are more sensitive to one extreme or the other (heat sensitive or cold sensitive). These mutations often define key molecular components of stress responses, such as the heat-shock proteins (discussed in Chapter 10).

Thermodynamic principles limit a cell's growth to a narrow temperature range. For example, heat increases molecular movement within proteins. Too much or too little movement will interfere with enzymatic reactions. We observe great biological diversity among microbes because different groups have evolved to grow within very different thermal ranges. A species grows within a specific thermal range because its proteins have evolved to tolerate that range. Outside that range, proteins will

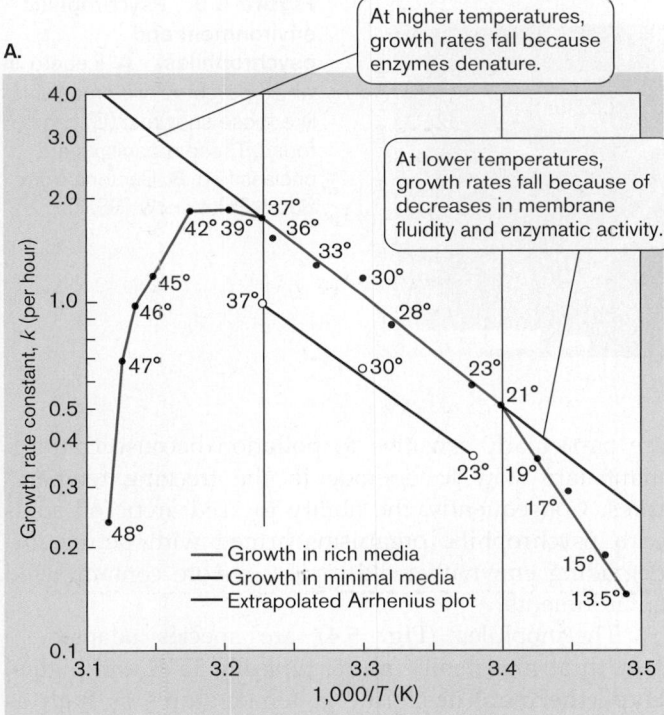

A.

At higher temperatures, growth rates fall because enzymes denature.

At lower temperatures, growth rates fall because of decreases in membrane fluidity and enzymatic activity.

— Growth in rich media
— Growth in minimal media
— Extrapolated Arrhenius plot

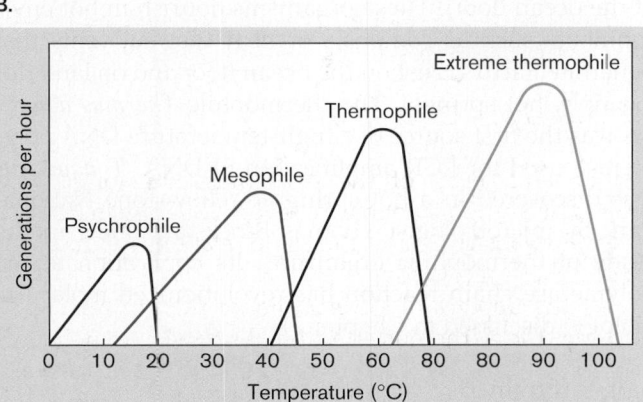

B.

Figure 5.2 Relationship between temperature and growth rate. A. Growth rate constant (k) of the enteric organism *Escherichia coli* is plotted against the inverse of growth temperature on the Kelvin scale ($1,000/T$ is used to give a convenient scale on the x-axis). This is a more detailed view of a mesophilic growth temperature curve. As temperature rises above or falls below the optimum range, growth rate decreases faster than predicted by the Arrhenius equation. **B.** Relationship between temperature and growth rates of different groups of microbes. Note that the growth rate increases linearly with temperature and obeys the Arrhenius equation. *Source*: B. Sherrie L. Herendeen et al. 1979. *J. Bacteriol.* **139**:185.

ranging from below 0°C to above 100°C. Temperatures over 100°C are usually found near thermal vents deep in the ocean. Vent water temperature can rise to 350°C but the pressure is sufficient to keep water liquid.

> **THOUGHT QUESTION 5.1** Why haven't cells evolved so that all their enzymes have the same temperature optimum? If they did, wouldn't they grow even more rapidly?

Microorganisms Are Classified by Growth Temperature

According to their ranges of growth temperature, microorganisms can be classified as mesophiles, psychrophiles, or thermophiles (**Fig. 5.2B**).

Mesophiles include the typical "lab rat" microbes, such as *Escherichia coli* and *Bacillus subtilis*. Their growth optima range between 20°C and 40°C, with a minimum of 15°C and a maximum of 45°C. Because they are easy to grow and because all human pathogens are mesophiles, much of what we know about protein, membrane, and DNA structure came from studying this group of organisms. Current advances, particularly in obtaining detailed three-dimensional (3-D) views of protein structures, are frequently based on studies of two other classes of organisms whose optimum growth temperature ranges flank that of the mesophiles—namely **psychrophiles** (on the low temperature side) and **thermophiles** (on the high temperature side). Because of their more stable folding structures, proteins from the thermophilic extremophiles are generally easier to crystallize than those of mesophiles or psychrophiles, so it is possible to determine their structures by X-ray crystallography (see Section 2.7).

Psychrophiles are microbes that grow at temperatures as low as 0°C, but their optimum growth temperature is usually around 15°C. Psychrophiles are prominent flora beneath icebergs in the Arctic and Antarctic (**Fig. 5.3**). In addition to true psychrophiles, whose growth optima are below 20°C, there are mesophiles that are cold resistant. For instance, the modern practice of refrigeration has selected for cold-resistant (psychrotolerant), yet mesophilic, pathogens such as *Listeria monocytogenes* (one cause of food poisoning and septic abortions).

Why do these organisms grow so well in the cold? One reason is that the proteins of psychrophiles are more flexible than those of mesophiles and require less energy (heat) to function. Of course, the downside to the increased flexibility of psychrophilic proteins is that they denature at lower temperatures than their mesophilic counterparts. As a result, psychrophiles grow poorly, if at all, when temperatures rise above 20°C. Another

denature or function too slowly for growth. The upper limit for protists is around 50°C, while some fungi can grow at temperatures as high as 60°C. Prokaryotes, however, have been found to grow at temperatures

A.

© Gerald & Buff Corsi/Visuals Unlimited

B.

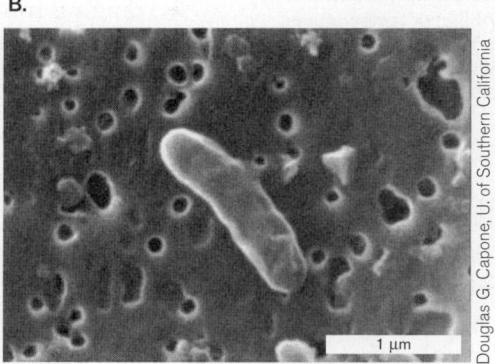

1 μm

Douglas G. Capone, U. of Southern California

Figure 5.3 Psychrophilic environment and psychrophiles.   **A.** Iceberg, in which psychrophilic organisms like those shown in (B) can be found. These organisms are unclassified. **B.** Bacteria from South Polar snow (SEM).

reason psychrophiles favor cold is that their membranes are more fluid at low temperature owing to the high proportion of unsaturated fatty acids present; at higher temperatures, their membranes are *too* flexible and fail to maintain cell integrity. Finally, bacteria and archaea that grow at 0°C in glaciers also contain antifreeze proteins and other cryoprotectants such as trehalose that can depress the freezing point by 2°C. So, although these organisms can grow in ice, they will not freeze. Interestingly, some psychrophilic and psychrotolerant bacteria actually stimulate ice formation in their surrounding environment (**Special Topic 5.1**).

Psychrophilic enzymes are of commercial interest because their ability to carry out reactions at low temperature has potential utility in food processing and bioremediation. Many food products require enzymatic processing. Enzymes help brew beer more quickly, break down lactose in milk, and can remove cholesterol from various foods. Production of foods at lower temperatures would be beneficial, since lower processing temperatures would minimize the growth of typical mesophiles that degrade and spoil food. Another goal is to find organisms that can safely degrade toxic organic contaminants (for example, petroleum) in the cold. Arctic environments

are particularly sensitive to pollution because contaminants are slow to degrade in the freezing temperatures. Consequently, the ability to seed arctic oil spills with psychrophilic organisms armed with petroleum-degrading enzymes could rapidly restore contaminated environments.

Thermophiles (**Fig. 5.4**) are species adapted to growth at high temperature, typically 55°C and higher. **Hyperthermophiles** grow at temperatures as high as 121°C, which occur under extreme pressure (for example, at the ocean floor). These organisms flourish in hot environments such as composts or near thermal vents that penetrate Earth's crust on the ocean floor and on land (for example, hot springs). The thermophile *Thermus aquaticus* was the first source of a high-temperature DNA polymerase used for PCR amplification of DNA. *T. aquaticus* was discovered in a hot spring at Yellowstone National Park by microbiologist Thomas Brock, a pioneer in the study of thermophilic organisms. Its application to the polymerase chain reaction has revolutionized molecular biology (discussed in Chapter 7).

www | Strain 21
| *microbiology2.com/links*

A.

©Momatiuk-Eastcott/Corbis

B.

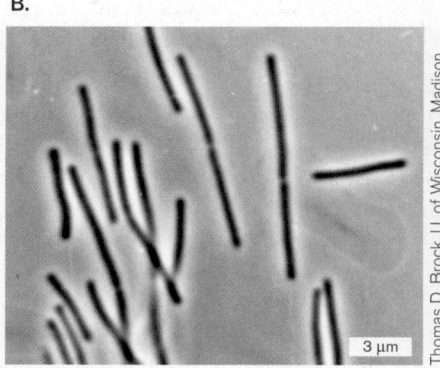

3 μm

Thomas D. Brock, U. of Wisconsin, Madison

C.

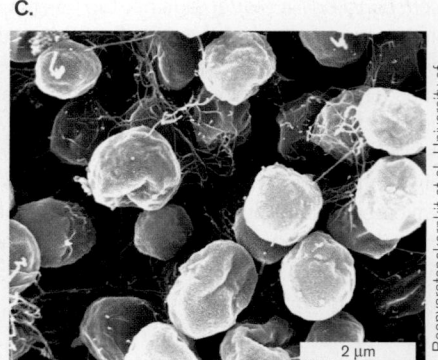

2 μm

B. Boonyaratanakornkit et al, University of California, Berkeley

Figure 5.4   Thermophilic environments and thermophiles.   A. Yellowstone National Park hot spring. **B.** *Thermus aquaticus*, a hyperthermophile first isolated at Yellowstone by Thomas Brock. Cell length varies from 3 to 10 μm. **C.** Thermophile *Methanocaldococcus jannaschii* grown at 78°C and 30 psi.

Special Topic 5.1   *It's Raining Bacteria*

How clouds, rain, and snow form has intrigued children and scientists for millennia. Little did anyone know that bacteria are major players in these processes. David Sands at Montana State University first hypothesized a link between rainfall and psychrotolerant bacteria in 1982. Though the scientific community initially scoffed at Sands' idea, he has now been proven right. Without bacteria (or some other tiny particles), clouds would never form because water vapor droplets are too small—about 250 nm (approximately one ten-thousandth of an inch) or less. A powerful surface tension forms the small curved surfaces on each droplet. Water vapor that goes into the liquid state must overcome this surface tension to form the larger droplets that make up clouds. Left alone, water vapor would never form clouds and there would never be rain.

There also would never be snow. Ice formation in clouds is required for snow and even most rainfall. At temperatures above −40°C, however, ice formation is not spontaneous. Tiny catalysts, known as ice nucleators, are required. Scientists knew for years that a protein in the outer membrane of certain bacterial plant pathogens, such as the psychrotolerant *Pseudomonas syringae* and *Erwinia* species, have the capacity to freeze pure water at temperatures as warm as −1°C. The protein binds water molecules in an ordered arrangement, providing a nucleating template that enhances ice crystal formation (**Fig. 1**). The ice crystals that form at this relatively warm temperature break the cell walls of the plants and release nutrients that the bacteria use as food. Ice-nucleating bacteria have been found at altitudes of several kilometers and have been documented in rain and snowfall.

A team led by Brent Christner at Louisiana State University in Baton Rouge has shown the ubiquity of these rainmaking microbes by looking at fresh snow collected at various mid- and high-latitude locations in North America, Europe, and Antarctica. They filtered the snow samples to remove particles, put those particles into containers of pure water, and slowly lowered the temperature, watching closely to see when the water froze. The higher the freezing temperature of any given sample, the greater the number of nuclei and the more likely they were to be biological in nature. To tease apart these two effects, the team treated the water samples with heat or chemicals to kill any bacteria inside, and again checked the freezing temperatures of the samples. In this way they found between 4 and 120 ice nucleators per liter of melted snow. Some 69%–100% of these particles appeared to be biological.

Why is this important? The ability to initiate freezing means that rainmaking bacteria can spur showers as a way of dispersing themselves worldwide. The bacteria facilitate cloud formation, clouds can move large distances in wind currents, and the resulting rain or snow will deposit the bacteria far afield from where they started. So, while climate can affect microbes, microbes can, in turn, affect the weather.

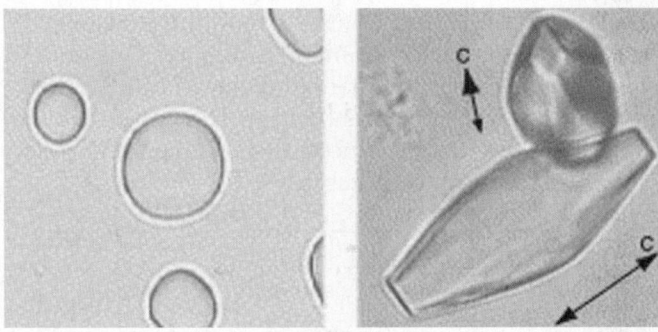

Figure 1   Ice crystallization by a bacterial ice nucleation protein. Photomicrographs of the ice crystal observed for **(A)** solvent buffer alone ([Tris−HCl] = 20 mM (pH 8.0), [EDTA] = 1 mM, [NaCl] = 0.5 M), and **(B)** approximately 400 μM of the ice nucleation protein INP96 from *Pseudomonas syringae* in the buffer. *Source:* Yoshihiro Kobashigawa et al. 2005. *FEBS Lett.* **579**:1493–1497. © 2005 Federation of European Biochemical Societies.

Extreme thermophiles often have specially adapted membranes and protein sequences. The thermal limits of these structures determine the specific high-temperature ranges in which various species can grow. Because enzymes in thermophiles (thermozymes) do not unfold as easily as mesophilic enzymes, they more easily hold their shape at higher temperatures. Thermophilic enzymes are stable, in part, because they contain relatively low amounts of glycine, a small amino acid that contributes to an enzyme's flexibility (glycines do not contain side chains, so they cannot form stabilizing intramolecular bonds). In addition, the amino termini of proteins in these organisms often are "tied down" by hydrogen bonding to other parts of the protein, making them harder to denature.

Like all microbes, thermophiles have chaperone proteins that help refold other proteins as they undergo thermal denaturation. Thermophile genomes are packed with numerous DNA-binding proteins that stabilize DNA. In addition, these organisms possess special enzymes that function to tightly coil DNA in a way that makes it more thermostable and less likely to denature (think of a coiled phone cord that has twisted and bunched up on itself). Special membranes also help give cells additional stability at high temperatures. Unlike the typical lipid bilayers of mesophiles, the membranes of thermophiles manage to "glue" together parts of the two hydrocarbon layers that point toward each other, making them more stable. They do this by incorporating more saturated linear lipids into their membranes. Saturated lipids form straight hydrocarbon tails that align well with neighboring lipids and form a highly organized structure stable to heat. The membranes of mesophiles are composed mostly of unsaturated lipids that bend against each other and align poorly. This property makes the membranes of mesophiles more fluid at lower temperatures.

The Heat-Shock Response

As insurance against extinction, most microorganisms possess elegant genetic programs that remodel their physiology to one that can temporarily survive inhospitable conditions. Rapid temperature changes experienced during growth activate batches of stress response genes, resulting in the **heat-shock response** (discussed in Chapter 10). The protein products of these heat-activated genes include chaperones that maintain protein shape and enzymes that change membrane lipid composition. The heat-shock response, first identified in *E. coli* by Yamamori and Yura in 1982, has since been documented in all living organisms examined thus far.

> **THOUGHT QUESTION 5.2** If microbes lack a nervous system, how can they sense a temperature change?

The appearance of different branches of life in the course of evolution reflects, in some ways, the narrowing of tolerance to heat. Different archaeal species, for example, can grow in extremely hot or extremely cold temperatures, and some can grow in the middle range. Bacteria, for the most part, tolerate a temperature range that bridges the archaeal extremes. Eukaryotes are less temperature tolerant than bacteria, with individual species capable of growth between 10°C and 65°C. Archaeal species have been found to grow at greater extremes than bacterial species, and bacteria at greater extremes than eukaryotes. As we will see, this evolutionary relationship holds for other environmental conditions.

TO SUMMARIZE:

- **Different species** exhibit different optimal growth values of temperature, pH, and osmolarity.
- **The Arrhenius equation** applies to growth of microorganisms: Growth rate doubles for every 10°C rise in temperature.

- **Membrane fluidity** varies with the composition of lipids in a membrane, which in turn dictates the temperature at which an organism can grow.
- **Mesophiles, psychrophiles, and thermophiles** are groups of organisms that grow at moderate, low, and high temperatures, respectively.
- **The heat-shock response** produces a series of protective proteins in organisms exposed to temperatures near the upper edge of their growth range.

5.3 Adaptation to Pressure

Living creatures at Earth's surface (sea level) are subjected to a pressure of 1 atmosphere (atm), which is equal to 0.101 megapascal (MPa) or 14 pounds per square inch (psi). At the bottom of the ocean, however, thousands of meters deep, hydrostatic pressure averages a crushing 400 atm and can go as high as 1,000 atm (101 MPa, or 14,000 psi) in ocean trenches (**Fig. 5.5**). Organisms adapted to grow at these overwhelmingly high pressures are called **barophiles** or **piezophiles**. From the curves in **Figure 5.6**, we can see that barophiles actually *require* elevated pressure to grow, while barotolerant organisms grow well over the range of 1–50 MPa, but their growth falls off thereafter.

Many barophiles are also psychrophilic because the average temperature at the ocean's floor is 2°C. However, barophilic hyperthermophiles form the basis of thermal vent communities that support symbiotic worms and giant clams (see Chapter 21).

How bacteria can survive pressures of 12,000 psi is still a mystery. It is known that increased hydrostatic pressure and cold temperatures similarly reduce membrane fluidity. Because fluidity of the cell membrane is critical to survival, the phospholipids of deep-sea bacteria commonly have high levels of polyunsaturated fatty acids to increase membrane fluidity. It is thought that in addition to these membrane changes, internal structures must also be pressure adapted. For example, ribosomes in the barosensitive organism *E. coli* dissociate at pressures above 60 MPa, so barophiles must contain uniquely designed ribosome structures that can withstand pressures even higher than that.

Specific applications for barotolerant proteins and pressure-regulated genes have not yet been developed. However, many food processes are carried out at high pressure to minimize bacterial contamination (destructive bacteria will not tolerate the pressure), so it is expected that barophiles will ultimately offer useful biotechnology products, such as enzymes that will carry out food processing at high pressure. High-pressure processing has been used on cheeses, yogurt, luncheon meats, and oysters to kill contaminating bacteria without destroying the flavor or texture of the food.

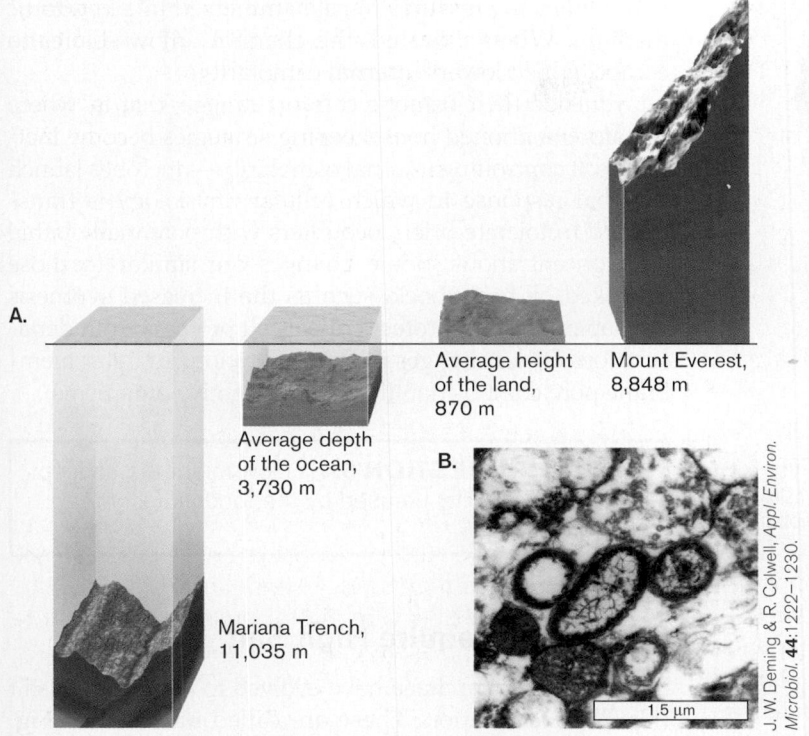

Figure 5.5   Barophilic environments and piezophiles.   A. Ocean depths. The deepest part of the ocean is at the bottom of the Mariana Trench, a depression in the floor of the western Pacific Ocean, just east of the Mariana Islands. The Mariana Trench is 1,554 miles long and 44 miles wide. Near its southwestern extremity, 210 miles southwest of Guam, lies the deepest point on Earth. This point, referred to as the "Challenger Deep," plunges to a depth of nearly 7 miles. **B.** Barophile *Shewanella violacea.* Length is approx. 1.5 μm.

J. W. Deming & R. Colwell, *Appl. Environ. Microbiol.* **44**:1222–1230.

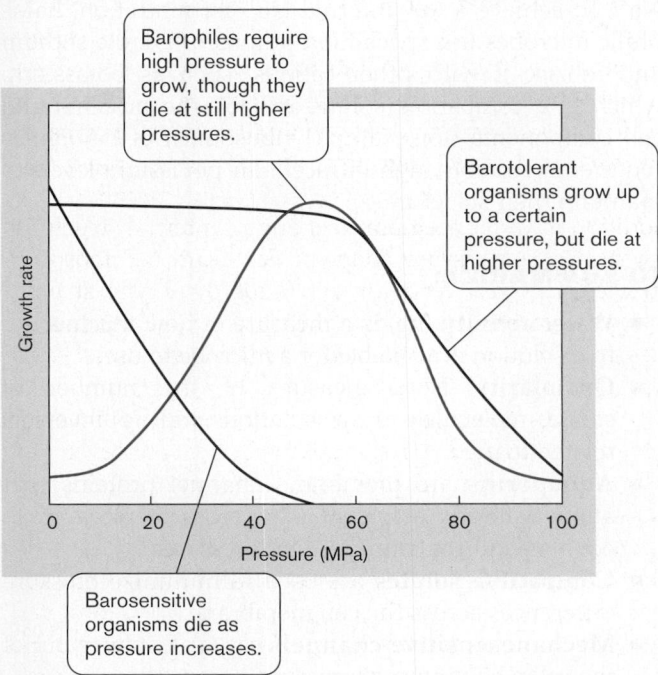

Figure 5.6   Relationship between growth rate and pressure.

TO SUMMARIZE:

- **Barophiles (piezophiles)** can grow at pressures up to 1,000 atm but fail to grow at low pressures.
- **Membrane fluidity** can be compromised at high pressures and cold temperatures. Specially designed membranes and protein structures are thought to enable the growth of barophiles.

THOUGHT QUESTION 5.3   What could be a relatively simple way to grow barophiles in the laboratory?

5.4   Water Activity and Salt

Water is critical to life, but environments differ in terms of how much water is actually available to growing organisms; microbes can only use water that is not bound to ions or other solutes in solution. Water availability is measured as **water activity** (a_w), a quantity approximated by concentration. Because interactions with solutes lower water activity, the more solutes there are in a solution, the less water is available for microbes to use for growth. Water activity is typically measured as the ratio of the solution's vapor pressure relative to that of pure water. A solution is placed in a sealed chamber and the amount of water vapor determined at equilibrium. If the air above the sample is 97% saturated relative to the moisture present over pure water, the relative humidity is 97% and the water activity is 0.97. Most bacteria require water activity to be greater than 0.91 (the water activity of seawater). Fungi can tolerate water activity levels as low as 0.86.

Osmotic Stress

Osmolarity is a measure of the number of solute molecules in a solution and is inversely related to a_w. The more particles there are in a solution, the greater the osmolarity and the lower the water activity (see Appendix 2). Osmolarity is important for the cell because it is related to water activity and also because a semipermeable membrane surrounds microbial cells, so osmolarity inside the cell can be, and often is, different from osmolarity outside. The principles of physical chemistry dictate that solute concentrations in two chambers separated by a semipermeable membrane will tend to equilibrate. Equilibrating osmolarity across a semipermeable cell membrane, which does not allow the movement of solutes, requires the movement of water. In hypertonic medium, where the

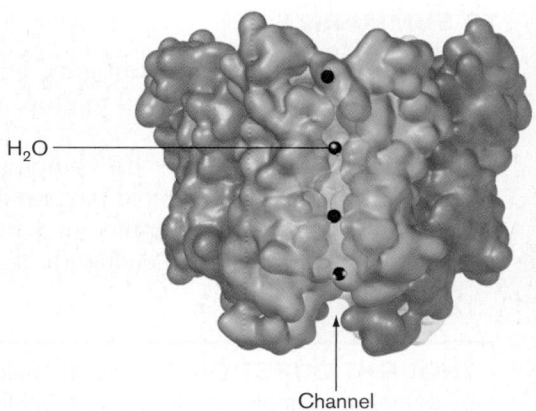

H₂O

Channel

Figure 5.7   Aquaporin.   Transverse view of the channel. Complementary halves of the channel are formed by adjacent protein monomers. The curvilinear, size-selective (~4 ± 0.5 nm) core of the channel (~18 nm) is lined primarily by hydrophobic residues. (PDB code: 1J4N)

external osmolarity is higher than the internal, water will try to *leave* the cell in an attempt to equalize osmolarity across the membrane. In contrast, suspension of a cell in a hypotonic medium (one of lower osmolarity than the cell) will cause an *influx* of water (see Appendix 2, Fig. A2.6).

The movement of water across cell membranes does not occur primarily by simple diffusion. Special membrane water channels formed by proteins called aquaporins enable water to traverse the membrane much faster than by diffusion and help protect cells against osmotic stress (**Fig. 5.7**). However, too much water moving in or out of a cell is detrimental. Cells may ultimately explode or implode, depending on the direction in which the water moves. Even bacteria with a rigid cell wall suffer. They may not explode like a human cell, but the forces placed on the cell wall are great.

Cells Minimize Osmotic Stress

In addition to moving water, microbes have at least two other mechanisms to minimize osmotic stress across membranes. When stranded in a hypertonic medium (higher osmolarity than the cell), bacteria try to protect their internal water by synthesizing or importing compatible solutes that increase intracellular osmolarity. Compatible solutes are small molecules that do not disrupt normal cell metabolism (as high levels of Na⁺ would do), even when present at high intracellular concentrations. Increasing intracellular levels of these compounds (for example, proline, glutamic acid, potassium, or betaine) elevates cytoplasmic osmolarity without any detrimental effects, making it unnecessary for water to leave the cell.

Cells also contain pressure-sensitive (mechanosensitive) channels that can be used to leak solutes out of the cell. It is believed that these channels are activated by

rising internal pressures in cells immersed in a *hypo*tonic medium. When activated, the channels allow solutes to escape, which lowers internal osmolarity.

Outside their osmotic comfort range—that is, where the aforementioned housekeeping strategies become ineffective at controlling internal osmolarity—microbes launch a global response in which cellular physiology is transformed to tolerate brief encounters with potentially lethal salt concentrations. Some changes are similar to those provoked by heat shock, such as the increased synthesis of chaperones that protect critical cell proteins from denaturation. Other changes include alterations in outer membrane pore composition (for Gram-negative organisms).

> **THOUGHT QUESTION 5.4**   How might the concept of water availability be used by the food industry to control spoilage?

Halophiles Require High Salt

Some species of archaea have evolved to require high salt (NaCl) concentration. These are called **halophiles** (**Fig. 5.8**). In striking contrast to most bacteria, which prefer salt concentrations from 0.1 to 1 M (0.2%–5% NaCl), the extremely halophilic archaea can grow at an a_w of 0.75 and actually require 2- to 4-M NaCl (10%–20% NaCl) to grow. For comparison, seawater is about 3.5% NaCl. All cells, even halophiles, prefer to keep a relatively low intracellular Na⁺ concentration. One reason for this is that some solutes are moved into the cell by symport with Na⁺. To achieve a low internal Na⁺ concentration, halophilic microbes use special ion pumps to excrete sodium and replace it with other cations, such as potassium, which is a compatible solute. In fact, the proteins and cell components (for example, ribosomes) of halophiles require remarkably high intracellular potassium levels to maintain their structure.

TO SUMMARIZE:

■ **Water activity** (a_w) is a measure of how much water in a solution is available for a microbe to use.
■ **Osmolarity** is a measure of the number of solute molecules in a solution and is inversely related to a_w.
■ **Aquaporins** are membrane channel proteins that allow water to move quickly across membranes to equalize internal and external pressures.
■ **Compatible solutes** are used to minimize pressure differences across the cell membrane.
■ **Mechanosensitive channels** can leak solutes out of the cell when internal pressure rises.
■ **Halophilic organisms** grow best at high salt concentration.

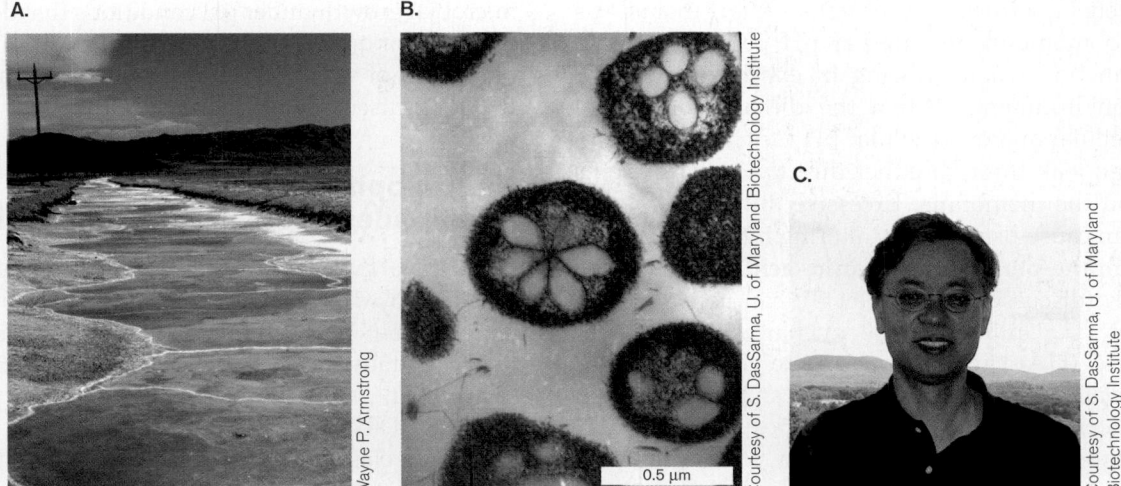

Figure 5.8　Halophilic salt flats and halophilic bacteria.　A. The halophilic salt flats along Highway 50, east of Fallon, Nevada, are colored pinkish red by astronomical numbers of halophilic bacteria. **B.** *Halobacterium* sp. (TEM). Cross section, cell width 0.5–0.8 μm. **C.** Shiladitya DasSarma and colleagues at the University of Maryland completed the genome sequence of *Halobacterium* species NRC-1. They demonstrated novel features of archaeal genetics, including intriguing similarities with molecular regulatory structures in eukaryotes.

5.5　Adaptation to pH Changes

As with salt and temperature, the concentration of hydrogen ions (H$^+$)—actually, hydronium (H$_3$O$^+$)—also has a direct effect on the cell's macromolecular structures. Extreme concentrations of either hydronium or hydroxyl ions (OH$^-$) in a solution will limit growth. In other words, too much acid or base is harmful to cells. Despite this sensitivity to pH extremes, living cells tolerate a greater range in environmental concentration of H$^+$ than of virtually any other chemical substance. *E. coli*, for example, tolerates a pH range of 2–9, a 10,000,000-fold difference. For a brief review of pH, refer to Appendix 1, Section A1.7.

pH Optima, Minima, and Maxima

The charges on various amino or carboxyl groups within a protein help forge the intramolecular bonds that dictate protein shape and thus protein activity. Because H$^+$ concentration affects the protonation of these ionizable groups, altering pH can alter the charges on these groups, which in turn changes protein structure and activity. The result is that all enzyme activities exhibit optima, minima, and maxima with regard to pH, much as they do for temperature. As we saw with temperature, groups of microbes have evolved to inhabit diverse niches, for which pH values can range from pH 0 to pH 11.5 (**Fig. 5.9**). However, species differences in optimum growth pH are *not* dictated by the pH limits at which critical cell proteins function.

Generally speaking, the majority of enzymes, regardless of the pH at which their source organism thrives, tend to operate best between pH 5 and 8.5 (which, if you think about it, is still a range in which the hydrogen ion concentration varies more than 1,000-fold). Yet many microbes grow in even more acidic or alkaline environments.

Unlike its temperature, the intracellular pH of a microbe, as well as its osmolarity, is not necessarily the same as that of its environment. Biological membranes

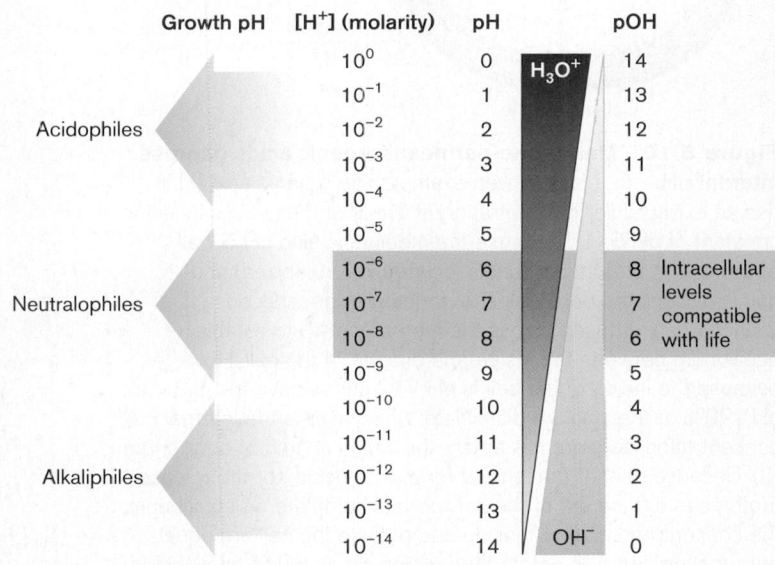

Figure 5.9　Classification of organisms grouped by optimum growth pH.

are relatively impermeable to protons—a fact that allows the cell to maintain an internal pH compatible with protein function when growing in extremely acidic or alkaline environments. When the difference between the intracellular and extracellular pH (ΔpH) is very high, protons can leak through either directly or via proteins that thread the membrane. Excessive influx or efflux of protons can cause problems by altering internal pH.

Membrane-permeant organic acids, also called weak acids (discussed in Chapter 3), can accelerate the leakage of H$^+$. Unlike H$^-$, the uncharged form of an organic acid (HA) can freely permeate cell membranes and dissociate intracellularly, releasing a proton that then acidifies internal pH (**Fig. 5.10**). The extent of the pH drop depends on the buffering capacity of the cell's proteins. This shuttling of protons can turn a relatively mild external pH level (say, pH 6) into a deadly acid stress. A naturally occurring example of organic acid stress is the lactic acid produced by lactobacilli during the formation of yogurt. The buildup of lactic acid limits the bacterial growth, leaving yogurt with plenty of food value. The food industry has taken advantage of this phenomenon by preemptively adding citric acid or sorbic acid to certain foods. This practice allows manufacturers to control microbial growth under pH conditions that do not destroy the flavor or quality of the food. **Figure 5.11** illustrates the pH values of various everyday items. Food microbiology is further discussed in Chapter 16.

Neutralophiles, Acidophiles, and Alkaliphiles Grow in Different pH Ranges

Cells have evolved to live under different pH conditions not by drastically changing the pH optima of their enzymes but by using novel pH homeostasis strategies that maintain intracellular pH above pH 5 and below pH 8, even when the cell is immersed in pH environments well above or below that range.

Three classes of organisms are differentiated by the pH of their growth range: neutralophiles, acidophiles, and alkaliphiles.

Neutralophiles generally grow between pH 5 and pH 8, and include most human pathogens. Many neutralophiles, including *E. coli* and *Salmonella enterica*, adjust their metabolism to maintain an internal pH slightly above neutrality, which is where their enzymes work best. They maintain this pH even in the presence of moderately acidic or basic external environments (**Fig. 5.12**). Other neutralophiles allow their internal pH to fluctuate with external pH but usually maintain a pH difference

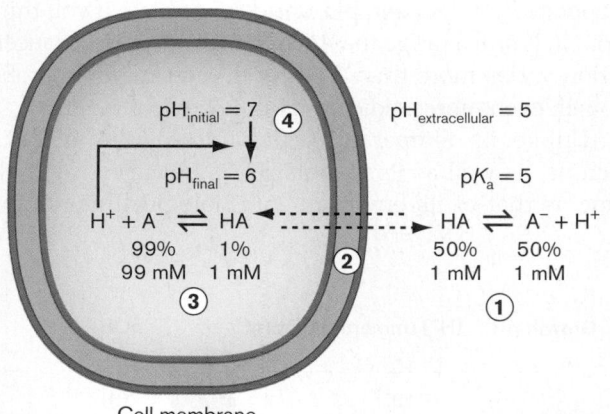

Figure 5.10 Membrane-permeant organic acids depress internal pH. In this contrived example, the organic acid (HA) has an extracellular concentration of 2 mM and has a dissociation constant of pH 5 (1). Because the medium is also pH 5, half of the acid is protonated (undissociated, or un-ionized) and half is dissociated (ionized). The un-ionized form, because it is uncharged, diffuses across the membrane (2) to establish equilibrium between the inside and outside of the cell. However, because the inside of the cell is pH 7 (2 units above the external pH), 99% of the acid will dissociate, which lowers the internal HA concentration, so more HA enters the cell in search of equilibrium (3). Because neither the ionized form of the acid nor the released proton can diffuse out of the cell, both accumulate. At equilibrium, the concentrations of HA inside and outside the cell are equal, but for every HA that enters the cell, ionization of HA has yielded 99 A$^-$ molecules and an equal number of protons. The protons lower the internal pH to an extent that depends on the buffering capacity of cellular proteins (4).

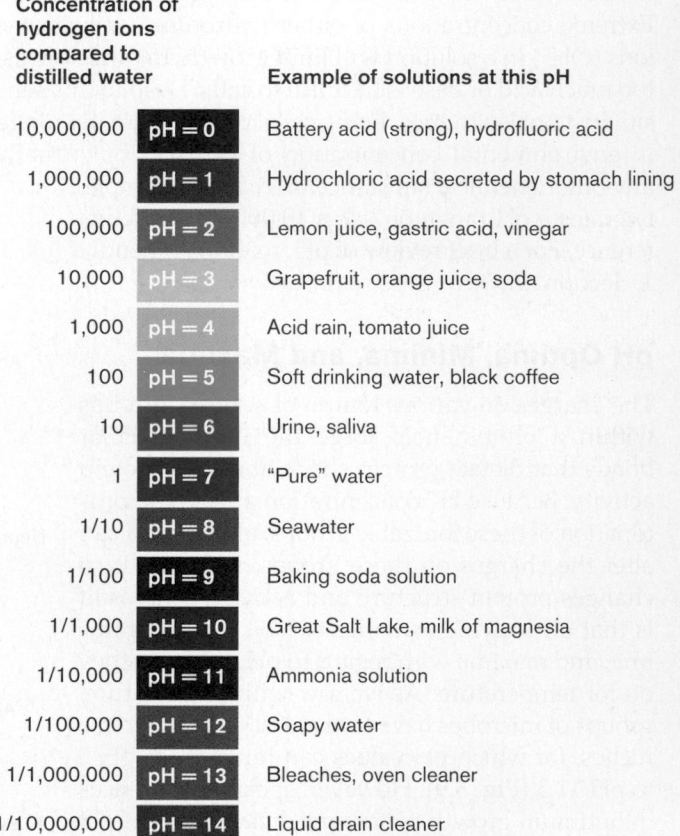

Concentration of hydrogen ions compared to distilled water		Example of solutions at this pH
10,000,000	pH = 0	Battery acid (strong), hydrofluoric acid
1,000,000	pH = 1	Hydrochloric acid secreted by stomach lining
100,000	pH = 2	Lemon juice, gastric acid, vinegar
10,000	pH = 3	Grapefruit, orange juice, soda
1,000	pH = 4	Acid rain, tomato juice
100	pH = 5	Soft drinking water, black coffee
10	pH = 6	Urine, saliva
1	pH = 7	"Pure" water
1/10	pH = 8	Seawater
1/100	pH = 9	Baking soda solution
1/1,000	pH = 10	Great Salt Lake, milk of magnesia
1/10,000	pH = 11	Ammonia solution
1/100,000	pH = 12	Soapy water
1/1,000,000	pH = 13	Bleaches, oven cleaner
1/10,000,000	pH = 14	Liquid drain cleaner

Figure 5.11 pH values of common substances.

(ΔpH) of about 0.5 pH unit across the membrane at the upper and lower limits of growth pH. The ΔpH value is an important component of the transmembrane proton potential, a source of energy for the cell (see Chapter 14).

> **NOTE:** The older term "neutrophile" used for this group of organisms is similar to the descriptor for a specific type of white blood cell ("neutrophil"). To avoid confusion, the term "neutrophil" should be reserved for the white blood cell and the term "neutralophile" used to designate microbes with growth optima near neutral pH.

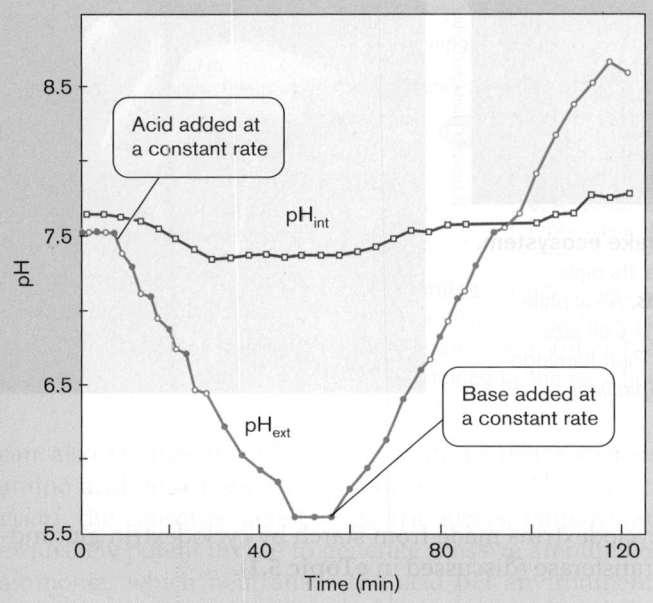

Figure 5.12 Maintaining internal pH (pH homeostasis) over a wide range of external pH. Internal pH of the neutralophile *Escherichia coli* measured following the addition of acid (HCl, at $t = 10$ min) to change external pH (circles) and subsequent addition of base (NaOH, at $t = 52$ min). In this experiment, internal pH (pH$_{int}$) was determined using nuclear magnetic resonance to measure changes in methyl phosphate (open circles/squares). The two phosphate species titrate over different pH ranges. *Source: Joan L. Slonczewski et al. 1981. PNAS* **78**:6271.

Acidophiles are bacteria and archaea that live in acidic environments. They are often chemoautotrophs (lithotrophs) that oxidize reduced metals and generate strong acids such as sulfuric acid. Consequently, they grow between pH 0 and pH 5. Acidophiles generally maintain an internal pH that is considerably more acidic than that of neutralophiles but still less acidic than their growth environment (**Fig. 5.13**). The ability to grow at this pH is due partly to altered membrane lipid profiles (high levels of tetraether lipids) that decrease proton permeability, as well as to ill-defined proton extrusion mechanisms. Often, an organism that is an extremophile with respect to one environmental factor is an extremophile with respect to others. *Sulfolobus acidocaldarius*, for example, is a thermophile and an acidophile (**Fig. 5.14**). It uses sulfur as an energy source and grows in acidic hot springs rich in sulfur.

Alkaliphiles occupy the opposite end of the pH spectrum, growing best at values ranging from pH 9 to pH 11. They are commonly found in saline soda lakes, which have high salt concentrations and pH values (as high as pH 11). Soda lakes, like Lake Magadi in Africa (**Fig.**

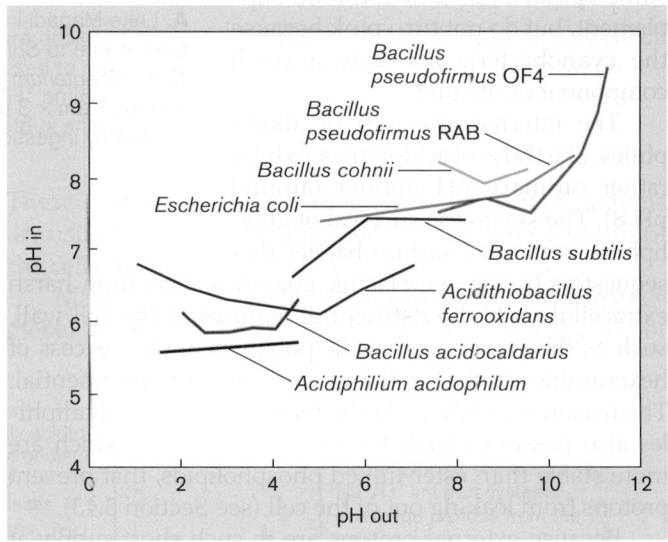

Figure 5.13 Cytoplasmic pH as a function of the external pH among acidophiles, neutralophiles, and alkaliphiles. *Source: Joan L. Slonczewski et al. 2009. Adv. Microb. Physiol.* **55**:1–79.

A.

B.

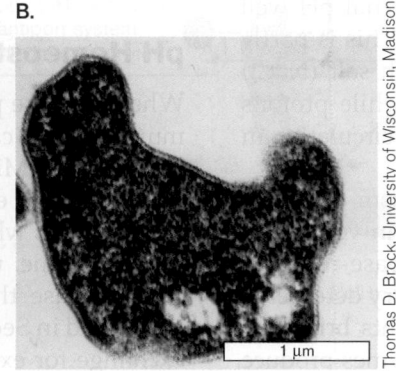

Figure 5.14 Sulfur Caldron acid spring and *Sulfolobus acidocaldarius*. A. Sulfur Caldron, in the Mud Volcano area of Yellowstone National Park, is one of the most acidic springs in the park. It is rich in sulfur and in *Sulfolobus*, a bacterium that thrives in hot, acidic waters with temperatures from 60°C to 95°C and a pH of 1–5. **B.** Thin-section electron micrograph of *S. acidocaldarius*. Under the electron microscope, the organisms appear as irregular spheres that are often lobed.

acid resistance. In a process analogous to the heat-shock response, bacterial physiology undergoes a major molecular reprogramming in response to hydrogen ion stress. The levels of a large number of proteins increase, while the levels of others decrease. Many of the genes and proteins involved in the acid stress response overlap with other stress response systems, including the heat-shock response. These physiological responses include modifications in membrane lipid composition, enhanced pH homeostasis, and numerous other changes with unclear purpose. Some pathogens, such as *Salmonella*, sense a change in external pH as part of the signal indicating that the bacterium has entered a host cell environment (see **eTopic 5.2**).

TO SUMMARIZE:

- **Hydrogen ion concentration** affects protein structure and function. Thus, enzymes have pH optima, minima, and maxima.
- **Microbes use pH homeostasis mechanisms** to keep their internal pH near neutral when in acidic or alkaline media.
- **Adding weak acids** to certain foods undermines bacterial pH homeostasis mechanisms, thereby preventing food spoilage and killing potential pathogens.
- **Neutralophiles, acidophiles, and alkaliphiles** prefer growth under neutral, low, and high pH conditions, respectively.
- **Acid and alkaline stress responses** result when a given species is placed under pH conditions that slow its growth. The cell increases the levels of proteins designed to mediate pH homeostasis and protect cell constituents.

5.6 Oxygen and Other Electron Acceptors

Many microorganisms can grow in the presence of molecular oxygen (O_2). Some even use oxygen as a terminal electron acceptor in the **electron transport chain**, a group of membrane proteins (cytochromes) that convert energy trapped in nutrients to a biologically useful form. The use of O_2 as the terminal electron acceptor is called **aerobic respiration** (see Chapter 14).

Oxygen Has Benefits and Risks

Electrons pulled from various energy sources (for example, glucose) possess intrinsic energy, an energy that cytochrome proteins can harness. They do this by extracting the energy in incremental stages and using it to move protons out of the cell. This unequal distribution of H^+ across the membrane produces a transmembrane electrochemical gradient, a sort of biobattery called the proton motive force (discussed in Section 14.2). Once the cell has drained as much energy as possible from an electron, that electron must be passed to a diffusible, final electron acceptor molecule floating in the medium. This clears the way for another electron to be passed down the electron transport chain (**Fig. 5.18**).

One such terminal electron acceptor is dissolved oxygen. However, oxygen and its breakdown products are dangerously reactive—a serious problem for all cells. As a result, different species have evolved to either tolerate or avoid oxygen altogether. **Table 5.2** gives examples of microbes that grow at different levels of oxygen.

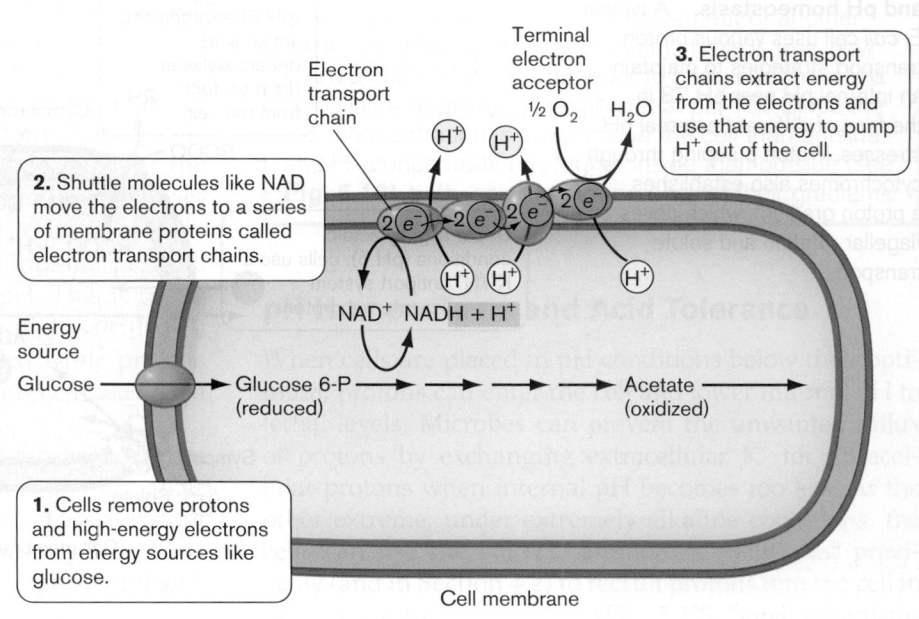

Figure 5.18 Role of oxygen as a terminal electron acceptor in respiration. This pumping of protons out of the cell by electron transport chains produces more positive charges outside the cell than inside, resulting in an electrochemical gradient (also called proton motive force). An electron from a high-energy source is passed from one member of the chain to the next. With each transfer of an electron to the next member of the chain, more energy is extracted and converted to proton motive force. At the end of the chain, the electron must be passed to a final or terminal electron acceptor, clearing the path for the next electron. This net process is called respiration.

2. Shuttle molecules like NAD move the electrons to a series of membrane proteins called electron transport chains.

3. Electron transport chains extract energy from the electrons and use that energy to pump H^+ out of the cell.

Electron transport chain

Terminal electron acceptor
$\frac{1}{2} O_2$ H_2O

NAD^+ $NADH + H^+$

Energy source

Glucose ⟶ Glucose 6-P (reduced) ⟶ ⟶ ⟶ Acetate (oxidized) ⟶

1. Cells remove protons and high-energy electrons from energy sources like glucose.

Cell membrane

Table 5.2 Examples of aerobes and anaerobes.

Aerobic	Facultative	Microaerophilic	Anaerobic

Neisseria spp.
  Causative organisms of meningitis, gonorrhea
	Escherichia coli
	  Normal GI flora; additional pathogenic strains
		Helicobacter pylori
		  Causative organism of gastric ulcers
			Azoarcus tolulyticus
			  Degrades toluene

Pseudomonas fluorescens
  Found in soil; degrade TNT and aromatic hydrocarbons
	Saccharomyces cerevisiae
	  Yeast; used in baking
		Lactobacillus
		  Ferments milk to form yogurt
			Bacteroides spp.
			  Normal GI flora

Mycobacterium leprae
  Cause of leprosy
	Bacillus anthracis
	  Cause of anthrax
		Campylobacter spp.
		  Causative organism of gastroenteritis
			Clostridium spp.
			  Soil microorganisms; causative agents of tetanus and botulism

Azotobacter
  Soil microorganisms; fix atmospheric nitrogen
	Vibrio cholerae
	  Cause of cholera
		Treponema pallidum
		  Causative organism of syphilis
			Actinomyces
			  Soil microrganisms; synthesize antibiotics

Rhizobium spp.
  Soil microorganisms; plant symbionts
	Staphylococcus spp.
	  Found on skin; causative agent of boils
			Desulfovibrio spp.
			  Reduce sulfate

Aerobes *vs.* Anaerobes

The relationships between microbes and oxygen are varied. **Figure 5.19** illustrates a test tube with growth medium. The top of the tube, closest to air, is oxygenated; the bottom of the tube has much lower levels of oxygen. Some microbes grow only at the top of the tube, while others prefer to grow at the bottom—the distributions based on each organism's relationship with oxygen. A **strict aerobe** is an organism that not only exists in oxygen but also uses oxygen as a terminal electron acceptor. In fact, a strict aerobe can *only* grow with oxygen present and possesses an aerobic metabolism (aerobic respiration). An aerobe will grow at the top of the tube shown in **Figure 5.19**. In contrast, a **strict anaerobe** dies in the least bit of oxygen. Strict anaerobes do not use oxygen as an electron acceptor, but this is not why they die in air. They die because this type of microbe is vulnerable to reactive oxygen molecules (also called reactive oxygen species, or ROS) produced by its own metabolism when it is exposed to oxygen. Anaerobes will grow at the bottom of the tube shown in **Figure 5.19**.

Do not confuse the ability of an organism to *exist* in the presence of oxygen with its ability to *use* oxygen as a terminal electron acceptor. Some aerobes can survive in oxygen but do not have the ability to use oxygen. Any organism that possesses NADH dehydrogenase 2—aerobe or anaerobe—will, in the presence of oxygen, inadvertently autooxidize the FAD (flavin adenine dinucleotide) cofactor within the enzyme and produce dangerous amounts of superoxide radicals ($^{\bullet}O_2^-$; **Fig. 5.20**). Superoxide will degrade to hydrogen peroxide (H_2O_2), another reactive molecule. Iron, present as a cofactor in several enzymes, can then catalyze a reaction with hydrogen peroxide to produce the highly toxic hydroxyl radical ($^{\bullet}OH$). All of these molecules seriously damage DNA, RNA, proteins, and lipids. Consequently, oxygen is actually an extreme environment in which survival requires special talents. Aerobes, but not anaerobes, destroy reactive oxygen species with the aid of enzymes such as superoxide dismutase (to remove superoxide) and peroxidase and catalase (to remove hydrogen peroxide). Aerobes also have resourceful enzyme systems that detect and repair macromolecules damaged by oxidation.

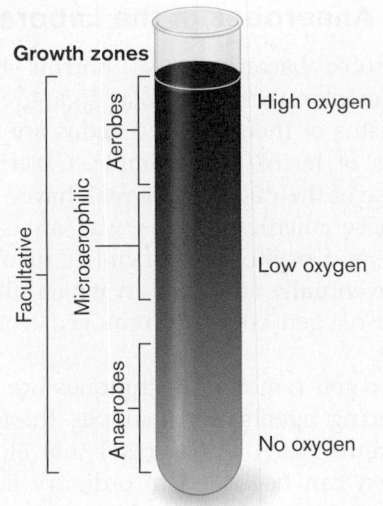

Growth zones

Aerobes — High oxygen

Microaerophilic — Low oxygen

Facultative

Anaerobes — No oxygen

Figure 5.19 Oxygen-related growth zones in a standing test tube.

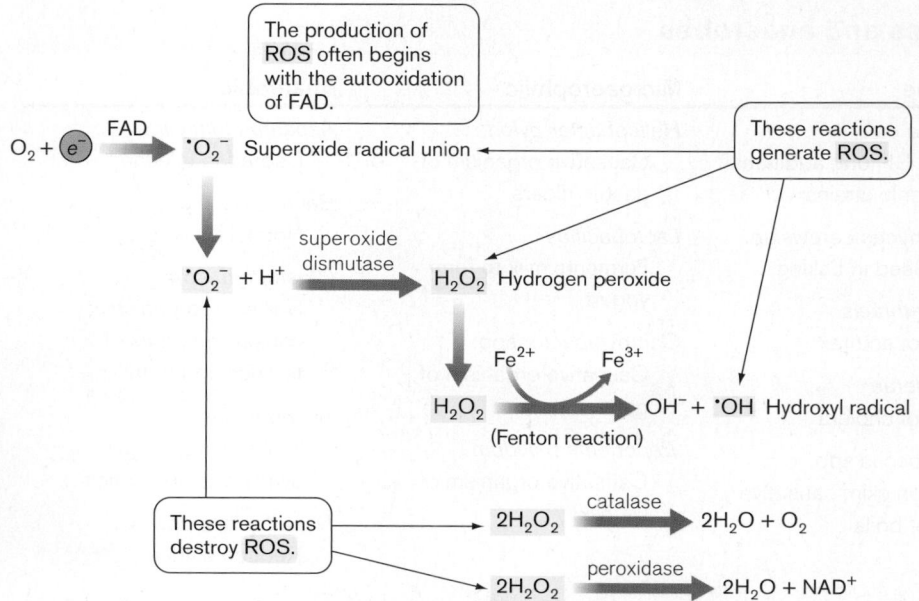

Figure 5.20 Generation and destruction of reactive oxygen species (ROS). The autooxidation of flavin adenine dinucleotide (FAD) and the Fenton reaction occur spontaneously to produce superoxide and hydroxyl radicals, respectively. The other reactions require enzymes. FAD is a cofactor for a number of enzymes (for example, NADH dehydrogenase 2). Catalase and peroxidase do not produce ROS but detoxify hydrogen peroxide.

Anaerobes Ferment and Respire without Oxygen

Anaerobic microbes fall into several categories. Some anaerobes actually do respire using electron transport systems, but instead of using oxygen, they rely on alternative terminal electron acceptors like nitrate to conduct **anaerobic respiration** and produce energy. Anaerobes of another ilk do not possess cytochromes, cannot respire, and so must rely on carbohydrate **fermentation** for energy (that is, they conduct **fermentative metabolism**). In fermentation, ATP energy is produced through substrate-level phosphorylation in a process that does not involve cytochromes.

Facultative organisms are microbes that can live with or without oxygen. They will grow throughout the tube shown in **Figure 5.19**. **Facultative anaerobes** (such as *E. coli*) also possess enzymes that destroy toxic oxygen by-products, but they have both fermentative *and* aerobic respiratory potential. Whether a member of this group uses aerobic respiration, anaerobic respiration, or fermentation depends on the availability of oxygen and the amount of carbohydrate present. **Aerotolerant anaerobes** only use fermentation to provide energy but contain superoxide dismutase and catalase (or peroxidase) to protect them from reactive oxygen species. These enzymes allow aerotolerant anaerobes to grow in oxygen while retaining a fermentation-based metabolism. Microorganisms that possess *decreased* levels of superoxide dismutase and/or catalase will be **microaerophilic**, meaning they will grow only at low oxygen concentrations.

The fundamental composition of all cells reflects their evolutionary origin as anaerobes. Lipids, nucleic acids, and amino acids are all highly reduced—which is why

our bodies are combustible. We never would have evolved that way if molecular oxygen had been present from the beginning. Even today, the majority of all microbes are anaerobic, growing buried in the soil, within our anaerobic digestive tract, or within biofilms on our teeth.

> **THOUGHT QUESTION 5.6** If anaerobes cannot live in oxygen, how do they incorporate oxygen into their cellular components?
>
> **THOUGHT QUESTION 5.7** How can anaerobes grow in the human mouth when there is so much oxygen there?

Culturing Anaerobes in the Laboratory

Many anaerobic bacteria cause horrific human diseases, such as tetanus, botulism, and gangrene. Some of these organisms or their secreted toxins are even potential weapons of terror (for example, *Clostridium botulinum*). Because of their ability to wreak havoc on humans, culturing these microorganisms was an early goal of microbiologists. Despite the difficulties involved, conditions were eventually contrived in which all, or at least most, of the oxygen could be removed from a culture environment.

Three oxygen-removing techniques are used today. Special reducing agents (for example, thioglycolate) or enzyme systems (such as Oxyrase) that eliminate dissolved oxygen can be added to ordinary liquid media. Anaerobes can then grow beneath the culture surface. A second, very popular way to culture anaerobes, especially on agar plates, is to use an anaerobe jar (**Fig. 5.21A**).

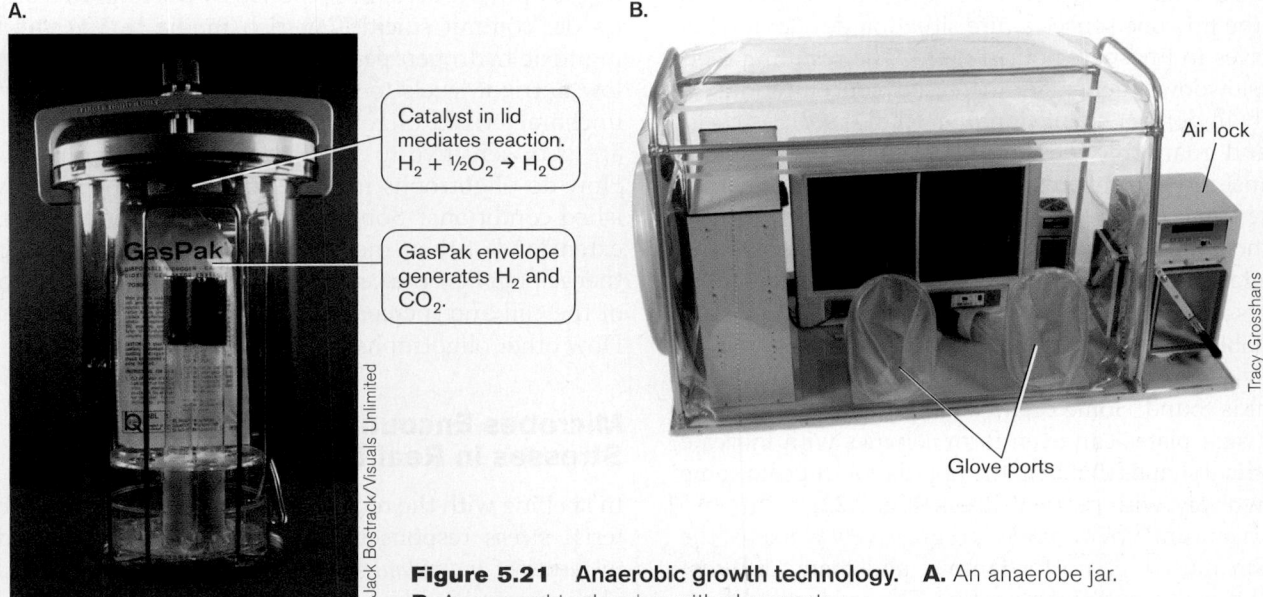

A.

Catalyst in lid mediates reaction.
$H_2 + \frac{1}{2}O_2 \rightarrow H_2O$

GasPak envelope generates H_2 and CO_2.

GasPak

©Jack Bostrack/Visuals Unlimited

B.

Air lock

Glove ports

Tracy Grosshans

Figure 5.21 Anaerobic growth technology.   A. An anaerobe jar.
B. An anaerobic chamber with glove ports.

Agar plates streaked with the organism are placed into a sealed jar with a foil packet that releases H_2 and CO_2 gases. A palladium packet hanging from the jar lid catalyzes a reaction between the H_2 and O_2 in the jar to form H_2O and effectively removes O_2 from the chamber. The CO_2 released is required by some reactions to produce key metabolic intermediates. Some microaerophilic microbes, like the pathogens *Helicobacter pylori* (the major cause of stomach ulcers) and *Campylobacter jejuni* (a major cause of diarrhea), require low levels of O_2 but elevated amounts of CO_2. These conditions are obtained by using similar gas-generating packets.

For strict anaerobes exquisitely sensitive to oxygen, even more heroic efforts are required to establish an oxygen-free environment. A special anaerobic glove box must be used in which the atmosphere is removed by vacuum and replaced with a precise mixture of N_2 and CO_2 gases (**Fig. 5.21B**).

THOUGHT QUESTION 5.8   What evidence led people to think about looking for anaerobes? *Hint*: Look up Spallanzani, Pasteur, and spontaneous generation on the Internet.

TO SUMMARIZE:

■ **Oxygen is a benefit to aerobes**—organisms that can use it as a terminal electron acceptor to extract energy from nutrients.

■ **Oxygen is toxic** to all cells that do not have enzymes capable of efficiently destroying the reactive oxygen species—for example, anaerobes.

■ **Anaerobic metabolism** can be either **fermentative** or **respiratory**. Anaerobic respiration requires the organism to possess cytochromes that can use compounds other than oxygen as terminal electron acceptors.

■ **Aerotolerant anaerobes** grow either in the presence or in the absence of oxygen, but use fermentation as their primary, if not only, means of gathering energy. These microbes also have enzymes that destroy reactive oxygen species, allowing them to grow in oxygen.

■ **Facultative anaerobes** grow with or without oxygen and have enzymes that destroy reactive oxygen species. In addition, they possess both the ability for fermentative metabolism and the ability to use oxygen as a terminal electron acceptor.

5.7 Nutrient Deprivation and Starvation

It is intuitively obvious that limiting the availability of a carbon source or other essential nutrient will limit growth. Not so obvious are the dramatic molecular events that cascade through a cell undergoing nutrient limitation and ultimately starvation. Optimizing growth rate at suboptimal nutrient levels is an important aim of free-living bacteria, given that intestinal, soil, and marine environments rarely offer nutrient excess.

Starvation Activates Survival Genes

Numerous gene systems are affected when nutrients decline (see Section 10.3 for details). Growth rate slows, and daughter cells become smaller and begin to experience what is termed a "starvation" response, in which the microbe senses a dire situation developing but still strives to find new nourishment. The resulting metabolic slowdown generates increased concentrations of critically important small signal molecules such as cyclic AMP and guanosine tetraphosphate (ppGpp) that globally transform gene expression. The highly soluble nature of these small molecules means they can quickly diffuse throughout the cell, promoting a fast response. During this metabolic retooling, transport systems for potential nutrients are produced even if the matching substrates are unavailable. Cells begin to make and store glycogen, presumably as an internal emergency store in case no other nutrient is found. Some organisms growing on nutrient-limited agar plates can even form colonies with intricate geometrical shapes that help the population cope, in some unknown way, with nutrient stress (**Fig. 5.22**).

As a cultural environment progressively worsens, the organism must prepare for famine, and many different stress survival genes become active. The products of these genes afford protection against stressors such as reactive oxygen radicals or temperature and pH extremes. No cell can predict the precise stresses it might encounter while incapacitated, so it is advantageous to be prepared for as many as possible. As described in Chapter 4, some species undergo elaborate developmental processes that ultimately produce dormant spores.

Oligotrophs

Most well-studied organisms thrive on organically rich media (more than 2 grams of carbon per liter) but struggle to grow at very low nutrient concentration (1–5 mg of carbon per liter). In natural ecosystems, however, the majority of existing microbes appear to be **oligotrophs**, organisms with a high rate of growth at extremely low organic substrate concentrations. Oligotrophs are well suited for growth in the nutrient-poor, or oligotrophic, environments found in certain lakes and streams. Some

oligotrophs are actually poisoned by concentrated organics or "commit suicide" in rich media by overproducing toxic hydrogen peroxide. Consequently, they *require* low nutrient levels to survive. For example, a variety of uncharacterized oligotrophic bacteria that exist in soil are very sensitive to NaCl and various L-amino acids. How do oligotrophs manage to grow in such impoverished conditions? Some oligotrophic bacteria have thin extensions of their membrane and cell wall called prothecaes (stalks) that essentially *expand* the surface area of the cell and increase capacity to transport nutrients. How other oligotrophs grow remains a mystery.

Microbes Encounter Multiple Stresses in Real Life

In keeping with the reductionist approach to science, bacterial stress responses have traditionally been studied in terms of *individual* stresses. *Escherichia coli*, for example, increases synthesis of a specific set of proteins when exposed to high temperature and a different set of proteins when exposed to high salt. Some members overlap the two sets; that is, several proteins may be highly expressed under both conditions, but each stress response also includes proteins unique to each stress.

Environmental situations in the world outside of the laboratory, however, can be quite complex, involving multiple, not just single, stresses. An organism could simultaneously undergo carbon starvation in a high salt, low pH environment. A classic study by Kelly Abshire and Frederich Neidhardt examined this situation using the pathogen *Salmonella enterica*, a cause of diarrhea. *S. enterica* invades human macrophage cells and survives in phagocytic vacuoles, where numerous stresses, such as low pH, oxidative stress, and nutrient limitations, are simultaneously imposed on the bacteria. Comparing the proteins synthesized by *Salmonella* growing in this compartment with the proteins synthesized under single stresses in the laboratory revealed an unexpected

A.

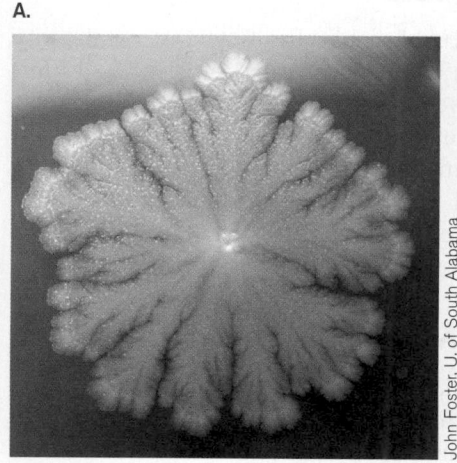

John Foster, U. of South Alabama

B.

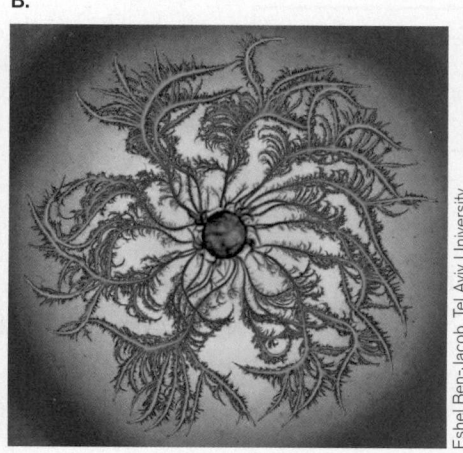

Eshel Ben-Jacob, Tel Aviv University

Figure 5.22   Effects of starvation on colony morphology.   A. Starving *E. coli* colony. **B.** *Pseudomonas dendritiformis* C morphotype, grown on hard agar (1.75%) under starvation conditions. The colony consists of branches with chiral twists (colored green), all with the same handedness. The added coloration indicates time of growth (yellow = oldest; red = most recent).

response pattern. Although many stress-related proteins were induced in the intracellular environment, no one set of stress-induced proteins was induced in its entirety. Furthermore, several bacterial proteins were induced by growth *only* within the macrophage phagolysosome, suggesting the presence of unknown intracellular stresses. Thus, caution is advised when trying to predict cell responses to real-world situations based solely on controlled laboratory studies that alter only single parameters.

Humans Influence Microbial Ecosystems

Natural ecosystems are typically low in nutrients (oligotrophic) but teem with diversity, so that numerous species compete for the same limiting nutrients. Maximum

diversity in a given ecosystem is maintained, in part, by the different nutrient-gathering profiles of competing microbes. Take a scenario in which microbe A is better than microbe B at gathering phosphate when phosphate levels are low, but microbe B is superior to A at culling limited quantities of nitrogen (**Fig. 5.23**). Neither organism dominates when both phosphate *and* nitrogen are in short supply. So even though B can harvest more nitrogen than A, it cannot outgrow A, since the phosphate concentration is low. However, if phosphate is suddenly increased so that it is no longer limiting for either organism, the one better adapted to low nitrogen (microbe B) will outgrow and overwhelm the other. The sudden infusion of large quantities of a formerly limiting nutrient, a process called **eutrophication**, can lead to a "bloom" of microbes. Organisms initially held in check by the limiting nutrient now exhibit unrestricted growth, consuming other nutrients to a degree that threatens the existence of competing species. The concept of limiting nutrients is covered in Chapter 21.

Humans have caused nutrient pollution in several ways. Runoff from agricultural fields, urban lawns, and golf courses is one source. Untreated or partially treated domestic sewage is another. Sewage was a primary source of phosphorus eutrophication of lakes in the 1960s and '70s, when detergents contained large amounts of phosphates. The phosphates acted as water softeners to improve cleaning action, but when washed into lakes they also proved to be powerful stimulants to algal growth (**Fig. 5.24**). The resulting algal "blooms" in many lakes led to oxygen depletion and consequent fish kills. Many

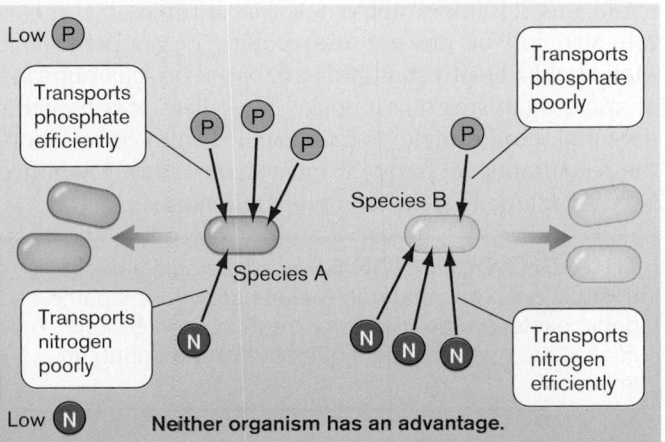
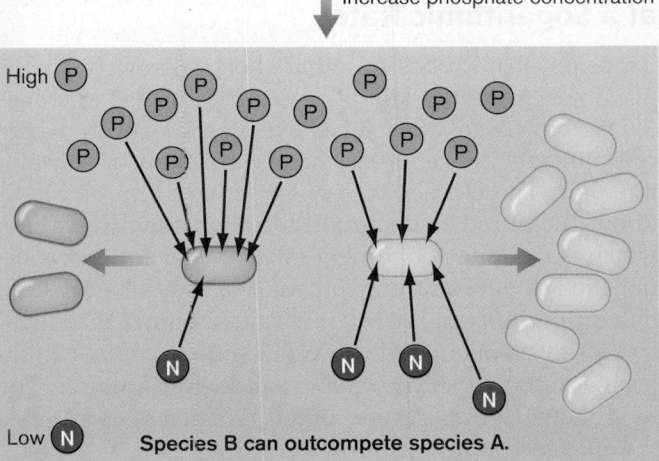

Figure 5.23 Maintaining microbial diversity through nutrient limitation. Species A and species B require both nitrogen and phosphate to grow. *Top*: Neither species dominates because each has a limiting nutrient (N = nitrogen; P = phosphate). *Bottom*: Following phosphate eutrophication, species B will outgrow species A because although both species have enough phosphate, B is better able to assimilate the limited quantity of nitrogen.

Figure 5.24 Eutrophication. Algal bloom in an experimental lake resulting from phosphate eutrophication. A divider curtain separates the lake and prevents mixing of water. The bright green color results from cyanobacteria growing on phosphorus added to the near side of the curtain.

native fish species disappeared, to be replaced by species more tolerant of the new conditions.

TO SUMMARIZE:

- **Starvation** is a stress that can elicit a starvation response in many microbes. Enzymes are produced to increase the efficiency of nutrient gathering and to protect cell macromolecules from damage.
- **The starvation response** is usually triggered by the accumulation of small signal molecules such as cyclic AMP or guanosine tetraphosphate.
- **Oligotrophs** are organisms that thrive in nutrient-poor conditions.
- **Human activities can cause eutrophication**, which damages delicately balanced ecosystems by introducing nutrients that can allow one member of the ecosystem to flourish at the expense of other species.

5.8 Physical, Chemical, and Biological Control of Microbes

Prevention, control, or elimination of microbes that are potentially harmful to humans is one of the primary goals of our health care system. Within the recent past, infectious disease was an imminent and constant threat to most of the human population. The average family in the United States prior to 1900 had four or five children, in part because parents could expect half of them to succumb to deadly infectious diseases. What today would be a simple infected cut in years past held a serious risk of death, and a trip to the surgeon was tantamount to playing Russian roulette with an unsterilized scalpel. Improvements in sanitation procedures and antiseptics and the advent of antibiotics have, to a large degree, curtailed the incidence and lethal effects of many infectious diseases. Success in this endeavor has played a major role in extending life expectancy and in contributing to the population explosion.

A variety of terms are used to describe antimicrobial control measures. The terms convey subtle, yet vitally important, differences in various control strategies and outcomes.

- **Sterilization** is the process by which *all* living cells, spores, and viruses are destroyed on an object.
- **Disinfection** is the killing, or removal, of *disease-producing* organisms from inanimate surfaces; it does not necessarily result in sterilization. Pathogens are killed, but other microbes may survive.
- **Antisepsis** is similar to disinfection, but it applies to removing pathogens from the surface of living tissues, like the skin. Antiseptic chemicals are usually

not as toxic as disinfectants, which frequently damage living tissues.

- **Sanitation**, closely related to disinfection, consists of reducing the microbial population to safe levels and usually involves both cleaning and disinfecting an object.

Antimicrobials can also be classified on the basis of the specific groups of microbes destroyed, leading to the terms "microbicide," "bactericide," "algicide," "fungicide," and "virucide." Furthermore, these agents can be classified either as "-static" (inhibiting growth) or "-cidal" (killing cells). For example, antibacterial agents are called **bacteriostatic** or **bactericidal**. Chemical substances that kill microbes are **germicidal** if they kill pathogens (and many nonpathogens). Germicidal agents do not necessarily kill spores.

Although these descriptions place emphasis on killing pathogens, it is important to note that antimicrobial agents can also kill or prevent the growth of nonpathogens. Many public health standards are based on *total* numbers of microorganisms on an object, regardless of pathogenic potential. For example, to gain public health certification, the restaurants we frequent must demonstrate low numbers of bacteria in their food preparation areas.

> **THOUGHT QUESTION 5.9**   Why would a bacteriostatic antibiotic that only inhibits growth of a pathogenic bacterium be useful for treating an infection? *Hint*: Is the human body a quiet bystander during an infection?

Cells Treated with Antimicrobials Die at a Logarithmic Rate

Exposure of microbes to lethal chemicals or conditions does not instantly kill all microorganisms. Microbes die according to a negative exponential curve, where cell numbers are reduced in equal fractions at constant intervals. The efficacy of a given lethal agent or condition is measured as **decimal reduction time (D-value)**, which is the length of time it takes that agent (or condition) to kill 90% of the population (a drop of 1 log unit, or a drop to 10% of the original value). **Figure 5.25** illustrates the exponential death profile of a bacterial culture heated to 100°C. The D-value is a little over 1 minute. The food industry uses several other parameters to evaluate the efficiency of killing.

Several factors influence the ability of an antimicrobial agent to kill microbes. These include the initial population size (the larger the population, the longer it takes to decrease it to a specific number), population composition (are spores involved?), agent concentration, and duration of exposure. Although the effect of concentration seems intuitively obvious, only over a narrow range is an

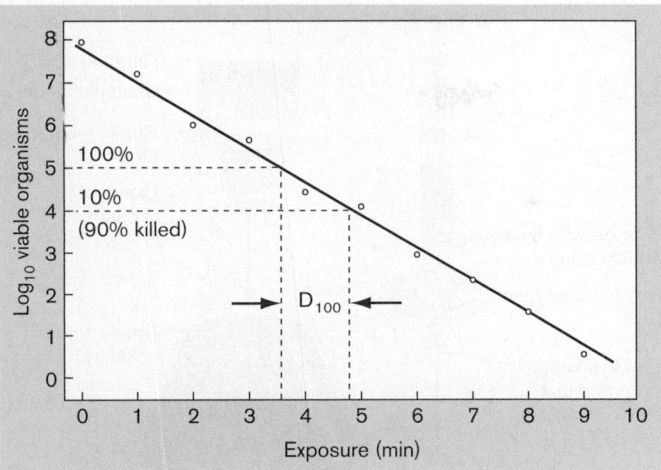

Figure 5.25 The death curve and the determination of D-values. Bacteria were exposed to a temperature of 100°C, and survivors were measured by viable count. The D-value is the time required to kill 90% of cells (that is, for the viable cell count to drop by one $\log_{10}$ unit). In this example, the D-value at 100°C (D_{100}) is approximately 1 minute.

increase in concentration matched by an increase in death rate. Increases above a certain level might not accelerate killing at all. For example, 70% ethanol is actually better than pure ethanol at killing organisms. The reason is that some water is needed to help ethanol penetrate cells. The ethanol then dehydrates cell proteins.

So, why is death a logarithmic function? Why don't all cells in a population die instantly when treated with lethal heat or chemicals? The reason is based, in part, on the random probability that an agent will cause a lethal "hit" in a given cell. Cells contain thousands of different proteins and thousands of molecules of each. Not all proteins and not all genes in a chromosome are damaged by an agent at the same time. Damage accumulates. Only when enough molecules of an *essential* protein or a gene encoding that protein are damaged will the cell die. Cells that die first are those that accumulate lethal hits early. Members of the population that die later have, by random chance, absorbed more hits on nonessential proteins or genes, sparing the essential ones.

Why, if 90% of a population is killed in 1 minute, aren't the remaining 10% killed in the next minute? It seems logical that all should have perished. Yet after the second minute, 1% of the original population remains alive. This phenomenon can also be explained by the random hit concept. Although there are fewer viable cells after 1 minute, each has the same random chance of having a lethal hit as when the treatment began. Thus, death rate is an exponential function, much like radioactive decay is an exponential function.

A final consideration is the overall fitness of individual cells. It is a mistake to assume that all cells in a population are identical. For example, at any given time, one cell may express a protein that another cell has just stopped expressing (for example, superoxide dismutase). In that instant, the first cell might contain a bit more of that protein. If the protein is essential or confers a level of stress protection (such as against superoxide), the cell can absorb more punishment before it is dispatched. The presence of lucky individuals expressing the right repertoire of proteins might also explain why death curves commonly level off after a certain point.

Physical Agents That Kill Microbes

Physical agents are often used to kill microbes or control their growth. Commonly used physical control measures are temperature extremes, pressure (usually combined with temperature), filtration, and irradiation.

High temperature and pressure. Even though microbes were discovered less than 400 years ago, thermal treatment of food products to render them safe has been practiced for over 5,000 years. Moist heat is a much more effective killer than dry heat, thanks to the ability of water to penetrate cells. Many bacteria, for instance, easily withstand 100°C dry heat but not 100°C boiling water. We humans are not so different, finding it easier to endure a temperature of 32°C (90°F) in dry Arizona than in humid Louisiana.

While boiling water (100°C) can kill most vegetative (actively growing) organisms, spores are built to withstand this abuse, and thermophiles prefer it. Killing spores and thermophiles usually requires combining high pressure and temperature. At high pressure, the boiling point of water rises to a temperature rarely experienced by microbes living at sea level. Even endospores quickly die under these conditions. This combination of pressure and temperature is the principle behind sterilization using the steam autoclave (**Fig. 5.26**). Standard conditions for steam sterilization are 121°C (250°F) at 15 psi for 20 minutes, a set of conditions that experience has taught us will kill all spores (except those of some thermophiles, which do not affect food or human health). These are also the conditions produced in pressure cookers used for home canning of vegetables.

Failure to adhere to these heat and pressure parameters can have deadly consequences, even in your own home. For instance, *Clostridium botulinum* is a spore-forming soil microbe that commonly contaminates fruits and vegetables used in home canning. The improper use of a pressure cooker while canning these goods will allow spores of this pathogen to survive. Once the can or jar is cool, the spores will germinate and begin producing their deadly toxin. All of this happens while the canned goods sit on a shelf waiting to be opened and consumed. Once ingested, the toxin makes its way to the nervous system

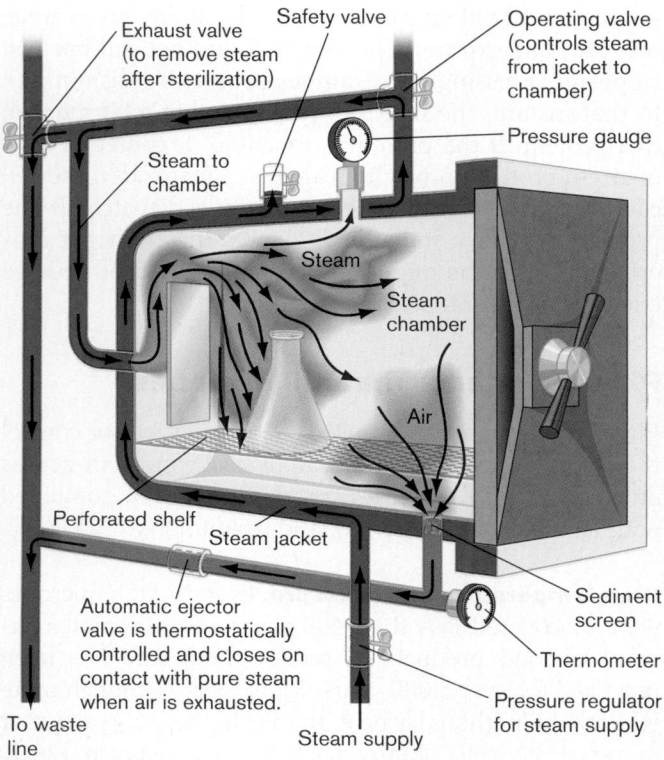

Figure 5.26   Steam autoclave.

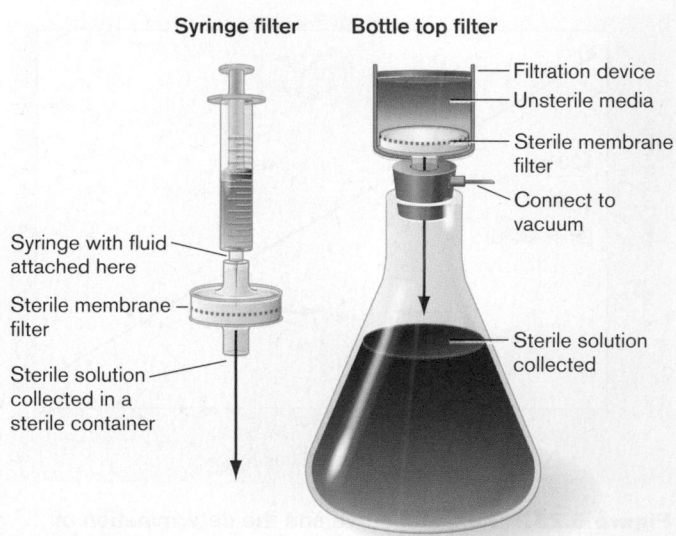

Figure 5.27   Membrane filtration apparatus.

and paralyzes the victim. Several incidents of this disease, called botulism, occur each year in the United States. (For more on food poisoning, see Chapter 16.)

THOUGHT QUESTION 5.10   How would you test the killing efficacy of an autoclave?

Pasteurization. Originally devised by Louis Pasteur to save products of the French wine industry from devastating bacterial spoilage, **pasteurization** today involves heating a particular food (such as milk) to a specific temperature long enough to kill *Coxiella burnetii*, the causative agent of Q fever, the most heat-resistant nonspore-forming pathogen known. Thus, pasteurization will also kill other disease-causing microbes. Many different time and temperature combinations can be used for pasteurization. The LTLT (low-temperature/long-time) process involves bringing milk to a temperature of 63°C (145°F) for 30 minutes. In contrast, the HTST (high-temperature/short-time) method (also called flash pasteurization) brings the milk to a temperature of 72°C (161°F) for only 15 seconds. Both processes accomplish the same thing: the destruction of *C. burnetii* and other bacteria.

Cold. Low temperatures have two basic purposes in microbiology: to temper growth and to preserve strains. Bacteria not only grow more slowly in cold, but also die

more slowly. Refrigeration temperatures (4°C–8°C, or 39°F–43°F) are used for food preservation because most pathogens are mesophilic and grow poorly, if at all, at those temperatures. One exception is the Gram-positive bacillus *Listeria monocytogenes,* which can grow reasonably well in the cold and causes disease when ingested.

Long-term storage of bacteria usually requires placing solutions in glycerol at very low temperatures (–70°C). Glycerol prevents the production of razor-sharp ice crystals that can pierce cells from without or within. This deep-freezing suspends growth altogether and keeps cells from dying. Another technique, called **lyophilization**, freeze-dries microbial cultures for long-term storage. In this technique, cultures are quickly frozen at very low temperatures (quick-freezing also limits ice crystal formation) and placed under vacuum, where the resulting sublimation process removes all water from the media and cells, leaving just the cells in the form of a powder. These freeze-dried organisms remain viable for years. Finally, viruses and mammalian cells must be kept at extremely low temperatures (–196°C), submerged in liquid nitrogen. Liquid nitrogen freezes cells so quickly that ice crystals do not have time to form.

Filtration. Filtration through micropore filters with pore sizes of 0.2 μm can remove microbial cells, but not viruses, from solutions. Samples from 1 ml to several liters can be drawn through a membrane filter by vacuum or can be forced through it using a syringe (**Fig. 5.27**). Filter sterilization has the advantage of avoiding heat, which can damage the material sterilized. Of course, since these filters do not trap viruses, the solutions are not really sterile.

Air can also be sterilized by filtration. This process forms the basis of several personal protective devices. A surgical mask is a crude example, while **laminar flow**

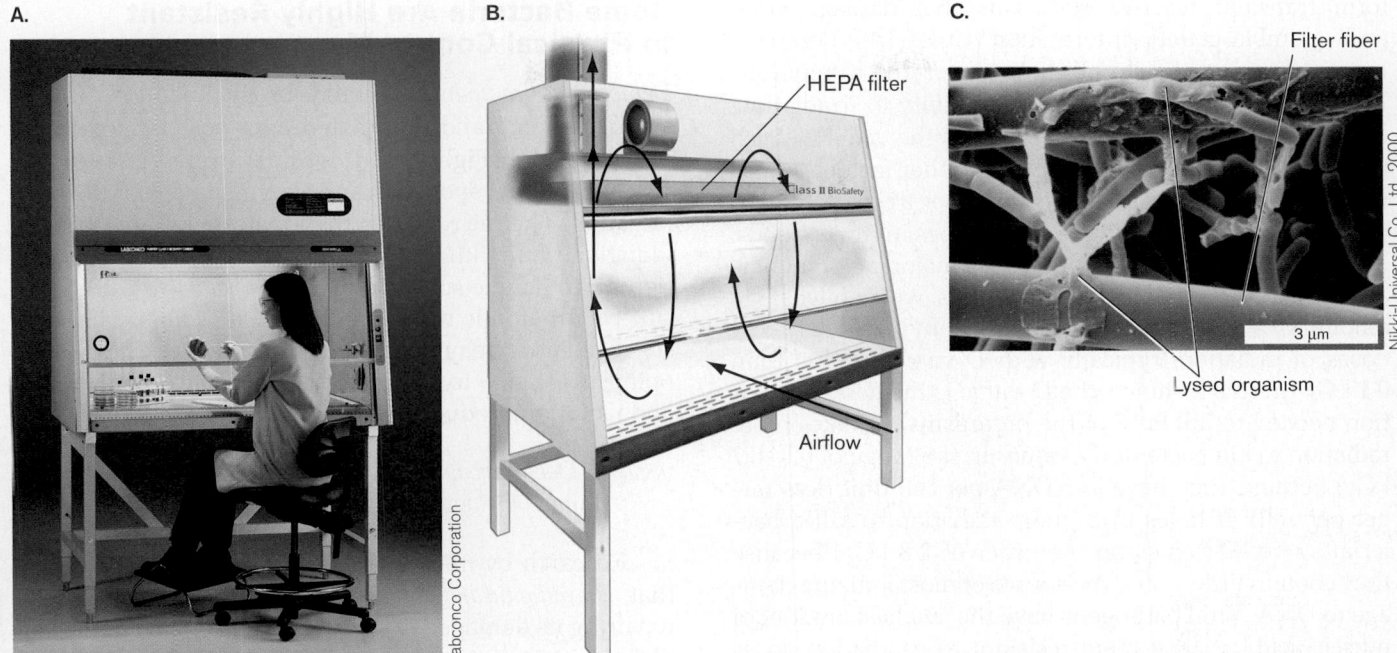

A. Labconco Corporation

B. HEPA filter — Class II BioSafety — Airflow

C. Filter fiber — Lysed organism — 3 µm — Nikki-Universal Co., Ltd., 2000

Figure 5.28 Biological safety cabinet. **A.** Worker examines sample under the hood. **B.** Air is drawn into the chamber and continually passed through the HEPA filter to remove hazardous agents. Filtration prevents the escape of aerosolized infectious agents. **C.** Immobilized enzyme filter. The primary function of this enzyme filter is to kill airborne microorganisms caught on the surface of the filter to protect against secondary contamination from microorganisms in air filtration systems. The photo shows lysed bacteria (*Bacillus subtilis*). The cell walls have been hydrolyzed by enzymatic action, and cell membranes are broken as a result of osmotic pressure pushing outward against the membrane.

biological safety cabinets are more elaborate (and more effective). These cabinets force air through high-efficiency particulate air (HEPA) filters and remove over 99.9% of airborne particulate material 0.3 µm in size or larger. Biosafety cabinets are critical to protect individuals working with highly pathogenic material (**Fig. 5.28A** and **B**). Newer technologies have been developed that embed antimicrobial agents or enzymes directly into the fibers of the filter (**Fig. 5.28C**). Organisms entangled in these fibers are not just trapped; they are attacked by the antimicrobials and lyse.

Irradiation. Public health authorities worldwide are increasingly concerned about food contaminated with pathogenic microorganisms such as *Salmonella* species, *E. coli* O157:H7, *Listeria monocytogenes*, and *Yersinia enterocolitica*. Irradiation, the bombardment of foods with high-energy electromagnetic radiation, has long been a potent, if politically sensitive, strategy for sterilizing food after harvesting. The food consumed by NASA astronauts, for example, has for some time been sterilized by irradiation as a safeguard against food-borne illness in space. Until recently, public pressure severely limited the use of this technology because of concerns about product safety and exposure of workers. However, the recent use of the U.S. mail service by a bioterrorist to disseminate

anthrax spores across the eastern United States has renewed interest in sterilization by irradiation. Foods do not become radioactive when irradiated, but there are concerns that reactive molecules potentially dangerous to humans are produced when high-energy particles are absorbed by the natural chemicals in the food and react to produce toxic substances.

Aside from ultraviolet light, which, owing to its poor penetrating ability, is useful only for surface sterilization, there are three other sources of irradiation: gamma rays, electron beams, and X-rays. Radiation dosage is usually measured in a unit called the gray (Gy), which is the amount of energy transferred to the food, microbe, or other substance being irradiated. A single chest X-ray delivers roughly half a milligray (1 mGy = 0.001 Gy). To kill *Salmonella*, freshly slaughtered chicken can be irradiated at up to 4.5 kilograys (kGy), about 7 million times the energy of a single chest X-ray. Recently the Food and Drug Administration also approved the use of irradiation (4 kGy) on lettuce and spinach, frequent sources of diarrhea-producing *Escherichia coli*. Water runoff from cattle pastures can carry fecal bacteria into adjacent fields. Post harvest, irradiation will kill microbes that can cause disease or shorten produce shelf-life.

When microbes present in food are irradiated, water and other intracellular molecules absorb the energy and form transient reactive chemicals that damage DNA and scramble genetic information. Unless the organism repairs this damage, it will die while trying to replicate. Microbes differ greatly in their sensitivity to irradiation, depending on the size of their DNA, the rate at which they can repair damaged DNA, and other factors. It also matters if the irradiated food is frozen or fresh, as it takes a higher dose of radiation to kill microbes in frozen foods.

The size of the DNA "target" is a major factor in radiation efficacy. Parasites and insect pests, which have large amounts of DNA, are rapidly killed by extremely low doses of radiation, typically with D-values of less than 0.1 kGy (in this instance, the D-value is the *dose* of radiation needed to kill 90% of the organisms). It takes more radiation to kill bacteria (D-values in the range of 0.3–0.7 kGy) because they have less DNA per cell unit (less target per cell). It takes even more radiation to kill a bacterial spore (D-values on the order of 2.8 kGy) because they contain little water, the source of most ionizing damage to DNA. Viral pathogens have the smallest amount of nucleic acid, making them resistant to irradiation doses approved for foods (viruses have D-values of 10 kGy or higher). Infectious agents that do not contain nucleic acids are an even bigger problem. Prions, for example, are misfolded brain proteins that "self-replicate" and cause neurodegenerative diseases (see Section 26.6). Because prions do not contain nucleic acids, the agent can be inactivated by irradiation only at extremely high doses. Thus, irradiation of food is effective in eliminating parasites and bacteria, but is woefully inadequate for eliminating viruses or prions.

NOTE: Electromagnetic radiation emitted by microwave ovens does not directly kill bacteria. However, the heat generated when electromagnetic radiation excites water molecules in an organism will kill the organism if the temperature attained is high enough.

Some Bacteria Are Highly Resistant to Physical Control Measures

Deinococcus radiodurans could be nicknamed "Conan the bacterium" and designated as a poster microbe for extremophiles (**Fig. 5.29**). It was discovered in 1956 in a can of meat that spoiled despite having been sterilized by radiation. The microbe has the greatest ability to survive radiation of any known organism. The amount of radiation it can handle suggests that *D. radiodurans* could even survive an atomic blast. The bacterium's ability to withstand radiation may have evolved as a side effect of developing resistance to extreme drought, since dehydration and radiation produce similar types of DNA damage.

www | *Deinococcus*
      microbiology2.com/links

Research by microbiologist John Battista has shown that *D. radiodurans* possesses an unusual capacity for repairing its damaged DNA, although the precise mechanisms remain a mystery. One possible mechanism is that a damaged chromosome in one member of the quartet shown in **Figure 5.29A** can restore itself using DNA from another member of the quartet. Michael Daly, a pathologist at the Uniformed Services University of the Health Sciences in Maryland, has shown that *Deinococcus* also protects its proteins, which are even more susceptible targets of radiation than is DNA, by accumulating high amounts of manganese. The large intracellular concentration of manganese removes the highly damaging free radicals generated by radiation.

On the basis of this research, *D. radiodurans* was genetically engineered to treat radioactive mercury-contaminated waste from nuclear reactors—a process called **bioremediation** (discussed in Chapter 14). The genes for mercury conversion were spliced from a strain of *E. coli* resistant to particularly toxic forms of mercury and inserted into *D. radiodurans*. The genetically altered

Figure 5.29 *Deinococcus radiodurans.*
A. The amount of radiation this organism can survive is equivalent to that of an atomic blast. The nature of the dark inclusion bodies in three of the four cells in the quartet is not known.
B. John Battista of Louisiana State University showed that *D. radiodurans* has exceptional capabilities for repairing radiation-damaged DNA.

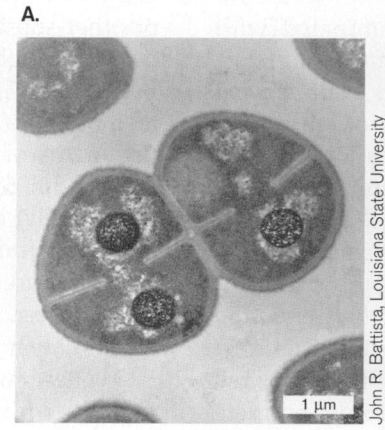

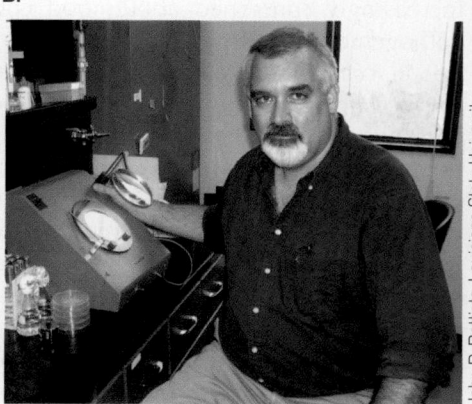

John R. Battista, Louisiana State University

superbug was able to withstand the ionizing radiation and transform toxic waste into forms that could be removed safely. Fortunately, there is little need to worry about its being a superpathogen, because the organism does not cause disease and is susceptible to antibiotics.

Chemical Agents

Disinfection by physical agents is very effective, but in numerous situations their use is impractical (kitchen countertops) or plainly impossible (skin). In these instances, chemical agents are the best approach. A number of factors influence the efficacy of a given chemical agent. These include:

- **The presence of organic matter.** A chemical placed on a dirty surface will bind to the inert organic material present, lowering the agent's effectiveness against microbes. It sometimes is not possible to clean a surface prior to disinfection (as in a blood spill), but the presence of organic material must be factored in when estimating how long to disinfect a surface or object.
- **The kinds of organisms present.** Ideally, the agent should be effective against a broad range of pathogens.
- **Corrosiveness.** The disinfectant should not corrode the surface or, in the case of an antiseptic, damage skin.
- **Stability, odor, and surface tension.** The chemical should be stable upon storage, possess a neutral or pleasant odor, and have a low surface tension so it can penetrate cracks and crevices.

The Phenol Coefficient. Phenol, first introduced by Joseph Lister in 1867 to reduce the incidence of surgical infections, is no longer used as a disinfectant, but its derivatives, such as cresols and orthophenylphenol, are still in use. The household product Lysol is a mixture of phenolics. Phenolics are useful disinfectants because they denature proteins, are effective in the presence of organic material, and remain active on surfaces long after application.

Today we know phenolics are toxic and should not be used on living tissues. Nevertheless, because of its potency and its history, phenol is the benchmark against which other disinfectants are measured. The **phenol coefficient test** consists of inoculating a fixed number of bacteria—for example, *Salmonella enterica* or *Staphylococcus aureus*—into dilutions of the test agent. At timed intervals, samples are withdrawn from each dilution and inoculated into fresh broth (which contains no disinfectant). The phenol coefficient is based on the highest dilution (lowest concentration) of a disinfectant that will kill all the bacteria in a test after 10 minutes of exposure, but leaves survivors after only 5 minutes of exposure. This

Table 5.3	**Phenol coefficients for various disinfectants.**	
Chemical agent	*Staphylococcus aureus*	*Salmonella enterica*
Phenol	1.0	1.0
Chloramine	133.0	100.0
Cresols	2.3	2.3
Ethyl alcohol	6.3	6.3
Formalin	0.3	0.7
Hydrogen peroxide	–	0.001
Lysol	5.0	3.2
Mercury chloride	100.0	143.0
Tincture of iodine	6.3	5.8

concentration is known as the maximum effective dilution. Dividing the reciprocal of the maximum effective dilution for the test agent (for example, ethyl alcohol) by the reciprocal of the maximum effective dilution for phenol gives the phenol coefficient (**Table 5.3**). For example, if the maximum effective dilution for agent X is $1/900$ and that of phenol is $1/90$, then the phenol coefficient of X is 900/90 = 10; the higher the coefficient, the higher the efficacy of the disinfectant.

Commercial Disinfectants

Ethanol, iodine, chlorine, and surfactants (for example, detergents) are all used to reduce or eliminate microbial content from commercial products (**Fig. 5.30**). The first three are compounds that damage proteins, lipids, and DNA. Highly reactive iodine complexed with an organic carrier forms an iodophor, a compound that is water-soluble, stable, nonstaining, and capable of releasing iodine slowly to avoid skin irritation. Wescodyne and Betadine (trade names) are iodophors used, respectively, for the surgical preparation of skin and for wounds. Chlorine is another highly reactive disinfectant with universal application. It is recommended for general laboratory and hospital disinfection and kills the HIV virus.

Detergents can also be antimicrobial agents. The hydrophobic and hydrophilic ends of detergent molecules (the coexistence of which makes the molecules amphipathic) will emulsify fat into water. Cationic (positively charged) but not anionic (negatively charged) detergents are useful as disinfectants because the cationic detergents contain positive charges that can gain access to the negatively charged bacterial cell and disrupt membranes. Anionic detergents are not antimicrobial but do help in the mechanical removal of bacteria from surfaces.

Low-molecular-weight aldehydes such as formaldehyde are highly reactive, combining with and inactivating proteins and nucleic acids. This characteristic makes them useful disinfectants.

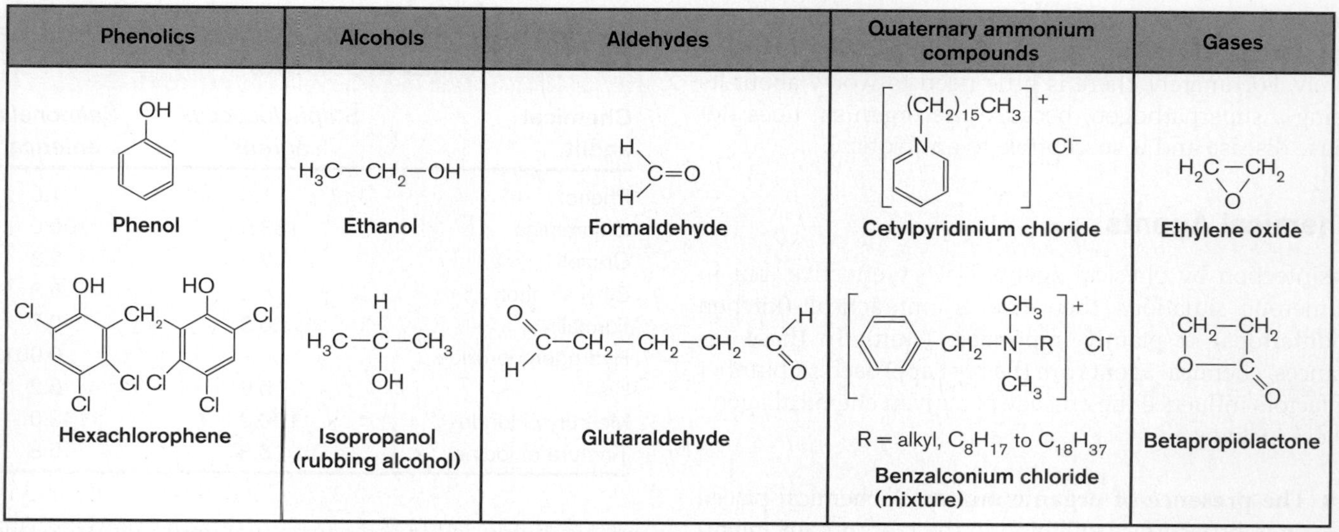

Phenolics	Alcohols	Aldehydes	Quaternary ammonium compounds	Gases
Phenol	Ethanol	Formaldehyde	Cetylpyridinium chloride	Ethylene oxide
Hexachlorophene	Isopropanol (rubbing alcohol)	Glutaraldehyde	R = alkyl, C_8H_{17} to $C_{18}H_{37}$ Benzalconium chloride (mixture)	Betapropiolactone

Figure 5.30 **Structures of some common disinfectants and antiseptics.**

Disposable plasticware like petri dishes, syringes, sutures, and catheters are not amenable to heat sterilization or liquid disinfection. These materials are best sterilized using antimicrobial gases. Ethylene oxide gas (EtO) is a very effective sterilizing agent; it destroys cell proteins, is microbicidal and sporicidal, and rapidly penetrates packing materials, including plastic wraps. Using an instrument resembling an autoclave, EtO at 700 milligrams per liter (mg/L) will sterilize an object after 8 hours at 38°C or 4 hours at 54°C if the relative humidity is kept at 50%. Unfortunately, EtO is explosive. A less hazardous gas sterilant is betapropiolactone. It does not penetrate as well as EtO, but it decomposes after a few hours, which makes it easier to dispose of than EtO.

A new procedure, known as gas discharge plasma sterilization, may replace EtO because it is less harmful to operators. Gas discharge plasma is made by passing certain gases through a radio frequency electric field to produce highly reactive chemical species that can damage membranes, DNA, and protein. It is not yet widely used.

Some attempts have been made to use more sophisticated microbicides as topical agents to prevent the spread of sexually transmitted infections. Several chemical microbicides have proved effective against the agents that cause gonorrhea, herpes, and chlamydia. Some are currently undergoing clinical trials as vaginal gels to determine if the compounds can quell the spread of human immunodeficiency virus (HIV). Although initial results are encouraging, there is, as yet, no definitive proof that any of them are effective.

Bacteria Can Develop Resistance to Disinfectants

It is widely known that bacteria can develop resistance to antibiotics used to treat infections. This is a serious concern in the medical community. So, one might wonder whether bacteria can also develop resistance against chemical disinfectants used to *prevent* infections. The answer is yes—and no. It is difficult for a bacterium to develop resistance to chemical agents that have multiple targets and can easily diffuse into a cell. Iodine, for example, has both of these characteristics. However, disinfectants that have multiple targets at *high* concentrations may have only a single target at lower concentrations—a situation that can foster the development of resistance. For instance, triclosan (a halogenated bisphenol compound used in many soaps and deodorants) targets several cell constituents, making it nicely bactericidal at high concentrations. However, at low concentrations triclosan only inhibits fatty acid synthesis and is merely bacteriostatic. Organisms have developed resistance to triclosan at low concentrations by altering the fatty acid synthesis protein normally targeted by triclosan.

Low-level resistance also can be achieved through membrane-spanning, multidrug efflux pumps (described in Chapter 4). For instance, the MexCD-OprJ efflux system of *Pseudomonas aeruginosa*, a Gram-negative bacterium that causes infections in burn and cystic fibrosis patients, can pump several different biocides, detergents, and organic solvents out of the cell, thereby reducing their efficacy. This finding and other reports of *Pseudomonas* gaining resistance to disinfectants have led many clinicians to advocate caution in the widespread use of certain chemical disinfectants.

Antibiotics Selectively Control Bacterial Growth

Antibiotics as made in nature are chemical compounds synthesized by one microbe that selectively kill other microbial species. Naturally occurring antibiotics act like tiny

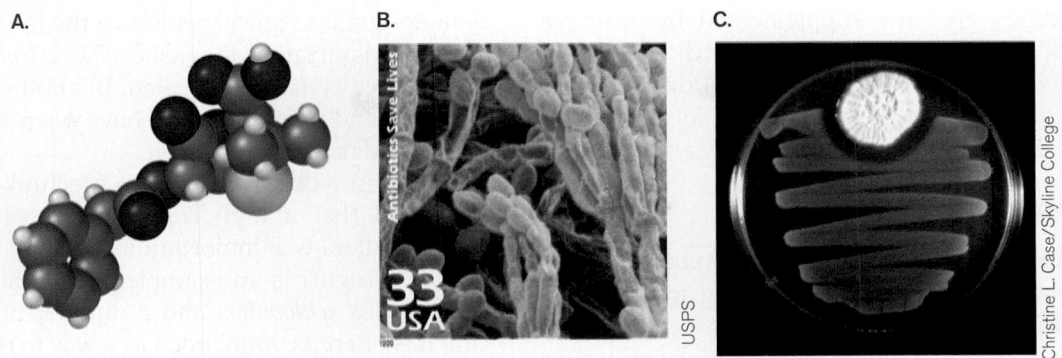

Figure 5.31 **Penicillin.** **A.** Penicillin G molecule produced by the *Penicillium* mold. **B.** U.S. postage stamp showing *Penicillium notatum*. **C.** *P. notatum* culture. The mold excretes penicillin, which inhibits the growth of *Staphylococcus aureus*.

molecular land mines. As a defense against competitors, some organisms secrete antibiotic compounds into their surrounding environment, where they remain until encountered by an intruder. If the interloper is susceptible to the antibiotic, the compound will target specific structures or proteins, disrupting their function. The target cell is either rendered helpless (by bacteriostatic antibiotics) or made nonviable (by bactericidal compounds). (See Chapter 27 for detailed modes of action.) When purified and administered to patients suffering from an infectious disease, these antibiotics can produce seemingly miraculous recoveries.

As we saw in Chapter 1, penicillin, produced by *Penicillium notatum*, was discovered serendipitously in 1929 by Alexander Fleming. **Figure 5.31A** shows this molecule, which mimics a part of the microbial cell wall. Because of this mimicry, penicillin binds to biosynthetic proteins involved in peptidoglycan synthesis and prevents cell wall formation. The drug is bactericidal because *actively growing* cells lyse without the support of the cell wall (**Fig. 5.32**). Other antibiotics target protein synthesis, DNA replication, cell membranes, and various enzyme reactions. These interactions are described throughout Parts 1–3 of this text.

WWW | Penicillin in Jmol
microbiology2.com/links

So how do antibiotic-producing microbes avoid suicide? In some instances, the producing organism lacks the target molecule. *Penicillium* mold, for instance, lacks peptidoglycan and is immune to penicillin by default. Some bacteria produce antimicrobial compounds that target other members of the same species. In this case, the producing organism can modify its own receptors so that they no longer recognize the compound (as with some bacterial colicins). Another strategy is to modify the antibiotic if it reenters the cell. This is the case with streptomycin produced by *Streptomyces griseus*. Streptomycin inhibits bacterial protein synthesis and does not discriminate between the protein synthesis machinery of

Streptomyces and that of others. However, while making enzymes that synthesize and secrete streptomycin, *S. griseus* simultaneously makes the enzyme streptomycin 6-kinase, which remains locked in the cell. If any secreted streptomycin reenters the cell, this enzyme renders the drug inactive by attaching a phosphate to it.

Because many microorganisms have become resistant to commonly used antibiotics, pharmaceutical companies are continually using a variety of drug discovery approaches to search for new antibiotics. Traditional procedures include scouring soil and ocean samples collected from all over the world for new antibiotic-producing organisms and chemically redesigning existing antibiotics so that they can bypass microbial resistance strategies.

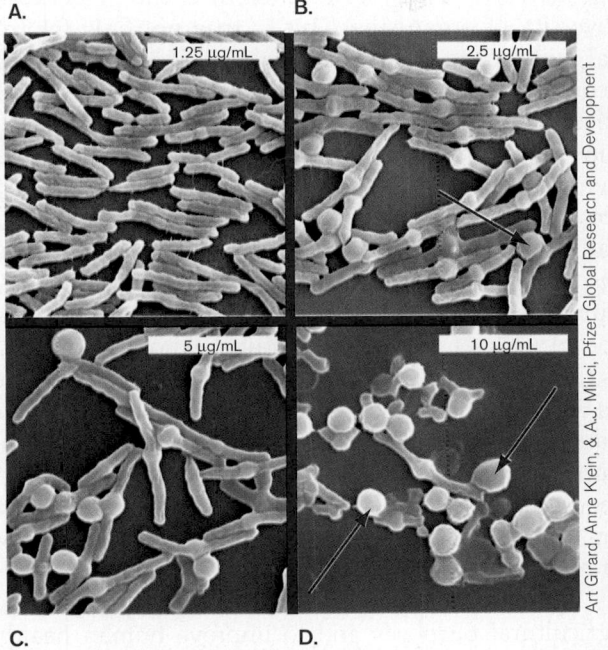

Figure 5.32 **Effect of ampicillin (a penicillin derivative) on *E. coli.*** Cells were incubated for 1 hour at the antibiotic concentrations shown. Swollen areas of cells in panels (B)–(D) reflect weakening cell walls. Cells shown are approx. 2 μm long.

These procedures are now supplemented by "mining the genomes" of microbes for potential drug targets. The frontiers of chemistry include rational drug design, which relies on computer-based methods for predicting the structure and function of potential new antibiotics.

TO SUMMARIZE:

- **Physical and chemical agents** kill microbes by denaturing proteins or DNA, or by disrupting lipid bilayers.
- **Sterilization** kills all living organisms.
- **Antisepsis** is the removal of potential pathogens from the surfaces of living tissues.
- **Antimicrobial** compounds can be **bacteriostatic** or **bactericidal.**
- **The D-value** is the time (or dose, for irradiation) it takes an antimicrobial treatment to reduce the numbers of organisms to 10% of the original value.
- **The autoclave** uses high pressure to achieve temperatures that will sterilize objects.
- **Food can be preserved** by pasteurization, refrigeration, filtration, and irradition.
- **Chemical disinfectants** are compared to one another on the basis of the phenol coefficient.
- **Antibiotics** are compounds produced by one living microorganism that kill other microorganisms.

Biological Control of Microbes

Pitting microbe against microbe is an effective way to prevent disease in humans and animals. One of the hallmarks of a healthy ecosystem is the presence of a diversity of organisms. This is true not only for tropical rain forests and coral reefs, but also for the complex ecosystems of human skin and the intestinal tract. In these environments, the presence of harmless microbial flora can retard the growth of undesired pathogens. The pathogenic fungus *Phytophthora cinnamomi*, for example, causes root rot in plants but is biologically controlled by fungi belonging to the genus *Myrothecium*. Naturally occurring *Staphylococcus* species on human skin produce short-chain fatty acids that retard the growth of pathogenic strains. Another illustration is the human intestine, which is populated by as many as 500 microbial species. Most of these species are non-pathogenic organisms that exist in symbiosis with their human host. Vigorous competition between members of the normal intestinal flora and the production of permeant weak acids by fermentation helps control the growth of numerous pathogens.

Microbial competition has been widely exploited for agricultural purposes and to improve human health in a process known as **probiotics**. In general, a probiotic is a food or supplement that contains live microorganisms and improves intestinal microbial balance. Newborn baby chicks, for instance, are fed a microbial cocktail of normal flora designed to quickly colonize the intestinal tract and prevent colonization by *Salmonella*, a frequent contaminant of factory-farmed chicken. In another example, *Lactobacillus* and *Bifidobacterium* have been used to prevent and treat diarrhea in children.

Russian Nobel laureate Ilya Mechnikov in 1908 first suggested that a high concentration of lactobacilli in intestinal flora was important for health and longevity in humans. Yogurt is an example of a probiotic containing *Lactobacillus acidophilus* and a number of other lactobacilli. It is often recommended as a way to restore a normal balance to gut flora (for example, after it has been disturbed by antibiotic treatment) and appears useful in the treatment of inflammatory bowel disease.

Phage therapy, another biocontrol method, was first described in 1907 by Félix d'Herelle at France's Pasteur Institute, long before antibiotics were discovered. Bacteriophages are viruses that prey on bacteria (discussed in Section 6.1). Each bacterial species is susceptible to a limited number of specific phages. Because a phage infection often causes bacterial lysis, it was considered feasible to treat infectious diseases with a phage targeted to the pathogen. At one time, doctors used phages as medical treatment for illnesses ranging from cholera to typhoid fever. In some cases, a liquid containing the phage was poured into an open wound. In other cases, phages were given orally, introduced via aerosol, or injected. Sometimes the treatments worked; sometimes they did not. When antibiotics came into the mainstream, phage therapy largely faded. Now that strains of bacteria resistant to standard antibiotics are on the rise, the idea of phage therapy has enjoyed renewed interest from the worldwide medical community.

TO SUMMARIZE:

- **Biocontrol** is the use of one microbe to control the growth of another.
- **Probiotics** contain certain microbes that, when ingested, aim to restore balance to intestinal flora.
- **Phage therapy** offers a possible alternative to antibiotics in the face of rising antibiotic resistance.
- **Disinfection** kills pathogens on inanimate objects.

Concluding Thoughts

Microbiology as a science was founded on the need to understand and control microbial growth. The initial impetus was to control the diseases of humans, as well as the diseases of plants and animals. But as we will see in later chapters, microbiology has developed into a science that has helped us understand the molecular processes of life. Concepts such as biological diversity, food microbiology, microbial disease, and antibiotics will be revisited in later chapters.

CHAPTER REVIEW

Review Questions

1. Explain the nature of extremophiles and discuss why these organisms are important.
2. What are the parameters that define any growth environment?
3. List and define the classifications used to describe microbes that grow in different physical growth conditions.
4. What do thermophiles have to do with the PCR reaction?
5. Why is water activity important to microbial growth? What changes water activity?
6. How do cells protect themselves from osmotic stress?
7. Why do changes in H$^+$ concentration affect cell growth?
8. How do acidophiles and alkaliphiles manage to grow at the extremes of pH?
9. If an organism can live in an oxygenated environment, does that mean that the organism uses oxygen to grow? If an organism can live in an anaerobic environment, does that mean it cannot use oxygen as an electron acceptor? Why or why not?
10. What happens when a cell exhausts its available nutrients?
11. List and briefly explain the various means by which humans control microbial growth. What is a D-value? What is a phenol coefficient?
12. How do microbes prevent the growth of other microbes?

Thought Questions

1. Given a natural lake environment with 100 species of bacteria, why does the species with the fastest generation time not overwhelm the others? Or does it?
2. *Escherichia coli* is a facultative species, able to grow with or without oxygen. What would it take to make this organism an anaerobe?
3. Two spore formers, *Bacillus stearothermophilus* and *Bacillus coagulans*, have D-values at 121°C of 5 min- utes and 0.07 minute, respectively. How could the spores from these organisms have such different D-values? *Hint*: Find the optimum growth temperatures for these organisms.
4. Phage therapy is touted by some as a solution to antibiotic resistance. Explain why phage therapy may not be able to solve this problem.

Key Terms

acidophile (161)
aerobic respiration (164)
aerotolerant anaerobe (166)
alkaliphile (161)
anaerobic respiration (166)
antibiotic (176)
antisepsis (170)
bactericidal (170)
bacteriostatic (170)
barophile (156)
bioremediation (174)
decimal reduction time (D-value) (170)
disinfection (170)
DNA microarray (150)
electron transport chain (164)
eutrophication (169)

extremophile (150)
facultative (166)
facultative anaerobe (166)
fermentation (fermentative metabolism) (166)
germicidal (170)
halophile (158)
heat-shock response (156)
hyperthermophile (154)
laminar flow biological safety cabinet (172)
lyophilization (172)
membrane-permeant organic acid (160)
mesophile (153)

microaerophilic (166)
neutralophile (160)
oligotroph (168)
osmolarity (157)
pasteurization (172)
phenol coefficient test (175)
piezophile (156)
probiotic (178)
psychrophile (153)
sanitation (170)
sterilization (170)
strict aerobe (165)
strict anaerobe (165)
thermophile (153)
water activity (157)

Recommended Reading

Atomi, Haruyuki. 2005. Recent progress towards the application of hyperthermophiles and their enzymes. *Current Opinion in Chemical Biology* **9**:163–173.

Blasius, Melanie, Ulrich Hubscher, and Suzanne Sommer. 2008. *Deinococcus radiodurans*: What belongs to the survival kit? *Critical Reviews in Biochemistry and Molecular Biology* **43**:221–238.

Christner, Brent C., Rongman Cai, Cindy E. Morris, Kevin S. McCarter, Christine M. Foreman, et al. 2008. Geographic, seasonal, and precipitation chemistry influence on the abundance and activity of biological ice nucleators in rain and snow. *Proceedings of the National Academy of Sciences USA* **105**:18854–18859.

Daly, Michael, Elena Gaidamakova, Vera Matrosova, Alexander Vasilenko, Min Zhai, et al. 2007. Protein oxidation implicated as the primary determinant of bacterial radioresistance. *PLoS Biology* **5**:769–779.

D'Amico, Salvino, Tony Collins, Jean-Claude Marx, Georges Feller, and Charles Gerday. 2006. Psychrophilic microorganisms: Challenges for life. *EMBO Reports* **7**:385–389.

Horikoshi, Koki. 1998. Barophiles: Deep-sea microorganisms adapted to an extreme environment. *Current Opinion in Microbiology* **1**:291–295.

Horikoshi, Koki. 1999. Alkaliphiles: Some applications of their products for biotechnology. *Microbiology and Molecular Biology Review* **63**:735–750.

Kota, Swathi, and Hari S. Misra. 2008. Identification of a DNA processing complex from *Deinococcus radiodurans*. *Biochemistry and Cell Biology* **86**:448–458.

Rhen, Mikael, and Charles J. Dorman. 2005. Hierarchical gene regulators adapt *Salmonella enterica* to its host milieus. *International Journal of Medical Microbiology* **294**:487–502.

Romero, Diego, Claudio Aguilar, Richard Losik, and Roberto Kolter. 2010. Amyloid fibers provide structural integrity to *Bacillus subtilis* biofilms. *Proceeding of the National Academy of Science* **107**:2230–2234.

Slonczewski, Joan, James A. Coker, and Shiladitya DasSarma. 2010. Microbial growth with multiple stressors. *Microbe* **5**:110–116.

Thomas, D. N., and G. S. Dieckmann. 2002. Antarctic Sea Ice—A habitat for extremophiles. *Science* **295**:641–643.

Yacoby, Iftach, and Itai Behar. 2008. Targeted filamentous bacteriophages as therapeutic agents. *Expert Opinion on Drug Delivery* **5**:321–329.

 For further study and review, please visit StudySpace at **microbiology2.com**.

Chapter 6

Virus Structure and Function

100 nm

All kinds of cells, including bacteria, eukaryotes, and archaea, can be infected by particles called viruses. Viruses are the smallest known units of reproduction, some with genomes of fewer than ten genes. A virus must infect a host cell and subvert the cell's machinery to reproduce more virus particles, usually killing the host cell. Alternatively, some viral genomes are copied into the host genome, where they replicate silently with the host. Degenerate viral genomes from ancient viral insertions make up a large part of the human genome.

Virus structure varies in complexity from single RNA molecules to spore-like packages containing multiple layers of protein and membranes. Some viruses resemble degenerate cells. Viral intracellular life cycles enable them to divert complex host cell processes to form new viruses. How can such particles, barely large enough to be detected by an electron microscope, consume entire cells and multicellular organisms? Alternatively, how can viruses integrate their genomes into the genome of their host, replicating with their host for future generations?

Current Research Highlight

Mimivirus, a cause of pneumonia, is the largest known virus; larger even than some bacteria. Its genome poses intriguing questions for evolution, such as, did viruses evolve from cells? This cryo-EM model of mimivirus was developed by Michael Rossmann and colleagues. The star-shaped vertex (colorized blue) opens to release viral DNA within the host cell. The coloring is based on radial distance from the center of the virus. Red, from 180 to 210 nm; and rainbow coloring from red to blue, between 210 and 250 nm. *Source: Chuan Xiao et al. 2009. PLoS Biology 7:e92.*

In April 2009, in Mexico and California, people fell ill from a mysterious strain of influenza virus. The strain was called "swine flu" because some of its genes revealed origin in influenza viruses infecting pigs. But pigs did not spread this new strain—people did. Travelers soon brought the strain to other countries, where healthy people succumbed. Because the strain had a new combination of surface proteins never experienced before, people lacked immunity. In some cities, schools were closed and sports stadiums were emptied. Some airports screened passengers thermally for signs of fever (**Fig. 6.1A**), although the measure fails to catch those incubating the virus before symptoms appear. It was feared that the strain could cause a death rate comparable to the 1918 pandemic, when millions of people died. Fortunately, the 2009 strain turned out to be mild, with a death rate of less than 1%, and prompt treatment was effective. Even so, every year seasonal strains of influenza kill half a million people. Yet the influenza virus is a small, noncellular particle (**Fig. 6.1B**) whose RNA genome has only eight genes encoding ten proteins. How could such a tiny biological entity cause such devastating illness?

Chapter 6 explores viral infections, as well as the many roles of viruses in biology. Some viruses generate frightening epidemics, from influenza to AIDS. Others fill essential niches in the environment, particularly in marine ecosystems, where they cycle carbon and curb toxic blooms of algae. Viral predation selects for much of the species diversity of marine protists and invertebrates.

In research, viruses have provided both tools and model systems for our discovery of the fundamental principles of molecular biology. The first genes mapped, the first regulatory switches defined, and the first genomes to be sequenced were all those of viruses. Vectors for gene cloning and gene therapy today continue to be derived from viruses.

In this chapter, we introduce major themes of virus structure and function and the fundamental challenges that all viruses face: genome packaging, cell attachment and entry, and the molecular strategies that enable viruses to divert the metabolism of their host cell. Viruses provide key tools and model systems for molecular biology. Our understanding of viruses, particularly bacterial viruses, called bacteriophages, provides a useful background for the molecular biology we will encounter in Part 2 of this book (Chapters 7–12). The molecular biology of viral life cycles is explored further in Chapter 11,

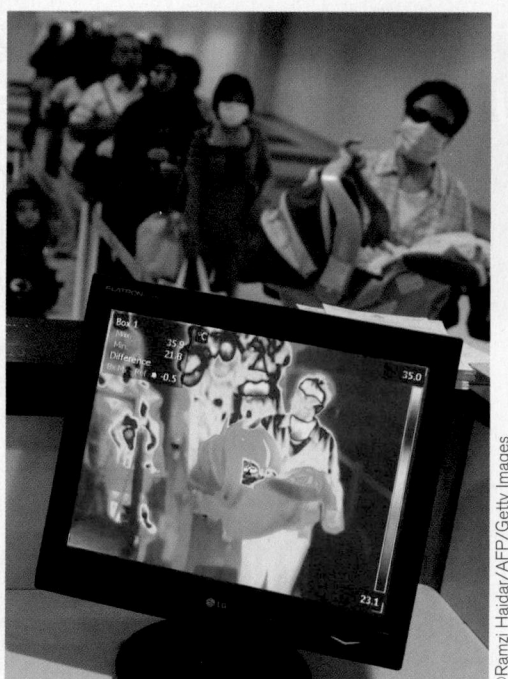

A.

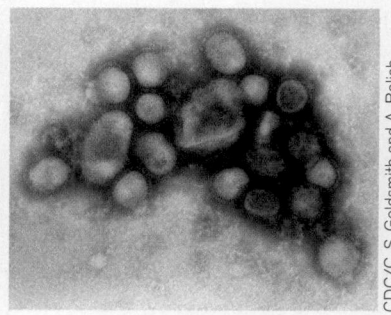

B.

©Ramzi Haidar/AFP/Getty Images

CDC/C. S. Goldsmith and A. Balish

Figure 6.1 **Screening for influenza.** **A.** Thermal scanning for fever in passengers at an airport. **B.** Influenza virions (transmission electron micrograph, TEM). Virion size, 100–200 nm.

and viral disease pathology and epidemiology are discussed in Chapters 25–27.

6.1 What Is a Virus?

A **virus** is a noncellular particle that must infect a host cell, where it reproduces. Our concept of a virus has developed over the past century of virology. The Dutch microbiologist Martinus Beijerinck (1851–1931) first proposed the existence of viruses as soluble infectious agents that passed through a filter too small to pass cells. The first virus to infect bacterial cells (bacteriophage) was observed in 1917 by French microbiologist Félix d'Herelle (1873–1949). In 1952, Alfred Hershey (1908–1997) and Martha Chase (1927–2003) tested the incorporation of ^{32}P-radiolabeled phosphate versus ^{35}S-radiolabeled amino acids into the genetic material of a bacteriophage, T2. When the phage T2 infected *E. coli*, only ^{32}P label was incorporated into the infected cells, thus confirming that DNA was the genetic material, not proteins (which would incorporate ^{35}S). Other viruses were shown to have genomes of RNA—for example, tobacco mosaic virus (TMV). The A-form structure of RNA was first solved for TMV by British crystallographer Rosalind Franklin (1920–1958). The first virus shown to cause tumors was an RNA retrovirus, Rous sarcoma virus (RSV), the discovery of which earned Peyton Rous (1879–1970) the Nobel Prize in Medicine in 1966.

The sequencing of viral genomes, and the discovery of viruses as large as cells, have deepened our understanding of viral origins. Some large viruses have evolved from bacteria, whereas other viruses may have evolved from host cellular components (discussed in Chapter 11).

The virus particle, or **virion**, consists of an infective nucleic acid (either DNA or RNA) contained within a protective shell made of protein, called the **capsid**. The capsid often has a molecular delivery device that enables transfer of the virion's genome into the host cell.

Viruses that infect bacteria are known as **bacteriophages** or **phages**. An example is bacteriophage T2, which infects *Escherichia coli* (**Fig. 6.2A**). The T2 and T4 phages have a capsid with a tail that inserts the viral genome into the host cell, where it directs the reproduction of progeny virions. Virions are released when the host cell lyses. As cells lyse, their disappearance can be observed as a **plaque**, a clear spot against a lawn of bacterial cells (**Fig. 6.2B**). Each plaque arises from a single virion or phage particle that lyses a host cell and spreads progeny to infect adjacent cells. Plaques can be counted as representing individual infective virions from a phage suspension.

An example of a virus that infects humans is measles virus. The measles virus has an envelope that is derived from the host cell plasma membrane during exit of the virus from the host cell. During entry of the virus into a new host cell, the envelope fuses with the host cell plasma membrane, releasing the viral contents into the cytoplasm of the cell. After replicating within the infected cell, newly formed measles virions become enveloped by host cell membrane as they bud out of the host cell (**Fig. 6.2C**). The spreading virus generates a rash of red spots on the skin of infected patients (**Fig. 6.2D**) and can be fatal (one in 500 cases).

Plants are also infected by viruses, such as tobacco mosaic virus (TMV). Within the plant cell, virions accumulate to high numbers (**Fig. 6.2E**) and travel through interconnections to neighboring cells. Infection by tobacco mosaic virus results in mottled leaves and stunted growth (**Fig. 6.2F**). Plant viruses cause major economic losses in agriculture worldwide.

Each species of virus infects a particular group of host species, known as the **host range**. Some viruses can infect only a single species; for example, HIV infects only humans. Close relatives of humans, such as the chimpanzee, are not infected, although they are susceptible to a closely related virus, simian immunodeficiency virus (SIV). In contrast, the West Nile virus, transmitted by mosquitoes, has a much broader host range, including many species of birds and mammals.

> **THOUGHT QUESTION 6.1** Which viruses do you know that have a narrow host range, and which have a broad host range?

WWW | American Society for Virology

WWW | International Committee on Taxonomy of Viruses
microbiology2.com/links

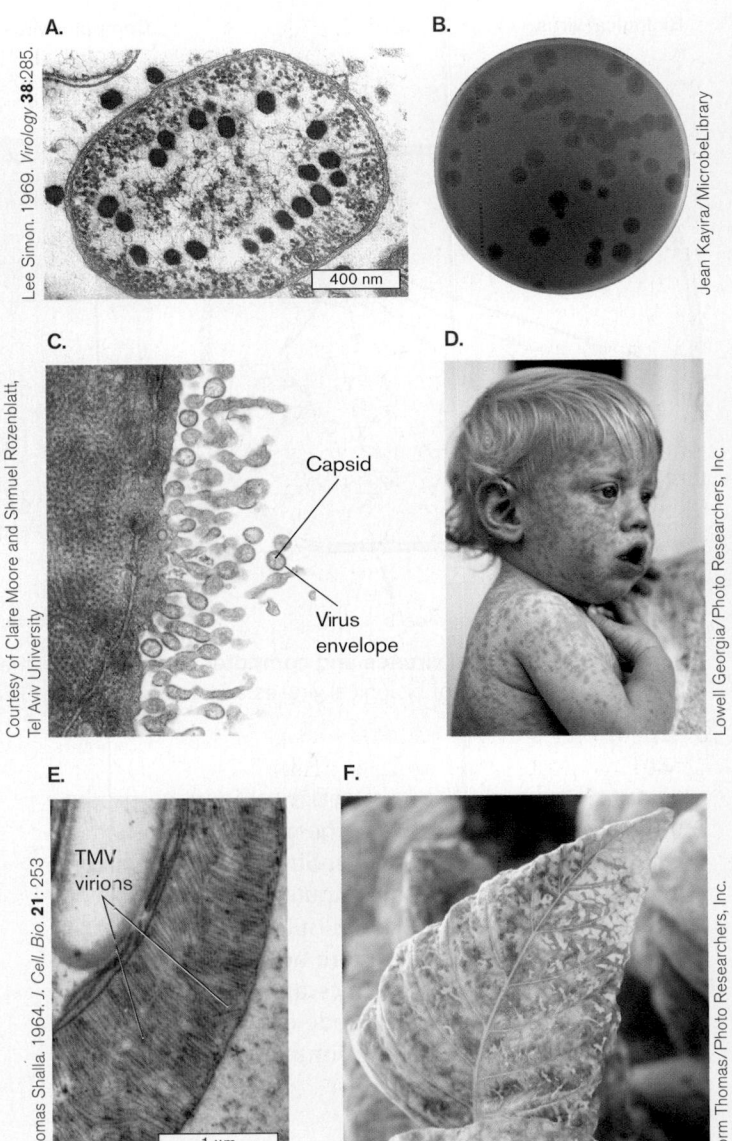

Figure 6.2 Virus infections. A. Bacteriophage T2 particles form a semicrystalline array within an *E. coli* cell (transmission electron micrograph, TEM). **B.** Bacteriophage infection forms plaques of lysed cells on a lawn of bacteria. **C.** Measles virions (diameter 200–600 nm) bud out of human cells in tissue culture (TEM). **D.** Child infected with measles shows a rash of red spots. **E.** Tobacco leaf section is packed with tobacco mosaic virus particles. **F.** Tomato leaf infected by tobacco mosaic virus shows mottled appearance.

Viruses Propagate Their Genomic Information

Viral propagation exemplifies the central role of information in biological reproduction. The propagation of viruses is mimicked by the spread of "computer viruses," whose information "infects" computer memory (**Fig. 6.3**). When a biological virus infects a host cell, the information in

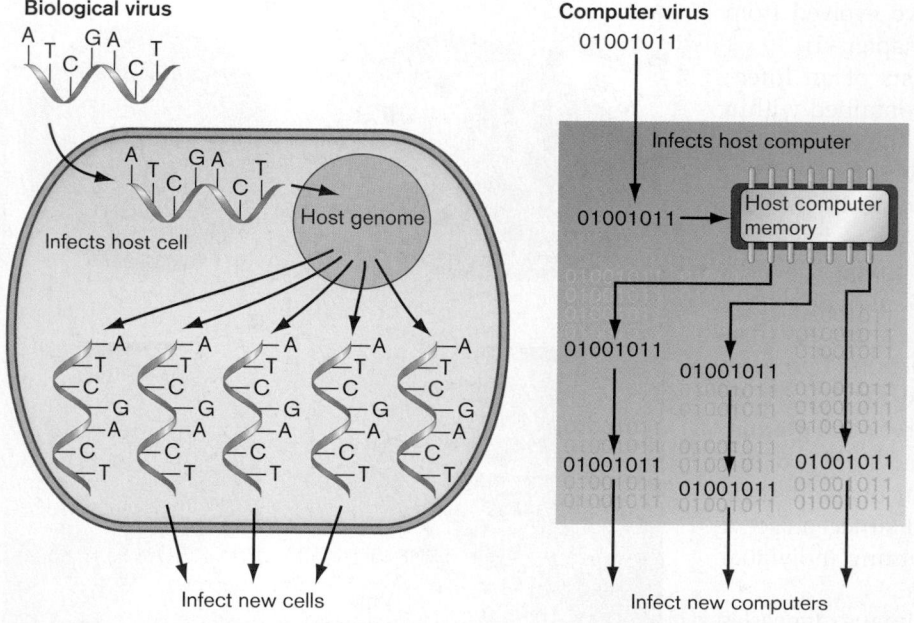

Figure 6.3 Biological viruses and computer viruses. From the standpoint of information, the behavior of biological viruses and computer viruses is analogous.

its genome subverts the host cell machinery to produce multiple copies of the virus; the multiple copies then escape to infect more host cells. Similarly, when a computer virus infects a host computer, its program code subverts the host to produce multiple copies of the virus, which then escape to infect more host computers. Computer viruses generate epidemics analogous to those of biological viruses. The virus's code can even be designed to "mutate" in order to foil the "immune system" of antivirus software.

Viruses Infect All Forms of Life

Viruses are ubiquitous, infecting every taxonomic group of organisms, including bacteria, eukaryotes, and archaea. In marine ecosystems, viruses act as major predators and sequester significant amounts of nutrients. For humans, viruses cause many forms of illness, whose influence on our history and culture would be hard to overstate. More people died of influenza in the global epidemic of 1918 than in the battles of World War I. In the past 30 years, the AIDS pandemic caused by HIV has killed 25 million people worldwide and continues to grow.

Viruses are part of our daily lives. The most frequent infections of college students are due to respiratory pathogens such as rhinovirus (the common cold) and Epstein-Barr virus (infectious mononucleosis), as well as sexually transmitted viruses such as herpes simplex virus (HSV) and papillomavirus (genital warts). Viruses also impact human industry; for example, bacteriophages (literally, "bacteria-eaters") infect cultures of *Lactococcus*

during the production of yogurt and cheese. Plant pathogens such as cauliflower mosaic virus and rice dwarf virus continue to cause substantial losses in agriculture.

In contrast to our vast arsenal of antibiotics (effective against bacteria), the number of antiviral drugs remains depressingly small. Because the machinery of viral growth is largely that of the host cell, viruses present relatively few targets that can be attacked by antiviral drugs without harming the host. However, an understanding of viral life cycles at the molecular level is now leading to the development of new antiviral drugs, such as AZT and protease inhibitors, which combat HIV.

Despite their lethal potential, viruses have made surprising contributions to medical research. A bacteriophage provides a protein that lyses the cell walls of anthrax bacteria. Other bacteriophages are used as **cloning vectors**, small genomes into which foreign genes can be inserted and cloned for gene technology. Even lethal viruses such as HIV are being developed as vectors for human gene therapy.

Some viruses introduce copies of their own genomes into the host's genome, a process that can mediate evolution of the host genome. Indeed, studies of molecular evolution reveal that viral genomes are the ancestral source of about a tenth of the human genome.

Viral Genomes

The genome of a virus can be small, encoding fewer than ten genes. In cauliflower mosaic virus, for example, the genome encodes only seven genes (**Fig. 6.4A**), which actually overlap each other in sequence. This overlap in sequence is made possible by the use of different **reading frames**, start positions to define the first base of the codon for translation to amino acid sequence. Many viral genomes, such as that of avian leukosis virus (**Fig. 6.4B**), are encoded by RNA. The RNA genome of avian leukosis virus has protein-encoding genes grouped by functional categories of core capsid, replicative enzymes, and envelope proteins (proteins embedded in the envelope phospholipid bilayer).

Larger viral genomes, such as that of herpes virus or of bacteriophage T4, have genes dispersed around the chromosome, similar to the genomes of bacteria. The giant mimivirus, which infects amebas and may cause human pneumonia, is as large as some bacteria (**Fig. 6.5**).

A. Cauliflower mosaic virus genome (DNA)

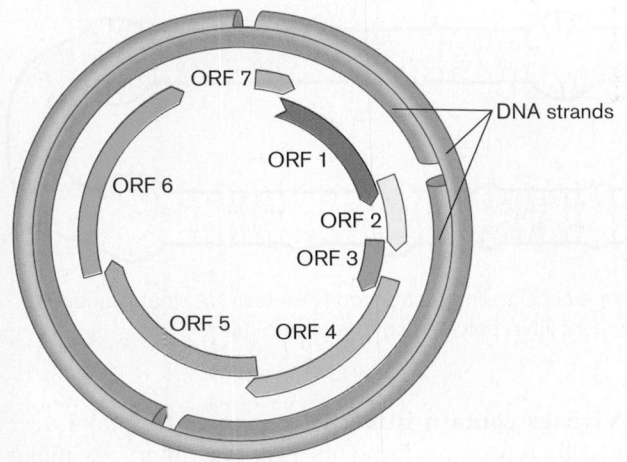

B. Avian leukosis virus genome (RNA)

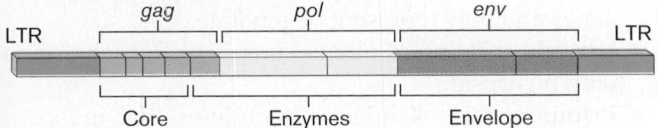

Figure 6.4 Simple viral genomes.   A. Cauliflower mosaic virus has a circular genome of double-stranded DNA, whose strands are interrupted by nicks. The genome encodes seven overlapping genes (open reading frames, ORFs). **B.** Avian leukosis virus: a single-stranded RNA retrovirus resembling eukaryotic mRNA. Three genes (*gag*, *pol*, and *env*) encode polypeptides that are eventually cleaved to form a total of nine functional products. LTR = long terminal repeat.

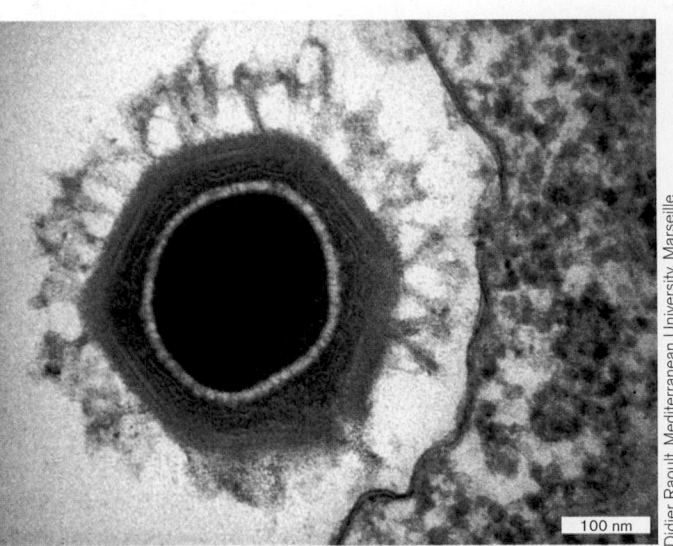

Figure 6.5   Mimivirus infecting an ameba.   About 300 nm across, the mimivirus is larger than some bacteria.

Mimivirus evolved by reductive evolution of an intracellular parasitic bacterium, and it still has a bacterium-sized genome of 1.2 million base pairs. The mimivirus virion conducts numerous cellular functions, including DNA repair and protein folding by chaperones. It probably represents a descendant of a cell that parasitized protist cells until most of its cellular traits were lost through degenerative evolution.

Viroids: Infective Genomes with No Capsid

Early in the twentieth century, viruses were believed to be the smallest particles capable of infecting hosts and propagating themselves. Then, even smaller virus-like infectious agents were discovered for which the nucleic acid genome is itself the entire infectious particle; there is no protective capsid. Such infectious agents are called **viroids**. Most viroids are RNA molecules that infect plants, including many kinds of fruits and vegetables. For example, citrus viroids have caused economic losses in the citrus industry of the United States, Australia, and Israel.

Most viroids are poorly understood, but a well-known example is the potato spindle tuber viroid (**Fig. 6.6A**). This viroid consists of a circular, single-stranded molecule of RNA that doubles back on itself to form base pairs interrupted by short unpaired loops. This unusual circularized form avoids breakdown by host RNase enzymes. The RNA folds up into a globular structure that interacts with host cell proteins. The RNA is replicated by host RNA polymerase. The host RNA polymerase normally requires a DNA template, but the replication process is modified by an unknown mechanism.

During viroid infection, the RNA-dependent RNA polymerase replicates progeny copies of the viroid, which encodes no products other than itself. Viroids can cause as much host destruction as "true viruses," and some authors, particularly in plant pathology, classify them as "viruses without capsids."

Some viroids have catalytic ability, comparable to enzymes made of protein. An RNA molecule capable of catalyzing a reaction is called a ribozyme (discussed in Chapter 8). Viral ribozymes may be able to cleave themselves or other specific RNA molecules. Their ability to cleave very specific RNA sequences has applications in medical research such as cleaving human mRNA involved in cancer.

Prions: Infection without Nucleic Acid?

A remarkable class of infectious agents is believed to consist of protein only. These agents, known as **prions**, are thought to be aberrant proteins arising from the host cell. Prions gained notoriety when they were implicated in brain infections such as Creutzfeldt-Jakob disease, a

Potato spindle tuber viroid—circular ssRNA

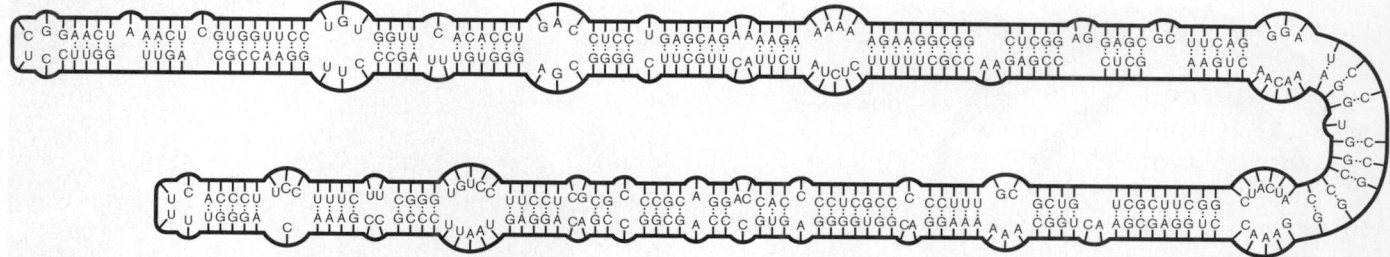

Figure 6.6 Viroids: infective RNA.   Potato spindle tuber viroid consists of a circular single-stranded RNA (ssRNA) that hybridizes internally. A viroid encodes no genes, but it hijacks the plant cell's RNA-dependent RNA polymerase to replicate itself.

variant form of which is known as "mad cow" disease because it may be transmitted through defective proteins in beef from diseased cattle. Other diseases believed to be caused by prion transmission include scrapie, a disease of sheep; and kuru, a degenerative brain disease found in a tribe of people who customarily consumed the brains of deceased relatives.

In prion-associated diseases, the infective agent is unaffected by treatments that destroy RNA or DNA, such as nucleases or UV irradiation. A prion is an aberrant form of a normally occurring cell protein that assumes an abnormal conformation or tertiary structure (**Fig. 6.7A**). The prion form of the protein acts by binding to normally folded proteins of the same class and altering their conformation to that of the prion. The multiplying prion then alters the conformation of other normal subunits, forming harmful aggregates in the cell and ultimately leading to cell death. In the brain, prion-induced cell death leads to tissue deterioration and dementia (**Fig. 6.7B**).

Prion diseases can be initiated by infection with an aberrant protein. More rarely, the cascade of protein misfolding can start with the spontaneous misfolding of an endogenous host protein. The chance of spontaneous unfolding is greatly increased in individuals who inherit certain alleles encoding the protein; thus, spontaneous prion diseases can be inherited genetically. Overall, prion diseases are unique in that they can be transmitted by an infective protein instead of by DNA or RNA; and they propagate conformational change of existing molecules without synthesizing entirely new infective molecules.

TO SUMMARIZE:

■ **Viruses** consist of noncellular particles that infect a host cell and direct its expression apparatus to produce virus particles.

■ A **virion** consists of a capsid enclosing a nucleic acid genome. Some viruses are further enclosed by an envelope of membrane with embedded proteins.

■ **All classes of organisms are infected by viruses.** Usually the hosts are limited to a particular host range of closely related strains or species.

■ **Viruses contain infective genomes** that take over a cell, reprogramming its cell machinery to make progeny virus particles (virions). Some viral genomes consist of fewer than 10 genes; others have 100 or 200 genes and may represent degenerate cells.

■ **Viroids** that infect plants consist of RNA hairpins with no capsid.

■ **Prions** consist of infectious proteins that induce a cell's native proteins to fold incorrectly and impair cell function.

A.

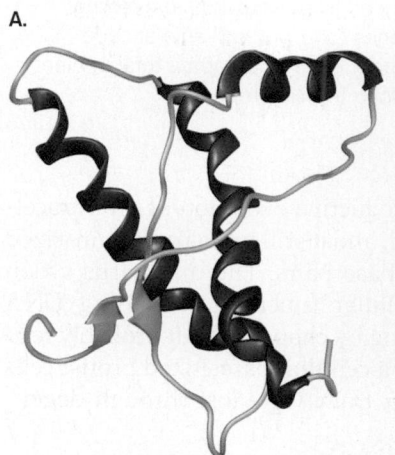

Normal conformation   Abnormal conformation

B.

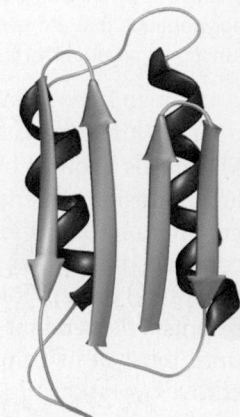

Ralph C. Eagle/Photo Researchers, Inc.

Figure 6.7 Prion disease. A. The normal conformation of a prion, compared to the abnormal conformation. The abnormal form "recruits" normally folded proteins and changes their conformation into the abnormal form. (PDB code: 1AG2) **B.** Section of a human brain showing "spongiform" holes typical of Creutzfeldt-Jakob disease.

6.2 Virus Structure

The packaged structure of a virus achieves two goals: It keeps the viral genome intact, and it enables infection of the appropriate host cell. First, the stable capsid protects the viral genome from degradation and enables it to be transmitted outside the host. Second, in order for the viral genome to reproduce, the virion must either insert its genome into the host cell or disassemble within the host. In the process, the original particle loses its stable structure and its own identity as such, but it generates numerous progeny virions.

> **THOUGHT QUESTION 6.2** What would happen if a virus particle remained intact within a host cell instead of releasing its genome?

Symmetrical Virus Particles

Different viral species make different forms of capsids. Virus particles may be symmetrical, in which case the capsid is one of two types: icosahedral or filamentous (helical). Each type of capsid exhibits geometrical symmetry. The advantage of symmetry is that it provides a way to form a package out of repeating protein units generated by a small number of genes and encoded by a short chromosomal sequence. The smaller the viral genome, the more genome copies can be synthesized from the host cell's limited supply of nucleotides. Nevertheless, some symmetrical viruses, such as herpes virus and mimivirus, have much larger genomes. Large genomes offer a greater range of functions for viral components.

Icosahedral viruses. Many viruses package their genome in an **icosahedral** (20-sided) **capsid**. Icosahedral viral capsids take the form of a polyhedron with 20 identical triangular faces. In the capsid, each triangle can be composed of three identical but asymmetrical protein units. An example of an icosahedral capsid is that of the herpes simplex virus (**Fig. 6.8A**). Each triangular face of the capsid is determined by the same genes encoding the same protein subunits. No matter what the pattern of subunits in the triangle is, the structure overall exhibits rotational symmetry characteristic of an icosahedron (**Fig. 6.8B**): threefold symmetry around the axis through two opposed triangular faces, fivefold symmetry around an axis through opposite points, and twofold symmetry around an axis through opposite edges. Capsid symmetry is important for structure determination and visualization and for the design of antiviral drugs.

> **THOUGHT QUESTION 6.3** Why do viral capsids take the form of an icosahedron instead of some other polyhedron?

Virus particles can be observed by standard transmission electron microscopy (TEM), but the details of capsid structure as in **Figure 6.8A** require visualization by digital reconstruction of cryo-EM (discussed in Section 2.6). Recall from Chapter 2 that in cryo-EM, the viral samples for TEM are prepared flash-frozen, preventing the formation of ice crystals. Flash freezing enables observation without stain. The electron beams penetrate the object; thus, images of individual capsids actually provide a glimpse of the virus's internal contents. By

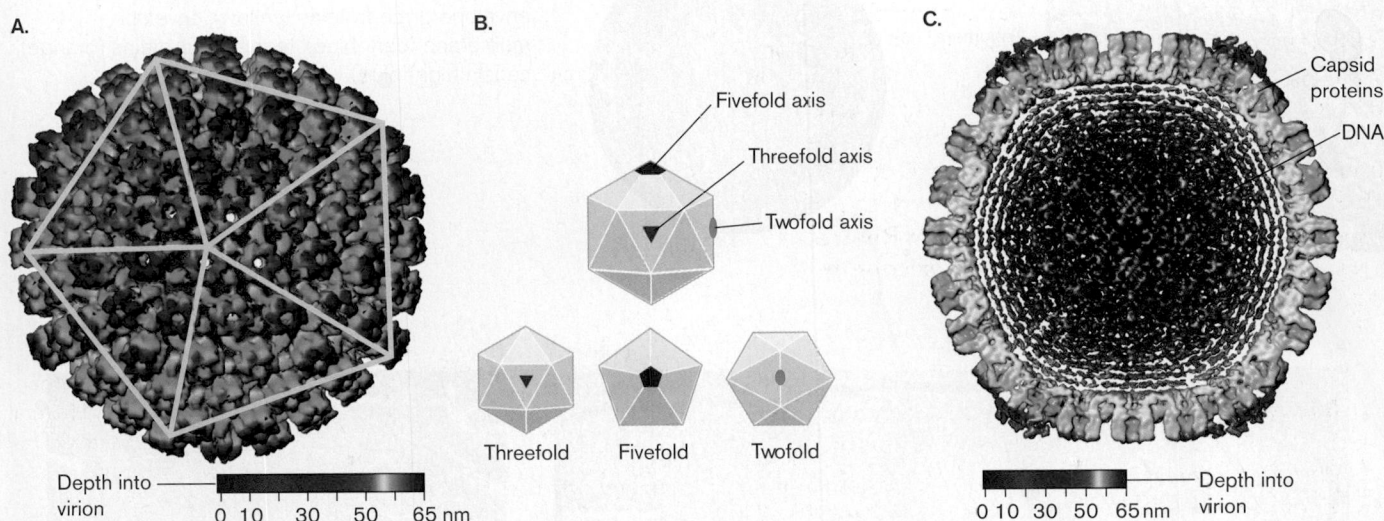

Figure 6.8 Herpes: icosahedral capsid symmetry. **A.** Icosahedral capsid of herpes simplex 1 (HSV-1), envelope removed. Imaging of the capsid structure is based on computational analysis of cryo–electron microscopy (cryo-TEM). Images of 146 virus particles were combined digitally to obtain this model of the capsid at 2-nm resolution. **B.** Icosahedral symmetry includes fivefold, threefold, and twofold axes of rotation. **C.** The icosahedral capsid contains spooled DNA. *Source:* (A.) Matthew L. Baker et al. 2003. *J. Mol. Biol.* **331**(2):447–56. (B.) Z. Hong Zhou et al. 1999. *J. Virol.* **73**:3210.

digitally combining and processing cryo-TEM images from a number of capsids, a three-dimensional reconstruction is built for the entire virus particle. Within the icosahedral capsid, the herpes genome is spooled and tightly packed (**Fig. 6.8C**). The double-stranded DNA of the herpes genome is packed under high pressure, whose release drives viral DNA into the host nucleus (discussed in Chapter 11).

In some icosahedral viruses, the capsid is enclosed in an **envelope** composed of plasma membrane from the host cell in which the virion formed. Other viruses, such as herpes virus, derive their envelope from intracellular membranes, such as the nuclear membrane or endoplasmic reticulum. The envelope and capsid contents of herpes virus are shown in **Figure 6.9A**. The mature envelope bristles with glycoprotein **spike proteins** that plug it onto the capsid. The spike proteins enable the virus to attach and infect the next host cell.

Between the envelope and the capsid, additional proteins may be found, called the **tegument** (**Fig. 6.9B**). Tegument proteins are expressed during infection of a host cell, and then packaged in the virion during envelope formation. Both viral and host proteins may be packaged. As the virus enters a new host cell, its envelope dissolves into the host cell membrane, and the tegument proteins are released to start the reproduction of new viruses.

Filamentous viruses. A second major category of virus structure is that of **filamentous viruses**. Filamentous viruses include bacteriophages such as phage M13 (**Fig. 6.10A**), as well as animal viruses such as Ebola virus, which causes a swiftly fatal disease of humans and related primates (**Fig. 6.10B**). Filamentous phages have applications in human medicine and industry. Phages similar to M13 infect the Gram-positive species *Propionibacterium freudenreichii*, a key fermenting agent for Swiss cheese. Another filamentous phage, CTXφ, integrates its sequence into the genome of *Vibrio cholerae*, where it carries the deadly toxin genes required for cholera. On the other hand, filamentous phages have been used in nanotechnology to nucleate the growth of crystalline "nanowires" for electronic devices.

The filamentous bacteriophage M13 (**Fig. 6.10A**) consists of a relatively simple capsid of protein monomers. The monomers are stacked around a coiled genome consisting of a circle of single-stranded DNA. At one end of

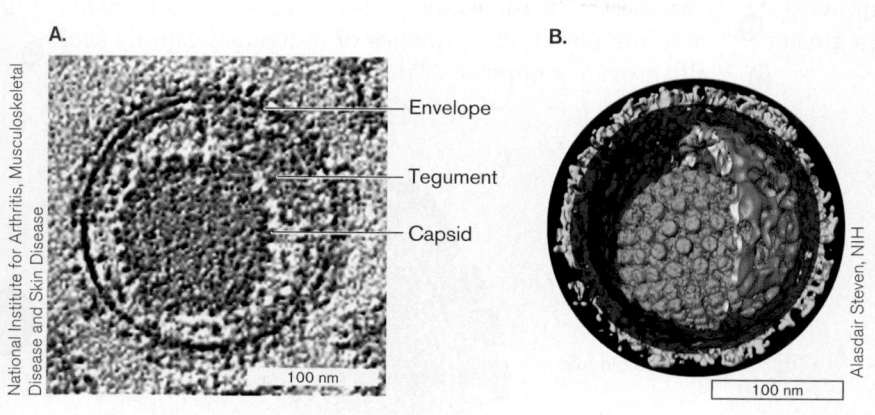

Figure 6.9 Envelope and tegument surround the herpes capsid.   A. Section showing envelope and tegument proteins surrounding capsid (cryo-TEM). **B.** Cutaway reconstruction: spike envelope glycoproteins (yellow), envelope membrane (dark blue), tegument proteins (orange), capsid (light blue).

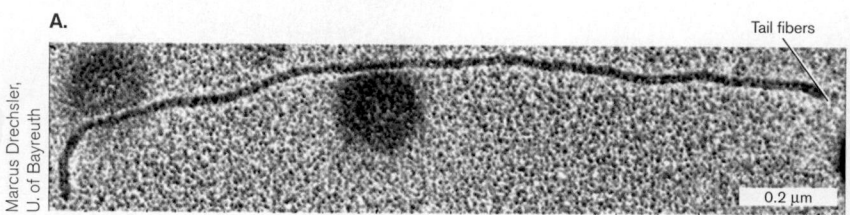
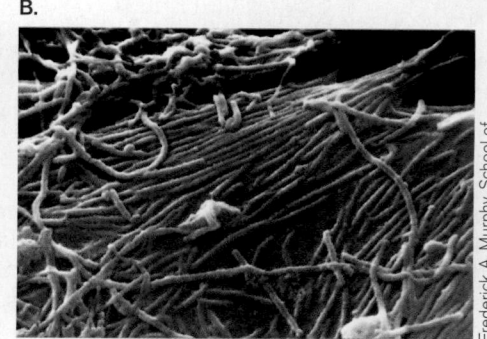

Figure 6.10 Filamentous viruses.   A. The filamentous bacteriophage M13 has a relatively simple helical capsid that surrounds the genome coiled within (TEM). **B.** Ebola virus filaments (SEM).

the filament, short tail fibers mediate specific attachment to the host. Attachment occurs at the F pilus of an *E. coli* bacterium containing an F plasmid that encodes pili.

Filamentous viruses show helical symmetry. The pattern of capsid monomers forms a helical tube around the genome, which usually winds helically within the tube. In a helical capsid, the genome is a single-stranded DNA (as in phage M13) or RNA (as in tobacco mosaic virus). **Figure 6.11A** shows how the RNA strand of tobacco mosaic virus winds in a spiral within a tube of capsid monomers laid down in a spiral array. Such a tube can be imagined as a planar array of subunits that coils around such that each row connects to the row above, generating a spiral (**Fig. 6.11B**). The length of the helical capsid may extend up to 50 times its width, generating a flexible filament.

Unlike the icosahedral capsid, which has a fixed size, the helical capsid can vary in length to accommodate different lengths of nucleic acid. Furthermore, some viruses package several genome segments into separate helical capsids. For example, influenza virus packages several different genome segments into separate helical packages of different sizes, contained together within a membrane envelope. Multiple helical packaging enables influenza virus to package different numbers of RNA segments into different virions. A surprising proportion of progeny virions are thus defective, but the process enables rapid evolution of new strains.

Complex and Asymmetrical Virus Particles

Some bacteriophages have complex multipart structures. Bacteriophages often supplement the icosahedral capsid or head coat with an elaborate delivery device. For example, bacteriophage T4 (**Fig. 6.12A**) has an icosahedral "head" containing the pressure-packed DNA, attached to a helical "neck" that channels the nucleic acid into the host cell. The neck has six jointed tail fibers that stabilize the structure on the host cell surface (**Fig. 6.12B**). The structure of phage T4 was first observed by microbiologists during the rise of the NASA space program, and its form was compared to that of the "lunar module" that landed on the moon (**Fig. 6.12C**). Indeed, in the 1960s, the "tailed phages," such as phage T4 and phage lambda, were to molecular biology what the lunar landings were to space exploration.

WWW T4 Bacteriophage Cell-Puncturing Device, a 3-D interactive tutorial
microbiology2.com/links

Other viruses lack a symmetrical capsid. In poxviruses, such as vaccinia and smallpox (**Fig. 6.13A**), the double-stranded DNA genome is stabilized by covalent connection of its two strands at each end (**Fig. 6.13B**). Instead of a capsid, the DNA is enclosed loosely by a core envelope studded

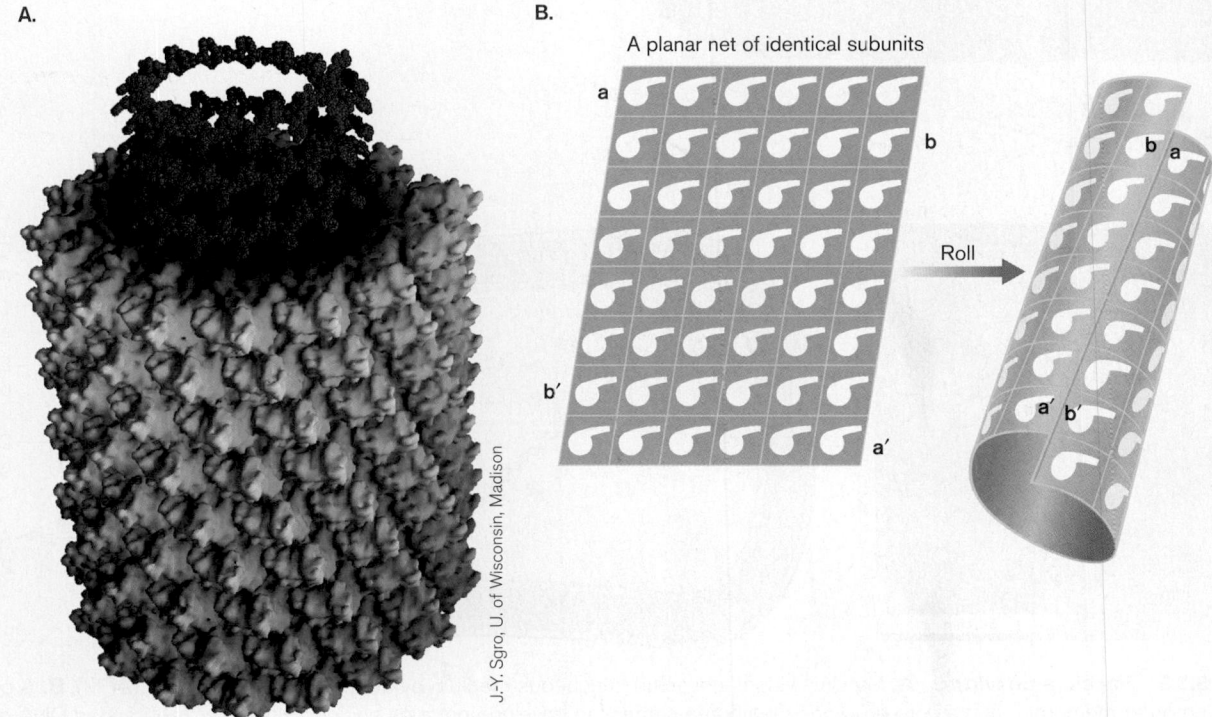

Figure 6.11 Tobacco mosaic virus: helical symmetry. A. The helical filament of tobacco mosaic virus (TMV) contains a single-stranded RNA genome coiled inside. Image reconstruction is based on X-ray crystallography. **B.** We can simulate an array corresponding to helical capsid structure by rolling up a planar array into a cylinder, while slanting the array along the vertical axis of the cylinder so that the horizontal elements are displaced by one unit.

A.

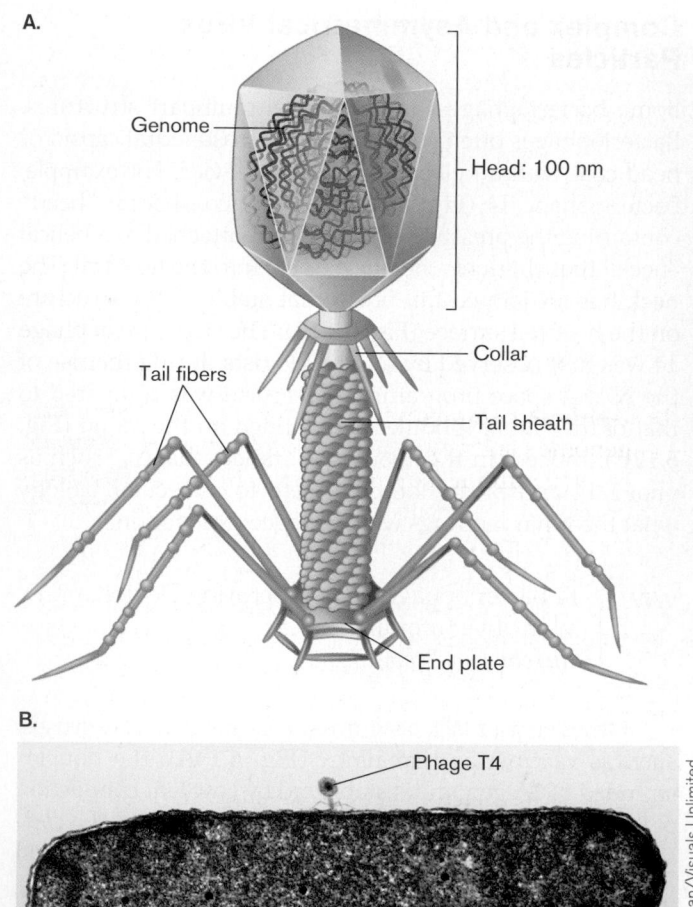

B.

C.

Figure 6.12 **Bacteriophage T4 capsid.** **A.** Phage T4 particle with protein capsid containing packaged double-stranded DNA genome. The capsid is attached to a sheath with tail fibers that facilitate attachment to the surface of the host cell. After attachment, the sheath contracts and the core penetrates the cell surface, injecting the phage genome. **B.** *E. coli* infected by phage T4 (colorized blue, TEM). **C.** The structure of phage T4 resembles an *Apollo* lunar module that landed on the moon.

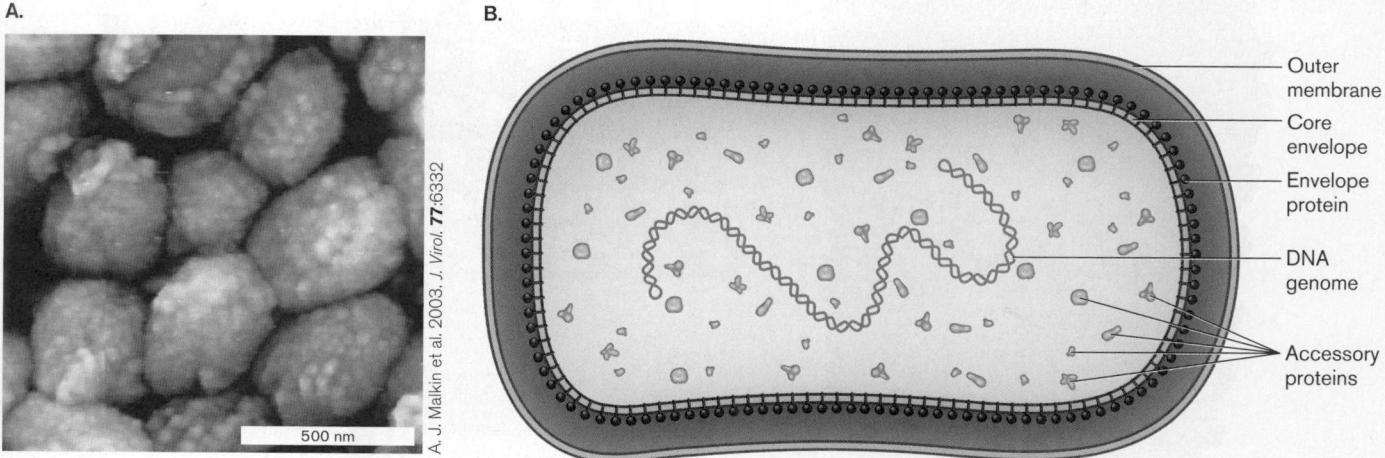

Figure 6.13 **Vaccinia poxvirus.** **A.** Vaccinia virions observed in aqueous medium by atomic force microscopy (AFM). **B.** A pox virion includes an outer membrane and a core envelope membrane containing envelope proteins enclosing the double-stranded DNA genome and accessory proteins. The DNA is stabilized by a hairpin loop at each end.

with spike proteins, surrounded by an outer membrane. The core envelope also encloses a large number of accessory proteins needed early in viral infection, such as initiation proteins for the transcription of viral genes and RNA-processing enzymes that modify viral mRNA molecules.

Asymmetrical DNA viruses usually contain a large number of accessory proteins. The first viruses to be studied, including TMV and poliovirus, had extremely simple structures that consisted only of nucleic acid and packaging proteins. These models led to the conclusion that a virion consists solely of a packaged genome. Further research, however, revealed that other kinds of viruses contain enzymes and regulatory proteins—encoded by the virus, its host, or both. The proteins may be found either inside the capsid or in the tegument between the capsid and the envelope. Examples include the reverse transcriptase and protease enzymes of HIV and the dozen different enzymes contained by vaccinia poxvirus. Large, asymmetrical viruses contain so many enzymes that they appear to have evolved from degenerate cells.

TO SUMMARIZE:

- **The viral capsid** is composed of repeated protein subunits—a structure that maximizes the structural capacity while minimizing the number of genes needed for construction.
- **The capsid packages the viral genome** and delivers it into the host cell.
- **Icosahedral capsids** have regular, icosahedral symmetry.
- **Filamentous (helical) capsids** have uniform width, generating a flexible filamentous virion.
- **Enveloped viruses** consist of a protein capsid and tegument proteins enclosed within phospholipid membrane derived from the host cell. The envelope includes virus-specific spike proteins.
- **Accessory proteins** are contained within the capsid or as tegument components between the capsid and envelope.

6.3 Viral Genomes and Classification

Living organisms today are classified by relatedness of their gene sequences. Genetic relatedness can also be used to compare closely related viruses, such as herpes viruses. The definition of a virus species, however, is problematic, given the small size and high mutability of viral genomes and the ability of different viruses to recombine their genome segments within an infected host cell. Furthermore, it is not clear whether all viral species are monophyletic—that is, descended from a common ancestor. In fact, it is more likely that different classes of viruses

arose from different sources—for example, from parasitic cells or from host cell components such as DNA replication enzymes. Viruses are classified by genome composition, virion structure, and host range.

The International Committee on Taxonomy of Viruses

For purposes of study and communication, a working classification system has been devised by the International Committee on Taxonomy of Viruses (ICTV). The ICTV classification system is based on several criteria:

- **Genome composition.** The nucleic acid of the viral genome can vary remarkably with respect to physical structure: It may consist of DNA or RNA; it may be single- or double-stranded; it may be linear or circular; and it may be whole or **segmented** (that is, divided into separate "chromosomes"). Genomes are classified by the Baltimore method (discussed next).
- **Capsid symmetry.** The protein capsid may be helical or icosahedral, with various levels of symmetry.
- **Envelope.** The presence of a host-derived envelope, and the envelope structure if present, are characteristic of related viruses.
- **Size of the virus particle.** Related viruses generally share the same size range; for example, enteroviruses such as poliovirus are only 30 nm across (about $1/30$ the size of bacteria such as *E. coli*), whereas poxviruses are 200–400 nm, as large as a small bacterium.
- **Host range.** Closely related viruses usually infect the same or related hosts. However, viruses with extremely different hosts can show surprising similarities in genetics and structure. For example, both rabies virus and potato yellow dwarf virus are enveloped, bullet-shaped viruses of the rhabdovirus family.

NOTE: In nomenclature, families of viruses are designated by Latin names with the suffix "-viridae": for example, *Papillomaviridae*. Nevertheless, the common forms of such family names are also used, for example, the "papillomaviruses." Within a family, a virus species is simply capitalized, as in "Papillomavirus."

The Baltimore Virus Classification

Of the classification criteria just described, many virologists consider genome composition the most fundamental; that is, viruses of the same genome class (such as double-stranded DNA) are more likely to share ancestry with each other than with viruses of a different class of genome (such as RNA). In 1971, David Baltimore proposed that the primary distinction among classes of viruses be the genome composition (RNA or DNA) and the route used to express messenger RNA (mRNA). Baltimore, together with Renato Dulbecco and Howard Temin, was awarded the Nobel Prize in Physiology or Medicine in 1975 for discovering how tumor viruses cause cancer.

All cells and viruses need to make messenger RNA to produce their fundamental protein components. The production of mRNA from the viral genome is central to a virus's ability to propagate its kind. Cellular genomes always make mRNA by copying double-stranded DNA. For viruses, however, different kinds of genomes require fundamentally different mechanisms to produce mRNA. The different means of mRNA production generate distinct groups of viruses with shared ancestry.

So far, the known mechanisms of replication and mRNA expression define seven fundamental groups of viral species (**Fig. 6.14**). These seven fundamental groups form the basis of the taxonomic survey of viruses in **Table 6.1**.

Group I. Double-stranded DNA viruses such as the herpes and smallpox viruses make their own DNA polymerase or use that of the host for genome replication. Their genes can be transcribed directly by a standard RNA polymerase, in the same way that a cellular chromosome would be transcribed. The RNA polymerase used can be that of the host cell, or it can be encoded by the viral genome.

A.

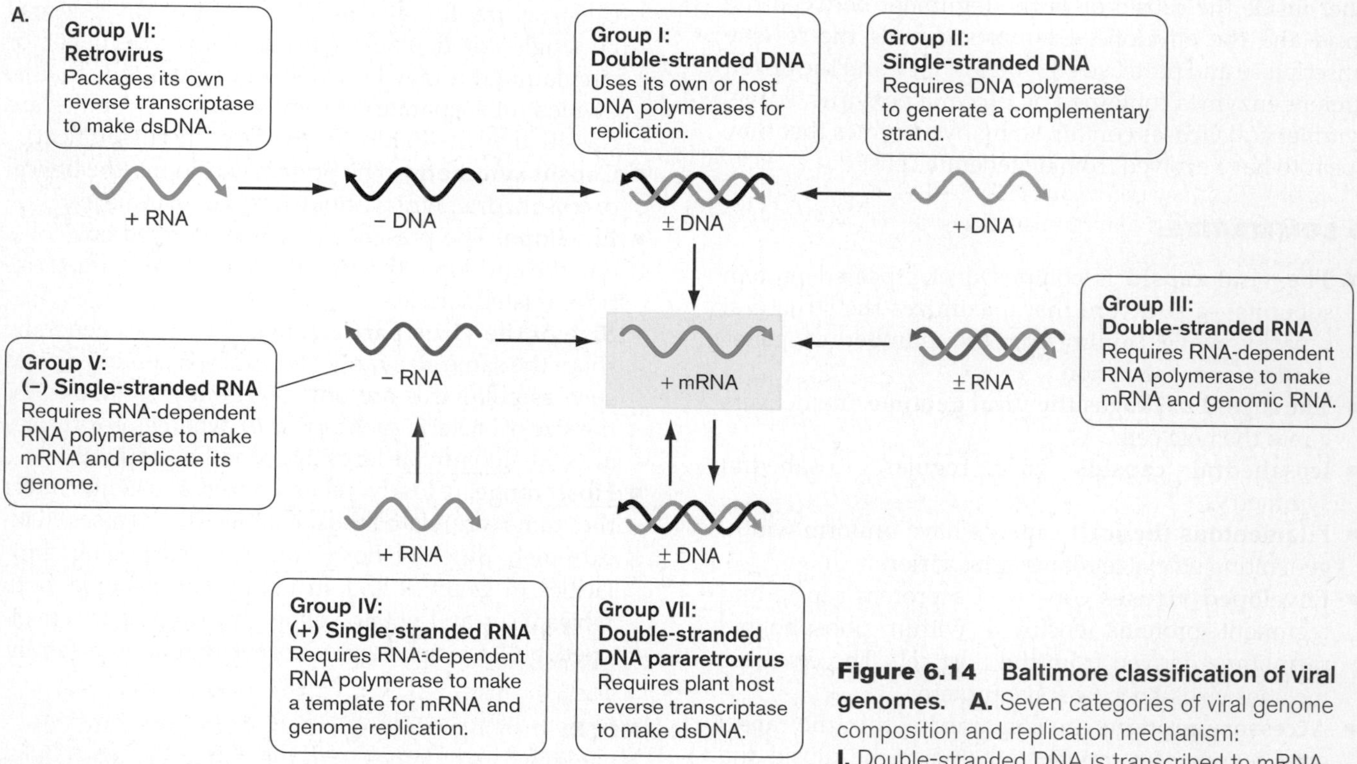

Group VI:
Retrovirus
Packages its own reverse transcriptase to make dsDNA.

Group I:
Double-stranded DNA
Uses its own or host DNA polymerases for replication.

Group II:
Single-stranded DNA
Requires DNA polymerase to generate a complementary strand.

+ RNA   − DNA   ± DNA   + DNA

Group V:
(−) Single-stranded RNA
Requires RNA-dependent RNA polymerase to make mRNA and replicate its genome.

− RNA   + mRNA   ± RNA

Group III:
Double-stranded RNA
Requires RNA-dependent RNA polymerase to make mRNA and genomic RNA.

+ RNA   ± DNA

Group IV:
(+) Single-stranded RNA
Requires RNA-dependent RNA polymerase to make a template for mRNA and genome replication.

Group VII:
Double-stranded DNA pararetrovirus
Requires plant host reverse transcriptase to make dsDNA.

B.

California Institute of Technology

Figure 6.14   Baltimore classification of viral genomes.   A. Seven categories of viral genome composition and replication mechanism:
I. Double-stranded DNA is transcribed to mRNA.
II. Single-stranded DNA generates a double-stranded form within the host cell, which is transcribed to mRNA.
III. Double-stranded RNA makes mRNA using RNA-dependent RNA polymerase.
IV. Single-stranded RNA (+) makes a complementary (−) strand, which is transcribed to mRNA.
V. Single-stranded RNA (−) is transcribed to mRNA.
VI. Single-stranded RNA (+) is reverse-transcribed to DNA, which is transcribed to mRNA.
VII. Double-stranded DNA (dsDNA) is transcribed to mRNA, which is reverse-transcribed to regenerate viral genomes for packaging into virions.
B. David Baltimore (left), with a graduate student at the California Institute of Technology. Baltimore won the 1975 Nobel Prize in Physiology or Medicine for his work on retroviruses; co-winners were Renato Dulbecco and Howard Temin.

Group II. Single-stranded DNA viruses require the host DNA polymerase to generate the complementary DNA strand. The double-stranded DNA can then be transcribed by host RNA polymerase. An example is canine parvovirus.

Group III. Double-stranded RNA viruses, such as the plant pathogenic reoviruses, require a viral **RNA-dependent RNA polymerase** to generate messenger RNA by transcribing directly from the RNA genome. Since the RNA polymerase is required immediately upon infection, such viruses usually package a viral RNA polymerase with their genome before exiting the host cell.

Group IV. (+) Sense single-stranded RNA viruses consist of a positive-sense (+) strand (the coding strand) that can serve directly as mRNA to be translated to viral proteins. Replication of the RNA genome, however, requires synthesis of the template (–) strand (complementary to the [+] strand) to form a double-stranded RNA intermediate. Positive-sense (+) RNA viruses are very common. They include the coronaviruses, the flavivirus that causes West Nile encephalitis, and hepatitis C virus.

Group V. (–) Sense single-stranded RNA viruses such as influenza virus have genomes that consist of template, or "negative-sense," RNA. Thus, they need to package a viral RNA-dependent RNA polymerase for transcribing (–) RNA to (+) mRNA. The (–) strand RNA viral genomes are often segmented; that is, they consist of more than one separate linear chromosome—a key factor in the evolution of killer strains of influenza (see Section 11.4).

Group VI. Retroviruses, or RNA reverse-transcribing viruses such as HIV and feline leukemia virus, have genomes that consist of (+) strand RNA. Instead of RNA polymerase, they package a **reverse transcriptase**, which transcribes the RNA into a double-stranded DNA (for details, see Section 11.5). The double-stranded DNA is then integrated into the host genome, where it directs the expression of the viral genes.

Group VII. Pararetroviruses, or DNA reverse-transcribing viruses, have a life cycle that requires reverse transcriptase. Animal viruses such as hepatitis B (a hepadnavirus) first copy their double-stranded DNA genomes into RNA, and then reverse-transcribe the RNA to progeny DNA using a reverse transcriptase packaged in the original virion. In contrast, plant pararetroviruses, such as cauliflower mosaic virus, generate an RNA intermediate that replicates using a reverse transcriptase made by the host cell. Many plant genomes include a gene for reverse transcriptase. Cauliflower mosaic virus is of enormous agricultural significance for its use as a vector to construct pesticide-resistant food crops.

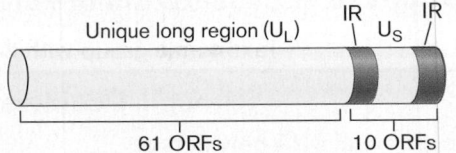

A. Varicella-zoster virus (VZV) genome (120 kb)

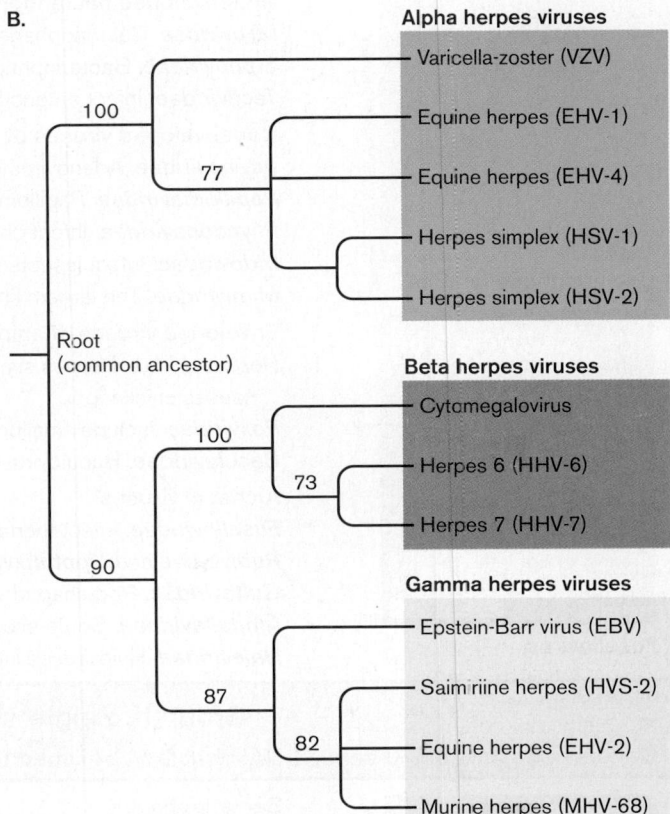

Figure 6.15 Phylogeny of herpes viral genomes. A. Genome structure of human varicella-zoster virus (VZV), the causative agent of chickenpox. **B.** Phylogeny of human and animal herpes viruses, based on whole-genome sequence analysis comparing clusters of orthologous groups of genes. Numbers measure genetic divergence.

Molecular Evolution of Viruses

The phylogeny, or genetic relatedness, of viruses can be determined within families. For example, the herpes family includes double-stranded DNA viruses that cause several human and animal diseases, such as chickenpox, oral and genital herpes infection, and respiratory and genital infections in horses. Herpes genomes consist of double-stranded DNA, 120–220 kilobases (kb) encoding about 70–200 genes; an example is that of varicella-zoster virus, the causative agent of chickenpox (**Fig. 6.15A**). The genome includes two "unique" segments of genes, one long and one short (U_L and U_S), joined by two inverted repeats (IRs). Other herpes genomes share similar structure, though they differ in gene order and IR position.

Table 6.1   Groups of viruses–Baltimore classification.

Viruses	Taxonomic group with key traits and examples

Group I. Double-stranded DNA viruses

Replicate using host or viral DNA polymerase.

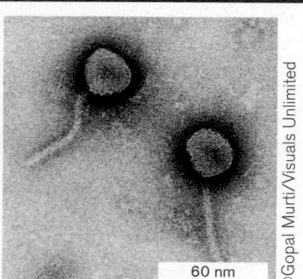

60 nm

©Gopal Murti/Visuals Unlimited

Bacteriophage lambda

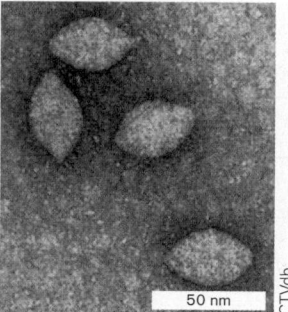

50 nm

ICTVdb

Fusellovirus

Nonenveloped bacteriophages (structure includes head and tail)
Myoviridae. Bacteriophage T4 infects *Escherichia coli.*
Siphoviridae. Bacteriophages lambda and mu infect *E. coli.* Others infect Gram-positive hosts.
Tectiviridae. Infect enteric bacteria.

Nonenveloped viruses of animals and protists
Adenoviridae. Adenovirus generates tumors in humans.
Papillomaviridae. Papillomavirus causes genital warts.
Phycodnaviridae. Infect chlorella, algal symbionts of paramecia and hydras.
Iridoviridae. Infect insects and amphibians.
Mimiviridae. The largest known viruses; infect *Acanthamoeba* and human macrophages.

Enveloped viruses of animals
Herpesviridae. Herpes simplex 1 and 2 cause oral and genital herpes; varicella-zoster virus causes chickenpox.
Poxviridae. Include smallpox and cowpox viruses.
Baculoviridae. Baculoviruses infect insects.

Archaeal viruses
Fuselloviridae. Infect thermoacidophiles (*Sulfolobus*) and halophiles (*Haloarcula*).
Rudiviridae and *Lipothrixviridae.* Infect *Sulfolobus* and *Thermoproteus.*
Guttaviridae. Rod-shaped viruses; infect *Sulfolobus.*
Ampullaviridae. Bottle-shaped viruses; infect *Acidianus.*
Haloviridae. Haloviruses infect haloarchea such as *Haloferax*, *Halobacterium*, and *Haloarcula.*

Group II. Single-stranded DNA viruses

Genome consists of (+) sense DNA; host DNA polymerase used; nonenveloped.

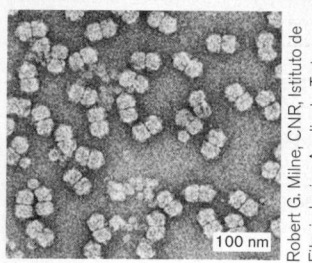

100 nm

Robert G. Milne, CNR, Istituto de Fitovirologica Applicata, Torino, Italy

Geminivirus

Bacteriophages
Inoviridae. Bacteriophage M13 infects *E. coli* and has a slow-release life cycle.
Microviridae. Bacteriophage φX174 infects *E. coli.*

Animal viruses
Parvoviridae. Cause various diseases in cats, pigs, and other animals.
Circoviridae. Infect pigs and birds, causing immunosuppression.

Plant viruses
Geminiviridae. Transmitted by aphids to tomato plants and other important crops. Their virions group in "twins," each member of the pair carrying one DNA circle with part of the genome.

Group III. Double-stranded RNA viruses

Require viral RNA-dependent RNA polymerase; usually package the polymerase before exiting host cell.

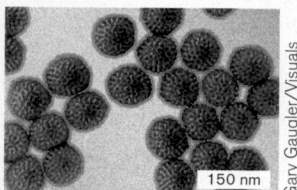

150 nm

Gary Gaugler/Visuals Unlimited

Nonsegmented, enveloped bacteriophages
Cystoviridae. Infect *Pseudomonas* species of bacteria.

Segmented, nonenveloped viruses of animals and plants
Birnaviridae. Infect marine and aquatic fish.
Reoviridae. Orthoreoviruses and rotaviruses infect humans and other vertebrates. Cypovirus infects insects. Rice dwarf virus (phytoreovirus) devastates rice crops worldwide.
Varicosaviridae. Infect plants.

Group IV. (+) Sense single-stranded RNA viruses

Require viral RNA-dependent RNA polymerase to generate (−) template for progeny (+) genome; usually nonsegmented.

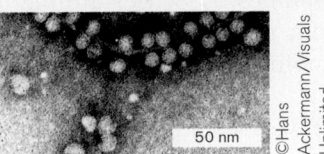

50 nm

©Hans Ackermann/Visuals Unlimited

Bacteriophage MS2

30 nm

Kenneth Eward/Photo Researchers, Inc.

Rhinovirus 14

Nonenveloped bacteriophages

Leviviridae. Bacteriophages MS2 and Qβ infect *E. coli.*

Nonenveloped animal and plant viruses

Bromoviridae. Infect many kinds of plants and are often carried by beetles.

Picornaviridae. Poliovirus causes poliomyelitis. Rhinovirus causes the common cold. Aphthovirus causes foot-and-mouth disease in cattle and other stock.

Tobamoviridae. Tobacco mosaic virus infects plants.

Potyviridae. Viruses such as plum pox virus infect fruits, peanuts, potatoes, and other plants.

Enveloped animal and plant viruses

Coronaviridae. Coronaviruses include SARS (severe acute respiratory syndrome) and animal viruses.

Flaviviridae. Infect humans; include West Nile virus, yellow fever virus, and hepatitis C virus.

Togaviridae. Include rubella virus and equine encephalitis virus.

Group V. (−) Sense single-stranded RNA viruses

Require viral RNA-dependent RNA transcriptase. Genome often segmented.

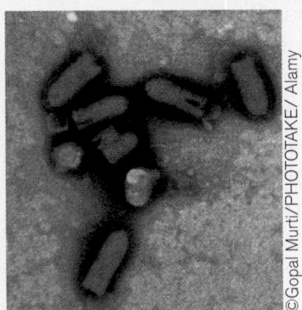

©Gopal Murti/PHOTOTAKE/ Alamy

Rabies virus

Segmented, enveloped viruses

Orthomyxoviridae. Influenza virus causes major epidemics among humans and animals.

Nonsegmented, enveloped viruses

Filoviridae. Ebola virus causes outbreaks among humans and chimpanzees.

Rhabdoviridae. Rabies virus infects mammals (Virion length 130–300 nm).

Paramyxoviridae. Infect humans and cause measles, mumps, and parainfluenza.

Segmented (+/−) strand, enveloped viruses

Arenaviridae. Spread by rodents; cause hemorrhagic fever and lymphocytic choriomeningitis.

Bunyaviridae. Hantaviruses are spread by rodents and infect humans. Tospoviruses are transmitted by thrips, infecting plants. Rift Valley fever virus infects livestock.

Group VI. Retroviruses (RNA reverse-transcribing viruses)

Viral reverse transcriptase copies RNA to DNA for integration into host chromosome.

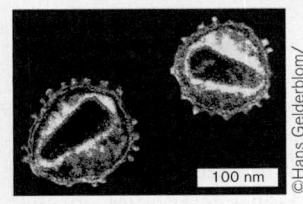

100 nm

©Hans Gelderblom/ Visuals Unlimited

HIV-1

Retroviridae. Simple retroviruses include oncogenic retroviruses: feline leukemia virus, Rous sarcoma virus, avian leukosis virus. Xenotropic murine leukemia–related virus (XMRV) is linked to prostate cancer and chronic fatigue syndrome. Lentiviruses (complex retroviruses) include human immunodeficiency virus (HIV, the cause of AIDS) and simian immunodeficiency virus (SIV).

Group VII. Pararetroviruses (DNA reverse-transcribing viruses)

DNA transcribed to RNA; reverse-transcribed to DNA using host reverse transcriptase or packaged viral reverse transcriptase.

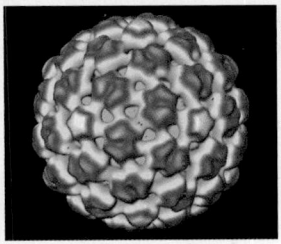

Encyclopedia of Virology. 1995 Academic Press Ltd.

Cauliflower mosaic virus

Nonenveloped plant viruses

Caulimoviridae. Transmitted by aphids; cauliflower mosaic virus and related viruses infect cauliflower, broccoli, groundnuts, soybeans, and cassava. The cauliflower mosaic virus promoter sequence is used to construct vectors to insert genes into transgenic plants.

Badnaviridae. Infect bananas, cocoa plants, citrus, yams, and sugarcane.

Enveloped animal viruses

Hepadnaviridae. Hepatitis B virus causes widespread disease of the human liver.

The relatedness of different herpes viruses can be measured by comparing their genome sequences. Comparison is based on **orthologous genes**, or **orthologs**. Orthologs are genes of common ancestry in two genomes that share the same function, a topic discussed in Chapter 8. Examples include the ribosomal RNA genes whose sequence is used to measure the relatedness of cellular organisms. Viruses have no ribosomal RNA, but closely related viruses share other orthologous genes. In pairs of orthologs, the amount of difference in sequence correlates approximately with the time following divergence from a common ancestor (a topic discussed in Chapter 17). Sequence comparison places herpes into three classes designated alpha, beta, and gamma (**Fig. 6.15B**). The alpha class includes human varicella-zoster virus and the oral and genital herpes viruses (HSV-1 and HSV-2), as well as equine herpes virus. The beta class includes cytomegalovirus, a common cause of congenital infections (present at birth), as well as two less known viruses. The gamma class includes Epstein-Barr virus (the cause of infectious mononucleosis), as well as several viruses of animals.

The more ancient evolution of viruses, however, is difficult to assess from DNA or RNA sequence. Unlike cells, viruses do not possess genes common to all species, such as the ribosomal RNA genes used to estimate times

of divergence of cellular organisms (discussed in Chapter 17). Furthermore, the genomes of viruses are highly mosaic—that is, derived from multiple sources. Mosaic genomes result from recombination between different viruses coinfecting a host.

In some cases, phylogeny is inconsistent with the fundamental chemical composition of the genome. For example, some DNA bacteriophages actually share closer ancestry with RNA bacteriophages than they do with DNA animal viruses. This is because two or more viruses can coinfect a cell and exchange genetic components. Thus, the genomic content of a virus can be influenced by its host range.

> **THOUGHT QUESTION 6.4** How can viruses with different kinds of genomes (RNA versus DNA) combine and exchange genetic information?

As more viral genomes are sequenced, classification methods have been devised to take advantage of sequence information without requiring gene products common to all species. One promising approach is that of **proteomics**, analysis of the **proteome**, the proteins encoded by genomes (discussed in Chapter 8). Proteins are identified through biochemical analysis of virus particles and through bioinformatic analysis of protein sequences encoded in the genomes. Proteomic analysis is useful for viruses because their small genomes encode a small number of proteins, which can be readily analyzed. Furthermore, because proteomic analysis is based on numerous gene products, it does not require a single sequence (such as rRNA) common to all species. Statistical comparison of all proteins generated by a set of viral species reveals underlying degrees of relatedness.

An example of viral classification based on proteomic analysis is that of the "proteomic tree" of bacteriophages proposed by Forest Rohwer and Rob Edwards (**Fig. 6.16**). Unlike earlier trees based on a single common gene sequence, the proteomic tree is based on the statistical comparison of phage protein sequences predicted by the genomic DNA of many different species of phages. The proteomic analysis predicts major evolutionary categories of phage species that share common host bacteria. For example, phages that infect Gram-negative hosts will show more

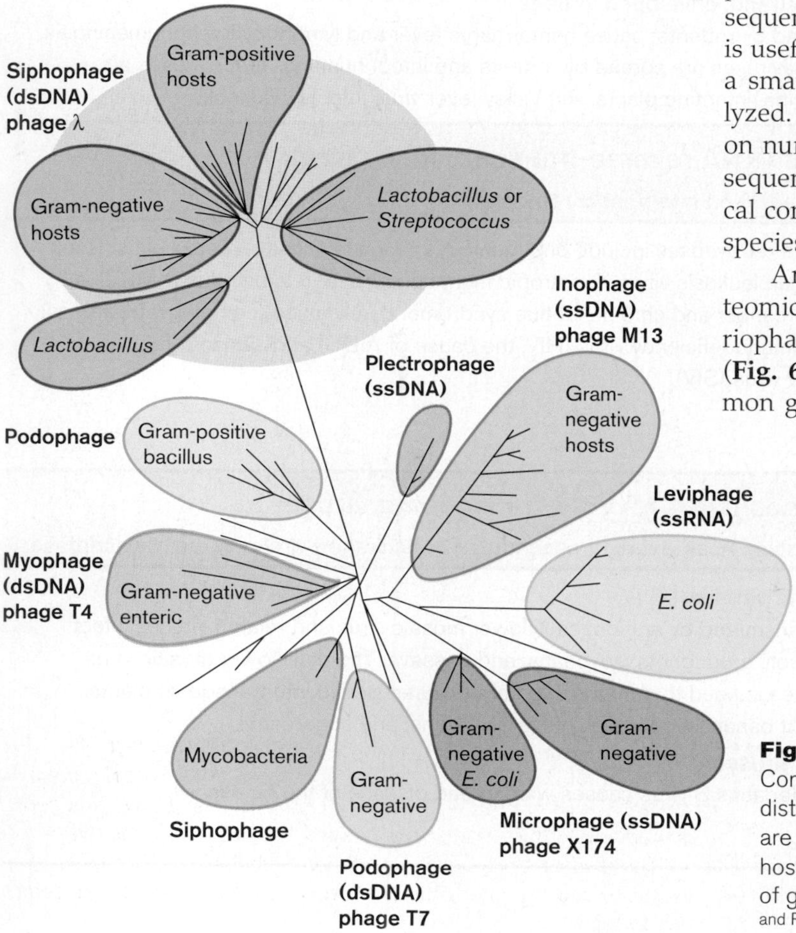

Figure 6.16 The bacteriophage proteomic tree. Comparison of all proteins encoded by each genome predicts distinct groups of bacteriophages. Within each group, there are subgroups of phages with shared hosts, since sharing of hosts facilitates genetic recombination and horizontal transfer of genes between different phages. *Source*: Based on Forest Rohwer and Rob A. Edwards. 2002. *J. Bacteriol.* **184**:4529.

genetic commonality with each other than with phages that infect Gram-positive hosts. Shared hosts have a significant impact on phage evolution because coinfecting a host enables phages to exchange genes.

TO SUMMARIZE:

- **Classification of viruses** is based on genome composition, virion structure, and host range.
- **The Baltimore virus classification** emphasizes the form of the genome (DNA or RNA, single- or double-stranded) and the route to generate messenger RNA.
- **Proteomic classification** includes information from all viral proteins. Statistical analysis reveals common descent of viruses infecting a common host.

6.4 Bacteriophage Life Cycles

All viruses require a host cell for reproduction. While viruses display a remarkable diversity of reproductive strategies, they all face these same needs for host infection:

- **Host recognition and attachment.** Viruses must contact and adhere to a host cell that can support their particular reproductive strategy.
- **Genome entry.** The viral genome must enter the host cell and gain access to the cell's machinery for gene expression.
- **Assembly of virions.** Viral components must be expressed and assembled. Components usually "self-assemble"; that is, the joining of their parts is favored thermodynamically.
- **Exit and transmission.** Progeny virions must exit the host cell, and then reach new host cells to infect. In the case of multicellular organisms, the virus must eventually reach other multicellular hosts (as discussed in Chapter 26).

Bacteriophages Attach to Host Cells

To commence an infectious life cycle, bacteriophages need to contact and attach to the surface of an appropriate host cell. Contact and attachment are mediated by **cell surface receptors**, proteins on the host cell surface that are specific to the host species and that bind to a specific viral component.

Receptor proteins. The cell surface receptor for a virus is actually a protein with an important function for the host cell, but the virus has evolved to take advantage of the protein. An extensively studied model system of virus-receptor binding is that of bacteriophage lambda (see **Table 6.1**, top row). The phage lambda virion attaches specifically to the maltose porin in the outer membrane of *Escherichia coli* (**Fig. 6.17**). Although the protein is often

called "lambda receptor protein," it actually evolved in the host as a way to obtain the sugar maltose to metabolize. Thus, natural selection maintains the maltose porin in *E. coli*, despite the danger of phage infection.

The precise domain of the maltose porin that binds to phage lambda was defined experimentally by mutations in *E. coli* that cause amino acid substitution in the protein. Some of the mutant *E. coli* strains were resistant to phage lambda infection. The mutations that conferred host resistance mapped to the domain of maltose porin that binds the phage capsid.

Phage Reproduction within Host Cells

Historically, the life cycles of bacteriophages have provided some of the most fundamental insights in molecular biology. In 1952, Alfred Hershey and Margaret Chase showed that the transmission of DNA by a bacteriophage to a host cell led to the production of progeny bacteriophages, thus confirming that DNA is the hereditary material. In 1950, André Lwoff and Antoinette Gutman showed that a phage genome could integrate itself within a bacterial genome—the first recognition that genes could enter and leave a cell's genome. Other fundamental concepts of the genetic unit and the basis of gene transcription came from experiments on bacteriophages, as discussed in Chapters 7–9.

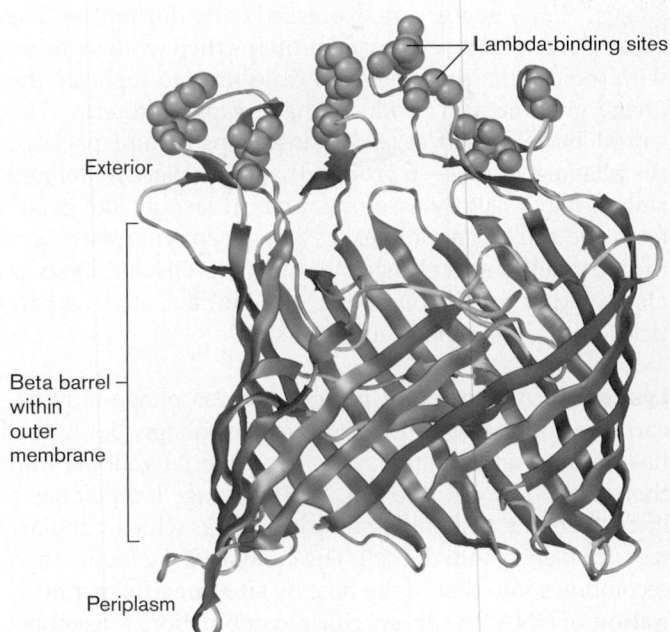

Figure 6.17 Bacteriophage lambda interacts with its host receptor. Phage lambda infects *E. coli* by binding to the maltose porin. The "beta barrel" of maltose porin (blue) is buried in the outer membrane. The phage binding sites (green) were identified by amino acid substitution mutations that prevent phage binding and confer host resistance to lambda. (PDB code: 1MAL)

Most bacteriophages (phages) insert only their genome into a cell through the cell envelope, thus avoiding the need for the capsid to penetrate the molecular barrier of the cell wall. For example, the phage T4 virion has a neck tube that contracts, bringing the head near the cell surface to insert its DNA (**Fig. 6.18A**). The pressure of the spooled DNA—as high as 50 atmospheres (atm)—is released, expelling the DNA into the cell. After the genome has been inserted, the phage capsid remains outside, attached to the cell surface. The empty capsid is termed a "ghost" because of its pale appearance in an electron micrograph.

The lytic cycle. In a **lytic cycle**, when a phage particle injects its genome into a cell, it immediately reproduces as many progeny phage particles as possible. The process of reproduction involves replicating the phage genome, as well as expressing phage mRNA to make enzymes and capsid proteins. Some phages, such as T4, digest the host DNA to increase the efficiency of phage production. Finally, the host cell lyses, releasing progeny phages.

Phage T4 reproduces entirely by the lytic cycle and thus is called a "virulent" phage. Other phages, such as phage lambda, have the options of reproducing by lysis or by lysogeny (**Fig. 6.18B**).

Lysis. After phage lambda inserts its DNA, its genes are expressed by the host cell RNA polymerase and ribosomes. "Early genes" are expressed early during the lytic cycle. Other phage-expressed proteins then work together with the cellular enzymes and ribosomes to replicate the phage genome and produce phage capsid proteins. The capsid proteins self-assemble into capsids and package the phage genomes—a process that takes place in defined stages, like a factory assembly line. At last, a "late gene" from the phage genome expresses an enzyme that lyses the host cell wall, releasing the mature virions. **Lysis** is also referred to as a burst, and the number of virus particles released is called the **burst size**.

Lysogeny. A **temperate phage**, such as phage lambda, can infect and lyse cells like a virulent phage, but it also has an alternative pathway: to integrate its genome into that of the host cell (see **Fig. 6.18B**). Phage lambda has a linear genome of double-stranded DNA, which circularizes upon entry into the cell. The circularized genome then recombines into that of the host by **site-specific recombination** of DNA. In site-specific recombination, a recombinase enzyme aligns the phage genome with the host DNA and exchanges the phosphodiester backbone links with those of the host genome. The process of exchange thus integrates the phage genome into that of the host. The integrated phage genome is called a **prophage**.

Integration of the phage genome as a prophage results in **lysogeny**, a condition in which the phage genome is replicated along with that of the host cell as the host reproduces (**Fig 6.14B**). Implicit in the term "lysogeny," however, is the ability of such a strain to spontaneously generate a lytic burst of phage. For lysis to occur, the prophage (integrated phage genome) directs its own excision from the host genome by an intramolecular process of site-specific recombination. The two ends of the phage genome exchange their phosphodiester backbone linkages so as to come apart from the host molecule. As the phage DNA exits the host genome, it circularizes and initiates a lytic cycle, destroying the host cell and releasing phage particles.

The "decision" between lysogeny and lysis is determined by proteins that bind DNA and repress the transcription of genes for virus replication (see Section 10.7). Exit from lysogeny into lysis can occur at random, or it can be triggered by environmental stress such as UV light, which damages the cell's DNA.

The regulatory switch of lysogeny responds to environmental cues indicating the likelihood that the host cell will survive and continue to propagate the phage genome. If a cell's growth is strong, it is more likely that the phage DNA will remain inactive, whereas events that threaten host survival will trigger a lytic burst. An analogous phenomenon occurs in animal viral infections such as herpes, in which environmental stress triggers reactivation of a virus that was dormant within cells (a latent infection). Reactivation of a latent herpes infection results in painful outbreaks of skin lesions.

Viruses transfer host genes. During the exit from lysogeny or during latent growth of animal viruses, the virus can acquire host genes and pass them on to other host cells. This process of transferring host genes is known as **transduction**. A transducing bacteriophage can pick up a bit of host genome and transfer it to a new host cell. In some forms of transduction, the entire phage genome is replaced by host DNA packaged in the phage capsid, resulting in a virus particle that transfers only host DNA. Host DNA transferred by viruses can become permanently incorporated into the infected host genome. The mechanisms of phage-mediated transduction (generalized and specialized) are discussed in Chapter 9.

The integration and excision of viral genomes make extraordinary contributions to the evolution of host genomes. Lysogenic prophages often express key toxins and virulence factors: For example, the CTXφ prophage encodes cholera toxin in *Vibrio cholerae*; the Shiga toxin prophage confers virulence on enteric pathogens *Shigella* and *E. coli* O157:H7; and prophages encode toxins for *Corynebacterium diphtheriae* (diphtheria) and *Clostridium botulinum* (botulism). In natural environments, phage transduction mediates much of the recombination of bacterial genomes. In the laboratory, the ability of phages to transfer genes provided vectors for recombinant DNA

A. Phage T4 DNA insertion

Phage T4 attaches to bacterium.

Sheath contracts and viral DNA enters bacterium.

B. Phage lambda reproductive cycle

Host genome

Phage attaches to host cell and inserts DNA.

Phage particle

Linear dsDNA cyclizes to circular DNA.

Lysogeny

Phage DNA integrates into host genome to form prophage.

Lytic cycle

Viral DNase cleaves host cell DNA. Cell synthesizes capsid proteins.

Phage recombines by re-joining the ends of its phosphodiester chain and enters the lytic cycle.

Cell replicates phage DNA. DNA is packaged into capsids.

Stress induces excision of phage DNA.

Integrated phage DNA reproduces with host genome.

Phage lyses cell, and progeny phages are released.

Integrated phage DNA replicates with host genome.

technology, the artificial construction of phages to carry genes from animals and plants for gene cloning.

When viral transfer of genes was first discovered in the 1960s, multicellular eukaryotes were thought to have a much lower tolerance for genome change. We have since learned, however, that much of the human genome also shows evidence of gene transfer mediated by ancient viruses, including possible retroviruses similar to HIV (discussed in Chapter 11).

Slow release. The slow-release cycle differs from lysis and lysogeny in that phage particles reproduce without destroying the host cell (**Fig. 6.19**). Slow release is performed by filamentous phages such as phage M13. In the slow-release life cycle, the single-stranded circular DNA of M13 serves as a template to synthesize a double-stranded intermediate. The double-stranded intermediate

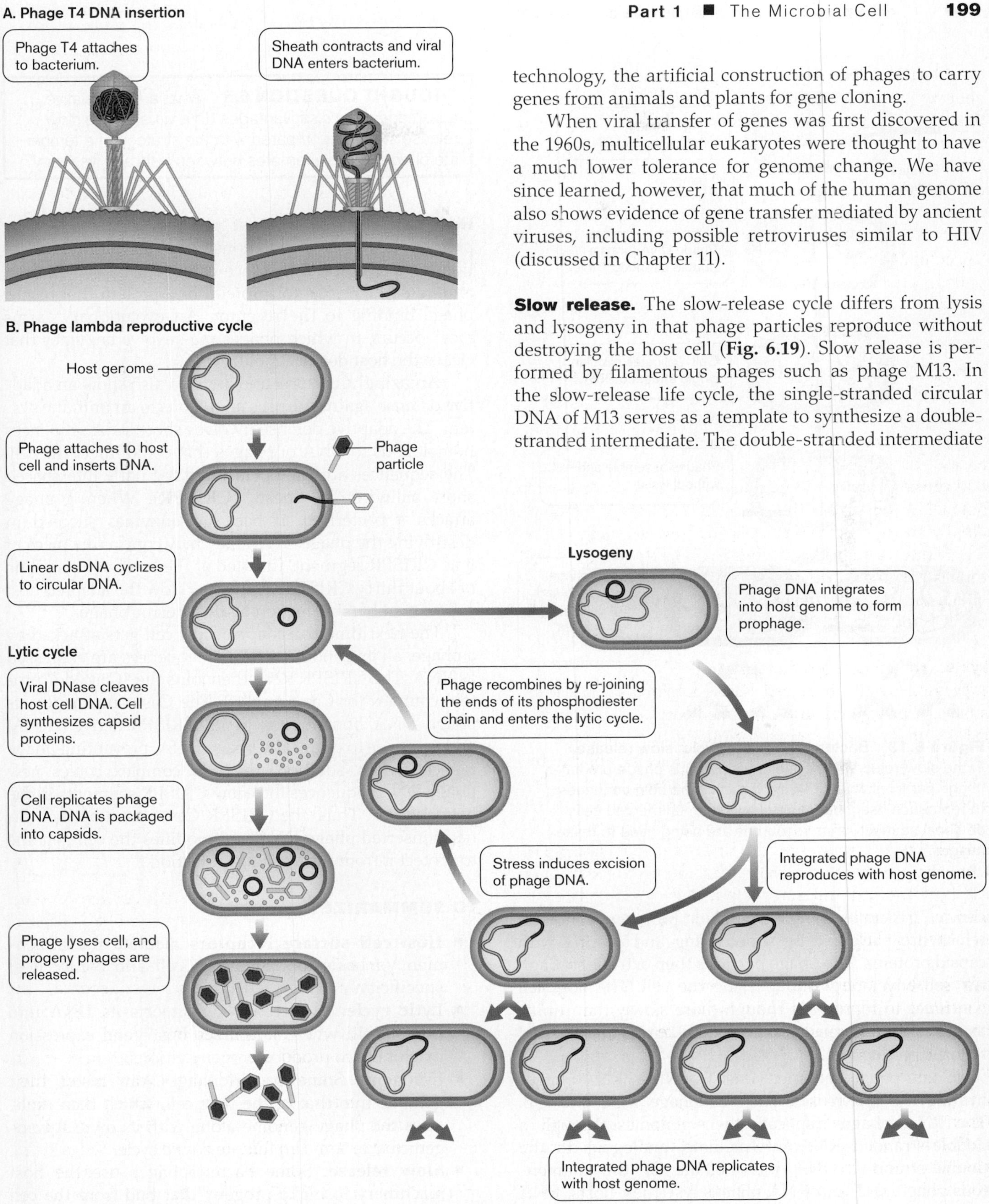

Figure 6.18 Bacteriophage reproduction: lysis and lysogeny. **A.** Phage T4 attaches to the cell surface by its tail fibers and then contracts to inject its DNA. The empty capsid remains outside as a "ghost." **B.** Lysis occurs when the phage genome reproduces progeny phage particles, as many as possible, and then lyses the cell to release them. In phage lambda, lysogeny can occur when the phage genome integrates itself into that of the host. The phage genome is replicated along with that of the host cell. The phage DNA, however, can direct its own excision by expressing a site-specific DNA recombinase. This excised phage chromosome then initiates a lytic cycle.
microbiology2.com/animations

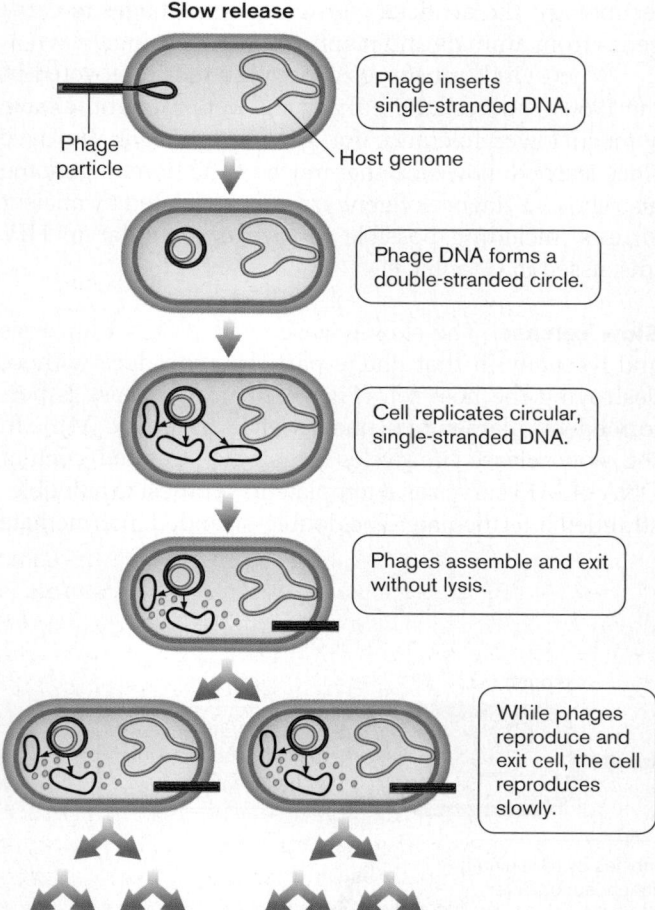

Slow release

Phage inserts single-stranded DNA.

Phage particle

Host genome

Phage DNA forms a double-stranded circle.

Cell replicates circular, single-stranded DNA.

Phages assemble and exit without lysis.

While phages reproduce and exit cell, the cell reproduces slowly.

Figure 6.19 Bacteriophage life cycle: slow release. In the slow-release life cycle, a filamentous phage produces phage particles without lysing the cell. The host continues to reproduce itself, but more slowly than uninfected cells do because much of its resources are being used to make phages.

slowly generates single-stranded progeny genomes, which are packaged by supercoiling and coating with capsid proteins. The phage particles then extrude through the cell envelope without lysing the cell. The host cell continues to reproduce, though more slowly than uninfected cells do because much of its resources are diverted to virus production.

Some phage genomes consist of RNA. RNA phages include the leviviruses, such as phage MS2 and Qβ. The RNA phages replicate their genomes through a double-stranded RNA intermediate, analogous to the double-stranded DNA intermediate of the filamentous phage M13; but RNA phages lyse their hosts. RNA phages have very simple genomes composed of only three or four genes. These genes encode a replicase, a coat protein, and a maturation factor. The RNA genomes of these phages are of historical interest as model systems to study ribosomal translation.

THOUGHT QUESTION 6.5 What are the relative advantages and disadvantages (to a virus) of the slow-release strategy, compared with the strategy of a temperate phage, which alternates between lysis and lysogeny?

Host cell defense against phages. Under attack by phages, host cell populations undergo natural selection for resistance. Resistance can occur by evolution of an altered host receptor protein, or a protein that blocks phage binding to the receptor. An evolutionary "arms race" occurs, in which phages may evolve enzymes that cleave the host defence molecules.

Amazingly, bacteria and archaea also show an adaptive defense against viruses analogous to an immune system. The adaptive defense involves short DNA sequences homologous to DNA of phages that could infect the cell. The sequences are called clustered regularly interspaced short palindromic repeats (**CRISPR**). When a phage attacks a bacterium, if bacterial enzymes succeed in destroying the phage DNA, they may copy a tiny piece of it as CRISPR segment, inserted at the head of a long line of about thirty CRISPR sequences. Now the adapted host cell "remembers" infection by the specific phage.

The next time the adapted host cell gets attacked by a phage, all its genomic CRISPR sequences are expressed as RNA. The CRISPR RNA then joins the "Cascade" protein complex (or Cas complex). The Cas complex detects phage DNA homologous to its CRISPR-derived RNA, and proceeds to cleave the phage DNA, preventing phage replication. In addition, the Cas complex copies new phage DNA sequences into new CRISPR segments in the host genome. Thus, the CRISPR/Cas system both eliminates inserted phage DNA and modifies the cell genome to protect it from future phage infection.

TO SUMMARIZE:

- **Host cell surface receptors** mediate the attachment of bacteriophages to a cell and confer host specificity.
- **Lytic cycle.** A bacteriophage injects its DNA into a host cell, where it utilizes host gene expression machinery to produce progeny virions.
- **Lysogeny.** Some bacteriophages can insert their genome into that of the host cell, which then replicates the phage genome along with its own. A lysogenic bacterium can initiate a lytic cycle.
- **Slow release.** Some bacteriophages use the host machinery to make progeny that bud from the cell slowly, slowing growth of the host without lysis.
- **CRISPR host defense.** Host bacteria and archaea contain CRISPR sequences homologous to phage DNA. CRISPR RNA combines with Cas complex to defend against phage infection.

6.5 Animal and Plant Virus Life Cycles

Animal and plant viruses solve problems similar to those faced by bacteriophages: host attachment, genome entry and gene expression, virion assembly, and virion release. The more complex structure of eukaryotic cells, however, leads to greater complexity and diversity of life cycles than is seen in phages. Viral reproduction may involve intracellular compartments such as the nucleus or secretory system and may depend on tissue and organ development in multicellular organisms. Study of animal virus life cycles reveals potential targets for antiviral drugs, such as protease inhibitors for HIV.

NOTE: The virus life cycles in Chapter 6 are simplified. For greater molecular detail, see Chapter 11.

Animal Viruses Show Tissue Tropism

Like bacteriophages, animal viruses evolve to bind specific receptor proteins on their host cell. An example of a human virus-receptor interaction is that of rhinovirus (see **Table 6.1**, Group IV), which causes the common cold. Rhinovirus attaches to ICAM-1, a human glycoprotein needed for intercellular adhesion (**Fig. 6.20A**). The rhinovirus binds to a domain of ICAM-1 essential for ICAM-1 to bind a lymphocyte protein called integrin.

The host receptors play a key role in determining the host range, the group of host species permitting infection. Within a host, receptor molecules can also determine the viral **tropism**, or ability to infect a particular tissue type. Some viruses, such as Ebola virus, exhibit broad tropism, infecting many kinds of host tissues, whereas others, such as papillomavirus, show tropism for only one type, the epithelial tissues. Tropism commonly depends on the availability of a protein on the host cell surface (receptor protein) that can bind a viral surface protein. For example, poliovirus infects only a specific class of human cells that display the immunoglobulin-like receptor protein PVR. Mice lack the PVR protein on their cell surfaces, and they are not normally infected by polio, but when transgenic mice were engineered to express PVR on their cells, the mice could then be infected.

Different strains of a virus may show radically different host tropisms based on their receptor specificity. An example is that of avian influenza strain H5N1. The H5N1 strain infects birds by binding to a glycoprotein (protein with sugar chains) receptor on cell surfaces of the avian respiratory tract. The H5N1 strain requires a receptor protein with a sialic acid sugar chain terminating in galactose linked at the C-3 position ($\alpha2, 3$), common in the avian respiratory tract. In humans, however, most nasal upper respiratory cells have receptors with galactose linked at the C-6 position ($\alpha2, 6$). Human cells with the C-3 linkage are more common in the lower respiratory tract. That is why avian influenza H5N1 infection of humans has been relatively rare. However, only a small mutation in the H5N1 envelope protein could enable it to bind to the ($\alpha2, 6$) receptor more effectively, allowing rapid transmission between humans.

THOUGHT QUESTION 6.6 How could humans evolve to resist rhinovirus infection? Is such evolution likely? Why or why not?

Genome uncoating. Most animal viruses, unlike bacteriophages, enter the host cell as virions. The internalized virons undergo **uncoating**, a process of virion disassembly in which the genome is released for replication and gene expression. A viral genome may be uncoated in several different ways. For example, measles virus, a paramyxovirus, enters the cell by binding host receptor proteins, which causes the viral envelope to fuse with the host cell membrane (**Fig. 6.20B**). The measles RNA genome is then uncoated and released directly into the cytoplasm. Alternatively, a flavivirus such as hepatitis C is taken up by **endocytosis** (**Fig. 6.20C**). In endocytosis, the cell membrane forms a vesicle around the virion and engulfs it, forming an endocytic vesicle. The endocytic vesicle fuses with a lysosome, whose acidity activates entry of the capsid into the cytoplasm. The capsid then comes apart, uncoating the viral genome. Still other species, such as adenovirus, enter by endocytosis but require transport to the nucleus (**Fig. 6.20D**). At the nuclear membrane, the virion docks at a nuclear pore and injects its DNA genome into the nucleus. The adenoviral DNA then has access to cellular DNA polymerase and RNA transcriptase.

Animal Virus Life Cycles

The primary factor determining the life cycle of an animal virus is the physical form of the genome. A viral genome made of DNA can utilize some or all of the host replication machinery. If, however, the genome consists of single-stranded RNA, it first needs to synthesize either an RNA-dependent RNA polymerase to generate an RNA template or a reverse transcriptase to generate a DNA template (in the case of retroviruses).

DNA virus life cycle. An example of a double-stranded DNA virus is human papillomavirus (HPV) (see **Table 6.1**, Group I), the cause of genital warts (**Fig. 6.21A**). HPV is the most common sexually transmitted disease in the United States and one of the most common worldwide. Certain strains infect the skin, whereas others infect the mucous membranes through genital or anal contact

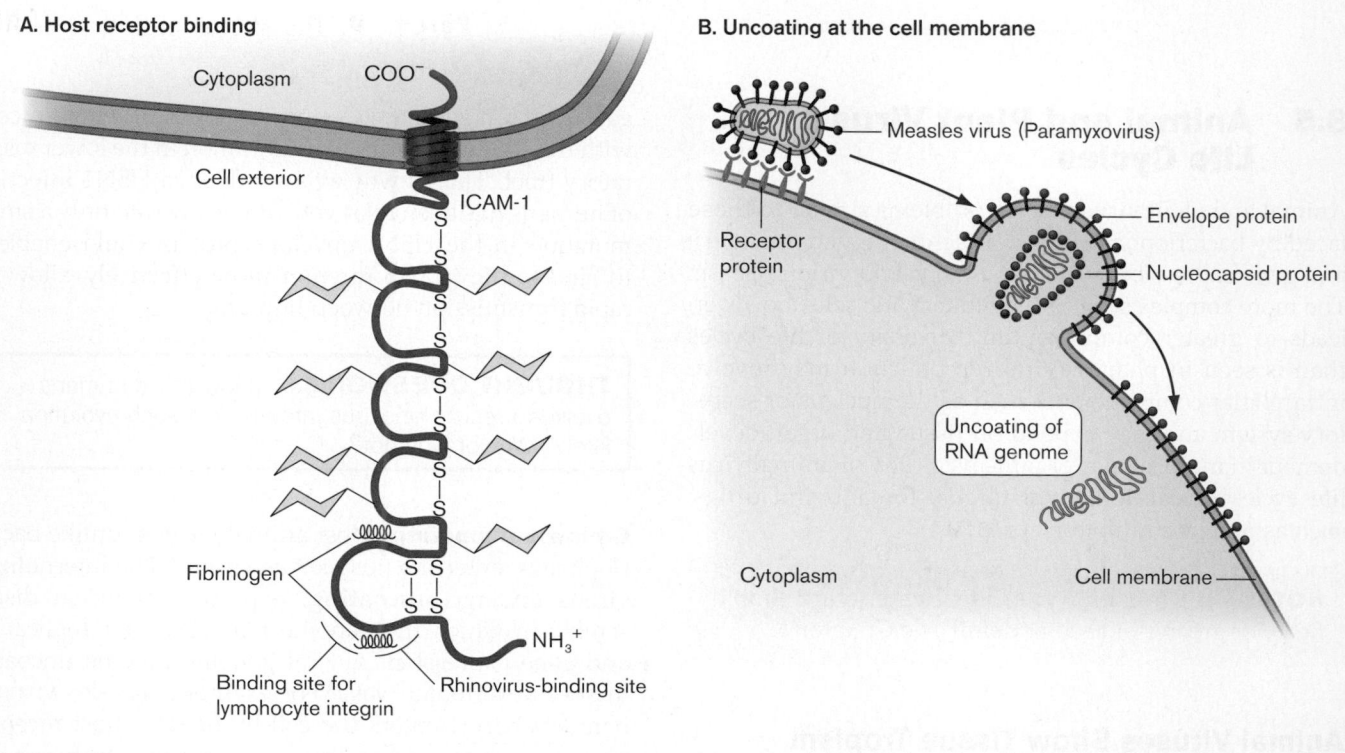

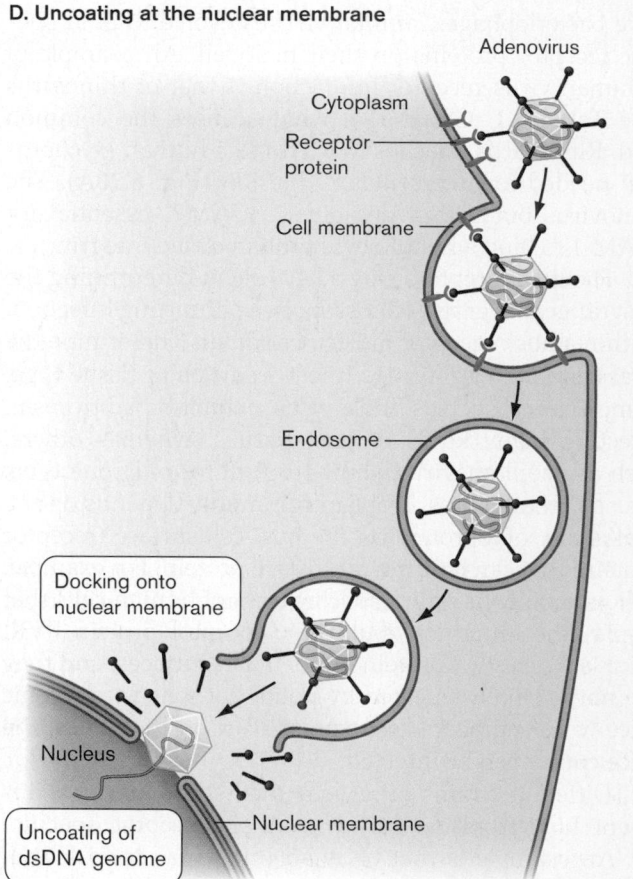

A. Host receptor binding

Cytoplasm

COO⁻

Cell exterior

ICAM-1

Fibrinogen

Binding site for lymphocyte integrin

Rhinovirus-binding site

NH₃⁺

B. Uncoating at the cell membrane

Measles virus (Paramyxovirus)

Receptor protein

Envelope protein

Nucleocapsid protein

Uncoating of RNA genome

Cytoplasm

Cell membrane

C. Uncoating within endosomes

Cytoplasm

Receptor protein

Cell membrane

Hepatitis C

Endosome

Lysosome fuses with and acidifies endosome.

Viral envelope fuses with endosome membrane.

Uncoating of RNA genome

H⁺ (acidity)

D. Uncoating at the nuclear membrane

Adenovirus

Cytoplasm

Receptor protein

Cell membrane

Endosome

Docking onto nuclear membrane

Nucleus

Uncoating of dsDNA genome

Nuclear membrane

Figure 6.20   Receptor binding and genome uncoating.   A. Rhinovirus attaches to the intercellular adhesion molecule (ICAM-1), a glycoprotein required by the host cells to bind a lymphocyte integrin, a cell surface matrix protein required for cell-cell adhesion. After binding a specific receptor on the host cell membrane, an animal virus enters the cell, where its genome is uncoated. **B.** Uncoating at the cell membrane. **C.** Uncoating within endosomes. **D.** Uncoating at the nuclear membrane. *Source:* Part A based on J. Bella et al. 1998. *PNAS* **95**:4140.

A.

B.

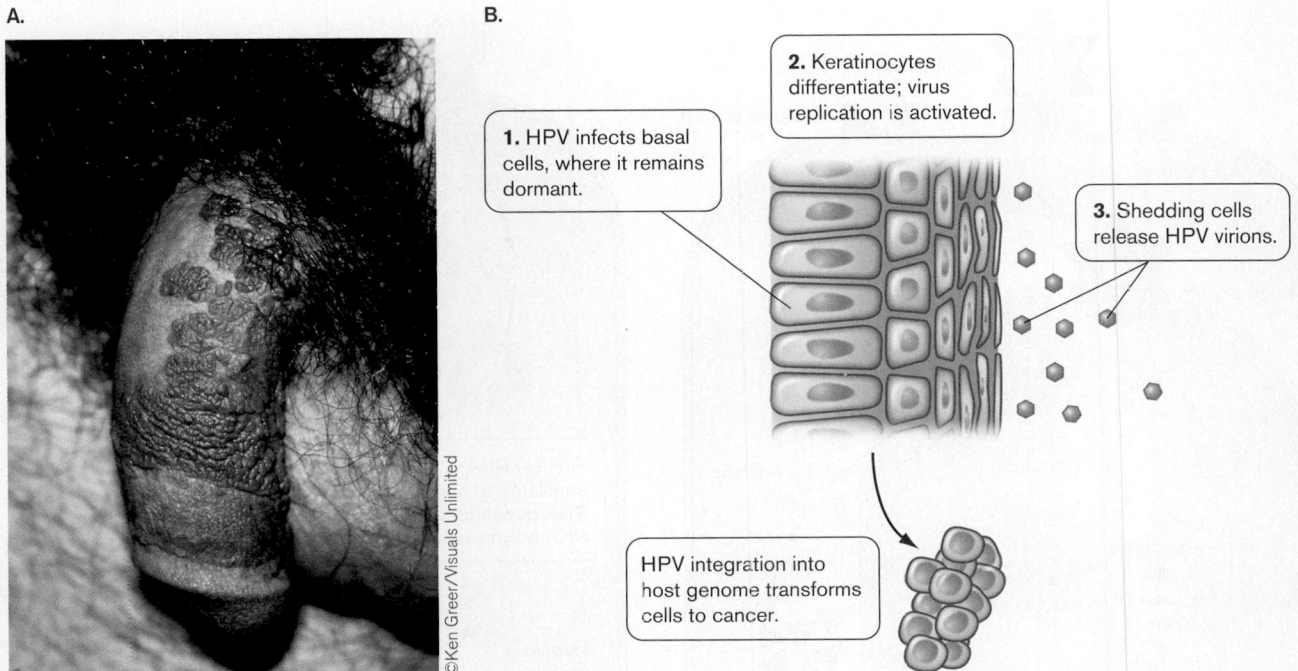

1. HPV infects basal cells, where it remains dormant.

2. Keratinocytes differentiate; virus replication is activated.

3. Shedding cells release HPV virions.

HPV integration into host genome transforms cells to cancer.

©Ken Greer/Visuals Unlimited

Figure 6.21   Human papillomavirus.   A. Certain strains of human papillomavirus (HPV) cause warts on the genitals or anus. **B.** HPV infects basal epithelial cells, where the DNA uncoats but remains dormant. As cells differentiate, new virions are synthesized and released by shedding cells. Some HPV proteins transform host cells to cancer cells.

(sexual transmission). Like phage lambda, papillomavirus has an active reproduction cycle and a dormant cycle in which the viral genome integrates into that of the host.

HPV initially infects basal epithelial cells (**Fig. 6.21B**). In the basal cells, papilloma virions enter the cytoplasm by receptor binding and membrane fusion. The virion then undergoes uncoating via disintegration of the protein capsid or coat, releasing its circular, double-stranded genome (**Fig. 6.22**). The uncoated viral genome enters the nucleus, where it gains access to host DNA and RNA polymerases.

The process of viral reproduction is complicated by the developmental progression of basal cells into keratinocytes and ultimately cells to be shed or sloughed off from the surface (see **Fig. 6.21B**). Viral replication is largely inhibited until the basal cells start to differentiate into keratinocytes (mature epithelial cells). Host cell differentiation induces the viral DNA to replicate and undergo transcription by host polymerases. The mRNA transcripts then exit the nuclear pores, as do host mRNAs, for translation in the cytoplasm. The translated capsid proteins, however, return to the nucleus for assembly of the virion. Nuclear virion assembly is typical of DNA viruses (with the exception of poxviruses, which replicate entirely in the cytoplasm).

As the keratinocytes complete differentiation, they start to come apart and are shed from the surface. The DNA virions are released from the cell during this shedding process. Unfortunately, in the basal cells HPV has an alternative pathway of integrating its genome into that of the host (analogous to phage lysogeny). The integrated genome can transform host cells into cancer cells. Transformation occurs through increased expression of viral oncogenes (cancer-causing genes). The oncogenes inhibit the expression of host tumor suppressor genes.

RNA virus life cycle. The picornaviruses include poliovirus as well as rhinovirus, which causes the common cold (see **Table 6.1**, Group IV). Picornavirus genomes contain (+) strand RNA, allowing viral reproduction to occur entirely in the cytoplasm, without any use of DNA.

Picornaviruses bind to a surface receptor, such as ICAM-1 for rhinovirus or the PVR receptor for poliovirus. The (+) strand RNA is uncoated by insertion through the cell membrane into the cytoplasm (**Fig. 6.23**). The role of endocytosis is debated; poliovirus requires no endocytosis, but rhinovirus genome uncoating may require endocytosis and low-pH induction.

After uncoating, a gene in the viral RNA is translated by host ribosomes to make RNA-dependent RNA polymerase. The RNA-dependent RNA polymerase uses the viral RNA template to make (−) strand RNA. The (−) strand RNA then serves as a template for other viral mRNAs, as well as progeny genomic RNA, which is replicated in virus-induced vesicles from the endoplasmic reticulum. Capsid proteins are translated by host ribosomes,

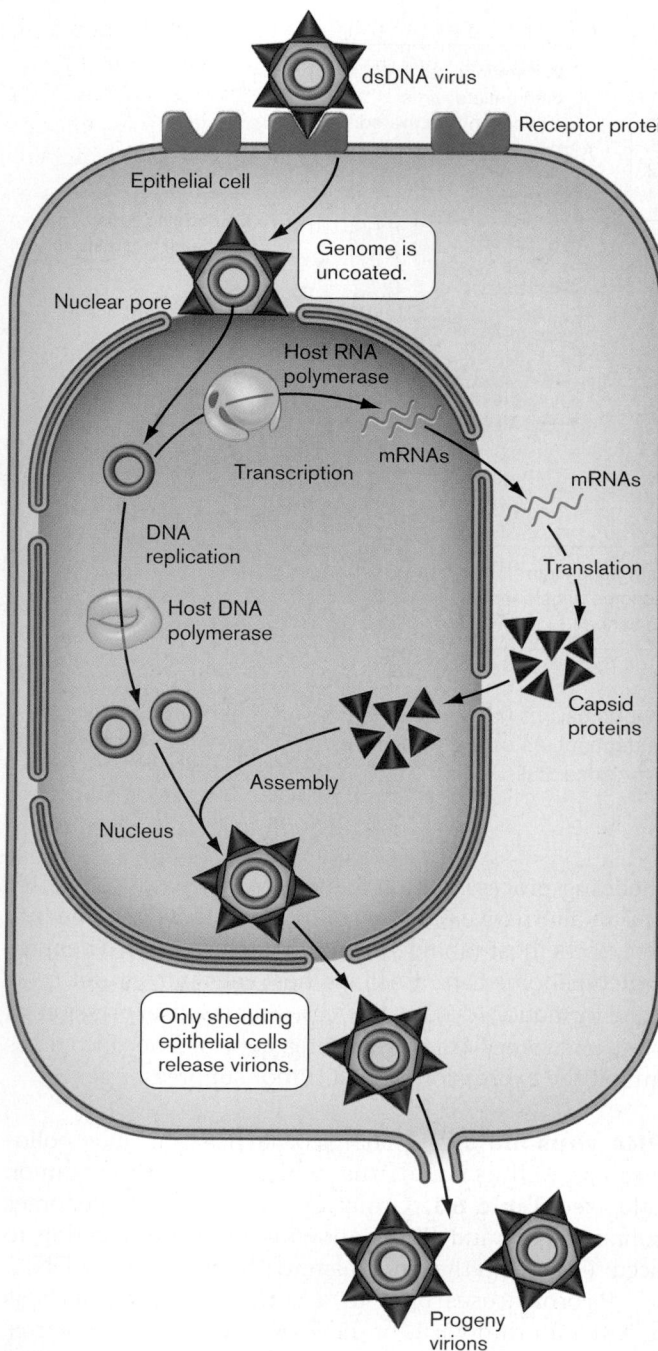

Figure 6.22 Papillomavirus life cycle. HPV, a double-stranded DNA virus, enters the cytoplasm, where the protein coat disintegrates. The viral DNA enters the nucleus for replication and transcription by host polymerases. Viral mRNA returns to the cytoplasm for translation of capsid proteins, which return to the nucleus for assembly of virions.

and the capsids self-assemble in the cytoplasm. Virions then assemble at the cell membrane and are released by subverting lysosomes, which attempt to digest them.

Note that other kinds of RNA viruses, such as influenza virus, encapsidate a (−) strand genome. In this case,

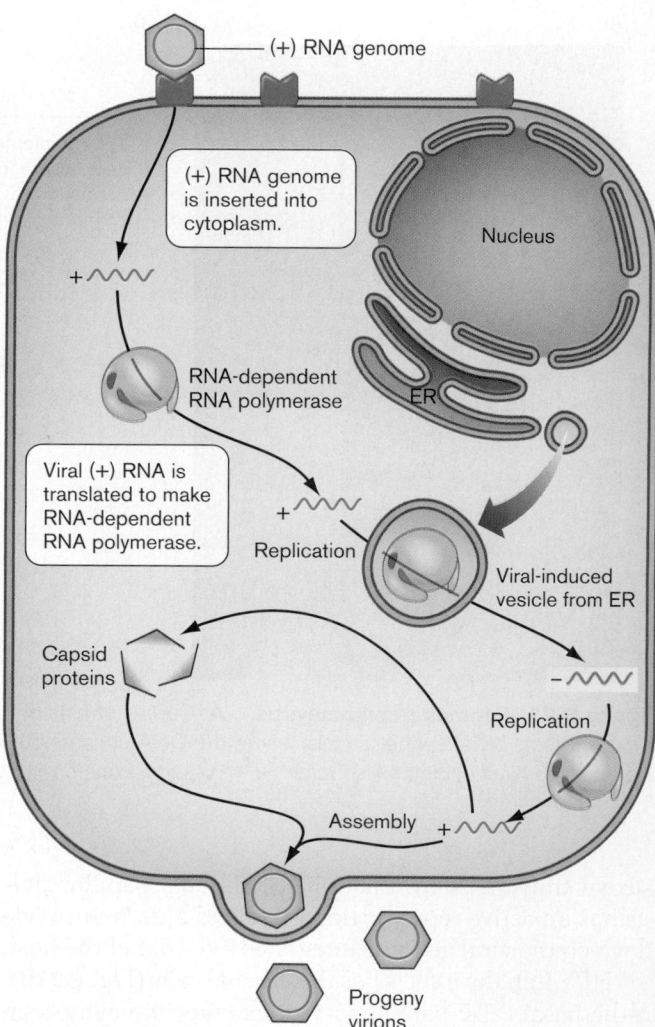

Figure 6.23 Picornavirus life cycle. A picornavirus inserts its (+) strand RNA into the cell. Reproduction occurs entirely in the cytoplasm. A key step is the early translation of a viral gene to make RNA-dependent RNA polymerase. The RNA-dependent RNA polymerase uses the picornavirus RNA template to make (−) strand RNA, which then serves as a template for other viral mRNAs, as well as progeny genomic RNA, which is replicated in virus-induced vesicles from the endoplasmic reticulum.

the (−) strand must serve as the template to generate mRNA as well as (−) strand progeny genomes. Influenza virus replication includes other interesting molecular complications, discussed in Chapter 11.

RNA retroviruses. Retroviruses include human immunodeficiency virus (HIV), the causative agent of AIDS; and feline leukemia virus (FeLV), a disease commonly afflicting domestic cats (see **Table 6.1**, Group VI). A retrovirus such as HIV uses a reverse transcriptase to make a DNA copy of its RNA genome (**Fig. 6.24**). Instead of being translated from an early gene, the reverse transcriptase is actually carried within the virion, bound to

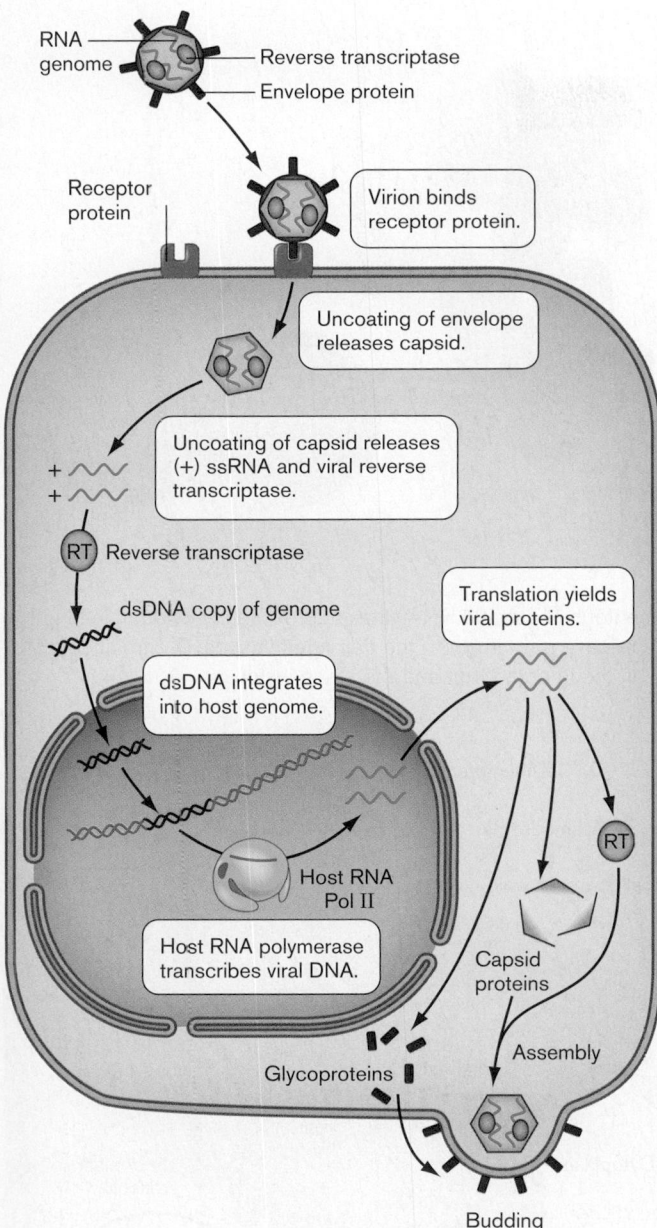

Figure 6.24 Retrovirus life cycle. A retrovirus such as human immunodeficiency virus (HIV) uses reverse transcriptase to copy its RNA into double-stranded DNA. The DNA then enters the nucleus to recombine in the host genome, where a host RNA polymerase generates viral mRNA and viral genomic RNA. The viral mRNA enters the cytoplasm for translation. Viral coat proteins are transported by the endoplasmic reticulum to the cell membrane, where virions assemble and bud out.

the RNA genome with a primer in place. The virion contains two copies of the HIV genome, each carrying its own reverse transcriptase. Virions infect T helper cells and other lymphocytes of the active immune system (discussed in Chapter 24). After uncoating in the cytoplasm, the viral RNA is copied into double-stranded DNA.

The DNA copy then enters the nucleus, where an integrase enzyme integrates the DNA into the host genome.

The viral genome then replicates silently in the host cell, generating only a small number of virions without apparent effect on the host. To generate virions, a host RNA polymerase transcribes viral mRNA and viral genomic RNA. The viral mRNA reenters the cytoplasm for translation to produce coat proteins and envelope proteins. The coat proteins are transported by the endoplasmic reticulum (ER) to the cell membrane, where virions self-assemble and bud out.

At a certain point, the host cell can suddenly begin to generate large numbers of virions, destroying the immune system. The cause of accelerated reproduction is poorly understood, although it involves cytokines released under stress conditions such as poor health or pregnancy. Activation occurs through protein regulators encoded by the virus. The full process of HIV reproduction is in Chapter 11.

Oncogenic viruses. Some retroviruses, such as feline leukemia virus (FeLV), carry **oncogenes**, which can transform host cells to cancer cells. Other cancer-causing viruses include DNA viruses such as HPV that can integrate their genomes into a host chromosome. The advantage of cancer transformation is that it expands the population of infected cells proliferating virus particles, or replicating a hidden viral genome.

In some cases, the transfer of host genes to abnormal chromosome locations can generate host-derived oncogenes, whose uncontrolled expression causes uncontrolled proliferation of the cell and ultimately cancer. Viruses that carry oncogenes are known as **oncogenic viruses**. This capacity for gene transfer may be manipulated artificially for gene therapy (discussed in Chapter 11).

> **THOUGHT QUESTION 6.7** What are the advantages and disadvantages to the virus of replication by the host polymerase, compared with using a polymerase encoded by its own genome?

Plant Virus Life Cycles

All kinds of plants are subject to viral infection. Plant viruses pose enormous challenges to agriculture, where the concentrated growth of a single strain (monoculture) provides ideal conditions for a virus to spread.

Plant virus entry to host cells. In contrast to animal viruses and bacteriophages, plant viruses infect cells by mechanisms that do *not* involve specific membrane receptors. The reason may be that plant cell membranes are covered by thick cell walls impenetrable to virion uptake or genome insertion. Thus, the entry of plant viruses usually requires **mechanical transmission**, nonspecific access through physical damage to tissues, such as

A.

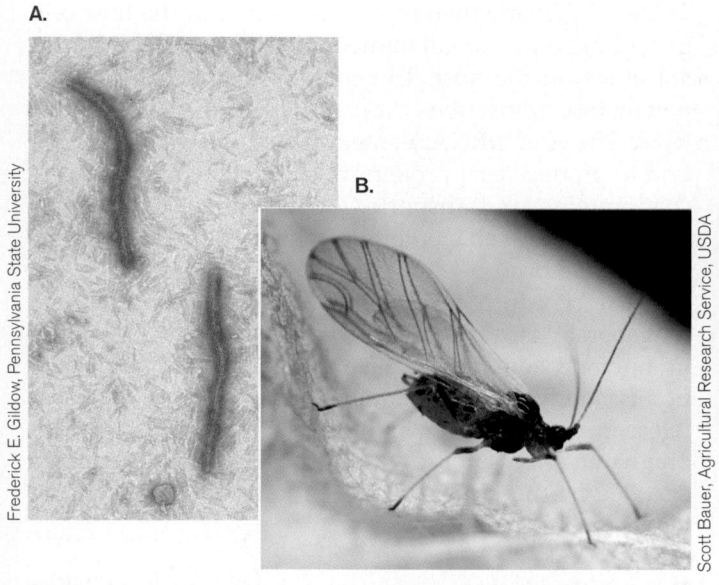

B.

C.

D.

Figure 6.25   Plum pox is caused by potyvirus.   A. Potyvirus, a filamentous (+) strand RNA virus, approximately 800 nm in length (TEM). **B.** Potyvirus is transmitted by aphids, which suck the plant sap and release potyvirus into the damaged tissues. **C.** Streaking of flowers caused by potyvirus infection. **D.** Ring-shaped pockmarks appear on the infected fruit and the stone inside.

abrasions of the leaf surface by the feeding of an insect. The means of mechanical transmission of plant viruses are limited by the cell wall and by the sessile nature of plants. Most plant viruses gain entry to cells by one of three routes:

- **Contact with damaged tissues.** Viruses such as tobacco mosaic virus appear to require nonspecific entry into broken cells.
- **Transmission by an animal vector.** Insects and nematodes transmit many kinds of plant viruses. For example, the geminiviruses are inoculated into cells by plant-eating insects such as aphids, beetles, or grasshoppers.
- **Transmission through seed.** Some plant viruses enter the seed and infect the next generation.

An economically important plant virus is the potyvirus called plum pox virus (see **Table 6.1**, Group IV), a major pathogen of plums, peaches, and other stone fruits. Plum pox virus, a (+) strand RNA virus, is transmitted by aphids (**Fig. 6.25**). Following infection, the spread of the virus generates streaked leaves and flowers, as well as ring-shaped pockmarks on the surfaces of the fruit and of the stone within.

Plant virus transmission through plasmodesmata. Within a plant, the thick cell walls prevent a lytic burst or budding out of virions. Instead, plant virions spread to uninfected cells by traveling through **plasmodesmata** (singular, **plasmodesma**). Plasmodesmata are membrane

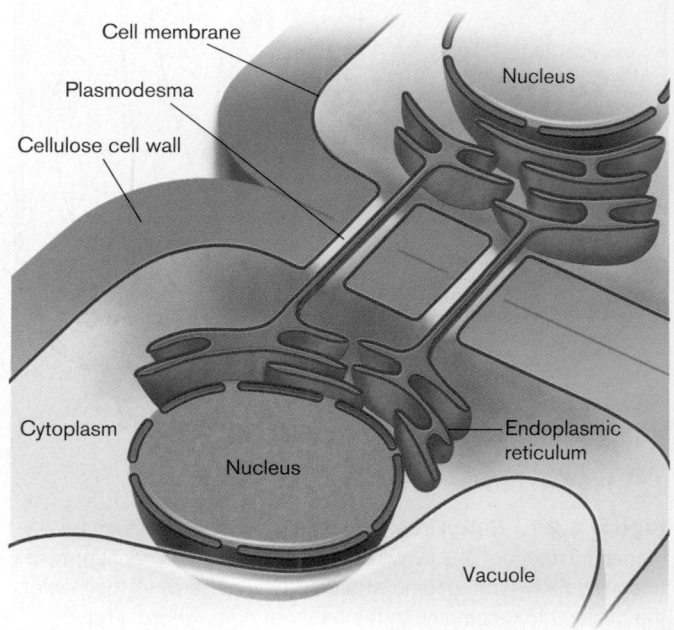

Figure 6.26   Plant cells connected by plasmodesmata. Plasmodesmata offer a route for plant viruses to reach uninfected cells.

channels that connect adjacent plant cells (**Fig. 6.26**). The outer channel connects the cell membranes of the two cells; the inner channel connects the endoplasmic reticulum.

Passage through the plasmodesmata requires action by movement proteins whose expression is directed by the viral genome. In some cases, the movement proteins transmit the entire plant virion; in other cases, only the nucleic acid itself is small enough to pass through. The independent movement and transmission of plant viral genomes suggest infective strategies in common with those of viroids, which lack capsids altogether.

DNA pararetroviruses. Pararetroviruses possess a DNA genome that requires transcription to RNA in the cytoplasm, followed by reverse transcription to form DNA genomes for progeny virions. Some pararetroviruses, such as hepadnavirus, infect humans, but the best-known pararetrovirus is cauliflower mosaic virus, or caulimovirus (see **Table 6.1**, Group VII). Caulimovirus is an important tool for biotechnology because it has a highly efficient promoter for gene transcription, allowing high-level expression of cloned genes. Vectors derived from caulimoviruses are used to construct transgenic plants.

Caulimovirus is transmitted by secretions from an insect whose bite damages plant tissues, providing access to the cytoplasm (**Fig. 6.27**). The caulimovirus genome moves from the cytoplasm to the cell nucleus through a nuclear pore. Within the nucleus, two promoters on its DNA genome direct transcription to RNA. The two RNA transcripts exit the nucleus for translation by host ribosomes to make viral proteins. A host reverse transcriptase, present in plant cells, copies the RNA into DNA viral genomes. After virions are assembled in the cytoplasm, movement proteins help transfer them through plasmodesmata into an adjacent cell.

A caulimovirus promoter sequence is commonly used in gene transfer vectors for plant biotechnology. The advantage of viral promoters is their highly efficient transcription of a transferred gene, such as one conferring pesticide resistance. In the field, 10% of cruciferous vegetables are typically infected with caulimovirus. Some critics of gene technology are concerned that the prevalence of the caulimovirus promoter in transgenic crops may lead to the evolution of new pararetroviruses.

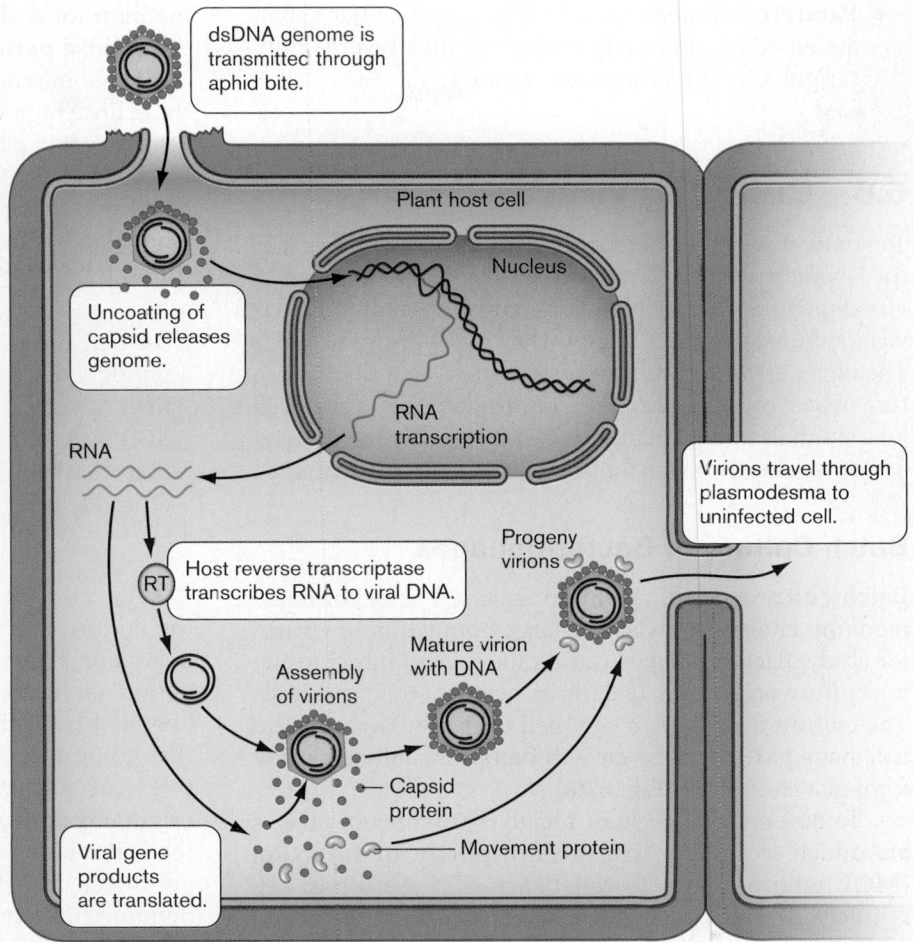

Figure 6.27 Caulimovirus life cycle. The cauliflower mosaic virus, or caulimovirus, uses host RNA polymerase to copy its DNA into RNA and uses host reverse transcriptase (RT) to make DNA copies. The DNA genomes are packaged into progeny virions that use virus-encoded movement proteins to travel through plasmodesmata to an adjacent uninfected cell.

TO SUMMARIZE:

- **Host cell surface receptors** mediate animal virus attachment to a cell and confer host specificity and tropism.
- **Animal DNA viruses** either inject their genome or enter the host cell by endocytosis. The viral genome requires uncoating for gene expression.
- **RNA viruses** use an RNA-dependent RNA polymerase to transcribe their messenger RNA.
- **Retroviruses** use a reverse transcriptase to copy their genomic sequence into DNA for insertion in the host chromosome.
- **Plant viruses** enter host cells by transmission through a wounded cell surface or an animal vector. Plant viruses travel to adjacent cells through plasmodesmata.

▪ **Pararetroviruses** contain DNA genomes but generate an RNA intermediate that requires reverse transcription to DNA for progeny virions.

6.6 Culturing Viruses

To study any living microorganism, we must culture it in the laboratory. Culturing viruses provides large numbers of virions and their components for analysis. A complication of virus culture is the need to grow the virus within a host cell. Therefore, any virus culture system must be a double culture of host cells plus viruses. Culturing viruses of multicellular animals and plants involves additional complications, as viruses show tropism for particular tissues or organs.

Batch Culture of Bacteriophages

Batch culture, or culture in an enclosed vessel of liquid medium, enables growth of a large population of viruses for study. Bacteriophages can be inoculated into a growing culture of bacteria, usually in a culture tube or a flask. The culture fluid is then sampled over time and assayed for phage particles. The growth pattern usually takes the form of a step curve (**Fig. 6.28**).

To observe one cycle of phage reproduction, phages are added to host cells at a **multiplicity of infection** (**MOI**, ratio of phage to cells) such that every host cell is infected. The phage particles immediately adsorb to surface receptors of host cells and inject their DNA. As a result, virions are virtually undetectable in the growth medium for a short period after infection; this is called the **eclipse period**.

For some species, it is possible to distinguish between the eclipse period and a **latent period**, which includes the eclipse period plus the time during which progeny viruses have been formed but remain trapped within the cell. The latent period is particularly significant in animal viruses, where large numbers of virions usually generate progeny through budding out of the host cell (**Fig. 6.29**).

> **NOTE:** The "latent period" of a lytic virus is the period between initial phage-host contact and the first appearance of progeny phage. This must be distinguished from the "latent infection" of a virus that maintains its genome within a host cell without reproducing virions.

As cells begin to lyse and liberate progeny viruses, the culture enters the **rise period**, in which virus particles are appearing in the growth medium. The rise period ends when all the progeny viruses have been liberated from their host cells. If the number of viruses that go on to reinoculate further host cells is small, then the virus concentration at the end point divided by the original concentration of inoculated phage approximates the **burst size**—that is, the number of viruses produced per infected host cell. We can estimate the burst size by dividing the concentration of progeny virions by the concentration of inoculated virions, assuming that all the original virions infect a cell.

Figure 6.28 One-step growth curve for a bacteriophage. After initial infection of a liquid culture of host cells, the titer of virus drops near zero as all virions attach to the host. During the eclipse period, progeny phages are being assembled within the cell. As cells lyse (the rise period), virions are released until they reach the final plateau. The infectious cycle is typically complete within less than an hour.

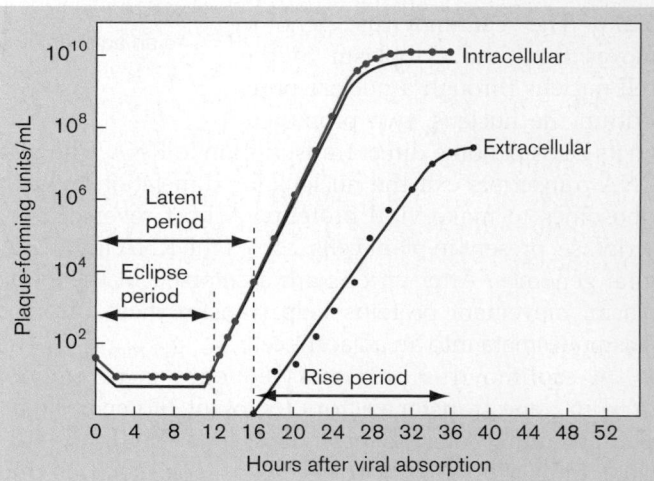

Figure 6.29 One-step growth curve for a virus. The titer of extracellular virus drops near zero during the latent period, as all virions adsorb to the host. Then progeny virions begin to emerge by budding out from the infected cell. The growth curve of a rapidly replicating animal virus typically takes hours to level off; the "burst" event is not defined as clearly as for phages.

The burst size, together with the cell density prior to lysis, determines the concentration of the resultant suspension of virus particles, called a **lysate**. In the case of bacteriophages, a lysate of phage particles can be extremely stable, remaining infective at room temperature for many years. Eukaryotic viruses tend to be less stable and need to be maintained in culture or deep freeze.

> **THOUGHT QUESTION 6.8** Why does bacteriophage reproduction give a step curve, whereas cellular reproduction generates an exponential growth curve? Could you design an experiment in which viruses generate an exponential curve? Under what conditions does the growth of cellular microbes give rise to a step curve?

Tissue Culture of Animal Viruses

In the case of animal and plant viruses, the multicellular nature of the host is an important factor in the pathology and transmission of the pathogen (discussed in Chapters 25 and 26). Animal viruses can be cultured within whole animals by serial inoculation, where virus is transferred from an infected animal to an uninfected one. Culture within animals ensures that the virus strain maintains its original **virulence** (ability to cause disease). But the process is expensive and laborious, involving the large-scale use of animals.

A historic event in 1949 was the first successful growth of a virus in tissue culture. Poliovirus, the causative agent of the devastating childhood disease poliomyelitis, was grown in human cell tissue culture (**Fig. 6.30**) by John F. Enders, Thomas J. Weller, and Frederick Robbins at the Children's Hospital in Boston. As heralded that year in *Scientific American*, "It means the end of the 'monkey era' in poliomyelitis research . . . Tissue-culture methods have provided virologists with a simple in vitro method for testing a multitude of chemical and antibiotic agents." Since then, tissue culture has remained the most effective way to study the molecular biology of animal and plant viruses and to develop vaccines and antiviral agents.

Some viruses can be grown in a tissue culture of cells growing confluently on a surface. The fluid bathing the tissue layer is sampled for virus concentration. As in the case of bacteriophage batch culture, we can define an eclipse period, a latent period before appearance of the first progeny virions in the culture fluid, and a rise period. The intracellular completion of virion assembly within the cell is important for transmission of viral infection because some viruses have the ability to spread their virions from cell to cell without ever existing as such outside. In this way, they can avoid detection by the immune system to which they might be vulnerable.

In tissue culture, the time course of animal virus replication is usually much longer (hours or days) than that of bacteriophages (typically less than an hour under optimal conditions). The burst size, however, of animal viruses is typically several orders of magnitude larger than that of phages. The reason for the larger burst size is probably that the volume of the eukaryotic host cell is usually much larger than that of a bacterial host, thus providing a larger supply of materials to build virions.

Not all animal viruses exhibit growth that can be represented by a step curve. Some species, called slow viruses, bud off virions relatively slowly, without immediately lysing the host cell. A well-known family of slow viruses is the **lentiviruses** (literally, "slow viruses"), a group of retroviruses that includes HIV. HIV and related retroviruses are known for their long incubation periods, in some cases many years, during which time extremely low concentrations of virions are produced by infected cells.

Plaque Isolation and Assay of Bacteriophages

For the investigation of cellular microbes, an important tool is the culturing of individual colonies on a solid substrate that prevents dispersal throughout the medium, as described in Chapters 1 and 4. Plate culture of colonies enables us to isolate a population of microbes descended from a common progenitor. But viruses cannot be isolated as "colonies." The reason is that although viruses can be obtained at incredibly high concentrations, they disperse in

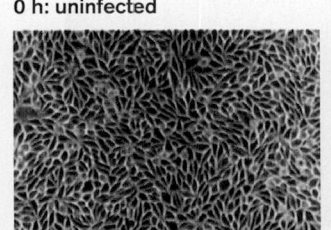 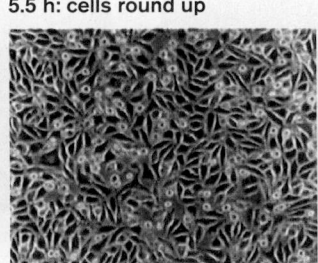 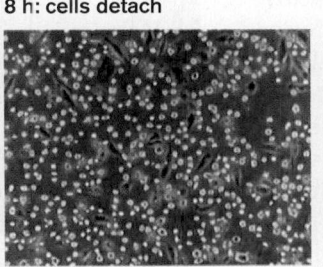 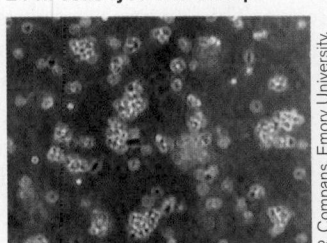

0 h: uninfected 5.5 h: cells round up 8 h: cells detach 24 h: cells lyse and clump

R. Compans, Emory University, School of Medicine

Figure 6.30 Poliovirus replication in human tissue culture. Before infection (0 hours), the cultured cells grow in a smooth layer. At 5.5 hours after infection, cells are starting to retract and round up. At 8 hours, infected cells detach from the culture dish. By 24 hours, cells have lysed or in some cases clumped with other cells. (cell length, approximately 50 μm)

suspension. Even on a solid medium, viruses never form a solid visible mass comparable to the mass of cells that constitutes a cellular colony.

In viral plate culture, viruses from a single progenitor lyse their surrounding host cells, forming a clear area called a **plaque**. The **plaque assay** for lytic bacteriophages was invented early in the twentieth century by the French microbiologist Félix d'Herelle, who was the first to recognize that bacterial cells can be infected by bacteriophages.

To perform a plaque assay of bacteriophages, a diluted suspension of bacteriophages is mixed with bacterial cells in soft agar, and the mixture is then poured over a nutrient agar plate (**Fig. 6.31**). Where no bacteriophages are present, the bacteria grow homogeneously as a "lawn," an opaque sheet over the surface (confluent growth). Where there is a bacteriophage, it infects a cell, replicates, and spreads progeny phages to adjacent cells, killing them as well (**Fig. 6.32**). The loss of cells results in a round, clear area seemingly cut out of the bacterial lawn. In contrast to the clear plaques produced by lytic phages (as shown in the figure), temperate (lysogenic) phages make cloudy plaques containing lysogenized viable cells. The prophage in the host genome protects the cells from subsequent infection and lysis by another phage.

Plaques offer a convenient way to isolate a recombinant DNA molecule contained in a bacteriophage vector. In **Figure 6.32B**, the blue plaques result from a phage vector carrying the gene encoding the enzyme beta-galactosidase. This enzyme converts a colorless compound into a blue dye. When the indicator gene is interrupted by an inserted recombinant gene, the phage produces white plaques, which indicate the successful production of recombinant DNA phages.

Plaques can be counted and used to calculate the concentration of phage particles, or **plaque-forming units (PFUs)**, in a given suspension of liquid culture. The liquid

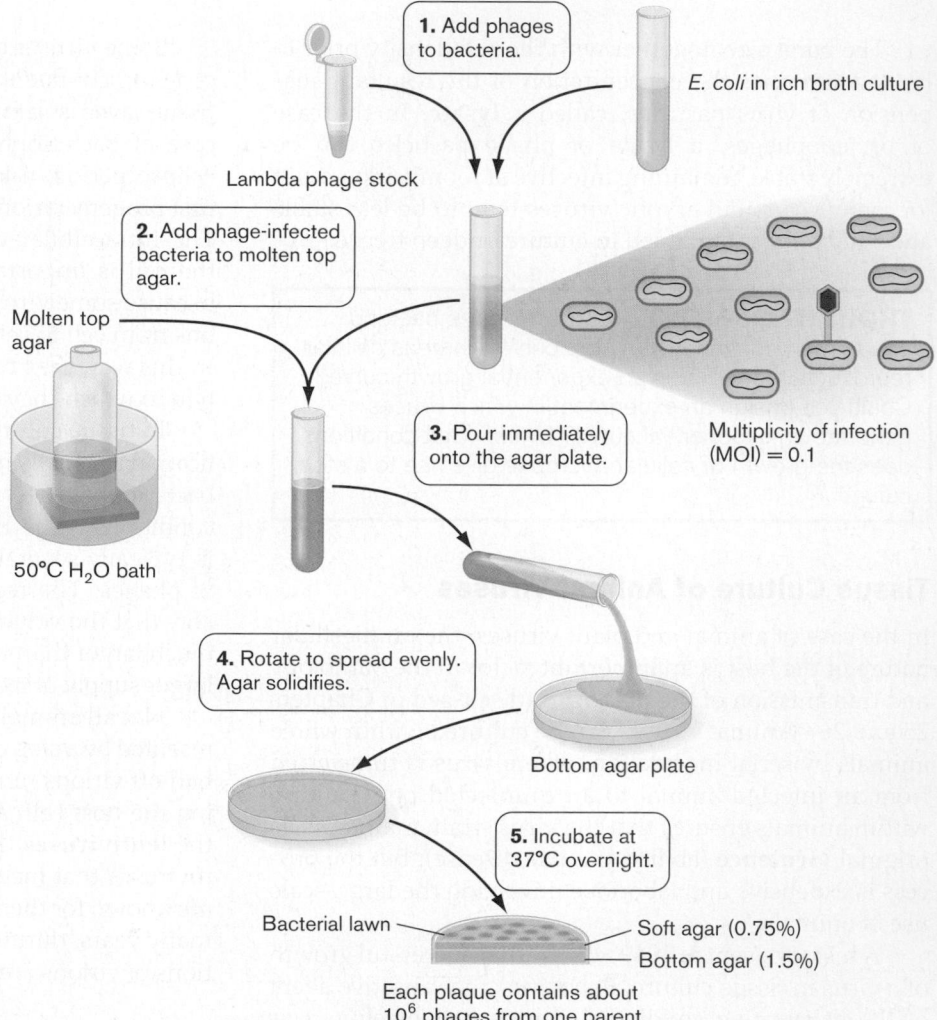

Figure 6.31   Plating a phage suspension to count isolated plaques.   A suspension of bacteria in rich broth culture is inoculated with a low proportion of phage particles (multiplicity of infection is approximately 0.1). This means that only a few of the bacteria become infected immediately, while the rest continue to grow. Each plaque arises from a single infected bacterium that bursts, its phage particles diffusing to infect neighboring cells.

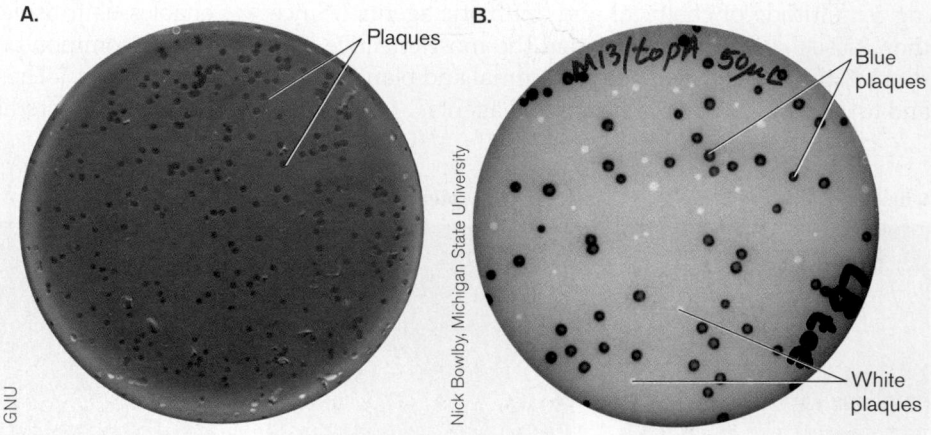

Figure 6.32   Phage plaques on a lawn of bacteria.   A. Phage lambda plaques on a lawn of *Escherichia coli* K-12. **B.** Plaques of recombinant phage M13 on *E. coli*. The original phage expresses beta-galactosidase, an enzyme that makes a blue product (blue plaques). White plaques are produced by phage particles whose genome is recombinant (contains a cloned gene interrupting the gene for beta-galactosidase).

culture can be analyzed by serial dilution in the same way one would analyze a suspension of bacteria.

Plaque Isolation and Assay of Animal Viruses

For animal viruses, the plaque assay has to be modified because it requires infection of cells in tissue culture. Tissue culture usually involves growth of cells in a monolayer on the surface of a dish containing fluid medium, which would quickly disperse any viruses released by lysed cells. To solve this problem, in 1952 Renato Dulbecco, at the California Institute of Technology, modified the tissue culture procedure for plaque assays (**Fig. 6.33A**). In Dulbecco's method, the tissue culture with liquid medium is first inoculated with virus. After sufficient time to allow for viral attachment to cells, the fluid is removed and replaced

by a gel medium. The gel retards the dispersal of viruses from infected cells, and as the host cells die, plaques can be observed. **Figure 6.33B** shows a plate culture of human coronavirus infection colon carcinoma cells.

Animal viruses that do not kill their host cells require a different kind of assay based on identification of a **focus** (plural, **foci**), a group of cells infected by the virus. In a **fluorescent-focus assay**, the infected cells are incubated for a sufficient period to allow the production of progeny virions. The plasma membranes of the cells are then made permeable by treatment with an organic solvent, and an antiviral antibody is added. Unattached antibodies are then washed away, and a second antibody is added that recognizes the first antibody molecule. The second antibody is conjugated to a fluorophore whose fluorescence reveals the foci of cells, each of which has arisen from a single virion in the original inoculum.

Another kind of fluorescent-focus assay uses green fluorescent protein (GFP). The viruses are genetically engineered to contain a GFP gene fusion—a labeling technique discussed in Chapter 3. The GFP fluorescence reveals foci of virus-infected cells (**Fig. 6.34A**).

A focus assay can also be used to isolate oncogenic viruses. Oncogenic, or cancer-causing, viruses actually **transform** their host cells into cancer cells. Cancer cells lose contact inhibition; they grow up in a pile instead of remaining in the normal monolayer. These piles of transformed cells, or transformed foci, can easily be

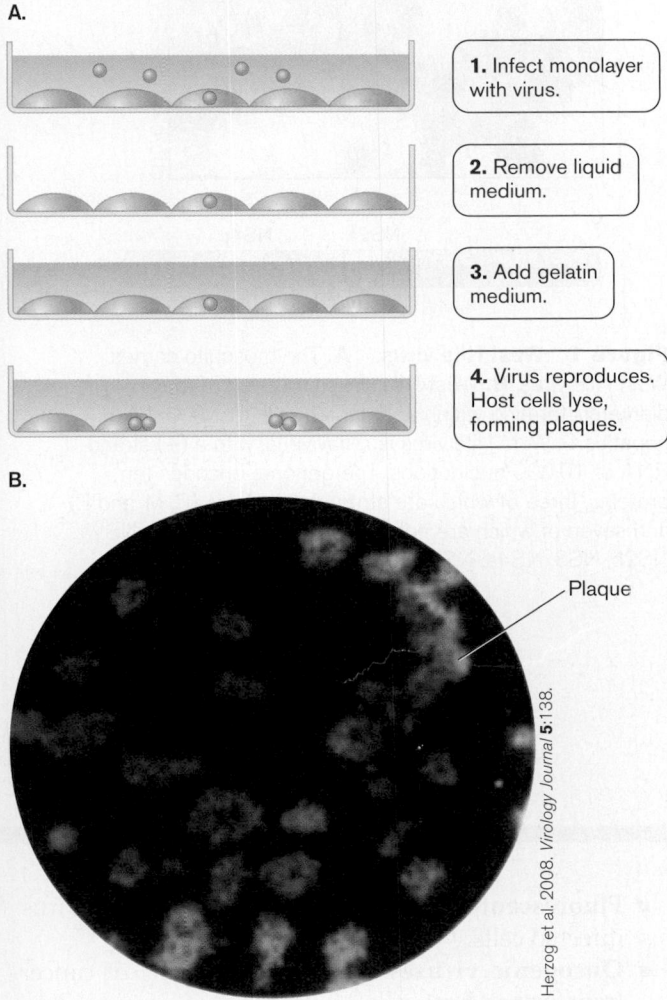

A.

1. Infect monolayer with virus.

2. Remove liquid medium.

3. Add gelatin medium.

4. Virus reproduces. Host cells lyse, forming plaques.

B.

Plaque

Herzog et al. 2008. *Virology Journal* **5**:138.

Figure 6.33   Plate culture of animal viruses.   A. Modified plaque assay for animal viruses. The gelled medium retards the dispersal of progeny virions from infected cells, restricting new infections to neighboring cells. The result is a visible clearing of cells (a plaque) in the monolayer. **B.** Plaque assay in which human coronavirus suspension was plated on a monolayer of colon carcinoma cells in tissue culture.

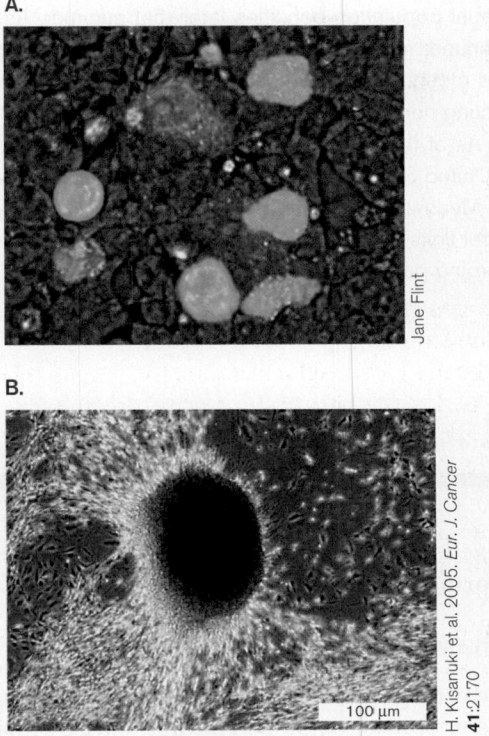

A.

Jane Flint

B.

100 μm

H. Kisanuki et al. 2005. *Eur. J. Cancer* **41**:2170

Figure 6.34   Focus assays of animal viruses.   A. GFP fluorescent-focus assay. **B.** Transformed-focus (cancer-forming) assay of an oncogenic virus. Transformed cells grow in an uncontrolled manner, which can produce a tumor.

Special Topic 6.1   West Nile Virus, an Emerging Pathogen

In the spring of 1999, New Yorkers noticed an unusual number of dead crows and other local birds. Then people began falling ill with an unfamiliar form of encephalitis (brain inflammation) caused by an avian virus never before seen in North America. The virus, a (+) single-stranded RNA virus, comes from the flavivirus family, which includes yellow fever and dengue (breakbone fever). These pathogens are all endemic to warmer environments, such as the Mediterranean and North Africa—hence the name, West Nile virus. Like yellow fever, a disease made infamous by its devastation of the southern United States in the nineteenth century, West Nile virus is transmitted between humans and animals by mosquitoes (**Fig. 1**).

Why the sudden appearance of West Nile virus in the United States? The answer is not known for certain, but some climatologists argue that West Nile and other emerging pathogens arise from unusual weather patterns related to global climate change (**Fig. 2**). West Nile virus is generally found in species of mosquitoes that fare poorly in the cold of winter. In 1998–99, however, the unusually warm winter enabled a few mosquitoes carrying the virus to survive until the spring. The spring that followed was drier than usual, forcing birds to spend more time at dwindling water pools, where mosquitoes were concentrated. The July heat wave then increased the rate of viral replication in the mosquitoes, which reached unusual population densities later that summer. As the cycle continued, mosquitoes reinfected birds, which then infected more mosquitoes. Eventually, the cycle included mosquitoes infecting humans.

As of this writing, West Nile virus has spread to most of the United States and appears to be endemic to North America. Meanwhile, research continues to determine whether clearer links can be established between climate changes and emerging pathogens.

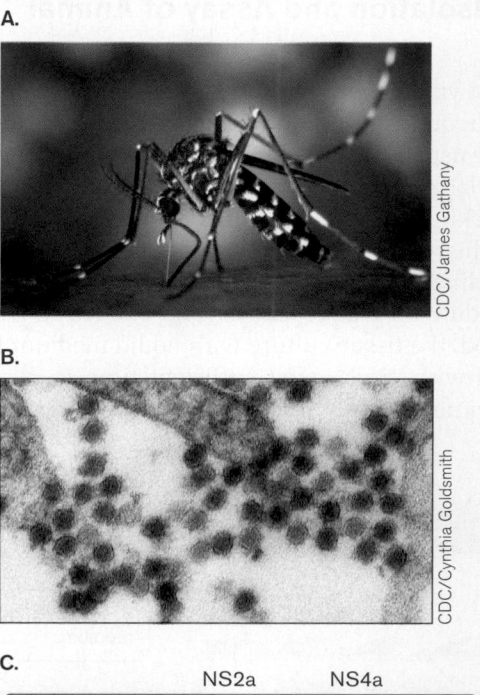

Figure 1 West Nile virus. A. The mosquito carries West Nile virus. **B.** West Nile virus particles (50 nm in diameter) forming within an infected cell (TEM). **C.** Like hepatitis C, West Nile virus is a flavivirus, with a (+) strand RNA of 10,000 nucleotides. The genome encodes ten proteins, three of which are structural proteins (C, M, and E) and seven of which are nonstructural proteins (NS1, NS2a, NS2b, NS3, NS4a, NS4b, and NS5).

---

visualized and counted. This procedure is known as the **transformed-focus assay** (**Fig. 6.34B**).

TO SUMMARIZE:

- **Culturing viruses** requires growth in host cells. Bacteriophages may be cultured either in batch culture or as isolated plaques on a bacterial lawn.
- **Batch culture** of viruses generates a step curve.
- **Animal viruses** are cultured within animals or as plaques in a tissue culture.

- **Fluorescent-focus assays** reveal foci of virus-infected cells.
- **Oncogenic viruses** are cultured as foci of cancer-transformed host cells.

6.7 Viral Ecology

Viruses exist naturally within host organisms in complex ecosystems. They play critical roles mediating their host population size. Viral persistence in natural ecosystems

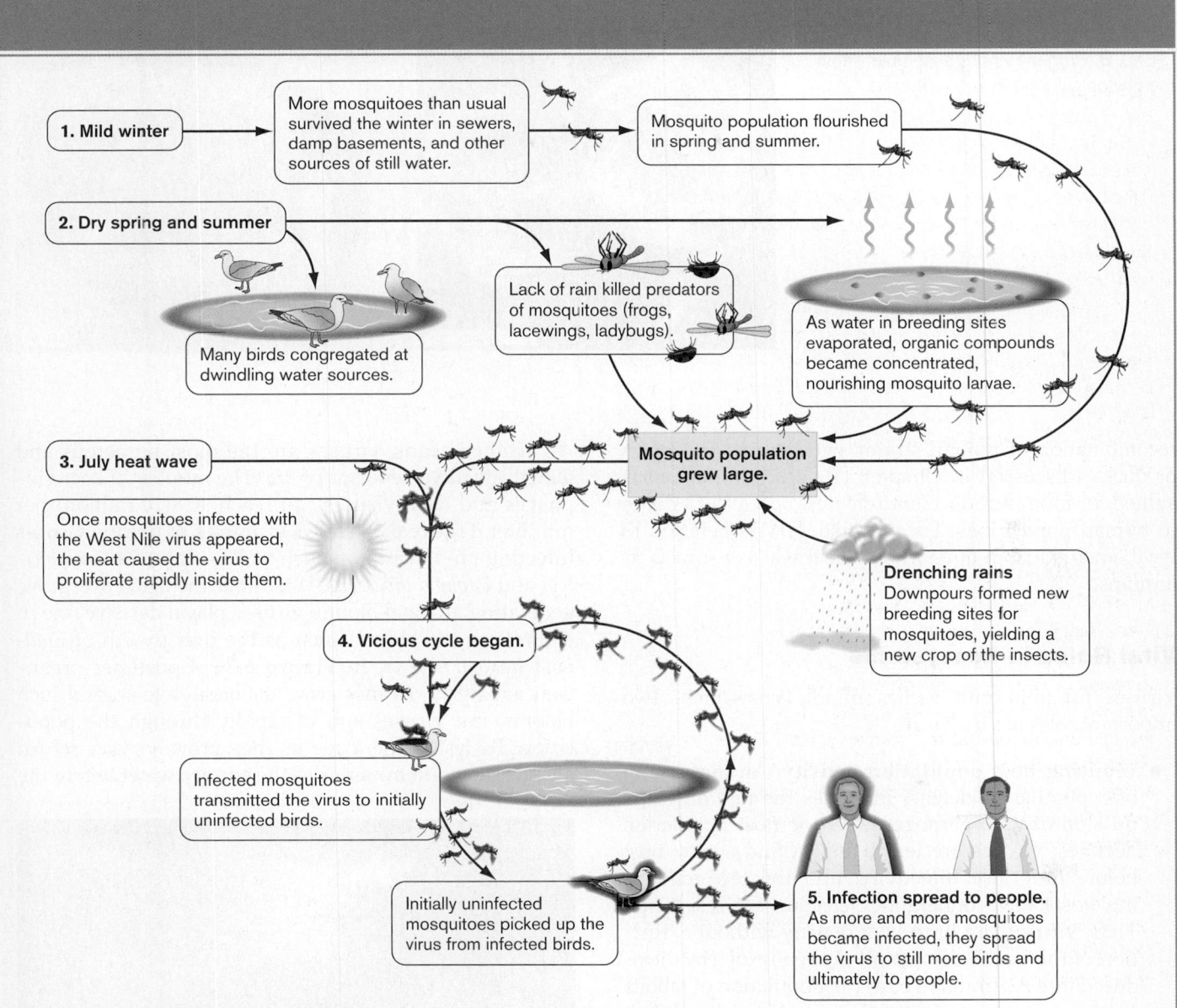

Figure 2 An emerging pathogen. In the changing global climate, an emerging pathogen threatens wildlife and people. West Nile virus, originally found in Africa and the Mediterranean, is now spreading through the Western Hemisphere, probably owing to an increase in warm weather. The spread of the virus involves transfer between mosquitoes and avian, animal, or human hosts. *Source*: Adapted from P. R. Epstein. 2000. *Sci. Am.* **283**:50–57.

impacts human health, and that of agricultural plants and animals. Persistence in wild populations provides a reservoir of infection that hinders eradication, especially when associated with an insect vector (carrier organism) such as a tick or a mosquito (see **Special Topic 6.1**).

Emergence of Viral Pathogens

Changing distribution patterns of insect vectors and animal hosts can generate new epidemics of a pathogen in regions where the virus could not spread before. Such changes in distribution can be brought about by many factors, including global climate change (**Special Topic 6.1**).

Some emerging viruses arise as variants of endemic milder pathogens. Viruses long associated with a host, such as the common-cold viruses (rhinoviruses), tend to have evolved a moderate disease state that provides ample opportunities for host transmission. A virus that "jumps" from an animal host, however, may cause a more acute syndrome with higher mortality. The best-known cases are the exceptionally virulent emerging strains of influenza, which generally result from intracellular

Figure 6.35 Viruses infect algae.
A. A virus attaches to the surface of a marine phytoplankton, *Emiliana huxleyi* (SEM). **B.** Progeny virions assemble within the phytoplankton *Pavlova vivescens*.

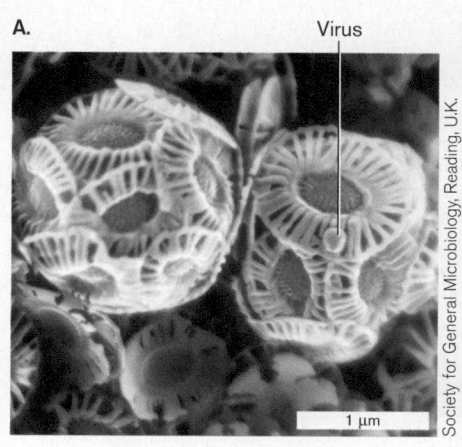

A.   Virus

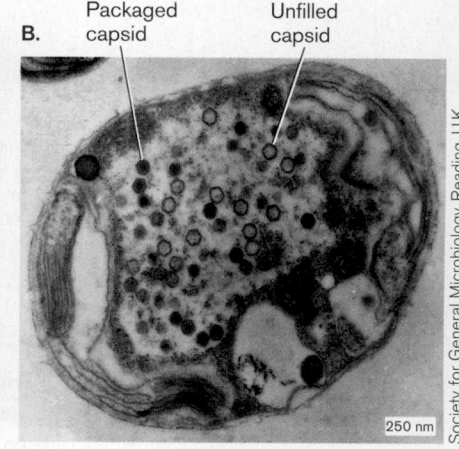

B.   Packaged   Unfilled
      capsid   capsid

recombination of human strains with strains from pigs or ducks (discussed in Chapter 11). Wild animals consumed for food introduce entirely new species of viruses to human populations. For example, HIV is believed to have emerged from apes whose meat was consumed by humans.

Viral Roles in Ecosystems

Viruses fill important niches in all ecosystems. Two important roles of viruses are

- **Limiting host population density.** An increase in host population density increases the rate of transmission of viral pathogens. As the host population declines, viruses are less likely to find a new host before they lose infectivity; and the few remaining hosts have undergone selection for resistance. Thus, viruses can limit host density without extinction of the host. An example is the myxovirus introduced into Australia to curb the population of rabbits (which had been previously introduced by British colonists). The myxovirus ultimately maintained the rabbits at a reduced density, less destructive to the ecosystem.

- **Selecting for host diversity.** Each viral species has a limited host range and requires a critical population density to sustain the chain of infection. The virus limits its host species to a population density far lower than that sustainable by the available resources. The resources then support other species resistant to the given virus (but susceptible to others). Thus, overall, viruses prevent the dominance of any one species. Viruses foster the evolution of many distinct host species. The role of viruses in diversity is exemplified in marine phytoplankton, which are highly susceptible to viruses and display an exceptional range of species diversity.

In the oceans, viruses are the most numerous and genetically diverse forms of life. The number of bacteriophages and algal viruses can reach 10^7 (10 million) per milliliter. **Figure 6.35** shows examples of marine viruses infecting phytoplankton such as the algae *Emiliana huxleyi* and *Pavlova vivescens*. When marine algae overgrow, generating an algal bloom, viruses play a decisive role in controlling the bloom, such as the overgrowth of *Emiliana huxleyi* shown in **Figure 6.36**. Consumer organisms apparently cannot grow fast enough to control such blooms, but viruses spread rapidly through the population. By lysing the algae as they grow, viruses return algal carbon and minerals to the surface water before the

Figure 6.36 Marine algal bloom controlled by viruses.
Bloom of the alga *Emiliana huxleyi* off Plymouth, England, detected by satellite remote sensing. The pale clouds in the water are the reflected light from billions of calcite plates, or "coccoliths," that coat each algal cell.

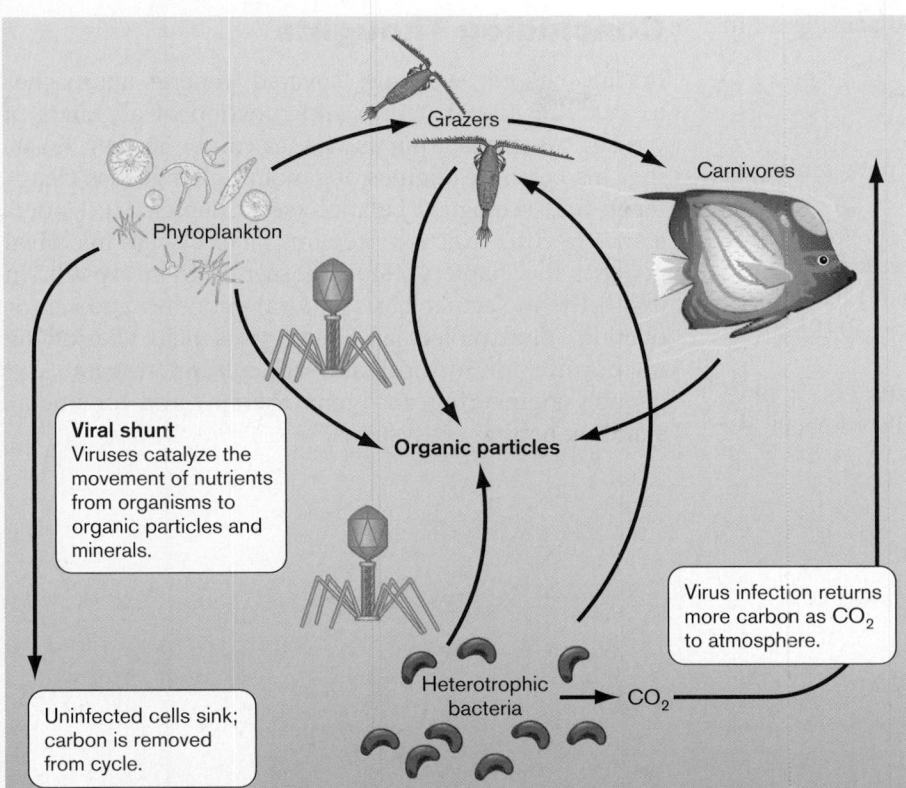

Viral shunt
Viruses catalyze the movement of nutrients from organisms to organic particles and minerals.

Uninfected cells sink; carbon is removed from cycle.

Virus infection returns more carbon as CO_2 to atmosphere.

algae starve to death and their bodies sink to less productive depths.

> **THOUGHT QUESTION 6.9** Suppose a certain virus depletes the population of an algal bloom. If some of the algae are selected for resistance, will they grow up again and dominate the producer community?

Viruses, particularly bacteriophages, are equally abundant in coastal and aquatic habitats, and they reach numbers tenfold higher in soil. In coastal waters, the populations of DNA viruses turn over daily, and they lyse 20%–100% of the bacteria, thus serving as major consumers. It was thought that biofilms on marine snow and sedimentary particles might be protected from virus infection by their polysaccharide matrix, but biofilm structures also provide receptors for viral recognition and attachment. In fact, viral infection rates for biofilm bacteria equal the infection rates of planktonic (floating) cells. Various species of viruses infect host organisms at all levels of the food chain; in a wetland, for example, besides bacteriophages infecting bacteria, viruses infect diatoms, insects, vertebrates (including fish and amphibians), and plants such as reeds and salt marsh grasses. Wetland plants may serve as reservoirs or as overwintering hosts for viruses harmful to agriculture.

A common measure of the influence of viruses in a given habitat is the virus-to-bacteria ratio. In marine water, the ratio has been estimated at values of 1–100 viruses per bacterium. Similar ratios are found in forested soil and in wetland soil. But in agricultural soil, the ratio can be as high as 1,000 viruses per bacterium. The reasons for this viral density in agricultural soil are not known, but one factor is poor soil quality. A statistical correlation shows that viral numbers in soil increase with dryness and low content of organic matter. In experimental tests, carbon- and nitrogen-containing nutrients are added to a model ecosystem. Following nutrient addition, the bacterial numbers are observed to increase rapidly, while viral increase lags. A possible explanation for the lag in virus production is that the conditions of ample water and nutrients favoring bacterial growth also favor bacteriophage lysogeny, in preference to lytic burst and release of phages.

On a global scale, viruses play a significant role in the carbon balance. Models of ocean carbon flux typically emphasize the consumption of phytoplankton by grazers and the consumption of grazers by carnivores (**Fig. 6.37**). At each level, some of the carbon fixed as biomass sinks to the ocean floor, removed from the carbon cycle. But at each level, viral infection and lysis converts the bodies of phytoplankton, grazers, and carnivores into detritus consisting of small organic particles. The organic material is consumed by bacteria, whose rapid respiration releases much CO_2. Thus, carbon is diverted from the ocean sink into CO_2 returning to the atmosphere—a factor that must be considered in models of global warming.

TO SUMMARIZE:

- **Environmental change** results in new emerging viral pathogens.
- **Host population density** is limited by virus infection, and host genetic diversity is increased.
- **Virus-to-bacteria ratios** range from 10–100 for marine and aquatic environments, and 10–1,000 for soil.
- **Marine viruses** are the major consumers of phytoplankton. Viral activity substantially impacts the global carbon balance.

Concluding Thoughts

In this chapter we have covered general approaches to studying the structure and function of all kinds of viruses. As devastating as viruses can be to their hosts, they also provide engines of genomic change (see Chapter 9) and ecological balance (see Chapter 21). Understanding virus function prepares us to discuss microbial genetics in Chapters 7–10. Chapter 11 then explores in depth the molecular basis of viral genomes and reproduction. Viral molecular biology is a field of growing importance for medical and agricultural research, for genetic engineering and gene therapy, and for understanding natural ecosystems.

CHAPTER REVIEW

Review Questions

1. Compare and contrast the form of icosahedral and filamentous (helical) viruses, citing specific examples.
2. How do viral genomes gain entry into cells in bacteria, plants, and animals?
3. Explain the structure and function of the seven Baltimore groups of viral genomes.
4. How do viral genomes interact with host genomes, and what are the consequences for host evolution?
5. Compare and contrast the lytic, lysogenic, and slow-release life cycles of bacteriophages. What are the strengths and limitations of each?
6. Compare and contrast the life cycles of RNA viruses and DNA viruses in animal hosts. What are the strengths and limitations of each?

7. Explain the plate titer procedure for enumerating viable bacteriophages. How must this procedure be modified to titer animal viruses? Oncogenic viruses?
8. Explain how a pure isolate of a virus can be obtained. How do the procedures differ from those used for isolating bacteria?
9. Explain the generation of the step curve of virus proliferation. Why is virus proliferation generally observed as a single step, or generation, in contrast to the life cycles of cellular microbes, outlined in Chapter 4?
10. Explain the key contributions of viruses to natural ecosystems. What may happen in an ecosystem where viruses are absent or fail to cause significant infection?

Thought Questions

1. Discuss the functions of different structural proteins of a virion, such as capsid, nucleocapsid, tegument, and envelope proteins. How do these functions compare and contrast with functions of cellular proteins?
2. Explain two different ways that viruses may cause cancer (oncogenesis). How can strongly oncogenic viruses be assayed in culture?

3. What are the relative advantages of the virulent phage life cycle of phage T4, the lysis-lysogeny options of phage lambda, and the slow-release life cycle of phage M13? Under what conditions might each strategy be favored over the others?
4. Given the basis of viral tropism, how might an animal evolve to become resistant to a virus infection?

Key Terms

bacteriophage (phage) (183)
batch culture (208)
burst size (198, 208)
capsid (183)
cell surface receptor (197)
cloning vector (184)
CRISPR (200)
DNA reverse-transcribing virus (193)
eclipse period (208)
endocytosis (201)
envelope (188)
filamentous virus (188)
fluorescent-focus assay (211)
focus (211)
host range (183)
icosahedral capsid (187)
latent period (208)
lentivirus (209)
lysate (209)

lysis (198)
lysogeny (198)
lytic cycle (198)
mechanical transmission (205)
multiplicity of infection (MOI) (208)
oncogene (205)
oncogenic virus (205)
orthologous gene (ortholog) (196)
pararetrovirus (193)
plaque (183, 210)
plaque assay (210)
plaque-forming unit (PFU) (211)
plasmodesma (206)
prion (185)
prophage (198)
proteome (196)
proteomics (196)
reading frame (184)
retrovirus (193)

reverse transcriptase (193)
rise period (208)
RNA-dependent RNA polymerase (193)
RNA reverse-transcribing virus (193)
segmented genome (191)
site-specific recombination (198)
spike protein (188)
tegument (188)
temperate phage (198)
transduction (198)
transform (212)
transformed-focus assay (212)
tropism (201)
uncoating (201)
viral envelope (188)
virion (183)
viroid (185)
virulence (209)
virus (182)

Recommended Reading

Anyamba, Asaf, Jean-Paul Chretien, Jennifer Small, Compton J. Tucker, Pierre B. Formenty, et al. 2009. Prediction of a Rift Valley fever outbreak. *Proceedings of the National Academy of Sciences USA* **106**:955–959.

Edwards, Robert A., and Forest Rohwer. 2005. Viral metagenomics. *Nature Reviews. Microbiology* **3**:504–510.

Grünewald, Kay, Prashant Desai, Dennis C. Winkler, J. Bernard Heymann, David M. Belnap, et al. 2003. Three-dimensional structure of herpes simplex virus from cryo–electron tomography. *Science* **302**:1396–1398.

Harris, Audray, Giovanni Cardone, Dennis C. Winkler, J. Bernard Heymann, Matthew Brecher, et al. 2006. Influenza virus pleiomorphy characterized by cryoelectron tomography. *Proceedings of the National Academy of Sciences USA* **103**:19123–19127.

Horvath, Phillippe and Randolphe Barrangou. 2010. CRISPR/Cas, the immune system of bacteria and archaea. *Science* **327**:167–170.

Hull, Roger. 2001. *Matthews' Plant Virology*. Academic Press, New York.

Moineau, S., D. Tremblay, and S. Labrie. 2002. Phages of lactic acid bacteria: From genomics to industrial applications. *ASM News* **68**:388–391.

Ptashne, Mark. 2004. *Genetic Switch: Phage Lambda Revisited*. Cold Spring Harbor Laboratory Press, Cold Spring Harbor, NY.

Raoult, Didier, Stéphane Audic, Catherine Robert, Chantel Abergel, and Patricia Renesto. 2004. The 1.2-megabase genome sequence of Mimivirus. *Science* **306**:1344–1350.

Simon J. Labrie, Julie E. Sampson, and Sylvain Moineau. Bacteriophage resistance mechanisms. 2010. *Nature Reviews. Microbiology* **8**:317–327.

Srinivasiah, Sharath, Jaysheel Bhavsar, Kanika Thapar, Mark Liles, Tom Schoenfeld, et al. 2008. Phages across the biosphere: Contrasts of viruses in soil and aquatic environments. *Research in Microbiology* **159**:349–357.

Suttle, Curtis A. 2007. Marine viruses—major players in the global ecosystem. *Nature Reviews. Microbiology* **5**:801–812.

Wilson, Willie. 2002. Giant algal viruses: Lubricating the great engines of planetary control. *Microbiology Today* **29**(November):180.

Xiao, Chuan, Yurii G. Kuznetsov, Siyang Sun, Susan L. Hafenstein, Victor A. Kostyuchenko, et al. 2009. Structural studies of the giant Mimivirus. *PLoS Biology* **7**:e1000092.

 For further study and review, please visit StudySpace at **microbiology2.com**.

Part 2

Genes and Genomes

AN INTERVIEW WITH

CHRISTINE JACOBS-WAGNER: THE THRILL OF DISCOVERY IN MOLECULAR MICROBIOLOGY

Christine Jacobs-Wagner has taught for 8 years at Yale University. With her students and postdocs, she has made striking discoveries by studying the bacterium *Caulobacter crescentus*, which generates two very different cell forms during each round of division. As a graduate student at the University of Liège, Belgium, she won the General Electric & *Science* Prize for her work on beta-lactam antibiotic resistance and cell wall sensing in Gram-negative bacteria. At Yale, she was designated an investigator of the Howard Hughes Medical Institute, one of the nation's largest medical philanthropic organizations.

Christine Jacobs-Wagner, Professor of Molecular, Cellular and Developmental Biology, Yale University; Investigator of the Howard Hughes Medical Institute.

Why did you decide to study microbiology?

Because microorganisms are fascinating creatures! The more I learn about them, the more I marvel. It often seems as if there isn't a thing microorganisms cannot do. They are extremely sensitive and responsive to their physical and chemical environment. They can multiply at a blazing speed and yet they replicate and segregate their genetic material with an exquisite temporal and spatial accuracy. Some of them can thrive under ridiculously harsh conditions, in Antarctic ice or boiling-hot deep-sea vents. Bacteria like *Deinococcus radiodurans* can withstand 500 times more gamma radiation than what could kill a human!

Microbes have been studied for a very long time and we have learned a great deal; yet they are still full of surprises. For example, bacteria are more than tiny mixed bags of molecules. They exhibit a remarkably elaborate internal organization, which has revolutionized the way we view and study bacteria. For instance, we have shown that *Caulobacter* localize key signaling proteins at opposite ends of the cell to monitor the progression of cell division. This can be important for cell cycle regulation and the initiation of different developmental programs in daughter cells. Protein localization at the cell ends can also be critical for many important bacterial processes, including chromosome segregation, cell division, motility, and pathogenesis, just to name a few.

How did the General Electric & *Science* Prize influence the development of your career?

This is a prize that recognizes the contribution of graduate students and it is wonderful that GE & *Science* and other foundations acknowledge their work in the form of a prize. More than anything, the GE & *Science* prize helped me realize I was doing work that others found important and interesting. It gave me confidence. My graduate career was unusual in that I spent time in five different labs in four different countries, each time learning the unique expertise that each lab offered; for example, biochemical characterization in Jean-Marie Frère's lab in Belgium, and transcriptional regulation in Staffan Normark's lab in Sweden.

How did you choose to study *Caulobacter crescentus*?

I was drawn to *Caulobacter* because of its tractability to study the bacterial cell cycle, given the ease of obtaining cell populations that are synchronized with respect to the cell cycle. The ability to synchronize cell populations is very important; it is a key reason why the budding yeast has contributed so much to our understanding of the eukaryotic cell cycle. Furthermore, *Caulobacter* is an excellent model to study both cellular polarization and development because its cell cycle is coordinated with a highly polarized developmental program that culminates with an asymmetrical division and the generation of daughter cells with different cell fates. Applied and environmental microbiologists are also developing *Caulobacter* as a living detector of various toxic compounds and as a multipurpose bioremediation agent.

Why do bacteria have a cytoskeleton? How does it relate to the cytoskeleton of animal cells?

I think that bacteria have a cytoskeleton for many of the same reasons that our own human cells do: to organize their cytoplasm, to govern their cell shape and size, and to generate force (for example, for DNA segregation). These activities are important for cellular life regardless of its origin, and a cytoskeleton seems like a good tool to perform these important functions.

Some elements of the bacterial cytoskeleton, such as actin and

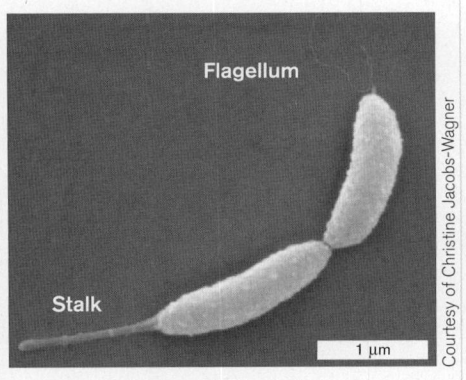

Caulobacter crescentus (SEM) undergoing cell division to form stalked cell and flagellated cell.

tubulin homologs, are evolutionarily linked to the animal cytoskeleton.

What current research question are you pursuing, and how do students participate?

We study how cellular organization is achieved and regulated in bacteria. In particular, we are interested in the spatiotemporal mechanisms involved in cell division, chromosome segregation, cell cycle regulation, and cell morphogenesis. Undergraduate and graduate students are an integral part of our research team. They each have an independent research project, which allows them to explore their ideas and use their creativity. Yet each student's project is sufficiently linked to others in the lab to foster interactions and intellectual exchange among lab members. One of the beauties of working with a highly tractable bacterial model is that students can learn many techniques, test numerous ideas, and explore different directions in a reasonable amount of time.

Last year, my postdoc Sebastian Poggio and my undergraduate student Constantin (Nick) Takacs took advantage of Nick's school break to spend a couple of weeks in Professor Waldemar Vollmer's lab at Newcastle University in England. Within that short amount of time, Nick and Sebastian, with the help of Professor Vollmer and one of his students, fully characterized the complex composition and structure of the *Caulobacter* peptidoglycan cell wall. They brought back their newly acquired expertise in high-performance liquid chromatography and mass spectrometry here to Yale so that now our lab can do this type of technically demanding work in-house. Everybody wins!

What genetic and molecular tools do you use? How does your research combine genetics and advanced microscopy?

We often cannot anticipate what approach will be most successful, so we try to use every tool at our disposal, at times simultaneously, to achieve our goal. For example, Nora Ausmees, as a postdoctoral fellow in my lab, led the discovery of bacterial intermediate filaments by visually screening *Caulobacter* mutants with cell shape defects. Intermediate filaments are major cytoskeletal components of human cells, and over 30 human diseases (such as progeria and some cardiomyopathies and lipodystrophies) have been associated with their malfunction. Yet little is known about their basic properties, such as their molecular function, assembly, and regulation within cells. *Caulobacter* now offers a highly tractable experimental model system to study intermediate-filament biology.

Does your work have implications for the developmental cycle of human pathogens?

Many of the developmental and cell cycle pathways that we and others study are essential for viability and are present in close relatives of *Caulobacter*, including dangerous pathogens (for example, *Brucella* and *Rickettsia* species, which cause a variety of diseases, including brucellosis and typhus). These pathogens are far less tractable, and *Caulobacter* research has stimulated exciting, ongoing research on how the pathways we study affect the cell cycle and development of these pathogens in their hosts.

What advice do you have for today's students?

I would strongly recommend that students spend some time in a microbiology lab. Whatever they aspire to do in the future, this experience will give them a better understanding of the world they live in. Microbes are

C. crescentus cells; intermediate filaments are labeled by pink fluorescence.

virtually omnipresent, and they are indispensable to our well-being and prosperity, even while some pose threats to our health and economy. Doing some microbiological research is also one of the best ways to learn hypothesis-driven reasoning and to develop skills in critical thinking. The speed at which microbes grow and can be manipulated allows students to design experiments and test hypotheses even when time is limited.

For students who already work in a lab, I suggest that they explore the possibility of visiting another lab with distinct expertise to learn a new approach that can directly help their research project. I did this when I was an undergraduate student, a graduate student, and a postdoctoral fellow. Each experience generated a huge boost in my research; enhanced my skill set; created new, long-lasting friendships; and generated connections to important people who have since supported my career.

How does your family relate to your work? Do you have interesting pursuits outside of science?

My mother has been a constant source of support throughout my life, which has helped a lot. The daily support comes from my husband, Matt, who is incredibly enthusiastic about my career. He is not a scientist (he is an e-learning developer), yet he plays an active role in my work that ranges from building and maintaining our lab website to being my top adviser on nonscientific issues, such as running a lab or raising money. Having someone close who listens and gives good input has been extremely valuable.

Chapter 7
Genomes and Chromosomes

A genome is all the genetic information that defines an organism. For over a hundred years, we have known that chromosomes consist of DNA, yet we have only recently been able to sequence complete genomes. Now we are in the postgenomic era in which biological research is driven, in large part, by knowledge of complete genome sequences.

Microbial genomes consist of one or more DNA chromosomes. Many bacteria have a single, circular chromosome, but some species have multiple chromosomes or even a mixture of linear and circular chromosomes. For example, the genome of *Sinorhizobium meliloti*, a major contributor to nitrogen fixation, consists of three circular chromosomes.

This chapter addresses basic as well as some advanced issues of DNA and genomes. What is DNA? How is it packaged in the cell? Because DNA is essential to life, nature must have devised a remarkable machine to replicate it. But how does this machine work? And do different species use different machines? Microbiology has played a leading role in answering these difficult questions—opening the door to postgenomic studies of all living things.

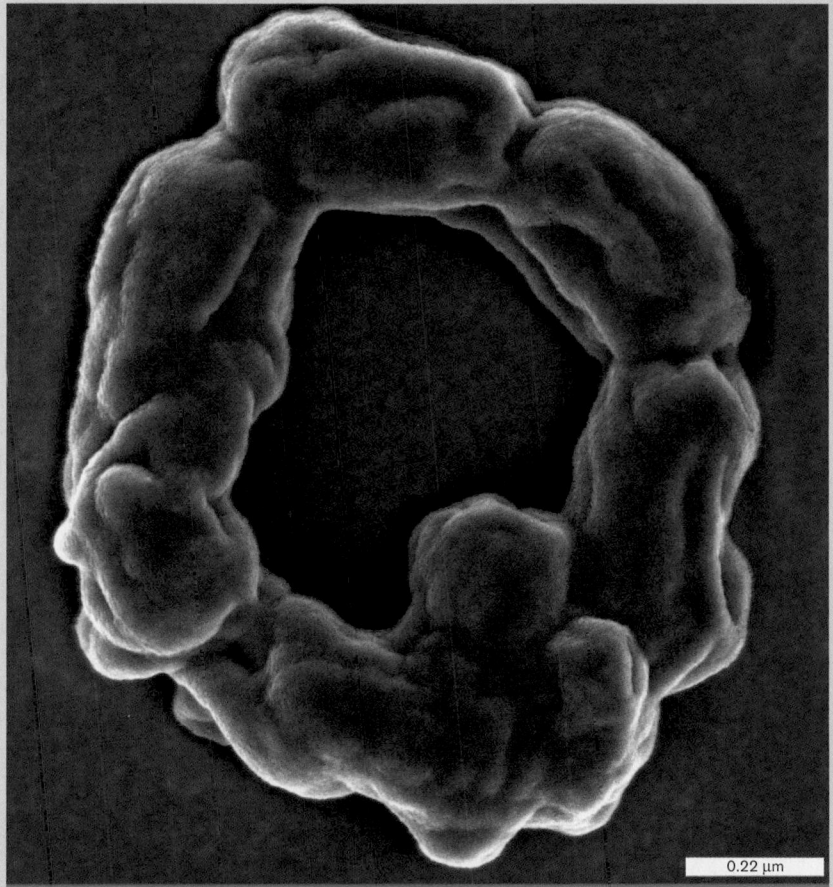

0.22 μm

Current Research Highlight

When cell division goes awry. Before a bacterial cell can divide, its genome must replicate completely. If you stop replication, you stop cell division. In contrast, the genome can continue to replicate even if cell division stops. For example, a defect in FtsZ which forms a ring at midcell to direct formation of the division septum will halt cell division but *not* DNA replication. Normally dividing bacteria appear as typical single cells. However, this pseudocolored scanning electron micrograph shows what happens to *Coxiella burnetii*, the cause of human Q fever, when FtsZ is defective. The genome continues to replicate, but the cell becomes filamented with multiple, incomplete division septa—in this case, forming the letter "Q." *Source:* ©Bryan Hansen, Rocky Mountain Laboratories Microscopy Unit from Paul A. Beare et al. 2009. *J. Bacteriol.* (cover image) **191**:1369–1381.

What makes each species unique is its genome, the sum total of its genes. The human genome is also supplemented by the genomes of all the microbes in the digestive tract. Collectively, intestinal microbes contain 100 times more genes than the human genome and provide key metabolic capabilities that humans lack, such as digesting complex plant materials. In the mid–twentieth century, one intestinal bacterium, *Escherichia coli*, became the focus of efforts to understand genes and genetics. The choice of *E. coli* was fortunate, as its genetics proved to be especially pliable.

In one remarkable experiment, a cell of *E. coli* was lysed, releasing its chromosome for electron microscopy. What spewed out of this single cell was a strand of DNA 1,500 times longer than *E. coli* itself. Scientists wondered how this enormous molecule could fit into a single cell, and how an enzyme could duplicate that much DNA (over 4.6 million base pairs) without the whole DNA molecule getting tangled up (**Fig. 7.1**). And how did it manage to do this in under 20 minutes, the doubling time of the organism? Half a century later, those questions continue to intrigue us.

Chapter 7 explores the nature of DNA, the structure of prokaryotic genomes, and the mechanisms used for their replication. We also examine how microbial enzymes that manipulate DNA were used to develop the fundamental techniques of present-day biotechnology, such as DNA restriction analysis, DNA sequencing, and the polymerase chain reaction (PCR). The expression of genes to make RNA and protein is discussed in Chapter 8. The remaining chapters of Part 2 present gene recombination and transmission (Chapter 9), the regulation of gene expression (Chapter 10), the specialized genetic mechanisms of viruses (Chapter 11), and the use of microbial genetic tools for research and biotechnology (Chapter 12).

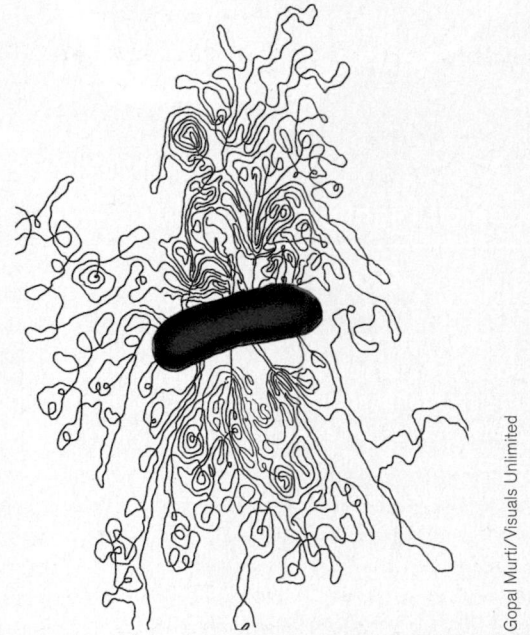

Gopal Murti/Visuals Unlimited

Figure 7.1 Osmotically disrupted bacterial cell with its DNA released.

From the advances of the last 50 years, a new molecular and genomic perspective of biology has emerged.

7.1 DNA: The Genetic Material

The ability of plants and animals to transfer genetic traits was known well before we knew that DNA was the carrier of genetic information. Early researchers understood the Mendelian model of gene inheritance and how it dictated the passage of traits from parent to offspring in eukaryotes. This type of gene transfer is known as **vertical transmission**, as from parent to child. Bacteria, however, appeared to operate differently, because in addition to vertical transmission, they are able to conduct **horizontal transmission**—the transfer of small pieces of genetic information from one cell into another. The study of horizontal gene transfer in bacteria ultimately led to the discovery of horizontal transfer throughout animals and plants, albeit on a slower timescale because of their longer generation times. Mechanisms of horizontal transmission are discussed in Chapter 9.

Our path to understanding DNA began in 1928 with the discovery by Fred Griffith (1881–1941) that some unknown substance from dead cells of a virulent strain of *Streptococcus pneumoniae* could transfer virulence to living but otherwise harmless strains when both strains were coinjected into mice. This form of horizontal gene transfer, called transformation, led to the discovery in 1944 that genetic information is embedded in the base sequence of deoxyribonucleic acid (DNA) (see Section 1.6). A more contemporary example of transformation is the toxic shock gene that can be moved horizontally among strains of *Staphylococcus aureus*. As a result of these and many other experiments, we now appreciate that chromosomes are made of contiguous packets of information, called genes.

Genes are units of information composed of a sequence of DNA nucleotides of four different types: adenine (A), guanine (G), thymine (T), and cytosine (C). (For review at the level of introductory biology, see Appendix 1.) A **structural gene** is a string of nucleotides that can be decoded by an enzyme to produce a functional RNA molecule. A structural gene usually produces an RNA molecule that in turn encodes a protein. A **DNA control sequence**, on the other hand, regulates the expression of a structural gene. DNA control sequences do not encode RNA but regulate the expression (RNA production) of an adjacent structural gene, for example, by binding a repressor protein (discussed in Chapter 8). DNA control sequences are not really genes, since they do not encode RNA or protein. Control sequences include the promoters that signal the start of RNA synthesis from a structural gene and the binding sites for regulatory proteins that can activate or inactivate that promoter.

As noted previously, the entire genetic complement of DNA in a cell that defines it as an organism is called its **genome**. Our primary goal in this chapter is to convey what is known about how genomes are maintained and replicated. Mining the informational content of DNA sequences and genes is described more fully in Chapters 8 and 10.

Bacterial Genomes

In the early twentieth century, the chromosomes of bacteria, unlike those of eukaryotes, could not be observed by light microscopy. The reason is that bacteria, unlike eukaryotes, do not undergo mitosis, a process in which chromosomes are condensed and thickened about 1,000-fold. Important clues to bacterial chromosome structure were, however, gleaned from painstaking genetic studies. In the 1950s, it was discovered that **conjugation**, a horizontal gene transfer mechanism requiring cell-to-cell contact, could transfer large segments of some bacterial chromosomes—not all at once, as in the established Mendelian model of plants and animals, but sequentially over a period of time (it takes 100 minutes to move the entire *E. coli* chromosome from one cell to another).

Thus, even though the bacterial chromosome could not be seen, conjugation allowed genes to be mapped relative to each other according to time of transfer. For example, a donor strain that can synthesize the amino acids alanine and proline can directly transfer the encoding genes to a recipient cell defective in those genes. The transfer process is nonspecific; any gene can be transferred in this way. Completion of transfer also requires **recombination**, in which the donor DNA fragment replaces the recipient DNA fragment. Successful transfer of the genes for amino acid synthesis enables the formerly defective recipient to form colonies on minimal media lacking either amino acid. But because transfer occurs over time from a fixed starting point (that is, not all genes are transferred at the same time), it might take one gene 10 minutes of cell contact to be transferred while the second gene, farther away from the starting point of DNA transfer, takes an additional 20 minutes.

Knowing that eukaryotic chromosomes are linear, scientists initially expected that bacterial chromosomes would be linear too. However, the early genetic maps drawn from conjugation experiments just would not fit together in a manner consistent with a linear model—for the simple reason that the bacterial chromosome in *E. coli*, the organism under study at the time, is circular. We now know that many microbes have circular chromosomes, although some, like the Lyme disease agent *Borrelia burgdorferi*, have linear chromosomes and others, like *Agrobacterium tumefaciens*, have a mixture of circular and linear chromosomes (**Table 7.1**).

It is instructive to compare genome sizes across the phylogenetic tree, from viruses to humans. The size range is enormous. In general, the simpler the organism, the smaller its genome. At some point, as DNA content is trimmed, the organism loses independence and *must* parasitize another organism, thus begging the question, What constitutes a "minimal" genome? How "lean" can a chromosome become and still support independent growth?

TO SUMMARIZE:

- **A genome** is all the genetic information that defines an organism.
- **Genomes** of bacteria and archaea are made up of chromosomes and plasmids consisting of DNA.
- **Chromosomes** of bacteria and archaea can be circular or linear, as can plasmids.
- **Functional units** of DNA sequences include structural genes and regulatory sequences.

7.2 Genome Organization

New techniques for constructing physical maps of genomes and determining the sequences of whole genomes have revealed tremendous diversity in the size and organization of prokaryotic genomes.

Genomes Vary in Size

Bacterial chromosomes range from 580 to 9,400 kilobase pairs (kb), and archaeal chromosomes range from 935 to 6,500 kb. For comparison, eukaryotic chromosomes range from 2,900 kb (microsporidia) to over 4,000,000 kb (humans).

NOTE: The designation "kb" can refer to the length of a double-stranded or single-stranded DNA molecule. A bacterial genome is, by definition, double-stranded. Some viral genomes can be single-stranded.

One of the smallest cellular genomes sequenced thus far is that of *Mycoplasma genitalium*. These pathogens rely on their host environment for many products but can still grow outside a host cell. The complete genome of *M. genitalium* consists of only 580 kb and encodes 480 proteins. It lacks the genes required for many biosynthetic functions. These organisms cannot synthesize amino acids or construct cell walls, and they do not have a functional tricarboxylic acid (TCA) cycle. In contrast, free-living bacteria that can grow in soil have larger genomes and dedicate many genes to the synthesis or acquisition of amino acids or TCA cycle intermediates. Even different strains of one species, such as *Salmonella enterica*, may vary considerably in gene distribution (**Fig. 7.2**).

NOTE: Current *Salmonella* nomenclature lumps many former species into one—*Salmonella enterica*—and downgrades former species names to serovar status. The serovar names are not italicized, and their first letters are capitalized (for example, *S. enterica* serovar Typhi, sometimes written *S.* Typhi).

Table 7.1 Genomes of representative bacteria and archaea.

Species (strain)	Genomic chromosome(s)[a] (kilobase pairs, kb)	Plasmid(s)[a] (kb)	Total (kb)
Bacteria	**Circular and linear**	**Circular**	
Mycobacterium tuberculosis Tuberculosis	4,400		4,400
Mycoplasma genitalium Normal flora, human skin	580		580
Burkholderia cepacia	3,870 + 3,217 + 876	93	8,056
Escherichia coli K-12 (W3110) Model strain for *E. coli* proteomics	4,600		4,600
Anabaena species (PCC 7120) Cyanobacteria: major photosynthetic producer of carbon source for aquatic ecosystems	6,370	110 + 190 + 410	7,080
Borrelia burgdorferi Lyme disease	911	21 plasmids with sizes between 9 and 58	>1,250
Agrobacterium tumefaciens Tumors in plants; genetic engineering vector	2,840 + 2,070	214 + 542	5,666
Archaea			
Methanocaldococcus jannaschii Methanogen from thermal vent	1,660	16 + 58	1,734
Haloarcula marismortui Halophile from volcanic vent	3,130 + 288	33 + 33 + 39 + 50 + 155 + 132 + 410	4,270
	1,000 kb	500 kb	

[a]Purple circles and lines indicate relative sizes of genomic elements and whether these are circular or linear. Size bars are provided under each column.

Another feature that distinguishes bacterial and archaeal genomes from those of eukaryotes is the amount of so-called noncoding DNA. Many, but not all, eukaryotes contain huge amounts of noncoding DNA scattered between genes. In some species (such as humans), over 90% of the total DNA is noncoding. Some noncoding regions include **enhancer** sequences needed to drive transcription of eukaryotic promoters and DNA expanses that separate enhancers. Enhancer sequences can function at large distances from the gene they regulate. A

promoter is the DNA sequence immediately in front of a gene that is needed to activate the gene's expression. Most of the noncoding spacers appear to be remnants of genes lost over the course of evolution and pieces of defunct viral genomes. Noncoding regions may, however, provide raw material for future evolution.

In contrast to many eukaryotes, bacteria tend to have very little noncoding DNA (typically less than 15% of the genome). Archaeal genomes have a few genes that contain noncoding DNA (introns) resembling introns of

Figure 7.2   The genome of *Salmonella enterica* serovar Typhimurium LT2.   The circular chromosome. Base pairs (bp) are indicated around the perimeter. The two outer multicolored circles indicate genes encoded on the separate strands, which means they are transcribed in opposite directions. Color codings represent homologies of eight compared species. Blue indicates genes present only in *Salmonella* Typhimurium LT2. Orange indicates that the gene has a close homolog in all eight genomes compared. Green indicates genes with a close homolog in at least one other *Salmonella* (*S.* Typhi, *S.* Paratyphi A, *S.* Paratyphi B, *S. arizonae*, or *S. bongori*) but not in *E. coli* K-12, *E. coli* O157:H7, and *Klebsiella pneumoniae*. Gray indicates other combinations. The black inner circle is the GC content of the LT2 DNA (peaks pointing outward indicate GC-rich areas).

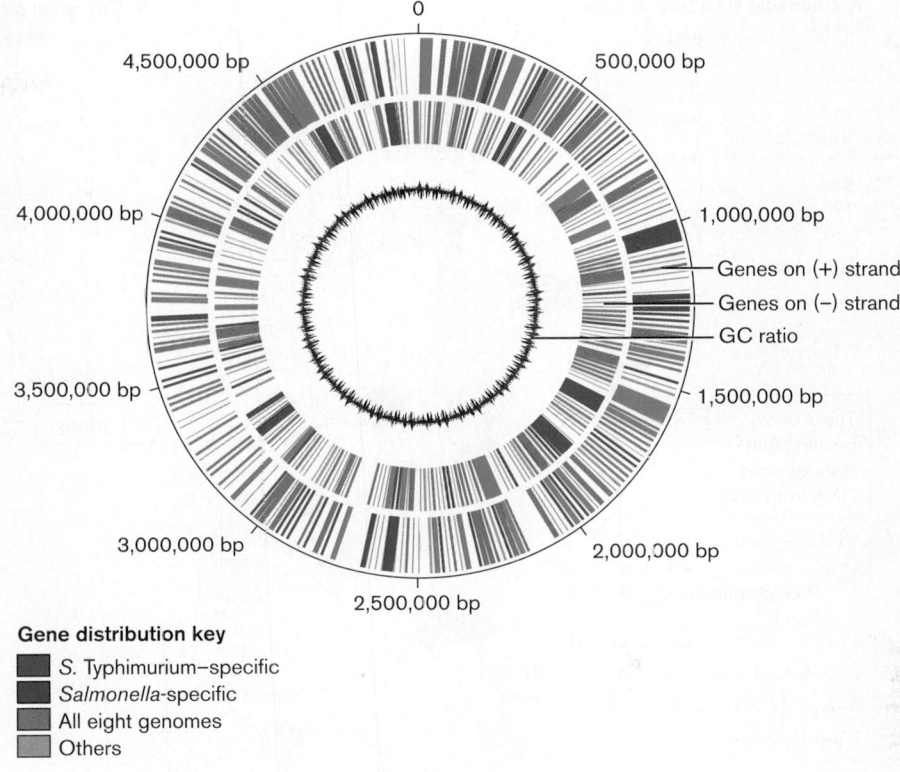

Gene distribution key
- ■ *S.* Typhimurium–specific
- ■ *Salmonella*-specific
- ■ All eight genomes
- ■ Others

eukaryotes, but overall they contain much less noncoding DNA. The completed genome sequence of the bacterium *Mycobacterium leprae*, however, has a substantial amount (about 50%) of DNA with no known or predicted function, and this noncoding DNA may represent degenerate or inactivated genes. Still, this is less than the percentage of noncoding DNA seen in most eukaryotes.

www | Genome sizes
microbiology2.com/links

Functional Units of Genes

In the simplest case, a gene can stand alone, operating independently of other genes. The RNA produced from a stand-alone gene is said to be monocistronic, coding for one protein. Alternatively, a gene may exist in tandem with other genes in a unit called an **operon** (**Fig. 7.3A**). All genes in an operon are situated head to tail on the chromosome and are controlled by a single regulatory sequence located in front of the first gene. The single RNA molecule produced from the operon contains all the information from all the genes in that operon.

On a functional level, a collection of genes and operons at different positions on the chromosome form a **regulon** when they have a unified biochemical purpose (such as amino acid biosynthesis) and are regulated by the same regulatory protein (**Fig. 7.3B**). The various

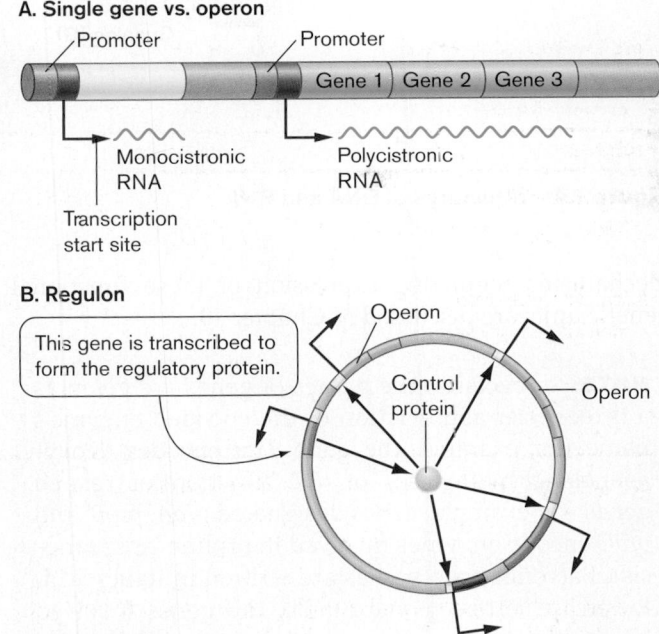

Figure 7.3   Gene organization.   A. Diagram of a segment of DNA containing a single gene producing a monocistronic message (codes for one protein) and an operon of three genes producing a polycistronic message (from which three different proteins are made, one corresponding to each gene). **B.** Diagram of a circular bacterial genome, containing genes and operons coordinately controlled by a regulatory protein.

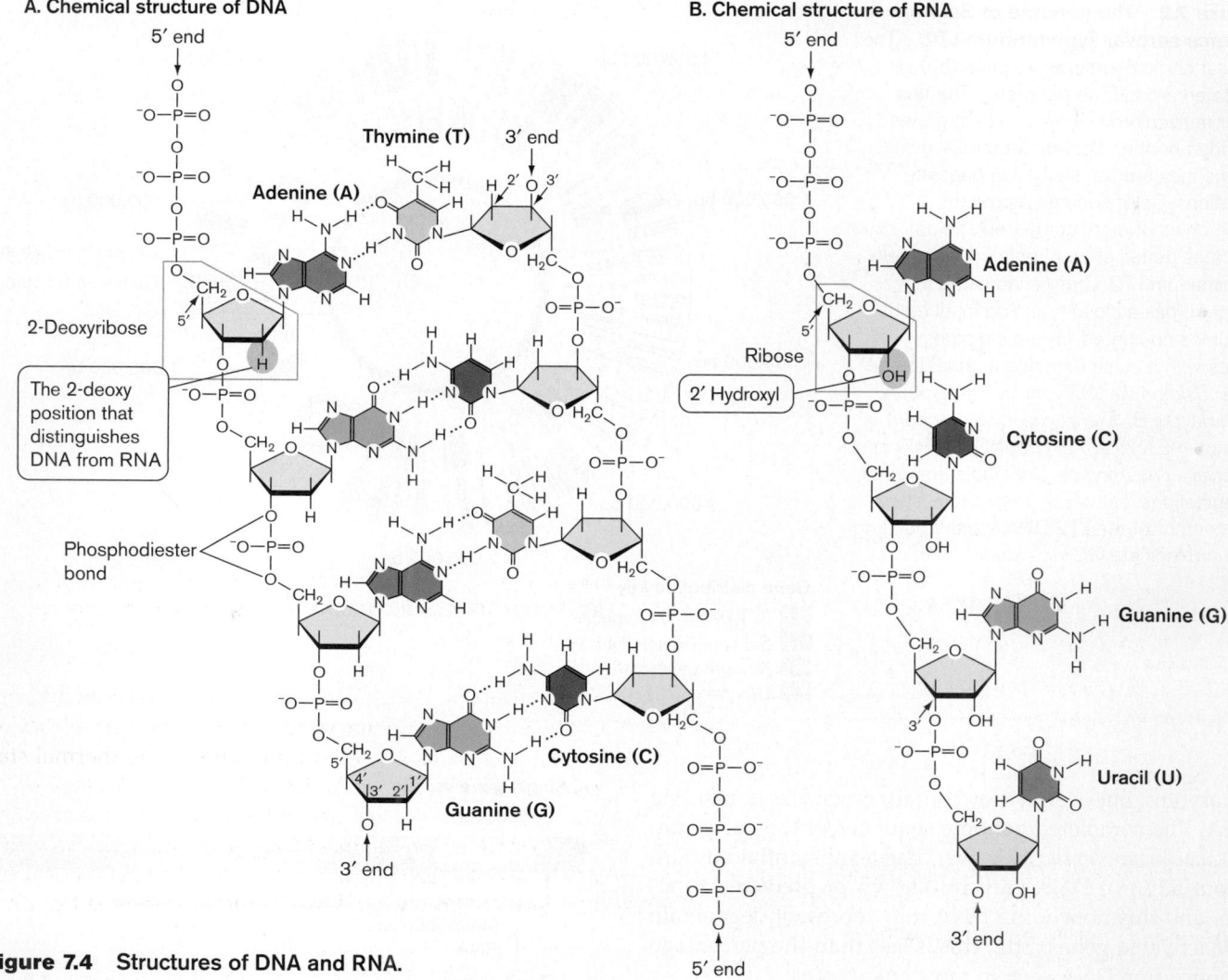

A. Chemical structure of DNA

2-Deoxyribose

The 2-deoxy position that distinguishes DNA from RNA

Phosphodiester bond

5′ end

Thymine (T)   3′ end

Adenine (A)

3′ end

B. Chemical structure of RNA

5′ end

Ribose

2′ Hydroxyl

Adenine (A)

Cytosine (C)

Guanine (G)

Uracil (U)

3′ end

5′ end

Figure 7.4   Structures of DNA and RNA.

mechanisms regulating expression of these functional genetic units are discussed in Chapter 10.

WWW | Microbial Genome Database
microbiology2.com/links

NOTE: In bacteria, the names of genes are given as a three-letter abbreviation of the encoded enzyme's name (for example, the gene *dam* encodes <u>d</u>eoxy-<u>a</u>denosine <u>m</u>ethylase) or the function of related genes (for example, genes designated *proA*, *proB*, and *proC* encode enzymes involved in <u>pro</u>line biosynthesis). Bacterial gene names are written in italics with lowercase letters. For example, the genes involved in catabolizing lactose are the *lac* genes. If several genes are involved in the pathway, a fourth letter, capitalized, is used. Thus, the three genes *lacZ*, *lacY*, and *lacA* are all associated with lactose catabolism. When speaking of a gene product, a nonitalic (roman) font is used, and the first letter is capitalized. Thus, *lacZ* is the gene, and LacZ is the protein product of that gene.

DNA Function Depends on Chemical Structure

DNA is composed of four different nucleotides linked by a phosphodiester backbone (**Fig. 7.4A**). Each nucleotide consists of a nitrogenous base attached through a ring of nitrogen to carbon 1 of 2-deoxyribose in the phosphodiester backbone. The 2-deoxy position that distinguishes DNA from RNA is circled in the figure. A **phosphodiester bond** (marked) joins adjacent deoxyribose molecules in DNA to form the phosphodiester backbone. Phosphodiester bonds link the 3′ carbon of one ribose to the 5′ carbon of the next ribose. The two backbones are **antiparallel** so that at either end of the DNA molecule, one strand ends with a 3′ hydroxyl group and the opposite strand ends with a 5′ phosphate. This antiparallel arrangement is necessary so that complementary bases protruding from

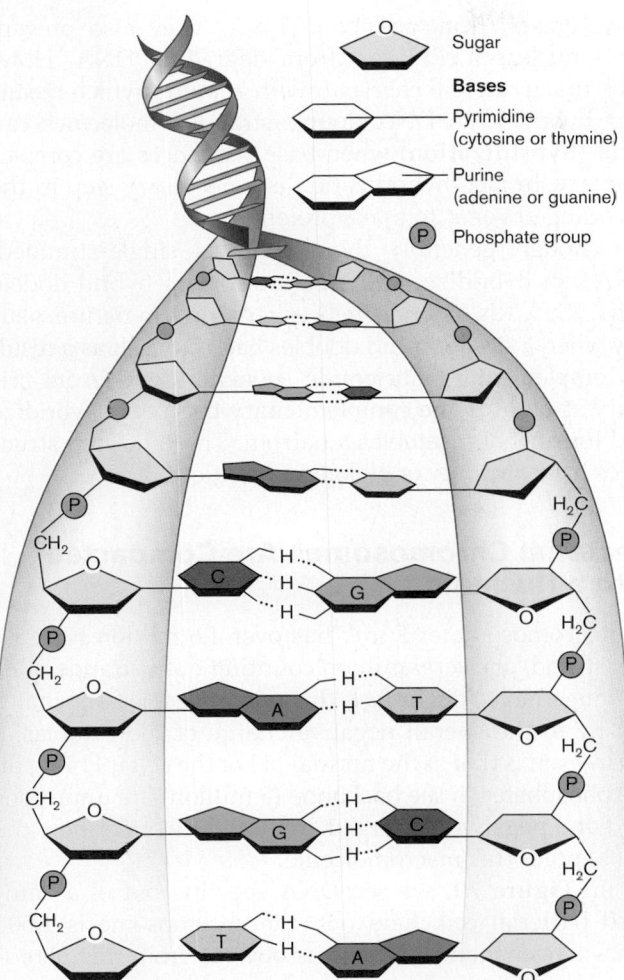

Figure 7.5   Progressive magnification of the DNA helix.
The top half of the molecule illustrates the typical double-helix structure that becomes magnified in the lower half to show individual bases and hydrogen bonding.

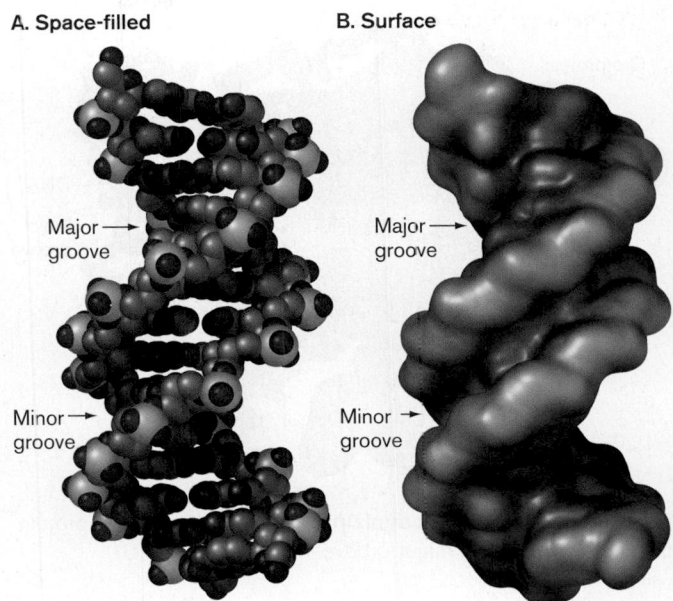

Figure 7.6   Models of DNA.   A. Space-filling model of DNA. **B.** DNA surface modeled using nuclear magnetic resonance. (PDB code: 1K8J)

the two strands can pair properly via hydrogen bonding. Base pairing is not possible if one tries to model DNA strands in a parallel arrangement.

The nitrogenous bases in DNA are planar hetero-aromatic structures stacked perpendicular to the phospho-diester backbone and parallel to each other (see **Fig. 7.4A**). **Purines** (bicyclic bases; adenine or guanine) pair with **pyrimidines** (monocyclic bases; thymine or cytosine). Under physiological conditions of salt (about 0.85%) and pH (pH 7.8), the hydrogen bonding of the bases permits adenine to pair only with thymine (via two hydrogen bonds) and likewise guanine with cytosine (via three hydrogen bonds). These complementary base interactions enable the two phosphodiester backbones to wrap around each other to form the classic double helix, or duplex.

The thousands of H-bonds that form between purines and pyrimidines along the interior of a DNA duplex (**Fig. 7.5**) make the bonding of the two complementary strands

of DNA highly specific, so that a duplex is formed only between complementary strands. Although H-bonds govern the specificity of strand pairing, the thermal stability of the helix is predominantly due to the stacking of the hydrophobic base pairs. Stacking enables interactions between these base pairs so that water and ions can interact with the negatively charged, hydrophilic phosphate backbone but not with the interior of the double helix.

> **THOUGHT QUESTION 7.1**   What do you think happens to two single-stranded DNA molecules isolated from *different* genes when they are mixed together at very high concentrations of salt? *Hint:* High salt concentrations favor bonding between hydrophobic groups.

As seen in the space-filling model of DNA in **Figure 7.6A** and the contour map in **Figure 7.6B**, the DNA double helix has grooves: a wide major groove and a narrower minor groove. The two grooves are generated by the angles at which the paired bases meet each other. These grooves provide DNA-binding proteins access to base sequences buried in the center of the molecule, so that proteins can interact with the bases without the strands being separated (**Fig. 7.7**).

At high temperatures (50°C–90°C), the hydrogen bonds in DNA break, and the duplex falls apart, or **denatures**, into two single strands. The temperature required to denature a DNA molecule depends on the GC/AT ratio of a sequence. More energy is required to break the three hydrogen bonds of a GC base pair than the two bonds of an AT base pair. Thus, DNA with a high GC content

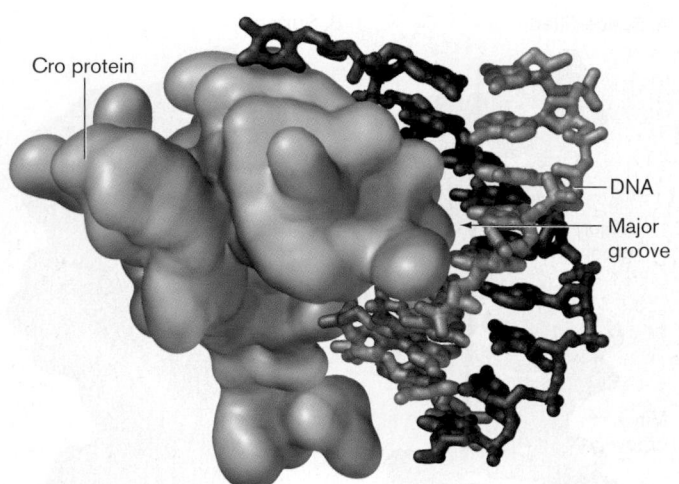

Cro protein

DNA

Major groove

Figure 7.7 **Protein recognizing DNA.** Cro repressor protein binds DNA within the major groove. (PDB code: 6CRO)

requires a higher denaturing temperature than does similar-sized DNA with a lower GC content. The black center ring in **Figure 7.2** illustrates how the GC content of bases can change around a single chromosome.

When DNA has been heated to the point of strand separation, lowering the temperature permits the two single strands to find each other and reanneal into a stable double helix. The kinetics of DNA denaturation is much faster than that of renaturation, since the latter is a random, hit-or-miss process of complementary sequences finding each other. This melting/reannealing property of DNA is exploited in a number of molecular techniques (see the discussion of the polymerase chain reaction in Section 7.6). Note, however, that bacteria and archaea growing at extreme pH or temperature protect their DNA from denaturation through the use of remarkable DNA-binding proteins. The role of DNA-binding proteins in microbial survival is the subject of considerable research.

> **THOUGHT QUESTION 7.2** How do you think the kinetics of denaturation and renaturation depend on DNA concentration?

RNA Differs Slightly from DNA

As we learned in Chapter 3, the growing cell makes temporary copies of its genes in the form of RNA (ribonucleic acid) molecules that direct the synthesis of proteins. DNA in a cell is usually double-stranded, whereas RNA is usually single-stranded. DNA and RNA are chemically similar, except that in RNA the sugar ribose replaces deoxyribose and the pyrimidine base uracil replaces thymine (**Fig. 7.4B**). Functionally, these two differences prevent enzymes meant to work on DNA, such as DNA

polymerases, from acting on RNA. They also prevent RNA nucleases (RNases) from degrading DNA. However, uracil can still base-pair with adenine, which means that hybrid RNA-DNA double-stranded molecules can form (**hybridization**) when base sequences are complementary. In fact, hybridization is a necessary step in the decoding of genes to make proteins.

Though generally thought of as single-stranded, RNA can hybridize with DNA to form a hybrid double helix. RNA-RNA double helices also form in nature, usually when a single strand doubles back on itself as a result of complementary nucleotide sequences within its primary structure. The complementary bases can hybridize and form what resembles a hairpin. These hairpin structures have a variety of biological functions.

Bacterial Chromosomes Are Compacted into a Nucleoid

The chromosome of *E. coli* has over 4.6 million bases in one strand, or over 9 million counting both strands. This is a huge molecule. In fact, this one molecule contributes greatly to the overall negative charge of the cytoplasm. The reason is that at the normal pH of the cell (pH 7.8), all the phosphates in the backbone (9 million) are unprotonated and negatively charged. You might wonder how the cell handles this macromolecule.

In **Figure 7.1**, we see DNA spewing out of a damaged bacterial cell. Laid out, the chromosome is 1,500 times longer than the cell. It is obvious from this photomicrograph that an intact, healthy cell must compact a huge bundle of DNA into a very small volume. DNA is the second largest molecule in the cell (only peptidoglycan is larger) and comprises a large portion of a bacterial cell's dry mass, about 3%–4%. While packaging 3% of a cell's dry weight may not seem like a challenge, realize that DNA is further confined only to ribosome-free areas of the cell, so the chromosome-packing density reaches about 15 mg/ml. In a test tube, DNA at 15 mg/ml is almost a gel, so how can anything move inside a cell? And how does all this DNA keep from getting hopelessly entangled?

As introduced in Chapter 3, cells pack their DNA into a manageable form that still allows ready access to DNA-binding proteins. Although bacteria lack a nuclear membrane, they pack their DNA into a series of protein-bound domains collectively called the **nucleoid** (see Section 3.5). Unlike the compact nucleus of eukaryotes, the bacterial nucleoid is distributed throughout the cytoplasm.

DNA Supercoiling Compacts the Chromosome

A nucleoid gently released from *E. coli* appears as 30–100 tightly wound loops (**Fig. 7.8**). The boundaries of each loop are defined by anchoring proteins called histone-like

Figure 7.8 Bacterial nucleoid.
Nucleoid showing domain loops after gentle release from cells. The single-strand nick unwinds (relaxes) only one loop.

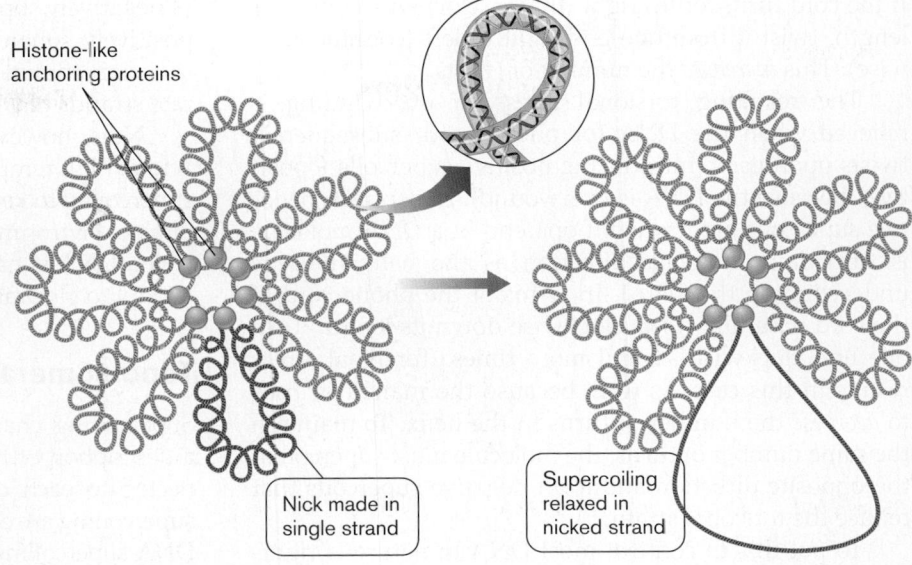

Histone-like anchoring proteins

Nick made in single strand

Supercoiling relaxed in nicked strand

proteins for their similarity to histones, the DNA-binding proteins of eukaryotes. The double helix within each domain is itself helical, or supercoiled. The easiest way to envision supercoiling is to picture a coiled telephone cord. After much use, a phone cord twists, or supercoils, upon itself. Note that supercoiled phone cords are quite compact, taking up less space than a relaxed cord. Circular DNA works the same way—a property used by the cell to pack its chromosome.

It is important to note that for DNA to form supercoils, its ends must be tethered. In a circular chromosome, the DNA ends are tethered to each other. Introducing an extra twist by breaking one or both strands, twisting one end, and then resealing the strands means that the increased or decreased torsional (twisting) stress is trapped in the final circular molecule. It cannot then spontaneously unwind. Details are discussed later.

Remarkably, the nucleoid with its 30–100 loops (or domains) can maintain different loops at different superhelical densities. The independence of supercoiled domains was demonstrated by introducing one single-strand nick in the phosphodiester backbone of one domain (see **Fig. 7.8**). You can do this by adding very small amounts of a nuclease (an enzyme that cleaves a nucleic acid). The ends of the broken strand, driven by the energy inherent in the supercoil, rotate about the unbroken complementary strand of the duplex and relax the supercoil. However, supercoils were removed from only the one domain. How is this possible if the chromosome is one circular molecule? The unaffected chromosomal domains remained supercoiled because they were constrained at their bases by anchoring proteins such as Hu and HNS (histone-like proteins) that prevent rotation.

How does DNA achieve the supercoiled state? The bacterial cell produces enzymes that can twist DNA into supercoils and relieve supercoils. A single twist introduced into a small (300-bp) circular DNA molecule forms a single supercoil as shown in **Figure 7.9 ▶**. An enzyme makes a double-strand break at one point in the circle, passes another part of the DNA through the break, and reseals it. The result is the same as if one end of the broken circle were twisted one full turn. Twisting in the opposite direction of the helical turn tightens the helix by adding more turns (overwinding). Think of the phone cord again. Look down the length of the cord from one end.

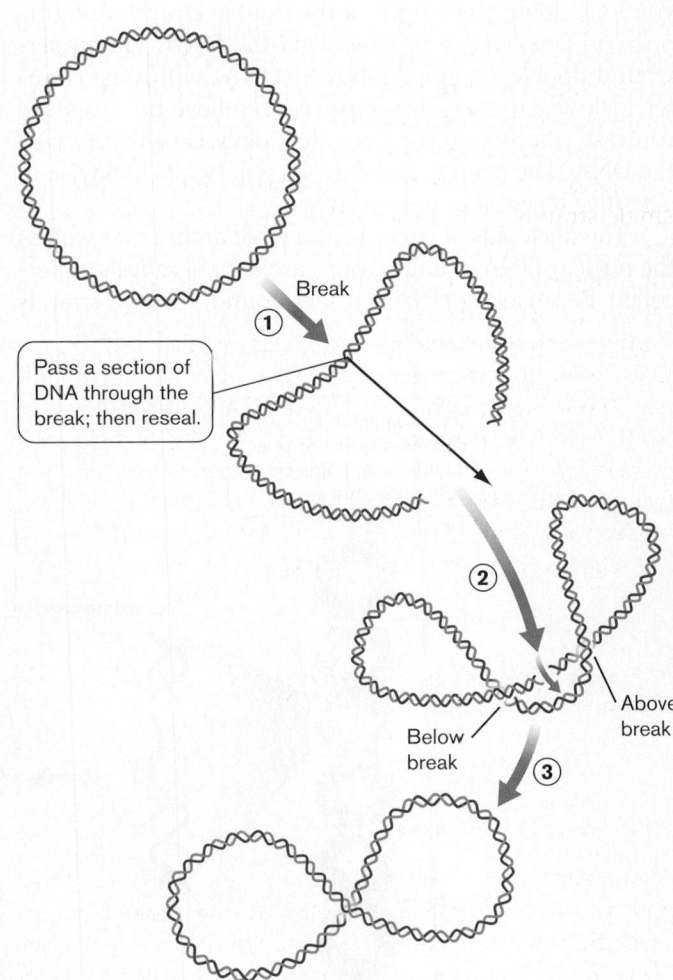

Break

① Pass a section of DNA through the break; then reseal.

②

Above break

Below break

③

Figure 7.9 Supercoiling of 300-bp circular DNA. A supercoil can be introduced into a double-stranded circular DNA molecule by (1) cleaving both strands at one site in the molecule, (2) passing an intact part of the molecule *between* ends of the cut site, and (3) reconnecting the free ends. ▶ *microbiology2. com/animations*

If the cord turns left to right (that is, clockwise) down its length, twist it from one end right to left (counterclockwise). This *increases* the number of twists.

The resulting torsional stress of overwinding is relieved when the DNA (or phone cord) subsequently twists upon itself, introducing positive supercoils ("positive" because the DNA is overwound). In contrast, negative supercoils are formed if one end of a DNA molecule is turned in the *same* direction as the helix (thereby underwinding the DNA). In terms of the phone cord, if the cord naturally turns clockwise down its length, turn one end clockwise several more times. Torsional stress results in this case, as well, because the maneuver tries to *decrease* the number of turns in the helix. To maintain the same number of turns, the molecule must supercoil in the opposite direction and form negative supercoils that reduce the torsional strain.

To put this in context, most DNA in nature is right-handed. Right-handed helical DNA turns clockwise when you look down the length of the double strand. Rotating one end clockwise will *underwind* the DNA. The underwound double-stranded DNA (dsDNA) will twist counterclockwise in a *negative supercoil* to relieve the stress. In contrast, rotating one end counterclockwise will *overwind* the DNA. The overwound dsDNA will twist clockwise in a *positive supercoil* to relieve stress.

The nucleoids of bacteria and most archaea, as well as the nuclear DNA of eukaryotes, are kept *negatively* supercoiled. Because the DNA is underwound, the two strands

of negatively supercoiled DNA are easier to separate than positively supercoiled DNA. This is important for transcription enzymes like RNA polymerase that must separate strands of DNA to make RNA.

Note, however, that some archaeal species living in acid at high temperature have nucleoids that are *positively* supercoiled to keep DNA double-stranded in these inhospitable environments (discussed below). Positively supercoiled DNA is harder to denature because it takes excess energy to separate overwound DNA.

Topoisomerases Supercoil DNA

Supercoiling changes the topology of DNA. Topology is a description of how spatial features of an object are connected to each other. Thus, enzymes that change DNA supercoiling are called **topoisomerases**. To maintain proper DNA supercoiling levels, a cell must delicately balance the activities of two types of topoisomerases. Type I topoisomerases cleave only one strand of a double helix, while type II enzymes cleave both strands. Type I enzymes are generally used to relieve or unwind supercoils, while type II enzymes use energy to introduce them. **Figures 7.10** and **7.11** illustrate the mechanisms used by type I and type II topoisomerases, respectively. Type I enzymes are usually single proteins, while type II enzymes have multiple subunits. An example of a type II topoisomerase is DNA gyrase, whose function is to introduce negative supercoils in DNA (see **Fig. 7.11**). The active gyrase complex is

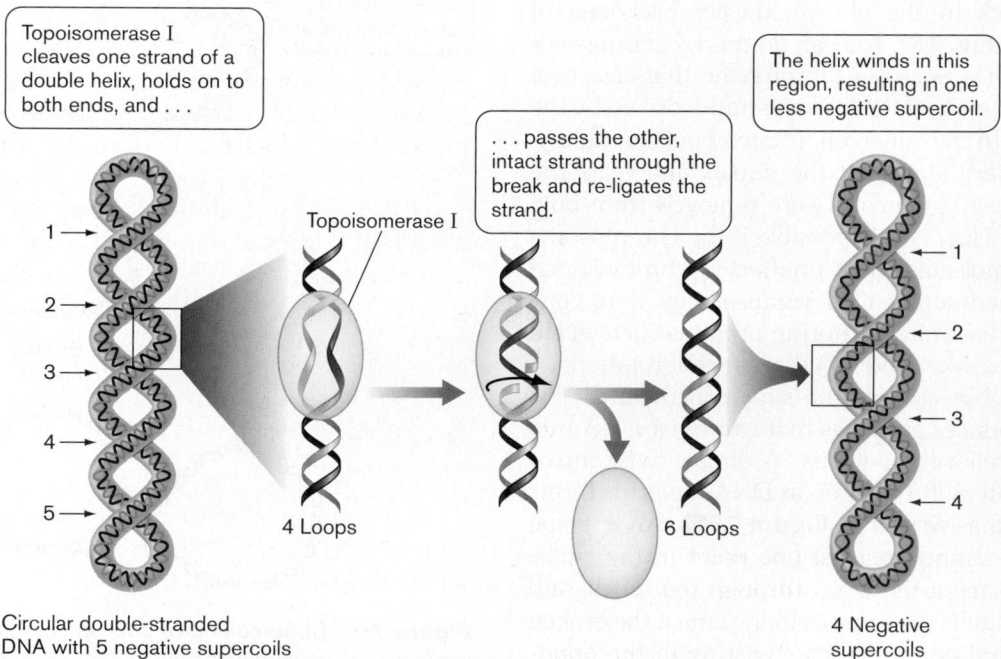

Topoisomerase I cleaves one strand of a double helix, holds on to both ends, and . . .

Topoisomerase I

. . . passes the other, intact strand through the break and re-ligates the strand.

The helix winds in this region, resulting in one less negative supercoil.

4 Loops

6 Loops

Circular double-stranded DNA with 5 negative supercoils

4 Negative supercoils

Figure 7.10   Mechanism of action for type I topoisomerases (topo I of *E. coli*).   Topoisomerase I relaxes a supercoiled DNA molecule. From left to right: Circular, supercoiled, double-stranded DNA is nicked in a single strand by topoisomerase I, unwound, and released with one less superturn. ●▶ *microbiology2.com/animations*

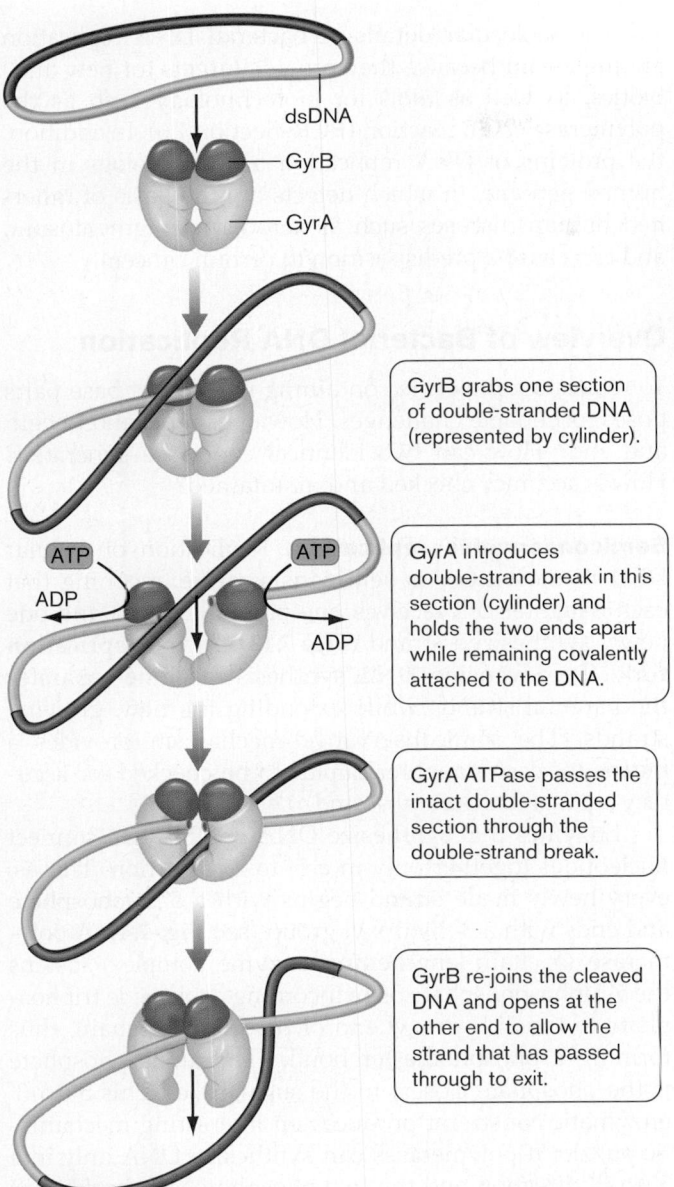

Figure 7.11 Mechanism of action for type II topoisomerases (DNA gyrase of *E. coli*). The gyrase enzyme grabs DNA and, in an ATP-dependent process, introduces a double-strand break, passes another part of the double helix through the break, and then reseals the break. The result is the introduction of a negative supercoil. ▶❚ *microbiology2.com/animations*

Labels in figure:
- dsDNA
- GyrB
- GyrA
- ATP
- ADP

GyrB grabs one section of double-stranded DNA (represented by cylinder).

GyrA introduces double-strand break in this section (cylinder) and holds the two ends apart while remaining covalently attached to the DNA.

GyrA ATPase passes the intact double-stranded section through the double-strand break.

GyrB re-joins the cleaved DNA and opens at the other end to allow the strand that has passed through to exit.

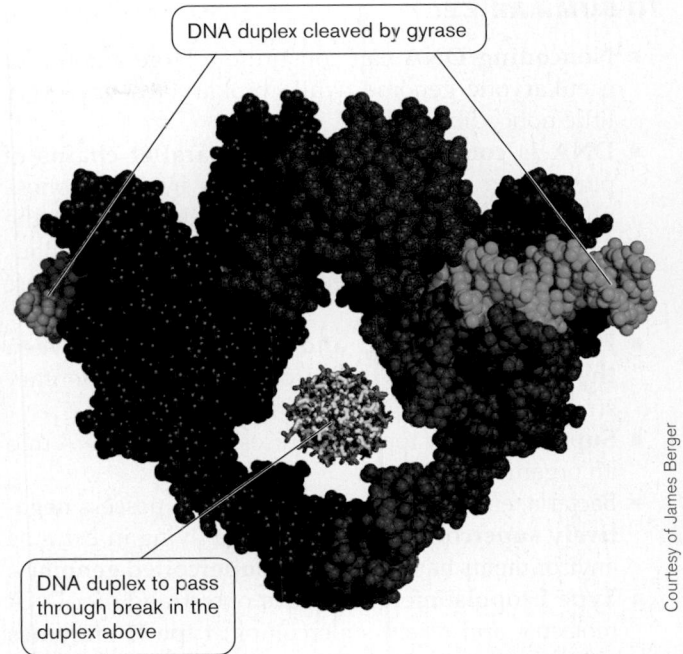

Figure 7.12 Three-dimensional representation of DNA gyrase. The 3-D model of gyrase was determined using data from X-ray crystallographic studies (blue and red regions). The gyrase complex is modeled in the process of gripping a broken DNA duplex (shown in green) and transporting a second duplex (the multicolored rosette).

Labels in figure:
- DNA duplex cleaved by gyrase
- DNA duplex to pass through break in the duplex above
- Courtesy of James Berger

a tetramer composed of two GyrA and two GyrB proteins. **Figure 7.12** shows a three-dimensional representation of DNA gyrase in the midst of generating a supercoil.

Enzymes that make or manage bacterial DNA, RNA, and proteins are common targets for antibiotics. For instance, the **quinolone antibiotics** specifically target bacterial type II topoisomerases. They do not affect eukaryotic topoisomerases. A modern quinolone, ciprofloxacin, was the treatment of choice for anthrax pneumonia during the 2001 anthrax attacks. The progenitors of this drug family, nalidixic and oxolinic acids, were used to map *gyrA* and *gyrB*, the first drug resistance genes identified in *E. coli*. The modern successors of these drugs, the fluoroquinolones, are among the most widely used antimicrobials in the world. These drugs do not block topoisomerase action but stabilize the complex in which DNA gyrase is covalently attached to DNA (see **Fig. 7.11**). The stuck complex forms a physical barrier in front of the DNA replication complex, and the bacterial cell dies.

Extreme thermophiles (hyperthermophilic archaea) possess an unusual gyrase called reverse DNA gyrase. In contrast to the DNA gyrase from mesophiles, reverse gyrase introduces positive supercoils into the chromosome. It is proposed that tightening the coil helps protect the chromosome against thermal denaturation. Because the DNA has *extra* turns, it takes more energy (heat) to separate the strands.

THOUGHT QUESTION 7.3 DNA gyrase is essential to cell viability. Why, then, are nalidixic acid–resistant cells that contain mutations in *gyrA* still viable?

THOUGHT QUESTION 7.4 Bacterial cells contain many enzymes that can degrade linear DNA. How, then, do linear chromosomes in organisms like *Borrelia burgdorferi* (the causative agent in Lyme disease) avoid degradation?

TO SUMMARIZE:

- **Noncoding DNA** can constitute a large amount of a eukaryotic genome, while prokaryotes have very little noncoding DNA.
- **DNA is composed of two antiparallel chains** of purine and pyrimidine nucleotides in which phosphate links the 5′ carbon of one nucleotide with the 3′ carbon of the next in the chain. The result is a double helix containing a deep major groove and a more shallow minor groove.
- **Hydrogen bonding and interactions between the stacked bases** hold together complementary strands of DNA.
- **Supercoiling** by topoisomerases compacts DNA into an organized nucleoid.
- Bacteria, eukaryotes, and most archaea possess **negatively supercoiled DNA**. Archaea living in extreme environments have **positively supercoiled genomes**.
- **Type I topoisomerases** cleave one strand of a DNA molecule and *relieve* supercoiling; **type II enzymes** cleave both strands of DNA and use ATP to *introduce* supercoils.

7.3 DNA Replication

Microbial DNA needs to replicate itself as accurately and as quickly as possible so that the organism can grow and compete with other species. Replication efficiency is one reason why bacterial pathogens such as *Salmonella* can cause disease so quickly after ingestion. In this respect, bacteria differ from multicellular organisms, which need to regulate cell division carefully within their tissues because unregulated growth within tissues leads to cancer. The process of bacterial replication involves an amazing number of proteins and genes coming together in a complex machine. A list of 20 DNA replication components can be found in **eTopic 7.1**. Operation of the replication complex is all the more remarkable considering that some bacteria, such as thermophilic *Bacillus* species that live in hot springs, can double in less than 15 minutes.

The molecular details of bacterial DNA replication are important because they provide targets for new antibiotics, as well as tools for biotechnology such as the polymerase chain reaction (PCR; Section 7.6). In addition, the proteins of DNA replication have homologs in the human genome, in which defects are the basis of inherited human diseases such as xeroderma pigmentosum, and can cause a predisposition to certain cancers.

Overview of Bacterial DNA Replication

To replicate a molecule containing millions of base pairs poses formidable challenges. How does replication begin and end? How can two identical copies be generated? How is accuracy checked and maintained?

Semiconservative replication. Replication of cellular DNA in most cases is **semiconservative**, meaning that each daughter cell receives one parental strand and one newly synthesized strand (**Fig. 7.13**). At the **replication fork**, the advancing DNA synthesis machine separates the parental strands while extending the new, growing strands. The semiconservative mechanism provides a means for each daughter duplex to be checked for accuracy against its parental strand.

Enzymes that synthesize DNA or RNA can connect nucleotides together only in a 5′-to-3′ direction. That is, every newly made strand begins with a 5′ triphosphate and ends with a 3′ hydroxyl group (see **Fig. 7.4**). A polymerase (a chain-lengthening enzyme complex) fastens the 5′ alpha-phosphate of an incoming nucleoside triphosphate to the 3′ hydroxyl end of the growing chain, thus forming a phosphodiester bond. (The alpha-phosphate is the phosphate closest to the sugar base.) This 5′-to-3′ enzymatic constraint produces an interesting mechanistic puzzle: If polymerases can synthesize DNA only in a 5′-to-3′ direction and the two phosphodiester backbones of the double helix are antiparallel, how are both strands of a moving replication fork synthesized simultaneously? One strand presents no problem, because it is synthesized in a 5′-to-3′ direction toward the fork; but synthesizing

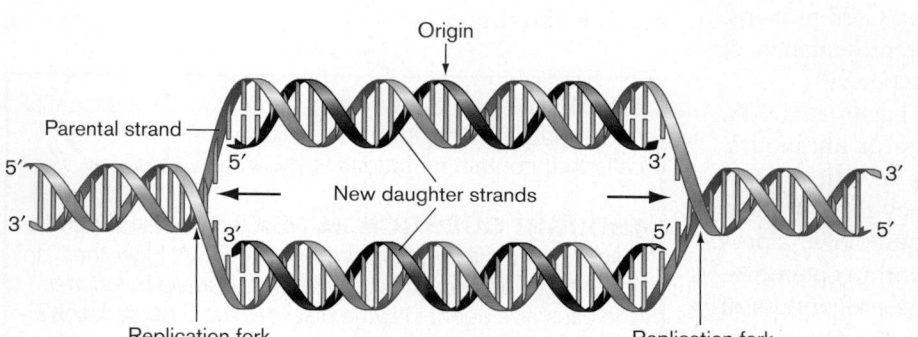

Figure 7.13 Semiconservative replication. A replication bubble with two replication forks. Parental strands are gray, and newly synthesized daughter strands are purple. Replication is called semiconservative because one parental strand is conserved and inherited by each daughter cell genome. It is called bidirectional because it begins at a fixed origin and progresses in opposite directions.

the other new strand in a 5′-to-3′ direction would seem to dictate that it move *away* from the fork (**Fig. 7.14**). How is this possible? The answer was revealed in Chapter 3 but is discussed in greater detail here.

> **THOUGHT QUESTION 7.5** Suppose you have the following capabilities: You can label DNA in a bacterium by growing cells in medium containing either nitrogen ^{14}N or the heavier isotope ^{15}N; you can isolate pure DNA from the organism; and you can subject DNA to centrifugation in a cesium chloride solution, a solution that forms a density gradient when subjected to centrifugal force, thereby separating the light (^{14}N) and heavy (^{15}N) forms of DNA to different locations in the test tube. Given these capabilities, how might you prove that DNA replication is semiconservative?

The process of DNA replication is divided into three phases: (1) initiation, which is the melting (unwinding) of the helix and the loading of the DNA polymerase enzyme complex; (2) elongation, which is the sequential addition of deoxyribonucleotides from deoxyribonucleoside triphosphates, followed by proofreading; and (3) termination, in which the DNA duplex is completely duplicated, the negative supercoils are restored, and key sequences of new DNA are methylated.

Replication from a Single Origin

Replication in bacteria begins at a single defined DNA sequence called the **origin** (*oriC*). Following initiation, a circular bacterial chromosome replicates *bidirectionally* (see **Fig. 7.14**, step 1) until it terminates at defined **termination** (*ter*) **sites** located on the opposite side of the molecule. Once the process has begun, the cell is committed to completing a full round of DNA synthesis. As a result, the decision of when to start copying the genome is critical. If it starts too soon, the cell accumulates unneeded chromosome copies; if it starts too late, the dividing cell's septum "guillotines" the chromosome, killing both daughter cells. Consequently, elaborate fail-safe mechanisms link the initiation of DNA replication with cell mass, generation time, and cellular health, making the timing of initiation remarkably precise.

> **NOTE:** Not all single-celled microbes have chromosomes with a single origin. The archaeon *Sulfolobus acidocaldarius*, for instance, has a single chromosome with three replication origins.

Fundamentals of DNA replication. The basic process of chromosome replication is outlined in **Figure 7.14**. After initiation of replication, a replication bubble forms

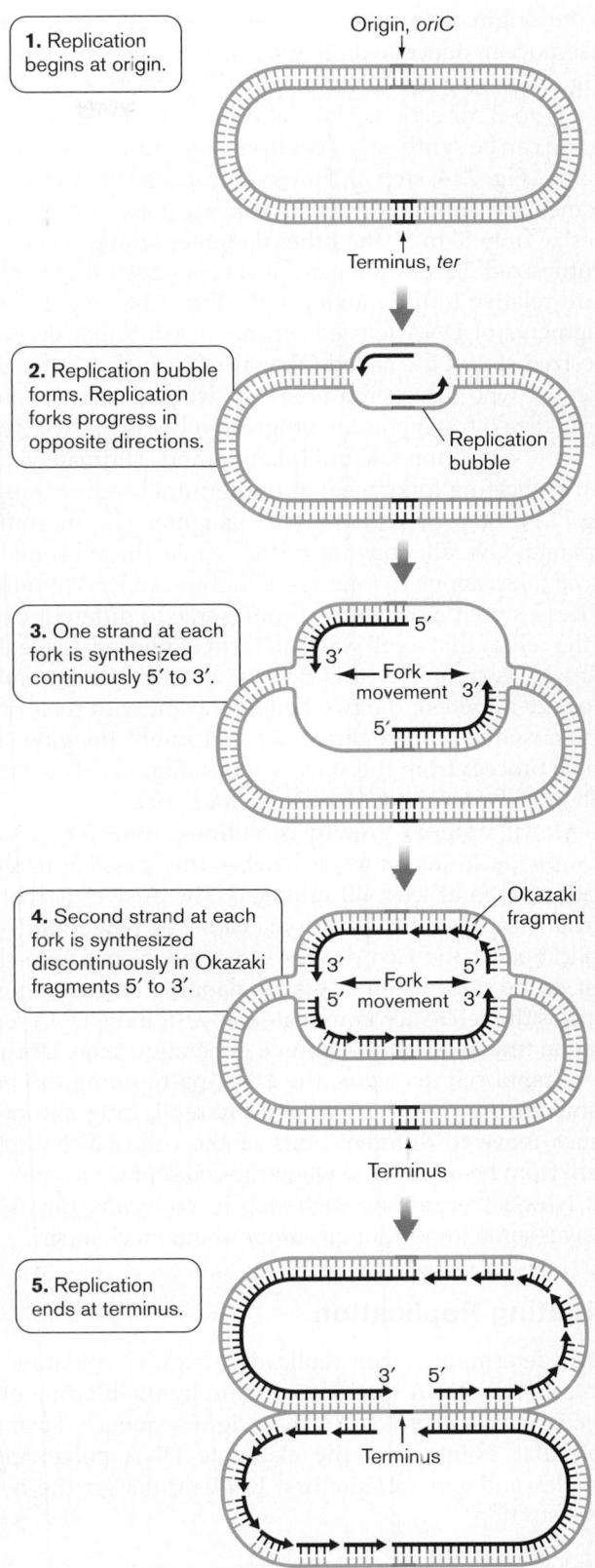

1. Replication begins at origin.

Origin, *oriC*

Terminus, *ter*

2. Replication bubble forms. Replication forks progress in opposite directions.

Replication bubble

3. One strand at each fork is synthesized continuously 5′ to 3′.

Fork movement

4. Second strand at each fork is synthesized discontinuously in Okazaki fragments 5′ to 3′.

Okazaki fragment

Fork movement

Terminus

5. Replication ends at terminus.

Terminus

Figure 7.14 **Comparing direction of fork movement with direction of DNA synthesis.**

at the origin. The bubble contains two replication forks that move in opposite directions around the chromosome (**Fig. 7.14**, step 2). DNA polymerases synthesize DNA in a 5'-to-3' direction. Thus, at each fork, one new DNA strand can be synthesized continuously until the terminus region (**Fig. 7.14**, step 3). However, because the two DNA strands are antiparallel and the DNA polymerases synthesize only 5' to 3', the other daughter strand has to be synthesized discontinuously, in stages—seemingly backward relative to the moving fork (**Fig. 7.14**, step 4). The fragments of DNA formed on this discontinuously synthesized strand are called **Okazaki fragments**, after the scientist who discovered them. As we will discuss later, the Okazaki fragments are progressively stitched together to make a continuous, unbroken strand. Ultimately, the two replicating forks meet at the terminal sequence (see **Fig. 7.14**, step 5) and the two daughter chromosomes separate. Overall, copying of the whole chromosome in *E. coli* takes about 40 minutes. Chromosome-partitioning processes then move each chromosome to different ends of the cell so that a cell wall can form at midcell. Once the cell wall is complete (this time is variable but generally about 20 minutes), the two daughter cells, with their new chromosomes, can separate. So you might imagine the whole process from the start of replication to cell separation would take about 60 minutes for *E. coli*.

Under optimal growth conditions, however, *E. coli* cells divide in 20 minutes. How is this possible if replication takes at least 40 minutes? The answer is that a partially replicated chromosome can start new rounds of replication at the two daughter origins even before the first round is complete. This overlapping of generations enables the cell to accommodate the 40-minute DNA replication time within a 20-minute generation time. During these rapid cell divisions, the DNA-partitioning mechanisms ensure that the two actively replicating chromosomes move to different ends of the cell, which keeps them from being severed when the cell septum forms.

Now let's examine each step in molecular detail to answer some important questions about mechanism.

Initiating Replication

What determines when replication begins? Initiation is controlled by DNA methylation, and by the binding of a specific initiator protein to the origin sequence. Further molecular events load the elaborate DNA polymerase complex and generate the first RNA primer for the new DNA strand.

DNA methylation controls timing. The chromosomal origin of *E. coli* is a sequence of 245 base pairs designated *oriC*. It is subject to a critical molecular control mechanism that dictates the precise timing of replication initiation. Initiation of replication at *oriC* is activated by one

protein, DnaA, and inhibited by another, SeqA. Immediately after a cell has divided, the level of active DnaA (DnaA bound to ATP) is low, and the inhibitor SeqA will bind to *oriC* and prevent ill-timed initiations (before the cell has grown enough to divide again).

How does SeqA know to bind just after the origin has replicated? The key is DNA methylation. *E. coli* uses the enzyme deoxyadenosine methylase (Dam) to attach a methyl group to the adenine residue at position N-6 (**Fig. 7.4A**; see the N of the NH_2 group attached to the six-membered ring). The methylated adenines appear in all GATC sequences. GATC sequences (the recognition sites of Dam methylase) are scattered throughout the chromosome and occur on both strands. Just after the origin has replicated, however, there is a short lag before the newly synthesized strand is methylated. As a result, the origin is temporarily hemimethylated, a situation in which only one of the two complementary strands is methylated. Because SeqA has a high affinity for hemimethylated origins, this inhibitor will bind most tightly immediately after the origin has been replicated and prevent another initiation event. Eventually, the Dam methylase will methylate the new strand and decrease SeqA binding.

The replication initiator protein, DnaA. The onset of the initiation phase is determined by the concentration of the replication initiator protein, DnaA. DnaA recognizes specific 9-bp repeats at *oriC*. As the cell grows, the level of active DnaA rises until it is sufficient to bind to a series of 9-bp repeats at *oriC* (**Fig. 7.15**⊙, step 1). DnaA actually binds as a complex with ATP (DnaA-ATP). This binding facilitates melting of DNA at the origin and initiates the assembly of a membrane-bound replication hyperstructure, a complex assembly of numerous proteins forming a functional unit, at midcell (at the cell equator).

The origin sequence (*ori*), after moving through the replication complex, cannot trigger another round of replication, because of inhibition by SeqA and decreasing levels of unbound DnaA-ATP. Another round of replication can begin only after (1) the origin becomes fully methylated, (2) SeqA dissociates, and (3) the DnaA-ATP concentration rises. The experiment in **Figure 7.16** shows that DNA in replicating cells of *E. coli* contains two origins and one terminator region, located 180° from the initial origin. Each site is labeled with a different fluorescent protein.

Initiation requires RNA polymerases. An unexpected feature of DNA replication is that its initiation actually requires RNA polymerases. The housekeeping RNA polymerase helps separate the strands of the DNA helix at the origin. A second RNA polymerase, DNA primase, produces the *primer*, or starter, fragments needed to synthesize new DNA. The housekeeping RNA polymerase used to make most of the RNA in the cell (discussed in Chapter 8) transcribes DNA at *oriC* to produce RNA that

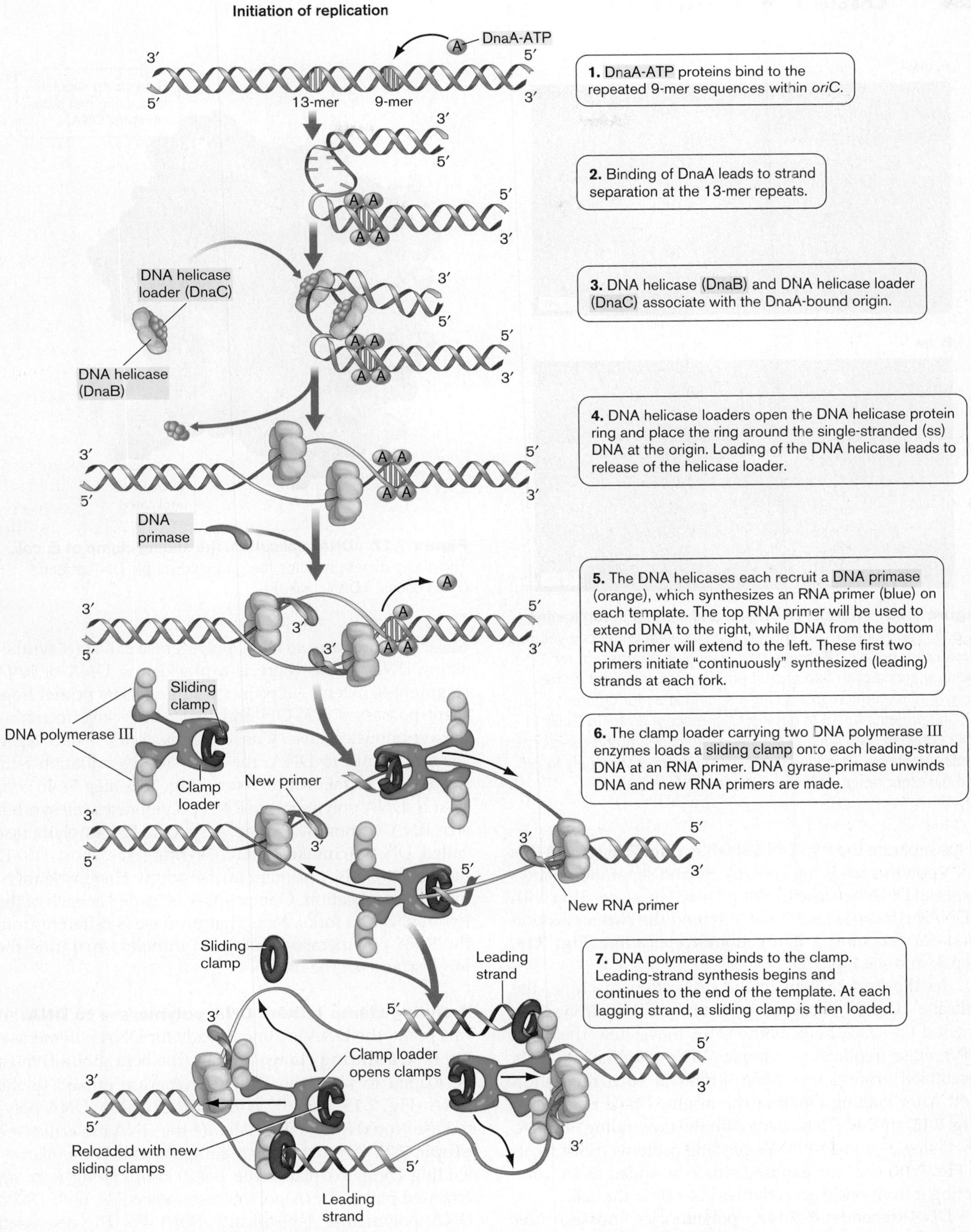

Initiation of replication

A — DnaA-ATP

3′ 5′
13-mer   9-mer
5′ 3′

1. DnaA-ATP proteins bind to the repeated 9-mer sequences within *oriC*.

3′
5′

A A
A A

5′

3′

2. Binding of DnaA leads to strand separation at the 13-mer repeats.

DNA helicase loader (DnaC)

DNA helicase (DnaB)

3′
5′

A A
A A

5′

3′

3. DNA helicase (DnaB) and DNA helicase loader (DnaC) associate with the DnaA-bound origin.

A A
A A

3′
5′

A A
A A

5′
3′

DNA primase

4. DNA helicase loaders open the DNA helicase protein ring and place the ring around the single-stranded (ss) DNA at the origin. Loading of the DNA helicase leads to release of the helicase loader.

A

3′
5′

3′

A
A A

5′
3′

5. The DNA helicases each recruit a DNA primase (orange), which synthesizes an RNA primer (blue) on each template. The top RNA primer will be used to extend DNA to the right, while DNA from the bottom RNA primer will extend to the left. These first two primers initiate "continuously" synthesized (leading) strands at each fork.

Sliding clamp

DNA polymerase III

Clamp loader

New primer

3′
5′

3′

5′

New RNA primer

3′
5′

5′
3′

6. The clamp loader carrying two DNA polymerase III enzymes loads a sliding clamp onto each leading-strand DNA at an RNA primer. DNA gyrase-primase unwinds DNA and new RNA primers are made.

Sliding clamp

Leading strand

7. DNA polymerase binds to the clamp. Leading-strand synthesis begins and continues to the end of the template. At each lagging strand, a sliding clamp is then loaded.

Clamp loader opens clamps

3′
5′

5′

5′
3′

Reloaded with new sliding clamps

3′

Leading strand

Figure 7.15 Initiation of DNA replication. The start of DNA replication is precisely timed and linked to the ratio of DNA to cell mass. In *E. coli*, the initiator protein DnaA accumulates during growth and then triggers the initiation of replication. It begins with DnaA-ATP complexes binding to 9-mer (9-bp) repeats upstream of the origin. The binding of DnaA-ATP (along with other proteins not shown) first causes the DNA to loop in preparation for being melted open by the helicase (DnaB). ▶❚❚ *microbiology2.com/animations*

A. oriC

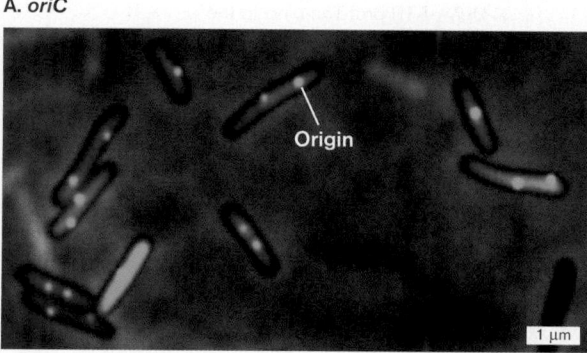

Origin

1 µm

B. ter

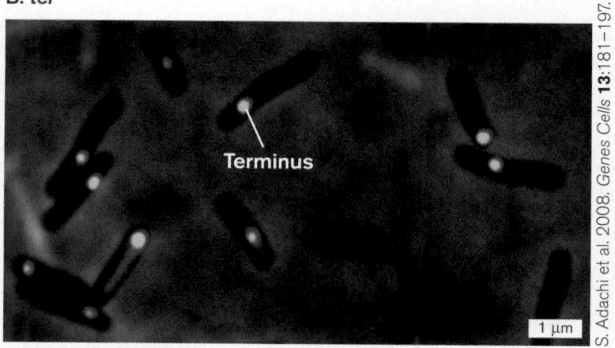

Terminus

1 µm

S. Adachi et al. 2008. *Genes Cells* **13**:181–197.

Figure 7.16   Number of origins and termini in replicating *E. coli*.   A. Fluorescent *oriC* foci (labeled with enhanced yellow fluorescent protein). **B.** Same cells with blue fluorescent *ter* foci (enhanced cyan fluorescent protein). The *oriC* and *ter* foci were engineered to contain DNA-binding sites for different binding proteins fused to different fluorescent proteins. Each cell contained both binding sites and expressed both fluorescent binding proteins. Thus, one fluorescent protein bound only to *ori* and the other bound only to *ter* in the same cell.

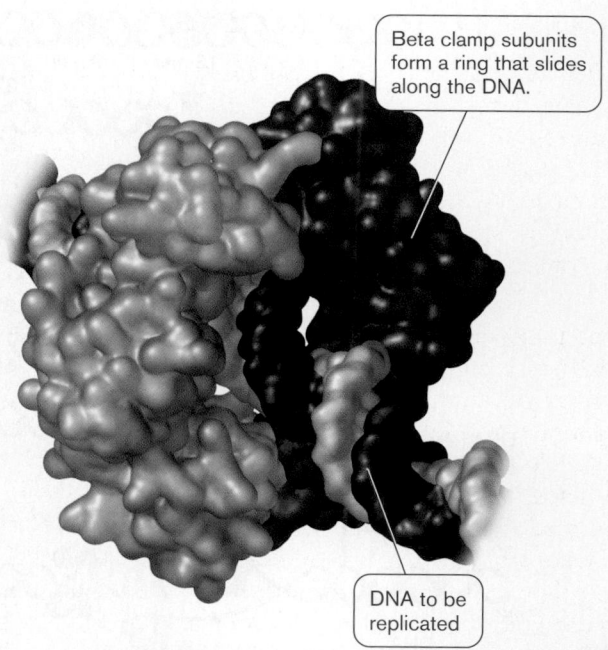

Beta clamp subunits form a ring that slides along the DNA.

DNA to be replicated

Figure 7.17   DNA replication: the sliding clamp of *E. coli*. The clamp dimer encircles the gray-and-purple DNA strand. (PDB codes: 1OK7, 1K8J)

helps separate the two DNA strands (**Fig. 7.15**, step 2; this RNA polymerase is not shown). Strand separation allows a special DNA helicase (protein DnaB) in association with a DNA helicase loader (DnaC) to bind the two replication forks formed during bidirectional replication (**Fig. 7.15**, step 3; and see **Fig. 7.17**).

As the lead protein of the replication machine, the helicase (DnaB) uses energy from ATP hydrolysis to unwind the DNA helix as the DNA moves into the DNA polymerase replicating complex. The ringlike DnaB is assembled around one DNA strand at each replication fork. After loading DnaB at the origin, DnaC is released (**Fig. 7.15**, step 4). Coincident with the unwinding of DNA, small single-stranded DNA–binding proteins (SSBs, seen in **Fig. 7.18**) coat the exposed single-stranded DNA, protecting it from nuclease activities patrolling the cell.

DNA-dependent DNA polymerases possess the unique ability to "read" the nucleotide sequence of a DNA template and synthesize a complementary DNA strand. The discovery of this activity earned Arthur Kornberg (1918–2007) a Nobel Prize in 1959. As remarkable as

these enzymes are, *no* DNA polymerase can start synthesizing DNA unless there is a preexisting DNA or RNA fragment to extend—a primer fragment. The primer fragment possesses a 3′ OH end that can receive incoming deoxyribonucleotides. Consequently, once DnaB (helicase) is bound to DNA, the next step in initiation is to make RNA primers at each fork (**Fig. 7.15**, step 5). In contrast to DNA polymerases, RNA polymerases *can* synthesize RNA without a primer. A specific RNA polymerase called DNA **primase** (DnaG) synthesizes short (10–12 nucleotides) RNA primers at the origin, thereby launching DNA replication. One primase is loaded at each of the two replication forks. Note that primase is different from the RNA polymerase involved in initially separating the DNA strands at the origin.

A sliding clamp tethers DNA polymerase to DNA. At this point, the DNA is almost ready for DNA polymerase. But first, a **sliding clamp** protein (the beta subunit) must be loaded to keep the DNA polymerase affixed to the DNA (**Fig. 7.15**, step 6). Without this clamp, DNA polymerase would frequently "fall off" the DNA molecule (see **eTopic 7.2**). A multisubunit complex (called the clamp-loading complex) places the beta clamp, along with an attached pair of DNA polymerase molecules, onto DNA. DNA polymerase (specifically DNA Pol III; discussed later) can then bind to the 3′ OH terminus of the primer RNA molecule and begin to synthesize new DNA (**Fig. 7.15**, step 7). **Figure 7.17** presents the molecular structure of the beta clamp loaded onto DNA.

Elongation of Replicating DNA

Escherichia coli contains five different DNA polymerase proteins, designated Pol I through Pol V. All polymerases catalyze synthesis of DNA in the 5'-to-3' direction. However, only the replication polymerases Pol III and Pol I participate directly in chromosome replication. The other polymerases conduct operations to rescue stalled replication forks and repair DNA damage.

DNA polymerase III. The main replication polymerase, Pol III, is a complex, multicomponent enzyme. Because of its complexity, it is often referred to as a "molecular machine." The DNA synthesis activity of Pol III is held in the alpha subunit of the complex, while other subunits are used for improving fidelity (accuracy of replication) and processivity (a measure of how long the polymerase remains attached to, and replicates, a template). The Pol III epsilon subunit (DnaQ), for example, contains a **proofreading** activity that prevents mistakes and improves fidelity.

Proofreading activities within DNA polymerases scan for mispaired bases that have been inappropriately added to a growing chain. Mispaired bases are detected by their increased mobility. A mispaired base mistakenly linked by a phosphodiester bond to a growing DNA chain is more mobile than the correct base because it does not hydrogen-bond to the template base. This motion halts DNA elongation by Pol III because the base is not properly positioned at the enzyme's active site. This stalling of Pol III activity triggers an intrinsic 3'-to-5' **exonuclease** activity in the epsilon subunit. Exonucleases degrade DNA starting from either the 5' end or the 3' end. The exonuclease activity of Pol III cleaves the phosphodiester bond, releasing the improperly paired base from the growing chain. Once the wayward mispaired base has been excised, Pol III activity can proceed.

Both DNA strands are elongated simultaneously. After initiation, each replication fork contains one elongating 5'-to-3' strand called the leading strand (look back at **Fig. 7.15**, step 7). But how is the opposite strand at each fork replicated? There are no known DNA polymerases capable of synthesizing DNA in the 3'-to-5' direction, which would seem to be needed if both strands are to be synthesized simultaneously. DNA synthesis of one strand continuing all the way back to the origin is not a solution, because it would leave the unreplicated strand at each fork exposed to possible degradation for too long and would double the time needed to complete DNA replication. The cell has solved this dilemma by coordinating the activity of *two* DNA Pol III enzymes in one complex—one for each strand. The two associated Pol III complexes together with DNA primase and helicase form

the **replisome**. As the dsDNA unwinds at the fork, the problem strand loops out and primase (DnaG) synthesizes a primer. The second Pol III enzyme binds to the primed section of the loop and synthesizes DNA in the 5'-to-3' direction (imagine the lower template strand in **Fig. 7.18** threading from left to right, through the polymerase ring). All the while, the second polymerase is moving along in tandem with the first polymerase (on the leading strand) relative to the fork (**Fig. 7.18**, step 1). Realize that in actuality, the replisome remains at a fixed, midcell location in the cell, probably attached to the membrane, and the template DNA threads through it (discussed later).

Note that simultaneous extension of the two strands requires that synthesis of the looped strand *lag* behind the leading strand and that new RNA primers be synthesized periodically by primase (DnaG), every thousand bases or so. Thus, the lagging strand is synthesized discontinuously, in pieces called Okazaki fragments, while the leading strand can be synthesized continuously. As the leading strand moves forward, advancing the fork, there remains a long stretch of lagging strand complementary to the already replicated leading strand. This lagging strand remains single-stranded but protected by single-stranded DNA–binding proteins (SSBs) (**Fig. 7.18**, step 2). After about 1,000 bases, DNA primase reenters and synthesizes a new RNA primer in anticipation of lagging-strand DNA synthesis (**Fig. 7.18**, step 3). At some point, the lagging-strand polymerase bumps into the 5' end of the previously synthesized fragment. This interaction causes DNA polymerase to disengage from that strand (**Fig. 7.18**, step 4), and the clamp loader loads a new clamp near the new RNA primer (**Fig. 7.18**, step 5). The DNA polymerase binds to that clamp and begins synthesizing another Okazaki fragment (**Fig. 7.18**, step 6). This process repeats every 1,000 bases or so around the chromosome.

DNA polymerase I. Discontinuous DNA synthesis results in a daughter strand containing long stretches of DNA punctuated by tiny patches of RNA primers. This RNA must be replaced with DNA to maintain chromosome integrity. To remove the RNA, cells typically use an RNase enzyme specific for RNA-DNA hybrid molecules (called RNase H). A DNA Pol I enzyme enters after the RNase and synthesizes a DNA patch using the 3' OH end of the preexisting DNA fragment as a priming site (**Fig. 7.19**). When DNA Pol I reaches the next fragment, the enzyme removes the 5' nucleotide and resynthesizes it. This process of replicating DNA increases accuracy and decreases mutations.

Once DNA Pol I stops synthesizing, it cannot join the 3' OH of the last added nucleotide with the 5' phosphate of the abutting fragment. The resulting nick in the phosphodiester backbone is repaired by **DNA ligase**, which

Elongation of DNA synthesis

1. The leading-strand DNA Pol III enzyme replicates the leading strand. SSBs cover and protect the unreplicated single strand. The DNA helicase remains on the lagging strand, unwinding the dsDNA moving into the replisome complex.

2. Lagging-strand DNA polymerase synthesizes the lagging strand, which loops out after passing through the polymerase.

3. After DNA helicase has moved approximately 1,000 bases, another RNA primer is synthesized on each lagging strand.

4. When the lagging-strand polymerase bumps into the 5′ end of a previously synthesized fragment, the DNA polymerase is released and the clamp is disengaged.

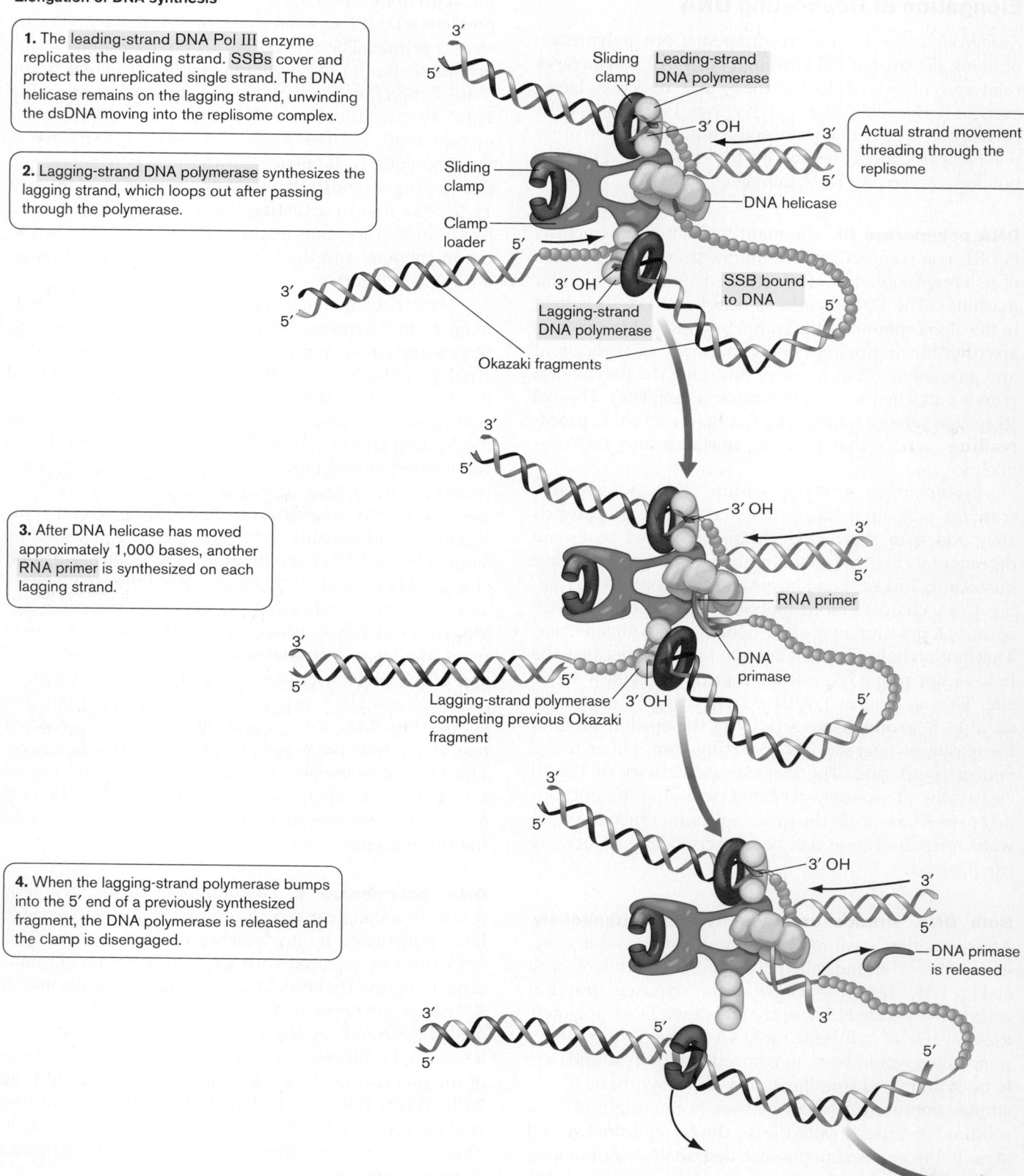

Figure 7.18 The DNA polymerase dimer acting at a replication fork. The leading and lagging strands are synthesized simultaneously in the 5′-to-3′ direction. For clarity, the beta clamp on the lagging strand is shown on the opposite side of Pol III as compared to its position on the leading strand. If the clamp were placed on the left side of the polymerase, as it is shown in many models, the lagging strand would have to completely loop around the polymerase and enter the polymerase from the left.

in *E. coli* and many other bacteria uses energy gained by cleaving nicotinamide adenine dinucleotide (NAD) to form the phosphodiester bond (see **Fig. 7.19**). NAD is not used in its usual way, as a reductant that oxidizes substrates. Energy inherent in the diphosphate bond of NAD is captured upon cleavage by DNA ligase and used to rejoin the 3′ OH and 5′ phosphate ends present at the nick. DNA ligase from eukaryotes and some other microbes uses ATP rather than NAD in this capacity.

THOUGHT QUESTION 7.6 How fast does *E. coli* DNA polymerase synthesize DNA (in nucleotides per second), given that the genome is 4.6 million base pairs, replication is bidirectional, and the chromosome completes a round of replication in 40 minutes?

DNA replication generates supercoils. As template DNA is threaded through the replisome, the helicase continually pulls apart the two strands of the DNA helix. As a result, the DNA ahead of the fork twists, introducing positive supercoils. (Try this yourself. Twist two pieces of string together, staple one end of the pair to a piece of cardboard, and then pull the two strands apart from the free end. Notice the supercoiling that takes place beyond, or downstream of, the moving fork.) The increasing torsional stress in the chromosome could stop replication by making strand separation more and more difficult. What prevents the buildup of torsional stress is the DNA gyrase (**Fig. 7.11**) that is located ahead of the fork, removing the positive supercoils as they form.

Another topological question, noted earlier, is how DNA polymerase maneuvers through the cell. Many

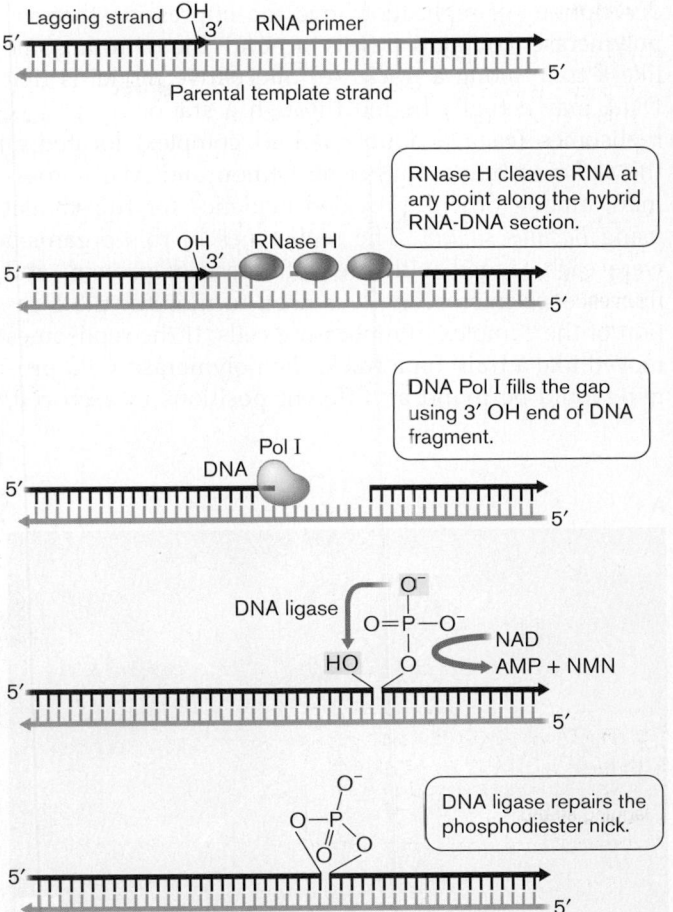

Figure 7.19 RNase H removing the RNA primer. RNase H cleaves the RNA primer (blue). DNA polymerase I uses the preexisting 3′ OH end of the DNA fragment to fill the gap. Finally, DNA ligase repairs the phosphodiester nick using energy derived from the cleavage of NAD.

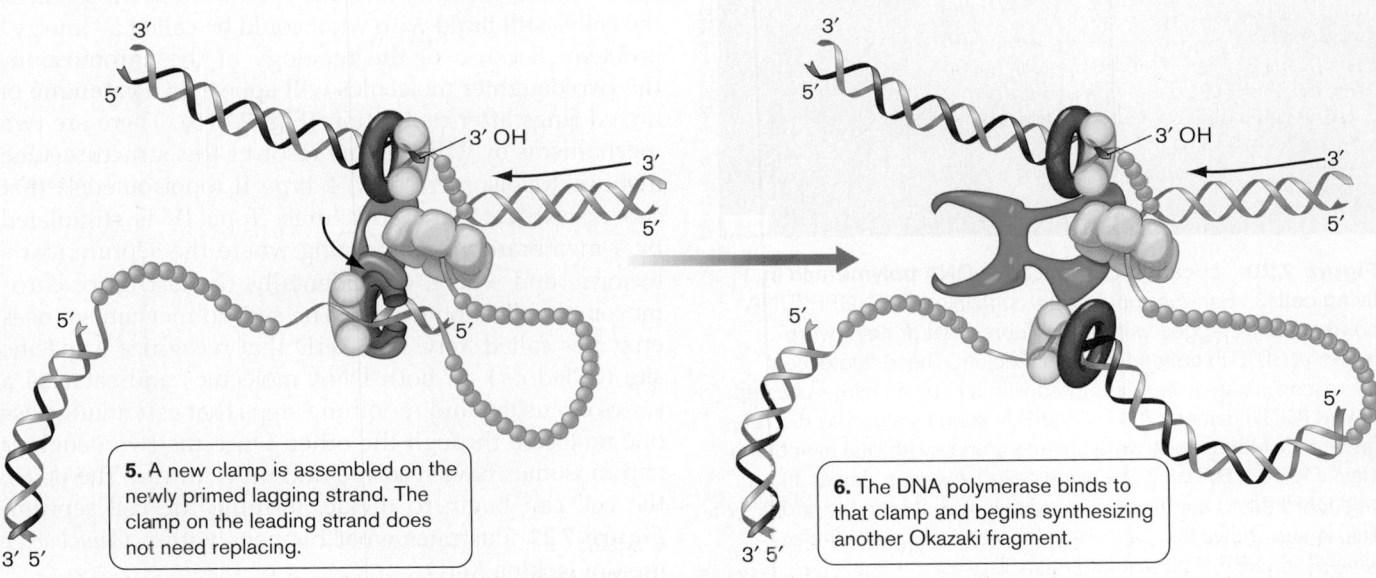

Figure 7.18 (*continued*) **The DNA polymerase dimer acting at a replication fork.**

descriptions of replication give the impression that the polymerase complex travels around the chromosome like a train along a track. An alternative model is that DNA may actually be fed through a stationary pair of replisomes (each a double Pol III complex) located at the cell membrane. Katherine Lemon and Alan Grossman, among others, provided evidence for this model using *Bacillus subtilis*. The replisomes in this organism were each tagged with green fluorescent protein, and fluorescence microscopy was used to monitor the location of the complex in replicating cells. If the replisomes moved like a train on a track, the polymerase-GFP protein would be found at different positions in each cell.

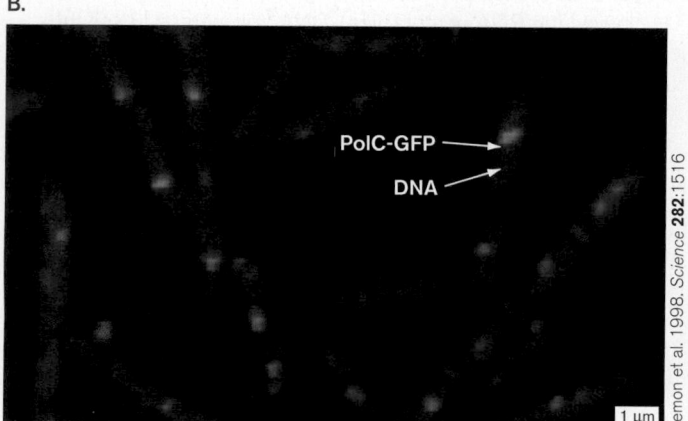

A.

PolC-GFP

1 μm

Lemon et al. 1998. *Science* **282**:1516

B.

PolC-GFP

DNA

1 μm

Lemon et al. 1998. *Science* **282**:1516

Figure 7.20 Location of replicative DNA polymerase in living cells. *Bacillus subtilis* cells containing PolC-GFP (DNA polymerase III tagged with green fluorescent protein) were grown at 30°C in defined minimal medium. These fluorescent microscopy images were captured with a cooled charge-coupled device (CCD) camera. **A.** PolC-GFP is seen localized as discrete green foci. Membranes were stained orange with vital membrane stain FM4-64. **B.** Using the same cells, DNA was stained blue with DAPI (blue) and the image overlaid with the one in part A. This image shows that DNA is distributed throughout the cell while PolC-GFP is localized to midcell.

Instead, however, in every replicating cell, replisomes were observed as distinct fluorescent foci located at or near midcell (**Fig. 7.20**). Cellular DNA stained with a blue fluorescent dye (DAPI) clearly occupied most of the cytoplasmic space. The "factory" model has recently been called into question, however. Studies using different fluorescently tagged replisome proteins suggest that the replisome starts at midcell but then splits and moves toward the two poles, more like a train on a track. The truth may lie somewhere between the two models. For now, the controversy continues.

Terminating Replication

Bidirectional replication of a circular bacterial chromosome results in the two membrane-associated replication machines attempting to replicate through the same DNA sequences 180° from the origin—that is, halfway around the chromosome. What tells the polymerases to stop? There are as many as ten terminator sequences (*ter*) on the *E. coli* chromosome that polymerases enter but rarely, if ever, leave (**Fig. 7.21A**). One set of terminators deals with the clockwise-replicating polymerase, while the other set halts DNA polymerases replicating counterclockwise relative to the origin. A protein called Tus (terminus utilization substance) binds to these sequences and acts as a counter-helicase when it comes in contact with an advancing helicase (DnaB). The bound Tus protein effectively halts polymerase movement. Multiple terminator sites ensure that the polymerase complex does not escape and continue replicating DNA. Which terminator site is used depends in part on whether replication of one fork has lagged behind replication of the other.

Once DNA polymerases have completed duplicating the chromosome and have been removed at the *ter* sites, the cell is still faced with what could be called a "knotty" problem. Because of the topology of the chromosome, the two daughter molecules will appear as a **catenane** of linked rings after replication (**Fig. 7.21B**). There are two mechanisms by which *E. coli* resolves this structure. One involves topoisomerase IV, a type II topoisomerase that can resolve catenane structures. Topo IV is stimulated by a membrane protein sitting where the septum starts to form, and where, coincidentally, the last bit of chromosome replication occurs. The second mechanism uses enzymes called XerC and XerD that recognize a specific site (called *dif*) on both DNA molecules and catalyze a series of cutting and re-joining steps that essentially pass one molecule through the other. Once the two daughter chromosomes have resolved and move toward the poles, the cell can begin to divide, forming the cell septum. **Figure 7.22** illustrates what happens with a *Caulobacter* mutant lacking XerD.

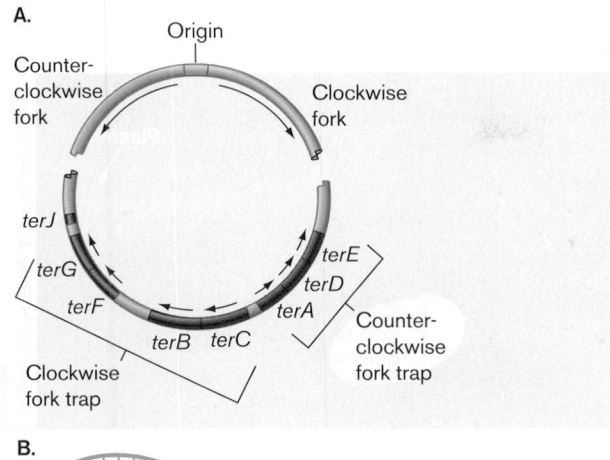

A.

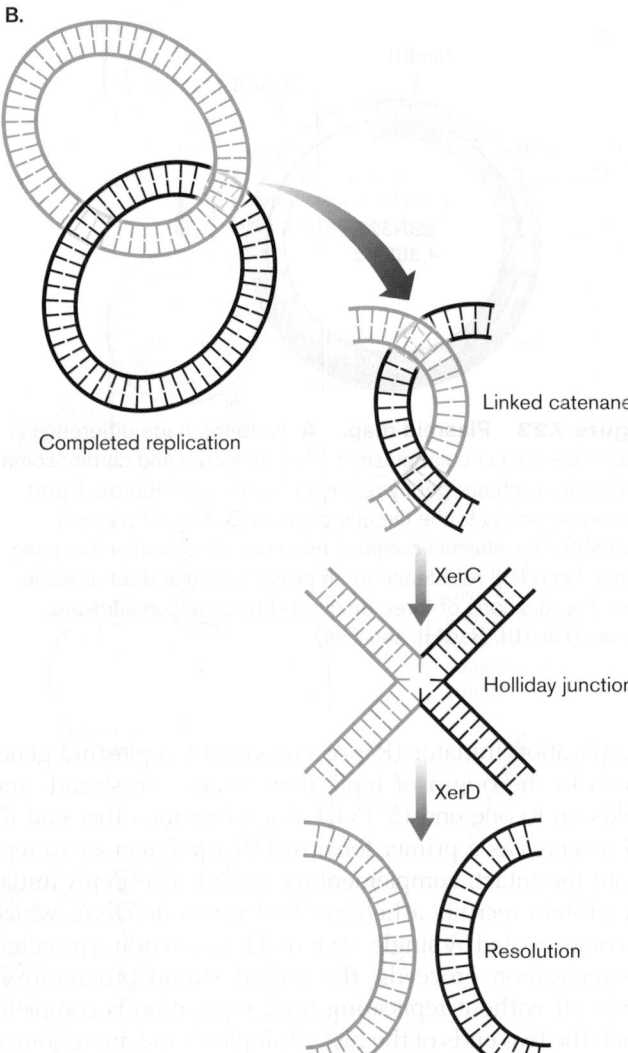

B.

Figure 7.21 Terminating replication of the chromosome. A. Terminator regions for DNA replication on the *E. coli* chromosome. Replication forks moving clockwise are trapped by *terJ*, *terG*, *terF*, *terB*, and *terC*. Counterclockwise-moving forks are trapped by *terA*, *terD*, and *terE*. **B.** Resolution of DNA replication catenanes. On the left is a linked-chromosome catenane. The highlighted area is enlarged to the right. XerC and XerD catalyze a breaking and re-joining that passes the chromosomes through each other, resolving the link.

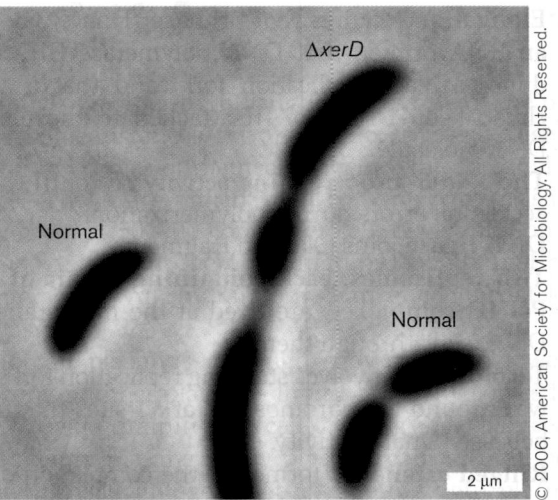

Figure 7.22 Effect of a *xerD* mutation on *Caulobacter* cell division. *Caulobacter* replicates by producing a motile cell at one end of a nonmotile, fixed, stalked cell. The *xerD* mutant cell is elongated with multiple constrictions because of a defect in chromosomal segregation. *Source: R. B. Jensen. 2006. J. Bacteriol.* **188**:6016–6019.

THOUGHT QUESTION 7.7 Individual cells in a population of *E. coli* typically initiate replication at different times (asynchronous replication). However, depriving the population of a required amino acid can synchronize reproduction of the population. What happens is that ongoing rounds of DNA synthesis finish, but new rounds do not begin. Replication stops until the amino acid is once again added to the medium—an action that triggers simultaneous initiation in all cells; that is, reproduction of the population becomes synchronized. Why?

THOUGHT QUESTION 7.8 The antibiotic rifampin inhibits transcription by RNA polymerase, but not by primase (DnaG). What happens to DNA synthesis if rifampin is added to a synchronous culture?

www | DNA replication in archaea
microbiology2.com/links

TO SUMMARIZE:

- **Replication is semiconservative**, with newly synthesized strands lengthening in a 5′-to-3′ direction. It involves initiation, elongation, and termination.
- **Replication is initiated** from a fixed DNA origin attached to the cell membrane. Initiation depends on the mass and size of the growing cell. It is controlled by the accumulation of initiator and repressor proteins and by methylation at the origin.

shortened remarkably over the past decade. It originally took many years to sequence the 4.6-million-bp genome of *E. coli*. Today it would take less than a month. The top DNA sequencing laboratories can sequence nearly 10 million bases per month.

How the information from sequences is mined will be discussed more in Chapter 8. Suffice it to say, many important discoveries have resulted from this technology. A particularly intriguing recent discovery was finding that the entire genome of the bacterium *Wolbachia* (which infects 20% of the world's insect population) is embedded within the genome of *Drosophila*, the common fruit fly. The evolutionary implications of this discovery are under intense investigation.

Metagenomics

Microorganisms in nature do not generally exist as pure cultures. They grow instead as complex consortia containing numerous species, most of which have never been grown in a laboratory. With newer and ever-faster DNA sequencing methodologies and PCR techniques, we can now "peer" deeper into the natural world. Like the explorers of old, scientists are finding new species never before seen and are beginning to understand the intricate ways bacterial species interact to form what could be called biological support groups.

Metagenomics is the use of modern genomic techniques to study microbial communities directly in their natural environments, bypassing the need for isolation and laboratory cultivation of individual species. The field began with the culture-independent retrieval of 16S rRNA genes from the environment by amplifying DNA directly from environmental samples using PCR protocols. Since genes encoding 16S rRNA from small ribosomal subunits (also called small-subunit rRNA) are highly conserved, it was possible to determine whether a given sequence of 16S rRNA represented a new species and to which known species it was most closely related. Since then, metagenomics has revolutionized microbiology by shifting focus even further away from clonal isolates capable of growth in the lab toward the estimated 99% of microbial species that cannot currently be cultivated. Since 2003 over 30 metagenomic projects have been carried out.

A typical metagenomic project begins with the construction of a clone library from DNA sequences retrieved from an environmental sample (soil, for example). Many sequence reads derived from a library remain as unassembled one-pass sequences because of the variety of sizes of environmental genomes and their abundance. Indeed, half of the reads for the Sargasso Sea data set and all of the reads for the Minnesota soil data set were unassembled sequences of roughly 700 bp.

Current metagenomic protocols generally use 16S rRNA sequence hits to deduce the various species of organisms present in a sample. The 16S information can also help determine how the many unknown organisms present in that sample relate to known species. But random sequencing will also identify most if not all of the other genes present in all the organisms that the consortium comprises. All of these genes make up what is known as the metagenome—that is, the combined yet unsorted genomic potential of all the organisms in the sample. Thus, metagenomics not only identifies the principal partners of a microbial consortium but also reveals many of their metabolic capabilities. Communities of microorganisms can actually form new metabolic pathways in which a useless by-product of one member can be used by another member in a reaction whose product can benefit a third member.

The science has even progressed to the point that if the sample contains a low number of species, whole-genome shotgun (WGS) sequencing methods such as pyrosequencing (see **Special Topic 7.1**) can reconstruct complete or nearly complete genomes for those species, whether bacteria, archaea, or viruses. The key is assembling hundreds of thousands of sequences generated by random sequencing into overlapping contigs. Powerful new genome-assembling computer algorithms allow simultaneous building of complete or nearly complete genomes from a mixture of microbes taken from a single sample.

A major strength of metagenomics is its potential for serendipitous discovery. One example where metagenomic data led to an unexpected finding came from studies of archaeal populations in acid mine drainage. DNA sequence information gained from these studies was used to show that genetic recombination occurs at a much higher frequency than previously predicted, and is the primary evolutionary force shaping these populations. In another example, data from the Pacific and Atlantic oceans revealed that the greatest variation within populations of *Prochlorococcus* (the most abundant photosynthetic organism in the ocean) occurs in genomic "islands." These islands are discrete regions in the genome that are believed to be hot spots for genomic innovation. They are derived in part from genes horizontally transferred from other organisms by viruses.

The power of metagenomics has also recently been directed toward solving the human microbiome. There are 100 trillion microbes (10 microbial cells for every one human cell) in the human body, but we do not know the identity of most of them and are ignorant as to how to grow them in the laboratory. How we interact with these organisms and they with us is a major question of modern microbiology—one that has led to the formation of the Human Microbiome Project, whose major focus is the gut. We already know that the composition of intestinal flora can influence blood chemistry and may contribute to irritable bowel syndrome (inflammation of the gastrointestinal tract). It is also suspected of influencing an individual's susceptibility to diabetes and obesity. Consequently, knowing the identity of these microbes and their collective metabolic potential will undoubtedly lead to new advances in the treatment and prevention of human diseases.

TO SUMMARIZE:

- **DNA restriction endonucleases** used for DNA analysis cleave DNA at specific recognition sequences, which are usually 4–6 bp in length and produce either a blunt or a staggered cut.
- **Agarose electrophoresis** will separate DNA molecules on the basis of their size.
- **Restriction endonuclease–digested DNA** molecules were first cloned into plasmids in the early 1970s.
- **The polymerase chain reaction (PCR)** will amplify one or two copies of a specific gene a millionfold.
- **The most commonly used DNA sequencing** technique (Sanger technique) uses nucleotides that lack both the 2′ OH and 3′ OH groups to terminate chain elongation.
- **Entire genomes** are sequenced by shotgun cloning of overlapping DNA fragments, each of which is sequenced. Overlapping regions are matched and joined by computer analysis until the entire genome is reconstructed.
- **Metagenomics** uses rapid DNA sequencing and other genomic techniques to study consortia of microbes directly in their natural environment.

Concluding Thoughts

We used to think of bacteria as simple, single-celled organisms. In fact, they are far from simple. Complexity is evident in the elegant mechanisms they use to organize and replicate their genomes. The next three chapters will expand on this idea and discuss the way bacteria make proteins, regulate genes, exchange DNA, and, in the process, evolve. Knowing how a bacterial cell replicates and genetically controls the repertoire of available proteins and enzymes provides a unique perspective from which to study the physiology of microbial growth, explored in Part 3 of this textbook. Furthermore, understanding the bacterial genome allows us to explore the physiology of pathogenic as well as nonculturable organisms and gain greater insight into the evolutionary diversity of species that we will discuss in Part 4.

CHAPTER REVIEW

Review Questions

1. What is the difference between vertical and horizontal gene transmission?
2. Explain the structural types of bacterial genomes. What is a structural gene?
3. Describe the three functional levels of gene organization.
4. What are the differences between DNA and RNA?
5. Explain DNA supercoiling. Why is it important to microbial genomes?
6. Discuss the mechanisms of topoisomerases. What drug targets a topoisomerase?
7. What are the basic mechanisms involved in DNA replication?
8. How does the bacterial cell regulate the initiation of chromosome replication?
9. What is the clamp loader? Primase? DNA helicase (DnaB)? Helicase loader (DnaC)? DNA proofreading?
10. How is the problem of replicating both strands at a replication fork solved?
11. What is a catenane? What does it have to do with DNA replication?
12. Explain the polymerase chain reaction.
13. What is rolling-circle replication?
14. What is DNA restriction and modification? Why is it important to bacteria? Why is it important to forensic scientists?
15. Explain the basic technique of DNA sequencing.

Thought Questions

1. When you sequence a genome, the DNA used is in fragments. How do you know where the base pairs in the genome are located?
2. During rapid growth, why would a bacterial cell die if an antibiotic drug formed "a physical barrier in front of the DNA replication complex"?
3. Why would bacteria, which replicate more quickly than eukaryotic organisms, lose pseudogenes more rapidly than eukaryotes do?

Key Terms

antiparallel (226)
catenane (240)
cloning (250)
conjugation (223)
contig (253)
denature (227)
DNA control sequence (222)
DNA ligase (237)
enhancer (224)
exonuclease (237)
genome (222)
genome library (249)
histone (245)
horizontal transmission (222)
hybridization (228)
intron (246)

metagenomics (254)
nucleoid (228)
Okazaki fragments (234)
operon (225)
origin (*oriC*) (233)
palindrome (247)
phosphodiester bond (226)
plasmid (242)
polymerase chain reaction (PCR) (249)
primase (236)
promoter (224)
proofreading (237)
pseudogene (246)
purine (227)
pyrimidine (227)

pyrosequencing (252)
quinolone antibiotic (231)
recombination (223)
regulon (225)
replication fork (232)
replisome (237)
restriction endonuclease (247)
restriction site (250)
semiconservative (232)
shuttle vector (249)
sliding clamp (236)
structural gene (222)
termination (*ter*) site (233)
topoisomerase (230)
vertical transmission (222)

Recommended Reading

Amick, Jean D., and Yves V. Brun. 2001. Anatomy of a bacterial cell cycle. *Genome Biology* 2:1020.1–1020.4.

Bates, David. 2008. The bacterial replisome: Back on track? *Molecular Microbiology* 69:1341–1348.

Boeneman, Kelly, and Elliott Crooke. 2005. Chromosomal replication and the cell membrane. *Current Opinion in Microbiology* 8:143–148.

Campbell, Christopher S., and R. Dyche Mullins. 2007. In vivo visualization of type II plasmid segregation: Bacterial actin filaments pushing plasmids. *Journal of Cell Biology* 179:1059–1066.

Cox-Foster, Diana L., Sean Conlan, Edward C. Holmes, Gustavo Palacios, Jay D. Evans, et al. 2007. A metagenomic survey of microbes in honey bee colony collapse disorder. *Science* 318:283–287.

Dillon, Shane, and Charles J. Dorman. 2010. Bacterial nucleoid-associated proteins, nucleoid structure, and gene expression. *Nature Reviews. Microbiology* 8:185–195.

Duggin, Ian G., R. Gerry Wake, Stephen D. Bell, and Thomas M. Hill. 2008. The replication fork trap and termination of chromosome replication. *Molecular Microbiology* 70:1323–1333.

Falkow, Stanley. 2001. I'll have chopped liver please, or how I learned to love the clone. *ASM News* 67:555.

Forterre, Patrick, Simon Gribaldo, Daniele Gadelle, and Marie-Claude Serre. 2007. Origin and evolution of DNA topoisomerases. *Biochimie* 89:427–446.

Giovannoni, Stephen J., H. James Tripp, Scott Givan, Mircea Podar, Kevin L. Vergin, et al. 2005. Genome streamlining in a cosmopolitan oceanic bacterium. *Science* 309:1242–1245.

Katayma, Tsutomu, Shogo Ozaki, Kenji Keyamuaa, Kazuyuki Fujimitsu. 2010. Regulation of the replication cycle: conserved and diverse regulatory systems for DnaA and oriC. *Nature Reviews. Microbiology* 8:163–170.

Matsunaga, Fujihiko, Kie Takemura, Masaki Akita, Akinori Adachi, Takeshi Yamagami, Yoshizumi Ishino. 2010. Localized melting of duplex DNA by Cdc6/Orc1 at the DNA replication origin in the hyperthermophilic archaeon *Pyrococcus furiosus. Extremophiles* 14:21–31.

Mott, Melissa L., and James M. Berger. 2007. DNA replication initiation: Mechanisms and regulation in bacteria. *Nature Reviews. Microbiology* 5:343–354.

Thanbichler, Martin, Patrick H. Viollier, and Lucy Shapiro. 2005. The structure and function of the bacterial chromosome. *Current Opinion in Genetic Development* 115:153–162.

Vologodskii, Alexander. 2010. DNA supercoiling helps unlink sister duplexes after replication. *Bioessays* 32:9–12.

For further study and review, please visit StudySpace at **microbiology2.com**.

Chapter 8

Transcription, Translation, and Bioinformatics

The cell accesses the vast store of data in its genome by reading a DNA template to make an RNA copy (transcription); and decoding the RNA to assemble protein (translation). The molecular machines that carry out these processes have been studied relentlessly, yet they still hold secrets.

After translation, each polypeptide must be properly folded and placed at the correct cellular or extracellular location. How does the cell do this? And what does the cell do with proteins it no longer needs? As we will discuss, elegant chaperone pathways help fold proteins, complex transport systems move proteins out of the cytoplasm, and regulated proteolytic systems properly dispose of unneeded or damaged proteins. What emerges is a picture of remarkable biomolecular integration, homeostatically controlled to maintain balanced growth and ensure survival.

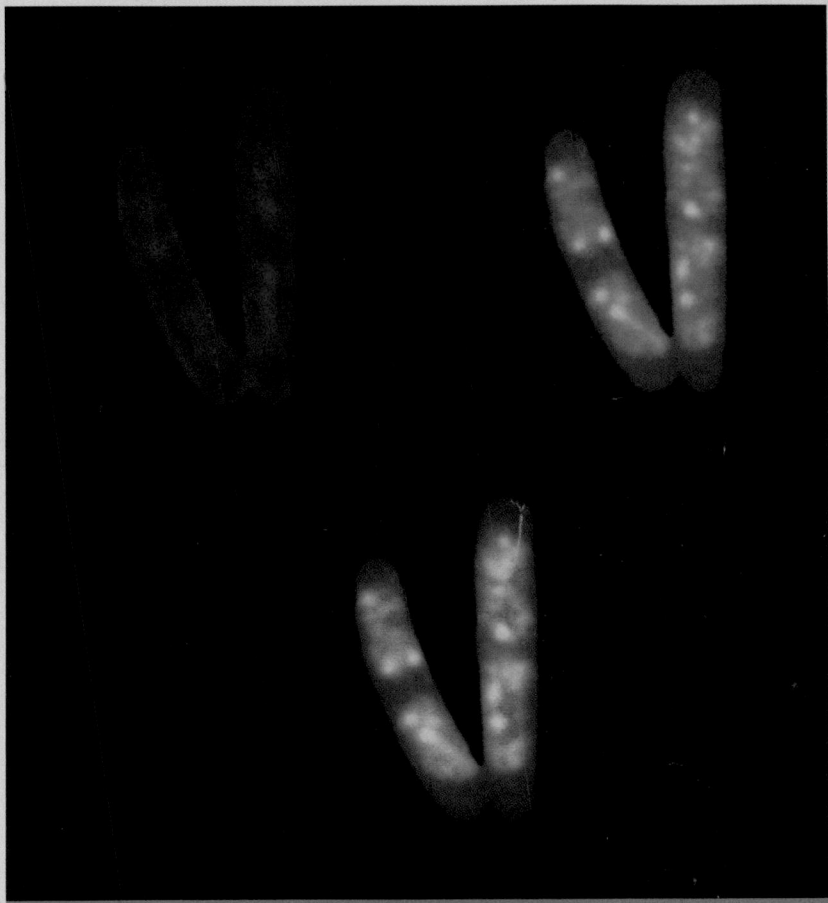

Current Research Highlight

RNA polymerase molecules transcribe DNA at multiple sites. This series of fluorescent images reveals multiple foci of RNA polymerase within growing cells, reflecting the simultaneous transcription of numerous genes. The first image shows the *Escherichia* coli nucleoid (red). The second image shows distribution of RNA polymerase (tagged with green fluorescent protein). The third image is a composite (overlay of the first two images). The article provides additional images showing the nucleoid elongating when transcription is stopped with rifamycin. The images provide evidence that active transcription helps drive nucleoid condensation. *Source*: Julio E. Cabrera et al. 2009. *J. Bacteriol.* **191** 4180–4185. © 2009, American Society for Microbiology. All Rights Reserved.

By 1960, it was clear that the genetic code is embedded in DNA, but the code itself remained a mystery, as did the mechanism by which the code produces protein. RNA was a suspected, but unproven, intermediate. Marshall Nirenberg and Heinrich Matthaei, a postdoctoral student of Nirenberg's at the National Institutes of Health (**Fig. 8.1A**), set out in 1959–60 to design a cell-free system (a cell lysate of *E. coli*) to test the hypothesis. Key to their investigation was the use of synthetic RNA molecules constructed by Maxine Singer (**Fig. 8.1B**). These contained simple, known, repeated sequences such as poly-A (consisting only of adenylic acid), poly-U (polyuridylic acid), poly-AAU, and poly-ACAC. These RNA molecules were tested in a cell-free protein-synthesizing system to see if the repetitive sequence might direct incorporation of a specific amino acid into a protein. Each synthetic polynucleotide was tested in the presence of a radiolabeled amino acid. If the radioactive amino acid was incorporated into a polypeptide, the polypeptide would also be radioactive. On the morning of May 27, 1960, the results of experiment 27Q indicated that the poly-U RNA specified

the assembly of radioactive polyphenylalanine. It was the first break in the genetic code.

Nirenberg reported the result at a conference in Moscow at the height of the Cold War. News that Nirenberg's poly-U experiment had determined the first "word" of the genetic code was an international media event. In January 1962, the *Chicago Sun-Times* announced: "No stronger proof of the universality of all life has been developed since Charles Darwin's *The Origin of Species* . . . In the far future, the . . . hereditary lineup will be so well known that science may deal with the aberrations of DNA . . . that produce cancer, aging, and other weaknesses of the flesh." This statement was truly prophetic, for understanding the genetic code was the first step in discovering the basis of many genetic diseases. For his work, Nirenberg, along with Har Gobind Khorara (who developed methods for making synthetic nucleic acids) and Robert Holley (who solved the structure of yeast transfer RNA), won the 1968 Nobel Prize in Physiology or Medicine.

Since Nirenberg's breakthrough, we have learned a tremendous amount about genes and proteins, and in this electronic age we can use the information in ways never before possible. A new discipline called bioinformatics uses powerful computer technologies to store and analyze gene and protein sequences. Bioinformatic programs can compare a new gene sequence with hundreds of thousands of known genes. These comparisons help predict for any given microbe what genes are present, what proteins are made, and even what food it consumes. Patterns in DNA sequences across species also allow us to pose new questions about microbial life, disease, and evolution.

In Chapter 8, we use *E. coli* as a paradigm to explore the way microbes interpret the nucleotide sequence of DNA and convert it to proteins. From there we look at what the cell does with those proteins once they are made. Specific proteins are targeted for movement into the periplasm, while other proteins must be inserted into membranes. We also show how damaged proteins are selectively degraded. We conclude with bioinformatics, now an essential tool of the new microbiology. Chapters 9 and 10 then discuss how microbes respond to their environment by controlling the expression of their genes.

Figure 8.1 Many scientists contributed to understanding the genetic code. **A.** Heinrich Matthaei (left) and Marshall Nirenberg (right) were first to crack the genetic code. An early hand-drawn model of what would eventually be known as translation is behind them. **B.** Maxine Singer, a key contributor to the genetic code experiments, also helped develop guidelines for recombinant DNA research.

8.1 RNA Polymerases and Sigma Factors

To survive and reproduce, every cell needs to form proteins using the information encoded within DNA. Chromosomal DNA is large and cumbersome, so the first step in the process is to make multiple copies of the information in snippets of RNA that can move around the cell and, like disposable photocopies, be destroyed once the encoded protein is no longer needed. This copying of DNA to RNA is called **transcription**. Much of what we have learned

about transcription has come from studying *E. coli*. However, bioinformatic analyses indicate that the mechanisms and components of transcription in other bacterial species are similar to those characterized in *E. coli*.

RNA Polymerase Transcribes DNA to RNA

An enzyme complex called RNA polymerase, also known as DNA-dependent RNA polymerase, carries out the process of transcription, making RNA copies (called **transcripts**) of a DNA template. The DNA **template strand** specifies the base sequence of the new complementary strand of RNA. Transcription involves (1) initiation, which is the binding of RNA polymerase to the beginning of the gene, followed by melting of the helix and synthesis of the first nucleotide of the RNA; (2) elongation, the sequential addition of ribonucleotides from nucleoside triphosphates; and (3) termination, in which sequences at the end of the gene trigger release of the polymerase and the completed RNA molecule.

RNA polymerase in bacteria consists of a "core polymerase," which is a protein complex containing the *essential* components required for elongation of the RNA chain, plus a **sigma factor**, a protein needed only for initiation of RNA synthesis, not for its elongation. Core polymerase plus sigma factor together are called "holoenzyme."

Core RNA polymerase is a complex of four different subunits: two alpha (α) subunits, one beta (β) subunit, and one beta-prime (β') subunit for every core enzyme (**Fig. 8.2**). The beta-prime subunit contains the Mg^{2+}-containing catalytic site for RNA synthesis, as well as sites for the ribonucleotide and DNA substrates and the RNA products. The three-dimensional structure of RNA polymerase resembles a hand, whose "fingers" consist of the beta and beta-prime subunits. DNA fits into a cleft formed by the beta and beta-prime subunits (**Fig. 8.2**). The alpha

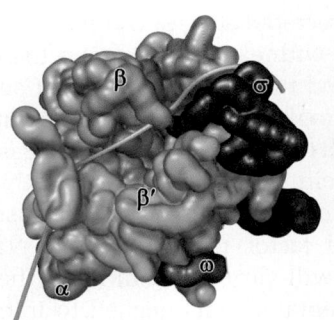

Figure 8.3   Three-dimensional structure of RNA polymerase.   RNA polymerase holoenzyme. Note the open channel able to accept DNA. To view the stereo image, locate code 1L9U in the RCSB Protein Data Bank on the Internet.

subunit assembles the other two subunits into a functional complex and communicates through physical "touch" with various regulatory proteins that can bind DNA. The resulting protein-protein communications instruct RNA polymerase in what to do after binding DNA.

Promoter binding. Where does a gene begin and end? The helical structure of a DNA sequence appears largely homogeneous compared to proteins, which bend and fold in complex ways. In fact, without sigma factor the core RNA polymerase binds and releases DNA at random. Yet there must be a mechanism that tells RNA polymerase where a gene starts, because random transcription is extremely wasteful.

The proteins that guide RNA polymerase to genes are the sigma (σ) factors (see **Fig. 8.3**). A sigma factor binds RNA polymerase through the alpha subunit and then helps the core enzyme detect a specific DNA sequence, called a **promoter**, which signals the beginning of a gene. We will discuss how sigma recognizes a promoter later.

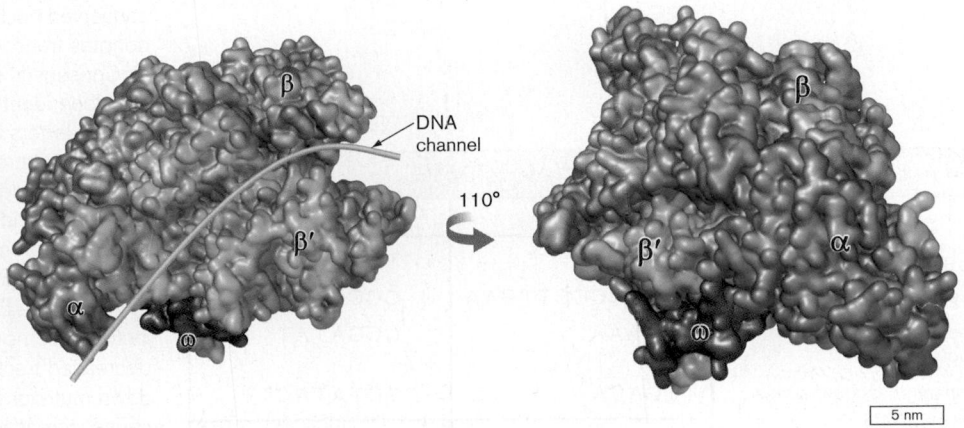

Figure 8.2   Subunit structure of core RNA polymerase.   Two views of RNA polymerase. The channel for the DNA template is shown by the yellow line. Subunits are color-coded. Omega (ω) is a subunit of unclear function sometimes associated with RNA polymerase. The molecule in (A) is rotated as shown to give the image in (B). To view stereo images of RNA polymerase, locate code 1HQM in the RCSB Protein Data Bank on the Internet. *Source:* Robert D. Finn et al. 2000. *EMBO J.* **19**:6833–6844.

A single bacterial species can make several different sigma factors, with each sigma factor helping core RNA polymerase find the start of a different subset of genes. However, a single core polymerase complex can bind only one sigma factor at a time. By acting as a sort of "seeing-eye" protein, sigma factors help RNA polymerase recognize different classes of promoter sequences. The specific sigma factor used to initiate transcription of a given gene will vary, depending on the gene and on the environmental signals needed to initiate transcription of that gene. But regardless of which sigma factor is used, core RNA polymerase is required for all gene transcription.

Sigma factors regulate major physiological responses. Individual sigma factors coordinately control genes involved in nitrogen metabolism, flagellar synthesis, heat stress, starvation, sporulation, and many other physiological responses. So far, over 100 sigma factor genes have been sequenced from numerous species. Although they all have different amino acid sequences, they also show sequence similarities that make them recognizable as sigma factors. We can determine the DNA sequence recognized by a given sigma factor by comparing known promoter sequences of different genes whose expression requires the same sigma. The DNA sequence similarities seen among these different promoters define a **consensus sequence** likely recognized by the sigma factor (**Fig. 8.4A** and **B**). A consensus sequence consists of the most likely base (or bases) at each position of the predicted promoter. Note that although the promoter is a double-stranded DNA

A. Strong *E. coli* promoters

```
                          -35                                              -10                  +1
tyr tRNA    TCTCAACGTAACACTTTACAGCGGCG··CGTCATTTGATATGATGC·GCCCCGCTTCCCGATAAGGG
rrn D1      GATCAAAAAAATACTTGTGCAAAAAA··TTGGGATCCCTATAATGCGCCTCCGTTGAGACGACAACG
rrn X1      ATGCATTTTTCCGCTTGTCTTCCTGA·GCCGACTCCCTATAATGCGCCTCCATCGACACGGCGGAT
rrn (DXE)₂  CCTGAAATTCAGGGTTGACTCTGAAA·GAGGAAAGCGTAATATAC·GCCACCTCGCGACAGTGAGC
rrn E1      CTGCAATTTTTCTATTGCGGCCTGCG·GAGAACTCCCTATAATGCGCCTCCATCGACACGGCGGAT
rrn A1      TTTTAAATTTCCTCTTGTCAGGCCGG··AATAACTCCCTATAATGCGCCACCACTGACACGGAACAA
rrn A2      GCAAAAATAAATGCTTGACTCTGTAG··CGGGAAGGCGTATTATGC·ACACCCGCGCCGCTGAGAA
λ PR        TAACACCGTGCGTGTTGACTATTTTA·CCTCTGGCGGTGATAATGG··TTGCATGTACTAAGGAGGT
λ PL        TATCTCTGGCGGTGTTGACATAAATA·CCACTGGCGGTGATACTGA··GCACATCAGCAGGACGCAC
T7 A3       GTGAAACAAAACGGTTGACAACATGA·AGTAAACGGTACGATGT·ACCACATGAAACGACAGTGA
T7 A1       TATCAAAAAGAGTATTGACTTAAAGT·CTAACCTATAGGATAGTACTA·CAGCCATCGAGAGGGACACG
T7 A2       ACGAAAAACAGGTATTGACAACATGAAGTAACATGCAGTAAGATAC·AAATCGCTAGGTAACACTAG
fd VIII     GATACAAATCTCCGTTGTACTTTGTT··TCGCGCTTGGTATAATCG·CTGGGGGGTCAAAGATGAGTG
```

B. Consensus sequences of σ⁷⁰ promoters

-35 region   -10 region
TTGACAT —17 ± 1 bp— TATAAT

C. Sequence of *lac* promoter

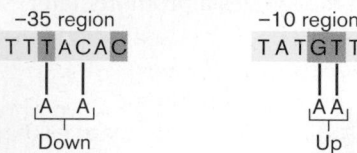

-35 region   -10 region
TTTACAC   TATGTT

A   A   A A
Down   Up

D.

Sigma factor	Promoter recognized	Promoter consensus sequence	
		-35 region	-10 region
RpoD σ⁷⁰	Most genes	TTGACAT	TATAAT
RpoH σ³²	Heat shock–induced genes	TCTCNCCCTTGAA	CCCCATNTA
RpoF σ²⁸	Genes for motility and chemotaxis	CTAAA	CCGATAT
RpoS σ³⁸	Stationary-phase and stress response genes	TTGACA	TCTATACTT
		-24 region	-12 region
RpoN σ⁵⁴	Genes for nitrogen metabolism and other functions	CTGGNA	TTGCA

Figure 8.4   -10 and -35 sequences of *E. coli* promoters. A. Alignment of sigma-70 (σ⁷⁰)–dependent promoters from different genes is used to generate a consensus sequence in panel (B) (yellow indicates conserved nucleotides; brown denotes transcript start site, +1). **B.** Consensus sequences of σ⁷⁰-dependent promoters (red-shaded letters indicate nucleotide positions where different promoters show a high degree of variability). **C.** Mutations in the *lac* promoter that affect promoter strength (*lac* genes encode proteins used to metabolize the carbohydrate lactose). Some mutations can cause decreased transcription (called down mutations), while others cause increased transcription (up mutations). **D.** Some *E. coli* promoter sequences recognized by different sigma factors.

(dsDNA) sequence, convention is to present the promoter as the single-stranded DNA (ssDNA) sequence of the sense strand, which has the same sequence as the RNA product.

The best consensus sequences are based on a large set of promoters that use the same sigma factor, or other regulator. Some positions in a consensus sequence are highly conserved—that is, the same base is found in that position in every promoter. Other, less conserved positions can be occupied by different bases. Few promoters will actually have the most common base at every position. Even highly efficient promoters usually differ from the consensus at one or two positions.

Microbes are guarded by an array of protein sensors that continually sample the internal and external environments. These sensors "look" for chemical deficiencies or dangers and, when triggered, direct the cell to increase synthesis or decrease destruction of the appropriate sigma factor. Accumulation of a specialty sigma factor dislodges other sigma factors from core polymerase and redirects RNA polymerase to the promoters of genes whose products can best solve the problem sensed. Regulation of multiple genes by a single sigma factor enables the cell to coordinately control those genes simply by regulating when a given sigma factor accumulates (Section 10.4). This control is critical to survival in constantly changing environments.

> **THOUGHT QUESTION 8.1**   If each sigma factor recognizes a different promoter, how does the cell manage to transcribe genes that respond to multiple stresses, each involving a different sigma factor?

Every cell has a "housekeeping" sigma factor that keeps essential genes and pathways operating. In the case of *E. coli* and other Gram-negative rod-shaped bacteria, that factor is sigma-70, so named because it is a 70-kilodalton (kDa) protein (its gene designation is *rpoD*). Genes recognized by sigma-70 all contain similar promoter consensus sequences consisting of two parts. The DNA base corresponding to the start of the RNA transcript is called nucleotide +1 (+1 nt). Relative to this landmark, the consensus promoter sequences are characteristically centered at –10 and –35 nt before the start of transcription (see **Fig. 8.4**). Other sigma factors typically recognize different consensus sequences at one or both of these positions (or at nearby locations in some cases); or they shape the overall polymerase complex to bind the promoter.

How do sigma factors, or any other DNA-binding proteins for that matter, recognize specific DNA sequences when the DNA is a double helix? The phosphodiester backbone is quite uniform, and the bases appear inaccessible owing to base pairing. Recognition of DNA sequences is possible, however, because the alpha-helical portions of DNA-binding proteins recognize base side groups (noncovalent binding) that protrude from the base into the major and minor grooves of DNA. **Figure 8.5A** illustrates how portions of sigma-70 from *E. coli* wrap around DNA, allowing certain parts of the protein to fit into DNA grooves.

A key to understanding how sigma factors recognize different promoters is their structure. Amino acid sequence comparisons indicate that sigma factors generally contain four highly conserved regions (**Fig. 8.5A and B**). Part of region 2 of the sigma-70 family recognizes –10 sequences, while region 4 recognizes the –35 sites. A part of region 1 helps separate the DNA strands in preparation for RNA synthesis. The structure of the holoenzyme RNA polymerase complex positioned at a promoter is shown in **Figure 8.5C**.

> **THOUGHT QUESTION 8.2**   With respect to two different sigma factors with different promoter recognition sequences, predict what would happen to the overall gene expression profile in the cell if one sigma factor were artificially overexpressed. Could there be a detrimental effect on growth?
>
> **THOUGHT QUESTION 8.3**   Why might some genes contain multiple promoters, each one specific for a different sigma factor?

TO SUMMARIZE:

- **RNA polymerase holoenzyme** consisting of core RNA polymerase and a sigma factor executes transcription of a DNA template strand.
- **Sigma factors** help core RNA polymerase locate consensus promoter sequences near the beginning of a gene. The sequences identified and bound by *E. coli* sigma-70 are located –10 and –35 bp upstream of the transcription start site.
- **Dynamic changes in gene expression** result when the relative levels of different sigma factors change.

8.2 Transcription of DNA to RNA

Sigma binding is the first step of transcription initiation. Once the promoter has been activated by binding with its sigma factor, RNA polymerase completes the process of initiation, in which the closed RNA polymerase complex becomes an open complex. In elongation, the RNA chain is extended. The final step is termination, in which RNA polymerase detaches from the DNA.

A.

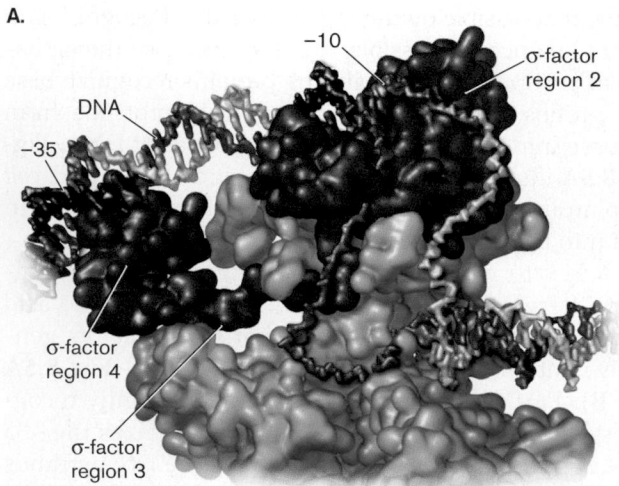

Figure 8.5 How sigma factors recognize specific DNA double-helix sequences. **A.** *E. coli* sigma factor binds to two sequences in a DNA promoter. Region 2 of sigma binds to the −10 region of promoters, while region 4 binds to the −35 promoter region. Cyan marks part of the core RNA polymerase. Purple on DNA denotes the −10 and −35 promoter regions. **B.** Sigma-70 amino acid homology and domain interactions with different regions of the promoter sequence. The primary amino acid sequences of different sigma factors are represented in linear form. Colored areas indicate conserved amino acid regions within examples of the sigma-70 family. **C.** Taq holoenzyme–promoter DNA complex. Double-stranded DNA is shown as atoms. The possible locations of the two alpha subunit C-terminal domains (drawn as green spheres labeled I and II) on the DNA UP elements are illustrated. The UP element is a DNA sequence that increases transcription. (PDB codes: 1L9Z, 1K8J) *Source*: Parts A and C from Seth A. Darst. 2001. *Curr. Opin. Struct. Biol.* **11**:155–162.

B.

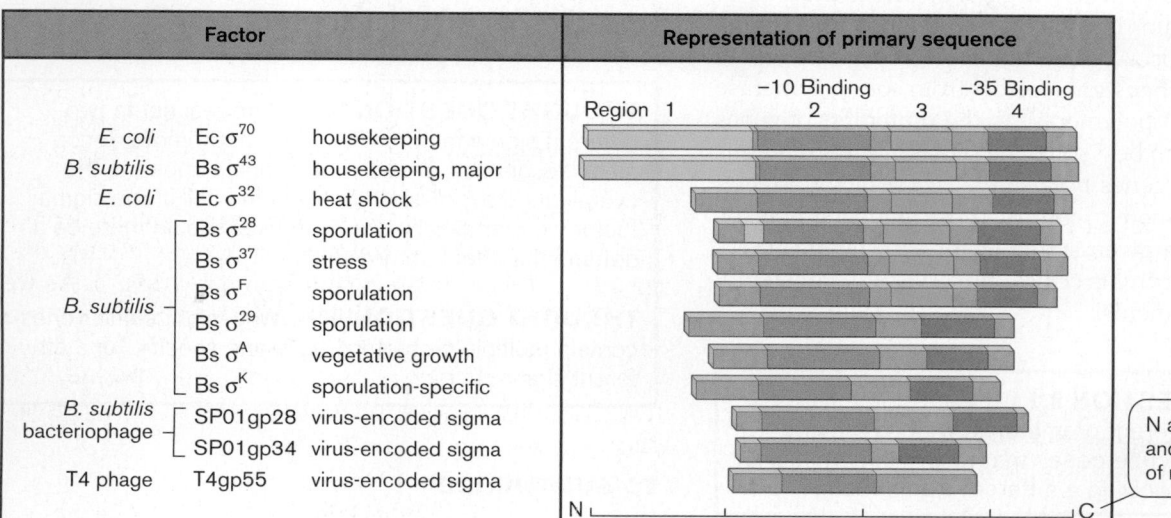

C.

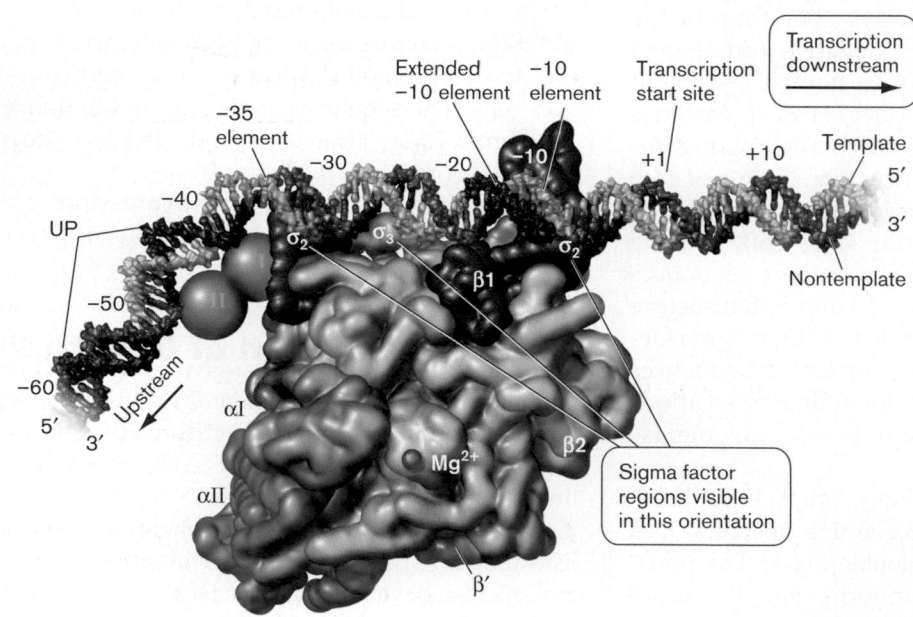

The Three Stages of Transcription

Initiation of transcription. RNA polymerase constantly scans DNA for promoter sequences (**Fig. 8.6**, step 1). It binds DNA loosely and comes off repeatedly. Once bound to the promoter, RNA polymerase holoenzyme forms a loosely bound, closed complex with DNA, which remains annealed and double-stranded (that is, unmelted) (step 2). To successfully transcribe a gene, this closed complex must become an open complex through the unwinding of one helical turn, which causes DNA to become unpaired in this area (step 3). After promoter unwinding, RNA polymerase in the open complex becomes tightly bound to DNA. Region 1 of the sigma factor is important for DNA unwinding, as is a "rudder" complex in beta-prime that separates the two strands (shown in **Fig. 8.7A**).

The open-complex form of RNA polymerase begins transcription. The first ribonucleoside triphosphate (rNTP) of the new RNA chain is usually a purine. It base-pairs to the position designated +1 on the DNA template, which marks the start of the gene. As the enzyme complex moves along the template, subsequent NTPs diffuse through a channel in the polymerase and into position at the DNA template (**Fig. 8.7B**). After the first base is in place, each subsequent ribonucleoside triphosphate transfers a nucleoside monophosphate to the growing chain while releasing a pyrophosphate:

$$\text{RNA—3'—OH} + \text{}^-\text{O—P—O—P—O—P—O—nucleoside—3'—OH}$$

$$\downarrow$$

$$\text{RNA—3'—O—P—O—nucleoside—3'—OH} + \text{}^-\text{O—P—O—P—O}^-$$

The energy released by cleaving the rNTP triphosphate groups (phosphoric anhydride bond) is used to form the phosphodiester link to the growing polynucleotide chain.

As seen in **Figure 8.7**, the beta-prime subunit forms part of the pocket, or cleft, in which the DNA template is read. Another part of the beta-prime cleft is involved in hydrolyzing ribonucleoside triphosphates. Deep at the base of the cleft is the active site of polymerization, defined by three evolutionarily conserved aspartate residues that chelate (bond with) two magnesium ions (Mg^{2+}). As we saw for DNA polymerases, these metal ions play a key role in catalyzing the polymerization reaction. (The Mg^{2+} metal ions lower the pK_a of water. A metal-bound hydroxyl ion then becomes a potent nucleophile that facilitates phosphoryl transfer.)

WWW | 3-D view of RNA polymerase interaction with DNA
microbiology2.com/links

Elongation of RNA transcripts. The sigma factor remains associated with the transcribing complex until about nine bases have been joined; then it dissociates (**Fig. 8.6**, step 4). The newly liberated sigma factor can then reassociate with an unbound core RNA polymerase to direct another round of promoter binding. Meanwhile, the original RNA polymerase continues to move along the template, synthesizing RNA at approximately 45 bases per second. The unwinding of DNA ahead of the moving complex forms a 17-bp transcription bubble. Because of this unwinding, positive supercoils are formed ahead of the advancing bubble. These positive supercoils are removed by enzymes that return the DNA to its normal negative supercoiled state (see Section 7.2).

Termination of transcription. We have learned how RNA polymerase recognizes the beginning of a gene and transcribes RNA, but how does it know when to stop? Again, the secret is in the sequence. All bacterial genes

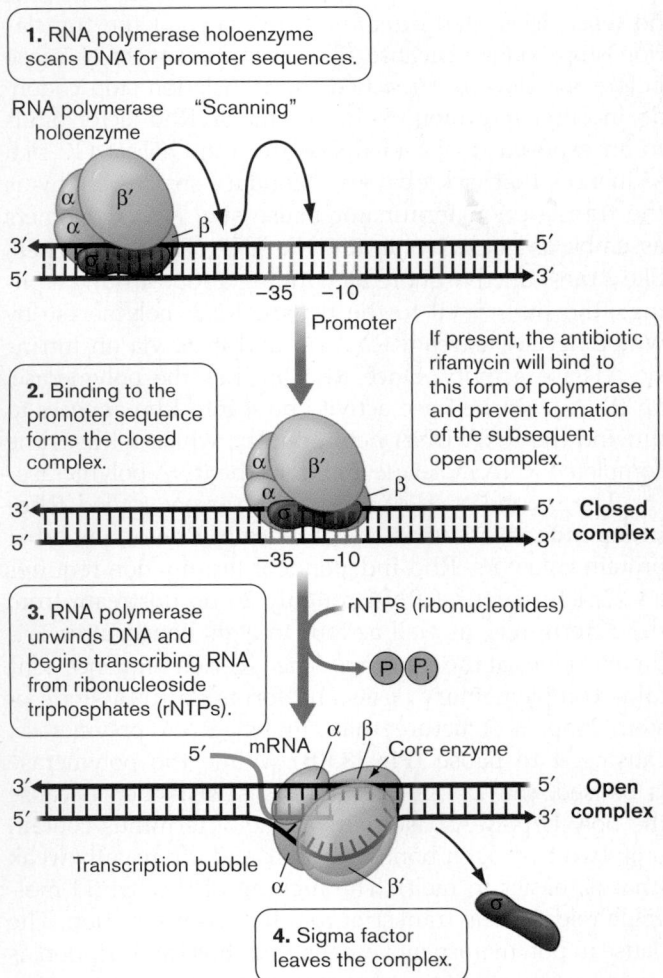

Figure 8.6 Transcription initiation.

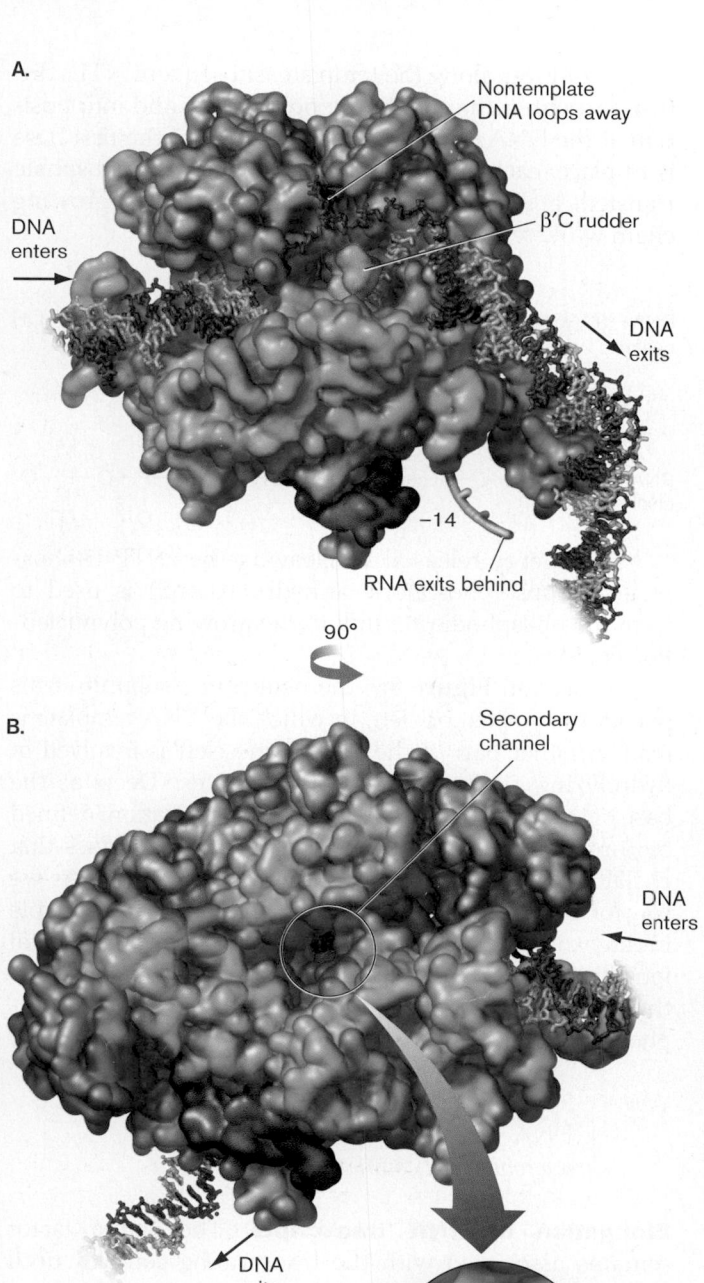

A.

Nontemplate
DNA loops away

DNA
enters

β'C rudder

DNA
exits

−14

RNA exits behind

90°

B.

Secondary
channel

DNA
enters

DNA
exits

Mg²⁺

Secondary channel

Figure 8.7 **Three-dimensional view of transcription by** *Thermus aquaticus* **(Taq) RNA polymerase showing DNA, RNA, and the position of the rudder.** The process is shown from two different angles. Transcription (polymerase movement on the DNA) is going from right to left in panel (A) and the opposite direction in panel (B). Color-coded molecular surfaces are as follows: beta subunit, green; beta G flap in red, beta-prime C rudder in gold; alpha and omega, not seen in this view; DNA template (template strand, purple; nontemplate strand, gray); RNA transcript, blue. The directions of the entering downstream DNA and exiting upstream DNA are indicated (black arrows). **A.** A view perpendicular to the main active-site channel, which runs roughly horizontal. Parts of the protein structure are colored. **B.** In this view, the structure is rotated 90° relative to view (A). With the entering DNA removed for clarity, Mg²⁺ is seen in the catalytic site; the secondary channel where nucleotides (orange) enter and a part of the RNA strand (blue) is also seen. (PDB codes: 1L9Z, 1K8J) Source: Seth A. Darst. 2001. *Curr. Opin. Struct. Biol.* **11**:155–162.

actually encodes a protein, to allow time for the ribosome to catch up to a transcribing RNA polymerase (discussed in Section 8.3). The transcription termination pause site, however, is located after the ORF, beyond the translation stop codon, because if transcription were to cease before the ribosome reached the translation stop codon, an incomplete protein would be made. Rho factor binds to an exposed region of RNA after the ORF at GC-rich sequences that lack obvious secondary structure. This is the transcription terminator pause site. Rho monomers assemble as a hexamer around the RNA (**Fig. 8.8A**). Then, like a raft pulled to shore by winding a rope around a piling, Rho pulls itself to the paused RNA polymerase by wrapping downstream RNA around itself via an intrinsic ATPase activity. Once Rho touches the polymerase, an RNA-DNA helicase activity built into Rho appears to unwind the RNA-DNA heteroduplex, which releases the completed RNA molecule and frees the RNA polymerase.

The second type of termination event, called **Rho-independent**, occurs in the absence of Rho or any other protein cofactors. Rho-independent termination requires a GC-rich region of RNA roughly 20 bp upstream from the 3′ terminus, as well as four to eight consecutive uridine residues at the terminus. The GC-rich sequence contains complementary bases and forms an RNA stem, or stem loop, a structure that "grabs" RNA polymerase, causing it to pause (**Fig. 8.8B**). While the polymerase is paused, the DNA-RNA duplex is weakened because the poly-U/poly-A base pairs at the 3′ terminus contain only two hydrogen bonds per pair and are usually weak (that is, easier to melt). The melting of the hybrid molecule releases the transcript and halts transcription. The pause in polymerase movement, and thus transcription, is important to prevent the formation of tighter base pairing downstream of the UA region.

use one of two known transcription termination signals. One termination mechanism, called **Rho-dependent**, relies on a protein called Rho and an ill-defined sequence at the 3′ end of the gene that appears to be a strong pause site. Pause sites are areas of DNA that slow or stall RNA polymerase movement. Some pause sites occur within the gene's **open reading frame (ORF)**, the part of a gene that

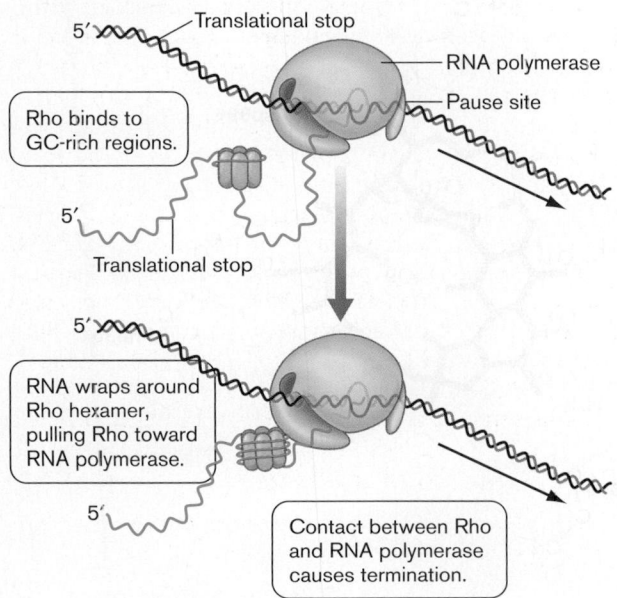

A. Rho-dependent termination

Translational stop

RNA polymerase

Pause site

Rho binds to GC-rich regions.

Translational stop

RNA wraps around Rho hexamer, pulling Rho toward RNA polymerase.

Contact between Rho and RNA polymerase causes termination.

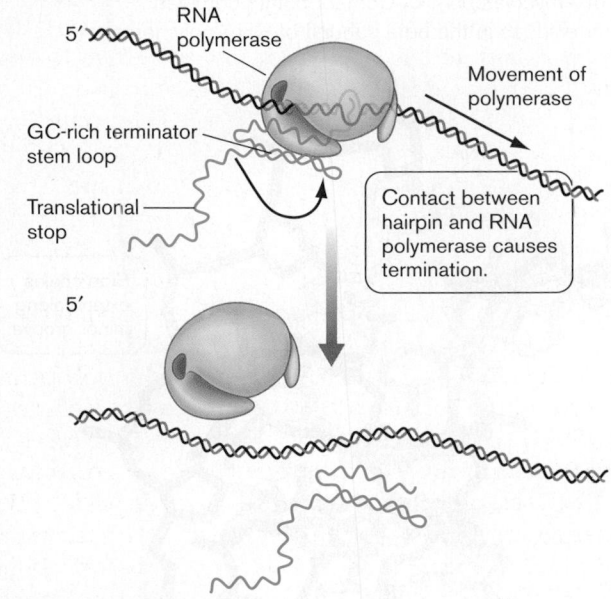

B. Rho-independent termination

RNA polymerase

Movement of polymerase

GC-rich terminator stem loop

Translational stop

Contact between hairpin and RNA polymerase causes termination.

Figure 8.8 Transcription termination.

Antibiotics That Affect Transcription

To be useful in medicine, all antibiotics must meet two fundamental criteria: They must kill or retard the growth of a pathogen, and they must not harm the host. Antibiotics used against bacteria, for instance, must *selectively* attack features of bacterial targets not shared with eukaryotes. Many antibiotics possess highly specific modes of action, binding to and altering the activity of one specific "machine" of one type of cell.

One example is the antibiotic rifamycin B (**Fig. 8.9A**), produced by the actinomycete *Amycolatopsis mediterranei* (**Fig. 8.9B**). Rifamycin B selectively targets bacterial RNA polymerase and binds to the bacterial RNA polymerase beta subunit near the Mg^{2+} active site, blocking the exit channel (**Fig. 8.9C**). RNA polymerase can carry out two or three polymerization steps, but then stops because the nascent RNA cannot exit. The antibiotic does not prevent RNA polymerase from binding to promoters, and it does not inhibit, or even bind to, complexes already transcribing DNA, because the exiting mRNA chain blocks rifamycin from the rifamycin-binding site within the channel. The semisynthetic derivative of this drug, rifampin, is used for treating tuberculosis and leprosy and is given to contacts (family members or roommates) of individuals with bacterial meningitis caused by *Neisseria meningitidis*.

Actinomycin D (**Fig. 8.10A**) is another antibiotic produced by an actinomycete. Its phenoxazone ring is a planar structure that "squeezes," or intercalates, between GC base pairs within DNA. The side chains then extend in opposite directions along the minor groove (**Fig. 8.10B**). Because it mimics a DNA base, actinomycin D blocks the elongation phase of transcription; but because it binds any DNA, it is not selective for prokaryotes. It can be used to treat human cancers because cancer cells replicate rapidly, but it has severe side effects because it inhibits DNA synthesis in normal cells. (Further discussion of antibiotics can be found in Chapter 27.)

Antibiotic Resistance Helped Reveal Components of RNA Synthesis Machines

How did scientists determine the details of RNA and protein synthesis? The processes were observed in cell-free systems. For example, transcription and translation were observed in the test tube using purified RNA polymerase and ribosomes. In addition, the various protein subunits of RNA polymerase were separated from each other and mixed together again to make an active complex.

One particularly clever strategy was based on the genetics of antibiotic resistance. We now know that many antibiotics target specific components of the bacterial transcription and translation machinery. Organisms such as *E. coli*, however, mutate at low frequency to become resistant to a given antibiotic. The resistance mutation often affects the part of the RNA or protein synthesis machine that binds to the antibiotic. The effect is a decreased affinity of that altered protein for the antibiotic.

The subunits that confer resistance to particular antibiotics were determined in cell-free systems. For example, rifamycin is an antibiotic that inhibits transcription.

A.

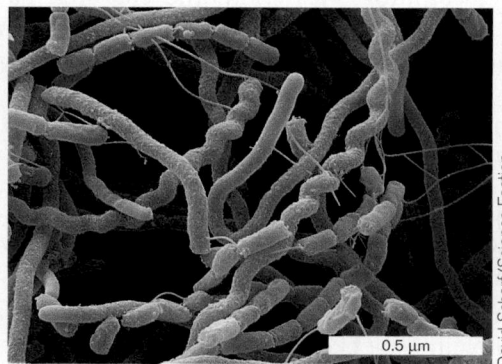

Rifamycin B $R_1 = CH_2COO^-$; $R_2 = H$

Rifampin $R_1 = H$; $R_2 = CH=N^+$⟨N–CH₃⟩

B.

David Scharf/Science Faction

0.5 µm

C.

Amino acid residue · Position in beta subunit

His406

Arg409

Asp396

Phe394

Ser411

Gln393

Rifamycin

Figure 8.9 **Structure and mode of action of rifamycin.** **A.** Structure of rifamycin. R groups indicated are added to alter the structure and pharmacology of the basic structure. **B.** Electron micrograph of *Amycolatopsis*. **C.** Contact points between rifamycin and residues in the beta subunit of RNA polymerase.

A.

Side chains were synthesized by addition of amino acids.

Methyl-val

Sarcosine

Pro

D-Val

Thr

Phenoxazone ring system

Actinomycin D

B.

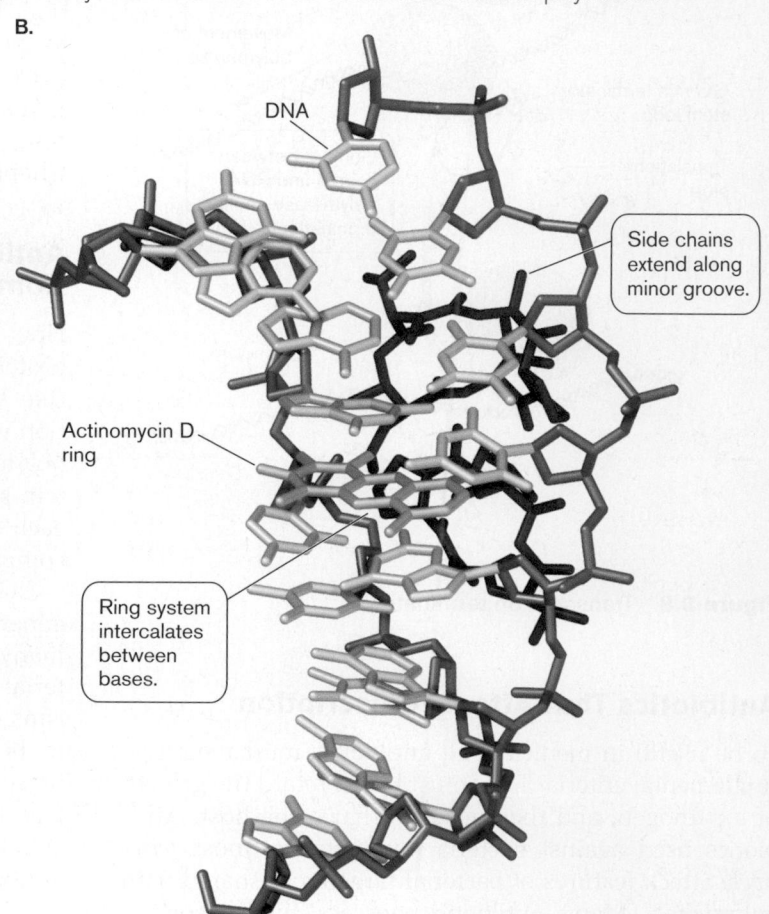

DNA

Side chains extend along minor groove.

Actinomycin D ring

Ring system intercalates between bases.

Figure 8.10 **Structure (A) and mode of action (B) of actinomycin D.** (PDB code for B: 1dsc)

To determine which RNA polymerase subunit was targeted by the drug, RNA polymerases from rifamycin-sensitive and -resistant strains were purified, the component parts of the polymerase were separated, and then a chimeric RNA polymerase was reassembled using all but one of the protein subunits from the sensitive strain. The missing subunit was supplied by the resistant strain and the polymerase preparation tested in vitro for an ability to make RNA in the presence of rifamycin. Resistance will happen only when the subunit conveying resistance has been added—in this case, the beta subunit. In subsequent in vitro assays and X-ray crystallography, investigators used this information to learn more about the enzymatic mechanism of this subunit.

Different Classes of RNA Have Different Functions

Now that we have shown how RNA is synthesized, it is important to note that not all RNA molecules are translated to protein. Six known classes of RNA are made. Each is designed for a different purpose (**Table 8.1**). RNA molecules that encode proteins are called **messenger RNA (mRNA)**. Molecules of mRNA average 1,000–1,500 bases in length but can be much longer or shorter, depending on the size of the protein they encode. Another class of RNA forms the scaffolding on which ribosomes are built and is called **ribosomal RNA (rRNA)**. As will be discussed later, rRNA also forms the catalytic center of the ribosome. **Transfer RNA (tRNA)** molecules ferry amino acids to the ribosome. A unique property of rRNA and tRNA is the presence of unusual modified bases not found in other types of RNA.

The fourth important class of RNA is called **small RNA (sRNA)**. Molecules of sRNA do not encode proteins but are used to *regulate* the stability or translation of specific mRNAs into proteins (discussed in Chapter 10). As will be discussed later in more detail, all mRNA molecules contain untranslated leader sequences that precede the actual coding region. Complementary sequences within some of these RNA leader regions snap back on themselves to form double-stranded stems and stem loop structures that can obstruct ribosome access and limit translation. The regulatory sRNAs are thought to disrupt or stabilize intrastrand stem structures by base-pairing to key regions within the mRNA molecule.

The final two classes of RNA are **tmRNA**, which has properties of both tRNA and mRNA (discussed later); and **catalytic RNA**. Catalytic RNA is usually found associated with proteins in which the enzymatic (catalytic) activity actually resides in the RNA portion of the complex rather than in the protein. All of these functional RNAs may represent remnants of the ancient "RNA world," where the earliest ancestral cells were built of RNA parts whose functions were later assumed by proteins.

RNA Stability

Once released from RNA polymerase, most prokaryotic mRNA transcripts are doomed to a short existence owing to their degradation by various intracellular RNases. This is especially true of transcripts that encode proteins (mRNA). The rationale is that the cell must be prepared to react rapidly to a changing environment, quickly halting synthesis of superfluous or even detrimental proteins. An effective way to do this is to rapidly destroy the

Table 8.1	Classes of RNA in *E. coli*.[a]				
RNA class	Function	Number of types	Average size	Approximate half-life	Unusual bases
mRNA (messenger RNA)	Encodes protein	Thousands	1,500 nt	3–5 minutes	No
rRNA (ribosomal RNA)	Synthesizes protein	Three	5S, 120 nt; 16S, 1,542 nt; 23S, 2,905 nt	Hours	Yes
tRNA (transfer RNA)	Shuttles amino acids	27 (86 genes)	80 nt	Hours	Yes
sRNA (small RNA)	Controls translation or transcription	20–30	<100 nt	Variable	No
tmRNA (properties of transfer and messenger RNA)	Frees ribosomes stuck on damaged mRNA	Roughly one per species	300–400 nt	3–5 minutes	No
Catalytic RNA	Carries out enzymatic reactions (e.g., RNase P)	??	Varies	3–5 minutes	No

[a]Six major classes of RNA exist in all bacteria. They differ in their function in the cell, number of types, average sizes, half-lives, and whether they contain modified bases.

mRNAs that encode those proteins. Rapid mRNA degradation makes it necessary to continually transcribe those genes as long as the need for their products exists.

RNA stability is measured in terms of half-life, which is the length of time the cell needs to degrade half the molecules of a given mRNA species. The average half-life for mRNA is 1–3 minutes but can be as little as 15 seconds. Other RNAs can be fairly stable, with half-lives of 10–20 minutes or longer. However, different kinds of RNAs differ drastically with respect to stability. The mRNAs are usually unstable compared with rRNAs and tRNAs, which contain modified bases less susceptible to RNase digestion. These more stable RNA molecules have half-lives of hours.

Recent discoveries show that components of the major RNA-degrading machine of *E. coli*—a four-protein complex including an RNase, an RNA helicase, and two metabolic enzymes—are organized in a helical cytoskeletal structure associated with the cell membrane. The findings suggest that the RNA degradosome is compartmentalized within the cell. Compartmentalization may facilitate regulation of RNA processing and degradation by preventing free access of the RNase to cytoplasmic RNA.

TO SUMMARIZE:

■ **A closed complex** occurs on initial binding of RNA polymerase to promoter DNA.

■ **Opening (melting) of the DNA** strands produces a "bubble" of DNA around the polymerase and forms the open complex.
■ **Rho-dependent and Rho-independent** mechanisms mediate transcription termination.
■ **Rifamycin B** is an antibiotic that selectively inhibits transcription initiation of bacterial RNA polymerase.
■ **Actinomycin D** is an antibiotic that nonselectively inhibits transcription elongation.
■ **Messenger RNA** molecules encode proteins.
■ **Ribosomal RNA** are stable molecules found in ribosomes and contain unusual modified bases.
■ **Transfer RNA** are stable molecules that bind amino acids and contain modified bases.
■ **Small RNA** molecules can regulate gene expression, have catalytic activity (catalytic RNA), or function as a combination of tRNA and mRNA (tmRNA).
■ **Half-lives of RNA** species vary within the cell.

8.3 Translation of RNA to Protein

Once a gene has been copied into mRNA, the next stage is **translation**, the decoding of the RNA message to synthesize protein. An mRNA molecule can be thought of as a sentence in which triplets of nucleotides, called **codons**, represent individual words, or amino acids

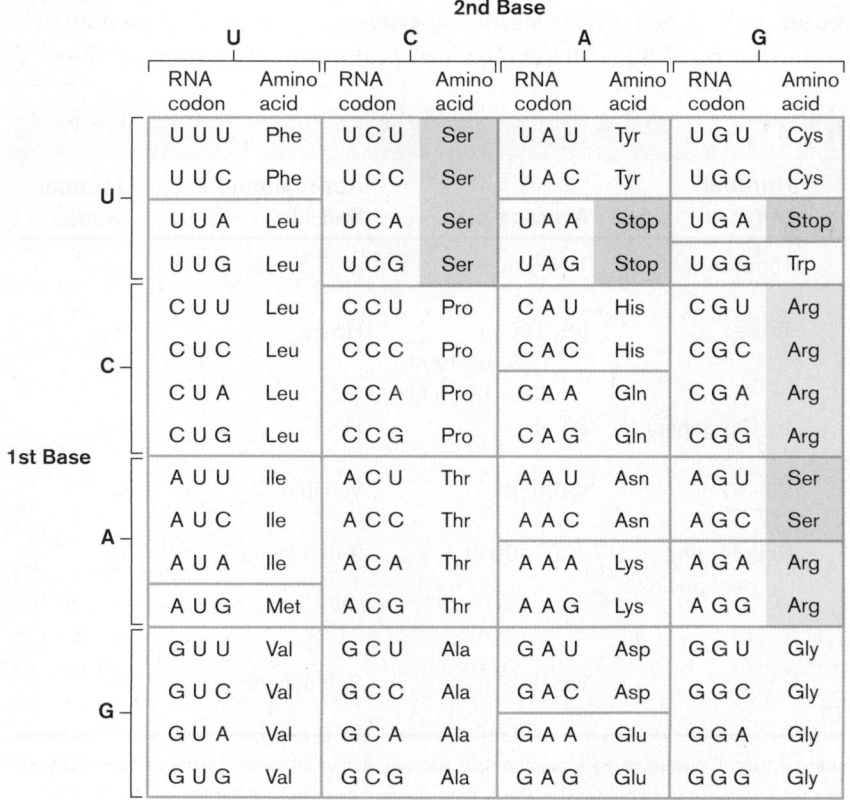

2nd Base

		U		C		A		G	
		RNA codon	Amino acid	RNA codon	Amino acid	RNA codon	Amino acid	RNA codon	Amino acid
U		U U U	Phe	U C U	Ser	U A U	Tyr	U G U	Cys
		U U C	Phe	U C C	Ser	U A C	Tyr	U G C	Cys
		U U A	Leu	U C A	Ser	U A A	Stop	U G A	Stop
		U U G	Leu	U C G	Ser	U A G	Stop	U G G	Trp
C		C U U	Leu	C C U	Pro	C A U	His	C G U	Arg
		C U C	Leu	C C C	Pro	C A C	His	C G C	Arg
		C U A	Leu	C C A	Pro	C A A	Gln	C G A	Arg
		C U G	Leu	C C G	Pro	C A G	Gln	C G G	Arg
A		A U U	Ile	A C U	Thr	A A U	Asn	A G U	Ser
		A U C	Ile	A C C	Thr	A A C	Asn	A G C	Ser
		A U A	Ile	A C A	Thr	A A A	Lys	A G A	Arg
		A U G	Met	A C G	Thr	A A G	Lys	A G G	Arg
G		G U U	Val	G C U	Ala	G A U	Asp	G G U	Gly
		G U C	Val	G C C	Ala	G A C	Asp	G G C	Gly
		G U A	Val	G C A	Ala	G A A	Glu	G G A	Gly
		G U G	Val	G C G	Ala	G A G	Glu	G G G	Gly

1st Base (left) / **3rd Base** (right)

Figure 8.11 The genetic code. Codons within a single box encode the same amino acid. Color-highlighted amino acids are encoded by codons in two boxes. Stop codons are highlighted pink. Often, single-letter abbreviations for amino acids are used to convey protein sequences: Ala (A); Arg (R); Asn (N); Asp (D); Cys (C); Gln (Q); Glu (E); Gly (G); His (H); Ile (I); Leu (L); Lys (K); Met (M); Phe (F); Pro (P); Ser (S); Thr (T); Trp (W); Tyr (Y); Val (V).

(**Fig. 8.11**). Ribosomes are the machines that read the language of mRNA and convert, or translate, it to protein. Numerous steps are involved in translation, the first of which is the search by ribosomes for the beginning of an RNA-coding region. Because the code consists of triplet codons, a ribosome must start translating at precisely the right base (in the right frame) or the product will be gibberish. Before we discuss how the ribosome finds the right reading frame, let's review the code itself and the major players in the translation process.

The Code and tRNA Molecules

When learning a new language, we need a dictionary that converts words from one language into the other, in this case from RNA to protein. Through painstaking effort, Marshall Nirenberg and colleagues broke the molecular code and found that each codon (a triplet of nucleotides) represents an individual amino acid. Remarkably, and with few exceptions, the code operates universally across species. (Some exceptions: UGA is normally read as a stop codon but encodes tryptophan in vertebrate mitochondria, *Saccharomyces*, and mycoplasmas; and *Candida albicans* reads CUG for serine instead of leucine.) On examining the code presented in **Figure 8.11**, we can see that most amino acids have multiple codon synonyms. Glycine, for example, is translated from four different codons, and leucine from six, but methionine and tryptophan each have only one codon. Because more than one codon can encode the same amino acid, the code is said to be degenerate or redundant. Notice that in almost every case, synonymous codons differ only in the *last* base. How the cell handles this degeneracy (redundancy) in the code will be explained later.

Only 61 out of a possible 64 codons specify amino acids. Four of these act as codons that can mark the beginning of the protein "sentence." In order of preference, they are AUG (90%), GUG (8.9%), UUG (1%), and CUG (0.1%). Because of its inefficiency in starting translation, the use of a rare start codon, such as CUG, limits the translation of any ORF in which it is found. Although these **start codons** are used to begin all proteins, they are not restricted to that role; they also encode amino acids found in the middle of coding sequences. Consequently, there must be another mRNA feature that signifies whether an AUG codon, for example, marks the start of a protein or resides internally, where it codes for an amino acid (discussed below).

The remaining three triplets (UAA, UAG, and UGA) are equally important, for they tell the ribosome when to stop reading a gene sentence. Called **stop codons**, they trigger a series of events (to be described later) that dismantle ribosomes from mRNA and release a completed protein.

> **THOUGHT QUESTION 8.4**   How might the redundancy of the genetic code be used to establish evolutionary relationships? *Hint*: Different species have different GC content.

The decoder (or adapter) molecules that convert the language of RNA (codons) to the language of proteins (amino acids) are the tRNAs. Their job is to travel through the cell and return to the ribosome with amino acids in tow. Approximately 80 bases in length, a tRNA laid flat appears as a clover leaf with three loops (**Fig. 8.12A** and **B**). In nature, however, the molecule folds into a characteristic "boomerang-like" three-dimensional structure (**Fig. 8.12C**). The very bottom or middle loop of all tRNA molecules (as depicted in **Fig. 8.12B** and **C**) harbors an **anticodon** triplet that base-pairs with codons in mRNA (**Fig. 8.13**). As a result, this loop is called the anticodon loop. Notice that codon and anticodon pairings are aligned in an antiparallel manner. Most tRNA molecules begin with a 5' G, and all end with a 3' CCA, to which amino acids attach (see **Figs. 8.12** and **8.13**). Because of this, the 3' end of the molecule is called the acceptor end.

Transfer RNA molecules contain a large number of unusual, modified bases, which accounts for the strange letter codes seen in **Figure 8.12A** (D, M, Y, T, and P). Some of the odd bases used are diagrammed in **Figure 8.14**. For example, wybutosine (with the symbol yW) has three rings instead of the two found in a normal purine. How do these unusual bases end up in tRNA? During transcription of the tRNA genes, normal, unmodified bases are incorporated into the transcript. Some of these are modified later by specific enzymes to make inosine and other odd bases. The remarkable stability of tRNA molecules is explained, in part, by these unusual bases, because they are poor substrates for RNases.

When looking at the cloverleaf structure of tRNA, two of the loops are named after modifications that are invariantly present in those loops. One is called the TPC (or TΨC) loop because this loop in every tRNA has the nucleotide triplet thymidine, pseudouridine (Ψ), and cytidine. The other loop always contains dihydrouridine and is called the DHU loop. These loops are recognized by the enzymes that match tRNA to the proper amino acid.

As noted previously, the code is redundant in that many amino acids have codon synonyms. Redundancy occurs primarily in the third position of the codon (for example, UU_U_ and UU_C_ both encode phenylalanine). This is the result of a "wobble" in the first position of the anticodon, which corresponds to the third position of the codon (remember, as with DNA, base-pairing between RNA strands is antiparallel). The wobble is due in part to the curvature of the anticodon loop and to the use of an unusual base (inosine) at this position in some tRNA molecules. The wobble structure allows one anticodon to pair with several codons differing only in the third position.

A.

5′ GCGGAUUUAGCUCAGDDGGGAGAGCMCCAGACUGAAYAUCUGGAGMUCCUGUGTPCGAUCCACAAUUCGCACCA 3′

B.

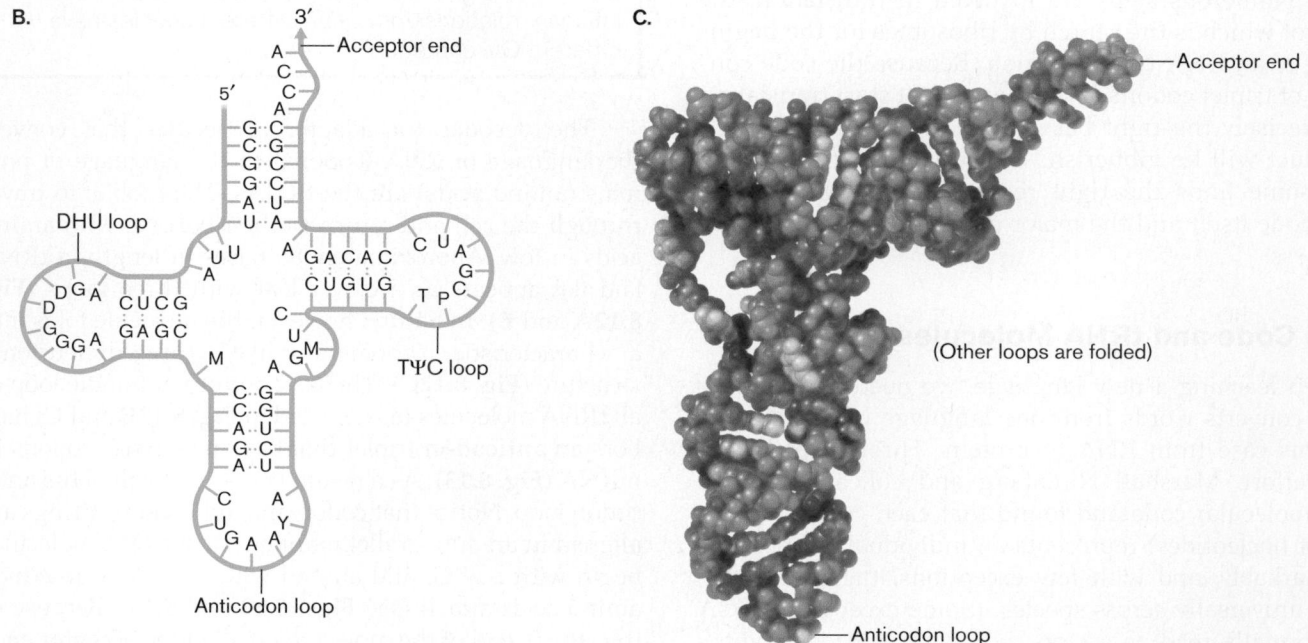

C.

(Other loops are folded)

Figure 8.12 Transfer RNA. A. Primary sequence. The letters D, M, Y, T, and P stand for modified bases found in tRNA. **B.** Cloverleaf structure. DHU (or D) is dihydrouracil, which occurs only in this loop; TΨC consists of thymine, pseudouracil, and cytosine bases that occur as a triplet in this loop. The DHU and TΨC loops are named for the modified nucleotides that are characteristically found there. **C.** Three-dimensional structures. The anticodon loop binds to the codon, while the acceptor end binds to the amino acid. (PDB code: 1GIX)

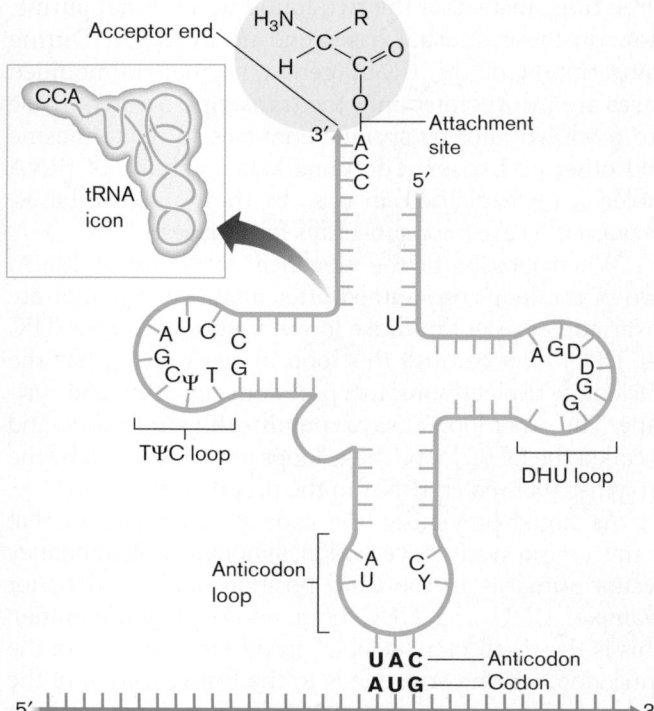

Figure 8.13 Codon-anticodon pairing. The tRNA anticodon consists of three nucleotides at the base of the anticodon loop. The anticodon hydrogen-bonds with the mRNA codon in an antiparallel fashion. This tRNA is "charged" with an amino acid covalently attached to the 3′ end.

Aminoacyl-tRNA Synthetases Attach Amino Acids to tRNA

When a tRNA GCG anticodon, for instance, pairs with a CGC codon, the ribosome has no way of checking that the tRNA is attached (that is, charged) to the "correct" amino acid (arginine in this case). Consequently, each tRNA must be charged with the proper amino acid *before* it encounters the ribosome. How are amino acids correctly matched to tRNA molecules and affixed to their 3′ ends? The charging of tRNAs is carried out by a set of enzymes called **aminoacyl-tRNA synthetases**. Each cell has generally 20 of these "match and attach" proteins, one for each amino acid. Some bacteria, however, have only 18, choosing instead to modify already charged glutamyl-tRNA and aspartyl-tRNA by adding an amine to make glutamine and asparagine derivatives. Each aminoacyl-tRNA synthetase enzyme has a specific binding site for its cognate (matched) amino acid. Each enzyme also has a site that recognizes the target tRNA and an active site that joins the carboxyl group of the amino acid to the 3′ OH (class II synthetases) or 2′ OH (class I synthetases) of the tRNA by forming an ester (**Fig. 8.15**). The tRNA synthetase disengages after the amino acid has been attached to the tRNA.

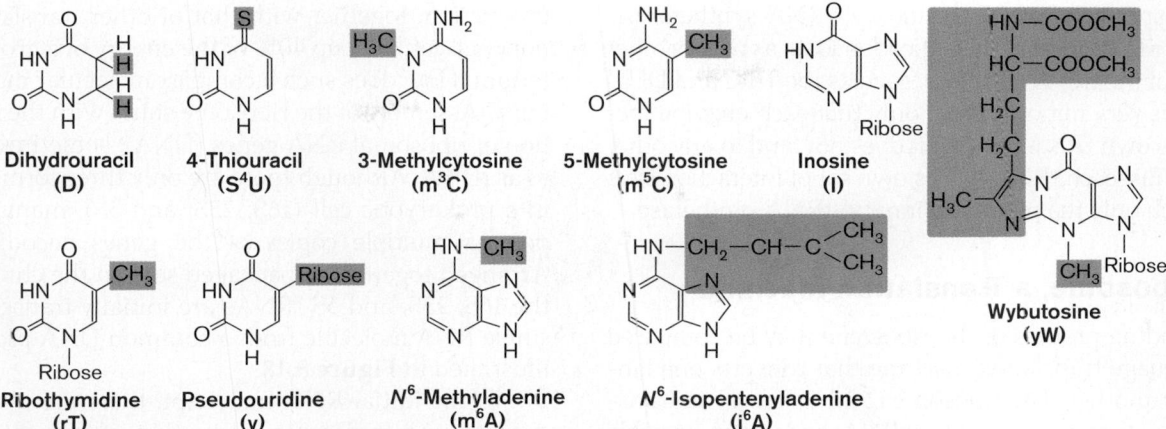

Figure 8.14 **Modified bases present in tRNA and rRNA molecules.** Modifications are highlighted red.

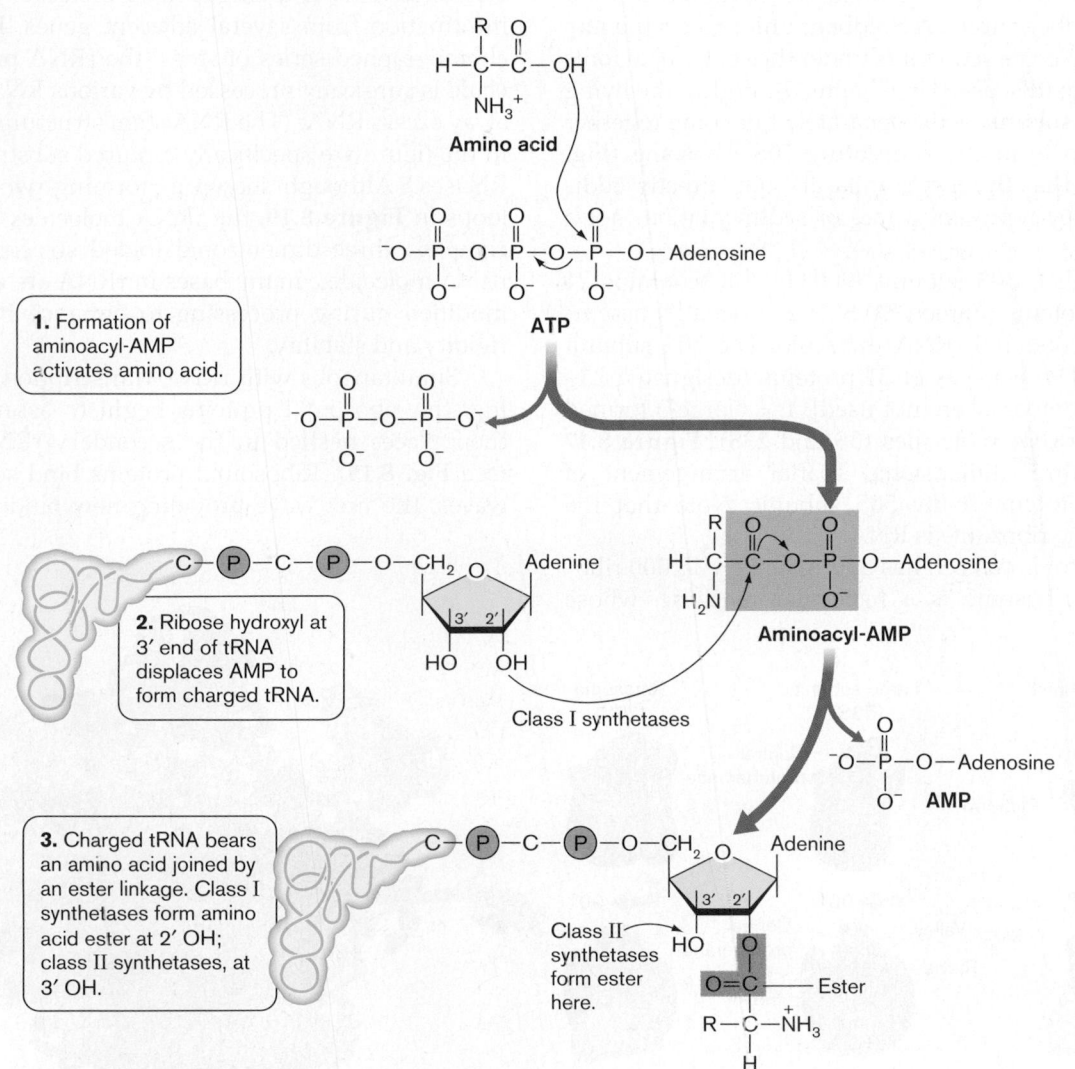

Figure 8.15 **Charging of tRNA molecules by aminoacyl-tRNA synthetases.**

The specificity of the aminoacyl-tRNA synthetase is based on recognition of the anticodon loop, as well as other features of the tRNA structure, such as the TΨC and DHU loops. It is very important not only that each enzyme recognize its own tRNA but that it does not bind to any other tRNA. Thus, each tRNA has its own set of interaction sites that match only the proper aminoacyl-tRNA synthetase.

The Ribosome, a Translation Machine

The decoding process of the ribosome may be compared to a language translation machine that converts one language to another. The ribosome can be viewed as translating the language of the mRNA code into sensible protein sequences that conduct the activities of the cell. Ribosomes are composed of two complex subunits, each of which includes rRNA and protein components. In prokaryotes, the subunits are named 30S and 50S for their "size" in Svedberg units. A Svedberg unit reflects the rate at which a molecule sediments under the centrifugal force of a centrifuge (discussed in Chapter 3). Within the living cell, the two subunits exist separately but come together on mRNA to form the translating 70S ribosome (**Fig. 8.16**). (Note that Svedberg units are not directly additive, since they represent a rate of sedimentation, not a weight.)

The smaller, 30S subunit (900,000 Da) contains 21 ribosomal proteins (named S1–S21; S = "small") assembled around one 16S rRNA molecule. The 50S subunit (1.6 million Da) consists of 31 proteins (designated L1–L34 [three numbers were not used]; L = "large") formed around two rRNA molecules (5S and 23S). **Figure 8.17** presents a three-dimensional spatial arrangement of rRNA and protein in the 50S subunit. Note that the majority of the ribosome is RNA.

The typical *E. coli* cell has approximately 18,000 ribosomes. The ribosome is a molecular machine, whose production, together with that of other translation components, can take up 40% of the energy of a growing bacterium. How does such a complex molecular machine get built? Assembly of the ribosome starts with the transcription of ribosomal RNA genes (DNA), sometimes referred to as rDNA. Although there are only three forms of rRNA in a prokaryotic cell (16S, 23S, and 5S), many microbes possess multiple copies of the genes encoding them. Arranged together as packaged sets on the chromosome, the 16S, 23S, and 5S rRNAs are initially transcribed as a single RNA molecule from a common DNA promoter, as illustrated in **Figure 8.18**.

In the initial RNA transcript, there are also leaders, spacers, and trailer sequences that often contain tRNA sequences. This polycistronic message is the starting point from which the ribosomes are built. **Cistron** is a name used to define a functional unit of RNA; a **polycistronic** RNA is a single RNA molecule that contains information from several adjacent genes. In a carefully choreographed series of steps, the rRNA precursor molecule is surgically processed by various RNases that trim away excess RNA. (The RNA stem structures represented in the figure are specifically required substrates for some RNases.) Although shown as forming two-dimensional loops in **Figure 8.19**, the rRNA molecules actually form complex, three-dimensional folded structures. As with tRNA molecules, many bases in rRNA are enzymatically modified during processing to enhance the molecule's rigidity and stability.

Simultaneous with rRNA transcription and processing, the ribosomal proteins begin to assemble, finding their places nestled in the secondary rRNA structures (see **Fig. 8.19**). Ribosomal proteins bind sequentially in waves, the first wave providing new binding sites for a

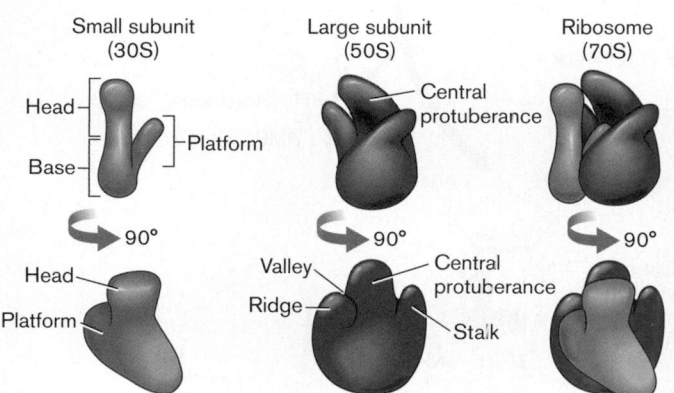

Figure 8.16 Schematic ribosome structure. Each panel shows two rotated views of the 30S subunit, 50S subunit, and 70S ribosome complex, respectively. Note that the platform of the 30S subunit fits into the valley of the 50S subunit when forming the 70S ribosome.

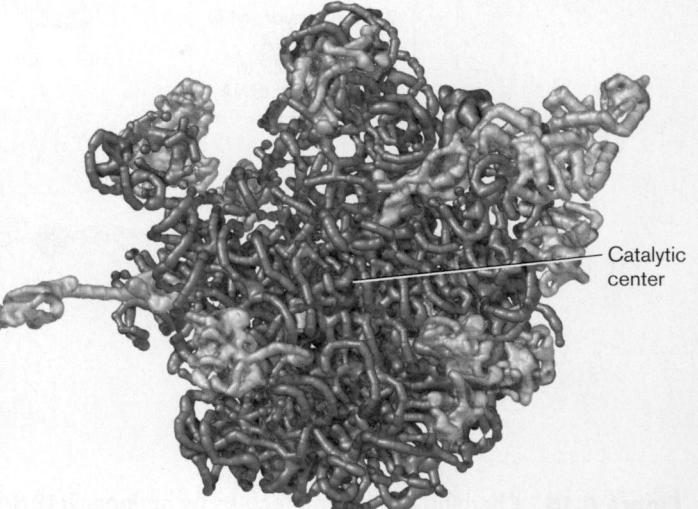

Figure 8.17 RNA-protein interfaces in the large (50S) ribosome subunit. rRNA is blue, proteins are gold. (PDB code 1GIY)

Figure 8.18 Processing of ribosomal RNA and tRNA. A. The genes encoding rRNAs in *E. coli* exist in operons that include several tRNA genes. *E. coli* contains seven such operons in its chromosome, all of which make the same rRNA molecules but differ in which tRNA molecules are present. **B.** Processing of precursor rRNA molecules. Specific RNases cleave the transcript at precise places to release 16S, 23S, and 5S rRNA and the interspersed tRNA molecules.

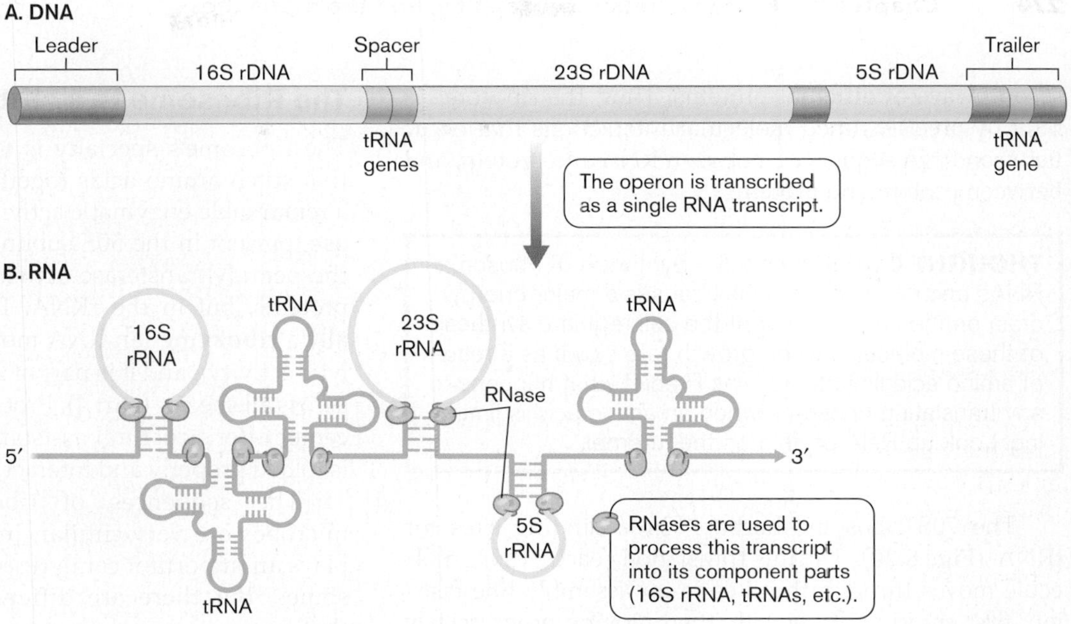

A. DNA

Leader | 16S rDNA | Spacer | 23S rDNA | 5S rDNA | Trailer

tRNA genes

tRNA gene

The operon is transcribed as a single RNA transcript.

B. RNA

16S rRNA   tRNA   23S rRNA   tRNA

RNase

5′   3′

tRNA

5S rRNA

RNases are used to process this transcript into its component parts (16S rRNA, tRNAs, etc.).

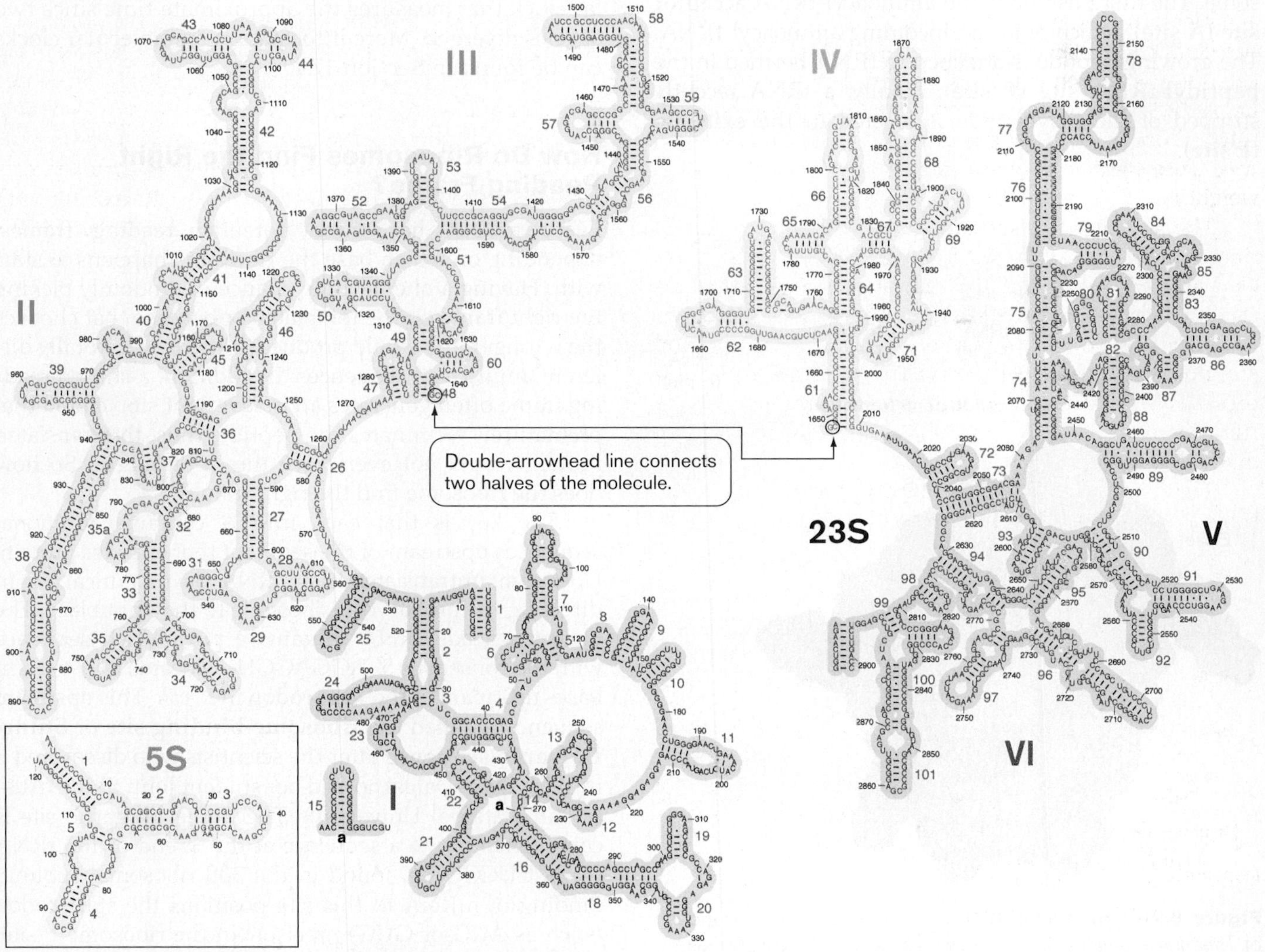

Double-arrowhead line connects two halves of the molecule.

23S

5S

Figure 8.19 Secondary structure of the 5S and 23S rRNA. Note the many double-stranded hairpin structures that fold into domains (roman numerals). Nucleotides are numbered in black; stem structures are numbered in blue. During the building of the ribosome, proteins bind to specific sites on the RNA molecule. The proteins bind to some of the secondary and tertiary rRNA structures and help form the key tertiary 23S structure. Domains where various ribosomal proteins bind are indicated by different colors. The 23S RNA is split where indicated to provide space for all secondary structures.

second wave of different proteins. Thus, the ribosome is built by precise, timed molecular interactions that occur between RNA sequences, between RNA and protein, and between protein and protein.

> **THOUGHT QUESTION 8.5** Synthesis of ribosomal RNAs and ribosomal proteins causes a major energy drain on the cell. How might the cell regulate synthesis of these molecules when growth rate slows as a result of amino acid limitation? *Hint*: Predict what happens to any translating ribosome when an amino acid is limiting. Look up RelA protein on the Internet.

The 70S ribosome harbors three binding sites for tRNA (**Fig. 8.20**). During translation, each tRNA molecule moves through the sites in an assembly-line fashion, first entering at one site, then moving progressively through the others before being jettisoned from the ribosome. The first position is the aminoacyl-tRNA **acceptor site (A site)**, which binds an incoming aminoacyl-tRNA. The growing peptide is attached to tRNA berthed in the **peptidyl-tRNA site (P site)**. Finally, a tRNA recently stripped of the polypeptide is located in the **exit site (E site)**.

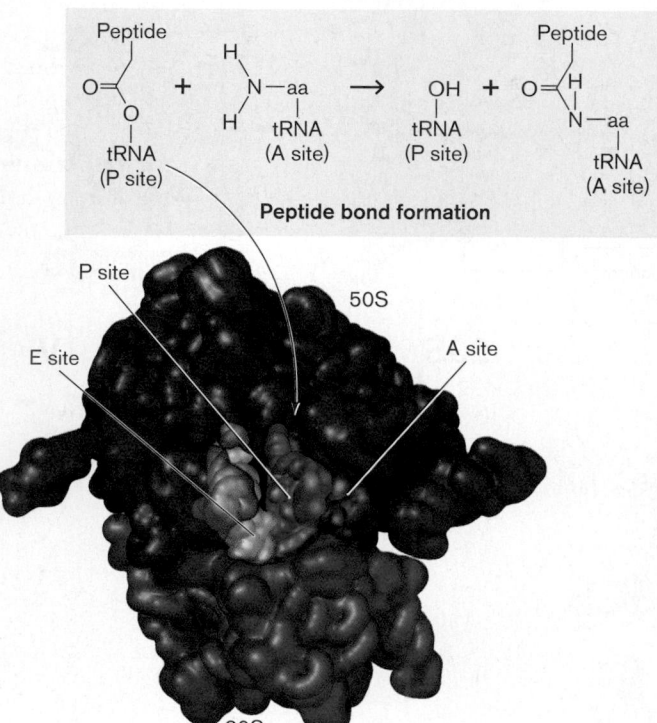

Figure 8.20 Binding of tRNA. X-ray crystallographic model of *Thermus aquaticus* ribosome with associated tRNAs. 50S is red, 30S is violet, and tRNAs in the A, P, and E sites are purple, green, and gold, respectively. (PDB codes: 1GIX, 1GIY) Inset shows the formation of a peptide bond between the peptidyl tRNA in the P site and amino acyl tRNA in the A site.

The Ribosome Is a "Ribozyme"

The ribosome's specialty is to make the peptide bonds that stitch amino acids together, which it does by using a remarkable enzymatic activity called **peptidyltransferase**, present in the 50S subunit. Contrary to expectation, the peptidyltransferase activity resides not in ribosomal proteins, but in the rRNA. Peptidyltransferase is actually a **ribozyme** (an RNA molecule that carries out catalytic activity), and it is part of 23S rRNA highly conserved across all species (part 4). Proteins surrounding this active center offer structural assistance to ensure that the RNA is folded properly and interact with tRNA substrates.

The sequences of ribosomal RNAs from all microbes are very similar, in large part because rRNA plays an important catalytic role in the activities of ribosomes. But there are differences in rRNA sequences that increase in relation to the evolutionary distance between species. As a result, rRNA serves as a molecular clock that measures the approximate time since two species diverged. More information on molecular clocks can be found in Section 17.4.

How Do Ribosomes Find the Right Reading Frame?

Every mRNA has three potential reading frames, depending on which base the ribosome happens to start with. Having a one in three chance of randomly picking the right frame on the template, a ribosome that chooses the wrong frame would produce proteins with totally different amino acid sequences. In addition, a shifted reading frame often generates an inadvertent stop codon that prematurely terminates the peptide; thus, the translated protein would not even have the right length. So how does the ribosome find the right frame?

The key is that each mRNA contains additional sequences upstream of the segment that encodes protein. Upstream, untranslated leader RNA plays a critical role in directing the prokaryotic ribosome to the right place. The upstream leader RNA contains a purine-rich sequence with the consensus 5′-AGGAGGU-3′ located four to eight bases upstream of the start codon in *E. coli*. This upstream sequence is called the **ribosome-binding site** or **Shine-Dalgarno sequence** after the scientists who discovered it in 1974, Lynn Dalgarno and her student John Shine (Australian National University). The Shine-Dalgarno site is complementary to a sequence at the 3′ end of 16S rRNA (5′-ACCUCCU-3′), found in the 30S ribosomal subunit. Binding of mRNA to this site positions the start codon (such as AUG or GUG) precisely in the ribosome P site, ready to pair with an *N*-formylmethionyl-tRNA.

Defining a gene. We have described the process of transcription and are about to translate a resulting mRNA.

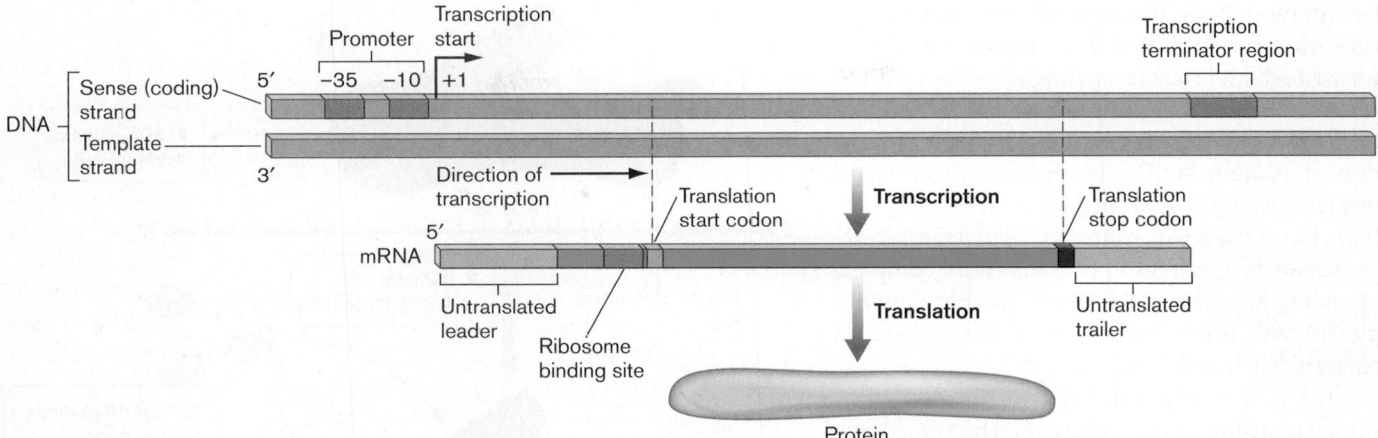

Figure 8.21 Alignment of a bacterial structural gene with its mRNA transcript. RNA polymerase binds to the −10 and −35 regions of DNA and, reading the template strand, begins RNA transcription at nucleotide +1. Messenger RNA has an untranslated region of variable length and then a ribosome binding site. The ribosome begins translating a few nucleotides downstream at the codon corresponding to the amino terminus of the translated protein.

It will help to illustrate the alignment between the DNA region comprising a structural gene (a gene encoding a protein), and the resulting mRNA transcript containing translation signals and the protein coding region. **Figure 8.21** shows the "sense" and "template" DNA strands of a gene. The sequence of the sense strand equals that of the mRNA transcript but with T's substituting for U's. The template strand is the strand actually "read" by RNA polymerase. Note that +1 marks the DNA base where the mRNA transcript starts. In the mRNA transcript, an untranslated "leader" sequence precedes the protein coding region, which is located between the translation start and stop signals (discussed later). Downstream of the translation stop signal lies an untranslated "trailer". The "leader" and "trailer" sequences help regulate gene expression. As you proceed in this chapter, refer back to this figure to clarify important connections between a gene, its transcript, and the resulting protein.

NOTE: Eukaryotic ribosomes have a different mechanism for finding the start codon. They generally start translating at the first AUG after the 5′ cap. The 5′ cap is a 7-methylguanosine added post-transcriptionally to the 5′ end of eukaryotic mRNA molecules. The normal role of the cap is thought to be as a signal to transport RNA out of the nucleus and/or to stabilize mRNA.

Although Dalgarno and Shine predicted the importance of the ribosome-binding site in 1974, experimental evidence that this mRNA region truly bound 16S rRNA was lacking. The evidence was supplied in 1975 by Joan Steitz in an elegant yet simple experiment (**Fig. 8.22**). The experiment demonstrated that a radiolabeled RNA fragment taken from the 5′ end of mRNA binds to the 3′ end of 16s rRNA (see **eTopic 8.1**).

The Three Stages of Protein Synthesis

Like transcription, peptide synthesis is a three-step process. Once the ribosome has been properly positioned on the message, peptide synthesis can begin. Peptide synthesis has three phases: initiation, which brings the two ribosomal subunits together, placing the first amino acid in position; elongation, which sequentially adds amino acids as directed by the mRNA transcript; and

Figure 8.22 Joan Steitz (second from left), at Yale University with some of her students. Steitz and her colleagues provided key evidence for the existence of the ribosome-binding site on mRNA.

termination. Note that several steps in the process can be inhibited by certain antibiotics (described later in this section).

Initiation of translation. In bacteria, initiation of protein synthesis requires three small proteins called initiation factors (IF1, IF2, and IF3). In archaea (for example, *Methanocaldococcus jannaschii*), initiation is much more complex, involving six different IF proteins. For simplicity, we will limit discussion to the process in bacteria, such as *E. coli*.

IF3 first brings mRNA and the 30S ribosome subunit together, allowing the ribosome-binding site to find its complementary site on 16S rRNA (**Fig. 8.23**). Next, IF1 binds and blocks the A site. IF2 bound to GTP then escorts the initiator *N*-formylmethionyl-tRNA (fMet-tRNA) to the start codon located at what will be the P site. *N*-formylmethionyl-tRNA binds to all start codons and is the only aminoacyl-tRNA to bind directly to the P site. Once the initiator tRNA is in place, IF3 is released. The 50S subunit then docks to the 30S subunit, GTP is hydrolyzed, and IF1 and IF2 are released. The ribosome is now "locked and loaded."

Elongation of the peptide. In elongation, three basic steps are repeated: (1) an aminoacyl-tRNA binds to the A (acceptor) site; (2) a peptide bond forms between the new amino acid and the growing peptide pre-positioned in the P site; (3) the message must move by one codon (**Fig. 8.24**). First, an elongation factor (EF-Tu) associates with GTP to form a complex (EF-Tu-GTP) that binds to charged aminoacyl-tRNAs (except for the initiator tRNA). Sequences within the 50S rRNA recognize features of the EF-Tu-GTP–aminoacyl-tRNA complex, guiding it into the A site (**Fig. 8.24**, step 1). Correct selection of tRNA is guided in large part by codon-anticodon pairing, but conformational changes sensed by various ribosomal proteins customize the fit. If the fit is not perfect, the tRNA is rejected.

Once an aminoacyl-tRNA is in the A site, a peptide bond is formed between the amino acid moored there and the terminal amino acid linked to tRNA in the P site (step 2). Simultaneously, a GTP is hydrolyzed and the resulting EF-Tu-GDP is expelled. Peptide bond formation effectively transfers the peptide from the tRNA in the P site to tRNA in the A site (step 3). Finally, for protein synthesis to continue, the ribosome must advance by one codon, which moves the peptidyl-tRNA from the A site into the

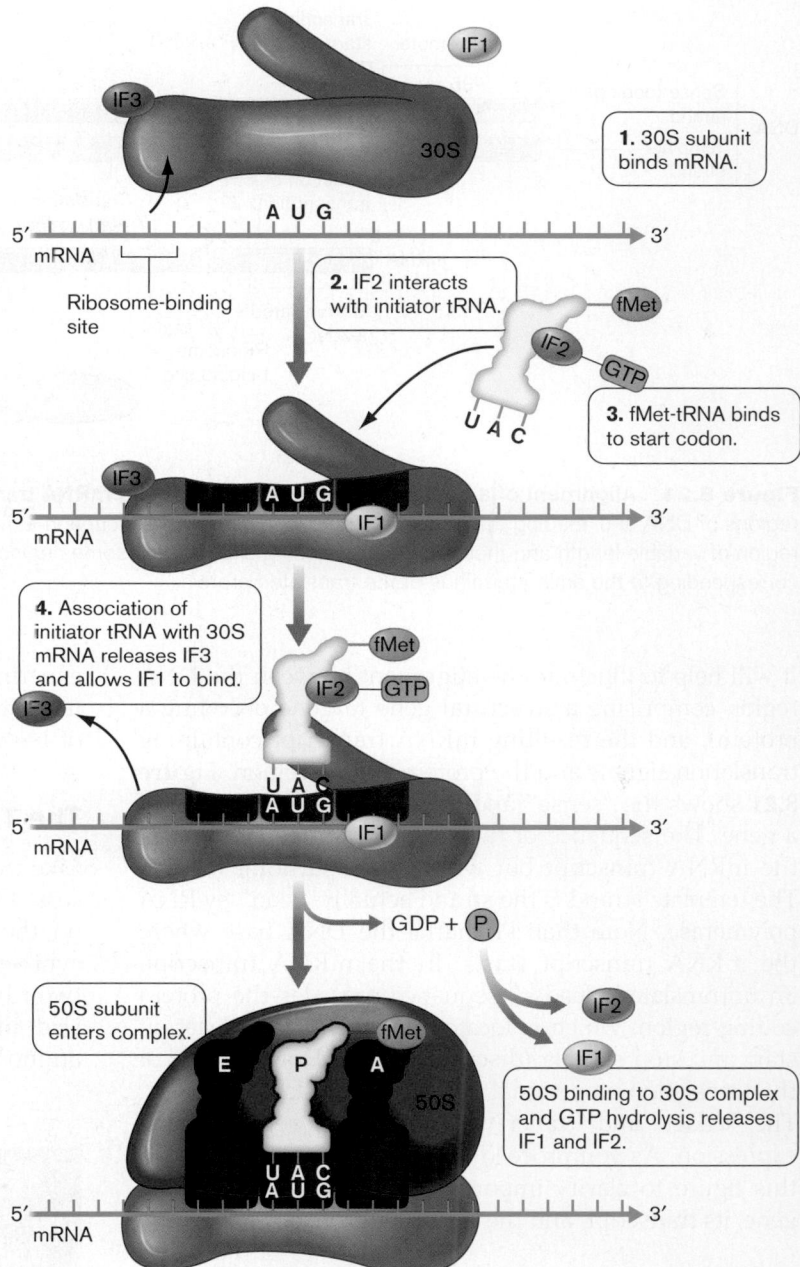

Figure 8.23   Translation initiation.   See text for details. ▶❚❙ *microbiology2. com/animations*

P site, leaving the A site vacant. The process, called **translocation**, involves another elongation factor, EF-G, associated with GTP. EF-G-GTP binds to the ribosome, GTP is hydrolyzed, and the 50S subunit ratchets ahead on the message by one codon (step 4). The 30S subunit then follows (step 5). This maneuver opens up the A site, moves the peptidyl-tRNA into the P site, and slides uncharged tRNA into the E (exit) site. The next aminoacyl-tRNA that enters the A site stimulates a conformational change in the ribosome that telegraphs through to the E site and ejects the uncharged tRNA.

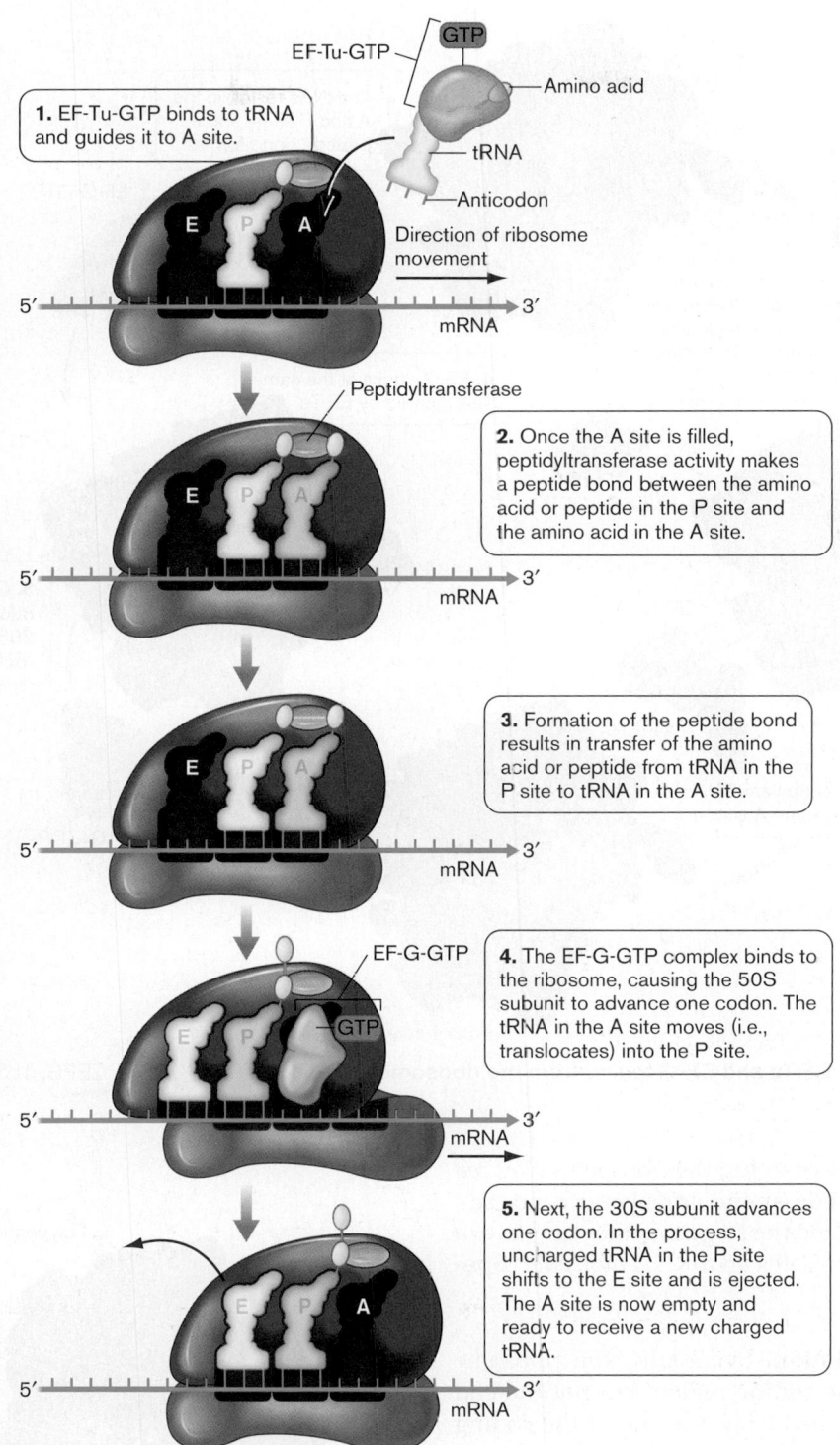

1. EF-Tu-GTP binds to tRNA and guides it to A site.

EF-Tu-GTP

GTP

Amino acid

tRNA

Anticodon

Direction of ribosome movement

5′ mRNA 3′

Peptidyltransferase

2. Once the A site is filled, peptidyltransferase activity makes a peptide bond between the amino acid or peptide in the P site and the amino acid in the A site.

5′ mRNA 3′

3. Formation of the peptide bond results in transfer of the amino acid or peptide from tRNA in the P site to tRNA in the A site.

5′ mRNA 3′

EF-G-GTP

GTP

4. The EF-G-GTP complex binds to the ribosome, causing the 50S subunit to advance one codon. The tRNA in the A site moves (i.e., translocates) into the P site.

5′ mRNA 3′

5. Next, the 30S subunit advances one codon. In the process, uncharged tRNA in the P site shifts to the E site and is ejected. The A site is now empty and ready to receive a new charged tRNA.

5′ mRNA 3′

Figure 8.24 **Elongation of peptide.** ▶ *microbiology2.com/animations*

Figure 8.25▶ presents a three-dimensional rendering of EF-Tu and EF-G cycling on and off the ribosome. Note that both factors bind to the same site. This reflects a structural similarity (or molecular mimicry) between EF-G-GTP and the EF-Tu-GTP–aminoacyl-tRNA complex. Because both factors bind to the same site, they cannot do so simultaneously, so they must cycle on and off the ribosome sequentially. **Figure 8.26**▶ illustrates the path of mRNA into the ribosome and the nascent (that is, incomplete) peptide emerging from it. While this

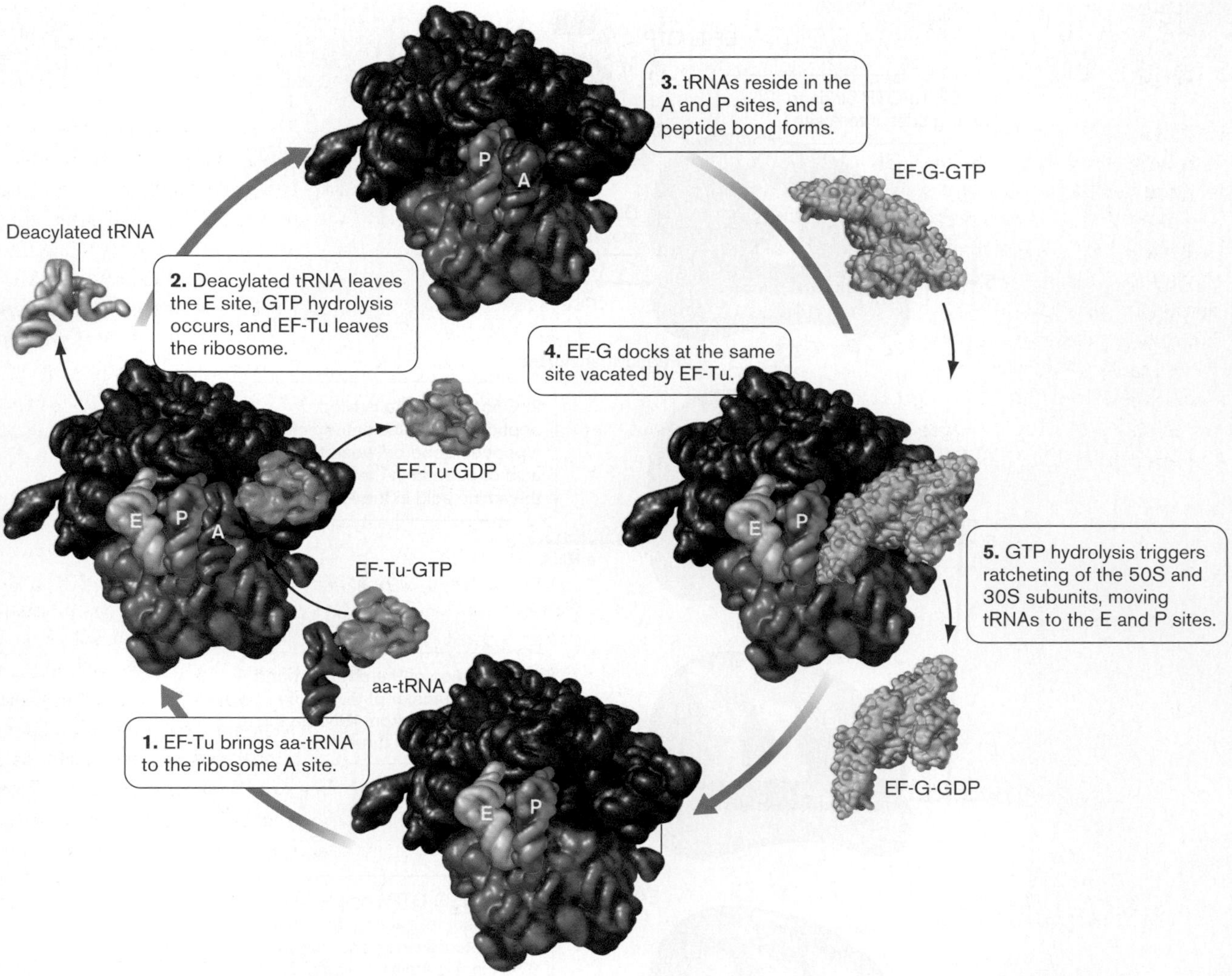

3. tRNAs reside in the A and P sites, and a peptide bond forms.
P
A
EF-G-GTP
Deacylated tRNA
2. Deacylated tRNA leaves the E site, GTP hydrolysis occurs, and EF-Tu leaves the ribosome.
4. EF-G docks at the same site vacated by EF-Tu.
EF-Tu-GDP
E
P
A
E
P
5. GTP hydrolysis triggers ratcheting of the 50S and 30S subunits, moving tRNAs to the E and P sites.
EF-Tu-GTP
aa-tRNA
1. EF-Tu brings aa-tRNA to the ribosome A site.
E
P
EF-G-GDP

Figure 8.25 Cycling of EF-Tu and EF-G to and from the ribosome. (PDB codes: 1GIX, 1GIY, 2EFG, 1LS2) *microbiology2.com/animations*

process seems incredibly complex, the ribosomes of *E. coli* manage to link together 16 amino acids per second. Scientists have now visualized the staggered movements of a single ribosome as it translates an mRNA molecule (**Special Topic 8.1**).

Termination of translation. Eventually, the ribosome arrives at the end of the coding region, but not the end of the RNA. As noted previously, the end of the coding region is marked by one of three stop codons. Formation of the last peptide bond and the subsequent translocation of mRNA leads to ejection of tRNA in the E site

Polypeptide
tRNAs
50S
E
P
A
Codon-anticodon site
30S
mRNA
5′
3′

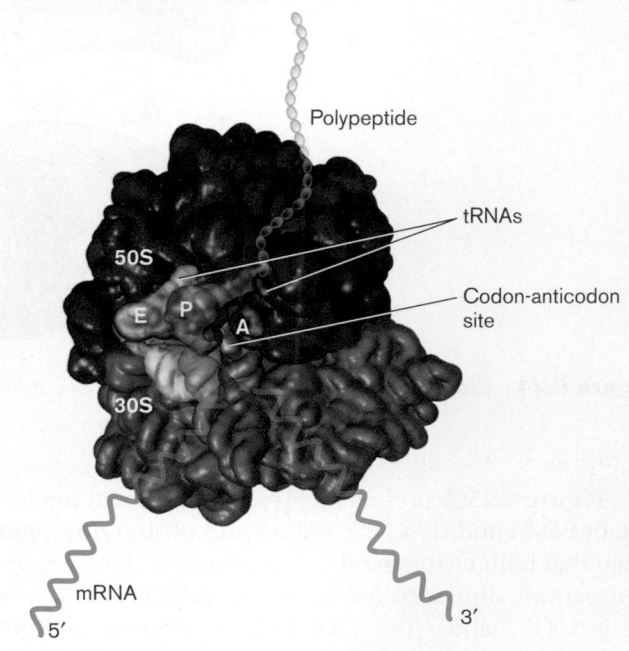

Figure 8.26 Orientation of tRNA molecules within the ribosome and tracks of mRNA and nascent polypeptide. Note that the mRNA travels along the 30S subunit, and the growing peptide exits from a channel formed in the 50S subunit. (PDB codes: 1GIX, 1GIY) *microbiology2.com/animations*

Special Topic 8.1   Stalking the Lone Ribosome

Recent advances in crystallography and cryo-electron microscopy have strengthened our understanding of ribosome structure and function. However, it has been extremely difficult to follow the steps of ribosomes during translational elongation in the test tube, where thousands of ribosomes are usually needed to see something happen. Because the dynamics of individual ribosomes are stochastic (random), the functions of thousands of ribosomes cannot be synchronized.

Recently Tinoco and colleagues used an optical tweezer to trap a single ribosome and map its progress along an mRNA codon by codon. **Figure 1** illustrates the trap and how it works. An RNA strand was made with a digoxigenin molecule affixed at the 3' end and a biotin molecule at the 5' end. Tiny polystyrene beads coated with streptavidin (which binds to biotin) or with antibodies to digoxigenin were used to trap separate ends of the RNA molecule. The bead attached to the 3' end was placed in a laser trap, and the bead attached to the 5' end was held in place by a micropipette tip. A single ribosome was stalled at the 5' side of the mRNA hairpin construct by leaving out components needed for translation. Once a translating mixture of appropriately charged tRNAs, translocation factors, and GTP was added, the progress of the unstalled, translating ribosome was monitored by measuring the change in size of the hairpin, which shortens as the ribosome moves through it.

Results are shown in **Figure 2**, which measures change in extension versus time. Strikingly, the extension shows a repeated step-pause pattern. Every arrow indicates a discrete step where the ribosome has progressed by one codon (three nucleotides), which, in turn, decreases the hairpin by six nucleotides as the hairpin unwinds. Each step took place in 0.078 second, with punctuating pauses lasting 2.8 seconds. The pauses are thought to represent the collective time it takes to introduce a new charged tRNA into the A site, catalyze peptide bond formation, and begin translocation. These studies have revealed for the first time that translation occurs not in a continuous manner, but as a series of translocation-pause events.

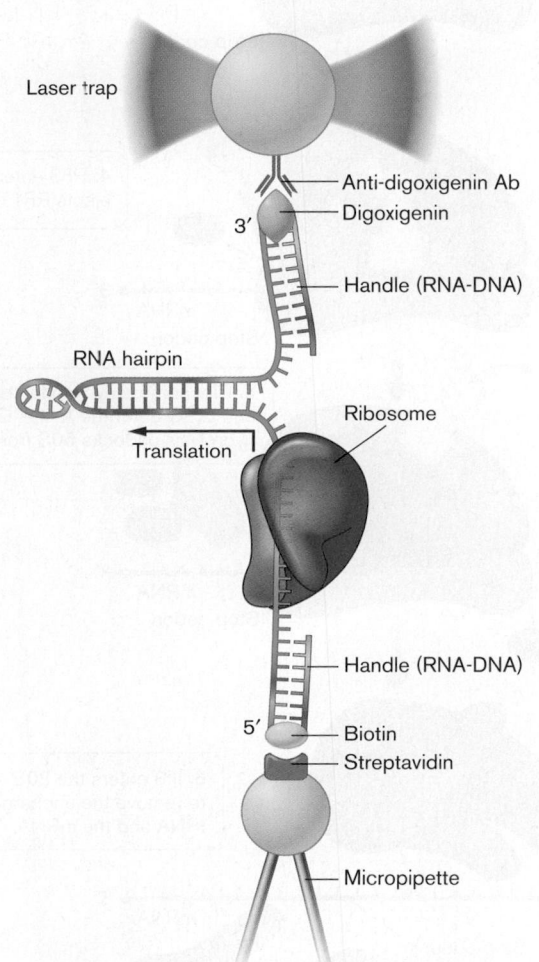

Figure 1   The ribosome trap concept.

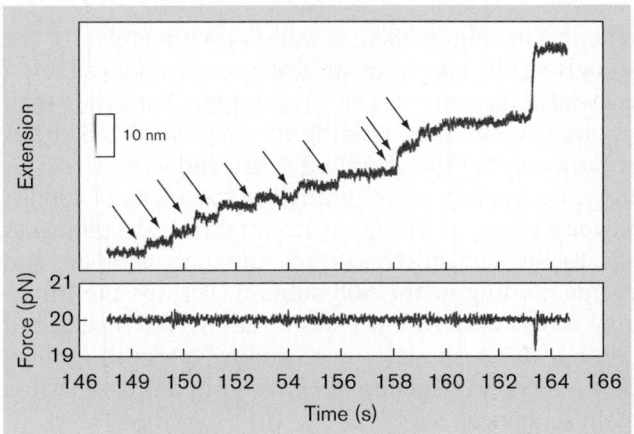

Figure 2   Results from a ribosome trap experiment.
Arrowheads indicate where the ribosome has moved by one codon, shortening the hairpin but lengthening the distance between the two beads. pN = piconewtons. *Source:* Jin-Der Went et al. 2008. *Nature* **452**:598–603.

(**Fig. 8.27**⏵ , step 1) and brings the stop codon into the A site (**Fig. 8.27**, step 2). No tRNA binds, but one of two **release factors** (RF1 or RF2) will enter (step 2) and activate the peptidyltransferase, thereby cutting the bond tethering the completed peptide to tRNA in the P site (step 3).

With the protein released, the ribosome must disassemble. RF3 causes RF1 or RF2 to depart the ribosome (step 4). Then, ribosome recycling factor (RRF), along with EF-G, binds at the A site, and an accompanying GTP hydrolysis undocks the two ribosomal subunits (step 5). IF3 then reenters the 30S subunit to replace the remaining uncharged tRNA and mRNA (step 6). The liberated ribosomal subunits are now free to diffuse through the cell, ready to bind yet another mRNA and begin the translation sequence anew.

> **THOUGHT QUESTION 8.6** While working as a member of a pharmaceutical company's drug discovery team, you find that a soil microbe snatched from the jungles of South America produces an antibiotic that will kill even the most deadly, drug-resistant form of *Enterococcus faecalis*, which is an important cause of heart valve vegetations in bacterial endocarditis. Your experiments indicate that the compound stops protein synthesis. How could you more precisely determine the antibiotic's mode of action? *Hint*: Can you use mutants resistant to the antibiotic?

Antibiotics That Affect Translation

Streptomycin (**Fig. 8.28A**), a well-known member of the aminoglycoside family of antibiotics, is produced by a species of *Streptomyces*. The drug targets bacterial small ribosomal subunits by binding to a region of 16S rRNA that forms part of the decoding A site and to protein S12, a protein critical for maintaining the specificity of codon-anticodon binding. At high concentrations, streptomycin binds to the 30S-mRNA-tRNA initiation complex and prevents binding of the 50S subunit. It stops the initiation of translation. At low concentrations, translation can go on but the A site becomes "sloppy," permitting illicit codon-anticodon matchups that result in a mistranslated protein sequence.

Bacteria usually become resistant to streptomycin by spontaneously mutating the S12 gene (*rpsL*), although mutations in 16S rRNA can also prevent binding. The altered S12 protein maintains function but will not bind streptomycin. Some bacteria gain resistance by acquiring an aminoglycoside phosphotransferase that modifies streptomycin so that it cannot bind to its target. Additional mechanisms are discussed in Chapter 27. Other therapeutically important aminoglycosides include gentamicin, kanamycin, and amikacin.

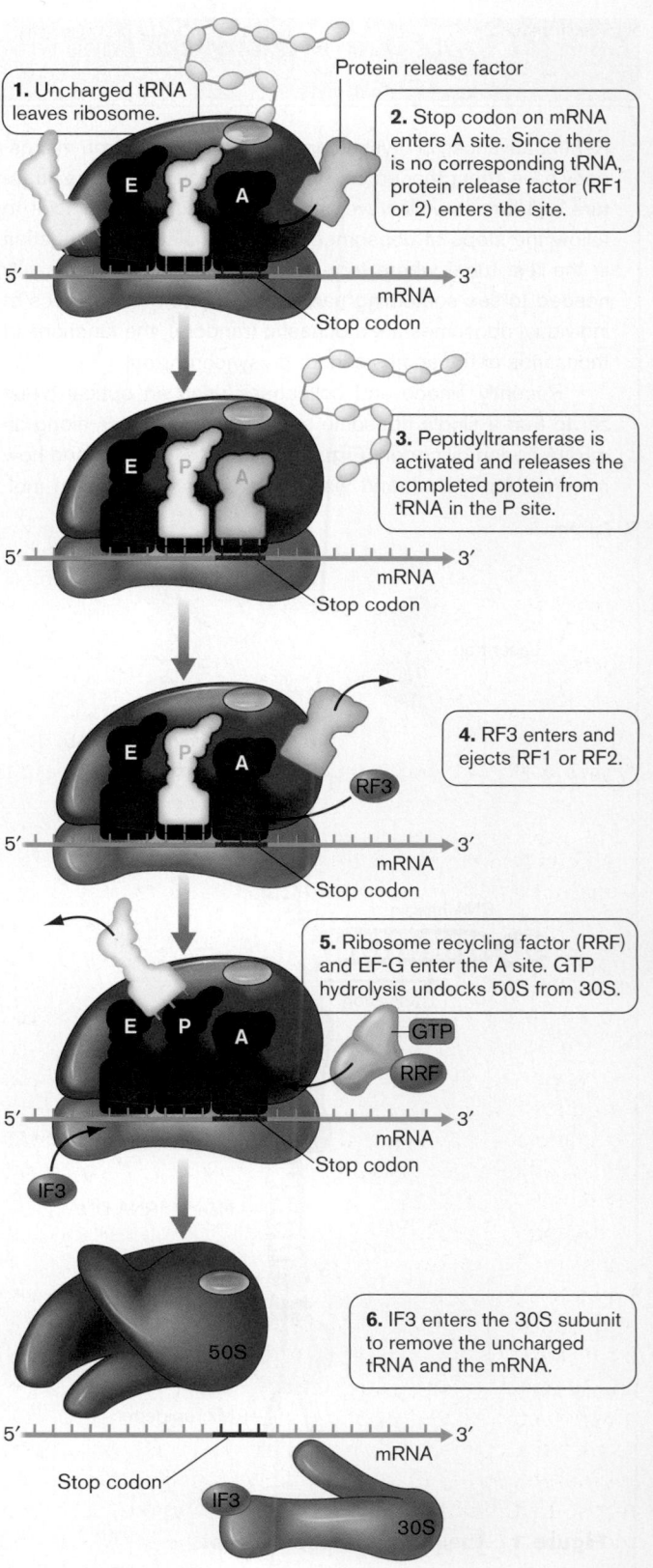

1. Uncharged tRNA leaves ribosome.

Protein release factor

2. Stop codon on mRNA enters A site. Since there is no corresponding tRNA, protein release factor (RF1 or 2) enters the site.

mRNA
Stop codon

3. Peptidyltransferase is activated and releases the completed protein from tRNA in the P site.

mRNA
Stop codon

4. RF3 enters and ejects RF1 or RF2.

RF3

mRNA
Stop codon

5. Ribosome recycling factor (RRF) and EF-G enter the A site. GTP hydrolysis undocks 50S from 30S.

GTP
RRF

IF3

mRNA
Stop codon

6. IF3 enters the 30S subunit to remove the uncharged tRNA and the mRNA.

50S

Stop codon

IF3

mRNA

30S

Figure 8.27 Translation termination. The completed protein is released, and the ribosome subunits are recycled. ⏵
microbiology2.com/animations

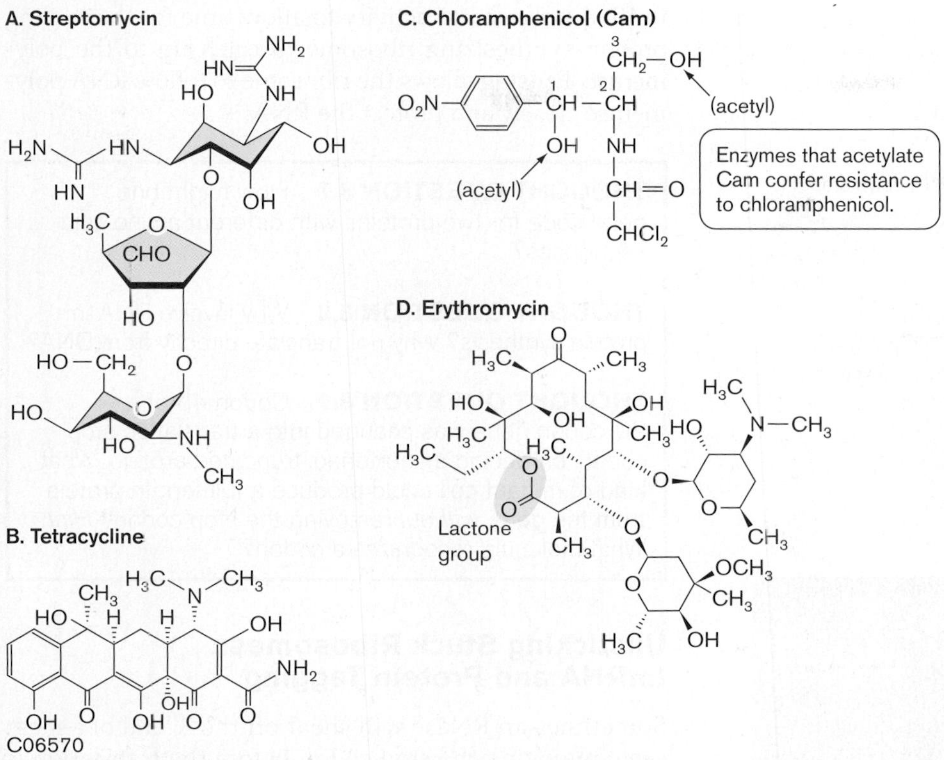

Figure 8.28 Antibiotics that inhibit protein synthesis in bacteria.

macrolide binding and confer resistance to these drugs. Other microbes reduce erythromycin binding by methylating the relevant area of 23S rRNA using an enzyme produced from another mobile genetic element.

Other translation-targeting antibiotics interfere with mRNA binding to the ribosome (kasugamycin), prevent translocation by targeting EF-G (fusidic acid), or use **molecular mimicry** to trick peptidyltransferase into action without having a bona fide tRNA in the A site (puromycin). We chronicle the discovery and use of antibiotics more completely in Chapter 27.

Prokaryotic Transcription and Translation Are Coupled

Many genes in bacterial chromosomes are arranged and transcribed in tandem and produce a polycistronic mRNA (see Section 7.2). Remember that the beginning of each gene in a polycistronic mRNA carries its own ribosome-binding site. This means that different ribosomes can bind simultaneously to the start of each cistron within a polycistronic mRNA.

Ribosomes bind to the 5' end of mRNA and begin translating protein even before RNA polymerase has finished transcribing the mRNA (**Fig. 8.29**). This is called coupled transcription and translation. Transcriptional-translational coupling in prokaryotes provides an opportunity to use translation as a means of regulating transcription, a process that is explained further in Chapter 10.

Eukaryotic microbes, on the other hand, use separate cellular compartments to carry out most of their transcription and translation. They transcribe genes in the nucleus and, after splicing and other forms of processing, transport mRNA to the cytoplasm, where the majority of translation occurs. However, a small amount of translation is carried out in the eukaryotic nucleus, where it is also coupled to transcription.

In prokaryotes, coupling of transcription and translation presents a potential problem. Ribosomes generally travel along mRNA more slowly than mRNA is generated by RNA polymerase, so RNA polymerase can potentially scoot ahead of the ribosome and leave large tracts of RNA unprotected and susceptible to nucleases. Unintentional cleavage of the RNA between the ribosome and polymerase would separate the two macromolecular

Tetracycline (**Fig. 8.28B**) also targets the 30S ribosomal subunit, but instead of preventing formation of the 70S complex (as with streptomycin), it outright prevents binding of aminoacyl-tRNA to the A site and stops protein synthesis. It binds to 16S rRNA, right above the A site for incoming tRNA. Resistance to tetracycline can be conferred by an efflux transport system that effectively removes the antibiotic from the bacterial cell. Genes encoding this resistance are usually carried on mobile genetic elements like plasmids that can be passed from cell to cell (see Section 9.2). Other resistance mechanisms are described in Chapter 27.

Another species of *Streptomyces* produces chloramphenicol (**Fig. 8.28C**), which attacks the 50S subunit. It binds to 23S rRNA at the peptidyltransferase active site and inhibits peptide bond formation. Resistance to this drug comes from an ability to synthesize an enzyme, chloramphenicol acetyltransferase, that destroys chloramphenicol activity by modifying the chemical.

Erythromycin, made by *Streptomyces erythraeus*, is one of a large group of related antibiotics called macrolides, whose hallmark is a large lactone ring of 12–22 carbon molecules (**Fig. 8.28D**). They all affix themselves to the 50S subunit by binding to protein L15 and 23S rRNA in the peptidyltransferase cavity. Binding triggers an abortive translocation step that evicts peptidyl-tRNA from the P site while preventing peptide bond formation. In some organisms, mutational alterations in L15 decrease

A.

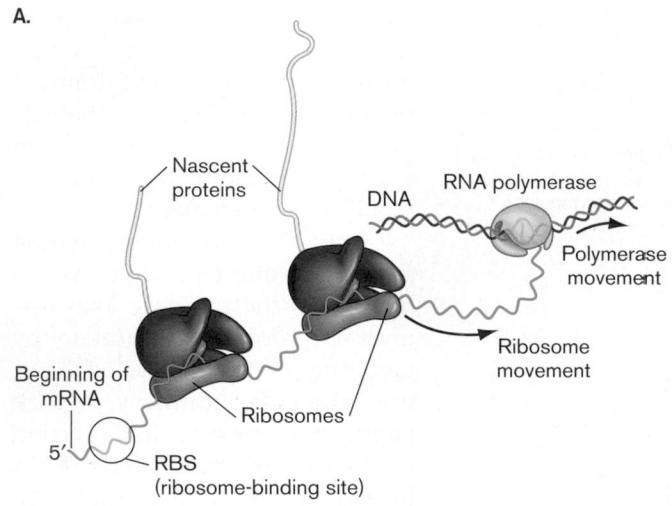

B.

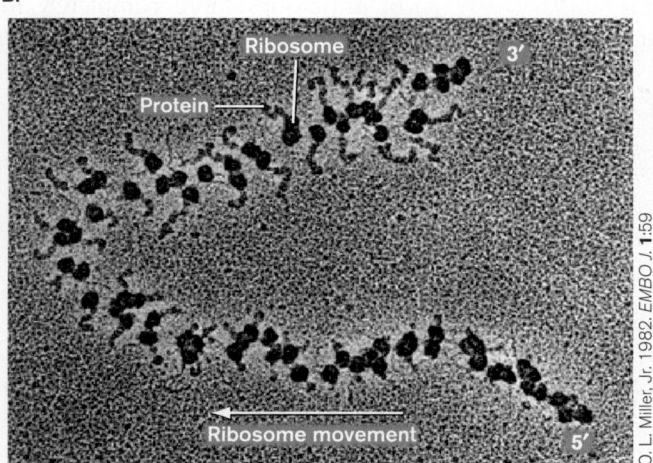

O. L. Miller, Jr. 1982. *EMBO J.* 1:59

Figure 8.29 Coupled transcription and translation in prokaryotes. **A.** Ribosomes attach at mRNA ribosome-binding sites and start synthesizing protein before transcription of the gene is complete. **B.** Electron micrograph of a polysome (ribosome is 21 nm across). Several ribosomes may translate a single mRNA molecule at the same time. The beginning (5′ end) of the mRNA is to the right (at the arrow), and the 3′ end is to the left. Note that the synthesized protein molecule grows longer and longer the closer the ribosome gets to the 3′ end of the mRNA. The protein molecule is seen most clearly at the end of the mRNA (to the upper right).

machines and destroy the message. The cell handles this problem by modulating transcriptional speed. The rate of RNA synthesis averages about 45 nt per second, which roughly equals the average rate of translation (16 amino acids per second). However, these speeds are not constant. For example, RNA polymerase pauses momentarily at sites rich in GC content (GC base pairs, with three hydrogen bonds, are harder to melt than AT pairs with only two hydrogen bonds). Once sigma factor exits the transcription complex, proteins called NusA and NusG enter the complex and can actually lengthen the pause

in RNA polymerase activity to allow time for the trailing protein-synthesizing ribosome to catch up to the polymerase. Pausing allows the ribosome to follow RNA polymerase closely and protect the RNA.

THOUGHT QUESTION 8.7   How might one gene code for two proteins with different amino acid sequences?

THOUGHT QUESTION 8.8   Why involve RNA in protein synthesis? Why not translate directly from DNA?

THOUGHT QUESTION 8.9   Codon 45 of a 90-codon gene was changed into a translation stop codon, producing a shortened, truncated protein. What kind of mutant cell could produce a full-length protein from the gene *without* removing the stop codon? *Hint:* What molecule recognizes a codon?

Unsticking Stuck Ribosomes: tmRNA and Protein Tagging

Sometimes an RNase will shear off the 3′ end of a message, removing the stop codon before the translation is complete. What happens after a ribosome has finished translating this damaged mRNA molecule? Without a stop codon, there is nothing to trigger ribosome release when the ribosome reaches the end of the message. So the ribosome is stuck at the end of the mRNA with a peptidyl-tRNA in the ribosome P site.

The answer to this problem is a translation rescue molecule, the previously mentioned tmRNA, which has the properties of both tRNA and mRNA. One end of the tmRNA molecule has an attached amino acid and a folded structure that resembles a tRNA. This aminoacyl end acts like tRNA, entering the unoccupied ribosome A site, where a peptide bond then forms between the stalled polypeptide (in the P site) and the emergency amino acid on the tmRNA. Another section of the tmRNA is then read as mRNA, which adds a tag of 12 or so amino acids to the now dislodged peptide (SsrA; **Fig. 8.30A**). These amino acids, called a proteolysis tag, predestine the protein for destruction. A stop codon present in tmRNA triggers peptide release and ribosome disassembly. A helper protein (SspB) recognizes the proteolysis tag and brings the useless aberrant polypeptide to the ClpXP protease (discussed shortly) for degradation. **Figure 8.30B** shows this process for tmRNA from *E. coli*.

TO SUMMARIZE:

■ **Triplet nucleotide codons** in mRNA encode specific amino acids. **Transfer RNA molecules** interpret the genetic code and bring specific amino acids to the A (acceptor) site in the ribosome.

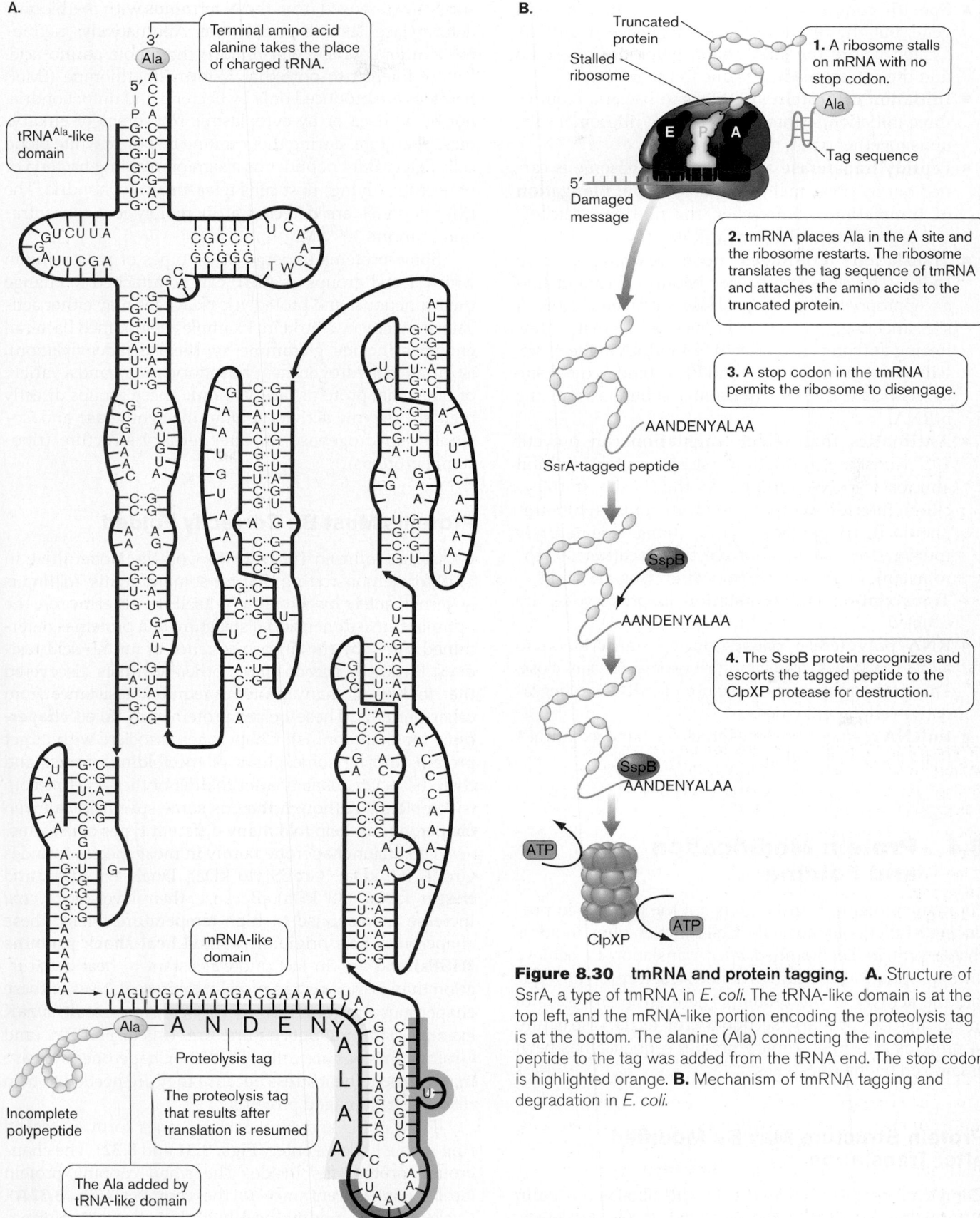

A. Structure of SsrA, a type of tmRNA in *E. coli*. The tRNA-like domain is at the top left, and the mRNA-like portion encoding the proteolysis tag is at the bottom.

Terminal amino acid alanine takes the place of charged tRNA.

tRNA^Ala-like domain

mRNA-like domain

Incomplete polypeptide

Proteolysis tag

The proteolysis tag that results after translation is resumed

The Ala added by tRNA-like domain

B.

Truncated protein

Stalled ribosome

Damaged message

Tag sequence

1. A ribosome stalls on mRNA with no stop codon.

2. tmRNA places Ala in the A site and the ribosome restarts. The ribosome translates the tag sequence of tmRNA and attaches the amino acids to the truncated protein.

3. A stop codon in the tmRNA permits the ribosome to disengage.

AANDENYALAA

SsrA-tagged peptide

AANDENYALAA

4. The SspB protein recognizes and escorts the tagged peptide to the ClpXP protease for destruction.

AANDENYALAA

ClpXP

Figure 8.30 tmRNA and protein tagging. A. Structure of SsrA, a type of tmRNA in *E. coli*. The tRNA-like domain is at the top left, and the mRNA-like portion encoding the proteolysis tag is at the bottom. The alanine (Ala) connecting the incomplete peptide to the tag was added from the tRNA end. The stop codon is highlighted orange. **B.** Mechanism of tmRNA tagging and degradation in *E. coli*.

- **Specific codons** mark the beginning and end of a gene, but the Shine-Dalgarno sequence in mRNA located before the start codon helps the ribosome find the correct reading frame in the mRNA.
- **Initiation of protein synthesis** in bacteria requires three initiation factors that bring the ribosomal subunits together on an mRNA molecule.
- **Peptidyltransferase** activity of the ribosome is carried out by ribosomal RNA, not protein. **Elongation of translation** occurs when the ribosome ratchets one codon length along the mRNA.
- **Translation terminates** upon reaching a stop codon. The ribosome pauses because it cannot find an appropriate tRNA. A release factor enters the A site and triggers peptidyltransferase activity, thus freeing the completed protein from tRNA in the A site.
- **Ribosome release factor** and EF-G bind to the A site to dissociate the two ribosomal subunits from the mRNA.
- **Antibiotics that affect translation** can prevent 70S ribosome formation (streptomycin), inhibit aminoacyl-tRNA binding to the A site (tetracycline), interfere with peptidyltransferase (chloramphenicol), trigger peptidyltransferase prematurely (puromycin), cause abortive translocation (erythromycin), or prevent translocation (fusidic acid).
- **Transcription and translation** in prokaryotes are coupled.
- **RNA polymerase pauses** during transcription to allow the slower-translating ribosomes to stay close. This pause minimizes exposure of mRNA to degradative cellular enzymes.
- **tmRNA** rescues ribosomes stuck on damaged mRNA that lacks a stop codon.

8.4 Protein Modification and Folding

For many proteins, translation is not the last step in producing a functional molecule. Cell function often requires that a protein be modified *after* translation to achieve an appropriate three-dimensional structure or to regulate the activity of the protein. There are many ways of modifying the primary, secondary, or tertiary structure of proteins after the primary protein sequence has been assembled by the ribosomes.

Protein Structure May Be Modified after Translation

Completed proteins released from the ribosome contain *N*-formylmethionine at the N terminus (as previously described). With some proteins, the *N*-formyl group is surgically removed from the N terminus with methionine deformylase, leaving methionine. Alternatively, methionyl aminopeptidase will remove the whole amino acid. It is of interest to note that *N*-formylmethionine (fMet) peptides are produced only by bacteria and mitochondria, not by archaea or by cytoplasmic ribosomes of eukaryotes. Therefore, during the immune response white blood cells detect fMet peptides as a sign of invading bacteria or of necrotic (dying) host cells releasing mitochondria. The fMet peptides are detected at incredibly low concentrations, around 10^{-12} M.

Some proteins undergo other types of processing in which acetyl groups or AMP can be attached to change their functions, and proteolytic cleavages may either activate or inactivate a protein. Examples of modified bacterial enzymes include glutamine synthetase (adenylylation), isocitrate dehydrogenase (phosphorylation), and a variety of ribosomal proteins (acetylation). These groups directly regulate enzyme activity (glutamine synthetase and isocitrate dehydrogenase) or alter tertiary structure (ribosomal proteins).

Proteins Must Be Correctly Folded

Christian Anfinsen (1916–1995) won the Nobel Prize in 1972 for demonstrating that for some proteins, folding is governed solely by the protein itself. In other words, the optimal three-dimensional structure of a protein is determined solely by the linear sequence of amino acid residues. But three decades later, other scientists discovered that folding of many proteins requires assistance from other proteins. These helper proteins are called **chaperones** (or chaperonins). Chaperones associate with target proteins during some phase of the folding process and then dissociate, usually after folding of the target protein is completed. Although there is some specificity, a given chaperone can help fold many different types of proteins.

The major chaperone family in most species includes GroEL (60 kDa), GroES (10 kDa), DnaK (70 kDa), and trigger factor (48 kDa). Because their levels in *E. coli* increase in response to high-temperature stress, these chaperones were originally named **heat-shock proteins (HSPs)** and are, in fact, more resistant to heat denaturation than is the average protein. Representatives of these chaperones are found in all species. As a result, DnaK examples throughout nature are called HSP70s, and GroEL homologs are called HSP60s. Chaperones increase in response to heat stress because they are needed to help refold heat-damaged proteins.

The GroEL and GroES chaperones form a stacked ring with a hollow center (**Figs. 8.31** and **8.32**). The chaperoned protein fits inside. The small capping protein GroES controls entrance to the chamber (**Fig. 8.32A**). Cycles of ATP binding and hydrolysis cause conformational changes within the chamber that can reconfigure target proteins. DnaK (HSP70) chaperones have a very different structure (**Fig. 8.32B**). They do not form rings like the GroEL and GroES chaperones, but can clamp down on a peptide to assist folding. Proteins emerging from a prokaryotic ribosome enter a folding pathway that involves a hierarchy of these chaperones.

TO SUMMARIZE:

- **Protein modifications** are made after translation.
- **The N-terminal amino acid** (*N*-formylmethionine) can be removed by methionine deformylase.
- **An inactive precursor protein** can be cleaved into a smaller active protein, or other groups can be added to the protein (for example, phosphate or AMP).
- **Chaperone proteins** help translated proteins fold properly.

8.5 Secretion: Protein Traffic Control

Microorganisms, especially Gram-negative bacteria, face an intriguing challenge in delivering proteins to different target locations in the cell. Recall that Gram-negative microbes are surrounded by two layers of membrane (the inner membrane, or cell membrane, and an outer membrane) between which lies a periplasmic space (see Section 3.4). Many proteins are specifically destined for one or another of these cell compartments. Other proteins are secreted completely out of the cell into the surrounding environment (for example, hemolysins that lyse red blood cells). But how do these diverse proteins know where to go? Protein traffic out of the cell is directed by an elaborate set of protein secretion systems. Each system selectively delivers a set of proteins originally made in the cytoplasm to various extracytoplasmic locations.

The term "secretion" is used to describe movement of a protein out of the cytoplasm. There are protein secretion systems that move proteins out of the cytoplasm into the cytoplasmic membrane and across the membrane to the periplasm, others that move proteins to the outer membrane, and still others that deliver proteins across both of the membranes and into the surrounding environment. An added complication of protein export is that periplasmic proteins are usually delivered unfolded into the periplasm and require another set of chaperones to fold properly in this cell compartment.

Protein Export Out of the Cytoplasm

Proteins destined for the bacterial cell membrane or envelope regions (periplasm, outer membrane, or extracellular spaces) require special export systems. These systems manage to move hydrophilic proteins through

Figure 8.31 Top view of the GroEL-GroES complex of *E. coli.* The heat-shock chaperone-10 (equivalent to GroES), shown in red, is tacked onto a GroEL ring. The chaperoned protein fits inside the open hole. (PDB code: 1PCQ)

A.

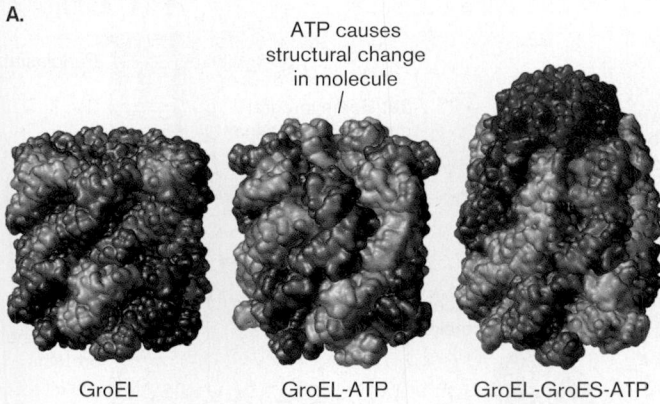

ATP causes structural change in molecule

GroEL   GroEL-ATP   GroEL-GroES-ATP

B.

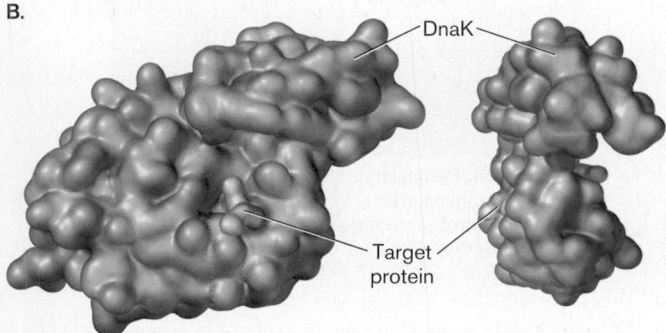

DnaK

Target protein

Figure 8.32 GroEL-GroES and DnaK (HSP70) structures. A. Three-dimensional reconstructions of GroEL, GroEL-ATP, and GroEL-GroES-ATP from cryo-EM. The GroES ring is seen as a disk above GroEL. (PDB codes: 1SS8, 2C7E, 1PCQ) **B.** DnaK clamping down on a peptide (yellow). (PDB code: 1DKX)

one or more hydrophobic membrane barriers. Proteins meant for the inner membrane (for example, cytochromes) are tagged, as part of the open reading frame, with very hydrophobic N-terminal **signal sequences** of 15–30 amino acids. Signal sequences tether nascent proteins to the membrane and confer conformations that allow the proteins to melt into the fabric of the membrane. Inner membrane proteins also contain hydrophobic transmembrane-spanning regions (20–25 amino acids) that aid in this insertion process. These hydrophobic regions are important because they are compatible with the hydrophobicity of the membrane itself. A nutrient transport protein often has 12 such membrane-spanning regions, which weave back and forth across the membrane.

One special export system begins with a complex called the **signal recognition particle (SRP)**, which targets proteins for inner membrane insertion. A second export mechanism uses a protein called trigger factor that assists the journey of proteins destined for the periplasm. (Trigger factor was mentioned earlier as a chaperone.) These two protein traffic pathways converge on a general secretion complex composed of three proteins, collectively called the SecYEG translocon, embedded in the cell (inner) membrane. Depending on the exported protein, SecYEG will assist export to the periplasm or insertion into the membrane.

Protein Export to the Cell Membrane

The pathway leading proteins to the inner (cell) membrane begins with SRP. In *E. coli*, SRP consists of a 54-kDa protein Ffh complexed with a small RNA molecule (*ffs*). SRP binds to the signal sequences of integral cell membrane proteins as they are being translated (**Fig. 8.33** ▶II) and halts further translation in the cytoplasm. The nascent protein with its paralyzed nontranslating ribosome is delivered to the membrane-embedded protein FtsY, where translation resumes. The partially translated protein is now subject to one of two fates: It may be cotranslationally inserted directly into the cell membrane, meaning that the protein is inserted even as it is still being translated. Alternatively, the protein may be completely synthesized, after which it is delivered to the SecYEG translocon for insertion. The route to membrane insertion depends on the protein.

Although much is known about how proteins are inserted into the cytoplasmic membrane, a machinery designed to insert proteins into the outer membrane is unknown, but highly anticipated.

Protein Export to the Periplasm: The Sec-Dependent General Secretion Pathway

The periplasm contains important proteins that bind nutrients for transport into the cell and other proteins that carry out enzymatic reactions. For example, one form of superoxide dismutase (SOD), an enzyme that degrades superoxide, is a periplasmic protein in *Salmonella enterica* and other Gram-negative bacteria. Many periplasmic proteins, such as SOD and maltose-binding protein (which imports the sugar maltose), are delivered to the periplasm by a common pathway called the general secretion pathway.

There are several steps in the general secretion pathway. First, the peptide is completely translated in the cytoplasm (**Fig. 8.34** ▶II, step 1). Trigger factor interacts with newly synthesized protein as it exits the ribosome and keeps pre-secreted proteins in a loosely folded conformation, awaiting interaction with the next component of the secretion machinery. The completed pre-secretion protein is then captured by a piloting protein called SecB (**Fig. 8.34**, step 2; and **Fig. 8.35**), which unfolds the pre-secretion protein and delivers it to SecA, a protein peripherally associated with the membrane-spanning SecYEG

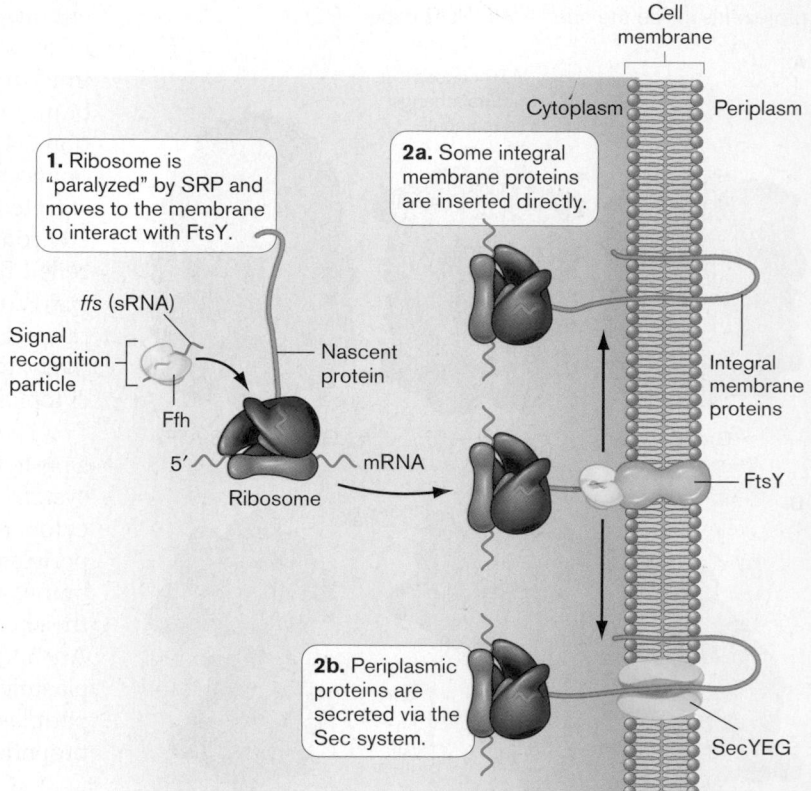

Figure 8.33 **SRP and cotranslational export.** A ribosome "paralyzed" by SRP does not resume translating protein until encountering FtsY in the membrane. Translation can then recommence. Some proteins designated for integral membrane location are inserted directly (top). Other integral membrane proteins, and proteins destined for the periplasm, are inserted or secreted via the Sec system (bottom). ▶II *microbiology2.com/animations*

translocon (step 3). Keeping a pre-secretion protein unfolded in the cytoplasm assists the secretion process because sliding an unfolded protein through a membrane is far easier than trying to deliver a folded one.

The SecA ATPase appears to act like a plunger (step 4). It inserts deep into the SecYEG channel, shoving about 20 amino acids of the target export protein into the channel. ATP hydrolysis causes SecA to release the protein and withdraw (step 5). At this point, SecA can bind fresh ATP, rebind the target protein, and reinsert, pushing another 20 amino acids through. Proteins needed in the periplasm have cleavable signal sequences at their amino-terminal ends. Immediately following translocation of the amino-terminal sequence into the periplasm, the sequence is snipped off by periplasmic signal peptidases (LepB is one of several examples in *E. coli*). This cleaving completes translocation and allows release of the mature protein into the periplasm (step 6). Signal peptidases will not, however, cleave signals from proteins destined to stay embedded within the membrane (integral membrane proteins).

NOTE: "Translocation" can refer to the movement of a ribosome along mRNA or describe movement of a protein from one cell compartment (cytoplasm) to another (periplasm).

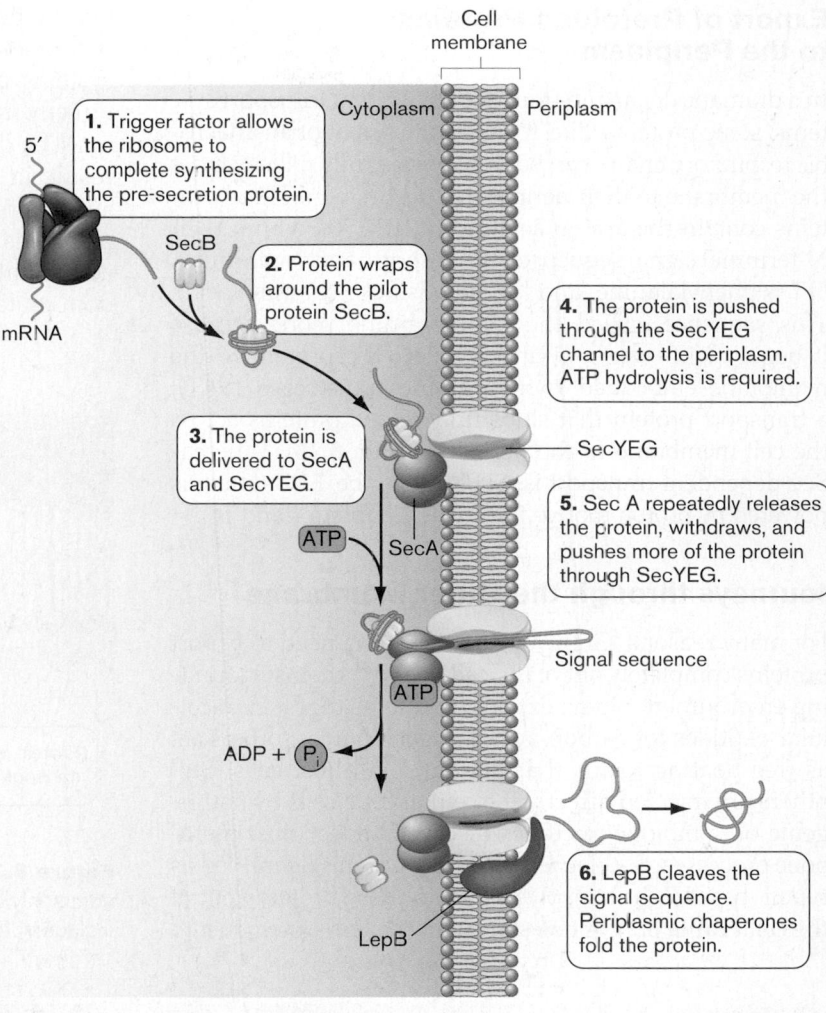

Figure 8.34 SecA-dependent general secretion pathway. ▶❙❙
microbiology2.com/animations

Periplasmic proteins delivered by the Sec system arrive unfolded and inactive. Because the folding chaperones mentioned earlier are cytoplasmic, periplasmic proteins need another set of dedicated periplasmic chaperones to guide their tertiary folding. Another problem with periplasmic proteins is that the oxidizing environment of aerobic cells can oxidize cysteines within a protein and produce inappropriate cysteine disulfide bonds that destroy enzyme function. Special periplasmic disulfide reductases are required to reverse these S—S bonds and form two SH groups. Many periplasmic proteins, however, need certain disulfide bonds to be active. The periplasm also contains a disulfide bond catalyst (DsbA) to make those bonds.

Eukaryotic microbes such as the yeast *Saccharomyces cerevisiae* also possess secretion systems that move proteins to the membrane and beyond. However, the eukaryotic Sec systems are more complex and are evolutionarily distinct from bacterial Sec systems. Archaeal secretion systems are actually more similar to those of eukaryotes.

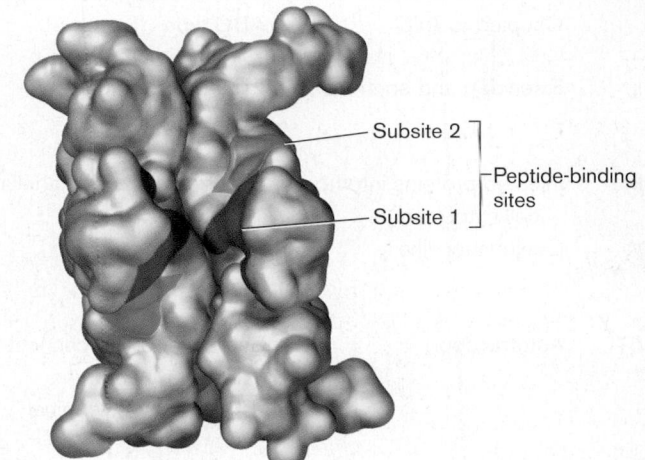

Figure 8.35 Proposed peptide-binding channel of SecB. The exposed surfaces encompassing the two proposed peptide-binding subsites are highlighted. The pre-secretion protein is wrapped around these sites in SecB. (PDB code: 1FX3)

Export of Prefolded Proteins to the Periplasm

In a dramatic departure from Sec-dependent transport systems, some proteins, like TorA, a component of an anaerobic respiratory chain, can be transported fully folded across the membrane to their periplasmic destination. These proteins contain the amino acid motif RRXFXK within their N-terminal signal sequence ("R" is shorthand for arginine, "F" is phenylalanine, and "X" stands for any amino acid). This sequence, called the twin arginine motif because it begins with two arginines, targets the protein to the membrane-embedded twin arginine translocase (TAT), a transport protein that ships fully folded proteins across the cell membrane to the periplasm (**Fig. 8.36**). Whereas Sec-dependent transport is ATP driven, the TAT system is powered by proton motive force (see Chapters 4 and 14).

Journeys through the Outer Membrane

For many reasons, Gram-negative bacteria need to export proteins completely out of the cell and into their surrounding environment. Some exported proteins digest extracellular peptides for carbon and nitrogen sources; others act as free-floating toxins that bind and kill host cells. Still others are injected directly into eukaryotic cells by pathogenic or symbiotic microbes to commandeer host metabolic processes. Six elegant secretion systems, identified as type I–type VI, have evolved to ship these proteins out of the cell (**Table 8.2**). A few start with the Sec system just to

get the protein into the periplasm, where dedicated outer membrane systems take over and complete export. Other systems provide nonstop service, delivering the protein directly from the cytoplasm to the extracellular space.

The diversity of system design is impressive. It is the result, in some instances, of selective evolutionary pressures appropriating established cell processes (for example, pilus assembly). New spheres evolve through the accidental duplication of one set of genes followed by random mutations that innovatively redesign the duplicated

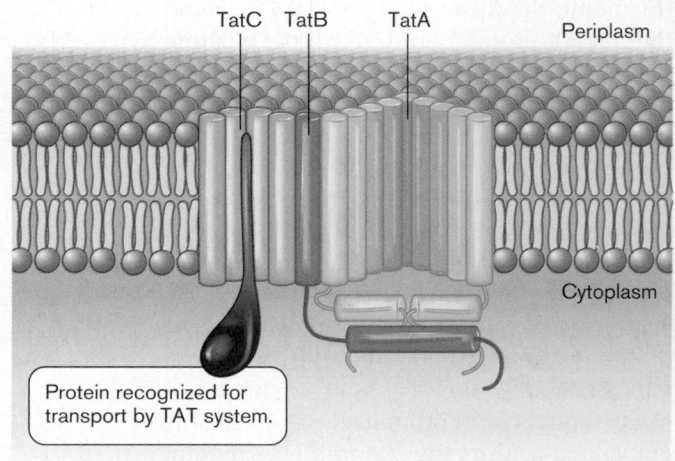

Figure 8.36   The twin arginine translocase (TAT). A commonly accepted model for the Tat protein translocase, which includes proteins TatA, TatB, and TatC.

Table 8.2   Comparison of mechanisms that secrete proteins across the outer membrane.[a]

Type	Mechanism	Structure	Location of protein substrate	Location of secretory signals	Number of components
I	Coupled to TolC	ABC type	Cytoplasm	N terminus, not cleaved	3
II	Extending and contracting	Pilus-like structure	Periplasm	Cleaved N terminus for Sec-dependent transport	12–16
III	Injected proteins into host cell cytoplasm	Syringe, related to flagella biogenesis	Cytoplasm	None	20
IV	Conjugation-like	Multicomponent	Some are cytoplasmic, others are periplasmic	None	8 or 9
V	Autotransport	Self-transporting channel in outer membrane formed by C-terminal domain	Periplasm	N terminus, cleaved	1
VI	Injecting proteins directly into host cytoplasm	Syringe-like, phage origin	Cytoplasm	None	12–20

[a]The six classes are grouped on the basis of their structure. Types II, III, and IV are evolutionarily related to mechanisms that assemble pili (type II) or flagella (type III) or that carry out conjugation (type IV). Type VI systems appear to be related to phage tail proteins. Some systems pick up protein substrates from the cytoplasm and transport them across both membranes. Substrates for other systems are collected in the periplasm. Substrates also differ as to the presence of N-terminal signal sequences.

set for a new purpose. We know this because the footprints of genetic divergence have been left behind in the DNA sequence. Type I secretion will be described here. Other systems will be covered during the discussion of pathogenesis in Chapter 25.

Type I Protein Secretion

Chapter 4 describes the family of ATP-binding cassette (ABC) influx transporters whose signature is an amino acid motif that binds ATP. (The term "cassette" refers to a sequence of amino acids that is conserved in many proteins with similar functions.) In addition to ABC influx transporters, similar ABC transporters function in the opposite direction to export various toxins, proteases, and lipases, as well as antimicrobial drugs (multidrug efflux transporters). These ABC transporters are the simplest of the protein secretion systems and make up what is called type I protein secretion (one example is the Hly system that secretes hemolysin; **Fig. 8.37◗**). Type I systems all have three protein components, one of which contains an ATP-binding cassette. One component is an outer membrane channel, the second is an ABC protein at the inner membrane, and the third is a periplasmic protein lashed to the inner membrane.

Proteins secreted through type I systems never contact the periplasm, because they pass through a continuous channel that extends from the cytoplasm to the outer membrane. The inner membrane and periplasmic subunits are generally substrate specific, but numerous ABC export systems share the channel protein TolC. TolC is an intriguing protein composed of a beta barrel channel embedded in the outer membrane and an alpha helix tunnel spanning the periplasm. **Figure 8.38** illustrates how the tunnel might open and close. The type I transport shown in **Figure 8.37** exports a hemolysin (HlyA) from *E. coli* that lyses red blood cell membranes. HylB and HylD are the ABC and periplasmic components, respectively.

Five other protein secretion systems are briefly summarized in **Table 8.2**. Some move proteins directly from the cytoplasm to the outside, similar to the type I system; others pick up proteins deposited in the periplasm by the Sec system. Note that they all play important roles in microbial pathogenesis and are more fully discussed in Chapter 25.

WWW | Chaperone-assisted protein folding
     microbiology2.com/links

TO SUMMARIZE:

- **Special protein export mechanisms** are used to move proteins to the inner membrane, the periplasm, the outer membrane, and the extracellular surroundings.
- **N-terminal amino acid signal sequences** help target membrane proteins to the membrane.

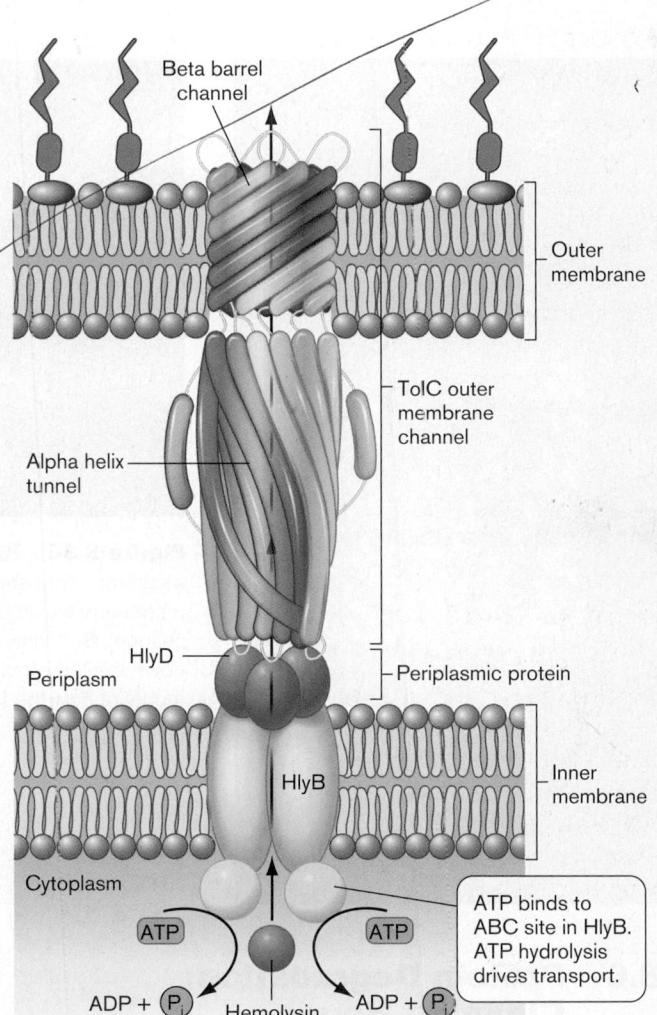

Figure 8.37   Type I secretion: the HlyABC transporter.
Hemolysin (HlyA) is transported directly from the cytoplasm into the extracellular medium through a multicomponent ABC transport system. The HlyB and D proteins are dedicated to HlyA transport. TolC is shared with other transport systems. Not drawn to scale. *Source*: Modified from Moat et al. 2002. *Microbial Physiology* 4th ed. Wiley-Liss.
◗ *microbiology2.com/animations*

- **The general secretory system** involving the SecYEG translocon can move unfolded proteins to the inner membrane or periplasm.
- **The signal recognition particle (SRP)** pauses translation of a subset of proteins that will be placed into the membrane.
- **SecB protein** binds to certain unfolded proteins that will eventually end up in the periplasm, and pilots them to the SecYEG translocon.
- **The twin arginine translocase (TAT)** can move a subset of already folded proteins across the inner membrane and into the periplasm.
- **Type I secretion systems** are ATP-binding cassette (ABC) mechanisms that move certain secreted proteins directly from the cytoplasm to the extracellular environment.

A.

B.

C.

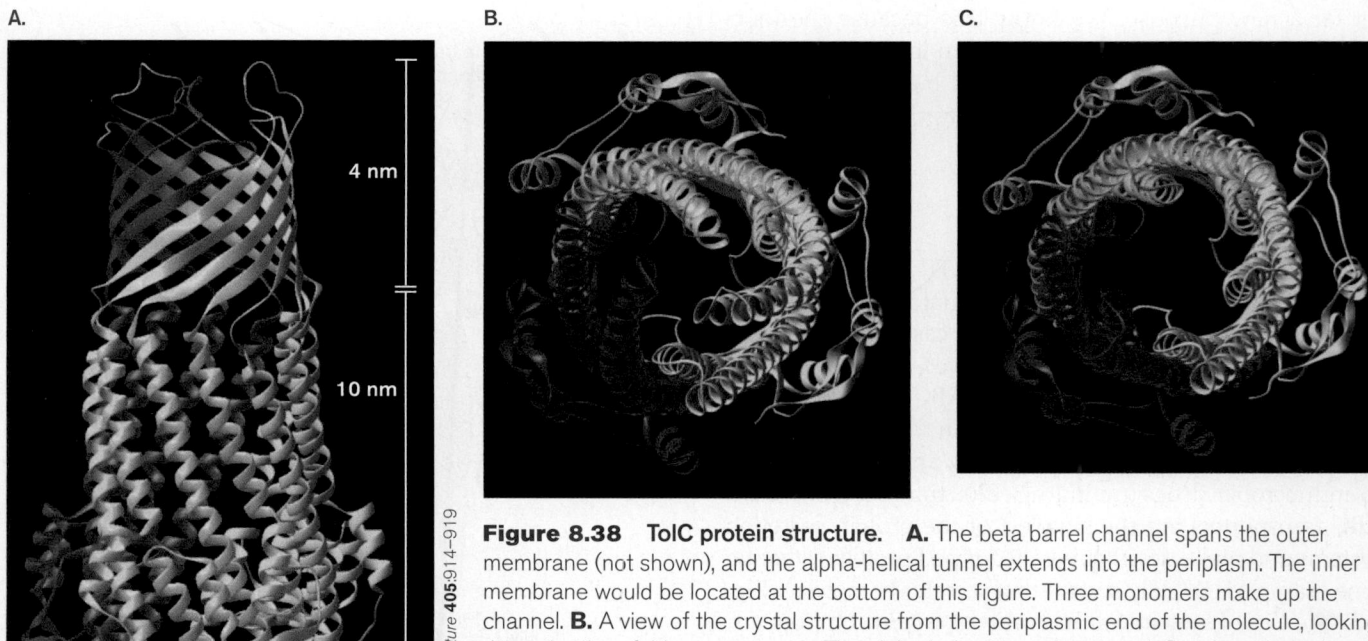

Vassilis Koronakis et al. 2000. *Nature* **405**:914–919

Figure 8.38   TolC protein structure.   A. The beta barrel channel spans the outer membrane (not shown), and the alpha-helical tunnel extends into the periplasm. The inner membrane would be located at the bottom of this figure. Three monomers make up the channel. **B.** A view of the crystal structure from the periplasmic end of the molecule, looking down the threefold symmetry axis. The alpha-helical tunnel is closed. **C.** A hypothetical model of how the tunnel might be opened, with the same orientation as in (B).

8.6   Protein Degradation: Cleaning House

What happens when a cell no longer needs a specific protein—or when a cell synthesizes a protein with incorrect amino acids? These useless or disabled proteins must be degraded to maintain cellular health. Because cellular needs are constantly changing, proteins no longer needed can produce deleterious effects on the cell. This is particularly true of regulatory proteins, whose concentrations must change with time or in response to alterations in the cellular condition.

Many normal proteins contain degradation signals called **degrons** that dictate the stability of a protein. The **N-terminal rule** describes one type of degron. The rule states that the N-terminal amino acid of a protein directly correlates with its stability. For example, proteins beginning with arginine, lysine, or phenylalanine experience a short half-life (2 minutes or less); while proteins with aspartic acid, glutamic acid, or cysteine in the lead position enjoy a much longer half-life (more than 10 hours). Why this correlation exists is not clear.

Abnormally folded proteins are recognized by proteases in part because regions that are normally buried within the protein's three-dimensional structure become exposed. The protein is progressively degraded into smaller and smaller pieces by a series of these proteases.

Initial cuts usually involving ATP-dependent endoproteases like Lon protein or ClpP are followed by digestion with tripeptidases and dipeptidases. Endopeptidases cleave proteins somewhere within the sequence but not from the ends of the sequence. Many peptidases use ATP hydrolysis to help unfold the target protein prior to digestion. Unfolding is necessary for the protein to slide into a barrel-shaped protease such as ClpAP, ClpXP, or the proteasomes of archaea (20S) and eukaryotic microbes (26S) (**Figs. 8.39** and **8.40**). Proteasomes are protein-degrading organelles of eukaryotes. The protein-degrading enzymes are classified as serine, cysteine, or threonine proteases, depending on the key residue in their active sites.

There is a striking resemblance between prokaryotic and eukaryotic ATP-dependent proteases (**Fig. 8.40**). The core particle of the eukaryotic proteasome contains two copies of 14 different proteins assembled in four stacked rings, each containing seven proteins (heteroheptameric). The core particle is capped at the top and bottom by the regulatory particles, also made from 14 proteins that are different from those in the core. Six of those proteins are ATPases, while others recognize a ubiquitin peptide tag (a 76-amino-acid peptide) placed on doomed proteins (see **eTopic 8.2**). The purpose of this cap is to unfold the target protein and inject it into the 20S proteasome barrel. Archaea also contain a proteasome, but it comprises 14 copies of two different proteins assembled in four stacked rings, each containing seven proteins (homoheptameric). Unlike eukaryotic proteasomes, there is no regulatory particle, and substrate proteins are not ubiquitinated. Bacterial proteins are not ubiquitinated either, as a rule. However, an exception to that rule has been found in *Mycobacterium tuberculosis*. These organisms place a

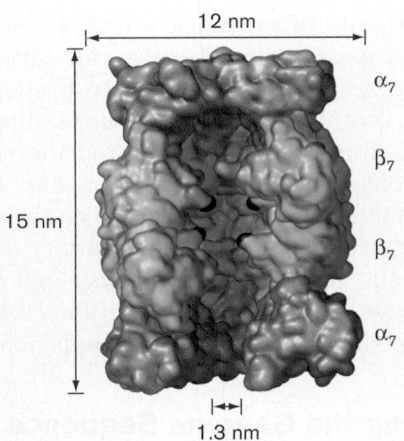

Figure 8.39 **Protein degradation machines.** Predicted structure of the 20S proteasome from the methanoarchaeon *Methanosarcina thermophila*. The active sites involved in peptide bond cleavage are indicated in red. (PDB code: 1GOU)

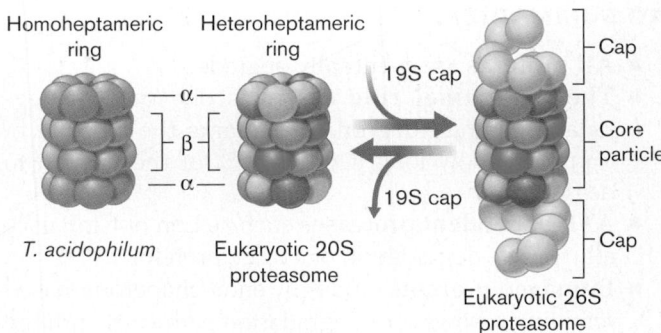

Figure 8.40 **Proteolysis machines from prokaryotes and eukaryotes.** Structural comparison of the 20S proteasome from an archaeon (*Thermoplasma acidophilum*), the 20S equivalent in a eukaryote, and the 26S eukaryotic proteasome. The 19S cap is a multisubunit regulatory structure.

26-amino-acid peptide (PUP) on proteins targeted for degradation. **Figure 8.40** shows a side-by-side comparison of the archaeal proteases and the eukaryotic 20S and 26S proteasomes.

Bacteria contain Clp proteases, which contain a proteolytic core made of two homoheptameric rings of the protein ClpP. These proteases also have interchangeable homohexameric ATPase caps made of ClpX, ClpA, ClpB, or ClpC, each of which recognizes different substrates. The accessory proteins recognize and present different substrate proteins to the ClpP protease and thereby regulate which proteins are degraded.

How Do Cells Cope with Stress-Damaged Proteins?

Microbes are constantly exposed to environmental insults such as high temperature or pH extremes that damage proteins and cause them to misfold. As an energy cost-saving device and to prevent interruption of protein function, injured proteins go through a kind of triage process that evaluates whether they are salvageable or must be destroyed before they can endanger the cell. Chaperones constantly hunt for misfolded (or otherwise damaged) proteins and attempt to refold them. But if the protein is released from a chaperone and remains misfolded, it can, by chance, either reengage the chaperone or bind a protease that destroys it (**Fig. 8.41**). This fold-or-destroy triage system is essential if a microbe is to survive environmental stress.

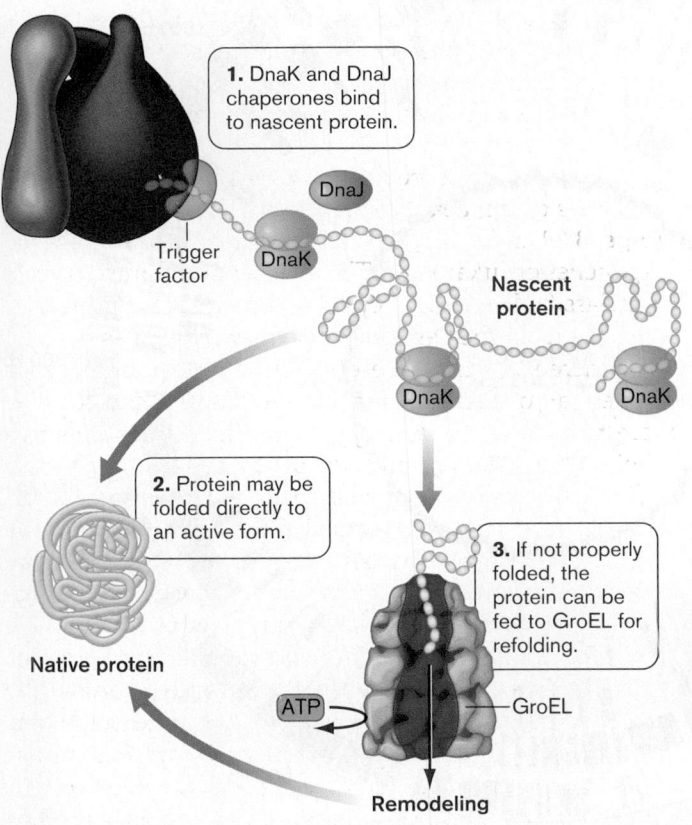

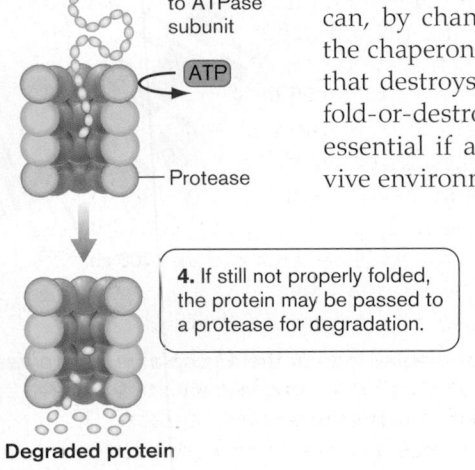

Figure 8.41 **Protein folding versus degradation triage pathways.** The figure depicts what can happen to a newly synthesized protein. However, a protein that unfolds as a result of environmental stress (for instance, heat) will undergo the same triage process.

TO SUMMARIZE:

- **All proteins** are eventually degraded.
- **The N-terminal rule** describes one type of degradation signal (degron) that marks the half-life of a protein (how long it takes 50% of the protein to degrade).
- **ATP-dependent proteases** such as Lon or ClpP usually initiate degradation of a large protein.
- **Damaged proteins** randomly enter chaperone-based refolding pathways or degradation pathways until the protein is repaired or destroyed.

8.7 Bioinformatics: Mining the Genomes

We have just described how the information within a genome is deciphered by the cell to produce proteins. But the experiments described cannot be performed in the vast majority of microbes, which are unculturable. In Chapter 7 we discussed how we can quickly and efficiently sequence entire genomes (see Section 7.6). This knowledge, combined with our understanding of protein structure (amino acid sequences) and the relationships between protein structure and function, has brought us to the postgenomic era. We can now call on the vast store of information gathered over the last century to make predictions about the genetics and physiology of microbes even when we cannot grow them in the laboratory. The following sections reveal how these predictions are made.

Annotating the Genome Sequence

Chapter 7 explained how the DNA sequence of a genome is obtained. Knowing the sequence is just a first step; meaningful information comes from **annotation** of the DNA sequence—that is, understanding what the sequence means. Annotation is analogous to identifying separate sentences and words in an unknown language.

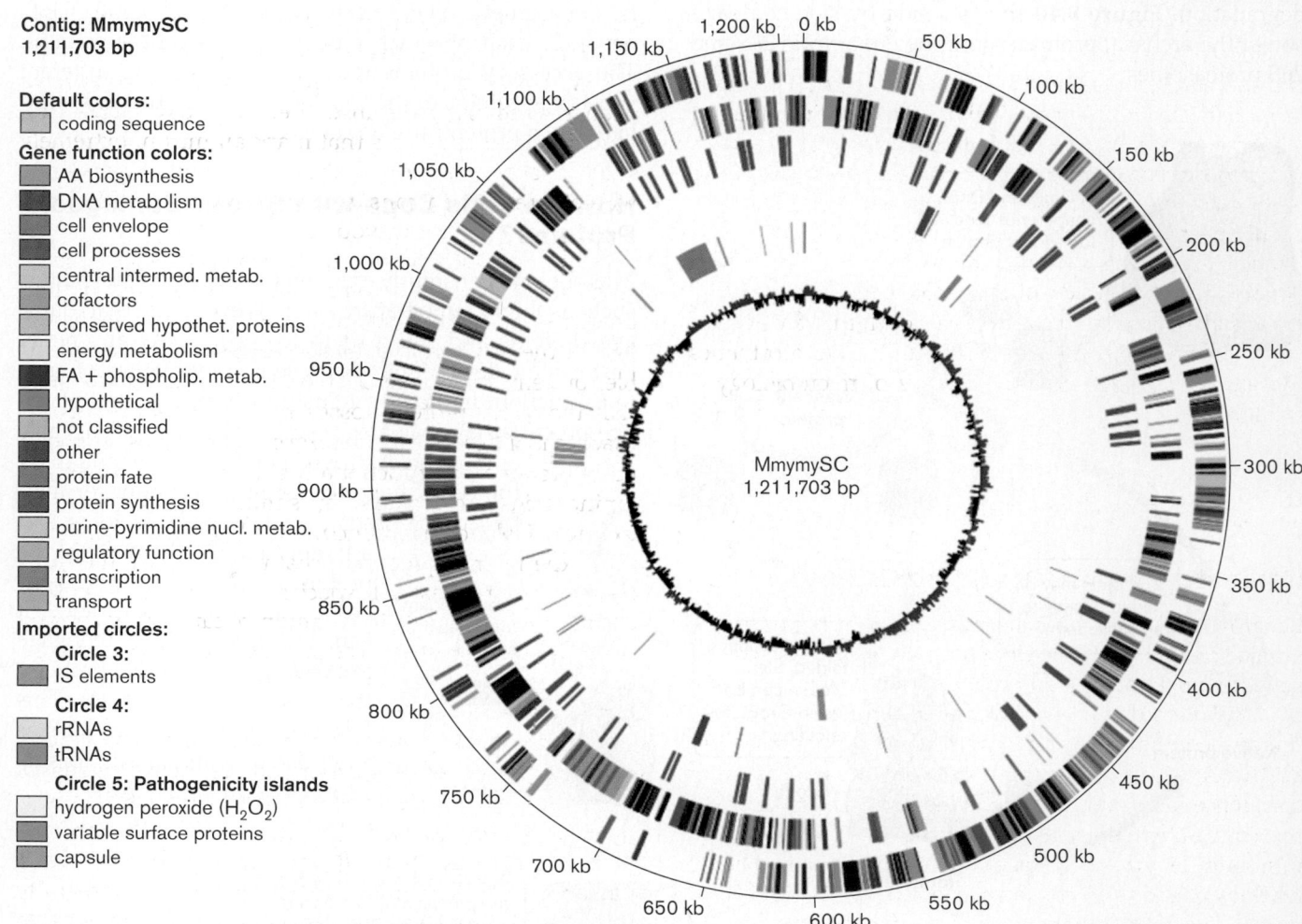

Figure 8.42   Circular display map of the *Mycoplasma mycoides* genome (1,211,703 bp).   Color code indicates gene clustering by function. The two outer rings represent genes transcribed from different DNA strands. The innermost circle uses peaks to show GC content levels above (red) or below (black) 50%.

For most languages, you would look for punctuation marks like periods or exclamation points to identify sentences and then scan for blank spaces between letters to signify the beginning and end of a word. You may not initially know what the sentence says, but defining what makes a sentence is the first step.

Annotation of a DNA sequence includes finding the start and stop sites of genes, as well as predicting the function of the gene product. Computers use established rules to mark potential genes—open reading frames (ORFs). Then, similarities are sought between the deduced amino acid sequences of those ORFs and the sequences of proteins with known functions. Similarities are used to infer the function of the unknown ORF. Genes encoding transfer RNA and ribosomal RNAs do not encode proteins but can be identified because of the conservation of these sequences across vast phylogenetic distances. The results of annotation can be represented in many different forms. **Figure 8.42** shows a commonly used method of presenting the results: a circular display map—in this case of the *Mycoplasma mycoides* genome—with genes color-coded by predicted function.

Since 1998, the complete genomes of over 225 microbial species have been published, and many others have been partially sequenced. This wealth of information has spawned a new discipline called **bioinformatics** dedicated to comparing genes of different species. Data from bioinformatics enable scientists to make predictions about an organism's physiology and evolutionary development, even if the organism cannot be grown in the laboratory. The magnitude of this achievement, and of the job ahead, can be appreciated if we recall that fewer than 1% of microorganisms can be cultured. Bioinformatics has forever changed how the science of microbiology is conducted.

www | Sequenced genomes
microbiology2.com/links

Computer Analysis and Web Science

Lengthy DNA sequences are analyzed by computer programs such as ORF Finder. The program uses the universal genetic code to deduce all possible protein sequences that could be formed in all reading frames on RNA molecules transcribed from either direction on the chromosome (**Fig. 8.43A**). An open reading frame (ORF), the equivalent of a sentence in our analogy, is defined as a DNA sequence that can potentially encode a string of amino acids of minimum length—say, for example, 50 residues. The 50 residues of the ORF could encode a 5,500-Da (5.5-kDa) protein, since the average weight of an amino acid is 110 Da. Each ORF begins with a translation start codon (usually ATG, or more rarely, GTG or TTG). A translation start codon marks where ribosomes start to read a messenger

RNA molecule. An ORF ends with a termination codon (in the DNA, these are TAA, TAG, or TGA), which, as UAA, UAG, or UGA in RNA, signals the ribosome to stop translation. In addition, the computer can identify an ORF by looking for potential ribosome-binding sites upstream of the start codon, but these ribosome-binding sites may differ between species, so finding one is not essential to declaring the presence of an ORF.

While the preceding methods work for prokaryotes, identifying ORFs in eukaryotes is more difficult. In eukaryotes, most genes contain **introns** (long, noncoding sequences that occur in the middle of genes), as do the initial mRNA products produced from those genes (**Fig. 8.43B**). The vast majority of known prokaryotic genes do not contain introns. Introns serve several regulatory functions in eukaryotes that determine whether a protein product is made. However, the eukaryotic cell must use special splicing mechanisms to remove the intron sequences and re-join the protein-encoding sequences, called **exons**, of the mRNA prior to translation. (In eukaryotic organisms, the mRNA transcribed directly from DNA is thus called the **preliminary mRNA transcript** or **pre-mRNA**; the "final" mRNA transcript is produced when the introns have been spliced out.) Because there are few identifying sequence characteristics that mark an intron, extremely sophisticated computer analysis is needed to determine where to remove introns in order to derive the ORF coding sequence of a eukaryotic gene.

Once an ORF is identified, computers use mathematical algorithms to determine whether the protein predicted by the ORF resembles any other protein deposited in the worldwide databases or, even better, resembles proteins of known function. By "resemble" we mean that the query protein possesses amino acid sequences that are identical or functionally similar to those found in other proteins. In a functionally similar sequence, certain amino acids are replaced by similar amino acids—for example, isoleucine for leucine. We know from considerable experimental precedent that proteins with the same function, regardless of species, usually have critical amino acid sequences in common because they evolved from the same ancestral sequence. Sequences common to two different protein or DNA molecules are called homologous sequences.

Figure 8.44 shows a sequence alignment among eight examples of dihydrolipoamide dehydrogenases from different species and illustrates how this alignment was used to partially characterize the function of a virulence gene in an important pathogen. Ordinarily, dihydrolipoamide dehydrogenase is integrally important to the tricarboxylic acid cycle, which extracts energy and carbon from carbohydrate food sources. Note the highlighted areas in the figure; these represent regions of identity or similarity at similar positions.

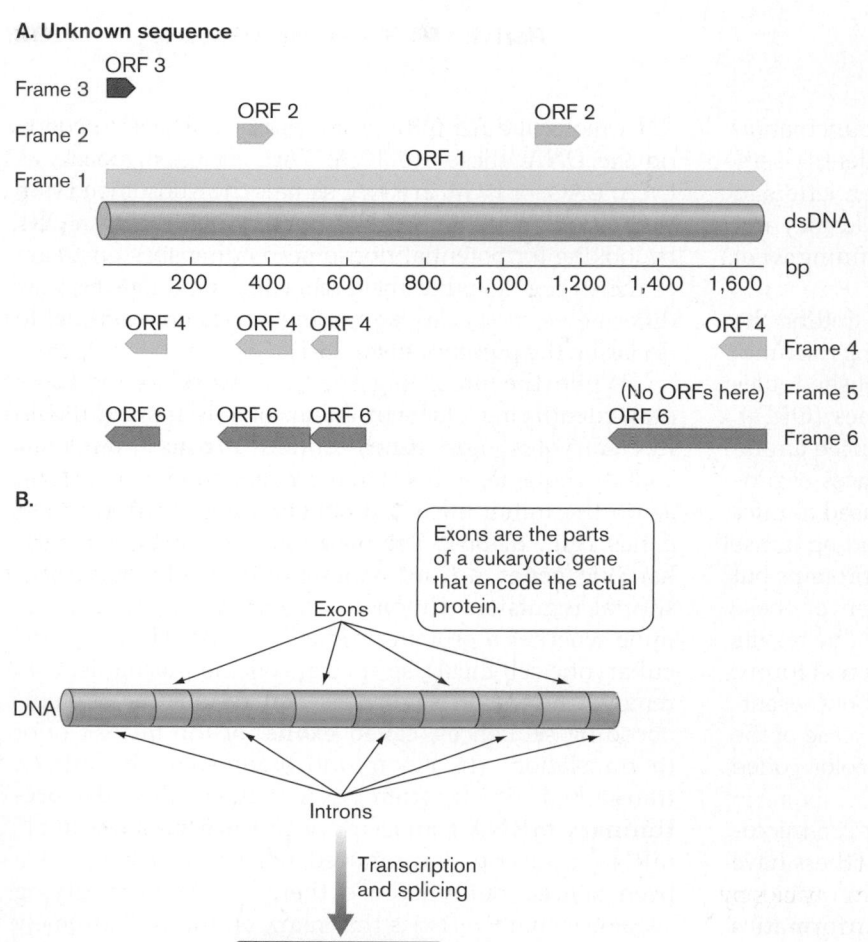

A. Unknown sequence

Figure 8.43 Predicting open reading frames (ORFs) in a DNA sequence. A. Each predicted ORF in this 1,600-bp bacterial sequence begins with AUG or GUG (potential translation start codons in mRNA) and ends with a translation terminator codon. The computer can translate in all six reading frames (three on top strand, three on bottom strand). Determining which is the real ORF requires additional information, such as potential ribosome-binding sites and a transcriptional start site. **B.** In eukaryotes, finding ORFs is complicated by the presence of noncoding DNA sequences called introns. The actual ORF (exons) is disrupted by introns, which must be spliced out of the primary RNA message before the final mRNA leaves the nucleus.

Exons are the parts of a eukaryotic gene that encode the actual protein.

```
                              FAD-binding domain                    Pyridine redox active site
C08 Streptococcus pneumoniae  GPAGYVAAIKAAQFGGKVALVEKSE------LGGTCLNRGCIPTKTYLHNAEIIENIGHAANRGIVIEN
C37 Streptococcus pneumoniae  --------------------------MYGGTCINIGCIPTKTLLVAAEKD---------------
Streptococcus pyogenes        GPAGYYAAIRGAQLGGKIAIVEKSE------FGGTCLNVGCIPTKTYLKNAEILDGIKIAAGRGINLAS
Clostridium magnum            GPGGYVAAIRAAQLGAKVTLIEKES------LGGTCLNVGCIPTKVLLHSSQLLTEMKEGDKLGIDIEG
Bacillus subtilis             GPGGYVAAIRAAQLGQKVTVVEKA------TLGGVCLNVGCIPSKALINAGHRYENAKHSDDMG--ITA
Azotobacter vinelandii        GPGGYVAAIKSAQLGLKTALIEKYKGKEGKTALGGTCLNVGCIPSKALLDSSYKFHEAHESFKLHG-IST
Escherichia coli              GPAGYSAAFRCADLGLETVIVERYN------TLGGVCLNVGCIPSKALLHVAKVIEEAKALAEHG--IVF
Neisseria meningitidis        GPGGYSAAFAAADEGLKVAIVERYK------TLGGVCLNVGCIPSKALLHNAAVIDEVRHLAANG--IKY


C08 Streptococcus pneumoniae  PNFTVDMEKLLETKSKVVNTLVGGVAGLLRSYGVTVHKGIGTITKDKNVLVNG--------------SEL
C37 Streptococcus pneumoniae  ----LSFEEVIATKNTITGRLNGKNYTTVAGTGVDIFDAEAHFLSNKVIEIQAGDE----------KQE
Streptococcus pyogenes        TNYTIDMDKTVDFKNTVVKTLTGGVQGLLKANKVTIFNGLGQVNPDKTVTIG---------------SQT
Clostridium magnum            -SIVVNWKHIQKRKKIVIKKLVSGVSGLLTCNKVKVIKGTAKFESKDTILVTKEDG---------VAEK
Bacillus subtilis             ENVTVDFTKVQEWKASVVNKLTGGVAGLLKGNKVGKEAYFVDSNSVRVMDEN----------SAQT
Azotobacter vinelandii        GEVAIDVPTMIARKDQIVRNLTGGVASLIKANGVTLFEGHGKLLAGKKVEVTAADG----------SSQV
Escherichia coli              GEPKTDIDKIRTWKEKVINQLTGGLAGMAKGRKVKVVNGLGKFTGANTLEVEGEN----------GKTV
Neisseria meningitidis        PEPELDIDMLRAYKDGVVSRLTGGLAGMAKSRKVDVIQGDGQFLDPHHLEVSLTAGDAYEQAAPTGEKKI

                                                                          NAD-binding domain
C08 Streptococcus pneumoniae  LETKKIILAGGSKVSKINVPGME-SPLVMTSDDILEMNEVPESLVIIGGGVVGIELGQAFMTFGSKVTVI
C37 Streptococcus pneumoniae  LTAETIVINTGAVSNVLPIPGLATSKNVFDSTGIQSLDKLPEKLGVLGGGNIGLEFAGLYNKLGSKVTVL
Streptococcus pyogenes        IKGRNVILATGSKVSRINIPGID-SKLVLTSDDILDLREMPKSLAVMGGGVVGIELGLVWASYGVDVTVI
Clostridium magnum            VNFDNAIIATGSMPFIPEIEGNK-LSGVIDSTGALSLESNPESIAIIGGGVIGVEFASIFNSLGCKVSII
Bacillus subtilis             YTFKNAIIATGSRPIELPNFKY--SERVLNSTGALALEIPKKLVVIGGGIIGTELGTAYANFGTELVIL
Azotobacter vinelandii        LDTENVILASGSKPVEIPPAPVD-QDVIDSTGALDFQNVPGKLGVIGAGVIGLELGSVWARLGAEVTVL
Escherichia coli              INFDNAIIAAGSRPIQLPFIPHE-DPRIWDSTDALELKEVPERLLVMGGGIIGLEMGTVYHALGSQIDVV
Neisseria meningitidis        VAFKNCIIAAGSRVTKLPFIPE--DPRIIDSSGALALKEVPGKLLIIGGGIIGLEMGTVYSTLGSRLDVV
```

Figure 8.44 Multiple-sequence alignment. The program ClustalW (available online) was used to align the amino acid sequence of dihydrolipoamide dehydrogenases from the *Streptococcus pneumoniae* C08 ORF with selected orthologs from other known species (only a portion of the alignment is shown). These proteins are called orthologs because they have similar sequences and carry out the same biochemical reaction. The top sequence in each tier is a different portion of the C08 sequence. Amino acid sequences of over 51% similarity are shaded. Highly conserved residues, where at least five residues are the same or similar amino acids, are shaded yellow. Very highly conserved residues, where at least seven out of eight are identical, are shaded brown. Deletions (sequences missing in one or more orthologs) are indicated as dashes.

"Identity" means the same amino acid is present in each protein; "similarity" means that chemically related amino acids are substituted (for example, valine is substituted by the similar leucine, or glutamate is substituted by aspartate). Also notice that the top sequence in the figure is an unknown ORF from the genome of *Streptococcus pneumoniae*, an organism that causes pneumonia. The computer program flagged the sequence similarities between this unknown ORF and the known proteins shown in the figure. On the basis of that homology, the ORF was deduced to encode a dihydrolipoamide dehydrogenase—a prediction confirmed by mutating the gene and by looking directly for the biochemical activity. However, *S. pneumoniae* is a strictly fermentative organism that does not have a complete TCA cycle. Nevertheless, deleting the gene decreased virulence of this pathogen in a mouse model, suggesting an important function for the enzyme unrelated to the TCA cycle. In this and many other instances, bioinformatics provides important insights into microbial physiology.

Alignments can be done either with base sequences in DNA or with protein sequences deduced from the DNA sequence. Because of the degeneracy of the code, however, it is usually easier to pick up evolutionary relationships between genes from distantly related species by comparing their protein products rather than the DNA sequence. Remember, one amino acid can be encoded by more than one DNA codon. So a DNA strand (nontemplate) encoding the peptide Ala-Leu-Ser could be 5′ GCU CCU UCC or 5′ GCA CCG UCA. It is easier to see the homology using the deduced peptide. On the other hand, not all genes encode proteins. Prime examples are the genes encoding stable ribosomal RNA molecules. The DNA sequences for these ribosomal RNAs are highly conserved across species, and their similarities reveal homology even among the three ancestral domains: Bacteria, Archaea, and Eukarya.

> **NOTE:** Do not confuse template and nontemplate strands of DNA for a given gene. The template strand, read 3′ to 5′ by RNA polymerase, is transcribed to make the 5′ to 3′ RNA transcript. The nontemplate DNA strand, also called the sense strand, has the *same* sequence as the RNA transcript, with the substitution of T's for U's.

Homologs, Orthologs, and Paralogs

Homologies found between genes or proteins suggest an evolutionary relatedness. Genes or proteins that are **homologous** probably evolved from a common ancestral gene. Homologous genes or proteins can be classified as either orthologous or paralogous. **Orthologous** genes have functions essentially the same but occur in

two or more different species. For example, the gene for glutamine synthase (*gltB*) in *E. coli* is orthologous to *gltB* in *Vibrio cholerae* and to *gltB* in the Gram-positive bacterium *Bacillus subtilis*. **Paralogous** genes arise by duplication within the same species (or progenitor) but evolve to carry out different functions (**Fig. 8.45**). For example, genes encoding type III protein secretion systems, which export virulence proteins, have evolved from paralogous genes encoding flagellar biogenesis. The two sets of genes share a common ancestor but have since evolved completely different functions (in this case secretion and motility). Paralogous genes are maintained in a microbial genome because their distinct functions contribute to the organism's adaptive potential. Duplicate genes of identical function usually result in loss of one copy or the other by degenerative evolution (see Chapter 9).

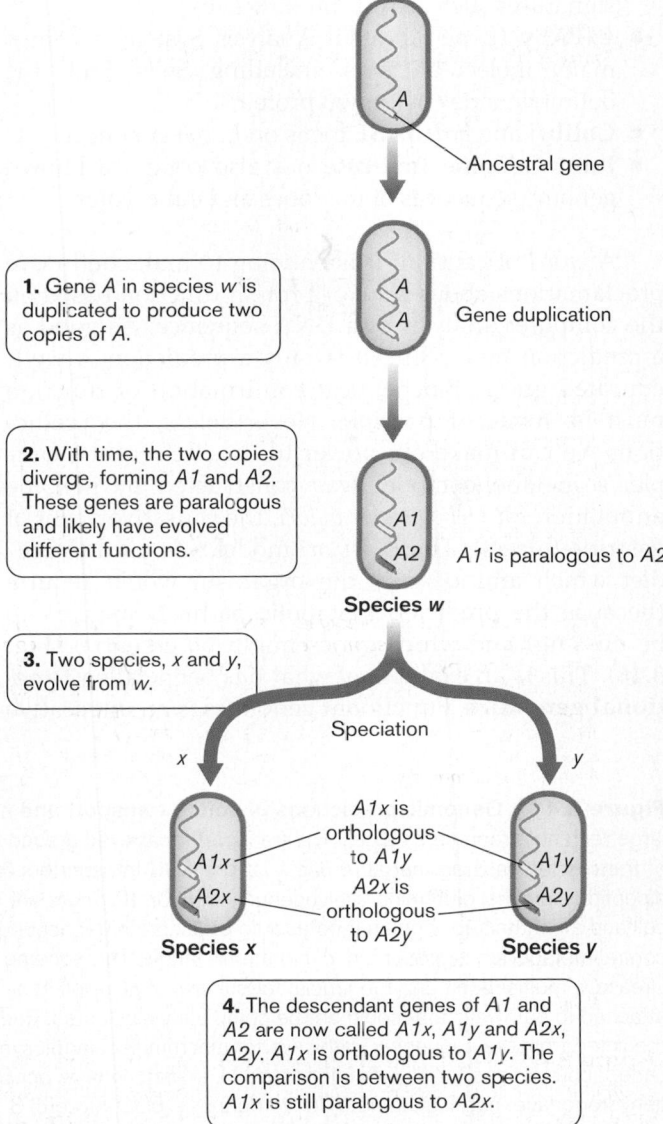

1. Gene *A* in species *w* is duplicated to produce two copies of *A*.

Ancestral gene

Gene duplication

2. With time, the two copies diverge, forming *A1* and *A2*. These genes are paralogous and likely have evolved different functions.

A1 is paralogous to *A2*

Species *w*

3. Two species, *x* and *y*, evolve from *w*.

Speciation

A1x is orthologous to *A1y*
A2x is orthologous to *A2y*

Species *x*   **Species** *y*

4. The descendant genes of *A1* and *A2* are now called *A1x*, *A1y* and *A2x*, *A2y*. *A1x* is orthologous to *A1y*. The comparison is between two species. *A1x* is still paralogous to *A2x*.

Figure 8.45 Paralogous versus orthologous genes. An ancestral gene can undergo a duplication to evolve an orthologous or paralogous gene.

Many computer programs and resources used to analyze DNA and protein sequences are freely available on the Web. Here are some useful programs and websites:

- **BLAST** (Basic Local Alignment Search Tool) compares a sequence of interest with all other DNA or protein sequences deposited in sequence databases.
- **Multiple Sequence Alignment** aligns sequences of genes identified as homologous by BLAST analysis.
- **KEGG** (Kyoto Encyclopedia of Genes and Genomes) outlines biochemical pathways in many sequenced organisms. Areas within this site graphically display reference biochemical pathways (for example, glycolysis) and then indicate which proteins in that pathway are predicted to occur in any organism with a sequenced genome.
- **Motif Search** searches DNA or proteins for sequence signatures such as ATP-binding sites.
- **ExPASy** (Expert Protein Analysis System) contains many molecular tools, including Swiss-Prot, the definitive index of known proteins.
- **Colibri** and *coli*BASE focus on *E. coli* genomics.
- **Joint Genome Institute** has also compiled known genome sequences of microbes and eukaryotes.

A word of caution: It is enticing to make definitive proclamations about gene or protein function based on the computer analysis of a DNA sequence. As good as a prediction may seem, it is only a prediction, a well-educated guess. Biochemical confirmation of function must be made, if possible. Nevertheless, the predictions we can make are powerful. In one recent example, a metabolic model was constructed by genome annotation for *Helicobacter pylori*, the causative agent of gastritis (ulcers). The *H. pylori* model was used to predict which amino acids the organism would require (because the predicted metabolic pathway appears to be missing) and which genes might be essential (**Fig. 8.46**). This is an example of what has been called **functional genomics**. Functional genomics is an integrative

process in which bioinformatic approaches allow scientists to make predictions about function for a set of genes and then test those predictions experimentally. In another application, some investigators are trying to predict the smallest number of genes required for a living cell (see **eTopic 8.3**).

THOUGHT QUESTION 8.10 An ORF 1,200 bp in length could encode a protein of what size and molecular weight? Hint: Find the molecular weight of an average amino acid.

The information age has spawned what could be called Web biology, an *in silico* science that has provided new insight into evolution, physiology, and pathogenesis. Bioinformatics has not only confirmed what we know in greater detail, but also revealed new information, such as the existence of ORFs with no known function. The sequencing of entire genomes has enabled a greater understanding of sequence similarity or homology between genes or proteins and has provided a foundation for understanding evolutionary relationships. For example, the sequencing of the alphaproteobacterium *Rickettsia prowazekii* revealed many similarities to genes within eukaryotic mitochondrial DNA. This finding led to the conclusion that eukaryotic mitochondria were derived from a rickettsial predecessor that had become an endosymbiont (an intracellular symbiotic organism), rather than food, for a eukaryotic cell.

TO SUMMARIZE:

- **Annotation** requires computers that look for patterns in DNA sequences. Annotation predicts regulators, ORFs, rDNA, and tRNA. Similarities in protein sequence (deduced from the DNA sequence) are used to predict protein structure and function.
- **An open reading frame (ORF)** is a sequence of DNA predicted by various sequence cues to encode an actual protein.

Figure 8.46 Genomic predictions of solute transport and metabolic pathways of *Helicobacter pylori* strain 26695. The large rectangle represents a cell. The transporters arrayed around the periphery were identified by sequence comparisons characteristic of transporters in Gram-negative bacteria. The ABC transporters (multicomponent) are predicted to transport oligopeptides, dipeptides, proline, glutamine, molybdenum, and iron III. There are also predicted P-type ATPases that extrude toxic metals from the cell and a glutathione-regulated potassium efflux protein (encoded by *kefB*). An integrated view of the main components of *H. pylori* central metabolism is presented within the rectangle. This scheme is based on using glucose as the sole carbohydrate source. Urease, a multisubunit enzyme crucial for survival of *H. pylori* at acid pH, is indicated as a complex (purple circle). A question mark is attached to pathways that could not be completely elucidated. Red arrows represent pathways or steps for which no enzymes were predicted from the sequence. Pathways for macromolecular biosynthesis (RNA, DNA, and fatty acids) were found but are not shown. Other abbreviations for *H. pylori* genes: *ackA*, acetate kinase; *acnB*, aconitase B; *aspC*, aspartate aminotransferase; *dld*, D-lactate dehydrogenase; *gdhA*, glutamate dehydrogenase; *gldD*, glycerol 3-phosphate dehydrogenase; *glnA*, glutamine synthetase; *gltA*, citrate synthase; *hydABC*, hydrogenase complex; *icd*, isocitrate dehydrogenase; NDH-1, NADH:quinone oxidoreductase complex; *pfl*, pyruvate formate lyase; *por*, pyruvate ferredoxin oxidoreductase; *ppc*, phosphoenolpyruvate carboxylase; *pps*, phosphoenolpyruvate synthase; *pta*, phosphate acetyltransferase.

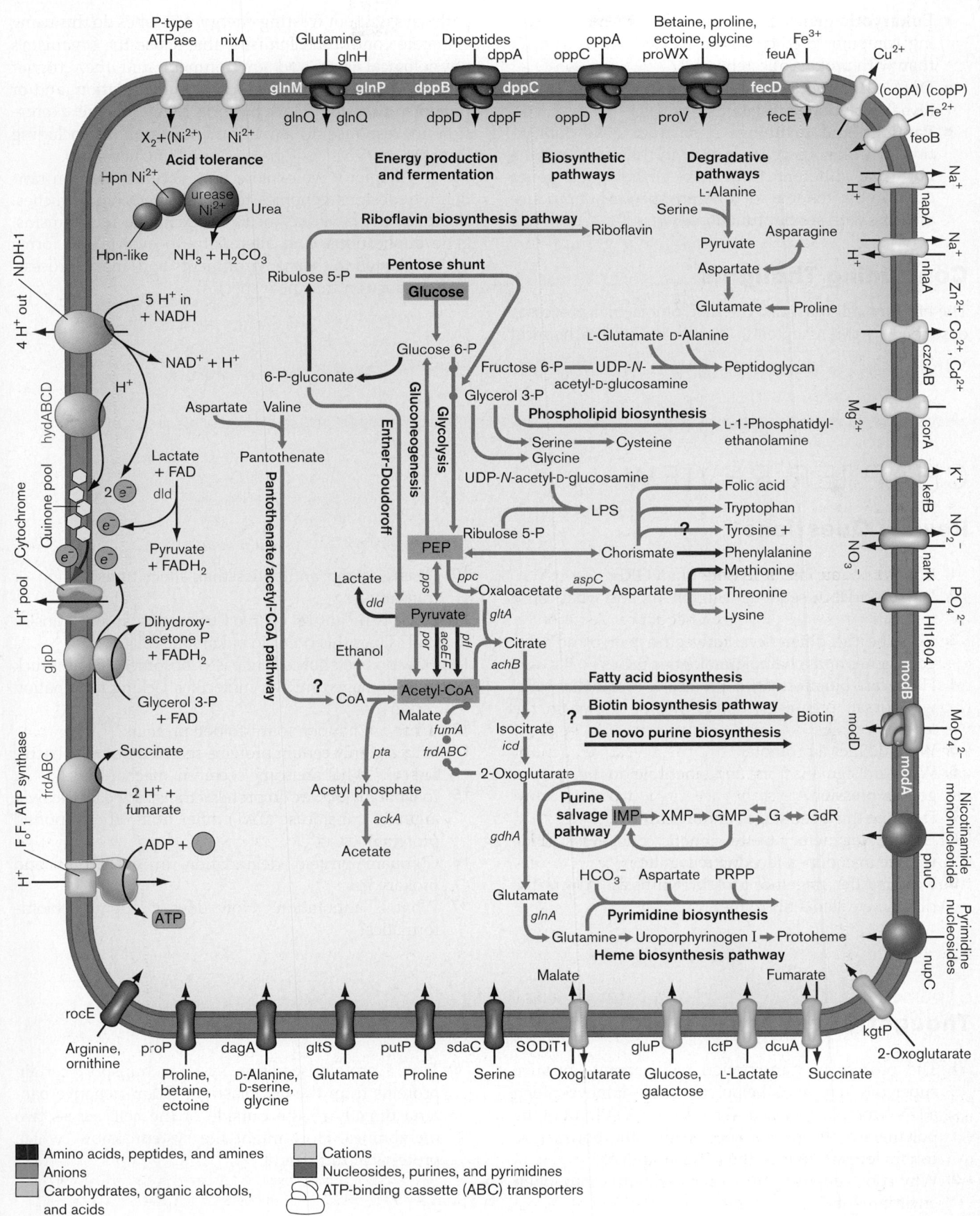

- **Eukaryotic genes** contain introns and exons, making computer predictions of an ORF more difficult than with prokaryotic genes.
- **DNA alignments** of similar genes or proteins can reveal evolutionary relationships.
- **Paralogs and orthologs** arise from gene duplications. Paralogous genes coexist in the same genome but have different functions. Orthologous genes occur in the genomes of different species but produce proteins with similar functions.

Concluding Thoughts

The efficient cell must use the macromolecular processes described in this chapter to synthesize its biochemical pathways without wasting energy. Microbes do this using elaborate control mechanisms that sense the organism's physiological state and environment and then trigger changes in replication, transcription, translation, and/or protein processing. How bacteria regulate gene expression in response to environmental stimuli, including threats to survival, will be discussed in Chapter 10.

In Chapter 9 we explore how natural selection randomly redesigns genomes to adapt to ecological niches. Microbes use a variety of DNA exchange mechanisms, gene duplications, and alterations to evolve into forms better adapted to their environments and, in the process, may produce entirely new species.

CHAPTER REVIEW

Review Questions

1. What are some characteristics of an ORF?
2. What is a DNA sequence alignment, and what can it tell you?
3. Describe the differences between a pair of orthologous genes and a pair of paralogous genes.
4. How can bioinformatics predict a metabolic pathway for an organism that cannot be grown in the laboratory?
5. What defines a promoter?
6. What are sigma factors, and what role do they play in gene expression?
7. Describe the three stages of transcription.
8. Explain degeneracy of the genetic code. What is the wobble in codon-anticodon recognition?
9. Describe the stages of protein synthesis. Why is the ribosome called a ribozyme?
10. Discuss some antibiotics that affect transcription or translation.
11. What is meant by coupled transcription and translation? Does this occur in eukaryotic cells?
12. How do bacterial cells release ribosomes that are stuck onto damaged mRNA molecules lacking termination codons?
13. What can happen to misfolded proteins?
14. Why are only certain proteins secreted from the bacterial cell? What are some secretion mechanisms?
15. In what major way do proteins transported by the twin arginine translocase (TAT) differ from other exported proteins?
16. Compare protein degradation in eukaryotes and prokaryotes.
17. What is annotation? How does it apply to bioinformatics?

Thought Questions

1. The process of transcription will generate positive supercoils in front of the polymerase as it moves along a DNA template. Why doesn't the DNA in front of the polymerase become so knotted that the polymerase can no longer separate the DNA strands?
2. Why do cells secrete some proteins into their environments?
3. Type I protein secretion systems transport certain proteins from the cytoplasm of Gram-negative bacteria directly to the outside of the cell, across two membranes. How might the system "know" which proteins to transport?

Key Terms

acceptor site (A site) (274)
aminoacyl-tRNA synthetase (270)
annotation (292)
anticodon (269)
bioinformatics (293)
catalytic RNA (267)
chaperone (284)
cistron (272)
codon (268)
consensus sequence (260)
degron (290)
exit site (E site) (274)
exon (293)
functional genomics (296)
heat-shock protein (HSP) (284)
homologous (295)
intron (293)

messenger RNA (mRNA) (267)
molecular mimicry (281)
N-terminal rule (290)
open reading frame (ORF) (264)
orthologous (295)
paralogous (295)
peptidyl-tRNA site (P site) (274)
peptidyltransferase (274)
polycistronic (272)
preliminary mRNA transcript
  (pre-mRNA) (293)
promoter (259)
release factor (280)
Rho-dependent (264)
Rho-independent (264)
ribosomal RNA (rRNA) (267)

ribosome-binding site (274)
ribozyme (274)
Shine-Dalgarno sequence (274)
sigma factor (259)
signal recognition particle (SRP) (286)
signal sequence (286)
small RNA (sRNA) (267)
start codon (269)
stop codon (269)
template strand (259)
tmRNA (267)
transcript (259)
transcription (258)
transfer RNA (tRNA) (267)
translation (268)
translocation (276)

Recommended Reading

Buttner, Daniela, and Ulla Bonas. 2002. Port of entry—the type III secretion translocon. *Trends in Microbiology* **10**:186–192.

Cascales, Eric. 2008. The type VI secretion toolkit. *European Molecular Biology Organization Reports* **9**:735–741.

Gowrishankar, Jayaraman, and Rajendran Harinarayanan. 2004. Why is transcription coupled to translation in bacteria? *Molecular Microbiology* **54**:598–603.

Haugen, Shanil P., Wilma Ross, and Richard L. Ghorse. 2008. Advances in bacterial promoter recognition and its control by factors that do not bind DNA. *Nature Reviews. Microbiology* **6**:507–519.

Kaczanowska, Magdalena, and Monica Rydén-Aulin. 2007. Ribosome biogenesis and the translation process in *Escherichia coli*. *Microbiology and Molecular Biology Reviews* **71**:477–494.

Murakami, Katsuhiko S., Shoko Masuda, Elizabeth A. Campbell, Oriana Muzzin, and Seth Darst. 2002. Structural basis of transcription initiation: RNA polymerase holoenzyme-DNA complex. *Science* **296**:1285–1290.

Proshkin, Sergey, A. Rachid Rahmouni, Alexander Mironov, and Evgeny Nudler. 2010. Cooperation between translating ribosomes and RNA polymerase in transcription elongation. *Science* **328**:504–508.

Pugsley, Antony P., Olivera Francetic, Arnold J. Driessen, and Victor de Lorenzo. 2004. Getting out: Protein traffic in bacteria. *Molecular Microbiology* **52**:3–11.

Santangelo, Thomas J., and John N. Reeve. 2006. Archaeal RNA polymerase is sensitive to intrinsic termination directed by transcribed and remote sequences. *Journal of Molecular Biology* **355**:196–210.

Schmeing, T. Martin, and Venkatraman Romakrishnan. 2009. What recent ribosome structures have revealed about the mechanism of translation. *Nature* **461**:1234–1242.

Taghbalout, Aziz, and Lawrence Rothfield. 2008. New insights into the cellular organization of the RNA processing and degradation machinery of *Escherichia coli*. *Molecular Microbiology* **70**:780–782.

Ward, Naomi, and Claire M. Fraser. 2005. How genomics has affected the concept of microbiology. *Current Opinion in Microbiology* **8**:564–571.

Wen, Jin-Der, Laura Lancaster, Courtney Hodges, Ana-Carolina Zeri, Shige H. Yoshimura, et al. 2008. Following translation by single ribosomes one codon at a time. *Nature* **452**:598–603.

 For further study and review, please visit StudySpace at **microbiology2.com**.

Chapter 9

Gene Transfer, Mutations, and Genome Evolution

DNA was once thought to be a monolithic "master molecule" governing all traits of an organism while remaining unchanged itself. We now know that DNA sequences change over generations through various mutations, DNA rearrangements, and gene transfers between species. Bacterial and archaeal genomes sometimes shuttle large clusters of genes between members of different taxonomic domains. All this inter- and intraspecies DNA traffic has led ecologists to expand the definition of the "microbial genome" to include DNA in the cell plus all the DNA out in the environment to which the microbe has potential access.

Important questions come to mind when we contemplate the consequences of DNA plasticity. For example, if the processes of adding, subtracting, and mutating genes all come at some cost, can the cell protect itself from undergoing too many drastic changes? Are some mutations good and others bad? Why would nature allow this much genetic uncertainty? This chapter explores these long-standing evolutionary questions and shows how microbial genomes continually change.

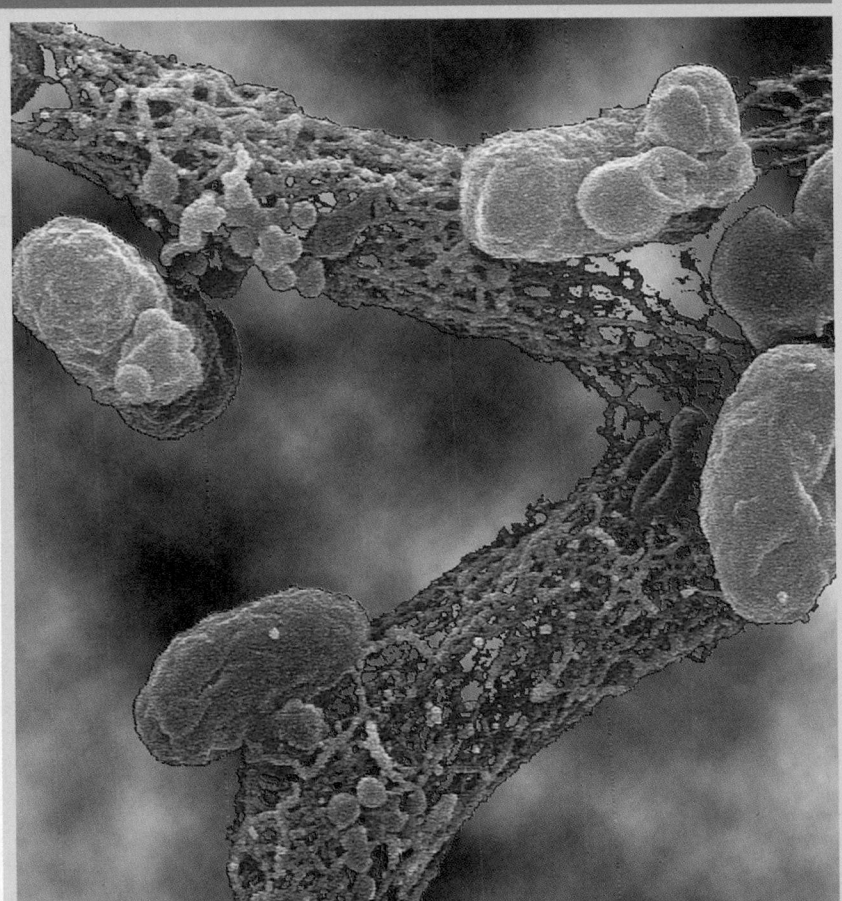

Current Research Highlight

Biofilm connections between *Escherichia coli* harboring a natural conjugative F plasmid. F plasmids mediate their own transfer from one cell to another using conjugation. Plasmid-encoded pili act as grappling hooks to pull plasmid donor and recipient cells together. F-factor DNA then passes through a membrane conjugation pore that forms between cells. Cells carrying F plasmids not only transfer DNA from strain to strain, but form complex mature biofilms. F-pilus connections between cells (seen in this scanning EM) activate production of colonic acid (a polysaccharide) and curli (a type of protein filament) to form a structured meshwork that supports very strong cell-to-cell and cell-to-surface interactions. The biofilm increases cell resistance to various stresses (e.g. antibiotics). Cell sizes approximately 1.5 µm. *Source*: Thithiwat May and Satoshi Okabe. 2008. *J. Bacteriol.* **190**:7479–7490. © 2008, American Society for Microbiology. All Rights Reserved.

We think of evolution as taking place over thousands if not millions of years. In New Zealand, however, John Sullivan and Clive Ronson witnessed evolution taking place within a mere decade. These scientists were studying the microbe *Mesorhizobium loti*, a bacterial species that establishes symbiotic relationships with plants by forming nitrogen-fixing nodules on the plants' roots. The genes encoding symbiosis are located as a group on the mesorhizobial chromosome. In addition to these symbiotic rhizobia, there are many nonsymbiotic species that are incapable of forging this relationship. In a remarkable experiment, Sullivan and Ronson inoculated a single strain of *M. loti* into an area of land devoid of natural nodulating rhizobia. Seven years later, they discovered that the area contained many genetically diverse symbiotic mesorhizobia now able to nodulate the flowering plant *Lotus corniculatus*. These microbes did not exist before the experiment. The scientists found that the 500-kb genome segment encoding symbiosis had somehow made its way from *M. loti* into these other bacteria—essentially generating new species. How did this happen so quickly?

In Chapter 9 we address the mechanisms that mediate transient as well as heritable DNA movements between species. Next we describe the competing processes of mutagenesis and mutation repair required for self-preservation and evolutionary change. The chapter closes with an examination of how all these processes collaborate over millennia to remodel genomes and build biological diversity.

9.1 The Mosaic Nature of Genomes

Genomic analysis has revealed that over the millennia, microbes have undergone extensive gene loss and gain. Archaea, for example, arose from a common eukaryotic/archaeal phylogenetic branch and, as a result, possess many traits in common with eukaryotes, such as the structure and function of their DNA and RNA polymerases. However, many archaeal genes whose products are involved with intermediary metabolism look purely bacterial. In fact, 37% of the proteins found in the archaeon *Methanocaldococcus jannaschii* are found in all three domains—Archaea, Eukarya, and Bacteria. Another 26% are otherwise found only among bacteria, while a mere 5% are confined to archaea and eukaryotes. This distribution clearly suggests that archaea enjoy a mixed heritage.

Another surprise arising from bioinformatic studies is the mosaic nature of the *E. coli* genome and, indeed, of all microbial genomes. Though we have intensively studied *E. coli* for over 100 years, we now find that the organism's DNA is rife with pathogenicity islands, fitness islands, inversions, deletions (when compared to similar species), paralogous genes, and orthologous genes. How did

all this genomic blending happen? The answer appears to involve heavy gene traffic between species (horizontal gene transfer), recombination events occurring within a species, and a variety of mutagenic and DNA repair strategies. All of these processes accelerated natural selection, where trial and error helped mold genomes.

Chapter 9 presents these mechanisms of gene transfer and alteration. The consequences of gene flow will be discussed in Chapter 17.

9.2 Gene Transfer

In 1928, a perceptive English medical officer, Fred Griffith (1879–1941), found that he could kill mice by injecting them with *dead* cells of a virulent pneumococcus (*Streptococcus pneumoniae*), a cause of pneumonia, together with *live* cells of a nonvirulent mutant. Even more extraordinary was the fact that the bacteria recovered from the dead mice were of the live, *virulent* type. Were the dead bacteria brought back to life? Unfortunately, Griffith was killed by a German bomb during an air raid on London in 1941 and never learned the answer.

In a landmark series of experiments published in 1944, Oswald Avery (1877–1955), Colin MacLeod (1909–1972), and Maclyn MacCarty (1911–2005) proved that Griffith's experiment was not a case of reviving dead cells, but involved the transfer of DNA released from the virulent, dead strain *into* the harmless living strain of *S. pneumoniae*—an event that transformed the live strain into a killer. The process of importing free DNA into bacterial cells is now known as **transformation**.

Transformation provided the first clue that gene exchange can occur in microorganisms. But why bacteria carry out this and other forms of gene transfer was not fully appreciated until much later. Recent genomic studies indicate that the fundamental purpose of bacterial gene transfer is to acquire genes that might be useful as the environment changes. Before we can discuss how bacterial genomes evolve, we need to understand the various gene exchange mechanisms available to them. Following the explanations of gene exchange, we will discuss the mechanisms that incorporate newly acquired DNA into genomes (for example, recombination).

Gene Exchange by Transformation

Many bacteria can import DNA fragments and plasmids released from nearby dead cells via the process of transformation. Natural transformation is conferred by specific protein complexes called **transformasomes**. Organisms in which transformation is a natural part of their growth cycles include Gram-positive bacteria like *Streptococcus* and *Bacillus*, as well as Gram-negative species such as *Haemophilus* and *Neisseria*.

Other bacteria, however, such as *E. coli* and *Salmonella*, do not possess the equipment needed to import DNA naturally. They require artificial manipulations to drive DNA into the cell. These laboratory techniques include perturbing the membrane by chemical ($CaCl_2$) and electrical (**electroporation**) methods. $CaCl_2$ alters the membrane, making these cells chemically competent and allowing DNA to pass. Electroporation, on the other hand, uses a brief electrical pulse to "shoot" DNA across the membrane.

Why do species such as *Neisseria* undergo natural transformation? First, species that indiscriminately import DNA may use the transformed DNA as food. Second, the more finicky species that transform only compatible DNA sequences may use DNA released from dead compatriots to repair their own damaged genomes. Finally, transformation may play an evolutionary role, enabling species to adjust to new environments by acquiring new genes from other species in a process called **horizontal gene transfer**. Horizontal gene transfer is distinguished from vertical transfer, the generational passing of genes from parent to offspring, as in cell division. If horizontal acquisition of a gene system improves the competitiveness of the cell, the new genes will be retained and the descendants will have evolved into a new kind of organism. Pathogens such as *Neisseria gonorrhoeae*, the cause of gonorrhea, may have used transformation to acquire new genes whose protein products now help the organism evade the host immune system.

Gram-positive organisms transform DNA using a transformasome complex. Natural transformation in Gram-positive organisms typically involves the growth phase–dependent assembly of a transformasome complex across the cell membrane (**Fig. 9.1**). The transformasome is composed of a binding protein that captures extracellular DNA floating in the environment, plus proteins that form a transmembrane pore. The complex also includes a nuclease that degrades one strand of a double-stranded DNA molecule while pulling the other strand intact through the pore and into the cell. Once inside, the strand can be incorporated into the chromosome by recombination—a process we will discuss in Section 9.3.

Once the transformasome is assembled, the cell is **competent**, meaning that it can import free DNA fragments and incorporate them into its genome. What triggers growth phase–dependent competence? For some Gram-positive bacteria, competence for transformation is generated by a chemical conversation (quorum sensing) that takes place between members of the culture. Every individual in a growing population produces and secretes a small, 15- to 20-amino-acid peptide generically called **competence factor** that accumulates in the medium until it induces a genetic program that makes the population

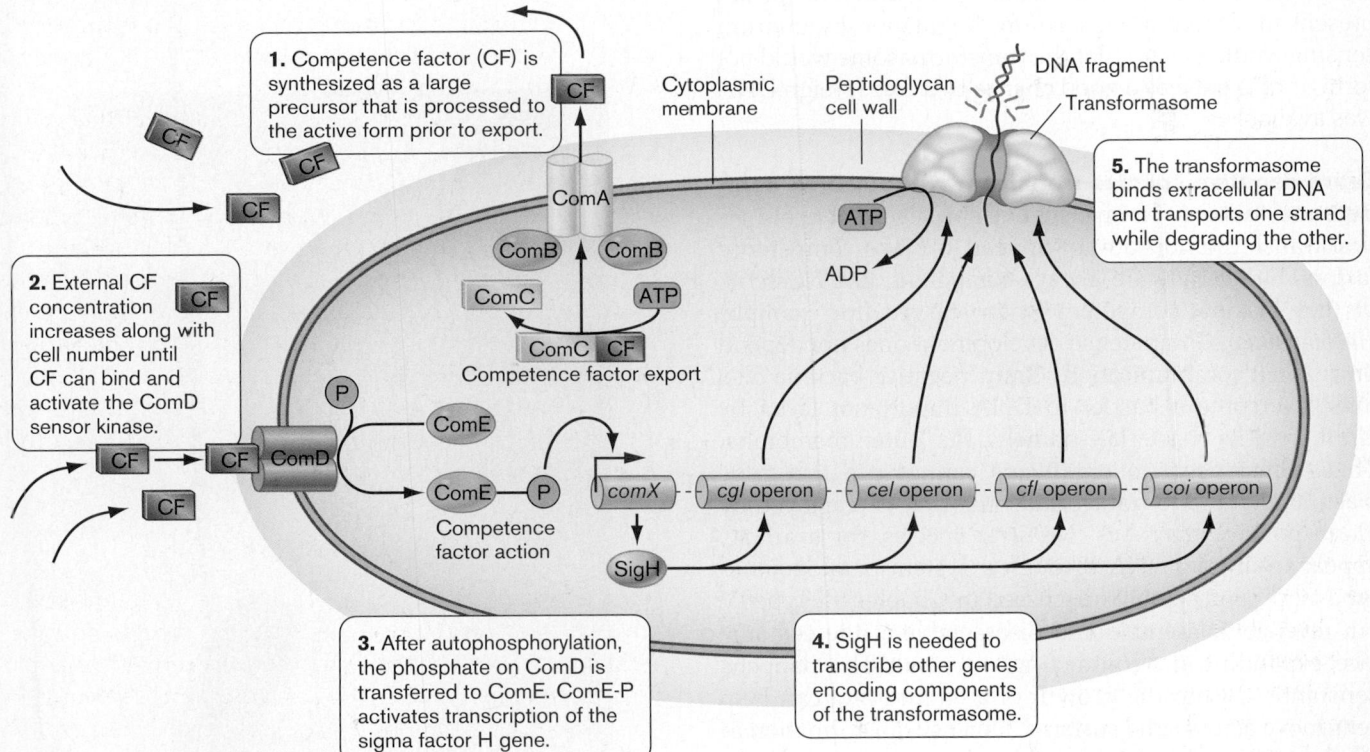

Figure 9.1   Transformation in *Streptococcus*. The process of transformation in this organism begins with the synthesis of a signal molecule (competence factor) and concludes with the import of a single-stranded DNA strand through a transformasome complex.

competent (**Fig. 9.1**, step 1). The sequence of the competence factor peptide is unique to each species, as are the specifics of the induction process. For *Streptococcus pneumoniae*, as the population increases—that is, as the cell density increases—so does the level of competence factor in the medium (step 2).

Above a certain concentration threshold, competence factor is able to bind to a sensory protein built into the cell membrane (ComD for *S. pneumoniae*). This binding begins what is called a phosphorylation cascade. In the phosphorylation cascade, the sensory protein uses ATP to autophosphorylate and then passes the phosphate to a cytoplasmic regulatory protein, ComE, that stimulates expression of a novel sigma factor, SigH (step 3). The resulting sigma factor is specifically used to transcribe genes encoding the transformasome (step 4). The protein products of these genes are assembled at the membrane, and the cell becomes competent (step 5).

Why would organisms regulate transformation competence by quorum sensing? One hypothesis holds that cells are unlikely to encounter stray DNA when growing in dilute natural environments such as ponds, where other bacteria are scarce. So, in this situation, why waste energy making the transformasome? However, when these same cells are growing at high density, as in a biofilm, they are more likely to encounter DNA released from dying neighbors. This is DNA they could use to repair their own damaged genomes, or use as food, or test for a new survival advantage should the DNA come from a different species present in a biofilm consortium. Regulation by quorum sensing would ensure that the transformasome would not form until there was a good chance that free foreign DNA was available.

Gram-negative species transform without DNA competence factors. Gram-negative species capable of natural transformation do not appear to make competence factors. Either they are always competent, like *Neisseria*, or they become competent when starved (for example, *Haemophilus*). Competence development does not depend on cell-cell communication. Gram-negative bacteria also must overcome a barrier to DNA import not faced by Gram-positive bacteria—namely, the outer membrane. Thus, Gram-negative organisms cannot use the cytoplasmic membrane transformasome that is employed by Gram-positive microbes. *Neisseria* species, for example, appear to import DNA through a system paralogous to type IV pilus assembly (discussed in Chapter 8). Type IV pili reversibly assemble and disassemble at the cell surface, expanding and contracting as a result. This happens constantly during the growth of a culture and can help cells move across solid surfaces. During transformation in species of *Neisseria*, disassembly of a pilus is thought to drag transforming DNA into the cell and across the two membranes.

In another departure from the Gram-positive example, transformation in species of *Haemophilus* and *Neisseria* is species and sequence specific, thereby limiting gene exchange between different genera. Specificity is due to particular sequences in DNA that are recognized by part of the uptake apparatus. However, not all Gram-negative competence systems display such specificity; *Acinetobacter calcoaceticus*, a lung pathogen, is able to take up DNA from any source at very high frequency. This is possible because the DNA-binding part of its transformation system does not need to recognize a specific nucleotide sequence.

Gene Transfer by Conjugation

Conjugation, or "bacterial sex," requires cell-cell contact typically initiated by a special pilus protruding from a donor cell (**Fig. 9.2**). Conjugation occurs in many species of bacteria and archaea, even in hyperthermophiles such as *Sulfolobus* species. In nature, mixed-species biofilms are thought to be a way in which even cross-species DNA transfer may occur, albeit at a low frequency. Some species, especially Gram-positives, use nonpilus attachment proteins. In the case of *E. coli*, the tip of the specialized plasmid-encoded sex pilus attaches to a receptor on the recipient cell and then contracts, drawing the two cells closer. The two cell envelopes fuse and generate a conjugation complex, similar to the transformation complex,

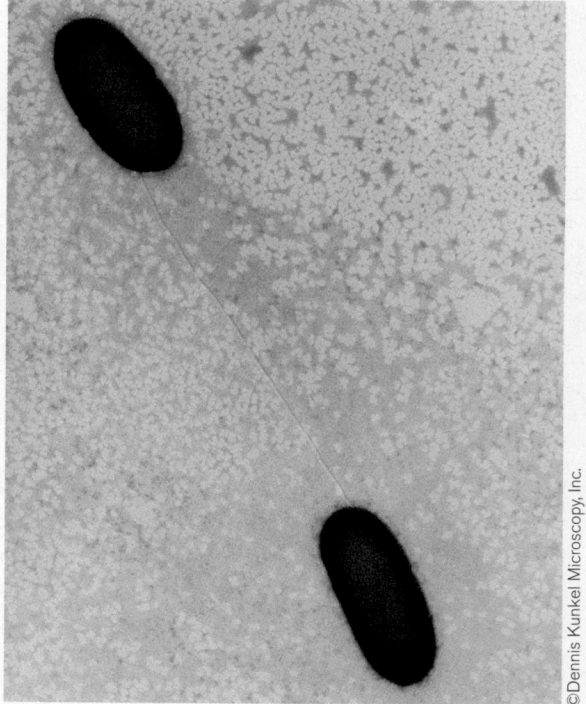

©Dennis Kunkel Microscopy, Inc.

Figure 9.2   Sex pilus connecting two *E. coli* cells.

through which single-stranded DNA passes from donor to recipient. The DNA passes across the membranes through a conjugation complex.

Bacterial conjugation requires the presence of special *transferable* plasmids that usually contain all the genes needed for pilus formation and DNA export. In some cases, plasmids can also contain genes encoding antibiotic resistance. A well-studied, transferable plasmid in *E. coli* is called **fertility factor (F factor)** (**Fig. 9.3**). Not only does this plasmid transfer itself to a recipient cell, but if it integrates into the chromosome of its host, it can also transfer host genes. F factor contains two replication

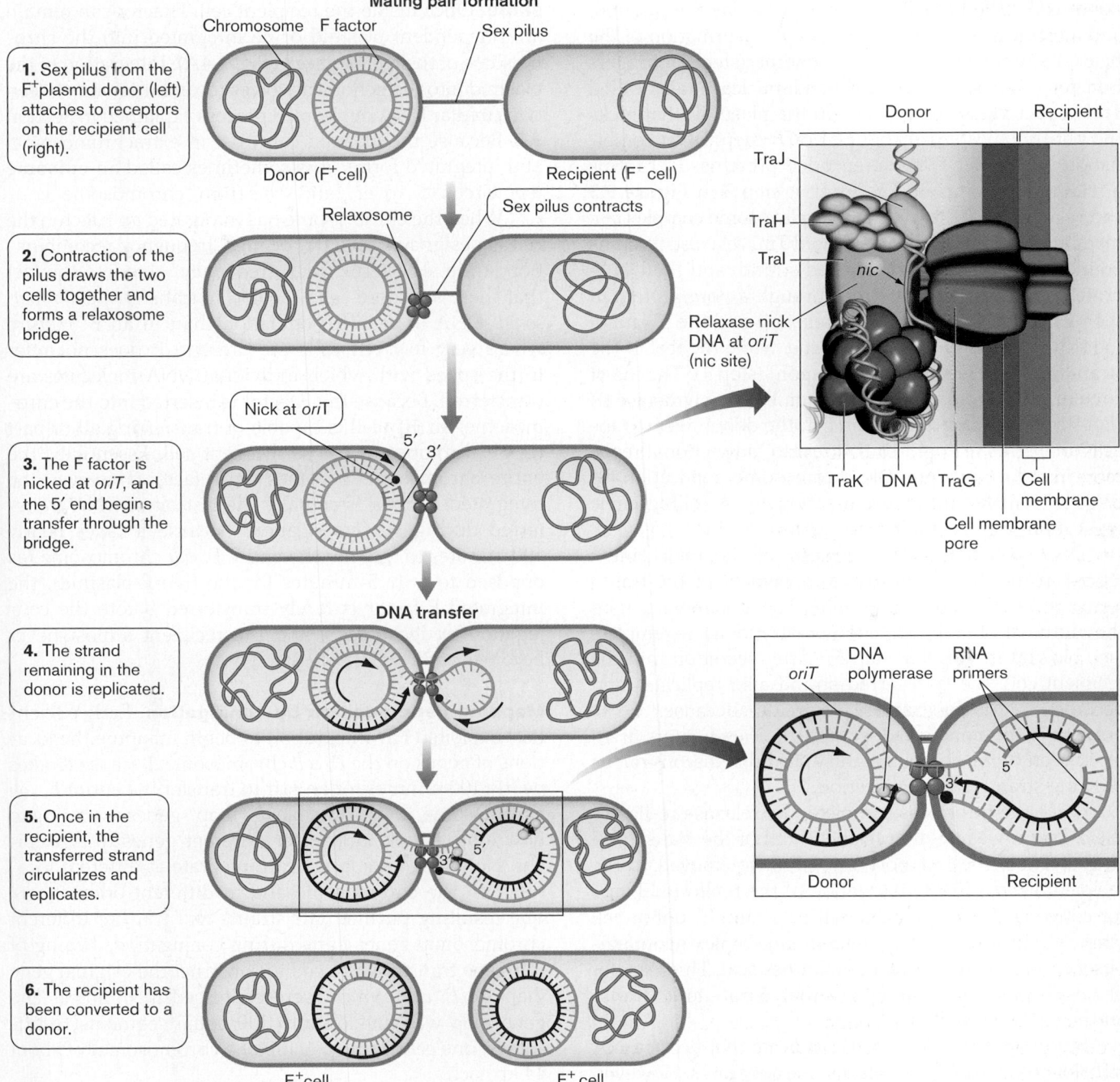

Mating pair formation

1. Sex pilus from the F⁺ plasmid donor (left) attaches to receptors on the recipient cell (right).

Chromosome   F factor   Sex pilus
Donor (F⁺ cell)   Recipient (F⁻ cell)

2. Contraction of the pilus draws the two cells together and forms a relaxosome bridge.

Relaxosome   Sex pilus contracts

3. The F factor is nicked at *oriT*, and the 5′ end begins transfer through the bridge.

Nick at *oriT*   5′ 3′

Donor   Recipient
TraJ   TraH   TraI   *nic*
Relaxase nick DNA at *oriT* (nic site)
TraK   DNA   TraG   Cell membrane
Cell membrane pore

DNA transfer

4. The strand remaining in the donor is replicated.

5. Once in the recipient, the transferred strand circularizes and replicates.

oriT   DNA polymerase   RNA primers
3′   5′
Donor   Recipient

6. The recipient has been converted to a donor.

F⁺ cell   F⁺ cell

Figure 9.3   The conjugation process.   Some plasmids, such as F factor in *E. coli*, can mediate cell-to-cell transfer of DNA. The plasmid is nicked at the *nic* site (step 3), and the 5′ end is transferred to the other cell. Purple arrowheads depict the 3′ end of replicating DNA. The inset at step 3 shows a current model of the DNA secretion apparatus. The relaxosome complex is composed of TraH, TraI (the helicase/endonuclease), TraJ, and TraK. The inset illustrates the step in which the TraI endonuclease is about to nick the donor plasmid. The 5′ end of the nick will move through the pore and remain attached to the membrane while the rest of the single-stranded DNA passes into the recipient. *Source*: Step 3 inset adapted from G. Schröder et al. 2002. *J. Bacteriol.* **184**:2767–2779. ▶ *microbiology2.com/animations*

origins, *oriV* and *oriT*, located at different positions on the plasmid. The origin *oriV* is used to replicate and maintain the plasmid in nonconjugating cells, whereas *oriT* is used only to replicate DNA during DNA transfer. The F factor also contains several *tra* genes whose protein products carry out the DNA transfer process.

Conjugation begins with cell-cell contact between a donor cell, called the **F⁺ cell**, which carries the plasmid, and a recipient **F⁻ cell** (**Fig. 9.3**, step 1). Formation of the fused membrane conjugation conduit (step 2) triggers F-factor synthesis of a helicase/endonuclease (also called TraI or relaxase, and encoded on the plasmid) that nicks the phosphodiester backbone at *oriT* (step 3). In combination with other F factor–encoded proteins, TraI forms a relaxosome complex. The inset at step 3 in **Figure 9.3** shows a model of the multifactor relaxosome complex at a membrane bridge. The translocated TraI relaxase remains bound to the 5′ end of the nicked strand, and the DNA-protein complex is transferred through a pore protein in the recipient. The translocated relaxase and the 5′ end of the strand remain in the membrane, while the rest of the strand is transferred through the pore (step 4). The intact circular strand remains in the donor. DNA polymerase III (Pol III) is then recruited to *oriT* in the donor, where replication begins (step 4). In contrast to bidirectional replication used to duplicate the chromosome, replication for conjugal transfer occurs unidirectionally by rolling-circle replication (step 5) (see Section 7.4).

DNA polymerase III synthesis of the replacement strand in the donor cell drives movement of the transferred strand through the pore. The polymerase uses the untransferred intact circular strand as a template (see blowup in **Fig. 9.3**, step 5). After secretion into the recipient cell, the transferred strand also replicates and becomes a double-stranded molecule. Because of the polarity of the transferred strand (the 5′ end enters first), replication of the complementary strand is discontinuous (lagging strand–type) replication.

Once the transfer is complete, the relaxase re-ligates the 5′ end it was holding to the 3′ tail of the transferred strand. Thus, the last portion of F factor moved to the recipient is *oriT*. Recircularization of the replicated F factor converts the F⁻ recipient cell to a new F⁺ donor cell (step 6). Ultimately, the conjugation complex spontaneously comes apart, and the membranes seal. This transfer process is very quick, taking less than 5 minutes to transfer the entire 110-kb F factor.

Studies with many bacteria indicate that donor DNA transfer occurs through a relaxosome apparatus. However, a recent paper reintroduced the old concept that transferred DNA moves through the hollow pilus itself. The evidence presented in that paper has yet to be confirmed.

Is it possible for two donors to exchange DNA with each other? The answer is generally no. Donors have mechanisms in place to prevent this. For example, they make a cell surface protein encoded by the F factor that inhibits formation of a conjugation complex with another donor possessing the same protein. This mechanism prevents the pointless transfer of a plasmid to a cell that already has that plasmid.

The F-factor plasmid can integrate into the chromosome. Once inside the recipient cell, F factor can remain an independent plasmid or be integrated into the chromosome of the recipient cell (**Fig. 9.4A**). Integration of the plasmid into the genome involves recombination between two circular DNA duplexes, a process explained in Section 9.3. Because the plasmid can exist in extrachromosomal and integrated forms, it is sometimes called an **episome** (*epi*, Greek for "over"; and *some* from "chromosome").

When the chromosome has integrated an F factor, the cell is designated an **Hfr**, or <u>h</u>igh-<u>f</u>requency <u>r</u>ecombination strain. The term "high frequency" refers to the fact that there are more cells capable of transferring *chromosomal* DNA in an Hfr population than in an F⁺ culture where very few Hfr cells are present. It does not refer to the speed with which individual DNA molecules are transferred. Because the F factor is inserted into the chromosome, an Hfr cell is capable of transferring all or part of the chromosome into a recipient cell. Essentially, the entire chromosome becomes the F factor. However, the integrated F factor is actually the last bit of DNA transferred during Hfr conjugation. Because it takes nearly 100 minutes to transfer the entire *E. coli* chromosome (as opposed to only 5 minutes for the free F plasmid), the integrated F factor is rarely transferred before the conjugation bridge breaks, and the recipient almost never becomes an F⁺ or Hfr cell.

Mapping gene position by conjugation. Earlier scientists exploited Hfr integration to begin mapping the locations of genes on the *E. coli* chromosome. Because it takes nearly 100 minutes for an Hfr to transfer the entire *E. coli* chromosome, it was possible to "map" genes according to how long it took to transfer different genes to a recipient. Note that an F factor can integrate at different locations on the chromosome and in different orientations. The resulting purified Hfr strains will transfer different chromosomal genes early during conjugation. Timing of Hfr gene transfer was used to construct the original gene maps of *E. coli*, which were expressed in minutes. The gene map was divided into 100 equal segments (centisomes; one centisome = 1/100 of a chromosome) of about 44 kb each.

The mapping process required mating an *E. coli* Hfr strain with an F⁻ recipient that contained a mutation in the gene to be mapped. It is important in any genetic cross that the mutation cause an observable phenotype. Say, for example, we wanted to map the gene *proA*, which is needed to synthesize the amino acid proline.

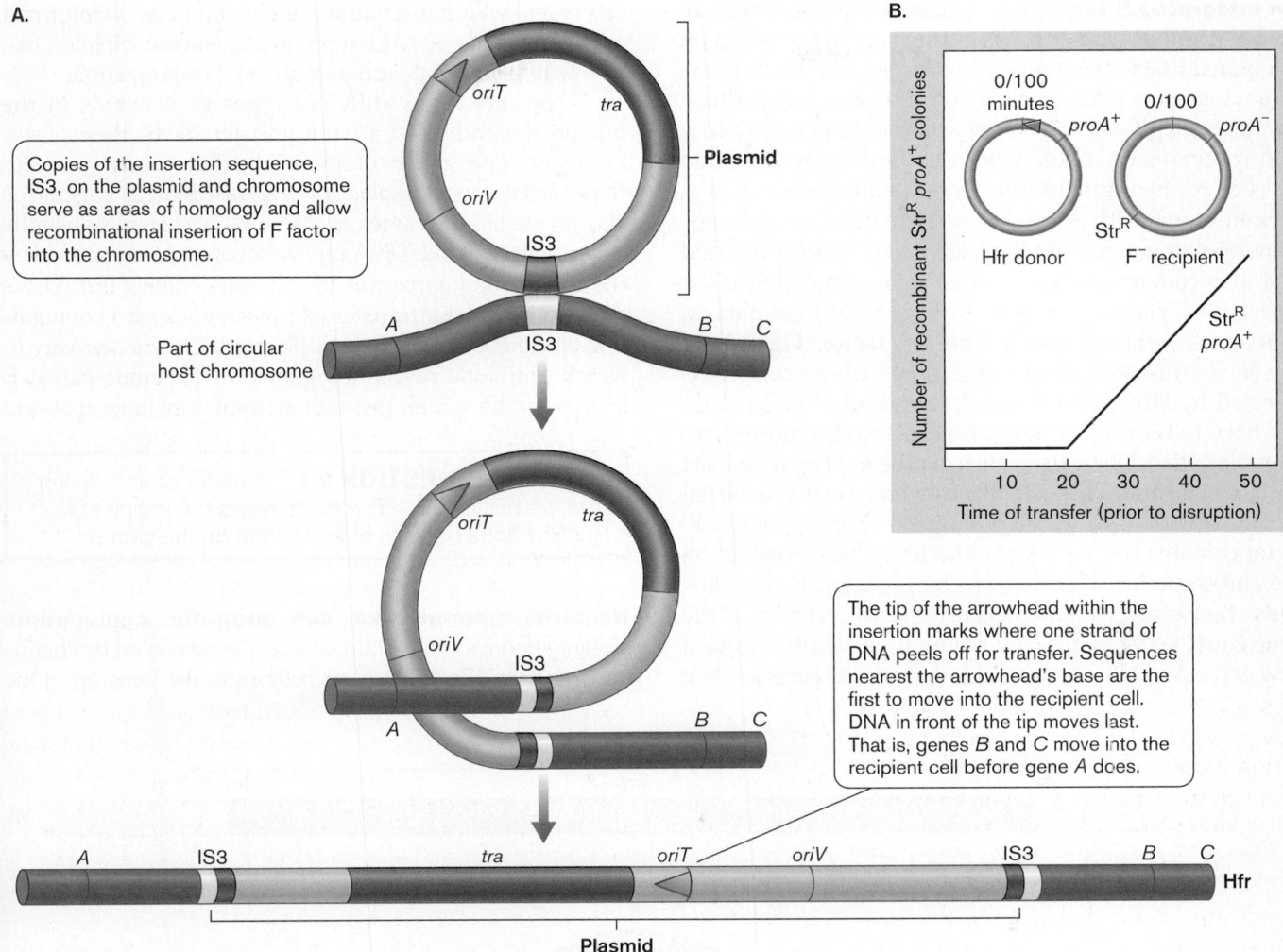

A.

Copies of the insertion sequence, IS3, on the plasmid and chromosome serve as areas of homology and allow recombinational insertion of F factor into the chromosome.

oriT
tra
Plasmid
oriV
IS3

Part of circular host chromosome

A
IS3
B C

oriT
tra
oriV
IS3
A
B C

The tip of the arrowhead within the insertion marks where one strand of DNA peels off for transfer. Sequences nearest the arrowhead's base are the first to move into the recipient cell. DNA in front of the tip moves last. That is, genes B and C move into the recipient cell before gene A does.

A
IS3
tra
oriT
oriV
IS3
B C
Hfr

Plasmid

B.

0/100 minutes
proA⁺
0/100
proA⁻

StrR

Hfr donor
F⁻ recipient

StrR
proA⁺

Number of recombinant StrR proA⁺ colonies

10 20 30 40 50
Time of transfer (prior to disruption)

Figure 9.4   Hfr formation and gene mapping.   A. The making of an Hfr by F-factor integration. Regions of sequence homology between an F factor and the host chromosome are recognized by host cell recombination proteins that then integrate F factor into the host chromosome. The transfer genes (*tra*), and replication origins are shown (vegetative = *oriV*; transfer = *oriT*). **B.** Conjugational mapping of genes. The figure shows Hfr donor and recipient chromosomes. The green arrowhead marks *oriT* (the transfer origin) at 0 minutes on the Hfr chromosome. Because it takes 100 minutes for conjugation to transfer the entire *E. coli* chromosome, the chromosome was arbitrarily divided into 100 units starting from 0 (located at the *thrA* gene) through 100. In the conjugation cross shown, the recipient requires proline added to media in order to form colonies. Mixing Hfr cells with recipient cells starts the conjugation "clock." The graph illustrates that it takes 20 minutes of conjugation to start seeing ProA⁺ recombinants.

Cells defective in *proA* will not grow on minimal medium unless proline is added. However, if the mutated *proA* gene is repaired by conjugation and subsequent recombination between the donor and recipient DNA molecules, the recombinant cell will grow and form a colony on agar medium *lacking* proline. It is also important that the F⁻ recipient be resistant to an antibiotic such as streptomycin, so the Hfr donor cells can be killed after mating.

An Hfr culture that is streptomycin sensitive is mixed with the streptomycin-resistant *proA* mutant culture (**Fig. 9.4B**). Then, at timed intervals, an aliquot (that is, a measured sample) of the mating mixture is removed and the conjugation bridges broken by blending in a blender. This is called interrupted mating. The longer the mating goes

before breaking the bridges, the more of the chromosome is transferred from the Hfr into the recipient. At each interval, the disrupted mixture is plated onto a selective agar lacking proline but containing streptomycin. The *only* cell that can grow on this medium is a recombinant in which the *proA⁺* gene transferred from the Hfr strain and recombined into the chromosome of the streptomycin-resistant F⁻ cell. Until the gene transfers, no colonies will form. Because the position of the integrated F factor on the chromosome is fixed, the farther the wild-type gene is from the insertion site, the longer it will take for the gene to move into the mutant. For example, in **Figure 9.4B** the *proA⁺* gene is located 20 map minutes away from the F-factor insertion.

An integrated F factor can excise from the chromosome. Another type of gene shuffling can occur when an integrated F factor subsequently excises from the chromosome. Once an F factor has been integrated into a host chromosome, it can also be excised via host recombination mechanisms. Usually, it is excised completely and restored to its original form. Occasionally, however, it is excised along with some DNA from the host chromosome, yielding a product that will be an F-factor plasmid that also contains some of the chromosomal DNA. The derivative F plasmid that contains host DNA is called an **F-prime (F′) plasmid** or **F-prime (F′) factor** (**Fig. 9.5**).

In contrast to chromosomal genes whose transfer is directed by Hfr, genes hitchhiking on an F′ plasmid do not have to recombine into the recipient chromosome to be maintained. The extra genes can be expressed as part of the F′ plasmid. This establishes what is called a partial diploid situation, in which a conjugal recipient of the F′ factor contains two copies of those few genes—one set on the chromosome, the other on the F′ factor. Partial diploids are useful to the cell because the second copy of the gene could serve as the raw material needed to evolve a new gene. And they are useful to geneticists because they can reveal whether a mutant allele of a gene is dominant over a normal, or wild-type, allele. Partial diploids can also help geneticists understand gene organization.

There are many different types of plasmids in the microbial world. Most are not transferable by themselves. However, one group of plasmids that cannot transfer themselves can be *mobilized* if a transferable plasmid is also present in the same cell. Mobilizable plasmids usually contain an *oriT*-like DNA replication origin recognized by the conjugation apparatus of the transferable plasmid. As a result, when the transferable plasmid begins conjugating, so does the mobilizable plasmid. This is one way in which antibiotic resistance genes on plasmids called R factors can be spread throughout a microbial population.

> **THOUGHT QUESTION 9.1** Transfer of an F factor from an F$^+$ cell to an F$^-$ cell converts the recipient to F$^+$. Why does transfer of an Hfr not do the same?

Bacterial pheromones can promote conjugation. Conjugation can, in certain species, be promoted by chemical communication between cells (quorum sensing). One

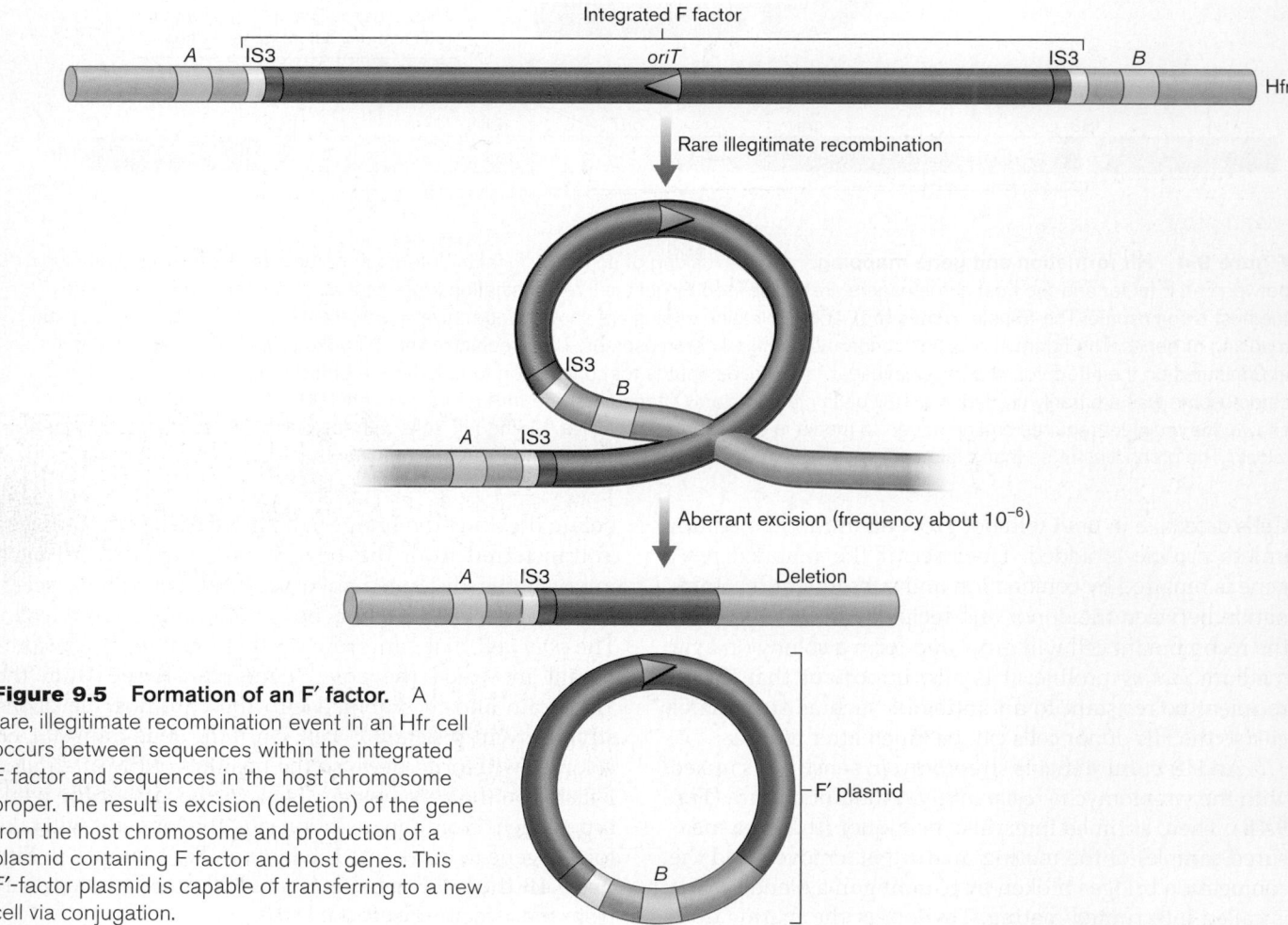

Figure 9.5 Formation of an F′ factor. A rare, illegitimate recombination event in an Hfr cell occurs between sequences within the integrated F factor and sequences in the host chromosome proper. The result is excision (deletion) of the gene from the host chromosome and production of a plasmid containing F factor and host genes. This F′-factor plasmid is capable of transferring to a new cell via conjugation.

example is plasmid pAD1 in *Enterococcus faecalis*. The pAD1 plasmid encodes an antimicrobial bacteriocin gene and produces a protein called aggregation substance, which catalyzes cell-cell contact and formation of the conjugation complex. Ordinarily, this gene and others on the plasmid are inactive. Potential recipients, however, try to entice donors to mate with them by producing small behavior-altering molecules called pheromones. The chemical signal molecules released by bacteria are small peptides that enter the donor cell through oligopeptide permeases present in the membrane. Once inside the donor cell, the pheromone stimulates transcription of the *pAD1* transfer genes. Another pheromone-responsive conjugation system was found in *Bacillus subtilis*, although this system involves the conjugative transfer of a transposon (see Section 9.6) rather than a plasmid. It should be noted that most known conjugation systems do not use pheromones to stimulate transfer.

Microbial transfer of genes into eukaryotes.

In eukaryotes, sexual exchange of genes usually occurs only within a single species. Microbes, on the other hand, are more promiscuous. Mating among different microbial genera is common and perhaps even desirable from an evolutionary viewpoint. For example, *Salmonella* can conduct interspecies mating with *E. coli*. Sharing genes between species allows each one to sample genes from the other and keep genes that increase fitness.

What may be surprising is that some bacteria can actually transfer genes across biological domains. One striking example is *Agrobacterium tumefaciens*, which causes crown gall disease in plants. This Gram-negative plant pathogen, common to the rhizosphere (the area around root surfaces), contains a tumor-inducing plasmid (Ti) that can be transferred via conjugation to plant cells. These bacteria detect and swim toward phenolic wound compounds released by damaged plant cells. The phenolic compounds signal where the microbe can gain access to plant cells. Subsequent transfer, integration, and expression of the Ti plasmid in the plant cell genome trigger the release of plant hormones that stimulate tumorous growth of the plant (**Fig. 9.6**). Plant cells within the tumor release amino acid derivatives called "opines" that the microbe can then use as a source of carbon and nitrogen.

The unique mode of action of *A. tumefaciens* has made this bacterium an indispensable tool for plant breeding. Any desired genes, such as insecticidal toxin genes (like those produced by *Bacillus thuringiensis*) or herbicide resistance genes, can be engineered into the bacterial plasmid DNA and thereby inserted into the plant genome (discussed in Chapter 16). The use of *Agrobacterium* not only shortens the conventional plant-breeding process, but also allows entirely new (nonplant) genes to be engineered into crops.

Scientists have recently discovered that interdomain transfer of DNA may be more common than was previously realized. There is evidence of massive DNA transfers from some bacteria to their eukaryotic hosts (**Special Topic 9.1**).

> **NOTE:** Conjugation in bacteria and archaea is mechanistically different from conjugation among eukaryotic microbes. Some eukaryotic microbes exchange nuclei through a structure, called a conjugation bridge, very different from that of bacteria (discussed in Chapter 20).

A.

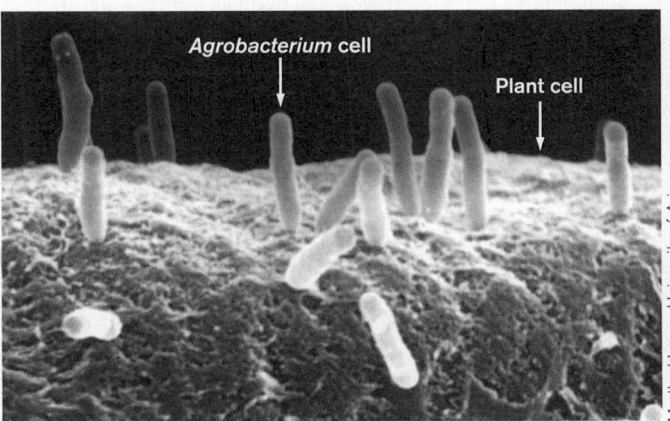

B.

Agrobacterium cell

Plant cell

Figure 9.6 An example of gene transfer between bacteria and plants. A. Crown gall disease tumor caused by the bacterium *Agrobacterium tumefaciens*. **B.** Electron micrograph of *A. tumefaciens* attached to plant cells.

Special Topic 9.1 | There's a Bacterial Genome Hidden in My Fruit Fly

In the 1986 movie *The Fly*, the DNA of a fly becomes mixed with the DNA of a man in a teleportation experiment gone horribly wrong. Predictably, the resulting "flyman" and his girlfriend have trouble coping with his horrible transmogrification. Originally thought to be solely the stuff of science fiction, we now know interdomain genome transfers do occur in nature. The transfer of Ti plasmid from *Agrobacterium tumefaciens* to plant cells is one well-documented example. However, there are other intriguing biological situations that provide opportunities for more frequent interdomain transfers. One of these situations involves bacterial endosymbionts that have evolved

to live within eukaryotic host cells. *Wolbachia pipientis* is a maternally inherited bacterial endosymbiont that infects a wide range of arthropods, including at least 20% of insect species, as well as filarial nematodes (discussed in Section 17.6). *Wolbachia* is passed from one generation of insect to the next because it lives inside the eggs of the insect (**Fig. 1**). Scientists recognized that the presence of this bacterium in developing gametes provided a unique opportunity for horizontal transfer from the bacterium to the eukaryotic chromosome.

Startling work has recently shown that *Wolbachia*-to-host genome transfer has actually occurred. *Wolbachia* DNA inserts have been discovered deep within the genomes of diverse invertebrate species, including wasps, fruit flies, and nematodes. **Figure 2** shows the presence of a particular *Wolbachia* gene embedded in a *Drosophila* (fruit fly) chromosome extracted from a fly cured of the bacterium by antibiotics. In fact, the entire *Wolbachia* genome was transferred, and many of the genes transferred were transcribed, though it is unclear whether they have any effect on fly metabolism. This finding is extremely important from an evolutionary viewpoint. Although the extent of bacterium-to-eukaryote gene transfer is not known, it is relevant to note that whole-eukaryote genome sequencing projects routinely exclude bacterial sequences on the assumption that these represent contamination. This approach may have to be reevaluated.

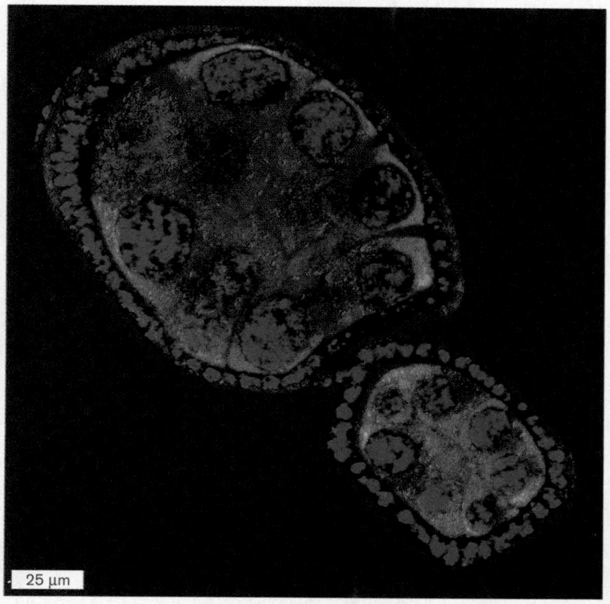

Figure 1 *Wolbachia* within the *Drosophila* female germ line. The intracellular alphaproteobacterium *Wolbachia* (red) accumulates in the germ-line cell that will form the fruit fly egg and embryo, and is thereby transmitted to the next host generation. Green indicates germ-line cytoplasm and blue indicates DNA in this fluorescent microscopy image.

Horacio M. Frydman and Eric Wieschaus

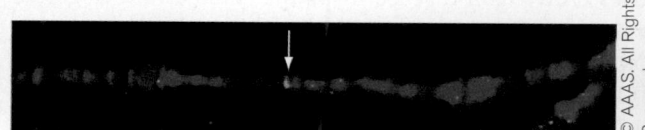

Figure 2 *Wolbachia* DNA in a *Drosophila* chromosome. A fluorescein-stained probe for the *Wolbachia* gene WD0484 is seen bound to a unique location (green, arrow) on chromosome 2L of *Drosophila ananassae* stained with propidium iodide (red). Chromosome 2L is 2Mbp. *Source*: Hotopp et al. 2007. *Science* **317**:1753–1756.

Gene Exchange by Phage Transduction

Bacteriophages harbor their own genomes, separate from that of their host cells. However, bacteriophages also can *accidentally* move bacterial genes between cells as an offshoot of the phage life cycle (see Section 6.4 and eTopic 7.3). The process in which bacteriophages carry payloads of host DNA from one cell to another is known as **transduction**. There are two basic types of transduction: generalized and specialized. **Generalized transduction**

can take any gene from a donor cell and transfer it to a recipient cell. In contrast, **specialized transduction** (also known as restricted transduction) can transfer only a few closely linked genes between cells.

Bacteriophages capable of generalized transduction have trouble distinguishing their own DNA from that of the host when attempting to package DNA into their capsids, so pieces of bacterial host DNA accidentally become packaged in the phage capsid. In the case of *Salmonella* P22 phage, the packaging system recognizes a certain DNA sequence on

P22 DNA called a *pac* site. During rolling-circle replication, P22 DNA forms long concatemers containing many P22 genomes arranged in tandem. The *pac* site defines the ends of the phage genome, marking where the packaging system cuts the P22 DNA and starts packaging DNA into the next empty phage head. However, certain DNA sequences on the *Salmonella* chromosome also "look" like *pac* sites. As a result, the packaging system sometimes mistakenly packages host DNA instead of phage DNA.

The result of this mistake is that 1% of the phage particles in any population of P22 do not contain phage DNA but carry host DNA plucked from around the chromosome (**Fig. 9.7**). The phages that carry host DNA are called transducing particles. Any one transducing particle will contain only one segment of host DNA, but different particles in the phage population will contain different segments of host DNA. When a transducing particle injects its DNA into a cell, no new phages are made; but the highjacked host DNA can recombine, or exchange, with sequences in the host chromosome of the newly infected cell, changing the genetic makeup of the recipient.

> **THOUGHT QUESTION 9.2** Would it be easier to demonstrate generalized transduction using a temperate phage (which generates a lysogen) or a lytic (virulent) phage?

Specialized or restricted transduction is a phage-mediated gene transfer mechanism that resembles the formation of the F′ factors previously described. In contrast to generalized transduction, specialized transduction can move only a limited number of host genes. *E. coli* phage lambda is the classic example of specialized transduction. The basic biology of lambda phage is discussed in Chapter 6. When lambda phage DNA first enters the cell, it is linear. It circularizes at cohesive ends called *cos* sites. Then a small DNA sequence in lambda called *attP* can specifically recombine with a small host DNA sequence (called *attB*) located between the *gal* (galactose catabolism) and *bio* (biotin synthesis) genes on the *E. coli* chromosome (**Fig. 9.8**, step 1). This process, carried out by the phage integrase protein, produces a chromosome with an integrated phage genome, referred to as a prophage, flanked by chimeric *att* sites called *attL* and *attR*. They are called chimeric because each one is made half from the bacterial and half from the phage *att* sites. Prophage DNA remains latent until something happens to the cell to activate it.

Specialized transduction begins with this prophage DNA. The majority of the time, when the prophage is reactivated (usually by DNA damage), recombination mechanisms involving phage proteins can excise the lambda DNA precisely, so that the chromosome and viral DNAs are restored to their native states. The viral DNA will then replicate and make more phage particles

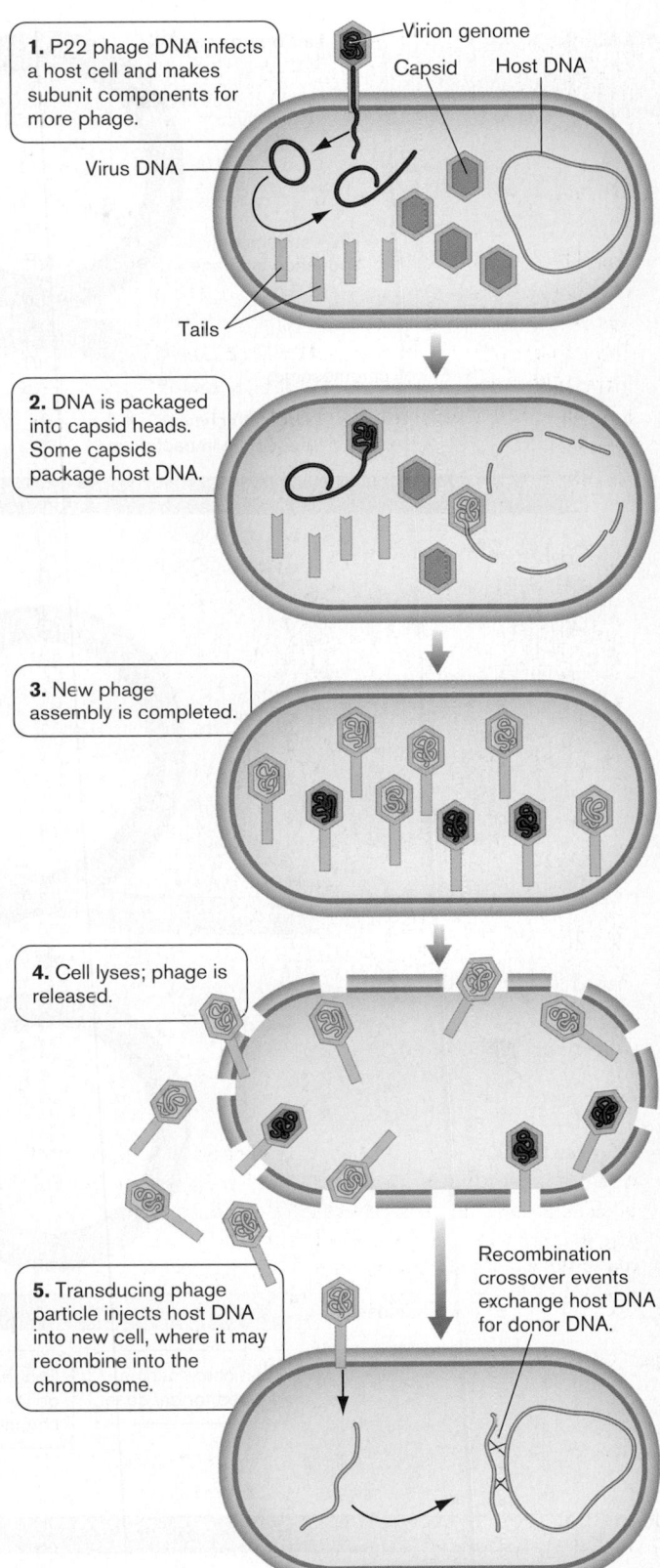

1. P22 phage DNA infects a host cell and makes subunit components for more phage.

Virion genome
Capsid   Host DNA
Virus DNA
Tails

2. DNA is packaged into capsid heads. Some capsids package host DNA.

3. New phage assembly is completed.

4. Cell lyses; phage is released.

5. Transducing phage particle injects host DNA into new cell, where it may recombine into the chromosome.

Recombination crossover events exchange host DNA for donor DNA.

Figure 9.7 Generalized transduction. Generalized transduction by phage vectors can move any segment of donor chromosome to a recipient cell. The number of genes transferred in any one phage capsid is limited, however, to what can fit in the phage head.

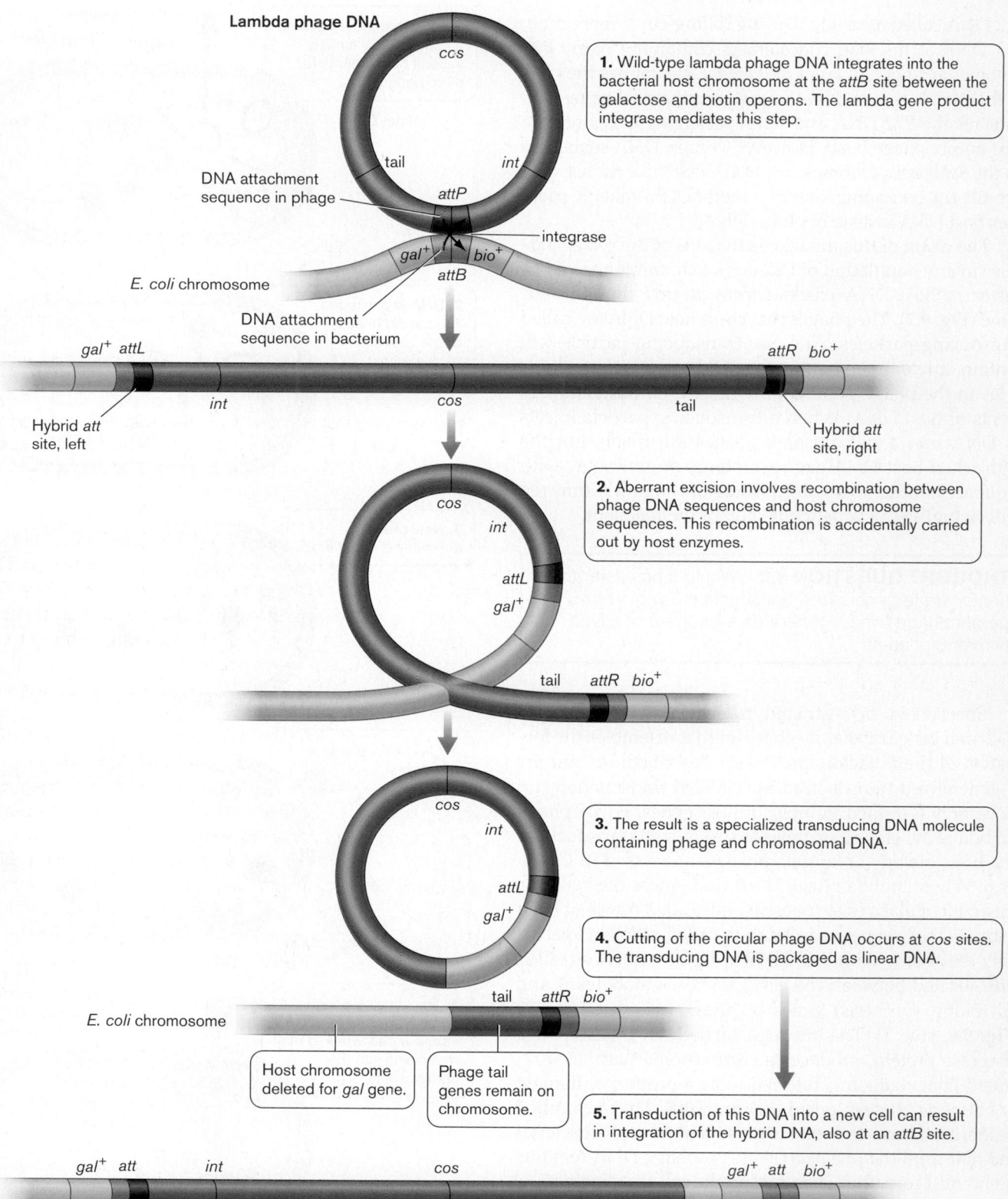

Figure 9.8 **Specialized transduction is restricted to moving host genes flanking the phage attachment site.** The number of genes transferred is limited by the size of the phage head. The resulting transductant chromosome becomes partially diploid for the transferred gene, in this case *gal*.

containing normal phage DNA. However, on rare occasions improper excision can take place between host DNA sequences that lie adjacent to the phage insertion site (*attB* in **Fig. 9.8**, step 1) and similar DNA sequences within the prophage. Improper, or aberrant, excision between virus and host sequences yields a virus that will lack a few viral genes (tail genes missing in step 3) but will include host genes lying adjacent to the phage attachment site (the galactose utilization gene *gal* in **Fig. 9.8**, step 3). Aberrant excision is accidentally carried out by host recombination enzymes. With some phages, the specialized transducing particles can replicate unaided. However, in **Figure 9.8** the result is a defective, specialized transducing phage DNA (lambda d*gal*, or λd*gal*). Specialized transducing phage particles that are defective cannot replicate by themselves but require the presence of a helper phage to supply missing gene products.

Once formed, specialized transducing phages can deliver the hybrid DNA molecule to a new recipient cell. This is the transduction process. Once in that cell, the phage DNA can integrate into the host *attB* site, carrying the donor host gene(s) with it (**Fig. 9.8**, steps 4 and 5). This will establish another partial diploid situation in which the new recipient contains two copies of a host gene: one originally present on its chromosome and one brought in by the transducing DNA.

Genes other than *gal* or *bio* can also be moved by specialized transduction if lambda is used to infect mutant host strains lacking the normal *att* site. Because the primary integration site is gone, lambda DNA can integrate at low frequency to other, secondary attachment sites somewhere else in the chromosome. Improper excision at one of those locations will package adjacent host DNA.

DNA Restriction and Modification

There are dangers to the cell associated with the indiscriminate transfer of DNA between bacteria. The most obvious risk involves bacteriophages whose goal is to replicate at the expense of target cells. A bacterium that can digest invading phage DNA while protecting its own chromosome has a far better chance of surviving in nature than do cells unable to make this distinction. Beyond the threat of overt destruction by phages, however, the unrestricted incorporation of foreign DNA can be an energetic drain on the cell. Genes encoding competing or useless products or products that lack regulatory restraints would squander resources and lower the overall fitness of the cell. As a result, bacteria have developed a kind of "safe-sex" approach to gene exchange. It is an imperfect approach, however, that still leaves room for beneficial genetic exchanges.

This protection system, called restriction and modification, involves the enzymatic cleavage (restriction) of alien DNA and the protective methylation (modification) of self DNA (**Fig. 9.9**). Most bacteria produce DNA

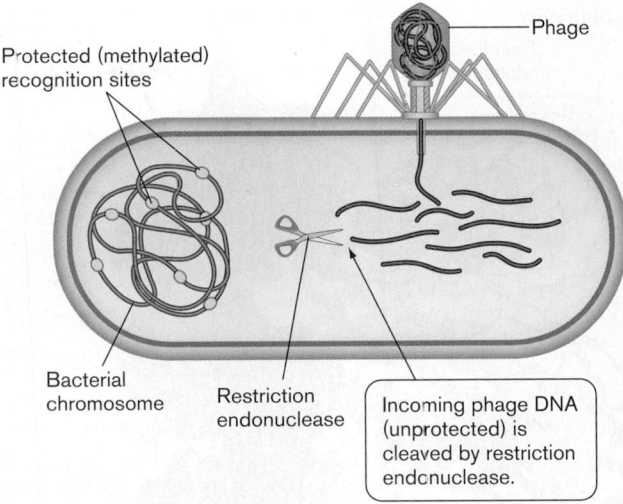

Figure 9.9   Restriction of invading phage DNA. Phage DNA is injected into a host where restriction endonucleases can digest it. Host DNA is protected because specific methylations of its own DNA prevent the enzymes from cutting it.

restriction endonucleases (also known as restriction enzymes), enzymes that recognize specific short DNA sequences (known as recognition sites) and cleave DNA at or near those sequences (**Fig. 9.10**). A structural model of the restriction enzyme *Eco*RI cleaving its target DNA sequence is shown in **Figure 9.11**.

Given that the producing organism's own chromosomes contain these same restriction sequences, how do bacteria avoid "committing suicide"? They protect themselves by producing matching modification enzymes that use *S*-adenosylmethionine to attach methyl groups to those same sequences. A three-dimensional model of a methylation enzyme interacting with a DNA recognition site is shown in **Fig. 9.12**. This modification makes the sequence invisible to the cognate (matched) restriction endonuclease (the "A" no longer looks like adenine to the restriction endonuclease). Only one strand of the sequence needs to be methylated to protect the duplex from cleavage;

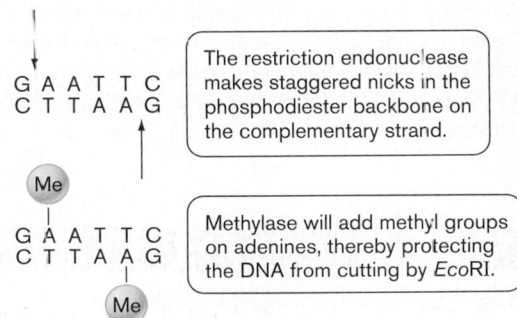

Figure 9.10   *Eco*RI restriction site. *Eco*RI is an example of a class II restriction-modification system. Shown here are cleavage (top) and methyl modification (bottom). The DNA sequence shown is specifically recognized by the endonuclease and methylase enzymes.

A. Side view

B. View from top

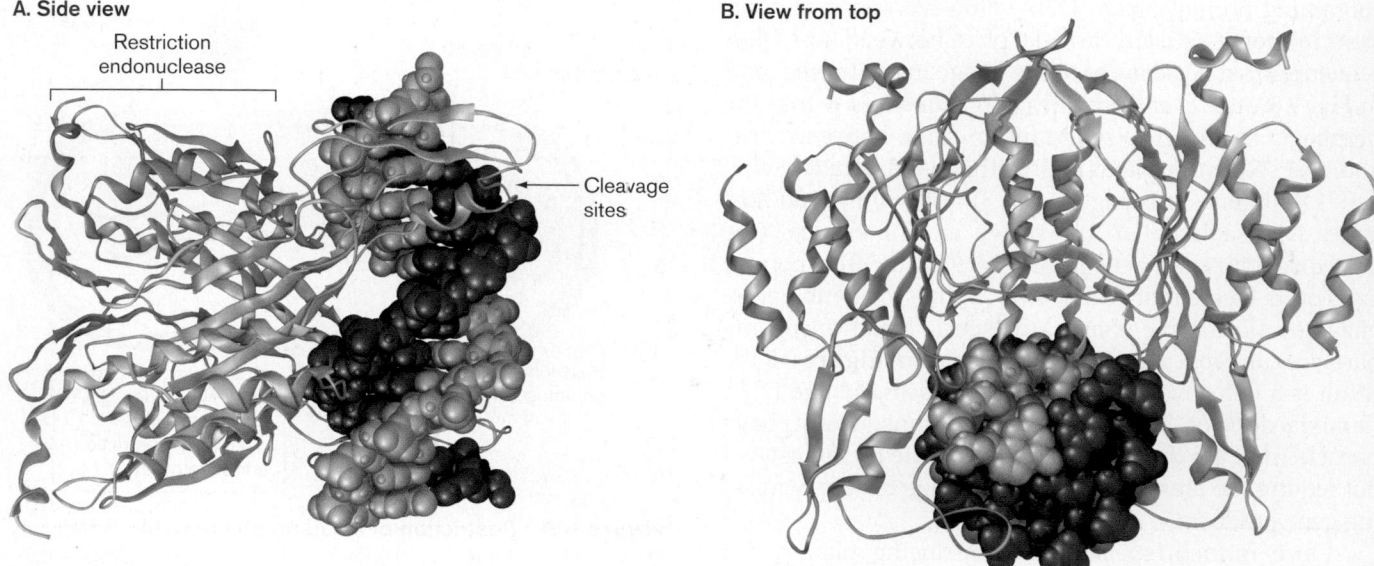

Figure 9.11   DNA restriction endonuclease attached to a DNA restriction sequence.   A. In the side view, the part of the protein shown in gold is about to cut the dark gray DNA backbone at the top. The light blue part of the protein will cut the light gray DNA at the bottom. **B.** Viewed from the top, the enzyme is seen to hold the DNA backbone. (PDB code: 1QRI)

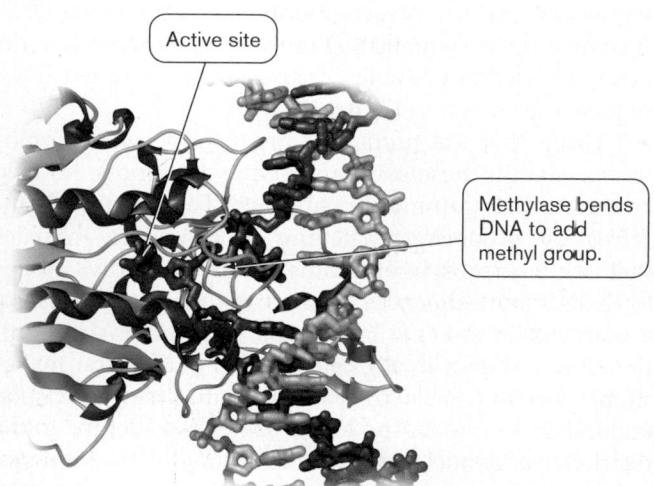

Figure 9.12   DNA methylase used for DNA modification. The DNA methylase adds a methyl group to a restriction site, changing the site's conformation so that it is no longer recognized by the restriction endonuclease. The methylase bends the DNA helix to bring the target base into the active site (left side). (PDB code: 1DCT)

thus, even newly replicated, and consequently hemimethylated (only one strand methylated), DNA sequences are invisible to the restriction endonuclease.

There are three main types of restriction endonucleases, based on how many subunits there are in the enzyme, what cofactors are required, and where the DNA cleavage sites are located relative to the recognition site (**Table 9.1**). Types I and III restriction endonucleases have

their restriction and modification activities in one multifunctional protein and cleave DNA some distance away from the recognition site. Type II restriction endonucleases, those used most often for cloning, only possess endonuclease activity. A separate modification protein carries out methylation of the restriction site. Type II restriction endonucleases generally recognize palindromic DNA sequences and cleave at those sites. **Figure 9.11** illustrates how the class II restriction endonuclease *Eco*RI holds the DNA as it makes a staggered cut at the cleavage (restriction) site.

THOUGHT QUESTION 9.3   How do you think phage DNA containing restriction sites evades the restriction-modification screening system of its host?

THOUGHT QUESTION 9.4   Each type I restriction-modification system includes separate proteins for restriction (HsdR), modification (HsdM), and sequence recognition (HsdS). Can a plasmid grown in a wild-type strain be used to transform a strain defective in *hsdR*? An *hsdS* mutant? Can a plasmid grown in an *hsdM* mutant strain be transformed into an *hsdR* defective strain?

TO SUMMARIZE:

- **Transformation** is the uptake by living cells of free-floating DNA from dead, lysed cells.
- **Competence** for transformation in some organisms is triggered by a genetically programmed physiological change.

Table 9.1 | **Main types of restriction-modification systems.**

Restriction-modification system	Restriction and modification activities	Number of subunits	Recognition site characteristics	Cleavage and modification sites	Examples
Type I	Present in one multifunctional protein	Three different subunits	5–7 bp, asymmetrical	Located 100 bp or more from recognition site	*Eco*K in *E. coli*; *Sty*LTIII in *Salmonella enterica*
Type II	Separate methylase and restriction enzymes	One or two identical subunits per activity	4–6 bp, palindromic	At or near recognition site	*Eco*RI in *E. coli*; *Hind*III in *Haemophilus influenzae*
Type III	Present in one multifunctional protein	Two different subunits	5–7 bp, asymmetrical	24–26 bp from recognition site	*Eco*571 in *E. coli*; *Bce*SI in *Bacillus cereus*

- **Conjugation** is a DNA transfer process mediated by a transferable plasmid that requires cell-cell contact and formation of a protein complex between mating cells.
- **Pheromone peptides** secreted by Gram-positive cocci such as *Enterococcus faecalis* activate transfer genes in nearby donor cells.
- **Some bacteria can transfer DNA across phylogenetic domains.** For example, *Agrobacterium tumefaciens* conjugates with plant cells.
- **Transduction** is the process whereby bacteriophages transfer fragments of bacterial DNA from one bacterium to another. In generalized transduction, a phage preparation can move any gene in a bacterial genome to another bacterium. In specialized transduction, a phage can move only a limited number of bacterial genes.
- **Restriction endonucleases** protect bacteria from invasion by foreign DNA. Restriction-modification enzymes methylate restriction target sites in the host DNA to prevent self-digestion.

9.3 Recombination

Once a new piece of DNA has entered a cell through one of the gene exchange systems, what happens to it? The answer to that question depends on the nature of the acquired DNA. If the DNA is a plasmid, capable of autonomous replication, it can coexist in the cell separate from the host chromosome. If the DNA is not capable of autonomous replication, it is usually degraded by nucleases. Alternatively, it may be incorporated into the chromosome through recombination.

Two different DNA molecules in a cell can recombine by one of several mechanisms. **Generalized**

recombination requires that the two recombining molecules have a considerable stretch of homologous DNA sequence. In contrast, the mechanism of **site-specific recombination** requires very little sequence homology between the recombining DNA molecules, but does require a short (10–20 bp) sequence recognized by the recombination enzyme.

Generalized Recombination and RecA

Enzymes participating in generalized recombination are able to find and align homologous stretches of DNA in two different DNA molecules and catalyze an exchange of strands. The result can be a **cointegrate** molecule in which a single recombination event (that is, a single crossover) joins the two participating DNA molecules. For example, when two circular DNA molecules are joined by a single recombination event, the two molecules are added together and form one large circular molecule (for an example, see **Fig. 9.4**). A second crossover separated by some distance from the first, however, will lead to an equal exchange of DNA in which neither molecule increases in length.

Why is recombination advantageous? There are three probable functions for generalized recombination in the microbial cell:

- Recombination probably first evolved as an internal method of DNA repair, useful to fix mutations or restart stalled replication forks. This role does not involve foreign DNA.
- Cells with damaged chromosomes use DNA donated by others of the same species to repair their damaged genes.
- Recombination is part of a "self-improvement" program that samples genes from other organisms for an ability to enhance the competitive fitness of the cell.

The central, but by no means only, player in generalized recombination is a protein called RecA. RecA molecules are also called synaptases because they are able to scan DNA molecules for homology and align the homologous regions, forming a triplex DNA molecule, or synapse (**Fig. 9.13**). Homologs of RecA are found in many other species.

The clearest model of RecA-mediated recombination comes from *E. coli* (**Figs. 9.13** and **9.14**). Before RecA can find homology between two DNA molecules, the donor double-stranded DNA molecule is converted to a single strand by the RecBCD enzyme. The RecBCD complex enters at the end of a DNA fragment and begins to unwind it (**Fig. 9.14**, steps 1 and 2). The complex changes activity when it encounters 8-bp sequences called Chi (crossover hot-spot instigator) that are scattered throughout most DNA molecules. RecBCD nicks DNA at a Chi site and continues unwinding the strand. RecA then loads onto the single strand as a filament (step 2).

When RecA finds homology between donor and recipient DNA (minimum required is about 50 bp), strand invasion occurs (step 3), in which the donor single-stranded DNA invades the homologous region in the double-stranded DNA recipient molecule and displaces its like strand. At this point a single-strand crossover has been made. The single-strand crossover produces a molecule with four double-stranded ends.

Other recombination proteins, RuvA and B, assemble at the crossover point and extend the invasion, in a process called strand assimilation or branch migration (**Fig. 9.14**, step 4 and inset). This process extends the base pairing between homologous donor and recipient strands.

Ultimately, the end of the displaced recipient strand is cleaved (step 5) and ligated to the donor strand (step 6). The resulting structure is known as a **Holliday junction**, after the scientist (Robin Holliday) who first proposed it (step 7; also seen in inset). The final step is resolution of the crossover Holliday junction. To visualize this, we mentally rotate the right half of the Holliday structure as shown in **Figure 9.14**, step 7. RuvC protein cleaves across the junction (step 8), and the products are ligated to form a complete crossover. The ligated product represents a true crossover. In each case, notice that there is a small **heteroduplex** region in which one strand comes from the donor and the complementary region comes from the recipient.

If we were viewing the end result of transduction, where the recipient is a circular chromosome and the donor is a double-stranded linear fragment, the recombinant molecule would appear as shown in **Figure 9.14**, step 9. A second crossover farther along the donor strand is then required to maintain circular integrity of the recipient (step 10). This double crossover means that any genes residing between the first and second crossovers are simultaneously exchanged.

One consequence of this type of double recombination is that genes clustered next to each other on a

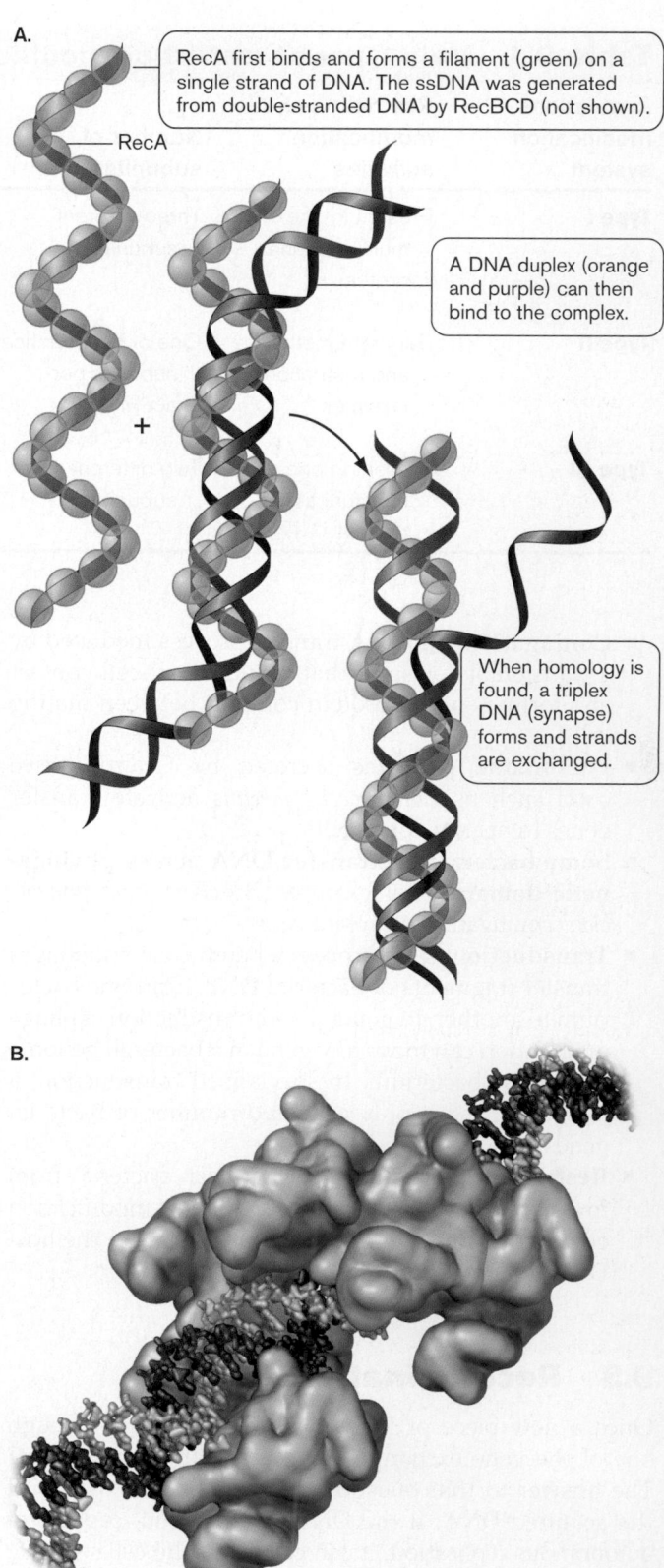

A.

RecA first binds and forms a filament (green) on a single strand of DNA. The ssDNA was generated from double-stranded DNA by RecBCD (not shown).

RecA

A DNA duplex (orange and purple) can then bind to the complex.

When homology is found, a triplex DNA (synapse) forms and strands are exchanged.

B.

Figure 9.13   RecA protein catalyzing recombination.
A. The standard mechanism of recombination. **B.** A molecular model of RecA protein filament catalyzing recombination. The DNA is light gray and dark gray. (PDB codes: 1N03, 1K8J)

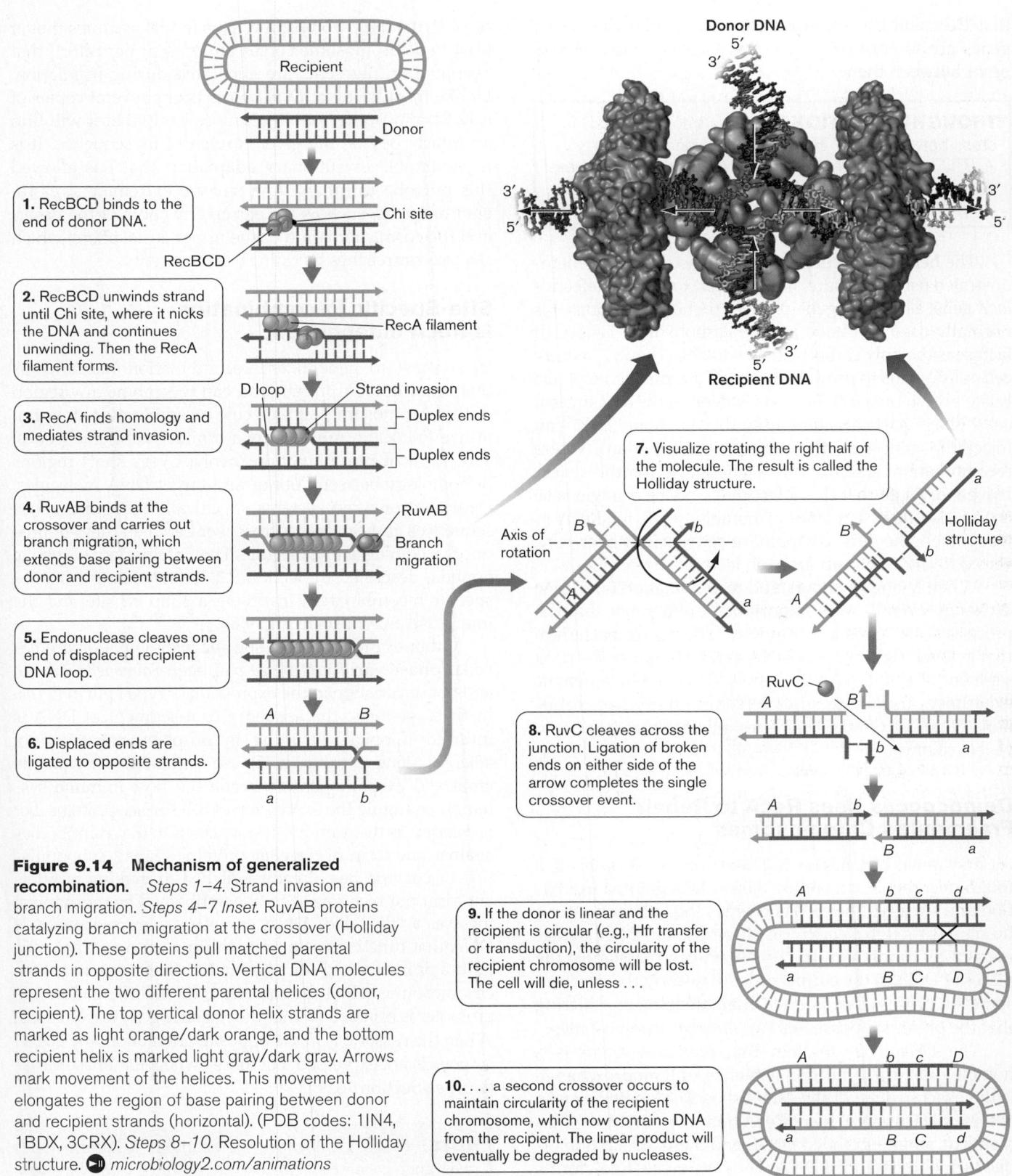

1. RecBCD binds to the end of donor DNA.

2. RecBCD unwinds strand until Chi site, where it nicks the DNA and continues unwinding. Then the RecA filament forms.

3. RecA finds homology and mediates strand invasion.

4. RuvAB binds at the crossover and carries out branch migration, which extends base pairing between donor and recipient strands.

5. Endonuclease cleaves one end of displaced recipient DNA loop.

6. Displaced ends are ligated to opposite strands.

7. Visualize rotating the right half of the molecule. The result is called the Holliday structure.

8. RuvC cleaves across the junction. Ligation of broken ends on either side of the arrow completes the single crossover event.

9. If the donor is linear and the recipient is circular (e.g., Hfr transfer or transduction), the circularity of the recipient chromosome will be lost. The cell will die, unless . . .

10. . . . a second crossover occurs to maintain circularity of the recipient chromosome, which now contains DNA from the recipient. The linear product will eventually be degraded by nucleases.

Figure 9.14 Mechanism of generalized recombination. *Steps 1–4.* Strand invasion and branch migration. *Steps 4–7 Inset.* RuvAB proteins catalyzing branch migration at the crossover (Holliday junction). These proteins pull matched parental strands in opposite directions. Vertical DNA molecules represent the two different parental helices (donor, recipient). The top vertical donor helix strands are marked as light orange/dark orange, and the bottom recipient helix is marked light gray/dark gray. Arrows mark movement of the helices. This movement elongates the region of base pairing between donor and recipient strands (horizontal). (PDB codes: 1IN4, 1BDX, 3CRX). *Steps 8–10.* Resolution of the Holliday structure. ▶❙ *microbiology2.com/animations*

chromosome can be recombined as a group into a recipient's genome. In the case of transduction, a piece of DNA from the donor bacterium replaces a segment of DNA in the recipient host. One gene in the segment is said to be cotransduced with another. The closer the two genes are physically on the chromosome, the higher the probability that they will be cotransduced, because the closer two genes are to each other, the less likely a crossover will occur between them.

> **THOUGHT QUESTION 9.5** In a transductional cross between an $A^+B^+C^+$ genotype donor and an $A^-B^-C^-$ genotype recipient, 100 A^+ recombinants were selected. Of those 100, 15% were also B^+, while 75% were C^+. Is gene B or C closer to gene A?

The benefit of recombination to bacteria can be demonstrated using a culture of E. coli that contains a defective lacZ gene. The lacZ$^+$ gene product (beta-galactosidase) is normally used to catabolize the carbohydrate lactose. If lactose is the only carbohydrate available, the lacZ mutant cell culture fails to grow. However, if a transducing phage lysate grown on a lacZ$^+$ strain is added to the lacZ mutant cells, phage particles containing the functional lacZ$^+$ can inject the gene into the mutant (transduction), where recombination systems can exchange it for the defective gene. (Note that the defect may only be a single base change, so there is plenty of homology left for RecA to recognize.) The new recombinant cell has now been converted to lacZ$^+$ and can grow on lactose.

When trying to understand recombination (and gene exchange overall), it is important to remember that the processes are generally random. The lucky bacterium that gets the right piece of DNA and recombines the right segment of that DNA will benefit. This is why a genetic experiment usually requires screening hundreds of millions of cells to find a few recombinants that grow into visible colonies.

Deinococcus Uses RecA to Repair Fragmented Chromosomes

As first noted in Chapter 5, Deinococcus radiodurans is a tough microbe. It can endure 3,000 Gy of gamma irradiation and survive. That is 3,000 times the lethal dose for a human. Although Deinococcus chromosomal DNA is fragmented by irradiation, the organism can reassemble the pieces of DNA in the correct order to re-form the chromosome—a process that is even more amazing considering that the organism possesses two different chromosomes.

The Deinococcus protein that mediates repair is a homolog of RecA, but the sequence of strand exchange is the exact inverse of the established RecA pathway (see previous discussion). Most RecA homologs first form filaments on single-stranded DNA, and then take up homologous double-stranded DNA. The Deinococcus RecA forms a filament on double-stranded DNA and then incorporates a homologous single-stranded molecule. Deinococcus RecA does not need ssDNA to bind. Instead, Deinococcus RecA is thought to bind the double-stranded fragments of irradiated DNA and splice homologous ends together to reconstruct the chromosome. This model assumes that at least two chromosome copies are present per cell so that overlapping fragments are generated during irradiation. Unlike most bacteria, Deinococcus keeps several copies of each chromosome, making it more likely that it will find an intact copy of any given stretch of its sequence. It is a remarkable evolutionary adaptation that has allowed this microbe to inhabit and survive extremely stressful environments, such as the extreme dryness of the desert and the cosmic radiation of the upper atmosphere, which Deinococcus reaches by riding wind currents.

Site-Specific Recombination Is RecA Independent

In contrast to generalized recombination mechanisms that require RecA protein and can recombine any region of the chromosome, site-specific recombination does not utilize RecA and moves only a limited number of genes. This form of recombination involves very short regions of homology between donor and target DNA molecules. Dedicated enzyme systems specifically recognize those sequences and catalyze a crossover between them to produce a cointegrate molecule. The integration of phage lambda, described in Section 9.2, is one example of site-specific recombination involving a 15bp att site and the intergrase enzyme (see **Fig. 9.8**).

Other examples of site-specific recombination are flagellar **phase variation** in the pathogen Salmonella enterica and phase variation in the expression of type I pili in E. coli. In these systems, the sequence of a segment of DNA is inverted (flipped), resulting in on-or-off regulation of adjacent gene expression. Phase variation is frequently employed by pathogens to evade the host immune system by changing the expression of cell surface proteins. For S. enterica, as the immune system starts to make antibodies against one form of flagellar protein, a small subpopulation of bacteria has already switched to making a different form not recognized by the antibody. The mechanism involves a "flippable" **DNA cassette** (a large sequence of DNA that functions as a unit within a genome) that contains a promoter and sits adjacent to a flagellar gene on the chromosome. When the cassette is in one orientation, the promoter is oriented so that the flagellar gene is expressed. When the cassette is in the opposite orientation, the flagellar gene is not expressed. For more on flagellar phase variation, see Section 10.6.

> **NOTE:** A DNA cassette—for example, a kanamycin resistance cassette—can be moved from one place to another, or invert within the genome. In the case of the kanamycin cassette, even the inverted form maintains function. Both in vivo and in vitro mechanisms can be used to move gene cassettes.

TO SUMMARIZE:

- **Recombination** is the process by which DNA sequences can be exchanged between DNA molecules.
- **General recombination** involves large regions of sequence homology between recombining DNA molecules.
- **RecA synaptase** mediates generalized recombination.
- **The extremely radiation-resistant** *Deinococcus radiodurans* is thought to use RecA protein to patch together homologous ends of fragmented DNA in a way that reconstructs the chromosome after extreme radiation damage.
- **Site-specific recombination** requires little homology between donor and recipient DNA molecules. Site-specific recombination enables phage DNA to integrate into bacterial chromosomes and is a process that can turn on or turn off certain genes, as in flagellar phase variation in *Salmonella*.

9.4 Mutations

Any permanent, heritable alteration in a DNA sequence, whether harmful, beneficial, or neutral, is called a **mutation**. There are two basic requirements to produce a heritable mutation: There must be a change in the base sequence, and the cell must fail to repair the change before the next round of replication. Repair mechanisms will be discussed in Section 9.5. Although it is tempting to think that all mutations destroy a protein's activity, this is not the case. A given mutation may or may not affect the informational content or phenotype of the organism.

Mutations come in several different physical and structural forms:

- A **point mutation** is a change in a single nucleotide (**Fig. 9.15A** and **B**). Among point mutations, changing a purine to a different purine or a pyrimidine to a different pyrimidine is called a **transition**, while swapping a purine for a pyrimidine (and vice versa) is a **transversion**.
- **Insertions** and **deletions** involve, respectively, the addition or subtraction of one or more nucleotides (**Fig. 9.15C** and **D**), making the sequence either longer or shorter than it was originally.
- An **inversion** occurs when a fragment of DNA is flipped in orientation relative to DNA on either side (**Fig. 9.15E**).
- A **reversion** occurs when a sequence altered by mutation returns to its original sequence.

Mutations can also be categorized into several informational classes (refer to Table 8.11 for the genetic code). Mutations that do not change the amino acid sequence of a translated open reading frame (ORF) are called **silent mutations**. For example, a point mutation changing

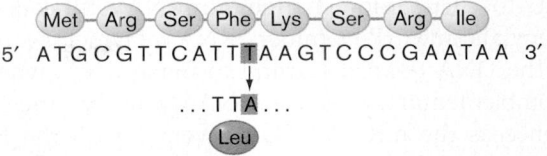

A. Missense point mutation

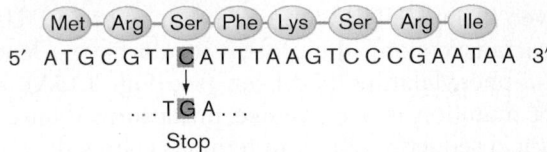

B. Nonsense point mutation

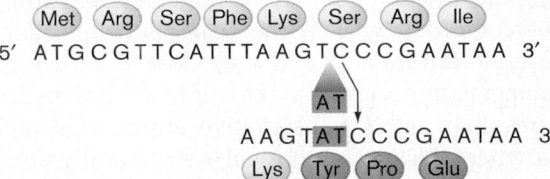

C. Insertion frameshift

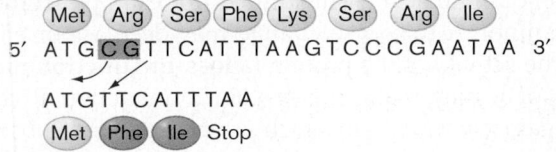

D. Deletion frameshift

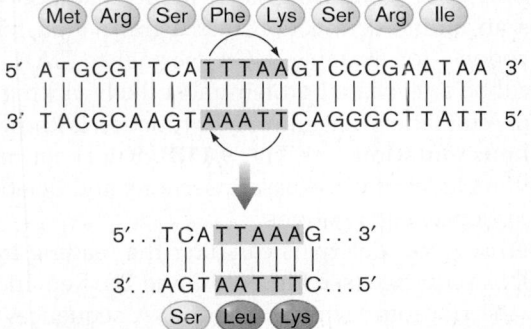

E. Inversion

Figure 9.15   Changes in a DNA sequence that result in different classes of mutations.   A. Missense mutation. The T-to-A transversion is a point mutation that converts the phenylalanine codon TTT to the leucine codon TTA. **B.** Nonsense point mutation. The C-to-G transversion converts the TCA codon, encoding serine, to the TGA stop codon. **C.** Insertion frameshift. The addition of AT into the middle of the TCC serine codon causes a shift in the reading frame that changes all the downstream amino acids. **D.** Deletion frameshift. Removing two nucleotides from the arginine codon also causes a frameshift. **E.** Inversion. Rotating a DNA sequence 180° relative to adjacent sequences changes the amino acids produced.

TTT to TTC in the sense DNA strand (corresponding to a UUU-to-UUC codon change in mRNA) still codes for a phenylalanine. (Remember, RNA polymerase reads from the DNA template strand to make RNA, whereas the complementary, or sense, DNA strand has the same sequence as the mRNA.) Thus, even though the DNA sequence has changed, the protein sequence remains the same. This is also called a synonymous substitution. However, if the UUU codon were changed to a UUA (a U-to-A transversion), the protein would have a leucine where a phenylalanine had been (see **Fig. 9.15A**). This type of mutation is a **missense mutation** because the amino acid sequence of the protein has changed.

The amino acid substitution resulting from a missense mutation may or may not alter protein function. The outcome depends on the structural importance of the original amino acid and how close in structure the replacement amino acid is to the original. Missense mutations result in either conservative amino acid replacements, where the new amino acid is structurally similar to the original (for example, leucine is substituted for isoleucine); or nonconservative replacements, in which a very different amino acid is substituted (for example, tyrosine for alanine). A missense change may decrease or eliminate the activity of the protein (a **loss-of-function mutation**), or it may make the protein more active. It could even gain a new activity, such as an expanded substrate specificity or a completely different substrate specificity (these are called **gain-of-function mutations**).

A **nonsense mutation** is a point mutation that changes an amino acid codon into a translation termination codon—for example, UCA (serine) to UAA. The result will be a truncated protein most likely lacking any function. A mutation that eliminates function is known as a **knockout mutation** (see **Fig. 9.15B**). Knockout mutations can include multiple-base insertions and deletions, as well as nonsense mutations.

Insertions and deletions can alter the reading frame of the DNA sequence (see **Fig. 9.15C** and **D**). Remarkable as it is, the ribosome simply reads RNA sequences one codon at a time, stringing amino acids together in the process. It translates each codon "word" but cannot understand the overall protein "sentence." It does not recognize when bases have been added or removed through mutation; instead, it keeps on reading the sequence in triplets. If the number of bases inserted or deleted is not a multiple of three, the ribosome will read the wrong triplets. The result is a **frameshift mutation**, which produces a garbled protein "sentence." This often causes the ribosome to encounter a premature stop codon, originally in a different reading frame. If the insertion or deletion involves multiples of three bases, the reading frame is not changed but one or more amino acids are added or removed.

An inversion mutation involves the flipping of a DNA sequence (see **Fig. 9.15E**). Imagine the highlighted sequence rotating 180° while the adjacent sequences (in black) remain stationary. The rotation would allow retention of 5'-to-3' polarity in the new molecule; but if the inversion occurred within a gene, it would likely change the codons in the area and alter the resulting protein. What is shown in the figure is a small inversion; however, inversions often involve large tracts of DNA encompassing several genes. If an entire gene with its promoter inverts, the gene will likely remain functional, and its encoded protein may very well still be made. Inversions occur within a genome as a result of recombination events between similar DNA sequences or as a consequence of mobile genetic elements jumping between different areas of a genome.

Mutations can affect both the genotype and the phenotype of an organism. The genotype of an organism reflects its genetic makeup. Regardless of whether a mutation causes a change in a biochemical trait, every mutation causes a change in the genotype (the makeup of the genome). In contrast to genotype, phenotype comprises only observable characteristics, such as biochemical, morphological, or growth traits. For instance, consider the arginine biosynthesis gene *argA*. The product of this gene allows an organism to make enough arginine to support growth in media devoid of arginine. Any mutation in *argA* is a genotypic change, but if the result is a synonomous amino acid substitution, there is no phenotypic change—the organism can still make its own arginine. However, if the mutation destroys the activity of the *argA* product, the microbe can no longer make arginine, and the amino acid must be supplied in the medium. This is a phenotypic change. (Note: A mutant that has lost the ability to synthesize a substance required for growth is called an auxotroph.)

It is also important to realize that small mutations, such as a single base substitution, can have large effects on phenotype. Conversely, large mutations—for instance, the insertion of a 5-kb transposon between two genes—may have little or no effect. To illustrate, consider that a single point mutation in the *hpr* gene of the phosphotransferase sugar transport system (discussed in Section 4.2) will render a bacterium incapable of growing on many sugars, but inserting 5 kb of DNA just past the *hpr* stop codon will have no effect on cell growth. For mutations, it's often not the size that counts; it's the location.

Mutations Arise by Diverse Mechanisms

Although DNA is quite stable, especially compared to mRNA, which is rapidly degraded, it is susceptible to damage inflicted by a variety of physical and chemical agents. For example, irradiation by X-rays can cause a massive number of DNA strand breaks. When this happens, the integrity of the chromosome is lost and the cell dies. Other agents can directly modify the bases in DNA while leaving its overall structure intact. In this case, the modified bases have altered hydrogen bond base-pairing properties that result in the incorporation of an inappropriate base during replication. When that happens, a mutation occurs.

Mutations can be caused by **mutagens**, chemical agents that can damage DNA (**Table 9.2**). Even in the absence of a mutagen, though, mutations arise spontaneously. Because DNA proofreading and repair pathways are so efficient, spontaneous mutations are rare, having a frequency of occurrence ranging from 10^{-6} to 10^{-8} per generation.

Spontaneous mutations in a genome arise for many reasons—for example, tautomeric shifts in the chemical structure of the bases as shown in **Figure 9.16**. Tautomeric shifts involve a change in the bonding properties

Table 9.2 Mutagenic agents and their effects.

Mutagenic agent	Effects
Chemical agent	
Base analog	Substitutes "look-alike" molecule for the normal nitrogenous base during DNA
Examples: caffeine, 5-bromouracil	replication: point mutation
Alkylating agent	Adds an alkyl group, such as methyl group ($-CH_3$), to nitrogenous base, resulting in
Example: nitrosoguanidine	incorrect pairing: point mutation
Deaminating agent	Removes an amino group ($-NH_2$) from a nitrogenous base: point mutation
Examples: nitrous acid, nitrates, nitrites	
Acridine derivative	Inserts (intercalates) into DNA ladder between backbones to form a new rung,
Examples: acridine dyes, quinacrine	distorting the helix: can cause frameshift mutations
Electromagnetic radiation	
Ultraviolet rays	Link adjacent pyrimidines to each other, as in thymine dimer formation, and thereby
	impair replication; lethal if not repaired
X-rays and gamma rays	Ionize and break molecules in cells to form free radicals, which in turn break DNA;
	lethal if not repaired

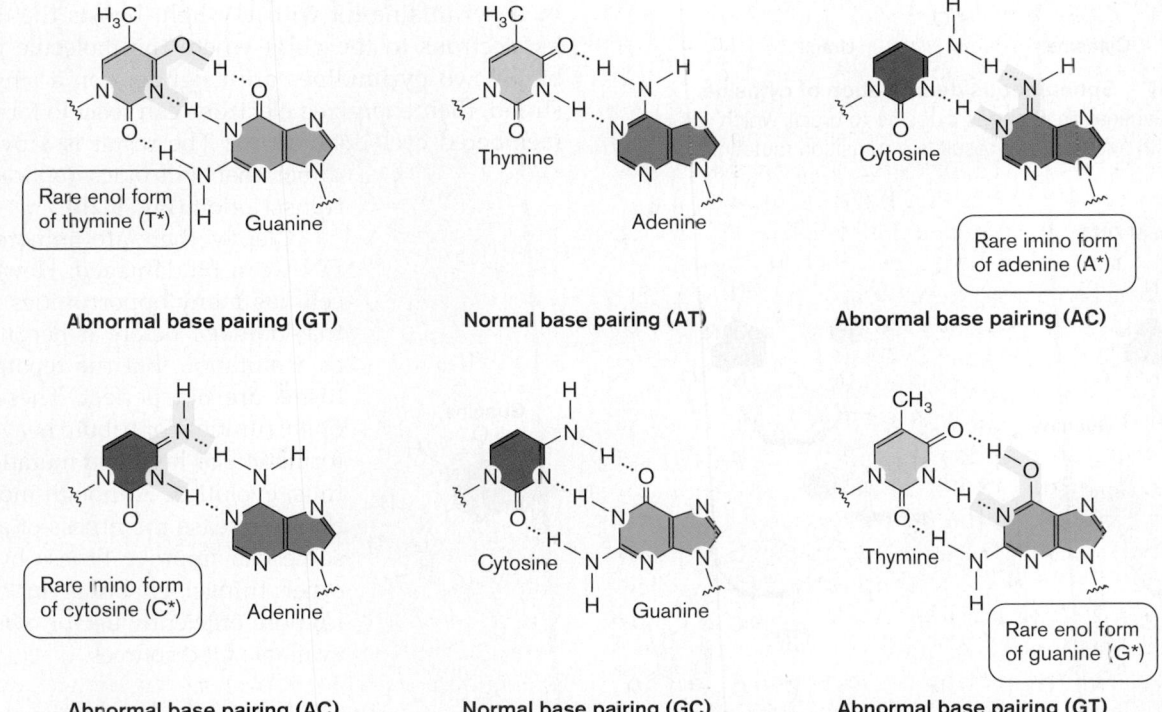

Figure 9.16 Rare tautomeric forms of bases have altered base-pairing properties. Cytosine, which normally base-pairs with guanine, will, in its imino form, pair with adenine. The imino form of adenine will also base-pair with cytosine, rather than with its normal complementary base thymine. Likewise, the enol form of thymine base-pairs with guanine, while the enol form of guanine binds thymine. These tautomeric transitions can lead to permanent mutations.

of amino (−NH₂) and keto (=O) groups. Normally, the amino and keto forms predominate (over 85%), but when an amino group, for example, shifts to an imino (=NH) group, base pairing changes. A cytosine, which normally base-pairs with guanine, will, in its rare imino form, base-pair with adenine. Tautomeric shifts that occur during DNA replication will increase the number of mutational events. Even though the replication apparatus is very accurate with various proofreading and repair functions, such as the 3′-to-5′ exonuclease activity of DNA polymerase III (see DnaQ in Section 7.3), mistakes do occur, at a very low rate.

Besides misincorporation mistakes, naturally occurring intracellular chemical reactions with water can damage DNA. These endogenous reactions are an important source of spontaneous mutations. For example, cytosine deaminates spontaneously to yield uracil (**Fig. 9.17**). The result is a GC-to-AT transition. In addition, purines are particularly susceptible to spontaneous loss from DNA by breakage of the glycosidic bond connecting the base to the sugar backbone (**Fig. 9.18**). The result of this loss is the formation of an **apurinic site** (one missing a purine

base) in the DNA. Lack of a purine would obviously hinder transcription and replication.

DNA can be damaged as a result of metabolic activities of the cell that produce intermediates such as hydrogen peroxide (H₂O₂), superoxide radicals (•O₂⁻), and hydroxyl radicals (•OH), collectively called reactive oxygen species. Even though bacteria have biochemical mechanisms to detoxify reactive oxygen species, sometimes the systems are overwhelmed. Oxidative damage causes the production of thymidine glycol or 8-oxo-7-hydrodeoxyguanosine in DNA (**Fig. 9.19**).

Naturally occurring intracellular methylation agents (for example, S-adenosylmethionine) can spontaneously methylate DNA to produce a variety of altered bases. The spontaneous methylation of the N-7 position of guanine, for example, weakens the glycosidic bond and results in spontaneous loss of the base (apurinic site formed) or opening of the imidazole ring, forming a methylformamide pyrimidine. In addition to mispairing, some of these spontaneous events can lead to major chromosomal rearrangements, such as duplications, inversions, and deletions. Mutagens tend to increase the mutation rate by increasing the number of mistakes in a DNA molecule, as well as by inducing repair pathways that themselves introduce mutations.

Ultraviolet (UV) light produces a striking structural alteration in DNA molecules because pyrimidines are highly susceptible to UV radiation. The energy absorbed by a pyrimidine hit with UV light boosts the energy of its electrons to the point where the molecule is unstable. If two pyrimidines are neighbors on a single DNA strand, their energized electrons can react to form a four-membered cyclobutane ring. The result is a pyrimidine dimer that will block replication and transcription (**Fig. 9.20**).

Clearly, there are numerous ways DNA can be damaged. However, the cell has many opportunities to repair that damage before it becomes fixed as a mutation. But the repair mechanisms are not perfect. These missed opportunities contribute heavily to the formation of heritable mutations and, thus, evolution. Although most mutations decrease the fitness of a species, some can improve fitness by, among other things, enabling an organism to more efficiently use or compete for available food sources.

Mutation Rates and Frequencies

The **mutation rate** for a given gene is defined as the number of mutations

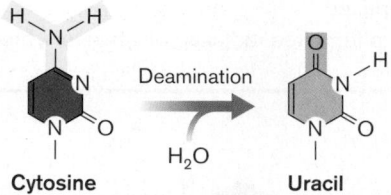

Figure 9.17   Spontaneous deamination of cytosine. Oxidative deamination changes cytosine to uracil, which will base-pair with adenine. The result is a transition mutation.

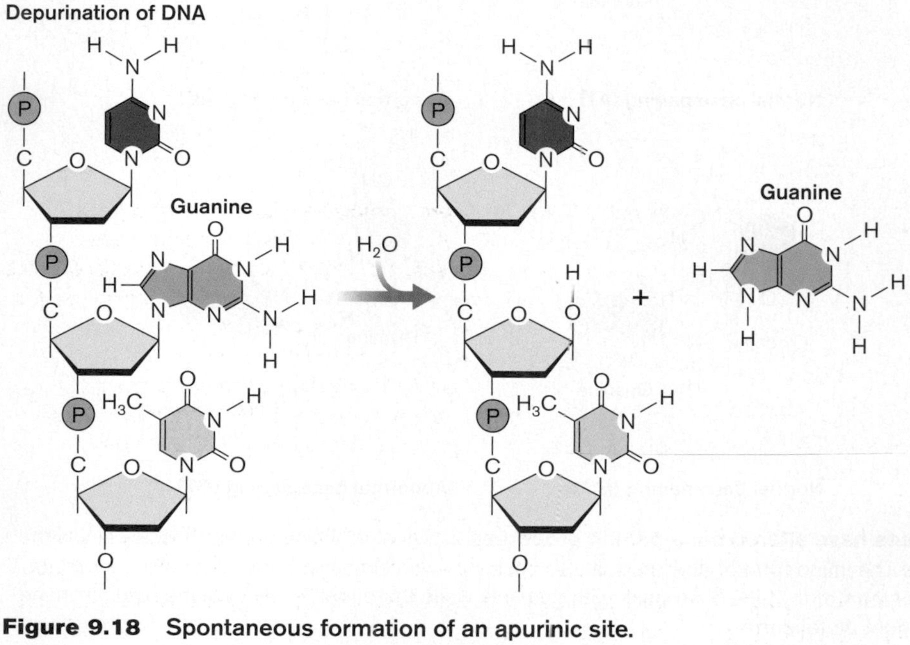

Cytosine   **Deamination**   **Uracil**

Depuration of DNA

Guanine   Guanine

Figure 9.18   Spontaneous formation of an apurinic site.

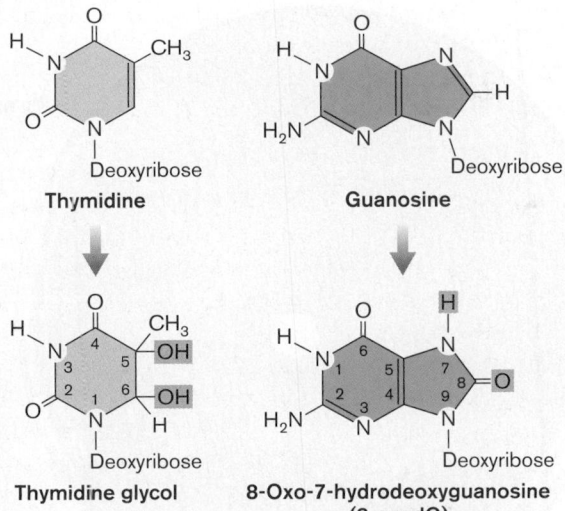

Figure 9.19 Examples of damage caused by reactive oxygen species. Growth in the presence of oxygen leads to the production of reactive oxygen species that can modify nucleotide residues. The modifications can interfere with polymerase function and stop replication or interfere with the transcription of affected genes. The blue-highlighted groups are modifications to thymidine and guanosine residues.

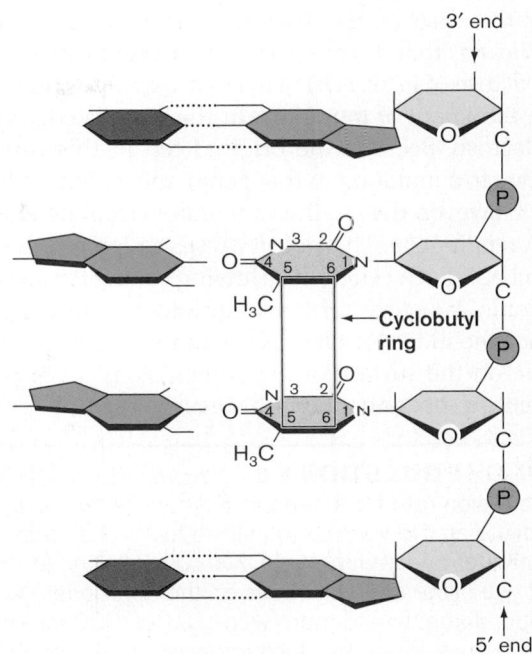

Figure 9.20 Production of a of pyrimidine dimer. The energy from ultraviolet light irradiation can be absorbed by pyrimidine molecules. The excited electrons of carbons 5 and 6 on adjacent pyrimidines can then be shared to form a four-membered cyclobutane ring between adjacent pyrimidines. The pyrimidine dimer acts as a block to replication and transcription.

formed per cell division. Since the number of cells in a typical culture starts out very low and grows to a very high density, the number of cell divisions is approximately equal to the number of cells in the culture. For example, to go from 1 cell to 8 cells requires 7 cell divisions (not generations). Mutation rate can be estimated according to the formula $a = m$ per number of cell divisions, where a is the mutation rate and m is the number of mutations that occur as the number of cells increases from N_0 to N. Thus, a culture growing from 10^3 cells to 10^7 cells requires approximately 10^7 cell divisions ($10,000,000 - 1,000 = 9,999,000$ cell divisions). If 10 mutants developed, then the mutation rate would be approximately $10/10^7 = 1 \times 10^{-6}$.

NOTE: The difference between the number of cell divisions and the number of generations can be appreciated by looking at the replication of 1,000 cells to give 2,000 cells. There are 1,000 cell divisions, but only 1 generation.

A less complicated calculation is **mutation frequency**. Mutation frequency tells us how many mutant cells are present in a population. It is calculated simply as the ratio of mutants per total cells in the population.

The colonies in **Figure 9.21** graphically illustrate the result of a high mutation rate. Single yeast cells placed on an agar surface multiplied to form the colonies shown (**Fig. 9.21A**). As the yeast cell replicates to form a colony, a spontaneous mutant may develop at some point. The

A.

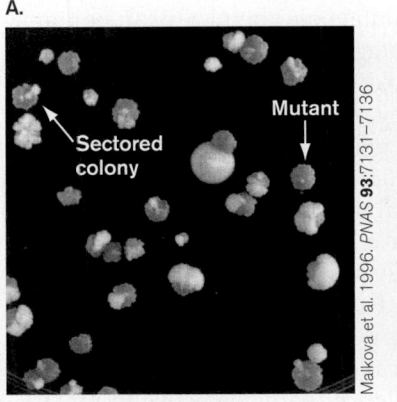

Sectored colony

Mutant

Malkova et al. 1996. *PNAS* **93**:7131–7136

B.

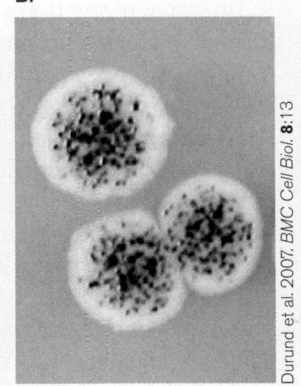

Durund et al. 2007. *BMC Cell Biol.* **8**:13

Figure 9.21 Sectored colonies of bacteria and yeast. A. Sectored colonies of yeast. On this medium, normal yeast cells form white colonies. The occasional mutant cell that develops as the colony grows will replicate as part of the colony but forms a reddish-orange sector. **B.** *E. coli* mutator strain. Cells are lactose-negative mutants that do not ferment lactose and do not turn the indicator blue. But because the strain has a mutation in one of the mutator genes, there is a high rate of reversion to the lactose fermentor strain, which shows up as blue papillae on the white colony.

mutant yeast is unable to metabolize a compound, which then accumulates and turns the cell red. As the mutant cell increases in number alongside normal cells, a sectored colony forms.

In the case of the bacteria, the medium contains an indicator that turns blue if the organism can ferment lactose (**Fig. 9.21B**). This is a useful tool to follow the development of mutations in the gene needed to ferment lactose. Because the original cell in this case was Lac⁻ (due to a mutation in that gene), most of the colony is white. However, the strain is a mutator strain defective in a DNA repair gene. This defect causes a high rate of reversion to Lac⁺. Each cell in the growing colony that reverted to become a lactose fermenter produced offspring that can turn the indicator blue. These mutants appear as blue papillae on the surface of the white colony. Each papilla represents a separate mutational event.

> **THOUGHT QUESTION 9.6** You want to calculate the mutation rate for a gene in *Salmonella enterica*, so you dilute an 18-hour broth culture to 1×10^4 cells per milliliter and let it grow to 1×10^9 cells/ml. At that point you dilute and plate cells on the appropriate agar medium and estimate there were 10,000,000 mutants in the culture. What you don't know is that the original dilution of 1×10^4 cells/ml already had 10 mutants defective in that gene. Does that affect your estimate of mutation rate? How?

Mutagens Can Be Identified Using Bacterial "Guinea Pigs"

In humans, many nonmutagenic chemicals can be transformed into potent mutagens by the liver. The liver is the chief organ for detoxifying the body—a task that its enzymes accomplish by chemically modifying foreign substances. In a world where new chemical entities are constantly being introduced, it is important to determine which ones are potential mutagens. Conventional methods require testing these compounds on large populations of animals. Bruce Ames and his colleagues invented a simple alternative that uses bacteria as a rapid initial screen, for which Ames received the National Medal of Science. The method, called the Ames test, relies on a mutant of *Salmonella enterica* that is defective in the *hisG* gene, whose product is involved in histidine biosynthesis (**Fig. 9.22**). The *hisG* mutant cannot grow on a minimal defined medium lacking histidine. However, if a reversion mutation occurs in the *hisG* gene and restores the gene to its original functional state, the new mutant cell will form a colony even in the absence of histidine. This method is called a reversion test.

Ames used this *his* reversion test to screen compounds for potential mutagenicity. If a chemical is able to mutate one gene (*hisG*), it has the potential to affect any gene. As shown in **Figure 9.22,** a mutagen-containing disk is placed in the middle of an agar plate spread with the original *his* mutant. As the mutagen diffuses into the medium and causes reversion mutations,

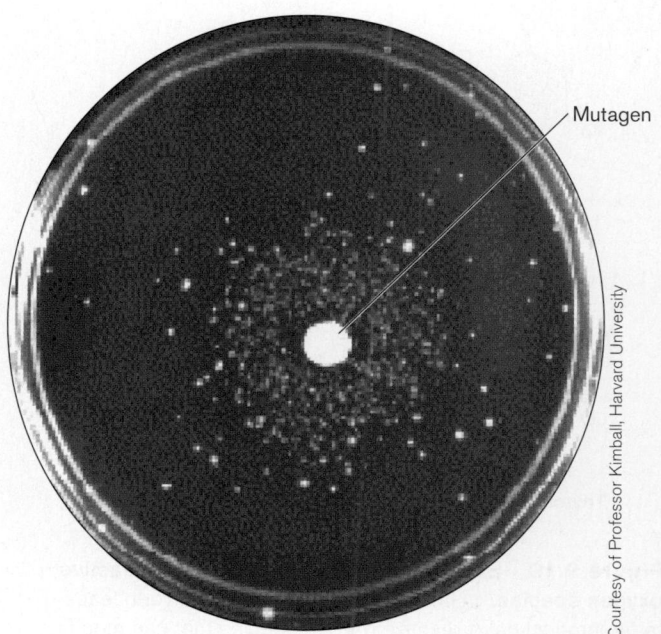

Figure 9.22 Basic Ames test for mutagenesis. A mutation in the *hisG* gene of *Salmonella enterica* produces a histidine auxotroph, which is a strain that requires histidine to grow. When plated onto medium lacking histidine, only a few spontaneous mutations (reversions) occur that reverse the mutation so the cell can make histidine again. These cells form colonies on medium lacking histidine. A mutagen in a paper disk placed in the center of the plate will diffuse and, over time, increase the number of prototrophic revertants seen orbiting the disk.

colonies start to appear, forming a ring around the disk. The longer the plate is incubated, the more revertants are produced.

How can this technique be used to screen for hidden mutagens that require processing by mammalian enzymes? The basic technique was modified by treating a potential mutagen with a rat liver extract before applying the mixture to bacteria for the *his* reversion test (**Fig. 9.23**). In the absence of liver enzymes, these compounds are not mutagenic. If the enzymes of the liver, the main detoxifying organ of the body, convert the compound to a mutagen (they do this accidentally—not on purpose), His⁺ revertant colonies will be seen. This method has helped accelerate the process of drug discovery by providing an inexpensive preliminary screen for weeding out mutagenic chemicals before the more expensive animal testing is undertaken.

The most recent assays for mutagenicity involve transgenic mice. Mice are engineered to have the *lacZ* gene, for instance, inserted into their chromosomes. The *lacZ* gene encodes an enzyme that is easily assayed, so it is useful as a reporter for mutagenic activity. This reporter gene will be distributed throughout the mouse in all organs. A potential mutagen is administered

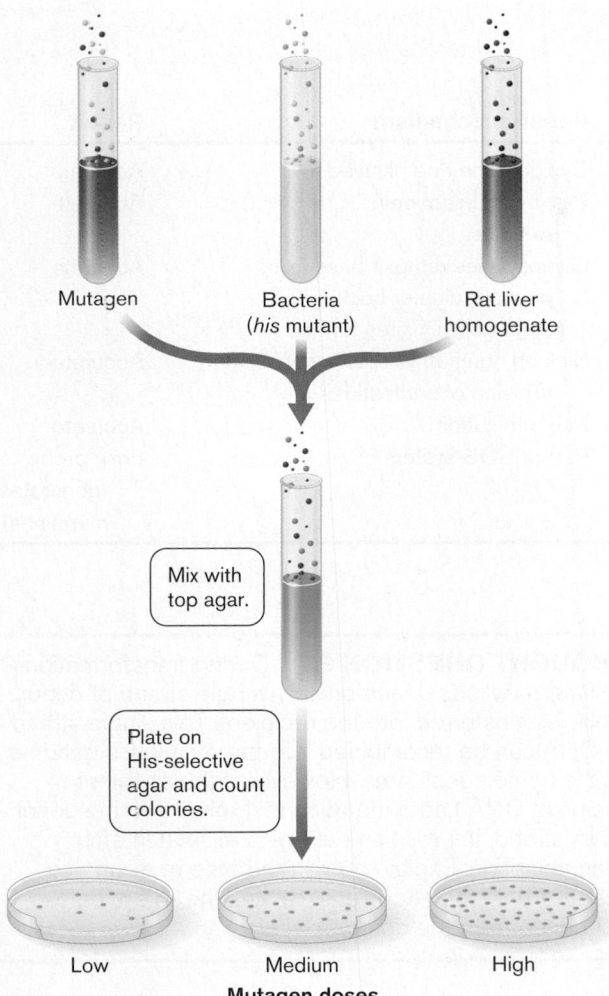

Figure 9.23 The modified Ames assay to test for the mutagenic properties of chemicals processed through the liver. The potential mutagen, *his*-mutant bacteria, and liver homogenate are combined and mixed with agar. The combination is poured into a petri plate. If enzymes from the liver extract modify the test compound and convert it to a mutagen, increasing numbers of His⁺ revertants will be observed with increasing doses of mutagen. If the compound is *not* mutagenic, few relevant colonies will be seen on any plate.

to the mouse, and after some time the mouse is sacrificed, organs are removed, and the DNA is extracted and examined for mutations arising in the *lacZ* gene. This approach allows scientists to track where the mutagen is distributed in the mouse and to determine whether certain organs convert harmless precursor chemicals to dangerous mutagenic compounds.

TO SUMMARIZE:

- **A mutation** is any heritable change in DNA sequence, regardless of whether there is a change in gene function.

- **Genotype** reflects the genetic makeup of an organism, whereas **phenotype** reflects its physical traits.
- **Classes of mutations** include silent mutations, missense mutations, nonsense mutations, point mutations, insertions, deletions, and frameshift mutations.
- **Spontaneous mutations** reflect tautomeric shifts in DNA nucleotides during replication, accidental incorporation of noncomplementary nucleotides during replication, or "natural" levels of chemical or physical (irradiation) mutagens in the environment.
- **Chemical mutagens** can alter purine and pyrimidine structure and change base-pairing properties.
- **Mutagenicity** of a chemical can be assessed by its effect on bacterial cultures.

9.5 DNA Repair

Microorganisms are equipped with a variety of molecular tools that repair DNA damage before the damage becomes a heritable mutation (see **Table 9.3**). The type of repair mechanism used (and when it is used) depends on two things: the type of mutation needing repair and the extent of damage involved. Some mechanisms excise whole fragments of DNA that contain damaged bases; others precisely excise the damaged bases or directly reverse the damage. After extensive damage, special "emergency" DNA polymerases are expressed that sacrifice replication accuracy to rescue the damaged genome. Whether damage is introduced by mutagens or by inaccurate DNA synthesis, microbial survival depends on the ability to repair DNA.

We will first cover the **error-proof repair** pathways that prevent mutations. These include methyl mismatch repair, photoreactivation, nucleotide excision repair, base excision repair, and recombinational repair. We will then discuss **error-prone repair** pathways. These pathways risk introducing mutations and operate only when damage is so severe that the cell has no other choice but to die.

Error-Proof Repair Pathways

Methyl mismatch repair. What happens if DNA polymerase simply makes a mistake and incorporates a normal but incorrect base? Although proofreading functions in DNA polymerases are designed to prevent the misincorporation of bases during replication, misincorporation does occur. Curiously, far fewer mutations arise than one would predict from the inherent error rate of DNA polymerase III (approximately one mistake per 10^6 bases synthesized, after proofreading). However, the mutation rate in a live cell is actually only 10^{-9} to 10^{-10} per base pair replicated. How can a cell repair a mutation after it has been introduced during replication? One method is

Table 9.3 **Basic types of DNA repair.**

System	Genes	Mutations recognized	Repair mechanism	Result
Photoreactivation	*phrB*	Pyrimidine dimers	Cyclobutane ring cleaved	Accurate
Nucleotide excision	*uvrABCD*	Helical destabilization, e.g., pyrimidine dimers	Patch of nucleotides excised	Accurate
Base excision	*fpg, ung, tag, mutY, nfo*	Various modified bases	Glycosylases remove base from phosphodiester backbone; apurinic (AP) sites formed	Accurate
Methyl mismatch	*mutHSL, dam*	Transitions, transversions	Nick on nonmethylated strand, excision of nucleotides	Accurate
Recombination	*recA*	Single-strand gaps	Recombination	Accurate
Translesion bypass synthesis	*umuDC*	Gaps	Part of SOS system	Error-prone (generates mutations)

methyl mismatch repair, which is based on recognition of the methylation pattern in DNA bases. As discussed in Section 7.3, many bacteria tag their parental DNA by methylating it at specific sites. In *E. coli*, for example, deoxyadenosine methylase (Dam) methylates the palindromic sequence GATC to produce GA^{ME}TC. The Dam methylase does this soon, but not immediately, after replication of a DNA sequence.

When DNA polymerase misincorporates a base during replication, a mismatch forms between the incorrect base in the newly synthesized but unmethylated strand and the correct base residing in the parental, methylated strand. Methyl-directed mismatch repair enzymes (MutS, MutL, and MutH) bind to the mismatch. MutS first binds to the mismatch and recruits MutL and MutH. MutL recognizes the methylated strand (GA^{ME}TC) and brings it in a loop to meet MutS and MutH; and then MutH cleaves the unmethylated strand containing the mutation, near the GATC (**Fig. 9.24**⬤). A DNA helicase called UvrD then unwinds the cleaved strand, exposing it to a variety of exonucleases. The result is a gap that is filled in by DNA polymerase I (Pol I) and sealed by DNA ligase.

The methyl-directed mismatch repair proteins (and genes) are called Mut (and *mut*) because a high mutation rate results in strains that are defective in one of these proteins. A bacterial strain with a high mutation rate is called a **mutator strain**.

NOTE: Not all DNA methylations serve as signals for methyl mismatch repair. DNA methylation by restriction-modification systems, for example, prevents cleavage by DNA restriction endonucleases but is not used to designate parental DNA after replication.

THOUGHT QUESTION 9.7 During transformation in *Streptococcus pneumoniae*, a single strand of donor DNA is transferred into the recipient. This single strand of DNA can be recombined into the recipient's genome at the homologous area. However, if the segment of recipient DNA had a mutation in it relative to the donor DNA strand, the result would be a mismatch after recombination. Explain how, in the face of mismatch repair, the recipient cell ever retains the wild-type sequence from the donor.

Photoreactivation and nucleotide excision. Other DNA repair mechanisms do not distinguish between parental and newly synthesized strands. Many microbes in the environment commonly encounter ultraviolet light. The pyrimidine dimers that form as a result of ultraviolet irradiation can be repaired by a light-activated mechanism called **photoreactivation**. In photoreactivation, the enzyme photolyase binds to the dimer and cleaves the cyclobutane ring linking the two adjacent, damaged nucleotides. The damage is repaired without any bases being excised.

While photolyase is specific for pyrimidine dimer repair and requires light, a system called **nucleotide excision repair (NER)** operates in the dark (or light) and is used to excise other kinds of damaged DNA as well as pyrimidine dimers. In this system, a three-subunit endonuclease (UvrABC) excises a patch of 12–13 nucleotides that includes the dimer (**Fig. 9.25**⬤). The basic mechanism involves UvrA, associated with UvrB, recognizing the damaged base (**Fig. 9.25**, steps 1–3). UvrA is ejected (step 4), and UvrB binds UvrC (step 5), which actually cleaves the damaged strand at two sites flanking the damaged base (step 6). The small fragment is removed and the gap repaired by DNA polymerase I (steps 7 and 8).

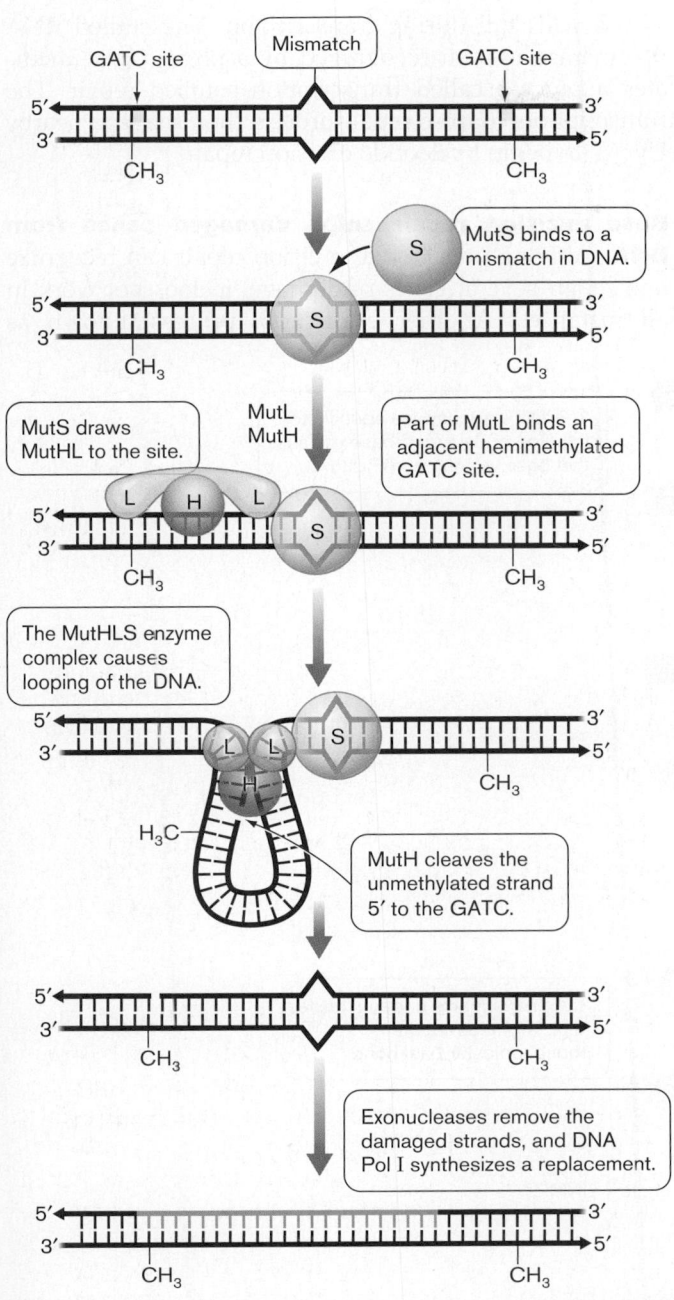

Figure 9.24 Methyl mismatch repair. The bacterial cell, in this case *E. coli*, can use specific methylations on DNA to recognize parental DNA strands for preferential DNA repair. Newly replicated strands are not immediately methylated. So when a mismatch is found, the mismatch repair system views the newly synthesized strand as suspect and replaces the section of unmethylated DNA encompassing the mismatch. The steps following cleavage are similar to what is shown in Fig. 9.25 (steps 7 and 8). ▶❙❙
microbiology2.com/animations

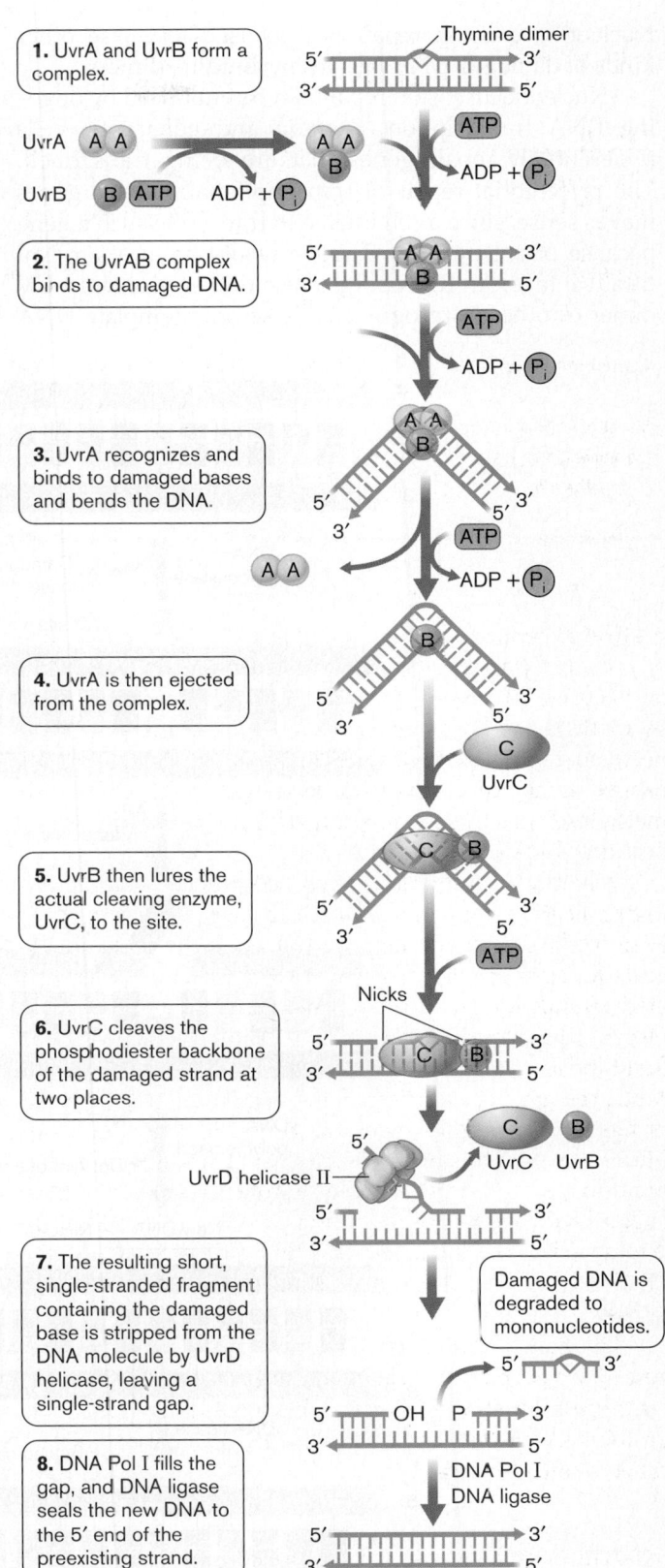

Figure 9.25 Nucleotide excision repair. This repair mechanism cuts out a single-stranded segment of DNA containing the mutation and uses DNA replication to repair the gap. ▶❙❙
microbiology2.com/animations

Nucleotide excision repair is also used to excise other kinds of damaged DNA besides pyrimidine dimers.

Nucleotide excision repair can be enhanced by ongoing RNA transcription. Bacteria and eukaryotic cells preferentially repair genes that are being transcribed. The preferential repair of transcriptionally active genes makes sense, since a cell unable to transcribe such a gene because of recent DNA damage would be at a survival disadvantage. An RNA polymerase that encounters a UV dimer or other unrecognizable base on a template DNA strand will stall during transcription. The stalled RNA polymerase is then recognized by a protein that mediates a process called transcription-coupled repair. The transcription-coupled repair protein then snares a nearby UvrAB to begin nucleotide excision repair.

Base excision repair snips damaged bases from DNA. Although nucleotide excision repair can recognize and repair several types of damage, it does not work in all instances. Another error-proof process, known as

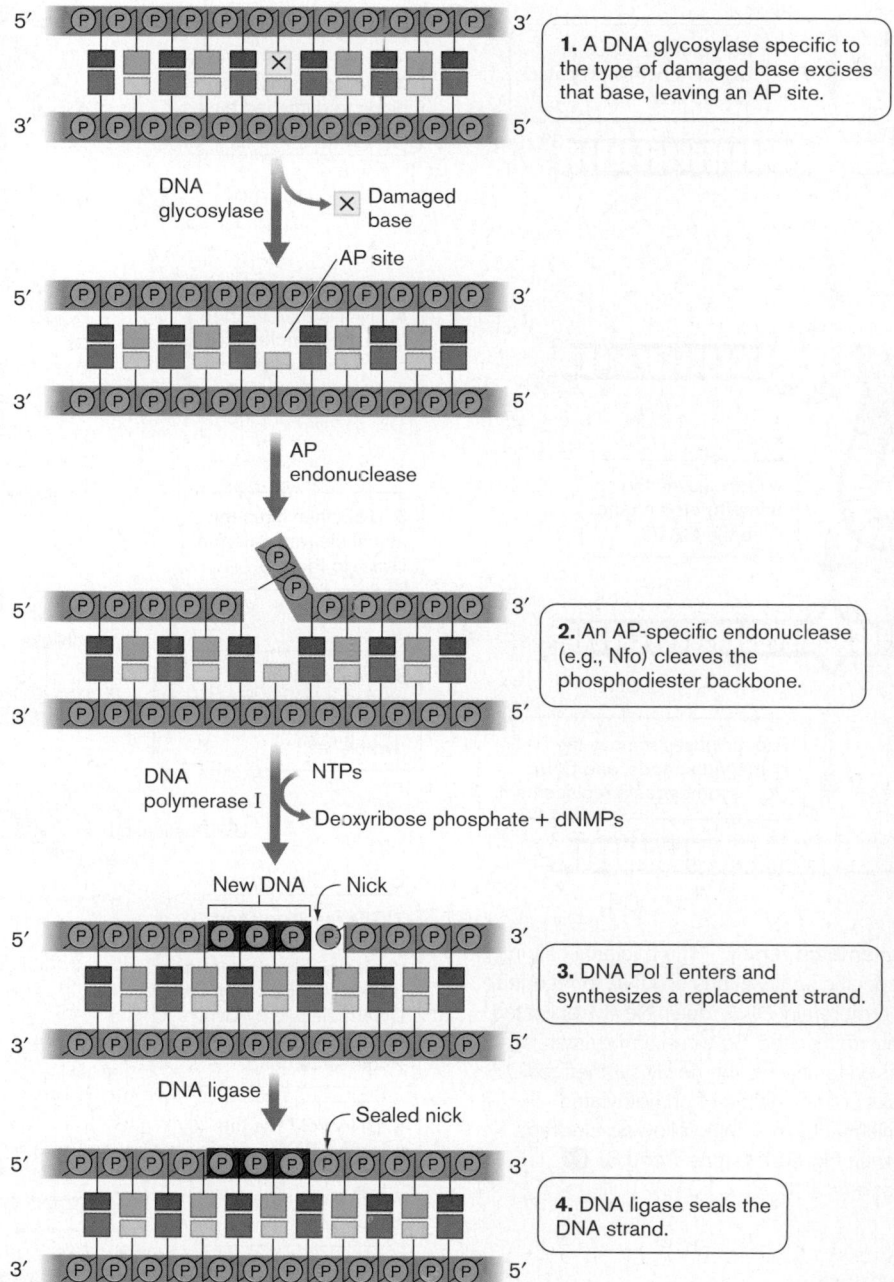

1. A DNA glycosylase specific to the type of damaged base excises that base, leaving an AP site.

2. An AP-specific endonuclease (e.g., Nfo) cleaves the phosphodiester backbone.

3. DNA Pol I enters and synthesizes a replacement strand.

4. DNA ligase seals the DNA strand.

Figure 9.26 Base excision repair. Specialized enzymes can recognize specific damaged bases and remove them from DNA without breaking the phosphodiester backbone. The result is an abasic site (AP; apurinic or apyrimidinic) that can be recognized and cleaved by a specific AP endonuclease. This AP site allows DNA polymerase I to synthesize a replacement strand containing the proper base. ▶❙

microbiology2.com/animations

base excision repair (BER), employs a battery of glycosylase enzymes that can recognize and clip certain damaged bases from the phosphodiester backbone. Uracil DNA glycosylase, hypoxanthine DNA glycosylase, and 3-methyladenine glycosylase recognize, respectively, uracil, hypoxanthine, and 3-methyladenine when these bases are present in DNA. Uracil can be found in DNA either as a spontaneous deamination product of cytosine (see **Fig. 9.17**) or as a result of inappropriate incorporation during replication. Spontaneous deamination of adenine residues produces hypoxanthine. Uracil and hypoxanthine have base-pairing properties very different from those of the original bases, so their formation can lead to mutations. Mutagens such as methyl methanesulfonate can alkylate adenine to produce the adduct 3-methyladenine. A 3-methyladenine residue will totally block DNA replication, making this a lethal form of damage. Consequently, repairing these mutations is vital for survival.

The glycosylases noted above cleave the bond connecting the damaged base to deoxyribose in the phosphodiester backbone (**Fig. 9.26**, step 1). The result is an intact phoshodiester backbone missing a base (an abasic site). The site is called an **AP site** because it is missing a purine (it is apurinic) or a pyrimidine (it is apyrimidinic). The next step in base excision repair involves **AP endonucleases** that specifically cleave the phosphodiester backbone at AP sites (step 2). The 5′-to-3′ exonuclease activity of the gap-filling DNA polymerase I will degrade the cleaved strand downstream of the AP site and at the same time synthesize in its stead a replacement strand containing the proper base (step 3). DNA ligase seals the remaining nick and the repair process is complete (step 4).

THOUGHT QUESTION 9.8 It has been reported that hypermutable bacterial strains are overrepresented in clinical isolates. Out of 500 isolates of *Haemophilus influenzae*, for example, 2%–3% were mutator strains having mutation rates 100–1,000 times higher than lab reference strains. Why might mutator strains be beneficial to pathogens?

All of the repair processes described thus far (mismatch repair, photoreactivation, nucleotide excision repair, and base excision repair) are error-proof pathways that rely on the presence of a good template strand opposite the damaged strand. Replication of the template strand is used to replace the damaged bases accurately. But what happens when both strands are damaged or when there is only one strand?

Recombinational repair. Figure 9.27 illustrates one repair mechanism that occurs when DNA replication takes place before a UV-induced dimer can be excised by nucleotide excision repair. Because DNA polymerase III (the main polymerase for chromosome replication)

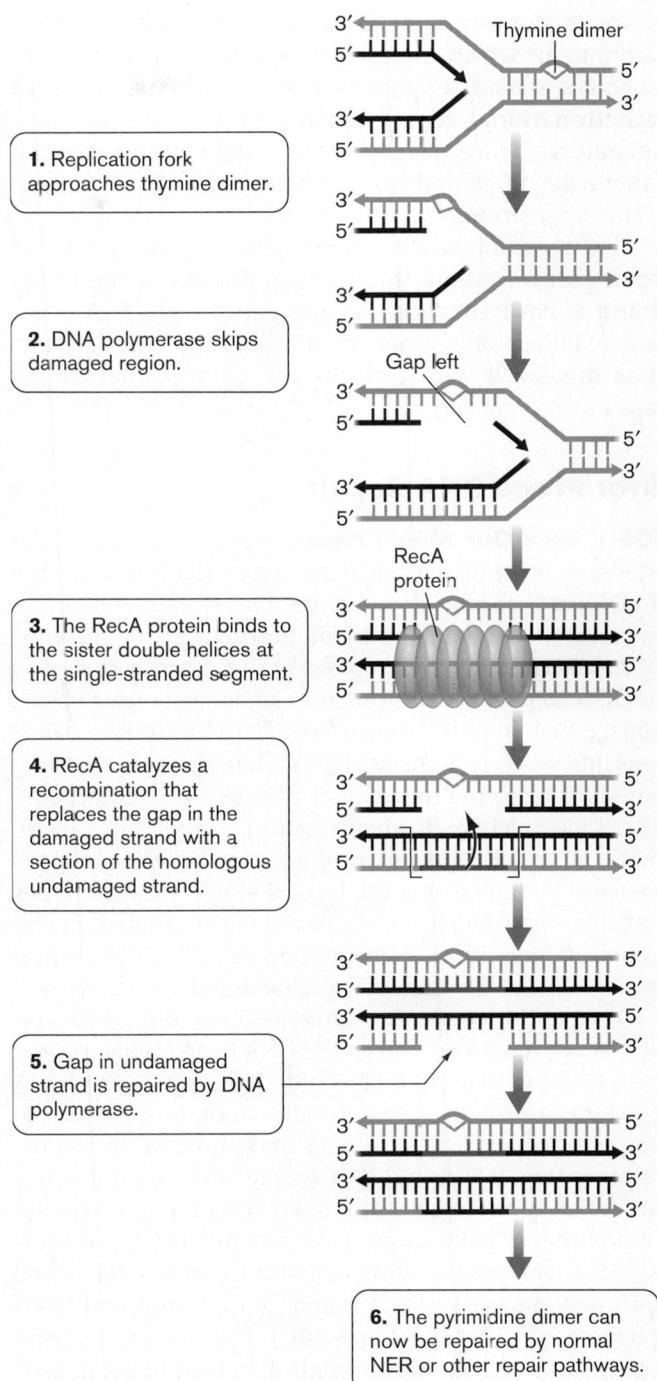

1. Replication fork approaches thymine dimer.

2. DNA polymerase skips damaged region.

3. The RecA protein binds to the sister double helices at the single-stranded segment.

4. RecA catalyzes a recombination that replaces the gap in the damaged strand with a section of the homologous undamaged strand.

5. Gap in undamaged strand is repaired by DNA polymerase.

6. The pyrimidine dimer can now be repaired by normal NER or other repair pathways.

Figure 9.27 Recombinational repair. Recombination can be used to repair DNA damage of replicating DNA when one daughter strand is undamaged. A single-stranded segment of the undamaged daughter strand can be used to replace a gap in the damaged daughter strand.

cannot decode a dimer, it skips over the damaged area and restarts with a new RNA primer, leaving a gap opposite the dimer. At this point, the Uvr complex cannot excise the UV dimer, because chromosome integrity would be lost. However, the RecA protein involved with recombination can bind to the gap and initiate a genetic exchange in which a piece of the undamaged, properly replicated strand is spliced into the gap. This is called **recombinational repair** and mostly occurs near replication forks. Once the gap is filled, the Uvr complex can remove the dimer and DNA polymerase I will fill the gap in the other strand.

The recombinational repair system is an error-proof repair pathway. Note that the gap formed in the donor strand is easily replaced by the gap-filling DNA polymerase I. Recombinational repair is not limited to pyrimidine dimers. It will work on any damage that causes gaps during replication.

Error-Prone DNA Repair

SOS ("Save Our Ship") repair. When DNA damage is extensive, none of the repair strategies that excise pieces of DNA or that rely on one of the two strands remaining undamaged can work without destroying the chromosome. The cell must take more drastic measures known as the **SOS response**, a coordinated cellular response to DNA damage that, in order to save the cell, can introduce mutations into severely damaged DNA. So in this case, "repair" refers to saving the integrity of the circular chromosome even if incorrect bases are introduced. In the SOS system, the RecA protein appears to sense when DNA damage is extensive by monitoring the level of single-stranded DNA available. For example, excessive ultraviolet irradiation will produce numerous ssDNA gaps because DNA polymerase cannot replicate through pyrimidine dimers.

Formation of RecA filaments (see **Fig. 9.13**) on ssDNA activates a second function of RecA, called coprotease activity, that stimulates autodigestion of the LexA repressor, a protein that normally binds to DNA at the promoters of DNA repair genes and prevents their transcription (**Fig. 9.28A** and **B**). Cleavage of LexA unleashes expression of these genes to make DNA repair enzymes. Among these enzymes are two "sloppy" DNA polymerases that lack proofreading activity: UmuDC, also called DNA polymerase V (Pol V); and DinB, also called DNA polymerase IV (Pol IV) (**Fig. 9.28C**). The UmuDC enzyme is perfect for replicating through damaged bases (a process called translesion replication) because it sacrifices accuracy for continuity. When the enzyme encounters an undecipherable damaged base, it will insert whatever nucleotide is available. This, of course, will lead to numerous mutations, but the benefit is that the cell has a chance to live if it can tolerate the mutations. This high mutation frequency is a calculated risk. However, the cell really has no other option, since housekeeping DNA polymerases like Pol III cannot move through damaged areas of chromosomal DNA. The choice really is "mutate or die." (Note that the term "housekeeping" is applied to proteins or enzymes that keep the cell running at all times.)

When the cell is severely compromised by mutation, it temporarily stops dividing. Like a pit stop during an auto race, this pause in cell division allows time for repair enzymes to fix the damage. The product of another SOS-regulated gene, called *sulA*, causes this pause by binding to the FtsZ cell division protein, keeping it from initiating cell division (see Section 3.6). Once the damage has been repaired, RecA coprotease is inactivated and LexA repressor accumulates, turning off all the SOS genes, including *sulA*. The lingering SulA protein is then degraded by a protease called Lon. The protein is called Lon because when it is missing, SulA inappropriately accumulates, inhibits cell division, and produces *long* filamentous cells.

Human homologs of bacterial repair genes. About 30% of *E. coli* genes have human homologs; the functions of the human genes may be similar to those of *E. coli*, or the human genes may have newly acquired functions. These genes often turn out to play key roles in human genetic diseases. An example is *mutS*, of the methyl mismatch DNA repair system. In humans, the ancestor of *mutS* evolved to have multiple repair functions, as well as to participate in postmeiotic segregation. Defects in this repair mechanism have been linked to increased risk of certain colon cancers. Numerous other human genetic diseases are caused by mutations in homologs of bacterial repair genes. For example, defective excision repair genes in humans cause xeroderma pigmentosa, a disease that causes blindness and skin cancers. A deficiency in transcription-coupled repair causes Cockayne syndrome in humans, a devastating neurodegenerative disease.

TO SUMMARIZE:

- **DNA repair pathways** in microorganisms include error-proof and error-prone mechanisms.
- **Methyl mismatch repair** uses methylation of the parental DNA strand to discriminate it from newly replicated DNA. The premise is that the parental strand will contain the proper DNA sequence.
- **Photoreactivation** cleaves the cyclobutane rings of pyrimidine dimers.
- **Nucleotide excision repair (NER)** clips out a patch of single-stranded DNA containing certain types of damaged bases.
- **Base excision repair (BER)** excises structurally altered bases without cleaving the phosphodiester backbone. The resulting AP (apurinic/apyrimidinic) site is targeted by AP nucleases.
- **Recombinational repair** takes place at a replication fork. A "good" strand of DNA is used to replace a homologous damaged strand.
- **Extensive DNA damage leads to induction of the SOS response**, producing increased levels of the error-proof repair systems, as well as error-prone

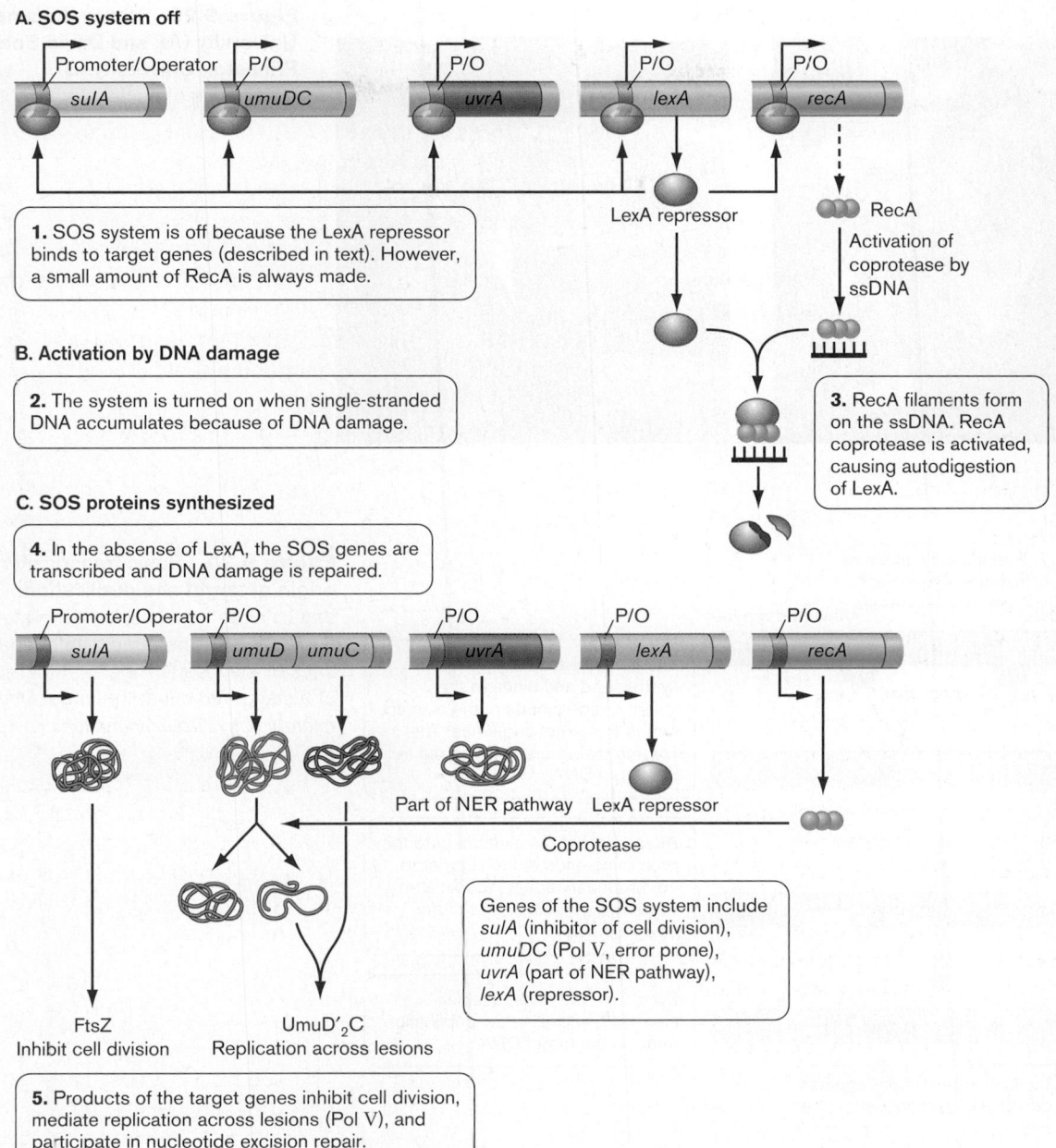

Figure 9.28 Regulation of the SOS response system. The emergency DNA repair system known as the SOS response is induced when there is extensive DNA damage. The system is not a single repair mechanism but a set of different mechanisms that collaborate to rescue the cell.

translesion bypass DNA polymerases that introduce mutations.

9.6 Mobile Genetic Elements

In 1948, Barbara McClintock (1902–1992) noticed that certain genetic traits of corn defied the laws of Mendelian inheritance. The genes encoding these traits, sometimes called "jumping genes," seemed to hop from one chromosome to another. Although McClintock's theories were provocative at the time, we now know that these types of genes, referred to as **transposable elements**, exist in virtually all life-forms and can move both within and between chromosomes.

Since McClintock's work, transposable elements have been studied most widely in bacteria. Nancy Kleckner, Russel Chan, Bik-Kwoon Tye, and David Botstein described the presence of transposable elements in bacteria in 1975 (**Fig. 9.29**). In this classic study, a transposable element carrying a tetracycline (Tc) resistance gene was found inserted into the P22 phage genome.

A.

©Marcus Halevi

B.

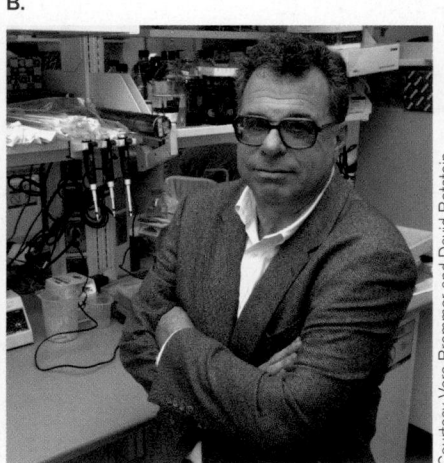

Courtesy Vera Bremmer and David Botstein

Figure 9.29   Nancy Kleckner, Harvard University (A); and David Botstein, Princeton University (B).   Two of the discoverers of bacterial transposition.

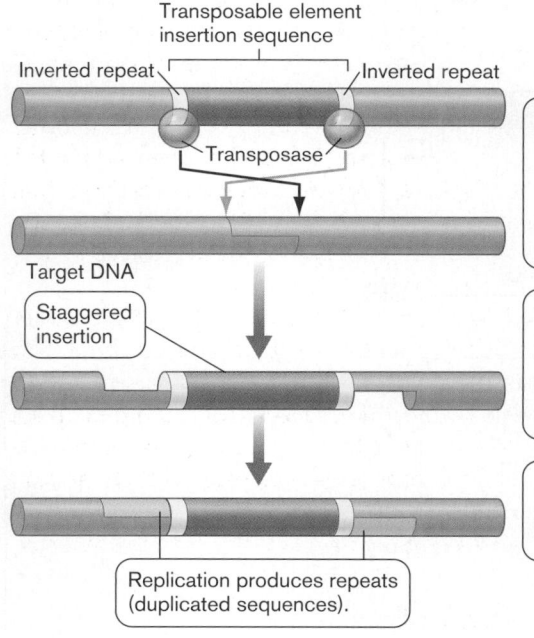

Transposable element insertion sequence

Inverted repeat   Inverted repeat

Transposase

Target DNA

Staggered insertion

Replication produces repeats (duplicated sequences).

Figure 9.30   Basic transposition and origin of target site duplication.   Enzymes that catalyze transposition generate duplications in the target site by ligating the ends of the insertion element to the long ends of a staggered cut at the target DNA site. ▶❚ *microbiology2.com/animations*

The transposase enzyme is synthesized and binds to the inverted-repeat ends of the element and to the target sequence. The enzyme makes a staggered cut in the target DNA.

Attachment of the element onto the protruding ends of the staggered cut produces repeats (duplicated sequences) at either end of the new insertion.

Every time the transposable element "jumps," a new duplication forms in the target DNA.

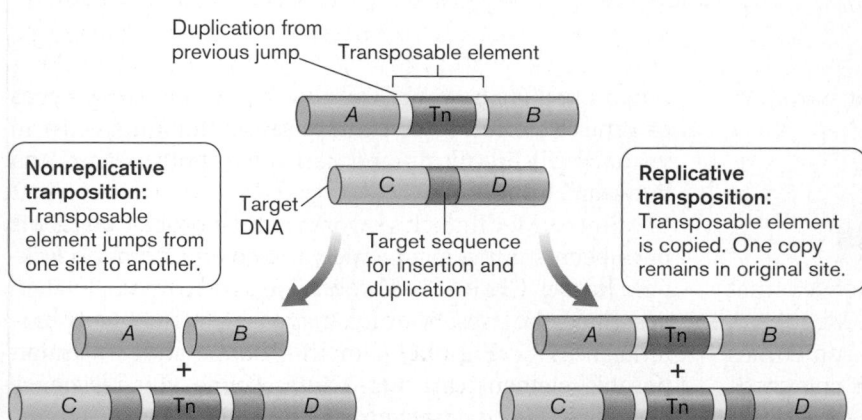

Duplication from previous jump   Transposable element

A   Tn   B

Nonreplicative tranposition:
Transposable element jumps from one site to another.

Target DNA

C   D

Target sequence for insertion and duplication

Replicative transposition:
Transposable element is copied. One copy remains in original site.

A   B
+
C   Tn   D

A   Tn   B
+
C   Tn   D

Figure 9.31   Products of nonreplicative and replicative transposition.   Nonreplicative transposition moves an insertion element from one DNA site to another without leaving a copy of the element at the original site. Replicative transposition leaves the element at the original site and moves a replicated copy to the new site. ▶❚ *microbiology2.com/animations*

P22 is a lysogenic phage that infects *Salmonella*. Recall that a lysogenic phage can integrate into the host chromosome and remain dormant for long periods of time.

The experiment used a defective P22 phage (carrying the Tc resistance gene) that could replicate in only certain strains of *Salmonella*. When this phage was used to infect nonpermissive strains of *Salmonella* (strains that would not allow the phage to grow or lysogenize), the *Salmonella* still became tetracycline resistant at a high frequency. Many of those tetracycline-resistant colonies also developed an auxotrophic requirement for an amino acid, nucleic acid base, or vitamin. The simple explanation was that the tetracycline resistance (Tc^R) element translocated ("jumped") out of the P22 DNA and into one of a number of locations on the *Salmonella* chromosome, sometimes inserting within a structural gene, thereby generating a mutation and phenotypic change. Today we know that these mobile DNA elements have contributed to remodeling of genomes throughout the evolutionary development of all species.

> **NOTE:** The symbol :: is a convention representing insertion. The element appearing after the two colons has been inserted into the element noted in front of the two colons. For example, *proA*::Tn*10* means Tn*10* (a transposable element) is inserted into the *proA* gene involved in proline synthesis.

Transposable elements are not autonomous; unlike plasmids, they are incapable of existing outside of a larger DNA molecule. They exist only as hitchhikers integrated into some other DNA molecule (**Fig. 9.30**●ll). The DNA sequence of a transposable element includes a gene encoding a **transposase**, an enzyme that catalyzes the transfer or copying of the element from one DNA molecule into another. A simple transposable element (700–1,500 bp) called an **insertion sequence (IS)** consists of a transposase gene flanked by short inverted-repeat sequences that are targets of the transposase (see **Fig. 9.30**).

> **NOTE:** An inverted repeat is a DNA sequence identical to another downstream sequence. The repeats have reversed sequences and are separated from each other by intervening sequences:
>
> 5'-AATCGAT . . . . ATCGATT-3'
>
> When no nucleotides intervene between the inverted sequences, it is called a palindrome.

Transposition is the process of moving a transposable element *within* or *between* DNA molecules. During transposition, a short DNA sequence on the target DNA molecule is duplicated so that one copy of the sequence will flank each end of the element (**Fig. 9.30**). Note that the transposase randomly selects one of many possible target sequences where it will move the insertion element.

IS elements transfer by one of two mechanisms: replicative or nonreplicative transposition (**Fig. 9.31**●ll). In nonreplicative transposition, the insertion sequence excises itself out of one host DNA while integrating into the next DNA. In replicative transposition, the sequence copies itself into the new host DNA while a copy remains within the original host.

Nonreplicative Transposition

The nonreplicative model of transposition is shown in **Figure 9.32**●ll. The transposase protein binds to the inverted-repeat ends of the transposable element and to the target DNA, forming a transpososome complex (**Fig. 9.32**, step 1). The transposase cuts the phosphodiester backbone (step 2), severing one strand at one end of the insertion sequence and the other strand at the other end. The 3′ OH ends of the IS element then attack the unnicked strands in a transesterification reaction that produces hairpin structures (step 3). In this instance, joining the ends of the double-stranded molecule produces a single strand with a hairpin.

After the host carrier DNA is ejected from the transpososome, the hairpin ends are renicked and the 3′ OH ends attack the target DNA molecule in a staggered manner (step 4). The element has successfully "jumped" from one molecule to another without replicating. Note that the target DNA segment replicates as a result of the insertion process, producing a copy of the target sequence at each end of the IS element (step 5).

Replicative Transposition

Figure 9.33 shows how a transposable element in a plasmid can bring about cointegration of the plasmid into a target DNA molecule during replicative transposition. As a result of replication of the IS element, the cointegrate molecule will contain two copies of the element flanking the integrated plasmid. Thus, transposition can provide an organism with a whole set of genes that might extend its metabolic capacity, provide drug resistance, or serve as the starting point for gene divergence through mutation. Do not be misled—transposition is not a process designed to help the host; rather, it evolved to ensure survival of the transposon. The benefit to the host is accidental.

Transposons are another type of transposable element. They are more complex than simple insertion elements, since they carry other genes in addition to those required for transposition. They come in different

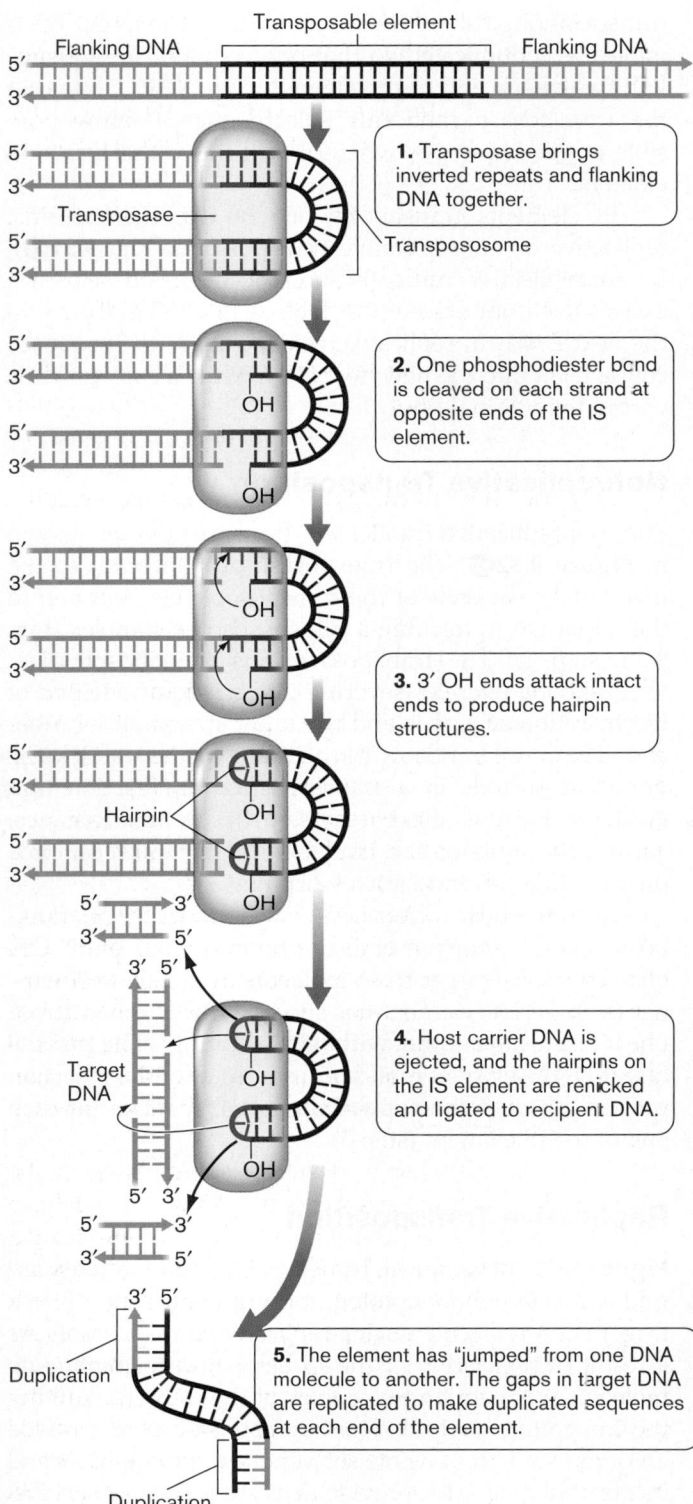

Figure 9.32 Nonreplicative transposition. The transpososome complex includes the transposase binding to the ends of the transposable element and the target DNA. The black DNA segments marked "Duplication" in the last panel correspond to the duplications represented in Figure 9.30. ▶ *microbiology2. com/animations*

The labelled steps in the figure:

1. Transposase brings inverted repeats and flanking DNA together.

2. One phosphodiester bond is cleaved on each strand at opposite ends of the IS element.

3. 3′ OH ends attack intact ends to produce hairpin structures.

4. Host carrier DNA is ejected, and the hairpins on the IS element are renicked and ligated to recipient DNA.

5. The element has "jumped" from one DNA molecule to another. The gaps in target DNA are replicated to make duplicated sequences at each end of the element.

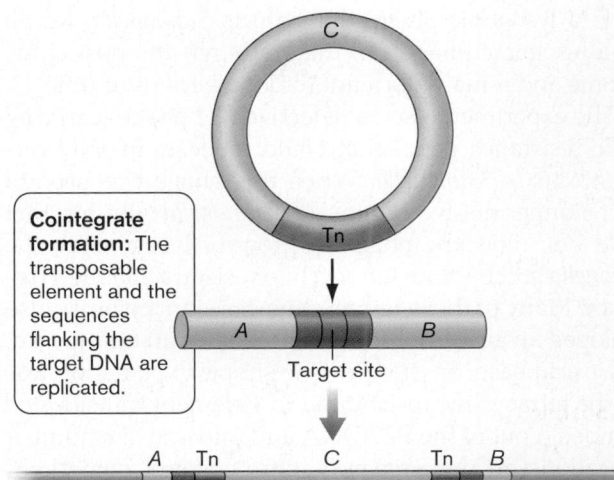

Cointegrate formation: The transposable element and the sequences flanking the target DNA are replicated.

Figure 9.33 Transposition-mediated cointegrate formation between two DNA elements. During replicative transposition, the insertion element is copied and simultaneously ligated to the target site. The result is a transient cointegrate molecule where donor DNA and target DNA become one molecule.

configurations. A **composite transposon** typically consists of two insertion sequences that flank an antibiotic resistance gene, or one or more catabolic genes (for example, genes for benzene catabolism). Often the interior inverted repeats of the IS elements have degenerated, so the transposase primarily acts on the two outermost inverted repeats, causing the whole transposon to move as one unit. Tn*10* is one such example (**Fig. 9.34A**). Some transposons also include regulatory genes, such as a repressor that inhibits transcription of the transposase.

Complex transposons have, as the name suggests, a more complex organization. Tn*3*, for instance, not only possesses transposase and antibiotic resistance genes, but includes a gene whose product, called resolvase, specifically resolves (unlinks) the cointegrate produced by replicative transposition (**Fig. 9.34B**). The transposase acts on the inverted-repeat ends of the element, causing simultaneous insertion and replication. Resolvase is a site-specific recombinase that mediates recombination at the internal Tn*3 res* sites. The result of this recombination is restoration of the two original circular replicons, each one now containing a copy of the transposon.

All of the transposition events mentioned so far involve transfer between DNA molecules within the same cell. Some transposons, called **conjugative transposons**, are able to transfer from one cell to another by conjugation. One example is the transposon Tn*916* of *Enterococcus*, which encodes tetracycline resistance. These conjugative transposons resemble plasmids in that they form a circular intermediate just prior to conjugation, but they do not replicate autonomously. They must be part of a larger self-replicating DNA entity (for example, plasmid, chromosome, or phage). The transfer mechanism of conjugative

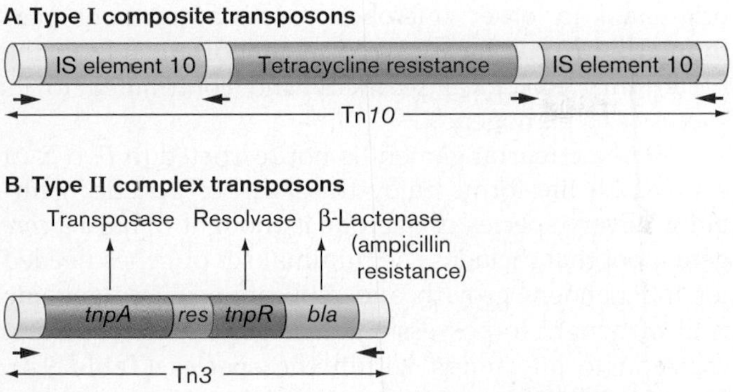

A. Type I composite transposons

| IS element 10 | Tetracycline resistance | IS element 10 |

Tn*10*

B. Type II complex transposons

Transposase Resolvase β-Lactenase
(ampicillin resistance)

| *tnpA* | *res* | *tnpR* | *bla* |

Tn*3*

Figure 9.34 Examples of composite and complex transposons. A. In Tn*10*, the yellow fragments represent inverted repeats and the arrows depict orientation. The internal repeats contain mutations that prevent the individual IS elements from jumping off on their own. **B.** TnpR is a repressor of the Tn*3* *tnpA* transposase gene and is the enzyme that recognizes the *res* site required to resolve cointegrates formed during transposition.

transposons also resembles that of temperate phage in that these elements excise from donor DNA and integrate into the recipient chromosomes. Tn*916* and some other conjugative transposons are designed to transfer when a population is facing danger. For example, in the absence of tetracycline, only a few cells in a population may have Tn*916*. However, the presence of just a small amount of tetracycline triggers conjugative transfer of the element, spreading it throughout the population.

The study of transposons may provide insight into the workings of the human immunodeficiency virus (HIV) that causes AIDS. The structure of transposase from transposon Tn*5* bears a striking similarity to the HIV integrase needed to embed the reverse-transcribed HIV viral DNA into the human genome (see **eTopic 9.1** and Section 11.5). Another type of mobile element, called an integron, has evolved in bacteria to capture DNA cassettes that encode antibiotic resistance genes (**eTopic 9.2**).

THOUGHT QUESTION 9.9 Diagram how a composite transposon like Tn*10* can generate inversions or deletions in target DNA during transposition. *Hint*: This happens when transposition within the chromosome occurs using the inverted repeats closest to the tetracycline resistance gene (see **Fig. 9.34A**). If you draw it right, you will see that in the end, the tetracycline resistance gene is lost.

TO SUMMARIZE:

- **Transposable elements and insertion sequences** ("jumping genes") move from one DNA molecule to another, usually without replicating separately (that is, they are not plasmids).

- **Transposase** is an enzyme that forms a transpososome complex with the transposable element and target DNAs.
- **Insertion sequences** are simple DNA transposable elements containing a transposase gene flanked by short inverted-repeat sequences.
- **Transposable elements move** by nonreplicative or replicative mechanisms.
- **Transposons** are complex transposable elements carrying additional genes (for example, drug resistance).
- **Composite transposons** have two duplicate insertion sequence elements that flank additional genes. Often the interior inverted repeats of these elements contain mutations.
- **Transposons can carry a variety of genes**, including antibiotic resistance genes.

9.7 Genome Evolution

"Nothing makes sense in biology except in the light of evolution," said biologist Theodore Dobzhansky (1900–1975). This statement, made in 1972, rings especially true now that we have access to rapid sequencing of whole genomes. A sense of evolution, of what went before, pervades efforts to interpret the results of the many genome projects. Microbial genomes are dynamic entities that continually evolve. Genes can be removed (deletion), added (insertion), rearranged (recombination), or divided to the point where the genome bears only a slight resemblance to what it once was. Evolution of a genome is a random process driven by natural selection. The process of microbial evolution is discussed in Chapter 17. Here we note the molecular mechanisms that drive genome change.

The result of genome evolution can be seen in the genomes of *Escherichia coli*. Several strains of *E. coli* have been sequenced. The first was the K-12 strain (called MG1655). It contains 4,639,221 base pairs (compared with the 3 billion base pairs of the human genome). Of this genome, 87.8% encodes proteins and 0.8% codes for stable RNAs such as tRNA and rRNA used in translation. DNA with no known function encompasses another 0.7%. Approximately 11% of the chromosome is involved with various forms of regulation. Even though *E. coli* is the best-studied organism on the planet, 28% of the 4,288 open reading frames still remain a mystery to us, having no known or annotated function.

Examining the sequences more closely reveals how the current organism we call *E. coli* came about through the evolution of an ancestral genome. Two basic processes are thought to contribute to genome remodeling: horizontal gene transfers, and duplications followed by functional divergence through mutation.

Horizontal Gene Transfer

Microbial genomes evolve by randomly assembling an eclectic array of genes from many sources. For example, *E. coli* strain O157:H7, the culprit in several fatal outbreaks of food-borne disease in the United States and Europe, contains 1,387 genes lacking in strain K-12. These additional genes represent about 25% of the O157:H7 genome and encode virulence factors, metabolic pathways, and prophages (phage genes integrated into host chromosomes), all of which were acquired from other species. Until 1990, it was thought that bacterial genomes were rather static; one strain of *E. coli* was thought to be very similar to any other strain. But because of genomic comparisons like this, we now know that enterobacteria are subject to a great deal more interspecies recombination than was previously thought.

Evidence for horizontal transfer. More evidence of gene shuffling comes from comparing the proportion of GC base pairs along the chromosome (also known as GC content). On average, the *E. coli* genome consists of 50.8% GC pairs. But 15% of the K-12 genome and 26% of the O157:H7 genome show GC proportions that differ significantly from the rest of the genome and show a different codon usage, suggesting that these genes came from other bacterial lines or species and were acquired by *E. coli* more recently. These horizontal gene transfers, occurring via conjugation, transduction, or transformation, are estimated to happen at the rate of about 16 kb, or 0.4% of the *E. coli* genome, every million years. A sudden change in the GC content within a chromosome is the equivalent of an evolutionary "footprint" marking a horizontal gene transfer.

Horizontal gene transfer appears to be a primary force in microbial evolution. Some estimates place the level of recombination between species to be about 100 times greater than the rate of mutation. In *E. coli*, it has been estimated that nearly 20% of its genome may have originated in other microbes. Gene acquisitions performed by *E. coli* have played an important role in the bacterium's ecological plasticity and contributed to its success as a pathogen.

Genome rearrangement is not restricted to *E. coli*, of course. All life-forms enjoy this form of genetic "gambling." Every species of microbe is thought to have a *core* gene pool that includes the minimal set of genes needed for independent growth and replication. Every genome is also thought to possess a *flexible* gene pool that is not common to all strains within the species (**Table 9.4**). Therefore, differences in genome size between strains are usually thought to reflect variations in the flexible gene pool brought about by loss (deletion) or gain (insertion) of chromosomal DNA.

Genomic islands are large, transferred genetic elements containing numerous genes that serve a common function. For example, a **pathogenicity island** is a genomic island that encodes virulence factors. They are found in bacteria pathogenic for plants and animals, but not in related nonpathogenic strains. *Salmonella enterica* serovar Typhimurium, for example, contains five pathogenicity islands, each with a role specific to a different stage of disease. Pathogenicity islands are often flanked by boundary regions such as direct repeats or insertion elements, usually found near tRNA genes (**Fig. 9.35**). The association of pathogenicity islands with tRNA genes may reflect the fact that some tRNA genes serve as attachment sites for the integration of some phage DNAs or transposons. In addition, some tRNA genes are dispensable, since there are often multiple copies of a given tRNA gene within a genome. The pathogenicity island as a whole is often genetically unstable, subject to further rearrangements or deletions. In addition to virulence genes, pathogenicity islands contain mobility genes encoding transposases, integrases, or other enzymes associated with recombination.

Another kind of genomic island helps a microbe enter into symbiotic relationships. This is called a mobile

Table 9.4	**Composition of bacterial genomes.**		
Core gene pool		**Flexible gene pool**	
DNA elements	**Encoded features**	**DNA elements**	**Encoded features**
Chromosomes and some plasmids	Ribosomes	Genomic islands	Pathogenicity
	Cell envelope	Genomic islets	Antibiotic resistance
	Key metabolic pathways	Bacteriophages	Secretion
	DNA replication	Plasmids	Symbiosis
	Nucleotide turnover	Integrons	Degradation
		Transposons	Secondary metabolism; restriction-modification; transposases/integrases

Source: U. Dobrindt et al. 2002. *Curr. Top. Microbiol. and Immunol.* **264**:157–175.

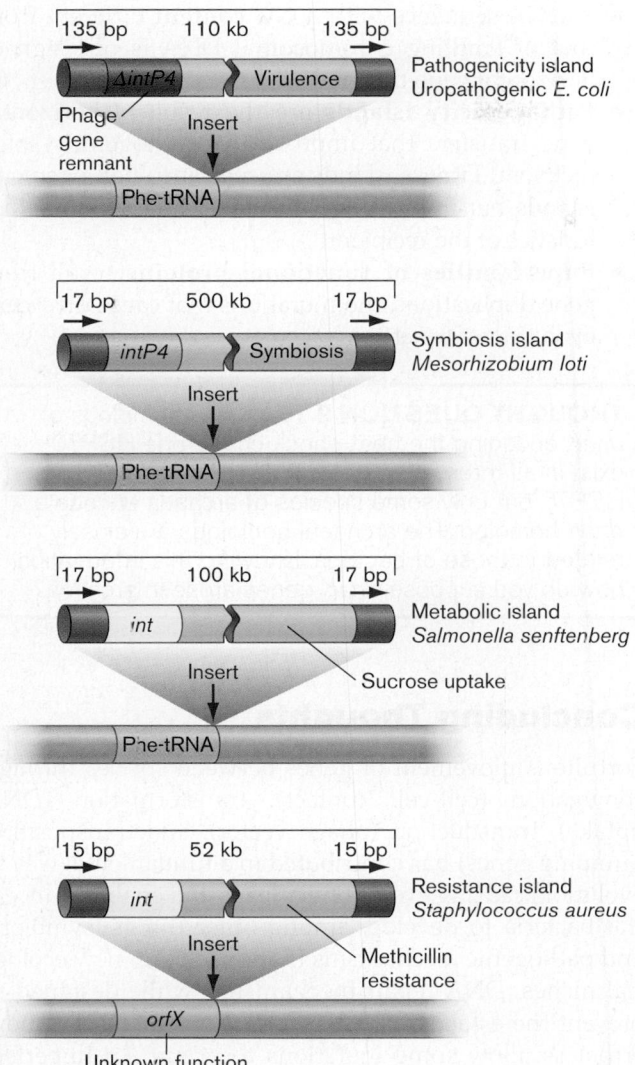

Figure 9.35 Genomic islands from different microorganisms. Pathogenicity islands are often found inserted adjacent to tRNA genes and have direct repeats at their ends (arrows). Functions for each island are identified.

symbiosis island. For example, *Mesorhizobium loti* is a symbiotic bacterium that lives in nodules on the roots of legumes. This root nodule bacterium derives nourishment from the plant in exchange for fixing nitrogen from the atmosphere that the plant can use. Several species of rhizobia carry the genes for nodulation and nitrogen fixation on large plasmids. Others, like *Bradyrhizobium* species, have symbiosis genes as part of their genomes, but clustered in a way that suggests they were once derived from horizontal transfer. These gene clusters appear to have lost whatever mobility they once had. The symbiosis genes of *M. loti,* however, are contained on a mobile symbiosis island. These were the genes horizontally transferred in the Sullivan and Ronson field experiment described in the introduction to this chapter.

Pathogenicity islands and symbiosis islands alike encode remarkable protein secretion systems (called type III and type IV secretion systems) capable of injecting effector proteins directly from the bacterium straight into a eukaryotic cell. Once inside the target cell, these effector proteins alter the function of eukaryotic cell proteins, making the plant, in the case of symbiosis, more accepting of the symbiont. Type III and IV secretion systems are discussed further in Chapter 25.

Other genomic islands are called fitness islands. Fitness islands encode functions that are important to survival in the environment. *E. coli*, for example, possesses an acid fitness island that enables the organism to survive at pH 2, a condition found in the human stomach (see Section 12.1).

Genome Reduction

It is important to note that evolution toward pathogenicity involves gene loss as well as gene acquisition. Large-scale loss of genes through evolution is known as genome reduction. For example, pathogenic shigellae (the cause of bacillary dysentery) exhibit chromosomal "black holes," regions lacking genes that occur in the closely related *E. coli*. Absence of these genes is required for a fully virulent phenotype (why is not clear). Similarly, the genome of the organism that causes whooping cough, *Bordetella pertussis*, is 100 kb smaller than the genome of the less pathogenic *Bordetella bronchiseptica*—another example of evolution by genome reduction. Genome reduction is discussed in Chapter 17.

Sometimes, evidence of genome reduction in the making can be observed. Many genomes contain pseudogenes, genes that by homology appear to encode an enzyme but are nonfunctional because a portion is missing through deletion. These appear to be the remnants of genes that were useful to an ancestral species but made superfluous when one of its descendants adapted to a new environmental niche. When evolutionary pressure to retain a functional form of the gene no longer exists, the imperative to repair mutations within that gene also evaporates. The pseudogene, therefore, is in the process of being eliminated.

Duplications and Divergence

Gene duplication is the most important mechanism for generating new genes and biochemical processes. Duplication frees a gene from its previous functional constraints and allows divergent evolution through mutation. Duplications can arise in several ways, including transposition of a transposable element and recombination after replication between multiple direct or inverted-repeat sequences located in the chromosome. **Figure 9.36** shows a duplication formed by recombination between direct repeats of chromosomal sequences. Superfamilies of

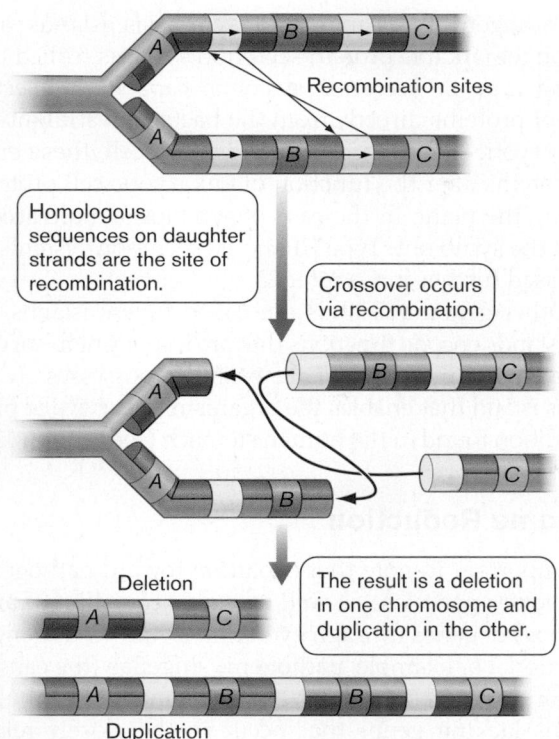

Recombination sites

Homologous sequences on daughter strands are the site of recombination.

Crossover occurs via recombination.

The result is a deletion in one chromosome and duplication in the other.

Deletion

Duplication

Figure 9.36 Duplication during replication.   Replicat ng DNA with three genes: *A*, *B*, and *C*. Yellow segments denote areas with sequence homology. Arrows represent orientation. Crossover lines indicate a recombination event between homologous areas in different places on daughter strands. Red arrows indicate where the strands reconnect after recombination.

proteins arising from divergent evolution share structural and functional features but may catalyze different reactions. One example is the ABC superfamily of ABC transporters. Homology between members of the ABC family may be as little as 10%, suggesting that once an enzyme adopts a new function, the sequence diverges rapidly. Thus, gene duplications and mutation-driven divergence are processes directly responsible for generating orthogous and paralogous genes discussed in Chapter 8.

TO SUMMARIZE:

■ **Horizontal gene transfers** between species are thought to occur by conjugation, transformation, and transduction.

■ A DNA sequence with a **GC content** different from that of flanking chromosomal DNA is one sign of horizontal gene transfer.

■ **Pathogenicity islands** are the result of horizontal gene transfers that improve the pathogenicity of a recipient. Fitness islands are similar to pathogenicity islands but improve environmental survival characteristics of the recipient.

■ **Superfamilies of functional proteins** result from gene duplications and mutations that cause divergent evolution of function.

THOUGHT QUESTION 9.10   Gene homologs of *dnaK* encoding the heat-shock chaperone HSP70 exist in all three domains of life. All bacteria contain HSP70 but only some species of archaea encode a *dnaK* homolog. The archaeal homologs are closely related to those of bacteria. Knowing this information, how do you suppose *dnaK* genes arose in archaea?

Concluding Thoughts

Fortuitous movement of genes between species through conjugation (cell-cell contact), transformation (DNA uptake), transduction (phage vector), and transposition (jumping genes) has contributed in a fundamental way to evolution, causing evolutionary leaps and enabling ancestral bacteria to develop, among other things, symbiotic and pathogenic mechanisms that opened up new ecological niches. DNA repair mechanisms, while designed to prevent the establishment of deleterious mutations, nevertheless allow some mutations to occur. An imperfect DNA repair mechanism offers the cell the opportunity to throw the genetic dice. The vast majority of mutations will be detrimental, but a key few may help the variant strain outcompete its neighbors in its ecosystem. A serious question is raised by our new awareness of genetic mobility: How do we now define a species? For instance, numerous strains of *Helicobacter pylori* show large differences in gene sequence and organization. Should each be called a different species? An arbitrary setting of ribosomal RNA gene divergence is used to assign species, but some wonder whether this is still a valid landmark. For further discussion of microbial evolution, see Chapter 17.

CHAPTER REVIEW

Review Questions

1. What are the basic ways microorganisms exchange DNA?
2. Discuss horizontal versus vertical gene transfer.
3. Describe competence and how it comes about in a population.
4. What is an F factor, and how does it (and other factors like it) contribute to gene exchange?
5. What is a merodiploid? Explain how a cell can become a merodiploid.
6. Compare the differences between the activities of competence factor made by *Streptococcus* and pheromones made by *Enterococcus*.
7. What does microbial gene exchange have to do with the plant disease called crown gall disease?
8. Compare specialized versus generalized transduction.
9. Discuss how bacteria protect themselves from invading bacteriophages.
10. What is the value of recombination to a species?
11. List the major proteins that contribute to recombination, and specify their roles in the process.
12. Define "Holliday structure," "strand assimilation," "cointegrate," and "cotransduction."
13. List and explain the different types of mutations.
14. How are genotype and phenotype different?
15. Describe six different DNA repair mechanisms. Which ones contribute to mutations?
16. Explain the basic process of transposition. Why are insertion sequences always flanked by direct repeats of host DNA? How are transposons different from plasmids?
17. What is a pathogenicity island? A fitness island? What are their characteristics?

Thought Questions

1. Calculate the mutation rate for a gene in which the number of mutations increases from 0 to 500 as cell number increases from 10^1 to 10^8.
2. Why does the competence factor for initiating synthesis and assembly of the transformasome have to be exported out of the cell to bind ComD? Why doesn't the molecule bind internally?
3. Rapidly growing bacteria can initiate a second round of replication before a previous round is completed. Would such rapid growth introduce more mutations due to replication errors than slower growth by the same species would?
4. *Agrobacterium tumefaciens* Ti plasmid has been used as a tool to genetically modify plants. Why would a plant biologist use a tumor-causing plasmid to breed new plants? Wouldn't the genetically altered plant develop a tumor?
5. Though we can identify sets of genes that have been horizontally transferred from one species of microbe to another, we rarely can identify the source species. What might account for this failure?
6. You have just isolated a new temperate bacteriophage for *Salmonella enterica*. How can you determine whether this phage mediates generalized or specialized transduction?

Key Terms

AP endonuclease (329)
AP site (329)
apurinic site (322)
base excision repair (BER) (329)
cointegrate (315)
competence factor (303)

competent (303)
complex transposon (334)
composite transposon (334)
conjugative transposon (334)
deletion (319)
DNA cassette (318)

electroporation (303)
episome (306)
error-prone repair (325)
error-proof repair (325)
F⁻ cell (306)
F⁺ cell (306)

Recommended Reading

Babic, Ana, Ariel B. Lindner, Marin Vulic, Eric J. Stewart, and Miroslav Radman. 2008. Direct visualization of horizontal gene transfer. *Science* **319**:1533–1536.

Bridges, Bryn A. 2005. Error-prone DNA repair and translesion synthesis: Focus on the replication fork. *DNA Repair* **4**:618–619.

Chen, Inês, Peter J. Christie, and David Dubnau. 2005. The ins and outs of DNA transfer in bacteria. *Science* **310**:1456–1460.

Claverys, Jean-Pierre, Bernard Martin, and Patrice Polard. 2009. The genetic transformation machinery: Composition, localization, and mechanism. *FEMS Microbiology Reviews* **3**:643–656.

Doyle, Marie, Maria Fookes, Al Ivens, Michael W. Mangan, John Wain, et al. 2007. An H-NS-like stealth protein aids horizontal DNA transmission in bacteria. *Science* **5**:654–666.

Fraser, Christophe, Eric J. Alm, Martin F. Polz, Brian G. Spratt, and William P. Hanage. 2009. The bacterial species challenge: Making sense of genetic and ecological diversity. *Science* **323**:741–746.

Frost, Laura S., Raphael Leplae, Anne O. Summers, and Ariane Toussaint. 2005. Mobile genetic elements: The agents of open source evolution. *Nature Reviews. Microbiology* **3**:722–732.

Hotopp, Julie C. Dunning, Michael E. Clark, Deodoro C. S. G. Oliveira, Jeremy M. Foster, Peter Fischer, et al. 2007. Widespread lateral gene transfer from intracellular bacteria to multicellular eukaryotes. *Science* **317**:1753–1756.

Lim, Yin Mei, Ad J. Groof, Mrinal K. Bhattacharjee, David H. Figurski, and Eric A. Schon. 2008. Bacterial conjugation in the cytoplasm of mouse cells. *Infection and Immunity* **76**:5110–5119.

Medini, Duccio, Claudio Donati, Hervé Tettelin, Vega Masignani, and Rino Rappuoli. 2005. The microbial pan-genome. *Current Opinion in Genetics and Development* **15**:589–594.

Thomas, Christopher, and Kaare M. Neilsen. 2005. Mechanisms of, and barriers to, horizontal gene transfer between bacteria. *Nature Reviews. Microbiology* **3**:711–721.

Tippin, Brigette, Phuong Pham, and Myron F. Goodman. 2004. Error-prone replication for better or worse. *Trends in Microbiology* **12**:288–295.

Truglio, James J., Deborah L. Croteau, Bennett Van Houten, and Caroline Kisker. 2006. Prokaryotic nucleotide excision repair: The UvrABC system. *Chemical Reviews* **106**:233–252.

For further study and review, please visit StudySpace at **microbiology2.com**.

Chapter 10

Molecular Regulation

A bacterial genome encodes thousands of different proteins needed to handle many different environmental contingencies. Some proteins, such as RNA polymerase, are always required for growth. Most proteins, however, are needed only under a limited set of conditions. Proteins that degrade the sugar lactose, for example, are useful only when lactose is present. Likewise, *Vibrio cholerae* needs to make cholera toxin only when it's growing inside the human body, not when it's living in the ocean. To compete successfully with others, the microbe will not waste energy making unneeded proteins. Cells achieve molecular efficiency by using elegant control systems that selectively increase or decrease gene transcription, mRNA translation, or mRNA degradation, as well as by degrading or sequestering regulatory proteins. But how does a cell "know" when to alter expression?

Chapter 10 describes the variety of mechanisms bacteria use to sense the environment and adjust gene expression.

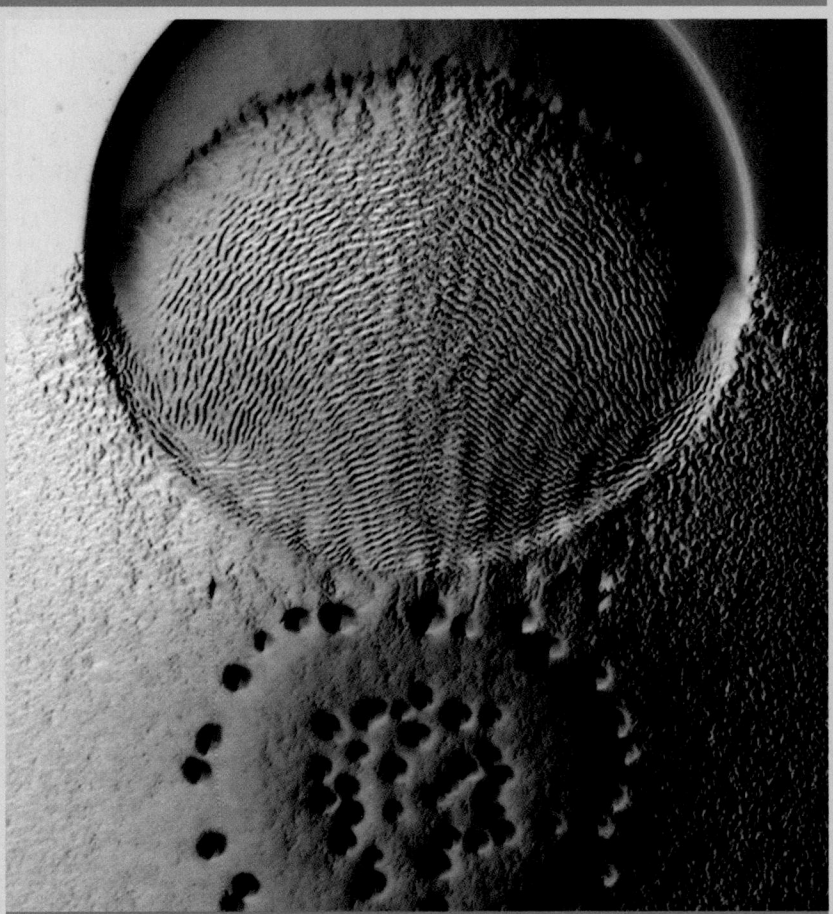

Current Research Highlight

Myxococcus uses chemotaxis to hunt for prey. Chemotaxis is the process by which bacteria sense and move toward favorable (or away from unfavorable) conditions. The process is highly regulated by sensory transduction mechanisms. In this image, the social bacterium *Myxococcus xanthus* (large lower colony) invades the upper *Escherichia coli* colony (diameter approx. 8 mm) and devours the cells. Movement of *Myxococcus* is regulated by a chemosensory signaling process that produces multicellular waves or "ripples" as the predator *Myxococcus* moves throughout the upper *E. coli* prey colony. The regulation of chemotaxis is similar in *Myxococcus* and *E. coli*. However, what attracts *Myxococcus* to the prey cells is unknown. (Small dots in the *Myxococcus* are fruiting bodies that contain spores.) Source: James E. Berleman et al. 2008. *PNAS* **105**:17127–17132 (cover image). Image courtesy of James E. Berleman and John R. Kirby. © 2008 by The National Academy of Sciences of the USA.

During the day, the bobtailed squid, *Euprymna scolopes*, remains buried in the sand of shallow reef flats around Hawaii. After sunset, the animal emerges from its hiding place and begins its search for food. As it swims in the moonlit night, its light organ projects light downward in an apparent attempt to camouflage the squid from predatory fish swimming below. Looking up, the fish do not see the squid, but instead see what appear to be stars. What does this have to do with microbiology? Inside the light organ are luminescent bacteria called *Vibrio fischeri*. The bacteria, not the squid, produce the light. However, these microbes do not *constantly* glow. They light up only when their cell density rises above a certain level, at which time the genes needed to make light are "turned on." This intricate level of genetic control by such a small, seemingly simple bacterium is truly amazing.

Microbes use numerous mechanisms to sense their internal and external environments and then translate that information into action. The cell's surface, for example, contains an array of sensing proteins that monitor osmolarity, pH, temperature, and the chemical content of the surroundings or that detect the presence of a host or competitor. Quorum sensing, in which bacteria secrete and sense chemical signal molecules, make it possible for members of microbial communities to communicate and cooperate.

All these mechanisms orchestrate the synthesis of waves of proteins whose concentrations change with changing environments. Even virulence genes have tapped into these sensing systems, using them to determine whether the microbe has entered a susceptible host. Chapter 10 explores the fundamental principles of gene regulation in microbes and discusses how individual systems are woven into a global regulatory network that interconnects many processes throughout the cell.

10.1 Regulating Gene Expression

A cell needs to monitor two different compartments in order to know when to alter gene expression and adjust

its physiology: the cytoplasm within the cell and the environment outside. Intracellularly, the concentrations of vitamins, amino acids, and nucleotides must be sufficient

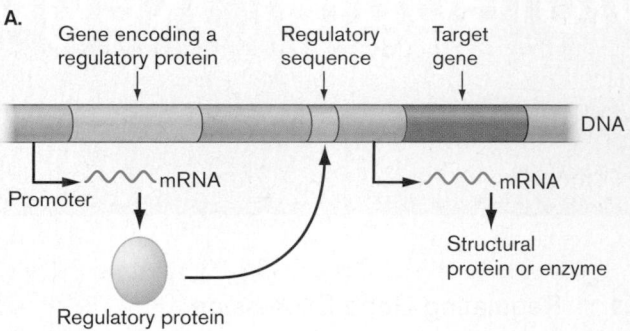

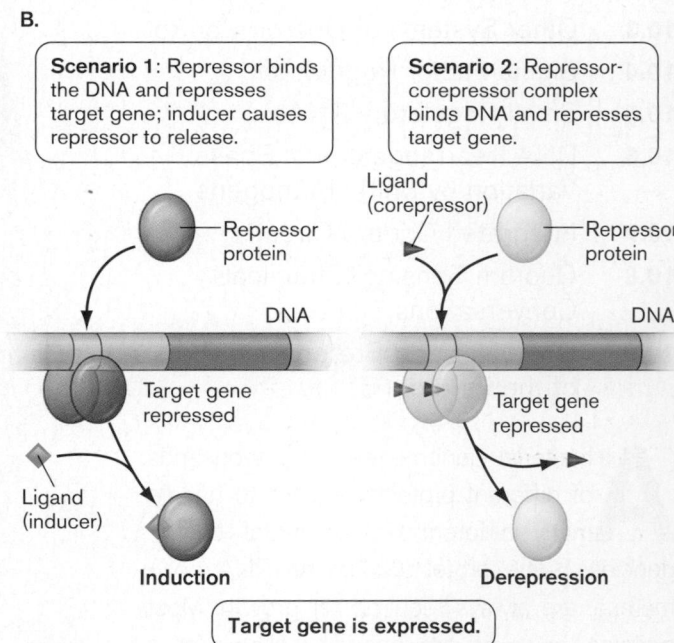

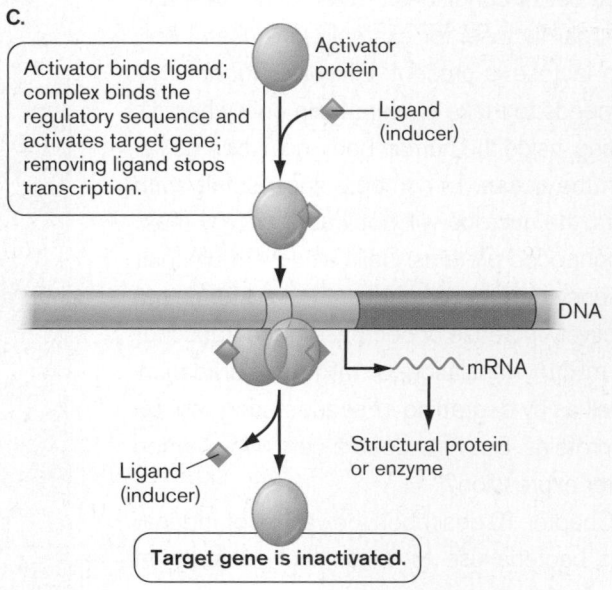

Figure 10.1   General aspects of transcriptional regulation by repressor and activator proteins.   A. Schematic drawing of a regulatory system. The regulatory gene may be located near or quite far away from the target gene on the chromosome. When it is located nearby, it is often, but not always, transcribed separately from the target gene. The product of the regulatory gene (the regulatory protein) binds to DNA sequences near the promoter of the target gene and controls whether transcription occurs. **B.** Repressor proteins bind to DNA sequences and prevent transcription. One type of repressor (scenario 1) releases from the target DNA sequence *after* binding a chemical ligand (called an inducer) that is specific for that repressor. The target gene is thereby induced. The second type of repressor (scenario 2) must bind a chemical ligand, called a corepressor, *before* it binds to DNA. **C.** Activator proteins generally bind to specific chemical ligands in the cytoplasm before the protein can bind to DNA sequences near target genes.

and balanced to supply the biosynthetic and energetic needs of the cell. To achieve balance, the cell must control de novo (new) synthesis of these compounds and "know" what carbon and energy sources are present in order to assemble the proper catabolic pathway.

The microbe also needs to know whether conditions outside the cell are hazardous, and it needs to distinguish, from the combination of chemicals present, whether it is floating in a pond of water, a gastrointestinal tract, or a mammalian host cell. Once the environment is sensed, the cell can change the repertoire of genes it expresses to meet its needs or to protect itself.

Cells use different mechanisms to sense and respond to conditions within and outside the cell membrane. Sensing conditions within the cell is relatively straightforward. **Regulatory proteins** bind specific small-molecular-weight compounds called ligands to determine their concentrations (**Fig. 10.1A**). Different regulatory proteins bind different ligands. For example, one regulator will bind a carbohydrate ligand and alert the cell that a new potential carbon source is available, while a different regulator will sense if there is enough of the amino acid tryptophan present in the cytoplasm to carry out protein synthesis. The ligand, once bound, then alters the ability of the regulatory protein to latch onto specific DNA regulatory sequences located near the promoters of target genes. DNA regulatory sequences are sometimes called operator sequences if binding downregulates expression of the target genes, or activator sequences if binding increases expression.

General Concepts of Transcriptional Control by Regulatory Proteins

Genes encoding regulatory proteins are usually, but not always, transcribed separately from the target gene (**Fig. 10.1A**). Regulatory proteins come in two forms: **repressors** and **activators** (not to be confused with activator *sequences*). Repressor proteins bind to regulator sequences and prevent the transcription of target genes, an event known as **repression**. However, repression happens in one of two ways, depending on the repressor. Some repressors bind DNA sequences by themselves and prevent transcription (**Fig. 10.1B**, scenario 1). Relief from repression requires that a specific ligand, called an **inducer**, bind to the repressor protein, causing it to release from the DNA sequence. Because a small inducer molecule is required, the increased expression of the target gene is called **induction**. The lactose operon, discussed later, is one example of an inducible system.

Other repressor proteins do not bind well to DNA regulatory sequences unless they first bind a small ligand called a **corepressor** (**Fig. 10.1B**, scenario 2). As the ligand disappears from the cell, it is no longer available to bind to the repressor protein. When this happens,

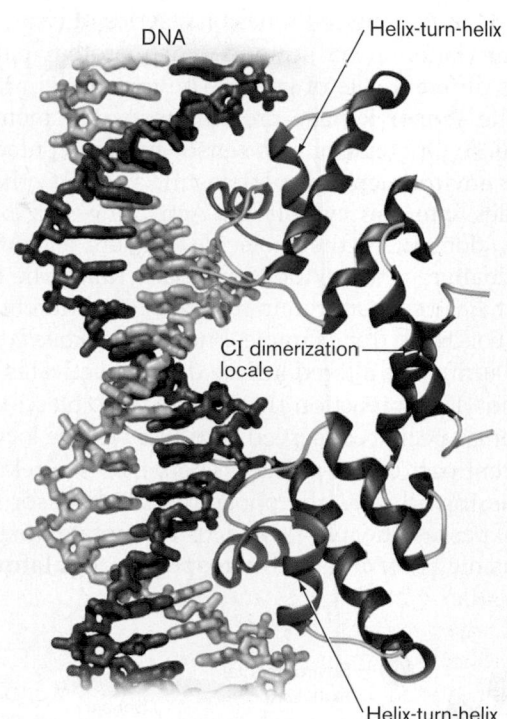

Figure 10.2 Binding of a repressor protein to DNA. Dimer of lambda CI repressor binding to DNA. Note the helix-turn-helix motif located in two successive major grooves. Helices are gold and purple; the turn is green. (PDB code: 1LMB)

the repressor releases from the DNA and the target gene is expressed. This process is called **derepression** rather than induction. The tryptophan operon, also discussed later, is an example of a repression/derepression system.

Activator proteins, on the other hand, generally bind poorly to DNA sequences unless an inducer is present (**Fig. 10.1C**). Most activator proteins, once bound to DNA, directly contact an RNA polymerase molecule stuck at the nearby promoter, spurring it to initiate transcription. When the intracellular concentration of inducer falls, the activator protein (without inducer) either leaves the DNA or moves to a nearby site from which it can no longer contact RNA polymerase. Thus, transcription of that target gene stops.

Figure 10.2 illustrates how one repressor protein, CI from lambda phage, binds to DNA. The protein forms a dimer, and one part of each molecule, called the DNA-binding domain, interacts with DNA in the major groove.

Sensing the Extracellular Environment

Sensing what goes on outside the cell is more challenging because intracellular regulatory proteins cannot reach through the membrane and touch what is outside. A common mechanism used by Gram-positive and Gram-negative organisms to transmit information from outside to inside the cell relies on a series of two-protein phosphorylation relay systems. Each protein pair will activate different sets of genes. The first protein in each relay, the **sensor kinase** protein, spans the membrane (**Fig. 10.3**). One end of this sensor protein contacts the outside environment (or periplasm), and the other end protrudes into the cytoplasm. Activating the external sensory domain of the molecule triggers a conformational change in the cytoplasmic part, called the kinase domain. Each sensor protein recognizes a different molecule or condition (for example, PhoQ in *Salmonella* senses magnesium). The altered kinase domain activates a self-phosphorylation reaction that uses ATP to place a phosphate on a specific conserved histidine residue located in a different part of the protein. Then, like two relay runners passing a baton, the phosphorylated sensor kinase protein passes the phosphate to a cognate (matched) cytoplasmic protein called a **response regulator**. This

transfer, called transphosphorylation, occurs at a specific aspartate residue within the response regulator. The phosphorylated response regulator commonly binds to regulatory DNA sequences in front of one or more specific genes and activates or represses expression. Regulatory relays of this type are called **two-component signal transduction systems**.

Microbes Control Gene Expression at Several Levels

Hundreds of intracellular and extracellular sensing mechanisms monitor the overall health of the cell and its environment. But how do these systems control gene expression?

It is important to note that expression of genes and operons and their products (mRNA, which produces the protein product, or small functional RNAs) can be controlled at various levels. Different levels of control offer different advantages to the cell. The major levels of control can be categorized as DNA sequence controls, transcriptional controls, translational controls, and post-translational modifications. In general, DNA sequence–level control is the most drastic and the least reversible, whereas control at the protein (translational/posttranslational) level offers the most rapid and reversible control. Examples of these control levels are summarized here and then discussed in detail later in the chapter.

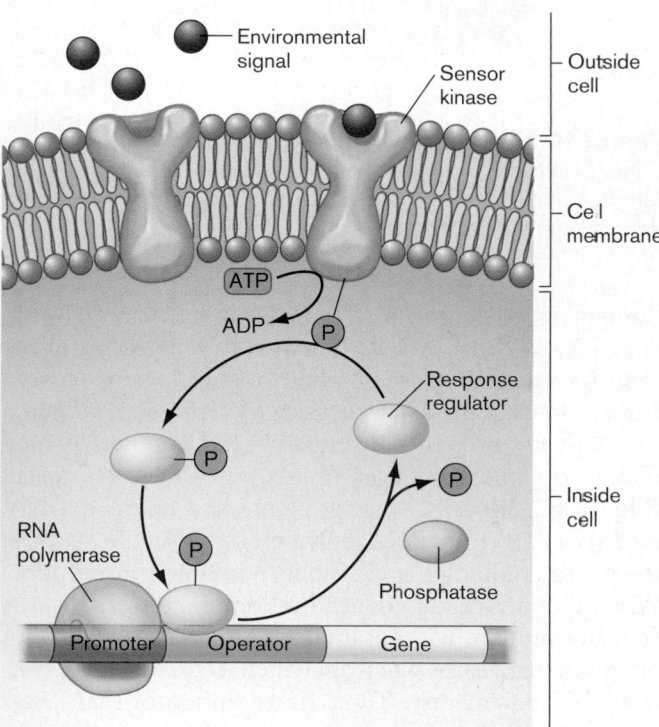

Figure 10.3 Two-component signal transduction systems sense the external environment. A transmembrane sensor kinase protein senses an environmental condition outside the cell (as in Gram-positive bacteria) or in the periplasm (as in Gram-negative bacteria). Binding of the environmental signal triggers autophosphorylation of the sensor kinase in some instances, as shown in the example, and in other instances prevents it. The phosphorylated sensor kinase will then transfer the phosphate to a cognate response regulator protein in the cytoplasm. The response regulator then binds to target operator or activator DNA sequences and inhibits or stimulates gene expression. A phosphatase will down-regulate the system by removing the phosphate.

- **Alterations of DNA sequence.** Some microbes can program the mutation of DNA so as to activate or disable a particular gene. An example of such a mechanism is phase variation, in which DNA rearrangement involving the reversible flipping of a DNA segment enables a pathogen to turn on or turn off expression of cell surface proteins to evade the host immune system.
- **Control of transcription and mRNA stability.** Many types of gene regulation occur at the level of transcription. The most common mechanisms of transcriptional control in prokaryotes involve operons. Recall that an operon is a string of two or more genes in a chromosome that are expressed from a common promoter in front of the first gene in the operon. Genes in an operon are coordinately regulated by protein repressors, activators, and sigma factors, as well as by small RNAs (sRNAs). Coordinate regulation means that the expression levels of all genes in the operon increase or decrease simultaneously.
- **Control of mRNA stability.** Levels of specific mRNA molecules are also regulated by RNase activity, which degrades some mRNA molecules as fast as they are transcribed. In some, but not all, instances, sRNA molecules bind RNA transcripts and help or hinder degradation.

- **Translational control.** Translation by ribosomes can be regulated by *translation initiation sequences* in the mRNA, which recognize specific translational repressor proteins. Translational control mechanisms are often coupled to transcriptional mechanisms, offering "fine-tuning" of operon control. Attenuation, for instance, is a mechanism that uses *translation* to sense the level of an amino acid in the cell and then increase or decrease transcription of the genes encoding the enzymes that synthesize the amino acid.
- **Posttranslational control.** Once proteins are made, their activity can be controlled by modifying their structure—for example, by protein cleavage, phosphorylation, methylation, or acetylation. These modifications can activate, deactivate, or even lead to destruction of the protein.

In recent years, studies of bacterial biology have revealed integrated control circuits in which all the previously described levels of control work together to coordinately regulate systems throughout the cell. We will discuss examples of integrated control circuits, such as the phage lambda lysis/lysogeny decision and the metabolic/genetic system of nitrogen regulation (see Section 10.7). But first we will explore model examples of control. Most of these examples are at the transcriptional level, which is the level best understood.

TO SUMMARIZE:

- **Regulatory proteins** help a cell sense changes in its internal environment and alter gene expression to match.

- **Repressor and activator proteins** bind to operator and activator DNA sequences, respectively, in front of target genes. Repressors prevent transcription, whereas activator proteins stimulate transcription.
- **Two-component signal transduction systems** also help the cell sense and respond to its environment, both inside and outside.
- **Gene expression** can be controlled at several levels: DNA sequence, transcription, translation, or by modifying the protein posttranslationally.
- **Integrated control circuits** connect many individual regulatory systems to coordinate regulation throughout the cell.

10.2 Paradigm of the Lactose Operon

In 1961, French scientists Jacques Monod (1910–1976) and François Jacob (1920–) proposed the revolutionary idea that genes could be regulated (**Fig. 10.4A** and **B**). They, among others, noticed that the enzyme used by *Escherichia coli* to consume the carbohydrate lactose was produced *only* when lactose was added to growth media. The term "induction" was coined to describe this phenomenon. The enzymes required to metabolize glucose, however, were different. They were always (that is, *constitutively*) present. The groundbreaking work of Jacob and Monod, and of André Lwoff (1902–1994) for his study of phage lysogeny (**Fig. 10.4C**), won them a Nobel Prize and launched the field of gene regulation, a scientific realm where discoveries continue to surprise us. Still, it took many years after Monod and Jacob's discovery to

  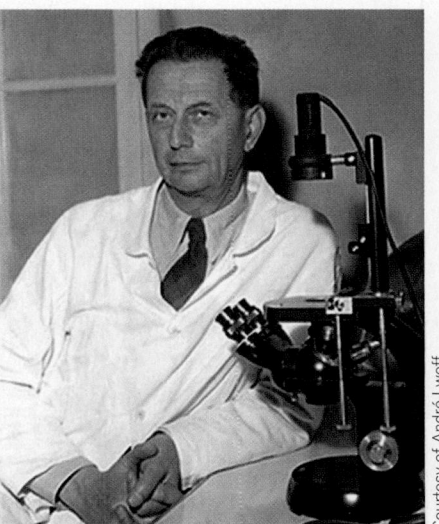

Figure 10.4 Discoverers of gene regulation. Jacques Monod (**A**), François Jacob (**B**), and André Lwoff (**C**) all worked at the Pasteur Institute and won the 1965 Nobel Prize in Physiology or Medicine for their groundbreaking work on induction and gene regulation.

learn exactly how the lactose-degrading enzyme beta-galactosidase is induced. Many of the concepts presented here hold for numerous bacterial gene systems.

Lactose Induces Expression of the *lacZYA* Operon

Figure 10.5A shows the genes in *E. coli* that encode the simple regulatory circuit for lactose catabolism. "Catabolism" is a general term to describe the degradation of an organic food source. The genes *lacZ*, *lacY*, and *lacA*

form an **operon**, which is a group of genes cotranscribed from a common promoter. The *lacY* gene encodes lactose permease (LacY), an integral membrane protein that imports (transports) lactose from the extracellular environment. The product of *lacZ*, beta-galactosidase (LacZ), intracellularly cleaves the disaccharide lactose into its component parts: glucose and galactose (**Fig. 10.6**). Both of these sugars are subsequently degraded by the enzymes of glycolysis (Section 13.5). To break down lactose in this way requires synthesis of both beta-galactosidase and lactose permease; without these two

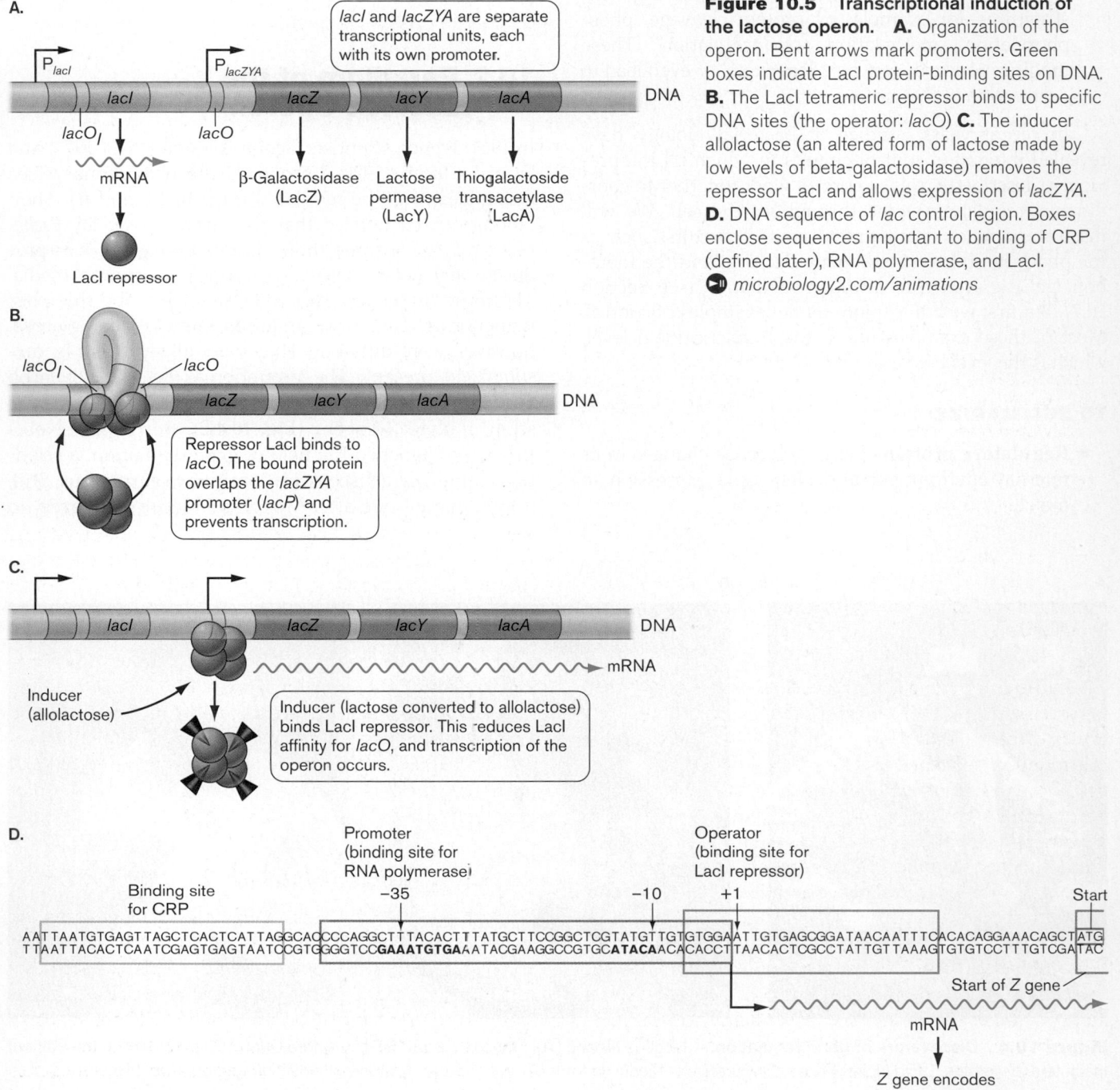

Figure 10.5 Transcriptional induction of the lactose operon. A. Organization of the operon. Bent arrows mark promoters. Green boxes indicate LacI protein-binding sites on DNA. **B.** The LacI tetrameric repressor binds to specific DNA sites (the operator: *lacO*) **C.** The inducer allolactose (an altered form of lactose made by low levels of beta-galactosidase) removes the repressor LacI and allows expression of *lacZYA*. **D.** DNA sequence of *lac* control region. Boxes enclose sequences important to binding of CRP (defined later), RNA polymerase, and LacI. ▶ⅠⅠ *microbiology2.com/animations*

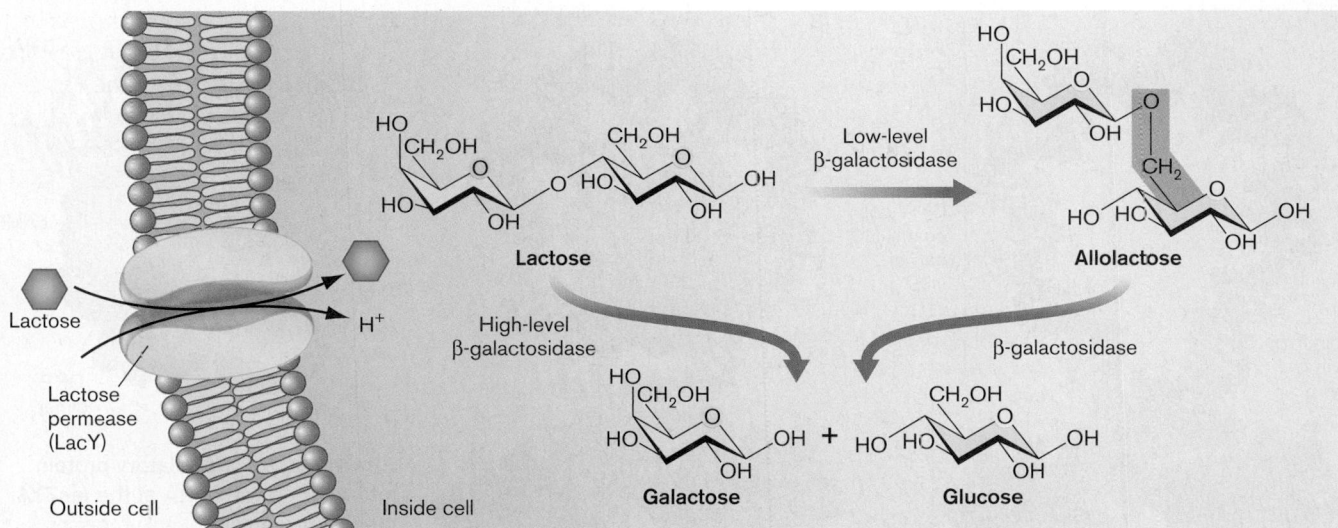

Figure 10.6   Lactose transport and catabolism.   A dedicated lactose permease uses proton motive force to move lactose (and a proton) into the cell. Once there, the enzyme beta-galactosidase can cleave the disaccharide into its component parts (galactose and glucose) or alter the linkage between the monosaccharides to produce allolactose, an important chemical needed to induce the genes that encode this pathway.

gene products, the catabolic energy of lactose is unavailable to the cell.

The role of *lacA*, encoding thiogalactoside trans-acetylase (LacA), is unclear, but it is not needed to ferment lactose. One function may be to detoxify a harmful by-product of lactose metabolism. As we can see, even well-studied systems such as the *lac* operon still pose unanswered questions.

> **NOTE:** "Lactose operon", "*lac* operon", and "*lacZYA* operon" all refer to the same system.

> **THOUGHT QUESTION 10.1**   If the gene *lacZ* has a nonsense mutation in its open reading frame, will *lacY* be translated?

In the absence of lactose, the *lac* operon is transcribed at extremely low levels, generating fewer than ten molecules of beta-galactosidase per cell. The reason is that transcription of *lacZYA* is repressed by the protein product of the regulator gene, *lacI*. The gene *lacI* is situated immediately upstream of *lacZYA* but is transcribed from a different promoter. A tetramer of LacI repressor protein forms in the cell and binds to two operator regions of DNA. One operator sequence is called *lacO*, which partially overlaps the *lacZYA* promoter (P_{lacZYA}). The second operator site occurs within *lacI* and is called *lacO_I* (**Fig. 10.5A ◑**). The *lac* operators control whether the three structural genes are transcribed. The LacI tetramer simultaneously binds the *lacO* and *lacO_I* operators (a dimer at each site). This causes the intervening DNA to loop out (**Figs. 10.5B and 10.7A**) and prevents RNA polymerase from continuing transcription into the structural

lacZYA genes. (A third binding site for LacI exists but is not important to our discussion here.)

Once lactose is added to the medium, the *lac* operon is expressed at 100-fold higher levels. How does lactose get into the cell to induce the operon if the operon encoding its transport protein is not expressed? It turns out that even when the *lacZYA* operon is uninduced, it is constitutively expressed at a low level. This means that lactose can be transported into the cell in low amounts because of constant low-level expression of the lactose transporter LacY. The tiny amount of beta-galactosidase (LacZ) that is expressed in uninduced cells is also important. At this low level, the enzyme does not completely *cleave* the glycosidic bond of lactose but rearranges it to form allolactose, the form of lactose that activates the operon (the structure of which can be seen in **Fig. 10.6**). Allolactose binds the repressor and "unlocks" the protein (by altering the conformation of LacI), so it is released from the operator (**Fig. 10.5C**). Once this happens, sigma D (the housekeeping sigma factor of *E. coli*, not shown in the figure) can find the *lac* promoter sequences and enable RNA polymerase to initiate transcription of the *lacZYA* structural genes.

Activation of Transcription by cAMP-CRP

There is another important mechanism that governs the level of *lacZYA* transcription in *E. coli*. This mechanism involves a small molecule called cyclic AMP (cAMP) that accumulates when a cell's energy state declines. Cyclic AMP is a derivative of AMP in which the 5' phosphate is linked to the 3' OH group of the ribose, making a cyclic bond. Cyclic AMP controls the expression of many genes by combining with a dimeric regulatory protein called

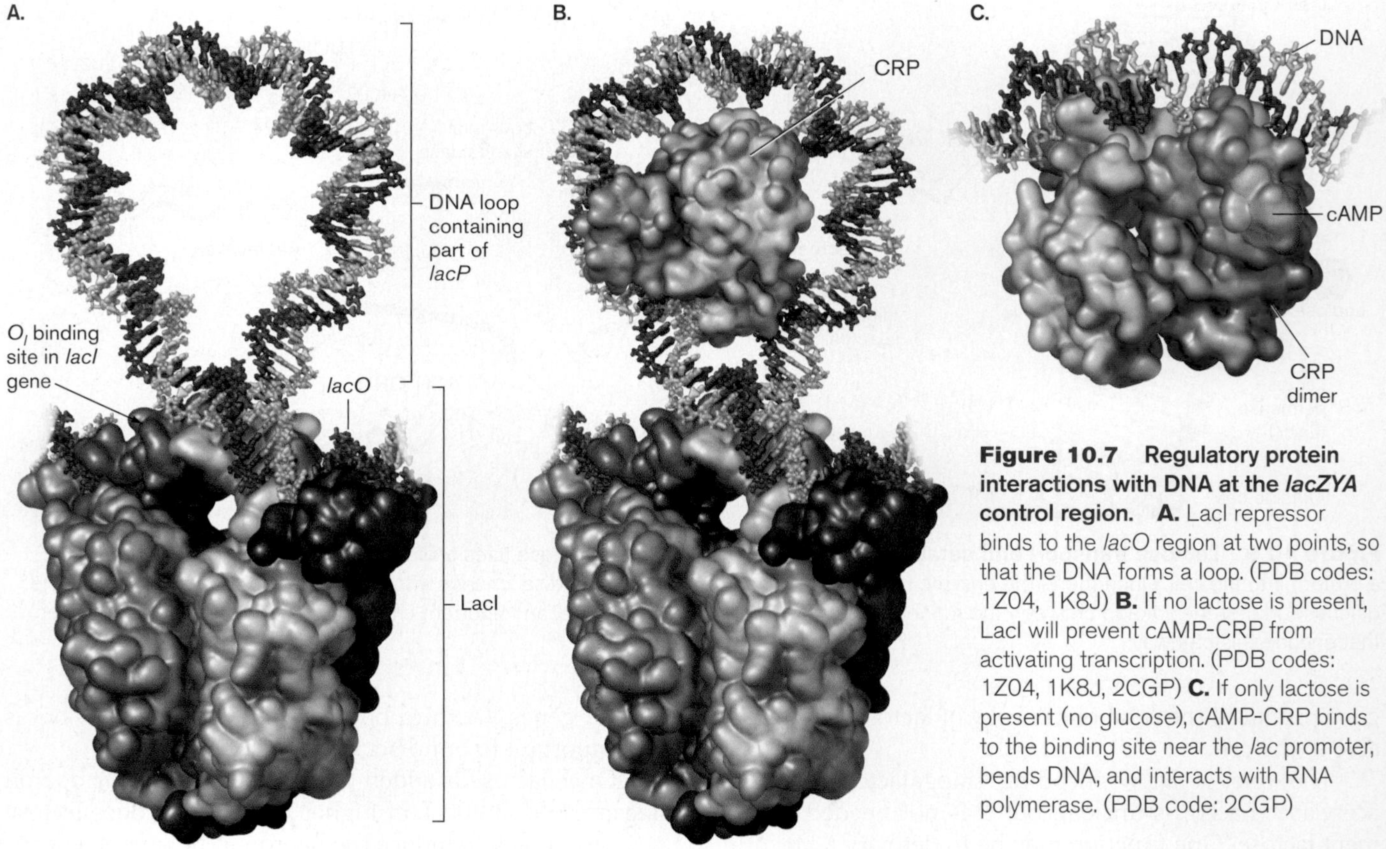

A.

DNA loop containing part of *lacP*

O_l binding site in *lacI* gene

lacO

LacI

B.

CRP

C.

DNA

cAMP

CRP dimer

Figure 10.7 Regulatory protein interactions with DNA at the *lacZYA* control region. **A.** LacI repressor binds to the *lacO* region at two points, so that the DNA forms a loop. (PDB codes: 1Z04, 1K8J) **B.** If no lactose is present, LacI will prevent cAMP-CRP from activating transcription. (PDB codes: 1Z04, 1K8J, 2CGP) **C.** If only lactose is present (no glucose), cAMP-CRP binds to the binding site near the *lac* promoter, bends DNA, and interacts with RNA polymerase. (PDB code: 2CGP)

cAMP receptor protein (CRP). The cAMP-CRP complex binds to specific DNA sequences located near many bacterial genes and modifies their transcription, usually acting as an activator. This is the case for the *lacZYA* operon (**Figs. 10.8** and **10.5C**).

> **NOTE:** The original name for CRP was "catabolite activator for protein" (CAP), given to reflect its role in catabolite repression. Though some investigators still use the CAP designation, the term "CRP" is generally preferred today. Thus, the gene encoding the protein is called *crp*.

How does the cAMP-CRP complex ultimately activate the expression of *lacZYA*? RNA polymerase cannot easily form an open complex (Section 8.2) and is essentially stuck at *lacP* and other CRP-dependent promoters, even in the absence of a LacI repressor. This problem is overcome when the cAMP-CRP complex binds to a 22-bp DNA sequence located −60 bp from the start of transcription (just upstream of the *lacZYA* operator), causing the DNA to curve (see **Figs. 10.5D** and **10.7C**). CRP can then directly interact with the alpha subunit of RNA polymerase bound at the *lac* promoter and activate transcription (**Fig. 10.9** and **eTopic 10.1**).

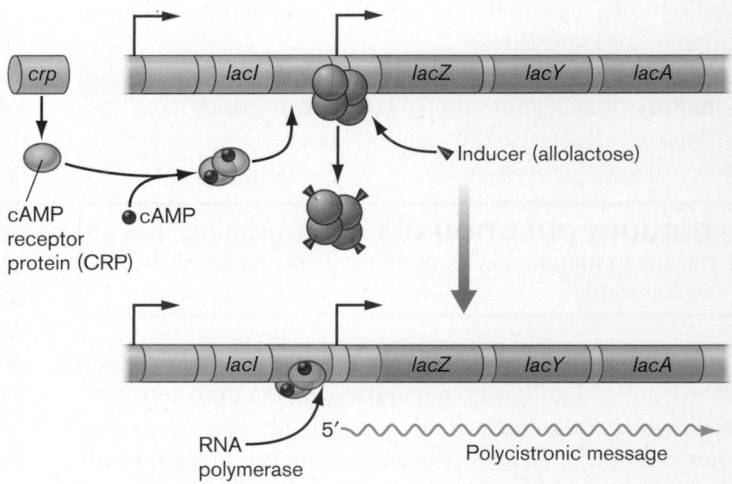

Figure 10.8 Activation of the *lacZYA* operon by cAMP and CRP. Although the inducer allolactose removes the repressor LacI and allows expression of *lacZYA*, maximum expression requires the presence of cAMP and cAMP receptor protein (CRP), which bind to a separate site at the *lacZYA* promoter. Bound CRP can interact with RNA polymerase and increase the rate of transcription *initiation*.

The C terminus of the RNA polymerase alpha subunit can bind DNA sequences near certain promoters and inhibit RNA polymerase movement. When cAMP-CRP touches regions in the C terminus of the alpha subunit,

Class I promoters

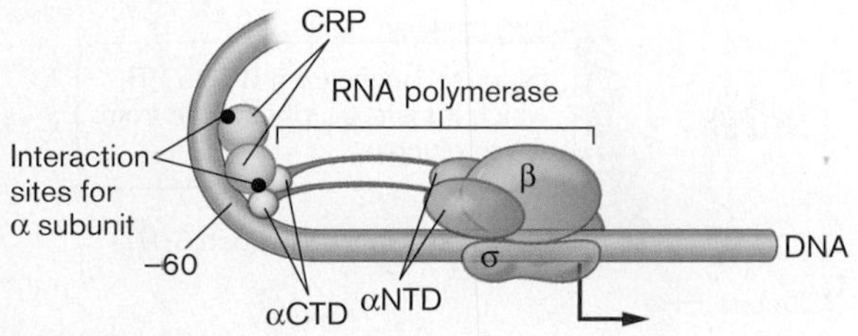

Abbreviations key

CRP: cAMP receptor protein
αCTD: C-terminal domain of RNA polymerase alpha subunit
αNTD: N-terminal domain of RNA polymerase alpha subunit

Figure 10.9 CRP interactions with RNA polymerase.
Promoters like the one at the *lacZYA* operon, possess a CRP-binding site positioned about −60 bp from the transcriptional start. CRP on these promoters interacts with the alpha subunit C-terminal domain (αCTD) of RNA polymerase. *Source*: Adapted from Moat et al. 2002. *Microbial Physiology*, 4th ed. Wiley-Liss, Delaware.

this inhibition is relieved. Contact between CRP and the alpha subunit is enough to stimulate class I promoters such as *lacP*. Notice in **Figure 10.7B** that when neither glucose nor lactose is present, cAMP-CRP cannot by itself activate *lacZYA* transcription, because it binds within the DNA loop formed by the LacI (repressor) tetramer, so there is no room for RNA polymerase to bind to DNA.

WWW | Biomolecules at Kenyon: catabolite activator protein
microbiology2.com/links

NOTE: In presenting a genotype, gene names written with a superscript + are considered wild-type genes. Gene names written *without* a superscript + are considered mutant genes. Also recall that a gene name written in roman (not italic) type and with a capital first letter is really the name of the protein product of that gene. Thus, *lacZ* is the gene, but LacZ is the protein.

THOUGHT QUESTION 10.2 Null mutations completely eliminate the function of a mutated gene. Predict the effect of the following null mutations on the induction of beta-galactosidase by lactose, and predict whether the *lacZ* gene is expressed at high or low levels in each case: *lacI, lacO, lacP, crp, cya* (the gene encoding adenylate cyclase). What effect will those mutations have on catabolite repression?

THOUGHT QUESTION 10.3 Predict what will happen to the expression of *lacZ* under the following partial diploid conditions (see Section 9.2). The genotypes of these strains are presented as follows: chromosomal genes/plasmid gene. (a) *lacI lacO$^+$P$^+$Z$^+$Y$^+$A$^+$/*plasmid *lacI$^+$*; (b) *lacO lacI$^+$P$^+$Z$^+$Y$^+$A$^+$/*plasmid *lacO$^+$*; (c) *crp lacI$^+$O$^+$P$^+$Z$^+$Y$^+$A$^+$/*plasmid *crp$^+$*.

Glucose Represses the *lac* Operon

What happens if, in addition to lactose, the medium contains an alternative carbon source, such as glucose? Enzymes for glucose catabolism (glycolysis) are always produced at high levels because glucose is the favored catabolite (carbon source), providing the quickest source of energy. Many carbohydrates, including lactose, must first be converted to glucose to be catabolized. So, should *E. coli* forestall induction of the *lac* operon while glucose is present, in the interest of greater efficiency? In fact, this is precisely what happens. In what is known as **catabolite repression**, an operon enabling catabolism of one nutrient is repressed by the presence of a more favorable catabolite (commonly glucose).

When glucose and lactose are both present in the medium, cells grow by breaking down glucose until the glucose is depleted; then growth stops. At this point, the cells sense a need for carbon and induce the *lacZYA* operon (because lactose is present). This allows them to begin consuming lactose and reinitiate growth. The biphasic curve of a culture growing on two carbon sources is often called **diauxic growth** (**Fig. 10.10A**). But if lactose was present from the start, why was *lacZYA* turned off? The answer is that glucose indirectly prevents the induction of *lacZYA*. **Figure 10.10B** illustrates that even when *lacZYA* is already induced, adding glucose stops (or represses) induction.

Glucose Causes Catabolite Repression via Inducer Exclusion

Failure of lactose to induce *lacZYA* during growth on glucose (**Fig. 10.10B**, dotted horizontal line) is due to mainly the fact that growth on glucose keeps lactose out of the cell. This phenomenon is known as **inducer exclusion**. If lactose cannot enter the cell, the *lacZYA* operon cannot be induced. The key to inducer exclusion is that a component of the glucose transport system (phosphotransferase system, or PTS) will bind to and inhibit LacY permease. Transport of some sugars through the PTS also affects adenylate cyclase activity, but this mechanism is not a major cause of glucose/lactose diauxic growth (see **eTopic 10.2**).

Many sugars, including glucose, are transported by the multicomponent phosphotransferase system (see Section 4.2). This system transfers a phosphate group from phosphoenolpyruvate (PEP), along a series of protein intermediates, ending with different membrane transport proteins that import specific carbohydrates. The carbohydrate-specific membrane proteins pass the phosphate to the sugar during transport. The glucose-specific PTS relay includes a cytoplasmic protein called IIA^{Glc} and the integral membrane protein IIBC^{Glc}. Protein IIA^{Glc} is the key to inducer exclusion.

NOTE: IIA^{Glc} and IIBC^{Glc} are also referred to as Enz IIA and Enz IIBC in the literature.

A.

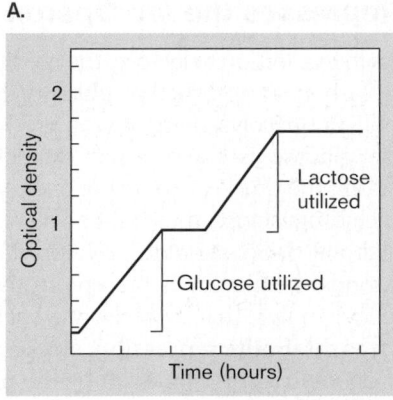

B.

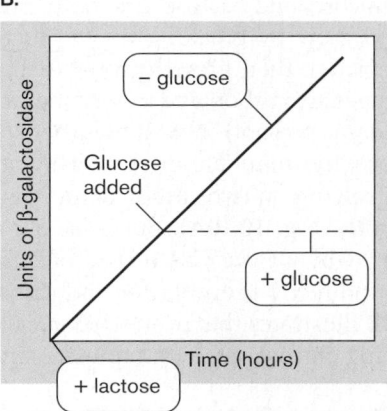

Figure 10.10   Catabolite repression of the *lacZYA* operon.
A. The diauxic growth curve of *E. coli* growing on a mixture of glucose and lactose. **B.** Glucose repression of LacZ (beta-galactosidase) production. Lactose was added at the beginning of the experiment to parallel cultures, and beta-galactosidase activity was measured. At the point indicated, glucose was added to one culture. The synthesis of beta-galactosidase continues to increase in the culture without glucose but stops in the culture with glucose.

A. Glucose present

B. Glucose absent

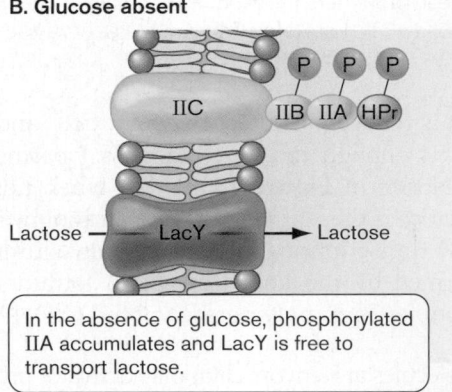

Figure 10.11   Glucose transport via the phosphotransferase system inhibits LacY (lactose permease).   A. When glucose is present, the level of phosphorylated IIA^{Glc} is low because glucose siphons off the phosphate. IIA^{Glc} then interacts with and inhibits LacY (lactose permease) activity, thereby excluding transport of inducer (inducer exclusion). **B.** In the absence of glucose, the phosphorylated forms of glucose-specific IIA^{Glc} and IIBC^{Glc} accumulate because they cannot pass the phosphate to substrate (there is no glucose). LacY functions in this situation to transport lactose.

Glucose is changed to glucose 6-phosphate during its transport through the PTS system. Thus, as the cell imports glucose, phosphate is continually removed from IIA^{Glc}, making unphosphorylated IIA^{Glc} the predominant form of the protein. This is important because the unphosphorylated form of the protein binds to and inhibits the activity of certain non-PTS carbohydrate permeases, such as LacY, the transporter for lactose. Shutting transport down keeps the inducer out, which, in the case of lactose, prevents *lacZYA* expression (**Fig. 10.11**). The precise nature of the communication between these two proteins is not well understood.

In contrast to Gram-negative bacteria, Gram-positive bacteria do not usually contain cAMP. Catabolite repression in these organisms is mediated by other mechanisms.

THOUGHT QUESTION 10.4   Researchers often use isopropyl-β-D-thiogalactopyranoside (IPTG) rather than lactose to induce the *lacZYA* operon. IPTG structure resembles lactose, which is why it can interact with the LacI repressor, but it is not degraded by beta-galactosidase. Why do you think the use of IPTG is preferred in these studies?

How Do We Study Protein-DNA Binding?

A long series of painstaking experiments led to the discovery of how the *lac* operon is regulated, but current techniques have made it much easier to study how DNA and proteins interact and how these interactions control gene expression. The search for DNA-binding sites often begins with bioinformatics.

DNA sites that bind regulatory proteins typically exhibit a sequence symmetry that allows unambiguous recognition. Dimers of a regulatory protein bind to the DNA, with each member of the duo binding half of the symmetrical DNA sequence. Symmetry usually involves an inverted repeat (**Fig. 10.12**). Once a potential regulatory protein and a presumed target gene have been identified, the DNA sequence extending 200–300 bp upstream of the target gene's transcriptional start site is examined using computer programs designed to reveal possible binding sites. Finding an inverted repeat within 100–200 bases of a promoter suggests the presence of a regulatory binding site. Finding very similar sequences in front of two or more different promoters controlled by the same regulatory protein is stronger evidence that, in fact, the sequences represent regulatory binding sites. Proof ultimately requires showing that the protein will bind a piece of DNA containing this sequence. Protein-DNA binding can be demonstrated in several ways, as described in Section 12.3.

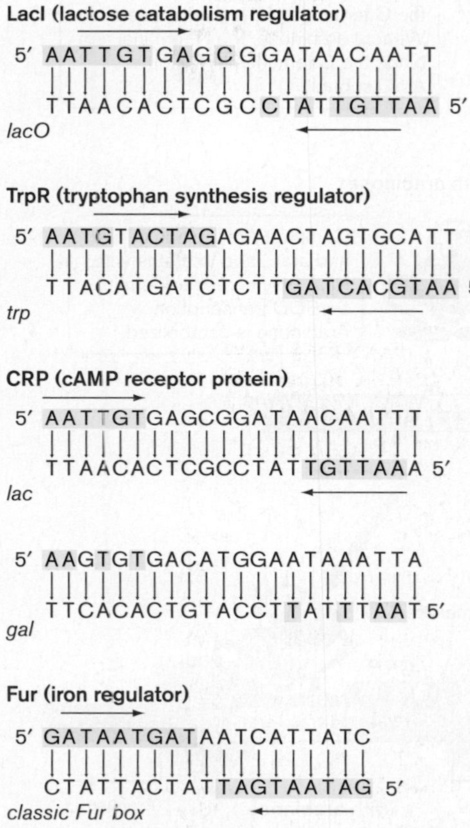

LacI (lactose catabolism regulator)

5′ AATTGTGAGCGGATAACAATT
   |||||||||||||||||||||
   TTAACACTCGCCTATTGTTAA 5′
lacO

TrpR (tryptophan synthesis regulator)

5′ AATGTACTAGAGAACTAGTGCATT
   ||||||||||||||||||||||||
   TTACATGATCTCTTGATCACGTAA 5′
trp

CRP (cAMP receptor protein)

5′ AATTGTGAGCGGATAACAATTT
   ||||||||||||||||||||||
   TTAACACTCGCCTATTGTTAAA 5′
lac

5′ AAGTGTGACATGGAATAAATTA
   ||||||||||||||||||||||
   TTCACACTGTACCTTATTTAAT 5′
gal

Fur (iron regulator)

5′ GATAATGATAATCATTATC
   |||||||||||||||||||
   CTATTACTATTAGTAATAG 5′
classic Fur box

Figure 10.12 Examples of DNA regulatory sequences. The sequences shown are located upstream of the genes noted in italics. Inverted repeats are shown in yellow. Note that in some inverted repeats, the occasional base is not repeated (only those bases that repeat are highlighted in the diagram). Arrows indicate direction of symmetry.

TO SUMMARIZE:

- **The lactose utilization *lacZYA* operon** of *E. coli* was the first gene regulatory system described.
- **LacI binds as a tetramer** to the operator region and represses the *lac* operon by preventing open complex formation by RNA polymerase.
- **Beta-galactosidase (LacZ)**, when at low concentration, cleaves and rearranges lactose to make the inducer allolactose.
- **Allolactose complexes to LacI**, reducing repressor affinity for the operator and allowing induction of the operon.
- **The cAMP-CRP complex** binds to the *lac* operator region and stimulates RNA polymerase through CRP interaction with the C-terminal domain of the RNA polymerase alpha subunit. cAMP-CRP regulates many types of operons.
- **In catabolite repression**, a preferred carbon source prevents the induction of an operon that enables catabolism of a different carbon source. One example is catabolite repression of the *lac* operon by glucose. Glucose transport through the PTS system causes catabolite repression by inhibiting LacY permease activity (inducer exclusion) and lowers cAMP levels.

10.3 Other Systems of Operon Control

The lactose operon is not the only model of regulation. Many other mechanisms that regulate genes and operons are utilized by bacteria. In this section, we discuss proteins that have dual regulatory functions—that perform as both activators (positive control) and repressors (negative control). We will also see how the coupling of transcription and translation is used to regulate gene expression by attenuation and how an idling ribosome sends a chemical signal that affects the expression of many genes and operons.

Repression of Anabolic (Biosynthetic) Pathways

Many different gene systems use repressor proteins to inhibit transcription. Repressing biosynthetic pathways is fundamentally different from repressing catabolic (degradative) systems. Repressor proteins that control catabolic pathways, such as lactose degradation, typically bind the initial substrate or a closely related product (for example, allolactose in the case of the *lac* operon). Binding the substrate decreases repressor protein affinity for operator DNA. Thus, increased concentration of the substrate or inducer actually removes the repressor from the operator and *derepresses* expression of the operon.

This derepression makes sense because the cell wants to make the enzymes that use the substrate as a carbon and energy source.

In contrast, genes encoding biosynthetic enzymes are regulated by repressors (called inactive aporepressors) that must bind the end product of the pathway (for example, tryptophan for the *trp* operon) to become active repressors. The end product that binds the aporepressor is called a corepressor. Binding of the corepressor (end product) to the repressor increases the repressor's affinity for the operator sequence upstream of the target gene or operon. As the concentration of end product, such as an amino acid or nucleotide, increases in the cell beyond

that needed to support growth, the cell will shut the biosynthetic pathway down and not waste energy making a superfluous pathway or compound.

The AraC Regulator Activates and Represses Transcription

The *lac* operon regulatory paradigm is a wonderful way to regulate gene systems involved in catabolism. What could be better than having a substrate induce expression of the operon needed to catabolize it? How about a regulator that can repress *or* activate gene expression, depending on whether substrate is available? This would

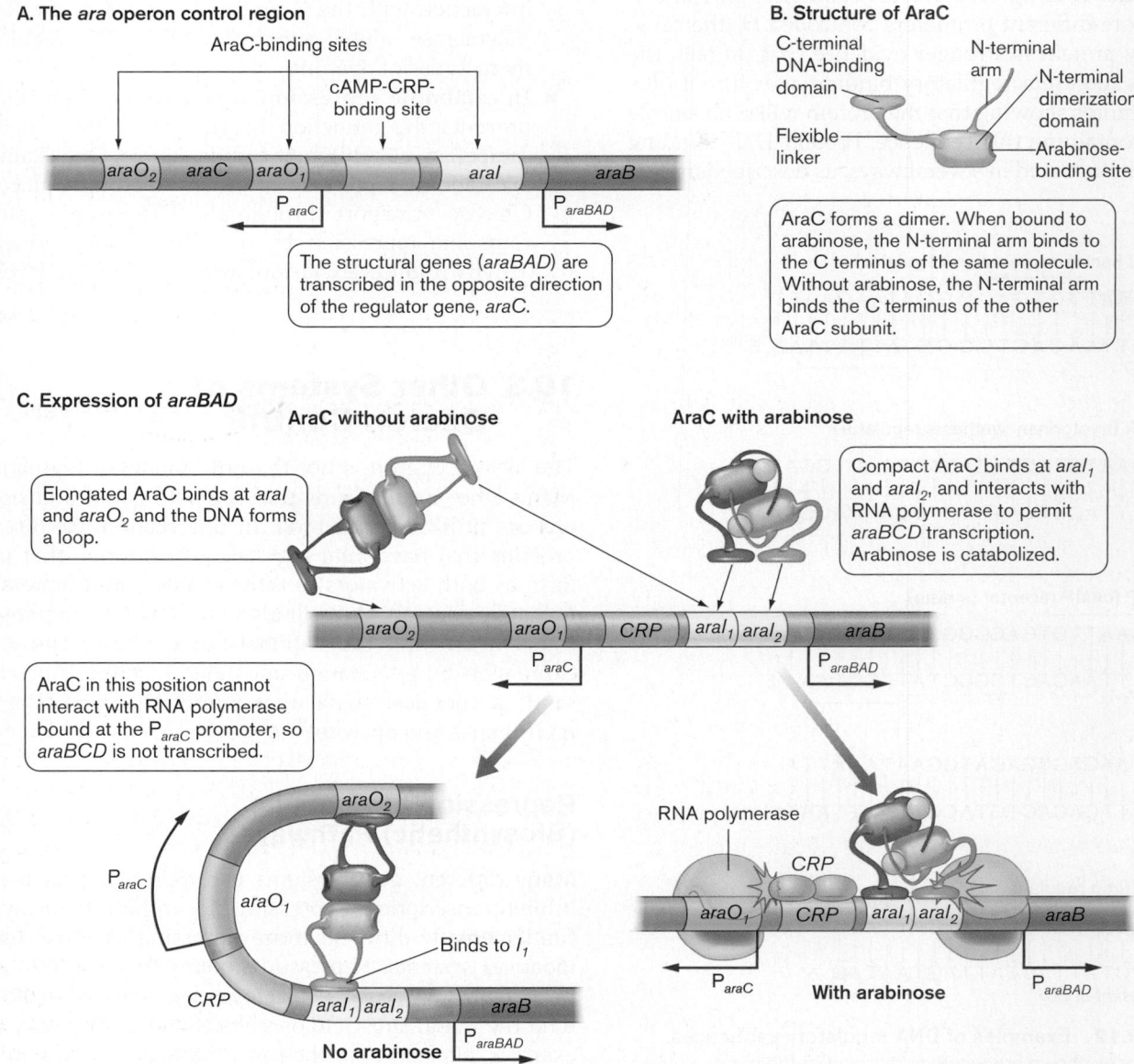

Figure 10.13   Regulation of *araBAD* operon by the AraC regulator.   A. Divergently transcribed *araC* and *araBAD* operons showing binding sites for the AraC and CRP proteins. The region shown is about 400 bp. **B.** View of AraC showing dimerization and DNA-binding domains. **C.** Alternative conformations of AraC dimers. The different conformations change the location of where the dimer can bind DNA. (Note that O_1 and O_2 are operators, and I_1 and I_2 are other DNA-binding sites that, when occupied by AraC, induce expression.)

provide very tight control over the synthesis of catabolic enzymes and enable an accelerated induction. One such regulator, called AraC, regulates the genes encoding arabinose catabolism. When arabinose is absent, AraC represses expression of the genes that break down arabinose; when arabinose is present, AraC activates these same genes. The products of these genes ultimately convert the five-carbon sugar L-arabinose to D-xylulose 5-phosphate, an intermediate in the pentose phosphate shunt, a pathway that provides reducing energy for biosynthesis (see Section 13.5). It turns out that many regulatory systems, including some virulence regulators, use AraC-type regulation.

The *ara* operon. Part of the operon encoding the arabinose degradation enzymes is shown in **Figure 10.13A**. Note that the *ara* operon is transcribed divergently from a central promoter region. The structural genes (*araBAD*) are transcribed in one direction and the regulator gene, *araC*, is transcribed in the opposite direction. The product of *araC* is a 33-kDa protein (AraC) with two domains: the C-terminal end and the N-terminal end (**Fig. 10.13B**). The C-terminal end contains a DNA-binding domain that contains two helix-turn-helix motifs. Helix-turn-helix motifs are made up of two alpha helices separated by a short amino acid chain that allows a bend. Usually, the helix closer to the C terminus fits into the major groove of the target DNA. The C-terminal DNA-binding domain is attached by a flexible linker peptide to an N-terminal dimerization domain. AraC forms a dimer in vivo that can assume one of two conformations, depending on whether arabinose is available. When arabinose is absent, the dimer exists in a rigid, elongated form that represses expression of the machinery for arabinose degradation (*araBAD*). When arabinose is present, it binds to the dimer, which then assumes a more compact form that activates expression of *araBAD* (**Fig. 10.13C**).

The flexible N-terminal arm of an AraC monomer is the key to determining what type of dimer forms. When arabinose is not present, the N-terminal arm of an AraC monomer binds to the C domain of the same molecule, forming a rigid, extended structure. However, when arabinose docks to its binding site in the AraC protein, the N arm can bend and bind to the C domain of its partner protein, producing a more compact structure. **Figure 10.13C** shows the alternative dimerization configurations of the AraC N-terminal domain in the absence and presence of L-arabinose.

So how does AraC act as both repressor and activator? First, unlike the *lac* repressor, AraC does not function by denying RNA polymerase access to the promoter. Instead, AraC represses *araBAD* expression by producing a looped DNA structure that limits activation by cAMP-CRP, which, as discussed in Section 10.2, is a global

regulator of many bacterial genes. Then, when placed in the right position on the DNA, AraC, as well as CRP, can activate *araBAD* transcription by physically contacting an RNA polymerase complex that is already bound to the promoter.

There are four DNA sites to which AraC can bind, as illustrated in **Figure 10.13C**. With arabinose absent, the elongated, repressor form of the AraC dimer (AraC$_2$) can bind only the widely separated sites I$_1$ and O$_2$, causing the intervening DNA to form a loop. The loop positions AraC too far from the *araBAD* promoter to ever interact with RNA polymerase. The loop also prevents CRP from interacting with RNA polymerase. However, the compact, inducer form of AraC$_2$, formed by the binding of arabinose, can bind to the DNA sites I$_1$ and I$_2$ (I represents inducer site), which are closer to the promoter. This places AraC closer to an RNA polymerase idling on the *araBAD* promoter and allows the two complexes to touch. This contact releases RNA polymerase from the promoter and allows it to move into and transcribe the *araBAD* structural genes. This position also allows bound CRP to activate expression. **Figure 10.14** illustrates the structural effect of arabinose on the N-terminal dimerization domain of AraC. The binding of arabinose to AraC engineers dimerization through interactions between coiled-coil alpha-helical regions of the protein.

Advantages of the activator-repressor strategy. This system offers an advantage over a simple repressor circuit. In the *lac* operon, LacI repressor must fully dissociate from operator DNA during induction and ends up dispersing throughout the cell, often nonspecifically binding to other DNA sequences. Reestablishing repression requires random diffusion of LacI protein back to the *lacO* DNA sequence. This takes some time because random diffusion of bulky proteins in the cytoplasm is slow. In contrast, AraC protein rarely leaves the vicinity of the *ara* operon. The response of AraC to varying levels of arabinose allows AraC to simply shuttle back and forth between repressor and activator sites, shortening the delay between induction and repression. The inducer arabinose is relatively small; thus, it can diffuse quickly throughout the cell to find AraC positioned at the promoter.

The family of AraC-like regulators. Computer analysis of numerous microbial genomes reveals a family of over 800 regulators that share a 100-amino-acid region of homology with AraC and the closely related XylS activator (xylose catabolism). This signature sequence, found at the C-terminal end of AraC, forms an independently folding domain that contains two helix-turn-helix DNA-binding motifs. The AraC/XylS family members are a diverse group of intracellular sensors. Collectively they regulate

A. Arabinose present

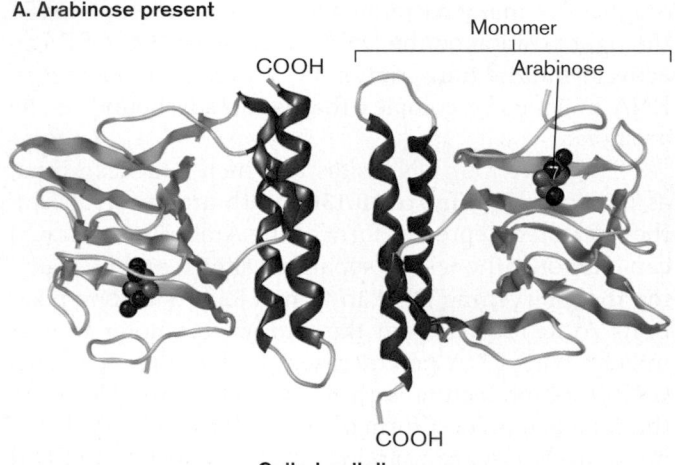

Coiled-coil dimer

B. No arabinose present

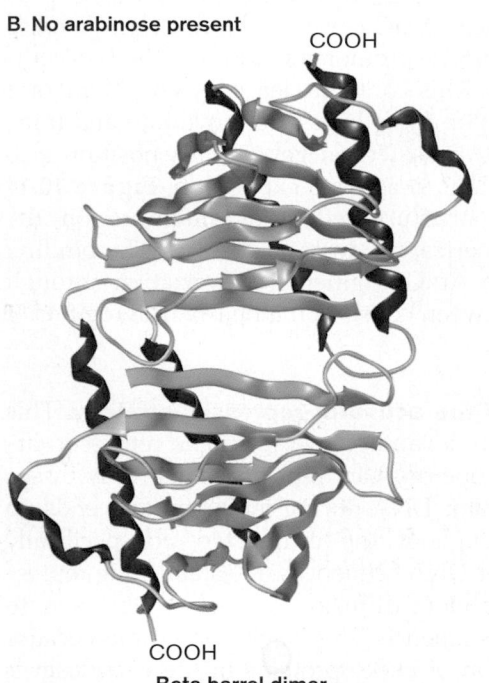

Beta barrel dimer

Figure 10.14 Molecular structure of the AraC-binding domain. Figures show alternative binding conformations of the dimerization domain only. **A.** Dimer formed by the interaction of coiled-coil regions in the presence of arabinose. (PDB code: 2ARC) **B.** Alternative dimer formed by the interaction between the beta barrel regions in the absence of arabinose. Note the helix-turn-helix motif (helices shown in red). This is a common structure present in many DNA-binding proteins. The helix-turn-helix is able to fit into the major groove of a DNA molecule. (PDB code: 2ARC)

a variety of different cell functions, including carbon metabolism and virulence, as well as many responses to environmental conditions.

Although family resemblance is strong at the DNA-binding domains of these AraC-like proteins, homology usually evaporates at the other end of the protein, known as the dimerization domain. The nonfamily domain is

where these proteins appear to bind or respond to some ligand (for example, arabinose). But for many, if not most, of the AraC/XylS family members, the identity of the ligand remains a mystery.

In Attenuation, Translation Regulates Transcription

As noted previously, many amino acid biosynthetic pathways are controlled by transcriptional repression in which a repressor protein binds to a DNA operator sequence to prevent transcription. For instance, when internal tryptophan levels exceed cellular needs, the excess tryptophan (acting as a corepressor) will bind to an inactive aporepressor protein, TrpR, converting it to an active holorepressor (**Fig. 10.15**). TrpR holorepressor then binds to an operator DNA sequence positioned upstream of the tryptophan (*trp*) operon, which encodes the enzymes required for tryptophan biosynthesis. Holorepressor bound to the *trp* operator represses expression by blocking RNA polymerase. But repression is not the whole story in regulating the *trp* operon. Many amino acid biosynthetic operons, including the *trp* operon, have adopted a second strategy for down-regulating tryptophan synthesis, which can be used alone or in conjunction with repression. This second mechanism is called **transcriptional attenuation**. Attenuation uses the ribosome as a sensor of amino acid levels. It is a transcriptional control mechanism in which the ability to *translate* part of an mRNA determines whether the RNA polymerase that began transcribing the operon is allowed to continue downstream transcription. Because attenuation halts transcription in progress, it affords an even quicker response to changing amino acid levels than simple repression.

Transcriptional attenuation was discovered by Charles Yanofsky and his colleagues at Stanford University. While examining the beginning of the *trp* operon in *E. coli*, they discovered an odd DNA region located between the *trp* operator and the first structural gene, *trpE*. This region, called the **leader sequence**, does not directly control the biosynthesis of tryptophan, but instead determines whether RNA polymerase, already authorized to begin transcription, is granted permission to proceed into the *trp* structural genes. The leader sequence encodes a peptide, but the peptide has no enzymatic function. The importance of the leader sequence lies in a pair of tryptophan codons (UGGUGG) embedded within it. The act of translating those codons couples the intracellular level of tryptophan (measured as charged tryptophanyl-tRNA) to transcription. If the level of tryptophan is sufficient to maintain a level of charged tryptophanyl-tRNA adequate to support growth at a rapid rate, the mRNA of the leader region jettisons RNA polymerase before it reaches *trpE*, the first structural gene involved in the biosynthesis of tryptophan.

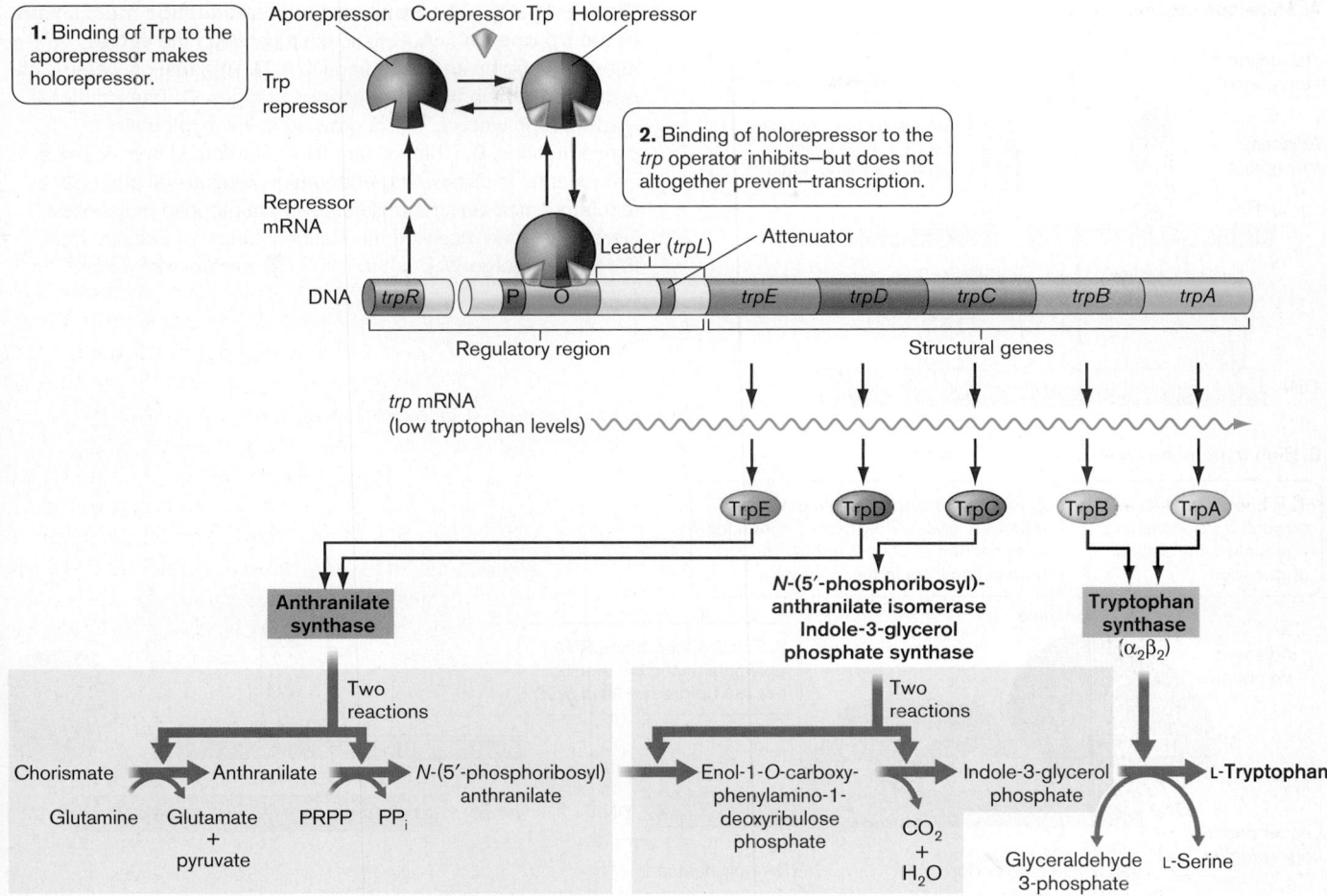

1. Binding of Trp to the aporepressor makes holorepressor.

2. Binding of holorepressor to the *trp* operator inhibits—but does not altogether prevent—transcription.

Figure 10.15   **The tryptophan biosynthetic pathway in *E. coli* and repression of the tryptophan operon.**   The tryptophan biosynthetic enzymes and their encoding genes. TrpR aporepressor (inactive repressor) binds excess tryptophan when intracellular concentration exceeds need. The holorepressor (active repressor) then binds to the *trp* operator and prevents transcription. Note the long polycistronic message in blue. Repression lowers expression about 100-fold. PRPP stands for phosphoribosylpyrophosphate.

The attenuation mechanism hinges on four complementary nucleotide stretches within the leader mRNA. These regions, numbered 1–4, can base-pair to form competing stem loop structures (**Fig. 10.16A** ⬤). Two of the stem loop structures are critical to the mechanism. These are the **anti-attenuator stem loop** formed by regions 2 and 3 and the **attenuator stem loop** (or terminator stem loop) formed by regions 3 and 4. If the 3:4 attenuator stem loop forms, then the RNA polymerase is ejected and transcription stops (**Fig. 10.16B**). Formation of the 2:3 anti-attenuator stem loop, however, prevents formation of the 3:4 stem loop because the 2:3 stem is longer than the 3:4 stem and, thus, more thermodynamically stable. The anti-attenuator stem allows RNA polymerase to transcribe into *trpE* (**Fig. 10.16C**). What controls which stem loop forms?

High tryptophan levels. The ribosome is very large and can barrel through RNA stem loop structures. When the cell is replete with charged tryptophanyl-tRNA and needs no more, the ribosome quickly translates through the key tryptophan codons but runs into a translation stop codon between regions 1 and 2. The ribosome stalls in this position, enveloping region 2 and preventing formation of the 2:3 stem. As a result, once RNA polymerase transcribes through region 4, the 3:4 attenuator stem snaps together. This structure interacts with the RNA polymerase ahead of it and halts transcription. As you would expect, the ribosome dissociates after reaching the region 2 stop codon, but because the 3:4 stem loop is already in place, a 2:3 stem loop does not form. Ribosome release, therefore, leads to formation of a 1:2 stem structure, precluding all possibility of regions 2 and 3 annealing.

Low tryptophan levels. However, if the level of charged tryptophanyl-tRNA is low, the ribosome following behind RNA polymerase stalls over the tryptophan codons. Because these codons occur right at the beginning of

A. Stem loop structures in attenuator region

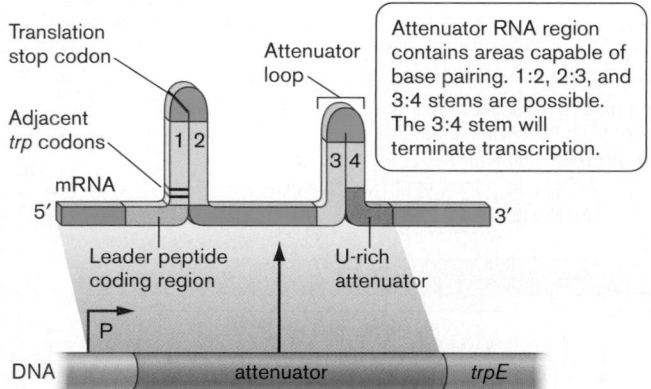

Translation stop codon

Attenuator loop

Adjacent *trp* codons

Attenuator RNA region contains areas capable of base pairing. 1:2, 2:3, and 3:4 stems are possible. The 3:4 stem will terminate transcription.

1 2

3 4

mRNA

5′ 3′

Leader peptide coding region

U-rich attenuator

P

DNA attenuator *trpE*

Figure 10.16 The transcriptional attenuation mechanism at the *trp* operon. A. Relationship between the mRNA attenuator region and encoding DNA. **B.** Attenuation when *E. coli* is growing in high tryptophan concentrations. **C.** Transcriptional read-through when *E. coli* is growing in low tryptophan concentrations. **D.** Charles Yanofsky (Stanford University) was instrumental in discovering attenuation and several other gene regulatory mechanisms while studying tryptophan metabolism. Here he is seen receiving the National Medal of Science from President George W. Bush in 2003. ▶❙ *microbiology2.com/ animations*

B. High tryptophan levels

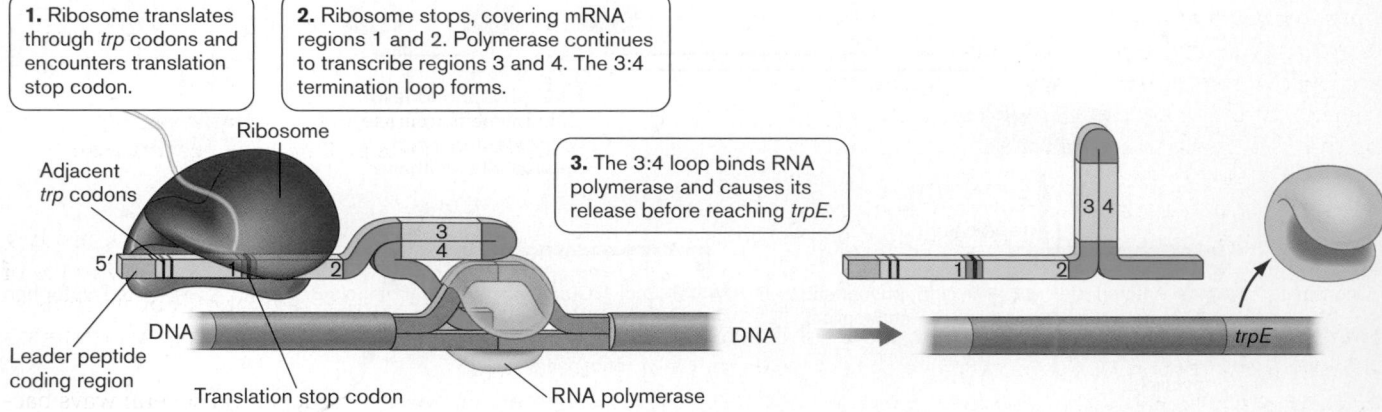

1. Ribosome translates through *trp* codons and encounters translation stop codon.

2. Ribosome stops, covering mRNA regions 1 and 2. Polymerase continues to transcribe regions 3 and 4. The 3:4 termination loop forms.

Ribosome

Adjacent *trp* codons

3. The 3:4 loop binds RNA polymerase and causes its release before reaching *trpE*.

3 4

5′

DNA

DNA

Leader peptide coding region

Translation stop codon

RNA polymerase

trpE

C. Low tryptophan levels

D.

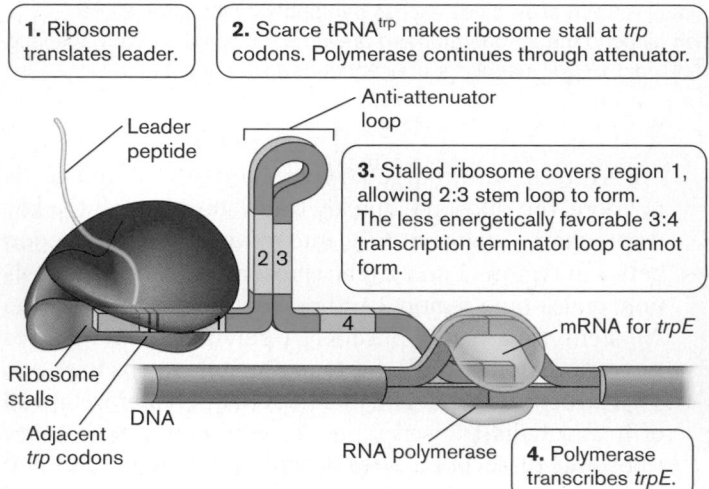

1. Ribosome translates leader.

2. Scarce tRNA^trp makes ribosome stall at *trp* codons. Polymerase continues through attenuator.

Anti-attenuator loop

Leader peptide

3. Stalled ribosome covers region 1, allowing 2:3 stem loop to form. The less energetically favorable 3:4 transcription terminator loop cannot form.

2 3

1

4

mRNA for *trpE*

Ribosome stalls

DNA

Adjacent *trp* codons

RNA polymerase

4. Polymerase transcribes *trpE*.

region 1, the ribosome does not cover and protect region 2. As soon as RNA polymerase transcribes region 3, the 2:3 anti-attenuator stem loop forms and precludes formation of the 3:4 attenuator stem loop. The result is that RNA polymerase can continue into the structural genes, and

ultimately more tryptophan is made. (A new ribosome binds to a ribosome-binding site at the *trpE* message.)

Realize that for the *trp* operon, attenuation is a fine-tuning mechanism. The holorepressor provides the majority of control. However, transcriptional attenuation is a common regulatory strategy used to control many operons that code for amino acid biosynthesis.

> **NOTE:** Even though translation is part of the attenuation control mechanism, attenuation is not considered translational control. The reason is that RNA polymerase, rather than the ribosome, is the target of the control. Translational control of gene expression will be discussed later.

When Energy Sources Dwindle, Ribosome Synthesis Slows

During transitions from nutrient-rich to nutrient-poor conditions, microbes must contend with dramatic fluctuations in growth rate. This presents a problem. When a cell is growing rapidly, its molecular machinery is geared for peak performance, and the synthesis of new ribosomes is frenetic, trying to keep pace with rapid cell division. The more ribosomes a cell contains, the faster that cell can make new proteins and the faster it can grow. But what happens when the party's over—when poor carbon and energy sources cannot supply enough energy to maintain rapid cell division? Without a way of curbing ribosome construction, cells would soon fill with idle ribosomes. Under these conditions, bacteria undergo a process called the **stringent response**. The stringent response causes a decrease in the number of rRNA transcripts made for ribosome assembly and alters the expression of numerous other genes.

In the stringent-response strategy, idling ribosomes produce a signal molecule called guanosine tetraphosphate (ppGpp) that interacts with RNA polymerase and lowers its ability to transcribe genes encoding ribosomal RNA (**Fig. 10.17**). How is ppGpp made? When an uncharged tRNA binds at the ribosome A site, which can happen during amino acid starvation, a ribosome-associated protein called RelA transfers phosphate from ATP to GTP to form ppGpp. This signal nucleotide interacts with the beta subunit of RNA polymerase and diminishes its recognition of promoters for operons producing rRNA and tRNA. The result is the down-regulation of rRNA and tRNA synthesis. The less rRNA scaffolding there is available for building ribosomes, the fewer ribosomes will be produced.

This raises another question. Even though the synthesis of ribosomal RNA has been curtailed, won't the cell continue to waste resources on the synthesis of ribosomal proteins? It turns out that some ribosomal proteins can bind to the mRNA that encodes them and inhibit translation. So when there is less rRNA in the cell, free ribosomal proteins accumulate in the cytoplasm, unassociated with ribosomes. These excess ribosomal proteins begin to bind to their own mRNA molecules and inhibit the translation of their own coding regions, as well as

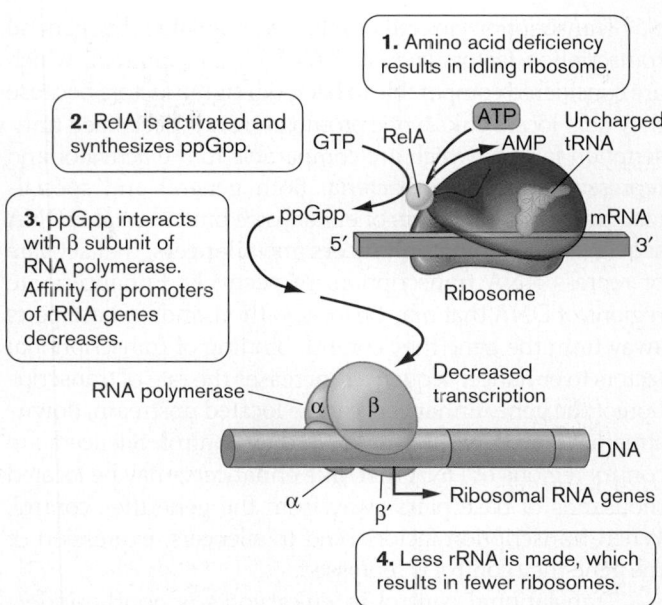

Figure 10.17 Ribosome-dependent synthesis of guanosine tetraphosphate and the stringent response.

the coding regions of other ribosomal proteins residing on the same polycistronic mRNA. This is an example of **translational control**, where regulation affects translation of an mRNA by ribosomes rather than transcription by RNA polymerase.

In this section we have touched on several ways bacteria control the expression of operons. There are many more examples, some of which (for example, riboswitches; see eTopic 15.2) are described in the metabolism section of this textbook.

In Eukaryotes, Most Genes Are Controlled Individually

There are a number of differences between gene regulation in prokaryotes and eukaryotes. In eukaryotes, most genes are present as isolated sequences encoding a single polypeptide, rather than as multigene operons. Eukaryotic microbes, like multicellular animals and plants, possess three different RNA polymerases: RNA polymerases I and III (pol I and pol III) to transcribe small functional RNAs and RNA polymerase II (pol II) for mRNA. Prokaryotes generally use a single RNA polymerase. Protein-coding sequences of eukaryotic microbes are interrupted by introns that do not encode protein. Special splicing mechanisms are required to remove the introns from mRNA prior to translating the exons that make up a single protein. Single-celled eukaryotic microbes generally have fewer introns than multicellular eukaryotes have. There are even fewer examples of introns in the prokaryotic world. Some introns occur in archaea—a feature of genetics that archaea share with eukaryotes.

Transcription in eukaryotes is regulated by general transcription factors such as TATA-binding protein, which are considered comparable to bacterial sigma factors because they help locate eukaryotic promoters; and specialized transcription factors, which are comparable to the activator and repressor proteins of bacteria. Both general and specialized transcription factors of eukaryotes bind regulator DNA sequences, including **enhancers** and **silencers**, that activate or repress RNA transcription, respectively. Enhancers are regions of DNA that may be located thousands of base pairs away from the gene they control. Binding of transcriptional factors to enhancer sequences increases the rate of transcription of the gene. Enhancers can be located upstream, downstream, or even within the gene they control. Silencers are control regions of DNA that, like enhancers, may be located thousands of base pairs away from the gene they control. When transcription factors bind to silencers, expression of the gene they control is repressed.

Translational control in eukaryotes is poorly understood. One form of control involving translation is nonsense-mediated mRNA decay, which terminates transcription in the nucleus. As the mRNA is being transcribed, nuclear ribosomes begin to translate the message. If a ribosome runs into a stop codon (resulting from faulty RNA transcription), transcription terminates and the faulty RNA is degraded.

TO SUMMARIZE:

- **DNA-binding proteins** often recognize symmetrical DNA sequences.
- **Many anabolic pathway genes** (for example, for amino acid biosynthesis) are repressed by the end product of the pathway (for example, the amino acid), which binds to a corepressor that inhibits transcription.
- **AraC-like proteins** are a large family of regulators, present in many bacterial species, that can activate or repress operons by assuming different conformations.
- **Attenuation** is a transcriptional regulatory mechanism in which translation of a leader peptide affects transcription of a downstream structural gene.
- **Idling ribosomes** synthesize the signal molecule ppGpp, which interacts with the beta subunit of RNA polymerase and decreases the affinity of RNA polymerase for ribosomal RNA and a variety of other genes needed for rapid growth. The result, called the **stringent response**, is lower transcription of these genes, reduced levels of rRNA, and a decrease in the number of ribosomes.

10.4 Sigma Factor Regulation

DNA repressor and activator proteins are great for controlling the expression of individual genes and operons and are even used to coordinate expression of regulons.

A **regulon** is a set of genes scattered around the chromosome that have related functions (such as tyrosine biosynthesis). In many cases, such as in the stringent response, bacteria need to coordinately activate large sets of genes and operons of seemingly disparate function that nevertheless collaborate to aid survival in particularly hostile environmental situations. One way to indirectly regulate expression of large sets of genes is to first regulate the synthesis or activity of the sigma factor that directs the expression of all those genes.

Many bacteria employ alternative sigma factors to direct transcription of distinct sets of genes (see Section 8.2). For example, many Gram-negative bacteria, such as *E. coli* or *Salmonella enterica*, use the sigma factor called sigma S (also called sigma-38, RpoS, σ^S, or σ^{38}) to initiate the transcription of a variety of genes associated with survival in stationary phase. The level of sigma S is low in exponentially growing cells but rises dramatically as cells enter stationary phase. There are myriad ways to control whether a sigma factor gene is expressed and, if expressed, whether its product accumulates or functions.

Sigma Factor Activity Can Be Regulated

The cell can regulate an enzymatic reaction by changing the amount of the enzyme in a cell or by changing the *activity* of an enzyme already present. The same is true for sigma factors. Some microbes use **anti-sigma factor** proteins to inhibit sigma factor activity. Anti-sigma factor proteins target specific sigma factors, blocking their access to core RNA polymerase (that is, blocking their activity). This strategy prevents expression of that sigma factor's target genes.

Salmonella uses an anti-sigma factor (FlgM) to time events involved in constructing flagella. The transcription factor sigma F (also called sigma-28, RpoF, σ^F, or σ^{28}) is required to synthesize proteins used in the last stages of flagellar biosynthesis. The anti-sigma factor FlgM, however, keeps sigma F function at bay until membrane assembly of the flagellar basal bodies is complete. Once completed, the basal body selectively secretes the anti-sigma factor from the cell. This frees intracellular sigma F to direct transcription of the final set of flagellar assembly genes.

But what happens when the blocked sigma factor is suddenly needed? How do cells counter anti-sigma factors? Some systems counter anti-sigma factors with **anti-anti-sigma factors** that bind the blocking protein, like a decoy, and free the sigma factor to join core RNA polymerase. Other systems link the liberation of a bound sigma factor with a cell cycle event or with the assembly of a structure. *Bacillus* species, for instance, produce spores. During the process of making a spore, the cell couples formation of a forespore cross-wall (see Section 4.7) to the activation of an anti-anti-sigma factor, which triggers the release of a whole cascade of new sigma factors needed only to complete spore formation (discussed later in this section).

Sigma Factor Translation Is Regulated

The synthesis or accumulation of sigma factors can also be regulated. Regulation of the heat-shock sigma factor sigma H (also called sigma-32, RpoH, σ^H, or σ^{32}) in *E. coli* illustrates how the synthesis of a sigma factor can be controlled at the level of translation (**Fig. 10.18**). Excessive heat, above 42°C for *E. coli*, will cause proteins to denature and membrane structure to deteriorate. All cells subjected to heat above their comfort zone (optimal growth range) will express a set of proteins called heat-shock proteins. These proteins include chaperones that refold damaged proteins and a variety of other proteins that affect DNA and membrane integrity. The transcription of many *E. coli* heat-shock genes requires the sigma factor sigma H. So one of the first consequences of growth at elevated temperature is an increase in the amount of sigma H present in the cell.

The level of sigma H is regulated by two temperature-dependent mechanisms; one of these controls the rate of sigma H synthesis, while the other determines its rate of proteolysis. The gene encoding sigma H is *rpoH*. At 30°C, *rpoH* mRNA adopts a secondary structure at the 5′ end that buries a ribosome-binding site, so *rpoH* mRNA is poorly translated. A sudden rise in temperature melts this secondary structure and exposes the ribosome-binding site, allowing translation to occur more readily. Thus, heat shock increases sigma H synthesis, which in turn increases transcription of the heat-shock genes whose products include chaperones and proteases.

Sigma Factor Levels Are Regulated by Proteolytic Degradation

Proteolysis is another mechanism used to control the level of certain sigma factors. It works by limiting sigma factor accumulation. For example, at 30°C, the *rpoH* message is poorly translated owing to secondary structure, as previously described, but some sigma

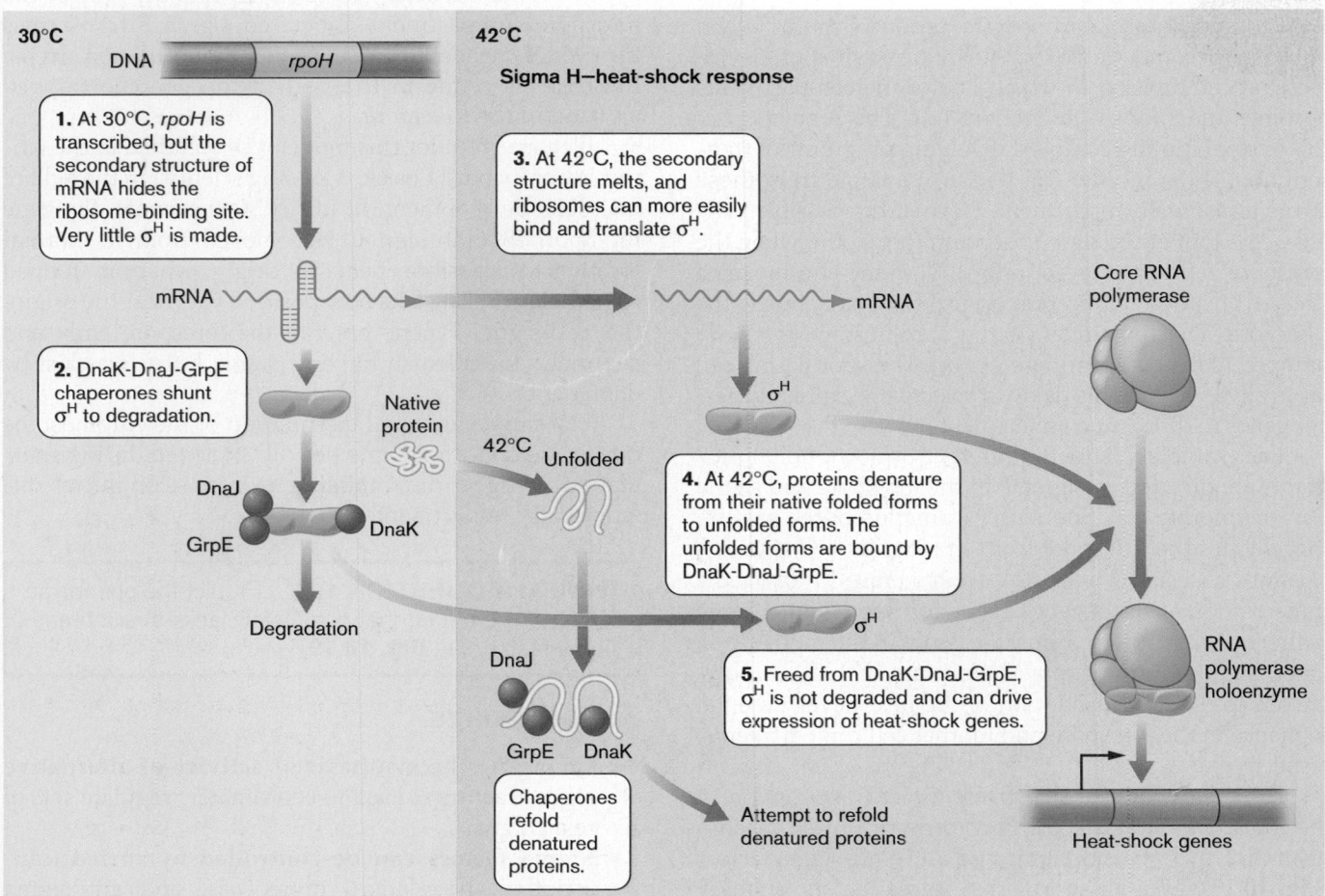

Figure 10.18   The heat-shock response of *E. coli*. Two mechanisms control sigma H levels. The small amount of sigma H that can be made at 30°C is met by the DnaK-DnaJ-GrpE chaperone system and shuttled toward degradation. At 42°C, however, misfolded cytoplasmic proteins siphon off the chaperone trio and release sigma H to direct transcription of the heat-shock genes.

H protein is made. To prevent inappropriate expression of the sigma H–dependent heat-shock genes at 30°C, the DnaK-DnaJ-GrpE chaperone system interacts with sigma H and shuttles it to various proteases for digestion (see **Fig. 10.18**). At 42°C, however, proteolysis of sigma H decreases, and sigma H is allowed to accumulate. Degradation decreases because at the higher temperature, the chaperones are siphoned away from sigma H by the many other heat-denatured proteins formed. The new goal of the chaperones is to refold and rescue those damaged proteins. Chaperone redeployment frees sigma H to transcribe the heat-shock genes, which include the chaperone genes *dnaK*, *dnaJ*, and *grpE*. Thus, as the temperature rises, the amount of sigma H is increased by two temperature-dependent mechanisms: One increases translation by exposing the ribosome-binding site, while the second redeploys chaperones that direct its proteolysis.

In Sporulation, Different Sigma Factors Are Activated in the Mother Cell and Forespore

Bacillus and *Clostridium* species produce spores when nutrients become scarce. Sporulation requires an asymmetrical cell division in which one of the compartments becomes the spore (see Section 4.7). The regulation of this system is quite complex, involving programs of transcription in the mother cell that are separate from those in the forespore compartment. How is this possible? The answer is that at the time of septum formation, when the forespore is first produced, only 30% of the chromosome (the part near the replication origin) is actually inside the forespore. The remainder of the chromosome is slowly pumped in over a 15-minute period. The sporulating cell takes advantage of this delay to selectively express different genes in different compartments.

For example, distinctly different transcriptional programs are directed by sigma F in the forespore and sigma E in the mother cell. Soon after formation of the septum that divides the forespore from the mother cell, sigma F becomes associated with RNA polymerase to direct the transcription of select genes in the forespore while sigma E directs transcription of specific genes in the mother cell. Later, sigma G and sigma K associate with RNA polymerase to direct transcription of the next developmental sequence in the forespore and mother cell compartments, respectively.

How does the cell selectively direct transcription in the forespore? It starts with an intricate timing mechanism that involves anti-sigma and anti-anti-sigma factors (**Fig. 10.19**). Like all sporulation sigma factors, sigma F is inactive when first synthesized. It is inactive because it binds to an anti-sigma F protein dubbed SpoIIAB that is cosynthesized with sigma F in the predivisional cell

(**Fig. 10.19A**, step 1). After division, both proteins become equally partitioned between the two compartments. However, in addition to an anti-sigma, there is an anti-anti-sigma F protein, called SpoIIAA (AA for "anti-anti"), which is also equally distributed (**Fig. 10.19A**, steps 2 and 3). Anti-anti factor frees sigma F from anti-sigma F, which is degraded by another protein complex, the ClpPC protease. This is not a problem in the mother cell because that compartment just makes more AB (anti-sigma), so sigma F is never active.

But all of the previous events occur in both compartments (forespore and mother cell). So why is sigma F activation limited to the forespore? Recall that at the time of septum formation, one chromosome is slowly pumped into the forespore. The gene encoding the AB anti-sigma factor is located far from the origin and, as such, is not in the part of the genome trapped in the forespore immediately after septal formation (**Fig. 10.19B**, step 2). Thus, no new AB (anti-sigma) is made in the forespore during this time, and what was there is inactivated by anti-anti factor and degraded by ClpPC (**Fig. 10.19B**, steps 3 and 4). This depletion of anti-sigma in the forespore efficiently frees sigma F to act only in the forespore compartment. Liberated sigma F transcribes the genes needed for subsequent sporulation steps; these genes reside in the part of the genome trapped early on in the forespore.

Elegant proof for this model was obtained in the laboratory of Richard Losick, a leading scientist in the field of microbial development. In Losick's experiment, the gene for SpoIIAB (anti-sigma) was moved from its normal location to a position near the origin, where it stopped sporulation. Because its new position was near the origin, the anti-sigma F gene entered the forespore early and continued to replenish the anti-sigma F that was lost by degradation.

The simple, unequal distribution of the chromosome during septal formation is heavily exploited during sporulation to trigger differential gene expression in two different cell compartments.

> **THOUGHT QUESTION 10.5** Predict the phenotype of a *spoIIAA* mutant that completely lacks this anti-anti-sigma factor (see **Fig. 10.19**).

TO SUMMARIZE:

- **Changing the synthesis or activity of alternative sigma factors** is used to coordinately regulate sets of related genes.
- **Sigma factors can be controlled** by altered transcription, translation, proteolysis, and anti-sigma factors.
- **Secondary structures** at the 5′ end of mRNA can obscure access to ribosome-binding sites. Conditions

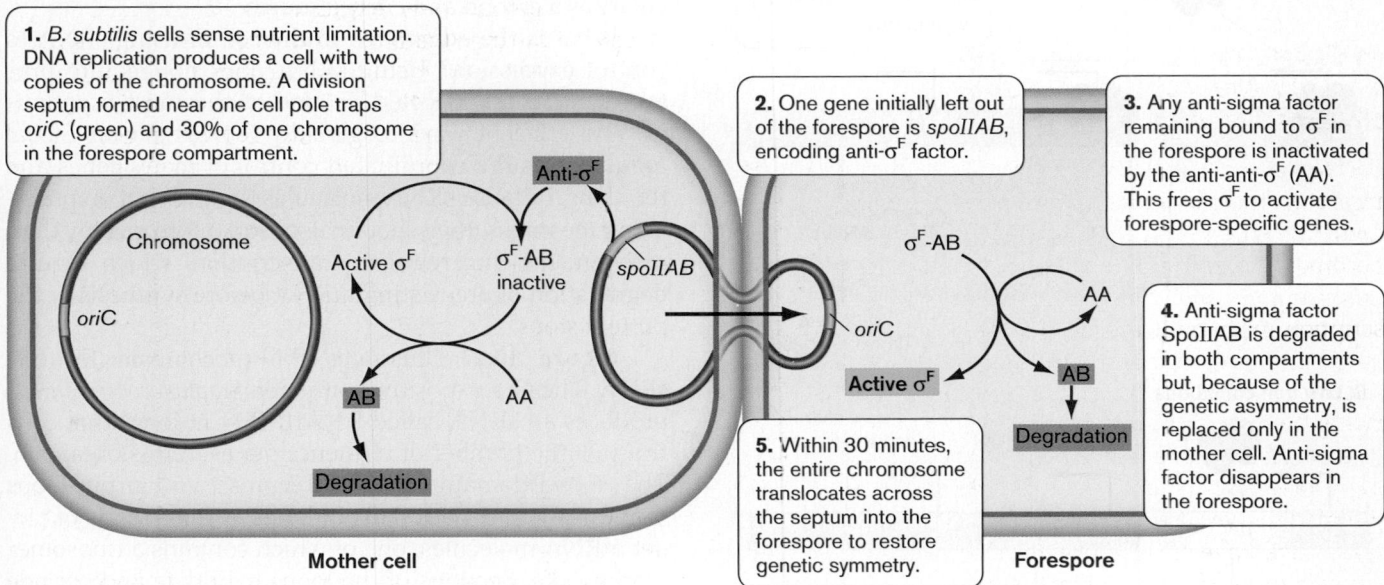

A. Vegetative cell—σ^F is inactive

Anti-σ^F is constantly replenished by *spoIIAB* transcription

1. σ^F is inactivated by anti-σ^F (AB).

Anti-σ^F (AB)

Active σ^F

σ^F-AB inactive

Anti-anti (AA)

2. Anti-anti-σ^F (AA) causes anti-sigma (AB) to dissociate from σ^F.

AB

Degradation

3. Anti-sigma (AB) is constantly degraded but replenished by transcription of *spoIIAB*.

Figure 10.19   Regulating sporulation by genetic asymmetry.   A. Summary of mechanisms that control sigma F activity in growing cells of *Bacillus subtilis*. The interactions of sigma F, anti-sigma F (AB), and anti-anti-sigma (AA) collectively keep sigma F inactive. **B.** Sigma F is activated in the forespore until the forespore chromosome is pumped into the forespore. Activation of sigma F leads to synthesis of sigma G in the forespore (not shown), which directs the next stage of spore development.

B. Forespore—σ^F is active

1. *B. subtilis* cells sense nutrient limitation. DNA replication produces a cell with two copies of the chromosome. A division septum formed near one cell pole traps *oriC* (green) and 30% of one chromosome in the forespore compartment.

2. One gene initially left out of the forespore is *spoIIAB*, encoding anti-σ^F factor.

3. Any anti-sigma factor remaining bound to σ^F in the forespore is inactivated by the anti-anti-σ^F (AA). This frees σ^F to activate forespore-specific genes.

Chromosome

oriC

Active σ^F

σ^F-AB inactive

Anti-σ^F

spoIIAB

oriC

σ^F-AB

AA

Active σ^F

AB

AB

AA

Degradation

4. Anti-sigma factor SpoIIAB is degraded in both compartments but, because of the genetic asymmetry, is replaced only in the mother cell. Anti-sigma factor disappears in the forespore.

5. Within 30 minutes, the entire chromosome translocates across the septum into the forespore to restore genetic symmetry.

Degradation

Mother cell

Forespore

that melt the secondary structures will increase translation of the sigma factor.

- **Chaperones** can direct certain sigma factors toward degradation by proteases. Conditions that draw those chaperones away from the sigma factor will allow the sigma factor to accumulate.
- **Sporulation** relies on the hierarchical activation of a series of different sigma factors.
- **Temporary asymmetrical distribution** of the chromosome between the forespore and the mother cell results in differential activation of compartment-specific alternative sigma factors.
- **Sigma F is a forespore-specific sigma factor** regulated by anti-sigma and anti-anti-sigma factors. Sigma F is preferentially activated by temporary

asymmetrical distribution of the chromosome during the early stages of sporulation.

10.5 Small Regulatory RNAs

A considerable fraction of a bacterial chromosome does not encode mRNA, rRNA, or tRNA. Many of these intergenic regions encode untranslated small RNA (sRNA) molecules that carry out a variety of biological functions. Examples of sRNA can be found in many different organisms, ranging from bacteria to mammalian cells. Many act as regulators of gene expression at a posttranscriptional level, either by interacting with proteins or by acting as **antisense RNAs** that bind to complementary sequences of target transcripts and stimulate or prevent translation. Regulatory RNAs are involved in the control of a variety of processes, such as plasmid replication, transposition in both prokaryotes and eukaryotes, phage development, viral replication, bacterial virulence, environmental stress responses, and developmental control in lower eukaryotes.

An example of regulation by an sRNA occurs within the Fur (ferric uptake regulator) regulon of *E. coli* (**Fig. 10.20**). (The genes that make up the Fur regulon, despite being scattered around the chromosome, are all controlled by Fur protein.) Iron is an important component of living systems. However, too much iron can be detrimental to the cell by increasing oxidative stress. Fur is a repressor

protein that regulates genes whose products scavenge iron from the environment or store iron in the cell (see Section 4.2). When intracellular iron levels are high, Fur represses expression of scavenging genes but induces production of iron storage proteins (**Fig. 10.20A**). Fur represses gene expression by directly binding to specific DNA sequences in front of target genes. But how does Fur activate other genes? One mechanism involves controlling production of a small RNA called RhyB (90 nt) (see **Fig. 10.20**). When iron levels are low, RhyB sRNA down-regulates production of several iron-storing and iron-using proteins (such as succinate dehydrogenase). In this way, the cell prevents the shuttling of limited iron resources in wasteful directions. The cell needs to use the available iron, not store it. When iron is plentiful, however, Fur will directly repress *rhyB* and indirectly promote expression of the storage genes. Succinate dehydrogenase, an iron-containing enzyme, can also be made, which enables the use of succinate as a carbon and energy source.

What is the advantage to the cell of using sRNA to control expression? Using sRNA does not require protein synthesis and could be one of the most economical and efficient ways to globally repress genes. Global regulation is the coordinated control of many genes and regulons. Because sRNA molecules typically act on preexisting messages, they should also work more quickly than mechanisms that regulate transcription, which require degradation of preexisting mRNA before synthesis of the protein stops.

Figure 10.21 illustrates the mechanism for one sRNA. The Gram-positive pathogen *Staphylococcus aureus* produces an sRNA called RNAIII (514 nt long) that controls a large number of virulence genes in this organism. The 3′ end domain of RNAIII forms two hairpin loops that can interact with hairpin loops at the 5′ ends of target mRNA molecules, one of which contains a ribosome-binding site. Sections of the loops hybridize and occlude ribosome access to the binding site. This antisense mechanism inhibits translation. The hybrid molecule also produces an alluring target for a ribonuclease that will degrade the message. RNAIII is expressed after the bacterial population reaches a certain density during infection (see the discussion of quorum sensing in Section 10.8). It then inhibits the translation of, and degrades, mRNAs required earlier in the infection (for example, those for some exotoxins and surface attachment proteins).

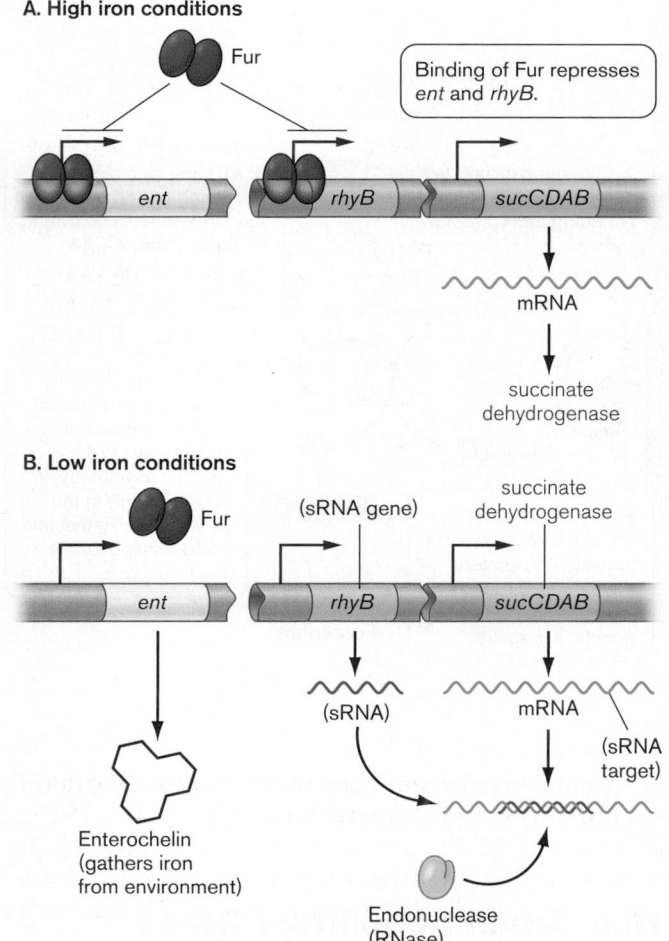

Figure 10.20 Activity of a small regulatory RNA molecule. **A.** When iron levels are high, Fur repressor protein binds to the *ent* and *rhyB* Fur box DNA sequence (a short specific DNA sequence in front of the genes regulated by Fur) and represses their expression. Enterochelin is no longer made, but the *sucCDAB* message encoding succinate dehydrogenase can be translated. **B.** Under low iron conditions, the small RNA *rhyB* is expressed. RhyB sRNA binds to the *sucCDAB* message and renders it susceptible to an RNase.

sRNA Genes Are Hard to Identify

How common are regulatory sRNA molecules? How can you look at a genome and predict what sequences encode sRNA genes? Identifying sRNA genes requires a different approach than identifying ORFs, which is relatively easy to do. ORFs have a translational start codon heading up a string of amino acid codons; sRNAs do not.

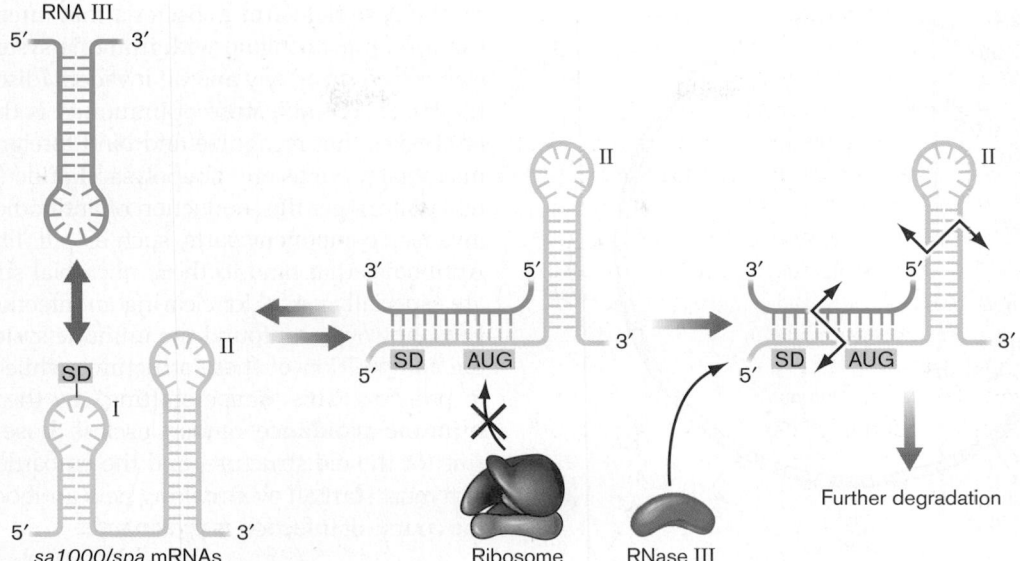

Figure 10.21   Model for small RNA function.   RNAIII binds to its target mRNAs (*sa1000/spa*) via loop-loop interactions. The initial base pairings extend, leading to the formation of an extended duplex. Binding of RNAIII hinders ribosome binding and promotes the access to RNase III. SD (Shine-Dalgarno ribosome-binding site) and AUG are in green, RNAIII is in red, and the mRNA target is in black. Broken arrows indicate degradation. *Source:* S. Boisset et al. 2007. *Genes Dev.* **21**:1353–1366.

A.

B.

Figure 10.22 Susan Gottesman (**A**, center) and Gisela Storz (**B**), both from the National Institutes of Health, played instrumental roles in establishing the importance of sRNA molecules in bacteria.

The translational start codon of an ORF is preceded by a potential ribosome-binding site, which is itself preceded by a potential –10/–35 (or similar) promoter site sequence. Genes encoding sRNA will not have the ribosome-binding site but should have a recognizable promoter (see Sections 8.1 and 8.2). The promoter is not, however, enough to mark a potential sRNA gene, because all genomes contain sequences that resemble promoters yet fail to bind RNA polymerase.

The laboratories of Susan Gottesman and Gisela Storz at NIH (**Fig. 10.22**) discovered that genes encoding known sRNA molecules always occur in intergenic regions between genes encoding ORFs and exhibit high degrees of homology between related species (greater than 80%). The researchers used this knowledge to compare the intergenic regions of *E. coli* with those of *Salmonella* and *Klebsiella*. After eliminating tRNAs, rRNAs, and other repetitive elements, they ended up with 295 possible sRNA candidates. Next, a microarray experiment (see Section 10.9) that included oligonucleotide probes to intergenic regions asked whether any of the predicted sRNA molecules were actually made. Using this strategy, 23 new sRNA species were identified (**Fig. 10.23**). It appears that sRNA genes are common and integral players in gene control.

Compared with prokaryotes, eukaryotic microbes possess an even broader array of small functional RNA molecules, including those of the spliceosome and of various cytoplasmic protein-RNA complexes. The genomes of eukaryotes such as yeast reveal thousands of potential

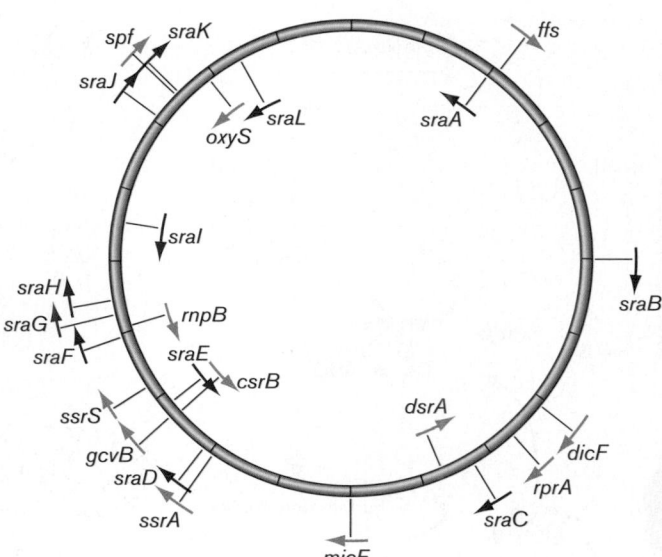

Figure 10.23 Predicted sRNA genes in *E. coli*. Location of various sRNA genes identified in the *E. coli* genome. Orange arrows depict sRNAs of proven function; red arrows indicate predicted sRNAs of unknown function. Arrowheads indicate direction of transcription.

sRNA sequences. The significance of sRNA discovery was highlighted by the decision of the journal *Science* to designate sRNA as the "Molecule of the Year" in 2002.

TO SUMMARIZE:

- **Small regulatory RNAs** found within bacterial intergenic regions regulate the transcription or stability of specific mRNA molecules.
- **The antisense nature** of sRNA allows these molecules to bind target mRNA. Binding can either stabilize the target mRNA or make it susceptible to degradation.
- **Identifying sRNAs** is more complex than identifying ORFs.
- **Eukaryotic microbes** use small interfering RNA to silence genes.

10.6 DNA Rearrangements: Phase Variation by Shifty Pathogens

All of the operon mechanisms previously described involve binding reactions between DNA, RNA, and proteins. More drastic means of control involve altering the DNA sequence itself.

Like a chameleon changing its color, some microbes use gene regulation to periodically change their appearance in a process called **phase variation**. Phase variation involves changing the amino acid composition of a particular protein on the bacterial surface. Bacteria that infect mammals, for example, must contend with immune systems whose mission is to destroy any and all invaders (discussed in Chapter 24). A central feature of immunity is the production of antibodies that recognize and bind foreign structures like microbial proteins and lipopolysaccharide (LPS). An infection will trigger the production of antibodies specific to the invader's component parts, such as pili, flagella, and LPS. Antibodies that bind to these microbial surface structures are especially useful for clearing an infection. Some pathogens, however, confound the immune system by changing the composition of these structures while the infection is in progress. This "shape-shifting" by the microbe, called **immune avoidance**, renders useless those antibodies specific for the old structure, and the embattled immune system must start all over making new antibodies. As a result, the course of infection is prolonged.

Gene Inversion: An On-Off Switch

Flagellar phase variation in the Gram-negative bacterium *Salmonella enterica* is a classic example of the regulatory genre known as gene inversion. Gene inversion is a recombinational event that flips the orientation of a gene or DNA segment in the chromosome, thereby turning that gene, or an adjacent gene, on or off. A major cause of diarrhea, *S. enterica* periodically changes the type of protein, called flagellin, used to make its flagella. Each organism has two genes, widely separated on the chromosome, that encode different forms of flagellin. The mechanism of the switch is a site-specific DNA recombination that turns off one gene while turning on the other. The target of the switch is a 993-bp DNA fragment (or cassette), called the H region, that contains an outwardly directed promoter and a gene called *hin*, whose product, Hin recombinase (also called Hin invertase), mediates the recombination (**Fig. 10.24**). In antigenic parlance, the term "H antigen" refers to flagella, so the acronym Hin stands for H inversion. The DNA cassette is flanked by short (26-bp) inverted repeats called *hixL* (left) and *hixR* (right).

NOTE: An **inverted repeat** is a sequence found in identical (but inverted) forms at two sites on the same double helix (for example, 5'-ATCGATCG-nnnnnnnnnnnnnnnCGATCGAT-3'). A **direct repeat** is a sequence found in identical form at two sites on the same double helix (for example, 5'-ATCGATCG-nnnnnnnnnnnnnnnATCGATCG-3'. A **tandem repeat** is a direct repeat without any intervening DNA sequence (for example, ATCGATCGATCGATCGATCGATCG).

WWW | Biomolecules at Kenyon: Hin recombinase molecular tutorial
microbiology2.com/links

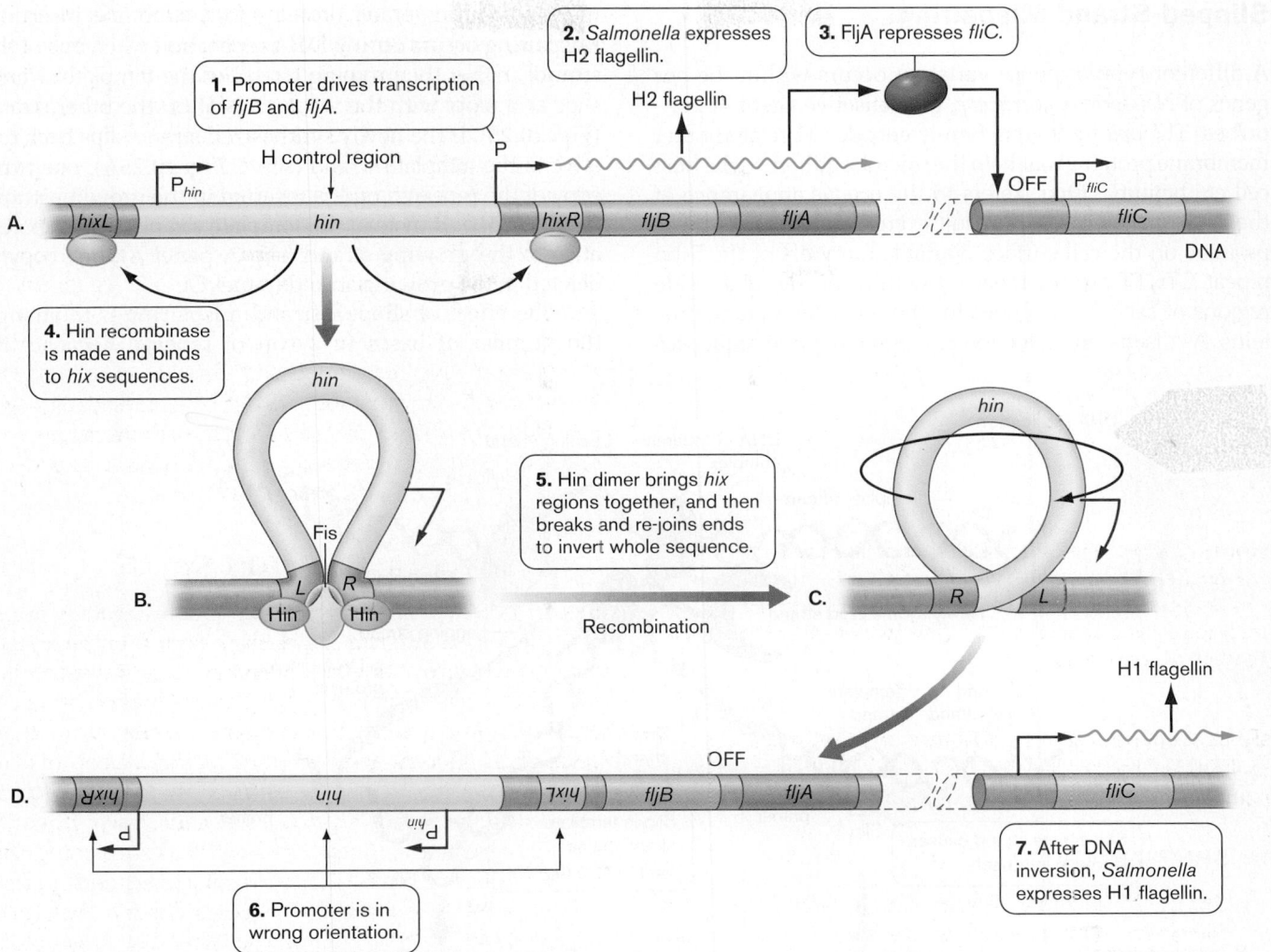

Figure 10.24   Phase variation of flagellar proteins in *Salmonella enterica*.   An invertible region containing a promoter controls the expression of two unlinked flagellar protein genes. In one orientation (**A**), the promoter drives synthesis of H2 flagellin (*fljB*) and a repressor (FljA) of the H1 flagellin gene (*fliC*). Action by the Hin recombinase causes the segment to invert (**B** and **C**), thereby removing the promoter. Because the repressor FljA is not formed, the gene for H1 can be expressed (**D**).

The Hin recombinase collaborates with other less specific DNA remodeling proteins, such as Fis, to engineer an association between the 26-bp left (*hixL*) and right (*hixR*) ends of the invertible DNA element. The two ends, each bound to a Hin monomer, are brought together by Hin-Hin protein interactions. DNA within the cassette then forms a loop. Hin cuts within the center of each *hix* site, producing staggered ends. An exchange of Hin subunits is thought to lead to strand inversion, so that the orientation of the DNA cassette is reversed relative to the flanking DNA on either side.

In one orientation, the outwardly directed promoter of the H region directs expression of H2 flagellin (encoded by *fljB*) and a repressor (FljA) that prevents transcription of the other flagellin gene, *fliC* (see **Fig. 10.24A**). After the switch, however, the promoter points in the wrong direction, so there is no production of H2 flagellin or FljA, the

repressor of *fliC* (see **Fig. 10.24D**). Lacking this repressor, the *fliC* flagellin gene is expressed. Thus, H1 flagellin (present in phase 1 cells) is synthesized instead of H2 flagellin, which is present in phase 2 cells. The amino acid sequences, and thus the antigenicities, of the two flagellar proteins are different, so this mechanism allows *Salmonella* to change its appearance to a host immune system. In each generation, the rate of the reversible switch varies from about one cell in 10^3 to one in 10^5. Note, however, that this switch would not accomplish much if the infecting population of bacteria started out mixed—that is, producing both types of flagella. The initial infection must be of one phenotype.

THOUGHT QUESTION 10.6   What is the phenotype of a *fljA* mutant? A *fliC* mutant?

Slipped-Strand Mispairing

A different type of phase variation occurs within the *opa* genes of *Neisseria gonorrhoeae*, the causative agent of gonorrhea. The *opa* multigene family encodes 11 related outer membrane proteins that help the microbe adhere to the host cell epithelium. "Opa" refers to the *opaque* appearance of the bacterial colonies that result from the presence of these proteins on the cell surface. Variable numbers of the 5-bp repeat CTCTT are interspersed within the signal peptide regions of each of the genes for the outer membrane proteins. As discussed in Section 8.5, signal peptide sequences

mark some membrane proteins for membrane insertion. Mispairing occurs during DNA replication when one of the strands, either the growing strand or the template strand, slips and pairs with the wrong repeat on the other strand (**Fig. 10.25**). If the newly synthesized strand slips back relative to the template strand (slip 1, **Fig. 10.25A**), one extra copy of the repeated unit is inserted in the growing strand (**Fig. 10.25B**). If instead the template strand slips back relative to the growing strand (slip 2, panel A), one copy is deleted in the growing strand (panel C).

The effect of slipped-strand mispairing is to change the number of bases in a run of repeats. Because the

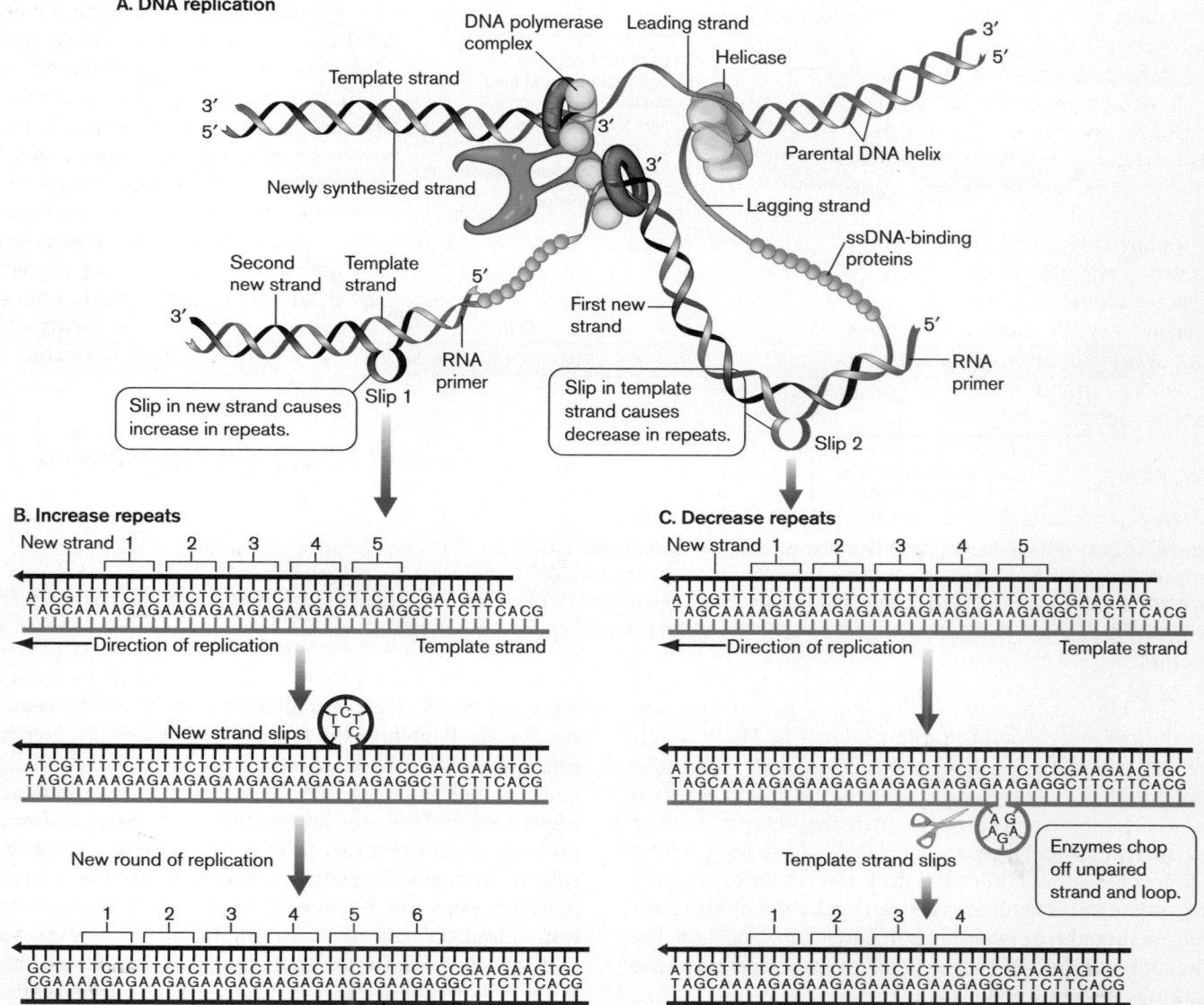

Figure 10.25 Slipped-strand mispairing. **A.** A DNA polymerase dimer moving along a replication fork. Slipping events (1 and 2) are portrayed on the lagging strand. **B.** The series of events that occur if the newly synthesized strand slips back relative to the template strand. When this new dsDNA, containing the looped-out slipped strand, is subsequently replicated, one of the daughter dsDNA molecules will have an extra repeat because of the slippage. **C.** What happens if, prior to replication, the original template strand slips back relative to the newly synthesized complementary strand? If repair enzymes get to the loop prior to replication, the loop is excised and a pentanucleotide repeat is removed. Subsequent daughter strands will be shortened.

length of the repeated sequence is not an even multiple of three, its insertion or deletion within the coding region of a gene generates a frameshift. The frameshift causes the gene to encode a run of amino acids completely different from that encoded by the original sequence (see Section 9.4). For *opa*, the number of CTCTT repeats in the amino-terminal portion of the gene determines whether the full-length protein is made (in which case the *opa* gene is considered on) or not (in which case the gene is off). One repeat more or less means a shift in the reading frame of the *opa* message so that a translation stop codon materializes, terminating production of that protein. The frequency of such slippage is very high; each time the bacteria divide, approximately one out of every 100 or 1,000 daughter cells will carry a mutation that changes the number of CTCTT repeats.

Each *opa* gene switches on or off independently. As a result, none, one, or several Opa polypeptides (produced from different *opa* genes) can be expressed simultaneously. Because no regulatory proteins are made, this is a wonderful low-maintenance way to evade the immune system.

Eukaryotic microbes, especially pathogenic sporozoa, possess elaborate mechanisms of phase variation. The trypanosome that causes "sleeping sickness" undergoes extensive genetic shuffling and mutation of its coat proteins over successive generations, essentially overwhelming the host immune system by presenting every possible form of antigen.

SUMMARY:

- **In phase variation**, the structure of a bacterial protein reversibly changes from one form to another via a genetic mechanism.
- **An invertible promoter switch** regulates two genes of *Salmonella enterica* encoding two alternative structural types of flagellin.
- **Slippage while replicating** through repetitive DNA sequences, a process called **slipped-strand mispairing**, can reversibly activate or inactivate a series of outer membrane proteins in *Neisseria gonorrhoeae*.

10.7 Integrated Control Circuits

The regulatory circuits outlined to this point have involved mostly single-switch mechanisms. But microbes build on these mechanisms by combining multiple switches with redundant feedback controls that can form a kind of integrated gene circuit. Integrated circuits can send a virus down alternate lifestyle paths (lysis or lysogeny), couple the genetic and biochemical control of a metabolic pathway (ammonia assimilation), or time the events of a developmental cycle such as sporulation. The deeper we delve into the regulatory workings that govern the expression of a cell's genome, the more clearly we see the hierarchy of control. Groups of simple control circuits are collectively regulated by other, more complex circuits, which, in turn, are governed by an even higher regulatory authority. One illustration of this hierarchy was already presented: the role of cAMP and CRP in regulating lactose and arabinose catabolism. The *lacZYA* and *araBAD* operons are each controlled separately by dedicated regulatory switches—LacI and AraC, respectively. At a higher level, however, *both* operons are coregulated by cAMP and CRP in response to energy status (whether or not glucose is present and utilized). The systems we are about to discuss represent increasingly complex examples of nature's inventiveness and need for control.

The first example we will discuss here actually has nothing to do with transcriptional/translational or post-translational control of protein production. It does illustrate the concept of two-component signal transduction, whereby one protein (the sensor kinase) senses something happening in the environment and then transmits a signal (via phosphorylation) to the second component in the system (the response regulator). Two-component signal transduction systems can be used to control transcription, but they can also be used to affect the physiology and behavior of a cell without changing transcription. The example we will use is chemotaxis, the ability of a bacterium to move toward environments favorable to growth and away from harsh environments.

Chemotaxis: How Bacteria Know Where They Are Going

Bacteria that use flagella to move do not simply dash about, "hoping" they will swim in a favorable direction. They have a sense about where they want to go and where not to go. Bacteria such as *E. coli* are actually attracted toward certain amino acids and carbohydrates present in media. They swim toward these chemicals by sensing their gradients. **Chemotaxis** is the ability of an organism to sense chemical gradients and modify its motility in response.

Before we discuss how these chemical gradients are sensed, let's first talk about how bacteria can change direction. You will recall that the flagellar motor can rotate clockwise or counterclockwise (Section 3.7, Fig. 3.44) and that on any one bacterium, all motors coordinate their rotations. Thus, all flagella on a single cell will rotate clockwise (or counterclockwise), and then suddenly and simultaneously switch to the opposite rotation. In the counterclockwise mode, all flagella sweep behind the cell, forming a rotating bundle that propels the organism forward in what is called smooth swimming or a "run." If the flagellar rotors all suddenly switch to a clockwise rotation, the bundle is disrupted and the bacterium "tumbles" in a random fashion. Once the flagellar rotors switch back to counterclockwise rotation, the bacterium will have reoriented itself in a different position, changing its swimming direction.

The key to chemotaxis is a mechanism that suppresses the number of tumbles an organism makes when it moves from a lower to higher concentration of an attractant chemical. For instance, an organism moving toward an attractant may tumble only twice in 5 seconds. In contrast, an organism moving in the wrong direction (that is, toward a lower concentration of attractant) may tumble eight times in 5 seconds. If, all of a sudden, the organism finds itself going in the right direction, the sensory transduction system will suppress the number of tumbles that occur, and the cell will continue moving in the right direction.

This is the same strategy you use to find a hamburger that you can smell cooking on a grill in your neighborhood but cannot see. You randomly walk around until the smell gets stronger, and then you keep moving in that direction until you happen upon it.

www | Video showing transition from bundled to dispersed flagella
microbiology2.com/links

How does the bacterium suppress tumble frequency? Let's start at the end of the system and work backward. The natural bias of the flagellar rotor is counterclockwise rotation (in other words, smooth swimming). In order to tumble, the cell phosphorylates a protein called CheY. CheY-P interacts with a rotor protein and causes the rotor to *reverse* direction to turn clockwise (**Fig. 10.26A**⬤, step 1). CheA protein is the kinase that phosphorylates CheY and is the "key" to chemotaxis. The more CheA kinase activity there is, the more CheY-P is made, resulting in a higher frequency of tumbling (reorienting) events. Another protein, CheZ, continually dephosphorylates CheY. When *E. coli* senses an attractant chemical like an amino acid at the cell surface, the result is *decreased* CheA kinase activity, which leads to lower CheY-P levels. The presence of fewer CheY-P molecules allows longer periods of counterclockwise rotation and, consequently, extended smooth swimming toward the attractant.

This makes sense, but how does a cell "know" that it has moved into an area with attractant? The answer begins with clusters of special membrane-spanning proteins called **methyl-accepting chemotaxis proteins** (**MCPs**, or chemoreceptors) located at the poles of *E. coli* and other bacteria (**Fig. 10.26A**, step 1). The periplasmic domains of different MCPs bind to different attractant molecules that flow into the periplasm as the cell moves. The cytoplasmic domain of each MCP binds to the CheA kinase noted earlier (via an intermediary protein, CheW) and controls CheA activity. When a chemoattractant chemical, such as serine, binds to the periplasmic side of

the MCP, the conformation of the cytoplasmic domain changes and *inhibits* CheA kinase activity (step 2). The result is that the CheY-P level decreases (because CheZ continues to dephosphorylate CheY-P) and smooth swimming ensues.

This, too, makes sense when an organism *first* encounters a chemoattractant. But how does the cell know to keep moving into even *higher* concentrations? As it happens, the conformational change in the MCP that inactivates CheA kinase activity also subjects the cytoplasmic side of the MCP to methylation (via *S*-adenosylmethionine) by CheR methylase (**Fig. 10.26A**, step 3). Methylation of glutamate residues in the cytoplasmic struts of the MCP reactivates the CheA kinase but desensitizes the MCP so that an even higher concentration of attractant must be present in order to bind the MCP and inhibit CheA kinase. So, if the organism is moving into a higher concentration of attractant, CheA is again inactivated, tumbling is suppressed, and the cell prolongs the run. If it is not moving into higher concentrations of attractant, CheY-P is made again and the organism will tumble. If the bacterium moves to a lower concentration of attractant, the cell will keep tumbling at a high frequency until it moves back into a higher concentration, when tumbling is suppressed to allow a smooth run.

When the cell moves away from attractant, the methylation switch is reset by another protein, CheB-P, which removes methyl groups so that the system is resensitized to attractant (**Fig. 10.26A**, step 4). The system is elegantly fine-tuned in that CheA kinase is also the protein that phosphorylates CheB. So, as the cell moves away from attractant, CheA kinase phosphorylates CheY to produce tumble and phosphorylates CheB to reset the sensitization switch.

MCP sensory proteins are not uniformly distributed around the cell surface, but assemble as clusters at both poles of *E. coli*. This clustering seems logical because the bacterium points with its pole in the direction it swims. **Figure 10.27** depicts one such MCP cluster (yellow) at one pole of the bacterium and illustrates the diffusion of CheY-P toward the opposite pole. CheY-P will interact with flagellar rotors all along the cell surface, causing tumble.

It is interesting to note that while the basic chemotactic mechanism is evolutionarily conserved in bacteria, different genera use it differently. For instance, in the Gram-positive organism *Bacillus subtilis*, ligand binding to an MCP *stimulates* CheA kinase, and CheY-P *stimulates* counterclockwise flagellar rotation and, thus, extended runs. This mechanism is the exact opposite of that used in the *E. coli* response.

Chemotaxis is important for some obvious and perhaps not so obvious reasons. It clearly provides a useful survival strategy in nature, keeping bacteria moving toward nutrient and away from trouble (for example, toxic compounds). For natural commensal bacteria or

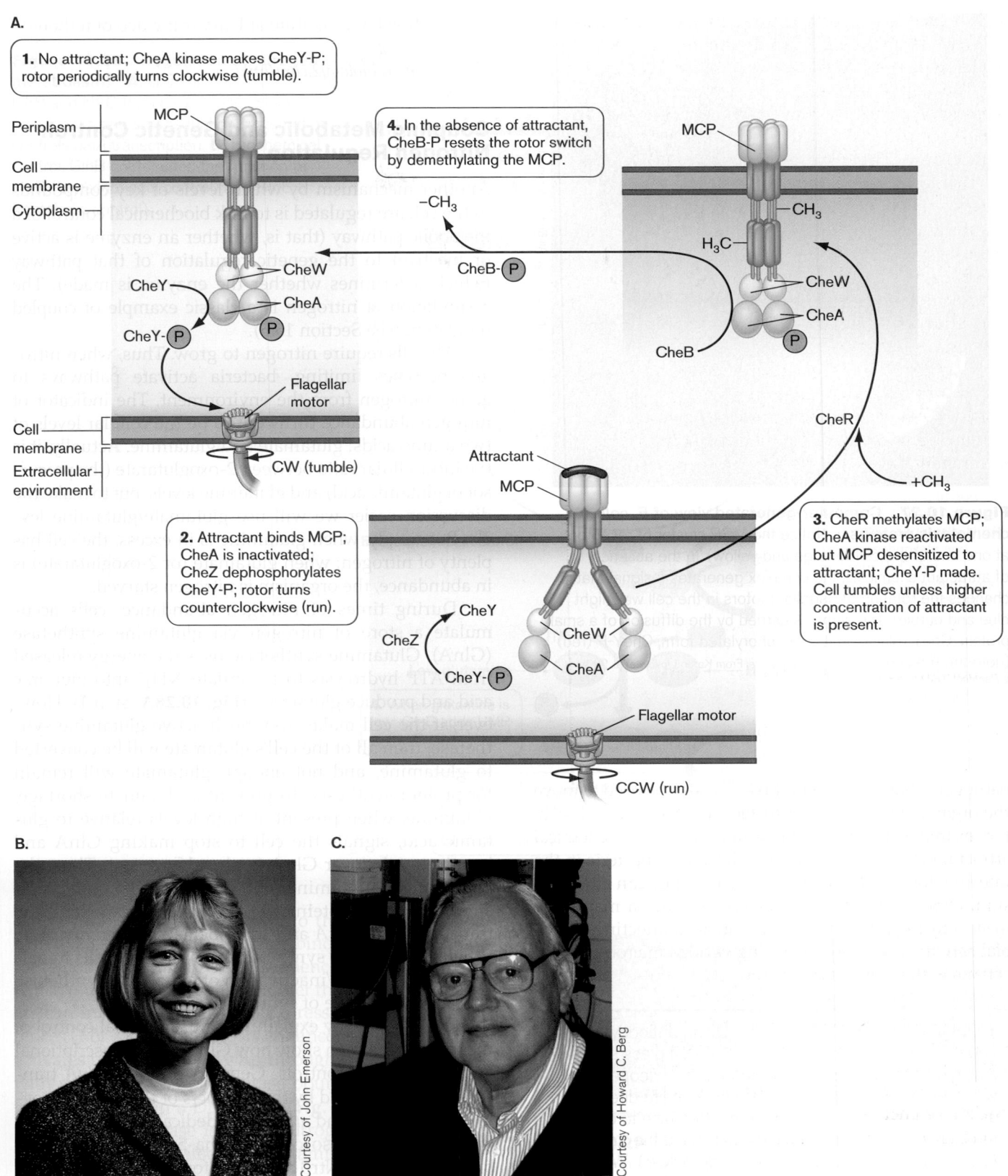

A.

1. No attractant; CheA kinase makes CheY-P; rotor periodically turns clockwise (tumble).

Periplasm

MCP

Cell membrane

Cytoplasm

CheW

CheY

CheA

CheY-(P) (P)

Flagellar motor

Cell membrane

Extracellular environment

CW (tumble)

4. In the absence of attractant, CheB-P resets the rotor switch by demethylating the MCP.

−CH₃

CheB-(P)

MCP

CH₃

H₃C

CheW

CheA

CheB

(P)

CheR

+CH₃

2. Attractant binds MCP; CheA is inactivated; CheZ dephosphorylates CheY-P; rotor turns counterclockwise (run).

Attractant

MCP

CheY

CheZ

CheY-(P)

CheW

CheA

Flagellar motor

CCW (run)

3. CheR methylates MCP; CheA kinase reactivated but MCP desensitized to attractant; CheY-P made. Cell tumbles unless higher concentration of attractant is present.

B.

Courtesy of John Emerson

C.

Courtesy of Howard C. Berg

Figure 10.26 Chemotaxis. A. Chemotaxis signaling pathway in *E. coli*. Ann Stock, Robert Wood Johnson Medical School (**B**), and Howard C. Berg, Harvard University (**C**), are major contributors to the fields of signal transduction, motility, and chemotaxis.
▶❙ *microbiology2.com/animations*

Special Topic 10.2 The Role of Quorum Sensing in Pathogenesis and in Interspecies Communications

Pseudomonas aeruginosa is a human pathogen that commonly infects patients with cystic fibrosis, a genetic disease of the lung. The organism forms a biofilm over affected areas and interferes with lung function. Key to the destruction of host tissues by *P. aeruginosa* are virulence factors such as proteases and other degradative enzymes. But these proteins are not made until cell density is fairly high, a point where the organism might have a chance of overwhelming its host. The organism would not want to make the virulence proteins too early and alert the host to launch an immune response. The induction mechanism involves two interconnected quorum-sensing systems called Las and Rhl, both composed of regulatory proteins homologous to LuxR and LuxI of *Vibrio fischeri*. Many pathogens besides *Pseudomonas* appear to use chemical signaling to control virulence genes. Genomic analysis has revealed homologs of known quorum-sensing genes in *Salmonella*, *Escherichia*, *Vibrio cholerae*, the plant symbiont *Rhizobium*, and many other microbes.

Some microbial species not only chemically talk among themselves, but appear capable of communicating with other species. *Vibrio harveyi*, for example, uses two different, but converging, quorum-sensing systems to coordinate control of its luciferase. Both sensing pathways are very different from the *V. fischeri* system. One utilizes an acyl homoserine lactone (AHL) as an autoinducer (AI-1) to communicate with other *V. harveyi* cells. The second system involves production of a different autoinducer (AI-2) that contains borate. Because many species appear to produce this second signal molecule, it is thought that mixed populations of microbes use it to "talk" to each other.

In the case of *V. harveyi*, specific membrane sensor kinase proteins are used to sense each autoinducer (**Fig. 1**). At low cell densities (no autoinducer), both sensor kinases initiate phosphorylation cascades that converge on a shared response regulator, LuxO, to produce phosphorylated LuxO. LuxOP appears to activate a repressor of the *lux* genes. Thus, at low cell densities the culture does not display bioluminescence. At high cell density, the autoinducers prevent signal transmission by inhibiting phosphorylation. The cell stops making repressor, thereby allowing another protein, LuxR (*not*

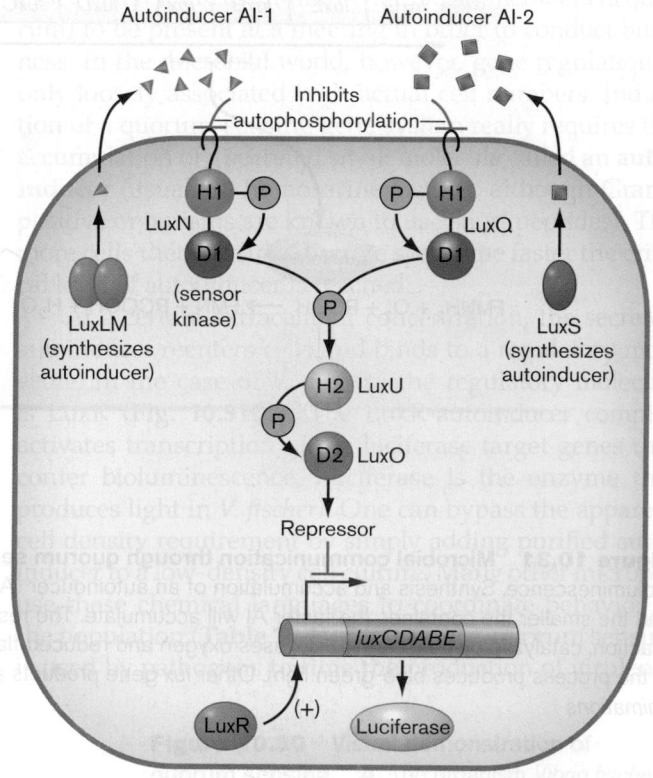

Figure 1 The two quorum-sensing systems of *Vibrio harveyi*. In the absence of autoinducers (AI-1 and AI-2), both sensor kinases trigger converging phosphorylation cascades that end with the phosphorylation of LuxO. Phosphorylated LuxO (LuxO-P) activates a repressor that inhibits expression of the luciferase genes. As autoinducer concentrations increase, they inhibit autophosphorylation of the sensor kinases and the phosphorylation cascade. As a result, repressor levels decrease, allowing the LuxR protein to activate the *lux* operon.

factors for optimum effect on the host. Quorum sensing can even be used to communicate between different species (**Special Topic 10.2**).

This knowledge begs the question, Why would a squid want to harbor this bioluminescent microbe? Buried in the sand by day, this animal emerges at night from its safe hiding place to hunt for food. In moonlight, the squid would appear as a dark silhouette from below, marking it as easy prey for predators. It is thought that

the squid camouflages itself by projecting light downward from its light organ, giving the appearance of celestial stars as perceived by predators below.

THOUGHT QUESTION 10.9 Genes encoding luciferase can be used as "reporters." What do you think would happen if the promoter for an SOS response gene were fused to the luciferase open reading frame?

a homolog of the *V. fischeri* LuxR), to activate the *lux* operon. The "lights" are turned on. Bonnie Bassler (**Fig. 2**) and Pete Greenberg (**Fig. 3**) are two of the leading scientists whose studies revealed the complex elegance of quorum sensing in *Vibrio* and *Pseudomonas* species. Other organisms, such as *Salmonella*, have been shown to activate the AI-2 pathway of *V. harveyi*, dramatically supporting the concept of cross-species communication.

A recent report by Ian Joint and his colleagues showed that bacteria can even communicate across the prokaryotic-eukaryotic boundary. The green seaweed *Enteromorpha* (a eukaryote) produces motile zoospores that explore and attach to *Vibrio anguillarum* bacterial cells in biofilms (**Fig. 4**). They attach and remain there because the bacterial cells produce acyl homoserine lactone molecules that the zoospores sense. Part of the evidence for this interdomain communication involved showing that the zoospores would even attach to biofilms of *E. coli* carrying the *Vibrio* genes for the synthesis of acyl homoserine lactone. The implications of possible interdomain conversations are staggering. Do our normal flora "speak" to us? Do we "speak" back?

For further discussion of molecular communication between prokaryotes and eukaryotes, see Chapter 21.

Figure 3 **Peter Greenberg, one of the pioneers of cell-cell communication research.** Peter Greenberg, first at the University of Iowa and now at the University of Washington, has studied quorum sensing in *Vibrio* species and various other pathogenic bacteria, such as *Pseudomonas*.

Figure 2 **Bonnie Bassler** (left) of Princeton University was instrumental in characterizing interspecies communication among bacteria.

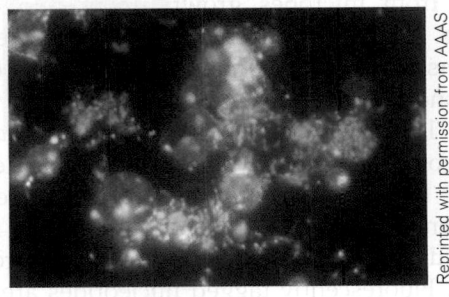

Figure 4 *Enteromorpha* zoospores (red), a type of algae, attach to biofilm-producing bacteria (blue) in response to lactones produced by the bacteria.

THOUGHT QUESTION 10.10 What do you think would happen if a culture were coinoculated with a *Vibrio fischeri luxI* mutant and a *luxA* mutant, neither of which produces light?

10.9 Genomics and Proteomics: Tools of the Future

Most of the studies exposing the regulatory systems discussed in this chapter were accomplished by examining only a small number of genes at any one time. Now, however, new technologies enable scientists examine thousands of genes or proteins in experiment.

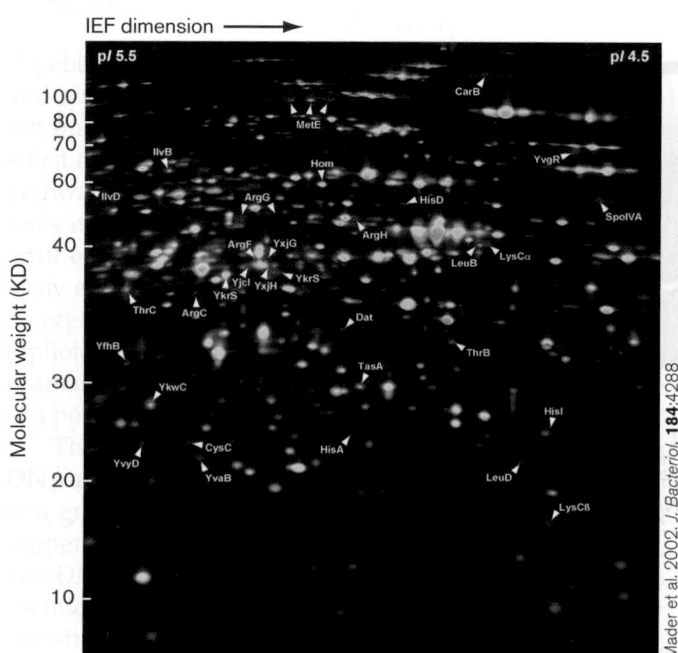

Figure 10.33 **Proteomic profile of *Bacillus subtilis* cells grown in minimal media with and without mixed amino acid supplementation (casamino acids).** The IEF gradient used in the first dimension was pH 4.5–5.5; this represents only a part of the entire proteome. The figure is the result of dual-channel image analysis of silver-stained gels. A computer assigns the color red to proteins expressed in minimal media and green to proteins expressed in amino acid–supplemented media. If the proteins are expressed under both conditions, the red and green colors combine to form yellow or orange.

conditions will reveal those proteins whose levels increase or decrease in response to the changing environment. **Figure 10.33** shows a dual-channel image analysis that compares *Bascillus subtilis* proteomes from cultures grown in a minimal glucose medium with or without amino acid supplementation. The different colors indicate whether the level of a protein is higher, lower, or the same in the two cultures.

Identifying Proteins Plucked from a Two-Dimensional Gel

The true power of proteomics becomes evident when it is linked to genomics. Knowing the complete sequence of an organism's chromosome makes it easier to identify proteins following growth under any experimental condition. For example, levels of certain proteins increase when a bioremediating microbe is grown on benzene instead of glucose. How does one determine which proteins are induced? Those proteins, and the genes encoding them, can be identified from a two-dimensional gel

if the organism's genome is sequenced. This procedure is shown schematically in **Figure 10.34**. The protein spot of interest is excised from the gel and digested into peptide fragments with a protease. The peptide fragment mixture is then analyzed by mass spectrometry, which determines the precise molecular weight of each fragment. In tandem mass spectrometry, each proteolytic fragment is then subfragmented by ionization. This produces progressively smaller secondary fragments missing one or more amino acids. Because the weight of each amino acid is distinct, tandem mass spec analysis will determine the amino acid sequence of the initial proteolytic fragment. Computer programs compare the amino acid sequence of each protein fragment with the predicted sequences of proteolytic fragments from all ORFs in a genome. Assigning a spot on a two dimensional gel to a specific protein comes from finding several peptides that match different regions of that protein.

> **THOUGHT QUESTION 10.11** Why might the mass of a protein excised from a gel fail to match any of the ORF molecular weights predicted by the genome sequence?

This technology has been used for many microbial applications. One fascinating example centers on the transcriptional events that occur during the developmental cycle of *Caulobacter crescentus*. This organism progresses from a nonreplicating swimming form (swarmer cell) to a sessile stalked form that can replicate to make more swarming offspring. DNA microarray and proteomic analysis have revealed several potential regulators of this complex developmental process.

TO SUMMARIZE:

- **The transcriptome and proteome** constitute all of a cell's mRNA molecules and proteins, respectively. Transcriptomes and proteomes change as environmental conditions change.
- **A DNA microarray chip** is used to monitor the levels of thousands of individual mRNA molecules made during growth.
- **Two-dimensional gels** separate proteins by isoelectric point and molecular weight. They offer a snapshot of the proteome at any given point of growth or under any given growth condition.
- **Mass spectrometry** can determine the exact molecular weight of a protein (or the sequence of a component peptide) taken from an otherwise anonymous spot on a two-dimensional gel. The exact molecular weight or peptide sequence is compared against the database of ORFs deduced from the genomic

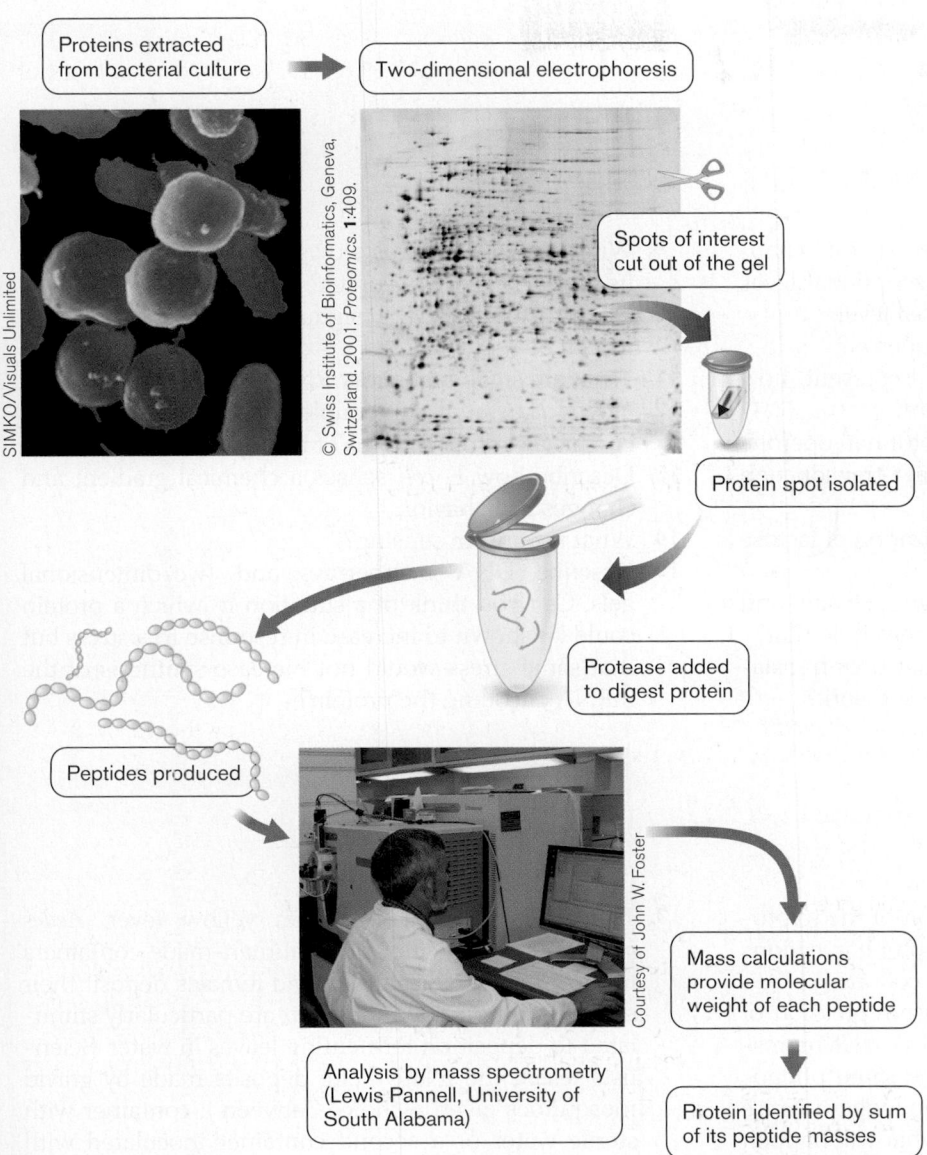

Proteins extracted from bacterial culture

Two-dimensional electrophoresis

© Swiss Institute of Bioinformatics, Geneva, Switzerland. 2001. *Proteomics.* 1:409.

SIMKO/Visuals Unlimited

Spots of interest cut out of the gel

Protein spot isolated

Protease added to digest protein

Peptides produced

Analysis by mass spectrometry (Lewis Pannell, University of South Alabama)

Courtesy of John W. Foster

Mass calculations provide molecular weight of each peptide

Protein identified by sum of its peptide masses

Figure 10.34 Proteomic identification of proteins in a two-dimensional gel. Proteins are extracted from a bacterial culture and subjected to two-dimensional electrophoresis. Spots of interest can be cut out of the gel and digested with trypsin. The resulting peptides are analyzed by mass spectrometry. In a process called MS/MS, the mass of each peptide is determined and then selected peptides are subjected to additional fragmentation by ion spray. Each resulting peptide fragment will differ in size by one or more amino acids. Knowing the mass of each amino acid and the masses of the different peptide fragments allows one to extrapolate the sequence of the original peptide. Then the MS/MS sequences obtained for all of the tryptic peptides are compared by computer to all the predicted ORFs in a microbial genome. If one ORF contains all the peptides, a match is declared and the protein is identified.

sequence of the organism. A successful match, therefore, determines the protein's identity.

Concluding Thoughts

The systems discussed in this chapter are but a small sampling of the known gene regulatory systems. Many more await discovery. How the cell coordinates all these systems is still a matter of conjecture. Scientists, like genetic cartographers, are now trying to map all the regulatory circuits of *E. coli*, as well as to identify novel regulatory strategies used by other organisms. The structural biology of regulatory protein-protein interactions and protein-DNA binding are certainly the research waves of the future. For though we know the "what and where" of many molecular interactions, we remain, for the most part, ignorant of *how* they occur structurally.

CHAPTER REVIEW

Review Questions

1. List regulatory mechanisms discussed in this chapter that work at the DNA level, transcriptional level, translational level, and posttranslational level.
2. How does lactose induce the *lacZYA* operon?
3. If *lacY* is not induced unless lactose is present, how does external lactose induce the system?
4. How does tryptophan repress the tryptophan operon?
5. Describe a two-component signal transduction system.
6. Discuss how glucose impacts the utilization of lactose as a carbon source.
7. Name a regulatory protein that can activate and repress an operon's expression. How does it do that?
8. What is the regulatory mechanism that uses translation to control transcription? How does it work?

9. When growth of *E. coli* slows, how does the cell slow its production of ribosomes?
10. Describe four ways that sigma factor production/activity can be regulated.
11. What are small regulatory RNA molecules?
12. How do bacterial cells couple genetic control and biochemical control in terms of nitrogen assimilation?
13. Describe how *E. coli* senses a chemical gradient and changes its behavior.
14. What is quorum sensing?
15. Describe DNA microarrays and two-dimensional gels. Can you think of a situation in which a protein could be shown to increase in response to a stress but that same stress would not increase synthesis of the mRNA encoding the protein?

Thought Questions

1. Explain what will happen to expression of the tryptophan operon if you replace the key tryptophan codons in the attenuator region with tyrosine codons.
2. Adding tryptophan to *E. coli* will cause repression of the *trp* operon genes. Mutations in the *trpR* repressor and the *trp* operator will have the same phenotype; that is, adding tryptophan will no longer repress expression of the *trp* genes. Explain what will happen to the phenotype if you transform each mutant with a plasmid carrying the wild-type *trpR* gene or the wild-type *trp* operator region.

3. The mosquito that transmits yellow fever, *Aedes aegypti*, requires water-filled human-made containers for egg laying. Normally, gravid females deposit their eggs in multiple containers but are particularly stimulated to deposit on fermenting leaves in water. Scientists tested the relative egg deposits made by gravid mosquitoes given a choice between a container with sterile water or a second container inoculated with 14 bacterial species isolated from a bamboo infusion (which was previously shown to be a favorite deposit medium). Ninety percent of the eggs were deposited on the surface of the bacterial infusion, and only 10% were deposited on the sterile water. How do you think the presence of the bacteria influenced the choice?

Key Terms

activator (343)
anti-anti-sigma factor (358)
anti-attenuator stem loop (355)
antisense RNA (361)
anti-sigma factor (358)
attenuator stem loop (355)
autoinducer (374)

catabolite repression (349)
chemotaxis (367)
complementary DNA (cDNA) (378)
corepressor (343)
derepression (343)
diauxic growth (349)
direct repeat (364)

DNA microchip (378)
enhancer (358)
immune avoidance (364)
inducer (343)
inducer exclusion (349)
induction (343)
inverted repeat (364)

isoelectric focusing (IEF) (378)

isoelectric point (378)

leader sequence (354)

methyl-accepting chemotaxis protein (MCP) (368)

operon (346)

phase variation (364)

proteome (378)

quorum sensing (374)

regulatory protein (343)

regulon (358)

repression (343)

repressor (343)

response regulator (344)

sensor kinase (344)

silencer (358)

stringent response (357)

tandem repeat (364)

transcriptional attenuation (354)

transcriptome (378)

translational control (357)

two-component signal transduction system (344)

Recommended Reading

Boisset, Sandrine, Thomas Geissmann, Eric Huntzinger, Pierre Fechter, Nadia Bendridi, et al. 2007. *Staphylococcus aureus* RNAIII coordinately represses the synthesis of virulence factors and the transcription regulator Rot by an antisense mechanism. *Genes and Development* **21**:1353–1366.

Bougdour, Alexandre, Christopher Cunning, Patrick J. Baptiste, Thomas Elliott, and Susan Gottesman. 2008. Multiple pathways for regulation of sigmaS (RpoS) stability in *Escherichia coli* via the action of multiple anti-adaptors. *Molecular Microbiology* **68**:298–313.

Camilli, Andrew, and Bonnie L. Bassler. 2006. Bacterial small-molecule signalling pathways. *Science* **311**:1113–1116.

Fang, Ferric C. 2005. Sigma cascades in prokaryotic regulatory networks. *Proceedings of the National Academy of Sciences USA* **102**:4933–4934.

Fozo, Elizabeth M., Matthew R. Hemm, and Gisela Storz. 2008. Small toxic proteins and the antisense RNAs that repress them. *Microbiology and Molecular Biology Reviews* **72**:579–589.

Fujita, Masaya, and Richard Losik. 2002. An investigation into the compartmentalization of the sporulation transcription factor σE in *Bacillus subtilis*. *Molecular Microbiology* **43**:27–38.

Henke, Jennifer M., and Bonnie L. Bassler. 2004. Bacterial social engagements. *Trends in Cell Biology* **14**:648–656.

Joint, Ian, Karen Tait, Maureen E. Callow, James A. Callow, Debra Milton, et al. 2002. Cell-to-cell communication across the prokaryotic-eukaryotic boundary. *Science* **298**:1207.

Jørgensen, Mikkel G., Deo P. Pandey, Milena Jaskolska, and Kenn Gerdes. 2009. HicA of *Escherichia coli* defines a novel family of translation-independent mRNA interferases in bacteria and archaea. *Journal of Bacteriology* **191**:1191–1199.

Lewis, Mitchell. 2005. The *lac* repressor. *Critical Reviews in Biology* **328**:521–548.

Lux, Renate, and Wenyuan Shi. 2004. Chemotaxis-guided movements in bacteria. *Critical Reviews in Oral Biological Medicine* **15**:207–220.

Magnusson, Lisa U., Anne Farewell, and Thomas Nyström. 2005. ppGpp: A global regulator in *Escherichia coli*. *Trends in Microbiology* **13**:236–242.

Majdalani, Nadim, Carin K. Vanderpool, and Susan Gottesman. 2005. Bacterial small RNA regulators. *Critical Reviews in Biochemistry and Molecular Biology* **40**:93–113.

Merino, Enrique, and Charles Yanofsky. 2005. Transcription attenuation: A highly conserved regulatory strategy used by bacteria. *Trends in Genetics* **21**:260–264.

Miller, Lance D., Matthew H. Russell, and Gladys Alexandre. 2009. Diversity in bacterial chemotactic responses and niche adaptation. *Advances in Applied Microbiology* **66**:53–75.

Parker, Christopher T., and Vanessa Sperandio. 2009. Cell-to-cell signalling during pathogenesis. *Cellular Microbiology* **11**:363–369.

Pichon, Christophe, and Brice Felden. 2007. Proteins that interact with bacterial small RNA regulators. *FEMS Microbiology Reviews* **31**:614–625.

Piggot, Patrick J., and David W. Hilbert. 2004. Sporulation of *Bacillus subtilis*. *Current Opinions in Microbiology* **7**:579–586.

Ponnusamy, Longanathan, Ning Xu, Satoshi Nojima, Dawn M. Wesson, Coby Schal, et al. 2008. Identification of bacteria and bacteria-associated chemical cues that mediate oviposition site preferences by *Aedes aegypti*. *Proceedings of the National Academy of Sciences USA* **105**:9262–9267.

Staron, Anna, and Thorsten Mascher. 2010. Extracytoplasmic function sigma factors come of age. *Microbe* **5**:164–170.

Stewart, Valley. 2008. The ribosome: A metabolite-responsive transcription regulator. *Journal of Bacteriology* **190**:4787–4790.

For further study and review, please visit StudySpace at **microbiology2.com**.

Chapter 11

Viral Molecular Biology

Viruses show many different forms. Their chromosomes range from a single nucleic acid coiled inside a protein tube, to a cell-like package composed of multiple layers of protein and membrane. Viral genomes may include fewer than 10 genes or more than 200. Their replication inside a host cell varies from the lytic program of bacteriophage T4 to the elaborated multiorganelle cycle of herpes viruses. Herpes and other latent viruses sequester their genomes within that of the host, causing devastating disease. Yet latent viral genomes contribute surprisingly to human evolution. And the viruses most deadly to humans can serve as vectors for gene therapy.

The molecular mechanisms of viruses raise intriguing questions: How do viruses trick host cell defenses into permitting their entry and replication? How did viruses originate: from free-living cells that took up predatory lifestyles or from host cell components? Viral molecular biology offers clues to these questions. The molecular mechanisms of viral infection reveal targets for antiviral drugs, and tools for gene therapy vectors.

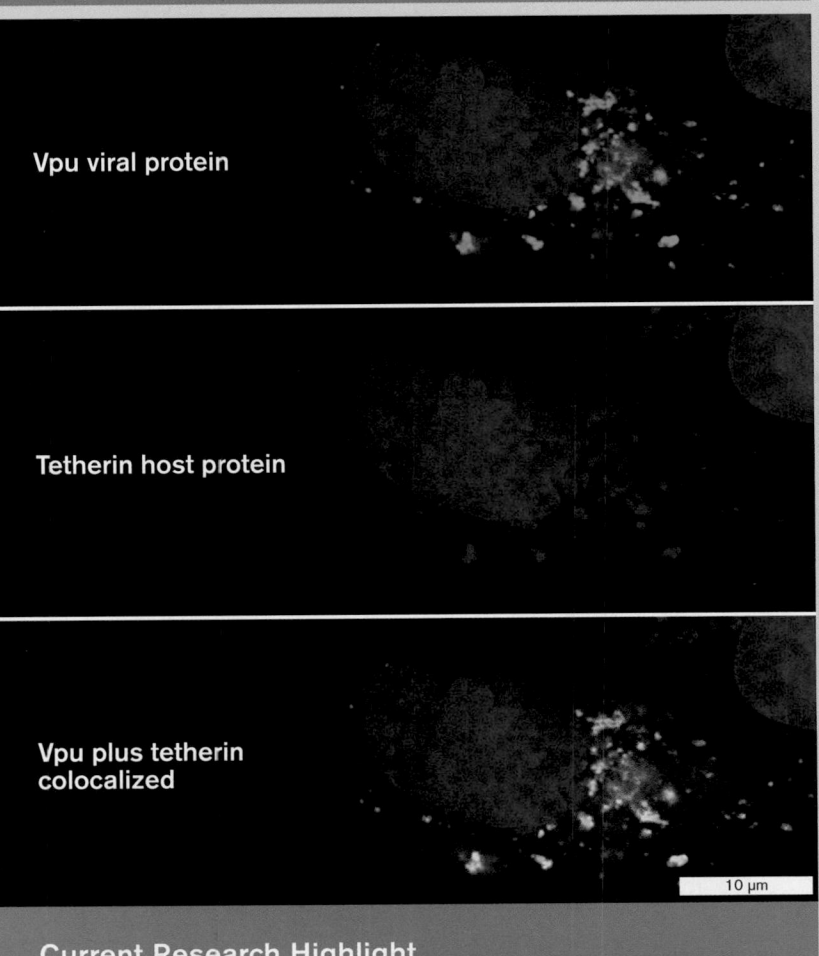

Vpu viral protein

Tetherin host protein

Vpu plus tetherin colocalized

10 μm

Current Research Highlight

Human immunodeficiency virus (HIV) causes AIDS. The molecular basis of HIV infection yields clues to possible drug therapies. This fluorescence micrograph shows an infected human cell producing HIV virions. The cell nucleus is labeled with a blue fluorophore. In the cytoplasm, the new virions are tethered to the cell surface by the host protein tetherin (red), requiring release by the HIV regulatory protein Vpu (green). The colocalization of tetherin and Vpu (yellow) supports the proposed role of Vpu in producing HIV virions from infected cells. *Source:* Stuart J. D. Neil et al. 2008. *Nature* **451**:425.

In the 1990s, as the AIDS epidemic spread around the globe, health care workers noticed certain individuals who remained free of HIV over many years of exposure. These people proved to have a defective gene that encodes a macrophage cell surface protein, CCR5 (**Fig. 11.1**). CCR5 enhances resistance to some pathogens, but the protein acts as a receptor for HIV. Individuals who carry a single copy of the gene defect, known as deletion 32 (CCR5-Δ32), possess only one functional copy of CCR5, encoded by the homologous chromosome. The CCR5-Δ32 heterozygotes show delayed onset of AIDS, and individuals with two copies of the CCR5 defect resist infection by the most common strains of HIV. CCR5-Δ32 is carried by about 10% of people of European descent; in other populations the allele is rare.

The basis of the CCR5-Δ32 resistance to HIV is that the virus envelope proteins need to recognize the specific protein sequence of CCR5 in order to attach and enter the host cell. The single CCR5 defect decreases the number of CCR5 receptor proteins and thus the efficiency of infection. The defect on both homologous chromosomes eliminates the protein and prevents virus attachment altogether. This protein-protein interaction offers hope for developing methods to prevent or treat AIDS. It may be possible to transfer the CCR5-Δ32 trait into the genome of a patient's bone marrow stem cells, which will then develop into macrophages that resist HIV infection. An alternative approach is to develop chemical blockers of CCR5 binding to HIV, thus decreasing cell infection. The first CCR5 blocker, maraviroc, was developed by Pfizer and approved for therapy in 2007.

The HIV receptor defect is just one example of viral molecular biology for medical research and biotechnology. In Chapter 10 we discussed how molecular mechanisms such as repressors and DNA inversions regulate the function of microbial cells. Chapter 11 presents the molecular mechanisms of viral infection of cells. We assume a fundamental understanding of the nature of viruses, as presented in Chapter 6. Here we explore in greater depth the molecular basis of infection. First we consider a classic bacteriophage whose virion assembly provides a key model for viral development at the molecular level. We then examine four viruses infecting humans, which include RNA and DNA genomes, as well as a retrovirus. We see that all viral infection processes share common themes, and unique molecular mechanisms as well.

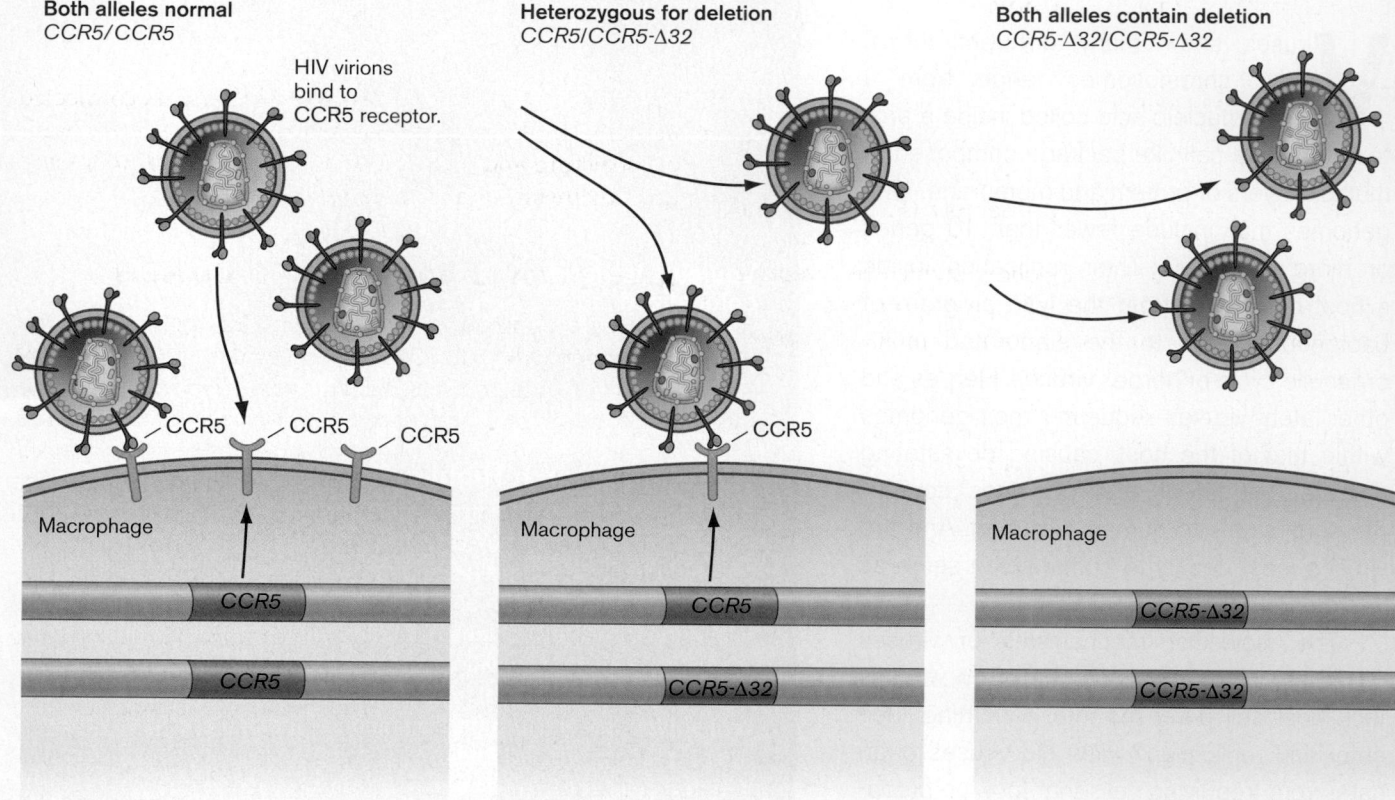

Figure 11.1   Resistance to HIV: a molecular mechanism.   HIV normally binds to the CCR5 surface protein of immune cells. Individuals heterozygous for a CCR5 deletion (CCR5-Δ32) show delayed onset of AIDS. Individuals homozygous for the CCR5 defect show full protection from the major strains of HIV, apparently because the virus fails to bind to its target host cell.

Finally, we see how the mechanisms of lethal viruses may actually be harnessed for lifesaving gene therapy.

This chapter emphasizes the fundamental biology of viral infection and propagation. The consequences of viral disease for the host organism are presented in Chapters 25 and 26.

www | The Viral Biorealm
microbiology2.com/links

11.1 Phage T4: The Classic Molecular Model

Bacteriophages are of interest for several reasons. Historically, phages provided the first living systems simple enough to dissect at the molecular level. Phages yielded fundamental discoveries in gene regulation, including the classic case of lambda lysogeny (discussed in eTopic 10.3). The first model for the genetic analysis of development was a phage system, the assembly of the "tailed phage" T4 (*Myoviridae*). Today, phages provide new tools for antibacterial agents and new ways to make vaccines. They are also used to assemble wires in microscopic devices. For an example, see **eTopic 11.1**.

Phage T4 Structure

Phage T4 may be the most complicated noncellular reproductive unit yet characterized. The intricate virion requires 20 gene products for assembly (**Fig. 11.2**). The T4 genome of 166 kilobase pairs (kb) specifies 170 genes, many of which encode metabolic enzymes that replace host cell functions (**Fig. 11.3**). This viral genome almost looks like a cell genome in miniature. Thus, T4 has served as a model to probe the fundamental molecular processes of transcription and gene recombination, as well as the global function of cells. The mutation strategies devised to dissect T4 function provided the basis for dissecting molecular development in animals.

www | 3-D tutorial on the T4 tail injector
microbiology2.com/links

The phage T4 virion has a **capsid**, a device carrying the genome, composed of specialized protein subunits. The capsid includes a polyhedral "head" containing its genome, a linear double strand of DNA (see **Fig. 11.2**) attached to a delivery device called the "tail"—hence the term **tailed phage**. The head is connected by a narrow neck tube to an internal tube that extends down within the sheath. The sheath plus the internal tube make up the

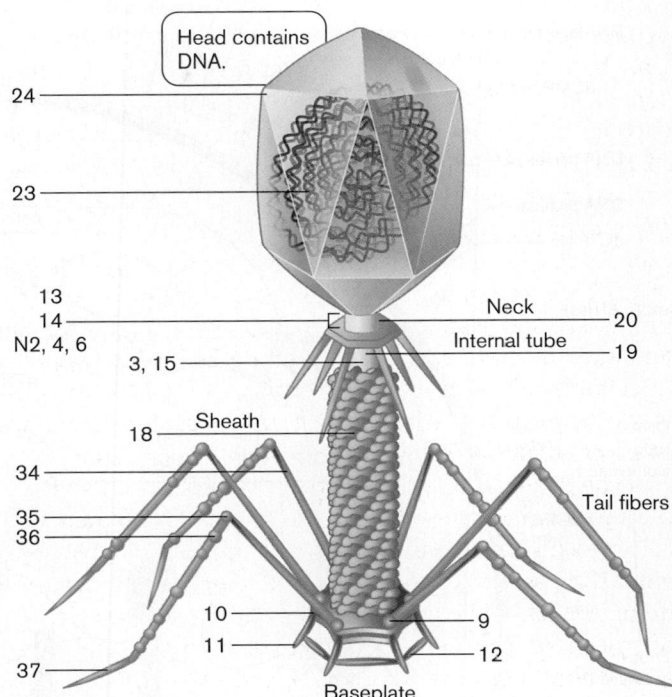

Figure 11.2   Structure of a phage T4 capsid (simplified). Proteins are assigned to specific genes (numbered) that are mapped on the chromosome in Figure 11.3.

tail. From the tail extend two sets of fibers, six connected at the neck and six at the baseplate. The elegant mechanical structure and function of this "molecular motor" was an early inspiration for nanotechnology.

Adsorption to Host and DNA Delivery

Viral infection requires a molecular "fit" between a viral component and one or more molecules of the host cell surface. Phage T4 infects strains of *Escherichia coli* whose outer membrane possesses lipopolysaccharide (LPS) and the porin OmpC. The phage adsorbs (attaches) to the cell surface by contact between its tail fibers and the outer membrane (**Fig. 11.4**). Once several of the tail fibers are anchored to the outer membrane by LPS, the phage baseplate makes contact. The contact between baseplate and cell outer membrane generates a conformational change in the tail. The outer tube of the tail, the sheath, is a helical tube of subunits of protein P18. The sheath contracts by shortening and widening, enabling the internal tube to push through, like a molecular syringe.

The internal tube of the tail is capped by the needle-like "injector," a unique trimer composed of proteins P27 and P5. The injector structure was first determined by Michael Rossman and colleagues at Purdue (**Fig. 11.5A**). Rossman's lab developed techniques of X-ray analysis

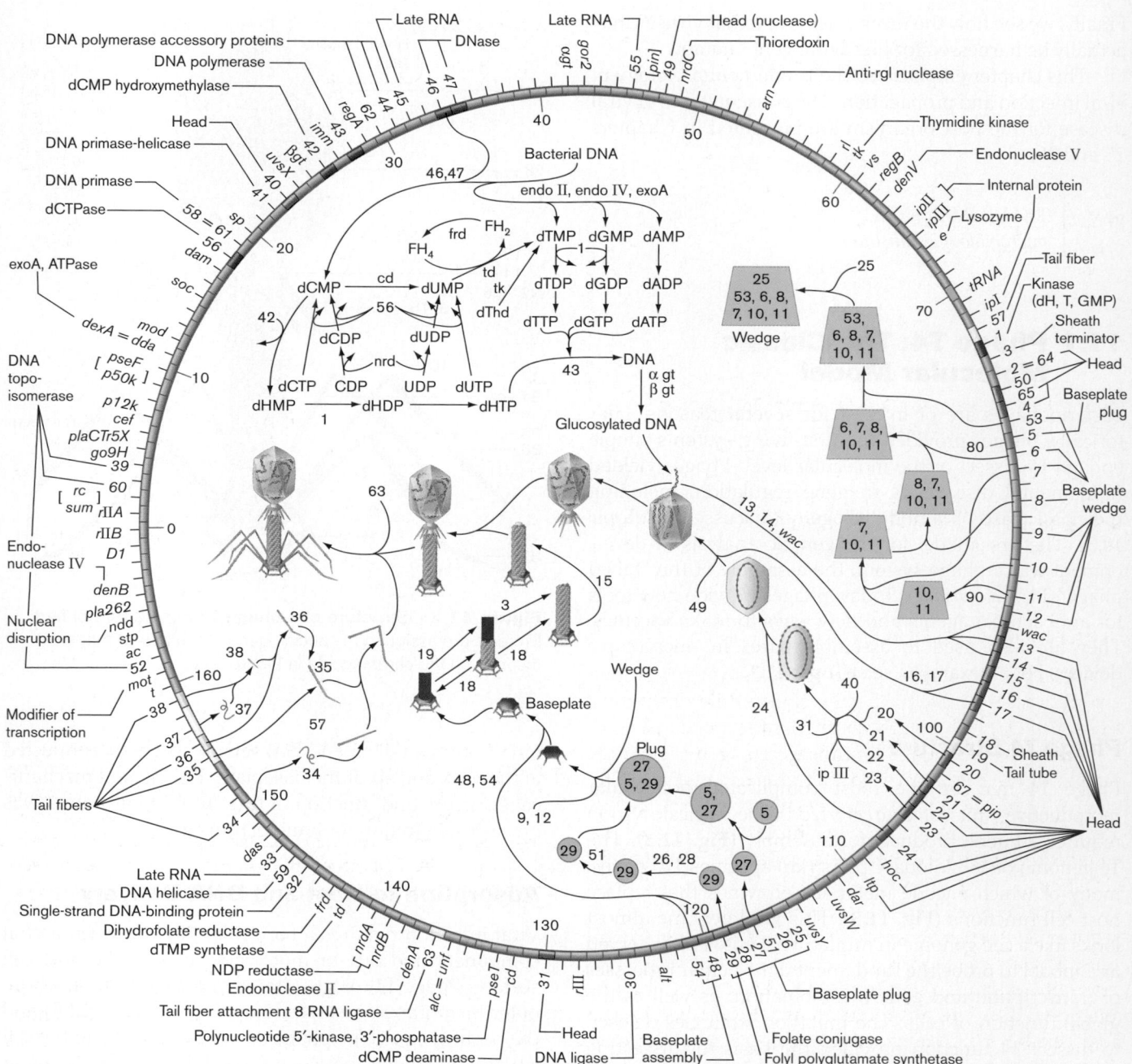

Figure 11.3 Phage T4 genome, showing functions of mapped genes. The gene numbers correspond to those assigned to steps of assembly in Figure 11.2. *Source:* Based on Matthews et al. 1983. *Bacteriophage T4*, ASM Press, Washington, D.C.

and cryo-EM to visualize many challenging viral structures, including whole virions such as rhinovirus (cold virus), dengue fever flavivirus, and canine parvovirus. These viral structures may now be used to test new antiviral drugs.

The P5 proteins form three intertwined strands in a "beta helix." As the sheath surrounding the internal tube contracts, the injector pokes through the host outer membrane and penetrates the cell wall (**Fig. 11.5B** and **C**). The

peptidoglycan surrounding the injector is then digested by protein P5, whose globular alpha-helical domain acts as lysozyme. The hole in the cell wall enables the entire tail to penetrate the cell wall and inner membrane. When the end of the tail tube penetrates the inner membrane, the phage DNA is released and forced out of the head under pressure. Double-stranded DNA viruses are packed at a pressure of up to 50 atmospheres (atm), some of the highest pressures found in living systems.

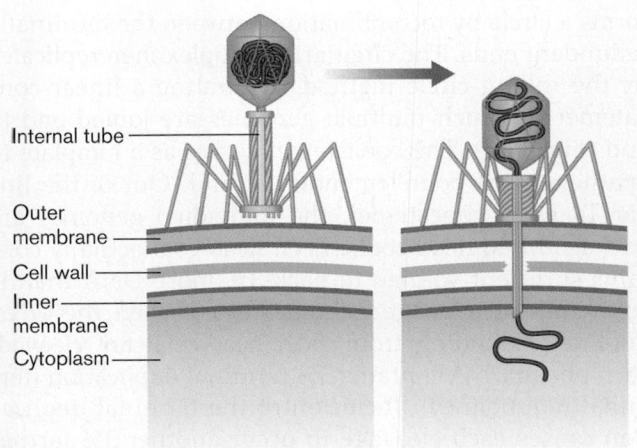

Figure 11.4   Phage T4 adsorption and DNA injection. First the tail fibers contact the outer membrane; then the baseplate completes the connection. The sheath contracts, the internal tube (composed of protein P19) penetrates the outer membrane, and then the peptidoglycan is digested. The tube extends to the inner membrane, and the head releases DNA under packing pressure, forcing it into the cytoplasm.

Virulent Replication

Phage T4 is fully virulent; that is, its only reproductive option is to assemble progeny virions while destroying a host cell (**Fig. 11.6A**). By contrast, other tailed phages, such as phage lambda, can undergo lysogeny, the integration of their genome into that of their host cell. The integrated phage genome then replicates passively with the host. Lysogeny is described in Chapter 6, and the genetic regulation of phage lambda is presented in eTopic 10.3.

When phage T4 DNA enters the host cytoplasm, a set of phage genes is activated for transcription. Phage gene activation involves DNA-binding regulators like those of bacterial gene regulation (discussed in Chapters 8 and 10). The phage genes are transcribed by the host cell RNA polymerase and translated by the host ribosomes. These host components, however, are supplemented by phage-encoded components such as tRNAs.

Phage genes expressed early in the infectious cycle are called **early genes**. The early-gene products include proteins needed to cleave host DNA, thus halting host macromolecular synthesis. (The phage's own DNA avoids cleavage because its cytosine bases are methylated.) In addition, several phage DNA synthesis enzymes replace key enzymes of the host. One T4 enzyme increases the rate of phage DNA replication tenfold. Another enzyme replaces cytosine with the modified base 5-hydroxymethylcytosine (**Fig. 11.6B**), which substitutes for cytosine throughout the phage

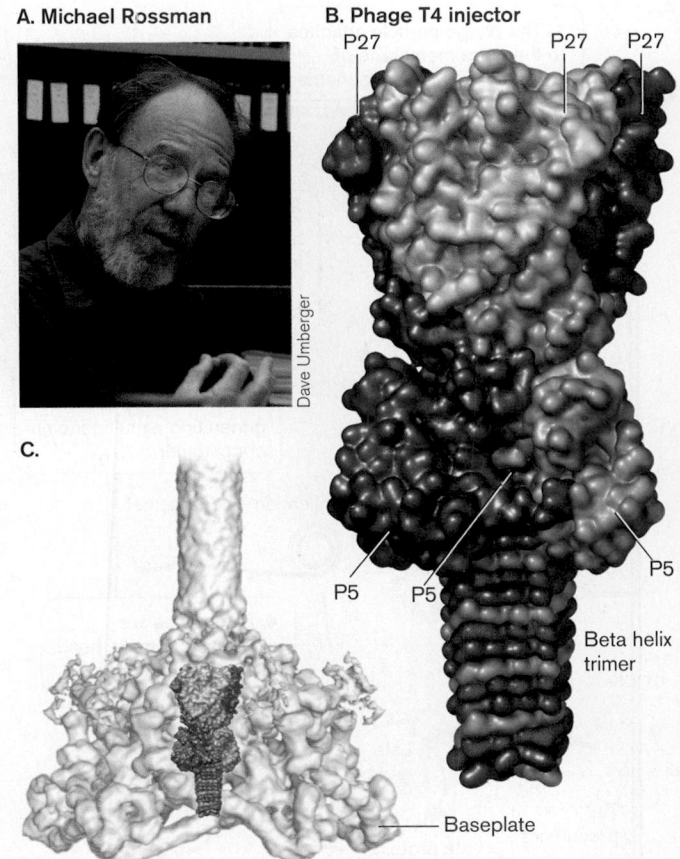

Figure 11.5   Injector device of phage T4.   A. Michael Rossman and colleagues at Purdue University deciphered the structure of the phage T4 injector. They have also solved the structures of numerous whole viruses, such as the first rhinovirus (cold virus), dengue fever flavivirus, and canine parvovirus. **B.** Molecular model of the injector for the phage DNA. (PDB code: 1K28) **C.** Position of the injector within the baseplate, using low-resolution X-ray data. *Source:* Based on Rossman et al. 2004. *Curr. Opin. Struct. Biol.* **14**:171–180.

chromosome. This modification of phage DNA prevents cleavage by viral and host endonucleases. The endonucleases (enzymes that cleave DNA) fail to "fit" DNA containing the modified cytosine.

> **THOUGHT QUESTION 11.1**   Why would phage T4 production require a tenfold higher rate of DNA replication than does the host cell? Why would the phage substitute all of its cytosine with an unusual base that requires greater energy to synthesize?

Phage T4 DNA is synthesized within the host cell by rolling-circle replication. The advantage of rolling-circle replication is that many genome copies are made quickly, without needing a special enzyme to complete linear ends. First the linear DNA duplex of phage T4

A.

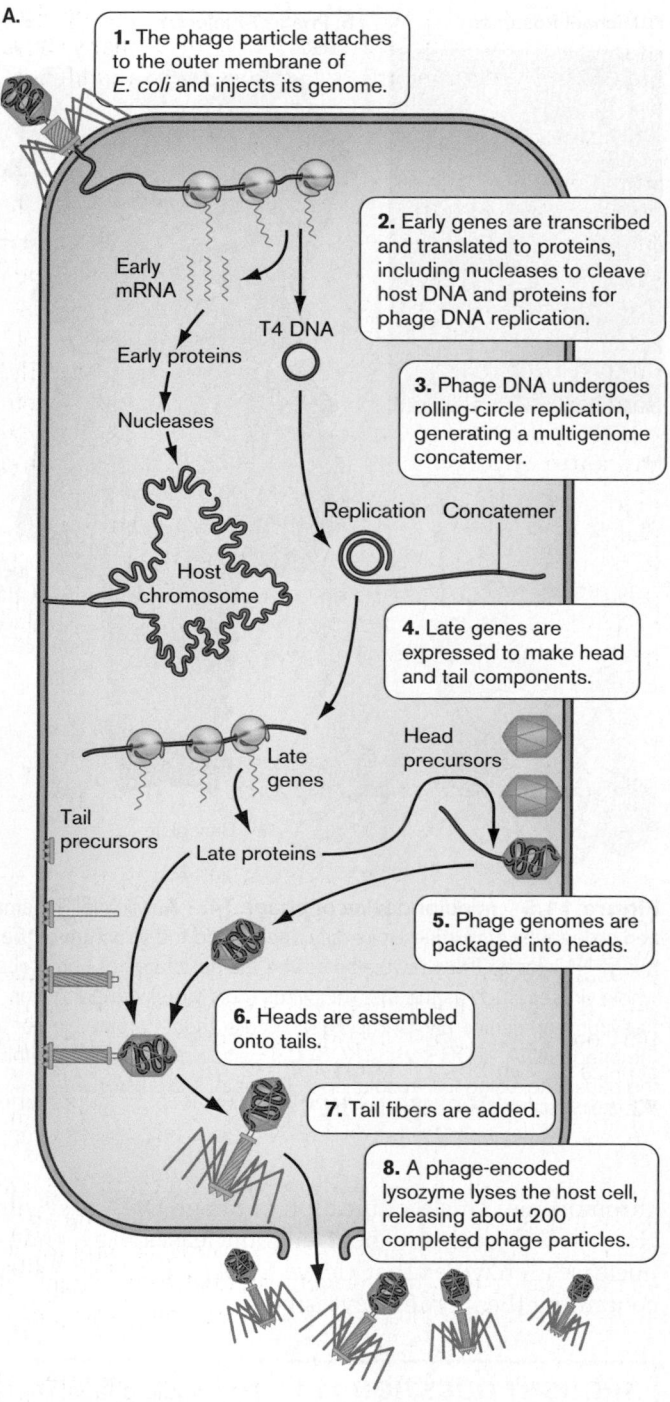

1. The phage particle attaches to the outer membrane of *E. coli* and injects its genome.

Early mRNA

T4 DNA

Early proteins

Nucleases

Host chromosome

2. Early genes are transcribed and translated to proteins, including nucleases to cleave host DNA and proteins for phage DNA replication.

3. Phage DNA undergoes rolling-circle replication, generating a multigenome concatemer.

Replication   Concatemer

4. Late genes are expressed to make head and tail components.

Late genes

Head precursors

Tail precursors

Late proteins

5. Phage genomes are packaged into heads.

6. Heads are assembled onto tails.

7. Tail fibers are added.

8. A phage-encoded lysozyme lyses the host cell, releasing about 200 completed phage particles.

B. 5-Hydroxymethylcytosine

Figure 11.6   Replicative cycle of phage T4.   A. The phage particle attaches to the surface of *E. coli* and injects its genome. The genome is reproduced and packaged into progeny virions that are released upon lysis of the host cell. **B.** 5-Hydroxymethylcytosine replaces cytosine during DNA replication by T4-encoded DNA polymerase.

forms a circle by recombination between the terminally redundant ends. The circularized duplex then replicates by the rolling-circle method, generating a linear concatemer in which multiple genomes are joined end to end (**Fig. 11.7**). The concatemer serves as a template to synthesize the complementary strand. Out of the linear T4 DNA concatemer, the individual genomes are packaged into head coats. Each head coat actually contains sufficient volume to pack 3% more DNA than is contained in a phage genome; thus, when the DNA duplexes extending from each head coat are cleaved, each phage DNA contains 3% terminal duplication (terminal redundancy). Inclusion of the terminal duplication causes each cleavage to occur another 3% farther along the genome. Thus, every T4 DNA molecule ends up with a different terminal repeat at some point throughout the genome.

Ultimately, the **late genes** are induced to produce the capsid and tail proteins that assemble on the membrane to make mature phage. During **assembly**, the phage genomes are stuffed into heads, the filled heads are attached to tails, and finally the various tail, collar, and capsid substructures are assembled into about 200 complete phage particles per cell. At last, a late gene encodes a lysozyme to digest the cell wall, releasing the phages into their surroundings. The signal for **lysis** and phage release remains unknown.

> **THOUGHT QUESTION 11.2**   The phage T4 chromosome is a linear piece of DNA, yet the genomic map of T4 (**Fig. 11.3**) is circular. Why?

The temporal separation of early genes and late genes is common to the replicative cycles of all kinds of phages as well as viruses of eukaryotes. Viruses differ, however, in the size of their genomes; some, such as human immunodeficiency virus, or HIV (discussed in Section 11.4), contain only a small number of genes, whereas others, such as phage T4, encode products that duplicate host functions of DNA synthesis, ribosomal proteins, tRNA, and biosynthetic enzymes such as dihydrofolate reductase. Some phage-encoded enzymes, such as the T4 DNA polymerase, modify a cellular function to favor phage reproduction.

One hypothesis proposed for the evolution of T4 is that the phage arose by degeneration of a cellular parasite that ultimately lost its cytoplasm but retained metabolic components. An alternative hypothesis is that the phage T4 genome acquired many host genes through recombination with scraps of the host genome. The acquired genes then evolved new functions that increased the efficiency of phage assembly; for example, dihydrofolate reductase is used to assemble folate into the baseplate of the T4 injector.

Phage Particles Self-Assemble

The assembly of phage T4 particles within the host cytoplasm offers exceptional opportunities to visualize a molecular pathway. Each phage particle is assembled by convergence of three pathways involving the head coat, the tail, and the tail fibers (**Fig. 11.8**). All of these stages

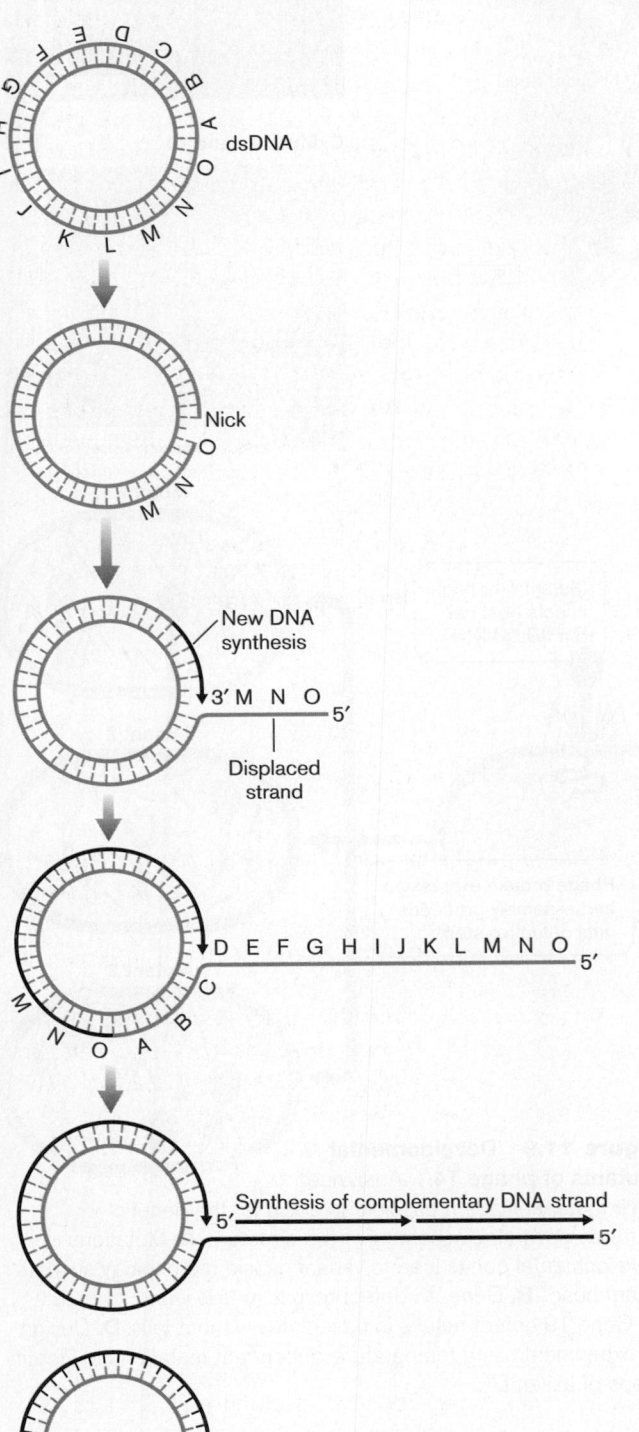

Nick

New DNA synthesis

3′ M N O 5′

Displaced strand

D E F G H I J K L M N O 5′

Synthesis of complementary DNA strand

5′

5′

ABC DEF——DEF GHI——GHI JKL——JKL MNO   MNO

can be isolated and observed within infected cells by electron microscopy (**Fig. 11.9**). But with all these many steps, how did we determine the order of assembly and identify the coding genes?

The stages of phage assembly were discovered using strains that carry mutations in various genes that encode proteins essential for development (see **Fig. 11.8**). In some cases, defective assembly leads to a bizarrely altered shape of the particle, such as a phage with a "giant head" (**Fig. 11.9A**). In other cases, a defective gene simply prevents progression in the pathway, halting assembly at an unfinished stage. For example, **Figure 11.9B** shows the phage particles found in a cell infected by a phage defective for gene 23. The unfinished particles consist of tail structures only. This defect occurs because gene 23 is required for assembly of the head, which must be completed before attachment to the tail. In **Figure 11.9C**, which shows phages defective for gene 19, we see only baseplates because the product of gene 19 is needed to build the tail upon the baseplate. In this case, assembled heads are found near the cell membrane.

> **THOUGHT QUESTION 11.3**   Which numbered genes from **Figure 11.8** might be mutated in each of the defective phage populations presented in **Figure 11.9D**?

An interesting problem arises in designing an experimental strategy: How do we grow populations of phages whose particles fail to complete assembly? The answer lies in using **conditional lethal mutations**—that is, mutations that cause a lethal defect under one growth condition but permit growth under a second condition. Two classes of conditional lethal mutant strains of phage T4 were obtained through mutagenesis by Robert Edgar and Richard Epstein in the early 1960s. These two classes are

- **Temperature-sensitive mutants.** Mutants that are temperature sensitive can grow at one temperature but fail to grow at another. For example, a point mutation may result in a protein product that is stable at 25°C but denatures at 42°C, so the phage can reproduce at 25°C but not at 42°C.
- **Nonsense mutations countered by suppressor tRNA.** A mutant that contains a nonsense codon (stop codon) in the middle of a gene prematurely

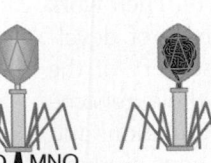

Figure 11.7 Rolling-circle replication of phage T4.   Initially, the linear genome circularizes; then it replicates as a concatemer. Out of the concatemer, the individual genomes are packaged into head coats and then cleaved. Each encapsidated DNA contains an end-duplicated 3% of its genome.

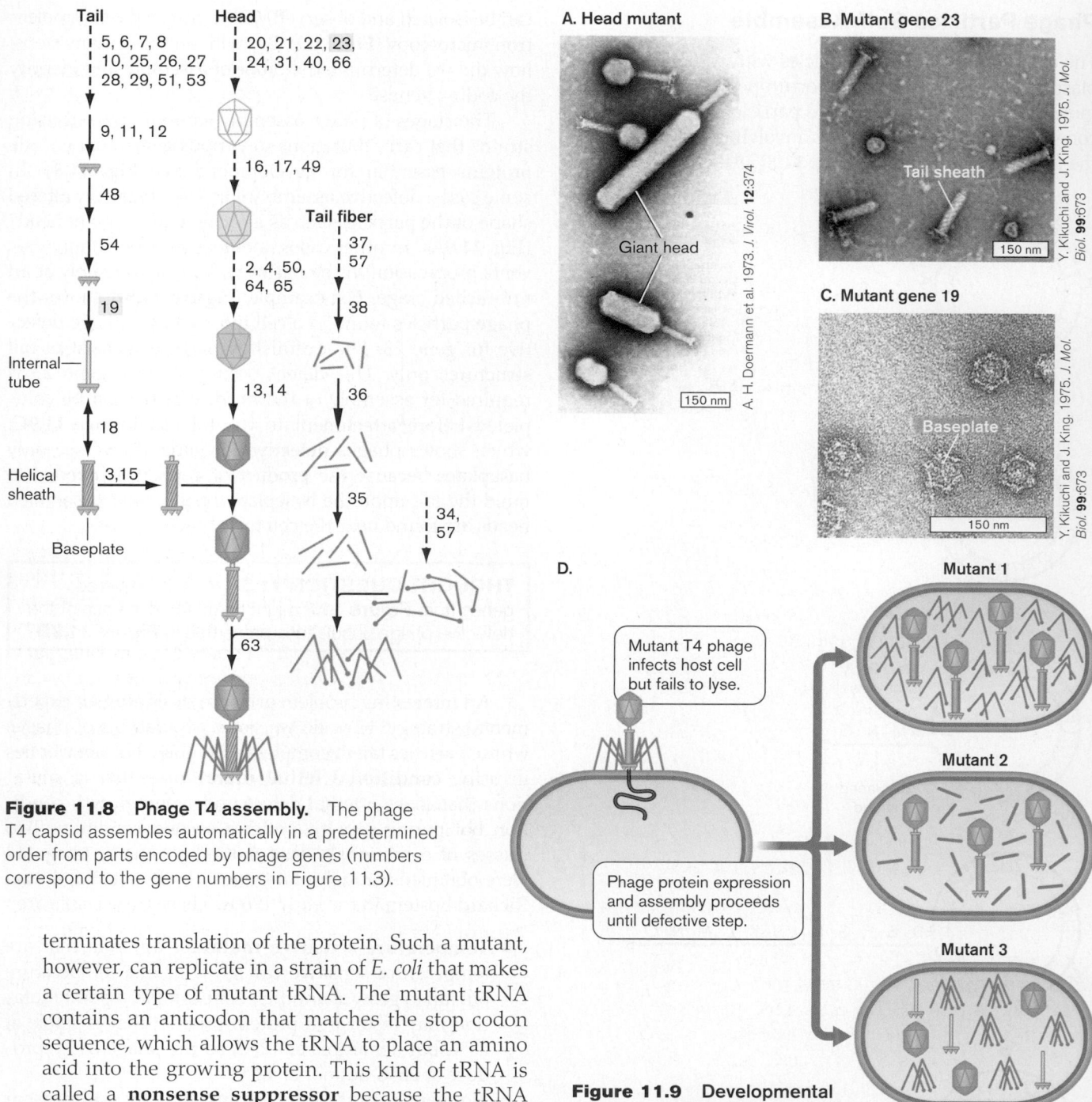

Figure 11.8 Phage T4 assembly. The phage T4 capsid assembles automatically in a predetermined order from parts encoded by phage genes (numbers correspond to the gene numbers in Figure 11.3).

Figure 11.9 Developmental mutants of phage T4. Analyzing such mutants made it possible to decipher the genetic pathway governing assembly of the structure. **A.** Mutations in developmental genes lead to variant capsid morphology such as "giant head." **B.** Gene 23 defect results in tails without heads. **C.** Gene 19 defect results in baseplates without tails. **D.** Design of experiment using temperature-dependent mutations to identify steps of assembly.

terminates translation of the protein. Such a mutant, however, can replicate in a strain of *E. coli* that makes a certain type of mutant tRNA. The mutant tRNA contains an anticodon that matches the stop codon sequence, which allows the tRNA to place an amino acid into the growing protein. This kind of tRNA is called a **nonsense suppressor** because the tRNA anticodon mutation suppresses the effect of the nonsense mutation in the phage.

Through many years of analysis of phage mutants, Edgar and Epstein and their colleagues showed that conditional lethal mutations could be used to dissect a developmental pathway as complex as that of phage T4. Their work laid the foundation for the molecular analysis of development in animals and plants and the discovery of the molecular basis of inherited diseases. With whole-genome sequences in hand, we use conditional lethal mutations and suppressor mutations to decipher how all the gene products

work together. For example, genetic analysis of development in the fruit fly *Drosophila* uses temperature-sensitive mutations to dissect mechanisms of gene regulation.

TO SUMMARIZE:

- **The phage T4 virion** consists of a head containing its DNA genome and accessory proteins, a tail composed of an internal tube with a sheath; and tail fibers.
- **T4 is adsorbed** by binding of the tail fibers to bacterial outer membrane receptors, followed by binding of the tail baseplate.
- **Injection of T4 DNA** through the internal tube leads to lytic infection. Early genes encode a DNA polymerase and a DNase to cleave host DNA.
- **Rolling-circle replication** generates progeny genomes linked in a concatemer. The concatemer is cut with an offset, so that each linear genome has a slight overlap, cut at a different position in the sequence.
- **Virions are self-assembled**, including packaging of DNA into the head, in arrays beneath the inner membrane.
- **Cell lysis** is caused by a lysozyme expressed by late genes in the T4 genome.
- **The phage T4 assembly process** was dissected in studies of temperature-sensitive mutants and nonsense mutants with suppressor tRNA. This strategy of genetic dissection was used as a model for experiments on animal development.

NOTE: The filamentous phage M13, including its replicative cycle and nanotechnology applications, is discussed in **eTopic 11.1**.

11.2 Poliovirus: (+) Strand RNA

Compared to bacteriophages, viruses that infect animal cells usually have the simpler task of crossing the eukaryotic cell membrane. Within the eukaryotic cell, however, the interior structure is generally more complex than that of prokaryotes. To use the cell's gene expression system, an animal virus must negotiate various cellular compartments, such as endocytic vesicles and the nucleus. Different viruses use very different strategies to manipulate the host cell.

A medically important group of viruses is the **picornaviruses** (**Table 11.1**), a class of icosahedral RNA viruses whose life cycle has some interesting parallels to that of phage T4. The picornavirus **poliovirus** causes the paralytic disease **poliomyelitis**, or **polio**. The host range for infection by poliovirus is limited to humans and our closest relatives, such as chimpanzees. Closely related picornaviruses, such as the **rhinoviruses** and coxsackievirus, cause the common cold. A major veterinary concern is aphthovirus, the cause of foot-and-mouth disease in cattle and swine. In 2001, an outbreak of foot-and-mouth disease in the United Kingdom caused major economic and societal disruption.

Poliovirus is famous historically for the outbreaks of poliomyelitis that swept the United States during the first half of the twentieth century (**Fig. 11.10**). The virus enters orally from fecal sources such as contaminated water. In most cases, poliovirus infection causes a mild gastrointestinal disease, but the virus can invade motor neurons of the spinal cord and lower brain, resulting in paralysis. Tens of thousands of Americans were infected, including many young children. Many had few or no symptoms but could transmit the disease to others. About 1% became paralyzed, of whom about 15% died. Decades later, paralyzed individuals may suffer post-polio syndrome, a condition in which muscles that were previously affected by the polio infection experience atrophy and pain. Today, polio infection has nearly been wiped out worldwide, although outbreaks still occur because of low levels of vaccination in isolated communities such as the Amish and in various countries in Asia and Africa.

President Franklin Roosevelt (**Fig. 11.10B**), himself a victim of the disease, established a national foundation called the March of Dimes to develop a vaccine. Support from the March of Dimes led Jonas Salk and colleagues at the University of Pittsburgh to develop the first vaccine against polio in 1952. The spectacular success of the March of Dimes set the pattern for future public support for research on more challenging diseases, such as cancer and AIDS.

Table 11.1   Picornavirus species (examples).

Genus	Virus	Disease	Host
Enterovirus	Coxsackievirus	Common cold, myocarditis	Humans
	Echovirus	Meningitis, paralysis, encephalitis	Humans
	Poliovirus (PV)	Meningitis, paralysis	Humans
Rhinovirus	Human rhinovirus (HRV)	Common cold	Humans
Aphthovirus	Foot-and-mouth disease virus (FMDV)	Foot-and-mouth disease	Cattle, swine
Hepatovirus	Hepatitis A virus (HAV)	Hepatitis	Humans
Cardiovirus	Encephalomyocardiovirus	Encephalitis, myocarditis	Mice
Erbovirus	Equine rhinotracheitis B virus (ERBV)	Rhinotracheitis (respiratory infection)	Horses

A. Quarantine notice

March of Dimes archive

B. President Franklin D. Roosevelt

Franklin D. Roosevelt Library

Figure 11.10   Polio in the United States.   A. Board of Health quarantine notice, around 1900, aiming to exclude polio carriers. **B.** The most famous polio victim, President Franklin Delano Roosevelt, with his granddaughter Ruthie Bie and his dog Fala. Roosevelt's legs were paralyzed permanently, and he used a wheelchair throughout his presidency.

Despite the success of the polio vaccine, questions remain. The vaccine is not perfect; could better vaccines avoid the one-in-a-million deaths actually caused by the vaccine? What is the cause of the post-polio syndrome that affects some former polio victims 30–40 years after their illness? Are there positive applications of poliovirus, perhaps as a gene therapy vector or even as a treatment for brain tumors? To address these questions, we need to understand the virus and its host cell at the molecular level.

Poliovirus Structure

The structure of poliovirus is so compact and symmetrical that it was crystallized early in the twentieth century. The three-dimensional structure was modeled by X-ray crystallography and by digital reconstruction based on cryo-EM. Like many viruses, the capsid is fundamentally icosahedral (20-sided) (discussed in Chapter 6). Its form can be visualized as 60 triangular faces, each composed of proteins VP1, VP2, and VP3 (**Fig. 11.11A**). The three proteins arise from one long "polyprotein," or polypeptide, synthesized as a unit and then cleaved into three pieces. The icosahedral form enables a small number of genes to encode a relatively large capsid. Because of this efficient coding, icosahedral capsids are common among animal and plant viruses, including, for example, the rhinoviruses (cold viruses), aphthovirus (the virus causing foot-and-mouth disease), herpes viruses, and rice yellow mottle virus.

In the poliovirus capsid, each group of three triangular faces forms a pentamer. Beneath the pentamers of VP1, VP2, and VP3 lie subunits of a fourth protein, VP4. VP4 subunits coat the interior, strengthen the structure, and help package the RNA genome.

The RNA genome of poliovirus consists of a (+) strand (the strand that directly encodes protein sequences). Its 3′ OH end has a poly-A tail, like eukaryotic mRNA, but its 5′ OH end is capped with a special viral protein, VPg. The VPg will serve as a primer for RNA-dependent RNA

A.

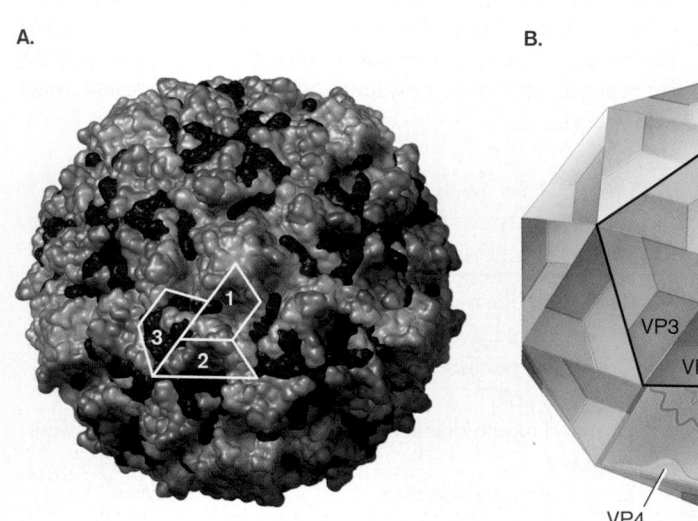

B.

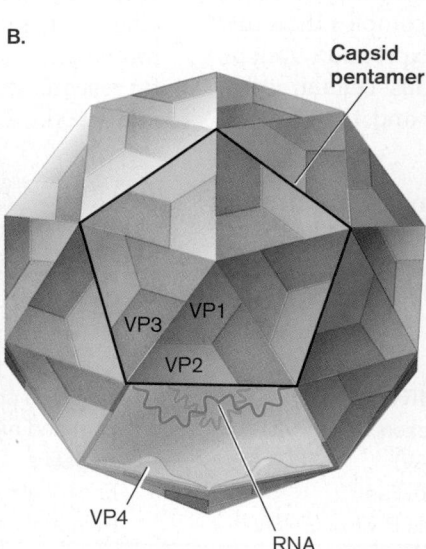

Capsid pentamer

VP1

VP3

VP2

VP4

RNA

Figure 11.11   Poliovirus structure. A. Digital reconstruction of the poliovirus capsid based on TEM. The icosahedral structure is composed of external capsid proteins VP1 (yellow), VP2 (blue), and VP3 (red). The three proteins form a structural unit, five of which form a pentamer. (PDB code: 1ASJ) **B.** Schematic of capsid with RNA genome and VP4 proteins inside.

synthesis. Unlike cellular polymerases (which use RNA primers exclusively), some viral polymerases are primed by proteins.

Virus Attachment and Genome Entry

The poliovirus attaches to the host at a species-specific receptor protein location. The virus can infect only those host cells that contain a **poliovirus receptor (PVR)**, a cell membrane glycoprotein, also known as CD155, found on cells of the intestinal epithelium. The normal function of CD155 is to help adhesion of adjacent epithelial cells, and to enhance the humoral immune response (antibody production). Unfortunately, poliovirus evolved to bind to this protein and gain entry into the cell. Poliovirus can bind only to the exact PVR sequence found in humans and chimpanzees. Thus, model animals such as mice cannot normally be infected. A transgenic mouse expressing a recombinant human gene for PVR can be infected by poliovirus. The polio mouse in 2003 became the first transgenic animal used to develop pharmaceuticals for a human disease. The mouse model was of limited use, however, because the mice develop only the intestinal symptoms but not the full neurological complications of poliomyelitis.

Poliovirus receptors diffuse freely within the host cell membrane. PVR can bind the polio virion at any vertex of the capsid pentamer (a pentamer is shown in **Fig. 11.11B**). PVR actually binds to adjacent VP2 and VP3 proteins surrounding the VP1 subunits, as shown in **Figure 11.12**. When poliovirus attaches to a cell, multiple receptor molecules diffuse near it and eventually bind to several faces of the capsid. As the poliovirus receptors bind, the membrane forms a cup around the capsid. Unlike other viruses, such as influenza virus, poliovirus requires only the earliest stage of endocytosis; the virion delivers its RNA to the cytoplasm immediately after internalization beneath the cell membrane. Thus, poliovirus infection is insensitive to drugs such as bafilomycin A1, which prevents infection by viruses that enter cells through endosomes.

As the poliovirus interacts with the poliovirus receptors, its pentamer structures shift so as to extrude the N-terminal ends of the five VP1 peptides (**Fig. 11.12**). The five VP1 molecules of the pentamer poke through the cell membrane together, generating a pore through which the chromosome passes into the cytoplasm. The transfer of the poliovirus chromosome parallels that of phage T4 in that the nucleic acid is injected across the cell membrane through a viral device while the capsid proteins are left outside.

The viral chromosome, a (+) strand RNA with a protein VPg attached to the 5′ end, is said to be **uncoated**. The uncoated genome is now ready to direct formation of progeny virions.

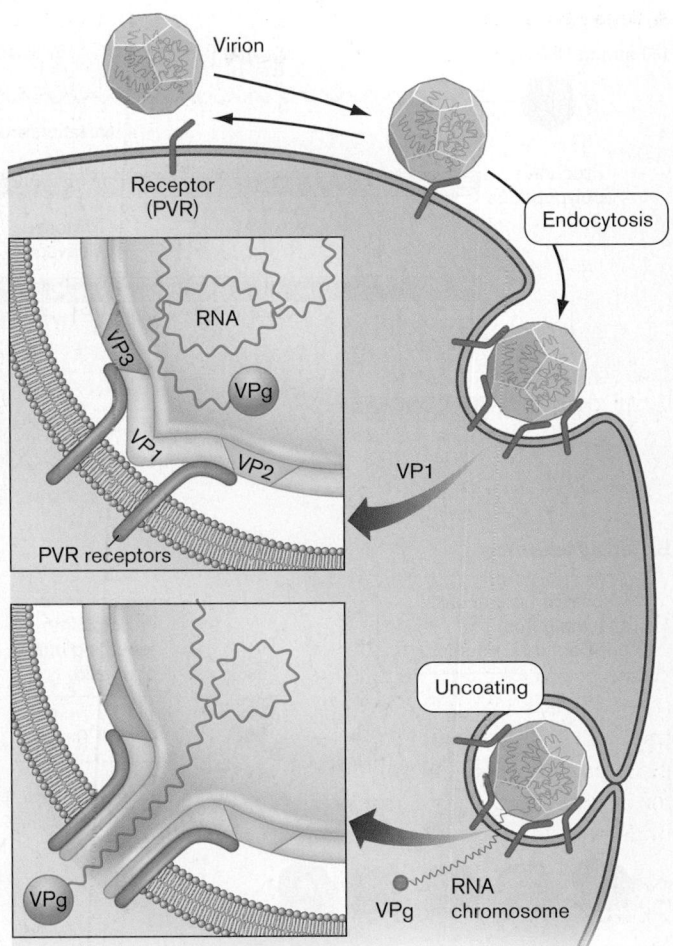

Figure 11.12 Poliovirus attachment to the host cell. The poliovirus binds PVR at the VP2 and VP3 subunits. A conformational change in the VP1 subunits allows insertion of the genome.

Gene Expression and Replicative Cycle

The uncoated viral RNA has two distinct roles in producing progeny virions: (1) mRNA translation by host ribosomes and (2) template replication of progeny RNA genomes. These two processes occur in separate compartments within the host cells.

Translation to make peptides. In the cytoplasm, the RNA is translated to make three large precursor peptides: P1, P2, and P3 (**Fig. 11.13A**). All three peptides are eventually cleaved by one of two viral proteases (products 2A and 3C) to generate a total of 11 different protein products. The P1 precursor forms the repeating unit of the isosahedral capsid; each unit is ultimately cleaved into four peptides. The P2 precursor is cleaved to form a two-subunit protease and an RNA-membrane interaction protein. The P3 subunit is cleaved to form the VPg primer for RNA synthesis (3B), another protease (3C), and RNA-dependent RNA polymerase (3D).

A. Polio genome

(+) strand RNA genome

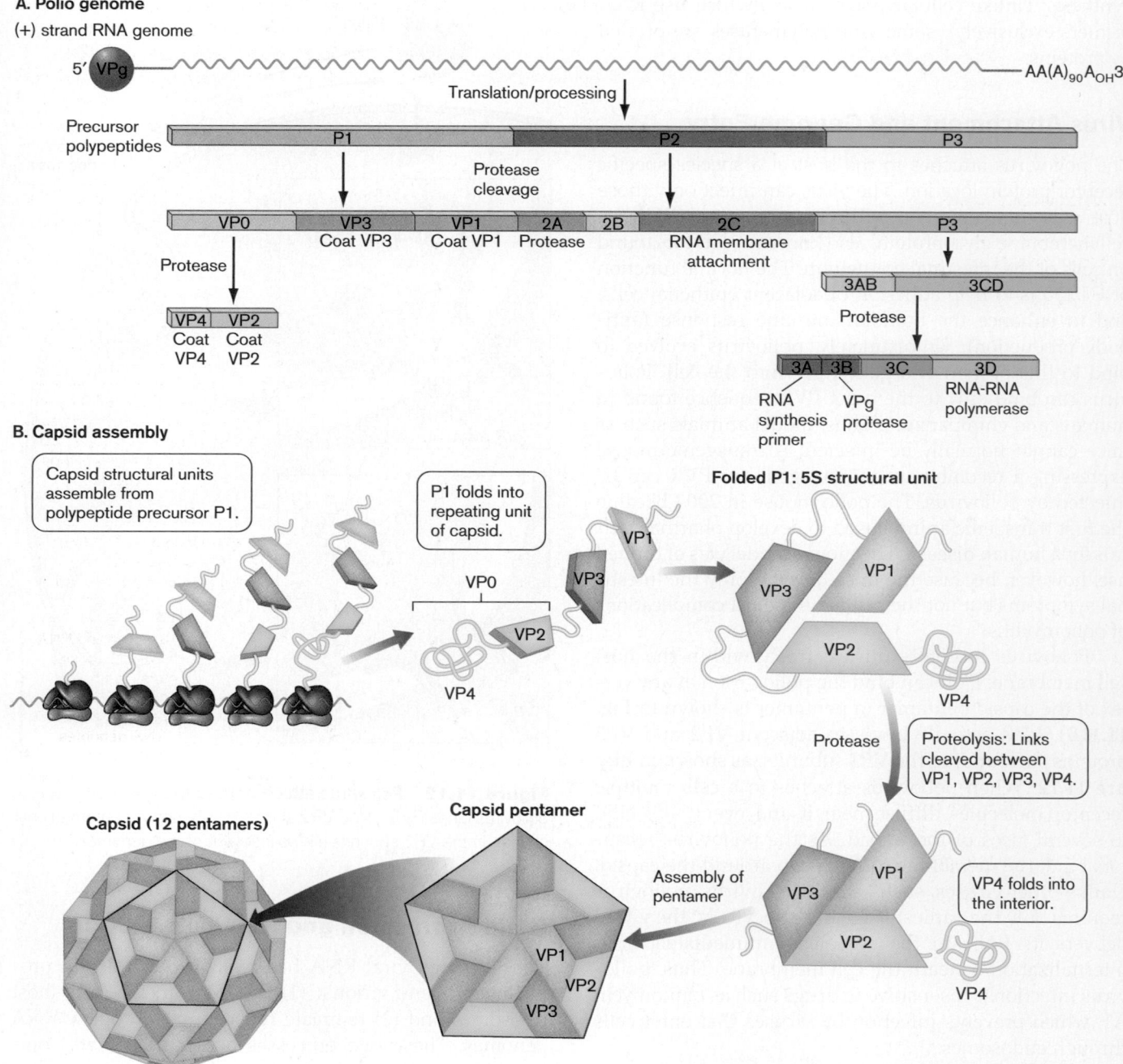

Figure 11.13 Polio genome and capsid assembly. A. The (+) strand RNA genome encodes three precursor polypeptides: P1, P2, and P3. These precursors are cleaved by proteases to generate the virus capsid proteins (VP1, VP2, and VP3), as well as the RNA-dependent RNA polymerase, proteases, and other assembly proteins. **B.** The capsid proteins are assembled together into the 5S structural unit before cleavage by protease. Five 5S units are packed into a pentamer, 12 of which assemble to build the icosahedral capsid.

The capsid proteins ultimately arise from cleavage of P1. Cleavage, however, does not occur immediately (**Fig. 11.13B**). Instead, the peptide P1 (including sequences VP4–VP2–VP3–VP1) first folds into the tertiary structure of the capsid's major repeating unit. The repeating unit is called the 5S structural unit, referring to its size in Svedberg units of sedimentation rate. After folding together, each 5S unit

then undergoes proteolysis at the junctions between the four subunits (VP1 through VP4). The 5S units then join together in groups of five to form pentamers, 12 of which ultimately assemble around a viral RNA chromosome.

RNA genome replication. In the overall cycle of replication (**Fig. 11.14**), the RNA chromosome serves as mRNA

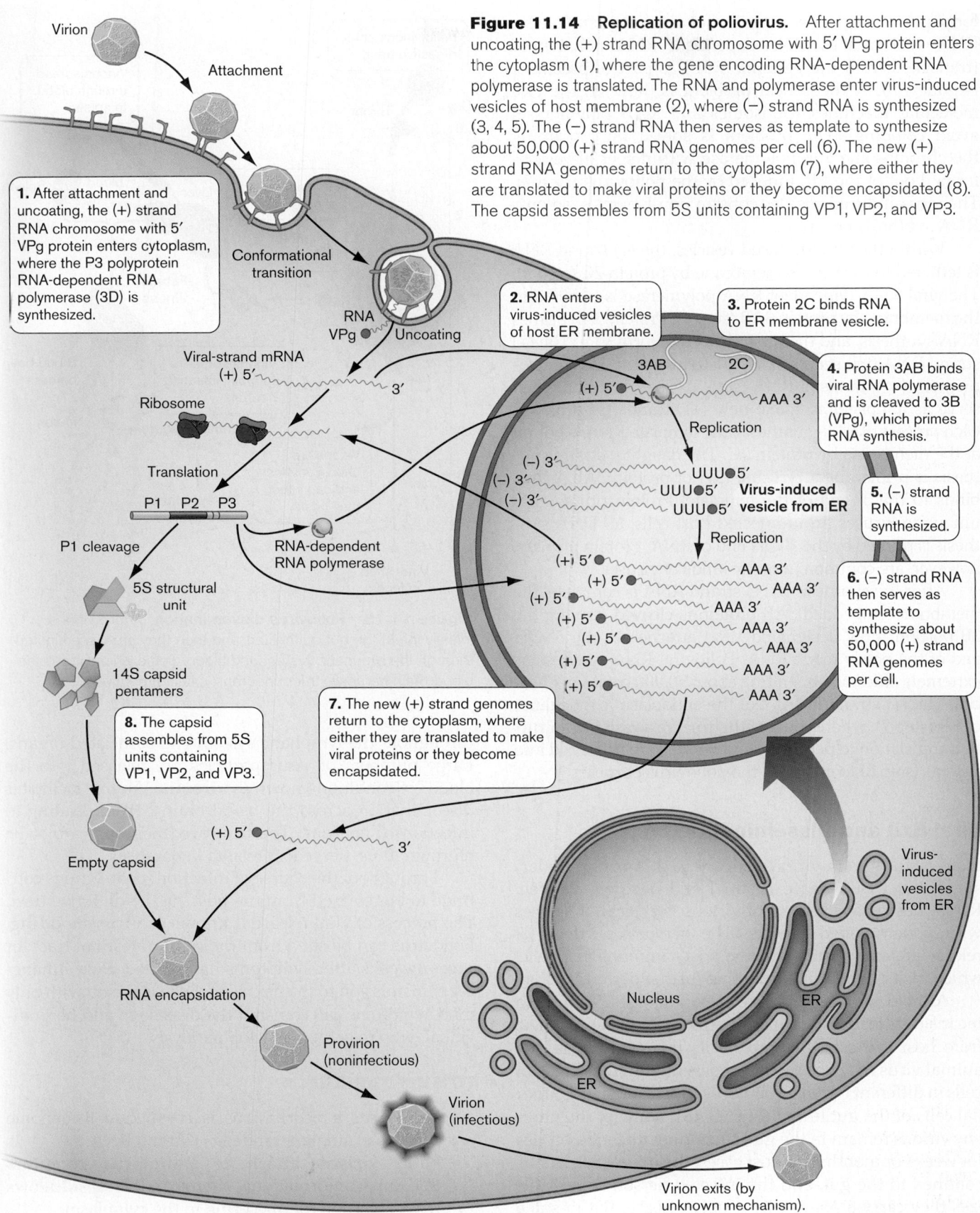

Virion

Attachment

Figure 11.14 Replication of poliovirus. After attachment and uncoating, the (+) strand RNA chromosome with 5' VPg protein enters the cytoplasm (1), where the gene encoding RNA-dependent RNA polymerase is translated. The RNA and polymerase enter virus-induced vesicles of host membrane (2), where (−) strand RNA is synthesized (3, 4, 5). The (−) strand RNA then serves as template to synthesize about 50,000 (+) strand RNA genomes per cell (6). The new (+) strand RNA genomes return to the cytoplasm (7), where either they are translated to make viral proteins or they become encapsidated (8). The capsid assembles from 5S units containing VP1, VP2, and VP3.

1. After attachment and uncoating, the (+) strand RNA chromosome with 5' VPg protein enters cytoplasm, where the P3 polyprotein RNA-dependent RNA polymerase (3D) is synthesized.

Conformational transition

RNA
VPg   Uncoating

Viral-strand mRNA
(+) 5'   3'

Ribosome

Translation

P1   P2   P3

P1 cleavage

5S structural unit

RNA-dependent RNA polymerase

14S capsid pentamers

8. The capsid assembles from 5S units containing VP1, VP2, and VP3.

Empty capsid

(+) 5'   3'

RNA encapsidation

Provirion (noninfectious)

Virion (infectious)

2. RNA enters virus-induced vesicles of host ER membrane.

3. Protein 2C binds RNA to ER membrane vesicle.

3AB   2C

(+) 5'   AAA 3'

Replication

4. Protein 3AB binds viral RNA polymerase and is cleaved to 3B (VPg), which primes RNA synthesis.

(−) 3'   UUU 5'
(−) 3'   UUU 5'   **Virus-induced**
(−) 3'   UUU 5'   **vesicle from ER**

Replication

5. (−) strand RNA is synthesized.

(+) 5'   AAA 3'
(+) 5'   AAA 3'
(+) 5'   AAA 3'
(+) 5'   AAA 3'
(+) 5'   AAA 3'
(+) 5'   AAA 3'
(+) 5'   AAA 3'

6. (−) strand RNA then serves as template to synthesize about 50,000 (+) strand RNA genomes per cell.

7. The new (+) strand genomes return to the cytoplasm, where either they are translated to make viral proteins or they become encapsidated.

Virus-induced vesicles from ER

Nucleus   ER

ER

Virion exits (by unknown mechanism).

for peptide translation (step 1) and also as a template for RNA synthesis by RNA-dependent RNA polymerase (translated from the genomic RNA as part of precursor peptide P3). A theoretical problem is, how do the two molecular machines (ribosomes and RNA polymerase) avoid colliding? The problem is solved by containing the chromosome and polymerase within special vesicles formed out of the ER, induced by the poliovirus (step 2). The virus-induced vesicles contain no ribosomes, so only RNA synthesis occurs.

Within the virus-induced vesicles, the (+) strand RNA is tethered to the vesicle membrane by protein 2C (step 3). The viral RNA-dependent RNA polymerase is tethered to the membrane by protein 3AB (step 4). Protein 3AB primes RNA synthesis, and then it is cleaved to form VPg. The (+) strand RNA first serves as template to synthesize complementary (−) strands of RNA (step 5). The (−) strands then serve as templates to make new (+) strands (step 6). For each process of RNA synthesis, the template RNA is bound at the membrane by protein 2C. The daughter strand synthesis is primed by a tyrosine-OH of protein 3AB, which binds to the 3′ end of the template. Protein primers are unique to viruses; in uninfected host cells, all DNA synthesis is primed by the 3′ OH end of RNA. Protein primers, however, are common in many kinds of viruses.

After the synthesis of (+) strand RNA is complete, the membrane-embedded 3AB protein is cleaved, leaving the 3B portion (now designated VPg) attached to the 5′ OH end of the newly made RNA. Polioviral RNA synthesis is extremely efficient, generating about 50,000 copies per host cell. The (+) strand RNAs exit the endoplasmic reticulum, or ER (step 7), where some of them serve as mRNA for further translation to peptides, while others are packaged into capsids (step 8) to complete the poliovirus particles.

Viral Exit and Dissemination

How mature polio virions exit the cell is poorly understood, but their release is rapid and destroys the cell. Rapid, destructive release of virions is typical of species with nonenveloped capsids. The cytopathic effects of release are compounded by the host's immune response, which includes inflammation that further damages cells.

Animal viruses, unlike bacteriophages, require a means of transmission not only from cell to cell but also from host to host. To complete its infectious cycle, the animal virus may need to infect several different kinds of cells in different tissues. Poliovirus initially infects epithelial cells of the gut lining (**Fig. 11.15**). Some of the progeny virions remain in the gut, sustaining infectious cycles for weeks or months. In most cases, the infection remains confined to the gut, and the infected person never realizes they carry a potentially deadly disease. But in some cases, virions move to the gut-associated lymphoid tissue, leading to **dissemination** (spread to different places)

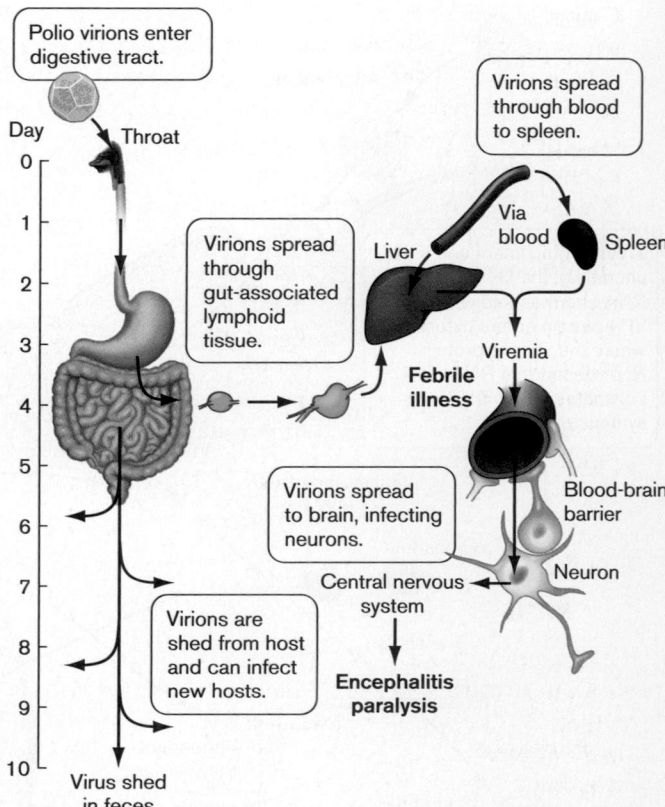

Figure 11.15 Poliovirus dissemination. Polio virions initially infect the gut epithelium, and then they are disseminated through the lymphatic system and blood. Some virions cross the blood-brain barrier to infect neurons, causing paralysis.

throughout the lymphatic system and associated organs. Large numbers of virus particles are released into the blood—a condition known as **viremia**. Viremia facilitates dissemination across the blood-brain barrier, leading to infection of neurons. The destroyed neurons cannot be regrown; thus, nerve loss results in paralysis.

Throughout the course of infection, polio virions continue to be released from the host via the digestive tract. The process of viral release is known as **virus shedding**. Poliovirus can be shed from the gastrointestinal tract for several weeks after symptoms have ended, thus enhancing transmission to the next host. Even a person with only mild symptoms can transmit the disease to another individual, who may then develop paralysis.

TO SUMMARIZE:

■ **Poliovirus**, a picornavirus, is icosahedral. Its genome consists of single-stranded (+) strand RNA.

■ **Polio virions attach to poliovirus receptors (PVRs).** Conformational change in the capsid allows DNA insertion for uncoating in the cytoplasm.

■ **Poliovirus RNA is replicated within ER-derived vesicles by a viral RNA-dependent RNA polymerase.**

Genome replication is primed by virus-encoded protein primer 3AB, which is cleaved to 3B (VPg).

■ **(+) strand RNA serves as a template for (–) strand RNA.** The (–) strand RNA then serves as template for (+) strand RNA, which is either translated to proteins or encapsidated into progeny virions.

■ **RNA for polio gene expression is exported** to the cytoplasm for translation by host ribosomes. Virions are packaged in the cytoplasm and exported by an unknown mechanism.

■ **Poliovirus enters the body through the gastrointestinal tract.** In severe cases, the virus disseminates through the lymph and blood, crossing the blood-brain barrier to infect neurons and cause paralysis.

11.3 Influenza Virus: (–) Strand RNA

Some viruses package a (–) strand RNA genome, whose replication involves a more complex replicative cycle than that of (+) strand RNA viruses. Influenza A virus, an orthomyxovirus, is one of the most common life-threatening viruses in the United States. Each year, influenza A infects approximately 10% of the U.S. population, causing about 36,000 deaths annually. The elderly are most susceptible, but in an epidemic year, mortality rises among young people.

Influenza shows a cyclic appearance of extremely virulent strains that cause pandemic mortality, such as the famous pandemic of 1918, which infected 20% of the world's population and killed more people than World War I did. The 1918 strain arose as a mutant form of an influenza strain infecting birds. In 2006, a similar avian influenza strain emerged as a potential source of the world's next influenza pandemic. As of this writing, however, the avian strain has not yet mutated to a form readily transmitted between humans. In 2009, a highly transmissible strain related to swine influenzas ("swine flu") spread rapidly around the world but caused relatively mild illness. A future strain might emerge combining high transmission with high human mortality.

Highly virulent strains of influenza usually result from reassortment between distantly related strains. The reassortment process is enhanced by a particular feature of their molecular biology—namely, the **segmented genome**. A segmented genome of a virus consists of multiple separate nucleic acids, like the multiple chromosomes of a eukaryotic cell. The influenza genome consists of eight separate linear (–) strands of RNA, which can reassort from different strains to generate a novel hybrid strain.

www | CDC page on influenza virus
microbiology2.com/links

Influenza Virion Structure and Genome

The influenza virion, unlike that of polio, has no geometrical capsid (**Fig. 11.16**). Its RNA chromosome segments are loosely contained by a shell of **matrix proteins** (M1). Matrix proteins enclose the core or capsid, supplementing its connection to the membrane envelope. The envelope derives from the phospholipid membrane of the host cell, which incorporates viral proteins such as hemagglutinin (HA) and neuraminidase (NA). These viral envelope proteins peg the membrane to the enclosed capsid, maintaining its structure. Neuraminidase can be blocked by the antiviral agent Tamiflu (oseltamivir), one of the main drugs available to treat influenza.

The matrix shell contains the eight RNA segments, which are noncoding (–) strands (**Fig. 11.17**). The RNA segments are coated with **nucleocapsid proteins (NPs)**. The term "nucleocapsid" refers generally to proteins coating a viral genome and packaged within or as part of the viral capsid. Each NP-coated RNA segment also possesses a bound RNA-dependent RNA polymerase complex. Thus, unlike poliovirus and most bacteriophages, the influenza virion contains more than its genome; it also contains its own RNA polymerase enzyme, bound to each RNA segment and poised for synthesis. During infection, each (–) strand RNA segment must be transcribed to a (+) strand mRNA for eukaryotic translation.

> **THOUGHT QUESTION 11.4** Why does influenza virus have to provide its polymerase ready-made, whereas poliovirus does not?

> **NOTE:** In figures of this chapter, AAA-OH$^{3'}$ represents a poly-A tail consisting of a variable number of adenine nucleotides with a 3' hydroxyl end.

The segmented genome of influenza poses a major problem for virus assembly: Given that viral infection requires all eight segments, how does the assembly mechanism package exactly eight segments, one of each? In fact, the segments are packaged at random, resulting in a vast majority of defective particles. The fraction of influenza particles observed to be capable of infection is consistent with random packaging of any eight segments into a capsid. The theoretical fraction of infectious virions would be given by the following equation:

$$\frac{8!}{8^8} = \frac{8 \times 7 \times 6 \times 5 \times 4 \times 3 \times 2 \times 1}{16,777,216} = \frac{1}{416}$$

One in 400 may sound tremendously inefficient, but if 10,000 virions are produced, there will be more than enough infective particles to propagate the virus.

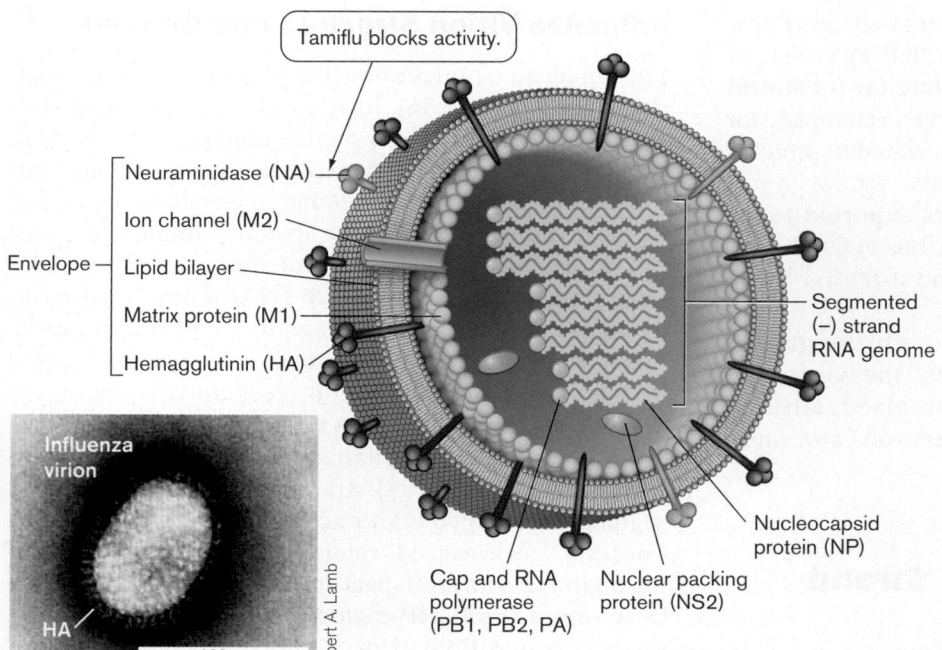

Figure 11.16 Structure of influenza A. Diagram of virion structure showing envelope (tan), envelope proteins, matrix protein (green), RNA segments with attached polymerase, and the nuclear packing protein NS2. The inset image shows influenza A virions (TEM). The brush-like border coating the envelope consists of glycoproteins, hemagglutinin (HA), and neuraminidase (NA).

Furthermore, animal viruses require a relatively small number of progeny to ensure infecting the neighboring cells of a multicellular tissue, as compared to phages or viruses of single-celled organisms. Thus, influenza virus may avoid the energetic expense of an accurate packaging mechanism for its segmented genome.

THOUGHT QUESTION 11.5
Explain why the expression $(8!/8^8)$ approximates the proportion of infective particles of influenza virus. What assumption might be changed to make the proportion greater or less?

The key advantage of a segmented genome is that it facilitates recombination between two strains coinfecting the same cell. Coinfection can generate an instant new strain that evades a host immune system. Even strains that infect animals such as ducks or swine may reassort their segments with those of a human virus (**Fig. 11.18**). Influx of genes from a distantly related strain can sharply increase virulence and mortality. Major epidemics of exceptionally

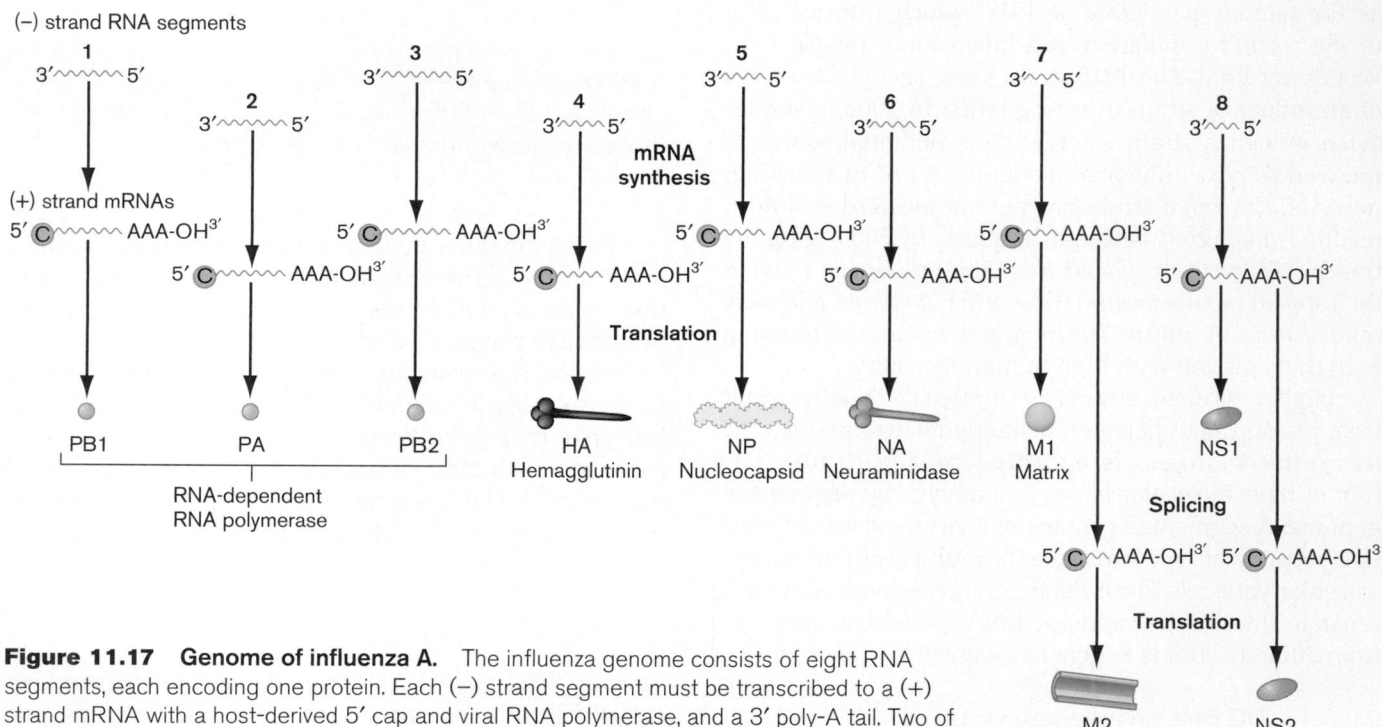

Figure 11.17 Genome of influenza A. The influenza genome consists of eight RNA segments, each encoding one protein. Each (−) strand segment must be transcribed to a (+) strand mRNA with a host-derived 5′ cap and viral RNA polymerase, and a 3′ poly-A tail. Two of the segments (7 and 8) are processed further to express two additional proteins (M2 and NS2).

virulent influenza arise from reassortment with genome segments from strains that evolved within ducks or swine, agricultural animals that live in close proximity to humans. In each genome, the "H" and "N" numbers designate alleles of the genes encoding hemagglutinin and neuraminidase, respectively. For example, the Hong Kong flu strain had alleles H3 and N2. The "avian" strain that emerged in 2006 had alleles H5 and N1. The "swine" strain in 2009 had alleles H1 and N1, similar to the pandemic 1918 strain.

Figure 11.18B shows how various combinations of human and avian genome segments led to new allelic combinations of the envelope proteins hemagglutinin ("H" numbers) and neuraminidase ("N" numbers).

These variant envelope proteins can foil the immune system; for example, a human immune system is unlikely to have developed antibodies against an avian envelope protein. Sequence analysis suggests that the 1918 pandemic virus derived all eight segments from an avian strain. Subsequently, the 1957 virulent strain incorporated avian segments 2, 4, and 5, whereas the 1968 virulent strain replaced segments 2 and 4 with those of yet another avian strain.

Overall, the opportunities for reassortment among influenza strains make it difficult to generate vaccines and predict which strains to vaccinate against in a given flu season. Nevertheless, alert clinicians and effective public health reporting enable physicians to predict lesser

A.

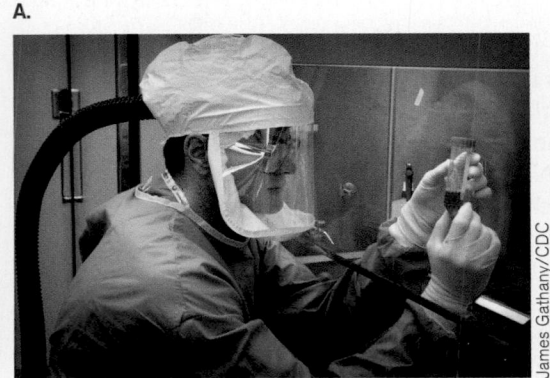

James Gathany/CDC

Figure 11.18 Reassortment between human and avian strains generates exceptionally virulent strains. A. The reconstructed strain of the 1918 pandemic influenza virus is studied by Dr. Terrence Tumpey, microbiologist at the Centers for Disease Control and Prevention, Atlanta. The 1918 strain may hold clues to combating the "avian influenza" strain today. **B.** Major epidemics of exceptionally virulent strains arise from reassortment with chromosomal segments from strains that evolved within ducks or swine, agricultural animals that live in proximity to humans. The segments are labeled as in Figure 11.17. The 1918 pandemic may actually have derived all eight segments from an avian strain. The 1957 strain incorporated avian segments 2, 4, and 5; subsequently, the 1968 strain replaced segments 2 and 4 with those of yet another avian strain. The numbers following "H" and "N" refer to different alleles of the genes encoding hemagglutinin and neuraminidase, respectively.

B.

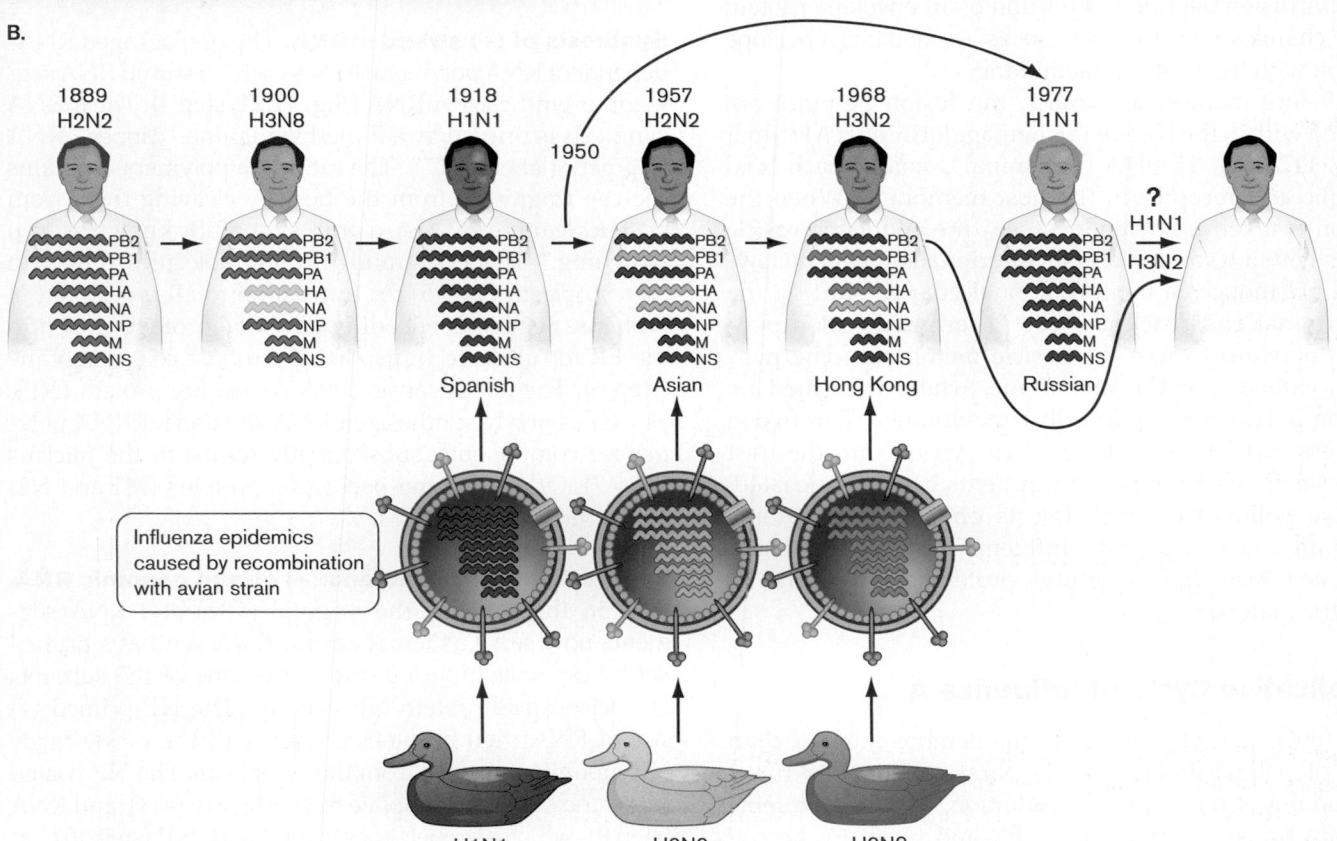

epidemics and protect many people by vaccination, as discussed in Chapter 28. The rarer pandemics involving exceptionally virulent avian-human recombinant strains remain a major challenge to public health.

Attachment and Entry to the Host Cell

An influenza virion attaches to a cell when its hemagglutinin envelope protein binds to a host cell receptor protein that contains polysaccharide terminating with sialic acid. The precise structure of the host receptor binding the hemagglutinin may determine whether a strain such as avian influenza H5N1 will spread directly between humans. For example, the sialic acid connection in the receptor polysaccharide can involve different OH groups of the sugar galactose: a linkage to the OH-3 (α2,3) or to the OH-6 (α2,6) (**Fig. 11.19**). The influenza strain H5N1 recognizes mainly the α2,3-linked protein, found in birds. In humans, the upper respiratory tract contains mainly α2,6-linked receptors; α2,3-linked receptors are found only deeper within the lungs. Thus, avian influenza strain H5N1 is rarely transmitted between humans. However, more rapid transmission could arise from a mutation in the avian hemagglutinin that binds the receptor, leading to an influenza pandemic.

The hemagglutinin complex consists of a trimer of subunits, each of which has an N-terminal **fusion peptide**. A fusion peptide is a portion of an envelope protein that changes conformation so as to facilitate envelope fusion with the host cell membrane.

Before membrane contact, the fusion peptides are buried within the core of the hemagglutinin (HA) trimer (**Fig. 11.20**). The HA C-terminal domains each bind a sialic acid receptor in the host membrane. When the virion is taken up by endocytosis, the endocytic vesicle fuses with a lysosome and its interior acidifies. The lowered pH induces a conformational change shifting the C-terminal ends back and the N-terminal fusion peptides outward to face the vesicle membrane. The peptides extend into the membrane, where they mediate fusion between viral and host membranes. The fusion process expels the contents of the virion into the host cytoplasm, where the RNA segments become uncoated. Unlike poliovirus, which injects only its chromosome with an attached peptide, influenza virus releases several enzymes and structural proteins along with its genetic material.

Replicative Cycle of Influenza A

The replication of influenza virus is more complex than that of poliovirus because the viral components travel in and out of the nucleus. In addition, envelope proteins require transport through the ER and Golgi to the cell

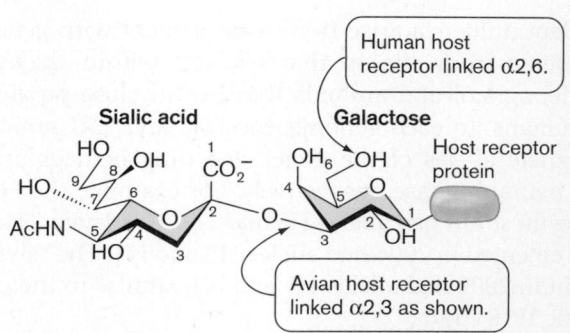

Figure 11.19   Sialic acid in influenza receptor. The avian host receptor polysaccharide contains sialic acid with an α2,3 bond to galactose, whereas the human receptor in the upper respiratory tract has an α2,6 bond.

membrane (**Fig. 11.21**). By contrast, for poliovirus no envelope is assembled.

After the influenza virion undergoes endocytosis (step 1) and release into the cytoplasm (step 2), all the viral (−) RNA segments with their prepackaged primers and polymerases are released. Each NP-coated RNA segment individually passes through the nuclear pores into the nucleus (step 3). In the nucleus, each viral polymerase attached to a genomic (−) RNA segment synthesizes (+) strand RNA for mRNA (step 4) or for templates to synthesize progeny (−) strand RNA segments (steps 8, 9).

Synthesis of (+) strand mRNA. The prepackaged RNA-dependent RNA polymerase uses each (−) strand RNA segment to synthesize mRNA (**Fig. 11.21**, step 4). The mRNA synthesis is primed by a 7-methylguanine-"capped" RNA fragment (labeled "C"). The influenza polymerase obtains the cap fragments from the host by cleaving them from host nuclear pre-mRNA, a process quaintly known as "cap snatching." The (+) strand mRNA molecules return to the cytoplasm (step 5) for translation to all types of viral proteins. Segments encoding envelope proteins attach to the ER for ultimate transport to the host cell membrane (step 6). The nucleocapsid RNA-packaging protein (NP), as well as newly synthesized RNA-dependent RNA polymerase components, subsequently return to the nucleus (step 7). Other genome-packaging proteins (M1 and N2) also return to the nucleus.

Synthesis of (+) strand and (−) strand genomic RNA. Back in the nucleus, the original (−) strand RNA segments now serve as templates for RNA synthesis primed *not* by cap snatching, but instead by one of the subunits of nucleocapsid protein NP (step 8). The NP-primed (+) strand RNA then becomes coated with the newly made NP subunits imported from the cytoplasm. The NP-coated (+) strand serves as template to synthesize (−) strand RNA (step 9), which also becomes coated with NP (step 10).

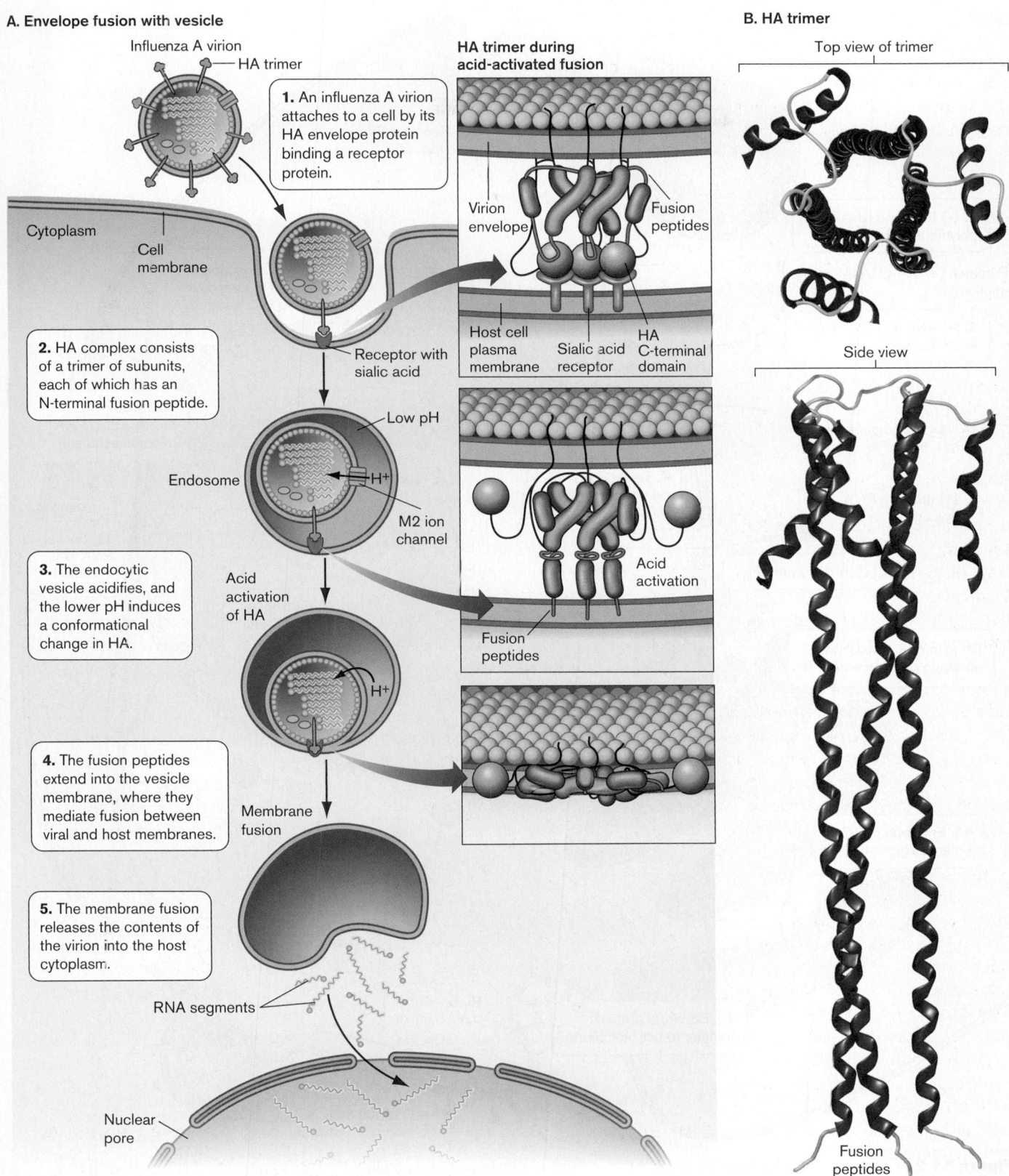

A. Envelope fusion with vesicle

Influenza A virion — HA trimer

1. An influenza A virion attaches to a cell by its HA envelope protein binding a receptor protein.

Cytoplasm

Cell membrane

Receptor with sialic acid

2. HA complex consists of a trimer of subunits, each of which has an N-terminal fusion peptide.

Low pH

Endosome

H+

M2 ion channel

3. The endocytic vesicle acidifies, and the lower pH induces a conformational change in HA.

Acid activation of HA

H+

4. The fusion peptides extend into the vesicle membrane, where they mediate fusion between viral and host membranes.

Membrane fusion

5. The membrane fusion releases the contents of the virion into the host cytoplasm.

RNA segments

Nuclear pore

HA trimer during acid-activated fusion

Virion envelope — Fusion peptides

Host cell plasma membrane — Sialic acid receptor — HA C-terminal domain

Acid activation

Fusion peptides

B. HA trimer

Top view of trimer

Side view

Fusion peptides

Figure 11.20   Entry and acid activation of influenza A virion.   A. The HA envelope protein of an influenza A virion attaches to a host cell (1) binding a sialic acid receptor protein. HA complex consists of a trimer of subunits, each of which has an N-terminal fusion peptide. When the virion is taken up by endocytosis, the interior of the endocytic vesicle acidifies, and the lower pH induces a conformational change in HA (2, 3). The receptor-binding domains fold back, and the fusion peptides extend into the vesicle membrane, where they mediate fusion between viral and host membranes (4). The membrane fusion releases the contents of the virion into the host cytoplasm (5). **B.** Model of the core portion of HA trimer (top view and side view). (PDB code: 1HTM) ▶❙❙ *microbiology2.com/animations*

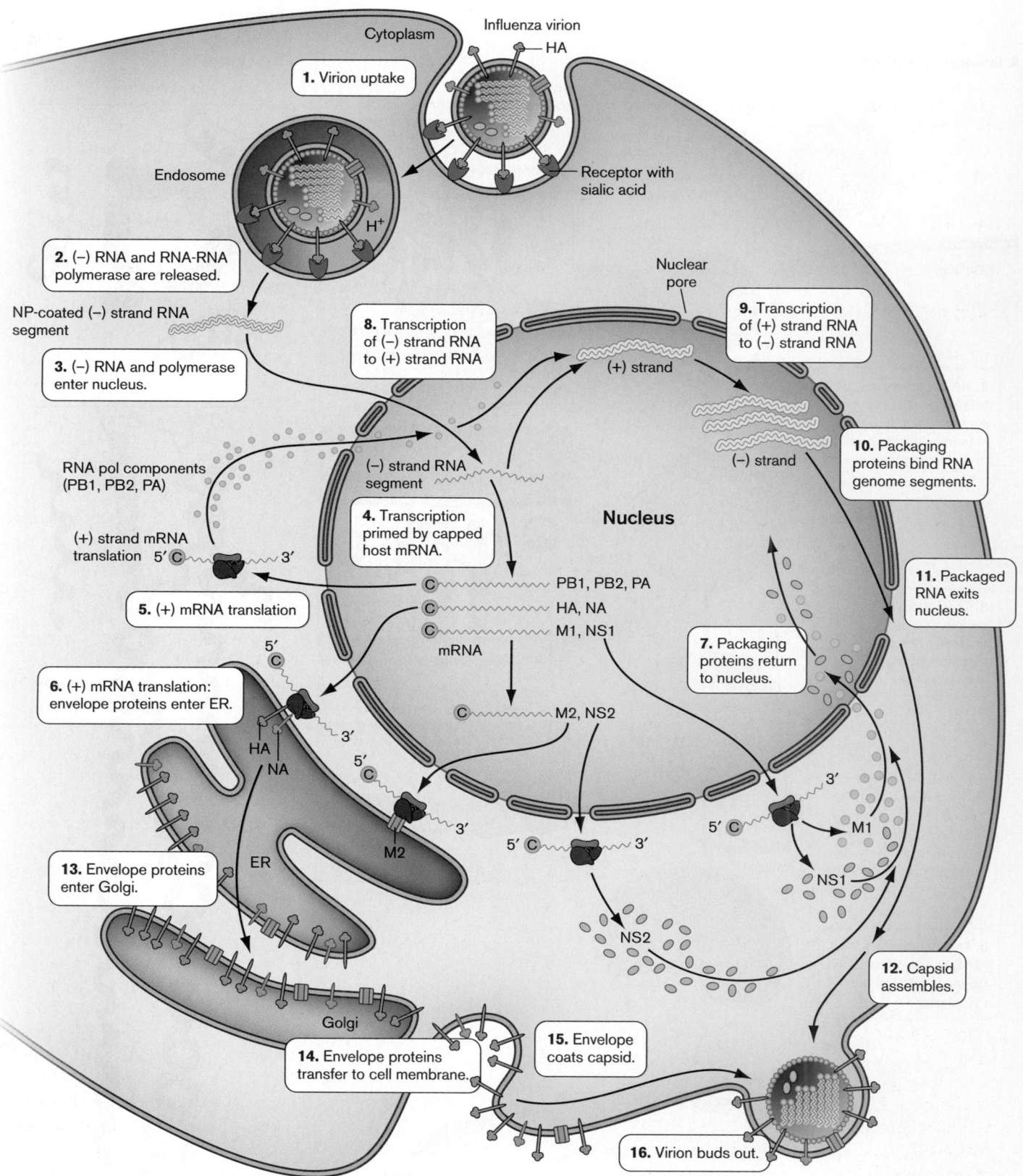

Figure 11.21 Replication of influenza virus. HA-mediated attachment and uptake (1) leads to release of NP-coated RNA segments in the cytoplasm (2). The (−) strands enter the nucleus (3), where the viral polymerase makes (+) strand RNA (4). RNA synthesis is primed by "capped" Ⓒ fragments of host mRNA, obtained by "cap snatching." The (+) strand RNA segments return to the cytoplasm for translation to viral proteins (5). Some viral proteins enter the ER (6); others return to the nucleus for packaging (7). Additional (+) strand RNA is made to serve as a template for (−) strands, primed by NP protein (8, 9). Proteins MI, NS1, and NS2, as well as newly synthesized RNA-dependent RNA polymerase components (PA, PB1, and PB2), return to the nucleus to bind the progeny (−) strand segments (10). Protein-packaged RNA segments then return to the cytoplasm (11), where the capsid assembles (12). Envelope proteins (M2, NA, and HA) are synthesized at the ER (6), where they are glycosylated by host enzymes and transferred to the Golgi (13) for export to the cell membrane (14). At the cell membrane, the packaged (−) RNA segments are enveloped by host membrane containing the envelope proteins (15). Mature virions bud out of the cell membrane (16). ▶‖ *microbiology2.com/animations*

The NP-coated (–) RNA associates with a newly made polymerase for a future cycle of viral replication. The RNA is then complexed with matrix protein (M1) and nuclear packaging protein (N2), proteins that were imported from the cytoplasm earlier. At last, the fully packaged (–) RNA segments exit the nucleus to the cytoplasm (step 11), where they approach the cell membrane for packaging into progeny virions (step 12).

Envelope synthesis and assembly. The envelope proteins synthesized at the ER include hemagglutinin (HA) and neuraminidase (NA). Within the ER lumen, these proteins are glycosylated by host enzymes and then transferred to the Golgi (step 13) for export to the cell membrane (step 14). Within the cell membrane, the envelope proteins assemble around a group of (–) RNA segments complexed with their matrix and packaging proteins, completing the virion particle (step 15). The virion buds out (step 16) and neuraminidase cuts the virion loose from host glycoproteins. Neuraminidase activity is inhibited by the drug oseltamivir (Tamiflu).

TO SUMMARIZE:

- **Influenza virus** causes periodic pandemics of respiratory disease. New virulent strains arise through reassortment of human and avian strains.
- **The influenza virus consists of segmented (–) strand RNA.** Each segment is packaged with nucleocapsid proteins. Segments from different strains recombine through coinfection.
- **Nucleocapsid and matrix proteins** enclose the RNA segments of the influenza virus. The matrix is enclosed by an envelope containing spike proteins.
- **Envelope HA proteins mediate virion attachment.** The HA protein includes a fusion peptide that undergoes conformational change to cause fusion between the viral envelope and host cell membrane. For influenza, the virion is internalized by endocytosis.
- **Lysosome fusion with endosomes** triggers viral envelope fusion with the endosome membrane. The viral genome and proteins are then released into the cytoplasm. Viral (–) strand RNA segments attached to RNA-RNA polymerase enter the nucleus.
- **Influenza mRNA synthesis is primed by capped RNA fragments**, cleaved from host mRNA. The viral mRNAs return to the cytoplasm for translation.
- **Genomic RNA synthesis is primed by the nucleocapsid protein (NP).** First, (+) strand RNA is synthesized as a template for (–) RNA strands, which are then packaged in newly made nucleocapsid protein and exported to the cytoplasm.
- **Envelope proteins** of influenza virus are synthesized at the ER for transport to the cell membrane.
- **Influenza virus is assembled** at the cell membrane, where capsid, matrix, and (–) strand RNA components are packaged into envelope.

11.4 Human Immunodeficiency Virus (HIV): Retrovirus

So far we have discussed (+) strand RNA viruses, which need to generate a (–) strand intermediate as a template both for (+) strand mRNA and for progeny genomes; and (–) strand viruses, which can generate mRNA directly but need to make a (+) strand template to generate progeny virions. Another major class of RNA viruses is **retroviruses**. Retroviruses are so called because they reverse the normal order of synthesis to copy their RNA into a double-stranded DNA, which is then integrated into the host genome. As introduced in Chapter 6, the RNA-to-DNA synthesis requires a key enzyme, reverse transcriptase (RT), the target of the major anti-AIDS drug azidothymidine (AZT).

www | Reverse transcriptase tutorial (Biomolecules at Kenyon)

www | UNAIDS
microbiology2.com/links

Retroviruses form a large family of viruses that infect animals; for examples, see **Table 11.2**. Retroviruses are divided into two major groups: the "simple retroviruses," whose genomes include just four genes; and the lentiviruses, with more complex genomes (discussed below). Simple retroviruses cause tumors and leukemias; for example, **Rous sarcoma virus (RSV)** was the first virus demonstrated to cause cancer.

The DNA insertion mechanism of retroviruses is ideally suited to alter host gene regulation, either by decoupling a host gene from its regulatory sites or by inserting an oncogene (an altered host gene that causes cancer). Another retrovirus, **feline leukemia virus (FeLV)**, is a major veterinary pathogen. FeLV remains America's number one killer of outdoor cats, despite the availability of an effective vaccine. Related to FeLV is primate T-lymphotrophic virus (PTLV-1), formerly called human T-cell leukemia virus (HTLV). PTLV-1 was the first retrovirus identified in humans, discovered by American virologist Robert Gallo in 1980.

The second group of retroviruses is the **lentiviruses**, or "slow viruses." Lentiviruses are retroviruses that cause infections that progress slowly over many years. Lentiviral genomes include the four key genes of retroviruses, plus genes for several regulatory proteins that modulate host interactions. The most famous lentivirus is **human immunodeficiency virus (HIV)**, the causative agent of **acquired immunodeficiency syndrome (AIDS)**. The virus HIV and its causative role in AIDS were discovered by French virologist Luc Montagnier, building on Gallo's studies of retroviruses. In 2008, according to the United Nations, 33 million people globally were estimated to be living with HIV. That year, the AIDS epidemic claimed 2

Table 11.2 Retroviruses of animals (examples).

Genus	Virus	Disease	Host
Simple retroviruses			
Alpharetrovirus	Avian leukosis virus (ALV)	Leukemia	Birds
	Rous sarcoma virus (RSV)	Sarcoma (tumor)	Birds
Betaretrovirus	Mouse mammary tumor virus (MMTV)	Mammary tumor	Mice
Gammaretrovirus	Feline leukemia virus (FeLV)	Lymphoma, immune deficiency	Cats
	Moloney murine leukemia virus (MMLV)	Leukemia	Mice
	Xenotropic murine leukemia virus–related virus (XMRV)	Associated with prostate cancer and chronic fatigue syndrome	Humans
Deltaretrovirus	Bovine leukemia virus (BLV)	Leukemia	Cattle
	Primate T-lymphotrophic virus (PTLV-1) (formerly human T-cell leukemia [HTLV])	Leukemia	Humans
Epsilonretrovirus	Walleye dermal sarcoma virus (WDSV)	Sarcoma	Fish
Lentiviruses	Human immunodeficiency virus (HIV-1, HIV-2)	AIDS	Humans
	Simian immunodeficiency virus (SIV)	Simian AIDS	Monkeys
	Equine infectious anemia virus (EIAV)	Anemia	Horses
	Maedi-Visna virus (MV)	Neurological disease	Sheep

million lives, and 3 million people became newly infected with HIV.

WWW | Feline leukemia virus, Cornell Feline Health Center

WWW | HIV InSite, University of California at San Francisco
microbiology2.com/links

HIV Causes AIDS

HIV is a lentivirus that evolved from viruses infecting African monkeys. Two major types are recognized: HIV-1, the cause of most infections at present; and HIV-2, another type that appears to have evolved from simian immunodeficiency virus (SIV). The virus is transmitted through blood and through genital or oral-genital contact. The HIV life cycle is that of a "slow virus" that can hide in the host cell for many years, with only gradual production of virions. Eventually, however, the virus destroys the body's T leukocytes, leaving the host defenseless against many organisms that normally would be harmless.

The first cases of AIDS in the United States were reported in 1981. Twenty-nine years later, HIV infects over a million Americans, and more than half a million have died of AIDS. Worldwide, HIV infects one in every 100 adults, equally among women and men. In the developed countries, treatment effectively prolongs the life of people with AIDS; but treatment is too expensive for the majority of those infected worldwide. Countries in southern Africa have already experienced population decline. In South Africa, approximately 20% of adults test positive for HIV. Despite all we know, AIDS continues to spread in

North America and Europe, and particularly high rates of increase are seen in eastern Europe and central Asia. As of this writing, there is no cure and no vaccine.

A historical view of AIDS in the United States emerges in the book *And the Band Played On* by Randy Shilts, adapted as an award-winning film in 1993. The book and film show how American society failed for many years to grasp the significance of AIDS because the syndrome first appeared in societal groups considered marginal (homosexual men and certain ethnic immigrants), although it spread to all social classes. In addition, the virus proved extremely difficult to detect and grow in culture. The discovery of HIV sparked controversy in the scientific community because Gallo failed to acknowledge his use of a virus-producing cell line from Montagnier, the first to isolate HIV-1. Since that time, the two scientists and many others have collaborated to develop a test for HIV-1 infection and to search for a vaccine. In 2008 the Nobel Prize in Physiology or Medicine was awarded to Montagnier and his associate Françoise Barré-Sinoussi for the discovery of HIV and its role in AIDS (**Fig. 11.22**).

The lack of a cure for AIDS should not surprise us, since only a small number of effective antiviral drugs have been found. But why can we find no vaccine when safe and effective vaccines were obtained long ago for major killers such as smallpox and polio and even for the feline retrovirus FeLV? The answers to this question are complex. They include

■ **High mutation rate.** The mutation rate of all retroviruses is exceptionally high because of the high error rate of reverse transcriptase. Even within one patient, the virus evolves into numerous genetically different

A. Françoise Barré-Sinoussi

B. Luc Montagnier and Robert Gallo

Figure 11.22 HIV discoverers. A. Françoise Barré-Sinoussi, at the Pasteur Institute, who worked with Luc Montagnier to discover the virus that causes AIDS. **B.** Luc Montagnier and Robert Gallo agree to collaborate on development of an AIDS vaccine, 2002.

strains, as different as different species. This collection of diverse strains within an individual is termed a **quasispecies**. Different strains of the quasispecies may involve different organs and predominate at different stages of the disease.

■ **Complex regulation of replication.** The lentiviruses express a greater number of regulator proteins than do the "simple" retroviruses. These complex regulatory options of a lentivirus may enable HIV to hide itself more effectively within host cells than other retroviruses, such as FeLV.

While a cure remains elusive, studying the molecular biology of HIV has led to successful treatments that greatly extend the life span and improve the quality of life for infected individuals. And the very traits that make HIV so insidious a pathogen also make it a promising vector for gene therapy (see Section 11.6).

HIV Structure and Genome

The structure of HIV as visualized by TEM consists of an electron-dense **core particle** (or capsid) surrounded by a phospholipid envelope (**Fig. 11.23A**). In some sections, the core appears round, whereas other sections cut longitudinally reveal an elongated shape, like a cone or cylinder. The thickening of the membrane around the core indicates the presence of **spike proteins**, which peg the membrane to the matrix, as in influenza virus. The envelope forms around the core by budding out of the host cell membrane (**Fig. 11.23B**)—a process that does not rapidly lyse the cell but has other devastating consequences for cell function. The conical core is composed of capsid subunits (CA) whose arrangement is partly icosahedral (**Fig. 11.24A**). The capsid subunits enclose two identical copies of the RNA genome with reverse transcriptase and other enzymes.

Each virion contains two separate single-stranded copies of the RNA genome (**Fig. 11.24A**). Each RNA genome is coated with nucleocapsid proteins (NC) similar in function to those of influenza virus. For priming, each RNA is complexed with a tRNA derived from the previously infected host cell. The primed and packaged RNA is contained within the capsid or core composed of capsid subunits (CA). The capsid also contains about 50 copies of reverse transcriptase (RT) and protease (PR), as well as a DNA integration factor (IN). Unique to type HIV-1, subunits of a host chaperone named cyclophilin A are incorporated into the structure, about one for every ten capsid subunits. An HIV-1 mutant that fails to incorporate cyclophilin A can attach to a host cell and insert its capsid, but the capsid fails to come apart, and infection is halted.

The capsid is surrounded by a matrix (MA subunits), which reinforces the host-derived phospholipid membrane. The membrane is pegged by spike proteins composed of the envelope subunits TM and SU. As in influenza virus, the spike proteins play crucial roles in host attachment and entry.

The genome of HIV was first cloned for molecular study by Flossie Wong-Staal, now at the University of California at San Diego (**Fig. 11.24B**). Born in China in 1947, Wong-Staal immigrated to the United States and then worked with Gallo on the early discoveries of HIV. Wong-Staal now pursues gene therapy approaches to combat HIV infection and is developing retroviral vectors for human gene therapy (discussed in Section 11.6).

The HIV genome includes three main open reading frames that are found in all retroviruses: *gag*, *pol*, and *env*. The *gag* sequence encodes capsid, nucleocapsid, and matrix proteins; *pol* encodes reverse transcriptase (RT), integrase, and protease; and *env* encodes envelope proteins. The *gag* and *pol* sequences overlap, but because they are translated in different reading frames (different ways

A. HIV virions

Briggs et al. 2006. *Structure* **14**:15–20

100 nm

B. Virion budding out

Edwin P. Ewing, Jr./CDC

100 nm

Figure 11.23   HIV-1 virus particles.   A. HIV virions (cryo–EM tomography) show the cone-shaped capsids (colorized red), proteases and host-derived proteins (yellow), and envelopes (blue). **B.** Budding virions show thickened membrane as spike proteins concentrate.

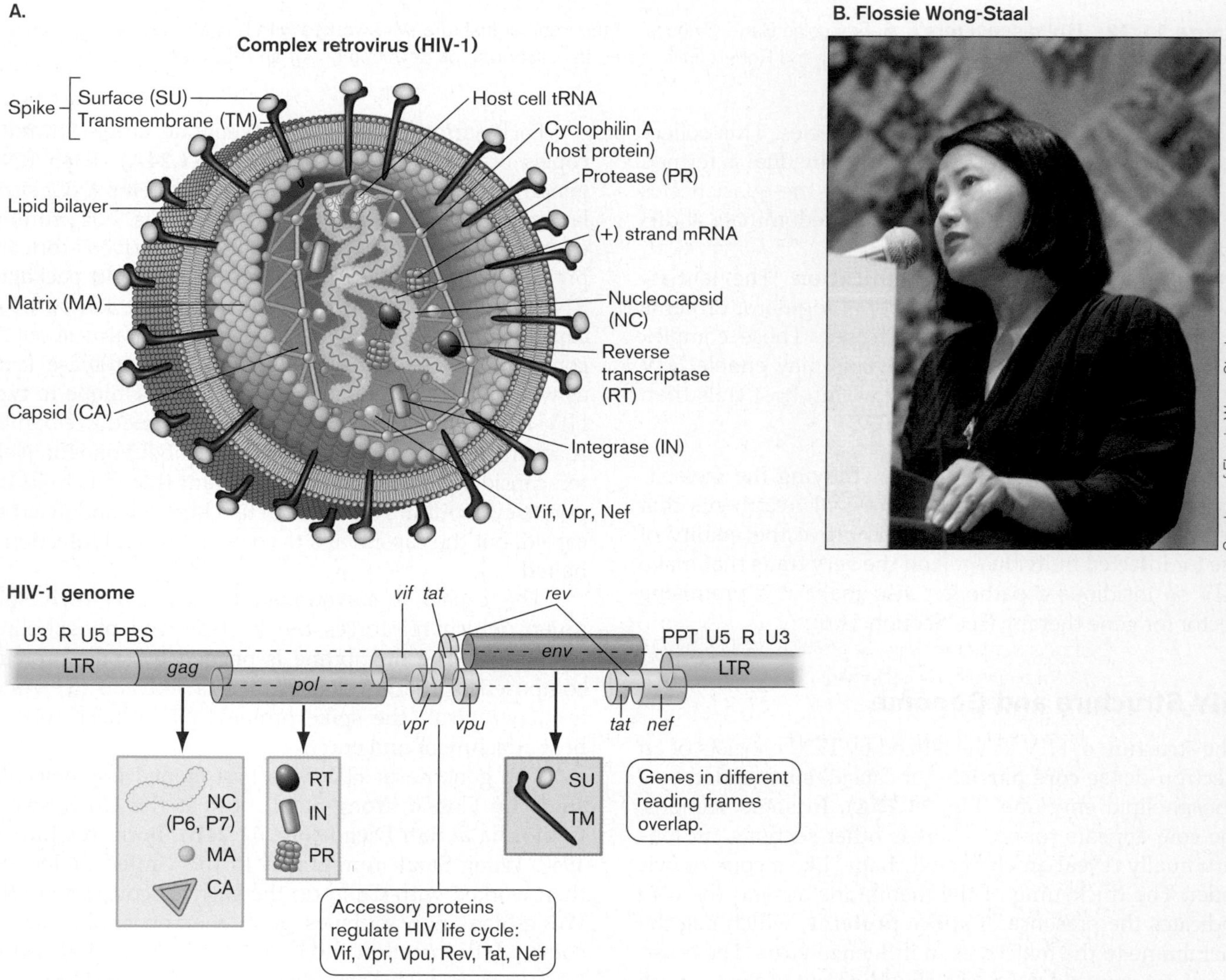

A.

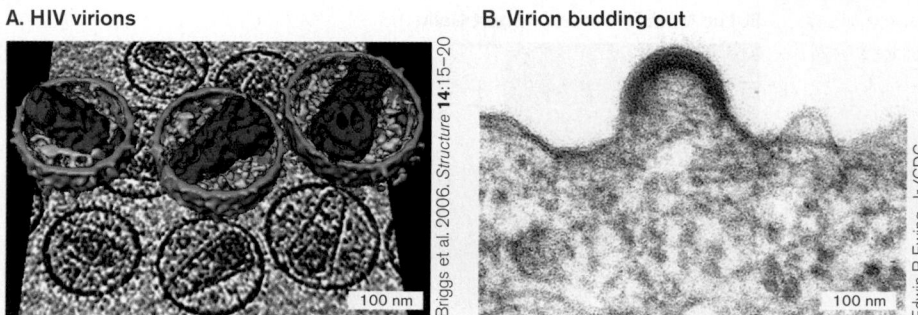

Complex retrovirus (HIV-1)

Spike [Surface (SU)
Transmembrane (TM)]

Lipid bilayer

Matrix (MA)

Capsid (CA)

Host cell tRNA

Cyclophilin A (host protein)

Protease (PR)

(+) strand mRNA

Nucleocapsid (NC)

Reverse transcriptase (RT)

Integrase (IN)

Vif, Vpr, Nef

B. Flossie Wong-Staal

Courtesy of Flossie Wong-Staal

HIV-1 genome

vif tat   *rev*

U3 R U5 PBS   PPT U5 R U3

LTR   *gag*   *pol*   *env*   LTR

vpr   vpu   *tat   nef*

NC (P6, P7)
MA
CA

RT
IN
PR

SU
TM

Genes in different reading frames overlap.

Accessory proteins regulate HIV life cycle: Vif, Vpr, Vpu, Rev, Tat, Nef

Figure 11.24   HIV-1 structure and genome.   A. Internal structure of virion, color-coded to match the genome. In the genome sequence, the staggered levels indicate three different reading frames. Each virion contains two copies of the RNA genome plus multiple copies of reverse transcriptase (RT) and protease (PR) enclosed within a conical capsid (CA subunits plus subunits of a host protein, cyclophilin A). The capsid is surrounded by a matrix (MA subunits), which reinforces the host-derived phospholipid membrane, pegged by spike proteins (SU, TM). The genome also encodes six accessory proteins that are expressed within the infected host cell and regulate the replicative cycle. **B.** Flossie Wong-Staal, pioneering AIDS researcher at the University of California at San Diego, was the first to clone the HIV genome. Wong-Staal now pursues gene therapy approaches to AIDS prevention and develops lentiviral gene vectors.
Source: Part A based on Briggs et al. 2006. *Structure* **14**:15–20.

to align the triplet code), the ribosome expresses each independently of the other. During infection, each reading frame is transcribed and translated as a "polyprotein"; then, at subsequent stages, the polyprotein is cleaved by proteases to form the mature products.

In HIV-1, the *gag* and *pol* sequences overlap, and *env* overlaps with genes encoding **accessory proteins**, a term for proteins that modify and regulate retroviral infection. Accessory proteins are unique to lentiviruses, mediating host response and maintaining the long-term slow progression of lentiviral disease. The accessory proteins are expressed within the infected host cell and regulate the replicative cycle (**Table 11.3**). For example, Tat protein activates transcription of the viral genome. The HIV-1 genome encodes at least six accessory proteins—a greater number than any other retrovirus. They are major targets for research and drug discovery aimed at preventing HIV proliferation.

HIV Attachment and Host Cell Entry

Like other viruses, HIV needs to recognize specific receptor molecules on the surface of its target cells. The primary receptor for HIV is the CD4 surface protein on CD4 T lymphocytes (T cells). The normal function of CD4 surface proteins is to connect the T cell with an antigen-presenting cell, which activates the T cell to turn on B-cell production of antibodies (discussed in Section 24.5). Disruption of this antibody production is the main cause of the AIDS-related susceptibility to opportunistic infections. Note, however, that CD4 proteins appear on many other cell types, such as microglia (macrophage-like cells in the central nervous system) and Langerhans cells (immune cells of the epidermis). Their presence may make other cells susceptible to infection by HIV.

Spike proteins mediate membrane fusion. The binding of HIV to CD4 receptors involves the envelope spike protein SU (**Fig. 11.25**). Spike proteins are the main external proteins accessible to the host immune system.

HIV attachment to the cell membrane requires a fusion peptide rearrangement similar to that of influenza virus. When SU binds to CD4, the spike transmembrane component (TM) unfolds and extends its fusion peptide into the host cell membrane (**Fig. 11.25**). In addition, SU binds to secondary receptors in the membrane called **chemokine receptors (CCRs)**. Chemokines are signal molecules for the immune system, but their receptor proteins can bind viruses that evolve to take advantage of them. After the spike protein SU binds receptors and the TM

Table 11.3 Accessory proteins of HIV-1.

Protein	Function	Effect of mutation
Vif	Virion component: • Protects reverse transcriptase from error-inducing host protein APOBEC3G. • Required for infectivity of progeny virions.	Virions produced are noninfective
Vpr	Virion component: • Transcription factor, activates HIV transcription during G_2 phase of cell cycle; arrests T-cell growth. • Imports DNA across nuclear membrane; avoids need to infect rapidly dividing cells in which mitosis dissolves the nucleus.	Lower production of virions
Nef	Virion component: • Internalizes and degrades CD4 receptors, to avoid superinfection by more HIV virions, and to lessen immune response to the infected cell. • Decreases expression of MHC proteins that stimulate cytotoxic T cells. • Enhances infectivity of virus particle.	Slower progression to AIDS
Vpu	Membrane protein: • Degrades CD4, releasing bound spike proteins. • Promotes virion assembly and release from cell surface tetherins.	Early death of host cell; lower production of virions
Rev	Nuclear phosphoprotein, combines with host cell proteins: • Stabilizes certain mRNAs in nucleus. • Exports mRNA out of nucleus into the cytoplasm. • By promoting mRNA export, induces shift from latent phase to virion-producing phase.	Failure of infection
Tat	Transcription factor: • Binds TAR site on nascent RNA to activate transcription. • Associates with histone acetylases and kinases to activate transcription of integrated viral DNA.	Failure of chromosome replication

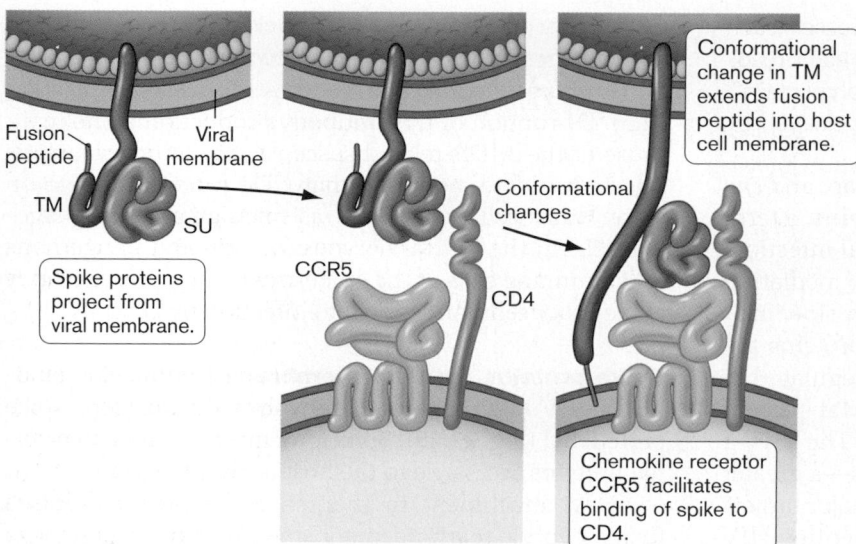

Figure 11.25   HIV attachment to host cell. The SU subunit of the spike protein complex attaches to the receptor, CD4 cell surface protein.

(Figure labels:)
Fusion peptide
Viral membrane
TM
SU
Spike proteins project from viral membrane.
CCR5
CD4
Conformational changes
Conformational change in TM extends fusion peptide into host cell membrane.
Chemokine receptor CCR5 facilitates binding of spike to CD4.

fusion peptide inserts, the HIV-1 envelope fuses with the plasma membrane. The CCR is also called a **coreceptor**, a protein acting with CD4 to bind the HIV spike proteins.

The requirement for CCR attachment varies among different types of HIV. Some chemokine receptors are found on neurons, and their involvement in HIV infection may mediate the neurological disorders seen in AIDS. As described at the beginning of the chapter, people with a defective gene for CCR5 are resistant to infection by HIV-1. Researchers used this knowledge to develop a transgenic therapy for HIV-1-positive volunteers. From the HIV-infected individuals, CD4 T cells were removed and placed into culture. The genome of the cultured T cells was engineered to contain a defective CCR5 gene; thus, the T cells no longer expressed CCR5 on their surface. The CCR5-negative T cells were then returned to the volunteers, where they produced a resurgence of HIV-1-resistant T cells in the blood. The transgenic therapy may prove useful in maintaining the immune systems of HIV-positive patients. It may not, however, prove useful for all types of HIV, and it does not cure the infection. Nevertheless, it represents hope for future molecular research to overcome AIDS.

After the HIV envelope fuses with the cell membrane, the HIV core directly enters the cytoplasm. The HIV entry mechanism differs from that of influenza virus, in which endocytosis and lysosome fusion are required to open the capsid and release the genome into the cell. The HIV core (composed of CA and the host-derived cyclophilin A) dissolves, releasing the twin RNA genomes along with associated viral enzymes into the cytoplasm.

> **THOUGHT QUESTION 11.6**   How do attachment and entry of HIV resemble attachment and entry of influenza virus? How do attachment and entry differ between these two viruses?

Retroviral RNA Genome

The twin RNA genomes each possess a 5′ "cap" and a 3′ poly-A "tail" that probably enable them to mimic host nuclear mRNA. Each RNA is hybridized to a tRNA primer for DNA synthesis. The primer is a lysine-specific tRNA from the previous infected cell. The primer might be expected to hybridize at the 3′ end of the template, where its 3′ OH "points" toward the opposite end, positioned to synthesize all the way down. Surprisingly, however, the 3′ OH end of the tRNA actually binds near the 5′ end, where initially it can generate only a brief sequence. These early sequences bind key regulatory factors for transcription and for DNA insertion into the host genome (discussed below).

> **THOUGHT QUESTION 11.7**   Compare and contrast the priming of chromosome replication in HIV with the priming mechanisms of poliovirus and influenza virus.

Reverse Transcriptase Copies RNA to DNA

A retrovirus, unlike other RNA viruses, must integrate its entire genome in the host genome in order to replicate viral genomes and produce new progeny virions. Thus, the RNA genome needs to serve as a template to synthesize a DNA complement; but then the original RNA template must be degraded and replaced by a DNA strand, for host integration. All these processes are accomplished by **reverse transcriptase (RT)**, the defining enzyme of a retrovirus. Reverse transcriptase is the source of the high error rate of retroviral replication—on average, one or two errors per copy of HIV. This high error rate generates the quasispecies of different strains within an HIV-infected person.

Reverse transcriptase is the target of the first clinically useful drug to treat HIV infection, the nucleotide analog **azidothymidine (AZT)**. AZT is incorporated into the growing DNA chain in place of a thymidine; but because its 3′ OH is replaced by an azido group (–N₃), no further nucleotides can be added.

The reverse transcriptase complex actually possesses three different activities:

- **DNA synthesis from the RNA template.** Synthesis of DNA is first primed by the host tRNA, which was hybridized to the chromosome within the virion.
- **RNA degradation.** After DNA synthesis, the template RNA is gradually removed through an RNase H activity of the RT complex. Removal of RNA enables replacement of the entire original RNA template by DNA.
- **DNA-dependent DNA synthesis.** To make the DNA complementary strand replacing the RNA, the RT needs to use the newly made DNA as its template. Thus, RT has the rare ability to use either DNA or RNA as template.

The RT complex is shown in **Figure 11.26A**. The RNA template with its short RNA primer is threaded through the RT complex between the "thumb" and "fingers"—a configuration typical of other RNA polymerases (discussed in Chapter 8). The RT complex adds successive deoxynucleotides from dNTPs starting at the 3′ OH end of the RNA primer. As DNA elongates, however, the RNA template is cleaved from behind by the RT complex. Thus, the new DNA actually replaces preexisting RNA sequence. This "destructive replication" is unique to retroviruses. The details are important, as they suggest possible targets for new antiviral drugs.

WWW | Reverse transcriptase tutorial (Biomolecules at Kenyon)
microbiology2.com/links

Initiation of reverse transcription. Figure 11.26B outlines the entire process of reverse transcription in HIV-1, from primed RNA template to integration of duplex DNA. First, the host-derived tRNA primer initiates synthesis (by reverse transcriptase, RT) of a DNA strand complementary to the RNA chromosome. DNA is elongated toward the 5′ end of the HIV chromosome, generating a short segment (step 1). The original RNA template for this short segment is then degraded by the RNase H activity of RT, leaving only the DNA extension off the tRNA primer (step 2).

The new DNA primes the second template. The original RNA template had repeated ends (labeled "r," lowercase for RNA), and the exposed DNA copy of the 5′ end has a complementary sequence ("R," uppercase for DNA). The "R" DNA from the second tRNA extension hybridizes to the 3′ end of the original RNA (step 3). The hybridized DNA elongates along the rest of the chromosome (step 4), up to the primer-binding site for mRNA transcription (pbs). DNA completion is followed by degradation of the remaining RNA template, except for occasional short fragments to serve as primers, such as the polypurine tract (ppt). A complementary DNA strand is then synthesized through the PPT primer, leaving a nick at U3 (step 5).

Interestingly, human host cells have evolved a mutator protein, APOBEC3G, that can be packaged into progeny virions. APOBEC3G deaminates cytosines in the viral genome and thus increases the error rate of RT. However, HIV has evolved an accessory protein, Vif (see **Table 11.3**), that prevents APOBEC3G from being packaged into virions.

DNA integration into the host genome. The final step of genome processing requires integration into host DNA. The double-stranded DNA copy of the HIV genome (plus a short repeat of the 5′ end) circularizes (step 6). The circular molecule then integrates into the host chromosome by site-specific recombination at a host target sequence (step 7), mediated by the HIV integrase protein (IN). This step forms an integrated viral genome, or **provirus**. Integration generates two copies of the provirus 5′ end (sequence U3–R–U5), which is called a **long terminal repeat (LTR)**. The provirus and LTR ends are also flanked by two copies of the host target sequence. Proviral sequences can now be expressed, directing production of progeny virions.

An alternative to virion production is that the integrated HIV genome can lie dormant, like the lambda prophage in *E. coli* (discussed in Chapter 6). The viral genome is replicated passively within the genome of its host cell, hiding for many years with only infrequent production of virions. The few virions shed by the patient, however, can infect an unsuspecting individual who shares sexual or blood contact.

Replicative Cycle of HIV

The steps of HIV replication are outlined in **Figure 11.27**. The main points of viral entry and replication are typical of retroviruses. HIV, however, has an exceptionally large number of accessory proteins that govern the level of virus production and the duration of the quiescent phase, when the integrated chromosome replicates with the host cell.

Synthesis of HIV mRNA and progeny genomic RNA. After the HIV virion attaches to the host receptors, its envelope fuses with the host membrane (step 1).

A. Reverse transcriptase

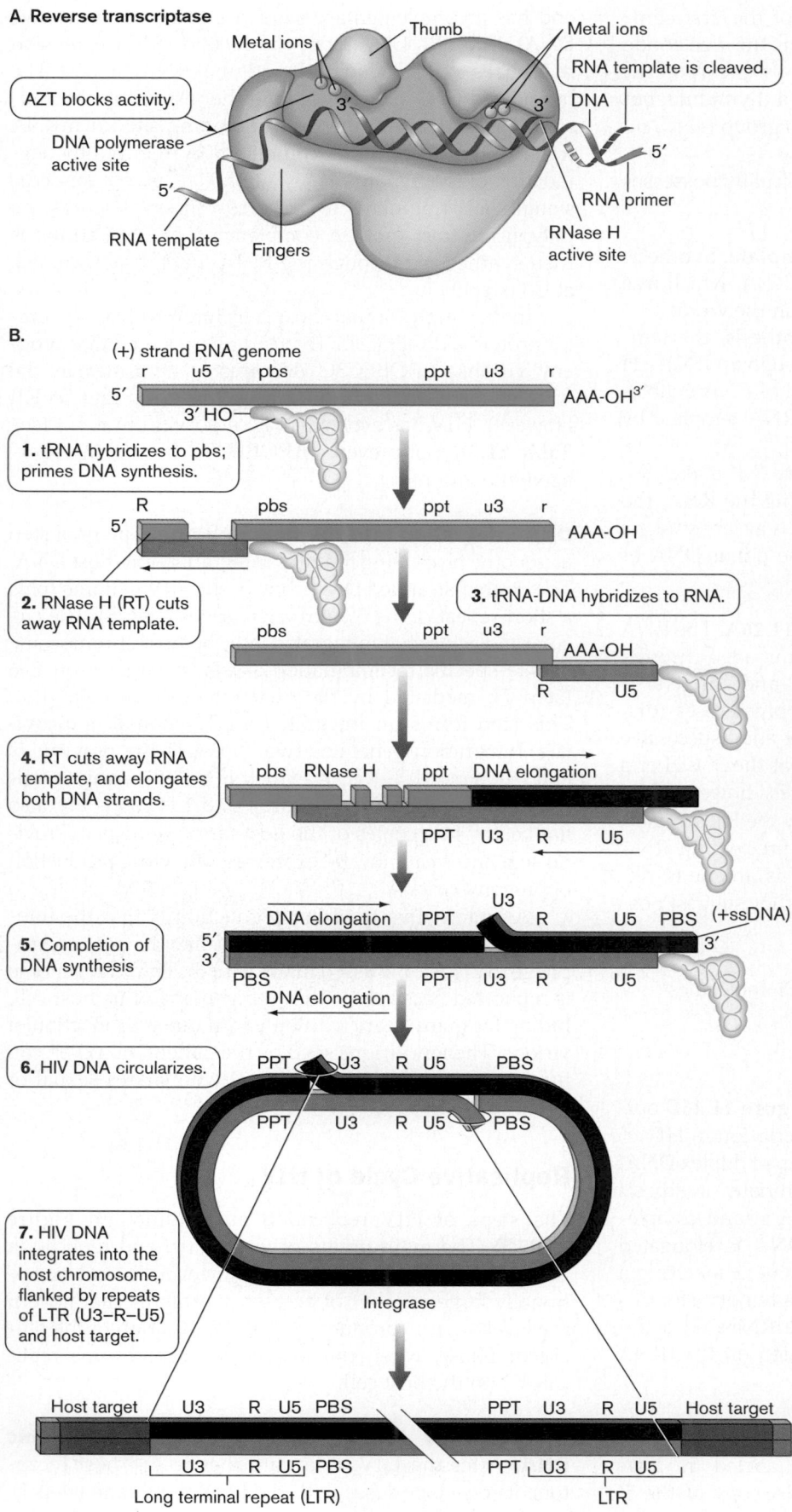

B.

1. tRNA hybridizes to pbs; primes DNA synthesis.

2. RNase H (RT) cuts away RNA template.

3. tRNA-DNA hybridizes to RNA.

4. RT cuts away RNA template, and elongates both DNA strands.

5. Completion of DNA synthesis

6. HIV DNA circularizes.

7. HIV DNA integrates into the host chromosome, flanked by repeats of LTR (U3–R–U5) and host target.

Figure 11.26 Reverse transcription of the HIV genome and integration into host DNA. A. The RNA template with its short RNA primer is threaded through the RT complex between the "thumb" and "fingers"—a configuration typical of other RNA polymerases. The RT complex adds successive deoxynucleotides from dNTPs (deoxyribonucleoside triphosphates), starting at the 3′ OH end of the RNA primer (as in regular DNA synthesis). As DNA elongates, the RNA template is cleaved from behind by "RNase H" activity. **B.** The tRNA primer (1) initiates a short sequence of DNA complementary to the 5′ end of the HIV chromosome (2). The corresponding template is then degraded. The original RNA template has repeated ends (r), so the exposed DNA end (extended tRNA primer of the second copy) then hybridizes to the 3′ end (3). DNA now elongates along the rest of the chromosome (4), up to the primer-binding site (pbs) for mRNA transcription. The remaining RNA template is degraded by RT and replaced by complementary DNA up to the PPT-U3 junction (5). The completed double-stranded DNA copy of the HIV genome, plus a short repeat of the 5′ end (U3–R–U5), circularizes (6) and integrates into the host chromosome (7), mediated by integrase (IN). The LTR (U3–R–U5) and "host target" sequences generate repeats, flanking the integrated HIV genome (provirus).

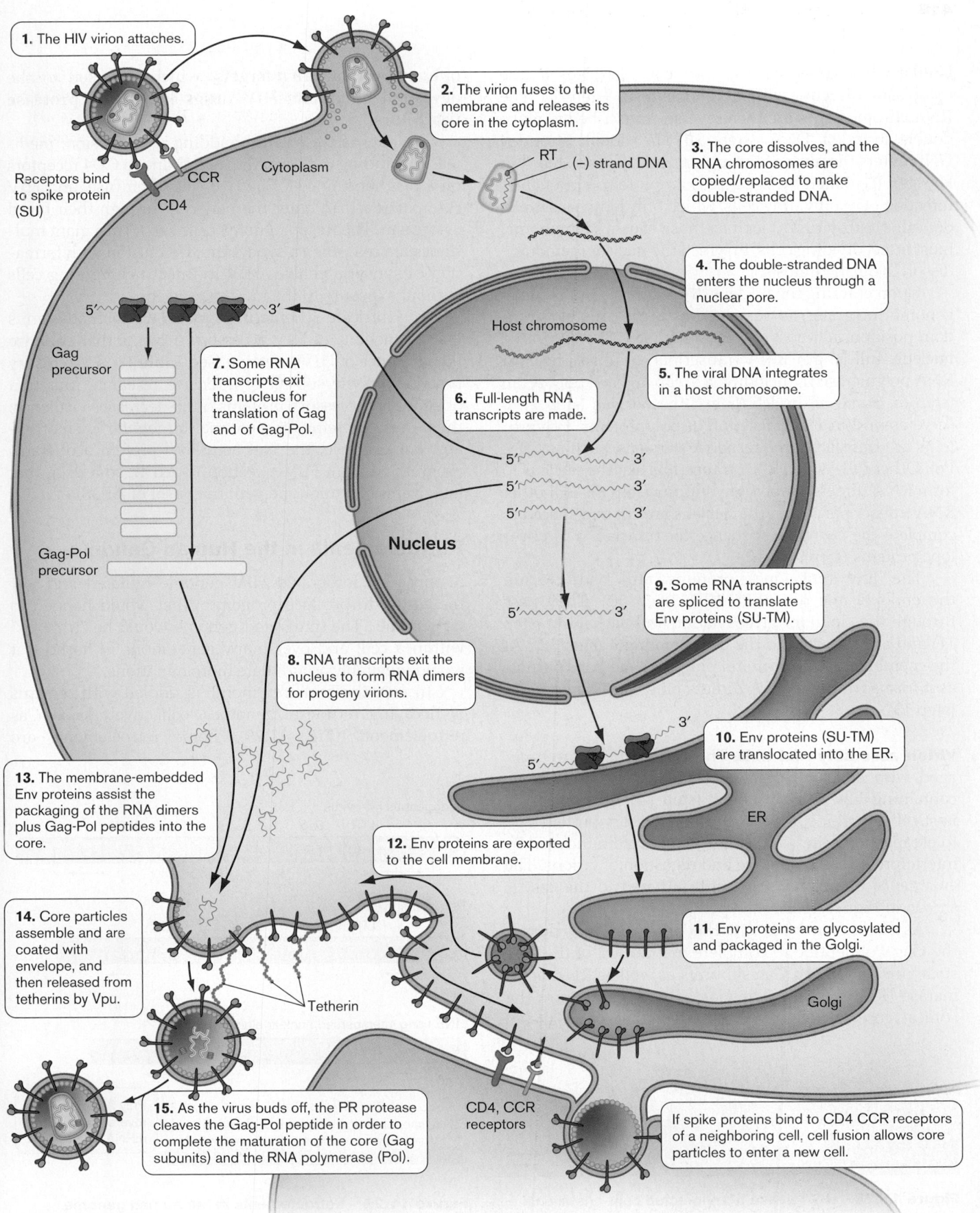

Figure 11.27 HIV replicative cycle. The HIV virion attaches its receptor and fuses with the host cell membrane, releasing its contents in the cytoplasm to undergo a replicative cycle. ▶ *microbiology2.com/animations*

1. The HIV virion attaches.

Receptors bind to spike protein (SU)

CCR

CD4

Cytoplasm

2. The virion fuses to the membrane and releases its core in the cytoplasm.

RT   (−) strand DNA

3. The core dissolves, and the RNA chromosomes are copied/replaced to make double-stranded DNA.

4. The double-stranded DNA enters the nucleus through a nuclear pore.

Host chromosome

5. The viral DNA integrates in a host chromosome.

5′   3′

Gag precursor

7. Some RNA transcripts exit the nucleus for translation of Gag and of Gag-Pol.

6. Full-length RNA transcripts are made.

Nucleus

5′   3′
5′   3′
5′   3′

Gag-Pol precursor

9. Some RNA transcripts are spliced to translate Env proteins (SU-TM).

5′   3′

8. RNA transcripts exit the nucleus to form RNA dimers for progeny virions.

5′   3′

10. Env proteins (SU-TM) are translocated into the ER.

ER

13. The membrane-embedded Env proteins assist the packaging of the RNA dimers plus Gag-Pol peptides into the core.

12. Env proteins are exported to the cell membrane.

11. Env proteins are glycosylated and packaged in the Golgi.

14. Core particles assemble and are coated with envelope, and then released from tetherins by Vpu.

Tetherin

Golgi

15. As the virus buds off, the PR protease cleaves the Gag-Pol peptide in order to complete the maturation of the core (Gag subunits) and the RNA polymerase (Pol).

CD4, CCR receptors

If spike proteins bind to CD4 CCR receptors of a neighboring cell, cell fusion allows core particles to enter a new cell.

Unlike influenza virus, the HIV core dissolves in the cytoplasm directly, without endocytosis (step 2). The RNA chromosomes are then reverse-transcribed to make double-stranded DNA (step 3). The double-stranded DNA enters the nucleus through a nuclear pore (step 4), a key step facilitated by Vpr accessory protein. Vpr enables infection of nondividing cells, which only lentiviruses can do; other retroviruses, such as those causing lymphoma, must infect dividing cells, in which the nuclear membrane dissolves during mitosis.

Upon entering the nucleus, the DNA copy of the HIV genome integrates its sequence as a provirus at a random position in a host chromosome (step 5). Within the nucleus, full-length RNA transcripts are made by host RNA polymerase II, including a 5′ cap and a 3′ poly-A tail (step 6). Some of the RNAs exit the nucleus (step 7) to serve as mRNA for translation of polyproteins. Polyproteins are translated in alternative versions, such as Gag-Pol. Other full-length RNA transcripts exit the nucleus to form RNA dimers for progeny virions (step 8). Still other RNA transcripts within the nucleus are cut and spliced to complete the *env* gene sequence for translation of envelope proteins (step 9).

The Env (envelope) proteins are made within the endoplasmic reticulum (ER) (step 10). They pass through the Golgi for glycosylation and packaging (step 11) and are exported to the cell membrane (step 12). At the membrane, Env proteins plug into the core particle as it forms from the RNA dimers plus Gag-Pol peptides (step 13).

Virion assembly and exit. The core particles are packaged with envelope derived from host cell membrane containing Env spike proteins (step 14). To escape the host cell, emerging virions require accessory protein Vpu to escape "tetherin," a host adhesion protein induced by interferon to cause reuptake and digestion of virions. The emergence of virions, some still tethered to the cell, is shown in **Figure 11.28**.

As the virion buds off, the protease (PR) cleaves the Gag-Pol peptide to complete maturation of the core structure containing Gag subunits as well as RNA polymerase (RT) (step 15). The Gag subunits now form the conical core structure. Proteases that cleave Gag-Pol

offer important drug targets, which have led to the development of anti-HIV drugs known as **protease inhibitors**.

An alternative to viral budding is cell fusion, mediated by binding of Env in the membrane to CD4 receptors on a neighboring cell. The two cells then fuse, and HIV core particles can enter the new cell through their fused cytoplasm. The fusion of many cells can form a giant multinucleate cell called a syncytium. Cell fusion with formation of syncytia enables HIV to infect neighboring cells without exposure to the immune system.

The intricate scheme in **Figure 11.27** actually omits many functions of HIV accessory proteins that enhance the virulence of HIV infection (see **Table 11.3**). Accessory proteins are the subject of intensive research. Mutation of genes for accessory proteins often decreases virulence; thus, these proteins are potential targets for chemotherapy. For example, the Nef accessory protein accelerates progression from HIV infection to AIDS, so a drug that inactivates Nef might prevent the onset of AIDS.

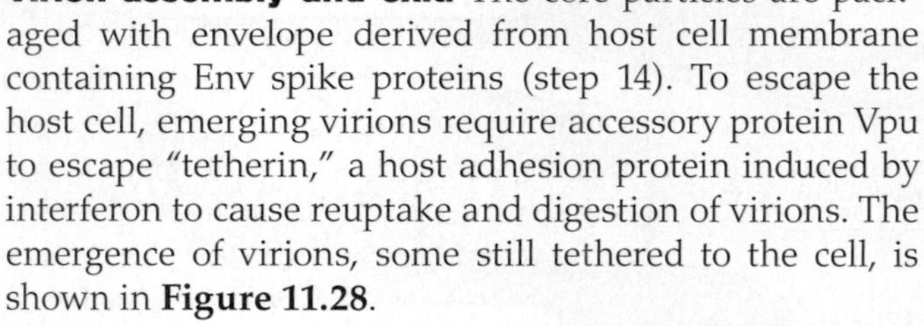

Retroelements in the Human Genome

Suppose an integrated HIV genome mutated and lost the ability to produce progeny. What would happen to its genome? The integrated genome would be "trapped" within a cell; and over many generations in its host, it would inevitably accumulate more mutations.

In fact, the human genome is riddled with remains of decaying retroviral genomes, collectively known as retroelements (**Fig. 11.29**). Some retroelements are

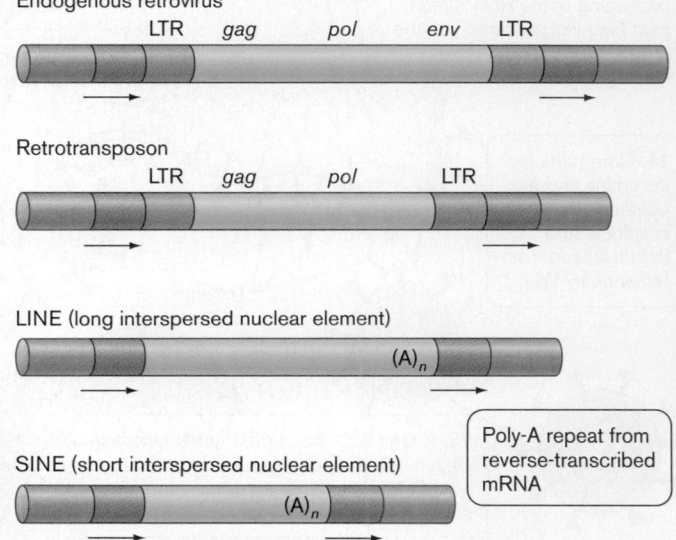

Figure 11.29   Retroelements in the human genome. Endogenous retroviruses and other retroelements in the human genome may arise from progressive degeneration of ancestral retroviruses, or they may be progenitors of new retroviruses.

Figure 11.28   HIV release from infected cell. As virions emerge, they remain tethered to the cell surface by host tetherins, requiring a release step mediated by accessory protein Vpu (TEM). Modified from Stuart J. D. Neil et al. 2008. *Nature* **451**:425, fig. 4e.

Neil, Stuart et al. 2008. *Nature* **451**:425.

endogenous retroviruses, sequences that contain all the genomic elements of a retrovirus, including *gag*, *env*, and *pol* genes, yet never generate virions. Presumably, they lost this ability by mutation. Other elements, known as **retrotransposons**, retain only partial retroviral elements but may maintain a reverse transcriptase to copy themselves into other genomic locations. An example of a retrotransposon is the well-known Alu sequence, a short sequence found in about a million copies in the human genome. In some cases, a retrotransposon such as Alu can interrupt a key human gene, leading to a genetic defect such as a defective lipoprotein receptor associated with abnormally high cholesterol level and heart failure. Still other retroelements, known as LINEs (long interspersed nuclear elements) and SINEs (short interspersed nuclear elements), show more vestigial remnants of retroviral genomes. Amazingly, about half the sequence of the human genome appears to have originated from viruses and retroelements. The origin of viruses, including their interaction with cellular evolution, is explored in **Special Topic 11.1**.

Could a virus as deadly as HIV be used to improve human health? In fact, the exceptional ability of lentiviruses to deliver their own DNA to human cells makes them promising candidates for gene therapy. Lentiviruses are particularly promising for their ability to transfer genes into nonmitotic cells (see Section 11.6).

TO SUMMARIZE:

- **Human immunodeficiency virus (HIV)** causes an ongoing pandemic of acquired immunodeficiency syndrome (AIDS). There is no cure, but molecular biology has led to drugs that extend life expectancy.
- **HIV is a retrovirus** whose RNA genome is reverse-transcribed into double-stranded DNA, which integrates into the DNA of the host cell.
- **The HIV core particle** contains two copies of its RNA genome, each bound to a primer (host tRNA) and reverse transcriptase (RT). The core is surrounded by an envelope containing spike proteins.
- **HIV binds the CD4 receptor** of T lymphocytes together with the chemokine receptor CCR5. Following virion-receptor binding and envelope-membrane fusion, HIV virions are released into the cytoplasm.
- **DNA is synthesized in the nucleus from the retroviral RNA**, primed by the tRNA, and synthesized by reverse transcriptase. **RNA degradation** allows formation of a double-stranded DNA. The retroviral **DNA integrates** into the host genome.
- **Retroviral mRNAs are exported to the cytoplasm** for translation. Envelope proteins are translated at the ER and exported to the cell membrane.

- **Retroviruses are assembled at the cell membrane**, where virions are released slowly, without lysis. Alternatively, the accumulation of retroviral proteins in the cell membrane leads to cell fusion, forming syncytia.
- **Accessory proteins regulate virion formation** and the latent phase, in which double-stranded DNA persists without reproduction of progeny virions.
- **Ancient retroviral sequences** persist within animal genomes, including the human genome.

11.5 Herpes Simplex Virus: DNA

Many important viruses of humans and other animals contain genomes of double-stranded DNA (**Table 11.4**). DNA viruses include the causative agents of well-known diseases such as smallpox, chickenpox, and infectious mononucleosis (mono). Most DNA viruses are considerably larger than RNA viruses and encode a wider range of viral enzymes; for example, the vaccinia genome encodes nearly 200 different proteins. The complexity of viruses such as vaccinia and herpes approaches that of small prokaryotic cells—a fact with interesting implications for viral evolution (see **Special Topic 11.1**).

Herpes viral DNA replicates by mechanisms similar to those used in the replication of prokaryotic and phage genomes, either bidirectionally from an origin of replication (as in bacteria) or by the rolling-circle method (as in phages such as T4). Like cellular genomes, herpes viral DNA replication requires more than a polymerase; enzymes such as helicase, primase, and single-strand binding proteins are also needed. Viral species differ as to the source of their replication components. Simian virus 40 (SV40) and Epstein-Barr virus rely entirely on cellular components, whereas poxviruses rely entirely on viral components for DNA replication. Other species, such as adenovirus and papillomavirus, combine viral and cellular components.

Herpes Simplex Virus Infects the Oral or Genital Mucosa

An important example of a DNA virus is herpes simplex virus (HSV). Strains HSV-1 and HSV-2 cause one of the most common infections in the United States. Approximately 60% of Americans acquire herpes simplex, usually HSV-1, in epithelial lesions commonly known as cold sores. About 30%–60% acquire genital herpes, usually HSV-2, through sexual contact (oral or vaginal). Genital herpes causes recurrent eruptions of infection in the reproductive tract (**Fig. 11.30**). Many of those infected are unaware of symptoms, but they can still transmit the disease to others.

Special Topic 11.1   How Did Viruses Originate?

The origin of viruses remains a mystery. Did viruses evolve within cells? Or could they possibly have predated cells? Unlike cellular organisms, viruses are too small to have left fossil traces that we can detect. Viral genomes show evidence of so much horizontal transfer that lineages are impossible to trace back over time. Unlike cells, which all possess ribosomal RNA genes with a common ancestor, all virus species share no single gene in common. No viral sequence data point to a first common ancestor, nor to a common means of origin.

Three theories are proposed to explain the origin of viruses (**Fig. 1**).

Reductive Evolution

Reductive evolution leads to gradual loss of unselected traits. For example, pathogens such as *Helicobacter* and *Myco-plasma* have lost many genes encoding metabolic functions and stress responses that are unnecessary when the host organism provides these functions. Suppose such a cell were

to become an intracellular parasite, increasingly dependent on host functions. Ultimately, it might lose all but a few metabolic genes, losing functions for genome replication with the rest of its functions provided by the host.

Some viruses show signs of degeneration from ancestral cells. For example, phage T4, the poxviruses, and herpes viruses possess large genomes of 80–200 genes. Their virions include enzymes and other proteins that provide some of the virus's own metabolic functions. The size of a poxvirus approaches half a micrometer—well within the size range of prokaryotic cells (**Fig. 2**). A vaccinia particle actually looks like a miniature cell, with its multilayered envelope containing a DNA chromosome surrounded by enzymes.

Recently, an even larger "cellular virus" was discovered, called the mimivirus because it mimics a cell (discussed in Chapter 6). Originally discovered within amebas, the mimivirus was at first mistaken for a bacterium. In fact, mimivirus consists of a giant icosahedral particle with a diameter of

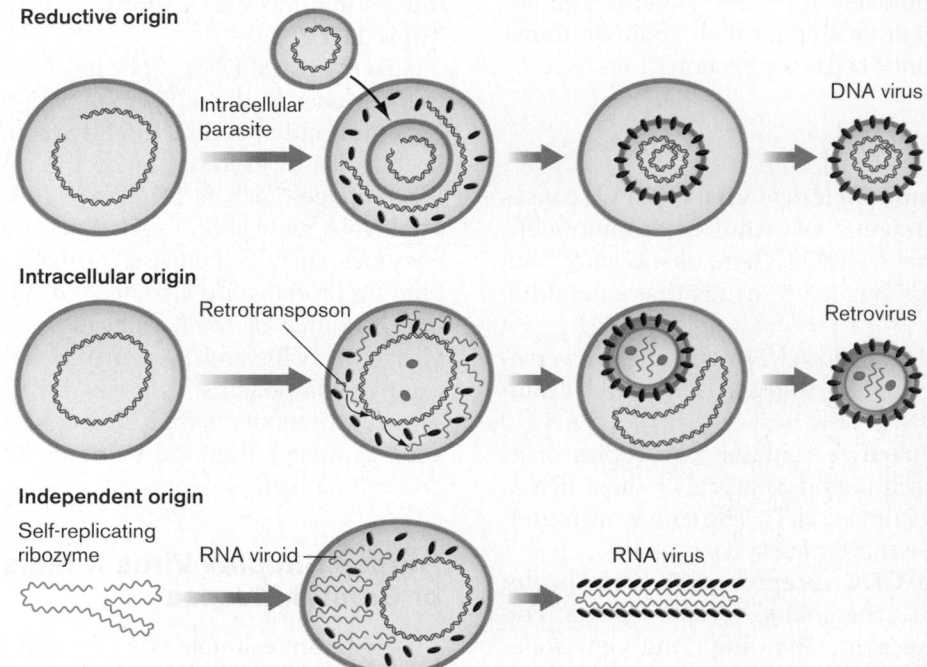

Figure 1   Theories of viral origin.   Three theories have been proposed to account for the origin of viruses. According to the theory of reductive origin, viruses evolved by degenerative evolution from intracellular parasitic cells. According to the theory of intracellular origin, viruses evolved from functional parts of cells that acquired an ability to reproduce themselves uncontrolled by the cell. The theory of independent origin holds that viroid nucleic acids could have evolved outside of cells during the "RNA world" and acquired the ability to infect cells.

400 nanometers (nm). Its genome is a circle of double-stranded DNA, about 1.2 megabases (Mb). The genome was sequenced by Jean-Michel Claverie and colleagues at the Institut de Biologie Structurale, Marseille. It includes 1,262 open reading frames—a greater number than the genomes of several prokaryotes, such as mycoplasmas. It is the first viral genome known to include protein translation components such as aminoacyl-tRNA synthetases and elongation and termination factors. Mimivirus may represent a "missing link" between bacteria and viruses.

Intracellular Origin

Other viruses, particularly retroviruses, support the idea of intracellular origin. Retroviruses require reverse transcriptase to convert their RNA information to DNA. When retroviruses were first discovered, reverse transcriptase was considered an anomaly because no such enzyme had been reported in host cells. We have since discovered specialized reverse transcriptases in both plant and animal cells. In retrospect, we now believe that a reverse transcriptase must have existed very early in the evolution of all cells, because something had to copy the information from the ancient "RNA world" cell into the DNA of modern genomes.

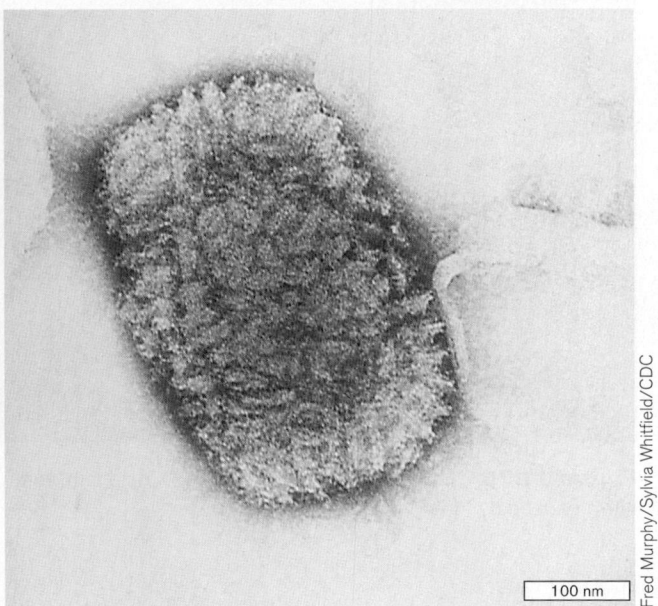

Fred Murphy/Sylvia Whitfield/CDC

100 nm

Figure 2   Large cell-like viruses.   Support for a reductive origin of some viruses can be seen in the size and complexity of viruses such as vaccinia, the infective agent of cowpox.

In animal cells, a native type of reverse transcriptase called telomerase replicates the ends of linear chromosomes. The sequence of telomerase shows homology with that of known reverse transcriptases of retroviruses. Mammalian genomes show repeated elements of DNA that have arisen from mistaken reverse copying of host RNA molecules back into the DNA genome. One can imagine that mutation of a cellular reverse transcriptase might lead to intracellular evolution of a retrovirus.

Independent Origin

Some of the simpler viral elements might have evolved by themselves in the RNA world, first as free-living entities, later as stowaways within host cells. Perhaps the best clue for such an origin is the infection of plant cells by viroid ribozymes. Viroids are self-contained RNA molecules that possess catalytic function. Present-day viroids rely entirely on host cell metabolism, but one could imagine that viroids might have evolved from an ancestral RNA molecule that managed its own self-replication.

An intriguing possibility is that some viruses not only originated as independent cells but were the first to evolve DNA, and they contributed DNA to all living cells. The viral origin of DNA is proposed by evolutionary biologists, including Patrick Forterre, at the Université Paris-Sud in Orsay, France. The earliest cells are believed to have had genomes of RNA, the simplest nucleic acid. These RNA-world cells may have been parasitized by viruses that evolved DNA chromosomes to avoid cleavage by host cellular enzymes. If a viral DNA chromosome became latent in a host cell (as do herpes viruses, for example), it might eventually acquire genes from the host genome through recombination events. Any host gene transferred to the DNA chromosome would be favored in evolution because DNA is more stable than RNA. Eventually, according to this hypothesis, the entire host genome would end up transferred to the DNA chromosome, and all cells would have genomes of DNA (as indeed cells do today). The hypothesis of a viral origin of DNA remains unproven but inspires exciting research to test its implications.

The origin of viruses has important implications for the principles of evolution, including origins of the earliest cells. In addition, the study of viral origins is important medically because it yields insights into virus function, suggesting pharmacological applications. For example, ribozymes could fight cancer by cleaving mRNA molecules expressed from oncogenes.

Table 11.4 **DNA viruses of animals (examples).**

Virus	DNA replication	Disease	Host
Adenoviruses (many strains)	Viral DNA polymerase, single-strand binding protein, and protein primer	Enteritis or respiratory diseases	Humans, other mammals, birds
Papovavirus—simian virus 40 (SV40)	Cellular DNA polymerases	Asymptomatic	Monkeys
Herpes viruses			
Herpes simplex virus 1 and 2	All viral components (DNA polymerase, primase, etc.)	Epithelial and genital lesions, latency in neurons	Humans
Varicella-zoster virus	Viral components	Chickenpox, shingles	Humans
Epstein-Barr virus	All cellular components (DNA polymerase, etc.)	Infectious mononucleosis, Hodgkin's lymphoma	Humans
Other strains		Epithelial lesions, cancer	Monkeys, cattle, horses
Papillomaviruses	Viral DNA helicase; cellular polymerase		
Human papillomaviruses (many strains)		Genital warts, cervical and penile cancer, skin warts	Humans
Other papillomaviruses		Warts, cancer	Rabbits, cattle, sheep
Poxviruses	All viral components		
Variola virus major		Smallpox	Humans
Vaccinia virus		Cowpox	Cattle, humans
Other poxviruses		Monkeypox	Monkeys, camels, birds

Herpes simplex virus typically infects cells of the oral or genital mucosa, causing ulcerated sores. The primary infection is epithelial, followed by latent infection within neurons of the ganglia. A common site of infection is the trigeminal ganglion, which processes nerve impulses between the face and eyes and the brain stem.

The latent infection of the ganglia later leads to new outbreaks of virus, often triggered by stress such as menstruation, sunlight exposure, or depression of the immune system. Progeny virions travel back down the dendrites to the epithelia, causing **lytic** infection. In the trigeminal ganglion, herpes reactivation can lead to eye disease or lethal brain infection. In most cases, herpes symptoms can be controlled by antiviral agents such as acyclovir (discussed in Chapter 27). There is no cure or prevention of future outbreaks. In pregnant women, HSV can be transmitted to the fetus, with serious complications for the child.

Herpes simplex virus is closely related to varicella-zoster virus (VZV), the cause of chickenpox, also an epithelial infection. Varicella, too, can hide in ganglial neurons, emerging decades later to cause skin lesions called shingles.

Herpes Simplex Virus Structure

The herpes virion comprises a double-stranded DNA chromosome packed within an icosahedral capsid (**Fig.**

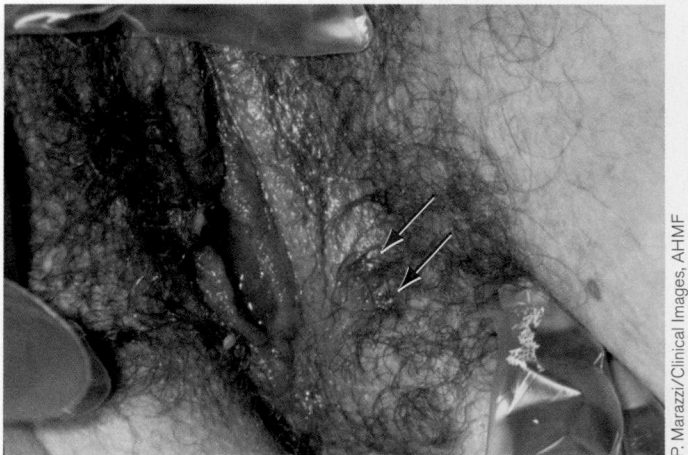

Figure 11.30 Genital herpes infection. Lesions of the vulva caused by HSV-2 infection.

P. Marazzi/Clinical Images, AHMF

11.31A). The capsid is surrounded by **tegument**, a collection of about 15 different kinds of virus-encoded proteins as well as proteins from the previous host. The tegument is contained within a host-derived membrane envelope with several kinds of spike proteins. The HSV-1 genome spans 152 kb, encoding more than 70 gene products (**Fig. 11.31B**). The sequence includes two unique segments, long (U_L) and short (U_S), each flanked by a terminal

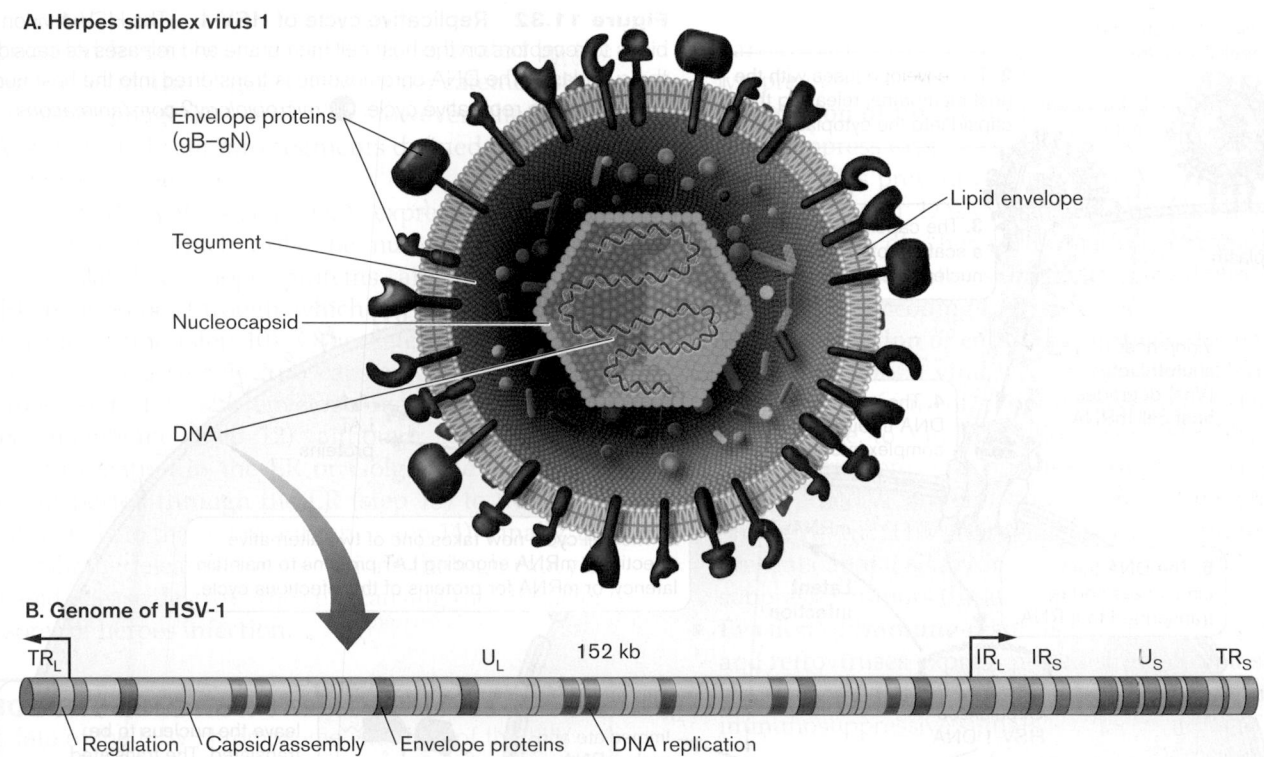

A. Herpes simplex virus 1

Envelope proteins (gB–gN)

Tegument

Nucleocapsid

DNA

Lipid envelope

B. Genome of HSV-1

TR$_L$ U$_L$ 152 kb   IR$_L$ IR$_S$ U$_S$ TR$_S$

Regulation   Capsid/assembly   Envelope proteins   DNA replication

Figure 11.31   Herpes simplex virus 1: virion and genome.   A. HSV-1 virion consists of a double-stranded DNA chromosome packaged within an icosahedral capsid. The capsid is surrounded by tegument, a collection of virus-encoded and host-derived proteins. The tegument is contained within a host-derived membrane envelope including several kinds of envelope proteins. **B.** The genome of HSV-1 spans 152,000 base pairs, encoding more than 70 gene products. The HSV sequence consists of two segments: long (U$_L$) and short (U$_S$). Each segment contains a unique region (U$_L$ or U$_S$) flanked by two inverted repeat regions, terminal (TR$_L$ or TR$_S$) and internal (IR$_L$ or IR$_S$), where the two segments meet.

repeat (TR$_L$ or TR$_S$) and an internal repeat (IR$_L$ or IR$_S$). Within the host, the genome circularizes, so the genetic linkage map appears circular.

Attachment and Entry of Herpes Simplex

Unlike poliovirus and HIV, the herpes virion can bind to several alternative receptor molecules in the host cell membrane, such as a homolog of tumor necrosis factor receptor called HveA or intercellular adhesion molecules called nectins (**Fig. 11.32**, step 1). As in the case of HIV, the entire herpes capsid enters the cytoplasm (step 2); but unlike HIV, whose capsid dissolves, the herpes capsid travels down a scaffold of microtubules (step 3) to the nuclear membrane. During this stage, the virion host shutoff factor (Vhs) degrades host mRNA, thus shutting off host protein synthesis. At a nuclear pore complex, the herpes capsid injects its DNA. The DNA is forced out of the capsid and into the nucleus by the high pressure of double-stranded DNA packing in the capsid, similar to the high-pressure injection of phage T4 DNA into a bacterial cell (step 4). The DNA then circularizes (step 5) to form a plasmid-like intermediate.

The herpes genome now takes one of two alternative directions: expression of mRNA for proteins of the infectious cycle, or expression of mRNA encoding LAT proteins to maintain latency (step 6). If the latent course is taken, the DNA circle can persist within the cell for decades before switching to lytic infection. Latent infection most commonly occurs in nerve cells, such as those of the trigeminal ganglion.

Replication of Herpes Simplex Virus

In the nucleus, herpes DNA is transcribed to mRNA by host RNA polymerase II (**Fig. 11.32**). If mRNA for lytic infection is produced, it exits the nucleus to be translated by ribosomes. Many different mRNAs are produced and exported, including those required for "immediate-early" and "early" stages of infection (step 7). The translated proteins return to the nucleus for packaging within capsids.

To generate progeny genomes, the circular DNA is replicated by viral enzymes, including DNA polymerase, single-strand binding protein, and a proofreading endonuclease (step 8). Additional enzymes

Recommended Reading

Barton, Erik S., Douglas W. White, Jason S. Cathelyn, Kelly A. Brett-McClellan, Michael Engle, et al. 2007. Herpesvirus latency confers symbiotic protection from bacterial infection. *Nature* **447**:326–329.

Belov, George A., Qian Feng, Krisztina Nikovics, Catherine L. Jackson, et al. 2008. A critical role of a cellular membrane traffic protein in poliovirus RNA replication. *PLoS Pathogens* **4**:e1000216.

Blesch, Armin, James Conner, Alexander Pfeifer, Mehdi Gasmi, Anthony Ramirez, et al. 2005. Regulated lentiviral NGF gene transfer controls rescue of medial septal cholinergic neurons. *Molecular Therapy* **11**:916.

Cartier, Nathalie, Salima Hacein-Bey-Abina, Cynthia C. Bartholomae, Gabor Veres, Manfred Schmidt, et al. 2009. Hematopoietic stem cell gene therapy with a lentiviral vector in X-linked adrenoleukodystrophy. *Science* **326**:818–823.

Cockrell, Adam S., and Tal Kafri. 2007. Gene delivery by lentivirus vectors. *Molecular Biotechnology* **36**:184–204.

Crotty, Shane, Maria-Carla Saleh, Leonid Gitlin, Oren Beske, and Raul Andino. 2004. The poliovirus replication machinery can escape inhibition by an antiviral drug that targets a host cell protein. *Journal of Virology* **78**:3378–3386.

Frampton, A. R., Jr., W. F. Goins, K. Nakano, E. A. Burton, and J. C. Glorioso. 2005. HSV trafficking and development of gene therapy vectors with applications in the nervous system. *Gene Therapy* **12**:891–901.

Girones, Rosina. 2006. Tracking viruses that contaminate environments. *Microbe Magazine* **1**:19–25.

Kane, Melissa, and Tatyana Golovkina. 2010. Common threads in persistent viral infections. *Journal of Virology* **84**:4116–4123.

Kanerva, Anna, and Akseli Hemminki. 2004. Modified adenoviruses for cancer gene therapy. *International Journal of Cancer* **110**:475–480.

Levine, Bruce L., Wendy B. Bernstein, Naomi E. Aronson, Katia Schlienger, Julio Cotte, et al. 2002. Adoptive transfer of costimulated CD4+ T cells induces expansion of peripheral T cells and decreased CCR5 expression in HIV infection. *Nature Medicine* **8**:47–53.

Nam, Ki Tae, Dong-Wan Kim, Pil J. Yoo, Chung-Yi Chiang, Nonglak Meethong, et al. 2006. Virus-enabled synthesis and assembly of nanowires for lithium ion battery electrodes. *Science* **312**:885–888.

Nicola, Anthony V., Anna M. McEvoy, and Stephen E. Straus. 2003. Roles for endocytosis and low pH in herpes simplex virus entry into HeLa and Chinese hamster ovary cells. *Journal of Virology* **77**:5324–5332.

Raoult, Didier, Stéphane Audic, Catherine Robert, Chantal Abergel, Patricia Renesto, et al. 2004. The 1.2-megabase genome sequence of mimivirus. *Science* **306**:1344–1350.

Rossman, Michael G., Vadim V. Mesyanzhinov, Fumio Arisaka, and Petr G. Leiman. 2004. The bacteriophage T4 DNA injection machine. *Current Opinion in Structural Biology* **14**:171–180.

Stevens, James, Ola Blixt, Terrence M. Tumpey, Jeffrey K. Taubenberger, James C. Paulson, et al. 2006. Structure and receptor specificity of the hemagglutinin from an H5N1 influenza virus. *Science* **312**:404–410.

van Riel, Debby, Vincent J. Munster, Emmie de Wit, Guus F. Rimmelzwaan, Ron A. M. Fouchier, et al. 2006. H5N1 virus attachment to lower respiratory tract. *Science* **312**:399.

 For further study and review, please visit StudySpace at **microbiology2.com**.

Chapter 12

Molecular Techniques and Biotechnology

12.1 Basic Tools of Biotech: A Research Case Study

12.2 Genetic Analyses

12.3 Molecular Analyses

12.4 "Global" Questions of Cell Physiology

12.5 Applied Biotechnology

Technologies that are developed to answer basic questions in biology often yield useful consumer products. This is especially evident in today's world of genetic engineering, where the term "biotechnology" has become ubiquitous. The science of biotechnology uses living organisms or their products to improve human health. Although we think of it as a new field, the earliest biotechnologists actually lived about 10,000 years ago. They learned to use yeast to make bread and alcohol. They also unwittingly used naturally existing bacteria to make cheeses and yogurts.

The field changed little over the millennia, until about 60 years ago, when the Scotsman Alexander Fleming discovered antibiotics. That watershed event, along with unraveling the structure of DNA and deciphering the genetic code, ushered in the modern high-tech era of biotechnology. We have since learned a great deal about the molecular biology of the microbes our ancestors used. Now, through gene-splicing techniques, we can force bacterial cells to produce human hormones; make fruits and vegetables; produce vaccines; and devise biochemical pathways to synthesize therapeutic molecules that never before existed.

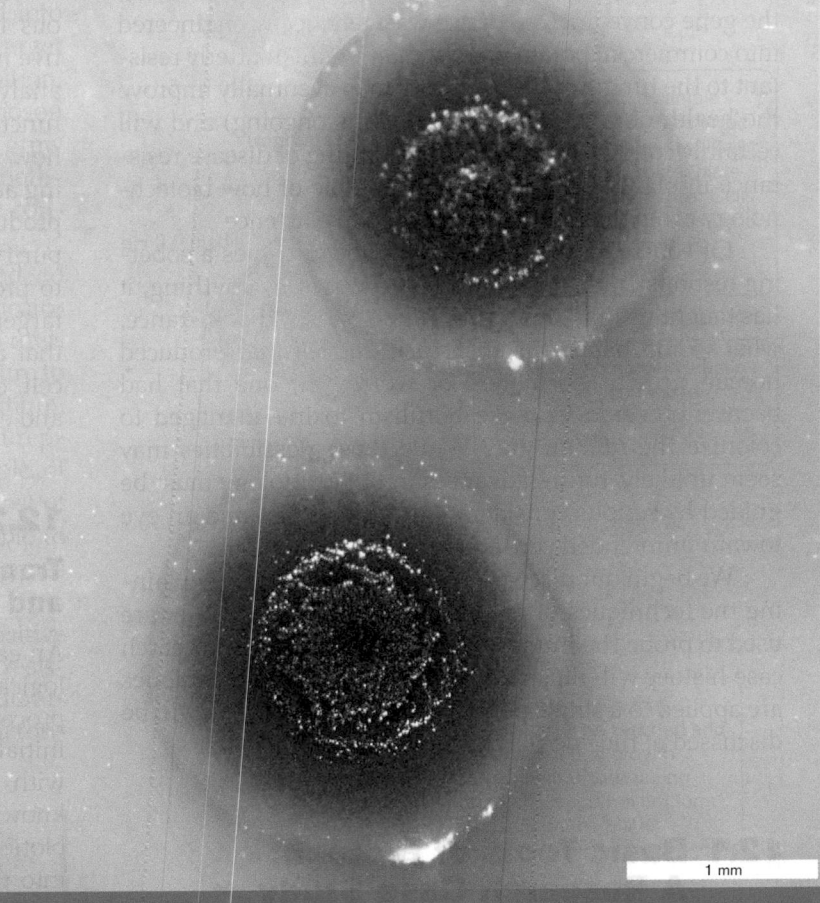

1 mm

Current Research Highlight

Complete genome transplants are now possible. We have been able to manipulate small pieces of bacterial genomes for many years via mutation, cloning, and recombination techniques, many of which will be described in this chapter. But manipulating an entire genome has been out of reach—until now. Scientists have taken the genome of one bacterial species, *Mycoplasma mycoides*, moved it to yeast (they added a yeast centromere), modified the genome using yeast genetics, and then transplanted the modified genome into a different species, *Mycoplasma capricolum*, replacing the *M. capricolum* genome. The resulting *M. mycoides* colonies are shown in this photo. This biotechnological advance is especially exciting because genetic manipulation of *M. mycoides* was previously impossible. Using this technique, we might redesign prokaryotic systems and even engineer new species. *Source:* Carole Lartigue et al. 2009. *Science* 325:1693–1696.

427

Chapter 13

Energetics and Catabolism

13.1 Energy and Entropy for Life

13.2 Energy and Entropy in Biochemical Reactions

13.3 Energy Carriers and Electron Transfer

13.4 Catabolism: The Microbial Buffet

13.5 Glucose Breakdown and Fermentation

13.6 The Tricarboxylic Acid (TCA) Cycle

13.7 Aromatic Pollutants

All living cells need energy to move and grow. Growth requires energy to incorporate nonliving substances into new cells. The energy to build cells comes from chemical reactions. For microbes, the variety of such reactions is limitless; virtually any kind of molecule in our biosphere—from hydrogen gas to chlorinated pollutants—can yield energy for some kind of microbe.

To capture and use energy, microbes must regulate their energy-yielding reactions and couple them to biosynthesis, using enzymes. Enzymes direct the transfer of energy onto carriers such as ATP.

Many energy-yielding reactions used by microbes breakdown complex molecules into smaller ones, a process called catabolism. Microbes catabolize chemicals within our own digestive tracts and all around us, in the soil and water. Collectively, microbes show astonishing potential to catabolize nearly any organic substance, including petroleum, hard woods, and synthetic polymers. The products of their catabolism range from ethanol to the generation of electricity in fuel cells.

Current Research Highlight

Deep within a South African gold mine, 2 kilometers (km) below Earth's surface, the energy for life comes from radioactive decay. Geomicrobiologist Lisa Pratt and her team have isolated live bacteria growing within rock, where they obtain energy from hydrogen gas. The hydrogen comes from water ionized by uranium decay. A vast underground ecosystem of such bacteria exists, powered by radioisotope decay instead of the solar energy that powers life aboveground. The photograph above shows Professor Tullis Onstott (Princeton University) opening a valve on a borehole in preparation for sampling microbial biomass from deep groundwater in the Witwatersrand Basin, South Africa. *Source*: Dylan Chivian et al. 2008. *Science* **322**:275. Photo: Courtesy of Lisa M. Pratt, Indiana University.

The U.S. Army manufactures billions of pounds of explosives such as trinitrotoluene (TNT). When a munitions plant shuts down, many acres of explosive-contaminated soil need to be remediated before the site is converted for civilian use. To clean up the site, the army, like many civilian communities, enlists the help of bacteria, through composting (**Fig. 13.1**). Compost bacteria consume the explosive molecules as food, gaining energy to grow new cells.

In composting, nitrogen-rich material (such as explosives or food wastes) is mixed with cellulose-rich yard waste or corn husks. The mixed compost materials are aerated to provide oxygen for oxidative breakdown by bacteria and fungi. The breakdown of complex molecules such as cellulose to smaller molecules, such as carbon dioxide, is called **catabolism**. Catabolism provides energy for **anabolism**, the reactions that build microbial cells. But some of the energy is always released as heat.

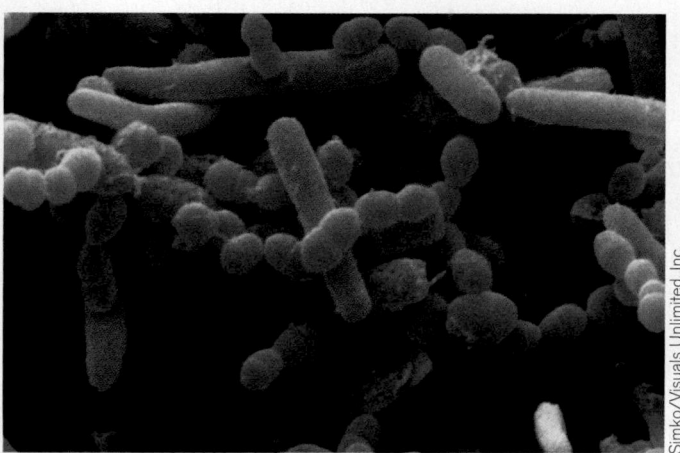

Figure 13.1   Microbial composting.   A. Composting soil contaminated by explosives at the Joliet Army Ammunition Plant. **B.** Rod-shaped (bacilli) and spherical (cocci) bacteria found in compost. Bacteria are approximately 0.5–5.0 μm in length.

In a compost pile, heat is produced faster than it dissipates, and the temperature typically rises to 60°C. As temperature rises, thermophilic species of *Bacillus* and *Thermus* take over. The thermophiles metabolize even faster, and the heat they release maintains the compost at a high temperature that favors their growth. The heat of compost can actually be used by other organisms, such as certain nesting birds that use compost to warm their eggs.

All microbes need energy from chemical reactions to "do the work" of growing new cells. The energy released from a reaction is transferred by enzymes to other reactions that build simple molecules into a complex cell. Chapter 13 explains how microbes gain energy from chemical reactions, by transfering electrons between molecules. This chapter emphasizes catabolism, the breakdown of complex organic molecules with electron transfer. Chapter 14 explores electron transport in greater depth, through pathways of organic respiration (oxidation of organic nutrients), lithotrophy (oxidation of *in*organic nutrients), and photosynthesis. The energy from all these pathways is used to build cells by pathways of anabolism (biosynthesis), the focus of Chapter 15. Finally, Chapter 16 explores commercial applications of microbial metabolism to produce food and beverages, as well as industrial products and pharmaceuticals.

The major reactions yielding energy for life are summarized in **Table 13.1**. Chapter 13 emphasizes reactions of organic compounds donating electrons to yield energy, known as **organotrophy**. For each class of metabolism, **Table 13.1** shows only a few examples. Microbes use many energy-yielding reactions unavailable to animals or plants. For example, *Pseudomonas* and related species of soil bacteria can gain energy by catabolizing camphor, benzene, or chlorinated aromatic pollutants. On the other hand, the archaeon *Pyrodictium occultum* gains energy by oxidizing hydrogen gas with sulfur—an inorganic reaction. Inorganic reactions yielding energy are covered in Chapter 14.

NOTE: Organotrophs use preformed organic compounds to yield <u>energy</u> (a process called **organotrophy**). Most organotrophs are also **heterotrophs**, organisms that use preformed organic compounds for <u>biosynthesis</u> (a process called **heterotrophy**).

13.1 Energy and Entropy for Life

Every form of life, from a composting microbe to a human body, uses energy. **Energy** is the ability to do work, such as flagellar propulsion or cell growth. Energy

Table 13.1 Energy acquisition in bacteria and archaea.

Compounds metabolized	Class of metabolism	Examples of energy-yielding reactions	Electron acceptor	Systems for energy acquisition
Organotrophy Organic compounds donate electrons	Fermentation Catabolism	$C_6H_{12}O_6 \rightarrow 2C_3H_6O_3$ (or other small molecules)	Organic	Glycolysis and other catabolism
	Organic respiration Catabolism with inorganic or small organic electron acceptor	$C_6H_{12}O_6 + 6H_2O + 6O_2 \rightarrow$ $6CO_2 + 12H_2O$	O_2	Glycolysis and other catabolism TCA cycle
		$C_6H_{12}O_6 + 6H_2O + 12NO_3^- \rightarrow$ $6CO_2 + 12H_2O + 12NO_2^-$	NO_3^-, SO_4^{2-}, Fe^{3+}, or other	Electron transport system
Lithotrophy Inorganic compounds donate electrons	Lithotrophy or chemoautotrophy	Electron donor for respiration is H_2, Fe^{2+}, H_2S, NH_4^+	O_2 or NO_3^-	Electron transport system
	Methanogenesis	Electron donor is H_2, CH_3OH, CH_3NH_2 $CO_2 + 4H_2 \rightarrow CH_4 + 2H_2O$	CO_2	Methanogenesis
Phototrophy Light absorption provides electrons	Photoautotrophy Light absorption drives CO_2 fixation	Photolysis of H_2O $6CO_2 + 12H_2O \rightarrow$ $C_6H_{12}O_6 + 6H_2O + 6O_2$		Photosystems I and II
		Photolysis of H_2S. HS^-, or Fe^{2+} $6CO_2 + 12H_2S \rightarrow$ $C_6H_{12}O_6 + 6H_2O + 12S$		Photosystem I or II
	Photoheterotrophy Catabolism with light absorption	Photolysis of H_2S, HS^-, or small organic molecules Light-driven H^+ pump or Na^+ pump		Photosystem I or II Bacteriorhodopsin or proteorhodopsin

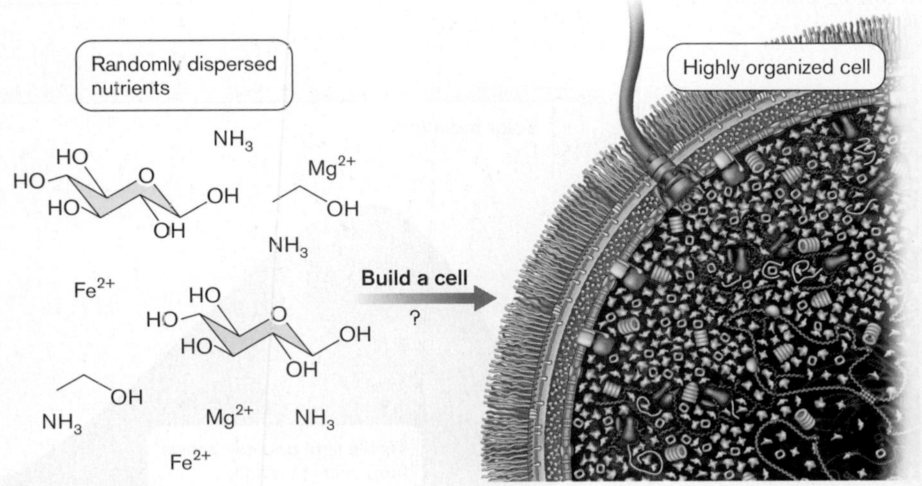

Figure 13.2 Living organisms acquire energy to build cells. Early-twentieth-century biologists wondered how the "spontaneous" growth of cells could obey the laws of conservation of energy and increase of entropy.

is used to organize proteins and to maintain ion gradients across the cell membrane. Yet the laws of thermodynamics tell us that systems tend to become less ordered and that **entropy**, the disorder, or randomness, of the universe, always increases. So how do cells assemble simple, disordered molecules into complex, ordered forms (**Fig. 13.2**)? How does life build order out of disorder?

Microbes Use Energy to Build Order

To explain how microbes build order requires a closer look at the thermodynamic relationship between energy and entropy. Gaining energy enables organisms to grow, thus increasing their structural order. As order increases, the cell is said to decrease in entropy, or disorder. But the decrease in entropy is local to the cell, and temporary.

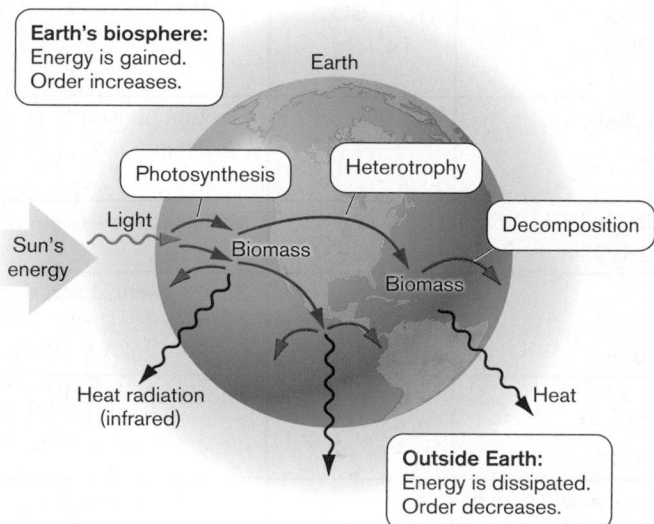

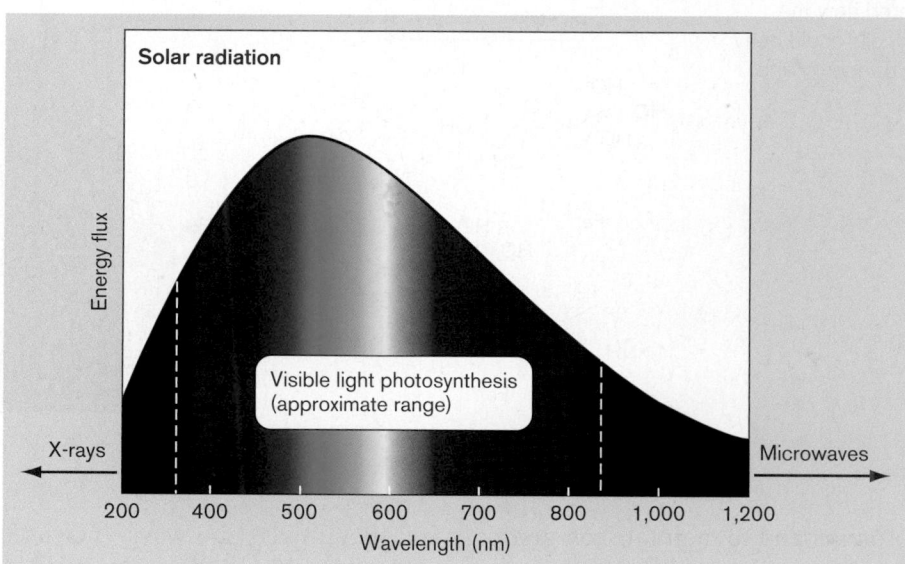

Figure 13.3 Solar energy. Solar radiation reaches Earth, where a small fraction is captured by photosynthetic microbes and plants. The microbial and plant biomass enters heterotrophs and decomposers, which convert a small fraction to biomass at each successive level. At each level, the majority of energy is lost, radiated from Earth as heat.

Ultimately, the cell's energy must be spent as heat, which radiates away, causing entropy to increase. In other words, the local, temporary gain of energy enables a cell to grow. Continued growth requires continual gain of energy and continual radiation of heat.

What is true of the cell holds as well for the entire biosphere. In Earth's biosphere, the total metabolism of all life must ultimately dissipate most energy as heat. Biological heat production is not always obvious, because soil and water provide a tremendous heat sink. But overall, Earth's biosphere behaves as a giant thermal reactor (**Fig. 13.3**). As solar radiation reaches Earth, a small fraction is captured by photosynthetic microbes and plants. The fraction captured is largely in the range of visible light (**Fig. 13.4**), the range of wavelengths in which photon energies are appropriate for the controlled formation and dissociation of molecular bonds. Some bacteria conduct photosynthesis using ultraviolet and near-infrared radiation. At shorter wavelengths (X-rays), chemical bonds are broken indiscriminately; at longer wavelengths (microwave and radio wave), the quantum energy is too low to drive chemical reactions.

Microbial and plant photosynthesis generates biomass, which is consumed by heterotrophs and decomposers. The consumers store a small fraction of their energy in biomass. At each successive level, the majority of energy is lost, radiated from Earth as heat. Despite growth of living organisms on Earth, the universe as a whole becomes more disordered.

> **THOUGHT QUESTION 13.1** Why can't photosynthesis be driven by microwaves or by X-rays?

Figure 13.4 The solar spectrum. The Sun radiates across the spectrum, but the intensity of solar radiation reaching Earth peaks in the range of visible light.

Gibbs Free Energy Change

To provide energy to a cell, a biochemical reaction must go forward from reactants to products. The direction of a reaction can be predicted by a thermodynamic quantity known as the **Gibbs free energy change**, Δ*G* (also known as free energy change or Gibbs energy change). The Δ*G* value of a reaction determines how much energy is potentially available to do work, such as to drive rotary flagella, to build a cell wall, or to store accurate information in DNA.

Δ*G* includes enthalpy and entropy. The free energy change Δ*G* has two components:

- Δ*H* = change in **enthalpy**, the heat energy absorbed or released as reactants become products at constant pressure. When reactants absorb heat from their surroundings as they convert to products, Δ*H* is positive. When, instead, heat energy is released, Δ*H* is negative. Release of heat (negative value of Δ*H*) can yield energy for the cell to use.
- Δ*S* = change in **entropy**, or disorder. Entropy is based on the number of states of a system, such as the number of possible conformations of a molecule. If a cellular reaction splits one molecule into two, all else being equal, entropy increases; the system is more disordered, and Δ*S* is positive. If, however, two molecules become one, then order is increased overall, and Δ*S* is negative.

The relationship between Δ*H* and Δ*S* is given by the Gibbs-Helmholtz equation:

$$\Delta G = \Delta H - T\Delta S$$

The value of free energy, Δ*G*, determines whether a process may go forward spontaneously. Negative values of Δ*G* mean that heat is released and/or order is decreased. If Δ*G* is negative, the process may go forward, whereas positive values mean that the reaction will go in reverse.

The overall sign of Δ*G* depends on its two components: Δ*H*, the absorption or release of heat energy, and −*T*Δ*S*, the product of entropy change (Δ*S*) and temperature (*T*). In living organisms, a sufficiently negative Δ*H* (energy lost as heat) often overrides −*T*Δ*S*, the term for increase in order (negative value of Δ*S*, positive value of −*T*Δ*S*). Thus, a living organism, whose development entails increasing order and decreasing Δ*S*, can grow as long as the sum of its metabolism has a sufficiently negative value of Δ*H*. The heat loss associated with Δ*H* is obvious in a compost pile, and it occurs as well for all living organisms and communities.

Negative Δ*G* Drives a Reaction Forward

An example of a thermodynamically favored reaction is the oxidation of hydrogen gas to form water. Hydrogen gas is oxidized for energy by many kinds of bacteria in soil and water. For example, hydrogenotrophic bacteria of the genus *Ralstonia* have been isolated from ultrapure water used for nuclear fuel storage, where radioactive ionization generates H_2. The chemical reaction of hydrogen oxidation is

$$2H_2 + O_2 \rightarrow 2H_2O$$

In this reaction, two molecules of hydrogen gas donate four electrons to oxygen, forming water. For this reaction, at standard temperature (298 kelvins, or K) and pressure (sea level), Δ*H* = −484 kilojoules per mole (kJ/mol). The Δ*H* is strongly negative (much heat is released) because the bonds of the product H_2O are much more stable than those of the substrates H_2 and O_2. However, entropy decreases because the three molecules are replaced by two, a more ordered state. Thus, Δ*S* is negative (Δ*S* = −0.0886 kJ/mol · K); and in the Gibbs equation the negative sign on the entropy term −*T*Δ*S* makes its contribution to Δ*G* positive, unfavorable for reaction. So which term wins: Δ*H* or −*T*Δ*S*?

$$\begin{aligned} \Delta G &= \Delta H - T\Delta S \\ &= -484 \text{ kJ/mol} - (298 \text{ K}) \, (-0.0886 \text{ kJ/mol} \cdot \text{K}) \\ &= -484 \text{ kJ/mol} + 27 \text{ kJ/mol} \\ &= -457 \text{ kJ/mol} \end{aligned}$$

Overall, Δ*G* is negative, and so bacteria with the appropriate enzyme pathways can use the reaction of hydrogen gas with oxygen to provide energy.

NOTE: The **joule (J)** is the standard SI unit to denote energy. 1 kilojoule (kJ) = 1,000 joules. Another unit commonly used is the kilocalorie (kcal). The conversion factor is: 1 kJ = 0.239 kcal.

TO SUMMARIZE:

- **Energy** enables cells to build ordered structures out of simple molecules from their environment.
- **The free energy change (Δ*G*)** includes enthalpy (Δ*H*), the heat energy absorbed or released; and −*T*Δ*S*, the product of temperature and entropy change (Δ*S*).
- **Negative values of Δ*G*** show that a reaction may drive the cell's metabolism. The sign of Δ*G* depends on the relative magnitude of Δ*H* and −*T*Δ*S*.

13.2 Energy and Entropy in Biochemical Reactions

The energy required for all growth processes is supplied by biochemical reactions. But which direction will a biochemical reaction go? The values of ΔH and $-T\Delta S$ determine which reactions will proceed and whether they can provide energy under actual conditions of the cell. The overall free energy change ΔG determines which "foods" a microbe can eat or, more precisely, which reactions between available molecules can be harnessed for microbial growth.

Enthalpy and Entropy in Metabolism

The relative contributions of ΔH and $-T\Delta S$ indicate the roles of chemistry and temperature in microbial metabolism.

■ **ΔH-driven reactions release heat.** The ΔH term dominates bioenergetic reactions involving strong oxidants such as O_2 (reactions such as glucose respiration and photosynthesis). ΔH-driven reactions release a lot of heat, causing, for example, the rise in temperature of an aerated compost pile.

■ **ΔS-driven reactions depend on temperature.** The entropy-driven component of Gibbs energy, $-T\Delta S$, grows larger as temperature increases. This term dominates reactions that generate a larger number of products than reactants but have a relatively small change in oxidation state. An example is glucose fermentation to ethanol and CO_2 during production of alcoholic beverages. Fermentation reactions produce less heat than oxidative catabolism, and their energy yield increases with environmental temperature.

An extreme case of entropy-driven metabolism is that of acetate conversion to methane and CO_2, conducted by soil archaea such as *Methanosarcina barkeri*. This methanogen actually cools its environment, absorbing heat to form products of higher chemical energy. Its growth is driven by the increase in entropy.

Measuring ΔH and ΔS. The actual ΔH and $-T\Delta S$ of biochemical reactions can be determined by **calorimetry**, measuring the heat released or absorbed during a reaction. One type of calorimetry is isothermal titration calorimetry, in which heat release is measured as the amount of energy needed to dissipate heat and keep the temperature constant. Observing the reaction at different fixed temperatures reveals the temperature dependence of the reaction and hence the relative contribution of the entropic term $-T\Delta S$. A common application of isothermal titration calorimetry is to study the ΔH and $-T\Delta S$ of a reaction in which a drug binds to its target receptor protein, such as an inhibitor binding to the protease

of HIV (human immunodeficiency virus). The ΔH and $-T\Delta S$ of the inhibitor-binding reaction indicate its binding strength, an important factor in predicting its efficacy in treating patients.

The Direction of a Reaction

The ΔG for a given reaction is determined by several factors:

■ **Molecular stability.** When reactants combine to form products with more stable bonds, the reaction has a negative ΔH. For example, the reaction of a sugar with oxygen has a negative ΔH because the relatively unstable oxygen molecules are reduced to H_2O.

■ **Entropy increase.** Reactions in which a complex molecule is broken down to a greater number of smaller molecules increase entropy (positive ΔS, negative $-T\Delta S$). Entropy also increases with formation of gaseous products, such as CO_2, or of carboxylic acids, such as lactic acid, in which H^+ dissociates from a carboxylate ion ($RCOO^-$).

■ **Concentrations and environmental factors.** The direction of reaction depends not only on the intrinsic properties of the reactants, but on environmental factors such as temperature, pressure, and reactant concentration. An excess of reactants over products makes ΔG more negative (favoring the forward reaction). An excess of product makes ΔG more positive (favoring the reverse reaction).

Standard reaction conditions. The thermodynamic values reported in data tables hold only for isolated reactions under "standard reaction conditions" defined for scientists to compare their work. Standard conditions differ greatly from the actual conditions of living cells. These conditions include temperature, ionic strength, and gas pressure (in the case of gaseous components, such as CO_2), as well as the concentrations of reactants and products. To enable comparison, thermodynamic values are commonly presented under standard conditions for temperature, pressure, and concentration. The standard conditions define a standard Gibbs free energy change, $\Delta G°$. The standard conditions for $\Delta G°$ are as follows:

■ The temperature is 298 K (25°C).
■ The pressure is 1 atm (standard atmospheric pressure).
■ All concentrations of substrates and products are 1 molar (M).

These standard conditions also apply to changes in enthalpy ($\Delta H°$) and entropy ($\Delta S°$), which contribute to $\Delta G°$. In biochemistry, an additional standard condition

Table 13.2 Effect of the concentration ratio on ΔG.

Initial ratio of products to reactants: $\dfrac{[C][D]}{[A][B]}$	Change in ΔG (kJ/mol) at standard temperature (298 K) and atmospheric pressure	Result of change from standard concentrations
10^{-4}	−23	Products increase
10^{-2}	−11	Products increase
1	0	ΔG = ΔG°
10^{2}	+11	Reactants increase
10^{4}	+23	Reactants increase

is that of pH 7, because living cells commonly maintain their cytoplasm within a unit of neutral pH. The thermodynamic terms in biochemistry are thus designated $\Delta G°'$, $\Delta H°'$, and $-T\Delta S°'$.

> **NOTE:** Remember to distinguish these forms of the Gibbs free energy term:
>
> ΔG = change in Gibbs free energy for a reaction under defined conditions.
> $\Delta G°$ = ΔG at standard conditions of temperature (298 K) and pressure (sea level), with all reactants and products at a concentration of 1 M.
> $\Delta G°'$ = ΔG at standard temperature, pressure, and concentrations for $\Delta G°$, plus the additional condition of pH 7 (H^+ concentration of 10^{-7} M) commonly specified by biochemists.

Concentrations affect ΔG. In living cells, the concentrations of reactants and products usually differ from 1 M; thus, the actual ΔG differs from $\Delta G°$ or $\Delta G°'$. Consider a reaction in which reactants A and B are converted to products C and D, as well as conversion in reverse:

$$A + B \rightleftharpoons C + D$$

Higher concentration of reactants (A or B) drives the reaction forward, whereas higher concentration of products drives it in reverse. So, ΔG includes the ratio of products to reactants:

$$\Delta G = \Delta G° + RT \ln \frac{[C][D]}{[A][B]}$$

$$= \Delta G° + 2.303\, RT \log \frac{[C][D]}{[A][B]}$$

where R is the gas constant ($R = 8.315$ J/mol) and T is the absolute temperature in kelvins (298 K for 25°C). The factor 2.303 converts the logarithm of the ratio of products to reactants from base e to base 10.

The effect of the concentration ratio on ΔG is shown in **Table 13.2**. In reactions at medium temperature (25°C–40°C), a hundredfold increase in the ratio of products to reactants adds about 11 kJ/mol to ΔG. ΔG is then less negative and the reaction less favorable. On the other hand, a hundredfold decrease in the concentration ratio makes ΔG more negative by −11 kJ/mol.

In some environments, a highly negative concentration term can override a positive $\Delta G°$, resulting in a reaction with negative ΔG that microbes can use for energy. For example, in an iron mine the high concentration of reduced iron favors iron-oxidizing microbes. Alternatively, a high temperature may increase the magnitude of the term 2.303 RT log [products]/[reactants] when it is negative, until it overrides a positive $\Delta G°$. This occurs in thermophiles such as *Sulfolobus*, which metabolize sulfur at 90°C by reactions unfavorable at lower temperatures.

Another way the direction of reaction may change is for a second metabolic process to remove one of the reaction's products (C or D) as fast as it is produced. The total ΔG for both reactions could then be negative, even if the first reaction had a positive ΔG. In **glycolysis**, the pathway of glucose breakdown to pyruvate, many of the individual reactions have $\Delta G°'$ values smaller than 5 kJ/mol. Their actual direction of reaction in the cell depends on concentrations of products and reactants. Glycolytic reactions with near-zero $\Delta G°'$ are reversible and can in fact participate in the biosynthetic pathway of gluconeogenesis (glucose biosynthesis).

Concentration Gradients

A growing cell needs to obtain essential molecules from outside, such as sugars and amino acids, and inorganic ions. The difference in concentration of a nutrient inside a cell versus outside generates a concentration gradient that stores energy.

In water solution, a dissolved substance diffuses by random movements until its distribution has the same concentration throughout (**Fig. 13.5A**). The random distribution of molecules at uniform concentration

A. Diffusion: positive ΔS

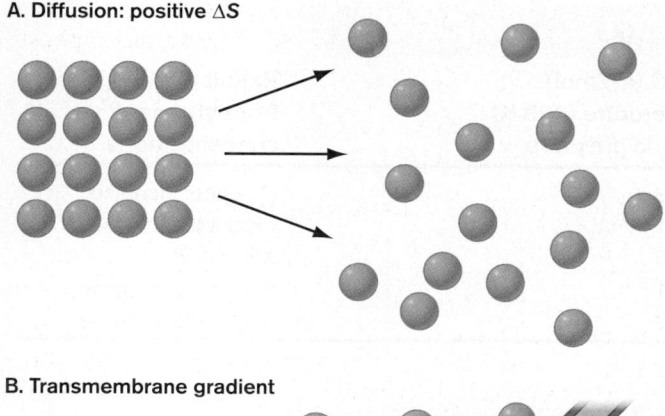

B. Transmembrane gradient

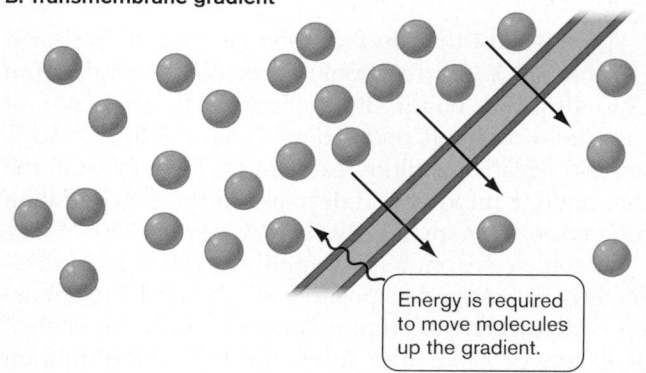

Energy is required to move molecules up the gradient.

Figure 13.5 Diffusion and transport. A. Water-soluble molecules diffuse to uniform concentration throughout the solution: ΔS is positive, −TΔS negative; the entropy term favors the process. **B.** If a membrane separating two compartments is permeable, molecules move from a compartment with high concentration to one with low concentration. Energy is required to move molecules up their concentration gradient.

represents the state of greatest entropy. Diffusion in the environment ultimately brings nutrients into contact with microbial cells, even cells that lack chemotactic motility to hunt for food. For example, sugars in the food we eat diffuse through our saliva to reach the bacteria growing in biofilms on our teeth.

The cell membrane contains transporters for useful molecules, allowing the nutrient molecules to cross. Entropy favors their movement from higher to lower concentration (**Fig. 13.5B**). In most environments, however, the nutrients are at lower concentrations than inside the cell. To obtain these molecules from outside, the bacterial cell must transport them against their gradient—that is, from lower to higher concentration—increasing the concentration difference and thus decreasing entropy. Uptake against a gradient requires an energy source to power transport proteins embedded in the membrane. The transporters must spend energy to compensate for the decrease in entropy when a cell transports the nutrient against its concentration gradient.

To form a chemical gradient requires expenditure of energy, so establishing a gradient stores energy. The energy stored in a gradient across a cell membrane can be interconverted with the energy of biochemical reactions such as sugar catabolism. For example, respiration generates a proton gradient (chemical gradient of H^+), which then drives the ATP synthase to synthesize ATP. The interconversion of gradient energy with chemical reaction energy, called the "chemiosmotic theory," earned Peter Mitchell the 1978 Nobel Prize in Chemistry (discussed in Chapter 14).

Within living cells, numerous other factors influence the Gibbs energy values of microbial growth. The Gibbs equation as presented assumes constant concentrations of reactants and products. In living cells, however, the reactant concentrations are continually changing. To account for changing concentrations requires a level of analysis more complex than we can present here.

An important source of concentration change is the coupling of energy-generating reactions to the energy-spending reactions of biosynthesis. The energy-spending reactions subtract from the observed free energy release of a growing microbe. A model proposed by Urs von Stockar and colleagues predicts that microbes optimize their coupling of catabolism with biosynthesis to maintain a net Gibbs energy of growth within a range of 300–450 kJ/mol per carbon of substrate catabolized. Different forms of catabolism, however, offer very different yields of biomass (cell mass synthesized per unit substrate). For example, the oxidative respiration of glucose yields five times the biomass produced by glucose fermentation, although respiring and fermenting cells show similar net Gibbs energy of growth.

A surprising discovery has been that many bacteria and archaea in natural environments grow extremely slowly, at values of ΔG approaching zero (that is, thermodynamic equilibrium). When the actual ΔG (under actual reaction conditions) equals zero, a reaction proceeds equally forward and in reverse, and there is no net change in energy. At equilibrium, the ratio of product and reactant concentrations exactly cancels ΔG° or ΔG°′:

$$\Delta G = 0 = \Delta G° + 2.303\ RT \log \frac{[C][D]}{[A][B]}$$

$$\Delta G° = -2.303\ RT \log \frac{[C][D]}{[A][B]}$$

Living cells can never grow exactly at equilibrium (ΔG = 0), but some gain energy from reaction pathways whose Gibbs free energy just exceeds equilibrium. Michael McInerney, Derek Lovley, and colleagues have identified soil bacteria and archaea that gain energy from metabolic pathways with ΔG values as small as −20 kJ/mol, less than one ATP's worth of energy per cycle. An example of an energy-yielding pathway with near-zero ΔG is CO_2 reduction to methane (a form of methanogenesis). How such pathways are coupled to biosynthesis is poorly understood, but the advantage of near-zero ΔG reactions is their greater efficiency of energy capture. The discovery of low-ΔG energetics has opened new possibilities for industrial microbiology previously thought impossible, such as the anaerobic digestion of complex organic pollutants in contaminated soil (Section 14.7).

THOUGHT QUESTION 13.2 Methanogens are a small proportion of the microbial community in soil, and the ΔG of methanogenesis is small. Yet methanogens produce large volumes of methane, large enough to contribute significantly as a greenhouse gas to global warming. Why would methanogens produce a relatively large quantity of waste product?

THOUGHT QUESTION 13.3 The soil bacterium *Geobacter* can metabolize acetic acid by the following reaction:

$$CH_3COOH + 2H_2O \rightarrow 2CO_2 + 4H_2$$

Yet the $\Delta G^{o'}$ is +95 kJ/mol. What could happen in the soil enabling *Geobacter* to grow?

TO SUMMARIZE:

- **The direction of reaction** is determined by the molecular stability of reactants and products and by entropy change associated with number of products, gaseous products, or products with multiple forms.
- **Concentrations of reactants and products affect ΔG.** The lower the concentration ratio of products to reactants, the more negative the value of ΔG.
- **A concentration gradient stores energy.** Solutes run down their concentration gradient unless energy is applied to reverse the flow.
- **Cellular thermodynamics involves** changing reactant and product concentrations and coupling energy-yielding reactions with energy-spending reactions.
- **Reaction pathways with ΔG near zero** can drive microbial growth.

13.3 Energy Carriers and Electron Transfer

Our ΔG equations show only the total energy of a reaction such as glucose oxidation. If all the energy were released at once, however, it would dissipate as heat without building biomass. In living cells, glucose oxidation never occurs in one step. Instead, the energy yield is divided among a large number of stepwise reactions with smaller energy changes. In this way, the cell can be thought of as "making change" by converting a large energy source to numerous smaller sources that can be "spent" conveniently for cell function and biosynthesis. The "spending" of energy is controlled by enzymes that couple all energy-providing reactions to specific energy-spending reactions.

Energy Carriers Gain and Release Energy

Many of the cell's energy transfer reactions involve **energy carriers**. Examples of energy carriers are ATP (adenosine triphosphate) and NADH (the reduced form of nicotinamide adenine dinucleotide). Energy carriers are molecules that gain and release small amounts of energy in reversible reactions. Energy carriers are used to transfer energy in a wide range of biochemical reactions.

Some energy carriers, such as NADH, transfer energy associated with electrons received from a food molecule. A molecule that transfers, or "donates," electrons to another molecule is called an **electron donor** or a reducing agent; a molecule that receives, or "accepts," electrons is called an **electron acceptor**. For example, during glucose catabolism, a molecule of glyceraldehyde 3-phosphate transfers a pair of electrons ($2e^-$) with a hydrogen ion to NAD^+, forming NADH. NAD^+ is an electron acceptor that receives the electrons; it then becomes the electron donor NADH. Electron donors such as NADH mediate electron transfer from reduced food molecules to a terminal electron acceptor such as oxygen. Energy carriers that transfer electrons are needed for all energy-yielding pathways and for biosynthesis of cell components such as amino acids and lipids.

Note that in living cells, all energy transfer reactions are coupled by enzymes to specific biochemical processes. Without enzymes, energy is given off as heat and thus would be lost from the living system.

ATP Carries Energy

Adenosine triphosphate, or **ATP** (**Fig. 13.6A**), is composed of a base (adenine), a sugar (ribose), and three phosphates. Note that adenine-ribose-phosphate (adenosine nucleotide) is equivalent to a nucleotide "link" of RNA. The base adenine is a fundamental molecule of life, one that forms spontaneously from methane and ammonia in experiments simulating the origin of life on early Earth. Like the sugar ribose, ATP is an ancient component of cells, found in all living organisms.

Under physiological conditions, ATP always forms a complex with Mg^{2+} (**Fig. 13.6B**). The magnesium cation partly neutralizes the negative charges of the ATP phosphates, stabilizing the structure in solution. Most enzyme-binding sites for ATP actually bind Mg^{2+}-ATP. This is one reason magnesium is an essential nutrient for all living cells.

A. ADP phosphorylation to ATP

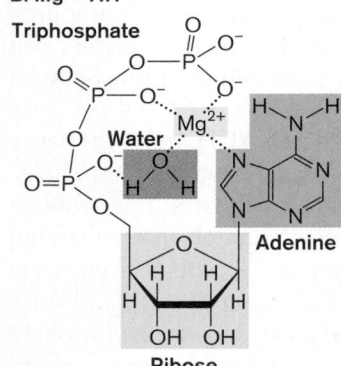

ADP

+ O⁻—P—OH + H⁺ →

P$_i$

→ ... + H_2O

ATP

B. Mg²⁺-ATP

Triphosphate

Water   Mg²⁺

Adenine

Ribose

Figure 13.6 ADP plus inorganic phosphate makes ATP. The reaction requires energy input (positive ΔG) because the negatively charged oxygens of the phosphates are forced into proximity. **A.** The chemical reaction phosphorylating ADP (adenosine diphosphate) to ATP (adenosine triphosphate). **B.** Model of Mg²⁺-ATP. The multiple negative charges of ATP are stabilized by being complexed with magnesium ion plus a water molecule.

ADP phosphorylation to ATP. During cell metabolism, ATP is generated by **phosphorylation**, the condensation of inorganic phosphate with adenosine diphosphate (ADP):

$$A-O-P-O-P-O^- + HO-P-O^- + H^+$$

$$\downarrow$$

$$A-O-P-O-P-O-P-O^- + H_2O \quad \Delta G^{\circ\prime} = 31 \text{ kJ/mol}$$

The phosphorylation of ADP to form ATP requires energy input (positive ΔG).

Why does ATP formation require energy? The inorganic phosphate has four oxygen atoms that share a negative charge. When phosphate reacts with another phosphate to form a bond, the charged oxygens of adjacent phosphates are forced into proximity, despite charge repulsion. The two phosphodiester bonds in ATP show charge repulsion, associated with a large negative ΔG of hydrolysis. Hydrolysis of each bond yields energy. The formation and hydrolysis of ATP can be shown as

$$A-P{\sim}P + H^+ + P_i \rightleftharpoons A-P{\sim}P{\sim}P + H_2O$$

where ~ designates each energy-storing bond and P$_i$ designates inorganic phosphate; or in more concise shorthand,

$$ADP + P_i \rightleftharpoons ATP + H_2O$$

Why do cells need to use different energy carriers, such as NADH and ATP, instead of just one kind?

- **Different amounts of energy.** The amount of energy carried by NADH transition is approximately three times the energy carried by ATP. If an ATP-driven enzyme were to use NADH instead, it would have to waste most of the energy available from NADH as heat.

- **Redox change.** The energy provided by NADH requires transfer of two electrons. Thus, NADH is used only for biochemical reactions involving substrate reduction.

ATP transfers energy. ATP can transfer energy to cell processes in three different ways: hydrolysis releasing phosphate; hydrolysis releasing pyrophosphate (diphosphate); and phosphorylation of an organic molecule. Each process serves different functions in the cell.

- **Hydrolysis releasing phosphate.** The **hydrolysis** of ATP at the terminal phosphate consumes H_2O to produce ADP and P$_i$. Hydrolysis of the phosphodiester bond releases energy. The energy released by ATP hydrolysis can be transferred to a coupled reaction of biosynthesis, such as building an amino acid. The two reactions are coupled by an enzyme with binding sites specific for ATP and the substrate.

- **Hydrolysis releasing pyrophosphate.** ATP can hydrolyze at the middle phosphate, releasing pyrophosphate (PP$_i$). The pyrophosphate usually hydrolyzes shortly afterward to make 2 P$_i$; thus, approximately twice as much energy is spent as in release of 1 P_i from ATP. The advantage of pyrophosphate release and subsequent hydrolysis over phosphate hydrolysis of ATP is that it drives a reaction strongly forward because twice as much energy would be required to reverse the reaction. Pyrophosphate is released in reactions that must avoid reversal—for example, the incorporation of nucleotides into growing chains of RNA.

■ **Phosphorylation of an organic molecule.** ATP can transfer its phosphate to the hydroxyl group of a molecule such as glucose to activate the substrate for a subsequent rapid reaction. No inorganic phosphate appears, and no water molecule is consumed:

$$\text{ATP} + \text{glucose} \rightarrow \text{ADP} + \text{glucose 6-phosphate}$$

Some enzymes catalyze ATP transfer of phosphate to activate sugar molecules for catabolism. Other enzymes couple the phosphorylation of a sugar to its transport across the cell membrane; consequently, these enzymes make up the **phosphotransferase system (PTS)**. The PTS enzymes play a critical role in determining which nutrients from the environment a microbe can acquire and catabolize (discussed in Chapter 4).

Note that besides ATP, other nucleotides carry energy. Guanosine triphosphate (GTP) provides energy for ribosome elongation of proteins. And the phosphodiester bonds of all four nucleotide triphosphates, as well as their corresponding deoxyribonucleotide triphosphates, carry energy for their own incorporation into RNA and DNA, respectively.

ATP produced by glucose catabolism. A large number of ATPs can be formed by coupling ATP synthesis to the step-by-step breakdown and oxidation of a food molecule such as glucose. In theory, complete oxidation of glucose through respiration can produce as many as 38 ATP molecules. The overall $\Delta G^{\circ\prime}$ of the coupled reactions is

$$C_6H_{12}O_6 + 6H_2O + 6O_2 \rightarrow$$
$$12H_2O + 6CO_2 \quad \Delta G^{\circ\prime} = -2{,}878 \text{ kJ/mol}$$
$$\underline{38\,[\text{ADP} + P_i \rightarrow \text{ATP} + H_2O] \quad \Delta G^{\circ\prime} = 38 \times (31 \text{ kJ/mol})}$$
$$\text{Net } \Delta G^{\circ\prime} = -1{,}700 \text{ kJ/mol}$$

The difference in $\Delta G^{\circ\prime}$ for the coupled reactions is the energy lost as heat and entropy, in this case $-1{,}700$ kJ/mol ($-2{,}978$ kJ/mol $+ 1{,}178$ kJ/mol). Thus, the maximal efficiency of energy capture by ATP is about 40%, a level that may be approached by highly efficient systems such as mitochondria. When ΔG values are corrected for cellular concentrations of reactants and products, the actual efficiency appears to be greater than 50%. For comparison, the efficiency of a typical machine, such as an internal combustion engine, is about 20%.

Under conditions such as low oxygen concentration, much smaller amounts of ATP are made per molecule of glucose. On the other hand, the vast majority of microbial catabolism in soil and water involves nonsugar substrates consumed by reactions at near-zero values of ΔG, which require multiple cycles to make a single molecule of ATP. These near-equilibrium reactions remain poorly understood, but their efficiency may approach 100%.

THOUGHT QUESTION 13.4 Linking an amino acid to its cognate tRNA is driven by ATP hydrolysis to AMP (adenosine monophosphate) plus pyrophosphate. Why release PP_i instead of P_i?

THOUGHT QUESTION 13.5 In the microbial community of the bovine rumen, the actual ΔG value has been calculated for glucose fermentation to acetate:

$$C_6H_{12}O_6 + 2H_2O \rightarrow 2C_2H_3O_2^- + 2H^+ + 4H_2 + 2CO_2$$
$$\Delta G = -318 \text{ kJ/mol}$$

If the actual ΔG for ATP formation is 44 kJ/mol and each glucose fermentation yields four molecules of ATP, what is the thermodynamic efficiency of energy gain? Where does the lost energy go?

NADH Carries Energy and Electrons

A major energy carrier that donates and accepts electrons is **nicotinamide adenine dinucleotide** (**NADH**, reduced; **NAD$^+$**, oxidized). NADH carries two or three times as much energy as ATP, depending on cell conditions. It is used to carry electrons from breakdown products of glucose. The oxidized form NAD$^+$ receives two electrons ($2e^-$) plus a hydrogen ion (H$^+$) from a food molecule; a second H$^+$ from the food molecule enters water solution. Overall, reduction of NAD$^+$ consumes two hydrogen atoms to make NADH:

$$\text{NAD}^+ + 2H^+ + 2e^- \rightarrow \text{NADH} + H^+ \quad \Delta G^{\circ\prime} = 62 \text{ kJ/mol}$$

For this reaction, $\Delta G^{\circ\prime}$ is positive; therefore, it requires input of energy from the food molecule. The reduced energy carrier NADH can then reverse this reaction by donating two electrons ($2e^-$) to another molecule, regenerating NAD$^+$.

NOTE: A hydrogen ion, or "proton" (H$^+$), does not exist free in solution. In water solution, H$^+$ combines with H_2O to form hydronium (H_3O^+), but for clarity we use H$^+$. An atom of hydrogen removed from a C–H bond consists of a proton (H$^+$) plus an electron (e^-). In a reaction, the proton and electron may be transferred to one molecule or to separate molecules.

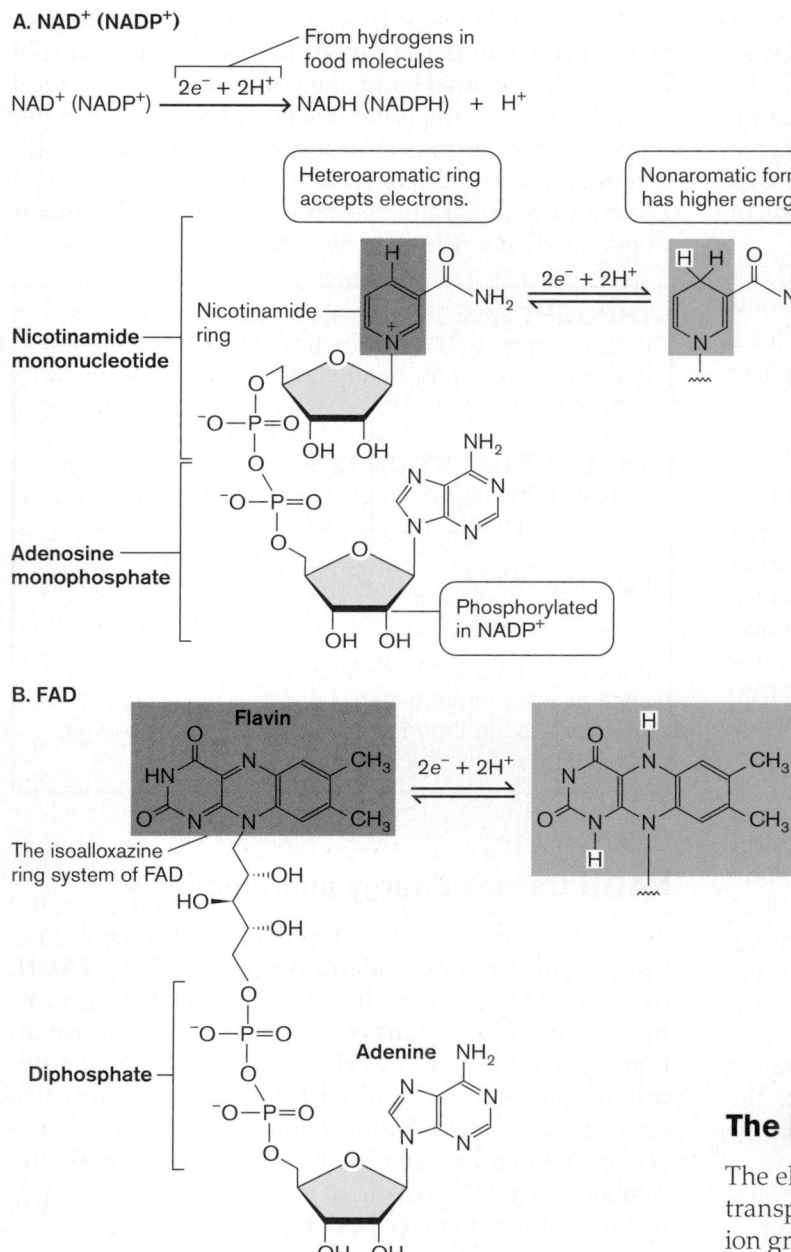

A. NAD⁺ (NADP⁺)

From hydrogens in food molecules

$$NAD^+ (NADP^+) \xrightarrow{2e^- + 2H^+} NADH (NADPH) + H^+$$

Heteroaromatic ring accepts electrons.

Nonaromatic form has higher energy.

$$2e^- + 2H^+$$

Nicotinamide mononucleotide

Nicotinamide ring

Adenosine monophosphate

Phosphorylated in NADP⁺

B. FAD

Flavin

The isoalloxazine ring system of FAD

$$2e^- + 2H^+$$

Diphosphate

Adenine

Ribose

Figure 13.7 Reduction of NAD⁺ and FAD.
A. In the reduction of NAD⁺, the nicotinamide ring (shaded pink) loses a double bond as two electrons are gained from an electron donor. Two hydrogen atoms are consumed; one bonds to NADH, while the other ionizes. **B.** In FAD, the flavin isoalloxazine ring system gains two electrons associated with two hydrogens.

A heteroaromatic ring is stable, but its disruption requires less energy than the disruption of benzene. Thus, it is possible to disrupt the ring by adding two electrons with H⁺ eliminating one double bond. The donation of electrons eliminates the ring's aromaticity, generating NADH, which carries energy in an amount useful for cell reactions.

The electrons transferred to NADH eventually must be put somewhere else, onto the next electron acceptor. If NADH builds up in a cell, no NAD⁺ remains to continue oxidizing food molecules. One way the energy stored by NADH can be spent is to transfer $2H^+ + 2e^-$ onto a product of catabolism. For example, in ethanolic fermentation to make wine or beer, NADH reduces pyruvate to ethanol. In this case, however, the energy is lost to the cell. Alternatively, NADH can transfer its electrons to one of a series of electron carrier molecules known as the **electron transport system (ETS)**.

The Electron Transport System (ETS)

The electron transport system (also known as the electron transport chain) stores energy from electron transfer as ion gradients across the membrane of the cell or an organelle. The ETS enables production of ATP. The ETS includes a series of proteins and small organic molecules that can be reduced and cyclically reoxidized. At the end of the series of oxidation-reduction transfers, the electrons are transferred to a **terminal electron acceptor** whose product leaves the cell. For example, molecular oxygen (O_2) as a terminal electron acceptor is reduced to H_2O.

The reaction of O_2 reduction to H_2O is coupled to oxidation of NADH or another reduced energy carrier:

$$NADH + H^+ \rightarrow$$
$$NAD^+ + 2H^+ + 2e^- \qquad \Delta G^{\circ\prime} = -62 \text{ kJ/mol}$$

$$\tfrac{1}{2}O_2 + 2H^+ + 2e^- \rightarrow H_2O \qquad \Delta G^{\circ\prime} = -158 \text{ kJ/mol}$$

$$NADH + H^+ + \tfrac{1}{2}O_2 \rightarrow$$
$$NAD^+ + H_2O \qquad \Delta G^{\circ\prime} = -220 \text{ kJ/mol}$$

NADH structure and function. NAD⁺ consists of an ADP molecule attached to a nicotinamide group instead of a third phosphate. The nicotinamide group, like a ribonucleotide, contains a nitrogenous base attached to a sugar phosphate. In NAD⁺, the dehydrogenated (oxidized) nicotinamide has a ring structure (shaded pink in **Fig. 13.7A**) that forms a stable cation.

NAD⁺ is a relatively stable structure because the ring electrons are **aromatic**; that is, the bonding electrons delocalize equally around the ring, as in benzene. Aromatic rings that contain noncarbon atoms are said to be **heteroaromatic**. Many biologically active molecules are heteroaromatic, including adenine and other nucleotide bases.

The total energy spent by NADH oxidation through the ETS is –220 kJ/mol (–62 kJ/mol – 158 kJ/mol). This energy is converted to transmembrane proton potential (composed of the H^+ concentration difference plus the charge difference across the membrane). The proton potential drives nutrient transport, motility, and synthesis of ATP. The ETS and the proton potential are discussed in detail in Chapter 14.

Many microbes can reduce terminal electron acceptors other than O_2, such as nitrate (NO_3^-) (see Table 13.1). Humans, unlike most microbes, have no alternative electron acceptors; thus, we need to breathe oxygen. When no terminal electron acceptor is available to accept electrons, NADH builds up to unfavorably high levels, leaving no more NAD^+ available to be reduced. When muscle tissues run out of oxygen, they transfer electrons from NADH back to the products of glucose catabolism, producing lactic acid. The buildup of lactic acid drives down pH, causing the muscle to cramp.

Other energy carriers that transfer electrons. Different steps of metabolism utilize different but related energy carriers. For example, NADPH differs from NADH only in its extra phosphate attached to the 2' carbon of adenine; the amount of energy carried is the same. Some enzymes can utilize both NADPH and NADH, whereas other enzymes use only one or the other. In many bacteria, NADPH is used for biosynthesis, whereas NADH feeds the ETS to yield energy. Other species, particularly phototrophs, use NADH and NADPH interchangeably.

Another related coenzyme is **flavin adenine dinucleotide (FAD)**, in which flavin substitutes for nicotinamide. The flavin nucleotide includes a ring structure whose aromaticity is eliminated by its receiving two electrons (**Fig. 13.7B**). The redox function of the flavin isoalloxazine ring system is similar to that of NADH:

$$FAD + 2H^+ + 2e^- \rightarrow FADH_2$$

Like NADH, $FADH_2$ donates $2e^-$ to an electron acceptor. $FADH_2$ is a weaker electron donor than NADH, but when combined with a strong electron acceptor such as O_2, electron transfer occurs and significant energy is released ($\Delta G^{\circ\prime} = -158$ kJ/mol).

For several reasons, different kinds of reactions use different energy carriers:

■ **Different amounts of energy.** Biochemical reactions yield different amounts of energy—that is, different values of ΔG. Suppose that a reaction can provide more than enough energy to generate ATP from ADP (31 kJ), but not quite enough to generate NADH from NAD^+ (62 kJ). An example is the conversion of succinate to fumarate in the TCA cycle (see Section 13.6). This reaction provides the energy to reduce FAD to $FADH_2$, whose oxidation by O_2 can yield two molecules of ATP. Thus, the use of $FADH_2$ enables the cell to make more efficient use of its food than if generation of ATP or NADH were the only choices.

■ **Different redox levels.** Food molecules may have more or fewer electrons (level of reduction/oxidation) than those associated with the cell structure. For example, lipids are more highly reduced than glucose. Thus, lipid catabolism requires a greater proportion of electron-accepting energy carriers (such as NAD^+ or $NADP^+$) than does glucose catabolism, and it makes relatively few ATP molecules directly. A combination of energy carriers that do or do not change redox state enables cells to balance their electrons while acquiring energy.

■ **Regulation and specificity.** Specific energy carriers can direct metabolites into different pathways serving different functions. For example, in many bacteria NADH is directed into the ETS, whereas NADPH, the 2'-phosphorylated form of NADH, is directed into biosynthesis of cell components such as amino acids and lipids.

The concentrations and reduction level of energy carriers provide much information on the state of a cell, such as the effects of environmental stress on cell metabolism. But energy carriers such as NADH undergo rapid turnover (interconversion with NAD^+). How can NADH and ATP concentrations be observed within living cells? One method uses nuclear magnetic resonance (NMR) spectroscopy (discussed in **eTopic 13.1**). The use of NMR to observe living cells led to the development of magnetic resonance imaging (MRI) to observe the entire human body.

Enzymes Catalyze Metabolic Reactions

In living cells, each reaction must occur only as needed, in the right amount at the right time. The rate of reaction is determined by the **activation energy** (E_a), the input energy needed to generate the high-energy transition state on the way to products (**Fig. 13.8**). Most biochemical reactions have an activation energy that exceeds the kinetic energy of the reactant molecules colliding. Thus, no matter how negative the ΔG, the reaction will proceed only when the activation energy is lowered by interaction with a catalyst, an agent that participates in a reaction without being consumed.

Biological reactions are catalyzed by **enzymes**, structures composed of protein (or, in some cases, RNA) that bind substrates of a specific reaction. The enzyme lowers the activation energy by bringing the substrates in proximity to one another and by correctly orienting them to react. In some cases, enzymes provide a reactive amino acid residue to participate in a transition state between

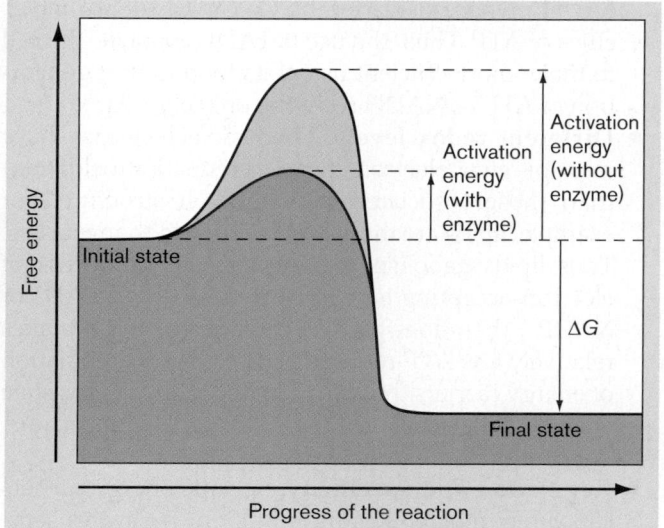

Figure 13.8   Enzyme lowers the activation energy of the transition state.   In the presence of an enzyme, the activation energy of the reactants is decreased, allowing rapid conversion of reactants to products.

reactants and products. Microbial enzymes have growing importance in industry; they are used for food production, fabric treatment, and drug therapies (discussed in Chapter 16).

Enzymes couple specific energy-yielding reactions (such as those of glucose breakdown) with the cell's reactions requiring energy, such as making ATP. An example of coupled reactions is shown in **Figure 13.9**. The substrate phosphoenolpyruvate (PEP), a phosphorylated

breakdown product of glucose, is converted to pyruvate while the phosphate is added to ADP to generate ATP:

Phosphoenolpyruvate

Pyruvate

The $\Delta G°'$ of phosphate cleavage from PEP is -62 kJ/mol, whereas the $\Delta G°'$ of ATP formation is only 31 kJ/mol. Thus, the net $\Delta G°'$ is negative (-31 kJ/mol), and the reaction goes forward to pyruvate. But a high activation energy makes the reaction extremely slow; it essentially never occurs on the timescale of life. The reaction occurs only when the two reacting substrates (PEP and ADP) are coupled by the enzyme pyruvate kinase (**Fig. 13.10**). Pyruvate kinase has specific binding sites for each of its substrates: PEP and ADP. ADP and PEP are brought together by the enzyme and positioned so as to lower the activation energy of phosphate transfer.

In addition to the catalytic site, the enzyme has an **allosteric site** (a regulatory site distinct from the substrate-

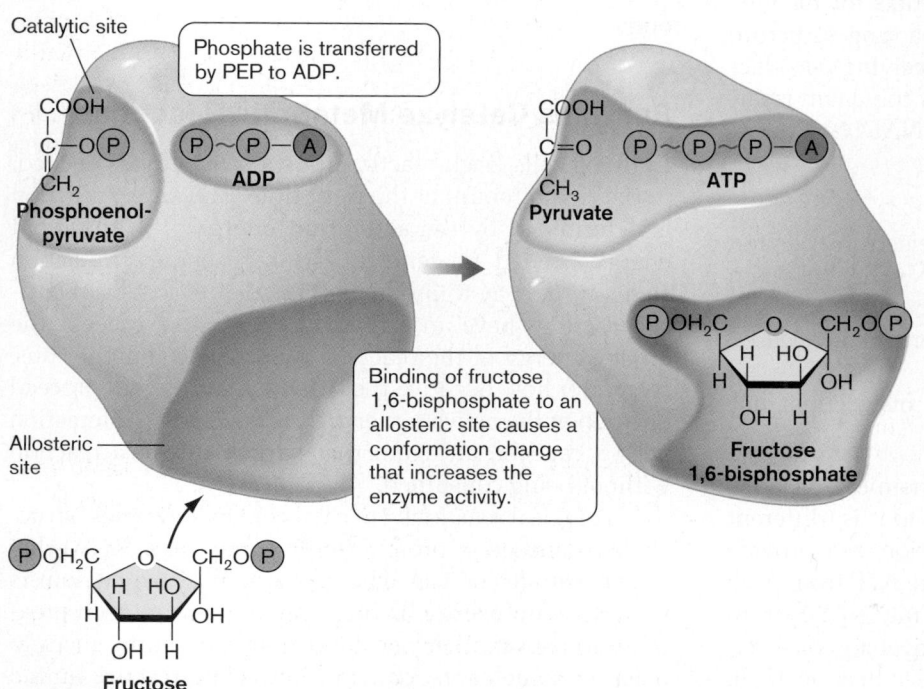

Figure 13.9   The enzyme pyruvate kinase.   Pyruvate kinase catalyzes the transfer of a phosphoryl group from PEP to ADP, generating pyruvate and ATP. The enzyme possesses separate binding sites for substrates and allosteric effectors.

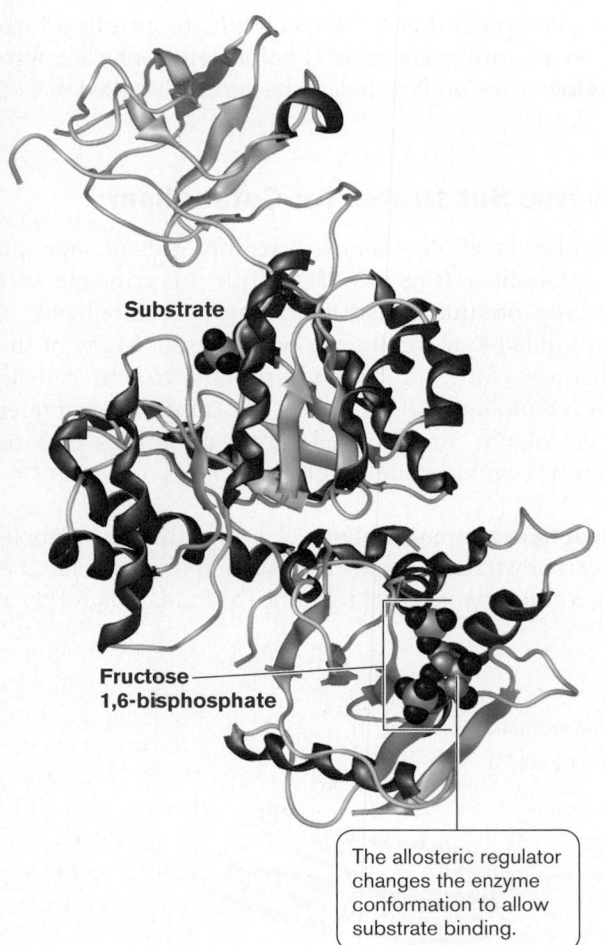

Substrate

Fructose
1,6-bisphosphate

The allosteric regulator changes the enzyme conformation to allow substrate binding.

Figure 13.10   Structure of pyruvate kinase. The molecular model of pyruvate kinase, based on a crystal structure of the enzyme bound to a substrate analog and an allosteric effector, fructose 1,6-bisphosphate. (PDB code: 1A3W)

binding site) for its activator fructose 1,6-bisphosphate. The difference between a substrate-binding site and an allosteric site can be seen in the molecular model based on X-ray crystallography (**Fig. 13.10**). The allosteric site is found at a distance from the substrate-binding site, but its interaction with the regulator, fructose 1,6-bisphosphate, alters the conformation of the entire enzyme, increasing the rate of reaction. As we shall see, fructose 1,6-bisphosphate is a central intermediate of glycolysis; thus, it makes sense that as this molecule builds up, it activates pyruvate kinase to remove products farther down the chain of catabolic reactions, maintaining the steady flow of metabolites.

THOUGHT QUESTION 13.6 What would happen to the cell if pyruvate kinase catalyzed PEP conversion to pyruvate but failed to couple this reaction to ATP production?

Note that pyruvate kinase is actually named for its reverse reaction, the use of ATP to phosphorylate pyruvate to PEP—perhaps because the activity of the purified enzyme was originally observed in the reverse direction. Enzymes can catalyze both forward and reverse reactions; the predominating direction depends on the concentrations of substrates and products, which determine ΔG, and on allosteric regulators.

TO SUMMARIZE:

■ **Catabolic pathways organize the breakdown of large molecules** in a series of sequential steps coupled to reactions that store energy in small carriers such as ATP and NADH.

■ **ATP and other nucleotide triphosphates store energy** in the form of phosphodiester bonds.

■ **NADH, NADPH, and FADH$_2$ each store energy** associated with an electron pair that carries reducing power.

■ **Enzymes catalyze reactions by lowering the ΔG** required to reach the transition state. They couple energy transfer reactions to specific reactions of biosynthesis and cell function.

13.4 Catabolism: The Microbial Buffet

In the early twentieth century, it was thought that microbes could catabolize a limited subset of naturally occurring organic molecules, such as sugars. Molecules not known to be catabolized were termed "xenobiotics," especially if they were "synthetic" products of human industry. We have since found that virtually any organic molecule can be catabolized by a microbe that has evolved the appropriate enzymes. Catabolism also plays key roles in microbial disease; for example, the causative agent of acne, *Propionibacterium acnes*, degrades skin cell components such as sialic acids, matrix molecules, lipids, and pore-forming factors.

Note that besides releasing energy, the breakdown of organic food molecules provides substrates for biosynthesis. Biosynthesis is covered in Chapter 15.

Classes of Catabolism

Major forms of catabolism are summarized in **Table 13.1**. Most catabolic pathways fall into one of three classes:

■ **Fermentation.** First identified by Louis Pasteur as *la vie sans air*, or "life without air," fermentation is the partial breakdown of organic food without net transfer of electrons to an inorganic terminal electron acceptor; thus, it can take place in the absence of oxygen. Fermentation has a favorable ΔG owing to breakdown of a large molecule to several smaller products, which are usually more stable as well. Types of fermentation pathways include ethanolic fermentation, producing alcoholic beverages; and lactic acid fermentation by lactic acid bacteria, producing cheese and yogurt.

■ **Respiration.** Respiration combines catabolic breakdown of organic molecules with electron transfer to a terminal electron acceptor such as oxygen. Respiration yields far more energy from catabolism than does fermentation. For humans, respiration is synonymous with breathing; but in the absence of O_2, many microbes use alternative electron acceptors, such as nitrate or sulfate. Thus, microbes—unlike humans and other animals—have the capacity for anaerobic respiration.

■ **Photoheterotrophy.** Some microbes conduct catabolism with a "boost" from light absorption. In photolytic catabolism, or photoheterotrophy, light absorption by chlorophyll drives the photolysis of an organic molecule such as succinate. Photoheterotrophy is common in marine and freshwater bacteria (discussed in Chapter 14).

Diverse Substrates for Catabolism

Microbes catabolize many different kinds of substrates, or catabolites (**Fig. 13.11**). While in principle virtually any organic constituent may be catabolized, certain kinds of substrates are widely used. Many of these substrates form products important to our nutrition and technology. Others, such as aromatic components of petroleum, are environmental pollutants that only microbes can degrade (**Section 13.7**).

Carbohydrate catabolism. Historically, the catabolism of carbohydrates (sugars and sugar polymers) has been important because of its relevance to human digestion and

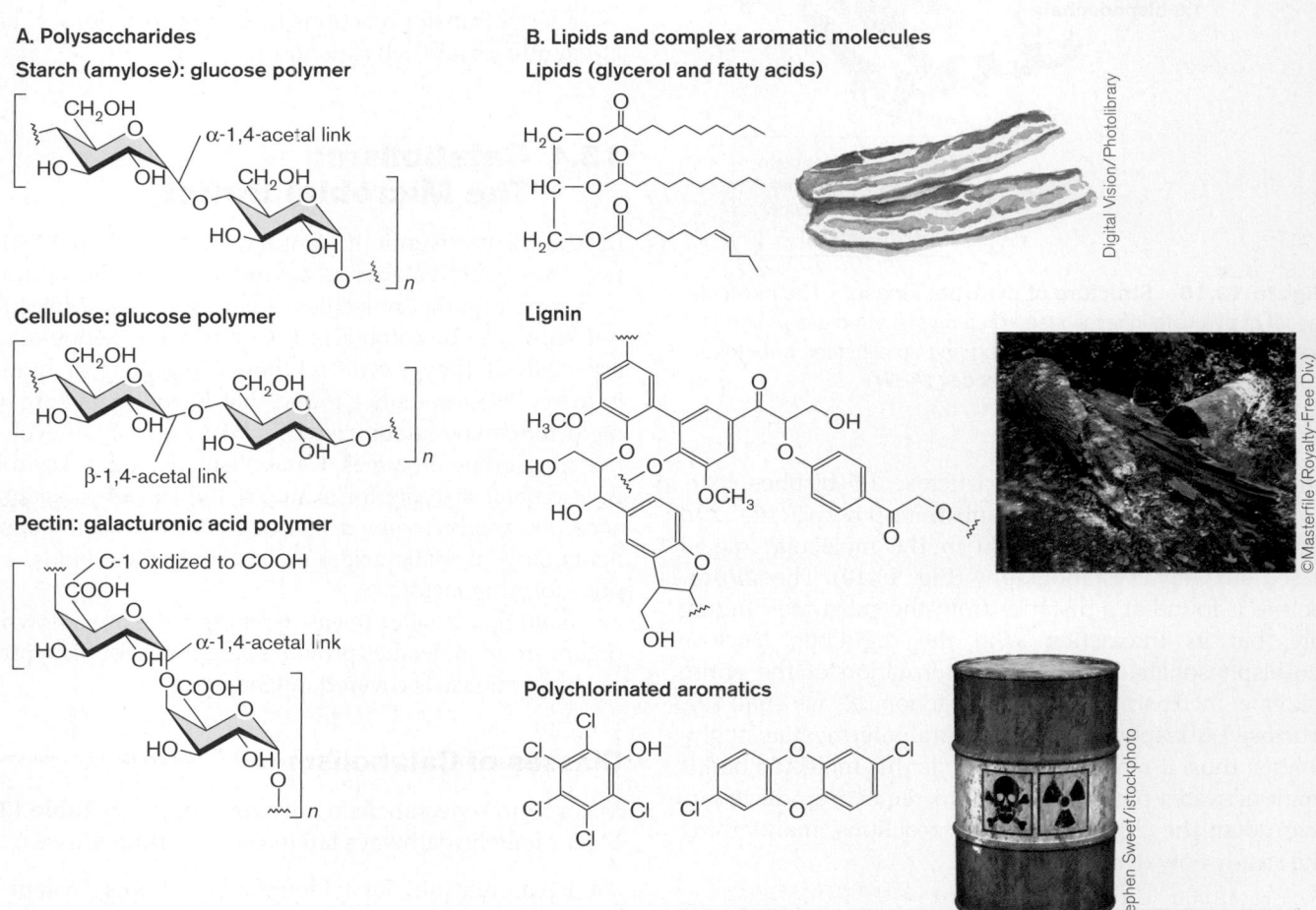

Figure 13.11 Carbon sources for catabolism. A. Polysaccharides such as starch, cellulose, and pectin are broken down by microbes to glucose, which undergoes glycolysis. **B.** Lipids and complex aromatic structures are broken down to acetate, which enters the TCA cycle. Halogenated aromatic pollutants are broken down by white rot fungi, a promising agent of bioremediation.

to the microbial production of food and drink (presented in Chapter 16). Glucose as such is rarely available to microbes, except to those pathogens growing within a host. But the pathways of glucose catabolism provide entry points for a diverse range of food molecules found in the environment, such as sugar chains, known as **polysaccharides**. For example, the cellulose of plant cell walls, the starch of potatoes, the pectin of fruit—all contain sugar chains that are broken down by microbial enzymes, first to short chains (oligosaccharides), then to two-sugar units (disaccharides), and then to monosaccharides such as glucose or fructose.

Even tiny differences between sugars can mean the difference between growth or starvation for a microbe. How can we possibly determine the difference between very similar substrates for catabolism?

One tool to reveal catabolite structure is nuclear magnetic resonance (NMR) (**Fig. 13.12A**). NMR spectroscopy gives detailed structural information about molecules containing atoms of odd atomic number or mass number, such as ^1H and ^{13}C. NMR detects magnetic spin transitions of these nuclei, which are affected by the electron clouds of the chemical bonds within the molecule. In many cases, the different hydrogen and carbon nuclei in a molecule can be linked to specific signals

in the NMR spectrum. **Figure 13.12B** shows ^{13}C NMR spectra of two monosaccharides, glucose and galactose, that differ in the position of the hydroxyl group on a single carbon, C-4. This small difference is important because different microbial enzymes (encoded by different genes) have evolved specifically to catabolize different sugars.

Polysaccharides are hydrolyzed to products that enter central catabolic pathways such as glycolysis (**Fig. 13.13**). In some sugars, the aldehyde is replaced by a hydroxyl (sorbitol, mannitol) or a carboxylate (gluconate, glucuronate). These sugar derivatives may be taken up by specific transporters and converted to glucose by specific enzymes. Alternatively, the sugar acids (carboxylate sugars) can be broken down by the Entner-Doudoroff pathway, an important variant of glycolysis. Pyruvate is commonly reduced to fermentation products such as lactate or ethanol. In the presence of a terminal electron acceptor such as oxygen, pyruvate releases an acetyl group to the tricarboxylic acid (TCA) cycle, also known as the Krebs cycle or citric acid cycle. The TCA cycle transfers electrons to the electron transport system, and ultimately the terminal electron acceptor. Glycolysis and the TCA cycle are presented in Sections 13.5 and 13.6, respectively.

A.   **B.**

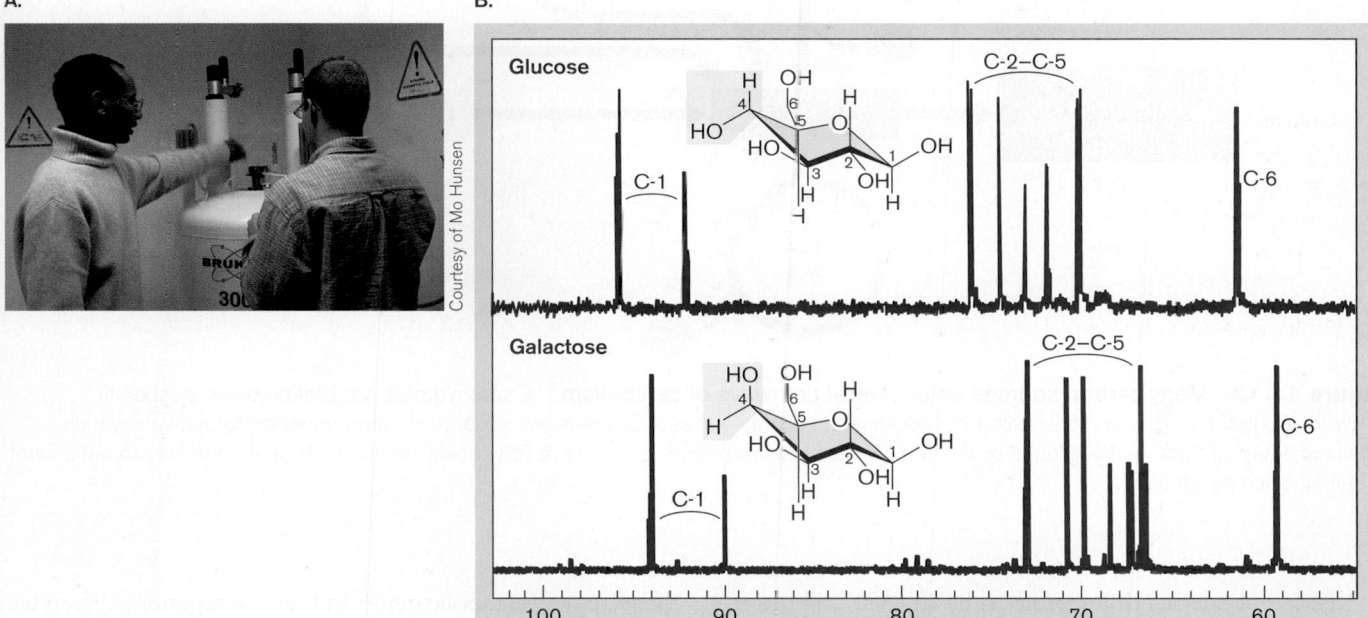

Figure 13.12 Different sugars distinguished by NMR.   A. Mo Hunsen (left), chemist at Kenyon College, demonstrates the use of the Bruker 300-MHz (megahertz) NMR. A sample is inserted into a superconducting magnet. In NMR spectroscopy, nuclei in a strong magnetic field absorb applied electromagnetic radiation, producing a signal. **B.** ^{13}C NMR spectra distinguish two highly similar sugars, glucose and galactose, which are carbon sources for microbial catabolism. Within each sugar, the signal of each ^{13}C nucleus has a "shift" in position based on its shielding by electrons in the molecule.

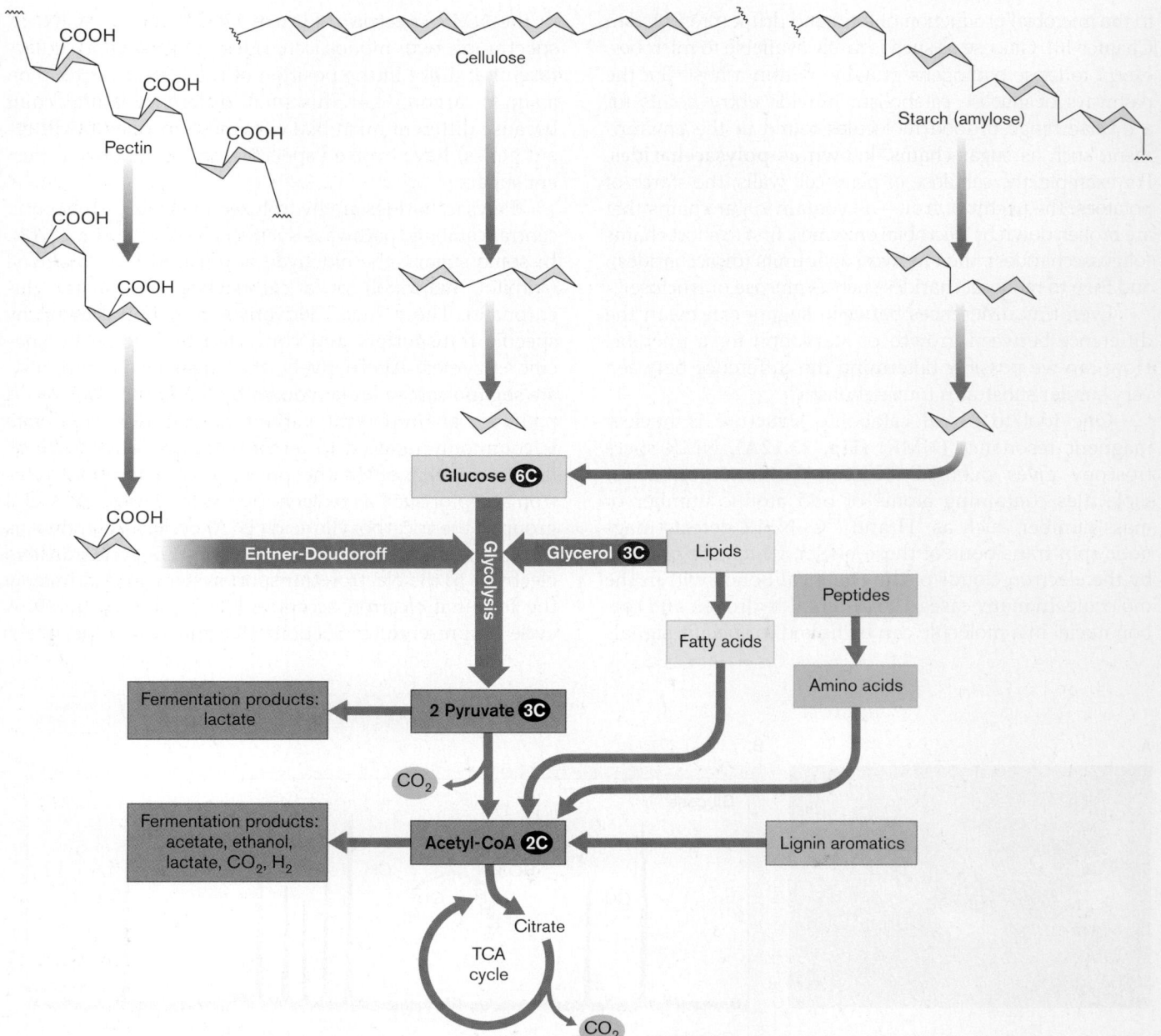

Figure 13.13   Many carbon sources enter central pathways of catabolism.   Carbohydrates are broken down by specific enzymes to disaccharides and then to monosaccharides such as glucose. Glucose and sugar acids are converted to pyruvate, which releases acetyl groups. Acetyl groups or acetate are also the breakdown products of fatty acids, amino acids, and complex aromatic plant materials such as lignin.

Different species differ profoundly in their abilities to digest particular polysaccharides. The starches (glucans) and pectins (pectic acids) are the most widely digested. Cellulose forms particularly stable fibers that are digestible by some bacteria and fungi, but not by animals. For this reason, animals such as cattle and termites require bacteria such as *Ruminococcus* and *Succinomonas* living within their digestive tracts to ferment cellulose from grasses or wood. By evolving a symbiosis with microbes, animals avoided the need to acquire new catabolic genes in their own genome. Even we humans derive about 15% of our caloric intake from catabolism of plant fibers by bacteria such as *Bacteroides thetaiotaomicron*. Microbiologists consider the human gut flora a functional part of the human body. The microbial genomes are functionally part of the human metagenome, the total sequence of genomes of a community of organisms. Medical conditions such as obesity may depend in part on the catabolic activities of gut flora.

In a complex environment such as the intestinal lumen, with numerous potential substrates for catabolism, organisms select substrates on the basis of their availability and energy efficiency. Substrates are selected through gene regulation, as discussed in Chapter 10. For example, in *E. coli* the sugar lactose induces transcription of genes that encode beta-galactosidase (*lacZ*) and lactose permease (*lacY*). But in the presence of glucose, a preferred carbon source, *lac* transcription is halted. Halting *lac* transcription enables preferential catabolism of glucose. The process of prioritized consumption of substrates is known as **catabolite repression**.

Catabolism of lipids and amino acids. Many bacteria respire lipids from sources such as milk, animal fats, and nuts; lipid oxidation causes the rancid odor of spoiled meat or butter. Lipids are catabolized by hydrolysis to glycerol and fatty acids (see **Fig. 13.13**). Glycerol, a three-carbon triol, can enter catabolism either as a three-carbon intermediate of glycolysis or through breakdown to acetate, entering the TCA cycle. Fatty acids undergo oxidative breakdown to acetyl groups.

Amino acids are the building blocks of proteins, but when present in excess of the cell's needs, they are catabolized to provide energy. Some pathogens catabolize specific amino acids as part of the disease process. For example, *Legionella pneumophila*, the cause of legionellosis, catabolizes threonine. Threonine catabolism is required for the pathogen to grow within macrophages (white blood cells). Thus, threonine catabolism may be a target for new drugs against *L. pneumophila*.

Specific enzymes catalyze the early steps in the degradation of each amino acid until products are formed that can enter common pathways of carbohydrate catabolism or the TCA cycle. The initial step of amino acid degradation is one of two kinds: decarboxylation (removal of CO_2) to produce an amine (**Fig. 13.14**); or deamination (removal of NH_3) to produce a carboxylic acid. Carboxylic acids are degraded through the TCA cycle. Amine products, such as cadaverine and putrescine, are often excreted, causing the noxious odor of decomposing flesh. In general, microbes favor decarboxylation during growth at low pH, because the amine products buffer the cell against acidity. Growth at higher pH favors deamination, because the acidic products buffer the cell against alkalinization.

The initial steps of amino acid catabolism are regulated genetically by multiple factors, including substrate availability and environmental conditions such as pH. In *E. coli*, low pH upregulates expression of several amino acid decarboxylases that release amines, raising pH. An example is the *cadBA* operon

encoding lysine permease and lysine decarboxylase (see **Fig. 13.14**). Expression of *cadBA* is upregulated by lysine binding to the membrane-embedded regulator LysP; by acid, which changes the conformation of regulator CadC; and by low oxygen, which increases transcription by unknown means.

Other reactions of amino acids generate small amounts of products with intense odors and flavors. Products of amino acid catabolism confer some of the distinctive flavors of fermented foods and beverages. For example, catabolites of aspartate and methionine generate the flavors of cheese (**Special Topic 13.1**). Microbial food production is discussed in Chapter 16.

Aromatic catabolism. Aromatic compounds are more difficult to digest than sugars because of the exceptional stability of aromatic ring structures. Yet many bacteria metabolize benzene derivatives, and even polycyclic aromatic molecules, either partly or all the way to CO_2. A particularly important aromatic substance found in nature is **lignin** (see **Fig. 13.11B**), which forms the key structural support of trees and woody stems. A discouragingly complex molecule, lignin is made from sugars oxidized to benzene rings, with ether connections that are difficult for enzymes to break

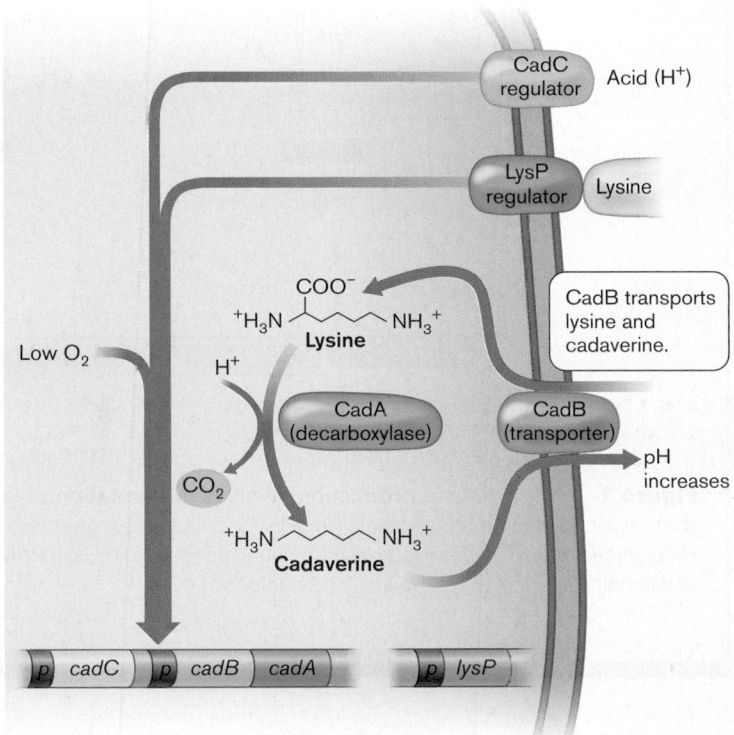

Figure 13.14   Lysine catabolism responds to the environment. Transcription of the operon encoding lysine transporter (CadB) and lysine decarboxylase (CadA) is induced by acid, lysine, and low oxygen concentration.

Special Topic 13.1 | Swiss Cheese: A Product of Bacterial Catabolism

Have you ever wondered how cow's milk can be used to make so many different-tasting kinds of cheese? The form and taste of cheese depend largely on microbial fermentation. Today, we can devise new kinds of cheese by cloning genes for new fermentation pathways and inserting them into cheese-producing bacteria. Swiss cheese is famous for its large, round holes, known as "eyes," as well as for its distinc-

tive flavor (**Fig. 1A**). Both the eyes and the flavor derive from common pathways of bacterial fermentation that occur during cheese production. The original Swiss cheese, or Emmentaler, was made in the Emmen Valley from cows pastured in the mountains. Today, related "Swiss-type" cheeses such as Jarlsberg are produced in large quantities throughout Europe and America.

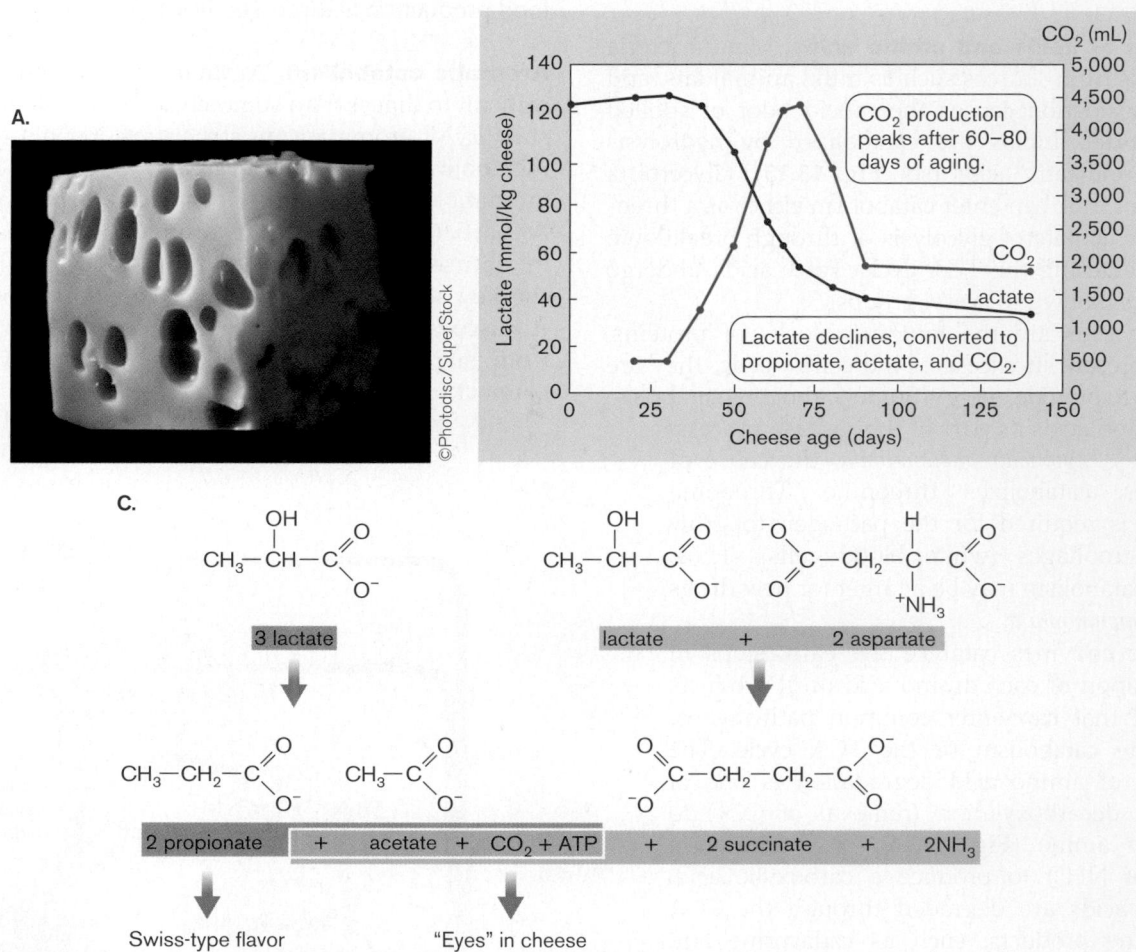

Figure 1   Swiss cheese production involves fermentation of lactate to propionate.   A. Swiss cheese (Emmentaler). **B.** As a Swiss cheese ages, lactate declines and CO_2 rises, gradually forming the "eyes" (holes). **C.** Lactate is fermented by *Propionibacterium freudenreichii* to propionate, acetate, and CO_2. Concurrent fermentation of lactate and aspartate generates additional CO_2, increasing the size and number of eyes.

down. Certain fungi and soil bacteria have evolved to digest lignin, ultimately directing its carbon into the TCA cycle.

Today, the environment contains increasing amounts of aromatics, including growing amounts of halogenated aromatics, such as polychlorinated biphenyls. These

compounds, produced for herbicides and other industrial uses, are the source of highly toxic dioxins. Aromatic pollutants concentrate in the food chain, reaching levels potentially dangerous to human consumers. Halogenated aromatics turn out to be catabolized by a number of soil microbes, which offer promising candidates for

Emmentaler cheese production involves two fermentation stages. The first stage, at high temperature, is dominated by thermophilic lactic acid bacteria such as *Lactobacillus helveticus* and *Streptococcus salivarius*. Lactic acid bacteria ferment the milk sugar lactose, a disaccharide of glucose and galactose, generating mainly lactic acid. The high temperature ensures rapid buildup of acid and retards growth of harmful species from the unpasteurized milk.

The second fermentation stage is unique to Swiss-type cheeses. The fermented milk curd is cooled and inoculated with *Propionibacterium freudenreichii*, a species that converts lactate to propionate, acetate, and CO_2 (**Fig. 1B** and **C**). This propionic fermentation generates another ATP, in addition to the ATPs of glycolysis. Propionic fermentation proceeds at cool temperatures, very slowly, over many months. During this period as the cheese is "aged," lactate declines while CO_2 increases. The gradual CO_2 production forms the eyes, or holes, for which Swiss cheese is known. The propionate contributes to the distinctive flavor.

The optimal production of eyes and flavor depends on several side pathways of fermentation of amino acids. Fermentation of aspartate plus lactate generates additional CO_2, thus increasing the size of the eyes, and it increases the overall efficiency of lactate fermentation. Thus, addition of aspartate, as well as use of a strongly aspartate-fermenting bacterial strain, can enhance the quality of the cheese.

The full flavor of a cheese results from a large number of minor fermentation pathways that produce trace amounts of aldehydes, esters, alcohols, and ketones. One substance that contributes to the Swiss-type flavor is methional, a derivative of methionine (**Fig. 2**). During the initial fermentation by lactic acid bacteria, small amounts of methionine are deaminated and decarboxylated to methional, a strong odorant. An alternative side reaction generates methanethiol, an odorant more characteristic of cheddar cheese. The basis for favoring one pathway or another remains unknown, but these side reactions generate many distinctive varieties of cheese.

Can new varieties of cheese be made with engineered bacteria? In 2009 Sean Hanniffy and colleagues cloned a gene for an enzyme of methionine catabolism from the bacterium *Brevibacterium linens*. The enzyme converts methionine to the cheddar flavor molecule, methanethiol. The gene was transferred into a cheese-producing strain of *Lactococcus lactis*, where it increased production of methanethiol. This engineered strain can now be tested for its effects on cheese production. Microbial food production is discussed further in Chapter 16.

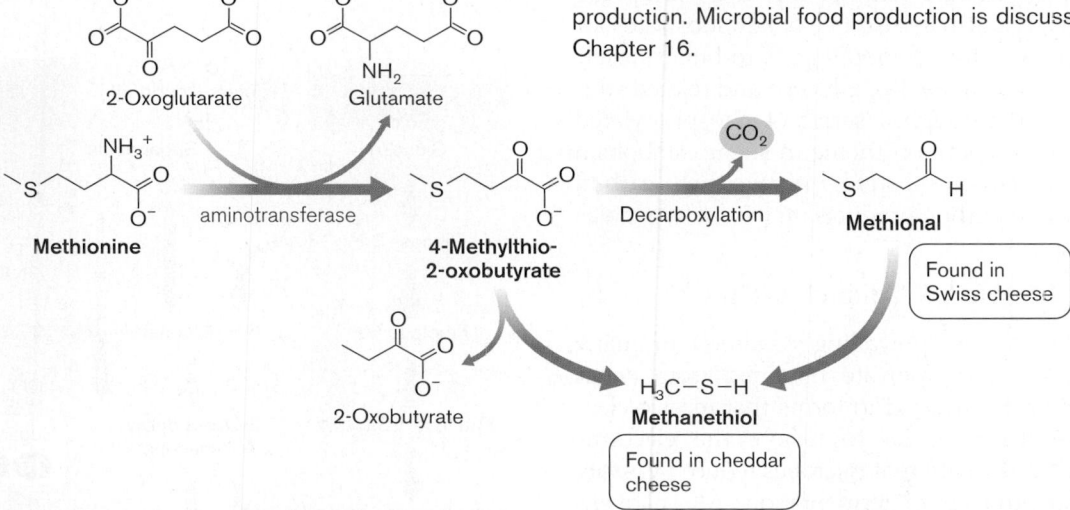

Figure 2 Methionine is fermented to flavor molecules. Methionine undergoes deamination and oxidation, followed by decarboxylation to methional, an aldehyde odorant characteristic of Swiss cheese. Alternative reactions generate methanethiol, more typical of cheddar.

bioremediation of polluted environments (described in Section 13.7).

A given species of bacterium or fungus may catabolize thousands of different carbon sources, each requiring specific transporters and enzymes for initial breakdown. In most cases, these products ultimately funnel into one of a few common pathways of metabolism. The remainder of this chapter presents key pathways in detail: glycolysis and other pathways of glucose catabolism, the TCA cycle, and the catechol pathway of benzoate catabolism. These pathways play key roles in medical microbiology and in industrial fields such as bioremediation.

TO SUMMARIZE:

■ **Fermentation, respiration, and photoheterotrophy** are forms of catabolism. In fermentation, the catabolite is broken down to smaller molecules without an inorganic electron acceptor. Respiration requires an inorganic terminal electron acceptor such as O_2 or nitrate. In photoheterotrophy, catabolism is supplemented by light absorption.

■ **Carbohydrates or polysaccharides** are broken down to disaccharides, and then to monosaccharides. Sugars and sugar derivatives, such as amines and acids, are catabolized to pyruvate.

■ **Pyruvate and other intermediary products of sugar catabolism** are fermented, or they are further catabolized to CO_2 and H_2O through the TCA cycle (in the presence of a terminal electron acceptor).

■ **Lipids and amino acids** are catabolized to glycerol and acetate, as well as other metabolic intermediates.

■ **Aromatic compounds such as lignin and benzoate derivatives** are catabolized to acetate through different pathways, such as the catechol pathway.

13.5 Glucose Breakdown and Fermentation

Glucose catabolism is important not only as a widespread source of energy, but also as a source of key substrates for biosynthesis, such as five-carbon sugars to build nucleic acids (discussed in Chapter 15). Glucose and related sugars are catabolized through a series of phosphorylated sugar derivatives. A common theme in sugar catabolism is the split of a six-carbon substrate into two three-carbon products. The three-carbon products may form two molecules of pyruvate:

$$C_6H_{12}O_6 \rightarrow 2C_3H_4O_3 + 4H \text{ (on NADH)}$$

Under anaerobiosis—the prevailing condition of many microbial habitats—the pyruvate obtained from sugar breakdown must be converted to forms that receive electrons from NADH, in order to restore the electron-accepting form NAD^+. Different microbes reduce pyruvate to different end products of fermentation. Alternatively, through anaerobic respiration, NADH may reduce an electron acceptor such as oxygen or nitrate, allowing pyruvate to feed into the TCA cycle (discussed in Section 13.6).

NOTE: The carboxylic acid intermediates of metabolism exist in equilibrium with their dissociated, or ionized, form, identified by the suffix "-ate." For example, lactic acid dissociates to lactate; acetic acid dissociates to acetate. We use the "-ate" terms for acids whose ionized form predominates under typical cell conditions (around pH 7).

To catabolize glucose, bacteria and archaea use three main routes to pyruvate (**Fig. 13.15**):

■ **Glycolysis,** or the **Embden-Meyerhof-Parnas (EMP) pathway,** in which glucose 6-phosphate isomerizes to fructose 6-phosphate, ultimately forming two molecules of pyruvate. EMP is used by many bacteria, eukaryotes, and archaea. From each glucose, the pathway generates net two ATP and two NADH.

■ **Entner-Doudoroff (ED) pathway,** in which glucose 6-phosphate is oxidized to 6-phosphogluconate, a phosphorylated sugar acid. Alternatively, sugar acids may be converted directly to 6-phosphogluconate. Sugar acids often derive from intestinal mucus, and the ED pathway is essential for enteric bacteria to colonize the intestinal epithelium. The ED pathway generates only one ATP, one NADH, and one NADPH.

■ **Pentose phosphate shunt (PPS),** in which glucose 6-phosphate is oxidized to 6-phosphogluconate, and then decarboxylated to a five-carbon sugar (pentose), ribulose 5-phosphate. The PPS produces sugars of three to seven carbons, which serve as precursors for biosynthesis or are converted to pyruvate as needed. The PPS generates one ATP plus two NADPH, the reducing cofactor most commonly associated with biosynthesis.

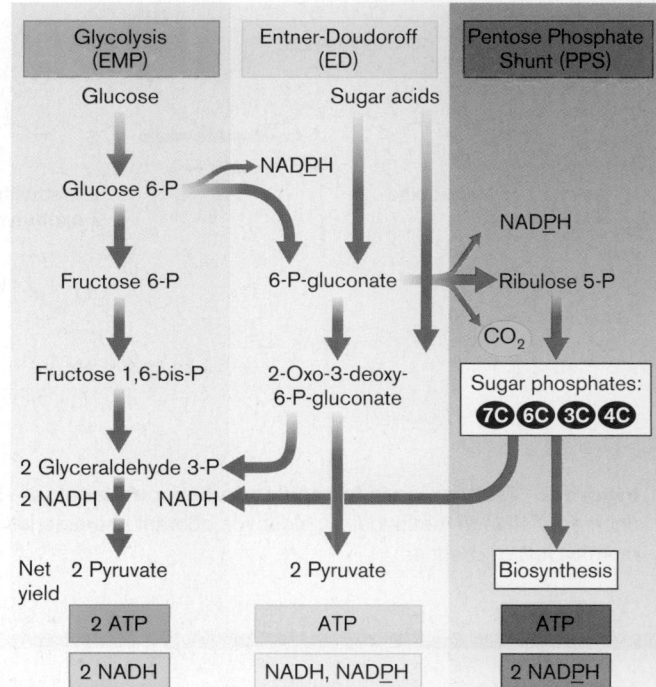

Figure 13.15   From glucose to pyruvate: three pathways. The Embden-Meyerhof-Parnas (EMP) pathway of glycolysis, the Entner-Doudoroff (ED) pathway, and the pentose phosphate shunt (PPS) each catabolize carbohydrates by related but different routes.

Glycolysis: The EMP Pathway

The **Embden-Meyerhof-Parnas (EMP) pathway**, or **glycolysis**, is the form of glucose catabolism most commonly studied in introductory biology; it is central for animals and plants, as well as many bacteria. In the EMP pathway, one molecule of D-glucose undergoes stepwise breakdown to two molecules of pyruvic acid (or its anion, pyruvate) (**Fig. 13.16**). Glucose breaks down in two stages. In the first stage, the glucose molecule is primed for breakdown by two steps of sugar phosphorylation by ATP. Each ATP phosphotransfer step requires input of Gibbs energy. The phosphoryl groups tag two sides of the glucose for splitting into two three-carbon molecules of glyceraldehyde 3-phosphate (G3P).

In the second stage, each glyceraldehyde 3-phosphate is oxidized by NAD^+ through steps leading to pyruvate. Each conversion of glyceraldehyde 3-phosphate to pyruvate forms two molecules of ATP, one from dephosphorylation of the substrate and one from addition of inorganic phosphate. Since 2 ATP were spent originally to phosphorylate glucose, the net gain of energy carriers is two molecules of NADH plus two molecules of ATP.

Most of the conversion steps are associated with a small change in energy—so small that the sign of ΔG depends on the concentrations of substrates or products; thus, some individual steps are reversible. In the cytoplasm, however, as intermediate products form they are quickly consumed by the next step, and so the pathway flows in one direction. The direction of flow is determined by the key irreversible reactions that consume ATP. These steps "prime" the pathway by spending energy.

Phosphorylation and splitting of glucose. In the first stage of the EMP pathway, the six-carbon sugar is activated by two phosphorylation steps (**Fig. 13.17**). The first phosphoryl group (phosphate) is added at carbon 6 of glucose. (In some species of bacteria, the first phosphoryl group is added by phosphoenolpyruvate instead of ATP, but the net effect is the same.) The next enzyme-catalyzed step, rearrangement of glucose 6-phosphate to fructose 6-phosphate, involves no significant change in energy but prepares the sugar to receive the second phosphoryl group, forming fructose 1,6-bisphosphate.

The sugar then splits into two three-carbon sugars (trioses), each tagged with one of the two phosphates. The splitting of this molecule has a favorable entropy change (ΔS), but its chemical change (ΔH) is unfavorable, largely canceling out the energy yield. The two triose phosphates, glyceraldehyde 3-phosphate and dihydroxyacetone phosphate, have nearly the same energy value, so an enzyme interconverts them reversibly. Interconversion is necessary because only glyceraldehyde 3-phosphate proceeds further in the pathway.

> **NOTE:** In Chapters 13–16, every substrate conversion shown requires catalysis by an enzyme. For the EMP pathway, the enzymes are shown, but for other pathways, the enzyme names are omitted.

ATP generation. In the second stage of the EMP pathway, each three-carbon glyceraldehyde 3-phosphate is directed into an energy-yielding pathway to pyruvate (**Fig. 13.17**). First, the aldehyde (R-CHO) is converted to the carboxylate (R-COO⁻ + H⁺). Conversion to a carboxylate releases a substantial amount of energy (negative ΔG). This oxidation of the aldehyde represents the major source of energy in glycolysis—the step at which the energy obtained is used to transfer a pair of electrons onto NAD^+, forming NADH with an ionized H⁺. In addition, sufficient energy is released to add a phosphoryl

Figure 13.16   Energy changes during the Embden-Meyerhof-Parnas pathway (glycolysis). Glucose is activated through two substrate phosphorylations by ATP. The breakdown of glucose to two molecules of pyruvate is coupled to net production of two ATP and two NADH. Phosphoryl groups are shown as (P).

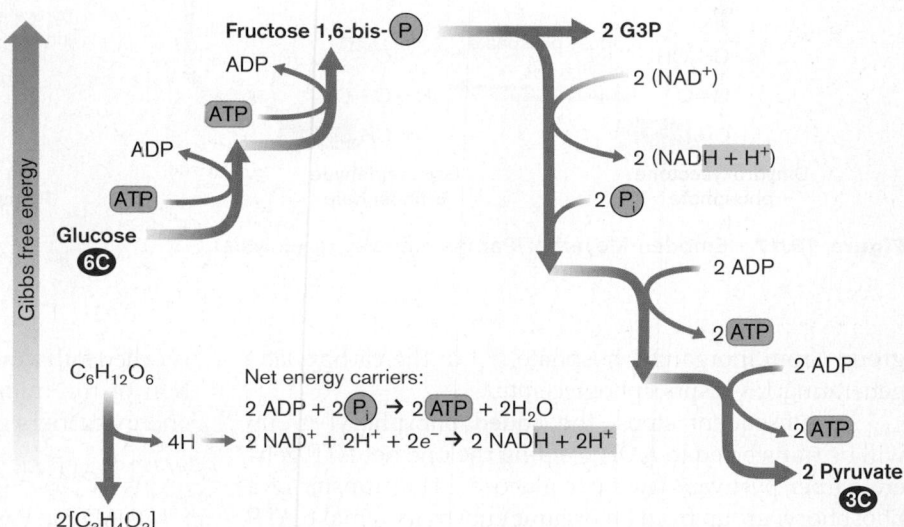

Net energy carriers:
2 ADP + 2 (P$_i$) → 2 (ATP) + 2H$_2$O
2 NAD⁺ + 2H⁺ + 2e⁻ → 2 NADH + 2H⁺

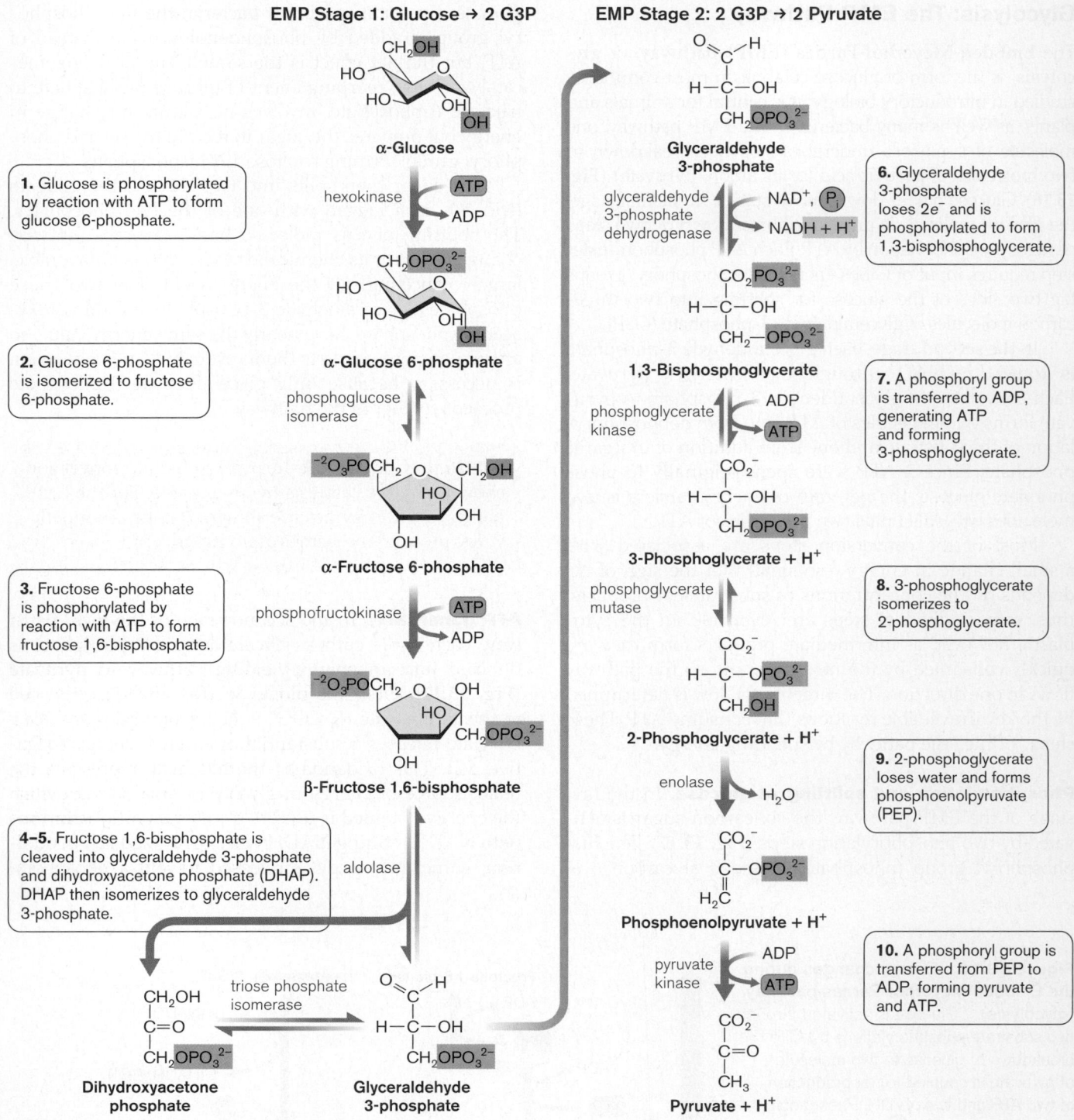

Figure 13.17 **Embden-Meyerhof-Parnas pathway (glycolysis).**

group (from inorganic phosphate, P_i) to the carboxylate, generating 1,3-bisphosphoglycerate.

In subsequent steps, the added phosphoryl group will be transferred to ADP, yielding the one net ATP generated per pyruvate (two per glucose). The transfer of a phosphoryl group from an organic substrate to make ATP

is called **substrate-level phosphorylation**. With subtraction of the initial two ATP molecules invested, the net energy carriers gained from each glucose are as follows:

$$2 \text{ NAD}^+ \rightarrow 2 \text{ NADH} + 2\text{H}^+$$

$$2 \text{ ADP} + P_i \rightarrow 2 \text{ ATP} + 2\text{H}_2\text{O}$$

> **THOUGHT QUESTION 13.7** In the EMP pathway, how are the two water molecules generated?

Regulation of glycolysis. Enzymes of catabolism are regulated at the level of transcription of the enzyme. In addition, the activities of certain enzymes in long pathways require allosteric regulation. Allosteric regulation by enzyme substrates and products ensures that excess intermediates do not build up and avoids release of more energy than the cell can use at a given time. Glycolysis is regulated so that its reactions go forward only when the cell needs energy, not when the cell is trying to synthesize glucose. The regulation occurs at steps where the products are consumed so rapidly that their forward reaction is effectively irreversible. Irreversible steps are shown as unidirectional arrows in **Figure 13.17**.

> **THOUGHT QUESTION 13.8** Some bacteria make an enzyme, dihydroxyacetone kinase, that phosphorylates dihydroxyacetone to dihydroxyacetone phosphate. Why would this enzyme be useful?

Enzymes that catalyze irreversible steps are regulated so as to maintain consistent levels of intermediates in the pathway. The most important irreversible reaction in glycolysis is the phosphorylation of fructose 6-phosphate to fructose 1,6-bisphosphate, mediated by the enzyme phosphofructokinase. This enzyme is activated allosterically by ADP and inhibited by ATP or by the alternative phosphoryl donor, phosphoenolpyruvate.

What happens when the cell needs to reverse glycolysis in order to make glucose? Most of the intermediate reactions are reversible, and so the same enzymes can be used for biosynthesis. Pathways that participate in both catabolism (breakdown) and anabolism (biosynthesis) are called **amphibolic**.

An amphibolic pathway such as glycolysis includes enzymes such as phosphoglucose isomerase that can run in either direction. The direction of the pathway at a given time is determined by key enzymes that operate only in the catabolic direction or in the anabolic direction. For example, in glycolysis the ATP phosphorylation of fructose 6-phosphate is catalyzed by the enzyme phosphofructokinase, whereas in biosynthesis this step is reversed by a different enzyme, fructose bisphosphatase. Instead of regenerating ATP, fructose bisphosphatase removes the second phosphoryl group as inorganic phosphate—a step yielding energy and thus driving the whole pathway in reverse (toward biosynthesis of sugar). The two enzymes are regulated differently; the catabolic enzyme phosphofructokinase is activated by ADP, a signal of energy need, whereas the biosynthetic enzyme is inhibited by such signals.

The Entner-Doudoroff Pathway

The **Entner-Doudoroff (ED) pathway** offers a slightly different route to catabolize sugars, as well as sugar acids (sugars with acidic side chains). The ED pathway was originally studied for its role in production of the Mexican beverage *pulque*, or "cactus beer," by *Zymomonas* fermentation of the blue agave plant. More recently, Tyrrell Conway and colleagues have found genes encoding the Entner-Doudoroff enzymes in the genomes of numerous bacteria and archaea. In the human colon, the ED pathway enables *E. coli* and other enteric bacteria to feed on mucus secreted by the intestinal epithelium (**Fig. 13.18**). Some gut flora, such as *Bacteroides thetaiotaomicron*, actually induce colonic production of the mucus that they consume. These bacteria that "farm" intestinal mucus may enhance human health by preventing colonization by pathogens.

The ED pathway probably evolved earlier than the EMP pathway, as it involves fewer substrate phosphorylation steps and produces less ATP, and it is found in a wider range of prokaryotes. As in the EMP pathway, glucose is phosphorylated to glucose 6-phosphate (**Fig. 13.19**). The next step, however, involves oxidation by NAD^+ at carbon 1, with loss of two hydrogens to form 6-phosphogluconate. Gluconate is a sugar acid found in intestinal mucus; it can be phosphorylated to enter the ED pathway.

The hydrogens and electrons from glucose 6-phosphate are transferred to $NADP^+$, instead of NAD^+ as in the EMP pathway. This step differs from the EMP pathway in two respects: The carrier used is $NADP^+$ instead of NAD^+, and the electron transfer occurs early, without a second ATP-consuming phosphorylation step. When the six-carbon substrate is eventually split into two three-carbon products, one of the three-carbon products is glyceraldehyde 3-phosphate, which enters the second stage of glycolysis (see **Fig. 13.17**). NADH is made,

A.   B.

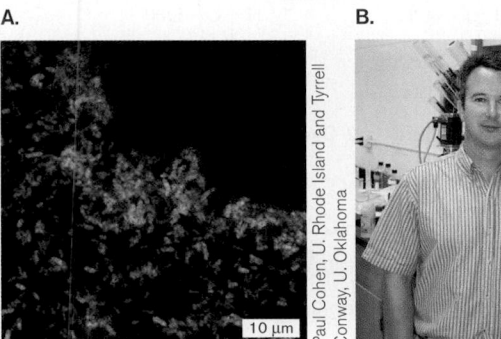

Paul Cohen, U. Rhode Island and Tyrrell Conway, U. Oklahoma

Courtesy of Tyrrell Conway

Figure 13.18 Intestinal bacteria use the Entner-Doudoroff pathway. A. Intestinal *E. coli* (orange) feed primarily on gluconate from mucus secretions (fluorescence micrograph). **B.** Tyrrell Conway, at the University of Oklahoma, used genomics and genetic analysis to dissect the role of the Entner-Doudoroff pathway in the enteric bacterial catabolism of sugar acids from intestinal mucus.

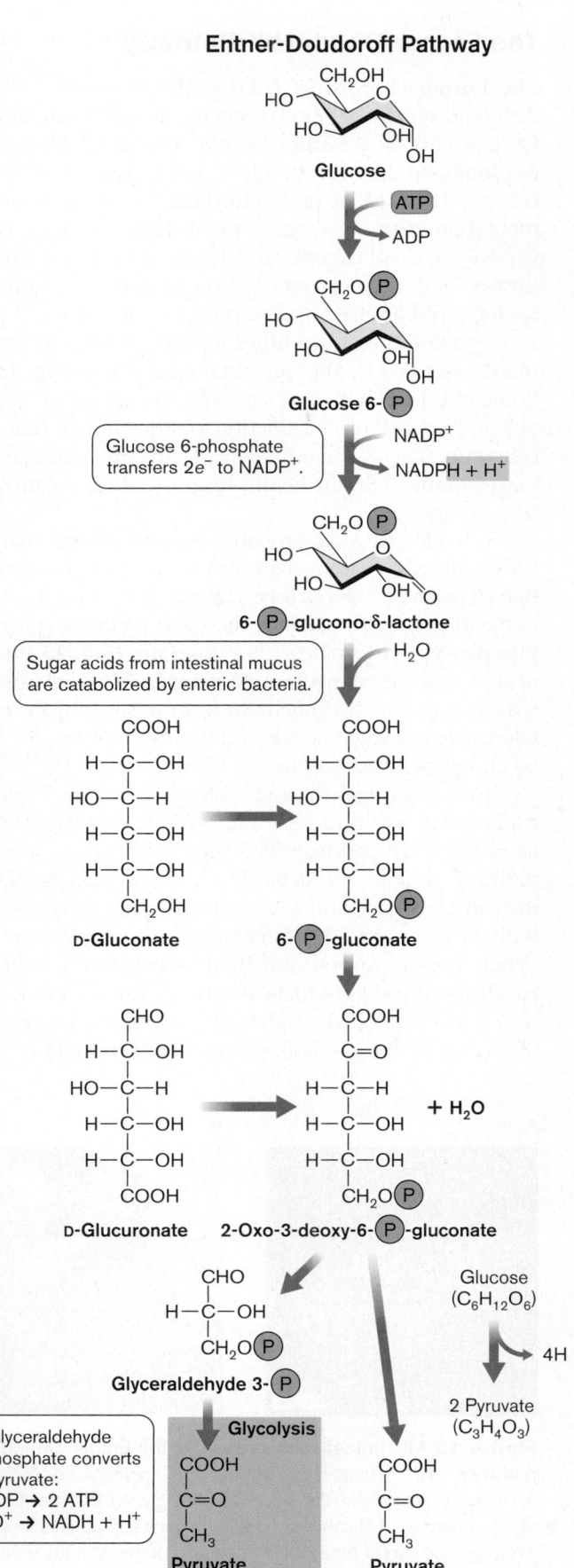

and two ATP are made by substrate-level phosphorylation. The remaining three-carbon product, however, is pyruvate. This one-step production of pyruvate short-circuits the catabolic pathway, missing the formation of an ATP. The unused potential energy is released as waste heat.

The net ATP gain from the Entner-Doudoroff pathway is only one ATP per molecule of glucose, half that of the EMP pathway (see **Fig. 13.15**). The electrons transferred, however, are equivalent: Instead of two molecules of NADH, the Entner-Doudoroff pathway generates one NADH and one NADPH.

> **THOUGHT QUESTION 13.9**   Explain why the ED pathway generates only one ATP, whereas the EMP pathway generates two.

The Pentose Phosphate Shunt

A third pathway of glucose catabolism is the **pentose phosphate shunt (PPS)**, which forms the key intermediate **ribulose 5-phosphate**, a five-carbon sugar. The pentose phosphate shunt generates one ATP with no NADH, but two NADPH for biosynthesis (**Fig. 13.20**). In addition, PPS generates a complex series of intermediates that can be redirected as substrates for biosynthesis of diverse cell components such as amino acids and vitamins.

The pentose phosphate shunt starts like the Entner-Doudoroff pathway: Glucose 6-phosphate gives up two electrons to form NADPH and is oxidized to 6-phosphogluconate. Thus, only one net ATP is gained from breakdown to pyruvate. The next step involves a second oxidation to NADPH, with loss of a carbon as CO_2. The loss of CO_2 generates the five-carbon sugar ribulose 5-phosphate (hence the pathway name "pentose phosphate"). In succeeding steps, pairs of sugars, such as sedoheptulose 7-phosphate and glyceraldehyde 3-phosphate, exchange short carbon chains, giving rise to sugar phosphates of various lengths—for example, ribose 5-phosphate and erythrose 4-phosphate, which are precursors of purines and aromatic amino acids, respectively. Alternatively, if these routes to biosynthesis are not taken, the

Net energy carriers:

$ADP + \boxed{P_i} \rightarrow \boxed{ATP} + H_2O$

$NADP^+ + 2H^+ + 2e^- \rightarrow NADPH + H^+$

$NAD^+ + 2H^+ + 2e^- \rightarrow NADH + H^+$

Figure 13.19   Entner-Doudoroff pathway. Glucose 6-phosphate is oxidized to 6-phosphogluconate, with one pair of electrons transferred to NADPH. The 6-phosphogluconate is dehydrated and cleaved to form one pyruvate plus one glyceraldehyde 3-phosphate that enters the EMP pathway to pyruvate.

Pentose Phosphate Shunt Pathway

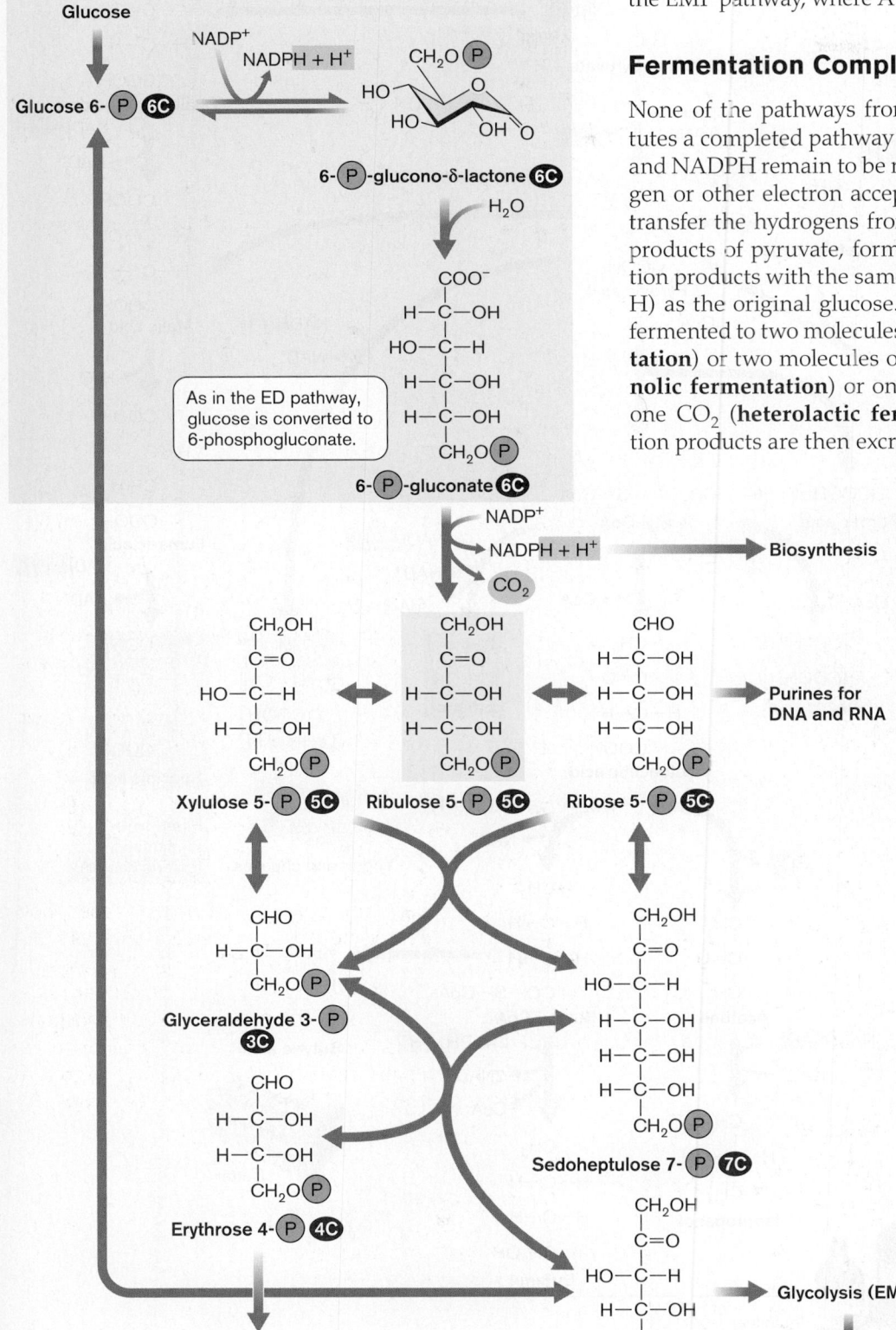

As in the ED pathway, glucose is converted to 6-phosphogluconate.

intermediates convert to fructose 6-phosphate and reenter the EMP pathway, where ATP and NADH are produced.

Fermentation Completes Catabolism

None of the pathways from glucose to pyruvate constitutes a completed pathway of catabolism, because NADH and NADPH remain to be recycled. In the absence of oxygen or other electron acceptors, heterotrophic cells must transfer the hydrogens from NADH + H$^+$ back onto the products of pyruvate, forming partly oxidized fermentation products with the same redox level (balance of O and H) as the original glucose. For example, glucose may be fermented to two molecules of lactic acid (**lactate fermentation**) or two molecules of ethanol plus two CO$_2$ (**ethanolic fermentation**) or one lactic acid, one ethanol, and one CO$_2$ (**heterolactic fermentation**). These fermentation products are then excreted from the cell.

Figure 13.20 Pentose phosphate shunt pathway of glucose catabolism. Like the Entner-Doudoroff pathway, the pentose phosphate shunt forms 6-phosphogluconate. One CO$_2$ is released, and one NADPH is produced for biosynthesis. The pathway can generate ribose 5-phosphate for purine synthesis or erythrose 4-phosphate to synthesize aromatic amino acids.

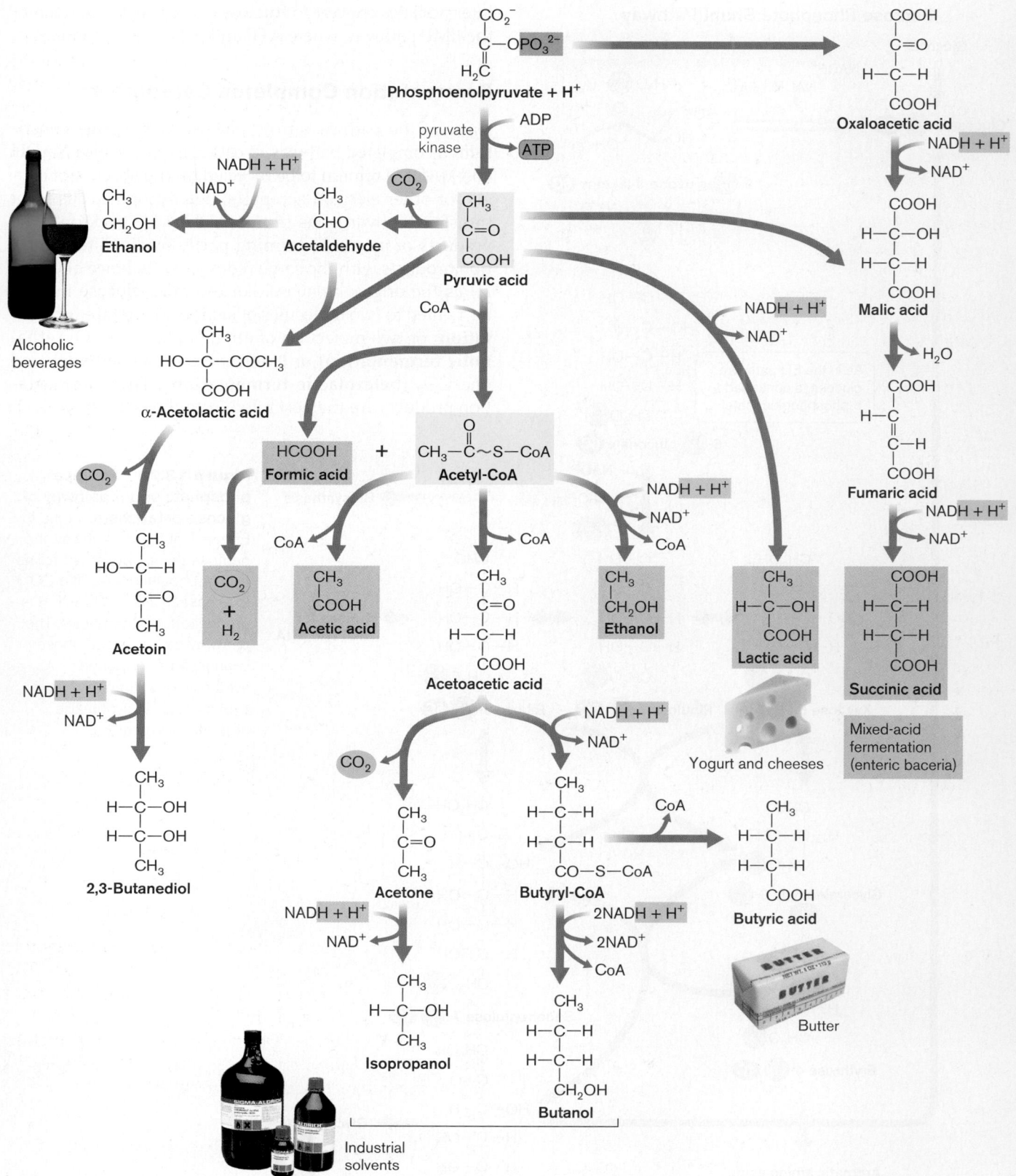

Figure 13.21 Fermentation pathways. Alternative pathways from pyruvate and phosphoenolpyruvate to end products, many of which we use for food or industry. Different species conduct different portions of the pathways shown. *Sources:* Wine: ©Food Collection/SuperStock; Butter: D. Hurst/Alamy; Solvents: Sigma-Aldrich Corporation; Cheese: ©Lee Hacker/Alamy.

The large quantities of substrate consumed in fermentation generate large amounts of waste products to be excreted. These bacterial wastes are actually useful to human "fermentation industries" such as the production of alcoholic beverages (ethanolic fermentation) or cheese (lactate fermentation).

Why do fermenting bacteria give up such large quantities of waste products that retain usable energy? The reason is that in the absence of oxygen (or another electron acceptor), the fermentation products cannot yield energy. Most fermentation pathways do not generate ATP beyond that produced by substrate-level phosphorylation, the direct transfer of a phosphate group from an organic phosphate to ADP. An example is the final step of glycolysis, catalyzed by pyruvate kinase (see **Fig. 13.17**). Much of the energy available from glucose remains unspent or is lost as heat. Nevertheless, fermentation is essential for microbes in environments such as anaerobic soil or animal digestive tracts. Even aerated cultures of bacteria start fermenting once their demand for oxygen exceeds the rate of oxygen dissolving in water. Microbes compensate for the low efficiency of fermentation by consuming large quantities of substrate and excreting large quantities of fermentation products.

Mixed-acid fermentation. Numerous different pathways have evolved to dispose of the waste in different forms (**Fig. 13.21**). *E. coli* ferments by a combination of routes known collectively as **mixed-acid fermentation**, forming acetate, formate, lactate, and succinate, as well as ethanol, H_2, and CO_2. Hydrogen (H_2) and CO_2 are the main gases passed by the human colon.

Colonic H_2 plus CO_2 can yield further energy for methanogenic microbes, through conversion to methane

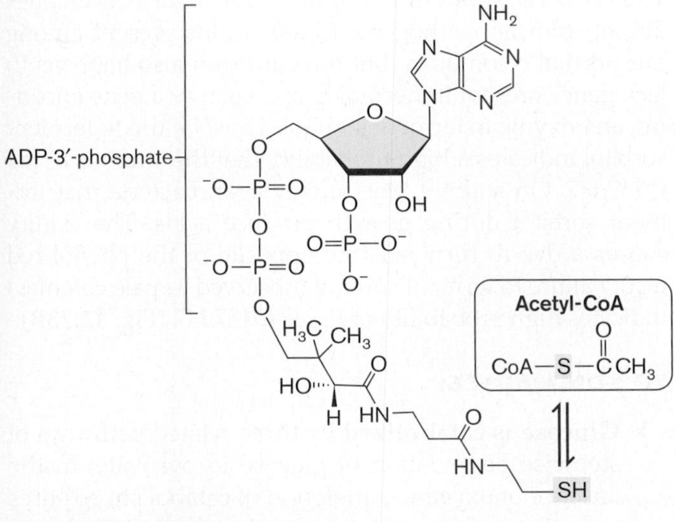

(discussed in Chapter 14.) Excess hydrogen and methane gases can cause problems for medical procedures such as colonoscopy. For colonoscopy, the colon may need to be flushed with a carbohydrate solution, which bacteria may ferment, releasing hydrogen. When a polyp is removed from the colon by electrocautery (high-frequency electric current), the gas may ignite, causing explosion. Colonic explosion can be avoided by flushing out the gases before electrocautery.

During mixed-acid fermentation, the proportions of products vary with pH. Low pH favors ethanol and lactate (which minimize acidification) in preference to formate and acetate. *Clostridium* species produce alcohols (butanol, isopropanol); *Porphyromonas gingivalis*, a cause of periodontal disease, produces short-chain acids (propionate, butyrate).

Many fermentation products share key intermediates, such as acetyl-CoA. Acetyl-CoA is a versatile two-carbon intermediate, the "Lego block" of metabolism. It consists of an acetyl group esterified to **coenzyme A** (**CoA**; **Fig. 13.22**), a famous coenzyme whose discovery won Fritz Lipmann the 1953 Nobel Prize in Physiology or Medicine with Hans Krebs. CoA has a thiol (–SH) that exchanges its hydrogen for an acyl group, thus activating the molecule for transfer in various metabolic pathways.

For example, the enzyme pyruvate formate lyase splits pyruvate to form acetyl-CoA plus formate:

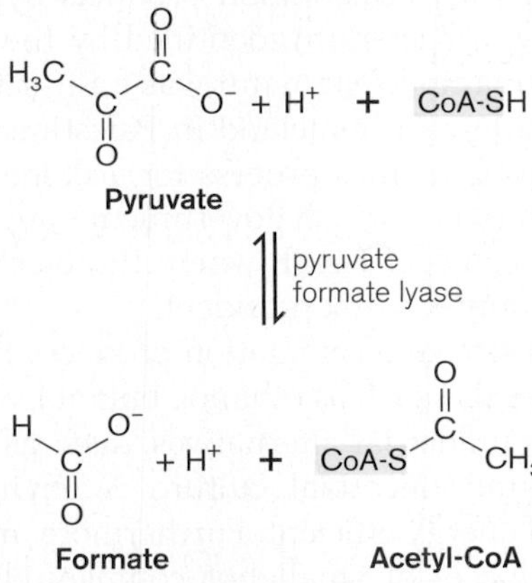

The SH group of CoA is indicated to emphasize its role in accepting the acetyl group to form acetyl-CoA. The hydrogen ($H^+ + e^-$) of the SH group is transferred onto the carboxyl carbon of pyruvate, generating formate.

Acetyl-CoA can be converted to various fermentation products by several different pathways. The simplest is water exchange with CoA, forming acetate:

$$CH_3CO-S-CoA + H_2O \rightleftharpoons$$
$$CH_3COO^- + H^+ + HS-CoA$$

Incorporation of water restores the hydrogen to the thiol of CoA, and the OH to acetate. The acetate thus formed may then be excreted by the cell.

Note that the excreted acids and alcohols are readily recovered by cells when an electron acceptor becomes

Figure 13.22 Structure of coenzyme A. The thiol (SH) forms an ester link with the COOH of acetic acid, generating acetyl-CoA.

available for oxidation or when these fermentation products are needed as building blocks for biosynthesis. Alternatively, the excreted "wastes" may be utilized by other species capable of metabolizing them further. For example, the H_2 released by gut bacteria through mixed-acid fermentation can be oxidized to water by the gastric pathogen *Helicobacter pylori*. Alternatively, H_2 plus CO_2 can be used by gut methanogens to yield energy with release of methane.

Food and industrial applications. The "waste products" of fermentation retain much of their organic structure and food value. Thus, fermentation products have proved extremely useful in human culture and technology. For thousands of years, ethanolic fermentation by yeast has been used to produce wine and beer, while lactate fermentation has been used to produce yogurt and cheese. The minor product butyric acid (butyrate) lends taste to butter. Another route from lactate to propionic acid (propionate) provides the distinctive flavor of Swiss cheese, while CO_2 gas generates the holes, or "eyes" (see **Special Topic 13.1**).

In the chemical industry, microbial fermentation produces industrial solvents such as butanol and acetone. Acetone production had historic impact during World War I, when Britain needed a source of acetone to manufacture gunpowder. Acetone and butanol were produced as fermentation products by *Clostridium acetobutylicum*, a bacterium identified by the biochemist Chaim Weitzmann. Weitzmann was a Russian-born Jew who sought a Jewish homeland in Palestine. At the end of the war, Weitzmann's process for acetone production helped earn him the British government's support for the national homeland of Israel, where the biochemist later became the country's first president.

Today, microbial fermentation produces many "commodity chemicals" such as ethanol, butanol, and glycerol. Compared to industrial alternatives, such as production from petroleum, microbial culture is environmentally desirable and energy efficient. Furthermore, microbes are "smart" producers of small but complex pharmaceuticals such as vitamins, amino acids, and antibiotics, molecules that would require many steps of organic synthesis. Applications of microbial biosynthesis are discussed further in Chapters 15 and 16.

Diagnostic applications. Another important application of fermentation lies in diagnostic microbiology. To quickly identify the microbe causing a disease and prescribe an effective antibiotic, hospitals use rapid and inexpensive biochemical tests. A general test for fermentation pathways is the **phenol red broth test** (**Fig. 13.23A**). Phenol red is a pH indicator, which is orange-red at neutral pH. It turns yellow in media acidified by fermentation acids (below pH 6.8) and red at higher pH (above pH 7.4). A culture of *Escherichia coli*, which ferments quickly, turns

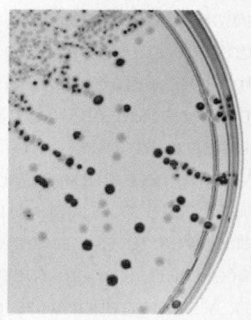

A. Phenol red broth test   **B. Sorbitol MacConkey agar**

Gas →

Dennis Strete, Fundamental Photographs, NYC

Troy Biological BD Worldwide

Figure 13.23   Clinical tests based on fermentation.
A. Phenol red broth test. Organisms are cultured in a tube of broth containing phenol red. The dye turns yellow when protonated at low pH. Phenol red is orange-red at neutral pH, yellow at lower pH (acid), and red at higher pH (base). A small inverted tube (Durham tube) collects gaseous fermentation products (CO_2 and H_2). Left to right: *Escherichia coli* gives acidic fermentation products (yellow) and gas in Durham tube; *Alcaligenes faecalis* does not ferment, tube turns red without gas; uninoculated control, red. **B.** Sorbitol fermentation test for pathogen *E. coli* O157:H7. White colonies (strain O157:H7) fail to ferment sorbitol, unlike red colonies (nonpathogenic *E. coli*).

phenol red to a medium yellow after 24 hours. *Alcaligenes faecalis*, meanwhile, ferments poorly on sugar but converts peptides in the broth to alkaline amines; this culture turns deep red.

More-specific tests depend on the microbe's ability to ferment specific sugars. Different species possess different enzymes able to convert different sugars and sugar derivatives to glucose, which then enters the common fermentation pathway. A well-known example is the use of sorbitol MacConkey agar to test for *E. coli* O157:H7, a lethal pathogen contaminating beef and vegetable produce. The O157:H7 strain of *E. coli* has a set of virulence genes absent from nonpathogenic *E. coli* strains present among our normal colon biota. But the pathogen also happens to lack genes present in normal *E. coli*, such as a gene encoding an enzyme to ferment sorbitol. Thus, failure to ferment sorbitol indicates a high probability that the strain is *E. coli* O157:H7. On sorbitol MacConkey agar, bacteria that ferment sorbitol during growth produce acids. The acidity causes a dye to turn red (the opposite of the phenol red test). Failure to ferment sorbitol (observed as pale colonies) indicates high probability of *E. coli* O157:H7 (**Fig. 13.23B**).

TO SUMMARIZE:

■ **Glucose is catabolized** by three related pathways of stepwise degradation of glucose to pyruvate. In the absence of oxygen, completion of catabolism requires generation of fermentation products.

■ **In the Embden-Meyerhof-Parnas (EMP) pathway**, glucose is activated by two substrate phosphorylations, and then cleaved to two three-carbon sugars. Both sugars eventually are converted to pyruvate. The pathway produces 2 ATP and 2 NADH.

- **In the Entner-Doudoroff (ED) pathway**, glucose is activated by one phosphorylation, and then dehydrogenated to 6-phosphogluconate. 6-Phosphogluconate is cleaved to pyruvate and a three-carbon sugar, which enters the EMP pathway to form pyruvate. The ED pathway produces 1 ATP, 1 NADH, and 1 NADPH.

- **In the pentose phosphate shunt**, 6-phosphogluconate is converted to a series of sugars, each containing three to seven carbons—including fructose 6-phosphate, which reenters the EMP pathway.

- **Intermediates of sugar catabolism may serve as substrates for biosynthesis** of amino acids and other cell components.

- **Glucose catabolism is reversible**, enabling cells to build glucose from small molecules. Glucose biosynthesis uses some enzymes of glycolysis in reverse, but also requires enzymes that bypass irreversible steps in the catabolic pathway.

- **Fermentation is the completion of catabolism** *without* the electron transport system and a terminal electron acceptor. The electrons from NADH are restored to pyruvate or its products in reactions that generate fermentation products, including alcohols and carboxylates, as well as H_2 and CO_2. Fermentation has applications in food, industrial, and diagnostic microbiology.

- **Acetyl-CoA is a key intermediate** in several fermentation pathways.

13.6 The Tricarboxylic Acid (TCA) Cycle

In the presence of O_2 or another terminal electron acceptor, the products of sugar breakdown can be catabolized to CO_2 and H_2O through the **tricarboxylic acid (TCA) cycle**. The TCA cycle is also known as the Krebs cycle, named for Hans Krebs (1900–1981), who shared the 1953 Nobel Prize in Physiology or Medicine with Fritz Lipmann. Krebs and his colleagues at Sheffield University, England, studied catabolism by observing the oxidizing activities of crude enzyme preparations from sources such as pigeon breast muscle, beef liver, and cucumber seeds. In all of these animal and plant tissues, the TCA cycle is conducted by mitochondria, using virtually the same process as their bacterial ancestors.

Glucose catabolism connects with the TCA cycle through pyruvate breakdown to acetyl-CoA and CO_2. Recall from Section 13.4 that acetyl-CoA is also generated from many other catabolic pathways, including those for breakdown of fatty acids, amino acids, and even benzoate derivatives. Regardless of its source, acetyl-CoA enters the TCA cycle by condensing with the four-carbon intermediate oxaloacetate to form citrate (**Fig. 13.24**). Citrate undergoes two steps of oxidative decarboxylation, in which CO_2 is released and two hydrogens with electrons are transferred to make NADH + H$^+$ or FADH$_2$. The TCA cycle, in whole or in part, is found in all microbial species except for degenerately evolved pathogens dependent on host metabolism.

We present first the connecting step between pyruvate and acetyl-CoA, followed by the details of the TCA cycle. Bacteria and archaea use at least ten known variations on

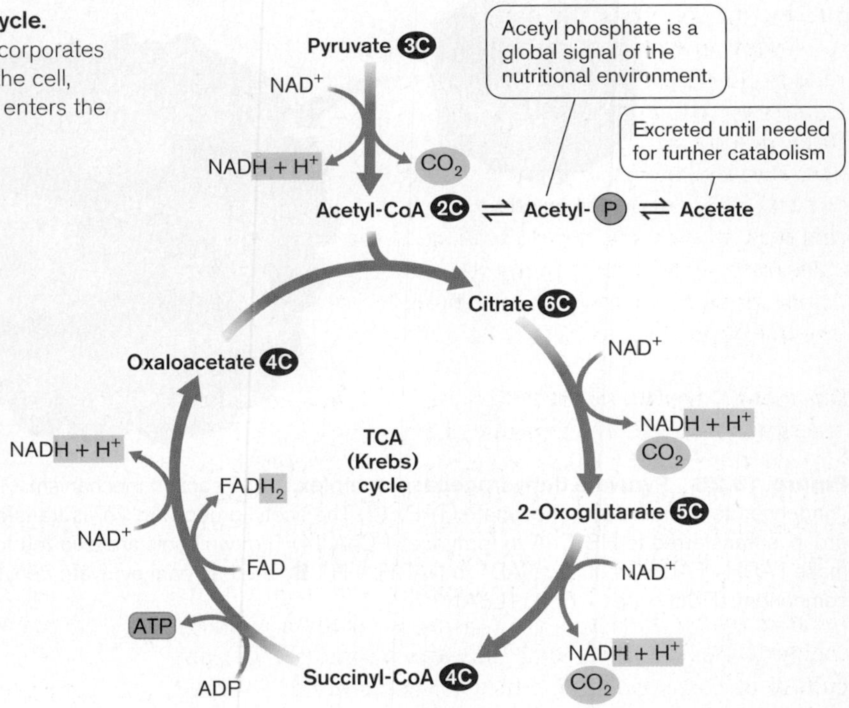

Figure 13.24   Acetyl-CoA feeds into the TCA cycle. Pyruvate undergoes oxidative decarboxylation and incorporates CoA to form acetyl-CoA. Depending on the state of the cell, acetyl-CoA is converted to acetate for excretion or it enters the TCA cycle.

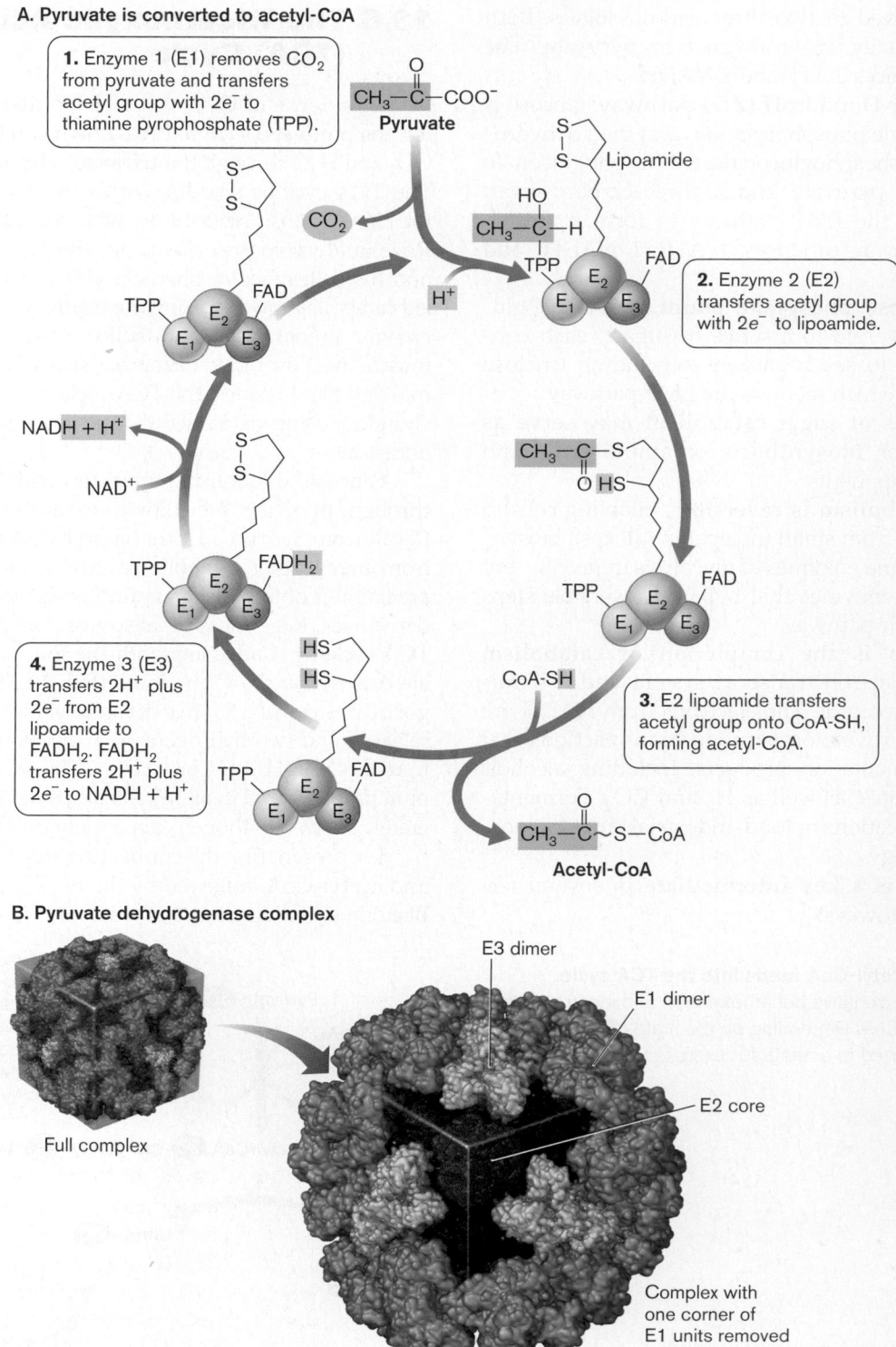

A. Pyruvate is converted to acetyl-CoA

1. Enzyme 1 (E1) removes CO_2 from pyruvate and transfers acetyl group with $2e^-$ to thiamine pyrophosphate (TPP).

2. Enzyme 2 (E2) transfers acetyl group with $2e^-$ to lipoamide.

3. E2 lipoamide transfers acetyl group onto CoA-SH, forming acetyl-CoA.

4. Enzyme 3 (E3) transfers $2H^+$ plus $2e^-$ from E2 lipoamide to $FADH_2$. $FADH_2$ transfers $2H^+$ plus $2e^-$ to NADH + H^+.

B. Pyruvate dehydrogenase complex

Full complex

E3 dimer

E1 dimer

E2 core

Complex with one corner of E1 units removed to reveal core

Figure 13.25 Pyruvate dehydrogenase complex. A. Reaction mechanism: (1) Enzyme 1 acquires the acetyl group of pyruvate, condensed to thiamine pyrophosphate (TPP). (2) The acetyl group with $2e^-$ is transferred to the lipoamide of enzyme 2. (3) The acetyl group is transferred to HS-CoA to form acetyl-CoA. (4) The two thiols are oxidized to disulfide, and their $2e^- + 2H^+$ reduce FAD to make $FADH_2$. $FADH_2$ reduces NAD^+ to NADH + H^+. **B.** The bacterial pyruvate dehydrogenase complex contains multiple copies of each component. (PDB codes: 1EAA, 1L8A)

the TCA cycle, conducted by diverse species under various environmental conditions. You will be relieved to hear that we present only one: the Krebs pathway found in most Gram-negative and Gram-positive bacteria, and in most eukaryotes. Variant pathways are detailed in online resources such as the KEGG Pathway Database.

www | KEGG
microbiology2.com/links

Pyruvate Dehydrogenase Connects Sugar Catabolism to the TCA Cycle

Pyruvate is converted to acetyl-CoA through removal of CO_2 and transfer of $2e^-$ onto NAD^+. The removal of CO_2 and transfer of two electrons is known as oxidative decarboxylation. The oxidative decarboxylation of pyruvate, coupled to CoA incorporation, is performed by an unusually large multisubunit enzyme called the **pyruvate dehydrogenase complex**, or **PDC** (**Fig. 13.25**). PDC is a key component of metabolism in bacteria and mitochondria, the first molecular player to direct sugar catabolism into respiration. In human mitochondria, defects in PDC affect organs with high metabolic rate, such as heart and brain, causing myocardial malfunction and heart failure, and neurodegeneration.

Reaction mechanism of PDC. The overall reaction catalyzed by PDC is

$$CH_3COCOO^- + H^+ + HS\text{-}CoA \rightarrow$$
$$CH_3CO\text{-}S\text{-}CoA + CO_2 + 2H^+ + 2e^-$$
$$NAD^+ + 2H^+ + 2e^- \rightarrow NADH + H^+$$

The removal of stable CO_2 yields energy for the electron transfer to NADH. The protons dissociated from pyruvate and thiol (SH) of CoA ($2H^+$ in total) are balanced by the net gain of protons by $NADH + H^+$.

The coupling of all the molecular transfers actually requires three distinct enzyme components within the complex, designated E1, E2, and E3 (see **Fig. 13.25**). Each of these enzymes possesses a distinct coenzyme to mediate electron transfer.

The bacterial pyruvate dehydrogenase complex contains multiple copies of all three enzymes (**Fig. 13.25B**). Central to the structure is a core containing 24 E2 proteins surrounded by 12 E1 dimers and 6 E3 dimers. Mitochondrial complexes are even larger and more intricate. The reason for this degree of complexity is unknown, but it is clear that organisms invest considerable energy and material in the construction and function of PDC.

Regulation of PDC. The activity of PDC is increased by high concentrations of its substrates, CoA and NAD^+; and inhibited by its products, acetyl-CoA and NADH. The product acetyl-CoA may enter one of several pathways.

In *E. coli*, when glucose is plentiful, acetyl-CoA is mostly converted to acetate via the intermediate acetyl phosphate (see **Fig. 13.24**). Acetyl phosphate is a global signal molecule that indicates to the cell the quantity and quality of carbon source available. As glucose decreases, the cell starts to take back acetate, converting it back to acetyl-CoA for entry into the TCA cycle.

At the level of gene expression, PDC responds to environmental conditions. As would be expected, PDC gene expression is repressed by carbon starvation and by low levels of oxygen. More surprising, expression is strongly induced both at high pH and at low pH, compared to growth at pH 7. The basis for extreme-pH induction is unknown.

> **THOUGHT QUESTION 13.10** Compare the reactions catalyzed by pyruvate dehydrogenase and pyruvate formate lyase (see Section 13.5). What conditions favor each reaction, and why?

Acetyl-CoA Enters the TCA Cycle

Acetyl-CoA enters the TCA cycle by condensing the acetyl group with oxaloacetate, a four-carbon dicarboxylate (double acid). The condensation forms citrate, a six-carbon tricarboxylate (**Fig. 13.26**). An advantage of intermediates with two or more acidic groups is that the concentration of the fully protonated form is extremely low; thus, the molecule is unlikely to be lost from the cell by diffusion across the membrane, as are monocarboxylic acids, such as acetate. Through the rest of the cycle, citrate loses two carbons as CO_2 by a series of reactions that transfer increments of energy to 3 NADH, $FADH_2$, and ATP. Each reaction step couples energy-yielding to energy-storing events.

Step 1. As the acetyl group condenses with oxaloacetate, the removal of CoA consumes a molecule of H_2O to restore HS-CoA. The removal of HS-CoA yields energy to incorporate acetate into oxaloacetate, forming citrate.

Step 2. Citrate undergoes two rearrangements with little energy change to form isocitrate. Isocitrate then undergoes oxidative decarboxylation. As we saw for pyruvate, removal of CO_2 yields energy to transfer $2H^+ + 2e^-$ to form $NADH + H^+$, producing 2-oxoglutarate (alpha-ketoglutarate).

Step 3. 2-Oxoglutarate undergoes decarboxylation to release CO_2 and make another $NADH + H^+$. In this case, CoA is incorporated, making succinyl-CoA.

Step 4. Succinyl-CoA releases CoA, providing energy to phosphorylate ADP to ATP. To form fumarate, $2H^+ + 2e^-$ are transferred to FAD to form $FADH_2$—a reaction involving negligible free energy change. FAD is reduced instead of NAD^+ because electron donation from succinyl-CoA does not yield enough energy to reduce NAD^+.

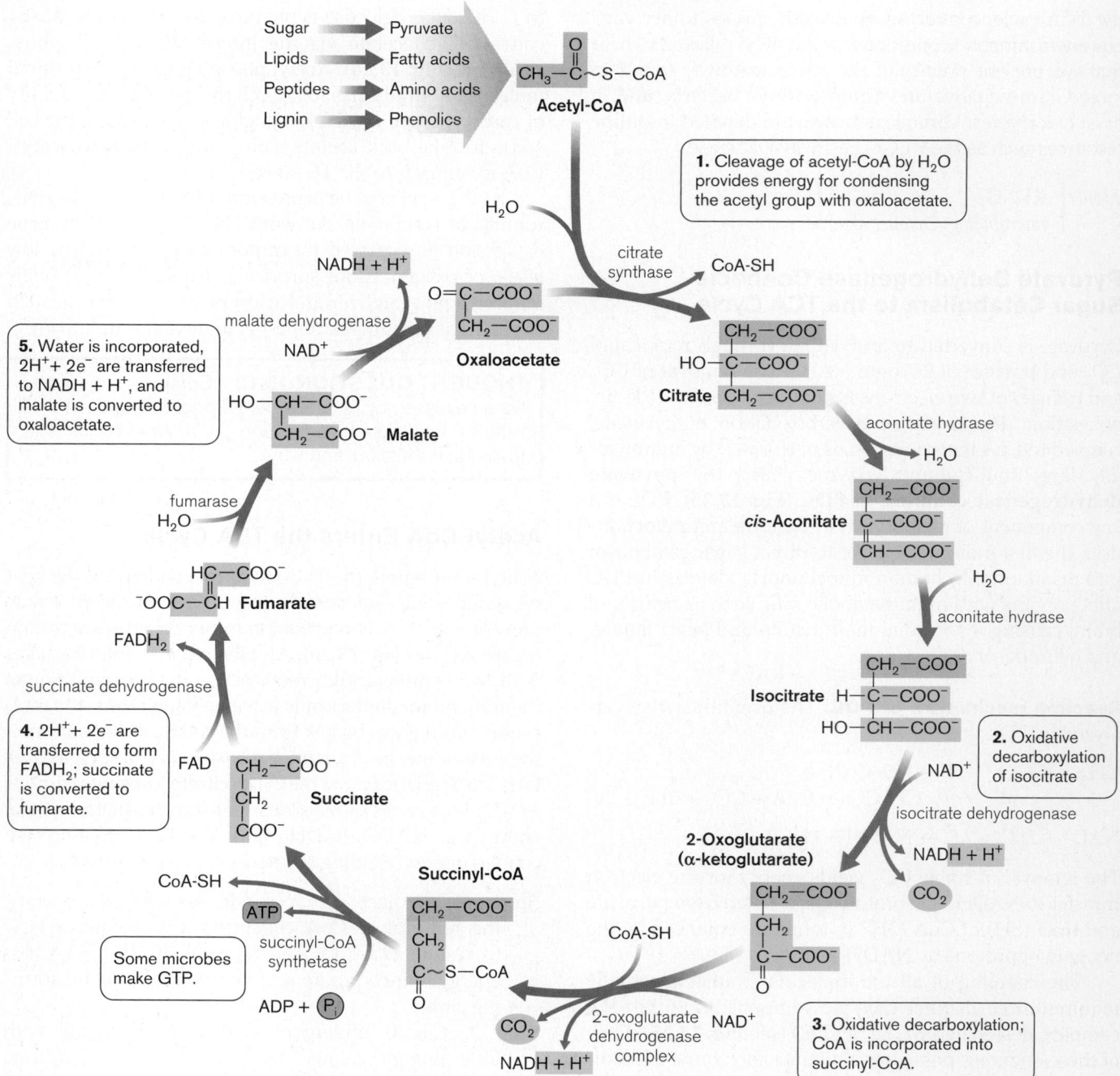

Figure 13.26 The tricarboxylic acid (TCA) cycle. Acetyl-CoA derived from pyruvate and other catabolic pathways enters the TCA cycle. Blue highlights show the fate of labeled acetate incorporated into the TCA cycle. (1) The acetyl group condenses with four-carbon oxaloacetate to produce citrate, a tricarboxylic acid. (2) Citrate rearranges to isocitrate, which is decarboxylated and transfers $2H^+ + 2e^-$ to form NADH + H$^+$. (3) 2-Oxoglutarate is decarboxylated and transfers $2H^+ + 2e^-$ to form NADH + H$^+$ while incorporating CoA to form succinyl-CoA. (4) Succinate is symmetrical; thus, the "old" and "new" carbons are now indistinguishable. Succinate transfers $2H^+ + 2e^-$ to form FADH$_2$, forming fumarate. (5) Water is incorporated, and $2H^+ + 2e^-$ are transferred to form NADH + H$^+$, forming oxaloacetate, ready to take up another acetyl group.

Step 5. Fumarate incorporates water across its double bond, forming the hydroxy acid malate. The increasing stability from fumarate to malate and from malate to oxaloacetate releases enough energy to form the final $NADH + H^+$. Oxaloacetate is the intermediate of lowest energy in the cycle, now ready to accept the next acetyl-CoA.

NOTE: Some textbooks state that succinyl-CoA synthetase, the enzyme catalyzing step 4 in the TCA cycle, phosphorylates GDP to GTP. According to the primary literature, ADP phosphorylation predominates in *E. coli* (Margaret Birney et al., 1996, *J. Bacteriol.* 178:2883), whereas in *Pseudomonas* species, various nucleotide diphosphates are phosphorylated (Vinayak Kapatral et al., 2000, *J. Bacteriol.* 182:1333). Human mitochondria have two enzymes, which form ATP and GTP, respectively (David Lambeth et al., 2004, *J. Biol. Chem.* 279:36621).

Observing the TCA cycle intermediates. How were all the TCA intermediates identified? The main experimental approach available to Krebs and his contemporaries was to guess at dozens of short-chain acids known to exist in cells, and then add each individually to an enzyme preparation and test for TCA cycle activity. In aerobic organisms, the TCA cycle is tightly tied to respiration, so an increase in uptake of oxygen signaled a TCA intermediate. A major experimental advance was the use of tracer isotopes such as ^{14}C, discovered by Martin Kamen in 1940.

Radioisotope tracers answered an important question about the TCA cycle. As the two acetyl carbons cycle through, which two carbons of each intermediate are removed as CO_2? Are the acetyl carbons lost first, or those of the original oxaloacetate? This question was answered by use of substrates radiolabeled with ^{14}C (**Fig. 13.26**, highlighted in blue). In the experiment, bacteria are fed ^{14}C-radiolabeled acetate, which enters the TCA cycle. The radiolabeled carbons are captured by the TCA intermediates and retained into the next cycle, whereas two COOH carbons from the original oxaloacetate are lost. Thus, the four-carbon intermediate does not recycle intact, but breaks down and re-forms each time it passes through the cycle.

Loss of the second CO_2 forms succinyl-CoA, which is converted to succinate. Succinate is a symmetrical molecule; thus, the former identities of the acetyl and oxaloacetate moieties are now erased, and the carbons are equally likely to disappear from either half of the molecule. Further conversion steps regenerate oxaloacetate—half its carbon from the acetyl group and half from the original oxaloacetate.

THOUGHT QUESTION 13.11 Suppose a cell is pulse-labeled with ^{14}C-acetate (fed the ^{14}C label briefly, then "chased" with unlabeled acetate). Can you predict what will happen to the level of radioactivity observed in isolated TCA intermediates? Plot a curve showing your predicted level of radioactivity as a function of the number of rounds of the cycle.

The TCA cycle and oxidative phosphorylation. In all, each acetate generates three NADH molecules, one $FADH_2$, and one ATP; and all the carbons from pyruvate (ultimately from glucose) have been released as waste CO_2. From the standpoint of the carbon skeleton, the glucose breakdown is now complete. But do we have a completed metabolic pathway? No, because all of the NADH and $FADH_2$ need to be recycled by donating their electrons onto a terminal electron acceptor.

The process of electron transfer from NADH and $FADH_2$ is mediated by a series of cell membrane proteins called the electron transport system (ETS) or electron transport chain (introduced in Section 13.3). The membrane proteins use the energy of electron transfer to pump protons, generating a gradient of hydrogen ions across the membrane (**Fig. 13.27**)—a process discussed in detail in Chapter 14. Assuming 3 ATP generated per NADH and 2 ATP per $FADH_2$, the hydrogen ion gradient then drives the membrane ATP synthase to synthesize as many as 34 ATP. Another 4 ATP came from glucose breakdown and the TCA cycle (38 total per glucose). The theoretical maximum yield is approached by mitochondria, but bacteria make less ATP; about 20 ATP per glucose are made by a well-aerated culture of *E. coli*. Mitochondria experience a relatively constant intracellular environment, whereas bacteria sacrifice efficiency for flexibility, spending energy to maintain a stable proton potential during extreme changes in external pH and redox levels.

The overall process of electron transport and ATP generation is termed **oxidative phosphorylation**. The overall process of oxidative catabolism from substrate breakdown to oxidative phosphorylation is called respiration. The overall equation for the respiration of glucose is

$$C_6H_{12}O_6 + 6H_2O + 6O_2 \rightarrow 12H_2O + 6CO_2$$

Glucose respiration can generate a relatively large number of ATPs per glucose, far more than fermentation. In bacteria, however, the actual number of ATPs generated varies widely with availability of carbon source and oxygen. For example, as oxygen decreases in the environment, the ability to oxidize NADH decreases, so the cell may make only one or two ATPs per NADH (discussed in Chapter 14).

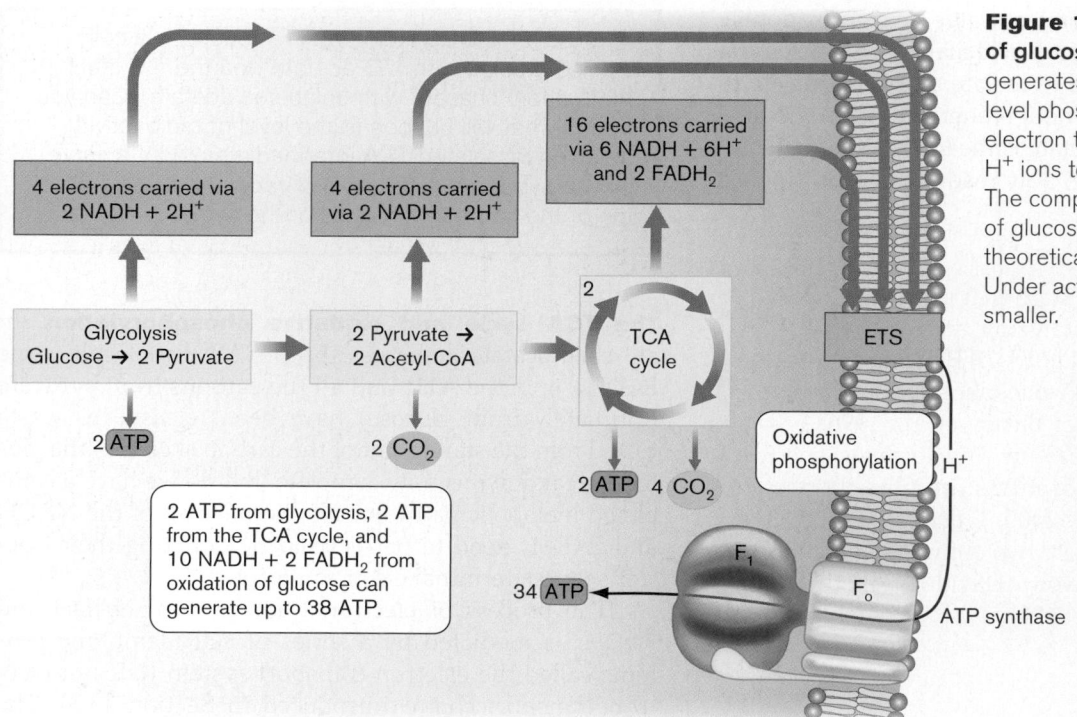

Figure 13.27 Complete oxidation of glucose. Glucose catabolism generates ATP through substrate-level phosphorylation and through the electron transport system pumping H^+ ions to drive the ATP synthase. The complete oxidative breakdown of glucose to CO_2 and H_2O could theoretically generate up to 38 ATP. Under actual conditions, the number is smaller.

Regulation of the TCA cycle. The enzymes of the TCA cycle are regulated extensively by substrate induction and product inhibition, and their expression is induced by high levels of oxygen and glucose. When glucose is absent, cells can catabolize acetate or fatty acids using a modified TCA cycle called the **glyoxylate bypass**. The glyoxylate bypass consists of two enzymes that divert isocitrate to glyoxylate and incorporate a second acetyl-CoA to form malate:

The glyoxylate bypass cuts out all loss of CO_2 and generation of ATP, $FADH_2$, and NADH, with the exception of one NADH from malate to oxaloacetate. Thus, limited energy is released; but, like the wheel of a bicycle in low gear, the cycle keeps turning under tough conditions. The glyoxylate bypass regenerates TCA cycle intermediates and provides substrates to build glucose and carbohydrates (discussed in Chapter 15).

TCA cycle intermediates as substrates for biosynthesis. Analysis of pathway evolution indicates that the TCA cycle originally evolved to provide substrates for biosynthesis. For example, the TCA cycle intermediate

2-oxoglutarate (alpha-ketoglutarate) is aminated to form glutamate, which leads to glutamine. Oxaloacetate is aminated to form aspartate, entering pathways to purines and pyrimidines as well. The TCA cycle, like glycolysis, is an amphibolic pathway that provides substrates for biosynthesis. Many bacteria use the TCA cycle and glycolytic enzymes to build their sugars and amino acids, as discussed in Chapter 15. Others, such as *Treponema pallidum*, the cause of syphilis, have lost the TCA cycle by reductive evolution. They must obtain amino acids from their host organism (see **Special Topic 13.2**).

TO SUMMARIZE:

■ **The pyruvate dehydrogenase complex (PDC)** removes CO_2 from pyruvate, generating acetyl-CoA. Two electrons are transferred to NAD^+, forming NADH + H^+. PDC activity is a key control point of metabolism, induced when carbon sources are plentiful and repressed under carbon starvation and low oxygen.

■ **The tricarboxylic acid cycle (TCA cycle)** converts the acetyl group to $2CO_2$ and $2H_2O$ in the presence of a terminal electron acceptor such as O_2 to receive the electrons associated with the hydrogen atoms.

■ **Acetyl-CoA enters the TCA cycle by condensing with oxaloacetate to form citrate.** A series of enzymes sequentially removes carbon dioxide and water molecules and generates 2 NADH, $FADH_2$, and ATP. Each reaction step couples energy-yielding to energy-storing events.

- **NADH and FADH$_2$ transfer electrons to the electron transport system** and ultimately the terminal electron acceptor, such as O$_2$. Electron transport generates a transmembrane proton potential that powers the membrane-embedded ATP synthase to make ATP.
- **Respiration** consists of substrate catabolism plus oxidative phosphorylation, the process of electron transport and ATP generation.

13.7 Aromatic Pollutants

Many carbon sources for catabolism contain aromatic forms such as the benzene ring. Natural sources of aromatic carbon include lignin from wood, and phenanthrenes (three fused rings) from petroleum. Aromatic components of petroleum cause much of the environmental damage during oil spills such as the Deepwater Horizon spill in 2010. Industrial aromatic compounds often pollute soil, such as nitrate explosives, aniline dyes, and the solvent toluene. To remove aromatic pollutants from water and soil, we depend on microbial catabolism. Aromatic molecules are notoriously difficult to catabolize because of the stability associated with the benzene ring. The benzene ring breaks down slowly, but over time, bacteria and fungi catabolize a wide range of aromatic molecules. These microbial processes have major industrial and environmental importance.

Benzoate Catabolism

Benzoate has a central role in aromatic catabolism, comparable to that of glucose in sugar catabolism. Bacteria such as *Pseudomonas* and *Rhodococcus* species degrade benzoate and related molecules aerobically or anaerobically. The aerobic pathways (**Fig. 13.28**) share key features:

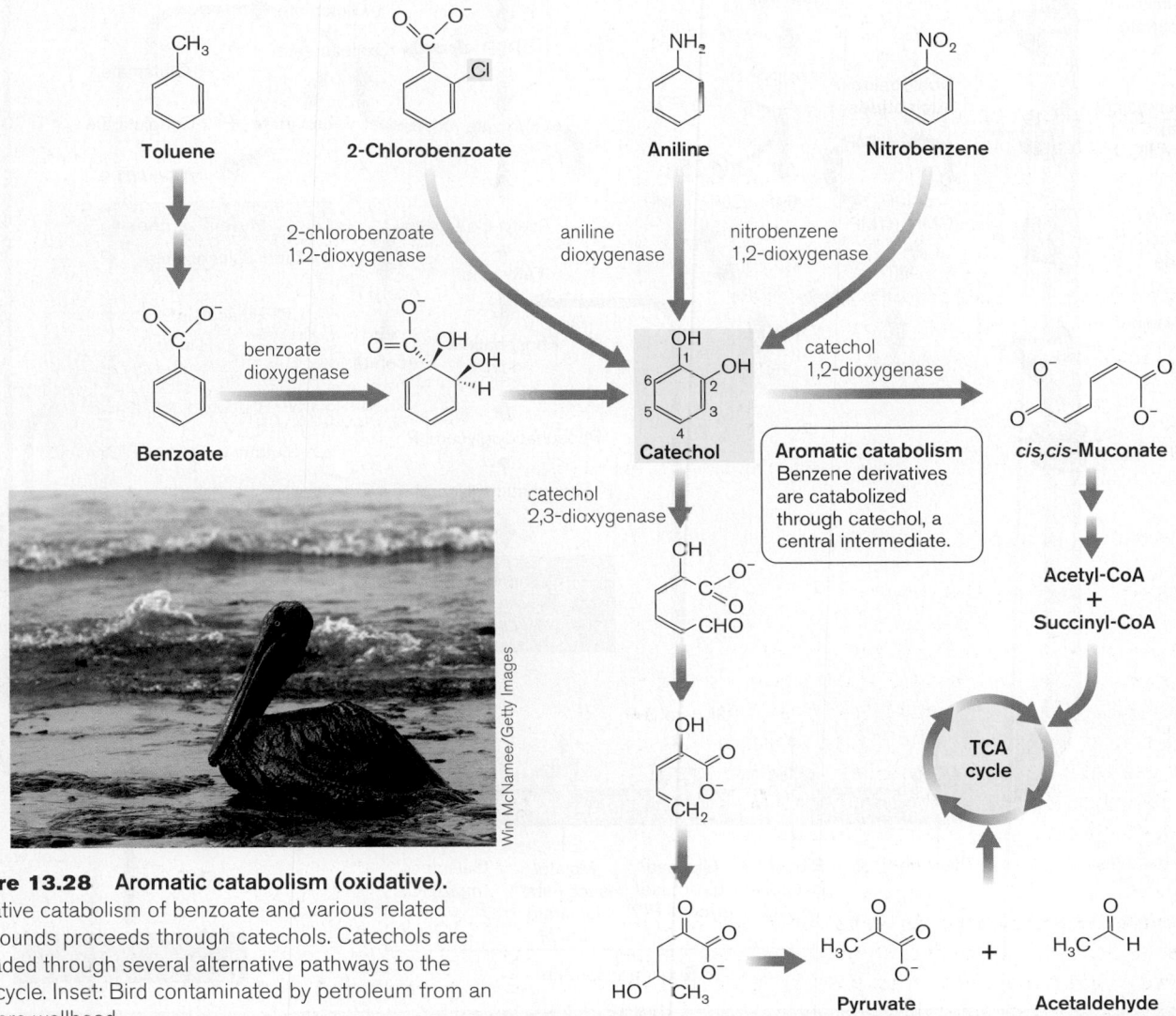

Figure 13.28 Aromatic catabolism (oxidative). Oxidative catabolism of benzoate and various related compounds proceeds through catechols. Catechols are degraded through several alternative pathways to the TCA cycle. Inset: Bird contaminated by petroleum from an offshore wellhead.

Special Topic 13.2   Genomic Analysis of Metabolism

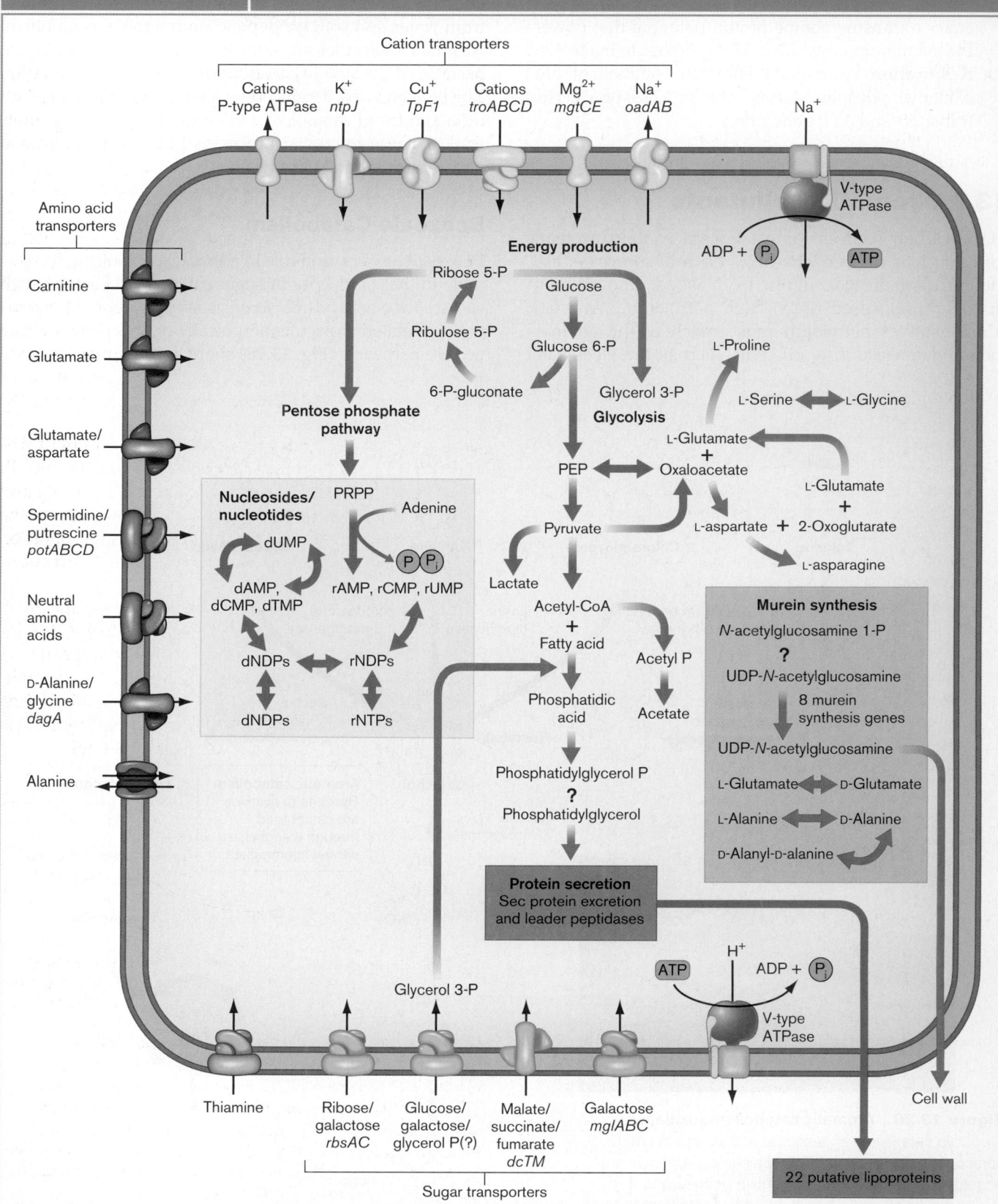

Dissection of the TCA cycle required decades of painstaking experiments in biochemistry and genetic analysis on model organisms such as *E. coli*. Now that so much metabolism is understood in model organisms, much of the metabolic capacity of other organisms can be deduced by comparison of their genomes. Genome analysis can be especially useful in characterizing pathogens that are difficult to culture and study, such as *Treponema pallidum*, the causative agent of syphilis.

The sequence of a microbial genome typically reveals a high percentage of genes whose function can be identified by homology with known genes in other species (discussed in Chapter 8). For example, the components of the pyruvate dehydrogenase complex show highly similar sequences in all organisms. Thus, it is possible to survey an annotated microbial genome and predict much of the microbe's metabolic capacities.

The genome of *T. pallidum* was sequenced by Claire Fraser-Liggett and colleagues at The Institute for Genomic Research (TIGR) (**Fig. 1**). The genome encodes key enzymes of glycolysis (red arrows, central), as well as those of fermentation to lactate or acetate. However, enzymes of the TCA cycle are completely absent, as are the proteins of electron transport. Thus, *T. pallidum* can generate ATP only by substrate-level phosphorylation, such as the ATP formation steps of glycolysis.

The genome does encode a small number of sugar transporters (green symbols) for entry to glycolysis, as well as amino acid transporters (red symbols). It does not encode enzymes of amino acid catabolism or biosynthesis. Thus, the organism can ferment a limited number of sugars but can neither catabolize nor form its own amino acids. It must depend on its host to supply many fundamental nutrients. The general loss of major metabolic systems from *T. pallidum* is typical of ancient pathogens that have evolved over a long period in obligate association with their host.

Figure 1 Genome of *Treponema pallidum* reveals metabolic pathways.   The *Treponema* genome encodes key enzymes of glycolysis and fermentation (red arrows, central). However, enzymes of the TCA cycle and electron transport are absent. Thus, *Treponema* is capable of little or no oxidative respiration. The genome does show sugar transporters (green symbols) and amino acid transporters (red symbols) but no enzymes for amino acid catabolic or biosynthetic pathways.

- Removal of substituents such as chlorides or nitrates
- Oxidation to form the dihydroxy derivative catechol
- Ring cleavage and catabolism to acetyl-CoA, which may enter the TCA cycle

Aerobic benzoate catabolism. Aerobic degradation typically proceeds through the key intermediate catechol, a benzene ring bearing two adjacent hydroxyl groups. Benzoate and various related compounds, such as toluene (methylbenzene) and nitrobenzene, are each converted to catechol by a specific dioxygenase, an enzyme that coordinately oxygenates two adjacent ring carbons.

The intermediate catechol then undergoes another key oxidation by a catechol dioxygenase, which adds two more oxygens while cleaving the ring. In different bacterial species, the enzyme may oxidize either the 1,2 positions, or the 2,3 positions as shown. The product is either a monocarboxylate or a dicarboxylate. In either case, the carbon skeleton is further oxidized, and its constituents are directed into the TCA cycle for full degradation to CO_2.

Genetic analysis of toluene catabolism. Many pathways of aromatic catabolism are still being discovered. To study unknown metabolism, we must identify the genes that encode the enzymes. These genes lie waiting in the sequenced microbial genomes, where a large proportion of genes remain unidentified. How are their functions discovered?

In bacteria, the enzymes for benzoate catabolism are often encoded on a large plasmid, of 100–450 kilobases (kb). The genes can be identified through use of clone libraries, as discussed in Chapter 12. For example, clone complementation was used by Hyung-Yeel Kahng and colleagues to reveal new genes encoding degradation of toluene and related soil pollutants (**Fig. 13.29**).

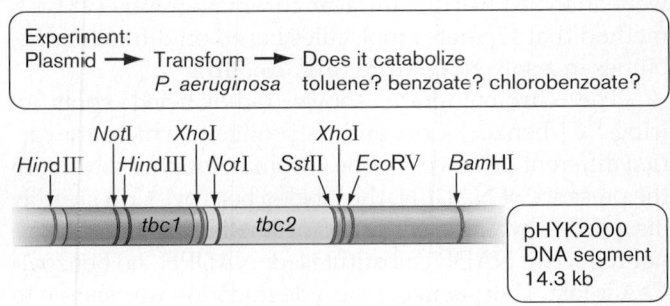

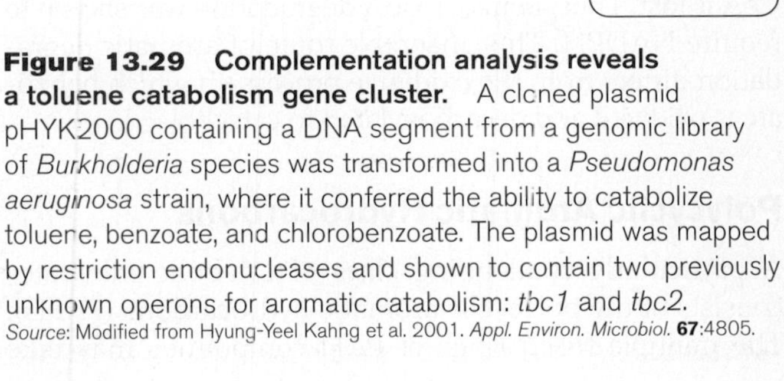

Figure 13.29 Complementation analysis reveals a toluene catabolism gene cluster.   A cloned plasmid pHYK2000 containing a DNA segment from a genomic library of *Burkholderia* species was transformed into a *Pseudomonas aeruginosa* strain, where it conferred the ability to catabolize toluene, benzoate, and chlorobenzoate. The plasmid was mapped by restriction endonucleases and shown to contain two previously unknown operons for aromatic catabolism: *tbc1* and *tbc2*.
Source: Modified from Hyung-Yeel Kahng et al. 2001. *Appl. Environ. Microbiol.* **67**:4805.

Kahng and co-workers obtained a strain of the soil bacterium *Burkholderia*, which degrades toluene, benzoate, and chlorobenzoate. A genomic library was generated to contain plasmids carrying cloned segments from throughout the *Burkholderia* genome. The library of clones was used to transform a strain of *Pseudomonas aeruginosa* that cannot degrade the toluene derivatives. Each transformed strain contained a plasmid with a different segment of the *Burkholderia* genome. One clone, pHYK2000, conferred the ability to catabolize toluene, benzoate, and chlorobenzoate. The plasmid in this clone was mapped by restriction endonucleases to delimit the range of DNA sequence needed for catabolism. Ultimately, the plasmid was shown to contain two previously unknown operons for aromatic catabolism: *tbc1* and *tbc2*. The functions of their gene products were tested subsequently by biochemical assays. These genes point to a potential new microbial route for biodegradation of soil pollutants.

Anaerobic benzoate catabolism. A challenge for biodegradation is that many pollution sources reach deep underground, where the soil is anoxic. Thus, oxygen is unavailable to conduct the conversions shown in **Figure 13.28**. How can microbes catabolize benzoate derivatives without oxygen? One example of an anaerobic pathway of benzoate catabolism was observed in the soil bacterium *Azoarcus evansii* (**Fig. 13.30**). In *Azoarcus*, under anaerobic conditions, benzoate undergoes reductive degradation instead of oxidation. First, the benzoate is activated by CoA, then it is reduced by NADPH. The NADPH reduction generates a C-C single bond, eliminating aromaticity.

The reduction of benzoyl-CoA by NADPH was demonstrated by Christa Ebenau-Jehle and colleagues at the University of Freiburg by the separation of radiolabeled intermediates. Ring carbons of benzoyl-CoA were labeled with ^{14}C, and reaction products from *Azoarcus* species were analyzed using thin-layer chromatography (TLC), a method that separates molecules based on different solubilities in a solvent mixture (**Fig. 13.30B**).

The chromatogram shows radiolabeled spots of [ring-^{14}C] benzoyl-CoA and its products, which are carried different distances by the chromatographic solvent. In the presence of NADPH, the labeled benzoyl-CoA steadily disappears, while products of catabolism emerge. If the nonreductive NADP$^+$ substitutes for NADPH, no benzoyl-CoA is lost. Thus, benzoyl-CoA degradation was shown to require NADPH. This anaerobic route of aromatic degradation differs from the oxidative process, in which benzoate is oxidized and decarboxylated to catechol.

Polycyclic Aromatic Hydrocarbons

A particularly challenging class of catabolic substrates consists of the polycyclic aromatic hydrocarbons (PAHs). The multiple fused rings of PAH compounds may take

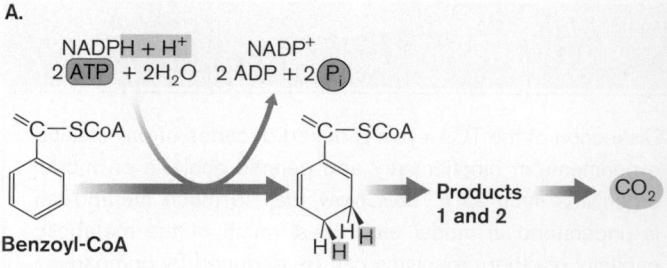

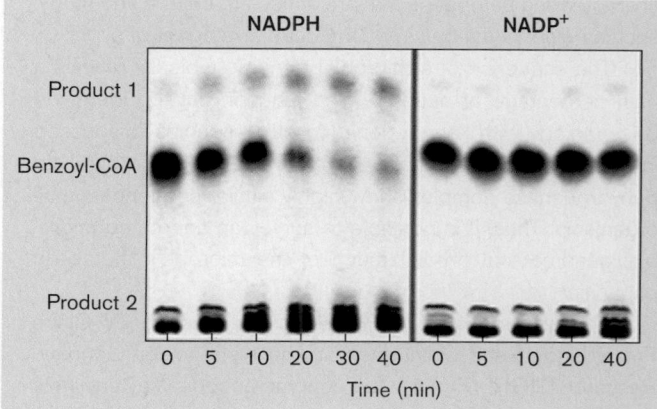

Figure 13.30 Anaerobic catabolism of benzoate: analysis by thin-layer chromatography.   A. Under anaerobic conditions, benzoate is activated by coenzyme A, and then reduced (hydrogenated) by NADPH. **B.** The reduction of benzoyl-CoA in *Azoarcus evansii* was observed by thin-layer chromatography (TLC). In the presence of the reducing agent NADPH, the spot due to benzoyl-CoA disappears, while those due to two products of catabolism increase. In the presence of NADP$^+$, with no reducing power, no reaction occurs. *Source*: Modified from Christa Ebenau-Jehle et al. 2003. *J. Bacteriol.* **185**:6119.

years to biodegrade in natural environments. PAHs are found in petroleum, as well as in coal tar products such as creosote, used to preserve wood. Creosote leaks from wood structures such as railway ties (**Fig. 13.31A** and **B**) and contaminates wetlands. Many PAHs are carcinogenic, and their increase in marine oil spills and wetlands is of growing concern.

The pathways and even the microbial species that degrade PAH in nature remain poorly known. In one study, the biodegradation of a PAH compound in creosote was observed by Sergey Selifonov and colleagues at the University of Minnesota. Selifonov used ^{13}C NMR to track the catabolism by *Pseudomonas* strain A2279 of ^{13}C-labeled acenaphthene, a tricyclic naphthalene derivative (**Fig. 13.31C**). The compound was labeled at a single aliphatic position so that its resonance would appear as a single peak separate from the range of aromatic carbon, where natural-abundance ^{13}C peaks appear. New peaks appeared, which were identified as various oxidized products of acenaphthene. The enzymes forming these products remain to be discovered. Furthermore, the completion of degradation remains unclear. Such complex

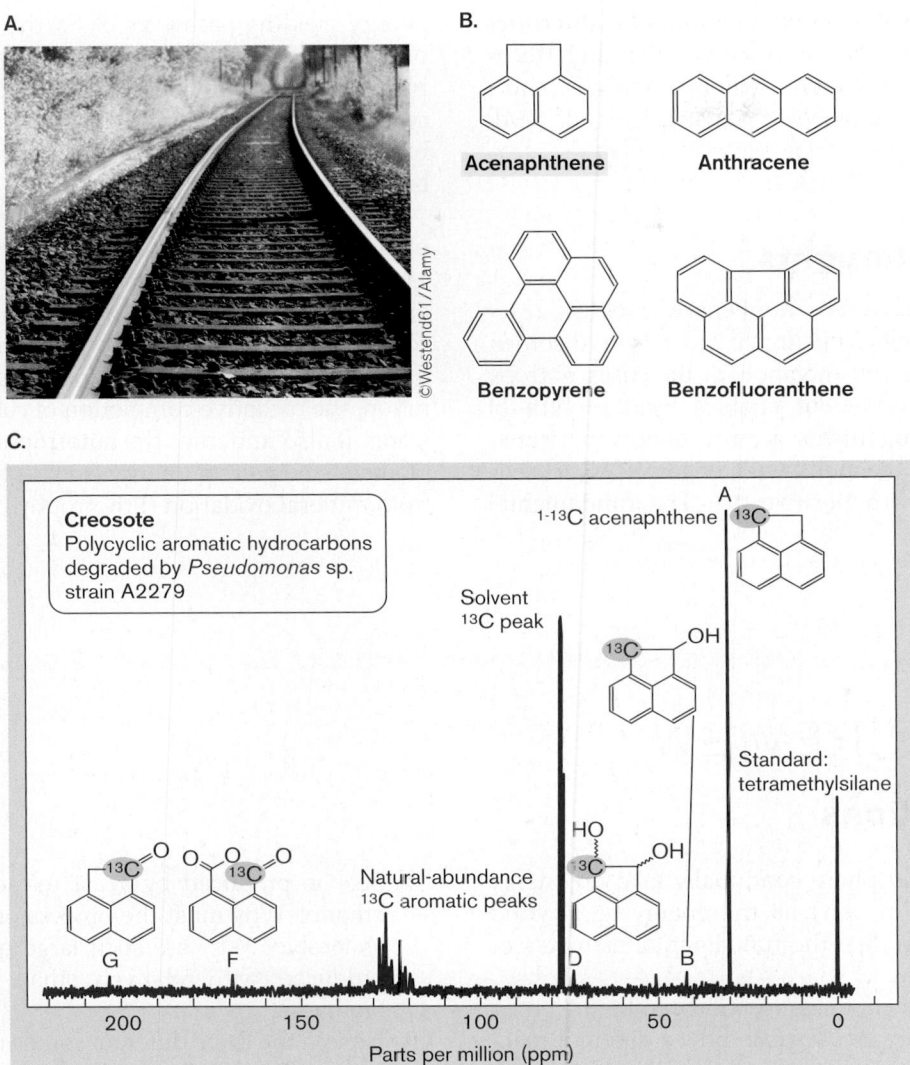

Figure 13.31   PAH biodegradation.   A. Railroad ties are a source of the pollutant creosote, which contains a mixture of polycyclic aromatic compounds. **B.** Typical structures of polycyclic aromatic hydrocarbons (PAHs). **C.** A ^{13}C-labeled polycyclic PAH, acenaphthene, is included in a creosote sample undergoing catabolism by *Pseudomonas* sp. strain A2279. After seven days, the oxidized digestion products are revealed by characteristic resonance peaks in the ^{13}C NMR spectrum. *Source: Part C modified from Sergey A. Selifonov et al. 1998. Appl. Environ. Microbiol.* **64**:1447.

molecules probably require a consortium of different species to complete their degradation to acetate.

A surprising discovery in marine water has been that many of the bacteria involved in degrading spilled oil are related to pathogens. The Deepwater Horizon spill showed oil degradation by bacteria of the genus *Vibrio*, such as *Vibrio parahemolyticus*, and *Vibrio vulnificus*, which cause serious human infections. Thus, a bacterium may have both negative consequences for humans as well as positive roles in our environment.

THOUGHT QUESTION 13.12   How might the enzymes that catalyze acenaphthene catabolism be discovered?

TO SUMMARIZE:

- **Catabolism of aromatic molecules** by bacteria and fungi recycles lignin and other important substances within ecosystems. Toxic pollutants are also degraded.
- **Benzoate undergoes aerobic catabolism to catechol.** The catechol ring is cleaved, generating acetyl-CoA, which enters the TCA cycle.
- **Anaerobic catabolism of benzoate** involves activation by HS-CoA and reduction of NADP$^+$ to NADPH.
- **Polycyclic aromatic hydrocarbons (PAHs)** are degraded slowly by microbes. PAH catabolism has exciting potential applications for bioremediation of toxic pollutants.

■ **Aromatic catabolism by environmental microbes** is an important field of research. Research techniques include genetic and genomic analysis, chromatography of radiolabeled substrates, and NMR spectroscopy.

Concluding Thoughts

In this chapter we have seen how energy-yielding reactions enable living cells to generate order from disorder. In fact, the result of cell metabolism increases entropy throughout the universe but enables local growth of complexity. Living organisms acquire energy by transferring energy from reactions with negative ΔG to cell-building reactions with positive ΔG. The fundamental energy-yielding pathways of Earth's biosphere are those of photosynthesis; and the biomass generated through photosynthesis provides materials for catabolism. Much of human civilization has been built on harnessing microbial catabolism for waste treatment, food production, and biotechnology.

All forms of metabolism involve chemical exchange of electrons through oxidation and reduction. In Chapter 14, we focus on the transport of electrons between donor and acceptor molecules, through membrane-embedded complexes that generate ion gradients across the membrane. This transport of electrons is fundamental to respiration, the oxidative completion of catabolism to CO_2 and water. It also underlies the autotrophic means of building biomass: the gain of energy from light (phototrophy) and from mineral oxidation (lithotrophy.)

CHAPTER REVIEW

Review Questions

1. Why must the biosphere continually take up energy from outside? Why can't all the energy be recycled among organisms, like the fundamental elements of matter?

2. Explain how a biochemical reaction can be driven by change in enthalpy, ΔH. Explain how a different reaction can be driven by change in entropy, ΔS. In each case, explain the role of the free energy change, ΔG.

3. Why do some biochemical reactions release energy only above a threshold temperature?

4. How do organisms determine which of their catabolic pathways to use? How does catabolism depend on environmental factors?

5. Beer is produced by yeast fermentation of grain to ethanol. Why must the process of beer production be anaerobic? Why are such large quantities of ethanol produced, with relatively small production of yeast biomass?

6. Explain the three different routes to catabolize glucose to pyruvate. Why is it necessary to start by spending one or two molecules of ATP?

7. Explain how the TCA cycle incorporates an acetyl group. How are the two CO_2 molecules removed?

8. Compare and contrast aerobic and anaerobic processes of benzoate catabolism.

Thought Questions

1. Why are glucose catabolism pathways ubiquitous, even in bacterial habitats where glucose is scarce? Give several reasons.

2. In glycolysis, explain why bacteria have to return the hydrogens from NADH back onto pyruvate to make fermentation products. Why can't NAD^+ serve as a terminal electron acceptor, like O_2?

3. Why does catabolism of benzene derivatives yield less energy than sugar catabolism? Why is benzene-derivative catabolism nevertheless widespread among soil bacteria?

4. Why do environmental factors regulate catabolism? For example, why are amino acids decarboxylated at low pH? Cite other examples.

Key Terms

activation energy (469)
adenosine triphosphate (ATP) (465)
allosteric site (470)
amphibolic (481)
anabolism (458)
aromatic (468)
calorimetry (462)
catabolism (458)
catabolite repression (475)
coenzyme A (CoA) (485)
electron acceptor (465)
electron donor (465)
electron transport system (ETS) (468)
Embden-Meyerhof-Parnas (EMP)
    pathway (478, 479)
energy (458)
energy carrier (465)
enthalpy (ΔH) (461)
Entner-Doudoroff (ED) pathway
    (478, 481)

entropy (ΔS) (459, 461)
enzyme (469)
ethanolic fermentation (483)
fermentation (471)
flavin adenine dinucleotide
    (FAD) (469)
Gibbs free energy change (ΔG) (461)
glycolysis (463, 478, 479)
glyoxylate bypass (492)
heteroaromatic (468)
heterolactic fermentation (483)
heterotrophy, heterotroph (458)
hydrolysis (466)
joule (J) (461)
lactate fermentation (483)
lignin (475)
mixed-acid fermentation (485)
nicotinamide adenine dinucleotide
    (NADH, NAD$^+$) (467)

organotrophy, organotroph (458)
oxidative phosphorylation (491)
pentose phosphate shunt (PPS)
    (478, 482)
phenol red broth test (486)
phosphorylation (466)
phosphotransferase system (PTS)
    (467)
photoheterotrophy (472)
polysaccharide (473)
pyruvate dehydrogenase complex
    (PDC) (489)
respiration (472)
ribulose 5-phosphate (482)
substrate-level phosphorylation (480)
terminal electron acceptor (468)
tricarboxylic acid (TCA) cycle (487)

Recommended Reading

Atlas, Ronald M. 1995. Petroleum biodegradation and oil spill bioremediation. *Marine Pollution Bulletin* **31**:178–182.

Bäckhed, Fredrik, Ruth E. Ley, Justin L. Sonnenburg, Daniel A. Peterson, and Jeffrey I. Gordon. 2005. Host-bacterial mutualism in the human intestine. *Science* **307**:915–920.

Brown, Stacie A., Kelli L. Palmer, and Marvin Whiteley. 2009. The host as a growth medium. *Nature Reviews. Microbiology* **6**:657–666.

Bunge, Michael, Lorenz Adrian, Angelika Kraus, Matthias Opel, Wilhelm G. Lorenz, et al. 2003. Reductive dehalogenation of chlorinated dioxins by an anaerobic bacterium. *Nature* **421**:357–360.

Chivian, Dylan, Eoin L. Brodie, Eric J. Alm, David E. Culley, Paramvir S. Dehal, et al. 2008. Environmental genomics reveals a single-species ecosystem deep within Earth. *Science* **322**:275–278.

Ebenau-Jehle, Christa, Matthias Boll, and Georg Fuchs. 2003. 2-Oxoglutarate:NADP$^+$ oxidoreductase in *Azoarcus evansii*: Properties and function in electron transfer reactions in aromatic ring reduction. *Journal of Bacteriology* **185**:6119–6129.

Fraser, Claire M., Steven J. Norris, George M. Weinstock, Owen White, Granger G. Sutton, et al. 1998. Complete genome sequence of *Treponema pallidum*, the syphilis spirochete. *Science* **281**:375–388.

Hanniffy, Sean B., Mark Philo, Carmen Peláez, Michael J. Gasson, Teresa Requena, et al. 2009. Heterologous production of methionine-γ-lyase from *Brevibacterium linens* in *Lactococcus lactis* and formation of volatile sulfur compounds. *Applied and Environmental Microbiology* **75**:2326–2332.

Head, Ian M., D. Martin Jones, and Wilfred F. M. Röling. 2006. Marine microorganisms make a meal of oil. *Nature Reviews. Microbiology* **4**:173–182.

Jackson, Bradley E., and Michael J. McInerney. 2002. Anaerobic microbial metabolism can proceed close to thermodynamic limits. *Nature* **415**:454–456.

Jeon, C. O., W. Park, P. Padmanabhan, C. DeRito, J. R. Snape, et al. 2003. Discovery of a bacterium, with distinctive dioxygenase, that is responsible for in situ biodegradation in contaminated sediment. *Proceedings of the National Academy of Sciences USA* **100**:13591–13596.

Kapatral, Vinayak, Xiaowen Bina, and A. M. Chakrabarty. 2000. Succinyl coenzyme A synthetase of *Pseudomonas aeruginosa* with a broad specificity for nucleoside triphosphate (NTP) synthesis modulates specificity for NTP synthesis by the 12-kilodalton form of nucleoside diphosphate kinase. *Journal of Bacteriology* **182**:1333–1339.

Ladas, Spiros D., George Karamanolis, and Emmanuel Ben-Soussan. 2007. Colonic gas explosion during therapeutic colonoscopy with electrocautery. *World Journal of Gastroenterology* **13**:5295–5298.

Maurer, Lisa M., Elizabeth Yohannes, Sandra S. BonDurant, Michael Radmacher, and Joan L. Slonczewski. 2005. pH regulates genes for flagellar motility, catabolism, and oxidative stress in *Escherichia coli* K-12. *Journal of Bacteriology* **187**:304–319.

Peekhaus, Norbert, and Tyrrell Conway. 1998. What's for dinner? Entner-Doudoroff metabolism in *Escherichia coli*. *Journal of Bacteriology* **180**:3495–3502.

Sonnenburg, Justin L., Jian Xu, Douglas D. Leip, Chien-Huan Chen, Benjamin P. Westover, et al. 2005. Glycan foraging in vivo by an intestine-adapted bacterial symbiont. *Science* **307**:1955–1959.

von Stockar, Urs, Thomas Maskow, Jingsong Liu, Ian W. Marison, and Rodrigo Patiño. 2006. Thermodynamics of microbial growth and metabolism: An analysis of the current situation. *Journal of Biotechnology* **121**:517–533.

Wolfe, Alan J. 2005. The acetate switch. *Microbiology and Molecular Biology Reviews* **69**:12–50.

Xu, Jian, Magnus K. Bjursell, Jason Himrod, Su Deng, Lynn K. Carmichael, et al. 2003. A genomic view of the human-*Bacteroides thetaiotaomicron* symbiosis. *Science* **299**:2074–2075.

For further study and review, please visit StudySpace at **microbiology2.com**.

Chapter 14

Respiration, Lithotrophy, and Photolysis

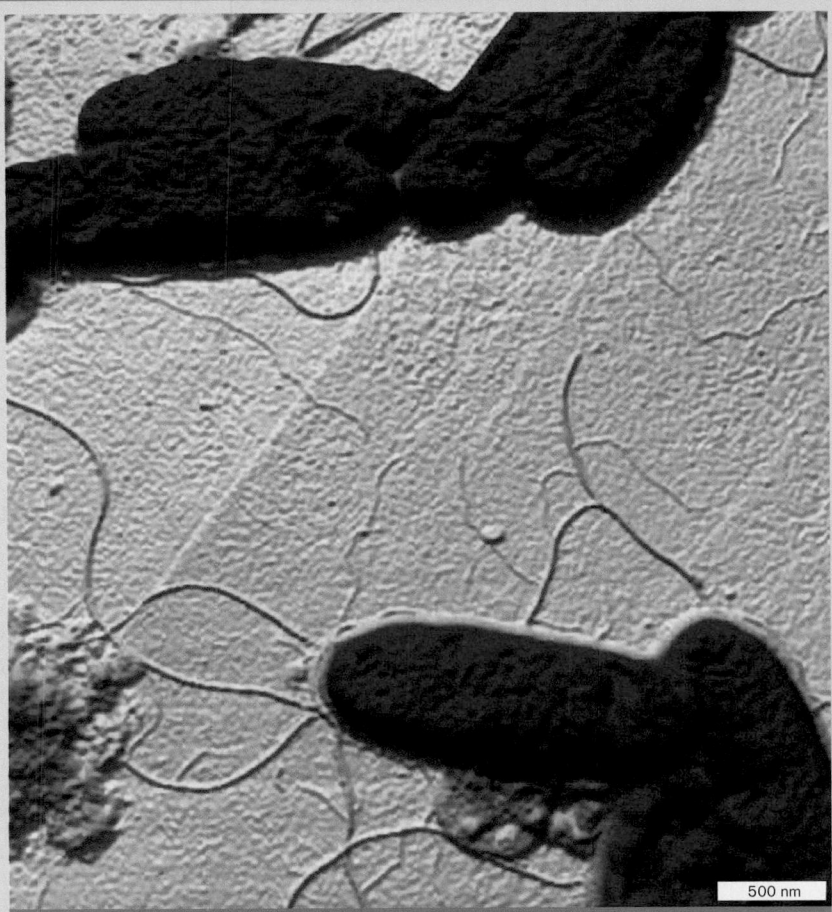

500 nm

Microbes transfer energy by moving electrons—the equivalent of an electric current. Electrons move from reduced food molecules onto energy carriers, from energy carriers onto membrane proteins called cytochromes, and from cytochromes onto oxygen or oxidized minerals. Some bacteria can actually donate electrons to an electrode and power a fuel cell. In the soil, bacteria donate electrons to metals such as iron, gold, and uranium. Bacteria in Earth's crust have deposited most of the iron mined today, and possibly the gold as well.

A special kind of electron current flows between proteins and cofactors of an electron transport system in a cell membrane. The electron transport system generates a "proton motive force" that drives protons across the membrane. The proton motive force stores energy to make ATP. Similar electron transport systems power our own mitochondria and the photosynthetic membranes of plant chloroplasts.

Current Research Highlight

Bacteria make electricity. Soil bacteria such as *Shewanella oneidensis* can transmit electrons through protein filaments called nanowires (AFM). These bacteria grow in biofilms on an electrode in a fuel cell. The fuel cell bacteria donate electrons from food, generating electricity to run small devices. *Source*: Moh El-Naggar et al. 2008. *Biophys. J.* **95**:L10–L12. Photo courtesy of Moh El-Naggar (University of Southern California, Los Angeles), Yuri Gorby (J. Craig Venter Institute, San Diego), and Irene Revenko (Asylum Research, Santa Barbara).

Electric current in living organisms has fascinated scientists ever since the eighteenth century, when Luigi Galvani (1737–1798) showed that a voltage potential caused a dead frog limb to flex. The idea of biological electricity inspired Mary Shelley's famous science fiction novel *Frankenstein: or, the Modern Prometheus*, in which a physician is imagined to "create life" by jolting body parts with an electric shock.

Today, electric currents and voltage potentials are known to be part of all cells. Nevertheless, it came as a surprise to discover that bacteria could directly power an electrical fuel cell. Bacteria commonly found in soil, such as *Geobacter metallireducens*, can oxidize organic nutrients using iron ions (ferric ion, Fe^{3+}) instead of oxygen. The organic molecules transfer electrons to the iron. But Fe^{3+} is usually present as insoluble grains of iron oxide, outside the cell; so how do cells reach the iron? Derek Lovley and colleagues at the University of Massachussetts, Amherst, wondered how bacteria gain access to iron oxide—and whether they might also transfer electrons to an electrode to make electricity (**Fig. 14.1**). Lovley found that *G. metallireducens* bacteria form a biofilm on a graphite electrode, where they transfer electrons and generate enough power to run a calculator.

The source of electrons for *Geobacter* metabolism can include hard-to-digest organic molecules, even aromatic compounds such as benzoates that often contaminate soil. Thus, bacteria may combine electricity generation with organic waste remediation. In fact, the bacterial community of our entire biosphere acts as a battery in which underground bacteria continually transfer electrons from Earth's reduced interior to the oxidizing atmosphere. Simply placing electrodes at appropriate depths in the ocean floor can yield useful electricity. Although the amount of power obtained at present is small, in the future microbial fuel cells may contribute significantly to our energy supply. The mechanism of a fuel cell will be presented later in **Special Topic 14.2**.

As we saw in Chapter 13, biochemical reactions can provide energy in different ways (see Table 13.1). Catabolism provides energy through breakdown of complex molecules (increasing entropy) and through electron transfer converting molecules to more stable forms (releasing heat energy). Electron transfer also yields energy from inorganic minerals (lithotrophy) and from light capture (phototrophy). Microbial electron transfer reactions have profound consequences for ecosystems; they acidify or alkalinize soil and water, and they can contribute or remove nitrogen and minerals.

Chapter 14 develops a unified view of electron transfer reactions, or oxidation-reduction reactions, in respiration, lithotrophy, and phototrophy. ("Reduction" of a molecule means the gain of an electron; "oxidation" means the loss of an electron.) In all kinds of metabolism, some of the energy from electron transfer is stored in the form of an electrochemical potential (voltage) across a membrane. The voltage potential includes a concentration gradient of ions (H^+ or Na^+) plus the charge difference across the membrane. Voltage potentials drive ATP synthesis and nutrient uptake, mediate pathogenesis of infective microbes, and shape the chemistry of our environment.

14.1 Electron Transport Systems

As we learned in Chapter 13, energy to support life is obtained through reactions that convert substrates to products of lower energy—that is, reactions in which ΔG (free energy change) is negative. Most such reactions involve transfer of electrons from a reduced **electron donor** to an oxidized **electron acceptor**. The simplest path of electron flow occurs in fermentation, where electrons from the fermented substrate, such as glucose, are transferred onto NAD^+ to make NADH (reduction), and

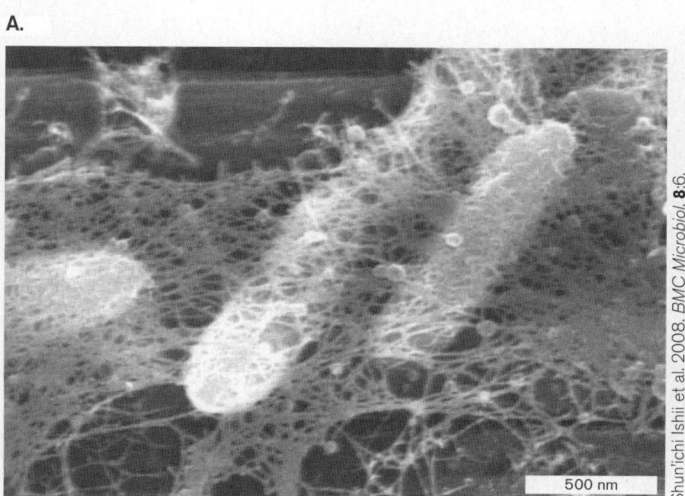

Figure 14.1   Electricity from iron-reducing bacteria.   A. *Geobacter metallireducens* biofilm on a graphite electrode (SEM). **B.** Bacterial fuel cell running a calculator.

then returned to the glucose breakdown products. More complex kinds of metabolism, such as aerobic respiration, transfer electrons through a series of membrane-soluble carriers called an **electron transport system (ETS)**, also known as an **electron transport chain**. A membrane-embedded ETS, unlike cytoplasmic redox reactions, can convert its energy into an ion potential or electrochemical potential between two compartments separated by the membrane. The ion potential is most commonly a proton (H^+) potential, also known as proton motive force (PMF). The proton motive force drives essential cell processes such as synthesis of ATP (discussed in Section 14.2).

Electron Donors and Acceptors

Different types of metabolism use electron transport systems with different components embedded in the bacterial membrane. Each ETS must accept electrons from an initial electron donor consisting of an organic "food" molecule, such as glucose, or a reduced mineral, such as Fe^{2+}. The ETS proteins and cofactors act sequentially as electron donors and acceptors, whose oxidation-reduction reactions are coupled to proton (H^+) transport across the membrane. Each ETS finally transfers its electrons to a terminal electron acceptor molecule, whose product may exit the cell.

Metabolism using an ETS is classified based on the nature of the initial electron donors and terminal electron acceptors. In this chapter we cover three major classes of prokaryotic energy acquisition involving an ETS (their relationships are summarized in Table 13.1):

- **Respiration** is the oxidation of electron donors such as sugars, lipids, or amino acids to CO_2 and H_2O. (Oxidation of minerals is a form of respiration called lithotrophy, discussed below.) Organic molecules donate electrons associated with hydrogen atoms. Additional hydrogen atoms are donated by water molecules. The oxidant (terminal electron acceptor) may be O_2 for aerobic respiration, or an anaerobic alternative for anaerobic respiration. The anaerobic electron acceptor may be inorganic (such as NO_3^-) or organic (such as fumarate, $^-O_2C-CH=CH-CO_2^-$.
- **Lithotrophy** (or **chemolithotrophy**) is the oxidation of inorganic electron donors, such as Fe^{2+} or H_2. The electron acceptor may be O_2 or an anaerobic alternative, such as NO_3^-. Some inorganic electron donors, such as H_2, can use organic electron acceptors. Many lithotrophs are autotrophs, fixing CO_2 for biosynthesis (discussed in Chapter 15).
- **Phototrophy** involves light capture by chlorophyll, usually coupled to splitting of H_2S or H_2O (photolysis) or organic molecules (photoheterotrophy). Photoautotrophs couple photolysis to CO_2 fixation for biomass (discussed in Chapter 15). Photoheterotrophs use light energy to supplement catabolism.

NOTE: Distinguish these prefix terms:

"Organo-"	Organic electron donors for energy
"Litho-"	Inorganic electron donors for energy
"Chemo-"	Chemical reactions yield energy
"Photo-"	Light absorption drives or supplements chemical reactions for energy
"Hetero-"	Organic molecules for carbon source
"Auto-"	Inorganic carbon source, usually CO_2

Energy Storage

In Chapter 13, we showed how the free energy change ΔG determines whether energy can be obtained from a given reaction of substrates to products. Similar considerations of ΔG govern the reactions of electron flow through an ETS and the storage of energy in transmembrane ion gradients.

The reduction potential. To obtain energy, biochemical reactions require a negative ΔG. In oxidation-reduction reactions (redox reactions), the ΔG values are proportional to the reduction potential (E) between the oxidized form of a molecule (electron acceptor) and its reduced form (electron donor). The reduction potential represents the tendency of a molecule to accept electrons, measured in volts (V) or millivolts (mV). A positive value of E has a negative ΔG, so the reaction can go forward.

The oxidized and reduced states of a molecule are called a **redox couple**. An example of a redox couple is the electron acceptor O_2 and the electron donor H_2O. Because O_2 is a strong electron acceptor, the redox couple O_2/H_2O has a high positive value of E. So, O_2 can be reduced to H_2O and yield a large amount of energy.

Standard values of E at equilibrium ($E°$) are known as the **standard reduction potential**. The reduction potential $E°'$ (assuming a concentration of 1 M for all components at pH 7) for H_2 formation from $2H^+ + 2e^-$ is -0.42 V. This large negative reduction potential means that it takes a lot of energy to add the two electrons to $2H^+$. Thus, the reverse reaction donating $2e^-$ from H_2 to an appropriate electron acceptor yields a lot of energy. Hydrogen-oxidizing bacteria grow in oral biofilms causing gum disease.

Reduction potentials represent a form of the standard free energy change, $\Delta G°'$. The value of $\Delta G°'$ (in kilojoules per mole, or kJ/mol) for an electron transfer reaction with a given reduction potential $E°'$ is given by

$$\Delta G°' = -nFE°'$$

where n is the number of electrons transferred, F is Faraday's constant ($F = 96.5$ kJ/V · mol) and $E°'$ is the reduction potential at 1-M concentration, 25°C, and pH 7. Note that the reaction is favored by <u>positive values of E</u>, which yield <u>negative values of ΔG</u>.

Table 14.1 "Electron tower" of standard reduction potentials.[a]

Electron acceptor	→	electron donor	$E^{\circ\prime}$ (mV)[b]
$CO_2 + 4H^+ + 4e^-$	→	[CH_2O] glucose + H_2O	−430
$2H^+ + 2e^-$	→	H_2	−420
$NAD^+ + 2H^+ + 2e^-$	→	$NADH + H^+$	−320
$S^0 + H^+ + e^-$	→	HS^-	−280
$CO_2 + 2H^+ + 3H_2 + 2e^-$	→	$CH_4 + 2H_2O$	−240
$SO_4^{2-} + 10H^+ + 8e^-$	→	$H_2S + 4H_2O$	−220
$FAD + 2H^+ + 2e^-$	→	$FADH_2$	−220[c]
$FMN + 2H^+ + 2e^-$	→	$FMNH_2$	−190
Menaquinone $+ 2H^+ + 2e^-$	→	Menaquinol	−74
Fumarate $+ 2H^+ + 2e^-$	→	Succinate	33
Ubiquinone $+ 2H^+ + 2e^-$	→	Ubiquinol	+110
$NO_3^- + 2H^+ + 2e^-$	→	$NO_2^- + H_2O$	+420
$NO_2^- + 8H^+ + 6e^-$	→	$NH_4^+ + 2H_2O$	+440
$MnO_2 + 4H^+ + 2e^-$	→	$Mn^{2+} + 2H_2O$	+460
$NO_3^- + 6H^+ + 5e^-$	→	$½N_2 + 3H_2O$	+740
$Fe^{3+} + e^-$	→	Fe^{2+} (at pH 2)	+770
$½O_2 + 2H^+ + 2e^-$	→	H_2O	+820

[a]Values taken from Rudolf K. Thauer, Kurt Jungermann, and Karl Decker. 1977. Energy conservation in chemotrophic anaerobic bacteria. *Bacteriological Reviews* 41:100–180.

[b]Voltage potentials for redox couples when all concentrations are 1 M at pH 7.

[c]This value is for free FAD; FAD bound to a specific flavoprotein has a different $E^{\circ\prime}$ that depends on its protein environment.

NOTE: A more positive $E^{\circ\prime}$ means that reducing the electron acceptor yields more energy. A more negative value of $E^{\circ\prime}$ means that oxidizing the electron donor yields more energy.

The standard reduction potential $E^{\circ\prime}$ gives the potential difference between the oxidized and reduced states at equilibrium, when both oxidized and reduced forms are at equal concentrations—namely, 1 M. Assuming equal concentrations at 1 M, we may compare $E^{\circ\prime}$ of various molecules available in the environment for microbial respiration. Such a comparison generates an "electron tower" representing the reduction potential $E^{\circ\prime}$ of molecules that may act as electron acceptors or donors (**Table 14.1**). The oxidized state of each redox couple in the table is shown in red, the reduced state in blue.

In the electron tower, the more negative values represent couples (half reactions) with a stronger electron donor (H_2, NADH), whereas the more positive values represent stronger electron acceptors (O_2, NO_3^-). For example, in the redox couple NAD^+/NADH + H^+, the oxidized form is NAD^+ and the reduced form is NADH. The reduction potential is strongly negative (−320 mV), so NADH is a strong electron donor; that is, the reverse reaction, $NADH + H^+ \rightarrow NAD^+$, releases a lot of energy (+320 mV).

A complete redox reaction combines two redox couples: one accepting electrons (arrow forward), the other donating electrons (arrow reversed). Redox couples with more negative values of $E^{\circ\prime}$ (higher in the tower) can provide electron donors (blue column in **Table 14.1**) for electron acceptors with more positive values of $E^{\circ\prime}$ (lower in the tower; red column). The $E^{\circ\prime}$ for the overall reaction is then given by adding the reversed reaction of the donor $E^{\circ\prime}$ to the acceptor $E^{\circ\prime}$. A positive value of $E^{\circ\prime}$ means that the reaction may proceed to provide energy. For example, the oxidation of NADH by O_2 results from combining two couples: O_2/H_2O (forward) and NADH/NAD^+ (reversed from **Table 14.1**). For the reversed couple, the sign of $E^{\circ\prime}$ changes:

Electron acceptor is reduced:
$$½O_2 + 2e^- + 2H^+ \rightarrow H_2O \qquad E^{\circ\prime} = 820\ mV$$

Electron donor is oxidized:
$$\underline{NADH + H^+ \rightarrow NAD^+ + 2e^- + 2H^+ \quad E^{\circ\prime} = 320\ mV}$$

$$NADH + H^+ + ½O_2 \rightarrow H_2O + NAD^+ \quad E^{\circ\prime} = 1{,}140\ mV$$

The aerobic oxidation of NADH pairs a strong electron donor (NADH) with a strong electron acceptor (O_2). Thus, NADH oxidation via the ETS provides the cell with a huge amount of potential energy (comparable to a 1-V battery cell) to make ATP and generate ion gradients. NADH is oxidized by many electron transport systems of bacteria and archaea, as well as mitochondria. Whenever we digest food, we make NADH for our mitochondria to oxidize.

Bacteria and archaea have evolved many alternative donor-acceptor systems; the one they use depends on what their environment provides. For example, in the human colon without oxygen, *Escherichia coli* can oxidize NADH with fumarate, which is reduced to succinate. From **Table 14.1**, we find that

$$NADH + H^+ + fumarate \rightarrow NAD^+ + succinate$$

$$E^{\circ\prime} = 320\ mV + 33\ mV = 353\ mV$$

When oxygen reappears in the environment, the succinate can donate electrons back to fumarate and water:

$$½O_2 + 2H^+ + succinate \rightarrow H_2O + fumarate$$

$$E^{\circ\prime} = 820\ mV - 33\ mV = 787\ mV$$

THOUGHT QUESTION 14.1 *Pseudomonas aeruginosa*, a cause of pneumonia in cystic fibrosis patients, oxidizes NADH with nitrate (NO_3^-) at neutral pH. What is the value of $E^{\circ\prime}$?

Concentrations of electron donors and acceptors. The standard reduction potential $E^{\circ\prime}$ assumes 1-M concentrations of reactants and products at pH 7. Actual values of E within cells depend on actual reactant concentrations. In Chapter 13, we saw that ΔG includes a term incorporating the ratio of product concentrations (C and D) to reactant concentrations (A and B):

$$\Delta G = \Delta G^{\circ\prime} + 2.303\, RT \log \frac{[\text{C}][\text{D}]}{[\text{A}][\text{B}]}$$

where R is the gas constant (8.315 joules per kelvin-mole, or J/Kmol), and T is the temperature in kelvins. The reduction potential E of a redox reaction depends on the ratio of reduced products to oxidized reactants:

$$E(\text{mV}) = \frac{-\Delta G}{nF}$$

$$= E^{\circ\prime} - \frac{2.303\, RT}{nF} \times \log \frac{[\text{C}][\text{D}](\text{reduced})}{[\text{A}][\text{B}](\text{oxidized})}\ \text{mV}$$

$$= E^{\circ\prime} - 60 \log \frac{[\text{C}][\text{D}](\text{reduced})}{[\text{A}][\text{B}](\text{oxidized})}\ \text{mV}$$

where n is the number of electrons transferred and F is the Faraday constant. As the ratio of reduced products to oxidized reactants increases, E decreases; in other words, there is less potential to do work. At moderate temperatures (25°C–40°C), the constants in the concentration term (2.303 RT/F) combine to yield approximately 60 mV per electron for a tenfold ratio of products to reactants. Thus, a tenfold ratio of products to reactants subtracts about 60 mV from ΔE (decreases energy yield), whereas a tenfold excess of reactants adds 60 mV to E (increases energy yield).

A high concentration of an electron donor can drive metabolism using reactions with small values of $E^{\circ\prime}$. For example, iron mine drainage releases high concentrations of reduced iron (Fe^{2+}). *Gallionella* bacteria oxidize Fe^{2+} using the reduction potential between Fe^{3+}/Fe^{2+} and O_2/H_2O. Overgrowth of *Gallionella* fills mine-polluted streams with thick orange biofilms of bacteria.

> **THOUGHT QUESTION 14.2** Could a bacterium obtain energy from succinate as an electron donor with nitrate (NO_3^-) as an electron acceptor?

An ETS Functions within a Membrane

An electron transport system transfers electrons onto membrane-embedded carriers in a series of increasing reduction potential, ending at a terminal electron acceptor such as O_2. Some electron transfer steps yield energy to pump ions across the membrane. To store this energy, the ETS must maintain an ion gradient across a membrane that fully separates two aqueous compartments. In bacteria, the cytoplasmic membrane separates the cytoplasm and external medium. Gram-negative bacteria such as *E. coli* have their ETS in the inner (cytoplasmic) membrane, which separates the cytoplasm from the periplasm. The Gram-negative outer membrane surrounding the periplasm is permeable to protons and other small molecules; thus, the outer membrane does not store energy. **Figure 14.2A** shows the inner membrane of *Helicobacter pylori*, a Gram-negative gastric pathogen. In *H. pylori*, the

A.

Inner (cytoplasmic) membrane

100 nm

Yanping Liu et al. 2006. *J. Gastroenterol.* **41**:569–574. © Springer-Verlag 2006.

B.

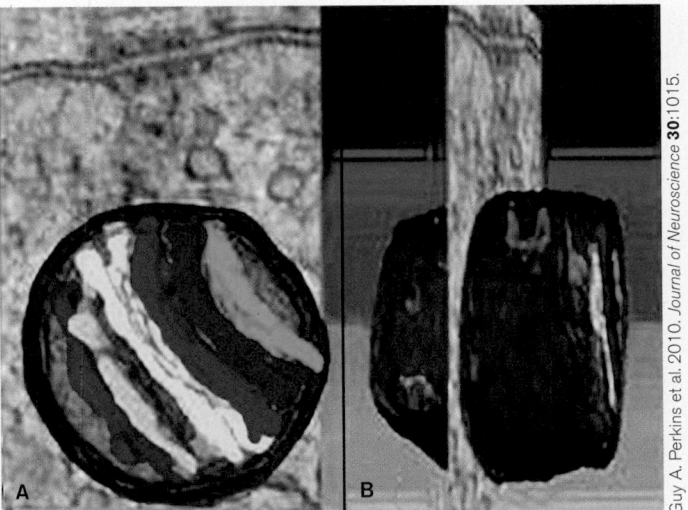

A   B

Guy A. Perkins et al. 2010. *Journal of Neuroscience* **30**:1015.

Figure 14.2 Respiratory membranes. A. Electron transport occurs in the inner (cytoplasmic) membrane of *Helicobacter pylori*. The cytoplasmic membrane and cell wall are surrounded by periplasm and outer membrane. **B.** Mitochondrion within a cat brainstem neuron, modeled by cryo-EM tomography, shown with one EM section through the cell. The mitochondrial outer membrane (blue) encloses inner membrane organized in pockets called cristae (other colors).

inner membrane contains all ETS complexes; the periplasm, cell wall, and outer membrane do not participate in respiration.

By contrast, bacteria that donate electrons to metals instead of to oxygen, such as *Shewanella* and *Geobacter* species, have extra cytochromes in their outer membrane. The outer membrane cytochromes help metal-reducing bacteria gain access to insoluble forms of metal—or to the surface of a fuel cell electrode (see **Special Topic 14.2**). In other bacteria, such as the nitrite oxidizer *Nitrospira*, the respiratory membranes form intracytoplasmic pockets called "lamellae."

Respiratory membranes similar to those of bacteria are found within our own mitochondria (**Fig. 14.2B**). Mitochondrial respiration uses only O_2 as a terminal electron acceptor. The ETS proteins are embedded in folds of the mitochondrial inner membrane called "cristae" (singular, crista). The inner membrane separates the inner mitochondrial space from the intermembrane space (between the inner and outer membranes). The mitochondrial inner membrane, including its electron transport proteins, evolved from the cell membrane of an endosymbiotic bacterial ancestor. Among modern bacteria, the closest relatives of mitochondria are the rickettsias, obligate intracellular parasites that cause serious diseases, such as Rocky Mountain spotted fever. Mitochondrial evolution is discussed further in Chapter 17.

The electron carrier molecules include proteins and small organic cofactors, molecules that are loosely or tightly bound to the proteins. The protein components of an ETS were first discovered in the 1930s at Cambridge University by Russian scientist David Keilin (1887–1963).

Keilin was an entomologist who studied insect mitochondria. The mitochondrial inner membranes contained proteins called **cytochromes**, which were named for their deep colors, typically red to brown.

In prokaryotes, cytochromes are often found in the cell membrane. **Figure 14.3** shows the absorbance spectrum of a cytochrome from the cell membrane of *Haloferax volcanii*, a halophilic archaeon isolated from the Dead Sea. In the reduced cytochrome, light absorption peaks in the blue range (440 nm) and in the red (607 nm). Upon oxidation, however, the cytochrome loses its absorption peak at 607 nm, and the 440-nm peak shifts to a shorter wavelength; thus, its reflected (unabsorbed) color shifts to the red. Different species of bacteria make many different cytochromes, with different absorption peaks, but all cytochromes show spectral change with change in reduction state.

A membrane ETS typically includes several different cytochromes with different reduction potentials. Keilin proposed that the cytochromes pass electrons sequentially from each protein complex to the next-stronger electron acceptor, with each step providing a small amount of energy to the organism (**Fig. 14.4**). The electron transport proteins are called **oxidoreductases** because they oxidize one substrate (removing electrons) and reduce another (donating electrons). Thus, they couple different half reactions in the electron tower (**Table 14.1**). Oxidoreductases consist of multiple-protein complexes that include cytochromes as well as noncytochrome proteins. The structure and function of ETS oxidoreductase complexes are discussed in Section 14.3.

TO SUMMARIZE:

■ **An electron transport chain (ETS)** consists of a series of electron carriers that sequentially transfer electrons to the carrier of next-higher reduction potential E (that is, the next-stronger electron acceptor). Electron flow through the ETS begins with an

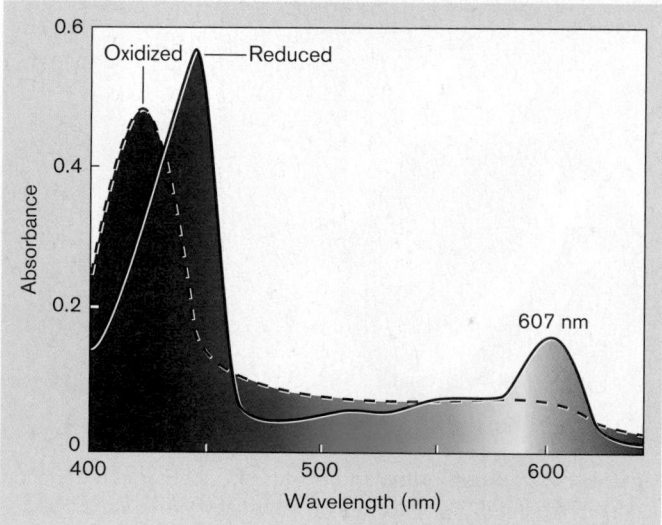

Figure 14.3 Light absorbance spectrum of a cytochrome. Absorption peaks shift between the oxidized and reduced forms of a cytochrome from the electron transport system of *Haloferax volcanii*, a halophilic archaeon.

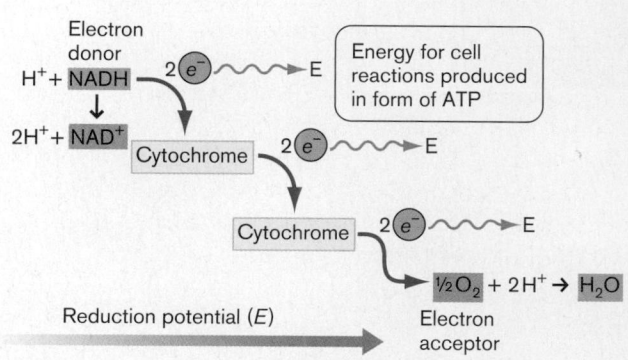

Figure 14.4 Electron transport system. Keilin's model for electron transfer through cytochromes. Each cytochrome in turn receives electrons from a stronger electron donor and transfers them to a stronger electron acceptor.

initial electron donor from outside the cell and ultimately transfers all electrons to a terminal electron acceptor that leaves the cell.

■ **The reduction potential E** for a complete redox reaction must be positive to yield energy for metabolism. The standard reduction potential $E^{\circ\prime}$ assumes all reactant concentrations equal 1 M, at pH 7.

■ **Concentrations of electron donors and acceptors** in the environment influence the actual reduction potential E experienced by the cell.

■ **The ETS is embedded in a membrane that separates two compartments.** Two aqueous compartments must be separate to maintain an ion gradient generated by the ETS.

■ **The ETS is composed of protein complexes and cofactors.** Protein complexes called oxidoreductases include cytochromes and noncytochrome proteins. Cytochromes are colored proteins whose absorbance spectrum shifts when there is a change in redox state.

14.2 The Proton Motive Force

The sequential transfer of electrons from one ETS protein to the next yields energy to pump ions (in most cases H^+) across the membrane. Proton pumping generates a **proton motive force** (or **proton potential**) composed of the H^+ concentration difference plus the charge difference across the membrane. The proton motive force (PMF) drives many different cell processes, including ATP synthesis, flagellar rotation, and nutrient transport, and flagellar rotation. Bacteria swim using rotary motors powered by a proton current.

The ETS Pumps Protons

The ion pumped by most ETS complexes is H^+. In this book, we use "hydrogen ions" and "H^+" interchangeably with "protons." In fact, we do not know the precise form of the ion that crosses the membrane. In water solution, H^+ never occurs as such, because it combines with a water molecule to form hydronium ion, H_3O^+. Current data, however, are consistent with the passage of hydrogen nuclei (protons) through the proton pumps of the electron transport system. Within the pump complex, the proton associates with one chemical group after another; for example, H^+ may combine with the amine of an amino acid (RNH_2) to form an ammonium ion (RNH_3^+).

The transfer of H^+ through a proton pump generates an H^+ concentration difference across the membrane. Since H^+ carries a positive charge, the proton transfer also generates a charge difference across the membrane. The H^+ concentration difference (ΔpH) plus the charge difference ($\Delta\psi$) make a proton potential (Δp), or proton motive force (PMF). The proton potential (in volts) equals the

electrochemical proton gradient (in joules) multiplied by Faraday's constant. The proton potential stores energy to drive protons back across the membrane through devices such as the ATP synthase and the flagellar motor. In pathogens, the proton potential drives drug efflux pumps that confer resistance to antibiotics such as tetracycline.

The discovery of the proton potential radically changed the field of biochemistry. Early in the twentieth century, Keilin and other scientists knew that the energy acquired by electron transport proteins was used to make ATP, but they did not know how. Most were convinced that electron transport was somehow directly coupled to ATP synthesis. The actual means of coupling was discovered by a student of Keilin's: Peter Mitchell (1920–1992). In 1961, Mitchell proposed an astonishing explanation for the coupling of electron transport to ATP synthesis (**Fig. 14.5**). The chemiosmotic theory states that the energy from electron transfer between membrane proteins is used to pump protons across the membrane, accumulating a higher H^+ concentration in the compartment outside. The pump generates the proton motive force Δp, which stores energy that can be used to make ATP (discussed in Section 14.3).

At first, many biologists found it difficult to understand how a proton potential could be coupled to ATP synthesis at an enzyme complex separate from the ETS, at a distant location on the membrane. The explanation is that the H^+ concentration gradient and charge difference exist everywhere on the membrane separating the

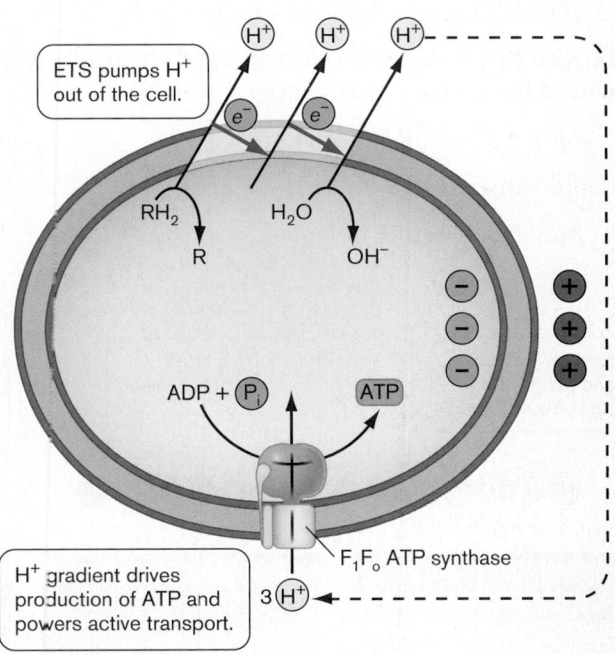

Figure 14.5 The chemiosmotic theory. The chemiosmotic theory states that the electron transport system pumps protons out of the cell. The resulting electrochemical gradient of protons (proton motive force) drives conversion of ADP to ATP through ATP synthase, as well as transport of substrates such as sugars.

two compartments. Because the membrane is impermeable to hydrogen ions, the H⁺ current can flow back only through a proton-driven complex such as the membrane ATP synthase (discussed in Section 14.3). In effect, the proton potential is a "proton battery," analogous to the electron potential of an electrical battery. The concept of cellular electricity proved highly controversial among scientists, who devised many experiments to test it before it was accepted (see **Special Topic 14.1**).

Δ*p* Includes Δψ and ΔpH

When protons are pumped across the membrane, energy is stored in two different forms—the separation of charge, or electrical potential (because H⁺ carries a positive charge), and the gradient of H⁺ concentration, or pH difference—as shown in **Figure 14.6**. Either form (or both) can drive Δ*p*-dependent cell processes. Both forms of energy contribute to the Δ*p* (proton potential):

- **The electrical potential** (Δψ) arises from the separation of charge between the cytoplasm (more negative) and the solution outside the cell membrane (more positive). For many bacteria, this "battery" potential is about –50 to –150 mV.
- **The pH difference** (ΔpH) is the log ratio of external to internal chemical concentrations of H⁺. For example, if the bacterial internal pH is 7.5 and the external pH is 6.5, the ΔpH is 1.0, and the ratio of $[H^+]_{out}$ to $[H^+]_{in}$ is 10. A ΔpH of 1.0 corresponds to a proton chemical concentration potential of approximately –60 mV.

The relationship between electrical and chemical components of the proton potential Δ*p* is given by

$$\Delta p = \Delta\psi - (2.3RT/F)\Delta pH \text{ mV}$$

or approximately

$$\Delta p = \Delta\psi - 60\Delta pH \text{ mV}$$

For cells grown at neutral pH, all three terms (Δ*p*, Δψ, and –60ΔpH) usually have a negative value, meaning that their force drives protons inward from outside.

In living cells, the relative contributions of Δψ and –60ΔpH vary depending on other sources of charge difference and protons, as shown in **Figure 14.6**. Both the Δψ and ΔpH components of Δ*p* are influenced by other factors besides ETS proton transport. For example

- **The charge difference Δψ** includes charges on other ions, such as K⁺ and Na⁺. These ions are pumped or exchanged by ion-specific membrane transport proteins.
- **The pH difference ΔpH** is affected by metabolic generation of acids, such as fermentation acids, or by pH changes outside the cell. Permeant acids can run down the gradient through the membrane, collapsing the ΔpH to zero.

Overall, the cell uses its various membrane pumps and metabolic pathways to adjust and maintain its proton motive force at a size sufficient to drive ATP synthesis but not so great as to disrupt the membrane.

> **THOUGHT QUESTION 14.3** What do you think happens to Δψ as the cell's external pH increases or decreases? What could happen to the Δ*p* of bacteria that are swallowed and enter the extremely acidic stomach?

How are the Δ*p* and its two components actually measured? In larger cells, such as those of eukaryotes, a microelectrode can be inserted into the cell to measure voltage potentials directly. Most bacterial cells are too small to have electrodes inserted into them, but Δψ and ΔpH can be measured by the uptake of molecules whose ability to permeate the membrane depends on their ionic state.

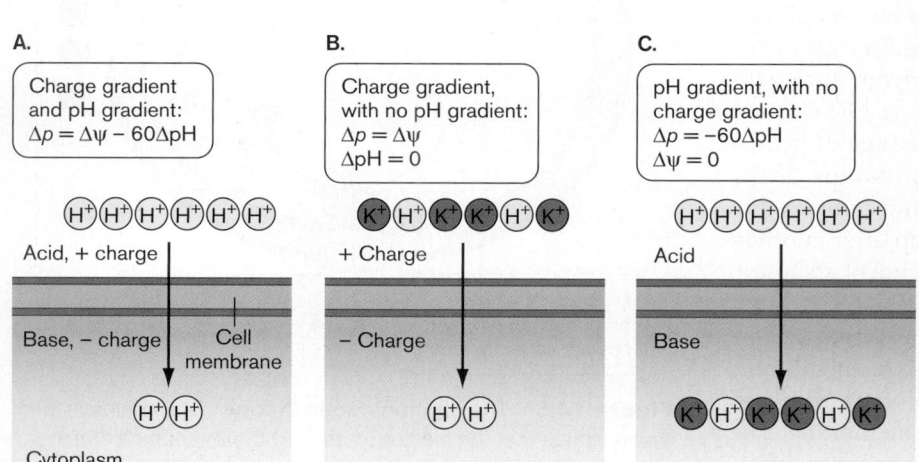

A.

Charge gradient and pH gradient:
Δ*p* = Δψ – 60ΔpH

H⁺ H⁺ H⁺ H⁺ H⁺ H⁺
Acid, + charge

Base, – charge   Cell membrane

H⁺ H⁺
Cytoplasm

B.

Charge gradient, with no pH gradient:
Δ*p* = Δψ
ΔpH = 0

K⁺ H⁺ K⁺ K⁺ H⁺ K⁺
+ Charge

– Charge

H⁺ H⁺

C.

pH gradient, with no charge gradient:
Δ*p* = –60ΔpH
Δψ = 0

H⁺ H⁺ H⁺ H⁺ H⁺ H⁺
Acid

Base

K⁺ H⁺ K⁺ K⁺ H⁺ K⁺

Figure 14.6 Electrical potential and pH difference. Proton motive force drives protons into the cell (arrow). **A.** The proton motive force Δ*p* is composed of the transmembrane electrical potential, Δψ (the charge difference); plus the transmembrane pH difference, ΔpH (the chemical concentration gradient of H⁺). **B.** If the pH inside and outside the cell is equal, then Δ*p* = Δψ. **C.** If the electrical charge inside and out is equal, then Δ*p* = –60ΔpH.

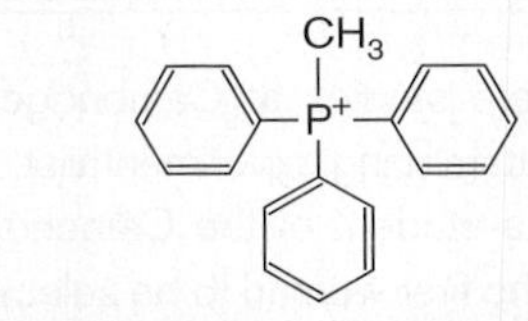

CH_3

Triphenylmethylphosphonium ion (TPMP)

Figure 14.7   A lipid-soluble cation.
Triphenylmethylphosphonium ion (TPMP) is soluble in the membrane because its positive charge on the phosphorus is surrounded by hydrophobic groups.

Lipid-soluble cations measure $\Delta\psi$. Most ions carrying positive charge (cations) cannot penetrate the hydrophobic membrane. However, certain cations contain multiple hydrophobic side chains whose lipid solubility overcomes the unfavorable interaction with the positive charge. An example is triphenylmethylphosphonium (TPMP) (**Fig. 14.7**). TPMP can penetrate and cross the membrane. Because no H^+ is carried, uptake of the cation is driven solely by the charge difference (usually negative inside the cell). A radiolabeled TPMP will accumulate in the cell in proportion to the charge difference ($\Delta\psi$). Thus, the cellular uptake of TPMP can be used to measure $\Delta\psi$.

Membrane-permeant weak acids measure ΔpH. Weak acids such as benzoic acid cross the membrane primarily in the uncharged form (HA) (**Fig. 14.8A**), as discussed in Chapters 3 and 4. Many fermentation acids, such as acetate and lactate, are membrane-permeant weak acids. At low concentration, the weak acid acts as a proton carrier, soluble in the membrane. Inside the cell, at higher pH, the HA dissociates to H^+ plus A^-. The acid accumulates inside until the pH inside falls to the level of the pH outside. Thus, uptake of a radiolabeled weak acid can be used to measure ΔpH.

Some weak acids cross the membrane in both charged and uncharged forms. The weak acid 2,4-dinitrophenol (DNP) dissociates to an anion whose charge is relatively evenly distributed around the molecule; thus, it remains overall hydrophobic enough to penetrate the membrane (**Fig. 14.8B**). Because both protonated (uncharged) and deprotonated (negatively charged) forms cross the membrane, DNP can cyclically bring protons into the cell and collapse both $\Delta\psi$ and ΔpH (**Fig. 14.8C**). Such molecules are called **uncouplers** because they uncouple electron transport from ATP synthesis through the membrane ATP synthase. Uncouplers are highly toxic to bacteria, and in the early twentieth century, they were considered for use as antibiotics. Unfortunately, uncouplers are also highly toxic to human cells. All cells need to maintain ion gradients and voltage potentials.

Δp Drives Many Cell Functions

Besides ATP synthesis, Δp drives many cell processes directly, such as the rotation of flagellar motors (**Fig. 14.9**). Proton transport is also coupled to transport of ions such as K^+ or Na^+ through parallel transport (symport) or oppositely directed transport (antiport), as discussed in Chapters 3 and 4. Ion flux then drives uptake of nutrients such as amino acids or efflux of molecules such as antibacterial drugs. Drug efflux pumps are a major problem in hospitals, where the pump proteins are encoded by genes carried on plasmids that spread among virulent strains.

TO SUMMARIZE:

■ **The ETS complexes generate a proton motive force (Δp).** The force is usually directed inward, driving protons into the cell.

■ **The proton potential drives ATP synthase** and other functions, such as ion transport and flagellar rotation.

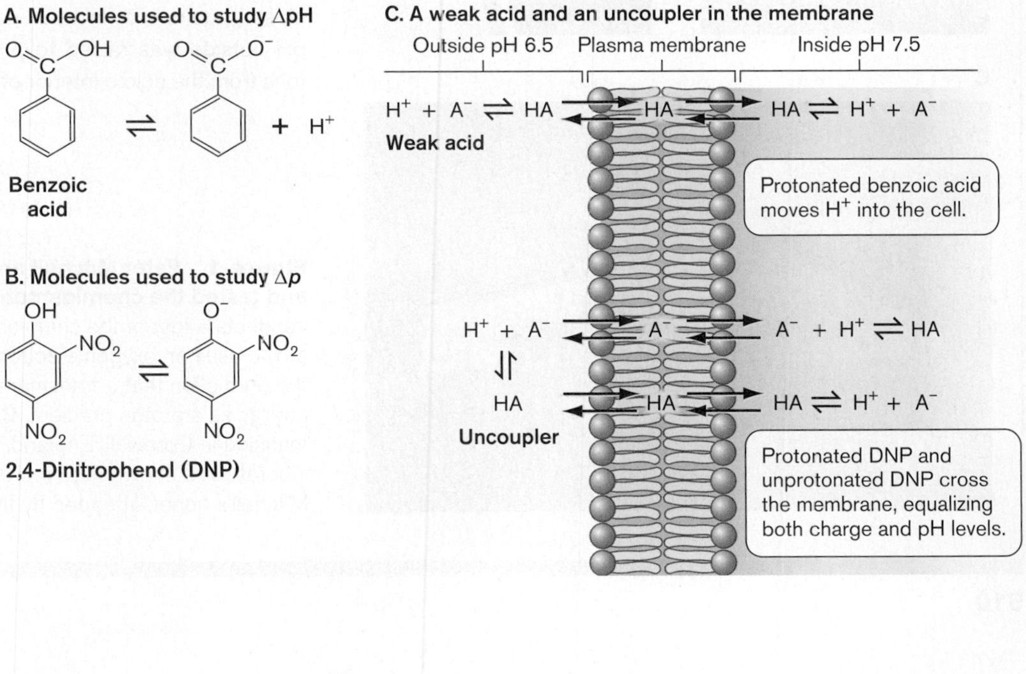

Figure 14.8   Membrane-permeant weak acids and uncouplers.   A. The transmembrane pH difference, ΔpH, drives uptake of a membrane-permeant weak acid such as benzoic acid, which dissolves in the membrane only in its protonated form. **B.** If an acid such as 2,4-dinitrophenol (DNP) crosses the membrane in both protonated and unprotonated forms, it cycles H^+ both ways until the entire proton motive force is dissipated. **C.** A weak acid and an uncoupler cross the membrane.

A. Molecules used to study ΔpH

Benzoic acid

B. Molecules used to study Δp

OH

2,4-Dinitrophenol (DNP)

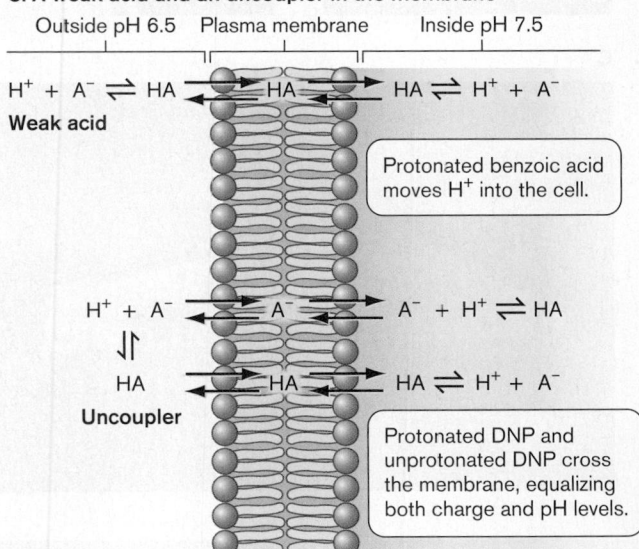

C. A weak acid and an uncoupler in the membrane

Outside pH 6.5   Plasma membrane   Inside pH 7.5

$H^+ + A^- \rightleftharpoons HA$ HA $HA \rightleftharpoons H^+ + A^-$
Weak acid

Protonated benzoic acid moves H^+ into the cell.

$H^+ + A^-$ A^- $A^- + H^+ \rightleftharpoons HA$

HA   HA $HA \rightleftharpoons H^+ + A^-$
Uncoupler

Protonated DNP and unprotonated DNP cross the membrane, equalizing both charge and pH levels.

In 1961, as a young researcher at the University of Edinburgh, Peter Mitchell (**Fig. 1A**) first developed the chemiosmotic theory. He developed the idea of vectorial metabolism—that metabolic energy could be stored in a directional form, in an ion gradient. The energy stored in a gradient of ions across a membrane could be interconverted with the energy of chemical interactions, such as oxidation of glucose. Through vectorial metabolism, the oxidation-reduction reactions of the electron transport system could drive the pumping of protons across a membrane, as well as transport of other ions and nutrients. The proton gradient could then drive protons back through the F_1F_o ATP synthase, driving synthesis of ATP.

But Mitchell's colleagues rejected his ideas. Researchers had spent many years discovering the molecular "intermediates" of metabolism, and they expected to find one or two more between the ETS and the ATP synthase. Furthermore, most biochemists lacked sufficient training in physics to appreciate Mitchell's ideas, and Mitchell's communication skills did not cross the gap. So Mitchell used his family wealth to leave Edinburgh and form the Glynn Research Foundation. His foundation supported a bioenergetics laboratory at Glynn House, a country estate in Cornwall that Mitchell restored and equipped (**Fig. 1C**). At Glynn, Mitchell developed a research program to test various principles and predictions of the chemiosmotic theory.

During his graduate studies at Cambridge, Mitchell had collaborated with an outstanding experimentalist, Jennifer Moyle (**Fig. 1B**). Moyle was a student of the Cambridge biochemist Marjorie Stephenson, the first woman to be selected as a Fellow of the Royal Society. Stephenson suggested that Moyle work with Mitchell, and it was Moyle who performed most of the early experiments testing the chemiosmotic theory. Moyle continued working with Mitchell at Edinburgh, and she moved with him to Glynn, where they worked together for the next 30 years.

In the 1960s, Mitchell and Moyle devised several experiments to test the chemiosmotic theory. A key requirement was to show that the ETS generates a proton potential. Moyle showed that respiration of mitochondria is associated with proton efflux: Mitochondria isolated from rat livers were exposed to oxygen, causing efflux of hydrogen ions. The H^+ efflux occurred as electrons were transferred across the ETS and hydrogen ions were expelled by the proton pumps. In Moyle's experiments, the number of protons extruded per electron transferred down the ETS was consistent with the chemiosmotic model. Similar results were obtained with vesicles made first from mitochondrial membranes and later from the membranes of chloroplasts and bacteria.

A greater challenge was to demonstrate that a proton motive force could, in fact, drive the ATP synthase and other ion transporters without any undiscovered intermediate. Many researchers attempted to show this, often with conflicting results. The first successful experiment was reported in 1966 by André Jagendorf, a colleague from Johns Hopkins University who had visited Mitchell at Glynn. Jagendorf and his postdoctoral fellow Ernest Uribe tested the effect of a proton gradient imposed on spinach chloroplasts (**Fig. 2A**). Chloroplasts contain ATP synthase directed outward (that is, driven by proton flow from inside to outside). The chloroplasts were partly opened by osmotic shock, and then suspended in medium containing concentrated hydrogen ions (pH 4). With the osmotic balance restored, the chloroplast membranes closed again, with increased [H^+] trapped inside. When the pH outside was raised to pH 8.3 (that is, lower [H^+]), protons from the acidic interior of the vesicles flowed out through

A.

Courtesy of Peter Rich, U. of London

B.

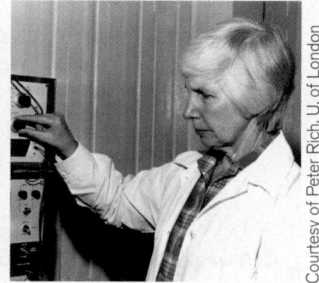

Courtesy of Peter Rich, U. of London

C.

Courtesy of Peter Rich, U. of London

Figure 1   Peter Mitchell and Jennifer Moyle proposed and tested the chemiosmotic theory.   A. Peter Mitchell conducts a test of the chemiosmotic theory. **B.** Jennifer Moyle uses an oxygen electrode for an experiment testing the prediction that a transmembrane voltage potential stores energy as a proton gradient. **C.** Glynn House, the Regency mansion in Cornwall, England, that Mitchell restored for his laboratory, is shown here in 1990 during a conference in Mitchell's honor, attended by five Nobel Prize winners.

the ATP synthase, and ADP and inorganic phosphate were converted to ATP. Jagendorf interpreted this result as consistent with Mitchell's chemiosmotic theory for the mechanism of phosphorylation.

A. A pH difference, ΔpH, drives ATP synthesis.

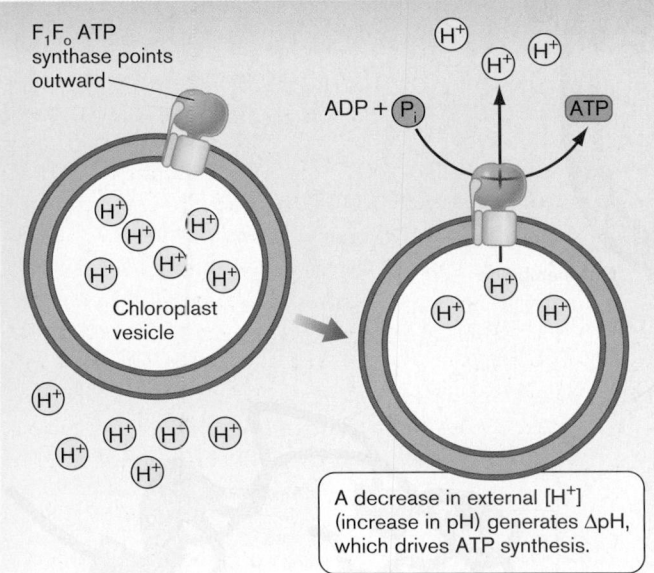

F_1F_o ATP synthase points outward

Chloroplast vesicle

ADP + P_i

ATP

A decrease in external [H^+] (increase in pH) generates ΔpH, which drives ATP synthesis.

B. A charge difference, Δψ, drives ATP synthesis.

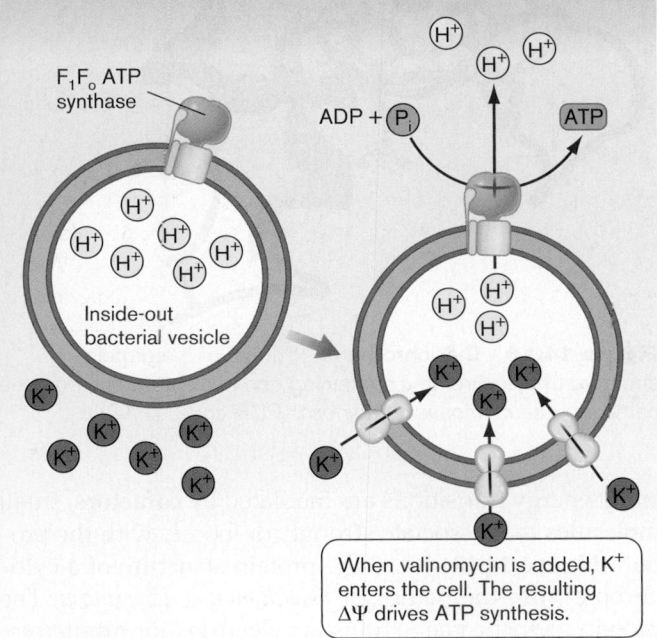

F_1F_o ATP synthase

ADP + P_i

ATP

Inside-out bacterial vesicle

When valinomycin is added, K^+ enters the cell. The resulting ΔΨ drives ATP synthesis.

Figure 2 **Either ΔpH or Δψ drives ATP synthesis.** **A.** A pH difference imposed across the chloroplast inner membrane drives a proton current through an outwardly directed ATP synthase. **B.** A charge difference generated by K^+ influx drives a proton current through ATP synthase.

Other researchers showed that ATP synthesis could be driven by a charge difference across a vesicle membrane (**Fig. 2B**). An artificial electrical potential (Δψ) was applied by loading vesicles with potassium ions (K^+). The potassium ions were conducted across the membrane by an ionophore, a small organic molecule that binds to an ion and solubilizes it in the membrane. The K^+ ionophore was the antibiotic valinomycin, a cyclic peptide of 12 amino acids produced by the bacterium *Streptomyces fulvissimus*. Valinomycin specifically binds K^+, while its hydrophobic side chains solubilize it in the membrane; thus the molecule conducts K^+ across the membrane down its concentration gradient (see **Fig. 2B**). In the experiment shown, the vesicles are "inside-out" bacterial membrane vesicles, in which the ATP synthase points outward instead of inward. The K^+ influx adds positive charge, which drives H^+ out through ATP synthase, catalyzing formation of ATP. Thus, Δψ drives formation of ATP.

These experiments, however, could not definitively rule out the existence of an undetected component of the membrane that somehow energized ATP synthase. A more compelling experiment was performed in 1975 by Efrem Racker and colleagues at Cornell University. Racker developed a process to make artificial vesicles called "liposomes," composed of purified phospholipids in the absence of any membrane proteins. He then purified a proton pump called bacteriorhodopsin from a halophilic archaeon, a *Halobacterium* species. Bacteriorhodopsin acts as a single-protein proton pump driven by light absorption. The bacteriorhodopsin was combined with phospholipids and with ATP synthase purified from mitochondria to obtain "reconstituted liposomes" with functional bacteriorhodopsin and mitochondrial ATP synthase. When the liposomes were exposed to light, the bacteriorhodopsin pumped protons and the ATP synthase made ATP. Thus, a proton pump and a proton-driven ATP synthase from two completely different organisms could work together in a membrane, connected solely by the proton motive force.

In 1978, Mitchell received the Nobel Prize in Chemistry, and the scientific community fully embraced the role of ion currents in metabolism. In later life, when a colleague disagreed with his views, Mitchell would turn off his hearing aid. Today the chemiosmotic principle is firmly established as the basis of energy transduction in respiration, lithotrophy, and photosynthesis. Interesting questions were addressed by Peter Rich (now at University College London), who succeeded Mitchell as director of research at Glynn in 1987. Rich uses spectrophotometry to investigate proton transport by specific ETS components, such as cytochrome oxidase. Other researchers study how extreme acidophiles and alkaliphiles manage to grow at pH values as low as pH 1 or as high as pH 14 while maintaining internal pH within two units of neutrality. Medical researchers investigate how pathogenic bacteria and tumor cells use proton-driven efflux pumps to expel antibiotics from the cell.

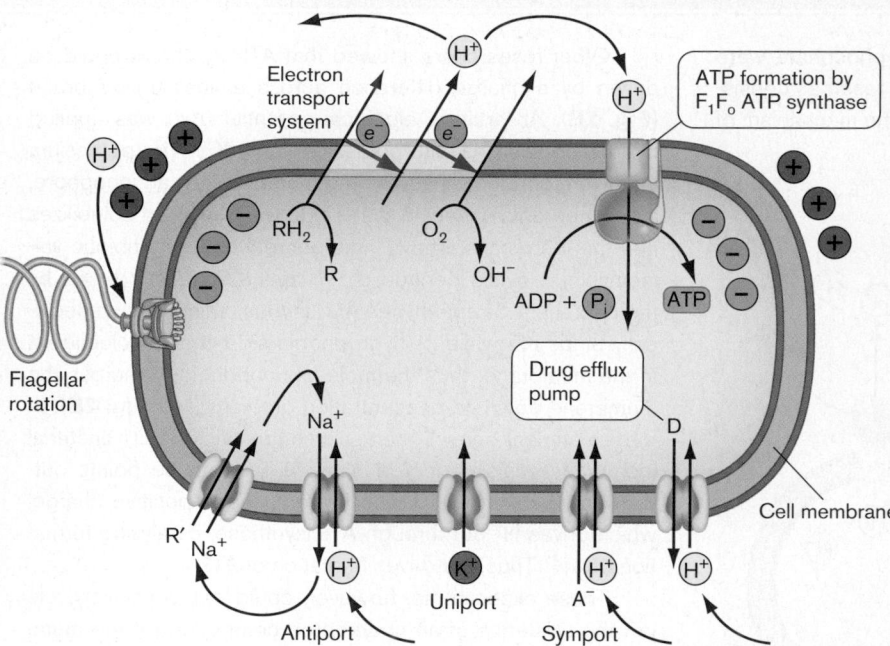

Figure 14.9 Processes driven by the proton motive force. Processes powered by proton potential include ATP synthesis through the F_1F_o ATP synthase, flagellar rotation, uptake of nutrients, and efflux of toxic drugs. "R" refers to an organic nutrient.

- **The proton potential (Δp, measured in millivolts)** is composed of the electrical potential ($\Delta\psi$) and the hydrogen ion chemical gradient (ΔpH): $\Delta p = \Delta\psi - 60\Delta$pH.

- **Uncouplers are molecules taken up by cells in both protonated and unprotonated forms.** Uncouplers can collapse the entire proton potential, thus uncoupling respiration from ATP synthesis.

THOUGHT QUESTION 14.4 Suppose that de-energized cells ($\Delta p = 0$) with an internal pH 7.6 are placed in a solution at pH 6. What do you predict will happen to the cell's flagella? What does this demonstrate about the function of Δp?

14.3 The Respiratory ETS and ATP Synthase

In the respiratory electron transport system (ETS), a series of carrier molecules harvests the reducing potential of electrons in small steps. The respiratory ETS most commonly presented is that of the mitochondrial inner membrane, in which NADH and $FADH_2$ transfer electrons ultimately to O_2, producing H_2O. Microbes also use many alternative electron donors and acceptors, including small organic molecules and minerals. Respiration is crucial for many bacterial pathogens.

Cofactors Allow Small Energy Transitions

ETS proteins such as cytochromes associate electron transfer with small, reversible energy transitions. The

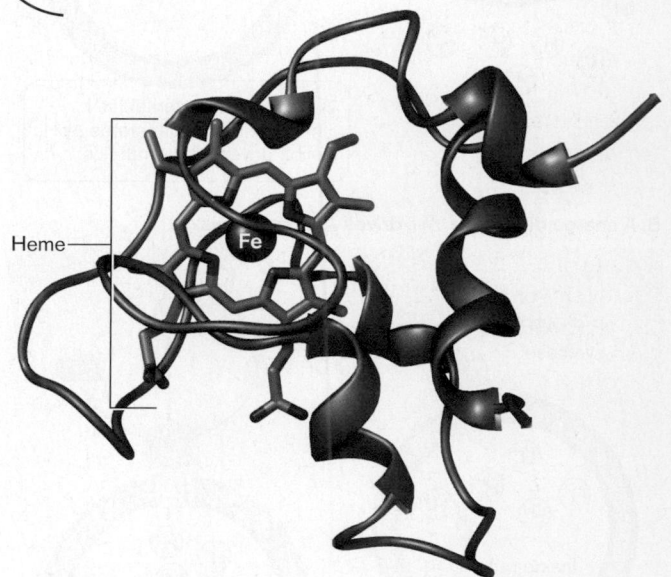

Figure 14.10 Cytochrome c. Portion of the protein structure of cytochrome c containing one heme cofactor, from the pathogen *Pseudomonas aeruginosa*. (PDB code: 2PAC)

small energy transitions are mediated by **cofactors**, small molecules that associate strongly or loosely with the protein. **Figure 14.10** shows the protein structure of a cytochrome from the pathogen *Pseudomonas aeruginosa*. The protein, cytochrome c, transfers electrons for nitrate respiration. Its reddish color derives from the buried cofactor called **heme**, a ring of conjugated double bonds surrounding an iron atom. The heme plays a key role in acquiring and transferring electrons, with an Fe^{2+}/Fe^{3+} transition. Metal-reducing bacteria such as *Geobacter* make hundreds of different types of cytochromes in their periplasm, where electrons accumulate until the bacterium finds an oxidized metal to accept them.

Cofactors such as heme allow small, reversible redox changes. Other examples include flavin mononucleotide (FMN) and ubiquinone (**Fig. 14.11**). The structure of each cofactor must allow transition of an electron between orbitals with closely spaced energy levels to avoid "spending it all in one place." If all energy were spent in one transition, most of it would be lost as heat instead of being converted to several small processes, such as pumping H^+ across the membrane.

Small energy transitions typically involve these kinds of molecular structures:

- **Metal ions such as iron or copper**, coordinated (and hence held in place) with amino acid residues. Iron is often coordinated by sulfur atoms of cysteine residues in the protein; examples shown in **Figure 14.11B** are [2Fe-2S] and [4Fe-4S]. Transition metals make useful electron carriers because their outer electron shell has several closely spaced energy levels, facilitating small energy transitions.
- **Conjugated double bonds and heteroaromatic rings**, such as the nicotinamide ring of NAD^+/NADH, also provide narrowly spaced energy transitions. Membrane-soluble carriers such as **quinones** (reduced to **quinols**) allow even smaller energy transitions than does NAD^+/NADH.

The major protein complexes of electron transport each contain one or more reaction centers containing either metal ions or conjugated double bonds or both. In FMN, the conjugated double bonds allow a small energy transition (**Fig. 14.11A**). In iron-sulfur clusters—[2Fe-2S] and [4Fe-4FS]—the metal atoms provide the site for a small energy transition, its size dependent on the cluster's connections within the associated protein (**Fig. 14.11B**). The heme group found in cytochromes and other oxidoreductases contains extensive conjugated double bonds, coordinated around the metal Fe^{3+} (**Fig. 14.11C**). Reduction by a transferred electron converts the iron to Fe^{2+}. The branching of side chains on the ring varies among different hemes, altering the magnitude of $E^{\circ\prime}$ for the redox couple Fe^{3+}/Fe^{2+}.

Some electron carriers are small molecules that associate loosely with a protein complex and then come off to diffuse freely throughout the membrane. Mobile electron carriers include quinones such as ubiquinone (**Fig. 14.11D**), which are reduced to quinols:

$$Q + 2H^+ + 2e^- \rightarrow QH_2$$

Quinols carry electrons and protons laterally within the membrane between the proton-pumping protein complexes of the ETS. After transferring their electrons to the next protein complex, they revert to quinones, capable of accepting electrons again.

> **NOTE:** *Dehydrogenases, reductases,* and *oxidases* are all **oxidoreductases**, which oxidize one substrate (remove electrons) and reduce another (donate electrons). Oxidoreductases that accept electrons from NADH or $FADH_2$ are also called dehydrogenases, because their reaction releases hydrogen ions.

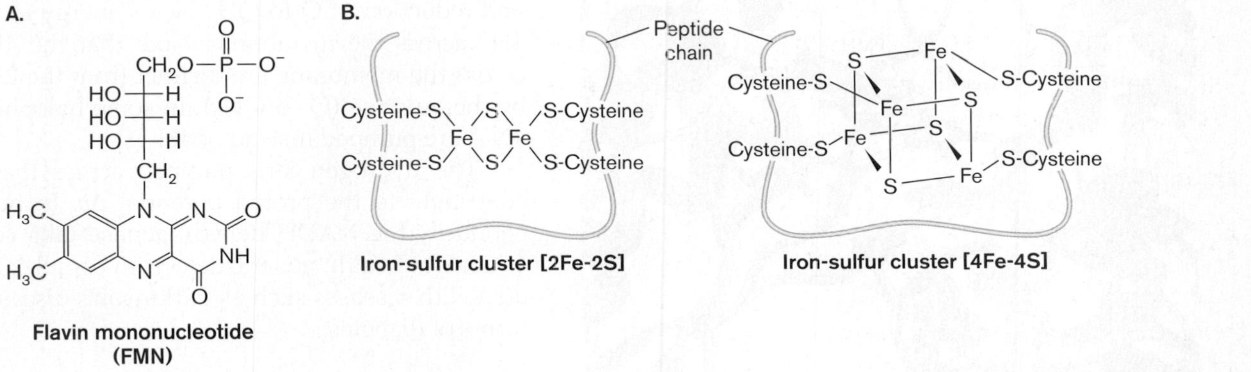

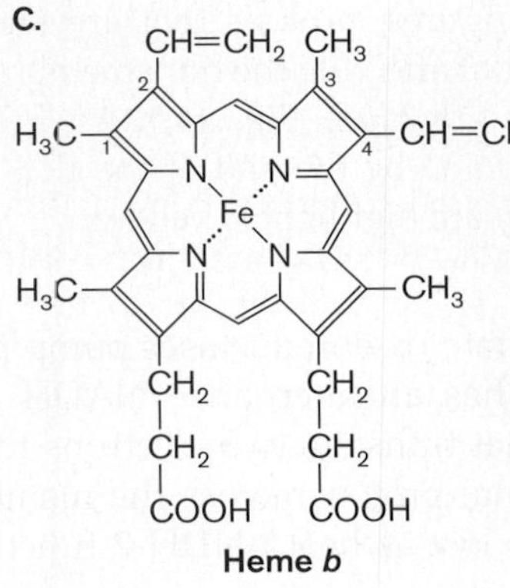

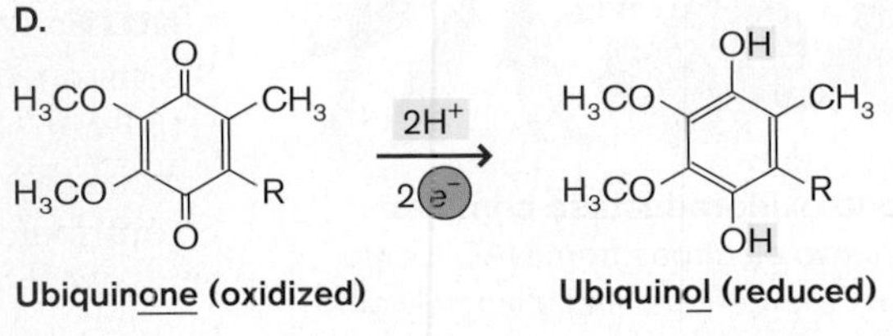

Figure 14.11   Reaction centers for electron transport.   A. Flavin mononucleotide (FMN). **B.** Iron-sulfur clusters: [2Fe-2S] and [4Fe-4S]. **C.** Heme *b*. The side chains of the ring vary among hemes, yielding different levels of redox potential. **D.** Ubiquinone, which is reduced to ubiquinol.

Oxidoreductase Protein Complexes

A respiratory electron transport system includes at least three functional components: an initial substrate oxido-reductase (or dehydrogenase), a mobile electron carrier, and a terminal oxidase. For each component, microbes make alternative versions that use alternative electron-donating substrates and terminal electron acceptors

A.

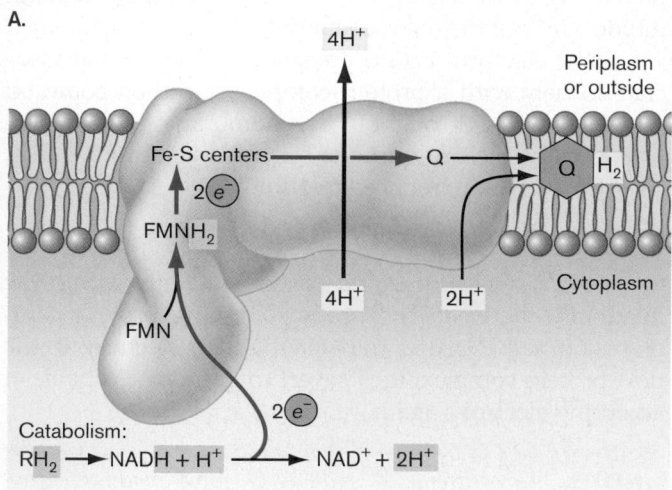

B.

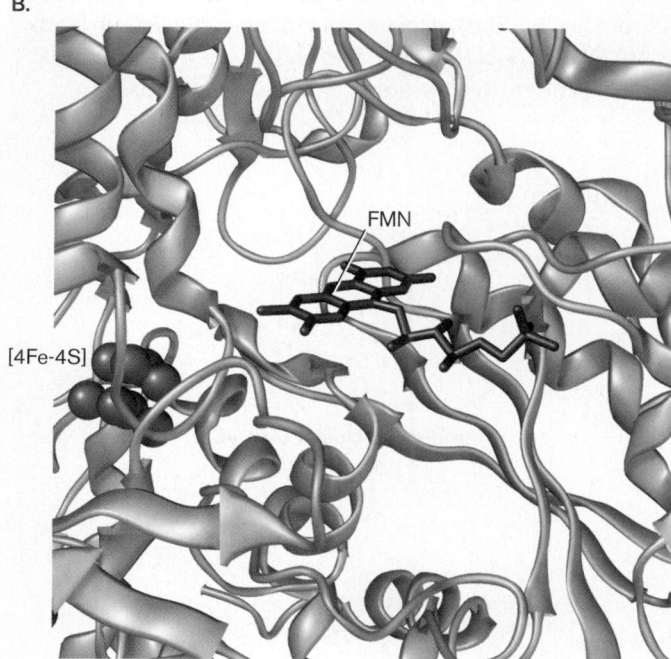

Figure 14.12   NADH:quinone oxidoreductase complex (NDH-1).   A. NDH-1 transfers two electrons from NADH onto FMN (FMNH$_2$), through several Fe-S centers, and ultimately onto a quinone, Q (forming quinol, QH$_2$). The energy from oxidizing NADH is coupled to pumping 4H$^+$ across the cell membrane. H$^+$ ions whose flow increases proton potential are highlighted yellow. **B.** Within the NDH-1 complex, FMN lies adjacent to the first iron-sulfur center, [4Fe-4S]. (PDB code: 1NOX) *Source*: B. L. A. Sazanov and P. Hinchliffe. 2006. *Science* **311**:1430–1436.

available in the environment. Here we present a typical bacterial ETS receiving electrons from NADH and transferring them to oxygen.

Initial substrate oxidoreductase. A respiratory ETS begins with an initial oxidoreductase that receives a pair of electrons from an organic substrate such as NADH. Note that NADH forms by receiving two electrons plus 2H$^+$ from an organic product of catabolism (designated RH$_2$) (**Fig. 14.12A**). The 2H$^+$ are ultimately balanced by 2H$^+$ from the cytoplasm combining with O$_2$ (or another terminal electron acceptor) at the end of the ETS. The two electrons (2e^-) from NADH enter an ETS protein complex embedded in the membrane.

NADH donates electrons to NADH dehydrogenase (NADH:quinone oxidoreductase, NDH-1) (**Fig. 14.12A**). In bacteria, the NDH-1 complex includes 14 different subunits; in mitochondria there are 46, forming one of the most elaborate membrane complexes known. A bacterial complex includes the cofactor FMN, as well as several iron-sulfur clusters—typically 7[4Fe-4S] and 2[2Fe-2S]. The cofactors "hand off" electrons to each other through adjacent connections; see, for example, the placement of FMN and the first [4Fe-4S] within the peptide coils of NDH-1 (**Fig. 14.12B**).

Each electron from NADH travels individually through FMN and the iron-sulfur series. At the end of the chain, the electrons and 2H$^+$ from solution are transferred to a quinone, which is thus reduced to a quinol. Quinones are designated Q; and quinols, QH$_2$.

Within the NDH-1 complex, the oxidation of NADH and reduction of Q to QH$_2$ yields energy to pump up to 4H$^+$ across the membrane. Note that the 4H$^+$ pumped across the membrane are distinct from the 2H$^+$ acquired by the quinone (Q → QH$_2$). In some halophilic bacteria, 4Na$^+$ are pumped instead of 4H$^+$.

The hydrogen ions pumped across the membrane contribute to the proton potential Δp. In human mitochondria, the NADH dehydrogenase (aka complex I) is critical for health; genetic defects in complex I are associated with diseases such as Parkinson's disease and some forms of diabetes.

> **NOTE:** In our figures, protons that are released and consumed without affecting the transmembrane potential are shaded gray (see **Fig. 14.12A**). Protons that cross the membrane by the end of the ETS (and thus contribute to Δp) are highlighted yellow.

Not all substrate oxidoreductases pump protons. For example, *E. coli* has an alternative NADH dehydrogenase (NDH-2) that transfers two electrons to Q without pumping additional protons across the membrane. (The unused energy is lost as heat.) NDH-2 functions during rapid growth, when the cell must limit its proton potential to avoid membrane breakdown. Other complexes, such as succinate dehydrogenase, transfer electrons from substrates that lack sufficient energy to pump extra protons.

Quinone pool. A quinone can receive $2e^-$ from the substrate oxidoreductase along with $2H^+$ from solution to balance the negative charges, yielding a quinol (**Fig. 14.12A**). The quinols diffuse within the membrane and carry reduction energy to other ETS components. After transferring $2e^-$ to the next protein complex, $2H^+$ are released. Usually the $2H^+$ released are on the opposite side of the membrane from where $2H^+$ were originally picked up (**Fig. 14.13**). Thus, besides electron transfer, a quinol may contribute two protons to the transmembrane proton potential. The reoxidized carriers then recycle back as quinones.

Each quinone can bind to a substrate dehydrogenase, pick up a pair of electrons and hydrogen ions, and then diffuse away and carry the electrons to a reductase. The quinones and quinols, referred to as the **quinone pool**, diffuse freely within the phospholipids of the membrane. Thus, the quinones/quinols are able to transfer electrons between many different redox enzymes.

Note that quinones with different side chains have different names, such as "ubiquinone" and "menaquinone." For clarity, we refer to all as quinones (Q), and their reduced forms as quinols (QH_2).

Terminal oxidase. A terminal oxidase complex receives electrons from a quinol (QH_2) and transfers them to a terminal electron acceptor, such as O_2 or NO_3^- (**Fig. 14.13**). The complex usually includes a cytochrome that accepts electrons from quinols. Cytochromes of comparable function are designated by letters, such as cytochrome b (*E. coli* and mitochondria) and cytochrome c (mitochondria). The cytochrome is bound to an oxidase complex containing a series of electron-transferring carriers: two iron-centered hemes and three copper atoms. This unique center couples electron transfer and proton pumping.

The *E. coli* cytochrome bo quinol oxidase consists of cytochrome b plus oxidase complex o. The cytochrome b subunit receives two electrons from a quinol ($QH_2 \rightarrow Q$) and releases the $2H^+$ out to the periplasm. Each electron from quinol travels through the two hemes of the oxidase complex. Thus, because $2H^+$ were consumed from the cytoplasm at the NADH oxidoreductase (**Fig. 14.12**) and $2H^+$ were released outside by a quinol, there is a net increase of $2H^+$ outside the cell. In addition, the transfer of $2e^-$ between the two oxidase hemes is coupled to pumping of $2H^+$ from the cytoplasm across to the periplasm (**Fig. 14.13**).

The second heme of the oxidase (heme o_3) receives an atom of oxygen from O_2. Each oxygen atom in turn receives two electrons and combines with two protons ($2H^+$) from the cytoplasm to form H_2O. The $2H^+$ consumed balances the $2H^+$ released by catabolism to make NADH + H^+.

Figure 14.13 Cytochrome *bo* quinol oxidase complex. Each quinol (QH_2) transfers two electrons to heme b. The two H^+ from each quinol are expelled to the periplasm (or outside the bacterial cell). The $2e^-$ from the quinols are transferred to heme o_3—a step coupled to pumping of $2H^+$ across the membrane. At heme o_3, $2e^-$ combines with $2H^+$ plus an oxygen atom from O_2 to form water. *Source:* Based on the structure determined by Jeff Abramson et al. 2000. *Nat. Struct. Biol.* **7**:910–917.

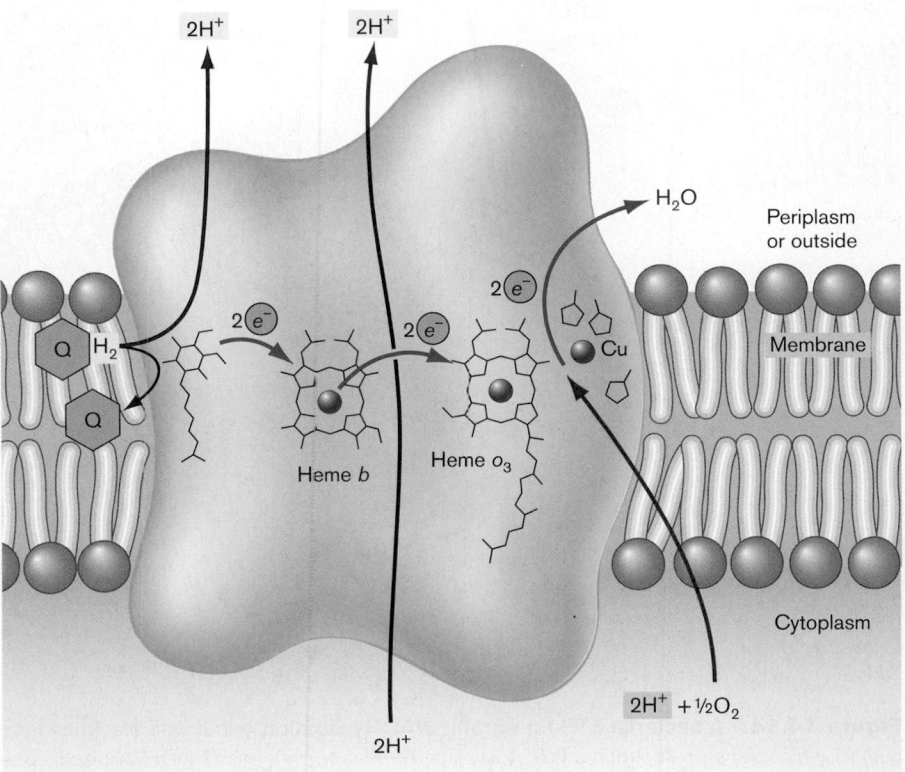

Note that the oxidase is conventionally shown as obtaining two electrons from the cytoplasm and donating them to $\frac{1}{2}O_2$, although, of course, $\frac{1}{2}O_2$ does not exist as such. A full reaction cycle of cytochrome *bo* quinol oxidase actually puts four electrons from two quinols (originally 2 NADH) onto O_2, taking up $4H^+$ from the cytoplasm to make two molecules of H_2O.

Besides cytochrome *bo* quinol oxidase, bacteria express different terminal oxidases that differ with respect to the ratio of cytoplasmic protons pumped to electrons transferred. For example, the alternative cytochrome *bd* oxidase reduces O_2 to water but pumps no extra protons. Although it pumps no protons, cytochrome *bd* oxidase can bind O_2 at much lower concentrations and donate electrons, completing the respiratory circuit. Thus, the *bd* oxidase enables *E. coli* to respire within low-oxygen habitats such as the mammalian intestine.

ETS Pathways for Respiration

As we have seen, a respiratory ETS includes at least three phases of electron transfer: (1) Electrons from an organic substrate are donated to an initial oxidoreductase; (2) the electrons are transferred to a quinone, which is reduced to a quinol; and (3) the quinol electrons are transferred to a terminal oxidase while its $2H^+$ are released outside. Both

enzymes and quinones show considerable complexity and diversity among different species in different environments. **Figure 14.14** summarizes one example of a complete ETS for oxidation of NADH by $\frac{1}{2}O_2$ in the inner membrane of *E. coli*.

The entering carrier NADH carries two electrons with protons obtained from catabolized food molecules. The two electrons ($2e^-$) from NDH-1 and two protons ($2H^+$) from solution are transferred onto Q (quinone), converting it to QH_2 (quinol). The electrons transferred from NADH include sufficient energy to pump up to $4H^+$ across the membrane. The exact number depends on cell conditions, such as the concentrations of NADH and the terminal electron acceptor.

The QH_2 diffuses within the membrane until it reaches a terminal oxidase complex, such as the cytochrome *bo* quinol oxidase. The $2H^+$ from QH_2 are released outside the cell, while the $2e^-$ enter the oxidase, reducing the two hemes. The two electrons join $2H^+$ from the cytoplasm, combining with an oxygen atom to make H_2O. The reaction is coupled to pumping of $2H^+$ across the membrane plus a net increase of $2H^+$ outside through redox reactions.

Overall, the oxidation of NADH exports about $8H^+$ per $2e^-$ transferred through the ETS to make H_2O. The export of protons generates a proton potential Δp.

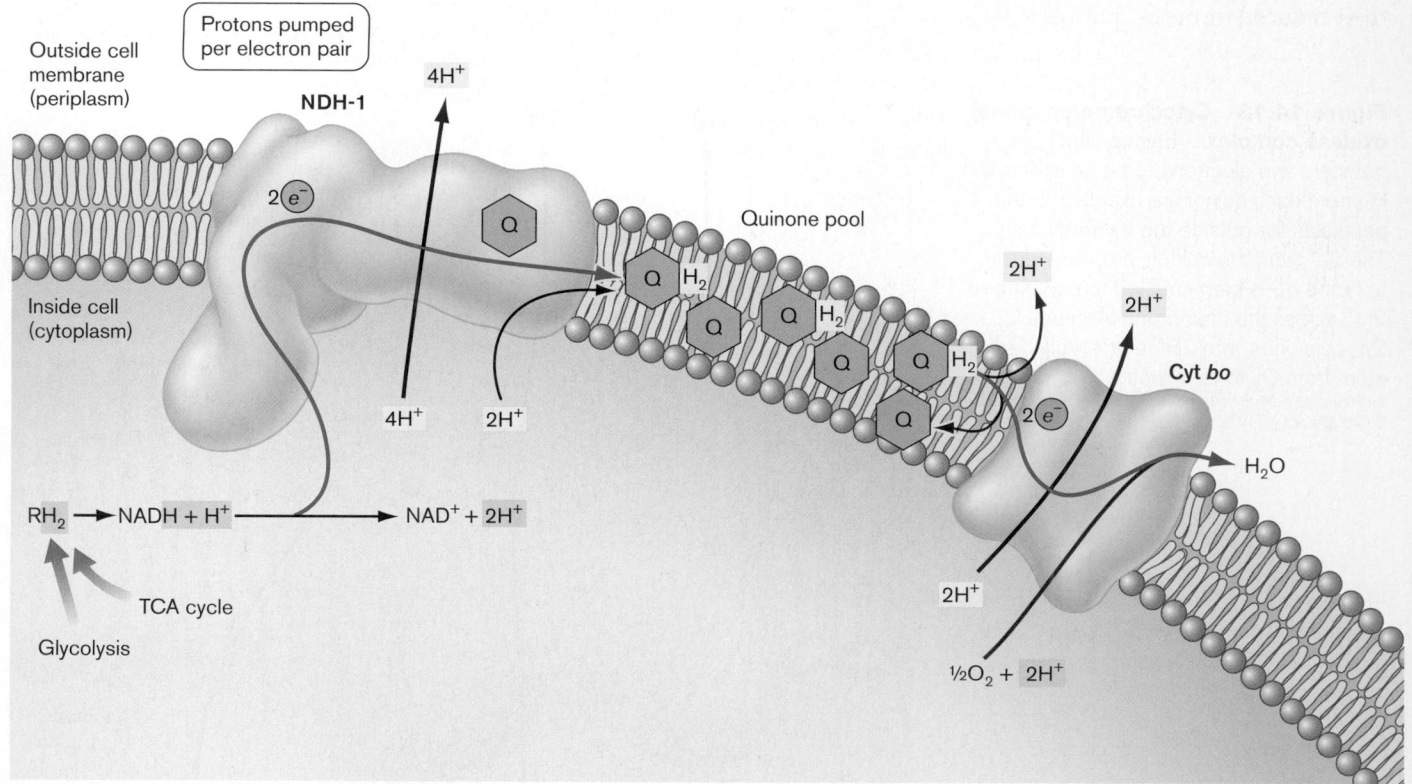

Figure 14.14   A bacterial ETS for aerobic NADH oxidation.   In *E. coli*, electrons from NDH-1 are transferred to quinones, generating quinols, which transfer electrons onto cytochrome *bo* (Cyt *bo*). For each NADH oxidized, up to $8H^+$ may be pumped across the membrane.
microbiology2.com/animations

NOTE: Protons (hydrogen ions, H^+) can have three different fates in the ETS:

1. Protons are pumped across the membrane by an oxidoreductase complex. Contributes Δp.
2. Protons are consumed from the cytoplasm by quinone/quinol, and then released across the membrane. Contributes Δp.
3. Protons are consumed by combining with the terminal electron acceptor (O_2). Balances protons released by catabolism; does not affect Δp.

THOUGHT QUESTION 14.5   What is the advantage of the oxidoreductase transferring electrons to a pool of mobile quinones, which then reduce the terminal reductase (cytochrome complex)? Why does each oxidoreductase not interact directly with a cytochrome complex?

THOUGHT QUESTION 14.6   Why are most electron transport proteins fixed within the cell membrane? What would happen if they "got loose" in aqueous solution?

Environmental modulation of the ETS. The ETS just described represents optimal conditions, when food and oxygen are unlimited. What happens in tough conditions, when food (that is, electron donors) or oxygen is scarce? Under varying environmental conditions, bacteria adjust the efficiency of their ETS by expressing alternative oxidoreductases. This adjustment depends on available concentrations of electron donors (such as NADH) and electron acceptors (such as O_2). Available concentrations of substrates determine the actual reduction potential E for NADH oxidation and hence the proton motive force Δp.

The overall equation for NADH oxidation by O_2 is

$$2\ NADH + 2H^+ + O_2 \rightarrow 2\ NAD^+ + 2H_2O$$

The reduction potential E depends on the concentrations of reactants and products:

$$E = E^{\circ\prime} - \frac{2.303\ RT}{nF} \times \log \frac{[NAD^+]}{[NADH]^2[H^+]^2[O_2]}$$

At low concentration of NADH or O_2 or at high pH (low H^+ concentration), the concentration term is negative, so the amount of energy released by the oxidation of NADH by O_2 is decreased. This means that fewer protons can be pumped per electron. Thus, it is advantageous for *E. coli* to have a range of options for its electron transport, pumping two to eight protons per NADH as needed.

The efficiency of the bacterial ETS is adjusted by making different NADH dehydrogenases and terminal oxidases under different environmental conditions (**Fig. 14.15**). In *E. coli*, the ETS components are regulated as follows:

- **High oxygen level.** When O_2 concentration is high, some electrons are donated to NDH-2 instead of NDH-1, allowing NAD^+ to be recycled without pumping extra protons. This is necessary to avoid generating too much negative potential and too high a pH in the cytoplasm.
- **Low oxygen level.** Cytochrome *bo* quinol oxidase requires a high concentration of O_2 to pump protons. At low $[O_2]$, as in the intestinal habitat, electrons are transferred instead from cytochrome *bd* quinol oxidase, which has a higher affinity for O_2. However, the *bd* oxidase cannot pump protons. So at low oxygen levels, expression of NDH-2 is repressed, allowing maximal proton pumping by NDH-1.
- **High external pH.** When external pH exceeds the pH of the cytoplasm, the cell needs to minimize proton export to maintain a constant internal pH. So the *bd* oxidase is expressed in preference to the proton-pumping *bo* oxidase.

THOUGHT QUESTION 14.7   Why would bacteria evolve membrane-embedded dehydrogenases in their ETS that fail to pump protons even when sufficient donor potential exists?

THOUGHT QUESTION 14.8   Experiments suggest that NADH dehydrogenase 2 is more often coupled with cytochrome *bo*, and dehydrogenase 1 more often with cytochrome *bd*. Why might this be the case?

Bacteria also have alternative oxidoreductases to serve different electron donors and acceptors. Some enzymes take electrons from donors lacking the potential

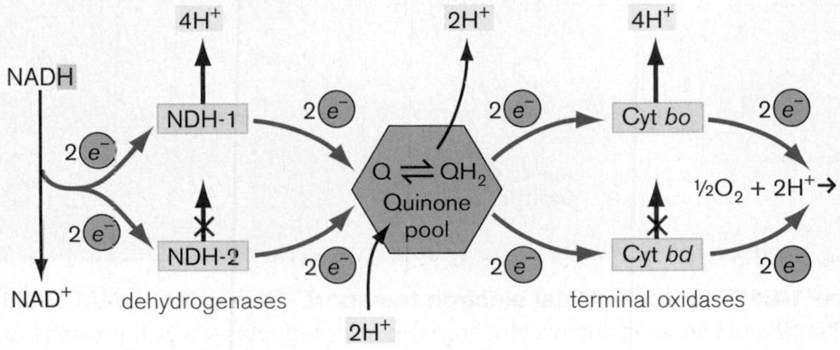

Figure 14.15 Alternative initial oxidoreductases and terminal oxidases.   NADH:quinone oxidoreductase complex (NDH-1) pumps $2H^+$, whereas NDH-2 pumps none. Cyt *bo* pumps an extra proton per electron, whereas Cyt *bd* only exports protons from QH_2 and consumes $2H^+$ to make H_2O.

of NADH; for example, succinate dehydrogenase catalyzes the one step of the TCA cycle that yields FADH$_2$, a step that provides not quite enough energy to produce NADH (discussed in Chapter 13). Succinate dehydrogenase is the only TCA enzyme embedded in the membrane as an ETS component. Other complexes donate electrons to anaerobic electron acceptors such as nitrate, or they accept electrons from inorganic electron donors (discussed in Sections 14.4 and 14.5).

Mitochondrial respiration. In contrast to bacteria, mitochondria have only a single ETS, optimized for a relatively uniform intracellular environment (**Fig. 14.16**). Protected by the eukaryotic cytoplasm, mitochondria do not need to use alternative versions of their ETS to respire under different conditions. Instead, a set of just four electron-carrying complexes has evolved so as to maximize the energy obtained from NADH, minimizing energy lost as heat. This highly efficient mitochondrial respiration is nearly universal throughout the cells of animals, plants, and most eukaryotic microbes.

The mitochondrial ETS has homologs (proteins encoded by genes with a common ancestor) of bacterial ETS components, including NADH dehydrogenase, succinate dehydrogenase, and cytochrome *c* oxidase. Nevertheless, the mitochondrial ETS differs from that of *E. coli* in these respects:

- **An intermediate cytochrome oxidase complex transfers electrons.** Besides NADH dehydrogenase (complex I) and cytochrome *c* oxidase (complex IV), mitochondria show an intermediate step of electron transfer via cytochrome *c* to ubiquinol:cytochrome *c* oxidoreductase (complex III, shown in **Fig. 14.16**). The intermediate electron transfer step pumps an additional 2H$^+$. Another 2e^- may come from succinate dehydrogenase (complex II), which forms FADH$_2$ through the TCA cycle.
- **The mitochondrial ETS pumps more protons per NADH.** As many as 10–12 protons may be pumped per NADH, in contrast to 2–8 protons in *E. coli*.
- **Homologous complexes have numerous extra subunits.** Mitochondria have evolved additional nonhomologous subunits specific to eukaryotes. For example, the homologous subunits of cytochrome *c* complex are enveloped by a series of eukaryotic-specific proteins.

The Proton Potential Drives ATP Synthesis

The proton potential drives synthesis of ATP by the membrane ATP synthase, also known as the F$_1$F$_o$ ATP synthase. The proton-driven synthesis of ATP completes the cycle of oxidative phosphorylation, in which hydrogen ions pumped by the ETS drive phosphorylation of ADP to ATP by the ATP synthase. The same ATP synthase can

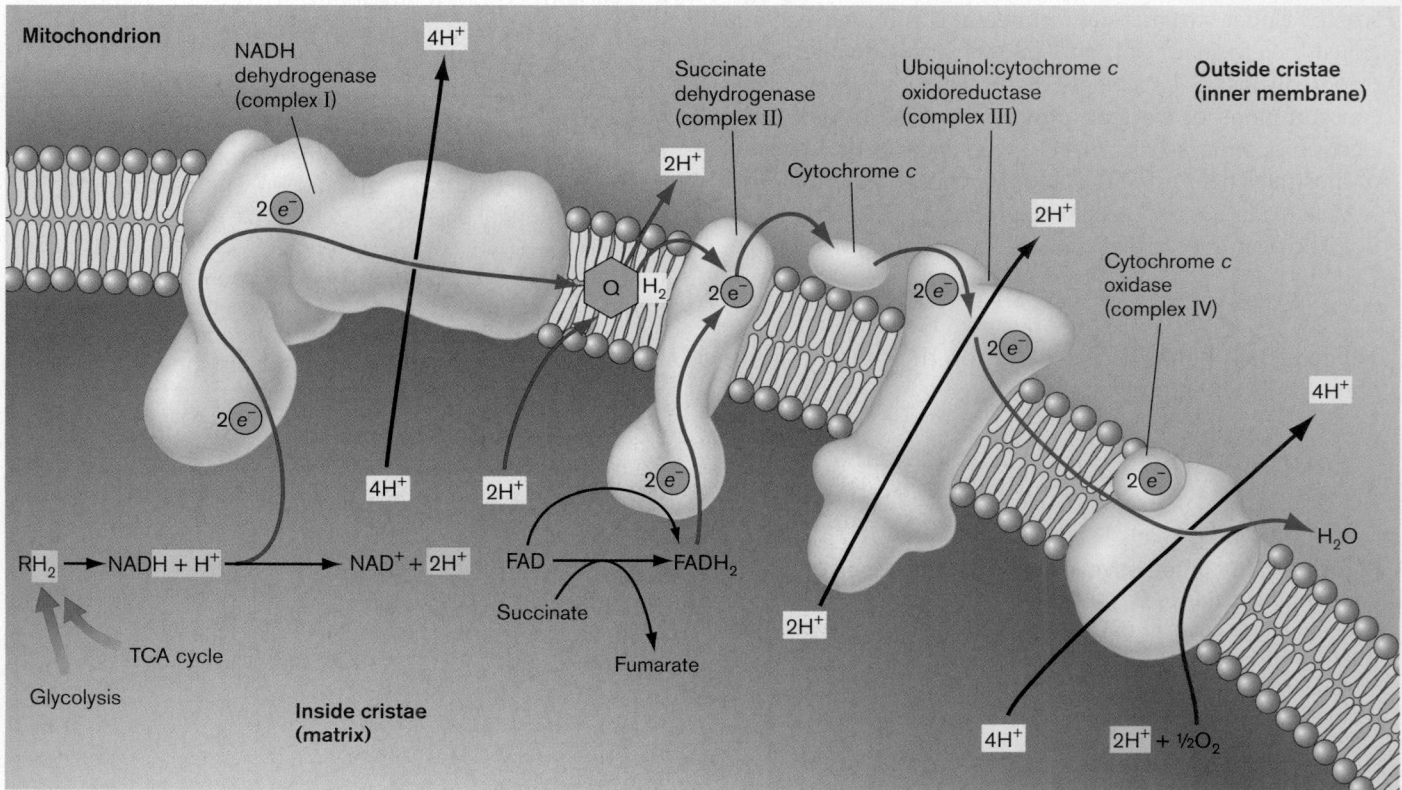

Figure 14.16   Mitochondrial electron transport. In addition to NADH dehydrogenase and a terminal cytochrome oxidase, mitochondria possess ubiquinol:cytochrome *c* oxidoreductase, which provides an intermediate electron transfer step. As a result, mitochondrial membranes export 10–12 H$^+$ per NADH.

use the proton potential generated by lithotrophy (Section 14.5) or by phototrophy (Section 14.6). Surprisingly, despite the homology of ATP synthase across all forms of life, the complex is a target for antibiotics. For example, the ATP synthase of *Mycobacterium tuberculosis*, the cause of tuberculosis, is inhibited specifically by the drug R207910, a diarylquinoline now in clinical trials.

The F_1F_0 ATP Synthase

The F_1F_0 ATP synthase is a protein complex highly conserved in the bacterial cell membrane, the mitochondrial inner membrane, and the chloroplast thylakoid membrane. An elegant molecular machine, the ATP synthase is composed of two rotating complexes: F_0 and F_1 (**Fig. 14.17**). The F_0 complex translocates protons across the membrane. Twelve c subunits form a cylinder embedded in the membrane, stabilized by subunits a and b. The F_1 complex consists of six alternating subunits of types alpha and beta surrounding a gamma subunit that acts as a drive shaft. Each "third" of the F_1 complex (an alpha plus a beta subunit) interconverts ADP + P_i with ATP + H_2O. The gamma subunit connects the tripartite "knob" of F_1 to the membrane-embedded F_0. Proton transport through F_0 drives ATP synthesis by F_1.

One proton at a time enters the subunit-a channel and moves into a c subunit of F_0 (**Fig. 14.18 ◐**). The proton potential directed inward ensures that protons more often enter from the outside than from the cytoplasm. Each entering proton causes a c subunit to rotate around the axle, with release of a bound proton to the cytoplasm. The flux of three protons through F_0 is coupled to forming one molecule of ATP by one alpha-beta-gamma unit of F_1. During each cycle generating ATP, the 12 c subunits of the F_0 rotor rotate in the membrane one-third of one turn relative to the F_1 in the cytoplasm.

Note that the generation of ATP is completely reversible, so ATP hydrolysis by F_1 can pump protons back through F_0 across the membrane. In the absence of a proton potential, a high ATP concentration can drive the F_0 in reverse, actually pumping protons to generate Δp. This reversal of ATP synthase is used, for example, by *Enterococcus faecalis*, intestinal Gram-positive cocci that generate ATP mainly by fermentation. *E. faecalis* can operate the membrane-embedded F_1F_0 ATP synthase in reverse, consuming ATP and thus generating a proton potential for nutrient uptake and ion transport.

The rotating ATP synthase motor is used as a model for nanoscale motor design. For example, James Martin and Wayne Frasch at Arizona State University study the mechanism of F_1 rotation by observing light reflected from a gold nanorod attached to the "drive shaft" (**Fig. 14.18C**).

How do alkaliphilic bacteria make ATP and maintain a proton potential when growing in environments up to pH 11, such as Lake Magadi, Kenya? At high external pH, the large inverted ΔpH would be expected to eliminate the proton potential; nevertheless, alkaliphiles make ATP. Terry Krulwich, at Mount Sinai School of Medicine, investigates the unusual properties of ATP synthase in alkaliphiles (discussed in **eTopic 14.1**).

> **THOUGHT QUESTION 14.9** Would *E. coli* be able to grow in the presence of an uncoupler that eliminates the proton potential supporting ATP synthesis?

Na$^+$ Pumps: An Alternative to H$^+$ Pumps

While a proton potential provides primary energy storage for most species, some bacteria generate an additional potential of sodium ions. A sodium motive force (ΔNa$^+$) is analogous to the proton motive force in that it includes the electrical potential $\Delta\psi$ plus the sodium ion concentration gradient (log ratio of the Na$^+$ concentration difference across the membrane). In extreme halophilic archaea, which grow in concentrated NaCl, the sodium potential entirely substitutes for the proton potential to drive ATP synthesis. These "haloarchaea" make use of the high external Na$^+$ concentration to store energy in the form of a sodium potential.

In some bacteria, an ETS oxidoreductase pumps Na$^+$ instead of H$^+$. For example, the proton-pumping NADH dehydrogenase can be supplemented by an NADH

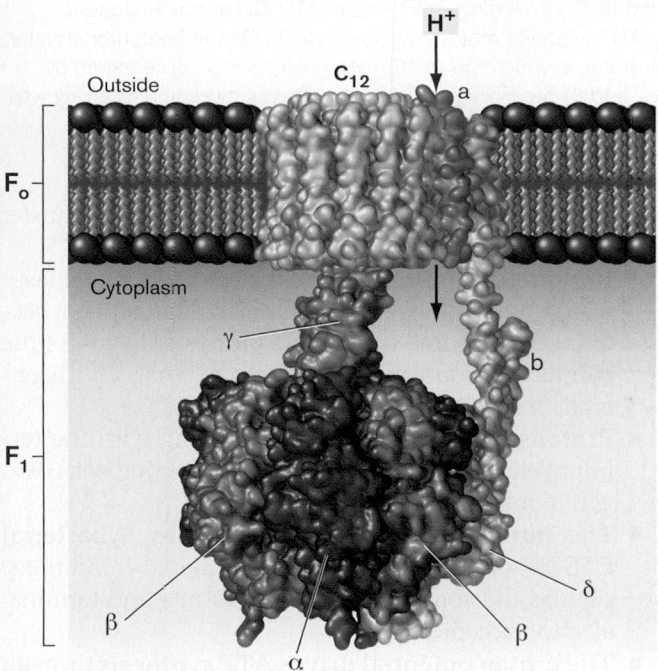

Figure 14.17 Bacterial membrane-embedded ATP synthase (F_1F_0 ATPase). The F_1F_0 complex plugged into the *E. coli* plasma membrane. The three pairs of alpha and beta subunits rotate around the gamma axle, catalyzing formation and hydrolysis of ATP. The ring of 12 c subunits rotates while translocating three protons. (PDB codes: 1B9U, 1C17, 1E79, 2CLY)

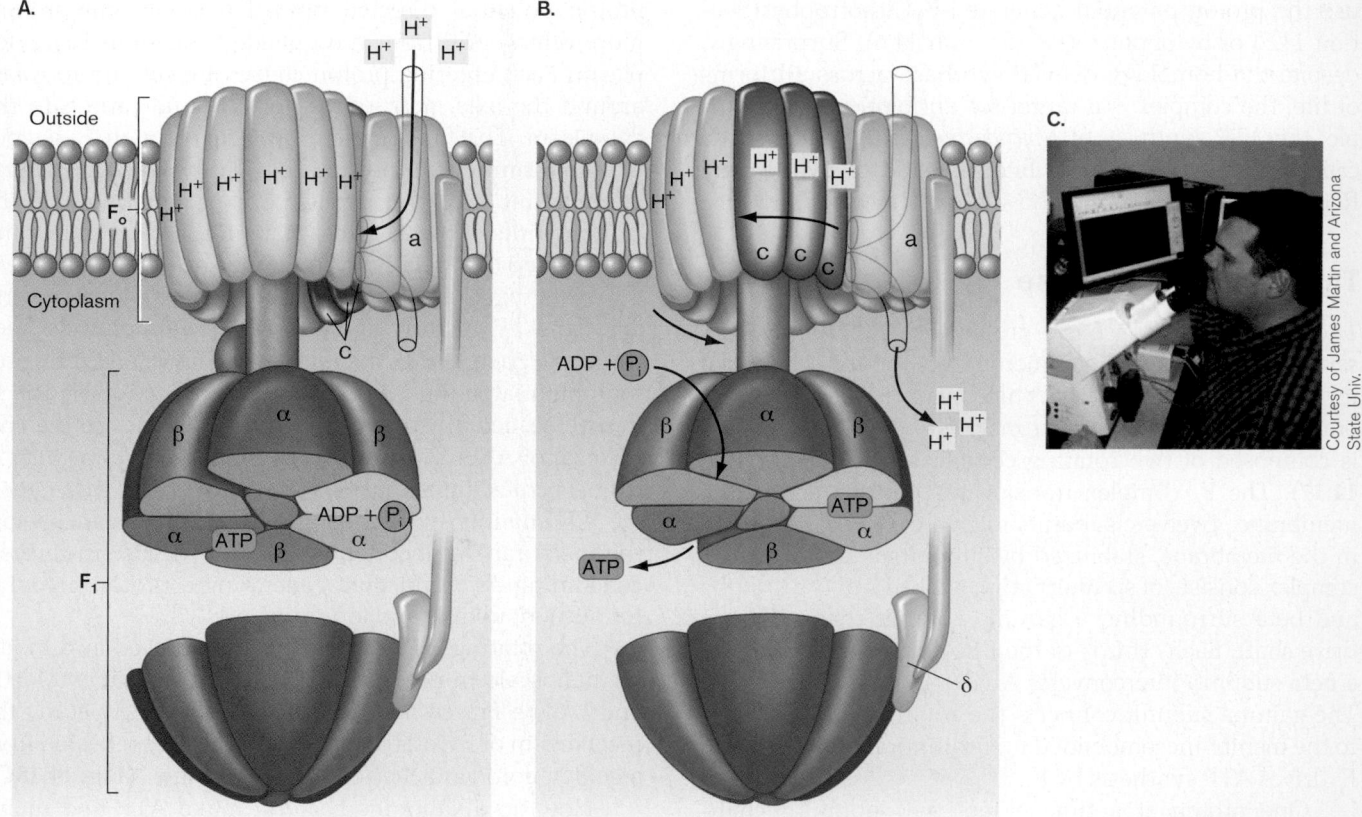

Figure 14.18   H$^+$ flux drives ATP synthesis.   A. Three protons enter c subunits of the F_o complex. **B.** The three c subunits rotate one-third turn relative to the F_1. Flux of three protons through F_o is coupled to F_1 converting ADP + P_i to ATP. **C.** Graduate student James Martin measures the rotation of single molecules of the *E. coli* F_1-ATP synthase motor powered by ATP. On the computer monitor, each spot of red light indicates a gold nanorod attached to the molecular motor, in which the rotational velocity can be determined by the flashing of red light from gold nanorods attached to the rotating drive shaft of the F_1 molecular motor. The single-complex assay was developed in the laboratory of Wayne Frasch, at the Arizona State University. ▶❙ *microbiology2.com/animations*

dehydrogenase that pumps Na$^+$ out of the cell. This primary sodium pump is found in many pathogens, including *Vibrio cholerae*, the cause of cholera, and *Yersinia pestis*, the cause of bubonic plague. These pathogens use the sodium-rich circulatory fluids of their hosts to store energy in a sodium potential.

TO SUMMARIZE:

- **Electron carriers** containing metal ions and/or conjugated double-bonded ring structures are used for electron transfer. For example, cytochrome *c* quinol oxidase has two hemes and three copper ions.
- **A substrate dehydrogenase** receives a pair of electrons from a particular reduced substrate such as NADH. NADH dehydrogenase (NADH:quinone oxidoreductase) typically has an FMN carrier and nine iron-sulfur clusters.
- **Quinones receive electrons** from the substrate dehydrogenase and become reduced to quinols. Typically, a quinol receives 2H$^+$ from the cytoplasm, and

then releases 2H$^+$ across the membrane upon transfer of 2e$^-$ to an electron acceptor complex.
- **Protons are pumped** by substrate dehydrogenases (oxidoreductases) and terminal oxidases. The bacterial cytochrome *bo* oxidase pumps 4H$^+$ across the membrane, coupled to transfer of 4e$^-$ onto O$_2$ to generate 2H$_2$O.
- **Protons are consumed** by combining with the terminal electron acceptor, such as combining with oxygen to make H$_2$O.
- **The number of protons pumped by a bacterial ETS** is determined by environmental conditions, such as the concentrations of substrate and terminal electron acceptor.
- **The proton potential drives ATP synthesis** through the membrane-embedded F_1F_o ATP synthase. Three protons drive each F_1F_o cycle, synthesizing one molecule of ATP.
- **A sodium potential (ΔNa$^+$) can drive ATP synthesis**, replacing or supplementing the proton potential for some bacteria and archaea.

14.4 Anaerobic Respiration

Nearly all multicellular animals and plants require electron transport to oxygen. Bacteria and archaea, however, sample an extraordinary menu of terminal electron acceptors, including metals, oxidized ions of nitrogen and sulfur, and chlorinated organic molecules. This **anaerobic respiration** generally occurs in environments where oxygen is scarce, such as in wetland soil and water and within the human digestive tract.

Electron Acceptors and Donors

Respiratory bacteria usually possess several different terminal oxidoreductases to reduce alternative electron acceptors, as shown for *E. coli* in **Figure 14.19**. These enzymes, although comparable to cytochrome quinol oxidases, are conventionally termed "reductases" to emphasize reduction of the alternative electron acceptor. Some of the electron acceptors are inorganic, such as nitrate (NO_3^-) reduced to nitrite (NO_2^-), or NO_2^- reduced to NO (nitric oxide). Others are organic products of catabolism; for example, the TCA cycle intermediate fumarate can be reduced to succinate. Organic electron acceptors play important roles in food decomposition. For example, the substance trimethylamine oxide, used by fish as an osmoprotectant against sea salt, is reduced by bacteria to trimethylamine—the main cause of the "fishy" smell.

At the top of the ETS, alternative dehydrogenases receive electrons from different organic electron donors, as well as from molecular hydrogen (H_2). The various dehydrogenases all "connect" to various terminal reductases through the pool of quinones. Note that the various electron donors differ greatly in their reduction potential and hence in their capacity to generate proton potential (see **Table 14.1**). In a given environment, the strongest electron donor and the strongest electron acceptor available are used. The best donor and best acceptor usually induce the expression of genes encoding their respective redox enzymes. For example, in the presence of nitrate, genes encoding nitrate reductase are expressed. At the same time, nitrate represses the expression of reductases for poorer electron acceptors, such as fumarate. (The mechanisms of gene induction and repression are discussed in Chapter 10.)

Nitrogen and Sulfur: Oxidized Forms

Oxidized forms of nitrogen and sulfur, usually anions such as nitrate or sulfate, can accept electrons from many species of soil and water bacteria. The nitrogen series offers an abundant source of strong electron acceptors. Reduction of oxidized states of nitrogen for energy yield is called **dissimilatory denitrification**. Dissimilatory nitrate reduction contributes to respiration, whereas assimilatory reduction of nitrate leads to ammonia for fixation into biomass (discussed in Chapters 15 and 22). Some pathogens are denitrifiers, such as *Neisseria meningitidis* (a cause of meningitis) and *Brucella* species that cause brucellosis in cattle, sheep, and dogs.

Oxidized forms of nitrogen. In the dissimilatory nitrogen redox series, a given oxidation state can serve as an acceptor in one redox couple but as a donor in the next. The redox couples are summarized here:

$$\overset{2e^-}{NO_3^-} \rightarrow \overset{e^-}{NO_2^-} \rightarrow \overset{e^-}{NO} \rightarrow \overset{e^-}{\tfrac{1}{2}N_2O} \rightarrow \tfrac{1}{2}N_2$$

| Nitrate | Nitrite | Nitric oxide | Nitrous oxide | Nitrogen gas |

In general, a single bacterial species expresses the terminal reductase for only one or two transformations in the series, such as reduction of nitrate and nitrite, while growing in communities that include species conducting the next steps. Each nitrogen state requires a specialized reductase, such as nitrate reductase or nitrite reductase, to receive electrons from the ETS. The full series of nitrate reduction to N_2 plays a crucial role in producing the nitrogen gas of Earth's atmosphere (discussed in Chapter 22).

Figure 14.19 Alternative electron donors and electron acceptors. *Escherichia coli* can oxidize various foods (electron donors) while reducing various electron acceptors. Each electron donor may utilize alternative dehydrogenases, depending on the environmental conditions. Each dehydrogenase sends protons through the quinol pool to the terminal reductases.

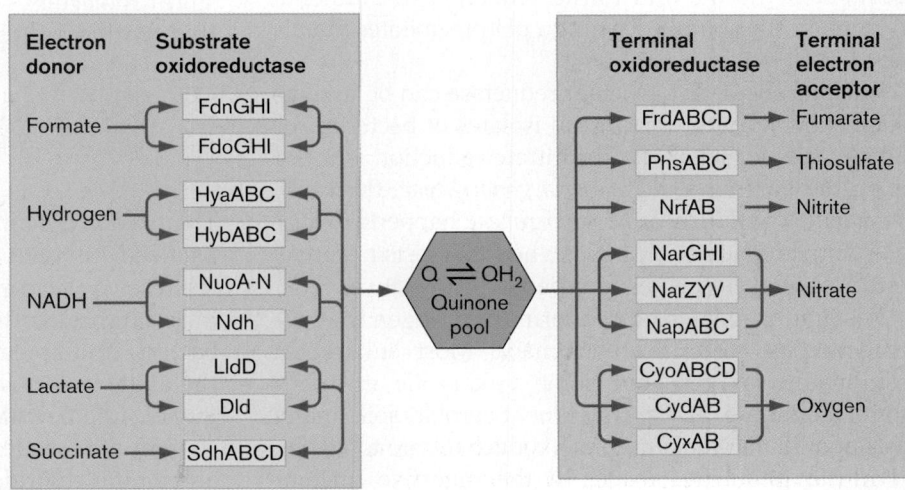

Special Topic 14.2 | Bacterial Electric Power

In natural biofilms, bacteria transfer electrons to metals and even between neighboring bacteria. What if we could harness bacterial electron transfer to power our own electrical devices? Kenneth Nealson (**Fig. 1A**) and students at the University of Southern California are trying to do just that.

The concept of bacterial electric power is surprisingly recent. Up to 20 years ago, it was thought that bacteria could reduce only soluble ions such as nitrate. But Nealson, like the famous nineteenth-century microbial ecologist Sergei Winogradsky, was fascinated by the metal transformations he observed in wetlands and lake sediments. Nealson was particularly intrigued by the high levels of reduced manganese (Mn^{2+}) found in Lake Oneida, New York. He reasoned that only microbial activity could account for so much reduced manganese, usually found in the oxidized form (Mn^{4+}, as the insoluble mineral MnO_2). In 1988, Nealson and his colleague Charles Myers discovered a bacterium that could respire anaerobically by donating electrons to MnO_2, releasing soluble Mn^{2+}. The bacterium, named *Shewanella oneidensis* MR-1 (for "manganese reducer") contains a large number of different cytochromes. The cytochromes help the bacterium donate electrons to different kinds of metals, even insoluble forms such as iron and cobalt embedded within clay.

How can bacteria reduce a substance outside their cell envelopes? Nealson and co-workers found evidence for several mechanisms, such as the shuttling of electrons from outer membrane cytochrome complexes, and the extension of electron-conducting "nanowires" made of protein. The nanowires help bacteria form a biofilm that connects them to each other and connects their cytochromes to the metal electron acceptor. In a fuel cell, the bacteria form a biofilm on the anode (electron-attracting electrode) (**Fig. 1B**). As the bacteria oxidize organic fuel, they cause charge separation between electrons and hydrogen ions. The electrons then pass in a current to the cathode.

The "fuel" for the cell can be a mixture of organic substances derived from any kind of food waste or sewage. Organic waste includes small organic molecules such as lactate, acetate, and even formaldehyde. The biofilm bacteria on the anode can remove hydrogens from these organic molecules, separating the electrons and hydrogen ions (**Fig. 1C**). The hydrogen ions migrate through a polymer membrane, whereas the electrons enter the anode leading to an electrical wire. The remaining carbon and oxygen atoms of the fuel are released as CO_2. The process is similar to natural respiration, except that instead of molecular oxygen, the electron acceptor is an electrode made of graphite. To complete the circuit, the electrons from the wire current ultimately react with oxygen and hydrogen ions to form water, as in aerobic respiration.

So far, microbial fuel cells have been able to generate milliamps of current, enough to drive small devices, such as clocks and marine data sensors. Microbial fuel cells offer a clean way to make electricity, with inexpensive fuel. The main challenge is generating larger currents. Many researchers are working to ramp up the currents and make the dream of bacterial electricity a reality.

An alternative option for many soil bacteria, such as *Bacillus* species, is to reduce nitrite to ammonium ion:

$$NO_2^- + 8H^+ + 6e^- \rightarrow NH_4^+ + 2H_2O$$

Dissimilatory nitrate and nitrite reduction to ammonia can increase the soil pH. The high pH precipitates metals such as iron.

The presence of a particular reductase can be used as a diagnostic indicator for clinical isolates of bacteria. For example, the chemical test for nitrate reduction is a key step in the diagnosis of *Neisseria gonorrhoeae*, the causative agent of gonorrhea. *N. gonorrhoeae* happens to lack the terminal reductase for nitrate; hence, it tests negative, whereas several closely related species test positive.

Respiration using oxidized forms of nitrogen is widespread among bacteria and archaea. Most eukaryotes must breathe oxygen, but some eukaryotic microbes respire on nitrate and nitrite. In anaerobic soil, many yeasts and filamentous fungi can reduce nitrate to nitrite, and nitrite to nitrous oxide. In the digestive tract of termites and of ruminant vertebrates, many fungi and protists grow anaerobically.

Oxidized forms of sulfur. In a similar series, oxidized sulfur molecules serve as electron acceptors for bacteria that synthesize appropriate reductases:

$$SO_4^{2-} \xrightarrow{2e^-} SO_3^{2-} \xrightarrow{2e^-} \tfrac{1}{2}S_2O_3^{2-} \xrightarrow{2e^-} S^0 \xrightarrow{2e^-} H_2S$$

Sulfate   Sulfite   Thiosulfate   Elemental sulfur   Hydrogen sulfide

Their redox potentials are generally lower than those for oxidized nitrogen forms. Nevertheless, sulfate and sulfite receive electrons from many kinds of electron donors, including acetate, hydrocarbons, and H_2. Sulfate-reducing bacteria and archaea are widespread in the ocean, from the arctic waters to submarine thermal vents. The ubiquity of sulfate reduction may be related to the high sulfate content of seawater, in which SO_4^{2-} is the most common anion after chloride.

A.

B.

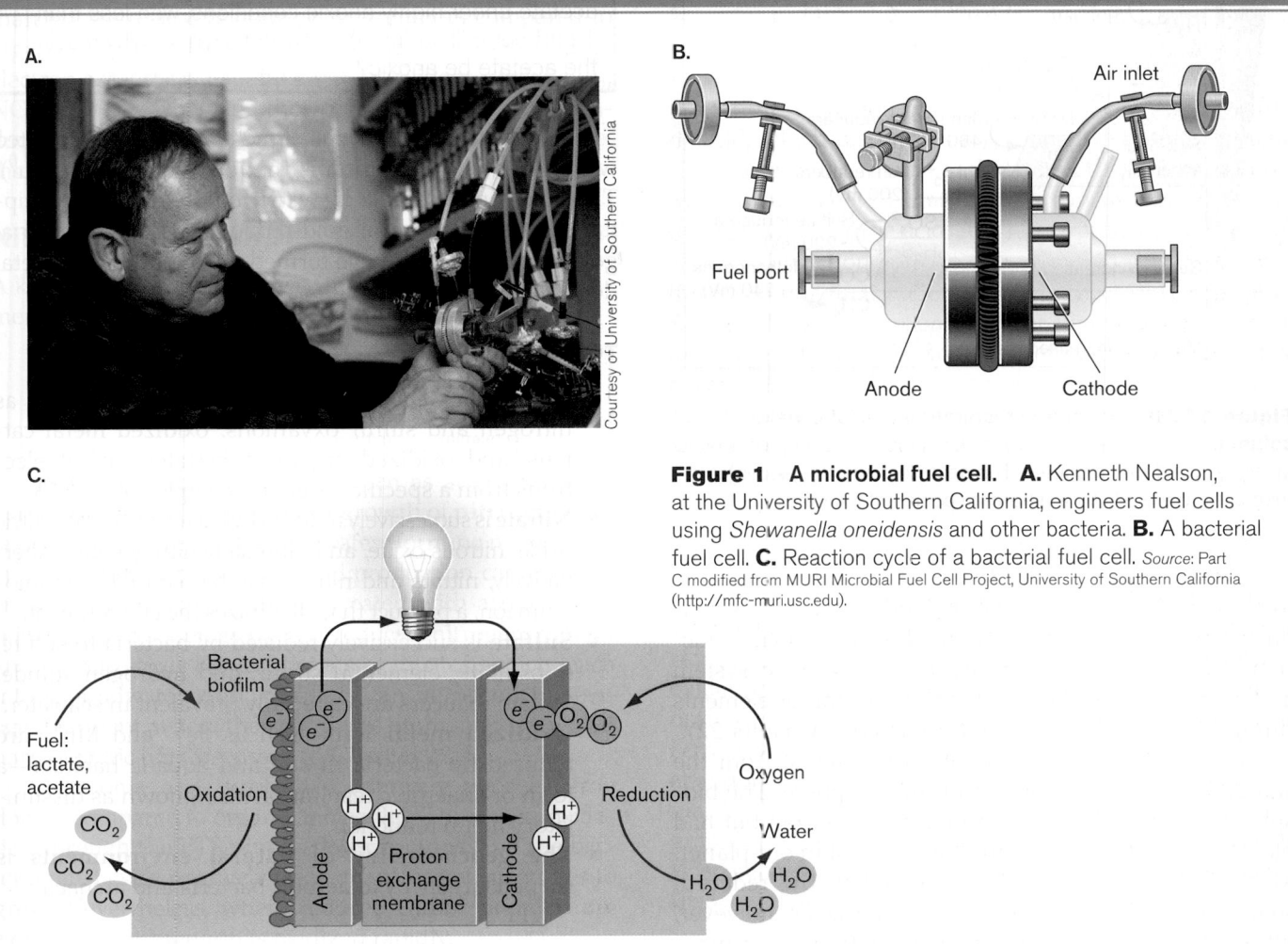

Figure 1 A microbial fuel cell. A. Kenneth Nealson, at the University of Southern California, engineers fuel cells using *Shewanella oneidensis* and other bacteria. **B.** A bacterial fuel cell. **C.** Reaction cycle of a bacterial fuel cell. *Source:* Part C modified from MURI Microbial Fuel Cell Project, University of Southern California (http://mfc-muri.usc.edu).

Dissimilatory Metal Reduction

An important class of anaerobic respiration involves the reduction of metal cations, or **dissimilatory metal reduction**. The term "dissimilatory" indicates that the metal reduced as a terminal electron acceptor is excluded from the cell. This is in contrast to minerals reduced for the purpose of incorporation into cell components (assimilatory metal reduction). Metal-reducing bacteria offer the intriguing prospect of making electricity in a fuel cell (see **Special Topic 14.2**). The metals most commonly reduced through anaerobic respiration are iron ($Fe^{3+} \rightarrow Fe^{2+}$) and manganese ($Mn^{4+} \rightarrow Mn^{2+}$), but virtually any metal with multiple redox states can be reduced or oxidized by some bacteria.

Metal electron acceptors require a specific oxidoreductase (or reductase). Metal ions in solution interact directly with reductases; but oxidized metals such as Fe^{3+} are often barely soluble in water. How does a bacterial reductase interact with an insoluble source of metal

outside the cell? Soil bacteria may "shuttle" electrons to extracellular metals using quinone-like degradation products of lignin, a complex aromatic substance that forms the bulk of wood and woody stems. Lignin degradation products are also called "humics" because of their presence in humus, the organic components of soil. Humics accumulate in anaerobic environments, where decomposition is slow. Alternatively, bacteria may contact insoluble metals using surface-bound cytochromes and even electron-conducting "nanowires" made of protein.

Anoxic or low-oxygen environments, such as the bottom of a lake or wetland sediment, offer a series of different electron acceptors (**Fig. 14.20**). The stronger electron acceptors will be consumed, in turn, by species that have the terminal oxidases to use them. For example, as oxygen grows scarce, nitrate is reduced to nitrogen gas ($NO_3^- \rightarrow N_2$) by nitrifying bacteria. As nitrate is used up, other species reduce manganese ($Mn^{4+} \rightarrow Mn^{2+}$), iron ($Fe^{3+} \rightarrow Fe^{2+}$), and sulfate ($SO_4^{2-} \rightarrow H_2S$). At the bottom of the lake or sediment, carbon dioxide is reduced

TO SUMMARIZE:

- **Bacteriorhodopsin** is a light-driven proton pump that supplements heterotrophy in haloarchaea. The homolog, proteorhodopsin, is found in marine proteobacteria.
- **The antenna complex** of chlorophylls and other photopigments captures light for transfer to the reaction center in chlorophyll-based photosynthesis.
- **Thylakoids** are folded membranes within phototrophic bacteria or chloroplasts. The membranes extend the area for chlorophyll light absorption, and they separate two compartments to form a proton gradient.
- **Photosystem I** separates electrons from H_2S or an organic electron donor. The electrons are ultimately transferred to form NADH or NADPH.
- **Photosystem II** separates an electron from bacteriochlorophyll and pumps H^+ to generate ATP. An electron ultimately returns to bacteriochlorophyll through cyclic photophosphorylation.
- **The oxygenic Z pathway in cyanobacteria and chloroplasts** includes homologs of photosystems I and II. Eight photons are absorbed and two electron pairs are removed from $2H_2O$, ultimately producing O_2.
- **Oxygenic photosynthesis generates 3 ATP + 2 NADPH** per $2H_2O$ photolyzed and O_2 produced. The ATP and NADPH are used to fix CO_2 into biomass.

Concluding Thoughts

Seemingly disparate biochemical means of nutrition, such as respiration and photosynthesis, share common mechanisms of electron flow and proton transfer. A recurring theme is that all forms of metabolism involve electron transfer reactions that yield energy for cell function. As molecules are rearranged by transfer of electrons from one substrate to another, energy is provided to form ion gradients and energy carriers. The energy carriers always need to be balanced between redox-neutral carriers, such as ATP, and reducing carriers, such as NADH or NADPH, for biosynthesis (as discussed in Chapter 15).

Whatever the means of obtaining energy, ultimately some energy must be spent for biosynthesis. Chapter 15 presents how microbes spend their energy to construct the most fundamental "nuts and bolts" of their cells.

CHAPTER REVIEW

Review Questions

1. Explain the source of electrons and the sink for electrons (terminal electron acceptor) in respiration, lithotrophy, and photolysis.
2. How do bacteria combine redox couples for a metabolic reaction that yields energy? Cite examples, calculating the reduction potential.
3. How do environmental conditions affect the reduction potential of a metabolic reaction?
4. Explain the role of cytochromes and redox cofactors in electron transport systems. What features of a molecule make it useful for redox biochemistry?
5. Explain how a proton potential is composed of a chemical concentration difference plus a charge difference. Explain how each component of Δp can drive a cellular reaction.
6. Explain the role of the substrate dehydrogenase (oxidoreductase), the quinones, and the cytochrome oxidase (oxidoreductase) in the respiratory ETS.

7. Compare the ETS function in lithotrophy with that in respiration.
8. Summarize the inorganic redox couples that can be used in anaerobic respiration and those that can be used in lithotrophy. What constraints determine whether a given molecule can serve as electron acceptor or as electron donor?
9. How do diverse forms of anaerobic respiration and lithotrophy contribute to ecosystems?
10. Explain the differences and common features of bacteriorhodopsin phototrophy and chlorophyll phototrophy.
11. Explain the differences and common features of photosystems I and II. Explain how the two photosystems combine in the Z pathway. Why can the Z pathway generate oxygen, whereas PS I and PS II cannot?

Thought Questions

1. The lung pathogen *Pseudomonas aeruginosa*, which also grows in soil, can respire aerobically or else anaerobically using nitrate. Under what conditions would *P. aeruginosa* use either form of metabolism? What part of its ETS would need to change?

2. What environments favor oxygenic photosynthesis, versus sulfur phototrophy? Explain.

3. In pathogens, which components of the ETS do you think would make good targets for new antibiotics, and why?

4. Devise a form of energy-yielding metabolism in which fumarate is converted to succinate; and a different form, in which succinate is converted to fumarate. Explain why the two reactions are reasonable, and try on the Internet to find actual organisms that obtain energy through these reactions.

Key Terms

anaerobic respiration (520)
anammox reaction (525)
antenna complex (533)
bacteriochlorophyll (533)
bacteriorhodopsin (529, 533)
carotenoid (533)
chlorophyll (532, 533)
chromophore (532)
cofactor (512)
cyclic photophosphorylation (536)
cytochrome (506)
dehalorespiration (527)
dissimilatory denitrification (521)
dissimilatory metal reduction (522)
electron acceptor (502)
electron donor (502)
electron transport system (ETS) (electron transport chain) (503)

ferredoxin (535)
heme (512)
hydrogenotrophy (527)
leaching (526)
lithotrophy (chemolithotrophy) (503, 524)
lumen (534)
methanogen (527, 529)
methanogenesis (527)
methanotrophy, methanotroph (529)
methylotroph (529)
nitrifier (525)
oxidoreductase (506, 513)
oxygenic Z pathway (535)
photoexcitation (530)
photoheterotrophy (530)
photoionization (530)
photolysis (530)

photosynthesis (530)
photosystem I (535)
photosystem II (535)
phototrophy (503)
proteorhodopsin (529)
proton potential (proton motive force) (507)
quinol (513)
quinone (513)
quinone pool (515)
reaction center (RC) (533)
redox couple (503)
respiration (503)
retinal (530)
standard reduction potential ($E°$) (503)
stroma (534)
thylakoid (533)
uncoupler (509)

Recommended Reading

Andries, Koen, Peter Verhasselt, Jerome Guillemont, Hinrich W. H. Göhlmann, Jean-Marc Neefs, et al. 2005. A diarylquinoline drug active on the ATP synthase of *Mycobacterium tuberculosis*. *Science* **307**:223–227.

Beatty, J. Thomas, Jörg Overmann, Michael T. Lince, Ann K. Manske, Andrew S. Lang, et al. 2005. An obligately photosynthetic bacterial anaerobe from a deep-sea hydrothermal vent. *Proceedings of the National Academy of Sciences USA* **102**:9306–9310.

Dinh, Hang T., Jan Kuever, Marc Mumann, Achim W. Hassel, Martin Stratmann, et al. 2004. Iron corrosion by novel anaerobic microorganisms. *Nature* **427**:829–832.

El-Naggar, Moh. 2009. The molecular density of states in bacterial nanowires. *Biophysical Journal* **95**:L10–L12.

Ferreira, Kristina N., Tina M. Iverson, Karim Maghlaoui, James Barber, and So Iwata. 2004. Architecture of the photosynthetic oxygen-evolving center. *Science* **303**:1831–1838.

Ishii, Shun'ichi, Takefumi Shimoyama, Yasuaki Hotta, and Kazuya Watanabe. 2008. Characterization of a filamentous biofilm community established in a cellulose-fed microbial fuel cell. *BMC Microbiology* **8**:6.

Jones, Shari A., Fatema Z. Chowdhury, Andrew J. Fabich, April Anderson, Darrel M. Schreiner, et al. 2009. Respiration of *Escherichia coli* in the mouse intestine. *Infection and Immunity* **75**:4891–4899.

Kashefi, Kazem, Jason M. Tor, Kelly P. Nevin, and Derek R. Lovley. 2001. Reductive precipitation of gold by dissimilatory Fe(III)-reducing bacteria and archaea. *Applied and Environmental Microbiology* **67**:3275–3279.

Lovley, Derek R. 2002. Dissimiliatory metal reduction: From early life to bioremediation. *ASM News* **68**:231–237.

Lower, Brian H., Liang Shi, Ruchirej Yongsunthon, Timothy C. Droubay, David E. McCready, et al. 2007. Specific bonds between an iron oxide surface and outer membrane cytochromes MtrC and OmcA from *Shewanella oneidensis* MR-1. *Journal of Bacteriology* **189**:4944–4952.

Newman, Dianne, and Jillian Banfield. 2002. Geomicrobiology: How molecular-scale interactions underpin biogeochemical systems. *Science* **296**:1071.

Philippot, Laurent. 2005. Denitrification in pathogenic bacteria: For better or worst? *Trends in Microbiology* **13**:191–192.

Reguera, Gemma, Kevin D. McCarthy, Teena Mehta, Julie S. Nicoll, Mark T. Tuominen, et al. 2005. Extracellular electron transfer via microbial nanowires. *Nature* **435**:1098–1101.

Rich, Peter R. 2003. The molecular machinery of Keilin's respiratory chain. *Biochemical Society Transactions* **31**:1095–1105.

Strous, Marc, Eric Pelletier, Sophie Mangenot, Thomas Rattei, Angelika Lehner, et al. 2006. Deciphering the evolution and metabolism of an anammox bacterium from a community genome. *Nature* **440**:790–794.

For further study and review, please visit StudySpace at **<u>microbiology2.com</u>**.

Chapter 15

Biosynthesis

How do microbes build their cells? Some bacteria and archaea build themselves entirely from carbon dioxide and nitrogen gas plus a few salts. From these simple molecules, microbial enzymes construct amino acids, the cell wall and envelope, and all the machinery of the cell. Some microbes produce complex secondary products with valuable properties, such as antibiotics. In the laboratory, microbial biosynthesis can be engineered to make cloned proteins, pesticides, and industrial reagents.

Once the key elements of carbon and nitrogen are incorporated into small molecules, more complex structures are built by many enzymes in intricate pathways. How do cells organize their biosynthesis to build precisely the forms they need? How do they avoid wasting energy on excess production? These questions are answered by the use of radioisotope tracers, genetic analysis of mutants, and the decoding of genomes. When a new species is discovered, its genome can reveal its capacities for biosynthesis, including hints of new pharmaceuticals.

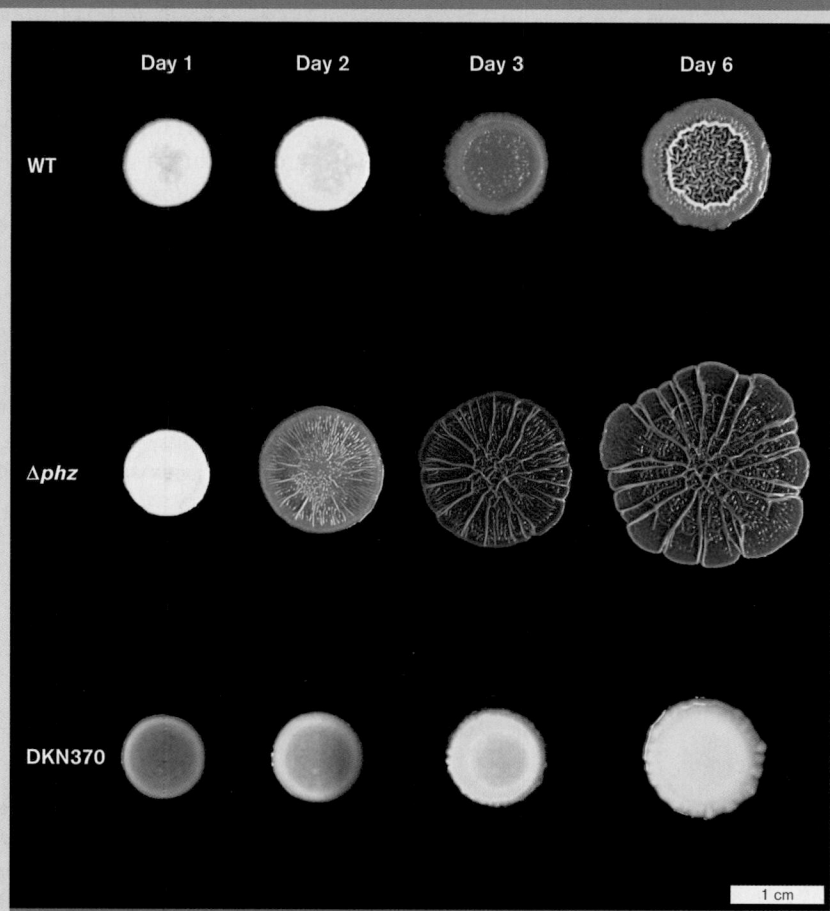

Day 1   Day 2   Day 3   Day 6

WT

Δ*phz*

DKN370

1 cm

Current Research Highlight

Bacteria make colored antibiotics. Colonies of the opportunistic pathogen *Pseudomonas aeruginosa* show different colors and textures depending on their biosynthesis of phenazines. Phenazines are aromatic molecules that can act as antibiotics; but they can also help the bacteria form biofilms while infecting the lungs. Biosynthesis of complex molecules plays key roles in microbial disease, and in microbial ecology. *Source*: Lars Dietrich et al. 2008. *Science* **321**:1203.

We have seen how microbes use chemical reactions to gain energy, storing it in ion gradients and in small molecules such as ATP. Autotrophs use this energy to build their entire cells from inorganic compounds such as water and CO_2. Pathogens use the energy to produce toxins and host penetration complexes. And helpful microbes build key cofactors that our own bodies have lost the ability to form yet still require for metabolism.

The cofactors we must obtain from our food are called vitamins, such as vitamin B_{12}, which is needed to make red blood cells. Virtually all our dietary B_{12} is produced by bacteria in the digestive tract of animals whose meat and dairy products we consume. Our own intestinal bacteria produce only small amounts of B_{12}, which are insufficient for our needs. Thus, strict vegetarians require vitamin B_{12} as a supplement, produced industrially by bacteria such as the Swiss cheese fermenter *Propionibacterium freudenreichii*. In vegan diets of traditional cultures, fermenting bacteria in unrefrigerated foods

provide this essential vitamin. Bacterial synthesis of vitamin B_{12} takes about 25 genes, each specifying a different enzyme to catalyze an essential step. The pathway is highly regulated to avoid wasting energy when the vitamin is already available.

Chapter 15 shows how microbes assimilate the key elements carbon and nitrogen to build complex biomolecules, and how they regulate biosynthesis to make only the products they need.

15.1 Overview of Biosynthesis

Biosynthesis is the building of complex biomolecules, also known as **anabolism**, the reverse of catabolism. **Figure 15.1** presents an overview of anabolism, or biosynthesis, and how it relates to catabolic pathways we have seen. Some species synthesize all their organic components,

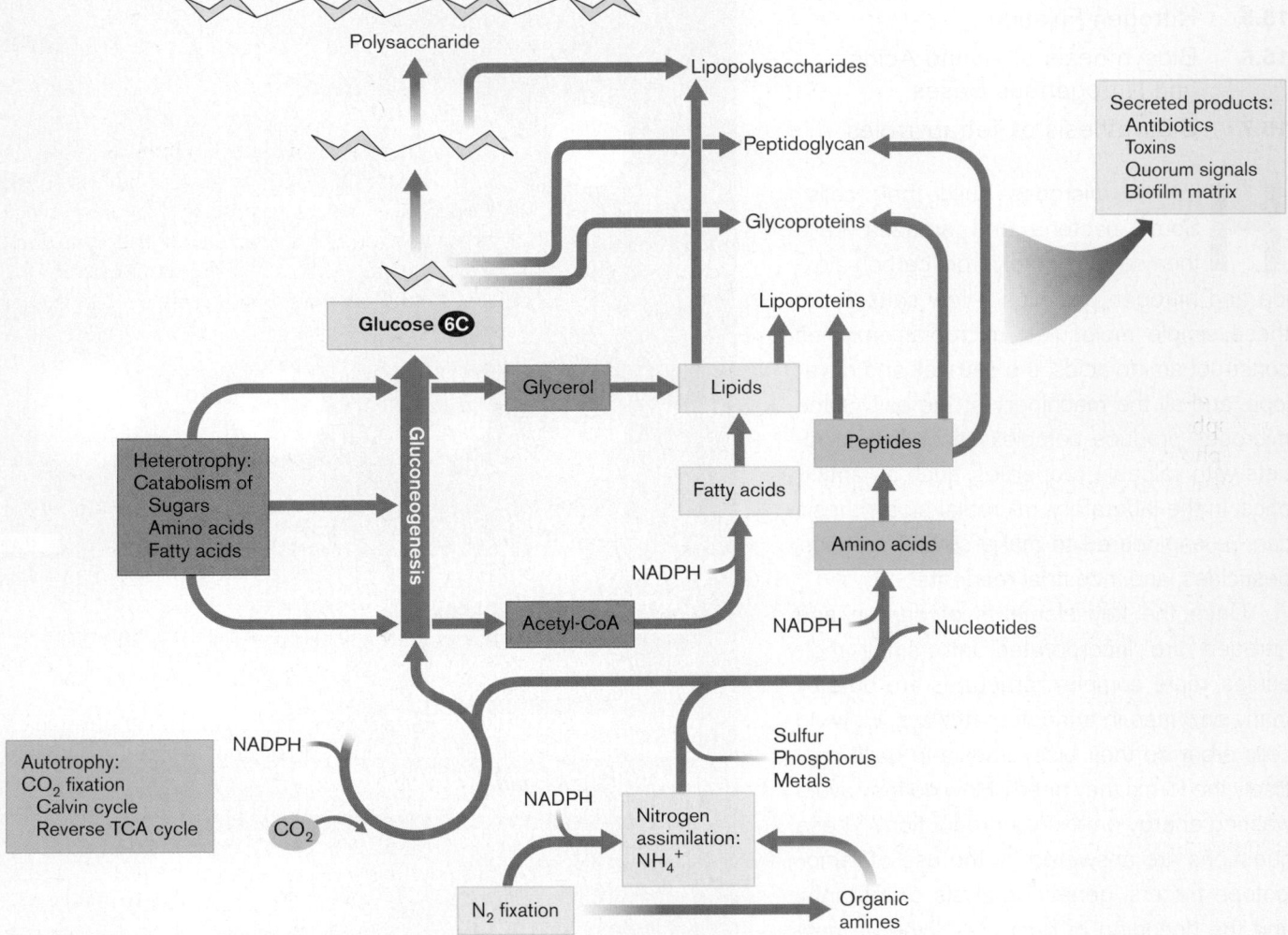

Figure 15.1 Biosynthesis: an overview. Microbes obtain carbon skeletons through CO_2 fixation, or through catabolism of compounds formed by other microbes. Nitrogen is obtained by N_2 fixation, or by uptake of nitrates or organic amines. All biosynthesis requires energy from ATP and from reducing cofactors such as NADPH.

while others must obtain essential molecules from their environment. Biosynthesis requires:

- **Essential elements.** Biosynthesis requires carbon, oxygen, hydrogen, nitrogen, and other essential elements. Carbon is obtained either through CO_2 fixation (autotrophy) or through acquisition of organic molecules made by other organisms. Autotrophs assemble carbon and water into small molecules such as acetyl-CoA that serve as substrates or building blocks for the cell. Heterotrophs break down molecules such as sugars and peptides to form acetyl-CoA and other key substrates. Besides carbon, biosynthesis must assimilate nitrogen and sulfur for proteins, phosphorus for DNA, and metals for metal-containing enzymes.
- **Reduction.** Many cellular structures require reduction of the substrate. Cell components such as lipids and amino acids are more reduced than the substrates available; so their biosynthesis must include reduction by a reducing agent such as NADPH.
- **Energy.** Building complex structures, with or without reduction, requires energy. Biosynthetic enzymes spend energy by coupling their reactions to the hydrolysis of ATP, the oxidation of NADPH, or the flux of ions down a transmembrane ion gradient.

Substrates for Biosynthesis

Many substrates for biosynthesis arise from glucose catabolism and the TCA cycle, central catabolic pathways discussed in Chapter 13 (**Fig. 15.2**). For example, the succinyl-CoA molecules used to build the tetrapyrrole ring of vitamin B_{12} come directly from the tricarboxylic acid (TCA) cycle, while the glycines derive from 3-phosphoglycerate, an intermediate of glycolysis. Glycerol 3-phosphate provides the glyceride backbone of lipids. Pyruvate forms the backbone of several amino acids with aliphatic side chains, whereas erythrose 4-phosphate contributes to the ring structures of aromatic amino acids. Other amino acids derive their carbon skeleton from TCA cycle intermediates oxaloacetate and 2-oxoglutarate, which incorporate nitrogen in the form of ammonium ion (NH_4^+).

Note that glucose catabolism and the TCA cycle are reversible; some autotrophs can synthesize entire sugar molecules, starting with CO_2 and working up through the reverse TCA cycle and reverse glycolysis. Thus, many common metabolites are available both to autotrophs and heterotrophs, as well as to microbes of mixed metabolism.

Biosynthesis Spends Energy

Biosynthesis costs energy in several ways. The organism's genome must maintain the DNA that encodes all the enzymes that catalyze all the steps of the pathway; a

mutation in any one enzyme may negate the entire process. Enzymes couple synthetic reactions to reactions releasing energy. The genomic and energetic costs lead microbes to evolve several strategies to control these costs:

- **Regulation.** Biosynthesis is regulated at several levels by the enzyme products. In this way, microbes avoid making more than they need under given environmental conditions. Regulation occurs at the levels of transcription and translation of enzymes, as well as through feedback inhibition of enzyme activity.
- **Genome degeneration.** Each species makes an evolutionary "choice" whether to maintain the expense of a particular assimilatory or biosynthetic pathway, such as N_2 fixation, or to lose the pathway and become dependent on other species in the environment. Parasitic microbes that grow only inside a host cell show extensive loss of biosynthetic genes.
- **Secondary products.** Free-living bacteria often produce secondary products that are not essential nutrients but enhance nutrient uptake or inhibit competitors. Many secondary products are antibiotics, such as streptomycin, produced by the actinomycete *Streptomyces coelicolor*.

First we present the pathways by which autotrophs assimilate carbon and nitrogen into simple building blocks, such as acetyl-CoA. Then we will present pathways of construction of the key parts of the cell, such as fatty acids and amino acids. Finally, we will consider exciting recent discoveries in the biosynthesis of secondary products such as antibiotics.

15.2 CO_2 Fixation: The Calvin Cycle

The fundamental significance of **carbon dioxide fixation** by green plants and algae was recognized in the early twentieth century. "Fixation" refers to the covalent incorporation of a small molecule into larger biochemical material. This incorporation of carbon requires tremendous energy input, as well as a large degree of reduction to incorporate hydrogen atoms.

The bulk of biomass on Earth consists of carbon fixed by chloroplasts and bacteria through the **reductive pentose phosphate cycle**, which recycles a pentose phosphate intermediate. The cycle is also known as the **Calvin cycle**, for which Melvin Calvin (1911–1997) was awarded the 1961 Nobel Prize in Chemistry. The mechanism of the pentose phosphate cycle was solved by Calvin with colleagues Andrew Benson and James Bassham, at the University of California at Berkeley. The importance of the Calvin cycle to the global ecosystem can scarcely be overestimated; it plays a major role in removing atmospheric CO_2.

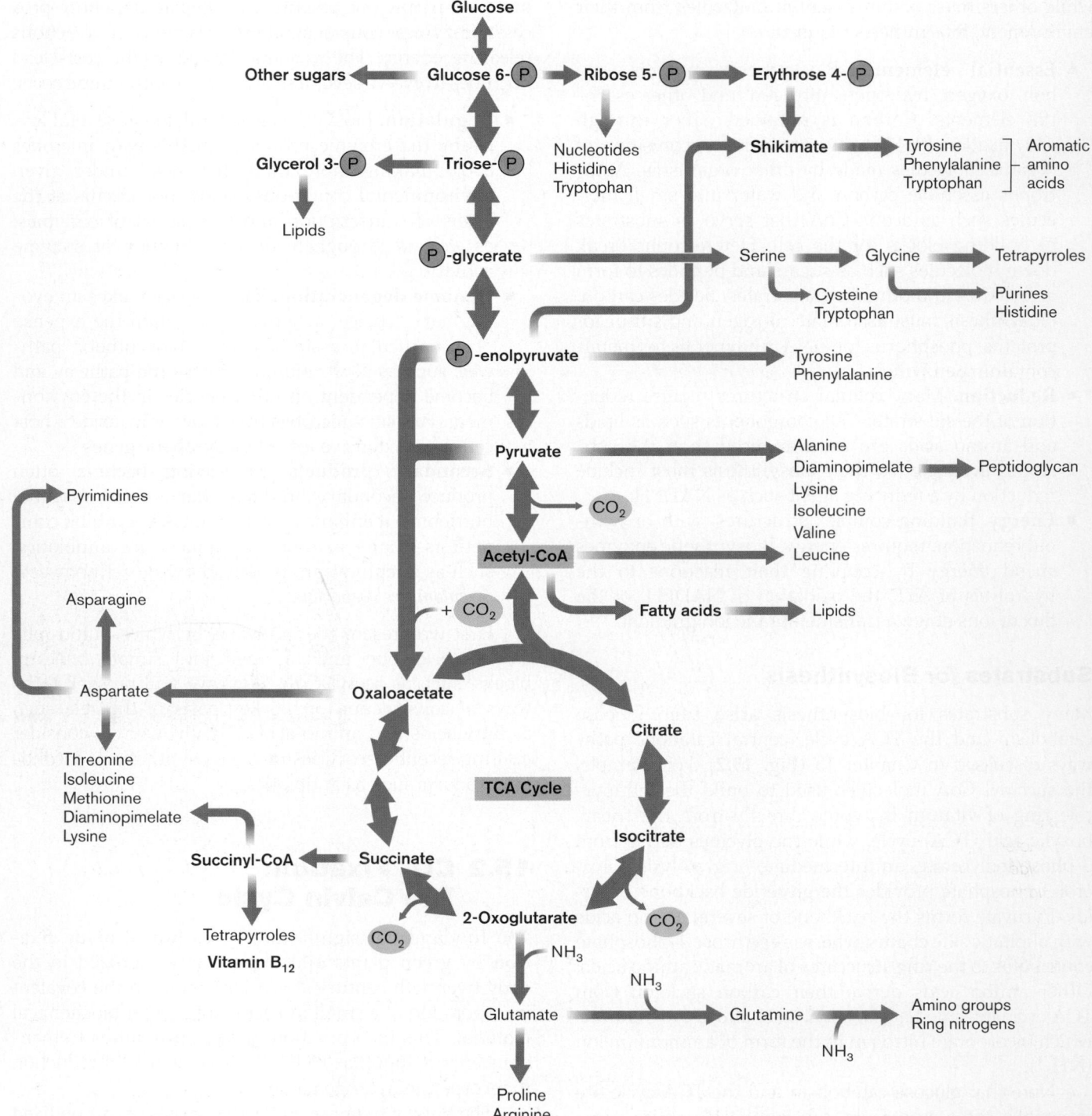

Figure 15.2 Substrates for biosynthesis. Substrates for biosynthesis of lipids and amino acids come from glucose catabolism and the TCA cycle. Acetyl-CoA links these two pathways and is a key substrate for biosynthesis.

The Calvin cycle's responsiveness to CO_2, temperature, and other factors must be considered in all models of global warming.

> **NOTE:** The Calvin cycle is also known as the **Calvin-Benson cycle**, the **Calvin-Benson-Bassham cycle**, or the **CBB cycle**. This book uses the terms "Calvin cycle" and "CBB cycle."

The Calvin cycle is performed by several categories of organisms (**Fig. 15.3**):

- **Oxygenic phototrophic bacteria**, mainly cyanobacteria, fix CO_2 by the Calvin cycle coupled to oxygenic photosynthesis. Cyanobacteria are believed to generate the majority of oxygen gas in Earth's atmosphere.
- **Chloroplasts** of algae and multicellular plants use the Calvin cycle as the "dark reaction," or "light-independent reaction," of oxygenic photosynthesis.

- **Facultatively anaerobic purple bacteria**, including sulfur oxidizers and photoheterotrophs such as *Rhodospirillum* and *Rhodobacter*, use the Calvin cycle to fix CO_2. These bacteria also obtain carbon through catabolism.
- **Lithotrophic bacteria** fix CO_2 through the Calvin cycle, using NADPH and ATP provided by the O_2 oxidation of minerals.

To date, the Calvin cycle has not been found among archaea and is largely absent in obligate anaerobic bacteria (**Table 15.1**). Archaea and anaerobes fix carbon by several different pathways, as described in Section 15.2.

Calvin Cycle Intermediates

Early in the twentieth century, biochemists tried to figure out the mechanism of CO_2 fixation, believing that agricultural photosynthesis could be made more efficient. With the tools then available, however, researchers had no hope of sorting out the intermediate products

A. **B.** **C.**

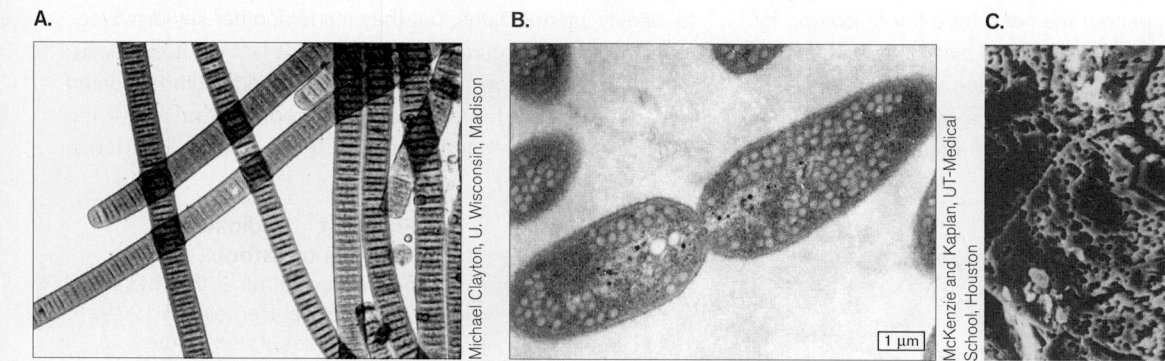
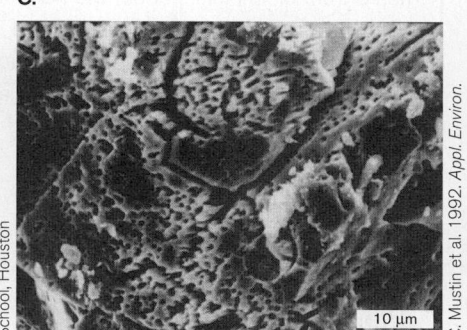

Figure 15.3 **Aerobic and facultatively anaerobic phototrophs use the Calvin cycle.** **A.** *Oscillatoria*, a filamentous cyanobacterium, fixes CO_2 through oxygenic photosynthesis. Filaments are about 8 μm wide (light micrograph). **B.** *Rhodobacter sphaeroides*, purple photoheterotrophs with spherical photosynthetic membranes (TEM). **C.** The iron and sulfur bacterium *Thiobacillus ferrooxidans* oxidizes the sulfur in pyrite (FeS_2), using the Calvin cycle to fix CO_2 (TEM).

Table 15.1 Carbon dioxide fixation pathways.

Pathway	Organisms in which pathways occur		
	Bacteria	Archaea	Eukaryotes
Aerobes and facultative anaerobes (Section 15.2)			
Calvin cycle	Cyanobacteria; purple phototrophs; lithotrophs	Rubisco homologs appear, but their function is unclear	Chloroplasts
Anaerobes and archaea (Section 15.3)			
Reductive (reverse) TCA cycle	Green sulfur phototrophs (*Chlorobium*); thermophilic epsilon-proteobacteria	Hyperthermophilic sulfur oxidizers (*Thermoproteus* and *Pyrobaculum*)	Anaplerotic reactions fix CO_2 to regenerate TCA intermediates
Reductive acetyl-CoA pathway	Anaerobes: acetogenic bacteria and sulfate reducers	Methanogens; other anaerobes	None known
3-Hydroxypropionate cycle	Green phototrophs (*Chloroflexus*)	Aerobic sulfur oxidizers (*Sulfolobus*)	None known

Special Topic 15.1 | The Discovery of ^{14}C

In 1937, a young physicist from the University of Chicago, Martin Kamen (1913–2002), arrived at the University of California at Berkeley to study nuclear reactions. Berkeley possessed one of the world's largest cyclotrons, an instrument in which charged particles are accelerated to high speeds by passage through a voltage potential (**Fig. 1**). One use of a cyclotron was to generate special isotopes of many elements by particle collision. A charged particle collides with a target atom, whose nucleus then undergoes a reaction to form a new isotope. For example, a deuteron (proton plus neutron, mass 2) can collide with ^{10}B (boron-10: five protons, five neutrons) to produce ^{11}C (six protons, five neutrons) plus a neutron. The radioactive isotope ^{11}C decays to ^{10}B with a half-life of 21 minutes.

Kamen and a colleague, chemist Samuel Ruben, were assigned the task of providing radioisotopes requested by physicians for use in cancer therapy—an arrangement that generated funding for the cyclotron. As a result, Kamen and Ruben heard from biochemists about the need for a tracer isotope for carbon. At that time, the only radioactive isotope of carbon that Kamen and Ruben knew how to produce was the short-lived ^{11}C. But it occurred to them that with a ready supply of ^{11}C at their cyclotron, they themselves could attempt tracer experi-

ments in physiology. They focused on a key question: the identity of the first compound to assimilate CO_2 during photosynthesis.

Kamen and Ruben designed experiments in which ^{11}C-radiolabeled CO_2 was assimilated by photosynthesis in *Chlorella*, a species of alga that is a food source for aquarium fish (**Fig. 2**). But the short half-life of ^{11}C made it difficult to complete an experiment while enough radioactivity remained to be detected. As the day's supply of ^{11}C was prepared by the cyclotron, the researchers in the biochemistry laboratory felt "like sprinters at the starting gate," according to Kamen: "Anyone looking in . . . when an experiment was in progress would have had the impression of three madmen hopping about in an insane asylum, what with the frenzied activity punctuated by loud classical music from the radio monitor, and Sam's yells to get on with it and hand him samples while he sat at the counter table, feverishly taking background and sample counts."

With their limited technology, Kamen and colleagues failed to identify intermediates, but they made another key discovery: Fixation of CO_2 occurs in organisms that lack photosynthesis. Colleagues had shown that an anaerobic bacterium fermented glycerol to succinate—a process that appeared to involve the addition of a fourth carbon from CO_2. So Kamen and Ruben

A.

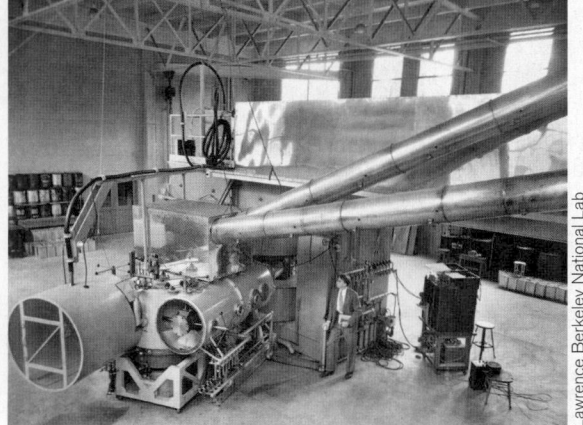

B.

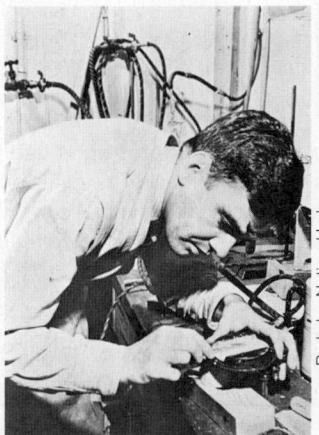

Lawrence Berkeley National Lab

Lawrence Berkeley National Lab

Figure 1 **Radioactive isotopes of carbon. A.** The 60-inch cyclotron at the Radiation Laboratory at Berkeley in 1939. Neutrons generated by the cyclotron bombarded stable atoms to generate radioactive isotopes. **B.** Martin Kamen at the cyclotron in 1941.

through which CO_2 was fixed. Elucidating the pathways of CO_2 fixation required development of two fundamental tools: tracer radioisotopes, which are specific compounds labeled with radioactivity; and paper chromatography, a means of separating labeled compounds based on differential migration in a solvent. Both isotope labeling and chromatographic separation remain key tools of biochemical analysis today, enhanced by modern developments.

Carbon isotope labeling. As a means of tracking the conversion of intermediates within a biochemical pathway, specific atoms within molecules can be "labeled" for detection. Atoms are labeled by substituting an uncommon isotope, a version of an element whose number of neutrons differs from that of the most common isotope found in nature. Chemically, the isotopes of an element behave nearly the same in biochemical reactions, although enzymes may show a slight preference for one isotope.

used ^{11}C to test for assimilation by various bacteria, by yeast, and even by rat liver and muscle. All of these organisms assimilated radioactive CO_2. It was ultimately established that all organisms—including humans—fix some CO_2 as part of their central metabolism, usually by reversal of steps in the TCA cycle.

Kamen and his colleagues sought a longer-lived radioisotope of carbon. Theoretical predictions suggested that a much longer half-life would be exhibited by the isotope ^{14}C. But ^{14}C had yet to be generated and isolated from the cyclotron. Kamen recalled observing unusual collision trails in a sample of nitrogen gas (^{14}N) exposed to neutrons from the cyclotron. Each collision of ^{14}N with a neutron formed a long trail characteristic of a positron (an emitted proton), attached to a short trail expected for a more massive product. The more massive product, Kamen guessed, was ^{14}C.

The isotope ^{14}C proved to have a half-life of 5,700 years. This stability made ^{14}C exactly the kind of tracer needed for experiments in biology. In 1945, Kamen reported the first biological experiment utilizing ^{14}C to investigate the assimilation of CO_2 by the bacterium *Clostridium thermoaceticum*. The bacteria were shown to assimilate two molecules of CO_2 to form acetic acid.

Kamen's own use of ^{14}C for tracer studies was cut short in 1944 by his dismissal from Berkeley stemming from charges that he was a security risk. During the war years, the cyclotron was devoted to weapons development, and all researchers were under tight scrutiny. Kamen faced a decade of accusations, including an appearance before the House Un-American Activities Committee, before he cleared himself of all charges. By that time, however, the fundamental questions of CO_2 fixation in oxygenic photosynthesis had been answered by Melvin Calvin and colleagues, using ^{14}C back at Berkeley. Kamen eventually discovered other kinds of CO_2 fixation in anaerobic bacteria, and he received the Enrico Fermi Award in 1996 for his original discovery of ^{14}C.

Today, short-lived radioisotopes such as ^{11}C find new applications in medical research. For human subjects, short-lived isotopes are desirable because their rapid decay limits exposure to harmful radiation. ^{11}C is particularly useful in positron emission tomography (PET), a technique in which a molecule labeled with ^{11}C is followed in the body by detection of emitted positrons. PET can be used to follow the fate of neurotransmitters and psychotropic substances in the brain.

Figure 2   The first carbon radioisotope experiments in biology.
A. Sam Ruben, chemist (seated), with plant physiologist Zev Hassid, demonstrating equipment for their ^{11}C studies of photosynthesis in *Chlorella*.
B. The alga *Chlorella*, in which the first ^{11}C-label studies were performed.

A.

Hansel Mieth/Time Life/Getty Images

B.

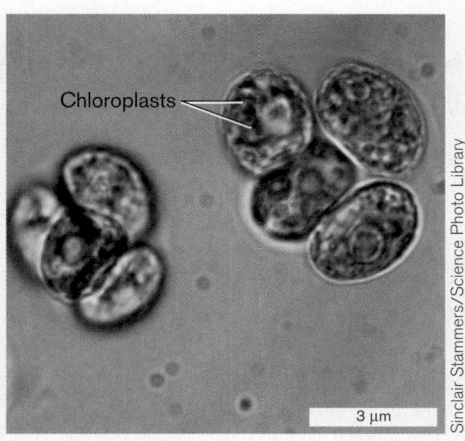

Chloroplasts

3 µm

Sinclair Stammers/Science Photo Library

The number of neutrons added to the number of protons (atomic number) yields the mass number of the isotope. For example, the predominant isotope of carbon, ^{12}C (carbon = 12), contains six protons and six neutrons, adding up to a mass number of 12. The next most abundant isotope, ^{13}C, contains seven neutrons. An isotopic label needs to be detected based on its physical properties, such as its mass (by mass spectroscopy) or its nuclear magnetic resonance (by NMR). The most

sensitive property for detection is radioactive decay. Radioactive decay requires an unstable isotope (radioisotope), such as ^{14}C.

In the 1930s, when Calvin started his research, few useful radioisotopes were known for biological molecules. The discovery of ^{14}C by Martin Kamen (1913–2002) in 1940 revolutionized biochemistry, enabling the discovery of all kinds of cellular metabolism (**Special Topic 15.1**).

Paper chromatography separates intermediates of CO_2 fixation. When ^{14}C-labeled substrates became available in the 1940s, a method was needed to isolate and characterize the unknown intermediates. The technique of **paper chromatography** was developed by Calvin for his studies of carbon fixation. In paper chromatography, a mixture of chemicals is spotted on a special paper, one end of which is immersed in a carrier solvent (**Fig. 15.4A**). The solvent is then drawn up through the paper by capillary action. Because the various chemical components of the solution differ in their solubility, they travel in the solvent at different rates. Afterward, the paper is dried and turned at a right angle. A second solvent is then added and drawn through in the crosswise direction. In the second solvent, the chemicals travel at different rates than in the first solvent, and thus are separated in two dimensions. The chance of two different kinds of molecules migrating the same distance in both solvents is very low.

Figure 15.4B shows one of Calvin's original chromatographic separations. The alga *Chlorella* was exposed to light so it could conduct photosynthesis. $^{14}CO_2$ was added. Then, samples of the alga were killed at specific times by being plunged into boiling alcohol. Five seconds after the addition of $^{14}CO_2$, the main spots showing radioactivity correspond to 3-phosphoglycerate (PGA) and glyceraldehyde 3-phosphate (G3P), early intermediates of CO_2

fixation. Thirty seconds after $^{14}CO_2$ was added, radiolabel is detected in subsequent products, such as amino acids.

The chromatography data illustrate the extraordinary speed of CO_2 fixation, while revealing a glimpse of its earliest intermediates. At even earlier times (2 seconds) only PGA shows radiolabel. **Figure 15.4C** shows the portion of the cycle deduced from the data: CO_2 is incorporated to form PGA, and then is hydrogenated to G3P by an unknown reducing agent (later shown to be NADPH). In this particular experiment, the key intermediate that first assimilates CO_2 remains undetected. The key intermediate was later identified as ribulose 1,5-bisphosphate, also known for its role in the pentose phosphate pathway of glucose catabolism (discussed in Section 13.5).

> **THOUGHT QUESTION 15.1**   Propose a simple experiment to reveal the key intermediate to receive CO_2.

Since Calvin's time, chromatographic separation has been developed into ever more sophisticated forms. High-performance liquid chromatography (HPLC) provides extremely high-resolution separation of chemical products through columns packed with beads of various physical properties.

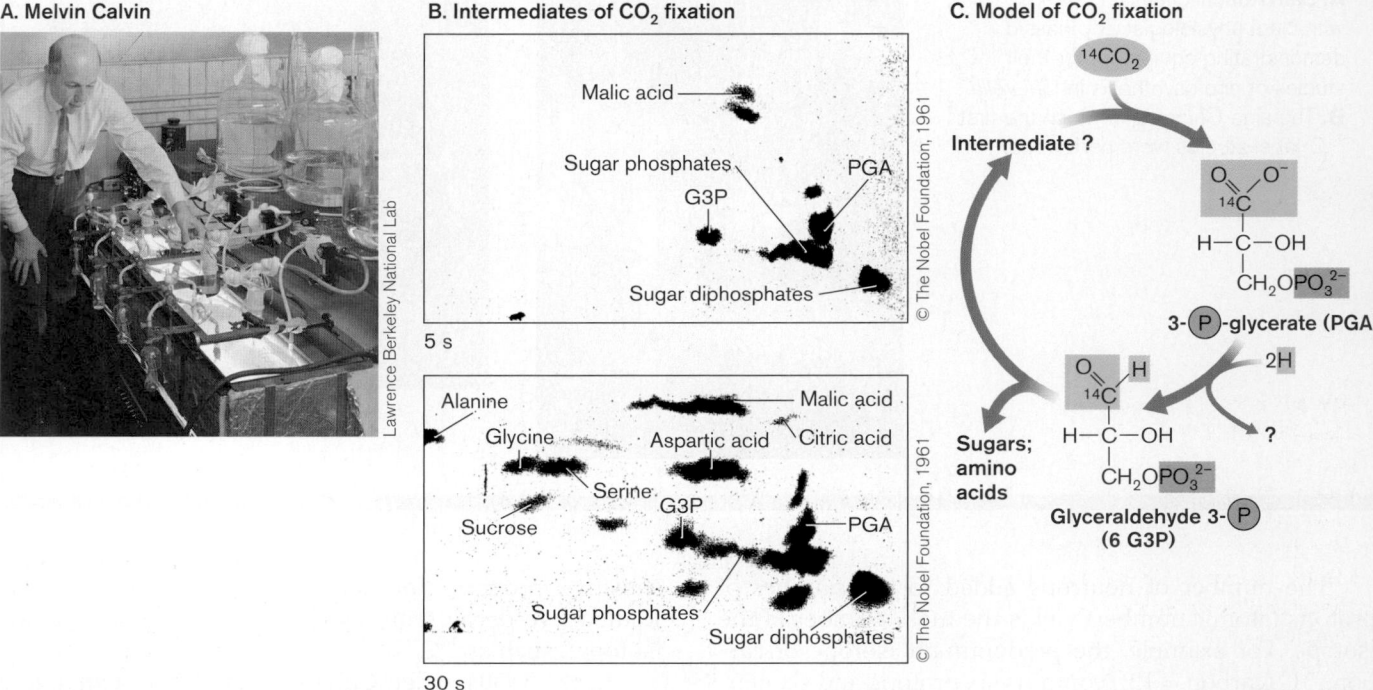

Figure 15.4   Discovery of the Calvin cycle.   A. Melvin Calvin won the Nobel Prize for elucidating CO_2 fixation through the Calvin cycle. **B.** Paper chromatography reveals early intermediate products. Five seconds after addition of $^{14}CO_2$ to photosynthesizing *Chlorella*, the main spots showing radioactivity contain 3-phosphoglycerate (PGA) and glyceraldehyde 3-phosphate (G3P). Thirty seconds after $^{14}CO_2$ addition, radiolabel has entered subsequent products, such as amino acids. **C.** Model of CO_2 fixation suggested by the chromatogram of ^{14}C-labeled intermediates.

Overview of the Calvin Cycle

Decades of biochemical and genetic experiments established the details of CO_2 fixation in the Calvin cycle. An overview is shown in **Figure 15.5**; greater detail appears in **Figure 15.7**.

In each "turn" of the cycle, one molecule of CO_2 is condensed (combined, forming a new C–C bond) with the five-carbon sugar ribulose 1,5-bisphosphate. The resulting six-carbon intermediate splits into two molecules of 3-phosphoglycerate (PGA). The fixed CO_2 ultimately ends up as a carbon of glyceraldehyde 3-phosphate (G3P), reduced by $2H^+ + 2e^-$ from NADPH + H⁺. Recall from Chapter 13 that NADPH is a phosphorylated derivative of NADH commonly associated with biosynthesis.

How does the fixed CO_2 become one "corner" of a glucose molecule? For every three turns of the cycle, fixing three molecules of CO_2, the cycle feeds one molecule of G3P ($C_3H_5O_3$–PO_3^{2-}) into biosynthesis:

$$3CO_2 + 6\ NADPH + 6H^+ + 9\ ATP + 9H_2O \longrightarrow$$
$$C_3H_5O_3\text{–}PO_3^{2-} + 6\ NADP^+ + 9\ ADP + 8\ P_i$$

The H_2O and the phosphoryl group of G3P are ultimately recycled during biosynthetic assimilation of G3P. Two molecules of G3P may condense (that is, form a new C–C bond) in a pathway to form glucose. The overall condensation of $6CO_2 \longrightarrow$ 2 G3P $\longrightarrow$ glucose yields:

$$6CO_2 + 12\ NADPH + 12H^+ + 18\ ATP + 18H_2O \longrightarrow$$
$$C_6H_{12}O_6 + 12\ NADP^+ + 18\ ADP + 18\ P_i$$

Alternatively, instead of forming glucose, G3P can enter biosynthesis of amino acids, vitamins, and other essential components of cells. Glyceraldehyde 3-phosphate is the fundamental unit of carbon assimilation into biomass.

Ribulose 1,5-Bisphosphate Is Reduced and Regenerated

Each "turn" of the Calvin cycle has three main phases: carboxylation and splitting into two three-carbon (3C) intermediates; reduction of two molecules of PGA to two molecules of G3P; and regeneration of ribulose 1,5-bisphosphate (see **Fig. 15.5**):

1. **Carboxylation and splitting: 6C $\longrightarrow$ 2[3C].** Ribulose 1,5-bisphosphate condenses with CO_2 and H_2O, mediated by ribulose 1,5-bisphosphate carbon dioxide reductase/oxidase, generally referred to by the acronym **Rubisco**. Rubisco generates a six-carbon intermediate, which immediately hydrolyzes (splits into

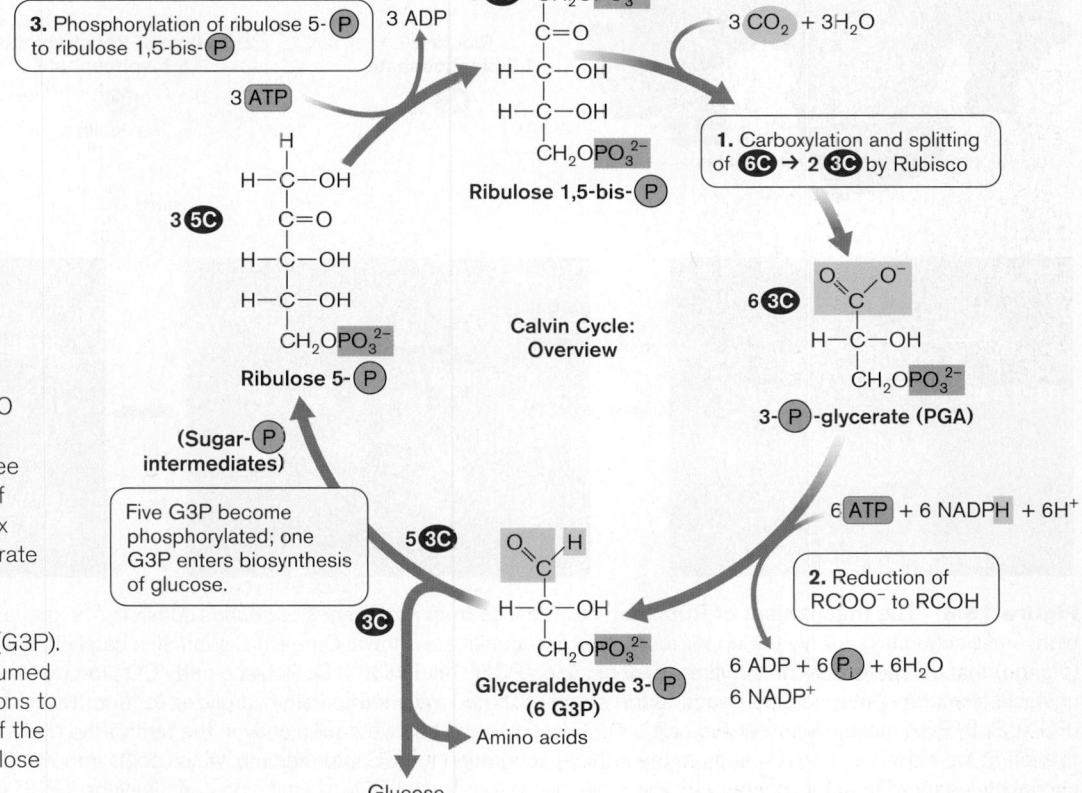

Figure 15.5   The Calvin cycle: overview.   The Calvin cycle condenses CO_2 and H_2O with the intermediate ribulose 1,5-bisphosphate. Overall, three molecules each of CO_2 and of H_2O are fixed and split into six molecules of 3-phosphoglycerate (PGA), which are reduced by 6 NADPH (with 6 ATP) to six glyceraldehyde 3-phosphate (G3P). An additional 3 ATP are consumed during sugar exchange reactions to regenerate three molecules of the recycled 5C intermediate ribulose 1,5-bisphosphate.

two parts by incorporating H_2O). The split produces two molecules of PGA, one of which contains the CO_2 fixed by this cycle.

2. **Reduction of PGA to G3P.** The carboxyl group of each PGA molecule is phosphorylated by ATP. The phosphorylated carboxyl group is then hydrolyzed and reduced by NADPH, forming G3P.

3. **Regeneration of ribulose 1,5-bisphosphate.** Of every <u>six G3P</u>, resulting from three cycles of fixing CO_2, <u>five G3P</u> enter a complex series of reactions (including hydrolysis of 3 ATP) to regenerate <u>three molecules of ribulose 1,5-bisphosphate</u>. The net conversion of five G3P molecules to three molecules of ribulose 1,5-bisphosphate forms $2H_2O$, restoring two of the $3H_2O$ fixed with $3CO_2$. <u>The remaining sixth G3P</u> exits the cycle, available to be used in the biosynthesis of sugars and amino acids. Thus, three fixed carbons lead to one three-carbon product.

Overall, each CO_2 fixed sends one carbon into biosynthesis and regenerates one ribulose 1,5-bisphosphate.

The full Calvin cycle occurs only in bacteria and in chloroplasts that evolved from bacteria. It has not been found in archaea or in the cytoplasm of eukaryotes. Thus, the Calvin cycle appears to have evolved after the divergence of the three domains of life. The relative uniformity of the pathway across species, compared with the diversity of anaerobic pathways of CO_2 fixation, supports the view of late emergence. Nevertheless, some archaeal genomes do show a homolog of Rubisco. The function of archaeal Rubisco is not yet understood.

Ribulose 1,5-Bisphosphate Fixes CO_2

Rubisco is found in large quantities in CO_2-fixing cells; it is one of the most prevalent proteins on Earth. Its characteristics determine the rate and efficiency of CO_2 fixation. The structure of Rubisco is highly conserved across bacterial and chloroplast domains. It consists of two types of subunits, designated small and large (**Fig. 15.6A**). The large (L) subunit contains the active (catalytic) site. The

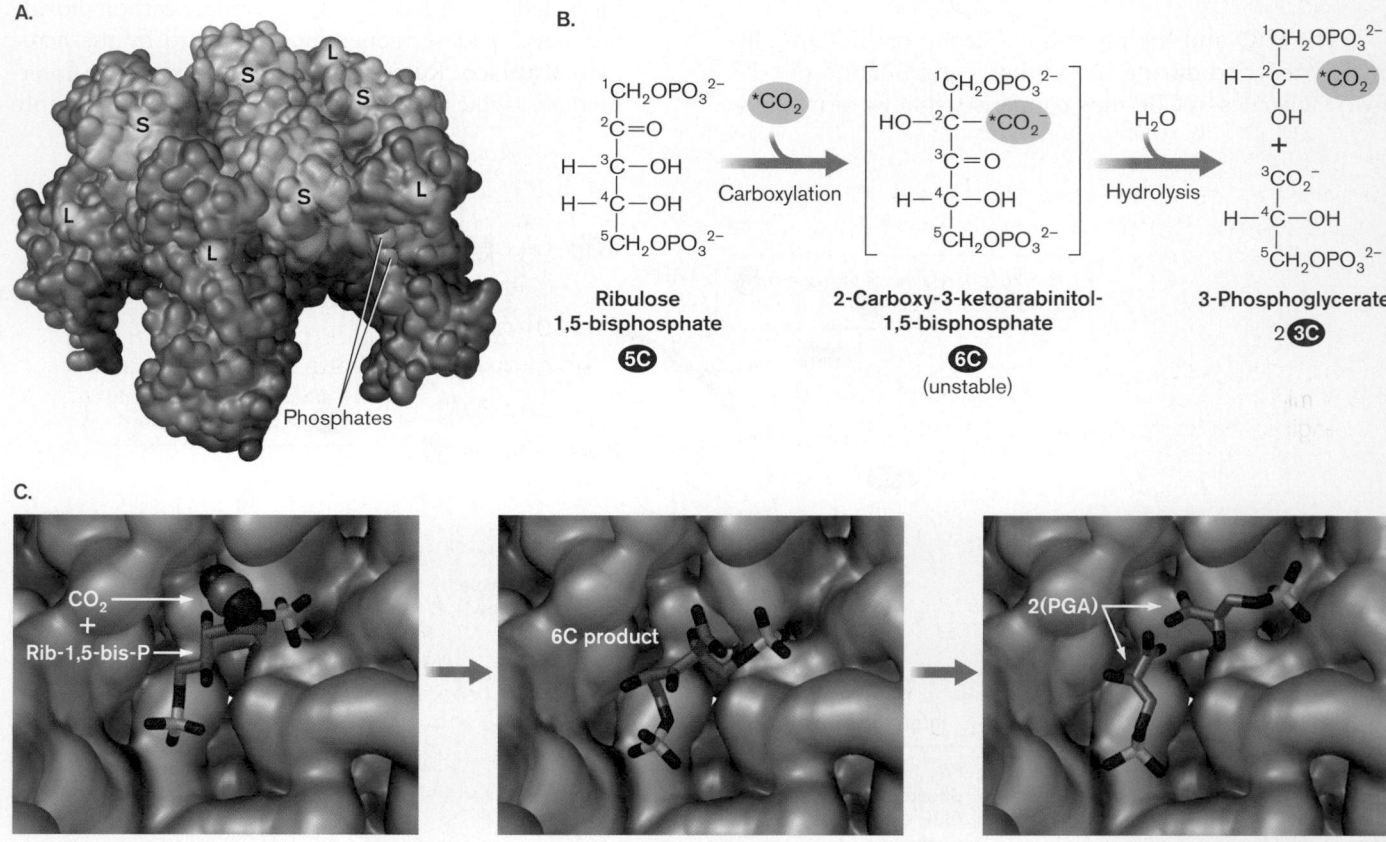

Figure 15.6 The mechanism of Rubisco. **A.** Rubisco from *Alcaligenes eutrophus* consists of eight large subunits (L) crowned by eight small subunits (S). Only the upper four L and S subunits are shown. On each L subunit, a catalytic site contains two phosphates (orange) that compete with the substrates for binding. (PDB code: 1BXN) **B.** Rubisco adds CO_2 to ribulose-1,5-bisphosphate to give an unstable six-carbon intermediate bound to the enzyme. The bound intermediate hydrolyzes to form two molecules of 3-phosphoglycerate (PGA). Both PGA molecules enter the cellular pool of PGA and behave equivalently in the rest of the cycle. **C.** The mechanism of CO_2 fixation at the catalytic site. CO_2 adds to the ketone, generating the 6C intermediate, which splits into two molecules of PGA. The two phosphates appear to act as anchors for the molecule(s) throughout CO_2 assimilation and cleavage. (PDB code: 1RUS)

function of the small (S) subunit remains unclear; some bacteria, such as *Rhodospirillum rubrum*, have a Rubisco with no small subunits. Different species contain different multiples of the small and large subunits: eight (in lithotrophs, green phototrophs, and some algae), six (in plant chloroplasts and some algae), or two (in purple phototrophs). Nevertheless, the fundamental mechanism of CO_2 fixation in all these organisms appears to be similar.

Mechanism of Rubisco. Initially, Rubisco adds CO_2 to ribulose 1,5-bisphosphate to give an unstable six-carbon intermediate that remains bound to the enzyme (**Fig. 15.6B** and **C**). The intermediate hydrolyzes into two molecules of PGA, each held at the active site by its phosphate. Only one molecule of PGA contains a carbon from the newly fixed CO_2; nevertheless, as both molecules dissociate from the enzyme, they enter the cellular pool of PGA and behave equivalently in the rest of the cycle.

Rubisco has a high affinity for CO_2, and the typical concentration of Rubisco active sites (one per large subunit, six per complex) in plant chloroplasts is 4 mM, about 500 times greater than the concentration of CO_2. Thus, considerable energy is invested in CO_2 absorption. Yet the efficiency of carbon fixation by Rubisco is lowered by the existence of a competing reaction with O_2 that leads to 2-phosphoglycolate instead of 3-phosphoglycerate. Researchers are trying to engineer a more efficient oxygen-resistant Rubisco enzyme to improve efficiency of crop yields.

> **THOUGHT QUESTION 15.2** Speculate on why Rubisco catalyzes a competing reaction with oxygen. Why might researchers be unsuccessful in attempting to engineer Rubisco without this reaction?

Regeneration of ribulose 1,5-bisphosphate. Overall, each glyceraldehyde 3-phosphate (G3P) arises from three rounds of CO_2 fixation by ribulose 1,5-bisphosphate. Thus, to maintain the cycle, for each G3P provided to biosynthesis, five other molecules of G3P must be recycled into three ribulose 1,5-bisphosphate. Details of regenerating the intermediate are shown in the lower half of **Figure 15.7**.

The regeneration pathway is summarized here:

- Two molecules of G3P condense to form the six-carbon sugar fructose 6-phosphate.
- Fructose 6-phosphate condenses with a third G3P. The nine-carbon molecule splits to form a five-carbon sugar (xylulose 5-phosphate) and a four-carbon sugar (erythrose 4-phosphate). Xylulose 5-phosphate rearranges to ribulose 5-phosphate.

- Erythrose 4-phosphate condenses with a fourth G3P (rearranged to dihydroxyacetone 3-phosphate) to make a seven-carbon sugar.
- The seven-carbon sugar condenses with the fifth G3P and splits into two molecules of the five-carbon sugar ribulose 5-phosphate (via xylulose 5-phosphate and ribose 5-phosphate).
- Each of the three five-carbon sugars receives a second phosphoryl group from ATP, generating ribulose 1,5-bisphosphate. Each ribulose 1,5-bisphosphate is now ready for Rubisco to fix another CO_2.

Why would a cycle have evolved requiring so many enzymatic steps to so many different intermediates? First, a cycle with many steps breaks down the energy flow into numerous reversible conversions with near-zero values of ΔG. The nearer to equilibrium, the more energy is conserved in the conversion. Second, the multiple different intermediates provide substrates for biosynthesis. For example, some molecules of erythrose 4-phosphate and ribose 5-phosphate are withdrawn from the cycle to build aromatic amino acids and nucleotides.

> **THOUGHT QUESTION 15.3** Why does ribulose 1,5-bisphosphate have to contain two phosphoryl groups, whereas the other intermediates of the Calvin cycle contain only one?
>
> **THOUGHT QUESTION 15.4** Which catabolic pathway (see Chapter 13) includes some of the same sugar-phosphate intermediates as those of the Calvin cycle? What might these intermediates in common suggest about the evolution of the two pathways?

Regulation of the Calvin Cycle

The Calvin cycle is tightly organized and regulated within cells. Expression of the CO_2-fixing enzymes varies with CO_2 concentration, light levels (for phototrophs), and temperature. The concentration of CO_2 is a special problem because CO_2 diffuses readily through phospholipid membranes. Thus, cells cannot concentrate this substrate across the cell membrane to reach the level needed to drive Rubisco. The gas concentration problem is solved by some species through enzymatic conversion of CO_2 to bicarbonate (HCO_3^-), which is trapped in the cytoplasm, unable to leak out of the cell membrane. This enzyme system is called the **carbon-concentrating mechanism (CCM)**. Other bacteria use alternative CO_2-fixing systems adapted to different CO_2 concentrations.

Figure 15.7   The Calvin cycle in detail.
The Calvin cycle assimilates <u>three</u> CO_2, forming <u>six</u> 3-phosphoglycerate (PGA) reduced to glyceraldehyde 3-phosphate (G3P), and converts <u>five</u> G3P into <u>three</u> ribulose 1,5-bisphosphate. Labels "3C," "5C," "6C," and so on, show the number of carbon atoms per molecule.

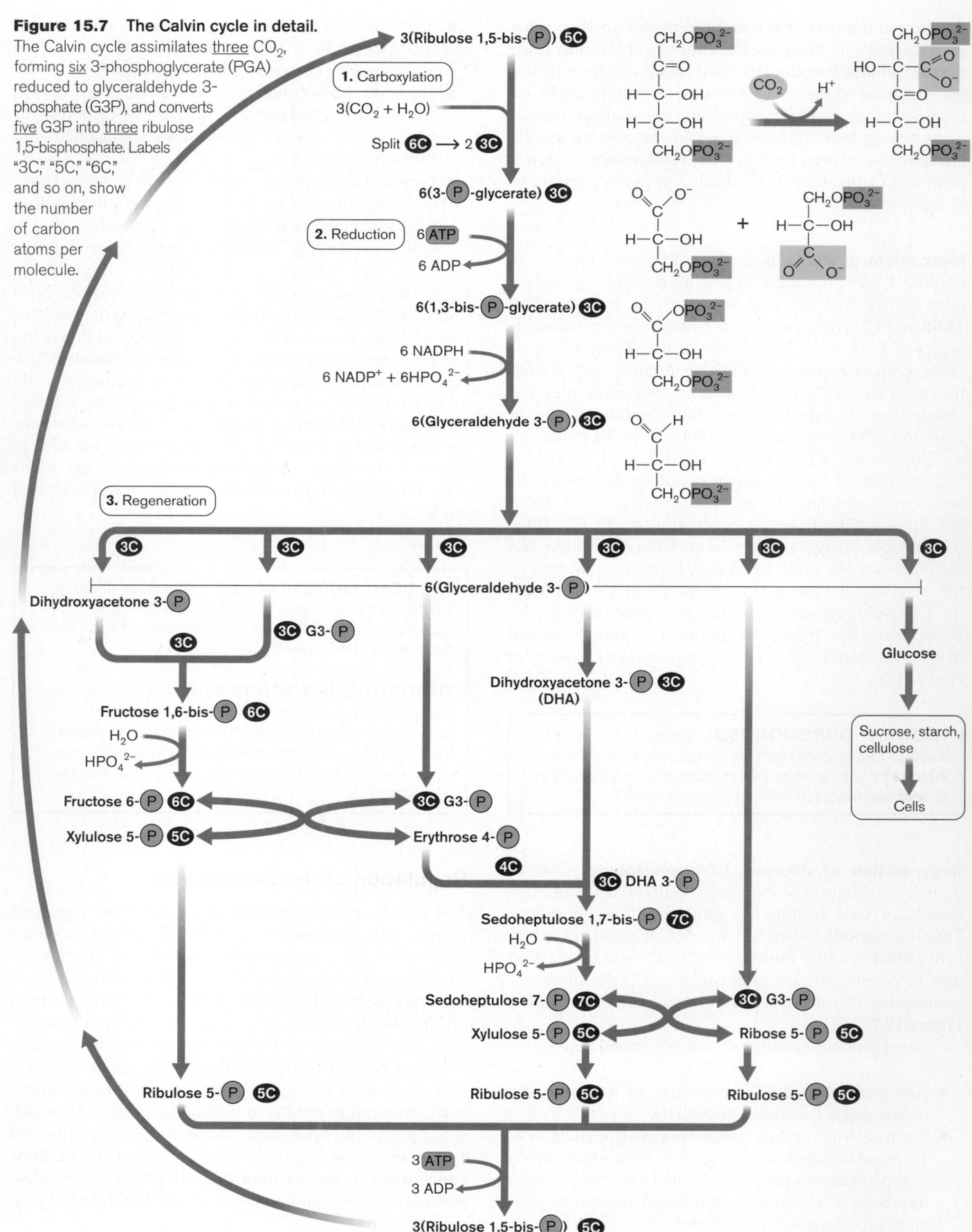

Carboxysomes contain Rubisco. Many organisms that fix CO_2 contain the Rubisco complex within subcellular structures called **carboxysomes** (**Fig. 15.8**). Carboxysomes are found within CO_2-fixing lithotrophs, as well as within cyanobacteria and chloroplasts. A carboxysome consists of a polyhedral shell of protein subunits surrounding packed molecules of Rubisco. The carboxysome takes up bicarbonate (converted from CO_2) by an unknown process. Once inside the carboxysome, the bicarbonate is immediately converted to CO_2 by the enzyme carbonic anhydrase. The CO_2 is then fixed by Rubisco to PGA—the first step of CO_2 fixation (shown in **Fig. 15.7**). The PGA exits the carboxysome to complete the Calvin cycle in the cytoplasm. Mutant strains of bacteria lacking carboxysomes can fix CO_2 only at high concentration (5%), much higher than the atmospheric CO_2 concentration (0.037%).

When CO_2 levels decline, genes are induced to express transmembrane transporters of CO_2 and bicarbonate (HCO_3^-) (**Fig. 15.9**). The inducible genes encode transport systems that participate in the carbon-concentrating mechanism. The CCM enables rapid concentration of HCO_3^- into the carboxysome, where it is dehydrated to CO_2 by the enzyme carbonic anhydrase.

The carbon-concentrating mechanism is regulated genetically. The CCM genes encode a low-affinity CO_2 transporter (NdhF4), a high-affinity CO_2 transporter (NdhF3), and two high-affinity HCO_3^- transporters (CmpABCD and SbtA). The low-affinity CO_2 transporter is effective only when CO_2 concentration is high. By contrast, the high-affinity transporters are induced by CO_2 starvation.

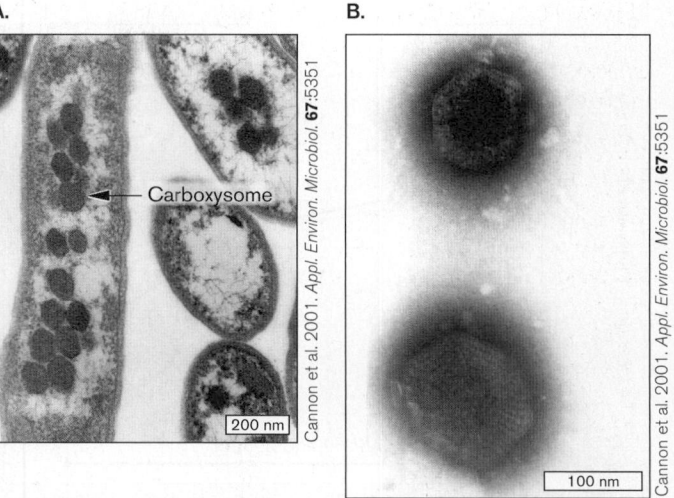

Figure 15.8 Carboxysomes. A. *Halothiobacillus neapolitanus*, a sulfur-oxidizing lithotroph; thin section (TEM) showing polyhedral carboxysomes. **B.** Carboxysomes isolated from *Synechococcus* species packed with Rubisco complexes (TEM).

Induction of transporters by low CO_2 levels is shown by reverse transcriptase PCR (RT-PCR), a technique for detection of minute quantities of mRNA using the enzyme reverse transcriptase followed by the polymerase chain reaction, a technique discussed in Chapter 12, Fig. 12.16. In RT-PCR, the RNA transcripts are reverse-transcribed to DNA; the DNA is then amplified by PCR to generate a detectable product. RNA transcripts are obtained for genes expressed before and

Figure 15.9 The carbon-concentrating mechanism (CCM) in cyanobacteria. CO_2 and HCO_3^- are brought into the cell by transport complexes in the inner membrane or thylakoid membranes of *Synechocystis* species. The HCO_3^- is concentrated in the carboxysome and converted to CO_2 by carbonic anhydrase. *Source:* Model from Dean Price and colleagues, Australian National University at Canberra; McGinn et al. 2003. *Plant Physiol.* **132**:218.

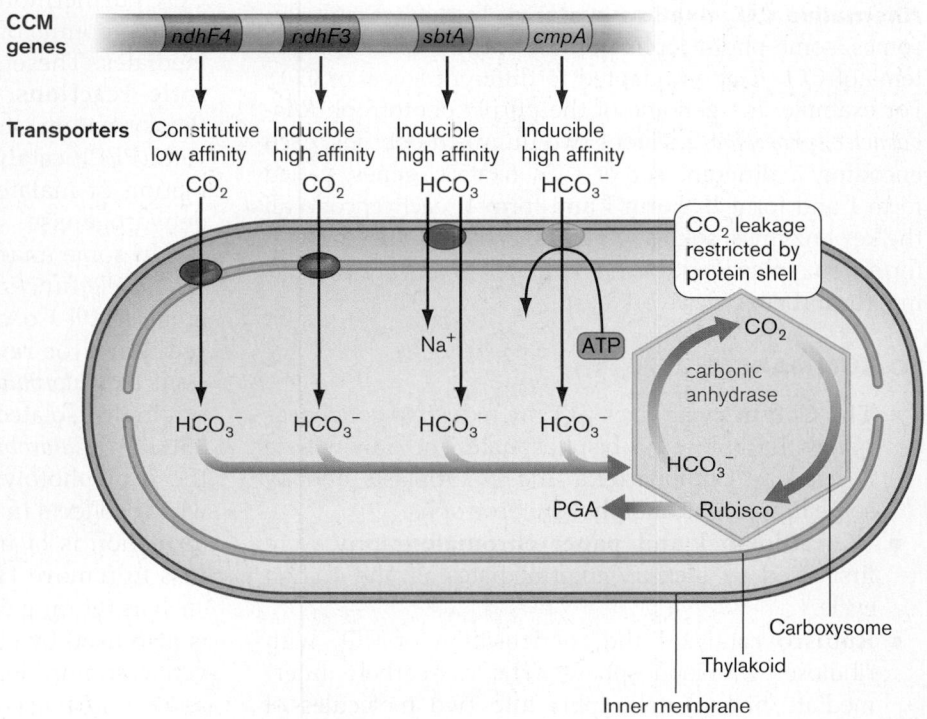

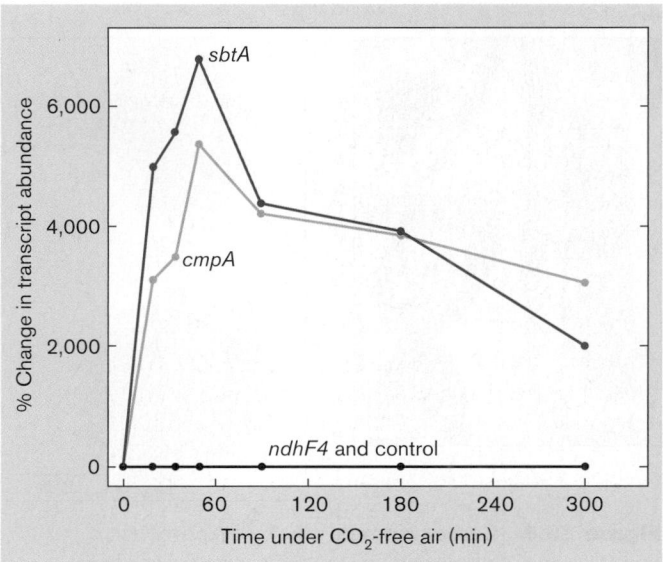

Figure 15.10   Time course of CCM gene transcription.
When CO_2 is limiting, *Synechocystis* genes encoding the carbon uptake complexes show induced transcription. Induction is measured by reverse transcriptase PCR (RT-PCR). The *sbtA* and *cmpA* genes show especially high induction, whereas the constitutive system (*ndhF4*) shows no induction.

after CO_2 limitation. The *sbtA* and *cmpA* transcripts, encoding HCO_3^- transporters, show particularly strong induction by CO_2 (see **Fig. 15.10**). By contrast, the constitutive low-affinity CO_2 transporter *NdhF4* shows no induction.

Alternative CO_2 fixation systems. Instead of carboxysomes, some phototrophs possess entire alternative systems of CO_2 fixation adapted to different levels of CO_2. For example, the genome of the purple phototroph *Rhodobacter sphaeroides* includes two unlinked operons, each encoding a different set of CO_2 fixation genes, called form I and form II. Form I and form II each encode all the key enzymes, such as Rubisco. When CO_2 is limiting, form I enzymes work best; when CO_2 levels are saturating, form II enzymes work best.

TO SUMMARIZE:

- **The Calvin cycle** fixes CO_2 by reductive condensation with ribulose 1,5-bisphosphate. The Calvin cycle is used by cyanobacteria and chloroplasts and by some lithotrophs and photoheterotrophs.
- **^{14}C radiolabel and paper chromatography** were first used to identify intermediates of the Calvin cycle.
- **Rubisco** catalyzes the condensation of CO_2 with ribulose 1,5-bisphosphate. The six-carbon intermediate immediately splits into two molecules of

3-phosphoglycerate (PGA), which are activated by ATP and reduced by NADPH to glyceraldehyde 3-phosphate (G3P).

- **One of every six G3P is converted to glucose** or amino acids. The other five molecules of G3P undergo reactions to regenerate ribulose 1,5-bisphosphate.
- **Carboxysomes sequester and concentrate CO_2** for fixation by the Calvin cycle. CO_2 levels regulate gene expression for carboxysome transporters and the Calvin cycle.

15.3 CO_2 Fixation in Anaerobes and Archaea

We have seen how aerobic organisms and facultative anaerobes fix CO_2 by the Calvin cycle. Some anaerobes, however, fix CO_2 by alternative means (see **Table 15.1**). The most ancient pathways, from the standpoint of evolution and phylogenetic distribution, are the reductive, or reverse, TCA cycle of anaerobic phototrophs and the reductive acetyl-CoA pathway of methanogens. Anaerobic CO_2 fixation plays an essential role in soil, aquatic, and wetland ecosystems, as well as in the digestive systems of animals such as cattle.

The Reductive, or Reverse, TCA Cycle

Portions of the TCA cycle (discussed in Section 13.6) are found in all major groups of organisms. Most of the individual reaction steps of the TCA cycle are reversible, enabling the assimilation of small amounts of CO_2. Furthermore, all organisms, including humans, fix small amounts of CO_2 that regenerate TCA cycle intermediates. These regeneration steps are called **anaplerotic reactions**. Common anaplerotic reactions are the formation of oxaloacetate from phosphoenolpyruvate (PEP), catalyzed by PEP carboxylase, and the formation of malate from pyruvate, catalyzed by malate dehydrogenase.

In some anaerobic bacteria and archaea, the entire TCA cycle functions in "reverse," reducing CO_2 to generate acetyl-CoA and build sugars (**Fig. 15.11A**). The **reductive**, or **reverse**, **TCA cycle** is used by bacteria such as *Chlorobium tepidum*, a green sulfur bacterium originally isolated from a New Zealand hot spring (**Fig. 15.11B**). *Chlorobium* conducts anoxygenic photosynthesis by photolyzing H_2S to produce elemental sulfur, which collects in extracellular granules. Sulfur granule formation is of interest for the development of a process to remove H_2S from sulfide-generating industries such as refining of coal and oil. The reductive TCA cycle is also used by epsilon-proteobacteria of hydrothermal vent communities and by sulfur-reducing archaea, such as *Thermoproteus* and *Pyrobaculum*.

A.

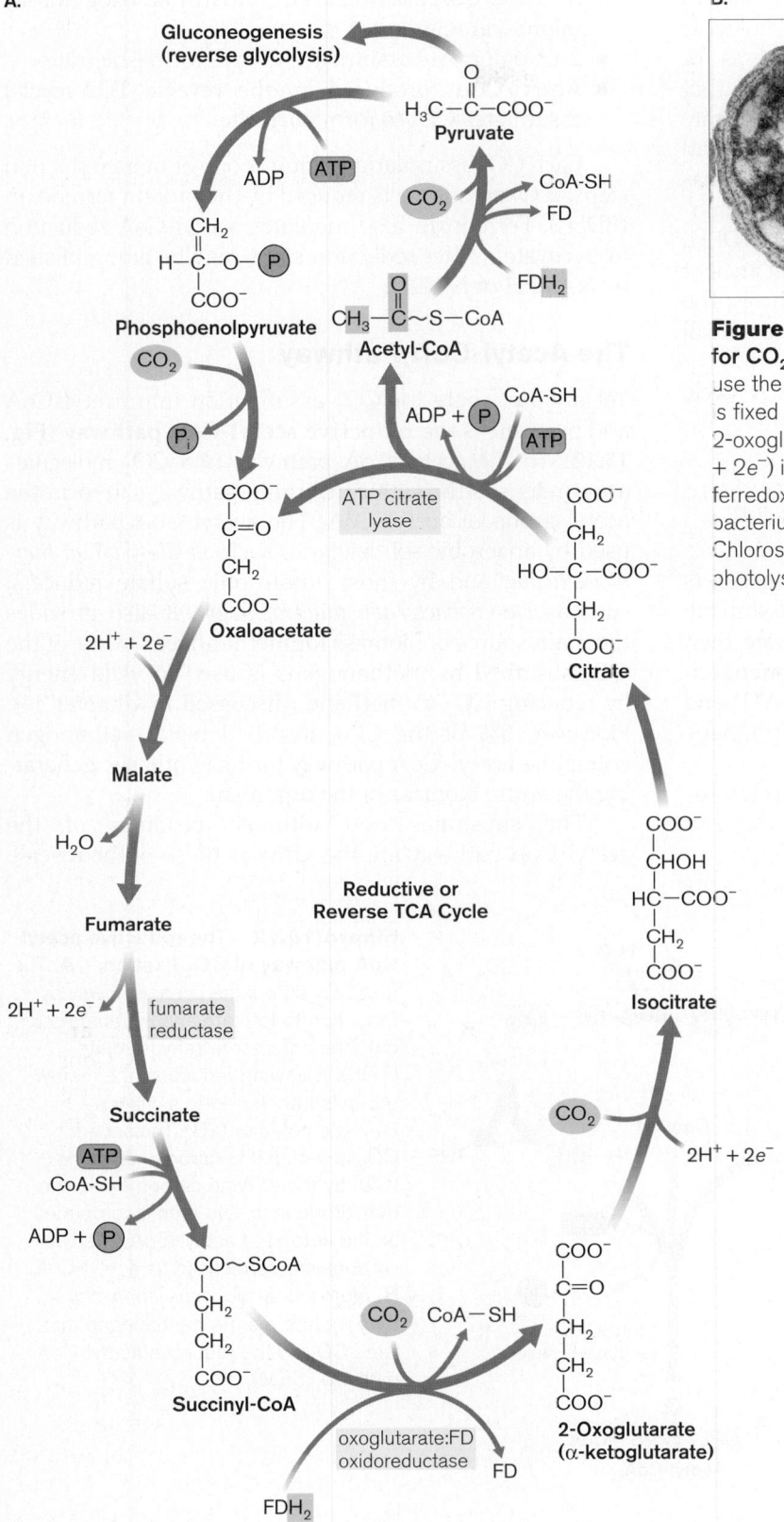

B.

Chlorosome

Niels-Ulrik Frigaard et al. 2002. *J. Bacteriol.* **184**:3368

100 nm

Figure 15.11   The reductive, or reverse, TCA cycle for CO_2 fixation.   A. Anaerobic phototrophs and archaea use the reverse TCA cycle to fix carbon into biomass. CO_2 is fixed by several intermediates, including succinyl-CoA, 2-oxoglutarate, and acetyl-CoA. Reduction (addition of $2H^+$ + $2e^-$) is performed by NADPH or NADH and by reduced ferredoxin (FDH_2). **B.** *Chlorobium tepidum*, a green sulfur bacterium that fixes CO_2 by the reductive TCA cycle (TEM). Chlorosome membranes contain the reaction centers of photolysis.

Reversal of the TCA cycle uses four or five ATP to fix four molecules of CO_2 and generates one molecule of oxaloacetate. The enzymes used are the same as for the "forward" cycle, except for three key enzymes that spend ATP to drive the reaction in reverse: ATP citrate lyase, 2-oxoglutarate:ferredoxin oxidoreductase, and fumarate reductase. For example, ATP citrate lyase catalyzes the cleavage of citrate into acetyl-CoA and oxaloacetate. CO_2 assimilation requires reduction by NADPH. The reductive TCA cycle is believed to be the most ancient means of CO_2 fixation and amino acid biosynthesis, the original cycle of biomass generation in the ancestors of all three living domains.

From CO_2 to acetyl-CoA, the reverse TCA cycle essentially reverses the overall cycle of catabolism:

$$2CO_2 + 2\,ATP + 8H^+ + 8e^- + \text{H-SCoA} \longrightarrow$$
$$\text{CH}_3\text{CO-SCoA} + 3H_2O + 2\,ADP + 2\,P_i$$

The coenzyme A is recycled as the acetyl group enters biosynthesis. For example, acetyl-CoA can assimilate another CO_2 with reduction to pyruvate. Pyruvate then builds up to glucose and related sugars by **gluconeogenesis** or reverse glycolysis, with expenditure of ATP and NADPH. Alternatively, acetyl-CoA can enter biosynthetic pathways to produce fatty acids or amino acids.

Unlike the Calvin cycle, the reverse TCA cycle provides several different molecular intermediates that can assimilate CO_2. Here are the key steps:

- Succinyl-CoA assimilates CO_2 to form 2-oxoglutarate (alpha-ketoglutarate).
- 2-Oxoglutarate assimilates CO_2 to form isocitrate.
- Acetyl-CoA (produced by the reverse TCA cycle) assimilates CO_2 to form pyruvate.

Each CO_2 assimilation requires one or more reduction steps. 2-Oxoglutarate is reduced by the protein ferredoxin (FDH_2). Ferredoxin also mediates acetyl-CoA reduction to pyruvate. Other reduction steps may be accomplished by NADPH or NADH.

The Acetyl-CoA Pathway

Yet another route for CO_2 assimilation into acetyl-CoA and pyruvate is the **reductive acetyl-CoA pathway (Fig. 15.12)**. In the acetyl-CoA pathway, two CO_2 molecules are condensed through converging pathways to form the acetyl group of acetyl-CoA. The acetyl-CoA pathway is used by anaerobic soil bacteria, such as *Clostridium thermoaceticum*, and by most autotrophic sulfate reducers, such as *Desulfobacterium autotrophicum*. It also provides the main source of biomass for methanogens. Most of the CO_2 absorbed by methanogens is used to yield energy by reducing CO_2 to methane (discussed in Chapter 14). However, 5% of the CO_2 absorbed by a methanogen enters the acetyl-CoA pathway for biosynthesis, generating the entire biomass of the organism.

The substrates and ultimate products of the acetyl-CoA pathway are the same as those of the reverse

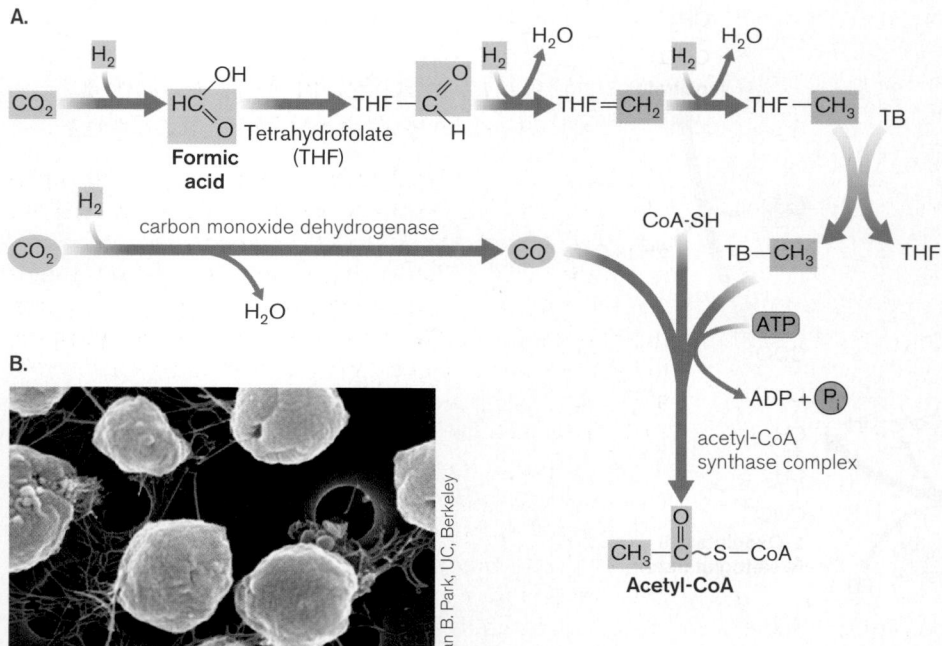

Figure 15.12 The reductive acetyl-CoA pathway of CO_2 fixation. A. The first CO_2 enters the linear pathway by reduction to formate and is transferred onto the cofactor tetrahydrofolate (THF). Following reduction, the methyl group is transferred to a vitamin B_{12}–like cofactor (TB). The second CO_2 is reduced to carbon monoxide (CO) by the enzyme carbon monoxide dehydrogenase, and then incorporated as the ketone of acetyl-CoA. The two carbons are condensed as acetyl-CoA. **B.** *Methanocaldococcus jannaschii*, a thermophilic marine methanogen that fixes CO_2 by the reductive acetyl-CoA pathway (SEM).

TCA cycle, except that the reducing agent is H_2 instead of NADPH. The acetyl-CoA pathway is slightly more efficient than the TCA cycle, requiring one less ATP:

$$2CO_2 + ATP + 4H_2 + H\text{-}SCoA \longrightarrow$$
$$CH_3CO\text{-}SCoA + 3H_2O + ADP + P_i$$

Nevertheless, the form of the pathway and its intermediate compounds are completely different from those of the TCA cycle. The reductive acetyl-CoA pathway is linear, with no recycled intermediates.

Reduction of the first CO_2. The first CO_2 enters the linear pathway by reduction to formate. The formate is transferred onto tetrahydrofolate (THF), a reduced form of **folate**. Folate (folic acid) is a heteroaromatic cofactor, comparable to NADH. Folate is required by many living organisms and is an essential vitamin for humans. It is particularly important for fetal brain development.

The carbon carried by reduced folate (THF) is reduced in successive steps to a methyl group ($-CH_3$). In methanogens, a THF methyl group is converted to methane for net energy acquisition. An occasional methyl group, however, is fixed into biomass. For incorporation into biomass, the methyl must be transferred from THF onto a different cofactor, a specialized tetrapyrrole (TB) related to vitamin B_{12}.

Reductive CO_2 incorporation into acetyl-CoA. The second CO_2 is reduced to carbon monoxide (CO) by the enzyme carbon monoxide dehydrogenase. This same enzyme is used in reverse reaction by some lithotrophs to gain energy from carbon monoxide as an electron donor. For fixation, the CO is condensed with the methyl group carried by TB to form acetyl-CoA. The acetyl-CoA then enters pathways of biosynthesis, as it does when formed by the reverse TCA cycle.

The fixation of CO_2 and release of methane by methanogens is a major concern for cattle farming. Within the rumen, bacteria convert feedstock into CO_2 and H_2, and methanogens divert a significant portion of these gases into methane. The methane then escapes through burping and flatulence. If methanogen growth could be prevented, the CO_2 and H_2 could be fixed by other microbes into fermentation acids assimilated by the rumen. Beef production would then release less of the greenhouse gas methane.

The 3-Hydroxypropionate Cycle

New pathways of carbon fixation continue to be discovered. In the **3-hydroxypropionate cycle**, CO_2 is fixed by acetyl-CoA into 3-hydroxypropionate (**Fig. 15.13A**). The cycle was discovered in the green photoheterotroph *Chloroflexus aurantiacus* (**Fig. 15.13B**). *Chloroflexus* bacteria are filamentous thermophiles that absorb light at longer wavelengths than cyanobacteria and often grow below cyanobacteria in thick mats of biofilm. The 3-hydroxypropionate cycle is also used by archaea, such as the thermoacidophile *Acidianus brierleyi* and the aerobic sulfur oxidizer *Sulfolobus metallicus*.

In the 3-hydroxypropionate cycle, acetyl-CoA condenses with hydrated CO_2 (bicarbonate ion, HCO_3^-) and is reduced by 2 NADPH to 3-hydroxypropionate. Condensation with a second molecule of HCO_3^- forms methylmalonl-CoA, followed by several intermediates to malyl-CoA. These intermediates were detected by [14C] acetate incorporation into the metabolism of *Chloroflexus* (**Fig. 15.13C**). Malyl-CoA releases a molecule of acetyl-CoA to renew the cycle, plus glyoxylate. Glyoxylate condenses with a third HCO_3^- to form pyruvate, which enters standard routes for biosynthesis of sugars and amino acids. Thus, in all, three molecules of CO_2 are fixed into one molecule of pyruvate, which serves as a substrate for biosynthesis.

TO SUMMARIZE:

- **The reductive, or reverse, TCA cycle** fixes CO_2 in some anaerobes and archaea.
- **Anaplerotic reactions** in all organisms regenerate TCA cycle intermediates by fixing CO_2.
- **The acetyl-CoA pathway** in anaerobic bacteria and methanogens fixes CO_2 by condensation to form acetyl-CoA.
- **The 3-hydroxypropionate cycle** in *Chloroflexus* and in some hyperthermophilic archaea fixes CO_2 in a cycle generating the intermediate 3-hydroxypropionate.

15.4 Biosynthesis of Fatty Acids, Polyesters, and Polyketides

The biosynthesis of cell parts begins with small, versatile substrates such as acetyl-CoA, a product of catabolic pathways such as the TCA cycle and benzoate catabolism. A key pathway of biosynthesis from acetyl-CoA generates fatty acids. Fatty acids condense with glycerol to form the phospholipids of cell membranes; they also form the lipid components of envelope proteins and are converted to enzyme cofactors. Fatty acid biosynthesis provides targets for antimicrobial drugs such as triclosan, an inhibitor of enoyl-ACP reductase. Related condensation pathways yield materials such as polyesters and polyketide antibiotics, such as tetracycline.

The repeating form of fatty acids, polyketides, and polyesters enables long-chain molecules to be synthesized with a relatively small number of different enzymes. By contrast, other cell components, such as amino acids, have more complex nonrepeating structures whose biosynthetic enzymes require long operons (presented in Section 15.6).

A.

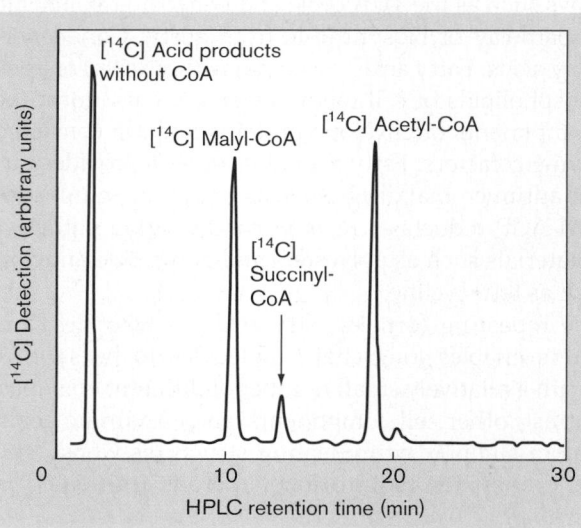

1. In the 3-hydroxypropionate cycle, acetyl-CoA condenses with CO₂ in the form of bicarbonate ion (HCO₃⁻), forming malonyl-CoA.

Pyruvate

Glyoxylate

Acetyl-CoA

Malonyl-CoA

2. Reduction by 2 NADPH yields the key intermediate 3-hydroxypropionate.

3-Hydroxypropionate

5. Malyl-CoA is cleaved to yield glyoxylate plus acetyl-CoA, which reenters the 3-hydroxypropionate cycle. Glyoxylate condenses with a third HCO₃⁻ to yield pyruvate.

L-Malyl-CoA

4. Methylmalonyl-CoA is converted to succinyl-CoA, then to malyl-CoA.

3-Hydroxypropionate Cycle

Succinyl-CoA

Propionyl-CoA

Methylmalonyl-CoA

3. Condensation with a third NADPH and a second HCO₃⁻ yields methylmalonyl-CoA.

B.

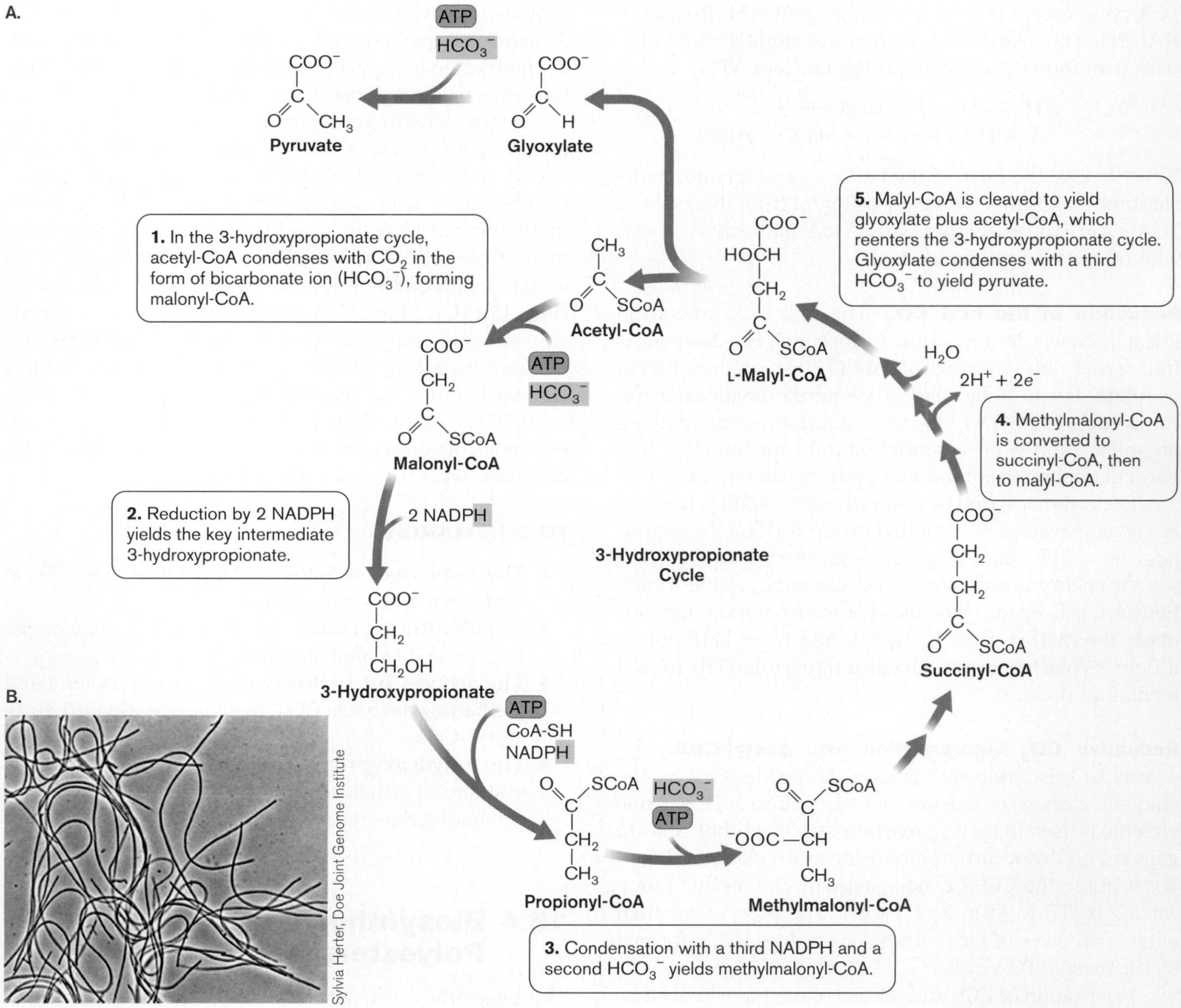

Sylvia Herter, Doe Joint Genome Institute

C.

[¹⁴C] Acid products without CoA

[¹⁴C] Malyl-CoA

[¹⁴C] Acetyl-CoA

[¹⁴C] Succinyl-CoA

[¹⁴C] Detection (arbitrary units)

HPLC retention time (min)

Figure 15.13   The 3-hydroxypropionate cycle of CO₂ fixation.
A. In the 3-hydroxypropionate cycle, acetyl-CoA condenses with CO₂ in the form of bicarbonate ion (HCO₃⁻), forming malonyl-CoA (1). Reduction by 2 NADPH forms the key intermediate 3-hydroxypropionate (2). Condensation with a third NADPH (3) and a second HCO₃⁻ forms methylmalonyl-CoA. Step 4 converts methylmalonyl-CoA to malyl-CoA. In step 5, malyl-CoA is cleaved to yield glyoxylate plus acetyl-CoA, which reenters the 3-hydroxypropionate cycle. Glyoxylate condenses with a third HCO₃⁻ to form pyruvate. **B.** *Chloroflexus*, a green phototroph that performs the 3-hydroxypropionate cycle. **C.** Products of [¹⁴C] acetate are incorporated into the 3-hydroxypropionate cycle. After 1 minute, samples were separated by HPLC (high-performance liquid chromatography). The nonpolar coenzyme A esters of each acid separate in the HPLC solvent. *Source: Part A modified from Silke Friedmann et al. 2006. J. Bacteriol.* **188**:2646. *Part C modified from Herter et al. 2002. J. Biol. Chem.* **277**:20277.

A.

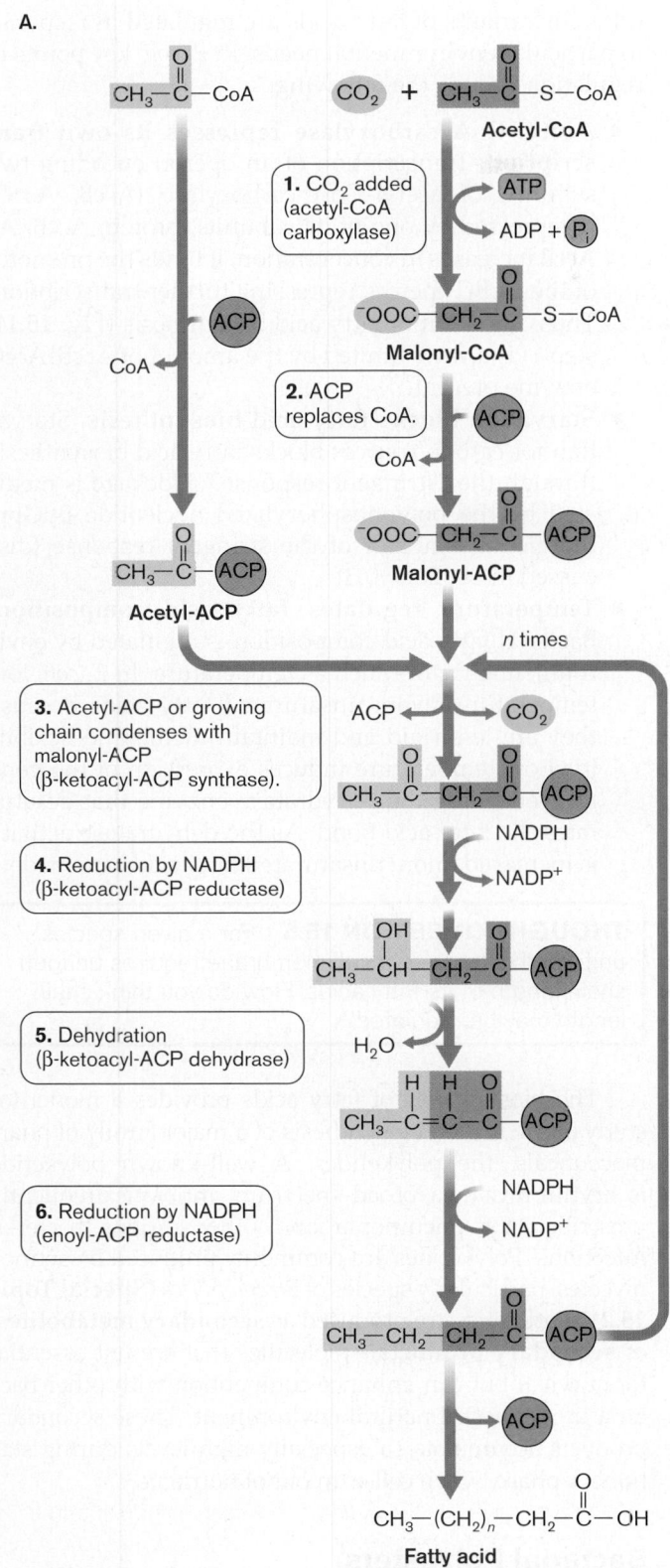

3. Acetyl-ACP or growing chain condenses with malonyl-ACP (β-ketoacyl-ACP synthase).

4. Reduction by NADPH (β-ketoacyl-ACP reductase)

5. Dehydration (β-ketoacyl-ACP dehydrase)

6. Reduction by NADPH (enoyl-ACP reductase)

B.

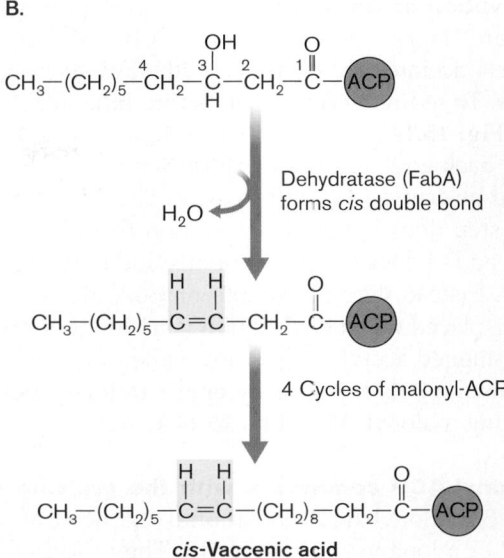

Dehydratase (FabA) forms *cis* double bond

4 Cycles of malonyl-ACP

cis-Vaccenic acid

Figure 15.14   Fatty acid biosynthesis. A. Stepwise elongation of a saturated fatty acid. Units of acetyl-CoA are carboxylated to malonyl-CoA, and then successively condense to form the long chain of a fatty acid. Each two-carbon unit requires reduction by 2 NADPH. **B.** An alkene "kink" can be left unsaturated during chain extension. A special dehydratase enzyme forms the double bond between the third and fourth carbons, rather than the second and third. In this position, the double bond escapes reduction as the chain lengthens through subsequent cycles of malonyl addition.

of molecules based on repeating units requires a cyclic pathway that feeds its products back repeatedly as substrates for further synthesis. The advantage of a cyclic process is that large polymers can be made with a limited number of enzymes; for example, the mycolic acids of *Mycobacterium tuberculosis* are extended to lengths of over 100 carbons.

While fatty acid structures used by cells include many different forms (discussed in Chapter 3), their synthesis always begins with the successive addition of two-carbon acetyl groups. Thus, most fatty acids contain an even number of carbon atoms.

The fatty acid synthase complex. The cyclic process of fatty acid synthesis is managed by the **fatty acid synthase complex**. The complex contains all the enzymes and component binding proteins bound together in proximity, so that all steps occur in one place without "losing" the unfinished molecule. The main bacterial version of this complex is designated FASII. All components are essential for viability and offer drug targets. Analogous multienzyme complexes are used by actinomycete bacteria to synthesize other long-chain products, such as polyketide antibiotics.

Fatty Acids Are Built from Repeating Units

Fatty acids exemplify the construction of cell components from repeating units (**Fig. 15.14**). The construction

Activation of acetyl-CoA. Most acetyl groups within the cell are "tagged" with coenzyme A (CoA), a cofactor that directs acetate into catabolic pathways such as the TCA cycle. To redirect acetyl groups into fatty acid biosynthesis (**Fig. 15.14**), acetyl-CoA molecules are first tagged at the "back end" by condensation with CO_2 (step 1), catalyzed by acetyl-CoA carboxylase. The addition of CO_2 in this step does not count as carbon fixation, because the CO_2 added does not permanently add to the carbon skeleton. Instead, like phosphate or CoA, the CO_2 exists to be displaced when its function is no longer required. The CO_2-tagged acetyl-CoA is now called malonyl-CoA. The coenzyme A is replaced by **acyl carrier protein (ACP)**, making malonyl-ACP (**Fig. 15.14A**, step 2).

Malonyl-ACP condenses with the growing chain. In step 3, malonyl-ACP hooks onto the head of an acetyl-ACP or a longer-growing chain. The process of hooking onto the new unit involves displacement of CO_2 from the back end of malonyl-ACP, which replaces ACP from the front end of the growing chain. The growing chain now contains a ketone from the former acetyl group. The ketone must be reduced to CH_2 by 2 NADPH, gaining $4H^+ + 4e^-$ (steps 4–6). The chain is now fully hydrogenated (saturated).

Once hydrogenated, the chain is ready to take on the next malonyl-ACP, which, as before, loses CO_2 and replaces the ACP from the front of the growing chain (return to step 3). Successive addition can continue many times to build a saturated fatty acid of indefinite length. The regulation of fatty acid chain length in bacteria is poorly understood.

Unsaturation. Certain fatty acids require unsaturated "kinks" in the chain to serve a structural purpose, such as to increase the fluidity of the membrane. Such a kink is formed by an unsaturated double bond between adjacent carbons (an alkene). Unsaturation can be generated during a cycle of elongation (**Fig. 15.14B**). During the first four cycles of acetyl addition, an alkene bond forms between the *second* and *third* carbons of the fatty acid. After the fifth two-carbon addition, however, a special dehydratase enzyme forms the alkene double bond between the *third* and *fourth* carbons. This alkene fails to be hydrogenated further. Instead, several additional malonyl units are added, generating a long-chain fatty acid with a kink of a *cis* double bond.

Regulation of fatty acid synthesis. The synthesis of fatty acids consumes enormous quantities of reducing energy; thus, it must be regulated closely to avoid waste. From a structural standpoint, production of fatty acids incorporated into membranes must be balanced with growth of the cytoplasm. Furthermore, the many

different variants of fatty acids are regulated in response to particular environmental needs. In *E. coli*, key points of regulation include the following:

■ **Acetyl-CoA carboxylase represses its own transcription.** Transcription of an operon encoding two subunits of acetyl-CoA carboxylase (AccB, AccC) is repressed by one of its subunits, protein AccB. As AccB increases in concentration, it binds the promoter of the *accBC* operon, repressing further transcription. Thus, initiation of fatty acid biosynthesis (**Fig. 15.14**, step 1) is always limited by the amount of AccB-AccC enzyme present.

■ **Starvation blocks fatty acid biosynthesis.** Starvation for carbon sources blocks fatty acid biosynthesis through the "stringent response." Blockage is mediated by the polyphosphorylated nucleotide ppGpp, the global regulator of the stringent response (discussed in Section 10.3).

■ **Temperature regulates fatty acid composition.** Bacterial fatty acid composition is regulated by environmental factors such as temperature. In *E. coli*, low temperature favors unsaturated fatty acids because they are less rigid and maintain membrane flexibility. Low temperature induces expression of the gene *fabA* encoding the dehydratase enzyme that desaturates the fatty acid bond. As the dehydratase activity is increased, more unsaturated fatty acids are made.

THOUGHT QUESTION 15.5   For a given species, uniform thickness of a cell membrane requires uniform chain length of its fatty acids. How do you think chain length may be regulated?

The biosynthesis of fatty acids provides a model for study of the related biosynthesis of a major family of pharmaceuticals, the polyketides. A well-known polyketide is erythromycin, a broad-spectrum antibiotic frequently prescribed for pneumonia and other serious bacterial infections. Polyketides are commonly produced by actinomycetes, particularly species of *Streptomyces* (**Special Topic 15.2**). Antibiotics are produced as **secondary metabolites**, or **secondary products**—molecules that are not essential for survival but can enhance competition with other bacteria in a crowded natural environment. These secondary products accumulate to especially high levels during stationary phase, when cells run out of nutrients.

Bacterial Polyesters

Cyclic pathways of elongation comparable to fatty acid biosynthesis are used to build polymers for energy storage. For example, many bacteria synthesize polyesters such as polyhydroxybutyrate (**Fig. 15.15**). The term

A.

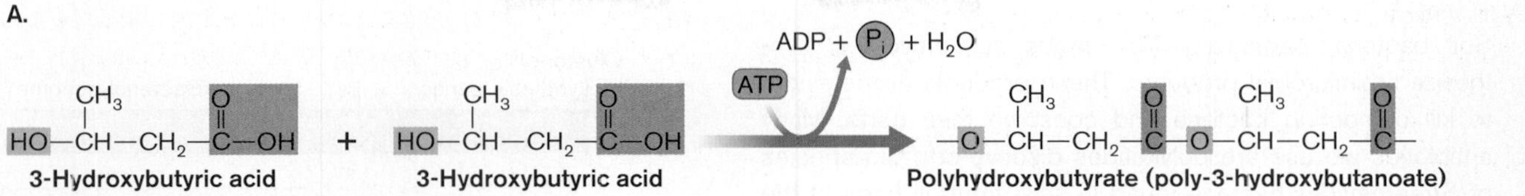

3-Hydroxybutyric acid + 3-Hydroxybutyric acid → Polyhydroxybutyrate (poly-3-hydroxybutanoate)

ATP → ADP + P_i + H_2O

B.

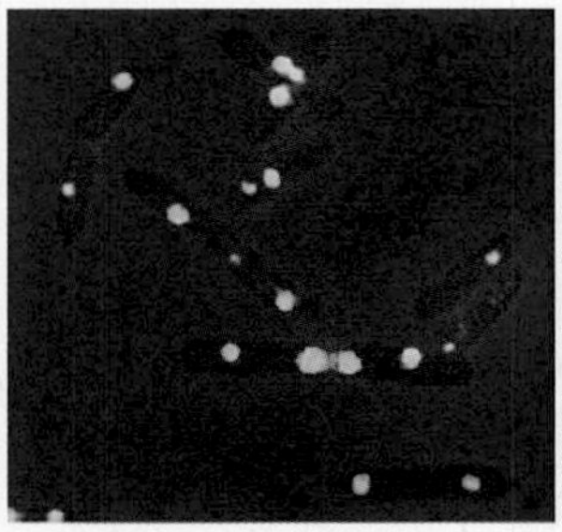

Figure 15.15   Microbial polyester: polyhydroxybutyrate.   A. Multiple units of 3-hydroxybutyric acid condense to form polyhydroxybutyrate, a polymer that stores energy. **B.** *Legionella pneumophila* bacteria form storage granules of polyhydroxybutyrate. *Source:* Stuart Mauchline et al. 1992. *J. Gen. Microbiol.* **138**:2371. Part B from Dieter Jendrossek. 2009. *Journal of Bacteriology* **191**:3195.

"polyester" indicates the multiple ester groups that are formed by repeated esterification of the carboxylic acid group of the chain with the hydroxyl group of a new alkanoate unit. Polyesters are insoluble in water, so they collect as storage granules within the bacterial cell, ready to release stored energy when needed. Polyester granules are formed by human pathogens such as *Legionella pneumophila* (**Fig. 15.15B**), the cause of legionellosis. Polyester storage helps *L. pneumophila* survive in water sources such as air-conditioning units. On the other hand, polyesters produced by soil bacteria such as *Ralstonia eutropha* are of interest for commercial development as biodegradable plastics. For example, surgical sutures made of polyhydroxybutyrate are resorbed by healed tissue.

Polyketide Antibiotics

The polyketide antibiotics are a diverse group of metabolites made by cyclic elongation. A prominent example is the broad-spectrum antibiotic erythromycin (see **Fig. 15.16**), originally isolated from *Streptomyces erythraeus*. Polyketides are synthesized by an enormous enzyme complex called a **modular enzyme**, which consists of multiple modules that add similar but nonidentical units to a growing chain. Modular enzyme biosynthesis offers exciting prospects for the design of new drugs.

Like fatty acids, polyketides are built by the successive condensation of malonyl-ACP units; but each malonyl group carries a unique extension, or R group. In erythromycin, the R groups are all methyl groups. In other polyketides, the R groups range from hydroxyl groups and amides to aromatic rings, some of which undergo secondary reactions and interconnections.

How is the order of different R groups determined? The polyketide is actually constructed on a modular enzyme (**Fig. 15.16**). A modular enzyme contains a series of active sites, or modules, each of which catalyzes the addition of one of a series of units. Each module contains all the enzyme activities needed to add the unit and recognizes a unit with one specific R group. The modular polyketide synthase may consist of a single protein with a series of domains, or it may consist of a complex of several protein subunits. Either the multidomain enzyme or the complex acts as an assembly line to generate the specific polyketide.

Figure 15.16A and **B** show the chain extension synthesis of a polyketide. The initial two domains of enzyme 1 are acyltransferase (AT) and an acyl carrier protein (ACP) domain, analogous to the ACP of fatty acid biosynthesis but specialized to accept one component of the polyketide. The acyltransferase transfers the acyl group (R_1-acetyl) from R_1-acetyl-CoA onto the ACP, with release of CoA (**Fig. 15.16A**). The R_1-acetyl is then transferred to a ketosynthase domain (KS).

Meanwhile, a malonyl group with its own R group (R_2-malonyl) has been transferred by a second AT onto the second ACP domain (**Fig. 15.16B**). The R_1-acetyl group then leaves KS and condenses with R_2-malonyl-ACP, releasing CO_2.

The R_1R_2 acyl chain subsequently undergoes a series of extensions by R-malonyl groups (**Fig. 15.16C**). For erythromycin, all R groups are methyl; other polyketides have R groups that are more complex. The initial AT and ACP domains constitute a "loading module" for the first acyl group, whereas subsequent sets of domains constitute "extender modules" that each add a different R-malonyl group. Extender modules can include secondary activities such as dehydratase (DH, removes OH) and ketoreductase (KR, hydrogenates a ketone to OH). In principle, the modular approach can construct a limitless range of products with diverse antibiotic properties.

Soil bacteria, particularly filamentous actinomycetes, synthesize antimicrobial products. These products are released to kill competing bacteria and consume their parts. Many antibiotics we use are polyketides discovered from species of *Streptomyces* bacteria found in soils from all parts of the world. For example, ivermectin (**Fig. 1**) is an antiparasitic drug used against worms that cause river blindness (onchocerciasis) and lymphatic filariasis. Ivermectin was isolated in 1975 from *S. avermitilis*, a bacterium found in a soil sample from a Japanese golf course. Another example, reported in 2002, is the experimental antitumor agent C-1027, isolated from *S. globisporus* (see **Fig. 1**).

Bacteria in nature produce numerous kinds of antibiotics we have yet to discover. One way to discover new polyketide antibiotics is by screening the genomes of new bacterial isolates for novel operons encoding polyketide synthase genes. The scanning of a large number of unsequenced genomes enables cloning of biosynthesis operons for hundreds of new secondary products. The pharmaceutical activities of these products can then be tested.

An even more ambitious approach is to screen a "metagenome," DNA isolated from numerous organisms in

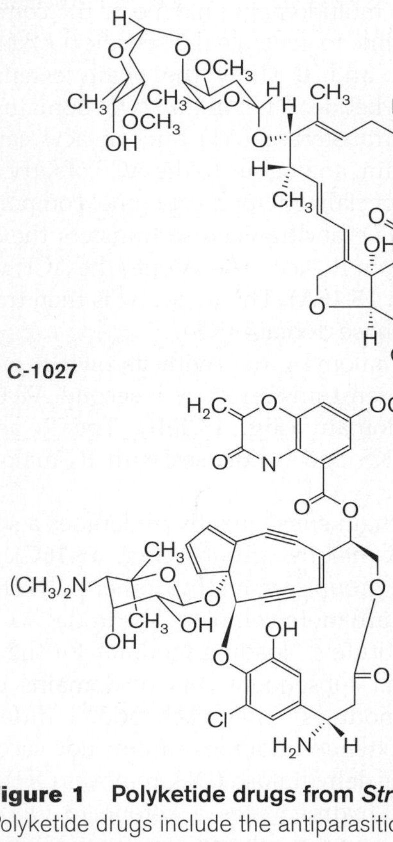

Ivermectin

C-1027

Figure 1 **Polyketide drugs from *Streptomyces* species.** Polyketide drugs include the antiparasitic drug ivermectin and the antitumor agent C-1027.

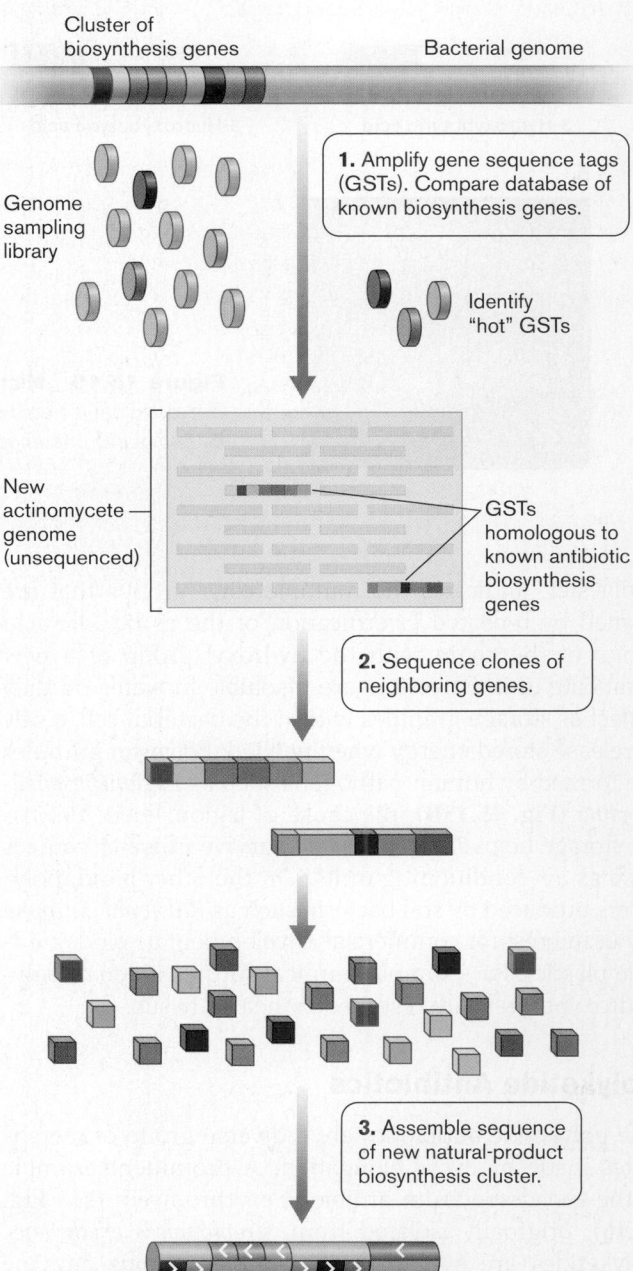

1. Amplify gene sequence tags (GSTs). Compare database of known biosynthesis genes.

Identify "hot" GSTs

GSTs homologous to known antibiotic biosynthesis genes

2. Sequence clones of neighboring genes.

3. Assemble sequence of new natural-product biosynthesis cluster.

Figure 2 **Screening genomes to identify a biosynthesis cluster for a new antibiotic.** Biosynthesis genes occur in clusters in a bacterial genome. To identify an unknown cluster whose genes are related to known clusters: (1) Amplify gene sequence tags (GSTs) from an unsequenced actinomycete genome. Compare GST sequences from the new genome with a database of known antibiotic biosynthesis gene clusters. Identify promising GSTs. (2) Hybridize GST sequences to clone plasmids containing sequences adjacent to the GST. Subclone and obtain the adjacent sequence, which may contain a new biosynthesis cluster. (3) Assemble the sequence of the new gene cluster for biosynthesis. *Source*: Based on E. Zazopoulos et al. 2003. *Nat. Biotechnol.* **21**:187.

a natural environment, without culturing. The metagenome of uncultured isolates may yield a hundred times as many potential drug sequences as are found in culturable isolates. An example of this approach was the screening of a sponge-associated bacterial metagenome by a research group at Kosan Biosciences. The Kosan group sequenced DNA equivalent to approximately 400 genomes from uncultured bacterial symbionts of the marine sponge *Discodermia dissoluta*. A clone library was generated, of which nearly 1% of clones contained genes encoding polyketide synthase modules.

How do we screen 400 genomes for unknown modular enzymes? The process of screening genomes requires two assumptions. First, genes for biosynthesis of secondary metabolites tend to be variants of genes for known products; that is, they share a common ancestral gene sequence. Second, the genes for synthesis of a particular product usually map together as a result of being transferred as a cluster from another organism. Thus, screening is designed to isolate an unknown cluster whose genes are related to known genes in other organisms. The result is valuable because often a variant of a known antibiotic proves to be active against pathogens resistant to the original substance, or the variant may have fewer side effects in the host.

To find a new antibiotic biosynthesis operon, the first step is to amplify gene sequence tags (GSTs) from an isolated bacterium (**Fig. 2**). A gene sequence tag is a short sequence of DNA that is amplified by PCR using primer ends homologous to known biosynthesis genes (step 1). The GSTs are amplified by PCR from plasmid clones in a genome library of a suspected drug-producing organism. The GSTs are compared with a database of known biosynthesis genes. The "hot" GSTs (those confirmed as biosynthetic homologs) are then used (step 2) to probe clones containing genes adjacent to the GST sequences. The adjacent genes are likely to be part of a gene cluster including the GST-defined genes, but they may encode enzymes with novel functions in biosynthesis. To identify these novel biosynthetic genes, DNA is sequenced upstream and downstream of the GST. Finally, the sequences are assembled to reveal the biosynthesis gene cluster (step 3). The sequence is annotated and deposited in the database for future genomic scanning.

The sponge symbiont metagenome revealed several multimodular polyketide synthase gene clusters. One large multimodular cluster is shown in **Figure 3**. The cluster includes a starter module (module 0) plus at least 14 extender modules. Each module includes a ketosynthase (tan) and an acyltransferase (red) to extend the chain by one ketide. Various modules contain "options" such as dehydratase (yellow), enoyl reductase (maroon), or ketoreductase (orange), depending on the fate of that segment of the polyketide. The antibiotic properties of this novel polyketide have yet to be determined.

Why would a sponge harbor so many bacteria encoding antibacterial agents? Presumably the sponge encourages growth of organisms that help protect it from microbial grazing on its tissues.

Polyketide synthesized by sponge symbiont

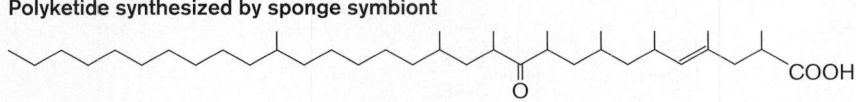

Open reading frames encoding modular enzymes

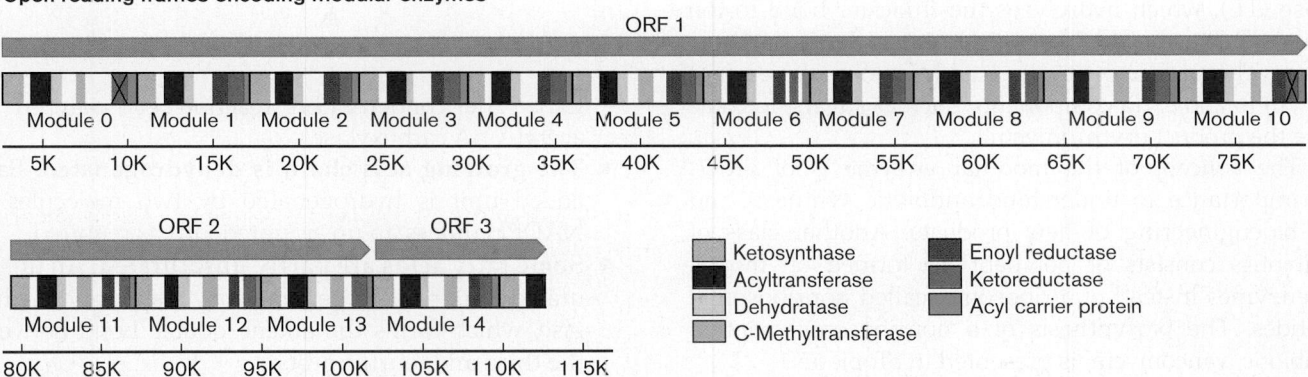

Figure 3 **Polyketide synthase gene cluster.** The sponge-symbiont gene cluster includes 14 complete modules, each encoding genes to extend and modify the polyketide chain.

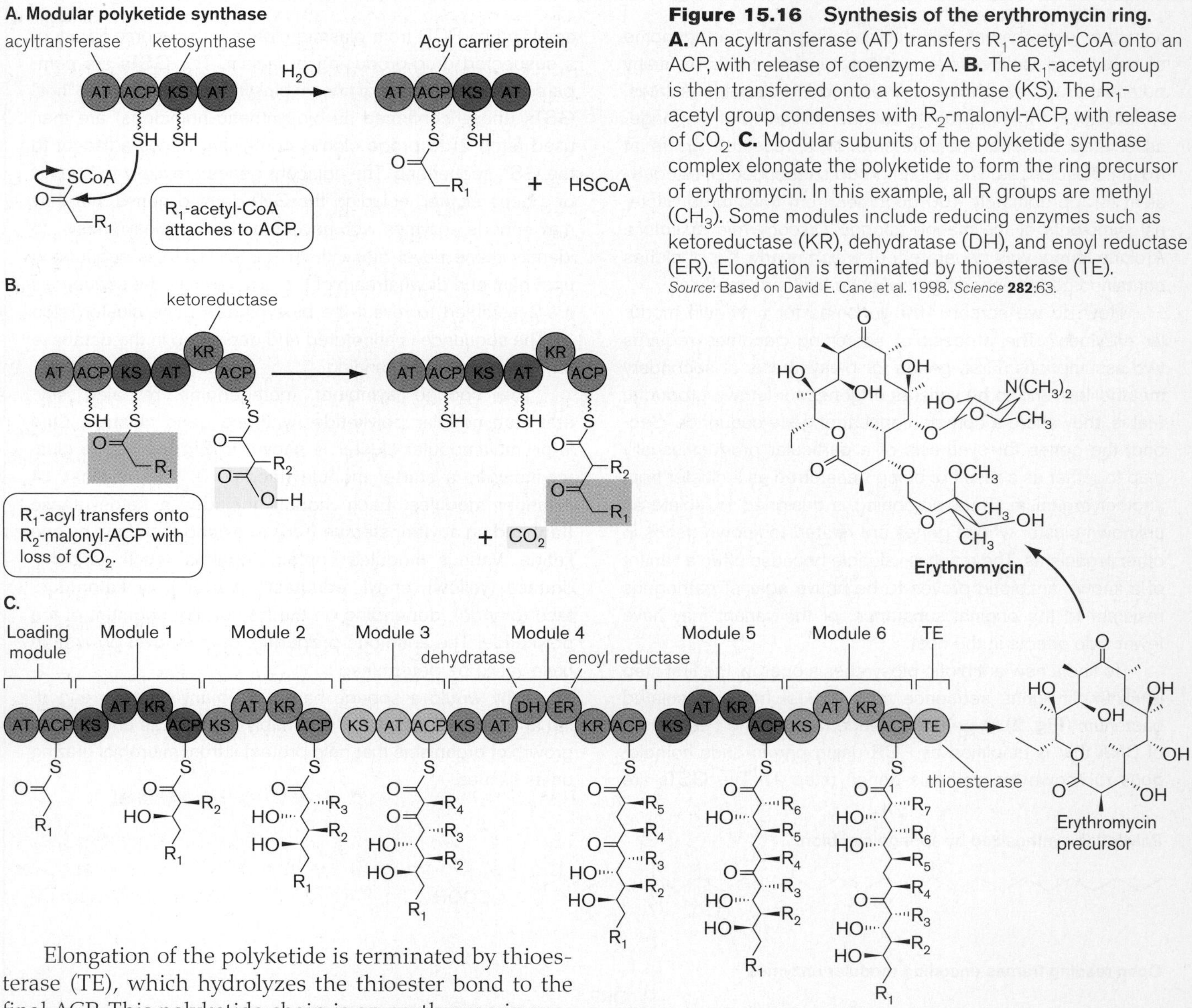

A. Modular polyketide synthase

B.

R₁-acyl transfers onto R₂-malonyl-ACP with loss of CO_2.

C.

Figure 15.16   **Synthesis of the erythromycin ring.**
A. An acyltransferase (AT) transfers R_1-acetyl-CoA onto an ACP, with release of coenzyme A. **B.** The R_1-acetyl group is then transferred onto a ketosynthase (KS). The R_1-acetyl group condenses with R_2-malonyl-ACP, with release of CO_2. **C.** Modular subunits of the polyketide synthase complex elongate the polyketide to form the ring precursor of erythromycin. In this example, all R groups are methyl (CH_3). Some modules include reducing enzymes such as ketoreductase (KR), dehydratase (DH), and enoyl reductase (ER). Elongation is terminated by thioesterase (TE).
Source: Based on David E. Cane et al. 1998. *Science* **282**:63.

Elongation of the polyketide is terminated by thioesterase (TE), which hydrolyzes the thioester bond to the final ACP. This polyketide chain is an erythromycin precursor. The precursor requires additional enzymes (not shown) to add extra components, such as sugars, to complete the product erythromycin.

The concept of the modular enzyme is of growing importance to understand antibiotic synthesis and the bioengineering of new products. Another class of antibiotics consists of polypeptides formed on modular enzymes instead of a ribosome, called nonribosomal peptides. The biosynthesis of a nonribosomal peptide antibiotic, vancomycin, is presented in eTopic 15.1.

TO SUMMARIZE:

■ **Fatty acid biosynthesis** involves successive condensation of malonyl-ACP groups formed from acetyl groups tagged with acyl carrier protein (ACP) and a carboxylate. Each successive malonyl group is transferred onto the growing acyl chain, with release of

CO_2. The condensation reaction is catalyzed by acetyl-CoA carboxylase.

■ **The growing acyl chain is dehydrogenated.** Each added unit is hydrogenated by two molecules of NADPH unless an unsaturated kink is required.

■ **Some fatty acids are partly unsaturated.** An unsaturated kink may be generated by a special dehydratase, which forms the alkene double bond between the third and fourth carbons.

■ **Fatty acid biosynthesis is regulated** by the levels of acetyl-CoA carboxylase and by the stringent response to carbon starvation. Bond saturation is regulated by temperature and other environmental factors.

■ **Polyesters** for energy storage are synthesized by cyclic elongation of polyalkanoates.

■ **Polyketide antibiotics are synthesized by modular enzymes.**

15.5 Nitrogen Fixation

The synthesis of amino acids, cell walls, and other cofactors requires an additional element we have not yet considered—namely, nitrogen. In principle, nitrogen should be more accessible to cells than carbon is, since nitrogen gas (N_2) comprises more than three-quarters of our atmosphere. In fact, however, N_2, with its triple bond, is one of the most stable molecules in nature. An enormous input of energy is required to reduce nitrogen to ammonia for assimilation into carbon skeletons. Early in evolution, all cells may have fixed their own N_2, but today only certain species of bacteria and archaea retain the ability. All other organisms depend on reduced or oxidized forms of nitrogen, which ultimately derive from N_2. Thus, all living organisms, directly or indirectly, depend on N_2-fixing prokaryotes.

> **NOTE:** Bacteria and archaea play essential roles in global cycling of nitrogen, sulfur, and phosphorus. The geochemical cycling of these elements is discussed in Chapter 22.

Nitrogen Assimilation

To assimilate into biomass, any form of nitrogen must be fully reduced to ammonia (NH_3), which at pH 7 is mostly protonated to ammonium ion, NH_4^+. Unlike carbon, nitrogen rarely appears in oxidized form in complex biomolecules. Inorganic forms, such as nitric oxide (NO), are used as defense mechanisms against invading pathogens (see Chapter 23) or as signaling molecules. In macromolecules, however, virtually all the nitrogen is reduced; organic compounds containing oxidized nitrogen are generally toxic.

While N_2 is the ultimate source and sink of biospheric nitrogen, several oxidized or reduced forms occur in the environment (**Fig. 15.17**). Most free-living bacteria can acquire nitrate (NO_3^-) or nitrite (NO_2^-) for reduction to ammonium ion, although they repress these energy-expensive pathways when ammonium ion is present. Even nitrogen-fixing legume symbionts, such as *Rhizobium*, can utilize nitrate.

In natural environments, most potential sources of nitrogen are subject to competition from dissimilatory metabolism in which the molecule is oxidized or reduced for energy, as discussed in Chapter 14. For example, anaerobic respirers convert nitrate and nitrite to N_2 (denitrification), whereas lithotrophs oxidize NH_4^+ to nitrite and nitrate (nitrification). An important consequence of nitrification is that most of the commercial ammonia fertilizer spread on agricultural fields is soon oxidized by lithotrophs to nitrates and nitrites. High concentrations of nitrites in water are harmful because they combine with hemoglobin, generating a form that cannot take up oxygen. When an infant drinks water with high nitrite, they may become ill with "blue-baby" syndrome.

Nitrogen Fixation: Early Discoveries

The first nitrogen fixers discovered by Martinus Beijerinck and colleagues in the late 1800s were soil and wetland bacteria such as *Beggiatoa* and *Azotobacter*. Species of

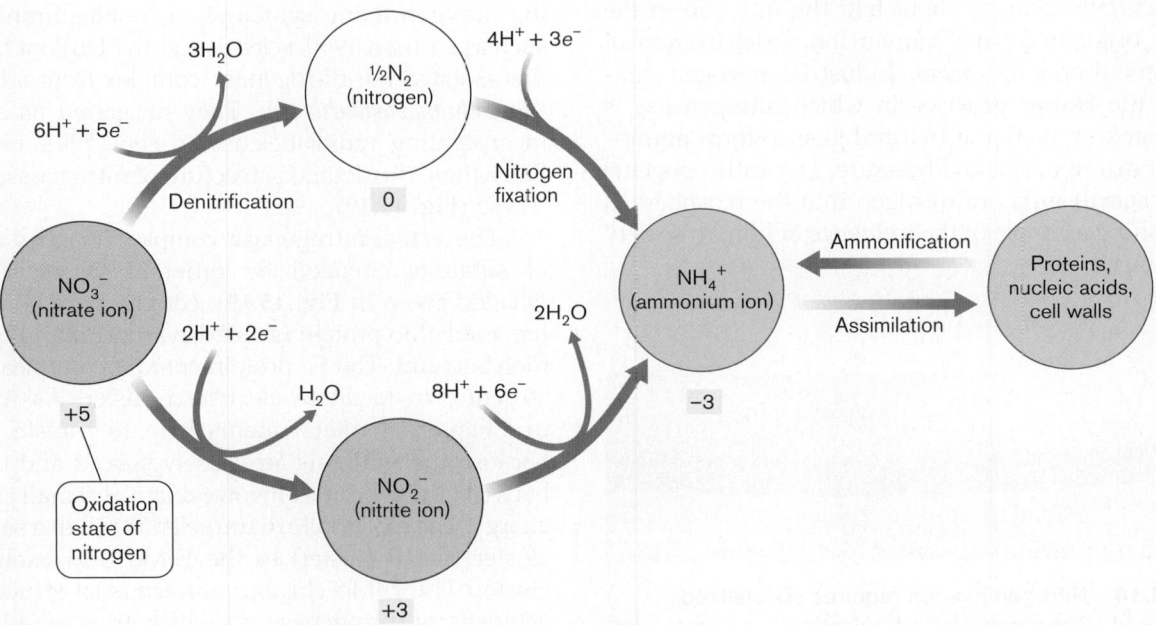

Figure 15.17 Nitrogen assimilation. Different oxidation states of nitrogen require different amounts of reducing energy for assimilation into biomass.

Rhizobium were discovered to fix nitrogen as endosymbionts of leguminous plants (see Chapter 21). For several decades, it was believed that N_2 was fixed only by a few special bacteria in the soil or in symbiotic plant bacteria. But in the 1940s, Martin Kamen and colleagues noticed that phototrophs such as *Rhodospirillum rubrum* produce hydrogen gas—a known by-product of nitrogen fixation. Nitrogen fixation was hard to demonstrate reliably in the laboratory because at that time researchers did not know that in *R. rubrum* nitrogen fixation is repressed by the presence of alternative nitrogen sources, such as ammonia. When traces of ammonia and other nitrogen sources were eliminated, Kamen's student Herta Bregoff found that photosynthetic bacteria actually fix nitrogen.

The fixation of nitrogen was confirmed by experiments showing uptake of the radioisotope ^{15}N into biomass. Nitrogen is now known to be fixed by most phototrophic bacteria (green and purple bacteria, as well as cyanobacteria) and by many archaea. Marine cyanobacteria fix a large proportion of both the nitrogen and carbon dioxide assimilated by our biosphere. Aquatic cyanobacteria such as *Anabaena* develop special cells called **heterocysts** to fix N_2, where photosynthesis is turned off to maintain anaerobic conditions (**Fig. 15.18**). Land ecosystems require nitrogen-fixing bacteria and archaea in the soil, often in mutualistic relationships with plants. For example, the bacterium *Sinorhizobium meliloti* associates with legume roots, causing them to form nodules. The nodules contain "bacteroids," forms of the bacteria that grow within the root cells. Nitrogen-fixing mutualism is discussed in Chapter 21.

Bacterial nitrogen fixation, however, has not sufficed to drive modern high-yield agriculture. By the end of the twentieth century, about half the nitrogen in the biosphere originated from human industrial fixation of N_2 for agricultural fertilizers. Industrial nitrogen fixation uses the **Haber process**, in which nitrogen gas is hydrogenated by methane (natural gas) to form ammonia, under extreme heat and pressure. The anthropogenic (human-caused) influx of nitrogen into the biosphere is an astonishing example of the influence of human society on our planet.

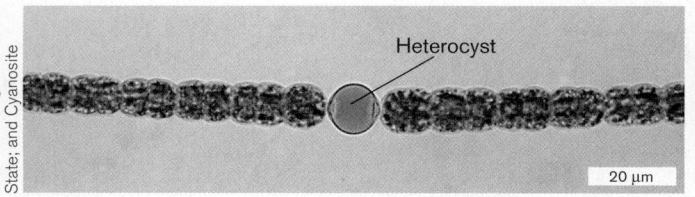

Figure 15.18   Nitrogen fixation requires specialized structures. *Anabaena spiroides*, a filamentous cyanobacterium, segregates N_2 fixation in heterocysts, cells that do not photosynthesize and thus maintain anoxic conditions.

The Mechanism of Nitrogen Fixation

In living cells, nitrogen fixation is an enormously energy-intensive process. The mechanism is largely conserved across species:

$$N_2 + 8H^+ + 8e^- + 16\,ATP \longrightarrow 2NH_3 + H_2 + 16\,ADP + 16\,P_i$$

The total energy investment includes approximately 3 ATP-equivalents per electron (assuming reduction by NADPH), plus 16 ATP molecules as shown. That makes $24 + 16 = 40$ ATP in all, more than that provided by complete oxidation of glucose. The production of H_2 is surprising, as it consumes extra ATP. Hydrogen loss results from initiating the cycle of nitrogen reduction (discussed shortly). Some bacteria have secondary reactions to reclaim the lost hydrogen with part of its lost energy.

> **NOTE:** Although nitrogen fixation is commonly represented as producing ammonia (NH_3), under most conditions of living cells the predominant form is protonated ammonium ion (NH_4^+).

Nitrogenase reaction mechanism. The overall conversion of nitrogen gas to two molecules of ammonia is catalyzed in four cycles by **nitrogenase**, an enzyme highly conserved in nearly all nitrogen-fixing species. Like Rubisco, nitrogenase probably evolved once in a common ancestor of all nitrogen-fixing organisms.

The mechanism of nitrogenase is of intense interest to agricultural scientists because of the potential benefits of improving efficiency of plant growth and of extending nitrogen-fixing symbionts to nonleguminous plants such as corn. In 1960, scientists at the DuPont Laboratory first isolated the nitrogenase complex from a bacterium, *Clostridium pasteurianum*. They measured its activity by incorporating radiolabeled nitrogen, $^{15}N_2$, into $^{15}NH_3$. Since then, the detailed structure of nitrogenase has been solved (**Fig. 15.19**).

The active nitrogenase complex includes two kinds of subunits encoded by different genes: Fe protein (shaded green in **Fig. 15.19**), containing a [4Fe-4S] center, and FeMo protein (shaded cyan), containing iron and molybdenum. The Fe protein contains a typical [4Fe-4S] structure to facilitate electron transfer. As we learned in Chapter 14, metal atoms help to transfer electrons because their orbitals are closely spaced and transitions between these orbitals involve relatively small amounts of energy. The electrons are funneled through a second Fe-S cluster (the P cluster) to the FeMo (iron-molybdenum) cluster. The FeMo cluster is an unusual structure characteristic of nitrogenase, in which trios of sulfur atoms alternate with trios of iron atoms. One end of the cluster is capped with another iron, and the other end is capped

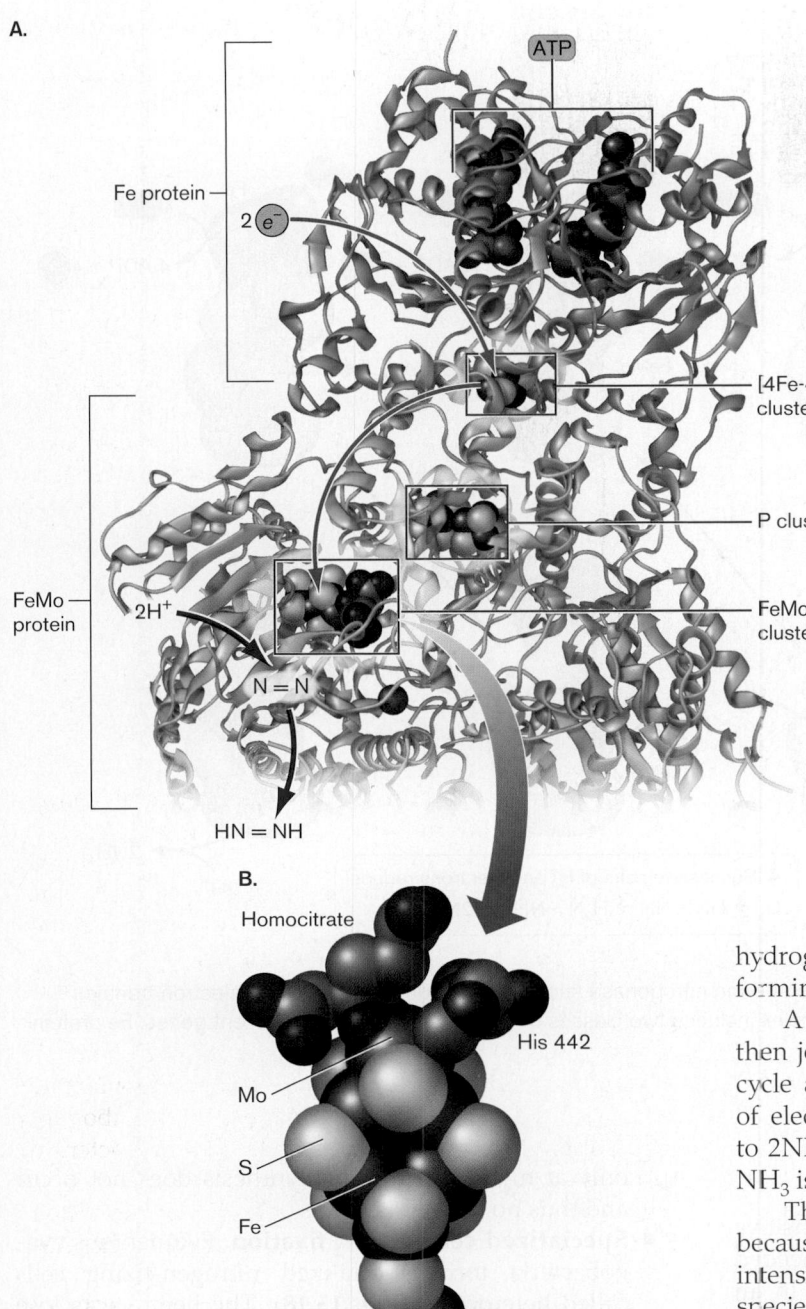

Figure 15.19   Structure of the nitrogenase complex.
A. In each active site of the Fe protein–FeMo protein complex, the Fe protein binds ATP and receives electrons from the electron donors. The electrons are subsequently channeled down through the [4Fe-4S] cluster and the P cluster (Fe-S cluster) to the FeMo cluster, where they reduce the N_2. (PDB code: 1N2C)
B. The metal cluster (Fe_7S_9Mo) is held in place by coordination with a molecule of homocitrate plus two amino acid residues of nitrogenase: His 442 and Cys 275.

with an atom of molybdenum (Mo). A consequence of the nitrogenase structure is that most nitrogen-fixing organisms (and their plant hosts, for leguminous symbionts) require the element molybdenum for growth. Some bacteria make an alternative nitrogenase that substitutes vanadium for molybdenum.

Four cycles of reduction. Nitrogen fixation requires four reduction cycles through nitrogenase (**Fig. 15.20**).

To initiate the first cycle of N_2 reduction, Fe protein acquires $2e^-$ from an electron transport protein such as a ferredoxin or a flavodoxin (step 1). Energy is provided by four molecules of ATP. The reduced Fe protein then transfers each electron to the FeMo center. The FeMo protein binds $2H^+$, which is reduced to H_2 (step 2). Only then does N_2 bind to the active site, by displacing the H_2 (step 3).

In the next reduction cycle, two electrons are transferred to Fe protein, where they reduce the iron. The reduced Fe protein transfers each electron to the FeMo center, near the binding site for N_2. The electron transfer requires expenditure of 2 ATP per electron, or 4 ATP per $2e^-$. Two hydrogen ions join the N_2 and receive the two electrons, forming HN=NH.

A third pair of electrons enters the Fe protein, which then joins another $2H^+$, reducing N_2 to $H_2N–NH_2$. The cycle again requires hydrolysis of 4 ATP. A final cycle of electron transfer and H^+ uptake reduces $H_2N–NH_2$ to $2NH_3$. At typical cytoplasmic pH values (near pH 7), NH_3 is protonated to NH_4^+.

The loss of H_2 during nitrogen fixation is puzzling because it represents lost energy in a highly energy-intensive process. The actual fate of the H_2 varies among species. *Klebsiella pneumoniae*, a Gram-negative organism in which nitrogen regulation has been much studied, gives off the H_2 without further reaction. On the other hand, *Azotobacter* uses an irreversible hydrogenase enzyme to convert H_2 back to $2H^+$ and recover the transferred electrons. In leguminous rhizobial symbionts such as *Sinorhizobium* species, H_2 recovery varies surprisingly among strains and can affect the efficiency of plant growth.

The process of nitrogen fixation is expensive, both in terms of protein synthesis and in terms of reducing energy and ATP. Even bacteria capable of nitrogen fixation repress the process in the presence of other nitrogen sources, such as nitrate (NO_3^-), nitrite (NO_2^-), ammonia (NH_3), or nitrogenous organic molecules acquired from dead cells.

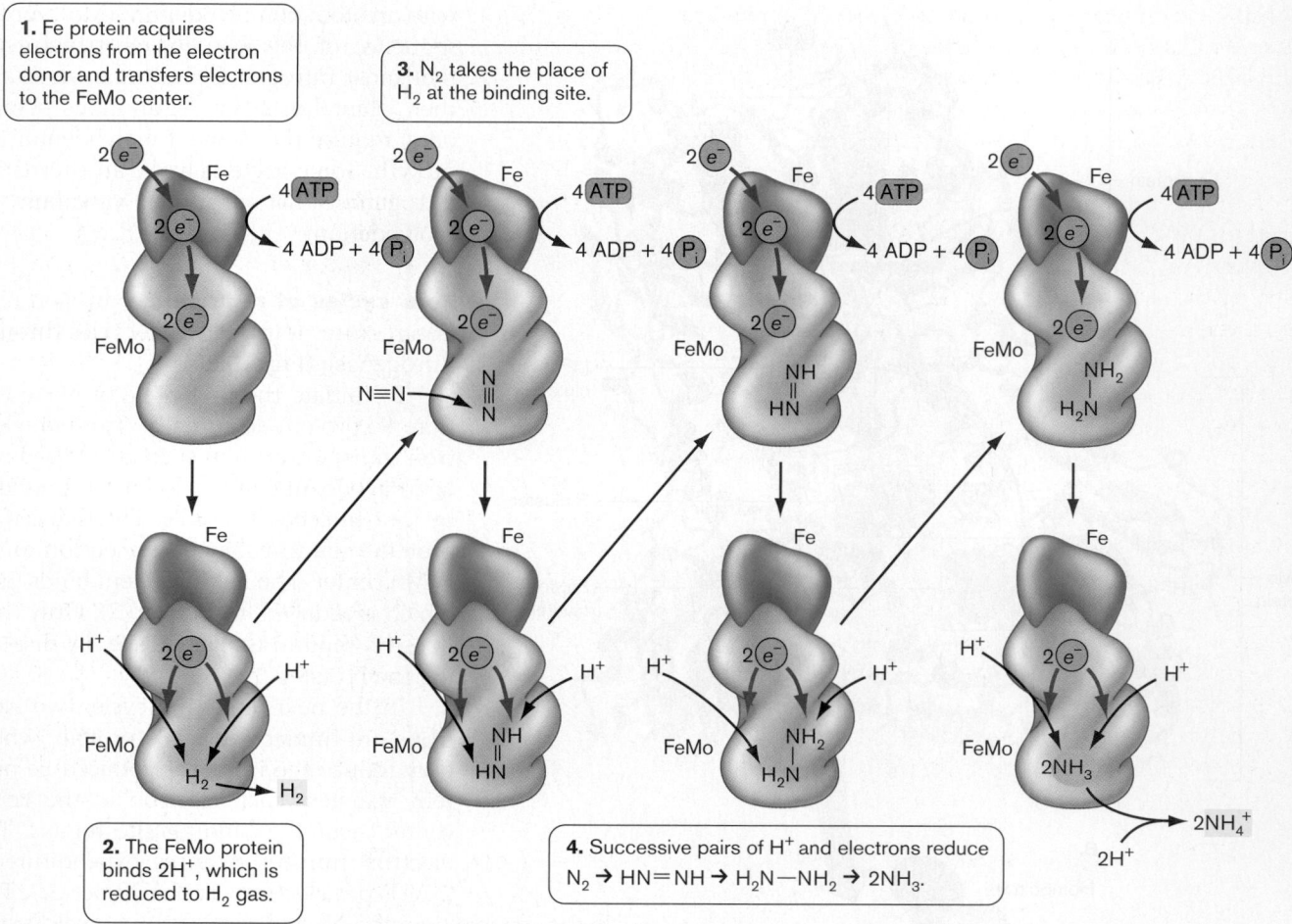

Figure 15.20 Nitrogen fixation by nitrogenase.   The enzyme nitrogenase successively reduces nitrogen by electron transfer, ATP hydrolysis, and H$^+$ incorporation. The nitrogenase complex includes two classes of subunits encoded by different genes: Fe protein (shaded green) and FeMo protein (shaded blue).

Anaerobiosis and N$_2$ Fixation

The reductive action of nitrogenase is extremely sensitive to oxygen because of the large reducing power needed to make NH$_4^+$. Thus, cells can fix nitrogen only in an anaerobic environment. This is no problem for anaerobes. But oxygenic phototrophs such as cyanobacteria face an obvious problem, as do bacterial symbionts of oxygenic plants. Aerobic and oxygenic organisms have developed several solutions to the problem of fixing nitrogen in aerobic environments.

- **Protective proteins.** Aerobic *Azotobacter* species synthesize protective proteins that stabilize nitrogenase and prevent attack by oxygen. For rhizobia, on the other hand, their plant hosts produce leghemoglobin (named for "legume" plants), a form of hemoglobin that sequesters oxygen away from the bacteria.
- **Temporal separation of photosynthesis and N$_2$ fixation.** Some species of cyanobacteria fix nitrogen only at night, when photosynthesis does not occur and thus no oxygen is made.
- **Specialized cells for N$_2$ fixation.** Filamentous cyanobacteria form specialized nitrogen-fixing cells called heterocysts (**Fig. 15.18**). The heterocysts lose their photosynthetic capacity entirely and specialize in nitrogen fixation. Heterocyst development is directed by a complex genetic program induced by nitrogen starvation. In natural environments, the heterocysts leak organic acids that attract heterotrophic bacteria, which use up all the oxygen around the heterocyst, generating ideal anaerobic conditions for nitrogen fixation.

Molecular Regulation of N$_2$ Fixation

Nitrogen fixation costs substantial energy, and therefore the process is highly regulated. Oxygen and NH$_4^+$ availability regulate expression of the *nif* genes (*nifHDKTY*) encoding nitrogenase and other nitrogen-fixation proteins. Regulation of *nif* is mediated by several molecular regulators, including a nitrogen starvation sigma factor (sigma-54) and the NtrB-NtrC two-component signal transduction system (discussed in Section 10.7). The NtrB-NtrC system has been dissected in *Klebsiella pneumoniae* by Sydney Kustu and colleagues at the University of California at Berkeley (**Fig. 15.21A**).

The NtrC protein regulates nitrogenase expression in response to NH_4^+ concentration (**Fig. 15.21B**). When cellular levels of NH_4^+ are high, no activators bind the *nifH-DKTY* promoter, and the nitrogen fixation genes are not expressed. When NH_4^+ is low, however, NtrC is phosphorylated by NtrB. The NtrC-P phosphoprotein now binds its DNA site to activate expression of *nifLA*, making NifL and NifA proteins. Expression of *nifLA* is coactivated by sigma-54. Sigma-54 and the NifA protein together activate expression of *nifHDKTY*. Thus, low NH_4^+ concentration turns on nitrogen fixation.

Nitrogen fixation cannot occur with oxygen, so high oxygen levels block nitrogenase expression. When oxygen is high, NifA binds NifL and is prevented from binding the nitrogenase promoter. Low oxygen allows NifA to bind the promoter and express nitrogenase to fix nitrogen. Overall, NtrC governs response to NH_4^+, whereas

NifA governs response to oxygen. Response to multiple environmental signals is typical of energy-expensive biosynthetic pathways.

TO SUMMARIZE:

- **Oxidized or reduced forms of nitrogen**, such as nitrate, nitrite, and ammonium ions, can be assimilated by bacteria and plants. Assimilation competes with dissimilatory reactions that obtain energy.
- **Nitrogen gas (N_2) is fixed into ammonium ion (NH_4^+)** only by some species of bacteria and archaea, never by eukaryotes.
- **Nitrogenase enzyme** includes a protein containing an iron-sulfur core (Fe protein) and a protein containing a complex of molybdenum, iron, and sulfur (FeMo protein). Electrons acquired by Fe protein (with energy from ATP) are transferred to FeMo protein to reduce nitrogen.
- **Four cycles of reduction by NADPH** or an equivalent reductant are required to reduce one molecule of N_2 to two molecules of NH_3. At neutral pH, NH_3 is protonated to NH_4^+.
- **Oxygen inhibits nitrogen fixation**. Bacteria and plants have various means of separating nitrogen

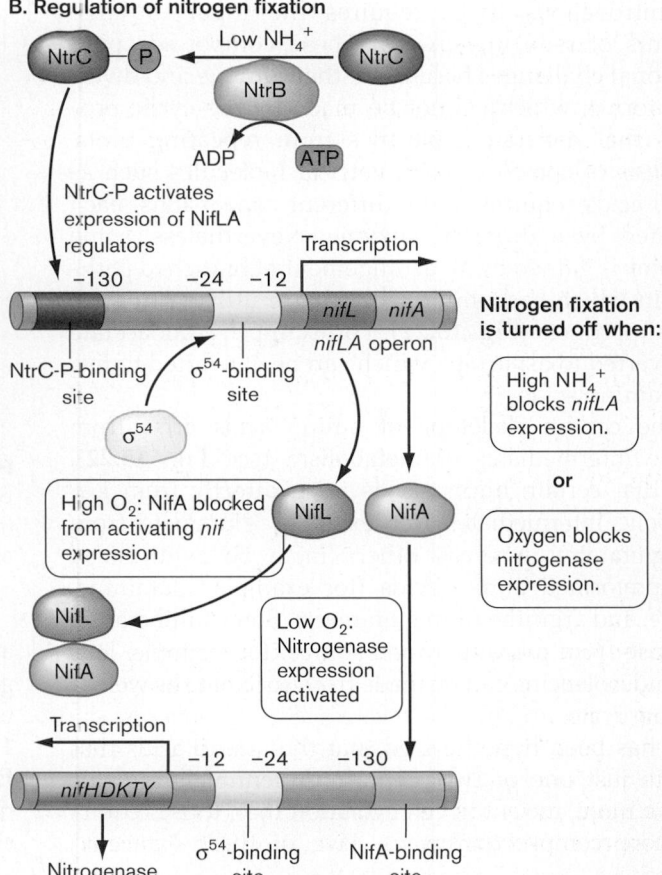

Figure 15.21 Regulation of nitrogen fixation.
A. Sydney Kustu has made important discoveries about NtrC and other forms of nitrogen regulation in the Gram-negative soil bacterium *Klebsiella pneumoniae*. **B.** When cellular levels of NH_4^+ are low, NtrC is phosphorylated and activates expression of *nifLA*. Expression is coactivated by the nitrogen starvation sigma factor, sigma-54. The NifA protein (and sigma-54) activates expression of *nif* genes encoding nitrogenase. When oxygen levels are high, however, nitrogen fixation cannot occur, so NifA is blocked by binding NifL. (Numbers refer to positions upstream of the translation site.)

fixation from aerobic respiration, such as heterocyst development or temporal separation.

■ **Regulation of nitrogen fixation** by nitrogen and oxygen levels occurs via NtrC and sigma-54 regulation of transcription.

15.6 Biosynthesis of Amino Acids and Nitrogenous Bases

Microbes need amino acids to make proteins and cell walls, as well as nitrogenous bases to form nucleic acids. When possible, they obtain these molecules from their environment through membrane-embedded transporters. But competition for such valuable nutrients is high, especially for free-living microbes in soil or water. Most free-living microbes and plants possess the ability to make all the standard amino acids and bases of the genetic code, as well as nonstandard variants used for cell walls and transfer RNAs. Microbial biosynthesis of amino acids "essential" for animals is used for industrial production of food supplements.

Amino Acid Synthesis

Like fatty acid biosynthesis, synthesis of amino acids and nitrogenous bases requires the input of large amounts of reducing energy. These compounds pose additional challenges because of their unique and diversified forms, which cannot be made by the cyclic processes that generate molecules from repeating units. Synthesis of complex, asymmetrical molecules such as amino acids requires many different conversions, each mediated by a different enzyme. Nevertheless, some economy is gained by an arrangement of branched pathways in which early intermediates are utilized to form several products (**Fig. 15.22**). For example, oxaloacetate is converted to aspartate, which can be converted to five other amino acids.

The carbon skeletons of amino acids arise from diverse intermediates of metabolism (see **Fig. 15.22**). Note that certain amino acids arise directly from key metabolic intermediates (for example, glutamate from 2-oxoglutarate), whereas others must be synthesized from preformed amino acids (for example, glutamine, proline, and arginine from glutamate). Some amino acids can arise from more than one source; for example, leucine and isoleucine can be made from succinate as well as from pyruvate.

It has been hypothesized that the amino acids that arise in just one or two steps from central intermediates are more ancient in cell evolution than those requiring more complex pathways. Five of these "ancient"

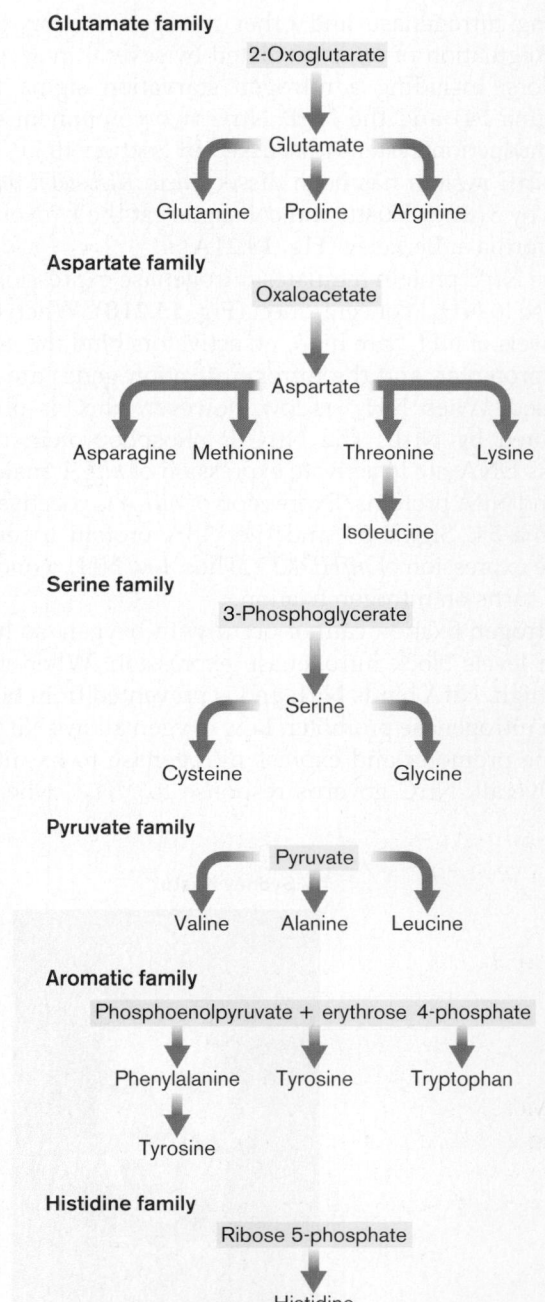

Figure 15.22   Major pathways of amino acid biosynthesis. Some amino acids arise from key metabolic intermediates (highlighted), whereas others must be synthesized out of other amino acids. Note alternative pathways (see Figure 15.2).

amino acids—glutamate, aspartate, valine, alanine, and glycine—are the same as those detected in meteorites, whose composition resembles that of prebiotic Earth (**Fig. 15.23**). These same five amino acids also appear in early-Earth simulation experiments in which methane, ammonia, and water are heated under reducing conditions and subjected to electrical discharge. Thus, we speculate that

Figure 15.23   The Murchison meteorite. Fragments of a meteorite that fell in Murchison, Australia, in 1969 were shown to contain the five fundamental amino acids of biosynthetic pathways (glutamate, aspartate, valine, alanine, and glycine).

the first amino acids that early cells evolved to make were the same as those that arose spontaneously in the prebiotic chemistry of our planet.

As in fatty acid biosynthesis, precursor molecules are channeled into amino acid biosynthesis by specialized cofactors and reducing energy carriers such as NADPH. Unlike fatty acids, however, amino acids must assimilate another key ingredient: nitrogen.

Assimilation of NH_4^+

The NH_4^+ produced by N_2 fixation or by nitrate reduction is the key source of nitrogen for biosynthesis. But NH_4^+ always exists in equilibrium with NH_3, which is toxic to cells. Moreover, NH_3 travels freely through membranes, making it difficult to store NH_4^+ within a cell. Deprotonation to NH_3 increases as pH rises; by pH 9.2, deprotonation reaches 50%. Even at neutral pH, a very small

equilibrium concentration of NH_3 can drain NH_4^+ out of the cell. Thus, cells avoid storing high levels of NH_4^+; instead, it is incorporated immediately into organic products.

2-Oxoglutarate and glutamate condense with NH_4^+.

In most bacteria, the main route for NH_4^+ assimilation is by condensing NH_4^+ with 2-oxoglutarate to form glutamate, or with glutamate to form glutamine (**Fig. 15.24**). Three key enzymes interconvert these substrates:

- **Glutamate dehydrogenase (GDH)**, actually named for its reverse activity, condenses NH_4^+ with 2-oxoglutarate to form glutamate. The condensation requires reduction by NADPH.
- **Glutamine synthetase (GS)** condenses a second NH_4^+ with glutamate to form glutamine, a process driven by spending one ATP.
- **Glutamate synthase (GOGAT, glutamine: 2-oxoglutarate amidotransferase)** converts 2-oxoglutarate plus glutamine into two molecules of glutamate. Different variants of this enzyme use different reducing agents: NADPH, NADH, or ferredoxin. The NADPH variant is shown.

All three substrates thus exchange amino groups readily. High nitrogen levels induce GDH to take up NH_4^+, but repress GS and GOGAT. GS has a higher affinity than GDH for NH_4^+. Low nitrogen levels induce GS (to make glutamine) and GOGAT (to convert some glutamine to glutamate as needed).

Both glutamate and glutamine contribute an amine, as well as their carbon skeletons, to the synthesis of other amino acids in the biosynthetic "tree." The transfer of ammonia between two metabolites such as glutamate and glutamine is called **transamination**. Transamination also occurs between many other pairs of amino acids and metabolic intermediates.

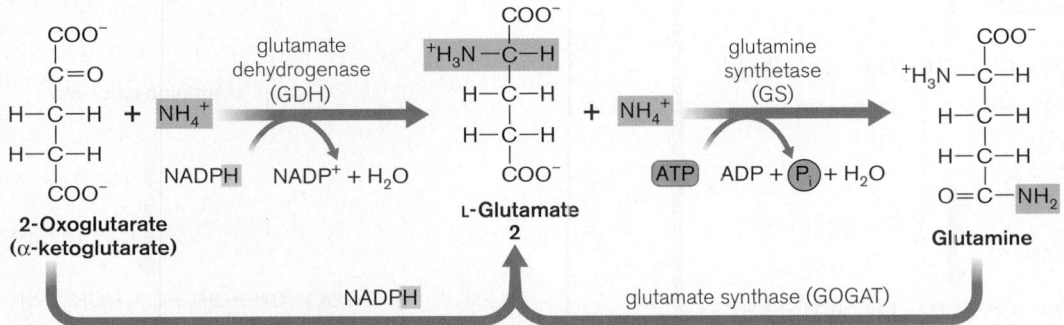

Figure 15.24   Assimilation of NH_4^+ into glutamate and glutamine. The key TCA cycle intermediate 2-oxoglutarate (alpha-ketoglutarate) incorporates one molecule of NH_4^+ at the ketone to form glutamate. Glutamine can combine with oxoglutarate to produce two molecules of glutamate. These reactions provide sources of nitrogen to feed other pathways of amino acid synthesis.

For example, glutamate transfers NH_3 to oxaloacetate, making aspartate and 2-oxoglutarate; the reaction can also be reversed. In another example, valine transfers an amine group to pyruvate, generating alanine and 2-ketoisovalerate.

THOUGHT QUESTION 15.6 Suggest two reasons why transamination is advantageous to cells.

The cellular levels of glutamate and glutamine act as indicators of nitrogen availability. When mineral forms such as ammonium and nitrate are absent, some microbes seek organic sources of nitrogen in their environment. During nitrogen starvation, NtrC is phosphorylated and induces expression of glutamine synthetase as well as nitrogenase. NtrC phosphate also induces a high-affinity ammonia transporter and transporters for organic sources of nitrogen, such as amino acids, oligopeptides,

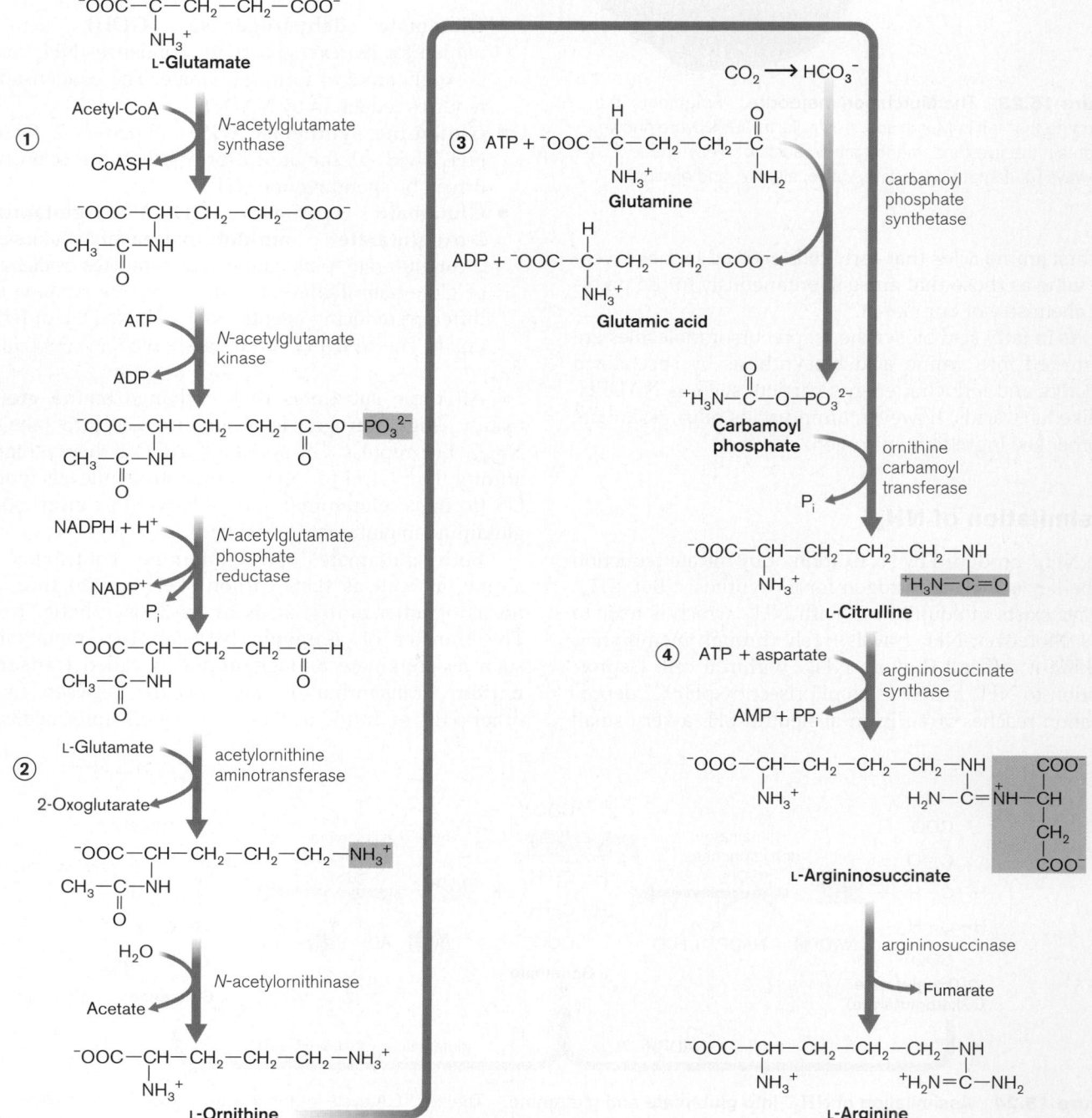

Figure 15.25 Arginine biosynthesis in *E. coli*. In arginine biosynthesis, two glutamate condense with acetyl-CoA, and nitrogen is donated by glutamate and glutamine.

and cell wall fragments containing amino sugars. All these molecules can be "scavenged" to obtain nitrogen for biosynthesis.

Building Complex Amino Acids

Seven "fundamental" amino acids show relatively simple biosynthetic pathways. Others require longer pathways involving numerous enzymes. An amino acid produced by a complex pathway is arginine (**Fig. 15.25**). Bacterial arginine biosynthesis generally involves about a dozen different enzymes distributed among four to eight operons. Operon expression is regulated by an arginine repressor that binds the promoter of each operon to prevent transcription in the presence of sufficient arginine for cell needs. In some species, the arginine repressor also activates enzymes of arginine catabolism, making the excess amino acid available as a carbon source.

> **THOUGHT QUESTION 15.7** Which energy carriers (and how many) are needed to make arginine from 2-oxoglutarate? (See **Fig. 15.25**.)

Arginine biosynthesis. Arginine synthesis begins with the condensation of glutamate with acetyl-CoA (**Fig. 15.25**, step 1). Three more amino groups (–NH₂) are transferred subsequently by glutamate, glutamine, and aspartate. A second glutamate transfers an amino group to the arginine precursor (step 2) while its own carbon skeleton cycles back as 2-oxoglutarate (alpha-ketoglutarate). This transfer of NH_4^+ between two organic intermediates is an example of transamination. The original acetyl group from acetyl-CoA is then hydrolyzed, producing ornithine. Ornithine is a central intermediate, used by cells to synthesize proline and various polyamines, as well as arginine.

Ornithine receives a third amino group (step 3) from glutamine via conversion of CO_2 to carbamoyl phosphate $[H_2N\text{-}CO(PO_4)^{2-}]$:

$$CO_2 + H_2O \longrightarrow H^+ + HCO_3^-$$

$$HCO_3^- + ATP + glutamine \longrightarrow$$
$$H_2N\text{-}CO(PO_4)^{2-} + ADP + glutamate$$

Carbamoyl phosphate provides a way to assimilate one nitrogen plus one carbon into a structure.

The fourth nitrogen is acquired from aspartate in a two-step process requiring ATP release of pyrophosphate (PP_i) (step 4). The release of fumarate, a TCA cycle intermediate, yields arginine.

Aromatic amino acids. The complexity of aromatic amino acids requires particularly energy-expensive biosynthesis. The number of different enzymes required, however, is minimized by the presence of a common core pathway branching to several different amino acids (**Fig. 15.26**).

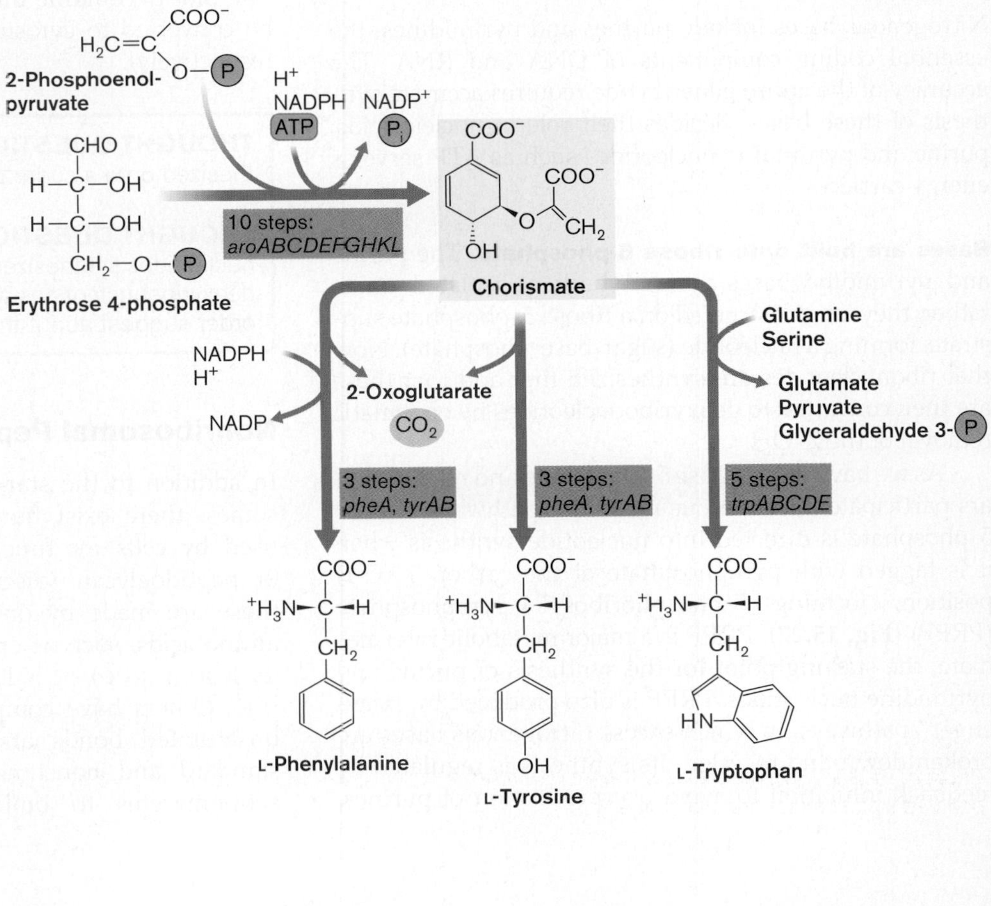

Figure 15.26 Biosynthesis of aromatic amino acids. Aromatic amino acids are assembled out of various carbon skeletons. In *E. coli*, the pathways to chorismate and tryptophan are encoded by operons of contiguous genes, *aro* and *trp*.

The three aromatic amino acids—phenylalanine, tyrosine, and tryptophan—each require a common precursor, chorismate. The pathway to chorismate starts with simple glycolytic intermediates (phosphoenolpyruvate and erythrose 4-phosphate), but its synthesis requires 10 enzymes encoded in a single operon, *aroABCDEFGHKL*. From chorismate to phenylalanine or tyrosine requires an additional three enzymatic steps. From chorismate to tryptophan, the most complex amino acid specified by the genetic code, requires another 5 enzymes, expressed by *trpABCDE*. Thus, a total of at least 15 enzymes encoded by different genes are required to make tryptophan out of simple substrates.

Not surprisingly, the presence of tryptophan in the cell effectively represses expression of its own biosynthetic enzymes. Repression of the *trp* operon includes two mechanisms: the Trp repressor, effective over moderate to high levels of tryptophan; and an RNA loop mechanism called attenuation, sensitive to lower levels of tryptophan (discussed in Chapter 10). Attenuation involves the destabilization of the transcription complex by formation of a specific stem loop on the nascent *trp* mRNA. The mechanism of attenuation in *E. coli* was discovered in 1981 by Charles Yanofsky, winner of the National Medal of Science. Several other amino acids commonly show attenuation in bacteria—particularly histidine, threonine, phenylalanine, valine, and leucine.

Purine and Pyrimidine Synthesis

Nitrogenous bases include purines and pyrimidines, the essential coding components of DNA and RNA. The accuracy of the entire genetic code requires accurate synthesis of these bases. Besides their role in nucleic acids, purine and pyrimidine nucleotides such as ATP serve as energy carriers.

Bases are built onto ribose 5-phosphate. The purine and pyrimidine bases are not built as isolated units; rather, they are constructed on a ribose 5-phosphate substrate, forming a nucleotide (sugar-base-phosphate). Note that ribonucleotides are synthesized first and that these are then converted to deoxyribonucleotides by enzymatic removal of the 2′ OH.

As we have seen, ribose 5-phosphate and related sugars participate in numerous metabolic pathways. Ribose 5-phosphate is directed into nucleotide synthesis when it is tagged with pyrophosphate at the carbon 1 (C-1) position, forming 5-phosphoribosyl-1-pyrophosphate (PRPP) (**Fig. 15.27**). PRPP is a major metabolic intermediate, the starting point for the synthesis of purine and pyrimidine nucleotides. PRPP is also produced by "scavenger" pathways, in which excess nitrogenous bases are broken down and recycled. Its synthesis is regulated by feedback inhibition to avoid overproduction of purines.

In humans, overproduction of PRPP leads to gout, a condition in which the purine breakdown product uric acid precipitates in the joints, causing painful swelling of wrists and feet.

Purine synthesis. The hydrolysis of PRPP releases pyrophosphate, thus spending two high-energy phosphate bonds. This irreversible reaction drives forward the subsequent reactions to make purine. To construct the purine, the pyrophosphate at carbon 1 (C-1) is replaced by an amine from glutamine. The purine is built up by addition of a series of single-carbon groups (formyl, methyl, and CO_2) alternating with nitrogens from glutamine and carbamoyl phosphate. The number of single-carbon assimilations, including assimilation of CO_2, is striking. It may reflect the ancient origin of the purine synthetic pathway, which evolved in CO_2-fixing autotrophs.

The first purine constructed is inosine, as inosine monophosphate. Inosine is the "wobble" purine found at the third position of transfer RNAs. Inosine monophosphate is subsequently converted to adenosine monophosphate (AMP) or guanosine monophosphate (GMP).

Pyrimidine synthesis. The pyrimidine is built by a slightly different route. First, the six-membered pyrimidine ring forms from aspartate plus carbamoyl phosphate. The pyrimidine ring then displaces the pyrophosphate of PRPP, attaching by a nitrogen to the ribosyl carbon 1. The first pyrimidine built is uracil (as UMP), which can be converted to cytosine or thymine (as CMP or TMP, respectively).

THOUGHT QUESTION 15.8   Why are purines synthesized onto a sugar phosphate base?

THOUGHT QUESTION 15.9   Why are the ribosyl nucleotides synthesized first, and then converted to deoxyribonucleotides as necessary? What does this order suggest about the evolution of nucleic acids?

Nonribosomal Peptide Antibiotics

In addition to the standard amino acids used by ribosomes, there exist hundreds of different amino acids used by cells for functions such as the cross-bridges of peptidoglycan (discussed in Chapter 3). Some of these are made by one-step modification of standard amino acids, such as epimerization (altering the chirality from L to D) or halogenation with chlorine or fluorine. Others have complex carbon skeletons, including unsaturated bonds and three-membered rings. Both standard and nonstandard amino acids are used by actinomycetes to build secondary metabolites with

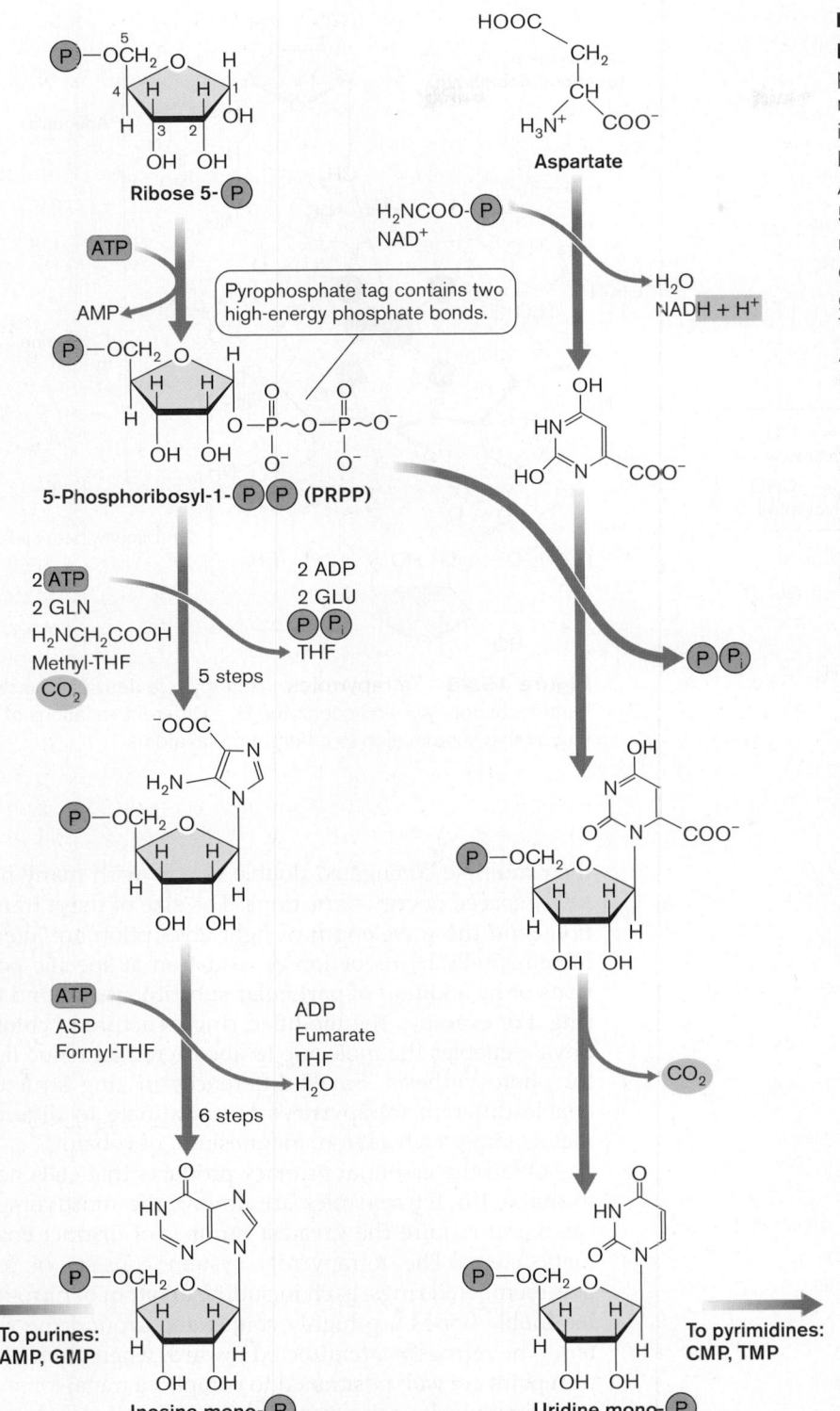

Figure 15.27 Biosynthesis of purines and pyrimidines. Both purines and pyrimidines are built onto ribose 5-phosphate. Ribose 5-phosphate is directed into purine and pyrimidine biosynthesis by the activating step of ATP $\longrightarrow$ AMP, converting the sugar to 5-phosphoribosyl-1-PP (PRPP). The purine ring is built up out of successive additions of amines, one-carbon units, plus a formyl group from formyl tetrahydrofolate (formyl-THF). The purine ring is modified to yield AMP and GMP. Likewise, the pyrimidine ring is modified to yield CMP or TMP.

hospital-borne pathogen *Clostridium difficile*. Vancomycin biosynthesis is presented in **eTopic 15.1**.

TO SUMMARIZE:

- **Amino acid biosynthesis** requires numerous different enzymes to catalyze many unique conversions. Structurally related amino acids branch from a common early pathway.
- **Metabolic intermediates** from glycolysis and the TCA cycle initiate amino acid biosynthetic pathways.
- **Ammonium ion is assimilated by TCA intermediates**, such as oxaloacetate into glutamate. Glutamate assimilates ammonium ion to form glutamine. Transamination is the donation of NH_4^+ from one amino acid to another, such as glutamine transferring ammonia to oxaloacetate to make aspartate.
- **Arginine biosynthesis** requires multiple steps of NH_3 transfer and carbon skeleton condensation.
- **Aromatic amino acids** are built from a common pathway that branches out. Their biosynthesis is regulated tightly at both transcriptional and translational levels.
- **Purines are built as nucleotides** attached to a ribose phosphate. Several single-carbon groups are assimilated, including CO_2—a phenomenon suggesting an ancient pathway. Pyrimidines are made from aspartate, and then added onto PRPP.
- **Nonribosomal peptide antibiotics** are built by modular enzymes analogous to those that build polyketides.

antimicrobial activity. These secondary metabolites are called **nonribosomal peptide antibiotics** because their peptide backbone is constructed by an enzyme without ribosomes. Nonribosomal peptide antibiotics are synthesized by modular "assembly-line" enzymes, analogous to those that build polyketides. An example is vancomycin, the main antibiotic against the lethal

Heme *b*

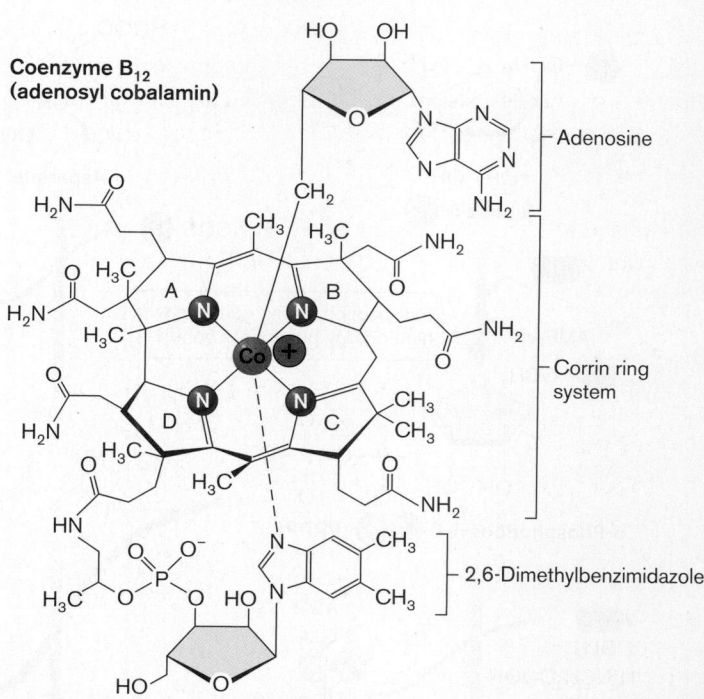

Chlorophyll

R = —CH₃ (chlorophyll *a*)
R = —CHO (chlorophyll *b*)

Coenzyme B₁₂ (adenosyl cobalamin)

Adenosine

Corrin ring system

2,6-Dimethylbenzimidazole

Figure 15.28 Tetrapyrroles. Tetrapyrrole derivatives include hemes, chlorophylls, and coenzyme B₁₂. Different variations of the ring enable coordination to different metal atoms.

15.7 Biosynthesis of Tetrapyrroles

The **tetrapyrrole** family includes some of the best-known and essential cofactors of energy transduction and biosynthesis, such as chlorophylls, hemes of cytochromes and hemoglobin, and coenzyme B₁₂ (**Fig. 15.28**). The versatility of tetrapyrroles for redox reactions comes from their multiple conjugated double bonds, with many narrowly spaced energy transitions. The size of these transitions and the wavelength of light absorption are altered incrementally by reduction or oxidation at specific positions or by addition of particular substituents around the ring. For example, the modified ring structure of chlorophyll *a* enables the molecule to absorb red and blue light for photosynthesis. Small differences in ring structure enable different tetrapyrroles to coordinate to different metal atoms, such as iron, magnesium, or cobalt.

Of all the essential primary products that cells need to make, the tetrapyrroles are among the most complicated and require the greatest number of distinct enzymatic steps. The tetrapyrrole system consists of four five-membered rings, each including an atom of nitrogen. Its double bonds are highly conjugated around the system. The nitrogens are directed inward, where their electron pairs are well positioned to complex a metal ion.

Tetrapyrroles arise from condensation of four pyrrole rings. The five-membered pyrrole ring has three substituent groups, one of which (–CH₂NH₂) serves as a linker in condensation of the tetrapyrrole. The individual pyrrole rings are generated by one of two different pathways (**Fig. 15.29**), found in different organisms: the glutamate pathway, in most bacteria as well as in plants; and the glycine-succinate pathway, in purple bacteria, animals, and yeast. The glutamate pathway is unusual in that it starts from a glutamine residue attached to a transfer RNA, generally used in protein synthesis. In either case, condensation and rearrangement generates the linear molecule 5-aminolevulinic acid (ALA). Two molecules of ALA cyclize to form the ring of porphobilinogen.

Once the pyrroles are formed, an enzyme condenses four of them in a chain, removing each amine as NH_4^+. The four linked pyrroles are conventionally labeled A through D. When the chain cyclizes, an unusual isomerization occurs in which the D pyrrole is inverted. The mechanism of the D-ring isomerization is unknown, but it imparts a critical asymmetry to the molecule as a whole. This first cyclized tetrapyrrole is called uroporphyrinogen III.

Uroporphyrinogen III forms the foundation of most biologically active tetrapyrroles. By inspecting functional tetrapyrroles, we can see the subtle modifications that distinguish them (see **Fig. 15.28**). Chlorophyll and coenzyme B_{12} (the active form of vitamin B_{12}, in which adenosine replaces cyanide) each have an extra-long side chain at the D ring, leading to completely different functional groups. The ring systems of both heme and chlorophyll are more unsaturated than the original ring system found in uroporphyrinogen III; the extra conjugated double bonds facilitate energy transitions in these electron carriers. The cobalamin core of coenzyme B_{12} has a more hydrogenated D ring and one less carbon (a carbon is lost in the connection between rings A and D).

The synthesis of tetrapyrroles is highly regulated by nutritional needs. For example, purple photoheterotrophs such as *Rhodospirillum rubrum* switch off synthesis of bacteriochlorophylls and other photopigments in the presence of oxygen, preferring aerobic respiration to phototrophy. Thus, when exposed to oxygen, the bacteria change color from purple-red to white.

Regulation of the biosynthesis of tetrapyrroles and other complex vitamins involves unusual mechanisms based on RNA structure, called "riboswitches." Riboswitches may derive from ancient processes that evolved early in life's history, in the "RNA world" before cells had DNA or protein. For more on riboswitches, see **eTopic 15.2**.

Figure 15.29 Biosynthesis of tetrapyrroles. The fundamental tetrapyrrole ring arises from simple TCA intermediates and amino acids. Two amino acids condense to form 5-aminolevulinic acid (ALA), which cyclizes to form porphobilinogen, a pyrrole ring with three substituents—two ending in COOH, one in NH_2. Four of the pyrroles link together, and then cyclize to form the corrin ring system. The final step of cyclization includes reversal of the D-ring linkage to yield uroporphyrinogen III, which enters various pathways to form different products. The products contain different metal ions (in parenthesis).

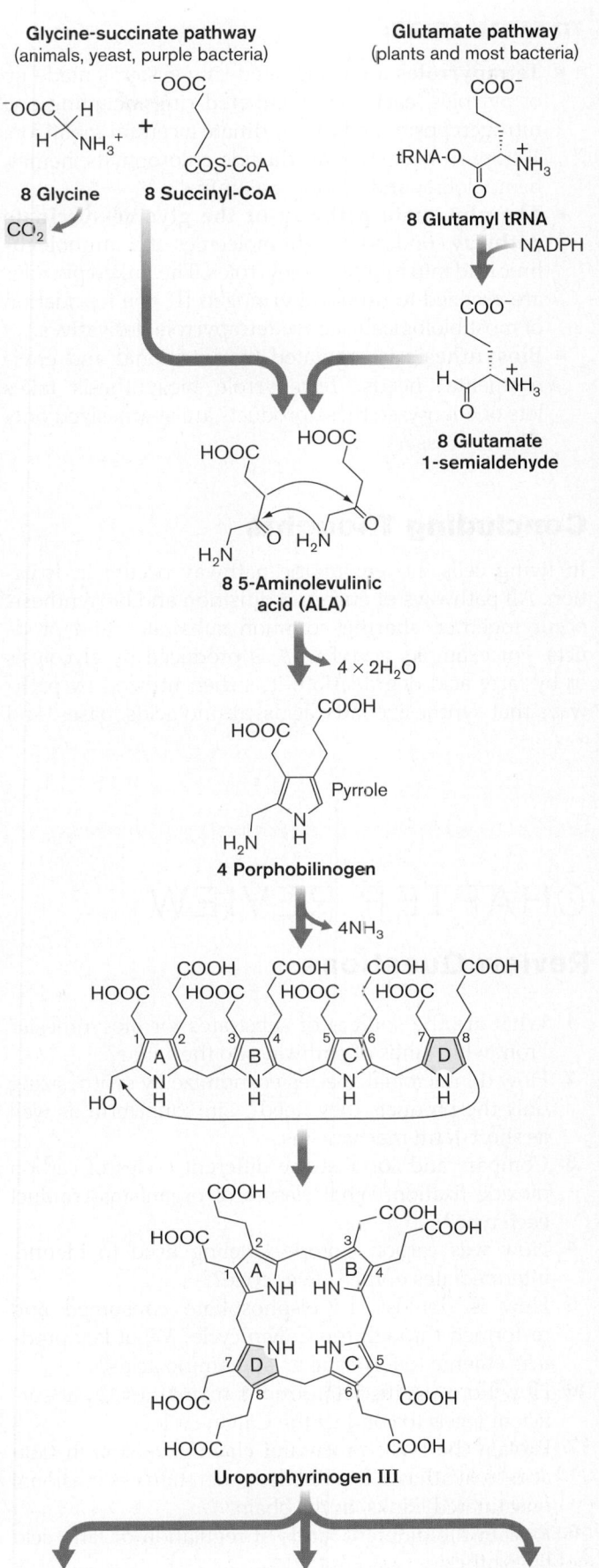

TO SUMMARIZE:

- **Tetrapyrroles** are conjugated ring systems made up of pyrroles, each five-membered ring including one nitrogen positioned to coordinate a central metal ion. Tetrapyrrole derivatives include chlorophylls, hemes, hemoglobin, and cyanocobalamin.
- **The glutamate pathway or the glycine-succinate pathway** condenses eight molecules of 5-aminolevulinic acid into four linked pyrroles. The linked pyrroles are cyclized to uroporphyrinogen III, the foundation of most biologically active tetrapyrrole derivatives.
- **Biosynthesis** is regulated by nutritional and environmental needs. Tetrapyrrole biosynthesis takes lots of energy, so these products are synthesized only when necessary.

Concluding Thoughts

In living cells, no enzymatic pathway occurs in isolation. All pathways of energy acquisition and biosynthesis occur together, sharing common substrates and products. For example, acetyl-CoA is produced by glycolysis or by fatty acid degradation; it is then utilized by pathways that synthesize fatty acids, amino acids, bases, and tetrapyrroles. Tetrapyrrole biosynthesis requires products of the TCA cycle. At the same time, the TCA cycle is needed to provide energy through catabolism. The microbial cell regulates these competing needs on an extremely rapid timescale; within seconds of the disappearance or appearance of a nutrient, such as an amino acid, its biosynthesis is turned on or shut off.

A given microbe needs to build as much biomass as it can by the most efficient means available. For a pathogen growing within a host, this means utilizing preformed host compounds for rapid growth and replication. Free-living microbes, however, face intense competition for organic nutrients; so, in addition, they need to fix essential elements into readily accessible forms. Many soil and marine microbes build the most complicated biological molecules out of single-carbon and single-nitrogen sources, although they scavenge preformed organic substrates when available.

Microbial metabolism has many applications in food preparation and preservation, and in industrial production of antibiotics and other pharmaceuticals. The principles of food and industrial microbiology are discussed in Chapter 16. Later, in Chapters 21 and 22, we see how microbial metabolism within populations and communities contributes to global cycles of Earth's biosphere.

CHAPTER REVIEW

Review Questions

1. What are the sources of substrates for biosynthesis? From what kinds of pathways do they arise?
2. How do microbial species economize by synthesizing only the products they need? Cite long-term as well as short-term mechanisms.
3. Compare and contrast the different cycles of carbon dioxide fixation. What classes of organisms conduct each type?
4. How was carbon isotope labeling used to identify intermediates of the Calvin cycle?
5. How is ribulose 1,5-bisphosphate consumed and re-formed through the Calvin cycle? What key products emerge to form sugars and amino acids?
6. How do oxygenic phototrophs maintain CO_2 at sufficient levels to conduct the Calvin cycle?
7. Explain the cyclic process of chain extension in fatty acid biosynthesis. Explain the generation of occasional unsaturated "kinks" in the chain.
8. Explain the different kinds of regulation of fatty acid biosynthesis.
9. What are the different environmental sources of nitrogen? Explain how and why microbes use these different sources.
10. Explain the process by which nitrogenase converts N_2 to $2NH_4^+$. Why is H_2 formed?
11. Explain the different ways that microbes maintain anaerobic conditions for nitrogenase.
12. Explain the molecular basis for regulation of nitrogen fixation and nitrogen scavenging.
13. Compare and contrast the general scheme of the biosynthesis of amino acids with that of fatty acids.
14. Outline the interconversions of 2-oxoglutarate, glutamate, and glutamine that provide nitrogen for amino acid biosynthesis.
15. Compare and contrast the processes of purine and pyrimidine biosynthesis. What is the role of the sugar ribose in each case?
16. How does the biosynthesis of tetrapyrroles generate diverse products from common small organic substrates?

Thought Questions

1. Why do some soil microbes fix N_2, whereas others depend on available nitrate, ammonium ion, or organic nitrogen? What environmental conditions would favor each strategy?

2. Disease-causing bacteria vary widely in their ability to synthesize amino acids. What kinds of pathogens would be likely to make their own amino acids, and what kinds would not?

3. *Mycoplasma genitalium*, an organism growing in human skin, lacks the ability to synthesize fatty acids. How do you think it makes its cell membrane? How could you test this?

Key Terms

acyl carrier protein (ACP) (562)
anabolism (544)
anaplerotic reaction (556)
biosynthesis (544)
Calvin cycle (Calvin-Benson cycle, Calvin-Benson-Bassham cycle, CBB cycle) (545, 547)
carbon-concentrating mechanism (CCM) (553)
carbon dioxide fixation (545)
carboxysome (555)
fatty acid synthase complex (561)

folate (559)
gluconeogenesis (558)
glutamate dehydrogenase (GDH) (573)
glutamate synthase (GOGAT) (glutamine:2-oxoglutarate amidotransferase) (573)
glutamine synthetase (GS) (573)
Haber process (568)
heterocyst (568)
3-hydroxypropionate cycle (559)
modular enzyme (563)

nitrogenase (568)
nonribosomal peptide antibiotic (577)
paper chromatography (550)
reductive acetyl-CoA pathway (558)
reductive pentose phosphate cycle (545)
reductive (reverse) TCA cycle (556)
Rubisco (551)
secondary metabolite (secondary product) (562)
tetrapyrrole (578)
transamination (573)

Recommended Reading

Eisen, Jonathan A., Karen E. Nelson, Ian T. Paulsen, John F. Heidelberg, Martin Wu, et al. 2002. The complete genome sequence of *Chlorobium tepidum* TLS, a photosynthetic, anaerobic, green-sulfur bacterium. *Proceedings of the National Academy of Sciences USA* **99**:9509–9514.

Gollnick, Paul, Paul Babitzke, Alfred Antson, and Charles Yanofsky. 2005. Complexity in regulation of tryptophan biosynthesis in *Bacillus subtilis. Annual Review of Genetics* **39**:47–68.

Hansen, S., V. B. Vollan, E. Hough, and K. Andersen. 1999. The crystal structure of rubisco from *Alcaligenes eutrophus* reveals a novel central eight-stranded beta-barrel formed by beta-strands from four subunits. *Journal of Molecular Biology* **288**:609–621.

Heath, Richard J., J. Ronald Rubin, Debra R. Holland, Erli Zhang, Mark E. Snow, et al. 1999. Mechanism of triclosan inhibition of bacterial fatty acid synthesis. *Journal of Biological Chemistry* **274**:11110–11114.

Herter, Sylvia, Jan Farfsing, Nasser Gad'On, Christoph Rieder, Wolfgang Eisenreich, et al. 2001. Autotrophic CO_2 fixation by *Chloroflexus aurantiacus:* Study of glyoxylate formation and assimilation via the 3-hydroxypropionate cycle. *Journal of Bacteriology* **183**:4305–4316.

Hügler, Michael, Carl O. Wirsen, Georg Fuchs, Craig D. Taylor, and Stefan M. Sievert. 2005. Evidence for autotrophic CO_2 fixation via the reductive tricarboxylic acid cycle by members of the ε subdivision of proteobacteria. *Journal of Bacteriology* **2187**:3020–3027.

Kamen, Martin. 1985. *Radiant Science, Dark Politics. A Memoir of the Nuclear Age.* University of California Press, Berkeley.

Liu, Wen, Steven D. Christenson, Scott Standage, and Ben Shen. 2002. Biosynthesis of the enediyne antitumor antibiotic C-1027. *Science* **297**:1170–1173.

Schirmer, Andreas, Rishali Gadkari, Christopher D. Reeves, Fadia Ibrahim, Edward F. DeLong, et al. 2005. Metagenomic analysis reveals diverse polyketide synthase gene clusters in microorganisms associated with the marine sponge *Discodermia dissoluta. Applied and Environmental Microbiology* **71**:4840–4849.

Valbuzzi, Angela, and Charles Yanofsky. 2001. *B. subtilis* regulatory protein TRAP. *Science* **293**:2057–2059.

Zazopoulos, Emmanuel, Kexue Huang, Alfredo Staffa, Wen Liu, Brian O. Bachmann, et al. 2003. A genomic-guided approach for discovering and expressing cryptic metabolic pathways. *Nature Biotechnology* 21:187–190.

Zhang, Yong-Mei, Stephen W. White, and Charles O. Rock. 2006. Inhibiting bacterial fatty acid synthesis. *Journal of Biological Chemistry* 281:17541–17544.

Zimmer, Daniel P., Eric Soupene, Haidy L. Lee, Volker F. Wendisch, Arkady B. Khodursky, et al. 2000. Nitrogen regulatory protein C-controlled genes of *Escherichia coli:* Scavenging as a defense against nitrogen limitation. *Proceedings of the National Academy of Sciences USA* 97:14674–14679.

For further study and review, please visit StudySpace at **microbiology2.com**.

Chapter 16

Food and Industrial Microbiology

Microbes have nourished humans for centuries, generating cheese, bread, wine and beer, tempeh, and soy sauce. And yet, from the moment of harvest, we humans compete with microbes for our food. Microbes from the food's surface or from the air colonize food, and their uncontrolled growth can render the food rancid or putrid. Historically, the need for food storage and preservation has led to practices such as drying, salting, smoking, and adding spices, all of which retard microbial growth.

The principles of food microbiology have been extended to a much broader field of industrial microbiology. Industrial microbiology includes the development of microbial products such as antibiotics and enzymes, as well as transgenic microbes that produce human proteins such as insulin and growth hormones. Many products derive from extremophiles, organisms adapted to extreme environments, whose enzymes can withstand industrial conditions. Industrial microbiology faces special challenges with regard to microbial growth conditions, scaled-up fermentation, genetic optimization of production, and safety testing.

Current Research Highlight

Chocolate production requires fermentation of cocoa beans by yeasts, lactic acid bacteria, and acetic acid bacteria. The beans must ferment immediately at the site of harvest. In 2009, two new species of lactic acid bacteria were isolated from fermenting cocoa beans in Ghana: *Lactobacillus fabifermentans* and *Lactobacillus cacaonum*. Lactic acid bacteria ferment citric acid from the cocoa fruit pulp and generate acetic acid, which helps the beans release flavors. *Source:* De Bruyne et al. 2009. *Int. J. Syst. Evol. Microbiol.* **59**:7. Photo: © PhotoAlto/Alamy.

The fermentation industry reaches back to 5000 BCE, the date of pottery jars excavated from a Neolithic mud-brick kitchen in the Zagros Mountains of modern Iran (**Fig. 16.1A**). The jars contain residue showing chemical traces of grapes fermented to wine. Besides making wine, Neolithic people leavened bread, another staple food requiring microbial growth. For thousands of years, microbial fermentation has been a daily part of human life and commerce.

Today, microbial growth drives companies such as Genencor International, now earning $400 million in annual revenues from engineering industrial enzymes. Genencor delivers 250 products ranging from contact lens cleansing agents to new fashion finishes for denim at manufacturing centers in the United States, Europe, Argentina, and China. Most of Genencor's product enzymes are made by bacteria grown in giant fermentation vessels (**Fig. 16.1B**). Genencor was founded as a joint venture by Genentech and Corning, and its success heralded a growing investment by biotech start-ups as well as traditional chemical companies in the fermentation industry: the use of microbes to produce industrial products.

The fermentation industry is based on long-standing microbial associations with human food. Until the relatively recent invention of steam-pressure sterilization, all foods contained live microbes. Microbial metabolism could spoil food; or it could improve the food by adding flavor, preserving valuable nutrients, and preventing growth of pathogens. As we learned in Chapter 1, one of the first great microbiologists, Louis Pasteur, began his career as a chemist investigating fermentation in winemaking. In Chapter 16, we see how the many kinds of microbial biochemistry presented in Chapters 13–15 give rise to the characteristics of food so familiar to us, from the taste of cheese and chocolate to the rising of bread dough and the physiological effects of alcoholic beverages. Industrial research continues to improve food through the fields of food biochemistry, discovering the molecular basis for the flavor and texture of microbial foods; food preservation, eliminating undesirable microbial decay by preservative methods; and food engineering, improving the quality, shelf life, and taste of microbial foods.

Beyond food, the commercial applications of microbes extend to industrial microbiology, the use of microbes to generate useful products of all kinds. Industrial microbiology generates enzymes, chemical feedstocks, fuels, and pharmaceuticals. New products are developed through genomic engineering and the technologies presented in Chapter 12. Increasingly, research microbiologists reach beyond the laboratory bench to invent new uses for microbial products—and start their own companies to develop and market them.

16.1 Microbes as Food

Certain kinds of microbial bodies have long been eaten as food, especially the fruiting bodies of fungi and the fronds of marine algae. Single-celled algae and cyanobacteria are also used as food supplements. These microbes can provide important sources of protein, vitamins, and minerals.

Edible Fungi

In the children's classic *Homer Price*, by Robert McCloskey, settlers on the Ohio frontier save themselves from starvation when they discover "forty-two pounds of edible fungus, in the wilderness a-growin'." Fungal **fruiting bodies**, multicellular reproductive structures that generate spores, are commonly known as mushrooms (see Chapter 20). Mushrooms offer a flavorful source of protein and minerals, albeit at the risk of consuming the deadly toxins produced by a few species. The protein content of edible mushrooms can be as high as 25% dry weight, comparable to that of whole milk (percent dry weight), and includes all essential dietary amino acids.

Mushrooms contributed to the survival of preindustrial humans, while killing those unlucky enough to consume varieties that were poisonous. Less than 1% of mushrooms are poisonous, but those few are deadly, such as the amanita, or "destroying angel," which produces toxins including the RNA polymerase inhibitor alpha-amanitin. People who ingest the amanita typically die of liver failure.

A.

Courtesy The University of Pennsylvania Museum

B.

Genencor International

Figure 16.1 The fermentation industry: then and now. A. A Neolithic jar that contained wine, from 5000 BCE. Excavated at Hajji Firuz Tepe, Iran. **B.** An industrial fermentation apparatus for growing microbes to produce enzyme products, at Genencor International.

Other kinds of mushrooms actually invite consumption in order to disperse their spores. For example, the underground truffles prized in European cooking produce odorant molecules that attract animals to dig them up. Truffle odorants include sex pheromones such as androstenol, found in human perspiration.

Many varieties of edible mushrooms are farmed and marketed for human fare. **Figure 16.2** shows the culturing of *Agaricus bisporus*, the mushroom variety most commonly sold in the United States as button mushrooms and portobellos. Button mushrooms are harvested at an early stage, whereas portobellos are harvested later, when the gills are fully exposed and some of the moisture has evaporated. The decrease in moisture in portobellos concentrates the flavor and gives them a dense, meaty texture; the mushrooms are often served in gourmet sandwiches as a vegetarian alternative to hamburger. *Agaricus* culture was first developed around 1700 in France, where the mushrooms were grown in underground caves. In modern mushroom farms, *Agaricus* mushrooms are cultured on composted horse manure or chicken manure in chambers controlled for temperature and humidity.

Other mushroom varieties from China and Japan are grown on logs or wooden blocks. Wood-grown mushrooms include the strong-flavored *Lentinula edodes* (black forest mushroom, or shiitake), *Pleurotus* (oyster mushrooms), and *Flammulina velutipes* (enoki mushrooms) with long thin white stalks and delicate flavor.

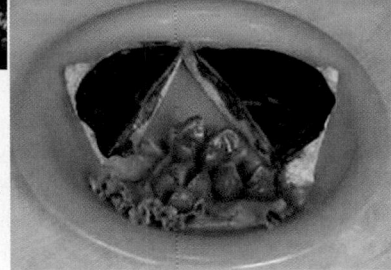

Figure 16.2   Commercial mushroom production.   A. Farming of *Agaricus bisporus* mushrooms on horse manure compost, studied by Penn State graduate student Kelly Ivors. **B.** Portobello mushroom sandwich, with lettuce and tomatoes, courtesy of Middle Ground Café, Gambier, Ohio.

Edible Algae

Several kinds of seaweed (marine algae) are cultivated, most notably in Japan. The red alga *Porphyra*, a eukaryotic "true alga" (discussed in Chapter 20), forms large multicellular fronds cultured for **nori** (**Fig. 16.3**). Nori is best known for its use in wrapping rice, fish, and vegetables to form sushi.

For nori production, the red algae are grown as "seeds," or starter cultures, in enclosed tanks. The starter cultures are then distributed on nets in a protected coastal area, usually an estuary. The cultures grow until they hang heavy from the nets, when they are harvested for processing into sheets. The sheets are toasted, turning dark green.

Another group of seaweeds, the "brown algae," or kelp, are grown for food or to extract food additives. Kelp of the genus *Laminari* are cultured as a source of the flavor additive monosodium glutamate (MSG). A different genus, the giant kelp *Macrocystis*, is grown off the coast of California for production of the polymer alginate, a common thickener for ice cream, salad dressings, and paints. Another additive produced from brown algae is carrageenan, an emulsifier (combiner of oil and water solutions) used in commercial pies and toothpastes.

Edible Bacteria and Yeasts

Since prehistoric times, consumption of single-celled fungi or yeasts has provided a supplemental source of protein and vitamins. Yeasts such as *Saccharomyces* species were grown to high concentration in fermented milks and grain beverages. Traditional beers contained only a low percentage of alcohol with a thick suspension of

Figure 16.3   Nori production for sushi.   A. Nori grows from nets in seawater. **B.** Toasted sheets of nori wrap rice, vegetables, and fish to make sushi.

A.

Courtesy of Bruce Russell

B.

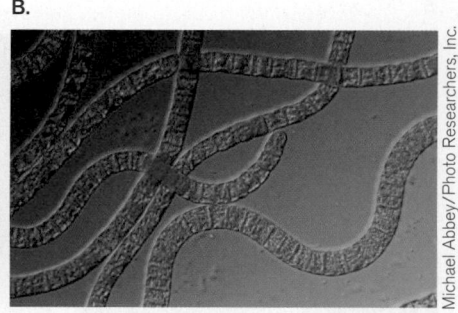

Michael Abbey/Photo Researchers, Inc.

Figure 16.4   Single-celled protein from *Spirulina*.   A. Farming *Spirulina*. **B.** *Spirulina* filaments are processed into protein-rich food supplements. (LM, 250×)

nutrient-rich yeasts. Especially important was the content of vitamin B_{12}, an essential substance for the human diet that cannot be obtained from plant sources.

Few bacteria are edible as isolated organisms, mainly because their small cells contain a relatively high proportion of DNA and RNA. The high nucleic acid content is a problem because nucleic acids contain purines, which the human digestive system converts to uric acid. Because we humans lack the enzyme urate oxidase, the uric acid cannot be metabolized. Consumption of more than 2 grams per day of nucleic acids causes uric acid to precipitate, resulting in painful conditions such as gout and kidney stones. For this reason, most bacteria cannot be consumed in quantity, except as a minor component of food mass, as in fermented foods.

An exception is the cyanobacterium *Spirulina*, whose purine content is low enough to include as a modest part of the human diet (**Fig. 16.4**). *Spirulina* consists of spiral-shaped cells that grow photosynthetically in freshwater. *Spirulina* is sold as a food additive, rich in protein, vitamin B_{12}, and minerals. It also contains antioxidant substances that may prevent cancer. *Spirulina* is grown with illumination in special ponds lined for food production. The final product is collected and vacuum-dried to form a dark green powder of flour-like consistency.

In the food industry, *Spirulina* is classified as a form of **single-celled protein**, a term for edible microbes of high food value. Other kinds of single-celled protein include eukaryotes such as yeast and algae. The twentieth century saw the development of single-celled protein as a food source for impoverished populations. The yeast *Saccharomyces cerevesiae* was grown for protein by Germany during World War I, using inexpensive molasses for culture; and during World War II, *Candida albicans* was grown on paper mill wastes. Single-celled protein was later promoted by Western countries as a food for rapidly expanding populations of developing countries. This idea inspired the science fiction film *Soylent Green* (1973), in which people of a future overpopulated Earth are forced to eat "Soylent," a food supposedly based on soybeans and single-celled protein, though its true source was recycled humans (**Fig. 16.5**).

Figure 16.5   *Soylent Green* (1973).   Science fiction film about a future Earth whose overgrown population is forced to eat a form of food called Soylent, supposedly based on soybeans and single-celled protein.

TO SUMMARIZE:

■ **Fungal fruiting bodies** such as mushrooms and truffles are consumed as a protein-rich food.

■ **Edible algae** include nori (toasted red algae), used to wrap sushi; and kelp (brown algae), which provides food additives.

■ *Spirulina* is an edible cyanobacterium, a source of single-celled protein. Most bacteria, however, are inedible because of their high concentration of nucleic acids.

■ **Yeasts** have been grown as an economical protein and vitamin supplement.

16.2 Fermented Foods: An Overview

Virtually all human cultures have developed varieties of **fermented foods**, food products that are modified biochemically by microbial growth. The purposes of food fermentation include the following:

- **To preserve food.** Certain microbes, particularly the lactobacilli, metabolize only a narrow range of nutrients before their waste products build up and inhibit further growth. Typically, the waste fermentation products that limit growth are carboxylic acids, ammonia (alkaline), or alcohol. Buildup of these substances renders the product stable for much longer than the original food substrate.
- **To improve digestibility.** Microbial action breaks down fibrous macromolecules and tenderizes the product, making it easier for humans to digest. Meat and vegetable products are tenderized by fermentation.
- **To add nutrients and flavors.** Microbial metabolism generates vitamins, particularly vitamin B_{12}, as well as flavor molecules such as esters and sulfur compounds.

Different societies have devised thousands of different kinds of fermented foods. Examples are given in **Table 16.1**. Fermented foods that are produced commercially include dairy products such as cheese and yogurt, soy products such as miso (from Japan) and tempeh (from Indonesia), vegetable products such as sauerkraut and kimchi, and various forms of cured meats and sausages. Alcoholic beverages are made from grapes and other fruits (wine), grains (beer and liquor), and cacti (tequila). Other kinds of foods require microbial treatment for special purposes, such as leavening by yeast (for bread), or cocoa bean fermentation (for chocolate). Besides commercial production, numerous fermented products are homemade by traditional methods thousands of years old. Such products are known as traditional fermented foods. Occasionally, a traditional fermented food enters commercial production and becomes widespread. For example, soy sauce, a traditional Japanese product, was marketed by the Kikkoman company and achieved global distribution in the twentieth century.

The nature of fermented foods depends on the quality of the fermented substrate, as well as on the microbial species and the type of biochemistry performed. Traditional fermented foods usually depend on **indigenous microbiota**—that is, microbes found naturally in association with the food substrate; or on starter cultures derived from a previous fermentation, as in yogurt or sourdough fermentation. Commercial fermented foods use highly engineered microbial strains to inoculate their cultures, although in some cases indigenous microbiota still participate. For example, wines and cheeses aged in the same caves for centuries often include fermenting organisms that persist in the air and the containers used.

Major classes of fermentation reactions are summarized in **Figure 16.6**. The most common conversions involve anaerobic fermentation of glucose, as discussed in Chapter 13. Glucose is fermented to lactic acid (**lactic acid fermentation**) in cheeses and sausages, primarily by lactic acid bacteria such as *Lactobacillus*. A second-stage fermentation of lactic acid to propionic acid (**propionic acid fermentation**) by *Propionibacterium* generates the special flavor of Swiss and related cheeses. Some kinds of vegetable fermentation, as in sauerkraut, involve production of lactic acid, ethanol, and carbon dioxide (**heterolactic fermentation**) by *Leuconostoc*. Fermentation to ethanol plus carbon dioxide without lactic acid (**ethanolic fermentation**) is conducted by yeast during bread leavening and production of alcoholic beverages.

In some food products, particularly those fermented by *Bacillus* species, proteolysis and amino acid catabolism generate ammonia in amounts that raise pH (**alkaline fermentation**). For example, alkaline fermentation forms the soybean product natto. Other products require the growth of mold, such as the mold-spiked Roquefort cheese and the soy product tempeh. Mold growth requires some oxygen for aerobic respiration. Respiration must be limited, however, to avoid excessive decomposition of food substrate and loss of food value.

Note that the conversions cited here include only the major reactions in achieving the food product. In addition, thousands of minor or secondary reactions occur, some of which produce tiny amounts of potent odorants and flavors.

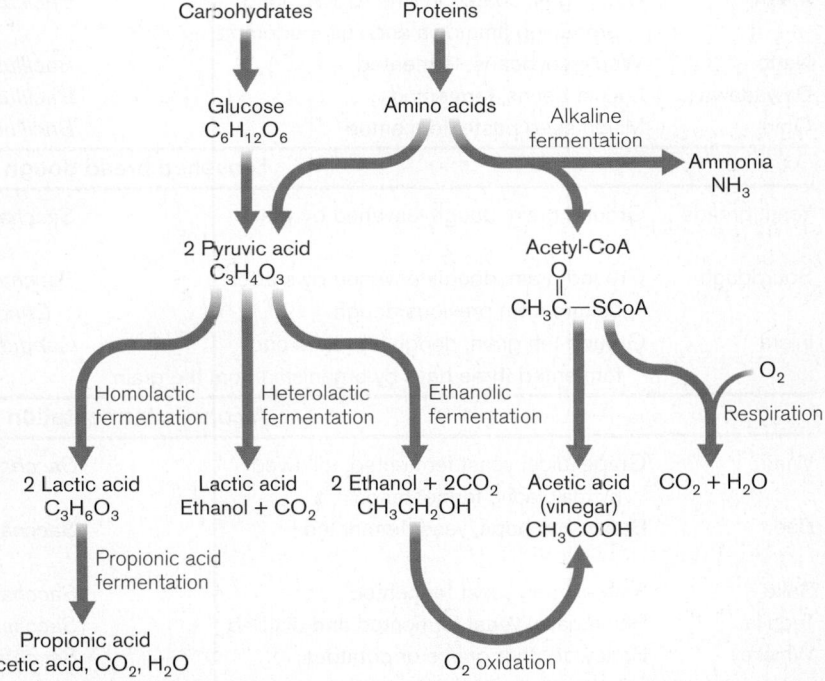

Figure 16.6 Major chemical conversions in fermented foods.

Table 16.1 Fermented foods and beverages.

Product	Description	Microbial genera	Geographic origin
Acidic fermentation of dairy products, meat, and fish			
Buttermilk	Bovine milk, lactic fermented	*Lactococcus*	Europe, Asia
Yogurt	Bovine milk, lactic fermented and coagulated	*Lactobacillus, Streptococcus*	Europe, Asia
Kefir	Bovine or sheep's milk, mixed fermentation, acidic with some alcohol	*Lactobacillus, Streptococcus,* yeasts, others	Eastern Europe, Russia
Sour cream	Bovine cream, lactic fermented	*Lactococcus*	Europe, Asia
Cheese (many kinds)	Milk (bovine, sheep, or goat), lactic fermented, coagulated, and pressed; in some cases cooked; mold ripened (spiked or coated)	Acid fermentation: *Lactobacillus, Streptococcus, Propionibacterium;* mold ripening: *Penicillium*	Europe, Asia
Sausage	Ground beef and/or pork encased with starter culture, lactic fermented, dried, or smoked	*Lactobacillus, Pediococcus, Staphylococcus,* others	Europe, Asia
Fermented fish	Many kinds of fish, mixed fermentation, acid and amines produced	Unknown	Africa, Asia
Acidic fermentation of vegetables			
Tempeh	Soybean cakes, fungal fermentation	*Rhizopus oligosporus*	Indonesia
Miso	Soy and rice paste, fungal fermentation	*Aspergillus*	Japan
Soy sauce	Extract of soy and wheat, fungal fermentation, brined, bacterial fermentation	*Aspergillus,* followed by halotolerant bacteria	China, Japan
Kimchi	Cabbage, peppers, and other vegetables, with fish paste, brined; container is buried	*Leuconostoc,* other bacteria	Korea
Sauerkraut	Cabbage, fermented, making lactic and acetic acids and CO_2	*Leuconostoc, Pediococcus, Lactobacillus*	Europe
Pickled foods	Cucumbers, carrots, fish; brined, then fermented	*Leuconostoc, Pediococcus, Lactobacillus*	Europe, Asia
Kenkey	Maize, fermented, wrapped in banana leaves and cooked	Unknown	Western Africa
Chocolate	Cocoa beans soaked and fermented before processing to chocolate	*Lactobacillus, Bacillus, Saccharomyces*	South America
Alkaline fermentation			
Pidan	Duck eggs, coated in lime (CaO), aged, producing ammonia and sulfur odorants	*Bacillus*	Japan, China
Natto	Whole soybeans, fermented	*Bacillus natto*	Japan, China
Dawadawa	Locust beans, fermented	*Bacillus*	Africa
Ogiri	Melon seed paste, fermented	*Bacillus*	Africa
Leavened bread dough			
Yeast breads	Ground grain, dough leavened by yeast	*Saccharomyces*	Europe, Asia, Americas
Sourdough	Ground grain, dough leavened by starter culture from previous dough	*Saccharomyces, Torulopsis, Candida*	Egypt
Injera	Ground teff grain, dough leavened and fermented three days by organisms from the grain	*Candida*	Ethiopia
Alcoholic fermentation			
Wine	Grape juice, yeast fermented, followed by malolactic fermentation	*Saccharomyces, Oenococcus*	Europe, Asia, Americas
Beer	Barley and hops, yeast fermented	*Saccharomyces*	Europe, Asia, Americas
Sake	Rice extract, yeast fermented	*Saccharomyces*	Japan
Tequila	Blue agave, yeast fermented and distilled	*Saccharomyces*	Mexico
Whiskey	Barley or other grains or potatoes, fermented and distilled	*Saccharomyces*	United Kingdom, Americas, Japan

While these flavor molecules have less nutritional consequence than the main fermentation products, they provide the complex, "sophisticated" taste for which fine cheeses, wines, and soy products are known.

THOUGHT QUESTION 16.1 Why do the lipid components of food experience relatively little breakdown during fermentation?

THOUGHT QUESTION 16.2 Why does oxygen allow excessive breakdown of food, compared with anaerobic processes?

TO SUMMARIZE:

- **Fermentation of food** enhances preservation, digestibility, nutrient content, and flavor.
- **Acidic fermentations** lead to organic acid fermentation products, such as lactate and propionate.
- **Alkaline fermentations** produce ammonia and break down proteins to peptides.
- **Ethanolic fermentation** produces ethanol and carbon dioxide.
- **Lipids are relatively stable** under anaerobic conditions of fermentation.

16.3 Acid- and Alkali-Fermented Foods

Many food fermentations produce acids or bases. An acid or base serves as an effective preservative because the pH change is unlikely to be reversed, and because animal or plant bodies grown at near-neutral pH are unlikely to support growth of acidophiles or alkaliphiles, which grow at extreme pH conditions.

Acidic Fermentation of Dairy Products

The conversion of milk to solid or semisolid fermented products dates far back in human civilization. The practice of milk fermentation probably arose among herders who collected the milk of their pack animals but had no way to prevent the rapid growth of bacteria. The milk had to be stored in a portable container such as the stomach of a slaughtered animal. After hours of travel, the combined action of lactic acid–producing bacteria and stomach enzymes caused the coagulation of milk proteins into **curd**. The curd naturally separated from the liquid portion, called **whey**. Both curds and whey can be eaten, as in the nursery rhyme "Little Miss Muffet." The curds, however, are particularly valuable for their concentrated protein content.

Curd formation. A **cheese** is any milk product from a mammal (usually cow, sheep, or goat) in which the milk protein coagulates to form a semisolid curd. Curd

formation results from two kinds of processes: acidification, usually as a result of the microbial production of lactic acid; and treatment by proteolytic enzymes such as rennet. The curd may then be separated and processed to varying degrees, depending on the type of cheese.

How does milk coagulate? The major organic components of cow's milk are milk fat (about 4% unless skimmed), protein (3.3%), and the sugar lactose (4.7%). Milk starts out at about pH 6.6, very slightly acidic. At this pH, the milk proteins are completely soluble in water; otherwise, they would clog the animal's udder as the milk came out. Fermentation generally begins with bacteria such as *Lactobacillus* and *Streptococcus* (**Fig. 16.7**). As bacteria ferment lactose to lactic acid, the pH starts to decline. The dissociation constant of lactic acid ($pK_a = 3.9$) allows greater deprotonation than with other fermentation products, such as acetate ($pK_a = 4.8$). Thus, lactic acid rapidly acidifies the milk product to levels that halt further growth of bacteria. Halting bacterial growth minimizes the oxidation of amino acids, thereby maintaining food quality.

Milk contains micelles (suspended droplets) of hydrophobic proteins called caseins. As the pH of milk declines below pH 5, the acidic amino acid residues of caseins become protonated, eventually destablilizing the tertiary structure. As the casein molecules unfold (or "denature"), they expose hydrophobic residues that regain stability by interacting with other hydrophobic molecules. The intermolecular interaction of caseins generates a gel-like network throughout the milk, trapping other substances, such as droplets of milk fat. This protein network generates the semisolid texture of **yogurt**, a simple product of milk acidified by lactic acid bacteria.

In most kinds of cheese formation, an additional step of casein coagulation is accomplished by proteases such as rennet. Rennet derives from the fourth stomach of a

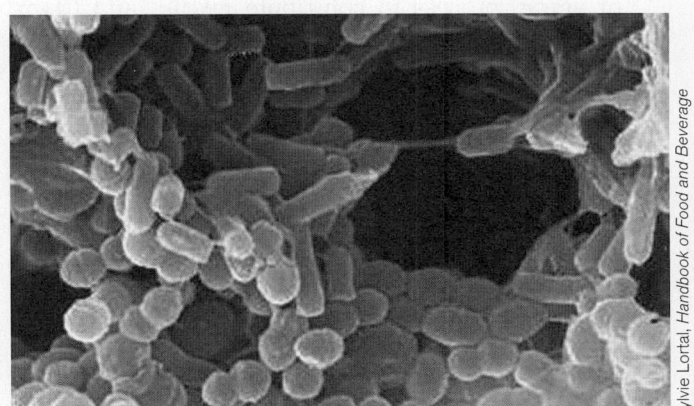

Figure 16.7 Bacterial community within Emmentaler cheese (SEM).   Bacterial species include *Lactobacillus helveticus* (rods, 2.0–4.0 μm in length) and *Streptococcus thermophilus* (cocci).

Sylvie Lortal, *Handbook of Food and Beverage Fermentation Technology*. 2004. CRC Press.

calf, although modern versions are made by genetically engineered bacteria. Calf rennet includes two proteolytic enzymes: chymosin and pepsin. Chymosin specifically cleaves casein into two parts, one of which is charged and water-soluble, the other hydrophobic. The hydrophobic portion forms a firmer curd than intact casein and results in the harder texture of solid cheeses. The water-soluble portion, about one-third of the total casein, enters the whey and is lost from the curd. Processing of some cheese varieties includes high temperature, which denatures even the whey protein so it is retained in the curd.

Varieties of cheese. An extraordinary number of cheese varieties have been devised (**Fig. 16.8**). These fall into several categories based on particular steps in their production.

- **Soft, unripened cheeses**, such as cottage cheese and ricotta, are coagulated by bacterial action, without rennet. The curd is cooked slightly, and the whey is partly drained, but their water content is 55% or greater. These cheeses spoil easily; there are no steps of aging, or **ripening**.
- **Semihard, ripened cheeses**, such as Muenster and Roquefort, include rennet for firmer coagulation, and the curd is cooked down to a water content of 45%–55%. The cheese is aged for several months.
- **Hard cheeses**, such as Swiss cheese and cheddar, are concentrated to even lower water content. Extrahard varieties, such as Parmesan and Romano, have a water content as low as 20%. These cheeses are aged for many months, even several years.
- **Brined cheeses**, such as feta, are permeated with brine (concentrated salt), which limits further bacterial growth and develops flavor. Harder cheeses, such as Gouda, may be brined at the surface.
- **Mold-ripened cheeses** are inoculated with mold spores that germinate and grow during the ripening, or aging, process to contribute texture and flavor. The mold may be inoculated on the surface, to form a crust (as in Brie and Camembert), or it may be spiked deep into the cheese (as in blue cheese or Roquefort).

Cheese production. Commercial production of cheese involves a standard series of steps (**Fig. 16.9**). At each of these steps, choices of treatment lead to very different varieties. Key steps are illustrated for the example of Gouda cheese in **Fig. 16.10**.

In the first step, the milk is filtered to remove particulate objects, such as straw, and microfiltered or centrifuged to remove potentially pathogenic bacteria and spores. Most modern production includes flash pasteurization (brief heating to 72°C), although some traditional cheeses continue to be made from unpasteurized milk. Unpasteurized milk in cheese has been linked to illness, particularly from *Listeria*, bacteria that grow at typical refrigeration temperatures.

The fermenting microbes are added as a **starter culture**. The starter was traditionally derived from a sample of previous fermentation, in which case the flora are undefined. Commercial cheese production now uses defined species. In all but the soft cheeses, bacterial coagulation and curd formation are supplemented by rennet or by genetically engineered proteases.

The solid curd is then cut, or **cheddared** (hence the name "cheddar" cheese). The finer the pieces, the more whey can be pressed out and the harder the cheese produced. Curd is then heat treated, with or without the whey; if whey is included, more protein is retained. Brining at this stage leads to a salty cheese, such as feta.

The pressed curd is then shaped into a mold, which determines the ultimate shape of the cheese. Before ripening (or aging), the cheese may be floated in brine to generate a rind; or it may be coated or spiked with a *Penicillium* mold. The ripening period then allows flavor to develop. Texture also changes; for example, where fermentation has produced CO_2, the trapped gas forms "eyes," or holes.

> **THOUGHT QUESTION 16.3** In an outbreak of listeriosis from unpasteurized cheese, only the refrigerated cheeses were found to cause disease. Why would this be the case?
>
> **THOUGHT QUESTION 16.4** Cow's milk contains 4% lipid (butterfat). What happens to the lipid during cheese production?

A.

B.

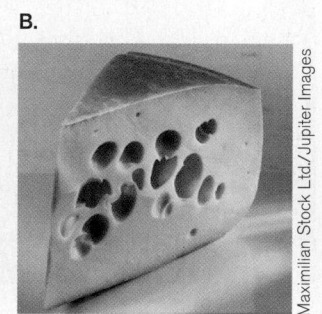

C.

D.

Figure 16.8 Cheese varieties. A. Cottage cheese, an unripened perishable cheese. **B.** Emmentaler Swiss cheese, with eyes produced by carbon dioxide fermentation. **C.** Feta cheese, a soft cheese from goat's milk, preserved in brine. **D.** Roquefort, a medium-hard cheese ripened by spiking with *Penicillium roqueforti*.

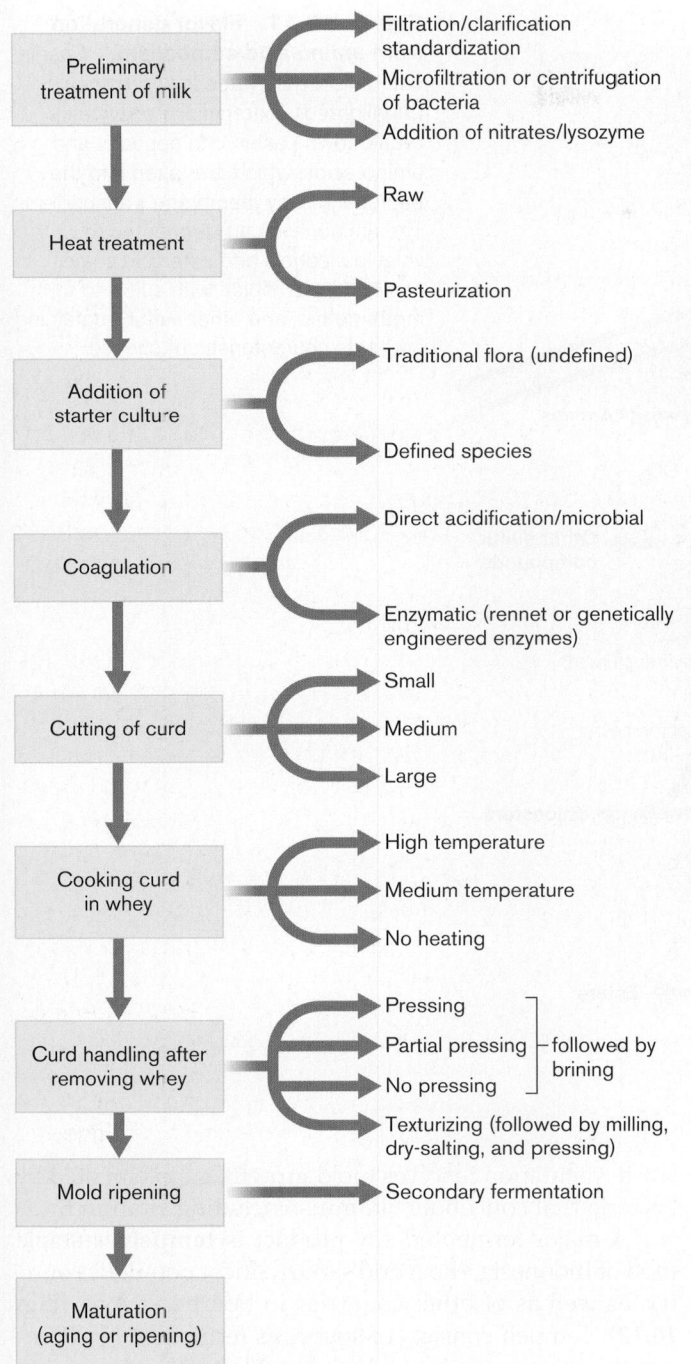

Figure 16.9 Flowchart for cheese production. The alternative procedures shown on the right produce different varieties.

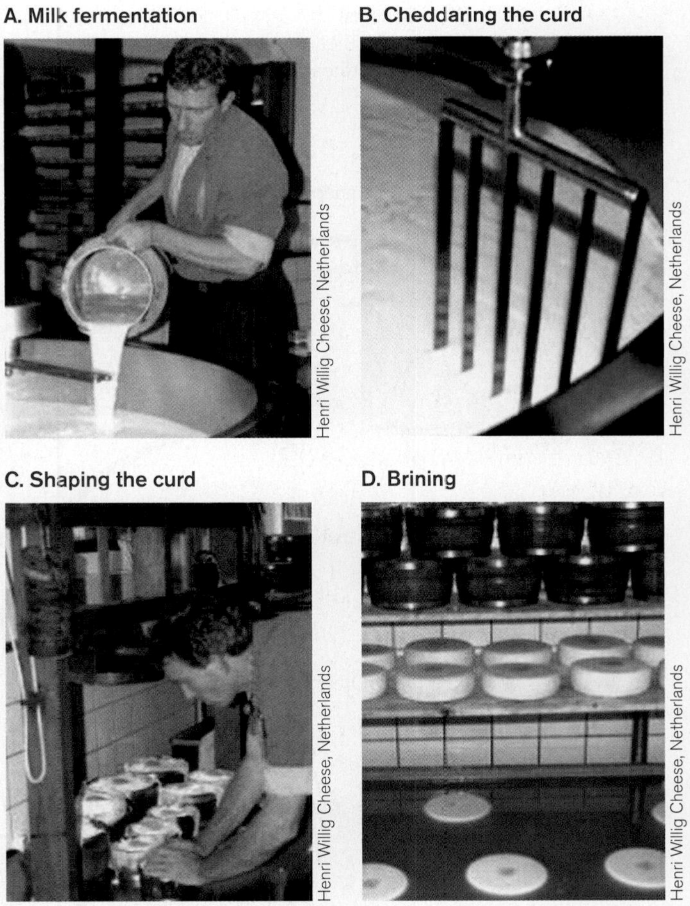

A. Milk fermentation

Henri Willig Cheese, Netherlands

B. Cheddaring the curd

Henri Willig Cheese, Netherlands

C. Shaping the curd

Henri Willig Cheese, Netherlands

D. Brining

Henri Willig Cheese, Netherlands

Figure 16.10 Cheese production. Gouda cheese is produced at the Henri Willig factory, the Netherlands. **A.** Milk is poured into a fermentation tub with a bacterial starter culture and rennet. **B.** The milk curd is cut, or "cheddared." **C.** The curds are shaped into round molds and then pressed to remove whey **D.** The solidified curds are floated in brine to form the characteristic rind of Gouda cheese. The cheese then dries and ripens on the shelf.

Flavor generation in cheese. In all fermented foods, microbial metabolism generates by-products that confer a characteristic aroma and flavor. In some cases, particular species confer distinctive flavors; for example, *Propionibacterium* ferments lactate or pyruvate to propionate, a distinctive flavor component of Swiss cheese. All bacteria generate a surprising range of side reactions, forming trace products that confer distinctive flavors. For example, while *Lactobacillus* converts most of the lactose to lactic acid, a small fraction of the pyruvate is converted to acetoin, acetaldehyde, or acetic acid, which contribute flavor. Most amino acids are retained intact; but traces are converted to flavorful alcohols, esters, and sulfur compounds (**Fig. 16.11**). For example, methanethiol (CH_3SH) contributes to the desirable flavor of cheddar cheese. Lipids are not significantly metabolized by *Lactobacillus*, but in mold-ripened cheeses such as Camembert, *Penicillium* oxidizes a small amount of the lipids to flavorful methylketones, alcohols, and lactones.

Acid Fermentation of Vegetables

Many kinds of vegetable products are based on microbial fermentation. Commercial products marketed globally include pickles, soy sauce, and sauerkraut. Other products provide staple foods for particular nations or regions, such as Indonesian tempeh and Korean kimchi.

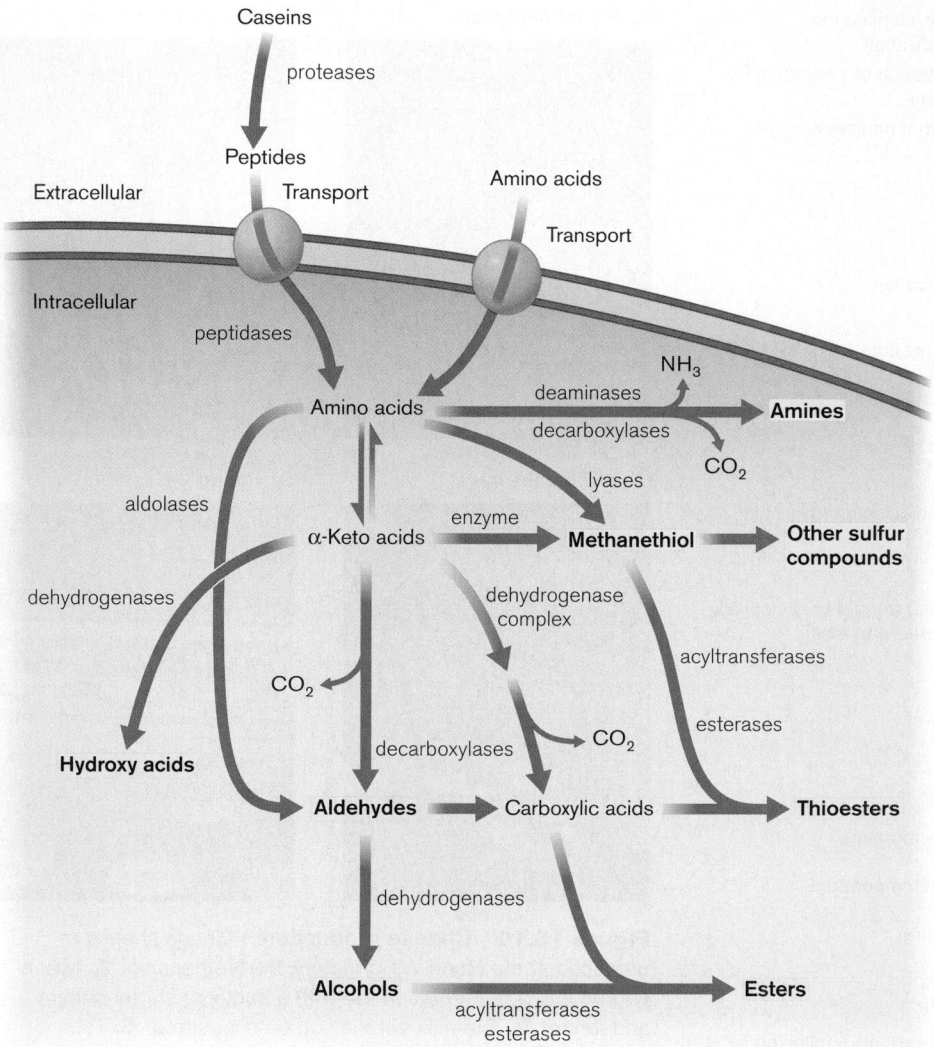

Figure 16.11 Flavor generation from amino acid catabolism. Casein catabolism generates flavor molecules (highlighted). Extracellular enzymes break down casein into peptides and amino acids, which are taken into the bacterial cell by membrane transporters. The amino acids are fermented to volatile alcohols and esters. In some cases, they combine with sulfur to form methanethiol and other sulfur-containing odorants characteristic of cheese.

Soy fermentation. Soybeans offer one of the best sources of vegetable protein and are indispensable for the diet of millions of people, particularly in Southeast Asia. In North America, soy products are important for vegetarian consumption and as a milk substitute, as well as for animal feed. But soybeans also contain substances that decrease their nutritive value. Phytate, or inositol hexaphosphate, chelates minerals such as iron, inhibiting their absorption by the intestine. Lectins are plant glycoproteins that bind to cell surface glycoproteins within the human body. At high concentration, lectins may upset digestion and induce autoimmune diseases. Protease inhibitors interfere with chymotrypsin and trypsin, thus decreasing the amount of protein that can be obtained from soy-based food.

All of these drawbacks of soybeans are diminished by microbial fermentation, while the protein content remains comparable to that of the unfermented bean (40%). A variety of fermented soy foods have been developed. Most soy fermentation involves mold growth, supplemented by bacteria that contribute vitamins, including vitamin B$_{12}$.

A major fermented soy product is **tempeh**, a staple food of Indonesia, the world's fourth most populous country, as well as of other countries in Southeast Asia (**Fig. 16.12**). Tempeh consists of soybeans fermented by *Rhizopus oligosporus*, a common bread mold. Besides decreasing the negative factors of soy, the mold growth breaks down proteins into more digestible peptides and amino acids. During World War II, tempeh was fed to American prisoners of war held by the Japanese. The tempeh was later credited with saving the lives of prisoners whose dysentery and malnutrition had impaired their ability to absorb intact proteins.

Tempeh is commonly produced in home-based factories in Indonesia. The soybeans are soaked in water overnight, allowing initial fermentation by naturally present lactic acid bacteria; in some cases, a crude "starter" may be introduced from the water of soybeans soaked previously.

Figure 16.12 Tempeh, a mold-fermented soy product. A. Fried tempeh. **B.** *Rhizopus oligosporus* mold, used to make tempeh.

A.

B.

This pre-fermentation allows bacterial generation of vitamins and produces mild acid that promotes growth of mold. The soaked beans are then dehulled, cooked, and cooled to room temperature for inoculation with *R. oligosporus* spores or with a previous tempeh culture. The inoculated beans are wrapped in banana leaves or in perforated plastic bags and then allowed to incubate for two days. The mold grows as a white mycelium that permeates the beans, joining them into a solid cake. The final product has a mushroom-like taste and is often served fried or grilled like a hamburger.

Other soy products undergo acidic fermentation by the mold *Aspergillus oryzae*. The Japanese condiment **miso** is made from ground soy and rice, salted and fermented for two months by *A. oryzae*. Soy sauce is made from jiang, a Chinese condiment similar to miso in which the rice starter culture is replaced by wheat. The fermentation generates glutamic acid, a flavor-enhancing compound known popularly in the form of its salt—monosodium glutamate, or MSG.

Fermentation of cabbage and other vegetables. Various leaf vegetables are fermented by traditional societies, originally as a means of storage over the winter months. In Europe and North America, the best-known fermented products include sauerkraut and pickles. Sauerkraut production involves heterolactic fermentation by *Leuconostoc mesenteroides*. In heterolactic fermentation, each fermented sugar molecule yields lactic acid, as well as ethanol and carbon dioxide. The culture is used to inoculate shredded cabbage, which is layered in alternation with salt. The salt helps limit the number of species and the extent of microbial growth. A similar brine-enhanced fermentation process is used to pickle cucumbers, olives, and other vegetables.

An important food based on brine-fermented cabbage is Korean **kimchi (Fig. 16.13).** Kimchi is prepared from Chinese cabbage, salted and layered with radishes, peppers, onions, and other vegetables. The vegetables are covered with a paste of fish, rice, and chili peppers. Pickled seafoods such as shrimp or oysters may be included. The entire mixture is stored in a pot, traditionally buried

underground for several months. The main fermentation organism is *Leuconostoc mesenteroides*, although *Streptococcus* and *Lactobacillus* species participate.

One of the most complex of all food fermentations is that of cocoa beans. Cocoa and coffee beans both require fermentation within the juice of the fruit before the beans are dried and processed. In the case of cocoa, at least three different kinds of fermentation must be performed by three different kinds of microorganisms (see **Special Topic 16.1**).

Alkaline Fermentation: Natto and Pidan

In Western countries, food-associated fermentation is almost synonymous with acidification. In Southeast Asia and in Africa, however, many food products involve *increased* pH. Such fermentations typically release small amounts of ammonia, which raises the pH to about 8, retarding growth of all but alkali-tolerant bacteria. The

A.

B.

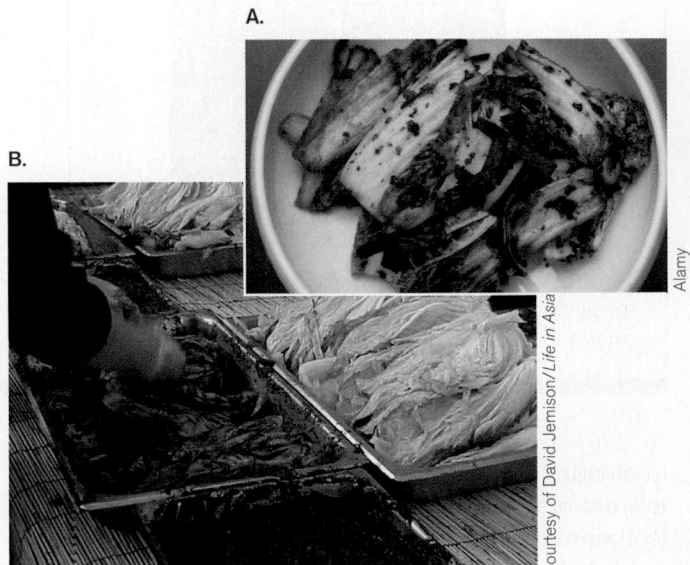

Figure 16.13 Kimchi. A. To make kimchi, cabbage leaves are layered alternating with chili paste containing salted fish and vegetables. **B.** The salted layers are packed to be buried and aged for two months.

Special Topic 16.1 | Chocolate: The Mystery Fermentation

Chocolate, the product of the cocoa bean *Theobroma cacao*, or "food of the gods," requires one of the most complex fermentations of any food. For all the commercialization of chocolate production, totaling 2.5 billion kilos per annum worldwide, no "starter culture" has yet been standardized to ferment the cocoa bean. Instead, the beans harvested in Africa or South America are heaped in mounds upon plantain leaves for fermentation by indigenous microorganisms, essentially the same way cocoa has been processed for

thousands of years (**Fig. 1**). The fermentation must occur immediately where the beans are harvested; they cannot be exported and fermented later.

The microbial fermentation actually occurs outside the cocoa bean, within the pulp that clings to the beans after they are removed from the cocoa fruit. The pulp contains 1% pectin, a complex branched polysaccharide, plus a rich supply of amino acids and minerals. These nutrients support growth of many kinds of microbes. Brazilian microbiologist Rosane Schwan,

Figure 1 Cocoa beans.
A. Cocoa fruit, showing beans encased in mucilage. **B.** Heap of beans covered by plantain leaves. Fruit pulp ferments, liquefies, and drains away, while the beans acidify and turn brown.

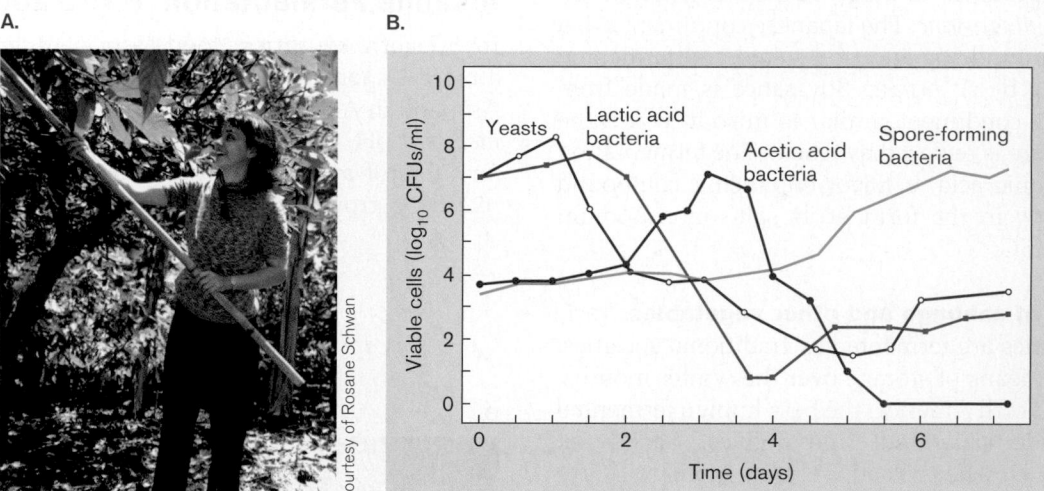

Figure 2 Microbial succession during cocoa pulp fermentation. **A.** Rosane Schwan harvests cocoa beans for her research on cocoa fermentation. **B.** Yeasts generate alcohol, which lactic acid bacteria convert to lactate and acetic acid bacteria oxidize to CO_2. CFUs = colony-forming units. *Source:* Part B from R. Schwan.

fermenting bacteria are usually *Bacillus*, aerobic species tolerant of moderate alkali and capable of extensive proteolysis and amino acid decomposition. The fermentation needs to be controlled to limit the loss of protein content, but the end result is a highly stable food product.

Natto. The Japanese soybean product **natto** is prepared by a process similar to that for tempeh. Soybeans are

washed, pre-fermented, and cooked briefly before incubation with the starter organism *Bacillus natto*. The fermenting beans are incubated in a shallow, ventilated container at a slightly raised temperature (40°C).

B. natto secretes numerous extracellular enzymes, including proteases, amylases, and phytases. These enzymes decrease the undesirable components of soy, such as phytates and lectins, while liberating more

of the University of Lavras, Brazil, analyzed the microbial community (**Fig. 2**). Schwan defined three stages of succession: yeasts, lactic acid bacteria, and acetic acid bacteria.

Yeast fermentation (anaerobic). The pulp initially is full of citric acid (pH 3.6). The acidity favors growth of yeasts, including *Candida, Kloeckera,* and *Saccharomyces.* The yeasts consume citric acid, increasing pH. They also degrade pectin into glucose and fructose, allowing the pulp to liquefy.

As the liquefied pulp drains, its remaining sugars are fermented to ethanol, CO_2, and acetate. These products eventually rise to levels that inhibit yeast growth. The ethanol and acetate penetrate the bean embryo, preventing germination and disrupting cell membranes. The disrupted cells release enzymes that generate key flavor molecules of chocolate by hydrolysis of proteins to produce hydrophobic peptides. Other products include psychoactive molecules, such as theobromine, a stimulant similar to caffeine, and 2-phenylethylamine, a neurotransmitter associated with the pleasure response (**Fig. 3**).

Lactic acid bacteria (anaerobic). The consumption of citric acid by yeast increases the pH to pH 4.2, encouraging growth of lactic acid bacteria. *Lactobacillus* species convert citric acid to acetate, CO_2, and lactic acid. As fermentable substrates disappear, however, lactic acid bacteria are inhibited.

Acetic acid bacteria (aerobic). After one or two days, the beans are turned over and mixed periodically to permit access to oxygen. Aerobic bacteria such as *Acetobacter* now convert the ethanol and acids to CO_2. The consumption of acids

neutralizes undesirable acidity. The bean is also penetrated by oxygen, which oxidizes key cocoa components, such as polyphenols. Polyphenol oxidation generates the brown color of cocoa and contributes flavor.

Oxidative respiration within the mound of beans generates heat faster than it can dissipate, causing the temperature to rise as high as 50°C, which halts fermentation. After fermentation and pulp drainage, the beans are dried and roasted. The roasting process completes the transformation of cocoa substances that contribute flavor. Without the preceding fermentation process, no flavor would develop. Cocoa liquor and cocoa butter are extracted from the beans and then recombined with sugar and other components to make "cocoa mass" (**Fig. 4**). The cocoa mass is stirred for several days to achieve a smooth texture; then it is molded into the decorative forms known as chocolate.

The quality of chocolate rests largely on the original process of pulp fermentation, whose details remain poorly understood. Schwan is working to develop a defined starter culture, a community of microbes to produce a predictable high quality of flavor. At present, however, one of the world's most refined and highly prized commercial food products still depends on the indigenous microbial fermentation of cocoa pulp.

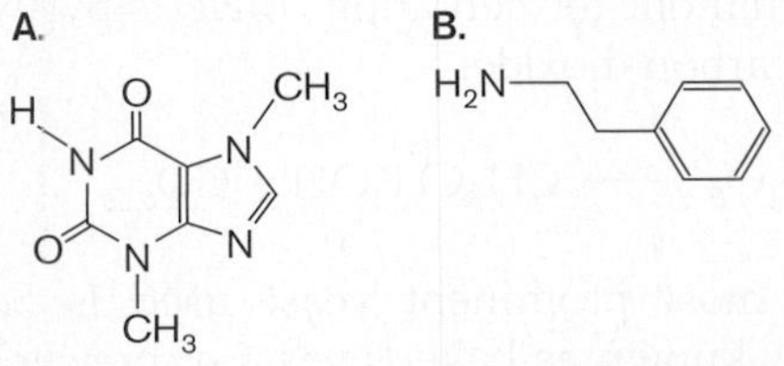

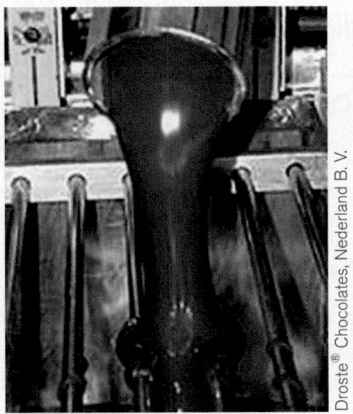

Droste® Chocolates, Nederland B. V.

Figure 3 **Psychoactive products of cocoa fermentation. A.** Theobromine, a stimulant. **B.** 2-Phenylethylamine, an antidepressant.

Figure 4. **Chocolate manufacture.**   Cocoa mass contains cocoa butter and liquor extracted from the cocoa beans, mixed with sugar and other ingredients.

easily digestible peptides and amino acids. Some of the amino acids are deaminated, generating ammonia; a well-ventilated natto chamber allows most of this gas to escape. In addition, *B. natto* synthesizes extracellular polymers such as polyglutamate (a peptide chain consisting exclusively of glutamic acid residues). Polyglutamate generates long elastic strings that bind the beans together. The stretching of these strings from

chopsticks is considered a sign of a good natto (**Fig. 16.14A**).

Alkali-fermented vegetables. In Africa, numerous vegetable products are based on alkaline fermentation predominantly by *Bacillus* species. An example is dawadawa, a paste of fermented locust beans common in western Africa. The locust beans are washed and supplemented with potash, or potassium hydroxide, originally obtained from wood ashes. This addition of alkali retards growth of bacteria other than *Bacillus* species, which predominate at higher pH. The beans are fermented by indigenous bacteria (bacteria already present in the beans). The fermented beans are sun-dried, releasing most of the ammonia, and pounded into cakes for storage. Similar alkali-fermented vegetables include ogiri, from melon seeds, and ugba, from oil beans.

Pidan. An ancient means of preserving eggs has produced the famous Chinese delicacy pidan, or "thousand-year egg," now a favorite at dim sum restaurants (**Fig. 16.14B**). To make pidan, duck eggs are covered with a mixture of brewed tea, lime (CaO), and sodium carbonate (Na_2CO_3). The lime and sodium carbonate react to form NaOH, which penetrates the eggshells, raising pH and coagulating the egg white proteins. The eggs are buried in mud for several months, during which time the combined action of alkali and *Bacillus* fermentation generates dark colors and interesting flavors. Hydrogen sulfide and acetaldehyde contribute to flavor generation.

A.

Courtesy of Matt Wegener

B.

Eleanor Nakama-Mitsunaga, Honolulu Star-Bulletin

Figure 16.14   Alkali-fermented foods. A. Natto consists of soybeans fermented by *Bacillus natto*. The fermentation generates long strings of polyglutamate. **B.** Pidan, or "thousand-year egg," consists of duck eggs coagulated by sodium hydroxide and fermented by *Bacillus* species. One egg is cut open, revealing the transformed yolk, which develops a greenish color.

> **THOUGHT QUESTION 16.5**   In traditional fermented foods, without pure starter cultures, what determines the kind of fermentation that occurs?

TO SUMMARIZE:

- **Milk curd** forms by lactic acid fermentation and rennet proteolysis, rendering casein insoluble. The cleaved peptides coagulate to form a semisolid curd. The main fermentative organisms are lactic acid bacteria.
- **Cheese varieties** include unripened cheese; semihard and hard cheeses that are cooked down and ripened; brined cheeses; and mold-ripened cheeses.
- **Cheese flavors** are generated by minor side products of fermentation, such as alcohols, esters, and sulfur compounds.
- **Soy fermentation** to tempeh and other products improves digestibility and decreases undesirable soy components such as phytates and lectins. The fermentative agent of tempeh is the bread mold *Rhizopus oligosporus*.
- **Vegetables** are fermented and brined to make sauerkraut and pickles. Cabbage and supplementary foods are fermented and brined to make kimchi.

- **Alkali-fermented vegetables** include the soy product natto, the egg product pidan, and the locust bean product dawadawa. The main fermenting organisms are *Bacillus* species.

16.4 Ethanolic Fermentation: Bread and Wine

Some of our most nutritionally significant and culturally important foods, including most bread and alcoholic beverages, derive from ethanolic fermentation by yeast fungi. Ethanolic fermentation converts pyruvate to ethanol and carbon dioxide:

$$C_3H_4O_3 \longrightarrow CH_3CH_2OH + CO_2$$

The most prominent yeast used is *Saccharomyces cerevisiae*, known as baker's yeast or brewer's yeast (**Fig. 16.15**). A hardy organism, *S. cerevisiae* easily survives on a grocery shelf for home use and is genetically tractable for fundamental research. The yeast has been studied since the time of Pasteur, who used it to prove the biological basis of fermentation. *S. cerevisiae* today is a major model system of cell biology, yielding the molecular secrets of human cancer and other diseases.

Bread making depends on carbon dioxide to generate air spaces that **leaven** the dough, making its substance easier to chew and digest. The small amount of ethanol produced is eliminated during baking. For alcoholic beverages, however, ethanol is the key product, accompanied by carbon dioxide bubbles for "fizz," known as carbonation.

Figure 16.15   Baker's yeast, the "champion" fermenter.   A. *Saccharomyces cerevisiae* cells budding; some show bud scars (SEM). **B.** *S. cerevisiae* is used to study the function of human proteins such as alpha-synuclein (green fluorescence), which plays a role in Parkinson's disease. Yeast cells engineered to express one copy of the gene (left panel) show the protein normally within their cell membrane. Two gene copies (right panel) cause the protein to clump and kill the cells.

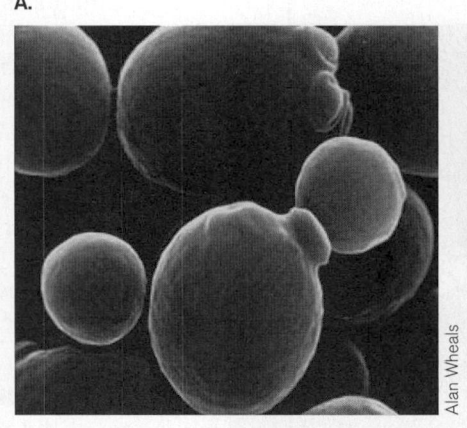
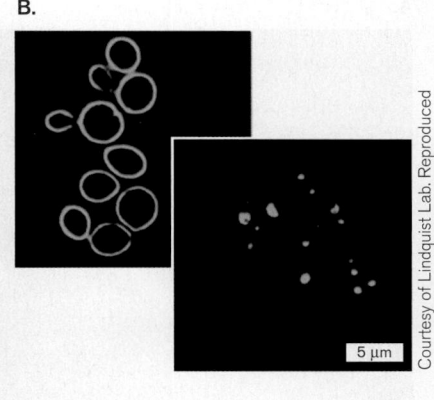

Bread Dough Is Leavened by Microbial CO$_2$

Bread is made in many different forms (**Fig. 16.16**) and from diverse kinds of flour, or ground grain. The earliest breads probably arose from grain mush naturally contaminated by yeast. Later, bread makers learned to include yeast left over from wine or beer production as a starter culture.

Yeast bread production. The preparation of all forms of yeast bread requires the same fundamental steps. A starter culture of yeast is included in the dough. The yeast can be commercial baker's yeast, or it can be **sourdough** starter, an undefined microbial population derived from a previous batch of dough. Analysis of sourdough shows mainly yeasts and lactobacilli, which release acids that favor the growth of the yeasts. The dough is kneaded to develop a fine network of air pockets and allowed to rise, expanding with production of the carbon dioxide gas (**Fig. 16.17**). The finest-textured breads are made from wheat flour, which contains gluten, a protein complex that forms a fine molecular network supporting the rising dough.

Modern bread sold commercially is produced on an industrial scale. The dough is made in huge vats and then cut into regular chunks that are placed into the bread pans. The loaves are baked under hot-air convection, a system that greatly decreases the baking time. In the United States, most commercial bread is sliced mechanically—an invention that dates back to 1912. Sliced bread requires preservative chemicals to prevent growth of mold on the exposed interior.

> **THOUGHT QUESTION 16.6**   Compare and contrast the role of fermenting organisms in the production of cheese and bread.

Injera: extended fermentation. Most kinds of bread involve only a short fermentation period, just long enough to produce enough gas for leavening. A prolonged fermentation, with more extensive microbial activity, occurs in the dough for an Ethiopian bread called **injera** (**Fig. 16.18**). The high microbial content provides a substantial source of vitamins not found in quick-rising breads.

Injera is made from teff, a grain with small, round kernels that grows in Ethiopia at high altitude. Teff contains no gluten, so it cannot rise as much as wheat flour,

Figure 16.16   Yeast bread.   Many varieties of bread are made.

Figure 16.17   Making bread.   As yeast fermentation generates carbon dioxide gas, the dough rises.

A.

Jim Sugar/Corbis

B.

Alamy

Figure 16.18   Injera.   A. After three days of fermentation, injera dough is baked in a ceramic pan upon a "Mirte" charcoal stove. **B.** Injera forms an edible tablecloth for a variety of Ethiopian foods.

but it makes a kind of flatbread. The dough is spread into a wide pancake, and the organisms present in the grain and air are allowed to ferment it for three days. The fermentation includes a succession of species, usually dominated by the yeast *Candida*. The extended fermentation generates a complex range of by-products that confer exceptional flavors in the baked product. Injera forms the basis of an entire Ethiopian meal, served with other food items placed upon it as an edible tablecloth. Diners wrap samples of each food in a fold of injera and consume them together.

Alcoholic Beverages: Beer and Wine

Ethanolic fermentation of grain or fruit was important to early civilizations because it provided a drink free of waterborne pathogens. Traditional forms of beer also provided essential vitamins in the unfiltered yeasts.

Ethanol is unique among fermentation products in that it provides a significant source of caloric intake, but it is also a toxin that impairs mental function. A modest level of ethanol enters the human circulation naturally from intestinal flora, equivalent to a fraction of a drink per day. The human liver produces the enzyme alcohol dehydrogenase, which detoxifies ethanol. This enzyme in a healthy liver can metabolize small amounts of alcohol without harm. However, excess alcohol consumption can overload the liver's capacity for detoxification and permanently damage the liver and brain.

Beer: alcoholic fermentation of grain. Beer production is one of the most ancient fermentation practices and is depicted in the statuary of ancient Egyptian tombs dated to 5,000 years ago (**Fig. 16.19A**). The earliest Sumerian beers were made from bread soaked in water and fermented. Today, most beer is produced commercially by fermenting barley using giant vats (**Fig. 16.19B**). Production of high-quality beer involves complex processing with many steps, including germination of barley grains, mashing in water and cooking, and introduction of hops for flavor (**eTopic 16.1**).

Before fermentation, the barley starch is broken down to maltose by enzymes from the grain. Most of the maltose is fermented by yeast to ethanol and carbon dioxide. However, minor side products contribute flavors—or unpleasant off-flavors if present in too great an amount (**Fig. 16.20**). Some of the off-flavors result from the presence of oxygen, which is needed because yeast is not a true anaerobe; it requires aerobic metabolism to synthesize some of its cell components. The presence of oxygen diverts some pyruvate into acetaldehyde, most of which converts to ethanol. Some of the acetaldehyde, however, remains unconverted, and some is converted to diacetyl. Both acetaldehyde and diacetyl cause off-flavors.

Most pyruvate undergoes ethanolic fermentation, but a small fraction is drawn off to make amino acids via TCA cycle intermediates, as discussed in Chapter 15. The 2-oxo acids of the TCA cycle are analogous to pyruvate, with the methyl group replaced by extended carbon chains (R group). A tiny amount of the 2-oxo acids is converted to long-chain alcohols, which add desirable flavor to beer.

> **THOUGHT QUESTION 16.7**   Compare and contrast the role of low-concentration by-products in the production of cheese and beer.

Wine: alcoholic fermentation of fruit. The fermentation of fruit gives rise to wine, another class of alcoholic products of enormous historical and cultural significance.

A.

Egyptian Museum, Cairo/Borromeo/
Art Resource, NY

B.

Courtesy of HowStuffWorks.com

Figure 16.19   Beer production: ancient and modern.   A. Making beer in ancient Egypt, circa 3000 BCE. The mash is stirred in earthen jars. **B.** Fermenters in a modern brewery.

yeast to begin fermenting immediately, with no need for preliminary breakdown of long-chain carbohydrates, as in the malting and mashing of beer.

Most modern wine production uses strains of the grape *Vitis vinifera*. The grapes are crushed to release juices, usually in the presence of antioxidants such as sulfur dioxide (**Fig. 16.21**). For white wine, the skins are removed before juice is fermented. For red wine, the skins are included in early fermentation to extract the red and purple anthocyanin pigments as well as phenolic flavor compounds. The first few days of fermentation are dominated by indigenous species of yeast naturally present on the grapes, such as *Kloeckera* and *Hanseniaspora* species. Commercial producers usually inoculate with standard *Saccharomyces cerevisiae*, whose population dominates the late stage of fermentation (6–20 days). Yeast growth ends once the ethanol level reaches about 15%; to achieve higher alcohol content, distillation is required.

Grapes produce the best-known wines, but wines and distilled liquors are also made from apples, plums, and other fruits. The key difference between fermentation of fruits and fermentation of grains is the exceptionally high monosaccharide content in fruits. Grape juice, for example, can contain concentrations of glucose and fructose as high as 15%. The availability of simple sugars allows

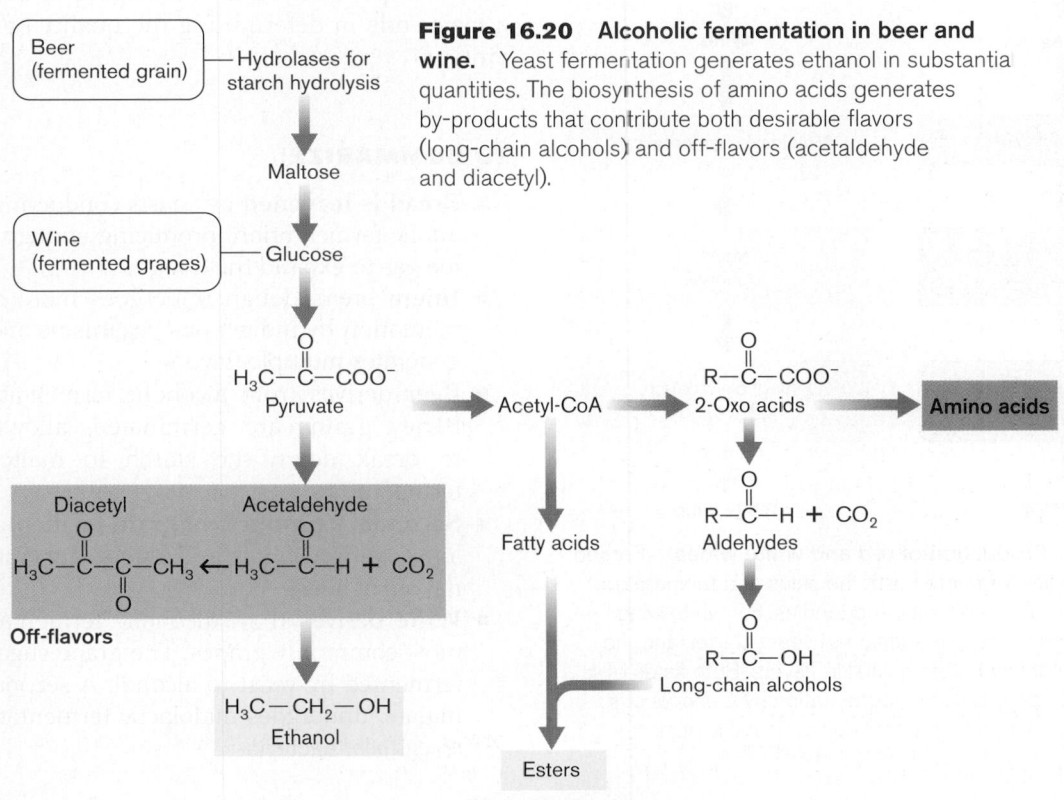

Figure 16.20   Alcoholic fermentation in beer and wine. Yeast fermentation generates ethanol in substantial quantities. The biosynthesis of amino acids generates by-products that contribute both desirable flavors (long-chain alcohols) and off-flavors (acetaldehyde and diacetyl).

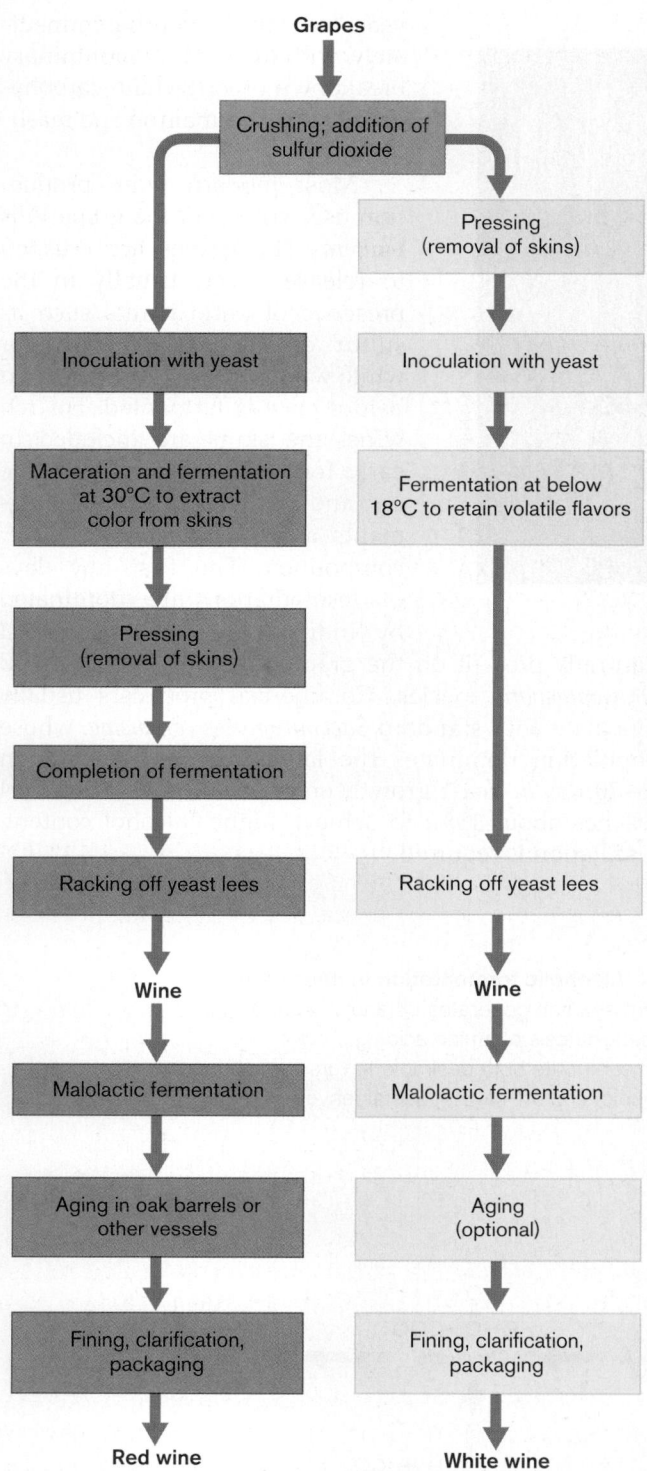

Figure 16.21   Production of red and white wines.   For red wine, the grapes are fermented with the skins at a temperature that increases extraction of color and tannins. For white wine, the skins are removed before addition of yeast starter, and the temperature is kept low to retain volatile flavors. Both kinds of wine usually undergo malolactic fermentation by *Oenococcus oeni*, special lactic acid bacteria that consume malic acid.
Source: G. H. Fleet. Wine. Chapter 37 in *Food Microbiology*, ASM Press.

After fermentation, the wine is drained, or "racked," from the sediment of grape and yeast material, the lees. The liquid may be further clarified by centrifugation. Then it is stored for two to three weeks in tanks or barrels. During storage, a second stage of fermentation may be performed, called **malolactic fermentation**. In malolactic fermentation, the wine is seeded with *Oenococcus oeni* bacteria, which ferment L-malate (deprotonated L-malic acid), a side product from glucose fermentation. The L-malate is decarboxylated to L- or D-lactate:

$$^-OOC-CH_2-CHOH-COO^- \longrightarrow$$
$$CH_3-CHOH-COO^- + CO_2$$

By removing a carboxylate group as CO_2, malolactic fermentation decreases acidity of the wine, improving flavor. It also removes a potential carbon source whose later fermentation by undesirable organisms could spoil the wine.

As in beer production, yeast fermentation of wine produces numerous minor products contributing flavor, such as long-chain alcohols and esters. At the same time, overgrowth of yeast or the growth of undesired species can produce excess amounts of these compounds, such as sulfides and phenolics, giving rise to off-flavors. Some undesired species require oxygen exposure, whereas others can grow during storage and bottling. The balance of microbial populations is challenging to control and has a major role in determining the quality of a given wine vintage.

TO SUMMARIZE:

- **Bread is leavened** by yeasts conducting limited ethanolic fermentation, producing enough carbon dioxide gas to expand the dough.
- **Injera** bread dough undergoes more extensive fermentation by indigenous organisms and, as a result, generates multiple flavors.
- **Beer** derives from alcoholic fermentation of grain. Barley grains are germinated, allowing enzymes to break down the starch to maltose for yeast fermentation.
- **Secondary products of grain fermentation**, such as long-chain alcohols and esters, generate the special flavors of beer.
- **Wine** derives from alcoholic fermentation of fruit, most commonly grapes. The grape sugar (glucose) is fermented by yeast to alcohol. A secondary product, malate, undergoes malolactic fermentation by *Oenococcus oeni* bacteria.

16.5 Food Spoilage and Preservation

We humans have always competed with microbes for our food. When early humans killed an animal, microbes commenced immediately to consume its flesh. Because meat perished so fast, it made economic sense to share the kill immediately and consume all as soon as possible. Vegetables might last longer, but eventually they succumbed to mold and rot. Later societies developed preservation methods, such as drying, smoking, and canning, that enabled humans to survive winters and dry seasons on stored food.

Modern food preservation depends on **antimicrobial agents**, chemical substances that either kill microbes or slow their growth, as well as physical preservative measures, such as treatment with heat and pressure. These general principles of microbial control are described in Chapter 5. Here we focus on microbial contamination and food preservation from the perspective of the food industry.

Food Spoilage and Food Contamination

After food is harvested, several kinds of chemical changes occur. Some begin instantly, whereas others take several days to develop. Some changes, such as meat tenderizing, may be considered desirable; others, such as putrefaction, render food unfit for consumption. The major classes of food change include:

- **Enzymatic processes.** Following the death of an animal, its flesh undergoes proteolysis by its own enzymes. Limited proteolysis tenderizes meat. Plants after harvest undergo other changes; for example, in harvested corn, the sugar rapidly converts to starch. That is why vegetables taste sweetest immediately after harvest.
- **Chemical reactions with the environment.** The most common abiotic chemical reactions involve oxidation by air—for example, lipid autooxidation, which generates rancid odors.
- **Microbiological processes.** Microbes from the surface of the food begin to consume it—some immediately, others later in succession—generating a wide range of chemical products. In meat, internal organs of the digestive tract are an important source of microbial decay.

Microbial activity can aid food production, but it can also have various undesirable effects. Two different classes of microbial effects are distinguished: food spoilage and food contamination with pathogens.

Food spoilage refers to microbial changes that render a product obviously unfit or unpalatable for consumption. For example, rancid milk or putrefied meats are unpalatable and contain metabolic products that may be deleterious to human health, such as oxidized fatty acids or organic amines. Even in these cases, however, the definition of spoilage partly depends on cultural practice. What is sour milk to one person may be buttermilk to another; what one society considers spoiled meat, another considers merely aged.

Different pathways of microbial metabolism lead to different kinds of spoilage. Sour flavors result from acidic fermentation products, as in sour milk. Alkaline products generate bitter flavor. Oxidation, particularly of fats, causes **rancidity**, whereas general decomposition of proteins and amino acids leads to **putrefaction**. The particularly noxious odors of putrefaction derive from amino acid breakdown products that often have apt names, such as the amines cadaverine and putrescine and the aromatic product skatole.

"Food contamination," or **food poisoning**, refers to the presence of microbial pathogens that cause human disease—for example, rotaviruses that cause gastrointestinal illness. Pathogens usually go unnoticed as food is consumed because their numbers are very low, and they may not even grow in the food. Even the freshest-appearing food may cause serious illness if it has been contaminated with a small number of pathogens.

How Food Spoils

Different foods spoil in different ways, depending on their nutrient content, the microbial species, and environmental factors such as temperature. **Table 16.2** summarizes common forms of spoilage.

Dairy products. Milk and other dairy products contain carbon sources, such as lactose, protein, and fat. In fresh milk, the nutrient most available for microbial catabolism is lactose, which commonly supports anaerobic fermentation to sour milk. Fermentation by the right mix of microbes, however, leads to yogurt and cheese production, as previously described.

Under certain conditions, bitter off-flavors may be produced by bacterial degradation of proteins. The release of amines causes a rise in pH. Protein degradation is most commonly caused by psychrophiles, species that grow well at cold temperatures, such as those of refrigeration.

Cheeses are less susceptible than milk to general spoilage because of their solid structure and lowered water activity. However, cheeses can grow mold on their surface. Historically, the surface growth of *Penicillium* strains led to the invention of new kinds of cheeses. But other kinds of mold, such as *Aspergillus*, produce toxins and undesirable flavors.

Meat and poultry. Meat in the slaughterhouse is easily contaminated with bacteria from hide, hooves, and intestinal contents. Muscle tissue offers high water content, which supports microbial growth, as well as rich

Table 16.2 Food spoilage (examples).

Food product	Signs of spoilage	Microbial cause
Dairy products		
Milk	Sour flavor	Lactic acid bacteria produce lactic and acetic acids.
Milk	Coagulation	Lactic acid bacteria produce proteases that destabilize casein and lower pH, causing coagulation.
Milk	Bitter flavor	Psychrophilic bacteria degrade proteins and amino acids.
Cheese	Open texture, fissures	Lactic acid bacteria produce carbon dioxide.
Cheese	Discoloration and colonies	Molds such as *Penicillium* and *Aspergillus* grow on the cheese.
Meat and poultry		
Meat and poultry	Rancid flavor	Psychrotrophic bacteria produce fatty acids that become oxidized.
Meat and poultry	Putrefaction	*Pseudomonas* and other aerobes degrade amino acids, producing amines and sulfides.
Meat and poultry	Discolored patches	Molds such as *Mucor* and *Penicillium* grow on the surface.
Eggs	Pink or greenish egg white	*Pseudomonas* and related bacteria grow on albumin, producing water-soluble pigments.
Eggs	Sulfurous odor	Bacterial growth on albumin releases hydrogen sulfide.
Seafood		
Fish	Fishy smell	Anaerobic psychrophiles such as *Photobacterium* convert trimethylamine oxide (TMAO) to trimethylamine.
Fish	Odor of putrefaction	*Pseudomonas* and other Gram-negative species degrade amino acids, producing amines and sulfides.
Shellfish	Odor of putrefaction	*Vibrio* and other marine bacteria decompose the protein.
Fruits, vegetables, and grains		
Plants before harvest	Rotting or wilting	Plant pathogens, most commonly fungi such as *Alternaria*, *Aspergillus*, and *Penicillium*.
Stored plant foods	Rotting or wilting	Molds or bacteria produce degradative enzymes, such as pectinases and cellulases.
Apples, pears, cherries	Geosmin off-flavor	*Penicillium* mold.
Peeled oranges	Discoloration and off-flavor	*Enterobacter* and *Pseudomonas* spp.
Pasteurized fruit juices	Medicine-like phenolic off-flavor	Acid- and heat-tolerant spore former, *Alicyclobacillus* sp., produces 2-methoxyphenol (guaiacol).
Bread	Ropiness	*Bacillus* spp. grow, forming long filaments.
Bread	Red discoloration	*Serratia marcescens*.

nutrients, including glycogen, peptides, and amino acids. The breakdown of peptides and amino acids produces the undesirable odorants that define spoilage (for example, cadaverine and putrescine).

Meat also contains fat, or adipose tissue, but the lipids are largely unavailable to microbial action because they consist of insoluble fat (triacylglycerides). Instead, meat lipids commonly spoil abiotically by autooxidation (reaction with oxygen) of unsaturated fatty acids, independent of microbial activity. Thus, when meats are exposed to air during storage, they turn rancid—particularly meats such as pork, which contains highly unsaturated lipids. Autooxidation can be prevented by anaerobic storage, such as vacuum packing, which also prevents growth of aerobic

microorganisms. The absence of the suppressed organisms, however, favors growth of lactic acid bacteria and anaerobes such as *Brochothrix thermosphacta*. These organisms generate short-chain fatty acids, which taste sour.

In industrialized societies, the most significant determinant of microbial populations in meat spoilage is the practice of refrigeration. Refrigeration prolongs the shelf life of meat because contaminating microbes are predominantly mesophilic (grow at moderate temperatures, as discussed in Chapter 5). But ultimately, the few psychrotrophs initially present do grow; typically, these are *Pseudomonas* species. The pseudomonads are also favored by the low pH of meat (pH 5.5–7.0), which results from the accumulation of lactic acid in the muscle.

Seafood. Fish and other seafood contain substantial amounts of protein and lipids, as well as amines such as trimethylamine oxide. Fish spoils more rapidly than meat and poultry for several reasons. First, fish do not thermoregulate, and they inhabit relatively low-temperature environments. Because fish grow in low-temperature environments, their surface microorganisms tend to be psychrotrophic and thus grow well under refrigeration. In addition, marine fish contain high levels of the osmoprotectant trimethylamine oxide (TMAO), which bacteria reduce to trimethylamine, a volatile amine that gives seafood its "fishy" smell. Finally, the rapid microbial breakdown of proteins and amino acids leads to foul-smelling amines and sulfur compounds, such as hydrogen sulfide and dimethyl sulfide.

THOUGHT QUESTION 16.8 Why would bacteria convert trimethylamine oxide (TMAO) to trimethylamine? Would this kind of spoilage be prevented by exclusion of oxygen?

Plant foods. Fruits, vegetables, and grains spoil differently from animal foods because of their high carbohydrate content and their relatively low water content. The low water content of plant foods usually translates into considerably longer shelf life than for animal-based foods. Carbohydrates favor microbial fermentation to acids or alcohols that limit further decomposition, and this microbial action can be managed to produce fermented foods, as described in Section 16.2.

Plant pathogens rarely infect humans but may destroy the plant before harvest. Most plant pathogens are fungi, although some are bacteria, such as *Erwinia* species. Historically, plant pathogens have caused major agricultural catastrophes, such as the Irish potato famine, caused by a fungus-like pathogen. Plant pathogens continue to devastate local economies and cause shortages worldwide; for example, the witches'-broom fungus *Crinipellis perniciosa* causes a fungal disease of cocoa trees that has drastically cut Latin American cocoa production.

After harvest, various molds and bacteria can soften and wilt plant foods by producing enzymes that degrade the pectins and celluloses that give plants their structure. In general, the more processed the food, the greater the opportunities for spoilage. For example, citrus fruits generally last for several weeks, but peeled oranges are susceptible to spoilage by Gram-negative bacteria.

Baked bread usually resists spoilage except for surface molds. In rare cases, however, improperly baked bread can show contamination. The appearance of red bread, caused by the red bacterium *Serratia marcescens*, is believed to have been the source of the "blood" observed in communion bread during a Catholic mass in the Italian town of Bolsena in 1263, an event that became known as the Miracle of Bolsena.

Pathogens Contaminate Food

Intestinal pathogens spread readily because many pathogens can be transmitted through food without any outward sign that the food is spoiled. The U.S. Centers for Disease Control and Prevention (CDC) estimates that there are 76 million cases of gastrointestinal illness a year in this country, usually spread through water or food. Thus, one in four Americans experiences gastrointestinal illness in a given year.

An example of food contamination is the 2008 outbreak of *Salmonella enterica* from peanut products. Peanuts contaminated at one processing plant led to an epidemic that sickened 700 people across the United States (**Fig. 16.22A**). The first cases of *Salmonella* infection were reported to the CDC on September 1, 2008. Most infected individuals developed diarrhea, fever, and abdominal cramps 12–72 hours after infection, and symptoms lasted four to seven days. Over the next six months, cases were reported from nearly all U.S. states. The curve of the outbreak (cases rising and then falling) followed the profile of a single-source epidemic, in which all infections are ultimately traced back to one source. (Epidemics are discussed in Chapter 28.)

The CDC researchers investigated the outbreak by comparing the food intake histories of ill persons against matched controls. They found a statistical association between illness and intake of peanut butter, eventually narrowed to a specific brand of peanut butter sold to institutions. As the epidemic grew, cases emerged in which the contaminated food product was crackers filled with peanut butter cream. Ultimately, the peanut butter and cream were traced back to peanuts from a single factory in Georgia. At the food plant, the source of *Salmonella* contamination could not be identified, but the plant records showed that product samples had tested positive for *Salmonella*. Instead of discarding the product, the plant had retested the samples until they "tested negative." Numerous health violations were cited, including gaps in the walls and dirt buildup throughout the plant.

All cases of illness showed a common strain of the pathogen *S. enterica* serovar Typhimurium. (A serovar is a strain whose surface proteins elicit a distinctive immune response.) The strain was identified by analyzing its genomic DNA cleaved by restriction endonucleases (see Section 7.6). Each restriction enzyme cleaves DNA at sequence-specific positions. Strains that differ at key restriction sites generate cleavage fragments of differing length, which are separated by pulsed-field electrophoresis (**Fig. 16.22B**). In electrophoresis, applied voltage causes DNA fragments to migrate different distances

A.

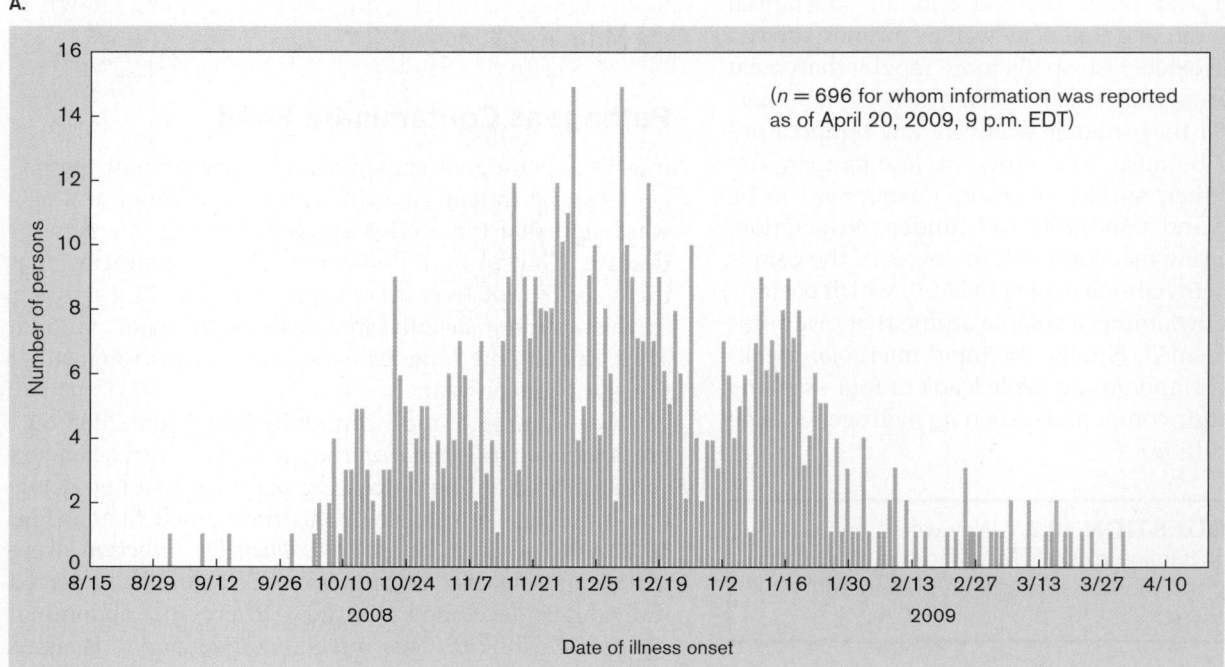

(n = 696 for whom information was reported as of April 20, 2009, 9 p.m. EDT)

Number of persons

8/15 8/29 9/12 9/26 10/10 10/24 11/7 11/21 12/5 12/19 1/2 1/16 1/30 2/13 2/27 3/13 3/27 4/10

2008   2009

Date of illness onset

B.

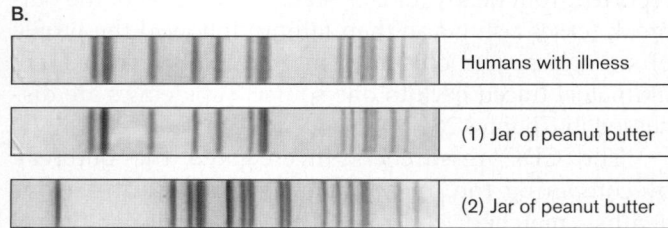

Humans with illness

(1) Jar of peanut butter

(2) Jar of peanut butter

Figure 16.22 *Salmonella enterica* **outbreak from contaminated peanut butter.** **A.** Infected individuals reported to the CDC from September 1, 2008, through April 20, 2009. **B.** Electrophoretic separation of restriction-digest DNA fragments from bacteria strains isolated from a human with illness. (1) Peanut butter containing the same strain; (2) peanut butter containing a different strain of *Salmonella enterica*. Source: www.cdc.gov.

according to size; the pulsed field optimizes separation of the largest sizes. The distance each fragment moves is visualized as a band in the gel. The band pattern, or "fingerprint," of *Salmonella* DNA from infected patients showed the same fragment lengths as *Salmonella* DNA from the peanut butter sample (labeled "(1)" in **Fig. 16.22B**). The band pattern from this peanut butter sample (1) was different from that of another infected peanut butter sample (2); the bacterium in (2) proved unrelated to the *Salmonella* outbreak.

This case illustrates several troubling features of food contamination in modern society. It shows the consequence of a food production plant's failure to follow regulations, and the failure of health inspection to enforce them. The contaminated product shipped out to a diverse array of institutions such as schools, and secondary producers such as cookie manufacturers, who incorporated the peanut butter cream ingredient. The bacteria then remained viable in contaminated food products for many months, sickening people long after the contamination event had occurred.

Food-Borne Pathogens Emerge from Environment and Agriculture

Food-borne pathogens typically arise from a range of sources; see, for example, the diagram of transmission of *Listeria monocytogenes*, a psychrotrophic pathogen that invades the cells of the intestinal epithelium, causing listeriosis (**Fig. 16.23**). (Psychrotrophic organisms grow optimally at moderate temperatures but also grow slowly at lower temperatures, typically 0°C–30°C.) *Listeria* can be transferred from soil and feed to cattle, whose manure then cycles it back to soil. From cattle, the pathogen contaminates milk and meat, where it can eventually infect human consumers. Because *Listeria* is a psychrotroph, it outcompetes other food-borne bacteria under refrigeration.

The U.S. Public Health Service judges the importance of food-borne illnesses based on their incidence and/or severity (**Table 16.3**). For example, the Norwalk-type viruses, or noroviruses, infect 180,000 Americans per year; their spread is very difficult to control, especially in

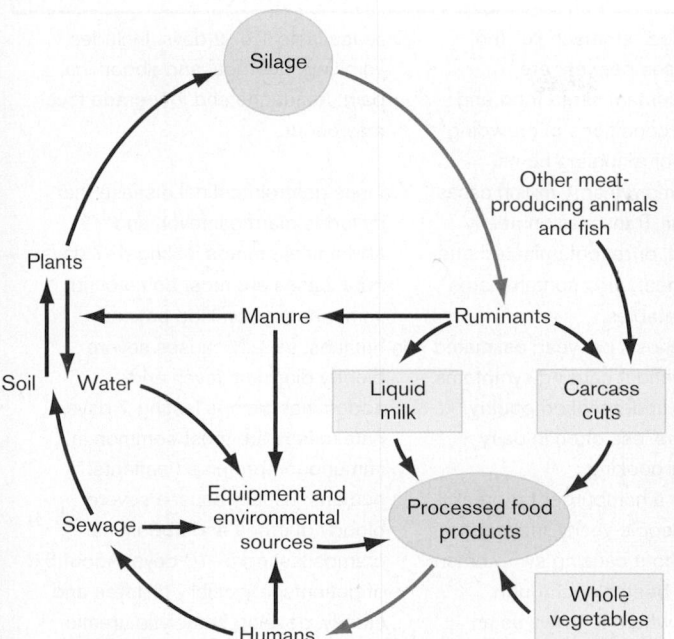

Figure 16.23 Transmission of *Listeria monocytogenes*. Transmission occurs through various routes, including passage through food products. Colored arrows indicate transmission of disease.

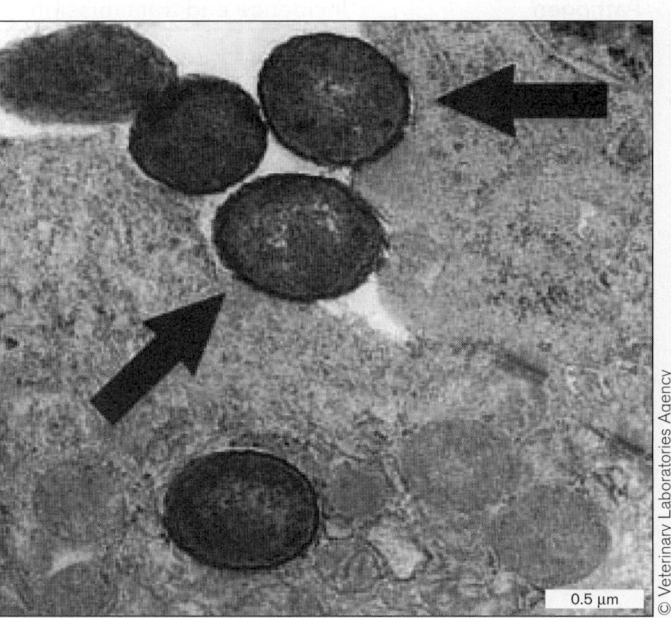

Figure 16.24 Intestinal crypt cells with adherent bacteria, strain *Escherichia coli* O157:H7. A gnotobiotic (germ-free) piglet was infected with the bacteria (arrows). A thin section was stained with toluidine blue.

close quarters, such as a cruise ship, where an outbreak can easily infect a large proportion of passengers. Fortunately, the course of illness, however unpleasant, is short, with few complications.

By contrast, the spore-forming pathogen *Clostridium botulinum* causes only about 100 cases of botulism per year. The incidence is relatively low, but if untreated, the fatality rate is 50%. In this case, the low incidence of botulism actually enhances its danger because the condition is likely to go undiagnosed.

What distinguishes a pathogen from a spoilage organism? Pathogens possess highly specific mechanisms for host colonization, as discussed in Chapter 25. **Figure 16.24** shows intestinal crypt cells covered with *Escherichia coli* O157:H7 bacteria, an emergent pathogen first recognized in 1982 in fast-food hamburgers. Since then, *E. coli* O157:H7 has also been found to contaminate spinach and other vegetables. The bacteria actually grow as **endophytes** within the plant transport vessels. By 2010, six lesser-known strains of *E. coli* had sickened people through contaminated lettuce or beef.

Bacterial factors that contribute to disease are often encoded together in the genome in a region known as a **pathogenicity island**. A pathogenicity island consists of a set of genes and operons that function coordinately (as discussed in Chapter 9). The colocalization of the genes enables transfer of virulence capability to other species

as pathogens evolve. **Figure 16.25** shows an example of a pathogenicity island in *Salmonella*. Four of its operons contribute to the type III secretion complex, which secretes toxins and host colonization factors (discussed in Chapter 25). Other genes encode outer membrane proteins that counteract host defenses, as well as regulators of virulence gene expression.

The most dangerous consequence of infection by food-borne pathogens is the production of a potentially fatal toxin. For example, *E. coli* O157:H7 infection of the intestines can be overcome, but the bacteria produce Shiga toxin, which can destroy the kidneys. In cases of adult botulism, *Clostridium botulinum* does not usually grow within the patient; the botulinum toxin comes from bacteria that grew previously in improperly sterilized food. The botulinum toxin has a highly specific effect, inhibiting synaptic vesicle fusion in the terminal of peripheral motor neurons (**Fig. 16.26**). Synaptic inhibition prevents activation of muscle cells, causing flaccid paralysis. Microbial toxins are discussed further in Chapter 25.

Food Preservation

Cultural practices and cuisines have long evolved so as to limit food spoilage. Such practices include cooking (heat treatment), addition of spices (chemical preservation), and fermentation (partial microbial digestion).

Table 16.3 Food-borne pathogens in the United States[a].

Pathogen	Incidence and transmission	Course of illness
Norovirus (Norwalk and Norwalk-like viruses)	Most common cause of diarrhea; also called "stomach flu" (no connection with influenza). 180,000 cases per year are estimated. Transmitted mainly by virus-contaminated food and water. Infection rates are highest under conditions of crowding in close quarters, such as inside a ship or a nursing home.	Disease lasts 1 or 2 days. Includes vomiting, diarrhea, and abdominal pain; headache and low-grade fever may occur.
Salmonella	Most common food-borne cause of death; more than 1 million cases per year; estimated 600 deaths per year. Transmission nearly always through food—raw, undercooked, or recontaminated after cooking, especially eggs, poultry, and meat; also contaminates dairy products, seafood, fruits, and vegetables.	Causes gastrointestinal disease that includes diarrhea, fever, and abdominal cramps lasting 4–7 days. Fatal cases are most common in immunocompromised patients.
Campylobacter	More than 1 million cases of campylobacteriosis per year; estimated 100 deaths per year. Grows in poultry without causing symptoms. Transmission is mainly through raw and undercooked poultry; contaminates half of poultry sold. Occurs less often in dairy products or in foods contaminated after cooking.	In humans, usually causes severe bloody diarrhea, fever, and abdominal cramps lasting 7 days. Fatal cases are most common in immunocompromised patients.
Escherichia coli O157:H7	An emerging pathogen, first recognized in a hamburger outbreak in 1982; now known to infect 73,000 people yearly, including 60 deaths per year. Grows in cattle without causing symptoms. Transmitted most often through ground beef; also through unpasteurized cider and from produce, where it grows as an endophyte.	In humans, usually causes severe bloody diarrhea and abdominal cramps lasting 5–10 days. About 5% of patients, especially children and elderly, develop hemolytic uremic syndrome, in which the red blood cells are destroyed and the kidneys fail.
Clostridium botulinum	Causes about 100 cases per year of botulism, with a 50% fatality rate if untreated. C. botulinum grows in improperly home-canned foods, more rarely in commercially canned low-acid foods and improperly stored leftovers such as baked potatoes. Spores occur in honey, endangering infants under 2 years of age.	Botulinum toxin from growing bacteria causes progressive paralysis, with blurred vision, drooping eyelids, slurred speech, difficulty swallowing, and muscle weakness. Infant botulism causes lethargy and impaired muscle tone, leading to paralysis.
Listeria monocytogenes	Listeria bacteria grow in animals without causing symptoms. Animal feces may contaminate water, which is then used to wash vegetables. Transmission occurs mainly through vegetables washed in contaminated water and through soft cheeses. Listeria is psychrotrophic, growing at refrigeration temperatures.	Listeriosis involves fever, muscle aches, and sometimes gastrointestinal symptoms. In pregnant women, symptoms may be mild but lead to serious complications for the unborn child.
Shigella	Shigella infects about 18,000 people a year in the United States; in developing countries, Shigella infections are endemic in most communities. Transmission occurs through fecal-oral contact or from foods washed in contaminated water.	Shigellosis involves gastrointestinal symptoms such as diarrhea, fever, and stomach cramps, usually lasting 7–10 days. Complications are rare.
Staphylococcus aureus	S. aureus is best known as the cause of skin infections transmitted through open wounds. However, S. aureus can also be transmitted through high-protein foods such as ham, dairy products, and cream pastries.	S. aureus causes toxic shock syndrome. Can also cause food poisoning via preformed toxins.
Toxoplasma gondii	T. gondii is a parasite believed to infect 60,000 people annually, most with no symptoms. In a few cases, serious disease results. T. gondii is transmitted through contact with feces of infected animals, particularly cats, or through contaminated foods such as pork.	Toxoplasmosis causes mild flu-like symptoms; but in pregnant women, its transmission to the unborn child can lead to severe neurological defects, including death. Neurological complications also occur in immunocompromised patients.
Vibrio vulnificus	V. vulnificus is a free-living marine organism that contaminates seafood or open wounds. About 40 cases per year are reported.	V. vulnificus can infect the bloodstream, causing septic shock. Mainly threatens people with preexisting conditions such as liver disease.

[a]Ten major food-borne pathogens highlighted by the U.S. Public Health Service (USPHS).

Pathogenicity island SPI2 in the *Salmonella* genome

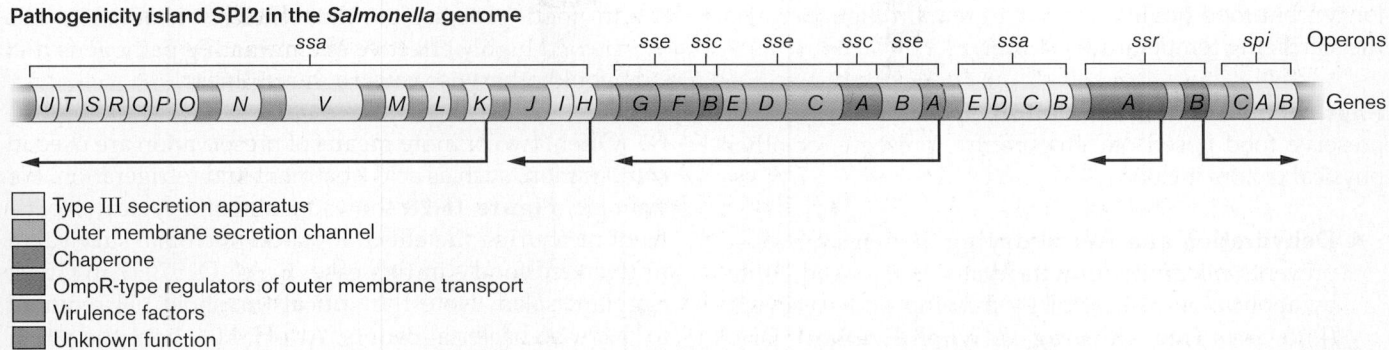

Figure 16.25   Virulence genes of *Salmonella* pathogenicity island SPI2.   Thirty-three virulence genes contained in nine contiguous operons help *Salmonella* bacteria grow within macrophages. *Source:* M. P. Doyle (ed). 2001. *Salmonella* species. Chapter 8 in *Food Microbiology*, ASM Press.

A.

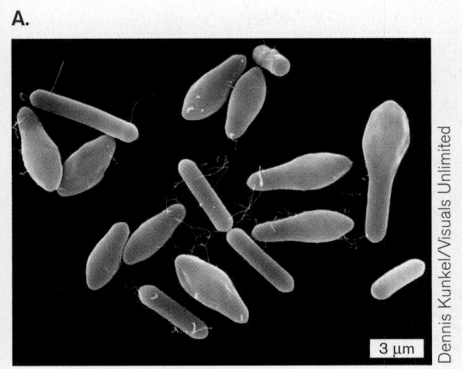

Figure 16.26   *Clostridium botulinum* produces botulinum toxin.   A. Club-shaped morphology of *C. botulinum* cells containing endospores. **B.** Botulinum toxin inhibits synaptic vesicle fusion in the terminal of a peripheral motor neuron, preventing activation of the muscle cell.

B.

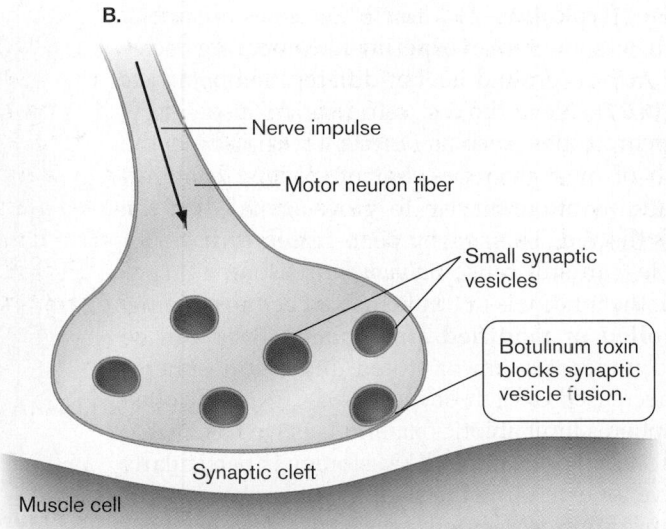

In modern commercial food production, spoilage and contamination are prevented by numerous methods based on fundamental principles of physics and biochemistry that limit microbial growth (discussed in Chapter 5).

The food industry uses several parameters to evaluate the efficiency of heat killing. The D-value (decimal reduction time) was described in Chapter 5. Two additional measures are 12D, the amount of time required to kill 10^{12} spores (or reduce a population 12 logs); and the z-value, the increase in degrees Celsius needed to lower the D-value to 1/10 of the time. If, for example, D_{100} (the D-value at 100°C) and D_{110} (the D-value at 110°C) for a given organism are 20 minutes and 2 minutes, respectively, then $12D_{100}$ equals 240 minutes (that is, 20 minutes × 12), and the z-value is 10°C (because a 10°C increase in temperature reduced the D-value to 1/10, from 20 minutes to 2 minutes). These measurements are determined empirically for each organism. The values are extremely important to the canning industry, which must ensure

that canned goods do not contain spores of *Clostridium botulinum*, the previously described anaerobic soil microbe that causes the paralyzing food-borne disease botulism (**Fig. 16.26**).

Because the tastes of certain foods suffer if they are overheated, z-values and 12D-values are used to adjust heating times and temperatures to achieve the same sterilizing result. Take the following example, where D_{121} is 10 minutes and $12D_{121}$ is 120 minutes. Sterilizing food at 121°C for 120 minutes might result in food with a repulsive taste, whereas reducing the temperature and extending the heating time may yield a more palatable product. The D- and z-values are used to adjust conditions for sterilization at a lower temperature. If D_{121} is 15 minutes (the time needed to kill 90% of cells) and the z-value is known to be 10°C (the temperature change needed to change D-value tenfold), then decreasing temperature by 10°C to 111°C will mean D_{111} is 150 minutes (10 × D_{121}). As a result, $12D_{111}$ is 1,800 minutes. Sterilization may take longer, but food quality is likely to remain high, because the sterilizing temperature is lower.

Physical means of preservation. Specific processes that preserve food based on temperature, pressure, or other physical factors include:

■ **Dehydration and freeze-drying.** Removal of water prevents microbial growth. Water is removed either by application of heat or by freezing under vacuum (known as **freeze-drying**, or **lyophilization**). Drying is especially effective for vegetables and pasta. The disadvantage of drying is that some nutrients are broken down.

■ **Refrigeration and freezing.** Refrigeration temperature (typically −2°C to 16°C) slows microbial growth, as shown in an experiment comparing bacterial growth in ground beef at different temperatures (**Fig. 16.27**). Nevertheless, refrigeration also selects for psychrotrophs, such as *Listeria*. Freezing halts the growth of most microbes, but preexisting contaminant strains often survive to grow again when the food is thawed. This is why deep-frozen turkeys, for example, can still cause *Salmonella* poisoning, especially if the interior is not fully thawed before roasting.

■ **Controlled or modified atmosphere.** Food can be packed under vacuum or stored under atmospheres with decreased oxygen or increased CO_2. Controlled atmospheres limit abiotic oxidation, as well as microbial growth. For example, CO_2 storage is particularly effective for extending the shelf life of apples.

■ **Pasteurization.** Invented by Louis Pasteur, pasteurization is a short-term heat treatment designed to decrease microbial contamination with minimal effect on food value and texture. For example, milk is commonly pasteurized at 63°C for 30 minutes, followed by quick cooling to 4°C. Pasteurization is most effective for extending the shelf life of liquid foods with consistent, well-understood microbial flora, such as milk and fruit juices.

■ **Canning.** In canning, the most widespread and effective means of long-term food storage, food is cooked under pressure to attain a temperature high enough to destroy endospores (typically 121°C). Commercial canning effectively eliminates microbial contaminants, except in very rare cases. The main drawback of canning is that it incurs some loss of food value, particularly that of labile biochemicals such as vitamins, as well as loss of desirable food texture and taste.

■ **Ionizing radiation.** Exposure to ionizing radiation, known as food irradiation, effectively sterilizes many kinds of food for long-term storage. The main concerns about food irradiation are its potential for unknown effects on food chemistry and the hazards of the irradiation process itself for personnel involved in food processing. Nevertheless, irradiation has proved highly effective at eliminating pathogens that would otherwise cause serious illness.

Often, two or more means of preservation are used in combination, such as acid treatment and refrigeration. For example, **Figure 16.28** shows results of a typical experiment measuring the effect of pH on microbial survival in refrigerated food—in this case, *E. coli* O157:H7 in Greek eggplant salad. Note the critical threshold pH required to decrease bacterial counts. At pH 4.0, about the pH of lemon juice, the bacteria show a steep exponential death curve, whereas at pH 4.5, the bacteria remain viable for many days.

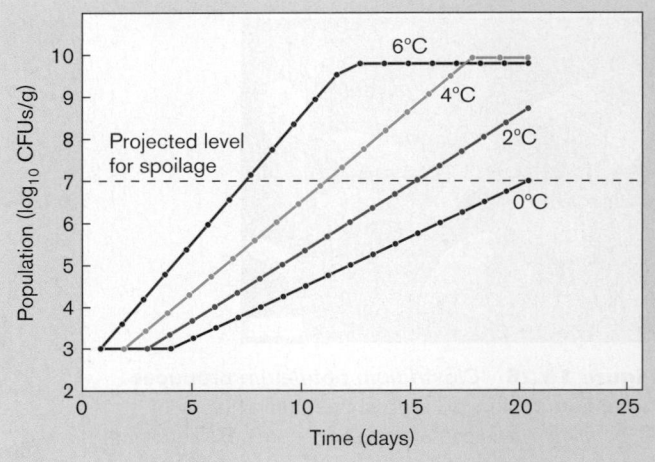

Figure 16.27   Bacterial growth in ground beef. The growth rate of total aerobic bacteria in ground beef declines at lower storage temperatures. CFUs = colony-forming units. *Source*: M. P. Doyle (ed.). 2001. Meat, poultry, and seafood. Chapter 5 in *Food Microbiology*, ASM Press.

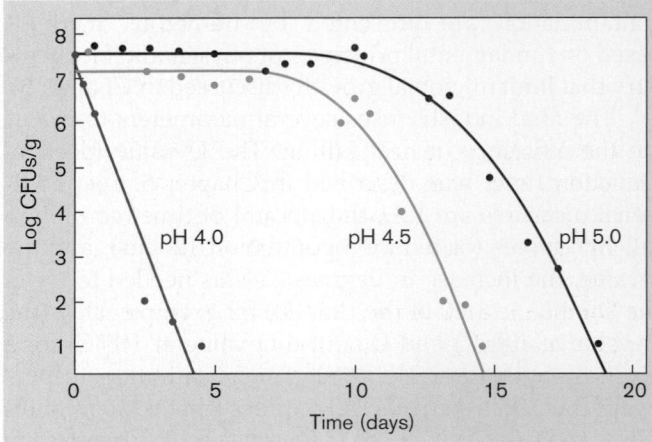

Figure 16.28   Bacteria die sooner at lower pH. Survival curves of *E. coli* O157:H7 in eggplant salad stored at 5°C at pH 4.0, pH 4.5, and pH 5.0. *Source*: Panagiotis N. Skandamis and George-John E. Nychas. 2000. *Appl. Environ. Microbiol.* **66**:1646.

Chemical means of preservation. Many kinds of chemicals are used to preserve foods. Major classes of chemical preservatives include:

- **Acids.** While microbial fermentation can preserve foods by acidification, an alternative approach is to add acids directly. Organic acids commonly used to preserve food include benzoic acid, sorbic acid, and propionic acid. The acids are generally added as salts: sodium benzoate, potassium sorbate, sodium propionate. These acids act by crossing the cell membrane in the protonated form and then releasing their protons at the higher intracellular pH. For this reason, they work best in foods that already have moderate acidity (pH 5–6), such as dried fruits and processed cheeses.
- **Esters.** The esters of organic acids often show antimicrobial activity whose basis is poorly understood. Examples include fatty acid esters and parabens (benzoic acid esters). They are used to preserve processed cheeses and vegetables.
- **Other organic compounds.** Numerous organic compounds, both traditional and synthetic, have antimicrobial properties. For example, cinnamon and cloves contain the benzene derivative eugenol, a potent antimicrobial agent.
- **Inorganic compounds.** Inorganic food preservatives include salts, such as phosphates, nitrites, and sulfites. Nitrites and sulfites inhibit aerobic respiration of bacteria, and their effectiveness is enhanced at low pH. These substances, however, may have harmful effects on humans; nitrites can be converted to toxic nitrosamines, and sulfites cause allergic reactions in some people.

THOUGHT QUESTION 16.9 Is it possible for physical or chemical preservation methods to completely eliminate microbes from food? Explain.

TO SUMMARIZE:

- **Food spoilage** refers to chemical changes that render food unfit for consumption. Food spoils through degradation by enzymes within the food, through spontaneous chemical reactions, and through microbial metabolism.
- **Food contamination, or food poisoning**, refers to the presence of microbial pathogens that cause human disease, or toxins produced by microbial growth. Food harvesting, processing, and shared consumption are all activities that spread pathogens.
- **Dairy products** can be soured by excessive fermentation or made bitter by bacterial proteolysis.
- **Meat and poultry** are putrefied by decarboxylating bacteria, which produce amines with noxious odors.

- **Fish and other seafood** spoil rapidly because their unsaturated fatty acids rapidly oxidize; they harbor psychrotrophic bacteria that grow under refrigeration; and their TMAO is reduced by bacteria to the fishy-smelling trimethylamine.
- **Vegetables** spoil by excess growth of bacteria and molds. Plant pathogens destroy food crops before harvest.
- **Food preservation** includes physical treatments, such as freezing and canning, as well as the addition of chemical preservatives, such as benzoates and nitrites.

16.6 Industrial Microbiology

The production and preservation of food is only one field of **industrial microbiology**, the commercial exploitation of microbes. Industrial microbiology is commonly understood to include a broad range of commercial products relevant to microbes, including vaccines and clinical devices; industrial solvents and pharmaceuticals; bioinformatic analysis of genomes; and genetically modified plants and animals using microbial vectors.

Other growing fields of industrial microbiology are wastewater treatment, bioremediation, and environmental management, covered in Chapters 21 and 22.

Industrial Microbiology Aims for Commercial Success

The practical application of a microbial product or device may arise out of an industrial laboratory, or it may be conceived by a research scientist with the aim of meeting a compelling need in society. For example, **Special Topic 16.2** profiles a biotechnology company formed to develop innovative products for the prevention, testing, and treatment of tuberculosis. In all cases, however, the key goal is success in the marketplace—that is, to generate a product that achieves adoption by customers in preference to alternative technologies. The product's sales must cover the costs of raw materials and production and (in a for-profit company) generate a profit for the shareholders. Success requires:

- **Identifying a useful product.** Possible products include small molecules, such as antibiotics; human proteins from cloned genes; or proteins from microbes with useful properties, such as thermostability.
- **Isolating a microbe to produce the product.** A novel product, such as an antibiotic, is generally discovered in a naturally occurring microbe. The genes encoding product biosynthesis are then cloned into an industrial vector.
- **Scaling up production in quantity.** The engineered microbial strain containing the vector with its cloned genes must be grown on an industrial scale, and the product must be isolated and purified.

Special Topic 16.2   Start-up Companies Take On Tuberculosis

In the 1990s, newspaper headlines announced the resurgence of tuberculosis (TB) in American cities. Carol Nacy was then a chief scientific officer at a small biotechnology firm, following a research career at the Walter Reed Army Institute of Research and a term as president of the American Society for Microbiology (**Fig. 1**). In 1996, Nacy was invited by the National Institutes of Health to assist in a review of tuberculosis research grant support to U.S. universities. She was surprised to discover the lack of attention to TB, a disease that is the world's number one killer of women aged 15–44 and the leading killer of men after traffic accidents. The United States spends $1 billion yearly to treat 13,000 incident cases; yet the standard antibiotics for tuberculosis were developed before 1970, and the main diagnostic test available (the tuberculin skin test) dates to 1880, the time of Robert Koch. The best available vaccine, the Bacille Calmette-Guérin (BCG) attenuated-live vaccine, is only 50% effective.

Despite the clear need, tuberculosis has been of little interest to industrial research or development. So Nacy decided to found two new companies to address TB: a nonprofit medical research company to develop vaccines and

Figure 1 Carol Nacy developed two companies to fight tuberculosis.

© Joanne Lawton/Washington Business Journal

conduct clinical trials and a for-profit company to develop innovative drugs and treatment devices.

A nonprofit company develops vaccines. The nonprofit company Aeras Global TB Vaccine Foundation, directed by Jerald Sadoff, has received over $200 million in grants from the Bill & Melinda Gates Foundation—more than doubling the amount spent worldwide on TB vaccines. Additional funding has come from the U.S. Centers for Disease Control and Prevention and from the government of Denmark. The Aeras foundation established a clinical trial site in Cape Town, South Africa. South Africa has the highest rate of pediatric TB in the world: over 500 cases per 100,000 persons. Aeras trained South African investigators and support staff in the art of clinical trials, while helping community health care workers to serve the local community. They helped test and counsel the community for HIV, a major risk factor for tuberculosis, and they vaccinated thousands of babies with BCG vaccine.

By 2009, Aeras had 12 different experimental TB vaccines in clinical and preclinical trials. The experimental vaccines are based on leading-edge principles of molecular biology, such as a recombinant vaccinia virus expressing antigenic proteins from *Mycobacterium tuberculosis*; recombinant fusion proteins from *M. tuberculosis*; and a mutant strain of the bacterium deleted for genes encoding two transcriptional sigma factors. An advanced "DNA vaccine" was developed consisting of a recombinant DNA vector designed to express antigenic proteins within host cells. The fundamental principles behind these experimental vaccines are introduced in Chapters 11 and 12.

A for-profit company develops drugs and devices. Nacy's for-profit company, Sequella, Inc., targeted innovative ideas with high risk but high potential to improve performance, rapidity, and safety of diagnosing and treating TB infections. Nacy and her cofounders scanned the academic community for novel ideas that had succeeded against TB in "proof of principle" animal models. The most promising ideas were developed for improved antibiotics, rapid and less invasive tests for TB exposure, and devices to measure the extent of pulmonary infection. Because any one idea had a high risk of failure, multiple prospects were pursued in each category.

■ **Developing a business plan.** The scientist-entrepreneur must obtain partners skilled at industrial management, finance, and marketing. Patents must be filed to protect the intellectual property rights.

■ **Safety and efficacy testing.** Human consumption or consumer use requires many levels of clinical testing for approval by government agencies.

■ **Effective marketing.** The benefits of the new product must be communicated effectively to convince customers of its superiority to current products or processes.

Failure at any of the preceding tasks spells doom for the product. Thus, a prudent business plan includes multiple alternative products in development. Although the

Several Sequella products have since reached advanced clinical trials. Their lead product is the Transdermal Patch test, a diagnostic transdermal patch to distinguish between active TB and prior TB infection or BCG vaccination (**Fig. 2**). Prior to the transdermal patch, it was difficult to determine whether an individual who tested positive for TB antibodies actually had active infection or had simply been exposed to the bacillus in the past. The transdermal product allows antigens to penetrate the skin without needle injection, and it produces results more reliable than the standard tuberculin skin test.

For treatment of TB, several possible new antibiotics are in the pipeline. The most promising antibiotic, SQ109, was discovered by high-throughput screening of a chemical library, in collaboration with Clifton Barry at the National Institutes of Health. The chemical library consisted of over 60,000 analogs of a known TB antibiotic, ethambutol (**Fig. 3**). Ethambutol is part of the current standard course of treatment for TB, whose six-month time course has a poor compliance rate. It is hoped that improved drugs will shorten the time course and improve compliance, thereby decreasing the appearance of drug-resistant strains.

The analog molecules were selected for their common diamine core, with different combinations of side chains. The 60,000 compounds in the library were subjected to combinatorial screening, a mathematically intensive analysis based on numerous tests. Of the compounds tested, 2,796 showed activity against *M. tuberculosis* in the test tube. The 69 best compounds were tested for cytotoxicity in tissue culture, activity in TB-infected macrophages, and activity in infected animals. The compound with the greatest efficacy and fewest side effects was SQ109, a molecule with an unusual cage-like side group of three fused rings. SQ109 is now in human clinical trials. With these products, Sequella has positioned itself to develop more effective vaccines, diagnostics, and treatments to alleviate the global burden of tuberculosis—and to make a handsome profit for investors.

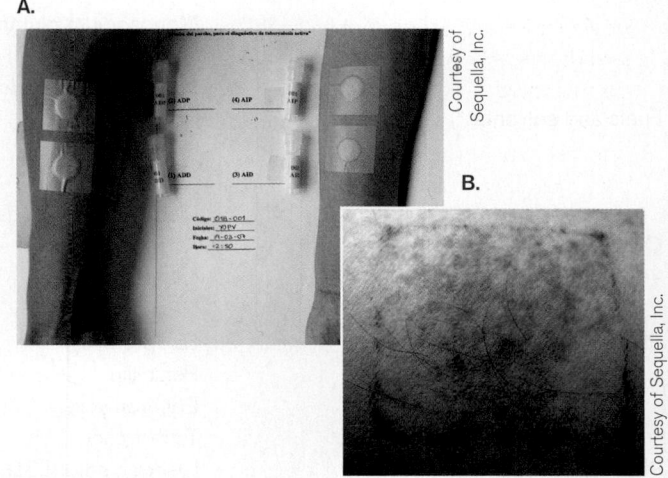

Figure 2   The transdermal patch reveals active TB infection.   A. The transdermal patch administers an antigen test reagent without requiring needle injection. **B.** A positive test result for active TB.

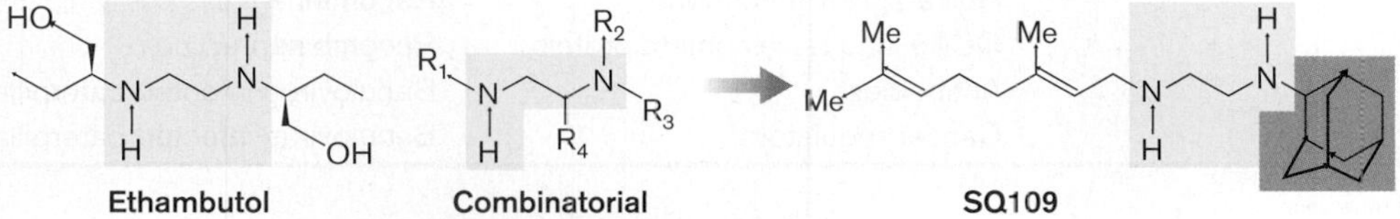

Ethambutol

Combinatorial library of 63,238 diamines

SQ109

Figure 3   A new antibiotic for TB is obtained by screening ethambutol analogs.   A combinatorial library of compounds containing the ethambutol diamine core (yellow) was screened for antibacterial effect against *Mycobacterium tuberculosis*. The most promising agent screened was SQ109, with an unusual carbon-cage side group (pink).

failure rate of new products is high, all the products we use had to overcome these risks.

Molecular Products from Human or Microbial Sources

Increasingly, microbes are grown to produce a commercially valuable chemical substance, such as a vitamin, an industrial solvent, or an enzyme (**Table 16.4**). Each product requires a gene or operon of genes encoding either the product itself or the enzymes for the product's biosynthesis. We may distinguish between two fundamentally different sources of products: cloned genes from human, animal, or plant sources; and native microbial products, often from newly discovered species in extreme environments. Cloned human genes typically encode a protein

Table 16.4 Commercial products from microbes.

Class of product	Product	Producer microbe[a]
Agricultural products	Gibberellin (plant hormone)	*Fusarium moniliforme* (fungus)
	Fungicides	*Coniothyrium minitans* (fungus)
	Insecticides (live pathogens)	*Bacillus thuringiensis*
Enzymes	Alpha-amylase	*Bacillus subtilis*
	Amyloglucosidase	*Aspergillus niger* (fungus)
	Lactase (beta-galactosidase)	*Kluyveromyces lactis* (fungus)
	Lipases	*Candida cylindraceae* (yeast)
	Alkaline protease	*Aspergillus oryzae* (fungus)
Food supplements	L-Lysine	*Brevibacterium lactofermentum*
	L-Tryptophan	*Klebsiella aerogenes*
	Monosodium glutamate (MSG)	*Corynebacterium ammoniagenes*
	Vitamin B_{12}	*Pseudomonas denitrificans*
	Vitamin C (ascorbic acid)	*Acetobacter suboxidans*
Fuels and solvents	Acetone	*Clostridium* spp.
	Butanol	*Clostridium acetobutylicum*
	Glycerol	*Zygosaccharomyces rouxii* (yeast)
	Methane (natural gas)	Methanogens (archaea)
Organic acids	Acetic acid	*Acetobacter xylinum*
	Citric acid	*Aspergillus niger* (fungus)
	Lactic acid	*Lactobacillus delbruckii*
Pharmaceuticals: antibiotics	Streptomycin	*Streptomyces griseus*
	Penicillin	*Penicillium chrysogenum* (fungus)
	Erythromycin	*Saccharopolyspora erythraea*
	Tetracycline	*Streptomyces aureofaciens*
Pharmaceuticals: other drugs	Lysergic acid (LSD, hallucinogen)	*Claviceps paspali* (fungus)
	Cyclosporin (immunosuppressant)	*Trichoderma polysporum* (fungus)
	Steroids	*Arthrobacter* spp.
Pharmaceuticals: cloned human proteins	Insulin	Recombinant *Escherichia coli*
	Interferon	Recombinant *E. coli* or *Saccharomyces cerevisiae* (yeast)
	Human growth hormone	Recombinant *E. coli*
	CC10 lung development protein	Recombinant *E. coli*
	Antibodies	Baculovirus-infected caterpillars
	Cancer regulators	Baculovirus-infected caterpillars

[a]Bacteria, unless stated otherwise

Source: M. J. Waites. 2001. *Industrial Microbiology*, Blackwell Science, Table 4.1, pp. 76–77.

of valuable function in the human body. For example, the Claragen company produces the recombinant human protein CC10, a lung development protein often deficient in the lungs of premature infants. The recombinant protein, produced and purified from a recombinant bacterium, can be provided to the infant to reduce lung inflammation.

Cloning of gene products in recombinant organisms is discussed in Chapter 12. Here we discuss obtaining products from native microbial sources. For example, the Danish company Novozymes markets over 700 microbial enzymes, including Stainzyme, a laundry detergent acting at lower temperatures, an oxidase to control bleaching of paper pulp, and proteases for leather treatment. Novozymes also produces numerous enzymes as biocatalysts

for green chemistry. "Green chemistry" refers to environmentally friendly procedures for reactions in organic chemistry, typically using water solution in place of petroleum-derived organic solvents.

Bioprospecting. To identify new microbial products, chemical companies screen thousands of microbial strains from diverse ecosystems. The search for organisms with potential commercial applications is called **bioprospecting**. Bioprospecting can be done anywhere, from one's backyard to Yellowstone National Park. Unique, endangered ecosystems are the most promising sources of previously unknown microbial strains with valuable properties. Extreme environments, such as the hot springs of Yellowstone, are particularly promising because their products may tolerate conditions of temperature or pH required for industrial use. Psychrophiles from extremely cold environments, such as Antarctica, are a useful source for enzymes that do not require heating for activity, such as the Stainzyme laundry detergent.

An important aspect of bioprospecting is "mining the genome." Once a promising source strain is obtained, its genome is sequenced to investigate the genetic sequences that encode the useful product and regulate its expression. The operons encoding the product (or enzymes for its production) can then be cloned for optimal production (see Chapter 12). The cloned genes are transferred into an **industrial strain**, a strain whose growth characteristics are well studied and optimized for industrial production. An industrial strain must possess the following attributes:

- **Genetic stability and manipulation.** The industrial strain must reproduce reliably, without major DNA rearrangements. It must also have an efficient gene transfer system by which vectors can introduce genes of interest into its genome.
- **Inexpensive growth requirements.** Industrial strains grow on low-cost carbon sources with minimal special needs, such as vitamins, and at easily maintained conditions of temperature and gases.
- **Safety.** Industrial strains are nonpathogenic and do not produce toxic by-products.
- **High level of product expression.** The strain or recombinant vector must possess an efficient gene

expression system to generate the desired product as a high proportion of its cell mass.

- **Ready harvesting of product.** Either the product must be secreted by the cell or, if the product is intracellular, the cells must be easily breakable to liberate the product.

Species for industrial strains include the bacterium *Bacillus subtilis*, the yeast *Candida utilis*, and the filamentous fungus *Aspergillus niger*. Each of these species is safe; grows to high density on inexpensive carbon sources, such as molasses; and expresses desired products at high concentration.

THOUGHT QUESTION 16.10 Why would different industrial strains or species be used to express different kinds of cloned products?

Fermentation Systems

Commercial success requires optimizing every detail of the fermentation system. "Fermentation" in industrial terms refers not just to anaerobic metabolism, but to all means of growth of microbes on an industrial scale. In an **industrial fermentor**, the growth vessel and all its environmental supports, such as temperature control and oxygenation, must be scaled up to thousands of liters (**Fig. 16.29**). This increase in scale generates many problems of quality control, such as maintaining even temperature, pH, and oxygenation throughout the vessel and minimizing foaming of the culture liquid.

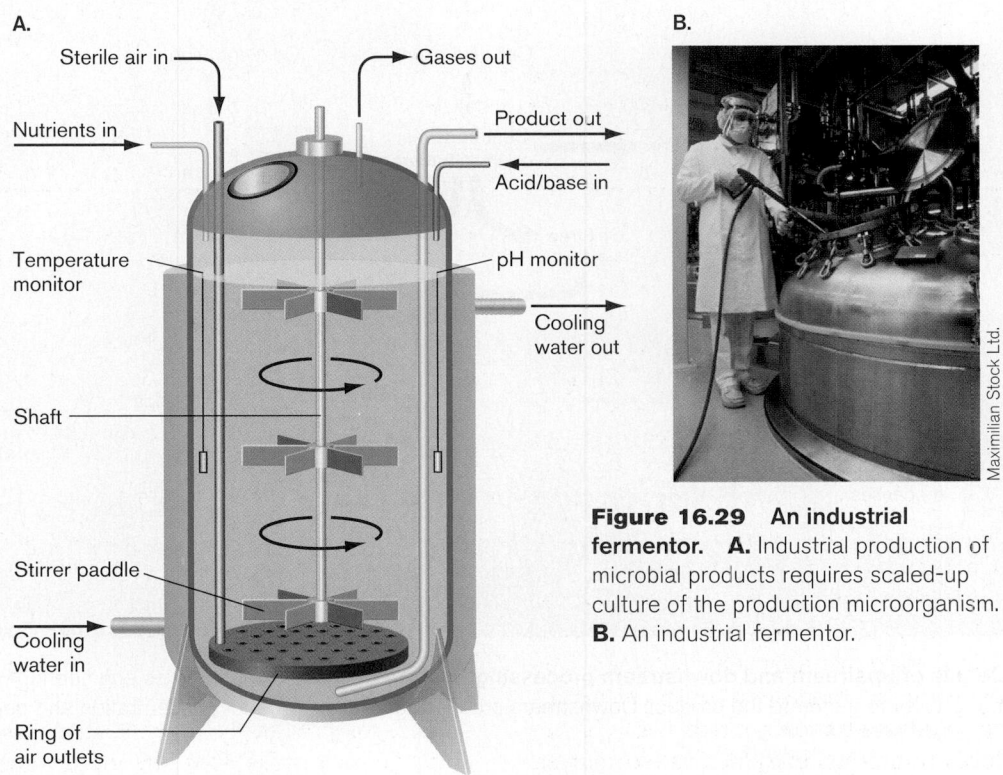

Figure 16.29 An industrial fermentor. A. Industrial production of microbial products requires scaled-up culture of the production microorganism. **B.** An industrial fermentor.

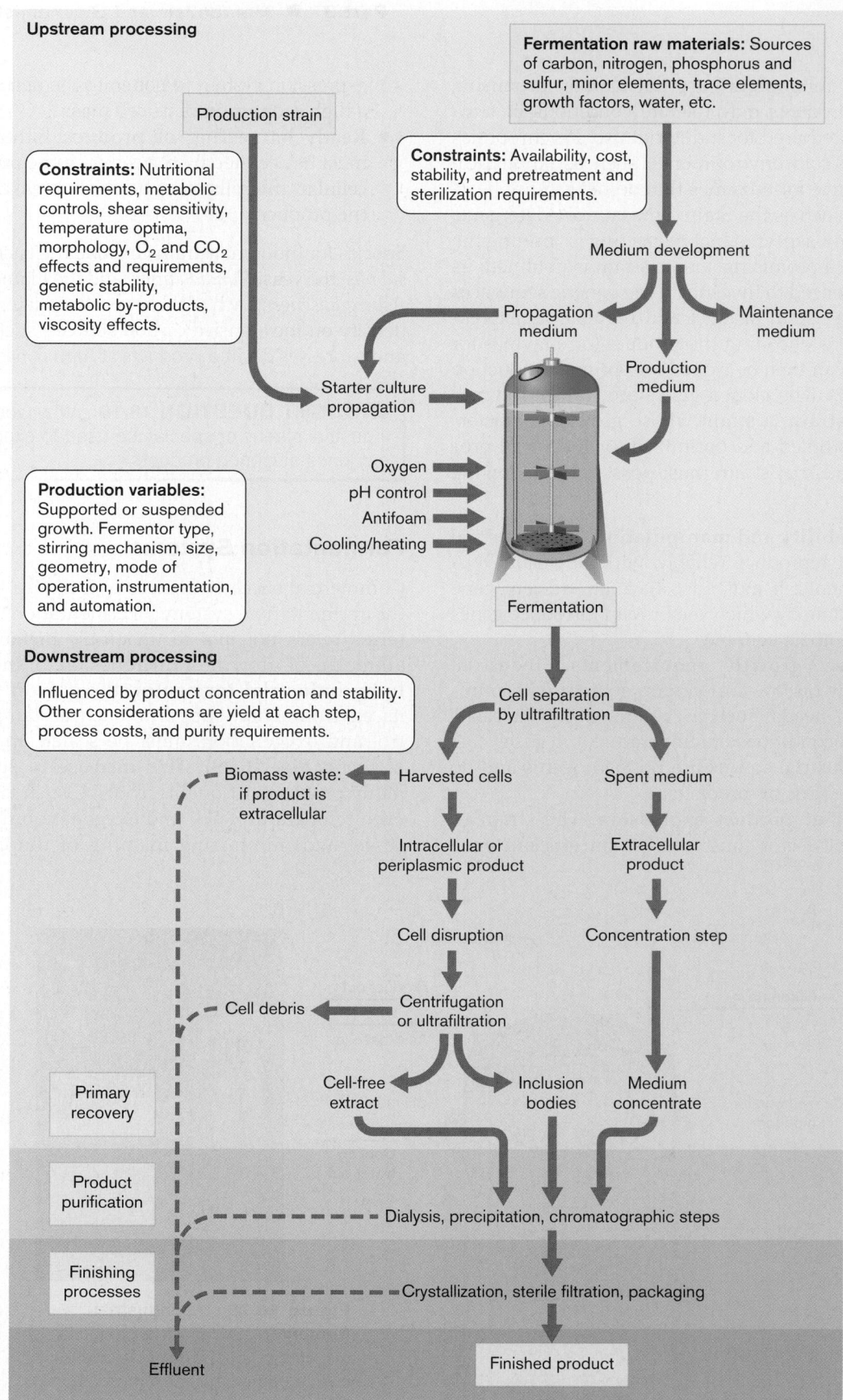

Figure 16.30 Details of upstream and downstream processing. Upstream processing involves engineering of the microbial strain and large-scale growth to generate the product. Downstream processing involves product concentration and purification. *Source*: M. J. Waites. 2001. *Industrial Microbiology*, Blackwell Science.

A small change in any of the growth factors can impact production costs and profit margin. Another major concern is to avoid contamination by other organisms.

The fermentor is the core of the first half of industrial production, known as **upstream processing (Fig. 16.30)**. "Upstream processing" refers to the culturing of the industrial microbe to produce large quantities of product. All aspects of the process must be controlled so as to maximize the final concentration of product, which in most cases peaks at a specific time in the microbial growth cycle. Following microbial growth, the culture must be harvested and the product purified. These processes constitute **downstream processing**. The first step of downstream processing is to separate the microbial cells from the culture fluid, by centrifugation or by filtration. Next, the **primary recovery** of product follows one of two different pathways, depending on whether the product is maintained within the cells or secreted into the culture fluid. Many kinds of subsequent purification and finishing steps are necessary before the product has acceptable quality for its desired use. Again, failure of any detail can render the entire product unusable.

Products designed for human internal use face formidable hurdles in clinical testing and, finally, approval by the appropriate regulatory agency, such as the U.S. Food and Drug Administration (FDA). After millions of dollars are invested in process development, the product may still fail this final test and never come to market. Not surprisingly, a pharmaceutical company must research thousands of potential products before achieving one that makes a profit. The consumer cost inevitably includes the development costs not only of the one successful product, such as recombinant insulin, but also of all the products that failed.

Production in Animal or Plant Systems

Some kinds of products require subtle kinds of processing that occur correctly only within eukaryotic cells. For example, protein products may require posttranslational modifications such as glycosylation (attachment of polysaccharide chains). In such cases, the expressed genes must be transferred from a bacterial or viral vector into an animal, fungal, or plant tissue culture or to a transgenic organism. The microbially transformed organism may itself be the industrial product. We present two examples of microbial production involving multicellular hosts: that of the plant vector host *Agrobacterium tumefaciens* and that of the baculovirus insect vector.

***Agrobacterium tumefaciens* engineers plants.** Bacteria and viruses provide the vectors for transferring genes into multicellular organisms, as discussed in Chapter 12. A bacterium of major industrial importance is *Agrobacterium tumefaciens*, a tumor-inducing plant pathogen that conducts natural genetic engineering on dicot plants **(Fig. 16.31 ⏵)**.

Long before scientists invented "recombinant DNA," *A. tumefaciens* had evolved a gene transfer system by which it induces infected plant cells to generate food molecules to feed the pathogen. This highly efficient gene transfer system is readily modified to insert genes conferring traits of interest, such as herbicide resistance, into plant genomes.

Tumorigenic strains of *A. tumefaciens* possess a special plasmid for engineering plant cells, called the **Ti plasmid** (tumor-inducing plasmid). The Ti plasmid is transferred into a plant cell through a process mediated by bacterial proteins, similar to conjugation (see Section 9.2). The Ti plasmid includes the *vir* operon, which encodes virulence genes, as well as T-DNA (transferable DNA), a sequence of genes that will be transferred to the host plant and recombined into its genome. T-DNA encodes tumor induction genes, as well as enzymes for biosynthesis of a carbon and nitrogen source called an opine. Opines are specialized amino acids made by a one-step synthesis from arginine, typically by amination of a central metabolite such as pyruvate or 2-oxoglutarate. A given strain of *A. tumefaciens* typically provides one type of opine synthesis enzyme and has the ability to metabolize the corresponding opine.

The *vir* gene products from the Ti plasmid detect the presence of a plant host and stimulate plasmid transfer (see **Fig. 16.31**). In the cell envelope, VirA protein detects a chemical signal from a wounded plant, which is capable of being infected. VirA then activates VirG to induce expression of other *vir* genes, encoding endonucleases VirD1 and VirD2. The VirD endonucleases cleave the left and right ends of the T-DNA and direct its transfer into the plant cell. Within the plant cell nucleus, the T-DNA becomes integrated into the plant genome, where it induces opine production. The opine-producing cells proliferate, forming a tumor.

For industrial use, a recombinant strain of *A. tumefaciens* has the Ti plasmid divided into two separate plasmids. One contains the *vir* operon conducting DNA transfer; the other contains T-DNA with its left and right ends intact but most of its genes substituted by a desired recombinant gene. Upon infection, the virulence system induces transfer of the T-DNA to the plant cell without any tumor-inducing genes, allowing genomic integration of the recombinant gene without tumor induction or opine production.

> **THOUGHT QUESTION 16.11** Why would an herbicide resistance gene be desirable in an agricultural plant? What long-term problems might be caused by microbial transfer of herbicide resistance genes into plant genomes?

Baculovirus in insects. Insect viruses such as baculovirus provide a surprisingly practical means of generating large amounts of product. The larvae, or caterpillars, grow rapidly on inexpensive substrates, accumulating large

Natural gene transfer

Industrial gene transfer

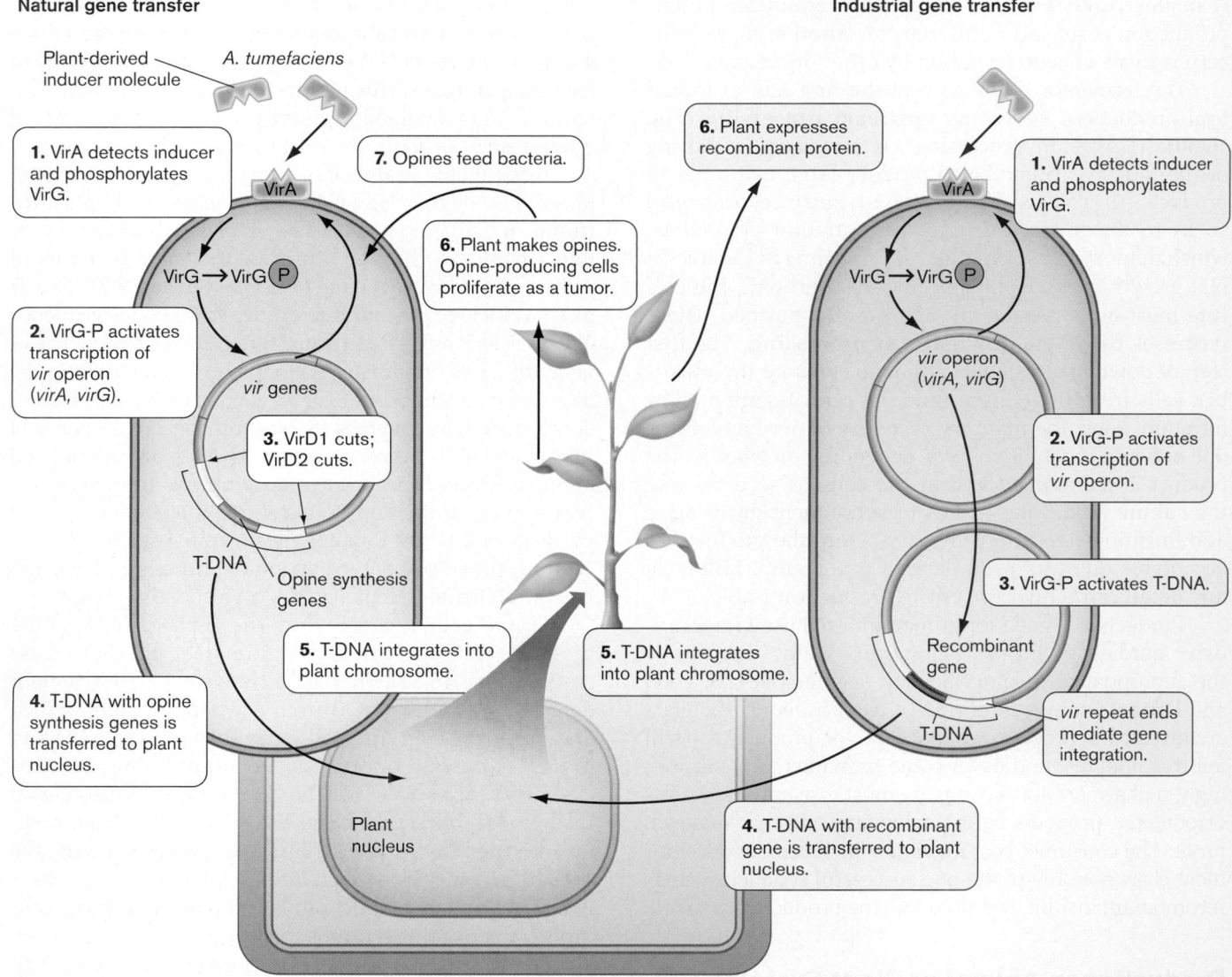

Figure 16.31 *Agrobacterium tumefaciens*: a natural gene transfer vector for plants.   Left: *A. tumefaciens* transfers T-DNA containing opine synthesis genes into a plant, which then produces opines to feed the bacteria. Right: *A. tumefaciens* can be engineered to transfer T-DNA containing a recombinant gene of interest into the plant genome. A recombinant strain of *A. tumefaciens* has the Ti plasmid divided into two separate plasmids: one containing the *vir* operon conducting DNA transfer, the other containing T-DNA with most of its genes substituted by a desired recombinant gene. ▶️ *microbiology2.com/animations*

quantities of biomass. The most effective host is generally a strain of an aggressive pest such as *Trichoplusia ni*, the cabbage looper, an insect pest that feeds on the underside of cabbage leaves. Virus-infected caterpillars are used to produce human proteins such as antibodies, growth factors, and antimicrobial peptides.

A baculovirus expression system (**Fig. 16.32**) was developed by insect virus researchers Max Summers and Gale Smith at Texas A&M University. This expression system involves two kinds of microbial vectors: a bacterial plasmid carrying an engineered product gene, and an insect virus. Baculoviruses are circular double-stranded DNA viruses that can infect a wide range of insects and insect cell lines in tissue culture. In the baculovirus

expression system, the gene of interest (encoding, for example, a vaccine antigen) is first cloned into a standard bacterial plasmid, called the transfer vector. The transfer vector contains an antibiotic resistance gene, a cloning site for gene insertion, and a promoter from the polyhedrin gene, a baculovirus gene promoter that induces high-level expression in infected insect cells. The transfer vector plus linearized baculovirus DNA are cotransfected (introduced) into insect cells in tissue culture. For the virus to reproduce, it must recombine with the transfer vector, restoring the circularity of its genome. All the progeny virus particles now contain the recombinant gene and, when replicated, they will express it to make high levels of recombinant protein.

Baculovirus expression system for eukaryotic gene products

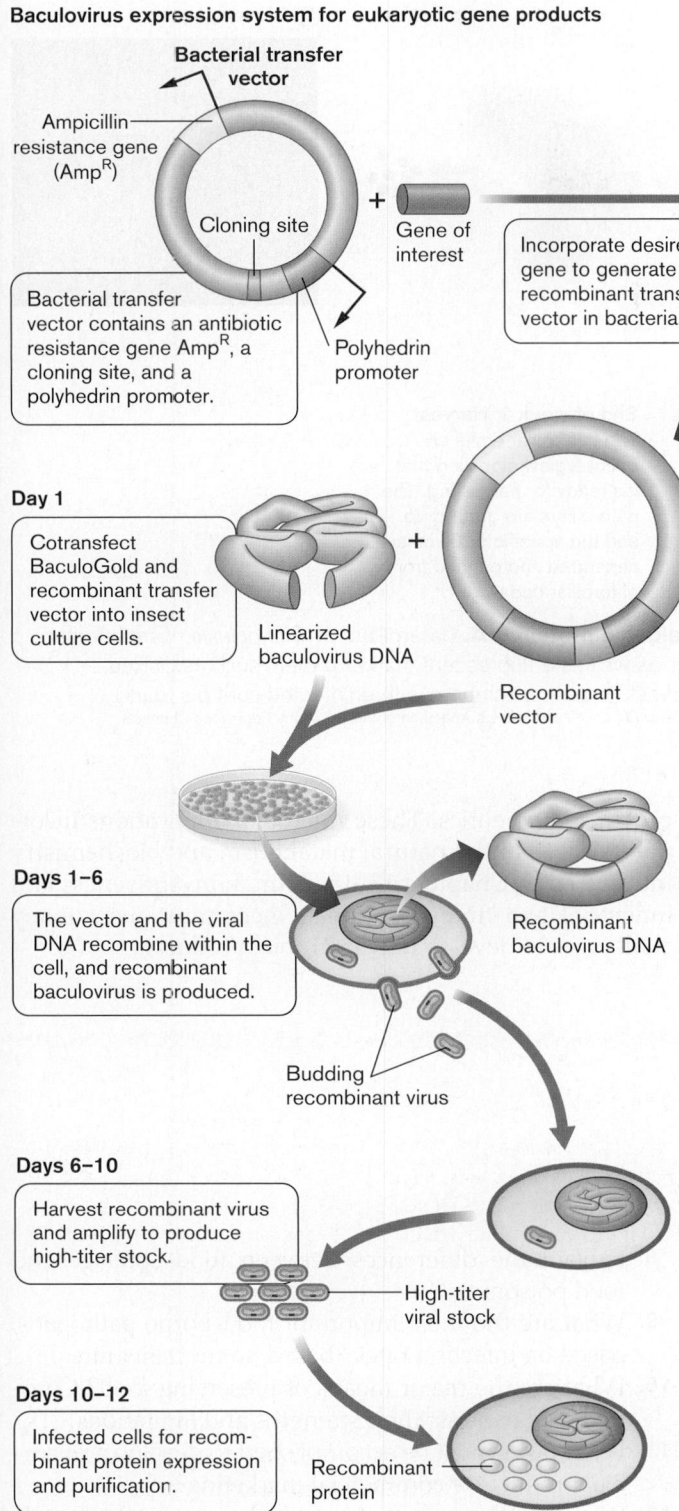

Day 1

Cotransfect BaculoGold and recombinant transfer vector into insect culture cells.

Days 1–6

The vector and the viral DNA recombine within the cell, and recombinant baculovirus is produced.

Days 6–10

Harvest recombinant virus and amplify to produce high-titer stock.

Days 10–12

Infected cells for recombinant protein expression and purification.

Figure 16.32 Baculovirus expression system for production in insects. A gene cloned in a bacterial transfer vector is recombined with an insect virus to express a protein product in an insect tissue culture. *Source*: BD Biosceinces.

product gene is sprayed into each chamber. The caterpillar eats the virus with its food, becoming infected. The recombinant virus then directs synthesis of its product, using the highly efficient promoter from the viral polyhedrin gene. In addition, a recombinant gene expresses a fluorescent marker protein such as DsRed. When the caterpillars glow with fluorescence, they are producing the desired protein and are ready for harvesting.

TO SUMMARIZE:

- **Industrial microbiology** includes the production of vaccines and clinical devices, industrial solvents and pharmaceuticals, and genetically modified plants and animals.
- **Microbial product molecules** may be indigenous, or they may be cloned from nonmicrobial sources. Thermophiles and psychrophiles are particularly important sources of new strains with potentially interesting new properties.
- **Microbial products must be competitive** with alternative technologies. Developing a competitive new molecular product requires identifying a useful molecule, isolating and engineering a strain to produce it, scaling up for production in quantity, developing a business plan, and testing for safety.
- **Bioprospecting** is the mass screening of new microbial strains for potentially valuable protein and small-molecule products.
- **Industrial strains**, commonly *Escherichia coli* or *Bacillus subtilis*, are used to incorporate the newly discovered genes into a host microbe.
- **Upstream processing** refers to the culturing of the industrial microbe to produce large quantities of product. **Downstream processing** involves product recovery and purification.
- **Posttranscriptional processing of human or plant genes** may be facilitated by use of a transgenic source or a virus-infected animal host.

Protein expression requires growth of the recombinant virus in insect caterpillars. A remarkably efficient system of caterpillar protein production was developed by the company Chesapeake PERL, Inc. (**Fig. 16.33**). In the protein production system, *Trichoplusia* caterpillars hatch from eggs and grow within chambers full of food substrate. A recombinant baculovirus containing the desired

A.

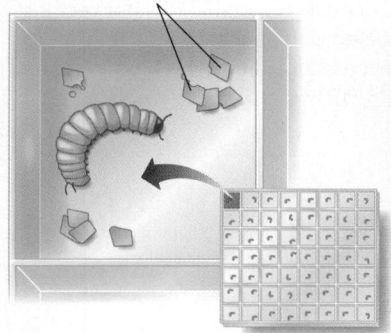
Food containing baculovirus

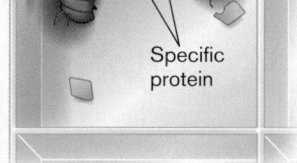
Specific protein

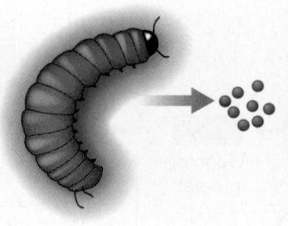

B. Glowing larvae
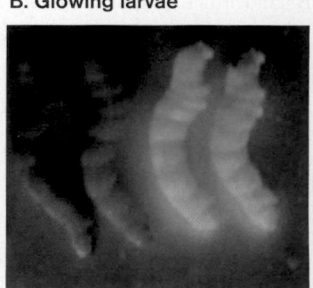
Chesapeake PERL, Inc.

Week 1: Infection
The caterpillar hatches from the egg. It sits amid food in a container made of 56 individual cells, each of which contains one caterpillar. At week's end, a recombinant insect virus encoding a specific protein is sprayed into each cell.

Week 2: Production
As the caterpillar grows, it eats the virus-infested food and becomes infected. The recombinant virus directs it to produce the specific protein, as well as a protein that causes it to glow.

End of week 2: Harvest
The caterpillar emits an intense glow signaling that it's ready for harvesting. The caterpillars are ground up, and the specific proteins are separated and purified from caterpillar cadavers.

Figure 16.33 Insect caterpillars are engineered to produce a valuable protein. A. Caterpillars of *Trichoplusia ni* are infected with recombinant baculovirus, which produces a desired protein product, as well as a fluorescent marker protein such as DsRed, indicating successful infection. **B.** The infected caterpillars ready for harvest fluoresce; compare with uninfected controls (dark).
Source: Photo courtesy of Chesapeake PERL, Inc., using a DsRed recombinant virus provided by Kevin O'Connell and Patricia Anderson of the U.S. Army Edgewood Chemical Biological Center.

Concluding Thoughts

Overall, we see how microbes from our environment contribute products for human use, from foods and vitamin supplements to industrial enzymes and highly specific protein therapeutics. These microbial applications incorporate much of the natural metabolism and biochemistry introduced in Chapters 13–15. Human inventiveness and industrial and financial management must meet many challenges to develop new microbial products.

CHAPTER REVIEW

Review Questions

1. What kinds of microbes are consumed as food? Why are most bacteria inedible for humans?
2. What are the advantages of fermentation for a food product?
3. What are the differences between traditional fermented foods and commercial fermented foods?
4. How do acidic fermentations contribute to the formation of different kinds of cheeses? What is the role of different kinds of metabolism performed by different microbial species?
5. Compare and contrast acidic and alkaline fermentation processes. What different kinds of foods are produced?
6. Compare and contrast the role of ethanolic fermentation in bread making and winemaking.
7. How do relatively minor fermentation reactions contribute to the flavor of food?
8. Explain the differences between food spoilage and food poisoning.
9. What are the most important food-borne pathogens, based on infection rates? Based on mortality rates?
10. What are the major means of preserving food? Compare and contrast their strengths and limitations.
11. What tasks must be accomplished to develop a microbial product for commercial marketing?
12. What are the sources of potential new microbial products? What is the difference between a source strain and an industrial strain?
13. Explain upstream processing and downstream processing.
14. Explain why genes encoding industrially useful products may be transferred into plant or animal systems for production. What are the roles of microbes in these systems?

Thought Questions

1. In cheese production, how do different kinds of fermenting microbes generate different flavors?
2. If you were a food safety regulator, which pathogen on the list in Table 16.3 would you consider your top priority? Defend your answer based on factors such as numerical incidence of infection, severity of disease, and economic losses due to illness.
3. Suppose you undertake industrial production of a recombinant glycoprotein for human therapy. Would you produce your product in bacteria, in yeast, or in caterpillars? Explain the advantages and limitations of your choice.

Key Terms

alkaline fermentation (587)
antimicrobial agent (601)
bioprospecting (612)
cheddared (590)
cheese (589)
curd (589)
downstream processing (615)
endophyte (605)
ethanolic fermentation (587)
fermented food (586)
food poisoning (601)
food spoilage (601)
freeze-drying (608)
fruiting body (584)

heterolactic fermentation (587)
indigenous microbiota (587)
industrial fermentor (613)
industrial microbiology (609)
industrial strain (613)
injera (597)
kimchi (593)
lactic acid fermentation (587)
leaven (596)
lyophilization (608)
malolactic fermentation (600)
miso (593)
natto (594)
nori (585)

pathogenicity island (605)
primary recovery (615)
propionic acid fermentation (587)
putrefaction (601)
rancidity (601)
ripening (590)
single-celled protein (586)
sourdough (597)
starter culture (590)
tempeh (592)
Ti plasmid (615)
upstream processing (615)
whey (589)
yogurt (589)

Recommended Reading

Centers for Disease Control and Prevention. 2008. Outbreak of *Listeria monocytogenes* infections associated with pasteurized milk from a local dairy—Massachusetts, 2007. *Morbidity and Mortality Weekly Report* **57**:1097–1100.

Centers for Disease Control and Prevention. 2009. Multistate outbreak of *Salmonella* infections associated with peanut butter and peanut butter–containing products—United States, 2008–2009. *Morbidity and Mortality Weekly Report* **58**:1–6.

Chang, Shu-Ting, and Philip G. Miles. 2004. *Mushrooms: Cultivation, Nutritional Value, Medicinal Effect, and Environmental Impact.* 2nd ed. CRC Press, New York.

Doyle, Michael P., and Larry R. Beuchat (eds.). 2007. *Food Microbiology: Fundamentals and Frontiers.* 3rd ed. ASM Press, Washington, DC.

Giraffa, Giorgio. 2004. Studying the dynamics of microbial populations during food fermentation. *FEMS Microbiological Reviews* **28**:251–260.

Hui, Y. H., Lisbeth Meunier-Goddick, Åse S. Hansen, Jytte Josephsen, Wai-Kit Nip, et al. (eds.). 2004. *Handbook of Food and Beverage Fermentation Technology.* Marcel Dekker, New York.

Marilley, L., and M. G. Casey. 2004. Flavours of cheese products: Metabolic pathways, analytical tools and identification of producing strains. *International Journal of Food Microbiology* **90**:139–159.

Mills, David A., Helen Rawsthorne, C. Parker, D. Tamir, and K. Makarova. 2005. Genomic analysis of *Oenococcus oeni* PSU-1 and its relevance to winemaking. *FEMS Microbiological Reviews* **29**:465–475.

Schwan, Rosane F. 1998. Cocoa fermentations conducted with a defined microbial cocktail inoculum. *Applied Environmental Microbiology* **64**:1477–1483.

Schwan, Rosane F., and Alan E. Wheals. 2004. The microbiology of cocoa fermentation and its role in chocolate quality. *Critical Reviews in Food Science and Nutrition* **44**:205–221.

Steinkraus, Keith H. (ed.). 1995. *Handbook of Indigenous Fermented Foods.* 2nd ed. Marcel Dekker, New York.

For further study and review, please visit StudySpace at **microbiology2.com**.

Part 4

Microbial Diversity and Ecology

Karl Stetter collecting samples from volcanic hot springs in Siberia.

AN INTERVIEW WITH

KARL STETTER: ADVENTURES IN MICROBIAL DIVERSITY LEAD TO PRODUCTS IN INDUSTRY

Karl Stetter, professor at the University of Regensburg in Germany, was the first person to discover living organisms growing at temperatures above 100°C. Hyperthermophiles grow in hot springs and at thermal vents in the ocean floor. Such organisms possess exceptionally stable enzymes with enormous potential uses in industry. In 2004, Stetter cofounded the Diversa Corporation (now the Verenium Corporation) to isolate and develop such enzymes. For example, Diversa developed an enzyme for bleaching paper from a microbe isolated at a geyser in Russia. Such industrial enzymes often replace polluting processes with processes that are more environmentally friendly.

Why did you decide to make a career in microbiology?
I had been fascinated about microbes since I was a little child, when my parents had given me a simple microscope.

What enabled you to discover the hyperthermophile *Pyrodictium occultum*?
Just believe the laws of physics and chemistry and never believe prejudices (such as, "If it would exist, other people would have discovered it already"). Finding life above 100°C was a big surprise. I had to be perfectly sure about my results and confirmed them several times. It finally paid off, and this topic became my first *Nature* paper.

Would you describe the discovery that led to your 1982 paper ("Ultrathin mycelia-forming organisms from submarine volcanic areas having an optimum growth temperature of 105°C," *Nature* 300[1982]:258–260)?

This paper raised a lot of public interest; I even had to give my first (completely unexpected!) press conference. Here is how my discovery came about: During my family holidays in 1981, in order to check my hypothesis of life possibly existing above 100°C (under overpressure to keep the water liquid), I did some scuba sampling at the shallow hot vents off the coast of Vulcano, Italy. This had been my first attempt in underwater sampling. To do this, I developed some special sampling gear, which I used and improved on all my later diving expeditions (electronic waterproof thermometers with sturdy, very long probes, syringes with enlarged inlets, and so on). During our holidays, my wife and my little daughter Sabine had fun supporting me in our little rubber boat, processing the still-boiling hot samples that I brought up to the surface. The samples were kept anaerobic by injecting a reducing agent, a mixture of sodium dithionite and sodium sulfide.

Back in the Regensburg lab, together with my technician, we began extensive cultivation attempts, which lasted for several months before the exciting bugs finally grew up. For the first time, from several samples at 100°C incubation temperature, disk-shaped cells grew up within two days. As we found out later, their fastest growth was in superheated water at 105°C, and their upper limit for growth temperature was even at 110°C!

What is it like for students working with extremophiles in the laboratory? How do you modify laboratory practice for the culture and observation of these organisms?
It's great fun for students to culture hyperthermophiles, as it is for me. It´s great to culture organisms that had been thought to be impossible. Of course, some safety briefing is essential to avoid lab accidents. Every culture bottle under pressure in my lab is housed in an explosion-safe cage. It's an incredible thrill being the first to see a novel bug growing!

What have we learned from the genome sequences of hyperthermophiles? What do they tell us about microbial evolution?
This question would require a whole book of answers. Here is my answer

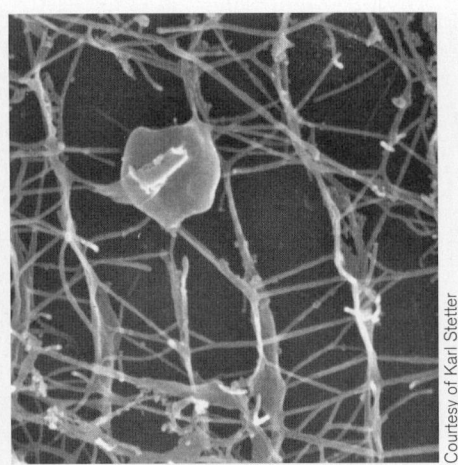

***Pyrodictium occultum*, the first hyperthermophile discovered.** The disk-shaped archaeal cells form complex networks of interconnecting filaments. (Cell diameter 2–4 μm.)

in a nutshell: Hyperthermophiles harbor rather small genomes. Interestingly, their DNA double helix is not stabilized by an elevated GC/AT content of the DNA. Instead, all hyperthermophiles possess reverse gyrase to stabilize DNA. The enzymes and other proteins are structurally related to their homologs in mesophilic (moderate-growth-temperature) organisms.

A lot of horizontal gene transfer goes on in hyperthermophiles. The information-processing machinery in hyperthermophiles (small-subunit rRNA, and so on) appears to be still rather primitive. In the small-subunit rRNA–based tree of life, hyperthermophiles include all the deep-branching lineages of organisms that diverged earliest.

You also discovered the unusual hyperthermophilic symbiont *Nanoarchaeum*, which lives at marine vents at 70°C–98°C associated with another archaeon, *Ignicoccus*. *Nanoarchaeum* cells are only 400 nm in diameter and lack cell walls, their cell membranes covered only by a protein S-layer. They require cell-cell contact with an actively growing *Ignicoccus* cell in order to grow.

What do we know about how these two organisms affect each other's growth?
Nanoarchaeum definitively needs *Ignicoccus* to grow. But *Ignicoccus* grows without *Nanoarchaeum*. This is all just under laboratory conditions. So far, we are pretty much away from a deeper understanding of this exciting symbiosis, which most likely has existed since the dawn of life.

Why is the Diversa company so interested in your extremophiles?
From its onset, Diversa has been interested in extremophiles, with the view of developing sturdy enzymes as industrial catalysts. Diversa has already developed several promising products, such as stock food additives, enzymes for bleaching paper pulp, and oil recovery enzymes.

What do you think are the most exciting areas for students entering microbiology today? What advice do you have for today's students?
I think our hyperthermophile research at present is only the tip of a damned hot iceberg! But truly, there are so many exciting questions open in all kinds of fields in microbiology. Students should follow their passion and try to do an excellent job in whatever field they choose. And if they don't mind burning their fingers once in a while . . . well, then hyperthermophiles may be the right choice for them!

After sharing your early discoveries, how does your family relate to your work today?
My wife Heidi is a microbiologist and a high school teacher of chemistry and biology. Our oldest daughter Sabine is now an attorney in criminal law. Our son Florian is a very successful actor, on the stage and in movies.

What are your pursuits outside science?
In addition to my number one hobby—cultivating hot bugs—I grow tropical orchids in my greenhouse (about 500 species) and drive a hot car (Ferrari) at 190 miles per hour on the German Autobahn.

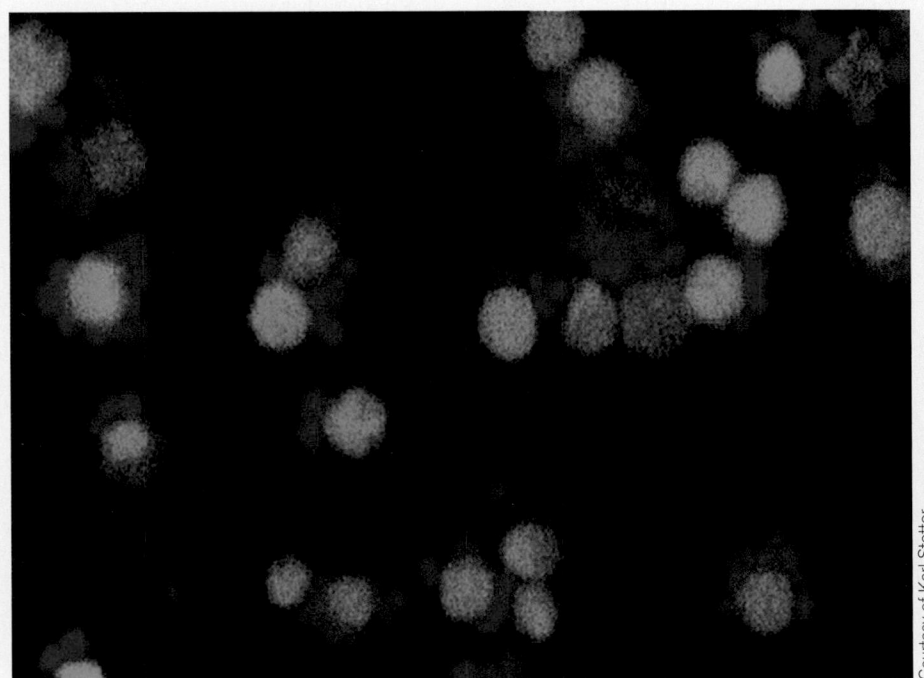

Symbiotic hyperthermophiles. This fluorescence micrograph reveals *Nanoarchaeum equitans* (red), the smallest archaeon known, as an obligate symbiont of another archaeon, *Ignicoccus* (green). These archaea were found at submarine thermal vents north of Iceland.

Chapter 17
Origins and Evolution

Microbial life may have appeared as early as 3.8 billion years ago, soon after our planet Earth formed out of dust from the young Sun. Since then, microbes have diverged into new forms adapted to diverse ways of life, from psychrophiles beneath the ice of Antarctica to anaerobes in the human colon. Descendants of early microbes include all living plants and animals, including ourselves. How did microbes originate on Earth? How did their metabolism shape the chemistry of Earth's crust and atmosphere?

DNA sequence–based taxonomy reveals vast numbers of previously unknown microbial species. How do we classify them? Some evidence, particularly the sequences of rRNA genes, supports a traditional branched tree of life. Other gene sequences, however, support a more complex picture, including gene transfer between distant branches of the tree. Many microbes have evolved mutualistic partnerships with other living things, such as bacteria that fix nitrogen within plant cells. Still other microbes evolve as pathogens whose adaptations enable them to live at the expense of their host and cause disease.

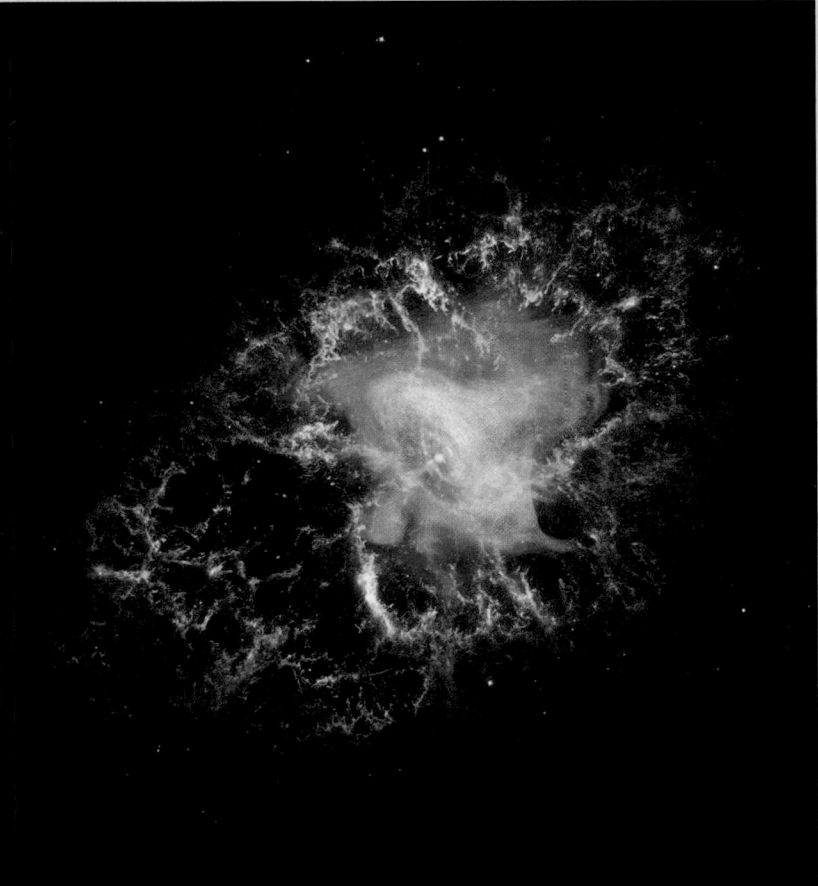

Current Research Highlight

Elements of life originated in exploding stars called supernovas, which left expanding gas clouds such as the Crab Nebula (shown here). The elements formed fundamental organic molecules of life, such as amino acids found in meteorites. Surprisingly, meteorite amino acids show the same left-handed chiral form that predominates in all life on Earth. Meteorite amino acids may have seeded Earth's prebiotic chemistry that led to the origin of life. *Source:* Daniel Glavin and Jason Dworkin. 2009. *PNAS* **106**:5487. X-ray: NASA/CXC/SAO/ F. Seward; Optical: NASA/ESA/ASU/J. Hester & A. Loll; Infrared: NASA/JPL-Caltech/Univ. Minn./ R. Gehrz.

For centuries, observers of the natural world have wondered where life came from. As early as 1802, the naturalist Erasmus Darwin, grandfather of Charles Darwin, wrote:

Organic life beneath the shoreless waves
Was born and nurs'd in ocean's pearly caves;
First forms minute, unseen by spheric glass,
Move on the mud, or pierce the watery mass;
These, as successive generations bloom,
New powers acquire and larger limbs assume.

Thus, nineteenth-century biologists developed the idea that all living organisms had evolved from microbes, perhaps even from cells too small to be seen with the "spheric glass" of a microscope. Even without modern tools of isotope analysis and genetics, thoughtful observers recognized the overwhelming commonalities among all living cells, such as the membrane-enclosed compartment of cytoplasm and common metabolic pathways such as sugar metabolism. Today, lines of evidence from geology, biochemistry, and genetics overwhelmingly support the microbial origin of life.

What did early life look like? The earliest forms of life for which we have clear fossil evidence are bacterial communities called **stromatolites** (**Fig. 17.1A**). A stromatolite is a bulbous mass of sedimentary layers of limestone (calcium carbonate, $CaCO_3$) accreted by microbes over years, even centuries. Within the outer layers, microbes grow as a "microbial mat," a kind of biofilm. The outermost layers of the mat contain oxygenic phototrophs, such as diatoms and filamentous cyanobacteria, that exude bubbles of oxygen. A few millimeters below the surface, red light supports bacteria photolyzing H_2S

to sulfate, which is then reduced by still lower layers of sulfate-reducing bacteria. Stromatolites today survive mainly in isolated pools whose high salt concentration excludes predators, as in Hamlin Pool, Shark Bay, Australia. But 3 billion years ago, without predators, stromatolites covered shallow seas all over Earth. Fossil stromatolites appear in ancient rock formations such as the 3.4-billion-year-old Strelley Pool Chert, a sedimentary formation in Pilbara Craton, Australia (**Fig. 17.1B**). Their layers preserve the wavy form of the microbial mats, remarkably similar to living stromatolites. Stromatolites are studied by NASA for clues as to how life may appear on other planets.

Questions about life's origin have long sparked controversy. In eighteenth-century Europe, the idea of "spontaneous generation" of microbes was considered so dangerous that priests and monks performed scientific experiments to disprove it (described in Chapter 1). The priests ultimately won their scientific point: All microbes today have "parents"—that is, preexisting microbes. But these experiments did not address the origin of the first living cells or how early life gave rise to modern species. This concept was so controversial that in the early twentieth century, laws forbade teaching it in American public schools. In 1929, such a law in Tennessee was put to the test by high school teacher John Scopes in what became the famous Scopes trial (**Fig. 17.2**). According to a 14-year-old student who testified at the trial, Scopes taught that "there was a little germ of one cell organism formed, and this organism kept evolving . . . and from this was man." A large body of evidence now supports the view that microbes living more than 3 billion years ago gave rise to all animals and plants, as well as to all modern microbes.

A.

B.

Figure 17.1 Stromatolites: an ancient life-form. A. Cyanobacteria form colonial stromatolites, the present-day structures believed to most closely resemble the earliest forms of life on Earth. Shark Bay, Western Australia. **B.** Section through a 3.4-billion-year-old stromatolite from the Strelley Pool Chert, Pilbara Craton, Australia.

Figure 17.2 Objecting to microbial ancestry. In 1925, outside the Scopes trial in Dayton, Tennessee, demonstrators opposed the teaching that all life—including humans—evolved from a microbe.

Chapter 17 explores the evidence from geochemistry and molecular biology for the nature of the earliest cells, as well as the challenges in interpreting data from so long ago. We explore how molecular techniques reveal deep similarities among all life-forms: the core macromolecular apparatus of DNA, RNA, and proteins, expressed by genomes that share a common set of ancestral genes. We focus on three major concerns of microbial evolution:

- The origin of life on Earth and the nature of the earliest cells.
- The divergence of microbes from common ancestors.
- Gene transfer and symbiosis as agents of evolution.

17.1 Origins of Life

Before the first cells could evolve, several fundamental conditions were required. These conditions included:

- **Essential elements.** Because all life on Earth is composed of molecules, the origin of life required the fundamental elements that compose organic molecules.
- **Continual source of energy.** The generation of life requires continual input of energy, which ultimately is dissipated as heat. The main source of energy for life is nuclear fusion reactions within the Sun.
- **Temperature range permitting liquid water.** Above 150°C, life's macromolecules fall apart; below the freezing point of water, metabolic reactions cease. Maintaining the relatively narrow temperature range conducive to life depends on the nature of our Sun,

our planet's distance from the Sun, and the heat-trapping capacity of our atmosphere.

Elements of Life

For life to arise and grow, elements such as carbon and oxygen needed to be available on Earth. The planet Earth coalesced during formation of the solar system 4.5 Gyr (gigayears, or billions of years) ago. Central to the solar system is our Sun, a "yellow" star of medium size and surface temperature (5,770 K). The Sun's surface temperature generates electromagnetic radiation across the spectrum, peaking in the range of visible light. As we learned in Chapter 13, the photon energies of visible light are sufficient to drive photosynthesis but not so energetic that they destroy biomolecules. Thus, the stellar class of our Sun makes organic life possible.

> **NOTE:** In geological description, a billion years (10^9) is a gigayear, or Gyr. A million years (10^6) is a megayear, or Myr.

The Sun's surface temperature and luminosity are generated by nuclear fusion reactions in which hydrogen nuclei fuse to form helium nuclei. (Be careful to distinguish nuclear reactions, involving nuclei, from chemical reactions, involving electrons.) Besides hydrogen and helium, 2% of the solar mass consists of heavier elements, such as carbon, nitrogen, and oxygen, as well as traces of iron and other metals—elements that compose Earth, including its living organisms. Where did these heavier elements come from? To answer this question, we must look to other stars in the universe at different stages of their development (**Fig. 17.3**).

Elements of life formed within stars. Throughout the universe, young stars such as our Sun fuse hydrogen to form helium. As stars age, they use up all their hydrogen. With hydrogen gone, the aging star contracts and its temperature rises, enabling helium nuclei to fuse, forming carbon (see **Fig. 17.3**). Carbon drives a cyclic nuclear reaction, the CNO cycle, to form isotopes of nitrogen and oxygen. Subsequent nuclear reactions generate heavier elements through iron (Fe). Thus, the major elements of biomolecules were formed within stars that aged before our solar system was born.

The later nuclear reactions of aging stars generate heavier nuclei up to iron. The aging star expands, forming a red giant (see **Fig. 17.3**). When a star of sufficient mass expands (a supergiant), it explodes as a **supernova**. The explosion of a supernova generates in a brief time all the heaviest elements and ejects the entire contents of the star

Life cycle of a massive star

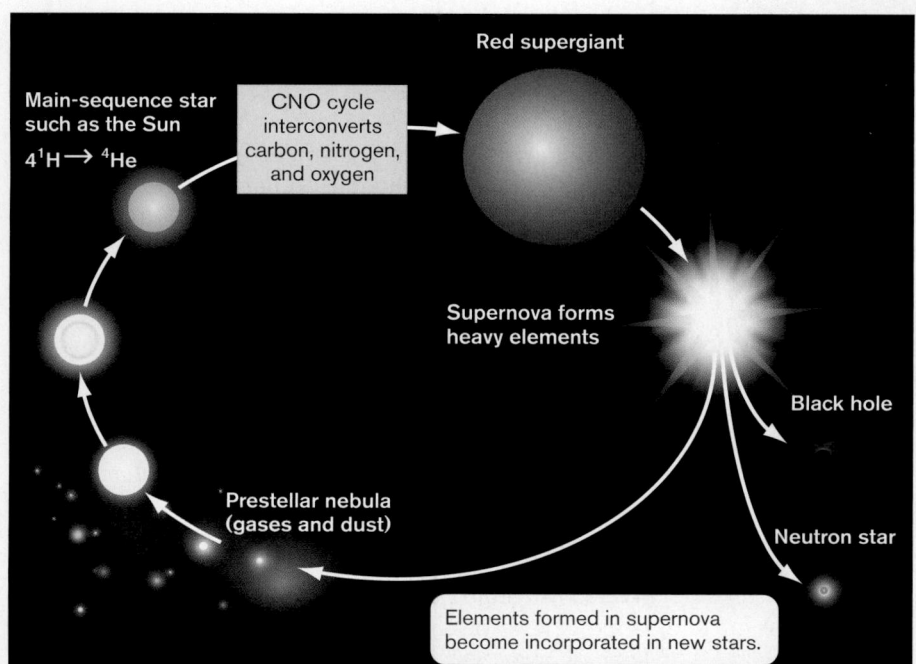

Figure 17.3   Stellar origin of atomic nuclei that form living organisms.   In young stars, hydrogen nuclei fuse to form helium. In older stars, fusion of helium forms carbon, nitrogen, oxygen, and all the heavier elements up to iron. Massive stars explode as supernovas, spreading all the elements of the periodic table across space. These elements are picked up by newly forming stars, such as our own Sun.

at near light speed. Billions of years before our Sun was born, the first stars aged and died, spreading all the elements of the periodic table across the universe. Some of these elements coalesced with our Sun and formed the planets of our solar system. In effect, all life on Earth is made of stardust, the remains of stars long gone.

> **THOUGHT QUESTION 17.1**   What would have happened to life on Earth if the Sun were a different stellar class, substantially hotter or colder than it is?

Elemental composition of Earth. When our solar system formed, individual planets coalesced out of matter attracted by the force of gravity. Because of Earth's small size, most of the hydrogen gas escaped Earth's gravity very early. The most abundant dense component of Earth was iron (**Fig. 17.4**). Much of Earth's iron sank to the center to form the core. The core is surrounded by a mantle, composed primarily of iron combined with less dense crystalline minerals such as silicates of iron and magnesium, $(Fe,Mg)_2SiO_4$. The mantle is coated by Earth's thin outer crust. The crust is composed primarily of silicon dioxide, SiO_2, also known as quartz or chert. Crustal rock includes smaller amounts of numerous minerals, including the carbonates and nitrates that provided the essential elements for life. Overall, the crust shows a redox gradient, reducing in the interior and oxidizing at the surface.

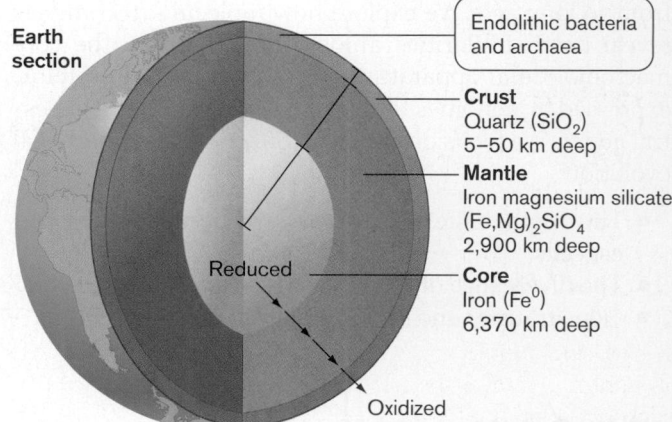

Figure 17.4   Geological composition of Earth.   The cross section of Earth shows the core, the mantle, and the thin outer crust. The core and mantle are rich in iron; oxygen content increases toward the crust. The crust is composed primarily of silicates such as quartz (SiO_2). Crustal rock supports endolithic bacteria and archaea.

The crust provides a habitat for microbes down to surprising depths, such as within gold mines excavated to 3 kilometers (km). Some endolithic microbes (microbes living within rock) metabolize by oxidizing electron donors generated through decay of radioactive metals. The discovery of endolithic organisms deep in the crust was of great interest to NASA scientists seeking life on Mars. Of the planets, Mars most closely resembles Earth in geology and distance from the Sun; and its crust might provide a habitat similar to Earth's.

The outer surface of the crust supports the remainder of the **biosphere**, the sum total of all life on Earth. The biosphere generates oxidants (electron acceptors), most notably O_2. Oxygen-breathing organisms can live only on the outer surface, where O_2 is produced by photosynthesis.

Earth's atmosphere. From the crust and the mantle of early Earth, volcanic activity released gases such as carbon dioxide and nitrogen, which formed Earth's first atmosphere, while volcanic water vapor formed the ocean. The composition of this first atmosphere, before life evolved, looked much like that of Mars: thin, about 1% as dense as that of Earth today, consisting primarily of CO_2. But unlike Mars, Earth developed living organisms that filled the atmosphere with gaseous N_2 and O_2 and that continue to produce these gases today. Organisms also produce CO_2, as well as fixing it into biomass. Some CO_2 and N_2 arise from geological sources such as volcanoes, but their contribution is small compared to that of biological cycles (discussed in Chapter 22). The overall composition of Earth's atmosphere is determined by living organisms, primarily microbes.

Temperature. Another important aspect of Earth's habitat, determined by the atmospheric density and composition, is temperature. Atmospheric gases absorb light and convert the energy to heat, raising the temperature of the surface and atmosphere. This rise in temperature is known as the **greenhouse effect**. Because carbon dioxide is an especially potent greenhouse gas, the CO_2-rich atmosphere of early Earth could have heated the planet to temperatures approaching that of Venus, eliminating the possibility of life. Instead, microbial consumption of CO_2 and generation of nitrogen and oxygen gases limited Earth's surface temperature to an average of 13°C. The cooling effect may have led to an ice age, possibly forestalled by rising methane from methanogens. One way or another, the history of Earth's atmosphere is intimately related to the history of microbial evolution.

Geological Evidence for Early Life

The first period of Earth's existence, ranging from 4.5 to 3.8 Gyr ago, is designated the **Hadean eon** (**Fig. 17.5**), named for Hades, the ancient Greek world of the dead. During the Hadean eon, repeated bombardment by meteorites vaporized the oceans, which then cooled and recondensed. Meteor bombardment may have killed off incipient life more than once before living microbes finally became established. Still, scientists speculate on whether some forms of life might have survived Hadean conditions, perhaps growing 3 km below the Earth's surface. Like the dead spirits imagined by the Greeks to have populated Hades, Earth's earliest cells might have reached deep enough within the crust that they were protected from the heat and vaporization at the surface.

In 2009 a belt of rock in Ungava, Québec, was dated by radioisotope decay at 4.3 Gyr, near the time the Earth formed. Some researchers find evidence for early life in this rock, but overall the evidence for life earlier than 3.8 Gyr ago remains speculative.

The Archaean eon. The earliest geological evidence for life that is generally accepted dates to 3.8–2.5 Gyr ago, in the **Archaean eon** (**Fig. 17.5**). In the Archaean, meteor bombardment was less frequent, and the Earth's crust had become solid. The Archaean marked the first period with stable oceans containing the key ingredient of life: liquid water. Water is a key medium for life because it remains liquid over a wide range of temperatures and because it dissolves a wide range of inorganic and organic chemicals. Rock strata dating to the Archaean eon reveal the first evidence of living organisms and their metabolic processes.

> **NOTE:** The term "Archaean" refers to the earliest geological eon when life existed, whereas "archaeal" is the adjective referring to the taxonomic domain Archaea. The domain Archaea (originally, Archaebacteria) was named by Carl Woese based on his theory that the species of this domain most closely resembled the earliest life-forms of the Archaean eon. In fact, early life may have encompassed diverse traits later associated with archaea, bacteria, and eukaryotes.

How and when did living cells arise out of inert materials? Without a time machine to take us back 4 Gyr, we must rely on evidence from Earth's geology. Interpreting geology is a challenge because most forms of evidence for early life are indirect and subject to multiple interpretations. The farther back in time, the more change has occurred to the rock strata and the greater the difficulties are. One way to meet this challenge, however, is to compare the results from different kinds of evidence (**Table 17.1**). If two or more kinds of evidence (such as microfossils and isotope ratios) point to life in the same location, the conclusion is strengthened.

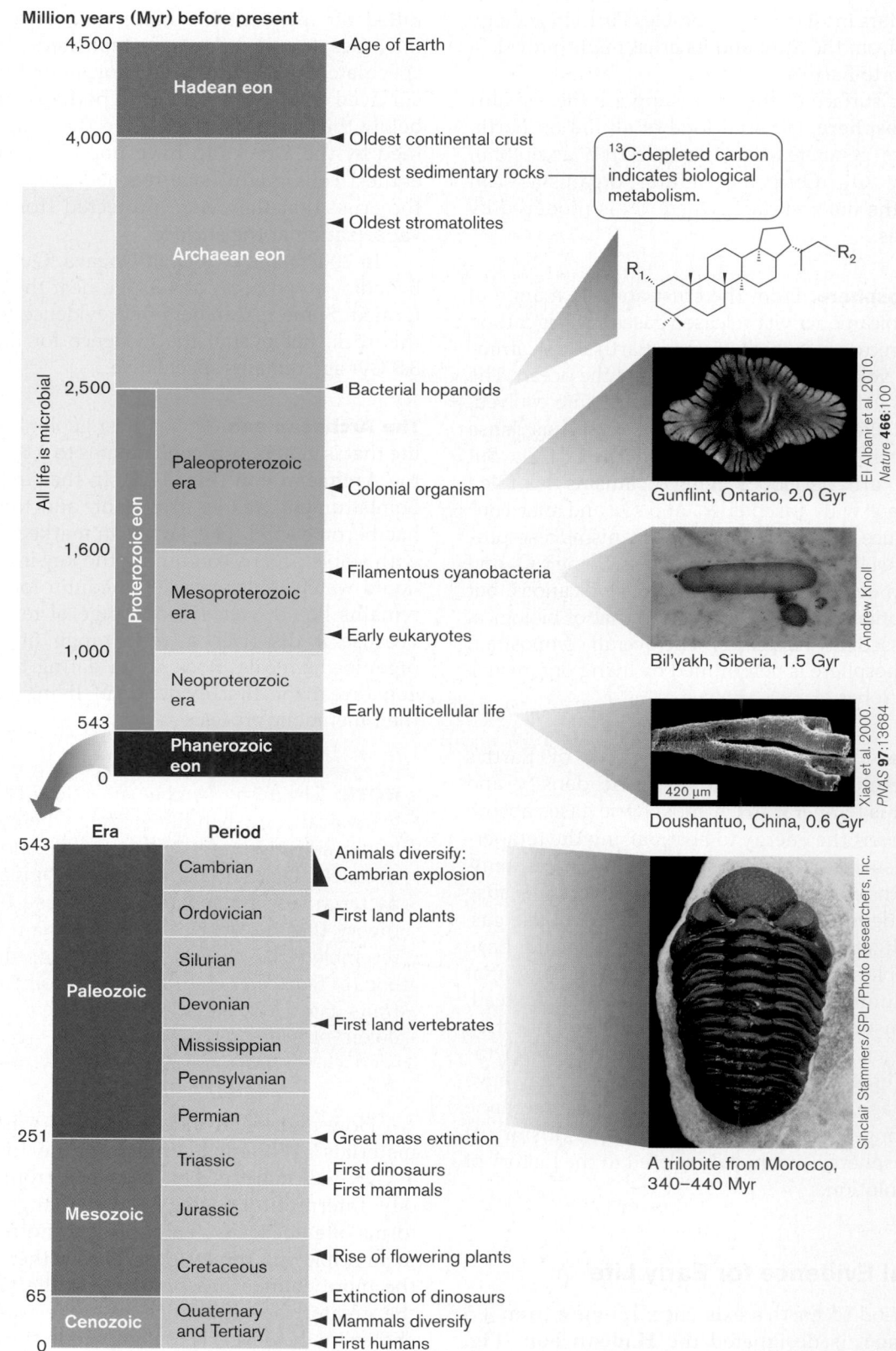

Million years (Myr) before present

4,500 — ◄ Age of Earth

Hadean eon

4,000 — ◄ Oldest continental crust
◄ Oldest sedimentary rocks
◄ Oldest stromatolites

Archaean eon

¹³C-depleted carbon indicates biological metabolism.

2,500 — ◄ Bacterial hopanoids

Paleoproterozoic era

◄ Colonial organism

Gunflint, Ontario, 2.0 Gyr

El Albani et al. 2010. *Nature* **466**:100

1,600 — ◄ Filamentous cyanobacteria

Mesoproterozoic era

◄ Early eukaryotes

Bil'yakh, Siberia, 1.5 Gyr

Andrew Knoll

1,000 —

Neoproterozoic era

◄ Early multicellular life

543 —

Phanerozoic eon

0 —

420 µm

Doushantuo, China, 0.6 Gyr

Xiao et al. 2000. *PNAS* **97**:13684

All life is microbial

Proterozoic eon

Era	Period	
543		
Paleozoic	Cambrian	◄ Animals diversify: Cambrian explosion
	Ordovician	◄ First land plants
	Silurian	
	Devonian	◄ First land vertebrates
	Mississippian	
	Pennsylvanian	
	Permian	
251		◄ Great mass extinction
Mesozoic	Triassic	◄ First dinosaurs / First mammals
	Jurassic	
	Cretaceous	◄ Rise of flowering plants
65		◄ Extinction of dinosaurs
Cenozoic	Quaternary and Tertiary	◄ Mammals diversify
0		◄ First humans

A trilobite from Morocco, 340–440 Myr

Sinclair Stammers/SPL/Photo Researchers, Inc.

Figure 17.5 Geological evidence for early life. The geological record shows evidence of microbial life early in Earth's history, 3 Gyr before the first multicellular forms.

Table 17.1 Geological evidence of early life.

Type of evidence	Advantages	Limitations
Stromatolites. Layers of phototrophic microbial communities grew and died, their form filled in by calcium carbonate or silica.	Fossil stromatolites can be observed in the oldest rock of the Archaean eon. Their distinctive shapes resemble those of modern living stromatolites.	Some layered formations attributed to stromatolites have been shown to be generated by abiotic (nonbiological) processes.
Microfossils. Early microbial cells decayed, and their form was filled in by calcium carbonate or silica. The size and shape of microfossils resemble those of modern cells.	Microfossils are visible and measurable under a microscope, offering direct evidence of cellular form.	Microscopic rock formations require subjective interpretation. Some formations may result from abiotic processes.
Isotope ratios. Microbes fix $^{12}CO_2$ more readily than $^{13}CO_2$. Thus, limestone depleted of ^{13}C must have come from living cells. Similarly, sulfate reduced by sulfate-respiring bacteria shows depletion of ^{34}S compared with ^{32}S.	Isotope ratios offer objective measurement of a highly reproducible physical quantity. They provide the best evidence for dating the earliest life. Isotope ratios generated by key biochemical reactions can calibrate the time line of phylogenetic trees.	We cannot prove absolutely that no abiotic process could generate a given isotope ratio. Isotope ratios tell us nothing about the form of early life or how it evolved.
Biosignatures. Certain organic molecules found in sedimentary rock are known to be formed only by microbes. These molecules are used as biosignatures.	Biosignatures such as hopanoids are complex molecules and highly specific to different life-forms.	A biosignature thought specific to one kind of organism may be discovered in others. In the oldest rocks, organic biosignatures are eliminated entirely by metamorphic processes.
Oxidation state. The oxidation state of metals such as iron and uranium indicates the level of O_2 available when the rock formed. Banded iron formations (BIF) suggest oxidation by microbial phototrophs that intermittently produced oxygen.	Oxidized metals offer evidence of microbial processes even in highly deformed rock.	It is hard to rule out abiotic causes of oxidation. If biological processes were the cause, the kind of metabolism is not revealed.

Stromatolites. Fossil stromatolites are layers of carbonate or silicate rock that resemble modern living stromatolites. Presumably, the fossils formed as layers of phototrophic microbial communities grew and died, their form filled in by calcium carbonate or silica. Fossil stromatolites accepted by geologists date as early as 3.4 Gyr ago (see **Fig. 17.1**). These rock formations appear remarkably similar to the layered forms of stromatolites today. The ancient rock, however, is too deformed to reveal the detailed structure of cells, and the biological origin of such fossils is questioned by some researchers.

Microfossils. The most convincing evidence for early microbial life is the visual appearance of **microfossils**, microscopic fossils in which minerals have precipitated and filled in the form of ancient microbial cells (**Fig. 17.6**). Microfossils are dated by the age of the rock formation in which they are found, which in turn is based on evidence such as radioisotope decay. Convincing microfossils need to show regular three-dimensional patterns of cells that cannot be ascribed to abiotic (nonbiological) causes.

The earliest convincing microfossils are dated at 2.0 Gyr. Microfossils dated to 2.0 Gyr include filamentous prokaryotes in the Gunflint Formation, Ontario, Canada (**Fig. 17.6A**). The Gunflint outcrops consist of chert, a kind of silicate formed by precipitation from an ancient sea. The sea was rich in carbonates and reduced iron, a good combination for redox metabolism. Some of the fossils resemble the form of filamentous iron-metabolizing bacteria today, such as *Leptothrix* species (**Fig. 17.6B**). Other microfossils dated at 2.0 Gyr resemble colonial cyanobacteria (compare **Fig. 17.6C** with **17.6D**). More recent strata, dated at 1.2 Gyr, contain larger fossil cells comparable to those of modern eukaryotes such as algae (**Fig. 17.6E** and **F**).

Microfossils

A. Filamentous prokaryotes

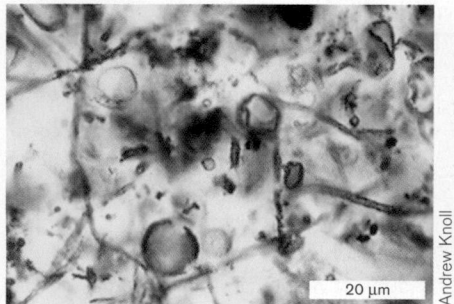

20 μm

Andrew Knoll

Modern species

B. *Leptothrix* sp.

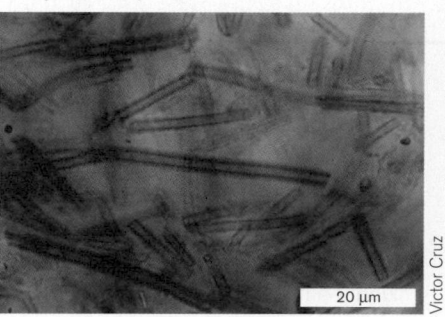

20 μm

Victor Cruz

Figure 17.6 **Microfossils compared with modern bacteria.** **A.** Filamentous prokaryotes, 2.0 Gyr, from Gunflint Formation, Ontario, Canada. **B.** Modern *Leptothrix* filamentous bacteria. **C.** Colonial cyanobacteria, about 2.0 Gyr, from Belcher Islands, Canada. **D.** Modern *Entophysalis* cyanobacteria. **E.** Filamentous algae, 1.2 Gyr, from arctic Canada. **F.** Modern red algae, a eukaryote, *Bangia* sp.

C. Colonial cyanobacteria

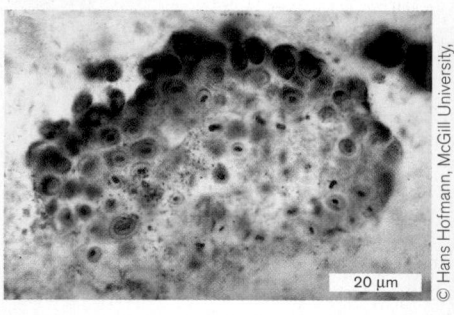

20 μm

© Hans Hofmann, McGill University, Montreal

D. *Entophysalis* sp.

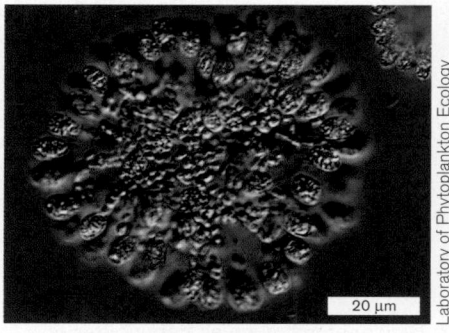

20 μm

Laboratory of Phytoplankton Ecology

E. Algae (eukaryote)

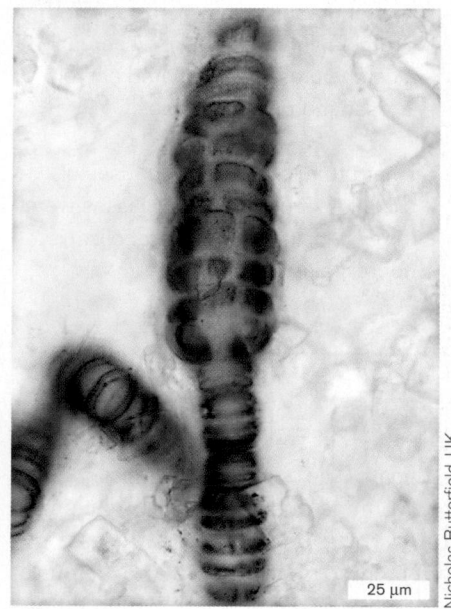

25 μm

Nicholas Butterfield, UK

F. *Bangia* sp., red algae

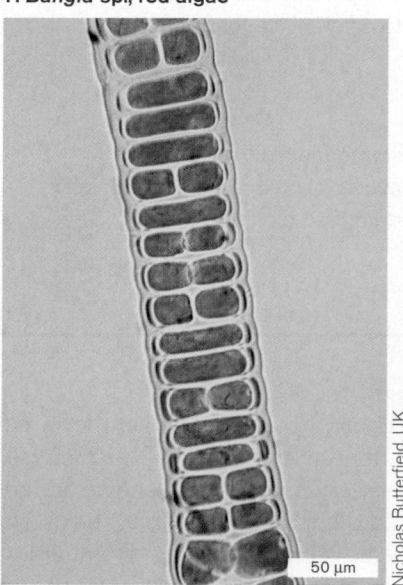

50 μm

Nicholas Butterfield, UK

If life existed in more ancient times, such as the Archaean eon, where are the microfossils? Archaean rock is metamorphic, greatly modified by temperature and pressure. The macroscopic contours of stromatolites can be identified, but microfossil interpretation is highly controversial. For example, microfossils of cyanobacteria dated to 3.85 Gyr by William Schopf in the early 1990s were accepted and described in many textbooks (**Fig. 17.7**). The 3.85-Gyr fossils have since been reinterpreted

by Martin Brazier and colleagues as nonbiogenic artifacts (caused by abiotic processes). The form of the proposed Archaean microfossils is less regular and convincing than the form of later specimens, particularly when observed at different angles not shown in the original publication.

Another kind of evidence is that of a **biosignature**, or **biological signature**, a chemical indicator of life. Biosignatures have been found that are even earlier than the oldest fossils. Their significance is limited, however, as it

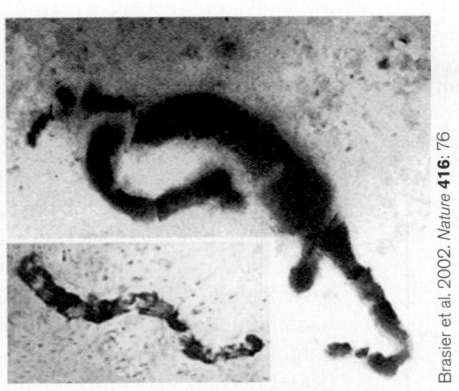

Figure 17.7 Microfossils or artifacts? Structures originally identified as cyanobacterial microfossils from Western Australia, dated to 3.85 Gyr. Further testing indicated that the structures are nonbiological artifacts.

is hard to rule out nonbiogenic explanations, so researchers seek additional, corroborating evidence based on independent principles.

Isotope ratios. An **isotope ratio** may serve as a biosignature if the ratio between certain isotopes of a given element is altered by biological activity. Enzymatic reactions, unlike abiotic processes, are so selective for their substrates that their rates may differ for molecules containing different isotopes. For example, the carbon-fixing enzyme Rubisco, found in chloroplasts, preferentially fixes CO_2 containing ^{12}C rather than ^{13}C. The carbon

dioxide fixed into microbial cells eventually is converted to calcium carbonate in sedimentary rock. Thus, the calcium carbonate deposited by CO_2-fixing autotrophs (such as cyanobacteria) shows lower ^{13}C content than calcium carbonate deposited by abiotic processes (**Fig. 17.8B**). The difference, $\delta^{13}C$, is defined by the fractional difference (in parts per thousand) between the $^{13}C/^{12}C$ ratios in a sample versus a standard inorganic rock:

$$\delta^{13}C = \frac{^{13}C/^{12}C \text{ (experimental)} - {}^{13}C/^{12}C \text{ (standard)}}{^{13}C/^{12}C \text{ (standard)}} \times 1,000$$

Typical $\delta^{13}C$ values are shown in **Figure 17.8B**. Organisms on land and sea show $\delta^{13}C$ values of −10 to −30 parts per thousand, a significant ^{13}C depletion through carbon fixation. A comparable $\delta^{13}C$ is observed in fossil fuels, which formed from plant and animal bodies decomposed by bacteria.

The most ancient mineral samples showing a substantial $\delta^{13}C$ (about −18 parts per thousand) are graphite granules in the Isua rock bed of West Greenland, dated at 3.7 Gyr. The graphite granules were analyzed by Minik Rosing, a native Greenlander at the Danish Lithosphere Centre (**Fig. 17.8A**). The graphite grains derive from microbial remains buried within sediment that subsequently metamorphosed, driving out the water content but leaving behind the telltale carbon. By contrast, carbonate rock of nonbiogenic origin from the same formation shows a $\delta^{13}C$ near zero.

A. Minik Rosing

BBC

Figure 17.8 Biosignatures of early life. A. Minik Rosing (left), in Greenland, shows the Isua rocks whose carbon isotope ratios indicate photosynthesis at 3.8 Gyr. **B.** ^{13}C isotope depletion (negative $\delta^{13}C$) occurs in biomass as a result of the Calvin cycle. Negative $\delta^{13}C$ is observed at 3.7 Gyr in sedimentary graphite, which may derive from sedimented phototrophs. Little or no isotope depletion is seen in carbonate rock, which has no biological origin. **C.** 2-Methylhopane, a biological signature of cyanobacteria, is found in rock strata dated at 2.5 Gyr. The 2-methyl group is highlighted.

B. ^{13}C **isotope depletion**

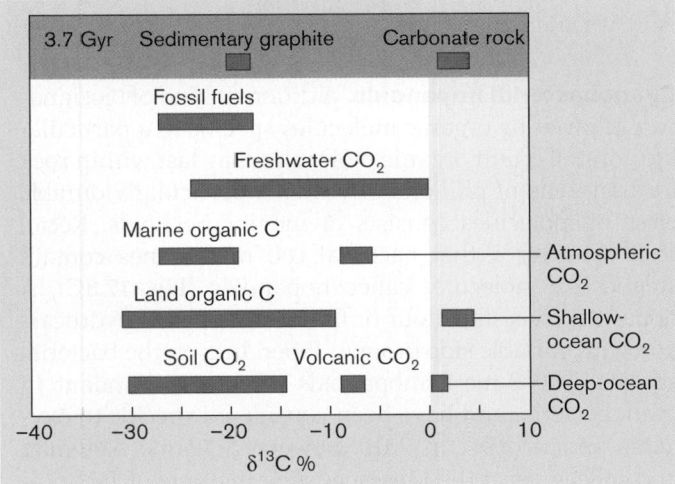

C. 2-Methylhopane

H_3C

A.

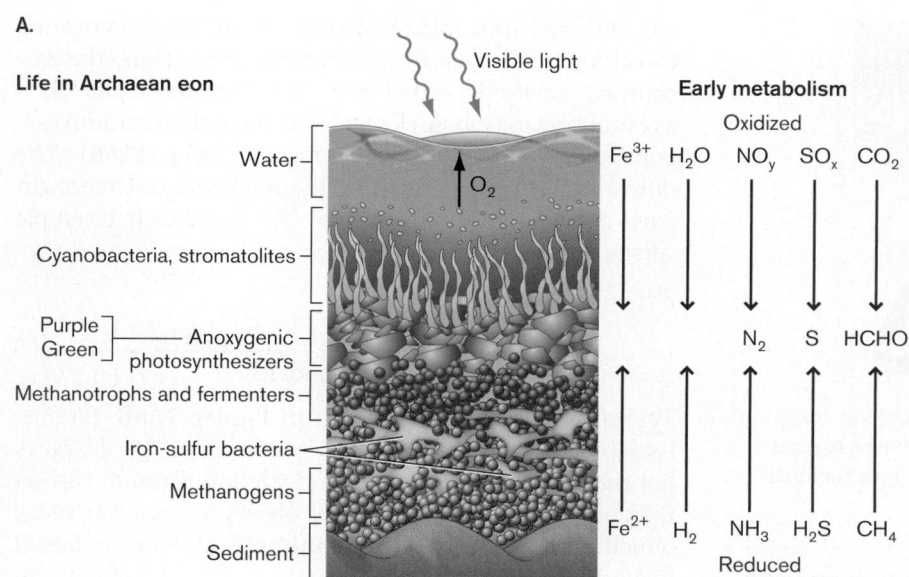

Life in Archaean eon

Water

Cyanobacteria, stromatolites

Purple / Green — Anoxygenic photosynthesizers

Methanotrophs and fermenters

Iron-sulfur bacteria

Methanogens

Sediment

Visible light

O_2

Early metabolism

Oxidized

Fe^{3+} H_2O NO_y SO_x CO_2

N_2 S HCHO

Fe^{2+} H_2 NH_3 H_2S CH_4

Reduced

Figure 17.9 Early metabolism.
A. Early metabolism could have been based on various reactions between oxidized minerals that diffuse down from the air and water and reduced minerals in the sediment, upwelling from hydrothermal vents. **B.** Early photosynthesis may have resembled that of haloarchaea, whose bacteriorhodopsin absorbs light in the central range of sunlight (yellow-green, 500–600 nm), in contrast to the chlorophyll of cyanobacteria and plants, which absorbs blue and red. *Source:* Part B adapted from Shil DasSarma, University of Maryland.

B.

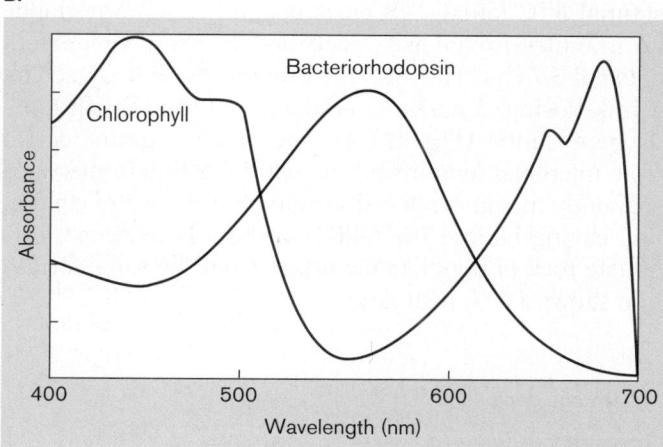

Cyanobacterial hopanoids. A different kind of biosignature is given by organic molecules specific to a particular life-form. Certain organic molecules may last within rock for hundreds of millions of years. A particularly durable class of molecules consists of membrane lipids. Recall from Chapter 3 that bacterial cell membranes contain steroid-like molecules called hopanoids (**Fig. 17.8C**). A hopanoid consists of four or five fused rings of hydrocarbon with variable side groups, depending on the bacterial species. The 2-methylhopanoids are most abundant in cyanobacteria, and have been considered specific to oxygenic phototrophs. In 2010, however, Dianne Newman and colleagues at the Massachusetts Institute of Technology reported finding 2-methylhopanoids in anoxygenic phototrophs such as the alphaproteobacterium *Rhodopseudomonas palustris*. Thus, the 2-methylhopanoids may be less specific a biomarker than previously thought.

The hopanoid derivative 2-methylhopane is found in sedimentary rock of the Hamersley Basin of Western Australia, dated to 2.5 Gyr. This biosignature offers evidence that either cyanobacteria or related species metabolism

existed by the end of the Archaean eon. The finding cannot, however, prove the early existence of oxygenic photosynthesis, since 2-methylhopanoids are found in species that do not produce oxygen. The study of 2-methylhopanoids demonstrates both the promise and pitfalls of biosignatures as evidence for the nature of early life.

Metabolism of the First Cells

How did the earliest life-forms metabolize without oxygen gas to respire and without the complex machinery of photosynthesis? The nature of the first metabolism is unknown, but geochemistry and modern metabolism suggest several possibilities (**Fig. 17.9A**):

- **Oxidation-reduction reactions.** The early oceans contained oxidized forms of nitrogen, sulfur, and iron that could interact with reduced minerals from the crust. For example, nitrate (NO_3^-) or sulfate (SO_4^{2-}) could be reduced by hydrogen gas to yield energy (hydrogenotrophy, discussed in Chapter 14). The oxidized molecules were generated by reactions driven by ultraviolet radiation, which penetrated the atmosphere in the absence of the ozone layer. Sulfur isotope ratios ($^{34}S/^{32}S$) suggest the growth of sulfate-reducing bacteria as early as 3.47 Gyr ago.
- **Light-driven ion pumps.** A simple light-driven pump, such as the bacteriorhodopsin of haloarchaea (halophilic archaea), could have conducted the first kind of phototrophy. The absorption spectrum of bacteriorhodopsin matches the peak of solar radiation reaching the upper layers of ocean (**Fig. 17.9B**), in contrast to the chlorophylls of cyanobacteria and plants, which absorb the outer ranges of blue and red. Perhaps cyanobacteria evolved in the presence of haloarchaea, filling the unexploited photochemical niches.
- **Methanogenesis.** Climate models of early Earth suggest a methane atmosphere, produced by methano-

genic archaea. Methanogenesis involves reaction of H_2 and CO_2 producing CH_4 and H_2O. Methanogens show highly divergent genomes—a finding that suggests early evolution of their common ancestor. Their biochemistry and evolution (discussed in Chapter 19) are consistent with proposed models of ancient life.

Oxygen from Cyanobacteria

An extraordinary event in the planet's history was the evolution of the first oxygenic phototrophs: cyanobacteria that split water to form O_2. The entry of O_2 into Earth's biosphere is often portrayed as a sudden event that would have been disastrous to microbial populations lacking defenses against its toxicity. In fact, geological evidence shows that oxygen arose gradually in the oceans, starting about 2 Gyr ago, and may have arisen and disappeared numerous times before reaching a high, steady-state level in our atmosphere. The mechanism of the oxygen fluctuation is unknown, but it must have occurred through cycles of aerobic and anaerobic microbial metabolism.

Banded iron formations. Evidence for oxygen in the biosphere comes from the oxidation state of minerals, particularly those containing iron. The bulk of crustal iron is in the reduced form (Fe^{2+}), which is soluble in water and reached high concentrations in the anoxic early oceans. Sedimentary rock, however, contains many fine layers of oxidized iron (Fe^{3+}), which is insoluble and forms a precipitate, such as iron oxide (Fe_2O_3). The layers of iron oxide suggest periods of alternating oxygen-rich and anoxic conditions. These layered iron minerals are called **banded iron formations (BIFs) (Fig. 17.10A)**. A common form of banded iron consists of gray layers of silicon dioxide (SiO_2) alternating with layers colored red by iron oxides and iron oxyhydroxides [$FeO_x(OH)_y$]. Banded iron formations are widespread around the world and provide our major sources of iron ore (**Fig. 17.10B**).

Banded iron formations are often found in rock strata containing signs of past life such as ^{13}C depletion and other biomarkers. For example, the Isua formations (Greenland) and Hamersley formations (Western Australia), which both show ^{13}C depletion, also contain extensive banded iron. Calculations show that the layers of oxidized iron could result from biological metabolism involving iron oxidation. One possibility is that the iron was oxidized by chemolithotrophs, using molecular oxygen produced by cyanobacteria. The Archaean and early Proterozoic eons experienced fluctuating levels of molecular oxygen in the atmosphere. These fluctuations could have led to oscillating levels of iron oxide, thus producing bands in the sediment, as microbes used up all the oxygen.

Alternatively, Dianne Newman, at the California Institute of Technology, proposes that the iron oxides arose directly from anaerobic photosynthesis, in which the reduced iron served as the electron donor (**Fig. 17.11**). In iron phototrophy, light excites an electron from Fe^{2+}, oxidizing the ion to Fe^{3+}, while the excited electron cycles through an electron transport system to yield energy (discussed in Chapter 14). Newman discovered iron phototrophy in modern purple bacteria such as *Rhodopseudomonas palustris*. In the ancient Earth, photosynthetic oxidation of Fe^{2+} to Fe^{3+}, or "photoferrotrophy," could have occurred in cycles until the marine iron was all oxidized, generating sedimentary layers of iron oxides and iron oxyhydroxides. Today, photoferrotrophs are abundant in deep anoxic lakes that resemble our model of the anoxic Archaean ocean.

By 2.3 Gyr, the prevalence of oxidized iron and other minerals indicates the steady rise of oxygen from photosynthesis in Earth's atmosphere. All the dissolved Fe^{2+} from the ocean floor was oxidized, leaving the oceans in the iron-poor state that persists today. Oxygen as the most efficient electron acceptor enabled the evolution of aerobic respiratory bacteria. Aerobic bacteria gave rise to mitochondria, which enabled the evolution of eukaryotes and ultimately multicellular organisms (**Fig. 17.12**).

Remarkably, modern cells are still composed primarily of reduced molecules, highly reactive with oxygen—a relic of the time when our ancestral cells evolved in the absence of oxygen. The conditions under which such cells may have evolved can be simulated in the laboratory—conditions under which some of life's most common molecules, such as adenine and simple amino acids, form spontaneously. These early-Earth simulation experiments can never prove the actual conditions under which life began, but they can suggest testable models with intriguing implications.

Figure 17.10 Banded iron formations. A. Banded iron formation in ancient sedimentary rock. Its main component is chert, a form of quartz (silicon dioxide, SiO_2) with layers colored red by iron oxide (Fe_2O_3). **B.** The BHP Billiton Iron Ore mine at Newman, Western Australia.

A.

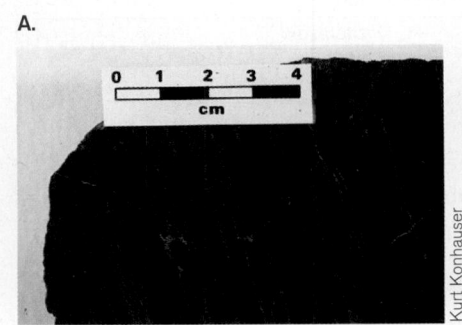

B.

Kurt Konhauser

A.

Courtesy of Dianne Newman

B.

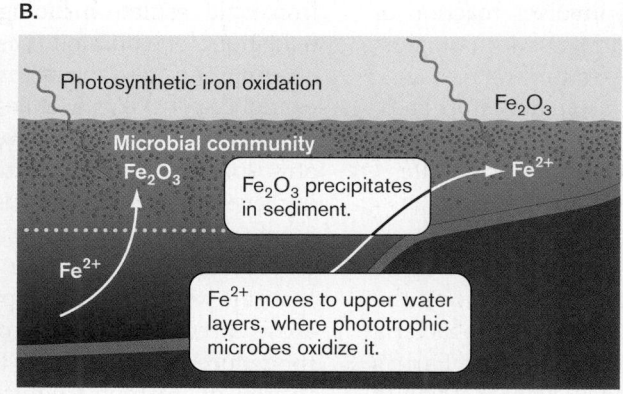

Photosynthetic iron oxidation

Fe_2O_3

Microbial community
Fe_2O_3

Fe_2O_3 precipitates in sediment.

Fe^{2+}

Fe^{2+}

Fe^{2+} moves to upper water layers, where phototrophic microbes oxidize it.

Figure 17.11 Iron phototrophy.
A. Dianne Newman at the California Institute of Technology proposes that early iron phototrophs caused the iron oxide deposition generating banded iron formations.
B. Photosynthetic oxidation of Fe^{2+} to Fe^{3+} may have generated sedimentary layers of Fe_2O_3 and $FeO_x(OH)_y$.

Origin and evolution of life on Earth

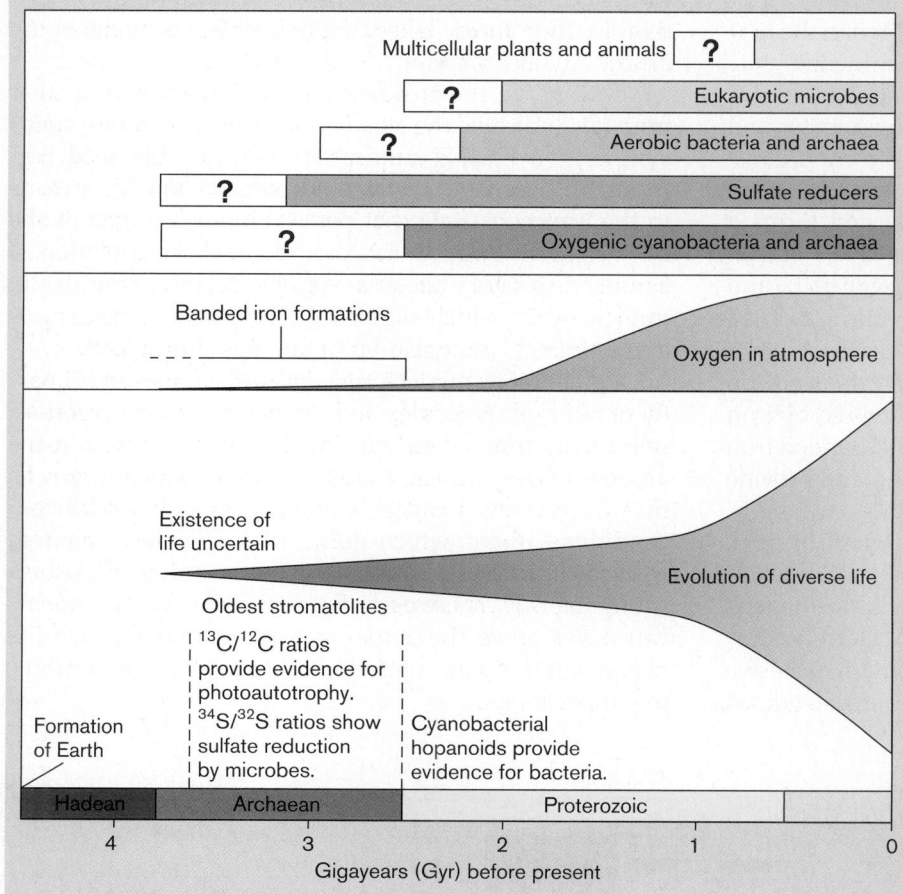

Figure 17.12 Proposed time line for the origin and evolution of life. The planet Earth formed during the Hadean eon (about 4.5 Gyr ago). The environment was largely reducing until cyanobacteria pumped O_2 into the atmosphere. When the O_2 level reached sufficient levels (about 0.6 Gyr ago), multicellular animals and plants evolved. Question marks designate periods when evidence for given life-forms is uncertain.

TO SUMMARIZE:

- **Elements of life** were formed through nuclear reactions within stars that exploded into supernovas before the birth of our own Sun.

- **Reduced molecules** compose Earth's interior. Oxidized minerals are found only near the surface. Early Earth had no molecular oxygen (O_2).

- **Archaean rocks show evidence for life** based on fossil stromatolites, isotope ratios, and chemical biosignatures. Fossil stromatolites appear in chert formations formed 3.4 Gyr ago. Isotope ratios for carbon indicate photosynthesis at 3.7 Gyr ago and sulfate reduction at 3.47 Gyr. Cyanobacterial hopanoids appear at 2.5 Gyr ago.

- **Microfossils of filamentous and colonial prokaryotes** date to 2.0 Gyr ago. At 1.2 Gyr, larger fossil cells resemble those of modern eukaryotes.

- **Early metabolism** involved anaerobic oxidation-reduction reactions. Likely forms of early metabolism include sulfate respiration, light-driven ion pumps, iron phototrophy, and methanogenesis.

- **Banded iron formations** reflect the cyclic increase and decrease of oxygen produced by cyanobacteria and consumed through reaction with reduced iron. After all the ocean's iron was oxidized, oxygen increased gradually in the atmosphere.

17.2 Models for Early Life

Various models have been proposed to explain how the first life-forms originated from nonliving materials and how they replicated and evolved. Models for early life attempt to address the following questions: In what kind of environment did the first cells form? What kind of metabolism did the first cells use to generate energy? What was their hereditary material?

Models for Origin of the First Cells

Models for early life include:

- **The prebiotic soup.** Organic building blocks of life could arise **abiotically** (in the absence of life) out of simple reduced chemicals such as ammonia and methane. This **prebiotic soup** could have generated complex macromolecules that eventually acquired the apparatus needed for self-replication and membrane compartmentalization.
- **Metabolist models.** The central components of intermediary metabolism, including the TCA cycle to generate amino acids, arose from self-sustaining chemical reactions based on inorganic chemicals. These abiotic reactions then acquired self-replicating macromolecules and membranes.
- **The RNA world.** Proposed by Francis Crick in the 1960s, the **RNA world** is a model of early life in which RNA performed all the informational and catalytic roles of today's DNA and proteins. The concept of an RNA world derives support from the emerging sequences of genomes, which reveal thousands of catalytic and structural RNAs.

The prebiotic soup. In the mid–twentieth century, biochemists Aleksandr Oparin, at Moscow University, and Stanley Miller and Harold Urey, at the University of Chicago, showed that organic building blocks of life such as amino acids could arise abiotically out of a mixture of water and reduced chemicals, including CH_4, NH_3, and H_2 (**Fig. 17.13**). The mixture was subjected to an electrical

discharge, similar to the lightning discharges that arise from volcanic eruptions, which would have been common in the late Hadean or early Archaean eon. The chemical reaction produced fundamental amino acids such as glycine and alanine. Similar experiments by Juan Oró, at the University of Houston, showed the formation of adenine by condensation of ammonia and methane.

The same amino acids and nucleic bases arising from early-Earth simulation are also found in meteorites, which are believed to retain the chemistry of the early solar system "frozen" in time. But unlike the chemical experiments, the meteorites show a remarkable predominance of left-handed mirror forms (L-enantiomers) of the amino acids—the same L-form utilized in the protein synthesis of all life on Earth. Some researchers hypothesize that organic compounds formed in outer space were restricted to the L-form by an unknown process and then "seeded" the propagation of L-form compounds in Earth's prebiotic soup.

The original model conditions for the prebiotic soup assumed that molecules in the early Archaean ocean were largely reduced, with little or no oxygen present in the atmosphere. More recent geochemical evidence suggests that the early ocean actually included oxidized forms of nitrogen, sulfur, and iron that arose through reactions driven by ultraviolet radiation, which penetrated the atmosphere in the absence of the ozone layer. These oxidized minerals could have reacted with the reduced crustal minerals, releasing energy to drive production of more complex biomolecules.

Forming the first cell, or "proto-cell," must have required enclosing the first biochemical reactants within a membrane-like compartment. Jack Szostak and colleagues at Harvard Medical School investigate how such compartments can arise spontaneously from fatty acid glycerol esters. The fatty acid derivatives are "amphipathic;" that is, they posses both hydrophobic portions that associate together and hydrophilic portions that associate with water. In water, the fatty acid derivatives collect in "micelles," small, round aggregates in which the hydrophobic portions associate in the interior and the hydrophilic portions associate with water. Under certain conditions, micelles can aggregate to form hollow vesicles

Figure 17.13 The prebiotic soup model for the origin of life. A. In the prebiotic soup, inorganic molecules could have reacted to form complex macromolecules that eventually acquired the apparatus for self-replication and membrane compartmentalization. **B.** Lightning accompanies eruption of the Galunggung Volcano in West Java, Indonesia, 1982. The first biomolecules may have formed as a result of lightning triggered by volcanic eruption.

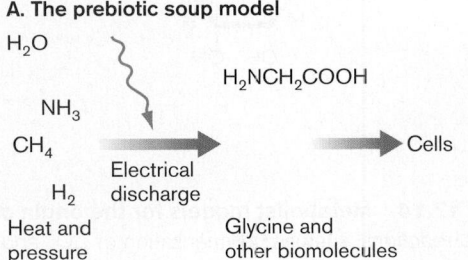

A. The prebiotic soup model

H_2O

NH_3

CH_4 → Electrical discharge → H_2NCH_2COOH → Cells

H_2

Heat and pressure

Glycine and other biomolecules

B.

of membrane. Szostak showed how the vesicles can take up molecules such as RNA, suggesting a primitive cell-like form. These spontaneous processes of membrane formation offer models for how the first living cells may have arisen.

Metabolist model. Other models attempt to explain the origin of biosynthesis on the basis of CO_2 fixation, the fundamental metabolism of all life today. Proponents of a **metabolist model**, including Harold Morowitz and Günter Wächterhäuser, at George Mason University, propose that CO_2-based metabolism originated through self-sustaining reactions (**Fig. 17.14A**). Simulation experiments suggest that such premetabolic reactions could have been catalyzed by metal sulfides prevalent in the early ocean. For example, abiotic polymerization of CO_2 into TCA cycle intermediates may be catalyzed by FeS.

How did simple inorganic reactions lead to biochemical cycles involving more complex biopolymers such as nucleic acids and proteins? A major challenge is to explain the genetic code by which nucleic acid codons are assigned to unique amino acids. The genetic code may have arisen from an earlier pretranslational mechanism of amino acid biosynthesis (**Fig. 17.14B**). In this mechanism, proposed by Morowitz and colleagues, each amino acid was originally synthesized from a TCA cycle acid complexed to a dinucleotide. The dinucleotide later evolved into the first two nucleotides of the codon specifying the amino acid. The proposed link between a specific dinucleotide and each amino acid would explain certain features of the genetic code, such as the fact that most amino acids specified by a codon starting with the same nucleotide are

synthesized from the same TCA cycle acid (discussed in Chapter 15).

The RNA world. Neither the prebiotic soup model nor the metabolist models account for the evolution of macromolecules that encode complex information, such as nucleic acids and proteins. A candidate for life's first "informational molecule" is RNA. RNA is a relatively simple biomolecule, with only 4 different "letters," compared to the 20 standard amino acids of proteins. Its purine base adenine arises spontaneously from ammonia and carbon dioxide under conditions believed to resemble those of the Archaean eon. Its ribose sugar is a fundamental building block of living cells, with key roles in numerous biochemical pathways, such as the Calvin cycle. Most exciting, in 2009, John Sutherland and co-workers discovered an organic reaction pathway that generates a sugar-base ribonucleotide in the absence of any enzyme. This discovery suggests a way that the first RNA molecules could have formed before there were living cells.

For several reasons, RNA is a better candidate than DNA for the earliest information molecule. Compared to DNA, RNA requires less energy to form and degrade. RNA's pyrimidine base uracil is formed early by biochemical pathways; only later is it transformed to the thymine used by DNA.

Most important, RNA molecules have been shown to possess catalytic properties analogous to those of proteins. Catalytic RNA molecules are called **ribozymes**. The first ribozyme, discovered by Nobel laureate Tom Cech in the protist *Tetrahymena*, can splice introns in mRNA. Other ribozymes actually catalyze synthesis of

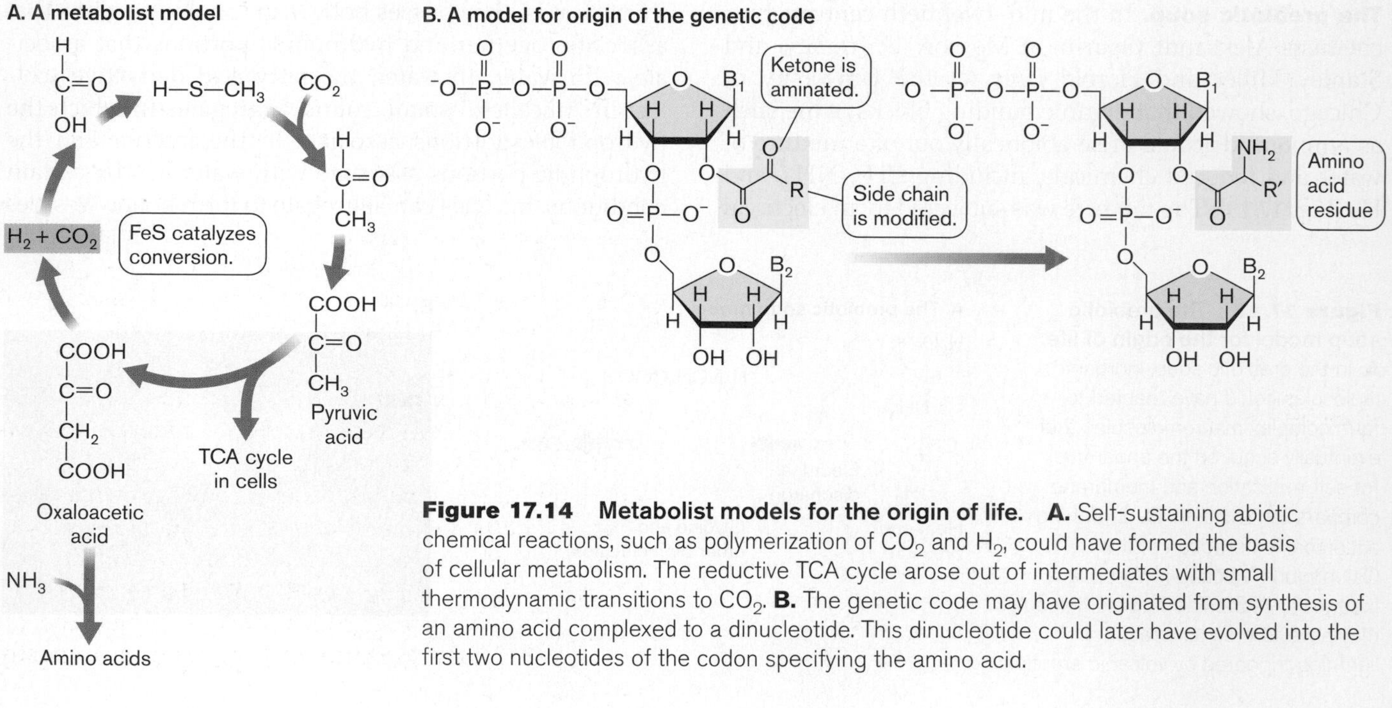

A. A metabolist model

B. A model for origin of the genetic code

Figure 17.14 Metabolist models for the origin of life. A. Self-sustaining abiotic chemical reactions, such as polymerization of CO_2 and H_2, could have formed the basis of cellular metabolism. The reductive TCA cycle arose out of intermediates with small thermodynamic transitions to CO_2. **B.** The genetic code may have originated from synthesis of an amino acid complexed to a dinucleotide. This dinucleotide could later have evolved into the first two nucleotides of the codon specifying the amino acid.

complementary strands of RNA, suggesting a model for early replication of RNA chromosomes. The most elaborate example of catalytic RNA is found in the ribosome. In the ribosome, X-ray crystallography reveals that the key steps of protein synthesis, such as peptide bond formation, are actually catalyzed by the RNA components, not proteins (discussed in Chapter 8). The ribosomal proteins possess relatively little catalytic function; their main role seems to be protection and structural support of the RNA.

Could RNA molecules have composed the earliest cells? In 2009, Tracey Lincoln and Gerald Joyce at the Scripps Research Institute devised a system in which two RNA ribozymes catalyze each other's synthesis. The ribozyme model system suggests how, in the earliest cells, RNA might have fulfilled the key functions that are today filled by DNA and proteins, including information storage, replication, and catalysis (**Fig. 17.15**). This model is known as the "RNA world."

The prominent function of RNA in the ribosome, one of life's most ancient and conserved molecular machines, suggests a model for the transition from an RNA world to the modern cell. In the ribosome, the actual steps of catalysis, such as forming the peptide bond, are conducted by RNA subunits acting as ribozymes. The ribozymes are stabilized by protein subunits. Tom Cech, Sidney Altman, and colleagues propose that the earliest RNA components of cells evolved by acquiring proteins to enhance stability. The proteins helped prevent the tendency of RNA to hydrolyze (come apart in reaction with water). As cells evolved, their peptide components increased through natural selection, and the RNA subunits may have shrunk by reductive evolution (the evolutionary loss of unneeded parts). A few complexes, such as the ribosome, still maintain their ribozymes; for others, perhaps all that remains are one or two nucleotides. Dinucleotide cofactors such as NADH persist in enzymes today, perhaps representing vestigial remnants of the

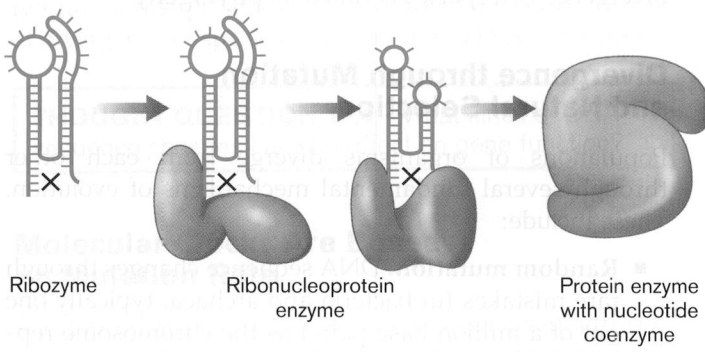

Ribozyme | Ribonucleoprotein enzyme | Protein enzyme with nucleotide coenzyme

Figure 17.15 From the RNA world to proteins. Earliest cells may have been composed of RNA enzymes (ribozymes). As the RNA cells evolved, ribozymes acquired protein subunits that eventually assumed most of their catalytic functions. Remnants of the original RNA may persist as nucleotide cofactors such as NADH.

RNA world. For more on RNA evolution, and surprising applications for medicine, see **eTopic 17.1**.

The RNA world model explains the central role of RNA throughout the history of living cells. Yet the role of RNA offers little clue as to the origins of cell compartmentalization and metabolism. No one model of life's origin yet addresses all the requirements for a living cell—metabolism, membrane compartmentalization, and hereditary material. Each model does, however, offer important insights into the evolutionary history and mechanisms of life today.

> **THOUGHT QUESTION 17.2** Outline the strengths and limitations of each model of the origin of living cells. Which aspects of living cells does each model explain?

Unresolved Questions about Early Life

Overall, geology and biochemistry provide compelling evidence that organisms resembling today's cyanobacteria lived on Earth at least 2.5 Gyr ago, possibly 3.7 Gyr ago, and that bacteria with anaerobic metabolism evolved as early or earlier. Many intriguing questions remain. We outline here three unresolved questions regarding the temperature of early Earth, the role of methane in the early atmosphere, and the actual source of Earth's first cells.

Thermophile or psychrophile? The apparent existence of life so soon after Earth cooled suggests a thermophilic origin. Thermophily is supported by the fact that in the domains Bacteria and Archaea, the deepest-branching species (that is, species that diverged the earliest from others in the domain) are thermophiles. Such organisms could have thrived at hydrothermal vents, which offer a continual supply of H_2S and carbonates.

On the other hand, after meteoric bombardment abated, early Earth should have become glacially cold. In the Archaean, solar radiation was 20%–30% less intense than it is today, and the thin CO_2 atmosphere was insufficient to increase the temperature by a greenhouse effect. A colder habitat would support psychrophiles. Psychrophiles might have had an advantage in an RNA world, given the thermal instability of RNA compared to DNA and proteins.

A world of methane? Among Earth's earliest lifeforms were methanogens (methane producers). Methanogens are one of the most widely divergent groups of organisms and persist today in environments ranging from anaerobic sediment to the human intestine. Methanogenesis requires only carbon dioxide and hydrogen gas, which would have been plentiful in early anoxic sediment. The production of methane, an extremely potent greenhouse gas, could have greatly

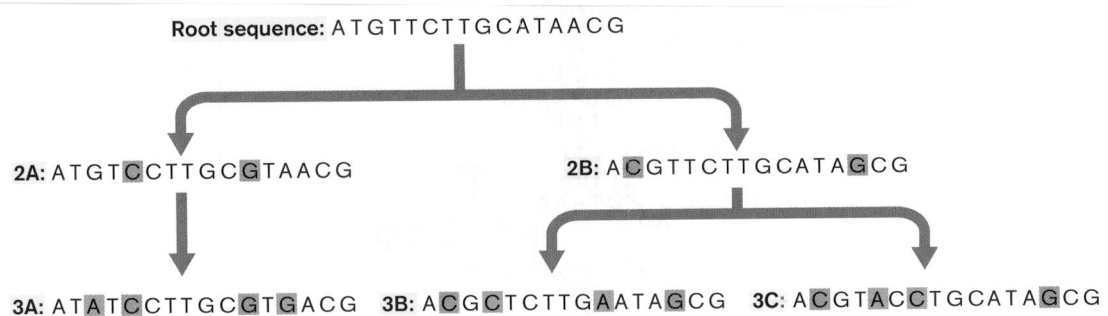

Root sequence: ATGTTCTTGCATAACG

2A: ATGTCCTTGCGTAACG   **2B:** ACGTTCTTGCATAGCG

3A: ATATCCTTGCGTGACG   **3B:** ACGCTCTTGAATAGCG   **3C:** ACGTACCTGCATAGCG

Figure 17.16 The molecular clock. As genetic molecules reproduce, the number of mutations accumulated at random is proportional to the number of generations and thus the time since divergence. In each sequence designation (for example, "2A"), the number indicates the generation and the letter identifies a specific strain.

increased Earth's temperature during the Archaean eon. Thus, methanogenesis could explain how Earth escaped the permanent freeze of Mars. Overheating would have been halted when the oxygen gas produced by cyanobacteria enabled growth of methanotrophs, bacteria that oxidize methane. The decline of methane and the rise of CO_2 then would have brought about relative thermal stability.

The debate over the temperature and climate of early Earth has interesting implications for Earth today, when we again face the prospect of massive global climate change. Human agriculture favors explosive growth of methanogens, which threaten to accelerate global warm

TO SUMMARIZE:

- **Prebiotic soup models** propose that the fundamental biochemicals of life arose spontaneously through condensation of reduced inorganic molecules.
- **Metabolist models** propose that components of intermediary metabolism arose from self-sustaining chemical reactions that connected nucleotides with amino acids, forming the basis of the genetic code.
- **The RNA world model** proposes that in the first cells, RNA performed all the informational and catalytic roles of today's DNA and proteins.

Special Topic 17.1   Phylogeny of a Shower Curtain Biofilm

The deep sea and the tropical forest receive much attention as habitats in which to discover new exotic life-forms. But novel microorganisms can be discovered closer to home—in the soil, on the roots of grass, in our own digestive tract. Even within our homes lurk microbial communities that include potentially dangerous pathogens. Such organisms were discovered by Norman Pace and colleagues at San Diego State University and at the University of Colorado, taking a break from their submarine studies to explore a domestic aquatic habitat: the "soap scum" biofilm on a shower curtain (**Fig. 1**).

Domestic water sources are well known as a source of pathogens such as *Legionella pneumophila*, the cause of legionellosis, a deadly form of pneumonia. Pathogens persist in biofilms growing within plumbing and air-conditioning lines. Lesser-known reservoirs for microbes are the biofilms that collect on household surfaces such as shower curtains, which are often neglected during cleaning. Viable cells of the shower curtain biofilm are revealed by fluorescence microscopy using the DAPI stain for DNA (**Fig. 1B**).

Pace's students Scott Kelley and Ulrike Thiesen obtained biofilm from a shower curtain and used it to extract DNA. The biofilm DNA was amplified for sequences transcribed to rRNA using the PCR technique. The PCR amplification was conducted with a pair of sequence primers based on portions of the SSU rRNA sequence that are extremely conserved across all known species of bacteria. The region between the primers, however, is known to vary among species; thus, the DNA amplified should show distinguishing features that identify the source organism.

The shower curtain biofilm revealed a wide range of species, including 117 unique sequences for SSU rRNA. The phylogeny shows how shower curtain samples relate to known

bacteria, offering clues as to how they might interact with humans. The most abundant groups of species include two genera of alphaproteobacteria: *Sphingomonas* and *Methylobacterium*. *Sphingomonas* is named for its outer membrane content of sphingolipids, lipids containing an amide link to fatty acids. The phylogeny of the shower curtain sphingomonads is shown in **Figure 2**. *Sphingomonas* species are known to grow in a wide range of soil and water habitats and are occasionally isolated as yellow colonies from natural water supplies. They catabolize complex carbon sources such as dibenzofuran and hexachlorocyclohexane and thus are of interest for bioremediation of environmental pollutants. Some species, however—particularly *S. paucimobilis* (highlighted in **Figure 2**)—cause opportunistic infections in immunocompromised individuals, including bacteremia, peritonitis, and abscesses.

Species of the next-most abundant genus, *Methylobacterium*, are versatile heterotrophs known for their ability to metabolize single-carbon sources such as methanol and methylamine. Species grow in soil and water, as well as within plant tissues; they are also isolated from automobile air-conditioning systems, printing paper machines, and dental unit water lines. Their pink color may be the source of the pink color commonly observed in shower biofilms. Some species, such as *M. extorquens* and *M. zatmanii,* cause illness in immunocompromised individuals, including pneumonia, skin ulcers, and bacteremia.

Growing numbers of immunocompromised individuals care for themselves at home, where they need to control their own microbiological exposure. Thus, it is of concern to discover opportunistic pathogens in a home setting. The authors conclude with a reminder that "exposure can be minimized by regular cleaning or by changing shower curtains."

A.   **B.**

Scott T. Kelley et al. 2004. *Appl. Environ. Microbiol.* **70**:4187

© Dena Digilio Betz

Figure 1   **Bacterial diversity on a shower curtain.**
A. Shower curtain "soap scum" consists of a biofilm.
B. Biofilm from a shower curtain is visualized by epifluorescence microscopy using DAPI stain.

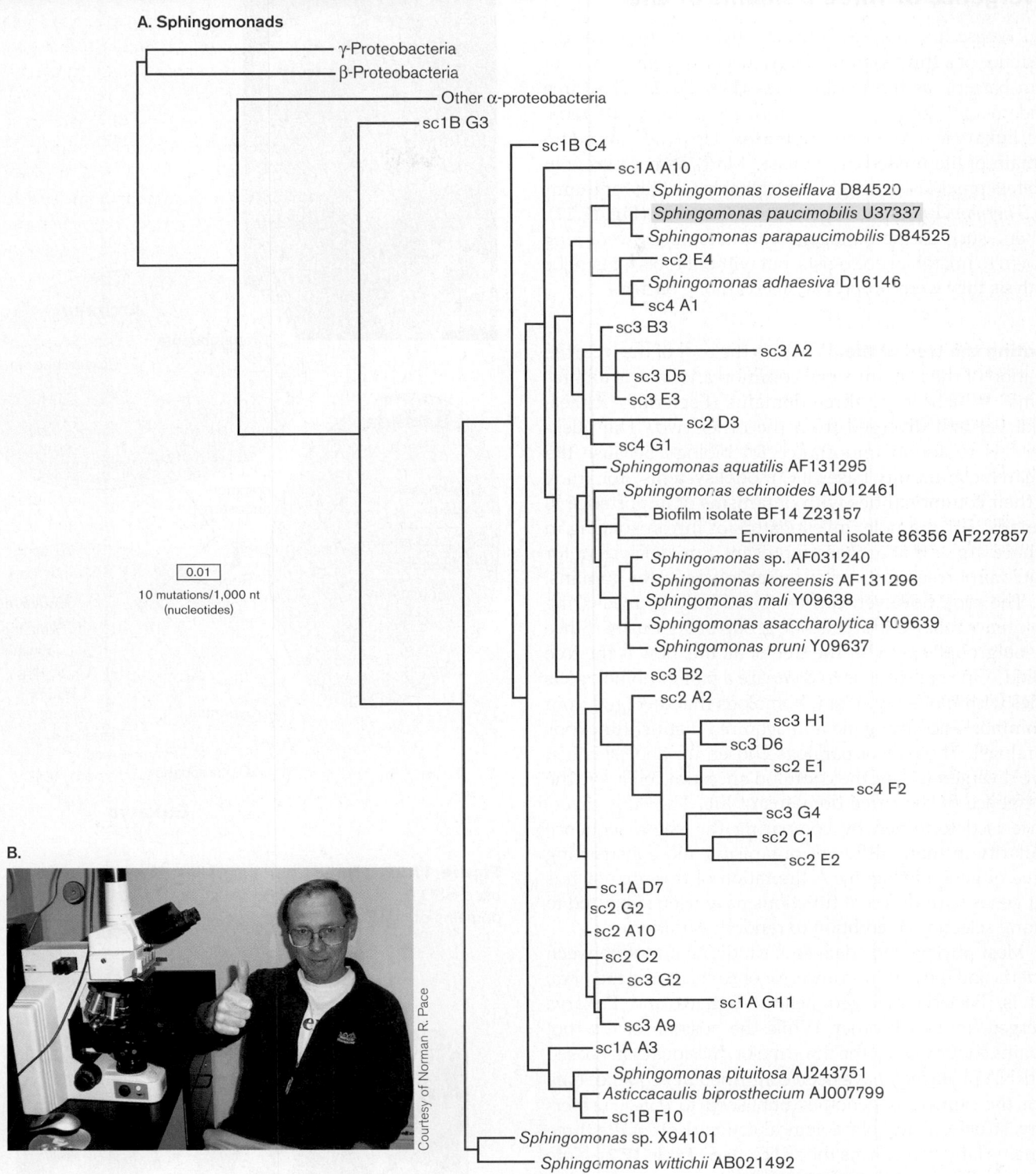

A. Sphingomonads

- γ-Proteobacteria
- β-Proteobacteria
- Other α-proteobacteria
- sc1B G3
- sc1B C4
- sc1A A10
- *Sphingomonas roseiflava* D84520
- *Sphingomonas paucimobilis* U37337
- *Sphingomonas parapaucimobilis* D84525
- sc2 E4
- *Sphingomonas adhaesiva* D16146
- sc4 A1
- sc3 B3
- sc3 A2
- sc3 D5
- sc3 E3
- sc2 D3
- sc4 G1
- *Sphingomonas aquatilis* AF131295
- *Sphingomonas echinoides* AJ012461
- Biofilm isolate BF14 Z23157
- Environmental isolate 86356 AF227857
- *Sphingomonas* sp. AF031240
- *Sphingomonas koreensis* AF131296
- *Sphingomonas mali* Y09638
- *Sphingomonas asaccharolytica* Y09639
- *Sphingomonas pruni* Y09637
- sc3 B2
- sc2 A2
- sc3 H1
- sc3 D6
- sc2 E1
- sc4 F2
- sc3 G4
- sc2 C1
- sc2 E2
- sc1A D7
- sc2 G2
- sc2 A10
- sc2 C2
- sc3 G2
- sc1A G11
- sc3 A9
- sc1A A3
- *Sphingomonas pituitosa* AJ243751
- *Asticcacaulis biprosthecium* AJ007799
- sc1B F10
- *Sphingomonas* sp. X94101
- *Sphingomonas wittichii* AB021492

0.01

10 mutations/1,000 nt
(nucleotides)

B.

Courtesy of Norman R. Pace

Figure 2 *Sphingomonas* **phylogeny from a shower curtain biofilm. A.** Phylogeny of sphingomonad bacteria was based on SSU RNA gene sequence comparison. **B.** Norman R. Pace characterized the first sequence phylogeny of thermophiles from high-temperature environments. His laboratory has since characterized the microbial diversity of other environments, such as a shower curtain biofilm.

Divergence of Three Domains of Life

Carl Woese first used SSU rRNA phylogeny to reveal the existence of a third kind of life, Archaea, roughly as distant from bacteria as from eukaryotes (**Fig. 17.21**). The three fundamental groups of life forms—Archaea, Bacteria, and Eukarya—are termed **domains**. How was an entire domain of life missed in the past? Many archaea grow in habitats previously thought inhospitable for life; for example, *Thermoplasma* species grow at 60°C at pH 2 (**Fig. 17.22**). Others, such as methanogens and halophiles, were long known to microbial ecologists, but without tools for genetic analysis they were simply classified among bacteria.

Rooting the tree of life. Where is the root of the tree, the position of the last universal common ancestor of all life-forms? Which of the three domains (Bacteria, Archaea, Eukarya) first diverged from the other two? The question has profound importance for biology because the research community bases its "model systems" for study on their commonalities with organisms of importance to humans. For example, investigators of intron splicing in archaea argue that archaea represent a model system for related processes in complex eukaryotes such as humans.

The root, however, can be found only by measuring divergence relative to an outside group of organisms. Since no "outgroup" exists for the tree of all life, how is the tree rooted? One approach is to compare a pair of homologous genes within one organism, homologs that diverged from a common ancestral gene and acquired distinct functions (paralogs). The pair of paralogs chosen for analysis must have diverged within the common ancestral cell *before* the divergence of the three domains of life. The early divergence is determined by comparing the pair's sequence similarity in many different organisms and constructing a tree of gene phylogeny. A limitation of this approach is that genes with different functions have been subjected to natural selection in addition to random variation.

Most phylogenetic data so far indicate a root between Bacteria and the common ancestor of Archaea and Eukarya; that is, Bacteria diverged before Archaea and Eukarya diverged from each other. While the position of the root remains controversial, the three major divisions of life based on rRNA phylogeny have been confirmed by sequence data from the numerous genomes published to date. Furthermore, structural and physiological comparison of the three domains largely confirms the rRNA tree (**Table 17.2**). Note first that all living cells on Earth share profound similarities. All cells consist of membrane-enclosed compartments that shelter the same fundamental apparatus of cell production, the DNA-RNA-protein machine. The fundamental components of this machine appear to have evolved before the three domains diverged from their last universal common ancestor. From a molecular standpoint, all cells on Earth appear more similar than they are different.

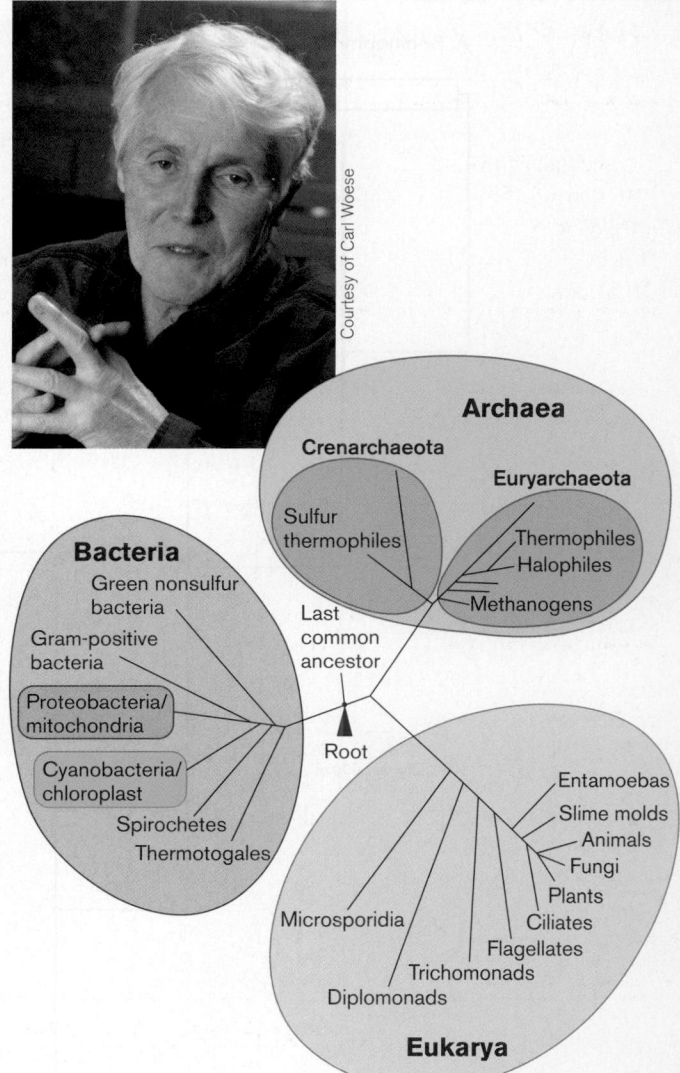

Figure 17.21   Three domains of life.   Carl Woese (inset) used SSU rRNA sequencing to reveal three equally distinct domains of life: Bacteria, Eukarya (eukaryotes), and Archaea.

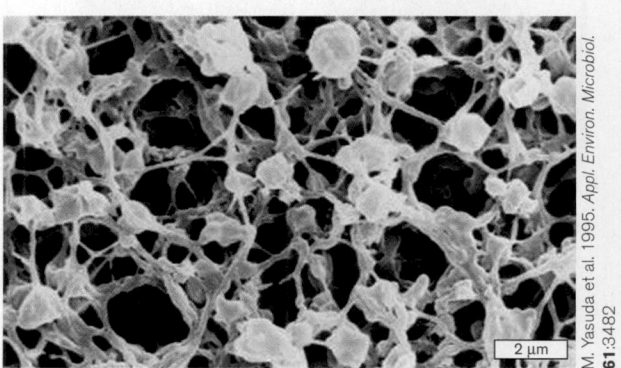

Figure 17.22   *Thermoplasma.*   This archaeon lives at 60°C at pH 2—with no cell wall, only a cell membrane.

Table 17.2 Three domains of life.

Characteristic	Traits of living organisms		
	All cells on Earth resemble each other in these traits:		
Chromosomal material	Double-stranded DNA		
RNA transcription	Common ancestral RNA polymerase		
Translation	Common ancestral rRNAs and elongation factors		
Protein	Common ancestral functional domains		
Cell structure	Aqueous cell compartment bounded by a membrane		
	Comparison of domains:		
	Bacteria	**Archaea**	**Eukarya**
	Archaea resemble bacteria in these traits:		
Cell volume	1–100 μm^3 (usually)		1–10^6 μm^3
DNA chromosome	Circular (usually)		Linear
DNA organization	Nucleoid		Nucleus with membrane
Gene organization	Multigene operons		Single genes
Metabolism	Denitrification, N$_2$ fixation, lithotrophy, respiration, and fermentation		Respiration and fermentation
Multicellularity	Simple		Simple or complex
		Archaea resemble eukaryotes in these traits:	
Intron splicing	Introns are rare	Introns are common	
RNA polymerase	Bacterial	Eukaryotic form	
Transcription factors	Bacterial	Eukaryotic form	
Ribosome sensitivity to chloramphenicol, kanamycin, and streptomycin	Sensitive	Resistant	
Translation initiator	Formylmethionine	Methionine (except mitochondria use formylmethionine)	
Cell wall	Peptidoglycan	Pseudopeptidoglycan or other polymer; or protein S-layer	
	Bacteria resemble eukaryotes and differ from archaea in these traits:		
Methanogenesis	No	Yes	No
Thermophilic growth	Up to 90°C	Up to 120°C	Up to 80°C
Photosynthesis	Many species; bacteriochlorophyll (proteorhodopsin derived from archaea)	Haloarchaea only; bacteriorhodopsin	Many species; chlorophyll (bacterial origin)
Chlorophyll light absorption	Red and blue	Green (central range of solar spectrum)	Red and blue (chloroplasts of bacterial origin)
Membrane lipids (major)	Ester-linked fatty acids	Ether-linked isoprenoid	Ester-linked fatty acids
Pathogens infecting animals or plants	Many pathogens	No pathogens	Many pathogens

Nonetheless, important differences emerge between each pair of domains; indeed, each domain shows distinctive traits absent or scarce in the other two. We summarize here the major features common to most members of each domain. Specific classes and species within each domain are explored in Chapters 18–20.

Archaea, bacteria, and eukaryotes. Of the three domains, the eukaryotes stand out as having a nucleus and other complex membranous organelles (see **Table 17.2**). Eukaryotic organelles include mitochondria and chloroplasts, which evolved from internalized bacteria. Bacteria and archaea possess no nucleus and have relatively simple intracellular membranes. Their size is limited by diffusion across the cell membrane, with occasional exceptions, such as the "giant bacterium" *Epulopiscium fisheisoni*. The larger size and complexity of eukaryotic cells generally require the most high-powered sources of energy, such as aerobic respiration and oxygenic photosynthesis, although some protists conduct fermentation.

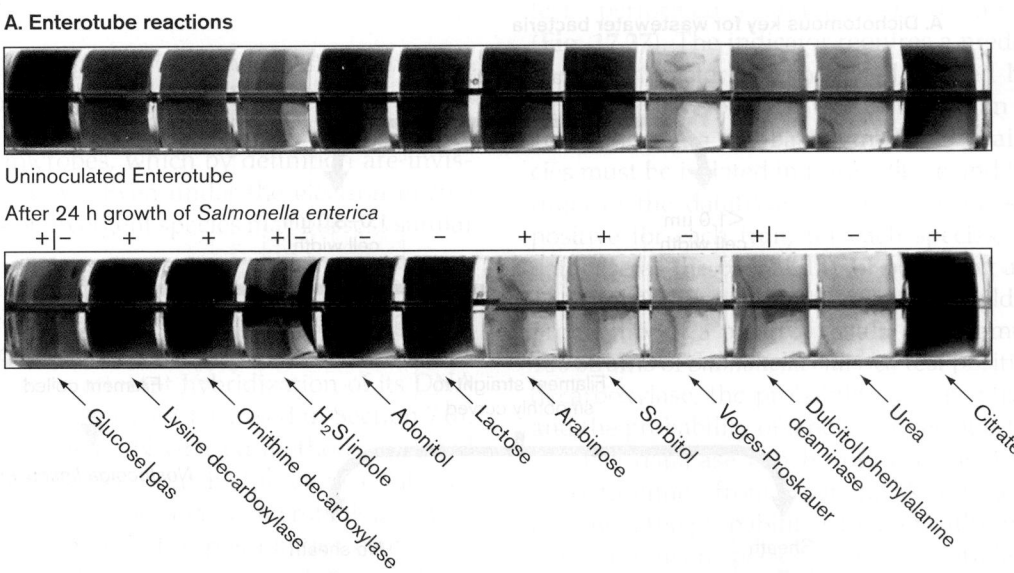

A. Enterotube reactions

Uninoculated Enterotube

After 24 h growth of *Salmonella enterica*

+|–   +   +   +|–   –   –   +   +   –   +|–   –   +

Glucose|gas · Lysine decarboxylase · Ornithine decarboxylase · H₂S|indole · Adonitol · Lactose · Arabinose · Sorbitol · Voges-Proskauer · Dulcitol|phenylalanine deaminase · Urea · Citrate

B.

Species	D-Glucose acid	Lysine decarbox.	Ornithine decarbox.	H₂S	Adonitol ferment	Lactose ferment	D-Sorbitol ferment	Dulcitol ferment	Score
Test Results:	+	+	+	+	–	–	+	–	
Salmonella enterica	0.99	0.98	0.97	0.95	0.01	0.01	0.95	0.96	3.3 E-2
Proteus mirabilis	0.99	0.01	0.99	0.98	0.01	0.02	0.01	0.01	9.2 E-5
Yersinia enterocolitica	0.99	0.01	0.95	0.01	0.01	0.05	0.99	0.01	8.7 E-5
Enterobacter aerogenes	0.99	0.98	0.98	0.01	0.98	0.95	0.99	0.05	8.9 E-6
Klebsiella pneumoniae	0.99	0.98	0.01	0.01	0.90	0.98	0.99	0.30	1.3 E-7

Figure 17.27   The Enterotube II, a probabilistic indicator to identify enteric Gram-negative pathogens.   A. The upper tube shows appearance of the uninoculated control. The lower tube shows test results for *Salmonella enterica*. **B.** A simplified indicator table in which the probabilities are multiplied, yielding a probability score for each organism in the database. *Source:* Joan Slonczewski and BD Diagnostics.

17.6 Symbiosis and the Origin of Mitochondria and Chloroplasts

So far, we have largely considered single species in isolation. In fact, however, all organisms evolve in the presence of other kinds of species, with whom they share interactions, both positive and negative. A major engine of evolution is **symbiosis**, the intimate association of two unrelated species. The physiology and behavioral adaptations of microbial symbiosis are discussed in Chapter 21; here we focus on the role of symbiosis in evolution.

The word "symbiosis" is popularly understood to mean **mutualism**, a relationship in which both partners

benefit and may absolutely require each other. Biologists, however, recognize **parasitism**, in which one partner is harmed, as a relationship equally as intimate as mutualism; both mutualism and parasitism are forms of symbiosis. Intimate relationships between species, either negative or positive, lead to **coevolution**, the evolution of two species in response to one another, showing parallel phylogeny. Coevolution commonly involves reductive evolution, where one partner species loses traits provided by the other partner. For example, the intracellular bacterium *Wolbachia*, found within nematodes that cause river blindness, has lost the ability to produce many nutrients, while retaining the ability to make purines and pyrimidines required by its host cell. Intracellular parasites such as chlamydias and

A.

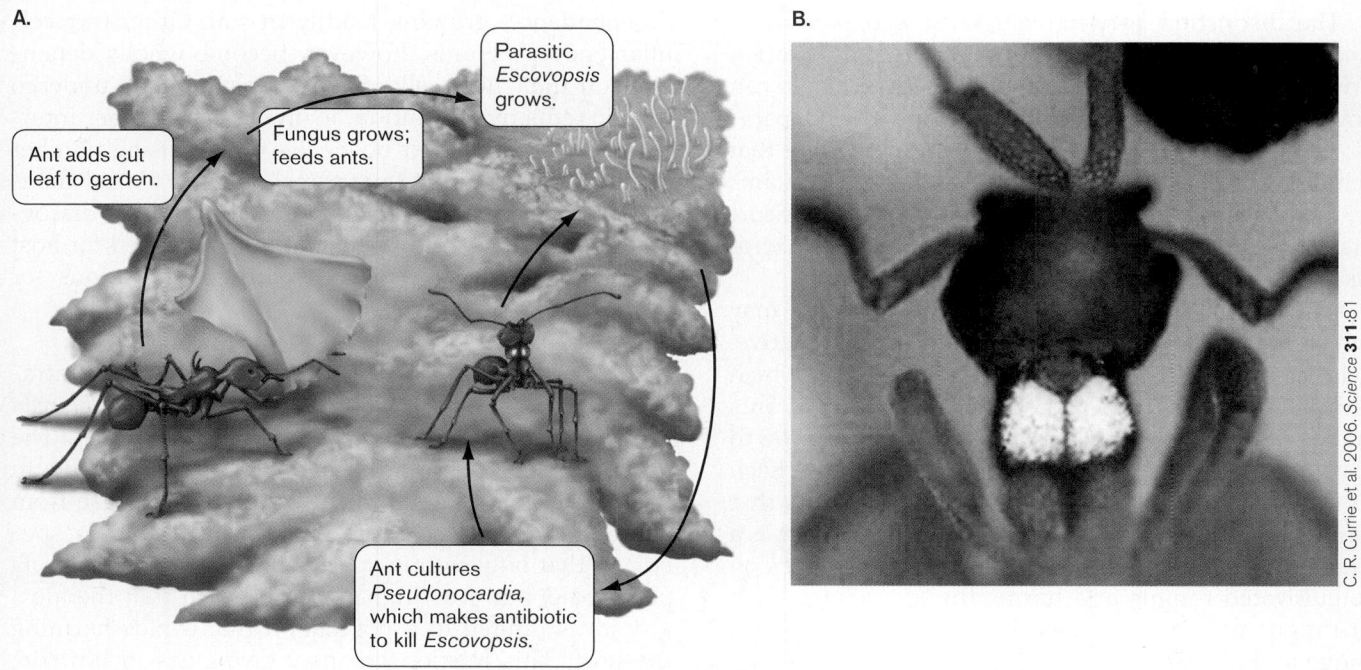

Ant adds cut leaf to garden.

Fungus grows; feeds ants.

Parasitic *Escovopsis* grows.

Ant cultures *Pseudonocardia*, which makes antibiotic to kill *Escovopsis*.

B.

C. R. Currie et al. 2006. *Science* **311**:81

C.

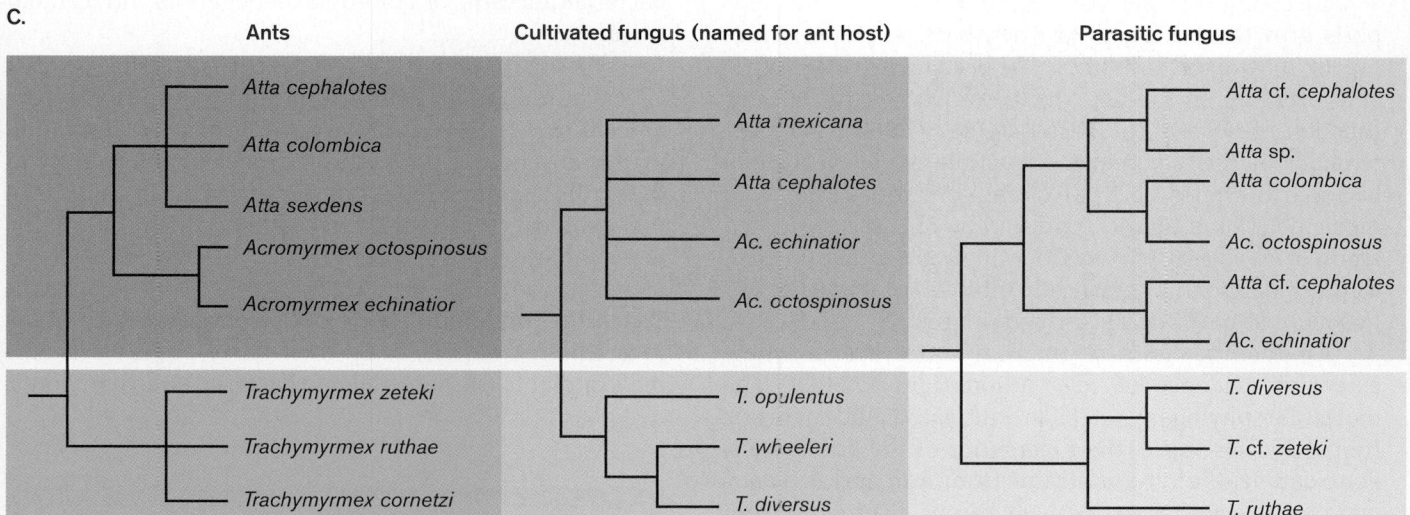

Ants	Cultivated fungus (named for ant host)	Parasitic fungus

Ants:
Atta cephalotes
Atta colombica
Atta sexdens
Acromyrmex octospinosus
Acromyrmex echinatior
Trachymyrmex zeteki
Trachymyrmex ruthae
Trachymyrmex cornetzi

Cultivated fungus (named for ant host):
Atta mexicana
Atta cephalotes
Ac. echinatior
Ac. octospinosus
T. opulentus
T. wheeleri
T. diversus

Parasitic fungus:
Atta cf. *cephalotes*
Atta sp.
Atta colombica
Ac. octospinosus
Atta cf. *cephalotes*
Ac. echinatior
T. diversus
T. cf. *zeteki*
T. ruthae

Figure 17.28 Ant-fungal gardens: a four-way symbiosis.   A. Attine ants cut leaves to cultivate mutualistic fungi, which feed the ants. When the parasitic fungus *Escovopsis* sp. grows, ants grow *Pseudonocardia* sp. on their bodies to produce antibiotics. **B.** Colonies of *Pseudonocardia* sp. on an ant cuticle. **C.** Coevolution of ants, cultivated fungi, and parasitic fungi. In most cases, the divergence of the ant species is paralleled by the divergence of their symbionts. Strain designations are based on the host ant gardens where they were isolated. *Source:* Parts A and C Cameron R. Currie et al. 2003. *Science* **299**:386.

rickettsias show similar reductive evolution, but they also evolve specialized traits enabling growth at host expense and evasion of the host immune system.

Microbial Interactions Lead to Coevolution

Interactions of microbial populations with different organisms include a range of positive through negative relationships (discussed in Chapter 21). A famous example of mutualism is that of bacterial nitrogen fixation, in which rhizobia form intracellular "bacteroids" within legume tissues. Both rhizobia and their plant hosts are highly evolved to respond to each other chemically and develop the nitrogen-fixing system. The plants provide protection and nutrients from photosynthesis while receiving fixed nitrogen from the bacteria. Another case of mutualism is that of lichens, multicellular systems composed of fungi and algae. These two cases and other forms of mutualism are discussed from the perspective of ecology in Chapter 21.

The distinction between mutualist and parasite is often subtle. Lichens consist of a mutualistic association between fungi and algae, but environmental change can convert the fungus to a parasite. On the other hand, parasitic microbes may coevolve with a host to the point that each depends on the other for optimal health. For example, the incidence of certain allergies and immune disorders, such as multiple sclerosis, appears to correlate with lack of exposure to pathogens and parasites.

In natural habitats, microbial symbiosis may involve multiple partners, both positive and negative. A classic example is that of the leaf-cutter ants, which cultivate fungal partners (**Fig. 17.28A**). Leaf-cutter ants arrange cut leaves in underground burrows for growth of a fungus that the ant is adapted to consume. Each ant species associates with one species of fungus that coevolved with the ants. A third "partner," however, is a parasitic fungus of the genus *Escovopsis*, which feeds on the cultivated fungus and harms the ant colony. Each ant-fungus pair is parasitized by an *Escovopsis* species unique to that pair.

To counteract the parasite, however, the ant supports growth of a fourth partner, bacteria that produce potent antifungal antibiotics. The bacteria grow on a specific structure on the ant cuticle, evolved for the specific function of hosting the bacteria (**Fig. 17.28B**). The bacteria, genus *Pseudonocardia*, are actinomycetes, a clade of bacteria that produce most of the antibiotics known to medicine. The antibiotic produced by one ant-associated strain of *Pseudonocardia* was identified as dentigerumycin, a cyclic peptide that specifically inhibits the parasitic fungus without harming the farmed fungus.

All four partners of the leaf-cutter system show extensive evidence of coevolution (**Fig. 17.28C**). The molecular phylogeny of both cultivated and parasitic fungal species tracks the divergence of the ant species. For example, the fungus strains from *Atta* and *Acromyrmex* host species show relatively recent divergence from each other, but they are highly diverged from the fungi of the more distantly related *Trachymyrmex* hosts. The antibiotic-producing bacteria also show host-related phylogeny.

Endosymbiosis Leads to Reductive Evolution and Obligate Association

The most intimate kind of symbiosis is **endosymbiosis**, in which one partner population grows within the body of another organism. Endosymbiosis includes communities of microbes within the digestive tract of animals, such as the human intestinal flora (discussed in Chapter 21). The internalized **endosymbiont** can also be intracellular, as in the case of rhizobial bacteroids within legume tissues. Rhizobia retain the genetic capacity for

independence, growing readily in soil. Other intracellular endosymbionts, however, become wholly dependent on their host cells. Such endosymbionts undergo drastic reductive evolution, evolving ever-deeper interdependence with their host cells. A surprising number of human invertebrate parasites, such as filarial nematodes and *Anopheles* mosquitoes, have been discovered to carry bacterial endosymbionts required for host growth. This discovery has exciting implications for treatment.

Microbial endosymbionts. A simple example of intracellular endosymbiosis is that of the alga *Chlorella* growing within *Paramecium bursaria* (**Fig. 17.29**). The algae conduct photosynthesis and provide nutrients to the paramecium, which in turn shelters the algae from predators and viruses. The relationship is highly specific in that only certain species of algae and paramecia participate, and it is highly controlled in that the algal growth is limited to a population that avoids harming the host. This relationship may give clues to how the bacterial ancestor of chloroplasts began its intracellular existence.

The algal symbiosis, however, is reversible in that *Chlorella* retains its ability to multiply outside the paramecium. Moreover, under conditions of starvation in the absence of light, the paramecium may start to digest its endosymbionts as prey. Thus, the nature of the symbiosis (mutualistic or predatory) depends on the environment.

Bacterial endosymbionts of invertebrates. Many invertebrate animals, often themselves parasites of animals or plants, possess obligate bacterial endosymbionts.

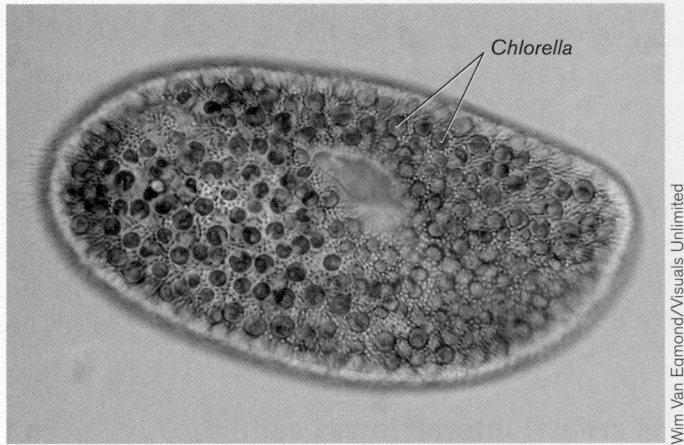

Figure 17.29 Endosymbiosis. *Paramecium bursaria*, a ciliate protist with endosymbiotic *Chlorella* algae. Cell length of *Paramecium*: 100–150 μm (LM).

Wim Van Egmond/Visuals Unlimited

In some cases the endosymbiont is a parasite, such as the bacterium *Wolbachia pipientis* that grows within cells of the fruit fly, *Drosophila*. *Wolbachia* bacteria are related to the rickettsias, obligate intracellular pathogens carried by arthropods and transmitted to humans. The *Wolbachia* strains that infect *Drosophila* cells are transmitted only through egg cells; they cannot exist outside the insect. The bacteria have evolved ways to manipulate *Drosophila* reproduction to enhance their own transmission; for example, they feminize infected males so as to produce eggs, which carry *Wolbachia* into the next generation of flies.

Other invertebrate endosymbionts are mutualists. In fact, 15% of insect species depend on intracellular bacteria to produce essential nutrients, such as certain amino acids or vitamins. In these mutualisms, both partners have lost essential traits by reductive evolution, and each now requires the partner species to provide the lost function. The degree of genome reduction is most extreme in the intracellular partner, which can be seen as heading in the same evolutionary direction that led to mitochondria and chloroplasts.

The discovery of intracellular bacteria in invertebrates has exciting medical implications for treatment of parasitic diseases. Invertebrate parasites such as filarial nematode worms invade human lymph nodes, causing forms of disease (filariasis) that are notoriously difficult to treat. Few antibiotics are sufficiently selective for worm metabolism versus human metabolism, since both are eukaryotic animals. A form of filariasis is elephantiasis, in which a limb expands with huge numbers of worms (**Fig. 17.30**). Filariasis afflicts more than 120 million people worldwide, largely in the Indian subcontinent and in Africa.

The filarial nematode *Brugia malayi* harbors *Wolbachia* endosymbionts (**Figs. 17.31**). These *Wolbachia* strains differ from those that parasitize insects. *Wolbachia* may have entered the nematode originally as a pathogen or parasite, and then persisted because of its metabolic contributions to the host. The nematode endosymbiont strains are mutualists; their presence is required for the nematode's embryonic development. The bacteria are found within tissue layers beneath the nematodes' skin and within the uterine tubes of females, where they enter the developing offspring (**Fig. 17.31**). When human patients infected by the nematodes are treated with antibiotics such as tetracycline, the bacteria disappear from worm tissues. The worm burden gradually decreases, and no offspring are produced. Antibacterial antibiotics eliminate the worms sooner and more completely than does treatment with anti-nematode agents.

The genome of a *Wolbachia* strain from a filarial nematode reveals extensive reductive evolution (**Fig. 17.32**). With barely a million base pairs, the *Wolbachia* genome has lost many metabolic pathways. It retains glycolysis and the TCA cycle but has lost the pathways for biosynthesis of all amino acids and most vitamins. It nonetheless retains pathways to make purines, pyrimidines, and the coenzymes riboflavin and FAD—essential pathways lost by its host nematode. Overall, *Wolbachia* appears to be evolving into an organelle of its host, like the ancestors of mitochondria and chloroplasts.

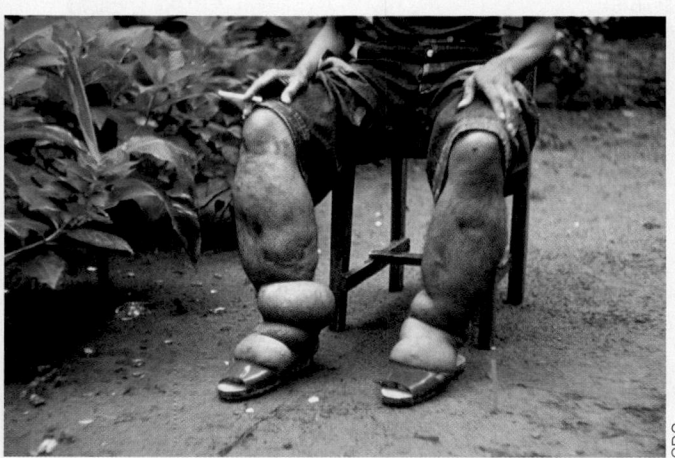

Figure 17.30 Filariasis. Patient suffering from filariasis, a form also known as elephantiasis.

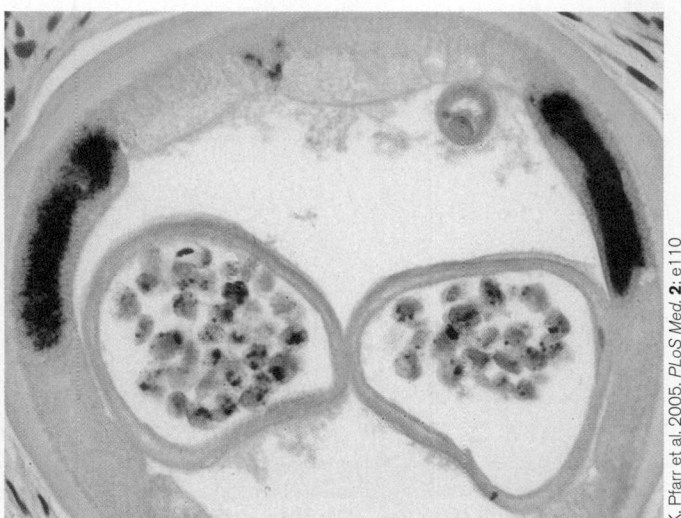

Figure 17.31 The filarial endosymbiont *Wolbachia*. Cross section of the nematode *Brugia malayi*, showing *Wolbachia* bacteria (stained pink) within the dermis and the uterine tubes.

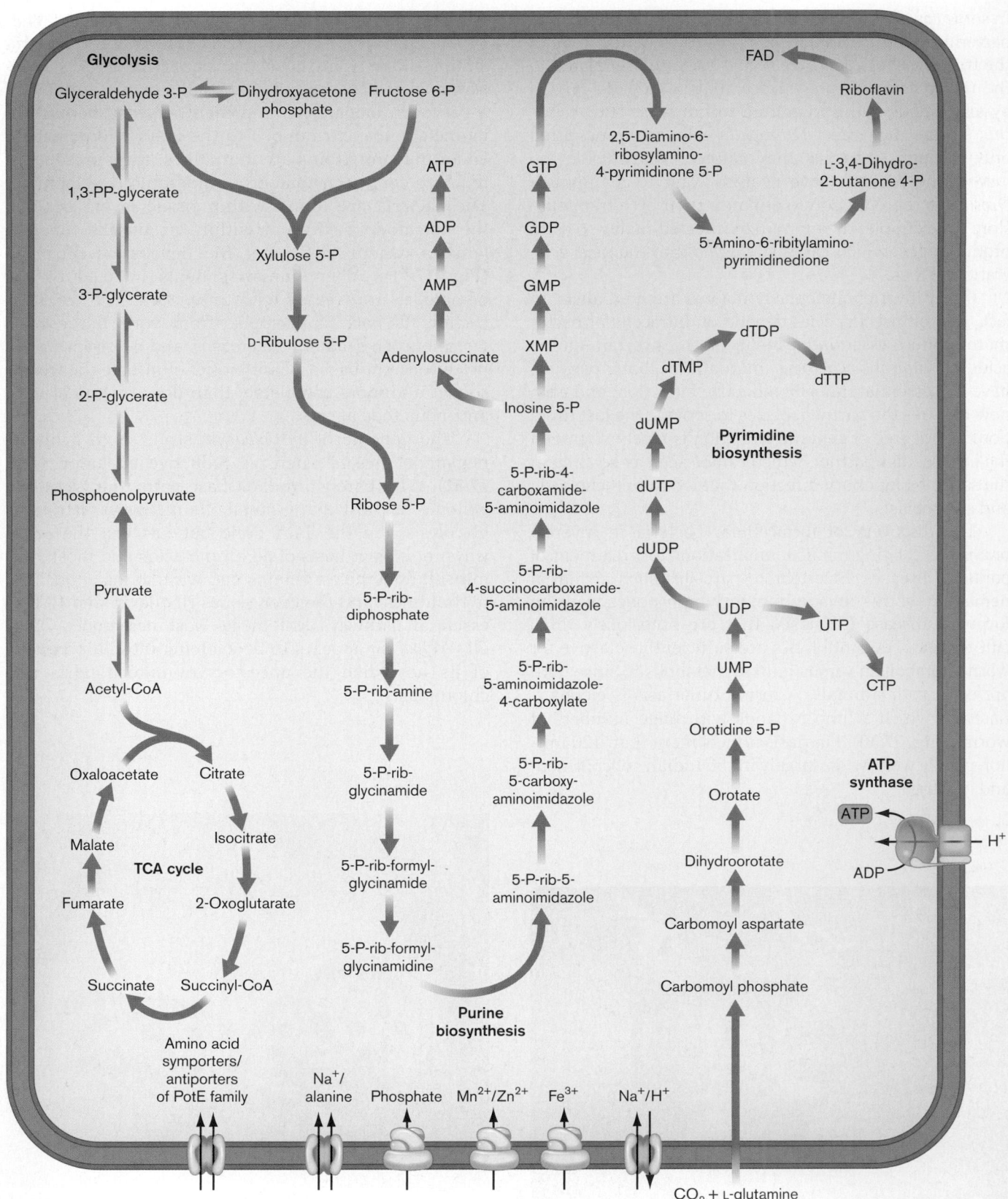

Figure 17.32 **Metabolic pathways retained in the *Wolbachia* genome.** Reductive evolution has eliminated much of the *Wolbachia* genome, but it retains metabolic pathways to make purines, pyrimidines, and vitamins such as riboflavin for its nematode host.

Intracellular endosymbionts are surprisingly common in arthropods, with exciting implications for medicine and agriculture. For further discussion, see Chapter 21.

Mitochondria and Chloroplasts

As Lynn Margulis and colleagues have shown, the assimilation of endosymbionts as mitochondria and chloroplasts played a central role in the evolution of eukaryotes (**Fig. 17.33**). Many similar endosymbioses are known today, such as the free-living algae *Chlorella* acquired by the protist predator *Paramecium bursaria* (see **Fig. 17.29**). Mitochondria, like *Wolbachia*, evolved from a bacterium related to the rickettsias. The mitochondrial ancestor must have entered the eukaryotic lineage as, or shortly after, the eukaryotes diverged from archaea, since all known eukaryotes retain mitochondria or vestigial remnants of mitochondrial genomes. Mitochondria provide the cell with the essential functions of electron transport and respiration. The electron transport system (ETS) is found in the mitochondrial inner membrane, believed to derive from the cell membrane of the ancestral bacterium. The outer membrane may derive from the invaginating membrane of the host cell that originally engulfed the endosymbiont.

Chloroplasts arose from cyanobacteria at some point before the divergence of red and green algae (discussed in Chapter 20). A model for cyanobacterial uptake can be seen in the protist *Glaucocystophyta*, which independently (and much later) took up cyanobacterial endosymbionts. The endosymbionts of *Glaucocystophyta* retain cell walls and some metabolism. Like mitochondria, chloroplasts possess inner and outer membranes, believed to derive from the ancestral endosymbiont and host, respectively. Photosynthetic complexes are located in the thylakoid membranes, similar to those of modern cyanobacteria.

The genomes of mitochondria and chloroplasts both show extreme reduction, even more extreme than that of any known endosymbiotic bacteria (**Fig. 17.34**). The few genes that remain include remnants of the central transcription-translation apparatus, such as rRNA and tRNAs, as well as a handful of genes whose products are essential for survival of the host cell: for respiration (mitochondria) or photosynthesis (chloroplasts). In human mitochondria, the essential products form the respiratory chain, including subunits of NADH dehydrogenase, cytochrome oxidase, and ATP synthase. Mutations in these key genes lead to serious diseases; for example, damage to mitochondrial genes of respiration is associated with motor neuron disease, parkinsonism, and forms of ataxia.

But thousands of genes encoding ETS subunits, as well as other essential parts of mitochondria, have migrated from the mitochondrion to the nucleus. The nuclear acquisition probably occurred through accidental

Figure 17.33 Endosymbiotic cells evolved into mitochondria and chloroplasts. Eukaryotic cells contain mitochondria and chloroplasts, organellar remnants of ancient endosymbioses. Inset: Lynn Margulis. *Source*: Margulis. 1993.

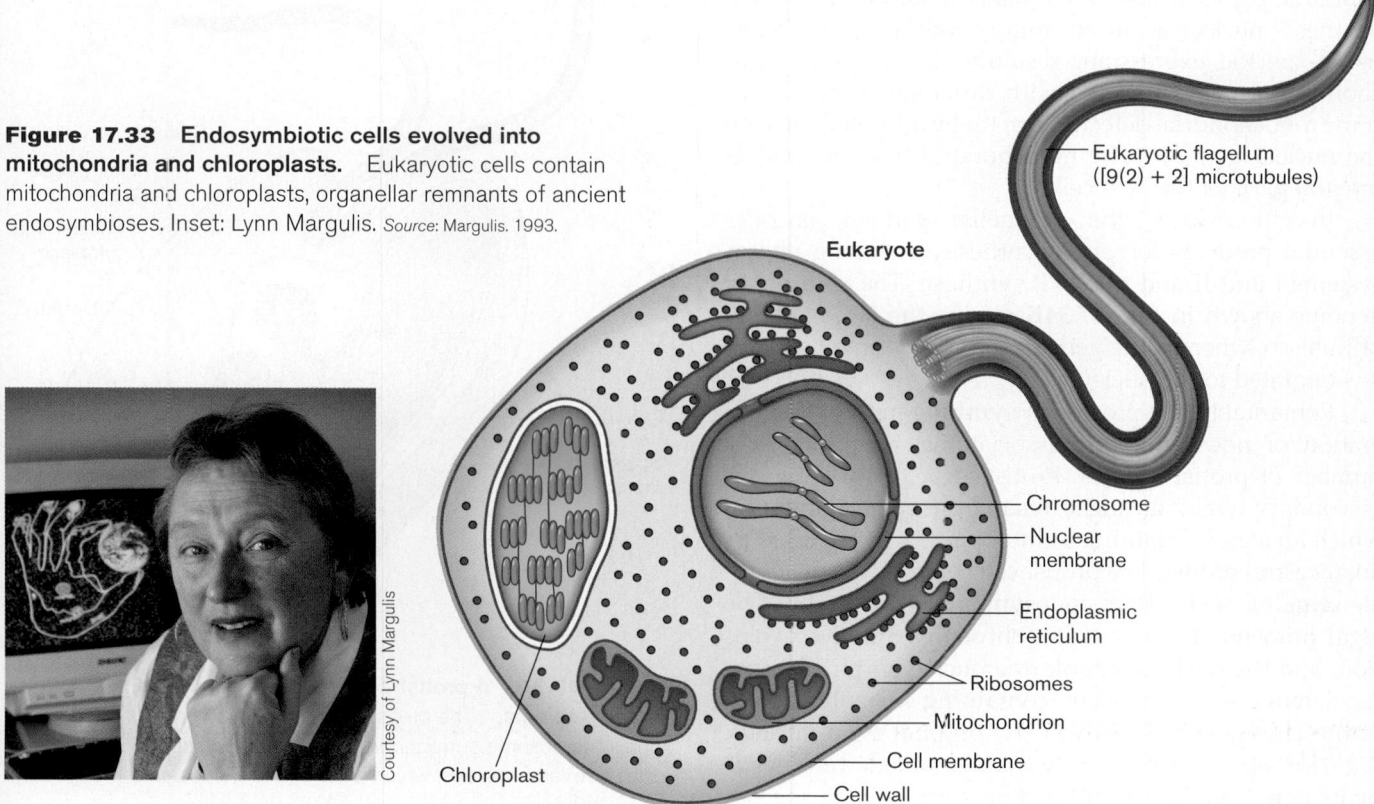

Courtesy of Lynn Margulis

Eukaryotic flagellum ([9(2) + 2] microtubules)

Eukaryote

Chromosome

Nuclear membrane

Endoplasmic reticulum

Ribosomes

Mitochondrion

Cell membrane

Cell wall

Chloroplast

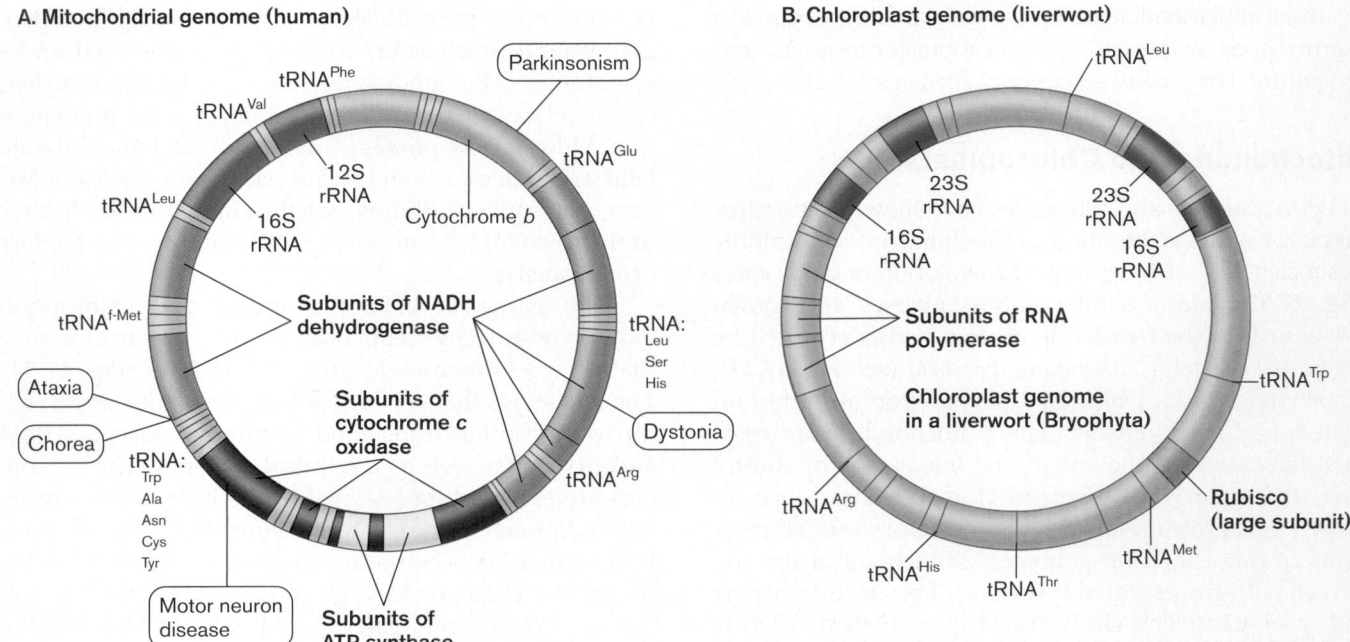

A. Mitochondrial genome (human)

B. Chloroplast genome (liverwort)

Figure 17.34   Genomes of mitochondria and chloroplasts.   A. The mitochondrial genome retains large- and small-subunit rRNAs, five tRNA genes, plus subunits of the respiratory electron transport chain. Bubbles indicate human diseases associated with mitochondrial defects. **B.** The chloroplast genome retains large- and small-subunit rRNAs (23S and 16S), several tRNA and RNA polymerase genes, plus Rubisco and components of photosystems I and II (dispersed around the circle, not labeled in figure).

copying of mitochondrial genes into the nuclear genome. Reductive evolution then occurred, faster in the mitochondrial copy because of the faster mutation rate. Some of these nuclear-acquired mitochondrial genes show tissue-specific expression, resulting in different mitochondrial types associated with different tissues. Thus, some mitochondrial defects are actually inherited through the nuclear genome. The mitochondria have evolved as integral parts of the host cell.

In chloroplasts, the organellar genome encodes essential products for photosynthesis, including photosystems I and II and the ATP synthase. The chloroplast genome shown in **Fig. 17.34B** retains the large subunit of Rubisco, whereas the gene encoding the small subunit has migrated to the nucleus.

Remarkably, the process of **symbiogenesis**, the generation of new symbiotic associations, continues in a number of protist species. Protist-algae, also known as "secondary symbiont" algae, result from symbiogenesis in which an alga (containing a chloroplast) was engulfed by an ancestral protist. The protist cell contains the degenerate remains of the algal endosymbiont (**Fig. 17.35**). The algal mitochondrion was lost through reductive evolution, and the nucleus shrank to a nucleomorph, the vestigial remains of a nucleus containing a small amount of the chromosomal DNA of the original algal nucleus. But the algal chloroplast was maintained, "enslaved" by its new host. The result is a new protist-alga species,

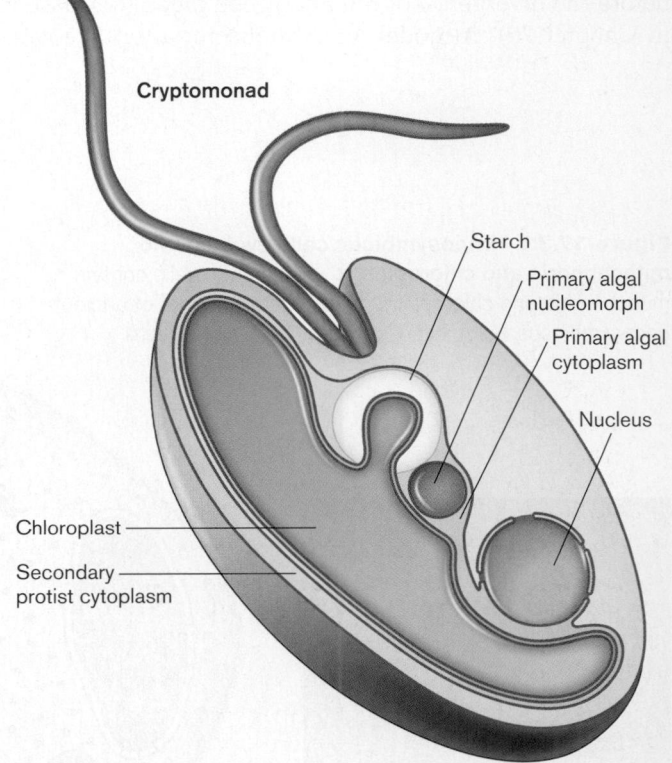

Figure 17.35   A protist-alga results from secondary endosymbiosis.   The protist *Guillardia theta* contains the primary algal chromosome surrounded by the remains of the primary algal cytoplasm and nucleus, which is shrunk to a nucleomorph.
Source: Modified from Paul R. Gilson. 2001. *Genome Biol.* **2**:1022.

Guillardia theta, capable of both phototrophy and heterotrophy. Its chloroplast has a double membrane derived from the original cyanobacterial ancestor of the chloroplast and from the algal ancestor, surrounded by another double membrane derived from the cell membranes of the alga and a secondary host.

Other secondary endosymbiont algae, such as kelps, diatoms, and dinoflagellates, are discussed in Chapter 20. In some species, tertiary symbiosis has been documented, in which a secondary-endosymbiont alga has been swallowed in turn by another protist.

> **THOUGHT QUESTION 17.8** Besides mitochondria and chloroplasts, what other kinds of entities within cells might have evolved from endosymbionts?

TO SUMMARIZE:

- **Symbiosis is the intimate association of two unrelated species.** A symbiosis in which both partners benefit is called **mutualism**. If one partner benefits while harming the other, this is called **parasitism**.
- **Symbiotic partners undergo coevolution**, the evolution of two species in response to one another. Coevolution involves reductive (degenerative) evolution, in which each partner species loses some functions that the other partner provides.

- **An endosymbiont lives inside a much larger host species.** Many microbial cells harbor endosymbiotic bacteria whose metabolism yields energy for their hosts.
- **Many invertebrates harbor endosymbiotic bacteria.** The bacteria are required for host survival and in some cases for pathology caused by a parasitic invertebrate.
- **Mitochondria evolved from endosymbionts.** The ancestor of mitochondria was an alphaproteobacterium related to rickettsias.
- **Chloroplasts evolved from endosymbionts.** The chloroplast ancestor was a cyanobacterium.

Concluding Thoughts

Over 3.85 Gyr ago, microbial communities had evolved, their strata building rock layers that persist today. From these or other microbes, all subsequent life evolved. It is hard to say which is more astonishing: the overall commonalities of all living cells, including membrane-enclosed support systems for genomes of 3.85 billion years of shared ancestry, or the subsequent evolution of organisms with vastly different adaptations to exploit every possible niche of our planet. The next three chapters explore these diverse adaptations: Chapter 18, bacterial diversity; Chapter 19, archaeal diversity; and Chapter 20, diversity among microbial eukaryotes including fungi, algae, and protists.

CHAPTER REVIEW

Review Questions

1. What was the composition of Earth's early crust and atmosphere? What processes changed their composition to that found today?
2. What kinds of evidence support the presence of life in the Archaean eon? What are the advantages and limitations of each kind of evidence?
3. What kinds of metabolism are believed to have existed in Archaean life? What kinds of evidence support their existence?
4. Compare and contrast three models for the origin of the first cells. Which features of life does each model explain, and which features are unexplained?
5. Explain the roles of classification, nomenclature, and identification for microbial taxonomy.
6. Why is the definition of species in bacteria and archaea more problematic than for eukaryotes? What is generally considered the present basis for defining prokaryotic species?

7. Discuss the roles of mutation, natural selection, and reductive evolution in the divergence of microbial species. Cite specific examples.
8. Explain the basis of a "molecular clock" for measuring microbial evolution. What fundamental properties must be met by a gene to function as a molecular clock? What are the limitations of a molecular clock?
9. Explain the basis of a phylogenetic tree. Why is the fundamental tree at the divergence of bacteria, archaea, and eukaryotes unrooted?
10. How does horizontal gene transfer determine genomic content? What kinds of genes are likely to undergo horizontal transfer?
11. Explain how endosymbiosis can lead to obligate association. Explain how reductive evolution and gene transfer lead to the evolution of organelles that are inseparable from host cells.

Thought Questions

1. How convincing are the microfossils in Figure 17.6? What criteria do you think would define a microfossil?
2. In the phylogeny shown here, where are the root and the outgroup? How does the outgroup organism differ from the others? Which two organisms are the most closely related? Which node represents the last common ancestor of *Neisseria* and *Haemophilus*? Which genome has evolved much faster than the others, and why?
3. Suppose you aim to use SSU rRNA gene sequences to sample all the species in a microbial community. How can you be sure of sampling everything? What are the limitations of this approach?

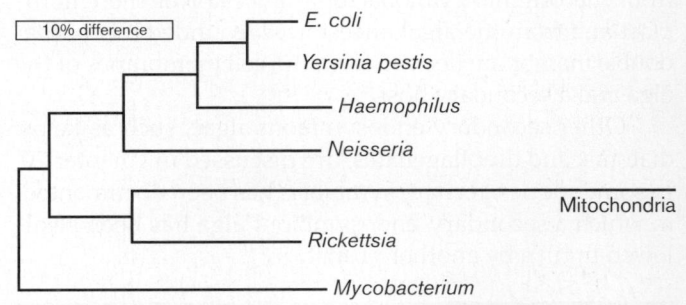

Key Terms

abiotic (635)
Archaean eon (627)
banded iron formation (BIF) (633)
biosignature (biological signature) (630)
biosphere (627)
candidate species (653)
clade (638)
classification (652)
coevolution (658)
core genome (652)
dichotomous key (656)
domain (646)
endosymbiont (660)
endosymbiosis (660)
genomic island (650)
genus name (652)
greenhouse effect (627)

Hadean eon (627)
horizontal gene transfer (648)
identification (652)
isolate (653)
isotope ratio (631)
metabolist model (635, 636)
metagenomics (653)
microfossil (629)
molecular clock (639)
monophyletic group (638)
mutualism (658)
node (642)
nomenclature (652)
pan-genome (652)
panspermia (638)
parasitism (658)
phylogenetic tree (641)

phylogeny (638)
prebiotic soup (635)
probabilistic indicator (656)
reductive evolution (639)
ribozyme (636)
RNA world (635)
root of a tree (642)
small-subunit (SSU) rRNA (641)
species (652)
species name (652)
stromatolite (624)
supernova (625)
symbiogenesis (664)
symbiosis (658)
taxon (652)
taxonomy (652)
vertical gene transfer (648)

Recommended Reading

Achtman, Mark, and Michael Wagner. 2008. Microbial diversity and the genetic nature of microbial species. *Nature Reviews. Microbiology* 6:431–440.

Bada, Jeffrey. 2004. How life began on Earth: A status report. *Earth and Planetary Science Letters* 226:1–15.

Barns, Susan M., Charles F. Delwiche, Jeffery D. Palmer, and Norman R. Pace. 1996. Perspectives on archaeal diversity, thermophily and monophyly from environmental rRNA sequences. *Proceedings of the National Academy of Sciences USA* 93:9188–9193.

Brasier, Martin D., Owen R. Green, Andrew P. Jephcoat, Annette K. Kleppe, Martin J. Van Kranendonk, et al. 2002. Questioning the evidence for Earth's oldest fossils. *Nature* 416:76–81.

Currie, Cameron R., Bess Wong, Alison E. Stuart, Ted R. Schultz, Stephen A. Rehner, et al. 2003. Ancient tripartite coevolution in the attine ant-microbe symbiosis. *Science* 299:386.

Fraser, Christophe, Eric J. Alm, Martin F. Polz, Brian G. Spratt, and William P. Hanage. 2009. The bacterial species challenge: Making sense of genetic and ecological diversity. *Science* 322:741–746.

Gevers, Dirk, Frederick M. Cohan, Jeffrey G. Lawrence, Brian G. Spratt, Tom Coenye, et al. 2005. Re-evaluating prokaryotic species. *Nature Reviews. Microbiology* **3**:733–739.

Gilson, Paul R. 2001. Nucleomorph genomes: Much ado about practically nothing. *Genome Biology* **2**:1022.1–1022.5.

Hanage, William P., Christophe Fraser, and Brian G. Spratt. 2005. Fuzzy species among recombinogenic bacteria. *BMC Biology* **3**:6.

Hanczyc, Martin M., Shelly M. Fujikawa, and Jack W. Szostak. 2003. Experimental models of primitive cellular compartments: Encapsulation, growth, and division. *Science* **302**:618–622.

Jiao, Yongqin, Andreas Kappler, Laura R. Croal, and Dianne K. Newman. 2005. Isolation and characterization of a genetically-tractable photoautotrophic Fe(II)-oxidizing bacterium, *Rhodopseudomonas palustris* strain TIE-1. *Applied and Environmental Microbiology* **71**:4487–4496.

Kasting, James F., and Janet L. Siefert. 2002. Life and the evolution of Earth's atmosphere. *Science* **296**:1066–1067.

Kelley, Scott T., Ulrike Theisen, Largus T. Angenent, Allison St. Amand, and Norman R. Pace. 2004. Molecular analysis of shower curtain biofilm microbes. *Applied and Environmental Microbiology* **70**:4187–4192.

Lapierre, Pascal, and J. Peter Gogarten. 2009. Estimating the size of the bacterial pan-genome. *Trends in Genetics* **25**:107–110.

Lincoln, Tracey A., and Gerald F. Joyce. 2009. Self-sustained replication of an RNA enzyme. *Science* **323**:1229–1232.

Moran, Nancy A., John P. McCutcheon, and Atsushi Nakabachi. 2008. Genomics and evolution of heritable bacterial symbionts. *Annual Reviews in Genetics* **42**:165–190.

Moser, Duane P., Thomas M. Gihring, Fred J. Brockman, James K. Fredrickson, David L. Balkwill, et al. 2005. *Desulfotomaculum* and *Methanobacterium* spp. dominate a 4- to 5-kilometer-deep fault. *Applied and Environmental Microbiology* **71**:8773–8783.

Ochman, Howard. 2005. Genomes on the shrink. *Proceedings of the National Academy of Sciences USA* **102**:11959–11960.

Salzberg, Steven L., Julie C. Dunning Hotopp, Arthur L. Delcher, Mihai Pop, Douglas R. Smith, et al. 2005. Serendipitous discovery of *Wolbachia* genomes in multiple *Drosophila* species. *Genome Biology* **6**:R23.

Shen, Yanan, Roger Buick, and Donald E. Canfield. 2001. Isotopic evidence for microbial sulphate reduction in the early Archaean era. *Nature* **410**:77–81.

Souza, Valerie, Amanda Castillo, and Luis Equiarte. 2002. Evolutionary ecology of *E. coli*. *American Scientist* **90**:332.

Tettelin, Hervé, David Riley, Ciro Cattuto, and Duccio Medini. 2008. Comparative genomics: The bacterial pan-genome. *Current Opinion in Microbiology* **12**:472–477.

Welander, Paula V., Maureen L. Coleman, Alex L. Sessions, Roger E. Summons, and Dianne K. Newman. 2010. Identification of a methylase required for 2-methylhopanoid production and implications for the interpretation of sedimentary hopanes. *PNAS* **107**:8537–8542.

Wernegreen, Jennifer J. 2005. Endosymbiosis: Lessons in conflict resolution. *PLoS Biology* **2**:307–311.

For further study and review, please visit StudySpace at **microbiology2.com**.

Chapter 18

Bacterial Diversity

Bacteria vary tremendously in their cell structure and metabolism. They include heterotrophs, phototrophs, and lithotrophs; some species can be classified as all three. There are obligate aerobes, anaerobes, and microaerophiles. Their cell shapes include rods, cocci, spirals, and budding forms. Ecologically, bacteria include mutualists, pathogens, and organisms that cannot be cultured in our laboratories. New species are continually discovered in soil and water, in our homes, and even within our own bodies.

How do we make sense of the thousands of different kinds of bacteria? This chapter introduces the major categories of bacteria about which we know the most. These include the Gram-positive rods and cocci, the Gram-negative proteobacteria, anaerobes, phototrophs, and spirochetes. There are cell wall–less chlamydias, planctomycetes, and deep-branching thermophiles. Even now, all the species we know represent but a tiny fraction of the diverse bacterial species growing in nature.

200 µm

Current Research Highlight

Biofilm of bacteria generating electricity. *Geobacter sulfurreducens* bacteria were cultured on the surface of a graphite electrode. The bacteria grow in towers with channels in between for nutrient flow. They donate electrons to the electrode, generating electricity. Image was obtained by laser confocal scanning microscopy; green fluorescence indicates metabolically active bacteria. *Source:* CDC/ Courtesy of Larry Stauffer, Oregon State Public Health Laboratory from Kelly P. Nevin et al. 2009. *PLoS One* 4:e5628.

Bacteria have evolved a bewildering array of life-forms that colonize every habitat on Earth, from the dunes of the Sahara desert to the ice-buried Lake Vostok of Antarctica. Every day we discover new species never before recorded, not just from exotic environments, but from habitats closer to home. Consider, for example, the digestive tract of a newborn infant (**Fig. 18.1**). Dutch researcher Antoon Akkermans and colleagues identified bacterial species from the feces of an infant during the first year of life. Infants acquire intestinal bacteria from their mother's skin at birth, and from all their surroundings thereafter; but which organisms persist and grow in the infant gut?

The bacteria were identified using polymerase chain reaction (PCR) to amplify small-subunit (SSU) rRNA gene sequences directly, without culturing. The rRNA sequences from the infant's fecal sample were separated by denaturing gradient gel electrophoresis (DGGE), a technique in which the cloned DNA products denature and form secondary structures that affect their migration in the gel. As each infant grows, the DGGE bands change as bacterial species disappear and new ones appear. The species distribution changes the most during months 4–6 as the infant is weaned from breast milk onto foods containing very different carbon sources.

Some of the DNA samples show close sequence similarity to well-characterized anaerobes such as *Bifidobacterium*, a Gram-positive rod that ferments without gas production. The lack of gas production by *Bifidobacterium* helps prevent colic (gas pains). Other Gram-positive rods were identified as clostridia, Gram-positive spore formers that include harmless species as well as pathogens causing tetanus and botulism. But more than half of the fecal isolates showed SSU rRNA with less than 97% similarity to any known sequence, a criterion for distinct species. Thus, even a human infant constitutes a microbial frontier full of unknown kinds of bacteria.

Characterizing bacteria is a formidable task. Even in familiar habitats the vast majority of species remain unknown and new phyla are discovered daily. At the same time, new fields of genomics, microscopy, and culture techniques clarify our view of major groups of bacteria with physiological, ecological, and medical significance. Chapter 18 presents these taxa in a format that emphasizes phylogeny (evolutionary relatedness), as well as key traits, such as the Gram-positive cell wall of Firmicutes and the oxygenic photosynthesis of cyanobacteria. Some microbial roles in communities are introduced. We explore microbial ecology in Chapter 21, and global cycling in Chapter 22.

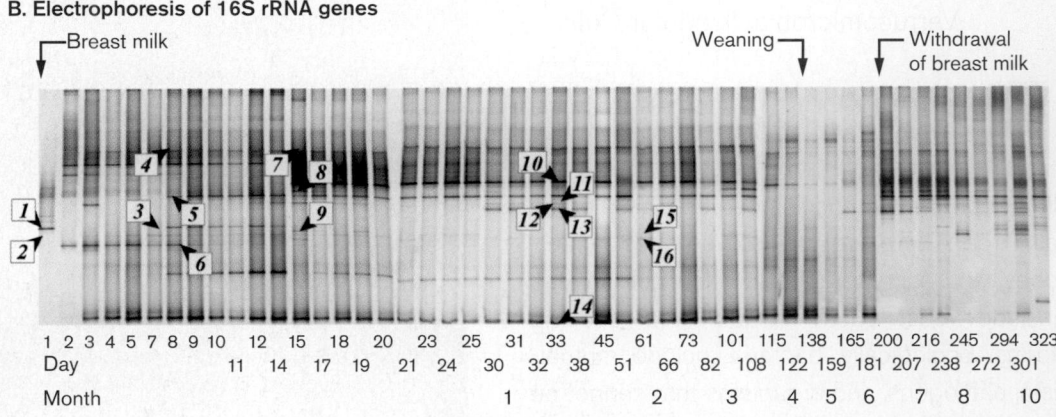

B. Electrophoresis of 16S rRNA genes

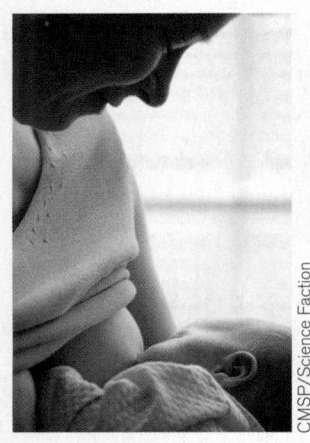
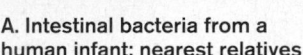

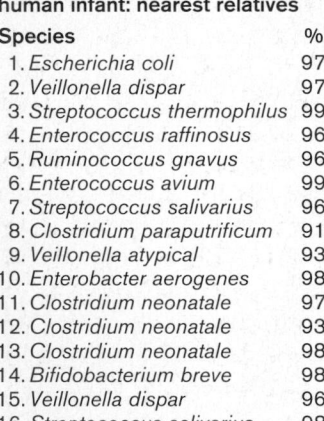

A. Intestinal bacteria from a human infant: nearest relatives

Species	%
1. *Escherichia coli*	97
2. *Veillonella dispar*	97
3. *Streptococcus thermophilus*	99
4. *Enterococcus raffinosus*	96
5. *Ruminococcus gnavus*	96
6. *Enterococcus avium*	99
7. *Streptococcus salivarius*	96
8. *Clostridium paraputrificum*	91
9. *Veillonella atypical*	93
10. *Enterobacter aerogenes*	98
11. *Clostridium neonatale*	97
12. *Clostridium neonatale*	93
13. *Clostridium neonatale*	98
14. *Bifidobacterium breve*	98
15. *Veillonella dispar*	96
16. *Streptococcus salivarius*	98

Figure 18.1 Bacterial succession in the digestive tract of a newborn baby. A. A nursing infant shows a unique combination of bacterial species. Comparative sequence analysis reveals the closest relative of each species isolated. **B.** From a newborn baby over time, fecal samples were tested. PCR-amplified sequences of 16S (SSU) rRNA genes were separated by denaturing gradient gel electrophoresis (DGGE). The patterns of bands change as species disappear and new ones appear, with the greatest changes occurring at weaning from breast milk to solid food. Band numbers refer to isolates for which 16S rRNA was sequenced (list at left). *Source:* Part B from Christine Favier et al. *Appl. Environ. Microbiol.* **68**:219.

For each major taxonomic group, we describe a few key species to represent the spectrum of diversity. For more information on particular species, consult the following public online resources:

■ **The National Center for Biotechnology Information (NCBI) Taxonomy Database**, funded by the U.S. government, includes the taxonomy of biological species reported in the literature and an ever-growing database of sequenced genomes.

www | National Center for Biotechnology Information (NCBI)
microbiology2.com/links

■ **MicrobeWiki (Microbial Biorealm)** at Kenyon College is a student-edited online resource for microbiology.

www | MicrobeWiki
microbiology2.com/links

18.1 Bacterial Diversity at a Glance

To survey bacterial diversity in one chapter is like touring all the countries of a continent in a single day. Like countries, bacterial taxa have complex traits and histories and often contested borders. At the same time, bacteria as a whole share major commonalities that offer an important perspective from which to appreciate their differences.

Common Traits of Bacteria

Chapter 17 summarized the differences and similarities of the three major domains—Bacteria, Archaea, and Eukarya (see Table 17.2). A profound commonality of bacteria is their central apparatus of gene expression, particularly their RNA polymerase and their ribosomal RNAs and translation factors. Bacterial gene expression complexes differ more from those of Archaea or Eukarya than those of Archaea or Eukarya differ from each other. This subtle point of molecular biology has a profound consequence for human medicine and agriculture: It underlies the selective activity of many antibiotics, such as streptomycin, that attack only bacterial pathogens, without harming the animal or plant host.

Another trait distinguishing bacteria from archaea and eukaryotes is that most bacterial cells possess a cell wall of peptidoglycan (discussed in Chapter 3). Peptidoglycan is composed of disaccharide-peptide chains that can cross-link in three dimensions; key enzymes that build the peptide links are blocked by antibiotics such as penicillin and vancomycin. Some archaea possess analogous sugar-peptide structures called "pseudopeptidoglycan" (discussed in Chapter 19), but their structural details

and antibiotic sensitivity differ fundamentally from bacterial peptidoglycan. Eukaryotes such as fungi and plants have cell walls of polysaccharides such as cellulose and chitin (discussed in Chapter 20).

In bacteria, variant forms of peptidoglycan distinguish different species. For instance, the Gram-positive pathogen *Staphylococcus aureus* has cell wall peptides cross-linked by pentaglycine (a chain of five glycine residues). Some species, such as mycoplasmas, lack peptidoglycan altogether, but they arose by reductive evolution from bacteria that possess it.

The Bacterial Tree

The phylogeny of known bacteria is presented in **Figure 18.2**. Not shown are numerous uncultivated bacteria that branch from all parts of this tree, known only from their DNA sequence. The names of well-studied phyla are lettered blue. A **phylum** is defined as a group of organisms sharing a common ancestor that diverged early from other bacteria based on SSU rRNA sequence (discussed in Chapter 17). Increasingly, whole-genome data are supplementing and refining our definition of bacterial clades. Phyla and other major divisions are also defined on the basis of historical convention and consensus of the research community.

A bacterial phylum comprises species that share key traits as well as ancestry. While sharing key traits, the member species often show remarkable diversity in other ways. Some phyla, such as Cyanobacteria, share a unique form of metabolism (oxygenic photosynthesis), yet have evolved many diverse cell shapes and grow in diverse habitats. Other phyla, such as Spirochetes, share a unique cell structure while diverging in habitat and metabolism.

Table 18.1 introduces the major groups of bacteria that appear most frequently in the literature. Bear in mind, however, that more than 100 other bacterial phyla are known to date and that deep-branching clades continue to be discovered. For example, in 2004 a new phylum TG-1 was described comprising many species of cellulose-digesting bacterial symbionts of protists within the termite gut. The TG-1 phylum is studied as a model for plant cellulose conversion to biofuels.

> **NOTE:** Note: Here is a way to approach Table 18.1:
> - Know the seven major categories and their traits in the shaded headings.
> - For each major category, select two or three species and learn their specific traits.

Deep-branching thermophiles. The most deeply branching bacterial phyla (that is, those that appear to have diverged earliest) include hyperthermophiles that share physiology and habitat with archaea. For example, *Aquifex* grows at 95°C and possesses traits typical

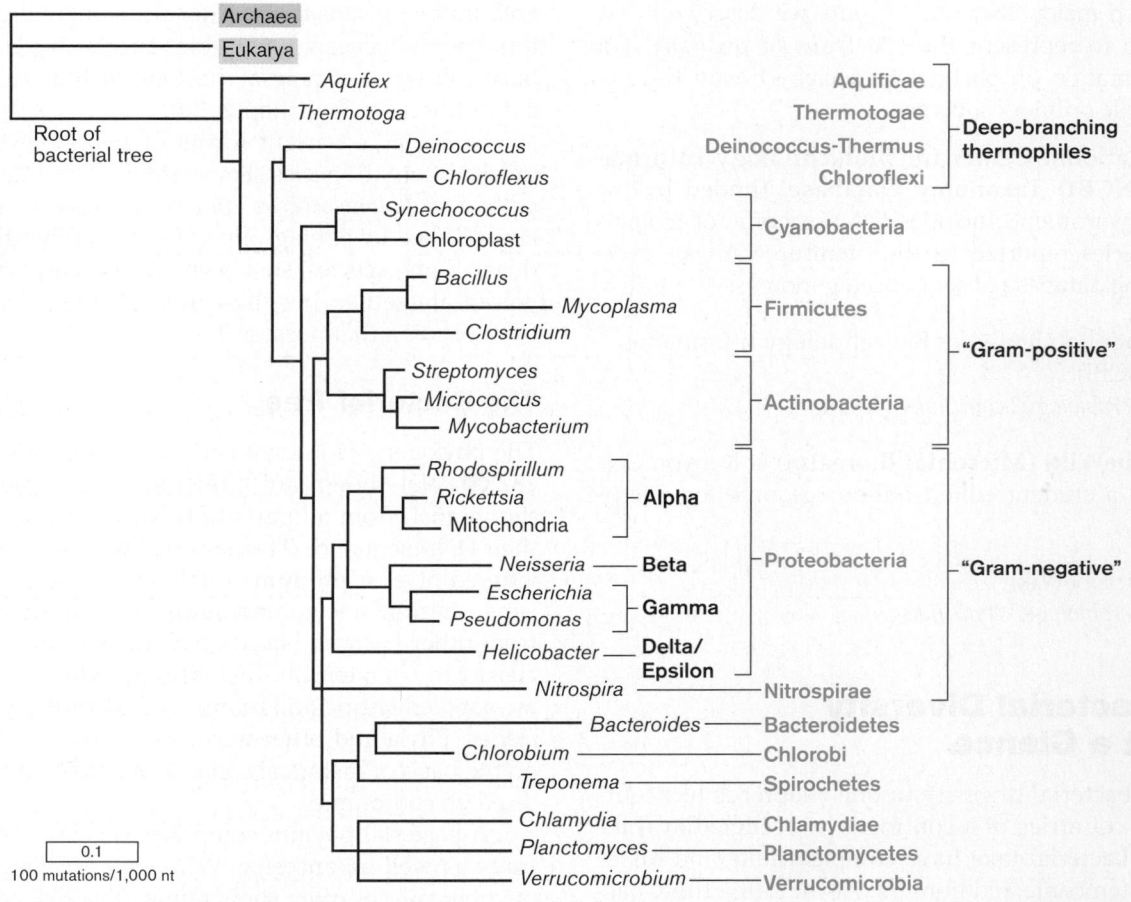

Figure 18.2 Bacterial phylogeny. A phylogenetic tree of representative Bacteria based on 16S rRNA sequence comparison. The tree is rooted with respect to Archaea and Eukarya. Bold labels correspond to major groups in Table 18.1. Phylum names are lettered blue. ⏵ᴵᴵ *microbiology2.com/animations*

of archaea, such as ether-linked membrane lipids. Some "archaeal" traits of bacteria arose from a common ancestor of bacteria and archaea, whereas others arose through lateral transfer of genes from archaea to bacteria sharing the high-temperature habitat. In other aspects of structure and physiology, the deep-branching phyla diverge from each other. Order Aquificales consists of Gramnegative rods that oxidize hydrogen, whereas order Thermotogales has sheathed Gram-negative rods that are obligate anaerobic heterotrophs. Deinococcus-Thermus includes moderately thermophilic Gram-negative heterotrophs, often filamentous, in the order Thermales. *Thermus aquaticus* was the first source of heat-resistant DNA polymerase for PCR. By contrast, members of the closely related order Deinococcales grow in the middle range of temperature and stain Gram-positive. *Deinococcus radiodurans* shows extreme resistance to ionizing radiation and desiccation (discussed in Chapters 5 and 9). The phylum Chloroflexi consists of thermophilic green phototrophs that oxidize H_2S or organic molecules. They often share habitat with thermophilic cyanobacteria.

Cyanobacteria. The **Cyanobacteria** are a deeply branching phylum of profound importance for all ecosystems. Cyanobacteria conduct photosynthesis by splitting water and releasing oxygen (O_2). Cyanobacteria may be poisoned by H_2S, but some species have adapted to use H_2S when available.

Only cyanobacteria and chloroplasts (eukaryotic organelles that evolved from cyanobacteria) can produce oxygen gas. They have a unique two-photosystem apparatus for oxygenic photosynthesis arranged in lamellar arrays of membranes called thylakoids. Unique to cyanobacteria and chloroplasts are the chlorophylls, most notably chlorophylls *a* and *b*; these are distinct from the "bacteriochlorophylls" used by nonoxygenic bacterial phototrophs. Cyanobacteria share a common metabolism, yet the cell shape and organization of cyanobacterial species show an immense range of different forms, including chains of cells, square arrays, and globular colonies. Many cyanobacteria grow in seawater or freshwater, whereas others grow in mutualism with a eukaryotic host.

Table 18.1 Representative groups of bacteria.

Note: Each trait applies to *most* members of the taxon described, but exceptions have evolved.
Blue-lettered terms indicate phyla.

• **Bulleted terms are representative orders** within the phylum (unless stated otherwise).

Deep-branching Thermophiles

Thermophilic bacteria that diverged early from the archaea and eukaryotes. Many genes transferred laterally from archaea.

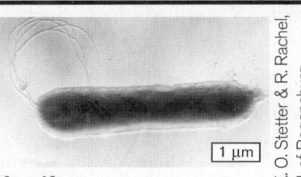

Aquifex

K. O. Stetter & R. Rachel, U. of Regensburg

Aquificae. Hyperthermophiles (70°C–95°C). Oxidize H_2, H_2S, or thiosulfate.
• **Aquificales.** Outer membrane; cells may be single with flagella or grow in filaments. *Aquifex, Thermocrinis.*

Chloroflexi. Filamentous green bacteria.
• **Chloroflexales.** Filamentous phototrophs; absorb light to oxidize organic compounds or H_2S. Most contain photosystem II in chlorosomes. *Chloroflexus aurantiacus* is thermophilic; forms mats in hot springs.

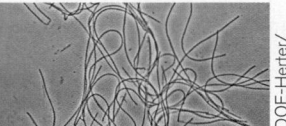

Chloroflexus aurantiacus

DOE-Herter/ Visuals Unlimited

Deferribacteres. Anaerobic vent thermophiles.
• **Deferribacterales.** Single rod or vibrio cells, flagellated. *Deferribacter autotrophicus* reduces Fe^{3+} with H_2; *D. abyssi* respires on small organic molecules.

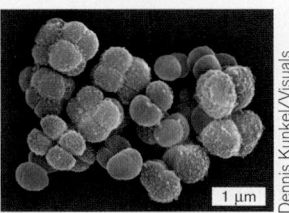

Deinococcus

Dennis Kunkel/Visuals Unlimited

Deinococcus-Thermus. Peptidoglycan contains ornithine. Diverse growth temperatures.
• **Deinococcales.** Thick envelope; stain Gram-positive. Not thermophilic, but extremely resistant to ionizing radiation and to desiccation. *Deinococcus.*
• **Thermales.** Species are filamentous clustered cells or single-celled. Grow at 70°C–75°C. *Thermus aquaticus* is the source of Taq polymerase for PCR.

Thermotogae. Thermophiles or hyperthermophiles (55°C–100°C).
• **Thermotogales.** Outer membrane with large periplasm; anaerobic heterotrophs. *Petrotoga, Thermotoga.*

Cyanobacteria

Oxygenic photoautotrophs with thylakoid membranes. Share ancestry with chloroplasts.

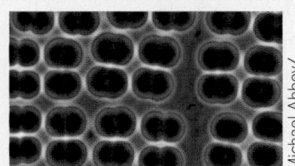

Merismopedia

Michael Abbey/ Visuals Unlimited

Cyanobacteria. Oxygenic photoautotrophs with thylakoid membranes. Fix CO_2 using Rubisco. Often mutualists. Share ancestry with chloroplasts.
• **Chroococcales.** Square colonies based on two division planes.
  Chroococcus. Single, double, or quartet of cells (10–20 μm per cell). Grow in pond sediment.
  Merismopedia. Platelike colonies of elliptical cells (5 μm).
  Synechococcus. Single or double cells. Marine producer.
• **Gloeobacterales.** Lack thylakoids; conduct photosynthesis in cell membrane. *Gloeobacter violaceus.*

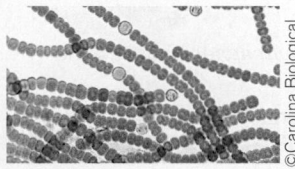

Anabaena

©Carolina Biological Supply/Visuals Unlimited

• **Nostocales.** Filamentous chains with N_2-fixing heterocysts. Often grow symbiotically with other microbes, corals, or plants.
  Anabaena. Aquatic. Grow in association with red water fern (*Azolla*).
  Nostoc. Aquatic. Grow independently; or mutualistically with fungi, as lichens; or as endosymbionts of *Gunnera* plant cells.

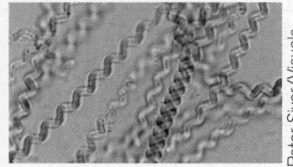

Spirulina

Peter Siver/Visuals Unlimited

• **Oscillatoriales.** Filamentous chains with motile hormogonia (short chains) (filament width, 50 μm).
  Oscillatoria. Aquatic or marine. Grow independently, or as sponge endosymbionts.
  Spirulina. Aquatic. Farmed as food supplement (filament width, 5 μm).
  Trichodesmium. Major marine producer. Forms "blooms" of overgrowth.
• **Pleurocapsales.** Globular colonies; reproduce through baeocytes. *Pleurocapsa, Myxosarcina.*
• **Prochlorales.** Tiny single cells, elliptical or spherical (1 μm).
  Prochlorococcus. Most abundant marine producer. One of the smallest cells of a free-living microbe, with a highly reduced genome.
  Prochloron. Tropical marine producer; endosymbiont of sea squirt.

(continued)

Firmicutes and Actinobacteria (Gram-positive)

Gram-positive. Peptidoglycan multiple layers, cross-linked by teichoic acids. Aerobes and facultative anaerobes.

Firmicutes. **Low-GC, Gram-positive rods and cocci.**

- **Bacillales.** Aerobic or facultative anaerobes (2–10 μm).

  Bacillus. Endospore-forming rods. Soil-growing: *B. subtilis*, *B. cereus*, *B. anthracis*, *B. thuringiensis*. Extremophiles: *B. alkalophilus*, *B. thermophilus*, *B. halodurans*.

  Listeria. Non-spore-forming rods; intracellular pathogens. *L. monocytogenes* causes listeriosis.

  Staphylococcus. Non-spore-forming cocci. Skin biota: *S. epidermidis*. *S. aureus* infects flesh.

- **Clostridiales.** Anaerobic rods.

  Carboxydothermus. Oxidize carbon monoxide (CO) to CO_2. Form endospores.

  Clostridium. Form endospores. *C. botulinum* and *C. difficile* are pathogens. *C. acetobutylicum* generates butanol.

  Dehalobacter. Non-spore-forming rods. Dechlorinate chloroethenes.

  Epulopiscium. Exceptionally large cells. Reproduce by "live birth."

  Heliobacterium, Heliophilum. Endospore-forming photoheterotrophs.

  Metabacterium. Exceptionally large cells; form multiple endospores. *M. polyspora*.

  Ruminococcus. Digestive flora of ruminant animals.

- **Lactobacillales.** Non–spore formers. Facultative anaerobes. Ferment, producing lactic acid.

  Enterococcus. Enteric cocci.

  Lactobacillus. *L. acidophilus* is used for dairy culture.

  Lactococcus. Used for dairy culture.

  Streptococcus. Chains of cocci. Group A streptococci cause "strep throat."

- **Mollicutes (class).** Lacks cell wall; requires an animal host. *Mycoplasma genitalium* has one of smallest known genomes. *M. pneumoniae* causes pneumonia.

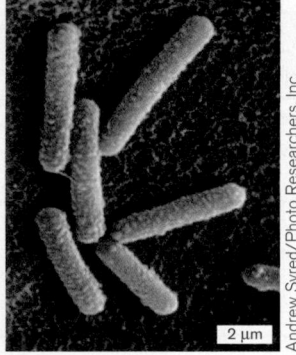

Bacillus subtilis

Andrew Syred/Photo Researchers, Inc.

2 μm

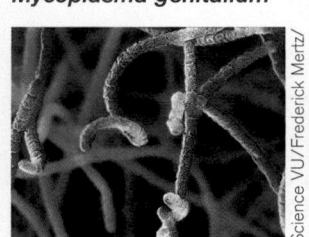

Mycoplasma genitalium

Don W. Fawcett/Photo Researchers, Inc.

Actinobacteria. High-GC, Gram-positive bacteria with moderate salt tolerance.

- **Actinomycetales.**

  Actinomycetaceae. Filamentous, producing aerial hyphae and spores (width, 1 μm).

  Actinomyces. *A. israelii* causes actinomycosis.

  Frankia. Saprophytes; grow on leaf litter. Fix nitrogen for plants.

  Salinispora. Sponge endosymbionts. Produce drug-like secondary products.

  Streptomyces. Produce many antibiotics. *S. coelicolor, S. griseus.*

  Corynebacteriaceae. Irregular rods. *Corynebacterium diphtheriae* causes diphtheria.

  Mycobacteriaceae. Exceptionally thick cell envelope holds acid-fast stain.

  Mycobacterium. *M. tuberculosis* causes tuberculosis; *M. leprae* causes leprosy.

  Micrococcaceae. Nonfilamentous soil bacteria; mostly obligate aerobes.

  Arthrobacter. Rods. Respire on oxygen or on chlorinated aromatics.

  Micrococcus. Aerobic cocci such as *M. luteus*. Grow in soil.

  Propionibacteriaceae. Propionic acid fermentation. *Propionibacterium acnes* causes acne; *P. freudenreichii* makes Swiss cheese.

- **Bifidobacteriales.** Ferment without gas. Enteric biota of breast-fed infants.

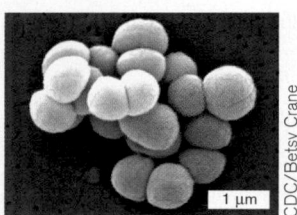

Streptomyces

Science VU/Frederick Mertz/Visuals Unlimited

Micrococcus luteus

CDC/Betsy Crane

1 μm

Proteobacteria and Nitrospirae (Gram-negative)

Gram-negative. Outer membrane contains LPS. Diverse metabolism. Share ancestry with mitochondria.

Proteobacteria. Gram-negative. Diverse cell forms and metabolism (1–10 μm).

Alphaproteobacteria (class). Heterotrophic rods or spirilla.

- **Caulobacterales.** Aquatic oligotrophs; alternate stalk and flagellum. *Caulobacter.*

- **Rhizobiales.** Plant mutualists and pathogens, methyl oxidizers, and animal pathogens.

  Agrobacterium. *A. tumefaciens* causes plant tumors; transgenic plant vector.

  Bartonella. *B. quintana* causes trench fever; *B. henselae* causes cat scratch fever.

  Brucella. Cause brucellosis in horses and sheep. *B. melitensis.*

  Nitrobacter. Nitrite-oxidizing lithotrophs. *N. winogradskyi.*

  Methylobacterium. Grow on single-carbon compounds.

  Rhizobium group. *Bradyrhizobium* and *Sinorhizobium* fix nitrogen within legumes.

  Rhodopseudomonas, Rhodomicrobium. Soil photoheterotrophs.

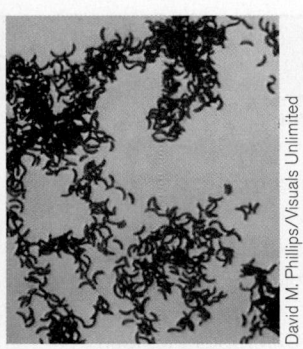

Rhodospirillum rubrum

David M. Phillips/Visuals Unlimited

- **Rhodobacterales, Rhodospirillales.** Flagellated photoheterotrophs: *Rhodobacter sphaeroides, Rhodospirillum rubrum, Roseobacter* (oxidizes CO). Nonphototrophs: *Acetobacter* makes vinegar.
- **Rickettsiales.** Includes intracellular parasites; lineage includes mitochondria.
  - ***Rickettsia.*** Intracellular parasites. *R. rickettsii* causes Rocky Mountain spotted fever; *R. prowazekii* causes typhus.
  - **SAR11 cluster.** Marine photoheterotrophs use proteorhodopsin. *Pelagibacter.*
- **Sphingomonadales.** Catabolize complex organics; used for bioremediation. Opportunistic pathogen: *Sphingomonas.* Aerobic photoheterotrophs: *Citromicrobium, Erythromicrobium.*

Betaproteobacteria (class). Phototrophs, lithotrophs, and pathogens.

- **Burkholderiales.** *Burkholderia pseudomallei* causes melioidosis in humans and farm animals.
- **Hydrogenophilales.** Lithotrophs. *Thiobacillus ferrooxidans* oxidizes iron and sulfur.
- **Neisseriales.** Mucous-membrane normal flora and pathogens. Aerobic or microaerophilic diplococci. *Neisseria gonorrhoeae* causes gonorrhea; *N. meningitidis* causes meningitis.
- **Nitrosomonadales.** Lithotrophs. *Nitrosomonas europaea* oxidizes ammonia.
- **Rhodocyclales.** Soil heterotrophs and photoheterotrophs. *Azoarcus evansii* catabolizes complex aromatic molecules. *Rhodocyclus* species are purple photoheterotrophs.

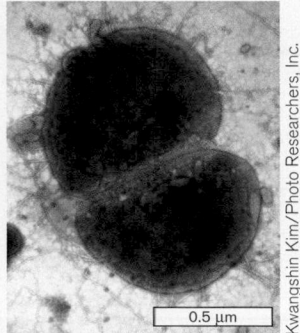

0.5 µm

Neisseria gonorrhoeae

Kwangshin Kim/Photo Researchers, Inc.

Gammaproteobacteria (class). Facultative anaerobes and lithotrophs.

- **Aeromonadales.** Aquatic heterotrophs. *Aeromonas hydrophila* infects fish and humans.
- **Alteromonadales.** Aquatic. *Shewanella oneidensis* reduces metals; used in fuel cells.
- **Chromatiales.** Lithotrophs: *Nitrococcus* oxidizes ammonia. Sulfur and iron phototrophs: *Chromatium.* Nitrite phototrophs: *Thiocapsa.*
- **Enterobacteriales.** Enterobacteriaceae. Facultative anaerobes; colonize the human colon.
  - ***Escherichia coli.*** Strains include normal flora and enteric pathogens.
  - ***Salmonella.*** *S. enterica* strains cause gastrointestinal infection and typhoid fever.
  - ***Yersinia.*** *Y. pestis* causes bubonic plague.
- **Legionellales.** *Legionella pneumophila* causes legionellosis pneumonia. *Coxiella burnetti* causes Q fever.
- **Oceanospirillales.** Marine heterotrophs, including degraders of petroleum. *Alcanivorax.*
- **Pseudomonadales.** Rods, aerobic or respire on nitrate; catabolize aromatics. *Pseudomonas aeruginosa* infects lungs in cystic fibrosis patients. *P. fluorescens* suppresses plant diseases.
- **Thiotrichales.** Lithotrophs and heterotrophs. *Beggiatoa alba* and *Thiomargarita namibiensis* oxidize sulfur. *Cycloclasticus* catabolizes polycyclic aromatic hydrocarbons in petroleum.
- **Vibrionales.** Marine heterotrophs. *Vibrio cholerae* causes cholera.
- **SAR86 cluster.** Marine photoheterotrophs use proteorhodopsin.

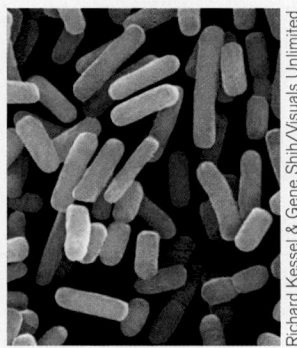

Escherichia coli

Richard Kessel & Gene Shih/Visuals Unlimited

Deltaproteobacteria (class). Lithotrophs and multicellular communities.

- **Bdellovibrionales.** Periplasmic predators. *Bdellovibrio bacteriovorus* consumes *E. coli.*
- **Desulfobacterales.** Reduce sulfate. *Desulfobacter* spp., *Desulfococcus.*
- **Desulfuromonadales.** Lithotrophs; reduce or oxidize sulfur or metals.
  - ***Geobacter metallireducens*** reduces iron oxide.
  - ***Desulfuromonas*** reduces elemental sulfur.
- **Myxococcales.** Gliding bacteria that form fruiting bodies. *Myxococcus* spp.

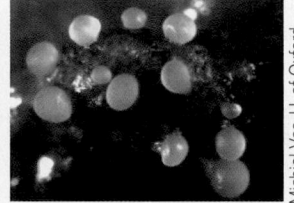

Myxococcus

Michiel Vos, U. of Oxford

Epsilonproteobacteria (class).

- **Campylobacterales.** Spirillar pathogens.
  - ***Campylobacter.*** *C. jejuni* causes food poisoning.
  - ***Helicobacter.*** *H. pylori* causes gastritis.
  - ***Nautilia, Hydrogenimonas, Sulfurimonas.*** Oxidize H_2 with sulfur or nitrate.
  - ***Thiovulum.*** Oxidizes sulfides, producing sulfur granules.

Nitrospirae. Nitrite oxidizers. Obligate aerobes.

- **Nitrospirales.** Tight spirilla. In soil and water, oxidize NO_2^- to NO_3^-. *Nitrospira, Leptospirillum.* Includes acidophilic Fe oxidizers.

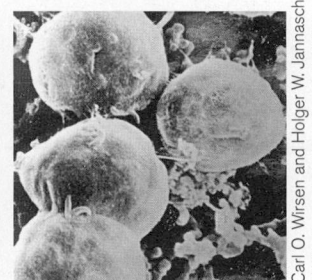

Thiovulum

Carl O. Wirsen and Holger W. Jannasch. 1978. *Journal of Bacteriology* **136**:765

Bacteroidetes and Chlorobi

Gram-negative. Obligate anaerobes and/or green sulfur phototrophs.

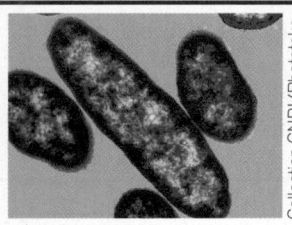

Bacteroides

Bacteroidetes. Obligate anaerobes. Heterotrophs that feed on diverse carbon sources.
- **Bacteroidales.** Obligate anaerobes, soil or human flora.
  - *Bacteroides.* Enteric. *B. thetaiotaomicron* digests complex plant carbohydrates in gut. *Bacteroides* species escaping the colon cause abscesses.
  - *Porphyromonas.* Includes gingival pathogens such as *P. gingivalis*.
- **Flavobacteriales.** Heterotrophs in soil and water. *Flavobacterium psychrophilum* infects fish.

Chlorobi. Green sulfur-oxidizing phototrophs.
- **Chlorobiales.** Anaerobic H_2S photolysis, absorbing red and infrared. *Chlorobium tepidum.*

Spirochetes (Spirochaeta)

Narrow, coiled cell encased by sheath, which encloses polar flagella.

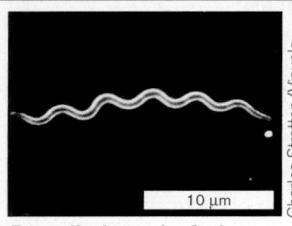

10 µm

Borrelia burgdorferi

Spirochetes. Narrow, coiled cell with axial filaments, encased by sheath. Polar flagella beneath sheath double back around cell. Often symbiotic or pathogenic.
- **Spirochaetales.** Aquatic, free-living or pathogenic (width, < 0.5 µm; length, 10–20 µm).
  - *Borrelia.* *B. burgdorferi* causes Lyme disease, transmitted by ticks.
  - *Hollandina.* Termite gut endosymbionts.
  - *Leptospira.* L-shaped animal pathogen; causes leptospirosis.
  - *Treponema.* *T. pallidum* causes syphilis.
  - *Spirochaeta.* Aquatic, free-living heterotrophs.

Chlamydiae, Planctomycetes, and Verrucomicrobia

Irregular cells lacking peptidoglycan, with subcellular structures analogous to those of eukaryotes.

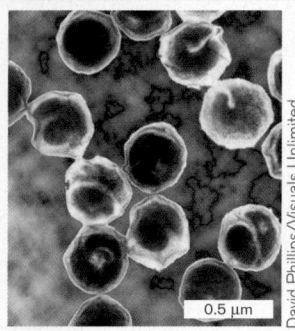

0.5 µm

Chlamydia trachomatis

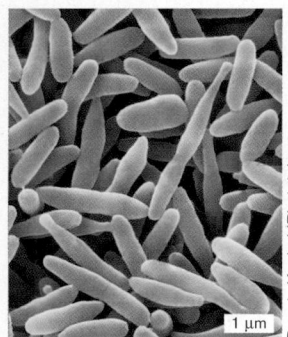

1 µm

Prosthecobacter fusiformis

Chlamydiae. Intracellular cell wall–less pathogens of animals or protists.
- **Chlamydiales.** Reticulate bodies form multiple spore-like transfer particles (elementary bodies) that infect the next host.
  - *Chlamydia.* *C. trachomatis* causes sexually transmitted disease and trachoma (eye infection).
  - *Chlamydophila.* *C. pneumoniae* causes pneumonia; *C. abortus* causes spontaneous abortion in animals; *C. psittaci* causes psittacosis in birds and humans.
  - *Parachlamydia.* Chlamydia-like species that infect free-living amebas.

Planctomycetes. Nucleoid has double membrane analogous to eukaryotic nuclear membrane and other intracytoplasmic subcellular membrane compartments.
- **Planctomycetales.** Aquatic or marine. Flexible cell shape.
  - *Brocadia.* Anaerobically oxidize ammonium and release N_2, using anammoxosomes.
  - *Gemmata.* Nuclear double membrane, surrounded by additional intracellular membrane.
  - *Pirellula.* Only one intracellular membrane. Has fimbriae and single flagellum. Marine habitat.
  - *Planctomyces.* One intracellular membrane. Halophile.
  - *Scalindua.* Conduct anammox in wastewater. Oxidize CO.

Verrucomicrobia. Stalk-like appendages contain actin filaments. Aquatic oligotrophs.
- **Verrucomicrobiales.**
  - *Prosthecobacter.* Each cell has a polar cytoplasmic extension called a prostheca.
  - *Verrucomicrobium.* Stellate cells have multiple prosthecae.

Gram-positive bacteria. Two major phyla, **Firmicutes** and **Actinobacteria**, are called the "Gram-positive bacteria" because most members stain Gram-positive. A few Firmicutes (the mycoplasmas) fail to stain Gram-positive because they lack cell walls, whereas some Actinomycetes possess a thick waxy coat that excludes the Gram stain. Most Firmicutes and Actinobacteria have an exceptionally thick cell wall, with several layers of peptidoglycan threaded by supporting molecules such as teichoic acids or mycolic acids. The thick, reinforced cell wall is what retains the Gram stain (discussed in Chapter 2). In addition, most Gram-positive species possess a well-developed S-layer of protein with glycan strands. By contrast, in other bacterial phyla the S-layer is either absent or present in diminished form.

Firmicutes and Actinobacteria differ genetically in their **GC content**—that is, the proportion of their genomes consisting of guanine-cytosine base pairs (as opposed to adenine-thymine). Firmicutes (low GC) generally grow as well-defined rods or cocci, isolated or in simple filaments consisting of cells that divide but remain attached end to end. Many Firmicutes form **endospores**, inert heat-resistant spores that can remain viable for thousands of years. Endospores are the most durable type of spore formed by bacteria. Actinobacteria have a relatively high GC content.

Actinobacteria (high GC) include the **actinomycetes** (order Actinomycetales), which undergo complex life cycles forming filamentous hyphae and arthrospores. Other groups closely related to actinomycetes grow as isolated rods and cocci, often with variable shape, such as the corynebacteria. Actinomycete relatives include the well-known causative agents of tuberculosis (*Mycobacterium tuberculosis*) and leprosy (*M. leprae*).

Gram-negative bacteria. Proteobacteria and Nitrospirae are called the "Gram-negative bacteria" because their single layer of peptidoglycan fails to retain the Gram stain. (Note, however, that members of other phyla also stain Gram-negative.) Species of Proteobacteria and of Nitrospirae have a complex outer membrane consisting of lipopolysaccharides (LPS) and porins. All Proteobacteria have in common the LPS outer membrane, but they show an immense range of shape (rods, cocci, spirals, and filaments) and metabolism (heterotrophy, lithotrophy, and anaerobic phototrophy). Most species are aerobic, facultative, or microaerophilic (requiring a low level of O_2; inhibited at higher levels).

Proteobacteria include five major classes, named Alphaproteobacteria, Betaproteobacteria, Gammaproteobacteria, Deltaproteobacteria, and Epsilonproteobacteria. (Some sources separate the Greek letter from the name Proteobacteria—for example, Alpha Proteobacteria, Beta Proteobacteria, and so on.) Proteobacteria include famous model organisms and pathogens, such as *Escherichia coli*, *Salmonella enterica*, and *Yersinia pestis*. Many are human or animal symbionts, either as mutualists or as pathogens—including *Rickettsia*, the genus most closely related to mitochondria. Members of the Gram-negative phylum Nitrospirae largely resemble proteobacteria in form, though they diverged earlier. Species of Nitrospirae (such as *Nitrospira* spp.) oxidize nitrite to nitrate—a lithotrophic conversion essential for ecosystems.

Bacteroidetes and Chlorobi. The phyla **Bacteroidetes** and **Chlorobi** stain Gram-negative, but nearly all their members are obligate anaerobes. Within Bacteroidetes, *Bacteroides* species ferment complex carbohydrates, serving as the major mutualists of the human gut. Chlorobi species are also obligate anaerobes, but they are "green sulfur" phototrophs that photolyze sulfides or H_2.

Spirochetes. The **Spirochetes** (Spirochaeta) exhibit a unique and complex cellular form of a flexible, extended spiral, resembling a telephone cord. The cytoplasm and cell membrane are contained within an outer membrane called the sheath. Between the sheath and the cell membrane extend flagella, doubled back from each pole. The rotation of these flagella is coordinated so as to twist and flex the helical body, generating motility and chemotaxis. Spirochetes include many free-living forms in aquatic systems, as well as digestive endosymbionts and pathogens.

Irregularly shaped bacteria. Three phyla of bacteria have unusual cell shapes: **Chlamydiae**, **Verrucomicrobia**, and **Planctomycetes**. Most members of these groups have lost their peptidoglycan cell walls, but they show complex structural adaptations and developmental forms. The Chlamydiae (best-known genus *Chlamydia*) are intracellular parasites that lose most of their cell envelope during intracellular growth. The replicating parasites generate multiple spore-like "elementary bodies" that escape to infect the next host.

Verrucomicrobia, by contrast, are free-living aquatic bacteria with wart-like protruding structures containing actin. Planctomycetes, also free-living, have stalked cells that reproduce by budding. Each planctomycete cell contains an extra double membrane surrounding its nucleoid, analogous to a eukaryotic nuclear membrane, though it evolved independently.

THOUGHT QUESTION 18.1 Which taxonomic groups in **Table 18.1** stain Gram-positive and which Gram-negative? Which group contains both Gram-positive and Gram-negative species? For which groups is the Gram stain undefined, and why?

THOUGHT QUESTION 18.2 Which groups of species share common structure and physiology within the group? Which groups show extreme structural and physiological diversity?

New Isolates: Unclassified and Uncultured Bacteria

The seven groups outlined in **Table 18.1** represent those bacteria of greatest historical interest and those best characterized by microbiologists. As new organisms emerge, they may be known initially only by their habitat and their small-subunit rRNA sequence. Such isolates require description of essential cell structure and metabolism before designation as new species. "Incompletely described" samples are designated as follows:

- An unclassified organism or uncultured organism is assigned to a taxonomic rank based on SSU rRNA sequence but not yet grown in pure culture—for example, an "uncultured actinomycete."
- An environmental sample is designated by its habitat and assigned a rank based on SSU rRNA. Examples of environmentally defined actinomycetes in the NCBI database include "oil-degrading bacterium AOB1" and "glacier bacterium FJS11."
- A species with some physiological characterization beyond DNA sequence may be published with a provisional status of **candidate species**, designated by the prefatory term "*Candidatus.*" For example, "*Candidatus* Nostocoida limicola" was published as "a filamentous bacterium from activated sludge."

Following our brief "one-day tour" of the bacterial domain, we now explore some of the diverse species within the major groups. Each section presents species from one of the seven major groups outlined in **Table 18.1**.

NOTE: The primary organizing principle used in Chapters 18–20 is that of phylogeny based on DNA relatedness. Characteristic traits described for each branch apply to the majority of its known species, though exceptions are discovered daily, such as members of Spirochetes that lack spiral form.

18.2 Deep-Branching Thermophiles

Several of the most deeply branching phyla show traits similar to those of archaeal extremophiles and share high-temperature habitats such as hot springs. The precise nature of their relationship with archaea is controversial, however, because these organisms also show the fastest doubling rates of all cells (in some cases less than 10 minutes) and high rates of mutation, which may have "accelerated" their molecular clock. Thus, the deep-branching thermophiles may appear to have diverged from other bacteria earlier than they in fact did.

Aquificales and Thermotogales: The Most Extreme Bacterial Hyperthermophiles

The order Aquificales includes bacteria growing at high temperatures, up to 95°C. Most members of this group are hydrogenotrophs, oxidizing hydrogen gas with molecular oxygen to make water. *Aquifex pyrophilus*, a flagellated rod (see **Table 18.1**), was first discovered by extremophile microbiologist Karl Stetter at Regensburg University in a submarine hydrothermal vent north of Iceland. Stetter named it *Aquifex*, Latin for "water maker," in reference to its metabolism of oxidizing hydrogen gas with O_2 to form water. *Aquifex* species are obligate autotrophs, fixing CO_2 into biomass using the reverse TCA cycle.

Other genera of Aquificales form filamentous mats in hydrothermal springs. *Thermocrinis ruber* was discovered by Thomas Brock at the University of Wisconsin–Madison as a mat of pink filamentous streamers ("pink filaments") growing at 82°C–88°C in the outflow channel of Octopus Spring, in Yellowstone National Park (**Fig. 18.3**). As the filaments grow, they break off flagellated single cells that grow new filaments. *T. ruber* gains energy by oxidizing hydrogen, thiosulfate, and elemental sulfur, as well as small organic molecules such as formate.

A.

B.

Figure 18.3 Deep-branching thermophiles. A. Mass of pink filamentous *Thermocrinis ruber* in a hot spring at near-boiling temperature. **B.** *T. ruber* filamentous cells (SEM).

Bob Lindstrom

R. Huber and W. Eber. *Prokaryotes.* 2002. Springer-Verlag

2 µm

Another deep-branching order of thermophiles (growing at 50°C–80°C) is Thermotogales. *Thermotoga maritima* was isolated from a geothermal vent in Volcano, Italy. Its cells have a loosely bound sheath, or "toga," for which the genus is named. It uses anaerobic respiration, reducing elemental sulfur, thiosulfate, and sulfite to hydrogen sulfide.

The genomes of *T. maritima* and of *Aquifex aeolicus* reveal a surprisingly high content of genes transferred horizontally from archaea, well after the bacteria diverged from their common ancestor. Nearly a quarter of the *T. maritima* genome appears to be of archaeal origin, even archaeal versions of enzymes fundamental to bacteria, such as glutamate synthase and inositol 1-phosphate synthase.

> **THOUGHT QUESTION 18.3** What taxonomic questions are raised by the apparent high rate of gene transfer between archaea and thermophilic bacteria?

Thermus and *Deinococcus*

The phylum Deinococcus-Thermus shares a unique structural trait, the substitution of L-ornithine for diaminopimelic acid in the peptidoglycan cross-bridge. *Thermus* species (growing at 70°C–75°C) are heterotrophs commonly isolated from hot tap water. *Deinococcus* species, however, are not thermophilic. *D. radiodurans* bacteria resist extremely high doses of ionizing radiation (discussed in Chapters 5 and 9). An otherwise ordinary heterotroph, *D. radiodurans* was originally isolated from cans of meat supposedly sterilized with a dose of several megarads. The natural habitat of *Deinococcus* species is not known, but the genus also resists extreme desiccation; it may have evolved as a soil organism capable of surviving drought. *Deinococcus* species, unlike *Thermus* species, have a thick cell wall that stains Gram-positive, as well as an S-layer.

Chloroflexi: Thermophilic Green Photoheterotrophic Bacteria

Chloroflexi bacteria are filamentous photoheterotrophs, absorbing light to split organic molecules. Photosystem II (PS II) generates ATP through cyclic electron flow (discussed in Chapter 14). Chloroflexi species conduct photoheterotrophy or H_2S photosynthesis. Most are moderate thermophiles, often found in hot springs in association with thermophilic cyanobacteria. *Chloroflexus aurantiacus* and thermophilic cyanobacteria form massive microbial mats in the hot springs of Yellowstone (**Fig. 18.4**).

Chloroflexi are informally called "green nonsulfur bacteria" to distinguish them from Chlorobi, a different phylum of green phototrophs that photolyze only H_2S and are strict anaerobes. Some chloroflexi appear green because their bacteriochlorophylls (Bchl) absorb primarily blue and red, allowing transmission of green (**Table 18.2**, top row). Other species, however, appear red or yellow owing to accessory pigments.

Chloroflexus species contain their photosynthetic apparatus within membranous organelles called **chlorosomes**. The chlorosome contains Bchl *c* and Bchl *d*, complexed with Bchl *a* at the plasma membrane. Other species of Chloroflexi, however, such as *Roseiflexus* and *Heliothrix*, lack chlorosomes. Chloroflexi are the main filamentous phototrophs outside the Cyanobacteria.

Table 18.2 summarizes key features of phototrophy in clades throughout the bacterial domain; we shall return to this table as a guide throughout the chapter. All light-harvesting complexes descend from one of two ancestral sources: the chlorophyll/bacteriochlorophyll with electron transport (PS I, PS II) or the proteorhodopsin proton pump. It is not clear whether one or both ancestral

A.

B.

Figure 18.4 *Chloroflexus*: a thermophilic phototroph. **A.** Hot-spring bacterial mat contains *Chloroflexus* and thermophilic cyanobacteria, Yellowstone National Park. **B.** *C. aurantiacus* filaments often combine with *Synechococcus* (cyanobacteria) to form a microbial mat. Red autofluorescence reveals the cyanobacteria.

Carol Pritchard-Martinez

*M. Ferris and D. J. Patterson, images used under license to MBL (micro*scope)*

Table 18.2 Phototrophic bacteria.

Taxon	Energy generation	Cell structure and photopigments	Absorption spectrum (whole cell)	Reaction center
Chloroflexi "Green nonsulfur" *Chloroflexus*	+O_2 Heterotrophy −O_2 Photoheterotrophy or use reduced sulfur	Chlorosome Bchl *a* Bchl *c/d*	400 600 800 1,000	PS II
Cyanobacteria "Blue-greens" *Anabaena*	+O_2 Oxygenic phototrophy −O_2 Photolithoautotrophy on reduced sulfur	Thylakoid Chl *a/b*	400 600 800 1,000	PS I and PS II
Firmicutes "Sun bacteria" *Heliobacterium*	+O_2 −O_2 Photoheterotrophy	Forms endospore Bchl *g*	400 600 800 1,000	PS I
Proteobacteria "Purple nonsulfur" (Alpha and Beta classes) Bchl *a* *Rhodospirillum*	+O_2 Heterotrophy −O_2 Photoheterotrophy	Bchl *a*	400 600 800 1,000	PS II
Bchl *b* *Blastochloris*	+O_2 Heterotrophy −O_2 Photoheterotrophy	Bchl *b*	400 600 800 1,000	PS II
"Purple sulfur" (Gamma class) *Chromatium*	+O_2 −O_2 Photolithoautotrophy on reduced sulfur	Bchl a/b S S S S	400 600 800 1,000	PS II
"Proteorhodopsin" (Alpha and Gamma classes) *Pelagibacter* (SAR11) SAR86	+O_2 Heterotrophy −O_2 Photoheterotrophy	Proteorhodopsin	400 600 800 1,000	PR
Chlorobi "Green sulfur" *Chlorobium*	−O_2 only Photolithoautotrophy on reduced sulfur	Chlorosome Bchl *a* S S S S S S Bchl *c/d/e* S S S S	400 600 800 1,000	PS I

Sources: J. Overman and F. Garcia-Pichel. 2001. The phototrophic way of life. In *Prokaryotes*. 2002. Springer-Verlag; Oded Béjà, et al. 2001. *Nature* **411**:786–789.

photosystems were present in the ancestor of all bacteria or whether extensive horizontal gene transfer distributed the photosystems among divergent clades.

NOTE: Distinguish phylum Chloroflexi (filamentous photoheterotrophs) from phylum Chlorobi (sulfur photoautotrophs of the Bacteroidetes and Chlorobi group). Both are phototrophs containing green photopigment complexes arranged in chlorosomes, but they are deeply divergent genetically.

TO SUMMARIZE:

- **Deep-branching thermophiles** such as *Aquifex* and *Thermotoga* species share traits and habitats with thermophilic archaea. They show extensive transfer of archaeal genes.
- **Deinococcus-Thermus** bacteria have ornithine in their peptidoglycan cross-bridges. *Deinococcus* bacteria are highly resistant to ionizing radiation, and they can be isolated from irradiated foods. *Thermus* bacteria are moderate thermophiles isolated from hot-water taps.
- **Chloroflexi** bacteria are deep-branching thermophilic photoheterotrophs, occurring as filamentous mats in association with thermophilic cyanobacteria in hot springs.

18.3 Cyanobacteria: Oxygenic Phototrophs

Cyanobacteria (or Cyanophyta) are the only bacteria that produce oxygen gas. The phylum is named for the blue phycocyanin accessory pigments possessed by some genera, giving them a bluish tint. Cyanobacteria commonly appear green because of the predominant blue and red absorption by chlorophylls *a* and *b* (see **Table 18.2**).Essentially the same chlorophylls are found in plant chloroplasts, which evolved from internalized cyanobacteria (discussed in Chapter 20). Some cyanobacteria, however, appear red because of the accessory pigment phycoerythrin, which absorbs blue-green light in a range missed by cyanobacterial chlorophylls. Cyanobacteria are the only prokaryotes that use both photosystems I and II, photolyzing water to produce oxygen (see **Table 18.2**; details of oxygenic photosynthesis are given in Chapter 14). Under anoxic conditions, most cyanobacteria can also photolyze hydrogen, reduced sulfur compounds, and organic compounds.

The success of oxygenic phototrophy is shown by its fundamental similarity across all members of Cyanobacteria, which account for about a quarter of all of Earth's biomass production and vary widely in cell form and behavior, as well as genetic diversity. Even two cyanobacterial genera such as *Synechococcus* and *Anabaena* share only 37% of their genes. Various cyanobacteria are found in all habitats, from tropical soils to Antarctica. Cyanobacteria include species edible for humans, in products such as *Spirulina* salad or *Nostoc* soup.

Cyanobacterial Cell Structure

The photosynthetic apparatus of cyanobacteria is organized within thylakoids, pockets of membrane resembling flattened spheres packed with reaction centers. The thylakoids may be distributed through the cell, as in filamentous genera such as *Nostoc* (**Fig. 18.5A**), or they may encircle the cell in concentric layers, as in the single-celled marine species *Prochlorococcus marinus* (**Fig. 18.5B**). *Prochlorococcus* is one of the smallest and most abundant oxygen producers in the biosphere. In both large cells and small, the thylakoids are completely separate from the plasma membrane, unlike the attached chlorosomes of Chloroflexi and the plasma membrane extensions of "purple" proteobacteria (see **Table 18.2**). Cyanobacterial thylakoids resemble the thylakoids of eukaryotic chloroplasts; they are the most complex and specialized form of photosynthetic apparatus.

Figure 18.5 Cyanobacterial cell structure. A. Intracellular organelles of *Nostoc*, a typical filamentous cyanobacterium (colorized TEM). **B.** Intracellular organelles of *Prochlorococcus*, a prochlorophyte cyanobacterium, the smallest known phototroph. *Prochlorococcus* accounts for 40%–50% of marine phototrophic biomass.

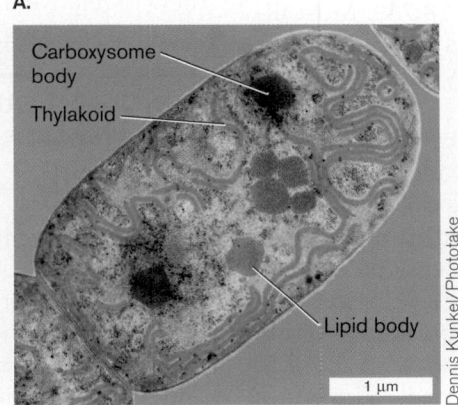

A.

Carboxysome body

Thylakoid

Lipid body

1 μm

Dennis Kunkel/Phototake

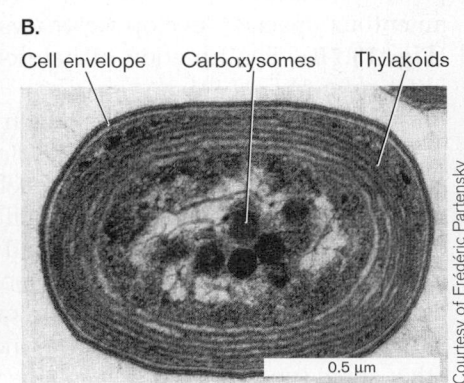

B.

Cell envelope   Carboxysomes   Thylakoids

0.5 μm

Courtesy of Frédéric Partensky

Cyanobacteria have several other subcellular structures. **Carboxysomes** (also known as polyhedral bodies) are rich in the enzyme Rubisco, and they fix CO_2. Cyanobacteria store energy-rich compounds in lipid bodies. To maintain height in the water column and thus access to sunlight, cyanobacteria have **gas vesicles** whose buoyancy enables cells to float. Their external structures include a thick peptidoglycan cell wall, similar to that of Gram-positive cells, plus several external layers that vary with different species. Many species move by "gliding," a form of motility poorly understood.

Besides fixing CO_2, most cyanobacteria fix N_2. Because nitrogen fixation requires the absence of oxygen, cyanobacteria have to solve the problem of maintaining anaerobic biochemistry while producing huge quantities of highly toxic O_2. Different species solve this problem in different ways:

- Formation of specialized nitrogen-fixing cells called **heterocysts**. Heterocysts exclude oxygen.
- Temporal separation, alternating between photosynthesis during daylight and nitrogen fixation at night.
- Accumulation of large aggregates of cells in which the interior becomes sufficiently anaerobic for nitrogenase to function, while the exterior continues oxygenic photosynthesis.
- Symbiosis with microbes or plants that consume oxygen or otherwise maintain anoxic conditions.

Filamentous and Colonial Cyanobacteria

Cyanobacteria have evolved several major categories of form. Single-celled forms include *Synechococcus* and *Prochlorococcus*, the most abundant phototrophs in the oceans. Other genera, such as *Oscillatoria*, generate long filaments of hundreds or even thousands of cells, which can associate in thick biofilms (**Fig. 18.6**). *Oscillatoria* cells are stacked like plates, wider than they are long. To disseminate their cells beyond the biofilm, the filaments produce **hormogonia**, short motile chains of three to five cells. Other filamentous genera, such as *Nostoc*, arrange their filaments in balls of mucilage, possibly to protect from grazing (**Fig. 18.7A**). Most filamentous species develop heterocysts to fix nitrogen (**Fig. 18.7B**). The function of heterocysts is discussed in Chapter 15.

Under environmental stress, such as light limitation or phosphate starvation, filamentous cyanobacteria such as *Anabaena* form specialized spore cells called **akinetes**. An akinete forms as a long, oval cell adjacent to a heterocyst, where it stores nitrogen and develops a thickened envelope. Like other types of spores, akinetes resist desiccation and remain viable for long periods. **Table 18.3** compares the properties of akinetes with those of other spore types (which are discussed under Firmicutes

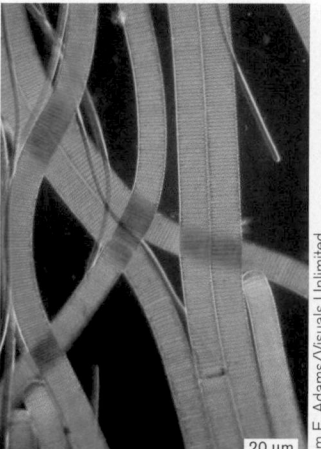 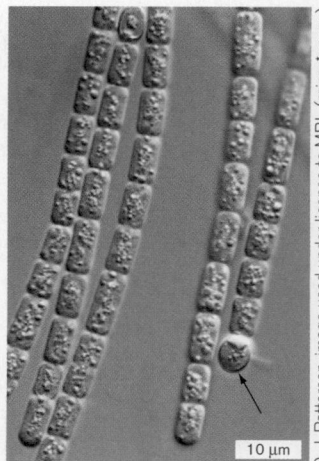

A. *Oscillatoria*   **B. Dome cell**

Figure 18.6   Filamentous cyanobacteria.   A. *Oscillatoria* forms large filaments whose cells are wider than their length. **B.** Cyanobacteria filament terminates in a dome cell (arrow).

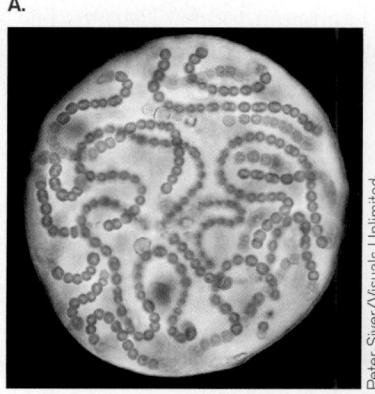 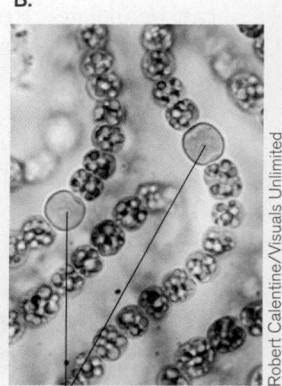

A.   B.

Heterocysts

Figure 18.7   Clumped filamentous cyanobacteria. A. *Nostoc* forms filaments that clump together in large balls of mucilage. **B.** *Nostoc* makes heterocysts (round cells approximately 5 μm in diameter).

and other taxa). Akinetes lie dormant but viable until improved conditions permit germination and growth of new vegetative filaments. In lake water, akinete germination may cause toxic blooms of *Anabaena*.

> **THOUGHT QUESTION 18.4**   What are the relative advantages and disadvantages of propagation by hormogonia, as compared with akinetes?

Nonfilamentous species divide to form small groups or larger colonies (**Fig. 18.8**). *Gloeocapsa* and *Chroococcus* form doublets or quartets, encased in a thick protective mucous slime (**Fig. 18.8A**). Others, such as *Merismopedia*, continue cell division in two planes, extending to form long square sheets of attached cells (**Fig. 18.8B**). Colonial

Table 18.3 Spore types in bacteria.

Spore type	Bacteria that produce the spore	Initiation of spore formation	Formation of the spore	Properties of the spore
Akinete	Filamentous cyanobacteria	Light limitation Cold temperature Phosphate starvation	An akinete develops next to a heterocyst, as a large oval cell with a multilayered envelope.	Desiccation and cold resistant Viable for decades
Arthrospore	Actinomycetes	Carbon starvation Phosphate starvation	At the tip of an aerial mycelium, cells undergo vegetative division and pinch off as arthrospores.	Desiccation resistant Heat resistant
Elementary body	Chlamydias	Completion of intracellular life cycle	Intracellular chlamydia reticular bodies replicate and then develop into elementary bodies with cross-linked outer membrane proteins. Elementary bodies survive outside the host cell.	Survives outside host Desiccation resistant
Endospore	Firmicutes	Carbon starvation Nitrogen starvation Phosphate starvation Low pH Peptide antibiotics	Individual cell develops motherspore and forespore. The forespore develops into an endospore with a spore coat reinforced by keratin and calcium dipicolinate.	Highly heat and desiccation resistant Viable for centuries
Myxospore	Myxobacteria	Nutrient starvation Heat shock Glycerol Dimethyl sulfoxide	Myxobacteria aggregate to form a fruiting body in which vegetative cell division forms a mass of myxospores.	Desiccation resistant UV resistant

A. *Gloeocapsa*

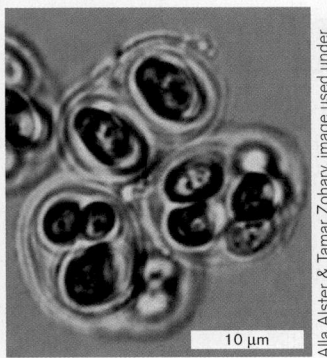

10 μm

Alla Alster & Tamar Zohary, image used under license to MBL (micro·scope)

B. *Merismopedia*

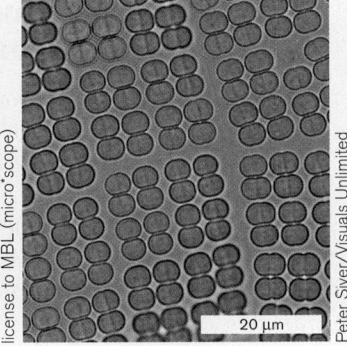

20 μm

Peter Siver/Visuals Unlimited

C. *Pleurocapsa*

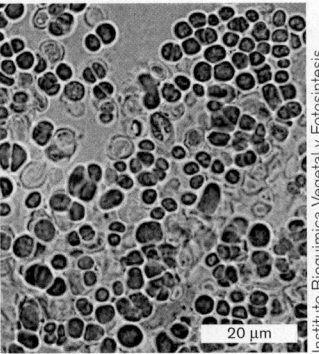

20 μm

Instituto Bioquímica Vegetal y Fotosíntesis

Figure 18.8 Single-celled and colonial cyanobacteria. **A.** *Gloeocapsa* is surrounded by mucus. Cells grow as single cells, doublets, or quartets. **B.** *Merismopedia* forms extended quartets, octets, and so on. **C.** *Pleurocapsa* forms enormous aggregates that release baeocytes.

genera such as *Myxosarcina* and *Pleurocapsa* reproduce by multiple fission, forming large cell aggregates (**Fig. 18.8C**). As the aggregate matures, some cells continue to divide and release single cells called baeocytes. Each baeocyte reproduces and develops into a new cell aggregate. The aggregate group maintains anoxic conditions at the center for nitrogen fixation.

> **THOUGHT QUESTION 18.5** What are the relative advantages and disadvantages of the different strategies for maintaining separation of nitrogen fixation and photosynthesis?

Cyanobacterial Communities

Cyanobacteria play key roles in many ecosystems. In the ocean, cyanobacteria are the main primary producers (discussed in Chapter 21). Environmental change causes the colonial cyanobacteria *Trichodesmium* to form giant blooms visible from outer space, covering many square kilometers of ocean surface (**Fig. 18.9**). Such a bloom can be triggered by an influx of iron carried by wind from a dust storm blowing off the Sahara desert. The concentrated *Trichodesmium* converts nitrogen to forms that promote growth of eukaryotic algae, which prove toxic to consumers such as fish and manatees.

Figure 18.9 Marine cyanobacterial bloom. A ship plows through an ocean bloom of *Trichodesmium*.

Cyanobacteria and diatoms
Purple sulfur proteobacteria
Long-wavelength purple sulfur bacteria

2.54 cm

Figure 18.10 Cyanobacteria in microbial mats. Cutaway through a multilayered microbial mat from the sand flats of Great Sippewissett Salt Marsh (Cape Cod, Massachusetts). Cyanobacteria and diatoms form the upper green layer, above layers of purple sulfur proteobacteria.

In salt marshes and sand flats, cyanobacteria participate in multilayered microbial mats, such as the section shown in **Figure 18.10**, cut from the sand flats of Great Sippewissett Salt Marsh, on Cape Cod, Massachusetts. The high concentration of sulfides in the sediment supports growth of high populations of sulfur phototrophs. Typically, cyanobacteria and eukaryotic algae such as diatoms form the upper green layer. Below, the purple layer consists of "purple sulfur" proteobacteria, whose photopigments (primarily Bchl *a*) absorb at longer wavelengths (discussed in Section 18.5). The pale-colored layer below the purple layer consists of proteobacteria with Bchl *b*, which absorbs farther into the infrared.

Cyanobacteria share many kinds of mutualism with animals, plants, fungi, and protists. Sponges growing on coral reefs may have endosymbiotic cyanobacteria that provide the sponge with nutrients from photosynthesis. The products of photosynthesis supplement the nutrients obtained by the sponge from its filter feeding, and these extra nutrients greatly augment the sponge growth rate within the competitive coral reef environment. Sponge symbionts produce many pharmaceutically active compounds.

TO SUMMARIZE:

■ **Cyanobacteria** are the only oxygenic prokaryotes. They conduct photosynthesis in thylakoids, fix CO_2 in carboxysomes, and maintain buoyancy using gas vesicles. They exhibit gliding motility.

■ **Single-celled cyanobacteria** such as *Prochlorococcus* are among the smallest and most abundant phototrophic producers in the oceans.

■ **Filamentous cyanobacteria** such as *Nostoc* and *Oscillatoria* are common in freshwater lakes. Filaments form heterocysts to fix nitrogen, and reproduce by hormogonia or by akinetes.

■ **Colonial cyanobacteria** such as *Myxosarcina* produce large cell aggregates with an anaerobic core for nitrogen fixation. The colonies reproduce through baeocytes.

■ **Symbiotic associations** of cyanobacteria are formed with animals, fungi, and plants.

18.4 Gram-Positive Firmicutes and Actinobacteria

Gram-positive bacteria were formerly all classified as Firmicutes, the "tough skin" bacteria, and older sources designate Actinobacteria within this phylum. Today, as shown in **Table 18.1**, the phylum Firmicutes is defined to contain "low-GC" species (having a low ratio of GC/AT base pairs), whereas Actinobacteria are the "high-GC" species. They comprise two distinct phylogenetic branches. Species in both groups have thick cell walls that retain the Gram stain and partly exclude antibiotics (discussed in Chapter 3). Their thick cell walls are reinforced by teichoic acids, cross-threading phosphodiester chains of glycerol and ribitol.

Many firmicutes, such as *Bacillus* and *Clostridium* species, survive unfavorable environmental conditions by forming durable endospores. Non–spore formers such as *Lactobacillus* and *Streptococcus* may have evolved from a common Firmicutes ancestor that formed endospores. In many cases, the machinery to form endospores is revealed in the sequenced genomes of organisms believed to be non–spore formers, such as *Carboxydothermus* sp., soil bacteria that oxidize CO (carbon monoxide) to CO_2. On the other hand, actinobacteria of the order Actinomycetales (actinomycetes), such as *Streptomyces*, do not form endospores, but they develop filaments dispersing arthrospores (see **Table 18.3**).

Firmicutes Include Endospore-Forming Rods

Endospore-forming bacteria are common in soil and air because their spore forms resist desiccation and can remain viable in a dormant state for thousands of years. The best-known orders are Bacillales (mainly aerobic respirers) and Clostridiales (obligate anaerobes). Both groups include species of environmental and economic importance, as well as causative agents of well-known diseases.

Bacillales, genus *Bacillus*. *Bacillus* was one of the first bacterial genera to be classified in the nineteenth century. Colonies soon appear on a nutrient agar plate exposed to air. *Bacillus* species can be isolated from soil or food by suspending a sample in water and heating at 80°C for half an hour. Vegetative cells (that is, cells undergoing binary fission) and non–spore–formers are killed at that temperature. The remaining endospores will germinate and grow on a beef broth agar plate at 25°C–30°C. *Bacillus* species isolated in this way include over a thousand

characterized strains, all but a few of them harmless to humans.

The large, rod-shaped **vegetative cells** (growing and replicating form) of *Bacillus* species are easily stained and visualized. A species of particular scientific importance is *Bacillus subtilis*, the best-studied Gram-positive organism, a "model system" for Firmicutes. The *B. subtilis* genome sequence reveals a large number of transporters for carbon sources and many secretory complexes for industrially important enzymes and drug resistance proteins. It also reveals several integrated prophages (integrated phage genomes). Phage genomes contribute to bacterial evolution by transferring genes between different strains and species, as discussed in Chapter 17.

B. subtilis is studied as a model system with respect to stress response. Enormous changes in protein expression accompany starvation and general stress conditions. As the vegetative cells run short of nutrients on an agar plate, they begin a program to "sporulate"—that is, develop inert endospores (**Fig. 18.11A**). The life cycle of endospore production involves a coordinated developmental plan, in which the cell divides near the pole instead of at the cell equator (**Fig. 18.11B**). The polar compartment develops as the **forespore**, directed by unique regulatory proteins such as SpoIIIE. The larger compartment, called the **motherspore**, provides DNA and nutrients to the growing forespore and disintegrates after release of the mature endospore. When the released endospore encounters favorable conditions of moisture and nutrients, it germinates and restarts vegetative growth. The full sporulation cycle is discussed in Chapter 4.

A.

Endospore

Couterstain   Empty capsule after spore released

CDC/Courtesy of Larry Stauffer, Oregon State Public Health Laboratory

Figure 18.11 ***Bacillus* species: Gram-positive endospore formers.** **A.** Gram-stained *Bacillus* species, sporulating culture. Endospores stain green. Malachite green stain. Cell length, 2–5 μm. **B.** Correlated fluorescence imaging of membrane migration, protein translocation, and chromosome localization during *B. subtilis* sporulation. Membranes were stained with red fluorescent FM4-64. Chromosomes were localized with the blue fluorescent nuclear counterstain DAPI. The small green fluorescent patches indicate the localization of a green fluorescent protein gene fusion to SpoIIIE, a protein essential for both initial membrane fusion and forespore engulfment. Progression of the engulfment is shown from left to right.

B.

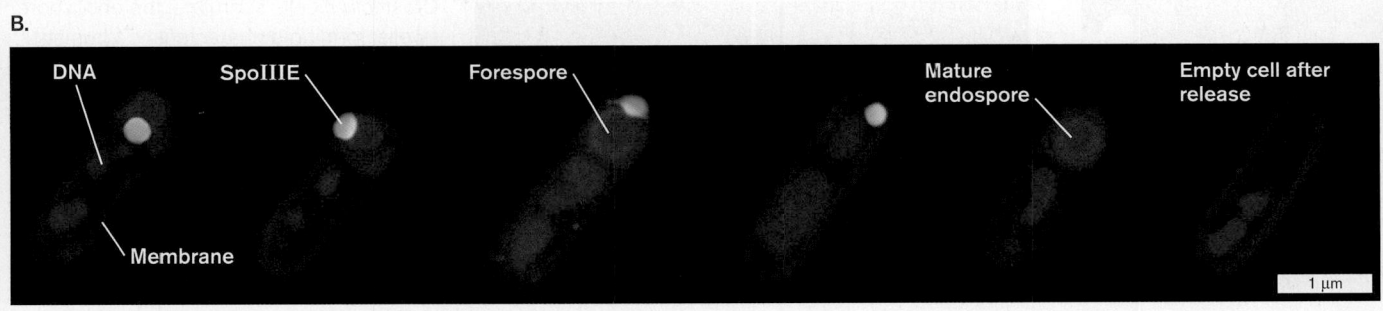

DNA   SpoIIIE   Forespore   Mature endospore   Empty cell after release

Membrane

1 μm

PNAS cover, Dec. 1999/Marc Sharp and Kit Pogliano, U. of California at San Diego

A.

Endospore   Crystalline inclusion

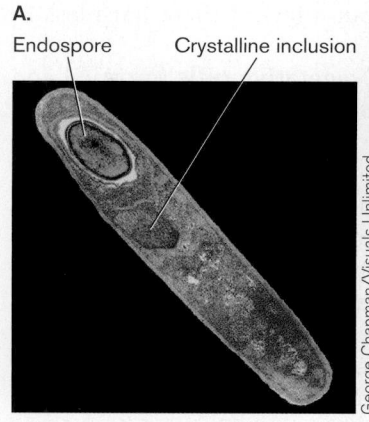

George Chapman/Visuals Unlimited

B.

1 μm

SciMAT/Photo Researchers, Inc.

Figure 18.12 *Bacillus thuringiensis*: **the biological insecticide.** **A.** Sporulating cell of *B. thuringiensis*, showing crystalline inclusion of insecticidal toxin (colorized TEM). **B.** Toxin crystals (colorized blue) embedded in *B. thuringiensis* colony.

A *Bacillus* species of great economic importance is *B. thuringiensis*, the most successful biological control agent yet produced (**Fig. 18.12**). *B. thuringiensis* was originally discovered in 1901 by Japanese bacteriologist Shigetane Ishiwata as a cause of disease in silkworms. The organism proved easy to culture on agar-based medium, and its spores are now applied as an insecticide against the gypsy moth caterpillar. During sporulation, *B. thuringiensis* generates a diamond-shaped crystal adjacent to its endospore (**Fig. 18.12A**). The crystal contains an insecticidal protein, known as delta endotoxin. The crystals are ultimately released from the cells (**Fig. 18.12B**). The toxin is activated only at high pH in the digestive tract of insect larvae; thus, it is completely safe for animals with acidic digestive tracts.

Nevertheless, controversy arose when the gene encoding the toxin was introduced transgenically into plants, with the aim of producing insect-resistant crops. The concern is that the *Bt* gene might escape into wild plants and endanger beneficial or harmless species of moths and butterflies.

Many *Bacillus* species are extremophiles, growing at high pH (*B. alkalophilus*), at high temperature (*B. thermophilus*), or in high salt (*B. halodurans*). Some species combine alkaliphily with thermophily, as in the case of the "alkalithermophile" *B. alkalophilus*. Genomic research has focused on the surprisingly subtle differences that

distinguish an extremophile from a closely related mesophile. For example, thermostability of proteins can be determined by comparing the protein sequences of a thermophile with those of a related mesophile. The thermophilic sequence shows specific patterns of amino acid residues that confer thermostability. Such amino acid substitutions provide useful information for industrial engineering of enzymes.

Clostridiales, genus *Clostridium*. The anaerobic family of spore formers includes the genus *Clostridium*, species of which cause botulism (*C. botulinum*) and tetanus (*C. tetani*) (**Fig. 18.13A**). The botulism toxin, botulinum, or "botox," is famous for its therapeutic use to relax muscle spasms and to smooth wrinkles in skin (**Fig. 18.13B**; discussed in Chapter 26). Other species, such as *C. acetobutylicum*, have economic importance as producers of industrial solvents such as butanol and acetone (for industrial microbiology, see Chapter 16). The butanol pathway is an example of the wide range of diverse fermentative strategies found among clostridia. Phylogenetically, the genus *Clostridium* is less tightly defined than *Bacillus*, whose members all show a highly similar DNA sequence, whereas *Clostridium* species diverge throughout the phylogeny of Firmicutes.

Clostridium cells sporulate in a form distinct from that of *Bacillus* (see **Fig. 18.13A**). The growing endospore swells the end of the cell, forming a "drumstick"

A. *Clostridium* sp.

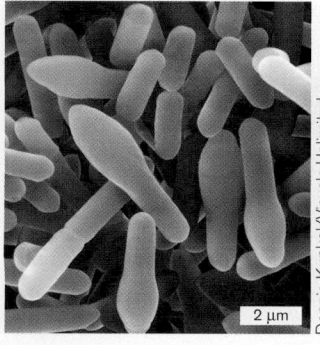

2 μm

Dennis Kunkel/Visuals Unlimited

B. Botox treatment

Before   After

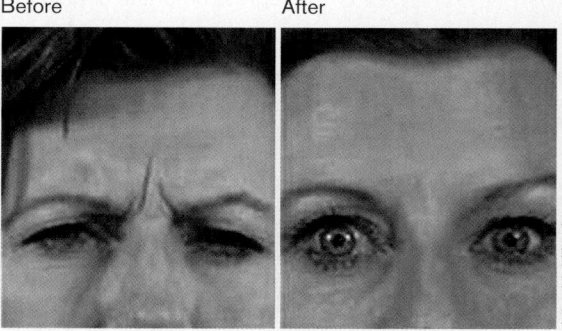

John M. Hilinski

Figure 18.13 *Clostridium* species: **spore-forming anaerobes.** **A.** As *Clostridium* cells sporulate, the endospore swells, forming a characteristic "drumstick" appearance. **B.** The deadly botulinum toxin (botox) from *C. botulinum* is used to relax muscle spasms. This woman was unable to open her eyelids fully (left frame)—a condition that was cured by injection of botox (right frame).

appearance. Mature clostridial spores are generally less heat resistant than the spores of *Bacillus* and thus more difficult to isolate in pure culture from natural environments. Nevertheless, *Clostridium* spores are found in soil and water, ready to germinate and grow when the environment becomes anoxic. Many harmless species are found in the human colon, particularly in infants (see **Fig. 18.1**). Pathogenic *C. botulinum* can grow within the colon of very young infants, and infant botulism has been implicated in some cases of sudden infant death syndrome. Another species, *C. difficile*, is emerging in hospitals as a life-threatening intestinal pathogen, resistant to most antibiotics. *C. difficile* grows in patients treated with antibiotics that eliminate normal enteric bacteria, allowing growth of the pathogen.

Related to *Clostridium* are the Heliobacteriaceae, or "Sun bacteria," the only known phototrophs in Firmicutes (see **Table 18.2**, third row). Their name derives from their yellowish color, caused by the shift of their red peak absorbance into the infrared, which allows transmission of red light plus green light (perceived as yellow). The heliobacteria are photoheterotrophs, with a PS I reaction center providing a modest boost of energy to supplement heterotrophic metabolism of simple organic compounds. Most heliobacterial cells are elongated rods with peritrichous flagella, and they form endospores—the only known endospore-forming phototrophs.

> **NOTE:** Distinguish *Heliobacterium* from *Helicobacter*, a helical-shaped species of the Gammaproteobacteria.

Variant sporulation and "live birth." The order Clostridiales includes species of exceptionally large bacteria that grow only within the digestive tract of specific animal hosts. These species have evolved intriguing variations of the endospore former life cycle, as revealed by Esther Angert and colleagues at Cornell University (**Fig. 18.14A**). *Metabacterium polyspora* (size 15–20 μm) grows throughout the digestive tract of guinea pigs (**Fig. 18.14B**). Endospores ingested from feces germinate in the upper intestine but rarely undergo binary fission. Instead, the growing cell forms forespores at both poles (**Fig. 18.14C**). The forespores actually multiply within the motherspore to form several endospores, which are released in the colon before defecation.

An even larger enteric endosymbiont is *Epulopiscium fishelsoni*, found in the digestive tract of surgeonfish (**Fig. 18.15**). These bacteria are large enough to be seen by eye, about the size of the period at the end of this sentence. As in *M. polyspora*, Angert showed that *E. fishelsoni* reproduction is synchronized with the digestive cycle of its host, but it has gone even further in transformation of the sporulation cycle. Binary fission is eliminated; the cell must fission at both poles. Each polar fission generates

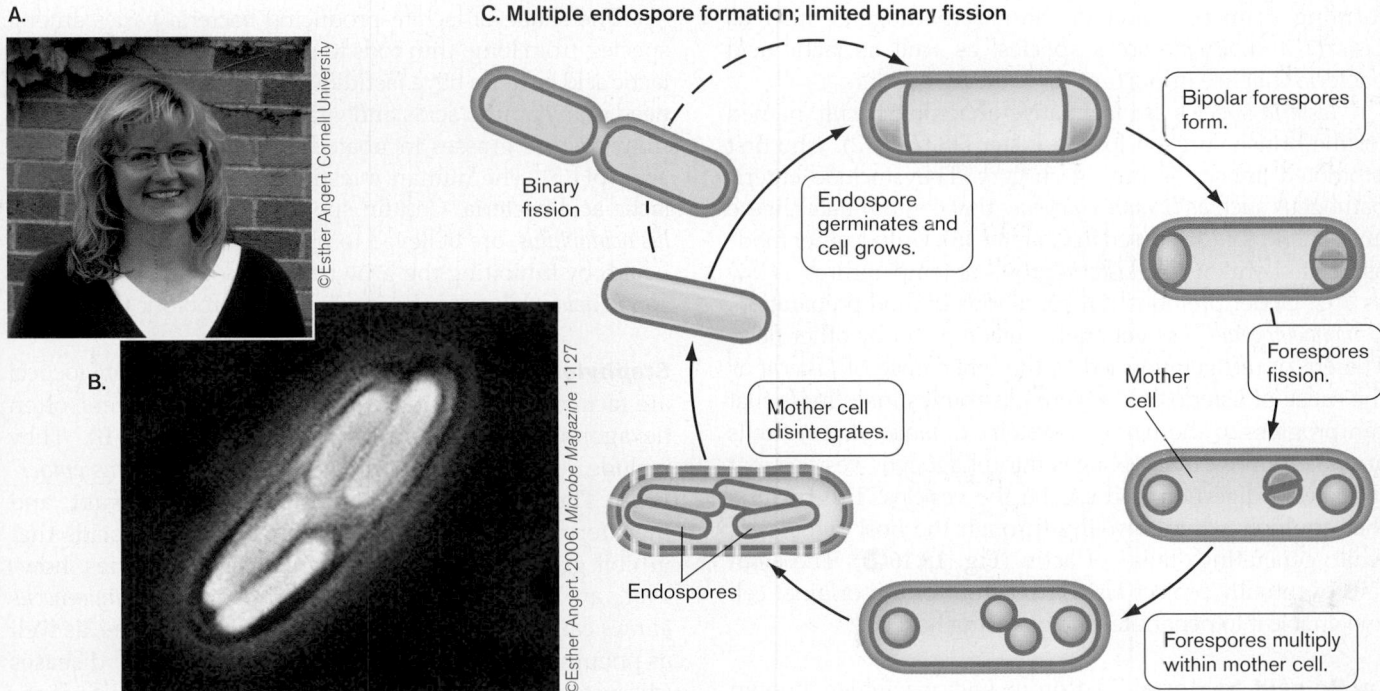

Figure 18.14   Multiple endospore formation.   A. Esther Angert, at Cornell University, characterized unusual forms of sporulation and reproduction in exceptionally large firmicute bacteria. **B.** *Metabacterium polyspora* forms multiple endospores (cell length 15–20 μm; phase-contrast LM). **C.** A forespore forms at each pole. Forespores fission and multiply within the motherspore and then are released. Germinated cells undergo limited or no binary fission.

A.

B. "Live birth" of offspring; no binary fission

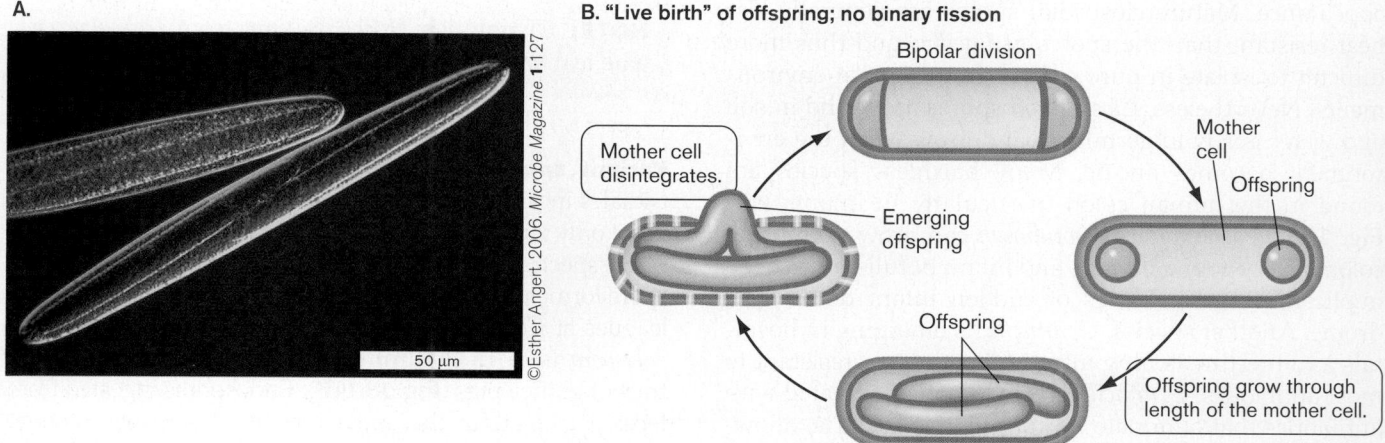

©Esther Angert. 2006. *Microbe Magazine* 1:127

50 μm

Figure 18.15 "Live birth" in *Epulopiscium*. **A.** *E. fishelsoni* forms offspring cells that grow internally. **B.** An offspring cell forms by fission at each pole. The cells grow internally until released. No binary fission occurs outside the mother cell.

an intracellular daughter cell that grows to nearly the full length of the mother cell. The two intracellular offspring ultimately emerge in "live birth" from the mother cell, which then disintegrates.

Non–Spore-Forming Firmicutes

Both Bacillales and Clostridiales, as well as other Gram-positive orders, include many non-spore-forming rods and cocci. In these taxa, the endospore-forming system was probably lost by reductive evolution. Non-spore-forming firmicutes include human pathogens such as *Listeria* and *Streptococcus* species, as well as lactic acid bacteria that are important for food production.

Listeria species are facultative anaerobic bacilli, named for the British surgeon Joseph Lister (1827–1912), who first promoted antisepsis during surgery. They include enteric pathogens such as *L. monocytogenes* that contaminate cheese and sauerkraut (discussed in Chapter 16). Unlike other food-associated organisms, *Listeria* grows at temperatures as low as 4°C. Under preindustrial conditions of food preparation, *L. monocytogenes* was generally outcompeted by other flora. The era of refrigeration led to the emergence of *Listeria* as the cause of listeriosis, a severe gastrointestinal illness that can progress to the nervous system. *L. monocytogenes* cells are taken up by macrophages into phagocytic vesicles, but they avoid digestion and escape the vesicles. The bacteria then multiply while traveling through the host cytoplasm, while generating "tails" of actin (**Fig. 18.16◉**). The actin tails eventually project *Listeria* cells out of the original cell and enable it to penetrate a neighboring host cell.

Lactic acid bacteria. The order Lactobacillales, known as lactic acid bacteria, are aerotolerant (capable of growth in the presence of oxygen), but most are obligate fermenters; that is, they generate ATP by substrate-level phosphorylation but cannot use oxygen to respire. They

ferment primarily by converting sugars to lactic acid (a fermentation pathway discussed in Chapter 13). As the acid builds up, the pH decreases until it halts bacterial growth; thus, the carbon source retains much of its food value for human consumption. This is the basis of yogurt and cheese production. *Lactococcus* and *Lactobacillus* species are extremely important for the dairy industry (discussed in Chapter 16). Other common genera of lactic acid bacteria include *Leuconostoc*, which often spoils meat, and *Pediococcus*, found in sauerkraut and fermented bean products, as well as meat products such as sausage.

The shape of lactate-producing bacteria varies among species, from long, thin rods to curved rods and cocci. Most lactic acid bacteria have fastidious growth requirements and need many amino acids and vitamins. They can be isolated from pasture grasses incubated anaerobically in moderate acid (pH 5). The human intestinal flora include species of lactic acid bacteria. Certain species, particularly *Lactobacillus acidophilus*, are believed to play a positive role in human health by inhibiting the growth of pathogens. For this reason *L. acidophilus* may be ingested as a probiotic therapy.

***Staphylococcus* and *Streptococcus*.** The staphylococci are facultative anaerobic cocci that grow in clusters, often hexagonal in arrangement (**Fig. 18.17A** and **B**). They include common skin flora such as *Staphylococcus epidermidis*. The staphylococci are generally salt tolerant, and their fermentation generates short-chain fatty acids that inhibit growth of skin pathogens. Certain species, however, are themselves serious pathogens. *Staphylococcus aureus* causes impetigo and toxic shock syndrome, as well as pneumonia, mastitis, osteomyelitis, and other diseases (discussed in Chapter 26). It is a major cause of nosocomial (hospital-acquired) infections, especially contamination of surgical wounds. The most dangerous strains, now resistant to most known antibiotics, are termed MRSA (methicillin-resistant *S. aureus*).

A.

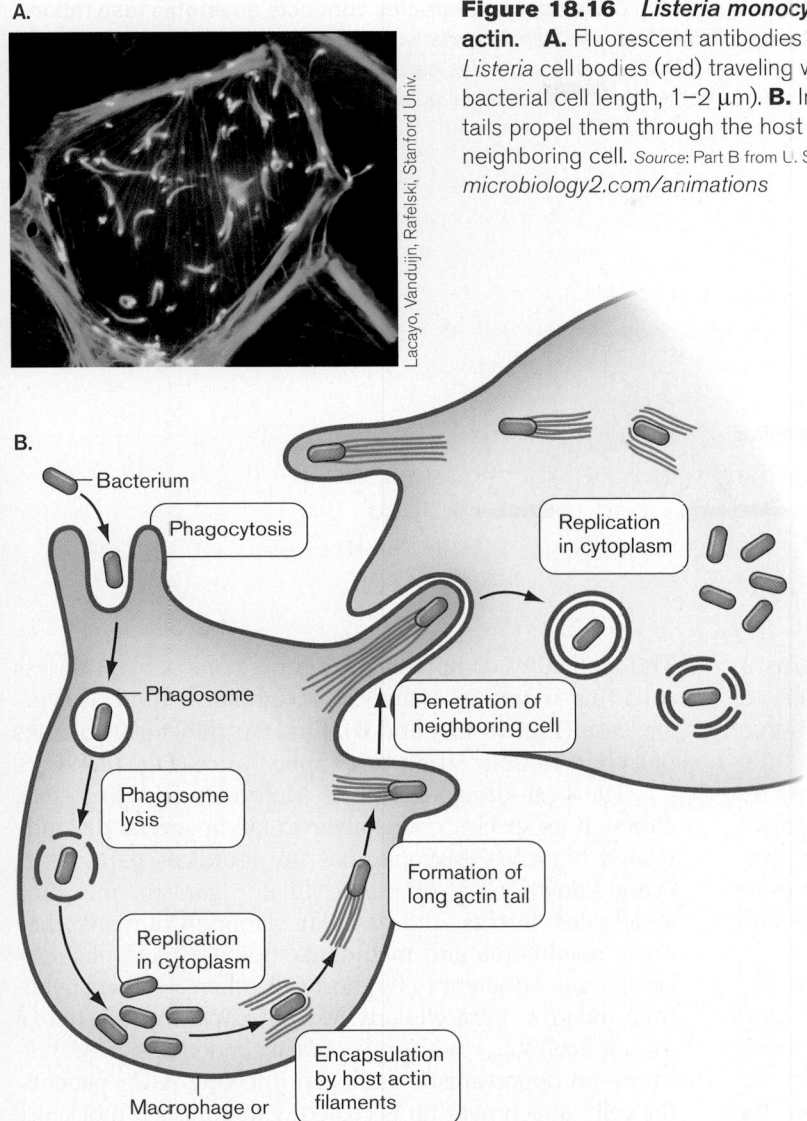

Lacayo, Vanduijn, Rafelski, Stanford Univ.

Figure 18.16 *Listeria monocytogenes*: intracellular pathogen travels on tails of actin. **A.** Fluorescent antibodies mark the tails of polymerized actin (green) behind the *Listeria* cell bodies (red) traveling within an infected macrophage (fluorescence micrograph; bacterial cell length, 1–2 μm). **B.** Invading bacteria encapsulate themselves in actin. Actin tails propel them through the host cytoplasm and out through the cell membrane to invade a neighboring cell. *Source*: Part B from U. South Carolina, Microbiology and Immunology On Line. ▶❚❚
microbiology2.com/animations

B.

- Bacterium
- Phagocytosis
- Phagosome
- Phagosome lysis
- Replication in cytoplasm
- Replication in cytoplasm
- Penetration of neighboring cell
- Formation of long actin tail
- Encapsulation by host actin filaments
- Macrophage or parenchymal cell

Streptococcus species generally form chains instead of clusters, as their cells divide in a single plane (**Fig. 18.17C** and **D**). They are aerotolerant (grow in the presence of oxygen) but metabolize by fermentation. Many live on oral or dental surfaces, where they cause caries (tooth decay). Their fermentation of sugars produces such high concentrations of lactic acid that the pH at the tooth surface can fall to pH 4. *Streptococcus* species cause many serious diseases, including pneumonia (*S. pneumoniae*), strep throat, erysipelas, and scarlet fever (*S. pyogenes* or group A strep).

The streptococci are less salt tolerant than *Staphylococcus* species, which tolerate as much as 5%–15% NaCl. Another genus whose size and fermentative metabolism resembles that of streptococci is *Enterococcus*. *E. faecalis* is a common member of the intestinal flora, and related strains are enteric pathogens.

Anaerobic dechlorinators. Some firmicutes from the soil show promising abilities to degrade chlorinated pollutants, such as dry-cleaning solvents that are biodegraded very slowly in the environment. The chlorinated molecules are reduced as alternative electron acceptors. *Dehalobacter restrictus*, a flagellated rod, was isolated as an anaerobe capable

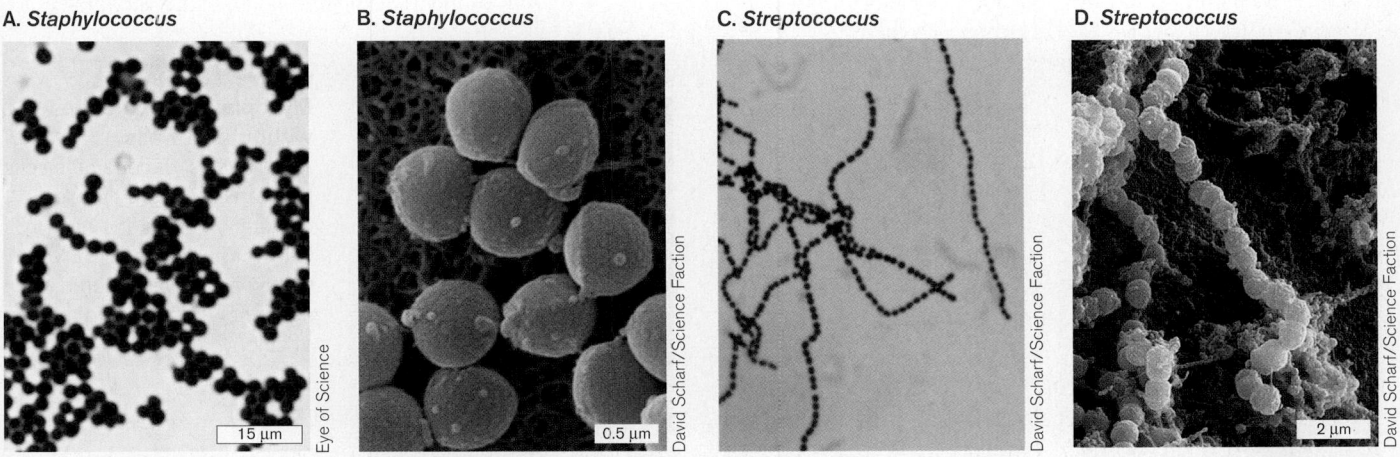

A. *Staphylococcus* — 15 μm — Eye of Science
B. *Staphylococcus* — 0.5 μm — David Scharf/Science Faction
C. *Streptococcus* — David Scharf/Science Faction
D. *Streptococcus* — 2 μm — David Scharf/Science Faction

Figure 18.17 Staphylococci and streptococci.   A. *Staphylococcus* species (cell size 0.5–1.0 μm; Gram stain). **B.** *Staphylococcus* species (colorized SEM). **C.** *Streptococcus* species (cell size 0.5–1.0 μm; Gram stain). **D.** *Streptococcus* species (colorized SEM).

A.

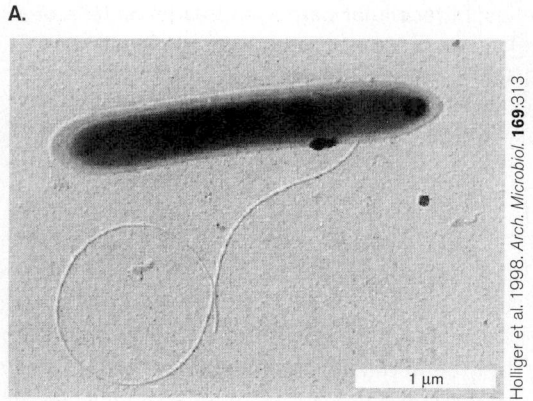

Holliger et al. 1998. *Arch. Microbiol.* **169**:313

1 µm

Figure 18.18 *Dehalobacter restrictus* **conducts anaerobic respiration by dechlorination.   A.** *Dehalobacter restrictus*, a flagellated rod related to clostridia. **B.** *D. restrictus* donates electrons to remove chlorine from tetrachloroethene, a major industrial pollutant. The ultimate product, ethene, is much less harmful in the environment.

B. Dehalogenation of tetrachloroethene by *Dehalobacter restrictus*

of respiring by donating electrons to chlorine atoms in tetrachloroethene (**Fig. 18.18**). Thus, the discovery of *D. restrictus* has promising potential for bioremediation of chlorinated pollutants.

While genetic analysis places *D. restrictus* among the clostridia, the bacterium actually stains Gram-negative, perhaps because its peptidoglycan layer is relatively thin. Nevertheless, the species shows no Gram-negative outer membrane. It does possess a thick S-layer of hexagonally tiled proteins, typical of Gram-positive bacteria.

Mollicutes lack a cell wall. Organisms of the class Mollicutes (named in Latin, "soft skin") have completely lost their cell wall and S-layer through reductive evolution, retaining only their cell membrane. Presumably, the loss of these energy-expensive structures enhanced the reproductive rate of cells in a protected host environment.

The Mollicutes comprise many genera of flexible wall-less cells that maintain a shape through some kind of cytoskeleton (**Fig. 18.19A** and **B**). On agar they form colonies of a characteristic "fried-egg" appearance (**Fig. 18.19C**).

The best-known genus of Mollicutes is *Mycoplasma*, although its species now appear to fall in several distantly related branches. Mycoplasmas are found as parasites of every known class of multicellular organism, including vertebrates, insects, and vascular plants; in humans, they cause pneumonia and meningitis. Some *Mycoplasma* species remain adherent to the host cell, whereas others penetrate and grow intracellularly. Most mycoplasma cells have a rounded cell shape with one or two extended tips. In *M. penetrans*, an opportunistic pathogen infecting AIDS patients, the cell's attachment tip is coated with adhesion molecules that enable attachment to a host cell surface (**Fig. 18.19A**). The attachment tip penetrates deep into epithelial tissues. By

A.

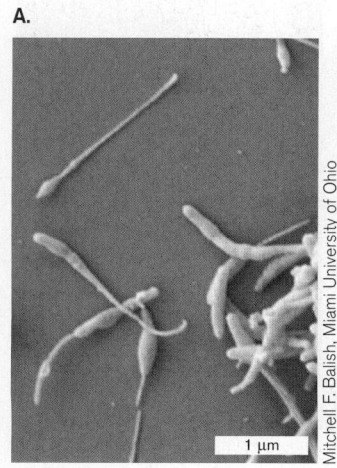

Mitchell F. Balish, Miami University of Ohio

1 µm

B.

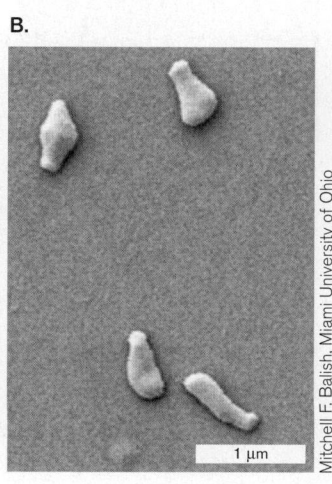

Mitchell F. Balish, Miami University of Ohio

1 µm

C.

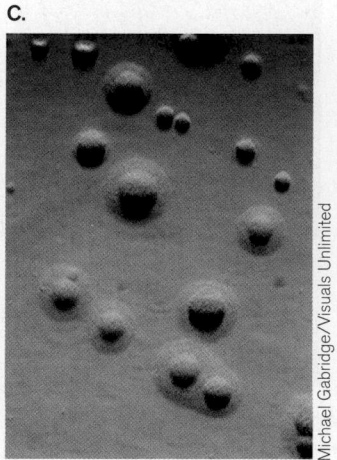

Michael Gabridge/Visuals Unlimited

**Figure 18.19
Mycoplasmas: parasites without cell walls.**
A. *Mycoplasma penetrans* cells have an elongated tip used for attachment to the host (SEM). **B.** *M. mobile* (SEM). **C.** Mycoplasmas cultured on agar show a "fried-egg" shape of colony.

contrast, the fish pathogen *M. mobile* has no attachment tip, but glides along a surface at a rate of seven cell lengths per second (**Fig. 18.19B**). The basis of mycoplasma motility is not understood, although it may involve gliding or cytoskeletal contraction. In some species, motility involves a specialized cell tip called the "terminal organelle," but species that lack this organelle are equally motile. Species of *Spiroplasma* have a spiral shape and undergo corkscrew motion; the basis for this motion is also unknown.

Mycoplasma genomes are among the smallest known cellular organisms, and they lack biosynthetic pathways for amino acids and phospholipids, which must be acquired from the host. Mycoplasmas have unique nutritional requirements, such as cholesterol, a membrane component typical of eukaryotes but rare for prokaryotes. For these reasons, mycoplasmas are difficult to grow in pure culture, although they readily infect tissue cultures. In fact, mycoplasma contamination of tissue culture is so prevalent that it has compromised major studies of cancer and AIDS.

Actinomycetes Form Multicellular Filaments

The phylum Actinobacteria, the "high-GC Gram-positives," comprises several orders, including Actinomycetales and Bifidobacteriales. Members of the order Actinomycetales are "actinomycetes." Actinomycetes can be identified under the microscope by the **acid-fast stain**, a procedure in which cells are penetrated with a dye that is retained under treatment with acid alcohol (discussed in Chapter 2). The acid-fast property is associated with unusual cell wall lipids, such as mycolic acids. Actinomycetes include filamentous spore formers such as *Streptomyces* and *Frankia*, as well as short-chain organisms such as *Mycobacterium*, and irregularly shaped cells such as *Corynebacterium*. An emerging group of marine actinomycetes such as *Salinispora* are isolated from sediment and from sponges. *Salinispora* species form numerous exotic secondary products that show promise as pharmaceutical agents.

***Streptomyces*: filamentous spore formers.** *Streptomyces* bacteria form multicellular filaments that generate dispersable spores. The streptomycete life cycle is discussed in Chapter 4. Streptomycetes play a major role in the ecosystems of soil (discussed in Chapter 21). Decaying *Streptomyces* cells produce the compound **geosmin**, which causes the characteristic odor of soil and can affect the taste of drinking water. In culture, the best-known species, *S. coelicolor* (Latin, "sky color"), forms strikingly blue colonies (**Fig. 18.20A**). The blue color derives from several pigments, including actinorhodin, a polyketide antibiotic. Other species produce filaments that are red, orange, green, or gray, depending on their distinctive products, many of which are antibiotics.

> **THOUGHT QUESTION 18.6** Why would *Streptomyces* produce antibiotics targeting other bacteria?

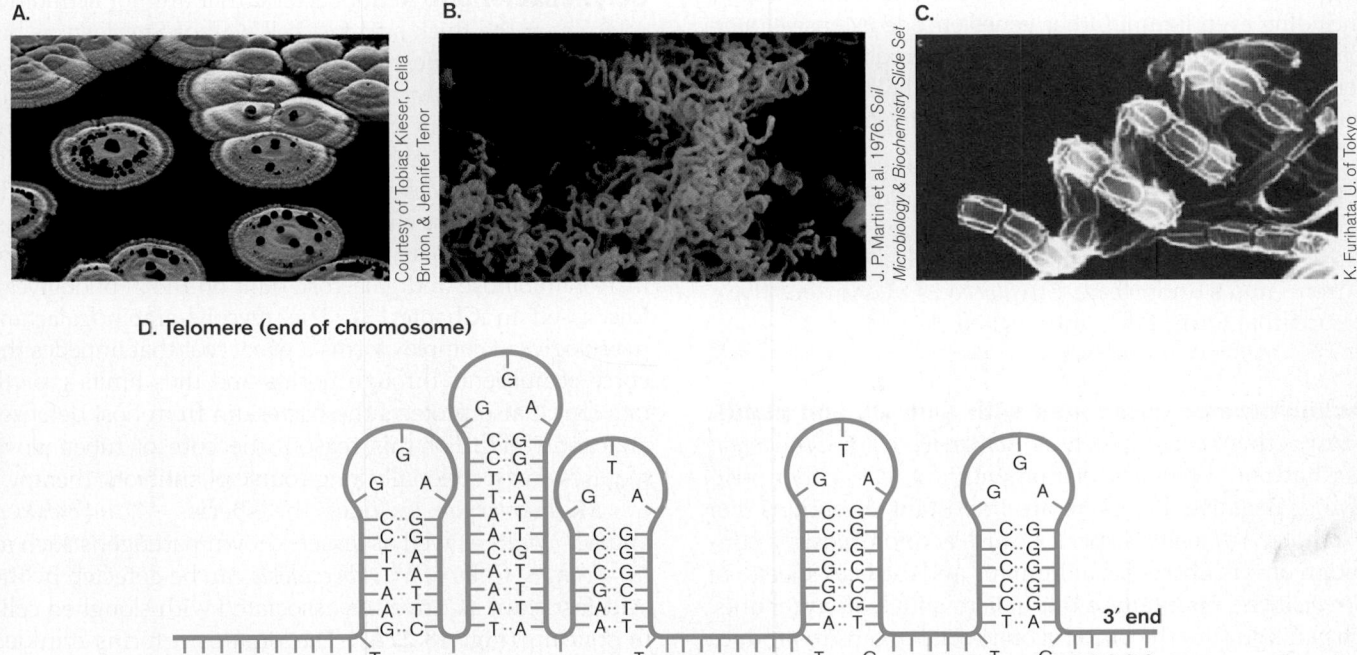

Figure 18.20 *Streptomyces* bacteria. **A.** Colonies of *S. coelicolor* show sky-blue mycelia. **B.** *Streptomyces* cells form coiled filaments (filament width, about 0.5 μm; SEM). **C.** Close-up of a coiled filament, showing individual cells (SEM). **D.** Hairpin-looped telomere end of the linear chromosome of *S. griseus*. *Source*: Part D from *Prokaryotes*, Springer.

Streptomyces species are obligate aerobes, requiring access to air to complete their life cycle. When a *Streptomyces* spore germinates, it extends branched filaments into the substrate called **vegetative mycelia**, branched filaments of actively dividing cells. Some of the filaments then grow upward into the air, where they develop into **aerial mycelia**. The aerial mycelia in some species grow in tightly coiled spirals (**Fig. 18.20B** and **C**). As the mycelial colony runs out of nutrients, older cells of the filament age and lyse, releasing nutrients absorbed by the younger cells. The nutrients also attract other scavengers, which may be killed by antibiotics produced by the aging streptomycete cells. The dead scavengers, too, release nutrients that feed the growing tip of the mycelium.

As mycelia mature, they fragment into smaller cells called **arthrospores**. Arthrospores are vegetative cells, not dormant like endospores (see **Table 18.3**). The arthrospores separate and are dispersed by the wind, enabling them to colonize a new location. Streptomycete mycelia can be obtained from natural habitats by burying a glass slide in soil and then waiting several days for spores to germinate, covering the slide with mycelia. They are challenging to isolate in pure culture, however, because their coiled filaments trap cells of other bacteria.

The *S. coelicolor* genome contains over 8 million base pairs—one of the largest prokaryotic genomes. Streptomycete chromosomes are linear with special "telomeres," single-stranded end sequences that double back to form hairpin loops (**Fig. 18.20D**). Much of the lengthy genome of a streptomycete encodes catabolism of a rich array of diverse organic components of decaying plant and animal matter, including even lignin. Other genes encode extensive operons for production of diverse secondary products (see Chapter 15), including antibiotics. More than half the antibiotics currently used in medicine derive from *Streptomyces* species.

> **NOTE:** Distinguish the order Actinomycetales from the genus *Actinomyces* within this order. Distinguish *Streptomyces*, filamentous rod-shaped actinomycetes, from nonactinomycete *Streptococcus*, Gram-positive cocci that form short, unbranched chains.

Actinomycetes associated with animals and plants.
Most actinomycetes, such as *Actinomyces* and *Streptomyces*, have no direct association with animals, either positive or negative, but there are important exceptions. For example, *Actinomyces* species cause actinomycosis, a condition of skin abscesses in humans and cattle. A species of *Streptomyces* maintains a mutualism with leaf-cutter ants. The ants culture the bacteria on special organs to produce antibiotics against parasites of their fungal gardens (discussed in Section 17.6).

Actinomycetes have many mutually beneficial associations with plants. For example, *Frankia* associates with the

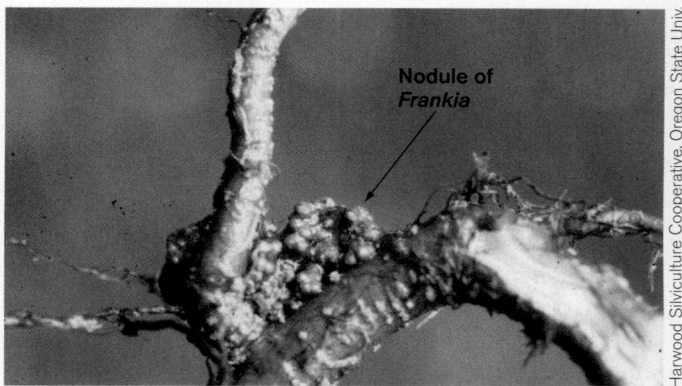

Figure 18.21 *Frankia* **species associate endophytically with plants.** Root of alder tree (*Alnus glutinosa*) bears the orange-yellow-colored nodules (arrowhead) containing *Frankia*.

alder tree, which develops orange-yellow-colored nodules on its roots to fix nitrogen (**Fig. 18.21**). *Frankia* species live as **endophytes**, endosymbionts of vascular plants. This nitrogen fixation mutualism is analogous to the better-known symbiosis between legumes and rhizobia (discussed under Alphaproteobacteria, in Section 18.5). Other *Frankia* species benefit wheat by conferring resistance to major fungal pathogens, such as the fungus that causes "take-all" disease, as well as to certain parasites and insects. Still other *Frankia* species grow as saprophytes, consuming dead leaf litter. A few actinomycetes cause plant diseases. For example, potato scab disease is caused by *Streptomyces scabies*.

Nonmycelial Actinobacteria: *Mycobacterium* and *Corynebacterium*. Actinobacteria that are not actinomycetes share the thick acid-fast cell wall of *Streptomyces* but lack the mycelial lifestyle. Two genera that cause dreaded diseases are *Mycobacterium* (**Fig. 18.22**) and *Corynebacterium*. Both genera have thick cell envelopes containing **mycolic acids** and phenolic glycolipids. The mycolic acids of *M. tuberculosis* are extremely diverse and include some of the longest-chain acids known, up to 90 carbons. The mycolic acids are linked to arabinogalactan, a polymer of arabinose and galactose built on the peptidoglycan (discussed in Chapter 3). The mycolyl-arabinogalactan-peptidoglycan complex forms a waxy coat that impedes the entry of nutrients through porins and thus limits growth rate; but it also protects the bacterium from host defenses and antibiotics. For this reason, the cure of tuberculosis requires an exceptionally long course of antibiotic therapy.

Mycobacterium includes the species *M. tuberculosis* and *M. leprae*, as well as lesser-known pathogens such as *M. ulcerans*. Cells of *M. tuberculosis* can be detected by the acid-fast stain as tiny rods associated with sloughed cells in sputum (**Fig. 18.22A**). The organism forms crinkled colonies after two weeks of growth on agar-based media (**Fig. 18.22B**).

The closely related species *M. leprae* causes the disfiguring disease leprosy (**Fig. 18.23A**). The species has

A.

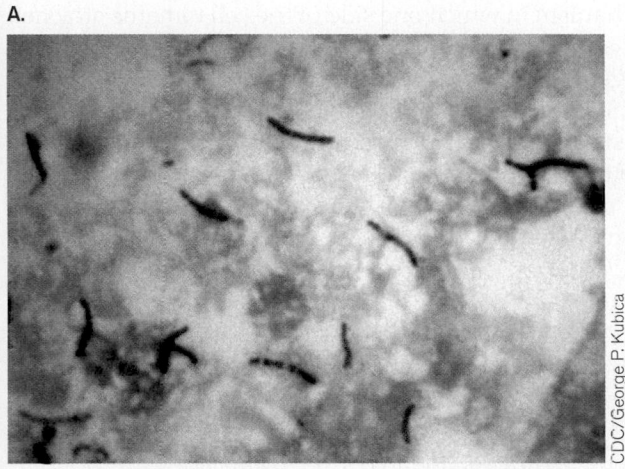

CDC/George P. Kubica

B.

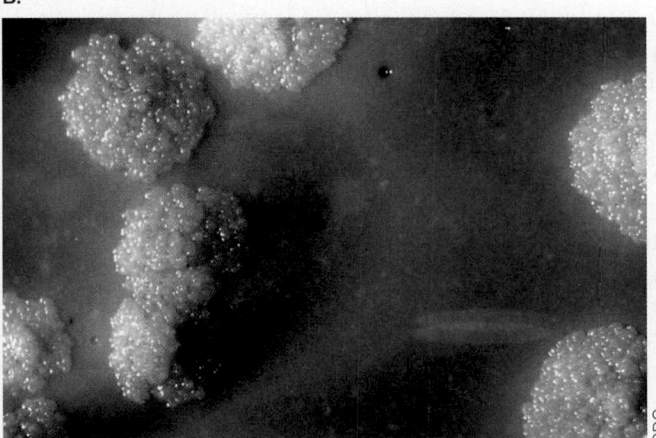

CDC

Figure 18.22 *Mycobacterium tuberculosis* **causes tuberculosis.** **A.** Acid-fast stain of tissue sample containing *M. tuberculosis* (chains of pink rods, 2–4 μm in length). **B.** Crinkled appearance of *M. tuberculosis* colonies. **C.** Mycolic acids coat the cell wall of *M. tuberculosis*.

C.

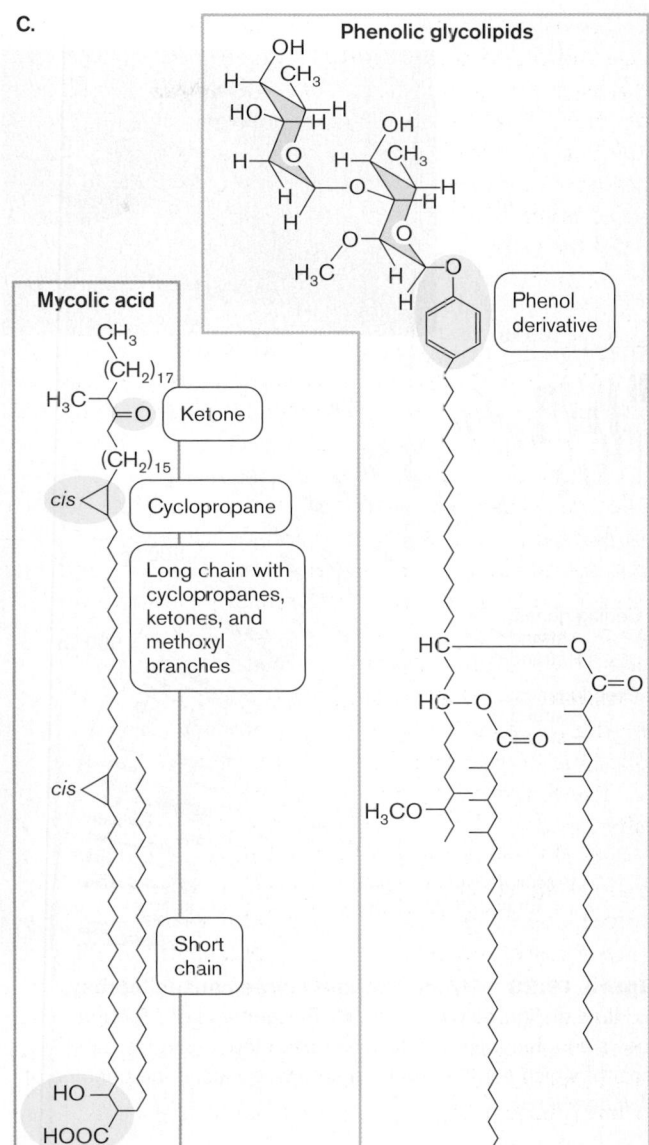

one of the longest known doubling times of any pathogen (about 14 days), and it can take a year to grow enough cells in the laboratory for observation. Lower temperature is required; for this reason, leprosy attacks the extremities (hands and feet, which have lower temperature). Culture on artificial media is impossible; the bacteria can be grown only within low-temperature animals, such as armadillos, or within genetically immunodeficient mice.

The genomes have been sequenced for both *M. tuberculosis* and *M. leprae*. *M. tuberculosis* has surprisingly few recognizable pathogenicity genes but a large number of environmental stress components, including 16 environmental sigma factors (discussed in Chapter 8), as well as 250 genes for its complex lipid metabolism. Over half the *M. leprae* genome consists of pseudogenes, homologs of *M. tuberculosis* genes undergoing reductive evolution (**Fig. 18.23B**). Thus, *M. leprae* appears to be an evolving

pathogen "caught in the act" of losing many genes no longer needed in its sheltered host environment. How it lost the need for so many genes preserved in *M. tuberculosis* remains a mystery.

Mycobacteria also include a much larger number of harmless commensals, such as *M. smegmatis*, isolated from human skin. Species of mycobacteria can be isolated from soil and water, as well as from various animal sources. Their culture is difficult because of their slow growth rates, but isolation can be enhanced by treatment with a base (NaOH or KOH) at concentrations that kill most other bacteria.

NOTE: Distinguish *Mycobacterium*, rods whose cell walls contain mycolic acids, from *Mycoplasma*, cell wall–less firmicutes related to *Bacillus* and *Clostridium*.

A.

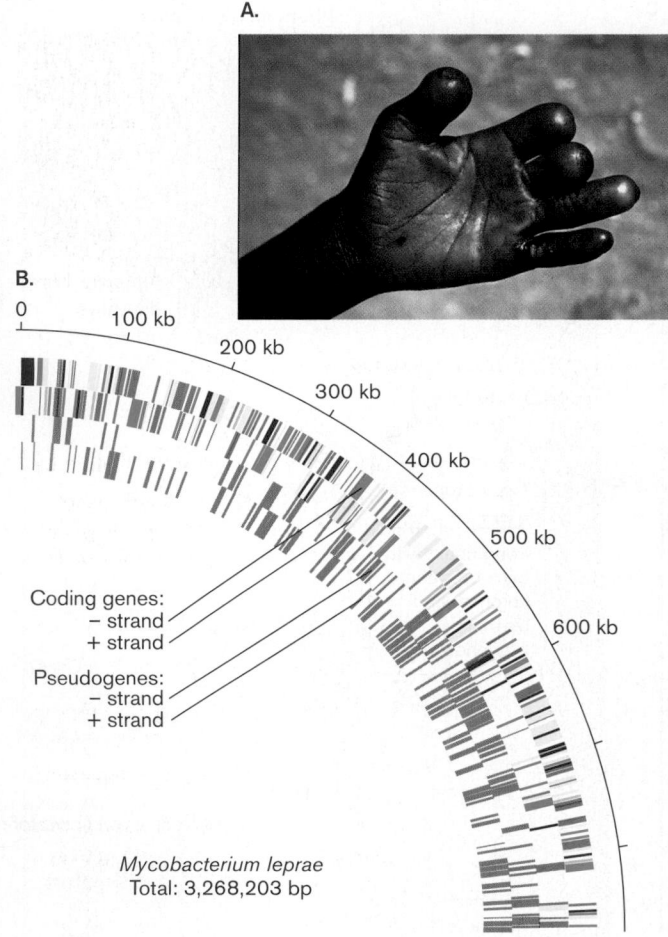

B.

Mycobacterium leprae
Total: 3,268,203 bp

Coding genes:
– strand
+ strand
Pseudogenes:
– strand
+ strand

C. James Webb/Phototake

Figure 18.23 *Mycobacterium leprae* **causes leprosy.**
A. Hand disfigured by leprosy. **B.** The genome of *M. leprae*
shows a high content of decaying pseudogenes (gray bars),
most of which correspond to functional genes in the genome of
M. tuberculosis.

Irregularly shaped actinomycetes. Several nonmycelial actinomycetes show unusual cell shapes. Members of
the genus *Corynebacterium* include soil bacteria, as well as
pathogens such as *C. diphtheriae*, the cause of the lung disease diphtheria. *Corynebacterium* species grow as irregularly shaped rods, which may divide by a "half-snapping"

mechanism in which one side of the cell remains attached
like a hinge (**Fig. 18.24A**). Related soil bacteria include
the genera *Nocardia* and *Rhodococcus*.

Soil bacteria of the genus *Arthrobacter* exhibit an
unusual cell cycle in which coccoid stationary-phase cells
sprout into rods, which eventually run out of nutrients
and revert to the coccoid form. The growing rods form
irregular branched filaments (**Fig. 18.24B**). An *Arthrobacter* species was discovered conducting anaerobic respiration by reduction of hexavalent chromium (Cr^{6+}), a
toxic metal pollutant, to a less toxic oxidation state. Now
Arthrobacter shows potential as an agent of bioremediation of hexavalent chromium.

Micrococcaceae. Relatives of *Arthrobacter* are nonfilamentous cocci such as *Micrococcus*. *M. luteus* is one of the
most ubiquitous of soil bacteria, appearing readily as yellow colonies on agar plates exposed to air. Historically,
the genus *Micrococcus* in the family Micrococcaceae was
classified with *Staphylococcus*, but its DNA sequence data
now place *Micrococcus* and most of the Micrococcaceae
within the actinomycetes.

Micrococci are aerobic heterotrophs, commonly isolated from air and dust, although their habitat of choice
is the human skin. Micrococci grow in square or cuboid
formations, dividing in two or three planes (**Fig. 18.25**);
cuboid clusters are known as **sarcinae**. *M. luteus* is harmless to humans and grows well at room temperature, so it
makes an excellent laboratory organism for observation
by students.

TO SUMMARIZE:

- **Firmicutes**, the low-GC Gram-positive bacteria,
include endospore-forming genera such as *Bacillus*
and *Clostridium*. The cycle of endospore formation
was probably present in the common ancestor of this
phylum.
- **Nonsporulating firmicutes** include pathogenic rods
such as *Listeria*, as well as food-producing bacteria
such as *Lactobacillus* and *Lactococcus*.
- *Staphylococcus* and *Streptococcus* are Gram-
positive cocci that include normal human flora, as

A. *Corynebacterium diphtheriae*   **B.** *Arthrobacter globiformis*

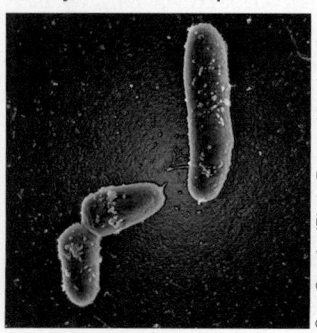

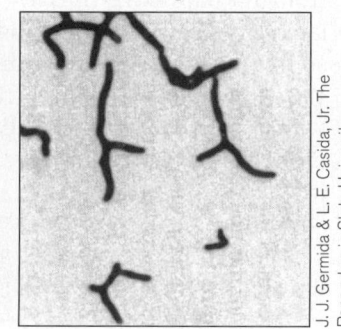

Gary Gaugler/Photo Researchers, Inc.

J. J. Germida & L. E. Casida, Jr. The Pennsylvania State University

Figure 18.24 Irregularly shaped actinomycetes:
Corynebacterium **and** ***Arthrobacter.*** **A.** *Corynebacterium
diphtheriae* (cells 1–2 µm in length) divide by snapping off
one side while remaining attached at the other; the result is
a typical V shape or "Chinese letter" arrangement (colorized
SEM). **B.** *Arthrobacter globiformis* cultures form coccoid cells in
stationary phase. With added nutrients, the coccoid cells grow
out as rods.

Figure 18.25 *Micrococcus* species grow in tetrads or sarcinae. **A.** *M. luteus* growing in tetrads (negative or indirect stain with nigrosin). **B.** *M. luteus* bacteria grow in clumps when cultured on a solid substrate (SEM).

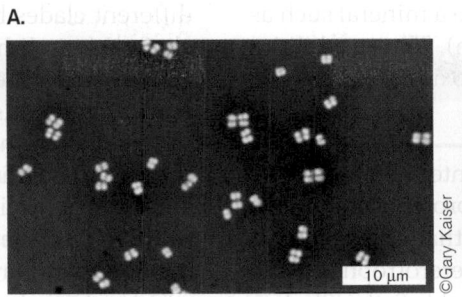

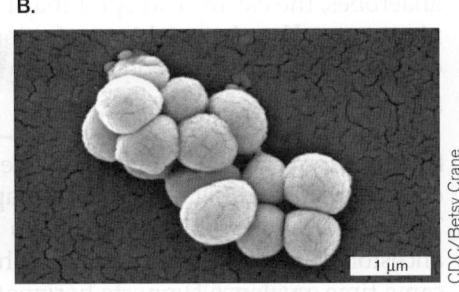

well as serious pathogens causing toxic shock syndrome, pneumonia, and scarlet fever.

■ **Mycoplasmas** belong phylogenetically to Firmicutes but lack the cell wall and S-layer. They have flexible cytoskeletons and show ameboid motility. Species cause diseases such as meningitis and pneumonia.

■ **Actinomycetes** (order Actinomycetales) include mycelial spore-forming soil bacteria, such as *Streptomyces*. Other actinomycetes have irregularly shaped cells, such as *Clostridium* species that cause tetanus and botulism. Actinomycetes stain acid-fast.

■ **Mycobacteria** are actinomycete rods whose cell envelope contains a diverse assemblage of complex mycolic acids. Mycobacterial species cause tuberculosis and leprosy. They stain acid-fast.

18.5 Gram-Negative Proteobacteria and Nitrospirae

The Proteobacteria, with all their diversity of form and metabolism, share a common structure—their triple-layered, Gram-negative cell envelope, which consists of outer membrane, peptidoglycan cell wall permeated by the periplasm, and inner membrane (plasma membrane) (discussed in Chapter 3). The outer membrane is packed with receptor proteins and porins, comprising two-thirds of the mass of the membrane. The outer membrane lipids contain long sugar polymer extensions (lipopolysaccharide, or LPS). In pathogens, LPS repels phagocytosis and has toxic effects when released by dying cells.

The Protean Metabolism of Proteobacteria

Proteobacteria, the "protean" or "many-formed" bacteria, comprise at least five classes, labeled Alpha through Epsilon (see **Table 18.1**). Even within each closely related group, we see nearly as wide a range of cell shape and metabolic strategies as we see in Proteobacteria as a whole.

A closer look at proteobacterial metabolism shows that processes that seem very different on the surface

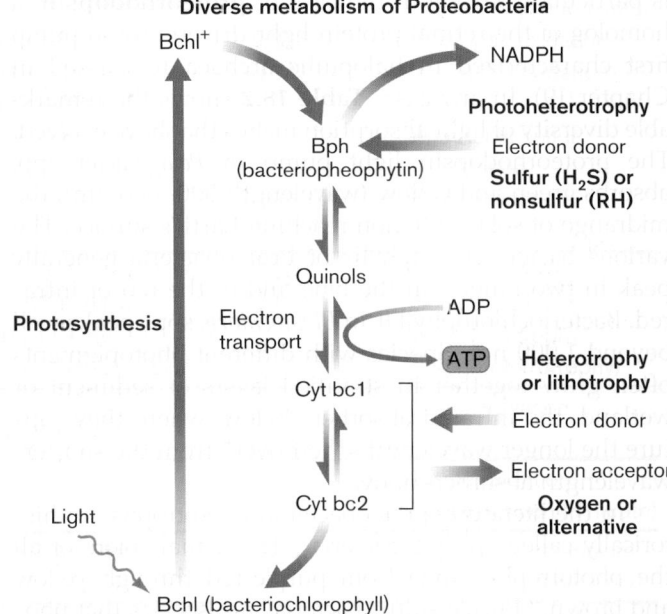

Figure 18.26   The protean metabolism of Proteobacteria. In many Gram-negative species, metabolic diversity arises through minor "add-ons" of biochemical modules such as light absorption by bacteriochlorophyll, use of sulfide or organic electron donors, and use of oxygen or alternative (anaerobic) electron acceptors.

actually connect through linked modules (**Fig. 18.26;** metabolism is discussed in detail in Chapter 14). Many proteobacteria can oxidize H_2 (hydrogenotrophy) and small organic acids, as well as complex organic molecules with aromatic rings. Some are "photoheterolithotrophs" capable of nearly all the fundamental classes of metabolism, depending on environmental conditions, such as availability of light, oxygen, and nutrients. An example is *Rhodopseudomonas palustris*, the species of Alphaproteobacteria characterized and sequenced by Caroline Harwood at the University of Washington. In such a "protean" organism, the core of all proteobacterial energy acquisition is a respiratory chain of electron donors and acceptors. Electrons may enter the chain from a photoexcited chlorophyll, from organic electron donors such as sugars or benzoates, or from a mineral electron donor such as reduced sulfur or iron (lithotrophy). In facultative

Special Topic 18.1 | Carbon Monoxide: Food for Bacteria?

Carbon monoxide (CO) is the most common cause of fatal poisoning in industrial countries. A product of incomplete combustion as in automobile emissions, inhaled CO kills by binding hemoglobin and preventing uptake of oxygen. But small amounts of CO are actually produced by human metabolism, as well as by plant photosynthesis and marine photochemistry. What happens to all this CO?.

In natural environments, CO serves as an electron donor, which some bacteria can oxidize for food:

$$CO + H_2O \longrightarrow CO_2 + 2H^+ + 2e^-$$

The CO is oxidized by water to CO_2, transferring two electrons to an ETS to yield energy. Thus, bacteria may play a previously unrecognized role in cycling CO and preventing toxic buildup in a local environment.

Where in nature do we find CO? Major sources are anthropogenic (human-made) combustion processes such as burning of fossil fuels, wood-burning stoves, and forest clearing for agriculture (**Fig. 1**). Volcanic eruptions and deep-sea hydrothermal vents also release substantial amounts. Small amounts of CO arise from living processes such as plant metabolism. Some bacteria that live in association with plant leaves or roots can assimilate the CO either as a sole energy source (lithotrophy) or as a lithotrophic supplement to organic respiration. The CO may donate electrons to molecular oxygen, or to anaerobic electron acceptors such as nitrate. Some nitrogen-fixing bacteria, such as *Bradyrhizobium* species, metabolize CO as part of their legume endosymbiosis: The intracellular bacteroids oxidize plant-produced CO to CO_2, which the plant can then fix into biomass.

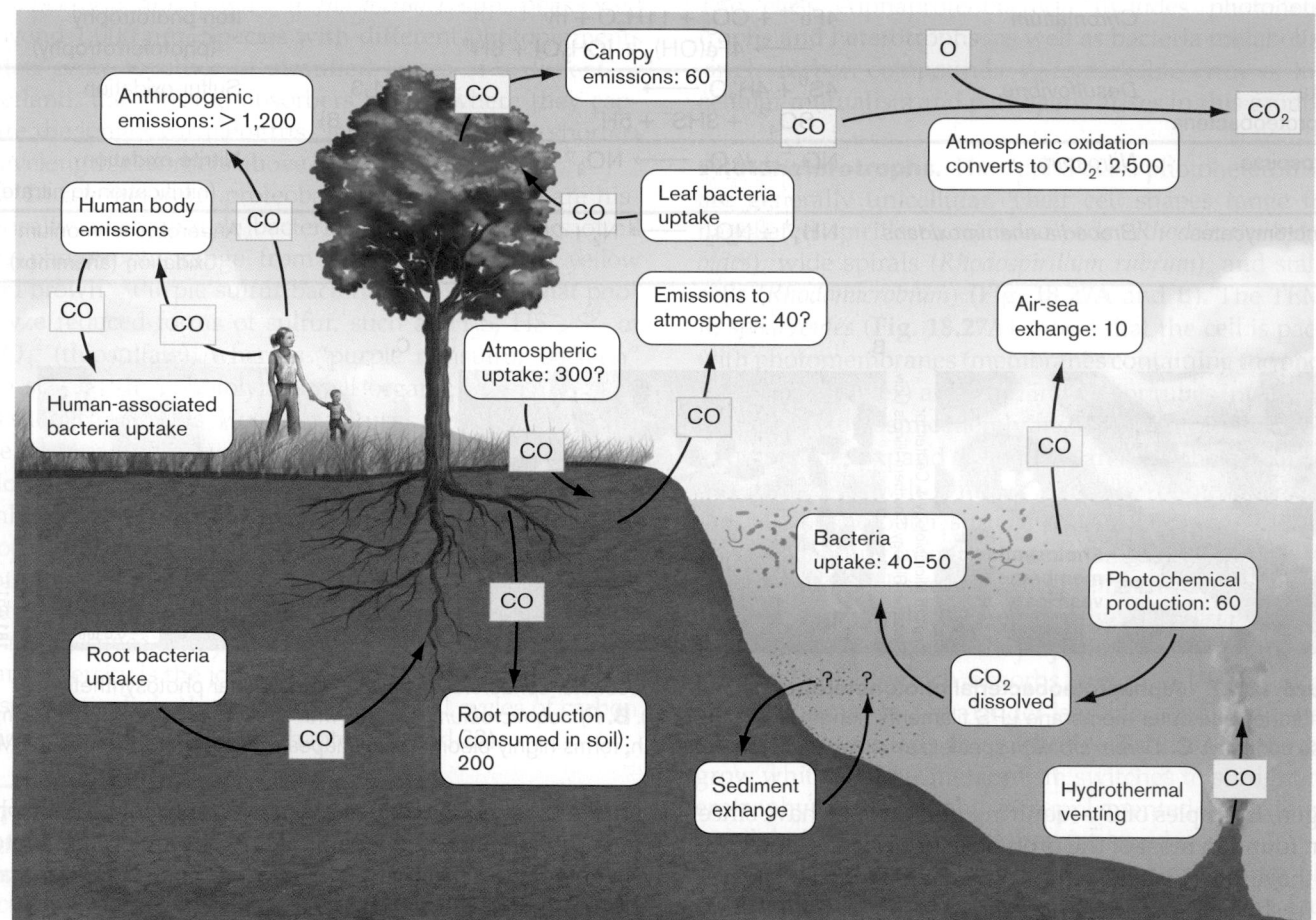

Figure 1 Carbon monoxide emission and uptake.   CO is produced in large amounts by oxidation in the atmosphere and by human burning of organic fuels. Biological sources of CO include plant and animal metabolism, and microbial processes. Much of this CO is consumed by bacteria. Human-associated bacteria consume CO from human metabolism. Units of CO are teragrams per year (10^{12} g/yr). *Source*: Gary M. King and Caroline F. Weber. 2007. *Nat. Rev. Microbiol.* **5**:107.

Human metabolism generates CO through breakdown of the heme portion of hemoglobin, generating a concentration of up to 5 parts per million, about five times the level of atmospheric CO. These low concentrations of CO serve as neurotransmitters and immune system regulators. But human CO may also supplement the metabolism of slow-growing pathogens such as *Mycobacterium tuberculosis*. Mycobacterial CO oxidation has been proposed as a target for antitubercular therapy.

In marine environments, a major source of CO emission is coastal water, where sunlight degrades dissolved organic matter from decayed plants and algae. These light-driven reactions, called photodegradation, yield CO as a by-product, building up surprisingly high levels in the water. More than 80% of this CO was found to be oxidized by microbial processes, but what organisms are involved? John D. Tolli and colleagues at the Woods Hole Oceanographic Institution investigated this question by isolating CO-oxidizing bacteria from the water at Vineyard Sound, Massachusetts. They used a technique called substrate-tracking microautoradiography (STAR) for uptake of radiolabeled CO by bacterial enzymes. The STAR technique was combined with fluorescent in situ hybridization (FISH), in which fluorescent-labeled DNA from known CO-oxidizing organisms is hybridized to DNA from newly isolated bacteria. The STAR-FISH technique allowed simultaneous assay of colonies for (1) ^{14}C-CO uptake (STAR), (2) total cellular DNA using the fluorophore DAPI, and (3) DNA hybridization (FISH) with known bacterial probes. The probe sequences used were for *Roseobacter*, a known CO-oxidizing bacterium of the Alphaproteobacteria clade. Thus, CO-oxidizing colonies were identified, and their relationship to *Roseobacter* was determined.

Figure 2 shows the phylogeny of some of the isolates Tolli found, based on SSU rRNA sequence comparison. Some isolates turned out to be known species of Proteobacteria; others showed novel sequences never before reported (those with prefix JT). The strongest CO oxidizers (JT-01, JT-08, JT-22) fell within the *Roseobacter* clade. *Roseobacter* species are oval-shaped bacteria that grow in rosettes and, in some cases, produce pink colonies. Abundant in coastal water, they are aerobic anoxygenic phototrophs; that is, they use only sulfide or organic electron donors for photosynthesis and do not produce O_2, but for unknown reasons their growth requires O_2. Four other JT isolates showed gammaproteobacterial DNA sequences distantly related to *Pseudomonas aeruginosa*, a soil bacterium and opportunistic pathogen. These diverse marine bacteria may play a key role in environmental CO removal, contributing to the health of the marine food chain.

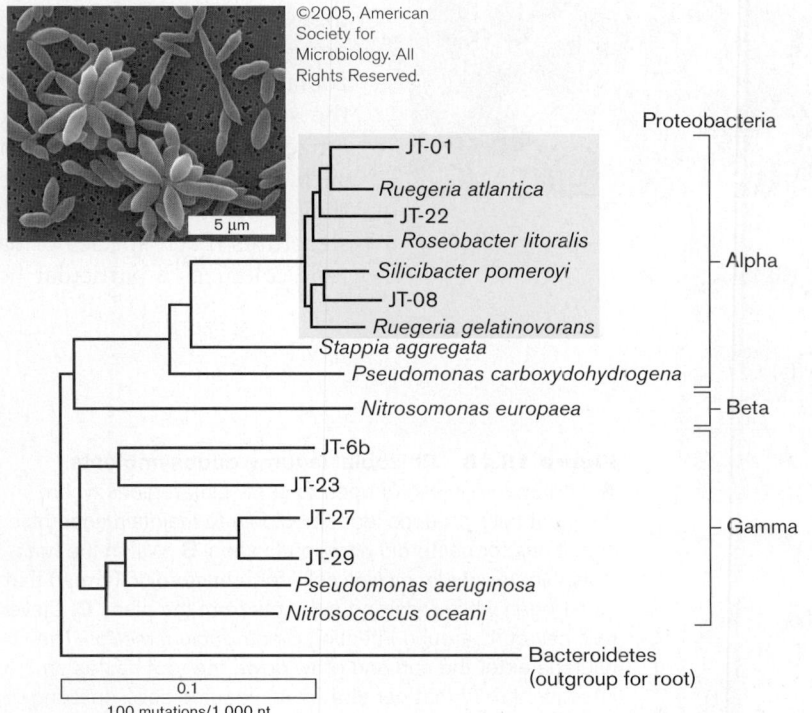

Figure 2   Phylogeny of coastal CO-oxidizing bacteria.   CO-oxidizing bacteria were isolated from Vineyard Sound, Massachusetts. The phylogenetic tree was calculated by comparing SSU rRNA sequences of newly described CO-oxidizing isolates (prefix JT) and known CO-oxidizing species. The *Roseobacter* clade is highlighted. Inset: *Roseobacter* species (SEM). Source: J. D. Tolli et al. 2006. *Appl. Environ. Microbiol.* **72**:1966. Inset from J. B. Bruhn et al. 2005. *Appl. Environ. Microbiol.* **71**:7263.

Oligotrophic bacteria often have unusual extended shapes enhancing nutrient uptake, such as the starlike cell aggregates of *Seliberia stellata*. In *Seliberia* species, the individual tightly coiled rods generate oval or spherical reproductive cells by a budding process. The budding reproductive cells germinate into rods, which then form new aggregates.

Some heterotrophs common in soil are pathogens. *Brucella* spp. are intracellular pathogens of animals that can also infect humans. The soil pathogen *Granulibacter bethesdensis* is associated with chronic granulomatous disease, an inherited disorder of the phagocyte oxidase system that leaves patients susceptible to infection. Other pathogens are carried by insects or animal hosts; for example, *Bartonella henselae* causes cat scratch fever.

Methylotrophy and methanotrophy. **Methylotrophy** is the ability of an organism to oxidize single-carbon compounds such as methanol, methylamine, or methane. The Alphaproteobacteria include several genera of methylotrophs, which are found in all environments, including soil, freshwater, and the ocean. Most methylotrophs can grow on both single-carbon and organic compounds. One such versatile genus, *Methylobacterium*, is equally at home in soil and water, on plant surfaces, and as a contaminant of facial creams and purified water for silicon chip manufacture. Other species are restricted to single-carbon compounds, incapable of metabolizing organic compounds with carbon-carbon bonds. Organisms that grow solely on methane (CH_4) are called **methanotrophs**.

Methane-oxidizing bacteria of both Alpha and Gamma classes contribute to aquatic ecosystems, serving as major food sources for zooplankton. They eliminate much of the methane produced by methanogens before it reaches the atmosphere, where it has a potent greenhouse effect (see Chapter 22).

Endosymbionts: mutualists and parasites. The Alphaproteobacteria include many highly evolved intracellular symbionts—some mutualists, others parasites. As isolated bacteria, they are generally rod-shaped with aerobic metabolism, but their shape is transformed within the host cell. Intracellularly, they need to solve special problems, such as the exclusion of oxygen from nitrogen fixation, a process requiring anaerobiosis—or, in the case of pathogens, the need to resist host defenses.

Nitrogen-fixing endosymbionts of plants include genera such as *Rhizobium*, *Bradyrhizobium*, and *Sinorhizobium*. The nomenclature has undergone many changes, but the species are generally referred to as rhizobia. Rhizobia can live freely in the soil, but they prefer to colonize plants, usually legumes such as peas or alfalfa, where they form distinctive nodule structures. Each species of bacteria colonizes a particular host

A.

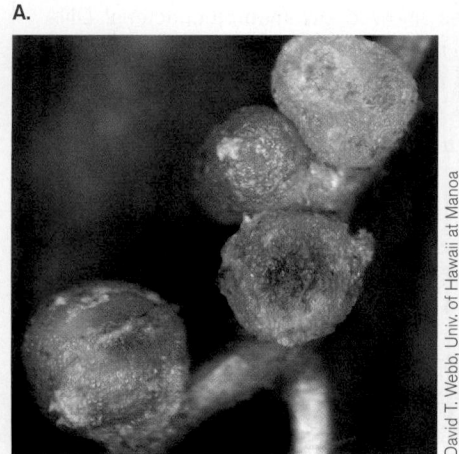

David T. Webb, Univ. of Hawaii at Manoa

B.

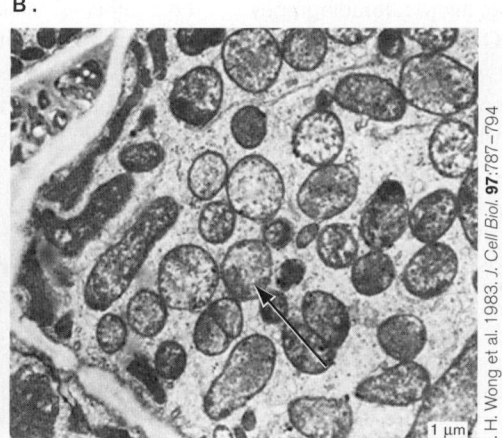

C. H. Wong et al. 1983. *J. Cell Biol.* **97**:787–794

C. *Rhizobium*-legume symbiosis early steps

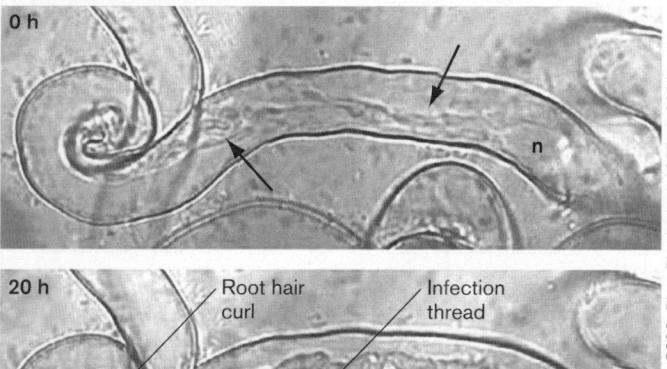

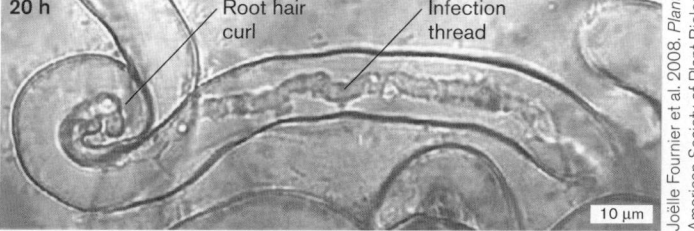

Joëlle Fournier et al. 2008. *Plant Physiol.* **148**:1985. © 2008 American Society of Plant Biologists

Figure 18.28   Rhizobia: legume endosymbionts. A. Legume nodules cut open to show pink regions where the plant cells produce leghemoglobin to maintain anaerobic conditions for bacteroid nitrogen fixation. **B.** Within the host cells, *Sinorhizobium meliloti* cells form bacteroids (arrow) that fix nitrogen while receiving nutrients from the plant. **C.** Clover root hair curls around infecting *Sinorhizobium meliloti*. The bacteria enter the curl and grow down the root hair as an infection thread that penetrates the legume cells, enabling the bacteria to colonize in the form of bacteroids. Micrographs show time 0 and 20 hours postinfection. n = nucleus.

range for infection. Complex chemosensory processes attract the bacteria to the surface of the host root, where their presence induces root hairs to form curls around the bacteria (**Fig. 18.28**). The curling of the root hairs enables the bacteria to form infection threads, chains of rod-shaped bacteria that invade the root cells. Within the host cells, the bacteria lose their cell wall and become rounded **bacteroids**, specialized for nitrogen fixation. The host plant cells provide the bacterioids with nutrients, as well as protective components such as **leghemoglobin**, an oxygen-binding protein that maintains anaerobiosis within infected cells. Leghemoglobin turns the nodule interior pink (**Fig. 18.28A**).

Agrobacterium. The genus *Agrobacterium* includes plant pathogens closely related to the rhizobia. *A. tumefaciens* is known for its ability to convert host cells to a form that produces tumors called galls (**Fig. 18.29**). The ability of *A. tumefaciens* to insert its DNA into plant genomes has made it a major tool for plant biotechnology (see Chapter 16).

Rickettsias: intracellular pathogens and mitochondria. Short coccoid rods, rickettsias lack flagella and can grow only within a host cell. The best-known rickettsia is *Rickettsia rickettsii*, the cause of Rocky Mountain spotted fever, a disease spread by ticks throughout the United States. *Rickettsia* species parasitize human endothelial cells (**Fig. 18.30**). The bacteria induce phagocytosis and then dissolve the phagocytic vesicle and escape into the cytoplasm. Some rickettsias propel themselves through the host cell by polymerizing cytoplasmic actin behind them—a process similar to that of *Listeria* (described under Firmicutes). The actin tails eventually project outward as filopodia, extensions of host cytoplasm and membrane that protect the bacteria from host defenses while enabling them to invade adjacent cells. *Coxiella burnetii*, the causative agent of human Q fever. Q fever is a

disease of livestock that can be transmitted to humans, causing flu-like symptoms.

Genetic analysis shows that the rickettsia clade includes mitochondria. All eukaryotic mitochondria appear to be descendants of an ancient rickettsial parasite whose respiratory apparatus ultimately became essential to power eukaryotic cells.

Genomic analysis of natural ecosystems continues to reveal surprises. We can now sequence a "community genome," or metagenome, of numerous organisms from a given microbial community. In 2004, Craig Venter and colleagues from several institutions sequenced a billion base pairs of the metagenome from the Sargasso Sea near Bermuda. Numerous uncultured species were revealed, including free-living rickettsias. Some of these rickettsias,

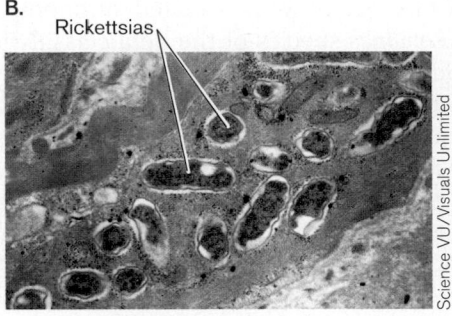

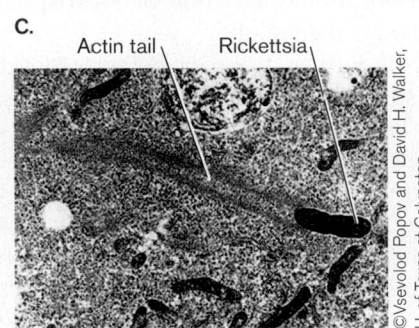

Figure 18.30   Rickettsias are obligate intracellular parasites. **A.** The Rocky Mountain tick carries *Rickettsia rickettsii*, the cause of Rocky Mountain spotted fever. **B.** *Rickettsia* species parasitize human endothelial cells (length, 1–2 μm; TEM). **C.** The rickettsias propel themselves through the host cell by polymerizing cytoplasmic actin behind them (TEM).

Figure 18.29   Plant gall induced by *Agrobacterium tumefaciens*.   Crown gall on chrysanthemum plant.

such as *Pelagibacter ubique* (isolate SAR11), contain proteorhodopsin, a bacterial version of the retinal-based light-driven proton pump found in halophilic archaea (discussed in Chapter 14). It was unexpected to find a nonparasitic marine rickettsia, let alone one with an archaeal form of photosynthesis.

Betaproteobacteria: Photoheterotrophs, Lithotrophs, and Pathogens

The Betaproteobacteria include photoheterotrophs such as *Rhodocyclus*, as well as a diverse range of lithotrophs (**Table 18.4**). Several important pathogens infect humans and other animals.

Lithotrophs: nitrifiers and sulfur oxidizers. An important group of nitrogen lithotrophs is the **nitrifiers**. Nitrifiers oxidize ammonia (NH_3) to nitrite (NO_2^-) (**Fig. 18.31A**) or nitrite to nitrate (NO_3^-). Typically, different species conduct the two reactions separately while coexisting in the soil and water. Nitrifiers are of enormous economic and practical importance for wastewater treatment because they decrease the reduced nitrogen content of sewage. Special systems have been developed to retain nitrifier bacteria behind filters as one stage of water treatment (**Fig. 18.31B**). In the system shown, nitrifying bacteria are encapsulated in pellets to retain them within the bioreactor while the treated water flows through a filter.

The most commonly isolated ammonia oxidizers are *Nitrosomonas* species of the Beta class. *Nitrosomonas* cells conduct electron transport through extensive internal membranes, either stacked or invaginated. Related genera, such as *Nitrosolobus* and *Nitrosovibrio*, have also been described, some with peritrichous or polar flagella. One ammonia-oxidizing genus of the Gamma group has been found: *Nitrosococcus*. Major nitrite oxidizers include the nonproteobacterial phylum Nitrospirae. Although genetically outside the Proteobacteria, the unusual spiral-shaped cells of Nitrospirae have a Gram-negative cell envelope.

Another important group of Betaproteobacteria consists of the sulfur and iron oxidizers, such as the genus *Thiobacillus* (**Fig. 18.32A**). Most are short rods or vibrios (comma-shaped). *Thiobacillus* and other sulfur-oxidizing genera can undergo a number of different reactions oxidizing H_2S to S^0 and S^0 to SO_4^{2-} (see **Table 18.4**). Sulfate production makes an environment acidic enough to erode stone monuments and the interior surface of concrete sewer pipes. Sulfur oxidation is often coupled to oxidation of iron, $Fe^{2+} \longrightarrow Fe^{3+}$. The bacterium *T. ferrooxidans* is known for its role in acidification of mine water, where it contributes to leaching of iron, copper, and other minerals (**Fig. 18.32B**).

Pathogens. The Beta class includes aerobic heterotrophic cocci such as *Neisseria*. The cocci of *Neisseria* species form distinctive pairs known as **diplococci** (shown in **Table**

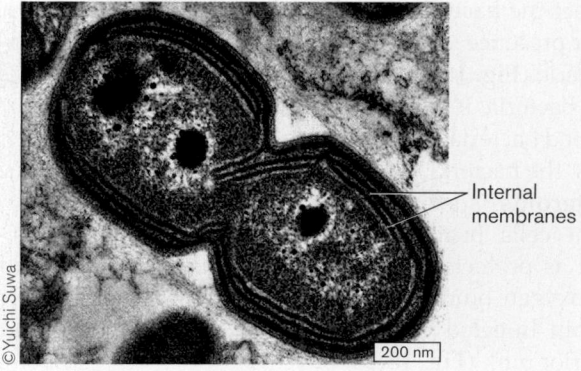

A. *Nitrosomonas europaea*

Internal membranes

200 nm

©Yuichi Suwa

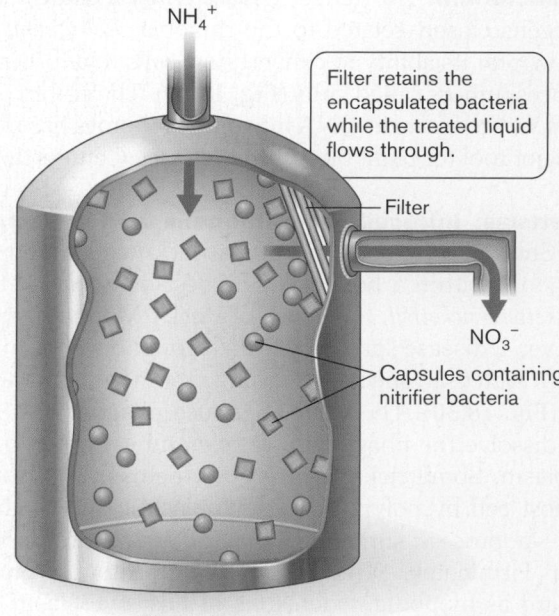

B. Encapsulated nitrifiers

NH_4^+

Filter retains the encapsulated bacteria while the treated liquid flows through.

Filter

NO_3^-

Capsules containing nitrifier bacteria

Bioreactor

Figure 18.31   Nitrifiers.   A. *Nitrosomonas europaea*, of the Betaproteobacteria, oxidizes ammonia to nitrite (cell length 2–4 μm; TEM). Internal membranes contain the electron transport complexes. **B.** Wastewater treatment uses nitrifier bacteria to remove ammonia.

18.1). Most *Neisseria* species are harmless commensals of the nasal or oral mucosa, but *N. gonorrhoeae* causes the sexually transmitted disease gonorrhea. *N. gonorrhoeae* is actually a microaerophile, requiring a narrow range of oxygen concentration; it has fastidious growth requirements, necessitating cultivation on special blood-based medium. A related organism, *N. meningitidis*, may be carried asymptomatically by a quarter of the human population, but occasionally it causes meningitis, which can be fatal. Other neisserias, such as *N. sicca*, are easily isolated from skin and often presented as an unknown for identification by undergraduates.

Other members of the Beta class are important animal and plant pathogens, such as *Burkholderia*. *B. cepacia*

A.

P. Romano et al. 2001. *J. Chem. Tech. Biotech.*
76:723

B.

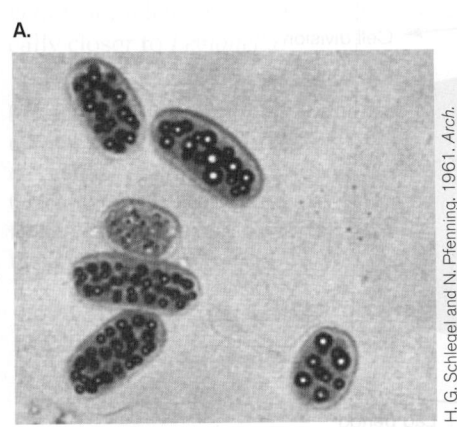

Photo courtesy of Kennecott Utah Copper, Salt Lake City, UT

Figure 18.32 *Thiobacillus*: **iron oxidizers. A.** *Thiobacillus ferrooxidans* leaches copper and iron from molybdenite ore (SEM). **B.** In a copper mine, *Thiobacillus* species can oxidize copper ores, leaching the copper into solution for retrieval.

was originally isolated from onions as a cause of bulb rot; it is now known to be a major opportunistic invader of the lungs of cystic fibrosis patients.

Gammaproteobacteria: Photolithotrophs, Enteric Flora, and Pathogens

The Gammaproteobacteria are well known for the Enterobacteriaceae family of facultative anaerobes found in the human colon, including important pathogens. In the environment, marine species catabolize complex organic pollutants, such as petroleum hydrocarbons. Gamma-proteobacteria also include unusual phototrophs, some of which oxidize iron and nitrite.

Sulfur and iron phototrophs. The Gamma group phototrophs, such as *Chromatium* species (**Fig. 18.33**), mainly utilize sulfide and produce sulfur, which is deposited as intracellular granules visible within the cytoplasm. Their phototrophy is entirely anaerobic. Some of the Gamma group are true autotrophs and do not use organic substrates. Some of these actually conduct phototrophy using iron (Fe^{2+}) to donate electrons. Iron phototrophy, or "photoferrotrophy," is considered an intriguing possibility for metabolism of early life (discussed in Chapter 17). *Thiocapsa* uses NO_2^- and was the first nitrogen-based phototroph discovered.

Sulfur lithotrophs. The sulfur-oxidizing genus *Beggiatoa* was one of the first kinds of lithotroph described by pioneer microbial ecologist Sergei Winogradsky (discussed in Chapter 1). *Beggiatoa* species oxidize H_2S to elemental sulfur, which collects as sulfur granules within the periplasm. *Beggiatoa* also stores carbon in cytoplasmic granules of polyhydroxybutyrate—a common strategy among proteobacteria. The cells of *Beggiatoa* grow as extended filaments with visible sulfur granules, forming biofilms on sulfide-rich sediment.

Enteric fermenters and respirers. The family Enterobacteriaceae, facultative anaerobes of the Gamma subdivision, include some of the most intensively studied species of all bacteria. Species are readily isolated from the

A.

B.

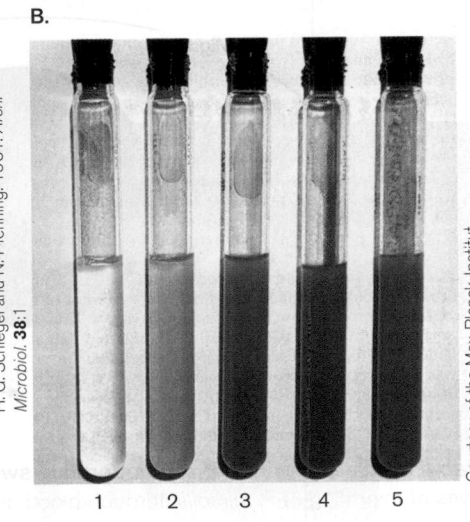

H. G. Schlegel and N. Pfenning. 1961. *Arch. Microbiol.* **38:1**

Courtesy of the Max-Planck-Institut

1   2   3   4   5

Figure 18.33 *Chromatium*: **sulfur and iron phototrophs. A.** *Chromatium* forms single-flagellated rods full of sulfur granules. **B.** Photoferrotrophy: Under illumination, color develops over time (1–5) as a *Chromatium* isolate oxidizes Fe^{2+} to Fe^{3+}.

A.
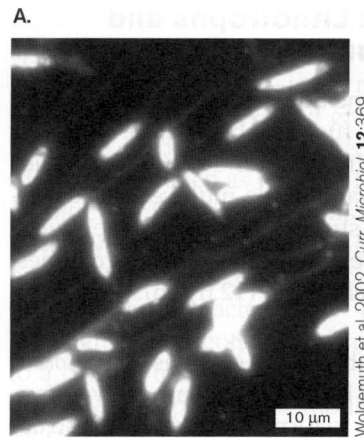

B.
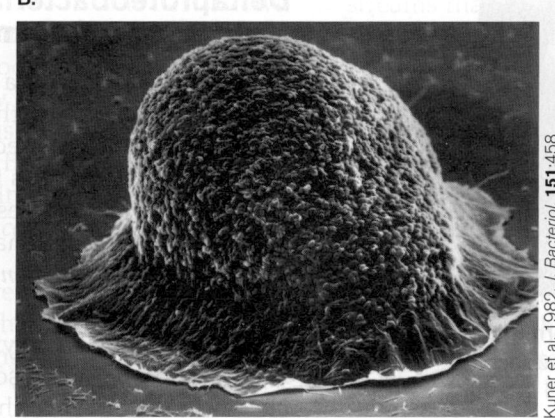

Wolgemuth et al. 2002. *Curr. Microbiol.* **12**:369

10 μm

Kuner et al. 1982. *J. Bacteriol.* **151**:458

Figure 18.37 Myxobacteria.
A. *Myxococcus xanthus* cells glide while producing trails of slime (dark-field LM). **B.** Upon starvation, cells come together to generate a fruiting body (100–200 μm across) packed with spherical myxospores (SEM).

Bdellovibrios parasitize bacteria. Bacteria can be parasitized or preyed on by smaller bacteria. The Delta class includes *Bdellovibrio* species, which attack proteobacterial host cells. The structure of the "attack cell" is a small comma-shaped rod with a single flagellum. The attack cell attaches to the envelope of its host and then penetrates into the periplasm, where it uses host resources to grow (**Fig. 18.39**). The growing cell produces enzymes that cross the inner membrane to degrade host macromolecules and make their components available to the bdellovibrio. The entire host cell loses its shape and becomes a protective incubator for the predator. This stage is called the bdelloplast.

Within the periplasm, the invading bdellovibrio elongates as a spiral filament while replicating several copies of its DNA. When most nutrients have been exhausted, the filament septates into multiple short cells. The cells develop flagella, and the bdelloplast bursts, releasing the newly formed attack cells.

Bdellovibrios can be isolated from sewage, soil, or marine water—any environmental source of Gram-negative prey bacteria. Different species infect *E. coli* and *Pseudomonas*, as well as *Agrobacterium* and *Rhizobium* in their free-living state. They can be cultured as plaques on a top-agar plate containing host bacteria—the same procedure used to isolate bacteriophages (discussed in Chapter 6).

Epsilonproteobacteria: Microaerophilic Helical Pathogens

Epsilonproteobacteria include the genera *Campylobacter* and *Helicobacter. Helicobacter pylori* is well known today as the causative agent of gastritis and stomach ulcers (**Fig. 18.40**). Yet until recently, most microbiologists believed that bacteria could not live in the acidic stomach. The discovery of *H. pylori* as the cause of gastritis earned the Nobel Prize in Medicine, in 2005, for Barry Marshall and J. Robin Warren, of the University of Western Australia.

Free-living cells

Spore dispersal

Starvation

Aggregation

Fruiting body

Figure 18.38 The life cycle of *Myxococcus xanthus.*
Starving myxobacteria glide toward each other to aggregate. The aggregation generates a fruiting body with bulges packed with small, spherical myxospores. *Source:* Dale Kaiser.

H. pylori and related species grow primarily on the stomach epithelium, at about pH 6, which is less acidic than the gastric contents (pH 2–4). The bacteria bury themselves in the epithelial layer and neutralize their acidic surroundings by secreting urease enzyme, which converts urea to ammonia and carbon dioxide. *Helicobacter* species form wide spiral cells (spirilla) with an unusual grouping of flagella at one end. The metabolism of *H. pylori* is microaerophilic, requiring a low level of oxygen. The bacteria can be isolated through biopsy of the gastric mucosa.

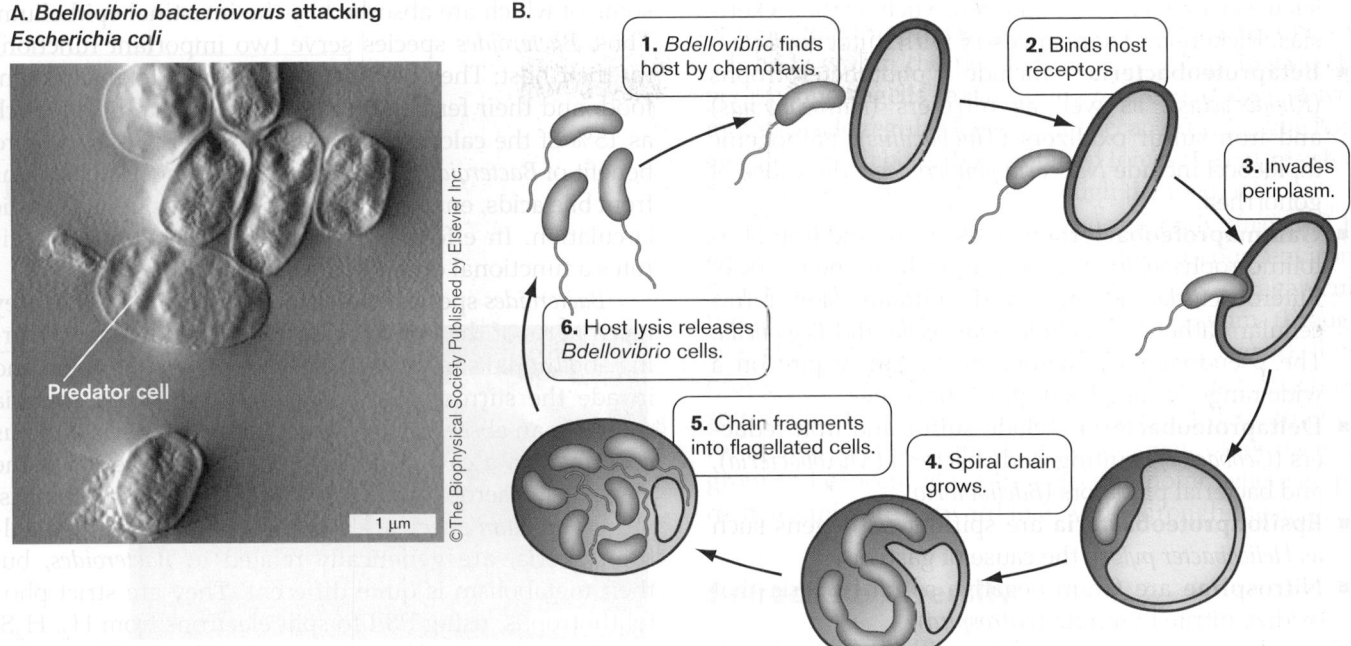

A. *Bdellovibrio bacteriovorus* attacking *Escherichia coli*

Predator cell

1 μm

©The Biophysical Society Published by Elsevier Inc.

B.

1. *Bdellovibric* finds host by chemotaxis.

2. Binds host receptors

3. Invades periplasm.

4. Spiral chain grows.

5. Chain fragments into flagellated cells.

6. Host lysis releases *Bdellovibrio* cells.

Figure 18.39   Parasites of bacteria: *Bdellovibrio*.   A. *Escherichia coli* under attack by *Bdellovibrio bacteriovcrus* (note predator cell within *E. coli* periplasm) (AFM). **B.** Life cycle of a bdellovibrio. A from Megan E. Núñez et al. 2003. *Biophys. J.* **84**:3379.

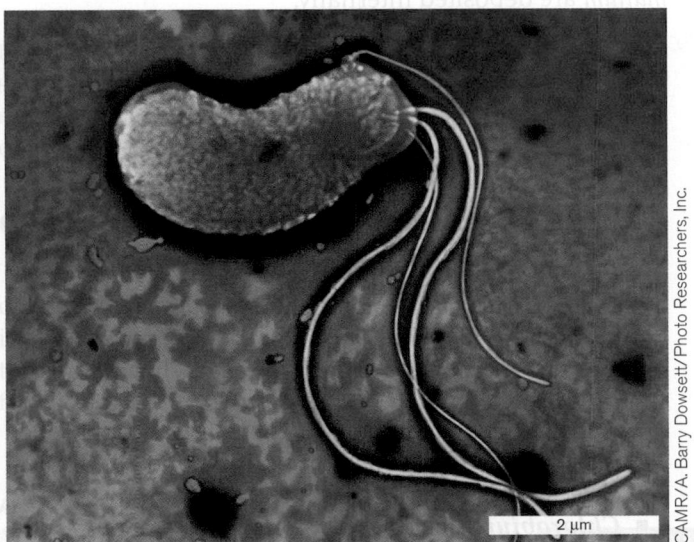

CAMR/A. Barry Dowsett/Photo Researchers, Inc.

2 μm

Figure 18.40   *Helicobacter*, a neutralophile growing within the acidic stomach.   *H. pylori*, a short spirillum with unusual knobbed flagella projecting from one end.

Several groups of related Epsilonproteobacteria are sulfur oxidizers and sulfur reducers, found in marine and aquatic habitats. *Thiovulum* species oxidize sulfides aerobically, using oxygen available in marine water. In deep-ocean sediment, near hydrothermal vents, hydrogenotrophs *Nautilia* and *Hydrogenimonas* oxidize H_2 using sulfur or nitrate. These bacteria enrich the marine habitat by cycling carbon, nitrogen, and sulfur.

Nitrospirae Oxidize Nitrite and Iron

The phylum Nitrospirae consists of Gram-negative spiral bacteria that oxidize nitrite ion to nitrate. Their cell structure resembles that of the Proteobacteria, although their phylogenetic branch is deep enough for assignment to a separate phylum. Most species, such as *Nitrospira* spp. (see **Table 18.1**), are true lithotrophs or autotrophs, fixing carbon in the form of carbon dioxide or carbonate using carboxysomes. *Nitrospira* species are generally found in freshwater or salt water. Their removal of excess nitrite makes a key contribution to aquatic ecosystems.

Another important genus is *Leptospirillum*, which includes acidophilic iron oxidizers. *Leptospirillum* spp. are also strict autotrophs, fixing carbon using Fe^{2+} as their electron donor and O_2 as the electron acceptor. Their metabolism generates acid, contributing to acid mine drainage in iron mines in Iron Mountain, California, where they grow in massive pink biofilms.

TO SUMMARIZE:

- **Proteobacteria** stain Gram-negative, with a thin cell wall and an LPS outer membrane. They show wide diversity of form and metabolism, including phototrophy, lithotrophy, and heterotrophy on diverse organic substrates.
- **Alphaproteobacteria** include photoheterotrophs (such as *Rhodospirillum*) and heterotrophs, as well as methylotrophs. They include intracellular mutualists

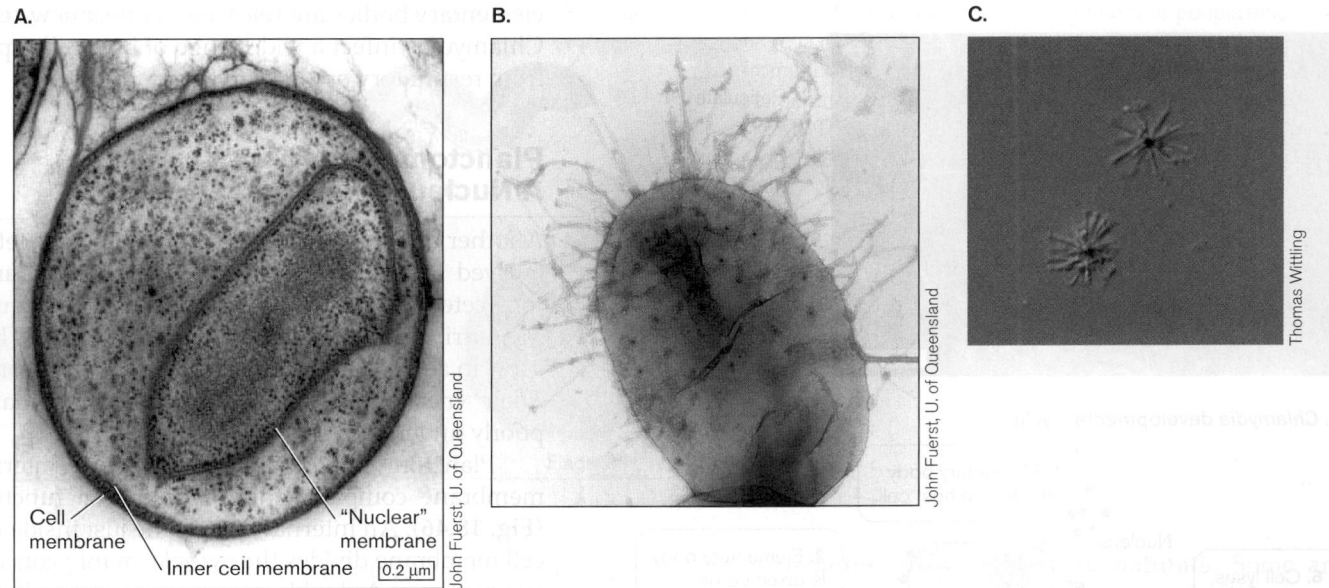

Figure 18.46 *Planctomyces*: bacteria with a "nuclear membrane." **A.** Section through *Gemmata obscuriglobus* showing membrane compartmentalization (TEM). The DNA is contained within a double membrane analogous to a eukaryotic nuclear membrane. **B.** A swarmer cell of *Planctomyces* sp. with multiple flagella. **C.** *Planctomyces bekefii* cells form stalks that attach to each other at the center to yield a starlike form. (Central cell size = 1–2 μm; stalk length = 1–5 μm.)

containing a poorly characterized cytoskeleton. The cytoskeleton appears to contain tubulin, a cytoskeletal protein previously believed to exist only in eukaryotes. In 2002, genes encoding tubulin were found in the partly sequenced genome of *Prosthecobacter dejongeii*, a free-living member of Verrucomicrobia. The genes appear so similar to those of eukaryotes that they must have undergone horizontal transfer from a eukaryotic genome. This horizontal transfer of a eukaryotic trait may be contrasted with the independent development of a nucleus-like structure in planctomycetes.

TO SUMMARIZE:

- **Chlamydiae** are obligate intracellular parasites that undergo a complex developmental progression, culminating in a spore-like form called an elementary body that can be transmitted outside the host cell. Chlamydiae lack cell walls.
- **Planctomycetes** lack cell walls, and they have evolved a membrane enclosing the nucleoid, analogous to the eukaryotic nuclear membrane.

- **Verrucomicrobia** have cell projections containing tubulin. Their tubulin genes are believed to have arisen through horizontal transfer from a eukaryote.

Concluding Thoughts

Several themes emerge in the diversity of the domain Bacteria. Some phyla, such as Proteobacteria and Actinobacteria, have evolved highly diverse metabolism and cell form. Others show remarkable uniformity in cell structure (Spirochetes) or in metabolism (Cyanobacteria). At the same time, our attempts to generalize about any given clade inevitably run up against the unexpected appearance of exceptions, such as the heliobacteria, photosynthetic endospore formers that unaccountably branch from clostridia. Given the range of bacterial phenotypes, it is hard to imagine that yet more diverse forms of microbial life exist; but they do, as we shall find among the archaea (Chapter 19) and the microbial eukaryotes (Chapter 20).

CHAPTER REVIEW

Review Questions

1. Which deep-branching phyla include hyperthermophiles? Why is the actual branch position of these groups controversial?
2. Compare and contrast Chloroflexi and Cyanobacteria with respect to habitat, cell structure, and means of photosynthesis.
3. Compare and contrast the colonial and filamentous cyanobacteria with respect to life cycle and means of nitrogen fixation.
4. Name and describe three genera of Firmicutes that form endospores. Discuss the difference between single and multiple spore formers. Explain the existence of non-spore-forming species of Firmicutes.
5. Compare and contrast firmicute endospores and actinomycete arthrospores with respect to their means of production, their resistance properties, and their dispersal mechanisms.

6. Describe the diverse kinds of metabolism available to different species of Proteobacteria. Explain how it is possible for some species to perform many different kinds of energy-gaining metabolism.
7. What do species of Bacteroidetes and Chlorobi have in common, and how do they differ?
8. Explain the structure and mechanism of motility typical in species of Spirochetes.
9. Discuss the properties of as many intracellular parasites as you recall from various phyla. What do they have in common, and how do they differ?
10. Describe the unique cell structure of *Planctomyces*. Explain how the traits of planctomycete cells appear analogous to aspects of eukaryotic cells.
11. How do microbiologists seek to find previously unknown kinds of bacteria? What techniques reveal unknown microorganisms?

Thought Questions

1. Why are the Proteobacteria metabolically diverse? How is it possible that so many closely related organisms use such different molecules to yield energy?
2. How do you think *Mycobacterium tuberculosis* manages to grow despite its thick envelope screening out most nutrients?

3. For motility, what are the relative advantages of external flagella versus the flexible spiral cells of spirochetes?
4. Why do different species of microbes grow together in layered biofilms (a) on a sand flat and (b) on the surface of human teeth?

Key Terms

acid-fast stain (691)
Actinobacteria (677)
actinomycetes (677)
aerial mycelium (692)
akinete (682)
arthrospore (692)
bacteroid (701)
Bacteroidetes (677)
candidate species (678)
carboxysome (682)
Chlamydiae (677)
Chlorobi (677)
chlorosome (679)
Cyanobacteria (672)
diplococcus (702)

elementary body (710)
endophyte (692)
endospore (677)
Firmicutes (677)
forespore (685)
fruiting body (705)
gas vesicle (682)
GC content (677)
geosmin (691)
heterocyst (682)
hormogonium (682)
leghemoglobin (701)
methanotroph (700)
methylotrophy (700)
mothospore (685)

mycolic acid (692)
myxospore (705)
nitrifier (702)
Nitrospirae (677)
phylum (671)
Planctomycetes (677)
Proteobacteria (677)
proteorhodopsin (696)
reticulate body (711)
sarcina (694)
Spirochetes (677)
swarming (704)
vegetative cell (685)
vegetative mycelium (692)
Verrucomicrobia (677)

Recommended Reading

Anderson, Gregory G., Joseph J. Palermo, Joel D. Schilling, Robyn Roth, John Heuser, et al. 2003. Intracellular bacterial biofilm-like pods in urinary tract infections. *Science* 301:105–107.

Andersson, Siv G. E., Alireza Zomorodipour, Jan O. Andersson, Thomas Sicheritz-Pontén, U. Cecilia M. Alsmark, et al. 1998. The genome sequence of *Rickettsia prowazekii* and the origin of mitochondria. *Nature* 396:133–140.

Angert, Esther. 2006. Beyond binary fission: Some bacteria reproduce by alternative means. *Microbe* 1:127–131.

Beatty, J. Thomas, Jörg Overmann, Michael T. Lince, Ann K. Manske, Andrew S. Lang, et al. 2005. An obligately photosynthetic bacterial anaerobe from a deep-sea hydrothermal vent. *Proceedings of the National Academy of Sciences USA* 102:9306–9310.

Bryant, Donald A., and Niels-Ulrik Frigaard. 2006. Prokaryotic photosynthesis and phototrophy illuminated. *Trends in Microbiology* 14:488–496.

Campbell, Barbara J., Annette Summers Engel, Megan L. Porter, and Ken Takai. 2006. The versatile ε-proteobacteria: Key players in sulphidic habitats. *Nature Reviews. Microbiology* 4:458–468.

Chouari, Rakia, Denis Le Paslier, Patrick Daegelen, Philippe Ginestet, Jean Weissenbach, et al. 2003. Molecular evidence for novel planctomycete diversity in a municipal wastewater treatment plant. *Applied and Environmental Microbiology* 69:7354–7363.

Daims, Holger, Jeppe L. Nielsen, Per H. Nielsen, Karl-Heinz Schleifer, and Michael Wagner. 2001. In situ characterization of *Nitrospira*-like nitrite-oxidizing bacteria active in wastewater treatment plants. *Applied and Environmental Microbiology* 67:5273–5284.

Dover, Lynn G., A. M. Cerdeno-Tarraga, M. J. Pallen, J. Parkhill, and Gurdyal S. Besra. 2004. Comparative cell wall core biosynthesis in the mycolated pathogens, *Mycobacterium tuberculosis* and *Corynebacterium diphtheriae*. *FEMS Microbiological Reviews* 28:225–250.

Ehrenreich, Armin, and Friedrich Widdel. 1994. Anaerobic oxidation of ferrous iron by purple bacteria, a new type of phototrophic metabolism. *Applied and Environmental Microbiology* 60:4517–4526.

Favier, Christine F., Elaine E. Vaughan, Willem M. De Vos, and Antoon D. L. Akkermans. 2002. Molecular monitoring of succession of bacterial communities in human neonates. *Applied and Environmental Microbiology* 68:219–226.

Frigaard, Niels-Ulrik, Asuncion Martinez, Tracy J. Mincer, and Edward F. DeLong. 2006. Proteorhodopsin lateral gene transfer between marine planktonic Bacteria and Archaea. *Nature* 439:847–850.

Geissinger, Oliver, Daniel P. R. Herlemann, Erhard Mörschel, Uwe G. Maier, and Andreas Brune. 2009. The ultramicrobacterium *"Elusimicrobium minutum"* gen. nov., sp. nov., the first cultivated representative of the termite group 1 phylum. *Applied and Environmental Microbiology* 75:2831–2840.

Giovannoni, Stephen J., H. James Tripp, Scott Givan, Mircea Podar, Kevin L. Vergin, et al. 2005. Genome streamlining in a cosmopolitan oceanic bacterium. *Science* 309:1242–1245.

Jenkins, Cheryl, Ram Samudrala, Iain Anderson, Brian P. Hedlund, Giulio Petroni, et al. 2002. Genes for the cytoskeletal protein tubulin in the bacterial genus *Prosthecobacter*. *Proceedings of the National Academy of Sciences USA* 99:17049–17054.

Jensen, Paul R., Philip G. Williams, Dong-Chan Oh, Lisa Zeigler, and William Fenical. 2007. Species-specific secondary metabolite production in marine actinomycetes of the genus *Salinispora*. *Applied and Environmental Microbiology* 73:1146–1152.

King, Gary M., and Carolyn F. Weber. 2007. Distribution, diversity and ecology of aerobic CO-oxidizing bacteria. *Nature Reviews. Microbiology* 5:107–118.

Schlieper, Daniel, María A. Oliva, José M. Andreu, and Jan Löwe. 2005. Structure of bacterial tubulin BtubA/B: Evidence for horizontal gene transfer. *Proceedings of the National Academy of Sciences USA* 102:9170–9175.

Sunenshine, Rebecca H., and L. Clifford McDonald. 2006. *Clostridium difficile*-associated disease: New challenges from an established pathogen. *Cleveland Clinic Journal of Medicine* 73:187–197.

Tolli, John D., S. M. Sievert, and C. D. Taylor. 2006. Unexpected diversity of bacteria capable of carbon monoxide oxidation in a coastal marine environment, and contribution of the *Roseobacter*-associated clade to total CO oxidation. *Applied and Environmental Microbiology* 72:1966–1973.

Venter, J. Craig, Karin Remington, John F. Heidelberg, Aaron L. Halpern, Doug Rusch, et al. 2004. Environmental genome shotgun sequencing of the Sargasso Sea. *Science* 304:66–74.

Wang, Jenny, Cheryl Jenkins, Richard I. Webb, and John A. Fuerst. 2002. Isolation of *Gemmata*-like and *Isosphaera*-like planctomycete bacteria from soil and freshwater. *Applied and Environmental Microbiology* 68:417–422.

Wu, Martin, Qinghu Ren, A. Scott Durkin, Sean C. Daugherty, Lauren M. Brinkac, et al. 2005. Life in hot carbon monoxide: The complete genome sequence of *Carboxydothermus hydrogenoformans* Z-2901. *PLoS Genetics* 1:e65.

For further study and review, please visit StudySpace at **microbiology2.com**.

Chapter 19
Archaeal Diversity

Archaea are the most ecologically diverse of the three domains. Species are found in all soil and water habitats, in association with animals and plants, and in "extreme" environments that exclude bacteria and eukaryotes. Archaea include hyperthermophiles inhabiting the hottest environments on Earth, as well as Arctic and Antarctic psychrophiles growing beneath sea ice and in subglacial lakes. Many hyperthermophiles experience extreme pressure or acidity, as well as extreme temperature. Haloarchaea grow in concentrated brine, coloring salt lakes red. Other species have forms of metabolism unique to archaea, such as methanogenesis. Their proposed resemblance to the earliest life-forms earned this group the name Archaea.

Archaea participate in many microbial communities, often growing with bacteria in multispecies biofilms. Marine archaea include free-living autotrophs, members of methane-eating syntrophies, and symbionts of fish and sponges. Methanogens grow in all kinds of anaerobic environments, from wetland soil to our own digestive tract. Yet surprisingly, the archaeal domain lacks pathogens. No archaeon has yet been shown to cause disease.

Current Research Highlight

Methanogenic archaea release methane, which can be collected as natural gas. Methanogens in a wastewater biofilm fluoresce red with a DNA probe for Methanomicrobiales, while other wastewater microbes fluoresce blue (DAPI stain). The filamentous methanogens shown are 7–10 μm in length. They entangle bacteria, forming "flocs" that can settle in the wastewater tank. The genome of a Methanomicrobiales species encodes enzymes for reducing formic acid and carbon dioxide with hydrogen gas, to make methane. Some bacteria, called methanotrophs, can then oxidize the methane to yield energy. *Source*: Iain Anderson et al. 2009. *PLoS One* **4**:e5797. Photo courtesy of Jose Luis Sanz/Universidad Autonoma de Madrid.

Archaea are best known for their dominance of super-heated habitats such as the hot springs of Yellowstone National Park and thermal vents at the ocean floor. Yet genetic surveys based on rRNA genes reveal archaeal species in the coldest habitats as well as the hottest. Ace Lake, Antarctica, is a well-studied cold habitat with temperatures in the range of 14°C–24°C and bottom anoxic layers never warmer than 2°C (**Fig. 19.1**). The lake supports a wide range of psychrophilic microbes, including methanogens and other archaea. Some of these psychrophilic archaea show surprising relatedness to the sulfur-metabolizing thermophiles found at thermal vents.

Overall, archaeal species grow in habitats at a wider range of temperature and other environmental factors than either bacteria or eukaryotes. Archaea grow in alkaline soda lakes, as well as highly acidic mine runoff streams; within oxygenated hot springs, as well as within

Figure 19.1   Ace Lake, Antarctica. Psychrophilic archaea and other microbes are studied here.

Table 19.1   Archaeal traits absent from bacteria and eukaryotes.

Traits of archaea	Archaea showing the trait	Alternative traits of bacteria and/or eukaryotes
Cell envelope		
Membrane lipids: isoprenoid L-glycerol ethers or diethers	All archaea	Membrane lipids: D-glycerol hydrocarbon diesters
Membrane lipid chains stiffen by covalent cross-links or by forming pentacyclic rings.	Most archaea	Chains stiffen by saturation.
S-layer of glycoprotein	Hyperthermophilic Crenarchaeota	Peptidoglycan (murein)
S-layer of protein, methanochondroitin, or sulfated polysaccharide	Methanogens and Haloarchaea	Peptidoglycan (murein)
Pseudomurein sacculus contains talosaminuronic acid; peptide bridges contain only L-amino acids.	Methanobacteriales, Methanopyrales	Peptidoglycan (murein)
Metabolism		
Nonphosphorylated intermediates of sugar catabolism and synthesis (EMP glycolysis in some cases)	All archaea	EMP pathway of glycolysis, with phosphorylation mediated by NAD or NADP (or Entner-Doudoroff)
Methanogenesis from H_2 and CO_2; or from CO, methanol, methyl sulfides, formate, or acetate	Methanogens	Anaerobic metabolism such as fermentation. No methane production.
Archaeal coenzymes: coenzymes F_{420}, F_{430}, and coenzyme M	Most methanogens	Flavin mononucleotide, coenzyme A, others
Retinal-associated light-driven membrane pumps for H^+ or Na^+	Haloarchaea	Chlorophyll-based photosynthesis
Nucleic acid structure and function		
Positive superturns generated by reverse gyrase protect DNA from extreme acid.	Hyperthermophilic archaea	Negative superturns generated by gyrase
Unique base structures in tRNA, such as the guanine analog archaeosine	Most archaea	tRNA bases, such as queuosine, found only in bacteria and eukaryotes

Sources: O. Kandler and H. König. 1998. *Cell. Mol. Life Sci.* **54**:305–308; C. Bullock. 2000. *Biochem. Mol. Biol. Ed.* **28**:186–191.

the anaerobic human digestive tract. Extremophiles are of interest to biotechnology companies because their exceptionally thermostable enzymes can catalyze reactions under industrially useful conditions, such as high salt or acidity or temperatures above the boiling point of water.

Archaea are also abundant in "moderate" habitats, such as the open ocean, the soil, and the surface of plant roots. Archaea often join bacteria within biofilms. Even in these shared habitats, species of archaea display unique traits, such as energy-yielding methanogenesis and cyclic diether membranes, found only in the archaeal domain, not in bacteria or eukaryotes. Perhaps the most compelling unique feature of archaea, from the human point of view, is the complete absence of known archaeal pathogens of animals or plants. A few reports find methanogens associated with periodontal disease, but even in these cases, no causal relationship has been demonstrated. This lack of pathogenicity is particularly striking given the intimate association of many archaea with animal digestive systems.

In Chapter 19, we explore the diversity of archaeal form and function. We emphasize the unique structures and metabolism of archaeal cells. While surveying their major taxonomic categories, we introduce research techniques used to study organisms in extreme environments and those with unique forms of metabolism such as methanogenesis.

As we saw for bacteria in Chapter 18, new species of archaea continue to be discovered daily. For further information, we recommend the online student site, MicrobeWiki, and the National Center for Biological Information (NCBI) Taxonomy Database.

19.1 Archaeal Traits

The Archaea show key features unique to this domain, as well as intriguing traits shared by eukaryotes, such as transcription factors required by eukaryotic RNA polymerase II. Archaea show distinctive forms, from intricate cell networks to square-shaped cells. Here we summarize key traits of archaea and then outline the major phylogenetic divisions.

Archaeal Cell Structure and Metabolism

Distinctive features of archaea, sometimes called "archaeal signatures," include components of the cell membrane and envelope and certain metabolic pathways to gain energy (**Table 19.1**).

Isoprenoid membranes and cell wall. The most distinctive structure of archaea is their membrane (**Fig. 19.2**). The membrane lipids of archaea differ

profoundly from those of bacteria and eukaryotes, with the exception of a few thermophilic bacteria that show evidence of horizontal transfer of genes conferring synthesis of archaeal-type lipids. Archaeal lipid structure includes:

■ **L-Glycerol.** Archaeal membrane lipids incorporate L-glycerol, rather than the mirror-symmetrical form D-glycerol, which is used by bacteria and eukaryotes. The two chiral forms show similar thermal stability, but their biochemistry requires different enzymes,

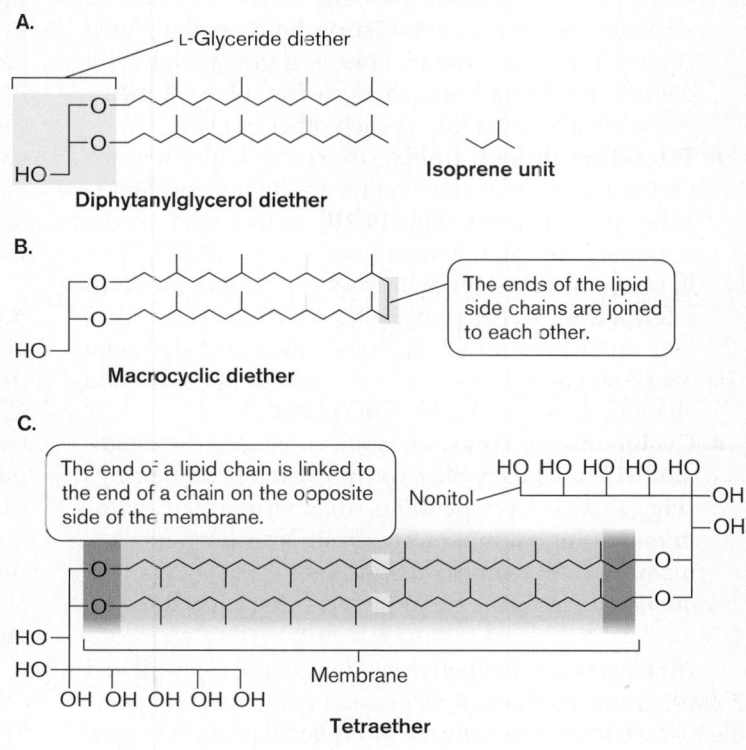

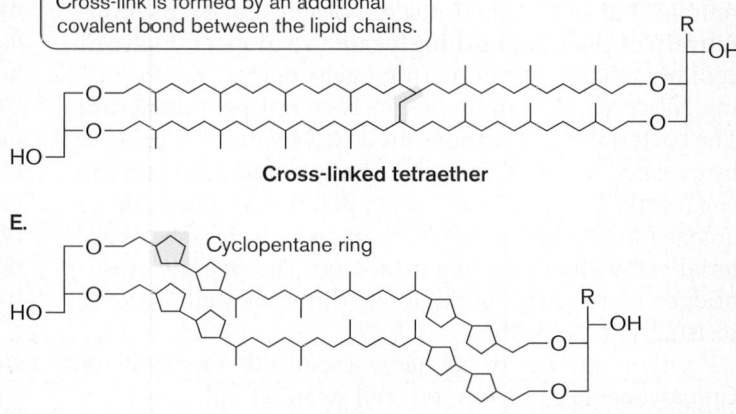

Figure 19.2 Branched-chain ether lipids are characteristic of archaeal membranes. Archaeal membranes show diverse forms of cyclization that offer different levels of protection from extremely high temperature or acidity. *Source:* R. M. Daniel and D. A. Cowan. 2000. *Cell. Mol. Life Sci.* **57**:250–264.

and thus they represent a deep divergence in ancestry. In some archaea, the glycerol is extended by six carbons, forming nonitol (nine OH groups).

- **Ether linkage.** The glycerol units are linked to side chains by ether links (R–O–R) instead of the ester links (R–COO–R) found in bacteria and eukaryotes (**Fig. 19.2A**). Ether links are much more stable than esters; in other words, breaking them requires more energy.
- **Isoprenoid chains.** The side chains of archaeal lipids are branched at every fourth carbon. The methyl branches arise by condensation (C–C bond formation) of units of isoprene (see **Fig. 19.2A**). Condensed isoprene chains are called **isoprenoid** or diphytanyl chains; thus, the overall lipid is diphytanylglycerol diether. Isoprenoid branched chains increase membrane stability by hooking each other in place.
- **Tetraether-linked lipids.** In some hyperthermophiles, the ends of side chains are linked covalently, either to each other (**Fig. 19.2B**) or to a lipid on the opposite side of the membrane (**Fig. 19.2C**). Two lipid chains cross-linked across the membrane form a **tetraether**, so called because the complex contains four ether links in all. In some cases, an additional covalent bond links the two linked pairs of side chains across the middle (**Fig. 19.2D**).
- **Cyclopentane rings.** In some archaea, the lipid's methyl branches cyclize, forming cyclopentane rings (**Fig. 19.2E**). Cyclopentane rings strengthen membranes at high temperature, by an unknown mechanism. At low temperature, archaea form lipids that are unsaturated (as bacteria and eukaryotes do).

Archaea show distinctive versions of the cell wall and S-layer. Some methanogens possess a cell wall comparable to the bacterial sacculus, a unimolecular cage of sugar chains linked by peptide bridges (discussed in Chapter 3). But the sacculus of methanogens is composed not of murein, but of a related macromolecule called **pseudomurein** or **pseudopeptidoglycan**. Pseudopeptidoglycan contains chains of alternating sugar derivatives including *N*-acetylglucosamine, as in bacterial peptidoglycan. The bacterial *N*-acetylmuramic acid, however, is replaced by a related sugar, *N*-acetyltalosaminuronic acid; and the sugar linkage is β(1,3) instead of β(1,4). As a result, these archaea are resistant to lysozyme, which degrades bacterial cell walls at the sugar linkage. The peptide cross-bridges of pseudopeptidoglycan differ as well, causing resistance to penicillin.

Other species of archaea, especially members of Crenarchaeota, possess no cell wall at all—only an S-layer composed of proteins. The shape of these cells is more flexible than that of most bacterial cells, which maintain turgor pressure against a sacculus. However, the S-layer proteins may be locked together in a structure as pressure resistant as the cell wall.

Unique metabolic pathways. Archaea show many distinctive metabolic pathways. Glucose is catabolized by several variants of the Entner-Doudoroff (ED) and Embden-Meyerhof-Parnas (EMP) pathways that rarely occur in bacteria (**Fig. 19.3**). For example, the sulfur thermophiles *Sulfolobus* and *Thermoplasma* convert glucose to gluconate without phosphorylation, ultimately generating pyruvate with no net production of ATP. On the other hand, halophilic archaea such as *Halobacterium* phosphorylate the dehydrated product of gluconate (2-oxo-3-deoxygluconate), thus channeling into the "standard" ED pathway. The ED pathway generates one molecule of pyruvate and one molecule of 3-phosphoglycerate, which produces one net ATP through the second stage of the EMP pathway. A unique variant of the EMP pathway is seen in the vent thermophile *Pyrococcus furiosus*, which oxidizes glyceraldehyde 3-phosphate using ferredoxin instead of NAD and avoids phosphorylation. The reduced ferredoxin is then used to reduce $2H^+$ to H_2 in an energy-yielding reaction.

The energy-yielding process of **methanogenesis** occurs only in archaea. Methane production by methanogenic archaea (methanogens) makes a growing contribution to global warming, as discussed in Chapter 22. Methanogenesis includes a unique pathway of carbon fixation, called the **carbon monoxide reductase pathway** because the key enzyme can fix CO as well as CO_2. A CO group is fixed into acetyl-CoA by condensation with a methyl group generated by methanogenesis. Acetyl-CoA then enters the TCA cycle. Methanogenesis also uses unique cofactors that largely replace NAD, FAD, and FMN (discussed in Section 19.4).

The only form of phototrophy in archaea is that of retinal-based ion pumps. Membrane protein complexes known as **bacteriorhodopsin** and halorhodopsin contain retinal and pumps for H^+ or Na^+ (discussed in Chapter 14). Bacteriorhodopsins are found in haloarchaea (formerly called halobacteria) such as *Halobacterium halobium*; the species was named before it was known to be an archaeon (discussed in Section 19.5). By contrast, the chlorophyll-based phototrophy found in bacteria and plants is completely unknown in archaea. Light-driven ion pumps were believed to be unique to archaea, until bacteriorhodopsin homologs were discovered in marine bacteria. These bacterial H^+ pumps, called proteorhodopsins, are believed to have evolved by horizontal gene transfer from archaea.

Nucleic acid structure. A distinctive feature of archaeal chromosome function is the "reverse gyrase" enzyme found in hyperthermophiles. All bacteria and eukaryotes have gyrase to maintain their DNA in an "underwound" negatively supercoiled state, but hyperthermophiles with reverse gyrase maintain positive supercoiling. The positive superturns "overwind" the DNA, enhancing

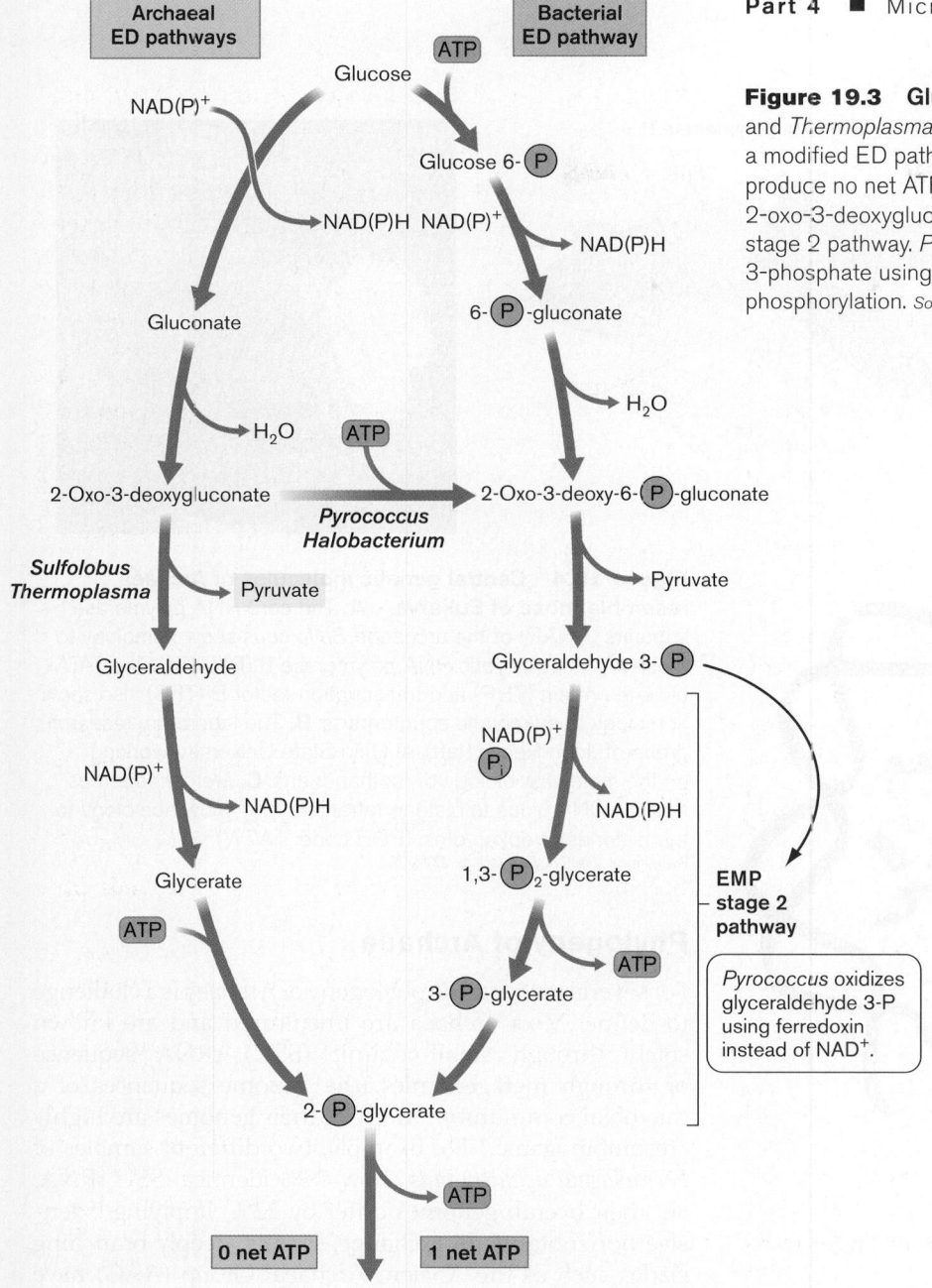

Archaeal ED pathways

Bacterial ED pathway

Figure 19.3   Glucose catabolism in archaea. *Sulfolobus* and *Thermoplasma* species catabolize glucose to pyruvate by a modified ED pathway without phosphorylating glucose and produce no net ATP. *Halobacterium* species phosphorylate 2-oxo-3-deoxygluconate and produce 1 net ATP by the EMP stage 2 pathway. *Pyrococcus furiosus* oxidizes glyceraldehyde 3-phosphate using ferredoxin instead of NAD$^+$ and avoids phosphorylation. *Source*: Hatim Ahmed et al. 2005. *Biochem. J.* **390**:529–540.

stability and preventing the helix from melting into separate strands at high temperature.

Archaea also have distinctive modified bases in their tRNA molecules. In particular, the guanosine analog archaeosine (7-formamidino-7-deazaguanosine) is used by nearly all archaea but no bacteria or eukaryotes. Other unusual tRNA bases, such as queuosine, are found only in bacteria and eukaryotes, not in archaea.

Archaeal Gene Regulation

The genomes of archaea generally resemble those of bacteria in size and gene density, and genes of related function are often arranged in operons. Certain tRNA gene sequences, however, are interrupted by nontranslated sequences called introns, similar to the tRNA introns found in eukaryotes. Furthermore, the archaeal apparatus for DNA and RNA polymerases, transcription factors, and protein synthesis show remarkable similarity to those of eukaryotes. **Figure 19.4A** compares the components of an archaeal RNA polymerase (from *Sulfolobus*) with those of eukaryotic RNA polymerase II. The archaeal polymerase possesses two transcription factors (regulatory protein components) that are found in eukaryotes: TATA-binding protein (TBP) and transcription factor B (TFIIB). By contrast, bacteria have no homologs of these factors. A consequence of the eukaryotic-like transcription and translation in archaea is that archaea

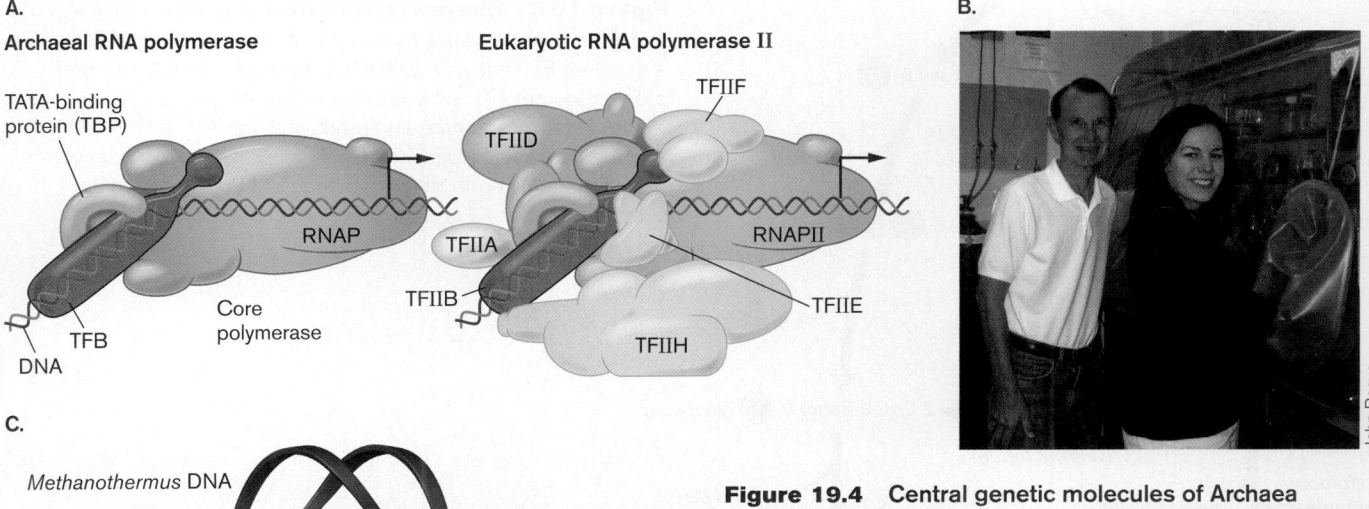

Figure 19.4 Central genetic molecules of Archaea resemble those of Eukarya. A. The core RNA polymerase subunits (RNAP) of the archaeon *Sulfolobus* show homology to those of the eukaryotic RNA polymerase II (RNAPII). The TATA-binding protein (TBP) and transcription factor B (TFB) also show homology to eukaryotic counterparts. **B.** The laboratory research group of John Reeve (left), at Ohio State University, working on the molecular biology of methanogens. **C.** *Methanothermus fervidus* DNA binds to histone tetramers that show homology to the histones of eukaryotes. (PDB code: 1A7W) *Source:* Kathryn A. Bailey et al. 2002. *J. Biol. Chem.* **277**:9293.

Phylogeny of Archaea

For several reasons, the phylogeny of Archaea is a challenge to define. Most archaea are uncultured and are known solely through small-subunit (SSU) rRNA sequence or through metagenomics (the genome sequences of a microbial community). Many of their genomes are highly "recombinogenic." For example, two different samples of *Ferroplasma acidarmanus* show 99% identical SSU rRNA, yet their overall genomes differ by 22%, implying extensive horizontal gene exchange. Finally, deeply branching clades such as the Ancient Archaeal Group (AAG) have such divergent SSU rRNA that their sequence fails to amplify with "standard" PCR primers based on the most highly conserved regions of the gene. We suspect that many deeply branching groups remain unknown, as yet undetected by our current molecular tools.

The domain Archaea includes two phyla, also called divisions, that have been extensively characterized (**Fig. 19.5**). These phyla are **Crenarchaeota** and **Euryarchaeota**. Major groups of species are outlined in **Table 19.2**.

NOTE: In Table 19.2, note that certain archaeal names contain a "bacteria" component—for example, the genus *Halobacterium*. Such organisms were known and named as bacteria before 1977, when the category "archaea" was first defined.

are resistant to antibacterial antibiotics that target transcription and translation.

Another eukaryotic structure for which archaeal homologs were discovered is that of **histones**, the fundamental packaging proteins of DNA. The histone complex found in eukaryotic chromosomes contains a histone $(H3 + H4)_2$ tetramer flanked by two histone $(H2A + H2B)$ dimers. The archaeal homologs form an $(H3 + H4)_2$ tetramer, with no $(H2A + H2B)$ homologs. John Reeve and colleagues at Ohio State University (**Fig. 19.4B**) showed that isolated DNA of specific sequences can be bound and curved around a histone tetramer from the methanogen *Methanothermus fervidus* (**Fig. 19.4C**). The DNA has AT-rich sequences that specifically fit the histone complex. The key sequences of the histones (the ones that bind to DNA) show homology to eukaryotic histones. Histones have since been found in many species of archaea.

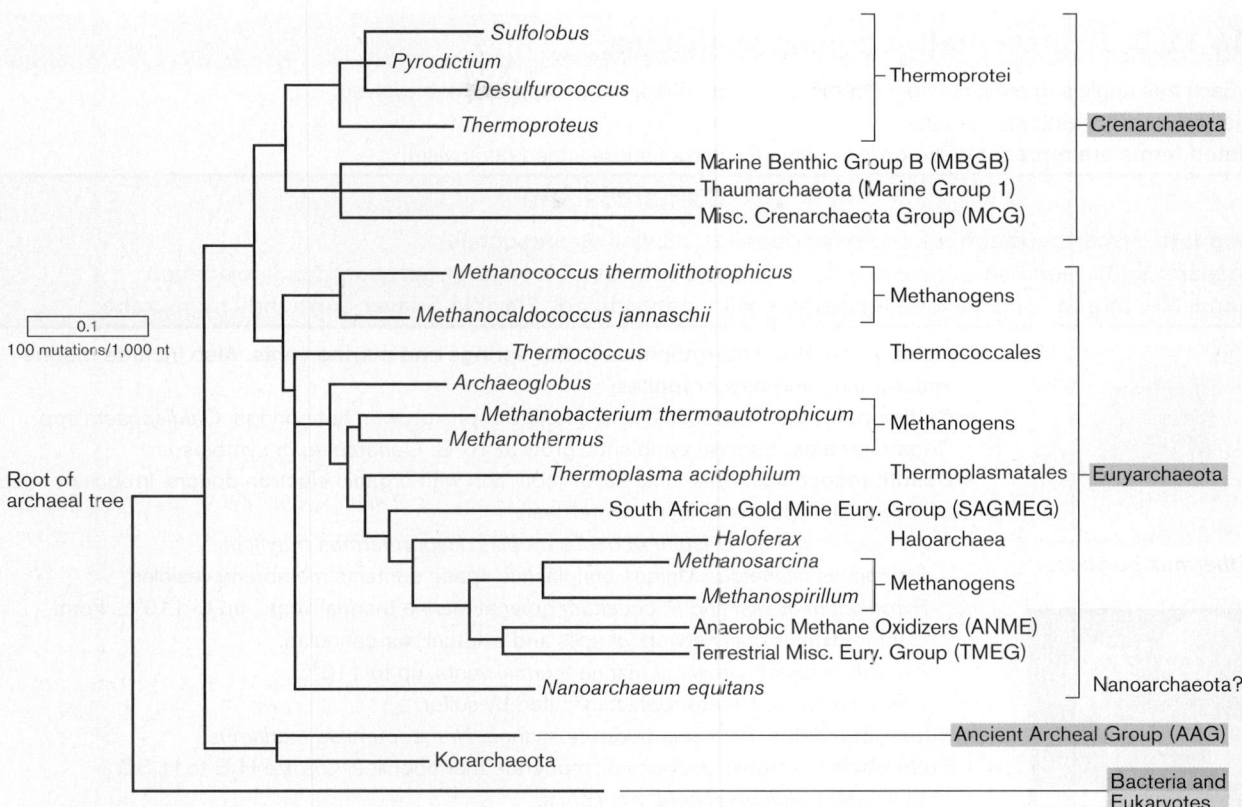

Figure 19.5   Archaeal divergence.   Major clades, divisions, and metagenomic groups of archaea. Divergence was based approximately on SSU rRNA and genomic sequences. *Sources:* Susan Barns et al.1996. *PNAS* **93**:9188; O. Nercessian et al. 2003. *Environ. Microbiol.* **5**:492; Christa Schleper et al. 2005. *Nat. Rev. Microbiol.* **3**:479; Andreas Teske and Ketil Sørensen. 2008. *ISME J.* **2**:3.

Crenarchaeota. The Crenarchaeota include a substantial proportion of soil, marine, and benthic (marine sediment) microbial communities, including associations with plants and animals. The few that are cultured are nitrifiers (ammonia oxidizers) or anaerobic heterotrophs. Understanding these organisms is important to assess their contribution to the global carbon cycle.

The better-studied crenarchaeotes are the thermophiles. Many thermophilic crenarchaeotes metabolize sulfur, either by anaerobic reduction (such as by H_2 to form H_2S) or by aerobic oxidation (by O_2 to form sulfuric acid). Anaerobic sulfur metabolizers include moderate thermophiles (growth range about 60°C–80°C) as well as hyperthermophiles (90°C–120°C). Many of the hyperthermophiles are also barophiles, growing under high pressure at hydrothermal vents on the ocean floor; an example is *Pyrodictium abyssi*. Other uncultured crenarchaeotes are psychrophiles, growing in Antarctic lakes. Thus, the Crenarchaeota encompass the widest range of growth temperature of any division of life.

Euryarchaeota. The phylum Euryarchaeota also includes members throughout soil and water, and associated with plants and animals. They show a greater range of metabolism than the Crenarchaeota. The most highly divergent group of Euryarchaeota is the methanogens. Methanogens serve a key energetic role in ecosystems by offering an anaerobic mechanism to remove excess H_2 and other small-molecule reductants. Despite their common energetic pathway, methanogens branch deeply among other euryarchaeotic groups, and they show a wide range of cell form and environmental adaptations. One methanogenic clade branches from the Haloarchaea, extreme halophiles that are the only form of life to grow in concentrated brine (NaCl). Most haloarchaea, such as *Halobacterium* species NRC-1, are photoheterotrophs that can supplement their metabolism with light-driven ion pumps. Still other genera, such as *Archaeoglobus*, show mixed physiology, including components of methanogenesis linked to other pathways, such as glycolysis. The euryarchaeotes include mesophiles as well as thermophiles, such as the order Thermococcales, but few psychrophiles have been found so far. With respect to pH, the euryarchaeotes show the widest range of any clade, from *Ferroplasma* (the order Thermococcales) growing at pH 0 to *Natronococcus* growing at pH 10.

Table 19.2 Representative groups of archaea.

Note: Each trait applies to *most* members of the taxon described, but exceptions have evolved.
Blue-lettered terms indicate classes.
• **Bulleted terms are representative orders** within the class (unless stated otherwise).

Crenarchaeota

Temperature: Hyperthermophiles and psychrophiles, as well as mesophiles
Metabolism: Sulfur and hydrogen oxidation, aerobic and anaerobic heterotrophy, ammonia oxidation
Envelope: Membrane lipids include tetraethers with crenarchaeol. Flexible S-layer surrounds membrane.

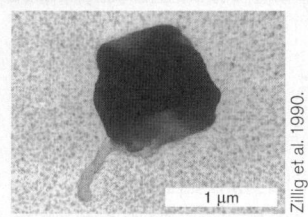

Hyperthermus butylicus

Zillig et al. 1990. *J. Bacteriol.* **172**:3959

Thermoprotei. Many thermophiles in hot springs and marine vents. Also include marine mesophiles and psychrophiles.
• **Caldisphaerales.** Thermoacidophilic heterotrophs grow in hot springs. *Caldisphaera* spp.
• **Cenarchaeales.** Sponge symbionts; grow at 10°C. *Cenarchaeum symbiosum.*
• **Desulfurococcales.** Anaerobic sulfur reduction with organic electron donors. Irregularly shaped cells with glycoprotein S-layer; no cell wall.
  Aeropyrum pernix, Desulfurococcus mobilis, Hyperthermus butylicus.
  Ignicoccus islandicus. Unique periplasmic space contains membrane vesicles.
  Pyrodictium abyssi and *P. occultum* grow at marine thermal vents, up to 110°C. Form three-dimensional network of cells and extracellular cannulae.
  Pyrolobus fumarii grows at marine thermal vents, up to 113°C.
  Thermosphaera. Heterotrophic; inhibited by sulfur.
• **Nitrosopumilales.** Ammonia oxidizers, marine. *Nitrosopumilus maritimus.*
• **Sulfolobales.** Aerobic acidophiles; moderate thermophiles. Oxidize H_2S to H_2SO_4.
  Sulfolobus, Sulfurisphaera, Acidianus.
• **Thermoproteales.** *Pyrobaculum, Thermoproteus, Vulcanisaeta.*

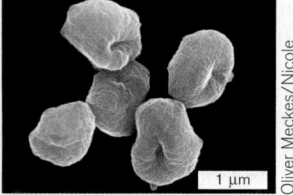

Sulfolobus sp.

Oliver Meckes/Nicole Ottawa

Uncultured groups. Revealed by metagenomic analysis.
• **Marine Benthic Group B (MBGB).** Anoxic marine sediments: deep-sea methane hydrates, hydrothermal vents, coastal regions. Metabolism uncertain.
• **Thaumarchaeota.** Proposed phylum including Nitrosopumilales, Marine Group 1 and other marine and soil archaea that oxidize NH_3 with O_2. Important marine source of nitrates for phytoplankton, and of methylphosphonate (CH_3-PO_3^{2-}), which bacteria convert to methane. *Nitrosopumilus maritimus.* Found in soil and associated with plant roots. Anaerobic heterotrophs.
• **Miscellaneous Crenarchaeota Group (MCG).** Terrestrial and marine, cold and hot, water and subsurface environments.
• **Psychrophilic marine crenarchaeotes.** Marine water, deep sea, and Antarctica. Anaerobic heterotrophs, sulfate reducers, nitrite-reducing methanotrophs.

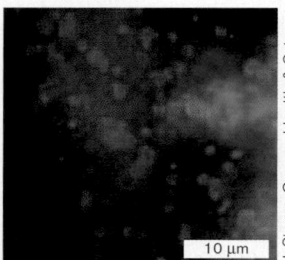

Crenarchaeotes (red fluorescence) from tomato root

H. Simon, Oregon Health & Science University, Beaverton, OR

Euryarchaeota

Temperature: Mesophiles and thermophiles; some psychrophiles
Metabolism: Methanogenesis, halophilic photoheterotrophs, and sulfur and hydrogen oxidizers; acidophiles and alkaliphiles
Envelope: Methanogens and halophiles have rigid cell walls of glycans, glycoproteins, or pseudopeptidoglycan.

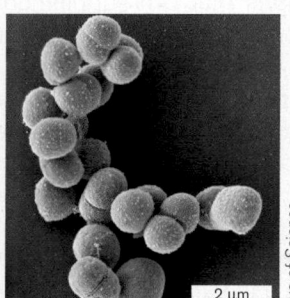

Halobacteriales

Eye of Science

Archaeoglobi. Hyperthermophiles; metabolize sulfur and use reverse methanogenesis pathway.
• **Archaeoglobales.** Sulfate oxidation of H_2 or organic hydrogen donors. *Archaeoglobus fulgidus.*

Haloarchaea. Halophiles; grow in brine (concentrated NaCl). Conduct photoheterotrophy, with light-driven H^+ pump and Cl^- pump.
• **Halobacteriales**
  Haloarcula, Halobacterium, Halococcus, Haloferax. Mesophiles at neutral pH. Grow in salterns and salt lakes. *Haloarcula marismortui; Halobacterium salinarum.*
  Haloquadra (Haloquadratum). Square-shaped mesophilic halophiles.
  Halorubrum lacusprofundi. Psychrophiles; grow in subglacial Antarctic salt lakes.
  Natronococcus, Natronomonas. Alkaliphiles; grow above pH 9 in soda lakes.

Table 19.2 Representative groups of archaea (*continued*)

Methanoculleus nigri

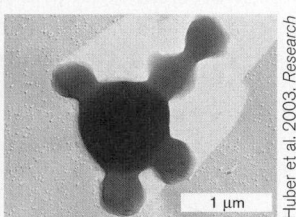

Methanosarcina mazei

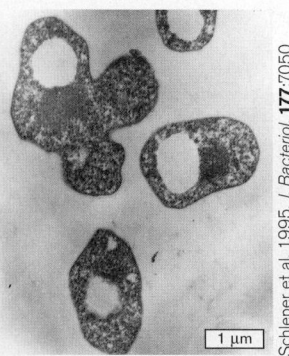

Nanoarchaeum equitans
(attached to surface of
Ignicoccus sp.)

Methanogens (four classes). Generate methane from CO_2 and H_2, formate, acetate, other small molecules. Strict anaerobes. Must associate with bacteria producing their substrates.
Cell walls of pseudopeptidoglycan or polysaccharides.
Wide variety of shapes: rods, filaments, cocci, and coccoid clusters.
Wide range of anaerobic habitats: wetlands, landfills, and intestinal tracts of animals. Psychrophiles in deep-ocean floor sediment generate methane hydrates.

- **Methanobacteriales.** Lack cytochromes; reduce CO_2, formate, or methanol with H_2.
  Methanobacterium spp. have peptidoglycan cell walls.
  Methanobrevibacter smithii and *Methanosphaera stadtmanae* inhabit human digestive tract.
    Increase caloric output of human food by reducing methanol, a breakdown product of pectin.
  Methanothermus fervidus. Marine vent thermophiles.

- **Methanomicrobiales, Methanococcales, and Methanopyrales.** Lack cytochromes; reduce CO_2, formate, or methanol with H_2.
  Methanocaldococcus jannaschii. Vent thermophile.
  Methanoculleus nigri.
  Methanogenium frigidum. Psychrophile.
  Methanopyrus kandleri. Marine vent thermophile.

- **Methanosarcinales.** Possess cytochromes; reduce methylamines and acetate (as well as CO_2 and formate) with H_2.
  Methanococcoides alaskense. Arctic marine habitat.
  Methanohalophilus. Halophile. Inhabits salt lakes, often at high pH.
  Methanosaeta concilii. Filamentous colonies.
  Methanosarcina acetivorans. Globular colonies; cell walls of sulfated polysaccharides.

Nanoarchaeota. Vent hyperthermophiles; obligate symbionts attached to *Ignicoccus*.
Nanoarchaeum equitans.

Thermococci. Hyperthermophiles (grow above 100°C) and barophiles (up to 200 atm pressure). Anaerobes; reduce sulfur.
- **Thermococcales.**
  Pyrococcus abyssi, P. furiosus. Use S^0 to oxidize H_2 or organic hydrocarbons. *Thermococcus* species are closely related.

Thermoplasmata. Extreme acidophiles; oxidize sulfur from pyrite (FeS_2), generating sulfuric acid. Mesophiles or moderate thermophiles.
- **Thermoplasmatales.**
  Ferroplasma acidiphilum and *F. acidarmanus* grow at 37°C–50°C. Oxidize sulfur from FeS_2, generating ambient pH as low as pH 0. No cell wall.
  Picrophilus torridus grows above 60°C. Oxidizes sulfur, generating acid. Possesses cell wall.
  Thermoplasma acidophilum grows at 59°C and pH 2.

Uncultured groups. Revealed by metagenomic analysis.
- **Anaerobic Methane Oxidizers (ANME).** Anoxic marine sediments. Oxidize methane from methanogens, in syntrophy with sulfate-reducing bacteria. Important control of greenhouse gas.
- **South African Gold Mine Euryarchaeotal Group (SAGMEG).** Terrestrial and marine deep subsurface. Anaerobic heterotrophs.
- **Terrestrial Miscellaneous Euryarchaeotal Group (TMEG).** Terrestrial deep subsurface and surface soils, marine and freshwater sediments.

Picrophilus oshimae

Deeply Branching Groups of Uncultured Archaea
Uncultured marine organisms, branching near root of archaeal tree.

Korarchaeota. Hyperthermophiles in Yellowstone hot springs and in deep-sea thermal vents.

Ancient Archaeal Group (AAG). Vent hyperthermophiles and benthic subsurface communities.

Marine Hydrothermal Vent Group (MHVG). Vent hyperthermophiles.

Emerging phyla. Besides the two major phyla Crenarchaeota and Euryarchaeota, metagenomic analysis continues to reveal new clades that branch near the root of the tree. The Ancient Archaeal Group (AAG) was discovered in hydrothermal vents and deep benthic sediment. Since these organisms could not be detected by PCR primers based on known SSU rRNA gene sequence, their detection required reverse transcription of rRNA from uncultured cells. The reverse transcripts showed that the rRNA sequence of Ancient Archaeal Group organisms differs by 20% from the most conserved regions of other known archaeal rRNA.

As more genomes have been sequenced, other proposed deeply branching lineages have since been assigned to known clades. The vent hyperthermophiles known as Korarchaeota appear to branch with the Ancient Archaeal Group. The **Nanoarchaeota** are tiny hyperthermophiles that grow as obligate symbionts of a crenarchaeote, *Ignicoccus*. Genomic analysis now suggests that the Nanoarchaeota are a fast-evolving lineage (like other symbionts with reduced genomes) and that they are euryarchaeotes, branching from the Thermococcales group.

> **THOUGHT QUESTION 19.1** If two deeply diverging clades each show a wide range of growth temperature, what does this suggest about the evolution of thermophily or psychrophily?

TO SUMMARIZE:

- **Archaeal membranes are composed of L-glycerol diether or tetraether lipids**, with isoprenoid side chains that may include cross-links or pentacyclic rings.
- **Many archaea have no cell wall—only an S-layer.** Some methanogens have cell walls of pseudomurein.
- **Glucose is catabolized by variants of the ED pathway.** Other metabolic pathways found in archaea include methanogenesis and retinal-associated light-driven ion pumps such as bacteriorhodopsin.
- **Central genetic functions of archaea resemble those of eukaryotes**, as seen in the structure of DNA and RNA polymerases and of histone-like DNA-binding proteins.
- **Two major phyla or divisions of archaea are Crenarchaeota and Euryarchaeota.** Crenarchaeota includes sulfur hyperthermophiles, mesophiles, and psychrophiles. Euryarchaeota includes methanogens, halophiles, and acidophiles growing at high and low temperatures.
- **Metagenomics reveals uncultured organisms.** We continue to discover deeply branching clades of archaea by designing new PCR primers, reverse-transcribing community rRNA, and sequencing metagenomes.

19.2 Crenarchaeota: Hyperthermophiles

The name Crenarchaeota means "scalloped archaea," derived from the amorphous scalloped cell shapes of the first hyperthermophiles discovered. Crenarchaeotes synthesize a distinctive tetraether lipid, called crenarchaeol, containing occasional six-membered cyclic rings (**Fig. 19.6**). Crenarchaeol is found in varying amounts in all known crenarchaeotes; thus, the molecule serves as a biosignature for this division.

The best-studied crenarchaeotes are members of the class Thermoprotei, which includes sulfur oxidizers, aerobic sulfur reducers, and heterotrophs (**Table 19.2**). The first members described were thermophiles; mesophiles and psychrophiles are now known as well.

Habitats for Thermophiles

Thermophiles and hyperthermophiles commonly grow in hot springs and geysers, such as those of Yellowstone National Park (**Fig. 19.7**) or the Solfatara volcanic area near Naples, Italy. A hot spring occurs where water seeps underground above a magma chamber, which heats the water to near boiling. The heated water expands and is forced upward through fissures, coming out in a heated spring. In a geyser, the water is heated under pressure.

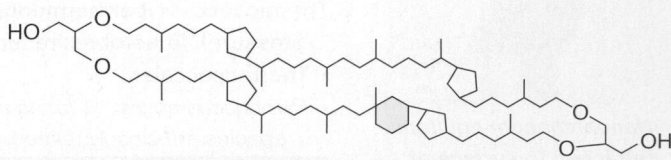

Figure 19.6 Crenarchaeol: a biosignature for crenarchaeotes. Crenarchaeol is a diphytanylglycerol diether containing a six-membered cyclic ring.

©Inga Spence/Visuals Unlimited

Figure 19.7 Thermophiles colonize the edge of a hot spring. Morning Glory Pool, a hot spring in Yellowstone National Park. Yellow-orange regions indicate growth of thermophiles.

As the water escapes upward, it turns into steam, which expands and jets upward, falling into a heated pool. These heated pools and their surrounding edges generate extreme ranges of temperature, mineral content, and acidity. They support a diverse range of microbial life, including thermophilic cyanobacteria and firmicutes, as well as archaea.

Several features of hot springs and geysers are important for thermophiles:

- **Reduced minerals.** The heated water dissolves high concentrations of sulfides and other reduced minerals. When the water emerges and cools, the minerals precipitate. These reduced minerals serve as rich energy sources for autotrophs.
- **Low oxygen content.** At higher temperatures, the oxygen concentration of water is decreased. Therefore, hyperthermophiles tend to be anaerobic, although there are important exceptions, such as *Sulfolobus*.
- **Steep temperature gradients.** The temperature of the water falls dramatically within a short distance from the source, forming a steep gradient. Different species of thermophiles are adapted to different temperatures and grow in separate patches at the different temperatures, causing a variegated pattern.
- **Acidity.** In some cases, hot-spring environments show extreme acidity. The acidity results from oxidation of sulfur or iron in reactions that generate strong inorganic acids such as sulfuric acid (H_2SO_4).

A major subcategory of volcanic hot-spring habitats is that of submarine hydrothermal vents on the ocean floor. So-called "vent" thermophiles must evolve adaptations to high pressure under several kilometers of ocean. Pressure increases by approximately 100 atm per kilometer of ocean depth. Organisms that grow only at high pressure are called **barophiles** (discussed in Chapter 5).

Desulfurococcales: Reducing Sulfur from Hot Springs

Species of the order **Desulfurococcales** show distinctive cell structures and forms of metabolism (**Fig. 19.8A**; **Table 19.3**). All possess elaborate S-layers but lack a cell wall; their diphytanylglycerol membranes contain a combination of diethers and tetraethers. Many take advantage of the high temperatures that increase the thermodynamic favorability of sulfur redox reactions. An example is *Desulfurococcus mobilis*, a flagellated coccoid cell isolated from hot springs; it grows optimally at 85°C (see **Fig. 19.8A**). *D. mobilis* respires anaerobically by reducing elemental sulfur (S^0) to sulfide (HS^-). Sulfur reduction is coupled to oxidizing small organic molecules such as sugars.

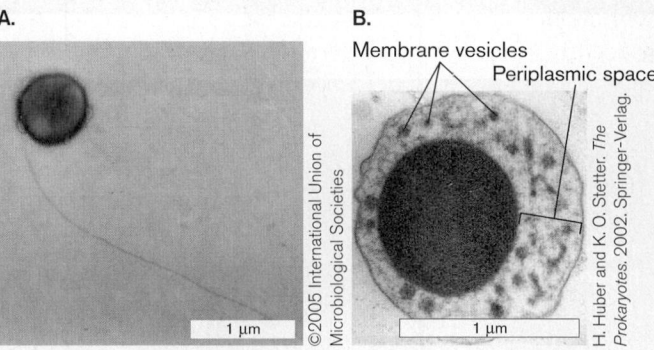

Figure 19.8 Hyperthermophilic crenarchaeotes. **A.** *Desulfurococcus mobilis* (shadow EM). **B.** *Ignicoccus islandicus* (TEM). Its unique periplasmic space contains membrane vesicles. *Source for A: A. Perevalova et al. 2005. Int. J. Syst. Evol. Microbiol.* **55**:995.

Species	Growth temperature (°C)	Growth pH	Cell shape	Metabolism
Aeropyrum pernix	70–100°	pH 5–9	Cocci	O_2 respiration
Desulfurococcus mobilis	78–87°	pH 6	Flagellated cocci	Anaerobic S^0 respiration or fermentation
Ignicoccus islandicus	70–98°	pH 5–7	Flagellated cocci with periplasmic space	Anaerobic lithotrophy, S^0 oxidation of H_2
Pyrodictium abyssi	80–110°	pH 5–7	Disks linked by cannulae	Anaerobic fermentation
Pyrodictium brockii	85–110°	pH 5–7	Disks linked by cannulae	Anaerobic oxidation of H_2 by S^0 or $S_2O_3^{2-}$ or fermentation
Sulfolobus solfataricus	50–87°	pH 2–4	Irregular cocci	O_2 respiration on S^0, producing H_2SO_4
Thermosphaera aggregans	65–90°	pH 5–7	Flagellated cocci in aggregates	Anaerobic fermentation

Table 19.3 Hyperthermophilic Crenarchaeota.

Another flagellated coccus, *Ignicoccus islandicus*, has an unusual cell architecture (**Fig. 19.8B**). *Ignicoccus* possesses an outer membrane surrounding its cytoplasmic membrane, with a large periplasmic space between them. The periplasmic space contains membrane-enclosed vesicles

of unknown function. No other species is known to have a periplasmic space containing vesicles. The evolution of this structure, similar to the extra membranes of the bacterial genus *Planctomyces* (see Chapter 18), suggests a model for an intermediate stage of evolution of the eukaryotic nucleus.

Ignicoccus islandicus, unlike *Desulfurococcus*, is a marine organism, growing at temperatures as high as 98°C. It is a lithotroph, oxidizing hydrogen with sulfur:

$$H_2 + S^0 \longrightarrow H_2S$$

Most of the cultured species of Desulfurococcales are obligate anaerobes. An exception is *Aeropyrum pernix*, one of the first archaea whose genome was sequenced. *A. pernix* is an aerobic heterotroph, respiring with O_2 on complex compounds during growth at 70°C–100°C.

Barophilic hyperthermophiles. The most extreme hyperthermophiles are barophiles adapted to grow near hydrothermal vents at the ocean floor. The high pressure beneath several kilometers of ocean allows water to remain liquid at temperatures above 100°C; the highest known temperature for growth of an organism is 121°C. Isolation and study of such organisms require specialized research apparatus (**Special Topic 19.1**).

A common feature of thermal vents is the **black smoker** (**Fig. 19.9**). A black smoker is a chimneylike structure resulting from the upwelling of seawater superheated by an undersea magma chamber. As in a geyser aboveground, the heated water is forced upward through a small opening. Because the thermal vent is under steam pressure, the water can reach temperatures of over 400°C, enabling it to dissolve high concentrations of minerals such as iron II sulfide (FeS). When the rising water escapes, however, it immediately cools, depositing iron sulfide around the edge of the vent chimney and precipitating iron sulfide particles that cloud the water—hence the term "black smoker." While no organism can grow at 400°C, various species of archaea are adapted to grow in the range of 100°C–120°C, where the vent stream meets the seawater and minerals precipitate (**Fig. 19.9B**).

Vent-adapted crenarchaeotes include *Pyrodictium abyssi* (**Fig. 19.10**), *P. occultum*, and *P. brockii*; the latter is named for Thomas Brock of the University of Wisconsin, Madison, a pioneering researcher of hyperthermophiles. For energy, *Pyrodictium* species reduce sulfur to H_2S, either with molecular hydrogen or with organic compounds. A membrane-embedded sulfur-reducing complex and a proton-translocating ATP synthase have been isolated from *P. abyssi*. The complexes are extremely heat stable, exhibiting a temperature optimum of 100°C.

Pyrodictium species grow as flat, disk-shaped cells that can be as thin as 0.1 μm. The cells contain a periplasm and outer membrane with an S-layer that is

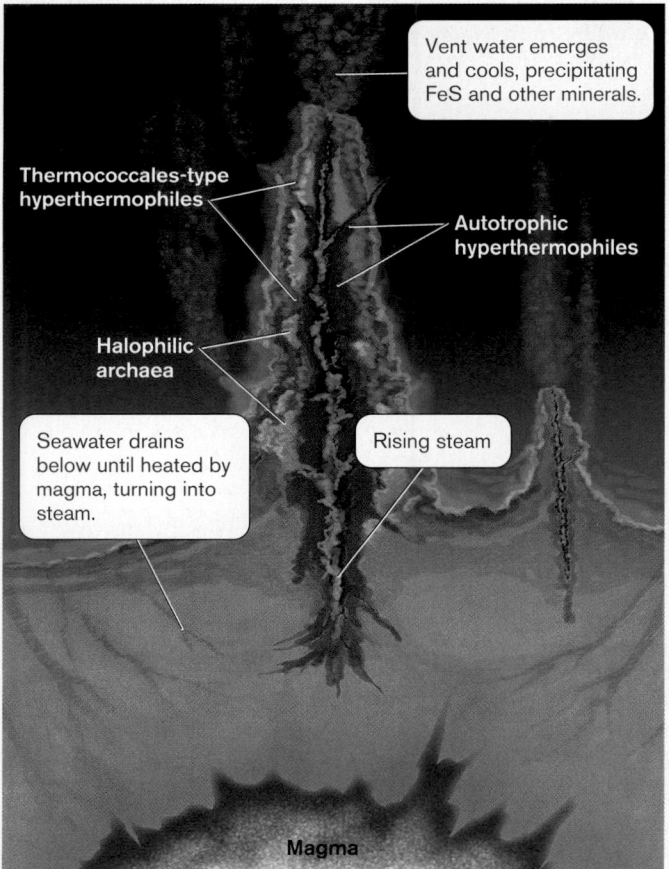

A.

Michael Perfit, U. of Florida, and NOAA VENTS program

B.

Vent water emerges and cools, precipitating FeS and other minerals.

Thermococcales-type hyperthermophiles

Autotrophic hyperthermophiles

Halophilic archaea

Seawater drains below until heated by magma, turning into steam.

Rising steam

Magma

Figure 19.9 Extreme temperature and pressure: black smoker vents.   A. Black smoker vents with steam escaping from "chimneys" of sulfide minerals at the Juan de Fuca Ridge on the ocean floor. **B.** Different parts of the smoker vent system support different classes of archaea.

Special Topic 19.1 | **Research on Deep-Sea Hyperthermophiles**

To study hyperthermophiles from black smoker vents requires specialized equipment and methods. The isolation of such organisms is a challenge because their habitats endanger our

Figure 1   Robotic sampling from a black smoker vent. Engineer Gene Massion, from the Monterey Bay Aquarium Research Institute, deploys the submersible Environmental Sample Processor with robotic collection arm at an ocean site off the coast of Maine. *Source*: Courtesy of Monterey Bay Aquarium Research Institute.

own survival. Undersea vent systems must be approached by a special submersible device with a robotic arm. An example is the Environmental Sample Processor from the Monterey Bay Aquarium Research Institute (**Fig. 1**). The robotic system samples temperature and other properties of fluid emerging from a black smoker hydrothermal vent. It can then sample organisms for study. An advanced version of this device can actually process the organism's DNA. Thus, the DNA can be obtained from vent-adapted microbes that could not survive transfer to a laboratory at sea level. The robotic sample processor is supported by NASA as a model for a future space probe to explore one of Jupiter's satellites, Europa, considered a possible source of extraterrestrial life.

Organisms that do survive transport to sea level must nonetheless be maintained at high pressure and temperature to ensure viability. In the laboratory, all devices for microscopy and cultivation (if possible) must also be maintained under pressure and at high temperature. **Figure 2A** shows a microscope system with a high-temperature and high-pressure cell installed, from the Marine-Earth Data and Information Department of the Japan Agency for Marine-Earth Science and Technology. Organisms are cultured in a pressurized cell such as the Deep Aquarium (**Fig. 2B**). The culture must be provided with reduced minerals and gases needed for growth of vent microbes.

A.

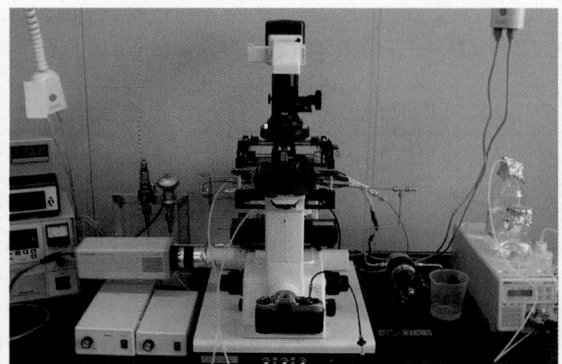

B.

Figure 2   Observation and culture of hyperthermophiles.   A. Microscope system with high-temperature and high-pressure cell installed. **B.** The "Deep Aquarium," a pressurized device for cultivation of vent organisms. *Source*: Images courtesy of Japan Agency for Marine-Earth Science and Technology.

coated with zinc sulfide, presumably precipitated from the vent minerals. The cell disks are interconnected by periplasmic extensions called **cannulae**. The cannulae can extend to more than 0.1 mm, forming complex networks of connections (**Fig. 19.10**); in liquid culture, the networks grow into white balls up to 10 mm in diameter. Cryo-electron tomography of a *Pyrodictium* cell shows that the cannulae bridge the periplasm between cells, but not the cytoplasm. The cannulae may enable

Pyrodictium cells to share nutrients and maintain a biofilm, while keeping their cellular identity distinct with separated cytoplasm.

What happens when a *Pyrodictium* cell divides? Cells of *P. abyssi* generate new cannulae as they undergo fission (**Fig. 19.11**). Some of the new cannulae form as loops connecting the two daughter cells while simultaneously pushing the two cells apart. In this fashion, the cell division process expands the cell network.

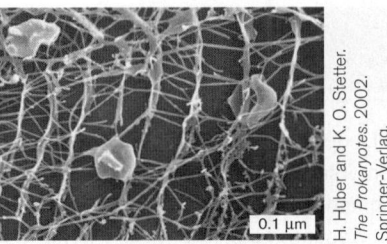

H. Huber and K. O. Stetter. 2002. *The Prokaryotes.* Springer-Verlag.

Figure 19.10 *Pyrodictium abyssi* **growing as networks of cells linked by cannulae.** SEM.

NOTE: Two genera of vent thermophiles have similar names but only distant genetic relatedness: *Pyrodictium abyssi*, a crenarchaeote; and *Pyrococcus abyssi*, a euryarchaeote.

The intricate cell network of *P. abyssi* is an example of a single-species biofilm. Other forms of single-cell and multicellular biofilms are found at hydrothermal vents. *Thermosphaera aggregans* forms colonies so tightly bound that they cannot be dissociated by protease treatment or sonication (**Fig. 19.12**). Multispecies biofilms of hyperthermophiles line the chimneys of black smoker vents.

THOUGHT QUESTION 19.2 What might be the advantages of flagellar motility for a hyperthermophile living in a thermal spring or in a black smoker vent? What would be the advantages of growth in a biofilm?

Sulfolobales: High Temperature and Extreme Acid

The crenarchaeote order **Sulfolobales** includes species that respire by oxidizing sulfur (instead of reducing it, like *Desulfurococcus*). These organisms, such as *Sulfolobus* species, grow at 80°C–90°C within hot springs and solfateras (volcanic vents that emit only gases). Ken Stedman and colleagues at Portland State University study *Sulfolobus solfataricus*, a species that grows at 80°C and pH 3 (**Fig. 19.13**). *S. solfataricus* oxidizes H₂S to sulfuric acid:

$$2S^0 + 3O_2 + 2H_2O \longrightarrow 2H_2SO_4 \longrightarrow 4H^+ + 2SO_4^{2-}$$

As a result of sulfuric acid production, the pH of the organism's surroundings falls to pH 2–3, effectively excluding all but acidophiles. While other archaea grow at higher temperatures (up to 120°C) or lower pH (below pH 0), *Sulfolobus* is of interest as a "double extremophile," requiring both high temperature and extreme acidity simultaneously.

Sulfolobus cells have a membrane composed mainly of tetraethers with cyclopentane rings (see **Fig. 19.2E**). Tetraether membranes are commonly seen in acidophilic thermophiles, probably because they are exceptionally impermeable to protons. Remarkably, *Sulfolobus* has no cell wall—only an S-layer of glycoprotein (**Fig. 19.14A**). Like all archaea, these organisms are nonpathogenic to animals and plants, yet they secrete toxins deadly to competitor strains of *Sulfolobus*.

Sulfolobus species are not obligate autotrophs; they can also grow heterotrophically on sugars or amino acids. In fact, many species are easily cultured in tryptone broth at 80°C, pH 3. Their internal pH is typically pH 6.5; thus, they maintain more than three units of pH difference across their membrane. The full metabolic potential of this organism is revealed by annotation of its genome, which reveals homologs of sugar and amino acid transporters, as well as the non-ATP-forming Entner-Doudoroff pathway of glucose catabolism. Genes are present for enzymes to oxidize HS⁻ and S₂O₃²⁻ as well as S⁰. The main redox carrier for respiration appears to be ferredoxin (instead of NADH, which is relatively unstable at high temperature).

THOUGHT QUESTION 19.3 What problem with cell biochemistry is faced by acidophiles that conduct heterotrophic metabolism?

Because *Sulfolobus* species are readily cultured, their metabolism and ecology have been studied extensively, revealing unexpected features. Some species show multiple origins of replication of their DNA; for example, *S. solfataricus* and *S. acidocaldarius* each use three active replication origins. The presence of multiple origins in some archaea may be related to the closer affinity of archaeal DNA management to that of eukaryotes.

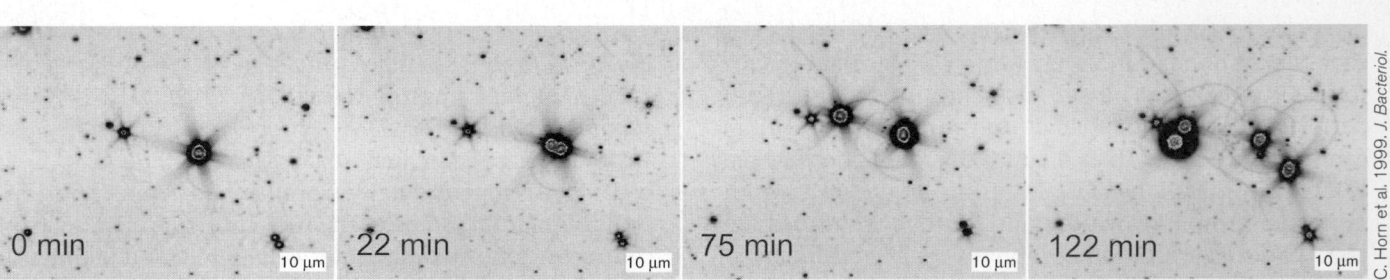

0 min   10 µm   22 min   10 µm   75 min   10 µm   122 min   10 µm

C. Horn et al. 1999. *J. Bacteriol.* **181**:5114

Figure 19.11 *Pyrodictium abyssi* **undergoing cell division.** Cells of *P. abyssi* generate new interconnecting cannulae as they divide.

Figure 19.12 Hyperthermal biofilms. Freeze-etched sample of *Thermosphaera aggregans* reveals a dense aggregate of cells (SEM).

H. Huber and K. O. Stetter. *The Prokaryotes.* 2002. Springer-Verlag.

Image courtesy of Ken Stedman

Figure 19.13 Isolating *Sulfolobus* at Yellowstone. A researcher holds a collecting tube at length to obtain samples from a steaming spring, Rabbit Creek, at Yellowstone National Park.

Another interesting discovery in *Sulfolobus* was that of archaeal viruses. *Sulfolobus* species are attacked by a number of viruses, including fuselloviruses (**Fig. 19.14B**). Fuselloviruses generally resemble bacteriophages in size and function, but their capsid is spindle-shaped, forming a cone at each end. Spindle-shaped capsids are common in archaeal viruses, but never seen in bacteriophages.

The process of a viral infection has been observed in *S. solfataricus* (**Fig. 19.15**). Cells were infected with *Sulfolobus* turreted icosahedral virus (STIV), a virus isolated from a boiling acid hot spring in Yellowstone National Park. Each icosahedral particle of STIV has 12 turret-like projections, as shown in the cryo-TEM image reconstruction (**Fig. 19.15A**). Mature virions contain a double-stranded DNA genome coated with lipid within the capsid. The thin section of an infected cell (**Fig. 19.15B**) shows progeny virions packed into a hexagonal array, portions of which poke through the host cell's S-layer in pyramidal forms. After lysis (**Fig. 19.15C**), the S-layer complex is all that remains of the empty cell. The S-layer appears surprisingly intact, showing that it provides a sturdy cell covering perhaps comparable in strength to a cell wall.

This lytic cycle resembles lytic and fast-release cycles of bacterial and eukaryotic viruses, although the capsid "turrets" and the pyramidal bulges of the lysing cell are unique to archaea. The structure of a capsid protein component shows surprising homology to both bacterial and eukaryotic viral proteins. This finding supports the hypothesis that modern viruses derive from an ancient reproductive form

predating the divergence of the three domains of life (discussed in Chapter 11).

Numerous other archaeal viruses have been discovered in high-temperature environments, including icosahedral, tailed, and filamentous forms. All have in common genomes consisting of double-stranded DNA, suggesting that only double-stranded DNA is stable enough for virus particles to persist at high temperature. Viruses of mesophilic archaea are only beginning to be explored.

THOUGHT QUESTION 19.4 What conclusions might be drawn if viruses of mesophilic archaea are found to have RNA genomes? What if they all have DNA genomes only?

A.

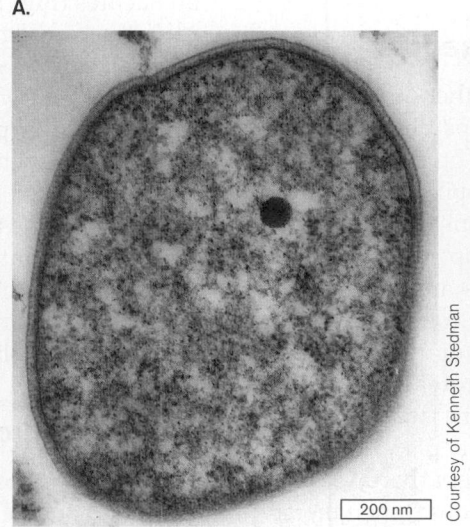

Courtesy of Kenneth Stedman

B.

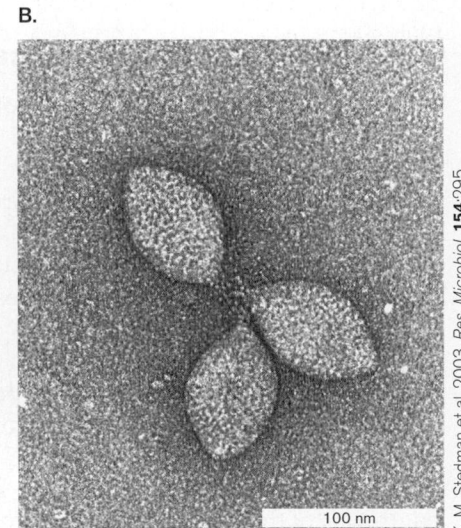

K. M. Stedman et al. 2003. *Res. Microbiol.* **154**:295

Figure 19.14 *Sulfolobus*. A. *Sulfolobus* cell with thick S-layer. **B.** Fusellovirus, a double cone–shaped virus that infects *S. solfataricus*.

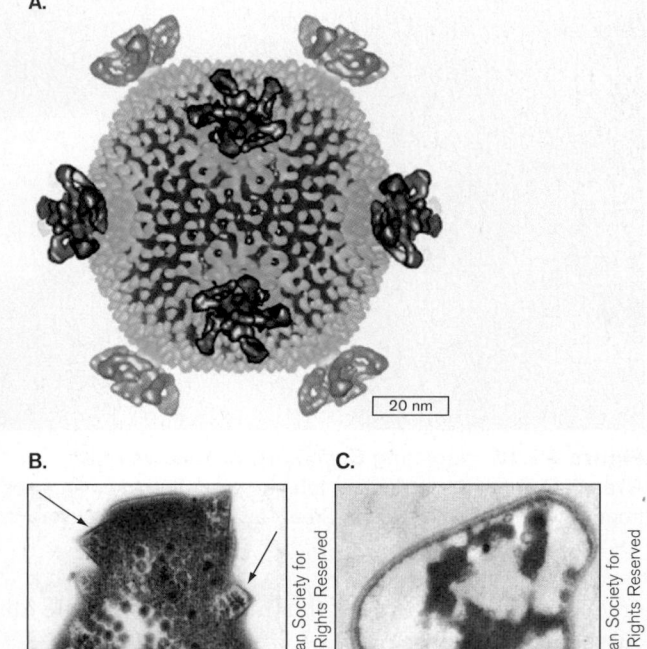

Figure 19.15 *Sulfolobus* **turreted icosahedral virus (STIV) infects** *Sulfolobus solfataricus.* **A.** STIV capsid with "turrets" (cryo-EM). Capsid diameter is 60 nm. **B.** A cell of *S. solfataricus* packed with hexagonal array of STIV particles. Arrows point to pyramidal bulges where virus arrays poke through a breach in the S-layer. **C.** Empty cell membrane and S-layer following lysis and viral release. Arrow points to released virus particles. *Sources:* Part A from George Rice et al. 2004. *PNAS* **101**:7716. Parts B and C from Susan K. Brumfield et al. 2009. *J. Virol.* **83**:5964.

Other Thermophilic Crenarchaeotes

The class Thermoprotei includes two other orders of thermophiles (as well as mesophiles and psychrophiles; see Section 19.3). Caldisphaerales is an order of thermoacidophiles first isolated from a Philippine hot spring at Mount Maquiling. Members of Caldisphaerales, such as *Caldisphaera*, typically grow at pH 3 up to 80°C. Unlike *Sulfolobus* species, *Caldisphaera* species are anaerobes or microaerophiles, tolerating only low concentrations of oxygen. They grow by fermentation or anaerobic respiration. Another order, **Thermoproteales**, includes hyperthermoacidophiles isolated from marine vents. *Thermoproteus* species grow up to 97°C at pH values lower than pH 3. Their rod-shaped cells are less than 0.3 μm in length, one of the smallest cell types known. Their metabolism is autotrophic, gaining energy by reducing sulfur with H_2 to H_2S. A related species, *Pyrobaculum islandicum*,

grows in temperatures up to 103°C, although at less acidic pH (pH 5–7). New thermophilic crenarchaeotes continue to be discovered.

TO SUMMARIZE:

- **Crenarchaeol** is a tetraether containing a six-membered ring, found in varying amounts in all known species of the division Crenarchaeota.
- **Habitats for hyperthermophiles** include hot springs and submarine hydrothermal vents. Vent organisms are barophiles as well as thermophiles.
- **Desulfurococcales includes diverse thermophiles.** Most are anaerobes that use sulfur to oxidize hydrogen or organic molecules.
- *Pyrodictium* **species show unusual networked cells.** Disk-shaped cells are interconnected by cytoplasmic bridges called cannulae.
- *Sulfolobus* **species are aerobic thermoacidophiles.** *Sulfolobus* species oxidize sulfur to sulfuric acid or catabolize organic compounds.
- **Caldisphaerales and Thermoproteales** include anaerobic hyperthermophilic acidophiles.

19.3 Crenarchaeota: Mesophiles and Psychrophiles

For two decades, crenarchaeotes were thought to be largely restricted to high temperatures. In 1992, however, fluorescent DNA probes based on rRNA sequences revealed mesophilic archaea growing throughout the ocean. Edward DeLong and colleagues first found archaea in the Pacific Ocean off the coast of North America; a survey in 2001 of marine archaea at the Hawaii Ocean Time-series Station found high numbers of crenarchaeotes (**Fig. 19.16**). The abundance of crenarchaeotes varied according to season and increased with depth, typically comprising 40% of the total microbial population at depths of 1,000 meters, where temperatures are cold. The high proportion of ribosomal RNA in these cells indicates that they are metabolically active, even at the relatively cold temperatures of the ocean. These uncultivated mesophiles and psychrophiles are probably the predominant species of Crenarchaeota on Earth. In addition, crenarchaeotes are now known to be present in many dryland environments as well, including soil and the surfaces of plant roots.

Psychrophilic Crenarchaeotes

Some of the psychrophilic crenarchaeotes found by DeLong and colleagues exhibit a remarkable symbiosis with marine animals such as the sponge (**Fig. 19.17**). The crenarchaeote *Cenarchaeum symbiosum* inhabits the

A.

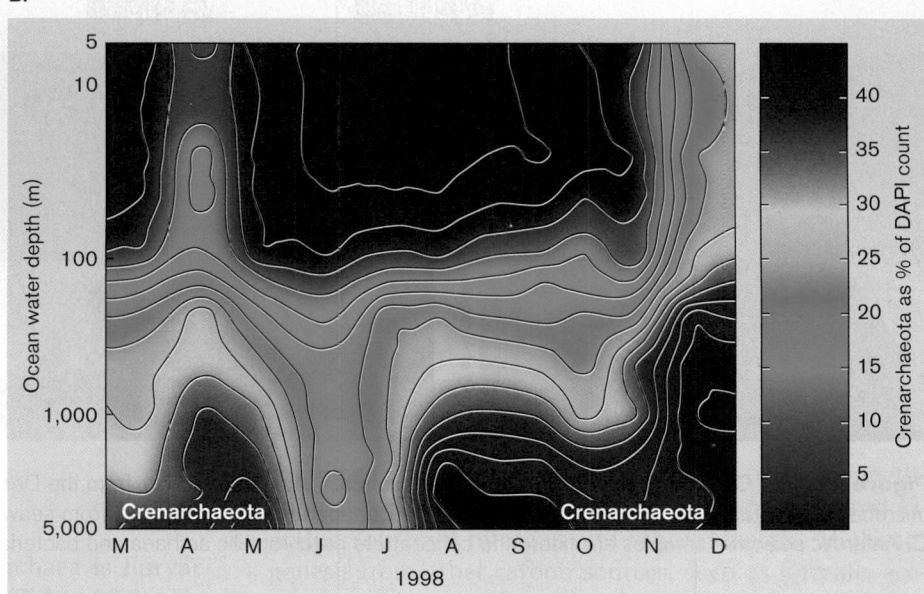

Figure 19.16   Crenarchaeota in the Pacific Ocean.   A. Ed DeLong, now at Massachusetts Institute of Technology, discovers marine crenarchaeotes. **B.** The proportion of Crenarchaeota (color profile) is shown as a function of depth and season. Crenarchaeotes were identified by RNA hybridization to an rRNA sequence DNA probe containing DAPI, a fluorophore. *Source:* Part B from Markus Karner et al. *Nature* **409**:507.

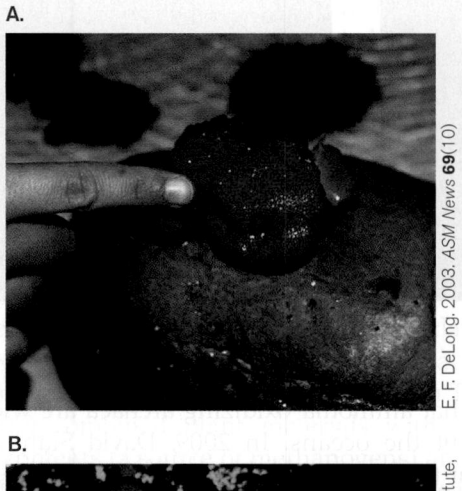

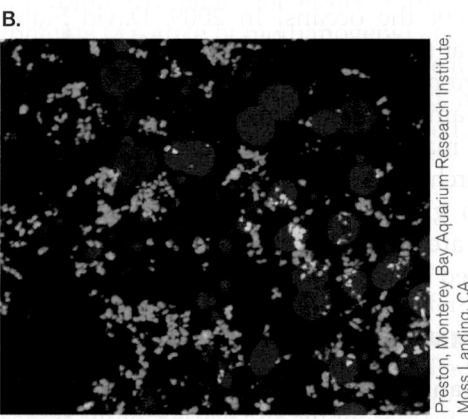

Figure 19.17   Symbiosis between a mesophilic crenarchaeote and a sponge.   A. *Cenarchaeum symbiosum* inhabits the sponge *Axinella mexicana*. **B.** Differential fluorescent staining of *C. symbiosum* present in sponge tissue reveals that bacterial DNA (stained with DAPI, green) is segregated from host cell ribosomes (stained with an rRNA probe, red).

sponge *Axinella mexicana*. Like other marine crenarchaeotes, *C. symbiosum* has yet to be grown in pure culture in the laboratory. The contributions of the microbe to its sponge host are unknown, but the sponge and its symbiotic crenarchaeotes can be cocultured in an aquarium for many years.

Even colder habitats were explored by Alison Murray and colleagues from the Desert Research Institute, who found psychrophilic crenarchaeotes growing in sea ice off Antarctica (**Figs. 19.18** and **19.19**). Archaea and bacteria were collected from ice and seawater at a temperature of –1.8°C; the water remains liquid below zero because of the high salt concentration. Cells were concentrated by filtration, and concentrated samples were analyzed to quantify the number of cells and classify them. The microbes collected were identified by fluorescence microscopy using probes of PCR-amplified rDNA, revealing both bacteria and archaea. Typically, 20% of the isolates are crenarchaeotes; occasional euryarchaeotes are found as well.

Most psychrophilic crenarchaeotes remain uncultivated, but the new technologies of genomics and micromanipulation now make it possible to study uncultivated organisms. The DNA of a single cell taken directly from the field can be amplified by PCR to generate a DNA library for genomic sequencing. Microarrays show which genes are expressed in these organisms and provide insights as to the molecular basis of life at cold temperatures.

Yet another important psychrophilic environment is the marine benthos, or seafloor sediment. At temperatures of about 2°C, seafloor strata show crenarchaeotes that are anaerobic heterotrophs, sulfate reducers, and anaerobic methanotrophs (methane oxidizers). Methanotrophs are

Methylamine:
$$4CH_3NH_2 + 2H_2O \longrightarrow 3CH_4 + CO_2 + 4NH_3$$

Dimethyl sulfide:
$$2(CH_3)_2S + 2H_2O \longrightarrow 3CH_4 + CO_2 + 2H_2S$$

In CO_2 reduction, the most common form of methanogenesis, carbon dioxide plus molecular hydrogen

Methanogens Show Diverse Cell Forms

Despite their metabolic similarity, methanogens display an astonishing diversity of form, perhaps as diverse as the entire domain of bacteria (**Fig. 19.20**). For example, *Methanocaldococcus jannaschii* cells grow as cocci with numerous flagella attached to one side, while *Methanosarcina barkeri* forms peach-shaped cocci lacking flagella. *Methanothermus fervidus* cells are short, fat rods without flagella,

A.   B.   C.

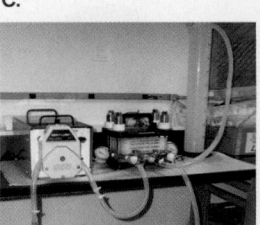

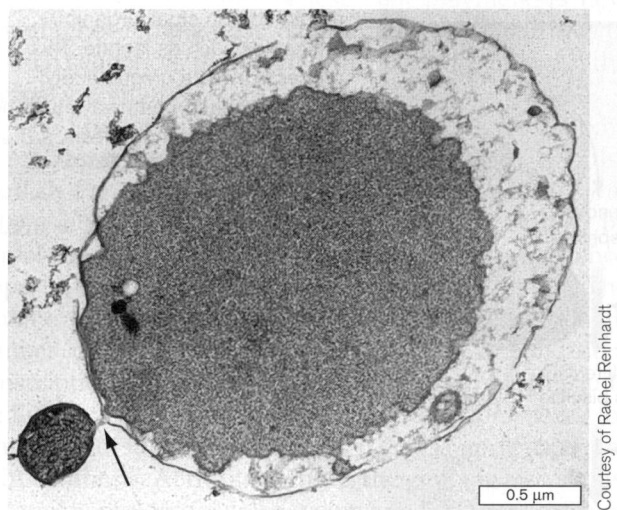

Courtesy of Rachel Reinhardt

0.5 µm

Figure 19.36 *Nanoarchaeum equitans*, a small obligate symbiont attached to *Ignicoccus hospitalis*. TEM.

equitans (**Fig. 19.36**). The organism consists of exceptionally small cells that are obligate symbionts of the crenarchaeote *Ignicoccus*. The *N. equitans* cell in **Figure 19.36** is attached to *I. hospitalis* by a membrane bridge (black arrow). The host cell may harbor up to four of the smaller cells. Both host and symbiont genomes have been sequenced, revealing extensive coevolution of the two. The *N. equitans* genome is exceptionally small (under 500 kb). The diminished genome is typical of a dependent organism that has lost numerous genes through degenerative evolution and dependence on a host. It remains unclear, however, whether *N. equitans* is parasitic on *Ignicoccus* or whether it makes a metabolic contribution to its host (which is able to grow without it).

TO SUMMARIZE:

- **Thermococcales includes hyperthermophiles and acidophiles.** Most species use sulfur to oxidize complex organic substrates.
- *Pyrococcus* and *Thermococcus* **are the source of vent polymerases** for polymerase chain reaction (PCR).
- **Tungsten is commonly used by enzymes** of Thermococcales species.
- *Archaeoglobus* **species reduces sulfate.** The methyl group of acetate is oxidized to CO_2 by reverse methanogenesis.

- **Thermoplasmatales includes extreme acidophiles.** *Ferroplasma* oxidizes iron pyrite ore (FeS_2) in a process that generates concentrated sulfuric acid, causing acid mine drainage.
- *Nanoarchaeum equitans* **is a tiny obligate symbiont that grows attached to** *Ignicoccus hospitalis.*

19.7 Deeply Branching Divisions

New archaeal species continue to be discovered through PCR-amplified rDNA probes. Some of their sequences diverge more deeply than the divergence between Crenarchaeota and Euryarchaeota. Most such strains are uncultivated, and little is known of them other than their rRNA sequence. Nevertheless, new techniques of genome sequencing enable us in some cases to read most of a genome from a single cell. The genomes of such mystery strains may provide clues as to their needs for cultivation.

A deeply branching division is the Ancient Archaeal Group (AAG) of hyperthermophiles. AAG includes the Korarchaeota vent hyperthermophiles isolated by Susan Barns and Norman Pace. The organisms have not been isolated in pure culture, but in 2008 the genome of one korarchaeote was sequenced from an enriched mixed culture at 80°C–90°C at Obsidian Pool, Yellowstone National Park. The organism, provisionally named *Korarchaeum cryptophilum*, grows in long, thin filaments less than 200 nm wide. Its genes suggest that it gains energy mainly from anaerobic peptide fermentation.

Concluding Thoughts

Overall, species of archaea inhabit a wider range of environments than either bacteria or eukaryotes, from extreme heat to extreme cold, from high pH to extreme acid. Their metabolism includes unique capacities, such as methanogenesis. Yet we probably know less about the actual scope of Archaea than we do about the other two domains because so many members remain uncultured. We recognize two major divisions: the crenarchaeotes, including sulfur-metabolizing hyperthermophiles as well as marine mesophiles and psychrophiles; and the euryarchaeotes, including methanogens, halophiles, and sulfur-metabolizing extremophiles. New deeply branching isolates continue to be discovered, potentially representing entire new divisions. Much of what we call Archaea remains to be explored.

CHAPTER REVIEW

Review Questions

1. What distinctive structures are seen in the archaeal cell membrane and envelope?
2. Which aspects of archaeal genetics resemble the genetics of bacteria, and which aspects have more in common with eukaryotes?
3. Compare and contrast the two major divisions of archaea: Crenarchaeota and Euryarchaeota.
4. Outline the genetic phylogeny and key traits of these groups of archaea: Haloarchaea, methanogens, Thermococcales, Thermoplasmatales.
5. What are some specific physiological adaptations found in hyperthermophiles? Psychrophiles? Extreme acidophiles?

6. Outline three different specific types of mutualism involving an archaeal symbiont.
7. Explain how different archaea contribute to cycling of nitrogen and sulfur in ecosystems.
8. Explain what is known and what is unknown about these groups of archaea: marine pelagic crenarchaeotes; marine benthic anaerobes; soil and plant root–associated crenarchaeotes. What kinds of experiments may reveal additional traits of these organisms?

Thought Questions

1. What approaches can you use to discover previously unknown deeply branching groups of archaea? Explain the strengths and limitations of each method.
2. Why do you think we have found no archaea that are pathogens of animals or plants?

3. Why do you think methanogens appear in many branches among different groups, whereas haloarchaea branch as a single group?

Key Terms

anaerobic methane oxidizers (ANME) (737)
bacteriorhodopsin (BR) (718, 742)
barophile (725)
black smoker (726)
cannula (727)
carbon monoxide reductase pathway (718)
Crenarchaeota (720)

Desulfurococcales (725)
Euryarchaeota (720)
gas vesicle (742)
Haloarchaea (739)
Halobacteriales (739)
halorhodopsin (HR) (742)
histone (720)
isoprenoid (718)
methane gas hydrate (737)

methanogen (733)
methanogenesis (718)
Nanoarchaeota (724)
pseudomurein (718)
pseudopeptidoglycan (718)
Sulfolobales (728)
syntrophy (733)
tetraether (718)
Thermoproteales (730)

Recommended Reading

Barns, Susan M., Charles F. Delwiche, Jeffrey D. Palmer, and Norman R. Pace. 1996. Perspectives on archaeal diversity, thermophily and monophyly from environmental rRNA sequences. *Proceedings of the National Academy of Sciences USA* **93**:9188–9193.

Biddle, Jennifer F., Julius S. Lipp, Mark A. Lever, Karen G. Lloyd, Ketil B. Sørensen, et al. 2006. Heterotrophic Archaea dominate sedimentary subsurface ecosystems off Peru. *Proceedings of the National Academy of Sciences USA* **103**:3846–3851.

Brochier, Céline, Patrick Forterre, and Simonetta Gribaldo. 2005. An emerging phylogenetic core of Archaea: Phylogenies of transcription and translation machineries converge following addition of new genome sequences. *BMC Evolutionary Biology* 5:36.

Brumfield, Susan K., Alice C. Ortmann, Vincent Ruigrok, Peter Suci, Trevor Douglas, et al. 2009. Particle assembly and ultrastructural features associated with replication of the lytic archaeal virus *Sulfolobus* Turreted Icosahedral Virus. *Journal of Virology* 83:5964–5970.

Cavicchioli, Ricardo. 2006. Cold-adapted archaea. *Nature Reviews. Microbiology* **4**:331–343.

DasSarma, Shiladitya. 2007. Extreme microbes. *American Scientist* **95**:224–231.

Edwards, Katrina J., Philip L. Bond, Thomas M. Gihring, and Jillian F. Banfield. 2000. An archaeal extreme acidophile involved in acid mine drainage. *Science* **287**:1796–1799.

Elkins, James G., Mircea Podar, David E. Graham, Kira S. Makarova, Yuri Wolf, et al. 2008. A korarchaeal genome reveals insights into the evolution of the Archaea. *Proceedings of the National Academy of Sciences USA* **105**:8102–8107.

Golyshina, Olga V., and Kenneth N. Timmis. 2005. *Ferroplasma* and relatives, recently discovered cell wall-lacking archaea making a living in extremely acid heavy metal-rich environments. *Environmental Microbiology* **7**:1277–1288.

Karner, Markus B., Edward F. DeLong, and David M. Karl. 2001. Archaeal dominance in the mesopelagic zone of the Pacific Ocean. *Nature* **409**:507–510.

Karr, Elizabeth A., Joshua M. Ng, Sara M. Belchik, W. Matthew Sattley, Michael T. Madigan, et al. 2006. Biodiversity of methanogenic and other archaea in the permanently frozen Lake Fryxell, Antarctica. *Applied and Environmental Microbiology* **72**:1663–1666.

Könneke, Martin, Anne E. Bernhard, José R. de la Torre, Christopher B. Walker, John B. Waterbury, et al. 2005. Isolation of an autotrophic ammonia-oxidizing marine archaeon. *Nature* **437**:543–546.

Lepp, Paul W., Mary M. Brinig, Cleber C. Ouverney, Katherine Palm, Gary C. Armitage, et al. 2006. Methanogenic Archaea and human periodontal disease. *Proceedings of the National Academy of Sciences USA* **101**:6176–6181.

Ng, Wailap V., Sean P. Kennedy, Gregory G. Mahairas, Brian Berquist, Min Pan, et al. 2000. Genome sequence of *Halobacterium* sp. NRC-1. *Proceedings of the National Academy of Sciences USA* **97**:12176–12181.

Pernthaler, Annelie, Anne E. Dekas, C. Titus Brown, Shana K. Goffredi, Tsegereda Embaye, et al. 2008. Diverse syntrophic partnerships from deep-sea methane vents revealed by direct cell capture and metagenomics. *Proceedings of the National Academy of Sciences USA* **105**:7052–7057.

Ruepp, Andreas, Werner Graml, Martha-Leticia Santos-Martinez, Kristin K. Koretke, Craig Volker, et al. 2000. The genome sequence of the thermoacidophilic scavenger *Thermoplasma acidophilum*. *Nature* **407**:508.

Samuel, Buck S., and Jeffrey I. Gordon. 2006. A humanized gnotobiotic mouse model of host–archaeal–bacterial mutualism. *Proceedings of the National Academy of Sciences USA* **103**:10011–10016.

Sapra, Rajat, Karine Bagramyan, and Michael W. W. Adams. 2003. A simple energy-conserving system: Proton reduction coupled to proton translocation. *Proceedings of the National Academy of Sciences USA* **100**:7545–7550.

Schleper, Christa, German Jurgens, and Melanie Jonuscheit. 2005. Genome studies of uncultivated archaea. *Nature Reviews. Microbiology* **3**:479–488.

Stams, Alfons J. M., and Caroline M. Plugge. 2009. Electron transfer in syntrophic communities of anaerobic bacteria and archaea. *Nature Reviews. Microbiology* **7**:568–577.

Teske, Andreas, and Ketil B. Sørensen. 2008. Uncultured archaea in deep marine subsurface sediments: Have we caught them all? *ISME Journal* **2**:3–18.

 For further study and review, please visit StudySpace at **microbiology2.com**.

Chapter 20

Eukaryotic Diversity

20.1 Phylogeny of Eukaryotes

20.2 Fungi and Microsporidia

20.3 Algae

20.4 Amebas and Slime Molds

20.5 Alveolates: Ciliates, Dinoflagellates, and Apicomplexans

20.6 Trypanosomes and Metamonads

The domain Eukarya encompasses a breathtaking range of size and form, from giant whales and sequoias to microbial fungi, algae, and protists. Some eukaryotic microbes are single cells as small as the smallest bacteria.

Fungi include multicellular forms such as mushrooms, as well as unicellular yeasts and filamentous *Penicillium* and *Neurospora*. Fungal filaments interconnect the roots of forest plants, forming a vast underground network of nutrients. Algae conduct photosynthesis using chloroplasts. Algae include broad sheets of kelp, as well as unicellular phytoplankton. Many algae turn out to be secondary endosymbionts—protists whose ancestors engulfed preexisting algae whole, only to assimilate them and utilize their prey's chloroplasts.

Protists include amebas, zigzag-swimming euglenas with their chloroplasts, paramecia with hundreds of cilia, and stalked vorticellae whose rings of cilia draw prey toward the mouth. Most protists are free-living, but some are deadly parasites, such as the trypanosome of sleeping sickness and the plasmodium of malaria.

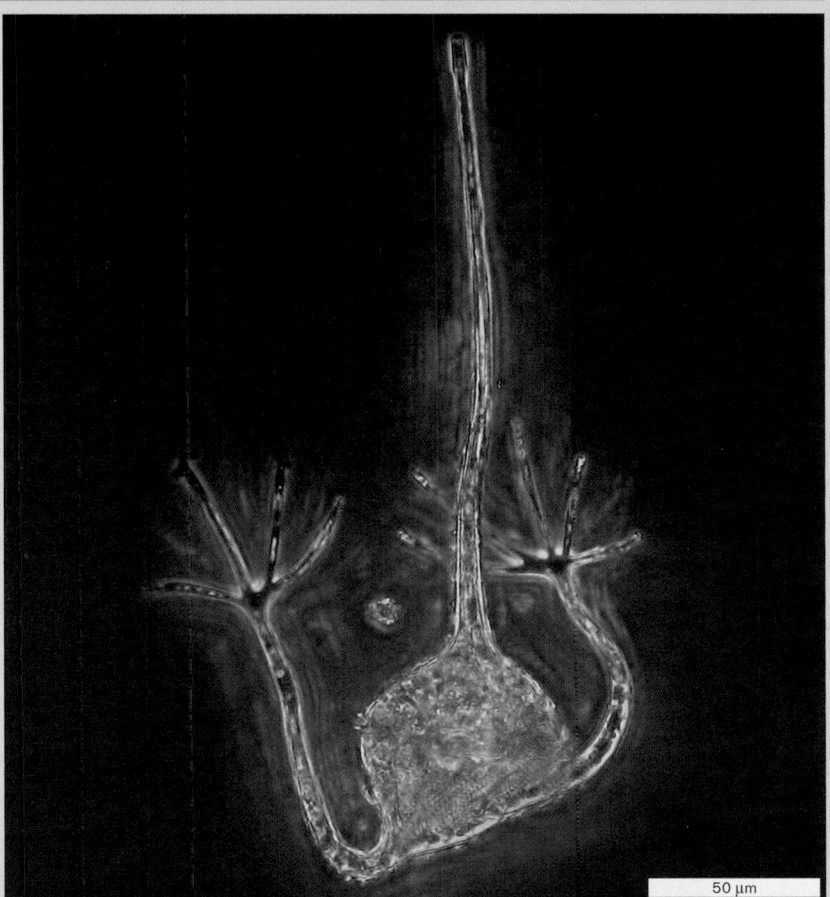

50 μm

Current Research Highlight

A tropical marine dinoflagellate, *Ceratium ranipes*, known for its remarkable hand-like extensions. The "fingers" of the phototroph grow during the day and then are reabsorbed at night. During the night, the fingerless cells swim with their flagella, seeking food. Upon illumination, the cells stop moving and begin to grow fingers. Dinoflagellates are protist algae that conduct a large part of marine photosynthesis throughout the oceans. Some dinoflagellates cause algal blooms such as red tides, releasing toxins that contaminate shellfish. Other species are mutualists of corals, providing photosynthetic products in exchange for a sheltered habitat.

Source: Marie-Dominique Pizay et al. 2009. *Protist* **160**:565–575. © 2009 Elsevier. All rights reserved.

When we think of eukaryotes, we think first of plants and animals consisting of complex multicellular bodies. Macroscopic multicellular eukaryotes provide most of the food we consume, from breakfast cereal to beef. But eukaryotes also include microbes, such as protists (algae and protozoa). While we rarely eat these directly, marine and aquatic protists form the core of a vast food web supporting fish and other metazoan seafood, a major food source of protein. A drop of pond water reveals filaments of algae full of green chloroplasts, and ciliated protists mating by conjugation (**Fig. 20.1**) Other protists include symmetrical diatoms, predators with engulfing mouths, and parasites that can infect our own bodies. Of the three domains of life, Eukarya shows the greatest range of size and shape.

In Chapter 20 we explore the diversity of eukaryotic microbes, including essential partners in ecosystems, as well as parasites that cause devastating pathology. Recall that the structure of eukaryotic cells is defined by the presence of the nucleus and other membrane-enclosed organelles, which enable eukaryotic cells to grow a thousandfold larger than those of prokaryotes (for review, see Appendix 2). Despite their extraordinary range of form and cell structure, the metabolism of eukaryotes is less diverse than that of either bacteria or archaea. Most eukaryotes are either oxygenic phototrophs or heterotrophs growing on complex carbon sources. All descended from an ancestral cell that incorporated a bacterial endosymbiont giving rise to mitochondria, the source of aerobic respiration.

This chapter presents the phylogeny of major groups of eukaryotes, including the branching of animals and plants from the microbial family tree. We explore the form and function of key microbes, including fungi, amebas and slime molds, algae, and various classes of parasites. As in Chapters 18 (Bacteria) and 19 (Archaea), we can show only representative examples of major clades of Eukarya (**Table 20.1**).

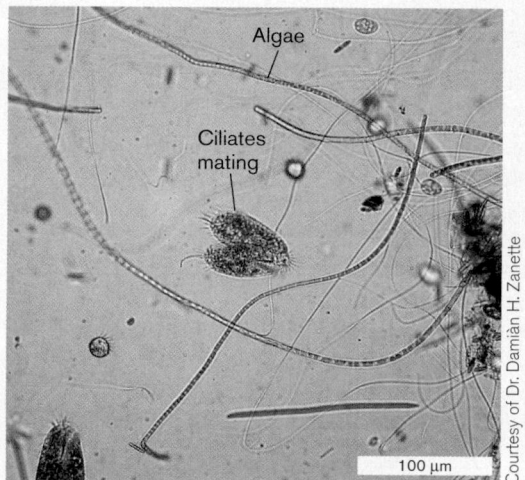

Figure 20.1 Eukaryotic microbes in a drop of pond water.
Algae fix CO_2 and produce biomass for consumers such as ciliated protists. From a freshwater pond in Southwestern Argentina.

20.1 Phylogeny of Eukaryotes

Our view of the domain Eukarya, including its relation to Bacteria and Archaea, has undergone several transitions over the past century, as discussed in Chapter 1. Classifying eukaryotes has always been a challenge, for several reasons. Complex eukaryotic cells frequently lose structures through reductive (degenerative) evolution. Thus, for example, a clade originally defined by possession of flagella often includes members that lack flagella. In addition, superficially similar forms of organisms have evolved independently in distantly related taxa; this is called convergent evolution. For example, the "water molds" that grow on aquarium fish superficially resemble fungi, but they actually evolved in a clade that includes brown algae and diatoms.

Another challenge for classification is the size and complexity of eukaryotic genomes, which has delayed the completion of genome sequences. Nevertheless, we have now completed the genomes of microbial eukaryotes representing most of the major clades. Researchers are approaching a consensus view of eukaryotic descent, including the branching of our own human species.

Historical Overview of Eukaryotes

For most of human history, life was understood in terms of macroscopic multicellular eukaryotes: animals (creatures that move to obtain food) and plants (rooted organisms that grow in sunlight). Fungi, which lack photosynthesis, were nonetheless considered a form of plant because they grow on the soil or other substrate. Thus, **mycology**, the study of fungi, was often included with botany, the study of plants. But the basis of fungal growth was poorly understood, and its mystery was often associated with magic, as seen in the Victorian painting in **Figure 20.2A**, depicting fairies upon a "fairy ring" of mushrooms.

In the eighteenth and nineteenth centuries, microscopists came to recognize microscopic forms of fungi such as hyphae (filaments of cells) and unicellular yeasts. Other unicellular life-forms, such as amebas and paramecia, were motile and appeared more like microscopic animals. These animal-like organisms were called **protozoa** (**Fig. 20.2B**). The cellular dimensions of protozoa were typically ten- to a hundredfold larger than those of bacteria, and their form and motility offered intriguing subjects for observation. So did single-celled phototrophs such as diatoms and dinoflagellates, which were called **algae**. Algae were thought of as unicellular plants, although simple multicellular forms were known. The unicellular and microscopic forms of fungi, protozoa, and algae came to be included in the subject of microbiology.

Discoveries in physiology led us to redefine these organisms. For example, the motile organisms defined

A.

Leicester Gallery

B.

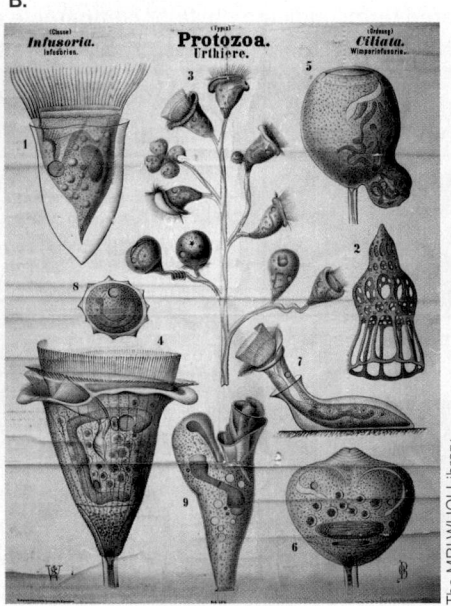

The MBLWHOI Library

Figure 20.2 Historical views of fungi and protozoa (protists). **A.** *A Fairy Ring*, a Victorian painting of fairies upon mushrooms, by Walter Jenks Morgan. **B.** A nineteenth-century depiction of ciliated protists, by Rudolf Leuckart.

as protozoa often contain chloroplasts and fit the classification of algae. On the other hand, slime molds, originally classified with fungi, show form and motility more typical of protozoa. By the mid–twentieth century, naturalists classified protozoa, unicellular algae, and undifferentiated colonial forms as **protists**. Researchers including Herbert Copeland, Robert Whittaker, and Lynn Margulis attempted to refine the definition of "protist" to better distinguish microbial life-forms. Today, molecular phylogeny shows that protists comprise several clades

equally distant from each other as they are from animals and plants (**Fig. 20.3**). In the terminology used today:

- **Protist** refers to single-celled and colonial eukaryotes other than fungi. Protists include many diverse algae and protozoa.
- **Protozoa (singular, protozoan)** are single-celled heterotrophic protists. They include environmental consumers, as well as medically important parasites such as *Giardia*.

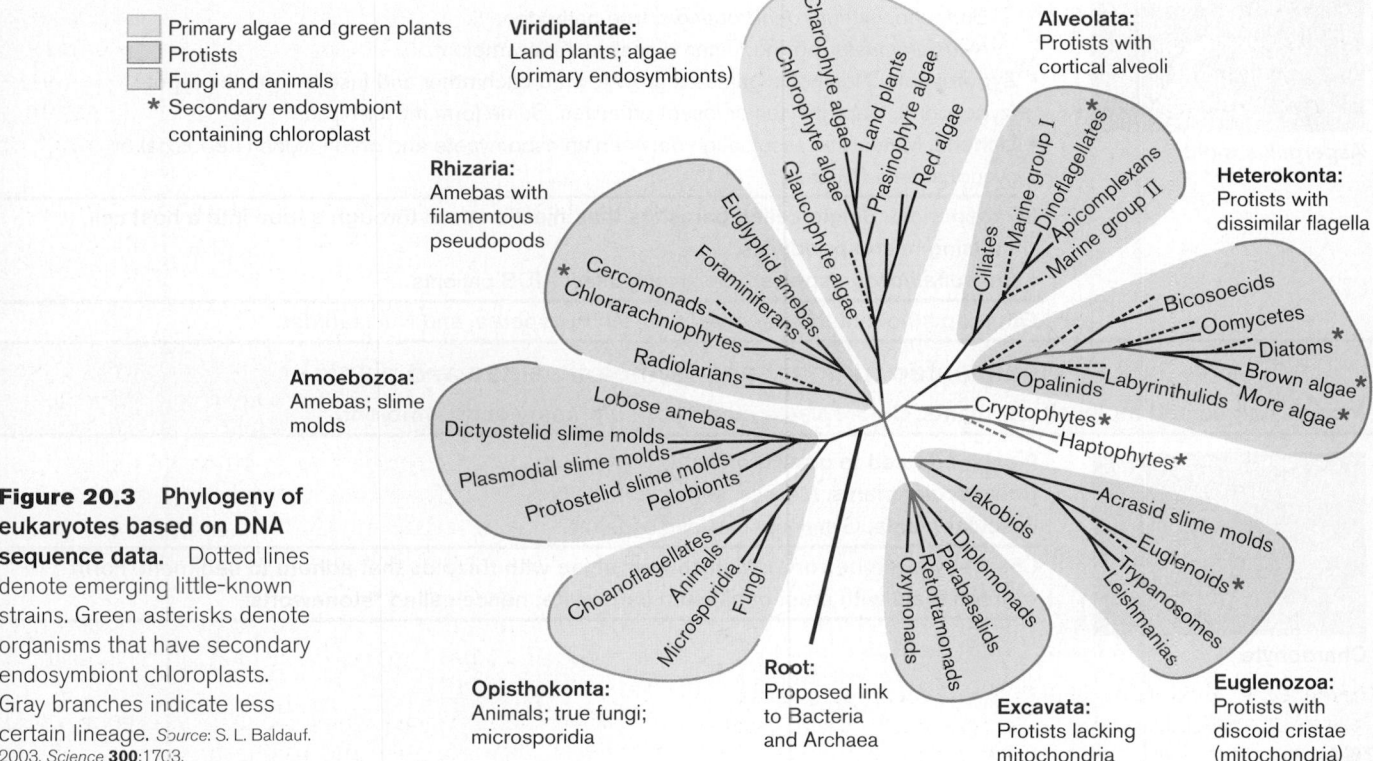

Figure 20.3 Phylogeny of eukaryotes based on DNA sequence data. Dotted lines denote emerging little-known strains. Green asterisks denote organisms that have secondary endosymbiont chloroplasts. Gray branches indicate less certain lineage. Source: S. L. Baldauf. 2003. *Science* **300**:1703.

Table 20.1 Representative groups of eukaryotes.

Note: Each trait applies to *most* members of the taxon described, but exceptions have evolved.
Blue-lettered terms indicate phyla or comparable rank.
• Bulleted terms are representative groups within the phylum.

Opisthokonta (fungi and metazoan animals)

Single flagellum on reproductive cells. Includes multicellular animals.

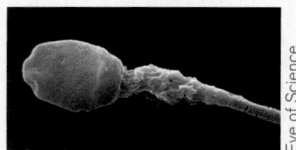

Human spermatozoan

Eye of Science

Choanoflagellate

©1994–2000 by C.J. O'Kelly and T. Littlejohn/U. of Montreal

Allomyces **zoospores**

J.C. Clark/California State Polytechnic Univ, Pomona

Aspergillus **mold**

©Dennis Kunkel/Visuals Unlimited

Animals. Metazoa. Multicellular organisms with motile cells and body parts.
• **Colonial animals.** Sponges, jellyfish.
• **Invertebrates.** Hydra, mollusks, arthropods, worms.
• **Vertebrates.** Fish, amphibians, reptiles, birds, mammals, including *Homo sapiens.*

Choanoflagellata. Single flagellum with collar of microvilli.
Resemble sponge choanocytes. Possible link to common ancestor of multicellular animals.

Fungi (Eumycota). Cells form hyphae with cell walls of chitin.
• **Ascomycota.** Fruiting bodies form asci containing haploid ascospores.
  Filamentous species. *Neurospora, Penicillium.*
  Aspergillus. Opportunistic pathogen; produces aflatoxin.
  Geomyces. Associated with bat mortality, white nose syndrome.
  Magnaporthe oryzae. Causes rice blast, the most serious disease of cultivated rice.
  Microsporum and *Trichophyton.* Cause ringworm skin infection.
  Trichoderma. Endophyte within plants; provides nutrients and induces plant defenses.
  Yeasts. Unicellular species have lost mycelial stages.
  Saccharomyces cerevisiae. Ferments beer and bread dough.
  Blastomyces dermatitidis, Candida albicans, and *Pneumocystis carinii.* Infect immunocompromised patients.
  Stachybotrys. Known as "black mold"; contaminates homes.
  Morels and truffles. Large fruiting bodies are highly valued foods.
• **Basidiomycota.** Basidiospores form primary and secondary mycelia; generate mushrooms.
  Lycoperdon is edible; *Amanita* is extremely poisonous; *Ustilago maydis* causes corn smut (a plant disease).
  Cryptococcus neoformans. Yeast-form basidiomycete; an opportunistic pathogen.
• **Chytridiomycota.** The deepest-branching fungal clade. Zoospores (motile gametes) with a single flagellum resemble the gametes of animals. Saprophytes or anaerobic rumen fungi. *Allomyces.*
  Batrachochytrium dendrobatidis, frog pathogen.
  Neocallomastix, bovine rumen digestive endosymbiont.
• **Zygomycota.** Nonmotile gametes grow toward each other and fuse to form the zygote (zygospore). Saprophytes or insect parasites. Some form mycorrhizae.
• **Lichens.** Mutualistic association between an ascomycete and green algae (*Trebouxia*) or cyanobacteria (*Nostoc*).

Microsporidia. Single-celled parasites that inject a spore through a tube into a host cell, causing microsporidiosis.
• *Encephalitozoon* species. Commonly infect AIDS patients.

Other opisthokonts: *Corallochytrium,* Ichthyosporea, and Nucleariidae

Viridiplantae (primary endosymbiotic algae and plants)

Includes algae and multicellular plants. Chloroplasts arose from a primary endosymbiont.

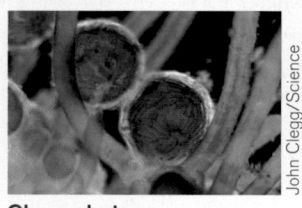

Charophyte

John Clegg/Science Photo Library

Plants. Adapted to growth on land.
Nonvascular plants. Mosses, ferns.
Vascular plants. Gymnosperms, angiosperms.

Charophyta (stoneworts). Multicellular algae with rhizoids that adhere to sediment. Form green mats with crust of calcium carbonate; hence called "stoneworts."

Table 20.1 **Representative groups of eukaryotes (*continued*)**

Cymopolia barbata

S. Berger/Heidelberger Institut

Chlorophyta (green algae). **Chlorophyll *a* confers green color. Inhabit upper waters.**

- **Unicellular with paired flagella.** *Chlamydomonas* is a unicellular green alga; a model system for research on algae. *Volvox* forms colonies of flagellated cells.
- **Multicellular.** *Ulva* grows in large sheets. *Spirogyra* forms chains of cells. *Cymopolia* forms calcified stalks with filaments.
- **Picoeukaryotes.** *Ostreococcus* and *Micromonas* are unicellular algae <3 μm in diameter.
- **Siphonous algae.** *Caulerpa* species consist of a single cell with multiple nuclei, growing to indefinite size.

Palmaria palmata

Edward Kinsman/Photo Researchers, Inc.

Rhodophyta (red algae). **Phycoerythrin obscures chlorophyll, colors the algae red. Absorption of blue-green light enables colonization of deeper waters.**

- *Porphyra* form sheets edible by humans.
- *Sebdenia* and *Plocamium* form branched fronds. *Palmaria*.
- *Mesophyllum* forms coralline algae, hardened by calcium carbonate crust; resembles coral.

Amoebozoa (amebas and slime molds)

Lobe-shaped (lobose) pseudopods driven by sol-gel transition of actin filaments.

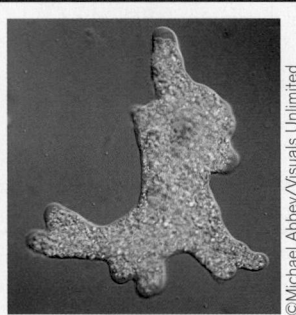

Chaos carolinense (size 1–2 mm)

©Michael Abbey/Visuals Unlimited

Amebas. **Unicellular. No microtubules to define shape. Life cycle is primarily asexual. Predators in soil or water.**

- *Amoeba proteus*, *Pelomyxa*, and *Chaos chaos* are giant free-living amebas in soil and water; they consume small invertebrates.
- *Entamoeba histolytica* is an intestinal parasite.
- *Acanthamoeba* is a soil predator, an opportunistic parasite causing keratitis and meningitis.

Mycetozoa. **Slime molds. Upon starvation, amebas aggregate to form a fruiting body, which undergoes meiosis and produces spores.**

- **Cellular slime molds.** Multicellular fruiting body alternates with single-celled amebas. *Dictyostelium discoideum*. Provides an important model system for multicellular development.
- **Plasmodial slime molds.** Multinucleate plasmodium alternates with single-celled amebas. *Physarum polycephalum*. In aqueous environment, amebas generate flagella.

Rhizaria (amebas with filament-shaped pseudopods)

Filament-shaped (filose) pseudopods. Some species have a test (shell) of silica or other inorganic materials.

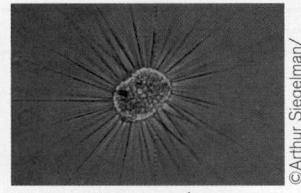

Actinophrys sol (cell size 30–90 μm)

©Arthur Siegelman/ Visuals Unlimited

Cercozoa. **Amebas with flagella and/or filamentous pseudopods.**

- **Euglyphida.** Amebas with a test.
- **Chlorarachniophyta.** Have algal endosymbionts.

Foraminifera. **Form spiral tests. Fossil "forams" are an indicator of petroleum deposits.**

Heliozoa. **Form thin pseudopods.** *Actinophrys sol*. Note: Phylogeny is disputed.

Radiolaria. **Form thin pseudopods (filopodia) reinforced by microtubules, radiating starlike from the center. Some possess inorganic skeletons or spicules.**

Alveolata (having cortical alveoli)

Cortex contains flattened vesicles called alveoli, reinforced below by lateral microtubules.

Vorticella (length 1 mm)

©Wim van Egmond/ Visuals Unlimited

Ciliophora. **Common aquatic predators. Reproduce by conjugation, in which micronuclei are exchanged; then regenerate macronucleus for gene expression. Macronuclear DNA may be cut into thousands of segments.**

Paramecium, *Spirostomum*, and *Oxytricha* are covered with cilia.
Didinium has two equatorial rings of flagella.
Vorticella and *Stentor* are stalked, with a mouth ringed by cilia.
Suctorians such as *Acineta* have knobbed tentacles.

(continued)

Table 20.1 Representative groups of eukaryotes (*continued*)

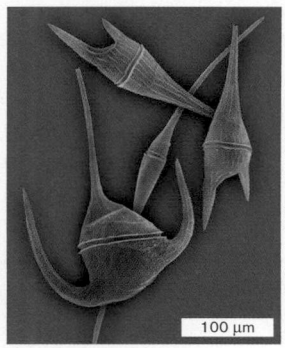

Ceratium tripos and Ceratium furca

Dinoflagellata. Secondary or tertiary endosymbiont algae, from engulfment of red algae or diatoms. Cortical alveoli contain stiff plates. Pair of flagella, one wrapped around the cell.

- **Free-living marine and aquatic dinoflagellates** such as *Peridinium* and *Ceratium* supplement their photosynthesis with predation. Some produce bioluminesence. Blooms of marine dinoflagellates generate the "red tide." Cause paralytic shellfish poisoning.
- **Zooxanthellae** such as *Symbiodinium* spp. are endosymbionts of corals, providing essential nutrition through photosynthesis. "Coral bleaching" is a serious condition in which the zooxanthellae are expelled under environmental stress. Zooxanthellae also inhabit clams, flatworms, mollusks, and jellyfish.

Apicomplexa (formerly sporozoa). Parasites with complex life cycles. Lack flagella or cilia; possess apical complex for invasion of host cells. Vestigial chloroplasts.

- *Plasmodium falciparum* causes malaria.
- *Toxoplasma gondii* causes feline-transmitted toxoplasmosis.
- *Cryptosporidium parvum* is a waterborne opportunistic parasite.

Heterokonta (having pair of dissimilar flagella)

Paired flagella of dissimilar form, one much shorter than the other.

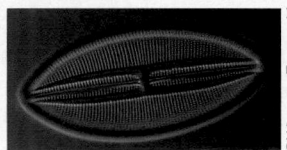

Diatom

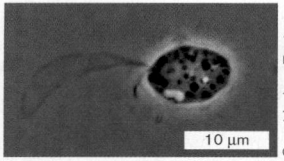

Paraphysomonas sp.

Secondary endosymbiont algae. Free-living in marine and aquatic systems. Mixotrophic; combine photrophy with heterotrophy.

- **Bacillariophyceae.** Diatoms. Possess intricate silicate shells with radial symmetry (centric) or bilateral symmetry (pennate).
- **Phaeophyceae.** "Brown algae" such as kelps and sargassum weed. Extend for many meters.
- **Chrysophyceae.** "Golden algae," pale-colored flagellates such as *Ochromonas* spp. and *Paraphysomonas* spp.
- **Prymnesiophyta (Haptophyta).** Coccolithophores. Covered with intricate plates of $CaCO_3$.
- **Xanthophyceae.** "Yellow-green algae" and other less-studied forms of phytoplankton.

Oomycetes. Water molds. Superficially resemble fungi; often infect fish or plants. *Phytophthora infestans* destroyed potato crops and caused the Great Irish Famine.

Euglenozoa or Discicristata (having disk-shaped cristae)

Usually possess a deep feeding groove. Disk-shaped cristae of mitochondria.

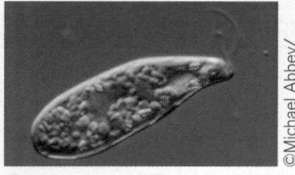

Euglena gracilis

Euglenida. Free-living flagellates. Some contain a secondary endosymbiotic chloroplast. *Euglena gracilis* is a common aquatic flagellate.

Jakobida. Free-living flagellates with lorica (stalk). *Reclinomonas americana.*

Trypanosomatidae. Parasites. *Trypanosoma brucei* causes sleeping sickness. *T. cruzi* causes Chagas' disease. *Leishmania* causes leishmaniasis.

Metamonada (lacking mitochondria)

Parasitic or symbiotic flagellates. Lost mitochondria and Golgi through degenerative evolution.

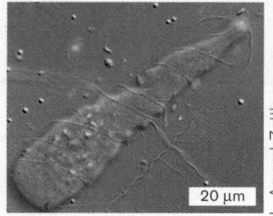

Pyrsonympha sp.

Diplomonadida and Retortamonadida. Human intestinal parasites. Highly degenerate cells, lacking mitochondria. *Giardia lamblia.* Cause of diarrhea from contaminated drinking water.

Oxymonadida and Parabasalia. Symbionts of termite gut. *Pyrsonympha.*

The phylogeny of micobial eukaryotes has only recently emerged from DNA sequences (**Fig. 20.3**). Eukaryotic genomes are harder to sequence because they are typically severalfold larger than those of bacteria and archaea, and 50%–90% of their DNA consists of noncoding sequences. Thus, eukaryotic genomes take longer to sequence and are more challenging to annotate than those of prokaryotes.

Furthermore, the evolution of eukaryotes includes multiple events of **endosymbiosis**, in which an engulfed cell evolved into an essential organelle. An endosymbiotic incorporation of a proteobacterium by the ancestor of all eukaryotes gave rise to mitochondria. A later incorporation of a cyanobacterium by the ancestor of plants and algae gave rise to chloroplasts. Much later, several lineages of protists took up chloroplast-bearing algae, which now show varying stages of evolution as organelles. The lineage of algae derived from a single endosymbiotic event are **primary endosymbionts** (the Viridiplantae in **Fig. 20.3**), while the protists that later incorporated algae are **secondary endosymbionts** (green asterisks in **Fig. 20.3**).

Although uncertainties remain, a consensus view of eukaryotic phylogeny has emerged based on a combination of DNA sequence comparison, protein trees, and the appearance of specific gene fusions and deletions. This consensus phylogeny is shown in **Figure 20.3**. The distinct clades of protists include amebas and slime molds (Amoebozoa); amebas with needllelike pseudopods (Rhizaria); ciliates and dinoflagellates (Alveolata); oomycetes, brown algae, and diatoms (Heterokonta); and a growing number of other groups not as well characterized (such as Excavata and Euglenozoa).

We will now summarize key traits of each major group. Diversity within each group is explored in the sections that follow. (compiled in **Table 20.1**).

NOTE: Medical textbooks also cover invertebrate animal parasites such as worms and mites as eukaryotic agents of disease, although they are not considered microbes.

Opisthokonts: Animals and Fungi

Where do humans and other multicellular animals fit into this taxonomy? The position of animals ("metazoa") among the eukaryotic microbial clades is of interest because it suggests which contemporary microbes most closely resemble our own cells. The degree of relatedness can help define microbial model systems for probing key questions of human cell biology, such as the basis of cancer and aging.

As we saw in Chapter 17, the first measure of species relatedness is based on DNA sequences encoding ribosomal RNA, particularly small-subunit rRNA. For eukaryotes, however, rRNA sequences yield ambiguous results, in part because of differing rates of change among different clades. So researchers have searched the genomes for stronger clues to relative divergence—that is, which clade diverged earliest from the others. A particularly strong clue would be the appearance of a gene insertion or deletion unique to a particular clade. Such a sequence indicates that all species containing the gene insertion or deletion must have diverged after their common ancestor branched from the outgroup.

Surprisingly, genomic analysis relates animals more closely to fungi than to motile single-celled eukaryotes such as paramecia or amebas. The true **fungi**, or **Eumycota**, are heterotrophs, either single-celled or growing in nonmotile filaments of cells called hyphae. Nevertheless, animals and fungi share several key gene insertions and deletions, as discovered by Sandra Baldauf and colleagues at the University of York. Baldauf focused on a short sequence insertion in a key gene encoding a protein translation factor, elongation factor 1α (EF-1α) (**Fig. 20.4**). The inserted DNA, encoding 12 amino acids, appears in all sequenced genomes of animals and fungi and is absent from all plants and most protists. It is present in a few organisms, such as microsporidians, that were previously considered protists. This gene sequence, as well as several others, supports the grouping of animals, fungi, and microsporidians together in one clade, the "opisthokonts" (see **Fig. 20.3**).

Animal and fungal cells also share a structural feature distinguishing them from protists: the presence of an unpaired flagellum. Both animals and fungi include species whose life cycle has a uniflagellar stage, in contrast to other microbial eukaryotes whose flagella are paired (for example, euglenas). In the case of humans, the uniflagellar stage is the spermatozoan. Similarly, some species of fungi generate uniflagellar reproductive cells called **zoospores**. As a whole, the clade including single-flagellum members is termed "opisthokont," based on the Greek words meaning "backward spear," referring to the appearance of the backward-pointing flagellum.

Among opisthokonts, the microbes that diverged most recently from animals (600 million years ago) appear to be the choanoflagellates. Genetic studies of choanoflagellates reveal several genes found only in animals. The prefix *choano-*, meaning "funnel," refers to the collar of filaments surrounding the flagellum. The collared cells of choanoflagellates closely resemble the choanocyte cells of colonial sponges, an ancient form of animal (**Fig. 20.5**). Furthermore, choanoflagellates form colonies that crudely resemble the colonial structure of sponges. Thus, choanoflagellates may represent a "missing link" between animals and the microbial eukaryotes.

NOTE: Eukaryotic flagella are whiplike organelles composed of microtubules and surrounded by a membrane; their action is powered by ATP along the entire filament. Prokaryotic flagella are rotary helical filaments composed entirely of protein subunits; their rotation is powered at the base by proton motive force.

EF-1α peptide sequence alignment

Group	Species	197	Insertion within EF gene	265
Opisthokonts: Animals	*Homo*	PISGWNGDNMLEPSAN-MPWFKG	WKVTRKDG-----NASG	TTLLEALDCILPPTRPTDKPLRLPL
	Drosophila	PISGWHGDNMLEPSTN-MPWFKG	WEVGRKEG-----NADG	KTLVDALDAILPPARPTDKALRLPL
	Acmaea	PISGYNGDNMLEKSPN-MPWYKG	WKVEQKDDKGNASTVTG	DTLTQALDSIQPPKRPTDKALRLPL
	Caenorhabditis	PISGFNGDNMLEVSSN-MPWFKG	WAVERKEG-----NASG	KTLLEALDSIIPPQRPTDRPLRLPL
	Hydra	PVSGWHGDNMIEPSPN-MSWYKG	WEVEYKDTG----KHTG	KTLLEALDNIPLPARPSSKPLRLPL
	Anemonia	PISGWHGDNMLEKSDK-MPWWNG	FELFNKSQG----SKTG	TTLFDGLDDINVPSRPTDKALRLPL
Opisthokonts: Fungi	*Sordaria*	PISGFNGDNMLEASTN-CPWYKG	WEKETKAG-----KSTG	KTLLEAIDAIEQPKRPTDKPLRLPL
	Triochoderma	PISGFNGDNMLTPSTN-CPWYKG	WEKETKAG-----KFTG	KTLLEAIDSIEPPKRPTDKPLRLPL
	Histoplasma	PISGFEGDNMIEPSPN-CTWYKG	WNKETASG-----KSSG	KTLLDAIDAIEPPTRPTDKPLRLPL
	Schizosacch.	PVSGFQGDNMIEPTTN-MPWYQG	WQKETKAG-----VVKG	KTLLEAIDSIEPPARPTDKPLRLPL
	Candida	PISGWNGDNMIEPSTN-CPWYKG	WEKETKSG-----KVTG	KTLLEAIDAIEPPTRPTDKPLRLPL
Opisthokonts: Microsporidians	*Glugea*	PISGYLGINIVEKGDK-FEWFKG	WKPV-SGA-----GDSI	FTLEGALNSQIPPPRPIDKPLRMPI
Plants	*Triticum*	PISGFEGDNMIERSTN-LDWYKG	-----------------	PTLLEALDQINEPKRPSDKPLRLPL
	Arabidopsis	PISGFEGDNMIERSTN-LDWYKG	-----------------	PTLLEALDQINEPKRPSDKPLRLPL
Protists	*Dictyostelium*	PISGWNGDNMLERSDK-MEWYKG	-----------------	RTLLEALDAIVEPKRPHDKPLRIPL
	Plasmodium	PISGFEGDNLIEKSDK-TPWYKG	-----------------	RTLIEALDTNQPPKRPYDKPLRIPL
	Entamoeba	PISGFQGDNMIEPSTN-MPWYKG	-----------------	PTLIGALDSVTPPERPVDKPLRLPL
	Trypanosoma	PISGWQGDNMIEKSEK-MPWYKG	-----------------	PTLLEALDMLEPPVRPSDKPLRLPL
	Euglena	PISGWNGDNMIEASEN-MGWYKG	-----------------	LTLIGALDNLEPPKRPSDKPLRLPL
	Giardia	PTSGWTGDNIMEKSDK-MPWYKG	-----------------	PCLIDAIDGLKAPKRPTDKPLRLPI

Figure 20.4 Alignment of peptide sequences for EF-1α reveals insertion in opisthokont species. A DNA insertion encoding 11–17 amino acid residues appears in the EF-1α sequence of animals, fungi, and microsporidians (known collectively as opisthokonts) but not in plants or most protists. This finding suggests that opisthokonts have a common ancestor that diverged from the eukaryote lineage before divergence of the plants and non-opisthokont protists. *Source:* S. L. Baldauf. 1999. *Am. Nat.* **154**:S178.

Fungi (Eumycota) consist of cells with chitinous cell walls that grow in chains called **hyphae** (singular, hypha). Fungi range from single-celled organisms, such as yeasts, to complex multicellular forms such as mushrooms. The deepest-branching clade of fungi, Chytridiomycota, produces the uniflagellar zoospores that define opisthokonts. Other fungi, however, generate nonmotile reproductive cells, a result of reductive evolution.

Several taxa that historically were grouped with fungi (for example, slime molds) are now classified genetically as protists. Slime molds generate populations of cells that migrate into a unified structure called a **fruiting body** to make reproductive cells. Slime molds are now grouped with amebas (Amoebozoa, discussed in Section 20.4). Water molds, plant and animal pathogens of the class Oomycetes, are now recognized as heterokont protists (see **eTopic 20.1**).

Algae Evolved by Engulfing Phototrophs

Algae are commonly defined as single-celled plants and simple multicellular plants lacking true stems, roots, and leaves. Algal cells contain **chloroplasts**, membrane-enclosed organelles of photosynthesis that evolved from

A. Choanoflagellate
Collar
Nucleus
Flagellum
Courtesy of D. Moran
B. Sponge body wall
Water flow
Pore
Choanocyte
Porocyte
Spicule
Amebocyte

Figure 20.5 Choanoflagellates resemble sponge choanocytes. A. A choanoflagellate, *Acanthocorbis unguiculata* (cell length 6 μm). **B.** Sponge choanocytes resemble choanoflagellates. Within the sponge, choanocytes assist the circulation of water and uptake of nutrients.

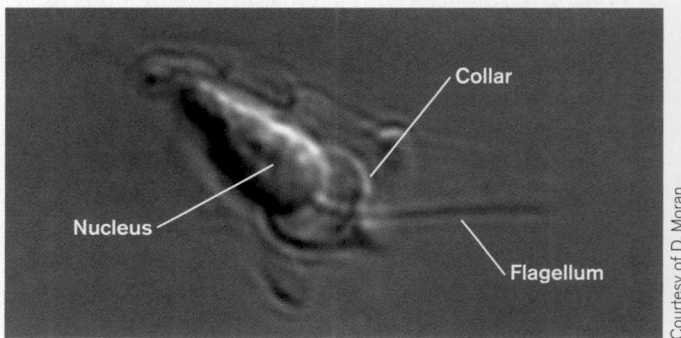

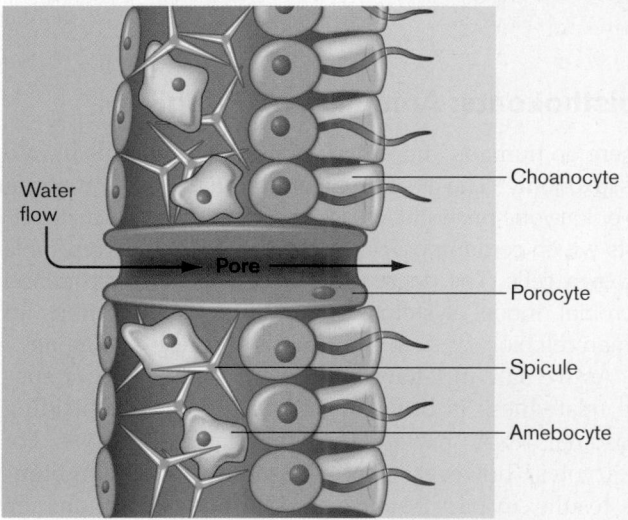

Figure 20.6   Chloroplast evolution: primary and secondary endosymbiosis.   A. Green algae (Chlorophyta) and red algae (Rhodophyta) contain chloroplasts (green) that evolved from engulfed cyanobacteria. Primary-host cytoplasm is colored yellow; primary-host nuclear membrane is colored brown. **B.** Cryptophyte algae contain chloroplast (green), vestigial primary-host cytoplasm (yellow), and vestigial nucleus or nucleomorph (brown) from the engulfed primary endosymbiont.

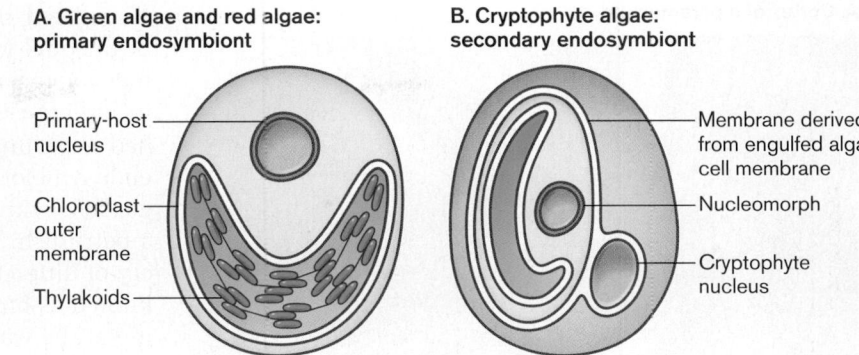

A. Green algae and red algae: primary endosymbiont

Primary-host nucleus

Chloroplast outer membrane

Thylakoids

B. Cryptophyte algae: secondary endosymbiont

Membrane derived from engulfed algal cell membrane

Nucleomorph

Cryptophyte nucleus

a cyanobacterium (**Fig. 20.6**). In Earth's biosphere, algae plus bacterial phototrophs feed all marine and aquatic ecosystems, producing the majority of oxygen and biomass available for Earth's consumers.

The "green plants" (Viridiplantae) include primary endosymbiont algae descended from a common ancestor containing a chloroplast. The chloroplasts of primary endosymbionts are enclosed by two membranes (**Fig. 20.6A**): the inner membrane (from the ancestral phototroph's cell membrane) and the outer membrane (from the host cell membrane as it enclosed its prey). Both **green algae (chlorophytes)** and **red algae (rhodophytes)** are primary endosymbionts. Their chloroplasts have diverged from their common ancestor to utilize pigments absorbing different ranges of the light spectrum. Genetically, the green algae are the most closely allied with plants.

Several major taxa traditionally considered to be algae turn out to be secondary endosymbionts derived from protist hosts. The symbiotic history is most evident in the cryptophyte algae (**Fig. 20.6B**), which still retain a vestigial nucleus, or **nucleomorph**, derived from the engulfed cell. Their chloroplast is surrounded by two extra membranes, one from the primary endosymbiont (engulfed alga) cell membrane and one from the secondary host. Other secondary endosymbiont algae include the **chrysophytes**, such as kelps and diatoms. The **dinoflagellates**, photoheterotrophs of the protist clade Alveolata, are secondary or tertiary endosymbionts, decended from a flagellate that consumed one or more types of algae. Dinoflagellates also engage in "kleptoplasty," or "chloroplast stealing," in which the chloroplast of a digested prey is retained long enough to derive some photosynthetic energy but ultimately consumed. The variety of endosymbiosis among protists provides clues as to how the original chloroplast evolved within the ancestral algae.

NOTE: "Algae" may refer to primary algae, or "true algae," as well as secondary endosymbiont algae that derive from protist clades. Fungi (the "true fungi," or Eumycota) are opisthokonts. Several fungus-like organisms have been reclassified within protist clades.

Protists Form Many Divergent Clades

The protists actually include several distantly related categories of eukaryotes (see **Table 20.1**). All protists are heterotrophs, commonly predators or parasites, although many also conduct photosynthesis as secondary endosymbionts (algae). Protists are important producers and consumers in marine, aquatic, and soil food webs. In ecology, phototrophic protists are termed phytoplankton and heterotrophs are termed zooplankton, although many in fact are "mixotrophs" that act as both producers and consumers.

Amebas are unicellular organisms of highly variable shape that form **pseudopods**, locomotory extensions of cytoplasm enclosed by the cell membrane. Their size can reach several millimeters and they can eat small invertebrates. There are two major groups of amebas: Amoebozoa and Rhizaria (see **Table 20.1**). The Amoebozoa, the most familiar kind of amebas, have lobed pseudopods, pseudopods that extend lobes of cytoplasm through cytoplasmic streaming. Most lobed amebas are free-living in aquatic habitats, but some cause human diseases such as meningitis. Lobed amebas also include slime molds, in which individual amebas converge to form a fruiting body. Amebas of the second group, Rhizaria, have thin, filamentous pseudopods, often radially arranged like a star, as in heliozoan amebas. Some Rhizaria, such as the foraminiferans, form inorganic shells called tests. Fossil foraminiferan tests are common in rock formations derived from ancient seas. Foraminiferan shells formed the white cliffs of Dover in Britain and the stone used to build the Egyptian pyramids.

NOTE: DNA sequencing led to reclassification of the filamentous amebas as Rhizaria (formerly Cercozoa, in our previous edition). Cercozoa now denotes a subgroup under Rhizaria.

Alveolates (the Alveolata) include ciliated protists (ciliates), dinoflagellates, and apicomplexans. Alveolates are known for their complex outer covering, or cortex. The cortex contains networks of vesicles called **cortical alveoli** (**Fig. 20.7A**). Alveoli store calcium ion and in some

A. Cortex of a paramecium

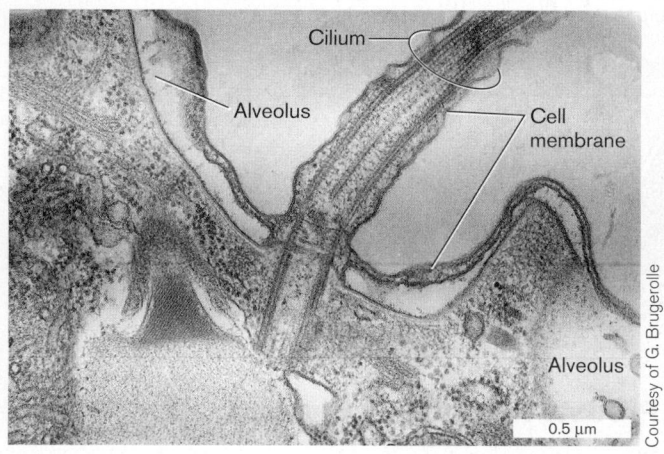

Courtesy of G. Brugerolle

B. Flagellum or cilium

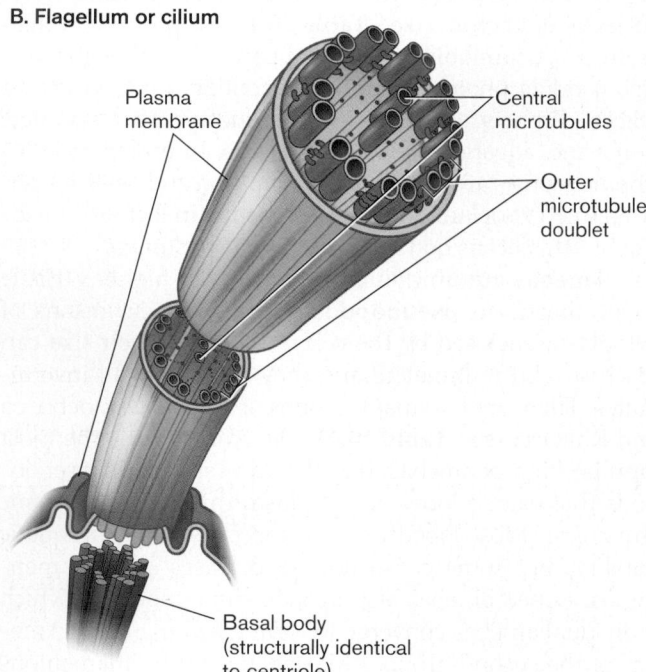

Figure 20.7 **Alveolates have alveoli in their cortex.**
A. Cortex of *Paramecium tetraurelia* with cilium and alveoli.
B. Flagella and cilia are composed of doublet microtubules and enveloped in cell membrane.

species grow protective plates. Organisms equipped with paired flagella or cilia are known as **flagellates** and **ciliates**, respectively. Flagella and cilia are essentially equivalent organelles composed of microtubules and enveloped by the cell membrane (**Fig. 20.7B**), although cilia are shorter and more numerous than flagella, and usually cover a broad surface. Alveolata also includes a major group of parasites that no longer possess flagella, the **apicomplexans**. A well-known apicomplexan parasite is *Plasmodium falciparum*, which causes malaria.

Heterokonts (the Heterokonta) are named for their pairs of differently shaped flagella (see **Table 20.1**).

Flagellated members of this clade possess two flagella of unequal length and different structure. They include many voracious zooplankton. Nonflagellated heterokonts include the oomycetes, or water molds (formerly classified with fungi), as well as diatoms and kelps (secondary endosymbiotic algae with chloroplasts). **Diatoms** are single cells with unique bipartite shells that fit together like a petri dish. The shells of diatoms form an infinite variety of different patterns for different species. **Kelps**, also known as brown algae, extend multicellular sheets floating at the water's surface. The kelp known as sargassum weed is famous for its growth in the Sargasso Sea.

Some protists, such as the Euglenozoa and the Excavata (excavates), show extensive evolutionary reduction. The Euglenozoa have mitochondria with distinctive disk-shaped cristae (membrane pockets). They include euglenas, free-living flagellates with a relatively simple cortex. Other Euglenozoa are parasites, such as the trypanosomes that cause sleeping sickness and Chagas' disease. The Excavata lack mitochondria altogether, although traces of mitochondrial genes persist in their nuclei. Excavata include the diplomonads, such as *Giardia lamblia*, a common contaminant of water supplies.

Newly Discovered Eukaryotes

Genes from natural communities continually reveal new species of microbial eukaryotes in previously unknown divisions. Many of the new isolates are single cells as small as bacteria, designated nanoeukaryotes (3–10 µm) and picoeukaryotes (0.5–3 µm). For example, the picoeukaryote *Ostreococcus tauri* is a green alga less than 2 µm across (**Fig. 20.8**). The tiny alga consists of a flat disk-shaped cell containing one or two of each type of organelle (a mitochondrion, a chloroplast, a stack of Golgi) packed tightly within the small volume.

The genome of *O. tauri* is also downsized; at only 8 Mb, it is barely twice that of the bacterium *Escherichia coli*. Genetic analysis shows that similar miniaturized eukaryotes branch deeply from all groups in the phylogenetic tree, potentially doubling the known number of major eukaryotic taxa.

Furthermore, we are finding eukaryotes in "extreme" environments previously believed to be restricted to bacteria and archaea, ranging from Antarctic sea ice to the hyperacidic Tinto River in Spain. The next decade may substantially reshape our overall understanding of the domain Eukarya.

TO SUMMARIZE:

- **Opisthokonta** includes true fungi (Eumycota) and multicellular animals (Metazoa), as well as certain kinds of protists.
- **Viridiplantae** includes green plants and primary endosymbiont algae.

A.

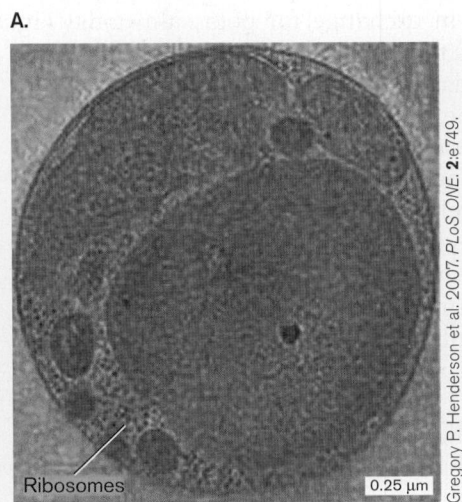

Ribosomes

0.25 μm

Gregory P. Henderson et al. 2007. *PLoS ONE.* **2**:e749.

B.

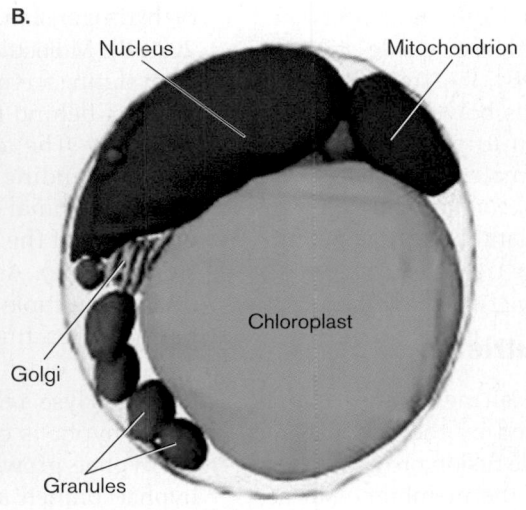

Nucleus

Mitochondrion

Golgi

Chloroplast

Granules

Figure 20.8 A tiny eukaryote still has organelles. *Ostreococcus tauri* visualized by cryo-EM (**A**) and in 3-D by tomography (**B**). *Source*: Gregory Henderson. 2007. *PloS One* **8**:e749.

■ **Protist groups** include Amoebozoa, the unshelled amebas; Rhizaria, shelled amebas; Alveolata, ciliates and flagellates with complex cortical structure; Heterokonta, the kelps, diatoms, and flagellates with nonequivalent paired flagella; and Euglenozoa and Excavata, primarily parasites.

■ **Many protists are phototrophs as well as heterotrophs**, based on secondary or tertiary endosymbiosis derived from engulfed algae.

20.2 Fungi and Microsporidia

Fungi provide essential support for all communities of multicellular organisms. Fungi recycle the biomass of wood and leaves, including substances such as lignin, which other organisms may be unable to digest. Underground fungal filaments called mycorrhizae extend the root systems of most plants, forming a nutritional "internet" that interconnects the plant community (discussed in Chapter 21). Mycorrhizae may have inspired the fictional underground tree network depicted in the film *Avatar* (2009).

Within animal digestive systems, fungi ferment plant materials. On the other hand, pathogenic fungi infect plants and animals, and they contribute to the death of immunocompromised human patients. Still other fungi produce antibiotics such as penicillin, as well as food products such as wine and cheeses (discussed in Chapter 16).

Subtle details of the fungal cell are exploited as targets for antifungal agents. For example, fungal membranes contain ergosterol, an analog of cholesterol not found in animals or plants. Inhibitors of ergosterol biosynthesis, such as the triazoles, are used to treat fungal infections.

The following traits are common to most fungal cells:

■ **Absorptive nutrition.** Most fungi cannot ingest particulate food, as do protists, because their cell walls cannot part and re-form, like the flexible pellicle of amebas and ciliates. Instead they secrete digestive enzymes and then absorb the broken-down molecules from their environment.

■ **Hyphae.** Most fungi grow by extending multinucleate cell filaments called hyphae (**Fig. 20.9A**). As a hypha extends, its nuclei divide mitotically without cell division, generating a multinucleate cell. Hyphae grow by cytoplasmic extension and branching. A branched mass of extending hyphae is called a **mycelium**.

A. Hyphae

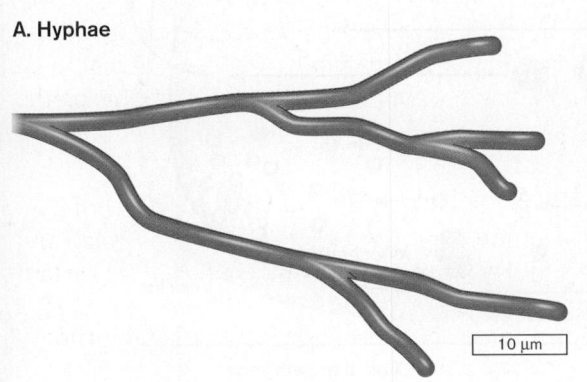

10 μm

B. Chitin

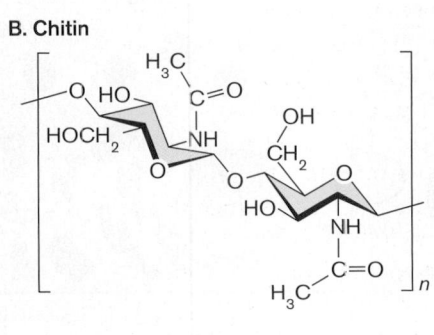

Figure 20.9 Fungi grow hyphae with cell walls of chitin. A. Fungal hyphae extend and form branches, generating a mycelium. **B.** Chitin consists of beta-linked polymers of *N*-acetylglucosamine.

■ **Cell walls containing chitin.** Chitin is an acetylated amino-polysaccharide of immense tensile strength, stronger than steel (**Fig. 20.9B**). Its strength derives from multiple hydrogen bonds between fibers. Chitinous cell walls enable fungi to penetrate plant or animal cells, including tough materials such as wood. Inhibitors of chitin synthesis, such as the polyoxins and nikkomycins, are used as antibiotics against fungal infections.

Fungal Hyphae Absorb Nutrients

At the growing tip of a hypha, the cell membrane expands by incorporating vesicles generated by the endoplasmic reticulum (**Fig. 20.10A**). The vesicle fusion provides phospholipids and proteins to extend the membrane surface area, allowing the cytoplasm to expand rapidly. Expansion of cell membrane and cytoplasm is driven by uptake

of hydrogen ions in exchange for potassium ions (**Fig. 20.10B**). Molecules that deregulate K^+/H^+ exchange, such as nystatin, serve as antifungal agents.

Just behind the hypha's growing tip lies its absorption zone. The absorption zone takes in nutrients from the surrounding medium, such as the cytoplasm of an invaded animal cell. Behind the absorption zone, the older part of the hypha collects and stores nutrients (the storage zone). As the storage zone expands, the nucleus divides multiple times. Septa form across the hypha, partly compartmentalizing the cytoplasm.

As the older part of the hypha ages, its tubular form begins to lyse, releasing cell constituents. This aging part of the hypha is called the senescence zone (not shown). As hyphae grow, branches extend from their sides. The hyphae branch and extend radially, forming the mycelium. The mycelium forms the characteristic round, fuzzy colony of a fungus or "mold." On a substrate such as wood or agar, mycelia grow in two forms: the aerial mycelium, which extends out into the air; and the surface mycelium, which grows into and along the surface of the substrate.

A.

Vesicles

Mitochondria

Golgi

Stephan Seiler et al. 1997. *EMBO J.* **16**:3025.

1 µm

Figure 20.10 Cellular basis of hyphal extension.
A. Section through growing tip of a hypha (TEM). Vesicles collect at the tip, where they fuse into the cell membrane, enabling extension. **B.** Hyphal extension is regulated by uptake of protons and extrusion of potassium ions at the tip of the apical growth zone. The absorption zone takes in nutrients, while the storage zone synthesizes and stores cell constituents.

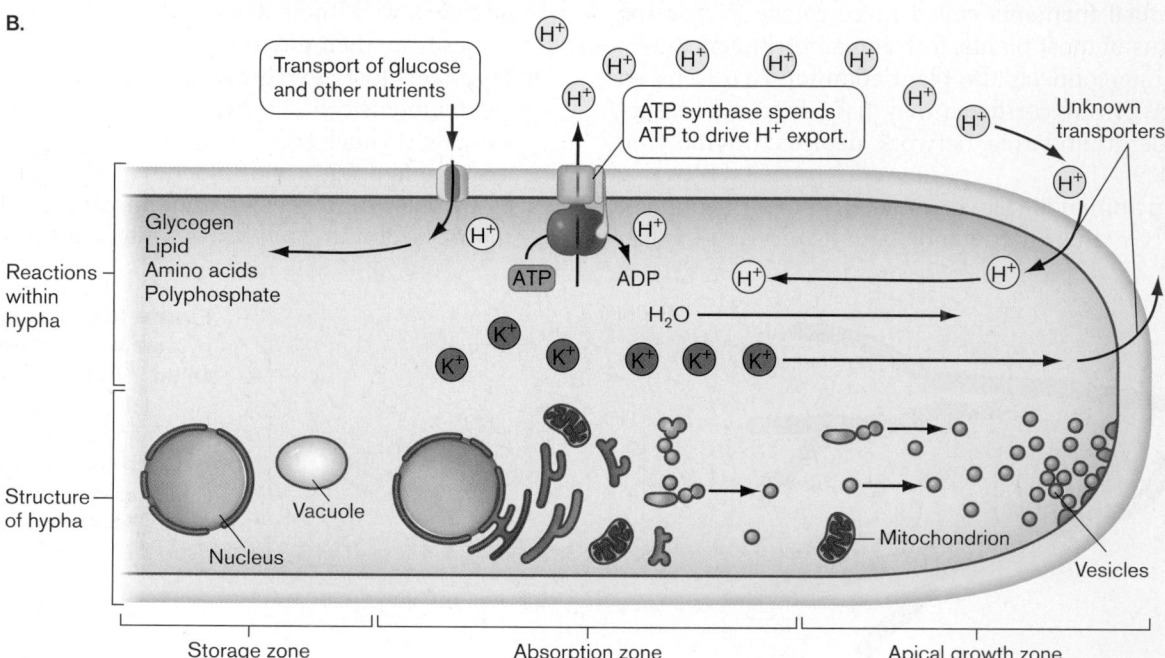

B.

Transport of glucose and other nutrients

ATP synthase spends ATP to drive H^+ export.

Unknown transporters

Reactions within hypha

Glycogen
Lipid
Amino acids
Polyphosphate

ATP ADP

H_2O

Structure of hypha

Vacuole

Nucleus

Mitochondrion

Vesicles

Storage zone   Absorption zone   Apical growth zone

Unicellular Fungi

Despite the advantages of multicellular hyphae for resource acquisition, some fungi are unicellular, known as **yeasts**. Yeast forms evolved in many different fungal taxa. Baker's yeast, *Saccharomyces cerevisiae* is used to leaven bread and to brew wine and beer (for more on food microbiology, see Chapter 16). Other yeasts, such as *Candida albicans* (**Fig. 20.11A**), are important members of human vaginal flora. Some are important opportunistic pathogens, occurring frequently in AIDS patients; for example, *Pneumocystis carinii* is a yeast-form ascomycete, whereas *Cryptococcus parvum* is a yeast-form basidiomycete (discussed shortly).

Some fungi, such as *C. albicans*, can grow either as yeast or as mycelia; these are known as "dimorphic" fungi. Another dimorphic fungal pathogen is *Blastomyces dermatitidis*, the cause of blastomycosis, a type of pneumonia. *B. dermatitidis* forms a mycelium in culture and in soil environments, but it grows as a yeast within the infected lung.

Yeasts such as *S. cerevisiae* provide major research subjects for eukaryotic biology. For example, the Nobel Prize in Physiology or Medicine in 2001 was won by yeast researchers Leland Hartwell and Paul Nurse (shared with Tim Hunt, studying sea urchins). Their work revealed key molecules of the eukaryotic cell cycle, relevant to concerns of human medicine such as cancer.

Some yeasts reproduce by **budding**, in which mitosis of the mother cell generates daughter cells of smaller size. The mother cell acquires a bud scar where the smaller one pinched off (**Fig. 20.11A**). After generating a limited number of buds, the mother cell senesces and dies. Thus, yeasts provide a unicellular model system for the process of aging.

> **THOUGHT QUESTION 20.1** Why would yeasts remain unicellular? What are the relative advantages and limitations of hyphae?

Some yeasts are asexual, whereas others can undergo sexual **alternation of generations** (**Fig. 20.11B**). This life cycle alternates between generation of a haploid population, with a single copy of each chromosome (n), and a diploid population, with a diploid chromosome number ($2n$). The haploid form develops gametes to fertilize each other, making a $2n$ zygote. After vegetative (nonsexual) divisions, the $2n$ form undergoes meiosis, regenerating the haploid form. (The process of meiosis is reviewed in Appendix 2.) Alternation of generations allows an organism to respond genetically to environmental change by reassorting its genes through meiosis, and by recombining them through fertilization. Gene reassortment and recombination provide new genotypes, some of which may increase survival in the changed environment.

In baker's yeast (*Saccharomyces cerevisiae*), haploid spores divide and proliferate by mitosis, forming a haploid mycelium. Under environmental signals such as starvation, mating factors induce the haploid cells to differentiate into gamete forms called "shmoos." Gametes of two different mating types fuse, and their nuclei combine to form a zygote. In the diploid generation, the zygote divides mitotically, generating a population of diploids that appear superficially similar to haploid cells. Under stress, particularly desiccation, the diploids

A. *Candida albicans*

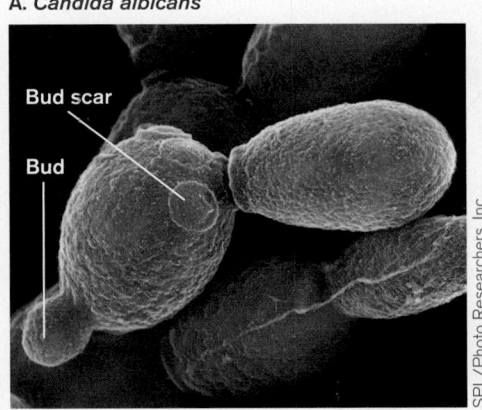

B. Yeast alternation of generations

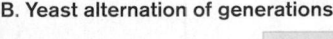
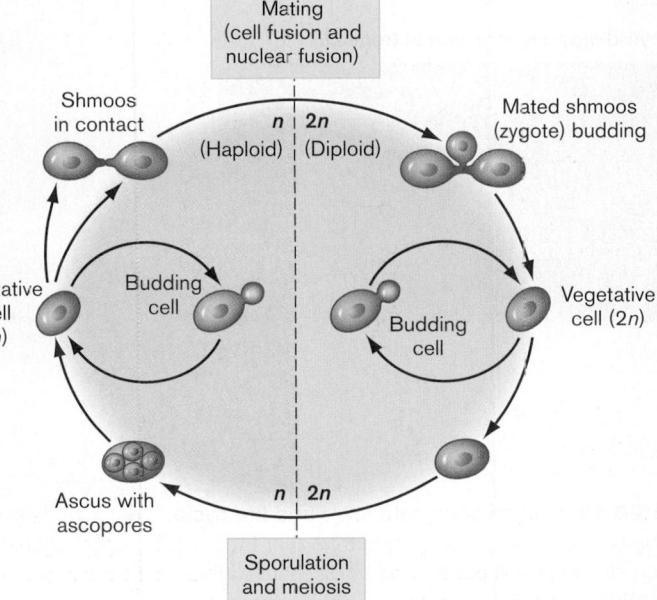

Figure 20.11 Yeasts are nonmycelial fungi.
A. *Candida albicans*, a common inhabitant of the human vaginal tract, reproduces by budding (SEM; cell size 3–6 μm). **B.** In the life cycle of baker's yeast (*Saccharomyces cerevisiae*), haploid cells reproduce many generations by budding.

undergo meiosis to reassort their genes for combinations that may better survive the changed environment. Meiosis generates an **ascus** that contains four haploid spores.

Many fungi and protists undergo modified versions of alternation of generations, utilizing a wide variety of haploid and diploid structures to accomplish essentially the same genetic tasks. In many fungi, the haploid form predominates; for example, ascomycetes such as *Aspergillus* and *Neurospora* form mainly haploid mycelia. In some fungi, no sexual reproduction has been observed, probably because the inducing conditions are unknown. Species that lack a known sexual cycle are called **mitosporic fungi**, also known as "imperfect" fungi. Mitosporic species are found in many different clades. An example is the famous *Penicillium* mold, an ascomycete, from which the antibiotic penicillin is derived.

THOUGHT QUESTION 20.2   Why would some fungi conduct asexual reproduction under most conditions? What are the advantages and limitations of sexual reproduction?

Mycelia, Mushrooms, and Mycorrhizae

Different species of fungi show vastly different forms, from the familiar mushrooms, fruiting bodies that can weigh several pounds, to the mycelia of pathogens and the symbiotic partners of algae in lichens. Major phyla of fungi include Chytridiomycota, Zygomycota, Ascomycota, and Basidiomycota.

NOTE: The major groups of fungi are also named with the alternative suffix *-etes*: Chytridiomycetes, Zygomycetes, Ascomycetes, Basidiomycetes.

Chytridiomycota: motile zoospores. The deepest-branching clade of fungi is Chytridiomycota (the chytrids), which share with animals and choanoflagellates the motile, flagellated reproductive form known as a zoospore. The zoospore form has been lost by other fungi.

Chytrid species include bovine rumen inhabitants whose hyphae penetrate tough plant material, facilitating digestion. An example is *Neocallomastix* species (**Fig. 20.12A**). Unlike most fungi, *Neocallomastix* is an

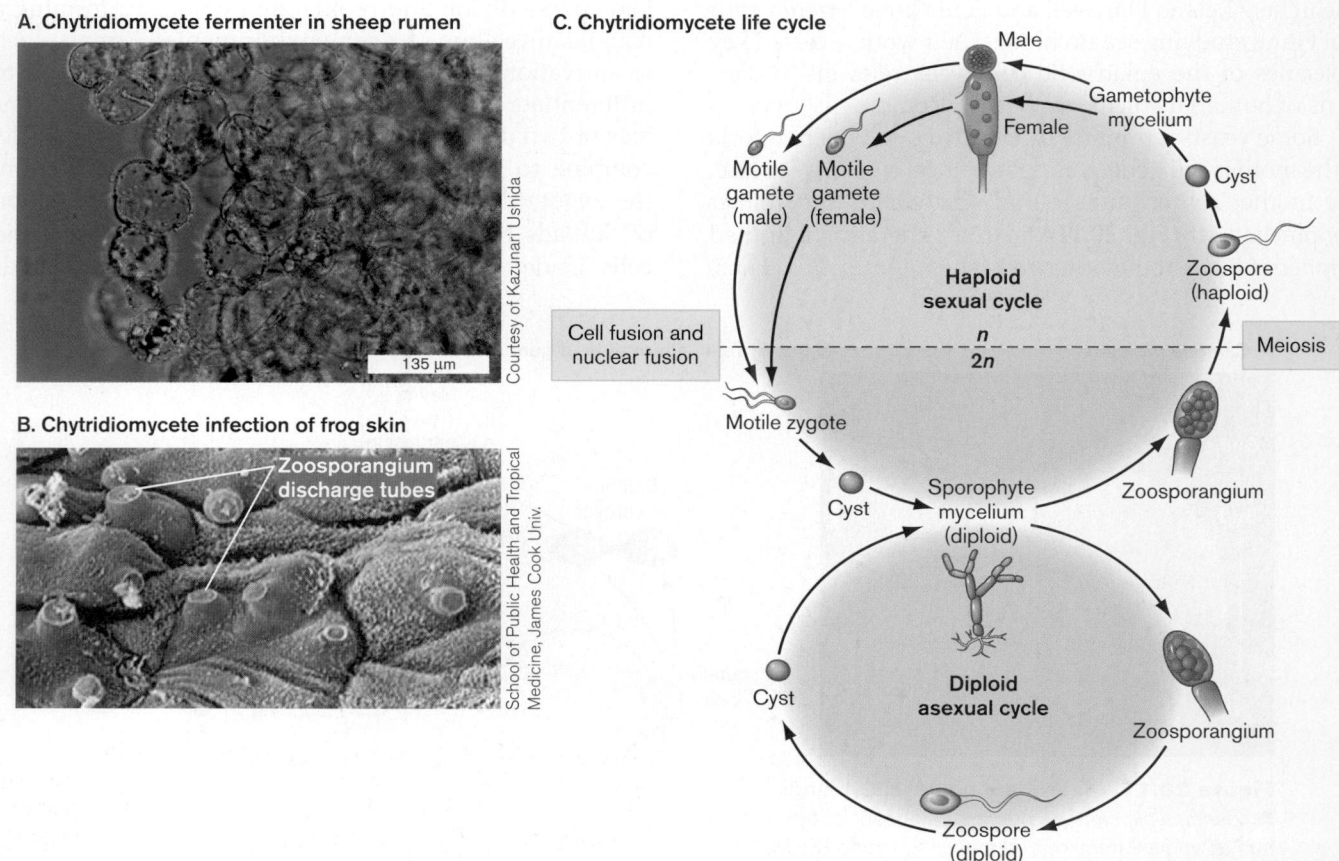

A. Chytridiomycete fermenter in sheep rumen

135 µm

Courtesy of Kazunari Ushida

B. Chytridiomycete infection of frog skin

Zoosporangium discharge tubes

School of Public Health and Tropical Medicine, James Cook Univ.

C. Chytridiomycete life cycle

Male
Female
Gametophyte mycelium
Motile gamete (male)
Motile gamete (female)
Cyst
Haploid sexual cycle
Zoospore (haploid)
Cell fusion and nuclear fusion
n
2n
Meiosis
Motile zygote
Cyst
Sporophyte mycelium (diploid)
Zoosporangium
Diploid asexual cycle
Cyst
Zoosporangium
Zoospore (diploid)

Figure 20.12   Chytridiomycete form and life cycle.   A. Chytrids such as *Neocallomastix* species ferment complex plant material within the bovine rumen, providing nutrition for the host. **B.** A pathogenic chytrid, *Batrachochytrium dendrobatidis*, infects the skin of a frog. Frog skin cells are penetrated by discharge tubes of diploid zoosporangia about to release zoospores. **C.** Life cycle of a chytrid. The diploid mycelium produces motile zoospores that form cysts in a poor environment. Alternatively, the diploid mycelium undergoes meiosis to form a haploid mycelium (gametophyte) that produces motile gametes.

obligate anaerobe whose mitochondria have evolved into **hydrogenosomes**, organelles that ferment carbohydrates in a pathway generating H_2. Hydrogenosomes are a unique adaptation of certain anaerobic fungi and protists.

Other chytrids are aerobic animal pathogens. **Figure 20.12B** shows the skin of a frog infected by the chytridiomycete *Batrachochytrium dendrobatidis*. *B. dendrobatidis* has caused a widespread die-off of frogs in Central and South America, in an epidemic associated with global warming. The mycelium of *B. dendrobatidis* grows within the frog skin, producing capsules full of diploid zoospores called zoosporangia. Each zoosporangium protrudes through the skin surface, ready to expel zoospores in search of a new host.

The life cycle of a chytrid includes both haploid (gametophyte) and diploid (sporophyte) mycelia (**Fig. 20.12C**). Haploid mycelia produce motile gametes that detect each other by sex-specific attractants. The gametes fuse to produce a motile zygote. The zygote forms a cyst, a cell with arrested metabolism that can persist for long periods. In a favorable environment, the cyst germinates to form a diploid mycelium, or sporophyte. The sporophyte generates zoosporangia full of zoospores. There are two alternative forms of zoosporangia: those that produce diploid zoospores, which form cysts and regenerate the diploid mycelium; and those that undergo meiosis to produce haploid zoospores. The haploid zoospores generate

a haploid mycelium (gametophyte) capable of producing haploid gametes.

Zygomycota: nonmotile sporangia. The **zygomycetes** and other nonchytridiomycete fungi generate nonmotile spores. Nonmotile spores require transport by air or water or ballistic expulsion (expulsion under pressure) from a spore-bearing organ, called the **sporangium**. A common zygomycete is the bread mold *Rhizopus* (**Fig. 20.13A**). Most zygomycetes, such as *Mucor* species, are soil molds that decompose plant material or other fungi or the droppings of animals (**Fig. 20.13B**). These modest molds fill important niches in all terrestrial ecosystems. Others, particularly *Glomus* species, form mutualistic associations with plant roots known as **vesicular-arbuscular mycorrhizae**. Mycorrhizae expand the roots' absorptive capacity, while obtaining plant sugars for the fungus (discussed in Chapter 21). Most trees and crop plants require mycorrhizae for optimal growth.

The life cycle of a zygomycete parallels that of a chytrid in its alternative options of haploid (n) and diploid ($2n$) forms (**Fig. 20.13C**). The mechanics differ, however, owing to the lack of motile gametes. The haploid spore (sporangiospore) is disseminated through air currents. A sporangiospore does not directly undergo sexual reproduction; it grows into a haploid mycelium. The haploid

A. Bread mold, *Rhizopus*

Gregory G. Dimijian

B. *Mucor* diploid hyphae form zygospores

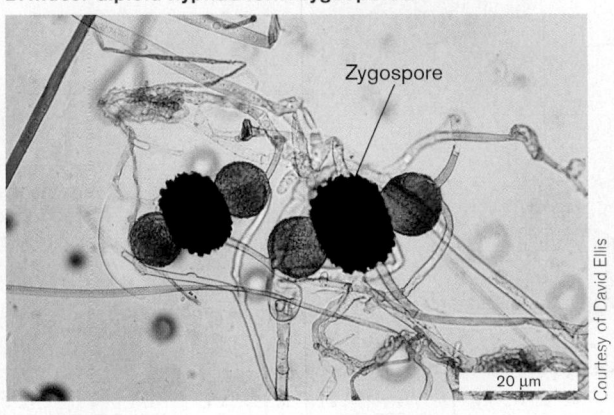

Zygospore

20 µm

Courtesy of David Ellis

C. Zygomycete life cycle

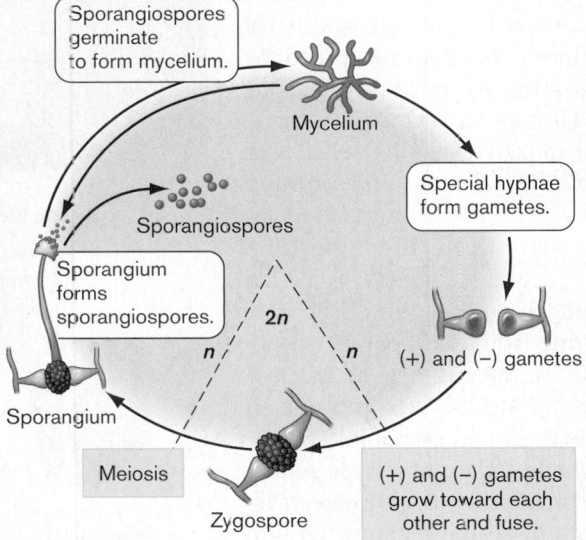

Figure 20.13   Zygomycete fungi form nonmotile sporangia. **A.** *Rhizopus* (bread mold) haploid sporangia contain sporangiospores. **B.** Diploid hyphae of *Mucor* species terminate in zygospores. **C.** The life cycle of zygomycetes involves primarily haploid mycelia. Special hyphae form gametes at their tips. Gametes of different mating types fuse to form the zygospore. The zygospore undergoes meiosis, regenerating haploid cells that form sporangia. The sporangia release sporangiospores, which germinate to form new mycelia.

mycelium then forms special hyphae whose tips differentiate into gamete cells. The gametes cannot separate from the filament; instead, two gamete-bearing hyphae must grow toward each other in order to fuse and form a **zygospore**. The zygospore undergoes meiosis and generates the sporangium, a haploid structure that releases **sporangiospores**.

THOUGHT QUESTION 20.3 What are the advantages and limitations of motile gametes, as compared to nonmotile spores?

Ascomycota: mycelia with paired nuclei. The **ascomycete** fungi are famous in the history of science, as well as in the culinary arts (**Fig. 20.14**). The bread mold *Neurospora* was used by George Beadle and Edward Tatum in the 1940s to formulate the one gene–one protein theory. In *Neurospora*, meiosis produces pods (asci) of **ascospores** aligned in rows that reflect the ordered tetrads of meiotic division (**Fig. 20.14B**). The tetrad patterns were used by geneticists to demonstrate the segregation and independent assortment of chromosomes. In other species, by contrast, the asci are packed in large mushroom-like fruiting bodies known as morels (*Morchella hortensis*) and truffles (*Tuber aestivum*). The ascospores of such fruiting bodies are spread by animals attracted by their delicious flavor. Human collectors traditionally use muzzled pigs to detect and unearth the famous underground truffles.

The ascomycete life cycle (**Fig. 20.14C**) includes a phase in which each cell possesses a pair of separate nuclei, one from each parent (chromosome number is designated $n + n$). The "dikaryotic" (paired-nuclei) phase is generated by haploid mycelia in which male and female reproductive structures fuse, followed by migration of all the male nuclei into

the female structure. The paired nuclei then undergo several rounds of mitotic division while migrating into the growing mycelium. In the mycelial tips, the paired nuclei finally fuse (becoming $2n$) and the mycelial tips develop into asci. Each ascus then undergoes meiosis in which the haploid products segregate in the same order that the meiotic chromosomes separated.

Some ascomycetes, such as *Aspergillus* and *Penicillium* species, form small asexual fruiting bodies called conidiophores for airborne spore dispersal (**Fig. 20.15**). *Penicillium* is known for producing penicillin, the first antibiotic in widespread use; forms of penicillin are still used today (discussed in Chapter 1). *Aspergillus* is a growing medical problem as an opportunistic pathogen of immunocompromised patients. *Aspergillus* can produce toxins (called mycotoxins) such as aflatoxin. Aflatoxin poisoning

A. Morel (an ascomycete fruiting body)   **B. Asci containing ascospores**

20 μm

Ed Reschke/Peter Arnold

C. Ascomycete life cycle

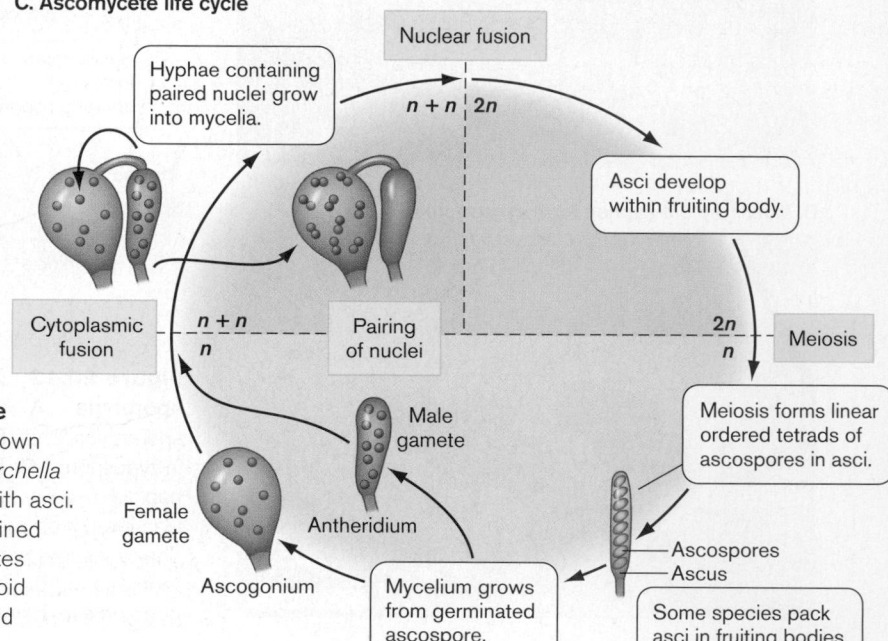

Figure 20.14 Ascomycetes produce large fruiting bodies.   A. The culinary delicacies known as morels are fruiting bodies of the species *Morchella hortensis*. The dark pits of the morel are lined with asci. **B.** Ascomycete asci containing ascospores (stained red). **C.** The life cycle of an ascomycete alternates between the diploid and haploid forms. The diploid mycelium produces asci, within which the haploid ascospores are formed.

A. **B.** **C.**

Figure 20.15 *Aspergillus.* **A.** *Aspergillus* forms a microscopic asexual fruiting structure called a conidiophore, containing spores in its spherical tip. **B.** Colony of *Aspergillus nidulans* on an agar plate. **C.** Mold grows above the flood line on a kitchen wall (shown by an undergraduate student volunteer) in New Orleans, 6 months after flooding caused by Hurricane Katrina.

commonly affects livestock, and in some cases agricultural workers; the toxin causes liver damage, immunosuppression, and cancer.

Conidiospore-forming ascomycetes are the major form of mold associated with dampness in human dwellings; for example, they caused massive damage to homes flooded in the wake of Hurricane Katrina (**Fig. 20.15C**). The flooding of homes full of drywall made ideal conditions for the growth of mold, which commonly consists of airborne ascomycete mycelia. Mold grew not only on materials submerged, but also on the surface above, exposed to water-saturated air, up to 3 feet above the flood line (the highest level submerged). Unfortunately, most home owner insurance policies covered damage only "up to the flood line."

Many ascomycetes are pathogens of animals or plants. For example, *Microsporum* and *Trichophyton* species cause ringworm skin infection, whereas *Magnaporthe oryzae* causes rice blast, the most serious disease of cultivated rice. Other species, however, are beneficial symbionts of plants, including crop plants such as beans, cucumbers, and cotton. *Trichoderma* species grow on the roots, or in some cases within the vascular tissue, of the plant. The fungi share nutrients with the plant, and they induce plant defenses against pathogens. *Trichoderma* even has a commercial use in cloth processing; the fungus is used to make "stonewashed jeans," as its cellulase enzymes partly digest the cotton.

Basidiomycota: cells with paired nuclei form mushrooms. The **basidiomycetes** form large, intricate fruiting bodies, known as "true" mushrooms (**Fig. 20.16**). Mushrooms produce some of the world's deadliest

A. *Amanita* **B.** *Aleuria*

Figure 20.16 **Mushrooms and other basidiomycetes form large, complex fruiting bodies.** **A.** *Amanita phalloides* makes one of the most dangerous toxins known: alpha-amanitin, an inhibitor of RNA polymerase II. **B.** *Aleuria* mushrooms releasing spores from their gills, to be carried on the wind.

A.

Peter von Dassow. 2009. *Nature* **459**:185–192.

Figure 20.26 Diatoms. A. Diatoms of various species. **B.** Life cycle of a centric diatom, *Thalassiosira* sp. Each vegetative cell division (lower right) requires formation of an in-fitting frustule half. Successive divisions result in progressive decrease in size (lower left). At a critical size limit, the diatom must undergo meiosis to form eggs and sperm (upper left). As gametes fuse, they form an auxospore (upper right), which regenerates a frustule of the original size.

Sperm exiting frustule

Peter von Dassow. 2009. *Nature* **459**:185–192.

5 µm

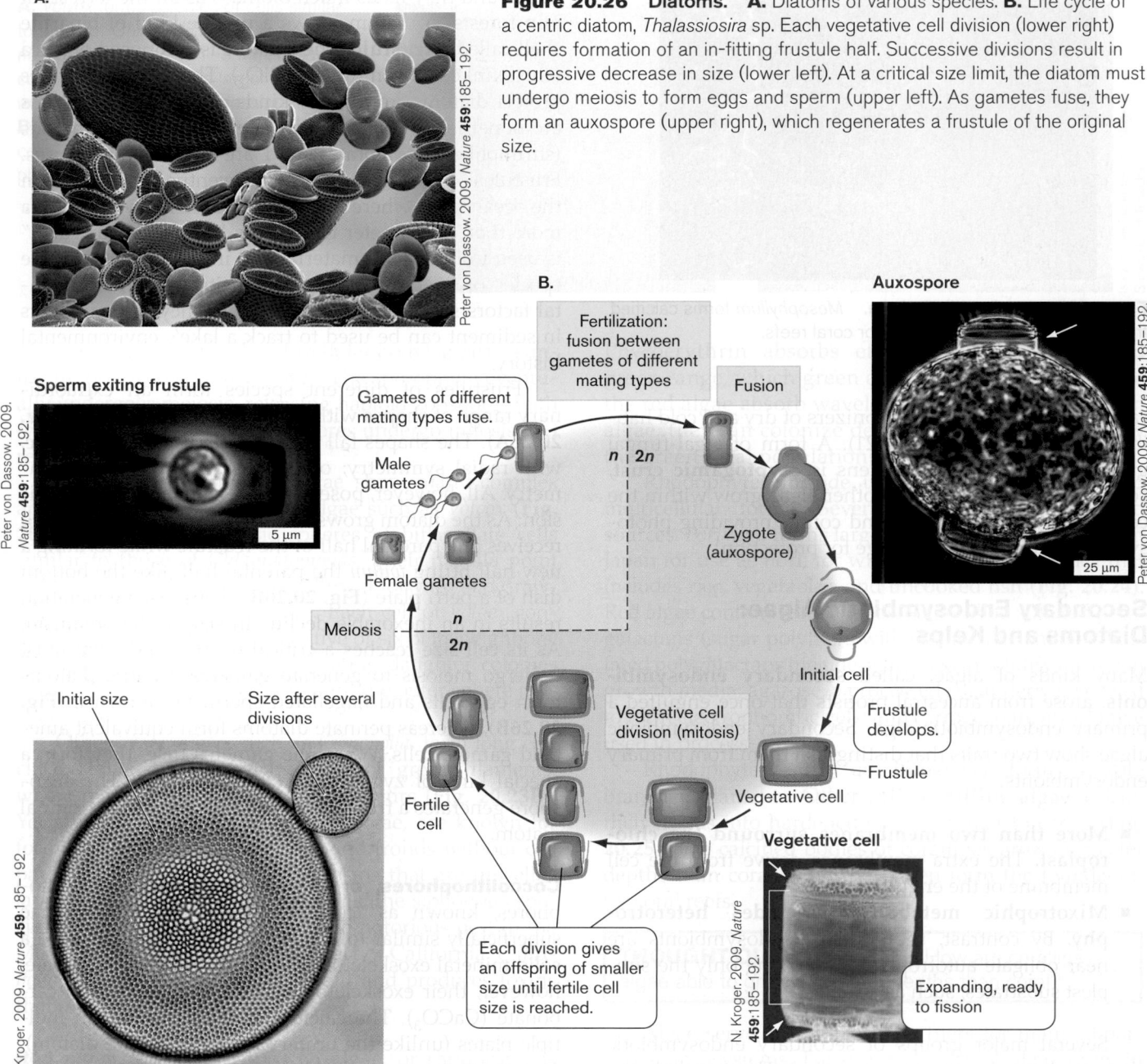

B.

Auxospore

Peter von Dassow. 2009. *Nature* **459**:185–192.

25 µm

Fertilization: fusion between gametes of different mating types

Gametes of different mating types fuse.

Fusion

Male gametes

Female gametes

Meiosis

n | 2*n*

n
2*n*

Zygote (auxospore)

Initial cell

Frustule develops.

Frustule

Vegetative cell division (mitosis)

Vegetative cell

Fertile cell

Each division gives an offspring of smaller size until fertile cell size is reached.

Vegetative cell

N. Kroger. 2009. *Nature* **459**:185–192.

Expanding, ready to fission

Initial size

Size after several divisions

N. Kroger. 2009. *Nature* **459**:185–192.

the surf rips the blades off and tosses them ashore. Kelps support important communities of multicellular organisms known as kelp forests. The most famous kelp forests are those of the unrooted **sargassum weeds**, which float upon the Sargasso Sea. Sargassum consists of stalks with photosynthetic blades and round gas bladders to keep the organism afloat (**Fig. 20.28**). Sargassum supports a complex food chain of invertebrate and vertebrate animals, including worms, crabs, and fish.

TO SUMMARIZE:

■ **Chlorophyta (green algae)** absorb red and blue light and grow near the top of the water column. Green algae include unicellular, filamentous, and sheet forms.

■ **Rhodophyta (red algae)** have the accessory photopigment phycoerythrin, which absorbs green and longer-wavelength blue light, enabling growth at greater depths. Red algae include species of diverse forms, many of which are edible for humans.

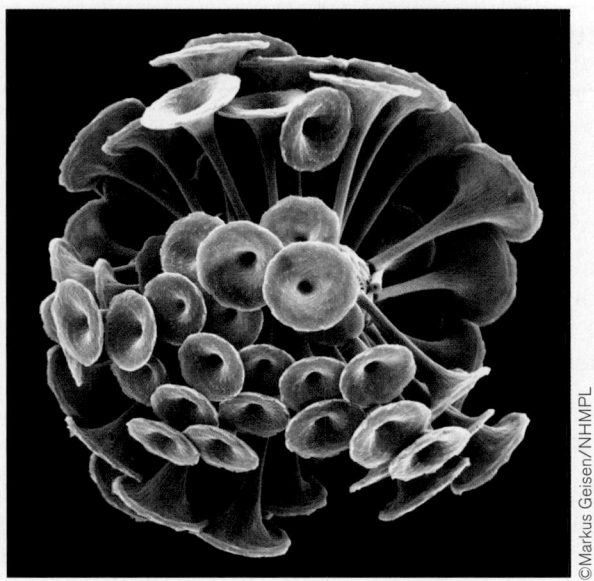

Figure 20.27   A coccolithophore, *Discosphaera tubifera.* The tiny cell is surrounded by a much larger volume of trumpet-shaped coccoliths. From the Alboran Sea, western Mediterranean.

©Markus Geisen/NHMPL

Sargassum weed

Figure 20.28   Kelp forests. *Sargassum natans* forms the basis of the Sargasso Sea. The brown alga forms stalks with leaflike blades and round gas bladders to keep the alga afloat.

Visuals Unlimited

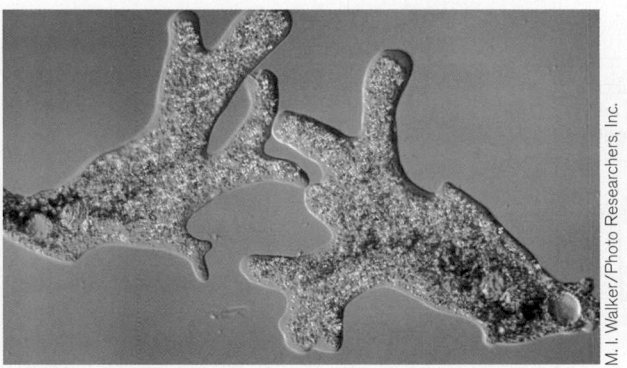

Figure 20.29   Amebas. *Amoeba proteus* moves by extending its pseudopods.

M. I. Walker/Photo Researchers, Inc.

- **Secondary endosymbiotic algae** are derived from protists that had engulfed primary symbiotic algae. They are mixotrophs, combining phototrophy and heterotrophy.
- **Diatoms are heterokonts with silicate shells called frustules.** Diatoms replicate by an unusual division cycle generating successively smaller frustules.
- **Kelps are heterokonts that grow in long sheetlike fronds.** Kelps play an important role in the ecology of the open ocean, as well as the ecology of marine beaches.

20.4 Amebas and Slime Molds

The ameba (alternative spelling, "amoeba") is familiar to most of us as an apparently amorphous form of microscopic life, capable of engulfing and consuming prey in a dramatic fashion (**Fig. 20.29**). While the ameba's shape is exceptionally variable, the pseudopods, or "false feet," that it extends, far from being amorphous, are complex structures that undertake highly controlled and specific movements. Most amebas are free-living predators in soil or water, engulfing prey by **phagocytosis**. They extend in size up to 5 mm, large enough to phagocytose bacteria, algae, ciliates, smaller amebas, and even invertebrates such as rotifers. A few are dangerous parasites of humans or animals. Furthermore, free-living amebas can harbor bacterial pathogens such as *Legionella pneumophila*, which contaminates water supplies and air ducts. The bacteria cause legionellosis, an often fatal form of pneumonia. The host ameba enables the pathogen's persistence and transmission to human hosts.

Free-living amebas such as *Acanthamoeba* species are common predators in the soil microbial community. They cause problems when they contaminate contact lens cleaning solutions, causing keratitis (infection of the cornea). Wearers of reusable contact lenses have a keratitis infection rate of two per thousand.

> **THOUGHT QUESTION 20.6**   What might happen when an ameba phagocytoses algae?

> **NOTE:** The taxonomy of amebas and slime molds remains problematic, with diverse views as to the number of clades, their relatedness, and their degree of divergence.

Pseudopod Motility

Ameba-like cell forms occur in many different species. Some species persist as an ameba throughout all or most of the life cycle. Others can convert to flagellates, particularly when the habitat fills with water, a low-viscosity condition favoring flagellar motility. Still other protist species, such as dinoflagellates, never become fully amebic, but they can extend a pseudopod to engulf prey.

Different kinds of amebas have different kinds of pseudopods. The "classic" ameba (phylum Amoebozoa) has lobe-shaped pseudopods (**Fig. 20.29**). Lobe-shaped pseudopods have the most variable shape. A different form is the sheetlike pseudopod, or **lamellar pseudopod**. Lamellar pseudopods are extended by dinoflagellates. Similar lamellar pseudopods are generated by human white blood cells such as leukocytes. Finally, needle-like pseudopods, or filopodia, are thin extensions reinforced by microtubules. Filopodia are made by the shelled amebas (phylum Rhizaria), such as Foraminifera (spiral shells) and Radiolaria (radial form).

The extension of lobe-shaped and lamellar pseudopods has been studied closely for its relevance to human white blood cells (for more on white blood cells as host defenses, see Chapter 23). The mechanism of pseudopod motility remains poorly understood, but it is known to involve a sol-gel transition between cortical cytoplasm (just beneath the cell surface) and the cytoplasm of the deeper interior (**Fig. 20.30**). The tip of a pseudopod contains a gel of polymerized actin beneath its cell membrane. From the center of the ameba, liquid cytoplasm (sol) containing actin subunits streams forward along microtubular "tracks," powered by ATP hydrolysis. The actin subunits stream into the pseudopod, where they polymerize, forming a gel. The gel region grows, pushing the membrane forward and extending the pseudopod. As the gel is pushed backward, it resolubilizes to continue the cycle.

Amebas can be uninucleate or multinucleate. They are usually haploid and reproduce asexually by nuclear mitosis, without dissolution of the nuclear membrane, followed by fission of the cytoplasm. Some species do have reproductive alternatives, such as cyst formation, gamete fusion and meiosis, and even growth of flagella in a favorable habitat.

> **THOUGHT QUESTION 20.7**   What kind of habitat would favor a flagellated ameba?

Ameba genetics is poorly understood, but at least one ameba genome has been sequenced—that of the intestinal parasite *Entamoeba histolytica*. The sequence contains 20 Mb of DNA in 14 chromosomes; some of these are linear, whereas others are circular. Closely related strains show considerable variation in organization, suggesting that ameba genomes (like their cytoplasm) undergo extensive rearrangement.

Slime Molds

Some amebas conduct a life cycle in which thousands of individuals (all members of one species) aggregate into a complex differentiated fruiting body (**Fig. 20.31A**). Such an organism is called a **slime mold**. Slime molds, as the name implies, were originally classified with fungi because their fruiting bodies superficially resemble fungal reproductive forms. There are two kinds of slime molds: **cellular slime molds**, in which the individual cells remain cellular; and **plasmodial slime molds**, in which the mass of aggregating cells becomes a multinucleate single cell.

A well-studied example of a cellular slime mold is *Dictyostelium discoideum*, historically an important model system for multicellular development (**Fig. 20.31**). *D. discoideum* amebas are relatively small, about 10 μm, but large enough to consume bacteria. They can be cocultured on a plate with *Escherichia coli*. As the haploid amebas consume bacteria, they divide asexually until their food runs out. At this point, a few amebas begin to emit the aggregation signal molecule cyclic AMP (cAMP). An ameba emitting cyclic AMP attracts other amebas nearby, which move toward the center and begin emitting cyclic AMP as well. Successive waves of cyclic AMP continue to attract thousands of amebas to the center, where they pile

Ameba cross section

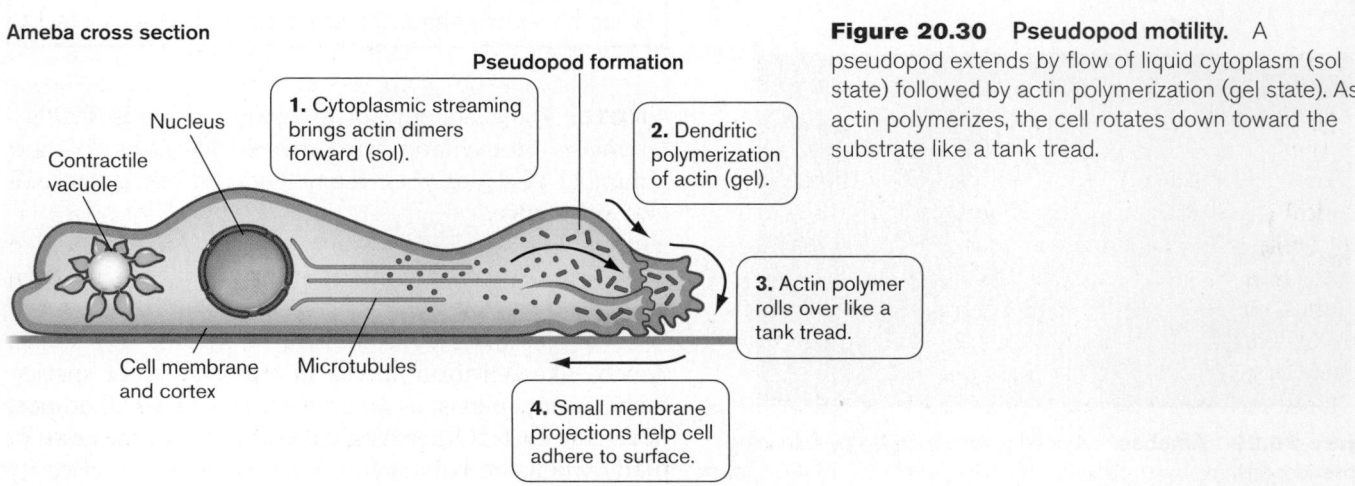

Pseudopod formation

Nucleus

Contractile vacuole

1. Cytoplasmic streaming brings actin dimers forward (sol).

2. Dendritic polymerization of actin (gel).

3. Actin polymer rolls over like a tank tread.

Cell membrane and cortex

Microtubules

4. Small membrane projections help cell adhere to surface.

Figure 20.30   Pseudopod motility.   A pseudopod extends by flow of liquid cytoplasm (sol state) followed by actin polymerization (gel state). As actin polymerizes, the cell rotates down toward the substrate like a tank tread.

A.

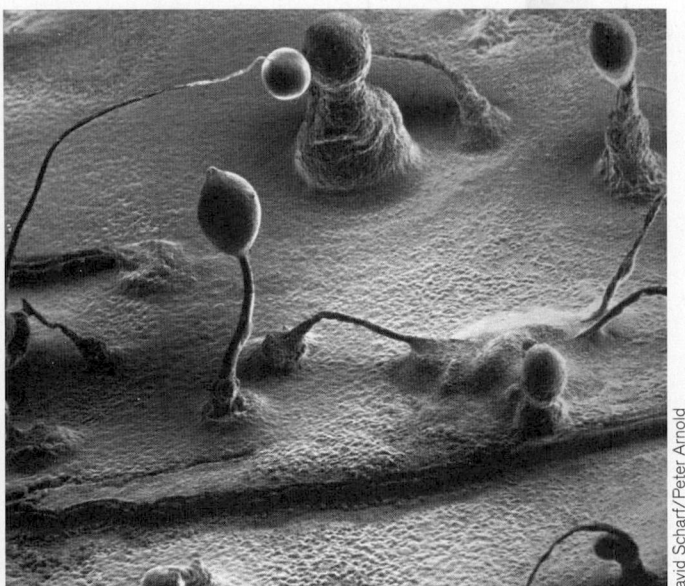

David Scharf/Peter Arnold

Figure 20.31   A cellular slime mold: *Dictyostelium discoideum*. A. Fruiting bodies of *D. discoideum* (composite light micrograph). **B.** Life cycle of *D. discoideum*.

B.

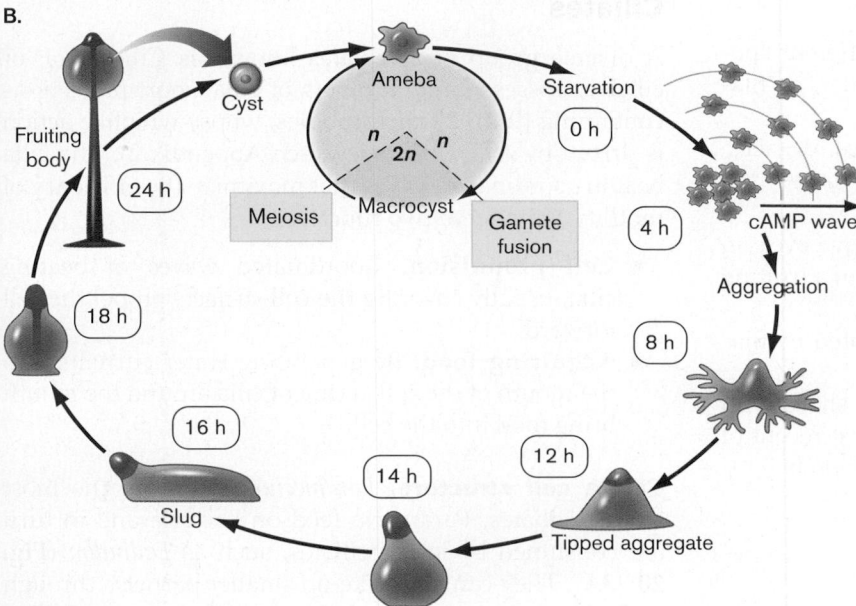

have a sexual alternative in which cells of opposite mating type can fuse to form a diploid zygote and then undergo meiosis, restoring haploid amebas.

By contrast, the amebas of a plasmodial slime mold, such as *Physarum polycephalum*, aggregate only following gamete fusion to form diploid cells. The diploids then aggregate and become transformed into **plasmodium**, a giant multinucleate cell that can spread over an area of many square centimeters. Out of the plasmodium arise fruiting bodies whose sporangia undergo meiosis, producing haploid spores. A large plasmodium can occasionally be seen as a yellow mass of slime spreading over decaying wood.

NOTE: Distinguish the term "plasmodium" (a large multinucleate cell) from the genus *Plasmodium* (an apicomplexan parasite, such as *Plasmodium falciparum*, which causes malaria).

Filamentous and Shelled Amebas

A distinct clade of amebas (the Rhizaria) form needlelike pseudopods. Most of these amebas are encased by mineral shells called tests (**Fig. 20.32A**). A major group of shelled amebas is the **radiolarians**, whose shells are made of silica perforated with numerous holes through which pseudopods appear to radiate in all directions. Many different kinds of radiolarians exist today, and many can be recognized in fossil rock. A second group of shelled amebas is the **foraminiferans** (**Fig. 20.32B**). The foraminiferans, or forams, generate shells of calcium carbonate as chambers laid down in helical succession. Their pseudopods all extend from one opening in the most recent chamber.

Radiolarians and forams grow in marine and aquatic habitats. Their shells make up a large part of reef formations, sedimentary rock, and beach sand. Forams, in particular, are used in geological surveys as indicators of petroleum deposits.

on top of each other to form a slug as long as 1 mm. The slug then migrates, attracted by light and warmth, to find an appropriate place to form a fruiting body and disperse its spores.

The slug at last differentiates into a fruiting body, a spherical sporangium supported upon a stalk of largely empty cells that emerges from a basal disk. The sporangium then releases spores (also called cysts), which are dispersed on air currents and can remain viable for several years. When a spore detects chemical signals from bacteria, it germinates an ameba to feed on them.

Note that the entire reproductive cycle just described is asexual; the amebas and their differentiated structures remain haploid throughout. *D. discoideum* amebas do

A. Radiolarian tests

0.25 mm

Dennis Kunkel Microscopy, Inc.

B. A foraminiferan

C. Hemleben/©Cushman Foundation for Foraminiferal Research, Inc.

Figure 20.32 **Filamentous and shelled amebas.**
A. Shells (tests) of radiolarians (colorized SEM). **B.** A live foraminiferan, *Globigerinella aequilateralis* (dark-field).

TO SUMMARIZE:

■ **Amebas** move using pseudopods. In different species, pseudopods are lobe-shaped, lamellar, or filamentous (filopodia).

■ **Cytoplasmic streaming** through cycles of actin polymerization and depolymerization drives the extension and retraction of pseudopods.

■ **Slime molds** show an asexual reproductive cycle in which individual amebas aggregate to form a fruiting body that produces spores.

■ **Radiolarians** have silicate shells penetrated by filamentous pseudopods.

■ **Foraminiferans** have calcium carbonate shells with helical arrangement of chambers, the most recent of which opens to extend filamentous pseudopods.

20.5 Alveolates: Ciliates, Dinoflagellates, and Apicomplexans

The Alveolata include voracious predators such as the ciliated protists (**Fig. 20.33A**). Alveolates are named for the flattened vacuoles called alveoli within their outer cortex (**Fig. 20.33B**; see also **Fig. 20.7A**). Some alveoli contain plates of stiff material, such as protein, polysaccharide, or minerals. Besides alveoli, most alveolate protists possess other kinds of cortical organelles, such as extrusomes for delivery of enzymes or toxins, bands of microtubules for reinforcement, and whiplike cilia or flagella. The alveolate cell form is highly structured, in contrast to the amorphous shape of amebas. Major groups of alveolates include ciliates, dinoflagellates, and apicomplexans.

Ciliates

A diverse group of alveolates known as Ciliophora, or ciliates, possess large numbers of cilia, short projections containing [9(2)+2] microtubules, whose whiplike action is driven by ATP (for review, see Appendix 2). The cilia beat in coordinated waves that maximize the efficiency of motility. Cilia serve two functions:

■ **Cell propulsion.** Coordinated waves of beating cilia, usually covering the cell surface, propel the cell forward.

■ **Acquiring food.** By generating water currents into the mouth of the cell, a ring of cilia around the mouth bring food into the cell.

Ciliate cell structure. *Paramecium* is one of the most studied ciliates. Paramecia feed on bacteria and in turn are consumed by larger ciliates, such as *Didinium* (**Fig. 20.33A**). They can also take up smaller particles through endocytosis by specialized pores in their cortex, called parasomal sacs (**Fig. 20.33B**).

The cell structure of *Paramecium* includes an **oral groove** for uptake of food driven by the beating cilia (**Fig. 20.33C**). Once ingested through the oral groove, a food particle travels within the digestive vacuole in a circuit around the cell. The digestive vacuole ultimately empties into the cytoproct, a specialized vacuole for the discharge of waste outside the cell. Paramecia maintain osmotic balance by means of a **contractile vacuole**, a vacuole that withdraws water from the cytoplasm to shrink it or contracts to expand it. Contractile vacuoles are widespread among protists and algae, but their mode of action has been most studied in paramecia.

Genetics and reproduction. Most ciliates have a complex genetic system involving one or more **micronuclei**

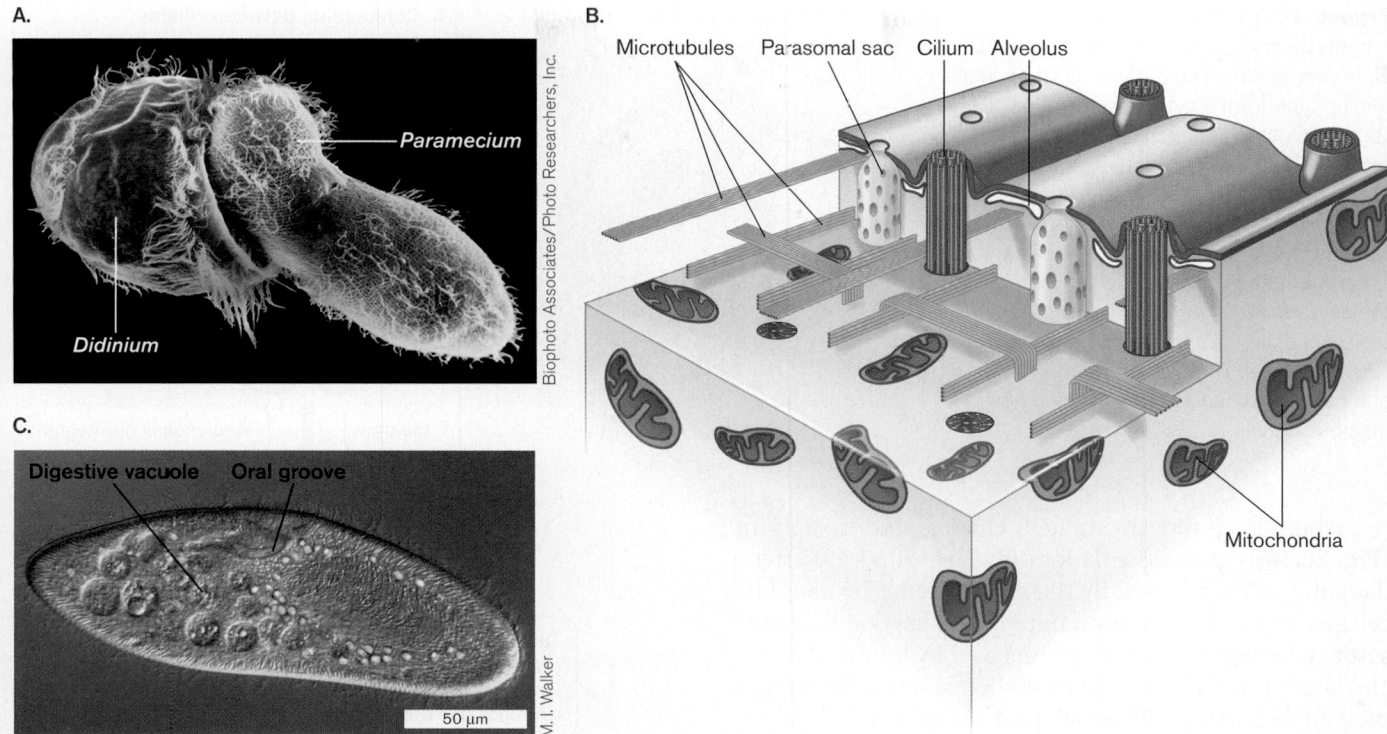

Figure 20.33   Ciliated protists.   A. *Didinium* consuming *Paramecium* (SEM). **B.** Cortical structure of a ciliate. Beneath the outer membrane lie flattened sacs of fluid called alveoli. The cilia, composed of [9(2)+2] microtubules, are rooted in a complex network of lateral microtubules. Parasomal sacs take up nutrients and form endocytic vesicles. **C.** A paramecium has digestive vacuoles and an oral groove for ingestion (colorized phase contrast).

and **macronuclei**. The micronucleus contains a diploid set of chromosomes that undergoes meiosis for sexual exchange (a process reviewed in Appendix 2). For gene expression, however, the micronucleus generates hundreds of copies of its DNA within a macronucleus. The DNA copies in the macronucleus are rearranged and fragmented to small segments. The small DNA segments generate a large number of "telomeres," chromosome ends. In humans, telomere maintenance is associated with aging; thus, ciliate macronucleus formation provides a model system for study of human aging (**eTopic 20.2**).

In ciliates, only macronuclear genes are transcribed to RNA and translated to protein. When a ciliate reproduces asexually, the micronucleus undergoes mitosis, whereas the macronucleus divides by a different mechanism that is poorly understood. Cell division occurs across the long axis, necessitating generation of a new oral groove for the posterior daughter cell and a new cytoproct for the anterior daughter cell—again, a process poorly understood.

Most ciliates are diploid and never produce haploid gamete cells. Instead, they reproduce sexually by exchanging haploid micronuclei. Micronuclei are exchanged through **conjugation**, a process in which two cells of opposite mating type join by a cytoplasmic bridge and exchange nuclear products of meiosis (**Fig. 20.34**). During fusion, the micronucleus of each cell undergoes

meiosis to form four haploid nuclei. Three out of four of the haploid nuclei disintegrate, as does the entire macronucleus. The haploid micronuclei then undergo mitosis, and one of each daughter nuclei is exchanged across the cytoplasmic bridge. Each transferred nucleus then fuses with its haploid counterpart, restoring diploidy. The two cells come apart, and each recombined micronucleus generates a new macronucleus.

THOUGHT QUESTION 20.8   Compare and contrast the process of conjugation in ciliates and bacteria (see Chapter 9).

THOUGHT QUESTION 20.9   For ciliates, what are the advantages and limitations of conjugation, as compared with gamete production?

Stalked ciliates. Some ciliates adhere to a substrate and use their cilia primarily to obtain prey. **Stalked ciliates** such as *Stentor* and *Vorticella* have a ring of cilia surrounding a large mouth (**Fig. 20.35A**). The ciliary beat is specialized to draw large currents of water and whatever prey it carries. Stalked ciliates are commonly found in pond sediment and in wastewater during biological treatment by microbial digestion, where they are attached to flocs of filamentous bacteria.

Figure 20.34 Conjugation. A. Two paramecia conjugating (light micrograph). **B.** In conjugation, two paramecia of opposite mating type form a cytoplasmic bridge. The 2n micronucleus of each cell undergoes meiosis. Each macronucleus, as well as three out of four meiotic products, disintegrates. The haploid micronuclei undergo mitosis, forming two daughter micronuclei. Daughter nuclei from each cell are exchanged across the cytoplasmic bridge and then fuse with their respective counterparts, restoring 2n micronuclei. The cells come apart, and each micronucleus generates a new macronucleus.

A. Conjugating paramecia

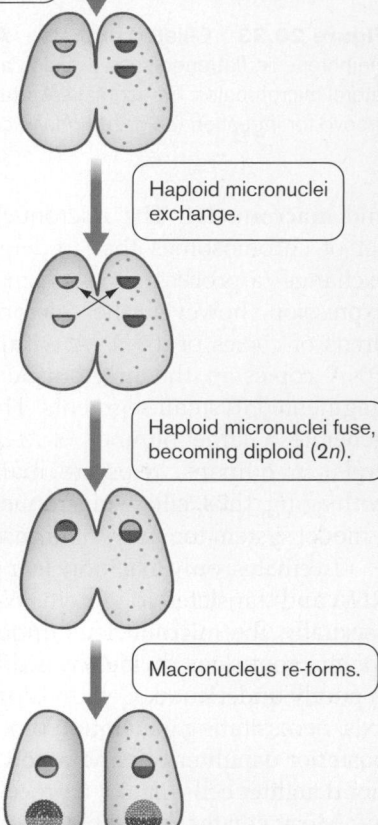

©Michael Abbey/Visuals Unlimited

B. Conjugation between ciliates

Micronucleus (2n)
Macronucleus

Conjugation

Meiosis   Macronuclei disintegrate.

Three of four meiotic products disintegrate, leaving one micronucleus (n).

Haploid micronuclei replicate and divide (mitosis).

Haploid micronuclei exchange.

Haploid micronuclei fuse, becoming diploid (2n).

Macronucleus re-forms.

Another group of stalked ciliates, the suctorians (**Fig. 20.35B**), possess cilia for only a short period after a daughter cell is released by the stalked cell. The daughter cell swims by ciliary motion until it finds a good habitat to settle, whereupon its cilia are replaced by knobbed tentacles similar to the filopodia of shelled amebas. Suctorians prey on swimming ciliates such as paramecia.

Dinoflagellates Are Phototrophs and Predators

The dinoflagellates (class Dinophyceae) are a major group of marine phytoplankton, essential to marine food chains. Like ciliates, they are highly motile, but instead of numerous short cilia, dinoflagellates possess just two long

A. *Stentor* (stalked ciliate)

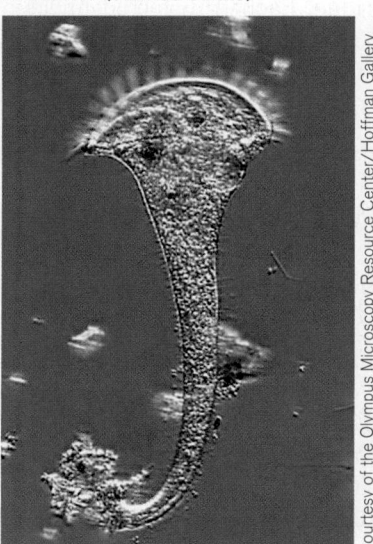

Courtesy of the Olympus Microscopy Resource Center/Hoffman Gallery

B. *Acineta* (suctorian)

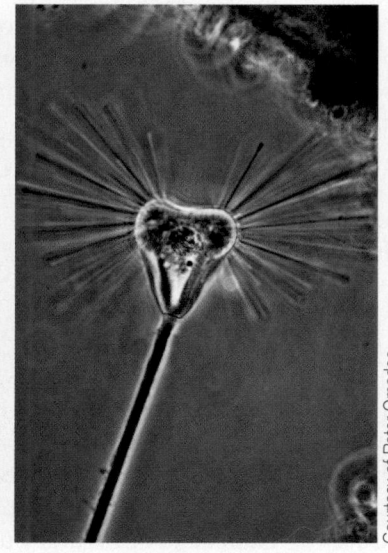

Courtesy of Peter Omodeo

Figure 20.35 Stalked ciliates. A. *Stentor*, a ciliate with a flexible stalk, 1.5–2.0 mm in length (phase contrast). The oral ring of cilia generates currents drawing food into the mouth. **B.** The suctorian *Acineta* replaces cilia with knobbed tentacles (light micrograph).

A. *Gymnodinium* (dinoflagellate)

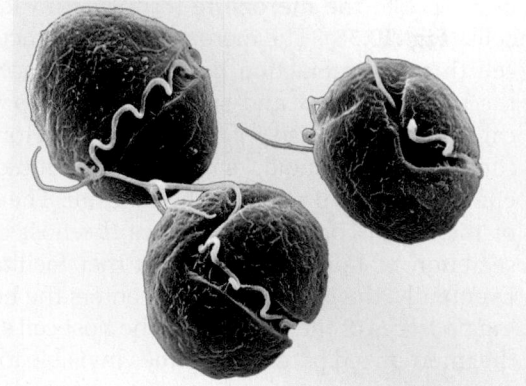

David M. Phillips/Visuals Unlimited

B. Red tide

Bill Bachman/Photo Researchers, Inc.

C. Dinoflagellate cell structure

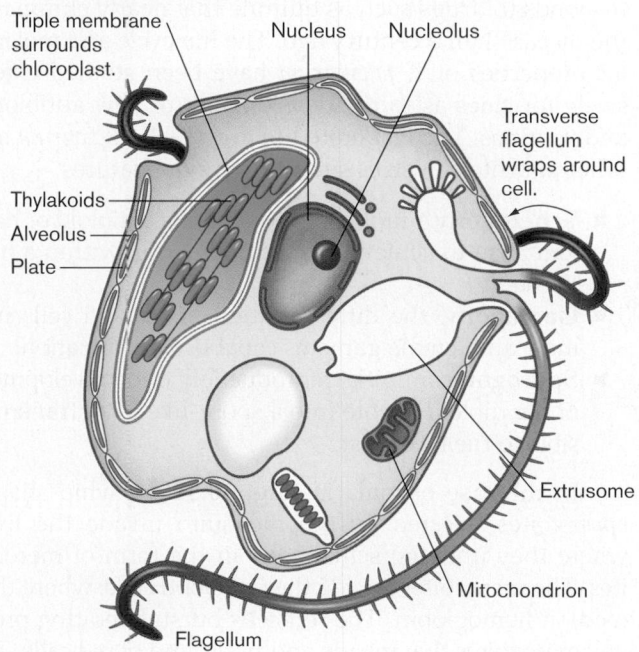

Triple membrane surrounds chloroplast.

Nucleus

Nucleolus

Thylakoids

Alveolus

Plate

Transverse flagellum wraps around cell.

Extrusome

Mitochondrion

Flagellum

Figure 20.36   Dinoflagellates.   A. *Gymnodinium* sp., a dinoflagellate, with flagellum wrapped around the cell (colorized SEM). **B.** "Red tide," caused by bloom of dinoflagellates. **C.** Diagram of a dinoflagellate. Protective plates of protein or calcified polysaccharides are formed within cortical alveoli. The chloroplast is surrounded by a triple membrane.

flagella, one of which wraps along a crevice encircling the cell (**Fig. 20.36A**). Dinoflagellates are secondary or tertiary algal endosymbionts. They have a chloroplast derived from a red alga, which in some species was later replaced by a heterokont alga, itself a secondary endosymbiont (**Fig. 20.36C**). Some dinoflagellates possess carotenoid pigments that confer a red color. Blooms of red dinoflagellates cause the famous red tide, which may have inspired the biblical story of the plague in which water turns to blood (**Fig. 20.36B**). Dinoflagellates release toxins that can be absorbed by shellfish, poisoning consumers months or years later.

The armor-plated appearance of a dinoflagellate results from its stiff alveolar plates, composed of proteins or calcified polysaccharides (**Fig. 20.36C**). The complex outer cortex includes various extrusomes (organelles that extrude a defensive substance) and endocytic pores, as well as a species-specific pattern of alveolar plates. Dinoflagellates

supplement their photosynthesis by predation, extending a lamellar pseudopod to engulf prey. Some dinoflagellates have evolved to lose their chloroplasts altogether, becoming obligate predators or parasites.

Some dinoflagellates inhabit other organisms as endosymbionts, providing sugars from photosynthesis in exchange for a protected habitat. Their hosts include shelled amebas, sponges, sea anemones, and, most important, reef-building corals. Coral endosymbionts, known as **zooxanthellae**, are vital to reef growth. The zooxanthellae are temperature sensitive, and their health is endangered by global warming. Rising temperatures in the ocean lead to coral bleaching (the expulsion of zooxanthellae), after which the coral dies.

Apicomplexans Are Specialized Parasites

Apicomplexans form a major group of parasites of humans and other animals. They are characterized by the presence of the **apical complex**, a highly specialized structure that facilitates entry of the parasite into a host cell. An example of an apicomplexan is *Toxoplasma gondii*, a parasite commonly carried by cats and transmissible to humans, where it can harm a developing fetus. Like the ciliates and dinoflagellates, apicomplexans possess an elaborate cortex composed of alveoli, pores, and microtubules. But as parasites, apicomplexans have undergone extensive reductive evolution, losing their flagella or cilia. They possess a unique organelle called the apicoplast, derived by genetic reduction from an endosymbiotic chloroplast. No capacity for photosynthesis remains, but the apicoplast provides one essential function in fatty acid metabolism.

The best-known apicomplexan is *Plasmodium falciparum*, the main causative agent of **malaria**, the most important parasitic disease of humans worldwide (**Fig. 20.37**). *P. falciparum* is carried by mosquitoes, which transmit the parasite to humans when the insect's proboscis penetrates the skin. The disease is endemic in areas inhabited by 40% of the world's population; it infects hundreds of millions of people and kills more than a million African children each year.

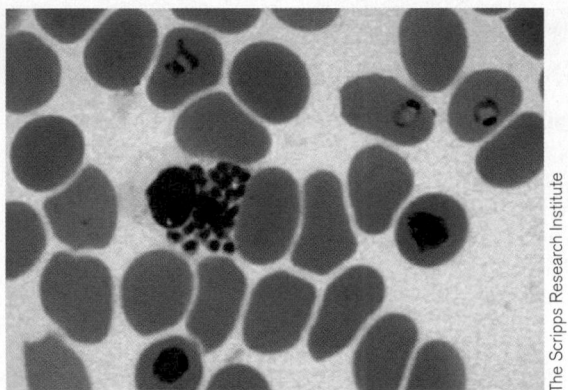

Figure 20.37 *Plasmodium falciparum*, **a cause of malaria.** Red blood cells infected with *P. falciparum*, which is stained purple with a dye that interacts with DNA (light micrograph). One red blood cell can be seen bursting, unleashing parasites on surrounding cells.

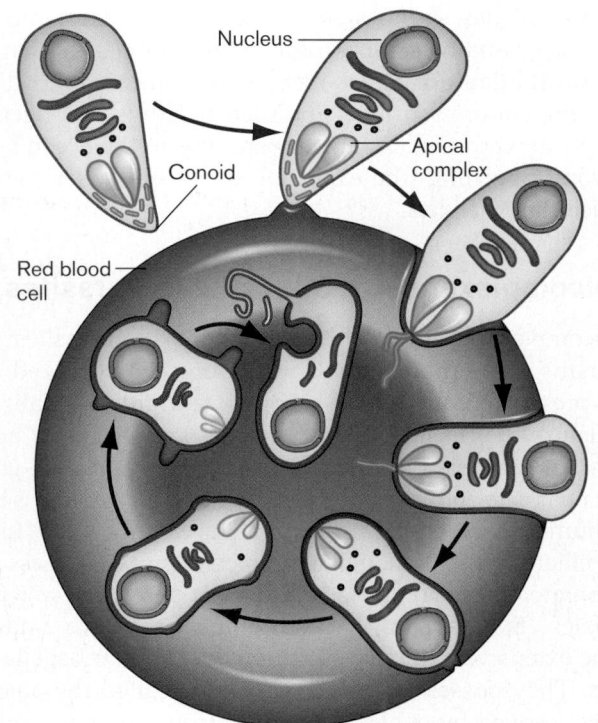

Figure 20.38 Merozoite form of *P. falciparum* invades a red blood cell. The apical complex facilitates invasion and then dissolves as the merozoite transforms into an intracellular form.

The transmitted parasites invade the liver and then develop into the **merozoite** form that invades red blood cells (**Fig. 20.38**). The merozoite first contacts a red blood cell through interaction between its surface proteins on the apical complex and specific host receptors. The apical complex consists of a pair of secretory organelles capped by the "conoid," a ring of microtubules that injects enzymes that aid entry of the parasite. The cone-like tip of the apical complex penetrates the host cell, enabling secretion of lipids and proteins that facilitate invasion. Eventually, the entire merozoite enters the host cell, leaving no traces of the parasite on the host cell surface. Thus, the internalized parasite becomes invisible to the immune system until its progeny burst out.

P. falciparum acquires resistance rapidly and no longer responds to drugs such as quinine that nearly eliminated the disease half a century ago. The life cycle and molecular properties of *P. falciparum* have been studied extensively for clues as to the development of new antibiotics and vaccines. The elaborate life cycle of *P. falciparum* and other parasites involves several common features:

■ **Schizogony**, mitotic reproduction of a diploid or haploid form to achieve a large population within a host tissue.

■ **Gamogony**, the differentiation of haploid cells into male and female gametes capable of fertilization.

■ **Sporogony**, mitotic reproduction and development of the diploid zygote into a spore-like form transmissible to the next host.

In the case of malaria (**Fig. 20.39**●), whip-shaped sporozoites injected by the mosquito invade the liver, where they undergo schizogony in the form of merozoites. The merozoites invade the red blood cells, where they feed on hemoglobin. The red cells burst, liberating progeny merozoites that invade another round of red cells. The bursting of red blood cells also releases cell fragments that trigger the cyclic fevers characteristic of malaria.

Some of the merozoites in the bloodstream undergo meiosis, generating pre-gamete cells, or gamonts. The gamonts multiply and mature (gamogony) and then they are acquired by bloodsucking mosquitoes. In the mosquito's midgut, the gamonts develop into female eggs and flagella-like male cells. The male cells fertilize the egg cells, and the resulting zygotes undergo mitosis (sporogony) and differentiate into sporozoites, which enter the salivary gland for transmission to the next human host.

The nuclear genome of *P. falciparum* consists of 24 Mb contained in 14 chromosomes. Sequence annotation and expression studies predict 5,300 protein-encoding ORFs, comparable to the number in a yeast genome. The parasite has lost many genes encoding enzymes and transporters while expanding its repertoire of proteins involved in antigenic diversity. In addition, the parasite contains two

Figure 20.39 Malaria: cycle of *Plasmodium falciparum* transmission between mosquito and human.

▶Ⅱ *microbiology2.com/animations*

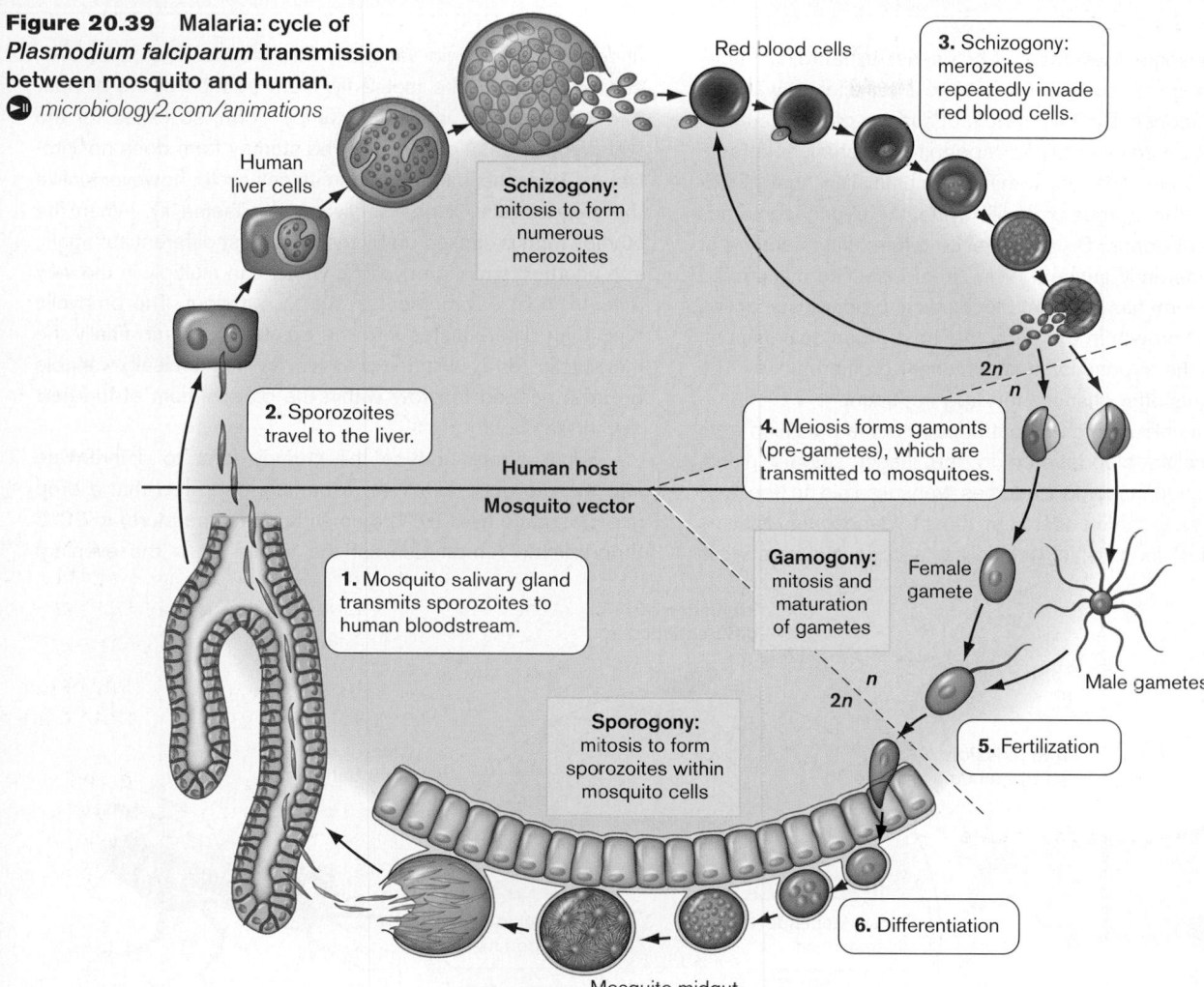

Red blood cells

3. Schizogony: merozoites repeatedly invade red blood cells.

Human liver cells

Schizogony: mitosis to form numerous merozoites

2. Sporozoites travel to the liver.

Human host

Mosquito vector

4. Meiosis forms gamonts (pre-gametes), which are transmitted to mosquitoes.

2n
n

1. Mosquito salivary gland transmits sporozoites to human bloodstream.

Gamogony: mitosis and maturation of gametes

Female gamete

Male gametes

n
2n

Sporogony: mitosis to form sporozoites within mosquito cells

5. Fertilization

6. Differentiation

Mosquito midgut

smaller nonnuclear genomes: that of its mitochondria and that of the chloroplast-derived apicoplast.

Unique metabolic and transport components emerge from its genome that offer promising targets for drug design. For example, the fatty acid biosynthesis occurring within the apicoplast is targeted by triclosan and other antibiotics. Other promising targets for antimalarial drugs are the unique proteases required to digest hemoglobin within the *P. falciparum* food vacuole.

TO SUMMARIZE:

- **Ciliates are covered with numerous cilia.** Cilia provide motility and help capture prey. Ciliates undergo complex reproductive cycles involving exchange of micronuclei through conjugation.
- **Dinoflagellates are phototrophic predators.** Dinoflagellates are tertiary endosymbiotic algae. Their alveoli contain calcified plates; they have paired flagella, one of which is used for propulsion. Predation occurs by extension of a lamellar pseudopod.

- **Apicomplexans are parasites that penetrate host cells.** The apicoplast is a specialized organ for cell penetration. Apicomplexans such as *Plasmodium falciparum* conduct complex life cycles within mammalian and arthropod hosts.

20.6 Trypanosomes and Metamonads

The group Euglenida includes flagellated protists such as *Euglena*, with chloroplasts arising from secondary endosymbiosis. Like other algal protists, the euglenas combine photosynthesis and heterotrophic nutrition. The Euglenida, however, include as well a group of obligate parasites called **trypanosomes**. Trypanosomes consist of an elongated cell with a single flagellum. The trypanosome has a unique organelle called the "kinetoplast," consisting of a mitochondrion containing a bundle of multiple copies of its circular genome, usually placed near the base of the flagellum.

The trypanosome *Trypanosoma brucei* has extraordinary abilities to change its surface proteins to indefinitely thwart the immune response. But this system of surface protein variation requires an enormous genetic repertoire, which must be activated only when needed, within the mammalian host bloodstream. At other stages of its life cycle, the trypanosome has very different needs. So the parasite differentiates among at least six differently shaped forms, from insect to mammalian host. Each form has different biochemical properties enabling survival and growth in the particular host organ environment. How does the trypanosome know when to differentiate? The control points offer chances for drug treatment.

An example of such a control point is the transition from the mammalian bloodstream to the insect salivary gland, where the trypanosome becomes transmissible to the next host (**Fig. 1**). In the bloodstream, most trypanosomes assume the "slender" form, which multiplies to large numbers while undergoing antigenic variation. The slender form, however, cannot survive in the tsetse fly. So a fraction of the slender forms differentiate into the "stumpy" form, so called for the wider appearance of the cell. The stumpy form does not proliferate, because it is arrested in its cell cycle; however, unlike the slender form, it can survive in the tsetse fly. When the stumpy form is sucked up by the fly, it must differentiate again, into another form ("procyclic"), which can multiply in the very different host environment of the fly's midgut. The procyclic form then differentiates into the epimastigote and finally the metacyclic form, which expresses the antigenically variable proteins needed to grow within the bloodstream of the next mammalian host.

What signals induce the stumpy form to differentiate into the procyclic form? An experiment showed that a drop in temperature from 37°C (human body temperature) to 20°C (approximate temperature of the tsetse fly in the evening)

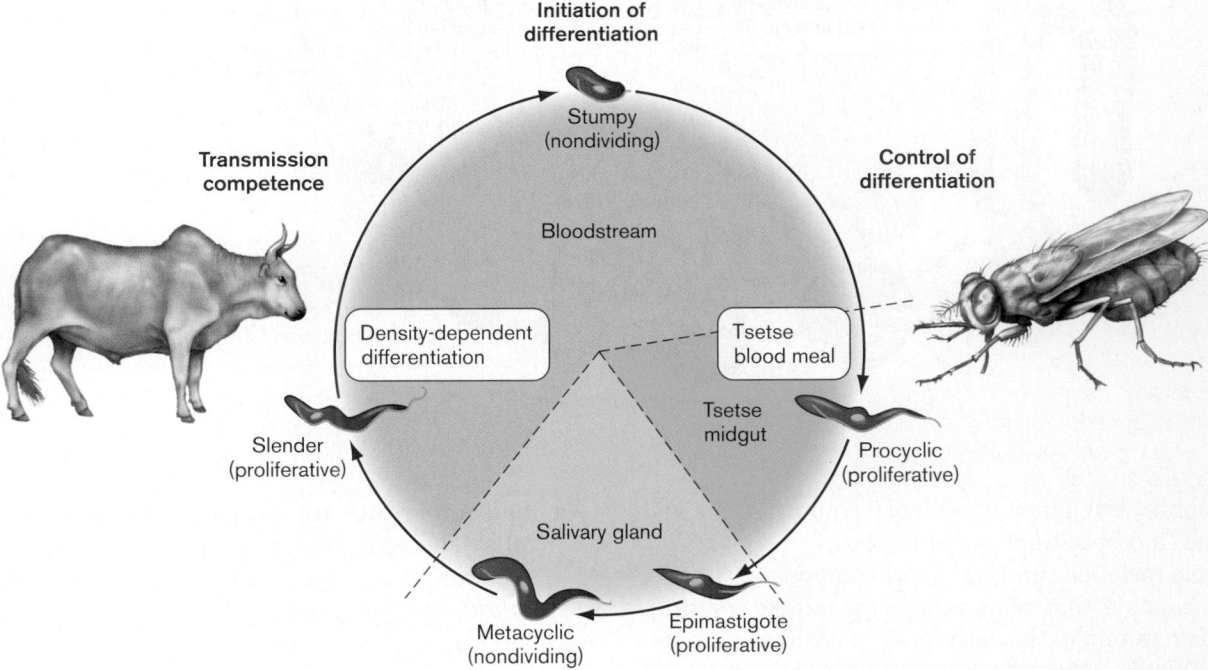

Figure 1 Stages of the trypanosome life cycle. The "slender" form of the trypanosome proliferates in the bloodstream, but only the "stumpy" form can survive in the tsetse salivary gland. The "stumpy" form must differentiate to the "procyclic" form and others that proliferate within the insect midgut.

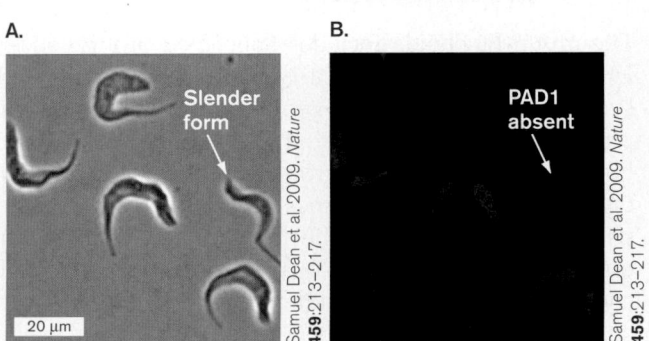

Figure 2 PAD1 expression in stumpy and slender forms. **A.** Under phase-contrast microscopy, the slender form is distinguished from the stumpy form. **B.** Immunofluorescence shows PAD1 protein (red) in the stumpy form but not in the slender form.

could induce differentiation of the stumpy form to the procyclic form. But what components of the stumpy form process the signal? This question was addressed by Keith Matthews and colleagues at the University of Edinburgh. Matthews compared the transcriptomes (the gene expression levels throughout the genome) of two different strains of trypanosome, one of which was defective for differentiation. The gene expression levels were compared by microarray analysis (discussed in Chapter 10). The microarray analysis revealed two genes expressed only in the differentiation-positive strain, encoding cell surface proteins PAD1 and PAD2, named for "proteins associated with differentiation."

The roles of the PAD1 and PAD2 genes in differentiation were tested by immunofluorescence microscopy. First, a mixture of stumpy and slender forms was prepared under phase contrast (**Fig. 2A**). The fixed forms were then treated with PAD1-specific antibody, followed by a fluorescent secondary antibody. Fluorescence (**Fig. 2B**) showed signal throughout the surface of the stumpy forms but not the slender forms. Thus, PAD1 protein is expressed only in the stumpy form; it may be required for differentiation of the slender form to the stumpy form.

What happens at the next transition, to the procyclic form within the tsetse fly? Other experiments showed that PAD1 disappeared in the procyclic form, but there was increased expression of a related protein, PAD2. Furthermore, the increased expression and cellular distribution of PAD2 specifically correlated with the temperature shift that activates differentiation. At the higher temperature of the mammalian bloodstream (37°C), PAD2 protein (green) was confined to the flagellar pocket of the trypanosome (**Fig. 3A**); whereas at the lower temperature of the insect environment (20°C), PAD2 was increased and distributed around the surface of the stumpy cell (**Fig. 3B**). The timing of PAD2 activation is consistent with its requirement for differentiation from the stumpy form to the procyclic form. Thus, PAD1 and PAD2 proteins play critical roles in trypanosome differentiation and suggest targets for antimicrobial therapy.

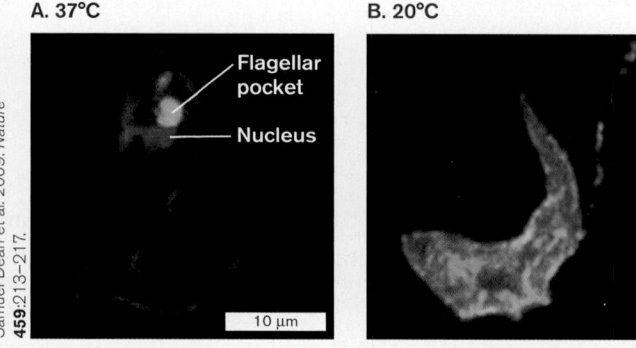

A. 37°C   **B. 20°C**

Flagellar pocket
Nucleus

10 μm

Samuel Dean et al. 2009. *Nature* **459**:213–217.

Samuel Dean et al. 2009. *Nature* **459**:213–217.

Figure 3   Temperature shift alters PAD2 expression.
A. At 37°C, PAD2 immunofluorescence (green) is confined to the flagellar pocket. DAPI fluorescence indicates nuclear DNA (blue), and alpha-tubulin immunofluorescence indicates the cytoplasm (red). **B.** At 20°C, PAD2 is increased and distributed throughout the cell.

Trypanosomes cause some of the most gruesome and debilitating conditions known to humanity; for example, *Leishmania major*, the agent of leishmaniasis, causes skin infections that may enter the internal organs. If untreated, leishmaniasis can lead to swelling and decay of the extremities and eventually death (**Fig. 20.40A** and **B**). Carried by sand flies, *Leishmania* infects 1.5 million people annually, in South America, Africa and the Middle East, and southern Europe. Leishmaniasis is a problem for Americans serving in Iraq; for this reason, returning veterans from Iraq are permanently restricted from blood donation.

Another major disease caused by trypanosomes is trypanosomiasis, also known as African sleeping sickness. A major killer of humans and livestock, in some sub-Saharan countries trypanosomiasis is the second biggest killer after AIDS. The parasite, *Trypanosoma brucei* (**Fig. 20.40C**), is carried by the tsetse fly. *T. brucei* multiplies in the bloodstream of the host animal, causing repeated cycles of proliferation and fever that ultimately lead to death if untreated. This trypanosome is known for its extraordinary degree of antigenic variation. Its genome includes 200 different active versions of its variant surface glycoprotein (VSG), the antigen inducing the immune response, as well as 1,600 different "silent" versions that can recombine with "active" VSG to make further variations. In effect, the trypanosome overwhelms the host immune system by continually generating new antigenic forms until the host repertoire of antibodies is exhausted. In order to infect its human host, the trypanosome needs to interconvert among several different forms of its life cycle. The molecular basis of conversion offers targets for drug therapy. An example of research on the trypanosome life cycle is described in **Special Topic 20.1**.

A related trypanosome, *T. cruzi*, is carried by reduviid bedbugs; it causes Chagas' disease, a debilitating infection of the heart and other internal organs. Chagas' disease is prevalent in South and Central America, and global warming is expected to expand its range north.

Metamonada, a highly degenerate group of parasites and symbionts, includes the intestinal parasite *Giardia lamblia*, an organism familiar in North America, Europe, and Russia (**Fig. 20.41**). *Giardia* is a frequent nemesis of day-care centers, as well as a contaminant of aquatic streams frequented by bears and other wildlife. *Giardia* occasionally contaminates community water supplies and has become endemic in some Russian cities. *Giardia* and other metamonads are noted for their anaerobic metabolism and their complete absence of mitochondria, reflecting their absolute dependence on the anaerobic intestinal environment of their hosts.

A. *Leishmania* infection

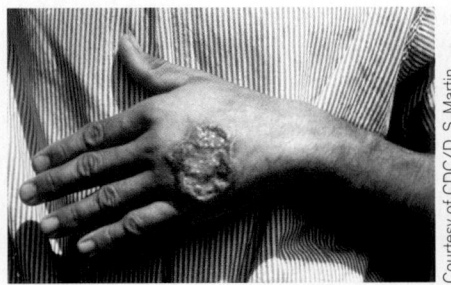

Courtesy of CDC/D. S. Martin

B. *Leishmania major*

©Dennis Kunkel

C. *Trypanosoma brucei*

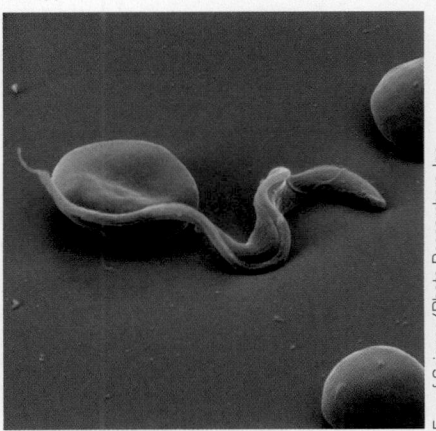

Eye of Science/Photo Researchers, Inc.

Figure 20.40 Trypanosomes. A. Patient suffering from *Leishmania* infection (leishmaniasis). **B.** Cluster of *L. major* undergoing schizogony within the sand fly. **C.** *Trypanosoma brucei*, cause of African sleeping sickness.

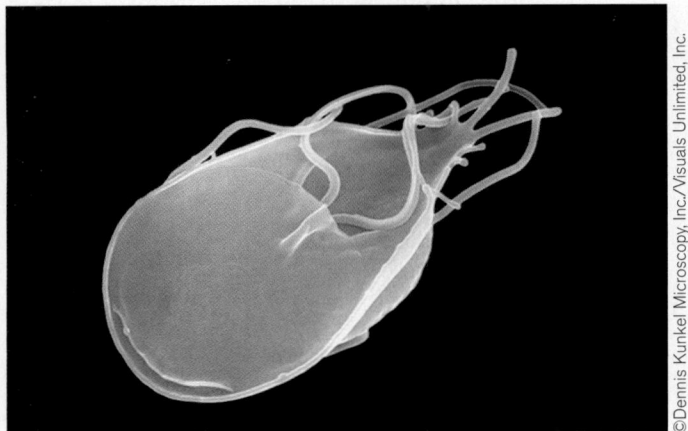

©Dennis Kunkel Microscopy, Inc./Visuals Unlimited, Inc.

Figure 20.41 A metamonad parasite. *Giardia lamblia*, a flagellate lacking mitochondria, and a common human intestinal parasite.

Concluding Thoughts

The microbial eukaryotes, including fungi, algae, and many kinds of protists, serve many diverse roles in our ecosystems. A survey of protists may leave an impression that most protist clades exist primarily to parasitize humans. However, recent genetic surveys based on DNA sequence analysis of environmental communities suggest the existence of twice as many protist clades in nature as we have studied to date. Most of these unknown clades have no direct connection with humans, yet they may fill crucial niches in the ecosystems on which human existence depends. The remaining chapters in Part 4 emphasize the interconnections among many kinds of microbes that form the foundations of Earth's biosphere.

CHAPTER REVIEW

Review Questions

1. Discuss the evidence for the branching of fungi and animals within one clade, the Opisthokonta, which is distinct from algae and protists.
2. How do primary symbiont algae differ from secondary and tertiary symbiont algae? Compare with respect to cell structure and nutritional options.
3. Compare and contrast the molecular basis of motility in amebas and ciliates. Cite particular species.
4. Compare and contrast the traits of these heterokont protists: kelps, diatoms, and dinoflagellates. Compare with respect to cell structure, colony organization, and nutritional options.
5. Summarize the key traits of fungi. What do fungi have in common with protists, and how do they differ?
6. Outline the life cycles of the major phyla of fungi: Chytridiomycota, Zygomycota, Ascomycota, and Basidiomycota. Explain their ecological significance.

7. Compare and contrast the traits of green and red algae (Chlorophyta and Rhodophyta, respectively).
8. Outline the life cycle of the slime mold *Dictyostelium discoideum*. Compare and contrast its features with that of fungi that produce fruiting bodies, such as basidiomycetes.
9. Outline the complex parasitic life cycles of an apicomplexan parasite and a trypanosome. Cite evidence of reductive evolution, as well as of evolution of elaborate specialized structures to facilitate the parasite's life cycle.

Thought Questions

1. Compare and contrast eukaryotic microbes that have inorganic shells or plates. What is their composition, and how do they grow?
2. Explain mixotrophy. Why are so many marine eukaryotes mixotrophs?
3. Why do eukaryotes show such a wide range of cell size? What selective forces favor large cell size, and what favors small cell size?
4. Do eukaryotic parasites have genomes that are larger or smaller than those of free-living organisms? Explain.

Key Terms

alga (752, 758)
alternation of generations (763)
alveolate (759)
ameba (759)
apical complex (781)
apicomplexan (760)
ascomycete (766)
ascospore (766)
ascus (764)
basidiomycete (767)
basidiospore (768)
budding (763)
cellular slime mold (776)
chloroplast (758)
chrysophyte (759)
ciliate (760)
conjugation (779)
contractile vacuole (778)
coralline alga (772)
cortical alveolus (759)
cryptogamic crust (773)
diatom (760)
dinoflagellate (759)
endosymbiosis (757)
Eumycota (757)

flagellate (760)
foraminiferan (777)
fruiting body (758)
frustule (773)
fungus (757)
gamogony (782)
green alga (chlorophyte) (759)
heterokont (760)
hydrogenosome (765)
hypha (758)
kelp (760)
lamellar pseudopod (776)
lichen (773)
macronucleus (779)
malaria (782)
merozoite (782)
micronucleus (778)
mitosporic fungus (764)
mycelium (761)
mycology (752)
nucleomorph (759)
oomycete (769)
oral groove (778)
phagocytosis (775)

phytoplankton (769)
plasmodial slime mold (776)
plasmodium (777)
primary endosymbiont (757)
protist (753)
protozoan (752)
pseudopod (759)
radiolarian (777)
red alga (rhodophyte) (759)
sargassum weed (774)
schizogony (782)
secondary endosymbiont (757, 773)
slime mold (776)
sporangiospore (766)
sporangium (765)
sporogony (782)
stalked ciliate (779)
trypanosome (783)
vesicular-arbuscular mycorrhizae (765)
yeast (763)
zoospore (757)
zooxanthella (781)
zygomycete (765)
zygospore (766)

Recommended Reading

Amaral-Zettler, Linda A., Felipe Gomez, Erik R. Zettler, Brendan G. Keenan, Ricardo Amils, et al. 2002. Eukaryotic biodiversity and physiology at acidic extremes: Spain's Tinto River. Eukaryotic diversity in Spain's River of Fire. *Nature* **417**:137.

Armbrust, E. Virginia. 2009. The life of diatoms in the world's oceans. *Nature* **459**:185–192.

Armbrust, E. Virginia, John A. Berges, Chris Bowler, Beverley R. Green, Diego Martinez, et al. 2004. The genome of the diatom *Thalassiosira pseudonana*: Ecology, evolution, and metabolism science. *Science* **306**:79–86.

Baldauf, Sandra L. 2003. The deep roots of eukaryotes. *Science* **300**:1703–1706.

Baum, Jake, Anthony T. Papenfuss, Buzz Baum, Terence P. Speed, and Alan F. Cowman. 2006. Regulation of apicomplexan actin-based motility. *Nature Reviews. Microbiology* **4**:621–628.

Deacon, Jim. 2006. *Fungal Biology*. 4th ed. Blackwell, Malden, MA.

Dean, Samuel, Rosa Marchetti, Kiaran Kirk, and Keith R. Matthews. 2009. A surface transporter family conveys the trypanosome differentiation signal. *Nature* **459**:213–217.

Gardner, Malcolm J., Shamira J. Shallom, Jane M. Carlton, Steven L. Salzberg, Vishvanath Nene, et al. 2002. The genome sequence of the human malaria parasite *Plasmodium falciparum*. *Nature* **419**:498–511.

Haldar, Kasturi, Sophien Kamoun, N. Luisa Hiller, Souvik Bhattacharje, and Christiaan van Ooij. 2006. Common infection strategies of pathogenic eukaryotes. *Nature Reviews. Microbiology* **4**:922–931.

Harman, Gary E., Charles R. Howell, Ada Viterbo, Ilan Chet, and Matteo Lorito. 2004. *Trichoderma* species—opportunistic, avirulent plant symbionts. *Nature Reviews. Microbiology* **2**:43–56.

Henderson, Gregory P., Lu Gan, and Grant J. Jensen. 2007. 3-D ultrastructure of *O. tauri*: Electron cryotomography of an entire eukaryotic cell. *PLoS One* **8**:e749.

Mochizuki, Kazufumi, Noah A. Fine, Toshitaka Fujisawa, and Martin A. Gorovsky. 2002. Analysis of a piwi-related gene implicates small RNAs in genome rearrangement in *Tetrahymena*. *Cell* **110**:689–699.

Parfrey, Laura W., and Laura A. Katz. 2010. Dynamic genomes of eukaryotes and the maintenance of genomic integrity. *Microbe* **5**:156–163.

Pounds, J. Alan, Martin R. Bustamante, Luis A. Coloma, Jamie A. Consuegra, Michael P. L. Fogden, et al. 2007. Widespread amphibian extinctions from epidemic disease driven by global warming. *Nature* **439**:161–167.

Wilson, Richard A., and Nicholas J. Talbot. 2009. Under pressure: Investigating the biology of plant infection by *Magnaporthe oryzae*. *Nature Reviews. Microbiology* **7**:185–196.

 For further study and review, please visit StudySpace at **microbiology2.com**.

Chapter 21

Microbial Ecology

Of all organisms, microbes grow in the widest range of habitats, from Antarctic lakes to the acid drainage of iron mines, as well as the air, water, and soil that surround us. Microbial communities form the foundation of Earth's biosphere, shaping the environments inhabited by plants and animals. In the ocean, vast quantities of microbes produce the biomass that ultimately feeds fish and humans. In forests and fields, microbes are actually the main consumers; they decompose the majority of plant material, generating fertile soil. Through their biochemistry, diverse microbes largely determine the quality of soil, air, and water for human life.

In all ecosystems, microbes associate with animals and plants. Microbes form elaborate symbiotic systems, such as mycorrhizae, underground networks of fungal hyphae that support and connect the roots of plants throughout forest and field. Rhizobia fix nitrogen for legumes, whereas algae photosynthesize for coral reefs. Anaerobes digest complex plant fibers for termites, cattle, and even humans. This chapter explores how microbes interact with each other and with their many diverse habitats on Earth.

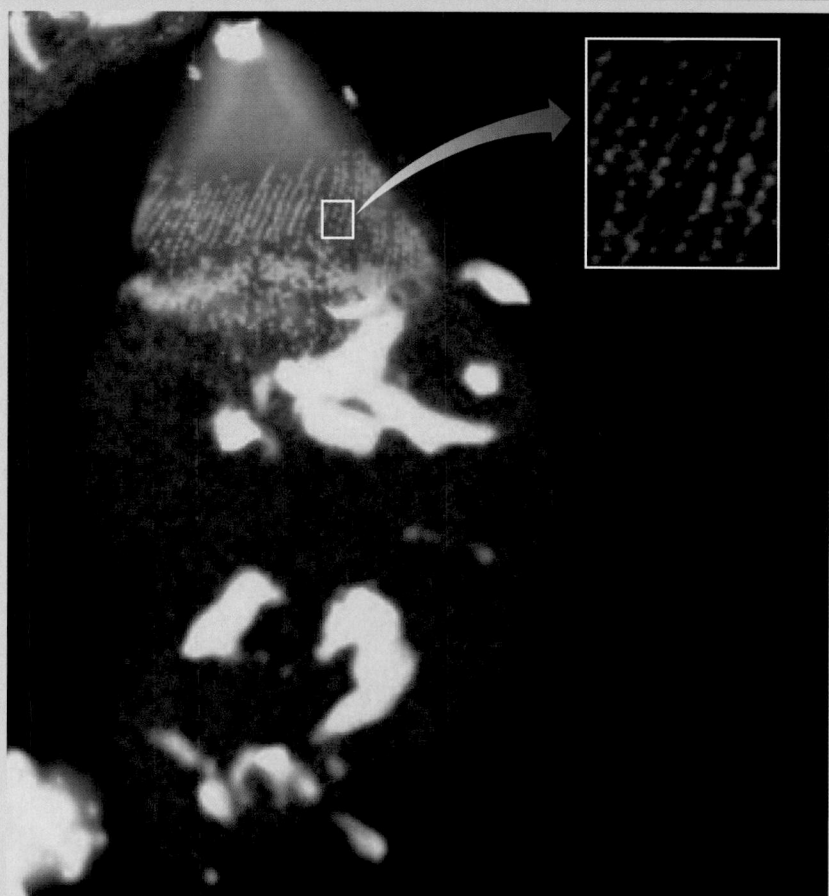

Current Research Highlight

A termite gut protist, *Nymphotricha agilis*, with multiple symbiotic bacteria that digest wood for the termite. Fluorescence microscopy showing wood particles ingested by the protist (yellow), intracellular bacteria known as Termite Group 1 (red), and sulfate-reducing *Desulfovibrio* bacteria that dot the protist's surface in rows (green). The intracellular bacteria catabolize cellulose from wood, while the surface-attached *Desulfovibrio* transfer electrons to sulfate. The microbial metabolism releases acetate for the termite, but also surprising amounts of hydrogen gas. The termite gut microbial community offers a model system for hydrogen biofuel.
*Source: Tomoyuki Sato et al. 2009. Environ. Microbiol. **11**:1007.*

The enteric disease cholera causes hundreds of thousands of deaths annually in India, Bangladesh, and other countries. The causative bacterium, *Vibrio cholerae*, inhabits both ocean and freshwater, contaminating the drinking water of millions of people too impoverished to boil it. Microbial ecologist Anwar Huq, from Bangladesh, worked with Rita Colwell, director of the U.S. National Science Foundation, to devise a simple way to decrease the incidence of cholera, based on the ecology of *V. cholerae*. Huq found that *V. cholerae* cells rarely exist in free suspension; rather, they colonize the surfaces of tiny invertebrates called copepods (**Fig. 21.1**). The copepods actually depend on these bacteria to eat through the chitin of their egg cases, releasing their young.

The copepods are much larger than individual bacteria—large enough to be excluded by a simple cloth filter. Most filter materials would be too expensive for villagers, but Huq found an alternative: the sari cloth worn by Bangladeshi women. A study showed that villagers could halve the incidence of cholera by filtering all their water through several layers of folded cloth from old

saris. Although the sari cloth could not remove isolated bacterial cells, it removed the copepods that *V. cholerae* inhabited. This case shows how public health can benefit from understanding microbial ecology.

Chapter 21 shows how microbes interact with each other in communities of organisms. Microbial communities critically impact other organisms in all habitats, from oceans to forests (**Fig. 21.2**). Microbes recycle organic material into aquatic and terrestrial ecosystems, providing resources for plants and animals. Our agricultural productivity depends in part on positive and negative effects of microbes in the soil and air. And deep below Earth's surface, microbial communities shape crustal rock. Following this chapter, Chapter 22 shows how microbial geochemistry shapes the global biosphere as a whole. Microbial cycling of elements through our soil, air, and water increasingly faces the impact of human technology. To preserve our planet's health, we will need to know far more about microbial ecology.

Discovery of microbial ecosystems is an interdisciplinary enterprise including methods presented throughout this book, such as fluorescence microscopy (Chapter 2), enrichment culture (Chapter 4), and molecular phylogeny and metagenomics (Chapters 12 and 17). This chapter presents examples of these methods and other approaches to microbial ecology:

- **Enumeration.** Counting organisms in populations, and mapping populations within habitats and over time.
- **Metabolic flux.** Measuring the transfer of carbon and nitrogen through trophic levels of a food web.
- **Behavior.** Observing the interactions among organisms on molecular, microscopic, and macroscopic levels.
- **Community gene expression.** Examining the function of microbial communities through expressed RNAs.

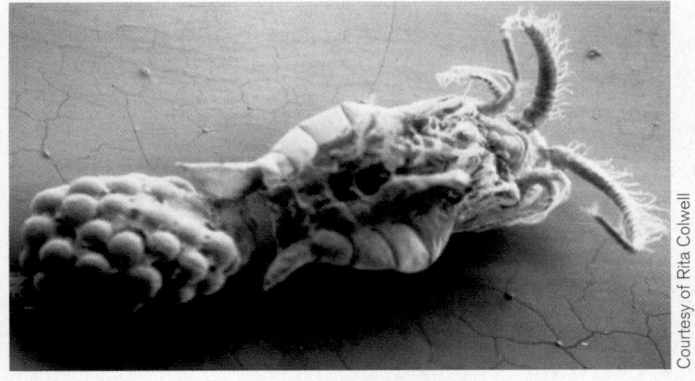

Figure 21.1 Copepods carrying *Vibrio cholerae*. Copepod with case of eggs, to be "hatched" by *V. cholerae* bacteria.

Courtesy of Rita Colwell

A.

Dave Powell, USDA Forest Service

B.

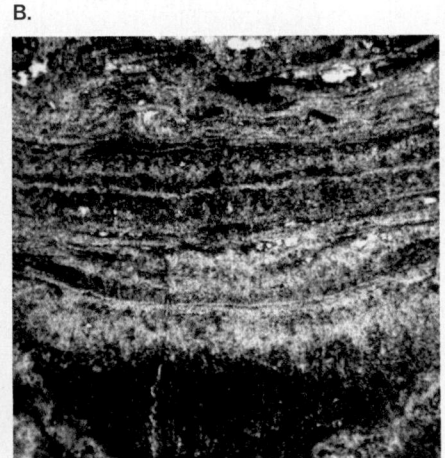

Courtesy of Kurt Konhauser

Figure 21.2 Microbes shape all environments. **A.** Microbes recycle organic material into aquatic and terrestrial habitats, providing resources for plants and animals. **B.** Microbes form crustal rock. Bacteria (dark layers) alternate with silica (white).

21.1 Microbes in Ecosystems

"No man is an island," nor is any microbe. All organisms exist within **ecosystems** consisting of populations of species plus their habitat or environment. A population is a group of individuals of one species living in a common location. Within an ecosystem, each population of organisms fills a specific **niche**. The niche is a set of conditions enabling an organism to grow and reproduce, including its habitat, resources, and relations with other species of the ecosystem. For example, the niche of *Synechococcus*, a cyanobacterium, is that of a free-living marine organism that fixes CO_2 into biomass while producing molecular oxygen utilized by swarms of heterotrophic bacilli. The habitat of *Synechococcus* is the upper water layer of the ocean; its biomass provides food for protist predators, which in turn feed invertebrates and fish.

Microbes Fill Unique Niches in Ecosystems

The essential role of microorganisms in all ecosystems was formulated by the Dutch microbiologist Cornelis B. van Niel (1897–1985). Van Niel was the first to show that bacteria in the soil and water can conduct photosynthesis without producing oxygen, using electron donors such as H_2S instead of H_2O. This surprising discovery revealed one of many kinds of metabolism unique to microbes, unknown in plants or animals. Other unique forms of microbial metabolism (discussed in Chapters 13–15) include nitrogen fixation by bacteria and archaea, and the degradation of lignin by bacteria and fungi. Moreover, microbial metabolism provides ecosystems with their sole source of key elements such as sulfur, phosphorus, and iron.

Unique microbial metabolism provides unique roles for microbes in ecosystems. Van Niel expressed this principle in two key postulates of microbial ecology:

- **Every molecule existing in nature can be used as a source of carbon or energy by a microorganism found somewhere in the biosphere.** Any cell component or product, such as amino acids and H_2S, can participate in some kind of energy-yielding reaction. We now extend this principle to include human-made compounds entering the environment. If an energy-yielding reaction exists, some microbe must have evolved to use it.
- **Microbes are found in every environment on Earth.** Every possible habitat for life supports microbes. In fact, the largest part of the biosphere (below Earth's surface) is inhabited solely by microbes.

The van Niel postulates imply an extraordinary variety of microbial species using different energy-yielding reactions. But what determines the contribution of each microbe? What a microbe gives to and takes from its ecosystem depends on two things:

- **The microbial genome.** Each genome provides a set of metabolic enzymes available to interconvert molecules. For example, the genomes of pseudomonads encode enzymes for catabolizing complex aromatic molecules, whereas genomes of cyanobacteria encode the apparatus of oxygenic photosynthesis. Microbial genomes share content through mobile genes (acquired by transfer mechanisms, discussed in Chapter 9).
- **Environmental factors.** The environment's physical and chemical factors, such as temperature, nutrient supplies, oxygen availability, and pH, determine the range of metabolic options. For example, if H_2 is available in the presence of O_2, *Rhodopseudomonas palustris* will oxidize hydrogen to water; on the other hand, if O_2 is absent but light is available, *R. palustris* will use anaerobic photosynthesis, splitting H_2S to make sulfur. Each organism's environment includes other organisms in the community. For example, the H_2 oxidized by *R. palustris* is provided by fermenting bacteria.

All organisms depend, directly or indirectly, on the presence of other organisms. Some populations cooperate by reciprocally supplying each other's needs. Cooperation by organisms may be incidental, as in the case of hydrogen-oxidizing bacteria using H_2 from fermenters; or it may involve mutualism, a highly evolved partnership in which two species have coevolved to support each other (discussed in Section 21.2).

Carbon Assimilation and Dissimilation: The Food Web

The interactions between microbes and their ecosystems include two common roles of metabolic input and output, often referred to as assimilation and dissimilation, respectively. Most of these metabolic processes were discussed in Chapters 13, 14, and 15, but ecology offers a different perspective.

Assimilation refers to processes by which organisms acquire an element, such as carbon from CO_2, to build into cells. When the environment lacks organic compounds containing an element such as nitrogen or phosphorus, microbes may assimilate the element from mineral sources. Common kinds of assimilation include carbon dioxide fixation and nitrogen fixation. Organisms that produce biomass from inorganic carbon (usually CO_2 or bicarbonate ions) are called **primary producers**. Producers are a key determinant of productivity for other members of the ecosystem.

Dissimilation is the process of breaking down organic nutrients to inorganic minerals such as CO_2 and

NO_2^-, usually through oxidation. Microbial dissimilation releases minerals for uptake by plants, and it provides the basis of wastewater treatment (discussed in Chapter 22). On the other hand, microbial dissimilation of an energy-rich nitrogenous fertilizer, such as ammonium ion, causes nitrogen to be lost from the soil.

This chapter covers microbial assimiliation and dissimilation of carbon and nitrogen in association with the plants and animals of an ecosystem. The cycles of other key elements, and their effects on the global biosphere, are explored in Chapter 22.

> **THOUGHT QUESTION 21.1**   From Chapters 13–15, give examples of microbial metabolism that fit patterns of assimilation and dissimilation.

The major interactions among organisms in the biosphere are dominated by the production and transformation of **biomass**, the bodies of living organisms. To obtain energy and materials for biomass, all organisms participate in the **food web** (**Fig. 21.3**). A food web describes the ways in which various organisms consume each other, as well as each other's products. Levels of consumption are called **trophic levels**. Organisms at each trophic level consume biomass of organisms from another level, usually closer to producers in the food web. At each trophic level, the fraction of biomass retained by the consumer is small; most is dissipated as CO_2 to provide energy.

Every food web depends on primary producers for two things:

■ **Absorbing energy from outside the ecosystem.** In most cases, the ultimate source of energy is sunlight, and the producers are phototrophs.
■ **Assimilating minerals into biomass.** The biomass of producers is then passed on to subsequent trophic levels.

The majority of carbon in Earth's biosphere is assimilated by oxygen-producing phototrophs such as cyanobacteria, algae, and plants. Certain important ecosystems are founded on carbon fixation by lithotrophs—for example, the hydrothermal vent communities, in which bacteria oxidize hydrogen sulfide to fix CO_2, capturing both gases as they well up from Earth's crust. The vent communities use oxygen generated by phototrophs from above.

In addition to producers, all ecosystems include **consumers** that acquire nutrients from producers and ultimately dissimilate biomass by respiration, returning carbon back to the atmosphere (**Fig. 21.3A**). Consumers constitute several trophic levels based on their distance from the primary producers. The first level of consumers, generally called **grazers**, directly feed on producers. Grazers usually convert 90% of the carbon back to CO_2 through respiratory metabolism, yielding energy. The next level of consumers, often called **predators**, feed on the grazers, again converting 90% back to atmospheric CO_2. In microbial ecosystems, the trophic relationships are often highly complex, because a given species may act as both producer and consumer.

At each trophic level, some of the organisms die, and their bodies are consumed by **decomposers**, returning carbon and minerals back to the environment for use by producers. Decomposers are all microbes (fungi or bacteria). Decomposers have particularly versatile digestive enzymes capable of breaking down complex molecules such as lignin. Without decomposers, carbon and minerals needed by phototrophs would be locked away by ever-increasing mounds of dead biomass. Instead, all biomass is recycled, and all the energy acquired by ecosystems is eventually converted to heat.

In the function of ecosystems, phylogenetic distinctions between

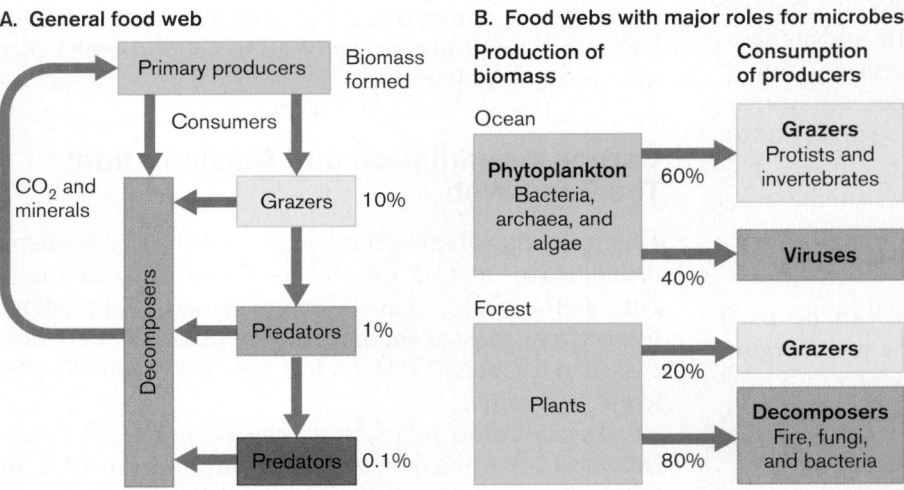

A. General food web

Primary producers — Biomass formed

Consumers

CO_2 and minerals

Decomposers

Grazers — 10%

Predators — 1%

Predators — 0.1%

B. Food webs with major roles for microbes

Production of biomass | **Consumption of producers**

Ocean

Phytoplankton Bacteria, archaea, and algae — 60% → **Grazers** Protists and invertebrates

40% → **Viruses**

Forest

Plants — 20% → **Grazers**

80% → **Decomposers** Fire, fungi, and bacteria

Figure 21.3   Microbes within food webs.   A. Biomass production. Primary producers fix CO_2 into biomass, grazers (primary consumers) that feed on primary producers, and predators consume the grazers and smaller predators. Decomposers consume dead bodies of organisms from every level, recirculating their minerals to the producers. Percentages indicate fraction of original CO_2 converted to biomass at each trophic level. **B.** In different habitats, the relative significance of the various trophic components varies. In marine ecosystems, the primary producers are bacteria, archaea, and algae. About 10%–40% of consumption is by viruses, which break down both producer and consumer microbes. In a forest, the major producers are trees, much of whose biomass is poorly digested by grazers; 80% of consumption occurs by fungal and bacterial decomposition. In some forests, the major decomposer is fire.

Table 21.1 Aerobic and anaerobic metabolism.

Element	Oxidized by O_2 (lithotrophy)	Reduced by CHO* or H_2 (anaerobic respiration)
Nitrogen	$NH_3 + O_2 \longrightarrow NO_3^-$	$NO_3^- + CHO \longrightarrow N_2$
Manganese	$Mn^{2+} + O_2 \longrightarrow Mn^{4+}$	$Mn^{4+} + CHO \longrightarrow Mn^{2+}$
Iron	$Fe^{2+} + O_2 \longrightarrow Fe^{3+}$	$Fe^{3+} + CHO \longrightarrow Fe^{2+}$
Sulfur	$H_2S + O_2 \longrightarrow SO_4^{2-}$	$SO_4^{2-} + CHO \longrightarrow H_2S$
Carbon	$CH_4 + O_2 \longrightarrow CO_2$	$CO_2 + H_2 \longrightarrow CH_4$

*CHO = organic material.

bacteria, archaea, and eukaryotes are of less significance than the biological and biochemical consequences of an organism's presence in the community—that is, what the organism produces or consumes. Thus, in this chapter we place greater emphasis on trophic roles than on phylogenetic distinctions.

The relative significance of microbial and multicellular producers and consumers varies considerably among different habitats. A major difference appears between marine and terrestrial ecosystems (**Fig. 21.3B**). In the oceans, most CO_2 fixation and biomass production are performed by the smallest inhabitants, phototrophic bacteria. The major consumers are protists and viruses; in terms of bulk biomass, multicellular organisms barely contribute. In terrestrial ecosystems, by contrast, the major primary producers and fixers of CO_2 are multicellular plants. Plants generate **detritus**, discarded biomass such as leaves and stems, that requires decomposition by fungi and bacteria.

The differences between the food webs of ocean and dry land explain why most of the food we harvest from the ocean consists of predators at the higher trophic levels (fish), whereas most food harvested on land consists of producers and first-level consumers (plants and herbivores). Fish depend on a vast invisible food web of microbes, and thus their numbers remain limited despite the seemingly huge volume of ocean. Thus, marine fish populations are especially vulnerable to overharvesting and to any environmental change that disrupts the food chain.

Oxygen and Other Electron Acceptors

The availability of oxygen and other electron acceptors is the most important factor that determines the kind of assimilation and dissimilation occurring in a given environment. Examples of aerobic and anaerobic metabolism are presented in **Table 21.1**. In aerobic environments, microbes use molecular oxygen as an electron acceptor to respire on organic compounds (abbreviated CHO) produced by other organisms (discussed in Chapters 13 and 14). Aerobic respiration on organic compounds is highly dissimilatory in that it tends to break compounds down to CO_2. Microbes also use oxygen to respire on reduced minerals such as NH_3, H_2S, and Fe^{2+} (lithotrophy). Lithotrophy is coupled to CO_2 fixation and is therefore assimilatory metabolism.

In anaerobic environments, such as deep soil, microbes use minerals such as Fe^{3+} and NO_2^- to oxidize organic compounds supplied by other organisms (anaerobic respiration). Other anaerobes use these electron acceptors to oxidize reduced minerals. Anaerobic environments allow much slower rates of assimilation and dissimilation than occurs in the presence of oxygen. The total volume of anaerobic microbial communities, however, far exceeds that of the oxygenated biosphere.

Temperature, Salinity, and pH

Other abiotic factors can profoundly affect the environment for microbes and other members of the food chain, either directly or by impacting other factors. The effects are most significant for **extremophiles**, species that grow in environments considered extreme by human standards (**Table 21.2**; discussed in Chapter 5). Temperature limits the rate of metabolism. Higher temperatures, particularly those found in hot springs, enable the fastest growth rates measured (doubling times as short as 10 minutes for some hyperthermophiles). On the other hand, temperature limits the oxygen concentration in water, so hyperthermophiles often use sulfur instead.

Table 21.2 Extremophiles.

Class of extremophile	Typical environmental conditions for growth
Acidophile	Acidic environments at or below pH 3
Alkaliphile	Alkaline environments at a range of pH 9–14
Barophile	High pressure, usually at ocean floor, from 200–1,000 atm
Endolith	Within rock crystals down to a depth of 3 km
Halophile	High salt, typically above 2-M NaCl
Hyperthermophile	Extreme high temperature, above 80°C
Oligotroph	Low carbon concentration, below 1 ppm
Psychrophile	Low temperature, below 15°C
Thermophile	Moderately high temperature, 50°C–80°C
Xerophile	Desiccation, water activity below 0.8

A. Lichen ground cover

Stephen & Sylvia Sharnoff

B. Blankets dyed with lichens

Stephen & Sylvia Sharnoff

Figure 21.5   Boreal forest ecosystems and native cultures depend on lichens.   A. Lichens cover most of the ground in boreal forests. **B.** Chilkat Tlingit dancing blankets are traditionally dyed with lichens.

Biological Soils

Figure 21.6   Cryptogamic crust.   Close-up view of cryptogamic crust just after rainfall.

Within the termite digestive organ (called the hind-gut), its cellulose-digesting bacteria form highly complex associations with protists such as *Mixotricha paradoxa* (**Fig. 21.7**), a large ciliate (as long as half a millimeter). The *Mixotricha* cell contains several kinds of bacterial endosymbionts. Intracellular bacteria digest cellulose from wood consumed by the termite. There are also organelles that

appear to be vestigial remnants of another endosymbiont, diminished by reductive evolution. On the protist's surface, four kinds of bacteria are attached. Two of the attached species are spirochetes, one significantly larger than the other. The spirochetes extend from the protist's cell membrane; they are flagellated, and their flagellar motility propels the protist cell. Two other species of "anchor bacteria" are Gram-negative rods attached to knobs of the protist surface; their function is unknown. All members of the partnership appear to have evolved an obligate relationship.

The relationship among microbial symbionts within the termite gut community generates a complex series of metabolic fluxes (**Fig. 21.8**). This kind of metabolic cooperation is called **syntrophy**, which means "feeding together." For syntrophy, the fluxes within the community must balance energetically with a negative value of ΔG, as they would for a single free-living organism (discussed in Chapter 13).

In the simplified model shown in **Figure 21.8**, the wood polysaccharides are converted by bacteria to fermentation acids, such as acetic acid, which is absorbed by the termite. (The termite's circulation then oxidizes the acetate to CO_2.) Other products of the termite bacterial fermentation include CO_2 and H_2, which can be converted to methane by methanogenic archaea. In some termites the H_2 builds up to levels as high as 30%. The termite microbial mutualism is being studied as a model system for production of hydrogen biofuel.

Digestive mutualisms are also important for mammals such as cattle (discussed in Section 21.6). Even we humans derive about 15% of our caloric intake from bacterial fermentation of plant fibers that we cannot otherwise digest.

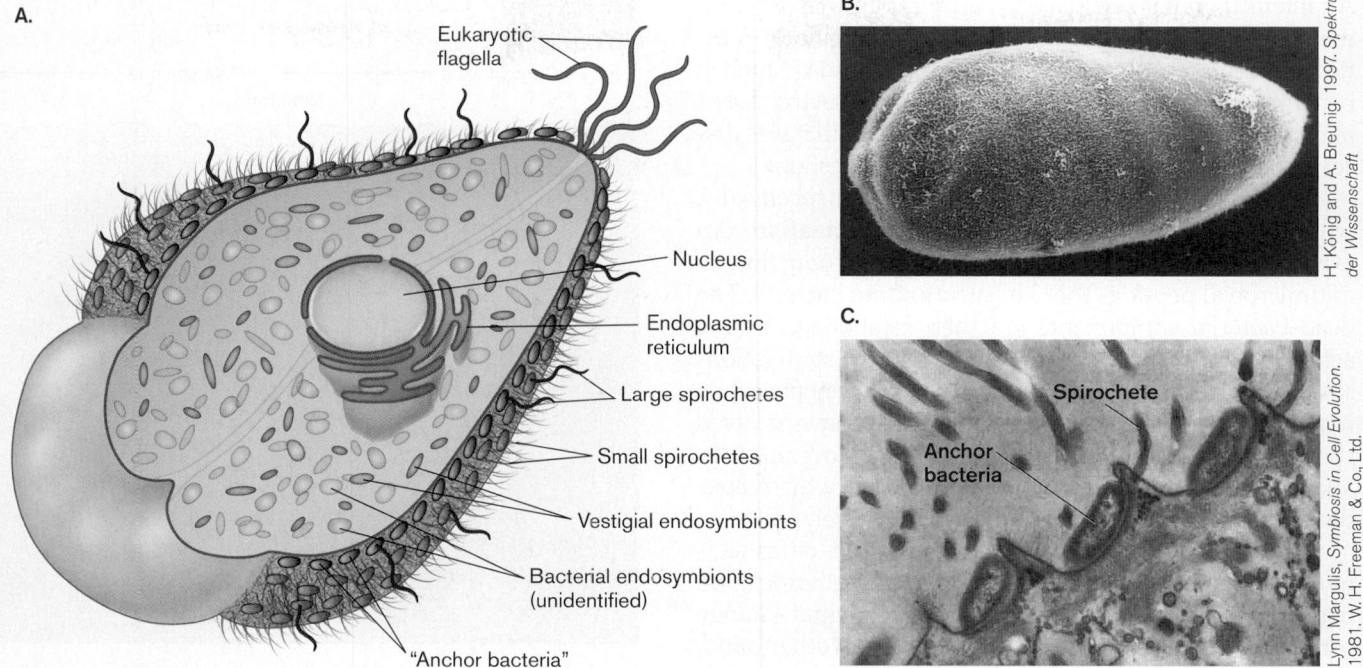

Figure 21.7 *Mixotricha paradoxa*: **a multiple symbiont.** **A.** The protist *Mixotricha paradoxa* possesses attached spirochetes (large and small species), "anchor bacteria," and two kinds of bacterial endosymbionts. **B.** *M. paradoxa* (length 300–500 μm; SEM). **C.** TEM section through pellicle including anchor bacteria and attached spirochetes. *Source*: Lynn Margulis, *Symbiosis in Cell Evolution*, 1993.

Symbiosis Involves Varying Degrees of Cooperation and Parasitism

Symbiosis between organisms involves a range of interdependence, from obligate mutualism (cooperation) to obligate parasitism (see **Table 21.3**). In a gut community, some of the microbial members may enhance each other's growth, but they can also grow independently. Their optional cooperation is called **synergism**, in which both species benefit but can grow independently and show less specific cellular communication. For example, human colonic bacteria produce fermentation products that colonic methanogens metabolize to methane. The methanogens gain energy, and the fermenting bacteria benefit energetically through the steady-state removal of their end products.

In other cases, one species derives benefit from another without return; for example, some wetland bacteria derive benefit from *Beggiatoa* because *Beggiatoa* bacteria oxidize H_2S, which inhibits growth of other species.

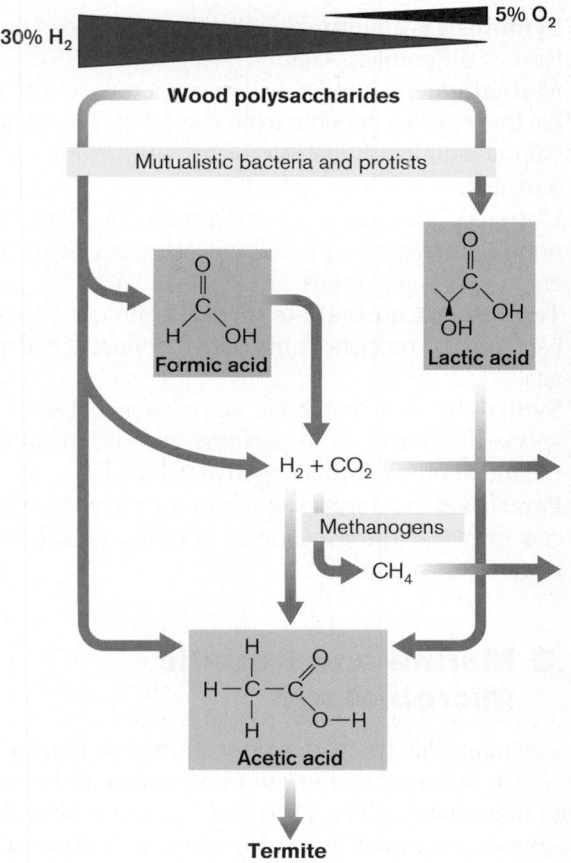

Figure 21.8 **Endosymbiont metabolic fluxes within termite hindgut.** Bacteria and their protist symbionts ferment wood polysaccharides to lactic acid, formic acid, acetic acid, H_2, and CO_2 within the hindgut of a termite, *Reticulitermes santonensis*. Some hydrogen is lost from the gut, some is converted to methane, and some is converted to acetic acid. Acetic acid is absorbed through the outer lining and feeds the termite.

An interaction that benefits one partner only is called **commensalism**. Commensalism is difficult to define in practice, since "commensal" microbes often provide a hidden benefit to their host. For example, gut bacteria such as *Bacteroides* species were considered commensals until it was discovered that their metabolism aids human digestion.

An interaction that harms one partner nonspecifically, without an intimate symbiosis, is called **amensalism**. An example of amensalism is actinomycete production of antimicrobial peptides that kill surrounding bacteria. The dead bacterial components are then catabolized by the actinomycete. Finally, **parasitism** is an intimate relationship in which one member (the parasite) benefits while harming a specific host. Many microbes have evolved specialized relationships as parasites, including intracellular parasitic bacteria such as the rickettsias, which cause diseases such as Rocky Mountain spotted fever.

The distinction between mutualist and parasite is often subtle. Lichens consist of a mutualistic association between fungus and algae, but environmental change can convert the fungus to a parasite. On the other hand, parasitic microbes may coevolve with a host to the point that each depends on the other for optimal health. For example, the incidence of certain allergies and immune disorders appears to correlate with lack of exposure to pathogens and parasites.

TO SUMMARIZE:

- **Symbiosis** is an intimate association between organisms of different species.
- **Mutualism** is a form of symbiosis in which each partner species benefits from the other. The relationship is usually obligatory for growth of one or both partners.
- **Lichens** are a mutualistic community of algae or cyanobacteria with fungi. Lichens are essential producers for dry soil habitats.
- **Termite gut mutualists** include cellulose-digesting bacteria and protists. A major by-product is hydrogen gas.
- **Syntrophy** is a metabolic association between two species, requiring both partners in order to complete the metabolism with a negative value of ΔG.
- **Parasitism** is a form of symbiosis in which one species grows at the expense of another, usually much larger, host organism.

21.3 Marine and Aquatic Microbiology

The oceans cover more than two-thirds of Earth's surface, reaching depths of several kilometers and forming an immense habitat (**Fig. 21.9**). Both oceans and freshwater support huge quantities of bacteria and algae, which

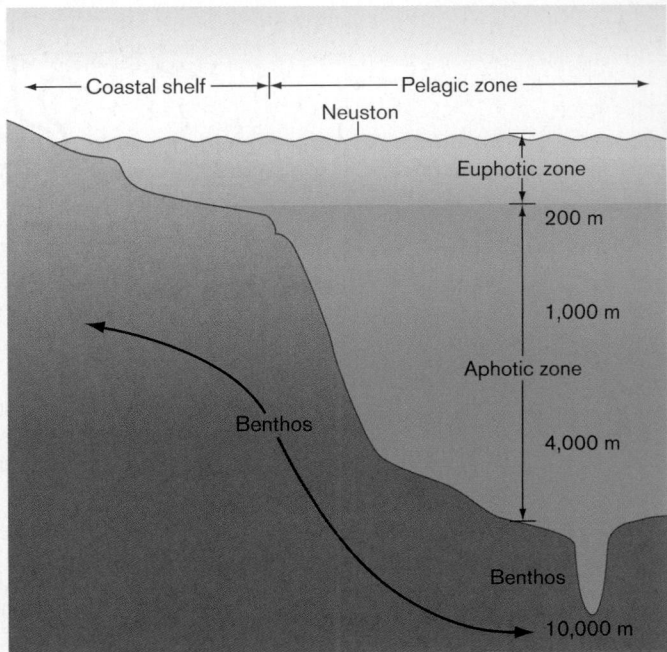

Figure 21.9 Regions of marine habitat. The marine habitat subdivides into several categories. The coastal shelf region is defined as the water extending from the shoreline out to a depth of 200 meters. The pelagic zone (open ocean) includes several depth regions: the neuston, the microscopic interface between water and air; the euphotic zone of light penetration, where phototrophs can grow, down to 100–200 meters; the aphotic zone, which supports only heterotrophs and lithotrophs; and the benthos, at the ocean floor.

drive vast ecosystems. Marine and freshwater microbes have enormous human impact, from the maintenance of fisheries to the production of biofuels. Algae are being engineered to produce petroleum, at rates 30-fold more energy-efficient than those of terrestrial biofuel plants. A science fiction scenario, *The Highest Frontier* (published in 2011) imagines an orbital space habitat powered by an outer shell of marine microbial phototrophs.

Marine water is distinguished by its salt concentration, averaging 3.5%. The major ions are Na^+ and Cl^-, with significant levels of sulfate, iodide, and bromide. Marine salt concentration is high enough to prevent growth of many aquatic and terrestrial bacteria, such as *E. coli*, although salt-tolerant organisms such as *Vibrio cholerae* grow well over a broad range of salt concentration.

Marine Habitats

Marine waters vary considerably with respect to temperature, pressure, light penetration, and concentration of organic matter. In the open ocean, the water column (known as the **pelagic zone**) is subdivided into distinct regions (see **Fig. 21.9**):

- **Neuston (about 10 μm).** The neuston is the air-water interface. Although extremely thin, the neuston layer contains the highest concentration of microbes. Many algae and protists have evolved so as to "hang" from the layer of surface tension.

- **Euphotic zone (100–200 m).** The **euphotic zone**, or **photic zone**, is the upper part of the water column that receives light for phototrophs. In the open ocean, the euphotic region extends down a couple of hundred meters, whereas at the **coastal shelf** (<200 meters to floor), a higher concentration of silt and organisms decreases the photic zone to as little as 1 meter.
- **Aphotic zone.** Below the reach of light, in the aphotic zone, only heterotrophs and lithotrophs can grow.
- **Benthos.** The benthos includes the region where the water column meets the ocean floor, as well as sediment below the surface. Organisms that live in the benthos, such as those of thermal vent communities, are called **benthic organisms**.

Another important determinant of marine habitat is the **thermocline**, a depth at which temperature decreases steeply and water density increases. A thermocline typically exists in an unmixed region. At the thermocline, a population of heterotrophs will peak, feeding on organic matter that settles from above.

Coastal regions show the highest concentration of nutrients and living organisms and the least light penetration. The open ocean is largely oligotrophic (extremely low concentration of nutrients and organisms). The concentration of heterotrophic microorganisms determines the **biochemical oxygen demand (BOD)**, the amount of oxygen removed from the water by aerobic respiration. Normally, the open ocean has such a low concentration of organisms that the BOD is extremely low; therefore, the dissolved oxygen content is high. This explains why enough oxygen reaches the ocean floor to serve lithotrophs and methanotrophs. The BOD rises, however, when there is excess sewage or petroleum such as that spilled by the Deepwater Horizon rig in 2010. Microbial consumption of these wastes at first deprives fish of oxygen, but the process is the only way the ocean recovers.

To nineteenth-century microbiologists, the oceans appeared virtually free of bacteria. Trained in the tradition of Robert Koch (discussed in Chapter 1), microbiologists attempted to isolate marine bacteria by plate culture, considered the definitive way to study a microbial species. But the colonies that grew on traditional plate media were extremely few. Then, in 1959, the German microbial ecologist Holger Jannasch (1927–1998; **Fig. 21.10A**) showed that many more bacteria could be seen by light microscopy than could be grown on plates. Later, with colleagues at the Woods Hole Oceanographic Institution, Jannasch went on to study life at the ocean floor, using the famous submersible vessel *Alvin* (**Fig. 21.10B**).

Jannasch was one of the first to show that most marine microorganisms could not be cultured in the laboratory. His observations sparked decades of controversy. How could the organisms be alive if they could not be cultured? We now know that the vast majority of bacteria alive in the biosphere are nonculturable under current laboratory conditions because we do not know their culture requirements. In some cases, microbial growth requires hidden mutualisms with other organisms in their natural environment. Such organisms are now termed **uncultured**, with the understanding that their culture requirements may be discovered in the future. Uncultured bacteria are identified by PCR amplification of their DNA and by fluorescence microscopy.

Uncultured organisms from marine communities are now characterized by **metagenomics**, the sequencing of "community DNA." For metagenomic analysis, the DNA is obtained from organisms taken directly from their habitat without culturing. In 2007, Craig Venter led a research team on the *Sorcerer II* Global Ocean Sampling expedition that sequenced over 6 billion base pairs of DNA from ocean water samples taken from around the globe. The researchers estimated that their sequences identified 25,000 different microbial species per liter of seawater. Their survey revealed many new gene sequences, doubling the number of known protein types for all living organisms.

So what kind of marine bacteria emerged out of Venter's DNA survey? Only 30% of the DNA sequenced showed

Figure 21.10 Holger Jannasch discovered unculturable marine bacteria. A. Holger Jannasch pioneered the study of marine microbes in the open sea and on the seafloor. **B.** A recent model of the submersible *Alvin* used for sampling deep marine organisms by researchers from the Woods Hole Oceanographic Institution.

A. Holger Jannasch

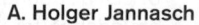

Photo by William Lambert, Woods Hole Oceanographic Institute.

B. Submersible *Alvin*

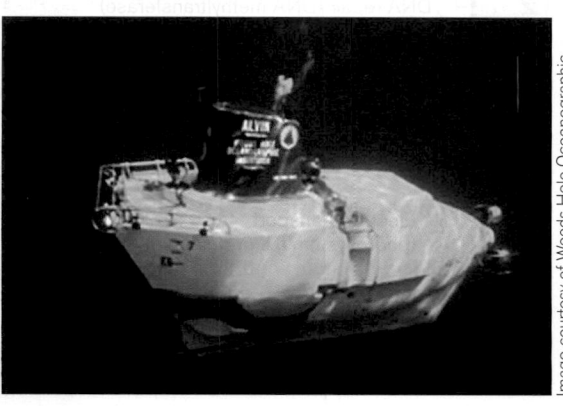

Image courtesy of Woods Hole Oceanographic Institutions.

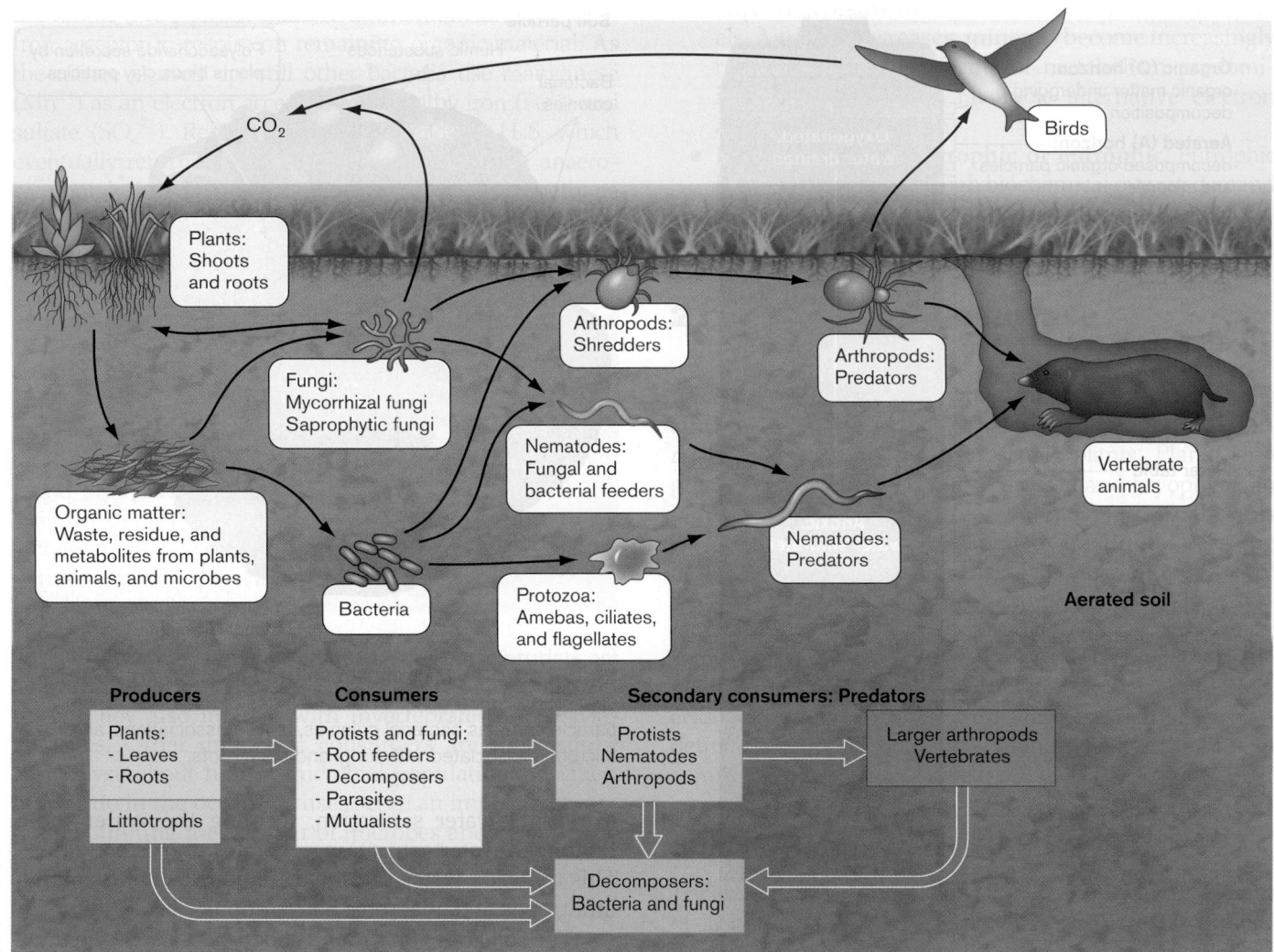

Figure 21.24   The soil food web. Plants are the major producers, although some production also occurs from lithotrophs such as ammonia oxidizers. Detritus from plants is decomposed by fungi and bacteria, which feed protists and small invertebrates such as nematodes. Protists and small invertebrates are consumed by larger invertebrates and vertebrate animals.

Plant matter is decomposed by many species of fungi, such as *Mycena* species, and by bacteria such as the actinomycetes. The actinomycetes include *Streptomyces*, a genus famous for the production of antibiotics and for generating chemicals whose odors give soil its characteristic smell (**Fig. 21.25A**). Besides fallen leaves, another source of organic matter from plants is the **rhizosphere**, the region of soil surrounding plant roots. The rhizosphere contains proteins and sugars released by roots, as well as sloughed-off plant cells. These materials feed large numbers of bacteria, which then cycle minerals back to the plant. Bacteria in the rhizosphere may also discourage growth of plant pathogens. Another habitat colonized by bacteria is the surface of fungi, whose filaments are considerably larger than bacterial cells (**Fig. 21.25B**).

At the next trophic level, bacteria feeding on leaf detritus and root exudates are then preyed on by protists and nematodes (**Fig. 21.25C**). Diverse predators such as nematodes, fungi, and protists exhibit different preferences for bacteria, fungi, or protists as prey. *Vampirella* protists drill holes in fungal hyphae to suck out their nutrients. Parasitic fungi prey on plants or invertebrates; some actually capture and strangle nematodes. Besides predation, there exist many complex forms of mutualism, such as lichens (discussed earlier). Another mutualistic interaction of fungi is that of **mycorrhizae**, fungal infections of plants in which the fungal hyphae extend the root surface area and increase the root's ability to absorb nutrients.

Microbes also cooperate to form complex communities such as microbial mats and biofilms. The microbes ultimately feed invertebrates, which then feed larger invertebrates and vertebrate predators. Some predators, such as earthworms and burrowing animals, enhance the soil quality by turning over the matter, aerating the soil particles and helping to mix the organic matter from above with the mineral particles from below.

A. *Streptomyces griseus*

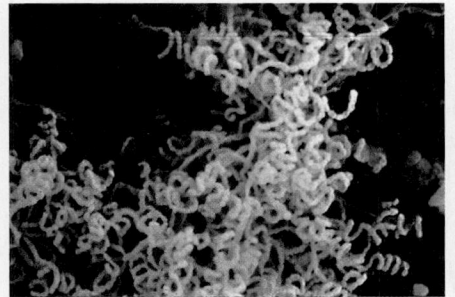

B. Bacteria grow on fungi

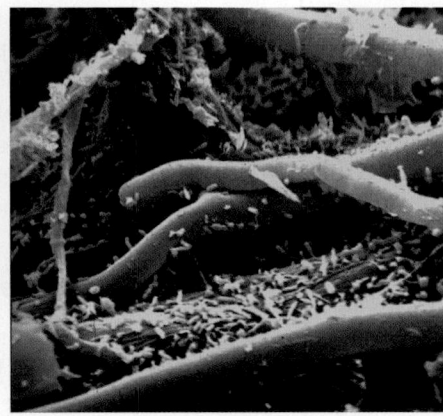

C. Nematode feeds on bacteria

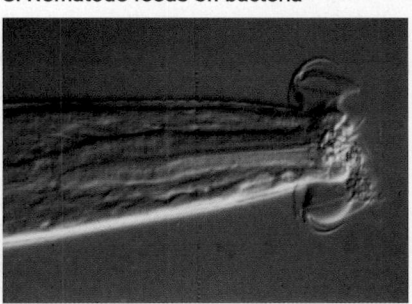

Figure 21.25 Soil microbes. A. Hyphae of actinomycete bacteria, *Streptomyces griseus* (filament width 1 μm; SEM). *Streptomyces* bacteria give the soil its characteristic odor. **B.** Bacteria grow on the surfaces of fungi (hyphal width 10–20 μm). **C.** A bacteria-feeding nematode has special mouthparts to consume bacteria. Other species of nematodes graze on plant roots or fungi. *Source:* Reprinted from *The Soil Biology Primer* with permission. ©2000 Soil and Water Conservation Society.

Decomposition of Lignin to Humus

A critical role of fungal decomposers (also known as **saprophytes**) is the breakdown of extremely complex structural components of vascular plants such as grass and trees. Trees, in particular, accumulate vast stores of biomass in forms that are difficult to digest, such as lignin. Fungal and bacterial decomposers possess enzyme systems to degrade lignin and other complex components of plants (presented in Section 13.7). Examples of decomposers include white rot fungi (**Fig. 21.26A**) and actinomycete soil bacteria. The prevalence of lignin is one reason that decomposition by fungi plays a much larger role in terrestrial ecosystems than in marine ecosystems.

Lignin is a highly complex and diverse covalent polymer composed of interlinked phenolic groups (benzene rings with OH or related oxygen-bearing side groups; **Fig. 21.26B**). These phenolic polymers can be broken down relatively rapidly to smaller units, some containing only a single phenol or benzoic acid. This first phase of microbial degradation (about 50%) occurs within a year of deposition in the soil. The remaining phenolics, however, may degrade less than 5% per year; and some samples dated by ^{14}C isotope ratios have been shown to last 2,000 years. These phenolic molecules are called **humic material** or **humus**. Because of its slow degradation, humic material provides a steady slow-release supply of nutrients for plant growth. But forests whose rate of microbial decomposition is particularly low—for example, the New Jersey Pine Barrens—depend on fire to clear the mounting layers of humus and return its minerals to the ecosystem.

Figure 21.26 Microbial formation of humus. A. *Mycena haematopus*, a white rot fungus, growing on a willow log. The fungus degrades lignin. **B.** Lignin is a complex organic polymer that is a component of wood and bark, one of the most abundant polymers on Earth.

A. White rot fungi

Duke University, Dept. of Biology

B. Lignin

Lignin is also a major constituent of newsprint and other kinds of paper disposed of by composting or in landfills. Because of the low degradation rate, particularly in anoxic soil, newsprint in landfills can maintain its structure for decades. Archaeology of landfills outside New York City has revealed readable headlines from papers dating to the 1940s.

Microbes Associated with Roots

The presence of plant roots provides yet another level of complexity to soil communities (**Fig. 21.27**). Plant roots influence the surrounding soil by taking up nutrients and by secreting organic substances and molecules that modulate their surroundings. The environment surrounding a plant root can be further subdivided into two categories: the **rhizoplane**, the root surface; and the rhizosphere, the region of soil outside the root surface but still influenced by plant exudates (materials secreted by the plant). Particular

bacterial species are adapted to these environments. For example, in anaerobic wetland soil, the rhizoplane and rhizosphere of plant roots provide oxygen for methanotrophs that oxidize methane produced by methanogens.

The rhizoplane and rhizosphere also provide the environment for symbiotic fungi that generate mycorrhizae. At least 80% of plants in nature, including 90% of forest trees, require mycorrhizae for growth.

Mycorrhizae: The Fungal Internet

The function and significance of mycorrhizae for plant growth is just beginning to be understood. Mycorrhizae (from *myco*, fungal, and *rhiza*, root) consist of fungal mycelia that associate intimately with the roots of plants,

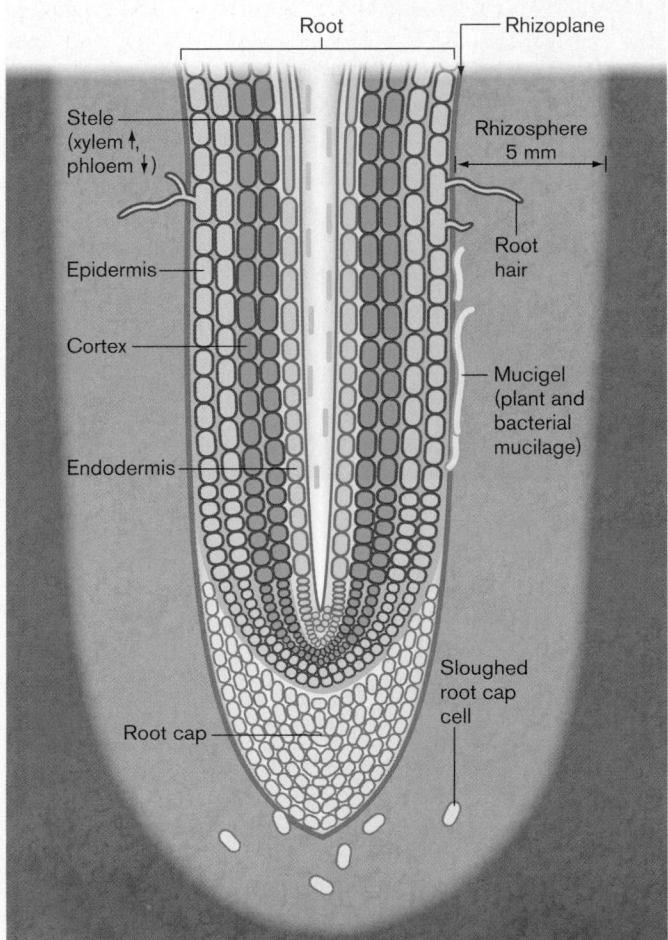

Figure 21.27   Plant roots offer special habitats for microbes. The rhizoplane is the region of soil directly contacting the plant root surface. The rhizosphere is the soil outside the rhizoplane that receives substances from the root, such as mucilage, sloughed cells, and exudates.

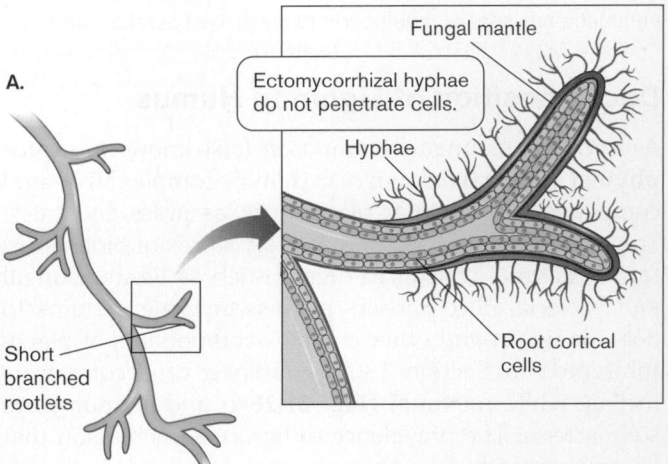

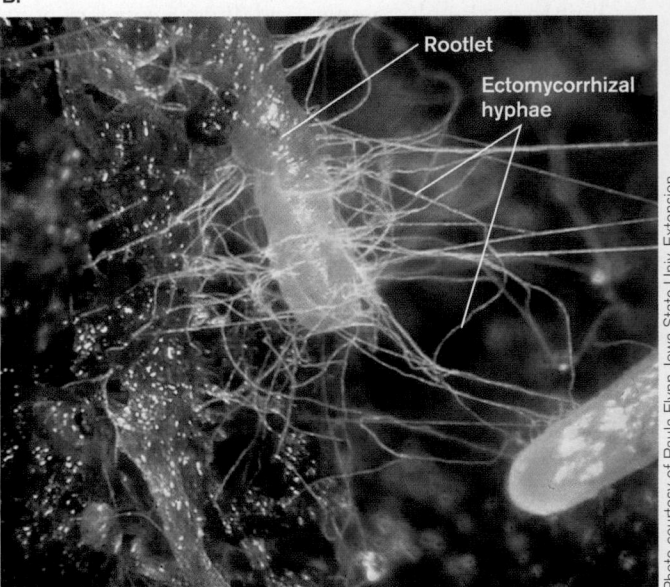

Figure 21.28   Ectomycorrhizae: fungi colonize the surface of the rootlet.   A. Plant root cross section. Ectomycorrhizae extend hyphae from the root surface. **B.** Ectomycorrhizal hyphae from a rootlet (light micrograph).

extending access to minerals while obtaining in return the energy-rich products of plant photosynthesis. Mycorrhizae were first discovered in the 1880s by German truffle hunters who sought to cultivate the prized delicacy, the fruiting body of an ascomycete (for review of fungi, see Chapter 20). The propagation of truffles was investigated by mycologist A. B. Frank at the Landwirtschaftliche Hochschule, Berlin. To Frank's surprise, he found that the truffles extended their mycelia far beyond the site of the fruiting body, and that the mycelia formed an impenetrable tangle with plant roots. Frank called the tangled mycelia "fungus-roots" or mycorrhizae. A century later, we are beginning to appreciate that these mysterious fungal-root tangles offer a vast interconnected network for exchange of nutrients among fungi and many different plants, like an internet connecting countless sites.

Two different kinds of mycorrhizae are observed: ectomycorrhizae and endomycorrhizae. **Ectomycorrhizae** colonize the rhizoplane, the surface of plant rootlets, the most distal part of plant roots (**Fig. 21.28**). The fungal mycelia never penetrate the root; instead, they form a thick mantle surrounding the root and then extend long mycelia away from the root to absorb nutrients. Numerous kinds of fungi form ectomycorrhizae, including ascomycetes, such as truffles; and basidiomycetes, known by their mushrooms, such as stinkhorns. Plants grown with ectomycorrhizae invest less of their body mass in roots and more in the aboveground stems and leaves—an important consideration for agriculture, in which the aboveground plant is usually the part harvested.

Endomycorrhizae form a more intimate association, in which the fungal hyphae penetrate plant cells deep within the cortex (**Fig. 21.29**). The penetrating hyphae form knobbed branches that resemble microscopic "trees," or arbuscules, within the root cells. Some of the hyphae form specialized vesicles within the plant that store nutrients. Another name for this kind of mycorrhizae is **vesicular-arbuscular mycorrhizae (VAM)**.

Endomycorrhizae are more specialized than ectomycorrhizae. They comprise a relatively small number of fungal species, such as members of the zygomycete genus *Glomus*, and they show obligate dependence on their host plants. Their presence in nature and their importance in the ecosystem, however, are actually greater. Endomycorrhizal species exist entirely underground (they do not form mushrooms), and they completely lack sexual cycles. They may acquire 25% of the photosynthetic product of their hosts in exchange for tremendous expansion of access to soil resources.

Mycorrhizae greatly enhance the plant's uptake of water, as well as minerals such as nitrogen and phosphorus. In addition, the hyphae sequester toxins, and they actually distribute organic substances from one plant to another. Mycorrhizae may join many different plants, even different species, in a vast nutrient-sharing network.

> **THOUGHT QUESTION 21.3**   Design an experiment to test the hypothesis that the presence of mycorrhizae enhances plant growth in nature.

A. Endomycorrhizae (plant root cross section)

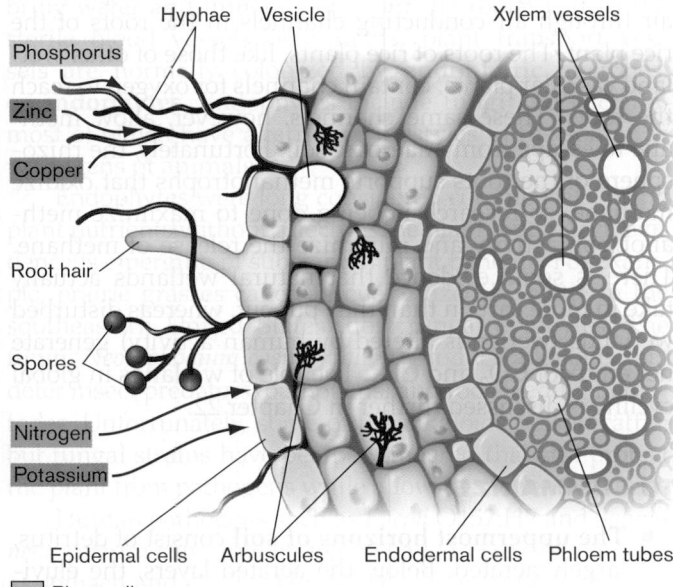

B. Arbuscules within root cells

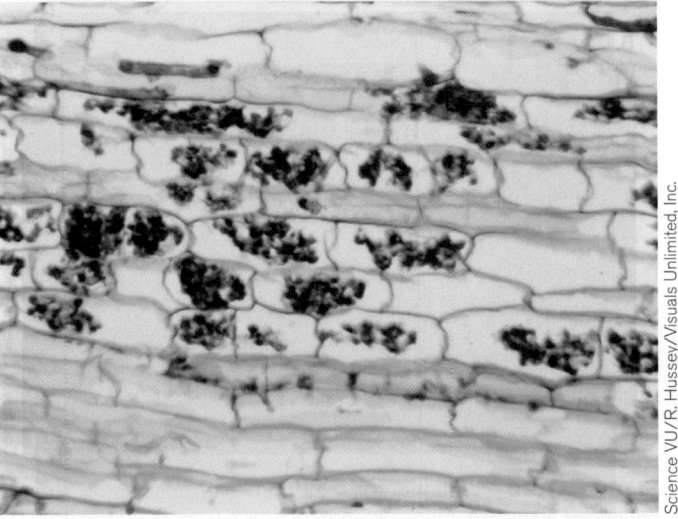

Science VU/R. Hussey/Visuals Unlimited, Inc.

Figure 21.29   Endomycorrhizae: fungi invade root cells, forming arbuscules.   A. Plant root cross section. Endomycorrhizal hyphae penetrate cells deep within the root cortex. **B.** The penetrating hypha forms an arbuscule within a root cell.

Figure 21.38 Endosymbiotic algae. A. Brain coral (*Diploria* sp.) containing zooxanthellae. **B.** *Symbiodinium* sp. are zooxanthellae (symbiotic dinoflagellate algae) of coral.

A.

Becky A. Dayhuff, National Oceanic and Atmospheric Administration/ Dept. of Commerce.

B.

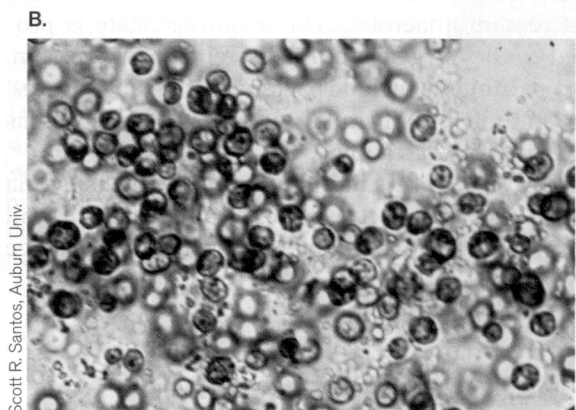

Scott R. Santos, Auburn Univ.

Sponge Communities

Animals often harbor extremely diverse communities of microbes whose specific contributions are poorly understood—for example, bacteria associated with marine sponges (**Fig. 21.39**). In 2001, Russell Hill and colleagues at the University of Maryland, Baltimore, characterized the bacterial species associated with the sponge *Rhopaloeides odorabile*, which grows off the Australian Great Barrier Reef. Using 16S rDNA probes, they identified novel species of deltaproteobacteria, planctomycetes, flavibacteria, green sulfur and nonsulfur bacteria, as well as deep-branching isolates of no known major clade.

Most interesting, they discovered a novel actinomycete that produces antimicrobial substances. Hill hypothesizes that the sponge harbors these bacteria to defend it from pathogens, which consume sponges that lack their own microbial community. Sponge bacteria, therefore, may offer a wealth of previously unknown antibiotics.

Digestive Communities

Highly complex and nutritionally important communities of microbes are found within the gut of vertebrate animals. Digestive chambers, such as the bovine rumen and the human colon, support thousands of species of bacteria, protists, and archaea. The human colon contains numerous fermenters and methanogens, some of which feed on intestinal mucus, whereas others digest complex plant fibers that our intestinal lining cannot, thus providing up to 15% of our caloric intake.

A. Sponge containing actinomycetes

Nicole Webster, Australian Institute of Marine Science.

B. Actinomycete isolated from sponge

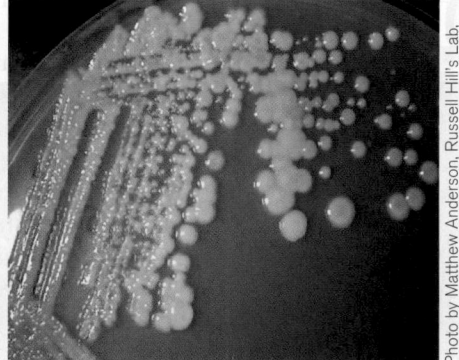

Photo by Matthew Anderson, Russell Hill's Lab, U. of Maryland Biotechnology Institute.

Figure 21.39 Sponge bacterial communities. A. The sponge *Rhopaloeides odorabile* contains actinomycetes that produce novel antimicrobial compounds. **B.** A sponge-associated actinomycete growing on an agar plate.

The bovine rumen microbial community. The most extensively studied digestive communities are those of ruminants, such as cattle, sheep, llamas, and caribou. Throughout most of human civilization, ruminants have provided protein-rich food, textile fibers, and mechanical work. A historical reference is the biblical injunction

A.

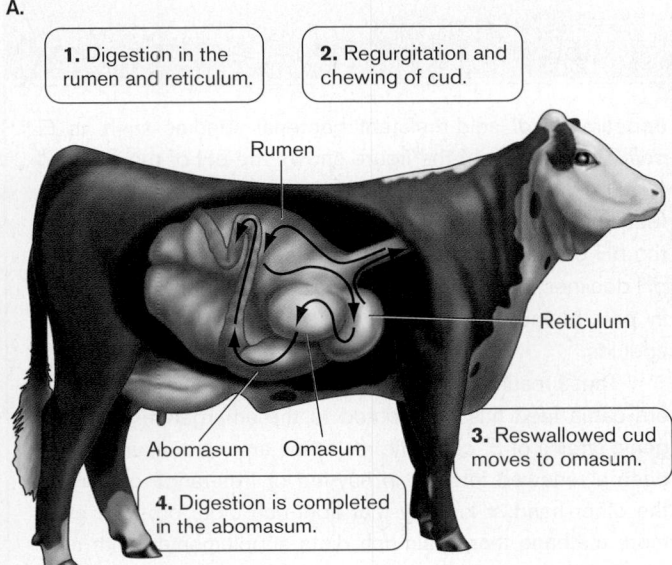

1. Digestion in the rumen and reticulum.

2. Regurgitation and chewing of cud.

Rumen

Reticulum

Abomasum   Omasum

3. Reswallowed cud moves to omasum.

4. Digestion is completed in the abomasum.

B.

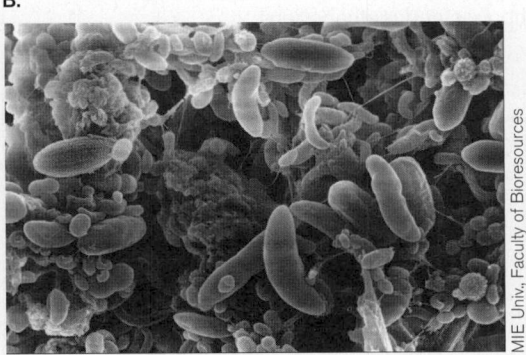

MIE Univ., Faculty of Bioresources

C.

Photo courtesy of James B. Russell, Cornell College of Veterinary Medicine.

Figure 21.40   The bovine rumen.   A. The rumen is the largest of four chambers in the bovine stomach. **B.** Bacteria growing within the rumen. **C.** Todd Callaway, Francisco Diez Gonzales, James B. Russell, and Minas Kizoulis study microbial diversity and metabolism of the bovine rumen.

to consume an animal that "is cleft-footed and chews the cud"—that is, "ruminates," or redigests its food in the fermentation chamber known as the **rumen (Fig. 21.40A and B).**

The microbial community of the rumen enables herbivores to acquire nutrition from complex plant fibers that the animal could not otherwise digest. From a genomic standpoint, such an arrangement makes evolutionary sense. If the animal had to digest all the diverse polysaccharide chains encountered in nature, its own genome would have to encode a wide array of different enzyme systems. Instead, the ruminant relies on diverse microbial species to conduct various kinds of digestion. Organisms that partly digest a substrate (for instance, converting sugars to lactate) provide a substrate such as short-chain acids that the host animal can absorb and digest to completion by aerobic respiration.

THOUGHT QUESTION 21.7   Why does ruminant fermentation leave food value for the animal host? How is the animal able to obtain nourishment from waste products that the microbes could not use?

The bovine gut system has four chambers (**Fig. 21.40A**). The rumen initially digests the feed and then passes it to the reticulum. The reticulum breaks the feed into smaller pieces and traps indigestible objects, such as stones or nails. After initial digestion, feed is regurgitated for rechewing and then returned to the rumen, by far the largest of the chambers. In the rumen, feed is broken down to small particles and fermented slowly by thousands of species of microbes. The partially digested feed passes to the omasum, which absorbs water and short-chain acids produced by fermentation. The abomasum then decreases pH and secretes enzymes to digest proteins before sending its contents to the colon for further nutrient absorption and waste excretion.

While rumen digestion has been studied since the 1830s, major advances in understanding rumen fermentation were first achieved in 1966 by Robert Hungate (1906–2004) at the University of California at Davis, who pioneered techniques of anaerobic microbiology. An example of Hungate's methods still in use today is that of obtaining anaerobic cultures from fistulated cattle—that is, cattle in which an artificial connection is made between the rumen and the animal's exterior (**Special Topic 21.1**). The cultures are analyzed by metagenomics and bioenergetics, as pursued by James Russell and colleagues at the Cornell College of Veterinary Medicine (**Fig. 21.40C**).

Different microbes fill different niches in ruminal metabolism (**Fig. 21.41**). Cattle grown on relatively poor forage (that is, forage high in complex plant content) show a high proportion of ruminal fungi, the chytridiomycetes (discussed in Chapter 20). Chytridiomycete mycelia

Special Topic 21.1 | A Veterinary Experiment: The Fistulated Cow

Ruminal ecology is vital to maximize dairy production and minimize contamination by deadly pathogens such as *E. coli* O157:H7. The rumen is observed by a simple surgical operation to form a fistula, a connecting hole between the rumen and the exterior of the animal (**Fig. 1**). The presence of the fistula, which can be capped and reopened at will, causes the animal no harm and ensures it several years of comfortable existence beyond the usual life of dairy cattle. The fistula can be used to insert experimental feed materials, such as grain or nitrogen supplements, and it can be used to test parameters of rumen fluid, such as pH and microbial content. Study of fistulated cattle is a common exercise for students of veterinary medicine.

An experiment using the fistula is shown in **Figure 2**. In the experiment shown, James Russell and colleagues tested the hypothesis that increasing proportions of grain in cattle feed result in increased acid production and favor the appearance of acid-resistant bacterial species such as *E. coli*. The top part of the figure shows the pH of ruminal fluid (obtained through the fistula) and of fluid from the colon. As feed includes higher proportions of grain (horizontal axis), the pH of the rumen declines, as does that of the colon. As pH declines, the number of *E. coli* bacteria increases greatly, in part because they are more acid resistant than other species.

These results suggest that the high grain content in modern cattle feed has contributed to the emergence of pathogenic strains of *E. coli*. One response among consumers has been a renewed interest in hay-fed or free-range cattle. On the other hand, it is likely that fiber-rich diets produce even more methane than grain-rich diets supplemented with antimethanogen agents. Further research on rumen flora will be needed to optimize cattle production while minimizing the growth of pathogens and methanogens.

A.

B.

Figure 1 **The fistulated cow.** **A.** A cow carries a surgical fistula, an artificial opening into the rumen. **B.** The fistula is uncapped, exposing the contents of the rumen.

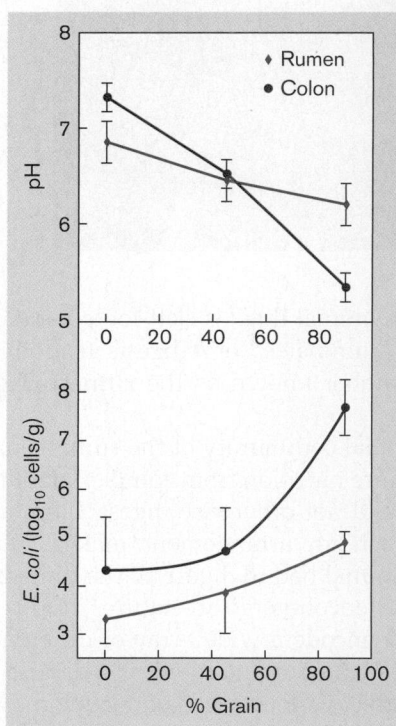

Figure 2 **Rumen pH and bacterial contents.** An experiment by James B. Russell and colleagues in which the fistula was used to test pH and *E. coli* bacterial content of the rumen. As the percentage of grain in feed rises, pH declines and *E. coli* bacteria increase. *Source*: Francisco Diez-Gonzalez et al. 1998. *Science* **281**:1666.

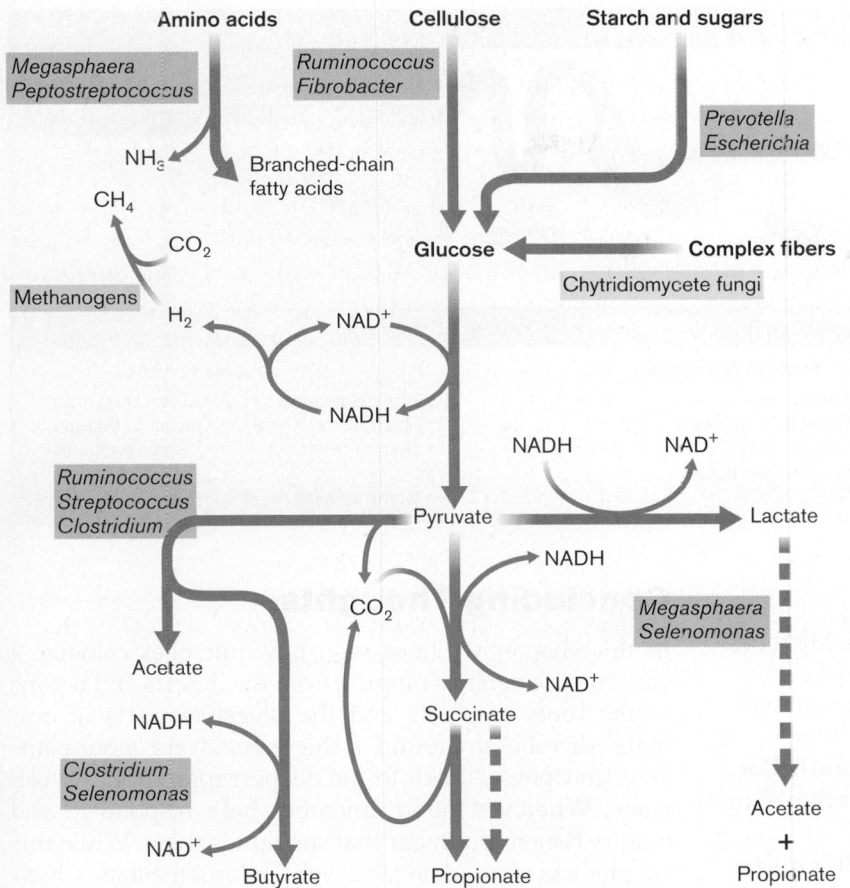

Figure 21.41   Ruminal metabolism. *Source*: Based on Russell and Rychlik. 2001. *Science* **292**:1119.

THOUGHT QUESTION 21.8 How do you think cattle feed might be altered or supplemented to decrease methane production?

A concern with modern cattle rearing is the shift in feed from hay to grain, whose higher content of starch leads to more rapid digestion and faster growth of the animal. Unfortunately, rapid starch digestion favors fermenters such as *Prevotella* species and *E. coli*, which generate higher acid levels and gases, leading to "starch bloat." Furthermore, rumen acidity selects for acid-resistant pathogens such as the *E. coli* strain O157:H7.

Other vertebrate digestive communities. Many kinds of vertebrate animals besides ruminants have gut microbial communities whose fermentation contributes to host nutrition (**Fig. 21.42**). Cattle are considered "foregut" fermenters in that their fermenting microbes receive the feed before the host's small intestine has digested and absorbed the best nutrients. Other foregut fermenters include colobine monkeys, and marsupials such as kangaroos. Kangaroos differ from cattle in that their foregut retention time is much shorter, so there is less time for action of slow-growing microbes such as methanogens. Thus, kangaroos produce little methane, in contrast to cattle. Switching from cattle to kangaroo meat has been proposed in Australia as a way to decrease greenhouse gas emissions.

Other animals, including humans, conduct microbial fermentation in the "hindgut" region (in humans, the colon). Hindgut fermentation favors bacteria capable of digesting complex plant materials that pass undigested through the small intestine. For example, the prominent *Bacteroides* species ferment mucopolysaccharides, pectin, and arabinogalactan among many others. *Bacteroides* species release oligosaccharides (short-chain sugar polymers) that are fermented further by *Bifidobacterium* species. In addition, bacterial biosynthesis may contribute essential amino acids absorbed by the colon.

Besides nutrition, human microbiota interact positively with our immune system and generate defenses against pathogens, as do plant endophytes. For example, *Bifidobacterium* species down-regulate human secretion of cytokines, molecules that cause inflammation in ulcerative colitis. Other colonic bacteria produce bacteriocins, molecules that prevent pathogens from adhering to the gut lining. Human microbiota are discussed further in Chapter 23.

appear on ruminal food particles, and their motile zoospores—formerly mistaken for protists—swim through rumen fluid. By contrast, cattle fed a high cellulose diet, such as hay, grow faster and show cellulolytic bacteria such as *Ruminococcus albus* and *Fibrobacter flavefaciens*.

Metabolism of cellulolytic bacteria requires the presence of small amounts of branched-chain fatty acids. The branched-chain acids turn out to be produced by amino acid fermenters such as *Megasphaera elsdenii* and *Peptostreptococcus anaerobius*. However, too much degradation of amino acids can lead to overproduction of ammonia, poisoning the animal; thus, the protein content of cattle feed must be limited.

Another problem with ruminal fermentation is the frequent production of H_2 and CO_2. Hydrogen production is hard to avoid because the quantity of electron donors (reduced food molecules) greatly exceeds that of the available electron acceptors. The H_2 and CO_2 from fermentation support methanogens, wasting valuable carbon from feed and contributing a substantial part of global methane emissions. So much methane is formed by the rumen that a cannula inserted into the rumen liberates enough of the gas to light a flame.

Figure 21.42   Vertebrate gut fermentation.   Vertebrate animals vary in their location of microbial fermentation. Fermenting microbes may be active mainly in the foregut (rumen), before the stomach, as in cattle; in the cecum, a digestive organ following the small intestine of rabbits and rodents; or in the colon (large intestine), as in humans. *Source:* William Karasov and Hannah Carey. 2009. *Microbe* **4**:323.

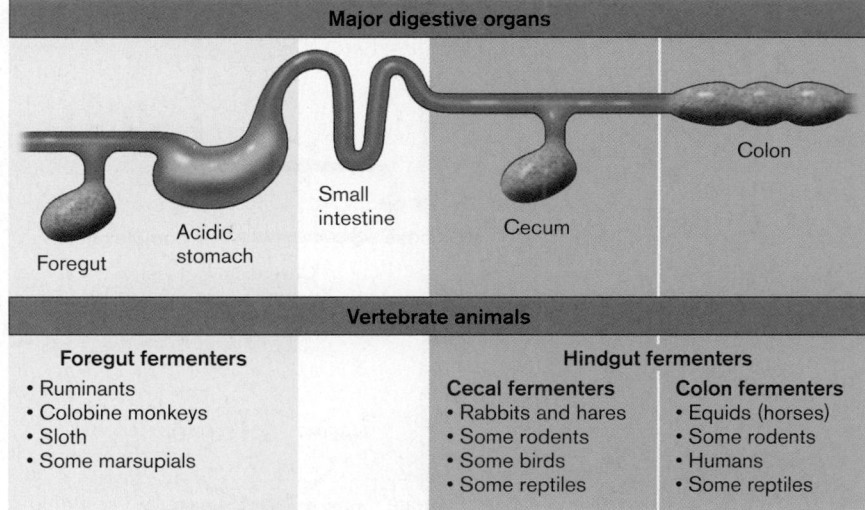

Major digestive organs				
Foregut	Acidic stomach	Small intestine	Cecum	Colon

Vertebrate animals

Foregut fermenters	Hindgut fermenters	
	Cecal fermenters	Colon fermenters
• Ruminants	• Rabbits and hares	• Equids (horses)
• Colobine monkeys	• Some rodents	• Some rodents
• Sloth	• Some birds	• Humans
• Some marsupials	• Some reptiles	• Some reptiles

TO SUMMARIZE:

■ **Animals harbor microbial communities.** Microbes populate the skin, where they provide protection from pathogens. Other species grow in the animal's digestive system, where they enhance nutrition.

■ **Corals and sea anemones harbor zooxanthellae,** algae that provide products of photosynthesis in exchange for a protected habitat.

■ **The rumen of ruminant animals is a complex microbial digestive chamber.** Rumen microbes, including bacteria, protists, and fungi, digest complex plant materials. The microbial digestion generates short-chain fatty acids that are absorbed by the intestinal lining.

■ **Foregut and hindgut fermentation contribute to digestion** in many vertebrate animals, including humans.

Concluding Thoughts

In this chapter we have seen how microbes colonize a vast array of habitats ranging from the deserts and oceans to the roots of plants and the digestive tracts of animals. Microbes are found in the upper layers of our planet's atmosphere, down to the deepest rock strata we can reach. Wherever found, microbes both respond to and modify the environment that surrounds them. While this chapter has focused on food webs in local habitats, Chapter 22 takes a global perspective of microbial ecology and its roles in the cycling of essential elements, such as nitrogen and iron, in Earth's biosphere.

CHAPTER REVIEW

Review Questions

1. What unique functions do microbes perform in ecosystems?

2. Explain the difference between carbon assimilation and dissimilation.

3. What kinds of microbial metabolism are favored in aerated environments? In anoxic environments?

4. What are examples of microbial producers in ecosystems? Include phototrophs as well as lithotrophs. Can a microbe be both a producer and a consumer? Explain.

5. Explain the microbial relationships in various forms of symbiosis, including mutualism, commensalism, and parasitism. Outline an example of each, detailing the contributions of each partner.

6. Compare and contrast the marine food web with the soil food web. What kinds of organisms are the producers and consumers? How many trophic levels are typically found?

7. Explain the role of biofilms in marine and soil habitats. Are these habitats typically uniform or patchy?
8. Compare and contrast the roles of microbes in photic and aphotic marine communities.
9. Compare and contrast the microbial activities in aerated and waterlogged soil.
10. Explain the different methods of quantifying microbial communities, including the advantages and limitations of each.
11. Explain how anaerobic microbial metabolism can enrich soil for plant cultivation.
12. What are mycorrhizae, and how are they important for plant growth?
13. Explain how bacterial mutualists fix nitrogen for plants.

Thought Questions

1. Explain with specific examples what mutualism and parasitism have in common, and how they differ. Explain an example of a relationship that combines aspects of both.
2. The photic zone and the benthic zone pose challenges and opportunities for marine microbes. What challenges do they have in common, and how do they differ?
3. How do we quantify the effects of a microbial population on an ecosystem? Discuss the roles of population size and turnover.

Key Terms

algal bloom (809)
amensalism (798)
assimilation (791)
bacteroid (817)
barophile (805)
benthic organism (799)
biochemical oxygen demand (BOD) (799)
biomass (792)
coastal shelf (799)
cold seep (806)
commensalism (798)
consumer (792)
coral bleaching (821)
cryptogamic crust (795)
decomposer (792)
detritus (793)
dissimilation (791)
ecosystem (791)
ectomycorrhiza (815)
endolith (811)
endomycorrhiza (815)
endophtye, endophytic (817)
euphotic (photic) zone (799)
eutrophic (808)

extremophile (793)
flavonoid (817)
food web (792)
grazer (792)
haustorium (821)
humic material (humus) (813)
hydric soil (816)
hydrothermal (thermal) vent (805)
infection thread (818)
leghemoglobin (818)
lichen (795)
lignin (813)
limiting nutrient (809)
littoral zone (808)
marine snow (801)
metagenomics (799)
microplankton (800)
mixotroph (805)
mutualism (795)
mycorrhiza (812)
nanoplankton (800)
niche (791)
oligotrophic (808)
parasitism (798)
pelagic zone (798)

phytoplankton (803)
picoplankton (800)
plankton (800)
predator (792)
primary producer (791)
psychrophile (805)
rhizobium (817)
rhizoplane (814)
rhizosphere (812)
rumen (823)
saphrophyte (813)
soil (810)
symbiosis (794)
symbiosome (818)
synergism (797)
syntrophy (796)
thermocline (799)
thermophile (805)
trophic level (792)
uncultured (799)
vesicular-arbuscular mycorrhiza (VAM) (815)
water table (811)
wetland (816)
zooxanthella (821)

Recommended Reading

Azam, Farooq, and Francesca Malfatti. 2007. Microbial structuring of marine ecosystems. *Nature Reviews. Microbiology* 5:782–792.

Brugerolle, Guy. 2004. Devescovinid features, a remarkable surface cytoskeleton, and epibiotic bacteria revisited in *Mixotricha paradoxa*, a parabasalid flagellate. *Protoplasma* 224:49–59.

DeLong, Edward F. 2009. The microbial ocean from genomes to biomes. *Nature* 459:200–212.

Diez-Gonzalez, Francisco, Todd R. Callaway, Menas G. Kizoulis, and James B. Russell. 1998. Grain feeding and the dissemination of acid-resistant *Escherichia coli* from cattle. *Science* 281:1666–1668.

Dubilier, Nicole, Claudia Bergin, and Christian Lott. 2008. Symbiotic diversity in marine animals: The art of harnessing chemosynthesis. *Nature Reviews. Microbiology* 6:725–740.

Fierer, Noah, and Robert B. Jackson. 2006. The diversity and biogeography of soil bacterial communities. *Proceedings of the National Academy of Sciences USA* 103:626–631.

Gage, Daniel J. 2004. Infection and invasion of roots by symbiotic, nitrogen-fixing rhizobia during nodulation of temperate legumes. *Microbiology and Molecular Biology Reviews* 68:280–300.

Huq, Anwar, Mohammed Yunus, Syed Salahuddin Sohel, Abbas Bhuiya, Michael Emch et al. 2010. Simple sari filtration is sustainable and continues to protect villagers from cholera in Matlab, Bangladesh. *mBio* 1:e00034-10.

Johnson, Zackary I., Erik R. Zinser, Allison Coe, Nathan P. McNulty, E. Malcolm, et al. 2006. Niche partitioning among *Prochlorococcus* ecotypes along ocean-scale environmental gradients. *Science* 311:1737–1740.

Karasov, William H., and Hannah V. Carey. 2009. Metabolic teamwork between gut microbes and hosts. *Microbe* 4:323–328.

Karl, David M. 2007. Microbial oceanography: Paradigms, processes, and promise. *Nature Reviews. Microbiology* 5:759–769.

Parniske, Martin. 2008. Arbuscular mycorrhiza: The mother of plant root endosymbioses. *Nature Reviews. Microbiology* 6:763–775.

Rusch, Douglas B., Aaron L. Halpern, Granger Sutton, Karla B. Heidelberg, Shannon Williamson, et al. 2007. The Sorcerer II global ocean sampling expedition: Northwest Atlantic through eastern tropical Pacific. *Public Library of Science Biology* 5(3):e77.

Russell, James B., and Jennifer L. Rychlik. 2001. Factors that alter rumen microbial ecology. *Science* 292:1119–1122.

Sato, Tomoyuki, Yuichi Hongoh, Satoko Noda, Satoshi Hattori, Sadaharu Ui, et al. 2009. *Candidatus* Desulfovibrio trichonymphae, a novel intracellular symbiont of the flagellate *Trichonympha agilis* in termite gut. *Environmental Microbiology* 11:1007–1015.

Sleator, Roy D., C. Shortall, and Colin Hill. 2008. Metagenomics. *Letters in Applied Microbiology* 47:361–366.

Walker, Jeffrey J., John R. Spear, and Norman R. Pace. 2005. Geobiology of a microbial endolithic community in the Yellowstone geothermal environment. *Nature* 434:1011–1014.

Young, I. M., and J. W. Crawford. 2004. Interactions and self-organization in the soil-microbe complex. *Science* 304:1634–1637.

 For further study and review, please visit StudySpace at **microbiology2.com**.

Chapter 22

Microbes and the Global Environment

22.1 Biogeochemical Cycles

22.2 The Carbon Cycle

22.3 The Hydrologic Cycle and Wastewater Treatment

22.4 The Nitrogen Cycle

22.5 Sulfur, Phosphorus, and Metals

22.6 Astrobiology

Microbes throughout the biosphere recycle carbon, nitrogen, sulfur, and other elements essential for all life. Through their biochemical transformations, diverse microbial activities largely determine the quality of soil, air, and water for human life. Today, all of these geochemical cycles are altered profoundly by human activity. The burning of fossil fuels, which were generated millions of years ago by subterranean microbes, releases quantities of carbon dioxide too great to be absorbed by marine bacteria and algae. Growing rice production increases the release of methane by methanogens that thrive in submerged rice paddies. Both carbon dioxide and methane are "greenhouse gases" that lead to global warming. Will we humans learn to manage the perturbations of our own biosphere, from pollution to global warming?

From local wastewater treatment to the control of greenhouse gases, microbes are our hidden partners on Earth. And Earth's microbial cycles lead us to wonder whether biospheres exist on other worlds. Our neighbor planet Mars shows tantalizing hints of water and the possibility of hidden microbial life.

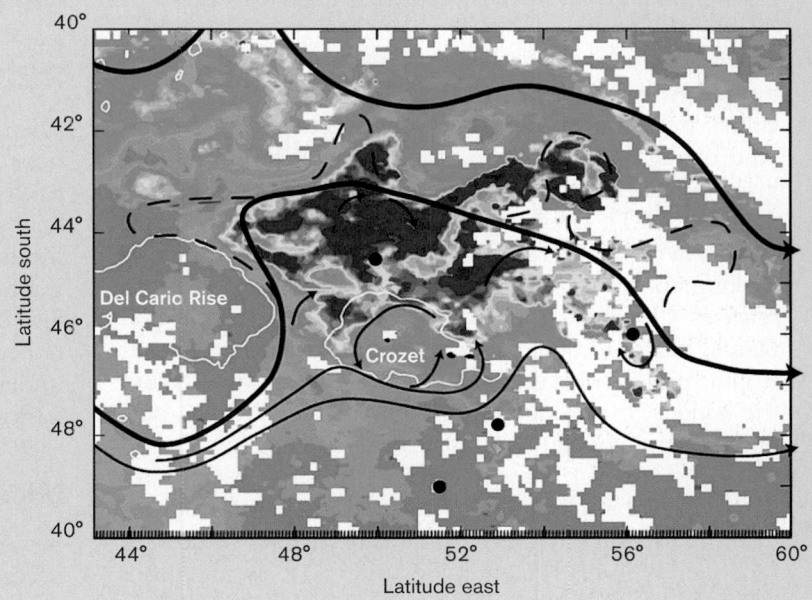

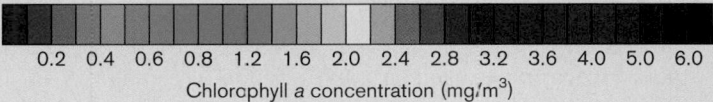

Chlorophyll *a* concentration (mg/m^3)

Current Research Highlight

Chlorophyll *a* concentration in algal bloom surrounding the Crozet Islands in the Southern Ocean between Africa and Antarctica. The island naturally releases iron, a limiting nutrient for phytoplankton (marine algae). Iron release causes an annual bloom of 120,000 km^2 in surrounding water. Marine algal blooms make a significant contribution to global carbon flux. Chlorophyll concentration is based on color measurements obtained by satellite using the Sea-viewing Wide Field-of-view Sensor (SeaWiFS) spectroradiometer, which measures bands of the visible light spectrum. Dark lines show water circulation, and dark spots indicate sampling stations for particulate carbon (biomass fixed by algae). Particulate carbon was measured to assess the ability of algal blooms to remove CO_2 from the atmosphere. *Source:* Raymond Pollard et al. 2009. *Nature* **457**:577. Photo courtesy of NASA.

Two billion years ago, ancient cyanobacteria acquired the ability to photolyze water and produce molecular oxygen. Oxygen is a powerful oxidant, lethal to most life in that anoxic era, when all living organisms were anaerobic microbes. Most anaerobic species must have gone extinct, until a lucky few species evolved mechanisms of antioxidant protection. Since then, microbes have shaped our biosphere by generating oxygen, by fixing gaseous nitrogen and returning it to the atmosphere, and by fixing and producing carbon dioxide. These and countless other microbial processes not only have made Earth's atmosphere what it is but continue to play a crucial role in its homeostasis.

During the past century, key aspects of global biospheric chemistry have been perturbed by human technology. Already nearly half the nitrogen in the biosphere comes from anthropogenic (human-generated) sources. And increases in atmospheric CO_2 and CH_4, important greenhouse gases, correlate with the rise of the industrial age (**Fig. 22.1**). These gases are introduced into the biosphere by bacteria, fungi, and protists, and increasingly by human technology. Human burning of petroleum releases CO_2 from the product of ancient bacteria and plants. Much of global carbon dioxide is fixed by phototrophic prokaryotes and algae, as well as land plants. But the added CO_2 production by human industry has outpaced the rate of plant and microbial CO_2 fixation.

Concern over the effect of increasing greenhouse gases on global climate change led the United Nations, meeting in Kyoto in 1997, to adopt the Kyoto Protocol for reduction of industrial emissions of CO_2, CH_4, and other gases that contribute to global warming. Over 160 nations adopted the protocol, although the United States declined, contending that it posed a disproportionate economic burden on developed nations. Will the human-induced global warming cause mass extinctions of a majority of Earth's species—as did the rise of ancient cyanobacteria? Or can we use our knowledge of microbial ecology to channel microbial activities into recovering the balance—for example, by increasing microbial CO_2 fixation?

Throughout most of this book, we present microbial biochemistry in the context of growth of individual organisms. In Chapter 22, we show how the collective metabolic activities of microbial populations contribute to global cycles of elements throughout Earth's biosphere. We consider the ways that we humans enlist microbes to manage our environment. Finally, we explore how our awareness of global microbiology has renewed our quest for life beyond Earth.

22.1 Biogeochemical Cycles

Microbes possess exceptional abilities to interconvert molecules. Nearly any kind of molecule can be metabolized by some species somewhere if the reaction provides energy or a useful nutrient. But even microbes face the limits of the actual elements themselves: No microbe, nor any other living creature, can convert one element to another. Living organisms conduct chemical reactions, not nuclear reactions—although microbes that live deep underground may obtain chemical energy from hydrogen ions generated by uranium decay.

Microbes Cycle Essential Elements

Because organisms cannot create their own elements, they need to get them from their environment. While small amounts of matter enter the biosphere from outer space, and gases continually escape Earth, the rate of these changes is tiny compared to the rates of living processes. So organisms acquire their elements either from nonliving components of their environment, such as by fixing atmospheric CO_2, or from other organisms, by grazing, predation, or decomposition. Furthermore, all organisms recycle their components back to the biosphere. The partners in this recycling include **abiotic** entities such as air, water, and minerals, as well as **biotic** entities such as predators and decomposers. Collectively, the metabolic interactions of microbial communities with the biotic and abiotic components of their ecosystems are known as **biogeochemistry** or **geomicrobiology**.

Which elements need to be recycled and made available for life? In most organisms, six elements predominate: carbon, oxygen, nitrogen, hydrogen, phosphorus, and sulfur (discussed in Chapter 4). **Table 22.1** summarizes the elemental composition typical of Gram-negative bacteria. In other microbes, elemental proportions vary with species and growth conditions. Some species include inorganic components such as the silicate shells of diatoms or the sulfur granules of phototrophic bacteria. All of these elements flow in **biogeochemical cycles** of nutrients throughout the biotic and abiotic components of the biosphere. The environmental levels of key elements can limit biological productivity. For example, the concentration of iron limits the

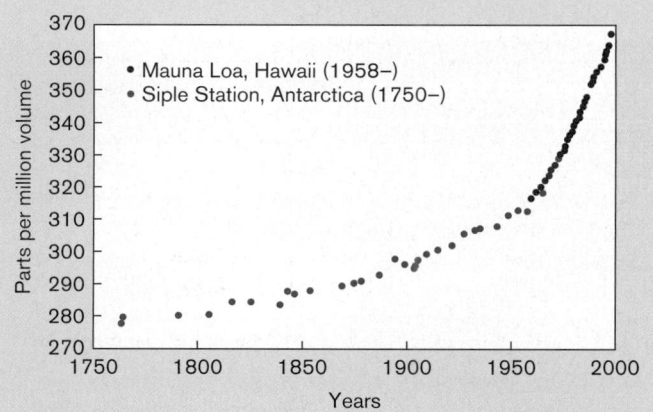

Atmospheric carbon dioxide (CO_2) concentrations (1750–present)

- Mauna Loa, Hawaii (1958–)
- Siple Station, Antarctica (1750–)

Figure 22.1 Global carbon dioxide levels. Atmospheric CO_2 levels since 1750 reveal the increase of this greenhouse gas accompanying the rise in industrial consumption of fossil fuels. Microbial CO_2 fixation helps limit CO_2 increase in the atmosphere. *Source:* World Resources Institute.

Table 22.1 Typical elemental composition of Gram-negative bacteria.

Element	Dry weight, %
Carbon	50
Oxygen	20
Nitrogen	14
Hydrogen	8
Phosphorus	3
Sulfur	1
Potassium	1
Sodium	1
Calcium	0.5
Magnesium	0.5
Chlorine	0.5
Iron	0.2
All others	0.3

marine populations of phytoplankton, which depend on iron supplied by wind currents. Other elements, such as zinc and copper, are **micronutrients**, nutrients required in still smaller amounts. Micronutrients, too, must be cycled, although their flux is challenging to measure.

Sources and Sinks of Essential Elements

Biogeochemical cycles include both biological components (such as phototrophs that consume CO_2 and heterotrophs that release CO_2) and geological components (such as volcanoes that release CO_2 and oceans that absorb CO_2). The major parts of the biosphere containing significant amounts of an element needed for life are called reservoirs of that element (**Table 22.2**). Each

reservoir acts both as a source of that element for living organisms and as a sink to which the element returns. For example, the ocean is an important reservoir for carbon (CO_2 in equilibrium with HCO_3^-, bicarbonate ion). The ocean's carbon cycles rapidly; thus, the ocean serves as both a source and a sink for carbon.

As elements cycle from sources to sinks, microbial metabolism generates a series of redox changes (discussed in Chapter 14). The major oxidation states of carbon, nitrogen, and sulfur are summarized in **Table 22.3**. Biospheric carbon can be found as CH_4 generated by methanogens (completely reduced, −4), as CO_2 produced by respiration and fermentation (completely oxidized, +4), or as one of various intermediate states of oxidation. Nitrogen is excreted by many organisms in its most reduced form, ammonia (NH_3), which is protonated to ammonium ion (NH_4^+). Ammonium is oxidized by lithotrophic bacteria through several stages to nitrite (NO_2^-) and nitrate (NO_3^-), which serve as terminal electron acceptors for anaerobic respiration. Sulfides (H_2S, HS^-) serve as electron donors for respiration and sulfur phototrophy, whereas oxidized forms, such as sulfite (SO_3^{2-}) and sulfate (SO_4^{2-}), serve in anaerobic respiration.

> **THOUGHT QUESTION 22.1** Why is oxidation state critical for the acquisition, usability, and potential toxicity of cycled compounds? Cite examples based on your study of microbial metabolism.

How do we study microbial cycling on a global scale? How do we figure out whether ecosystems are net sources or sinks of CO_2? Does microbial activity enhance or limit availability of nitrogen? Rates of flux of elements in the biosphere are very difficult to measure, yet the questions have enormous political and economic implications.

Table 22.2 Global reservoirs of carbon, nitrogen, and sulfur.[a]

Reservoir	Carbon	Rate of cycling	Nitrogen	Rate of cycling	Sulfur	Rate of cycling
Atmosphere	700 (CO_2)	Fast	3,900,000 (N_2)	Slow	0.0014 (SO_2, H_2S)	Fast
Ocean						
Biomass	4	Fast	0.5	Fast	0.15	Fast
Organic molecules	2,100	Fast	300	Fast	–	–
Inorganic molecules	38,000 (HCO_3^-, CO_3^{2-})	Fast	20,000 (N_2) / 690 (NO_3^-, NO_2^-, NH_4^+)	Slow / Fast	1,200,000 (SO_4^{2-})	Slow
Land						
Biomass	500	Fast	25	Fast	8.5	Fast
Organic matter (soil)	1,200	Fast	110	Slow	16	Fast
Crust (below land and ocean)	120,000,000	Slow	770,000	Slow	18,000,000	Slow
Fossil fuel (oil, coal, natural gas)	13,000	Fast				

[a]Units are 10^9 metric tons, or 10^{12} kg.

Source: R. Maier et al. 2000. *Environmental Microbiology*. Academic Press: San Diego, CA.

Table 22.3 Oxidation states of cycled compounds.

Oxidation state	Carbon		Nitrogen		Sulfur	
−4	CH_4	Methane				
−3			NH_3, NH_4^+	Ammonia, ammonium ion		
−2	CH_3OH, $(CH_2)_n$	Methanol, hydrocarbon	H_2N-NH_2	Hydrazine	H_2S, HS^-	Sulfides
−1			NH_2OH	Hydroxylamine		
0	$(CH_2O)_n$	Carbohydrate	N_2	Nitrogen	S^0	Elemental sulfur
+1			N_2O	Nitrous oxide		
+2	HCOOH	Formic acid	NO	Nitric oxide	$S_2O_3^{2-}$	Thiosulfate
+3			HNO_2, NO_2^-	Nitrous acid, nitrite ion		
+4	CO_2, HCO_3^-	Carbon dioxide, bicarbonate ion	NO_2	Nitrogen dioxide	SO_3^{2-}	Sulfite
+5			HNO_3, NO_3^-	Nitric acid, nitrate ion		
+6					H_2SO_4, SO_4^{2-}	Sulfuric acid, sulfate

To measure environmental carbon, nitrogen, and other elements, various methods are used. These methods fall under the following categories:

■ **Chemical and spectroscopic analysis.** Bulk quantities of CO_2, nitrates, and other chemicals can be determined by sophisticated chemical instrumentation. Atmospheric CO_2 is measured by infrared absorption spectroscopy, applied to samples from towers such as those of the NASA FLUXNET study (**Fig. 22.2**). Gas chromatography is used to separate and quantify various gases, including oxygen, nitrogen, sulfur dioxide, and carbon monoxide. Mass spectroscopy detects extremely small quantities of different molecules, even distinguishing between elemental isotopes.

■ **Radioisotope incorporation.** The influx and efflux of CO_2 can be measured by the uptake of ^{14}C-labeled substrates in a small, controlled model ecosystem called a **mesocosm**. Alternatively, CO_2 flux can be measured with radioisotope tracers in the field using a field chamber.

■ **Stable isotope ratios.** Some enzyme reactions show a preference for one isotope over another, such as ^{14}N versus [15]N. For example, denitrifiers (bacteria that metabolize nitrate) strongly prefer the ^{14}N isotope, leaving behind nitrate enriched in [15]N. The ^{14}N/^{15}N ratio is measured using mass spectroscopy. Measuring nitrogen isotope ratios can indicate whether denitrifiers could have conducted metabolism in the sample.

Figure 22.2 Measuring flux of elements in the biosphere. An American FLUXNET tower at Austin Cary Memorial Forest is used for atmospheric CO_2 sampling as part of a global effort to monitor carbon flux.

TO SUMMARIZE:

- **Microbes cycle essential elements in the biosphere.** Many key cycling reactions are performed only by bacteria and archaea.
- **Elements cycle between organisms and abiotic sources and sinks.** The most accessible source of carbon and nitrogen is the atmosphere. The Earth's crust stores large amounts of key elements, but their availability to organisms is limited.
- **Environmental flux of elements is measured through chemistry.** Methods include infrared and mass spectroscopy, gas chromatography, radioisotope incorporation, and the measurement of isotope ratios.

22.2 The Carbon Cycle

The foundation of all food webs involves influx and efflux of carbon. The major reservoirs of carbon are shown in **Table 22.2**. In theory, carbonate rock forms the largest

reservoir of carbon. But Earth's crust is the source least accessible to the biosphere as a whole. Crustal rock provides carbon only to organisms at the surface and to subsurface microbes that grow extremely slowly. Thus, subsurface carbon turnover is very slow. The carbon reservoir that cycles most rapidly is that of the atmosphere, a source of CO_2 for photosynthesis and lithotrophy. The atmosphere also acts as a sink for CO_2 produced by heterotrophy and by geological outgassing from volcanoes.

The atmospheric reservoir is much smaller than other sources, such as the oceans, crustal rock, and fossil fuels. For this reason, the industrial burning of fossil fuels has perturbed the balance between atmospheric CO_2 and larger reservoirs, such as the ocean. The ocean actually absorbs a good part of the extra CO_2, where the carbon is eventually converted to carbonates. In addition, marine phototrophs such as diatoms and coccolithophores trap a substantial amount of carbon in biomass. A portion of their biomass sinks to the ocean floor through the weight of their silicate or carbonate exoskeletons. But despite these buffering effects of the ocean, atmospheric CO_2 continues to increase at a rate of about 1% per year. The CO_2 traps solar radiation as heat—a process known as the **greenhouse effect**. CO_2 is one of several greenhouse gases contributing to global warming, the overall rise in temperature of our biosphere over the past hundred years.

Note, however, that the fact that CO_2 is a greenhouse gas does not make it inherently "bad" for the environment. In fact, if heterotrophic production of CO_2 were to cease altogether, phototrophs would run out of CO_2 in roughly 300 years, despite the vast quantities of carbon present in the ocean and crust. Thus, both CO_2 fixers and heterotrophs need each other.

Carbon Cycles Depend on Oxygen

The global cycle of carbon in the biosphere is closely linked to the cycles of oxygen and hydrogen, elements to which most carbon is bonded (**Fig. 22.3**). Overall, carbon cycles between carbon dioxide (CO_2) and various reduced forms of carbon, including biomass (living material). Note that the results of carbon cycling differ greatly, depending on the presence of molecular oxygen.

Aerobic carbon cycling. Aerated ecosystems include the photic zone of oceans and the oxygenated surface of terrestrial habitats (discussed in Chapter 21). In an aerated (or oxic) habitat, such as the marine photic zone, the

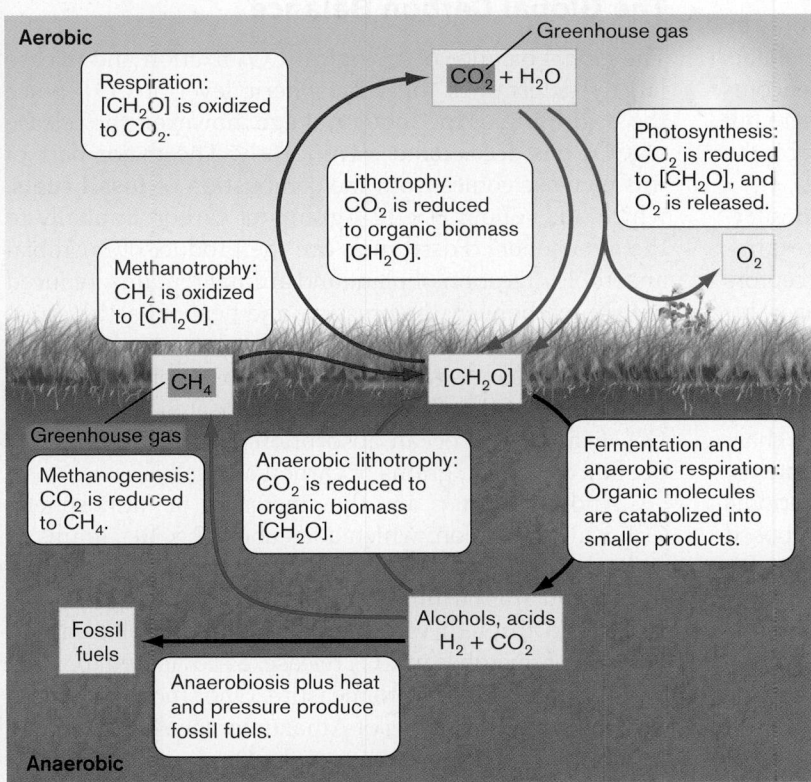

Figure 22.3 The carbon cycle: aerobic and anaerobic. Aerobic and anaerobic conversions of carbon. Blue = reduction of carbon; red = oxidation of carbon; black = fermentation; [CH$_2$O] = organic biomass. In an aerobic environment, photosynthesis generates molecular oxygen (O$_2$), which enables the most efficient metabolism by heterotrophs and lithotrophs. In an anaerobic environment, photosynthesis generates only oxidized minerals, which support limited anaerobic respiration. Fermentation generates organic carbon products, as well as CO$_2$ and H$_2$. In the absence of oxygen, methanogens convert carbon dioxide (CO$_2$) and molecular hydrogen (H$_2$) to methane (CH$_4$), one of the most potent greenhouse gases.

ecosystem absorbs enough light for the rate of photosynthesis to exceed the rate of heterotrophy. Microbial and plant photosynthesis fixes CO_2 into biomass, designated by the shorthand $[CH_2O]$. Phototrophs include bacteria and protists, as well as plants. Aerobic CO_2 fixation is accompanied by release of O_2. The O_2 is then used by heterotrophs (such as bacteria, protists, and animals) to convert $[CH_2O]$ back to CO_2. In the presence of light, a net excess of O_2 is released.

Biomass is also produced through lithotrophy—the oxidation of hydrogen, hydrogen sulfide, ferrous iron (Fe^{2+}) and other reduced minerals, and even carbon monoxide. Lithotrophy is especially prominent in soil and in weathered areas of crustal rock. Lithotrophy is performed solely by bacteria and archaea, essential microbial partners in these ecosystems.

Anaerobic carbon cycling. Anoxic environments support lower rates of biomass production than do oxygen-rich environments because they depend on oxidants of lower redox potential and limited quantity, such as Fe^{3+}. Anaerobic conversion of CO_2 to biomass is done mainly by bacteria and archaea. Vast, permanently anaerobic habitats extend several kilometers below Earth's surface, encompassing greater volume than the rest of the biosphere put together. In these habitats, endolithic bacteria inhabit the interstices of rock crystals (discussed in Chapter 21).

In soil and water, anaerobic metabolism includes fermentation of organic carbon sources, as well as respiration and lithotrophy with alternative electron acceptors such as nitrate, ferric iron (Fe^{3+}), and sulfate. Anaerobic decomposition by microbes is one stage in the formation of fossil fuels such as oil and natural gas (primarily methane). In soil, anoxic conditions (extremely low in O_2) favor incomplete breakdown of organic material. This characteristic of anaerobic soil actually enriches ecosystems, particularly those of wetlands, which undergo periodic cycles of aeration and hydration. The partly decomposed matter becomes available for further decomposition with oxygen. Anoxic environments near the surface also favor production of methane from the H_2, CO_2, and other fermentation products of anaerobes.

A major source of concern for global greenhouse gases is the methane hydrates accumulating in deep marine sediments, generated by huge benthic communities of methanogens (discussed in Chapters 19 and 21). Warming of methane hydrate releases gaseous methane, which quickly rises to the atmosphere. Geological evidence suggests that rapid methane release accompanied the retreat of the glaciers during ice age transitions. A rapid methane release today could accelerate global warming.

Another surprising source of marine methane, discovered recently, is the aerobic ammonia-oxidizing archaea, such as *Nitrosopumilus* species (discussed in Chapter 19). These archaea conduct reactions that degrade toxic phosphonates, organic compounds containing a direct carbon-phosphorus bond. The reactions release methylphosphonate, a compound that many kinds of bacteria convert to methane in order to acquire the scarce phosphate for phospholipids and nucleic acids. The methane is then released into the atmosphere.

At present, 80% of methane hydrates are oxidized by syntrophic microbial mats of sulfate-reducing bacteria and anaerobic methane-oxidizing archaea (ANME, discussed in Section 19.4). The overall reaction of the syntrophic community (or consortium) is:

$$CH_4 + SO_4^{2-} \longrightarrow HCO^{3-} + HS^- + H_2O$$

where methane is the initial electron donor oxidized by the ANME partner, and sulfate is the terminal electron acceptor reduced by the bacteria. The high sulfate concentration of marine water drives this reaction in benthic and deep-sea regions where all O_2 has been consumed.

The Global Carbon Balance

The global balance of biological CO_2 fixation and release largely determines the atmospheric level of CO_2. Since the beginning of the industrial age, however, the release of CO_2 has accelerated significantly. The major part of this increase comes from the combustion of **fossil fuels**, which adds about 6×10^{15} grams of carbon annually to the atmosphere. Fossil fuels are the product of microbial anaerobic digestion of plant and animal remains, reduced to hydrocarbons by the pressure and heat of Earth's crust. When burned as fuel, carbon that had accumulated over millions of years is rapidly returned to the atmosphere as CO_2. Some of the CO_2 flux is compensated by increased CO_2 fixation and ocean absorption, but there remains a net flux of 4×10^{15} grams of carbon annually.

Another factor in the increase of atmospheric CO_2 is deforestation, which adds about 2×10^{15} grams of carbon annually to the atmosphere, or approximately a third as much as burning fossil fuels. As forests are cut, CO_2 is released through wood burning and microbial decay. The role of microbes in CO_2 release, as compared to wood burning, is challenging to measure. One study focused on carbon flux in the forests of Amazonia, the Amazon River watershed. The Amazon watershed may account for as much as 10% of global carbon flux from deforestation. In this study, researchers mapped various processes of carbon uptake and efflux throughout the region (**Fig. 22.4**). Processes that generate CO_2 include burning of trees (red line) and microbial decay of cut trees (yellow line). CO_2 is removed from the atmosphere by regrowth of trees (green line). Since 1978, the major contribution to total net carbon efflux has been microbial decomposition of felled trees. Thus, more CO_2 has been released by tree cutting and microbial decay than by burning.

Carbon flux out of Amazonian forest

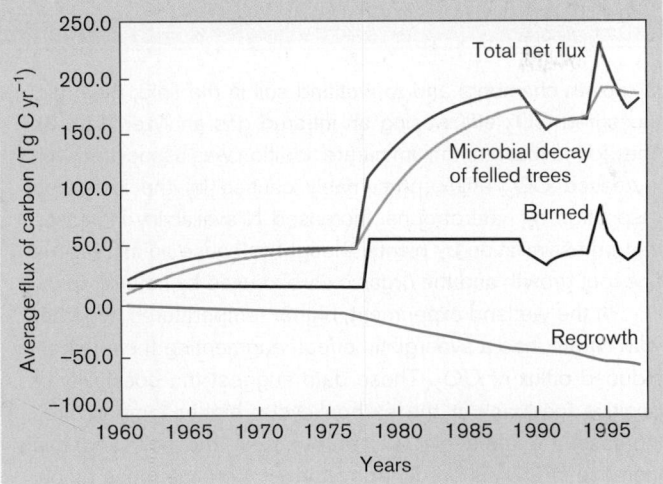

Figure 22.4   Carbon flux from Amazonian forest during 1962–1998. Carbon flux (CO_2 efflux to the atmosphere and uptake by forests) was measured from the Brazilian Amazon forest watershed. Total net carbon flux includes contributions from microbial decay, burning of trees, and regrowth of trees. Tg C yr^{-1} = teragrams of carbon per year (1 Tg = 10^{12} g).
Source: Based on R. A. Houghton et al. 2000. *Nature* **403**:301.

The contribution of ocean ecosystems to carbon flux is even more difficult to measure. Metagenomic surveys are revealing previously unrecognized communities of marine phototrophs, such as submicroscopic cyanobacteria, the prochlorophytes (*Prochlorococcus* species). In the deep ocean, similar studies reveal new methane-oxidizing archaea (ANME) that play a critical role in removing benthic methane hydrates.

Studies of carbon and oxygen flux based on isotope ratios suggest a much greater level of biological production than earlier estimates suggested. The oceans conduct more than half of the global biological uptake of carbon from the atmosphere. This kind of data is crucial for predicting the global effects of changes in atmospheric CO_2.

From a political standpoint, ecosystems such as forests that act as **carbon sinks** by fixing carbon into stable biomass are considered desirable because they lessen the rate of CO_2 input into the atmosphere. Ecosystems that act as sources of carbon dioxide may be viewed with disfavor because they contribute to global warming. But what if CO_2-generating ecosystems provide other environmental benefits? For example, wetlands are among Earth's most productive ecosystems, supporting vast amounts of plant and animal life, but they may also release significant amounts of CO_2 and methane (see **Special Topic 22.1**).

TO SUMMARIZE:

■ **The most accessible reservoir for carbon is the atmosphere (CO_2).** Atmospheric carbon is severely perturbed by the burning of fossil fuels.

■ **Cyanobacteria and other phytoplankton cycle much of the CO_2 in the biosphere.** Oxygen released by phototrophs is used by aerobic heterotrophs and lithotrophs.

■ **Carbon cycling is linked to the cycling of hydrogen and oxygen.**

■ **Anaerobic environments cycle carbon through bacteria and archaea.** Bacteria conduct fermentation and anaerobic respiration. Methanogens release methane, much of which is removed by methane-oxidizing bacteria and archaea.

■ **Microbial decomposition returns CO_2 to the atmosphere.** Microbial decomposition is a major contributor to the accelerated CO_2 flux associated with deforestation.

22.3 The Hydrologic Cycle and Wastewater Treatment

The fate and distribution of complex carbon compounds are largely functions of the **hydrologic cycle**, or **water cycle**, the cyclic exchange of water between atmospheric water vapor and Earth's ecosystems (**Fig. 22.5A**). A vast reservoir of water is supplied by the ocean. In the hydrologic cycle, water precipitates as rain, which is drawn by gravity into groundwater, rivers, lakes, and ultimately the ocean. All along this route, of course, evaporation returns water to the air. Human communities interact with the hydrologic cycle by drawing water for drinking and other purposes, and by returning wastewater. Wastewater requires cleansing of organic contaminants, including key forms of treatment by microbes.

Biochemical Oxygen Demand

A major problem in aquatic ecosystems is organic contamination. Water passing through the ground and aquatic ecosystems carries organic carbon material from humus, sewage, and fertilizer runoff. A sudden influx of rich carbon substrates accelerates respiration by aquatic microbes. Microbial respiration then competes with that of fish, invertebrates, and amphibians for the limited supply of oxygen dissolved in water, raising the **biochemical oxygen demand (BOD)**, also called the biological oxygen demand (BOD). The higher the concentration of organic substances, the higher the BOD arising from microbial oxygen consumption.

High BOD can cause a massive die-off of fish and other aquatic animals. Thus, a routine part of monitoring the health of lakes and streams is the measurement of BOD. A standard value of BOD is defined by measuring the rate of oxygen uptake in a water sample by a defined set of heterotrophic bacteria (**Fig. 22.5B**). Oxygen uptake is observed in a BOD analyzer (**Fig. 22.5C**), which

Special Topic 22.1 | Wetlands: Disappearing Microbial Ecosystems

Ecologist Siobhan Fennessy (**Fig. 1**) studies wetland ecosystems in Ohio—more than 90% of which have already been destroyed or substantially disturbed by human development. Fennessy and her undergraduate students at Kenyon College compare disturbed and undisturbed wetlands to investigate species diversity and the role that wetlands play in carbon and nitrogen cycling. Wetlands sequester huge amounts of carbon; globally they cover only 4%–5% of Earth's land area, but they contain up to 450 gigatons C, or approximately 20% of the carbon in the terrestrial biosphere. Much of the carbon and nitrogen is cycled by anaerobic bacteria and archaea such as methanogens.

Students Kanmani Venkateswaran and John Tisdale addressed the relationship between nitrogen loading and carbon efflux from a wetland (**Fig. 2**). Excess nitrogen loading into wetlands results from industrial and agricultural runoff of nitrogenous fertilizers. Does this influx of nitrogen to a nitrogen-limited ecosystem affect the rate of carbon flux? The students added ammonium nitrate fertilizer to wetland soil cores in growth chambers and to wetland soil in the field; then they measured CO_2 efflux using an infrared gas analyzer (**Fig. 3**). They found that ammonium nitrate addition was associated with increased CO_2 efflux, presumably caused by the increased respiration by heterotrophs. Increased N availability increases carbon assimilation by plants, thought to cause an increase in fine root growth and the organic carbon used by heterotrophs.

In the wetland experiment, higher temperatures, together with nitrate, had a synergistic effect, augmenting the fertilizer-induced efflux of CO_2. These data suggest the possibility of positive feedback in the carbon cycle: higher temperatures increase the rates of CO_2 output from the soil, and this increase in CO_2 is predicted to further increase mean global temperatures, accelerating soil CO_2 efflux. Soil carbon flux is one of the largest fluxes in the global carbon cycle, so even small changes in soil flux can lead to substantial carbon inputs to the atmosphere. Thus, the rise in greenhouse gases may lead to significant changes in carbon dynamics.

Figure 1 Ecologist Siobhan Fennessy.

Figure 2 **Field monitoring stations in an Ohio wetland.** The stations are used to study carbon flux and water levels. The white PVC pipes are wells with attachments for CO_2 flux measurements using a portable infrared gas analyzer.

detects dissolved oxygen in water. The rate of decrease of dissolved oxygen measured by the BOD analyzer is approximately proportional to the amount of dissolved organic matter available for respiration. Note, however, that the BOD in a natural environment will depend on the microbes actually present, as well as the plants and animals competing for oxygen and other resources.

Until recently, BOD was considered a local issue, affecting the health of lakes and rivers in a community. Today, however, large regions of ocean have become

Today's land developers are required to replace lost wetlands with "created wetlands." But are natural wetlands replaceable? Student Dan Gustafson addressed the survival of mycorrhizae (symbiotic fungal extensions of plant roots) in the anaerobic soil of well-established wetlands. Mycorrhizae are important for the productivity of natural ecosystems, as well as human agriculture. Gustafson proposed that mycorrhizae require oxygen provided by the oxygenated rhizosphere surrounding the roots of wetland plants. This hypothesis was tested by cutting off the oxygen moving down through the root channels. Oxygen levels were measured by determining soil redox potentials, which decrease when oxygen is depleted. In low oxygen, mycorrhizae failed to grow. Thus, the experiment showed that the mycorrhizae indeed require oxygen provided by the plant roots, even in anaerobic soil. Subsequent field studies showed that "created" wetlands fail to support as much mycorrhizae as natural wetlands do. This delicate mutualism between fungus and plant requires a long-established wetland habitat, a system that could take many decades to form in a "created wetland."

Figure 3   Measuring carbon flux. Kanmani Venkateswaran, undergraduate at Kenyon College, uses an infrared gas analyzer to measure carbon flux at a field monitoring station.

Courtesy of Kanmani Venkateswaran

which includes inputs from the Ohio and Missouri rivers, the Mississippi builds up high levels of organic pollutants, as well as nitrates from agricultural fertilizer. When these nitrogen-rich substances flow rapidly out to the Gulf in the spring, they lift the nitrogen limitation on algal growth and feed massive algal blooms. The algal population then crashes, and their sedimenting cells are consumed by heterotrophic bacteria. The heterotrophs use up the available oxygen, causing **hypoxia**. Hypoxia kills off the fish, shellfish, and crustaceans over a region equivalent in size to the state of New Jersey.

In 2010 the Gulf of Mexico dead zone was expanded by the unprecedented spill of oil from the Deepwater Horizon oil well blowout (**Fig. 22.6B**). The offshore oil platform exploded, releasing millions of barrels of oil into the Gulf over three months. The leaked oil killed wildlife throughout the Gulf, causing unprecedented environmental damage. Workers tried to contain the spill, but ultimately most of the oil was consumed by marine bacteria. Only bacteria and archaea possess the enzymes needed to degrade the complex organic mixture of petroleum, which includes waxy aromatic compounds such as paraffins.

While consumption by oil-eating bacteria was important for breaking down the pollution, their respiration raised BOD levels throughout the gulf, thus lowering oxygen available to wildlife. Furthermore, studies of earlier oil spills such as the *Exxon Valdez* oil spill in 1989, show that certain oil components linger in the environment for decades, including toxins such as polycyclic aromatic hydrocarbons (PAH). Further research is needed to increase the microbial degradation of pollutants.

Dead zones now occur along the coasts of industrial and developing countries throughout the world, including, for example, India and Australia. They deplete habitat for much marine life—for example, forcing sharks to swim out of hypoxic regions and nearer the shore. Dead zones contribute to the crash of fisheries worldwide, removing a critical food resource for human populations. Plans to clean up dead zones face daunting costs.

To avoid such dead zones, release of industrial pollutants such as petroleum must be prevented. In addition, all the communities throughout the river drainage areas must treat their wastes to eliminate nitrogenous wastes before disposal. There are two common approaches to community wastewater treatment, both of which involve microbial partners: wastewater treatment plants and wetland filtration. Both approaches depend on microbes to remove organic carbon and nitrogen from water before it returns to aquatic systems and ultimately the ocean.

Wastewater Treatment

In industrialized nations, all municipal communities use some form of **wastewater treatment** (**Fig. 22.7**). The purpose of wastewater treatment is to decrease the

dead zones, or **zones of hypoxia**, devoid of most fish and invertebrates. A well-known dead zone is a region in the Gulf of Mexico off the coast of Louisiana where the Mississippi River releases about 40% of the U.S. drainage to the sea (**Fig. 22.6A**). Over its long, meandering course,

A. Hydrologic cycle

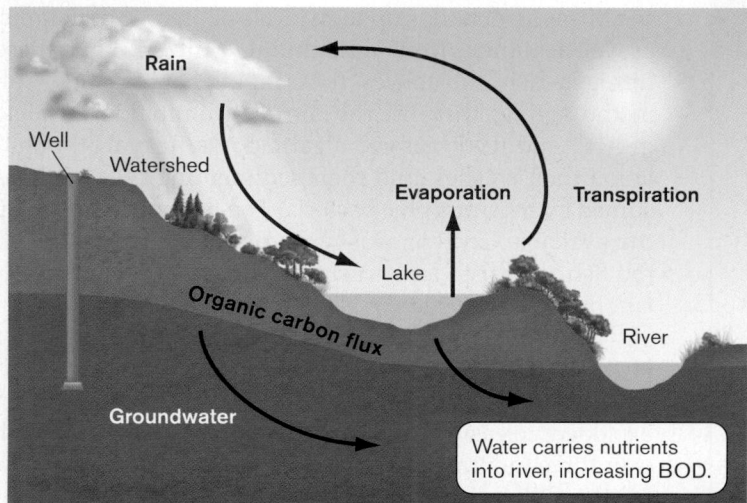

Figure 22.5   The hydrologic cycle interacts with the carbon cycle.   A. The hydrologic cycle carries bacteria and organic carbon into groundwater and aquatic systems. **B.** Bottled water samples are measured for dissolved oxygen over time; the rate of decrease of dissolved oxygen indicates biochemical oxygen demand (BOD). The rate of decrease of dissolved oxygen in water samples is approximately proportional to the concentration of organic matter available for respiration. **C.** A microprocessor-controlled BIOX-1010 BOD analyzer measures rate of respiration. Water samples are mixed with a concentrated microbial biomass, and a dissolved-oxygen (DO) sensor measures small rates of oxygen decrease over time.

B. BOD measurement

Joel Gordon, Pittsfield wastewater treatment plant

C. BOD analyzer

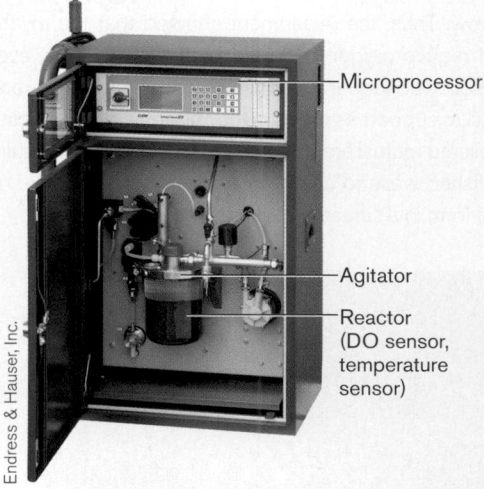

Endress & Hauser, Inc.

BOD and the level of human pathogens before water is returned to local rivers. The treatment process includes microbial metabolism. A modern wastewater plant can convert sewage into water that exceeds all government standards for humans to drink.

The wastewater treatment plant is the final destination for all household and industrial liquid wastes passing through the municipal sewage system. A typical plant includes the following stages of treatment:

■ **Preliminary treatment** consists of screens that remove solid debris, such as sticks, dead animals, and feminine hygiene items.
■ **Primary treatment** includes fine screens and sedimentation tanks that remove insoluble particles. The particles eventually are recombined with the solid products of wastewater treatment (known as **sludge**). The sludge ultimately is used for fertilizer or landfill.
■ **Secondary treatment** consists primarily of microbial ecosystems that decompose the soluble organic content of wastewater, by aerobic and anaerobic respiration. The microbes form particulate flocs of biofilm.

The flocs are sedimented as sludge, also known as activated sludge, owing to their microbial activity.
■ **Tertiary (advanced) treatment** includes chemical applications such as chlorination to eliminate pathogens.

The microbial ecosystems of secondary treatment require continual aeration to maximize breakdown of molecules to carbon dioxide and nitrates. The microbes typically include bacilli such as *Zoogloea*, *Flavobacterium*, and *Pseudomonas*, as well as filamentous species such as *Nocardia* (**Fig. 22.8A**). Optimal treatment depends on the ratio of filamentous to single-celled bacteria: enough filaments to hold together the flocs for sedimentation, but not so many as to trap air and cause flocs to float and foam, preventing sedimentation.

Besides bacteria, the ecosystem of activated sludge includes filamentous methanogens (discussed in Chapter 19), which metabolize short-chain molecules such as acetate within the anaerobic interior of flocs. Thus, wastewater treatment generates methane, often in quantities that can be recovered as fuel.

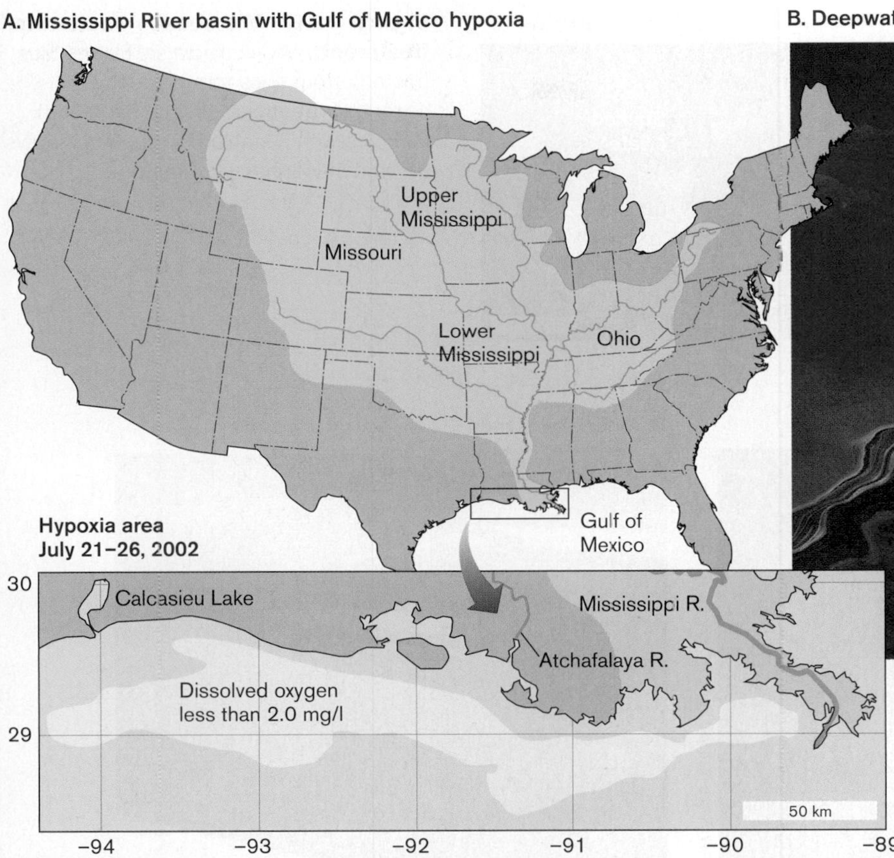

A. Mississippi River basin with Gulf of Mexico hypoxia

B. Deepwater Horizon oil spill

Upper
Mississippi

Missouri

Lower
Mississippi

Ohio

Gulf of
Mexico

**Hypoxia area
July 21–26, 2002**

30

Calcasieu Lake

Mississippi R.

Atchafalaya R.

Dissolved oxygen
less than 2.0 mg/l

29

50 km

−94   −93   −92   −91   −90   −89

REUTERS/Daniel Beltra

Figure 22.6   The dead zone in the Gulf of Mexico.   A. Map of the Mississippi River drainage, which empties into the Gulf of Mexico. Every summer, drainage high in organic carbon and nitrogen causes algal blooms, leading to hypoxia and death of fish. **B.** Aerial view of the oil leaked from the Deepwater Horizon oil wellhead in the Gulf of Mexico, 2010. *Source*: Part A from the National Oceanographic and Atmospheric Administration.

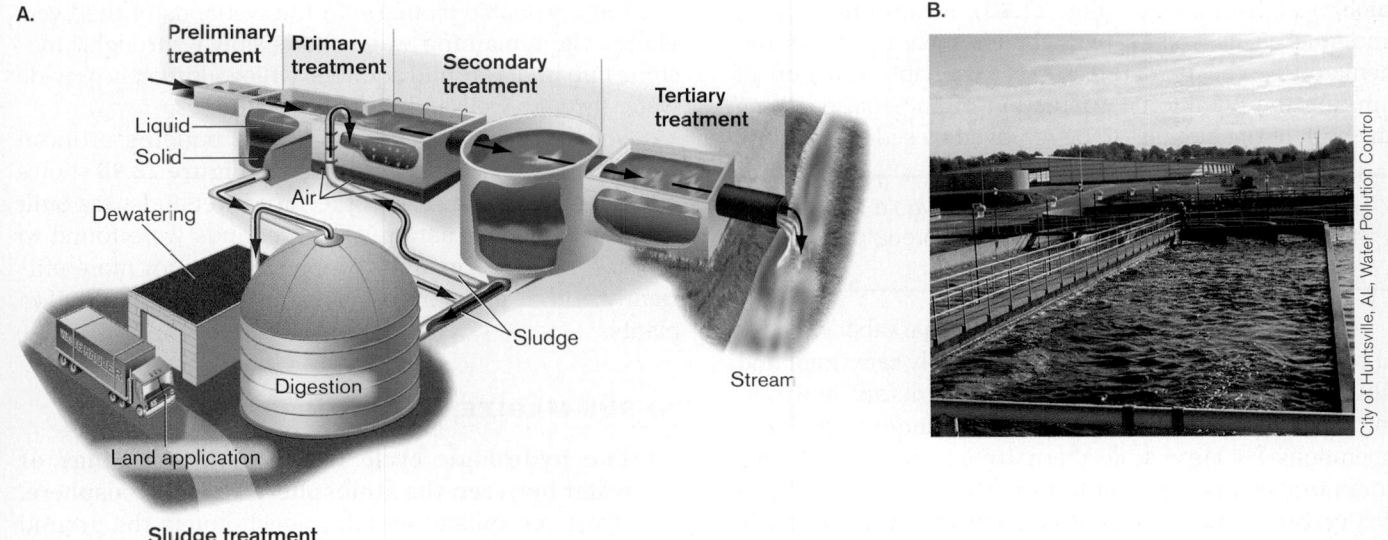

A.

Preliminary
treatment

Primary
treatment

Secondary
treatment

Tertiary
treatment

Liquid

Solid

Air

Dewatering

Sludge

Digestion

Stream

Land application

Sludge treatment

B.

City of Huntsville, AL, Water Pollution Control

Figure 22.7   Wastewater treatment plant.   A. In a typical municipal treatment plant, wastewater undergoes primary treatment (filtering and settling), secondary treatment (microbial decomposition), and tertiary treatment (chlorination or other chemical treatments). **B.** Aeration basin for secondary treatment.

A. *Nocardia* sp.

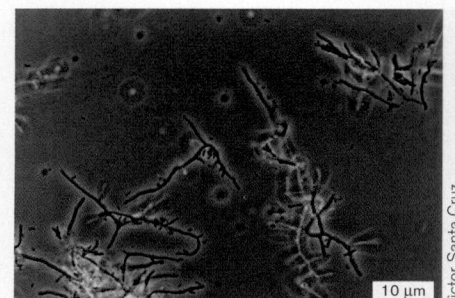

B. Flocs

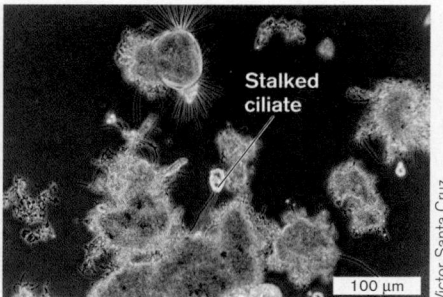

Stalked ciliate

Figure 22.8 Microbes in wastewater treatment. A. *Nocardia* sp. filamentous bacteria from flocs formed during secondary treatment (light micrograph). **B.** Flocs with a stalked ciliate, which preys on bacteria (light micrograph).

A. Everglades

B. Artificial wetlands

Figure 22.9 Water filtration by wetlands. A. The marshes of the Everglades act as natural filters for the aquifers of southern Florida. **B.** A filtering system installed by Steve Kerns on his hog farm in Taylor County, Iowa. A series of hillside terraces form wetlands containing bacteria that purify hog manure and wastewater.

In addition, the bacteria are preyed upon by protists such as stalked ciliates (**Fig. 22.8B**), swimming ciliates, and amebas, as well as invertebrates such as rotifers and nematodes. The predators serve a valuable function of limiting the planktonic population of bacteria, enabling the bulk of the biomass to be removed by sedimentation.

> **THOUGHT QUESTION 22.2** What would happen if wastewater treatment lacked microbial predators? Why would the result be harmful?

Wastewater treatment plants are remarkably effective at converting human wastes to ecologically safe water and ultimately human drinking water. The plants are, however, impractical for purifying the runoff from large agricultural operations. For large-scale alternatives to treatment plants, communities and agricultural operations are looking to wetland restoration. Much of our current water supply is already filtered and purified by natural wetlands, such as the Florida Everglades (**Fig. 22.9A**). Wetlands remove nitrogen through the action of denitrifying bacteria. In wetlands, rainwater and river water trickle slowly through vast stretches of soil, where microbial conversion acts as the foundation of a macroscopic ecosystem including trees

and vertebrates. Thus, much of the carbon and nitrogen is fixed into valuable biomass. In the wetlands of the Everglades, the remaining water filters slowly through limestone into underground aquifers, which ultimately provide water through wells to human communities.

Some agricultural operations are building artificial wetlands to replace treatment plants. **Figure 22.9B** shows a hog farm where a series of terraced wetlands was built to drain liquefied manure. The wetlands were found to produce fewer odors and to remove organics more efficiently and at a lower cost than do traditional filtration plants.

TO SUMMARIZE:

- **The hydrologic cycle is the cyclic exchange of water between the atmosphere and the biosphere.** Water precipitates as rain, which enters the ground and ultimately flows to the oceans. Along the way, water evaporates, returning to the atmosphere.
- **Water carries organic carbon that generates biochemical oxygen demand (BOD).** High BOD accelerates heterotrophic respiration and depletes oxygen needed by fish.

- **Wastewater treatment cuts down BOD.** Secondary treatment involves microbial communities that decompose the soluble organic content.
- **Wetlands filter water naturally.** Wetland filtration helps purify groundwater entering aquifers.

22.4 The Nitrogen Cycle

Besides carbon, oxygen, and hydrogen, another major element that cycles largely by microbial conversion is nitrogen (**Fig. 22.10**). The nitrogen cycle is notable in two respects:

- **Many oxidation states.** A larger number of oxidation states exist for biological nitrogen than for any other major biological element (see **Table 22.3**). Conversion between these oxidation states requires metabolic processes such as N_2 fixation, lithotrophy, and anaerobic respiration.
- **Dependence on prokaryotes.** Many steps of the nitrogen cycle require bacteria and archaea. Without bacteria and archaea, the nitrogen cycle would not exist.

> **THOUGHT QUESTION 22.3** How many kinds of biomolecules can you recall that contain nitrogen? What are the usual oxidation states for nitrogen?

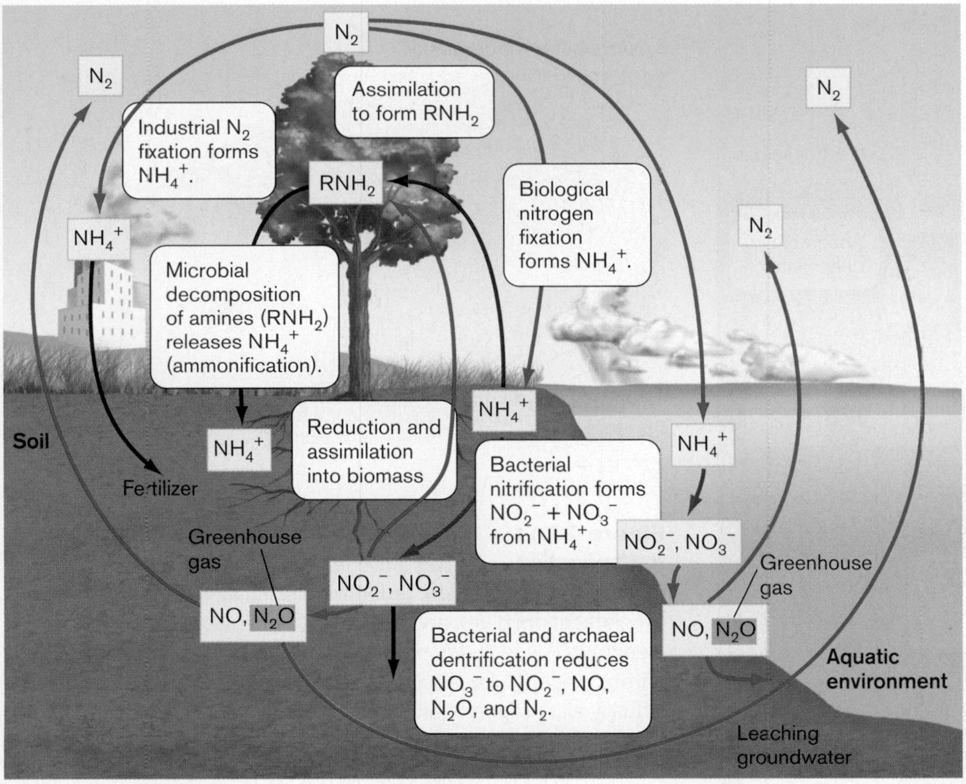

Sources of Nitrogen

Where is nitrogen found on Earth? A significant amount of nitrogen is found in Earth's crust, in the form of ammonium salts in rock (see **Table 22.2**). This form is largely inaccessible to microbes. The major accessible source of nitrogen is the atmosphere, of which dinitrogen gas (N_2) constitutes 79%. However, N_2 is a highly stable molecule that requires enormous input of reducing energy before assimilation is possible (see Chapter 15). Thus, for many natural ecosystems, and most forms of agriculture, nitrogen is the limiting nutrient for primary productivity.

Until recently in Earth's history, nitrogen was fixed entirely by nitrogen-fixing bacteria and archaea. In the twentieth century, however, the Haber process was invented for artificial nitrogen fixation to generate fertilizers for agriculture. The process was devised by German chemist Fritz Haber, who won the 1918 Nobel Prize in Chemistry. In the Haber process, N_2 is hydrogenated by methane under extreme heat and pressure. Today, the Haber process for producing fertilizers accounts for approximately 30%–50% of all nitrogen fixed on Earth. Other human activities, such as fuel burning and use of nitrogenous fertilizers, contribute to oxidized nitrogen pollutants such as nitrous oxide (N_2O), a potent greenhouse gas.

The nitrogen cycle is the most perturbed of the major biogeochemical cycles. The ultimate effects of perturbation are not yet clear. Some models show that increased nitrogen fixation could fertilize CO_2 fixation by marine and terrestrial ecosystems, partly decreasing net CO_2 emissions. The amount, however, will be too small to prevent global warming. Other models show that accelerated nitrogen use could limit nitrogen availability in the global biosphere, with unknown consequences.

The "Nitrogen Triangle"

The numerous oxidation states available for microbial metabolism of nitrogen generate a complex cycle of conversions. One way to organize the complexity is to envision a "nitrogen triangle" whose corners include three key forms: atmospheric N_2; the reduced forms NH_3 and NH_4^+; and the oxidized forms NO_2^- and NO_3^- (**Fig. 22.11A**). At the base of the triangle, both reduced and oxidized forms of nitrogen are assimilated into biomass.

Figure 22.10 The global nitrogen cycle. Prokaryotic conversions of nitrogen occur throughout the biosphere. Blue = reduction; red = oxidation; black = redox-neutral.

We shall consider in turn the three "legs" of the nitrogen triangle:

- **N₂ fixation** to ammonium (NH_4^+), a form assimilated into biomass by microbes and plants.
- **Ammonia oxidation**, in which ammonia is oxidized aerobically to nitrite (NO_2^-) and nitrate (NO_3^-), which are also assimilated by microbes and plants. Aerobic oxidation to nitrite or nitrate is called **nitrification**. In the absence of oxygen, anaerobic ammonium oxidation by nitrite generates N₂—a reaction called **anammox**.
- **Denitrification** of NO_2^- and NO_3^- back to N₂ (or, in carbon-rich habitats, reduction to NH_3).

Nitrogen fixation. The main avenue for entry of nitrogen into the biosphere is **nitrogen fixation**—specifically, fixation of dinitrogen, or nitrogen gas (N₂) into NH_3, protonated to NH_4^+. Ammonium ion is rapidly assimilated by bacteria and plants, typically through combination with TCA cycle intermediates to form key amino acids, such as glutamate (discussed in Chapter 15). The ability to assimilate NH_4^+ into nitrogenous organic molecules is found in virtually all primary producers.

Fixation of nitrogen requires enormous energy because the triple bond of N₂ is exceptionally stable. Breaking the triple bond to generate ammonia requires a series of reduction steps involving high input of energy:

Dinitrogen

$$N{\equiv}N \xrightarrow[2H^+ + 2e^-]{} HN{=}NH \xrightarrow[2H^+ + 2e^-]{} H_2N{-}NH_2 \xrightarrow[2H^+ + 2e^-]{} 2NH_3$$

All the intermediate reactions of nitrogen fixation are tightly coupled, so the intermediate compounds rarely become available for other uses. The final product, ammonia, exists in ionic equilibrium with ammonium ion:

$$NH_3 + H_2O \rightleftharpoons NH_4^+ + OH^-$$

Fixation of nitrogen is catalyzed by the enzyme nitrogenase (discussed in Chapter 15). The highly reductive reactions of nitrogenase require exclusion of oxygen; yet they also require tremendous input of energy, usually provided by aerobic metabolism. Therefore, nitrogen-fixing bacteria generally separate anaerobic nitrogen fixation from aerobic metabolism by one of several mechanisms, such as the heterocysts of cyanobacteria (discussed in Chapter 15).

Given the energy expense and the need to exclude oxygen, many species of bacteria, as well as all eukaryotes, have lost the nitrogen fixation pathway by degenerative evolution. But all ecosystems, both marine and terrestrial, include some species of bacteria and archaea that fix N₂ into ammonia. Nitrogen-fixing bacteria in the soil include obligate anaerobes, such as *Clostridium* species, and facultative Gram-negative enteric species of *Klebsiella* and *Salmonella*, as well as obligate respirers such as *Pseudomonas*.

A. The nitrogen triangle

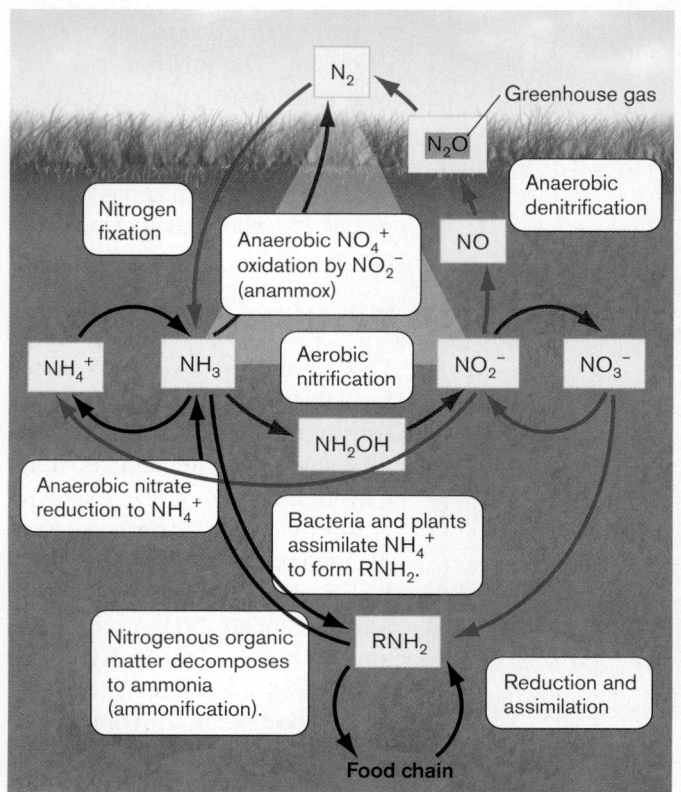

B. *Nitrobacter winogradskyi*

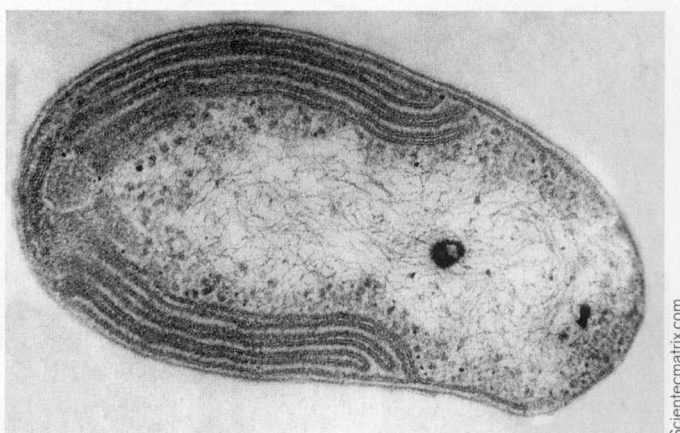

Figure 22.11   The nitrogen cycle: fixation, nitrification, denitrification.   A. The "nitrogen triangle" involves nitrogen fixation and assimilation, reductive dissimilation of nitrate (<u>deni</u>trification; blue), and oxidation (<u>nitri</u>fication; red). Denitrification includes production of the potent greenhouse gas nitrous oxide (N₂O). Assimilation into biomass is often reductive; virtually all nitrogen in biomolecules is highly reduced. Oxidation of ammonia generates nitrites and nitrates, whose runoff can pollute water supplies. **B.** *Nitrobacter winogradskyi* oxidizes nitrite to nitrate (TEM). Folded layers of membrane contain the electron transport complexes. (cell length, 0.5–1.0 μm.)

In ocean and in freshwater systems, cyanobacteria are the major nitrogen fixers. After fixation, all these organisms assimilate the reduced nitrogen into essential components of their cells. Within an ecosystem, nitrogen fixers ultimately make the reduced nitrogen available for assimilation by nonfixing microbes and plants, either directly through symbiotic association (such as that of rhizobia and legumes) or indirectly through predation (marine cyanobacteria) and decomposition (soil bacteria).

If nitrogen fixation is ubiquitous in soil and water, then why is nitrogen limiting? Nitrogen fixation is extremely energy intensive; thus, the rate of fixation usually fails to meet the potential demand of other members of the ecosystem. The one exception is legume symbiosis with rhizobia, which provide ample nitrogen for their hosts.

In agriculture, symbiotic nitrogen fixation by rhizobial bacteria increases the yield of crops such as soybeans (**Fig. 22.12A**). Soybeans can derive as much as 50% of their nitrogen from microbial symbionts. To enhance colonization by nitrogen fixers, farmers apply molecules called isoflavonoids, which mimic the natural plant-derived attractants for rhizobia.

Nitrification. Free ammonia in soil or water is quickly oxidized for energy by **nitrifiers**, bacterial species that possess enzymes for oxidation of ammonia to nitrite (NO_2^-), or of nitrite to nitrate (NO_3^-). This process is called **nitrification**. Nitrification of ammonia is a form of lithotrophy, an energy generation pathway involving oxidation of minerals. The nitrification pathway generates the red base of the triangle in **Figure 22.11A**. The pathway of nitrification includes:

$$\begin{array}{ccccc}
\text{Ammonia} & & \text{Nitrite} & & \text{Nitrate} \\
NH_3 & \longrightarrow NH_2OH \longrightarrow & NO_2^- & \longrightarrow & NO_3^- \\
+\ \tfrac{1}{2}O_2 & +\ O_2 & + & +\ \tfrac{1}{2}O_2 & + \\
& & H_2O + H^+ & & H^+
\end{array}$$

The pathway includes two separate energy generation mechanisms, utilized by different species: (1) ammonia through NH_2OH to nitrite, and (2) nitrite to nitrate. Typically, soil contains both kinds of microbes living together, a collaboration ensuring complete conversion to nitrate. Nitrifying genera include *Nitrosomonas*, which oxidizes ammonia to nitrite; and *Nitrobacter* (**Fig. 22.11B**) and *Nitrospira*, which oxidize nitrite to nitrate. Note that production of both nitrite and nitrate generates acid, which can acidify the soil.

THOUGHT QUESTION 22.4 In the laboratory, which bacterial genus would likely grow on artificial medium including NH_2OH as the energy source: *Nitrosomonas* or *Nitrobacter*?

Nitrate produced in the soil is assimilated by plants and bacteria nearly as quickly as ammonium ion, although extra

A. Soybeans grow with symbiotic rhizobia

B. Nitrate contamination of groundwater

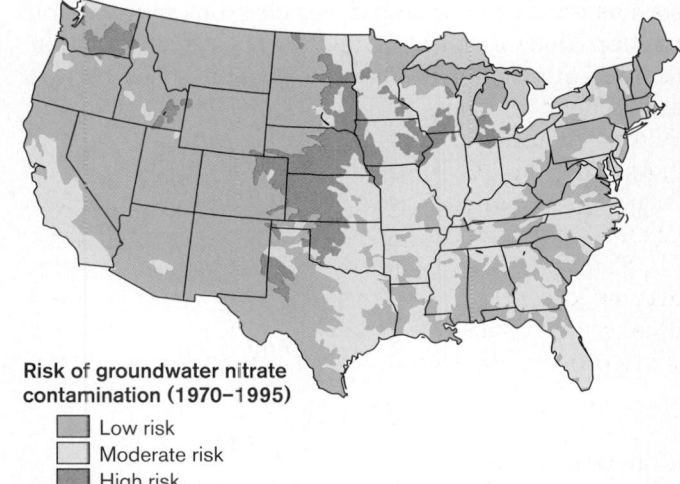

Risk of groundwater nitrate contamination (1970–1995)

- Low risk
- Moderate risk
- High risk

Figure 22.12 Agricultural benefits and consequences of microbial nitrogen metabolism. A. Soybean field in Maine. Rhizobial nitrogen fixation enhances growth of soybeans and other major crops. **B.** Nitrate in drinking water is especially prevalent in agricultural regions of the United States. Ammonification of fertilizer, followed by nitrification, generates nitrate and nitrite. Nitrate and nitrite runoff from oxidized nitrogenous fertilizers pollutes streams and groundwater.

energy is needed to reduce nitrate to NH_4^+ for incorporation into biomass. Nitrate assimilation to biomass is called **assimilatory nitrate reduction**. In agriculture, intensive fertilization generates a large excess of ammonia resulting from **ammonification**, the breakdown of organic nitrogen releasing ammonia. Ammonia is oxidized rapidly by lithotrophic bacteria (nitrifiers). The excess ammonia leads to a buildup of nitrites and nitrates, which are highly soluble in water, and they readily diffuse into aquatic systems. Aquatic nitrate reacts with organic compounds to form toxic nitrosamines. Nitrate influx also relieves the nitrogen limit on algae, causing algal blooms and raising BOD. Chronic nitrate influx leads to eutrophication and die-off of fish.

Consumption of nitrate in drinking water can lead to methemoglobinemia, a blood disorder in which hemoglobin is inactivated. Methemoglobinemia occurs in infants whose stomachs are not yet acidic enough to inhibit growth of bacteria that convert nitrate to nitrite. Nitrite oxidizes the iron in hemoglobin, eliminating its capacity to carry oxygen. The failure to carry oxygen leads to a bluish appearance, one cause of "blue baby syndrome." Nitrite-induced blue baby syndrome is a problem in intensively cultivated agricultural regions, such as Kansas and Nebraska (**Fig. 22.12B**).

Recently, another role was discovered for nitrite: as an electron donor for photosynthesis. Certain gamma-proteobacteria, related to *Thiocapsa*, can use light energy to oxidize nitrite to nitrate. The extent of nitrogen-based photosynthesis in global nitrogen cycles is unknown.

Denitrification. N_2 is regenerated by anaerobic respiration (**Fig. 22.11A**), in which an oxidized form of nitrogen such as nitrate or nitrite receives electrons from organic electron donors. Bacteria and archaea reduce nitrate through a series of decreased oxidation states back to atmospheric nitrogen:

$$
\begin{array}{ccc}
\text{Nitrate} & \text{Nitrite} & \text{Nitric oxide} \\
2NO_3^- \longrightarrow & 2NO_2^- \longrightarrow & 2NO \longrightarrow \\
4H^+ + e^- & + \quad 4H^+ + 2e^- & + \quad 2H^+ + 2e^- \\
& 2H_2O & 2H_2O
\end{array}
$$

$$
\begin{array}{cc}
\text{Nitrous oxide} & \text{Dinitrogen} \\
N_2O \longrightarrow & N_2 \\
2H^+ + 2e^- & + \\
& H_2O
\end{array}
$$

In terms of elemental flux through ecosystems, nitrate and nitrite reduction are known as **denitrification** or **dissimilatory nitrate reduction**. (In contrast, assimilatory nitrate reduction incorporates the nitrogen into biomass.) Anaerobic respirers in soil or water use an oxidized form of nitrogen as an alternative electron acceptor in the absence of O_2 (discussed in Chapter 14). All types of nitrogen-based anaerobic respiration are repressed in the presence of oxygen, a more favorable electron acceptor; therefore, denitrification is limited to anoxic habitats.

In undisturbed environments, the products of nitrate respiration rarely build up to levels that harm the ecosystem. Heavy fertilization, however, causes buildup of excess nitrate and so increases the environmental rate of denitrification. During this process, some of the nitrogen escapes as nitrous oxide gas (N_2O). A highly potent greenhouse gas, N_2O generates 200 times the warming effect of CO_2; thus, relatively small amounts of N_2O can make a disproportionate contribution to global warming. Furthermore, N_2O in the upper atmosphere reacts catalytically with ozone, depleting the ozone layer. Thus, the atmospheric effects of bacterial denitrification are a serious concern in agricultural and waste treatment processes.

Nitrous oxide also builds up in marine dead zones. Syed Wajih Naqvi and colleagues at the National Institute of Oceanography (Goa, India) investigated denitrification in the zone of hypoxia off the Indian coast (**Fig. 22.13A**). Under anoxic conditions, some bacteria and archaea respire with nitrate. The study revealed unexpectedly high levels of N_2O production at a series of stations within the zone. In an experiment, introduction of nitrate into water samples from the dead zone led quickly to production of nitrite and N_2O (**Fig. 22.13B**), a process almost certainly caused by denitrifying bacteria. Thus,

Figure 22.13 N_2O production from a coastal dead zone.
A. Zone of hypoxia off the coast of India. Circles represent stations for sample collection. **B.** Levels of NO_3^-, NO_2^-, and N_2O following addition of NO_3^- to a water sample from the zone of hypoxia. The sequential rise of NO_2^- and N_2O indicates metabolism of denitrifying bacteria. *Source:* Based on S. W. A. Naqvi et al. 2000. *Nature* **408**:346.

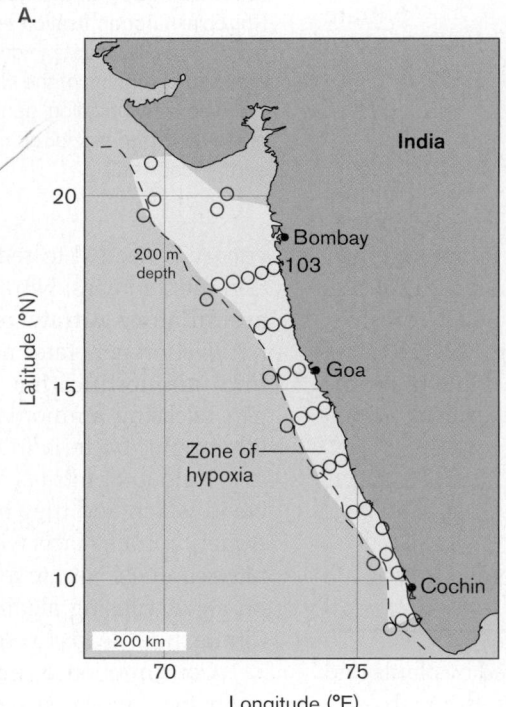

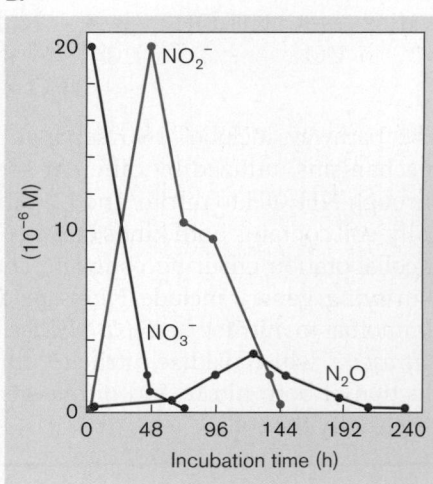

bacterial denitrification in polluted ocean waters may contribute significantly to global warming.

Dissimilatory nitrate reduction to ammonia. Most environmental nitrate and nitrite are reduced via N_2O as described previously. Certain conditions, however, favor the alternate route of dissimilatory nitrate reduction to ammonia (DNRA) (see **Fig. 22.11A**). Nitrate reduction to ammonia is a form of anaerobic lithotrophy or hydrogenotrophy in which nitrate serves as an electron acceptor and hydrogen gas (H_2) is the electron donor:

$$NO_3^- + 4H_2 + H^+ \longrightarrow NH_3 + 3H_2O$$

Bacteria reduce nitrate to ammonia mainly in anaerobic environments rich in organic carbon and H_2 generated by fermentation, but low in reduced nitrogen, such as sewage sludge and stagnant water. Another carbon-rich habitat favoring this pathway is the rumen, the digestive tract of cattle, goats, and other ruminant animals (see Chapter 21). Ruminants consume grasses whose cellulose requires digestion by microbial endosymbionts. Ruminants also depend on their microbial endosymbionts to assimilate nitrate into ammonia and synthesize amino acids.

Anaerobic ammonium oxidation (anammox). For many years, denitrification was considered the main way that nitrogen compounds return N_2 to the atmosphere. In 2003, two research groups from Europe and Costa Rica, led by Tage Dalsgaard and Marcel Kuypers, showed that in anoxic deep-sea water the major source of N_2 is anaerobic ammonium oxidation by nitrite (the anammox reaction):

$$NH_4^+ + NO_2^- \longrightarrow N_2 + 2H_2O$$

In anammox, ammonium ion serves as electron donor, and nitrite as anaerobic electron acceptor—an unusual combination of two different oxidation states of the same key element, nitrogen.

Anammox by bacteria and archaea has now been observed in all kinds of anaerobic habitats, including terrestrial soil and aquatic sediment. The reaction accounts for a majority of all N_2 returned to the atmosphere. Among bacteria, the main anammox contributors are planctomycete genera such as *Brocadia* and *Scalindua* (**Fig. 22.14**). Planctomycetes (discussed in Chapter 18) are unusual cell wall–less bacteria with special membranes for the anammox reaction. More recent metagenomic studies show comparable anammox contributions by archaea, primarily crenarchaeotes. Promising anammox microbes have been identified from wastewater sludge, in the hope of using them for more effective removal of excess nitrogen from wastewater.

> **THOUGHT QUESTION 22.5** The nitrogen cycle has to be linked with the carbon cycle, since both contribute to biomass. How might the carbon cycle of an ecosystem be affected by increased input of nitrogen?

TO SUMMARIZE:

- **Nitrogen in ecosystems is found in a wide range of oxidation states.** Interconversion of most of these states requires prokaryotes.
- **The main source and sink of nitrogen is the atmosphere.** Nitrogen gas is fixed into ammonium ion by some bacteria and archaea. Denitrifying bacteria reduce NO_3^- successively back to N_2 and return it to the atmosphere.
- **Nitrogen fixation is conducted by symbiotic bacteria in association with specific plants.** Legume-associated nitrogen fixation is critical for agriculture.

Figure 22.14 Anammox bacteria.
Phylogeny of anammox bacteria obtained from anaerobic natural environments and wastewater sludge. Based on SSU rRNA sequences as described in Chapter 17; bar represents 2% sequence divergence.
Source: Modified from Jun Nakajima et al. 2008. *Appl. Microbiol. Biotechnol.* **77**:1159.

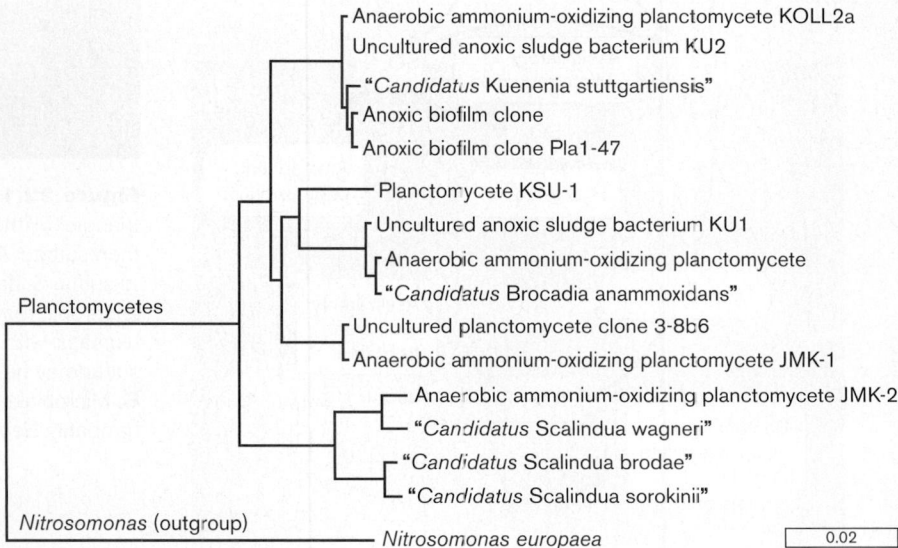

- **Nitrification is aerobic oxidation of ammonia to nitrite and nitrate.** Nitrification yields energy for lithotrophic bacteria in soil and water.
- **Nitrate can be reduced to ammonia under anoxic conditions.** Especially in the deep ocean, NO_3^- may be reduced by hydrogen gas to ammonia.
- **Ammonium is oxidized by the anammox reaction.** Anaerobic ammonium oxidation accounts for a large part of all N_2 returned to the atmosphere.

22.5 Sulfur, Phosphorus, and Metals

Besides carbon and nitrogen, many other elements participate in biochemical cycles that have important consequences for the biosphere, as well as for human environments. Sulfur undergoes a "triangle" of redox conversions analogous to that of nitrogen. Phosphorus, unlike other biological elements, is generally assimilated in the oxidized state, and phosphate is often a limiting nutrient for plants. Iron cycles in complex interactions with sulfur and phosphate. Beyond these macronutrients, many toxic metals in the environment, such as mercury and arsenic, are either bioactivated or detoxified by microbes.

The Sulfur Cycle

Sulfur is a major component of biomass, including proteins and cofactors. Like carbon and nitrogen, sulfur can be assimilated in either mineral or organic form. At the same time, assimilation competes with dissimilation. Reduced and oxidized forms of sulfur (comparable to those of nitrogen; see **Table 22.3**) offer electron donors and acceptors for dissimilatory reactions that generate energy. Reduced forms of sulfur, such as H_2S, can also participate in photolysis through reactions analogous to those of H_2O. Oxidized forms, such as sulfate, serve as anaerobic electron acceptors. As with nitrogen, most of these redox reactions are performed solely by bacteria and archaea. The biochemistry of sulfur cycling has important environmental consequences, such as the corrosion of concrete and iron.

In the ocean, sulfate is the second most common anion after chloride. Marine sulfate turns over slowly and constitutes an essentially limitless supply. Sulfur is similarly plentiful in freshwater sediments. Thus, sulfur is rarely a limiting nutrient.

While the amount of sulfur in ocean water is relatively large compared to that of nitrogen or carbon, the amount in the atmosphere, mainly sulfur dioxide, is small. Nevertheless, these small amounts of sulfur compounds

A. The sulfur triangle

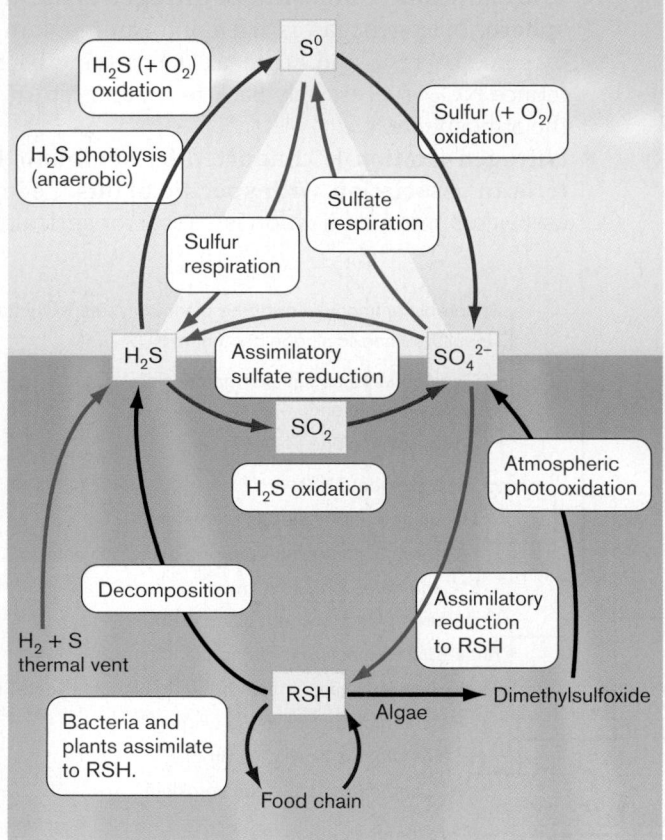

B. White sulfur bacteria

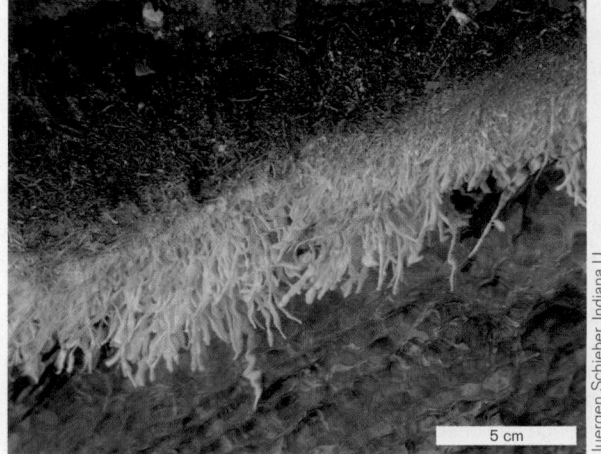

Juergen Schieber, Indiana U.

Figure 22.15 The sulfur cycle. A. The "sulfur triangle." With oxygen, H_2S is oxidized to sulfur dioxide, then sulfate. Anaerobically, H_2S may be photolyzed to sulfur. Sulfate serves as an electron acceptor for anaerobic respiration, or it may be assimilated into biomass, with reduced sulfur groups (RSH). Elemental sulfur may be oxidized to sulfate by lithotrophs.
B. Microbial mat of white sulfur-oxidizing bacteria (probably *Beggiatoa*) growing at a sulfide spring.

generate toxic effects, as well as acidic pollution. Thus, both biochemical and industrial sources of atmospheric sulfur are of concern.

Competing assimilatory and dissimilatory sulfur reactions in the biosphere form a "sulfur triangle" between H_2S, S^0, and SO_4^{2-} (**Fig. 22.15A**). The oxidation and reduction pathways are analogous to those of nitrogen. Sulfur pathways include further options of anaerobic phototrophy, such as H_2S photolysis. In an aquatic system, H_2S arises from spring waters welling up from the sediment and from decomposition of detritus. In decomposition, anaerobic respirers such as *Desulfovibrio* species convert sulfate to sulfur, then to H_2S. As H_2S rises to the oxygenated surface water, it is readily oxidized by sulfur-oxidizing bacteria such as *Thiobacillus* and *Beggiatoa* (**Fig. 22.15B**). *Beggiatoa* species are known as "white sulfur bacteria" because they form white mats, their appearance due to the sulfur granules generated by sulfide oxidation. Microbial sulfur oxidation is helpful for environments because it removes H_2S, which is highly toxic to most nonsulfur bacteria and plants. Because of the toxicity of H_2S, autotrophs more readily assimilate SO_4^{2-} into biomass, despite the extra energy needed to reduce SO_4^{2-}.

If light is available, an alternate route for H_2S oxidation is photolysis by anaerobic phototrophs such as *Rhodopseudomonas* species. Some phototrophic bacteria further oxidize the S^0 generated to sulfate. In some sulfur-rich lakes, such as the Russian Lake Sernoye, underground springs

pump so much H_2S that most of the sulfur is photolytically converted to S^0, which forms up to 5% of the sediment. This elemental sulfur can be mined commercially.

In thermal vent ecosystems, several sulfur-based reactions drive metabolism. For example, thermal vent archaea such as *Pyrodictium* species use sulfur to oxidize hydrogen gas to H_2S:

$$H_2 + S^0 \longrightarrow H_2S$$

This type of sulfur-based hydrogenotrophy is enhanced by the vent conditions of extreme pressure and temperature (100°C), conditions under which elemental sulfur exists in a molten state more accessible to microbes than solid sulfur.

Decomposition of biomass generates various organic sulfur compounds, many of which are volatile. Odors of certain microbial sulfur products contribute to the smell of rotting eggs, but others enhance the taste of cheeses (discussed in Chapter 16). In marine environments, algae excrete substantial quantities of dimethyl sulfide, which escapes to the atmosphere, providing part of the characteristic "smell of the sea." Dimethyl sulfide is believed to be the major biogenic source of atmospheric sulfur. Dimethyl sulfide and other organic sulfur compounds in the atmosphere react abiotically with light to SO_2 and SO_4^{2-}.

Sulfur is reduced by sulfur-reducing bacteria in many subsurface environments, such as beneath deposits of oil and coal (**Fig. 22.16A**). Oil and coal contain sulfur in

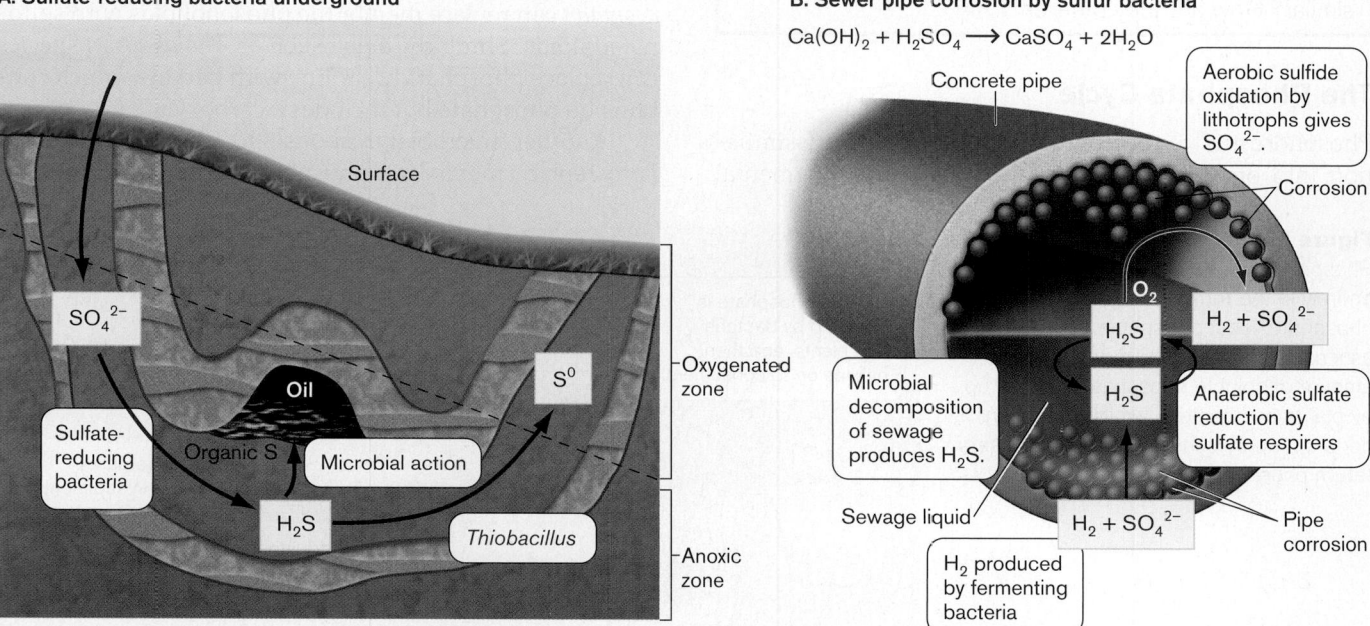

Figure 22.16 Consequences of the sulfur cycle. A. Oil- or coal-bearing geological strata provide rich electron donors for sulfate-reducing bacteria (sulfate respirers). Eventually, various forms of sulfur contaminate the oil or coal, which, when burned as fuel, generate first SO_2 and ultimately sulfuric acid (acid rain). **B.** In a sewer pipe, sulfate-reducing bacteria (sulfate respirers) generate H_2S. Sulfur-oxidizing bacteria then oxidize H_2S to sulfate in the form of sulfuric acid. The sulfuric acid reacts with calcium hydroxide in the concrete, thus corroding the interior surface of the pipe.

the form of SO_4^{2-} and provide a carbon source for sulfate respirers (sulfate-reducing bacteria that respire using sulfate as a terminal electron acceptor). The products of sulfate respiration include S^0 and organic sulfur. When the fuel is burned, these forms of sulfur cause severe pollution. Before burning, however, the S^0 and organic sulfur can be oxidized by sulfur-oxidizing bacteria. Fuel processors are now turning to sulfur-oxidizing microbes for experimental use in "desulfuration," the removal of sulfur from coal.

A habitat that exhibits the entire range of sulfur oxidation states is that of a sewer pipe. In a concrete sewer pipe, the alternation between anaerobic and oxygenated sulfur biochemistry causes severe corrosion (**Fig. 22.16B**). The microbial decomposition of sewage yields large quantities of toxic H_2S, which then volatilizes to high levels that endanger sewer workers. The H_2S is then oxidized to sulfuric acid by *Thiobacillus ferrooxidans*, a bacterium that colonizes the surface of the concrete. The sulfuric acid (H_2SO_4) decreases the pH at the concrete surface to pH 2. In the concrete surface, sulfuric acid converts calcium hydroxide to soluble calcium sulfate. Over several years, this corrosion can eat away half the thickness of a sewer pipe.

Sulfur metabolism shows important connections with metabolism of metals. For example, sulfur-oxidizing bacteria such as *Thiobacillus ferrooxidans* oxidize iron as well (discussed shortly).

THOUGHT QUESTION 22.6 Compare and contrast the cycling of nitrogen and sulfur. How are the cycles similar? How are they different?

The Phosphate Cycle

Phosphorus is a macronutrient required for assimilation into organic biomass. Phosphate is a fundamental component of nucleic acids, phospholipids, and phosphorylated proteins. Unlike sulfur and nitrogen, which are found in several different oxidation states, phosphorus is cycled almost entirely in the fully oxidized state of phosphate (**Fig. 22.17**). The absence of fully reduced phosphorus in ecosystems may be due to the fact that reduced phosphorus (phosphine, PH_3) undergoes spontaneous combustion in the presence of oxygen. Nevertheless, some anaerobic decomposers use phosphate as a terminal electron acceptor, reducing it to phosphine. In marshes and graveyards, where extensive decomposition occurs, phosphine emanates from the ground, where it ignites with a green glow. Microbial respiration of phosphate might be the cause of such "ghostly" apparitions.

Although phosphate is abundant in Earth's crust, its availability in ecosystems is limited by its tendency to precipitate with calcium, magnesium, and iron ions. Thus, dissolved phosphate in water and soil is often a limiting nutrient for productivity. In natural ecosystems, the available phosphate is taken up rapidly by bacteria and phytoplankton, and then consumed by grazers and predators and dispersed by decomposers.

Marine water is extremely limited for phosphate, because of the distance from sediment minerals. Thus, the genomes of marine phototrophs encode many systems for acquiring phosphate from organic sources. In addition, marine phototrophs show an unusual ability to substitute nonphosphorus lipids for phospholipids, thus cutting their phosphorus needs by half. For example, the cyanobacteria *Prochlorococcus*, *Synechococcus*, and *Trichodesmium* can replace membrane phospholipids with sulfonated lipids. Similarly, algae such as *Thalassiosira* species can replace phosphatidylcholine with betaine, which contains no phosphate but includes a carboxylate group.

Another microbial response to phosphate scarcity is to replace membrane phosphates with phosphonates,

Figure 22.17   The phosphate cycle. Phosphorus in the biosphere occurs entirely in the form of inorganic or organic phosphate. Most phosphate precipitates as insoluble salts in sediment. The small amount of soluble phosphate is taken up by plants and bacteria, which may then be taken up by consumers. Decomposers return phosphate to the environment.

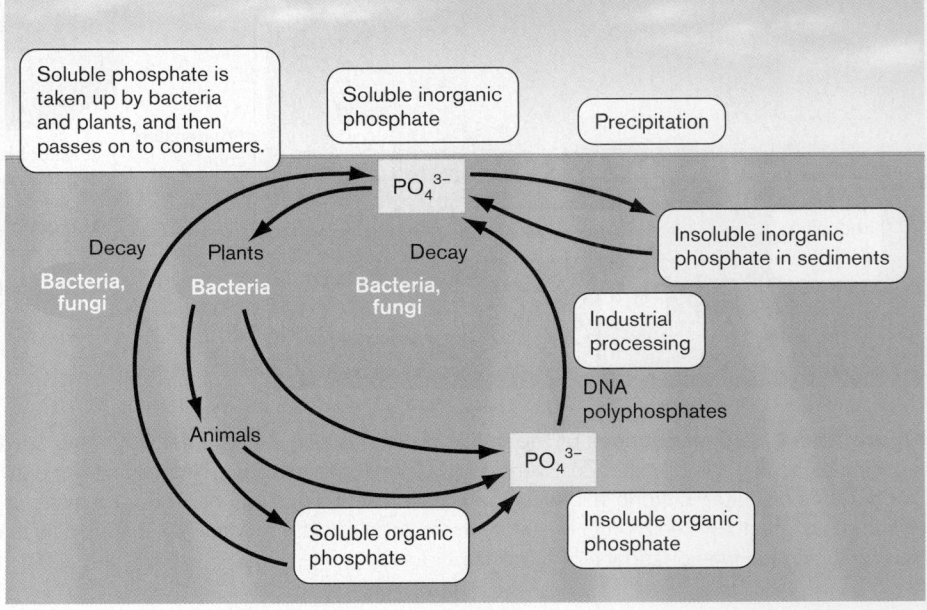

organic molecules in which the phosphorus atom bonds directly to carbon, instead of the easily hydrolyzed phosphoester bond. These phosphonate compounds, however, are cleaved by marine archaea, the ammonia oxidizers such as *Nitrosopumilus*. The archaea release methylphosphonate, which many bacteria can cleave to obtain phosphate. Phosphate scavenging from methylphosphonate releases methane, thus interacting with the carbon cycle.

In agriculture, phosphate is often added as a fertilizer. Phosphate fertilizer is obtained by treating calcium phosphate rock with sulfuric acid, producing calcium sulfate (gypsum) and phosphoric acid. Excess phosphate from fertilizer or industry may drain into streams and lakes where phosphorus is the limiting element. The sudden influx of phosphate causes an algal bloom. The overgrowth of algae leads to overgrowth of heterotrophs, depletion of oxygen, and destruction of the food chain.

> **THOUGHT QUESTION 22.7** Compare and contrast the cycling of nitrogen and phosphorus. How are the cycles similar? How are they different?

The Iron Cycle

Iron is a micronutrient, an element that forms a negligible part of biomass but is essential for growth of organisms (discussed in Chapter 4). Nearly all organisms require iron, which acts as a cofactor for enzymes and is an essential component of oxygen carrier molecules such as hemoglobin. Other common micronutrients required for enzymes and cofactors are zinc, copper, and selenium.

Iron in soil and sediment. Iron is a major component of Earth's crust, and substantial quantities are present in most soil and aquatic sediment. Yet the availability of iron to organisms is limited by its extremely low solubility in the oxidized form. In microbial biochemistry, the oxidized form, ferric iron (Fe^{3+}), interconverts with the reduced form, ferrous iron (Fe^{2+}). In the presence of oxygen, iron metal (Fe^0) "rusts" and is therefore available to organisms mainly as Fe^{3+}. Ferric iron is especially insoluble at high pH, precipitating with hydroxide ions as ferric hydroxide [$Fe(OH)_3$] or with phosphate ions as ferric phosphate ($FePO_4$) (**Fig. 22.18A**).

In iron mines, where pyrite (FeS_2) is exposed to air, spontaneous oxidation releases sulfuric acid:

$$2FeS_2 + 7O_2 + 2H_2O + 4H^+ \longrightarrow 2Fe^{2+} + 4H_2SO_4$$

This reaction drives the growth of lithotrophs such as *Thiobacillus ferrooxidans*, which greatly increases the rate of sulfuric acid production, leading to acid mine drainage. Where acid mine drainage enters an aquatic system, the

A. The iron cycle

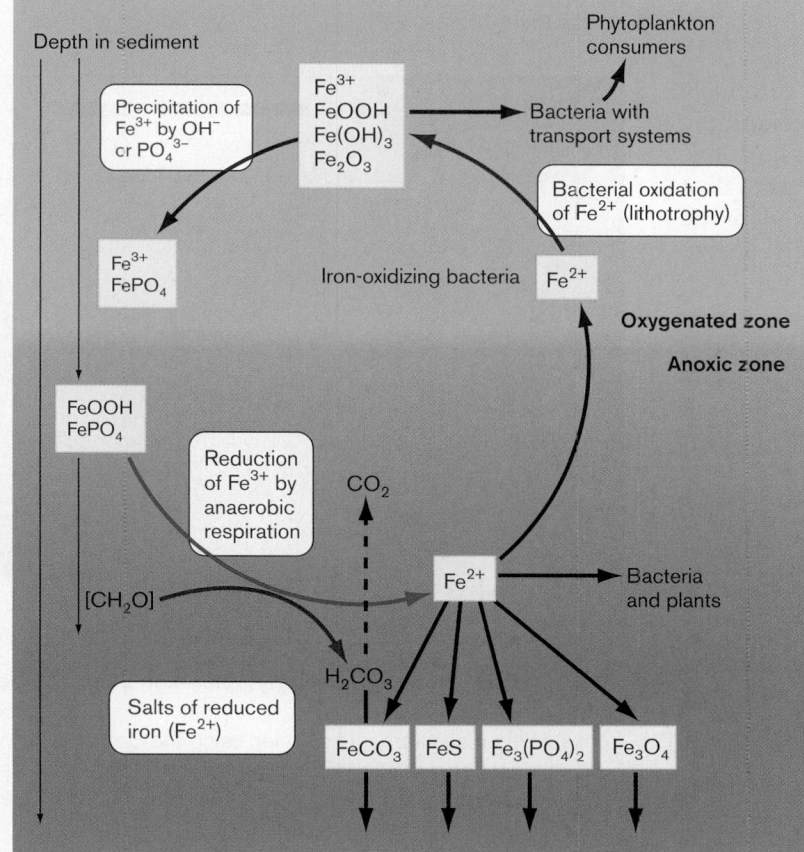

B. Iron precipitation due to lithotrophy

Bryan Allen

Figure 22.18   The iron cycle.
A. Ferric iron (oxidized iron, Fe^{3+}) precipitates with hydroxide or phosphate. Only bacteria can assimilate Fe^{3+}. In anoxic sediment, bacterial respiration reduces Fe^{3+} to Fe^{2+}, a more soluble form available to plants. **B.** Home plumbing shows signs of lithotrophic oxidation of Fe^{2+} to Fe^{3+} (orange precipitate).

reduced iron is oxidized by lithotrophs to ferric hydroxide, forming an orange precipitate, seen in home plumbing (**Fig. 22.18B**). Iron mine drainage causes severe pollution of streams (**Fig. 22.19**).

In anoxic habitats, such as benthic sediment, Fe^{3+} is largely reduced to Fe^{2+} by anaerobic respiration. Reduction leads to loss of the reddish color, generating gray-colored sediment, as in wetland soil (discussed in Chapter 21). The reduction of Fe^{3+} to Fe^{2+} is one of the enriching contributions of anaerobic sediment to aquatic wetlands and coastal estuaries, because reduced iron is more available to plants and bacteria. Oxidized iron, in aerobic soils, can be taken up only by bacteria. Bacteria synthesize special iron uptake systems, including molecules called **siderophores** that bind ferric ion outside the cell (see Section 4.2). Bacterial iron then becomes available to consumers in the ecosystem.

Iron cycling often connects with the sulfur cycle in ways that can prove unfortunate for human engineering. The most serious problem is anaerobic corrosion of iron. Iron corrodes spontaneously in the presence of oxygen. In anoxic environments, however, little spontaneous corrosion occurs. Instead, iron is corroded by sulfate-reducing bacteria (**Fig. 22.20**). Bacteria such as *Desulfovibrio* and *Desulfobacter* can grow on the iron surface, forming an anaerobic biofilm. Within the biofilm, iron is oxidized to Fe^{2+}, and sulfate dissolved in the water is reduced to ferrous sulfide:

$$4Fe^0 + SO_4^{2-} + 4H_2O \longrightarrow FeS + 3Fe^{2+} + 8OH^-$$

The FeS flakes away, exposing more iron to react with water, resulting in cyclic corrosion.

Marine iron. In most ocean water, iron is extremely scarce. This is because the benthic sediment containing iron is so distant that its iron is largely inaccessible. The main source of marine iron is eolian (wind-borne) dust from dry land. For example, a major source of iron for the North Atlantic is windstorms from the Sahara desert.

Most of the wind-borne iron is particulate, oxidized, and unavailable to eukaryotic phytoplankton (algae). Thus, bacteria that acquire and reduce ferric iron provide the main entry of iron into the ecosystem. This may explain why so many marine algae are mixotrophs: Although their photosynthesis can fix plenty of carbon for biomass, they must consume bacteria as a source of iron.

Iron is thus a limiting nutrient for marine phytoplankton. As discussed in Chapter 21, when higher quantities of a limiting nutrient enter an ecosystem, the limited populations rapidly increase. The rise of algal populations following iron addition was demonstrated in an "iron fertilization" experiment conducted by Phillip Boyd and colleagues at the University of Otago, New Zealand (**Fig. 22.21**). Several thousand kilograms of ferrous sulfate ($FeSO_4$) were

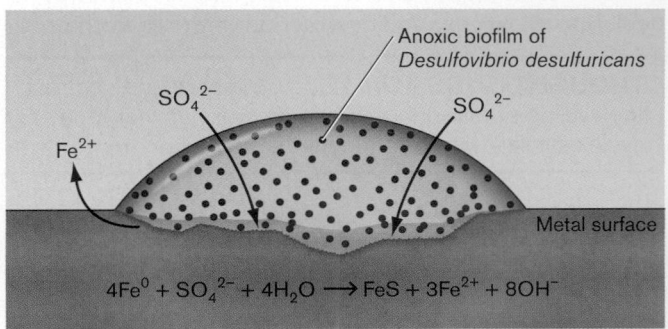

Figure 22.20 Anaerobic corrosion. Iron-sulfur bacteria convert Fe^0 to FeS. FeS flakes off, exposing iron to further oxidation and rapid corrosion. *Source:* H. T. Dinh et al. 2004. *Nature* **427**:829–832.

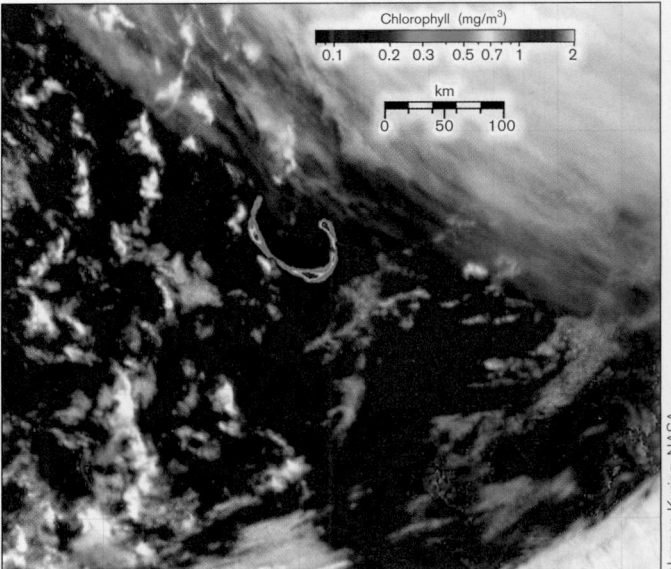

Figure 22.21 Iron-induced phytoplankton bloom in the Southern Ocean. Chlorophyll concentration was determined from satellite measurements of color, using the Sea-viewing Wide Field-of-view Sensor (SeaWiFS). The phytoplankton bloom was observed several days after iron sulfate was introduced. White regions indicate cloud cover. *Source:* Phillip W. Boyd et al. 2007. *Science* **315**:612.

Figure 22.19 Mine drainage enters a stream. Lithotrophic oxidation of reduced iron yields the orange material polluting this stream in Preston County, West Virginia.

released in the Southern Ocean, near Antarctica. Within a week after the iron release, there was a bloom of phytoplankton. The dominant phytoplankton in the bloom were diatoms of the species *Fragilariopsis kerguelensis*. The diatom bloom was so large that it was detected by a NASA satellite as a region of increased color reflected by chlorophyll.

The dramatic effects of iron fertilization led some researchers to propose that increasing iron throughout the oceans could cause blooms of diatoms removing CO_2 from the atmosphere. If a sufficiently large proportion of the diatoms were to escape predation and fall to the ocean floor, they would effectively remove their carbon from circulation. An alternative outcome, however, might be eutrophication, causing an ecological collapse comparable to a dead zone.

In 2009, a major test of iron fertilization was conducted by the LOHAFEX project of the National Institute of Oceanography of India and the Alfred Wegener Institute for Polar and Marine Research of Germany. The LOHAFEX experiment (*loha* meaning "iron" in Hindi; *fex*, "fertilization experiment") was opposed by many environmental groups, who considered it a form of pollution, but it was ultimately carried out. In the experiment, 20,000 kg of iron sulfate were introduced in the Southern Ocean, causing a bloom across 300 km^2. The bloom consisted of coccoliths, algae with plates of calcium carbonate instead of silicate (discussed in Chapter 20). But the coccoliths attracted amphipods, shrimplike predators that soon consumed the algae, recycling most of their carbon back to CO_2.

While iron fertilization is unlikely to meet our need for CO_2 removal, microbiology offers other approaches to offset the perturbation of Earth's carbon cycle by human technology. For example, bacteria and algae may be engineered to provide biofuels, such as hydrogen gas, whose use yields energy without releasing CO_2.

Our view of Earth's biosphere increasingly is moving toward environmental management—the concept that wilderness as such no longer exists, only a biosphere to be managed for better or worse. Microbes will have many roles as partners in environmental management.

Other Metals in the Environment

Another concern for environmental management is that of toxic metals such as mercury and arsenic. Besides iron, numerous trace metals interact with bacterial species as either electron donors or acceptors (**Table 22.4**). Bacterial conversion of metals can either produce toxic species or remove toxic metals from ecosystems. For example, chromium-6 [Cr(VI)] is an extremely toxic pollutant that can be reduced by soil bacteria to the much less toxic Cr(III). Microbial metal cycling is discussed further in **eTopic 22.1**.

Multiple Limiting Factors Modulate Complex Ecosystems

A common assumption of terrestrial and aquatic ecology is that a single nutrient, such as phosphorus or iron, is limiting for a given ecosystem. The requirement for more than one limiting factor is known as **resource colimitation**. In highly oligotrophic marine water, populations often depend on multiple limiting factors. For example, in

Table 22.4 Microbial metabolism of metals.

Metal	Major conversions	Microbial genera (examples)	Effects on environment
Manganese, Mn	$Mn^{2+} \longrightarrow Mn^{4+}$ $Mn^{4+} \longrightarrow Mn^{2+}$	*Hyphomicrobium, Arthrobacter* *Geobacter, Pseudomonas*	Mn is a trace element required for enzymes Anaerobic respiration in sediment
Mercury, Hg	$Hg^{2+} \longrightarrow Hg^0$ $Hg^{2+} \longrightarrow (CH_3)Hg^+$	*Thiobacillus* *Desulfovibrio*	Hg^0 volatilizes; little harm $(CH_3)Hg^+$ is a severe neurotoxin; accumulates at higher trophic levels, such as fish
Arsenic, As	$AsO_2^- \longrightarrow AsO_4^{3-}$	*Alcaligenes, Pseudomonas*	AsO_2^- (arsenite) poisons metabolism; used as terminal e^- acceptor
	$AsO_4^{3-} \longrightarrow AsO_2^-$	*Bacillus, Chrysiogenes, Pyrobaculum*	AsO_4^{3-} (arsenate) poisons metabolism; used as terminal e^- acceptor
	$AsO_4^{3-} \longrightarrow (CH_3)_3As$	*Candida* (a fungus), *Scopulariopsis* (a fungus)	Methylarsines. Poisonous; inhalation of moldy wallpaper with arsenic pigment
Chromium, Cr	$CrO_4^{2-} \longrightarrow Cr^{3+}$	*Aeromonas, Arthrobacter, Desulfovibrio*	CrO_4^{2-} [Cr(VI)] is mutagenic and carcinogenic; as e^- acceptor, reduced to Cr^{3+} [Cr(III)], less toxic
Vanadium, V	$VO_3^- \longrightarrow VO(OH)$	*Veillonella, Desulfovibrio, Clostridium*	V is a trace element for some nitrogenases and invertebrate blood pigments VO_3^- (vanadate) is oxidized as an e^- donor
Selenium, Se	$Se^0 \longrightarrow SeO_3^{2-}$	*Bacillus, Micrococcus*	Se is a trace element; higher doses toxic
Uranium, U	$UO_2^{2+} \longrightarrow UO_2$	*Veillonella, Shewanella, Geobacter*	UO_2^{2+} [U(VI)], soluble; is respired to uranium dioxide, UO_2 [U(IV)], insoluble; used for cleanup of radioactive uranium

the Baltic Sea both nitrogen and phosphorus are so low that they must be added together in order to stimulate a phytoplankton bloom. In another example, the North Atlantic is limited for both phosphorus and iron. Addition of both P and Fe stimulates growth of nitrogen-fixing cyanobacteria.

In some cases, different species are limited by different nutrients. In the LOHAFEX experiment, the Southern Ocean diatoms were limited for both iron and silicate; but coccoliths, whose shells are calcium carbonate rather than silicate, required only iron addition to bloom.

TO SUMMARIZE:

- **Sulfate is abundant in marine water.**
- **Oxidized and reduced forms of sulfur are cycled in ecosystems.** Sulfate and sulfite serve as electron acceptors for respiration. Hydrogen sulfide serves as an electron donor. Sulfur cycling participates in acid mine drainage and pipe erosion.
- **Phosphorus cycles primarily in the fully oxidized form (phosphate).** Phosphate limits growth of phototrophic bacteria and algae in some aquatic and marine systems.
- **Iron cycles in oxidized and reduced forms.** Oxidized iron (Fe^{3+}) serves as a terminal electron acceptor in anaerobic soil and water. Reduced iron (Fe^{2+}) from rock is oxidized through weathering or mining. Bacterial lithotrophy accelerates iron oxidation, leading to acidification.
- **Metal toxins can be metabolized by bacteria.** Bacterial metabolism may either increase or decrease toxicity.
- **Marine habitats show resource colimitation.** Multiple resources may be limiting for the phytoplankton community or for different populations.

22.6 Astrobiology

As we come to appreciate the ubiquitous contributions of microbes to shaping our planet, in all its diverse habitats, increasingly we wonder whether microbes exist on worlds beyond Earth. Is Earth unique in supporting life, or have living cells evolved as well on Mars or Venus or on Jupiter's planet-sized moons? Astrobiology is the study of life in the universe, including its origin and possible existence outside Earth. The discovery of life beyond Earth would arguably be the most significant advance in science since a human set foot on the moon.

If the same physical and chemical laws govern the universe everywhere, then it is hard to suppose that only one of the billions of stars would have a planet supporting life. On the other hand, we have no idea how many planets have been capable of developing and sustaining a biosphere. As Isaac Asimov said, "There are two possibilities. Maybe we're alone. Maybe we're not. Both are equally frightening."

If life exists elsewhere, is it built on the same fundamental elements as ours? Many lines of evidence suggest that the biochemistry of life elsewhere would resemble that of Earth. Terrestrial life is founded on macroelements in the first two rows of the periodic table, including carbon, nitrogen, oxygen, phosphorus, and sulfur (see **Table 22.1**; for periodic table, see Appendix 1). The valence numbers of these elements from the middle of the periodic table enable them to form complex molecular structures with strong covalent bonds. The same fundamental molecules that appear in early-Earth simulation experiments, such as adenine and glycine, also appear in meteorites. Thus, we suspect that the fundamental building blocks for biochemistry are universal. On the other hand, if life could indeed be founded on some other basis, how would we recognize it?

If life is found on other planets, it will almost certainly include microbes. In fact, a case can be made that the majority of biospheres in our galaxy would consist entirely of microbial life, since microbes inhabit a wider range of conditions than do multicellular plants and animals. Even on Earth itself, the largest bulk of our biosphere—including deep sediments and rock strata—consists of microbial ecosystems.

Did Mars Once Support a Biosphere?

For several reasons, the most studied candidate for extraterrestrial life has been the planet Mars:

- **Geology.** Of all the solar planets, Mars seems the most similar to Earth in its topography; indeed, some areas of Mars remarkably resemble desertscapes on Earth. Martian rock contains the fundamental elements needed for life.
- **Day and year length.** Mars has a day length similar to that of Earth, and a year only twice as long as ours.
- **Temperature.** The average temperature on Mars is 220 K (–53°C), too cold for most biochemistry on Earth, but its temperature rises above freezing at the equator. By contrast, the torrid heat of Venus (460°C) would exclude stable macromolecules.
- **Atmosphere.** The overall atmospheric pressure on Mars, 6 mbar, is barely a hundredth that of Earth (1,013 mbar) and lacks molecular oxygen. Thus, aerobes could not grow. But the Martian atmosphere does include carbon dioxide, actually at 20 times the CO_2 content on Earth, so there would be plenty for photoautotrophic production of biomass.
- **Water.** Surface water freezes out of the Martian atmosphere, and mineral formations suggest liquid water in the past. Liquid water may yet exist deep underground, supporting life-forms similar to the endoliths of Earth's crustal rock.

The existence of liquid water is a key question because on Earth, wherever liquid water exists—even brine (concentrated salt) at –20°C—there we find microbial life. Evidence for liquid water, past or present, was a priority for NASA's highly successful Phoenix Lander mission of 2008. Water was observed to condense and freeze out of the atmosphere. Ice clouds and snow were observed for the first time on Mars. The Martian soil composition showed high levels of perchlorate, a chemical that attracts water to form liquid solution and can support metabolism of some Earth microbes.

Other evidence for water on Mars comes from geological formations surveyed by the Mars Reconnaissance Orbiter in 2007. The orbiter mapped sedimentary deposits in Mars's Jezero Crater (**Fig. 22.22**). The crater formations reveal flow patterns typical of a river delta flowing into a lake. Clay-like mineral deposits (false-colored green in **Fig. 22.22**) may have trapped organic compounds needed for life. The layered patterns are best explained by a model involving fluid flow, such as the flow of water.

If life once existed on Mars, what became of it? Two main possibilities are considered:

- **Life developed and existed until the planet froze.** Under this scenario, life originated much as it did on Earth, during a time of heavy bombardment from space and outgassing of nitrogen and carbon dioxide. Unfortunately, however, life failed to generate sufficient atmosphere for a greenhouse effect to sustain temperate conditions. No oxygenic organisms produced molecular oxygen and an ozone layer.

Figure 22.22   Evidence of past water on Mars.
Sedimentary deposits in the Mars Jezero Crater, mapped by the Mars Reconnaissance Orbiter in 2007. The flow patterns resemble a river delta, best explained by a model involving the flow of liquid water. Green indicates clay-like mineral deposits.

Bethany Ehlmann, NASA/JPL/JHUAPL/MSSS/Brown University

- **Life developed and still exists underground.** In the absence of an ozone layer, the Martian surface is sterilized by cosmic radiation. Nevertheless, microbes may yet exist deep underground, similar to the endolithic prokaryotes on Earth. Endoliths metabolize within interstitial water and are protected from cosmic radiation.

Do Biosignatures Indicate Life?

If microbes do exist on Mars, they should be detectable. But the detection of unknown life, even on Earth, remains a challenge. The advent of PCR sequence detection of species based on ribosomal genes has revealed thousands of unknown species, most of which cannot be grown or recognized by other methods. Other species may well be missed if their rRNA sequences fail to amplify with our probes. On Mars, assuming life evolved independently of Earth life-forms, we would have no way to define sequence probes for detection.

Instead, researchers try to define **biosignatures**, chemical and physical signs that could only have been formed by life. Most of the proposed biosignatures are based on types of evidence for life on Earth, either as fossils of ancient life (discussed in Chapter 17) or as signs of current life in extreme habitats. Proposed biosignatures include the following:

- **Microfossils.** The mineralization of microbial cells leads to formation of structures that can be visualized under a microscope. The cell fossils must be sufficiently distinct to establish that no abiotic process could have formed them.
- **Isotope ratios.** Certain biochemical reactions preferentially use one isotope of an atom over another; for example, Rubisco, the key enzyme of carbon dioxide fixation, uses ^{12}C in preference to ^{13}C. Thus, photosynthetic use of ^{12}C can decrease the $^{12}C/^{13}C$ ratio in subsequent carbonate deposits. Isotope ratios of nitrogen, oxygen, and sulfur are also used as biosignatures.
- **Mineral deposits.** Certain mineral formations are observed to be caused only by microbial activity. For example, insoluble manganese oxides such as Mn_2O_3 are almost always the result of microbial oxidation of reduced manganese. Reduced manganese ions are very stable, and their abiotic oxidation rate is extremely slow.
- **Metabolic activity.** Samples of soil can be incubated with radioactive tracer substances such as $^{14}CO_2$ and tested for metabolic conversion or incorporation into biomass. It must be established that the conversion could not have occurred abiotically and that no organisms from Earth were present.

Various kinds of evidence for life on Mars have been reported, such as possible microfossils within a Martian meteorite that landed in Antarctica. Metabolic activity was tested in samples obtained by the NASA Viking Lander in 1975. As of this writing, however, no evidence has proved conclusive for active or past living microorganisms on Mars.

Could Mars Be Terraformed to Support Earth Life?

If no life exists—or if it exists only in the form of microbes deep underground—should we consider human intervention, or **terraforming**, to make Mars habitable for life from Earth? Scenarios for terraforming, long explored in science fiction, are now receiving serious thought among space scientists. Terraforming Mars would require increasing the temperature and air pressure. The temperature of the atmosphere might be increased by release of greenhouse gases. In addition, sufficient carbon dioxide and nitrogen gases must be made available from the Mars surface rock. In principle, microbes from Earth could be seeded to grow and generate an atmosphere containing nitrogen, oxygen, and CO_2.

The dilemmas and consequences of terraforming are depicted in the novel *Red Mars* (1993) by Kim Stanley Robinson. In favor of terraforming, it is argued that Mars offers enormous natural resources of potential benefit for humanity, especially as terrestrial resources are used up. Human settlements on Mars would be a major step forward for space exploration. On the other hand, it is argued that the planet Mars is a natural monument, a place with its own right to exist as such. It should be allowed to remain in its natural state for future generations to appreciate. As a practical matter, terraforming remains unfeasible for the near future. For example, the amount of chlorofluorocarbons calculated to raise Martian temperature is 100 times greater than our global capacity to produce such substances.

Does Europa Have an Ocean?

Farther out in the solar system, surprising candidates for microbial life are the moons of Jupiter. In 2000, the Galileo space probe passed several of Jupiter's moons, including Ganymede, Callisto, and Europa (**Fig. 22.23**). While their distance from the Sun results in extreme cold, these bodies receive extra heat from friction generated by tidal forces from the giant planet Jupiter. In the case of Europa, the tidal forces have been calculated to provide enough heat to liquefy water without boiling it off. Furthermore, measurement of Europa's magnetic field suggests that its composition includes a dense iron core surrounded by 15% water. Most of the water must be locked in ice, but tidal heating could melt enough water for an underlying ocean of brine. On Earth, similar brine lakes beneath the ice of Antarctica harbor halophilic archaea that grow at –20°C.

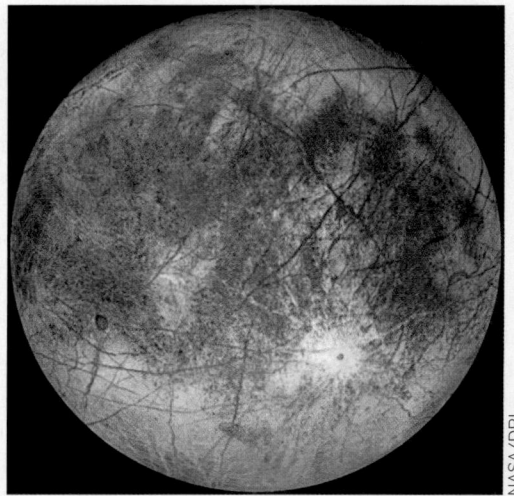

Figure 22.23 Jupiter's moon Europa. Does life exist beneath Europa's ice, in a salty sea?

If such oceans do exist, how could they support life without photosynthesis? Photosynthesis is impossible at Jupiter's distance from the Sun. However, an alternative source of chemical energy might be the influx of charged particles accelerated in Jupiter's magnetic field. Charged particles entering Europa's ice can react with water to form hydrogen peroxide (H_2O_2). The H_2O_2 then breaks down, releasing molecular oxygen. Oxygen reaching the brine layer below could combine with electron donors from crustal vents and power metabolism. Alternatively, molecular hydrogen (H_2) could be generated from water ionized by decay of radioisotopes. A similar source of H_2 for life has been proposed to occur on Earth in rock strata several kilometers below the surface, where it may support lithotrophs. The hydrogen could then combine with oxygen gas or other oxidants to power life. These schemes are highly speculative—but tantalizing enough to encourage further NASA missions to take a closer look at Europa and its sister moons.

Does Life Exist on Planets of Distant Stars?

The past decade of astronomy has seen an extraordinary growth in our knowledge of solar systems beyond our own. Now that we know so many other stars possess planets, we can only wonder whether they, too, possess biospheres of life-forms we would recognize.

In recent years, several hundred "extrasolar" planets have been detected around distant stars, including some of low mass comparable to that of Earth. Could we ever hope to detect signs of life on an extrasolar planet? A possible means of detection is suggested by William Sparks and colleagues at the Space Telescope Science Institute, Baltimore, and collaborating institutions. They show that

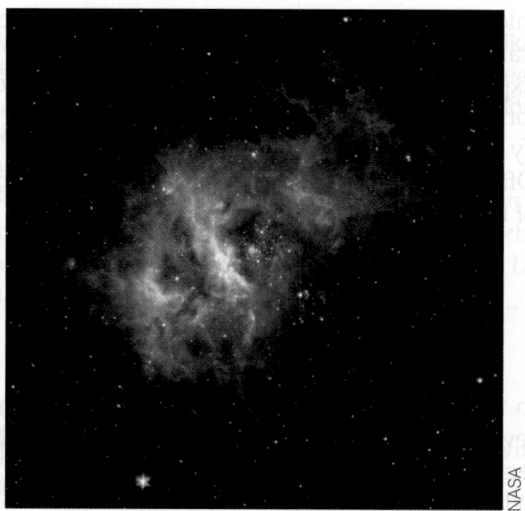

Figure 22.24   A stellar nursery.   Nebula RCW-49 contains more than 300 newborn stars, from which NASA's Spitzer Space Telescope detected spectroscopic signals of common organic constituents of life (infrared photograph).

light scattered by marine phytoplankton exhibits a property called circular polarization. Circular polarization arises from substances that are homochiral—that is, present in only one of two mirror forms. Homochirality is a strong biosignature, typical of proteins and metabolites. If someday we have telescopes capable of detecting light from distant planets, circular polarization might provide evidence of life.

Figure 22.24 shows an infrared photograph through the Spitzer Space Telescope of nebula RCW-49, a gaseous cloud full of newborn stars. Recall from Chapter 17 that as stars form, they take up dust of supernovas that includes all the elements needed to form biomolecules. Spectroscopic observation of RCW-49 reveals stars surrounded by disks coalescing into planets. The planetary disks contain icy particles full of organic molecules such

as methanol, glycine, and ethylene glycol, a reduced form of sugar. Could there be biospheres in the making?

TO SUMMARIZE:

- **Astrobiology is the study of life in the universe, including possible habitats outside Earth.**
- **The search for extraterrestrial life is based on methods similar to those used to seek early life on Earth.** Evidence includes chemical and physical biosignatures, isotope ratios, microfossils, and metabolic activity.
- **Mars is the planet whose geology most closely resembles that of Earth.** Geological features strongly support the past existence of flowing water, a prerequisite for microbial life.
- **Jupiter's moon Europa is proposed as another possible site for life.** Europa is bathed in a sea of brine similar to terrestrial habitats for halophiles.

Concluding Thoughts

Our observations of distant molecules, as well as our analysis of meteorites (discussed in Chapter 17), suggest that the biomolecules of Earth fit into a universal pattern of interstellar chemistry. Whether life exists elsewhere or we are alone, we must remember that our Earth is the only place we know of at this time that can support humans and the forms of life we require for our own survival. The function of our entire biosphere depends on our microbial partners to cycle key elements and acquire energy to drive the food web. For the first time in history, human technology now rivals the ability of microbes to alter fundamental cycles of biogeochemistry. But to manage and moderate our alterations—for our own survival and that of the biosphere—our fate still depends on the microbes.

CHAPTER REVIEW

Review Questions

1. Explain the major sources and sinks of carbon, nitrogen, and sulfur. Which sources recycle rapidly, and why?
2. Explain how the carbon cycle differs in oxygenated and anoxic environments.
3. Explain two different chemical methods of measuring the environmental levels of carbon and nitrogen.

4. Outline the hydrologic cycle. Explain the role of biochemical oxygen demand (BOD) in water quality and how it may be perturbed by human pollution.
5. Outline the function of a wastewater treatment plant. Include the phases of primary, secondary, and tertiary treatment. Explain the roles of microbes in these phases of water treatment.

Imagine two vacationing tourists, unknown to each other, each visiting the Alabama Gulf of Mexico coast at the same time. Both decide to take a swim in the Gulf. One, a 12-year-old girl, has a cut on her leg caused by the sharp edge of a scooter. The other vacationer, a man 66 years of age, recently cut his arm on a nail protruding from a wooden handrail. Within moments of entering the Gulf, several small Gram-negative microbes invade both of their bodies through those cuts. Four days later, the young girl is heading back to Birmingham in the back of the family car, oblivious to the battle recently waged in her bloodstream. At the same time, the man lies dead of an aggressive blood infection.

Both swimmers were attacked by the same pathogen, *Vibrio vulnificus*. Why did one live and the other die? As is the case for most healthy people, a variety of nonspecific, innate immune factors present in the girl's body killed the invading pathogen before it could multiply. The man, on the other hand, was an alcoholic with liver disease. He was deficient in several of those defense mechanisms—mechanisms that made the difference between life and death. What are these powerful innate factors? How can they be nonspecific, able to attack many different microbes without ever previously encountering any of them? And why do they kill the invaders and not our own cells?

Before exploring these questions of immunity and self-defense, we must first understand that each of us is a self-contained ecosystem, home to numerous microbes inhabiting all sorts of body niches. This ecological relationship helped shape our immune system. The day-to-day presence of these microbial guests also sharpens our immunity, but it is a benefit that comes with considerable risk.

23.1 Human Microbiota: Location and Shifting Composition

From the moment of our birth to the time of our death, we are constantly exposed to microbes. As a result, humans—and all other higher life-forms—are heavily populated with bacteria. In fact, our bodies carry ten times as many bacterial cells as human cells. Colonization typically occurs where our body interfaces with the external environment (for example, mouth, skin, and parts of the genitourinary tract). The more sequestered body sites are sterile. For example, most internal organs, the blood, and the cerebrospinal fluid should not harbor any bacteria. Microbes discovered at these sites usually have deleterious effects on the host. Their presence means an infection is under way. Bacteria normally found at various nonsterile body sites (such as the intestine) are called **commensal organisms** (*commensal*, from the Latin "to share a table"). By strict definition, a commensal organism derives benefit from the host but does not harm the host. However, many organisms originally defined as

commensal were later shown to benefit the host in relationships that are called mutualistic (see Section 17.6). The organisms, however, are still called commensal, especially in human hosts. A remarkable variety of bacterial species populate these sites, coexisting in delicate, balanced ecosystems.

We are now coming to understand that bacteria colonizing our bodies may be as important collectively as a kidney or liver. For instance, Gary Siuzdak and his colleagues at the Scripps Research Institute recently showed that our body cells are constantly bathed in metabolites produced by gut microbes. It is now thought that these metabolites, which circulate in our blood, significantly and favorably influence human health and development. The consortium of colonizing microbes has been dubbed the human **microbiota** or **microbiome**. Metagenomic strategies (described in Chapter 7) are under way to sequence all the genomes of our resident microbes. Remember, the majority of species that make up our microbiota are unknown and have never been grown in the lab. The collective genome of all microorganisms living in or on the human body is called a metagenome. We expect to gain tremendous insight into host-microbe relationships from the various microbiome projects under way. For instance, comparisons of microbiomes from the guts of lean and obese mice have shown vast differences in microbial species. The data suggest that the different microbiotas influenced the efficiency of calorie harvesting from the diet and altered how the derived energy was stored. Indeed, a microbiome study of sets of human twins in which one from each set was obese and the other lean found dramatic differences in their gut microbiomes.

WWW | Human Microbiome Project
     | *microbiology2.com/links*

In this section we describe some of the well-known members of the human microbiota and why they populate different body sites. **Figure 23.1** illustrates various human body sites colonized by bacteria and provides examples of the bacteria found there. Realize that the various species comprising normal microbiota at a given body site exist there as a biofilm consortium, chemically communicating with each other and their host.

First we will consider microbe-human interactions that take place in the host's outermost layers—the skin and the eye—followed by the environments of the nose and throat. We will then take up the relatively commensal-free respiratory tract and the commensal-rich digestive tract. Finally, normal microbiota of the genitourinary tract, with its widely diverse environments, will be described. **Table 23.1** lists the four prominent ecosystems of humans, along with the number of microorganisms that typically inhabit them (called the **bioburden**). The table includes the ratio of aerobes to anaerobes and the initial origins of resident species.

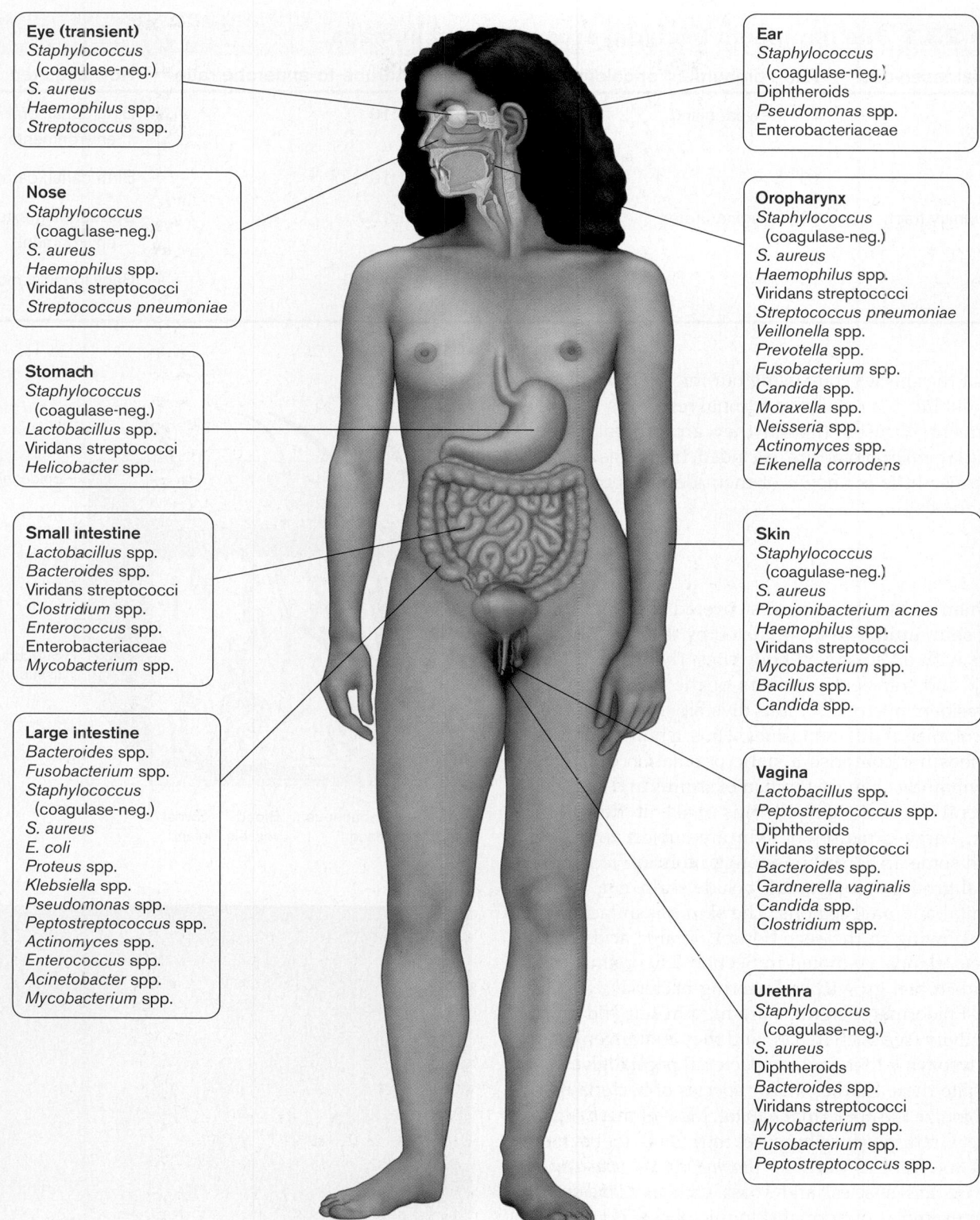

Eye (transient)
Staphylococcus
  (coagulase-neg.)
S. aureus
Haemophilus spp.
Streptococcus spp.

Nose
Staphylococcus
  (coagulase-neg.)
S. aureus
Haemophilus spp.
Viridans streptococci
Streptococcus pneumoniae

Stomach
Staphylococcus
  (coagulase-neg.)
Lactobacillus spp.
Viridans streptococci
Helicobacter spp.

Small intestine
Lactobacillus spp.
Bacteroides spp.
Viridans streptococci
Clostridium spp.
Enterococcus spp.
Enterobacteriaceae
Mycobacterium spp.

Large intestine
Bacteroides spp.
Fusobacterium spp.
Staphylococcus
  (coagulase-neg.)
S. aureus
E. coli
Proteus spp.
Klebsiella spp.
Pseudomonas spp.
Peptostreptococcus spp.
Actinomyces spp.
Enterococcus spp.
Acinetobacter spp.
Mycobacterium spp.

Ear
Staphylococcus
  (coagulase-neg.)
Diphtheroids
Pseudomonas spp.
Enterobacteriaceae

Oropharynx
Staphylococcus
  (coagulase-neg.)
S. aureus
Haemophilus spp.
Viridans streptococci
Streptococcus pneumoniae
Veillonella spp.
Prevotella spp.
Fusobacterium spp.
Candida spp.
Moraxella spp.
Neisseria spp.
Actinomyces spp.
Eikenella corrodens

Skin
Staphylococcus
  (coagulase-neg.)
S. aureus
Propionibacterium acnes
Haemophilus spp.
Viridans streptococci
Mycobacterium spp.
Bacillus spp.
Candida spp.

Vagina
Lactobacillus spp.
Peptostreptococcus spp.
Diphtheroids
Viridans streptococci
Bacteroides spp.
Gardnerella vaginalis
Candida spp.
Clostridium spp.

Urethra
Staphylococcus
  (coagulase-neg.)
S. aureus
Diphtheroids
Bacteroides spp.
Viridans streptococci
Mycobacterium spp.
Fusobacterium spp.
Peptostreptococcus spp.

Figure 23.1 **Examples of normal microbiota present at various colonizing sites.** *Staphylococcus* species are differentiated by their differing abilities to produce coagulase, an enzyme that coagulates serum. *S. aureus* is generally coagulase-positive (there are some coagulase-negative mutants), whereas other species of staphylococci normally found on skin are coagulase-negative. Some organisms—*Streptococcus pneumoniae*, for example—are pathogens but can be found as normal microbiota in some individuals.

Table 23.1 The prominent bacterial ecosystems of humans.

Microbial reservoirs	Total "bioburden" or colony-forming units	Aerobe-to-anaerobe ratio	How acquired
Skin	10^{4-6}/sweat gland	1:10	Birth canal; oral environments
Mouth	10^{6-8}	1:10	Birth canal; caregiver
Genitourinary tract	10^{8-9}/vagina/ureter	1:100	Surrounding external environment
Intestine	10^{11}/cm^3	1:1,000	Baby formula; mother; caregiver

Note that the species composing our microbiota will vary throughout life as a result of continual reseeding from the environment. Viruses, though they are present among our cellular microbiota, are excluded from this discussion because little is known of their identities or impact on human health.

Skin

The human adult, on average, is covered with 2 m^2 (over 9 ft^2) of skin (**epidermis**) populated by 10^{12} microorganisms. As with all colonized body sites, there are resident (normal) and transient members of the microbiota. But even a resident microbe exhibits diversity in that different strains colonize at different times. Thus, a human's microbiome does not comprise a static population but represents a vibrantly changing milieu of strains and species.

Several features of epidermis make it difficult to colonize. Large expanses of skin are subject to drying, although some areas harbor enough moisture to support microbial growth; moist areas include scalp, ear, armpit, and genital and anal regions. The skin has an acidic pH (pH 4–6) owing to the secretion of organic acids by oil and sweat glands. As noted in Section 5.5, organic acids inhibit microbial growth by lowering bacterial cytoplasmic pH. Epidermal secretions are high in salt and low in water activity (see Section 5.4), and they contain enzymes such as lysozyme that degrade bacterial peptidoglycan.

Despite these hurdles, many species of bacteria manage to colonize the epidermal habitat. Most of these organisms are Gram-positive because they tend to be more resistant to salt and dryness. *Staphylococcus epidermidis*, various *Bacillus* species, and yeast such as *Candida* are common examples of normal skin microbes. Though normal, they are not always innocuous. One member of the skin's resident microbiota, the Gram-positive, anaerobic rod *Propionibacterium acnes*, causes acne, a very visible plague of adolescence. Curiously, this is the same genus (but different species) that makes Swiss cheese (see Section 16.2). Increased hormonal activity during the teen years stimulates oil production by the sebaceous glands

A.

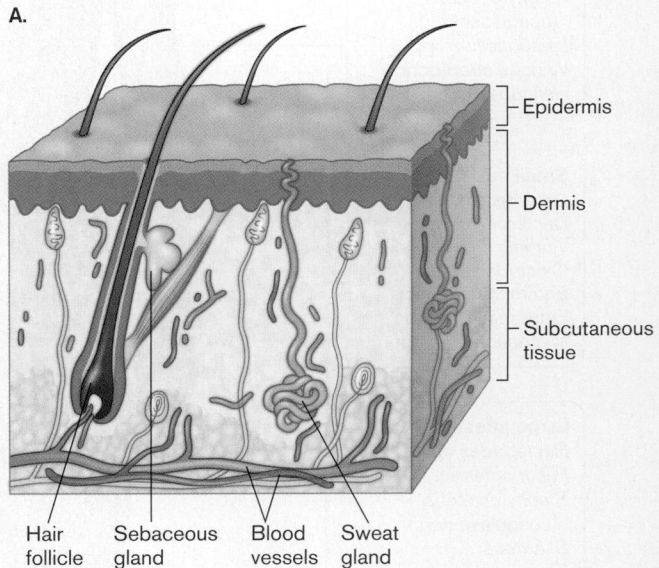

Epidermis

Dermis

Subcutaneous tissue

Hair follicle   Sebaceous gland   Blood vessels   Sweat gland

B.

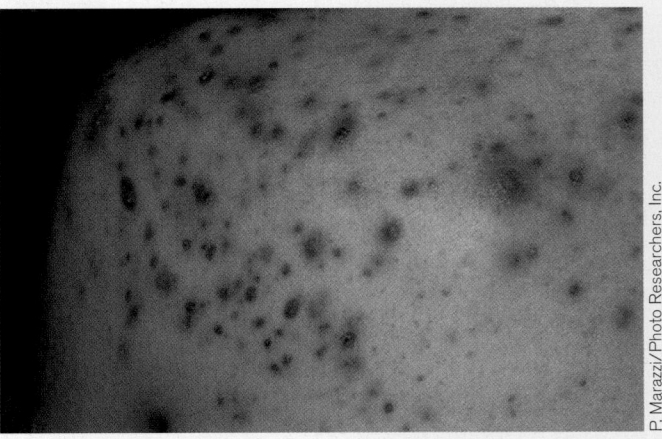

P. Marazzi/Photo Researchers, Inc.

Figure 23.2 Microbiology of skin and development of acne.
A. Location of sebaceous glands in skin. **B.** Acne.

(**Fig. 23.2**). *P. acnes* readily degrades the triglycerides in this oil, turning them into free fatty acids that then promote inflammation of the gland. One consequence of the inflammatory response is the formation of a blackhead, a plug of fluid and keratin that forms in the gland duct. The result is the typical skin eruptions of acne. Because of its microbial basis, treatments for acne include tetracycline to kill the microbe.

Eye

Because the eye is exposed to the outside environment, one might expect it to be heavily colonized. Actually, this is not the case; colonization is inhibited by the presence of antimicrobial factors such as lysozyme in the tears that continually rinse the eye surface (conjunctiva). Despite this, a few transient commensal bacteria can be found on the conjunctiva. Skin flora such as *Staphylococcus epidermidis* and diphtheroids (Gram-positive rods that look like clubs), as well as some Gram-negative rods such as *Escherichia coli*, *Klebsiella*, and *Proteus*, manage to at least temporarily make the eye their home without causing damage. Occasionally, bacteria such as *Streptococcus pneumoniae* and *Haemophilus influenzae*, as well as some viruses, can cause an ocular disease known as pinkeye, in which the eye becomes reddened with a watery discharge.

Oral and Nasal Cavities

Within hours of birth, a human infant's mouth becomes colonized with nonpathogenic *Neisseria* species (Gram-negative cocci), *Streptococcus*, *Actinomyces*, *Lactobacillus* (all Gram-positive), and some yeasts. These organisms come from the environment surrounding the newborn, such as the mother's skin and garments. As teeth emerge in the newborn, the anaerobic space between teeth and gums supports the growth of anaerobes such as *Prevotella* and *Fusobacterium* (**Fig. 23.3B** and **C**). Whatever the organism, colonizers of the oral cavity must be able to adhere to surfaces, like teeth and gums, to avoid mechanical removal and flushing into the acidic stomach. The teeth and gingival crevices are colonized by over 500 species of bacteria.

Organisms like *Streptococcus mutans*, which attaches to tooth enamel, and *Streptococcus salivarius*, which binds gingival surfaces, form a glycocalyx that enables them to firmly adhere to oral surfaces and to each other. They are but two of the microbes that lead to dental plaque formation. The acidic fermentation products of these organisms demineralize teeth and cause dental caries (tooth decay).

Important microbial habitats of the throat include the **nasopharynx**, which is the area leading from the nose to the oral cavity, and the **oropharynx**, which lies between the soft palate and the upper edge of the epiglottis (**Fig. 23.3A**). Organisms like *Staphylococcus aureus* and

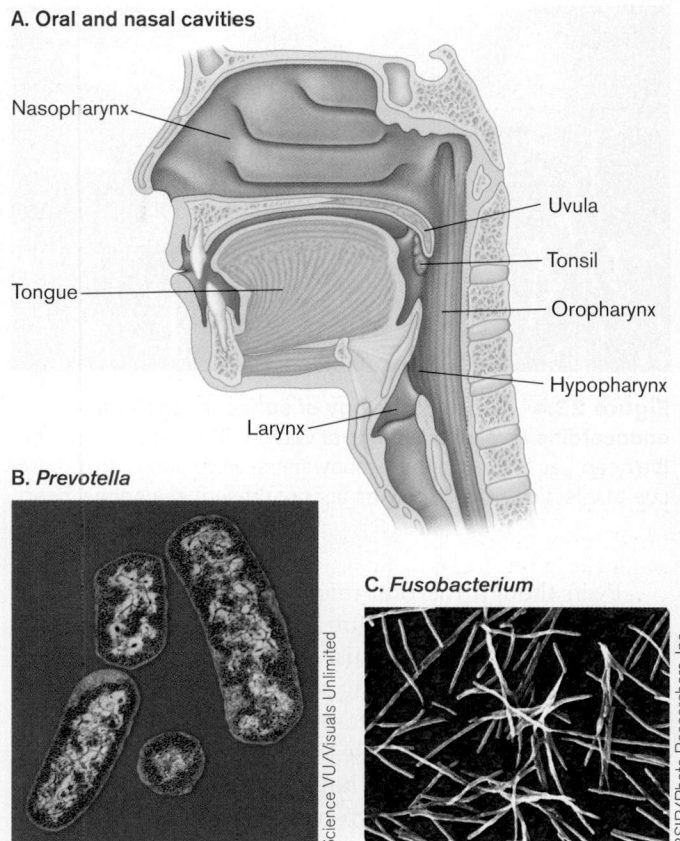

A. Oral and nasal cavities

Nasopharynx

Tongue

Larynx

Uvula

Tonsil

Oropharynx

Hypopharynx

B. Prevotella

Science VU/Visuals Unlimited

C. Fusobacterium

BSIP/Photo Researchers, Inc.

D. Periodontal disease

Biophoto Associates/Photo Researchers, Inc.

Figure 23.3   Examples of anaerobic oral microbiota and periodontal disease.   A. Structures of the oral and nasal cavities. **B.** *Prevotella* (colorized TEM, each cell approx. 2 μm long). **C.** *Fusobacterium* (SEM, each cell approx. 1–5 μm long). **D.** Symptoms of periodontal disease include red swollen gums, bleeding gums, gum shrinkage, and teeth drifting apart.

S. epidermidis populate these sites. The nasopharynx and oropharynx can also harbor relatively harmless streptococci such as *Streptococcus salivarius* and *S. oralis*, as well as *S. mutans*, a major cause of tooth decay. Other oropharyngeal organisms include a large number of diphtheroids and the small Gram-negative rod *Moraxella catarrhalis*. Within the tonsillar crypts (small pits along the tonsil surface) lie anaerobic species such as *Prevotella*, *Porphyromonas*, and *Fusobacterium*.

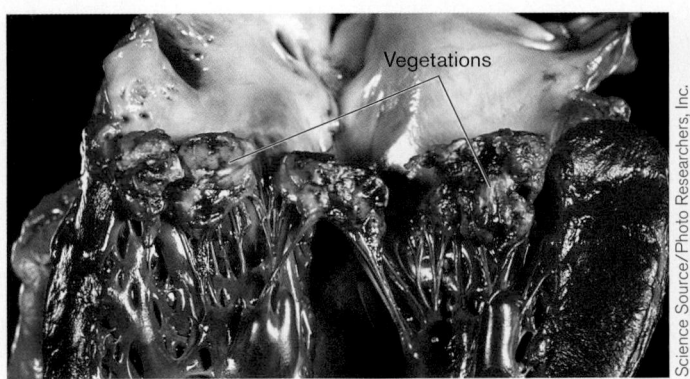

Figure 23.4 **Gross pathology of subacute bacterial endocarditis involving the mitral valve.** The left ventricle of the heart has been opened to show mitral valve fibrin vegetations due to infection. These growths are not present in a normal heart.

Figure 23.5 **Mucociliary elevator.** Movement of these hairlike cilia ushers particles up and out of the trachea and lungs (colorized SEM, diameter between 0.5 and 1 μm).

Even though the oral microbiota is normally harmless, it does hold potential for disease. Dental procedures, for instance, will often cause these organisms to enter the bloodstream, producing what is called **bacteremia** (infection of the bloodstream). Normal immune mechanisms typically clear these transient bacteremias quite easily; but in patients who have a heart mitral valve prolapse (heart murmur), microbes can become trapped in the defective valve and form bacterial vegetations. Vegetations consist of a large number of bacterial cells encased within glycocalyx, a polysaccharide or peptide polymer secreted by the organism, and fibrin, produced by clotting blood. Because the onset of disease is often insidious (slow), it is called subacute bacterial endocarditis (**Fig. 23.4**). Once ensconced within a vegetation, the microbes are extremely difficult to kill with antibiotics.

> **THOUGHT QUESTION 23.1** How can an anaerobic microorganism grow on skin or in the mouth, both of which are exposed to air?
>
> **THOUGHT QUESTION 23.2** What can a person with mitral valve prolapse do to prevent formation of subacute bacterial endocarditis when visiting the dentist?

Respiratory Tract

The lungs and trachea do not harbor normal microbiota. Many organisms entering the nasopharynx are trapped in the nose by cilia that beat toward the pharynx. Their ultimate destination is the acidic stomach and death. Microorganisms that make it into the trachea are trapped by mucus produced by ciliated epithelial cells that line the airways leading to the lung, and the cilia themselves usher the microbes up and away from the lungs. The ciliated mucous lining of the trachea, bronchi, and

bronchioles makes up the **mucociliary elevator** (**Fig. 23.5**). The mucociliary elevator constantly sweeps foreign particles up and out of the lung. This action is extremely important for preventing respiratory infections. When the mucociliary elevator fails or is overwhelmed by inhalation of too many infectious microbes, infections such as the common cold (for example, rhinovirus) or pneumonia (for example, *Streptococcus pneumoniae*) can result.

Stomach

We have known for a hundred years that the stomach contents are acidic and that gastric acidity can kill bacteria. Just how important that acidity is for protection against microbes is illustrated by the infection caused by *Vibrio cholerae*, the causative agent of cholera. Cholera is a severe diarrheal disease endemic to many of the poorer countries of the world. The toxin produced by the organism acts on the intestinal lining to cause voluminous diarrhea—as much as 10 liters a day. Death can occur rapidly as a result of dehydration and shock. Although cholera actually affects the intestines, not the stomach, the bacteria must survive passage through the stomach to reach the intestines.

In the middle of the last century, the U.S. government was concerned that troops dispatched to endemic countries could develop cholera and become incapacitated in large numbers, so the U.S. Army decided to test potential cholera vaccines for their efficacy. The vaccines were living cultures of genetically weakened cholera bacteria that could not cause serious disease. Living strains were preferred because they were expected to grow in the gastrointestinal tract and stimulate natural mucosal immunity (discussed later and in Section 24.3). But after huge numbers of virulent organisms were administered to healthy volunteers (as a control group), very few contracted cholera. This result was confusing because hundreds of thousands of malnourished people in India easily contract the disease.

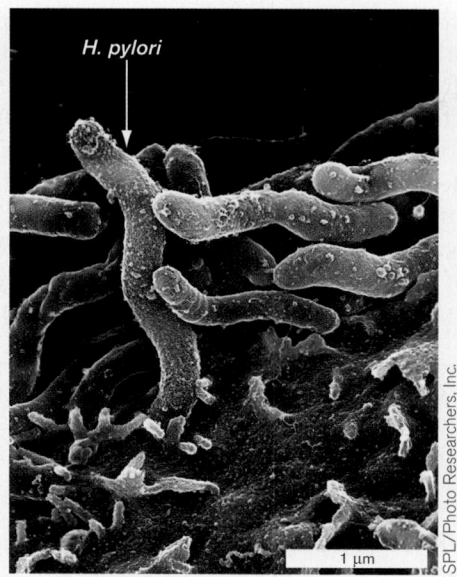

Figure 23.6 *Helicobacter pylori* **defies stomach acidity.**
Helicobacter pylori (colorized SEM) growing in the mucus on
stomach epithelium. *H. pylori* bacteria attach to gastric epithelial
cells and induce specific changes in cellular function, such as
an increase in the expression of laminin receptor 1, a protein
associated with malignancy.

Why did volunteers given live cholera generally fail
to develop the disease? Variations in resistance to cholera
were discovered to be related to differences in stomach
acidity. The organism *V. cholerae* is extremely sensitive to
low pH; even a pH of 4 will readily kill it. Because the pH
in the stomachs of healthy volunteers was well below 4,
it easily killed *V. cholerae*. Malnourished people, however,
suffer from hypochlorhydria (decreased stomach acid).
This permissive environment gives ingested microbes
time to enter the less acidic intestine, where they can
thrive and cause disease. Cholera epidemics in which
thousands of people die, even today, typically occur
among poor, malnourished populations. Because the U.S.
populace is better nourished, cholera is not a major problem in the United States.

Although the contents of the stomach are very acidic,
the mucous lining of the stomach is much less so. It is
there that some bacteria can take refuge. In fact, the
stomach harbors a diverse microbiota (as detected using
cultural and molecular techniques such as polymerase
chain reaction [PCR] identification of 16S rDNA). A classic stomach pathogen is *Helicobacter pylori*. Although this
organism has a remarkable ability to resist acid pH (it survives at pH 1 using the enzyme urease to generate ammonia), it will not grow at that pH. It can, however, grow and
divide in the mucous lining of the stomach, where the pH
is closer to 5 or 6 (**Fig. 23.6**). The U.S. Centers for Disease
Control and Prevention (CDC) estimates that *Helicobacter*
colonizes the stomachs of half the world's population.

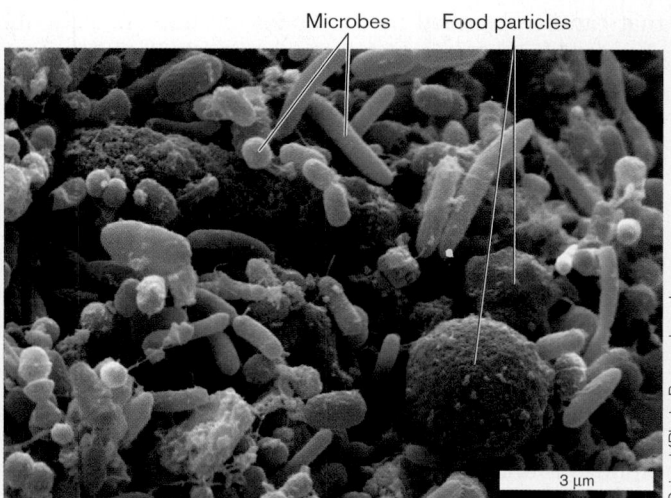

Figure 23.7 **Human feces showing food particles and
resident bacteria (colorized SEM).**

Most of the time, *Helicobacter* does not cause any apparent
problem, but on occasion, the organism can produce gastric ulcers and even cancer.

Human Intestine

The gastrointestinal tract beyond the stomach consists
of an extremely long tube made of several sections, each
of which provides a uniquely different environment that
supports the growth of different bacterial species (see
Fig. 23.1). Pancreatic secretions (at pH 10) enter the intestine at a point just past the stomach and raise the intestinal pH to about pH 8, which immediately relieves the
acid stress placed on microbes. The relatively high pH and
bile content of the duodenum and jejunum allow colonization by only a few resident, mostly Gram-positive, bacteria (enterococci, lactobacilli, and diphtheroids). These
particular Gram-positive organisms possess a bile salt
hydrolase that helps them grow in the presence of intestinal bile salts. The distal parts of the human intestine
(ileum and colon) have a slightly acidic pH (pH 5–7) and a
lower concentration of bile salts—conditions that support
a more vibrant ecosystem. The intestine actually contains
roughly 10^9–10^{11} bacteria per gram of feces (**Fig. 23.7**).
This generally anaerobic environment is populated by
both anaerobic and facultative microbes in a ratio of 1,000
anaerobes to one facultative organism. Why is the intestine anaerobic? The small amount of oxygen that diffuses
from the intestinal wall into the lumen is immediately
consumed by the facultative microbes, which render the
environment anaerobic.

One such anaerobe, *Bacteroides thetaiotaomicron*, provides us with a great benefit. Much of this organism's
genome is dedicated to taking many of the complex carbohydrates we eat and breaking them down into products

that can be absorbed by the body. We humans actually absorb 15%–20% of our daily caloric intake in this way.

Only recently have we really begun to ascertain what microbes inhabit our intestines (mainly because we cannot culture them). Through the use of culture-independent techniques (like ribotyping, bacterial classification based on 16S RNA sequences), we now have a sense of the identity, diversity, and variation among species of the human microbiota. For example, of 395 unique "phylotypes" recently identified in humans, 60% were unknown at the species level, 80% have never been cultured, and huge differences in the population of microbes were observed between individuals.

All organisms normally residing in the intestine are innocuous and considered normal biota. But if they accidentally escape into nearby tissues, they can cause serious disease. Intestinal bacteria, encompassing 400–500 different species, have special talents that allow them to bind and colonize the intestinal cell wall. One reason why such a large and eclectic mix of species is supported is that different bacteria attach to different host cell receptors. Another reason is that many different food sources are available to support diverse groups of microbes.

Organisms that inhabit the intestine include facultative anaerobes in the family Enterobacteriaceae, of which *E. coli* is a member, and many anaerobes, such as *Bacteroides* (Gram-negative rods), *Peptostreptococcus* (Gram-positive cocci), and *Clostridium* (Gram-positive, spore-forming rods). In fact, the vast majority of our intestinal biota consists of the phyla Bacteroidetes and Firmicutes, of which *Bacteroides* species and *Clostridium* species, respectively, are members. Besides bacteria, other common and innocuous inhabitants are the yeast *Candida albicans* and protozoa such as *Trichomonas hominis* and *Entamoeba hartmanni*.

The large intestine is both rich and diverse in terms of nutrient availability. The majority of intestinal bacteria require a fermentable carbohydrate for growth, so it has generally been assumed that carbohydrate metabolism is necessary for colonization by most species. In the large intestine, carbohydrates are derived from partially digested food and from host secretions. Naturally occurring sugar acids, such as galacturonate from pectin and gluconate and ketogluconate from muscle tissues, are present in the foods we eat. The mucus layer covering epithelial tissues is also recognized as an important source of carbohydrates in the intestine. Mucus is a complex gel of glycoproteins and glycolipids. The sugar substituents of mucus include *N*-acetylglucosamine, *N*-acetylgalactosamine, galactose, fucose, sialic acids, and lesser amounts of glucuronate and galacturonate. Interestingly, *E. coli* will grow rapidly in mucus by catabolizing gluconate, a component of secreted mucus, using the Entner-Doudoroff pathway (see Section 13.5). The organism grows very poorly in feces itself. Thus, *E. coli* doesn't grow on what we eat but on what we *secrete*.

Despite this buffet-like assortment of food, many species still end up vying for the same nutrient, so competition for food is fierce. Perturbing the balance of power with antibiotics, diet, or stress can lead to disease.

Although we have considerable knowledge of the gastrointestinal microbiota, there remains much to learn. For example, many of the species inhabiting the intestine have never been cultured in the laboratory. Molecular techniques such as PCR are now being used to identify the nonculturable organisms in this environment.

> **THOUGHT QUESTION 23.3** Why do many Gram-positive microbes that grow on the skin, such as *Staphylococcus epidermidis*, grow poorly or not at all in the gut?
>
> **THOUGHT QUESTION 23.4** How might normal biota escape the intestine and cause disease at other body sites?

Probiotics. As already noted, the microbial composition of the intestine is complex but balanced. It is complex because many species reside there. It is balanced because these species manage to coexist without eliminating each other. The rival organisms help prevent infection by competing with pathogenic species for nutrients and can actually contribute to nutrition (for example, *E. coli* makes vitamin B_{12}, whereas other organisms help digest complex substrates). Emotional stress, a change in diet, or antibiotic therapy can throw off that balance and lead to poor digestion or disease, such as pseudomembranous enterocolitis (**Fig. 23.8**). Some people favor restoring the natural microbial balance by orally ingesting living microbes such as the lactobacilli present in yogurt. Taking these supplements, called **probiotics** (from the Greek meaning "for life"), is thought to restore balance to the microbial community and return the host to good health. The most commonly used probiotic genera are *Lactobacillus* and *Bifidobacterium* (**Fig. 23.9**). The potential mechanisms by which they improve intestinal health include competitive bacterial interactions (normal biota prevent growth of pathogenic bacteria), production of antimicrobial compounds, and immunomodulation (an ability to change the activity of cells of the immune system). The emerging use of probiotics in several gastrointestinal disorders (including inflammatory bowel disease, IBS) has led to increased interest in their use in patients. Early results are promising, but much of the research remains controversial.

> **THOUGHT QUESTION 23.5** Why can the colon be considered a fermenter?

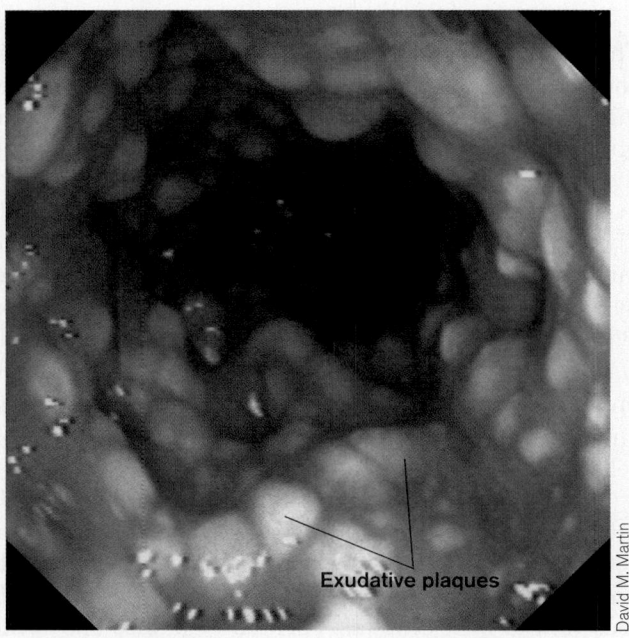

Figure 23.8 Pseudomembranous enterocolitis caused by *Clostridium difficile.* The organism is often a part of the normal intestinal microflora, but antibiotic treatment, especially following surgery, can kill off competing microbes, leaving *C. difficile* to grow unabated. A toxin produced by the organism growing at the epithelial surface damages and kills host cells, causing exudative plaques to form on the intestinal wall. The small plaques eventually coalesce to form a large pseudomembrane that can slough off into the intestinal contents.

Genitourinary Tract

Much of the genitourinary tract is normally free from microbes. These areas include the kidneys, which remove waste products from the blood; ureters, which remove urine from the kidneys; and the urinary bladder, which holds urine until it is excreted. The distal urethra, however, because of its proximity to the outside world, normally contains *Staphylococcus epidermidis*, *Enterococcus* species, and some members of Enterobacteriaceae. Some of these organisms can cause bladder disease (known as urinary tract infection, or UTI) if they make their way into the bladder (for example, via catheterization).

The large surface area and associated secretions of the female genital tract make it a rich environment for microbes. Composition of the vaginal microbiota changes with the menstrual cycle, owing to changing nutrients and pH. The mildly acidic nature of vaginal secretions (approximately pH 4.5) discourages the growth of many microbes. As a result, the acid-tolerant *Lactobacillus acidophilus* is among the most populous vaginal species. As with all microbial ecosystems associated with the human body, the balance between competing species is crucial to preventing infection. Antibiotic therapy to treat an infection can also cause the loss of normal biota. In the vagina, this allows infection by *Candida albicans* (yeast infection), which is not susceptible to antibiotics designed to kill bacteria.

TO SUMMARIZE:

- The **normal microbiota** present on body surfaces constantly changes.
- **Normal microbiota can cause disease** if microbes breach the surfaces they colonize and gain access to the circulation or deeper tissues.
- **Normal microbiota help prevent disease** by inhibiting colonization by pathogens.
- **Colonization of skin** by microbes is difficult owing to surface dryness, an acidic pH, high salinity, and the presence of degradative enzymes.
- **Skin microbiota** consists primarily of Gram-positive microbes, including *Propionibacterium acnes*, which can cause acne.
- **Microbe-free areas of the body** are the lungs, cerebrospinal fluid, and bladder. The eyes are generally microbe-free, although some microbes can be present for short periods.
- **Oral and nasal surfaces** are colonized by aerobic and anaerobic microbes.
- **The intestine** is populated by 10^9–10^{11} microbes per gram of feces in a ratio of approximately 1,000 anaerobes to one aerobe.

Figure 23.9 Commonly used probiotic microorganisms. Colorized SEMs of **A.** *Bifidobacterium* (note the "Y" shape of some cells) and **B.** *Lactobacillus acidophilus.*

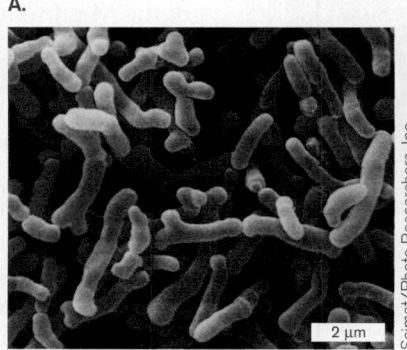

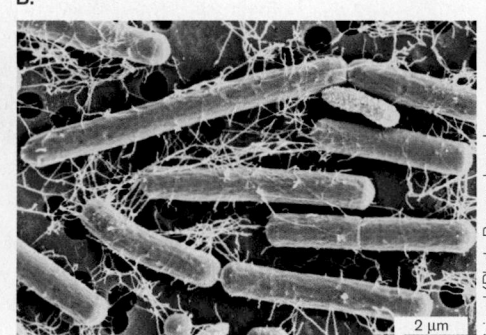

23.2 Risks and Benefits of Harboring Microbial Populations

It is enticing to portray colonizing bacteria as hostile armies encamped at our body's gates (also called portals of entry), just waiting for the opportunity to invade. But bacteria do not possess hostile intent, only the need to find food. The microbes of our normal microbiota, unlike true pathogens, have not evolved mechanisms to specifically breach human defenses. For the most part, the needs of commensal organisms are met at the colonizing site. However, accidental penetration beyond these sites by normal microbiota can cause disease. A cancerous lesion in the colon, for example, can provide a passageway for microbes to enter deeper tissues, where they can begin to grow unabated. *Bacteroides fragilis*, a harmless anaerobe in the gut, can invade tissues through surgical wounds, causing abscesses that persist (and even lead to gangrene) after abdominal surgery (**Fig. 23.10**).

By and large, the body's barriers of defense work well to prevent incursions by microbiota or, in the event of an incursion, to kill the invader. These defenses sometimes break down in persons with an immune system compromised by medical treatments (such as anticancer drugs) or by disease (such as a deficiency in complement factors, discussed in Section 23.3). Such a person is described as a **compromised host** and can be repeatedly infected by normal biota. Organisms causing disease in this situation are called **opportunistic pathogens**.

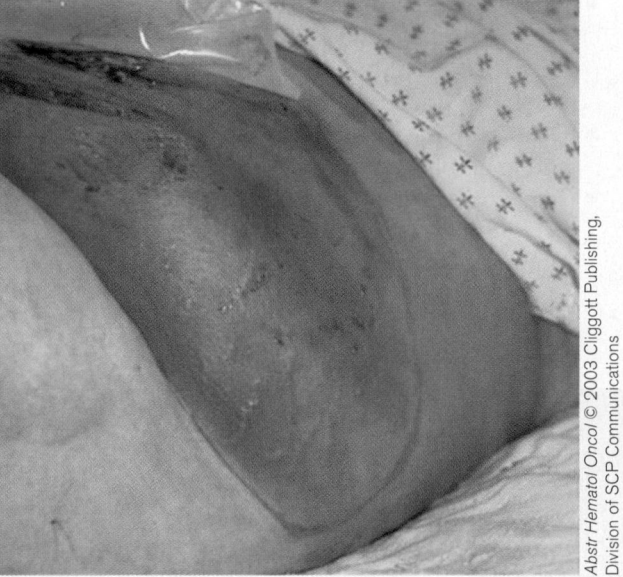

Abstr Hematol Oncol © 2003 Cliggott Publishing, Division of SCP Communications

Figure 23.10   Anaerobic gas gangrene of the abdominal wall caused by *Bacteroides*.   This infection was an unfortunate result of bowel surgery. Organisms escaped from the intestine and initiated infection in the abdominal wall.

Commensal species are often beneficial (mutualistic). They can interfere with the colonization of pathogens by competing for attachment receptors on host cells, competing for food sources, and synthesizing antimicrobial compounds such as fermentation end products. For example, the acidic fermentation products produced by lactobacilli in the female genitourinary tract help maintain a low pH. The low pH dissuades colonization by various pathogenic microbes. In another example, a polysaccharide produced by *B. fragilis* stimulates production of an anti-inflammatory cytokine that will prevent gastrointestinal colitis caused by *Helicobacter hepaticus*. **Cytokines** are small secreted host proteins that bind to various cells of the immune system and regulate the extent and duration of responses by those cells. (Cytokines are discussed in Sections 23.5 and 24.6.)

Commensal organisms also enhance function of the immune system. Bacterial proteins such as catalase can act as **immunomodulins**. Immunomodulins made by normal biota growing on mucosal surfaces modify the secretion of host proteins, such as cytokines and tumor necrosis factor, which influence the immune response.

Some toxic bacterial products are also double-edged swords. **Enterotoxins**, for example, are proteins produced by some Gram-negative pathogens that damage the small intestine of the host and cause diarrhea. But they may also protect against colorectal cancer by activating membrane calcium channels in intestinal epithelial cells. Increasing membrane conductance for calcium turns on an antiproliferative pathway that provides resistance to colon cancer.

A recently discovered benefit of commensal organisms has been their development as vaccine delivery systems. Select commensal species can be genetically engineered to produce proteins from pathogenic species on their cell surfaces. Colonization of a host by the modified commensal strain will elicit an immune response to the cloned protein and provide immunity to the pathogen. Commensal microbes such as *Streptococcus salivarius* (part of the oral and nasopharyngeal microbiota), *Lactococcus* species (residents of the intestinal tract), and *Lactobacillus* species (present in the urogenital and rectal tracts) are being exploited in this way. For example, if one wished to provide protection against *Streptococcus pyogenes*, the causative agent of strep throat, an oral commensal organism could be engineered to present the conserved portion of the M protein (the pilus that *S. pyogenes* uses to attach to human cells) to the oral mucosa. The resulting local immune response could protect the individual from strep throat—at least a strep throat caused by that M protein–bearing strain of *S. pyogenes* (there are at least 80 different M protein–type strains). This approach can be adapted, in theory, to virtually any pathogen that enters through or colonizes a given mucosal surface on any human or animal. It is important, however, to choose proteins from the pathogen that will not also confer pathogenicity on the commensal strain in which it is expressed.

The benefit that commensal microbiota offer to their hosts is especially apparent in gnotobiotic animals. A **gnotobiotic animal** is an animal that is germ-free or one in which *all* the microbial species present are *known*. Developing a gnotobiotic colony of animals involves delivering offspring by cesarean section under aseptic conditions in an isolator. The newborn is moved to a separate isolator where all entering air, water, and food are sterilized. Once gnotobiotic animals are established, the colony is maintained by normal mating between the members. Germ-free animals, however, often have poorly developed immune systems, lower cardiac output, and thin intestinal walls, and they are more susceptible to infection by pathogens. The lesson from these animals is that the presence of microbiota continually challenges the immune system and keeps it active.

TO SUMMARIZE:

- **Opportunistic pathogens** infect only compromised hosts.
- **Benefits of commensal microbes** include interfering with pathogen colonization, production of immunomodulatory proteins, and potential use as vaccine delivery vehicles.
- **Gnotobiotic animals** are germ-free or colonized by a known set of microbes.

23.3 Overview of the Immune System

Because we are surrounded by and host numerous bacteria, how is it we are not constantly infected? Here we begin our discussion of the many layers of protection designed to prevent infection and disease. As you will see, numerous physical and chemical barriers provide an effective first line of defense against infection by invading microbes. As such, they play a critical role in managing our resident ecosystems. But these barriers are not unbreachable. Organisms can still slip through. Consequently, humans, as well as other mammals, have a more aggressive defense called the immune system. The **immune system** is an integrated system of organs, tissues, cells, and cell products that differentiates self from nonself and neutralizes potentially pathogenic organisms or substances. This complex collection of cells and soluble proteins is capable of responding to nearly any foreign molecular structure.

Innate and Adaptive Immunity

There are two broad types of immunity: **nonadaptive immunity**, often called **innate immunity**; and **adaptive immunity**. Innate and adaptive immunity have several key differences. Innate immunity includes physical barriers such as skin, chemical barriers such as stomach

acid, and relatively nonspecific cellular responses to infection that engage if the physical and chemical barriers are breached. The cellular innate responses are triggered by microbial structures such as peptidoglycan. Innate immunity is essentially "hardwired" into the body. It is present at birth, so that mechanisms of innate immunity exist before the body ever encounters a microbe. The protection they afford is nonspecific, capable of blocking or attacking many different types of foreign substances and organisms.

In contrast to innate immunity, adaptive immunity is designed to react to very specific structures called **antigens**. An antigen is any chemical, compound, or structure foreign to the body that elicits an immune response. Adaptive immune responses to a specific antigen do not occur until the body "sees" that antigen. Once activated, adaptive immune mechanisms can recognize at least 10^{10} different antigenic structures and specifically launch a directed attack against each one. Once such an attack has been activated, the organism keeps a "memory" of the exposure in the form of specific memory cells, and an encounter with the microbe years later will reactivate the memory cells specific to the antigen.

The two types of immunity are illustrated by the response of the immune system to infection by the microorganism *Neisseria gonorrhoeae*, which causes the sexually transmitted disease gonorrhea. A component of the innate immune response is **complement**, composed of several soluble protein factors constantly present in the blood. Within moments of an initial infection, complement attacks the bacterial membrane. Complement proteins form holes in bacterial membranes, thereby killing the microbe (see Section 23.8). On the other hand, the adaptive immune response will generate specific antibodies made to the cells of *N. gonorrhoeae* that escaped the innate mechanisms. These antibodies, however, are not made until well after the organism infects a person. Together, innate and adaptive immunity can help ward off disease caused by this organism.

It is important to know that the innate and adaptive immune systems do not work completely independently of each other. Antibodies made by the adaptive immune system will trigger parts of the innate immune system, such as the complement cascade we will discuss later. Likewise, activation of innate resistance mechanisms will cause the release of small immunomodulatory peptides that influence the type and strength of adaptive immunity brought to bear. In military terms, it is similar to the army coordinating its actions with those of the air and naval forces.

Infection versus Disease

Contact with an infectious agent does not guarantee that a person will actually contract the disease. If the number of infecting organisms is small and the immune system (innate and adaptive) is effective, the individual may not develop disease (see chapter introduction), although a person known to have been exposed to certain microorganisms will be treated with antibiotics as a preventive measure. Chapter 26 will more fully discuss the difference between being infected and having a disease.

Any microbe that launches a successful attack on a human or other animal and causes disease must first breach the host's physical and chemical barriers to gain

entrance to the body. It must then survive the innate defense mechanisms and begin to multiply. Finally, the microbe must surmount the last line of defense, adaptive immunity, which begins to respond as the microbe struggles to overcome innate immune defenses. The rest of this chapter will discuss the various innate defense mechanisms. Adaptive immunity is discussed in Chapter 24.

Although innate and adaptive immunity are often treated as separate entities, certain kinds of cells and organs play a role in both types of immunity. We will thus introduce here the various cells and organs of the immune system as a whole before focusing on innate immunity. Our coverage of innate immunity will include physical and chemical barriers to infection, inflammation and nonspecific killing through phagocytosis, interferon, natural killer cells, and complement.

Cells of the Immune System

Blood is composed of red blood cells, white blood cells (also generally known as leukocytes), and platelets (**Fig. 23.11**). The many types of white blood cells are formed by differentiation of stem cells produced in bone marrow (**Fig. 23.12**). Among these white blood cells are various components of innate immunity that differentiated from myeloid stem cells. These differentiated cells include:

- Polymorphonuclear leukocytes (PMNs)
- Monocytes
- Macrophages
- Dendritic cells
- Mast cells

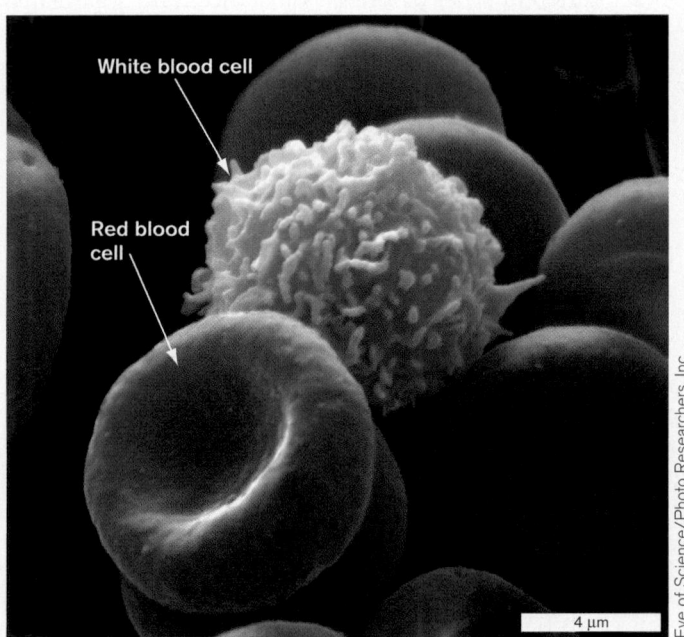

Figure 23.11 Red and white blood cells. This scanning electron micrograph illustrates the relative sizes and three-dimensional morphologies of these cell types.

Eye of Science/Photo Researchers, Inc.

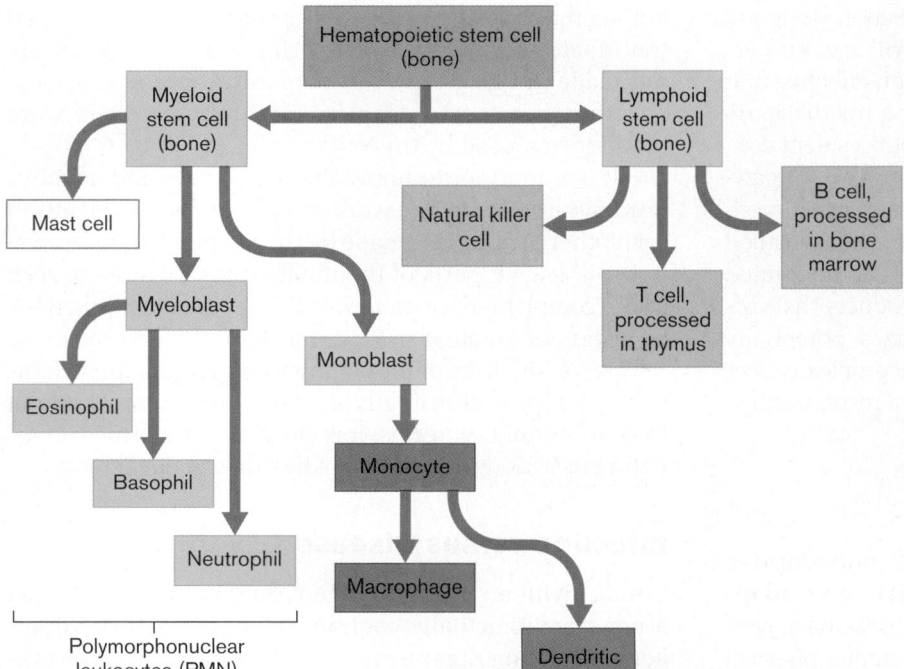

Figure 23.12 Development of white blood cell components of the immune system. Pluripotent stem cells in bone marrow divide to form two lineages. One lineage consists of the myeloid stem cells, which develop into polymorphonuclear leukocytes (PMNs) and monocytes. These cells primarily function as part of innate immunity. The other branch consists of the lymphoid stem cells, which ultimately form natural killer cells, B cells, and T cells. Final maturation into B cells and T cells (the principal cells involved in adaptive immunity) occurs in the bone marrow and thymus, respectively. Colors indicate a group of differentiated cells that arise from the same progenitor.

PMNs, also called granulocytes, have multilobed nuclei, differentiate from an intermediate cell called the myeloblast, and contain enzyme-rich lysosome organelles. PMNs are of several types named for their different staining characteristics. Each cell type has a different function. **Neutrophils** (**Fig. 23.13A**▶‖), making up the vast majority of white cells in the blood, can engulf microbes by phagocytosis and kill them by fusing enzyme-gorged lysosomes with the **phagosomes** containing the organism (**Fig. 23.14**). Phagosomes are the vacuoles formed around a microorganism as it is engulfed by a host cell. Enzymes spilling from the lysosome into the phagosome will destroy various components of the microbe and, ultimately, the microbe itself. **Basophils**, which stain with basic dyes, and **eosinophils** (**Fig. 23.13B**▶‖), which stain with the acidic dye eosin, do not phagocytose microbes but release products, such as major basic protein, that are toxic to the microbe. These blood cells also release chemical mediators (called vasoactive agents) that affect the diameter and permeability of blood vessels (the significance of which is discussed in Section 23.5). **Mast cells** are similar to basophils in structure but differentiate in a lineage separate from PMNs. Unlike PMNs, mast cells are residents of connective tissues and mucosa and do not circulate in the bloodstream. Basophils and mast cells also contain high-affinity receptors for a class of antibody called immunoglobulin E (IgE) associated with allergic responses (detailed in Section 24.8).

> **NOTE:** Because neutrophils comprise the vast majority of PMNs, the terms "neutrophil" and "PMN" are often used interchangeably.

Monocytes (**Fig. 23.13C**▶‖) are white blood cells with a single nucleus (not multilobed like a PMN); they engulf (phagocytose) foreign material. Monocytes circulating in the blood can migrate out of blood vessels into various tissues and differentiate into **macrophages** and **dendritic cells** (**Fig. 23.15**▶‖). Macrophages are phagocytic and form a major part of the amorphous **reticuloendothelial system**, which is widely distributed throughout the body. The reticuloendothelial system is a group of cells with the ability to take up and sequester particles. In addition to macrophages, the system is composed of specialized endothelial cells lining the sinusoids of the liver, spleen, and bone marrow, as well as reticular cells of lymphatic tissue (macrophages) and of bone marrow (fibroblasts). The function of macrophages and the reticuloendothelial system is to phagocytose microorganisms.

Macrophages are the cells most likely to make first contact with invading pathogens. They have two functions.

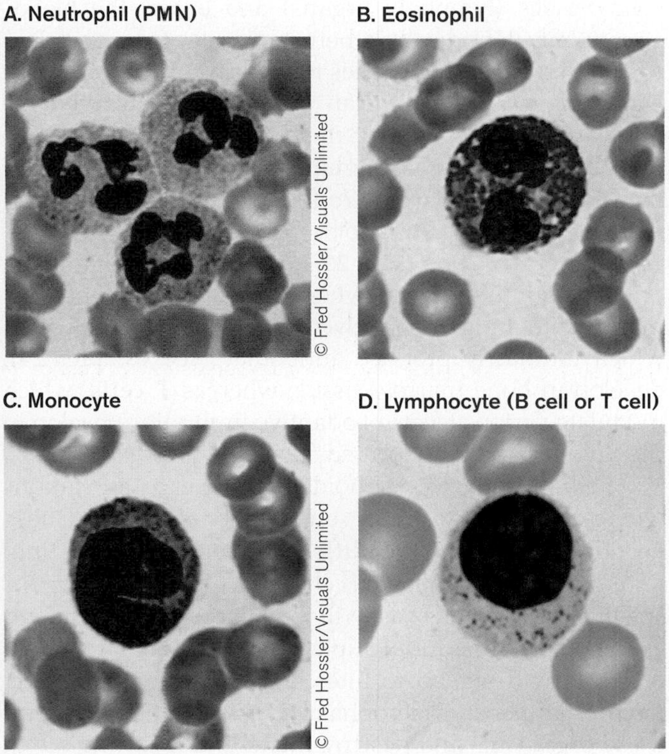

A. Neutrophil (PMN)

© Fred Hossler/Visuals Unlimited

B. Eosinophil

© Fred Hossler/Visuals Unlimited

C. Monocyte

© Fred Hossler/Visuals Unlimited

D. Lymphocyte (B cell or T cell)

© John D. Cunningham/Visuals Unlimited

Figure 23.13   Types of white blood cells.   A. Neutrophil (PMN). **B.** Eosinophil. **C.** Monocyte. **D.** Lymphocyte (B cell or T cell). ▶‖ *microbiology2.com/animations*

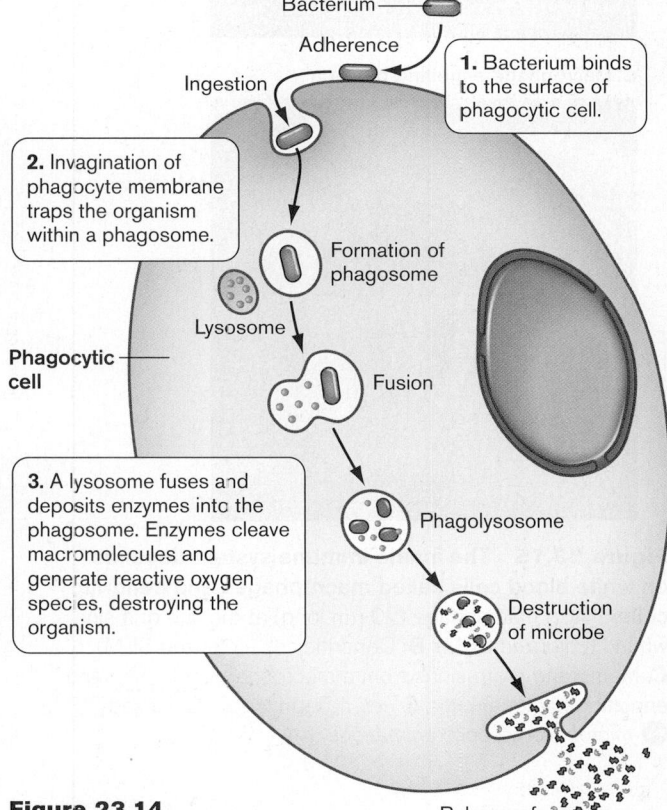

1. Bacterium binds to the surface of phagocytic cell.

2. Invagination of phagocyte membrane traps the organism within a phagosome.

3. A lysosome fuses and deposits enzymes into the phagosome. Enzymes cleave macromolecules and generate reactive oxygen species, destroying the organism.

Bacterium

Adherence

Ingestion

Phagocytic cell

Formation of phagosome

Lysosome

Fusion

Phagolysosome

Destruction of microbe

Release of microbial debris

**Figure 23.14
Phagosome-lysosome fusion.**

A. Macrophage

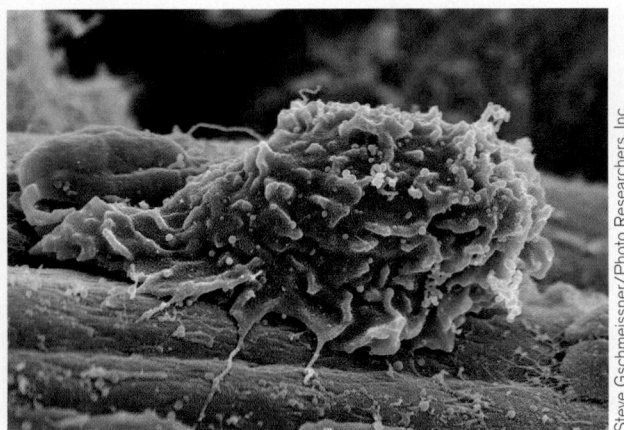

Steve Gschmeissner/Photo Researchers, Inc.

B. Dendritic cell

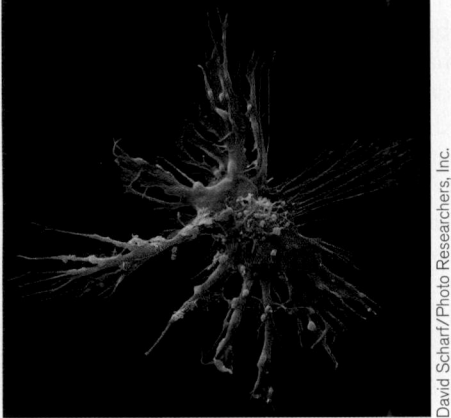

David Scharf/Photo Researchers, Inc.

C. Macrophage engulfing bacteria

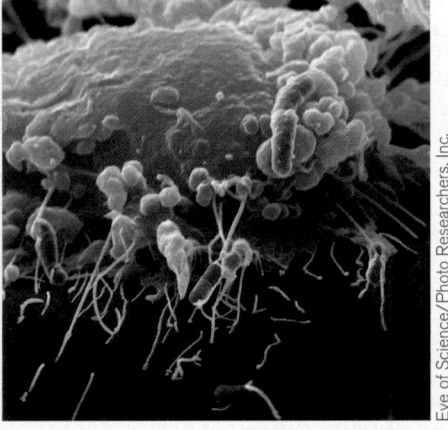

Eye of Science/Photo Researchers, Inc.

Figure 23.15 **The innate immune system depends on white blood cells called macrophages and dendritic cells.** **A.** A macrophage (20 μm long) at the site of a skin wound (colorized SEM). **B.** Dendritic cell (colorized SEM). **C.** Membrane protrusions from a macrophage detecting and engulfing bacteria (pink; *E. coli*, 1.5 μm long). (Colorized SEM.)
▶ *microbiology2.com/animations*

As part of innate immunity, they kill invaders directly. Protrusions from the macrophage surface extend and clasp nearby bacteria, pulling them into the cell (see **Fig. 23.15C**). Once ingested by the macrophage, the bacteria are destroyed. Subsequently, as the first step in adaptive immunity, the remnants are processed (degraded) into smaller peptides (called antigens), which are then presented on the macrophage cell surface. Specific white blood cells called T cells (a type of lymphocyte) can bind the displayed antigens and become activated as part of the adaptive immune response. Thus, macrophages are considered **antigen-presenting cells** (see Chapter 24).

Dendritic cells are also antigen-presenting cells. Present in the spleen and lymph nodes, they, like macrophages, can take up, process, and present small antigens on their cell surface. Dendritic cells are different from macrophages in structure and in the fact that dendritic cells primarily take up small soluble antigens from their surroundings rather than through phagocytosis and degradation of whole bacteria.

Lymphoid Organs

Lymphocytes (**Fig. 23.13D**), which are the main participants in adaptive immunity, are present in blood at about 2,500 cells per microliter, accounting for over one-third of all peripheral white blood cells. However, an individual lymphocyte spends most of its life within specialized solid tissues (lymphoid organs) and enters the bloodstream only periodically, where it migrates from one place to another, surveying tissues for possible infection or foreign antigens. No more than 1% of the total lymphocyte population can be found in the blood at any one time.

The tissues of the immune system where the great majority of lymphocytes are found are classified as primary or secondary lymphoid organs or tissues, depending on their function (**Fig. 23.16**). The primary lymphoid organs and tissues are where immature lymphocytes mature into antigen-sensitive B cells and T cells. **B cells**, which ultimately produce antibodies (see Chapter 24), develop in <u>b</u>one marrow tissue, whereas **T cells**, which modulate various facets of adaptive immunity, develop in the <u>t</u>hymus, an organ located above the heart.

The secondary lymphoid organs serve as stations where lymphocytes can encounter antigens. These encounters lead to the differentiation of B cells into antibody-secreting plasma cells, and T cells into antigen-specific helper cells, as will be discussed in Chapter 24. The spleen is an example of a secondary lymphoid organ. It is designed to filter blood directly and detect microorganisms. Macrophages in the spleen engulf these organisms and destroy them, and then migrate to other secondary lymphoid organs and present pieces of the microbe (called antigens) to the B and T cells, which then become activated.

The **lymph nodes** are another kind of secondary lymphoid organ. Lymph nodes are arranged to trap organisms from local tissues, not from blood. Lymph nodes are situated at various sites in the body where lymphatic vessels converge (for example, under the armpits). There are also lymphoid tissues in the mucosal regions of the gut and respiratory tracts (for example, Peyer's patches and gut-associated lymphoid tissue, or GALT, discussed later). Other secondary lymphoid organs are the tonsils, adenoids, and appendix.

TO SUMMARIZE:

- **The immune system** consists of both innate and adaptive mechanisms that recognize and eliminate pathogens.
- **Innate immunity** includes physical barriers and some cellular responses to various microbial structures.
- **Adaptive immunity** is a cellular response to specific structures (antigens) in which a memory of exposure is produced.
- **Myeloid bone marrow stem cells** differentiate to form cells of the innate immune system—phagocytic PMNs, monocytes, macrophages, antigen-presenting dendritic cells, and mast cells.
- **Lymphoid stem cells** differentiate into natural killer cells (part of the innate immune system) and lymphocytes (cells of the adaptive immune system). Lymphocytes are classified as B cells, which ultimately produce antibodies; and T cells, which regulate adaptive immunity.

23.4 Barbarians at the Gate: Innate Host Defenses

When defending a castle in medieval times, the first lines of defense included physical barriers (the castle wall), chemical barriers (boiling oil tossed onto invaders trying to scale the wall), and finally hand-to-hand combat once the wall was breached. Similarly, the body's initial defenses against infectious disease are composed of physical, chemical, and cellular barriers designed to prevent a pathogen's access to host tissues. Although generally described as nonspecific, some innate defense systems are more specific than others.

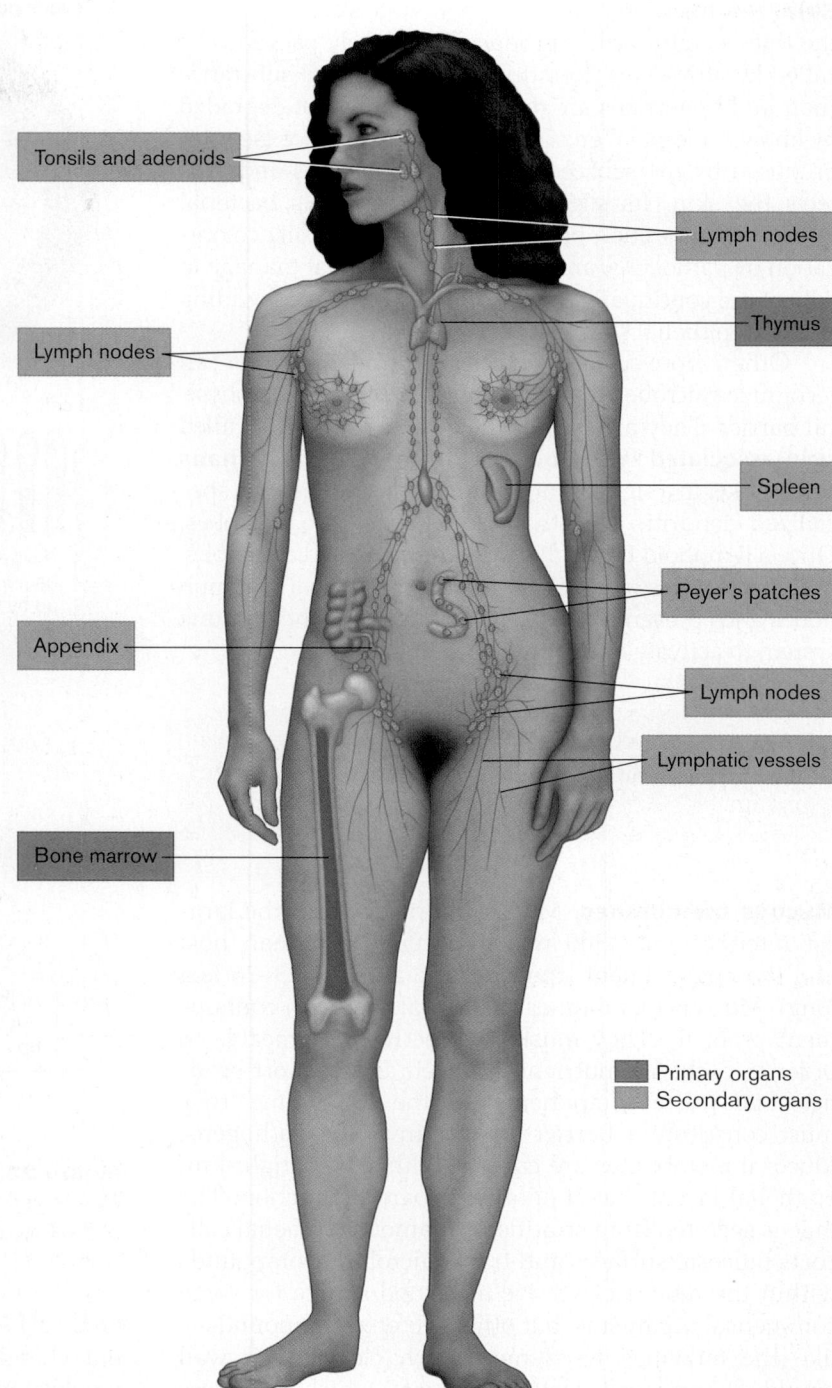

Tonsils and adenoids

Lymph nodes

Thymus

Lymph nodes

Spleen

Peyer's patches

Appendix

Lymph nodes

Lymphatic vessels

Bone marrow

■ Primary organs
■ Secondary organs

Figure 23.16 Lymphoid organs.

Physical Barriers to Infection

The first line of defense against any potential microbial invader (either commensal or pathogenic) occurs where parts of the body interface with the environment. These interfaces (skin, lung, gastrointestinal tract, genitourinary tract, and oral cavities) have similar defense strategies, although each has unique characteristics.

Skin. Few microorganisms can penetrate skin, because of the thick keratin armor produced by closely packed cells called keratinocytes. Keratin protein is a hard substance (hair and fingernails are made of it) that is not degraded by known microbial enzymes. An oily substance (sebum) produced by the sebaceous glands also covers and protects the skin. Its slightly acidic pH inhibits bacterial growth. Competition between species also limits colonization by pathogens, and microorganisms that manage to adhere are continually removed by the constant shedding of outer epithelial skin layers.

Other, more specialized cells just under the skin can recognize microbes managing to slip through the physical barrier. They are part of a consortium of cells called **skin-associated lymphoid tissue (SALT). Langerhans cells** make up a significant portion of SALT. They are specialized dendritic cells that can phagocytose microbes. Once a lymphoid Langerhans cell has ingested a microbe, the cell migrates by ameboid movement to nearby lymph nodes and "presents" parts of the microbe to the immune system to activate antimicrobial immunity.

> **NOTE:** Do not confuse phagocytic Langerhans cells with the pancreatic "islets of Langerhans" that secrete insulin.

Mucous membranes. Mucosal surfaces form the largest interface (200–300 m^2) between the human host and the environment (the intestine alone is 23–26 feet long). Mucous membranes in general present a containment problem. They must be selectively permeable in order to exchange nutrients, as well as to export products and waste components. At the same time, they must constitute a barrier against invading pathogens. Mucosal membranes are covered with special tight-knit epithelial layers that support this barrier function. The mucus secreted from stratified squamous epithelial cells coats mucosal surfaces and traps microbes. Compounds within the mucus can serve as a food source for some commensal organisms, but other secreted compounds—like the enzymes lysozyme, which cleaves cell wall peptidoglycan, and lactoperoxidase, which produces superoxide radicals—can kill an organism trapped in the mucus.

Semispecific innate immune mechanisms are also associated with mucosal surfaces. Host cells, even epithelial cells, in mucosa have evolved mechanisms to distinguish harmless compounds and commensal microorganisms from dangerous pathogens. Patterns of conserved structures on pathogenic microbes, called **pathogen-associated molecular patterns (PAMPs)**, are recognized by host cell surface receptors such as various Toll-like receptors (discussed in Section 23.7) and CD14.

A. Peyer's patch

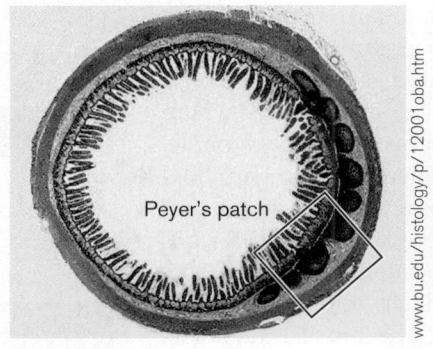

www.bu.edu/histology/p/12001oba.htm

B. M cell

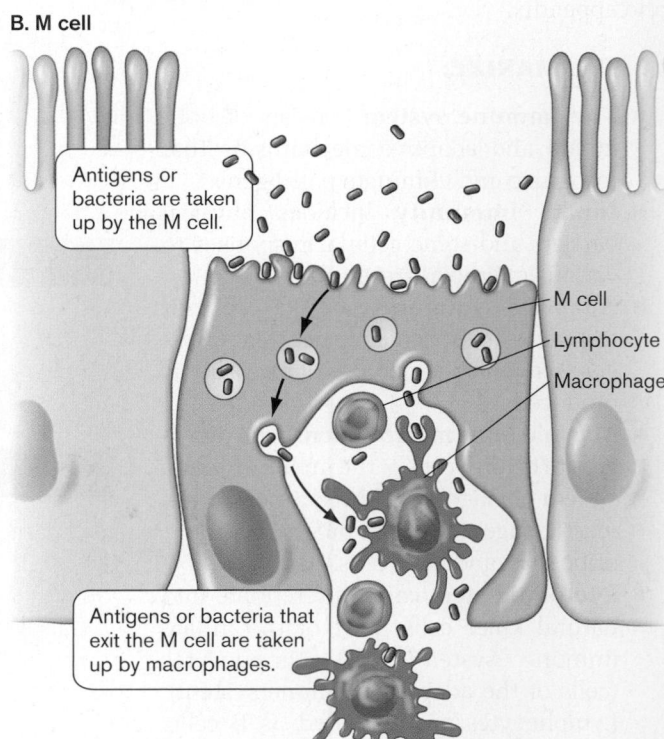

Figure 23.17 Gut-associated lymphoid tissue (GALT).
A. A Peyer's patch located on the intestine. **B.** Diagram of an M cell (microfold cell).

Once a PAMP has been recognized, the host cell sends out chemicals that can activate immune system cells involved with innate and adaptive immune mechanisms.

Like skin, the gastrointestinal system possesses an innate mucosal immune system, in this case called **gut-associated lymphoid tissue (GALT)**. GALT includes tonsils, adenoids, and Peyer's patches (**Fig. 23.17A**). These tissues include specialized **M cells** that dot the intestinal surface and are wedged between epithelial cells. "M" stands for "microfold," which describes their appearance (**Fig. 23.17B**). These are fixed cells that take up microbes from the intestine and release them, or pieces of them, into a pocket formed on the opposite, or basolateral, side of the cell. Other cells of the innate immune system, such as macrophages (a type of white blood cell that migrates through tissues), gather here and collect the organisms that emerge. Macrophages engulf and try to kill the organism and then place small, degraded components of the organism on the macrophage cell surface where other immune system cells can recognize them. As a result, M cells are extremely important for the development of mucosal immunity to pathogens. However, M cells can also serve as a portal for some pathogens to gain entry to the body.

The lungs. The lungs also have a formidable defense. In addition to the respiratory ciliary elevator discussed in Section 23.1, microorganisms larger than 100 μm become trapped by hairs and cilia lining the nasal cavity and trigger the forceful expulsion of air from the lungs (a sneeze). The sneeze is designed to clear the organism from the respiratory tract. Organisms that make it to the alveoli are met by phagocytic cells called **alveolar macrophages**. These cells can ingest and kill most bacteria.

Lung epithelia can also internalize and destroy organisms. This mechanism is thought to explain why *Pseudomonas aeruginosa* does not typically infect healthy lungs but readily infects the lungs of cystic fibrosis patients, whose cells cannot bind and kill the bacteria.

Cystic fibrosis is a genetic disease in which a membrane chloride channel called the **cystic fibrosis transmembrane conductance regulator (CFTR)** is defective. CFTR is important for regulating chloride movement across the membrane, an essential part of hydrating mucus in healthy individuals. Localization of the protein at the lung cell surface is depicted in **Figure 23.18**. In addition to

its role in regulating chloride channels, CFTR appears to serve as a receptor for internalizing *P. aeruginosa* by lung epithelial cells, which helps clear the organism from the lung. While normal lung epithelial cells prevent *Pseudomonas* infection, cells defective in CFTR cannot bind and clear the organism, leading to life-threatening disease. The *P. aeruginosa* strains posing the most serious threat produce a thick slimy material (alginate) that retards other lung clearance mechanisms (**Fig. 23.19**).

Chemical Barriers to Infection

Some examples of chemical barriers were mentioned previously, such as the acid pH of the stomach, lysozyme in tears, and generators of superoxide. In addition, a variety of human cells generate small antimicrobial, cationic (positively charged) peptides called **defensins** (**Table 23.2**). These antimicrobial peptides are important components of innate immunity against microbial infections.

Defensins. Defensins range in length from 29 to 47 amino acids and are found in mammals, birds, amphibians, and plants. They kill by destroying the invader's cytoplasmic membrane and are effective against Gram-positive and Gram-negative bacteria, fungi, and even some viruses (those with membranes, such as HIV). To kill Gram-negative bacteria, the peptides must first bind outer membrane lipopolysaccharides, which are negatively charged, and move into the periplasm. The defensins then insert into the cytoplasmic membrane of either Gram-positive or Gram-negative bacteria. Insertion is driven by the transmembrane electrical potential, which in bacteria is about −150 mV (cell interior more negative than exterior).

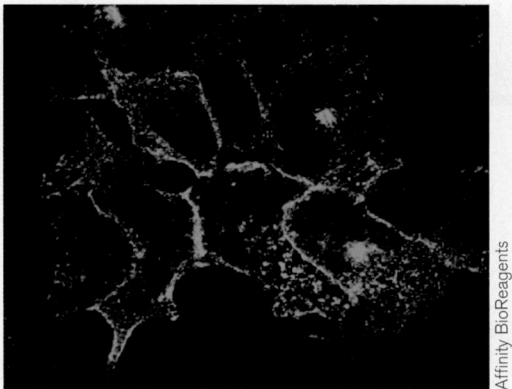

Affinity BioReagents

Figure 23.18 Plasma membrane location of the cystic fibrosis gene product (CFTR). Laser scanning confocal fluorescence microscopy was used to image living epithelial cells of the human airway. The cells were transfected with DNA expressing a cystic fibrosis transmembrane conductance regulator (CFTR) protein tagged with green fluorescent protein (GFP-CFTR). Note that the fluorescence is heaviest at the edge of the cell, indicating that the GFP-CFTR protein is localized at the cell surface.

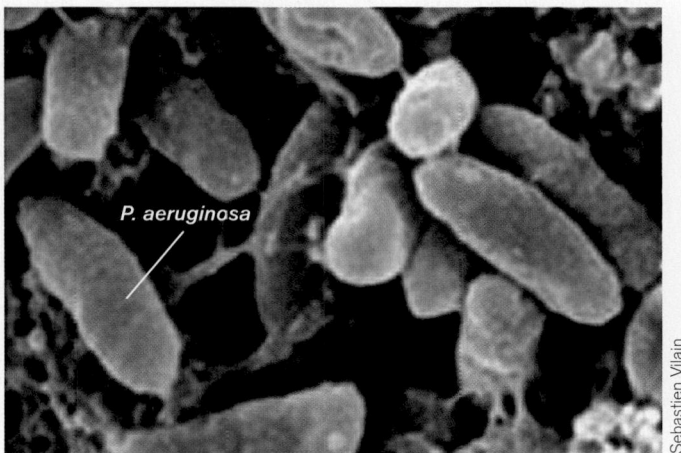

P. aeruginosa

Sebastien Vilain

Figure 23.19 Mucoid strains of *Pseudomonas aeruginosa* that infect cystic fibrosis patients. The organism is shown growing in a biofilm with strings of dehydrated exopolysaccharide (EPS) connecting cells. This EPS material gives colonies of these strains a very mucoid appearance. (Cells 2–3 μm long, SEM.)

Table 23.2 Categories of natural antimicrobial peptides.

Class	Examples	Major presence in humans	Source
Alpha defensins	-1, -2, -3, -4[a]	Neutrophils	Stored in granules
Alpha defensins	-5, -6	Paneth cells, small intestine	Stored in granules
Cathelicidins	LL-37, hCAP-18	Neutrophils	Secreted
Histatins		Saliva	Secreted
Beta defensins	HBD-1, -2	Epithelia	Secreted
Other species			
Maganins		Frogs	
Protegrins		Pigs	
Indolicidin		Cattle	

[a]Alpha defensins are named alpha defensin-1, alpha defensin-2, etc.

The negative charge pulls the small cationic peptides into the membrane, where they assemble into channels that destroy the cytoplasmic membrane barrier, killing the bacterial cell. Defensins generally do not affect eukaryotic cells, because these cells have a very low membrane potential (–15 mV) across their plasma membrane. Defensins are produced by many human cells, including cells of the skin, lungs, genitourinary tract, and gastrointestinal tract (**Fig. 23.20**). Variation in the defensins produced by different animals may partially explain pathogen-host specificity (see **Special Topic 23.1**).

There are two classes of vertebrate defensins: alpha (α) and beta (β). The alpha defensins are stored in membrane-enclosed granules within neutrophils and in Paneth cells in the small intestine (**Fig. 23.20A**). When stimulated, these cells **degranulate** (release their granule contents) by fusing their granule membranes to cytoplasmic or vacuolar membranes, dumping their contents into the surroundings or into phagocytic vacuoles, where the alpha defensins can destroy engulfed microbes (**Fig. 23.20B**). In contrast, the beta defensins are not stored in cytoplasmic granules. The synthesis of beta defensins is activated only after contact with bacteria or their other products. Other cationic antimicrobial peptides are listed in **Table 23.2**. Cathelicidins are discussed in **eTopic 23.1**.

THOUGHT QUESTION 23.6 Why do defensins have to be so small? Do defensins kill normal microbiota?

A.

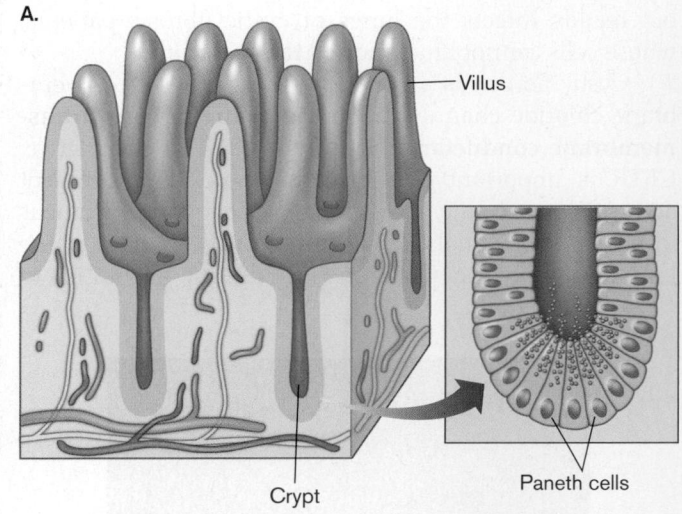

B.

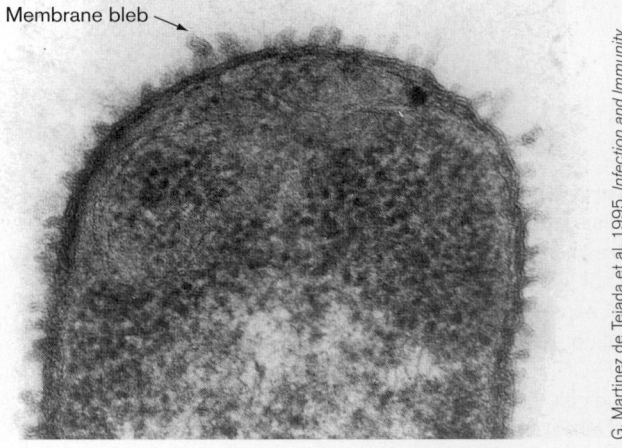

Figure 23.20 Defensins. A. Certain defensins are produced in the crypts of the intestine. The crypts contain granule-rich Paneth cells (inset) that discharge their granules into the crypt lumen in response to the entry of bacteria or as a result of food-related stimulation by acetylcholine. **B.** Effect of cationic peptides on *E. coli* 0111. Polymyxin B is a small cationic peptide antibiotic that mimics the action of defensins. In this micrograph, polymyxin B causes blebs of membrane to ooze from the surface of the cell. *Source:* A. C. L. Bevins, Research Institute, and D. Schumick, Department of Medical Illustration, Cleveland Clinic Foundation. ©2000, Cleveland Clinic Foundation.

G. Martinez de Tejada et al. 1995. *Infection and Immunity* **63**:3054. © American Society for Microbiology

Special Topic 23.1 Do Defensins Help Determine Species Specificity for Infection?

Differences in cationic peptides between host species may explain why some bacteria cause devastating disease in certain animals and only mild disease in others. *Salmonella enterica* serovar Typhimurium, for example, usually kills mice, but most humans survive. Insects, completely lacking an immune system, resist invading pathogens altogether. Could these differences be due to the type of defensin made?

Studies suggest that differences in immunity across species may indeed reflect species-specific defensins. Scientists placed the gene for human defensin-5 (made in human intestinal Paneth cells) into mice, which do not normally produce this peptide. When exposed to high doses of *S.* Typhimurium,

the engineered (transgenic) mice did not get sick, while normal (wild-type) mice became profoundly ill and died (**Fig. 1**).

Studies of defensins and their selectivity offer the promise of a new class of antibiotics. Frog skin, for example, is rich with antimicrobial peptides called maganins. The effectiveness of these peptides is impressive. Frog skin maganins are so powerful that frogs do not become infected when put into natural pond water teeming with bacteria. Harnessing these peptides as antibiotics would be a tremendous advance in medicine. The potential to discover new antibiotics is another reason for protecting our rain forests, which harbor many undiscovered species, including frogs whose unique chemical defenses have yet to be studied.

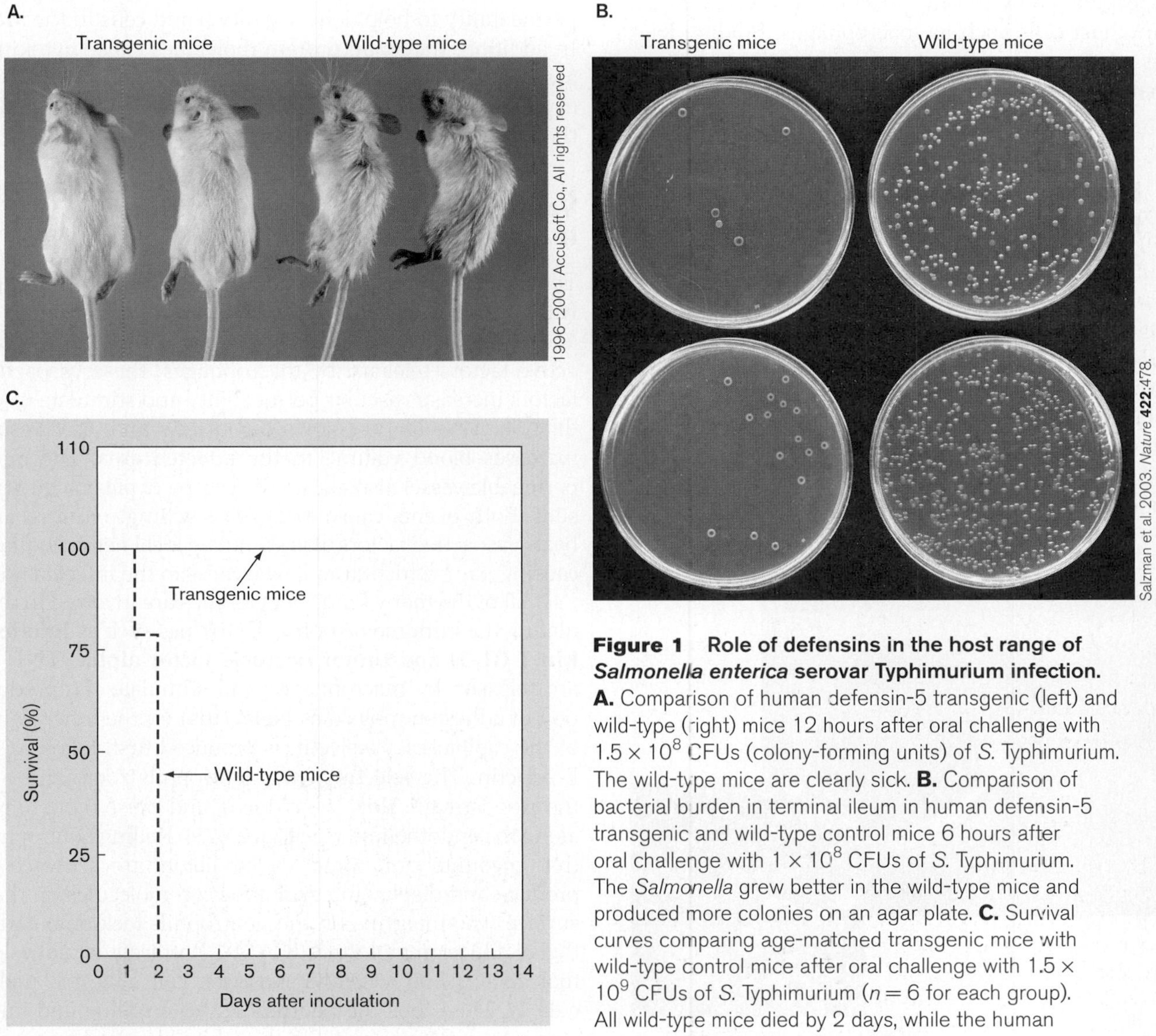

A.

Transgenic mice   Wild-type mice

B.

Transgenic mice   Wild-type mice

Salzman et al. 2003. *Nature* **422**:478.

C.

Figure 1 **Role of defensins in the host range of** *Salmonella enterica* **serovar Typhimurium infection.**
A. Comparison of human defensin-5 transgenic (left) and wild-type (right) mice 12 hours after oral challenge with 1.5×10^8 CFUs (colony-forming units) of *S.* Typhimurium. The wild-type mice are clearly sick. **B.** Comparison of bacterial burden in terminal ileum in human defensin-5 transgenic and wild-type control mice 6 hours after oral challenge with 1×10^8 CFUs of *S.* Typhimurium. The *Salmonella* grew better in the wild-type mice and produced more colonies on an agar plate. **C.** Survival curves comparing age-matched transgenic mice with wild-type control mice after oral challenge with 1.5×10^9 CFUs of *S.* Typhimurium ($n = 6$ for each group). All wild-type mice died by 2 days, while the human defensin-5 mice lived at least 14 days.

TO SUMMARIZE:

- **Skin defenses** against invading microbes include closely packed keratinocytes and a SALT lymphoid system made up largely of phagocytic Langerhans cells.
- **Mucous-membrane defenses** involve secreted enzymes, cytokines, and GALT tissues, such as Peyer's patches, that contain phagocytic M cells.
- **M cells** in gut-associated lymphoid tissues sample bacterial cells at their surface and release pieces of them to immune system cells.
- **Phagocytic alveolar macrophages** inhabit lung tissues, contributing to nonspecific defense.
- **Chemical barriers against disease** include cationic defensins, acid pH in the stomach, and superoxide produced by certain cells.
- **Pathogen-associated molecular patterns (PAMPs)** are recognized by Toll-like receptors (TLRs) found on many host cells such as macrophages. Binding triggers release of chemical signal molecules that activate innate and adaptive immune mechanisms.

23.5 Innate Immunity: The Acute Inflammatory Response

The boil shown in **Figure 23.21** is an inflammatory response triggered by infection with the organism *Staphylococcus aureus*. Inflammation is a critical innate defense in the war between microbial invaders and their hosts. It provides a way for phagocytic cells (such as neutrophils) normally confined to the bloodstream to gain access to infected sites within tissues. Movement of these cells out of blood vessels is called **extravasation**, which we discuss shortly. Once at the infection site, the neutrophils

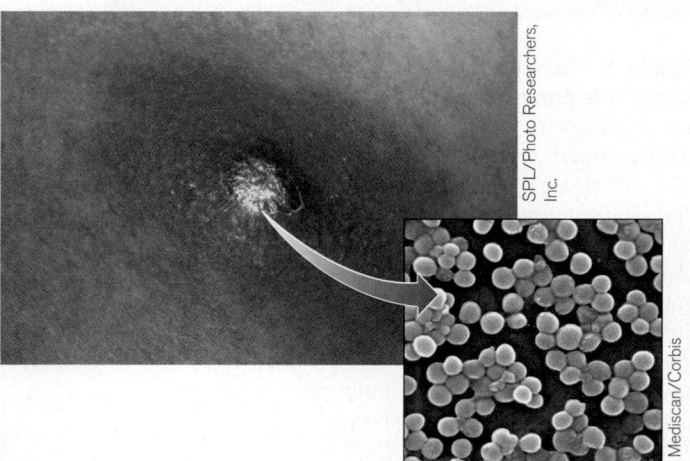

Figure 23.21   Inflammation caused by infection.   Boil resulting from infection of a hair follicle by *Staphylococcus aureus* (size 0.5–1.0 µm, colorized SEM).

SPL/Photo Researchers, Inc.

© Mediscan/Corbis

begin engulfing microbes. The white pus associated with an infection is teeming with these white blood cells. The five cardinal signs of inflammation, first described over 2,000 years ago, are: redness, warmth, pain, swelling, and altered function at the affected site.

Although many things can trigger inflammation, our focus is how microbes cause the response. The process begins with the infection itself. Microorganisms introduced into the body—for example, on a wood splinter—will begin to grow and produce compounds that damage host cells (**Fig. 23.22**●)). Resident macrophages that wander into the infected area engulf these organisms and then release inflammatory mediators (chemoattractants) that "call" for more help. These mediators include **vasoactive factors** such as leukotrienes, platelet-activating factor, and prostaglandins, which act on blood vessels of the microcirculation, increasing blood volume and capillary permeability to help deliver white blood cells to the area. In addition, the small protein molecules called cytokines are secreted, diffusing to the vasculature and stimulating expression of specific receptors (selectins) on the endothelial cells of capillaries and venules.

Vasoactive Factors and Cytokines Initiate Extravasation

The eventual passage of neutrophils through vascular walls (extravasation; **Fig. 23.23**) requires a relaxation of endothelial cell adhesion. The process is initiated by vasoactive factors released by macrophages; these vasoactive factors increase vascular permeability and stimulate vasodilation. Vasodilation slows blood flow and, as a result, increases blood volume in the affected area. The more permeable vessel also allows the escape of plasma into tissues. Both events cause localized swelling, redness, and heat. Vasoactive factors also stimulate local nerve endings, causing pain, which draws awareness to the infected area.

All of the many kinds of cytokines are involved in regulating the immune response. Cytokines such as **interleukin 1 (IL-1)** and **tumor necrosis factor alpha** (TNF-α) are released by macrophages and stimulate the production of adhesion molecules (**selectins**) on the inner lining of the capillaries. P-selectin is produced first, followed by E-selectin. The selectins snag neutrophils zooming by in the bloodstream, slow them down, and cause them to roll along the endothelium (see **Fig. 23.23**). Rolling neutrophils that encounter inflammatory mediators are activated to produce and display integrin adhesion molecules on their surface. The integrins on the neutrophils lock onto endothelial adhesion molecules ICAM-1 (intracellular adhesion molecule 1) and VCAM-1 (vascular cell adhesion molecule 1). This stops the neutrophils from rolling and initiates extravasation, in which the white blood cells squeeze through the endothelial wall and into the tissues, where they can help macrophages attack the invading microbes.

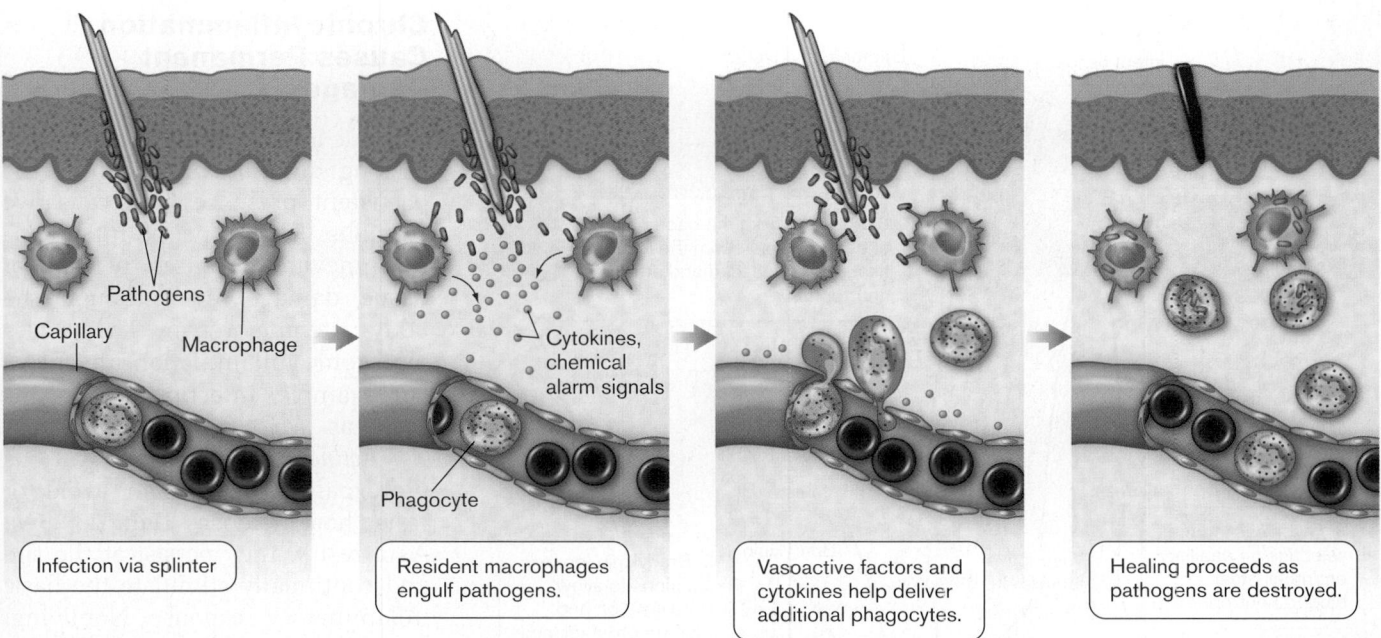

Figure 23.22 Basic inflammatory response. Neutrophils (a type of phagocyte) circulate freely through blood vessels and can squeeze between cells in the walls of a capillary (extravasation) to the site of infection. They then engulf and destroy any pathogens they encounter. ▶❚❚ *microbiology2.com/animations*

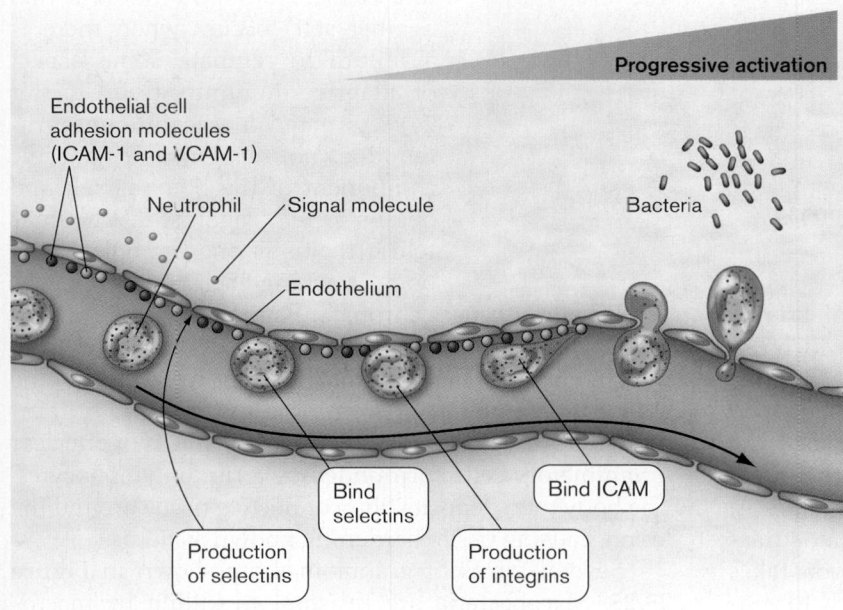

Figure 23.23 Mechanism of extravasation. Extravasation is the process by which leukocytes move from the bloodstream into surrounding tissues. Signal molecules produced by damaged tissue cells induce the production of selectins (produced early in the process) on the surfaces of endothelial cells and integrins on the surface of the white blood cell (selectins and integrins not shown). Selectins capture leukocytes traveling through blood vessels. Leukocytes begin to roll along the vessel wall, and the integrins on their surface lock onto the endothelial cell's adhesion molecules (ICAM-1 or VCAM-1). The leukocytes are progressively activated while rolling. The neutrophil ultimately squeezes through the wall between endothelial cells (extravasation).

Damaged tissue cells in the area of inflammation will release **bradykinin**, a nine-amino-acid polypeptide that helps loosen the tight junctions between endothelial cells to promote extravasation (**Fig. 23.24**). Bradykinin molecules also bind to mast cells in the area, causing a calcium influx that triggers degranulation. The histamine released from mast cells further loosens the endothelial cell junctions, allowing more fluid and cells to move out (increased vascular permeability), adding to fluid accumulation (edema). Bradykinin will attach to capillary cells and induce synthesis of prostaglandins that in turn stimulate nerve endings, causing pain in the area. A key enzyme involved in prostaglandin synthesis is cyclooxygenase (COX). Aspirin and the popularly prescribed anti-inflammatory agents Vioxx and Celebrex are COX inhibitors that prevent the synthesis of prostaglandins and thus reduce inflammatory pain.

Once neutrophils have passed through the vascular wall, chemotactic factors are needed to lure them to the proper location. One of these, fMet-Leu-Phe peptide, is made by the bacterium itself. Many bacterial proteins have fMet (*N*-formylmethionine) as their N-terminal

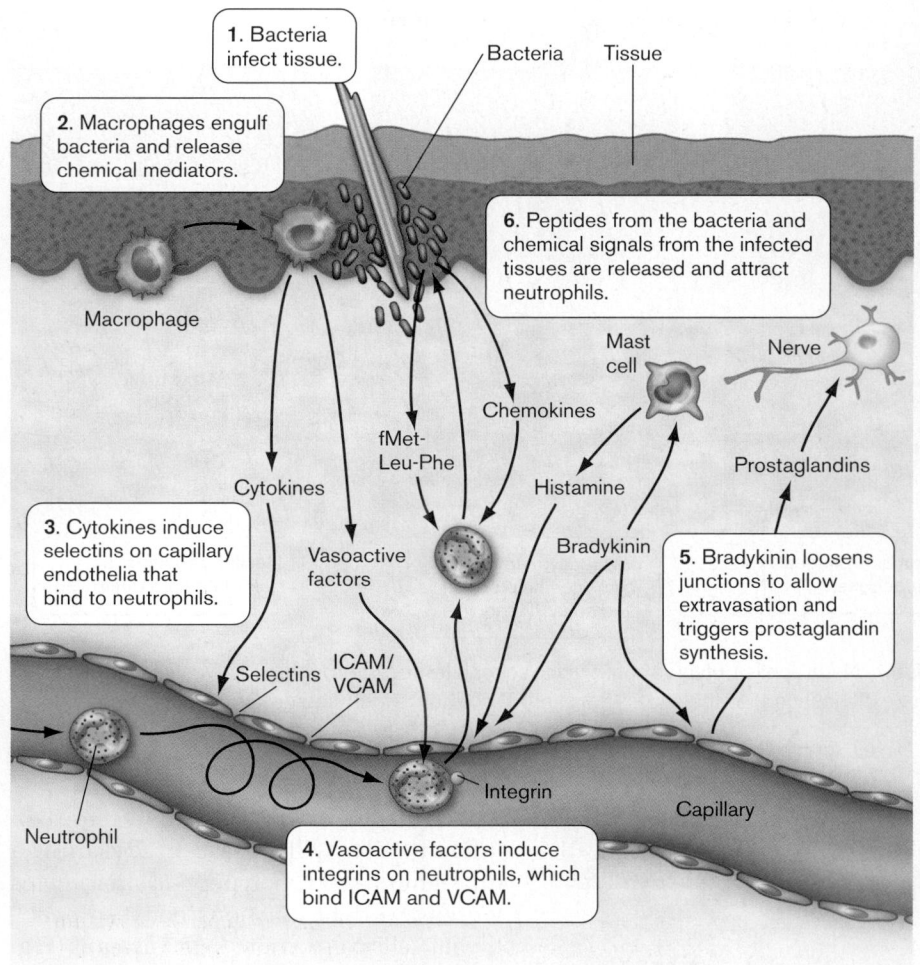

1. Bacteria infect tissue.

2. Macrophages engulf bacteria and release chemical mediators.

Bacteria   Tissue

Macrophage

6. Peptides from the bacteria and chemical signals from the infected tissues are released and attract neutrophils.

Mast cell   Nerve

Chemokines

fMet-Leu-Phe

Cytokines   Histamine   Prostaglandins

3. Cytokines induce selectins on capillary endothelia that bind to neutrophils.

Vasoactive factors   Bradykinin

5. Bradykinin loosens junctions to allow extravasation and triggers prostaglandin synthesis.

Selectins   ICAM/VCAM

Integrin

Neutrophil   Capillary

4. Vasoactive factors induce integrins on neutrophils, which bind ICAM and VCAM.

Figure 23.24   Summary of the inflammatory response.

amino acid, but mammalian proteins, in general, do not. Bacteria will often cleave off the fMet peptide, which can then diffuse away from the bacterium. The fMet peptide binds to neutrophil receptors and stimulates pseudopod projections aimed toward the microbe. This causes the white blood cell to migrate in the direction of the infection. In addition, **chemokine** peptides (IL-8 and MCP-1) produced by damaged tissues can serve as chemoattractants for these white blood cells. Thus, much of what takes place at the site of an infection is not due directly to substances produced by the microbe but is the result of the body's reaction to the presence of the intruding organism. Once the cascade of inflammatory events delivers phagocytic cells to the site of infection, they begin devouring the microbes in an attempt to cure the infection.

> **THOUGHT QUESTION 23.7**   What happens to all the neutrophils that enter a site of infection once the infection has resolved?

Chronic Inflammation Causes Permanent Damage

Chronic inflammation is a response of long duration provoked by the persistent presence of a causative stimulus. This type of inflammation inevitably causes permanent tissue damage, even though the body attempts repair. The causes of chronic inflammation are many. For example, infectious organisms such as *Mycobacterium tuberculosis*, *Actinomyces bovis*, and various protozoan parasites can avoid or resist host defenses (**Fig. 23.25A**). As a result, they persist at the site and continually stimulate the basic inflammatory response. Nonliving, irritant material like wood splinters, inhaled asbestos particles, or surgical implants can also lead to chronic inflammatory responses. Autoimmune diseases are another important cause. Autoimmunity (reaction against self) occurs when there is a failure to regulate some aspect of adaptive immunity (see Chapter 24). As a result, the immune system does not recognize one specific component of the body as self and begins to react against it. Rheumatoid arthritis is one example of the body attacking itself.

Whatever causes chronic inflammation, there is continual recruitment of macrophages and lymphocytes from the circulation. The body may attempt to "wall off" the site of inflammation by forming a **granuloma**. A granuloma begins as an aggregation of the mononuclear inflammatory cells surrounded by a rim of lymphocytes. The body then deposits fibroconnective tissue around the lesion, causing tissue hardening known as fibrosis.

Several forms of granulomas are shown in **Figure 23.25**. Mycobacteria are resistant to killing by macrophages and can be found living within these cells (**Fig. 23.25A**). This resistance to host defense can lead to longterm chronic infections and the production of granulomas. For example, skin infections caused by *Mycobacterium marinum* will produce skin granulomas, while *M. tuberculosis* can cause liver granulomas (**Fig. 23.25B** and **C**). Crohn's disease, which commonly manifests as abdominal pain, frequent bowel movements, and rectal bleeding, has been attributed to an autoimmune reaction possibly activated in some way by intestinal microbiota. In this

A. *Mycobacterium tuberculosis*   B. Fish tank granuloma   C. Liver granuloma

D. Healthy colon   E. Crohn's disease

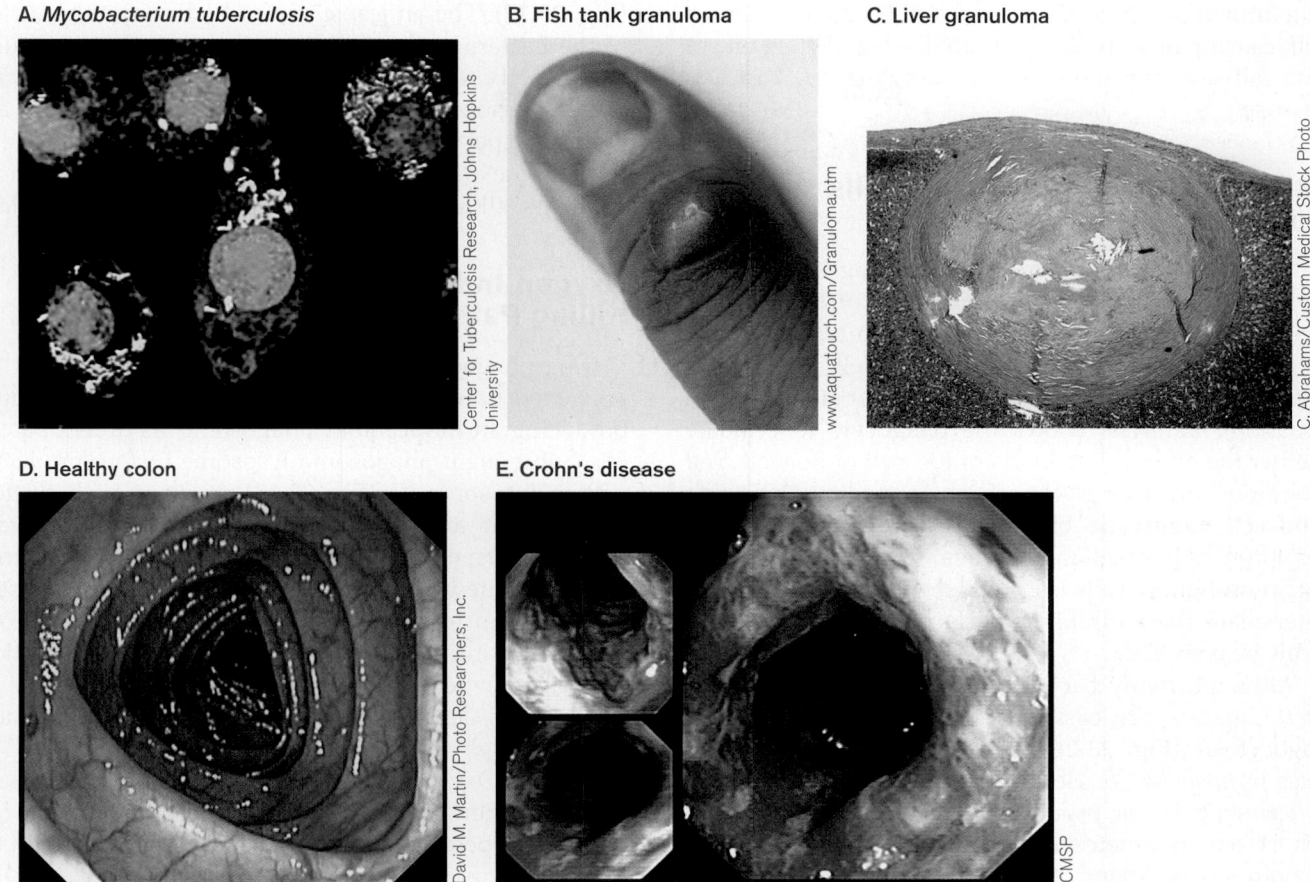

Figure 23.25   Chronic inflammation.   A. The fluorescent orange organisms shown here are *Mycobacterium tuberculosis* (2 μm long, fluorescence microscopy) within macrophages in a tuberculosis abscess. A thick, waxy cell wall protects mycobacteria against the mechanisms used by macrophages to destroy microorganisms. As a result, they survive for prolonged periods within macrophages. The continual stimulation of an inflammatory response leads to chronic inflammation. **B.** Fish tank granuloma. *Mycobacterium marinum* can cause tuberculosis-like infection in fish. The organism can accidentally enter the human body through an open wound or abrasion during cleaning of an aquarium containing infected fish. The infection is first noted as a lesion that heals very slowly on the hand or forearm. Once it does heal, it often forms a granuloma at the site that contains live organisms. **C.** Cross section of liver showing a necrotizing granuloma caused by *M. tuberculosis* that spread through the bloodstream to this site after escaping the lung. **D.** Healthy colon. **E.** Intestinal granulomas of Crohn's disease.

case, the intestinal bacteria are thought to cause a chronic inflammation resulting in characteristic granulomas (**Fig. 23.25D** and **E**).

TO SUMMARIZE:

- **Inflammation** begins with a mechanism (extravasation) that moves neutrophils from the bloodstream into infected tissues.
- **Macrophages** in tissues engulf microorganisms and release vasoactive factors that increase vascular permeability, cytokines that stimulate production of selectin receptors, and chemoattractant molecules that call in neutrophils.
- **Extravasation** is the process by which neutrophils pass between cells of the endothelial wall. Once out

of the circulation, the neutrophils travel to the site of infection.
- **Bradykinin** causes the release of prostaglandins, which produce pain in the affected area.
- **Chronic inflammation** results from the persistent presence of a foreign object.

23.6 Phagocytosis

Phagocytosis is an effective nonspecific immune response. However, it would be disastrous if white blood cells indiscriminately phagocytosed host as well as nonhost structures. Fail-safe controls built into our immune defenses prevent this from happening. Without these restraints, the mammalian immune system would constantly attack itself, causing more problems than it solves. When one of these fail-safe mechanisms fails, autoimmune diseases can arise.

Phagocytes Recognize Alien Cells and Particles

For phagocytosis to proceed, macrophages and neutrophils must first recognize the surface of a particle as foreign (**Fig. 23.26**). When a phagocyte surface interacts with the surface of another body cell, the phagocyte becomes temporarily paralyzed (inept at pseudopod formation). Paralysis allows the phagocyte to evaluate whether the other cell is friend or foe, self or nonself. Self recognition involves glycoproteins located on the white blood cell membrane binding inhibitory glycoproteins present on all host cell membranes. The inhibitory glycoprotein on human cells is called CD47. Because invading bacteria lack these inhibitory surface molecules, they can readily be engulfed.

Although many bacteria, such as *Mycobacterium* or *Listeria* species, are easily recognized and engulfed by phagocytosis (**Fig. 23.26A–C**), others, such as *Streptococcus pneumoniae*, possess polysaccharide capsules that are too slippery for pseudopods to grab (**Fig. 23.26D**). This is where innate immunity and adaptive immunity join forces. Adaptive immunity produces anticapsular antibodies that aid the innate immune mechanism of phagocytosis through a process known as **opsonization**

(**Fig. 23.27**). The anticapsular antibodies coat the surface of the bacterium, leaving the tail end of the antibodies, called the Fc region (see Section 24.3), pointing outward. The Fc regions of these antibodies are recognized and bound by specific receptors on phagocyte cell surfaces. As a result, the antibodies bind the bacteria to phagocytic Fc receptors, allowing the phagocyte to engulf the invader.

Oxygen-Independent and -Dependent Killing Pathways

During phagocytosis, the cytoplasmic membrane of the phagocyte flows around, and then engulfs, the bacterium, producing an intracellular phagosome, as described earlier. Subsequent phagosome-lysosome fusion (producing a phagolysosome; see **Fig. 23.14**) results in both oxygen-independent and oxygen-dependent killing pathways. Oxygen-independent mechanisms include enzymes like lysozyme to destroy the cell wall, compounds such as lactoferrin to sequester iron away from the microbe, and defensins, small cationic antimicrobial peptides (described in Section 23.4).

Oxygen-dependent mechanisms are probably activated through the Toll-like receptors (discussed in Section 23.7). Oxygen-dependent mechanisms kill through the production of various oxygen radicals. NADPH oxidase, myeloperoxidase, and nitric oxide synthetase in the phagosome membrane are extremely important. NADPH oxidase yields superoxide ion ($^\bullet O_2^-$), hydrogen peroxide, and, ultimately, hydroxyl radicals ($^\bullet OH$) and ions (OH^-).

A. White blood cell attacking bacteria

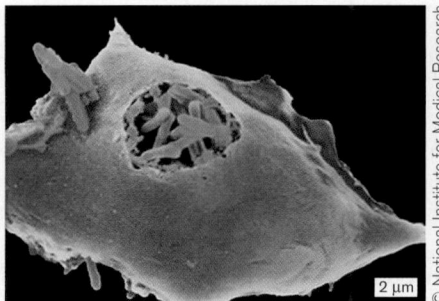

© National Institute for Medical Research

B. Contacts between phagocyte and target microbe

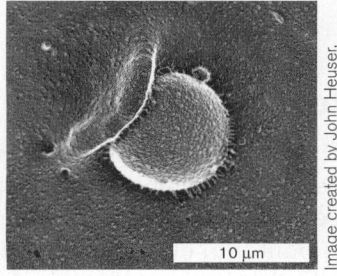

Image created by John Heuser, Washington Univ. School of Medicine

C. Macrophage engulfing bacterium

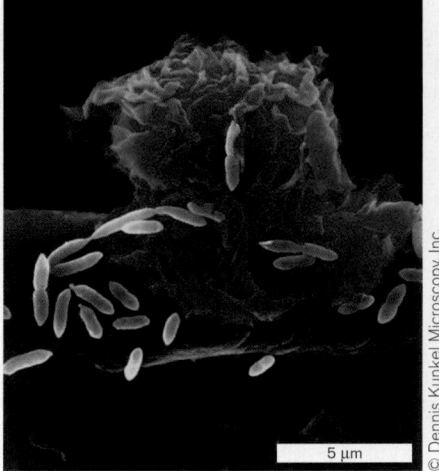

© Dennis Kunkel Microscopy, Inc.

D. *Streptococcus pneumoniae* and capsule

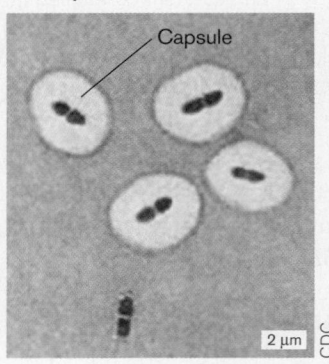

Capsule

CDC

Figure 23.26   Images of phagocytosis.   A. A white blood cell attacking bacteria. *Mycobacterium* (green, 2 μm long) being phagocytosed by a white blood cell (colorized SEM). **B.** Contacts between phagocyte and target microbe, illustrating how phagocytes "grab" the target bacterium. **C.** A macrophage engulfing bacteria on the outer surface of a blood vessel (SEM, magnification 1,315×). **D.** *Streptococcus pneumoniae* and capsule (India ink preparation). The slippery nature of the polysaccharide capsule makes phagocytosis more difficult.

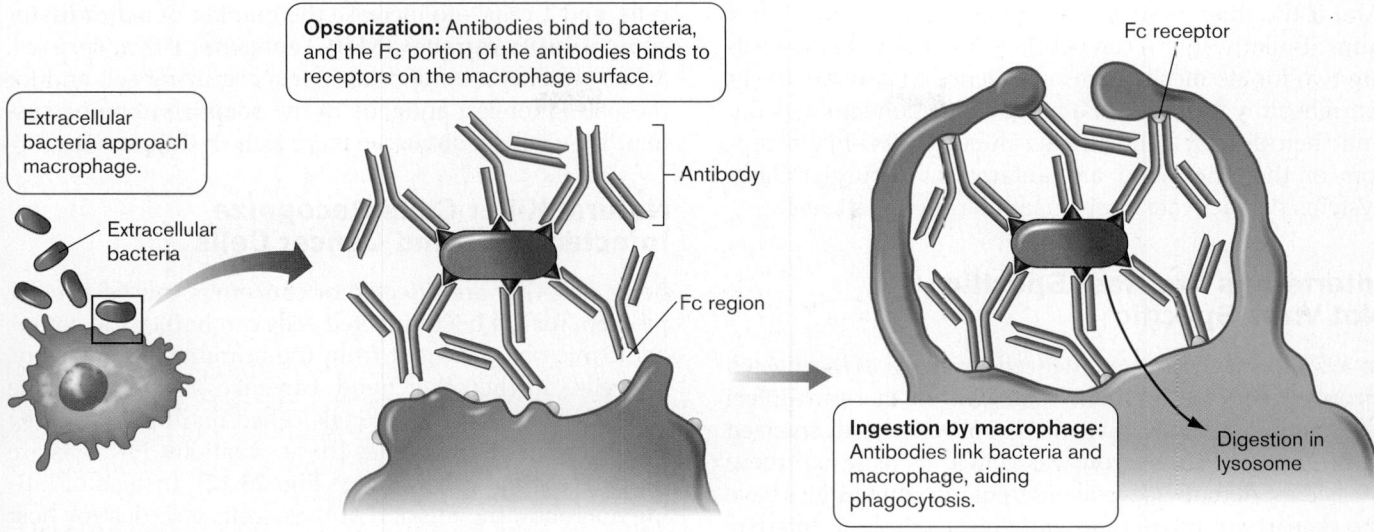

Figure 23.27 Opsonization. Opsonization is a process that facilitates phagocytosis. Here, macrophage Fc receptors bind to the Fc region of antibodies binding to bacteria.

$$\text{NADPH} + 2O_2 \xrightarrow{\text{NADPH oxidase}} \text{NADP}^+ + H^+ + 2\ {}^{\bullet}O_2^{-} \xrightarrow{\text{superoxide dismutase}}$$

$$H_2O_2 + Cl^- \xrightarrow{\text{myeloperoxidase}} OH^- + HOCl$$

Myeloperoxidase, present only in neutrophils, converts hydrogen peroxide and chloride ions to hypochlorous acid (HOCl).

Macrophages, mast cells, and neutrophils also generate reactive nitrogen intermediates that serve as potent cytotoxic agents. Nitric oxide (NO) is synthesized by NO synthetase. Further oxidation of NO by oxygen yields nitrite (NO_2^-) and nitrate (NO_3^-) ions. All of these reactive oxygen species attack bacterial membranes and proteins. These mechanisms cause the large increase in oxygen consumption noted during phagocytosis, called the **oxidative burst**. The reactive chemical species formed during the oxidative burst do little to harm the phagocyte because the burst is limited to the phagosome and because the various reactive oxygen species, such as superoxide, are very short-lived. Although these phagocytes are very good at clearing infectious agents, many bacteria have developed ways to outsmart this aspect of innate immunity. One example involving *Salmonella* is described in **eTopic 23.2**.

Autophagy and Intracellular Pathogens

Intracellular pathogens that grow in eukaryotic cytoplasm can be a serious problem for the host. Many pathogens, such as *Mycobacterium tuberculosis*, the cause of tuberculosis, enter the host cell in ways that bypass endosome formation or, if they do enter via an endosome, can escape from that compartment. These pathogens block normal host cell clearance pathways. To circumvent this problem, eukaryotic host cells (not just phagocytes) have taken a

cell function normally used to degrade damaged organelles (called **autophagy**) and have adapted it to clear themselves of intracellular pathogens. During the process of autophagy, the cell constructs a double membrane around the organism (or damaged organelle). This structure, called the autophagosome, sequesters the microbe from the nutrient-rich cytosol. Lysosomes then fuse with the autophagosome, depositing degradative enzymes that digest the organism. Ever-adapting intracellular microbes, however, have found ways to suppress autophagy and survive. **eTopic 23.2** discusses some ways that bacterial pathogens can avoid the innate immune system.

TO SUMMARIZE:

- **Phagocytosis** is selective for particles recognized as foreign to the body.
- **Oxygen-independent and oxygen-dependent mechanisms of killing** are initiated by fusion between a lysosome and a bacteria-containing phagosome.
- **The oxidative burst**, a large increase in oxygen consumption during phagocytosis, results in the production of superoxide ions, nitric oxide, and other reactive oxygen species.
- **Autophagy** is a process by which intracellular bacteria can be sequestered from the cytoplasm (via an autophagosome) and killed following fusion with a lysosome.

23.7 Innate Defenses: Interferon, Natural Killer Cells, and Toll-Like Receptors

When a community is threatened by a thief, a neighbor who has been robbed alerts others to take precautions.

And if the thief is discovered, police are called to arrest him. Similarly, interferon peptides and natural killer cells are two innate mechanisms of defense that, respectively, warn healthy host cells of a nearby infection and seek out and then destroy cells already infected. Toll-like receptors, on the other hand, are tantamount to burglar alarm systems that activate upon encountering an intruder.

Interferons Are Host Specific, Not Virus Specific

In 1957, it was discovered that cells exposed to inactivated viruses produce at least one soluble factor that can "interfere" with viral replication when applied to newly infected cells. The term **interferon** was coined to represent these molecules. Actually, several different macromolecules have the property of interfering with viral replication. Interferons are low-molecular-weight cytokines (14–20 kDa) produced by many eukaryotic cells in response to intracellular infection. The action of interferons is usually species specific (interferon from mice will not work on human cells) but virus nonspecific (human interferon will help protect against both poliovirus and influenza virus, for example).

There are two general types of interferons, which differ in the receptors they bind and the responses they generate (**Table 23.3**). Type I interferons have high antiviral potency; they consist of IFN-alpha (IFN-α), IFN-beta (IFN-β), and IFN-omega (IFN-ω). Type II interferon, IFN-gamma (IFN-γ), has more of an immunomodulatory function that will be discussed in the next chapter.

Type I interferons can bind to specific receptors on uninfected host cells and render those cells resistant to viral infection. The host cell becomes resistant because interferon induces the intracellular production of two classes of proteins. One class encompasses double-stranded RNA-activated endoribonucleases that can cleave viral RNA. Proteins in the second class are protein kinases that phosphorylate and inactivate eukaryotic initiation factor eIF2, which is required to translate viral RNA. These mechanisms affect both RNA and DNA viruses because protein synthesis is required for the propagation of all viruses. However, RNA viruses are better than DNA viruses at inducing interferon. Type I interferons are used to treat certain viral infections (for example, hepatitis C).

Type II interferon functions by activating various white blood cells—for example, macrophages, natural killer cells, and T cells—to increase the number of **major histocompatibility complex (MHC)** antigens on their surfaces. MHC proteins are important for recognizing self and for presenting foreign antigens to the adaptive immune system. They will be discussed more fully in Chapter 24.

Natural Killer Cells Recognize Infected Cells and Cancer Cells

Body cells that are infected or cancerous can be a major problem for the host. Infected cells can harbor the pathogenic microbe, hiding it from the immune system. Cancer cells, on the other hand, can take over and kill the host. A class of lymphoid cells called natural killer (NK) cells identifies and handles these situations. NK cells are formed in bone marrow (see **Fig. 23.12**). Instead of killing microbes, the mission of these cells is to destroy host cells that harbor microorganisms or that have been transformed into cancer cells (**Fig. 23.28**). Natural killer cells recognize changes in cell surface proteins of infected or cancer cells (MHC class I molecules; see Section 24.6) and then degranulate to release chemicals that kill those cells.

Natural killer cells recognize their targets in two basic ways. One involves MHC class I molecules, and the other utilizes Fc receptors. A normal host cell displays two classes of MHC molecules on the outside of the cell membrane. MHC I is an indicator of "self." (MHC II molecules will be discussed in Chapter 24.) NK cells have specific receptors that bind to self MHC I molecules on

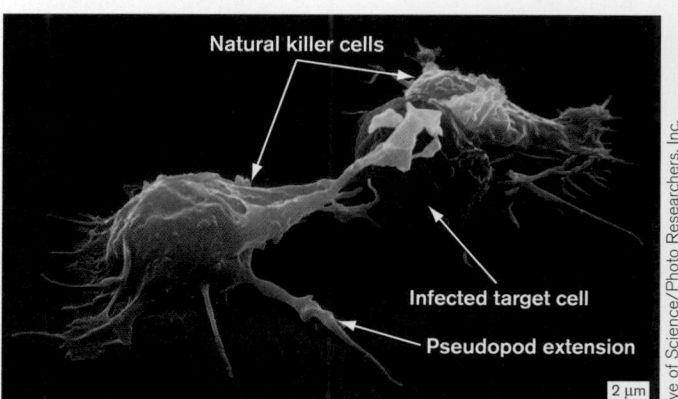

Figure 23.28   Natural killer cells.   Natural killer (NK) cells attack eukaryotic cells infected by microbes, not the microbes themselves. Perforin produced by the NK cell punctures the membrane of target cells, causing them to burst.

Table 23.3   Classes of interferons.

Type	Examples	Function	Mechanism of action
Type I	IFN-alpha IFN-beta IFN-omega	Antiviral	Induce dsRNA endonuclease, eIF2 kinase
Type II	IFN-gamma	Immunomodulatory	Triggers signal cascades in macrophages, NK cells, and T cells; increases MHC II on cell surfaces

the surfaces of other cells in the body. An NK cell that "touches" a self MHC I–containing cell from the same person will not attack that cell. However, if a host cell lacks MHC class I molecules, NK cells perceive the target as foreign and a potential threat. Host cells can lose their MHC molecules during infection or as a result of malignant transformation. When an NK cell encounters a host cell lacking these markers, the NK cell inserts a pore-forming protein (**perforin**) into the membrane of the target cell, through which cytotoxic enzymes are delivered.

Natural killer cells also contain Fc receptors on their cell surface. The second killing mechanism, called **antibody-dependent cell-mediated cytotoxicity (ADCC)**, is activated when the Fc receptor on the NK cell links to an antibody-coated host cell. The part of an antibody that does not bind to a target molecule (antigen) is called the Fc region. Why would host cells be coated with antibodies? During their replication, many viruses place viral proteins in the membrane of the infected cell. Antibodies to those viral proteins will coat the compromised cell, tagging it for ADCC. Once the compromised cell has been targeted, it is killed by the NK cell in the same way as described for cells lacking MHC, by insertion of a perforin molecule. This killing mechanism is another example of cooperation between innate (NK cells) and adaptive immunity (antibody-producing lymphocytes).

> **THOUGHT QUESTION 23.8** If NK cells can attack infected host cells coated with antibody, why won't neutrophils?

Toll-Like Receptors Recognize Pathogen-Associated Molecular Patterns

The faster the body can detect the presence of pathogens, the more quickly it can begin to deal with them. The more quickly it deals with them, the better the outcome of an infection. However, it takes time for the adaptive immune system to make antibody specific for a microbe (see Chapter 24). All the while, the pathogen can grow and cause disease. Fortunately, bacteria and viruses possess unique structures that immediately tag them as foreign. Structures such as peptidoglycan, flagellin, and lipoteichoic acids are not normally present in tissues unless bacteria or viruses are present. These structures have pathogen-associated molecular patterns (PAMPs) that can be recognized by Toll-like receptors present on various host cell types (**Table 23.4**).

First discovered in insects and named Toll receptors, **Toll-like receptors (TLRs)** are evolutionarily conserved cell surface glycoproteins present on the cells of many eukaryotic genera. They are transmembrane proteins with an extracellular Toll/interleukin 1 receptor domain (TIR domain). Humans have numerous Toll-like receptors, each of which recognizes different PAMPs present on pathogenic microorganisms, making them an innate defense mechanism with some degree of specificity. For example, TLR2 binds to lipoarabinomannan from mycobacteria, zymosan from yeast, lipopolysaccharide (LPS) from spirochetes, and peptidoglycan. TLR4, on the other hand, binds LPS, as well as host proteins released at sites of infection (for example, heat-shock protein 60). CD14, another host cell surface protein, serves as a coreceptor for LPS. Note that these receptors bind fragments of structures after they are released from the microbe. They don't interact with the whole organism.

> **NOTE:** The term "Toll gene" came from Christiane Nüsslein-Volhard's 1985 exclamation, "That's weird" (in German, "Das war ja toll") when shown the underdeveloped posterior of a mutant fruit fly. Thus the gene was dubbed "Toll." Toll-like receptors in mammals display sequence similarities to the Toll genes involved with insect embryogenesis.

Once bound to their ligands, the TLRs trigger an intracellular regulatory cascade via their TIR domain, causing the host cell to release proteins called cytokines that diffuse away from the site, bind to receptors on various cells of the immune system (see **Table 23.4**),

Table 23.4 Examples of Toll-like receptors and PAMPs.

Receptor	PAMPs recognized	Source	Host cells	Location
TLR1	Lipopeptides	Bacteria	Monocytes/macrophages, dendritic cells, B cells	Cell surface
TLR2	Glycolipids, lipoteichoic acids, peptidoglycan	Bacteria	Monocytes/macrophages, dendritic cells, mast cells	Cell surface
TLR3	Double-stranded RNA	Viruses	Dendritic cells, B cells	Cell compartment
TLR4	Lipopolysaccharide, heat-shock proteins	Bacteria	Monocytes/macrophages, dendritic cells, mast cells, intestinal epithelium	Cell surface
TLR5	Flagellin	Bacteria	Monocytes/macrophages, dendritic cells, intestinal epithelium	Cell surface
TLR6	Diacyl lipopeptides	Mycoplasma	Monocytes/macrophages, mast cells, B cells	Cell surface
TLR9	Unmethylated CpG residues in DNA	Bacteria	Monocytes/macrophages	Cell compartment

and direct them to engage the invader. Cytokines are discussed further in Chapter 24. The cells that respond to cytokines can be part of innate immunity, adaptive immunity, or both. TLR recognition of PAMPs can also trigger autophagy in infected cells.

TO SUMMARIZE:

- **Interferons are species-specific molecules** that can nonspecifically interfere with viral replication (type I) and modulate the immune system (type II).
- **Natural killer (NK) cells** are a class of white blood cell that destroy cancer cells or cells harboring microorganisms.
- **Natural killer cells target host cells** that have lost MHC class I receptors as a result of infection or cancer, or host cells that are coated with antibody (antibody-dependent cell-mediated cytotoxicity, ADCC).
- **Natural killer cells kill** by inserting perforin pores into the membranes of target cells.
- **Toll-like receptors (TLRs)** on host cells recognize different pathogen-associated molecular patterns (PAMPs).

23.8 Complement's Role in Innate Immunity

Bacterial invaders can also be attacked by a mechanism known as the complement cascade. Complement was first discovered as a heat-labile component of blood that enhances (or complements) the killing effect of antibodies on bacteria. The complement system consists of up to 20 proteins, several of which are proteases that sequentially cleave each other. (The liver is the main source of complement proteins.) Once a complement cascade is triggered, a number of things can happen. Pores are inserted into membranes and cause cytoplasmic leaks; in addition, pieces of the complement system can attract white blood cells and also facilitate phagocytosis (opsonization).

Complement Activation Pathways

The three routes to complement activation are officially known as the classical pathway, the alternative pathway, and the lectin pathway. The classical complement pathway depends on antibody, so it is part of adaptive rather than innate immunity and is discussed in Chapter 24 (Section 24.7). The lectin pathway requires synthesis of mannose-binding lectin by the liver in response to macrophage cytokines. This lectin coats the surface of invading microbes and activates complement without antibody. The lectin pathway connects to the classical complement pathway. We focus in this chapter on the alternative pathway because it is a well-characterized part of innate resistance. Like the lectin pathway, the alternative pathway is

a nonspecific defense mechanism (does not require antibody for activation). The alternative complement pathway can attack invading microbes long before a specific immune response can be launched.

One goal of the complement cascade is to insert pores into target microbial membranes. The pores destroy membrane integrity, which kills the cell. The alternative complement pathway begins with the complement factor C3 (**Fig. 23.29**). In blood, C3 slowly cleaves into C3a and C3b (step 1). C3b, under normal circumstances, is rapidly degraded— a process that thwarts inadvertent complement activation. However, if C3b meets LPS on an invading Gram-negative microbe, the bound C3b becomes stable and binds another factor, designated factor B (step 2), and makes factor B susceptible to cleavage by yet another protein, factor D (step 3). The resulting complex, called C3bBb, is changed into what is called C5 convertase by properdin, another serum protein (step 4). C5 convertase cleaves C5 in serum to C5a and C5b (step 5). (C3bBb is also called C3 convertase because it can quickly cleave more C3 and amplify the cascade.) C5b now starts to form a prepore complex by binding to C6 and C7 (step 6). The resulting C5bC6C7 complex binds to membranes. Finally, C8 and C9 factors join in to form the **membrane attack complex (MAC)**, becoming a destructive pore in the membrane of the target cell (step 7).

If the target of complement is a eukaryotic cell, Na^+ and H_2O enter and the cell osmotically lyses. Eukaryotic cells infected with viruses and coated with antibody can activate complement via the classical complement pathway (see Section 24.7). When complement affects Gram-negative cells, lysozyme (present in serum) enters through outer membrane pores and cleaves peptidoglycan, making the cell membrane more susceptible to the membrane attack complex. Teichoic acid on Gram-positive cell walls can induce the complement cascade but is less effective than LPS. Nevertheless, Gram-positive bacteria are resistant to complement because they lack an outer membrane (no LPS to efficiently start the cascade) and have a thick peptidoglycan layer that hinders access of complement components. However, even in the absence of LPS there are ways to activate complement involving antigen-antibody complexes (see Section 24.7).

Other Roles for Complement Peptides in Innate Immunity

Factor C3b, in addition to initiating the complement cascade, is a potent opsonin. An **opsonin** is any factor that can promote phagocytosis of an organism. PMNs (neutrophils) have specific C3b receptors on their surface. Thus, when C3b binds to a bacterial cell surface, it tags that cell and makes it easier for PMNs to grab and engulf the organism.

The complement fragments C5a (which forms part of the prepore complex) and C3a have many roles in immune function beyond their contribution to MAC formation. They are also **anaphylatoxins**, proteins that can trigger

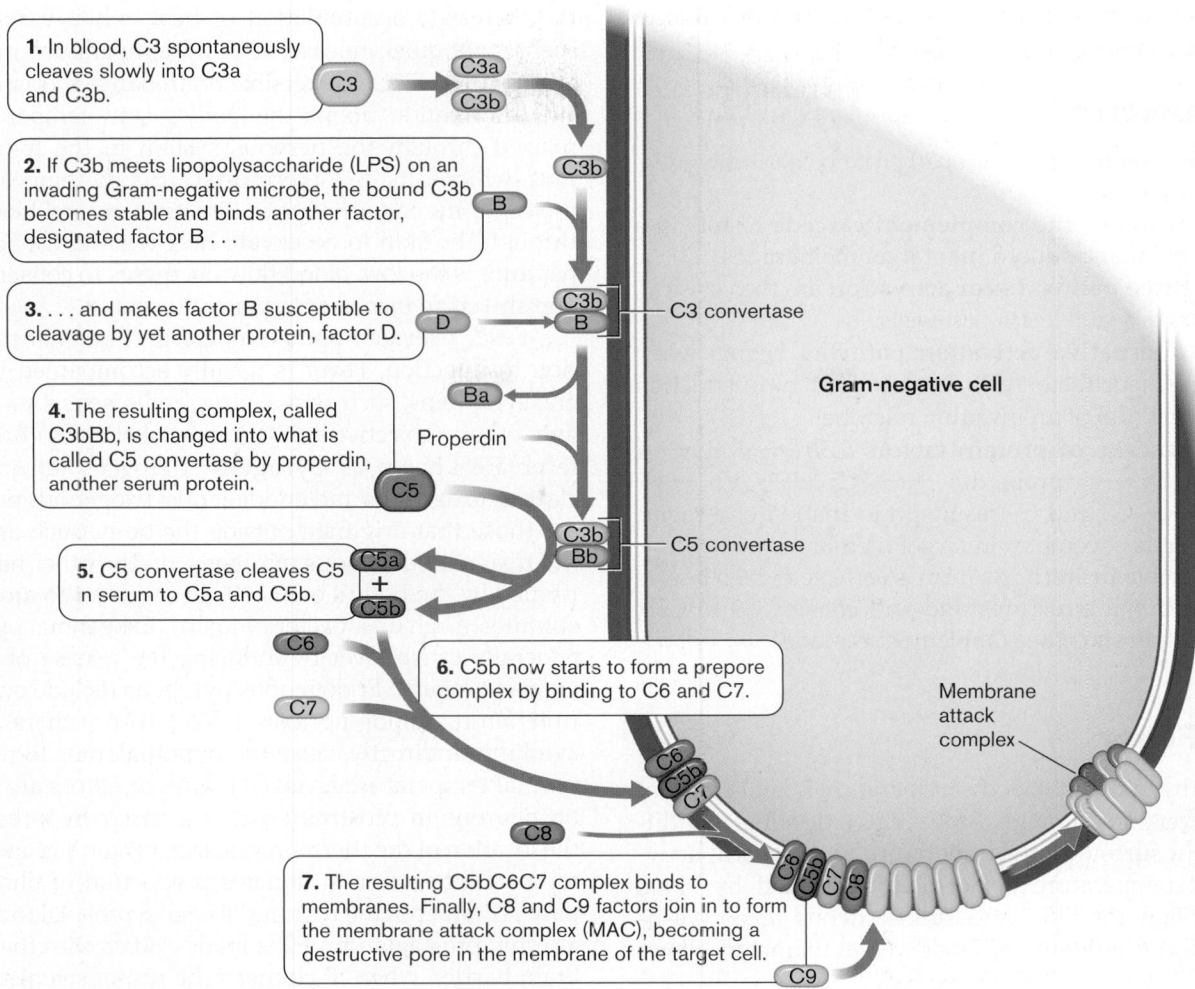

1. In blood, C3 spontaneously cleaves slowly into C3a and C3b.

2. If C3b meets lipopolysaccharide (LPS) on an invading Gram-negative microbe, the bound C3b becomes stable and binds another factor, designated factor B . . .

3. . . . and makes factor B susceptible to cleavage by yet another protein, factor D.

4. The resulting complex, called C3bBb, is changed into what is called C5 convertase by properdin, another serum protein.

5. C5 convertase cleaves C5 n serum to C5a and C5b.

6. C5b now starts to form a prepore complex by binding to C6 and C7.

7. The resulting C5bC6C7 complex binds to membranes. Finally, C8 and C9 factors join in to form the membrane attack complex (MAC), becoming a destructive pore in the membrane of the target cell.

C3 convertase

C5 convertase

Properdin

Gram-negative cell

Membrane attack complex

Figure 23.29   The alternative complement pathway.   Although called "alternative," this complement cascade is part of the first-line innate defense.

degranulation of vasoactive factors such as histamine from endothelial cells, mast cells, or phagocytes. They can also stimulate chemotaxis of immune cells. C5a and C3a bind to separate receptors on plasma membranes and activate signal cascade pathways. C5a triggers Ca^{2+} release from intracellular stores and stimulates the actin polymerization needed for cell migration. Conversely, C3a triggers Ca^{2+} influx, which facilitates extravasation. These peptide mediators also stimulate the release of certain cytokines, such as IL-4, from monocytes and mast cells that prepare capillary endothelium for rolling and adhesion of neutrophils. C5a will also up-regulate P-selectin and ICAM-1. Thus, some complement factors not only help directly destroy target bacterial cells, but also facilitate phagocytosis and contribute directly to inflammation.

Acute-Phase Reactants, Complement, and Heart Disease

As noted earlier, inflammation is associated with the production of various cytokines by macrophages. Some of

these cytokines (such as IL-1, TNF-α, and IL-6) travel to the liver, where they stimulate synthesis of several so-called acute-phase reactant proteins, including **C-reactive protein.** Named for its ability to activate complement, C-reactive protein is an acute-phase reactant that will bind to components of bacterial cell surfaces but not to host cell membranes (**Fig. 23.30**). Once tied to the bacterial cell surface, C-reactive protein will bind complement factor C1q of the classical complement pathway (Chapter 24) ultimately converting factor C3 to C3b, propagating the complement cascade. C-reactive protein accelerates C3b production at the bacterial surface, where it can do the most damage.

Elevated levels of C-reactive protein have been linked to an increased risk of cardiovascular disease. The hypothesis is that fatty deposits in the arteries trigger inflammation in the area of deposition. Inflammation then triggers an increase in C-reactive protein, which can be measured as an indicator of cardiovascular disease. Taking an aspirin a day may reduce the risk of heart attack by preventing inflammation and, in turn, reducing levels of C-reactive protein. Whether C-reactive protein has a direct role in cardiovascular disease is not known.

TO SUMMARIZE:

- **Complement** is a series of 20 proteins naturally present in serum.
- **Activation of the complement cascade** results in a pore being introduced into target membranes.
- **The three pathways for activation** are the classical, alternative, and lectin pathways.
- **The alternative activation pathway** begins when complement factor C3b is stabilized by interaction with the LPS of an invading microbe.
- **The cascade of protein factors** C3b $\longrightarrow$ B $\longrightarrow$ factor D $\longrightarrow$ properdin $\longrightarrow$ C5 $\longrightarrow$ C6 $\longrightarrow$ C7 $\longrightarrow$ C8 and C9 results in formation of a membrane attack complex in target membranes.
- **C-reactive protein** in serum is activated when bound to microbial structures and will convert C3 to C3b, which can start the complement cascade.

23.9 Fever

In a healthy individual, body temperature is kept constant within a very small range (36°C–38°C), despite large differences in surrounding temperature and physical activity. Body temperature is normally regulated by blood flow through the skin and subcutaneous areas. Vasoconstriction (tightening of blood vessel diameter) allows

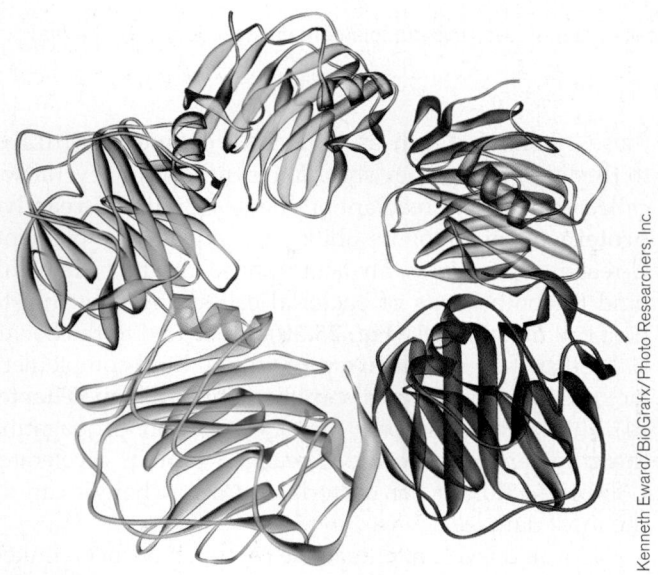

Figure 23.30 C-reactive protein. The three-dimensional structure of human C-reactive protein (CRP) is unique. CRP is synthesized as a 206-amino-acid polypeptide that folds to form a flattened jelly roll structure, which then assembles into a radially symmetrical pentamer that circulates in serum. This structure can bind bacterial cell surfaces and C3 complement factor.

Kenneth Eward/BioGrafx/Photo Researchers, Inc.

the increased accumulation of heat, while vasodilation secures its quick release. If body temperature increases above 42°C–43°C, irreversible brain damage occurs.

Information about the body's skin temperature is relayed through the nervous system to the hypothalamus, which acts as a thermostat. If body temperature is too high, the hypothalamus directs increased blood flow through the skin to accelerate heat release. If body temperature is too low, blood flow decreases to conserve heat and shivering increases to generate heat.

Fever (elevated body temperature) is a natural reaction to infection. Fever is usually accompanied by general symptoms, such as sweating, chills, sensation of cold, and other subjective sensations. Substances that cause fever are known as **pyrogens**. Pyrogens fall into two classes: exogenous and endogenous. Exogenous pyrogens are those that originate outside the body, such as bacterial toxins. Endogenous pyrogens, on the other hand, are formed by the body's own cells in response to an outside stimulus (such as a bacterial toxin). Exogenous pyrogens generally cause fever by inducing the release of endogenous pyrogens. Endogenous pyrogens include cytokines (interferon, tumor necrosis factor, IL-6, others). These cytokines indirectly cause the hypothalamus to reset the normal temperature level. Cytokine receptors are located on neurons in proximity to the anterior hypothalamus, the location of the thermoregulatory center. The cytokine-receptor interaction stimulates production of phospholipase A2, an enzyme required to make prostaglandins. As a result, prostaglandin E2 is made and crosses the blood-brain barrier, where it changes the responsiveness of the thermosensitive neurons that make up the thermoregulatory center. In other words, it turns up the thermostat.

Because the ideal growth temperature for many microbes is 37°C, elevated temperature can place the organism outside its "comfort zone" of growth. There is also evidence that fever reduces iron availability to bacteria (cytokine release causes an increase in iron storage protein). Slower growth of the pathogen allows the body's immune system time to subdue the infection before it is too late. Consequently, interventions that reduce a moderate fever caused by infection may be counterproductive to a speedy recovery.

> **THOUGHT QUESTION 23.9** If increased fever limits bacterial growth, why do bacteria make pyrogenic toxins?

TO SUMMARIZE:

- **The hypothalamus** acts as the body's thermostat.
- **Exogenous and endogenous pyrogens** elevate body temperature by stimulating prostaglandin production.
- **Prostaglandins** change the responsiveness of thermosensitive neurons in the hypothalamus.

Concluding Thoughts

The human body plays host to many species of microbes (our microbiota). Many of which contribute to our health and well being. The various "hardwired" innate immune mechanisms described in this chapter keep our microbiota at bay and provide an effective first line of defense against potential pathogens. The next chapter addresses what happens when microbes breach these innate defenses. Unlike the general protective mechanisms discussed in this chapter, the system of adaptive immunity generates a molecular defense specifically tailored to a given pathogen.

CHAPTER REVIEW

Review Questions

1. Name some sterile body sites.
2. What body sites are colonized by normal microbiota?
3. Under what circumstances can commensal organisms cause disease?
4. Why are commensal organisms beneficial to the host?
5. Name and describe various types of innate immunity.
6. What are probiotics? How do they help maintain health?
7. How do the lungs avoid being colonized?
8. What is a gnotobiotic animal?
9. Describe GALT and SALT.
10. Describe some chemical barriers to infection.
11. Discuss the different types of white blood cells.
12. What is a lymphoid organ?
13. Outline the process of inflammation.
14. Explain why phagocytes do not indiscriminately phagocytose body cells.
15. What is interferon?
16. Describe antibody-dependent cell-mediated cytotoxicity.
17. How does complement kill bacteria?
18. Why might fever be helpful in fighting infection?

Thought Questions

1. Is it common for microbial pathogens to pass the placental barrier (called transplacental transmission) and infect the fetus? Consider papillomavirus, *Listeria monocytogenes*, *Escherichia coli*, HIV, *Treponema pallidum*, *Neisseria gonorrhoeae*, and *Staphylococcus aureus*.

2. The vagina contains competing, commensal microbes that contribute to the health of the organ. So, is "cleaning" the vagina by douching actually unhealthy and likely to lead to an increased chance of infection (vaginosis)?
3. Why have microbes not adapted to avoid being recognized by Toll-like receptors?

Key Terms

adaptive immunity (871)
alveolar macrophage (877)
anaphylatoxin (888)
antibody-dependent cell-mediated cytotoxicity (ADCC) (887)
antigen (871)
antigen-presenting cell (874)
autophagy (885)
B cell (874)
bacteremia (866)
basophil (873)
bioburden (862)
bradykinin (881)
C-reactive protein (889)

chemokine (882)
commensal organism (862)
complement (871)
compromised host (870)
cystic fibrosis transmembrane conductance regulator (CFTR) (877)
cytokine (870)
defensin (877)
degranulate (878)
dendritic cell (873)
enterotoxin (870)
eosinophil (873)
epidermis (864)
extravasation (880)

gnotobiotic animal (871)
granuloma (882)
gut-associated lymphoid tissue (GALT) (876)
immune system (871)
immunomodulin (870)
innate immunity (871)
interferon (886)
interleukin 1 (IL-1) (880)
Langerhans cell (876)
lymph node (875)
lymphocyte (874)
M cell (876)
macrophage (873)

major histocompatibility complex
  (MHC) (886)
mast cell (873)
membrane attack complex (MAC) (888)
microbiome (862)
microbiota (862)
monocyte (873)
mucociliary elevator (866)
nasopharynx (865)
neutrophil (873)
nonadaptive immunity (871)

opportunistic pathogen (870)
opsonin (888)
opsonization (884)
oropharynx (865)
oxidative burst (885)
pathogen-associated molecular pattern
  (PAMP) (876)
perforin (887)
phagosome (873)
probiotic (868)

pyrogen (890)
reticuloendothelial system (873)
selectin (880)
skin-associated lymphoid tissue (SALT)
  (876)
T cell (874)
Toll-like receptor (TLR) (887)
tumor necrosis factor alpha (TNF-α)
  (880)
vasoactive factor (880)

Recommended Reading

Barton, Gregory M., and Jonathan C. Kagan. 2009. A cell biological view of Toll-like receptor function: Regulation through compartmentalization. *Nature Reviews. Immunology* **9**:535–542.

Delgado, Monica A., and Vojo Deretic. 2009. Toll-like receptors in control of immunological autophagy. *Cell Death and Differentiation* **16**:976–983.

Ganz, Tomas, and Robert I. Lehrer. 1998. Antimicrobial peptides of vertebrates. *Current Opinion in Immunology* **10**:41–44.

Gunn, Jon S. 2000. Mechanisms of bacterial resistance and response to bile. *Microbes and Infection* **2**:907–913.

Haas, Pieter-Jan, and Jos van Strijp. 2007. Anaphylatoxins: Their role in bacterial infection and inflammation. *Immunological Research* **37**:161–175.

Hathaway, Lucy J., and Jean-Pierre Kraehenbuhl. 2000. The role of M cells in mucosal immunity. *Cellular and Molecular Life Science* **57**:323–332.

Hewetson, J. T. 1904. The bacteriology of certain parts of the human alimentary canal and of the inflammatory processes arising therefrom. *British Medical Journal* **2**:1457–1460.

Hornef, Mathias W., Mary J. Wick, Mikael Rhen, and Staffan Normark. 2002. Bacterial strategies for overcoming host innate and adaptive immune responses. *Nature Immunology* **3**:1033–1040.

Menezes, Juscilene da Silva, Daniel de Sousa Mucida, Denise C. Cara, Jaqueline I. Alvarez-Leite, Momtchilo Russo, et al. 2003. Stimulation by food proteins plays a critical role in the maturation of the immune system. *International Immunology* **15**:447–455.

Niedergang, Florence, and Jean-Pierre Kraehenbuhl. 2000. Much ado about M cells. *Trends in Cell Biology* **10**:137–141.

Ogawa, Michinaga, and Chihiro Sasokawa. 2006. Bacterial evasion of the autophagic defense system. *Current Opinion in Microbiology* **9**:62–68.

Orvedahl, Anthony, and Beth Levine. 2009. Eating the enemy within: Autophagy in infectious diseases. *Cell Death and Differentiation* **16**:57–69.

Peekhaus, N., and Tyrrell C. Conway. 1998. What's for dinner? Entner-Doudoroff metabolism in *E. coli*. *Journal of Bacteriology* **180**:3495–3502.

Salzman, Nita H., Dipankar Ghosh, Kenneth M. Huttner, Yvonne Paterson, and Charles L. Bevins. 2003. Protection against enteric salmonellosis in transgenic mice expressing a human intestinal defensin. *Nature* **422**:522–526.

Stetson, Daniel B., and Ruslan Medzhitov. 2006. Type I interferons in host defense. *Immunity* **25**:373–381.

Stuart, Lynda M., and R. Alan B. Ezekowitz. 2005. Phagocytosis: Elegant complexity. *Immunity* **22**:539–550.

Sullivan, Ann, and Carl E. Nord. 2002. The place of probiotics in human intestinal infections. *International Journal of Antimicrobial Agents* **20**:313–319.

Tlaskalova-Hogenova, H., L. Tuckova, R. Lodinova-Zadnikova, R. Stepankova, B. Cukrowska, et al. 2002. Mucosal immunity: Its role in defense and allergy. *International Archives of Allergy and Immunology* **128**:77–89.

Turnbaugh, Peter J., Micah Hamady, Tanya Yatsunenko, Brandi L. Cantarel, Alexis Duncan, et al. 2009. A core gut microbiome in obese and lean twins. *Nature* **457**:480–484.

Wikoff, William R., Andrew T. Anfora, Jun Liu, Peter G. Schultz, Scott A. Lesley, et al. 2009. Metabolomics analysis reveals large effects of gut microflora on mammalian blood metabolites. *Proceedings of the National Academy of Sciences USA* **106**:3698–3703.

For further study and review, please visit StudySpace at **microbiology2.com**.

Chapter 24

The Adaptive Immune Response

Memory does not reside solely in the brain. Adaptive immunity also depends on a kind of memory that does not rely on nerves and neurons. Our immune system's memory relies on special lymphocytes in the circulation called memory B cells and T cells that form during an infection and "remember" it for years. Although a single member of these cells can recall exposure to only one part of an infecting agent, collectively memory cells remember every microbe that has managed to breach our innate defenses. Armed with this knowledge, they circulate throughout the body like tiny sentries, ready to detect and quickly respond to a second attack.

From birth, the immune system is able to adapt and recognize billions of possible foreign antigens. However, a system this flexible can also attack itself. This chapter will demystify how adaptive immunity develops, regulates itself and neutralizes potential threats.

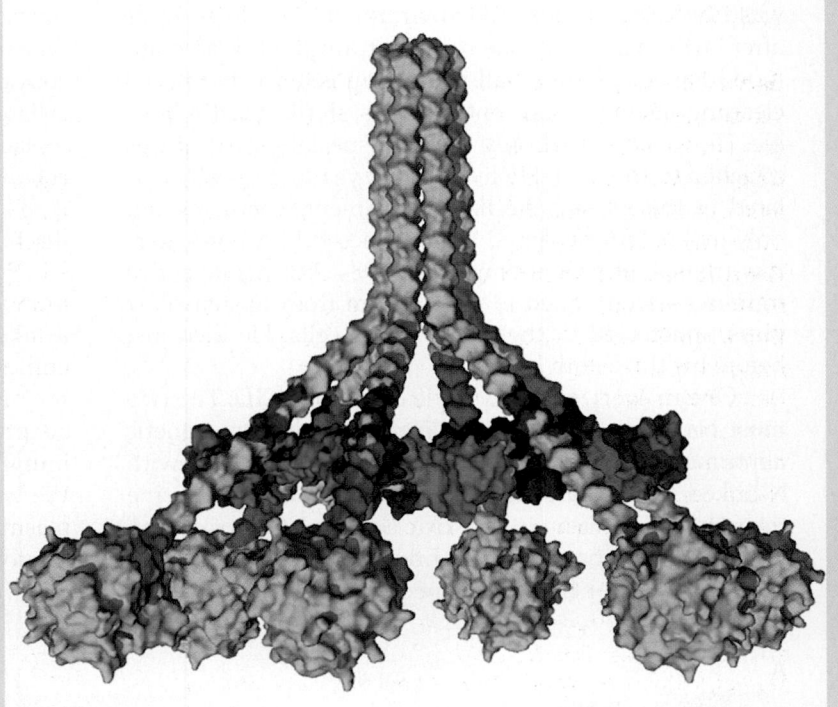

Current Research Highlight

Three-dimensional structure of serum complement component C1, a complement factor that links the adaptive and innate immune responses. The gray globular structures of C1 bind to antibodies attached to bacterial cells. The binding of C1 to antibody activates the C1 protease. The multicolored C1 protease lies in a horizontal plane relative to the six C1 stalks (gray). The protease domains in the center of the complex (not shown) cleave other complement components to initiate the complement cascade. The result is formation of the membrane attack complex that kills the bacterial cell. *Source*: A. E. Phillips et al. 2009. *J. Immunol.* **182**:7708–7717 (cover photo). © 2009 by The American Association of Immunologists, Inc.

David Philip Vetter was born in 1971 without an immune system. Better known as the "bubble boy," he was the only human known to live in a plastic, germ-free bubble for his entire 12 years of life (1971–1984; **Fig. 24.1A**). His predicament stemmed from a genetic disease known as severe combined immunodeficiency (SCID). In the most serious form of this disease, the patient harbors no T cells and has dysfunctional (or sometimes no) B cells, so the body cannot launch a meaningful immune defense against any invading microbe—whether pathogen or normal microbiota. Because David's older brother had died of SCID before David was born, physicians were alerted that David might have the disorder too. Consequently, David was transferred to a sterile environment within seconds after birth to await a bone marrow transplant. Water, air, food, diapers, clothes—all were disinfected with special cleaning agents before entering his sterile plastic bubble. He was handled only through special plastic gloves attached to the wall. He lived for 12 years physically isolated in this plastic, sterile environment, venturing out only in a NASA-designed sterile space suit. A bone marrow transplant was attempted in 1984, but his defective immune system failed to protect him from Epstein-Barr virus undetected in the transplanted cells. He died just before his thirteenth birthday.

One in every million people develops SCID. The two most common forms of the disease result from genetic abnormalities linked to the X chromosome. Patients with X-linked SCID either lack an enzyme called adenosine deaminase (resulting in the toxic accumulation of deoxyguanosine triphosphate) or cannot produce an important cytokine receptor that T cells need for their development and to communicate with B cells. Cytokines were introduced in Section 23.2.

A dramatic therapeutic attempt to cure the disease in 2004 used stem cells. Stem cells are undifferentiated, but are capable of changing into many different cell types, including B cells and T cells (see Section 23.3). A team of scientists headed by Dr. Adrian Thrasher of University College London removed stem cells from the bone marrow of four SCID children who lacked the gene for the cytokine receptor. They inserted the gene encoding the normal cytokine receptor into a severely defective leukemia virus that can infect cells without causing disease. The debilitated leukemia virus served as a gene vector to deliver the normal cytokine receptor gene into the patient's stem cells. Once the genetic material entered the nucleus, the healthy copy of the gene began to function and the corrected stem cells were reintroduced into the patients. After this gene replacement therapy, all four children started making T cells with the correct receptor and produced functional B cells. The patients developed an immune system, were discharged, and are now living at home.

SCID dramatically illustrates the importance of our immune system and how fragile its development is. All it takes is a small defect in a single gene to subvert the entire process.

Chapter 24 begins by describing the two types of adaptive immunity and the factors that influence the immunogenicity of foreign proteins, lipids, and so on. We will explore how B cells ultimately differentiate into plasma cells and make antibodies (defining what is called humoral, or circulating, immunity) and how certain T-cell lymphocytes develop to directly kill infected host cells in what is called cell-mediated or cellular immunity. You will also learn that another type of T cell controls the balance between humoral and cell-mediated responses to a given infection. Ultimately, you will appreciate that adaptive immunity is a major reason the human race still exists.

24.1 Adaptive Immunity

As Chapter 23 points out, the immune system has both nonadaptive and adaptive mechanisms. Nonadaptive (innate, or nonspecific) immune mechanisms are present from birth. Adaptive immunity, in contrast, develops as the need arises. For instance, adaptive immunity against malaria is not developed until the individual has encountered the plasmodial parasite that causes the disease. The **adaptive immune response** is a complex, interconnected, and cross-regulated defense network.

A. **B.**

Courtesy of Texas Children's Hospital

Courtesy of UCL News

Figure 24.1 **Living without an immune system.** **A.** David Vetter, the bubble boy, inside his environmental bubble. The tube behind him is a port that was used to introduce sterile food, clothes, etc. **B.** Adrian J. Thrasher, University College London, conducted a successful gene therapy trial on children with severe combined immunodeficiency (SCID).

Two types of adaptive immunity are recognized: humoral immunity and cell-mediated immunity. In **humoral immunity**, **antibodies** are produced that directly target microbial invaders. The term "humoral" means "related to body fluids." Thus, antibodies are proteins that circulate in the bloodstream and recognize foreign structures called antigens. An **antigen** (also called an **immunogen**) is any molecule that will, when introduced into a person, elicit the synthesis of antibodies that specifically bind the antigen. Antigens stimulate B cells (B lymphocytes) to differentiate into antibody-producing cells. **Cell-mediated immunity**, the second type of adaptive immunity, employs teams of T cells (T lymphocytes) that recognize antigens and then destroy host cells infected by the microbe possessing the antigen. In truth, the humoral and cellular immune responses are intertwined, each relying on some facet of the other to work efficiently. T cells serve a central role in adaptive immunity by determining whether humoral or cell-mediated mechanisms predominate in response to a specific antigen.

Adaptive immunity develops over a 3- to 4-day period after exposure to an invading microbe. The immune system does not recognize the *whole* microbe, but innumerable tiny *pieces* of it. Each small segment of an antigen that is capable of eliciting an immune response is called an **antigenic determinant** or **epitope**. Many single-protein antigens are recognized when the larger antigen is broken into smaller segments upon being phagocytosed. Even distinct tertiary (three-dimensional) shapes within a protein may be counted as antigenic determinants if they produce a specific response. This happens when two or more stretches of amino acids are far removed from each other in a protein's primary sequence yet protein folding aligns them side by side in three-dimensional space (**Fig. 24.2**). Such a three-dimensional structure may be recognized by the immune system as a single entity or antigen. Besides proteins, other structures in the cell, such as complex polysaccharides, can have linear and three-dimensional epitopes. So the immune response to a microbe is really a composite of responses to different epitopes by thousands of individual B cells, which are the cells that produce antibodies. The response to each individual epitope is **clonal**; that is, it gives rise to a population of cells that originate from a single cell. This means that each clone of immune host cells will target a unique epitope.

As mentioned previously, the humoral response starts when an antigen triggers the differentiation of B cells into antibody-producing cells. In contrast, the cell-mediated immune response occurs when certain types of T cells become activated by microbial antigens placed on infected host cell surfaces or on the surface of phagocytic cells that have engulfed the microbe. In cellular immunity, the activated T cell can directly kill the infected host cell in an attempt to kill the invading microbe. In

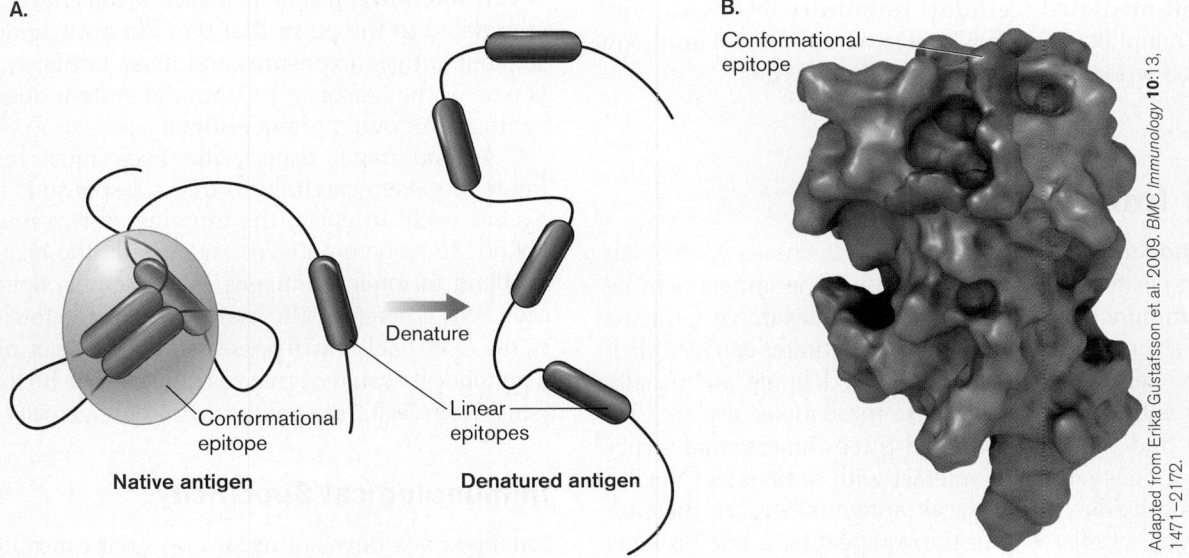

Adapted from Erika Gustafsson et al. 2009. *BMC Immunology* **10**:13, 1471–2172.

Figure 24.2 Antigens and epitopes. A. Native proteins fold into a three-dimensional shape, where several regions separated in the linear sequence can reside next to each other to form a conformational epitope. Denaturing the protein with the detergent sodium dodecyl sulfate (SDS) and reducing agents like dithiothreitol (to remove disulfide bonds) will unfold the protein and separate the various amino acid stretches that formed the conformational epitope. **B.** A three-dimensional protein structure showing, in red, four amino acids that form a conformational epitope.

addition, T cells synthesize soluble growth factors called cytokines (introduced in Section 23.2) that incite nearby macrophages to indiscriminately attack cells in the local area. Cytokines are discussed in Section 24.6. Thus, cellular immunity, in general, is critical for dealing with intracellular pathogens such as viruses, whereas humoral immunity is most effective against extracellular bacterial pathogens like *Streptococcus pneumoniae*, one cause of pneumonia.

As we proceed through the chapter, we will reveal how the immune system functions in layers, with each layer building on the previous one. We will periodically return to a particular aspect of the immune response—for example, B-cell differentiation into plasma cells—to integrate seemingly distinct parts of the immune system into a unified concept of immunity.

> **THOUGHT QUESTION 24.1**   Two different stretches of amino acids in a single protein form a three-dimensional antigenic determinant. Will the specific immune response to that three-dimensional antigen also respond to one of the two amino acid stretches alone?

TO SUMMARIZE:

- **An antigen** can elicit an antibody response. An antigen is usually made of many different epitopes (antigenic determinants), each of which binds to a different, specific antibody.
- **Humoral immunity** against infection is the result of antibody production originated by B cells.
- **Cell-mediated (cellular) immunity** involves a type of lymphocyte called T cells, which control antibody production and can directly kill host cells.

24.2 Immunogenicity

Immunogenicity measures the effectiveness by which an antigen elicits an immune response. One antigen can be more immunogenic than another. For example, proteins are the strongest antigens, but carbohydrates can also elicit immune reactions. Nucleic acids and lipids are usually weaker antigens, in part because these molecules are very flexible and present a variable three-dimensional structure that does not easily interact with antibodies. Nucleic acids and lipids are also weak antigens because they are both made of relatively uniform repeating units. Proteins are more effective antigens for three reasons: They form a variety of shapes, they maintain their tertiary structure, and they are made of many different amino acids that can be assembled in many different combinations. These features provide stronger interactions with antibodies in the

bloodstream and enable better recognition by lymphocytes, the cellular workhorses of the immune system.

Several other factors contribute to the immunogenicity of proteins (see **eTopic 24.1**). For example, the larger the antigen, the more likely it is that phagocytic cells will "see" and engulf it. This is important because for an immune response to occur, phagocytic cells such as macrophages and dendritic cells must first engulf large antigens and degrade them, presenting the epitopes on their cell surface. Phagocytic cells that degrade larger antigens and thus expose antigenic determinants are called **antigen-presenting cells (APCs)**. Many types of cells can be antigen-presenting. They include "professional" APCs, such as macrophages (monocytes), mast cells, or dendritic cells; and "nonprofessional" APCs (for example, fibroblasts) under the right circumstances. The immune response begins when a B cell binds to a foreign peptide and a T lymphocyte binds to the same foreign peptide displayed on the surface of an antigen-presenting cell.

Presentation of antigens on APCs requires that the antigen be placed on a membrane surface protein structure called the **major histocompatibility complex**, or **MHC** (discussed later). The more tightly an antigen can bind to these MHC surface proteins, the more immunogenic it is. The stronger the binding, the easier it is for T cells to recognize the complex.

Each specific antigen shows a different **threshold dose** needed to generate an optimal response. A dose higher or lower than that threshold will not generate as strong an immune response. Lower doses activate only a few B cells. Exceedingly high doses of antigen can cause **B-cell tolerance**, a state in which B cells have been overstimulated to the point that they do not respond to subsequent antigen exposures and make antibody. Tolerance is part of the reason your immune system does not react against your own protein antigens.

As you might expect, the body must regulate the immune system carefully so that a response is not leveled against itself. In effect, the immune system must become "blind" to its own antigens; as a result, the host will often be blind to foreign antigens that resemble epitopes of its own cells. Therefore, the more complex the foreign protein is, the more likely it will possess antigenic determinants that a lymphocyte can recognize as nonself. The further an antigen is from "self," the greater its immunogenicity will be.

Immunological Specificity

Smallpox is a devastating disease that caused enormous suffering and killed millions of people in the seventeenth and eighteenth centuries (**Fig. 24.3**). There was no cure, and the only available preventive treatment was to take dried material from lesions of a previous smallpox sufferer, place it on a healthy person, and hope the person

A. Smallpox patient

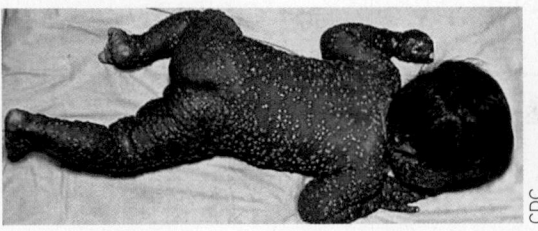

Figure 24.3   Immunological specificity is the basis of vaccination.   A. Photo of a smallpox patient, showing the white pox pustules. **B.** The smallpox virus, variola major (300 nm long, TEM). The photo shows the dumbbell-shaped, membrane-enclosed nucleic acid core. **C.** The vaccinia virus that causes cowpox (360 nm long, electron micrograph). Edward Jenner recognized the similarity between the deadly smallpox and less severe cowpox diseases and used cowpox scrapings to vaccinate humans against smallpox. *Source for C:* M. Hollinstead et al. 1999. *J. Virol.* **73**:1503.

B. Variola major

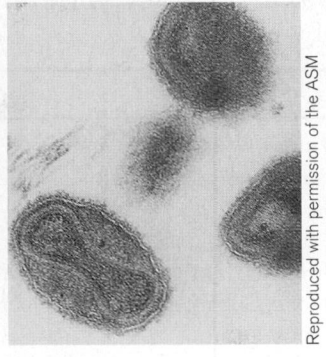

Reproduced with permission of the ASM MicrobeLibrary

C. Vaccinia

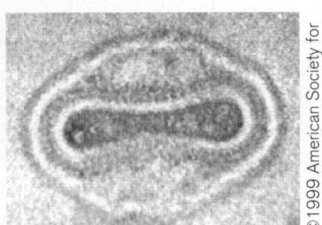

©1999 American Society for Microbiology

survived. Those who survived were protected from subsequent bouts of smallpox but were still susceptible to other diseases. This early observation gave rise to the idea of **immunological specificity**, which means that an immune response to one antigen is not effective against a different antigen. In other words, the immune response to smallpox will not protect someone against the plague bacillus (*Yersinia pestis*), which is antigenically different from the smallpox virus.

While immunological specificity is important, it is not absolute. As described in Chapter 1, an English country physician named Edward Jenner (see Fig. 1.18) in the late eighteenth century (long before viruses were discovered) learned to protect townsfolk from deadly smallpox disease by inoculating them with scrapings from lesions produced by a tamer disease, cowpox. By this process, Jenner unwittingly transferred the vaccinia virus (which we now know causes cowpox; see **Fig. 24.3C**) to villagers susceptible to smallpox (caused by the related but more dangerous variola virus). The resulting immune reaction to vaccinia produced an effective cross-protection against variola and thus prevented smallpox. This illustrates that an immune reaction against one organism or virus may be sufficient to protect against an antigenically related, if not identical, organism. The technique of exposing individuals to "tame" microbes to protect them against pathogens, now generally called **vaccination**, has been used to protect humans against many bacterial and viral pathogens (**Table 24.1**). Most vaccinations today involve administering crippled (attenuated) strains of the pathogenic microbe or inactivated microbial toxins (for example, diphtheria toxin).

Cross-protection, in which immunization against one microbe protects against a second, will work only if two proteins critical to the pathogenesis of the two different microorganisms share key antigenic determinants. No cross-protection occurs if these determinants differ significantly. A good example is the common cold, which is caused by hundreds of closely related rhinovirus strains (rhinitis, a runny nose, is one of the symptoms of this viral disease). Infection with one strain will not immunize the victim against a second strain. The reason is that the structures of the viral proteins used to attach to the ICAM-1 surface protein on host cells differ dramatically between different strains of rhinovirus (**Fig. 24.4A**). Antibodies called neutralizing antibodies, which bind to the attachment protein on one strain of rhinovirus, will prevent infection by the same virus strain (**Fig. 24.4B**) but will not bind a similar but antigenically distinct ICAM-1 receptor protein from a different strain. A key to one lock will not work on a different lock.

THOUGHT QUESTION 24.2   How does a neutralizing antibody that recognizes a viral coat protein prevent infection by the associated virus?

Antigens and Immunogens

A basic understanding of the ABO blood group system illustrates key concepts of immune specificity. Karl Landsteiner (1868–1943) discovered the ABO blood group system in 1901, for which he received the 1930 Nobel Prize in Physiology or Medicine (**Fig. 24.5A**). He observed that type A red blood cells (RBCs) remained intact when introduced into another type A individual, and type B RBCs were undamaged when introduced into a type B individual. However, RBCs from a type A individual, whose RBCs contained the A antigen, were destroyed (lysed) when transfused into a type B person, whose RBCs contained the B antigen (**Fig. 24.5B**). This is the phenomenon of blood group incompatibility, which occurs when specific antibodies in the serum of an individual of one blood type bind to antigens on red blood cells of a different type (foreign RBCs). We now know that type B individuals carry specific anti-A antibodies in their bloodstream, while type A individuals carry specific anti-B antibodies. (Type AB people carry neither

Table 24.1 Vaccines against viral and bacterial pathogens.

Disease	Vaccine	Vaccination recommended for:
Viral		
Chickenpox	Attenuated strain (will still replicate)	Children 12–18 months
Hepatitis A	Inactivated virus (will not replicate)	Travelers to endemic areas
Hepatitis B	Viral antigen	Medical personnel; children 1–18 months
Influenza	Inactivated virus or antigen	Adults over 65 years
Measles, mumps, rubella	Attenuated viruses; MMR combined vaccine	Children 15–19 months
Polio	Attenuated (oral, Sabin); inactivated (injection, Salk)	Children 2–3 years
Rabies	Inactivated virus	Persons in contact with wild animals
Yellow fever	Attenuated virus	Military personnel
Bacterial		
Anthrax	*Bacillus anthracis*, components of toxin; unencapsulated strain	Agricultural and veterinary personnel; key health care workers
Cholera	Killed *Vibrio cholerae,* toxin components	Travelers to endemic areas
Diphtheria	Toxoid (inactivated toxin)	Children 2–3 months
Pertussis	Acellular *Bordetella pertussis*	Children 2–3 months
Tetanus	Toxoid	Children 2–3 months
Haemophilus influenzae type b (meningitis)	Bacterial capsular polysaccharide	Children under 5 years
Lyme disease	*Borrelia burgdorferi*, lipoproteins OspA and OspC surface antigens	Canines, human vaccine discontinued
Meningococcal disease	*Neisseria meningitidis*, bacterial capsular polysaccharides	Military personnel; high-risk individuals
Pneumococcal pneumonia	*Streptococcus pneumoniae*, bacterial capsular polysaccharides	Adults over 50 years
Tuberculosis (*Mycobacterium tuberculosis*)	Attenuated *Mycobacterium bovis* (BCG vaccine)	Exposed individuals
Typhoid fever	Killed *Salmonella* Typhi	Individuals in endemic areas
Typhus fever	Killed *Rickettsia prowazekii*	Medical personnel in endemic areas; scientists

Figure 24.4 Antibodies prevent rhinovirus attachment to cell receptors. A. This figure illustrates the complexity of the rhinovirus capsid and shows the attachment of the virus to the cell surface molecule ICAM-1 (intercellular adhesion molecule, shown in reddish brown). (PDB code: 1rhi) **B.** This figure shows rhinovirus coated with protective (neutralizing) antibodies (green) that block the ICAM-1 receptors on the virus. As a result, the virus fails to attach to and infect the host cell. (PDB code: 1RVF)

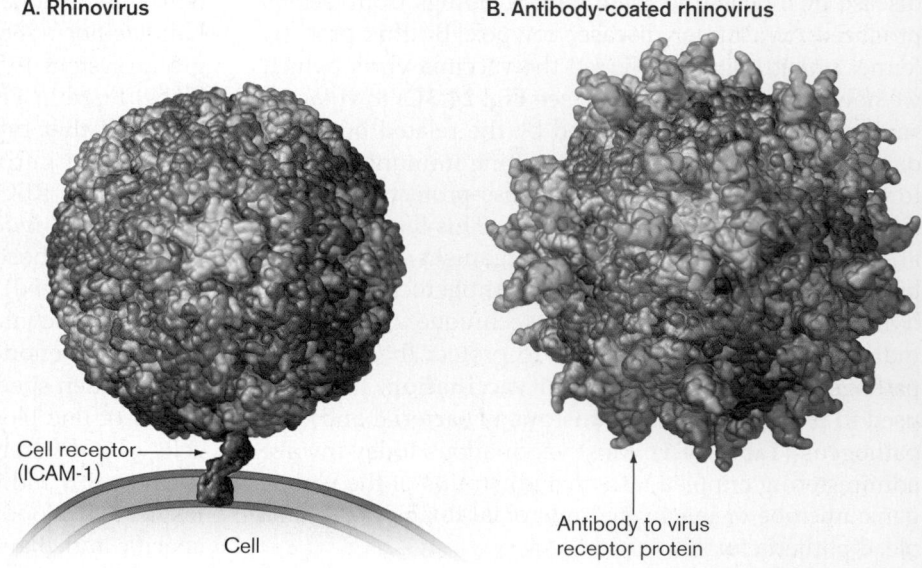

A. Rhinovirus

B. Antibody-coated rhinovirus

Cell receptor (ICAM-1)

Cell

Antibody to virus receptor protein

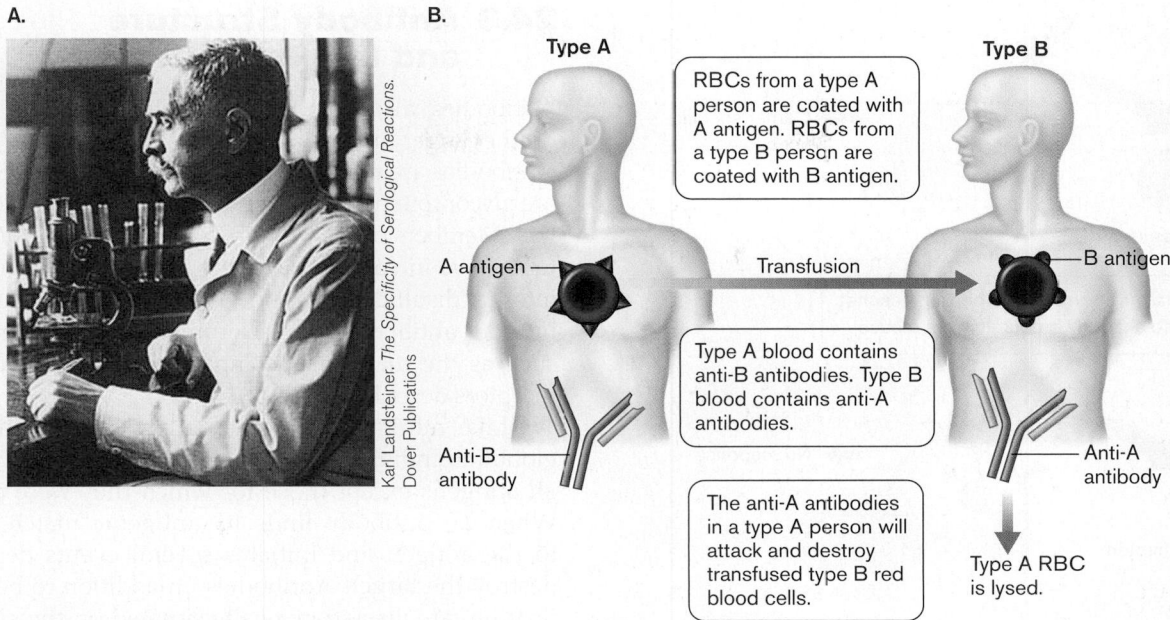

A. *Karl Landsteiner. The Specificity of Serological Reactions. Dover Publications.*

B.

Type A

RBCs from a type A person are coated with A antigen. RBCs from a type B person are coated with B antigen.

Type B

A antigen

Transfusion

B antigen

Type A blood contains anti-B antibodies. Type B blood contains anti-A antibodies.

Anti-B antibody

Anti-A antibody

The anti-A antibodies in a type A person will attack and destroy transfused type B red blood cells.

Type A RBC is lysed.

Figure 24.5   Karl Landsteiner and ABO blood groups.   A. Karl Landsteiner discovered ABO blood groups. **B.** Individuals with type A antigens on their red blood cells (RBCs) possess antibodies to B antigen, while type B individuals have B antigens on RBCs and carry anti-A antibodies. Transfusing RBCs from a type B individual into a type A individual will cause the anti-B antibodies to attack and destroy the type B red blood cells.

antibody; and type O individuals, whose RBCs contain neither A nor B antigen, carry both anti-A and anti-B antibodies.) The discovery by Landsteiner defined the basic concept of immunological specificity and explains why a type A person cannot donate blood to a type B individual, and vice versa. It also explains why a type O person (called a universal donor) can donate to type A, B, AB, or O individuals.

> **THOUGHT QUESTION 24.3**   Blood group type A individuals do not have B antigen, yet they make anti-B antibodies. If antibody production comes about in response to the presence of a specific antigen, what stimulates the production of anti-A or anti-B antibodies?

Antigens that, by themselves, can elicit antibody production are called **immunogens**. Landsteiner further defined the concept of immunological specificity using very simple molecules that were too small to elicit antibodies. Molecules of molecular weight less than 1,000 are generally not immunogenic (we now know this is because they do not bind MHC molecules). However, these small molecules, called **haptens** (from the Greek word meaning "to fasten" and the German word for "stuff"), will elicit production of specific antibodies if they are covalently attached to a larger carrier protein or other molecule (**Fig. 24.6**). Haptens can be thought

of as small, incomplete antigens. An example of a hapten is the antibiotic penicillin, a serious cause of immune hypersensitivity reactions in some individuals (see Section 24.8).

A protein carrier like bovine serum albumin (BSA) injected into a mouse would elicit antibodies that react against BSA (**Fig. 24.6**). Because BSA antigen elicits an immune response to itself, it is called an immunogen. In contrast, a mouse injected with the hapten benzene-sulfate fails to produce antibodies to the hapten. However, when the hapten is attached to BSA and the hapten-BSA complex is injected, the mouse produces antibodies that react to the carrier (BSA), as well as other antibodies that react against the benzene hapten. The reason for this carrier effect will be evident later as we describe how antigens are processed by cells of the immune system. Thus, antigens include immunogens that elicit an immune response by themselves, and haptens that must be attached to an immunogen in order to generate an immune response.

> **THOUGHT QUESTION 24.4**   The attachment proteins of different rhinovirus strains all bind to ICAM-1. How can all these proteins be immunologically different if they find the same target (ICAM-1)? Why won't antibodies directed against one rhinovirus strain block attachment of other rhinovirus strains?

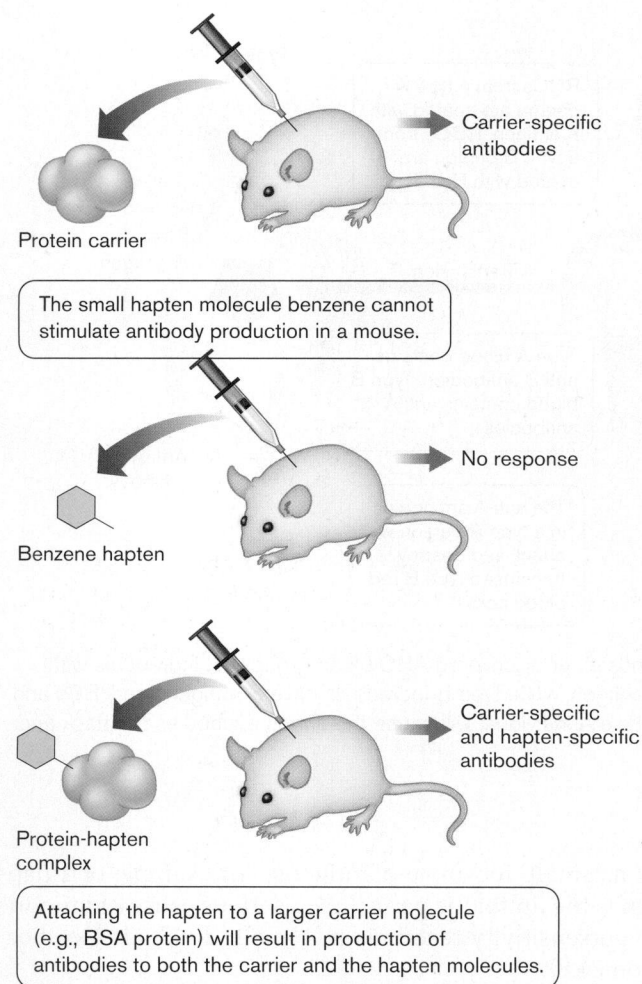

The small hapten molecule benzene cannot stimulate antibody production in a mouse.

Attaching the hapten to a larger carrier molecule (e.g., BSA protein) will result in production of antibodies to both the carrier and the hapten molecules.

Figure 24.6 Basic schematic showing how haptens can elicit antibody production.

TO SUMMARIZE:

■ **Proteins are better immunogens** than nucleic acids and lipids because proteins have more diverse chemical forms.

■ **Antigen-presenting cells** such as phagocytes, degrade microbial pathogens and present distinct pieces on their cell surface MHC proteins.

■ **Immunological specificity** means that antibody made to one epitope will not bind to different epitopes (although some weak cross-binding to similar epitopes can happen; for example, antibody to cowpox virus will bind to a similar epitope on smallpox virus).

■ **A hapten** is a small compound that must be conjugated to a larger carrier antigen to elicit production of an antibody.

24.3 Antibody Structure and Diversity

Antibodies, also called **immunoglobulins**, are members of the larger immunoglobulin superfamily of proteins. Antibodies are the keys to immunological specificity. They are glycoproteins made by the body in response to an antigen. Members of the immunoglobulin superfamily of proteins have in common a 110-amino-acid domain with an internal disulfide bond. The immunoglobulin superfamily includes antibodies and other important binding proteins, such as the major histocompatibility proteins and B-cell receptors described later.

Like miniature "smart bombs," antibody immunoglobulins individually circulate through blood, ignoring all antigens except those for which they were designed. When an antibody finds its antigenic match, it binds to the antigen and initiates several events designed to destroy the target. Antibodies, in addition to being free-floating, are also strategically situated on the surfaces of B cells, where they enable these lymphocytes to recognize specific antigens.

A typical antibody consists of four polypeptide chains. There are two large **heavy chains** and two smaller **light chains** (**Fig. 24.7**). The four polypeptides combine to form a Y-shaped tetrameric structure held together by disulfide bonds. Two bonds connect the two identical heavy chains to each other. One light chain is then attached near its carboxyl end to the middle of each heavy chain by a single disulfide bond. The antigen-binding sites are formed at the amino-terminal ends of the light and heavy chains. One antibody molecule possesses two identical antigen-binding sites, one on each "arm" of the molecule.

Antibody Structure Enables Immunoprecipitation

A typical antibody molecule has two antigen-binding sites, each of which bind identical antigens. Because of the ability to bind to more than one antigen molecule, antibodies can cross-link antigens in solution, ultimately forming complexes that are too large to remain soluble and thus fall out of solution (**Fig. 24.8**). The phenomenon, called **immunoprecipitation**, is normally observed only in vitro, where the concentration of antigen and antibody can be manipulated experimentally.

Immunoprecipitation occurs only with appropriate ratios of antigen and antibody molecules. Too many antigen molecules (antigen excess; **Fig. 24.8A**) or too few antigen molecules (antibody excess; **Fig. 24.8B**) result in complexes too small to immunoprecipitate. Large complexes are formed only at an appropriate antigen/antibody ratio called **equivalence** (**Fig. 24.8C**). Equivalence is the point where the number of antigenic sites is roughly equal to the number of antigen-binding sites.

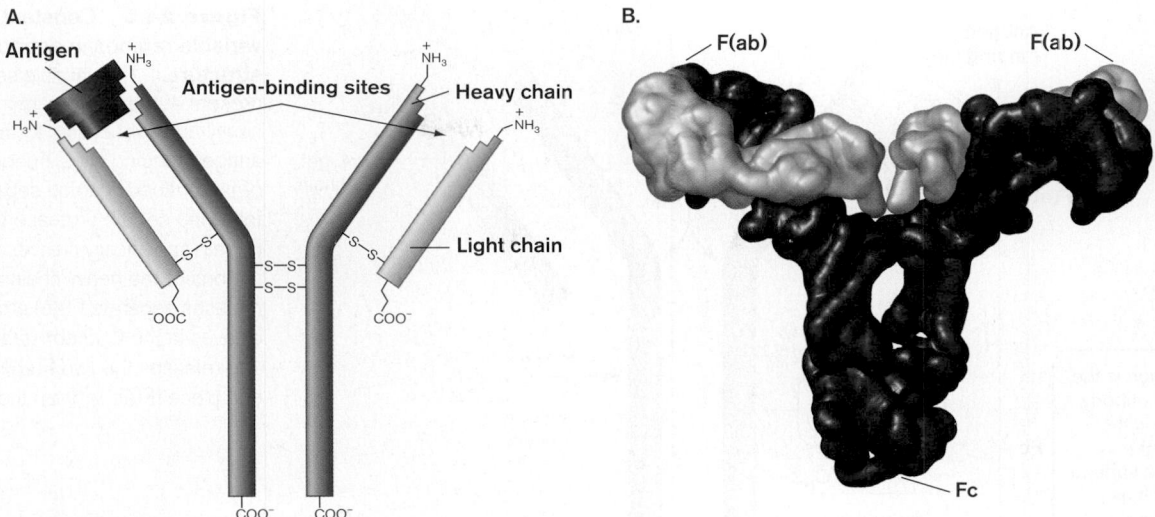

Figure 24.7   Basic antibody structure.   A. Each antibody contains two heavy chains and two smaller, light chains held together by disulfide bonds. The Y-shaped structure contains two antigen-binding sites, one at each arm of the molecule. The two antigen-binding sites are formed by the amino-terminal regions of the heavy- and light-chain pairs. **B.** Three-dimensional structure of an antibody. The heavy chains are shown in blue, the light chains in yellow. The F(ab) regions represent the antigen-binding sites. The Fc portion points downward and is used to attach the antibody to different cell surface molecules. (PDB code: 1R70)

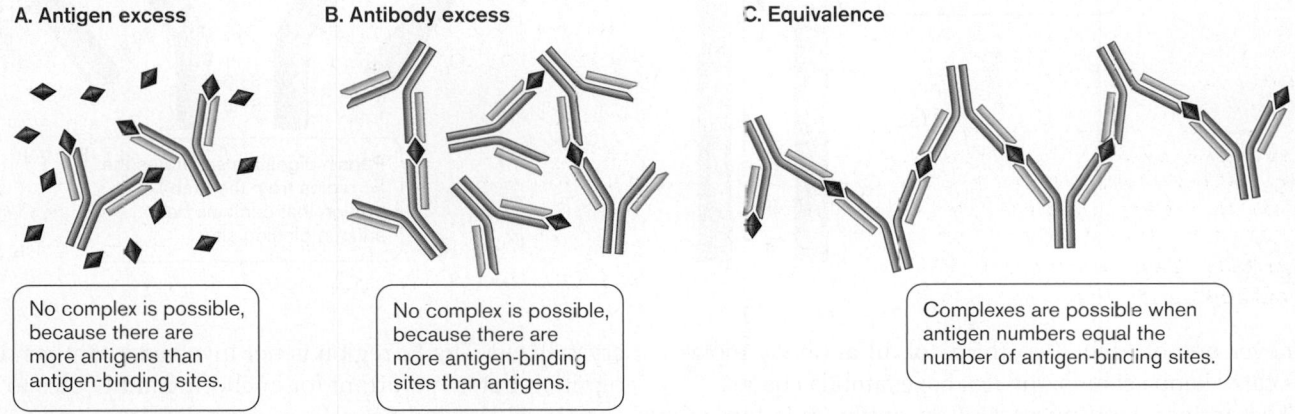

Figure 24.8   Basis of immunoprecipitation.   Only when the numbers of epitopes and antigen-binding sites are roughly equivalent will a large complex form and fall out of solution.

Immunoprecipitation is the basis for many experimental immunological techniques. For example, the concentration of an antigen can be determined in vitro, or antibodies can be used to identify and remove specific antigens from a complex mixture because the specific antigen-antibody complex falls out of solution. **eTopic 24.2** illustrates some experimental uses for immunoprecipitation, such as radial immunodiffusion and Western blotting.

Antibodies Have Constant and Variable Regions

There are five classes of antibodies, defined by five different types of heavy chains called alpha (α), mu (μ),

gamma (γ), delta (δ), and epsilon (ε). The heavy-chain classes are distinguished one from another by regions of highly conserved amino acid sequences, known as **constant regions** (denoted C_H or C_L for heavy and light chains, respectively; **Fig. 24.9**). Antibodies containing gamma heavy chains are called IgG; those with alpha, delta, mu, and epsilon heavy chains are called IgA, IgD, IgM, and IgE, respectively. Each antibody class serves a specific purpose in the immune system.

In contrast to heavy chains, there are only two classes of light chains—kappa (κ) and lambda (λ)—which are defined by their own constant regions (C_L). A single antibody of any heavy-chain class (for example, IgG) may contain two kappa light chains or two lambda light chains,

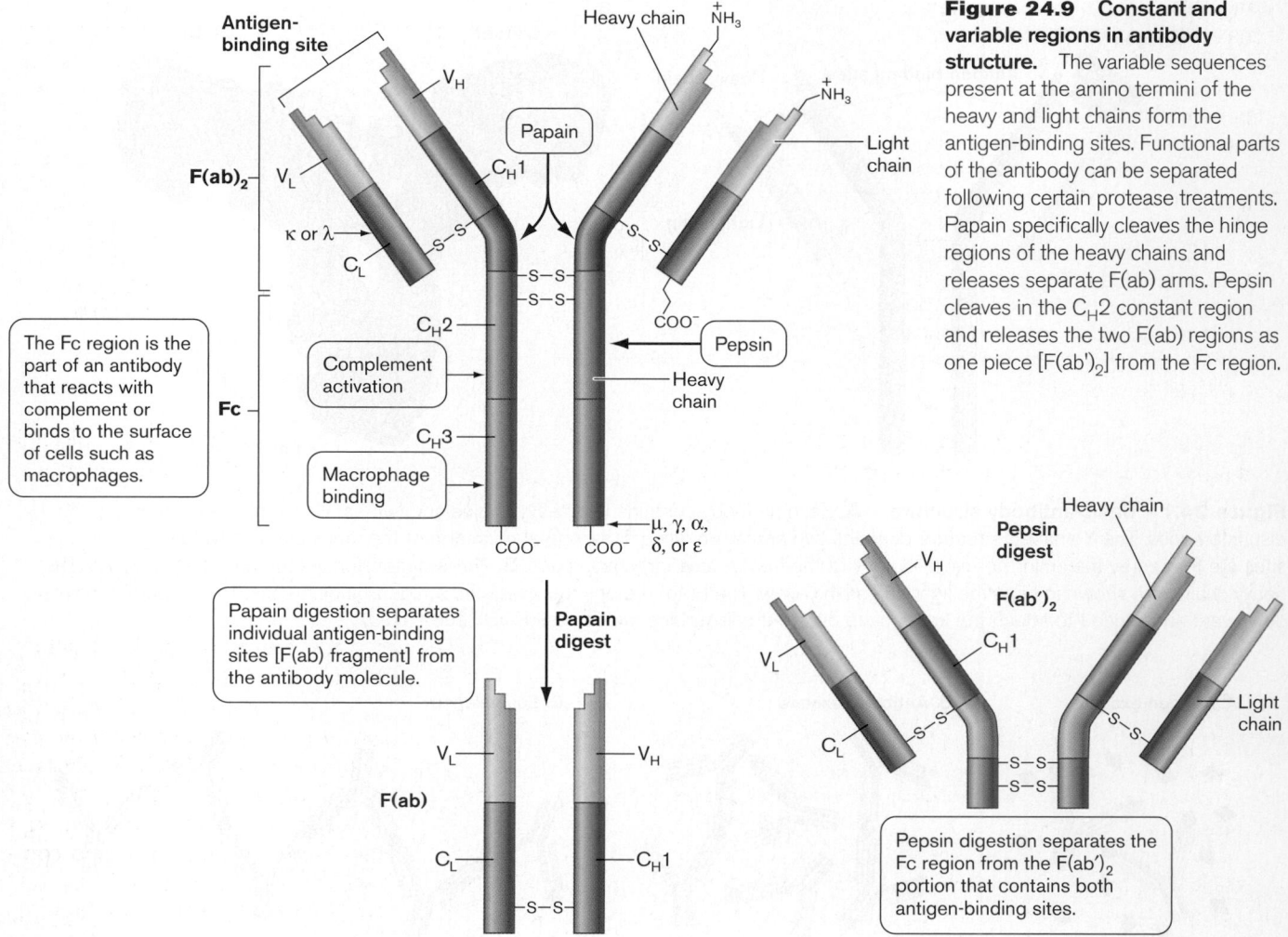

Figure 24.9 Constant and variable regions in antibody structure. The variable sequences present at the amino termini of the heavy and light chains form the antigen-binding sites. Functional parts of the antibody can be separated following certain protease treatments. Papain specifically cleaves the hinge regions of the heavy chains and releases separate F(ab) arms. Pepsin cleaves in the C_H2 constant region and releases the two F(ab) regions as one piece [F(ab')$_2$] from the Fc region.

The Fc region is the part of an antibody that reacts with complement or binds to the surface of cells such as macrophages.

Papain digestion separates individual antigen-binding sites [F(ab) fragment] from the antibody molecule.

Pepsin digestion separates the Fc region from the F(ab')$_2$ portion that contains both antigen-binding sites.

but never one of each. Two-thirds of all antibody molecules carry kappa chains; the rest have lambda chains.

The antigen-binding part of an antibody is formed by highly variable amino acid sequences situated at the amino-terminal ends of the light and heavy chains. These **variable regions** are referred to as the V_L and V_H regions (**Fig. 24.9**). The rest of each immunoglobulin chain is composed of the highly conserved constant regions. C_H1, C_H2, and C_H3 denote the three different constant regions in each heavy chain. Section 24.5 discusses the genes that code for these regions and the combinatorial possibilities underlying the formation of different antibody molecules.

In addition to the two "arms" that bind antigen, every antibody contains a "tail" that serves a different function. Intact antibodies can be dissected into their two functional parts through digestion with the protease papain. This endoprotease cleaves the molecule at the hinge region, which releases the tail of the antibody (called the **Fc region**) from the antigen-binding portion of the molecule (called the **F(ab')$_2$ region**; see **Fig. 24.9**). The "c" in Fc refers to the ease with which this fragment can be crystallized. The Fc region is not involved in antigen recognition but is important for anchoring antibodies to the surface of certain host cells and for binding components of the complement system.

Isotypes, Allotypes, and Idiotypes

As noted earlier, antibodies are classified according to variations in amino acid sequences in regions of the light and heavy chains. Certain differences in the constant region give rise to **isotypes**, which are common to all members of a particular species. IgG, IgM, IgA, IgD, and IgE are antibody isotypes (**Fig. 24.10**). The IgG isotype is identical among humans but is different from the IgG isotype of monkeys.

Within an isotype there may be "allotypic" differences, amino acid differences in the constant region that are shared by some, but not all, members of a species. As shown in **Figure 24.10**, allotypic differences in an amino acid sequence usually occur in the constant regions of light chains. For example, John possesses circulating IgG

Figure 24.10 Isotype, idiotype, and allotype differences on antibodies. Colors indicate different amino acid sequences (or epitopes). Isotype differences occur in heavy-chain constant regions within the same individual (any IgE *vs.* any IgG from Sherrie in this case). Idiotype differences occur in the antigen-binding regions in antibodies directed against different antigens (differences in Sherrie's IgG molecules directed against antigen A *vs.* antigen B). Allotype differences occur in the light or heavy chain constant regions of different people (e.g. differences in anti-B IgG antibodies taken from John compared to Sherrie).

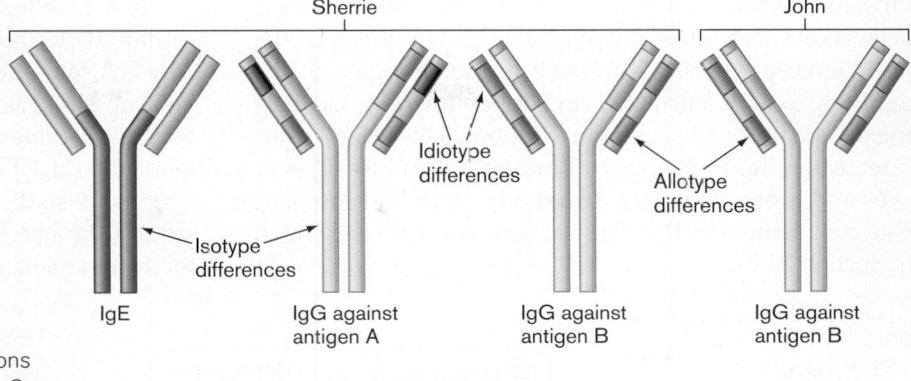

molecules that have allotypic differences from the IgG antibodies circulating in Sherrie. (John and Sherrie have different IgG **allotypes**.)

On another level, alterations in *hyper*variable regions within a single antibody class in a single person are referred to as "idiotypic" differences. These differences occur in the antigen-binding sites of the various antibodies of a single individual. Thus, the IgG molecules in Sherrie that bind to epitope A possess idiotypic differences from Sherrie's IgG molecules that bind to epitope B. In sum, the human IgG isotype is composed of different allotypes (constant-region differences between individuals), and each allotype is composed of different **idiotypes** (variable-region differences within an individual).

Note that antibodies are proteins and, as such, are themselves antigens. Thus, the amino acid differences found within a single class of antibody—IgG, for example—also represent different antigenic *epitopes* within that class. Isotypic, allotypic, and idiotypic differences, discussed above, define epitopes in antibody molecules that can be detected using immunological techniques as described in **eTopic 24.2**.

Antibody Isotype Functions and "Super" Structures

All antibody isotypes have the same basic structure. However, each isotype has a unique "super" structure (for example, monomer, dimer), and each is designed to carry out a different task. Some key properties of the five different immunoglobulin classes are listed in **Table 24.2**. **IgG** is the simplest and most abundant antibody in blood and tissue fluids. It is made as a monomer but has four subclasses. Each subclass varies in its amino acid composition and the number of interchain cross-links. IgG molecules carry out several missions for the immune system. First, they bind and **opsonize** microbes; that is, they make microbes more susceptible to phagocytes.

Table 24.2 Properties of human immunoglobulins.

Property	IgG				IgA		IgM	IgD	IgE
	IgG1	IgG2	IgG3	IgG4	IgA1	IgA2			
Mol. wt. (kDa)	146	146	165	146	160		970	184	188
% Carbohydrate		3			7–11		9–12	9–11	12
Serum half-life (days)	21	20	7	21	6		10	3	2
% Total serum Ig		70%			15–20%		5–10%	0.2%	0.002%
Avg. concentration (mg/ml)	9	3	1	0.5	3	0.5	1.5	0.03	5×10^{-5}
Ag-binding sites		2			2–4		5–10	2	2
Heavy chain	$\gamma 1$	$\gamma 2$	$\gamma 3$	$\gamma 4$	$\alpha 1$	$\alpha 2$	μ	δ	ε
Light chain		$\kappa\lambda$			$\kappa\lambda$		$\kappa\lambda$	$\kappa\lambda$	$\kappa\lambda$
Produced by fetus		Poor, if at all			Poor, if at all		Yes	?	Poor, if at all
Transmitted across placenta		Yes			No		No	No	No
Bind complement		Yes			No		Yes	No	No
Opsonizing		Yes			No		No	No	No
Bind mast cells		No			No		No	No	Yes

Opsonizing IgG antibodies coat the microbe with their Fc portions protruding outward. Phagocytes possess surface Fc receptors that can attach to the Fc region of the antibody to gain a firmer "grip" on the microbe, facilitating phagocytosis (discussed in Section 23.6). IgG can also directly neutralize viruses by binding to virus attachment sites and is one of only two antibody types that can activate complement by the classical pathway (to be described in Section 24.7).

IgA is secreted across mucosal surfaces and is most commonly found as a dimer (**Fig. 24.11A**). This explains why IgA can bind four molecules of antigen (each monomer can bind two molecules of antigen). The components of the IgA dimer are linked by disulfide bonds to a protein called the J chain, which joins two IgAs by their Fc regions. A sixth protein, the secretory piece, is wrapped around the IgA dimer during the secretion process. The secreted molecule, now called sIgA (secretory IgA), is

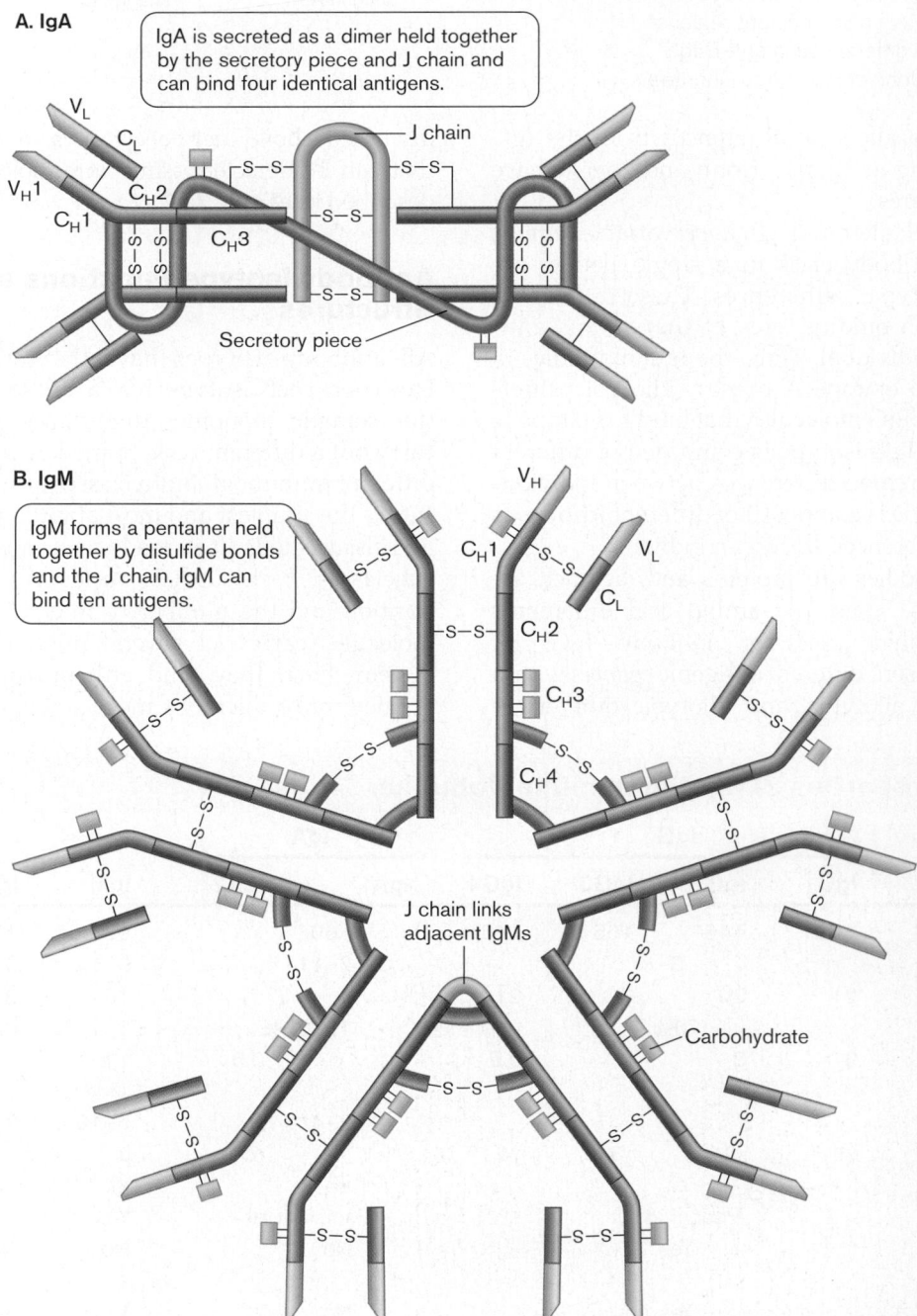

Figure 24.11 Structures of IgA and IgM. The antibodies are made as multimers of two (IgA) or five (IgM) immunoglobulin molecules.

found in tears, breast milk, and saliva and on other mucosal surfaces. The molecule sIgA is important for mucosal immunity against pathogens.

Circulating **IgM** is a huge, Ferris wheel–shaped molecule formed from five monomeric immunoglobulins tethered together by the J-chain protein (**Fig. 24.11B**). It can also be found in monomeric form on the surfaces of B cells, where it forms part of the B-cell receptor. IgM is the first antibody isotype detected during the early stages of an immune response. Unlike the smaller IgG immunoglobulins, IgM is so large that it cannot cross the placenta (see **Table 24.2**).

Two other antibody isotypes are present at very low levels in the blood. **IgD** is a monomer that can neither bind complement nor cross the placenta. IgD molecules are, however, abundantly found on the surface of B cells. Attached to the cell surface by its Fc region, IgD, along with monomeric IgM, can bind antigen and signal B cells to differentiate and make antibody.

IgE is also present in trace amounts in the blood, but it is found more prominently bound to the surface of mast cells and basophils, where it has potent biological activity. Mast cells and basophils contain granules loaded with inflammatory mediators. The primary role of IgE is to amplify the body's response to invaders. Once secreted into serum, IgE attaches to mast cells (**Fig. 24.12A and B**), again by way of its Fc region, and, like a Venus flytrap, waits until its matched antigen binds to its antigen-binding site. When surface IgE molecules of two mast cells are cross-linked by antigen, a signal is sent internally that triggers degranulation (see Section 24.8). The subsequent release of histamine and other pharmacological mediators from these granules helps orchestrate the acute inflammation that takes place during early host responses to microbial infection (that is, while the

antibody response is gearing up). The system also causes severe allergic hypersensitivities, such as anaphylaxis, and milder forms like hay fever (**Fig. 24.12C**).

TO SUMMARIZE:

- **Antibodies, or immunoglobulins**, are members of the immunoglobulin superfamily.
- **Antibodies are Y-shaped** molecules that contain two heavy chains and two light chains.
- **There are five classes (isotypes) of heavy chains.** Each antibody isotype is defined by the structure of the heavy chain.
- **Each antibody molecule contains two antigen-binding sites.** Each binding site is formed by the hypervariable ends of a heavy- and light-chain pair.
- **The Fc portion** of an antibody can bind to specific receptors on host cells. This binding is antigen independent.

24.4 Primary and Secondary Antibody Responses

After a lag period of several days following primary immunization or natural infection, antibodies begin to appear in the **serum** (the fluid that remains after blood clots). During the lag period, a series of molecular and cellular events causes a distinct subset of B cells to proliferate and differentiate into antibody-secreting **plasma cells** and **memory B cells**. Each B cell is genetically programmed to make antibodies to one antigen or epitope—a process called the **primary antibody response**, the events of which will be discussed later. A secondary exposure to the antigen, which can take place months or years after

A. Mast cell (SEM)   **B. Mast cell (TEM)**   **C. Hay fever**

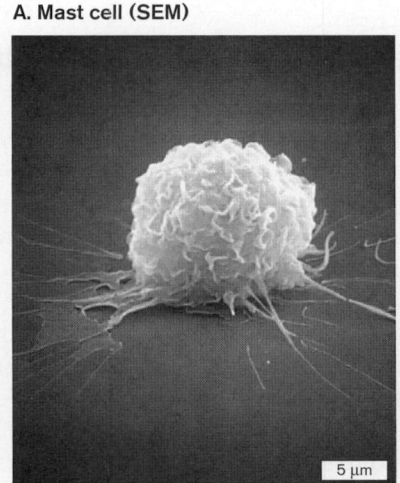
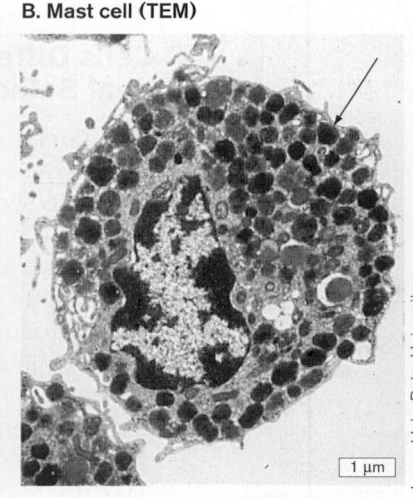
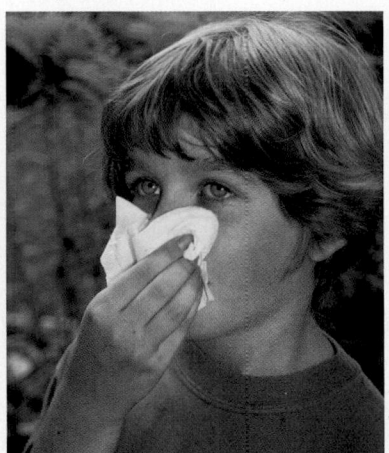

Laura Hale, Duke University   Laura Hale, Duke University   Mark Clarke

Figure 24.12   Mast cells.   A. Scanning EM of a mast cell. **B.** Electron transmission EM showing granules (arrow). **C.** Hay fever is the result of degranulation of IgE-coated mast cells, which release histamine and other pharmacological mediators.

the initial encounter, will trigger a rapid, almost instantaneous increase in the production of antibodies and is called the **secondary antibody response** (**Fig. 24.13**). This quick response occurs thanks to the memory B cells formed during the primary response. Once stimulated, memory B cells rapidly differentiate into plasma cells and secrete antibody. Plasma cells are much larger than B cells because of an enormous increase in protein synthesis and secretion machinery.

The net result of the primary antibody response is the early synthesis and secretion of IgM molecules specifically directed against the antigen, also called the immunogen. Later, during the primary response, a process known as **isotype switching** (**class switching**) occurs, by which the predominant antibody type produced becomes IgG rather than IgM (discussed shortly). Antibodies made during this primary phase, while specific for the immunogen, are actually *not* of the highest affinity. Mechanisms to increase antibody affinity occur later.

As the immunogen is cleared from the body, the levels of both IgG and IgM decline because the plasma cells that produced them die. Plasma cells have an average life span of only 100 days. However, the immune system has been primed to respond more aggressively to the immunogen should it encounter it again. This is because the memory B cells produced as a result of the primary response react to antigen more quickly and take less time to make antibody than naive B cells that had no prior exposure to the antigen.

B cells are maintained in the body because, unlike plasma cells, a subpopulation of B cells maintained in the lymph nodes continues to divide. If a person later encounters the same antigen, memory B cells quickly proliferate and differentiate into plasma cells with no lag phase. Thus, memory B cells quickly initiate the secondary antibody response (or anamnestic response, from the Greek *anamnesis*, meaning "remembrance"). Memory B cells comprise approximately 40% of the circulating B-cell population. During the secondary response, copious amounts of IgG antibody are secreted from plasma cells made from memory B cells that have undergone isotype switching. These antibodies have a higher specificity for the antigen than do the antibodies produced during the primary response (see "Making Memory B Cells" in Section 24.5). Small amounts of IgM are also produced from the few memory cells that did not undergo isotype switching during the primary response. (Most memory B cells undergo isotype switching and produce IgG rather than IgM, but a few do not.)

The speedy production of antibody during the secondary response is the basis of immunization. However, before a secondary response is possible, the host must undergo a primary response following initial exposure to a pathogen. But pathogens can do considerable harm during the lag phase of the primary response. To avoid this harm, an innocuous version of a pathogen, or a harmless piece of it, can be injected into a person to trigger the primary response without producing disease (or, at worst, producing only a mild form of the illness). Immunization thus primes the immune system to respond efficiently and without delay upon encountering the real pathogen. **Table 24.1** lists a variety of viral and bacterial diseases for which immunizations are available.

> **THOUGHT QUESTION 24.5** The mother of a newborn was found to be infected with rubella, a viral disease. Infection of the fetus could lead to serious consequences for the newborn. How could you determine whether the newborn was infected while in utero?

B Cells Differentiate into Plasma Cells by Clonal Selection

As mentioned earlier, each B cell circulating throughout the body or ensconced in a lymphoid organ is programmed to synthesize antibody that reacts with a *single* epitope. In a process called **clonal selection**, an invading antigen will inadvertently select which B-cell clone will proliferate to large numbers and differentiate into antibody-producing plasma cells or memory B cells. In this way, large amounts of antibody specific for the antigen are made. The mechanism of clonal selection begins with the antigen binding to a matching B cell (a B cell preprogrammed to bind to that antigen) (**Fig. 24.14**).

Mature naive B cells (those that have not previously encountered antigen) can produce only IgM and IgD,

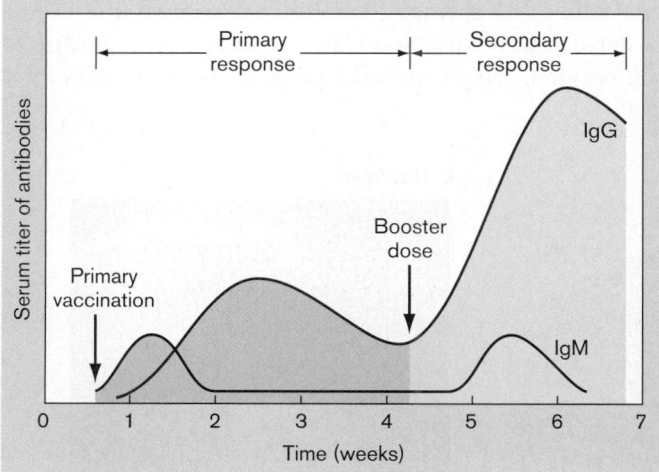

Figure 24.13 Primary versus secondary antibody response. Primary vaccination or infection leads to the early synthesis of IgM followed by IgG. Reinfection or a second, booster dose of a vaccine results in a more rapid antibody response consisting mainly of IgG due to memory B cells formed during the primary response. Note that the time course and level of Ab made vary with the immunogen and the host.

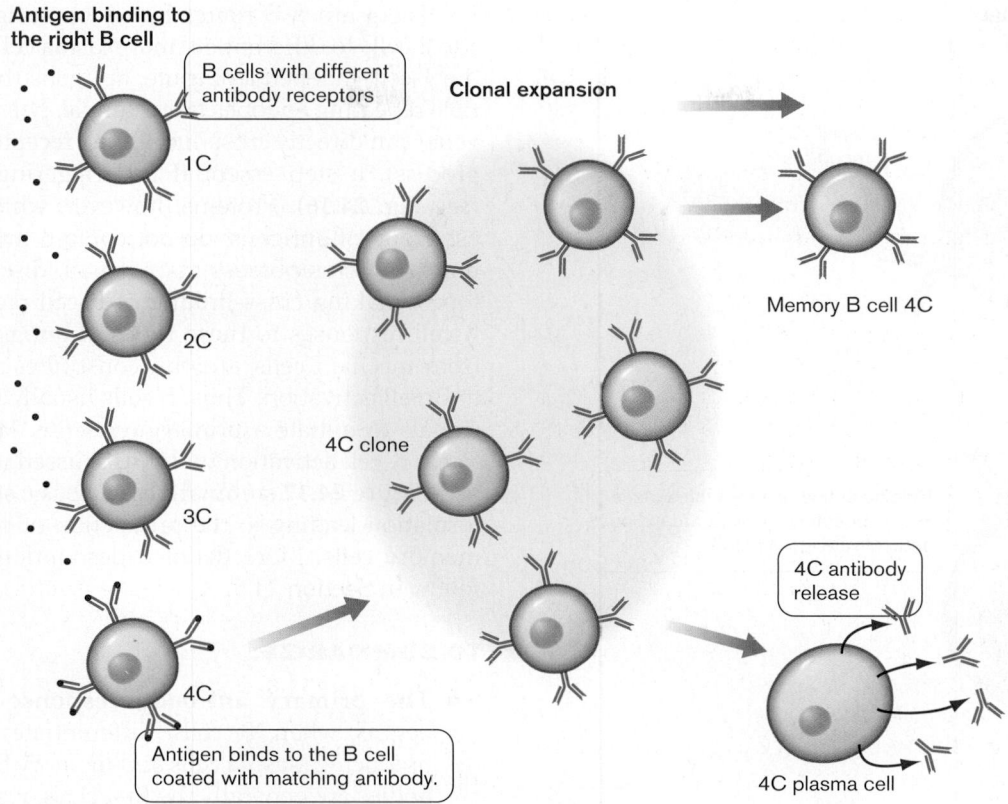

Figure 24.14   Clonal selection theory.   The B-cell population is composed of individuals that have specificity for different antigens. When a B cell contacts its cognate antigen, an intracellular signal is generated, leading to proliferation and differentiation of that clone (clonal expansion). Plasma cells and memory B cells result.

which have identical antigen specificities. These two antibody classes are displayed like tiny satellite dishes on the B-cell surface, anchored by their Fc regions through hydrophobic transmembrane segments. These surface antibodies are the keys to stimulating the proliferation and differentiation of B cells into antibody-producing plasma cells or memory B cells. Upon binding to its corresponding antigen via these surface antibodies, the B cell is said to become **activated**, whereby it multiplies and differentiates into a plasma cell that ultimately synthesizes only one antibody isotype (for example, IgG1 or IgA2). Clonal selection has begun. (Remember that some of the activated B cells become memory B cells and do not become antibody-secreting plasma cells.)

Each antibody bound to a B-cell membrane is associated with two other membrane proteins, called Igα and Igβ (these are not immunoglobulins but are designated Ig because they *associate* with the surface antibody). The complex is called the **B-cell receptor (Fig. 24.15)**. Each B cell may have upwards of 50,000 B-cell receptors. When B cells bind antigen, they often, but not always, differentiate into plasma cells (some become memory cells).

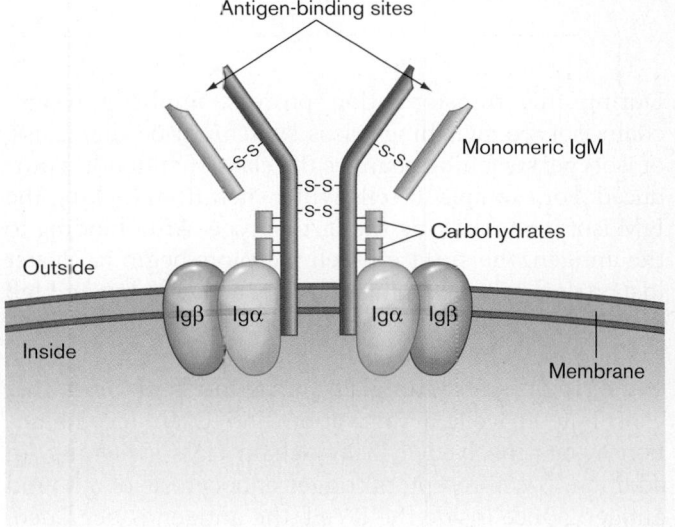

Figure 24.15   B-cell receptor.   The B-cell receptor is formed as a complex between a monomeric IgM and Igα and Igβ in the membrane. Igα and Igβ are not immunoglobulins.

Mature B cell

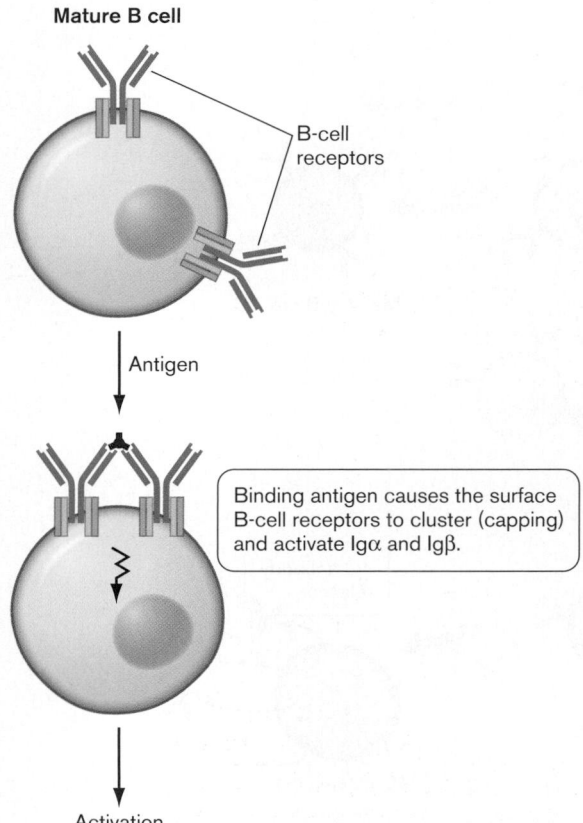

Binding antigen causes the surface B-cell receptors to cluster (capping) and activate Igα and Igβ.

Figure 24.16 Capping and activation of the B cell. The capping process initiates a signal cascade that activates differentiation and proliferation of the B cell independent of T-cell help. Antigens with repeating epitopes, such as polysaccharides, can directly cross-link B-cell receptors.

During this transformation process, antibody heavy-chain isotype switching (class switching) occurs. Class, or isotype, switching changes the class of antibodies produced. For example, B cells will switch from making the IgM isotype to making the IgA isotype. After binding to the antigen, the surface B-cell receptors begin to cluster in a process called capping. Capping activates Igα and Igβ to initiate a phosphorylation signal cascade directed into the nucleus (**Fig. 24.16**). The transcription factors at the end of the cascade stimulate transcription of genes that contribute to cellular activation and DNA recombination events involved in heavy-chain class switching. In addition, B-cell receptors trigger endocytosis of a bound antigen. Once inside the B cell, the antigen is degraded, processed, and repositioned on the B-cell membrane—a step that will ultimately enable T cells to directly bind and communicate with the B cell, giving the B cell "permission" to become an antibody-secreting plasma cell. How these antigens are placed on the B-cell membrane will be discussed in Section 24.6.

There are two routes by which antigens can stimulate B cells to differentiate into plasma cells. In one, called the T cell–independent route, antigens that possess multiple repeating epitopes (for example, polysaccharide antigens) can directly cross-link B-cell receptors (the capping process), a step essential for triggering differentiation (see **Fig. 24.16**). Proteins, however, which are the largest group of antigens, do not contain multiple repeating units. Proteins possess many small, discrete, single epitopes, making cross-linking of B-cell receptors difficult. B-cell responses to these types of antigens require help from specific T cells, and this constitutes the second route to B-cell activation. Thus, B cells usually require multiple signals to initiate a primary response. How T cells help foster B-cell activation will be discussed in Section 24.6.

Figure 24.17 summarizes the basic steps of antibody formation leading to the production of plasma cells and memory cells. More detailed descriptions of the events follow in Section 24.5.

TO SUMMARIZE:

- **The primary antibody response** to an antigen begins when B cells differentiate into antibody-producing plasma cells and memory B cells. IgM antibodies are generally the first class of antibodies made during the primary response.
- **Isotype switching** occurs during the primary response when a subpopulation of B cells switches during differentiation from making IgM to making other antibody isotypes.
- **The secondary antibody response** occurs during subsequent exposures to an antigen and arises because memory B cells are activated. IgG is the predominant antibody made.
- **Clonal selection** is the rapid proliferation of a subset of B cells during the primary or secondary antibody response.
- **A B-cell receptor** consists of a membrane-embedded antibody in association with the Igα and Igβ proteins. Binding of antigen to the B-cell receptor triggers B-cell proliferation and differentiation.

24.5 Genetics of Antibody Production

It is estimated that each human can synthesize 10^{11} different antibodies. Given that each B cell displays antibodies to only one antigenic determinant, it follows that there are 10^{11} different B cells in the body. We have learned, however, that each person possesses only about 1,000 genes or gene segments involved in antibody formation. How are 10^{11} different antibodies made from only 10^3

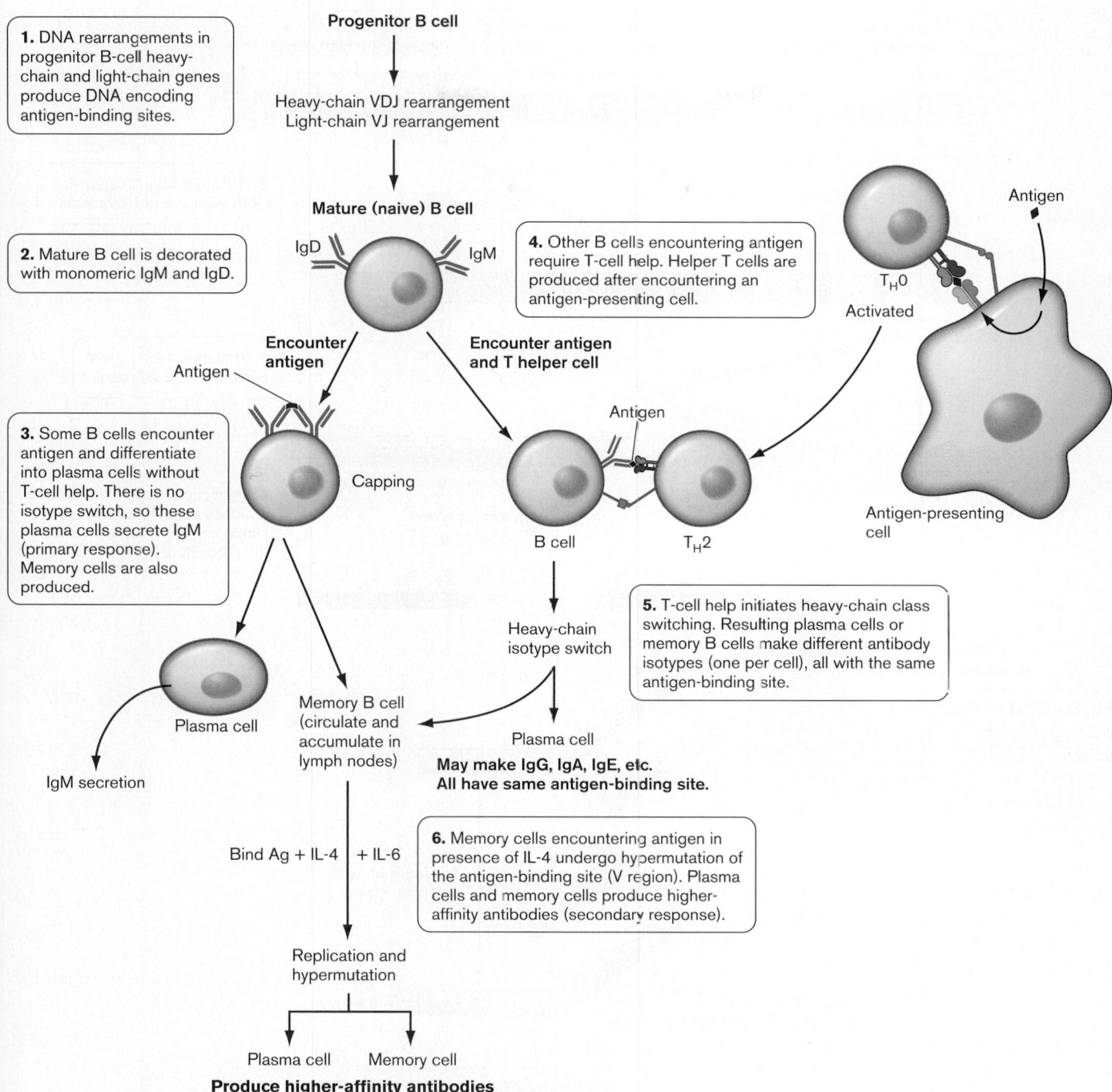

1. DNA rearrangements in progenitor B-cell heavy-chain and light-chain genes produce DNA encoding antigen-binding sites.

Progenitor B cell

Heavy-chain VDJ rearrangement
Light-chain VJ rearrangement

Mature (naive) B cell

2. Mature B cell is decorated with monomeric IgM and IgD.

IgD   IgM

4. Other B cells encountering antigen require T-cell help. Helper T cells are produced after encountering an antigen-presenting cell.

Antigen

T_H0

Activated

Encounter antigen

Antigen

Encounter antigen and T helper cell

3. Some B cells encounter antigen and differentiate into plasma cells without T-cell help. There is no isotype switch, so these plasma cells secrete IgM (primary response). Memory cells are also produced.

Capping

Antigen

Antigen-presenting cell

B cell T_H2

5. T-cell help initiates heavy-chain class switching. Resulting plasma cells or memory B cells make different antibody isotypes (one per cell), all with the same antigen-binding site.

Heavy-chain isotype switch

Plasma cell

Memory B cell (circulate and accumulate in lymph nodes)

Plasma cell

IgM secretion

May make IgG, IgA, IgE, etc. All have same antigen-binding site.

Bind Ag + IL-4 | + IL-6

6. Memory cells encountering antigen in presence of IL-4 undergo hypermutation of the antigen-binding site (V region). Plasma cells and memory cells produce higher-affinity antibodies (secondary response).

Replication and hypermutation

Plasma cell   Memory cell

Produce higher-affinity antibodies

Figure 24.17   Steps in antibody formation.

genes? Susumu Tonegawa was awarded the 1987 Nobel Prize in Physiology or Medicine for discovering that antibody genes can move and rearrange themselves within the genome of a differentiating cell. Three steps are involved: rearrangement of antibody gene segments (or cassettes), the random introduction of somatic mutations, and the generation of different codons during antibody gene splicing. In humans, antibody diversity is generated constantly over a lifetime.

DNA Rearrangements and Hypermutation Generate Antibody Diversity

The first step in making a specific antibody occurs during the formation of a B cell from a progenitor stem cell (progenitor B cell) in bone marrow. Immunoglobulin genes in a bone marrow progenitor B cell consist of many gene segments that can rearrange in many possible combinations. During differentiation of a progenitor cell into

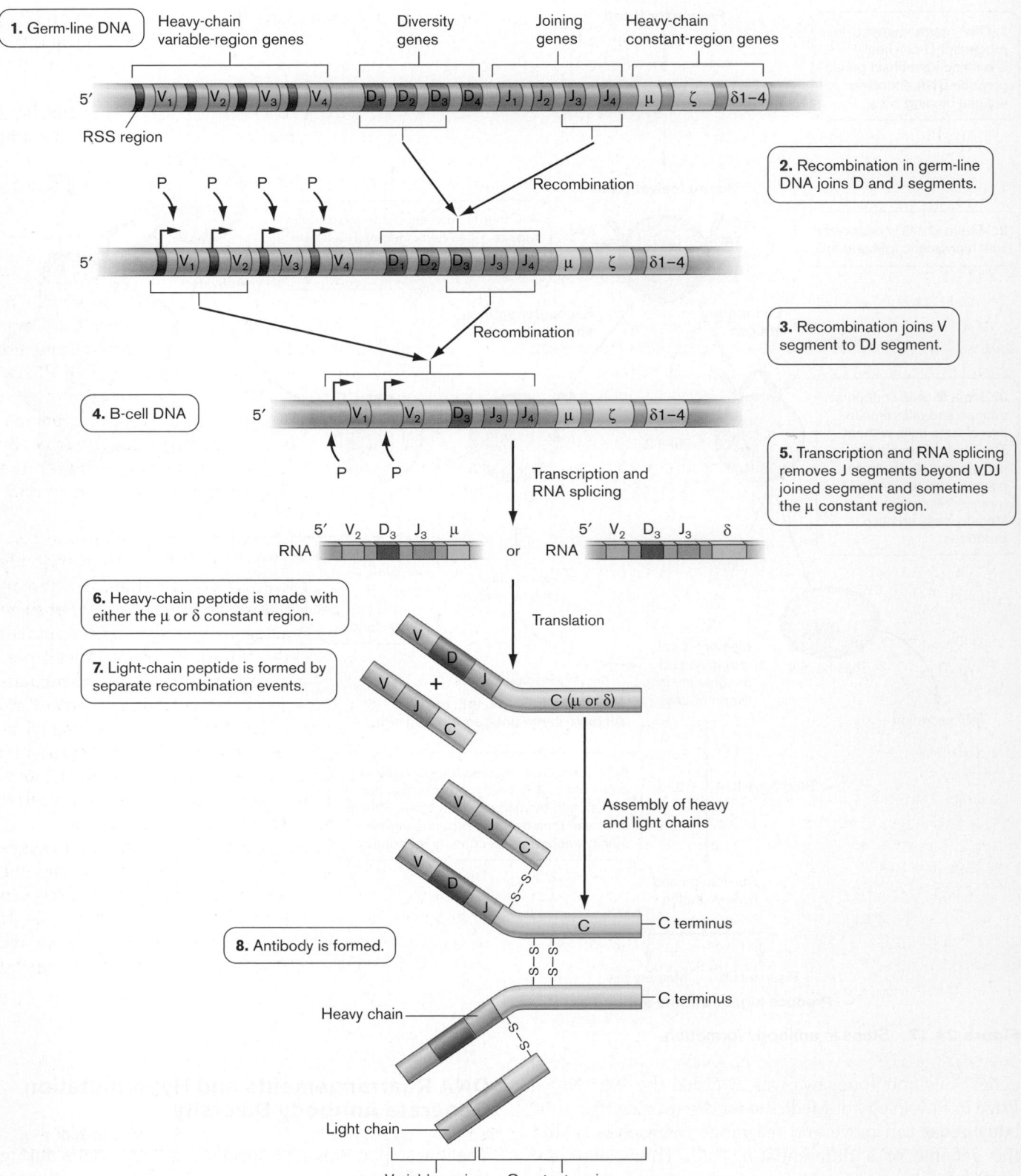

Figure 24.18 Formation of the VDJ regions of heavy chains. Note that only a small subset of the V, D, and J genes listed in Table 24.3 is actually shown in this model. RSS = recombination signal sequences.

a mature B cell, DNA segments are deleted in a process called **gene switching**, which decreases the number of gene segments in the mature B-cell DNA. The process starts at the 5′ end of an immunoglobulin gene cluster, which corresponds to the variable (V) end of the ultimate peptide (**Fig. 24.18**). The 5′ end encodes the N-terminal end of the protein that ultimately binds antigen (the antigen-binding site).

In both the heavy- and light-chain genes are a number of tandem gene cassettes encoding potential variable regions separated by **recombination signal sequences (RSS)**. These sequences allow recombination to bring two widely separated gene segments together. There are approximately 170 V gene segments for the heavy and light chains. The light-chain variable-region gene cluster lies upstream of a cluster of J (joint or joining)-region genes, which are ultimately used to join the variable region to the light-chain constant regions (C), whose genes reside farther downstream in the DNA. The arrangement of the heavy-chain genes is slightly more complex. In this case, the heavy-chain V cluster is followed by a D (diversity) cluster and then the J region.

> **NOTE:** Do not confuse the J-region gene segments used to make heavy- and light-chain proteins with the J-chain protein that holds together IgM and IgA multimers. They are completely different and unrelated. The J-region gene segments do not encode the J chain.

The processes leading to antibody formation in B cells are summarized in **Figure 24.18**. Antibody formation begins with a recombination event between RSS sites at one D and one J segment, which deletes all the intervening D and J segments. This is followed by the joining of the new DJ region to one of the V segments, deleting all of the intervening V and D segments. The result is a joined VDJ DNA sequence, at which point transcription can occur. Each V segment has its own promoter. However, if extra V segments remain upstream of the rearranged VDJ sequence on the DNA, the only promoter that will fire is the one immediately upstream of the VDJ rearranged

segment. The primary RNA transcript will then undergo RNA splicing to remove any J-segment RNA sequences that remain downstream of the VDJ RNA sequence. The result is a mature (naive) B cell.

A similar sequence of events occurs for the light chains, except that the product is VJ. **Table 24.3** illustrates the amount of antibody diversity in humans that can be achieved simply by this combinatorial re-joining (a total of about 2.5×10^6 antigens can be recognized).

Where does the rest of the diversity arise? After recombination, the V regions of the germ lines are susceptible to high levels of somatic mutation, resulting in the hypervariable regions. Hypermutation happens every time a memory B cell is exposed to the antigen: The memory B cells divide and the hypervariable regions mutate. Additional diversity comes from the junctions of VJ and VDJ, where recombinational splicing can occur between different nucleotides. Each gene splice event can generate additional codons, so the resulting peptides will differ by one or more amino acids. The interactions between the light- and heavy-chain hypervariable regions in an antibody form the antigen-binding sites.

In sum, a combination of gene splicing and random mutations gives us the remarkable level of antibody diversity we each possess. As noted earlier, the human body is capable of responding to 10^{11} antigens; yet there are only 10^8 estimated antigens in nature. This apparent overkill suggests that the immune system is well prepared to cope with any possible antigen it could encounter. Unfortunately for humans, enterprising microbes, such as the trypanosomes that cause sleeping sickness, can stay one step ahead of the immune system by changing the structure of key surface antigens. Changing the antigenic structure of a protein renders useless those antibodies made to the previous structure.

It isn't hard to understand why multicellular organisms, like humans, need the capacity to make any one of billions of different antibodies quickly. Pathogens can undergo many generations of growth within the single life span of a human host. So humans have to generate recombinant clones of cells quickly to overcome rapidly dividing pathogens.

Table 24.3 Antibody diversity attributed to combinatorial joining in the human germ line.

Chain type	Number of:			Number of combinations
	V regions	D regions	J regions	
λ Light chains	30	0	4	$30 \times 4 = 120$
κ Light chains	40	0	5	$40 \times 5 = 200$
Heavy chains	100	27	6	$100 \times 27 \times 6 = 16{,}200$
Number of possible antibodies	16,200 heavy-chain combinations $\times$ 120 λ-chain combinations $= 1.94 \times 10^6$			
	16,200 heavy-chain combinations $\times$ 200 κ-chain combinations $= 3.24 \times 10^6$			
	$1.94 \times 10^6 + 3.24 \times 10^6 = 5.18 \times 10^6$ combinations			

Isotype Class Switching Is the Result of DNA Recombination

We just described how a progenitor B cell becomes a mature (naive) B cell (see **Fig. 24.17**). The mature B cell (a B cell that has not yet "seen" the antigen but has already assembled its immunoglobulin VDJ-binding site) produces both IgM and IgD B-cell receptors (the naive B cell is referred to as IgM$^+$ IgD$^+$). The next step in B-cell differentiation occurs during the primary immune response.

The primary immune response is launched when a mature (naive) B-cell receptor finds its matched antigen. As mentioned, in the early stages of the response plasma cells produced from these B cells secrete only IgM, and an antigen-activated mature B cell will make IgM and IgD (IgM$^+$ IgD$^+$). As indicated in the summary diagram shown in **Fig. 24.17**, if this activated B cell also receives appropriate signals from a certain type of T cell known as a **helper T cell (T$_H$ cell)**, further immunoglobulin isotype switching will occur; the switched B cell may then make IgG, IgA, or IgE each with the same antigen recognition domain (variable region). The type of switch is influenced by small peptides called **cytokines** that are secreted by helper T cells.

How does the switch occur? Notice in **Figure 24.19** that the constant-region gene segments defining different antibody heavy-chain classes are arranged in tandem *after* a VDJ region. The mechanism by which a B cell switches to make IgG (C regions gamma 1–4), IgE (C-epsilon), or IgA (C-alpha 1 and 2) is very similar to VDJ formation. Each constant segment, except delta, contains a repeating DNA base sequence called a **switch region**. Recombination between these switch regions will delete the intervening DNA between the VDJ region and one of the constant regions. The primary RNA transcript produced after this recombination event has all of its introns spliced out before translation. Because the VDJ region is the same

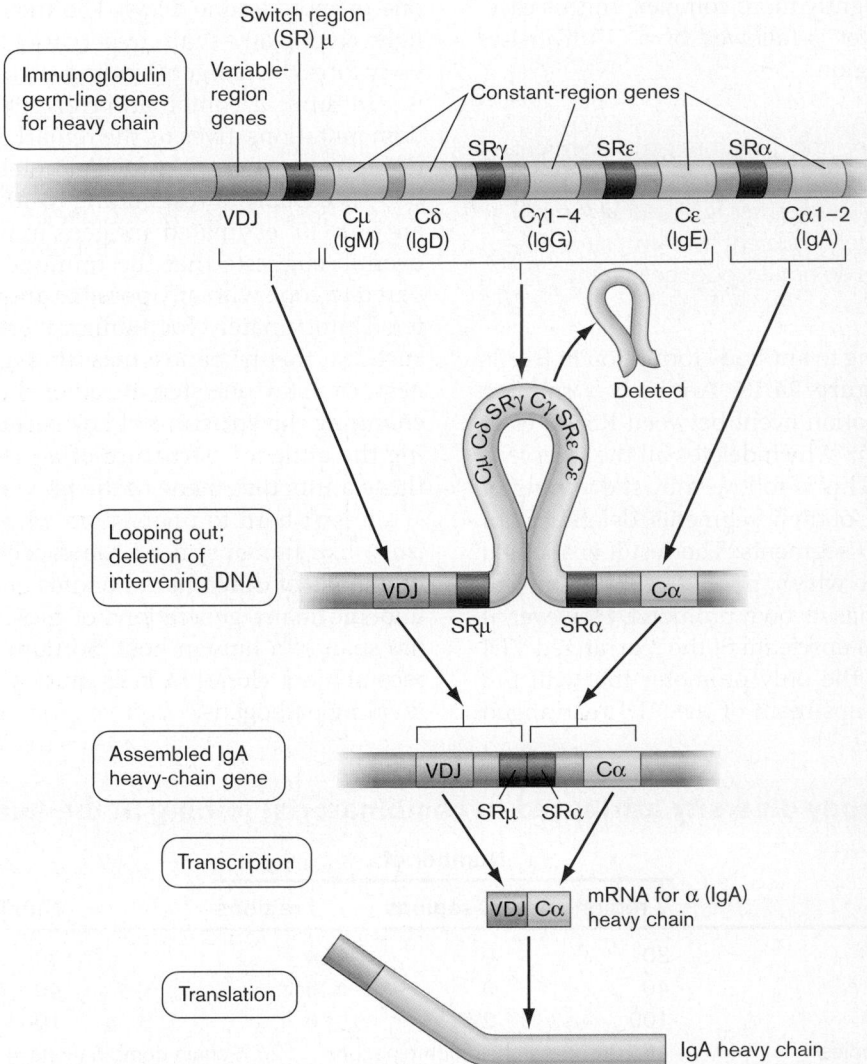

Figure 24.19 Heavy-chain class switching. As a B cell becomes activated, a switch in antibody isotype will occur. The switch involves recombination between isotype cassettes that brings one heavy-chain constant region (Cα in the example, for IgA) in tandem with a VDJ sequence.

regardless of which C_H gene is selected, whatever antibody is produced will have the same antigenic specificity as the original IgM. However, the heavy-chain switch selection process is *not* random. The type of cytokine present at the time of the switch will influence which C_H gene is selected.

Note that any heavy-chain peptide can combine with any light chain. However, a single, mature B cell can make only one type of heavy chain and one type of light chain.

THOUGHT QUESTION 24.6 B cells in early stages have both IgM and IgD surface antibodies, but the delta region has no switch region. Why does the delta region have no switch region?

THOUGHT QUESTION 24.7 Why do individuals with type A blood have anti-B and not anti-A antibodies?

Making Memory B Cells

As illustrated in **Figure 24.14**, an antigen-activated mature B cell may become a memory B cell. Memory B cells are B cells that have already undergone class switching (that is, they are committed to making IgG) but are very long-lived. Normally, chromosome ends (called telomeres) in every cell progressively shorten with each round of DNA replication. When they become too short, the cell dies. Production of memory B cells is associated with an increase in telomerase activity. Recall that telomerase is a reverse transcriptase that shares common ancestry with the reverse transcriptase of the virus HIV. Telomerase activity maintains the length of the telomeres and prolongs the life of the cell.

Memory B cells hypermutate the VDJ and VJ regions of antibody genes (as already described). These mutations increase the affinity of the antibody produced during the secondary response. Hypermutation helps fine-tune the immune response, making it even more specific toward a given microbe. Note, too, that as an immune response progresses, antigen becomes scarce. This means that only B cells with the highest-affinity antibodies as part of their B-cell receptors will remain activated. The process is known as affinity maturation.

THOUGHT QUESTION 24.8 Why do immunizations lose their effectiveness over time?

TO SUMMARIZE:

- **Progenitor B cells** can make any antibody isotype but are programmed to bind only one epitope.
- **Mature (naive) B cells** have undergone VDJ rearrangement and make IgM and IgD surface antibodies (part of B-cell receptors).
- **Experienced B cells** have completed isotype switching, which occurs after naive B cells bind their target

antigen. The B-cell receptors are made of a single antibody isotype that specifically binds one epitope.
- **During the primary antibody response**, IgM is the predominant class of antibody secreted.
- **Memory B cells** rapidly proliferate and differentiate into antibody-secreting plasma cells upon a second encounter with an antigen/epitope. Most antibody made during this secondary response is IgG.
- **Antibody diversity (idiotype)** occurs via a complex series of splicing events between adjacent DNA cassettes, as well as mutational events in DNA sequences encoding the hypervariable regions of heavy and light chains.

24.6 T Cells, Histocompatibility, and Antigen Processing

In this section we describe how antigens are recognized by T cells and how T cells then influence both humoral and cellular immunities. In the process, we will expand on how B cells become activated once they bind antigen, and we'll begin to explain why a host usually does not react against itself.

The T Cell Links Humoral and Cell-Mediated Immunity

Recall that there are two general types of immune response: humoral (antibody-based) immunity and cell-mediated immunity. Different types of infection tilt the immune response toward one or the other. B cells are clearly linked to the humoral immune system because they ultimately become plasma cells that make antibodies. T cells, on the other hand, serve as a nexus, or link, between humoral and cell-mediated responses. They play integral roles in both antibody production and cell-mediated immunity. Although derived from the same progenitor stem cell as B cells, T cells develop in the thymus (rather than in the bone marrow, where B cells develop) and they contain surface antigens different from those of B cells. In general, T cells can be divided into two broad groups: helper T cells (already mentioned) and **cytotoxic T cells (T_C cells)**. The two T-cell groups are differentiated by the type of cellular differentiation (CD) proteins present on their cell surfaces (**Table 24.4**). Helper T cells display the surface antigen CD4, while cytotoxic T cells display CD8.

Helper T cells come in three types: precursor T_H0 cells that can differentiate into the other two types; T_H1 cells, which assist in the activation of cytotoxic T cells; and T_H2 cells, which stimulate B-cell differentiation into plasma cells. Cytotoxic T (T_C) cells are the "enforcers" of the cell-mediated immune response. They destroy the membranes of host cells infected with viruses or bacteria. The ratio between the number of T_H1 cells and T_H2 cells produced during an infection affects whether the

Table 24.4 Major classes of T cells.

T-cell type	Coreceptor	MHC restriction	Cytokines produced	Major function
T_H0 cell Helper T cell	CD4	Class II	Wide range	Differentiate into T_H1 or T_H2
T_H1	CD4	Class II	IFN-γ, IL-2, TNF-β	Activate cytotoxic T cells
T_H2	CD4	Class II	IL-4, IL-5, IL-6	B-cell helper
Cytotoxic T cell (Tc)	CD8	Class I	IFN-γ, TNF	Kill virus-infected and cancer cells

immune response will lean more toward humoral (antibody, T_H2) or cell-mediated (T_H1) immune mechanisms. As will be described, different cytokines produced during infection influence whether an activated T_H0 cell develops into a T_H1 or T_H2 cell.

Why are T_H and T_C cells different? The reasons are many, but it all starts with their relative abilities to recognize two different protein complexes that are present on antigen-presenting cells. These complexes are called the major histocompatibility proteins.

The Major Histocompatibility Complex Is Critical to the Immune System

The factors affecting immune responsiveness are many, but the major histocompatibility complex (MHC)

proteins encoded by a set of genes on a single chromosome play a particularly important role. MHC proteins differ between species and among individuals within a species. They help determine whether a given antigen is recognized as coming from the host (a self antigen) or from another source (a foreign antigen), in a phenomenon called histocompatibility (hence the name "major histocompatibility complex"). There are two classes of MHC molecules found on cell surfaces. Both classes belong to the immunoglobulin family of proteins, but they are not immunoglobulins. Class I MHC receptors are found on all nucleated cells, while class II MHC molecules have a more limited distribution, principally on antigen-presenting cells (**Fig. 24.20A**). They are always expressed by the antigen-presenting dendritic cells and B cells, for example, but can also be made by

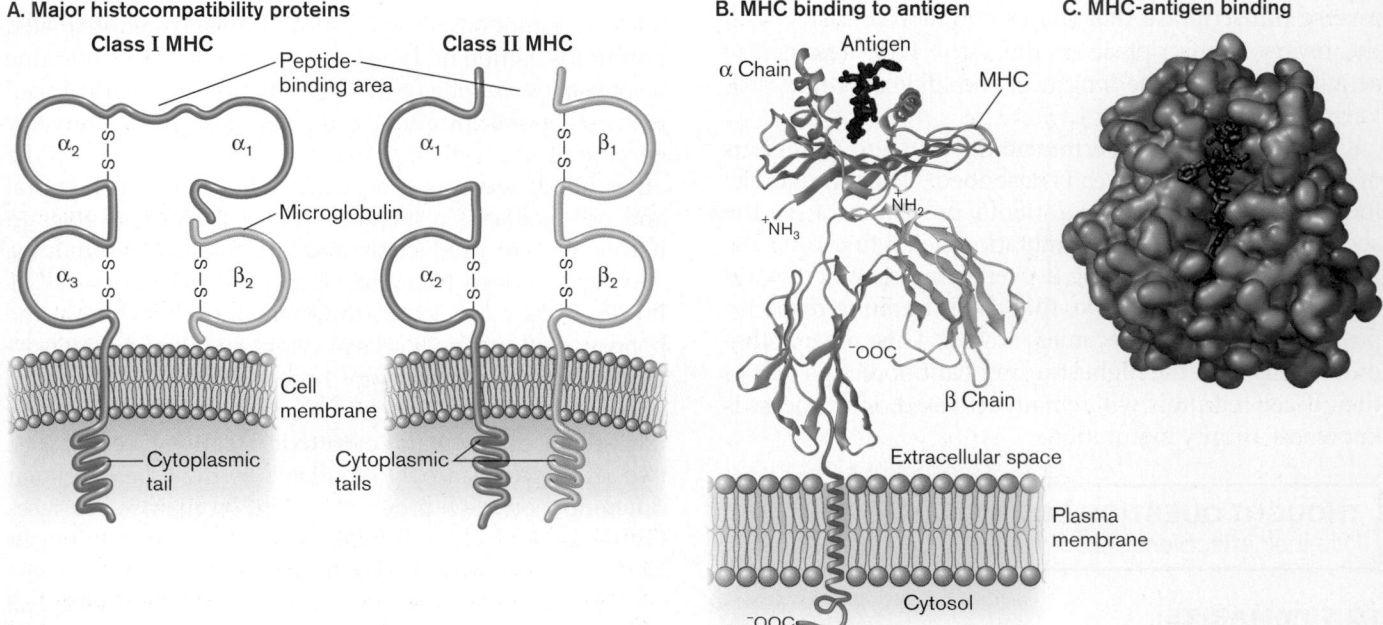

Figure 24.20 Major histocompatibility proteins. A. MHC class I molecules are composed of a 45-kDa chain and a small peptide called β_2-microglobulin (12 kDa). MHC class II molecules contain an alpha chain (30–34 kDa) and a beta chain (26–29 kDa). The peptide-binding regions of both classes show variability in amino acid sequence that yields different shapes and grooves. Peptide antigens nestle in the grooves and are held there awaiting interaction with T-cell receptors. CD8 T cells recognize antigen peptides associated with class I molecules, while CD4 T cells recognize peptides bound to class II molecules. **B.** Antigen binding to an MHC. **C.** Top view of antigen (red) nestled in the MHC peptide-binding site. (PDB code: 1BII)

macrophages under the influence of cytokines such as interferon-gamma (IFN-γ). The salient feature of MHC molecules is that they bind antigen, presenting it on the cell surface. The antigen-binding clefts of different MHC molecules differ markedly and have distinct binding affinities for different antigenic peptides (**Fig. 24.20B** and **C**).

MHC molecules are critical to the immune system because T cells only recognize antigens associated with MHC proteins; they do not recognize free-floating antigens. (B cells, on the other hand, *do* recognize free-floating antigens.) Once a T cell recognizes an antigen-presenting cell with a foreign antigen attached to an MHC molecule, the T cell becomes activated and in turn activates the immune system. It is important to note that not all potentially antigenic peptides can be recognized in one animal. An animal will not respond to an antigen if its antigen-presenting cells lack the MHC molecule needed to bind that antigen. This is the basis of tolerance to the self antigens present on an animal's own tissues. Tolerance means your immune system does not launch an immune response to your own antigens.

T-Cell Education and Deletion

Greek mythology tells of Narcissus, a young man punished by the gods for scorning the women who fell in love with him. One day Narcissus saw his beautiful reflection in a pool of water, fell in love, fell into the pool, and drowned. His inability to recognize himself is analogous to the danger posed by an immune system. Immune systems must be able to distinguish what is self (meaning antigens present in our own tissues) from what is not self. Innate immunity accomplishes this, in part, through the use of pattern recognition proteins such as the Toll-like receptors (see Section 23.7). How adaptive immune mechanisms avoid recognizing self is equally important. At the outset of development, the immune system is fully capable of reacting against self.

One mechanism the body uses to avoid attacking itself is to delete any T cells that react strongly against self antigens. In humans, this occurs throughout life as new T cells are made. In this process, T cells undergo a two-stage selection or "education" process in the thymus to recognize self versus nonself. This education process involves the use of self antigens, not foreign antigens. T cells bearing T-cell receptors (TCRs) that *weakly* recognize self MHC proteins displayed on thymus epithelial cells are allowed to survive (**positive selection**). These T cells leave the thymus to seed secondary lymphoid organs such as the spleen. T cells in the thymus that recognize self MHC peptides *too strongly*, however, are killed, or deleted from the population (**negative selection**). Almost 95% of T cells entering the thymus die during these positive and negative selection processes.

The positive selection process is needed because T cells must be able to recognize self MHC to "see" the associated antigen. Antigen bound to MHC actually increases the affinity to the matched TCR. If our T-cell repertoire included cells that bound self MHC too tightly, then our T cells would constantly react to our own MHC molecules, regardless of what antigen peptides were attached.

In contrast to negative selection for B cells, which occurs in bone marrow, T-cell education is limited to the thymus. But, if the thymus expresses only thymus antigens, how can T cells that respond to antigens expressed on other host cells (for example, heart cells) be removed? The answer is a special gene activator in thymus cells that allows them to synthesize *all* human proteins in small amounts. This expression is necessary to complete T-cell education within the thymus.

You might ask how someone who has had the thymus removed (a treatment for myasthenia gravis) can live if the organ is critical for T-cell maturation and for deleting self-reactive T cells. Actually, within a few years after birth, the thymus begins to lose its utility, such that very little function remains in adults. Research has shown that some T-cell maturation can occur extrathymically in secondary lymphoid tissues, although not as efficiently as in the thymus. This is at least part of the reason why children and adults with thymectomies are able to live relatively normal lives. The situation is more serious with newborns who have had the thymus removed. Newborns have not had sufficient time to populate their secondary extrathymic lymphoid organs with T cells.

Two Paths of Antigen Processing and Presentation

How are foreign antigens placed, or presented, on host cell surfaces? In the initial stages of an immune response, microbes either infect (as with viruses) or are engulfed by (as with bacteria) antigen-presenting cells. Inside the APCs, foreign proteins are degraded into smaller pieces (for example, converted to peptides). These smaller pieces are placed within MHC-binding clefts and transported back to the cell surface. Whether an antigen peptide binds to class I or class II molecules generally depends on how the antigen initially entered the cell (**Fig. 24.21**). Antigens synthesized within the cytoplasm of an APC (called endogenous antigens), as would occur during infections with viruses and intracellular bacteria, will attach to class I MHC molecules on the endoplasmic reticulum and are moved to the cell surface. Antigens produced outside of the APC (called exogenous antigens), as are most bacterial antigens, will enter the cell via phagocytosis and attach to class II MHC molecules transported to the acidic phagosome or lysosome (**Fig. 24.21**, right side). The MHC class II–peptide complex is

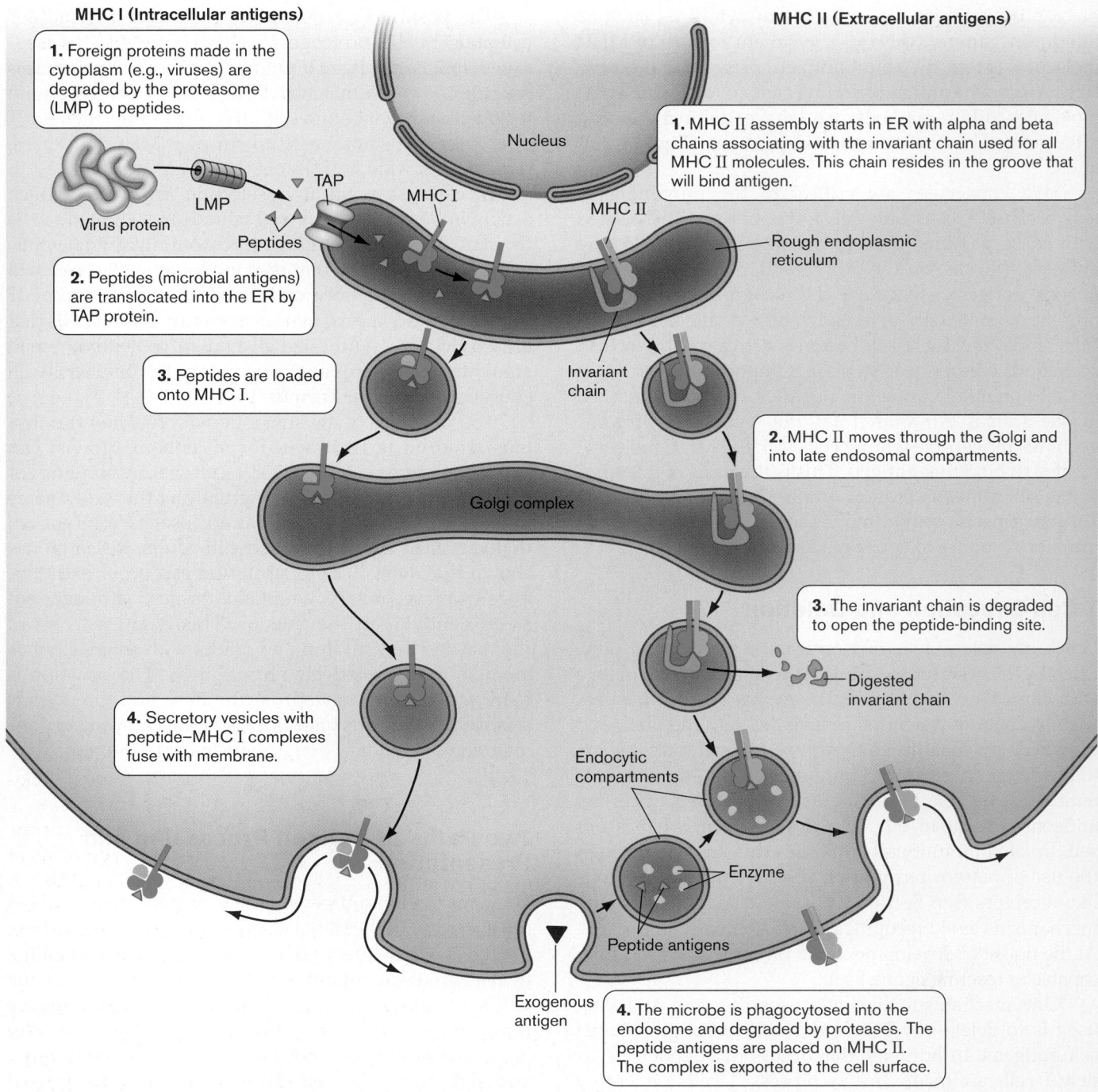

MHC I (Intracellular antigens)

1. Foreign proteins made in the cytoplasm (e.g., viruses) are degraded by the proteasome (LMP) to peptides.

2. Peptides (microbial antigens) are translocated into the ER by TAP protein.

3. Peptides are loaded onto MHC I.

4. Secretory vesicles with peptide–MHC I complexes fuse with membrane.

MHC II (Extracellular antigens)

1. MHC II assembly starts in ER with alpha and beta chains associating with the invariant chain used for all MHC II molecules. This chain resides in the groove that will bind antigen.

2. MHC II moves through the Golgi and into late endosomal compartments.

3. The invariant chain is degraded to open the peptide-binding site.

4. The microbe is phagocytosed into the endosome and degraded by proteases. The peptide antigens are placed on MHC II. The complex is exported to the cell surface.

Nucleus

LMP

Virus protein

TAP

Peptides

MHC I

MHC II

Rough endoplasmic reticulum

Invariant chain

Golgi complex

Digested invariant chain

Endocytic compartments

Enzyme

Peptide antigens

Exogenous antigen

Figure 24.21   Processing and presentation of antigens by antigen-presenting cells on class I and class II MHC proteins. Microbial proteins made in the host cytoplasm (upper left) are degraded and peptides are placed on MHC class I molecules in the endoplasmic reticulum (ER). Microbial proteins made outside the cell (lower center) are endocytosed, degraded in the endosome, and placed on MHC class II molecules. LMP = low-molecular-mass polypeptide component of the proteasome; TAP = transporter of antigen peptides.

then carried to the cell surface. Once the antigen is presented on the APC cell surface, T cells can interact with it via the T-cell receptors. Antigen-presenting cells interact with naive T cells within the lymph nodes, spleen, or Peyer's patches in the gut. So the APCs must make their way to those locations. How T cells then distinguish between MHC I and MHC II presentation is explained later.

T-Cell Receptors Bind Antigens on Antigen-Presenting Cells (APCs)

T-cell receptors (TCRs) are antigen-binding molecules (not immunoglobulins) present on the surfaces of T cells (**Fig. 24.22**). These receptors do not bind *soluble* antigen. A TCR, in general, will bind only to antigens attached to MHC surface proteins that are present on antigen-presenting cells. However, the TCRs of cytotoxic T cells can also bind viral antigens present on any virus-infected cell, whether or not that cell is normally considered an APC.

The T-cell receptor is composed of several transmembrane proteins. The part used to recognize antigen is composed of two molecules: alpha and beta. Much like the immunoglobulins, the alpha and beta proteins of the TCR are formed from gene clusters that undergo gene rearrangements. The diversity of TCR antigen receptors comes from random recombinations of various V, D, and J segments (analogous to, but different from, the immunoglobulin genes) and to variability in the precise joining of segments. There is no hypermutation, however.

The TCR alpha and beta proteins are found in a complex with four other peptides; together these form the

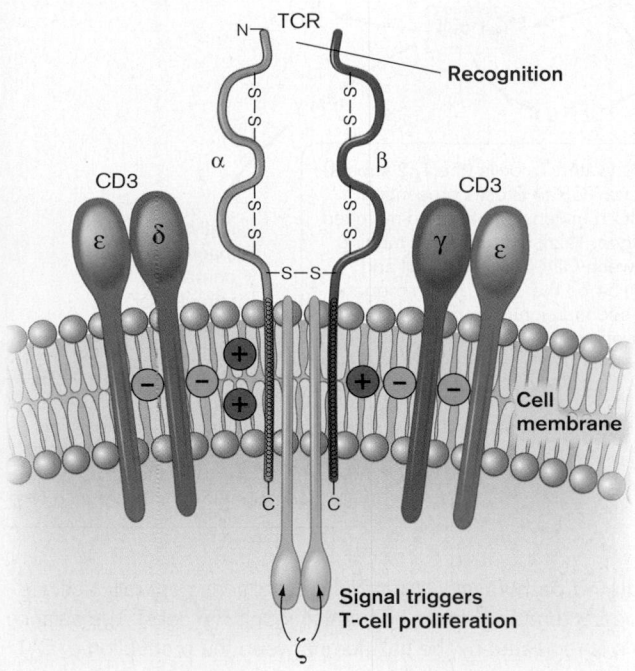

Figure 24.22   The T-cell receptor (TCR) and CD3 complex. The T-cell receptor proteins are associated with CD3 proteins at the cell surface. Antigen binds to the alpha and beta subunits. The positive and negative charges holding the complex together come from amino acids in the peptide sequences. Once bound to antigen, the complex transduces a signal into the cell. This signal triggers T-cell proliferation.

CD3 complex. When stimulated, these ancillary CD3 complex proteins recruit and activate intracellular protein kinases and launch a phosphorylation cascade that triggers proliferation of the T cell.

Activation of T_H0 Helper Cells

Figure 24.23 summarizes the steps that lead to T-cell activation in the lymph node and highlights how activated T cells influence the two arms of adaptive immunity: humoral and cellular. We will refer to **Figure 24.23** often in the sections that follow.

Two molecular signals are required to activate T cells. In the first step, regardless of the T-cell type, T-cell receptors are cross-linked by antigen-MHC proteins on antigen-presenting cells (**Fig. 24.23**◗, step 1). In the case of T_H0 cells, this step involves class II MHC complexes. But this first step alone is not enough to activate a T cell. A second signal is required by T_H cells, which involves the binding of a CD28 molecule present on the T-cell surface to a molecule called B7 protein on the APC cell surface (**Fig. 24.23**, step 2). Different cytokines produced during infection will then convert activated T_H0 cells to T_H1 or T_H2 cells (**Fig. 24.23**, steps 3 and 4). Activation of cytotoxic T cells (T_C cells) will be examined later.

B-Cell Activation Revisited

As with T cells, B cells need two signals to become activated. The first signal, described earlier, occurs when antigen cross-links B-cell receptors on B-cell membranes (in the process called capping). During the early phase of the primary response, the second activation signal for one type of B cell in the lymph node is binding to complement C3 on the bacteria or virus particle (see Section 23.8), or having a Toll-like receptor on the B cell interact with a PAMP (Section 23.7). This activation does not result in heavy-chain class switching; only IgM is secreted by the resulting plasma cell.

To become activated during the later primary immune response, most B cells require helper T cells. But not just any helper T cell can stimulate a B cell; specificity is involved. How does a B cell gain specific T-cell help? T cells (primarily the T_H2 type) "know" which B cells to help via their T-cell receptors. As part of the B-cell receptor capping mechanism, some antigen bound by B-cell receptors becomes internalized, to be processed and presented back on the B cell's surface MHC receptor (see 24.23, step 4 under "MHC II"). T_H2 cells that are act by dendritic cells presenting antigen A, for inst use their T-cell receptor to bind to antigen A on the matched B cells. This contact allows B cell to bind CD154 on the T cell—an i triggers the second intracellular signal activation (**Fig. 24.23**, step 5).

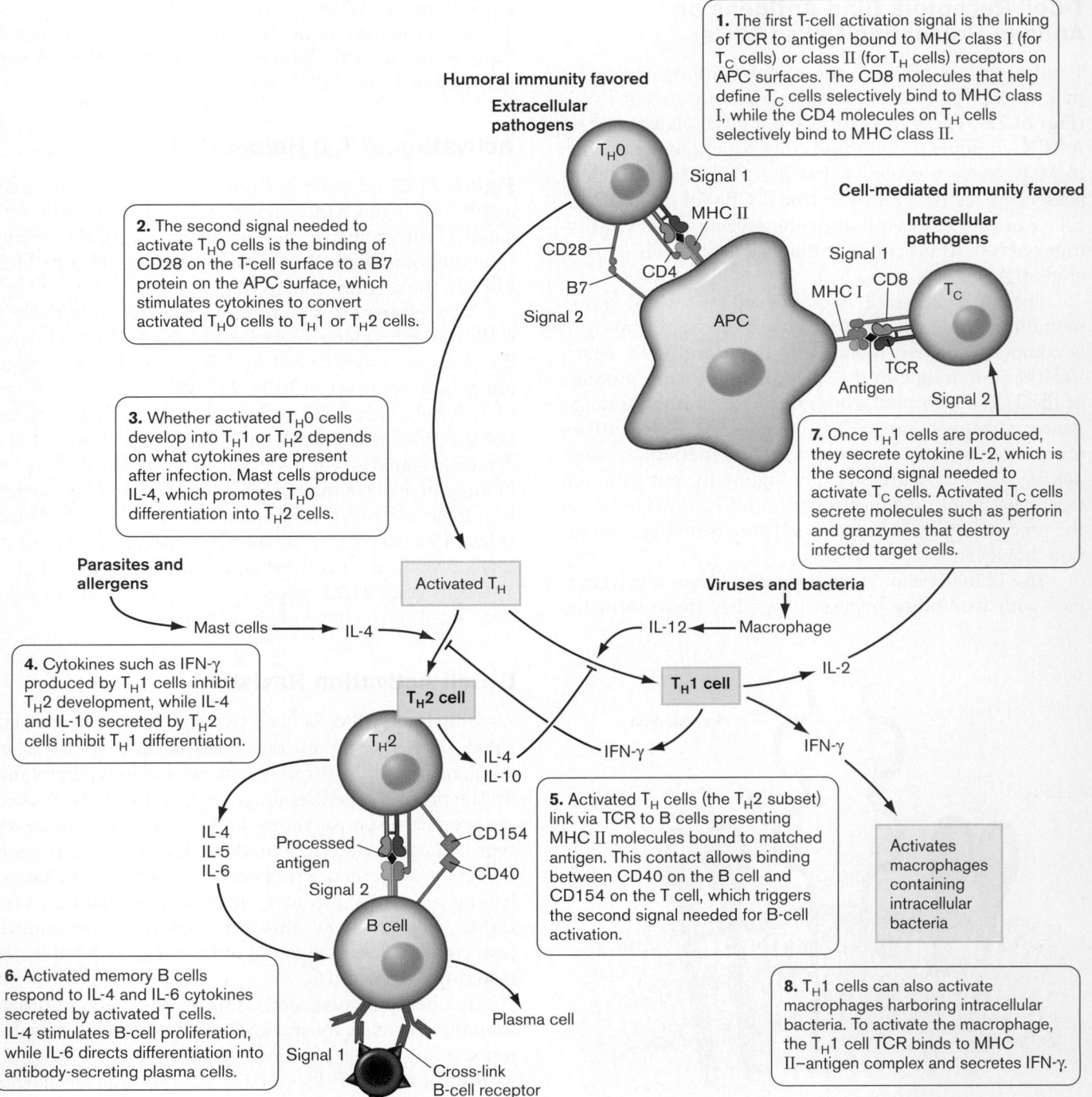

1. The first T-cell activation signal is the linking of TCR to antigen bound to MHC class I (for T_C cells) or class II (for T_H cells) receptors on APC surfaces. The CD8 molecules that help define T_C cells selectively bind to MHC class I, while the CD4 molecules on T_H cells selectively bind to MHC class II.

Humoral immunity favored

Extracellular pathogens

Cell-mediated immunity favored

Intracellular pathogens

2. The second signal needed to activate T_H0 cells is the binding of CD28 on the T-cell surface to a B7 protein on the APC surface, which stimulates cytokines to convert activated T_H0 cells to T_H1 or T_H2 cells.

3. Whether activated T_H0 cells develop into T_H1 or T_H2 depends on what cytokines are present after infection. Mast cells produce IL-4, which promotes T_H0 differentiation into T_H2 cells.

7. Once T_H1 cells are produced, they secrete cytokine IL-2, which is the second signal needed to activate T_C cells. Activated T_C cells secrete molecules such as perforin and granzymes that destroy infected target cells.

Parasites and allergens

Viruses and bacteria

4. Cytokines such as IFN-γ produced by T_H1 cells inhibit T_H2 development, while IL-4 and IL-10 secreted by T_H2 cells inhibit T_H1 differentiation.

5. Activated T_H cells (the T_H2 subset) link via TCR to B cells presenting MHC II molecules bound to matched antigen. This contact allows binding between CD40 on the B cell and CD154 on the T cell, which triggers the second signal needed for B-cell activation.

Activates macrophages containing intracellular bacteria

6. Activated memory B cells respond to IL-4 and IL-6 cytokines secreted by activated T cells. IL-4 stimulates B-cell proliferation, while IL-6 directs differentiation into antibody-secreting plasma cells.

8. T_H1 cells can also activate macrophages harboring intracellular bacteria. To activate the macrophage, the T_H1 cell TCR binds to MHC II–antigen complex and secretes IFN-γ.

Figure 24.23 Summary of the activation of humoral and cell-mediated pathways. Intracellular pathogens generally activate cell-mediated immunity by stimulating cytotoxic T cells. Extracellular pathogens tend to activate humoral immunity (B cells). The balance between cell-mediated and humoral immune responses to a given infection is regulated by the balance between the production of T_H1 (cell-mediated) versus T_H2 (antibody) helper cells. This balance is influenced by whether the foreign antigen was made by intracellular pathogens (T_H1-favored) or extracellular pathogens (T_H2-favored), and whether antigen presentation (by an APC) occurred through a macrophage (T_H1-favored) or a mast cell (T_H2-favored). T_H1 cells will encourage activation of cytotoxic T cells (cell-mediated immunity; best for killing intracellular pathogens), while T_H2 promotes antibody production (humoral immunity; best for attacking extracellular pathogens). See text for more details. ▶ *microbiology2.com/animations*

A.

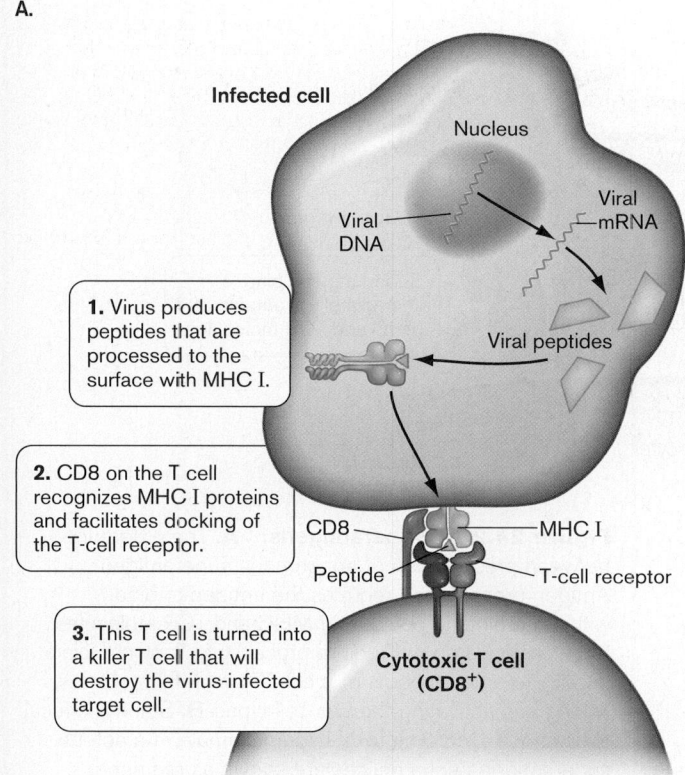

Infected cell

Nucleus

Viral DNA

Viral mRNA

1. Virus produces peptides that are processed to the surface with MHC I.

Viral peptides

2. CD8 on the T cell recognizes MHC I proteins and facilitates docking of the T-cell receptor.

CD8 —— MHC I

Peptide —— T-cell receptor

3. This T cell is turned into a killer T cell that will destroy the virus-infected target cell.

Cytotoxic T cell (CD8⁺)

B.

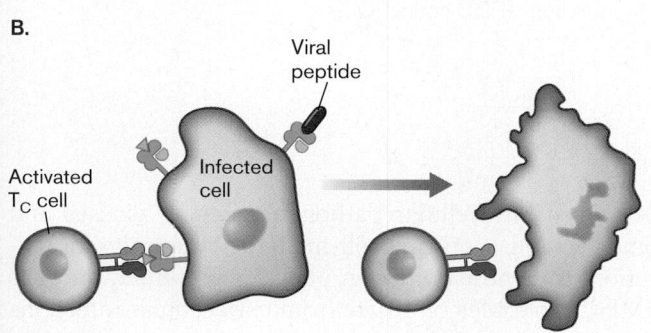

Viral peptide

Activated T_C cell

Infected cell

Figure 24.24 Presentation of a viral antigen to a T cell and cytotoxic T-cell action. A. CD8 protein on a T cell directs the interaction between a T-cell receptor and a viral antigen bound to a class I MHC protein on an antigen-presenting cell. **B.** The activated T_C cell then leaves the lymph node and migrates to the site of infection, where it can recognize the viral peptide presented on the MHC class I receptor on an infected cell. This interaction authorizes the cytotoxic T cell to kill the infected cell.

During the secondary response, B cells still need help to become plasma cells but do not need direct contact with helper T cells. Memory B cells that have antigen bound to their B-cell receptors can respond to the soluble IL-4 and IL-6 cytokines secreted by activated helper T cells without having direct contact with the T_H cell. IL-4 stimulates B-cell proliferation, while IL-6 directs differentiation into antibody-secreting plasma cells (**Fig. 24.23**, step 6).

Activation of Cytotoxic T Cells

As with helper T cells, the first signal needed to activate cytotoxic T cells is the binding of the TCR to an antigen-MHC complex (in this case, MHC class I) on an antigen-presenting cell (see **Fig. 24.23**, step 1). **Figure 24.24A** shows a closer look at the interaction between an antigen-presenting cell and a CD8 cytotoxic T cell. This interaction occurs in a lymph node. As with the activation of helper T cells, this interaction alone will not activate the cytotoxic T cell. The second signal needed here is actually a cytokine called IL-2 produced by the T_H1 class of helper T cells (**Fig. 24.23**, step 7). The first signal (the TCR–antigen–MHC I complex) initiates the synthesis of IL-2 receptors on the T_C-cell surface. Upon subsequent binding of IL-2, the T_C cell gains cytotoxic activity, after which it can leave the lymph node, travel to the site of infection, and kill any cell bearing the same peptide–MHC class I complex (for example, cells infected with the same virus that triggered T_C production) (**Fig. 24.24B**). The B7-CD28 interaction that was needed to activate helper T cells is not required for T_C-cell activation but can help.

WWW | Activation of cytotoxic T cells
microbiology2.com/links

Superantigens Do Not Require Processing to Activate T Cells

Normal antigens require processing by antigen-presenting cells. Each peptide produced by antigen processing is ultimately recognized in the context of self MHC molecules by a specific cognate T-cell receptor. However, when an antigen is introduced into a host, there are only a few T cells with the proper TCR needed to recognize that antigen—in the range of 1–100 cells in a million. Proliferation of the T cells through antigen-dependent activation increases that number and increases the immune response to that antigen.

Some proteins, called **superantigens**, stimulate T cells much more rapidly by bypassing the normal route of antigen processing. In fact, recognition as an antigen is not even involved. Certain microbial toxins, such as staphylococcal toxic shock syndrome toxin (TSST), are actually superantigens. These proteins, as illustrated in **Figure 24.25**, simultaneously bind to the outside of T-cell receptors on T cells and to the MHC molecules on antigen-presenting cells (for example, macrophages). This joining of the T cell and macrophage activates more T cells than does a typical immune reaction, and stimulates release of massive amounts of inflammatory cytokines from both cell types.

The effect of superantigens can be devastating because cytokines like tumor necrosis factor can overwhelm the host immune system regulatory network and cause severe damage to tissues and organs. The result is

A.

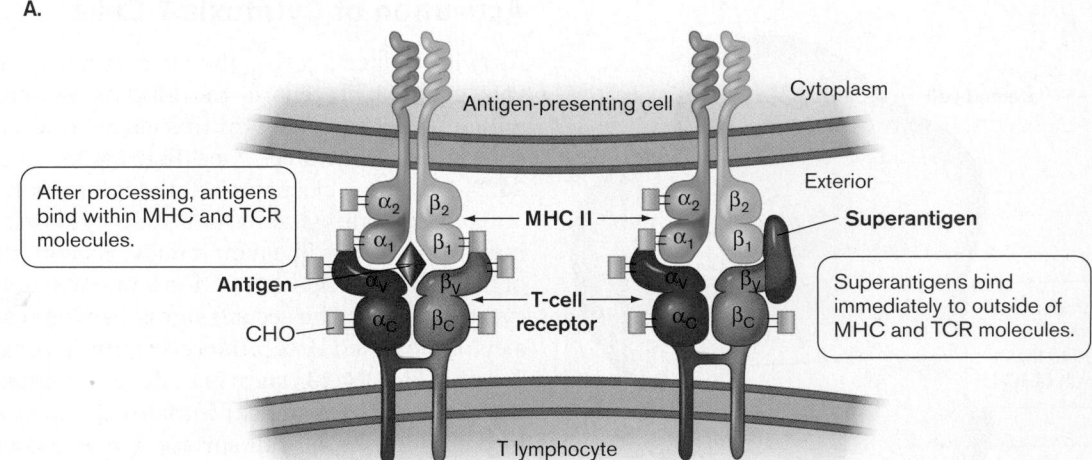

After processing, antigens bind within MHC and TCR molecules.

Antigen-presenting cell

Cytoplasm

Exterior

MHC II

Superantigen

Antigen

T-cell receptor

CHO

Superantigens bind immediately to outside of MHC and TCR molecules.

T lymphocyte

B.

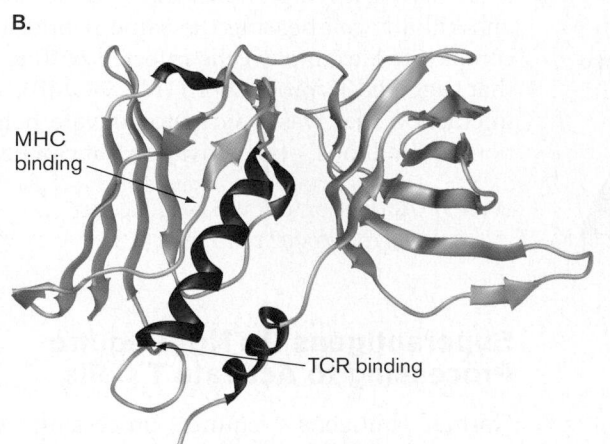

MHC binding

TCR binding

Figure 24.25 **Superantigens.** **A.** The difference between presentation of antigen and superantigen. Antigen presentation requires the antigen to bind within the binding pockets of MHC and TCR molecules. Superantigens do not require processing. They can bind directly to outer aspects of the TCR and MHC proteins, linking and activating the two cell types. **B.** Staphylococcal toxic shock syndrome toxin is one example of a potent superantigen. Alpha helices are shown as red ribbons, while sections of the beta sheet are shown as blue ribbons. (PDB code: 2QIL)

disease and sometimes death. Jim Henson, the creator of Kermit the Frog and other Muppet characters, died in 1990 from complications of pneumonia caused by a potent superantigen produced by *Streptococcus pyogenes*. This example is but one of many ways that overreaction by the immune system causes morbidity (disease) and mortality (death)—effects more pronounced than the direct effects of the pathogen involved.

The CD4 and CD8 Proteins Help T Cells Distinguish MHC I from MHC II

One question not yet addressed is, How do the TCRs on T_C and T_H0 cells "know" to respectively bind MHC class I and class II antigen complexes? The answer is based on the different clusters of cellular differentiation (CD) molecules displayed on T_C and T_H0 cell surfaces. The surface CD proteins CD8 and CD4 help T cells distinguish between MHC class I and class II molecules on antigen-presenting cells. The CD8 molecules on T_C cells selectively bind MHC class I, while CD4 molecules on T_H cells selectively bind MHC class II (see **Fig. 24.23**, step 1). Antigens presented on class I MHC molecules generally

arise from intracellular pathogens such as viruses and some bacteria, requiring cell-mediated immunity for resolution. In contrast, antigen peptides presented on class II MHC molecules originate from extracellular infections and are recognized only by CD4 T_H cells.

Once activated, cytotoxic CD8 T_C cells secrete molecules such as **perforin** and **granzymes** that destroy infected target cells (similar to natural killer cells; discussed in Section 23.7). Perforin forms a pore in target cells that is used to deliver toxic enzymes (granzymes). Because they directly attack host cells, CD8 T_C cells are the major "enforcers" of the cellular immune system, along with macrophages and NK cells.

Why does the presentation of antigens on MHC I molecules activate cytotoxic T cells rather than B cells and antibody production? A major reason is that intracellular pathogens (for example, viruses) hide inside host cells, where they are protected from antibody. Consequently, these pathogens are best killed when the harboring host cell is also sacrificed (via cellular immunity). Because class I MHC molecules are found on all nucleated cells, any infected cell can potentially activate T_C cells, converting them to cytotoxic T cells that ultimately kill the

Table 24.5 Select cytokines that modulate the immune response.

Cytokine[a]	Sample sources	General functions
IL-1	Many cell types, including endothelial cells, fibroblasts, neuronal cells, epithelial cells, macrophages	Affects differentiation and activity of cells in inflammatory response; central nervous system; acts as endogenous pyrogen
IL-2	T_H1 cells	Stimulates T-cell and B-cell proliferation
IL-3	T cells, mast cells, keratinocytes	Stimulates production of macrophages, neutrophils, mast cells, others
IL-4	T_H2 cells, mast cells	Promotes differentiation of CD4 and T cells into T_H2 helper T cells; promotes proliferation of B cells
IL-5	T_H2 cells	Acts as a chemoattractant for eosinophils; activates B cells and eosinophils
IL-6	T_H2 cells, macrophages, fibroblasts, endothelial cells, hepatocytes, neuronal cells	Stimulates T-cell and B-cell growth; stimulates production of acute-phase proteins
IL-8	Monocytes, endothelial cells, T cells, keratinocytes, neutrophils	Chemoattracts PMNs; promotes migration of PMNs through endothelium
IL-10	T_H2 cells, B cells, macrophages, keratinocytes	Inhibits production of IFN-γ, IL-1, TNF-α, IL-6 by macrophages
IFN-α/β	T cells, B cells, macrophages, fibroblasts	Promotes antiviral activity
IFN-γ	T_H1 cells, cytotoxic T cells, NK cells	Activates T cells, NK cells, macrophages
TNF-α	T cells, macrophages, NK cells	Exerts wide variety of immunomodulatory effects
TNF-β	T cells, B cells	Exerts wide variety of immunomodulatory effects

[a]IL = interleukin; IFN = interferon.

infected cell (T_C cells are not actually cytotoxic until they are activated). The infected host cell is sacrificed for the good of the whole animal, human or otherwise.

Whereas antigens presented on class I MHC molecules elicit only cell-mediated immunity, antigens presented on class II molecules can actually stimulate either arm of the immune system. In contrast to CD8 T_C cells, CD4 helper T cells do not directly kill host cells. Instead, activated CD4 T_H cells release various cytokines called lymphokines that, among other things, attract additional white cells to the area. **Table 24.5** lists a fraction of the many different cytokines produced by various cell types and their influence over the immune system. Activated T_H1 cells secrete cytokines that stimulate the cytotoxic T_C cells of the cellular immune system (see **Fig. 24.23**, step 7), whereas activated T_H2 cells directly interact with the class II MHC–antigen peptide complexes on B cells, as described earlier, and stimulate those cells to differentiate and produce antibody (see **Fig. 24.23**, step 5).

AIDS (acquired immunodeficiency syndrome) provides a vivid and tragic illustration of the importance of CD4 T cells in immunity. The human immunodeficiency virus (HIV) binds to CD4 molecules and thus is able to invade and infect CD4 T cells. As the disease progresses, the number of CD4 T cells declines below its normal level of about 1,000 per microliter. When the number of CD4 T cells drops below 400 per microliter, the ability of the patient to mount an immune response declines dangerously. The patient not only becomes hypersusceptible to infections by pathogens, but also becomes susceptible to infections by commensal organisms. Most AIDS patients actually die from these opportunistic infections.

The MHC Restriction Rule Limits T-Cell Reactivity

As noted previously, T cells can only recognize antigens complexed to an MHC molecule. Actually, a T cell can only recognize a self MHC molecule. The MHC molecules of each person are unique. Thus, T cells from one individual normally do not recognize peptides placed on antigen-presenting cells from another individual—a property known as **MHC restriction**. CD4 T_H cells are class II MHC restricted, while CD8 T_C cells are class I restricted. Rolf Zinkernagel (University of Zurich) and Peter Doherty (St. Jude Children's Research Hospital, Memphis) discovered this phenomenon in 1976 with a simple yet profoundly insightful experiment. They demonstrated that mouse cytotoxic T cells would kill only virus-infected target cells that possessed the same MHC surface protein. In contrast, if the infected target cell came from a different strain of mouse containing a different MHC protein, the cell was spared. Thus, T-cell recognition has two components: self and nonself. The self component comes from the MHC receptor, while the nonself component comes from the pathogen. This concept won Zinkernagel and Doherty the 1996 Nobel Prize in Physiology or Medicine.

Chapter 25

Microbial Pathogenesis

Mammals have elaborate physical, chemical, and immunological defenses that protect against disease-causing microbes, or pathogens. Yet every fortress has its weakness. Pathogenic microbes exploit those weaknesses. The result is disease.

How is a pathogen different from a commensal organism? The answer varies, depending on the pathogen. Some avoid phagocytosis by host cells, whereas others actively encourage it. Even more mysterious are pathogens that develop a latent, undetectable stage in the host, then later emerge to cause disease. Some pathogens efficiently slay their hosts. Others persist for many years without killing. Which is the more successful pathogenic strategy?

The fundamental question of microbial pathogenesis is how an organism too small to be seen with the naked eye can kill a human that is 1 million times larger. This chapter explores the strategies different bacterial and viral pathogens use to accomplish this feat.

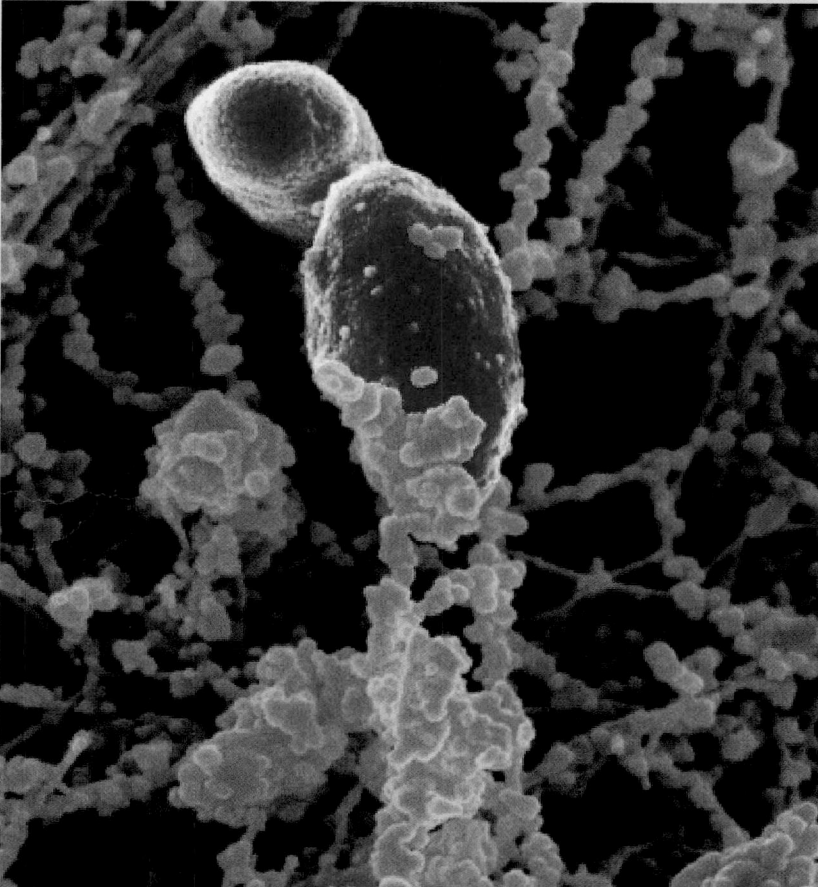

Current Research Highlight

Scanning electron micrograph of the obligate intracellular pathogen *Rickettsia rickettsii* (blue), the cause of Rocky Mountain spotted fever, showing its association with host actin (gold). Several pathogens propel themselves though host cytoplasm by polymerizing host actin at one pole of the bacterial cell. *Source*: Damon W. Ellison et al. 2008. *Infect. Immun.* **76**:542–550. Photo courtesy of E. R. Fischer and T. Hackstadt.

A 23-year-old Hispanic mother brought her 3-year-old daughter into the emergency room. The child was lethargic, had a fever of 40°C (104°F), and was having difficulty breathing. The mother explained that the family had arrived in the United States from El Salvador the previous week. The attending physician noted an extreme swelling of the child's cervical lymph nodes, giving the girl a thick, "bull-neck" appearance. She also noticed the beginnings of a membranous growth at the back of the child's throat that was beginning to obstruct the trachea. It was grayish in color and bled when scraped. When asked, the distraught mother admitted that the child had not received any vaccinations before arriving in New Mexico. Suspecting the nature of the child's illness, the physician granted immediate admission to the hospital and ordered administration of penicillin and a specific antitoxin. The results of a throat swab sent to the microbiology lab confirmed the physician's suspicion. The root of the child's disease was *Corynebacterium diphtheriae*, which causes diphtheria.

C. diphtheriae is a deadly pathogen able to kill humans by attacking the respiratory and cardiac systems. When untreated, death often comes by suffocation when the gray membrane completely covers the trachea. This pathogen is a Gram-positive, non-spore-forming rod identical in appearance to many common commensal throat microbes, such as the more docile *C. striatum*. So, two Gram-positive rods, identical in appearance, are both found in the human throat. Why is one a killer and the other not? The child with diphtheria most likely came in contact with the pathogen before leaving El Salvador, where vaccinations in some areas are difficult to obtain. The question we ask in this chapter is not how the disease could have been prevented, but what genetic distinctions differentiate closely related disease-causing pathogens from innocuous nonpathogens. In the case of diphtheria, the difference between friend and foe is a bacteriophage genome embedded in the genome of the organism. This phage carries the gene for diphtheria toxin, whose properties will be described later.

Pathogens, such as *C. diphtheriae*, that kill their hosts or only transiently reside there must also be prepared for life outside the host. These microbes possess an alternate physiology that allows survival in nonhost environments such as a lake or soil. Other pathogens, however, are not as versatile and die when separated from their host. To survive, they must be passed directly from person to person. If the host dies before transmission, the pathogen dies with it. Thus, to kill or not to kill is an important question each pathogen must address. Chapter 25 will discuss various relationships that occur between pathogens and their hosts and the factors that contribute to **pathogenesis**, the process by which microbes cause disease in a host. The degree of harm that is caused depends on the mechanisms the pathogen has at its disposal.

25.1 Host-Pathogen Interactions

Parasites, in the broadest sense, include bacteria, viruses, fungi, and protozoa that colonize and harm their hosts. However, the term **pathogen** is typically used to refer to bacterial, viral, and fungal agents of disease. Disease-causing protozoa and worms, on the other hand, are normally called parasites. Pathogens and parasites infect their animal and plant hosts in a variety of ways and enter into a variety of host-pathogen relationships, depending on the site of colonization. For example, organisms that live on the surface of a host are called **ectoparasites**. The fungus *Trichophyton rubrum*, one cause of athlete's foot, is an ectoparasite (**Fig. 25.1**). *Wuchereria bancrofti*, the worm parasite that causes elephantiasis, is an **endoparasite** because it lives inside the body (**Fig. 25.2**).

Before examining the mechanisms microbes use to cause disease, it is helpful to know the terminology of pathogenesis. An **infection** occurs when a pathogen or parasite enters or begins to grow on a host. Be aware that

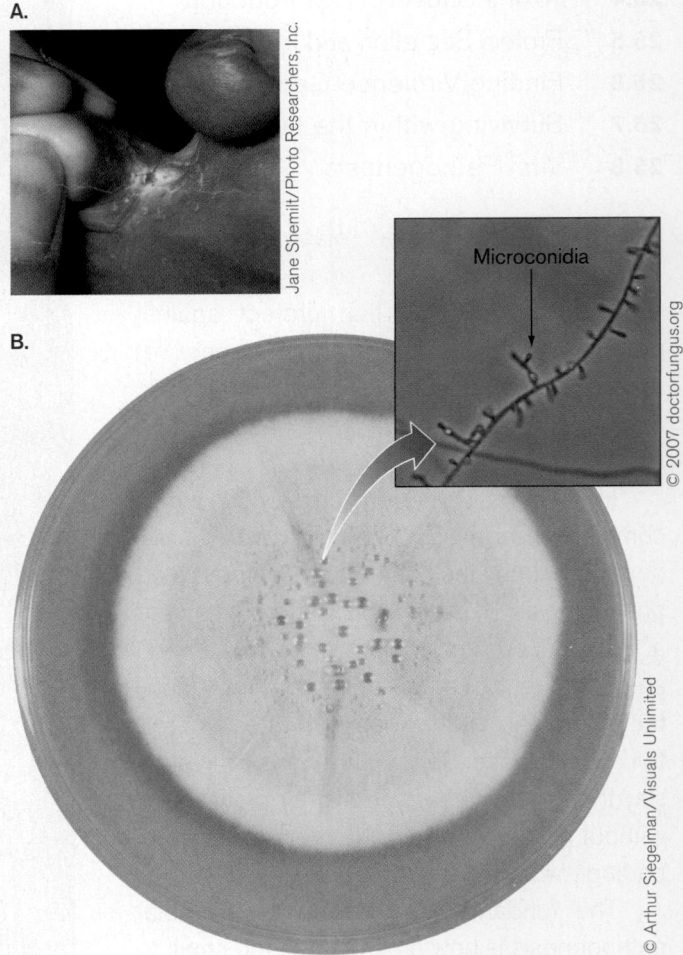

Figure 25.1 An ectoparasite. A. Athlete's foot is caused by the fungus *Trichophyton rubrum*. **B.** Colony morphology and microscopic, branching conidia (blowup) of *T. rubrum*. Conidia are asexual spores that grow on stalks called conidiophores (see Chapter 20).

the term "infection" does not necessarily imply overt disease. Any potential pathogen growing in or on a host is said to cause an infection, but that infection may be only transient because immune defenses kill the pathogen

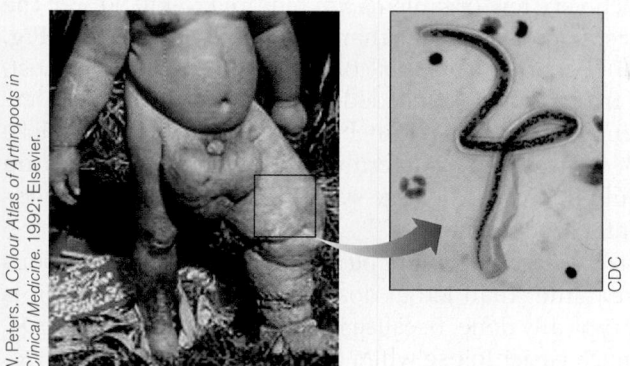

Figure 25.2   An endoparasite.   The disease filariasis, commonly known as "elephantiasis" for obvious reasons, is caused by the worm *Wuchereria bancrofti* (blowup), which enters the lymphatics and blocks lymphatic circulation. Adult worms are threadlike and measure 4–10 cm in length. The young microfilariae (shown) are approximately 0.5 mm in length. Though not a problem in the United States, *W. bancrofti* and elephantiasis are found throughout middle Africa, Asia, and New Zealand.

before noticeable disease results. Indeed, most infections go unnoticed. For example, every time you have your teeth cleaned by a dentist your gums bleed and your resident oral microbes transiently enter the bloodstream, but you rarely suffer any consequences.

Primary pathogens are disease-causing microbes possessing the means to breach the defenses of a healthy host. For example, *Shigella flexneri*, the cause of bacillary dysentery, is a primary pathogen. When ingested, it can survive the natural barrier of an acidic (pH 2) stomach, enter the intestine, and begin to replicate. **Opportunistic pathogens**, on the other hand, cause disease only in a compromised host. *Pneumocystis jirovecii* (previously *P. carinii*) is an opportunistic pathogen that causes life-threatening infections in AIDS patients, whose immune systems have been eroded (**Fig. 25.3A**). Some microbes even enter into a **latent state** during infection, in which the organism cannot be found by culture. Herpes virus, for instance, can enter the peripheral nerves and remain dormant for years, and then suddenly emerge to cause cold sores (**Fig. 25.3B**). The bacterium *Rickettsia prowazekii* causes epidemic typhus, but it can also enter a latent phase and months or years later cause a disease relapse called recrudescent typhus.

The term **pathogenicity** refers to an organism's ability to cause disease. It is defined in terms of how easily an organism causes disease (infectivity) and how severe that disease is (virulence). Pathogenicity, overall, is shaped by the genetic makeup of the pathogen. In other words, an organism is more—or less—pathogenic, depending on the tools at its disposal (such as toxins) and their effectiveness.

Virulence is a measure of the degree or severity of disease. For instance, Ebola virus and the closely related Marburg virus have case fatality rates near 70%. This means that they are highly virulent (**Fig. 25.4**). On the other hand, rhinovirus, the cause of the common cold, is very

A.

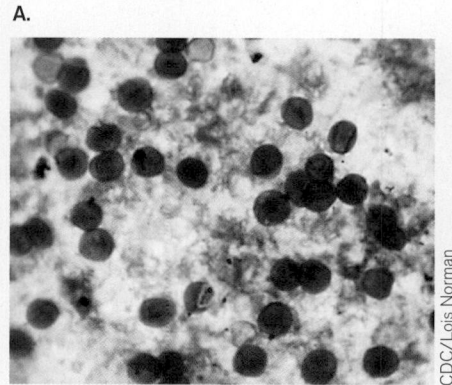

B.

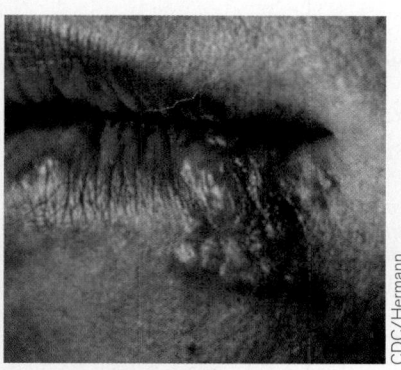

Figure 25.3   Opportunistic and latent infections.   **A.** *Pneumocystis jirovecii* cysts (5–10 μm in diameter) in bronchoalveolar material. Notice how the fungi look like crushed Ping-Pong balls. **B.** Cold sore produced by a reactivated herpes virus hiding latent in nerve cells.

Figure 25.4   Highly virulent viruses.
A. Ebola virus (approx. 1 μm long, TEM).
B. The body of a victim of Marburg virus is placed in a coffin for safe burial in Angola. Marburg and Ebola cause hemorrhagic infections in which patients bleed from the mouth, nose, eyes, and other orifices. They have a 70%–80% mortality rate.

A.

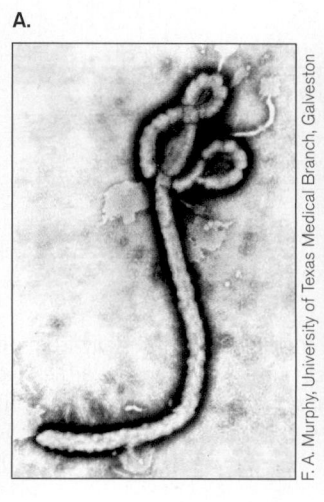

B.

effective at causing disease but almost never kills its victims. So it is highly infective but has a low virulence. Both organisms are pathogenic, but with one you live and the other you die.

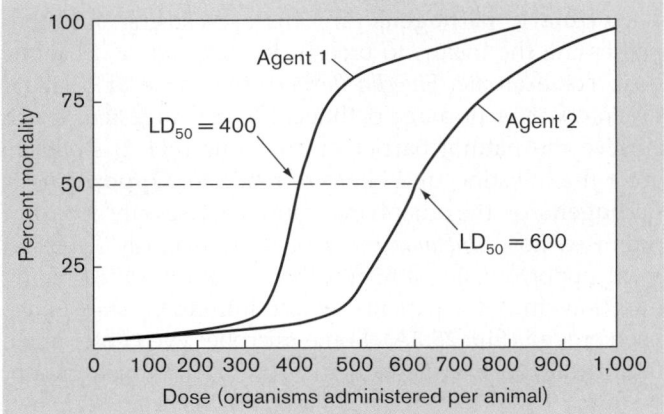

Figure 25.5 Measurement of virulence. Each LD_{50} measurement requires infecting small groups of animals with increasing numbers of infectious agent and viewing how many animals die. The number of microbes that kill half the animals is called the LD_{50} dose. In this example, agent 1 is more virulent than agent 2.

One way to measure virulence is to determine how many bacteria or virions are required to kill 50% of an experimental group of animal hosts. This is called the **LD_{50}** (lethal dose 50%). An organism with a low LD_{50}, in which very few organisms are required to kill 50% of the hosts, is more virulent than one with a high LD_{50} (**Fig. 25.5**). For organisms that colonize but do not kill the host, the **infectious dose** needed to colonize 50% of the experimental hosts (ID_{50}) can be measured. Infectious dose 50% is measured by determining how many microbes are required to cause disease symptoms in half of an experimental group of hosts.

Although one might be able to measure the infectious dose, rather than lethal dose, for a lethal pathogen, it is not typically done. Because it gives a clear end point, LD_{50} is much easier to use when trying to determine the effectiveness of a given treatment (an antibiotic, for example) or quantify the role of a given gene in pathogenesis.

THOUGHT QUESTION 25.1 Is a microbe with an LD_{50} of 5×10^4 more or less virulent than a microbe with an LD_{50} of 5×10^7?

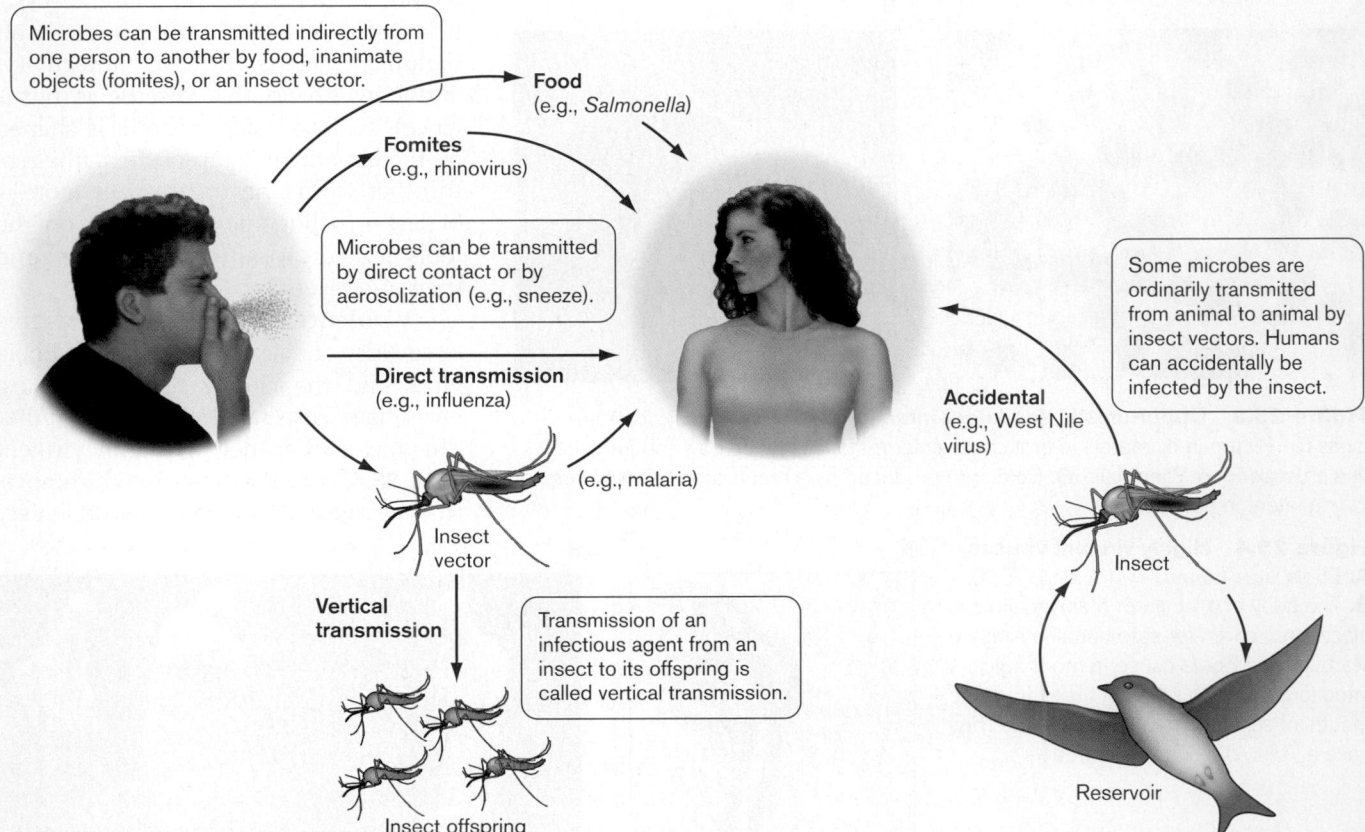

Figure 25.6 Infection cycles. Infectious agents can be transmitted horizontally from one member of a species to another by a variety of means: food, fomites, aerosolization, direct contact, or arthropod (insect or tick) vector. Vertical transmission is passage from parent to offspring during birth, while accidental transmission happens when a host that is not part of the normal infectious cycle unintentionally encounters that cycle.

Infection Cycles Can Be Direct or Indirect

We tend to separate medical and environmental considerations when discussing pathogens. However, greater insight can be gained from studying pathogens when both perspectives are considered. Most pathogens maintain a significant presence outside the human host. Some find sanctuary in the natural environment, as is the case with the fungus *Histoplasma* (found in soil) or the bacterium *Legionella pneumophila* (found in natural water). Many pathogens sustain their numbers in animals or insects (as in the case of eastern equine encephalitis virus or avian flu virus, for example). Keep this in mind as we discuss various infectious diseases because many virulence factors actually evolved as adaptations within diverse nonhuman environments. Environment and evolution are integral factors to consider in the study of infectious disease.

Somehow, pathogens must pass from one person or animal to another if a disease is to spread. The route an organism takes to accomplish this is called the **infection cycle**. A cycle of infection can be simple or complex (**Fig. 25.6**). Organisms that spread directly from person to person, such as the rhinovirus or *Shigella*, have simple infection cycles. A person can also transmit some infectious agents to another person by contaminating food during preparation. Inanimate objects through which pathogens can be relayed to hosts are called **fomites** (for example, the tissue used to stiffle a sneeze). More complex cycles often involve **vectors**, usually insects or ticks (arthropods), as intermediaries. Vectors serve to carry infectious agents from one animal to another. A mosquito vector, for example, transfers the virus causing yellow fever from infected to uninfected individuals (**Fig. 25.7**) in what is called **horizontal transmission**. The mosquito can also bequeath this particular virus to its offspring via infected eggs in a form of **vertical transmission** called **transovarial transmission**. Although yellow fever is not a problem in the United States today, West Nile virus, a flavivirus closely related to yellow fever, currently claims several victims each year in this country.

Because insects and ticks are instrumental in transmitting pathogens, killing these arthropod vectors is an important way to halt the spread of disease. Interventions include spraying insecticide in a community during egg-hatching season or using other microbes as assassins "trained" to kill the vector. For example, *Bacillus thuringiensis* will kill many types of insects that carry infectious agents. More recently, an insect virus called baculovirus has been developed that kills the *Culex* mosquito vectors carrying the West Nile virus. The advantage of these vector-targeting microbes is that they do not kill other insects or animals, as do many chemical insecticides.

Another critical factor in an infection cycle is the "reservoir" of infection. A **reservoir** is an animal, bird, or insect that normally harbors the pathogen. In the case of

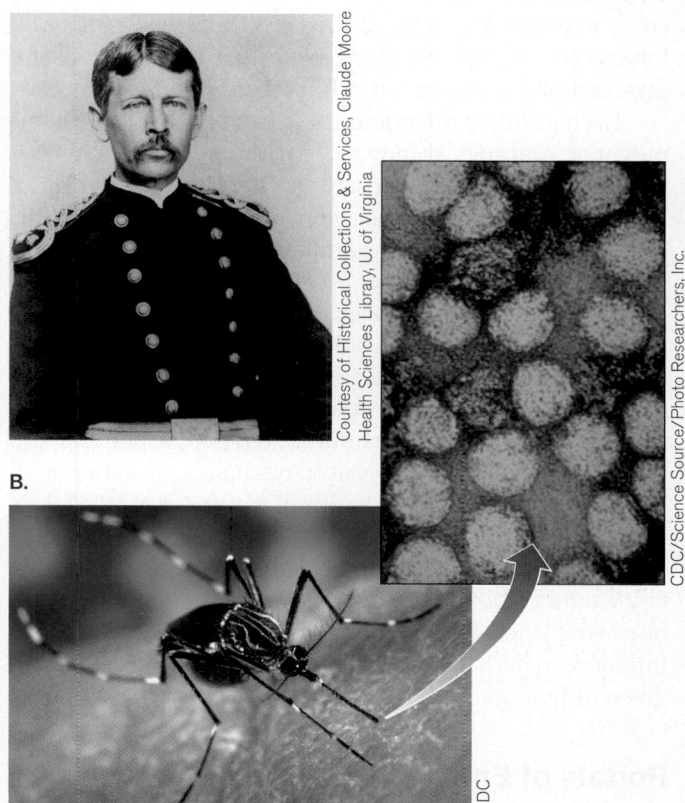

Figure 25.7 Insect vector and yellow fever. A. Walter Reed, a member of the University of Virginia School of Medicine class of 1869, proved in 1901 that the mosquito *Aedes aegypti* is the vector of transmission for yellow fever, a disease named for the jaundice produced by liver damage. **B.** *A. aegypti*. Yellow fever is caused by a flavivirus (inset, TEM) carried by this mosquito. The virus varies in size from 50 to 90 nm. Yellow fever remains endemic in the northern part of South America and in central Africa.

yellow fever, the mosquito is not only the vector, but the reservoir as well, because the insect can pass the virus to future generations of mosquitoes through vertical transmission. The virus causing eastern equine encephalitis (EEE), however, uses birds as a reservoir. The microbe is normally a bird pathogen and is transmitted from bird to bird via a mosquito vector. The virus does not persist in the insect, but transmission by the insect vector keeps the virus alive by passing it to new avian hosts. Humans or horses entering geographic areas harboring the disease (called endemic areas) can also be bitten by the mosquito. When this happens, they become accidental hosts and contract disease. The virus does not replicate to high titers in mammals, which means that horses and humans are poor reservoirs for it. EEE virus does, however, replicate to high numbers in the avian host. Reservoirs are critically important for the survival of a pathogen and as a source of infection. If the eastern equine encephalitis virus had to rely on humans to survive, the virus would cease to exist because of limited replication potential and limited access to mosquitoes. Note that the reservoir of a given pathogen might not exhibit disease.

Even a simple infectious cycle can become more complex. For example, rhinovirus can be spread person-to-person by a sneeze (airborne) or through the sharing of inanimate objects (fomites) such as contaminated utensils (fork, pen), towels, cloth handkerchiefs, and doorknobs. Handshaking is also an efficient means of transferring some pathogens. Imagine that one person in a city of 100,000 people has a cold and sneezes on her hands. Then, without washing her hands, she goes through the day shaking hands with ten people, and each of those people shakes hands with another ten people per day, and so on. If there were no repeat handshakes, and if none of the contacts washed their hands, it would take only 4 days to spread the virus throughout the population.

In this example, eventually the entire populace of the city would come in contact with the virus, but not everyone would actually contract disease. Additional factors influence whether the virus successfully replicates in a given individual.

Portals of Entry

How do infectious agents gain access to the body? Each organism is adapted to enter the body in different ways. Food-borne pathogens (for example, *Salmonella*, *E. coli*, *Shigella*, and rotavirus) are ingested by mouth and ultimately colonize the intestine. They have an oral portal of entry. Airborne organisms, in contrast, infect through the respiratory tract. Some microbes enter through the conjunctiva of the eye, others through the mucosal surfaces of the genital and urinary tracts. Agents that are transmitted only by mosquitoes or other insects enter their human hosts via the parenteral route, meaning injection into the bloodstream. Wounds and needle punctures can also serve as portals of entry for many microbes. For instance, shared needle use between drug addicts has been an important factor in the spread of HIV.

Immunopathogenesis of Infectious Diseases

Although we focus in this chapter on the mechanisms microbes use to cause disease (toxins, for example), often it is "friendly fire" by our immune system reacting to a pathogen that causes major tissue and organ damage. The immune response to any infection involves activating a complex network of cell types and soluble factors (discussed in Chapters 23 and 24) that may inadvertently damage the host to such a degree that it causes illness and even death. This collateral damage or "immunopathology" is a calculated risk taken by the host in its haste

to eradicate the pathogen. The term **immunopathogenesis** applies when the immune response to a pathogen is a contributing cause of pathology and disease.

The disease dengue hemorrhagic fever is a case in point. Caused by the dengue virus and transmitted by the *Aedes* mosquito, dengue fever manifests as a severe headache, muscle and joint pain, fever, and rash. The symptoms can also include abdominal pain, nausea, and vomiting. However, these symptoms are more a consequence of immunopathogenesis than they are a direct result of viral replication. Replication of the virus in host cells will lead to a massive activation of T cells (CD4$^+$ and CD8$^+$). The process triggers a cytokine cascade (some call it a "storm") that targets vascular endothelial cells and produces an endothelial "sieve" effect leading to fluid and protein leakage. Cytokines such as TNF-α, IL-1, and IFN-gamma (discussed in Chapters 23 and 24), along with many others that are released, contribute to inflammation and disease symptoms. So, to fully understand any infectious disease you must be aware of the pathogenic mechanisms wielded by the pathogen, but also realize that the disease symptoms may be due to immunopathogenesis.

TO SUMMARIZE:

- **Infection** with a microbe does not always lead to disease.
- **Primary pathogens** have mechanisms that help the organism circumvent host defenses.
- **Opportunistic pathogens** cause disease only in a compromised host.
- **Pathogenicity** refers to the mechanisms a pathogen uses to produce disease and how efficient the organism is at causing disease, whereas virulence is a measure of disease severity.
- **Diseases can be spread** by direct or indirect contact between infected and uninfected persons/animals or by insect vectors.
- **Pathogens use portals of entry** best suited to their mechanisms of pathogenesis.
- **Immunopathogenesis** occurs when the immune system's response to an infection damages host cells and tissues.

25.2 Virulence Factors and Pathogenicity Islands: The Tools and Toolkits of Microbial Pathogens

Pathogens can be distinguished from their avirulent counterparts by the presence of **virulence factors** that help establish the organism in the host and that alter host functions to cause disease. Virulence factors, which are encoded by virulence genes, include toxins, attachment proteins, capsules, and other devices used to avoid host innate and adaptive immune systems. All of these factors enhance the disease-producing capability, or pathogenicity, of the pathogen.

Extensive sequencing efforts have allowed us to compare genomes of many pathogens and expose some "footprints" of their evolution. For example, in bacterial pathogens, most chromosomes are dotted with clusters of pathogenicity genes that encode virulence functions. These gene clusters, called **pathogenicity islands**, can be considered the toolboxes of pathogens (originally discussed in Section 9.7). Many virulence genes reside in pathogenicity islands, although many others do not. Some virulence genes reside on plasmids (for example, the genes for the diarrhea-producing labile toxin of certain *E. coli* strains) or in phage genomes (such as the genes encoding the diphtheria toxin of *Corynebacterium diphtheriae*).

Many of the genes in pathogenicity islands were originally inherited through horizontal transmission from other organisms by, for example, conjugation or transduction (discussed in Section 9.2). But what do the pathogenicity genes do? Some genes encode molecular "grappling hooks," such as pili that attach to host cells. Once attached, microbes can secrete toxins that injure the host cell. Other bacteria wall themselves off to prevent damage by host inflammatory responses. Some bacterial pathogens are even capable of what could be called "host cell reprogramming." These organisms inject proteins directly into the host cell to disrupt normal signaling pathways. This reprogramming causes the target cell to engulf the bacterium, "commit suicide" (undergo apoptosis), or provide an even more intimate attachment platform at the cell surface.

Pathogenicity Islands Are Different from the Rest of the Genome

How do new pathogens evolve? DNA sequencing efforts suggest that gene transfer followed by divergent evolution is a major force in the development of emerging pathogens. Horizontal gene transfers move whole blocks of DNA (more than 10 kb) from one organism to another, placing the blocks directly in the chromosome in what is called a **genomic island** (see Chapter 9). If the island increases the "fitness" (virulence) of a microorganism (pathogen) that interacts with a host, it is called a pathogenicity island. Genomic islands generally reveal themselves by several anomalies that they possess with respect to the rest of the host genome:

■ Genomic islands are often linked to a tRNA gene and generally have a GC/AT ratio very different from that of the rest of the chromosome (**Fig. 25.8A**). For example, a plot of GC content along the length of a

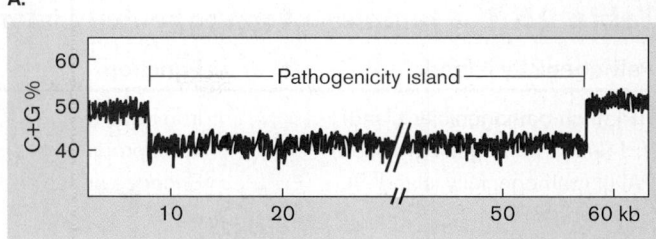

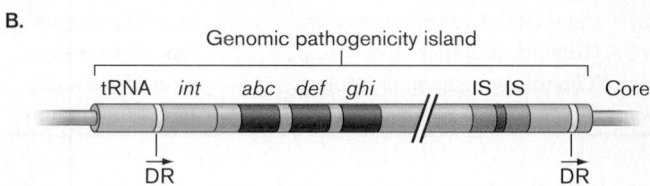

Figure 25.8 Model pathogenicity island. A. The guanine + cytosine (GC) content of the island is different from that of the core genome. **B.** Schematic model of a pathogenicity island. The DNA block is linked to a tRNA gene and flanked by direct repeats (DR) that may be "footprints" of a transposon or viral-mediated transfer. The integrase gene (*int*) and insertion sequences (IS) may also be remnants of transposition.

chromosome may reveal that most of the genome has a 50% GC content. But somewhere in the middle, a 50-kb region sticks out on the graph, showing a content of 40%. This probably reflects the GC content of the microbe that donated the island. The reason tRNA genes are often the targets for insertion of pathogenicity islands is not known. One hypothesis is that the conserved secondary structure of tRNA genes provides a structural motif that facilitates integration by an integrase.

■ Genomic islands are typically flanked by genes with homology to phage or plasmid genes (**Fig. 25.8B**). This is thought to reflect the transfer vector used to move the island from one organism to another.

■ Pathogenicity islands carry gene clusters with specific functions, such as protein export systems that secrete toxins (for example, type III secretion systems that inject toxic proteins directly into target host cells; see Section 25.5). These are the operons that contribute to the fitness of the organism in pathogenic circumstances.

Table 25.1 lists several pathogenicity islands and their functions. It provides examples of pathogenicity islands present in different bacteria, names a key function of the products of the island, and gives the disease caused. Various pathogenicity island functions will be described as the chapter proceeds.

WWW | Virulence factors of pathogenic bacteria
microbiology2.com/links

Table 25.1 Examples of pathogenicity islands.

Pathogenicity island	Function	Organism	Disease
HPI (high-pathogenicity island)	Iron uptake	*Yersinia* spp.	Plague, enterocolitis
VPI (*Vibrio* pathogenicity island)	Toxin production	*Vibrio cholerae*	Cholera
PAI III (pathogenicity island III)	Encodes adhesins	Uropathogenic *E. coli*	Urinary tract infection
SPI-1 and SPI-2 (*Salmonella* pathogenicity islands)	Type III secretion	*Salmonella enterica*	Gastroenteritis
SHI-1 and SHI-2 (*Shigella* islands)	Type III secretion	*Shigella flexneri*	Bloody diarrhea
YSA (*Yersinia* secretion apparatus)	Type III secretion	*Yersinia* spp.	Plague, enterocolitis
cag PI (cytotoxin-associated gene)	Type IV secretion	*Helicobacter pylori*	Gastric ulcers, gastric cancer
icm/dot (intracellular multiplication)	Type IV secretion	*Legionella pneumophila*	Legionnaires' disease

Shigella and *Escherichia* Evolved through Horizontal Transmission

Figure 25.9 shows a schematic comparison of the circular *Shigella flexneri* genome with the genomes of *E. coli* K-12 (a commonly used avirulent lab strain) and *E. coli* O157:H7 (a virulent strain of enterohemorrhagic *E. coli*). All three Gram-negative rods are closely related but differ greatly in terms of pathogenic potential and mechanisms. *S. flexneri* and *E. coli* O157:H7 cause bloody diarrhea, while *E. coli* K-12 has a commensal origin. DNA sequence analysis has revealed chromosome regions that are common among these organisms, as well as genomic islands specific to individual species and strains (**Fig. 25.9**). The core genes needed for sustaining growth (that is, for transcription,

translation, replication, and so on) are common to all three organisms, whereas other genes may be present in only one. These unique genes and islands are thought to be the result of horizontal gene transfers originating from widely different genera. Within the unique DNA segments of *S. flexneri* and *E. coli* O157:H7 are genes encoding host cell attachment, toxin secretion, and the toxins themselves—all of which are absent from K-12. The degree of gene shuffling that was required to separate these otherwise similar bacteria is remarkable.

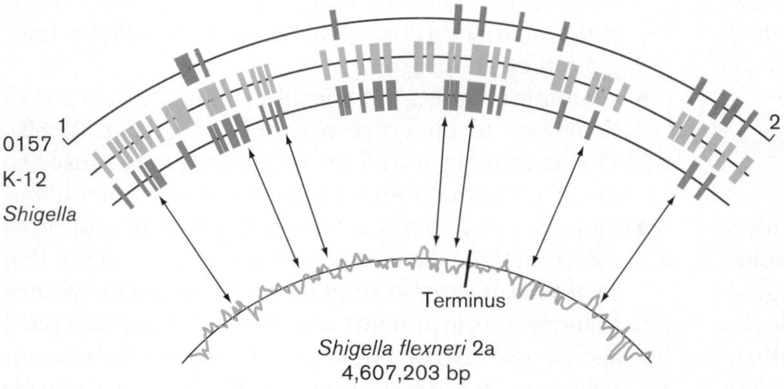

Caught in the Act: *Streptococcus agalactiae* Evolved through Conjugation

The Gram-positive organism *Streptococcus agalactiae*, also called group B streptococcus for its capsular antigen type, is the leading neonatal pathogen in the developed world. It is part of the normal vaginal microbiota of some women and can be transferred to a neonate passing through the birth canal. The organism is a major cause of newborn septicemia, which can lead to numerous long-term symptoms and even death. At least five clusters of strains are recognized, which vary in the severity of disease they can cause. Philippe Glaser and colleagues at the Pasteur Institute recently discovered how these different strain clusters may have evolved.

The scientists demonstrated that different *S. agalactiae* strains exchange large fractions of their chromosome and apparently mobilize them via conjugative elements located at different positions around the chromosome. The lengths of the exchanged regions are very large, ranging from 20 to 330 kb. The authors demonstrated that two strains of *S. agalactiae* will transfer large segments of DNA in vitro via conjugation. They used a donor strain whose genome was tagged at 30 different places with an antibiotic resistance marker. Recipient strains that became resistant to the drug after conjugation were isolated, and the position

Figure 25.9 Comparison of the *Shigella flexneri* 2a chromosome with chromosomes of *E. coli* K-12 and O157:H7 (EDL933). Segments of the three genomes between 1 Mb and 2 Mb are shown. Gray line indicates DNA sequences shared among the organisms (O157, top arc; K-12, second arc; *Shigella*, third arc). Colored boxes depict genomic islands (including pathogenicity islands) present in each organism. The bottom arc illustrates the GC content of the *Shigella* genome. Each data point along the graph indicates GC content relative to AT content as averaged over a sliding 10-kb window. Note how major differences in *Shigella* GC content shown above or below the center line (indicating 50% GC content) often correlate with the genomic islands depicted in the third arc. Arrows indicate some islands with obvious correlation to GC content.

of the marker was identified by DNA sequence analysis. The length of the recombined segment was then determined by looking for single nucleotide polymorphisms (SNPs) in flanking DNA, which would denote differences between the two sequenced genomes.

The scientists then compared genome sequences of eight clinical isolates using BLAST analysis (**B**asic **L**ocal **A**lignment **S**earch **T**ool), a method for rapidly searching nucleotide and protein databases. Since the BLAST algorithm detects both local and global alignments, regions of similarity embedded in otherwise unrelated proteins can be detected. A direct way to expose historical recombination events is to study SNPs present in the core genomes of related isolates. Nucleotide substitutions between strains were counted in 500-bp windows across the previously defined core chromosome. Strikingly, each pairwise comparison revealed an alternating pattern of highly conserved regions and less conserved regions, suggesting the occurrence of recombinational exchanges. The abrupt transition between conserved and divergent blocks allowed the investigators to predict recombination end points involving DNA exchanges of up to hundreds of kilobases.

Conjugative transfer and the replacement of hundreds of kilobases of bacterial chromosomes had already been described under laboratory conditions, but their impact on natural populations remained largely unknown. This work demonstrated that large conjugal exchanges have contributed significantly to the genome dynamics of *S. agalactiae* and strengthen the role of integrative conjugative elements in the dynamics of bacterial chromosomes.

Staphylococcus aureus Swaps Pathogenicity Islands by Transduction

Another pathogen caught in the act of transferring virulence genes is *Staphylococcus aureus*, a cause of boils, toxic shock syndrome, and many other infections. The *S. aureus* pangenome possesses a large, 15- to 17-member family of chromosomal pathogenicity islands (SaPIs), most of which encode superantigens that cause superantigen-type diseases such as toxic shock syndrome (superantigens are discussed in Chapter 24). What is remarkable is that most SaPIs transfer between different strains with relative ease. Because of this mobility, SaPIs are widely distributed in the genomes of *S. aureus*, with many strains possessing two or more. Interestingly, the genome organization of these elements is similar to that of a temperate phage. However, the SaPI phages are defective and do not form phage-like particles unless the strain is infected by certain types of helper phages that supply missing proteins. Once that happens, the elements are packaged into particles composed of phage proteins. The SaPI particles are released from the donor cell and can "infect" another

strain of *S. aureus*, where the pathogenicity island is then incorporated into the recipient's genome. This is a clear example of horizontal gene transfer happening every day somewhere on the planet. Even more startling is that these phage particles can also transfer the SaPI genes with equal ease to *Listeria monocytogenes*, a completely different genus that causes food-borne infections.

The following sections describe some of the specific tools that pathogens use to undermine the integrity of the body. From attachment, to toxins, to intracellular invasion, the infection process is like a chess match, with each side, human and microbe, trying to outmaneuver the other.

TO SUMMARIZE:

- **Pathogenicity islands** are DNA sequences within a species that are acquired by horizontal gene transfer from a different species.
- **Virulence genes** encode genes whose products enhance the disease-causing ability of the organism. Many virulence genes can be found within pathogenicity islands, but some are located outside of an obvious genomic island or reside in plasmids.
- **Pathogenicity islands** contain distinct features, such as GC content and the remnants of phages or plasmids, that mark them as being different from the rest of the genome.
- **Gene transfer mechanisms** will, by horizontal transfer, move virulence factor genes and pathogenicity islands among bacterial strains and species.

25.3 Virulence Factors: Microbial Attachment

Regardless of the disease, pathogens must reach a colonization site either through their own motility or by hitchhiking with a vector. Once at the site, the pathogen needs attachment mechanisms to stay there.

The human body has many ways to exclude pathogens. The lungs use a mucociliary elevator (see Fig. 23.5) to rid themselves of foreign bodies, the intestine uses peristaltic action to ensure that its contents are constantly flowing, and the bladder uses contraction to propel urine through the urethra with tremendous force. How do bacteria ever manage to stay around long enough to cause problems? Like a person grasping a telephone pole during a hurricane, successful pathogens moving through the body manage to grab on to host cells and tenaciously hold on. Thus, the first step toward infection is attachment, also called adhesion. An **adhesin** is the general term for any microbial factor that promotes attachment.

Viruses attach to the host through their capsid or envelope proteins, which bind to specific host cell receptors discussed in Chapters 6 and 11. Bacteria have a variety of similar strategies. They can use hairlike appendages called **pili** (also called **fimbriae**), whose tips contain receptors for mammalian cell surface structures. Or they can use a variety of adherence proteins or other molecules (adhesins) that are not part of a pilus. Sometimes they use both. **Table 25.2** summarizes bacterial attachment strategies. Note that different pili can impart tissue specificity for attachment (for example, uropathogenic versus diarrheagenic *E. coli*).

Types of Pili

Different pili from different bacterial species have been classified historically by phenotype. In some cases the phenotype classes have turned out to be inconsistent with sequence-based homologies. We will consider three groups in this chapter.

Type I pili are a group that, in general, adhere to mannose residues on host cell surfaces. Because adding

free mannose will inhibit attachment of most type I pili, this binding is called mannose sensitive. Another group of pili is called mannose resistant because the pili do not bind to mannose residues. There are at least two types of mannose-resistant pili. Members of one type, sometimes called **type III pili**, bind to red blood cells treated with tannic acid. The other, more commonly studied type is called **type IV pili**. Unlike type I pili, which simply stick out from the cell surface, type IV pili are more dynamic and are assembled on the cell surface through a very different pathway.

There are other types of attachment pili that do not fall neatly into one of these three groups. The primary classification of pili is now based largely on protein sequence information (deduced from DNA sequence) and may contradict earlier phenotype-based schemes. For instance, the pyelonephritis-associated pili (Pap) of the uropathogenic *E. coli* are mannose resistant, since they bind to a digalactoside on host surfaces (called the P blood group antigen). However, amino acid sequence

Table 25.2 Specific attachments of bacteria to cell or tissue surfaces.

Bacterium	Adhesin	Host receptor	Attachment site	Disease
Streptococcus pyogenes	Protein F	Amino terminus of fibronectin	Pharyngeal epithelium	Sore throat
Streptococcus mutans	Glucan	Salivary glycoprotein	Pellicle of tooth	Dental caries
Streptococcus salivarius	Lipoteichoic acid	Unknown	Buccal epithelium of tongue	None
Streptococcus pneumoniae	Cell-bound protein	N-Acetylhexosamine galactose disaccharide	Mucosal epithelium	Pneumonia
Staphylococcus aureus	Cell-bound protein	Amino terminus of fibronectin	Mucosal epithelium	Various
Neisseria gonorrhoeae	N-Methylphenylalanine pili	Glucosamine galactose carbohydrate	Urethral/cervical epithelium	Gonorrhea
Enterotoxigenic *E. coli*	Type I fimbriae (pili)	Species-specific carbohydrate(s)	Intestinal epithelium	Diarrhea
Uropathogenic *E. coli*	Type I fimbriae (pili)	Complex carbohydrate	Urethral epithelium	Urethritis
Uropathogenic *E. coli*	P pili (pyelonephritis-associated pili)	P blood group	Upper urinary tract	Pyelonephritis
Bordetella pertussis	Pili ("filamentous hemagglutinin")	Galactose on sulfated glycolipids	Respiratory epithelium	Whooping cough
Vibrio cholerae	N-Methylphenylalanine pili	Fucose and mannose carbohydrate	Intestinal epithelium	Cholera
Treponema pallidum	Peptide in outer membrane	Surface protein (fibronectin)	Mucosal epithelium	Syphilis
Mycoplasma	Membrane protein	Sialic acid	Respiratory epithelium	Pneumonia
Chlamydia	Unknown	Sialic acid	Conjunctival or urethral epithelium	Conjunctivitis or urethritis
Corynebacterium diphtheriae	Pili	Unknown	Pharyngeal epithelium	Diphtheria

homology suggests that Pap is very similar to type I pili. Note that pili, or other adhesins, can be virulence factors on some organisms but not on others.

Pilus Assembly

How bacteria assemble pili on their cell surfaces is an engineering marvel. The shafts of pili are cylindrical structures composed of identical pilin protein subunits. Several different proteins adorn the tip, including one at the very apex that binds to host receptors (**Fig. 25.10A**). In addition to these structural components, numerous other proteins work together as a machine to assemble the structure. Genes encoding a given pilin protein and the associated assembly apparatus are typically arranged on the chromosome as an operon (**Fig. 25.10B**).

Figure 25.11 illustrates the assembly of a type I pilus using uropathogenic *E. coli* Pap as a model. The mechanism is representative of other type I pili; only the names of the proteins will differ for each system. Protein components are secreted into the periplasm by the SecA-dependent general secretory system (discussed in Section 8.5). Once in the periplasm, the subunits are chaperoned by PapD to the membrane site of assembly, which is marked by the presence of the usher protein PapC. PapC proteins form channels in the outer membrane large enough to accommodate individual pilus subunits and, like an usher in a theater, direct the subunits to their proper places. Chaperoning of the pilin building blocks by PapD is necessary to prevent pilin subunits from inadvertently assembling in the periplasm. As

illustrated in **Figure 25.11**, appropriate assembly of pili at the usher site starts with the tip protein, PapG, which will ultimately bind to carbohydrates on host membranes after the pilus is complete. After PapG, the ushers add PapF and PapE, forcing PapG farther away from the surface. Then, identical PapA pilin subunits are strung together in a series to form the shaft. The PapA subunits assemble by swapping domains, essentially linking themselves together like pieces of a jigsaw puzzle.

Interestingly, shear forces generated by fast-moving urine in the urethra may actually tighten bacterial adhesion to target cells. The tip protein, FimH, which is analogous to PapG but for a different type I pilus, contains two

A. FimH adhesion tip

16 nm

C. Hal Jones et al. 1995. *PNAS USA* **92**:2081

B. Type I pilus gene cluster

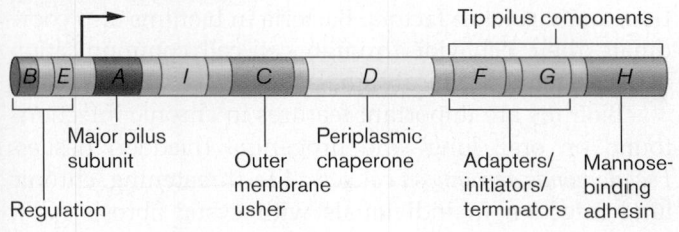

Figure 25.10   Attachment pili and encoding operon.   A. High-resolution micrograph showing a type I pilus (TEM). The FimH adhesin at the tip is the protein that binds to the cell receptor. **B.** Genetic organization of the type I gene cluster, which includes genes involved in pilus assembly. The genes are designated *fim A–I*.

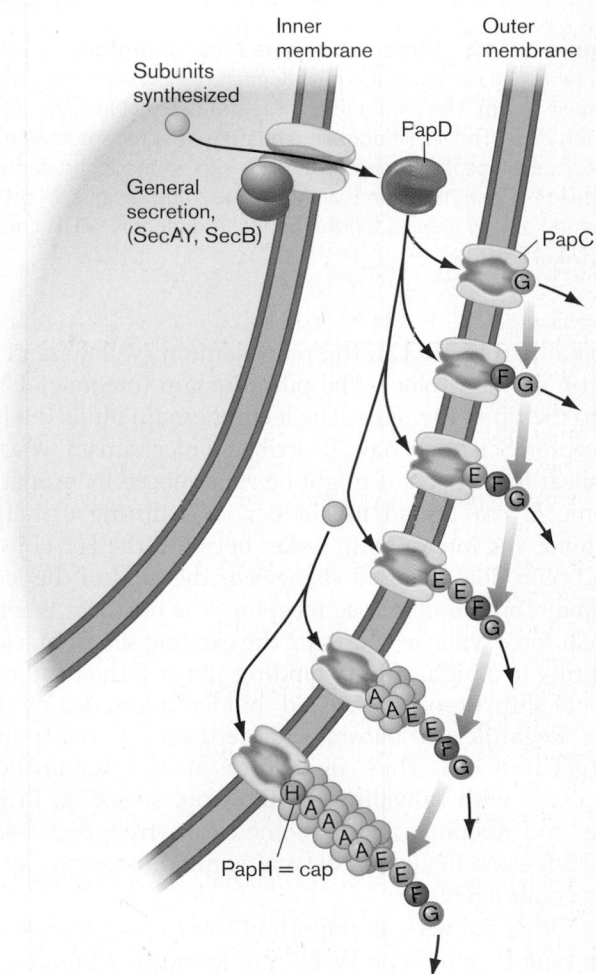

Figure 25.11   Assembly of type I pili.   The figure illustrates pyelonephritis-associated pilus (Pap) assembly but is representative of other type I pili such as in Fim in Fig. 25.10. Only the names of the proteins will differ. Proteins are secreted by the Sec system to the periplasm, where they are chaperoned by PapD to the site of assembly. PapC, also called the usher, assembles the individual proteins in the proper order. Assembly starts with the tip protein, PapG, marked at the far right, which ultimately binds to carbohydrates on host membranes. The subunits fit together like pieces of a jigsaw puzzle. The arrow at the head of the elongating pilus indicates the direction of pilus growth.

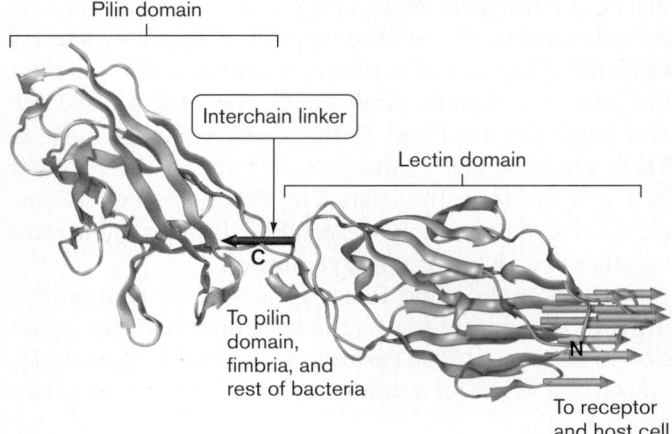

Pilin domain

Interchain linker

Lectin domain

C

To pilin
domain,
fimbria, and
rest of bacteria

N

To receptor
and host cell

Figure 25.12   Structure of type I pili tip protein.   FimH contains two domains, the pilin domain (yellow) and the lectin domain (blue). The pilin domain integrates FimH into the tip of the pilus. The lectin domain binds the host receptor. When shear develops, the receptor-binding site on the lectin domain is pulled in one direction (yellow arrows), and residue T158 (orange arrow) is pulled in the opposite direction. (PDB codes: 2CO4, 1TR7)

domains (**Fig. 25.12**): the pilin domain (yellow) and the lectin domain (blue). The pilin domain integrates FimH into the tip of the pilus. The lectin domain binds the host receptor. Scientists have described a mechanism whereby a shear force, such as might be experienced by uropathogenic *E. coli* bound to bladder cells during urination, extends the interdomain linker between the FimH lectin and pilin domains and *strengthens* the hold of the lectin domain on the mannose receptor. It is not clear whether shear force ends up changing the existing site from a low-affinity to a high-affinity binding site or if the conformational shift exposes a second, hidden, mannose-binding site. Regardless, it becomes harder to pry *E. coli* from its target host cell. Thus, fluid forces in the human body, as occur with salivating, swallowing, sneezing, urinating, and weeping, may in some cases strengthen bacterial adhesion instead of detaching and flushing away the infectious agents.

Other pili with an important role in pathogenesis are the type IV pili. Type IV pili are found in a broad spectrum of Gram-negative bacteria and share amino acid homology in their major pilin structure (**Fig. 25.13**). All type IV pili use similar secretion and assembly machinery involving at least a dozen proteins. One major difference between the assembly of type IV and type I pili is that type IV pilus proteins are never free in the periplasm; they are transported directly from the cytoplasm through a channel in the outer membrane (**Fig. 25.13A**). Thus, type IV pilus assembly is SecA independent. As described here, the blueprints for this mechanism were adapted and modified through evolution to secrete other proteins via

what is generally called type II protein secretion (discussed in Section 25.5). Species with type IV pili include *Vibrio cholerae*, *Pseudomonas aeruginosa* (**Fig. 25.13B**), certain pathogenic strains of *E. coli*, *Neisseria meningitidis* (**Fig. 25.13C**), and *N. gonorrhoeae*.

Type IV pili can actually make cells move because the assembly process involves the reiterative elongation and retraction of the pili. The process, called "twitching motility," occurs when the pilus elongates, attaches to a surface, and then depolymerizes from the base, which shortens the pilus and pulls the cell forward. This mechanism is akin to using a grappling hook to scale a building. The gliding motility of the slime mold *Myxococcus xanthus* is also due to type IV pili. The type IV pili of *Neisseria meningitidis*, shown in **Figure 25.13C**, are essential for crossing the blood-brain barrier and causing bacterial meningitis.

Bacteria also carry afimbriate adhesins (proteins that aid in attachment but do not form pili) that mediate binding to host tissues (**Fig. 25.14**). Some examples include *Bordetella* pertactin (which binds to integrin), *Streptococcus* protein F (binds to fibronectin), *Streptococcus* M protein (binds to fibronectin and complement regulatory factor H), and intimin of enteropathogenic *E. coli* (binds to Tir; discussed in Section 25.5). Fimbriae (pili) often mediate the initial binding between bacterium and host, after which a more intimate attachment is formed by an afimbriate attachment protein. In the case of *Neisseria gonorrhoeae*, once the type IV pilus has attached to the surface of the mucosal epithelial cell, the filamentous pilus contracts, pulling the bacterium down onto the host cell membrane. Tight secondary interactions are then mediated by the neisserial Opa membrane proteins, another example of an afimbriate adhesin (Opa is named because of the *opa*city it adds to colony appearance).

Biofilms and Infections

As first discussed in Section 4.6, bacteria in most environments form organized, high-density communities of cells called biofilms that are embedded in self-produced exopolymer matrices. Biofilm development is an ancient prokaryotic adaptation that allows microorganisms to adhere to any surface, living or nonliving, and facilitates survival in hostile environments. Within a single biofilm you can find localized differences in the expression of surface molecules, antibiotic resistance, nutrient utilization, and virulence factors. Bacteria in biofilms also coordinate their behavior through cell-cell communication using secreted chemical signals.

Biofilms are important features in chronic infections found on oral, lung, and urogenital (bladder) tissues. *Pseudomonas aeruginosa* causes a life-threatening, chronic lung infection in individuals with cystic fibrosis (CF).

A.

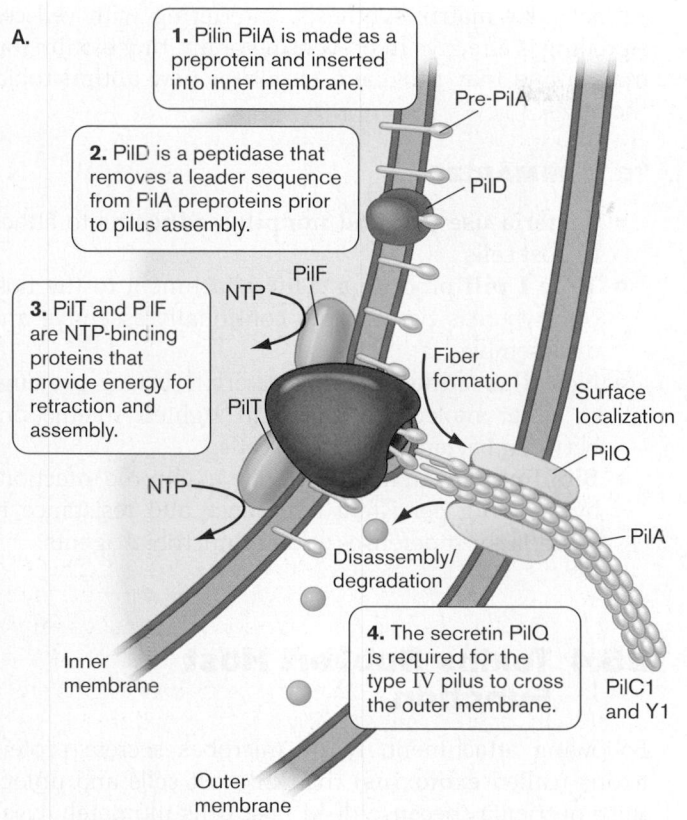

1. Pilin PilA is made as a preprotein and inserted into inner membrane.

2. PilD is a peptidase that removes a leader sequence from PilA preproteins prior to pilus assembly.

3. PilT and PilF are NTP-binding proteins that provide energy for retraction and assembly.

4. The secretin PilQ is required for the type IV pilus to cross the outer membrane.

Pre-PilA

PilD

PilF

NTP

PilT

NTP

Fiber formation

Surface localization

PilQ

PilA

Disassembly/ degradation

PilC1 and Y1

Inner membrane

Outer membrane

Figure 25.13   Type IV pili.   A. Model of pilus assembly and disassembly. In this example, PilA is the pilin protein, and PilC1 and Y1 form the attachment tip. Diameter of the filament is approximately 6 nm. Note that assembly/disassembly requires hydrolysis of nucleoside triphosphate and takes place at the inner membrane, not in the periplasm. **B.** Photographic evidence of type IV pilus extension and retraction in cells of *Pseudomonas aeruginosa*. Filament b retracts; then filament d extends at 6 seconds and retracts. Filament c attaches briefly at its distal tip (note straightening at 24 seconds) and then begins to retract. Time (*t*) in seconds. Fluorescent microscopy. **C.** Type IV pili are essential for the interaction of *Neisseria meningitidis* with brain endothelial cells. Type IV pili are green in this SEM. Diplococcal cells are approximately 1.6 μm in diameter. *Sources:* A. Bardy et al. 2003. *Microbiol.* **149**:295–304; B. J. M. Skerker and H. C. Berg. 2001. *PNAS* **98**:6901–6904.

B. Extension and contraction of Type IV pili

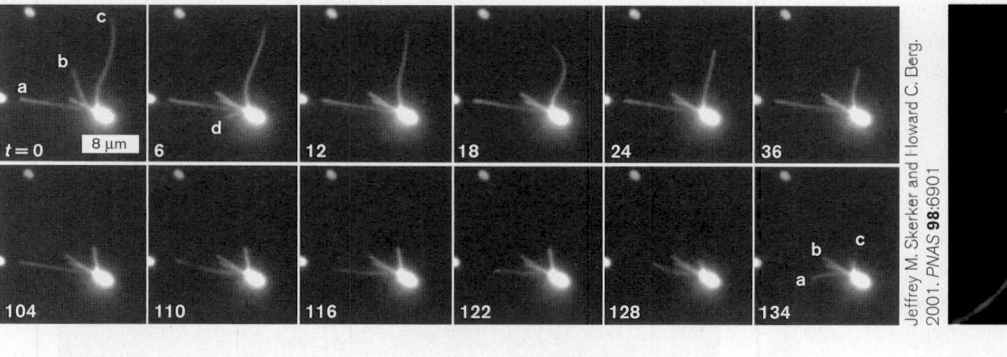

Jeffrey M. Skerker and Howard C. Berg. 2001. *PNAS* **98**:6901

C. Type IV pili of *Neisseria meningitidis*

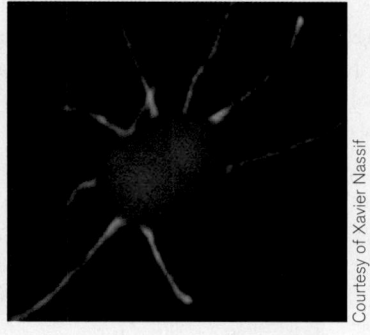

Courtesy of Xavier Nassif

A. M protein

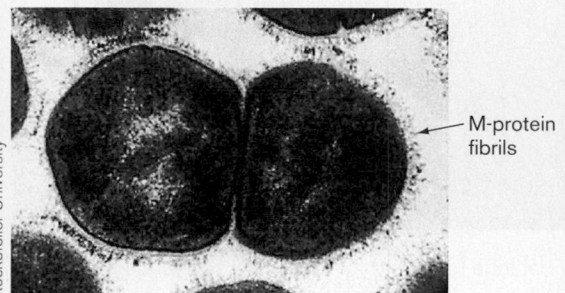

Maria Fazio and Vincent A. Fischetti; Rockefeller University

M-protein fibrils

B. *B. pertussis* colonizing the trachea

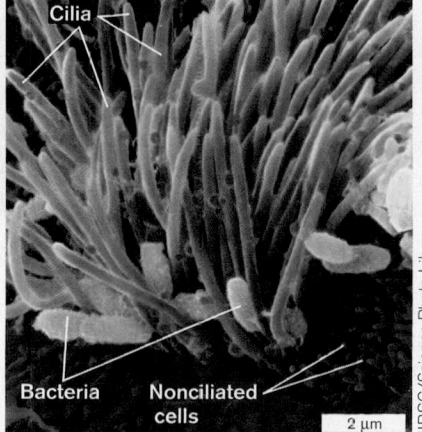

Cilia

Bacteria

Nonciliated cells

NIBSC/Science Photo Library

Figure 25.14   Nonpilus adhesins.   A. M-protein surface fibrils on *Streptococcus pyogenes* (TEM). Cell size 0.5–1 μm in diameter. **B.** Colonization of tracheal epithelial cells by *Bordetella pertussis* (SEM). This organism uses a surface protein called pertactin, as well as a pilus called filamentous hemagglutinin (FHA) to bind bronchial cells.

This microbe has been found growing as aggregates enclosed in a matrix within mucus from CF patients. It is thought that insufficient mucociliary clearance contributes to *P. aeruginosa* biofilm formation. Biofilms are also important in periodontitis (gum disease), indwelling catheter infections, infections of artificial heart valves, chronic urinary tract infections, recurrent tonsillitis, rhinosinusitis, chronic otitis media (middle ear infection), chronic wound infections, and osteomyelitis (bone infection). **Figure 25.15A** shows a scanning EM of a chronic infection of tonsil tissue from a pediatric patient. The fluorescent confocal image in **Figure 25.15B** shows a three-dimensional view of a biofilm on adenoid tissue with live cells stained green and dead cells stained red. The chronic presence of these organisms in biofilms will continually stimulate innate immune mechanisms through interactions with Toll-like receptors and cause chronic inflammation as a result. A biofilm infection may linger for months, years, or even a lifetime. If associated with an implanted medical device, the device may have to be replaced. Today, vascular catheter-related bloodstream infections are the most serious and costly health care–associated infections.

Biofilm infections are very important clinically because bacteria in biofilms exhibit tolerance to antimicrobial compounds and persistence in spite of sustained host defenses. Thus, biofilm infections are hard to cure. Tolerance to antibiotics may be caused by poor nutrient penetration into the deeper regions of the biofilm, leading to a stationary phase–like dormancy. Bacterial factors important to biofilm formation include type IV pili, quorum-sensing structural genes and regulators, and

extracellular matrix synthesis. Interfering with cell-cell signaling is effective in preventing or limiting biofilm formation and may provide a target for new antimicrobial therapies.

TO SUMMARIZE:

- **Bacteria use pili and nonpilus adhesins** to attach to host cells.
- **Type I pili** produce a static attachment to the host cell, whereas type IV pili continually assemble and disassemble.
- **Nonpilus adhesins** are bacterial surface proteins, or other molecules, that can tighten interactions between bacteria and target cells.
- **Biofilms** play an important role in chronic infections by enabling persistent adherence and resistance to bacterial host defenses and antimicrobial agents.

25.4 Toxins Subvert Host Function

Following attachment, many microbes secrete protein toxins (called **exotoxins**) that kill host cells and unlock their nutrients (because dead host cells ultimately lyse). Bacterial pathogens have developed an impressive array of toxins that take advantage of different key host proteins or structures. All Gram-negative bacteria also possess a nonprotein yet toxic compound called **endotoxin** that can hyperactivate host immune systems to harmful levels.

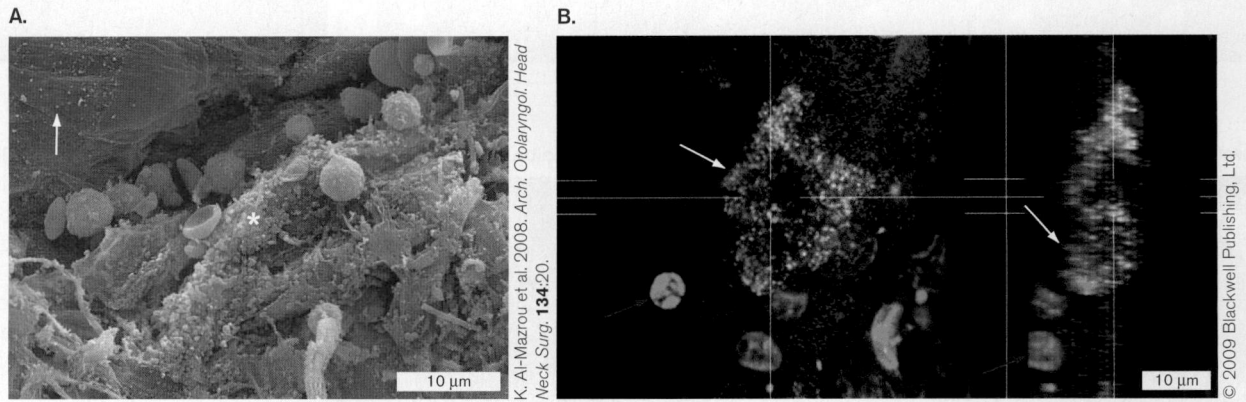

Figure 25.15   A bacterial biofilm infection.   A. Scanning EM showing mixed bacteria (asterisk) and adherent biofilm on the surface epithelium (arrow) of an infected tonsil. **B.** Confocal micrograph showing biofilm clusters (white arrows) consisting of rods and cocci on the mucosa of a pediatric adenoid. The adenoid was removed as routine treatment of recurrent otitis media (middle ear infection). Specimens were treated with nucleic acid stains using the LIVE/DEAD BacLight Bacterial Viability Kit, in which live bacteria stain green and dead bacteria stain red. Host inflammatory cells (red arrows) were also stained but were readily distinguished on the basis of size and nuclear morphology. The mucosal surface (blue) was imaged using reflected light. *Source for B:* Luanne Hall-Stoodley and Paul Stoodley. 2009. *Cell Microbiol.* **11**:1034–1043.

Microbial Exotoxins Have Many Modes of Action

Microbial exotoxins fall into five broad categories based on their mechanisms of action (**Table 25.3**). Three of these classes are illustrated in **Figure 25.16**.

- **Cell membrane disrupters.** Members of the first class, exemplified by the alpha (α) toxin of *Staphylococcus aureus,* disrupt the cell membrane and cause leakage of cell constituents (**Fig. 25.16A**).
- **Protein synthesis disrupters.** A second class, exemplified by diphtheria and Shiga toxins, targets eukaryotic ribosomes and destroys protein synthesis (**Fig. 25.16B**).
- **Second messenger pathway disrupters.** The third broad mechanism of action involves the toxin subverting host cell second messenger pathways. Cholera toxin and *E. coli* ST (stable toxin), for instance, cause runaway synthesis of cAMP (described later) and cGMP (**Fig. 25.16C**), respectively, in target cells. Elevated cAMP or cGMP levels, in turn, trigger critical changes in ion transport and fluid movement.
- **Superantigens.** The fourth class includes toxins that act as superantigens and activate the immune system without being processed by antigen-presenting cells (discussed in Section 24.6). The pyrogenic toxins of *S. aureus* (toxic shock syndrome toxin) and *Streptococcus pyogenes* are examples of superantigenic toxins.
- **Proteases.** The fifth class of toxins consists of proteases. One example is tetanus toxin, a protease that cleaves host components involved in nerve signal transmission.

This section focuses on the key concepts of microbial toxin activity. Section 25.5 discusses how bacteria secrete these toxins.

A common structural theme among many, but not all, bacterial toxins is that they have two subunits, usually called A and B. These two-subunit complexes are termed AB toxins. The actual toxic activity in AB toxins resides within the A subunit. The role of the B subunit is limited to binding host cell receptors. Thus, the B subunit for each toxin delivers the A subunit to the host cell. Many AB toxins have five B subunits arranged as a ring, in the center of which is nestled a single A subunit (**Figs. 25.16B** and **25.17A**).

A major AB toxin subclass comprises toxins that have **ADP-ribosyltransferase** enzymatic activity. These enzyme toxins transfer the ADP-ribose group from an NAD molecule to a target protein (**Fig. 25.17B**). The ADP-ribosylated protein has an altered function. Sometimes the function is destroyed (for example, protein synthesis is destroyed by diphtheria toxin); other times, the protein is locked into an active form insensitive to regulatory feedback control (for example, cAMP synthesis continues unchecked in the presence of cholera toxin).

Mechanisms of selected toxins representing the various classes are described in this section.

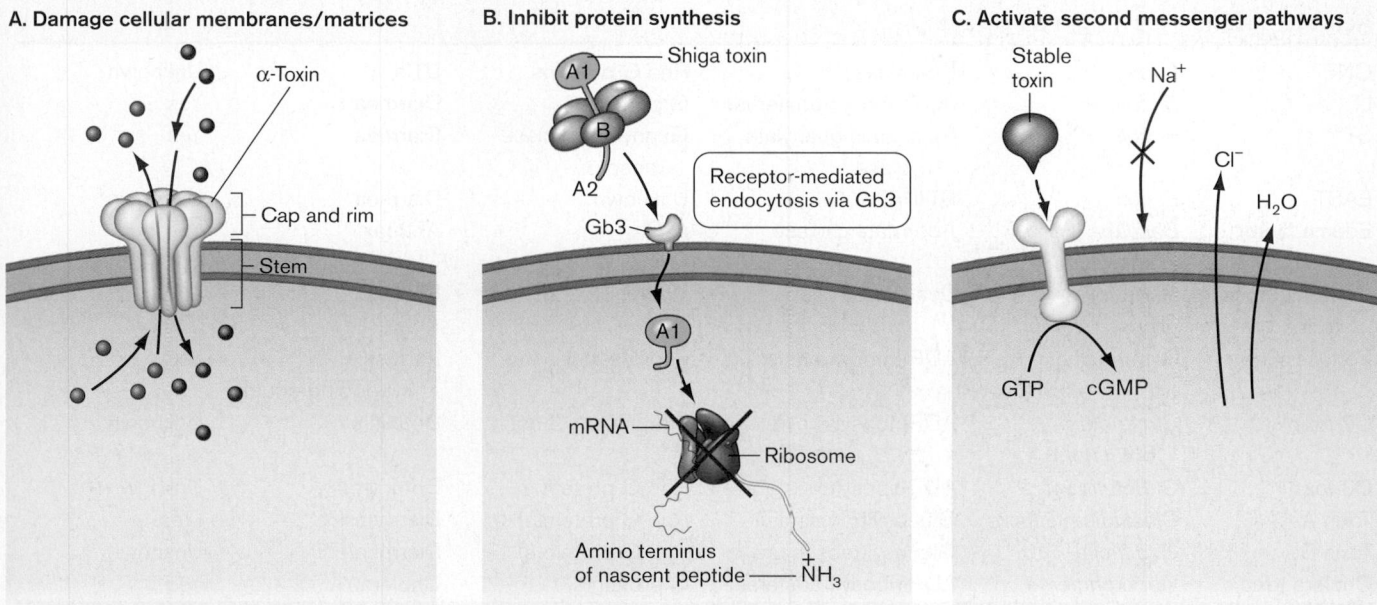

A. Damage cellular membranes/matrices

α-Toxin
Cap and rim
Stem

B. Inhibit protein synthesis

Shiga toxin
A1
B
A2
Receptor-mediated endocytosis via Gb3
Gb3
A1
mRNA
Ribosome
Amino terminus of nascent peptide — $\overset{+}{N}H_3$

C. Activate second messenger pathways

Stable toxin
Na^+
Cl^-
H_2O
GTP cGMP

Figure 25.16 Three classes of microbial exotoxins. These classes are defined by mode of action. **A.** Pore-forming toxins assemble in target membranes and cause leakage of compounds into and out of cells. **B.** Shiga toxin attaches to ganglioside Gb3, enters the cell, and cleaves 28S rRNA in eukaryotic ribosomes to stop translation. **C.** Enterotoxigenic *E. coli* heat-stable toxin affects cGMP production. The result is altered electrolyte transport: inhibition of Na^+ uptake and stimulation of Cl^- transport. In response to the resulting electrolyte imbalance, water leaves the cell.

Table 25.3 Characteristics of bacterial exotoxins.[a]

Toxin	Organism	Mode of action	Host target	Disease	Toxin implicated in disease[b]
Damage membranes					
Aerolysin	*Aeromonas hydrophila*	Pore former	Glycophorin	Diarrhea	(Yes)
Perfringolysin O	*Clostridium perfringens*	Pore former	Cholesterol	Gas gangrene[c]	Unknown
Hemolysin[d]	*Escherichia coli*	Pore former	Plasma membrane	UTIs	(Yes)
Listeriolysin O	*Listeria monocytogenes*	Pore former	Cholesterol	Food-borne systemic illness, meningitis	Yes
Alpha toxin	*Staphyloccocus aureus*	Pore former	Plasma membrane	Abscesses[c]	(Yes)
Panton-Valentine leukocidin	*Staphyloccocus aureus*	Pore former	Plasma membrane	Abscesses, necrotizing pneumonia	(Yes)
Pneumolysin	*Streptococcus pneumoniae*	Pore former	Cholesterol	Pneumonia[c]	(Yes)
Streptolysin O	*Streptococcus pyogenes*	Pore former	Cholesterol	Strep throat, scarlet fever	Unknown
Inhibit protein synthesis					
Diphtheria toxin	*Corynebacterium diphtheriae*	ADP-ribosyltransferase	Elongation factor 2	Diphtheria	Yes
Shiga toxins	*E. coli/Shigella dysenteriae*	*N*-Glycosidase	28S rRNA	HC and HUS	Yes
Exotoxin A	*Pseudomonas aeruginosa*	ADP-ribosyltransferase	Elongation factor 2	Pneumonia[c]	(Yes)
Activate second messenger pathways					
CNF	*E. coli*	Deamidase	Rho G proteins	UTIs	Unknown
LT	*E. coli*	ADP-ribosyltransferase	G proteins	Diarrhea	Yes
ST[d]	*E. coli*	Stimulates guanylate cyclase	Guanylate cyclase receptor	Diarrhea	Yes
EAST	*E. coli*	ST-like?	Unknown	Diarrhea	Unknown
Edema factor	*Bacillus anthracis*	Adenylate cyclase	ATP	Anthrax	Yes
Dermonecrotic toxin	*Bordetella pertussis*	Deamidase	Rho G proteins	Rhinitis	(Yes)
Pertussis toxin	*B. pertussis*	ADP-ribosyltransferase	G protein(s)	Pertussis (whooping cough)	Yes
C2 toxin	*Clostridium botulinum*	ADP-ribosyltransferase	Monomeric G-actin	Botulism	Unknown
C3 toxin	*C. botulinum*	ADP-ribosyltransferase	Rho G protein	Botulism	Unknown
Toxin A	*Clostridium difficile*	Glucosyltransferase	Rho G protein(s)	Diarrhea/PC	(Yes)
Toxin B	*C. difficile*	Glucosyltransferase	Rho G protein(s)	Diarrhea/PC	Unknown
Cholera toxin	*Vibrio cholerae*	ADP-ribosyltransferase	G protein(s)	Cholera	Yes

Table 25.3 Characteristics of bacterial toxins[a] (*continued*)

Toxin	Organism	Mode of action	Host target	Disease	Toxin implicated in disease[b]
Activate immune response					
Enterotoxins	*S. aureus*	Superantigen	TCR and MHC II	Food poisoning[c]	Yes
Exfoliative toxins	*S. aureus*	Superantigen (and serine protease?)	TCR and MHC II	Scalded skin syndrome[c]	Yes
Toxic shock syndrome toxin	*S. aureus*	Superantigen	TCR and MHC II	Toxic shock syndrome[c]	Yes
Pyrogenic exotoxins	*S. pyogenes*	Superantigens	TCR and MHC II	Toxic shock syndrome	Yes
				Scarlet fever	Yes
Protease					
Lethal factor	*B. anthracis*	Metalloprotease	MAPKK1/MAPKK2	Anthrax	Yes
Neurotoxins A–G	*C. botulinum*	Zinc metalloprotease	VAMP/synaptobrevin, SNAP-25/syntaxin	Botulism	Yes
Tetanus toxin	*Clostridium tetani*	Zinc metalloprotease	VAMP/synaptobrevin	Tetanus	Yes

[a]**Abbreviations:** CLDT, cytolethal distending toxin; CNF, cytotoxic necrotizing factor; EAST, enteroaggregative *E. coli* heat-stable toxin; HC, hemorrhagic colitis; HUS, hemolytic uremic syndrome; LT, heat-labile toxin; MAPKK, mitogen-activated protein kinase kinase; MHC II, major histocompatibility complex class II; PC, antibiotic-associated pseudomembranous colitis; SNAP-25, synaptosomal-associated protein; ST, heat-stable toxin; TCR, T-cell receptor; UTI, urinary tract infection; VAMP, vesicle-associated membrane protein.

[b]Yes, strong causal relationship between toxin and disease; (Yes), role in pathogenesis has been shown in animal model or appropriate cell culture.

[c]Other diseases are also associated with the organism.

[d]Toxin is also produced by other genera of bacteria.

Source: C. K. Schmidt, K. C. Meysick, and A. O'Brien. 1999. Bacterial toxins: Friends or foes? *Emerging Infectious Diseases* **5**:224–234.

Figure 25.17 AB toxins.
A. A typical AB toxin consists of an A subunit and a pentameric B subunit joined noncovalently.
B. Many AB toxins are ADP-ribosyltransferase enzymes that modify protein structure and function.

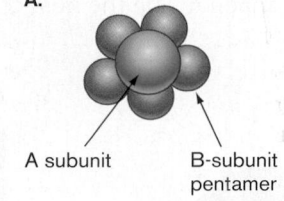

A subunit   B-subunit pentamer

NAD

ADP-ribosyltransferase

Target protein (active)

Target protein

ADP-ribosyl protein (inactive)

Nicotinamide

Alpha toxin. The hemolytic alpha toxin is produced by *Staphylococcus aureus*, an organism that causes boils and blood infections. Alpha toxin forms a transmembrane, oligomeric (seven-member) beta barrel pore in target cell membranes. It is easy to see how the resulting leakage of cell constituents and influx of fluid cause the target cell to burst. To form the pore, hydrophobic areas of each monomer face the lipids of the membrane, and hydrophilic residues face the channel interior. A completed pore and a cutaway view exposing the channel are illustrated in **Figure 25.18A** and **B**. Diagnostic microbiology laboratories visualize hemolysins (proteins that lyse red blood cells) such as alpha toxin by inoculating bacteria onto agar plates containing sheep

A. Alpha hemolysin

B. Cross section of alpha hemolysin

C. Hemolysis by *S. aureus*

Figure 25.18   Alpha hemolysin of *Staphylococcus aureus*.   A. Three-dimensional figure of the pore complex comprising seven monomeric proteins. (PDB code: 7AHL) **B.** Cross section showing the channel. **C.** A blood agar plate inoculated with *S. aureus*. The alpha toxin is secreted by the organism and diffuses away from the producing colony. It forms pores in the red blood cells embedded in the agar, causing them to lyse, which is visible as a clear area surrounding each colony.

red blood cells (**Fig. 25.18C**). The clear, yellow zones around the *S. aureus* colonies growing on blood agar indicate that the microbe secretes a hemolysin.

Cholera and *E. coli* labile toxins. *Vibrio cholerae* (**Fig. 25.19A**) produces a severe diarrheal disease called cholera that generally afflicts malnourished people populating poor countries like Bangladesh, or countries in which access to clean water has been disrupted by war

or natural disasters. This microbe produces a gastrointestinal enterotoxin nearly identical to one produced by some strains of *E. coli* associated with what is known as "traveler's diarrhea." Enterotoxins specifically affect the intestine. The *E. coli* enterotoxin is called **labile toxin (LT)** because it is easily destroyed by heat. Cholera toxin (CT) and labile toxin are both AB toxins with identical modes of action, which is to increase the level of cAMP made inside the host cell.

A. *Vibrio cholerae*

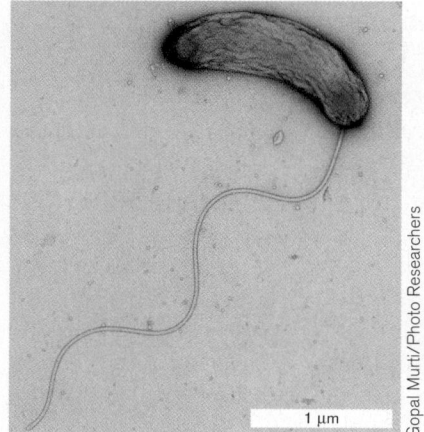

B. Cholera toxin

A subunit

B₅ subunits

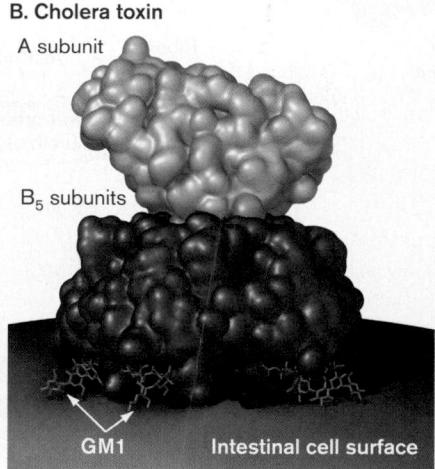

GM1   Intestinal cell surface

C. Brush border of intestine

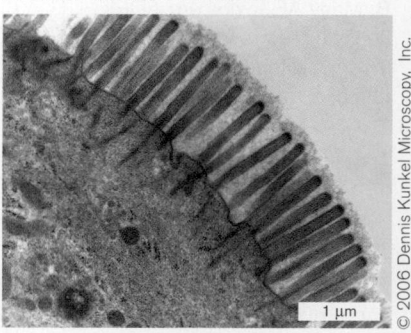

D. *V. cholerae* attachment

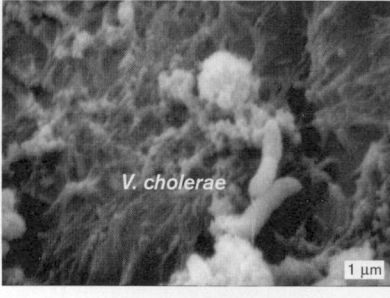

V. cholerae

Figure 25.19   Pathogenesis of cholera.   A. *Vibrio cholerae* (SEM). Note the slight curve of the cell and the presence of a single polar flagellum. **B.** Three-dimensional structure of cholera toxin binding ganglioside GM1 on the intestinal cell surface. (PDB code: 1S5F) **C.** Brush border of intestine (TEM). *V. cholerae* binds to the fingerlike villi on the apical surface. **D.** View of *V. cholerae* binding to the surface of a host cell (SEM). Note that *V. cholerae* does not invade the host cell.

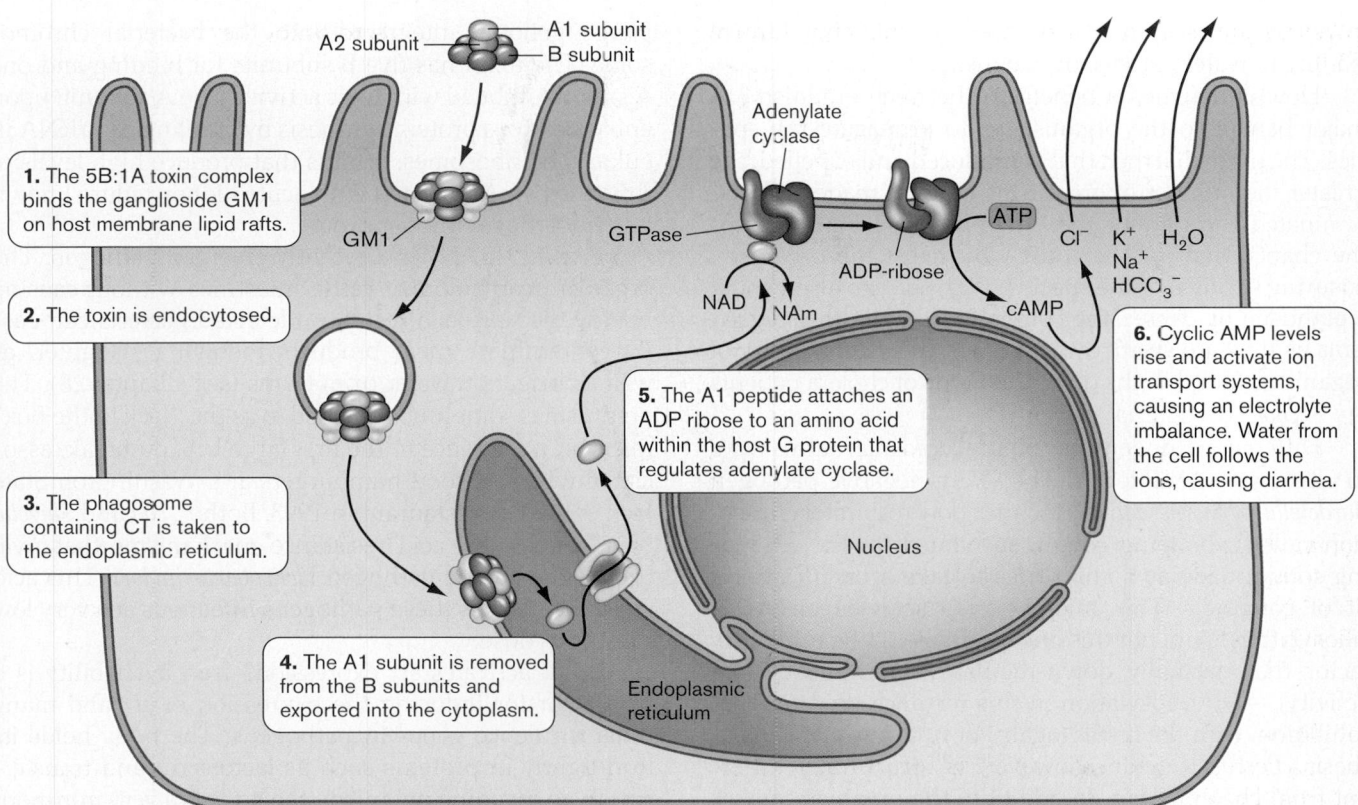

Figure 25.20   Cholera toxin mode of action.   Delivery of cholera toxin into target cells and deregulation of adenylate cylase activity. NAm = nicotinamide. ⏯ *microbiology2.com/animations*

After the bacteria attach to the cells lining the intestinal villi (**Fig. 25.19C** and **D**), they secrete AB toxins, the structure of which is shown in **Figure 25.19B**. Both toxins have five B subunits arranged as a ring around a single A subunit. The B subunits bind to ganglioside GM1 on eukaryotic cell membranes and deliver the A subunit to the target cell (**Fig. 25.20**⏯, steps 1–4). The A subunit possesses the toxic part of the molecule, an ADP-ribosyltransferase, which must be activated by the host.

The binding of CT or LT to GM1 triggers endocytosis and the formation of a toxin-containing vacuole. The phagosome is then transported to the endoplasmic reticulum. During this time, the A subunit is cleaved by a host protease into two fragments called A1 and A2, which are still held together by a disulfide bond. The reducing environment in the vacuole reduces that bond and frees the A1 peptide containing active ADP-ribosyltransferase into the endoplasmic reticulum, which exports the toxin into the cytoplasm.

The mission of the A1 peptide is to modify (ADP-ribosylate) a membrane-associated GTPase (called a G protein or G factor) that binds to adenylate cyclase and controls its activity (**Fig. 25.20**, step 5). To understand how the toxin produces diarrhea, it is helpful to know that the intestinal epithelium has both absorptive and secretory functions. Transport is normally characterized by a

net absorption of NaCl, short-chain fatty acids (SCFAs), and water, allowing extrusion of a feces containing very little water and salt. In addition, the epithelium secretes mucus, bicarbonate, and KCl. Cholera toxin initiates a cascade of events leading to secretion of large amounts of chloride and other ions.

How are G factors and cAMP involved? Human cells have two types of G-factor complexes that stimulate (G_s) or inhibit (G_i) adenylate cyclase, respectively, when bound to GTP (see **eTopic 25.1** for details). An intrinsic GTPase in each G factor hydrolyzes GTP and prevents continual stimulation of adenylate cyclase by G_s, or its inhibition by G_i. Cholera toxin (and *E. coli* labile toxin) ADP-ribosylate G_s and thereby inhibit the intrinsic GTPase activity (**Fig. 25.20**, step 5). As a result, adenylate cyclase is constantly stimulated to produce cAMP. The increased level of cAMP stimulates a host protein kinase that activates various ion transport channels, including the cystic fibrosis transmembrane conductance regulator (CFTR), so named because a defect in this protein manifests as the lung disease cystic fibrosis. CFTR controls chloride transport in several cell types, including intestinal epithelia (discussed in Section 23.4). As a result of CFTR activation, chloride, sodium, and other ions leave the cell and, in an attempt to equilibrate osmolarity, water leaves as well. Because the affected cells line the intestine, the escaping water enters the intestinal lumen, leading to watery stools, or diarrhea.

How is diarrhea a benefit to the microorganism? A major benefit to the organism is to propagate the species. The more diarrhea that is produced and expelled, the greater the number of organisms that are made and disseminated throughout the environment. This increases the chance that another host will ingest the organism, ensuring survival of the species. Diarrhea can also benefit a pathogen by decreasing competition with other organisms as they are swept away. In fact, the vast majority of organisms found in the diarrheal fluids of cholera patients are *V. cholerae* bacteria.

Different pathogens have discovered alternative ways to alter host cAMP levels. The Gram-negative pathogen *Bordetella pertussis* causes a childhood respiratory infection called whooping cough, so named for the whooping sound made as a child tries to take a breath after a fit of coughing. This microbe also secretes an ADP-ribosylating toxin, but this one modifies G_i (the inhibitory factor that normally down-regulates adenylate cyclase activity). ADP-ribosylation in this instance prevents that inhibition, with the result (again) of runaway cAMP synthesis. Pertussis toxin, however, is structurally different from cholera toxin. In addition, the organism makes a "stealth" adenylate cyclase that is secreted from the bacterium but remains inactive until it enters host cells, where it becomes active by binding the calcium-binding protein calmodulin. The resulting increase in cAMP levels causes inappropriate triggering of certain host cell signaling pathways.

Many other bacterial toxins utilize ADP-ribosylation to target different host proteins. A classic example, discussed in **eTopic 25.2**, is diphtheria toxin, produced by *Corynebacterium diphtheriae*. This toxin kills cells by ADP-ribosylating protein synthesis factor EF2. *C. diphtheriae* is a good example of a pathogen whose toxin gene (*dtx*) is part of a prophage genome integrated into the bacterial chromosome.

THOUGHT QUESTION 25.2   Antibodies to which subunit of cholera toxin will best protect a person from the toxin's effects?

THOUGHT QUESTION 25.3   How might you experimentally determine whether a pathogen secretes an exotoxin?

Shiga toxin: *Shigella* and *E. coli* O157:H7. *Shigella flexneri* and *E. coli* O157:H7 (also known as enterohemorrhagic *E. coli*) cause food-borne diseases whose symptoms include bloody diarrhea. These organisms produce an important toxin known as Shiga toxin (or Shiga-like toxin). The gene (*stx*) encoding Shiga toxin is part of a

phage genome integrated into the bacterial chromosome. The toxin has five B subunits for binding and one A subunit imbued with toxic activity. The A subunit, upon entry, destroys protein synthesis by cleaving 28S rRNA in eukaryotic ribosomes. Strains that produce high levels of this toxin are associated with acute kidney failure known as hemolytic uremic syndrome.

E. coli O157:H7 is a recently emerged pathogen. The organism can colonize cattle intestines without causing bovine disease; as a result, undetected bacteria can easily contaminate meat products following slaughter, as well as irrigation water from farms (see Chapter 28). The organism is sometimes referred to as the "Jack in the Box" microbe, a reference to the first large U.S. outbreak, associated with fast-food hamburgers at a Washington State Jack in the Box restaurant in 1993. Both *E. coli* and *Shigella* have remarkable acid resistance mechanisms that rival that of the gastric pathogen *Helicobacter pylori.* This acid resistance makes these pathogens infectious at a very low infectious dose.

What activates *stx* expression? Iron availability is a key factor for inducing the expression of *stx* and many other virulence genes in pathogens. The body holds its iron tightly in proteins such as lactoferrin and transferrin. To an invading organism, the body is a very iron-poor environment. Shiga toxin offers a way to rob the host of its iron stores.

An intriguing question often pondered by scientists is, What roles do virulence factors play in the natural ecology of these bacteria? Surely factors such as Shiga toxin did not evolve after *Shigella* started infecting humans. William Lainhart, Gino Stolfa, and Gerald Koudelka from the University at Buffalo predicted that Shiga toxin was actually a natural defense against *Tetrahymena thermophila*, a ciliated protist that grazes on bacteria. Predatory ciliates are a major source of bacterial mortality in the environment. Confirming their hypothesis, Lainhart and his colleagues found that *Tetrahymena* was killed when cocultured with Shiga toxin–producing bacteria. They propose that reactive oxygen species produced by *Tetrahymena* induce the SOS response in *Shigella*, which then activates phage replication and the expression of *stx* (SOS induction is discussed in Section 9.5). This is another example of altruistic behavior by members of a bacterial species in which some individuals sacrifice themselves (as a result of phage lysis) to save the population.

THOUGHT QUESTION 25.4   Would patients with iron overload (excess free iron in the blood) be more susceptible to infection?

Anthrax. A century ago, anthrax (caused by *Bacillus anthracis*; **Fig. 25.21A**) was mainly a disease of cattle

A.

B.

Single PA protein

Heptamer

C.

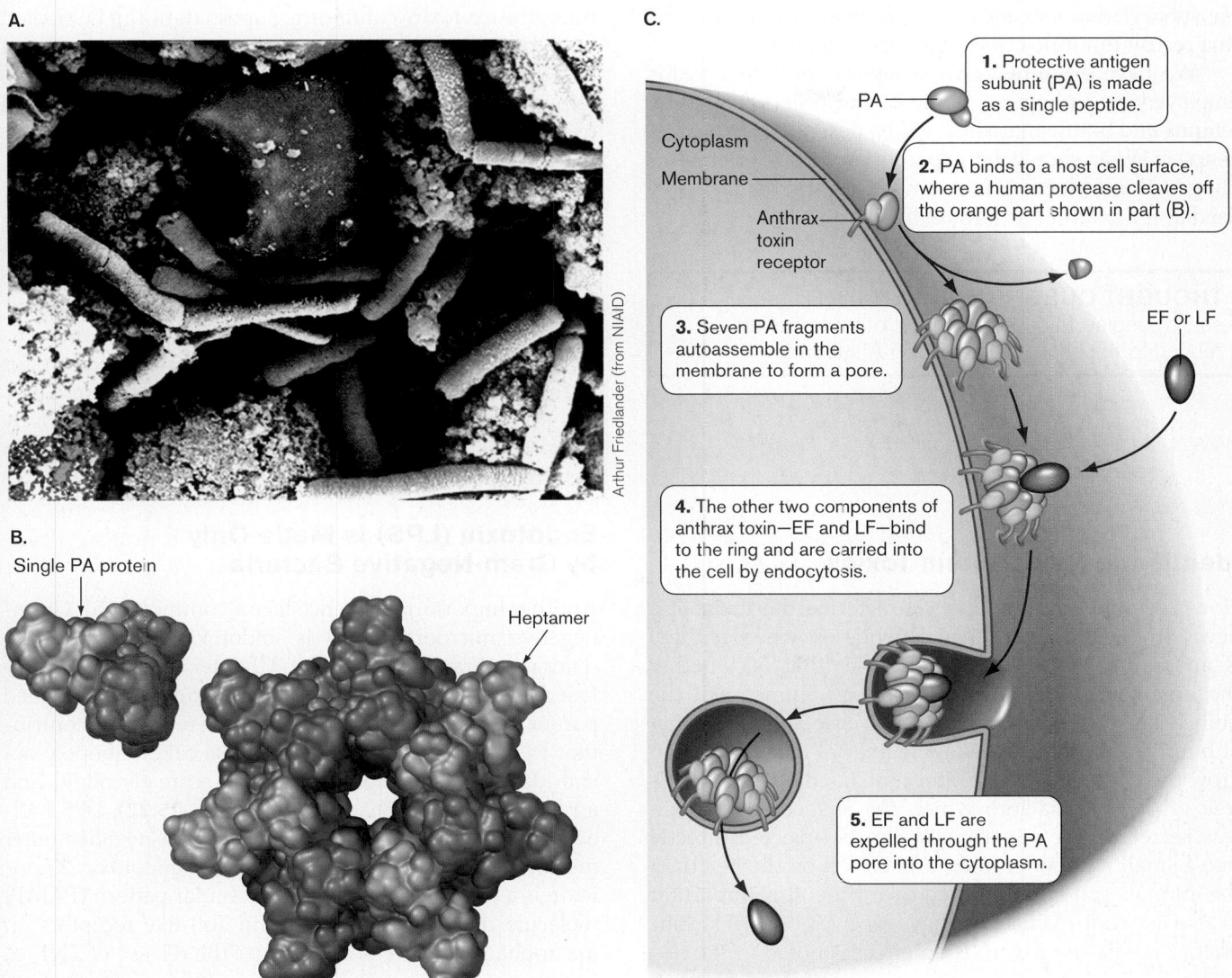

Figure 25.21 *Bacillus anthracis* **and anthrax toxin.** **A.** *B. anthracis* (SEM; approx. 2 μm in length). Splenic tissue from a monkey. Spores are not visible. **B.** Single subunit and heptamer of protective antigen. (PDB code: 1TZO) **C.** Mechanism of toxin entry.

and sheep. Humans acquired the disease only accidentally. Today we fear the deliberate shipment of *B. anthracis* through the mail or its dispersion from the air ducts of heavily populated buildings. What makes this Gram-positive, spore-forming microbe so dangerous? In large part, its lethality is due to the secretion of a plasmid-encoded tripartite toxin. The core subunit of the toxin is called **protective antigen (PA)** because immunity to this protein *protects* hosts from disease. Protective antigen is made as a single peptide but then binds to the host cell surface, where a human protease cleaves off a fragment (**Fig. 25.21B**). The remaining part of PA can autoassemble in the membrane to form a heptameric (seven-membered) pore. The other two components of anthrax toxin—**edema factor (EF)** and **lethal factor (LF)**— bind to separate rings and are carried into the cell (**Fig. 25.21C**). The complex is endocytosed, and the two

proteins carried in are passed through the pore into the host cytoplasm.

Edema factor and lethal factor are the toxic parts of anthrax toxin. Both are enzymes that attack the signaling functions of the cell. Edema factor is an adenylate cyclase that remains inactive until entering the cytoplasm, where it binds calmodulin. This binding activates adenylate cyclase, resulting in a huge production of cAMP, and inactivates calmodulin from its normal function in the cell.

Lethal factor is actually a protease that cleaves several host protein kinase kinases, each of which is part of a critical regulatory cascade affecting cell growth and proliferation. A protein kinase kinase is an enzyme that phosphorylates, and thereby activates, another protein kinase that can then phosphorylate one or more subsequent target proteins. One consequence of subverting these phosphorylation cascades is a failure to produce signals that recruit immune cells to fight the infection.

We have examined only a few of the many toxins employed by pathogens. Some of the others, including tetanus and botulism toxins, will be described in the next chapter. What should be apparent from our brief sampling is the evolutionary ingenuity that pathogens have used to try to tame the human host.

THOUGHT QUESTION 25.5 Use an Internet search engine to determine what other toxins are related to the cholera enterotoxin A subunit.

WWW | Anthrax tutorial
microbiology2.com/links

Identifying New Protein Toxins

How does one identify and characterize the toxin of a new pathogen? In part that depends on whether there is an animal model for the disease (that is, whether the organism will infect a laboratory animal like the guinea pig or mouse and cause disease similar to that of humans). If there is a suitable animal model, one can grow the pathogen in laboratory media and harvest cell-free growth media into which the suspected exotoxin was secreted. The supernatant can be injected directly into a small number of mice, and effects on the health of the animals can be monitored over time. It is important to also use control mice that have received only fresh laboratory media (no toxin). If an effect is noted in the test

mice, the exotoxin-containing supernatant can be treated with proteinase. If the toxin is protein, the treatment will destroy toxic activity. Subsequently, a variety of protein purification techniques, such as ion-exchange and molecular-sieve chromatography, can be used to purify the protein.

One can also utilize tissue culture cells to test for the presence of a cell-free toxin in growth media. In using this method, one thing to keep in mind is that the effect may be tissue specific: The toxin may work on one tissue but not another. Another concern is whether the laboratory medium used to grow the bacterial cells is sufficient to elicit toxin production. A high-iron medium, for example, will prevent synthesis of diphtheria toxin. Thus, it is important to mimic the host environment as closely as possible when coaxing a pathogen to make exotoxin in vitro.

Endotoxin (LPS) Is Made Only by Gram-Negative Bacteria

Another important virulence factor common to all Gram-negative microorganisms is endotoxin present in the outer membrane (discussed in Chapter 3). Not to be confused with secreted exotoxins, endotoxin is an embedded part of the bacterial cell surface and an important contributor to disease. Endotoxin, otherwise called lipopolysaccharide (LPS), is composed of lipid A, core glycolipid, and a repeating polysaccharide chain (**Fig. 25.22**). LPS molecules form the outer leaflet of the Gram-negative outer membrane. As bacteria die, they release endotoxin. Endotoxin is a pathogen-associated molecular pattern (PAMP) molecule that can bind to certain Toll-like receptors on macrophages or B cells and trigger the release of TNF-α,

A. LPS membrane

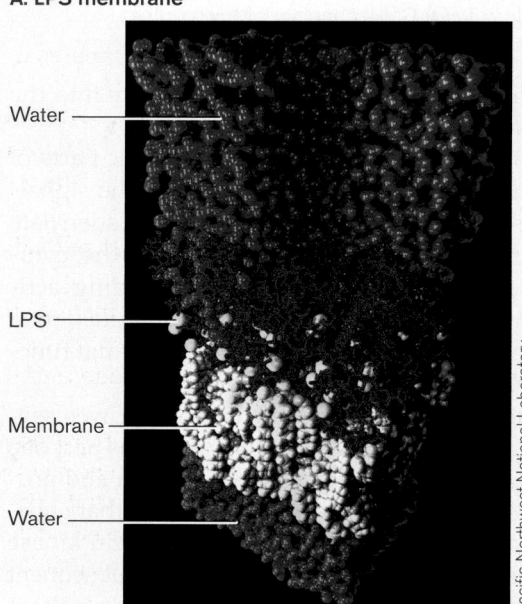

Water

LPS

Membrane

Water

Pacific Northwest National Laboratory

B. Gram-negative bacterial endotoxin (lipopolysaccharide, LPS)

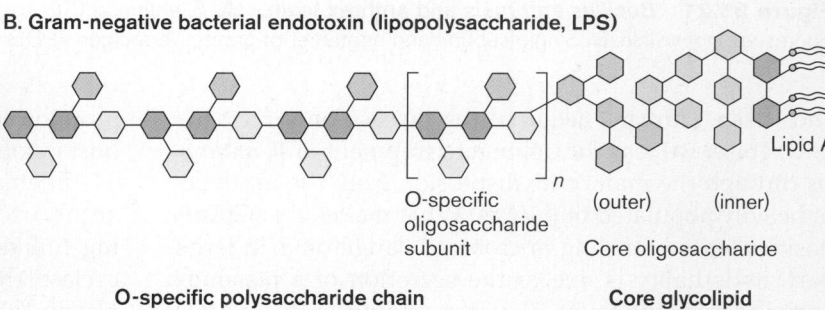

Lipid A

O-specific oligosaccharide subunit

(outer)   (inner)

Core oligosaccharide

O-specific polysaccharide chain   **Core glycolipid**

Figure 25.22 Endotoxin. A. Model of a lipopolysaccharide (LPS) membrane of *Pseudomonas aeruginosa*, consisting of 16 lipopolysaccharide molecules (red) and 48 ethylamine phospholipid molecules (white). **B.** Basic structure of endotoxin, showing the repeating O-antigen side chain that faces out from the microbe and the membrane proximal core glycolipid and lipid A (contains endotoxic activity).

interferon, IL-1, and other cytokines (PAMPs, Toll-like receptors, and cytokines are discussed in Chapters 23 and 24). The release of these active agents causes a variety of symptoms, such as:

- Fever
- Activation of clotting factors, leading to disseminated intravascular coagulation
- Activation of the alternative complement pathway
- Vasodilation, leading to hypotension (low blood pressure)
- Shock due to hypotension
- Death when other symptoms are severe

The lipid A moiety of LPS possesses endotoxic activity (**Fig. 25.22**).

NOTE: Lipopolysaccharides are also the outer membrane structures called O antigens that are used to classify different strains of *E. coli* (for example, *E. coli* O157 versus *E. coli* O111), as well as other Gram-negative organisms.

The role of endotoxins can be seen in infections with the Gram-negative diplococcus *Neisseria meningitidis* (**Fig. 25.23A**), a major cause of bacterial meningitis. *N. meningitidis* has, as part of its pathogenesis, a septicemic phase in which the organism can replicate to high numbers in the bloodstream. The large amount of endotoxin present causes a massive depletion of clotting factors that leads to internal bleeding, most prominently displayed to a physician as small pinpoint hemorrhages called **petechiae** on the patient's hands and feet (**Fig. 25.23B**). Capillary bleeding near the surface of the skin causes petechiae. One danger of treating massive Gram-negative sepsis with antibiotics is that the enormous release of endotoxin from dead bacteria could well kill the patient. Untreated Gram-negative sepsis is, however, almost always fatal, so its treatment, albeit risky, is imperative.

A recently proposed approach to prevent endotoxic shock is based on the knowledge that LPS must bind to Toll-like receptor TLR4 to cause endotoxic shock. What if we could neutralize TLR4 by antibody and prevent it from binding LPS during an infection? In a recent proof-of-principle experiment, antibody raised to TLR4 (anti-TLR4) was injected into mice. The antibody successfully blocked TLR4 receptors and protected the mice from *E. coli*–induced septic shock.

TO SUMMARIZE:

- **There are five categories of protein exotoxins** based on mode of action. These include toxins that

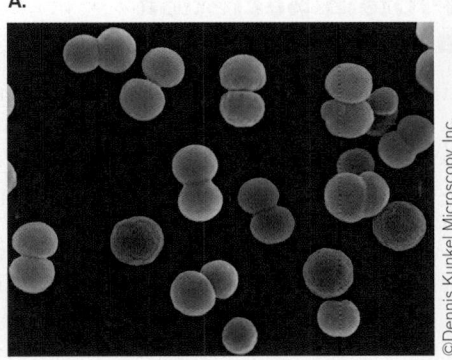
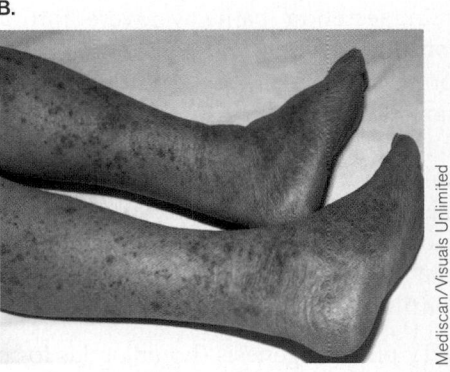

Figure 25.23 Effect of *Neisseria meningitidis* endotoxin. **A.** *N. meningitidis* (cell 0.8–1 µm diameter, SEM). **B.** Petechial rash caused by *N. meningitidis*.

disrupt membranes, inhibit protein synthesis, or alter host cell signal molecule synthesis, as well as superantigens and target-specific proteases.
- **AB-subunit toxins** are common. The B subunit promotes penetration through host cell membranes, and the A subunit has toxic activity.
- ***Staphylococcus aureus* alpha toxin** forms pores in host cell membranes.
- **Cholera toxin, *E. coli* labile toxin, and pertussis toxin** are AB toxins that alter host cAMP production by adding ADP-ribose groups to different G-factor proteins.
- **Shiga toxin** is an AB toxin that cleaves host cell 28S rRNA in host cell ribosomes.
- **Anthrax toxin** is a three-part AB toxin with one B subunit (protective antigen) and two different A subunits that affect cAMP levels (edema factor) and cleave host protein kinases (lethal factor).
- **Lipopolysaccharide (LPS)**, known as endotoxin, is an integral component of Gram-negative outer membranes and an important virulence factor that triggers massive release of cytokines from host cells. The indiscriminate release of cytokines can trigger fever, shock, and death.

25.5 Protein Secretion and Pathogenesis

A recurring theme among bacterial pathogens is the secretion of proteins that destroy, cripple, or subvert host target cells. The bacterial toxins described in Section 25.4 are secreted into the surrounding environment, where they float randomly until chance intervenes and they hit a membrane-binding site. However, many pathogens attach to a tissue and inject bacterial proteins (called effectors) directly into the host cell cytoplasm. The proteins may not kill the cell but redirect host signaling pathways in ways that benefit the microbe.

Protein secretion pathways were introduced in Section 8.5, focusing on ATP-binding cassette (ABC) proteins as a model. Additional secretion models are described here in their critical role of delivering pathogenicity proteins such as toxins. A particularly interesting aspect of these secretory systems is that many of them evolved from, and bear structural resemblance to, other more innocuous systems. Other molecular processes that are evolutionarily related to secretion include:

- Type IV pilus biogenesis (homologous to type II protein secretion)
- Flagellar synthesis (homologous to type III protein secretion)
- Conjugation (homologous to type IV protein secretion)

Table 25.4 lists features of the six export systems and examples of associated virulence effector proteins. Three of the systems will be described in more detail.

Type II Secretion Resembles Pilus Assembly

Cholera toxin, discussed in Section 25.4, is a well-known example of a toxin secreted by a type II secretion system. **Type II secretion** offers a clear example of how nature has modified the blueprints of one system to do a very different task. DNA sequence analysis has revealed that the genes used for type IV pilus biogenesis (see Section 25.3) were duplicated at some point during evolution and redesigned to serve as a protein secretion mechanism. Type IV pili have the unusual ability to extend and retract from the outer membrane, a property that produces the gliding motility of *Myxococcus* (see Section 4.7) and the twitching motility of *Neisseria* and *Pseudomonas*. As you might guess, assembly/disassembly of

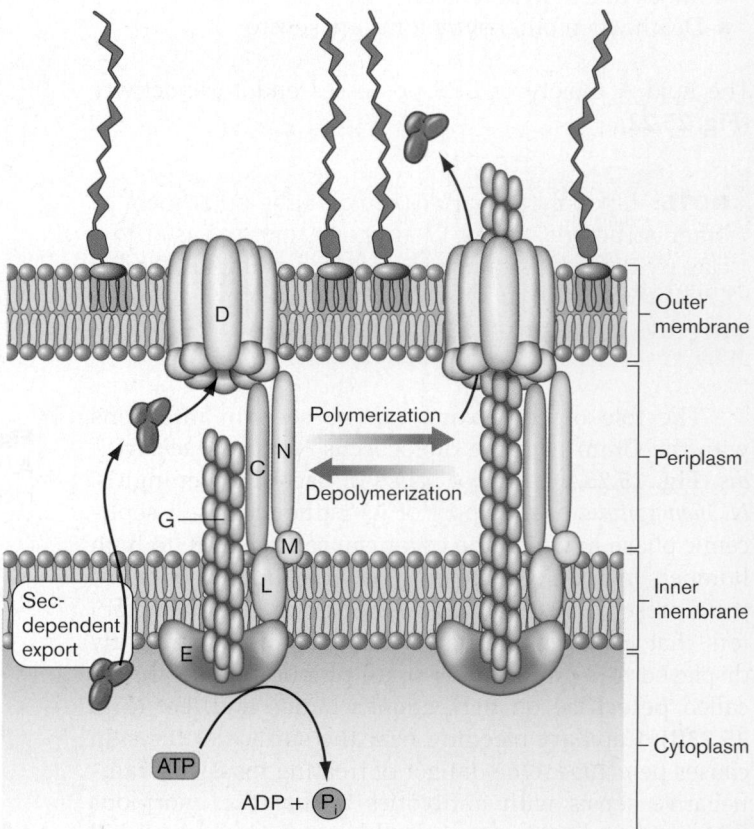

Figure 25.24 Type II secretion. C, D, E, G, L, M, and N are protein components of the secretion system. *Source:* Modified from Moat, Foster, and Spector. 2002. *Microbial Physiology*, 4th ed: Wiley-Liss.

Table 25.4 Secretion systems for bacterial toxins.

Secretion type	Features	Examples
I	SecA dependent, one effector per system	*E. coli* alpha hemolysin, *Bordetella pertussis* adenylate cyclase
II	SecA dependent, similar to type IV pili	*Pseudomonas aeruginosa* exotoxin A, elastase, cholera toxin
III	SecA independent, syringe, related to flagella, secrete multiple effectors	*Yersinia* Yop proteins, *Salmonella* Sip proteins, enteropathogenic *E. coli* (EPEC) EspA proteins, TirA
IV	Related to conjugational DNA transfers, secrete multiple effectors	*B. pertussis* toxin, *Helicobacter* CagA
V	Autotransporter, SecA dependent to periplasm, self-transport through outer membrane, one effector per system	Gonococcal and *Haemophilus influenzae* IgA proteases
VI	Related to phage tails, single effector	*Burkholderia* and *Vibrio cholerae* VgrG

Figure 25.25 The needle complex of the *Salmonella enterica* serovar Typhimurium type III secretion system. **A.** Unlike other secretion systems, the type III mechanism injects proteins directly from the bacterial cytoplasm into the host cytoplasm. The proteins in these systems are related to flagellar assembly proteins. Shown here are TEMs of osmotically shocked *S.* Typhimurium with needle complexes (arrows) visible in the bacterial envelope. **B.** Purified needle complexes (electron micrograph). **C.** Schematic representation of the *S.* Typhimurium needle complex and its putative components. *Source:* Modified from Moat, Foster, and Spector. 2002. *Microbial Physiology,* 4th ed. © Wiley-Liss, Inc.

▶▮ *microbiology2.com/animations*

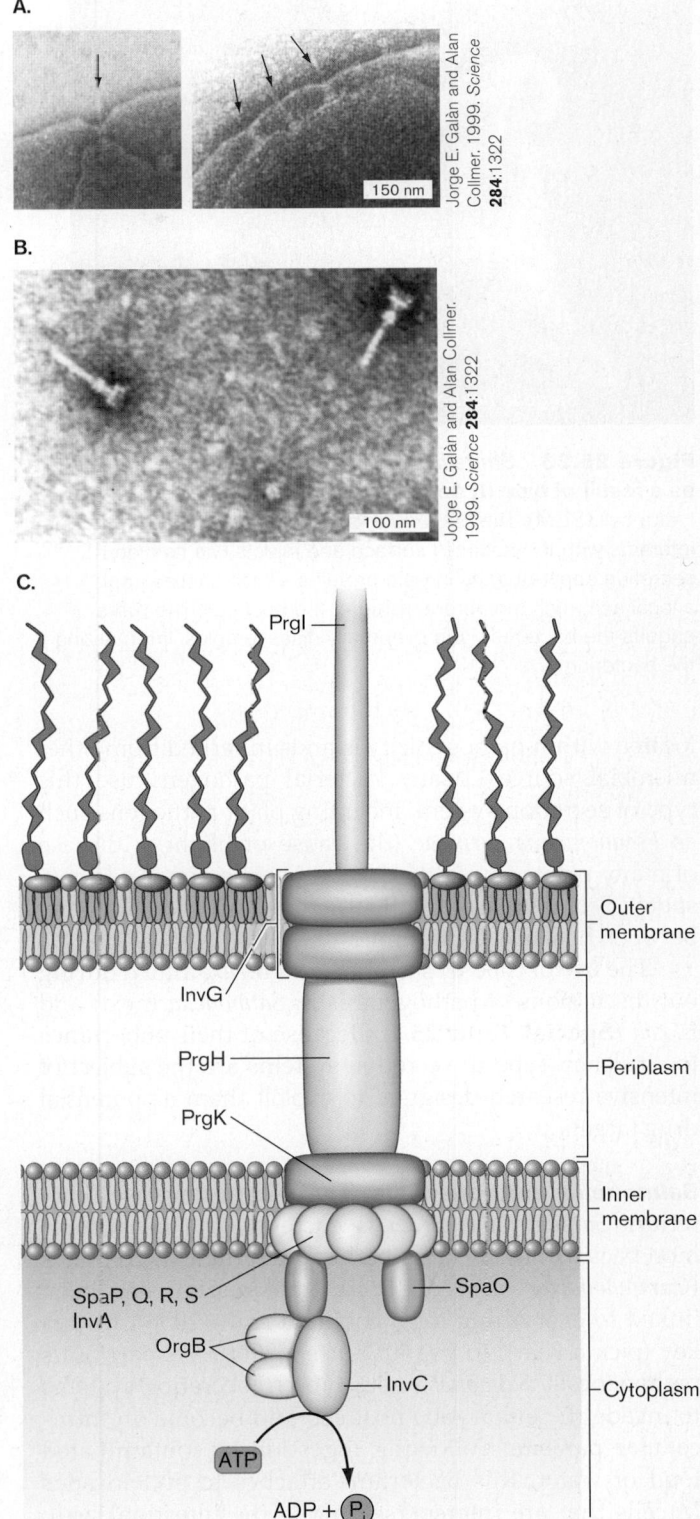

these appendages is quite complex. Type II protein secretion mechanisms mirror this complexity. Proteins to be secreted first make their way, via the Sec-dependent general secretion pathway, to the periplasm, where they then encounter the appropriate type II secretion system. Because of this periplasmic "layover," the proteins are folded *before* secretion. Type II secretion systems use the pilus-like structure as a piston to ram the folded proteins through an outer membrane pore structure and into the surrounding void (**Fig. 25.24**). Piston action occurs via cyclic assembly and disassembly of pilus-like proteins, driving the pilus-like structure through an outer membrane pore and then retracting.

Type III Secretion Injects Effector Proteins into Host Cells

In the 1990s, it was discovered that the etiological agents of Black Death and various forms of diarrhea (caused by species of *Yersinia, Salmonella,* and *Shigella*) could somehow take the bacterial virulence proteins made in their cytoplasm and drive them directly into the eukaryotic cell cytoplasm without the protein ever getting into the extracellular environment. Direct delivery is a good idea because it eliminates the dilution that happens when a toxin is secreted into media. Another advantage of this strategy is that it avoids the need to tailor the toxin to fit a preexisting host receptor.

What kind of molecular machine can directly deliver cytoplasmic bacterial proteins into target cells? Research has shown that some microbes use tiny molecular syringes embedded in their membranes to inject proteins directly into the host cytoplasm; this mechanism of delivery is called **type III secretion. Figure 25.25** shows electron micrographs of actual type III secretion needles and a model of the complex. Genes encoding type III systems are actually related to flagellar genes, whose products export the flagellin proteins through the center of a growing flagellum (discussed in Section 3.7). It appears that flagellar genes were evolutionarily reengineered to act more like molecular syringes (**Fig. 25.25C▶▮**). The bacterial virulence proteins secreted by type III systems subvert normal host cell signaling pathways, some of which cause dramatic rearrangements of host cytoskeleton at the cell membrane that lead to engulfment of the microbe (**Fig. 25.26**). The genes encoding type III systems are usually

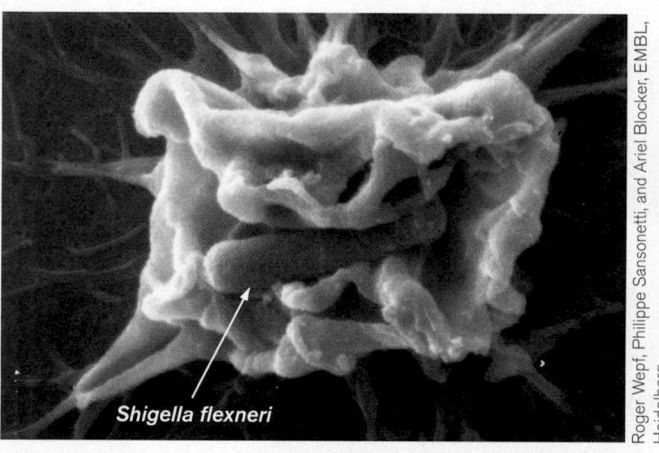

Roger Wepf, Philippe Sansonetti, and Ariel Blocker, EMBL, Heidelberg

Shigella flexneri

Figure 25.26   *Shigella* invades a host cell ruffle produced as a result of type III secretion.   *Shigella flexneri* entering a HeLa cell (SEM). The bacterium (small diameter approx. 1 μm) interacts with the host cell surface and injects (via its type III secretion apparatus) its invasin proteins, which choreograph a local actin-rich membrane ruffle at the host cell. The ruffle engulfs the bacterium and eventually disassembles, internalizing the bacterium.

located within pathogenicity islands inherited from other microbial sources. Many bacterial pathogens use this type of secretion system, including plant pathogens such as *Pseudomonas syringae* (the cause of blight, a disease of many plants in which leaves or stems develop brown spots). Secretion is normally triggered by cell-cell contact between host and bacterium.

The use of type III secretion will be examined during our discussions of pathogenesis in *Salmonella* (next) and *E. coli* (**Special Topic 25.1**). Because of their importance to virulence, type III secretion systems are the subject of intensive research designed to exploit them as potential drug targets.

Salmonella **pathogenesis.** The Gram-negative bacterium *Salmonella enterica* is currently the most common bacterial food-borne pathogen in the United States (*Campylobacter* is second). Its transmission has been linked to everything from cookie dough (2008), to turkey (pick a year), to peanut butter (2009). As part of its pathogenesis, *Salmonella* uses type III secretion systems to invade the eukaryotic host cell and become an intracellular parasite. Following ingestion of contaminated food or water, this bacterium attaches to and invades M cells that are interspersed along the intestinal wall. M cells are specialized intestinal epithelial cells (see Fig. 23.17B) that sample normal intestinal microbes and transfer pathogens across the epithelial barrier for recognition by the immune system. *Salmonella* subverts the normal function of M cells and causes an inflammatory response that leads to diarrhea. During its evolutionary journey toward becoming a pathogen, *Salmonella*

has acquired at least 5 and as many as 12 pathogenicity islands that are absent from related, but harmless, *E. coli* strains. Only two will be discussed here. *Salmonella* pathogenicity island 1 (SPI-1) encodes a type III protein secretion system that delivers a cocktail of at least 13 different protein toxins (called effector proteins) directly into the cytosol of host epithelial cells in the gut (**Fig. 25.27**). Inside epithelial cells, these effector proteins interfere with signal transduction cascades and modulate the host response. One mission of these effectors is to induce cytoskeletal rearrangements that cause ruffling of the eukaryotic membrane around the microbe (see **Fig. 25.26**). The membrane ruffling starts the process of engulfment. *Salmonella* induces this response as a way to avoid the normal endocytic process.

Once *Salmonella* enters an epithelial cell or a macrophage, it finds itself in a vacuole. In the normal course of events, an enzyme-packed lysosome would then fuse with the phagosome and release its contents in an effort to kill the invader. *Salmonella*, however, possesses a second pathogenicity island, called SPI-2, that subverts this host response. SPI-2 uses another type III secretion system to inject proteins that alter vesicle trafficking, thereby reducing phagosome-lysosome fusion so that the intracellular bacteria are spared. Insofar as there are 12 recognized pathogenicity islands present in the genome, clearly the pathogenesis of *Salmonella* is even more complex than just outlined.

WWW | The type III protein secretion system
     | *microbiology2.com/links*

Type IV Secretion Resembles Conjugation Systems

As Chapter 9 describes, many bacteria can transfer DNA from donor to recipient cells via a cell-cell contact system known as conjugation (discussed in Section 9.2). The conjugation systems of some pathogens have, through evolution, been modified into new systems that transport proteins, or proteins plus DNA, directly into target cells. *Agrobacterium tumefaciens*, for example, uses its Vir system to transfer the tumor-producing Ti plasmid and some effector proteins into plant cells. The result is a plant cancer called crown gall disease. The bacterium that causes whooping cough in humans, *Bordetella pertussis*, also uses a type IV secretion system, to export pertussis toxin; but it simply exports the toxin without injecting it into the host (**Fig. 25.28**). Type IV systems also differ with respect to whether the protein is taken directly from the cytoplasm, like CagA from *Helicobacter*, or from the periplasm, as with pertussis toxin. In the latter case, the SecA-dependent general secretory system first delivers the toxin to the periplasmic space.

Figure 25.27 Schematic overview of *Salmonella* pathogenesis. Effector proteins injected by *Salmonella* into a host M cell affect the activity of host proteins that trigger cell death (apoptosis), influence gene expression (Cdc42 and JNK), influence electrolyte movements, and induce actin rearrangement. Actin rearrangement produces the ruffling of host membrane to engulf the organism.

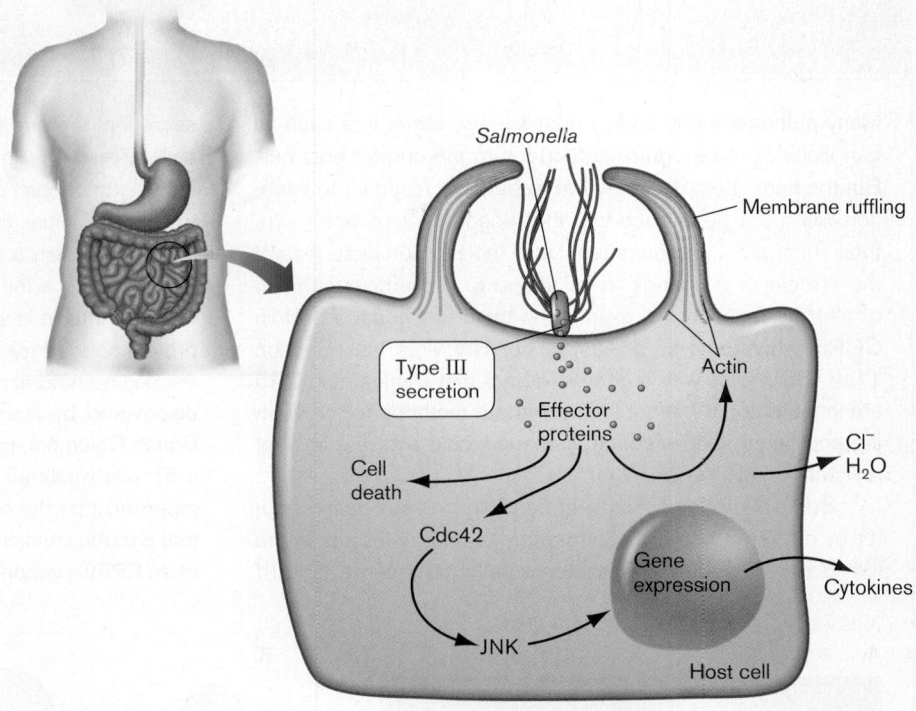

THOUGHT QUESTION 25.6   Protein and DNA have very different structures. Why would a protein secretion system be derived from a DNA-pumping system?

TO SUMMARIZE:

- **Many pathogens** use specific protein secretion pathways to deliver toxins.
- **Type II secretion** systems use a pilus-like extraction/retraction mechanism to push proteins out of the cell.
- **Type III secretion** uses a molecular syringe to inject proteins from the bacterial cytoplasm into the host cytoplasm.
- **Type IV secretion** utilizes an entourage of proteins that resemble conjugation machinery to secrete proteins from either the cytoplasm or the periplasm.

25.6 Finding Virulence Genes

All pathogens must enter a host; find their unique niche; avoid, circumvent, or subvert normal host defenses; multiply; and eventually be transmitted to a new susceptible host. Although certain common pathogenic tactics have come to be appreciated, each microbe has a unique "pathogenic signature" that permits survival and leads to its freedom to multiply. When an emerging pathogen causes disease, how do we figure out its mechanisms of virulence? Virulence constitutes a measurable

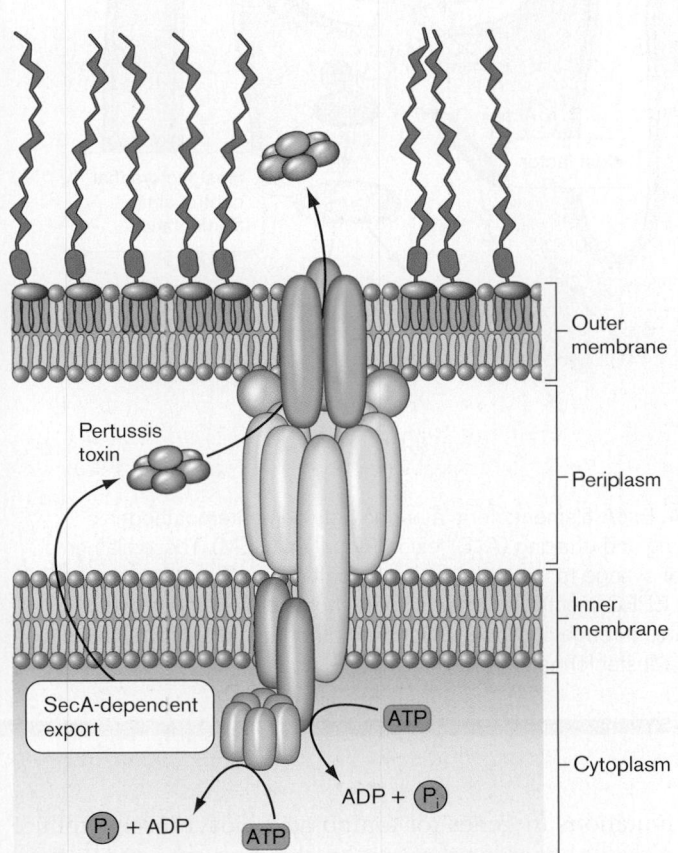

Figure 25.28 Type IV secretion of pertussis toxin. Evolutionarily related to conjugation systems, this type IV system in *Bordetella pertussis* takes pertussis toxin from the periplasm (moved there by SecA-dependent transport) across the outer membrane.

Special Topic 25.1	The Bacterial Trojan Horse: Bacteria That Deliver Their Own Receptor

Many pathogens rely on key host surface structures such as gangliosides to recognize and attach to the correct host cell. But the host species can evolve to become resistant to infection when the gene encoding the receptor (or receptor synthesis) mutates. The mutation could lead to complete loss of the protein or a change in its shape to prevent recognition or alter its function. An example is the T-cell surface protein CCR5, which acts as a receptor for HIV virus (see Section 11.4). Individuals with a genetic defect that eliminates CCR5 are immune to HIV infection, so without methods for preventing and curing HIV infection, humans would evolve a level of resistance to HIV.

Some microbes circumvent possible loss of a host receptor by not relying on the natural array of host receptors in the first place. Instead, these bacterial pathogens use a type III secretion system to insert their own receptors into target cells (**Fig. 1A**). One such group of enterprising pathogens is enteropathogenic *E. coli* (EPEC). Although EPEC uses pili to form an initial, loose attachment, the bacterium must ultimately establish a more intimate attachment. EPEC achieves this using an adhesion molecule on the bacterial cell called **intimin**. Intimin is a 94-kDa integral outer membrane bacterial protein needed for intimate adherence to host cells. However, there is no natural receptor for intimin on host membranes. As discovered by Brett Finley and his colleagues in Vancouver, British Columbia, EPEC must deliver its own intimin receptor, a 57-kDa bacterial protein called Tir (for translocated intimin receptor) that the bacteria inject into the host cell. The genes that encode intimin, Tir, and the secretion apparatus are all part of an EPEC pathogenicity island.

A.

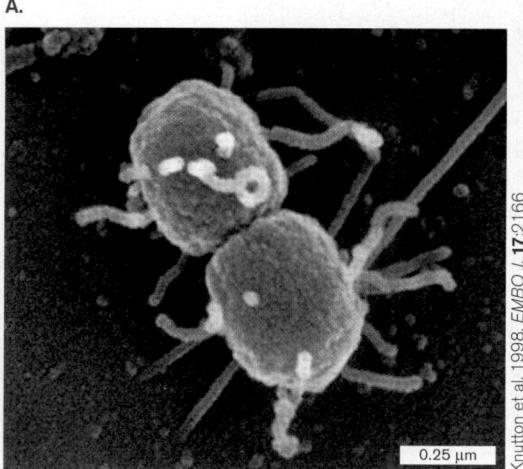

Knutton et al. 1998. *EMBO J.* **17**:2166

B.

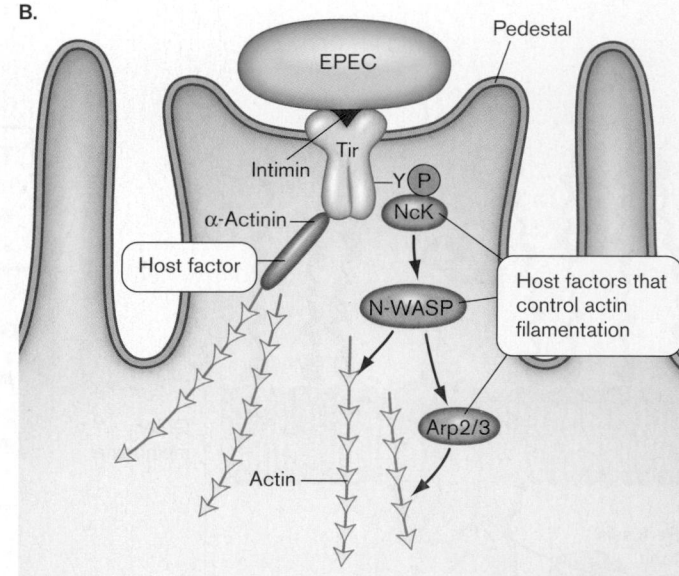

Figure 1   *E. coli* type III secretion and cell-cell interaction.   A. EspA filaments form a bridge between enteropathogenic *E. coli* (EPEC) and an epithelial cell during the early stages of attaching and effacing (A/E) lesion formation (SEM). These filaments are part of the type III secretion apparatus, functioning as a molecular syringe to inject proteins from pathogenic *E. coli* into the host cell. **B.** Model of cytoskeletal components within the EPEC pedestal. EPEC injects Tir protein into the host cell, where it moves to the membrane and acts as a receptor for intimin. Tir also communicates through phosphorylation with other host proteins to cause a change in actin cytoskeleton, which leads to pedestal formation. **C.** Pedestal formation (SEM).

phenotype of an organism. Genes encoding virulence factors are expressed just like those for other traits (for example, expression of the *lac* operon confers the ability to metabolize lactose; discussed in Section 10.2). For metabolic or biosynthetic pathways, identifying the requisite genes is relatively straightforward. For instance,

mutations in genes for amino acid biosynthesis produce a clear phenotype that can be observed on an agar plate. If the pathway is defective, the amino acid is not made and must be added to the medium or the mutant will not grow. Because virulence is a consequence of growth in a host, there is no way to identify a phenotype for virulence

Once injected and placed in the host membrane, Tir binds intimin on the bacterial surface (**Fig. 1B**). Think of Tir as a wall anchor you poke into a board in order to attach something to it. The result is the more intimate adherence between the bacterium and host cell surface required for infection to proceed. In addition, host protein kinases phosphorylate Tir at tyrosine residue 474. Phosphorylated Tir directly triggers a remarkable reorganization of host cellular cytoskeletal components (actin, alpha-actinin, ezrin, talin, and myosin light chain) such that a membrane "pedestal" is formed, raising the microbe up (**Fig. 1C**). The result of this attachment is the characteristic attaching and effacing (A/E) lesion, characterized by destruction of the microvilli and pedestal formation.

WWW | *E. coli* infection
microbiology2.com/links

C.

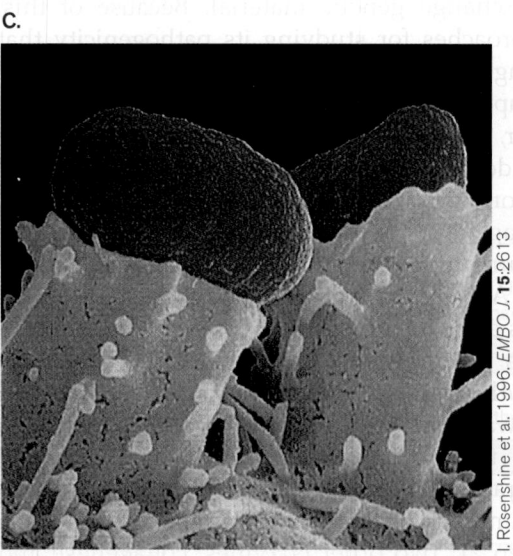

I. Rosenshine et al. 1996. *EMBO J.* **15**-2613

gene mutants in vitro. Virulence mutants fail to survive or fail to cause disease in animals. So how are these genes identified?

If the genome has been sequenced and a gene bears similarity to a known virulence gene present in another organism, that gene can be selectively mutated and the

resulting mutant strain tested for virulence, usually by measuring LD$_{50}$. But what of genes not assigned an annotated function? They, too, could encode virulence factors. And what of genes that are expressed only during an infection? How do we find those?

In Vivo Expression Technologies Can Identify Virulence Genes

Bacterial genes that are only expressed by a pathogen during growth within an animal host can now be discovered using what is called **in vivo expression technology (IVET)**. The first step in IVET is to cripple a pathogen so that it cannot grow in the host animal unless a "rescue" gene that produces the missing factor is expressed. Next, the rescue gene is placed without its promoter into a plasmid to make what is called a promoter trap plasmid. The promoter of the plasmid-borne rescue gene is then replaced with random promoters cloned from the microbial genome. If one of the cloned promoters drives expression of the bacterial rescue gene after the organism is injected into the animal, the organism will flourish and kill the host. The trick is to find the promoters that express in vivo but that do not express, or that very poorly express, in laboratory media. Promoters that fit these criteria respond only to unknown in vivo conditions within the host and can be used to determine what those in vivo conditions are.

The IVET procedure was first described in 1995 by J. Mekalanos and colleagues at Harvard University using a promoter trap plasmid in which random fragments of *Salmonella* DNA were inserted (that is, cloned) upstream of a promoterless *cat-lacZ* fusion in the plasmid (**Fig. 25.29**). The *cat* gene encodes chloramphenicol acetyltransferase, an enzyme that inactivates chloramphenicol, making the strain resistant to the drug. The collection of random clones was then injected into mice that were immediately injected with chloramphenicol. When a promoter allowed in vivo expression of the *cat* gene, the microbe became resistant to the drug, grew in the animal, and traveled to the spleen.

The researchers knew that some of the in vivo–expressed promoters would also be expressed in vitro (for example, promoters that drive genes encoding ribosome proteins). However, the goal was to identify promoters expressed *only* in the host—a characteristic that would indicate the presence of a downstream gene dedicated to virulence. So, the next task was to identify which of the in vivo–expressed promoters was *not* expressed in vitro. To do this, the bacteria extracted from the spleen were plated on artificial medium containing X-Gal (an indicator of beta-galactosidase enzyme made from the *lacZ* gene). A colony that was blue indicated that the promoter was expressed in vitro. A white colony indicated that the promoter was *not* expressed in vitro, but since the strain grew in the mouse, that same promoter must be expressed

A.

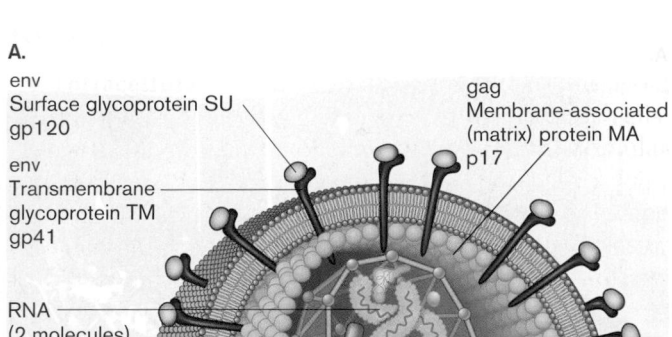

env
Surface glycoprotein SU
gp120

env
Transmembrane
glycoprotein TM
gp41

RNA
(2 molecules)

gag
Capsid CA
(core shelf)
p24

pol
Protease PR p9
Polymerase RT
RNase H RNH p66
Integrase IN p32

gag
Membrane-associated
(matrix) protein MA
p17

B.

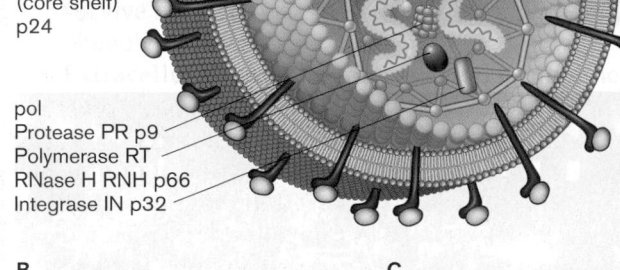

CDC/C. Goldsmith, P. Feorino,
E. L. Palmer, W. R. McManus

C.

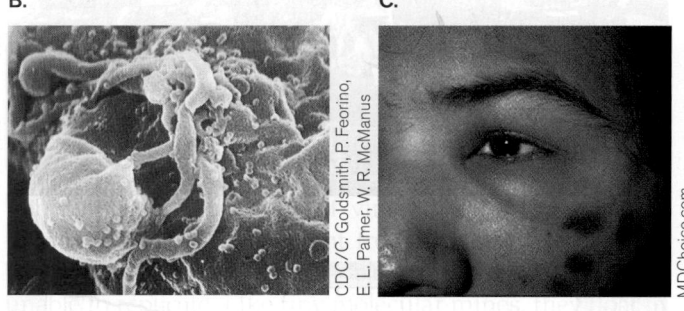

MDChoice.com

D.

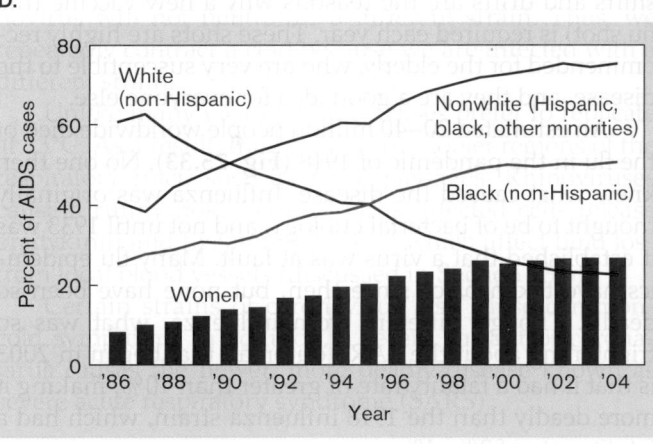

Figure 25.34   HIV and AIDS.   A. HIV. **B.** HIV-1 budding from cultured lymphocyte (SEM). **C.** Kaposi's sarcoma (oval spots) and periorbital cellulitis infection. **D.** AIDS incidence in the United States by race and age.

hinders apoptosis. Nef is one of several HIV proteins that affect pathogenesis (**Table 25.5**). As a result, the virus can replicate and bud from these lymphocytes (**Fig. 25.34B**).

Despite the presence of Nef, however, the lymphocytes eventually do die. Because CD4$^+$ T cells are the primary target of HIV, the levels of these T cells decline dramatically in an infected patient and cause an immunodeficiency, particularly in immune regulation. Monitoring T-cell numbers, therefore, is a critical diagnostic tool.

Comparing HIV infection and AIDS, the disease it causes, underscores the difference between infection and disease. Detection of HIV in an individual does not equate with that person's having overt disease. The individual can go for years without developing symptoms. But even if infected individuals are not showing signs of disease, they are fully able to transmit the virus to another person sexually, through blood transfusion, or through the use of shared hypodermic needles.

As with many viral diseases, AIDS begins with flu-like symptoms—fever, headache, fatigue, sore throat, and sometimes a rash. Consequently, symptomology is never

HIV can block chemokine binding and disconnect communication between the target cell and the immune system. You might still expect that the infection would be stopped because another aspect of the viral infection should, on its own, trigger apoptosis of the infected cell. Viral protein gp120 cross-links CD4 surface molecules on T cells, which, coupled with *antigen* binding to the T-cell receptor, should trigger apoptosis. The ensuing programmed cell death would simultaneously kill the infecting virus. However, the virus makes a protein (negative factor, Nef) that

Table 25.5	**HIV proteins that affect pathogenesis.**
Proteins	**Role in pathogenesis**
Tat (transcriptional trans-activator)	Secreted from infected cells; accelerates HIV gene transcription by host polymerase, has chemokine-like properties, acts as growth factor for endothelial cells, alters gene expression in target cells
Vpr (viral protein R)	Induces G$_2$ cell cycle arrest and nuclear import of the preintegration complex
Nef (negative factor)	Down-regulates cell surface CD4 and MHC class I (important for virus release); enhances virion infectivity by inhibiting apoptosis
Vif (virion infectivity factor)	Counteracts a host cellular protein that naturally inhibits HIV replication
Vpu (viral protein U)	Enhances HIV release from infected cells and CD4 degradation by targeting CD4 to the proteasome

a reliable way to diagnose HIV infection. The detection of antibodies is a good initial screen to gauge whether a person has been exposed to the virus. Once exposure has been confirmed by a serum test (a change in test result from negative to positive is called seroconversion), it is important to use PCR techniques to measure actual HIV viral load in a patient's blood. The higher the viral load, the faster the disease will progress. Thus, for proper disease management, it is important to know viral load in addition to monitoring T-cell numbers.

The symptoms of AIDS itself begin to manifest once T-cell numbers fall to about 300 per microliter of blood (the normal range is 500–1,500). It can often take 4 or 5 years (or more) before a patient starts to experience this decline. With this immunocompromised state come secondary infections such as shingles (herpes zoster virus), *Pneumocystis* pneumonia, tuberculosis, and oral thrush (a fungal infection of the mouth caused by the yeast *Candida albicans*). Late-stage AIDS is defined by $CD4^+$ counts below 200, at which point a malignancy called **Kaposi's sarcoma** can develop (**Fig. 25.34C**). Kaposi's sarcoma is a malignancy originating in endothelial or lymphatic cells, and tumors can arise anywhere—gastrointestinal tract, mouth, lungs, skin, or brain.

An important cause of AIDS symptoms is the HIV Tat protein (trans-activator of the HIV promoter). A trans-activator is a protein that can diffuse through the cytoplasm to bind and regulate a target gene sequence. The Tat protein is needed for viral replication. In addition, it is released from infected monocytes and enters other cells, where it alters host regulatory cascades and gene expression. The result is an increase in the production of nuclear factor kappa beta (NFκB), which helps further disassemble the immune system; and activation of endothelial cell growth, which can lead to Kaposi's sarcoma. Some believe that a vaccine using inactivated Tat will be useful in preventing AIDS. Other HIV proteins important to the pathogenesis of HIV and AIDS are presented in **Table 25.5**.

The relative incidence of AIDS by race since 1986 is examined in **Figure 25.34D**. Since 1990, the overall incidence, approximately 40,000 annual cases, has declined in the United States, mainly as a result of improvements in antiviral therapies. However, the decrease is also due to a drop in the number of white males afflicted. The numbers for women, African Americans, and Hispanic minorities continue to rise.

Human Papillomavirus

Human papillomavirus (HPV) is a very common virus that causes warts, an abnormal growth of skin tissue (**Fig. 25.35A** and **B**). HPV can produce warts on the feet, hands, vocal cords, mouth, and genital organs. Over 60 types of HPV have been identified so far. Each type infects different parts of the body. Some strains of HPV have also been associated with cervical and penile cancer. Genital HPV is one of the most common sexually transmitted diseases among college students. HPV replication is discussed in Section 6.5.

How does this virus cause excessive growth of tissues? Several mechanisms are involved. In one, HPV synthesizes a protein called E7 (**Fig. 25.35C**). E7 binds to a host protein, pRb, that normally prevents cells from completing a cell cycle. Protein pRb inhibits cell division by binding to a transcription factor called E2F, preventing it from activating the oncogenes that encode the c-Myc and cyclin E proteins. The c-Myc protein is needed for mitosis, and cyclin E is needed for cell cycle progression toward mitosis. So blocking their transcription limits cell division and controls growth. The HPV protein E7 targets pRb for destruction by the cell proteasome (discussed in Section 8.6). With pRb thus removed, c-Myc and cyclin E proteins are made continually. As a result, cell division proceeds unabated to produce warts or, in the worst-case scenario, cervical or penile cancer. After the initial infection has resolved, the virus may remain latent in tissues and can reactivate if immune system function is impaired.

An effective HPV vaccine that will protect against cervical or penile cancer has been developed, although guidelines for its use remain controversial. The problem is that it must be administered before a girl or boy becomes sexually active (and potentially exposed) in order to work. The FDA and CDC have recommended its use in males and females between ages 9 and 26.

Herpes Virus Micro-RNAs Affect Pathogenesis and Survival in Host Cells

Herpes viruses cause a variety of diseases, including cold sores (herpes simplex type 1, HSV-1), genital herpes (herpes simplex type 2), mononucleosis, and some cancers (Epstein-Barr virus). A key feature of herpes virus infections is the development of a latent state, most noticeable as recurrent cold sores around the mouth or genital herpes lesions. Part of how these viruses become latent involves shutting down viral replication and integrating viral DNA into a host genome. However, what has not been clear is why the presence of the virus does not trigger apoptosis in the host cell.

Apoptosis is a process triggered in human cells that are undergoing difficulties such as in an infection. It is a protective measure because, while it kills the host cell, it also destroys the virus in the process. So it is in the best interest of the virus to keep its host cell alive. What has recently been discovered is that herpes viruses produce a number of small RNA molecules called micro-RNAs (miRNA) that can interfere with the apoptosis program. One miRNA produced by herpes simplex 1 virus prevents the translation of two host cell proteins required for cell death. The miRNA binds to regions in the target mRNAs encoding these proteins and promotes their degradation. Because the latent virus does not produce any viral proteins, the immune system of the infected individual cannot detect the infected cell.

A.

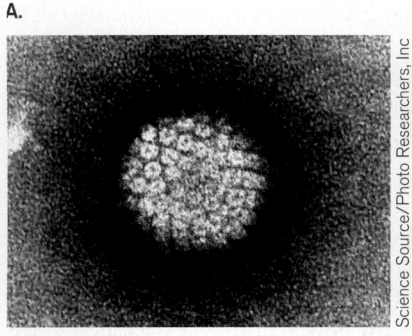

Science Source/Photo Researchers, Inc

B.

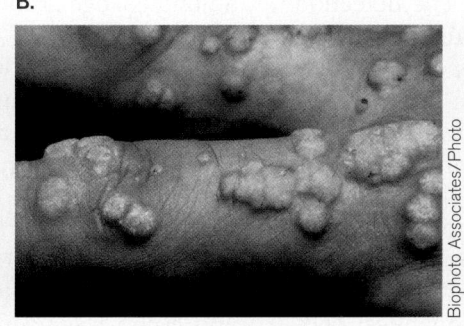

Biophoto Associates/Photo Researchers, Inc

**Figure 25.35   Human papillomavirus.
A.** Human papillomavirus (40–50 nm diameter, TEM). **B.** View of hand warts. **C.** Part of the strategy HPV uses to deregulate cell division in infected cells. Protein pRb normally helps limit cell division by inactivating E2F. Virus E7 protein, however, removes pRb, so the cell will begin to divide uncontrollably and become cancerous.

C.

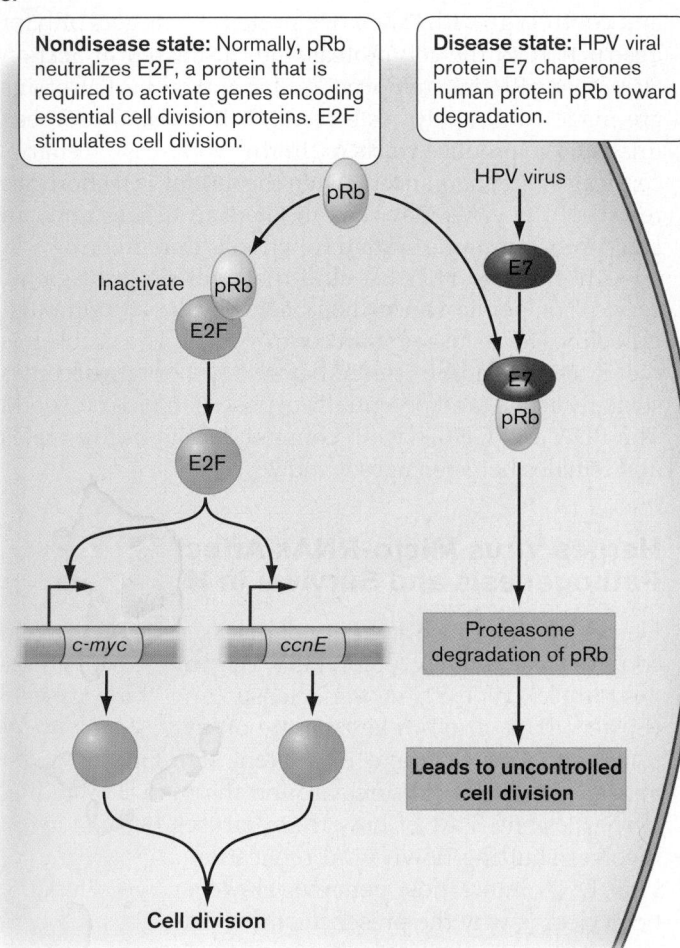

minor sequence differences in attachment proteins (for example, capsid). Because of these differences, neutralizing antibodies made during a previous infection are ineffective against the new strain.

■ **Viral infection** can increase a patient's susceptibility to other, less virulent microbes.
■ **Antigenic shifts** in a viral antigen (as in the hemagglutinin of influenza) can lead to a new pandemic of the disease.
■ **Animals coinfected with two viruses** can serve as incubators for antigenic shifts or the evolution of new viruses.
■ **Binding to host cells** can involve a single host receptor (as in rhinovirus) or multiple receptors (as in HIV).
■ **Virus-infected cells** can secrete proteins that enter uninfected cells and disrupt signaling pathways.

Concluding Thoughts

This chapter has only scratched the surface of all that is known about microbial pathogenesis. It should be clear at this point that attachment, immune avoidance, and subversion of host signaling pathways are common goals of most successful pathogens, whether bacterial, viral, or parasitic. But even with all we know, there remains much to learn.

The remaining chapters deal with the basic principles used to diagnose disease and eradicate offending pathogens, but many pathogens remain hard to detect and difficult, if not impossible, to kill. So the more we know about the mechanisms microbes use to cause disease, the better we will be at developing effective countermeasures.

As you will learn in the coming chapters, new pathogens are constantly emerging. Over the last few decades, we have seen the development of HIV, SARS, avian flu, hantavirus, West Nile virus, *E. coli* O157:H7, and the reemergence of flesh-eating streptococci, to name but a few. Can we ever stop pathogens from emerging? Probably not. For every countermeasure we develop, nature designs a counter-countermeasure. Our hope is that continuing research into the molecular basis of pathogenesis and antimicrobial pharmacology will keep us one step ahead.

Current treatments of HSV-1 rely on antiviral agents such as acyclovir that target viral polymerase and inhibit DNA replication. They are effective against replicating virus but do not target latent virus; thus, cold sores will return throughout the lifetime of an affected individual. Finding this anti-apoptosis miRNA offers the first potential antimicrobial target against latent infection.

TO SUMMARIZE:

■ **Multiple reinfection** of the same person by a virus can occur because different strains of the virus have

CHAPTER REVIEW

Review Questions

1. Describe the differences between infection versus disease; pathogenicity versus virulence; LD_{50} versus ID_{50}.
2. What is meant by direct versus indirect routes of infection?
3. What are the characteristics of a good reservoir for an infectious agent?
4. Name the various portals of entry for infectious agents, and a disease associated with each.
5. Describe the basic features of a pathogenicity island.
6. Explain various ways in which bacteria can attach to host cell surfaces.
7. Describe the basic steps by which pili are assembled on the bacterial cell surface. How do type I and type IV pili differ?
8. Explain the five broad categories of toxin mode of action.
9. What is ADP-ribosylation and how does it contribute to pathogenesis?
10. Explain the differences between exotoxins and endotoxins.
11. Explain the mechanisms of secretion carried out by type II and type III protein secretion systems. What are the paralogous origins of these systems?
12. Describe the key features of *Salmonella* pathogenesis.
13. How can genomic approaches help identify pathogens in an infection?
14. What different mechanisms do intracellular pathogens use to survive within the infected host cell?
15. Describe different molecular strategies that microbes use to avoid the immune system.
16. How do bacteria determine whether or not they are in a host environment?
17. What is antigenic shift?
18. How is attachment of influenza virus different from that of HIV? Why does HIV produce an immune deficiency while influenza does not?
19. Explain the basics of how human papillomavirus can trigger cancer.

Thought Questions

1. Why do you think new versions of swine and avian flu often originate in Asia?
2. How can you modify Koch's postulates to prove that a bacterial gene is a virulence factor?
3. How would you approach determining whether a particular pilus on group A streptococci (GAS) is required for the organism's pathogenesis? Use a tissue culture model.
4. Why have humans not developed resistance to microbial toxins?
5. *Vibrio cholerae* is an acid-sensitive organism. *E. coli*, on the other hand, is very acid resistant and able to survive stomach acidity for long periods of time. Is there an ethical issue to consider when trying to move the acid resistance system from *E. coli* into *V. cholerae*?

Key Terms

adhesin (943)
ADP-ribosyltransferase (949)
antigenic shift (970)
ectoparasite (936)
edema factor (EF) (955)
endoparasite (936)
endotoxin (948)
exotoxin (948)
facultative intracellular pathogen (966)
fimbria (944)
fomite (939)
genomic island (941)
horizontal transmission (939)
immunopathogenesis (940)

in vivo expression technology (IVET) (963)
infection (936)
infection cycle (939)
infectious dose (938)
intimin (962)
intracellular pathogen (966)
Kaposi's sarcoma (973)
labile toxin (LT) (952)
latent state (937)
LD_{50} (938)
lethal factor (LF) (955)
opportunistic pathogen (937)
parasite (936)

pathogen (936)
pathogenesis (936)
pathogenicity (937)
pathogenicity island (941)
petechia (957)
pilus (944)
primary pathogen (937)
protective antigen (PA) (955)
protein A (967)
reservoir (939)
type I pilus (944)
type II secretion (958)
type III pilus (944)
type III secretion (959)

type IV pilus (944)
vector (939)

vertical (transovarial) transmission
(939)

virulence (937)
virulence factor (940)

Recommended Reading

Aktories, Klaus, and Joseph T. Barbieri. 2005. Bacterial cytotoxins: Targeting eukaryotic switches. *Nature Reviews. Microbiology* **3**:397–410.

Barel, Monique, Ara G. Hovanessian, Karin Meibom, Jean-Paul Briand, Marion Dupuis, et al. 2008. A novel receptor–ligand pathway for entry of *Francisella tularensis* in monocyte-like THP-1 cells: Interaction between surface nucleolin and bacterial elongation factor Tu. *BMC Microbiology* **8**:145.

Brochet, Mathieu, Christophe Rusniok, Elisabeth Couvé, Shaynoor Dramsi, Claire Poyart, et al. 2008. Shaping a bacterial genome by large chromosomal replacements, the evolutionary history of *Streptococcus agalactiae*. *Proceedings of the National Academy of Sciences USA* **105**:15961–15966.

Burrows, Lori L. 2005. Weapons of mass retraction. *Molecular Microbiology* **57**:878–888.

Casadevall, Arturo. 2008. Evolution of intracellular pathogens. *Annual Review of Microbiology* **62**:19–33.

Chen, John, and Richard P. Novick. 2009. Phage-mediated intergeneric transfer of toxin genes. *Science* **323**:139–141.

Dean, Paul, Marc Maresca, and Brendan Kenny. 2005. EPEC's weapons of mass subversion. *Current Opinion in Microbiology* **8**:28–34.

Dietrich, Guido, Sebastian Kurz, Claudia Hübner, Christian Aepinus, Stephanie Theiss, et al. 2003. Transcriptome analysis of *Neisseria meningitidis* during infection. *Journal of Bacteriology* **185**:155–164.

Dobrindt, Ulrich, Bianca Hochhut, Ute Hentschel, and Jörg Hacker. 2004. Genomic islands in pathogenic and environmental microorganisms. *Nature Reviews. Microbiology* **2**:414–424.

Groisman, Eduardo, and Josep Casadesús. 2005. The origin and evolution of human pathogens. *Molecular Microbiology* **56**:1–7.

Hall-Stoodley, Luanne, and Paul Stoodley. 2009. Evolving concepts in biofilm infections. *Cell Microbiology* **11**:1034–1043.

Hamon, Mélanie, Hélène Bierne, and Pascale Cossart. 2006. *Listeria monocytogenes*: A multifaceted model. *Nature Reviews. Microbiology* **4**:423–434.

Hensel, Michael. 2004. Evolution of pathogenicity islands of *Salmonella enterica*. *International Journal of Medical Microbiology* **294**:95–102.

Hsiao, Ansel, and Jun Zhu. 2009. Genetic tools to study gene expression during bacterial pathogen infection. *Advances in Applied Microbiology* **67**:297–314.

Jin, Qi, Zhenghong Yuan, Jianguo Xu, Yu Wang, Yan Shen, et al. 2002. Genome sequence of *Shigella flexneri* 2a: Insights into pathogenicity through comparison with genomes of *Escherichia coli* K12 and O157. *Nucleic Acid Research* **30**:4432–4441.

Lainhart, William, Gino Stolfa, and Gerald B. Koudelka. 2009. Shiga toxin as a bacterial defense against a eukaryotic predator, *Tetrahymena thermophila*. *Journal of Bacteriology* **191**:5116–5122.

Mota, Luís Jaime, Isabel Sorg, and Guy R. Cornelis. 2005. Type III secretion: The bacteria-eukaryotic cell express. *FEMS Microbiology Letters* **252**:1–10.

Olsen, Randall J., Samuel A. Shelburne, and James M. Musser. 2009. Molecular mechanisms underlying group A streptococcal pathogenesis. *Cell Microbiology* **11**:1–12.

Pallen, Mark J., and Brendan W. Wren. 2007. Bacterial pathogenomics. *Nature* **449**:835–842.

Renesto, Patricia, Nicolas Crapoulet, Hiroyuki Ogata, Bernard La Scola, Guy Vestris, et al. 2003. Genome-based design of a cell-free culture medium for *Tropheryma whipplei*. *Lancet* **362**:447–449.

Schilling, Joel D., Matthew A. Mulvey, and Scott J. Hultgren. 2001. Structure and function of *Escherichia coli* type 1 pili: New insight into the pathogenesis of urinary tract infections. *Journal of Infectious Diseases* **183**:S36–S40.

Smith, Kristina M., Yigong Bu, and Hiroaki Suga. 2003. Induction and inhibition of *Pseudomonas aeruginosa* quorum sensing by synthetic autoinducer analogs. *Chemistry & Biology* **10**:81–89.

Taubenberger, Jeffrey K., Ann H. Reid, and Thomas G. Fanning. 2005. Capturing a killer flu virus. *Scientific American* **292(1)**:48–57.

Taubenberger, Jeffrey K., Ann H. Reid, A. E. Krafft, K. E. Bijwaard, and Thomas G. Fanning. 1997. Initial genetic characterization of the 1918 "Spanish" influenza virus. *Science* **275**:1793–1796.

Thierry, Roger, Céline Froidevaux, Didier Le Roy, Marlies Knaup Reymond, Anne-Laure Chanson, et al. 2009. Protection from lethal Gram-negative bacterial sepsis by targeting Toll-like receptor 4. *Proceedings of the National Academy of Sciences USA* **106**:2348–2352.

Thomas, Wendy E., Elena Trintchina, Manu Forero, Viola Vogel, and Evgeni V. Sokurenko. 2002. Bacterial adhesion to target cells enhanced by sheer force. *Cell* **109**:913–923.

Thompson, Lucinda J., and Hilde de Reuse. 2002. Genomics of *Helicobacter pylori*. *Helicobacter* **7**:1–17.

Wiles, Siouxsie, William P. Hanage, Gad Frankel, and Brian Robertson. 2006. Modelling infectious disease—Time to think outside the box? *Nature Reviews. Microbiology* **4**:307–312.

For further study and review, please visit StudySpace at **microbiology2.com**.

Chapter 26

Microbial Diseases

In the U.S. Civil War, twice as many people died from infections as from battle wounds themselves. Clearly, developments in medicine had not kept pace with innovations in rifle accuracy. Today, antibiotics can quell most wound infections. Yet millions of people still die as a result of dangerous microbes that spread by food, air, body fluids, and even insects. How do clinicians recognize these many different types of infections? What are the key symptoms or laboratory tests that help determine whether a person suffers from a viral or bacterial disease?

Microbial diseases are with us daily and continue to be a major contributor to global mortality and morbidity. The emergence of new pathogens, increasing drug resistance, and threats of bioterrorism have heightened the need for investigations into microbial disease mechanisms and the body's ability to combat infectious agents. Along with those basic studies, effective diagnostic algorithms are needed to quickly identify infectious diseases and prevent their spread. This chapter explores the general classes and causes of microbial diseases and introduces the art of diagnosis.

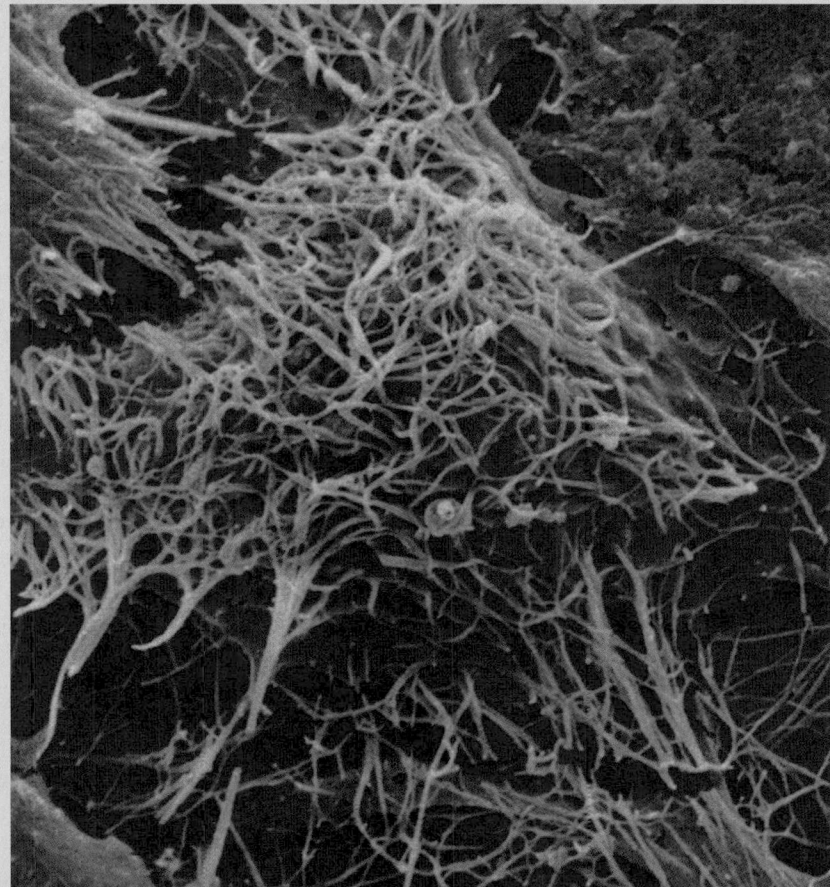

Current Research Highlight

Threadlike Ebola virions budding from a cell (center). Ebola virus is an extremely aggressive and lethal pathogen. Infected individuals bleed from multiple orifices and typically die within days of infection. Part of the reason the disease progresses so rapidly is that the virus is released from infected host cells in huge numbers. The reason for this was not known until recently. Work by Rachel Kaletsky and colleagues has demonstrated that a host protein called tetherin, which is normally used to limit virus release from infected cells, is disabled by Ebola virus. The result is the rapid release of massive numbers of virus particles that can then quickly spread infection to other organs and tissues. *Source:* Rachel L. Kaletsky et al. 2009. *PNAS* **106**:2886–2891. Photo courtesy of Paul Bates/Univ. of Penn. School of Med.

A visibly ill 16-year-old girl was taken to her pediatrician, where she complained of diarrhea, high fever (39.9°C; 103.8°F), and vomiting. She told the physician she had been healthy until two days prior. Upon physical exam, the doctor noticed that the girl had a much lower than normal blood pressure (76/48 mm Hg), a rapid heart rate (120 beats per minute), and an erythematous (red) rash on her trunk. Because of her deteriorating condition, the patient was admitted to the pediatric intensive care unit at the local hospital, and cultures were obtained. She was immediately given intravenous fluids and IV antibiotics, yet she died within a week. Patient history taken upon admission revealed that the girl had started her menstrual period 4 days before becoming ill, providing an important clue as to the cause of the disease.

This tragedy was part of a larger story that emerged in the late 1970s and early 1980s, when women started dying from this dangerous, new (emerging) disease. Now known as toxic shock syndrome, scientists and physicians learned that it is caused by certain strains of *Staphylococcus aureus* that produce a superantigen type of toxin (toxic shock syndrome toxin, or TSST; see Section 24.6). Why was the disease not seen in previous decades? The answer turned out to be the use of one brand of superabsorbent tampons (since removed from the market). The tampons produced a rich growth environment for *S. aureus*. So if the patient was colonized by a TSST-producing strain, huge amounts of toxin were released to circulate in the bloodstream. Today we recognize that toxic shock syndrome is a possible consequence of any *S. aureus* infection and can occur in both men and women. We now have tools to diagnose it quickly.

In Chapter 25, we discussed the arsenal of weapons that microbes use to cause disease (including TSST). But what of the diseases themselves? Microbial diseases are often presented from the microbe's point of view—for example, where in the body does *S. aureus* cause disease? Another approach, which we take in Chapter 26, is to look at infections from the vantage point of the infected organ. We might ask, for example, what organisms cause vaginal infections versus lung infections, and how are they differentially diagnosed? Are the patient's complaints consistent with gastrointestinal tract disease or a lung infection? This approach is more attuned to the practices of health care workers and the fundamentals by which a physician interacts with a patient. An individual who goes to a clinician complaining of a fever, cough, and chest pain does not typically report having encountered a particular bacterium. The physician must determine from the patient that the disease is localized to the chest and then intuit that it may be a respiratory tract infection. Appropriate samples are then collected and sent to a clinical microbiology laboratory, where data are collected to confirm or dispute the presence of a specific etiological agent, such as *Streptococcus pneumoniae* or other lung pathogens.

We do not present an exhaustive compendium of microbial illnesses in this chapter, but a representative sampling of infections that illustrate key aspects of microbial disease. The material is arranged and presented so as to integrate your knowledge of microbiology and immunology within the framework of the practice of medicine and the study of infectious disease.

26.1 Characterizing and Diagnosing Microbial Diseases

Although this chapter will classify representative microbial diseases by organ system, there are other very instructive ways to group and study microbial infections. These include the organism and portal-of-entry approaches. Each has clear benefits and pitfalls. The organism approach is useful when examining the different ways a given species can cause disease. For instance, *E. coli* can cause genitourinary tract infections, gastrointestinal infections, and meningitis. How do the various strains causing these diseases differ from one another? Are virulence factors present in the uropathogenic strain that are absent or different in the diarrhea-producing strain? (Virulence factors are discussed in Chapter 25; see Sections 25.2 and 25.3.)

Pathogens are also commonly classified by their route of infection (or portal of entry; see Chapter 23). In this approach, pathogens are classified as food-borne, airborne, blood-borne, or sexually transmitted. The portal-of-entry approach has its merits, but it sometimes seems at odds with the disease itself. *Salmonella enterica* serovar Typhi, for example, is categorized as a food-borne pathogen (the cause of typhoid fever). Food-borne pathogens typically have a fecal-oral route of infection and cause pronounced gastrointestinal disease. Yet *S.* Typhi does not actually cause gastrointestinal disease until very late in the disease process. After ingestion, it quickly enters the bloodstream from the intestine to produce a serious septicemia (blood infection) called enteric (typhoid) fever. At this stage, the organism cannot be found in feces. (Much later the organism can re-enter the intestine via the gallbladder and cause mild gastrointestinal symptoms.) The obvious value of knowing the route of infection is that society can design effective measures to avoid disease. Hand washing and proper food preparation, for instance, will effectively prevent the septicemia of typhoid fever. But clearly, not all organisms entering via the gastrointestinal tract cause gastrointestinal disease. Note that *S.* Typhi is different from *Salmonella enterica* serovar enteritidis, a major cause of diarrhea. *S. Enteritidis* is also a food-borne pathogen but rarely enters the bloodstream.

The organ system approach has its drawbacks as well, given that organisms that initially affect one organ system can disseminate to infect other organ systems. *Neisseria gonorrhoeae*, for example, initially infects the genitourinary tract but can disseminate via the bloodstream, causing a blood infection. *Yersinia enterocolitica*, a cause of gastroenteritis, can disseminate from the intestine and cause abscesses elsewhere in the body. Yet we favor the organ system approach because it is illuminating to know how microbial diseases present themselves to physicians treating patients. The physician's examination usually begins not by taxonomic evaluation of the agent or knowledge of the portal of entry, but with symptoms emanating from the organ system that has been affected. You will benefit, however, from studying infectious disease from various perspectives, and you should keep in mind factors such as the portal of entry and the taxonomy of the infectious organism.

Patient Histories Provide Important Diagnostic Clues

A problem often faced by clinicians is that many infectious diseases display similar symptoms, making diagnosis difficult. *Vibrio cholerae* and enterotoxigenic *E. coli*, for instance, both produce diarrheal diseases characterized by cramps, lethargy, and liters of watery stool each day. Cholera, however, is not commonly seen in the United States, Canada, or other developed countries. Nevertheless, a clinician might suspect cholera if a patient recently traveled to or arrived from an endemic area (a specific geographic locale such as India or Bangladesh) where the disease is regularly observed. This travel information is not gained by examining the patient but comes from taking a "patient history."

Knowledge of the patient's hobbies can be equally important for diagnosing a disease. This is another critical part of taking a patient's history. For example, a person who is suffering from enlarged glands, fever, and headaches and who was recently rabbit hunting may have been exposed to the Gram-negative rod *Francisella tularensis*, an intracellular pathogen that infects various wild animals and is the cause of tularemia, an illness also known as "rabbit fever" (**Fig. 26.1A**). A hunter with these symptoms could have accidentally infected a cut while cleaning the kill. Tularemia is an example of a **zoonotic disease**, an infection that normally affects animals but can be transmitted to humans.

How can knowing a person's occupation be important? Consider the case of a man with acute pneumonia. The clinician discovers that the man is a sheep farmer. With this knowledge, *Coxiella burnetii*, a pathogenic bacterial species that infects sheep and is shed in large quantities in placental materials, would be considered one of the possible sources of infection. The organism grows intracellularly only (it is an obligate intracellular pathogen). It infects sheep, cattle, and goats but does not usually cause clinical disease in these animals. *C. burnetii* is secreted in body fluids and shed in high numbers in the animal's amniotic fluids and placenta, heavily contaminating the soil. When the contaminated soil dries, the microbes become aerosolized whenever the dirt is disturbed. Humans (such as farmers) that inhale the dried particles can develop the lung infection called Q fever, another type of zoonotic disease (**Fig. 26.1B**).

One more example of an "occupational hazard" is illustrated by a woman with severe respiratory disease who works in a leather factory, handling animal hides. Knowing her occupation, a clinician would investigate whether she might have contracted anthrax (woolsorter's disease) from inhaling spores present on the animal hide.

Because patient histories are so helpful in diagnosing infectious diseases, we will present several case histories in the following sections and segue into discussions of various microbes that can infect each organ system. Key aspects of infectious diseases will be revealed as we proceed.

It should be stressed that further insight into pathogenesis and new innovations in diagnosis will come from the remarkable number of microbial genomes whose sequences have been completed, as well as from those that are still under way. A sampling of the pathogens whose genomes have already been sequenced is presented in **eTopic 26.1**. Many of the ORFs identified have

Figure 26.1 Examples of bacteria that cause zoonotic diseases. A. *Francisella tularensis* (colorized SEM), the cause of tularemia, a highly infectious disease spread usually via a tick vector but also through cuts. **B.** *Coxiella burnetii* (colorized TEM), the cause of Q fever. This irregularly shaped organism undergoes developmental stages that range in size from 0.2 to 1 μm.

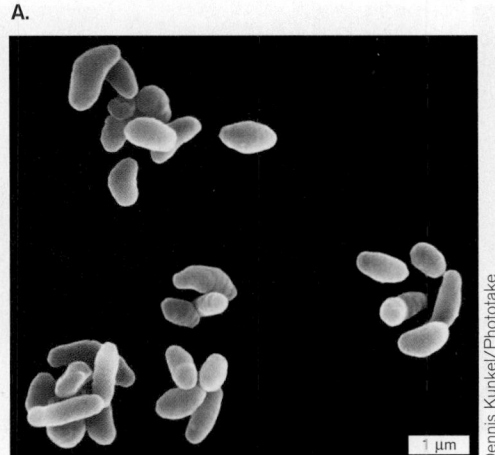
A.

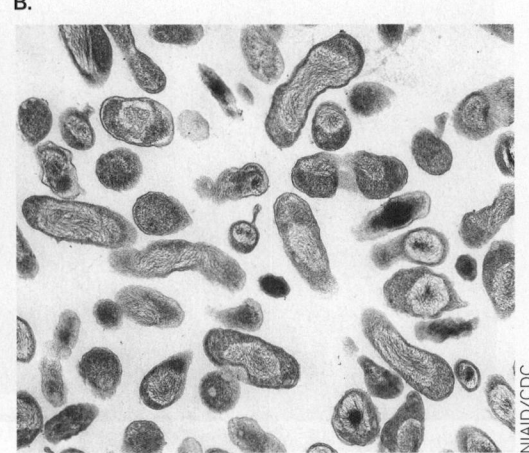
B.

no known function and may eventually explain differences in host and tissue specificity and reveal new toxins or other pharmacologically active molecules produced by virulent microbes.

TO SUMMARIZE:

- **Pathogens can be classified** as food-borne, air-borne, blood-borne, or sexually transmitted.
- **Understanding infectious disease** requires knowledge of the organ system, the portal of entry, and the infectious organism.
- **Patient histories** are vital in diagnosing microbial diseases.
- **Zoonotic diseases** are animal diseases accidentally transmitted to humans.

26.2 Skin and Soft-Tissue Infections

Skin infections range from simple boils to severe, complicated so-called flesh-eating diseases (**Table 26.1**). Recall that the integrity of the skin, as well as the presence of normal skin microbiota, prevent infection. However, even minor insults to the skin (such as a paper cut) can result in infections, most of which are caused by the Gram-positive pathogen *Staphylococcus aureus*. Healthy individuals develop infections of the skin only rarely; however, people with underlying immunosuppressive diseases such as diabetes are at much higher risk.

Boils

Staphylococcus aureus (**Fig. 26.2A**) is a common cause of the painful skin infections called boils, or carbuncles. This Gram-positive organism, often a normal inhabitant of the nares (nostrils), can infect a cut or gain access to the dermis via a hair follicle. It possesses a number of enzymes that contribute to disease, including coagulase, which helps coat the organism with fibrin, thereby walling off the infection from the immune system and antibiotics. As a result, boils generally require surgical drainage, as well as antibiotic therapy. As noted earlier, some strains of *S. aureus* also produce toxic shock syndrome toxin, a superantigen that can lead to serious systemic symptoms (see Chapters 24 and 25).

A particularly dangerous strain of *S. aureus*, called methicillin-resistant *S. aureus* (MRSA), has recently emerged. *S. aureus* infections are commonly treated with penicillin-like drugs, such as methicillin. MRSA has developed resistance to methicillin and many other penicillin-like drugs through a mutation that alters one of the proteins (called a penicillin-binding protein, PBP) involved in cell wall synthesis. Because methicillin is normally used as a first line of defense against staphylococcal infections, treatment failure of MRSA can be life-threatening, and alternative drugs, such as vancomycin, need to be used. MRSA first appeared as an agent of nosocomial infections—infections that occur after a patient enters a hospital. However, these antibiotic-resistant organisms are no longer contained only in the hospital. Individuals who have not been in a hospital are being infected with MRSA in what are called community-acquired infections. This is occurring at an epidemic rate in the United States (an incidence of approximately 20 per 100,000 population). Seventy percent of staphylococcal skin infections are now caused by MRSA. Thus, a physician can no longer assume that a patient walking into the office with a staphylococcal infection will respond to methicillin. The doctor must assume that it may be MRSA. As a result, treatment regimens around the country, and the world, are being forced to change. Today, vancomycin and linezolid are the antibiotics typically used to treat staph infections. Antibiotics are discussed in Chapter 27.

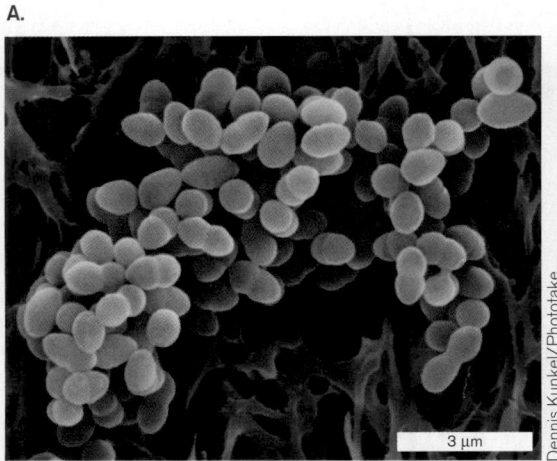

A.

Dennis Kunkel/Phototake

3 μm

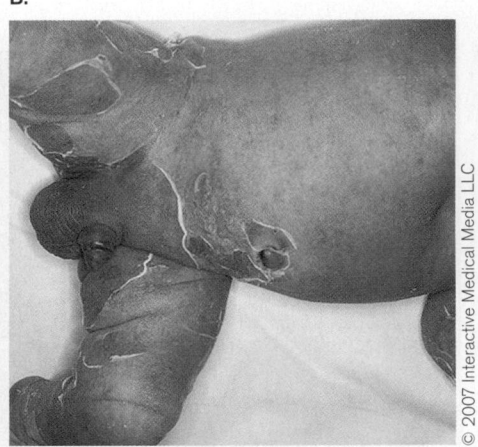

B.

© 2007 Interactive Medical Media LLC

Figure 26.2
Staphylococcus aureus.
A. *S. aureus* (colorized SEM). **B.** Exfoliative toxin from some strains of *S. aureus* cause scalded skin syndrome.

Table 26.1 Common infectious diseases of the skin.

Disease	Symptoms	Etiological agent	Virulence factors
Bacterial			
Folliculitis	Boils	*Staphylococcus aureus* (G+ cocci); fibrin wall around abscess renders it poorly accessible to antibiotics	Coagulase, protein A, TSST, leukocidin, exfoliative toxin
Scalded skin syndrome	Peeling skin on infants, systemic toxin		
Impetigo	Skin lesions on the face, mostly children		
Scarlet fever	Sore throat, fever, rash	*Streptococcus pyogenes* (G+ cocci)	M-protein pili, C5a peptidase, hemolysin, pyrogenic toxins, others
Erysipelas	Skin lesions, usually facial, that spread to cause systemic infection		
Necrotizing fasciitis	Rapidly progressive cellulitis		
Viral			
Rubella[a]	Discolored, pimply rash; mild disease unless congenital	Rubella virus [ssRNA(+)]	
Measles[a]	Severe disease, fever, conjunctivitis, cough, rash	Rubeola virus [ssRNA(−)]	V protein (interferes with interferon signaling)
Chickenpox[a]	Generalized discolored lesions	Varicella-zoster (dsDNA)	Glycoprotein B (fusion of viral and cellular membranes)
Shingles	Pain and skin lesions, usually on trunk in adults		Glycoprotein E (required for cell-cell fusion)
Smallpox[b]	Raised, crusted skin rash, highly contagious	Variola major (dsDNA)	SPICE (smallpox inhibitor of complement enyzmes)
Warts[a]	Rapid growth of skin cells	Papillomaviruses (dsDNA)	E6 and E7 oncoproteins (Chapter 25)
Fungal			
Dermatomycosis	Dry, scaly lesions like athlete's foot	Dermatophytes	
Sporotrichosis	Granulomatous, pus-filled lesions; can disseminate to lungs or other organs	*Sporothrix schenckii*	
Blastomycosis	Granulomatous, pus-filled lesions; can disseminate to lungs or other organs	*Blastomyces dermatitidis*	BAD1 adherence
Candidiasis	Patchy inflammation of mouth (thrush) or vagina; can disseminate in immunocompromised patients	*Candida albicans*; *Candida glabrata*	Proteinase, phospholipase, Ssn6/Tup1 regulators, others
Aspergillosis	Infected wounds, burns, cornea, external ear	*Aspergillus* spp.	PacC/FOS1 regulators, gliotoxin
Zygomycosis	Oropharyngeal infections; mainly affects diabetic patients; can rapidly disseminate	*Mucor* and *Rhizopus* spp.	

[a] Vaccine is available against this agent.

[b] Vaccine is no longer in use, because disease has been eradicated.

Other staphylococcal diseases are caused by toxin-producing strains in which the organism remains localized but the toxin disseminates. We have already mentioned TSST, but there are other toxins. For example, some strains of *S. aureus* produce a toxin called exfoliative toxin that causes a blistering disease in children called staphylococcal scalded skin syndrome (**Fig. 26.2B**). Exfoliative toxin, like TSST, is a superantigen, but it also cleaves a skin cell adhesion molecule that, when inactivated, results in blisters.

Table 26.1 presents other common infections of the skin and soft tissues.

Case History: Necrotizing Fasciitis by "Flesh-Eating" Bacteria

*One weekend in June, Cassi was camping with her three children. She suffered a minor cut on her finger, which she bandaged properly. She also injured the left side of her body while playing sports with her kids. Not thinking much of either of her minor injuries, she went to bed. Two days later, Cassi was extremely ill. Her symptoms included vomiting, diarrhea, and a fever. She was also in severe pain where she had injured her side, and the area had begun to bruise (the skin was not broken). By the next day she could barely get out of bed, and by the end of the night she was breathing with difficulty and could not see. Her side began to leak fluid and blood. Cassi was admitted to the hospital in shock, with no detectable blood pressure. An infectious disease specialist diagnosed the problem as necrotizing fasciitis, and she was rushed into surgery. In an effort to save her life, about 7% of her body surface was removed. Because the large wound infection in her side would need to resolve before a skin graft could be performed to repair it, the hole in Cassi's body was left wide open (**Fig. 26.3B**). After nearly 3 months and several operations, Cassi recovered.*

What kind of organism can cause this type of devastating disease? The disease **necrotizing fasciitis**, also known as flesh-eating disease, is rare and is often caused by the Gram-positive coccus *Streptococcus pyogenes* (**Fig. 26.3A**), a microbe normally associated with throat infections (pharyngitis). Although sometimes described as a recently emerging infectious disease, necrotizing fasciitis was first discovered in 1783, in France. Its incidence may have risen recently owing to the increased use of non-steroidal, anti-inflammatory drugs (e.g. ibuprofen), which increase a person's susceptibility to infection. In this case history, Cassi probably had this organism on her skin when the injury to her side occurred. The injured area probably suffered an

invisible microabrasion, providing a good growth environment for the organism, leading to the secretion of potent toxins and death of surrounding tissues.

Rapid, aggressive antibiotic treatment is required in these extreme cases, even before the clinical microbiology lab has had time to identify the organism. Therapy can include several antibiotics, such as clindamycin and metronidazole, which act against anaerobes and Gram-positive cocci, and gentamicin, a drug particularly effective against Gram-negative microbes. (Chapter 27 further discusses these and other antibiotics.) Often, however, antibiotic treatment of patients with necrotizing fasciitis is ineffective because of insufficient blood supply to the affected tissues.

> **THOUGHT QUESTION 26.1** Why would treatment of some infections require multiple antibiotics?

WWW | Antibiotic treatment
    | *microbiology2.com/links*

Is There an Immunological Predilection to Necrotizing Fasciitis?

Streptococcus pyogenes is best known for causing sore throats and for the immunological sequelae that can develop, such as rheumatic fever and glomerulonephritis. **Sequelae** are secondary diseases that develop after resolution of a primary infection. They are often the result of a cross-reactivity between bacterial and host antigens. The immune system, activated by the bacterial antigen, begins to attack the cross-reacting self antigens and damages tissues (see Section 24.8). Why some patients infected with *S. pyogenes* develop necrotizing fasciitis while others do not may be due more to the immunological status of the

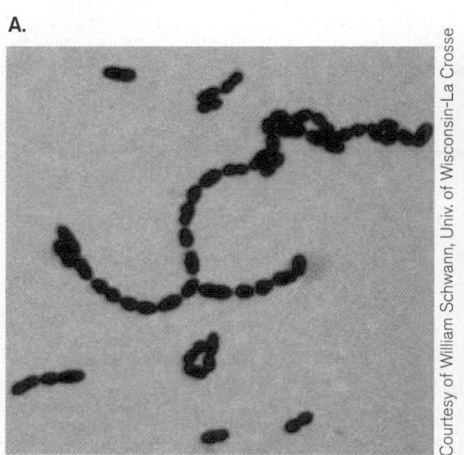

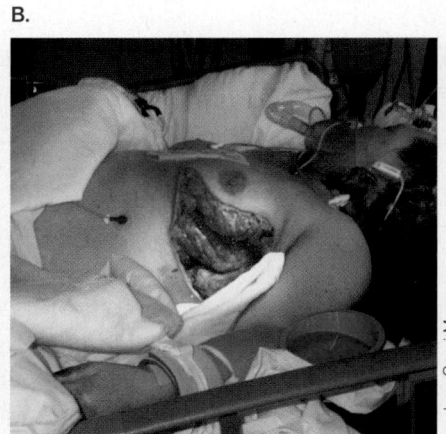

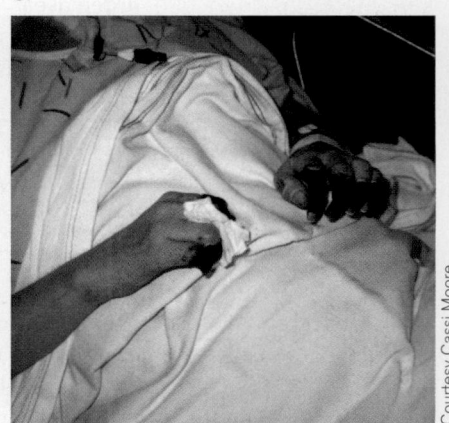

A. Courtesy of William Schwann, Univ. of Wisconsin-La Crosse
B. Courtesy Cassi Moore
C. Courtesy Cassi Moore

Figure 26.3 Flesh-eating *Streptococcus pyogenes.* A. Gram stain of *Streptococcus pyogenes*. **B.** Flesh removed from a patient in an effort to stop the spread of necrotizing fasciitis. **C.** Gangrene of the fingers caused by the same infection in the same patient.

patient than to the virulence arsenal of the microbe, but its arsenal is vast nonetheless.

The sources of many established putative virulence factor genes in *S. pyogenes* are the numerous prophages present in its genome. Prophages constitute approximately 10% of the organism's genome. One study found that a soluble factor produced by human pharyngeal cells can facilitate activation of at least some of these phages and cause horizontal transfer of the associated virulence factors between strains of this pathogen.

Viral Diseases Causing Skin Rashes

Several viruses can produce skin rashes, although their route of infection is usually through the respiratory tract. Measles, for example (see Section 6.1), is a highly contagious viral infection caused by a paramyxovirus, whose hallmark symptom is skin rash (Fig. 6.2). Also known as rubeola, the first signs are fever, cough, runny nose, and red eyes occurring 9–12 days after exposure. A few days later, spots in the mouth appear (Koplik's spots) along with a sore throat. Then a skin rash develops that typically starts on the face and spreads down the body. The virus replicates in the lymph nodes and spreads to the bloodstream (viremia), where it can infect endothelial cells of the blood vessels. The rash occurs when T cells begin to interact with these infected cells.

Although skin rash is the main symptom of measles, infection can also cause respiratory symptoms and complications, including pneumonia, bronchitis, croup, and even a fatal encephalitis in immunocompromised patients. In the United States, measles has been almost completely eliminated by the measles, mumps, and rubella (MMR) multivalent vaccine (one exception is the AIDS patient, in which measles can be fatal). Worldwide, however, measles is still a serious problem in places where vaccinations are not routine.

Rubella virus, a togavirus, causes a maculopapular rash known as German measles, or 3-day measles. The rash is similar but less red than that of measles (**Fig. 26.4**). German measles is an infection that affects primarily the skin and lymph nodes and is usually transmitted from person to person by aerosolization of respiratory secretions. It is not dangerous in adults or children; the virus can, however, cross the placenta in a pregnant woman and infect her fetus. If the virus crosses the placenta within the first trimester, the result is congenital rubella syndrome, which can cause death or serious congenital defects in the developing fetus.

Other viruses affecting the skin, such as chickenpox, the related disease shingles, and smallpox, are discussed in Chapters 6 and 11 and included in **Table 26.1**.

TO SUMMARIZE:

■ *Staphylococcus aureus* and *Streptococcus pyogenes* are common bacterial causes of skin infections. The organisms usually infect through broken skin.

■ **Methicillin-resistant *S. aureus* (MRSA)** has become an important cause of community-acquired staphylococcal infections.

■ **Necrotizing fasciitis** is usually caused by *S. pyogenes* but can be the result of other infections.

■ **Infections of the skin** can disseminate via the bloodstream to other sites in the body.

■ **Rubeola and rubella viruses** infect through the respiratory tract, but their main manifestation is the production of similar maculopapular skin rashes.

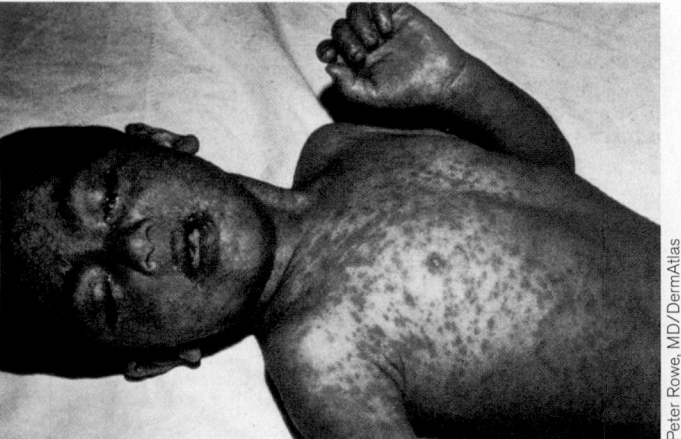

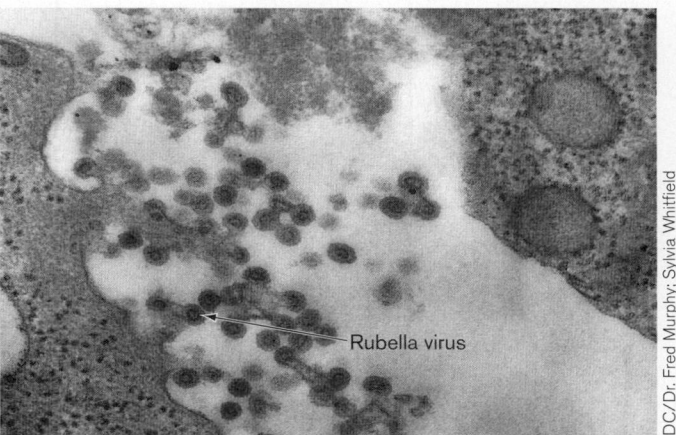

Figure 26.4   German measles.   A. Skin rash caused by rubella virus. **B.** Rubella virus budding from cell surface to form an enveloped virus particle (approx. 50–70 nm; TEM).

26.3 Respiratory Tract Infections

Lung and upper respiratory tract infections are among the most common diseases of humans. Many different bacteria, viruses, and fungi are well adapted to grow in the lung. Successful lung pathogens come equipped with appropriate attachment mechanisms and countermeasures to avoid various lung defenses (such as alveolar macrophages). One re-emerging bacterial pathogen, *Bordetella pertussis*, the cause of whooping cough, inhibits the mucocilliary elevator by binding to lung cilia (**eTopic 26.2**). Although many microbes can cause lung infection, most respiratory diseases are of viral origin, and most viral infections (such as the common cold) do not spread beyond the lung. Fortunately, viral diseases by and large are self-limiting and typically resolve within 2 weeks; however, the damage caused by a primary viral infection can lead to secondary infections by bacteria.

Bacterial infections of the lung, whether of primary or secondary etiology, require intervention. Today this means antibiotic therapy. Before the advent of antibiotics, the only recourse was to insert a tube into the patient's back to drain fluid accumulating in the pleural cavity around the lung (a pathological process known as pleural effusion). Unless released, the pressure on the lung will collapse the alveoli and make breathing difficult.

Bacterial infections of the lung that arise secondarily to viral disease occur, in part, because the patient dehydrates. The resulting increase in mucus viscosity hampers the mucociliary elevator (described in Section 23.1), as will growth of the bacterium itself. Because a compromised mucociliary elevator makes it difficult to expel the microbe, the patient's susceptibility to secondary bacterial infection increases. Hence, cold sufferers are advised to drink plenty of fluids, which help decrease the viscosity of mucus and consequently improve mucociliary elevator function.

Case History: Bacterial Pneumonia

*In March, an 80-year-old resident of a New Jersey nursing home had a fever accompanied by a productive cough with brown sputum (mucous secretions of the lung that can be coughed up). He reported to the attending physician that he had pain on the right side of his chest and suffered from night sweats. Blood tests revealed that his white blood cell (WBC) count was 14,000/μl (normal is 5,000–10,000/μl) with a makeup of 77% segmented forms (polymorphonuclear leukocytes, PMNs) and 20% bands (immature PMNs). The chest radiograph revealed a right upper lobe infiltrate with cavity formation (**Fig. 26.5A**). From this information, the clinician made a diagnosis of pneumonia. Microscopic examination of the patient's sputum revealed Gram-positive cocci in pairs and short chains surrounded by a capsule (**Fig. 26.5B**). Bacteriological culture of his sputum and blood yielded* Streptococcus pneumoniae.

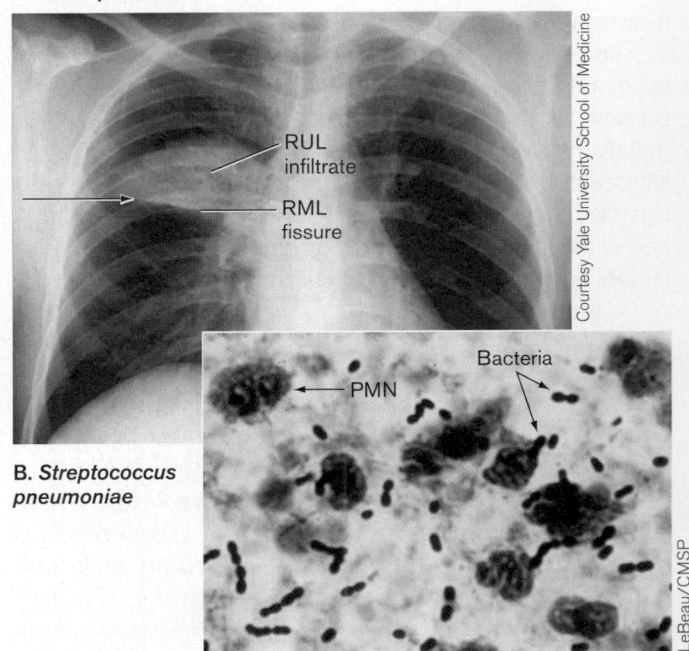

A. Lobar pneumonia

B. *Streptococcus pneumoniae*

C. Relative incidence of pneumonia

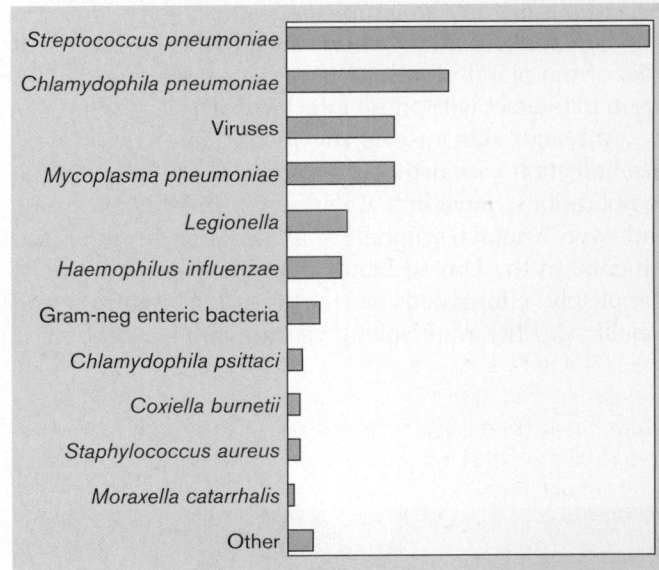

Figure 26.5 Pneumonia caused by *Streptococcus pneumoniae.* **A.** X-ray view of a patient with lobar pneumonia. Infiltrate in the right upper lobe (RUL) is caused by *S. pneumoniae*. The sharp lower border represents the upper boundary of the right middle lobe (RML) fissure (arrow). **B.** Micrograph of *S. pneumoniae*. Sputum sample showing numerous PMNs and extracellular diplococci in pairs and short chains. Bacteria range from 0.5 to 1.2 μm in diameter. **C.** Relative incidence of pneumonia caused by various microorganisms.

Note that pneumonia is a disease, not a specific infection. Many different microbes can cause pneumonia (**Table 26.2**). The pneumococcus *S. pneumoniae* accounts for about 25% of community-acquired cases of pneumonia, but pneumococcal pneumonia occurs mostly among the elderly and immunocompromised, including smokers, diabetics, and alcoholics. A breakdown of pneumonia cases by causative organism is shown in **Figure 26.5C**.

The noses and throats of 30%–70% of a given population can contain *S. pneumoniae*. The microbe can be spread from person to person by sneezing, coughing, or other close personal contact. Pneumococcal pneumonia may begin suddenly, with a severe shaking chill usually followed by high fever, cough, shortness of breath, rapid breathing, and chest pains.

After being aspirated into the lung, the microbe will grow in the nutrient-rich edema fluid of the alveolar spaces. Neutrophils and alveolar macrophages then arrive to try to stop the infection. They are called into the area from the circulation by chemoattractant chemokines released by damaged alveolar cells. The thick polysaccharide capsule of the pneumococcus, however, makes phagocytosis very difficult. In an otherwise healthy adult, pneumococcal pneumonia usually involves one lobe of the lungs; thus, it is sometimes called lobar pneumonia. The infiltration of PMNs and fluid lead to the typical radiological findings of diffuse cloudy areas. In contrast, infants, young children, and elderly people more commonly develop an infection in other parts of the lungs, such as around the air vessels (bronchi), causing bronchopneumonia.

The white cell count in the case history is telling. The patient had an elevated WBC count (normal is 5,000–10,000/μl) and elevated band cells (normal is 0%–8%). These increases are indicative of a bacterial, not viral, infection. Neutrophils (PMNs), the front-line combatants against infection, rise in response to bacterial infections and are first released from bone marrow as immature band cells, whose presence is a sure sign of bacterial infection.

Several outbreaks of pneumococcal pneumonia have occurred over recent years in nursing homes, where numerous residents have been affected. This underscores the importance of elderly people receiving the pneumococcal polysaccharide vaccine (PPV) as a hedge against infection. While there are over 80 antigenic types of pneumococcal capsular polysaccharides, the injected vaccine contains only the 23 types that are most often associated with disease. A vaccine that is formulated to respond to multiple antigens is termed a multivalent vaccine. The immune response that results will protect vaccinated individuals against infection by those antigenic types. The vaccine is recommended for individuals over 65, as well as for those who are immunocompromised. The patient in this case history failed to receive the vaccine.

In addition to causing serious infections of the lungs (pneumonia), *S. pneumoniae* can invade the bloodstream (bacteremia) and the covering of the brain (meningitis). The death rates for these infections are about one out of every 20 who get pneumococcal pneumonia, about two out of 10 who get bacteremia, and three out of 10 who get meningitis. Individuals with special health problems, such as liver disease, AIDS (caused by HIV), or organ transplants, are even more likely to die from the disease, because of their compromised immune systems.

An emerging infectious disease problem throughout the United States and the world is the increasing resistance of *S. pneumoniae* to antibiotics. At least 30% of the strains isolated are already resistant to penicillin, the former drug of choice for treating the disease. Chapter 27 discusses why antibiotic resistance is on the rise for this and other microbes.

Case History: Disseminated Disease from a Fungal Lung Infection

*A 35-year-old male boxer named Tyrrell, recently admitted to a Maryland hospital, was in good health until 6 months ago, when he developed a chronic cough that produced blood-tinged white sputum. He also experienced flu-like symptoms, a decrease in appetite, and weight loss. One month prior to admission, he became so short of breath that he could no longer continue boxing. At that time, an X-ray taken in the emergency room showed right upper lobe infiltrate, indicating pneumonia (**Fig. 26.6A**). A tuberculosis skin test was negative. He was given a prescription for the antibiotic azithromycin (a macrolide antibiotic commonly used to treat bacterial infections of the respiratory tract) and discharged. Despite antibiotic treatment, the cough never diminished and he developed several painless subcutaneous nodules. The largest nodule was on his left leg, contained pus, caused pain, and eventually hindered walking (**Fig. 26.6B**). This prompted his current admission to the hospital. The patient's history revealed that he installed home insulation for a living and had not traveled outside the area for the past year. He complained of fevers, chills, night sweats, and a 10-kg (22-lb) weight loss. His right tibia was tender to the touch, indicating bone involvement. He denied having prior pneumonia, sinus infection, arthritis, hematuria (blood in the urine), numbness, or muscle weakness. He had no history of intravenous drug use and had been in a monogamous relationship for 4 years. His white cell count was 10,500/μl with a normal differential of PMNs and band cells.*

In this case, the expression of frankly purulent material (pus) from the left-leg nodule suggested that this was an infectious process. The infection in this patient probably started in the lung (clued by the cough), after which the organism disseminated throughout the body via

Table 26.2 Selected microbes that cause respiratory tract diseases.

Disease	Symptoms	Etiological agent
Bacterial		
Inhalation anthrax	Fever, muscle aches, hypotension, respiratory failure	*Bacillus anthracis* (G+ rod)
Diphtheria	Tracheal pseudomembrane	*Corynebacterium diphtheriae* (G+ rod)
Whooping cough	Fever, runny nose, sneezing, violent cough followed by inhalation "whoop"	*Bordetella pertussis* (G− coccobacillus) (see **eTopic 26.2**)
Tuberculosis	Fever, chills, cough, bloody sputum, fatigue, weight loss	*Mycobacterium tuberculosis* (acid-fast bacillus)
Pneumonia	Infects cystic fibrosis patients; fever, chills, cough, chest pain	*Pseudomonas aeruginosa* (G− rod)
	Fever, chills, cough, chest pain	*Streptococcus pneumoniae* (G+ diplococcus)
	Fever, sore throat, chest pain, runny nose (also associated with cardiovascular disease)	*Chlamydophila pneumoniae*
Psittacosis	Fever, sore throat, chest pain, runny nose	*Chlamydophila psittaci*
"Walking pneumonia"	Fever, sore throat, nonproductive cough, chills	*Mycoplasma pneumoniae* (wall-less microbe)
Legionnaires' disease (legionellosis)	Fever, chest pain, chills, cough, muscle pain, vomiting, diarrhea	*Legionella pneumophila* (G− rod)
Viral		
CMV disease	Fever, chills, cough, chest pain	Cytomegalovirus (CMV); dsDNA
RSV disease	Fever, chills, cough, chest pain	Respiratory syncytial virus (RSV); ssRNA (−)
Influenza	Fever, chills, cough, chest pain, sore throat, muscle pain	Influenza and parainfluenza viruses; ssRNA (−), segmented
	Sore throat, fever, runny nose	Adenovirus; dsDNA
Severe acute respiratory syndrome	Fever, chills, cough, chest pain	SARS virus; ssRNA (+)
Fungal		
Aspergillosis	Affects lungs and sinuses; fever, chills, breathing difficulty	*Aspergillus* spp.
Histoplasmosis	Flu-like; pulmonary infiltrate	*Histoplasma capsulatum*
Coccidioidomycosis	Flu-like; pulmonary infiltrate	*Coccidioides immitis*
Blastomycosis	Chest pain, cough, skin lesion, pulmonary infiltrate	*Blastomyces dermatitidis*
Pneumocystosis	Infects AIDS patients; cough, fever, weight loss	*Pneumocystis jirovecii* (formerly *carinii*)

[a]Bacillus Calmette-Guérin (a weakened strain of the bovine tuberculosis strain).

Virulence properties	Source	Treatment	Vaccine
Peptide capsule, PA, LF, and EF toxins	Airborne	Ciprofloxacin	+, Military
Diphtheria toxin	Airborne; localizes to nasopharynx	Penicillin	+
Tracheal cytotoxin, adenylate cyclase toxin, filamentous hemagglutinin (adhesin)	Respiratory droplets	Erythromycin	+
Cord factor, wax D intracellular growth	Humans	Combination therapy (rifampin, isoniazid, ethambutol, pyrazinamide)	+ BCG[a]
Exotoxin A, phospholipase C, exopolysaccharide, others	Water, soil	Quinolones, aminoglycosides	
Capsule, pneumolysin	Inhalation	Macrolides, quinolones, ceftriaxone	Multivalent from capsule antigens
Obligate intracellular growth; prevents phagolysosome fusion	Person-to-person	Quinolones	
Obligate intracellular growth; prevents phagolysosome fusion	Bird droppings, dust; inhalation	Tetracycline, erythromycin	
Adhesin tip	Inhalation, person-to-person	Erythromycin and other macrolides	
Intracellular growth, hemolysin, cytotoxin, protease	Inhalation	Erythromycin	
	Saliva, tears, breast milk	Ganciclovir	Experimental
	Respiratory droplets	Treat symptoms; ribavirin	
	Respiratory droplets	Zanamivir (Relenza)	+
	Respiratory droplets	Treat symptoms	+, Military
	Respiratory droplets		
	Inhalation	Amphotericin B, voriconazole	
	Inhalation; bird, chicken, bat droppings	Amphotericin B, itraconazole	
	Inhalation	Amphotericin B, fluconazole	
	Inhalation	Amphotericin B, itraconazole	
	Inhalation	Trimethoprim sulfamethoxazole	

A. Pneumonia infiltrate

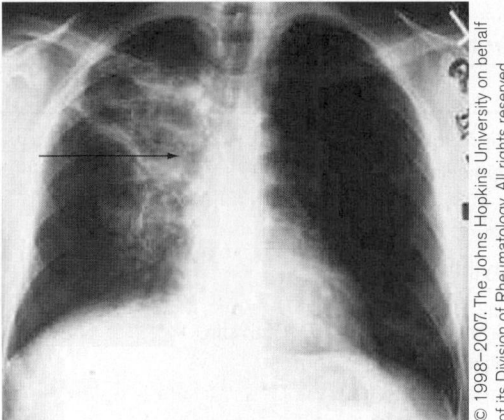

B. Leg lesion

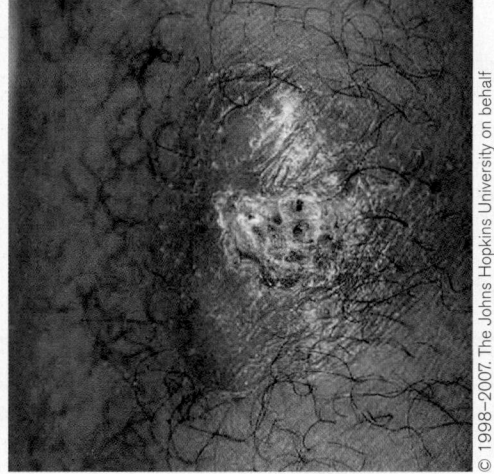

C. Colony of *Blastomyces dermatitidis*

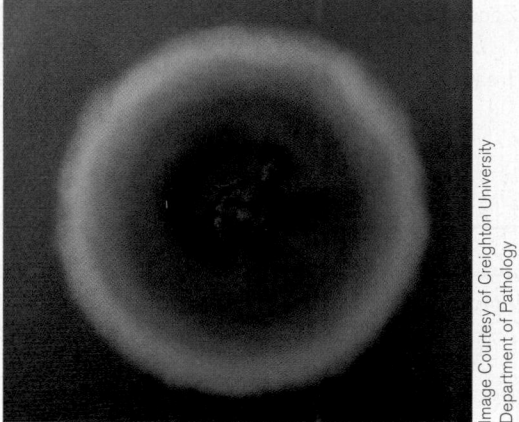

Figure 26.6   Pneumonia and metastatic disease caused by *Blastomyces dermatitidis*.   A. Diffuse infiltrate in the right lung (arrow). **B.** Metastatic leg lesion at tibia. **C.** Fungal colony of *B. dermatitidis*.

the bloodstream. Fungus is a probable cause, given the chronic nature of the patient's symptoms. An alternative possibility would be tuberculosis, caused by the bacterium *Mycobacterium tuberculosis*. *M. tuberculosis* also causes a chronic lung infection and might have been suspected, except for the negative TB skin test. The tuberculin skin test involves injecting a small amount of mycobacterial antigen called PPD (purified protein derivative) under the skin of the lower arm. A person who has been infected with *M. tuberculosis* will exhibit a localized delayed-type hypersensitivity reaction at the site of injection, although this does not equate to currently active disease.

The most likely fungal causes of infection in this case history are the endemic mycoses, such as histoplasmosis, blastomycosis, and coccidioidomycosis. This patient had never traveled to the western United States, where coccidioidomycosis is endemic, so exposure to *Coccidioides* was ruled out. Histoplasmosis most commonly presents as a flu-like pulmonary illness, with erythema nodosum (tender bumps on skin) and arthritis (swollen joints) or arthralgia (joint pain), none of which the patient had. Blastomycosis can disseminate to the lung, skin, bone, and genitourinary tract, consistent with the pattern of organ involvement seen in this patient. *Cryptococcus*, an encapsulated yeast, is not the likely cause, since it typically requires an immunocompromised host to cause disease— not the case in this instance. (*Cryptococcus*, which causes cryptococcosis, is an opportunistic pathogen that commonly infects AIDS patients.) The most prevalent clinical form of cryptococcosis is meningoencephalitis, although disease can also involve the skin, lungs, prostate gland, urinary tract, eyes, myocardium, bones, and joints.

Amphotericin B, an antifungal agent, was finally given to this patient. His fever lowered almost immediately, and the skin nodules diminished. After 2 weeks, a fungus was found in the cultures of the nodule biopsy, bronchoalveolar lavage (washes), and urine (**Fig. 26.6C**). This fungus was identified by a DNA probe as *Blastomyces dermatitidis*, confirming the diagnosis of blastomycosis.

B. dermatitidis is a dimorphic fungus that resides in the soil of the Ohio and Mississippi river valleys and the southeastern United States. The portal of entry is the respiratory tract, and infection is usually associated with occupational and recreational activities in wooded areas along waterways, where there is moist soil with a high content of organic matter and spores. The incubation period ranges from 21 to 106 days. This patient most likely inhaled conidia (fungal spores) from the soil while crawling underneath houses installing insulation. The physician learned that he used only a T-shirt to cover his mouth and nose— not an effective method of keeping spores from entering the respiratory tract. He should have worn a respirator.

Several critical features of this case help differentiate it from the preceding case of pneumococcal pneumonia. First, the initial macrolide antibiotic, azithromycin, should have killed most bacterial sources of infection. Second, the X-ray finding of diffuse infiltrate is more indicative of fungal lung infection than bacterial infection, which in a patient of this age would likely be confined to one lobe. The patient was young and in good health prior to the infection, making it unlikely to be pneumococcal pneumonia. The blood count was also a clue. Fungal infections do not usually cause an increase in WBCs or an increase in band cells. Finally, the **metastatic lesions** (infectious lesions that develop at a secondary site away from the initial site of infection) on the leg were in no way consistent with *S. pneumoniae*. They arise when the organism moves through the bloodstream from the primary site of infection to another body site, where it can begin to grow. Many infectious diseases start out as a localized infection but end up disseminating throughout the body to cause metastatic lesions.

> **NOTE:** The term "metastasis" means the spread of disease from one organ to another noncontiguous organ. Only microbial infections and malignant cancer cells can metastasize.

Tuberculosis as a Reemerging Disease

Until recently, tuberculosis, caused by the acid-fast bacillus *Mycobacterium tuberculosis* (**Fig. 26.7**, inset), was considered of passing historical significance to physicians practicing in the developed world. In 1985, however, owing primarily to the newly recognized HIV epidemic and a growing indigent population, TB resurfaced, especially in inner-city hospitals. In 1991, highly virulent multidrug-resistant (MDR) strains of *M. tuberculosis* were reported. These strains not only produced fulminant (rapid onset) and fatal disease among patients infected with HIV (the time between TB exposure to death is 2–7 months), but also proved highly infectious. Tuberculin skin test conversion rates of up to 50% are reported in exposed health care workers. A positive tuberculin skin test is seen as a delayed-type hypersensitivity reaction to *M. tuberculosis* proteins (called purified protein derivative [PPD]) injected under the skin. More information on mycobacteria can be found in Section 18.4.

M. tuberculosis primarily causes a respiratory infection, but it can disseminate through the bloodstream to produce abscesses in many different organ systems. Disseminated disease is called miliary TB because the size of the infected nodules, called tubercles, approximates the size of millet seeds. Tubercles in the lung are filled with *M. tuberculosis* cells and drain into bronchial tubes and the upper respiratory tract. The organisms are spread from person to person (no animal reservoir) through aerosolization of respiratory secretions. Once in the lung, the bacteria are phagocytosed by macrophages and survive ensconced within modified phagolysosomes. A delayed-type hypersensitivity response results, and small, hard tubercles form. Over time, the tubercles develop into caseous lesions that have a cheese-like consistency and can calcify into the hardened Ghon complexes seen on typical X-ray findings (**Fig. 26.7**).

After a period of 4–12 weeks, patients with primary disease exhibit a productive cough generating sputum and experience fever, night sweats, and weight loss. They also become tuberculin test positive owing to delayed-type hypersensitivity. However, note that a positive tuberculin skin test does not signify *active* disease, but only that the person was infected at one time. The bacterium may have been killed by the immune system without having caused disease.

The majority of patients recover from the primary TB episode. However, live bacteria can remain dormant for long periods of time, only to be reactivated months or years after initial infection—a condition called secondary or reactivation TB. Secondary TB commonly occurs in immunocompromised people. The tubercles described here form during secondary TB. The symptoms of secondary TB are more serious than those of primary TB and include severe coughing, greenish or bloody sputum, low-grade fever, night sweats, and weight loss. The gradual wasting of the body led to the older name for tuberculosis: consumption.

Treatment of active disease is aggressive and involves a four-drug regimen including isoniazid, rifampin, pyrazinamide, and ethambutol given over a course of several months. MDR strains are defined as being resistant to two or more first-line drugs. Because of this resistance, MDR strains are treated with a nine-drug regimen. Even more dangerous are extensively drug-resistant tuberculosis

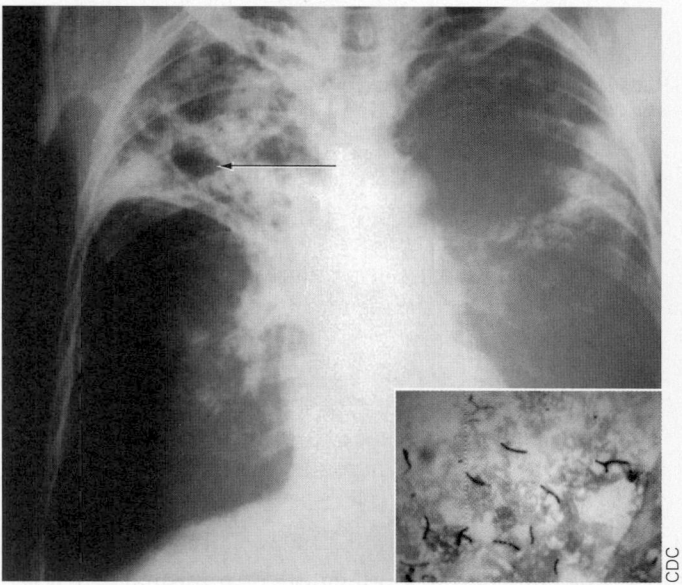

Figure 26.7 Calcified Ghon complex of tuberculosis. The arrow points to a complex in a patient's right upper lobe. Note the difference in appearance compared to Figure 26.6A. Inset is an acid-fast stain of *M. tuberculosis*.

(XDR-TB) strains, which are resistant to three or more second-line drugs in addition to two or more first-line drugs. These strains are almost untreatable.

> **THOUGHT QUESTION 26.2** Explain why patient noncompliance (failure to take drugs as directed) is thought to have led to XDR-TB.

Viral Diseases of the Lung

Numerous viruses can cause lung infections (see **Table 26.2**). Influenza, rhinovirus, and SARS are discussed in Chapters 6, 11, and 25. An important viral lung infection not discussed elsewhere is respiratory syncytial disease, caused by respiratory syncytial virus (RSV). A negative-sense, single-stranded RNA, enveloped virus, RSV is the most common cause of bronchiolitis and pneumonia among infants and children under 1 year of age. Illness begins most frequently with fever, runny nose, cough, and sometimes wheezing. RSV is spread from respiratory secretions through close contact with infected persons or by contact with contaminated surfaces or objects. Infection can occur when the virus contacts mucous membranes of the eyes, mouth, or nose and possibly through the inhalation of droplets generated by a sneeze or cough. Unlike rubella or rubeola, which infect the respiratory tract and disseminate though the body, RSV remains localized in the lung.

The majority of children hospitalized for RSV infection are under 6 months of age. RSV can cause repeated infections throughout life, usually associated with moderate to severe cold-like symptoms. Severe lower respiratory tract disease may occur at any age, especially among the elderly or people with compromised cardiac, pulmonary, or immune systems. As yet, a vaccine to control this disease is not available.

Table 26.2 presents many other bacterial, fungal, and viral microbes that can cause respiratory tract infection. Be aware that very different diseases can produce similar symptoms. For instance, people constantly confuse influenza (the flu) with the common cold. Symptomatically, they may start out similarly, but there are telling differences. Influenza is characterized by fever, myalgia (muscle aches), pharyngitis (sore throat), and headache. A runny nose is *not* one of the symptoms. The common cold, however, manifests as a runny nose, nasal congestion, sneezing, and throat irritation. No myalgia. The observant clinician will note the difference.

TO SUMMARIZE:

- **The mucociliary elevator** is a primary defense mechanism used by the lung to avoid infection.
- **Pneumonia** is caused by many microorganisms.
- **An elevated white cell count** in blood is an indicator of bacterial infection.
- **Pneumococcal vaccine** should be administered to the elderly because they are often immunocompromised.
- **Fungal agents** commonly cause long-term, chronic infections.
- **Tuberculosis** is an ancient yet reemerging bacterial disease with an increasing mortality rate caused by the development of multidrug-resistant strains, the susceptibility of HIV patients, and an increasing indigent population.
- **Localized infections in the lung can disseminate** via the bloodstream to form metastatic lesions at other body sites.
- **Respiratory syncytial virus** is one of several viruses that can cause lung disease but rarely disseminates.

26.4 Gastrointestinal Tract Infections

Nearly everyone has experienced diarrhea—a condition characterized by frequent loose bowel movements accompanied by abdominal cramps. Hundreds of millions of cases occur each year in the United States and are a major cause of death in developing countries. The loose stools usually result from inflammation (called gastroenteritis) due to viral growth, bacterial growth, or toxin production, causing large amounts of water and electrolytes to leave the intestinal cells and enter the intestinal lumen. As a result, the patient not only suffers diarrhea, but can become dangerously dehydrated. As with respiratory tract infections, most diarrheal disease is viral in origin, with rotavirus being the primary culprit. Among the bacteria, the Gram-negative curved bacillus *Campylobacter* is the most frequent cause of self-limiting diarrheal disease.

The main symptoms of gastroenteritis, whether bacterial or viral, are watery diarrhea and vomiting. A more severe form of gastroenteritis is called dysentery, whose symptoms include abdominal pain, a persistent desire to empty the bowels, and diarrhea with passage of blood or mucus. Bacterial causes of dysentery include several *Shigella* species and some strains of *E. coli*. An ameba, *Entamoeba histolytica*, can also cause a form of the disease, called amebic dysentery. Features of bacterial dysentery are discussed in more detail later in this section.

Remarkably, *Staphylococcus aureus* causes gastrointestinal disease without ever producing infection. Everyone has heard of the local church picnic where scores of people become violently ill within hours of eating unrefrigerated potato salad. *S. aureus* is the usual cause of these disasters. Not all strains of *S. aureus* cause food poisoning, but certain strains can secrete enterotoxins into tainted foods such as pies, turkey dressing, or potato salad. After ingestion, the toxin travels to the intestine, where it enters the bloodstream and stimulates nerves leading to the vomit center in the brain. Because the toxin is preformed, symptoms occur quickly after ingestion. Within 2–6 hours, the poisoned patient will begin vomiting and may also experience diarrhea. The disease, though violent, is not life-threatening and usually resolves spontaneously within 24–48 hours. In contrast, diarrhea caused by infectious agents, such as *Salmonella enterica*, that must first grow in the victim do not occur until 12–24 hours after ingestion, sometimes longer. A clinician noting quick onset of symptoms in a patient will immediately suspect staphylococcal food poisoning. Obviously, antibiotic treatment is not needed for staph food poisoning but may be indicated for other gastrointestinal infections.

Antibiotics Are Often Inappropriate When Treating Gastroenteritis

Although antibiotic treatment of infectious gastroenteritis seems intuitive, it is rarely used. Most gastrointestinal infections are viral (e.g. Norovirus, Norwalk virus, and Rotavirus), so antibiotics are ineffective. Likewise, gastroenteritis caused by bacteria usually resolves spontaneously without antibiotic treatment. Of course, severe systemic disease stemming from gastroenteritis can develop, often in the young or elderly. Diseases such as typhoid fever (*Salmonella* Typhi) or bacillary dysentery (*Shigella dysenteriae*) respond well to antibiotics.

In some cases, antibiotic treatment can actually *trigger* gastrointestinal disease. For example, the antibiotic clindamycin can kill most normal intestinal bacteria except the naturally resistant Gram-positive anaerobe *Clostridium difficile*, the causative agent of pseudomembranous enterocolitis. Unrestrained by microbial competition, *C. difficile* will grow in the intestine and produce specific toxins that can damage intestinal cells. The organism's growth leads to inflammation and the formation of pseudomembrane structures along the intestinal wall (refer to Fig. 23.8). Because they block the intestinal mucosa, pseudomembranes cause malabsorption of nutrients and water, which results in diarrhea. As the pseudomembrane enlarges, it begins to slough off and pass into the stool. Diagnosis of this disease involves PCR identification of the organism or immunological identification of the toxin in fecal samples.

Case History: Diarrhea and Dysentery

In April, a 6-year-old girl from Montgomery County, Pennsylvania, arrived at the ER with bloody diarrhea, a temperature of 39°C (102.2°F), abdominal cramping, and vomiting. Hospital admission occurred 5 days after a kindergarten field trip to the local dairy farm. When the parents were questioned about the child's activities during the trip, they said the child had purchased a snack while at the farm. Upon laboratory analysis, a fecal smear was positive for leukocytes, and isolation of organisms confirmed the presence of Gram-negative rods that produced Shiga toxins 1 and 2. Subsequent testing of the isolate by pulsed-field gel electrophoresis was indistinguishable from E. coli O157:H7. By this time, the child developed further problems. Symptoms included puffy face and hands, as well as neurological abnormalities. Initial examination suggested renal failure, and laboratory analyses supported this diagnosis with thrombocytopenia (reduced blood platelet count) and hemolytic uremic syndrome (HUS; renal failure). The child was treated by fluid and electrolyte replacement (intravenous). Antibiotics were not administered.

In this case history, the presence of leukocytes in a fecal smear is a sign that the intestinal pathogen has invaded the epithelial mucosa of the intestine. Breaching this barrier sends out a chemical call to neutrophils, which then enter the area. *Shigella dysenteriae, Salmonella enterica,* and enteroinvasive *E. coli* (EIEC) are invasive; because they actually enter enterocytes, they are considered intracellular pathogens. Enterohemorrhagic *E. coli* (EHEC), which also produces leukocytes and blood in stools, is not an intracellular parasite (it is not invasive) but causes damaging attachment and effacing lesions, described in Special Topic 25.1, which cause destruction of the mucosal epithelium. The resulting inflammation, in conjunction with damage caused by Shiga toxin, leads to blood and white cells in the stool. *E. coli* O157:H7, the etiological agent in the case history, is a common serotype of EHEC.

There are at least six different classes of pathogenic *E. coli* that differ in their repertoire of pathogenicity islands, plasmids, and virulence factors. They include EIEC and EHEC, already mentioned; enterotoxigenic *E. coli* (ETEC) and uropathogenic *E. coli* (UPEC), both described in Chapter 25; as well as enteropathogenic *E. coli* (EPEC) and enteroaggregative *E. coli* (EAEC). All but UPEC cause gastrointestinal disease. To tell them apart, each group has telltale O and H antigens that can be identified using serology.

NOTE: "O antigen" is part of the bacterium's LPS, while "H antigen" is flagellar protein. Thus, "O157:H7" denotes the specific version of LPS (O157) and flagellar protein (H7) found on *E. coli* O157:H7. Other pathogenic strains of *E. coli* have different O and H antigens.

Shiga toxin. *Shigella* and EHEC, the agent in the preceding case history, both produce toxins called Shiga toxin 1 and 2 that are encoded by genes of bacteriophage genomes embedded in the bacterial chromosome. These toxins inhibit host protein synthesis and, in the process, damage endothelial cells in the kidney and brain. Endothelial damage triggers the formation of platelet-fibrin microthrombi (clots) that occlude blood vessels in the various organs, leading to two major syndromes: hemolytic uremic syndrome (HUS) and thrombotic thrombocy-topenic purpura (TPP). HUS occurs when the micro-thrombi are limited to the kidney. The microclots clog the tiny blood vessels in this organ and cause decreased urine output, ultimately leading to kidney failure and death. In TPP, the clots occur throughout the circulation, causing reddish skin hemorrhages called petechiae and purpuras. Neurological symptoms (for example, confusion, severe headaches, possibly coma) then arise from microhemor-rhages in the brain. The hemorrhaging occurs because platelets needed for normal clotting have been removed from the circulation as they form the microthrombi. The decreased number of platelets is called thrombocytopenia.

The toxins, which are absorbed through the intes-tine and disseminated via the bloodstream, have five B subunits used to bind to target cell membranes and one A subunit imbued with toxic activity (see Section 25.4). The A subunit, upon entry, destroys protein synthe-sis by cleaving an adenine from 28S rRNA in eukaryotic ribosomes.

Enterohemorrhagic *E. coli* (EHEC). *E. coli* O157:H7 is a recently emerged pathogen that can colonize cattle intes-tines without causing disease and as a result can contam-inate meat products following slaughter. The organism is sometimes referred to as the "Jack in the Box" microbe, a reference to the first documented large-scale U.S. out-break in 1993, linked to fast-food hamburgers purchased at a Washington State Jack in the Box restaurant. As many as 600 people were sickened in that outbreak, and three children died. Though 1993 marked the first large-scale outbreak in the United States, the first report of the dis-ease in this country occurred 11 years earlier, in 1982.

E. coli O157:H7 rarely affects the health of the res-ervoir animal. But when an infected steer is slaughtered, fecal contamination of the carcass can happen, despite manufacturers' considerable efforts to prevent it. Grind-ing the tainted meat into hamburger distributes the microbe throughout. Cooking burgers to 160°C is essen-tial to kill any existing EHEC. Also be aware that cross-contamination between foods is possible. Using the same cutting board to prepare meat and salad is a great way to contaminate the salad, which will not be cooked.

Despite EHEC's common association with ham-burger, vegetarians are not safe from this organism. Dur-ing heavy rains, waste from a cattle farm can easily wash into nearby vegetable fields unless precautions are taken. If the cattle waste contains *E. coli* O157:H7, the crops are contaminated. One such outbreak occurred in 2006, when spinach from certain areas of California became contami-nated with this pathogen, prompting a nationwide recall of bagged spinach and a month without spinach salad.

Early on, the remarkably low infectious dose of *E. coli* O157:H7 mystified researchers. However, we have since learned that *E. coli* has an impressive level of acid resis-tance rivaling that of the gastric pathogen *Helicobacter*

pylori. Acid resistance mechanisms permit survival of *E. coli* in the acidic stomach and enable a mere 10–100 individual organisms to cause disease.

As already noted, EHEC strains produce two toxins that are identical to the Shiga toxins produced by *Shigella* species. The toxins cleave host ribosomal RNA, thereby halting translation. As discussed earlier, one consequence of Shiga toxin is HUS. The development of HUS, as in the case history described, is a common consequence of *E. coli* O157:H7 infection. Unfortunately, HUS can be treated only with supportive care, such as blood transfusions and dialysis throughout the critical period until kidney func-tion resumes. Antibiotic treatment can increase the release of Shiga toxins from the organisms and actually trigger HUS. Thus, antimicrobial therapy is not recommended.

In contrast to the case just described, many gastro-intestinal infections do not produce fecal leukocytes or blood in the stool. Diarrheal diseases caused by *Vibrio cholerae* (cholera) or enterotoxigenic *E. coli* (ETEC), which produces a cholera-like disease, do not involve invasion of the intestinal lining by the microbe and yield copious amounts of watery diarrhea. In these two toxin-driven diseases, the bacteria attach to cells lining the intestine and secrete toxins that are imported into the target cells (see Section 25.4).

Epidemiology of EHEC. Epidemiology is the study of factors and mechanisms involved in the spread of dis-ease. The father of epidemiology was the British physician John Snow, who, even before there was a clear connec-tion between bacteria and disease, managed to trace the source of a cholera epidemic in London in 1854 to a single well in the city. He found that the addresses of all the cholera patients clustered around just one of the city's numerous water pumps, which had, as it turns out, been contaminated with human feces. Sealing the well ended the epidemic.

The U.S. Centers for Disease Control and Prevention (CDC) used the same basic strategy to identify the risk factors associated with the case study presented earlier. Fifty-one infected patients and 92 controls (children who visited the farm but did not become ill) were interviewed. Infected patients were more likely than controls to have had contact with cattle, an important reservoir for *E. coli* O157:H7. All 216 cattle on the farm were sampled by rectal swab, and 13% yielded *E. coli* O157:H7 with a DNA restric-tion pattern indistinguishable from that isolated from the patients. This finding indicated that the cattle were the source of infection. Activities that promoted hand-mouth contact, such as nail biting and purchasing food from an outdoor concession were more common among the chil-dren who contracted disease (fecal-oral route of infec-tion). Furthermore, separate areas were not established for eating and interactions with farm animals. Visitors could touch cattle, calves, sheep, goats, llamas, chickens, and a pig while eating and drinking. Hand-washing facilities were unsupervised and lacked soap, and disposable hand towels were out of the children's reach. All of these situations provided opportunity for infection.

How can we prevent disease caused by enterohemorrhagic *E. coli*? Industry approaches include thorough washing of carcasses before processing, maintaining cold temperatures, and testing for possible contamination. In addition, the use of gamma irradiation to sterilize beef, spinach, and lettuce has been approved, although this remains a controversial process.

Type III secretion and diarrhea. Type III secretion was first described in Section 25.5, where we discussed the pathogenesis of *Salmonella enterica*, but these secretion systems are present in numerous Gram-negative pathogens, such as EHEC in our case history. Recall that type III protein secretion systems directly inject proteins from the cytoplasm of a bacterial pathogen into the cytoplasm of a target eukaryotic host cell. The system delivers proteins across three membranes—two for the Gram-negative bacterial pathogen and one for the target cell.

In addition to stimulating bacterial entry into host cells, bacterial proteins injected by type III transport systems cause host cells to secrete pro-inflammatory cytokines. The cytokines then "call in" inflammatory cells and alter ion transport through the epithelial membrane. Excessive export of ions such as chloride causes water to leave the cell in an attempt to equilibrate the internal and external ionic concentrations. The water entering the intestine results in diarrhea. We are just beginning to understand how the type III–translocated effector proteins induce these host cellular responses and how this leads to intestinal inflammation during an infection.

Although most gastrointestinal disease is intestinal in locale, specialized microbes can also target the stomach, with its harsh acidic environment.

Case History: Ulcers—It's Not What You Eat

Gary is a 34-year-old accountant who immigrated to Nebraska from Poland 7 years ago. Since his teenage years, he has been bothered periodically by episodes of epigastric pain (pain around the stomach), nausea, and heartburn. Antacids usually alleviated the symptoms. Over the years, he received several courses of treatment with Tagamet or Pepcid to reduce acid secretion and provide relief. Recently, an upper-GI endoscopy was performed, in which a long, thin tube tipped with a camera and light source was inserted into Gary's mouth and into his stomach. The view through the endoscope showed some reddened areas in the antrum (bottom part) of the stomach. The endoscope was also equipped with a small clawlike structure that obtained a small tissue sample from the lining of Gary's stomach. A urease test performed on the antral biopsy turned positive in 20 minutes. Histological examination of the biopsy confirmed moderate chronic active gastritis (inflammation of the stomach lining) and revealed the presence of numerous spiral-shaped organisms. Cultures of the antral biopsy were positive for Helicobacter pylori.

Painful and sometimes life-threatening gastric ulcers were for many years blamed on spicy foods and stress. These factors were believed to cause increased acid production that ate away at the stomach lining, even though the gastric mucosa is normally well protected from stomach acid, which can fall as low as pH 1.5. This protection argued against the model but was ignored. In the 1980s, after discovering odd, helical bacteria present in the biopsies of gastric ulcers, Australians Robin Warren and Barry Marshall (a medical intern at the Royal Perth Hospital at the time; **Fig. 26.8A**) proposed that bacteria, not pepperoni, cause ulcers (**Fig. 26.8B**). Their hypothesis was viewed with skepticism and declared as heresy by the established medical community. Faced with disbelief bordering on ridicule, the young intern drank a vial of the helical organisms and waited. A week later he began vomiting and suffered other painful symptoms of gastritis. Barry Marshall could not have been happier. He had proved his point. We now know that this curly microbe causes the vast majority of stomach ulcers.

The discovery of *Helicobacter pylori* and its association with gastric ulcer disease led to an upheaval in gastroenterology. Prior to this

A.   **B.**

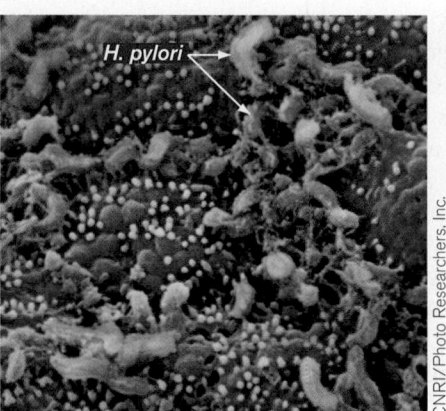

Figure 26.8 A bacterial cause of gastric ulcers. A. Australian physician Barry Marshall was so sure he was right about the cause of stomach ulcers, he swallowed bacteria to prove his point. **B.** View of *Helicobacter pylori* in stomach crypts (colorized SEM). Bacteria are approximately 2 μm in length.

discovery, treatment focused on suppressing acid production, which did not provide long-term relief. Within one year after acid-suppressive therapy, up to 80% of patients suffer a relapse of their ulcer. Therapy now includes antimicrobial treatment to kill the bacteria and acid suppression therapy to prevent further inflammation while the ulcer heals. Warren and Marshall, who recovered from his gastritis, received the 2005 Nobel Prize in Physiology or Medicine for their groundbreaking work.

The exact mechanism by which the organism causes gastric ulcers is not known, although a variety of virulence factors have been described (see Section 25.6). The urease enzyme of *H. pylori* is a characteristic virulence feature. Urease converts urea, produced by the gastric lining, to ammonia and carbon dioxide. The ammonia neutralizes acid in and around the organism and allows the microbe to survive in the stomach. In an animal model, the urease enzyme of *H. pylori* was found to be important for bacterial colonization in the stomach. Treatment with urease inhibitors such as acetohydroxamic acid, however, does not lead to eradication of *H. pylori*. This failure is likely due to protection afforded the organism once it reaches the mucus layer that blankets the gastric epithelium (see **Fig. 26.8B**). The pH of this environment is closer to neutral, so urease may no longer be needed. Tools useful for diagnosing *H. pylori* include histology, rapid urease testing, and serology (for example, enzyme-linked immunosorbent assay (ELISA) to detect antibody to the CagA antigen).

Enzyme-linked immunosorbent assay (ELISA). ELISA is a common immunological tool to detect the presence, in serum, of antibodies to a specific organism—an indication of infection. The assay is performed by coating wells in a plastic dish with an antigen (for example, *Helicobacter* CagA). Serum from the patient is then added to the well. If antibodies to CagA are present, they will bind to the CagA antigen. Unbound antibody is removed by washing, and a secondary antibody that binds human IgG is added to the well. These anti-antibodies have an enzyme linked to them. A sandwich is formed as follows: [plastic dish]–[CagA protein]–[anti-CagA antibody]–[anti-IgG antibody]–[enzyme]. When the appropriate enzyme substrate is added to the well, the enzyme acts on it to produce light or a chromogenic (colored) product that can be detected. The more anti-CagA antibody present in the serum, the more light or product that is produced by the linked enzyme. Applications of ELISA and further details of the methodology involved are described in Section 28.2.

***Helicobater pylori* infection, cancer, and bad breath.** Problems caused by *H. pylori* are not limited to gastric ulcers. The microbe has also been associated with gastric cancer. The evidence is not yet conclusive, but many individuals with gastric cancer are also colonized by *H. pylori*. The most compelling experimental proof is that gerbils infected with this organism develop gastric cancer. In addition, when the CagA protein of *H. pylori* is injected into gastric epithelial cells, it is phosphorylated on a tyrosine residue and activates a regulatory cascade that causes the gastric cell to proliferate (see Fig. 25.31). The connection between *H. pylori* and gastric cancer is sobering when you consider that *H. pylori* can be detected in about two-thirds of the world's population, especially in impoverished countries.

There is also the unusual case of a patient with a 60-year history of halitosis (extreme bad breath) that was resistant to all standard therapies but was cured by a triple-drug therapy regimen used for *H. pylori*. This case is an illustration of how microorganisms can cooperate to cause disease. *H. pylori* did not cause the smell. It appears that the patient did not produce much stomach acid. As a result, the urease produced by *H. pylori* easily neutralized the remaining stomach acid, allowing anaerobes ingested with food to start putrefying the stomach contents, leading to the foul-smelling breath. Other examples of disease caused by microbial cooperation will be described later.

Rotavirus Is the Single Greatest Cause of Gastroenteritis

Many people wrongly think that most cases of diarrhea are caused by a bacterial agent. Actually, a virus, **rotavirus**, causes more intestinal disease than any bacterial species (Table 6.1; dsRNA viruses). Rotavirus is highly infectious, spreading by the fecal-oral route; it is endemic around the globe; and it affects all age-groups, although children between 6 and 24 months are most severely affected. It is estimated that by age 3, all children have had a rotavirus infection.

The incubation period is approximately 2 days, after which the victim suffers frequent watery, dark green, explosive diarrhea. All of this may be accompanied by nausea, vomiting, and abdominal cramping. Severe dehydration and electrolyte loss due to the diarrhea will cause death unless supportive measures, such as fluid replacement, are undertaken. There is no cure, but most patients recover if rehydrated properly. Few deaths from rotavirus occur in the United States, but each year more than 600,000 children worldwide die from this viral diarrhea. However, the mortality and incidence of this disease will now decrease, because of a new, safe, and effective vaccine.

Protozoa Are Another Major Cause of Diarrheal Diseases

Most students of biology are familiar with protozoa (also called protists) such as paramecia and amebas, but many are surprised to learn that some species of protists cause serious human diseases. For instance, *Entamoeba histolytica* and *Cryptosporidium parvum* cause the diarrheal diseases amebic dysentery and cryptosporidiosis, respectively. In

Figure 26.9 *Giardia lamblia.* This protist is a major cause of diarrhea in the world. **A.** Cysts (7–14 μm) present in fecal matter (colorized SEM). **B.** Trophozoite form (colorized SEM, 5–15 μm in length).

A.

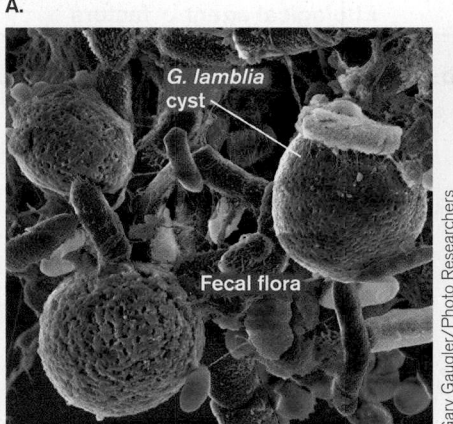

G. lamblia cyst

Fecal flora

Gary Gaugler/Photo Researchers

B.

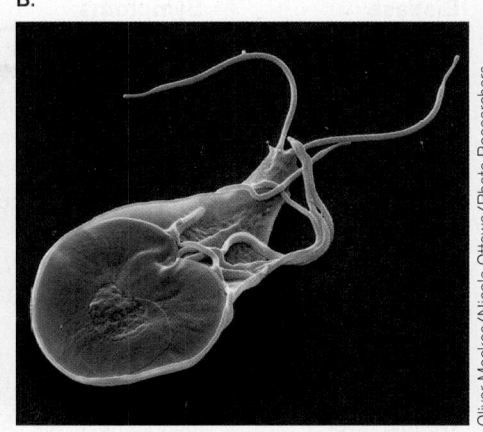

Oliver Meckes/Nicole Ottawa/Photo Researchers

2005, the CDC tallied 2,640 cases of cryptosporidiosis, a reportable disease in the United States. Two other amebas, *Naegleria* and *Acanthamoeba*, cause amebic meningoencephalitis. Because some species of *Acanthamoeba* can infect the eye, soft-contact wearers should take precautions to prevent contamination of their lenses.

The flagellated protozoan *Giardia lamblia* is a major cause of diarrhea throughout the world. In the United States alone, *G. lamblia* caused over 11,000 reported cases of giardiasis diarrhea in 2005 and likely caused thousands more that were not reported. *G. lamblia* enters a human or other host as a cyst present in drinking water contaminated by feces (**Fig. 26.9A**). Aside from humans, *G. lamblia* can be found in various rodents, deer, cattle, and even household pets. It is very infectious. Ingestion of as few as 25 cysts can lead to disease. Following ingestion, the hard, outer coating of the cyst is dissolved by the action of digestive juices to produce a trophozoite, which attaches itself to the wall of the small intestines and reproduces (**Fig. 26.9B**). Offspring quickly encyst and are excreted out of the host's body.

Asymptomatic carriers of *G. lamblia* are common—it has been estimated that anywhere from 1% to 30% of children in U.S. day-care centers are carriers. Disease usually manifests as greasy stools alternating between a watery diarrhea, loose stools, and constipation. However, some patients will experience explosive diarrhea. Diagnosis usually comes from observing the cysts or trophozoite forms of the protozoan in feces. Metronidazole is a drug often used to cure the disease. As for prevention, proper treatment of community water supplies is essential.

We have examined only a handful of the microbes that cause gastrointestinal infection. Others are presented in **Table 26.3** and described in Chapter 20.

TO SUMMARIZE:

- **Diarrhea** leads to dehydration, for which fluid replacement is a critical treatment. Antibiotic treatment is usually not recommended.

- **Staphylococcal food poisoning** is not an infection. It is a toxigenic disease.
- **Antibiotic treatments** can sometimes cause gastrointestinal disease (for example, pseudomembranous enterocolitis by *Clostridium difficile*).
- **Bacteria that invade intestinal epithelial mucosal cells** lead to the presence of red and white blood cells in fecal contents. This occurs with intracellular pathogens such as *Shigella*, *Salmonella*, and EIEC.
- **Bacteria that do not invade intestinal cells** usually produce watery diarrhea. EHEC is an exception because the attachment and effacing lesions it produces result in bloody stools.
- **Bacterial toxins** produced by bacterial enteric pathogens can cause systemic symptoms.
- **John Snow** founded the science of epidemiology while studying a cholera outbreak in London.
- **The bacterium *Helicobacter pylori*,** a common cause of gastric ulcers, lives in the stomach and is highly acid resistant.
- **Rotavirus** is the single greatest cause of diarrhea worldwide.
- *Giardia lamblia* is a major protozoan cause of diarrhea worldwide.

26.5 Genitourinary Tract Infections

Although the genital and urinary tracts are different organ systems, their close association in the body has led them to be grouped together when discussing infections.

Urinary Tract Infections

The urinary tract includes the kidneys, ureters, urinary bladder, and urethra. Infections anywhere along this route are called urinary tract infections (UTIs). Urinary tract infections are the second most common type of

Table 26.3 Selected microbes that cause diseases of the gastrointestinal tract.

Disease	Symptoms	Etiological agent	Virulence factors	Source	Treatment
Bacterial					
Staphylococcal food poisoning	Symptoms within 4 h of ingestion; nausea, vomiting, diarrhea	*Staphylococcus aureus* (G+)	Enterotoxin	Preformed toxin in foods	Supportive
Botulism	Symptoms begin quickly; flaccid paralysis	*Clostridium botulinum* (G+, anaerobe)	Neurotoxin	Preformed toxin in foods	Antiserum
Salmonellosis	Symptoms after 18 h; abdominal pain, diarrhea; invade intestinal M cells	*Salmonella enterica* [Gram-negative (G–)]	Type III secretion, intracellular growth	Chickens, other animals; fecal-oral route	Oral rehydration; antibiotics if severe
Typhoid fever	Headache, fever, chills, abdominal pain, rash (rose spots), hypotension, diarrhea in late stages	*Salmonella enterica* serovar Typhi (G–)	Type III secretion, intracellular growth, PhoPQ regulators, Vi antigen capsule	Human carriers (gallbladder reservoir), food, water	Quinolones
Traveler's diarrhea	Watery diarrhea	Enterotoxigenic *E. coli*	Labile and stable toxins	Humans; food, water	Oral rehydration
Gastroenteritis	Bloody diarrhea, HUS	Enterohemorrhagic *E. coli*	Intimin, Tir, type III secretion, Shiga toxin	Contaminated foods (hamburger) and crops	Oral rehydration; antibiotics if severe
Shigellosis	Bloody diarrhea, HUS	*Shigella* spp. (G–)	Shiga toxin, type III secretion, intracellular growth, actin-based motility, escape phagosome	Human fecal-oral route	Oral rehydration; antibiotics if severe
Cholera	Watery diarrhea	*Vibrio cholerae* (G–)	Cholera toxin, TCP pili, ToxR regulator	Human waste–contaminated water	Oral rehydration, antibiotics
Gastroenteritis	Diarrhea, blood in stool	*V. parahaemolyticus* (G–)	Enterotoxin	Raw seafood	Self-limiting
Gastroenteritis	Watery diarrhea, nausea	*Clostridium perfringens* (G+)	Alpha toxin	Soil, food	Self-limiting
Pseudomembranous enterocolitis	Fever, abdominal pain, diarrhea, pseudo-membrane in colon	*C. difficile* (G+)	Cytotoxin, antibiotic resistance	Animals, normal microbiota	Vancomycin
Gastroenteritis	Fever, muscle pain, watery diarrhea, blood in stool, headache	*Campylobacter jejuni* (G–)	Cytotoxin, enterotoxin, adhesin	Poultry, unpasteurized milk	Erythromycin
Gastric ulcers	Abdominal pain, bleeding, heartburn	*Helicobacter pylori* (G–)	Adhesin, urease CagA, vacuolating toxin	??	Triple drug (omeprazole, clarithromycin, metronidazole)
Viral					
Stomach "flu"	Nausea, vomiting, diarrhea	Noroviruses (Norwalk virus)	??	Fecal-oral route	Oral rehydration
		Rotavirus (most common cause)	??	Fecal-oral route	Oral rehydration

bacterial infection in humans, ranking in frequency just behind respiratory infections such as bronchitis or pneumonia. Along with numerous office visits, bladder infections and other urinary tract infections result in 100,000 hospital admissions and $1.6 billion in medical expenses each year in the U.S. alone. Most sufferers are women. One in five women will get a urinary tract infection at some point in her life, and 20%–40% of those infected will develop recurrent infections.

Urine, as produced in the kidneys and stored in the bladder, is normally sterile. Microorganisms must be introduced into the bladder to cause infection. Active infection of the urinary tract occurs in one of three basic ways:

- **Descending infection from the kidney.** Descending infection occurs when an infected kidney sheds bacteria that descend via the ureters into the bladder. Kidney infections arise when microorganisms are deposited in the kidneys from the bloodstream.
- **Ascending infection to the kidney.** In ascending infection, an established infection in the bladder ascends along the ureter to infect the kidney.
- **Infection from the urethra to the bladder.** Bacteria residing along the superficial urogenital membranes of the urethra can ascend to the bladder. This is more common in women than men. Microorganisms can also be introduced into the bladder by means of mechanical devices such as catheters or cystoscopes that are passed through the urethra into the bladder.

THOUGHT QUESTION 26.3 Why do you think most urinary tract infections occur in women?

Urine is bacteriostatic to most of the commensal organisms inhabiting the perineum and vagina, such as *Lactobacillus*, *Corynebacterium*, diphtheroids, and *Staphylococcus epidermidis*. In contrast, many Gram-negative organisms thrive in urine. As a result, most urinary tract infections are caused by facultative Gram-negative rods from the GI tract. The most common etiological agents of UTIs are:

- Certain serotypes of *E. coli* that comprise the uropathogenic *E. coli* (75% of all UTIs)
- *Klebsiella, Proteus, Pseudomonas, Enterobacter* (20%)
- *S. aureus, Enterococcus, Chlamydia,* fungi, *Staphylococcus saprophiticus,* other (5%)

Case History: Classic Urinary Tract Infection

Lashandra is 24 years old and has been experiencing back pain, increased frequency of urination, and dysuria (painful urination) over the past 3 days. She consulted her general practitioner, who requested that a midstream specimen of urine be examined. This was the first time Lashandra had ever suffered from persisting

dysuria. Upon microscopic examination, the urine was found to contain more than 50 leukocytes per microliter and 35 red blood cells per microliter. No epithelial squamous cells were seen. The urine culture plated on agar medium yielded more than 10^5 colonies per milliliter (meaning more than 10^5 organisms per milliliter in the urine) of a facultative anaerobic Gram-negative bacillus capable of fermenting lactose.

The first question to ask in this case is whether the patient had a significant UTI. The purpose of the midstream urine collection is to provide laboratory data to make this determination. Even though urine in the bladder is normally sterile, urine becomes contaminated with normal microbiota that have adhered to the urethral wall. In a midstream collection, the patient urinates briefly, stops to position a collection jar, and resumes urinating to collect a midstream sample. This minimizes the number of organisms in the sample by washing away organisms clinging to the urethra before actually collecting the sample. Nevertheless, the collected sample will still contain low numbers of organisms representing normal microbiota of the urethra. A diagnosis of UTI is made when the number of bacteria in the sample becomes greater than 10^5/ml. (However, that number can be as low as 1,000/ml in symptomatic women.) The patient in the case history is symptomatic and has sufficient numbers of bacteria in her urine to indicate a UTI.

The laboratory found the organism to be a Gram-negative bacillus that ferments lactose, suggesting *E. coli* as the likely culprit. Given that the normal habitat of *E. coli* is the gastrointestinal tract and this is the first UTI suffered by the patient, the infection is likely the result of an inadvertent introduction of the microbe into the urethra. The organism makes its way up the urethra and into the bladder. In contrast to ascending infections, organisms from an infected kidney can descend along a ureter to cause recurring bladder infections in some patients (also see **Special Topic 26.1**).

Gram-negative rods not only thrive in urine but may be adapted to cause urinary tract infections through specialized pili. These pili have terminal receptors for glycolipids and glycoproteins present on urinary tract epithelial cells (**Fig. 26.10A**). Uropathogenic strains of *E. coli*, for example, typically have P-type pili, with a terminal receptor for the P antigen (a rather appropriate name for a bladder-specific virulence factor). The P antigen is a blood group marker found on the surface of cells lining the perineum and urinary tract; it is expressed by approximately 75% of the population. These individuals are particularly susceptible to UTIs. The P antigen can also be shed and found in vaginal and prostate secretions. When this happens, the secreted P antigens can be protective by acting as a decoy. They bind to the bacterial receptor, preventing binding of the organism to the surface epithelium. Individuals most susceptible to UTIs are those who express P antigen on their cells but lack P antigen in their secretions.

Doctors used to think that women suffering from recurrent bladder infections were repeatedly infected as a result of sexual activity or poor hygiene—a presumption many women with chronic infections found frustrating and offensive. New evidence indicates that bacteria may actually hide inside bladder cells, forming a reservoir for reinfection.

Specialized colonies of *E. coli* have been found living in the surface layer of cells lining the bladders of infected mice (**Fig. 1A**). Though the mouse served as a model system, the bacteria used were originally isolated from humans. These strains of *E. coli* cause about 80% of all urinary tract infections. The bacteria use pili to latch on to proteins coating the bladder cells. Those proteins form a protective substance, known as uroplakin, that strengthens the epithelial cells and shields them from toxins that may build up in the urine (**Fig. 1B**). The bacteria can slip under the protective barrier and penetrate superficial cells lining the bladder. Once inside the cell, the microbes begin to multiply, and the epithelial cell fills with bacteria also coated with uroplakin, taking on the appearance of a pod (**Fig. 1C**).

The pod-like *E. coli* biofilms we describe here are notoriously resistant to antibiotic treatment and attacks by the immune system. Living inside host cells and surrounded by uroplakin, the *E. coli* in the bladder pods have devised a clever mechanism for surviving. Bacteria living on the edges of the biofilm may break free, leading to subsequent rounds of infection. Understanding the cycle of infection and pod formation may help researchers design drugs to prevent new urinary tract infections or stop established ones.

Although pods have so far been found only in mice, researchers are tracking women with persistent infections to see whether pods are also present in humans. Biofilms hiding inside cells elsewhere in the body also may act as reservoirs for other hard-to-treat chronic or recurring infections, such as ear infections.

Source: G. G. Anderson, J. J. Patermo, J. D. Schilling, R. Roth, J. Heuser, and S. J. Hultgren. 2003. Intracellular bacterial biofilm-like pods in urinary tract infections. *Science* 301:105–107.

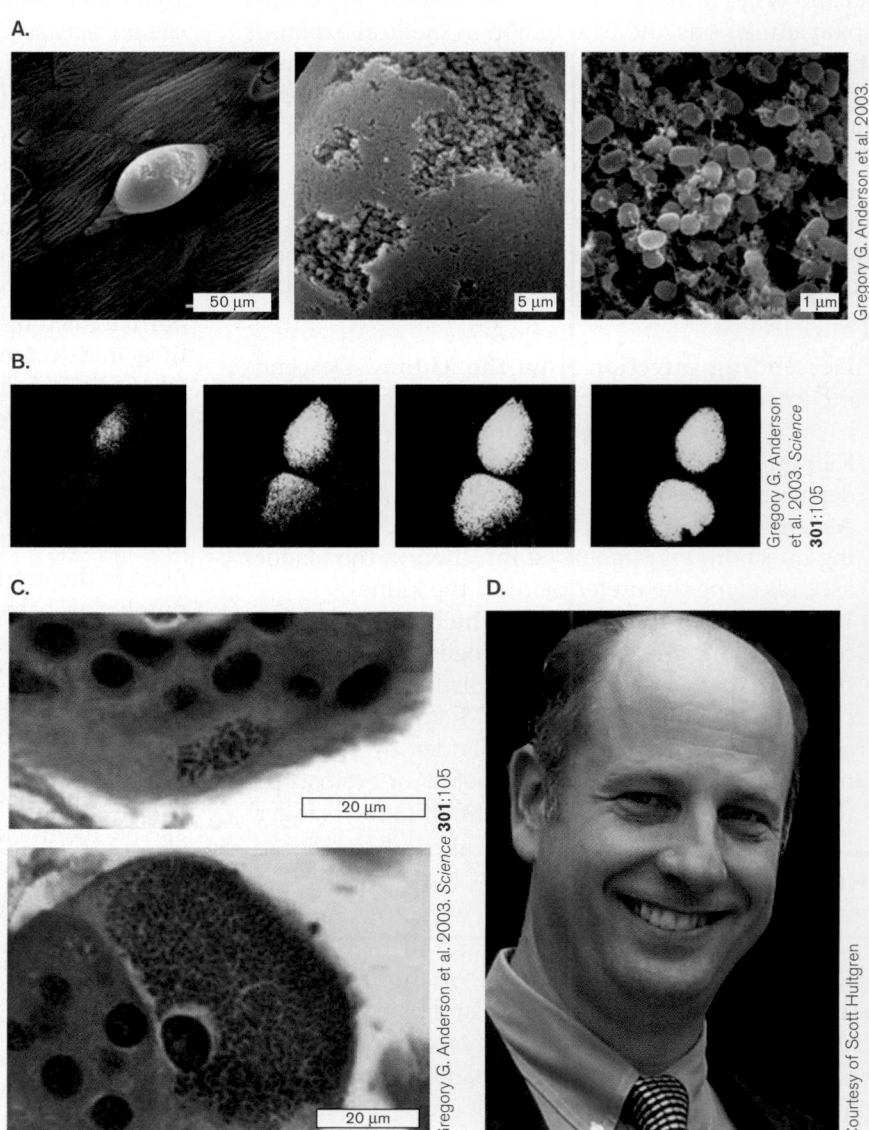

Figure 1 Bladder pods of uropathogenic *E. coli*. Scott Hultgren's laboratory (Washington University in St. Louis) has discovered that intracellular bacterial communities extend like pods into the bladder lumen. **A.** Increasing magnifications of large intracellular communities of uropathogenic *E. coli* (UTI89) inside pods on the surface of a mouse bladder infected for 24 hours (SEM). **B.** An epithelial pod of *E. coli* within a protective uroplankin shell. Confocal photos of a whole-mounted bladder infected with UTI89 expressing green fluorescent protein from the plasmid pcomGFP. Uroplakin on the surface of the two pods is revealed by treatment with antibody to uroplakin (primary antibody) and tetramethyl rhodamine isothiocyanate–labeled secondary antibody (red). The series starts on the left with the luminal surface and progresses toward the right in sections that move downward through the epithelium. Optical section thickness 1 μm. **C.** Hematoxylin- and eosin-stained sections of UTI89-infected mouse bladders show a bacterial factory 6 hours after inoculation (top panel) and a pod 24 hours after inoculation (bottom panel). Bacteria in the pod were densely packed and shorter, and completely filled the host cell. **D.** Scott Hultgren.

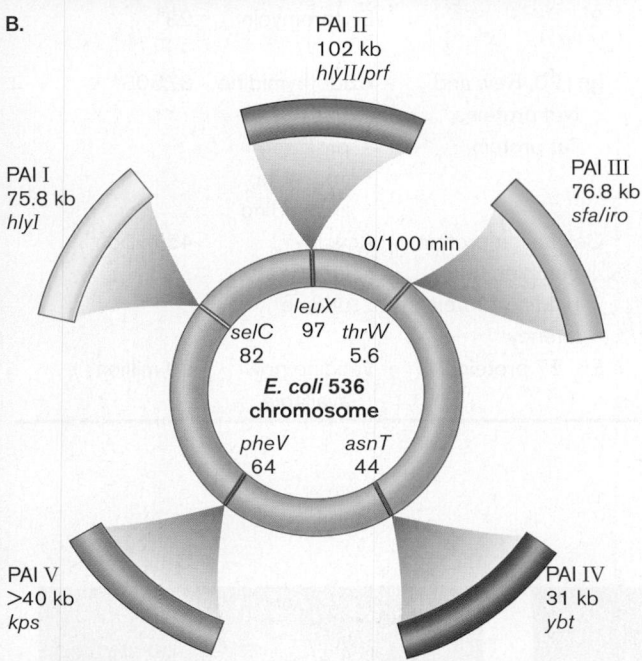

antibiotic therapy is recommended. In older patients, UTIs frequently show atypical symptoms, including delirium, which disappears when the UTI is treated.

> **THOUGHT QUESTION 26.4**   Urine samples collected from six hospital patients were placed on a table at the nurses' station awaiting pickup from the microbiology lab. Several hours later, a courier retrieved the samples and transported them to the lab. The next day, the lab reported that four of the six patients had UTIs. Would you consider these results reliable? Would you start treatment based on these results?

What makes uropathogenic *E. coli* different from other *E. coli*? This is a question still under investigation, but genomic analysis has exposed five pathogenicity islands unique to these strains (**Fig. 26.10B**). The functions of these pathogenicity islands are still under investigation.

Sexually Transmitted Diseases

Sexually transmitted diseases (STDs) are defined as infections transmitted primarily through sexual contact, which may include genital, oral-genital, or anal-genital contact. The organisms or viruses involved are generally very susceptible to drying and require direct physical contact with mucous membranes for transmission. Because sex can take many forms in addition to intercourse, these microbes can initiate disease in the urogenital tract, rectum, or oral cavities. Examples of common sexually transmitted diseases are given in **Table 26.4**.

Case History: Secondary Syphilis

A pregnant 18-year-old woman came to the county urgent-care clinic with a low-grade fever, malaise, and headache. She was sent home with a diagnosis of influenza. She again sought treatment 7 days later, after she discovered a macular rash (flat, red) developing on her trunk, arms, palms of her hands, and soles of her feet. Further questioning of the patient when serology results were known revealed that one year earlier, she had had a painless ulcer on her vagina that healed spontaneously.

The vaginal ulcer, the long latent period, and secondary development of rash on the hands and feet described in the case history are classic symptoms of syphilis. Syphilis was recognized as a disease as early as the sixteenth century, but the organism responsible, the spirochete *Treponema pallidum*, was not discovered until 1905 (**Fig. 26.11A**; Chapter 18 describes spirochete structure). The illness has several stages. The disease has often been called the great imitator because its symptoms in the second stage, as exhibited in the case history, can mimic many other diseases. The incubation stage can last from 2 to 6 weeks after transmission, during which time the organism multiplies and spreads throughout the body.

Figure 26.10   **Uropathogenic *E. coli*.**   **A.** Bladder cell with adherent uropathogenic *E. coli* (SEM). Bacteria are approximately 1 μm in length. **B.** Distribution of pathogenicity islands (PAI) in uropathogenic *E. coli*. The location of each insert is given in map units within the circle representing the genome. The 0–100 map units are called centisomes. Each centisome is approximately 44 kb of DNA. Zero is arbitrarily placed at the *thr* (threonine) gene. The origin of replication on this map is near 82 centisomes. A chromosomal gene flanking the insert is also provided. The size of each island is provided above the insert. A key virulence gene for each island is listed.

Urinary tract infections are among those most frequently acquired during a hospital stay (so-called nosocomial infections). In these cases, the causal organism is less likely to be *E. coli* and more likely to be another Gram-negative bacterium or *Staphylococcus*. Many UTIs resolve spontaneously, but others can progress to destroy the kidney or, via Gram-negative septicemia, the host. As a result,

Table 26.4 Microbes that cause sexually transmitted diseases.

Disease	Symptoms	Etiological agent	Virulence factors	Treatment	Reported cases
Gonorrhea	Purulent discharge, burning urination; can lead to sterility	*Neisseria gonorrhoeae* (G–)	Type IV pili, phase variation	Ceftriaxone	356,000[a]
Syphilis	1°: chancre; 2°: joint pain, rash; 3°: gummata, aneurism, CNS damage	*Treponema pallidum* (spirochete)	Motility	Penicillin	41,000[a]
Nongonococcal urethritis	Watery or mucoid urethral discharge, burning urination	*Chlamydia trachomatis*	Intracellular growth; prevents phago-lysosome fusion	Azithromycin	1.1 million[a]
Trichomoniasis	Vaginal itching, painful urination, strawberry cervix	*Trichomonas vaginalis* (protozoan)	Cytotoxin	Metronidazole	7.4 million (estimated)[a]
Chancroid	Painful genital lesion	*Haemophilus ducreyi* (G–)	?	Erythromycin	23[a]
Acquired immunodeficiency syndrome (AIDS)	Fever, diarrhea, cough, night sweats, fatigue, opportunistic infections	HIV	gp120, Rev, and Nef proteins; Tat protein	Azidothymidine (AZT), protease inhibitors, zidovudine	37,503[a]
Genital herpes	Painful ulcer on external genitals, painful urination	Herpes simplex 2	Cell fusion protein, complement-binding protein, latency	Acyclovir, iododeoxy-uridine	45 million[b]
Genital warts	Warts on external genitals	Human papillomavirus	E6, E7 proteins	Vaccine now available	20 million[b]

[a] CDC, reported cases for 2007.

[b] Total current cases.

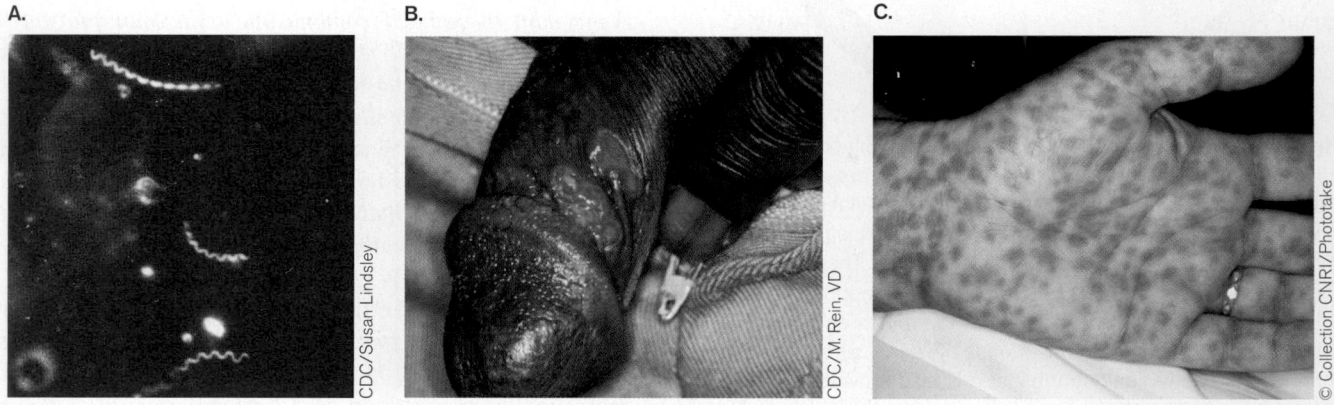

Figure 26.11   Syphilis.   A. *Treponema pallidum* (dark-field microscopy). Organisms are 10–25 μm long. **B.** Chancre of primary syphilis. **C.** Rash of secondary syphilis.

Primary syphilis is an inflammatory reaction at the site of infection called a **chancre** (**Fig. 26.11B**). About a centimeter in diameter, the chancre is painless and hard, and it contains spirochetes. Patients are usually too embarrassed to seek medical attention and, because it is painless, hope that it will "go away." It does go away after several weeks, and without scarring. The disease has now entered the primary latent stage. Over the next 5 years, symptoms may be absent, but at any time, as described in the case history, the infected person can develop the rash typical of **secondary syphilis** (**Fig. 26.11C**). The rash can be similar to rashes produced by many different diseases, which contributes to the "great imitator" label. The patient remains contagious in this stage. Some patients eventually progress over years to **tertiary syphilis**, in which a great number of symptoms can develop, affecting mainly the cardiovascular and nervous systems. The patient can develop dementia and eventually dies from the disease.

The presence of the organism in tissues can be detected with fluorescent antibody, but the initial screen is usually serological (that is, patient serum is tested for antibodies). Antibiotics are useful for eradicating the organism, but there is no vaccine, and cure does not confer immunity.

The disease is particularly dangerous in pregnant women. The treponeme can cross the placental barrier and infect the fetus to cause **congenital syphilis**. At birth, infected newborns will have notched teeth (visible on X-rays), perforated palates, and other congenital defects. Women should be screened for syphilis as part of their prenatal testing to prevent these congenital infections.

Columbus and the New World theory of syphilis. An outbreak of syphilis that spread throughout Europe soon after Christopher Columbus and his crew returned from the Americas (1493) led to the theory that Columbus brought the treponeme to Europe from the New World. However, the theory that Columbus brought syphilis to Europe has been difficult to prove.

Scientists advocating either the New World origin or the Old World origin of syphilis have relied mainly on bone evidence to support their views. Treponemal diseases, such as syphilis and Yaws (a milder form of disease caused by *T. palidum* subspecies *pertenue*) leave characteristic marks on skeletons, usually in the tertiary phase of disease. So, if bones with signs of treponemal disease were found in Native Americans who died *before* Columbus arrived in America but were not found in pre-Columbian European remains, case closed. Columbus did it. The problem is that paleontologists have reported evidence of treponemal diseases in the pre-Columbian skeletons of Native Americans and in those of Europeans.

The answer to this 500-year-old debate may be in a recent study by Kristin Harper and colleagues who used a phylogenetic approach to address the question. The group sequenced 26 geographically disparate strains of pathogenic *Treponema*. Of all the strains examined, the sexually-transmitted syphilis-causing strains originated most recently and were more closely related to the non sexually-transmitted Yaws-causing strains from South America, supporting the New World theory of syphilis. Thus, it seems the crew of the Columbus voyages brought smallpox and measles to America and returned to Europe with syphilis.

The Tuskegee experiment. Unfortunately, much of what we know of syphilis is the result of the infamous Tuskegee experiment conducted in the 1930s in Alabama. The study was entitled "Untreated Syphilis in the Negro Male." Through dubious means and deception, a group of African-American males was enlisted in a study that promised treatment but whose real purpose was to observe how the disease progressed without treatment. Today, such experiments are barred, thanks to strict institutional review board (IRB) oversights in which human subjects must sign informed consent forms. An interesting treatise on the Tuskegee experiment can be found on the Internet at the National Center for Case Study Teaching in Science (search for "bad blood tuskegee").

Chlamydial Infections Are Often Silent

Chlamydia is the most frequently reported sexually transmitted infectious disease in the United States, according to the Centers for Disease Control and Prevention, but many people are completely unaware that they are infected. Three-fourths of infected women, for instance, have no symptoms.

The chlamydia are unusual Gram-negative organisms with a unique developmental cycle (Chapter 18 describes chlamydial morphology). They are obligate intracellular pathogens that start as a small, nonreplicating, infectious elementary body that enters target eukaryotic cells. Once inside vacuoles, they begin to enlarge into replicating reticulate bodies (**Fig. 26.12**). As the vacuole fills, the reticulate bodies divide to become new nonreplicating elementary bodies. *Chlamydia trachomatis* and *Chlamydophila pneumoniae* can both cause STDs, as well as other diseases, such as trachoma of the eye or pneumonia.

People most at risk of developing genitourinary tract infections with chlamydia are young, sexually active men and women; anybody who has recently changed sexual partners; and anybody who has recently had another sexually transmitted disease. The astute clinician knows that when one STD is discovered, others may also be present. A recent report indicates that *Chlamydia* can bind to sperm in a process called hitchhiking. It is thought that this interaction helps the organism spread to females.

Left untreated, chlamydia can cause serious health problems. In women, the organism can produce pelvic inflammatory disease, a damaging infection of the uterus and fallopian tubes that can be caused by several different microbial species. The damage produced can lead to infertility, tubal pregnancies, and chronic pelvic pain. Men left untreated can suffer urethral and testicular infections and a serious form of arthritis.

Case History: Gonorrhea

A 22-year-old mechanic saw his family doctor for treatment of painful urination and urethral discharge. The patient was sexually active, with three regular and several "one time–good time" partners. Physical examination was unremarkable except for prevalent urethral discharge. The discharge was Gram-stained and sent for culture. The Gram stain revealed many pus cells, some of which contained numerous phagocytosed

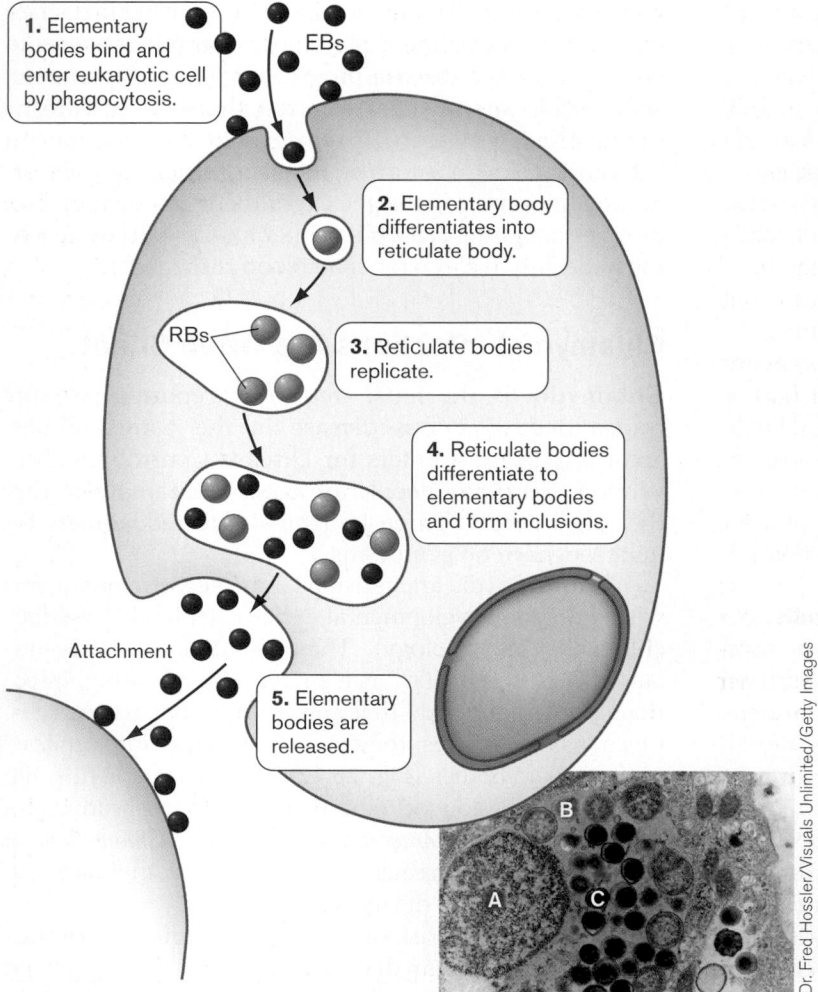

Figure 26.12 **Replication cycle of *Chlamydia*.** The inset is an EM micrograph of a *C. trachomatis*–containing vacuole in an infected cell, showing a reticulate body (A), an intermediate form between the elementary body and the reticulate body (B), and an elementary body (C).

Gram-negative diplococci (**Fig. 26.13A**). *Blood was drawn for syphilis serology, which proved negative. The patient was given an intramuscular injection of ceftriaxone (250 mg), and oral tetracycline (500 mg, four times a day) was prescribed for 7 days. The bacteriology lab was able to recover the bacteria seen in the Gram-stained smear of the urethral discharge. The organism produced characteristic colonies on chocolate agar (agar plates containing heat-lysed red blood cells that turn the medium chocolate brown;* **Fig. 26.13B**). *The case was subsequently reported to the state public health department. Upon his return visit, the patient's symptoms had resolved, and a repeat culture was negative. The patient confirmed that all three of his regular sexual contacts had been seen and evaluated at the STD clinic.*

The disease here is classic gonorrhea caused by *Neisseria gonorrhoeae*. A characteristic distinguishing *Neisseria* infections from *Chlamydia* infections is that bacterial cells are seen in gonorrheal discharges but not in chlamydial discharges. Gonorrhea has been a problem for centuries and remains epidemic in this country today. Symptoms generally occur 2–7 days after infection but can take as long as 30 days to develop. Most infected men exhibit symptoms; only about 10%–15% do not. Symptoms include painful urination, yellowish white discharge from the penis, and in some cases swelling of the testicles and penis. The Greek physician Galen (AD 129–199) originally mistook the discharge for semen. This led to the name *gonorrhea*, which means "flow of seed."

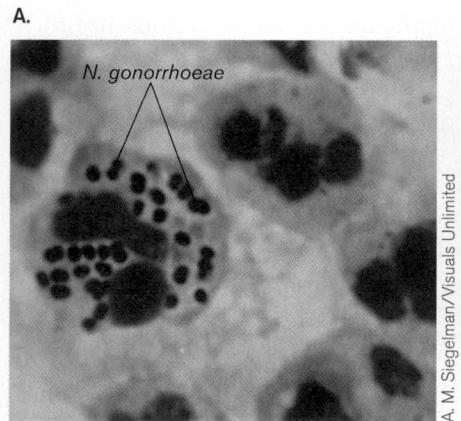

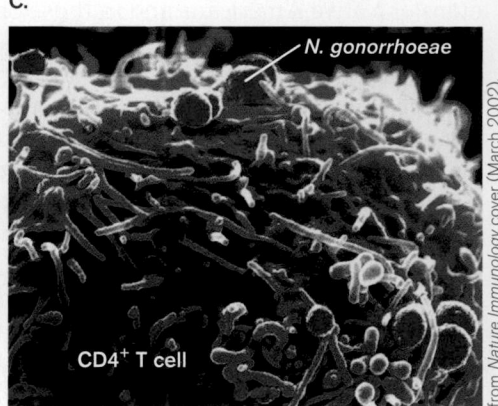

Figure 26.13 ***Neisseria gonorrhoeae.*** **A.** Within pus-filled exudates, the Gram-negative diplococci are found intracellularly inside PMNs. The intracellular bacteria (approx. 0.6–1 μm in diameter) in this case are no longer viable, having been killed by the antimicrobial mechanisms of the white cell. **B.** Colonies of *N. gonorrhoeae* growing on chocolate agar (agar plates containing heat-lysed red blood cells that turn the medium chocolate brown). **C.** *N. gonorrhoeae* binding to CD4⁺ T cells (colorized SEM), inhibiting T-cell activation and proliferation, which may explain the ease of reinfection.

In contrast to men, most infected women (80%) do not exhibit symptoms and constitute the major reservoir of the organism. If they are asymptomatic, they have no reason to seek treatment and thus can spread the disease. When symptoms are present, they are usually mild. A symptomatic woman will experience a painful burning sensation when urinating and will notice vaginal discharge that is yellow or occasionally bloody. She may also complain of cramps or pain in her lower abdomen, sometimes with fever or nausea. As the infection spreads throughout the reproductive organs (uterus and fallopian tubes), pelvic inflammatory disease occurs (see earlier discussion of chlamydia). It is important to note that there is no serological test or vaccine for gonorrhea, because the organism frequently changes the structure of its surface antigens.

Although *N. gonorrhoeae* is generally serum sensitive owing to its sensitivity to complement, certain serum-resistant strains can make their way to the bloodstream and carry infection throughout the body. As a result, both sexes can develop purulent arthritis (joint fluid containing pus), endocarditis, or meningitis. An infected mother can also infect her newborn during parturition (birth), leading to a serious eye infection called ophthalmia neonatorum. Because of this risk and because most infected women are asymptomatic, all newborns receive antimicrobial eyedrops at birth.

Because adults engage in a variety of sexual practices, *N. gonorrhoeae* can also infect the anus or the pharynx, where it can develop into a mild sore throat. These infections generally remain unrecognized until a sex partner presents with a more typical form of genitourinary gonorrhea. Because no lasting immunity is built up, reinfection with *N. gonorrhoeae* is possible. This is due in part to the phase variation in various surface antigens that takes place and because the organism can apparently bind to CD4$^+$ T cells, inhibiting their activation and proliferation to become memory T cells (**Fig. 26.13C**).

THOUGHT QUESTION 26.5 Considering that *Neisseria gonorrhoeae* is exquisitely sensitive to ceftriaxone, why do you suppose the patient in this case was also treated with tetracycline? And why can one person be infected repeatedly with *N. gonorrhoeae*?

HIV Causes Sexually Transmitted and Blood-Borne Disease

Though HIV (human immunodeficiency virus) is believed to have originated around the year 1900, it was not discovered until 1981, when the virus caused the greatest pandemic of the late twentieth century. HIV remains a serious problem today, especially in Africa, where it is estimated that 10%–30% of the population is infected with HIV (prevalence in the United States is less than 1%). HIV has claimed the lives of more than 22 million people worldwide, over half a million in the United States alone. The molecular biology and virulence of HIV are discussed in Chapters 11 and 25. This section focuses on the disease that HIV causes: acquired immunodeficiency syndrome (AIDS).

HIV, a lentivirus in the retroviral family, is a prominent example of viruses that can be transmitted either sexually (vaginally, orally, anally, homosexually, or heterosexually) or through direct contact with body fluids, such as occurs with blood transfusion or the sharing of hypodermic needles by intravenous drug users. HIV is *not* transmitted by kissing, tears, or mosquito bites. It can, however, be transferred from mother to fetus through the placenta (transplacental transfer). **Figure 26.14A** illustrates the incidence of AIDS cases by year. The drop-off beginning in 1990 corresponds to the development of more effective treatment regimens. Starting in 1999,

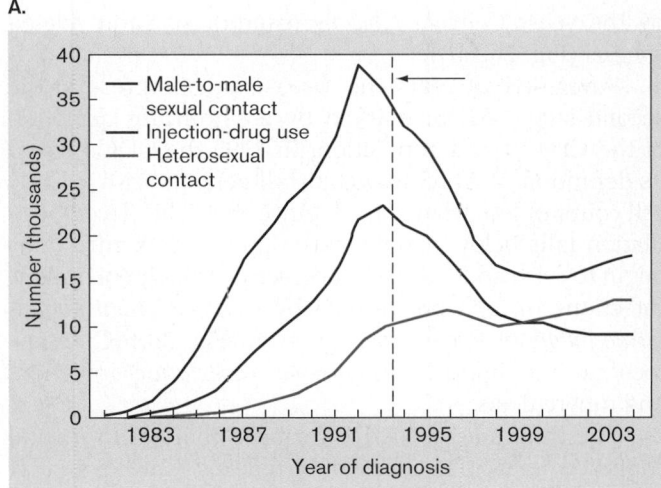

A.

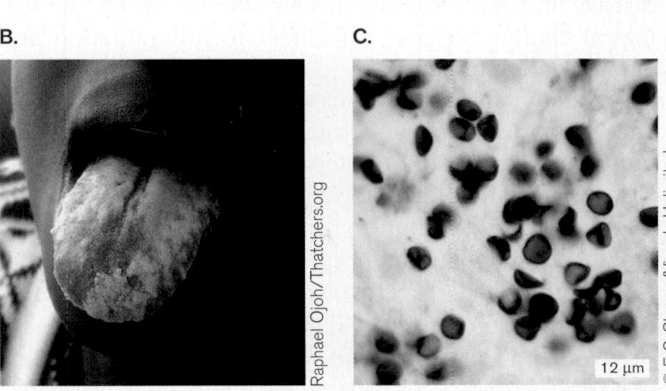

B.   C.

Figure 26.14   Acquired immunodeficiency syndrome.
A. Number of AIDS cases in the United States by major transmission category and year. In 1993, government agencies developed a specific definition of HIV infection that is used for public health surveillance only (vertical dotted line). It is estimated that 250,000–300,000 HIV-infected individuals are unaware of their HIV infection. **B.** Oral candidiasis. The white patches are caused by secondary infection by the yeast *Candida albicans*. **C.** *Pneumocystis jirovecii* infection of the lung. Note the cuplike appearance of the fungus, almost like crushed Ping-Pong balls. Organisms range from 2 to 6 μm in diameter.

however, the trend began climbing upward once again. The proportion of AIDS cases by sex and ethnicity is described in Chapter 25 (see Fig. 25.34).

Having entered the bloodstream, HIV infects CD4$^+$ T cells and replicates very rapidly, producing a billion particles per day. The subsequent decrease in CD4$^+$ T cells leads to the disease symptoms collectively called AIDS. Though the symptoms of AIDS may take years to develop (the average time between infection and AIDS is 11 years), nearly all HIV-infected individuals eventually become ill and die from the disease.

The first stage of the disease, previously known as AIDS-related complex (ARC), can include fever, headache, macular rash, weight loss, and the appearance of antibodies to HIV in serum. These early, relatively mild symptoms can manifest within a few months of infection, resolve within a few weeks, and then may recur. The debilitated immune system leads to secondary infections by the yeast *Candida albicans* (candidiasis) and related species (**Fig. 26.14B**).

After several years the disease can progress to the second stage, AIDS, marked by a significant depletion of the CD4$^+$ T-cell population. In 1993 the CDC revised its definition of AIDS to include all persons with a CD4$^+$ cell count of less than 200/μl. Once the CD4$^+$ T-cell population falls below 400 cells/μl, opportunistic infections begin to arise and disease processes begin. Opportunistic infections include pneumonia by *Mycobacterium avium-intracellulare* or *Pneumocystis jirovecii* (**Fig. 26.14C**), cryptococcal meningitis, *Histoplasma capsulatum* infection, and tuberculosis.

The third stage of AIDS includes changes in mental cognition, muscular action, and reflexes. Cardiovascular disease and brain tumors are also common. The neurological changes appear coincident to inflammation and demyelination of neurons.

The fourth stage of disease is marked by the appearance of various cancers triggered when the depressed immune system cannot detect and destroy cancers initiated by secondary agents. For instance, Kaposi's sarcoma (Fig. 25.34C) is a common cancer seen in AIDS patients that is caused by human herpes virus type 8 (HHV8).

Diagnosis of AIDS involves detecting anti-HIV antibodies and determining CD4$^+$ cell count in a patient. Assay for HIV to determine viral load is done via quantitative PCR to detect HIV-specific genes such as *gag, nef,* or *pol*. Remember, a person who is HIV-positive does not necessarily have AIDS; the disease may take years to develop. A vaccine is not yet available to prevent AIDS, in part because the envelope proteins of the virus (see Fig. 11.25) typically change their antigenic shape. However, progression of the disease can be controlled by antimicrobials that inhibit two HIV enzymes critical to the replication of the virus—reverse transcriptase and protease; this is discussed further in Chapter 27.

> **THOUGHT QUESTION 26.6**   Human immunodeficiency virus was discussed in this section and in Chapter 25. Like the plague, it is a blood-borne disease. Why, then, do fleas and mosquitoes fail to transmit HIV?

The Protozoan *Trichomonas vaginalis* Causes a Common Vaginal Infection

Trichomonas vaginalis is a flagellated protozoan (**Fig. 26.15**) that causes an unpleasant sexually transmitted vaginal disease called trichomoniasis. Approximately 2–3 million infections occur each year in the United States. It can occur in men or women; however, men are usually asymptomatic. Even among infected women, 25%–50% are considered asymptomatic carriers.

There is no cyst in the life cycle of *T. vaginalis*, so transmission is via the trophozoite stage only (the form of a protozoan in the feeding stage). The female patient with trichomoniasis may complain of vaginal itching and/or burning and a musty vaginal odor. An abnormal vaginal discharge also may be present. Males will complain of painful urination (dysuria), urethral or testicular pain, and lower abdominal pain.

Owing to colonization by lactobacilli (which produce large amounts of acidic lactic acid), the normal, healthy vagina has a pH of less than 4.5. However, since *T. vaginalis* feeds on bacteria, the pH of the vagina rises as the numbers of lactobacilli decrease. Definitive diagnosis requires demonstrating the flagellated protozoan in secretions by microscopy. PMNs, which are the primary host defense against the organism, are also usually present. As with giardiasis, this disease is treated with metronidazole.

David M. Phillips/Visuals Unlimited

Figure 26.15 *Trichomonas vaginalis.* *T. vaginalis* is a protozoan that causes a common sexually transmitted disease (SEM, size 7 μm × 10 μm).

TO SUMMARIZE:

- **Urinary tract infections (UTIs)** can result from ascending (to the kidney) or descending (from the kidney) routes of infection. The most common route leading to bladder infection, however, is through the urethra.
- ***E. coli*** is the most common cause of UTI.
- **Syphilis, gonorrhea, and chlamydia** are the most common sexually transmitted diseases.
- **A patient with one sexually transmitted disease** often has another sexually transmitted disease.
- **Complement sensitivity** prevents dissemination by *Neisseria gonorrhoeae*. In contrast, *N. meningitidis*, a cause of meningitis, frequently disseminates in the bloodstream because it is complement resistant.
- **HIV depletion of CD4⁺ T cells** results in lethal secondary infections and cancers.
- ***Trichomonas vaginalis*** is a flagellated protozoan that causes a sexually transmitted vaginal disease. The reservoirs for this organism are the male urethra and female vagina.

26.6 Central Nervous System Infections

Microbes cannot gain easy access to the brain in large measure because of the **blood-brain barrier**, a filter mechanism that allows only selected substances into the brain. The blood-brain barrier works to our advantage when harmful substances, such as bacteria, are prohibited from entering. However, it works to our disadvantage when substances we want to enter the brain, such as antibiotics, are kept out. The barrier is not a single structure but is a function of the way blood vessels, especially capillaries, are organized in the brain. Furthermore, the endothelial cells in those vessels have tight junctions that do not allow most compounds or microbes to cross. And yet, brain infections do occur.

Case History: Meningitis

In April 2001, a 4-month-old infant from Saudi Arabia was hospitalized with fever, tender neck, and purplish spots (purpuric spots) on her trunk (**Fig. 26.16A**). *Suspecting meningitis, the clinician took a cerebrospinal fluid (CSF) sample and examined it by Gram stain. The smear revealed Gram-negative diplococci inside PMNs. The CSF was turbid with 900 leukocytes per microliter, and Neisseria meningitidis was confirmed by culture. The child was treated with cefotaxime and made a full recovery. Her father, the person who brought her in, was clinically well. However, meningococcus was isolated from his oropharynx, as well as from the throat of the patient's 2-year-old brother. Isolates from the patient, her father, and her brother were positive by agglutination with meningococcus A, C, Y, W135 polyvalent reagent. The father's vaccination certificate confirmed that he had received a quadrivalent meningococcal vaccine. All three isolates were sent to the WHO (World Health Organization) Collaborating Centre, which confirmed meningococcus serogroup W135. The three isolates were examined by pulsed-field gel electrophoresis and were found to be indistinguishable (that is, they had identical DNA restriction patterns).*

Meningitis is an inflammation of the meninges, the membranes that surround the brain and spinal cord. Meningitis can be either bacterial or viral in origin. Sinus and ear infections can extend directly to the meninges, while septicemic spread requires passage through the blood-brain barrier. Viral meningitis is serious but rarely fatal in people with a normal immune system. The symptoms generally persist for 7–10 days and then completely resolve. Bacterial meningitis is usually caused by

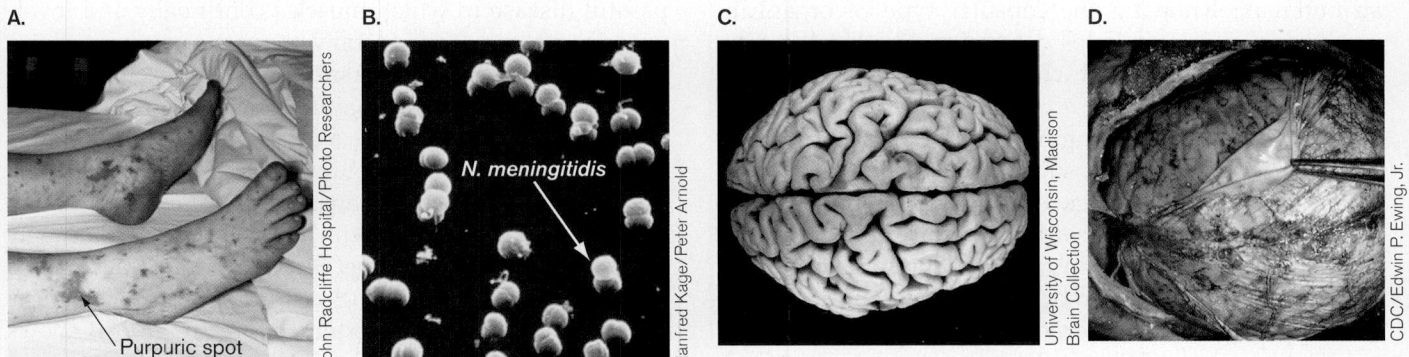

Figure 26.16   Bacterial meningitis.   Meningococcal disease is dreaded by parents and medical practitioners alike for its rapid onset and the difficulty of obtaining a timely and accurate diagnosis. **A.** Purpuric spots produced by local intravascular coagulation due to *Neisseria meningitidis* endotoxin. The rash in meningitis typically has petechial (small) and purpuric (large) components. **B.** *N. meningitidis* (SEM, approx. 1 μm diameter). **C.** Normal brain. **D.** Autopsy specimen of meningitis due to *Streptococcus pneumoniae*. Note the greening of the brain, compared with the pink normal brain.

Streptococcus pneumoniae, Neisseria meningitidis, or *Haemophilus influenzae.* Symptoms of bacterial meningitis can include sudden onset of fever, headache, neck pain or stiffness, painful sensitivity to strong light (photophobia), vomiting (often without abdominal complaints), and irritability. Prompt medical attention is extremely important because the disease can quickly progress to convulsions and death.

The meningococcus *N. meningitidis* (**Fig. 26.16B**) can colonize the human oropharynx, where it causes mild, if any, disease. At any given time, 10%–20% of the healthy population can be colonized and asymptomatic. The organism spreads directly by person-to-person contact or indirectly via droplet nuclei from sneezing or fomites. The problem arises when this organism enters the bloodstream. Unlike *N. gonorrhoeae,* the cause of gonorrhea, *N. meningitidis* is very resistant to complement, owing to its production of a polysaccharide capsule. The capsule allows the microbe to produce a transient blood infection (bacteremia) and reach the blood-brain barrier.

N. meningitidis uses type IV pili to adhere to these endothelial cells and becomes intracellular. Then, utilizing the process of **transcytosis**, in which a cell moves from one side of a host cell to another, the microbe moves to the brain side of the capillary, where it exits. Once in the cerebrospinal fluid, microbes can multiply almost at will. **Figure 26.16C** and **D** show the remarkable damage that *N. meningitidis* and other microbes, such as *S. pneumoniae,* cause in the brain.

Several antigenic types of capsules, called type-specific capsules, are produced by different strains of pathogenic *N. meningitidis:* types A, B, C, W135, and Y. Types A, C, Y, and W135 are usually associated with epidemic infections seen among people kept in close proximity, such as college students or military personnel. Type B meningococcus is typically involved in sporadic infections. Antibodies to these capsular antigens are used to classify the capsular types of the organisms causing an outbreak. Knowing the capsular type of organism involved in each case helps determine whether the disease cases are related and where the infection may have started.

Meningococcal meningitis is highly communicable. As a result, close contacts, such as the parents or siblings of any patient with meningococcal disease, should receive antimicrobial prophylaxis within 24 hours of diagnosis; a single dose of ciprofloxacin (a quinolone antibiotic, discussed in Section 27.4) can be given to adults, and 2 days of rifampicin given to children.

Highly susceptible populations can be immunized with a vaccine containing the four capsular structures. Generally, large numbers of people housed together are susceptible to this disease. For instance, it is a significant problem in college dormitories, prompting several colleges to request that incoming students be vaccinated.

> **THOUGHT QUESTION 26.7**   Normal cerebrospinal fluid is usually low in protein and high in glucose. The protein and glucose content does not change much during a viral meningitis, but bacterial infection leads to greatly elevated protein and lowered glucose levels. What could account for this?

Case History: Botulism—It Is What You Eat

In June, a 47-year-old resident of Oklahoma was admitted to the hospital with rapid onset of progressive dizziness, blurred vision, slurred speech, difficulty swallowing, and nausea. Findings on examination included drooping eyelids, facial paralysis, and impaired gag reflex. He developed breathing difficulties and required mechanical ventilation. The patient reported that during the 24 hours before onset of symptoms, he had eaten home-canned green beans and a stew containing roast beef and potatoes. Analysis of the patient's stool detected botulinum type A toxin, but no Clostridium botulinum *organisms were found. The patient was hospitalized for 49 days, including 42 days on mechanical ventilation, before being discharged.*

Bacterial Neurotoxins That Cause Paralysis

Imagine a disease that causes complete loss of muscle function. Using secreted exotoxins, two microbes cause lethal paralytic diseases. In one instance, the victim suffers a flaccid paralysis in which the muscles go limp, as in the case history just given, causing paralysis and respiratory difficulty. Voluntary muscles fail to respond to the mind's will because botulinum toxin interferes with neural transmission. The disease (called **botulism**) is typically food-borne and is caused by an anaerobic, Gram-positive, spore-forming bacillus named *Clostridium botulinum* (Chapter 18).

In striking contrast to botulism is tetanus, a very painful disease in which muscles continually and involuntarily contract (called tetany or spastic paralysis). Tetanus is caused by **tetanospasmin**, a potent exotoxin made by another anaerobic, Gram-positive spore-forming bacillus, called *Clostridium tetani* (**Fig. 26.17A**). Tetanospasmin interferes with neural transmission, but in contrast to botulism toxin, it causes *excessive* nerve signaling to muscles, forcing the victim's back to arch grotesquely while the arms flex and legs extend. The patient remains locked this way until death. Spasms can be strong enough to fracture the patient's vertebrae. **Figure 26.17B** shows the result of injecting a mouse's hind leg with just a tiny amount of tetanus toxin. In both botulism and tetanus, death can result from asphyxiation.

Botulism is typically caused by ingesting preformed toxin, although infected wounds or germination of ingested spores can also occasionally produce disease.

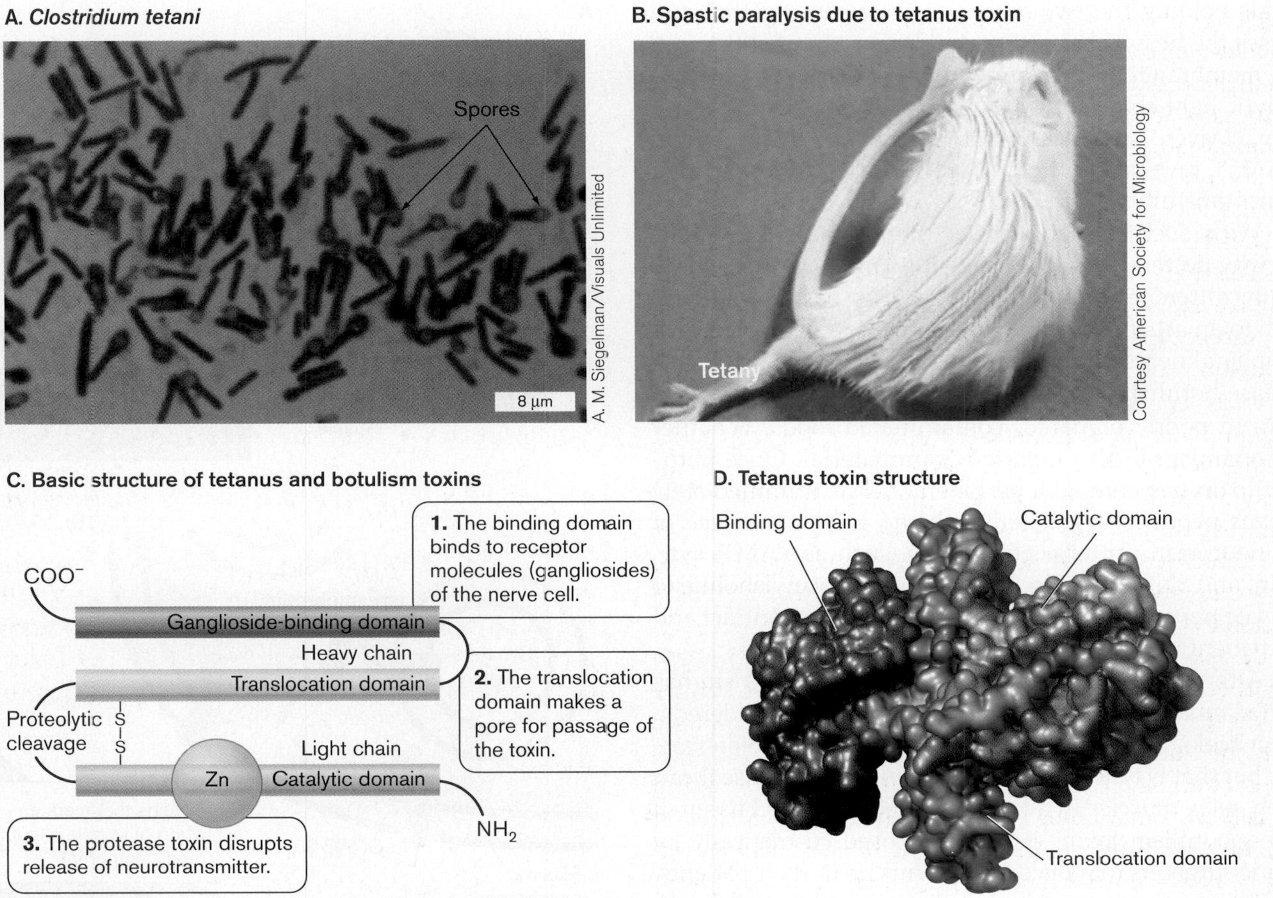

A. *Clostridium tetani*

B. Spastic paralysis due to tetanus toxin

C. Basic structure of tetanus and botulism toxins

1. The binding domain binds to receptor molecules (gangliosides) of the nerve cell.

2. The translocation domain makes a pore for passage of the toxin.

3. The protease toxin disrupts release of neurotransmitter.

D. Tetanus toxin structure

Figure 26.17   Tetanus and botulism toxins.   A. Photomicrograph of *Clostridium tetani* (cell length 4–8 μm). **B.** Mouse injected with tetanus toxin in left hind leg. **C.** Schematic diagram of tetanus and botulism toxins. **D.** Three-dimensional representation of the tetanus neurotoxin with the domains marked. (PDB code: 3BTA)

The microbe germinates in the food and produces toxin. Because the organism is an anaerobe, an anaerobic environment must be present. Home-canning processes are designed to remove oxygen in the canned food, as well as sterilize the food; but if the sterilization process is not complete, surviving spores germinate and produce toxin. The toxin is susceptible to heat but will remain active in improperly cooked food. After ingestion, the toxin is absorbed from the intestine. As in the case history given here, it is not unusual for the organism to be absent from stool samples. This is why serological identification of the toxin is important.

Note that a rare form of botulism, called infant botulism or "floppy head syndrome," can occur when infants are fed honey. Honey can harbor *C. botulinum* spores that can germinate in the gastrointestinal tract after which the growing vegetative cells will secrete toxin.

THOUGHT QUESTION 26.8   Knowing the symptoms of tetanus, what kind of therapy would you use to treat the disease?

Toxin structure. Botulism and tetanus toxins share 30%–40% identity and have similar structures and nearly identical modes of action. Botulism and tetanus toxins are each composed of two peptides—a large, or heavy, fragment analogous to a B subunit of AB toxins and a small, or light, fragment analogous to an A subunit. Both toxins are initially made as single peptides (about 150 kDa) that are cleaved after secretion to form two fragments (heavy and light chains) that remain tethered by a disulfide bond (**Fig. 26.17C**). Each heavy chain includes the binding domain, which binds to receptor molecules (gangliosides) on the nerve cell membrane, and a translocation domain that makes a pore in the nerve cell through which the toxin passes. The light chains (catalytic domains) are proteases that disrupt the movement of exocytic vesicles containing neurotransmitters needed for contraction or relaxation. **Figure 26.17D** shows a three-dimensional rendition of tetanus toxin with the three domains marked.

Mechanism of action. Both toxins bind to target membranes and are brought into the cell by endocytosis. The low pH that forms in the endosome reduces the disulfide bonds holding the two halves of the toxin together and causes the heavy chain to assemble as a channel through the membrane. That channel allows release of the proteolytic light chain into the cytoplasm. The toxic subunit then cleaves key host proteins such as synaptobrevin (VAMP), involved in the exocytosis of vesicles containing neurotransmitters (**Fig. 26.18**).

With such a high degree of mechanistic similarity, why do tetanus and botulism toxins have such drastically different effects? The answer is based on where each toxin acts in the nervous system. Both toxins enter at peripheral nerve endings where nerve meets muscle. The disease botulism results from the ingestion of preformed toxin in poorly prepared, contaminated foods. Whether the organism is also ingested is immaterial. Once botulism toxin has entered a peripheral nerve, it immediately cleaves peptides associated with the exocytic release of the neurotransmitter acetylcholine, such as VAMP, syntaxin, and SNAP-25 (**Fig. 26.18**). Without acetylcholine to activate nerve transmission, muscles will not contract and the patient is paralyzed.

Although botulism disease is now rare, the toxin has gained renewed interest in recent years. It is considered a select biological toxic agent of potential use to bioterrorists, but that is not where most interest lies. Because it can safely relax muscles in a localized area if injected in small doses, botulism toxin, or Botox, is used cosmetically by plastic surgeons to reduce facial wrinkles in their patients.

Tetanus, in contrast to botulism, is not a food-borne disease. *Clostribium tetani* spores are introduced into the body by trauma (such as by stepping on a dirty, rusty nail). Necrotic tissue then provides the anaerobic environment required for germination. Growing, vegetative cells release tetanospasmin, which enters the peripheral nerve cells at the site of injury. But rather than cleaving targets here, the toxin travels up axons in the direction opposite to nerve signal transmission until it reaches the spinal column, where it becomes fixed at the presynaptic inhibitory motor neuron. There the toxin also cleaves proteins like VAMP, but the function of vesicles in these nerves is to release *inhibitory* neurotransmitters that dampen nerve impulses. Tetanus toxin blocks release of the inhibitory neurotansmitters GABA and glycine into the synaptic cleft, leaving nerve impulses unchecked. As a result, impulses come too frequently and produce the generalized muscle spasms characteristic of tetanus. **Figure 26.19** illustrates the mechanism of action of tetanus toxin.

THOUGHT QUESTION 26.9 How do the actions of tetanus toxin and botulinum toxin actually help the bacteria colonize or obtain nutrients?

THOUGHT QUESTION 26.10 If *Clostridium botulinum* is an anaerobe, how might botulism toxin get into foods?

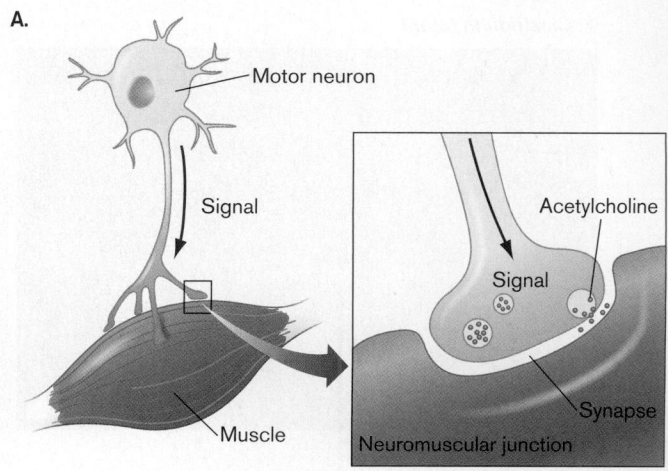

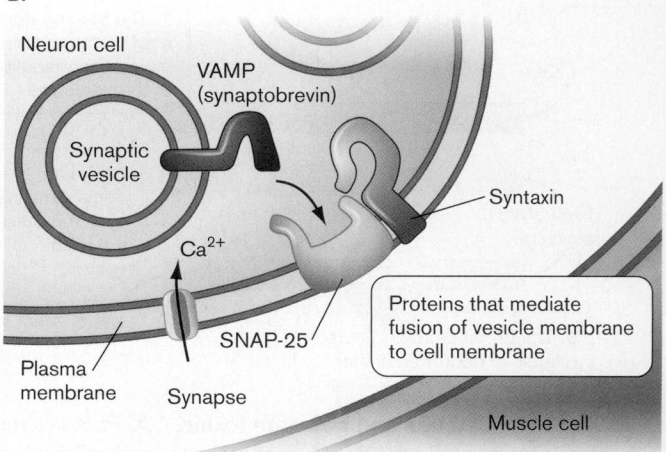

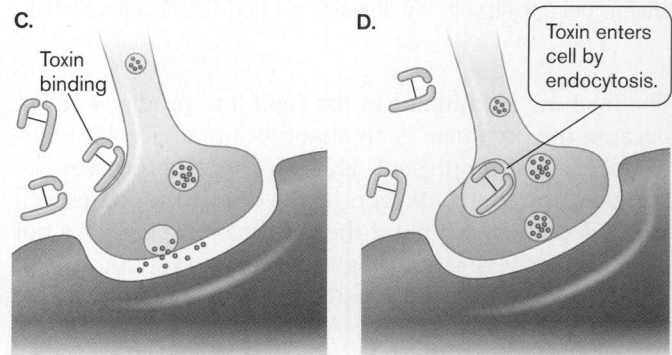

Figure 26.18 Mechanism of action of botulism toxin. **A.** The neuromuscular junction. Inset shows vesicles filled with neurotransmitters. **B.** A series of proteins within the nerve is needed to allow synaptic vesicles to bind to the nerve endings. Fusion of the membranes releases acetylcholine into the neuromuscular junction. Botulism toxin types A and E cleave SNAP-25. Botulism toxins B, D, F, and G cleave VAMP. Botulism toxin C1 cleaves syntaxin and SNAP-25. **C, D.** Toxin binds via the heavy chain and is endocytosed into the nerve terminal. Once the toxin cleaves its target, the nerve terminal is no longer able to release acetylcholine.

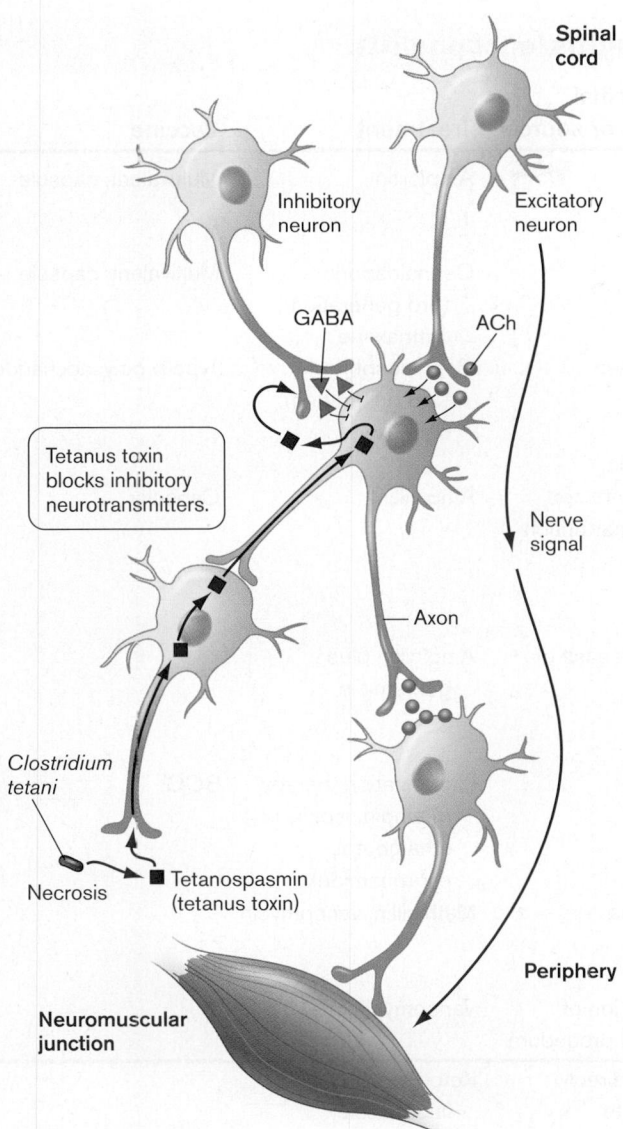

Figure 26.19 Retrograde movement of tetanus toxin to an inhibitory neuron. Tetanus toxin enters the nervous system at the neuromuscular junction and travels retrogradely up the axons until reaching an inhibitory neuron located in the central nervous system. There it cleaves VAMP protein (Fig. 26.18B) associated with exocytosis of vesicles containing inhibitory neurotransmitters. GABA = gamma-aminobutyric acid; ACh = acetylcholine. ⏯ *microbiology2.com/animations*

Case History: Eastern Equine Encephalitis

In August, Mr. C brought his 21-year-old son, Rich, to a New Jersey emergency room. Rich appeared dazed and had trouble responding to simple commands. When questioned about his son's activities over the past few months, Mr. C told the physician that Rich planned to enter veterinary school in the fall and had spent the month of July relaxing and sunning himself on New Jersey beaches. Aside from the shore, his favorite locale

was a pond in a wooded area near a horse farm. On the afternoon prior to admission, Rich became lethargic and tired. He returned home and went to bed. That evening, his father woke him for supper but Rich was confused and had no appetite. By 11 p.m., Rich had a fever of 40.7°C (103.5°F) and could not respond to questions. A few hours later, when his father had trouble rousing him, he brought Rich to the ER. Over the next week, Rich's condition worsened to the point where his limbs were paralyzed. Two weeks later he died.

Serum samples taken upon entering the hospital and a few days before he died showed a sixfold rise in antibody titer to eastern equine encephalitis (EEE) virus. Brain autopsy showed many small foci of necrosis in both the gray and white matter.

The EEE virus, a member of the family *Togaviridae*, is transmitted from bird to bird by mosquitoes. Horses contract the disease in the same way. Rich contracted it inadvertently while visiting the pond. The disease, an encephalitis, is often fatal (35% mortality) but fortunately rare. One reason human disease is rare is that the species of mosquito that usually transmits the virus between marsh birds does not prey on humans. But sometimes a human-specific mosquito bites an infected bird and then transmits EEE virus to humans. Another reason human disease is rare is that the virus generally does not fare well in the body. Many persons infected with EEE virus have no apparent illness because the immune system thwarts viral replication. However, as already noted, mortality is high in those individuals that do develop disease.

An interesting point in the case history is that diagnosis relied on detecting an increase in antibody titer to the virus. Typically, at the point that disease symptoms first appear, the body has not had time to generate large amounts of specific antibodies. After a week or so, often when the patient is nearing recovery (convalescence), antibody titers have risen manyfold. The rule of thumb is that a greater than fourfold rise in antibody titer between acute disease and convalescence (or in this case death) indicates that the patient has had the disease. While this knowledge could not help the patient in this case, it was valuable in terms of public health and prevention strategies.

Several bacterial, fungal, and viral causes of encephalitis and meningitis are given in **Table 26.5**.

Prion Diseases Are Caused by Proteins

An unusual infectious agent called the prion has been implicated as the cause of a series of relatively rare but invariably fatal brain diseases (**Table 26.6**). Prions are infectious agents that do not have a nucleic acid genome. It seems that a protein alone can mediate an infection. The **prion** has been defined as a small proteinaceous infectious particle that resists inactivation by procedures that modify nucleic acids. The discovery that proteins alone can transmit an infectious disease came as

Table 26.5 Selected microbes that cause meningitis/encephalitis.

Type of meningitis	Etiological agent	Virulence factors	Typical initial infection or source	Treatment	Vaccine
Bacterial (septic)	*Streptococcus pneumoniae* (G+)	Capsule, pneumolysin	Lung	Ampicillin	Multivalent, capsule
	Neisseria meningitidis (G–)	Capsule IgA protease, endotoxin	Throat	Cephalosporin (3rd generation), ceftriaxone	Multivalent, capsule
	Haemophilus influenzae type B (G–)	Polyribitol capsule, IgA protease, endotoxin	Ear infection	Cephalosporin (3rd generation), ceftriaxone	Type b polysaccharide
	Other G– bacilli	Endotoxin	Septicemia		
	Group B streptococci (G+)	Sialic acid capsule, streptolysin, inhibition of alternate complement path	Neonate infected during parturition	Ampicillin	Capsular
	Listeria monocytogenes (G+)	Intracellular growth, PrfA regulator, actin-based motility	Mild GI disease of mother	Ampicillin plus gentamicin	
	Mycobacterium tuberculosis	Cord factor, wax D, intracellular growth	Lung	Combination therapy (rifampin, isoniazid, ethambutol, pyrazinamide)	BCG
	Staphylococcus aureus (G+)	Coagulase, protein A, TSST, leukocidin	Septicemia	Methicillin, vancomycin	
	Staphylococcus epidermidis (G+)	Biofilm, slime	Complication of surgical procedure	Vancomycin, methicillin	
Aseptic[a]	Fungi (e.g., *Coccidioides*, *Cryptococcus*)		Sinusitis, direct spread to meninges	Ketoconazole, fluconazole	
	Amebas (e.g., *Naegleria*)		Swimming in contaminated waters	Amphotericin B	
	Treponema pallidum		Syphilis		
	Mycoplasmas	Adhesin tip	Respiratory	Erythromycin	
	Leptospira	Burrowing motility	Septicemia, water contaminated with animal urine	Erythromycin	Killed whole cell, animals only
Viral	Viruses (90% caused by enteroviruses)			Self-limiting	
Eastern equine encephalitis	EEE virus	?	Mosquito bite	None; often fatal	
West Nile disease	West Nile virus	?	Mosquito bite	Supportive therapy[b]	

[a]Aseptic meningitis is an inability to isolate bacterial sources by ordinary means.

[b]Supportive therapy means hospitalization, intravenous fluids, airway management, respiratory support, prevention of secondary infections.

Table 26.6 Prion diseases.

Disease	Susceptible animal	Incubation period	Disease characteristics
Creutzfeldt-Jakob (sporadic, familial, new variant [vCJD])	Human	Months (vCJD) to years	Spongiform encephalopathy (degenerative brain disease)
Kuru	Human	Months to years	Spongiform encephalopathy
Gerstmann-Straussler-Scheinker syndrome	Human	Months to years	Genetic neurodegenerative disease
Fatal familial insomnia	Human	Months to years	Genetic neurodegenerative disease with untreatable insomnia
Mad cow disease	Cattle	5 years	Spongiform encephalopathy
Wasting disease	Deer	Months to years	Spongiform encephalopathy

a surprise to the scientific community. Diseases caused by prions are especially worrying, since prions resist destruction by many chemical agents and remain active after heating at extremely high temperatures. There have even been documented cases in which sterilized surgical instruments, originally used on a prion-infected person, still held infectious agent and transmitted the disease to a subsequent surgery patient.

How can a nonliving entity without nucleic acid be an infectious agent? Prions associated with human brain disease are thought to be aberrantly folded forms of a normal brain protein. The theory is that when a prion is introduced into the body and manages to enter the brain, it will cause normally folded forms of the protein to refold incorrectly (**Fig. 26.20A**). The improperly folded proteins fit together like Lego blocks to produce damaging aggregated structures within brain cells.

Some investigators suggest that living, infectious agents, such as *Spiroplasma* (a spiral-shaped genus whose members lack cell walls), can somehow precipitate a conformational change in the normal brain protein, converting it to an infectious prion protein. This proposal is not, however, universally accepted and has yet to be proved.

Prion diseases are often called **spongiform encephalopathies** because the postmortem appearance of the brain includes large, spongy vacuoles in the cortex and cerebellum. These are visible in a brain sample from a victim of one of these diseases, called Creutzfeldt-Jakob disease (CJD) (**Fig. 26.20B** and **C**). Most mammalian species appear to develop these diseases.

Since 1996, mounting evidence points to a causal relationship between outbreaks in Europe of a disease in cattle called bovine spongiform encephalopathy (BSE, or "mad cow disease") and a disease in humans called variant Creutzfeldt-Jakob disease (vCJD). Both disorders are invariably fatal brain disorders. They have unusually long incubation periods (measured in years) and are caused by an unconventional transmissible agent.

As of this writing, only four cases of BSE have been detected in the United States. Because of aggressive surveillance efforts in the United States and Canada (where only nine cases have been found), it is unlikely, but not impossible, that BSE will be a food-borne hazard to humans in this country. The CDC monitors the trends and current incidence of typical and variant CJD in the United States (see **eTopic 26.3**). The worldwide incidence of CJD is one per 1 million population.

A.

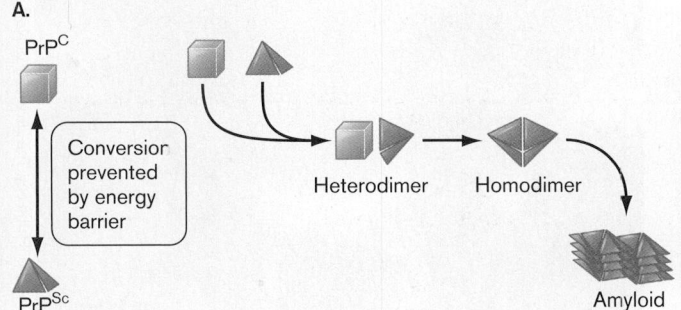

B.

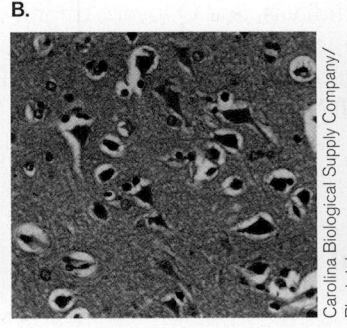

C.

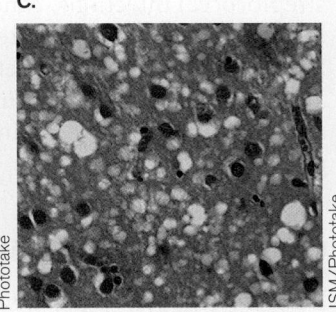

Figure 26.20 Spongiform encephalopathies. A. Refolding model of prion diseases. PrP is a brain protein that can take two forms: the normal form (PrP^C), which is a natural brain protein; and the prion form (PrP^{Sc}). When a prion form gains access to the brain (following ingestion or other means), it can cause the slow refolding of normal PrP^C proteins to form aggregates that damage the brain. **B.** Normal brain section. **C.** Section of brain taken from a CJD victim. Note the "Swiss cheese" appearance, indicating brain damage.

TO SUMMARIZE:

- ■ *Neisseria meningitidis* is resistant to serum complement because it produces a type-specific capsule. This allows the organism to reach and then cross the blood-brain barrier.
- ■ **A vaccine for** *N. meningitidis* is available, but none exists for *N. gonorrhoeae*.
- ■ **Tetanospasmin** causes spastic paralysis.
- ■ **Botulism toxin** causes flaccid paralysis.
- ■ **Serological diagnosis of an infectious disease** is possible if specific pathogen antibody titer rises fourfold between the acute and convalescent stages.
- ■ **Spongiform encephalopathies** are believed to be caused by nonliving proteins called prions.

26.7 Cardiovascular System Infections

Infections of the cardiovascular system include septicemia, **endocarditis** (inflammation of the heart's inner lining), pericarditis (inflammation of the heart's outer lining) and, possibly, atherosclerosis (the deposition of fatty substances along the inner lining of arteries; **eTopic 26.4**). These are all life-threatening diseases.

Septicemia is, by strict definition, the presence of bacteria or viruses in the blood. The presence of viruses is a condition called **viremia**, while bacteria in the circulation is more specifically called **bacteremia**. In practice, however, the terms "septicemia" and "bacteremia" are often used interchangeably. Septicemia can develop from a local infection situated anywhere in the body, although blood factors such as complement can nonspecifically kill many types of bacteria that enter the blood. Nevertheless, Gram-positives, Gram-negatives, aerobes, and anaerobes can all produce septicemia under the right conditions.

Endocarditis can be either viral or bacterial in origin. It can be a consequence of many bacterial diseases, such as brucellosis, gonorrhea, psittacosis, staphylococcal and streptococcal infections, candidiasis, and Q fever. Among the many viral causes are coxsackievirus, echovirus, Epstein-Barr, and HIV. Bacterial infections of the heart are always serious. Viral infections, although common, are rarely life-threatening in healthy individuals and are usually asymptomatic.

Case History: Bacterial Endocarditis

Elizabeth is 38 years old and has a history of mitral valve prolapse (a common congenital condition in which a heart valve does not close properly). She was recently admitted to the hospital complaining of fatigue, intermittent fevers for 5 weeks and headaches for 3 weeks, symptoms the physician recognized as possible indications of endocarditis. Elizabeth reported having a dental procedure a few weeks prior to the onset of symptoms. A sample of her blood placed in a liquid bacteriological medium grew Gram-positive cocci, which turned out to be Streptococcus mutans, a member of the viridans streptococci. As a result, the diagnosis of bacterial endocarditis was confirmed. The patient began a one-month course of intravenous penicillin G and gentamicin therapy and eventually recovered to normal health.

Endocarditis (inflammation of the heart) is traditionally classified as acute or subacute, depending on the pathogenic organism involved and the speed of clinical presentation. Subacute bacterial endocarditis (SBE) has a slow onset with vague symptoms. It is usually caused by bacterial infection of a heart valve (**Fig. 26.21**). Subacute bacterial endocarditis infections are usually (but not always) caused by a viridans streptococcus from the oral micrbiota (for example, *Streptococcus mutans*, a common cause of dental caries). "Viridans streptococci" is a general term used for commensal streptococci whose colonies produce green alpha hemolysis on blood agar ("viridans" from the Greek *viridis*, "to be green"). Most patients who develop infective endocarditis have mitral valve prolapse (90%), although this is frequently not the case when patients are intravenous drug abusers or have hospital-acquired (nosocomial) infections.

As in the case history presented here, bacterial endocarditis often begins at the dentist's office. Following a dental procedure (such as tooth restoration), oral bacteria can transiently enter the bloodstream and circulate. *S. mutans*, which is not normally a serious health problem, can become lodged onto damaged heart valves, grow as a biofilm, and secrete a thick glycocalyx coating that encases the microbes and forms a vegetation on the valve, damaging it further. If untreated, the condition can be fatal within 6 weeks to a year.

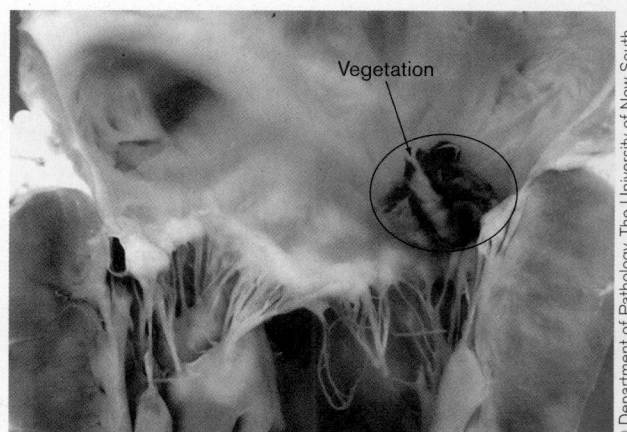

© Department of Pathology, The University of New South Wales, Sydney Australia

Figure 26.21   View of bacterial endocarditis.   Close-up of mitral valve endocarditis, showing vegetation (arrow).

When more virulent organisms, such as *Staphylococcus aureus*, gain access to cardiac tissue, a rapidly progressive (acute) and highly destructive infection ensues. Symptoms of acute endocarditis include fever, pronounced valvular regurgitation (backflow of blood through the valve), and abscess formation.

Most patients with subacute bacterial endocarditis present with a fever that lasts several weeks. They also complain of nonspecific symptoms, such as cough, shortness of breath, joint pain, diarrhea, and abdominal or flank pain. Endocarditis is suspected in any patient who has a heart murmur and an unexplained fever for at least one week. It should also be considered in an intravenous drug abuser with a fever, even in the absence of a murmur. In either case, definitive diagnosis requires blood cultures that grow bacteria. Blood cultures involve taking samples of a patient's blood from two different locations (such as two different arms). Growth of the same organism in cultures taken from two body sites rules out inadvertent contamination with skin microbes, which would likely yield growth in only one culture. The blood is then added to liquid culture medium and incubated at 37°C. Incubation should be done aerobically and anaerobically.

Curing endocarditis is difficult because the microbes are usually ensconced in a nearly impenetrable glycocalyx. Consequently, eradicating microorganisms from the vegetations almost always requires hospitalization, where high doses of intravenous antibiotic therapy can be administered and monitored. Antibiotic therapy usually continues for at least a month, and in extreme cases, surgery may be necessary to repair or replace the damaged heart valve.

Although at increased risk, patients with heart valve prolapse should not shy away from the dentist. Prophylactic treatment with penicillin or erythromycin taken an hour before a procedure will kill any oral bacteria that enter the bloodstream and prevent the development of endocarditis.

Viruses such as adenovirus and some enteroviruses can also cause endocarditis, as well as a condition known as myocarditis, an inflammation of the heart muscle.

> **THOUGHT QUESTION 26.11**   A patient presenting with high fever and in an extremely weakened state is suspected of having a septicemia. Two sets of blood cultures are taken from different arms. One bottle from each set grows *Staphylococcus aureus*, yet the laboratory report states that the results are inconclusive. New blood cultures are ordered. What might explain this result?

Malarial Parasites Feed on Blood

Malaria is the most devastating infectious disease known. Each year 300–500 million people develop malaria worldwide, and 1–3 million of these people, mostly children, die. Fortunately, the disease is relatively rare in the United States; only about 1,000 cases occur annually, and almost all are acquired as a result of international travel to endemic areas (**Fig. 26.22A**). Once again, this illustrates the diagnostic importance of knowing a patient's history.

The disease is caused by four species of *Plasmodium*: *P. falciparum* (the most deadly), *P. malariae*, *P. vivax*, and *P. ovale*. The life cycle of *Plasmodium*, discussed in detail in Chapter 20, is complex and involves two cycles: an asexual erythrocytic cycle in the human and a sexual cycle in the mosquito (discussed in Chapter 20; Fig. 20.39).

In the erythrocytic cycle, the organisms enter the bloodstream through the bite of an infected female *Anopheles* mosquito (the mosquito injects a small amount

A.

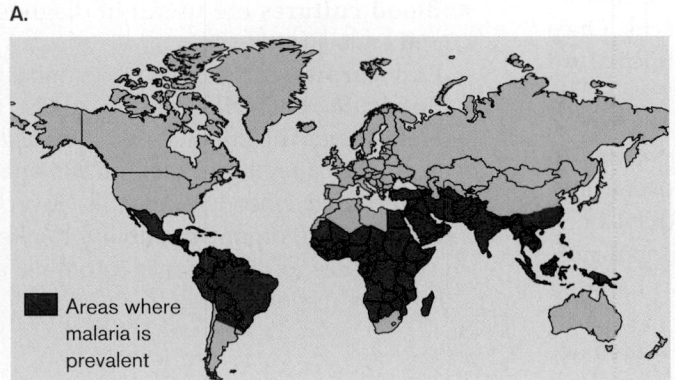

Areas where malaria is prevalent

B.

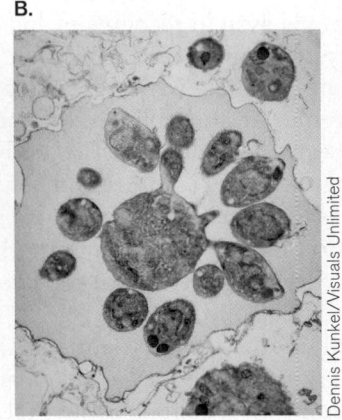

Dennis Kunkel/Visuals Unlimited

Figure 26.22   Malaria is a major disease worldwide.   A. Endemic areas of the world where malaria is prevalent. **B.** *Plasmodium falciparum* (TEM). Schizont after completion of division. A residual body of the organism (yellow-green) is left over after division. The erythrocyte has lysed and only a ghost cell remains; no cytoplasm is seen surrounding the merozoites just being released. Free merozoites are seen outside the membrane.

of saliva containing *Plasmodium*). The haploid sporozoites travel immediately to the liver, where they undergo asexual fission to produce merozoites. Released from the liver, the merozoites attach to and penetrate red blood cells, where *Plasmodium* consumes hemoglobin and enlarges into a trophozoite. The protist nucleus divides, so that the cell, now called a schizont, contains up to 20 or so nuclei. The schizont then divides to make the smaller, haploid merozoites (**Fig. 26.22B**). The glutted red blood cell eventually lyses, releasing merozoites that can infect new red blood cells (see Fig. 20.38).

WWW | Malaria animation
      | *microbiology2.com/links*

Sudden, synchronized release of the merozoites and red cell debris triggers the telltale symptoms of malaria: violent, shaking chills followed by high fever and sweating. The erythrocytic cycle, and thus the symptoms, repeats every 48–72 hours. After several cycles, the patient goes into remission lasting several weeks to months, after which there is a relapse.

Much of today's research focuses on why malarial relapse happens. Why does the immune system fail to eliminate the parasite after the first episode? When *Plasmodium* invades the red blood cells, it lines the blood cells with a protein, PfEMP1, that causes the parasite to stick to the sides of blood vessels. This removes the parasite from circulation, but the protein cannot protect the parasite from patrolling macrophages, which eventually detect the invader and recruit other immune cells to fight it. So, during a malarial infection a small percentage of each generation of parasites switches to a different version of PfEMP1 that the body has never seen before. In its new disguise, *Plasmodium* can invade more red blood cells and cause another wave of fever, headaches, nausea, and chills.

These sticky surface proteins are the antigens that the body's immune system recognizes and attacks. Once the immune system fights off one version of falciparum malaria, the parasite alters which gene is expressed. The resulting antigenic variation blinds the immune system, allowing a new wave of illness. The body now has to repeat the recognition and attack responses all over again. The parasite has 60 cloaking genes, called *var*, that can be turned on and off individually, changing the organisms' antigenic structure, like a criminal repeatedly changing his disguise to elude police.

In April 2005, Australian scientists Alan Cowman, Brendan Crabb, and colleagues (Walter and Eliza Hall Institute of Medical Research) showed that *var* genes are regulated by chromosome packaging, which unwraps one gene to be expressed at a time and literally packs away the inactive genes. DNA can be encased so securely by some proteins that other proteins cannot access the

nucleic acid for transcription—a process known as epigenetic silencing. Becoming immune to all the types of malaria can take upwards of 5 years and requires constant exposure; otherwise the immunity is lost. Many children do not live long enough to gain immunity to malaria in all its forms.

Diagnosis involves microscopic demonstration of the protist within erythrocytes (Wright stain) or through serology to identify antimalarial antibodies. Treatment regimens include chloroquine or mefloquine, which kill the organisms in their erythrocytic asexual stages, and primaquine, effective in the exoerythrocytic stages. The chloroquine family of drugs acts by interfering with the detoxification of heme generated from hemoglobin digestion. Malaria parasites accumulate the hemoglobin released from red blood cells in plasmodial lysosomes, where digestion occurs. The parasites use the amino acids from hemoglobin to grow but find free heme toxic. To prevent eating themselves to death (from accumulating too much heme), the organism detoxifies heme via polymerization, which produces a black pigment. Many antimalarial drugs, such as chloroquine, prevent polymerization by binding to the heme. As a result, the increased iron level (from heme) kills the parasite.

Chloroquine is given prophylactically to persons traveling to endemic areas. Unfortunately, *Plasmodium* has been developing resistance to these drugs, forcing development of new ones. The antigenic shape-shifting carried out by this parasite has so far stymied development of an effective vaccine.

Bear in mind, we have described only selected organisms that cause cardiovascular infections—there are many more (for example, *Rickettsia typhi*). There is even evidence that *Chlamydophila pneumoniae* may have a role in coronary artery disease (see **eTopic 26.4**).

TO SUMMARIZE:

- **Blood cultures** are useful in diagnosing septicemia and endocarditis.
- **Endocarditis** can have acute or subacute onsets.
- **Subacute bacterial endocarditis** is usually an endogenous infection caused by *Streptococcus mutans*.
- **Malaria, caused by *Plasmodium* species**, manifests as repeated episodes of chills, fever, and sweating owing to the organism's ability to alter the antigenic appearance of its surface proteins and evade the immune response.

26.8 Systemic Infections

Many pathogenic bacteria can produce a septicemia as a way to disseminate throughout the body and infect other organs. These organisms cause what are considered systemic infections.

Case History: The Plague

*A 25-year-old New Mexico rancher was admitted to an El Paso hospital because of a 2-day history of headache, chills, and fever (40°C; 104°F). The day before admission, he began vomiting. The day of admission, an orange-sized, painful swelling in the right groin area was noted (**Fig. 26.23A**). A lymph node aspirate and a smear of peripheral blood were reported to contain Gram-negative rods that exhibited bipolar staining (**Fig. 26.23B**). The patient's white blood cell count was 24,700/μl (normal is 5,000–10,000/μl), and platelet count was 72,000/μl (normal is 130,000–400,000/μl). In the two weeks prior to becoming ill, the patient had trapped, killed, and skinned two prairie dogs, four coyotes, and one bobcat. The patient had cut his left hand shortly before skinning a prairie dog. PCR and typical biochemical testing of a Gram-negative rod isolated from blood cultures identified the organism as* Yersinia pestis, *the organism that causes plague. The patient received an antibiotic cocktail of gentamicin and tetracycline. He eventually recovered, after 6 weeks in intensive care.*

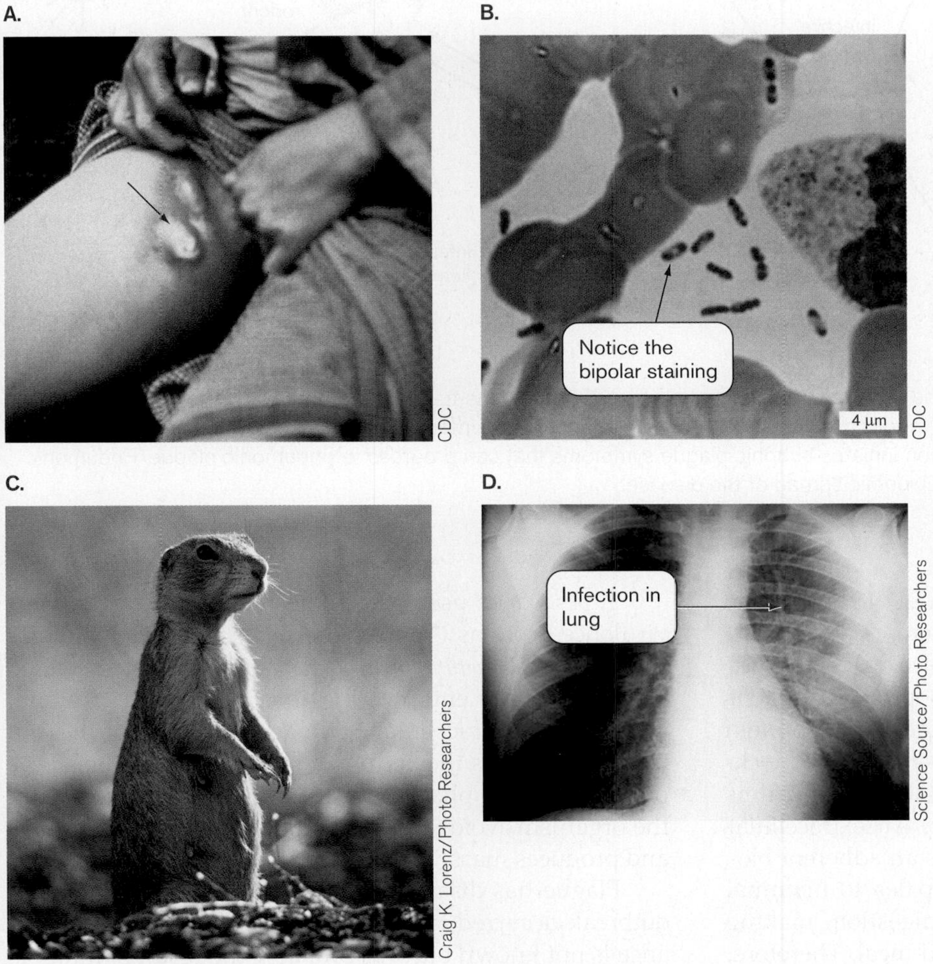

Plague is caused by the bacterium *Yersinia pestis*, which can infect both humans and animals. During the Middle Ages, the disease, known as the Black Death, decimated over a third of the population of Europe. Such was the horror it evoked that invading armies would actually catapult dead plague victims into embattled fortresses. This was probably the first case of biowarfare.

Contrary to popular belief, *Y. pestis* is present in the United States. The organism is endemic in 17 western states. It is normally transmitted from animal to animal, typically rodents like rats and even prairie dogs (**Fig. 26.23C**) by the bite of infected fleas. **Figure 26.24** illustrates the various infective cycles of the plague bacillus. Humans are not typically part of the natural infectious cycle. However, in the absence of an animal host, the flea can take a blood meal from humans and thereby transmit the disease to them. During the Middle Ages, urban rats venturing back and forth to the countryside became infected by the fleas of wild rodents that served as a reservoir. Upon returning to the city, the rat flea passed the organism on to other rats, which then died in droves. The rat fleas, deprived of their normal meal, were forced to feed on city dwellers, passing the disease on to them.

Individuals bitten by an infected flea or accidentally infected through a cut while skinning an infected animal first exhibit the symptoms of **bubonic plague**. Bubonic plague emerges as the organism moves from the site of infection to the lymph nodes, producing characteristically enlarged nodes called buboes (see **Fig. 26.23A**). From the lymph nodes, the pathogen can enter the bloodstream, causing **septicemic plague**. In this phase, the patient can go into shock from the massive amount of endotoxin in the bloodstream. Neither bubonic nor septicemic plague is passed from person to person. As the organism courses through the bloodstream, however, it will invade the lungs and produce **pneumonic plague**, which can be easily transmitted from person to person through aerosol droplets generated by coughing (**Fig. 26.23D**). Pneumonic plague is the most dangerous form of the disease because it can kill quickly and spread rapidly through a population. Pneumonic plague is so virulent

Figure 26.23 The plague. A. Classic bubo (swollen lymph node) of bubonic plague. **B.** *Yersinia pestis*, bipolar staining (length 1–3 μm). **C.** Prairie dogs are often hosts to fleas that carry plague bacilli. **D.** X-ray of pneumonic plague, showing bilateral pulmonary infection.

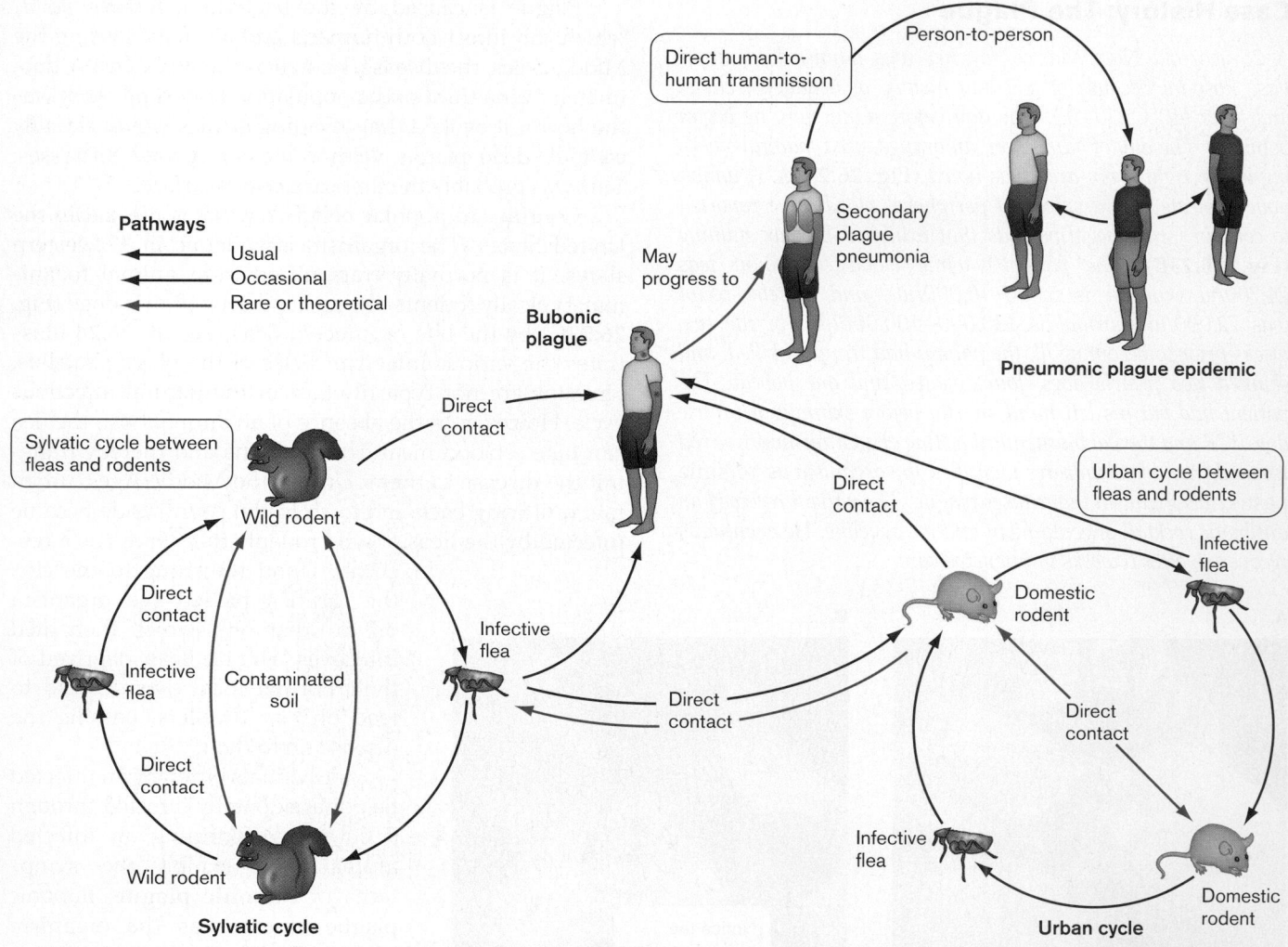

Pathways

◄──── Usual
◄──── Occasional
◄──── Rare or theoretical

Direct human-to-human transmission

Person-to-person

Secondary plague pneumonia

May progress to

Pneumonic plague epidemic

Bubonic plague

Direct contact

Sylvatic cycle between fleas and rodents

Wild rodent

Direct contact

Infective flea

Contaminated soil

Direct contact

Wild rodent

Sylvatic cycle

Infective flea

Direct contact

Direct contact

Direct contact

Urban cycle between fleas and rodents

Infective flea

Domestic rodent

Direct contact

Infective flea

Domestic rodent

Urban cycle

Figure 26.24   The cycles of plague.   The sylvatic cycle occurs in the wild, where fleas transmit the organism between rodents. An accidental interaction with urban rats can trigger a similar urban cycle. Humans can be infected through contact with infected fleas coming from either cycle. Flea bite transmission initiates bubonic plague symptoms that can progress to pneumonic plague. Pneumonic plague is highly infectious, which can cause epidemic spread of the disease.

that an untreated patient can die within 24–48 hours. The organism is usually identified postmortem.

The organism has numerous virulence factors, including so-called V and W cell surface lipoprotein antigens that inhibit phagocytosis. Another factor, the F1 protein capsule surface antigen, is partly responsible for blocking phagocytosis in mammalian hosts. Certain biofilms formed by *Y. pestis* are also important. An extracellular matrix synthesized by *Y. pestis* produces an adherent biofilm in the flea midgut that contributes to flea-to-mammal transmission. The biofilm blocks flea digestion, making the flea feel "starved" even after a blood meal. Therefore, the flea jumps from host to host in a futile effort to feel full. As the flea tries to take a bloodmeal, the blockage causes the insect to regurgitate bacteria into the wound. This curious effect of *Y. pestis* on the insect vector is another unique aspect of how plague spreads so quickly.

Y. pestis also uses type III secretion systems to inject virulence proteins (YopB and YopD) into host cell membranes. Unlike *Salmonella*, which uses type III–secreted proteins to gain entrance into host cells, *Y. pestis* is not primarily an intracellular pathogen. Injection of the Yop proteins disrupts the actin cytoskeleton and so helps the organism evade phagocytosis. By evading phagocytosis, the organism avoids triggering an inflammatory response and produces massive tissue colonization.

Plague has disappeared from Europe; the last major outbreak occurred in 1772. The reason for its disappearance is not known but was probably the result of multiple factors, not the least of which was human intervention. Although it wasn't until the nineteenth century that doctors understood how germs could cause disease, Europeans recognized by the sixteenth century that plague was contagious and could be carried from one area to another.

Beginning in the late seventeenth century, governments created a medical boundary, or *cordon sanitaire*, between Europe and the areas to the east from which epidemics came. Ships traveling west from the Ottoman Empire were forced to wait in quarantine before passengers and cargo could be unloaded. Those who attempted to evade medical quarantine were shot.

Case History: Lyme Disease

*Brad, a 9-year-old from Connecticut, developed a fever and a large (8-cm) reddish rash with a clear center (**erythema migrans**) on his trunk (**Fig. 26.25A**). He also had some left facial nerve palsy. Brad had returned a week previously from a Boy Scout camping trip to the local woods, where he did a lot of hiking. When asked by his physician, Brad admitted finding a tick on his stomach while in the woods but thinking little of it. The doctor ordered serological tests for Borrelia burgdorferi (the organism that causes Lyme disease), Rickettsia rickettsii (which produces Rocky Mountain spotted fever), and Ehrlichia equi (which causes ehrlichiosis). The ELISA test for B. burgdorferi came back positive, confirming a diagnosis of Lyme disease. The boy was given a 3-week regimen of doxycycline (a tetracycline derivative), which led to resolution of the rash and palsy.*

Lyme arthritis was first reported in Lyme, Connecticut, in the 1970s, but the causative organism, *Borrelia*

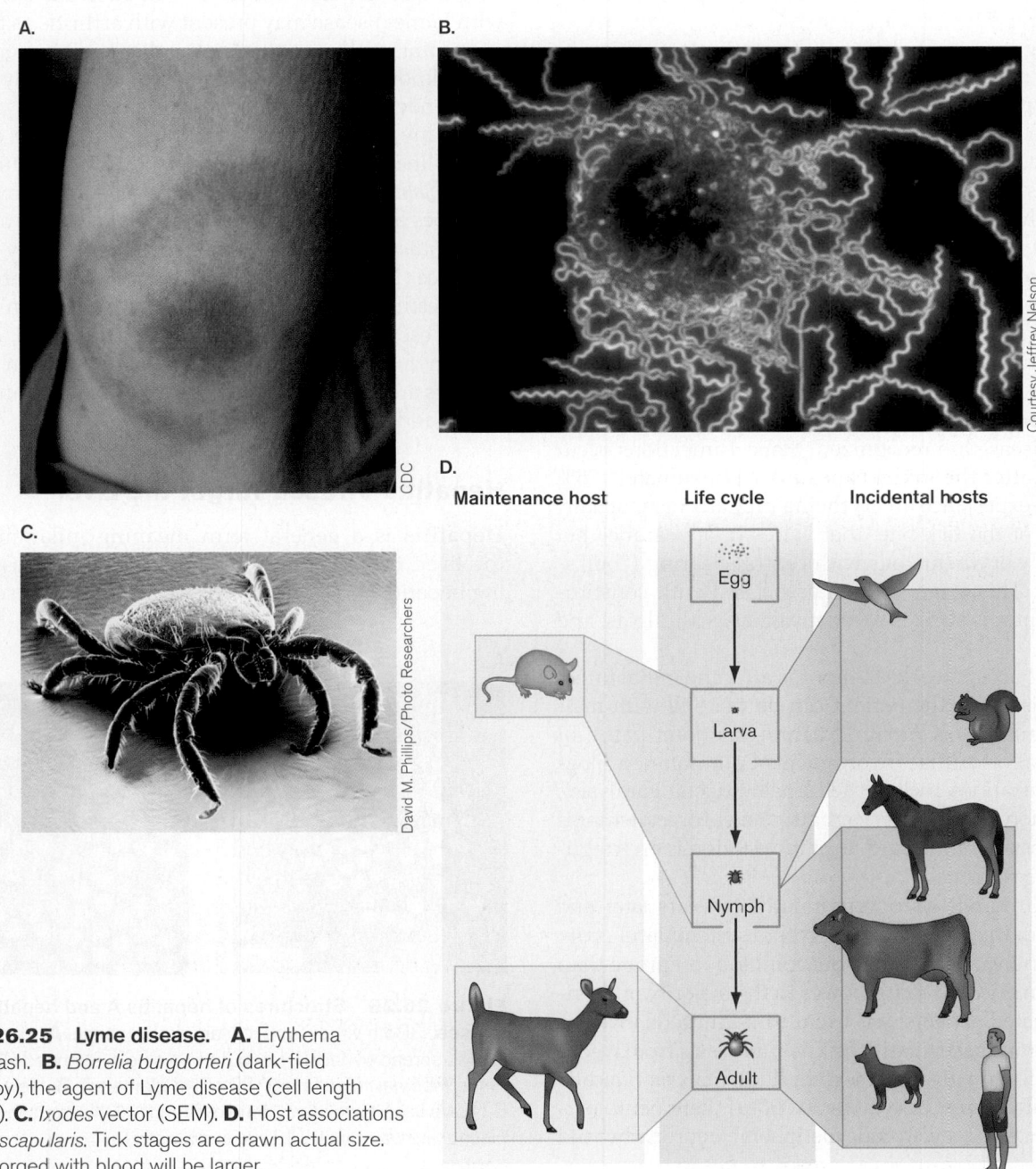

Figure 26.25 Lyme disease. A. Erythema migrans rash. **B.** *Borrelia burgdorferi* (dark-field microscopy), the agent of Lyme disease (cell length 5–30 μm). **C.** *Ixodes* vector (SEM). **D.** Host associations of *Ixodes scapularis*. Tick stages are drawn actual size. Ticks engorged with blood will be larger.

burgdorferi, was not identified until 1982. Since then, Lyme disease (aka **borreliosis**) has become the most common vector-borne illness in the United States and is considered an emerging infectious disease. The main endemic areas are the northeastern coastal area from Massachusetts to Maryland, Wisconsin and Minnesota, and northern California and Oregon. Lyme disease is also common to parts of Europe, such as Sweden, Germany, Austria, Switzerland, and Russia.

B. burgdorferi is a spirochete (**Fig. 26.25B**) transmitted to humans by ixodid ticks (hard ticks; **Fig. 26.25C** and **D**). In the northeastern and central United States, where most cases occur, the deer tick *Ixodes scapularis* transmits the spirochete, usually during the summer months. In the western United States, *I. pacificus* is the tick vector.

During its nymphal stage (the stage after taking its first blood meal), *I. scapularis* is the size of a poppy seed. Its bite is painless, so it is easily overlooked. Infection takes place when the tick feeds because the spirochete is regurgitated into the host. However, the organism grows in the tick's digestive tract and takes about 2 days to make its way to the tick's salivary gland, so if the tick is removed before that time, the patient will not be infected. Once it is transferred to the human, the microbe can travel rapidly via the bloodstream to any area in the body, but it prefers to grow in skin, nerve tissue, synovium (joint lining), and the conduction system of the heart.

Lyme disease has three stages. Three general stages of Lyme disease are recognized. Stage 1 infections occur 3–30 days after the initial exposure. Approximately 75% of patients experience an erythema migrans rash, usually at the site of the tick bite, that varies in appearance but is classically erythematous with central clearing ("bull's-eye" rash). This stage is often associated with constitutional symptoms such as fever, myalgias, arthralgias, and headache.

Stage 2 occurs weeks to months after the initial infection. In this stage, the patient can be quite ill with malaise, myalgias (muscle pain), arthralgias (joint pain), or neurological or cardiac involvement. Common neurological manifestations include Bell's palsy (facial paralysis), inflammation of spinal nerve roots, and chronic meningitis. The most common cardiac manifestation is an irregular heart rhythm.

Stage 3 borreliosis occurs months to years later and can involve the synovium, nervous system, and skin, though skin involvement is more common in Europe than in North America. Arthritis occurs in the majority of previously untreated patients; it is usually intermittent, involves the large joints, particularly the knee, and lasts from weeks to months in any given joint. Joint fluid analysis typically shows a WBC count of 10,000–30,000/μl. Late neurological involvement may include peripheral neuropathy and

encephalopathy, manifested as memory, mood, and sleep disturbances.

Treatment with antibiotics is recommended for all stages of Lyme disease but is most effective in the early stages. Treatment for early Lyme disease (stage 1) is a course of doxycycline for 14–28 days. Lyme arthritis is typically slow to respond to antibiotic therapy. Despite antimicrobial drug treatment, patients with persistent active arthritis and persistently positive PCR tests may have incomplete microbial eradication and may be the most likely to benefit from repeated treatment with injected antibiotics.

Curiously, 25% of people infected with *B. burgdorferi* never experience erythema migrans, and many infected individuals are also unsure of tick bites. Hence, patients with Lyme disease may present with arthritis as their first complaint. Although this makes diagnosis extremely difficult, knowing that the patient lives or recently traveled to an endemic area can provide a critical clue.

Many other bacteria can cause septicemia and systemic illness (**Table 26.7**). Gram-negative organisms like *E. coli*, *Salmonella* Typhi, and *Francisella*, and Gram-positive microbes like *Staphylococcus aureus*, *Enterococcus*, and *Bacillus anthracis*, can grow in the bloodstream if they can gain entrance. Even anaerobes that are normal inhabitants of the intestine (for example, *Bacteroides fragilis*) can be lethal if they escape the intestine and enter the blood, as might happen following surgery. This is one reason surgical patients are given massive doses of antibiotics immediately before and after surgery.

Hepatitis Viruses Target the Liver

Hepatitis is a general term meaning inflammation of the liver. Hepatitis is caused by several viruses, including hepatitis A, B, and C viruses. Although these viruses

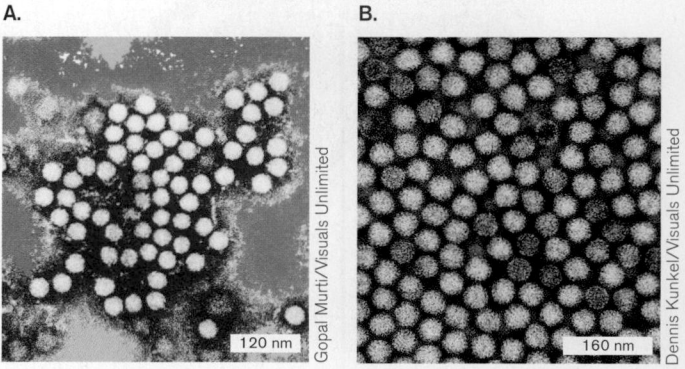

A.   **B.**

Figure 26.26   Structures of hepatitis A and hepatitis B viruses.   Both viruses are icosahedral in shape. **A.** Hepatitis A (TEM), spread by fecal-oral route, is a single-stranded RNA virus in the *Picornaviridae* family (30 nm in diameter). **B.** Hepatitis B (TEM) is an enveloped, double-stranded DNA virus in the *Hepadnaviridae* family (40 nm in diameter).

Table 26.7 Selected microbes that cause systemic disease.

Disease	Symptoms	Etiological agent	Virulence properties	Source	Treatment	Vaccine
Lyme disease	Stage 1: rash; stage 2: chills, headache, malaise, systemic involvement; stage 3: neurological changes	*Borrelia burgdorferi* (spirochete)	Antigenic variation, OspE (binds complement)	Deer tick	Penicillin, tetracycline	Yes
Brucellosis	Fever, weakness, sweats, splenomegaly, osteomyelitis, endocarditis, others	*Brucella abortis* (G– rod)	Intracellular, growth in monocytes	Animal products, unpasteurized milk	Doxycycline	Yes, for animals
Leptospirosis	Fever, photophobia, headache, abdominal pain, skin rash, liver involvement, jaundice	*Leptospira interrogans* (spirochete)	Burrowing motility	Urine of infected animals	Erythromycin, penicillin	Yes, for animals
Epidemic typhus	Chills, fever, headache, muscle pain, splenomegaly, coma	*Rickettsia prowazekii* (G– rod)	Obligate intracellular growth, escapes phagosome	Human louse, flying-squirrel flea	Tetracycline, chloramphenicol	
Tularemia	Fever, chills, headache, muscle pain, rash, bacteremia	*Francisella tularensis* (G– rod)	Intracellular	Rabbits, rodents, insect vectors	Gentamicin, streptomycin	Yes, but not currently in the U.S.
Typhoid fever	Septicemia, chills, fever, hypotension, rash (rose spots)	*Salmonella* Typhi (G– rod)	Type III secretion, intracellular growth, PhoPQ regulators, Vi antigen capsule	Gallbladder of human carrier	Ciprofloxacin, ceftriaxone	Vi antigen
	Septicemia, chills, fever, hypotension	*Salmonella choleraesuis* (G–) rod)	Intracellular growth, invasin	Animals, poultry	Ceftriaxone	
Vibriosis	Serious with immunocompromised patients; fever, chills, multi-organ damage, death	*Vibrio vulnificus* (G– curved rod)	Cytolysin, capsule	Seawater, raw oysters	Tetracycline plus aminoglycoside	
Bubonic plague	Buboes (swollen lymph glands), high fever, chills, headache, cough, pneumonia, septicemia	*Yersinia pestis* (G– rod)	Intracellular growth, type III secretion of YOPs (*Yersinia* outer proteins), phospholipase D, toxin	Rodents, rodent fleas, human respiratory aerosol, potential bioterrorism agent	Streptomycin or tetracycline	Yes, but not available in the U.S.

are members of very different families, they all target the liver. We include them in the section on systemic infections because their infectious routes take them to the bloodstream before arriving at the liver.

Hepatitis A virus (HAV) is a single-stranded RNA picornavirus (**Fig. 26.26A**) that causes an acute infection spread person-to-person by the fecal-oral route, but hepatitis A can also result from eating undercooked shellfish collected from contaminated waters. The virus replicates in the intestinal endothelium and is disseminated via the bloodstream to the liver. After replicating in hepatocytes, the progeny enter the bile and are released into the small intestine, explaining why stools are so infectious. Though the virus has an early viremic stage after leaving the intestine, it is rarely transmitted by transfusion because the viremic stage is transient, ending after development of liver symptoms. In contrast, hepatitis B and C viruses produce persistent viremia and are readily transmitted by transfusion.

Many people who are infected with HAV are asymptomatic or exhibit very mild symptoms that include nausea, vomiting, diarrhea, low-grade fever, and fatigue. As the virus attacks the liver, patients may become jaundiced (from accumulation of bilirubin in the skin), and their urine will turn dark brown. There is no specific treatment, but the disease usually lasts for only a few months and then resolves without establishing a carrier state. Disease can, however, be prevented if immunoglobulin is given to someone who has had contact with an infected individual. There is an inactivated hepatitis A vaccine available called HepA vaccine, which is administered after 1 year of age. For those not vaccinated, frequent hand washing is important for preventing the spread of the disease because it interrupts the fecal-oral cycle.

In contrast to HAV, hepatitis B virus (HBV) is a partially double-stranded circular DNA virus (family *Hepadnaviridae*) that causes diseases of varying severity. These include acute and chronic hepatitis, cirrhosis, and hepatocarcinoma. The virus also wears a membrane envelope donned when progeny viruses are released from infected cells. The virion coat protein, a surface antigen, is called HBsAg. The virus makes an excess amount of HBsAg, so it is sometimes extended as a tubular tail on one side of the virus particle and is often found in the blood of infected individuals in the form of noninfectious filamentous and spherical particles (**Fig. 26.26B**). The presence of HBsAg in blood is an indicator of HBV infection.

HBV is transferred primarily via blood transfusions, contaminated needles shared by IV drug users, and any human body fluid (saliva, semen, sweat, breast milk, tears, urine, feces). It can be transferred transplacentally to a fetus and can be sexually transmitted. Infection by HBV has two stages: a short-term acute phase and a long-term chronic phase that, if it extends beyond 6 months, may never resolve. Symptoms resemble those of the flu but with jaundice and brown urine. Liver damage caused by HBV infection is in large part due to an efficient cell-mediated immune response. Cytotoxic T cells and natural killer cells cause immune lysis of infected liver cells. Over the long term, chronic hepatitis will lead to a scarred and hardened liver (cirrhosis), the only recourse being a liver transplant. Fortunately, about 90% of those infected are able to fight off infection and never proceed to the chronic stage. A HepB vaccine (made from recombinant HBsAg) is available. Its administration is recommended after birth, followed by booster shots administered by 2 months and 18 months of age.

Hepatitis C virus (HCV) causes another form of hepatitis. HCV is a single-stranded linear DNA virus with a lipid coat; it is a member of the *Flaviviridae* family. It is transmitted by blood transfusions and causes 90% of transfusion-related cases of hepatitis. It can also be transmitted by needle sticks and, less frequently, by

sex. Over 100 million people worldwide are infected with HCV. Most HCV-infected individuals (80%) do not exhibit symptoms, and in those that do, symptoms may not appear for 10–20 years. At least 75% of patients that exhibit symptoms ultimately progress to chronic hepatitis requiring a liver transplant. Fortunately, infection can be detected using ELISA. Liver biopsies of HCV patients are used to determine the extent of liver damage, which in turn helps establish the stage of disease.

Prevention of HBC or HCV infection for health care personnel includes avoiding inadvertent needle sticks and, if one should occur, the administration of immunoglobulin within 7 days. Chronic hepatitis can be treated with interferon but only with modest success. Though vaccines have been developed for HAV and HBV, no vaccine is yet available for HCV. It is important to note that since hepatitis viruses can be spread via contaminated blood products, all blood donations collected by the Red Cross and other agencies are tested for the presence of these viruses, as well as for HIV. Thus, the blood supply is safe.

Ebola–The Perfect Pathogen or Too Deadly for Its Own Good?

How would you define the perfect pathogen? Would it be an organism that can kill its host with terrifying ease and quickness? If so, Ebola virus would fit the description. Ebola virus, a lipid-enveloped, threadlike RNA virus (*Filoviridae*; Fig. 25.4), was first associated with an outbreak of 318 cases of a hemorrhagic disease in Zaire. Of the 318 people who contracted the disease, 280 of them died within days. The disease was characterized by acute (rapid) onset of fever, severe muscle pains, horrible bleeding from multiple orifices (nose, mouth, anus, and vagina), and ultimately death. In that same year (1976), 284 people in Sudan were infected with the virus, and 156 of them died. The Ebola virus has a frightening reputation. It spreads like wildfire through the body after infection, causing severe hemorrhagic fever, and typically kills 90% of its victims. Internal bleeding results in shock and acute respiratory distress, leading to death.

The symptoms of Ebola (and of a related disease caused by the Marburg virus) reflect subversion of the innate immune system coupled with uncontrolled viral replication, particularly in macrophages and dendritic cells. Ebola virus infection of these cells enhances production of pro-inflammatory cytokines, such as TNF-α, and inhibits stimulation of T-cell maturation by dendritic cells. Thus, Ebola infections stimulate inflammatory processes leading to tissue damage but shut down early immune responses and prevent activation of adaptive immune responses, which allows unfettered viral replication.

Ebola viral proteins and their locations in the virion are shown in **Figure 26.27**. Ebola VP35 protein is a component of the viral RNA polymerase complex, but it is

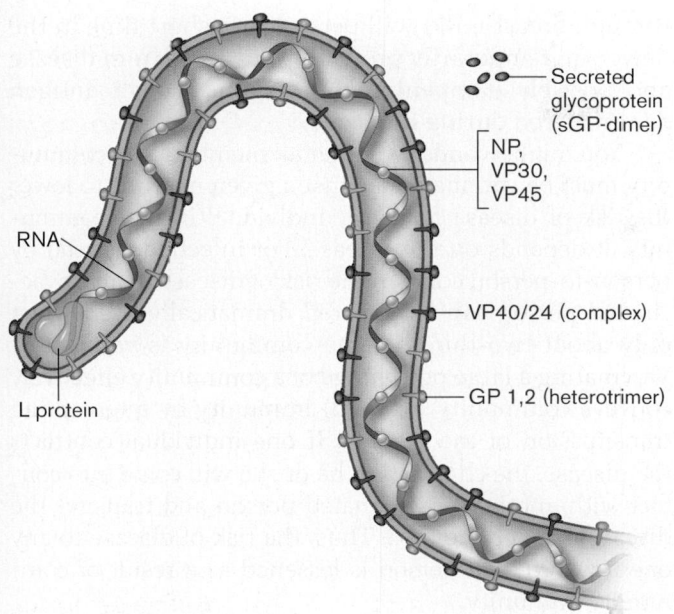

RNA

L protein

Secreted
glycoprotein
(sGP-dimer)

NP,
VP30,
VP45

VP40/24 (complex)

GP 1,2 (heterotrimer)

Figure 26.27   Ebola virion.   Composition of the virus. The ribonucleoprotein complex consists of the nucleoprotein (NP), the structural proteins VP30 and VP35, and the virion-associated RNA-dependent RNA polymerase (L). The glycoprotein (GP-sGP) is an integral membrane protein that can be secreted.

also a potent inhibitor of host interferon (IFN) production. The cellular response to whatever IFN is made is inhibited by VP24, which blocks the nuclear accumulation of a regulatory protein called STAT1. STAT1 is critical to IFN-stimulated gene expression. These and other strategies allow rapid replication of the virus.

After replicating, Ebola offspring sprout from the cell surface in a mass of tangled threads (as seen in the chapter opener photo). These new virions go on to attack new cells, riddling blood vessels and organs with damage as they go. Rapid release of new virions involves subverting another host mechanism, tetherin, designed to slow viral spread. Paul Bates and his colleagues at the University of Pennsylvania discovered that the cellular protein tetherin essentially *tethers* mature virus particles inside a cell so that they are unable to spread. Tetherin is IFN induced and can restrict the spread of structurally diverse enveloped viruses, including HIV (thus, it is part of the innate response). Ebola glycoprotein, however, counteracts tetherin so that nothing slows down viral spread.

The outlook for a patient infected with Ebola is dire. The incubation period is 4–16 days, and death occurs within 7–16 days. There are no effective drugs or vaccines. Administering blood plasma from people who have recovered, anticoagulation agents to reduce hemorrhaging, and interferon have had limited success. Chemical inhibitors of the host cathepsin proteases, which are required for viral replication, have been suggested as one possible antiviral therapy. Ebola enters a cell when the virus

membrane glycoprotein attaches to host membranes. The virus is then taken up in an endosome. Cathepsins in the endosome cleave the viral glycoprotein and allow the virus membrane to fuse with the endosome membrane, releasing the uncoated virus into the cytoplasm. Cathepsin inhibitors prevent release and, thus, replication. There is currently some hope that an attenuated, replication-deficient Ebola virus may serve as a vaccine.

Ebola epidemics result from person-to-person contact or from inadvertent laboratory exposures. Fortunately, Ebola outbreaks are self-contained because the viruses kill their victims quickly. Death comes before the virus can be transmitted to a new host. Some scientists therefore argue that efficient and quick killing is not the mark of a perfect pathogen. The better pathogen lets its host linger to ensure a home for itself and more opportunity to disseminate. So where does Ebola go when it is not infecting humans? The natural ecology of these viruses is largely unknown, although an association with monkeys and/or bats as possible reservoirs is suggested.

Global health authorities recently have become very concerned about a report of Ebola virus transmission from a pig to a farmer in the Philippines. Given that the particular Ebola strain in question, called Ebola-Reston, has never caused disease in humans (the pig farmer developed antibodies to the strain, not disease), why should the World Health Organization be so concerned? While Ebola was thought to be carried only in humans and monkeys, it now appears that pigs represent a new reservoir for the virus. In addition, animals often play an important ecological role in the evolution of new viruses that can infect humans. For instance, different influenza viruses mix their genomes in birds (avian flu) and pigs (swine flu). Finding an Ebola virus in pigs, albeit one that is innocuous to humans, raises the possibility that a new emerging strain of Ebola-Reston could jump to humans and cause disease.

TO SUMMARIZE:

- **Septicemia** is caused by many Gram-positive and Gram-negative bacterial pathogens. It can start with the bite of an infected insect, introduction via a wound, escape from an abscess, or penetration of the mucosal epithelium by the pathogen (as through the intestine or vagina); it can lead to disseminated, systemic disease.
- **Plague** has sylvatic and urban infection cycles involving transmission between fleas and rats.
- ***Yersinia pestis*–infected flea bites** lead to bubonic plague. Bubonic plague can progress to septicemic and pneumonic stages.
- **Aerosolized respiratory secretions** will directly spread *Y. pestis* pneumonic plague from person to person (no insect vector).

- **Lyme disease** is caused by the spirochete *Borrelia burgdorferi*, which is transmitted from animal reservoirs to humans by the bite of *Ixodes* ticks.
- **There are three stages of Lyme disease**: stage 1, a bull's-eye rash (erythema migrans); stage 2, joint, muscle, and nerve pain; stage 3, arthritis with WBCs in the joint fluid.
- **Hepatitis** is caused by several unrelated viruses; among them, HAV, HBV, and HCV account for most disease.
- **HAV** is transmitted by the fecal-oral route, does not establish chronic infection, and can be prevented by a vaccine.
- **HBV and HCV** can be transmitted by blood products (such as transfusions) and shared hypodermic needles and can lead to chronic hepatitis.
- **Vaccine for HBV**, but not HCV, is available.
- **Ebola virus** spreads from human to human and kills its victims quickly. Its viral proteins alter cytokine production and facilitate virus release from infected cells.

26.9 Immunization

In our discussion of various infections caused by bacteria and viruses, we often noted the availability of vaccines that prevent disease. In Chapter 24, we also listed many of the available vaccines and their basic makeup (see Table 24.1). As noted in that table, some vaccines utilize killed organisms (examples are the hepatitis A vaccine or the Salk inactivated polio vaccine), while others contain attenuated microbes (BCG for tuberculosis or the Sabin live polio vaccine) or consist of purified components of an infectious agent (*Streptococcus pneumoniae* and *Haemophilus influenzae* type b capsuluar antigens). Some vaccines are injected in combination as polyvalent vaccines to control multiple diseases (measles, mumps, and rubella), while others are given individually but may change from year to year (influenza viral capsid proteins).

Multiple vaccines can be given simultaneously and safely. Most vaccines are administered during childhood, when the diseases can be most devastating. **Table 26.8** provides the current immunization schedule recommended for children and adolescents by the Centers for Disease Control and Prevention. As you will notice, all of these vaccines are given in multiple doses, called booster doses. The exception to this rule is the influenza vaccine, which is given in a single dose but changes every year. The reason for multiple doses, as noted in Chapter 24, is that secondary exposure to an antigen provides a more robust and long-lasting immunity. However, most vaccines are not administered until 2 months of age because maternal antibody crossing through the placenta to the fetus (or to a newborn

through breast milk) will persist for a short time in the newborn, temporarily protecting the baby from disease and possibly dampening the response to an antigen administered during that time.

You might wonder whether all members of a community must be vaccinated against a given microbe to lower the risk of disease for every individual in that community. It depends on the disease. For infections spread by person-to-person contact, the risk of disease to an unvaccinated person can be lowered dramatically even when only about two-thirds of the community is vaccinated. Vaccinating a large percentage of a community effectively conveys community (or herd) immunity by interrupting transmission of the disease. If one individual contracts the disease, the chance that he or she will come into contact with another unvaccinated person and transmit the disease is much reduced. Thus, the risk of disease to any one unvaccinated person is lessened as a result of community immunity.

Herd immunity works well for diseases such as diphtheria, whooping cough (pertussis), measles, and mumps. However, herd immunity will not lower the risk of an unvaccinated person's contracting tetanus, which is not spread by person-to-person contact. *Clostridium tetani*, the agent whose toxin causes tetanus, is a ubiquitous soil organism transmitted through punctured skin. The risk of tetanus for an unvaccinated person doesn't change even if every other person in the community is vaccinated against tetanus.

The vast majority of people who receive vaccines suffer no, or only mild, reactions, such as fever or soreness at the injection site. Very rarely do more serious side effects, such as allergic reactions, occur. Vaccines are *extremely* safe. Unsettling, erroneous information circulates on the Internet purporting a link between vaccinations and other diseases, such as diabetes or autism, but no well-controlled scientific study supports these claims. Clearly, the risks, including death, associated with these preventable infectious diseases are far greater than the minimal risk associated with being vaccinated against them. As proof of this point, the undervaccination of children in the United States in the 1980s and 1990s has led to an increase in cases of whooping cough (25,616 in 2005), caused by *Bordetella pertussis* (cases did drop to 13,278 in 2008). Successful vaccination programs carried out in the United States have come close to eradicating many diseases once feared, such as polio, rubella, and diphtheria, and have dramatically lowered the morbidity and mortality of many others.

We acknowledge that the current risk of infection for some diseases (because they are so rare) is lower than the risk of an adverse reaction to immunization. However, failing to vaccinate, as just mentioned, will result in a population susceptible to the microbe and a resurgence of serious disease.

Table 26.8 Recommended childhood and adolescent immunization schedule, by vaccine and age—United States, 2009.

Vaccine Age →	Birth	1 month	2 months	4 months	6 months	12 months	15–18 months	24 months	4–6 years	11–12 years
Hepatitis B[a]	HepB	HepB			HepB			HepB series		
Rotavirus			Rota	Rota	Rota					
Diphtheria, tetanus, pertussis[b]			DTaP	DTaP	DTaP		DTaP		DTaP	Tdap
Haemophilus influenzae type b[c]			Hib	Hib	Hib[c]	Hib				
Inactivated poliovirus			IPV	IPV	IPV				IPV	
Measles, mumps, rubella						MMR			MMR	MMR
Varicella						Varicella			Varicella	
Meningococcal[d]										MCV4
										MCV4
Pneumococcal[e]			PCV	PCV	PCV	PCV		PCV		PPV
Influenza[f]					Influenza (yearly)			Influenza (yearly)		
Hepatitis A[g]						HepA series		HepA series		

This schedule indicates the recommended ages for routine administration of currently licensed childhood vaccines, as of December 1, 2005.

▓ Indicates age groups that warrant special effort to administer those vaccines not previously administered.

▓ Range of recommended ages   ▓ Catch-up immunization   ▓ Assessment at age 11–12 years

[a]Hepatitis B vaccine (HepB). *At birth:* All newborns should receive monovalent HepB, administered soon after birth and before hospital discharge.

[b]Diphtheria and tetanus toxoids and acellular pertussis vaccine (DTaP). Tdap is a modified vaccine with lower doses of diphtheria and pertussis toxoids.

[c]*Haemophilus influenzae* type b conjugate vaccine (Hib).

[d]Meningococcal vaccine (MCV4). Meningococcal conjugate vaccine (MCV4) should be administered to all children at age 11–12 years, as well as to unvaccinated adolescents at high school entry (age 15 years). Vaccine contains four types of capsules.

[e]Pneumococcal vaccine. The heptavalent pneumococcal conjugate vaccine (PCV) is recommended for all children 2–23 months and for certain children aged 24–59 months. The final dose in the series should be administered at age ≥12 months. Pneumococcal polysaccharide vaccine (PPV) is recommended in addition to PCV for certain high-risk groups. PPV contains 23-capsule antigens.

[f]Influenza vaccine. Influenza vaccine is recommended annually for children aged ≥6 months with certain risk factors (including, but not limited to, asthma, cardiac disease, sickle-cell disease, HIV infection, diabetes, and conditions that can compromise respiratory function or handling of respiratory secretions or that can increase the risk of aspiration), health care workers, and other persons (including household members) in close contact with persons in groups at high risk (see *Morbidity and Mortality Weekly Report* 2005; 54[No. RR-8]).

[g]Hepatitis A vaccine (HepA). HepA is recommended for all children at age 1 year (i.e., 12–23 months).

Source: www.cdc.gov/vaccines/recs/schedules

THOUGHT QUESTION 26.12 How might you design/construct a more effective vaccine for an antigen (for instance, the *Yersinia pestis* F1 antigen) that would harness the power of a Toll-like receptor (TLR)? *Hint:* TLR5 from Table 23.4.

TO SUMMARIZE:

- **Vaccines** can be made from live attenuated organisms, killed organisms, or purified microbe components.
- **Herd immunity** can help protect unimmunized persons from diseases transmitted person-to-person.
- **Serious side effects** very rarely result from immunizations.

Concluding Thoughts

This chapter described key concepts of infectious disease using just a small sample of disease-causing pathogens. Other diseases are considered elsewhere in the book or on the accompanying website. These include anthrax, cholera, the common cold (rhinovirus), dengue fever, diphtheria, gastric ulcers (*Helicobacter pylori*), herpes, acquired immunodeficiency syndrome (HIV), warts (human papillomavirus), influenza, salmonellosis, *Streptococcus agalactiae* infections, West Nile virus, and Whipple's disease (*Tropheryma whipplei*). The principal goal with this and the previous chapter was to illustrate how microbial metabolism can undermine the physiology of a human host. The next chapter will describe how humans fight back using pharmacology to sabotage the physiology of the infecting microbes.

CHAPTER REVIEW

Review Questions

1. How are pathogens classified by portal of entry?
2. Discuss some common skin infections.
3. What causes boils?
4. What are the symptoms of necrotizing fasciitis?
5. What is the difference between primary and secondary infections?
6. How do pneumococci avoid engulfment by phagocytes?
7. What causes the clouding seen in X-rays of infected lungs?
8. What are the key features of the pneumococcal vaccine?
9. Name the common fungal causes of lung disease.
10. What is a metastatic lesion?
11. Why is diarrhea watery?
12. What is the most common microbial cause of diarrhea?
13. List common bacterial agents that cause diarrhea.
14. Would you suspect *Salmonella* infection in a cluster of nauseated patients rushed to the hospital directly from a church picnic? Why or why not?
15. What is the significance of finding leukocytes in stool?
16. What is one reservoir of *E. coli* O157:H7?
17. How is a UTI diagnosed?
18. What is an important virulence determinant of uropathogenic *E. coli*?
19. What is the most common sexually transmitted disease?
20. How is gonorrhea different in men and women?
21. Why will *Neisseria gonorrhoeae* not usually disseminate in the bloodstream, while *N. meningitidis* will?
22. Name the major causes of bacterial meningitis. What are the two routes of infection?
23. If tetanus and botulism toxins have the same mode of action, why do they cause opposite effects on muscles?
24. What are some virulence factors of *Yersinia pestis*?

Thought Questions

1. Chickenpox is a disease of children and young adults caused by herpes virus 3 (varicella). It is characterized by a rash of fluid-filled vesicles that eventually become crusty. The rash starts on the trunk and spreads to the extremities. Illness usually resolves in 7–10 days, and the patient becomes immune to the disease. However, some individuals later in life (>60 years old) develop a painful disease called shingles caused by the same virus—even if they have never been reexposed to the virus. The lesions are tender, persistent vesicles that form on the skin. If such patients were immune when younger, hypothesize about how they contracted shingles and why the lesions are painful.

2. A 5-year-old male was brought to the emergency room by his grandmother, who had found the boy on the floor of her apartment covered in bloody, loose feces. Patient history revealed the boy attended day care regularly. The diagnostic laboratory determined the etiological agent to be a Gram-negative rod. This organism is a facultative intracellular pathogen that escapes the host cell vacuole and moves in and between cells by actin polymerization. From your readings in Chapters 25 and 26, what is the most likely genus and species? Why is the fact that the boy attended day care significant?

3. Why are urinary tract infections among the most commonly acquired nosocomial (hospital-acquired) infections?

4. Visit the MMWR (Morbidity and Mortality Weekly Report) page of the CDC website that provides the "Summary of Notifiable Diseases" (the URL as of this writing is www.cdc.gov/mmwr/mmwr_nd). Study the table that summarizes the monthly incidence of infections in the United States and compare the data for gonorrhea and Lyme disease. Explain the differences in monthly incidence of these two diseases and then view the incidence of those diseases in your state.

Key Terms

bacteremia (1012)
blood-brain barrier (1005)
borreliosis (1018)
botulism (1006)
bubonic plague (1015)
chancre (1000)
chlamydia (1001)
congenital syphilis (1001)
endocarditis (1012)

erythema migrans (1017)
hepatitis (1018)
metastatic lesion (989)
necrotizing fasciitis (982)
pneumonic plague (1015)
primary syphilis (1000)
prion (1009)
rotavirus (994)
secondary syphilis (1000)

septicemia (1012)
septicemic plague (1015)
sequela (982)
spongiform encephalopathy (1011)
tertiary syphilis (1001)
tetanospasmin (1006)
transcytosis (1006)
viremia (1012)
zoonotic disease (979)

Recommended Reading

Broudy, Thomas, and Vincent Fischetti. 2003. In vivo lysogenic conversion of Tox⁻ *Streptococcus pyogenes* to Tox⁺ with lysogenic streptococci or free phage. *Infection and Immunity* **71**:3782–3786.

Clark, Ian A., Alison C. Budd, Lisa M. Alleva, and William B. Cowden. 2006. Human malarial disease: A consequence of inflammatory cytokine release. *Malaria Journal* 5:85.

Duerr, Ann, Judith N. Wasserheit, and Lawrence Corey. 2006. HIV vaccines: New frontiers in vaccine development. *Clinical Infectious Diseases* 43:500–511.

Hizume, Masaki, Atsushi Kobayashi, Kenta Teruya, Hiroaki Ohashi, James W. Ironside, et al. 2009. Human prion protein (PrP) 219K is converted to PrPSc but shows heterozygous inhibition in variant Creutzfeldt-Jakob disease infection. *Journal of Biological Chemistry* 284:3603–3609.

Kaletsky, Rachel L., Joseph R. Francica, Caroline Agrawal-Gamse, and Paul Bates. 2009. Tetherin-mediated restriction of filovirus budding is antagonized by the Ebola glycoprotein. *Proceedings of the National Academy of Sciences USA* 106:2886–2891.

Kim, Kwang Sik. 2006. Microbial translocation of the blood-brain barrier. *International Journal for Parasitology* 36:607–614.

McGowan, Sheena, Corrine J. Porter, Jonathan Lowther, Colin M. Stack, Sarah J. Golding, et al. 2009. Structural basis for the inhibition of the essential *Plasmodium falciparum* M1 neutral aminopeptidase. *Proceedings of the National Academy of Sciences USA* **106**:2537–2542.

Mizel, Steven B., Aaron H. Graff, Nammalwar Sriranganathan, Sean Ervin, Cynthia J. Lees, et al. 2009. Flagellin-F1-V fusion protein is an effective plague vaccine in mice and two species of nonhuman primates. *Clinical and Vaccine Immunology* 16:21–28.

Perrin, Agnès, Stéphane Bonacorsi, Ettiene Carbonnelle, Driss Talibi, Philippe Dessen, et al. 2002. Comparative genomics identifies the genetic islands that distinguish *Neisseria meningitidis*, the agent of cerebrospinal meningitis, from other *Neisseria* species. *Infection and Immunity* 70:7063–7072.

Petersen, Andreas M., and Karen A. Krogfelt. 2003. *Helicobacter pylori*: An invading microorganism? A review. *FEMS Immunology and Medical Microbiology* 36:117–126.

Russell, David G. 2007. Who puts the tubercle in tuberculosis? *Nature Reviews. Microbiology* 5:39–47.

Trevejo, Rosalie T., Margaret C. Barr, and Robert Ashley Robinson. 2005. Important emerging bacterial zoonotic infections affecting the immunocompromised. *Veterinary Research* 36:493–506.

Zhou, Dongsheng, and Ruifu Yang. 2009. Molecular Darwinian evolution of virulence in *Yersinia pestis*. *Infection and Immunity* 77:2242–2250.

Zinkernagel, Annelies S., and Victor Nizet. 2007. *Staphylococcus aureus*: A blemish on skin immunity. *Cell Host & Microbe* 1:161–162.

 For further study and review, please visit StudySpace at **microbiology2.com**.

Chapter 27

Antimicrobial Chemotherapy

The discovery of antibiotics about 80 years ago has played a major role in increasing life expectancy throughout the world. Prior to 1918, the average life expectancy in the United States was 45–50 years. That number is now between 70 and 75 years, thanks in part to antibiotics. But antibiotics may soon become useless. Over several decades, antibiotics have been used indiscriminately to treat patients, and they are often included in animal feed—not to treat animals but to grow larger ones. These abuses, combined with the ability of bacteria to become antibiotic resistant, have led experts to predict an impending crisis in which the human race is vulnerable to infectious diseases we once thought had been conquered.

Important questions about chemotherapeutic agents will be addressed in this chapter, including: Why do antimicrobials inhibit the growth of bacteria, but not humans or animals? What is antibiotic resistance? How do clinicians know which antibiotic to use to treat an infection? We will also discuss what makes a good antibiotic target, and how new antibiotics are discovered.

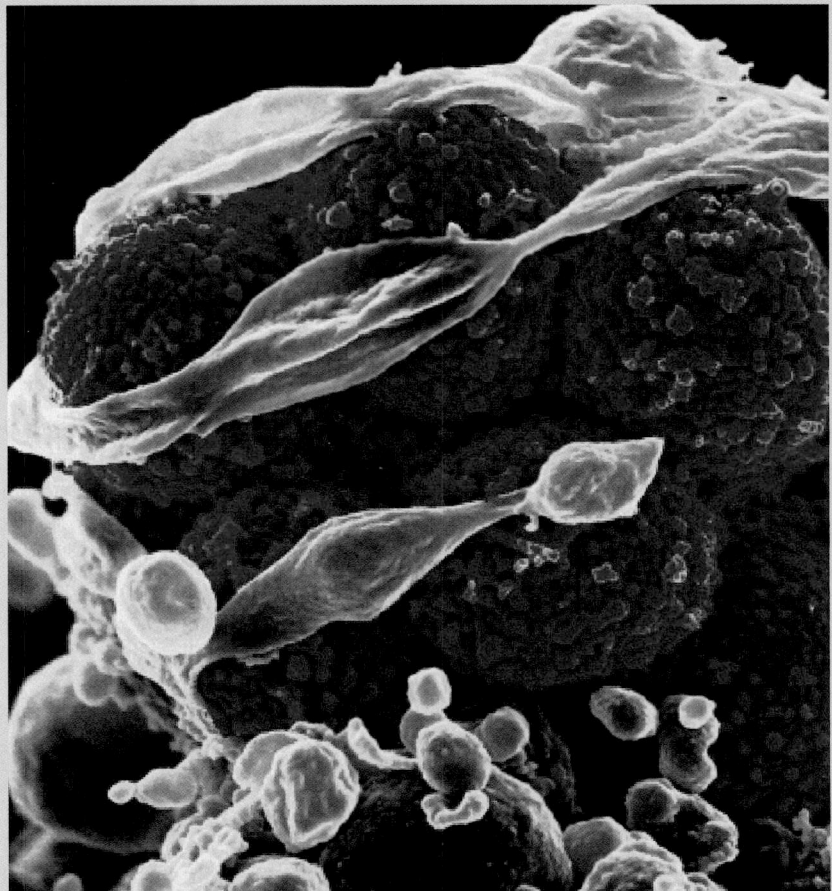

Current Research Highlight

The "superbug," methicillin-resistant *Staphylococcus aureus* (MRSA). MRSA is a highly pathogenic strain of *S. aureus* that once caused only hospital-acquired infections. MRSA has now been unleashed into the community, causing many serious, life-threatening infections. The photomicrograph (SEM) depicts MRSA cells (shown in red) destroying a human neutrophil trying to kill the bacteria. As discussed in this chapter, methicillin resistance is the result of an altered cell wall biosynthesis enzyme that can no longer bind methicillin. As serious as MRSA is, it has become apparent that different strains of *S. aureus* vary in the severity of disease they cause. Scientists recently used comparative whole-genome sequencing to determine that community MRSA isolates emerged from a single strain that recently underwent clonal expansion and diversified into a variety of strains possessing different levels of virulence. *Source:* Adam D. Kennedy et al. 2008. *PNAS* cover image 105(4). © 2008 by The National Academy of Sciences of the U.S.

Antibiotics undeniably have been a benefit to modern society. We live longer and contribute more to society than in the past, in large measure because of antibiotics. Infections we consider minor today often killed their victims just 50 or 60 years ago. But there has been a downside. Consider the case of a 56-year-old man with diabetes who entered the hospital for a heart transplant. The operation went well, but one week after the surgery, he developed a severe chest wound infection notable for exuded pus. Treatment with methicillin, a penicillin derivative commonly used to treat infections, failed and the patient became comatose. The diagnostic microbiology laboratory ultimately identified the agent as a methicillin-resistant *Staphylococcus aureus* (MRSA), prompting the surgeon to immediately place the patient on intravenous vancomycin for 6 weeks. Vancomycin is an antibiotic, structurally different from methicillin, that can usually kill methicillin-resistant bacteria. Fortunately, this patient recovered; too often, they do not.

Where did the infection come from? Surprising as it may seem, this drug-resistant pathogen was a resident of the hospital itself. To find the source, nasal swabs were taken of all hospital personnel and screened for the presence of *S. aureus.* The results revealed that several members of the surgical team actually harbored this organism as part of their resident microbiota. But which individual was the actual source of the patient's infection? In what could be described as an example of forensic microbiology, each strain was subjected to pulsed-field gel electrophoresis, and the resulting genomic restriction patterns were compared with that of the isolate from the infected patient. A match pointed to the source. The unwitting culprit turned out to be the perfusionist who manipulated the tubing used for cardiopulmonary bypass.

Hospital-acquired, or **nosocomial**, infections are not unusual. As many as 5%–10% of all patients admitted to acute-care hospitals acquire nosocomial infections, resulting in over 80,000 deaths each year. The deaths are due, in part, to the poor health of the patient and in part to the antibiotic-resistant nature of bacteria lurking in hospitals. In fact, as much as 60%–70% of staph infections that develop in a hospital setting are the result of methicillin-resistant *S. aureus* (MRSA). A foreboding report from the United Kingdom finds that one in four nursing-home residents is colonized by MRSA. A recent study in the U.S. found that the noses of nearly one in five non-hospitalized people are also colonized by MRSA. This problem will only get worse.

We begin Chapter 27 with a discussion of the golden age of antibiotic discovery (1940–1960) and move on to describe the basic concepts of antibiotics and their use and misuse. We will also delve into the ways genomic and proteomic approaches broaden our ability to search for new antibiotics and identify new antibiotic targets— all part of our attempt to stay one step ahead of evolving, antibiotic-resistant pathogens.

27.1 The Golden Age of Antibiotic Discovery

Antibiotics (from the Greek meaning "against life") are compounds produced by one species of microbe that can kill or inhibit the growth of other microbes. While we think of antibiotics as being a recent biotechnological development, their use has historical precedent. Ancient remedies called for cloths soaked with organic material to be placed on wounds to allow them to heal faster. This organic material likely contained "natural antibiotics" that killed bacteria and prevented further infection. The medicinal properties of molds were also recognized for centuries. Historical accounts refer to the ancient Chinese successfully treating boils with warm soil and molds scraped from cheeses, and in England a paste of moldy bread was a home remedy for wound infections up until the beginning of the twentieth century.

The modern antibiotic revolution began with the discovery of **penicillin** in 1928 by Sir Alexander Fleming (1881–1955). This discovery was actually a rediscovery, and was arguably one of the greatest examples of serendipity in science. Although Fleming generally receives the credit for discovering penicillin, a French medical student, Ernest Duchesne (1874–1912), originally discovered the antibiotic properties of *Penicillium* in 1896.

Duchesne observed that Arab stable boys at the nearby army hospital kept their saddles in a dark and damp room to encourage mold to grow on them. When asked why, they told him the mold helped heal saddle sores on the horses. Intrigued, Duchesne prepared a solution from the mold and injected it into diseased guinea pigs. All recovered. Although he submitted his work as a dissertation to the Pasteur Institute, it was ignored because of his young age and because he was unknown.

Penicillium was forgotten in the scientific community until Fleming rediscovered it one day in the late 1920s. Petri dishes were glass in those days and could be rewashed and sterilized. Fleming was preparing to wash a pile of old petri dishes he had used to grow the pathogen *Staphylococcus aureus*. He opened and examined each dish before tossing it into a cleaning solution. He noticed that one dish had grown contaminating mold, which in and of itself was not unusual in old plates, but all around the mold the staph bacteria had failed to grow (**Fig. 27.1A**). Fleming took a sample of the mold and found it was from the penicillium family, later identified as *Penicillium notatum*. The mold appeared to have synthesized a chemical, now known as penicillin (**Fig. 27.1B**), that diffused through the agar, killing cells of *S. aureus* before they could form colonies. Fleming (**Fig. 27.1C**) presented his findings in 1929, but they raised little interest, since penicillin appeared to be unstable and would not remain active in the body long enough to kill pathogens.

A.

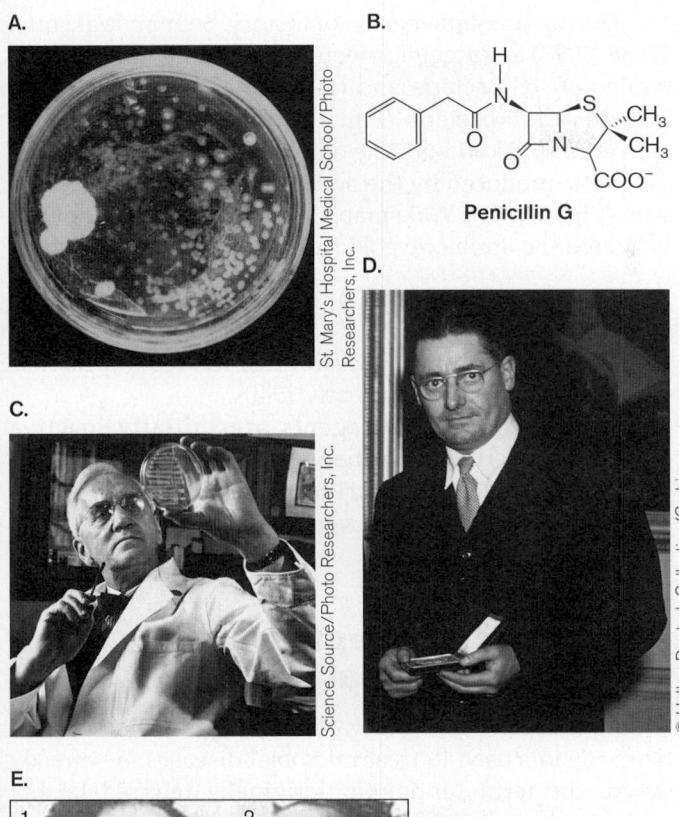

St. Mary's Hospital Medical School/Photo Researchers, Inc.

B.

Penicillin G

C.

Science Source/Photo Researchers, Inc.

D.

© Hulton-Deutsch Collection/Corbis

E.

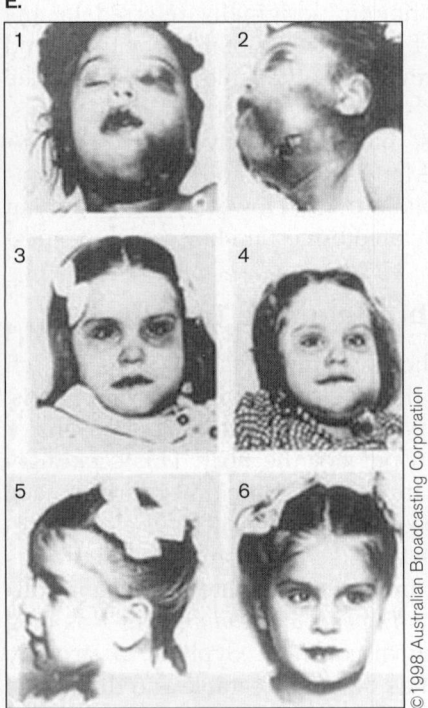

©1998 Australian Broadcasting Corporation

Figure 27.1 The dawn of antibiotics. A. Alexander Fleming's photo of the dish with bacteria and penicillin mold. **B.** The chemical structure of penicillin G. **C.** Alexander Fleming at work in his laboratory. **D.** Howard Florey. **E.** Pictures taken in 1942, shortly after the introduction of penicillin, show the improvement in a child suffering from an infection 4 days (panel 2) and 9 days (panel 4) after treatment. Panels 5 and 6 show her fully recovered.

As World War II began, an Oxford professor named Howard Florey (**Fig. 27.1D**), together with his colleague Ernst Chain, rediscovered Fleming's work, thought it held promise, and set about purifying penicillin. To their amazement, when the purified penicillin was injected into mice infected with staphylococci or streptococci, the majority of mice survived. Subsequent human trials also proved successful (**Fig. 27.1E**), and penicillin gained wide use, saving countless lives during the war. As a fitting tribute to serendipity, Fleming, Florey, and Chain received the 1945 Nobel Prize in Physiology or Medicine for their work. Duchesne's work, however, went unrecognized by the scientific community for decades.

The next landmark discovery in antibiotics was made by Gerhard Domagk (1895–1964), a German physician at the Bayer Institute of Experimental Pathology and Bacteriology who investigated antimicrobial compounds in the 1930s (**Fig. 27.2A**). In 1935, Domagk's 6-year-old child was afflicted with a serious streptococcal infection induced by an innocent pinprick to the finger. The infection spread to the lymph nodes under her arm and became so severe that lancing and draining the pus 14 times did little to help. The only remaining alternative was to amputate the arm. Unfortunately, even this drastic measure would probably not save her life. Frustrated, Gerhard Domagk took what would appear to be drastic measures. He administered a dose of an innocuous red dye he was investigating that, on agar plates (the usual medium for testing antibiotics), had shown absolutely no ability to inhibit the growth of streptococcus. Nevertheless, Domagk's daughter recovered completely.

How did Domagk conceive of such a therapy when the conventional method of screening for antibiotic activity on agar plates indicated that this compound (Prontosil) was useless? The answer lay in his prior use of laboratory animals to study the drug. When administered to mice, Prontosil was very effective at preventing infection. Had he not used live animals, Domagk would never have discovered that Prontosil is metabolized by the body into another compound, sulfanilamide, clearly lethal to the streptococcus. This finding led to an entire class of drugs, the **sulfa drugs**, which saved hundreds of thousands of lives, including that of Domagk's own child. Sulfanilamide is an analog of para-aminobenzoic acid (PABA), a precursor of folic acid, a vitamin necessary for nucleic acid synthesis (**Fig. 27.2B**). Sulfanilamide and other sulfur drugs bind to and inhibit the enzyme that converts PABA to folic acid. Without folic acid to make nucleic acid precursors, the cell stops growing.

Folic acid is not synthesized by humans but is a dietary supplement. Furthermore, bacteria do not transport folic acid, which explains why the sulfa drugs can inhibit bacterial growth in humans without affecting human cells.

A.

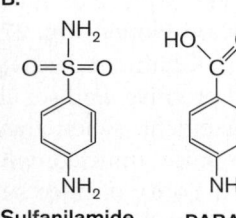

B.

Sulfanilamide   PABA

C.

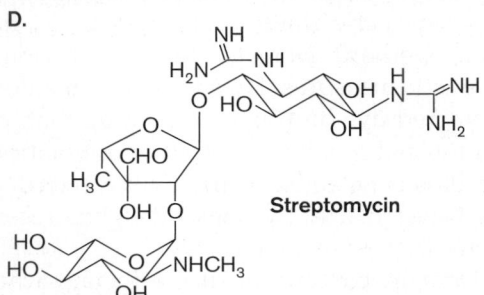

D.

Streptomycin

Figure 27.2   The discoverers of sulfanilamide and streptomycin.   A. Gerhard Domagk discovered sulfanilamide. **B.** Chemical structure of sulfanilamide, an analog of para-aminobenzoic acid (PABA), a precursor of the vitamin folic acid, necessary for growth. Sulfanilamide inhibits one of the enzymes that converts PABA into folic acid. **C.** Selman Waksman discovered streptomycin in 1944. **D.** Chemical structure of streptomycin.

In a dark turn, the pressing need for effective antibiotics during World War II dictated a change in the in vivo screening methods used by Domagk's German employer. Animal testing was abandoned in favor of human testing, but the humans were not volunteers. They were concentration camp prisoners intentionally infected with bacterial diseases, such as gangrene, and then treated with new chemical compounds. Domagk's possible participation in these experiments, even if unwilling, was a source of controversy that haunted him long after the war ended. Yet his contributions to medicine continued as he also developed the first effective chemotherapy for tuberculosis via the thiosemicarbazones and isoniazid, still used today.

During the same period of history, Selman Waksman (1888–1973) at Rutgers University began screening 10,000 strains of soil bacteria and fungi for an ability to inhibit growth or kill bacteria (**Fig. 27.2C**). In 1944, this herculean effort paid off with the discovery of streptomycin, an antibiotic produced by the actinomycete *Streptomyces griseus* (**Fig. 27.2D**). Waksman's discovery of streptomycin triggered the antibiotic gold rush that is still under way.

TO SUMMARIZE:

■ **The importance of antibiotics** in treating disease was recognized in the early 1940s.
■ **Some antimicrobial agents are initially inactive** until converted by the body to an active agent.
■ **Florey and Chain purified penicillin** and capitalized on Fleming's discovery of penicillin.

27.2   Basic Concepts of Antimicrobial Therapy

Antibiotics comprise the vast majority of chemotherapeutic agents used to treat microbial diseases. As already noted, the term "antibiotic" originally referred to any compound produced by one species of microbe that could kill or inhibit the growth of other microbes. Today, the term is also used for synthetic chemotherapeutic agents, such as sulfonamides, that are clinically useful but chemically synthesized. Many natural and synthetic compounds affect microbial growth. However, their utility in a clinical setting is dictated by certain key characteristics.

Antibiotics Exhibit Selective Toxicity

As early as 1904, the German physician Paul Ehrlich (1854–1915) realized that a successful antimicrobial compound would be a "magic bullet" that selectively kills or inhibits the pathogen but not the host. This seemingly obvious premise was innovative at the time. Ehrlich made several discoveries based on this concept, the most celebrated of which was the arsenical compound known as Salvarsan. Salvarsan proved to be quite effective in killing the syphilis agent *Treponema pallidum* (this was long before penicillin was discovered). Syphilis, a sexually transmitted disease, had been untreatable and the source of considerable long-term suffering. Ehrlich's "magic bullet" concept is now known as **selective toxicity**.

Selective toxicity is possible because key aspects of a microbe's physiology are different from those of eukaryotes. For example, suitable bacterial antibiotic targets include peptidoglycan, which eukaryotic cells lack, and ribosomes, which are structurally distinct between the Bacteria and Eukarya. Thus, chemicals like penicillin, which prevents peptidoglycan synthesis, and tetracycline, which binds to bacterial 30S ribosomal subunits, inhibit bacterial growth but are essentially invisible to host cells, since they do not interact with them at low doses.

While their intended targets are bacterial cells, some antibiotics, particularly at high doses, can interact with elements of eukaryotic cells and cause side effects that harm the patient. For example, chloramphenicol, a drug that targets bacterial 50S ribosomal subunits, can interfere with the development of blood cells in bone marrow—a phenomenon that may result in aplastic anemia (failure to produce red blood cells). The toxicity of an antibiotic can also depend on the age of the patient. Tetracycline, for instance, can cause defects in human bone growth plates and should not be administered to children. Problems can even arise if the drug does not directly impact mammalian physiology. For example, many people develop an extreme allergic sensitivity to penicillin, in which case the treatment of an infection may end up being worse than the infection itself. Physicians must be aware of these allergies and use alternative antibiotics to avoid harming their patients.

As a student of microbiology, it is important that you properly use the terms drug "susceptibility" and drug "sensitivity." A microbe is susceptible to the drug's action, but a human can develop an allergic sensitivity to the drug.

Antimicrobials Have a Limited Spectrum of Activity

No one antimicrobial drug affects all microbes. As a result, antimicrobial drugs are classified by the type of organisms they affect. Thus, we have antifungal, antibacterial, antiprotozoan, and antiviral agents. The term "antibiotic" is usually reserved for antimicrobial compounds that affect bacteria. Even within a group, one agent might have a very narrow **spectrum of activity**, meaning it affects only a few species, while another antibiotic inhibits many species. For instance, penicillin has a relatively narrow spectrum of activity, primarily killing Gram-positive bacteria. However, ampicillin is penicillin with an added amino group that allows the drug to more easily penetrate the Gram-negative outer membrane. As a result of this chemically engineered modification, ampicillin kills Gram-positive and Gram-negative organisms and is described as having a spectrum of activity broader than that of penicillin. There are antimicrobials as well that exhibit extremely narrow activities. One example is isoniazid, which is clinically useful only against *Mycobacterium tuberculosis*, the agent of tuberculosis. **eTopic 27.1** explores the spectrum of activity of select antibiotics.

As discussed in several prior chapters, we are increasingly aware that our natural microbiota contribute in important ways to human health and development. However, few studies have explored the possible impact of antibiotic use on host-microbiota interactions. We have known for decades that antibiotics—especially broad-spectrum antibiotics—can destroy the ecological balance between bacterial species in the gut and lead to gastrointestinal disease. Pathology can result when one species resistant to the antibiotic gains a growth advantage over various drug-susceptible species that ordinarily keep the pathogen in check (see the discussion of *Clostridium difficile* in Section 26.4). But what other effects might there be when one disturbs the microbial balance of power in the intestine? We may ultimately find that directing the toxicity of an antibiotic to a single bacterial species (while leaving all others alone) has benefits to human health we do not currently understand.

Antibiotics Are Classified as Bacteriostatic or Bactericidal

Patients typically believe that all antibiotics kill their intended targets. This is a misconception. Many drugs simply prevent growth of the organism and let the body's immune system dispatch the intruding microbe. Thus, antimicrobials are also classified on the basis of whether or not they kill the microbe. An antibiotic is **bactericidal** if it kills the target microbe; it is **bacteriostatic** if it merely prevents bacterial growth.

TO SUMMARIZE:

- **Antimicrobial agents** may be produced naturally or artificially.
- **Selective toxicity** refers to the ability of an antibiotic to attack a unique component of microbial physiology that is missing or distinctly different from eukaryotic physiology.
- **Antibiotic side effects** on mammalian physiology can limit the clinical usefulness of an antimicrobial agent.
- **Antibiotic spectrum of activity** refers to the range of microbes that a given drug affects.
- **Bactericidal antibiotics** kill microbes; **bacteriostatic antibiotics** inhibit microbial growth.

27.3 Measuring Drug Susceptibility

One critical decision a clinician must make when treating an infection is what antibiotic to prescribe for the patient. There are several factors to consider, including:

- The relative effectiveness of different antibiotics on the organism causing the infection. This includes learning whether the organism isolated from a specific patient has developed resistance to the drug.

■ The average attainable tissue levels of each drug. An antibiotic may appear to work on an agar plate, but the concentration at which it affects bacterial growth may be too high to be safe in the patient. In fact, an important aspect of designing new antibiotics is to enhance the pharmacological activity of an existing drug—for example, modifying it so that the body does not break it down or quickly secrete it in urine.

Minimal Inhibitory Concentration Reflects Antibiotic Efficacy

The in vitro effectiveness of an antimicrobial agent is determined by measuring how little of it is needed to stop growth. This is classically measured in terms of an antibiotic's **minimal inhibitory concentration (MIC)**, defined as the lowest concentration of the drug that will prevent the growth of an organism. But the MIC for any one drug will differ among different bacterial species. For example, the MIC of ampicillin needed to stop the growth of *Staphylococcus aureus* will be different from that needed to inhibit *Shigella dysenteriae*. The reasons that a drug may be more effective against one organism than another include the ease with which the drug penetrates the cell and the affinity of the drug for its molecular target.

So how do we measure MIC? As shown in **Figure 27.3**, an antibiotic is serially diluted along a row of test tubes containing nutrient broth. After dilution, the organism to be tested is inoculated at low, constant density into each tube, and the tubes are usually incubated overnight. Growth of the organism is seen as turbidity. Note in **Figure 27.3** that the tubes with the highest concentration of drug are clear, indicating no growth. The tube containing the MIC is the tube with the *lowest* concentration of drug that shows no growth. However, the MIC does not indicate whether a drug is bacteriostatic or bactericidal.

> **THOUGHT QUESTION 27.1** The drug tobramycin is added to a concentration of 1,000 µg/ml in a tube of broth from which serial twofold dilutions were made. Including the initial tube (tube 1), there are a total of ten tubes. Twenty-four hours after all the tubes are inoculated with *Listeria monocytogenes*, turbidity is observed in tubes 6–10. What is the MIC?
>
> **THOUGHT QUESTION 27.2** What additional test performed on an MIC series of tubes will tell you whether a drug is bacteriostatic or bactericidal?

MIC determinations are very useful for estimating a single drug's effectiveness against a single bacterial pathogen isolated from a patient but not very practical when trying to screen 20 or more different drugs. Dilutions take time—time that the technician, not to mention the patient, may not have. The time required to evaluate antibiotic effectiveness can be reduced by using a strip test (like the Etest shown in **Fig. 27.4**) that avoids the need for dilutions. The strip, containing a gradient of antibiotic,

µg/ml

0.06 0.125 0.25 0.5 1.0 2.0 4.0 8.0

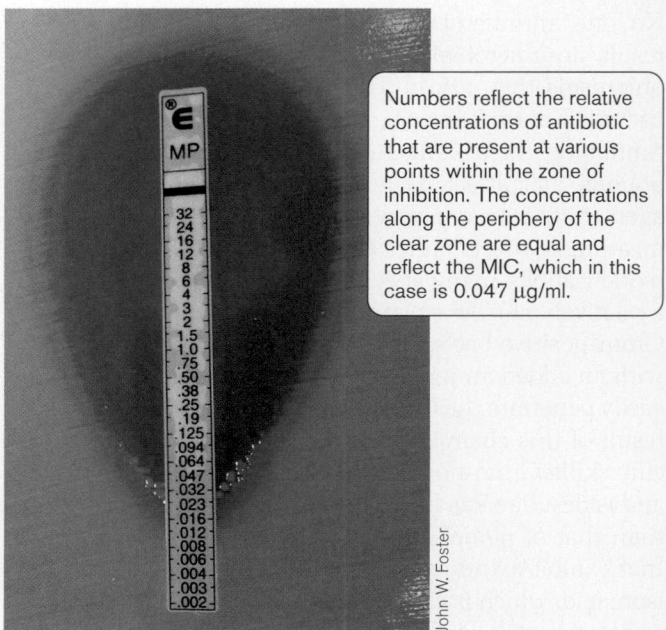

Numbers reflect the relative concentrations of antibiotic that are present at various points within the zone of inhibition. The concentrations along the periphery of the clear zone are equal and reflect the MIC, which in this case is 0.047 µg/ml.

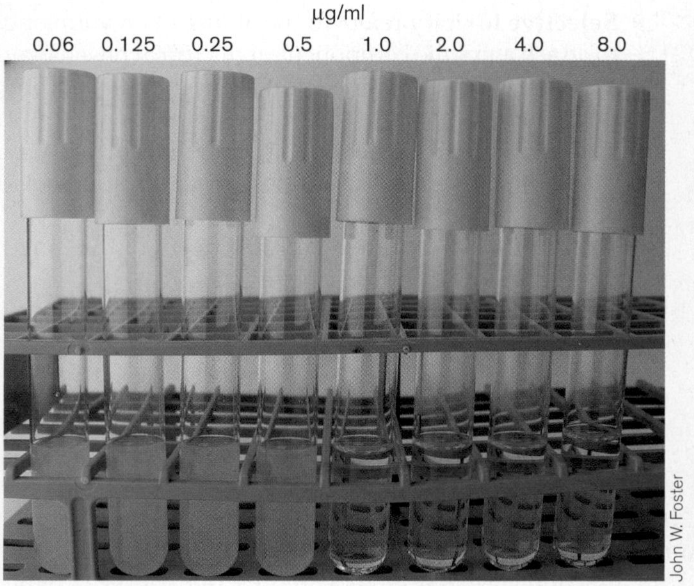

Figure 27.3 Determining minimal inhibitory concentration (MIC). In this series of tubes, tetracycline was diluted serially starting at 8 µg/ml (far right tube). Each tube was then inoculated with an equal number of bacteria. Turbidity indicates that the antibiotic concentration was not sufficient to inhibit growth. The MIC in this example is 1.0 µg/ml.

Figure 27.4 An MIC strip test. The Etest (AB Biodisk) is a commercially prepared strip that produces a gradient of antibiotic concentration (µg/ml) when placed on an agar plate. The MIC corresponds to the point where bacterial growth crosses the numbered strip.

is placed onto an agar plate freshly seeded with a dilute lawn of bacteria. While the bacteria are trying to grow, the drug diffuses out of the strip and into the media. Drug emanating from the more concentrated areas of the strip will travel faster and farther through the agar than will drug from the less concentrated areas of the strip. Thus, the higher concentrations will kill or inhibit the growth of cells farther from the strip than the lower concentrations. The result is a **zone of inhibition** where the antibiotic has stopped bacterial growth. The MIC is the point at which the elliptical zone of inhibition intersects with the strip.

Kirby-Bauer Disk Susceptibility Test

Although the strip test eliminates the time and effort needed to make dilutions, it would take 20 or more plates to test an equal number of antibiotics for just one bacterial isolate. Clinical labs can receive up to 100 or more isolates in one day, so individual MIC determinations are not practical. A simplified agar diffusion test, however, which can test 12 antibiotics on one plate, makes evaluating antibiotic susceptibility a manageable task.

Named for its inventors, the **Kirby-Bauer assay** uses a series of round filter paper disks impregnated with different antibiotics. A dispenser (**Fig. 27.5**) delivers up to 12 disks simultaneously to the surface of an agar plate covered by a bacterial lawn. Each disk is marked to indicate the drug used. During incubation, the drugs diffuse away from the disks into the surrounding agar and inhibit growth of the lawn to different distances. The zones of inhibition vary in width, depending on which antibiotic is used, the concentration of drug in the disk, and the susceptibility of the organism to the drug. The diameter of the zone correlates to the MIC of the antibiotic against the organism tested.

Correlations between MIC values and Kirby-Bauer zone sizes are made empirically. Every disk containing a given antibiotic is impregnated with a standard concentration of drug, and every antibiotic has a quantifiable MIC that will differ when tested against different bacterial species and strains. The outermost ring of the no-growth zone in a Kirby-Bauer disk test must, by definition, contain the minimal concentration of drug needed to prevent growth on agar. Thus, if species A and B have MIC values for penicillin of 4 μg/ml and 40 μg/ml, respectively, then species A will exhibit a proportionally larger zone of inhibition than species B in the disk test. A graph plotting MIC on one axis and zone diameter on the other provides the correlation.

A.

Figure 27.5 The Kirby-Bauer disk susceptibility test.
A. Device used to deliver up to 12 disks to the surface of a Mueller-Hinton plate. The device is placed over the plate, and the plunger is depressed to deliver the disks. **B.** Disks impregnated with different antibiotics are placed on a freshly laid lawn of bacteria and incubated overnight. The clear zones around certain disks indicate growth inhibition. Shown are the results with normal *Staphylococcus aureus.* **C.** Methicillin-resistant *S. aureus* (MRSA). Note the lack of inhibition by the oxacillin disk (arrow). This strain is resistant to both methicillin and oxacillin. **D.** *Streptococcus pneumoniae.* The brownish tint of the blood agar plates outside the zones of bacterial inhibition is caused by a hemolysin secreted by the lawn of pneumococci. C, chloramphenicol; CC, clindamycin; CZ, cefazolin; E, erythromycin; NOR, norfloxacin; OX, oxacillin; P, penicillin; RA, rifampin; SAM, sulbactam-ampicillin; SXT, sulfatrimethoprim; TE, tetracycline; VA, vancomycin.

B.

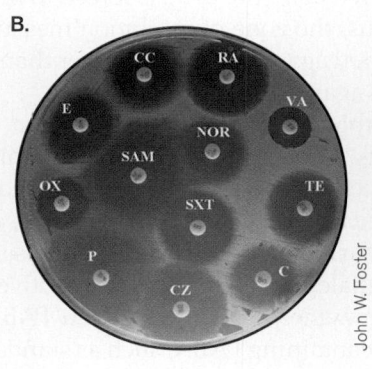

C.

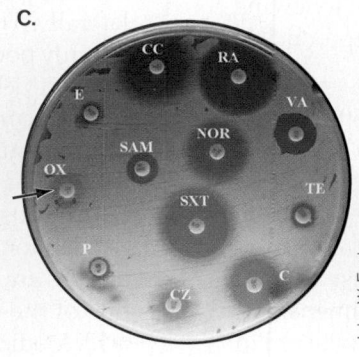

D.

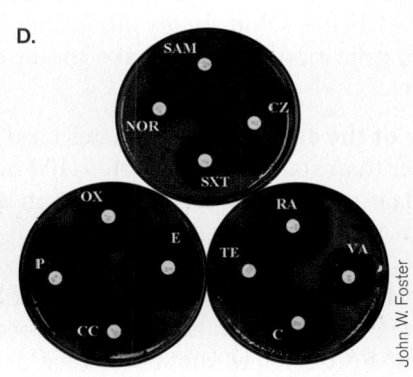

Table 27.1 Susceptibility results for *Staphylococcus aureus*.

Antibiotic	Quantity in disk (μg)	Zone of inhibition diameter (mm)		
		Resistant	Intermediate	Susceptible
Ampicillin	10	<12	12–13	>13
Chloramphenicol	30	<13	13–17	>17
Erythromycin	15	<14	14–17	>17
Gentamicin	10	≤12.5		>12.5
Streptomycin	10	<12	12–14	>14
Tetracycline	30	<15	15–18	>18

After incubating agar plates in the Kirby-Bauer test, the diameters of the zones of inhibition around each disk are measured, and the results are compared with a table listing whether a zone is wide enough (meaning the MIC is low enough) to be clinically useful. **Table 27.1** shows susceptibility data for *Staphylococcus aureus*. The concentration of antibiotic used and the zone size that is considered clinically significant are correlated with the average attainable tissue level for each antibiotic. For the antibiotic to remain effective in vivo, it is important that the tissue concentration of the drug remain above the MIC; otherwise, invading bacteria will not be affected.

Correlating antibiotic MIC with tissue level. The average attainable tissue level for a drug depends on how quickly the antibiotic is cleared from the body via secretion by the kidney or destruction in the liver. It also depends on when side effects of the drug start to appear. As shown in **Figure 27.6**, as long as the concentration of the drug in tissue or blood remains higher than the MIC, the drug will be effective. The concentration can be kept at sufficient levels either by administering a higher dose, which runs the risk of side effects, or by giving a second dose at a time when the levels from the first dose have declined. This is why patients are told to take doses of some antibiotics four times a day and other antibiotics only once a day.

To ensure reproducibility, the Kirby-Bauer test was standardized a half century ago. Reproducibility means that results from a laboratory in California will match those in Alabama, Ohio, or any other state. The following are standardizations used to make the test reproducible and easier.

- **Size of the agar plate.** The plates used (150 mm) are larger than standard agar plates (100 mm) to accommodate more disks and to maintain sufficient distance between disks so that zones of inhibition do not overlap.
- **Depth of the media.** Antibiotics diffuse out of impregnated disks in not two, but three dimensions. Because diffusion cannot occur very far downward in

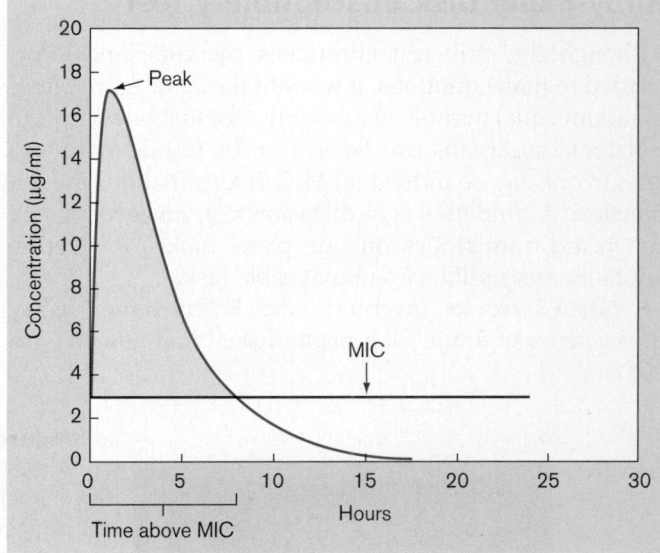

Figure 27.6 Correlation between MIC and serum or tissue level of an antibiotic. This graph illustrates the serum level of ampicillin over time. The important consideration here is how long the serum level of the antibiotic remains higher than the MIC. Once the concentration falls below the MIC, owing to destruction in the liver or clearance through the kidneys and secretion, the infectious agent fails to be controlled by the drug—in this case about 7–8 hours after the initial dose. To maintain a serum level higher than the MIC, a second dose would be taken. The shaded area of the curve represents time above MIC.

a thinly poured agar, the drug is forced to move more laterally. Thus, the zone of inhibition measured from a thinly poured agar plate will be larger than the zone from a thick agar plate.

- **Media composition.** Media composition can also affect results. For example, most common laboratory media contain para-aminobenzoic acid (PABA), which is used by the cell to make folic acid, a vitamin needed for purine and pyrimidine synthesis. Sulfonamides are analogs of PABA that competitively inhibit one of the enzymes needed to convert PABA to folic acid. Media containing PABA, such as standard nutrient agar, will flood the bacterial cell with PABA and limit the ability of the sulfa drugs to compete for the enzyme. Thus, even though the drug might be effective in vivo, on the agar plate there would be no zone of inhibition. The standardized media used for the Kirby-Bauer test, called **Mueller-Hinton agar**, contains no PABA.

■ **The number of organisms spread on the agar plate.** The number of organisms placed on an agar surface is inversely proportional to the size of the zone of inhibition. The more organisms there are, the smaller the zone. This phenomenon is observed because there is a time lag between dropping a disk onto a plate and the diffusion of the antibiotic. The more organisms there are on a plate, the less time it takes for them to form visible growth; so by the time the antibiotic gets to them, visible growth has already formed. As a result, a standard optical density solution of each organism is prepared, and a cotton swab is used to spread the entire agar surface.

■ **Size of the disks.** A standard diameter of 6 mm means that all antibiotics start diffusing into agar at the same point.

■ **Concentrations of antibiotics in the disks.** The higher the concentration of antibiotic in a disk, the faster the drug can diffuse through the plate and kill bacteria (or at least inhibit growth) before replicating cells are able to form visible growth. To avoid differences between labs, the concentration of each given drug impregnating a disk has been standardized.

■ **Incubation temperature.** Incubation temperature will not affect growth and diffusion equally. To avoid differences, a temperature of 37°C is standard.

THOUGHT QUESTION 27.3 You are testing whether a new antibiotic will be a good treatment choice for a patient with a staph infection. The Kirby-Bauer test using the organism from the patient shows a zone of inhibition of 15 mm around the disk containing this drug. Clearly, the organism is susceptible. But you conclude from other studies that the drug would not be effective in the patient. What would make you draw this conclusion?

TO SUMMARIZE:

■ **The spectrum of an antibiotic and the susceptibility of the infectious agent** are critical points of information required before prescribing antibiotic therapy.

■ **Minimal inhibitory concentration (MIC)** of a drug, when correlated with average attainable tissue levels of the antibiotic, can predict the effectiveness of an antibiotic in treating disease.

■ **MIC is measured** using tube dilution techniques but can be approximated using the Kirby-Bauer disk diffusion technique.

27.4 Mechanisms of Action

As noted earlier, selective toxicity of an antibiotic depends on enzymes or structures unique to the bacterial target cell. The following aspects of a microbe's physiology are classic targets:

■ Cell wall
■ Cell membrane
■ DNA synthesis
■ RNA synthesis
■ Protein synthesis
■ Metabolism

Table 27.2 summarizes the general targets of common antibiotics. Chapters 3, 7, and 8 describe these cellular components and provide the tools needed to understand how antibiotics work.

Cell Wall Antibiotics

Bacterial cell walls are an obvious structure that could be the basis for selective toxicity because peptidoglycan does not exist in mammalian cells; thus, antibiotics that target the synthesis of these structures should selectively kill bacteria. The following case history illustrates the use of two cell wall–targeting antibiotics and also reveals how bacteria can evolve to escape destruction.

Table 27.2 Mechanisms of action of antimicrobial agents.

Target	Antibiotic examples
Cell wall synthesis	Penicillins, cephalosporins, bacitracin, vancomycin
Protein synthesis	Chloramphenicol, tetracyclines, aminoglycosides, macrolides, lincosamides
Cell membrane	Polymyxin, amphotericin, imidazoles vs. fungi
Nucleic acid function	Nitroimidazoles, nitrofurans, quinolones, rifampin; some antiviral compounds, especially antimetabolites
Intermediary metabolism	Sulfonamides, trimethoprim

Case History: Meningitis

A 3-year-old child was brought to the emergency room crying, with a stiff neck and high fever. Gram stain of cerebrospinal fluid revealed Gram-positive cocci, generally in pairs. The diagnosis was meningitis. The physician immediately prescribed intravenous ampicillin. Unfortunately, the child's condition worsened, so antibiotic treatment was changed to a third-generation cephalosporin (which will cross the blood-brain barrier). The patient began to improve within hours and was released after 2 days. A report from the clinical microbiology laboratory identified the organism as Streptococcus pneumoniae.

Both of the antibiotics used in this case kill bacteria by targeting cell wall synthesis. Synthesis of peptidoglycan (introduced in Chapter 3) is a complex process but is represented simply in **Figure 27.7**. Basically, sugar molecules

called *N*-acetylglucosamine (NAG) and *N*-acetylmuramic acid (NAM) are made by the cell and linked together by a **transglycosylase** enzyme into long chains assembled at the cell wall. *N*-acetylmuramic acid contains a short side chain of amino acids that is assembled enzymatically, not by a ribosome. Rigidity of the macromolecular structure, essential for maintaining cell shape, is achieved by cross-linking the side chains from adjacent strands. The enzyme **transpeptidase** (D-alanyl-D-alanine carboxypeptidase/transpeptidase) catalyzes the cross-link. **Figure 27.7** also indicates where several antibiotics target various stages of this assembly process.

Peptidoglycan is assembled outside the cell membrane. Before we can explain how cell wall antibiotics work, it is important to know how the cell wall is made. Synthesis of peptidoglycan starts in the cytoplasm with a uridine diphosphate (UDP)–NAM molecule. The

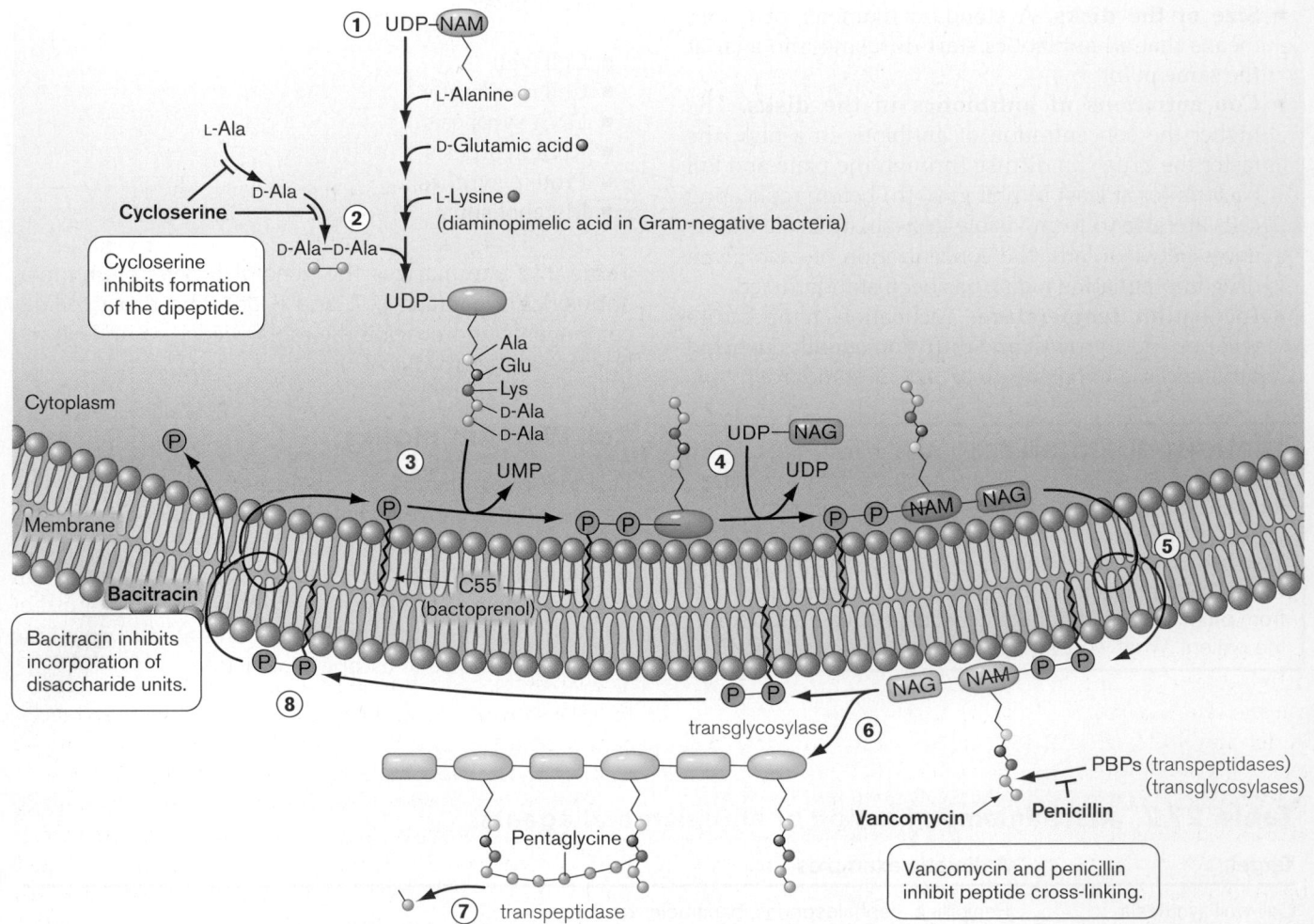

Figure 27.7 Peptidoglycan synthesis and targets of antibiotics. A series of small-molecular-weight compounds are joined to form a disaccharide unit that will be added to preexisting chains of this unit. Steps in the synthesis are described in the text. Red lines indicate inhibition. Cycloserine inhibits ligation of the two D-alanines (step 2). Bacitracin inhibits linking of the disaccharide units, while vancomycin and the beta-lactams, such as penicillin, inhibit the peptide cross-linking of peptidoglycan side chains.

amino acids L-alanine, D-glutamic acid, and L-lysine (or diaminopimelic acid in Gram-negative organisms) are individually and sequentially added to NAM (**Fig. 27.7**, step 1); and then a dipeptide of D-alanine is attached (step 2). Next, the NAM-pentapeptide is transferred to a membrane-situated, 55-carbon, lipid molecule called bactoprenol (step 3) and UMP is released. Another sugar molecule, NAG, is then linked to NAM, once again through a UDP intermediate (step 4). All of this takes place on the cytoplasmic side of the membrane. Bactoprenol carrying NAM-NAG then moves to the outer side of the cytoplasmic membrane (step 5), where transpeptidases and transglycosylases (two so-called **penicillin-binding proteins**, or **PBPs**) bind to the D-Ala-D-Ala part of the pentapeptide. Transglycosylase attaches the new disaccharide unit to an existing peptidoglycan chain (step 6). Transpeptidase then links two peptide side chains with a pentaglycine cross-link (in *Staphylococcus aureus*). The pentaglycine connects L-lysine on one side chain and the penultimate D-Ala on the other side chain (step 7). The terminal D-Ala is removed in the process. Other bacteria do not use a pentaglycine cross-link but directly form a peptide bond between L-lysine (or DAP) and the penultimate D-Ala. Cross-linking, which strengthens the cell wall, can take place within the same strand or between adjacent strands. In the last part of the cycle, one of the phosphates on the now liberated bactoprenol is removed and the lipid moves back to the cytoplasmic side of the membrane, ready to pick up and taxi another unit of peptidoglycan to the growing chain (step 8).

Penicillin and other beta-lactam antibiotics target penicillin-binding proteins. Penicillin is an antibiotic derived from cysteine and valine, which combine to form the beta-lactam ring structure shown in **Figure 27.8A**. Different R groups can be added to the basic ring structure to change the antimicrobial spectrum and stability of the derivative penicillin (**Fig. 27.8B**). Penicillin itself chemically resembles the D-Ala-D-Ala piece of peptidoglycan. This molecular mimicry allows the drug to bind transpeptidase and transglycosylase (which is why they are called penicillin-binding proteins), preventing their activities and halting synthesis of the chain. The consequence is a disaster for bacteria that are trying to grow larger and larger. Eventually, the growing cell bursts for lack of cell wall restraint. Penicillin, then, is a bactericidal drug (unless the treated organism is suspended in an isotonic solution). Penicillin is most effective against Gram-positive organisms because the drug has difficulty passing through the Gram-negative outer membrane. Ampicillin, which was used in the case history, is a modified version of penicillin that more easily penetrates this membrane and is more effective than penicillin against Gram-negative microbes. Thus, ampicillin has a broader spectrum of activity than penicillin.

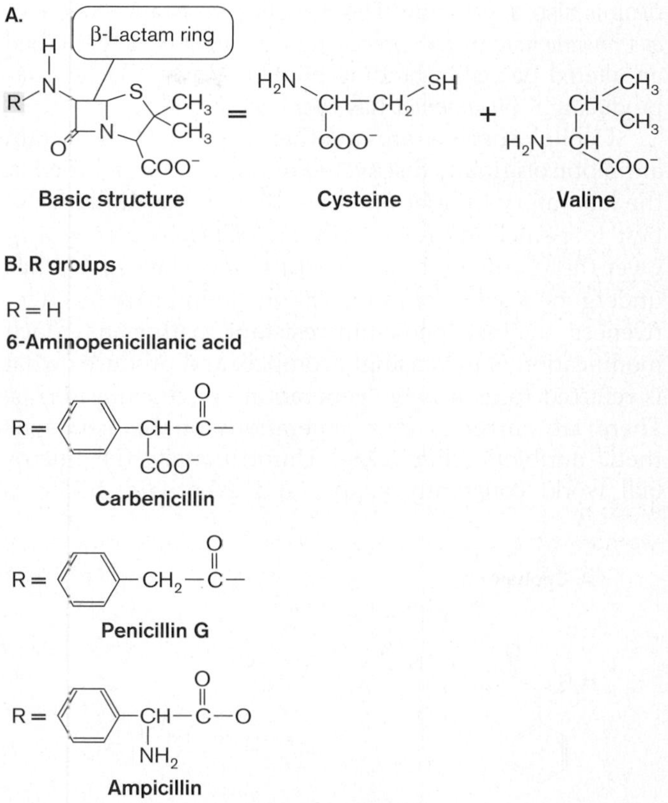

Figure 27.8   The structure of penicillins.   A. Penicillanic acid (R = H) is derived from cysteine and valine. **B.** The R group marked in (A) can be any one of a number of different groups, some of which are shown here. Modifying this group changes the pharmacological properties and antimicrobial spectrum Methicillin is penicillin G with $-OCH_3$ groups on the 2nd and 6th carbon of the benzene ring.

As noted earlier, antibiotic resistance is a growing problem throughout the world. Bacteria develop resistance to penicillin in two basic ways. The first is through inheritance of a gene encoding one of the beta-lactamase enzymes, which cleave the critical ring structure of this class of antibiotics. Beta-lactamase is transported out of the cell and into the surrounding media (for Gram-positives) or the periplasm (for Gram-negatives), where it can destroy penicillin before the drug even gets to the cell. Bacteria that produce beta-lactamase are still susceptible to certain modified penicillins and cephalosporins engineered to be poor substrates for the enzyme. Methicillin, for example, works well against beta-lactamase-producing microbes.

The second way a microbe can become resistant is through mutations in genes encoding key penicillin-binding proteins. Resistance occurs when the mutated gene produces an altered protein that no longer binds to the antibiotic. Methicillin-resistant bacteria use this strategy. Hospitals take special interest in methicillin-resistant *Staphylococcus aureus* (MRSA) because very few drugs can kill it. One of the few remaining antibiotics effective against MRSA is vancomycin. Unfortunately, resistance to this drug is also developing. The penicillin-resistant *Streptococcus pneumoniae* in the preceding case history actually had an altered penicillin-binding protein. No beta-lactamase-producing *S. pneumoniae* has ever been found.

Cephalosporins are another type of beta-lactam antibiotic originally discovered in nature but modified in the laboratory to fight microbes that are naturally resistant to penicillins (especially *Pseudomonas aeruginosa*). Over the years, the basic structure of cephalosporin has undergone a series of modifications to improve its effectiveness against penicillin-resistant pathogens. Each modification is increasingly complex and produces what is referred to as a new "generation" of cephalosporins. There are currently four generations of this semisynthetic antibiotic (**Fig. 27.9**). Unfortunately, the microbial world constantly adapts and eventually becomes resistant to new antibiotics. In the case of the cephalosporins, new beta-lactamases evolve that can attack the sterically buried beta-lactam rings in these molecules. It is also important to note that since the core feature of these drugs is the beta-lactam ring, persons who are sensitive to penicillins may also suffer a hypersensitivity reaction to cephalosporins.

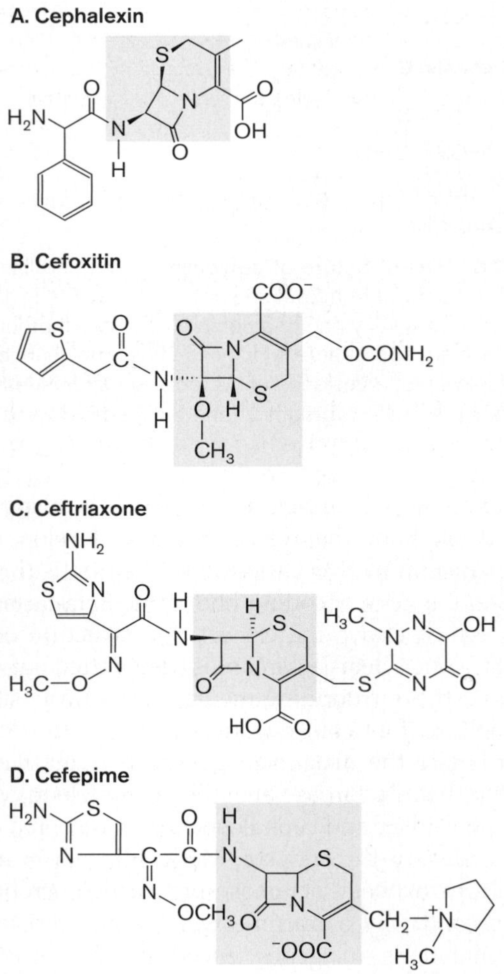

Treatment note: In the preceeding case history, the infecting strain of *S. pneumoniae* turned out to be resistant to ampicillin. Had the patient been an adult, a fluoroquinolone (see Section 7.2) might have been the best secondary drug of choice because its target, a type II topoisomerase, is unrelated to cell wall synthesis. Quinolones are not recommended for children, as in this case, because of potential side effects. Other beta-lactam antibiotics, such as the third-generation cephalosporins, may still work on penicillin-resistant *S. pneumoniae*, since the modified antibiotic often can still bind the altered PBP. Nevertheless, vancomycin or an oxazolidinone are the best last choices, since cephalosporin-resistant strains of *S. pneumoniae* are now appearing.

> **NOTE:** Archaeal peptidoglycan contains talosaminuronic acid instead of muramic acid and lacks the D-amino acids found in bacterial peptidoglycan. Archaea are thus insensitive to penicillins, which interfere with bacterial transpeptidases. This natural resistance is not a problem, since there are no known archaeal pathogens.

Figure 27.9 Cephalosporin generations. Representative examples. **A.** First generation: cephalexin (Keflex). **B.** Second generation: cefoxitin. **C.** Third generation: ceftriaxone. **D.** Fourth generation: cefepime. Note that with each successive generation, the side groups become more complex. Highlighted areas indicate the core structure of each of the cephalosporins, with beta-lactam rings.

Cell wall antibiotics target other steps in peptidoglycan synthesis. Another antibiotic that affects cell wall synthesis is **bacitracin**, a large polypeptide molecule produced by *Bacillus subtilis* and *Bacillus licheniformis* (**Fig. 27.10A**). The antibiotic inhibits cell wall synthesis by binding to the bactoprenol lipid carrier molecule that normally transports monomeric units of peptidoglycan across the cell membrane and to the growing chain (see **Fig. 27.7**). Bacitracin binds to and inhibits dephosphorylation of the carrier, thereby preventing the carrier from accepting a new unit of UDP-NAM. Resistance to bacitracin can develop if the organism can rapidly recycle the phosphorylated lipid carrier molecule through dephosphorylation or if the organism possesses an efficient drug export system (discussed in Section 27.6). Normally, bacitracin is used only topically because of serious side effects, such as kidney damage, that can occur if it is ingested.

Cycloserine (made by *Streptomyces garyphalus*) is one of several antimicrobials used to treat tuberculosis (**Fig. 27.10B**). Relative to bacitracin, it acts at an even earlier step in peptidoglycan synthesis. Cycloserine inhibits the two enzymes that make the D-Ala-D-Ala dipeptide. As a result, the complete pentapeptide side chain on *N*-acetylmuramic acid cannot be made (see **Fig. 27.7**).

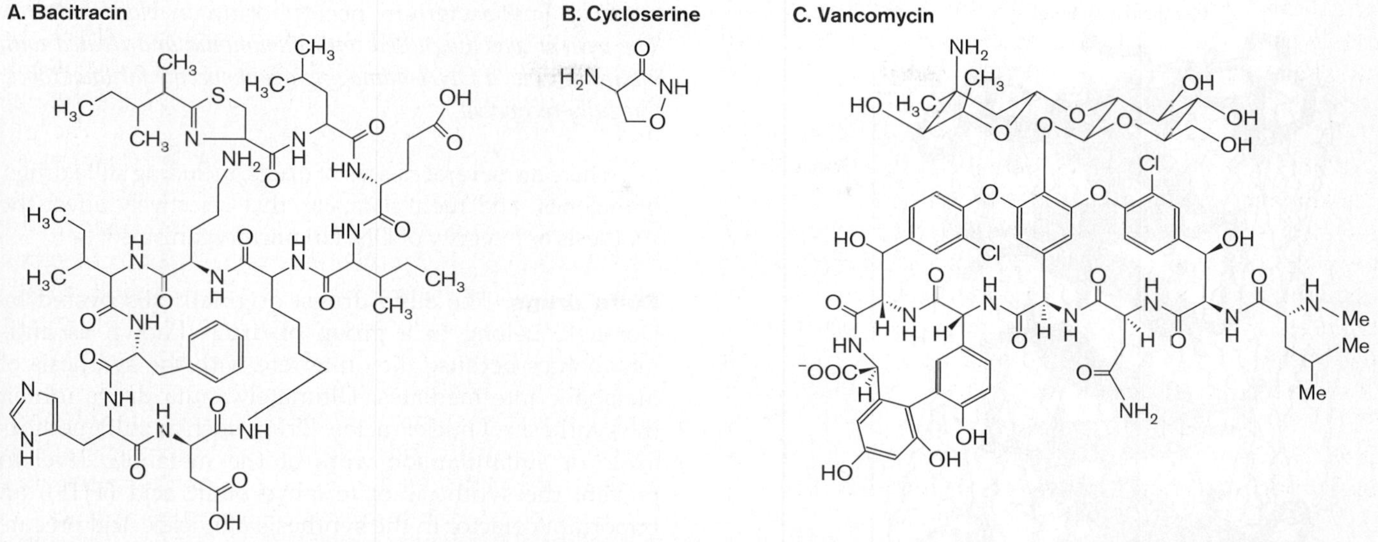

A. Bacitracin

B. Cycloserine

C. Vancomycin

Figure 27.10   Other antibiotics that affect peptidoglycan synthesis.   A. Bacitracin is produced by *Bacillus subtilis*. It is generally used only topically to prevent infection. **B.** Cycloserine, an analog of D-alanine, is one of several drugs used to treat tuberculosis. **C.** Vancomycin is a cyclic polypeptide made by *Amycolatopsis orientalis*, previously classified as a streptomycete. These antibiotics, especially bacitracin and vancomycin, are synthesized by exceedingly complex biochemical pathways in the producing organisms. Me = methyl.

Without these alanines, cross-linking cannot occur and peptidoglycan integrity is compromised.

Vancomycin, a very large and complex glycopeptide produced by the streptomycete *Amycolatopsis orientalis* (**Fig. 27.10C**), binds to the D-Ala-D-Ala terminal end of the disaccharide unit and prevents the action of transglycosylases and transpeptidases (see **Fig. 27.7**). The mechanism of resistance is very different for vancomycin and penicillin. This difference makes vancomycin particularly useful against penicillin-resistant bacteria. To prevent development and spread of vancomycin-resistant bacteria, this antibiotic is typically used only as a drug of last resort. Resistance can develop when products from a cluster of *van* genes collaborate to make D-lactate and incorporate it into the ester D-Ala-D-lactate, to which vancomycin cannot bind. Another enzyme in the *van* gene cluster prevents the accumulation of D-Ala-D-Ala; as a result, the D-Ala-D-Lactate replaces D-Ala-D-Ala in peptidoglycan. Peptidoglycan containing D-Ala-D-lactate functions just fine, but since vancomycin cannot bind the D-lactate form, the organism is resistant to the antibiotic.

Note that antibiotics targeting cell wall biosynthesis generally kill only *growing* cells. These drugs do not affect static or stationary-phase cells, because in this state the cell has no need for new peptidoglycan.

THOUGHT QUESTION 27.4   When treating a patient for an infection, why would combining a drug such as erythromycin with a penicillin be counterproductive? Erythromycin is described in Section 8.3.

Drugs That Affect Bacterial Membrane Integrity

Poking holes in a bacterial cytoplasmic membrane is an effective way to kill bacteria. There are a few compounds useful in this regard, among them a group called the peptide antibiotics, of which **gramicidin** is an example. Gramicidin, produced by *Bacillus brevis*, is a cyclic peptide composed of 15 alternating D- and L-amino acids. It inserts into the membrane as a dimer, forming a cation channel that disrupts membrane polarity (**Fig. 27.11**). Polymyxin (from *Bacillus polymyxa*), another polypeptide antibiotic, has a positively charged polypeptide ring that binds to the outer (lipid A) and inner membranes of bacteria, both of which are negatively charged. Its major lethal effect seems to be to destroy the inner membrane, much like a detergent. These antibiotics are used only topically to treat or prevent infection. Because they can also form channels across human cell membranes, they should never be ingested. Polymyxin has been fused to some bandage materials used to treat burn patients who are particularly susceptible to Gram-negative infections (for example, *Pseudomonas aeruginosa*).

Drugs That Affect DNA Synthesis and Integrity

Bacteria generally make and maintain their DNA using enzymes that closely resemble those of mammals. Thus, you might consider it impossible to selectively target bacterial DNA synthesis, but as you will see, this is possible.

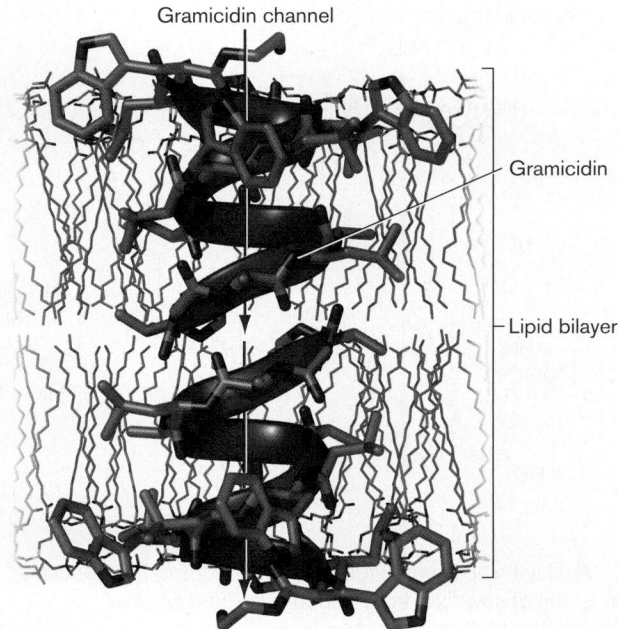

Gramicidin channel

Gramicidin

Lipid bilayer

Figure 27.11   Gramicidin is a peptide antibiotic that affects membrane integrity.   Gramicidin forms a cation channel across cell membranes through which H^+, Na^+, or K^+ can freely pass. (PDB code: 1GRM)

Case History: Pneumonia Due to a Gram-Negative Anaerobe

A 23-year-old woman arrived at the emergency room by ambulance with fever, chills, and severe muscle aches. She developed a nonproductive cough, had difficulty breathing, had pleuritic chest pain, and became hypotensive (low blood pressure). An X-ray showed lower-lobe infiltrate in the lungs, and the clinical laboratory reported the presence of the Gram-negative anaerobe Fusobacterium necrophorum in blood cultures. The patient was diagnosed with pneumonia and treated with metronidazole, a DNA-damaging agent specific for anaerobes. She fully recovered.

There are several classes of drugs, including sulfa drugs, quinolones, and metronidazole, that selectively affect the synthesis or integrity of DNA in microorganisms.

Sulfa drugs. The sulfa drugs, originally discovered by Domagk, belong to a group of drugs known as antimetabolites because they interfere with the synthesis of metabolic intermediates. Ultimately, sulfa drugs inhibit the synthesis of nucleic acids. Drugs such as sulfamethoxazole or sulfanilamide work at the metabolic level to prevent the synthesis of tetrahydrofolic acid (THF), an important cofactor in the synthesis of nucleic acid precursors (**Fig. 27.12**). All organisms use THF to synthesize nucleic acids, so why are the sulfa drugs selectively toxic to bacteria? The selectivity occurs because mammalians do not synthesize a precursor of THF called folic acid. Higher mammals generally rely on bacteria and green leafy vegetables as sources of folic acid. Bacteria make folic acid from the combination of PABA, glutamic acid, and pteridine. Sulfanilamide (SFA), a structural analog of PABA, competes for one of the enzymes in the bacterial folic acid pathway and inhibits both folic acid and THF production (**Fig. 27.12C**). Because humans lack that pathway, sulfa drugs are selectively toxic toward bacteria.

Quinolones. Another group of drugs inhibits DNA synthesis by targeting microbial topoisomerases such as DNA gyrase and topoisomerase IV (discussed in Section 7.2). Because these enzymes are structurally distinct from

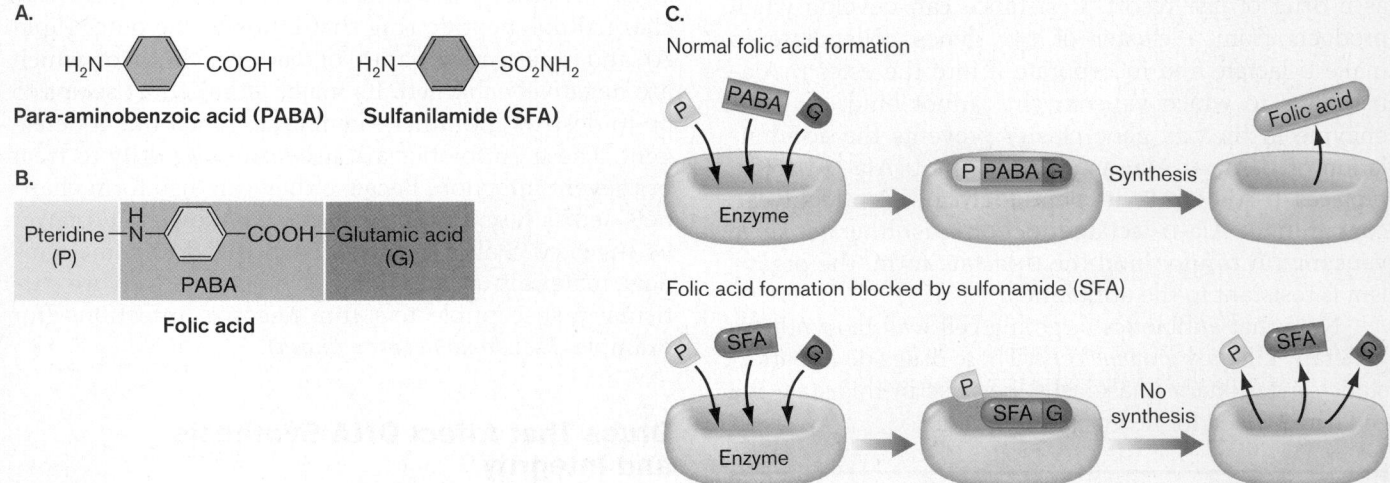

Figure 27.12   Mode of action of sulfanilamides.   A. The structures of PABA and sulfanilamide are very similar. **B.** PABA, pteridine, and glutamic acid combine to make the vitamin folic acid. **C.** Normal synthesis of folic acid requires that all three components engage the active site of the biosynthetic enzyme. The sulfa drugs replace PABA at the active site. The sulfur group, however, will not form a peptide bond with glutamic acid, and the size of sulfanilamide sterically hinders binding of pteridine, so folic acid cannot be made.

their mammalian counterparts, drugs can be designed to selectively interact with them without interfering with mammalian DNA metabolism. In 1963, one such drug, nalidixic acid, was discovered as a by-product of the synthesis of chloroquine, an antimalarial drug. Nalidixic acid, which targets DNA gyrase, has a very narrow antimicrobial spectrum, covering only a few Gram-negative organisms. However, various chemical modifications, such as adding fluorine and amine groups, have increased its antimicrobial spectrum and its half-life in the bloodstream. The result is the class of drugs known as the **quinolones**. (The mode of action of quinolones and fluoroquinolones was discussed in Section 7.2.)

Metronidazole. Also known as Flagyl, metronidazole is an example of a drug that is harmless until activated, known as a prodrug. Metronidazole is activated after it receives an electron (is reduced) from the microbial protein cofactors flavodoxin and ferredoxin, found in microaerophilic and anaerobic bacteria such as *Bacteroides* (**Fig. 27.13**). Once activated, the compound begins nicking DNA at random, thus killing the cell. Because the etiological agent in our case history was an anaerobe, metronidazole was an effective therapy. Metronidazole is also effective against protozoa such as *Giardia, Trichomonas,* and *Entamoeba.* Aerobic microbes, although in possession of ferredoxin, are incapable of reducing metronidazole, presumably because oxygen is reduced in preference to metronidazole.

> **THOUGHT QUESTION 27.5** The enzyme DNA gyrase, a target of the quinolone antibiotics, is an essential protein in DNA replication. The quinolones bind to and inactivate this protein. Research has proved that quinolone-resistant mutants contain mutations in the gene encoding DNA gyrase. If the resistant mutants contain a mutant DNA gyrase and DNA gyrase is essential for growth, why are these mutations not lethal?

RNA Synthesis Inhibitors

Antibiotics that inhibit transcription, such as rifampin and actinomycin D (**Fig. 27.14**), were described in Chapter 8. These drugs are considered bactericidal in action and are most active against growing bacteria. The tricyclic ring of actinomycin D binds DNA from any source. As a result, it is not selectively toxic and not used to treat infections. Rifampin (also called rifampicin), on the other hand, does exhibit

selective toxicity and is often prescribed for treatment of tuberculosis or meningococcal meningitis. Curiously, because rifampin is reddish orange, it turns bodily secretions, including breast milk, orange. The astute physician will warn the patient of this highly visible but harmless side effect to avoid unnecessary anxiety when the patient's urine changes color.

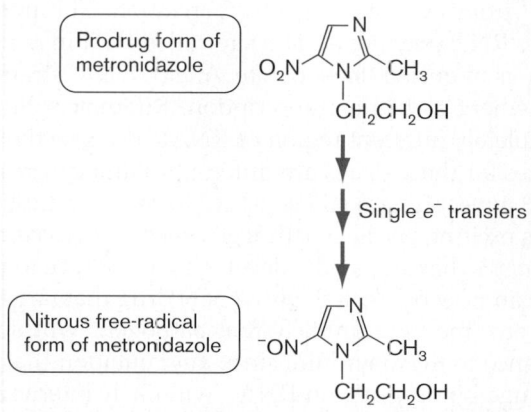

Figure 27.13   Activation of metronidazole. Single-electron transfers are made by ferredoxin and flavodoxin from anaerobes. Ferredoxin and flavodoxin are reducing agents capable of reducing oxidized molecules such as thioredoxin.

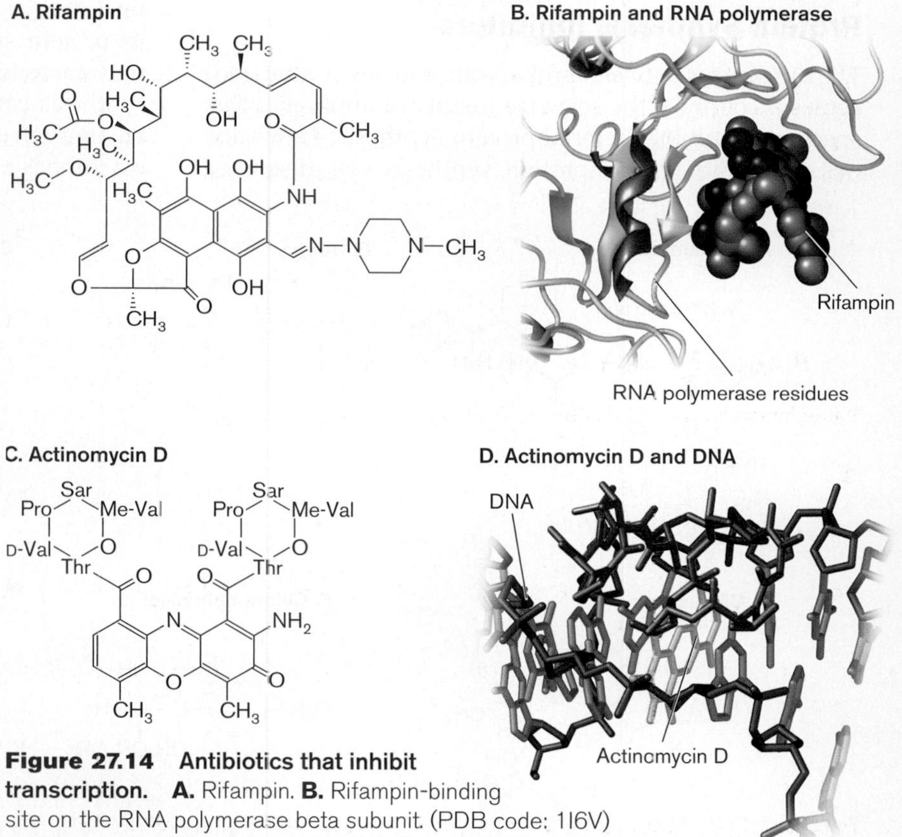

Figure 27.14   Antibiotics that inhibit transcription. A. Rifampin. **B.** Rifampin-binding site on the RNA polymerase beta subunit. (PDB code: 1I6V) **C.** Actinomycin D. **D.** Actinomycin D (yellow and red) interacting with DNA. Covalent intercalation of actinomycin interferes with DNA synthesis and transcription. (PDB code: 1DSC)

Until recently, rifampicin was the only antibiotic known to directly target bacterial RNA polymerase (RNAP). Rifampicin binding to RNAP prevents transcription only after the enzyme has started polymerization (it obstructs the exit tunnel for nascent RNA, as described in Chapter 8). In an exciting development, a new class of antibiotics (pyronins), represented by myxopyronin (produced by *Myxococcus fulvis*), has been discovered that prevents RNAP from ever starting polymerization. The pyronins bind to RNA polymerase at a site called the hinge region, which is required to separate (melt) DNA strands—a requirement to begin transcription. Rifampicin binds to a completely different region of RNAP. Because the binding sites for these drugs are different, rifampicin-resistant RNAP molecules are still sensitive to the new antibiotics. This is exciting because pathogens such as *Mycobacterium tuberculosis* that are rapidly developing resistance to rifampicin can now be treated with a new drug that targets the same enzyme. It is unlikely that pathogens will develop resistance to myxopyronin, since any mutation that alters the hinge binding site on RNAP will likely interfere with the enzyme's ability to perform an essential task, melting DNA. Thus, any attempt to become resistant would likely kill the bacterium anyway. Representatives of these new antibiotics are undergoing clinical trials.

Protein Synthesis Inhibitors

The differences between prokaryotic and eukaryotic ribosomes account for the selective toxicity of antibiotics that specifically inhibit bacterial protein synthesis. How various antibiotics inhibit protein synthesis was discussed in Section 8.3. We recommend reviewing Chapter 8 for details. Recall that protein synthesis inhibitors can be classified into several groups based on structure and function (**Fig. 27.15**). Note that most of these antibiotics work by binding and interfering with the function of bacterial rRNA, which differs from eukaryotic rRNA. Also recall that protein synthesis inhibitors are, by and large, bacteriostatic.

Case History: Erysipelas in a Penicillin-Sensitive Patient

Sixteen-year-old Jamal arrived at the emergency room after 2 days of fever, malaise, chills, and neck stiffness. His most notable symptom was a painful, red, rapidly spreading rash covering the right side of his face. The rash covered his entire cheek, which was swollen, and extended into his scalp. About 7 days earlier, Jamal had had a severe sore throat. However, because it subsided in 2 days he was not clinically evaluated. Throat cultures taken on admission revealed group A Streptococcus pyogenes, suggesting that the rash was a case of erysipelas caused by this organism. Although penicillin would be the drug of choice, Jamal is known to be allergic to this antibiotic.

When a patient is known to be immunologically sensitive to the usual drug of choice, a structurally distinct drug is best. Often, that drug will be one that inhibits protein synthesis; in this case, the drug chosen was the macrolide azithromycin. Drugs that inhibit protein synthesis can be subdivided into different groups based on their structures and on what part of the translation machine is targeted.

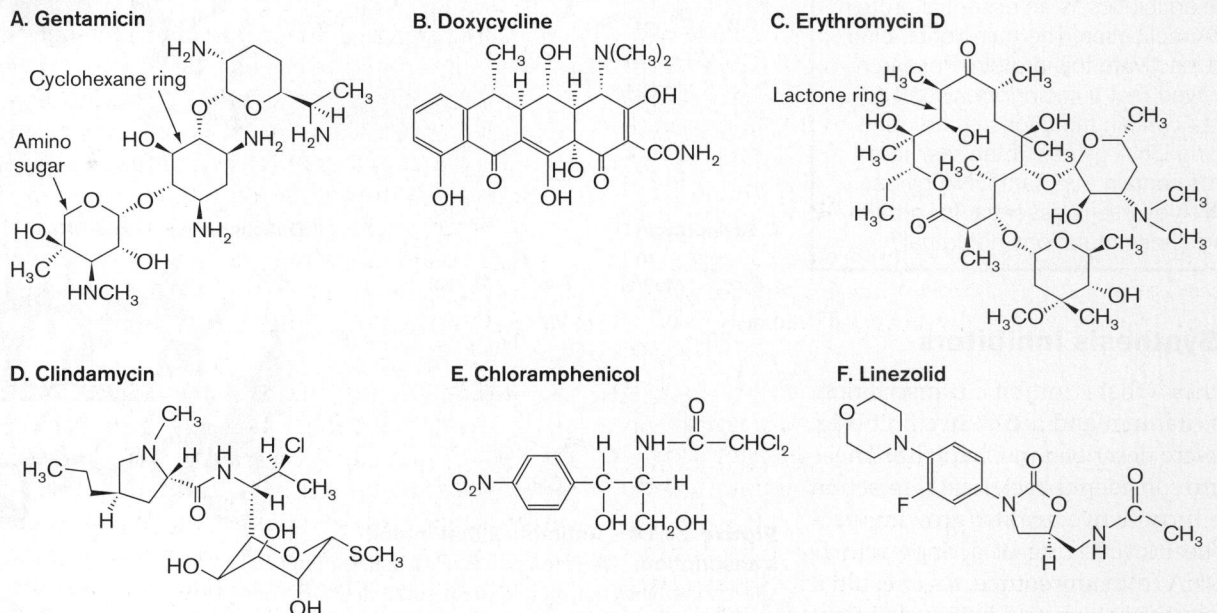

Figure 27.15   Protein synthesis inhibitors.   A. The aminoglycoside gentamicin. **B.** The tetracycline doxycycline. **C.** The macrolide antibiotic erythromycin. **D.** The lincosamide antibiotic clindamycin. **E.** Chloramphenicol. **F.** The oxazolidinone linezolid.

Drugs That Affect the 30S Subunit

The classification of antibiotics affecting protein synthesis is initially based on the bacterial ribosomal subunit targeted. Thus, one class of antibiotics interferes with 30S subunit function, and the other scrambles 50S subunit activities.

Aminoglycosides. There is considerable variation in structure among different aminoglycosides, but all contain a cyclohexane ring and amino sugars (**Fig. 27.15A**). The aminoglycosides are unusual among protein synthesis inhibitors in that they are bactericidal rather than bacteriostatic. Most of them bind 16S rRNA and cause translational misreading of mRNA, which is why these drugs are bactericidal. The resulting synthesis of jumbled peptides wreaks havoc with physiology and kills the cell. Streptomycin and gentamicin (see **Fig. 27.15A**) are two widely used drugs in this class. Ototoxicity (hearing damage) is a major, but uncommon, side effect of these antibiotics (approximately 0.5%–3% of patients treated with gentamicin suffer from this toxicity). Hearing is generally affected at frequencies above 4,000 Hz.

Tetracyclines. Tetracycline antibiotics are characterized by a structure with four fused cyclic rings—hence the name. **Figure 27.15B** shows one frequently used example, called doxycycline. Tetracyclines are bacteriostatic and work by binding to and distorting the ribosomal A site that accepts incoming charged tRNA molecules. Doxycycline is used to treat early stages of Lyme disease (*Borrelia burgdorferi*), acne (*Propionibacterium acnes*), and other infections. An important adverse side effect of tetracyclines is that they can interfere with bone development in a fetus or young child. Tetracycline use by pregnant mothers will also cause yellow discoloration of the infant's teeth. As a result, this drug is not recommended for pregnant women or nursing mothers.

Drugs That Affect the 50S Subunit

Five classes of drugs subvert translation by binding to the 50S ribosomal subunit. Most of these drugs were discussed in Chapter 8 and will be recapped here only briefly.

- **Macrolides**, all of which contain a 14- to 16-member lactone ring (**Fig. 27.15C**), inhibit translocation of the growing peptide (bacteriostatic action). Commonly prescribed examples are erythromycin and azithromycin. Azithromycin was the antibiotic used to treat the *Streptococcus pyogenes* infection in our case history, although other drugs could have been used. Because it is structurally dissimilar to any of the beta-lactam antibiotics, such as penicillin, it can be used safely in patients who are penicillin sensitive.

- **Lincosamides** (**Fig. 27.15D**), such as clindamycin, are similar to macrolides in function but have a different structure.
- **Chloramphenicol** (**Fig. 27.15E**) inhibits peptidyltransferase activity (bacteriostatic). Bone marrow depression leading to aplastic anemia is the most common serious side effect and limits its clinical use.
- **Oxazolidinones** (**Fig. 27.15F**) are a recently discovered class of synthetic antibiotics effective against many antibiotic-resistant microbes. In fact, this was the first new class of antibiotics discovered since the "golden age" of antibiotic discovery over 35 years ago. Oxazolidinones such as linezolid bind to the 23S rRNA in the 50S subunit of the prokaryotic ribosome and prevent formation of the protein synthesis 70S initiation complex. This is a novel mode of action; other protein synthesis inhibitors either block polypeptide extension or cause misreading of mRNA. Linezolid binds to the 50S subunit near where chloramphenicol binds, but it does not inhibit peptidyltransferase. Resistance is limited because most bacterial genomes have multiple operons encoding 23S rRNA. Usually more than one of these genes must mutate to confer high-level resistance. The more unmutated 23S rRNA genes there are, the more ribosomes susceptible to the antibiotic will be present. Oxazolidinones are useful primarily against Gram-positive bacteria. Gram-negative bacteria are intrinsically resistant because of multidrug efflux pumps (Section 27.6) and decreased permeability due to the outer membrane.
- **Streptogramins** (**Fig. 27.16**), produced by some *Streptomyces* species, fall into two groups, A and B. Streptogramins belonging to group A have a large nonpeptide ring. Streptogramin B members are cyclic peptides. The two groups differ in their modes of action, although both inhibit bacterial protein synthesis by binding to the peptidyltransferase site. Group A streptogramins bound to the peptidyltransferase site distort the ribosome to prevent binding of tRNA to the ribosome A site. In contrast, group B streptogramins are thought to narrow the peptide exit channel, preventing exit of the peptide and thereby blocking translocation.

Natural streptogramins are produced as a mixture of A and B, the combination of which is more potent than either individual compound alone (an example of synergy). In tribute to this synergistic action, the drug combination is marketed under the name Synercid. Synergy between the two drugs occurs because the A-type streptogramin alters the binding site for the B-type drug, increasing its affinity. Bacteria can develop resistance through ribosomal modification (the modification in 23S rRNA is the same one that provides resistance to

A. Streptogramin A

B. Streptogramin B

Figure 27.16   The streptogramins.
A. Streptogramin A is a large nonpeptide ring
structure. **B.** Streptogramin B is a cyclic peptide.

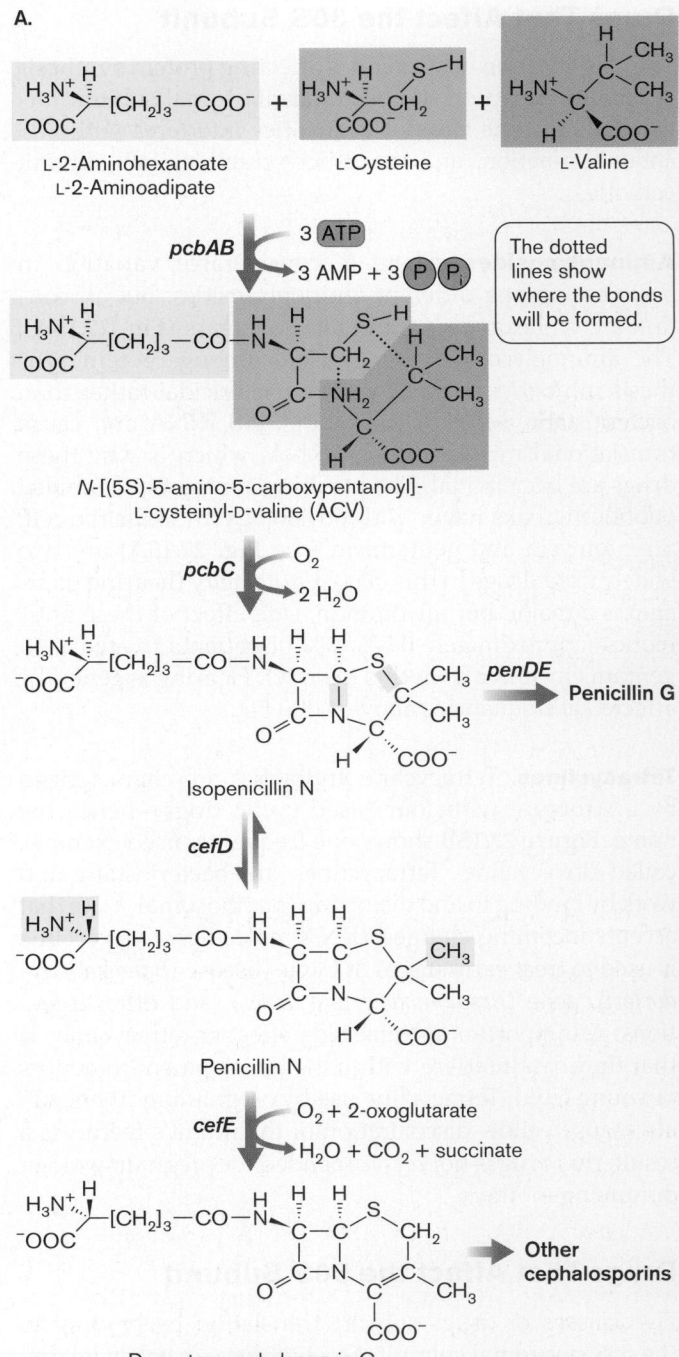

A.

L-2-Aminohexanoate
L-2-Aminoadipate   L-Cysteine   L-Valine

pcbAB   3 ATP
          3 AMP + 3 P P_i

The dotted
lines show
where the bonds
will be formed.

N-[(5S)-5-amino-5-carboxypentanoyl]-
L-cysteinyl-D-valine (ACV)

pcbC O_2
          $2 H_2O$

Isopenicillin N   **penDE** ➔ Penicillin G

cefD

Penicillin N

cefE O_2 + 2-oxoglutarate
          $H_2O + CO_2$ + succinate

➔ Other
cephalosporins

Deacetoxycephalosporin C

**Figure 27.17   Core biosynthetic pathway for penicillin and
the cephalosporins.   A.** Penicillin and cephalosporins share
a biochemical pathway. The *pcbAB*, *pcbC*, and *penDE* genes
encode ACV synthetase, IPN synthase, and IPN acyltransferase,
respectively. CefD is required to make penicillin N, a precursor
of the cephalosporins; CefE is a synthase that makes the first
cephalosporin structure in the biosynthetic sequence.

---

macrolides), via the production of inactivating enzymes,
or by active efflux of the antibiotic.

TO SUMMARIZE:

- **Antibiotic specificity** for bacteria can be achieved by
  targeting a process that occurs only in bacteria, not
  host cells; by targeting small structural differences
  between components of a process shared by bacteria
  and hosts; or by exploiting a physiological condition
  such as anaerobiosis present only in certain bacteria.
- **Antibiotic targets** include cell wall synthesis, cell
  membrane integrity, DNA synthesis, RNA synthesis,
  protein synthesis, and metabolism.
- **Antibiotics targeting the cell wall** bind to the
  transglycosylases, transpeptidases, and lipid carrier
  proteins involved with peptidoglycan synthesis and
  cross-linking.
- **Antibiotics interfering with DNA** include the anti-
  metabolite sulfa drugs that inhibit nucleotide synthesis, quinolones that inhibit DNA topoisomerases,
  and a drug, metronidazole, that, when activated, randomly nicks the phosphodiester backbone.
- **Inhibitors of RNA synthesis** target RNA polymerase (rifampin and pyronins) or bind DNA and
  inhibit polymerase movement (actinomycin D).
- **Aminoglycosides and tetracyclines** bind the 30S
  subunit of the prokaryotic ribosome.
- **A variety of antibiotics** bind the 50S ribosomal
  subunit and inhibit translocation (macrolides, lincosamides), peptidyltransferase (chloramphenicol),

B.

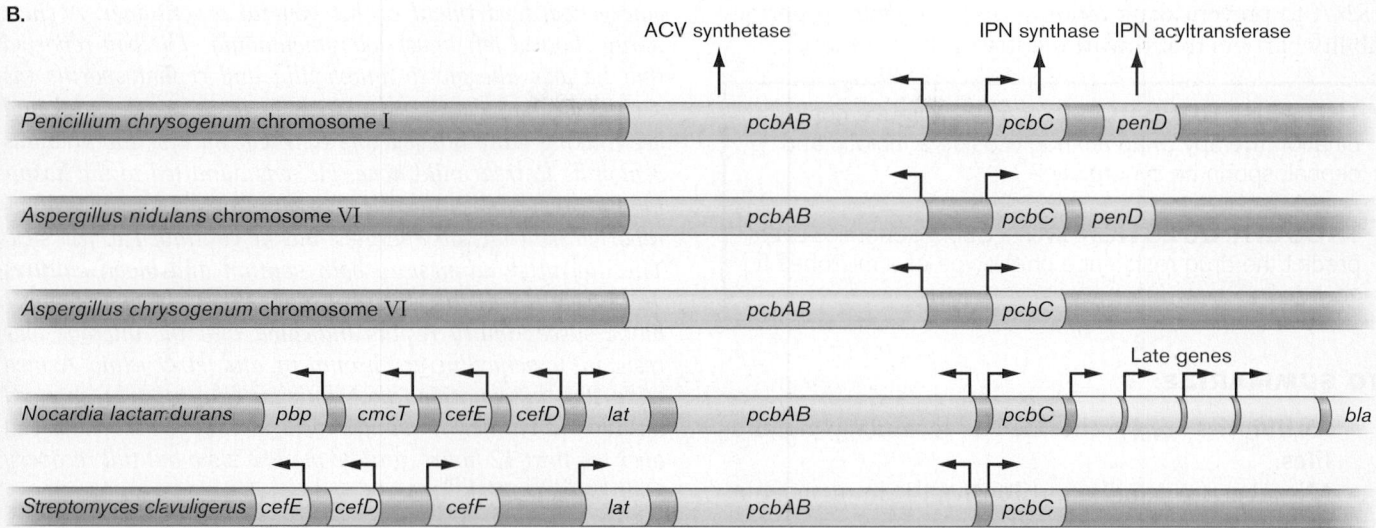

Figure 27.17 (continued)   B. Penicillin gene clusters in various fungi that synthesize penicillins and cephalosporins. The arrows indicate the orientation of the genes. Note that the *pcbAB-pcbC* promoter region is transcribed bidirectionally. Genes involved in cephalosporin synthesis are labeled *cef*. The *bla* gene encodes a beta-lactamase that protects the producing organism *Nocardia lactamdurans* and appears to serve as an ancestral homolog to *bla* genes found in clinical pathogens.

formation of 70S complex (oxazolidinones), or peptide exit through the ribosome exit channel (streptogramins).

27.5 Antibiotic Biosynthesis

Antibiotics are considered **secondary metabolites** because they often have no apparent use in the producing organism. This most likely means a purpose has yet to be identified. Today, antibiotic production might help one microbe compete with another in nature. Antibiotic production also can be useful in mutualistic relationships in which an antibiotic produced by a microbe protects the organism it colonizes from pathogens (as *Streptomyces* does for leaf-cutter ants; see Sections 17.6 and 18.4). Whatever its use today, growth inhibition may not have been the original purpose of secondary metabolite production. The relative complexity of the biosynthetic pathways suggests a more immediate purpose—for example, cell-cell signaling—that evolved into cross-species inhibition. Antibiotic biosynthetic pathways are sometimes relatively simple (as in penicillin) and sometimes quite complex. **Figure 27.17A** outlines the relatively straightforward biosynthetic pathway used to make penicillin and the related cephalosporins, all of which contain a beta-lactam ring. Note that the ring results from the joining of two amino acids: cysteine and valine. Several different types of fungi can produce these drugs. All use similar pathways and evolutionarily related genes (**Fig. 27.17B**).

Polyketide antibiotics such as erythromycin (**Fig. 27.15C**) are synthesized in a modular fashion by modular polyketide synthase enzymes (discussed previously, in Section 15.4). The synthesis strategy is reminiscent of the way fatty acids are made. Malonyl-ACP units containing different R groups are successively condensed by modular subunits of the polyketide synthase complex until elongation is terminated by the thioesterase. The repeating forms of polyketides enable these long-chain molecules to be synthesized by a small number of enzymes.

Curiously, peptide antibiotics such as bacitracin or the gramicidins are not synthesized using ribosomes. Complex biosynthetic pathways (enzyme factories) are involved that do not rely on mRNA. Nonribosomal synthesis of peptide antibiotics is discussed in Section 15.6 and eTopic 15.1.

One theory of why certain microbes make antibiotics is that they prevent the growth of competing organisms, but how do microbes that make antibiotics avoid committing suicide? Fungi that make penicillin do not face any consequence for having done so, because the organism does not contain peptidoglycan. Actinomycetes that produce compounds such as streptomycin or chloramphenicol, however, could be susceptible to their own secondary metabolite. Ribosomes isolated from *Streptomyces griseus*, for example, are fully sensitive to the streptomycin produced by this organism. *S. griseus* avoids killing itself in two ways. First, the organism synthesizes an inactive precursor of streptomycin, 6-phosphorylstreptomycin, which is secreted from the cell and, once outside the mycelium, becomes activated by a specific phosphatase. In addition, this streptomycete has an enzyme that inactivates any streptomycin that may leak back into the mycelium. Other organisms protect themselves by methylating key residues on their rRNA to prevent drug binding or by setting up permeability barriers that thwart reentry of the antibiotic.

THOUGHT QUESTION 27.6   Why might a combination therapy of an aminoglycoside antibiotic and cephalosporin be synergistic?

THOUGHT QUESTION 27.7   Could genomics ever predict the drug resistance phenotype of a microbe? If so, how?

TO SUMMARIZE:

- **Antibiotics are synthesized as secondary metabolites.**
- **Microbes may make antibiotics** to eliminate competitors in the environment.
- **Antibiotic producers prevent self-destruction** by means of various antibiotic resistance mechanisms.

27.6 Challenges of Drug Resistance

In this section we discuss the various ways microbes can become resistant to antibiotics and explore the evolutionary road to resistance. In addition, we address how resistance moves between species and how we can deal with the spread of resistance throughout the world.

Case History: Multidrug-Resistant Pneumonia

A 14-year-old boy with fever (39°C; 102.2°F), chills, and left-sided pleuritic chest pain was referred to a hospital emergency department by his general practitioner. A chest X-ray showed left lower-lobe pneumonia. The boy reported that he was allergic to amoxicillin and cephalosporins (as a child he had developed a rash to these agents) and had been taking daily doxycycline (tetracycline) for the previous 3 months to treat mild acne. He was admitted to the hospital and treated with intravenous erythromycin because of his reported beta-lactam allergies, but he continued to feel sick. The day after admission, both sputum and blood cultures grew Streptococcus pneumoniae. *After 48 hours, antibiotic susceptibility results indicated that the microbe was resistant to penicillin, erythromycin, and tetracycline. Armed with this information, the clinician immediately changed antibiotic treatment to vancomycin. The boy's fever resolved over the next 12 hours, and he made a slow but full recovery over the next week.*

Unfortunately, the scenario presented in this case is far too common and has become an extremely serious concern. **Figure 27.18** illustrates the rapid rise of penicillin resistance among *Streptococcus pneumoniae* in the world. Another instance of emerging antibiotic resistance is unfolding in Europe and the Far East. The non-Enterobacteriaceae Gram-negative rod *Acinetobacter baumanii* is increasingly seen as a dangerous cause of nosocomial infections. It commonly colonizes hospitalized patients, particularly those in intensive care units. Before 1998, there were almost no cases of multidrug-resistant *A. baumanii*. The rate is now as high as 8%. The organism is resistant to drugs as diverse as ciprofloxacin (a quinolone), amikacin (an aminoglycoside), penicillins, third-generation cephalosporins, tetracycline, and chloramphenicol. Imipenem, one of a relatively new class of beta-lactam drugs, is currently useful, but resistance to it is also likely to develop.

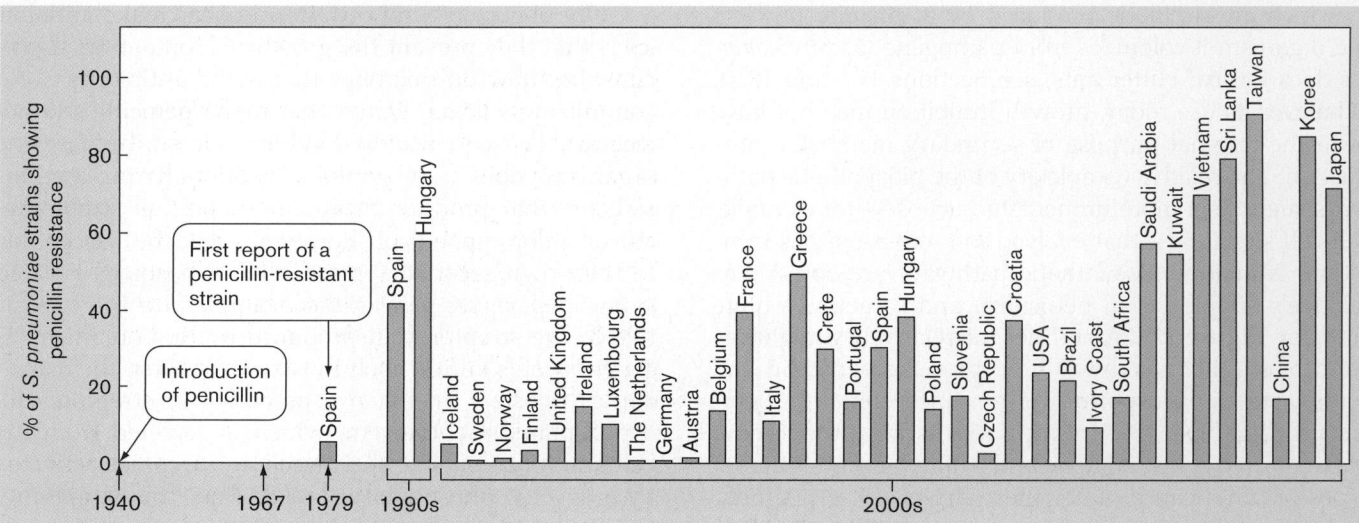

Figure 27.18   **The rise of penicillin-resistant *Streptococcus pneumoniae* throughout the world.**   Numbers reflect the number of penicillin-resistant strains among clinical isolates (strains of disease-causing bacteria isolated from patients from different countries). No resistance among clinical isolates was noted until after 1967.

There are four basic forms of antibiotic resistance. The resistant organism can:

- **Modify the target so that it no longer binds the antibiotic.** Mutations in key penicillin-binding proteins and ribosomal proteins, for instance, can confer resistance to methicillin and streptomycin, respectively. These mutations occur spontaneously and are not typically transferred between organisms.
- **Destroy the antibiotic before it gets into the cell.** One example is the enzyme beta-lactamase (or penicillinase), which is made exclusively to destroy penicillins. The sites of ring cleavage and structure of the enzyme are illustrated in **Figure 27.19**.
- **Add modifying groups that inactivate the antibiotic.** For instance, there are three classes of enzymes that modify and inactivate aminoglycoside antibiotics. The results of these types of enzyme modifications are illustrated for kanamycin in **Figure 27.20**.
- **Pump the antibiotic out of the cell using specific (for example, tetracycline export) and nonspecific transport proteins.** This strategy works because the pumps bail drugs out of the cell faster than the drugs can get in. Some are single-component pumps present in the cytoplasmic membrane of Gram-negative and Gram-positive bacteria (for example, NorA in *Staphylococcus aureus*, PmrA in *Streptococcus pneumoniae*, and the TetA and B proteins in Gram-negatives). Other drug efflux pumps are

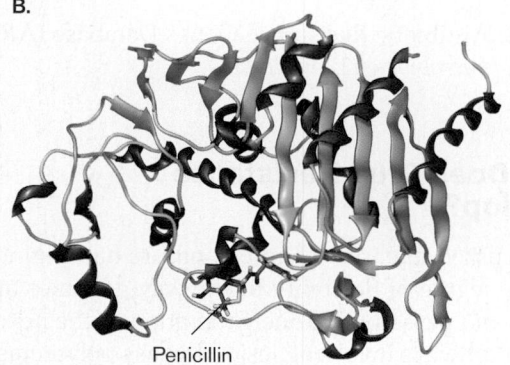

A.

Enzyme-Ser—OH +

β-Lactam

B.

Penicillin

Figure 27.19 Destroying penicillin. A. Beta-lactamase (or penicillinase) cleaves the beta-lactam ring of penicillins and cephalosporins. There are two types of penicillinases, based on where the enzyme attacks the ring. In either type, a serine hydroxyl group launches a nucleophilic attack on the ring. **B.** Structure of a beta-lactamase and location of the penicillin-binding site. (PDB code: 1XX2)

multicomponent systems present in Gram-negative bacteria only (discussed shortly). Efflux in either case is usually energized by proton motive force.

> **THOUGHT QUESTION 27.8** Fusaric acid is a cation chelator that normally does not penetrate the *E. coli* membrane, which means *E. coli* is typically resistant to this compound. Curiously, cells that develop resistance to tetracycline become *sensitive* to fusaric acid. Resistance to tetracycline is usually the result of an integral membrane efflux pump that pumps tetracycline out of the cell. What might explain the development of fusaric acid sensitivity?
>
> **THOUGHT QUESTION 27.9** Mutations in the ribosomal protein S12 (encoded by *rpsL*) confer resistance to streptomycin. Given a cell containing both $rpsL^+$ and $rpsL^R$ genes, would the cell be streptomycin resistant or sensitive? (Recall that genes encoding ribosome proteins for the small subunit are designated *rps*. A + indicates the wild-type allele, while R indicates a gene whose product is resistant to a certain drug.)

A particularly dangerous type of drug resistance is mediated by what are called **multidrug resistance (MDR) efflux pumps (Fig. 27.21)**. A single pump in this class can export many different kinds of antibiotics with little regard to structure. MDR pumps of Gram-negative microbes are similar to the ABC export systems described in Section 4.2. They include three proteins: an inner membrane pump protein (fueled by proton motive force—a distinction from true ABC exporters), an outer membrane channel connected to the pump protein, and an accessory protein that may link the other two proteins. For instance, the ArcB transporter, shown in **Figure 27.22**, almost indiscriminately binds antibiotics in a large central cavity (a promiscuous binding site) and uses proton motive force to move those compounds through a pore and out of a funnel that connects to an outer membrane channel, TolC.

Antibiotic efflux pumps are now believed to contribute significantly to bacterial antibiotic resistance because of the very broad variety of substrates they recognize and because of their expression in important pathogens. Strains of the pathogen *Mycobacterium tuberculosis*, for instance, have developed multidrug-resistant phenotypes in part because of MDR pumps. Approximately 2 million people die from tuberculosis annually, mostly in developing nations. What is even more alarming is that an increasing number of *M. tuberculosis* strains isolated from patients exhibit multidrug resistance. Although most of the antibiotic resistance in most *M. tuberculosis* multidrug-resistant strains is due to the accumulation of independent mutations in several genes, MDR pumps are thought to increase the level of resistance. Chemists typically try to tweak the structure of an antibiotic to overcome a specific type of resistance mechanism; but the MDR pumps act on an

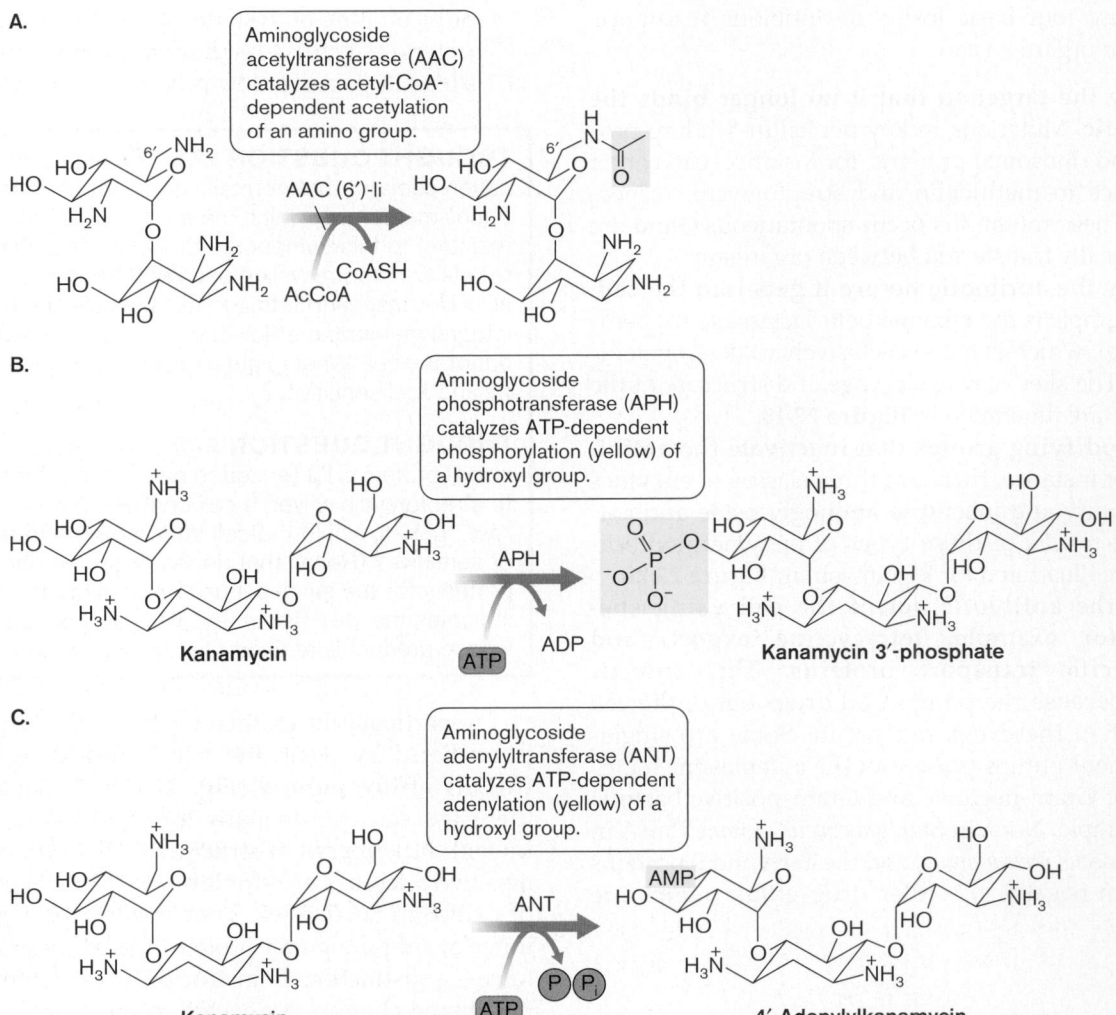

A.

Aminoglycoside acetyltransferase (AAC) catalyzes acetyl-CoA-dependent acetylation of an amino group.

AAC (6')-li

CoASH
AcCoA

B.

Aminoglycoside phosphotransferase (APH) catalyzes ATP-dependent phosphorylation (yellow) of a hydroxyl group.

Kanamycin

APH

ATP → ADP

Kanamycin 3'-phosphate

C.

Aminoglycoside adenylyltransferase (ANT) catalyzes ATP-dependent adenylation (yellow) of a hydroxyl group.

Kanamycin

ANT

ATP → P Pᵢ

4'-Adenylylkanamycin

Figure 27.20 Aminoglycoside-inactivating enzymes. Different enzymes can inactivate aminoglycoside antibiotics.

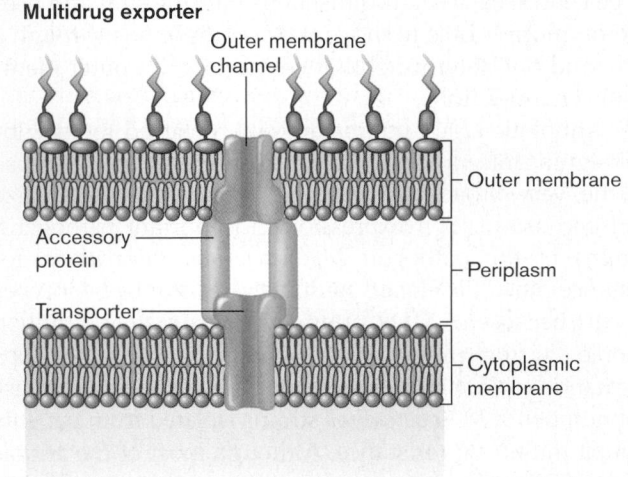

Multidrug exporter

Outer membrane channel

Outer membrane

Accessory protein

Periplasm

Transporter

Cytoplasmic membrane

Figure 27.21 Basic structure of a multidrug resistance efflux pump in Gram-negative bacteria. These efflux systems have promiscuous binding sites that can bind and pump a wide range of drugs out of the bacterial cell.

exceptionally wide range of antibiotics, almost without regard to structure.

WWW | Antibiotic Resistance Genes Database (ARDB) *microbiology2.com/links*

How Does Drug Resistance Develop?

As discussed in earlier chapters, nature has engineered a certain degree of flexibility in the way genomes are replicated and passed from one generation to the next. DNA repair pathways involving lesion bypass polymerases (for example, UmuDC; see Section 9.5) are thought to play a large role in randomized adaptive and evolutionary processes. For instance, at some point during evolution, gene duplication and mutational reshaping generated a gene product able to cleave the beta-lactam ring, producing an organism resistant to penicillin.

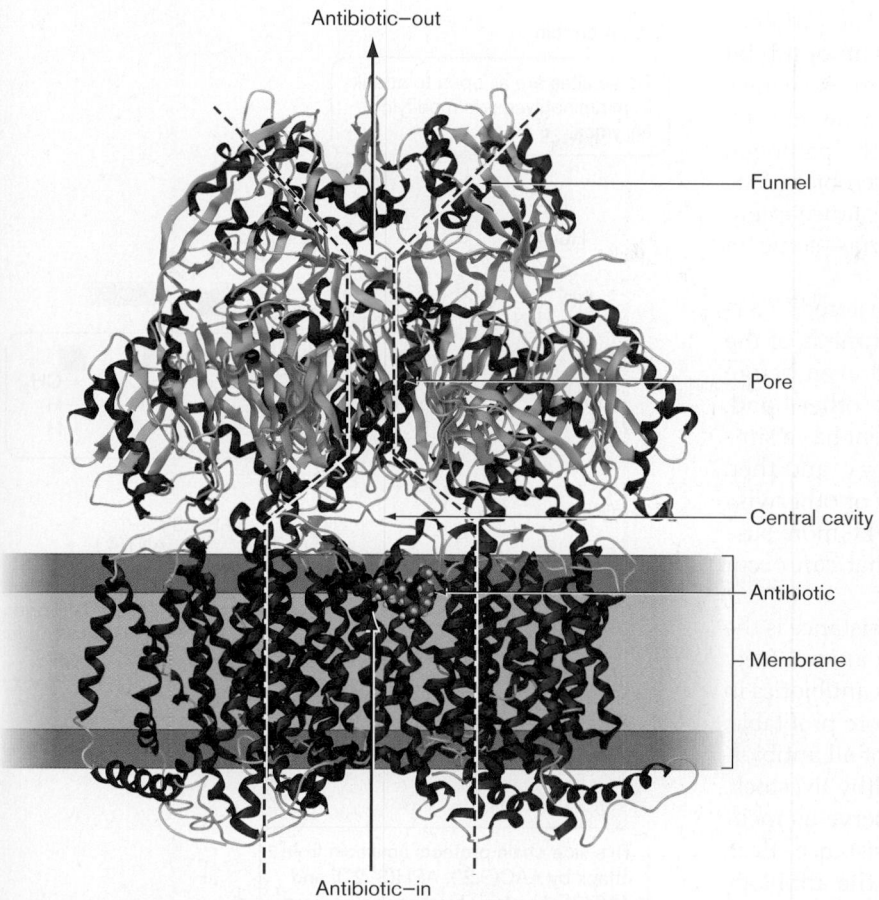

Figure 27.22 Structure of the *E. coli* ArcB multidrug resistance efflux pump. The structure is trimeric and driven by proton motive force. Dotted lines outline the central cavity, pore, and efflux funnel through which the drug is pumped. Connection to TolC in the outer membrane is not shown. (PDB code: 2GIF)

However, de novo antibiotic resistance through gene duplication and/or mutation does not occur in all species. Why reinvent the wheel—or, in this case, drug resistance? Gene transfer mechanisms such as conjugation, described in Chapter 9, can move antibiotic resistance genes from one organism to another and from one species to another. In fact, several drug resistance genes found in pathogenic bacteria actually had their start in the chromosomes of the drug-producing organisms and were passed on through gene swapping. For instance, *Streptomyces clavuligerus* produces a beta-lactamase (encoded by the *bla* gene) that protects this organism from the penicillin it produces. Transfer of a drug resistance gene is particularly evident if the gene has been incorporated into a plasmid and that plasmid is found in a new species.

An interesting case study of antibiotic resistance is provided by the Gram-positive bacterium *Enterococcus faecalis*, a natural inhabitant of the mammalian gastrointestinal tract that can cause life-threatening disease if granted access to other body sites (as in subacute bacterial endocarditis; see Section 26.7). *E. faecalis* is naturally resistant to numerous antibiotics, making disease treatment particularly challenging.

Vancomycin is one of the last lines of defense for treating serious *E. faecalis* infections. Unfortunately, increasing numbers of vancomycin-resistant strains have arisen in recent years. The completed genome sequence of one vancomycin-resistant strain illustrates the reason. The organism has an incredible propensity for incorporating mobile genetic elements that encode drug resistance. About a quarter of the genome consists of mobile or exogenously acquired DNA, including 7 probable phages, 38 insertion elements, numerous transposons, and integrated plasmid genes. One such mobile element encodes vancomycin resistance in the sequenced strain.

Recently, multidrug resistance in various microbes (for example, *Salmonella enterica*) has also been attributed to the presence of integrons. Integrons are gene expression elements that account for rapid transmission of drug resistance because of their mobility and ability to collect resistance gene cassettes (see eTopic 9.2).

We should note that the development of antibiotic resistance is not without consequence to the bacterium. Altering one aspect of an organism's physiology for the better may weaken another area. Bacteria may become resistant to a certain antibiotic, but that resistance comes at a price. For example, the altered DNA gyrase that affords resistance to quinolones may not function as well as the "normal" gyrase does. Thus, when both resistant and susceptible organisms cohabit the same environment, the wild-type (sensitive) strain may grow faster and eventually overwhelm the mutant strain—unless fluoroquinolone is present, of course.

How Did We Get into This Mess?

Consider the following case: A young mother brings her 4-year-old child to the physician. The child is screaming because he has an extremely painful sore throat. Simply looking at the throat is not diagnostic. The raw tissue could mean the child is suffering from a bacterial infection, in which case antibiotics are needed. Alternatively, a virus could be the cause—a situation in which antibiotics do nothing but pacify the parent. More often than not, the clinician will prescribe an antibiotic without ever knowing the cause of disease. The problem is this: The more an antibiotic is used, the more opportunities there are to select for an antibiotic-resistant organism. The presence of drug does not *cause* resistance but will kill off or inhibit the growth of competing bacteria that are sensitive, thereby allowing the resistant organism to grow to high numbers. Resistance can form not only in the pathogen itself, but also in a member of the normal microbiota. The danger, then, is that the gene imparting resistance might be horizontally transferred to other bacteria—some of them pathogens.

When *should* antimicrobials be administered? Certainly, in life-threatening situations where time is of the essence, antibiotics should be administered even before knowing the cause of the infection. On the other hand, the most prudent course to take when a patient has a simple infection is to confirm a bacterial etiology, and then prescribe. An exception may be in an elderly or otherwise immunocompromised individual, who may be more susceptible to secondary bacterial infections that can occur subsequent to viral disease.

Another proposed source of antibiotic resistance is the widespread practice of adding antibiotics to animal feed. No one quite knows why, but giving animals antibiotics in their food makes for larger, and therefore more profitable, animals. Some estimates suggest that 70% of all antibiotics used in the United States are fed to healthy livestock. The consequence is that the animals may serve as incubators for the development of antibiotic resistance. Even if the resistance develops in nonpathogens, the antibiotic genes produced can be transferred to pathogens. Efforts have been under way since the 1960s to curb this practice.

www | Video: "Antibiotics: Is a Strong Offense the Best Defense?"
microbiology2.com/links

Fighting Drug Resistance

Several strategies are being used to stay one step ahead of drug-resistant pathogens. In some instances, dummy target compounds that inactivate resistance enzymes have been developed. Clavulanic acid, for example, is a compound sometimes used in combination with penicillins such as amoxicillin. Clavulanic acid, a beta-lactam compound with no antimicrobial effect, competitively binds to beta-lactamases secreted from penicillin-resistant bacteria. Because the enzyme releases bound clavulanic acid very slowly, the amoxicillin remains free to enter and kill the bacterium. Another strategy is to alter the structure of the antibiotic in a way that sterically hinders the access of modifying enzymes. **Figure 27.23** illustrates how adding a side chain to gentamicin, converting it to amikacin, blocks the activity of various aminoglycoside-modifying enzymes. Of course, we are now seeing resistance to amikacin develop (this resistance is ribosome based, involving mutational alteration of the S12 protein or 16S rRNA).

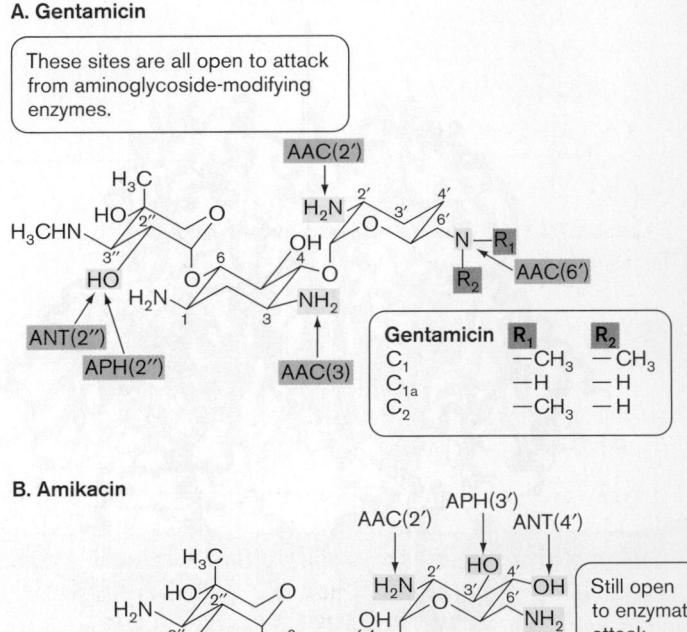

A. Gentamicin

These sites are all open to attack from aminoglycoside-modifying enzymes.

Gentamicin	R_1	R_2
C_1	—CH_3	—CH_3
C_{1a}	—H	—H
C_2	—CH_3	—H

B. Amikacin

Still open to enzymatic attack

This side chain protects amikacin from attack by AAC(3,2'), APH(3',2''), and ANT(2'') by steric hindrance.

Figure 27.23 Fighting drug resistance. A. Sites where gentamicin is vulnerable to enzymatic inactivation. AAC = aminoglycoside acetyltransferase; ANT = aminoglycoside adenylyltransferase; APH = aminoglycoside phosphotransferase. Inset shows the R groups for different gentamicin compounds: C_1, C_{1a}, and C_2. **B.** Gentamicin can be chemically modified at the highlighted sites to prevent loss of activity due to enzyme action. The side groups block access to enzyme active sites by steric hindrance (that is, the added group prevents the active site from interacting with its target structure) but do not inactivate the antibiotic.

Linking antibiotics is another strategy currently used to limit resistance. Recent advances have been made in linking a quinolone to an oxazolidinone to form a hybrid antibiotic with dual modes of action (**Fig. 27.24**). Because it has two modes of action, this hybrid antibiotic may limit the development of antibiotic resistance. Here's why: The rate of spontaneous resistance to a given antibiotic is roughly one out of 10^7 cells. For spontaneous resistance to develop to two antibiotics, that probability rises to one out of 10^{14}, making it very unlikely that an organism can become doubly drug resistant. However, multidrug resistance efflux pumps, integron cassettes, and plasmids carrying multiple antibiotic resistance genes can overwhelm that approach.

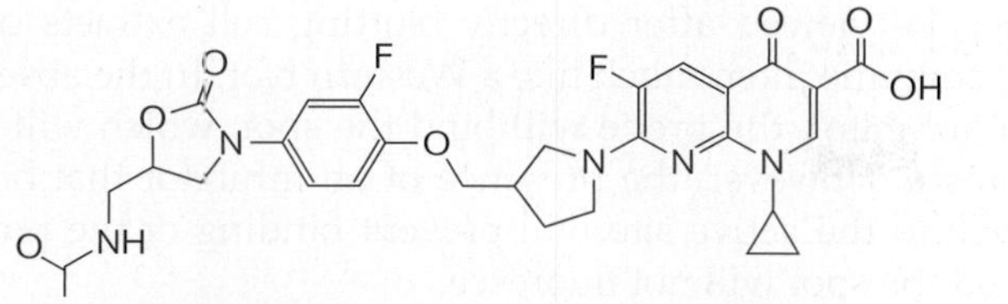

Figure 27.24   Quinolone-oxazolidinone hybrid.   Physically combining two different antibiotics may help reduce the emergence of drug-resistant bacteria.

Biofilms, Persisters, and the Mystery of Antibiotic Tolerance

A well-recognized puzzle of microbiology, unsolved for decades, is the problem of persister cells, cells that neither grow nor die in the presence of bactericidal agents. Joseph Bigger in 1944 noticed that penicillin would lyse a growing culture of *Staphylococcus aureus* but a small number of persister cells always survived. These persisters were not mutants made permanently resistant through mutation. They acted as though dormant.

In any biofilm or population of late exponential-phase cells, one can find persister cells that tolerate antibiotic treatment. It is now thought that persisters help explain antibiotic treatment failures and even latent bacterial infections such as recrudescent typhus or latent tuberculosis.

How might dormancy explain antibiotic tolerance? If persisters are dormant and have little or no cell wall synthesis, translation, or topoisomerase activity, then even if antibiotics can bind to them, their function cannot be corrupted. Tolerance enables antibiotic resistance at the price of not growing. Thus, an antibiotic can kill all susceptible bacteria in an infection, but the remaining persister cells serve as a source of population regrowth once the antibiotic is removed. You might say persister cells are the microbial equivalent of medieval monks whose simple but critically important role was to keep and protect learned texts. Persister cells, therefore, may be considered monk-like "keepers of the genome." So far, the molecular mechanisms that produce persister cells remain a mystery.

TO SUMMARIZE:

- **Antibiotic resistance** is a growing problem worldwide.
- **Mechanisms of antibiotic resistance** include modifying the antibiotic, destroying the antibiotic, altering the target to reduce affinity, and pumping the antibiotic out of the cell.
- **Multidrug resistance efflux pumps** use promiscuous binding sites to bind antibiotics of diverse structure.

- **Antibiotic resistance can arise spontaneously** through mutation, can be inherited by gene exchange mechanisms, or can arise de novo through gene duplication and mutational reengineering.
- **Indiscriminate use of antibiotics** has significantly contributed to the rise in antibiotic resistance.
- **Measures to counter antibiotic resistance** include chemically altering the antibiotic, using combination antibiotic therapy, and adding a chemical decoy.

27.7   The Future of Drug Discovery

Is humankind doomed to a future in which antibiotics will no longer work? The general consensus is no. Through prudent use of current antibiotics and innovative approaches for finding new ones, humans should continue to effectively control evolving bacterial pathogens.

How does one screen for new drugs that target specific proteins? Certainly the classic approach in which microbes, plants, and even animals collected from around the world are screened for their abilities to make new antibiotics is still valid and remains the most fruitful source of new drugs. However, the ability to search for novel bacterial drug targets or validate theoretical targets has been revolutionized by genome sequence analysis and associated genetic techniques. Powerful computational and bioinformatic methods are required in the initial identification and selection of molecular targets, followed by a series of postgenomic approaches to validate and characterize the targets, devise screens for effective inhibitor molecules, and pursue structure-based drug design to handle the inescapable emergence of future resistance. An example of a metagenomic approach to drug discovery is described in **Special Topic 15.2**.

The modern drug discovery process can be outlined as follows:

1. Identify new targets (such as unique essential enzymes) based on genomics.
2. Find or design compounds that inhibit the target in vitro and show that the inhibitory compound has actual antibacterial activity.
3. Show that the target within the bacterial cell is the same as the in vitro target.
4. Optimize the MICs against susceptible species by altering the compound's structure.
5. Examine the compound's spectrum of activity (Gram-positive, Gram-negative, aerobe, anaerobe, and so on) and the rate at which antibiotic resistance develops in pathogens.
6. Determine the new drug's toxicity to animals and humans, and its pharmacological properties (for instance, how long it stays at therapeutic levels in a patient).

The functions of genes that constitute potential new drug targets fall into three broad categories: those required only for growth of bacteria in laboratory media (in vitro expressed genes), those required only for bacterial infection in vivo (in vivo induced genes, discussed in Section 25.6), and those required for bacterial growth both in vitro and in vivo (housekeeping genes). Products of novel and essential genes falling into the second and third categories constitute new potential drug targets.

Take, for example, the case of *Streptococcus pneumoniae*. Twenty years ago, *S. pneumoniae* was uniformly sensitive to penicillin. Today, a growing number of strains have become resistant (as illustrated in the multidrug-resistant pneumonia case history presented earlier; also see **Fig. 27.18**). The discovery of proteins within this microbe that could potentially be targeted by new antibiotics would offer hope of future treatments. Recently, a computer-assisted strategy has helped identify several potential drug targets in this pathogen. The search consisted of scanning the pneumococcal genome for sequence motifs commonly found in cell surface–exposed or virulence-related proteins of other bacteria.

One such motif, called the choline-binding domain (CBD), is used by a variety of bacteria to interact with host cells. These domains consist of repeats of two to ten amino acids. The rationale used to find potential drug targets in the pneumococcus was to look for pneumococcal proteins that contain these domains. It was already known that various streptococcal species had one such protein, CbpA, which functions as a surface adhesin and plays an important role in nasopharyngeal colonization. By performing genome-wide searches with the C-terminal choline-binding region of *cbpA*, Khoosheh Gosink and his collaborators at St. Jude Children's Research Hospital (Memphis, Tennessee) identified six genes predicted to encode CBD-containing proteins ranging from 20 to 80 kDa (CbpD, E, F, G, I, and J). They constructed mutations in each gene and tested the mutants for pathogenicity. Mutants defective in CbpG adhered poorly to nasopharyngeal cells and were less virulent in a mouse infection model. Therefore, CbpG appears to play an important role in invasion and infection of the mucosa and the bloodstream. Its structural similarity to proteases suggests that it may be an excellent candidate for target-based drug development because numerous chemical inhibitors of proteases already exist.

Another advance in drug design utilizes a combination of genomic and proteomic approaches. The basis of this technique is to develop fluorescent molecules that bind only to *active* proteins of a given family of proteins. The probes can be devised by rational design in which knowledge of the active site structure allows chemists to engineer chemicals that interact with that site. A true inhibitor of that protein, and thus a potential antibiotic, will prevent binding of the tagged probe—an event that can be viewed after directly blotting cell extracts on a membrane filter, much like a Western blot. In the absence of inhibitor, the probe will bind the spot, which will fluoresce. However, the presence of an inhibitor that binds well to the active site will prevent binding of the probe, and the spot will not fluoresce.

Alternatively, if the probe binds to more than one protein, the same test can be performed using gel electrophoresis separation techniques. Proteins are separated on a nondenaturing gel so that the protein in question will retain its shape and its ability to bind probe. Tagged probe, with or without a potential inhibitor, is added to duplicate gels or a blot of those gels. The protein band will fluoresce in the absence of inhibitor or in the presence of a poor inhibitor. The band will not light up if an effective inhibitor is present.

The inhibitors are often made using more or less random combinatorial chemistry techniques, in which random variations of a molecule are made and tested. A library of these related compounds, which could number in the thousands, is robotically screened for inhibitory activity.

Genomic and combinatorial chemistries are certainly the cutting edge of antibiotic discovery. However, the old brute-force screening of natural products was recently used in a novel way to discover a promising new class of antibiotic. Merck scientists screened 250,000 natural product extracts for an ability to specifically inhibit bacterial fatty acid biosynthesis. Fatty acid biosynthesis is an attractive target for antibiotic development because the process and proteins involved are different from those of eukaryotes. The strategy specifically targeted the protein FabF because it is an essential component of fatty acid synthesis and is conserved among key pathogens such as *Staphylococcus aureus*.

Scientists engineered a strain of *S. aureus* to contain a gene expressing an antisense RNA to *fabF* mRNA. When the antisense RNA was induced, it bound to *fabF* mRNA and prevented efficient translation. As a result, the level of FabF protein in the cell decreased. The strain could still grow but would be exquisitely sensitive to any compound that targeted the remaining FabF protein. Fewer molecules of FabF present per cell means that fewer molecules of an active ingredient in a natural product are needed to stop fatty acid synthesis. Thus, zones of inhibition on an agar plate will be wider than they would be if cells produced normal amounts of FabF.

This novel targeted screening method led to the discovery of the antibiotic platensimycin, made by *Streptomyces platensis*, an organism isolated from South African soil. Platensimycin was subsequently shown to bind FabF and exhibit bacteriostatic, broad-spectrum activity, acting on Gram-positive and Gram-negative bacteria. It is only the third entirely new antibiotic developed in the last four decades. The novel chemical structure of platensimycin and its unique mode of action provide a great opportunity to develop a new class of critically needed antibiotics—a class that selectively targets fatty acid biosynthesis. More information on fatty acid biosynthesis and inhibitors can be found in Chapter 15.

Other intriguing ideas may lead to novel antimicrobial therapies. One involves using nanotubes to poke holes in bacterial membranes (**eTopic 27.2**). Another idea involves using photosensitive chemicals that can penetrate the microorganism and generate toxic reactive oxygen species (such as superoxide) when exposed to specific wavelengths of visible light (obviously good only for topical use). A third idea involves generating antimicrobials that "cork" the type III secretion apparatus that delivers virulence proteins into host target cells. Another clever approach is to interfere with quorum-sensing mechanisms of pathogens (**Special Topic 27.1**).

TO SUMMARIZE:

- **Potential targets for rational antimicrobial drug design** include proteins expressed only in vivo or proteins expressed both in vivo and in vitro.
- **Candidate antimicrobial compounds** can be designed to interact at the active site of a known enzyme and inhibit its activity.
- **Combinatorial chemistry** is used to make random combinations of compounds that can be tested for enzyme inhibitory activity and antimicrobial activity.

27.8 Antiviral Agents

A father pleading with a physician to give his child antibiotics when the infant is suffering with a cold is an all too common dilemma faced by the general practitioner, but there is nothing of substance the physician can do. The common cold is caused by the rhinovirus, and no antibiotic designed for bacteria can touch it. So why are there so few antiviral agents in the clinician's arsenal? The reason is that applying the principle of selective toxicity is much harder to achieve for viruses than it is for bacteria. Viruses routinely usurp host cell functions to make copies of themselves. Thus, a drug that hurts the virus is likely to also harm the patient. Nevertheless, there are several useful antiviral agents in which selective viral targets have been found and exploited. Some of these agents are listed in **Table 27.3**. Select examples are discussed in this section. Note that all of the molecular mechanisms of viruses presented in Chapter 11 are studied as potential drug targets.

Antiviral Agents That Prevent Virus Uncoating or Release

Membrane-coated viruses are vulnerable at two stages: when the virus is invading the host cell and after viral propagation, when the progeny viruses release from the host cell. The flu virus presents a good example of both.

Case History: Antiviral Treatment of Infant Influenza

A 9-month-old infant arrived at the Johns Hopkins Hospital with an acute onset of fever, cough, regurgitation from his gastrostomy feeding tube, and dehydration. This illness occurred following a series of chronic problems, including bronchiolitis (infection and inflammation of the bronchioles) caused by respiratory syncytial virus, and neonatal group B streptococcal sepsis. [Neonatal sepsis is often caused by Lancefield group B Streptococcus agalactiae, described in Chapter 28.]

Table 27.3 Examples of antiviral agents.

Virus	Agent	Mechanism of action	Result
Influenza	Amantadine	Inhibits viral M2 protein	Prevents viral uncoating
	Zanamivir	Neuraminidase inhibitor	Prevents viral release
Herpes simplex virus and varicella-zoster virus (shingles)	Acyclovir	Guanosine analog	Halts DNA synthesis
	Famciclovir	Prodrug of penciclovir, a guanosine analog	Halts DNA synthesis
Cytomegalovirus	Ganciclovir	Similar to acyclovir	Halts DNA synthesis
	Foscarnet	Analog of inorganic phosphate	Binds and inhibits virus-specific DNA polymerase
Respiratory syncytial virus and chronic hepatitis C	Ribavirin	RNA virus mutagen	Causes catastrophic replication errors
HIV	Zidovudine (AZT)	Nucleoside analog, resembles thymine	Inhibits reverse transcriptase
	Nevirapine	Binds to allosteric site	Inhibits reverse transcriptase
	Nelfinavir	Protease inhibitor	Prevents viral maturation
	Raltegravir	Integrase inhibitor	Prevents integration into host genome
	Maraviroc	CCR5 entry inhibitor	Prevents virus entry into host cells

Special Topic 27.1 | Disrupting Cell-Cell Communications to Prevent Infection: Good Idea?

Biofilms bolster the infectious process in many ways. From guarding against innate and adaptive immune mechanisms, to secreting toxins and producing antibiotic tolerance, biofilms serve as sturdy, infectious fortresses. **Figure 1** shows, for instance, a biofilm of *Pseudomonas aeruginosa* being built on a suture. This biofilm will constantly seed the wound with bacteria and will be hard to kill with antibiotics.

Since cell-cell signaling (quorum sensing) is critical to biofilm formation and function, jamming these signals may help resolve infections more quickly and prevent chronic infections from developing. As a result, research designed to destroy acyl homoserine lactones (AHLs, the chemical signal molecules) or interfere with their activities is under way. Some scientists are trying to use paraoxonases, natural enzymes that inactivate quorum-sensing signal molecules. Others are attempting to devise catalytic antibodies that can selectively

degrade the signal molecules. Underlying the latter approach is the idea that an antibody raised against a molecule resembling the transition form of an AHL undergoing hydrolysis will actually catalyze AHL hydrolysis.

An intriguing twist to these approaches was recently published by Vanessa Sperandio and colleagues. Her group studied the two-component signal transduction system consisting of QseC (a sensor kinase) and QseB (a response regulator). The QseCB system, present in enterohemorrhagic *E. coli* (EHEC), *Salmonella*, *Francisella*, and other important bacterial pathogens, recognizes an enigmatic bacterial signal molecule (autoinducer 3), but also appears to sense host adrenergic molecules (for example, epinephrine). Upon sensing any of these signals, QseCB activates transcription of key virulence genes. Sperandio's group screened 150,000 randomly synthesized small organic molecules for any that would interfere with QseC-dependent gene activation, the idea being that such a molecule would prevent induction of virulence genes. One compound was selected for its potency as compared with other molecules, minimal toxicity toward bacterial and human cell lines, and potential for chemical modification: LED209, or *N*-phenyl-4-{[(phenylamino) thioxomethyl] amino}-benzenesulfonamide (**Fig. 2**). When tested for effects

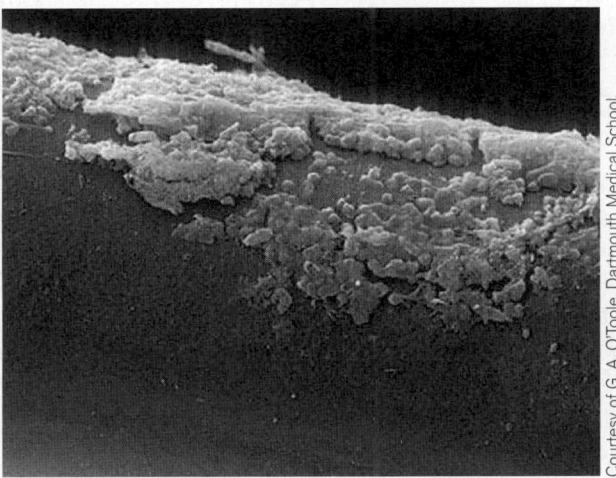

Courtesy of G. A. O'Toole, Dartmouth Medical School.

Figure 1 *Pseudomonas aeruginosa* **biofilm forming on a suture.** *Source*: T. R. de Kievit. 2009. *Environ. Microbiol.* **11**:279–288.

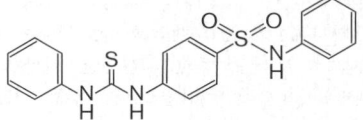

Figure 2 **The structure of LED209.** *Source*: D. A. Rasko et al. 2008. *Science* **321**:1078–1080.

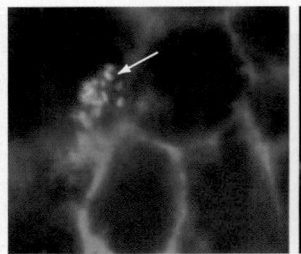 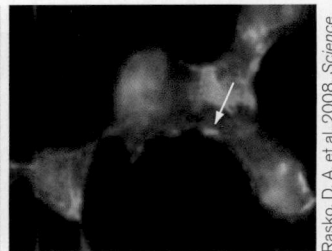

Rasko, D. A. et al. 2008. Science 321:1078–1080.

Figure 3 **LED209 inhibits pedestal formation by EHEC.** Cell nuclei and bacterial cells are stained red (with propidium iodide) and the cytoskeleton is stained green (with fluorescein-isothiocyanate phalloidin). **A.** EHECs are perched atop the green-stained, stemlike pedestals. The white arrow points to bacterial cells raised off the surface by short, bright-green pedestals. **B.** Treatment with 5-μM LED209 prevented pedestal formation. The white arrow points to a bacterial cell sitting on the host surface without a bright-green pedestal lifting it up. *Source*: D. A. Rasko et al. 2008. *Science* **321**:1078–1080.

Physical exam revealed fever, a severe cough resulting in respiratory distress, a rapid heart rate, and moderate dehydration. Nasopharyngeal aspirate was positive for influenza A antigen. The patient was treated with amantadine when influenza was diagnosed. He gradually improved and was discharged home 4 days after admission.

An unusually severe form of influenza spread across the United States in 2003. Most states reported a higher-than-normal number of influenza-related deaths of children and young adults. Several factors, however, helped keep the outbreak from becoming an epidemic of larger proportions. First, the administration of flu vaccine

on pathogenicity, the compound interfered with pedestal formation of EHEC—a key virulence feature of this organism (**Fig. 3**)—and reduced the virulence of *Francisella tularensis* in mice (**Fig. 4**).

While Sperandio's research sounds incredibly promising, we must remain cautious. There is a dearth of knowledge about interactions between bacterial signals and host signals. As recent studies seem to indicate, hosts may gauge their immune response by monitoring the signal molecules produced by pathogens and by their own microbiota. The danger of these therapies lies in developing antagonistic molecules that interfere with host and bacterial receptors or that destroy the signal molecules themselves, not knowing the consequence to the host. The possibility of unintended consequences must be recognized but does not diminish the exciting promise of this research. Further studies designed to unravel quorum-sensing networks that regulate virulence may lead to the discovery of new virulence factors and novel targets for vaccine and drug development.

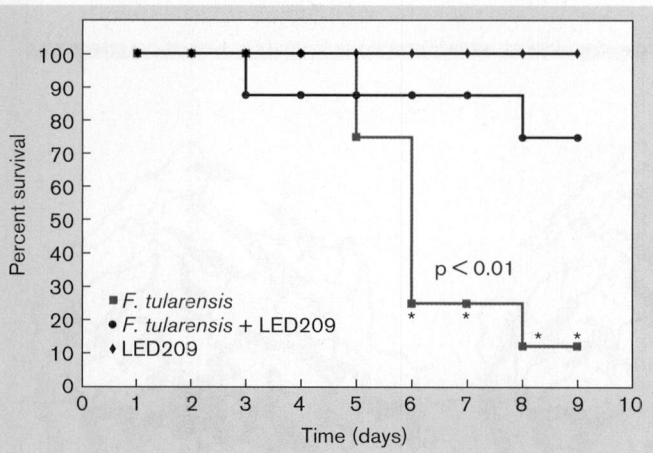

Figure 4   LED209 reduces virulence of *Francisella tularensis*.   Survival plot of mice infected intranasally with *F. tularensis* upon oral treatment with LED209 (20 mg/kg mouse weight). Asterisks indicate results that are significantly different from controls ($p < 0.01$). Source: Adapted from D. A. Rasko et al. 2008. *Science* **321**:1078–1080.

afforded the population what is called herd immunity (discussed in Section 26.9). Herd immunity occurs even when only a fraction of a population is immunized. This fraction of individuals will not become infected and so cannot spread the disease to others. At the very least, this slows the progression of infection throughout the

population. (It is impossible to immunize all humans against any given disease. However, it is estimated that immunizing two-thirds of a population can eliminate the disease by cutting off transmission. Even this level of immunization, however, is rarely achieved.)

The second factor that prevented an influenza pandemic was the availability of antiviral agents that can limit the disease course. As a result of studying the molecular biology of influenza, scientists discovered two selective targets. Influenza virus (200 nm) is encased in a membrane envelope donned when the virion buds from an infected cell. As described in Section 11.3, the envelope contains the viral proteins neuraminidase (NA) and hemagglutinin (HA). Spikes of hemagglutinin bind to glycoprotein receptors on the host cell and trigger receptor-mediated endocytosis. The virus ends up inside the resulting endosome. After the endosome is formed, proton pumps in the membrane acidify endosomal contents. The drop in pH changes the structure of hemagglutinin on the viral membrane so it can now bind to receptors on the endocytic membrane. The result is fusion of the two membranes and release of the virion into the cytoplasm. For this change in hemagglutinin structure to occur, it is essential that the *interior* of the enveloped virion also be acidified, and this is facilitated by the formation of a channel by the virus-encoded M2 envelope protein. Amantadine (**Fig. 27.25A**), a drug that has been used to treat some strains of the flu virus, is a specific inhibitor of the influenza M2 protein; however, it is useful against the influenza type A viral strains only. It is also effective against the avian H5N1 strain that has occasionally infected humans and is feared as the source of a future influenza pandemic. Unfortunately, amantadine-resistant strains are circulating in Asia in part because of the widespread use of amantadine by Chinese poultry farmers.

The second target of the influenza virus is the envelope protein neuraminidase. The newer antiflu drugs, such as zanamivir (Relenza), are **neuraminidase inhibitors** that act against types A and B influenza strains (**Fig. 27.25B** and **C**). Neuraminidase on the viral envelope allows virus particles to leave the cell in which they were made. Neuraminidase inhibitors prevent this release and cause the virus particles to aggregate at the cell surface, reducing the number of virus particles released. The contributions of different NA and HA genes to the severity of influenza are discussed in **Special Topic 27.2** and **eTopic 27.3**.

Amantadine and the neuraminidase inhibitors, when used within 48 hours of disease onset, decrease shedding and reduce the duration of influenza symptoms by approximately one day. However, flu symptoms generally last only from 3–10 days. While this does not sound like a substantial benefit, in the elderly, shortening the course of the flu can minimize damage to the lungs, which in

Special Topic 27.2	**Resurrecting the 1918 Pandemic Flu Virus**

As Chapter 25 describes, the influenza virus that swept the globe from 1918 to 1919 took the lives of 20–40 million people, making it a greater killer than the First World War. Why was the 1918 flu strain so much more virulent than any strain since? This question may now have been answered.

Teams of scientists working in fortress-like biohazard research facilities used forensic molecular biology techniques to re-create that deadly virus. Viral genomic RNA gene pieces were fished from tissues of a 1918 victim frozen in the Alaskan permafrost and from Formalin-treated archived tissues. Once sequenced and resynthesized, the eight viral genes were reassembled into a plasmid vector and sent to the laboratory of Terrence Tumpey at the Centers for Disease Control and Prevention. The plasmid was introduced into eukaryotic tissue culture cells that allowed the viral genes to function as a unit and produce infectious 1918 influenza virus.

Scientists then analyzed the ability of the resurrected strain to infect a variety of cell types and cause disease. An important difference noted between seasonal influenza strains and the 1918 version is that replication of typical flu strains is generally limited to lung cells, whereas the 1918 flu seemed able to replicate in many different types of cells. What could explain the more promiscuous replication of the 1918 strain? Previous work by Christopher Basler and colleagues revealed that the *specific combination* of hemagglutinin (HA) and neuraminidase (NA) from the 1918 strain was critical to virulence (**eTopic 27.3**). The researchers all knew that proteolytic cleavage of HA is required for the fusing of viral and host cell membranes and that most flu strains infect only those host cells that contain high levels of a particular protease typically found in lung cells. Experiments with the resurrected 1918 strain, however, revealed that the NA produced by the 1918 strain imparts an HA-cleaving ability to proteases found in many different cell types. Consequently, the 1918 flu strain can damage many other tissues beyond lung. It was the combination of HA and NA proteins from the 1918 strain that made the flu disaster possible.

Might today's antiviral drugs prevent a disaster if the 1918 flu strain reemerges? The answer, fortunately, is yes. Research has shown that typical M2 inhibitors such as amantadine will inhibit replication of reconstructed 1918 flu strains, affording some level of comfort as we await the arrival of the next great flu pandemic.

A. Amantadine   **B. Zanamivir**   **C.**

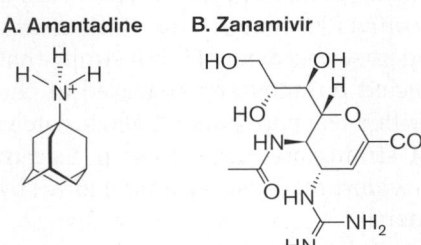
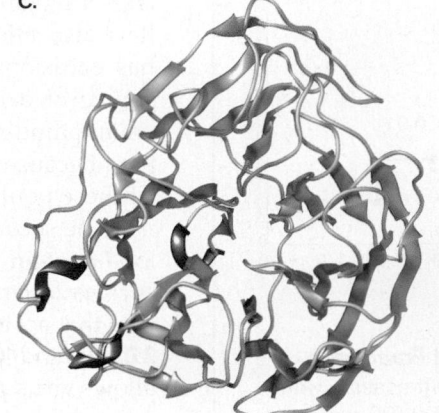
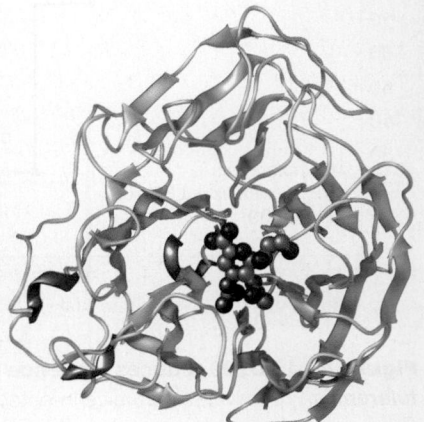

Figure 27.25 Inhibitors of influenza proteins. A. Amantadine inhibits the M2 protein. **B.** Neuraminidase inhibitor zanamivir. **C.** Neuraminidase without (left) and with (right) bound inhibitor. (PDB: 2HTQ)

turn reduces the chance of developing life-threatening secondary bacterial infections such as pneumonia and bronchitis.

Antiviral DNA Synthesis Inhibitors

Most antiviral agents work by inhibiting viral DNA synthesis. These drugs chemically resemble normal DNA nucleosides, molecules containing deoxyribose and analogs of adenine, guanine, cytosine, or thymine. Viral enzymes then add phosphate groups to these nucleoside analogs to form DNA nucleotide analogs. The DNA nucleotide analogs are then inserted into the growing viral DNA strand in place of a normal nucleotide. Once inserted, however, new nucleotides cannot attach to the nucleoside analogs, and DNA synthesis stops (**Fig. 27.26**). These DNA chain-terminating analogs are selectively toxic because viral polymerases are more prone to incorporate nucleotide analogs into their nucleic acid than are the more selective host cell polymerases. Antiviral DNA

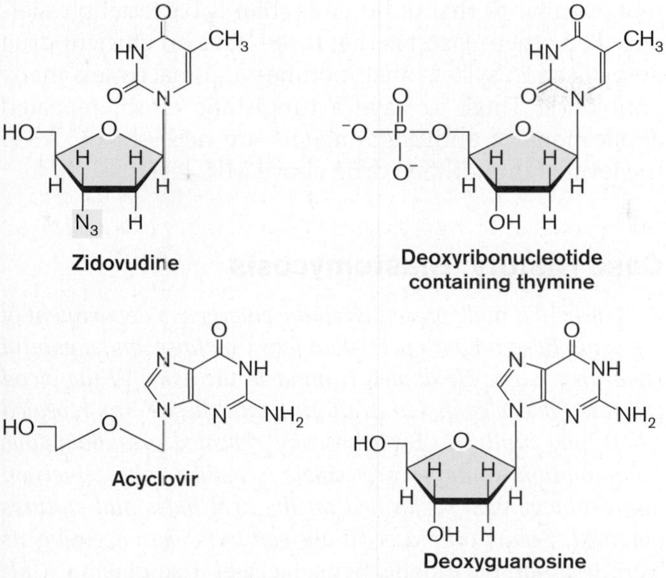

Figure 27.26   Antiviral inhibitors that prevent DNA synthesis. Zidovudine (AZT) and acyclovir are analogs of thymine and guanine nucleotides, respectively. Because the analogs have no 3′ OH to which another nucleotide can add, chain elongation ceases.

synthesis inhibitors work on DNA viruses or retroviruses, but not on viruses such as influenza with its RNA genome.

Case History: Treatment of HIV

A married couple come to the community clinic for prenatal care. He is 20 years old. She is 19 and reportedly 2 months pregnant with her first child. She denies intravenous (IV) drug use or a history of other sexual partners and has no history of sexually transmitted disease; however, a routine prenatal HIV antibody screen is reported as positive for HIV-1. Careful questioning of the patient and her husband elicits from him a history of IV drug use 5 years earlier. An HIV antibody screen for him is also positive. The laboratory results indicate that the wife may not yet require therapy (she has a low viral load—that is, less than 1,000 copies per milliliter of blood—and a high CD4 T-cell count), but since she is pregnant, a short course of antiretroviral therapy (AZT) will be helpful in preventing transmission of the virus to her child. The husband has an HIV viral load of 10,000 copies per milliliter and is started on combination antiretroviral therapy including a protease inhibitor.

Nucleoside and Nonnucleoside Reverse Transcriptase Inhibitors

Human immunodeficiency virus (HIV) is an RNA retrovirus that uses a reverse transcriptase to make DNA that then integrates into host nuclear DNA to form a provirus (discussed in Section 11.4). The antiretroviral drug zidovudine (abbreviated ZDV or AZT), which was used in the case history, is a nucleoside analog recognized by

reverse transcriptase. Once incorporated into a replicating HIV DNA molecule, the DNA chain-terminating property of AZT prevents further DNA synthesis. Because HIV transmission from the mother to the neonate can occur at delivery or by breast-feeding, treatment of the mother and the child are important steps to prevent transmission.

There are also nonnucleoside reverse transcriptase inhibitors. For example, the drug delavirdine binds directly to reverse transcriptase and allosterically inactivates the enzyme.

Protease Inhibitors

To make optimum use of its limited provirus DNA sequence, HIV makes long nonfunctional polypeptide chains that are proteolytically cleaved to make the actual proteins and enzymes used to replicate and produce new virions. For example, the *gag* and *pol* genes reside next to each other in the HIV genome and are transcribed as a single mRNA molecule (see Section 11.4). The Gag and Pol open reading frames overlap but are offset by one base. This mRNA produces two polyproteins, called Gag and Gag-Pol, the latter being the result of a shift in reading frames that takes place during translation. Once made, both polyproteins are cleaved by HIV protease. The Gag protein is proteolytically cleaved to make different capsid components (p17, p24, and p15, which is further cleaved to make nucleocapsid protein p7), while Gag-Pol is cleaved to make reverse transcriptase and integrase (**Fig. 27.27A**). Protease inhibitors such as Viracept and Lopinavir belong to a powerful new class of drugs that block the HIV protease (**Fig. 27.27B**). When the protease is inactivated, even though new virus particles are made the polyproteins remain uncleaved and the virus cannot mature. Because immature HIV particles cannot infect other cells, progress of the disease stalls. Note that protease inhibitors do not cure AIDS; they can only decrease the number of infectious copies of HIV.

Antiviral therapy with protease inhibitors is recommended for patients with symptoms of AIDS and for asymptomatic patients with viral loads above 30,000 copies per milliliter. Treatment should be considered even for patients with viral loads above 5,000 copies per milliliter, as for the husband in the case history presented earlier.

Because HIV can mutate rapidly and become resistant to single-drug therapies, treatment today involves administering combinations of three or more antiretroviral drugs. This therapeutic strategy is called Highly Active Antiretroviral Therapy (HAART). Most current HAART regimens include three drugs; usually two nucleoside reverse transcriptase inhibitors plus a protease inhibitor, a nonnucleoside reverse transcriptase inhibitor or an integrase inhibitor. Integrase inhibitors, first approved in 2007 (U.S.), block the enzyme needed to insert viral DNA into the host genome.

A.

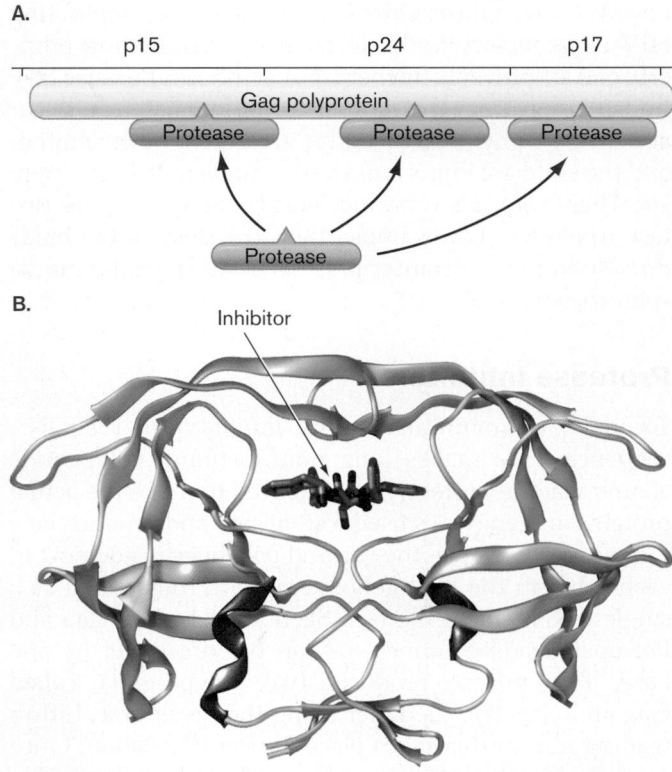

B.

Inhibitor

Figure 27.27   HIV protease inhibitor.   A. Representation of HIV protease cleavage of a single Gag polyprotein into multiple, smaller proteins. **B.** The protease enzyme is shown as a ribbon structure, while the protease inhibitor BEA 369 is shown as a stick model. (PDB code: 1EBY)

TO SUMMARIZE:

■ **Fewer antiviral agents** than antibacterial agents are available because it is harder to identify viral targets that provide selective toxicity.

■ **Preventing viral attachment to, or release from, host cells** is a mechanism of action for antiviral agents, such as amantadine and zanamivir, used to treat influenza virus.

■ **Inhibiting DNA synthesis** is the mode of action for most antiviral agents, although they work only for DNA viruses and retroviruses.

■ **HIV treatments** include reverse transcriptase inhibitors that prevent synthesis of DNA and protease inhibitors that prevent maturation of viral proteins into active forms.

27.9 Antifungal Agents

Fungal infections are much more difficult to treat than bacterial infections, in part because fungal physiology is more similar to that of humans than is bacterial physiology. The other reason is that fungi have an efficient drug detoxification system that modifies and inactivates many antibiotics. Thus, to have a fungistatic effect, repeated applications of antifungal agents are necessary to keep the level of unmodified drug above MIC levels.

Case History: Blastomycosis

A 37-year-old male presented to the emergency department of a Florida hospital with persistent fever, malaise, and a painful right-arm mass. He denied trauma to the arm. White blood cell count was elevated at 27,000/μl, and chest X-ray revealed a left-lung infiltrate. Bronchoscopy revealed granulomatous inflammation containing a single yeastlike mass. Incision and drainage was performed on the arm mass and cultures obtained. Serum cryptococcal antigen tests were negative, as were tests for Bartonella henselae *and* Toxoplasma. *Cultures from the right arm grew out a fungal form similar to that identified from the bronchoscopy specimens. A tentative diagnosis of* Blastomyces dermatitidis *was confirmed using PCR. The patient was placed on amphotericin B, and his fevers and leukocytosis subsequently subsided. His medication was changed to fluconazole for a recommended duration of 6 months.*

Superficial mycoses (fungal infections), such as athlete's foot, and systemic mycoses, such as blastomycosis, require very different treatments. Imidazole-containing drugs (clotrimazole, miconazole) are often used topically in creams for superficial mycoses (**Fig. 27.28A**). Others, such as itraconazole, are administered orally. Superficial mycoses include infections of the skin, hair, and nails, as well as *Candida* infections of moist skin and mucous membranes (for example, vaginal yeast infections). The imidazole-containing drugs appear to disrupt the fungal membrane by inhibiting sterol synthesis. Lamisil (a terbinafine compound) is a different class of agent that selectively inhibits ergosterol synthesis by fungi but not humans. It is currently a popular antifungal agent used to treat superficial mycoses. More chronic dermatophytic infections typically require another antifungal agent, called **griseofulvin**, produced by a *Penicillium* species (**Fig. 27.28B**). Griseofulvin disrupts the mitotic spindle and derails cell division (called metaphase arrest). This does not kill the fungus, but as the hair, skin, or nails grow and are replaced, the fungus is shed.

Vaginal yeast infections caused by *Candida* are often treated with nystatin, a polyene antifungal agent synthesized by *Streptomyces* that forms membrane pores (**Fig. 27.28C**). The name "nystatin," by the way, came about because two of the people who discovered it worked for the New York State Public Health Department.

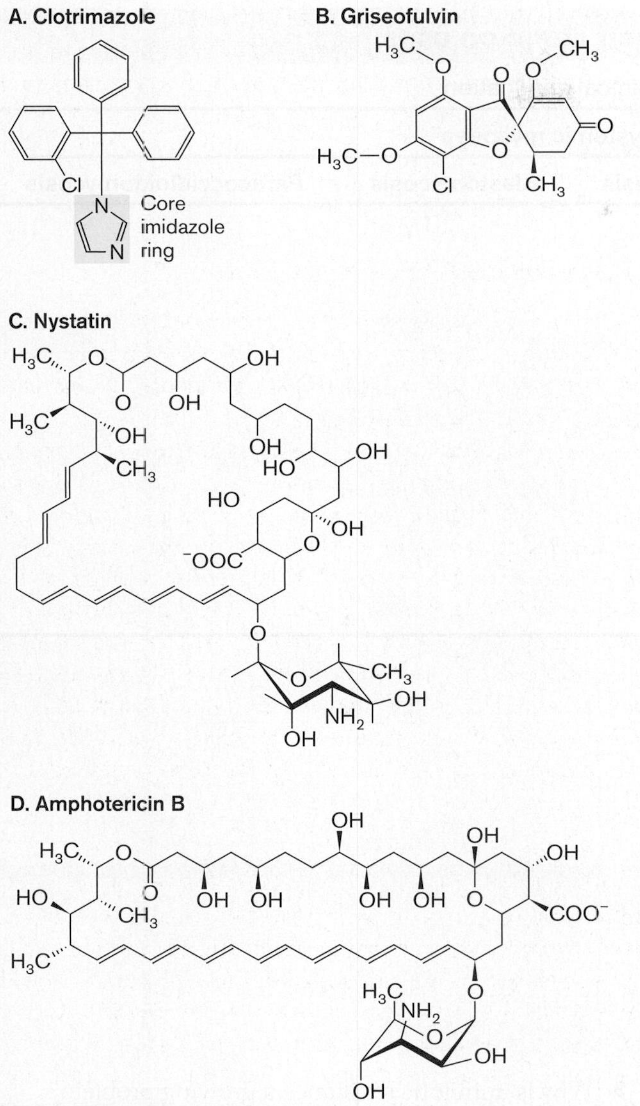

Figure 27.28   Examples of antifungal agents.
A. Clotrimazole belongs to the group of imidazole antifungals, so named because they all contain an imidazole ring. **B.** Griseofulvin is produced by *Penicillium griseofulvum*. **C.** Nystatin is a polyene macrolide produced by *Streptomyces noursei*. **D.** Amphotericin B is a polyene produced by *S. nodosus*.

The serious, sometimes fatal consequences of systemic mycoses require more aggressive therapy. The drugs used in these instances include **amphotericin B** (produced by *Streptomyces*; see **Fig. 27.28D**) and fluconazole. Amphotericin B binds to the sterols in fungal membranes and destroys membrane integrity. It has a high affinity for ergosterol, which is prevalent in fungal, but not mammalian, membranes. Fluconazole, on the other hand, inhibits the synthesis of ergosterol. Thus, fungal cells grown in the presence of fluconazole make defective membranes. Typically, curing systemic fungal infections requires long-term treatment to prevent disease relapse. **Table 27.4** lists a number of other commonly used antifungal agents.

TO SUMMARIZE:

- **Fungal infections** are difficult to treat because of similarities in human and fungal physiologies.
- **Imidazole-containing** antifungal agents inhibit sterol synthesis.
- **Griseofulvin** inhibits mitotic spindle formation.
- **Nystatin** produces membrane pores.
- **Amphotericin B** binds to membranes and destroys membrane integrity.

Concluding Thoughts

Antibiotics have done much to improve the health and well-being of people and animals around the globe. Although they have successfully kept at bay infectious diseases dreaded for centuries (such as the plague and tuberculosis), a crisis of antibiotic resistance looms because of the irresponsible use of antimicrobial agents. The fact that very few new, clinically useful antimicrobials have been discovered over the last quarter century should give us pause. Redoubling efforts to find new drugs, combined with the responsible use of existing antibiotics, is required to maintain our advantage over constantly evolving pathogens.

Table 27.4 **The major antifungal agents and their common uses.**

| | Clinical application | | | |
| | Systemic mycoses | | | |
Drug	Coccidioidomycosis	Histoplasmosis	Blastomycosis	Paracoccidioidomycosis
Polyenes				
Amphotericin B	+	+	+	+
Nystatin	–	–	–	–
Pimaricin	–	–	–	–
Imidazoles				
Clotrimazole	–	–	–	–
Miconazole	–	–	–	–
Ketoconazole	+	+	+	+
Triazoles				
Itraconazole	+	+	+	+
Fluconazole	+	?[b]	?	?
Antimetabolite				
5-Fluorocytosine[c]		–	–	–

[a] mc = mucocutaneous but not systemic candidiasis.

[b] Insufficient data.

[c] Used only in combination with amphotericin B.

CHAPTER REVIEW

Review Questions

1. What is selective toxicity? Provide examples.
2. Explain the difference between antibiotic susceptibility and antibiotic sensitivity.
3. What does the term "spectrum of antibiotic activity" mean?
4. Provide examples of bacteriostatic and bactericidal antibiotics.
5. What is the Kirby-Bauer test? Does it indicate whether a drug is bacteriostatic or bactericidal?
6. Give examples of drugs that target cell wall synthesis; RNA synthesis; protein synthesis; DNA replication. What are their modes of action?
7. What mechanism do producing organisms use to synthesize peptide antibiotics?
8. How do antibiotic-producing microorganisms prevent "suicide"?
9. Why is antibiotic resistance a growing problem?
10. What are the four basic mechanisms of antibiotic resistance?
11. Explain the basic concept of an MDR efflux pump.
12. Discuss the current concepts of the origin of antibiotic resistance.
13. What are some mechanisms used to combat the development of drug resistance?
14. Why are there few antiviral agents available to treat disease?
15. What is herd immunity?
16. How does amantadine inhibit influenza?
17. Discuss the general modes of action of antifungal agents.

Aspergillosis	Candidiasis	Cryptococcosis	Dermatophytosis	Other
+	+	+	–	
–	mc[a]	–	–	
–	–	–	–	Mycotic keratitis
–	mc	–	+	
–	mc	–	+	
–	+	–	+	
+	mc	+	+	Sporotrichosis
–	+	+	+	Sporotrichosis
+	+	+	–	Phaeohyphomycosis

Table header: Clinical application — Opportunistic mycoses

Thought Questions

1. Design a summary figure of a bacterium that illustrates the common targets of antimicrobial therapy.
2. A patient presented to the emergency room complaining of a nonproductive cough (no sputum) that had persisted for 6 weeks. The clinician prescribed a 7-day course of cephalosporin. After day 7 the patient returned, no better than when he started. Laboratory tests later showed the infection was caused by *Mycoplasma pneumoniae*. Explain why the antibiotic did not work.
3. How would you determine the MIC of an obligate intracellular pathogen such as *Rickettsia prowazekii*, the cause of typhus?

Key Terms

amphotericin B (1059)
antibiotic (1028)
bacitracin (1038)
bactericidal (1031)
bacteriostatic (1031)
chloramphenicol (1043)
cycloserine (1038)
gramicidin (1039)
griseofulvin (1058)
Kirby-Bauer assay (1033)
lincosamide (1043)

macrolide (1043)
minimal inhibitory concentration (MIC) (1032)
Mueller-Hinton agar (1035)
multidrug resistance (MDR) efflux pump (1047)
neuraminidase inhibitor (1055)
nosocomial (1028)
oxazolidinone (1043)
penicillin (1028)
penicillin-binding protein (PBP) (1037)

quinolone (1041)
secondary metabolite (1045)
selective toxicity (1030)
spectrum of activity (1031)
streptogramin (1043)
sulfa drug (1029)
transglycosylase (1036)
transpeptidase (1036)
vancomycin (1039)
zone of inhibition (1033)

Recommended Reading

Dagan, Ron. 2009. Impact of pneumococcal conjugate vaccine on infections caused by antibiotic-resistant *Streptococcus pneumoniae*. *Clinical Microbiology and Infection* **15**(Suppl.3):16–20.

David, Michael Z. and Robert S. Daum. 2010. Community-associated methicillin-resistant *Staphylococcus aureus*: epidemiology and clinical consequences of an emerging epidemic. *Clinical Microbioliological Reviews* **23**:616–687.

Davies, Julian C., and Diana Bilton. 2009. Bugs, biofilms, and resistance in cystic fibrosis. *Respiratory Care* **54**:628–640.

Fernandez-Lopez, Sara, Hui-Sun Kim, Ellen C. Choi, Mercedes Delgado, Juan R. Granja, et al. 2001. Antibacterial agents based on the cyclic D,L-α-peptide architecture. *Nature* **412**:452–455.

Fernandez-Tornero, Carlos, Rubens Lopez, Ernesto Garcia, Guillermo Gimenez-Gallego, and Antonio Romero. 2001. A novel solenoid fold in the cell wall anchoring domain of the pneumococcal virulence factor LytA. *Nature Structural Biology* **8**:1020–1024.

Freiberg, Christoph, and Heike Brötz-Oesterhelt. 2005. Functional genomics in antibacterial drug discovery. *Drug Discovery Today* **10**:927–935.

Gorski, Andrzej, Ryszard Miedzybrodzki, Jan Borysowski, Beata Weber-Dabrowska, Malgorzata Lobocka, et al. 2009. Bacteriophage therapy for the treatment of infections. *Current Opinion in Investigational Drugs* **10**:766–774.

Gosink, Khoosheh K., Elizabeth R. Mann, Chris Guglielmo, Elaine I. Tuomanen, and H. Robert Masure. 2000. Role of novel choline binding proteins in virulence of *Streptococcus pneumoniae*. *Infection and Immunity* **68**:5690–5695.

Hosaka, Takeshi, Mayumi Ohnishi-Kameyama, Hideyuki Muramatsu, Kana Murakami, Yasuhisa Tsurumi, et al. 2009. Antibacterial discovery in actinomycetes strains with mutations in RNA polymerase or ribosomal protein S12. *Nature Biotechnology* **27**:462–464.

Jayaraman, R. 2008. Bacterial persistence: Some new insights into an old phenomenon. *Journal of Bioscience* **33**:795–805.

Johnson, Kirk W., Denene Lofland, and Heinz E. Moser. 2005. PDF inhibitors: An emerging class of antibacterial drugs. *Current Drug Targets—Infectious Disorders* **5**:39–52.

Kloss, Patricia, Liqun Xiong, Dean L. Shinabarger, and Alexander S. Mankin. 1999. Resistance mutations in 23S rRNA identify the site of action of the protein synthesis inhibitor linezolid in the ribosomal peptidyl transferase center. *Journal of Molecular Biology* **294**:93–101.

Kohanski, Michael A. Daniel J. Dwyer, and James J. Collins. 2010. How antibiotics kill bacteria: from targets to networks. *Nature Reviews Microbiology* **8**:423–435.

Kohanski, Michael A., Daniel J. Dwyer, Boris Hayete, Carolyn A. Lawrence, and James J. Collins. 2007. A common mechanism of cellular death induced by bactericidal antibiotics. *Cell* **130**:797–810.

Marquez, Beatrice. 2005. Bacterial efflux systems and efflux pumps inhibitors. *Biochimie* **87**:1137–1147.

Paulsen, I. T., L. Banerjee, G. S. Myers, K. E. Nelson, R. Seghadri, et al. 2003. Role of mobile DNA in the evolution of vancomycin-resistant *Enterococcus faecalis*. *Science* **299**:2071–2074.

Shah, P. M. 2005. The need for new therapeutic agents: What is the pipeline? *Clinical Microbiology and Infection* **11**(Suppl.3):36–42.

Tumpey, Terrence M., and Jessica A. Belser. 2009. Resurrected pandemic influenza viruses. *Annual Review of Microbiology* **63**:79–98.

Turnidge, John, and David J. Paterson. 2007. Setting and revising antibacterial susceptibility breakpoints. *Clinical Microbiology Reviews* **20**:391–408.

Walsh, Christopher. 2003. *Antibiotics: Actions, Origins, Resistance*. ASM Press, Washington, DC.

Wang, Jun, Stephen M. Soisson, Katherine Young, Wesley Shoop, Srinivas Kodalil, et al. 2006. Platensimycin is a selective FabF inhibitor with potent antibiotic properties. *Nature* **441**:358–361.

 For further study and review, please visit StudySpace at **microbiology2.com**.

Chapter 28

Clinical Microbiology and Epidemiology

Throughout history, infectious diseases have killed more people than all of our wars combined. Our present success in controlling the spread of disease is due in large part to worldwide surveillance agencies that are equipped to detect outbreaks quickly, before major epidemics develop. These agencies rely on smaller clinical microbiology laboratories scattered throughout the world that help clinicians diagnose infectious diseases.

This chapter discusses the principles of clinical microbiology and epidemiology that are used to identify, treat, and contain outbreaks. How do clinical laboratories know what organisms to suspect in a given case? And what tests will unequivocally identify the right pathogen? How are the *real* pathogens found among the normal microbiota? Finding the *source* of an outbreak is also critical. How do epidemiologists identify "patient zero" and recognize and contain emerging diseases? Last but not least, we'll consider bioterrorism. Clinical scientists and epidemiologists are trained to detect bioterrorist attacks using the same principles they employ to detect naturally occurring infectious diseases.

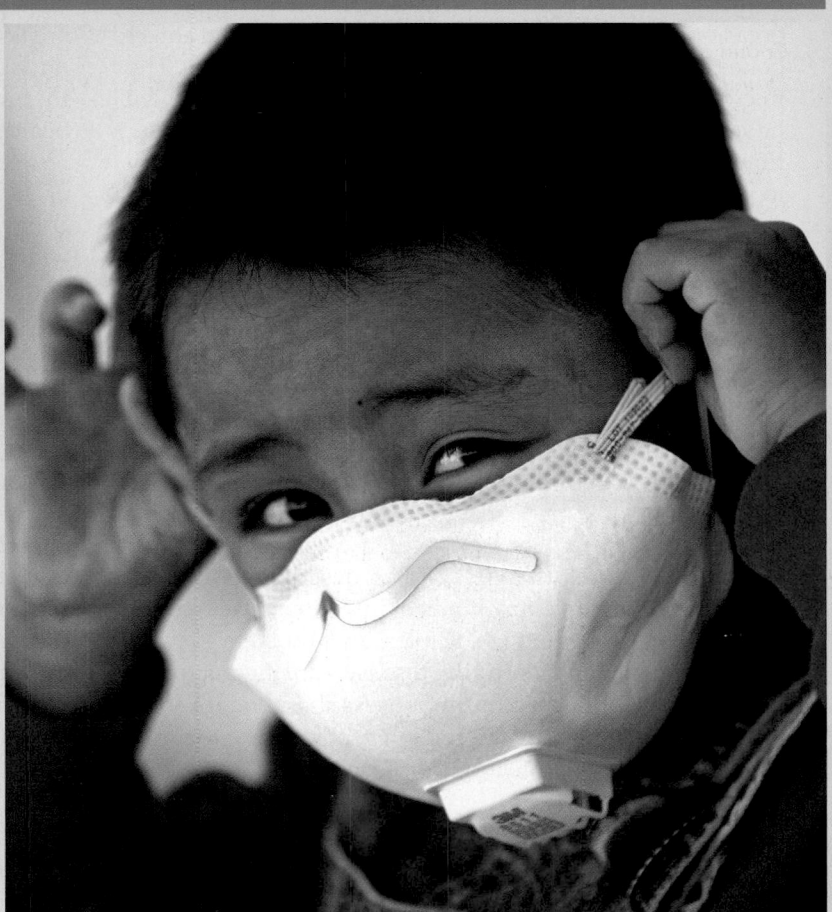

Current Research Highlight

The 2009 global outbreak of influenza A (H1N1) in humans appears to have originated in Mexico from infected pigs. The portrait is of "patient zero," the earliest documented human case, from La Gloria, Veracruz. The patient is now well. He lives near a pig farm but a direct connection to his infection has not been established. Initial epidemiological analyses of the pandemic suggested a case fatality ratio of 0.4% and indicated that transmission frequency was higher than in typical seasonal flu strains. Children were infected twice as often as adults. Medical detectives used the microbial detection and tracking strategies outlined in this chapter to trace the spread of the pandemic around the globe. The information gathered led to the implementation of numerous control measures, including vaccinations, designed to limit the global impact of this virus in terms of illnesses and deaths. Source: Christophe Fraser et al. 2009. *Science* **324**:1557–1561. Photo courtesy of Erich Schlegel/Rapport/Newscom.

Wilfrid Saintilus lay immobile on a straw mat when first seen by Dr. Paul Farmer in 1996. The 33-year-old peasant farmer had been sent home without hope by health care workers in a small clinic near his village in Haiti. He was pronounced paralyzed from the waist down; it was said that nothing more could be done. Farmer, however, quickly realized that Wilfrid's problem was not paralysis but a long-term *Salmonella* infection, very likely contracted from an unclean drinking-water supply—a problem for 80% of Haiti's population. Farmer realized that, while normally a gastrointestinal pathogen, *Salmonella* can occasionally disseminate via the bloodstream and infect bones, causing a painful disease called osteomyelitis. The microbe had infected Wilfrid's hip, causing pain so severe he could not move. His leg muscles had atrophied over time, and he now weighed under 100 pounds. Contrary to prediction, Wilfrid did not die. Dr. Farmer's diagnostic skills and antibiotics saved his life.

Thousands of stories like this, most with more tragic outcomes, underscore the integral roles that clinical microbiology and epidemiology have in health care. In Wilfrid's case, a simple laboratory test would have revealed a fully treatable *Salmonella* infection and prevented his enormous physical and emotional pain.

As in Chapters 24–27, we use case histories in Chapter 28 to introduce various strategies and methodologies used by modern medicine to diagnose infectious diseases. The approach is, by necessity, selective. Not all techniques, nor all diseases, can be described. Our goal is not to catalog the many kinds of infectious diseases but to demonstrate general principles and problem-solving approaches used in identifying disease-causing microbes. It will become apparent in the process that modern tools of clinical microbiology are indispensable in the diagnosis and treatment of infectious disease.

28.1 Principles of Clinical Microbiology

As in any good detective mystery, the first step in investigating an infectious disease is to identify the most likely suspects. This can be accomplished, in part, from observing the disease symptoms in a patient and knowing what organisms typically produce those symptoms. It also helps if the physician is aware of a similar disease outbreak under way in the community. Beyond these clues, the etiological agent must be identified through biochemical, molecular, serological, or antigen detection strategies.

Why Take the Time to Identify an Infectious Agent?

This is the first question many students ask when contemplating the effort and expense required to identify the genus and species of an organism causing an infection. Why not simply treat the patient with an antibiotic and be done with it? While this approach sounds appealing, there are several compelling reasons for identifying an infectious agent.

Many bacteria are resistant to certain antibiotics. As discussed in Chapter 27, antibiotic resistance is an increasingly serious global problem. Characterizing a microbe's antibiotic resistance profile is thus part of any microbial identification process. Understanding which antibiotics are effective in treating an infectious agent will help the physician avoid prescribing an inappropriate drug. Furthermore, knowing which drugs are ineffective enables health organizations to track the spread of antibiotic-resistant strains. For example, before 1970, most strains of *Neisseria gonorrhoeae* were susceptible to penicillin. Today, most are penicillin resistant, in part because of the widespread use of penicillin to treat gonorrhea during and after the Vietnam War, when U.S. soldiers first contracted penicillinase-producing *N. gonorrhoeae* (PPNG).

There are strain-specific disease complications. Many diseases have serious complications that are common to a given organism or strain of organism. For example, children whose sore throats are caused by certain strains of *Streptococcus pyogenes* can develop serious complications affecting the heart and kidney long after the infection has resolved. These complications are the immunological consequence of bacterial and host antigen cross-reactivity; they are called **sequelae** because they occur *after* the infection itself is over. Life-threatening sequelae such as rheumatic fever and acute glomerular nephritis caused by certain strains of *S. pyogenes* produce severe damage to the heart and kidney, respectively. Knowing early on that *S. pyogenes* has caused a child's sore throat allows the physician to prescribe antibiotics that will quickly eradicate the infection and prevent development of the sequelae.

> **NOTE:** Penicillin or a penicillin-like antibiotic is the usual treatment for *Streptococcus pyogenes* infection. Contrary to what you might expect, *S. pyogenes* has not developed any penicillin-resistant strains. It is not understood why.

Tracking the spread of a disease can lead to its source. Consider a situation in which ten infants scattered throughout a city develop bloody diarrhea. The clinical laboratory identifies the same strain of *Shigella sonnei*, a Gram-negative bacillus, as the cause in each case. Finding the same strain in all cases suggests they probably originated from the same source. Shigellosis is transferred from person to person by what is called the fecal-oral route. Carriers of *Shigella* will shed this organism in their feces. Inadequate hand washing after defecation will leave bacteria on the hands, which can then transfer the pathogen to foods or utensils or to another person by touching. Uninfected persons can become infected after eating the contaminated food or placing their fingers in their mouth.

Public health officials, armed with the knowledge that all the patients have the same strain of *Shigella*, then question the parents and learn that all of the children attend the same day-care center. By testing the other children and workers in that center, officials can stop the infection from spreading. This investigative process, called **epidemiology**, is covered more completely later in this chapter.

TO SUMMARIZE:

Identifying a pathogen enables us to:

- Use appropriate **antibiotics**, if necessary.
- Anticipate possible **sequelae**.
- **Track the spread** of the disease.

28.2 Approaches to Pathogen Identification

Pathogens can be identified on the basis of a variety of observations, including the patient's symptoms, the presence of organisms in stained clinical specimens, biochemical clues, and serology (the presence in a patient's serum of antibodies reactive against a specific microbe).

Bacterial Pathogen Identification Requires Knowledge of Microbial Physiology and Genetics

Thousands of bacterial species are capable of causing disease. How can a laboratory quickly, sometimes within hours, identify which one causes a given infection? The solution, in part, comes from knowing the biochemical and enzymatic features of these bacteria. Because no two species have the same biochemical "signature," the clinical lab can look for reactions, or combinations of reactions, that are unique to a given species.

The same is now true at the genetic level. In the current genomics-dominated era, DNA sequences of many pathogens have been completely elucidated. This has enabled design of polymerase chain reaction (PCR) primers that rapidly detect species-specific genes or even genes that are unique to highly pathogenic strains within a species. One example of such a gene is the attaching and effacing locus (*eae*) that is present in enterohemorrhagic, but not commensal, strains of *E. coli*. Finding *E. coli* in a fecal stool sample is not unusual. Finding *E. coli* with the *eae* gene, however, indicates that it is a pathogen.

Case History: Medical Detective Work

A 38-year-old woman with no significant previous medical history came to the emergency room complaining of a mild sore throat persisting for 3 days. Her symptoms included arthralgia (joint pain), myalgia (muscle pain), and low-grade fever. The day before, she had had a severe headache with neck stiffness, nausea, and vomiting. She was not taking any medications, had no known drug allergies, and did not smoke. She lived with her husband and two children, all of whom were well. Cerebrospinal fluid (CSF) was collected from a spinal tap. The CSF appeared cloudy (it should be clear) and contained 871 white blood cells per microliter (normal is 0–10/µl); the glucose level was 1 mg/dl (normal is 50–80); and the total protein level was 417 mg/dl (normal is under 45). Gram stain of a CSF smear revealed Gram-negative rods. The CSF sample was sent to the diagnostic laboratory for microbial identification.

As discussed in Thought Question 26.7, low glucose and elevated protein levels in CSF are indicators of *bacterial* infection, not viral. The increase in white blood cells revealed that the woman's immune system was trying to fight the disease. The presence of Gram-negative rods in the CSF smear confirmed a diagnosis of bacterial meningitis, since CSF should be sterile. Now it was up to the clinical laboratory to determine the etiological agent.

Problem-Solving Algorithms to Identify Bacteria

How does a clinical microbiology laboratory handle incoming specimens such as the one considered here? What are the first steps toward identifying the etiological agent of a disease? Over the years, clinical microbiologists have developed algorithms (step-by-step problem-solving procedures) that expose the most likely cause of a given infectious disease. For instance, there are only a limited number of microbes known to cause meningitis. The microbiologist poses a series of binary yes/no questions about the clinical specimen in the form of biochemical or serological tests. Typical questions in this case might include: Is an organism seen in the CSF of a patient with symptoms of meningitis? Is the organism Gram-positive or Gram-negative? Does it stain acid-fast? Answers to a first round of questions will then dictate the next series of tests to be used.

Because speed is of the essence in deciding how to treat the patient, a series of tests is not always carried out sequentially. To save time, a slew of tests are carried out simultaneously, but the results are interpreted sequentially using the algorithm. In our case history, for instance, consider the most common causes of bacterial meningitis: *Neisseria meningitidis, Streptococcus pneumoniae, Haemophilus influenzae,* and *Escherichia coli.* The CSF sample was Gram-stained by a microbiologist and simultaneously plated onto three media: chocolate agar, blood agar, and Hektoen agar. Chocolate agar is an extremely rich medium that looks brown owing to the presence of heat-lysed red blood cells (**Fig. 28.1A** and **B**). Because it is so nutrient-rich, all four organisms will grow on chocolate agar. However, nutritionally fastidious organisms such as *N. meningitidis* and *H. influenzae* will not grow well, if at all, on ordinary blood agar because these bacteria cannot lyse red blood cells and release required nutrients. Less fastidious organisms, such as *S. pneumoniae* and *E. coli*, will grow on blood agar, but only *E. coli* can grow on Hektoen agar (**Fig. 28.1C–E**), which is a selective and differential medium for enteric Gram-negative rods. Differential and selective media are described in Section 4.3.

In our case history, the Gram stain of the CSF revealed Gram-negative rods, which ruled out *Neisseria* (a Gram-negative diplococcus) and *Streptococcus pneumoniae* (a Gram-positive diplococcus). The organism in CSF did grow on blood agar, which eliminated *Haemophilus influenzae* (a Gram-negative, nonenteric rod) as a candidate. It also grew on Hektoen, where it produced orange lactose-fermenting colonies. Thus, the organism was a Gram-negative, enteric rod and likely *E. coli*. Additional biochemical tests confirming the identity of the organism had to be carried out, but this simple example shows how simultaneous tests can be interpreted.

Identifying Gram-negative bacteria. The Gram-negative bacterium in this case was subjected to a battery of biochemical tests packaged as 20 separate chambers in a patented analytical profile index (API 20E) strip that can be used for pathogen identification (**Fig. 28.2**). The strip requires overnight incubation and tests whether

A. Chocolate agar

B. Colonies of *N. gonorrhoeae*

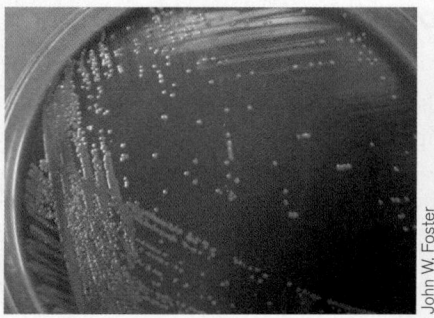

C. Hektoen agar

D. *E. coli* **colonies on Hektoen**

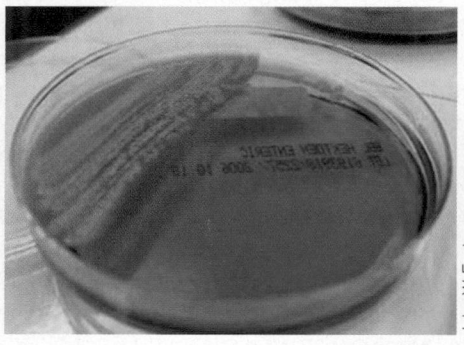

E. *S. enterica* **colonies on Hektoen**

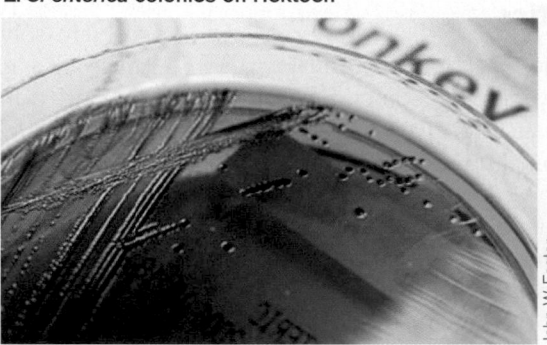

Figure 28.1   Chocolate agar and Hektoen agar: two widely used clinical media.   A. Uninoculated chocolate agar. Its color is due to gently lysed red blood cells that provide a rich source of nutrients for fastidious bacteria. **B.** Chocolate agar inoculated with a species of *Neisseria*. This organism will not grow well on typical blood agar, because important nutrients remain locked within intact red blood cells. **C.** Uninoculated Hektoen agar, which contains lactose, peptone, bile salts, thiosulfate, an iron salt, and the pH indicators bromothymol blue and acid fuchsin—the bile salts prevent growth of Gram-positive microbes. **D.** Hektoen agar inoculated with *Escherichia coli*. This organism ferments lactose to produce acidic fermentation products that give the medium an orange color owing to the pH indicators acid fuchsin and bromophenol blue. **E.** Hektoen agar inoculated with *Salmonella enterica*. This organism does not ferment lactose but instead grows on the peptone amino acids. The resulting amines are alkaline and produce a more intense blue color with bromothymol blue. *Salmonella* species also produce hydrogen sulfide gas from the thiosulfate. Hydrogen sulfide reacts with the medium's iron salt to produce an insoluble, black iron sulfide precipitate visible in the center of the colonies.

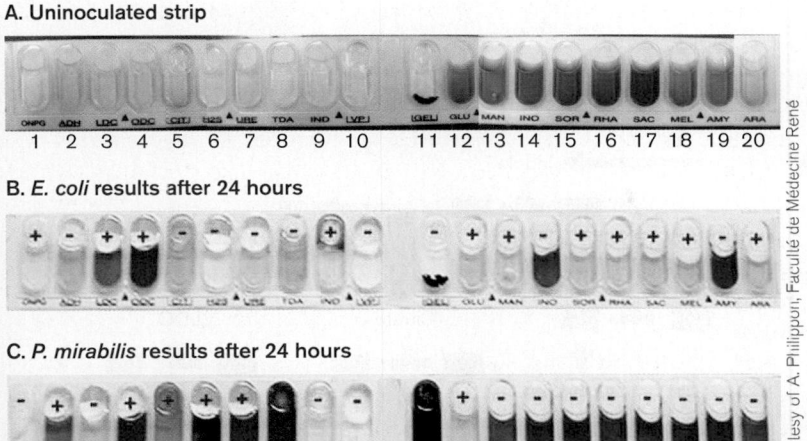

A. Uninoculated strip

| ONPG | ADH | LDC | ODC | CIT | H2S | URE | TDA | IND | LVP | | GEL | GLU | MAN | INO | SOR | RHA | SAC | MEL | AMY | ARA |
| 1 | 2 | 3 | 4 | 5 | 6 | 7 | 8 | 9 | 10 | | 11 | 12 | 13 | 14 | 15 | 16 | 17 | 18 | 19 | 20 |

B. *E. coli* results after 24 hours

C. *P. mirabilis* results after 24 hours

Courtesy of A. Philippon, Faculté de Médecine René Descartes

Figure 28.2 API 20E strip technology for the biochemical identification of Enterobacteriaceae. A. Uninoculated API strip. Each well contains a different medium that tests for a specific biochemical capability. The numbers correspond to Table 28.1. The color of the media after 24-hour incubation indicates a positive or negative reaction (see Table 28.1). **B.** API results for *E. coli*. Plus (+) and minus (−) indicate positive and negative reactions, respectively. **C.** API results for *Proteus mirabilis*.

the organism can ferment a series of carbon sources. It also provides evidence of specific end products produced as a result of fermentation. The results appear as different-colored reactions in each chamber and are scored as positive or negative, depending on the color (**Table 28.1** and **eTopic 28.1**). For example, in the indole

Table 28.1 Reading the API 20E.

	Tests	Reaction tested
1	ONPG	Beta-galactosidase
2	ADH	Arginine dihydrolase
3	LDC	Lysine decarboxylase
4	ODC	Ornithine decarboxylase
5	CIT	Citrate utilization
6	H_2S	H_2S production
7	URE	Urea hydrolysis
8	TDA	Deaminase
9	IND	Indole production
10	VP	Acetoin production
11	GEL	Gelatinase
12	GLU (glucose)	Fermentation/oxidation
13	MAN (mannitol)	Fermentation/oxidation
14	INO (inositol)	Fermentation/oxidation
15	SOR (sorbitol)	Fermentation/oxidation
16	RHA (rhamnose)	Fermentation/oxidation
17	SAC (sucrose)	Fermentation/oxidation
18	MEL (melibiose)	Fermentation/oxidation
19	AMY (amylose)	Fermentation/oxidation
20	ARA (arabinose)	Fermentation/oxidation

chamber (ninth well from the left in **Fig. 28.2B**), a red reaction at the top of the tube is positive and indicates that the organism can produce indole from tryptophan. A colorless chamber would be a negative result. Similar API strips are available for the identification of Gram-positive bacteria and yeasts.

The results of the API chambers can be interpreted in two ways. First, the results can be used as a dichotomous key, with lab personnel making a stepwise interpretation that begins with a key reaction, such as lactose fermentation. The reaction is read as positive or negative depending on the color. Then, following a printed flowchart, the technician goes to the next key reaction, indole production, and reads it as positive or negative. If the organism is a lactose fermenter and the indole test is positive, the choices have been narrowed to *E. coli* or *Klebsiella* species. Another reaction is read to distinguish between the remaining choices. The process continues until a single species is identified. A simplified example is shown in **Figure 28.3** using a limited number of enteric Gram-negative species. In reality, many more reactions than the 11 shown have to be used to make a definitive identification, because of species differences with respect to a single reaction. For example, **Figure 28.3** shows that *K. pneumoniae* and *K. oxytoca* exhibit opposite indole reactions, even though they are of the same genus. Even within a given species, only a certain percentage of strains might be positive for a given reaction. The inherent danger in using a dichotomous key is that one anomalous result can lead to an incorrect identification.

In a second approach, the API strips can be used to generate a seven-digit number that identifies the bacterium. To generate this number, individual reactions are given numerical values based on whether they give a positive or negative result, and then the number values from three reactions are added to produce a single digit. Because there are 21 reactions (the twenty-first reaction is an oxidase test performed on colonies), the end result will be a seven-digit number. The example in **Figure 28.4A** shows how this is done. Once the number is obtained, it is compared to a database of numbers generated by computer. The database was compiled by taking all possible test results and applying them to all species of Enterobacteriaceae. Each species has a known probability of being positive for a given reaction. As discussed in Section 17.5, taking each species and multiplying the probabilities for the results generated by the test organism will generate a probability score. The test organism is identified as the species with the highest score. This probability approach is considerably more accurate than the dichotomous key.

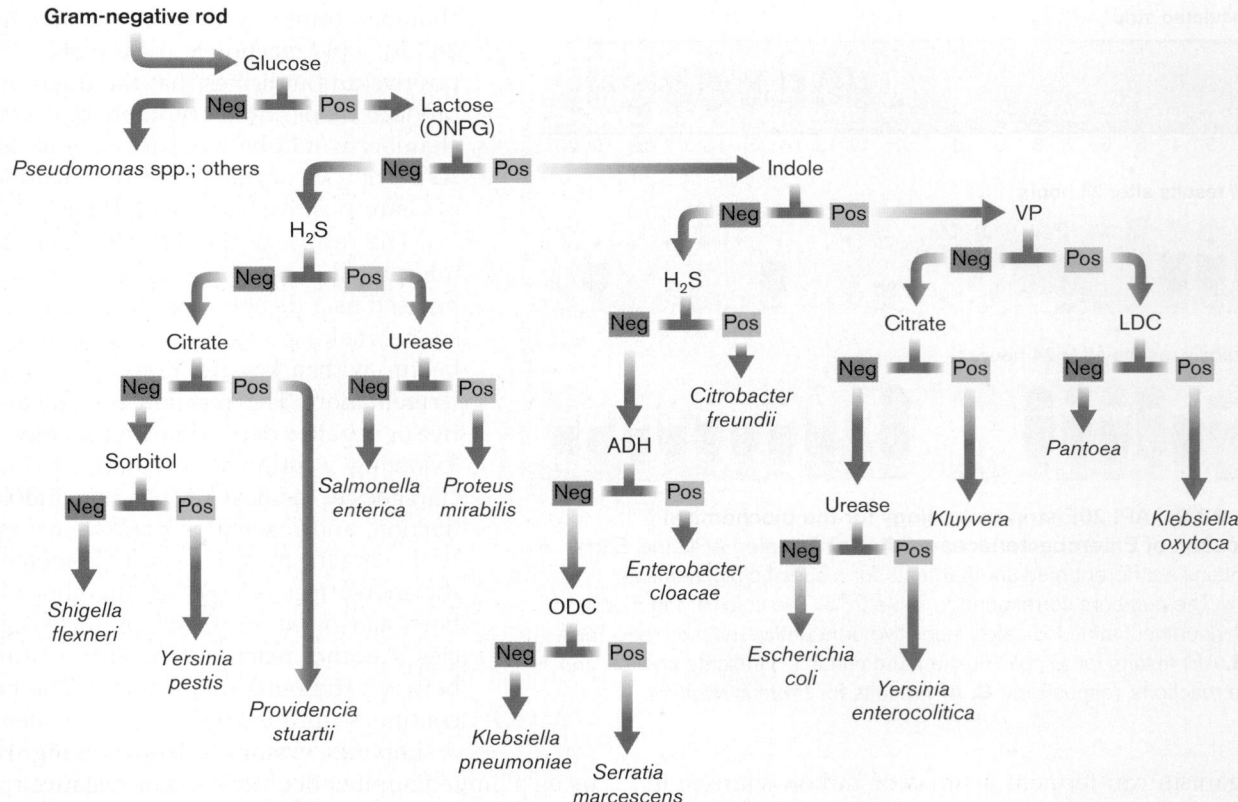

Figure 28.3 **Simplified biochemical algorithm to identify Gram-negative rods.** The diagram presents a dichotomous key using a limited number of biochemical reactions and selected organisms to illustrate how species identifications can be made using biochemistry. <u>Abbreviations and reactions:</u> **ADH** (arginine dihydrolase): stepwise degradation of arginine to citrulline and ornithine; **Citrate**: citrate utilization as a carbon source; **H₂S**: production of hydrogen sulfide gas; **Indole**: indole production from tryptophan; **Lactose (ONPG)**: lactose fermentation to produce acid (orthonitrophenyl beta-D-galactoside, cleaved by beta-galactosidase); **LDC** (lysine decarboxylase): cleavage of lysine to produce CO_2 and cadaverine; **ODC** (ornithine decarboxylase): cleavage of ornithine to make CO_2 and putrescine; **Sorbitol**: sorbitol fermentation to produce acid; **Urease**: production of CO_2 and ammonia from urea; **VP** (Voges-Proskauer test): production of acetoin or 2,3-butanediol. (*Note*: "Neg" for *Pseudomonas* in terms of glucose indicates an inability to ferment glucose. *Pseudomonas* can still use glucose as a carbon source.)

Most clinical laboratories in the United States and Europe now use automated identification systems that are even more sophisticated than API. These include bioMérieux's VITEK, Siemens Healthcare Diagnostics' WalkAway, and Becton Dickinson's Phoenix systems (**Fig. 28.4B**). All of them use cards or plates with 30 or more biochemical or enzymatic reactions. As with the API chambers, the results appear as colored reactions that can be read by computer. The BD Phoenix system, for example, will convert the results into a ten-digit code that is used to identify genus and species.

www | API 20E strip animation
microbiology2.com/links

Identifying nonenteric Gram-negative bacteria. The procedures we have outlined will accurately identify members of Enterobacteriaceae, but pathogenic Gram-negative bacilli can also be found in other phylogenetic families. One possibility in the preceding case history is that the Gram-negative bacillus seen in CSF smears would *not* grow on blood agar or on the other selective media, but would grow as small, glistening colonies on chocolate agar. This result would implicate the Gram-negative rod *Haemophilus influenzae*. Meningitis caused by *H. influenzae* was a major problem prior to 1988, before the introduction of vaccinations with *H. influenzae* type b capsule material. Growth of *H. influenzae* requires hemin (X factor) and NAD (V factor), so confirming the identity of *H. influenzae* involves growing the organism on agar medium containing hemin and NAD. This can be done by placing small filter paper disks containing these compounds on a nutrient agar surface (not blood agar) that has been covered with the organism. *H. influenzae* will grow only around a strip containing both X and V factors. Alternatively, X and V factors can be incorporated into Mueller-Hinton agar, as shown in **Fig. 28.5A**. The organism grows on chocolate medium because the lysed

A.

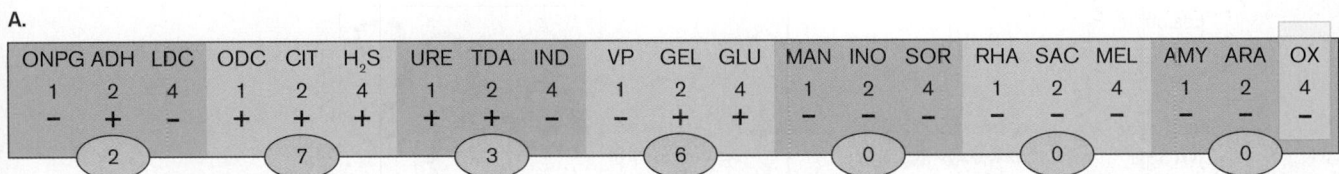

ONPG	ADH	LDC	ODC	CIT	H₂S	URE	TDA	IND	VP	GEL	GLU	MAN	INO	SOR	RHA	SAC	MEL	AMY	ARA	OX
1	2	4	1	2	4	1	2	4	1	2	4	1	2	4	1	2	4	1	2	4
−	+	−	+	+	+	+	+	−	−	+	+	−	−	−	−	−	−	−	−	−

2   7   3   6   0   0   0

B.

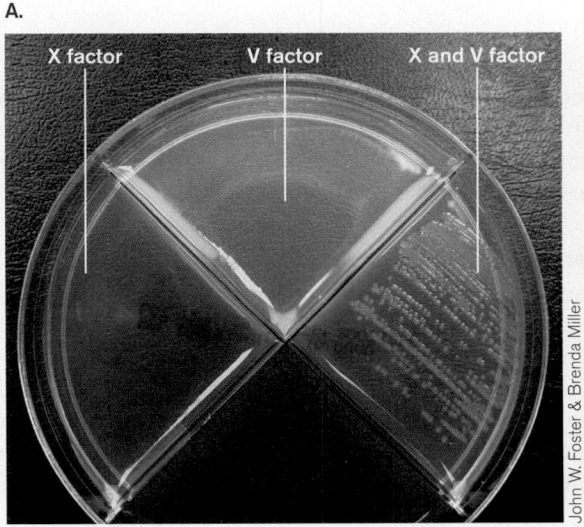

Courtesy and © Becton, Dickinson and Company

Figure 28.4 Generating an identification number from the API 20E strip. A. Results from Figure 28.2C were used to generate an identification number. The 20 reactions in the API plus an oxidase reaction performed separately are divided into seven reaction triplets. Each reaction within a triplet is given a value (1, 2, or 4). If the reaction is positive, that value is added to the value of any other positive reaction in that group, and the resulting number is placed in the oval. The result is a seven-digit number that will be unique to a given genus and species of enteric microorganism. Note that because of biochemical diversity within different strains of a species, several different numbers may be generated; but the test's design ensures that the numbers will still be unique to the species. **B.** Automated microbiology system. The BD Phoenix system uses plates with numerous reaction wells and a computerized plate reader to automatically identify pathogenic bacteria. This type of instrument automatically generates and evaluates the numbers.

red blood cells release these factors. Although XV growth phenotype is still used for identification, fluorescent antibody staining is more specific (discussed shortly).

A completely different identification scheme would have been used in the preceding meningitis case if the laboratory had discovered the organism to be a Gram-negative diplococcus (rather than a Gram-negative rod). A Gram-negative diplococcus would suggest *Neisseria meningitidis*. *N. gonorrhoeae* is also possible, but less likely because the gonococcus lacks the protective capsule that

meningococci use to survive in the bloodstream. Without this capsule, *N. gonorrhoeae* is not as resistant to serum complement and cannot disseminate to the meninges.

The first test to determine whether the organism is a species of *Neisseria* is the cytochrome oxidase test (**Fig. 28.5B**). In this test, a few drops of the colorless reagent *N,N,N′,N′*-tetramethyl-*p*-phenylenediamine dihydrochloride are applied to the suspect colonies. The reaction, which takes place only if the organism possesses both cytochrome oxidase and cytochrome *c*, turns the *p*-phenylenediamine

A.

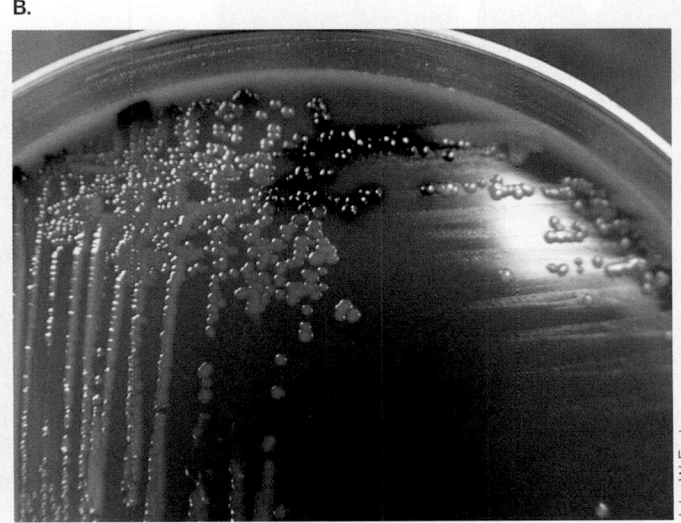

John W. Foster & Brenda Miller

B.

John W. Foster

Figure 28.5 *Haemophilus influenzae* **growth factors and** *Neisseria meningitidis* **oxidase reaction.** **A.** *H. influenzae* will grow on an agar plate (here, Mueller-Hinton agar) only when the medium has been fortified with both X factor (hemin) and V factor (NAD), but not either one alone. **B.** Oxidase-positive reaction for *Neisseria meningitidis*. Oxidase reagent (which is colorless) was dropped onto colonies of *N. meningitidis* grown on chocolate agar. The test is called the cytochrome oxidase test, but it really tests for cytochrome *c*.

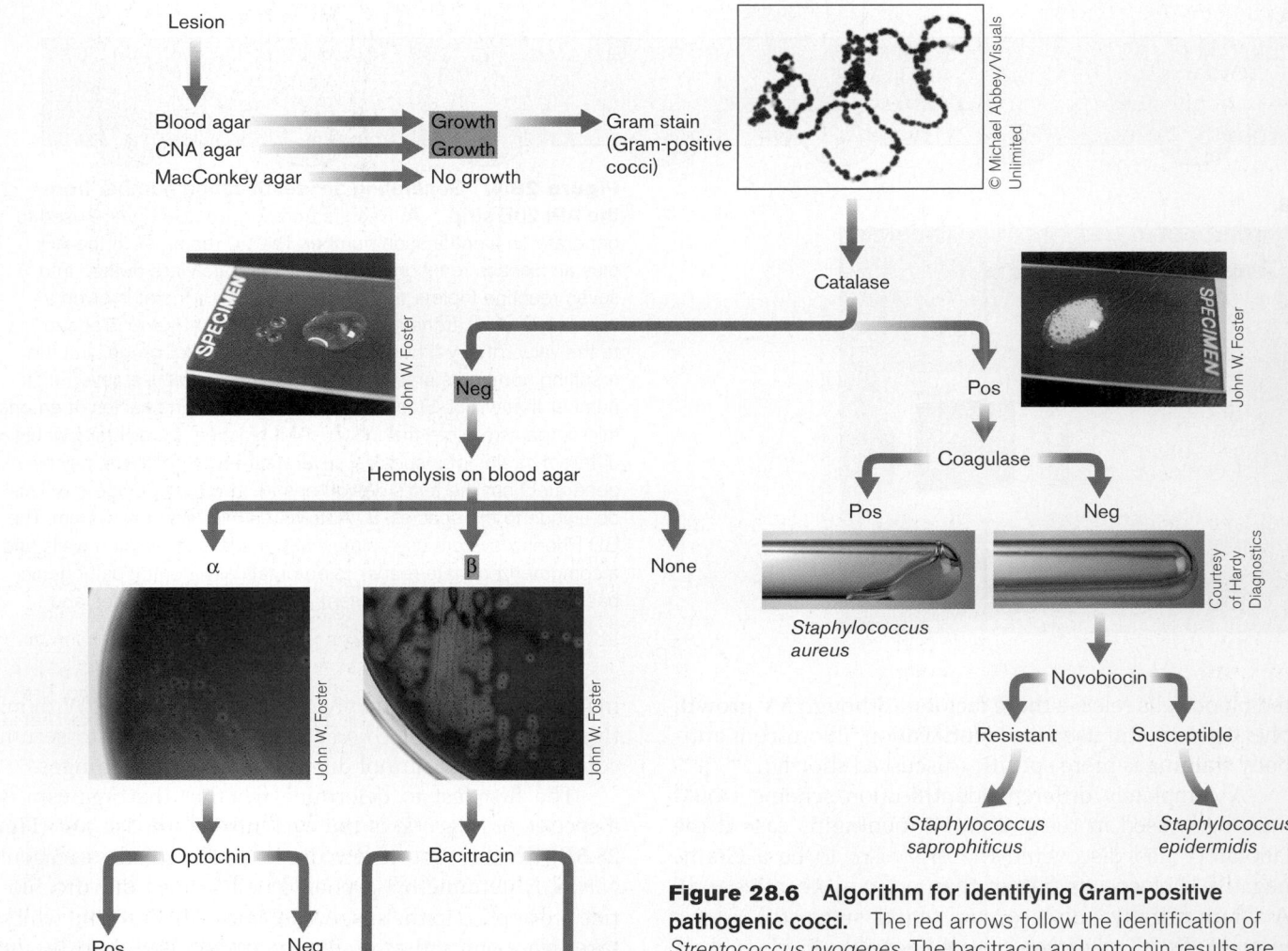

Figure 28.6   Algorithm for identifying Gram-positive pathogenic cocci.   The red arrows follow the identification of *Streptococcus pyogenes*. The bacitracin and optochin results are designated "positive" if the organism is susceptible and "negative" if the organism is resistant to the agent.

reagent (and the colony) a deep purple/black. While many bacteria possess cytochrome oxidase, *Neisseria* is one of only a few genera that also contain cytochrome *c* in their membranes. Oxidase-positive organisms use cytochrome oxidase to oxidize cytochrome *c*, which then oxidizes *p*-phenylenediamine. Other oxidase-positive bacteria include *Pseudomonas*, *Haemophilus*, *Bordetella*, *Brucella*, and *Campylobacter* species—all of them Gram-negative rods. None of the Enterobacteriaceae, however, are oxidase-positive, because they lack cytochrome *c*.

An oxidase-positive, Gram-negative diplococcus is very likely a member of *Neisseria*. Differentiation between

species of *Neisseria* is based on their ability to grow on certain carbohydrates; or it can be determined using immunofluorescent antibody staining tests that test for the presence of different capsule antigens in *N. meningitidis*.

Identifying Gram-positive pyogenic cocci. Recall the case of the woman with necrotizing fasciitis (see Section 26.2). How did the laboratory determine that the etiological agent was *Streptococcus pyogenes*? A sample algorithm, or flowchart (**Fig. 28.6**), shows how this is done. The physician sends a cotton swab containing a sample from a lesion to the clinical laboratory. The laboratory technician streaks the material onto several media: (1) blood agar (which will grow both Gram-positive and Gram-negative organisms); (2) blood agar containing the inhibitors colistin and naladixic acid (called a CNA plate, this agar will grow *only* Gram-positives); and (3) MacConkey agar (which will grow only Gram-negative organisms (see Section 4.3). The suspect organism in this case grows on the CNA and blood plates. Because it grows in the presence of the Gram-negative inhibitory compounds in CNA, one would immediately suspect the organism to be Gram-positive—an assumption borne out by the Gram stain.

> **NOTE:** Though the skin is normally populated by many different microorganisms, samples from an infected lesion are overwhelmingly populated by the etiological agent. This occurs because the pathogenic microbe outgrows normal microbiota. Using selective media to isolate the infectious agent will further simplify diagnosis by decreasing growth of any normal microbiota that may still be present.

The algorithm tells the laboratory technician that since the organism is a Gram-positive coccus, the next step is to test for catalase production. Catalase, which converts hydrogen peroxide (H_2O_2) to O_2 and H_2O, clearly distinguishes staphylococci from streptococci (remember, Gram stain morphology alone is insufficiently reliable to make that distinction). The catalase test is performed by mixing a colony with a drop of H_2O_2 on a glass slide. Effusive bubbling due to the release of oxygen indicates catalase activity (see **Fig. 28.6**). Staphylococci are catalase-positive; streptococci are catalase-negative. Note that many other organisms possess catalase activity, including the Gram-negative rod *E. coli.* However, based on the algorithm, *E. coli* would not be considered, since it does not grow on CNA agar and is not Gram-positive.

> **NOTE:** When performing a catalase test from colonies grown on blood agar, be sure not to transfer any of the agar, since red blood cells also contain catalase.

Having established that the organism is catalase-negative, the technician examines the blood plate for evidence of hemolysis. Three types of colonies are possible: nonhemolytic, alpha-hemolytic, and beta-hemolytic. Nonhemolytic streptococci do not produce any lytic zone. Alpha-hemolytic strains produce large amounts of hydrogen peroxide that oxidize the heme iron within intact red blood cells to produce a green product. As a result, alpha-hemolytic streptococci produce a green zone around their colonies called alpha hemolysis—even though the red blood cells remain intact. (For example, *Streptococcus mutans*, a cause of dental caries and subacute bacterial endocarditis, is alpha-hemolytic.) Still other streptococci produce a completely clear zone of true hemolysis surrounding their colony. This is called beta hemolysis. Complete hemolysis of red blood cells occurs owing to the export of enzymes, called hemolysins, that lyse red-cell

Marine Biological Laboratory

Figure 28.7   Rebecca Lancefield.   In 1918, Dr. Lancefield joined the Rockefeller Institute for Medical Research in New York City, where she studied the hemolytic streptococci, known then as *Streptococcus haemolyticus.* She was the first to use serum precipitation methods to classify *S. haemolyticus* into groups according to differences in cell wall carbohydrate antigens. The basic technique is still used today and is known as the Lancefield classification scheme in her honor.

membranes. The flowchart in **Figure 28.6** indicates that the organism from the case history was beta-hemolytic.

The final relevant test in this flowchart involves susceptibility to the antibiotic bacitracin, which identifies the most pathogenic group of beta-hemolytic streptococci. The beta-hemolytic streptococci are subdivided into many different groups, known as the Lancefield groups, based on differences in the composition of their carbohydrate peptidoglycans. These cell wall differences are distinguished from each other immunologically and divide the streptococci into Lancefield groups A–U. Rebecca Lancefield, for whom the classification scheme is named, was the first to use immunoprecipitation to group the streptococci (**Fig. 28.7**). The vast majority of streptococcal diseases are caused by group A beta-hemolytic streptococci (also called GAS), defined as the species *Streptococcus pyogenes.*

Unfortunately, the Lancefield classification procedure is somewhat time-consuming and not readily amenable as a rapid identification method. However, the group A beta-hemolytic streptococci are uniformly susceptible to the antibiotic bacitracin. Thus, a simple antibiotic disk susceptibility test can be used to indicate the group A beta-hemolytic streptococci (that is, *S. pyogenes*). But beware—many bacteria are bacitracin sensitive, so like the catalase test, the bacitracin test must be used in conjunction with an algorithm to be useful for identification. The technician must follow the appropriate algorithm before assigning importance to this or any other test result. It is irrelevant, for instance, if an alpha-hemolytic organism is bacitracin susceptible. Some of these may exist, but they are not associated with disease. The organism in our case of necrotizing fasciitis, however, was beta-hemolytic, so bacitracin susceptibility indicated that the organism was *S. pyogenes*.

The other tests named in **Figure 28.6** are equally important for identifying Gram-positive infectious agents. For example, *Streptococcus pneumoniae* is an important cause of pneumonia. Like *S. pyogenes*, *S. pneumoniae* is a catalase-negative, Gram-positive coccus; but unlike *S. pyogenes*, it is alpha-hemolytic. Optochin susceptibility is a property closely associated with *S. pneumoniae*, while other alpha-hemolytic strains of streptococci are resistant to this compound. Thus, an optochin susceptibility disk test is a useful tool for identifying *S. pneumoniae*.

Coagulase is a key reaction used to distinguish the pathogen *Staphylococcus aureus*, which causes boils and

bone infections, from other staphylococci, such as the normal skin species *S. epidermidis*. To conduct a coagulase test, a tube of plasma is inoculated with the suspect organism. If the organism is *S. aureus*, it will secrete the enzyme coagulase. Coagulase will convert fibrinogen to fibrin and produce a clotted, or coagulated, tube of plasma. Coagulase-negative staphylococci can still be medically important, however. *S. saprophiticus*, for instance, is an important cause of urinary tract infections. It can be distinguished from *S. epidermidis* by resistance to novobiocin.

> **THOUGHT QUESTION 28.1** Use **Figure 28.6** to identify the organism from the following case: A sample was taken from a boil located on the arm of a 62-year-old man. Bacteriological examination revealed the presence of Gram-positive cocci that were also catalase-positive, coagulase-positive, and novobiocin resistant.

Table 28.2 Some DNA-based detection tests.

Test	Intended use	Specimen for PCR	Transport conditions
Detection of HIV proviral DNA	Diagnosis of HIV infection in newborns; to resolve indeterminate serological results	>2 ml whole blood in EDTA tube. EDTA prevents coagulation by chelating (binding) ions.	Room temperature, within 24 hours after collection
Detection of *Bordetella pertussis*	Diagnosis of whooping cough (pertussis)	Nasopharyngeal swab	Room temperature, within 24 hours after collection
Detection of *Borrelia burgdorferi*	Diagnosis/surveillance of Lyme disease	Tick	Wrap in moist tissue in small plastic bag
		Skin biopsy; >1 ml spinal fluid; or urine	4°C (cold packs), within 24 hours after collection
Detection of *Rickettsia rickettsii*	Diagnosis/surveillance of Rocky Mountain spotted fever	Tick	Wrap in moist tissue in small plastic bag
		>2 ml whole blood in EDTA tube	Blood: room temperature
		Skin biopsy	Skin lesion biopsy within 24 hours after collection; skin: 4°C
Detection of *Ehrlichia chaffeensis*	Diagnosis/surveillance of human monocytic ehrlichiosis	Tick	Wrap in moist tissue in small plastic bag
		>2 ml whole blood in EDTA tube	Room temperature, within 24 hours after collection
Detection of Norwalk and other small round-structured viruses	Diagnosis of viral gastroenteritis; investigation of food-borne and waterborne outbreaks	>1 ml diarrheal stool in sterile container	4°C (cold packs), within 72 hours after collection
Typing of *Mycobacterium tuberculosis*	Determining relatedness of *M. tuberculosis* isolates; investigation of suspected outbreaks	Sputum, urine	Room temperature
Typing of various Gram-negative and Gram-positive bacteria	Determining relatedness of isolates; investigation of suspected nosocomial or food-borne outbreaks	Food, feces	4°C (cold packs), within 24 hours after collection (if *Campylobacter* suspected, transport at room temperature)

Pathogen Identifications Based on Molecular Genetics

Many diagnostic laboratories have implemented DNA-based methods to detect and type bacteria and viruses. Molecular detection methods for bacteria are often more rapid than the traditional culture-based methods described earlier. DNA detection (for example, PCR) takes only a few hours, while culture-based identification takes days to weeks. DNA/RNA detection methods are especially useful for viruses, which otherwise require elaborate electron microscopy to view morphology or serology to detect an increased presence of antiviral antibodies. The problem with serology is that by the time these antibodies become detectable in blood, the patient is already recovering from the disease. Molecular detection techniques may provide the *only* means of identifying newly recognized, or emerging, pathogens.

The polymerase chain reaction (PCR) is the most widely used molecular method in the clinical laboratory's diagnostic arsenal. DNA primers that bind to unique genes in a pathogen's genome can be used to specifically amplify DNA or RNA present in a clinical specimen. Successful amplification is visualized as an appropriately sized fragment in agarose gels following electrophoresis.

Why is PCR needed to detect the presence of these nucleic acids? Without the PCR amplification steps, clinical samples usually provide too little nucleic acid from infecting microorganisms to be detected. For example, *Mycobacterium tuberculosis*, the cause of tuberculosis, can take weeks to grow on standard bacteriological media. Detecting the presence of its DNA in the specimen, however, would allow a very rapid diagnosis. Unfortunately, the amount of bacterial DNA present in the sample is infinitesimal, too little to detect by standard techniques such as Southern blot. Amplifying *M. tuberculosis* nucleic acid by PCR, however, turns one copy of DNA into billions of copies.

PCR is very quick. Preparing a clinical specimen for PCR by extracting the DNA or RNA usually takes less than an hour. PCR itself is completed in 2–3 hours. Detection of the PCR-amplified DNA by DNA gel electrophoresis takes an additional 1–2 hours. So, what might take 2–3 days (or sometimes weeks) using biochemical algorithms may take less than a day by molecular strategies. Most clinical laboratories are now equipped with thermocyclers (the instruments used to precisely and rapidly cycle temperature for PCR reactions) and can carry out this type of identification protocol for organisms for which specific primers are available.

Table 28.2 lists several instances in which DNA detection tests are useful. A specific example presented in **Figure 28.8** illustrates the use of PCR to type different strains of the anaerobic pathogen *Clostridium botulinum*, the cause of food-borne botulism. These organisms

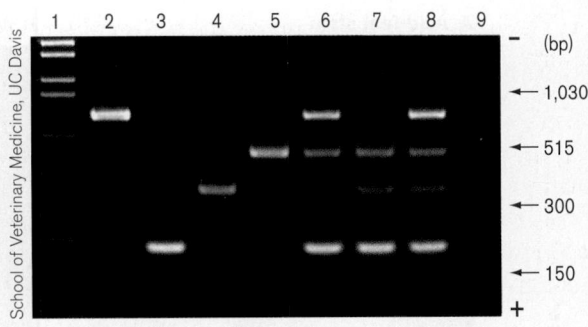

Figure 28.8   Multiplex PCR identification of *Clostridium botulinum*. *C. botulinum* cells are serologically typed not on the basis of surface antigens, but by which toxin genes a strain possesses. Isolates can be typed in a single PCR reaction that includes primer pairs specific for each of the four major toxin genes: types A, B, E, and F. Because multiple products are sought in a single reaction, this is called multiplex PCR. Each lane in the agarose gel was loaded with multiplex products from different isolates of *C. botulinum* and subjected to electrophoresis. The slower-moving fragments (toward top of gel) are larger than those moving farther down the gel toward the positive pole. Lane 1, DNA size markers; lane 2, type A (*cntA*); lane 3, type B (*cntB*); lane 4, type E (*cntE*); lane 5, type F (*cntF*); lane 6, types A, B, and F; lane 7, types B, E, and F; lane 8, types A, B, E, and F.

are not typed serologically, as is the case for *Streptococcus pneumoniae*. *C. botulinum* is divided into different types based on the neurotoxin genes they possess. In this example, **multiplex PCR** was used to simultaneously search for these toxin genes. Multiplex PCR uses multiple sets of primers, one pair for each gene, combined in a single tube with a specimen. Care must be taken to be sure that the primers chosen make different-sized products, do not interfere with each other, and do not produce artifactual products that can confuse interpretation. Multiplex PCR can help identify sets of specific genes present in a single species or can screen for the presence of multiple pathogens in a clinical sample. In the latter, primer sets are designed to amplify genes unique to each pathogen.

Case History: Outbreak of Hospital-Acquired Wound Infections in New Delhi

*From August through December, 45 patients in the pediatric surgery unit of a New Delhi hospital developed postoperative wound infections. Of these, 42 were outpatients, and 3 were inpatients who had undergone major surgery. The diseases ranged from chronic ear infections to bacteremia associated with the use of hemodialysis equipment. Thirty-two clinical samples of pus and wound exudates were tested for acid-fast bacilli by the Ziehl-Neelsen method (**Fig. 28.9A**). The same smear samples were cultured on Lowenstein-Jensen agar (**Fig. 28.9B**) and examined for growth over a course of several weeks. Biochemical tests indicated that the organism was a mycobacterium in each case. The slow-growing*

A. Acid-fast stain

B. *M. tuberculosis* colonies on Lowenstein-Jensen agar

C. Genetic fingerprint

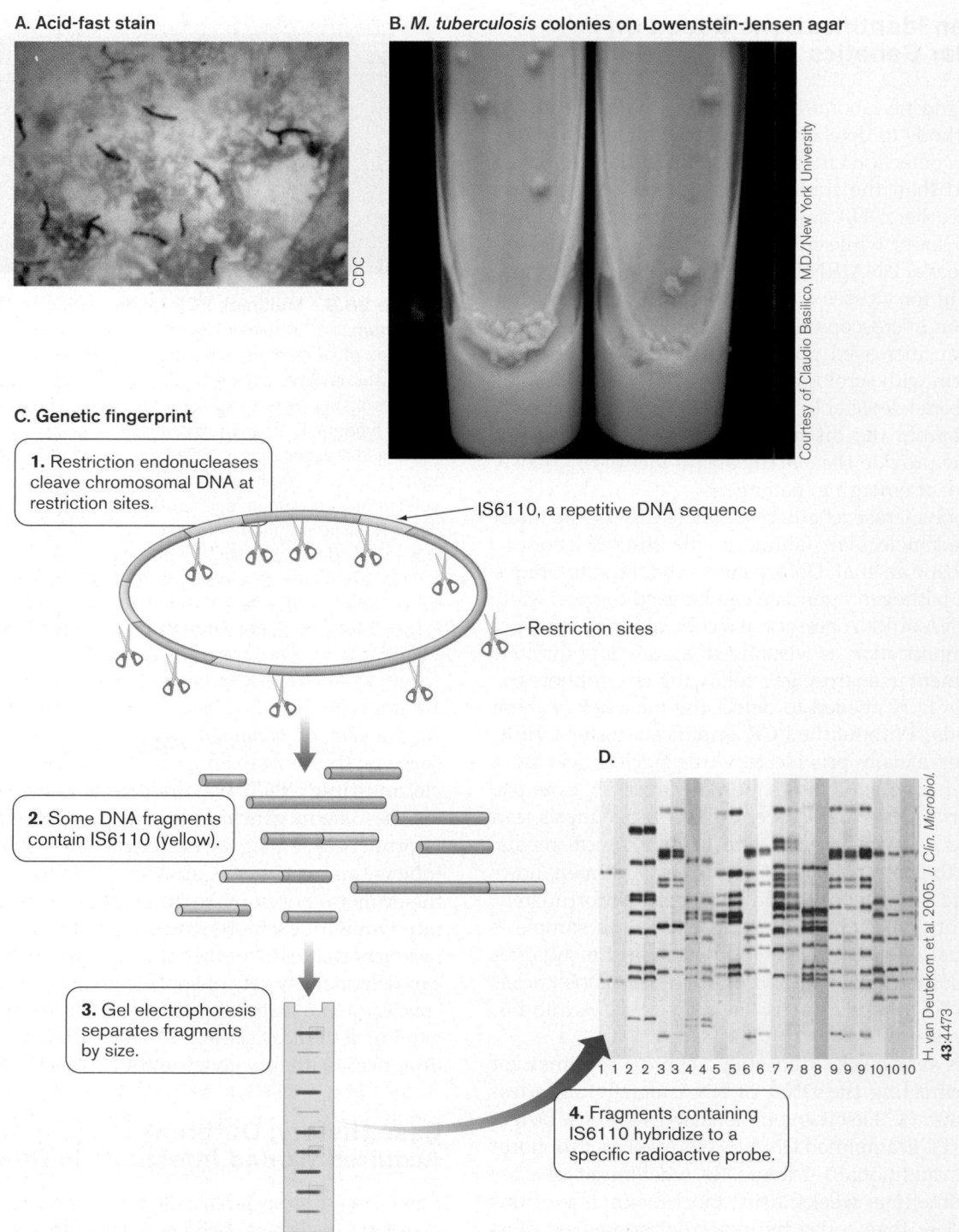

1. Restriction endonucleases cleave chromosomal DNA at restriction sites.

IS6110, a repetitive DNA sequence

Restriction sites

2. Some DNA fragments contain IS6110 (yellow).

D.

3. Gel electrophoresis separates fragments by size.

1 1 2 2 3 3 4 4 5 5 6 6 7 7 8 8 9 9 9 101010

4. Fragments containing IS6110 hybridize to a specific radioactive probe.

CDC

Courtesy of Claudio Basilico, M.D./New York University

H. van Deutekom et al. 2005. *J. Clin. Microbiol.* **43**:4473

Figure 28.9 Morphology, growth, and DNA fingerprinting of *Mycobacterium tuberculosis*. A. Acid-fast Ziehl-Neelsen stain of *M. tuberculosis*. **B.** Lowenstein-Jensen medium enables growth of mycobacterial species, some of which grow extremely slowly. The colonies have a "bread crumb–like" appearance. **C.** Genetic fingerprinting of *M. tuberculosis* isolates. Fragments containing IS6110 are marked yellow. **D.** Fragments containing IS6110 hybridize to the specific radioactive probe. A characteristic banding pattern (fingerprint) appears for each isolate. Isolates with similar banding patterns are assigned the same number in this example.

organism, Mycobacterium abscessus, *was identified by PCR. Based on the DNA fingerprint, the source of the outbreak was traced to the tap water in the operating room and to a defective autoclaving process (the result of a leaking vacuum pump and faulty pressure gauge in the autoclave).*

This case history points out an underlying problem with some biochemical identifications. It took weeks to grow enough organisms to perform these biochemical tests. An alternative PCR test that probed for mycobacterium-specific genes encoding 16S rDNA could have been performed directly on the clinical specimen and would have confirmed the diagnosis immediately, much sooner than by biochemical methods. The PCR test that was done was not performed until *after* the cells were grown in the laboratory.

M. tuberculosis can be presumptively diagnosed in the clinical laboratory by the acid-fast stain, a technique first described in 1882 (called the Ziehl-Neelsen stain) that is used to find the tubercle bacillus in a patient's sputum (**Fig. 28.9A**). The acid-fast stain enables a technician to visualize bacteria such as *Mycobacterium* species that are *not* stained by the Gram stain. Mycobacteria have a very waxy outer coat composed of mycolic acid that resists penetration by most dyes—an obstacle the acid-fast stain was designed to overcome. The original Ziehl-Neelsen acid-fast stain used phenol and heat to drive carbolfuchsin (a red dye) into mycobacterial cells on glass slides. Destaining with an acid alcohol solution removes the stain from all cell types *except* mycobacteria. The slide is subsequently counterstained with methylene blue, after which the mycobacteria will be seen as curved, red rods (called acid-fast bacilli), while everything else will appear blue. A more modern version of the acid-fast stain uses the fluorochrome auramine O to stain the mycolic acid. This dye also resists removal by an acid alcohol wash, so when observed under a fluorescence microscope, the mycobacteria will fluoresce bright yellow.

Although the acid-fast stain is very useful, sometimes organisms cannot be found in a sputum sample; and even if they are found, confirmatory tests are needed for a definitive diagnosis. Relying on growth of the organism means that diagnosis may not be clear for several weeks. For our case history, a rapid DNA-based method would have made the job of determining species much easier. This test probes a 383-bp DNA sequence located at the end of a heat-shock gene that is highly conserved among all mycobacterial species. As a result, the sequence can be amplified by PCR regardless of the strain. However, fragments amplified from different species will have somewhat different DNA sequences. The alterations in DNA sequence can be exposed by restriction digestion. Sequences from different species will produce unique digestion fragment patterns that can then be used for identification. The clinical laboratory compared the restriction patterns produced from mycobacteria isolated from different areas throughout the hospital in the case history. The results allowed the infection control staff to trace the source of all the infections to the tap water in the operating room.

In addition to PCR, restriction fragment length polymorphisms (RFLPs), a form of DNA fingerprinting, can be used to track a specific strain of bacteria (not just species) during the course of an epidemic. This technique relies on small differences in sequence that occur between strains of a single species. Because of these small differences in sequence, the number and location of restriction endonuclease sites in the chromosome will differ. In the example in **Figure 28.9**, DNA from a strain of *M. tuberculosis* is cut with restriction endonucleases, and the digested fragments are separated by gel electrophoresis (**Fig. 28.9C** and **D**). The gel is then probed with a radioactively labeled fragment of insertion sequence IS6110 (a fluorescent probe can also be used). IS6110 is a repetitive DNA sequence present at multiple sites in the chromosome. The radioactive probe can hybridize only to DNA fragments that contain IS6110 sequences. Strains of *M. tuberculosis* obtained from different outbreaks would likely display different hybridization patterns. Strains isolated from the same outbreak would look very similar, if not identical.

Case History: West Nile Virus

A 55-year-old man was admitted to a local hospital complaining of headache, high fever, and neck stiffness. The man appeared confused and disoriented. He also complained of muscle weakness. History indicated he had received several mosquito bites approximately 2 weeks previously. A blood specimen was sent to the laboratory. The report the following day indicated that the patient was suffering from West Nile virus.

West Nile virus is primarily an infection of birds and culicine mosquitoes (a group of mosquitoes that can transmit human diseases), with humans and horses serving as incidental, dead-end hosts. Replication of virus in this bird-mosquito-bird cycle begins when adult mosquitoes emerge in early spring and continues until fall. Among humans, the incidence of disease peaks in late summer and early fall. Birds provide an efficient means of geographic spread of the virus. As a result, over the past several years the virus has spread throughout much of the United States.

Isolating a disease-causing virus is extremely challenging. Most laboratories are not equipped for the special tissue culture techniques required to grow viruses. Consequently, most viral infections, including human West Nile virus infections, are usually diagnosed by measuring the antibody response of the patient. For instance, the presence of West Nile virus–specific IgM in cerebrospinal fluid is a good indicator of current West Nile virus infection, but it is indirect and not conclusive. Real-time PCR is a molecular test that can quickly reveal the presence of the virus itself.

Real-time quantitative PCR (RTQ-PCR, or qPCR) is used routinely for the high-throughput diagnosis of viral pathogens such as West Nile virus. Because West Nile virus (*Flaviviridae* family) contains single-stranded RNA, its RNA must be converted to DNA using reverse transcriptase before PCR can be attempted. The quantitative advantage of RTQ-PCR is that you can estimate the number of virus particles present in the sample by the number of viral RNA molecules there.

The basic technique is as follows: RNA is first extracted from the sample. A sub-sequence of viral RNA is then converted to cDNA using a single primer and reverse transcriptase (**Fig. 28.10**). The more virus particles there are in the sample, the more viral RNA will be present and the more cDNA product will be made. The cDNA is then amplified by PCR using two specific primers and Taq polymerase (see Section 12.3; Fig. 12.16). The trick for quantitation is in the method used to detect amplification. In one method, a third, fluorescent oligonucleotide (called the probe) is added to the PCR reaction (see Fig. 12.16A). The probe contains a

fluorescent dye at the 3′ end and a chemical dye at the 5′ end that quenches (absorbs) energy emitted from the fluorescent dye. As long as the two chemicals are kept in close proximity by the intact probe, no light is emitted. The probe is designed to anneal to a sequence between the binding sites of the two other primers (modifications on the ends of the probe prevent it from being used as a primer). So in a successful amplification, Taq polymerase will synthesize DNA from the two outside primers and degrade the probe oligonucleotide as it passes through that area. This cleavage separates the dye from the quencher, and the dye begins to fluoresce. The greater the amount of cDNA there was to begin with, the fewer cycles it takes to register a fluorescence increase over background (**Fig. 28.11**).

WWW | Molecular-beacon PCR
microbiology2.com/links

> **THOUGHT QUESTION 28.2** Why does finding IgM to West Nile virus indicate current infection? Why wouldn't finding IgG do the same?

Identifications Based on Serology

Case History: Ebola

Between October and November 2000, 62 residents of a small village north of Gulu, a town in Uganda, became ill with high fever, diarrhea, headache, vomiting, and gastrointestinal bleeding from the rectum. As many as 36 patients died. By January, 425 cases were reported, of whom 224 died. The presumptive diagnosis was Ebola. Laboratory confirmation tests included viral antigen detection and antibody ELISA tests. Laboratory-confirmed Ebola patients were defined as patients who were positive for either Ebola virus antigen or Ebola IgG antibody. Once identified, rigorous quarantine mechanisms were implemented to limit spread of the disease to other villages.

What are these tests?

The ELISA test detects antibodies indirectly or antigens directly. Enzyme-linked immunosorbent assay (ELISA) can detect antigens or antibodies present in nanogram and picogram quantities. One form of ELISA detects serum antibodies. It is carried out in a 96-well microtiter plate, which allows multiple patient serum samples to be tested simultaneously. An antigen from the virus (Ebola in this case) is attached (adsorbed) to the plastic of the wells (**Fig. 28.12**). Albumin or powdered milk is used to block the remaining sites on the plastic that could result in false positives. Patient serum is then added. Ebola-specific antibodies present in the serum will react with the antigen attached to the microtiter plate. The antigen-antibody complex is then reacted with goat-antihuman IgG to which an enzyme has been attached, or conjugated (for

Conversion of mRNA to cDNA by reverse transcription

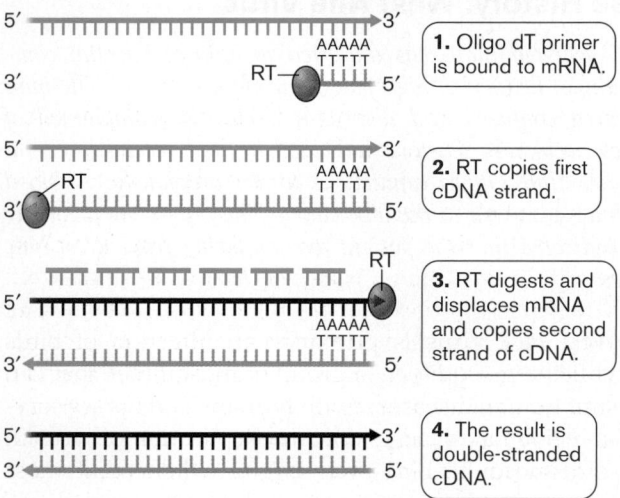

1. Oligo dT primer is bound to mRNA.

2. RT copies first cDNA strand.

3. RT digests and displaces mRNA and copies second strand of cDNA.

4. The result is double-stranded cDNA.

Figure 28.10 Reverse transcriptase synthesis of DNA. Reverse transcriptase (RT) uses mRNA as a template to make DNA. The DNA can then be amplified using standard PCR methods. Molecules of eukaryotic and viral mRNA usually contain poly-A tails at their 3′ ends. An oligonucleotide poly-T primer added to the reaction tube will anneal to the poly-A tail and allow RT to synthesize the first strand of a complementary DNA (cDNA). A second primer specific to the viral gene is then used to prime RT synthesis of the opposite strand (second-strand synthesis), using the first DNA strand as template. Subsequent amplification by PCR requires the addition of a thermostable DNA polymerase (Taq) that can withstand the denaturing and DNA synthesis temperatures required for amplifying the cDNA.

A.

Cycle number	Amount of DNA
0	1
1	2
2	4
3	8
4	16
5	32
6	64
7	128
8	256
9	512
10	1,024
11	2,048
12	4,096
13	8,192
14	16,384
15	32,768
16	65,536
17	131,072
18	262,144
19	524,288
20	1,048,576
21	2,097,152
22	4,194,304
23	8,388,608
24	16,777,216
25	33,554,432
26	67,108,864
27	134,217,728
28	268,435,456
29	536,870,912
30	1,073,741,824
31	1,400,000,000
32	1,500,000,000
33	1,550,000,000
34	1,580,000,000

B.

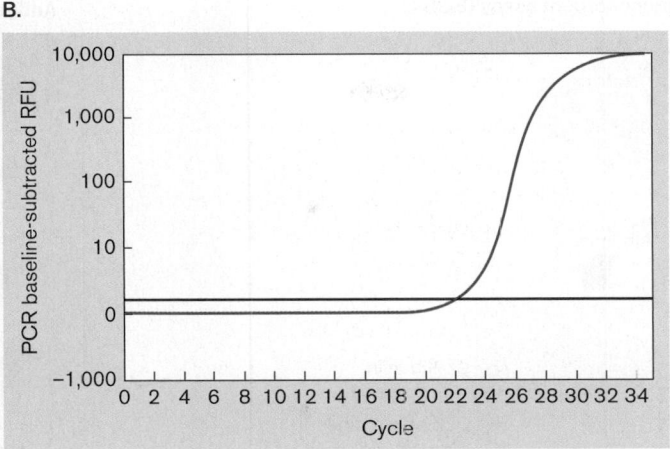

Figure 28.11   Results of real-time PCR.   A. The exponential increase in PCR products after each cycle of hybridization and polymerization. The switch to yellow indicates the point where the increase in product plateaus because the primers have been exhausted. **B.** Increase in relative fluorescence units (RFUs) as the fluorescent dye is released from the dual-labeled probe during real-time PCR. The blue curve representing PCR product is flat for the first 22 cycles because the amount of DNA made, and therefore the level of fluorescent dye released, remains below background level (red line). The more starting DNA there is, the sooner RFU values will increase over background (that is, the fewer cycles will be needed to see the increase over background). The slope eventually decreases because the fluorescent probe has become limiting.

example, horseradish peroxidase). This forms an antibody "sandwich" that attaches the enzyme to the well. The chromogenic substrate for the enzyme is added next (for example, tetramethylbenzidine). If enzyme-conjugated antibody has bound to any human IgG, the enzyme will convert the substrate to a colored product (blue for tetramethylbenzidine). Enzyme activity can be measured with an ELISA plate reader. The amount of colored product formed, detected as absorbance with a spectrophotometer, will be an indication of the amount of anti-Ebola antibody present in the patient sample.

> **THOUGHT QUESTION 28.3** Why does adding albumin or powdered milk prevent false positives in ELISA?

Antigen capture is another ELISA technique, but in this instance, anti-Ebola antibody, not viral antigen, is adsorbed to the wells of a microtiter plate (**Fig. 28.13**). Patient serum is then added to the wells. If the serum contains Ebola antigen, the antigen will be captured by the antibody in the well. Then a second, enzyme-conjugated antibody against the Ebola antigen is added. The more antigen there is in the serum, the more enzyme-linked antibody will affix to the well. Addition

of the appropriate chromogenic substrate will produce a colored product that can be measured.

Antibody against Ebola may be easier to detect than viral antigen because antibodies will be present at higher levels than the virus itself. But because there is a delay between the time when the virus is first present in serum and when the body manages to make antibody, a speedier diagnosis can be made by directly detecting viral antigen.

While bacterial infections are commonly diagnosed by growing the infecting organism on artificial medium in the clinical laboratory, viral diseases are usually diagnosed by immunological means. Viruses are more difficult to grow than bacteria and do not exhibit the biochemical diversity so useful for identifying different bacterial species. In addition to immunological tests, viral diseases can be diagnosed using molecular approaches, such as PCR, to identify viral DNA or RNA sequences.

> **THOUGHT QUESTION 28.4** Specific antibodies against an infecticus agent can persist for years in the bloodstream, long after the infection resolves. So how is it possible that antibody titers can be used to diagnose recently acquired diseases such as infectious mononucleosis? Couldn't the antibody be from an old infection?

Enzyme-linked immunosorbent assay (ELISA)

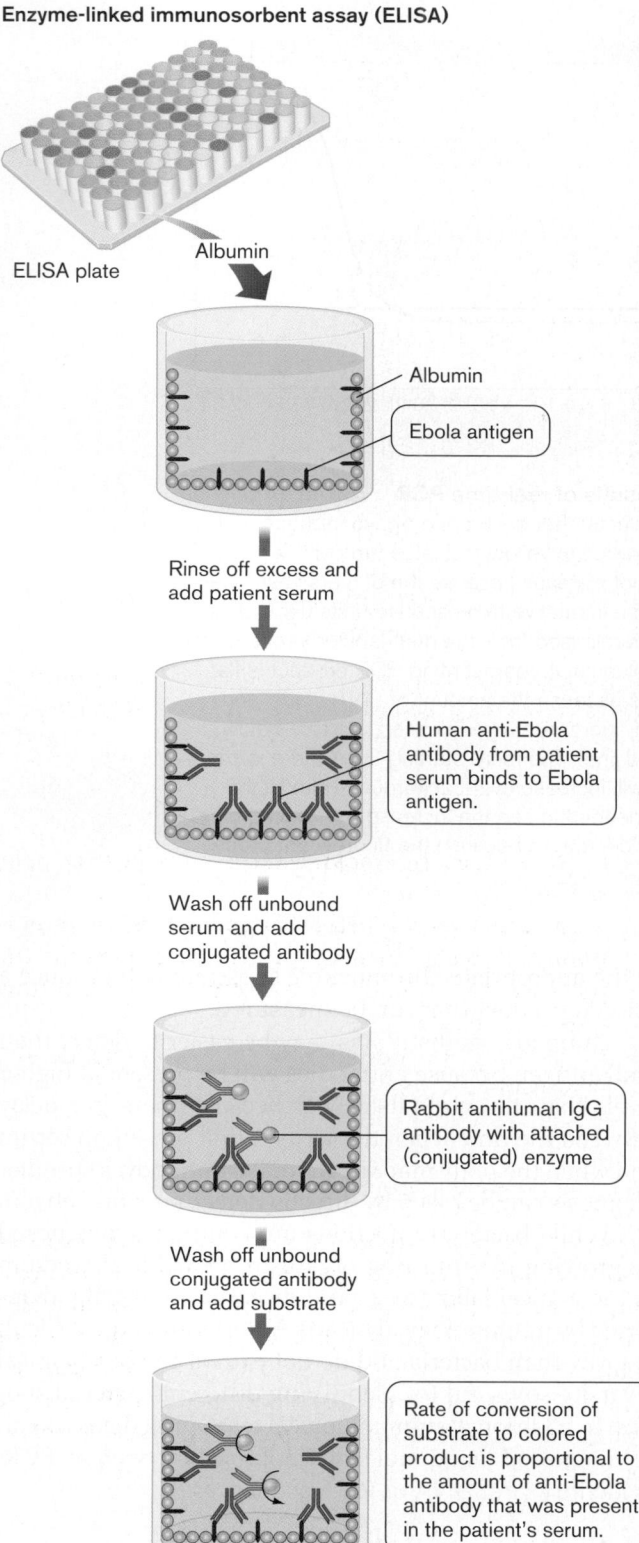

ELISA plate

Albumin

Albumin

Ebola antigen

Rinse off excess and add patient serum

Human anti-Ebola antibody from patient serum binds to Ebola antigen.

Wash off unbound serum and add conjugated antibody

Rabbit antihuman IgG antibody with attached (conjugated) enzyme

Wash off unbound conjugated antibody and add substrate

Rate of conversion of substrate to colored product is proportional to the amount of anti-Ebola antibody that was present in the patient's serum.

Figure 28.12 Enzyme-linked immunosorbent assay (ELISA). ELISA to detect anti-Ebola antibodies circulating in patient serum. The 96-well plate can be used to make dilutions of a single patient's serum to more precisely determine the amount of anti-Ebola antibody, or it can be used to test samples from multiple patients.

Antigen capture

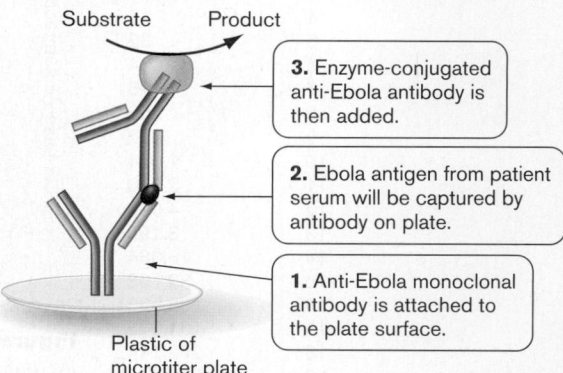

4. If Ebola antigen is present, the conjugated antibody will be captured by the complex. The addition of substrate will lead to production of a colored product.

Substrate   Product

3. Enzyme-conjugated anti-Ebola antibody is then added.

2. Ebola antigen from patient serum will be captured by antibody on plate.

1. Anti-Ebola monoclonal antibody is attached to the plate surface.

Plastic of microtiter plate

Figure 28.13 Antigen capture ELISA. ELISA technique to capture Ebola antigens circulating in patient serum.

Fluorescent Antibody Staining

Chapter 26 presents a case history involving an 80-year-old nursing-home resident who contracted pneumonia caused by *Streptococcus pneumoniae*. The laboratory diagnosis was probably made using the biochemical algorithm previously described. However, there are over 80 serological types of *S. pneumoniae*, each one containing a different capsular antigen. How can the lab identify which antigenic type has caused the infection? One way is to stain the organism with antibodies.

Figure 28.14A illustrates the result of staining a smear of the isolated streptococcus with fluorescently tagged antibodies directed against a specific antigenic type of capsule. Viewed under a fluorescence microscope, the organism is "painted" green when the right antibody binds to the capsule. In the case of pneumonia, this knowledge probably will not help in treatment of the individual patient, but its broader value is in determining whether a *single type* of organism is responsible for an outbreak of pneumonia, which in turn is of epidemiological value for identifying the source of the bacterium.

On the other hand, fluorescent antibody staining techniques are critically important for rapidly identifying organisms that are difficult to grow. Infected tissues can be subjected to direct fluorescent antibody staining. **Figure 28.14B**, for example, shows a direct fluorescent antibody stain of pleural fluid from a patient with Legionnaires' disease.

Other Microbes

Space does not permit a complete listing of the various diseases and the methods used to identify the

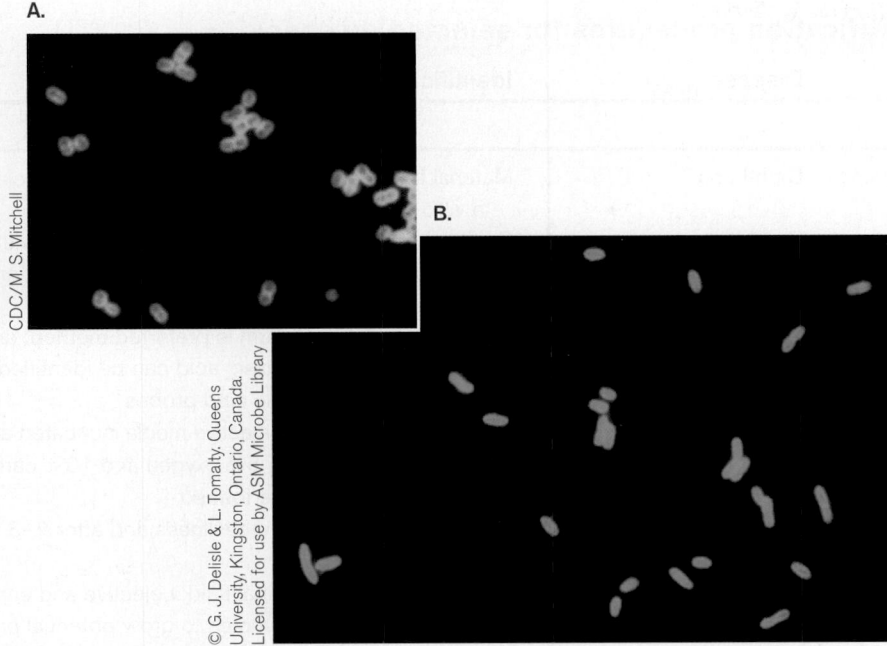

Figure 28.14   Fluorescent antibody stain.   A. *Streptococcus pneumoniae* capsule (cells approx. 0.8 μm, fluorescence microscopy). The capsule is the green halo. The center of the halo is the cell. **B.** *Legionella pneumophila* (approx. 1 μm in length) from a respiratory tract specimen.

etiological agents, but **Table 28.3** presents some additional examples. In general, bacteria that are easily cultured are grown in the laboratory, after which biochemical tests are performed. Bacterial, viral, and fungal species that are difficult to grow are typically identified using immunological techniques. These tests either identify antigen from the microbe in infected tissues or measure a rise in antibody titer. As noted earlier, DNA-based methodologies are gaining increasing acceptance. Eukaryotic microbial parasites such as *Plasmodium* species (the cause of malaria), *Giardia lamblia* (which causes giardiasis, a diarrheal disease), and *Entamoeba histolytica* (the cause of amebic dysentery) can be identified via their telltale morphologies under the microscope, making biochemical tests unnecessary.

TO SUMMARIZE:

- **Selective media** are used to inhibit growth of one group of organisms while permitting growth of others (such as Gram-positive bacteria versus Gram-negative bacteria). This technique is often used to prevent growth of normal microbiota while permitting growth of pathogens.
- **Differential media** exploit the unique biochemical properties of a pathogen to distinguish it from similar-looking nonpathogens.
- **Bacterial species can be identified** with biochemical analyses, molecular techniques (for example,

PCR), and/or immunological methods (for example, ELISA).

- **Viral diseases are often diagnosed using immunological tests,** such as ELISA, that measure the presence of antibody or antigen, or by real-time quantitative PCR.
- **Fluorescent antibody staining** can rapidly identify organisms or antigens present in tissues.

28.3 Specimen Collection

Physicians must collect, and the laboratory must process, a wide variety of clinical specimens. Types of specimens range from simple cotton swabs of sore throats, in which the swab is placed into a liquid transport medium before being sent to the laboratory, to urine and fecal samples that are transported directly. **Tables 28.2** and **28.3** outline some of these samples and the techniques used to collect them.

Case History: Abdominal Abscess

A 4-year-old boy was admitted to the hospital for evaluation and treatment of persistent pain in the rectal area. His problem had begun about one week earlier with ill-defined pain in the same area. He had a white blood cell count of 24,900 with 87% granulocytes. An abdominal computed tomography scan revealed an abscess adjacent to his rectum.

Table 28.3 Identification procedures for selected diseases.

Agent	Disease	Identification
Bacteria		
Corynebacterium diphtheriae	Diphtheria	Material from nose and throat cultured on a special medium; in vivo or in vitro tests for toxin
Bordetella pertussis	Pertussis (whooping cough)	Smears of nasopharyngeal secretions stained with fluorescent antibody; ELISA for toxin in respiratory secretions; culture on special media
Legionella pneumophila	Legionellosis, Legionnaires' disease, Pontiac fever	Culture on special medium is preferred method; antigen detection by ELISA; *Legionella* nucleic acid can be identified in clinical material using PCR and nucleic acid probes
Campylobacter spp.	Campylobacteriosis	Isolation of bacteria on selective media incubated at 42°C in atmosphere of nitrogen containing 5% oxygen and 10% carbon dioxide; then biochemical testing performed
Leptospira interrogans	Leptospirosis	Serological tests early in the illness and after 2–3 weeks to detect rise in antibody titer
Listeria monocytogenes	Listeriosis	Culture of blood and spinal fluid; selective and enrichment cultures performed on food samples to grow potential pathogens; DNA probe for rapid identification of colonies
Chlamydia trachomatis	Chlamydial genital infections	Identification of *C. trachomatis* antigen in urine or pus using monoclonal antibody; nucleic acid probes
Treponema pallidum	Syphilis	Direct fluorescent antibody staining; serological tests
Francisella tularensis	Tularemia	Cultures using cysteine-containing media; fluorescent antibody stain of pus; detection of rise in antibody titer
Yersinia pestis	Plague	Identification of capsular antigen using fluorescent antibody or ELISA
Viruses		
Rhinovirus	Common cold	Strain identification requires use of specific antibodies
Influenza virus	Flu	Tests comparing influenza antibody levels in blood samples taken during acute and convalescent stages of illness
Hantavirus	Hantavirus pulmonary syndrome	Antigen detection in tissues using electron microscopy or monoclonal antibody; ELISA and Western blot tests for IgG and IgM antibodies in victim's blood
Herpes simplex virus	Different strains cause cold sores, ocular lesions, and genital lesions	Identifying the viral antigen in clinical material using fluorescent antibody or DNA probes
Mumps virus	Mumps	Rise in antibody titer, or presence of IgM antibody to mumps virus in patient's blood
Rotavirus	Diarrhea	Electron microscopy or ELISA of diarrheal stool for virus
HIV	AIDS	Detection of antibody to HIV-1 in patient's blood
Fungi		
Coccidioides immitis	Coccidioidomycosis, infection of lung; can disseminate to almost any tissue	Observation of large, thick-walled, round spherules from clinical specimens; PCR identification
Histoplasma capsulatum	Histoplasmosis, intracellular infection of lung; sometimes disseminates	Stained material from pus, sputum, tissue, etc., examined for intracellular *H. capsulatum* yeast phase; blood tests for antibody to the organism

A needle aspiration drained 20 ml of yellowish, foul-smelling fluid from the abscess. Aerobic cultures of this specimen plated on blood and MacConkey agars were negative. Why didn't the infectious agent grow?

The problem in this instance is related to specimen collection and processing. Internal abscesses located near the gastrointestinal tract are often anaerobic infections, in this case caused by the Gram-negative rod *Bacteroides fragilis*, a strict anaerobe (**Fig. 28.15A**). Intestinal microbes, the majority of which are anaerobic, can sometimes escape the intestine if the organ is damaged in

A.

CDC/Don Stalons

B.

1. Remove tube from package.

2. Remove plunger with attached swab. Collect sample.

3. Reinsert swab and press plunger through stopper so that inner tube drops to bottom of outer tube.

4. Mix by swirling. Transport to laboratory.

- Plastic foil package
- Plunger
- Stopper
- Swab
- Inner tube
- Platinum catalyst
- Anaerobic indicator

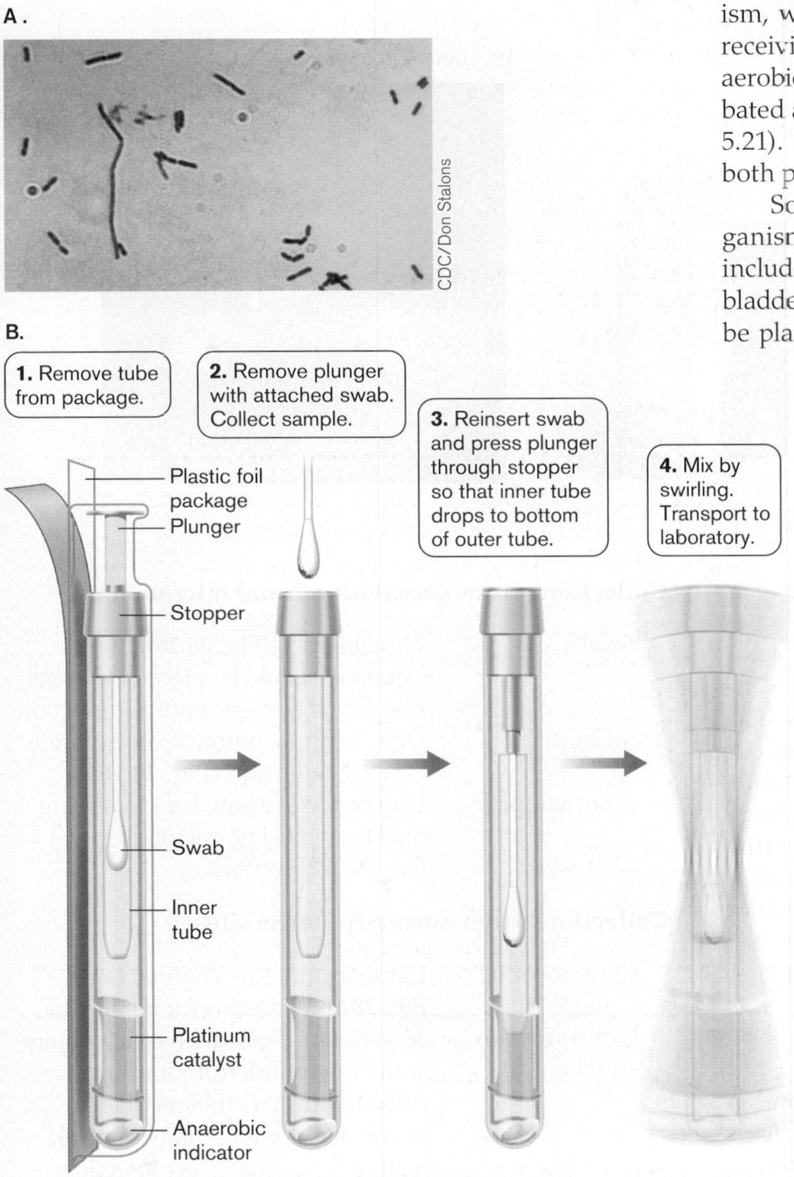

Figure 28.15 Anaerobic infection. A. Gram stain of *Bacteroides fragilis* (1.5–4 μm in length). **B.** Vacutainer anaerobic specimen collector. Plunging the inner tube to the bottom will activate a built-in oxygen elimination system. The anaerobic indicator changes color when anaerobiosis has been achieved.

some way. The specimen in this instance should have been collected under anaerobic conditions by aspiration into a nitrogen-filled tube prior to transport to the clinical laboratory. Alternatively, a swab of the abscess material can be inserted into a special transport tube that has a built-in oxygen elimination system (**Fig. 28.15B**). Because the specimen was collected, transported, and handled in air, many anaerobic microbes were probably killed by the oxygen.

Because *B. fragilis* has a stress response system that permits survival of this anaerobe for one or two days in oxygen, some of the bacteria may have survived transport. The laboratory still had a chance to find the organism, which raises the second problem in the case. After receiving the specimen, the lab cultured it only under aerobic conditions. The laboratory should have also incubated a series of plates anaerobically (see Section 5.6; Fig. 5.21). This case, therefore, illustrates the importance of both proper specimen collection and proper processing.

Some body sites should not contain any microorganisms when collected from a healthy individual. These include blood, cerebrospinal fluid, and urine from the bladder. Because these sites are sterile, specimens can be plated onto nonselective agar media as well as selective media. Nonselective media, such as blood or chocolate agars, can be used because any organism found in these specimens is considered significant.

Urine collection can be problematic, however. When collected from a catheterized patient, urine should be sterile (**Fig. 28.16A**). Catheterization involves passing thin, sterile tubing through the urethra and directly into the bladder. (Catheterization is used primarily to assist urination by immobilized patients, but it also provides a convenient way to collect urine for bacteriological examination.) Collections made from the tube should be sterile unless an infection is present. Unfortunately, the simple process of inserting the catheter through the nonsterile urethra can sometimes introduce organisms into the bladder and precipitate an infection. In addition, the urine should be collected from the catheter, never from the collection bag. Urine may sit for hours in the collection bag, so organisms initially present at low numbers have time to replicate to high numbers even if the patient does not have a urinary tract infection (UTI).

When a catheter is not in place, urine is most commonly collected by what is called a midstream clean-catch technique, which is performed by the patient. In this procedure, the external genitals are first cleaned with a sterile wipe containing an antiseptic. The patient

Figure 28.16 Specimen collection. A. Urinary catheter showing placement in the urethra. **B.** Throat swab. **C.** Sputum collection. A TB patient has coughed up sputum and is spitting it into a sterile container. (The patient is sitting in a special sputum collection booth that prevents the spread of tubercle bacilli. The booth is decontaminated between uses.) **D.** Lumbar puncture to obtain cerebrospinal fluid.

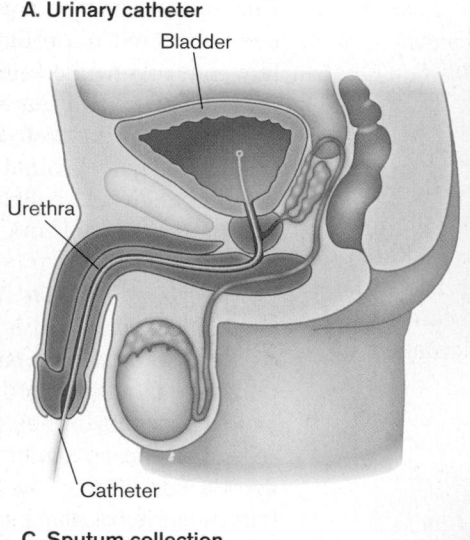

A. Urinary catheter

B. Throat swab

C. Sputum collection

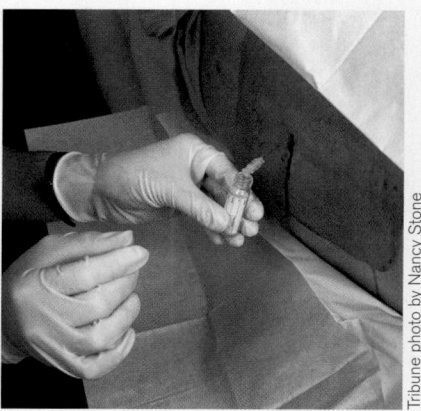

D. Lumbar puncture

then partially urinates to wash as many organisms as possible out of the urethra and then collects 5–15 ml of the midstream urine in a sterile cup. This urine sample will usually not be sterile, because of urethral contamination, but the number of bacteria will be low. The clinical laboratory determines how many organisms per milliliter are present in the midstream catch and informs the physician as to whether an infection is present. Finding more than 100,000 organisms per milliliter of urine from a midstream clean catch is considered indicative of an infection even in an asymptomatic patient; 10,000 or fewer is considered normal. If the patient actually has symptoms of a UTI, however, then any count greater than 1,000 colony-forming units (CFUs) per milliliter of a single species is considered significant and should be treated.

Identifying pathogens present at sites that contain normal microbiota is more challenging. A stool or fecal sample, for instance, is normally teeming with microbes. These specimens are typically plated onto selective media (for example, MacConkey, Hektoen, CNA agar) to eliminate or decrease the number of normal microbiota that might contaminate the specimen. The following represent techniques used to collect specimens from sterile and nonsterile body sites:

Collections from sites with normal microbiota:

Swabs	Throat swabs (**Fig. 28.16B**), for example, should be placed in specialized liquid nutrient transport medium
Sputum	Deep lung secretions expectorated for oral collection (**Fig. 28.16C**)
Stool samples	Cup or rectal swab; for identifying diarrhea-causing microbes
Abscesses	Needle aspirations

Collections from normally sterile sites:

Cerebrospinal fluid	Lumbar punctures (spinal taps; **Fig. 28.16D**); testing for meningitis
Urine samples	Midstream clean catch (will contain some microflora from urethra) or collected from catheters placed in the bladder (should be sterile); testing for urinary tract infections
Blood samples	Generally taken by syringe from two body sites and placed in liquid media for aerobic and anaerobic culture. The same organism isolated from blood samples taken from the two sites is considered the likely etiological agent.

THOUGHT QUESTION 28.5 Two blood cultures, one from each arm, were taken from a patient with high fever. One culture grew *Staphylococcus epidermidis*, but the other blood culture was negative (no organisms grew out). Is the patient suffering from septicemia caused by *S. epidermidis*?

THOUGHT QUESTION 28.6 A 30-year-old woman with abdominal pain went to her physician. After the examination, the physician asked the patient to collect a midstream urine sample that they would send to the lab across town for analysis. The woman complied and handed the collection cup to the nurse. The nurse placed the cup on a table at the nurses' station. Three hours later, the courier service picked up the specimen and transported it to the laboratory. The next day, the report came back "greater than 200,000 CFUs/ml; multiple colony types; sample unsuitable for analysis." Why was this determination made?

TO SUMMARIZE:

- **Special collection precautions** must be taken when collecting specimens from abscesses of suspected anaerobic etiology.
- **Common specimens** include blood, pus, urine, sputum, throat, stool, and cerebrospinal fluid.
- **Specimens** from sites containing normal microbiota must be handled differently than specimens taken from normally sterile body sites.

28.4 Biosafety Containment Procedures

Case History: Fatal Meningitis

On July 15, an Alabama microbiologist was taken to the emergency room with acute onset of generalized malaise, fever, and diffuse myalgias. She was given a prescription for oral antibiotics and released. On July 16, she became tachycardic and hypotensive and returned to the hospital. She died 3 hours later. Blood cultures were positive for Neisseria meningitidis serogroup C. Three days before the onset of symptoms, the patient had prepared a Gram stain from the blood culture of a patient subsequently shown to have meningococcal disease; the microbiologist had also handled agar plates containing cerebrospinal fluid (CSF) cultures from the same patient. Coworkers reported that fluids were aspirated from blood culture bottles at the open laboratory bench. No biosafety cabinets, eye protection, or masks were used for this procedure. Testing at CDC indicated that the isolates from both patients were indistinguishable. The laboratory at the hospital infrequently processed isolates of N. meningitidis and had not processed another meningococcal isolate during the previous 4 years.

Medical and laboratory personnel are exposed to extremely dangerous pathogens on a daily basis.

The microbiologist in this case did not take appropriate measures to protect herself and ended up with a laboratory-acquired infection leading to meningitis. The CDC has published a series of regulations designed to protect workers at risk of infection by human pathogens. Infectious agents are ranked by the severity of disease and ease of transmission. On the basis of this ranking, four levels of containment are employed (**Table 28.4**).

Category I organisms have little to no pathogenic potential and require the lowest level of containment (biosafety level 1). Standard sterile techniques and laboratory practices are sufficient. Category II agents have greater pathogenic potential but are not generally transmitted by the respiratory route. They require more rigorous containment procedures, such as limiting laboratory access when experiments are in progress and using biological laminar flow cabinets if aerosolization is possible (biosafety level 2). Category III pathogens can be transmitted via the respiratory route and present a high risk of infection. To safely handle these organisms, level 2 procedures are supplemented with a lab design ensuring that ventilation air flows only *into* the room (negative pressure) and that exhaust air vents directly to the outside. Negative pressure will keep any organism that may aerosolize from escaping into hallways. In addition, access to the lab is strictly regulated and includes double-door air locks at the entrance (biosafety level 3).

By law, extremely dangerous pathogens such as the Ebola virus may be studied only at a biosafety level 4 containment facility. Practices here dictate that lab personnel wear positive-pressure lab suits connected to a separate air supply (**Fig. 28.17**). The positive pressure ensures that if the suit is penetrated, organisms will be blown away from the breach and not sucked into the suit. As reasonable as these regulations may seem, they were not always in effect. Prior to 1970, scientists had an almost cavalier approach toward handling pathogens. For instance, culture material was routinely transferred from one vessel to another by mouth pipetting (essentially using a glass or plastic pipette as a straw). This is now forbidden for obvious reasons.

TO SUMMARIZE:

- **Various levels** of protective measures are used in handling potentially infectious biological materials.
- **Category I** agents are generally not pathogenic and require the lowest level of containment.
- **Category II** agents are pathogenic but not typically transmitted via the respiratory tract. Laminar flow hoods are required.
- **Category III** agents are virulent and transmitted by respiratory route. They require laboratories with special ventilation and air-lock doors.
- **Category IV** agents are highly virulent and require the use of positive-pressure suits.

Table 28.4 Biological safety levels and select agents.

	Containment level			
	Level 1	**Level 2**	**Level 3**	**Level 4**
Class of disease agent	**Category I** Agents not known to cause disease	**Category II** Agents of moderate potential hazard; also required if personnel may have potential contact with human blood or tissues	**Category III** Agents may cause disease by inhalation route	**Category IV** Dangerous and exotic pathogens with high risk of aerosol transmission; only six such labs in the United States
Recommended safety measures	Basic sterile technique; no mouth pipetting	Level 1 procedures plus limited access to lab; biohazard safety cabinets used; hepatitis vaccination recommended	Level 2 procedures plus ventilation providing directional airflow into room, exhaust air directed outdoors; restricted access to lab (no unauthorized persons)	Level 3 procedures plus one-piece positive-pressure suits; lab completely isolated from other areas present in the same building or is in a separate building
Representative organisms in class	*Bacillus subtilis* *E. coli* K-12 *Saccharomyces* spp.	*Bordetella pertussis* *Burkholderia mallei* (glanders) *Campylobacter jejuni* *Chlamydia* spp. *Clostridium* spp. *Corynebacterium diphtheriae* *Cryptococcus neoformans* *Cryptosporidium parvum* Dengue Diarrheagenic *E. coli* *Entamoeba histolytica* *Francisella tularensis* (tularemia) *Giardia lamblia* *Haemophilus influenzae* *Helicobacter pylori* Hepatitis *Legionella pneumophila* *Listeria monocytogenes* *Mycoplasma pneumoniae* *Neisseria* spp. Pathogenic *Vibrio* spp. *Salmonella* spp. *Shigella* spp. *Staphylococcus aureus* *Toxoplasma* *Yersinia enterocolitica* *Yersinia pestis*	*Bacillus anthracis* (anthrax) *Brucella* spp. (brucellosis) California encephalitis *Coxiella burnetii* (Q fever) EEE (eastern equine encephalitis) Japanese encephalitis virus La Crosse encephalitis LCM (lymphocytic choriomeningitis) *Mycobacterium tuberculosis* Rabies *Rickettsia prowazekii* (typhus fever) Rift Valley fever SARS (severe acute respiratory syndrome) Variola major (smallpox) and other poxviruses VEE (Venezuelan equine encephalitis) West Nile virus Yellow fever	Ebola Guanarito virus Hantavirus Junin virus Kyasanur Forest disease virus Lassa fever Machupo virus Marburg virus Tick-borne encephalitis viruses

Organisms in blue are on the list of CDC select agents that are considered possible agents of bioterrorism.

Figure 28.17 Biosafety level 4 containment. Dr. Kevin Karem at the CDC performs viral plaque assays to determine the neutralization potential of serum from smallpox vaccination trials. He is protected by a positive pressure suit working in a biosafety level 4 laboratory. The airflow into his suit is so loud he must wear earplugs to protect his hearing.

28.5 Principles of Epidemiology
Case History: Inhalation Anthrax

On October 16, a 56-year-old African-American U.S. Postal Service worker became ill with a low-grade fever, chills, sore throat, headache, and malaise. This was followed by minimal dry cough, chest heaviness, shortness of breath, night sweats, nausea, and vomiting. On October 19, he arrived at a local hospital, where he presented with a normal body temperature and normal blood pressure. He was not in acute distress but had decreased breath sounds and rhonchi (dry sounds in lungs due to congestion). No skin lesions were observed, and he did not smoke. Total WBC count was normal, but there was a left shift in the differential— that is, more polymorphonuclear leukocytes (PMNs; see Section 26.3). A chest X-ray showed bilateral pleural effusions (accumulation of fluids in the lung) and a small right lower-lobe air space opacity. Within 11 hours, blood cultures taken upon admission grew Bacillus anthracis. *Ciprofloxacin, rifampin, and clindamycin antibiotic treatments were initiated, and the patient recovered. His job at the post office was to sort mail.*

From October 4 to November 2, 2001, the Centers for Disease Control and Prevention (CDC) and various state and local public health authorities reported 10 confirmed cases of inhalational anthrax and 12 confirmed or suspected cases of cutaneous anthrax in persons who worked in the District of Columbia, Florida, New Jersey, and New York. Many of them were postal workers. It was clear that a biological attack was in progress.

The word "epidemiology" is derived from the Greek meaning "that which befalls man." In scientific parlance, **epidemiology** examines the distribution and determinants of disease frequency in human populations. Put more simply, epidemiologists determine the source of a disease outbreak and the factors that influence how many individuals will succumb to the disease. Epidemiological principles are also used to determine the effectiveness of therapeutic measures and to identify new syndromes, such as SARS (severe acute respiratory syndrome) and Lyme disease. Some of the basic concepts of epidemiology were already covered in Chapter 25 when we discussed infection cycles. Now we will explore how those principles are used to track disease.

Epidemiological early-warning systems require an extensive organization that coordinates information from many sources. In the United States, that duty falls to the Centers for Disease Control and Prevention (CDC). On the world stage, it is the World Health Organization (WHO). Any disease considered highly dangerous or infectious is first reported to local public heath centers, usually within 48 hours of diagnosis. The local centers forward that information to their state agencies, which then report to the CDC in Atlanta. This is how authorities in 2001 quickly recognized that an outbreak of anthrax was under way.

The terms "endemic" and "epidemic" are often used when referring to disease outbreaks. A disease is **endemic** if it is always present in a population at a low frequency. For example, Lyme disease, caused by the spirochete *Borrelia burgdorferi*, is endemic to the northeastern United States because the organism has found a reservoir in deer and ticks. Recall that a **reservoir** is an animal, bird, or insect that harbors the infectious agent and is indigenous to a geographic area. Humans become infected only when they come in contact with the reservoir. Thus, the disease incidence is low but relatively constant. A disease is **epidemic**, on the other hand, when larger than normal numbers of individuals in a population become infected over a short time. This is due, in part, to rapid and direct human-to-human transmission.

Figure 28.18A illustrates the difference in the frequency of cases observed between endemic and epidemic disease. An endemic disease can become epidemic if the population of the reservoir increases, allowing for more frequent human contact; or if the infectious agent evolves to spread directly from person to person, bypassing the need for a reservoir. This is the concern with the H5N1 avian flu virus, which is endemic in animals and birds in Asia (see Sections 11.3 and 25.8). A **pandemic** is an epidemic that occurs over a wide geographic area, usually the world. Pandemics may be long-lived, such as the bubonic plague pandemic in the fourteenth century and the AIDS pandemic in the late twentieth and early twenty-first centuries; or they may be short-lived, as with the 1918 flu pandemic.

A. Endemic versus epidemic

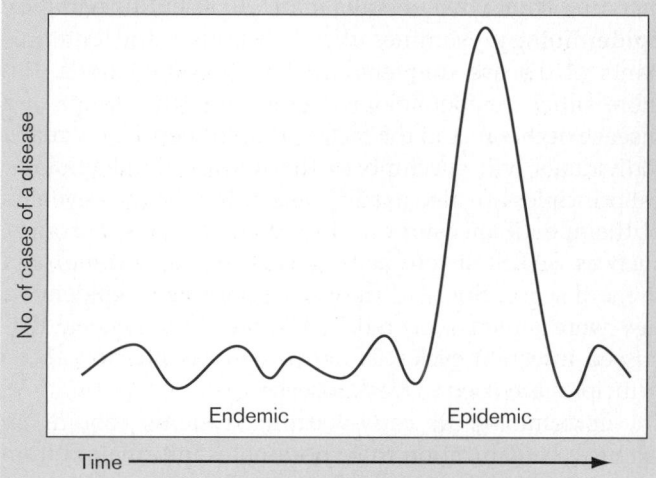

A.

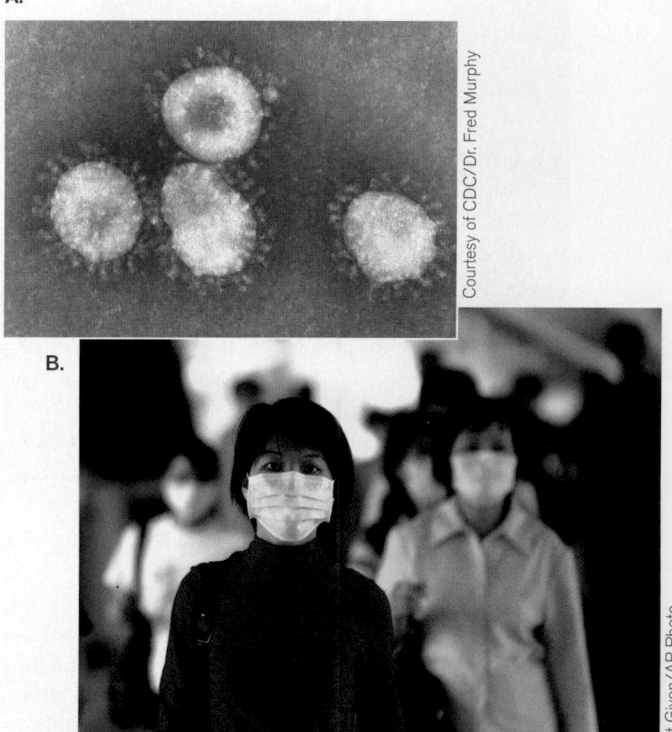

Courtesy of CDC/Dr. Fred Murphy

B.

Anat Givon/AP Photo

Figure 28.19 Severe acute respiratory syndrome (SARS).
A. The coronavirus that causes SARS (TEM). **B.** Citizens of China, including the military, donned surgical masks in 2003 to slow the spread of SARS.

B.

AP Photo/Jean-Marc Bouju

Figure 28.18 The difference between endemic and epidemic disease. A. An endemic disease is continually present at a low frequency in a population. A sudden rise in disease frequency constitutes an epidemic. **B.** A health care worker stands outside a quarantined area housing Ebola patients in the Ivory Coast. Epidemics can be minimized if infected persons are kept segregated from the general population (quarantined) to avoid spread of the infectious agent.

Finding Patient Zero

When trying to contain the spread of an epidemic, it is vital to track down the first case of the disease (known as the **index case** or patient zero) and then identify everyone who has had contact with that individual so that they can be treated or separated from the general population (that is, **quarantined**). When a new disease arises, the epidemiological search for the index case starts only after a number of patients have been diagnosed and a new disease syndrome declared. This is what happened with AIDS in the 1970s.

Identifying an index case within a specific community is easier if the disease syndrome is already recognized, as was the case with the 2003 SARS outbreak in Singapore. According to the World Health Organization, a suspected case of SARS is defined as an individual who has a fever greater than 38°C (100.4°F), who exhibits lower respiratory tract symptoms, and who has traveled to an area of documented disease or has had contact with a person afflicted with SARS. The index case in Singapore was a 23-year-old woman who had stayed on the ninth floor of a hotel in Hong Kong while on vacation. A physician from southern China who stayed on the same floor of the hotel during this period is believed to have been the source of her infection, as well as that of the index patients who precipitated subsequent outbreaks in Vietnam and Canada.

During the last week of February, the woman, who had returned to Singapore, developed fever, headache, and a dry cough. She was admitted to Tan Tock Seng Hospital, Singapore, on March 1 with a low white blood cell count and patchy consolidation in the lobes of the right lung. Tests for the usual microbial suspects (*Legionella*, *Chlamydia*, *Mycoplasma*) were negative. Electron microscopy of nasopharyngeal aspirations showed virus particles with widely spaced club-like projections (**Fig. 28.19A**). At the time of her admission to the hospital, the clinical features and highly infectious nature of SARS were not known. Thus, for the first 6 days of hospitalization, the patient was in a general ward, without barrier infection control measures. During this period, the index patient infected at least 20 other individuals, including hospital staff, nearby patients, and visitors.

Within weeks, the WHO named the disease in China SARS and issued travel alerts (discussed further in Section 28.6). These alerts allowed Singapore health officials to rapidly identify the index patient and her contacts. As a result, they were able to limit spread of the illness. When all was said and done, SARS killed fewer than 1,000 victims worldwide, although thousands more became ill and recovered (**Fig. 28.19B**). Today, SARS remains a threat but one of lesser concern. Methicillin-resistant *Staphylococcus aureus* (MRSA), H5N1 avian flu, and H1N1 swine flu are considered more pressing dangers.

Identifying Disease Trends Depends on Physician Notification of Health Organizations

How do epidemiologists first recognize that an epidemic is under way, and then identify the agent and its source?

Certain diseases, because of their severity and transmissibility, are called reportable, or notifiable, diseases (**Table 28.5**). Physicians are required to report instances of these diseases to a central health organization, such as the CDC in the United States and the WHO. As a result, the incidences of certain diseases within a population can be tracked and upsurges noted. An emerging disease not on the list of notifiable diseases can be detected as a cluster of patients with unusual symptoms or combinations of symptoms. This is possible because diseases of unknown etiology are also reported to health authorities. A new disease could manifest with common symptoms (for example, the cough and fever of SARS) that cannot be linked to a known disease agent by clinical tests. An upsurge in cases, of either a reportable disease or an emerging disease, will set off institutional "alarms" that initiate epidemiological efforts to determine the source and cause of the outbreak.

John Snow (1813–1858), the Father of Epidemiology

The first case in which the source of a disease outbreak was methodically investigated took place in the

Table 28.5 Notifiable infectious diseases.

Bacterial

Anthrax	Hansen's disease (leprosy)	Shigellosis
Botulism	Legionellosis	Staphylococcal enterotoxin
Brucellosis	Leptospirosis	Streptococcal invasive disease
Campylobacter infection	Listeriosis	*Streptococcus pneumoniae*, invasive disease
Chlamydia infection	Lyme disease	Syphilis
Cholera	Meningitis, infectious	Tuberculosis
Diphtheria	Pertussis	Tularemia
Ehrlichiosis	Plague	Typhoid fever
E. coli O157:H7 infection	Psittacosis	Vancomycin-resistant *Staphylococcus aureus* (VRSA)
Gonorrhea	Q fever	
Haemophilus influenzae, invasive disease	Rocky Mountain spotted fever	
	Salmonellosis (nontyphoid fever types)	

Viral

Dengue fever	Mumps	Varicella (chickenpox), fatal cases only (all types)
Hantavirus infection	Poliomyelitis	Yellow fever
Hepatitis, viral	Rabies	
HIV infection	Rubella and congenital rubella syndrome	
Measles	Smallpox	

Fungal

Coccidioidomycosis		

Parasitic

Amebiasis	Giardiasis	Trichinosis
Cryptosporidiosis	Malaria	
Cyclosporiasis	Microsporidiosis	

mid-nineteenth century. During a serious outbreak of cholera in London in 1854, John Snow (**Fig. 28.20A**) used a map to plot the locations of all the diarrheal cases he learned about. The source of the infection was unknown, but Snow thought that if the cases clustered geographically, he might gain a clue as to its location. Water in that part of London was pumped from separate wells located in the various neighborhoods. Snow's map revealed a close association between the density of cholera cases and a single well located on Broad Street (**Fig. 28.20B**). Simply removing the pump handle of the Broad Street well put an end to the epidemic, proving that the well water was the source of infection (known as the "point source" since all infections originated from that point). This approach succeeded brilliantly, even though the infectious agent that causes cholera, *Vibrio cholerae*, was not recognized until 1905, over 50 years later. Identifying clusters of patients afflicted with a given disease is still used to locate potential sources of infectious disease outbreaks.

THOUGHT QUESTION 28.7 Methicillin, a beta-lactam antibiotic, is very useful in treating staphylococcal infections. The development of methicillin-resistant strains of *Staphylococcus aureus* (MRSA) is a very serious development because few antibiotics can kill these strains. Imagine a large metropolitan hospital in which there have been eight serious nosocomial infections with MRSA and you are responsible for determining the source of infection so that it can be eliminated. How would you accomplish this using common bacteriological and molecular techniques?

Genomic Strategies Help Identify Nonculturable Pathogens

Microbiologists have successfully developed many strategies to identify the causes of infectious diseases. Robert Koch in the late nineteenth century devised a set of postulates that, when followed, can identify the agent of a new disease (discussed in Section 1.3). One important tenet of Koch's postulates is that the suspected organism must be grown in pure culture. However, there are bacterial diseases for which the agent cannot be cultured. How might they be identified?

Case History: Whipple's Disease

In April, a 50-year-old man had an abrupt onset of watery diarrhea with a stool frequency of up to ten times every 24 hours. This was the latest episode in a 6-year history of illness beginning with recurring fevers, flu-like symptoms, profuse night sweats, and painful joint swelling. The current bout of diarrhea was associated with gripping lower abdominal pain, especially after meals. No blood or mucus was found in the stools. Blood and stool cultures tested negative for known infectious agents. Serology was also negative for syphilis, brucellosis, toxoplasmosis, and leptospirosis. His weight fell rapidly from 83 to 73 kg within 4 weeks of onset. A flexible sigmoidoscopy showed only diffuse mild erythema in the bowel. However, the appearance of the small bowel was consistent with malabsorption, a disease in which nutrients are poorly absorbed by the intestine. A duodenal biopsy to look for the organisms of Whipple's disease showed large macrophages in the lamina propria (a layer of loose connective tissue beneath

---

A.

Figure 28.20 Early epidemiology. A. John Snow. **B.** The map of London that Snow used to pinpoint the source of the cholera outbreak in 1854. The Broad Street well is found within the red circle. Each black bar represents a death from cholera.

B.

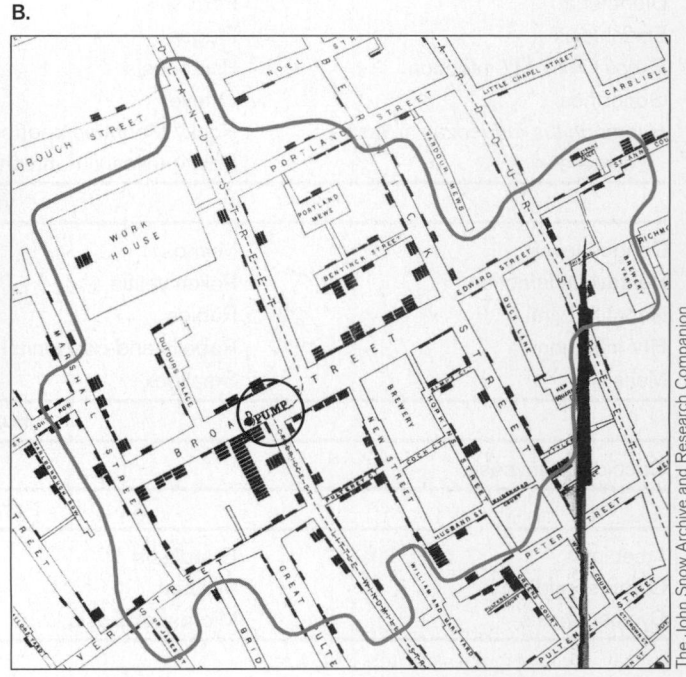

The John Snow Archive and Research Companion

the epithelium of an organ). Bacterial rods characteristic of Whipple's disease were seen with electron microscopy. PCR analysis of tissue samples confirmed the diagnosis.

First diagnosed in 1907 by George Whipple, the symptoms of this disease are malabsorption, weight loss, arthralgia (joint pain), fevers, and abdominal pain. Any organ system can be affected, including the heart, lungs, skin, joints, and central nervous system. The cause of Whipple's disease went undiscovered for 85 years but was suspected to be of bacterial etiology, even though an organism was never successfully cultured. The agent was finally identified in 1992 not by culturing, but by blindly amplifying 16S rDNA sequences from biopsy tissues.

All bacterial 16S rRNA genes have some sequence regions that are highly homologous across species and other sequences that are unique to a species. Tissues from numerous patients diagnosed with Whipple's disease were subjected to PCR analysis using the common 16S rDNA primers. If a bacterial agent was present, it was predicted that PCR should successfully amplify a DNA fragment corresponding to the agent's 16S rRNA gene. All tissues produced such a fragment, indicating that there were bacteria in the tissues. DNA sequence analysis of these fragments indicated that the organism was similar to actinomycetes but was unlike any of the known species.

The organism is actually a Gram-positive soil-dwelling actinomycete that has been named *Tropheryma whipplei* in honor of the physician who first recognized the disease. Because of its bacterial etiology, Whipple's disease can be treated with antibiotics, usually trimethoprim sulfamethoxazole (Bactrim, Septra, Cotrim).

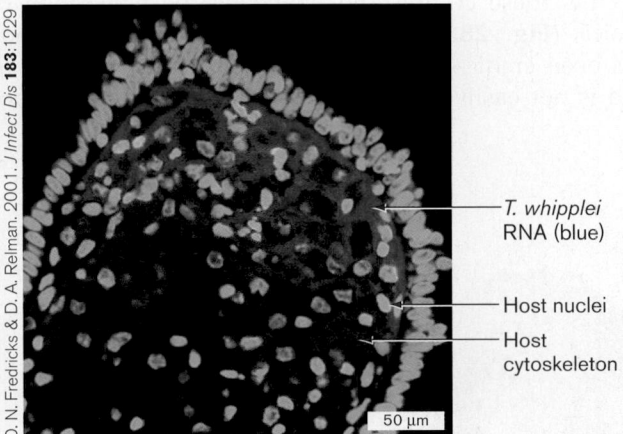

D. N. Fredricks & D. A. Relman. 2001. *J Infect Dis* **183**:1229

T. whipplei RNA (blue)

Host nuclei

Host cytoskeleton

50 μm

Figure 28.21 Whipple's disease. Fluorescent in situ hybridization of a small intestinal biopsy in a case of Whipple's disease (laser scanning confocal microscopy). In this test, a fluorescently tagged DNA probe that specifically hybridizes to *Tropheryma whipplei* RNA is added to the tissue. Other fluorescent probes are used to visualize host nuclei and cytoskeleton. *T. whipplei* rRNA is blue, nuclei of human cells are green, and the intracellular cytoskeletal protein vimentin is red. Magnification approx. 200×.

Figure 28.21 shows an in situ hybridization for *T. whipplei* RNA in a tissue biopsy.

This story reveals that Koch's postulates (see Figure 1.17) must sometimes be modified when identifying the cause of a new disease. In this instance, the organism could not be cultured in pure form in the laboratory but was found, via molecular techniques, in all instances of the disease. Note that more recent studies have successfully cultured *T. whipplei* in vitro. The medium used was formulated on the basis of nutritional requirements deduced from knowing the DNA sequence of the *T. whipplei* genome (discussed in Section 25.6).

> **THOUGHT QUESTION 28.8** What are some reasons why some diseases spread quickly through a population while others take a long time?

Molecular Approaches Can Be Used for Disease Surveillance

A worldwide pandemic of pulmonary tuberculosis currently affects over 2 billion people. Many of the *Mycobacterium tuberculosis* infections are caused by multidrug-resistant strains that are difficult, if not impossible, to kill with existing antibiotics (see Section 26.3). This problem is especially serious among refugee populations attempting to flee war-torn countries. As a result, it is important to screen these refugees as they enter neighboring countries with low incidences of tuberculosis. Although chest X-rays are mandatory in many cases, a positive image will be obtained only if the disease is at a relatively advanced stage. Actively infected individuals who have not developed the characteristic lung tubercles seen on X-ray will not be identified.

Unfortunately, the acid-fast staining of sputum samples (discussed in Section 28.2) also fails to detect individuals at an early stage of infection. Studies have shown, however, that PCR techniques are much more sensitive for detecting these individuals. As time progresses, PCR surveillance strategies will be used more often to track the worldwide ebb and flow of microbial diseases. PCR and restriction fragment length polymorphism (RFLP) strategies are already used for epidemiological purposes to type (that is, determine the relatedness of) different microbial isolates by generating a complex DNA profile that is specific for a particular strain (Section 28.2). For example, DNA profiles were used to link an outbreak of over 2000 cases of *salmonellosis* in 2010 to a single strain of *S. enterica* serovar Enteritidis. The results, conducted by a national network of public health agencies (PulseNet), led to the recall of over a half billion eggs.

Bioterrorism

"A wide-scale bioterrorism attack would create mass panic and overwhelm most existing state and local systems within a few days," said Michael T. Osterholm, director of the Center for Infectious Disease Research & Policy at the University of Minnesota. "We know this from simulation exercises." These words were spoken in October 2001.

Less than a month after the September 11 attack on the World Trade Center, a biodefense scientist working at the U.S. Army Medical Research Institute of Infectious Diseases (Fort Detrick, Maryland) sent weapons-grade anthrax spores through the U.S. mail. Thankfully, only 5 persons died and a mere 25 became ill. But even though the efficiency of the attack was poor, the impact was enormous. Over 10,000 people took a 2-month course of antibiotics after possible exposure, and mail deliveries throughout the country were affected. The simple act of opening an envelope suddenly became a risky endeavor.

As a result of this attack and other events, the CDC and the National Institutes of Health (NIH) assembled a list of select agents (marked in blue in **Table 28.4**) that could potentially be used as bioweapons. A bioweapon is considered to be any infectious agent or toxin that has high virulence and/or mortality rate. Microorganisms considered bioweapons can be used to conduct biowarfare, with the intent of inflicting massive casualties; or bioterrorism, which may result in only a few casualties but cause widespread psychological trauma.

Although the list of select agents is recent, biowarfare is not new. In the Middle Ages, victims of the Black Death (plague caused by *Yersinia pestis*) were flung over castle walls using catapults; during the French and Indian War, in the eighteenth century, Jeffrey Amherst distributed smallpox-infected blankets to Native Americans; and during World War II, the Imperial Japanese Army experimented with infectious disease weapons using Chinese prisoners as guinea pigs. Even the United States has participated through the development of weapons-grade anthrax spores. It was originally thought that the strain of *Bacillus anthracis* used in the 2001 anthrax attack was the same as the strain developed by the military. We now know that the strain used in the attack had a more ordinary pedigree.

The first documented act of bioterrorism in the United States occurred in 1984, when followers of the cult leader Bhagwan Shree Rajneesh tried to control a local election in Dalles, Oregon, by infecting salad bars with *Salmonella*. Over 900 people became ill. Rajneesh was sentenced to 20 years in prison.

How effective are bioweapons? The method by which a biological agent is dispersed plays a large role in its effectiveness as a weapon. Only a few people became ill during the 2001 anthrax attack, not only because of the epidemiological surveillance but because anthrax is inherently difficult to disperse. Thousands of spores must be inhaled to contract disease, which means that effective dispersal of the spores is critical for the use of anthrax as a weapon. Once spores hit the ground, the threat of infection is limited. Weapons-grade spores are very finely ground so that they stay airborne longer. But as we saw, the letter-borne dispersal system was not very effective in terms of generating large numbers of victims. Nevertheless, the potential threat of weapons-grade anthrax on the battlefield led the U.S. military in 2006 to resume vaccinating all soldiers serving in Iraq, Afghanistan, and South Korea.

An effective bioweapon would be one that capitalizes on person-to-person transmission. In an easily transmitted disease, one infected person could disseminate disease to scores of others within one or two days. So in terms of generating massive numbers of deaths, anthrax was a poor choice. But the goal of most terrorists is not to kill large numbers of people, but to terrorize them. In that regard, the anthrax attack succeeded (**Fig. 28.22A**).

The most effective bioweapon in terms of inflicting death (biowarfare) would have a low infectious dose, be easily transmitted between people, and be one to which a large percentage of the population is susceptible. Smallpox fits these criteria and would be the bioweapon of choice (**Fig. 28.22B**). Fortunately, however, smallpox has been eradicated (almost) from the face of the Earth and is not easily obtained. Two laboratories still harbor

A.

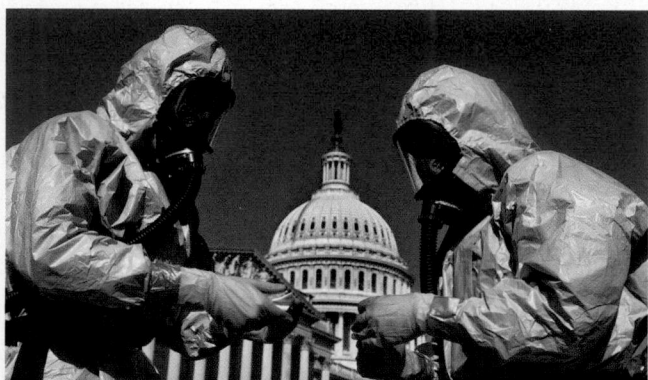

Kenneth Lambert/AP

B.

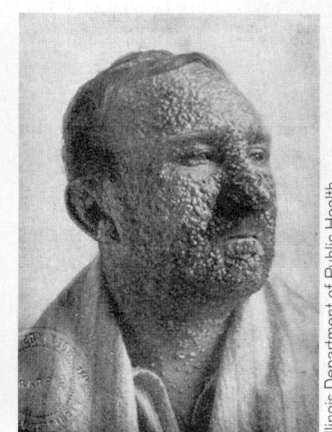

Illinois Department of Public Health

Figure 28.22 Dealing with bioterrorism. A. Members of a hazardous materials team near Capitol Hill during the anthrax attacks in 2001. **B.** An Illinois man suffering from smallpox, 1912.

the virus—one in the United States and one in Russia. It is believed that the virus has been destroyed in all other laboratories. Since smallpox is the perfect biowarfare agent, it is imperative that the last two smallpox repositories remain secure.

While the good news is that smallpox disease has been eradicated, the bad news is that no individual born after 1970 has been vaccinated (actually some military and a few laboratory personal are vaccinated). As a result, anyone under 35 years of age is susceptible to smallpox. Even those of us who received the smallpox vaccination over 35 years ago are at risk, since our protective antibody titers have diminished. A terrorist attack with smallpox would cause terrible numbers of deaths. The vaccine does, however, exist. Were a smallpox attack to be launched, the vaccine would be rapidly administered to limit the spread of disease. Nevertheless, the economic and psychological impact of a smallpox epidemic would be devastating.

Research with organisms considered to be select agents is tightly regulated. Because *Yersinia pestis*, for instance, is a select agent, laboratory personnel working with it must now possess security clearance with the Department of Justice (even though the organism itself can be handled under biosafety level 2 conditions; see **Table 28.4**). The laboratory must also register with the CDC to legally possess this pathogen, and access to the lab and the organism must be tightly controlled.

Much has improved since Michael Osterholm offered his dire assessment of a wide-scale bioterrorism attack. Education and surveillance procedures have been bolstered, and new detection technologies are being developed (**Special Topic 28.1**; see also **eTopic 28.2**). We will probably never be fully protected from attack, biological or otherwise, but recent efforts have improved the situation.

WWW | Bioterrorism preparedness act
microbiology2.com/links

TO SUMMARIZE:

- **John Snow** founded the discipline of epidemiology.
- **Epidemiology** examines factors that determine the distribution and source of disease.
- **Endemic, epidemic, and pandemic** are terms for different frequencies of disease in different geographic areas.
- **Finding patient zero (the index case)** is important for containing the spread of disease.
- **Molecular approaches using PCR and nucleic acid hybridization** are used to identify nonculturable pathogens and to track disease movements.
- **Bioweapons**, when they have been used, typically kill few people but incite great fear.
- **The CDC** has assembled a list of select agents with bioweapon potential.

28.6 Detecting Emerging Microbial Diseases

Case History: SARS

A 48-year-old man was hospitalized in a Dutchess County, New York, hospital with a 38.3°C (101°F) fever, headache, and body aches. He also had difficulty breathing. He had just returned from a business trip to China.

SARS: An Epidemiological Success Story

Within 7 weeks in early 2003, epidemiologists armed with global technologies and rapid DNA sequencing techniques tracked, named, identified, completely sequenced, and contained a newly emerging disease with a scary death rate: severe acute respiratory syndrome (SARS). This was a remarkable feat, especially when we consider that after the first cases of AIDS appeared in 1981, it took over 3 years just to identify the virus; and it took 7 years to track down the Lyme disease spirochete, *Borrelia burgdorferi*, after Lyme disease was first recognized in 1975.

We now know that SARS first developed in Asia, and then began to spread by jet to other countries, such as Canada. While the death rate seemed modest at about 5%, it was higher than the 4% reached during the 1918 flu epidemic that killed 20–40 million people worldwide. SARS clearly posed a formidable threat. Unprecedented cooperation between the WHO, the CDC, and numerous other health organizations around the world played a major role in tracking and containing the disease. Patient information, case histories, and possible treatment regimens flooded into a secure WHO website. The most exciting posting was the 30,000-bp sequence of the SARS genome, taken from one-millionth of a gram of genetic material isolated from a Toronto patient. The following 2003 time line illustrates the remarkable speed with which all this took place:

February 14—China first reports 305 cases of atypical pneumonia to WHO.

March 15—WHO issues emergency travel alert; names illness SARS.

March 17—Leading epidemiologists join WHO in conference call; agree to unprecedented cooperation.

April 4—U.S. President George Bush authorizes quarantine of SARS patients.

April 12—Canadian scientists in Vancouver post the completed genome on the Internet.

April 16—WHO announces that SARS is caused by a pathogen never before seen in humans, a new coronavirus.

Special Topic 28.1 | **What's Blowing in the Wind? Quick Pathogen Detection Systems Guard against Bioterrorism**

In April 1979, workers at a secret Soviet biological weapons facility neglected to replace a filter in a laboratory ventilation system, and a cloud of highly weaponized *Bacillus anthracis* spores quickly spewed into the air outside. The deadly plume drifted downwind, infecting humans and cattle across a wide area. The resulting outbreak ultimately killed more than 64 people. Had a rapid pathogen detection system been available and deployed at the facility, all those lives could have been saved. Unfortunately, 30 years ago such technology was only the stuff of science fiction. Today, that technology exists in small, portable forms.

Several multipathogen molecular detection platforms have been developed. Some involve machines that detect antigen-antibody interactions; others use automated PCR to amplify specific pathogen genes and detect the products through hybridization to immobilized oligonucleotides. The wet chemistries for all of these detection systems are carried out in a roughly 2-inch square called lab-on-a-chip.

An interesting addition to the field of quick pathogen detection is the PANTHER sensor (PAthogen Notification for THreatening Environmental Releases), a device developed in 2008 that can detect and identify a pathogen in as little as 3 minutes—significantly faster than traditional methods requiring isolation and growth of the pathogen (which can take 48 hours or longer). The PANTHER sensor (**Fig. 1**) uses a cell-based technology that can detect as few as a dozen particles of a pathogen per liter of air. Currently, it can detect 24 pathogens, including the potential bioterrorism agents of anthrax, plague, smallpox, and tularemia.

The device uses an array of B cells, each displaying antibodies specific to a particular bacterium or virus. The cells are engineered to emit photons of light when they detect

their target pathogen. More specifically, the B cells have been bioengineered to express the gene that encodes aequorin, a calcium-sensitive bioluminescent protein. When a pathogen surface antigen cross-links the appropriate B-cell surface antibodies, a signal transduction cascade produces a rapid influx of calcium (**Fig. 2**). Aequorin in the B cell will luminesce within seconds of calcium influx, and the light emitted is detected by a photon detector. The device then displays a list of any pathogens found.

Figure 1   The PANTHER detection system. *Source:* Rider et al. 2003. *Science* **301**:213–215.

Reprinted with permission of MIT Lincoln Laboratory, Lexington, MA

Technology Helps New Infectious Agents Emerge and Spread

Despite all we know of microbes and despite the many ways we have to combat microbial diseases, our species, for all its cleverness, still lives at the mercy of the microbe. Lyme disease, MRSA, SARS, Ebola, *E. coli* O157:H7, HIV, "flesh-eating" streptococci, hantavirus, swine flu—all of these and many other new diseases have emerged over the last 30 years. Worse yet, forgotten scourges, such as tuberculosis, have reappeared. Yet in the 1970s, medical science was claiming victory over infectious disease. What happened?

Part of the equation has been progress itself. Travel by jet, the use of blood banks, and suburban sprawl have

all opened new avenues of infection. People unwittingly infected by a new disease in Asia or Africa can, traveling by jet, bring the pathogen to any other country in the world within hours. A person may not even show symptoms until days or weeks after the trip. This means that diseases can spread faster and farther than ever before. In addition, newly emerged blood-borne pathogens can spread by transfusion. This was a major problem with HIV before an accurate blood test was developed to screen all donated blood.

Although human encroachments into the tropical rain forests have often been blamed for the emergence of new pathogens, one need go no farther than the Connecticut woodlands to find such developments. *Borrelia*

Quick pathogen detection technologies are constantly being improved in the effort to defend against bioterrorism. Small quick pathogen detection devices could be used in buildings, subways, and other public areas. Eventually, such technology is expected to supplant the classical clinical microbiology practices that require growth of the microbe to achieve identification. There is hope, for example, that such devices can be used on farms or in food-processing plants to test for contamination by *E. coli*, *Salmonella*, or other food-borne pathogens. Another potential application is in medical diagnostics, where the technology could be used to test patient samples, giving rapid results without having to send samples to a laboratory. Wherever it is used, successful quick pathogen detection should provide the clinician with better information to quickly prescribe the appropriate course of treatment.

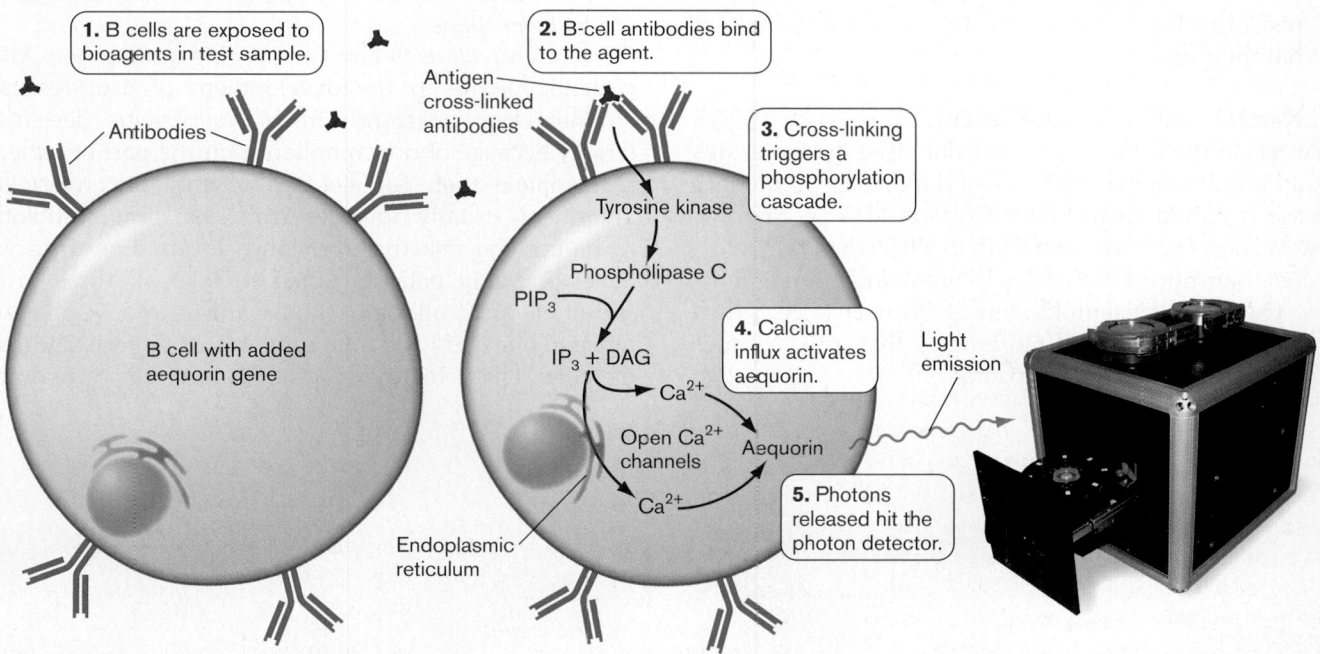

Figure 2   **Principle of the bioelectronic sensor used in the PANTHER system.**   DAG = diacylglycerol; IP_3 = 1,4,5-inositol trisphosphate; PIP_3 = phosphatidylinositol 3,4,5-trisphosphate. *Source*: Martha S. Petrovick et al. 2008. *Lincoln Lab. J.* **17**(1).

burgdorferi, the spirochete that causes Lyme disease, lives on deer and white-footed mice and is passed between these hosts by the deer tick (Fig. 26.25)—an infectious cycle that has been going on for years. Humans crossed paths with these animals long before the disease erupted in our communities. Why have we suddenly become susceptible? The answer appears to be suburban development. In the wild, foxes and bobcats hunt the mice that carry the Lyme agent. These predators disappear as developers clear land and build roads and houses, leaving the infected mice and ticks to proliferate. Humans in these developed areas are more likely to be bitten by an infected tick and contract the disease than in prior decades. Luckily, many diseases that successfully leap from animals to humans find the new host to be a dead end, unable to spread the disease to others.

There are numerous examples in which technology and progress have had the unintended consequence of breeding disease. Here are just a few:

■ **Mad cow disease.** Modern farming practices (North America and Europe) of feeding livestock the remains of other animals helped spread transmissible spongiform encephalopathies similar to Creutzfeldt-Jakob disease associated with prions. Because the prion is infectious, the brain matter from one case of mad cow disease could end up infecting hundreds of other cattle, which in turn increases the chance that the disease could spread to humans.

- **Lyme disease.** Suburban development in the northeastern United States destroys predators of the mice that carry *Borrelia burgdorferi*.
- **Hepatitis C.** Transfusions and transplants spread this blood-borne disease.
- **Influenza.** Live poultry markets in Asia serve as breeding grounds for avian flu viruses that can jump to humans.
- **Enterohemorrhagic *E. coli* (for example, *E. coli* O157:H7).** Modern meat-processing plants can accidentally grind trace amounts of these acid-resistant, fecal organisms into beef while making hamburgers.

Natural environmental events can also trigger upsurges in the incidence of unusual diseases. For example, an unprecedented outbreak of hantavirus pulmonary disease occurred in the Four Corners area of Arizona, New Mexico, Colorado, and Utah in 1993 when rain led to greater-than-normal increases in plant and animal numbers. The resulting tenfold increase in deer mice, which carries the virus, made it more likely that infected mice and humans would come in contact.

Detecting Emerging and Reemerging Pathogens

A world map showing the general locations of emerging and reemerging diseases is shown in **Figure 28.23**. A reemerging disease is one thought to be under control but whose incidence has risen. For example, the incidence rate of tuberculosis dropped sharply in the 1950s, but the number of recent cases has increased just as dramatically. Tuberculosis is a reemerging disease. The trigger for its reemergence was the AIDS pandemic beginning in 1980. Immunocompromised AIDS patients are highly susceptible to infection by many organisms, including *Mycobacterium tuberculosis*.

M. tuberculosis is also reemerging among non-AIDS patients, owing to the development of drug-resistant strains. Drug resistance in *M. tuberculosis* developed largely because of noncompliance on the part of patients to complete their full courses of antibiotic treatment. Treatment usually involves three or more antibiotics to reduce the risk that resistance to any one drug will develop. Many patients failed to take all three drugs simultaneously, allowing the organism to develop resistance to one drug at a time until it became resistant to all of them. These highly drug-resistant strains are almost

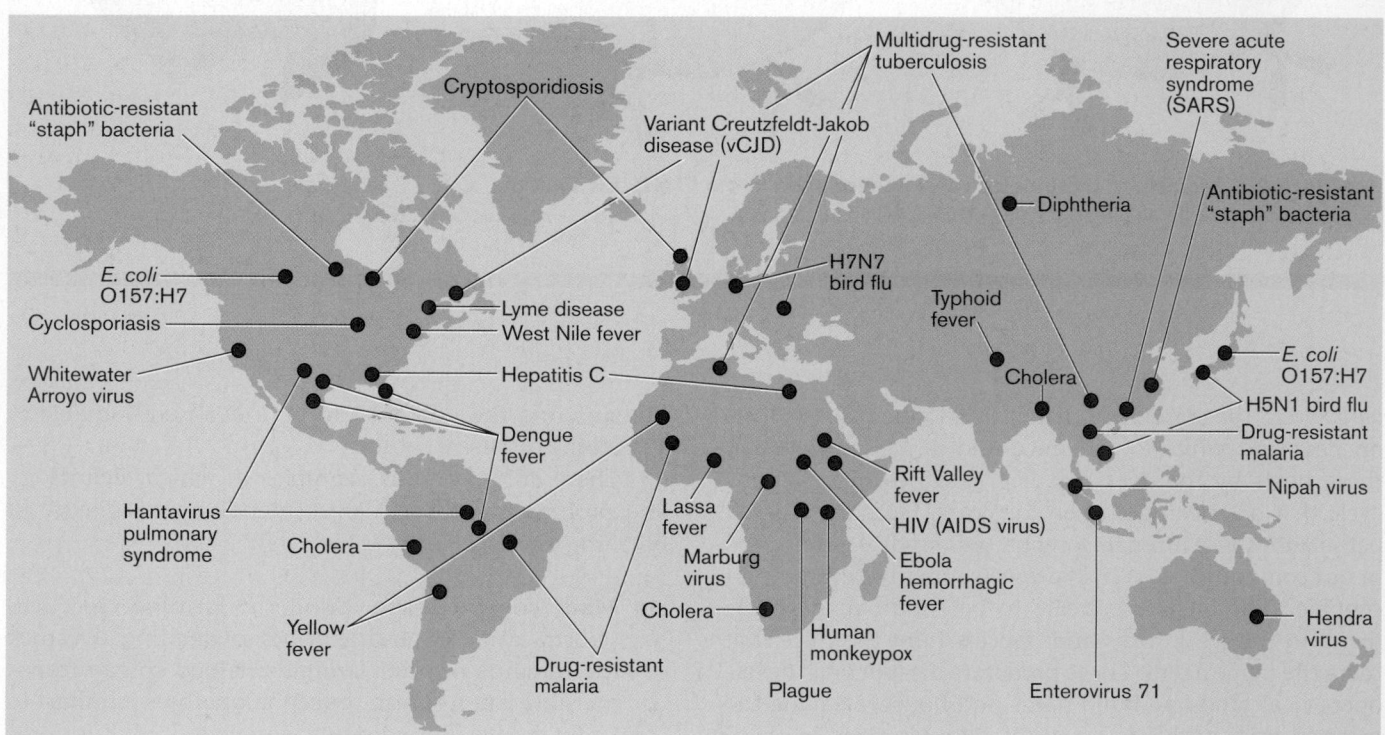

Figure 28.23 Locations of some emerging and reemerging infectious diseases. The examples given represent extreme increases in the reported cases. Many of these diseases, such as HIV and cholera, are widespread but show alarming increases in the areas indicated. Cases of H1N1 swine flu have increased around the globe.

impossible to kill. The link between noncompliance and the development of drug resistance is the primary reason for the current requirement that tuberculosis patients be in the presence of a medical staff member when taking the multiple antibiotics prescribed.

As described in Section 28.5, highly infectious diseases are identified using aggressive epidemiological surveillance. Communication among local, national, and world health organizations is critical and helped expose the rise in tuberculosis. To see for yourself how emerging diseases are monitored, search the Internet for a website titled ProMED-mail. There you will find daily reports posted from around the world that describe new outbreaks of infectious diseases.

For example, a posting on March 6, 2009, reports finding "skin pustules on an African elephant calf and 3 of its caretakers" at the San Diego Zoo. The cause of the infection was methicillin-resistant *Staphylococcus aureus* (MRSA). Further investigation revealed that infection of the elephant calf likely came from a caretaker colonized with MRSA. This is the first reported case of MRSA in an elephant and of suspected MRSA transmission from a human to a zoo animal. Another posting, on August 21, 2009, from Uganda, reports a case of hemorrhagic fever possibly due to Ebola virus. Definitive diagnosis was awaiting viral confirmation, but the patient was persistently bleeding from his mouth, nose, and ears, indicating a hemorrhagic fever of some sort. Given these signs of a new Ebola outbreak, medical staff at Mbale Hospital were on high alert, and the patient was placed in isolation. The site also posts reports of undiagnosed illnesses that *may* be the initial signs of emerging diseases. For instance, on August 15, 2009, there was a report of six cases of hemorrhagic diarrhea in dogs from Sarasota County, Florida, five of which died. The cause was unknown, but *E. coli* O157:H7 was suspected. If this had been proven correct, another possible source for human spread of the disease would have been exposed. In contrast to cattle, which show little ill effect of harboring *E. coli* O157, canines show severe disease. Tracking these reports allows trends of disease spread to be determined.

THOUGHT QUESTION 28.9 On the ProMED-mail web page (www.promedmail.com), click on **Links** to view outbreaks recorded by WHO. What outbreaks happened throughout the World during the current year?

Ecology Influences the Epidemiology and Evolution of Pathogens

We have provided numerous examples of how the natural environment can harbor pathogens and foster the emergence of new ones. Knowledge that pathogens have reservoirs in animals, arthropods, and plants has given rise

to a new collaborative effort among clinicians, scientists, veterinarians, and ecologists called the One Health Initiative. The goal of the One Health Initiative is to control human health through animal health and vice versa. For example, vaccinating wild rodents can decrease Lyme disease in humans. The idea has recently been extended to plant pathology because some bacteria are pathogens of both plants and humans (for example, *Pantoea agglomerans*, formerly *Enterobacter agglomerans*). In addition, some enteric bacteria can live within a plant vascular system (*E. coli*, *Klebsiella*, *Salmonella*).

An excellent example of how multidisciplinary collaboration resulted in better understanding of an infectious outbreak occurred in 2006 when approximately 200 people in 26 states were diagnosed with a particularly virulent case of *E. coli* O157:H7. Nearly half of the cases were hospitalized, and many suffered from hemolytic uremic syndrome (HUS). The source of the infection was contaminated spinach traced to the Salinas Valley of California. It turned out the organisms were contained within the vascular system of the spinach, so washing the spinach would not remove the pathogen. Had this outbreak been viewed through only the narrow lens of human health, efforts would have focused on morbidity, mortality, outbreak investigation, laboratory diagnosis, and clinical treatment. The origin of the disease would have remained a mystery.

Working together, epidemiologists and veterinarians found a genetically identical organism in cattle close to where the spinach was produced and in wild hogs that ran through the same fields. Ecologists and hydrologists understood that the groundwater and surface water in this region were being mixed because of a drought followed by heavy rains, and that irrigation systems were strained in the effort to keep up with intensified agricultural production. Eventually, the same *E. coli* strain was found in one of the water ditches close to the spinach fields in the area. Scientists pondering these facts within the One Health framework deduced that cattle harboring the *E. coli* had defecated in a field, thereby contaminating wild hogs. The wild hogs, by running through the spinach fields, contaminated those fields with their feces, and the irrigation water then swept the pathogen into the plant vasculature. Only by integrating our knowledge of the environment and ecology could this investigation be completely understood and appropriate intervention and prevention strategies implemented. This outbreak exemplifies the fact that human health and animal health are inextricably linked and that a holistic approach is needed to understand, protect, and promote the health of all species.

TO SUMMARIZE:

- **Emerging diseases** can spread quickly around the world as a result of air travel.
- **Modern technology** and urban growth have provided opportunities for new diseases to emerge.

■ **Multidisciplinary collaboration** among ecologists, veterinarians, clinicians, and other scientists is necessary for devising appropriate strategies aimed at disease intervention and epidemic prevention.

Concluding Thoughts

This chapter has examined the basic principles used to collect, detect, and track pathogenic microorganisms.

As you can see, the task of controlling the spread of disease is daunting and ever changing because microorganisms continue to evolve. Known pathogens change, eluding eradication efforts by the immune system and antibiotics, while new pathogens keep emerging. The world is depending on the next generation of microbiologists to face the growing threat of these evolving microbes.

CHAPTER REVIEW

Review Questions

1. Why is it important to identify the genus and species of a pathogen?
2. What is an API strip, and what is its use in clinical microbiology?
3. Describe three examples of selective media.
4. If a colony on a nutrient agar plate is catalase-positive, does this mean it is made up of Gram-positive microorganisms? Why or why not?
5. Describe the types of hemolysis visualized on blood agar.
6. What is the clinical significance of a group A, beta-hemolytic streptococcus?
7. How does one distinguish *Staphylococcus aureus* from *Staphylococcus epidermidis*?
8. Why are PCR identification tests preferable to biochemical approaches?
9. How is RTQ-PCR performed?
10. Describe an ELISA.
11. Name some sterile and nonsterile body sites.
12. List seven common types of clinical specimens collected for bacteriological examination.
13. Describe key features of the four levels of biological containment.
14. How is a pandemic different from an epidemic?
15. How can genomics help identify nonculturable pathogens?
16. List and briefly describe four emerging diseases.
17. Name four select agents and the diseases they cause.

Thought Questions

The first four thought questions below are based on the following case history: *An infectious disease physician in Florida telephoned the Centers for Disease Control and Prevention to report two possible cases of botulism. The two male patients presented with drooping eyelids, double vision, difficulty swallowing, and respiratory problems. The physician had drawn sera and collected stool specimens from the men to test for botulinum toxin, but no results were available.*

1. What are the major concerns raised by these two possible cases of botulism?
2. How might you go about swiftly determining if there was a link between the two cases and if there are other cases of botulism?
3. The two patients ate only one food item in common: a cured and fermented fish called "moloha." How would you determine whether the men were indeed suffering from botulism and whether the fermented fish was the source of disease?
4. How could this anaerobic pathogen grow and make toxin in this fish product?
5. Five small outbreaks of Ebola occurred within days of one another in five different cities in the United States and Canada. Patient histories revealed that one person from each city had been in the Atlanta airport at the same time 2 days prior to becoming seriously ill. None had left the country recently; none flew on the same jet or even crossed paths while in the airport. As an epidemiologist, conjure up a scenario, excluding bioterrorism, that could account for this scattered outbreak. You are aware that an outbreak recently occurred in the Congo. (This is a hypothetical case, by the way.)

Key Terms

endemic (1085)

epidemic (1085)

epidemiology (1065, 1085)

index case (1086)

multiplex PCR (1073)

pandemic (1085)

quarantined (1086)

reservoir (1085)

sequela (1064)

Recommended Reading

Baker, Edward L., Margaret A. Potter, Deborah L. Jones, Shawna L. Mercer, Joan P. Cioffi, et al. 2005. The public health infrastructure and our nation's health. *Annual Review of Public Health* **26**:303–318.

Bengis, R. G., F. A. Leighton, J. R. Fischer, M. Artois, T. Morner, et al. 2004. The role of wildlife in emerging and re-emerging zoonoses. *Reviews in Science and Technology* **23**:497–511.

Bravo, Lulette T., and Gary W. Procop. 2009. Recent advances in diagnostic microbiology. *Seminars in Hematology* **46**:248–258.

Espy, Mark, James Uhl, Lynne Sloan, Seanne Buckwalter, Mary Jones, et al. 2006. Real time PCR in clinical microbiology: Application for routine laboratory testing. *Clinical Microbiology Reviews* **19**:165–256.

Fournier, Pierre-Edouard, Michel Drancourt, and Didier Raoult. 2007. Bacterial genome sequencing and its use in infectious diseases. *Lancet Infectious Diseases* **7**:711–723.

Fraser, Christophe, Christl A. Donnelly, Simon Cauchemez, William P. Hanage, Maria D. Van Kerkhove, et al. 2009. Pandemic potential of a strain of influenza A (H1N1): Early findings. *Science* **324**:1557–1561.

Gaydos, Charlotte A., Mellisa Theodore, Nicholas Dalesio, Billie J. Wood, and Thomas C. Quinn. 2004. Comparison of three nucleic acid amplification tests for detection of *Chlamydia trachomatis* in urine specimens. *Journal of Clinical Microbiology* **42**:3041–3045.

Kuiken, Thijs, Ron Fouchier, Guus Rimmelzwaan, and Albert Osterhaus. 2003. Emerging viral infections in a rapidly changing world. *Current Opinion in Biotechnology* **14**:241–246.

Madoff, Lawrence C., and John P. Woodall. 2005. The Internet and global monitoring of emerging diseases: Lessons from the first 10 years of ProMED-mail. *Archives of Medical Research* **36**:724–730.

Molinari, J. A. 2003. Diagnostic modalities for infectious diseases. *Dental Clinics of North America* **47**:605–621.

Pejcic, Bobby, Roland De Marco, and Gordon Parkinson. 2006. The role of biosensors in the detection of emerging infectious diseases. *Analyst* **131**:1079–1090.

Raman, Lakshmy Anantha, Noman Siddiqi, Mohammed Shamim, Monorama Deb, Geeta Mehta, et al. 2000. Molecular characterization of *Mycobacterium abscessus* strains from a hospital outbreak. *Emerging Infectious Diseases* **6**:561–562.

Relman, David A., Thomas M. Schmidt, Richard P. MacDermott, and Stanley Falkow. 1992. Identification of the uncultured bacillus of Whipple's disease. *New England Journal of Medicine* **327**:283–301.

Rhoads, J. Mark, Nicole Y. Fatheree, Johana Norori, Yuying Liu, Joseph F. Lucke, et al. 2009. Altered fecal microflora and increased fecal calprotectin in infants with colic. *Journal of Pediatrics* **155**:823–828.

Shi, Pei-Yong, Elizabeth Kauffman, Ping Ren, Andy Felton, Jennifer Tai, et al. 2001. High throughput detection of West Nile virus RNA. *Journal of Clinical Microbiology* **39**:1264–1271.

Shvartzman, Pesach, and Yussuf Nasri. 2004. Urine culture collected from gel-based diapers: Developing a novel experimental laboratory method. *The Journal of the American Board of Family Practice* **17**:91–95.

Soto, S. M. 2009. Human migration and infectious diseases. *Clinical Microbiology and Infection* **15**(Suppl.1):26–28.

For further study and review, please visit StudySpace at **microbiology2.com**.

Appendix 1

Biological Molecules

This appendix reviews information typically covered in an introductory biology course. We first cover the chemical bonding principles needed to understand biological molecules, with an emphasis on the special properties of water. We then discuss organic molecules, paying particular attention to four important classes of organic biomolecules: proteins, carbohydrates, nucleic acids, and lipids. Finally, we explore common chemical principles, such as concentrations, thermodynamics, equilibrium, pH, and oxidation-reduction reactions.

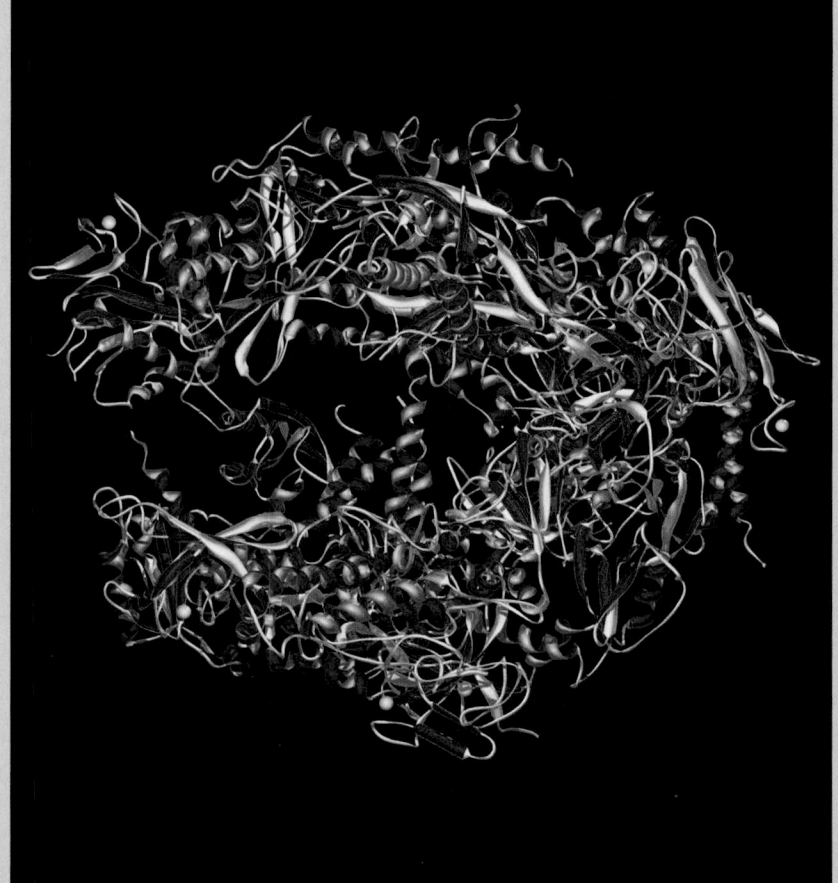

Current Research Highlight

The enzyme RNA polymerase II (computer model). The molecule comprises 12 subunits. This enzyme synthesizes a complementary mRNA strand from a strand of DNA during a process called transcription. It recognizes a start sign on the DNA strand and then moves along the strand building the mRNA until it reaches a stop sign. mRNA is the intermediary between DNA and its protein product. *Source:* Mark J. Winter/Photo Researchers, Inc.

Living cells are remarkably complex machines, able to integrate and respond to multiple stimuli, to catalyze reactions, and to replicate themselves. Yet despite all the various tasks that cells perform, 98% of the mass of living organisms consists of only six elements: hydrogen (H), oxygen (O), nitrogen (N), phosphorus (P), sulfur (S), and carbon (C); and 90% of the mass is accounted for by just C, H, and O. Of the compounds formed from these elements, the most abundant in cells is water. The remainder of the cell consists, for the most part, of just four different kinds of organic (carbon-based) macromolecules: proteins, nucleic acids, carbohydrates, and lipids. Cells can be thought of as compartments that orchestrate chemical reactions between these organic molecules. It is clear that to understand life, we must understand the properties of water and organic molecules and of their building blocks, the chemical elements.

A1.1 Elements, Bonding, and Water

Cells consist mostly of water and organic molecules. These cellular components are formed from atoms of various elements. An atom consists of a positively charged nucleus that contains protons and neutrons, surrounded by negatively charged electrons. The hydrogen nucleus consists of a single proton. Protons, neutrons, and electrons differ in their mass and charge as summarized in **Table A1.1**.

The elements can be organized into a periodic table as in **Figure A1.1**, indicating each element's **atomic number** (the number of protons) and **atomic mass** (the mass, in grams, of 1 mole of the element). The defining characteristic of an element is the atomic number. For example, all carbon atoms have six protons in their nucleus. To maintain neutrality, atoms have negatively charged electrons equal in number to the positively charged protons.

The atomic mass is the sum of the number of protons and neutrons. The atomic mass is an average that takes into account the relative abundance of each isotope. **Isotopes** are atoms of an element that differ in the number of neutrons. For example, the most abundant isotope of carbon is carbon-12 (with six neutrons), but there are naturally occurring isotopes of carbon-13 (seven neutrons) and carbon-14 (eight neutrons). Because carbon-13 and carbon-14 are rare, the average atomic mass is close to but not exactly 12. While some isotopes are stable, others decay at a known rate and give off radioactivity. Carbon-14 has a **half-life** (the amount of time it takes for half of a sample to decay) of 5,700 years and is used in radiocarbon dating to determine the age of organic material. Carbon-14 and shorter-lived isotopes such as tritium (hydrogen with a mass of 3—one proton and two neutrons) are used by scientists as tracers to follow specific atoms in metabolic pathways.

Bonds between Atoms Form Molecules

Atoms combine by sharing electrons to form molecules. For example, two atoms of oxygen combine to form molecular oxygen, O_2. Molecules may also contain more than one kind of element; an example is water, H_2O. The symbols O_2 and H_2O are examples of **molecular formulas**, a shorthand notation indicating the number and type of atoms present in a molecule.

Each column (group) in the periodic table (**Fig. A1.1**) contains elements of similar reactivity as a result of the similarity in their electronic configurations, particularly of electrons in the outermost shell. The shell closest to the nucleus can hold a maximum of two electrons, and the next shell a maximum of eight electrons. **Figure A1.2A** shows all the electrons for hydrogen, carbon, nitrogen, and oxygen. Each unpaired electron is capable of participating in a bond. If we examine **Figure A1.2A**, it is clear that H can form one bond; C, four bonds; N, three bonds; and O, two bonds. For example, carbon has four unpaired electrons in its outermost shell; each can form a bond with an unpaired electron from another atom. The two atoms involved in this bond "share" the two electrons. This sharing of electrons is termed a **covalent bond**. Covalent bonds are very strong and difficult to break. In methane (CH_4), carbon forms four covalent bonds with four hydrogen atoms (**Fig. A1.2B**). Methane is stable because both carbon and hydrogen have filled their outer shells. Atoms can also share more than one pair of electrons with another atom, forming double or triple bonds. In carbon dioxide, CO_2, each oxygen shares two pairs of electrons with carbon; and in diatomic nitrogen, N_2, the nitrogen atoms each share three pairs of electrons (see **Fig. A1.2B**). The bonding in molecules can be represented in a shorthand representation by a **structural formula**, in which covalent bonds are shown as a line between two atoms (**Fig. A1.2C**).

Another way atoms can obtain full outer shells is by gaining or losing electrons. A complete transfer of electrons can occur between two atoms that have a large

Table A1.1	The mass and charge of atomic particles.	
Particle	**Mass (atomic mass unit)**	**Charge (electronic charge unit)**
Proton	1	+1
Neutron	1	0
Electron	0.0005	−1

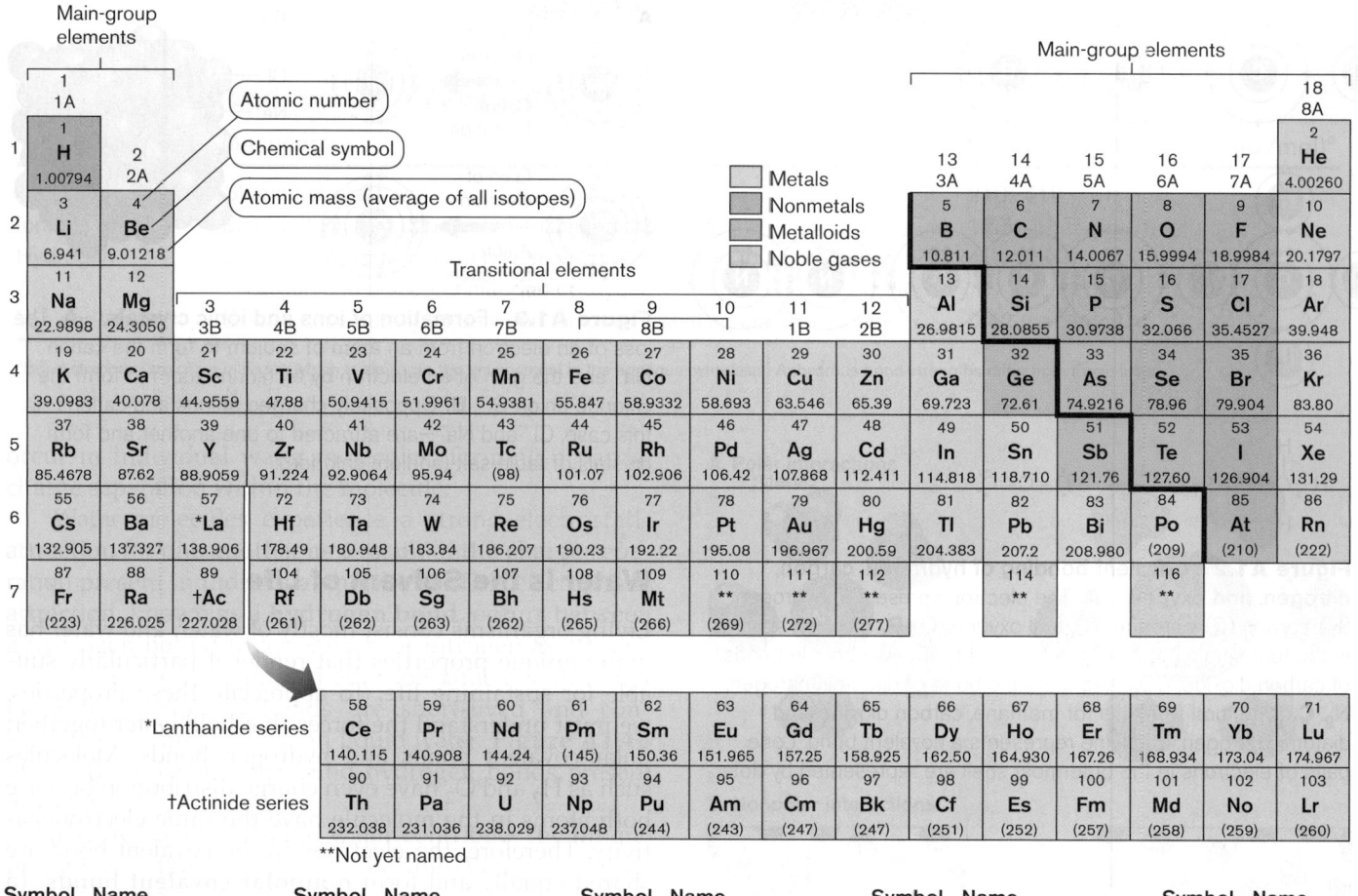

Figure A1.1   Periodic table of the elements.   The atomic number (number of protons) and atomic mass are shown for each element.

difference in **electronegativity**, a measure of the affinity of an atom for electrons. A large electronegativity indicates a strong attraction for electrons. Of the elements listed in **Table A1.2**, oxygen has the greatest attraction for electrons and sodium the weakest. If two elements

with greatly different electronegativities come into close contact, one element can "steal" an electron from the other. For example, sodium (Na) and chlorine (Cl) interact to form table salt, NaCl. The electronegative Cl strips an electron away from Na, and both Cl^- and Na^+ now

A value of 298 K is often used for temperature (this is roughly 25°C). The relationship between K_{eq} and $\Delta G°$ is usually shown with the base 10 logarithm instead of the natural log:

$$\Delta G° = -2.303\ RT \log K_{eq}$$

This equation defines a relationship between $\Delta G°$ and K_{eq} that is discussed in Chapter 13. When K_{eq} is 1 (neither products nor reactants favored), $\Delta G°$ (defined as the change in free energy when concentrations of all substances are at 1 molar) is zero because the reaction is at equilibrium at that point. A K_{eq} of less than 1 (reactants favored) corresponds to a positive $\Delta G°$ because at 1-molar concentrations of reactants and products, the reaction will move to the left. A K_{eq} of more than 1 (products favored) corresponds to a negative $\Delta G°$ because at 1-molar concentrations, the reaction will move to the right.

Free Energy in Cells and the Law of Mass Action

Although the standard changes in free energy are useful for comparing the energetics and equilibria of different reactions, they do not reflect what is happening in the cell, where concentrations of substances are probably not 1 molar. While K_{eq} and $\Delta G°$ are constants and unique for a particular reaction, ΔG depends on the actual concentrations of products and reactants as shown in the equation relating ΔG to $\Delta G°$:

$$\Delta G = \Delta G° + 2.303\ RT \log ([C][D]/[A][B])$$

In this equation, the concentrations of products and reactants are not their equilibrium concentrations, but the actual concentrations present at a given point in time. Because the concentrations of products and reactants will change as a reaction proceeds, the value of ΔG will change over time.

At equilibrium, where the total energy on each side of the reaction is equivalent, ΔG is zero. For an exergonic reaction at equilibrium, the total energy of reactants and products is equivalent because the amount of products is greater than the amount of reactants. If, at equilibrium, reactants (A, B) are added or products (C, D) are removed, ΔG will become negative. A negative ΔG causes the reaction to move to the right to reestablish equilibrium. As A and B convert to C and D, ΔG becomes less and less negative until it reaches zero again at equilibrium. If, at equilibrium, products are added or reactants are removed, ΔG will become positive and the reaction will proceed to the left to reestablish equal energy on both sides. This **law of mass action** is the tendency of a reaction to reestablish equilibrium after perturbations in the concentrations of products or reactants.

The law of mass action means that a reaction with a positive $\Delta G°$ can be made spontaneous and driven to the right by keeping the concentration of reactants high or the concentration of products low. The cell employs this strategy in metabolic pathways, where the product of one reaction is constantly removed by a subsequent reaction.

The Rate of a Reaction Depends on the Activation Energy

It is important to realize that although a reaction with a large positive K_{eq} and negative $\Delta G°$ is spontaneous, it may be slow. The value of $\Delta G°$ says nothing about the rate of a reaction. An everyday example of a spontaneous but slow reaction is the rusting of metal. A reaction will be slow, even though it is spontaneous, if it must pass through an unstable, high-energy transition state on the way to forming products. The **activation energy** (E_a) is the energy needed to reach this transition state (**Fig. A1.22**). For reactions with low activation energies, random collisions between reactants may provide enough energy to boost them up and over the E_a. For reactions with high activation energies, collisions between molecules may not provide enough energy for them to reach the transition state.

Enzymes are biological catalysts that can speed up reaction rates by stabilizing transition states and lowering the activation energy. Most enzymes are proteins that specifically bind reactants and provide an environment that facilitates product formation. Although enzymes increase reaction rates, they do not change the $\Delta G°$ or

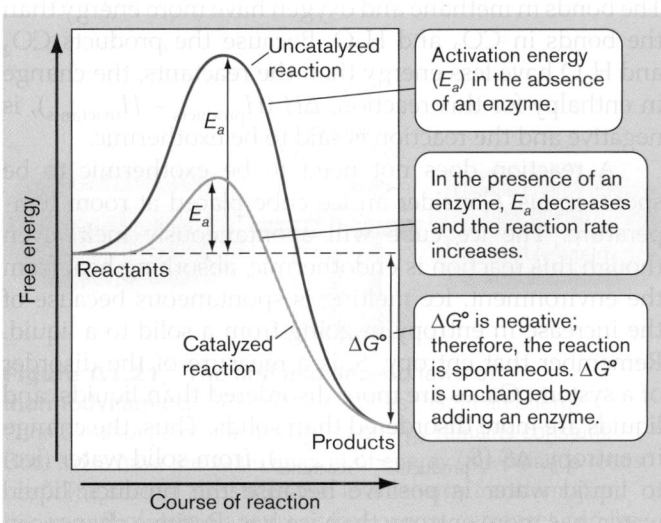

Figure A1.22 **Enzymes, activation energies, and reaction rates.**

K_{eq} of a reaction. **Figure A1.22** illustrates how the difference in free energy between products and reactants is unchanged even in the presence of an enzyme that lowers the activation energy.

Biological Processes Depend on pH

Many biological processes occur within only a narrow range of hydrogen ion concentrations. Hydrogen ions (H^+) are also referred to as protons, because they have lost their single electron and consist of only a proton. Hydrogen ion concentration is reported using a pH (power of hydrogen) scale, where pH is the negative logarithm of the hydrogen ion concentration: $pH = -\log_{10}[H^+]$. In pure water, which is considered neutral, $[H^+]$ is 1×10^{-7} molar (M), a pH of 7. Because the pH scale is logarithmic, every pH unit corresponds to a tenfold change in hydrogen ion concentration. A solution with a pH of 6 has a hydrogen ion concentration of 1×10^{-6} M, ten times the hydrogen ion concentration at pH 7. Hydrogen ion concentration $[H^+]$ multiplied by hydroxide ion concentration $[OH^-]$ always equals 1×10^{-14}. Hence, the concentrations of H^+ and OH^- are reciprocally related (if one goes up, the other goes down). If the pH is less than 7, the solution is acidic (more H^+, less OH^-). If the pH is greater than 7, the solution is basic, or alkaline (less H^+, more OH^-).

Acids (for example, carboxyl groups) release protons, and bases (for example, amino groups) bind protons (**Fig. A1.23**). At intracellular pH (near 7), the carboxyl and amino groups of amino acids are both ionized (charged), the carboxyl group carrying a negative charge and the amino group a positive charge. The ionization of these groups can change if the pH changes. At lower pH (more acidic solution) protons move back onto the ionized carboxylic acid. For example, the side chain of an acidic amino acid such as glutamate (see **Fig. A1.9**) is ionized at normal cell pH but may regain a proton and become glutamic acid at a lower pH. At higher pH values (lower proton concentrations), amino groups lose protons to become uncharged. Disturbances in the ionization state of carboxyl or amino groups on the side chains of amino acid residues can disrupt ionic bonds between these groups and lead to protein denaturation. This effect of pH on protein tertiary structure is one reason why cells can tolerate only a narrow range of intracellular pH.

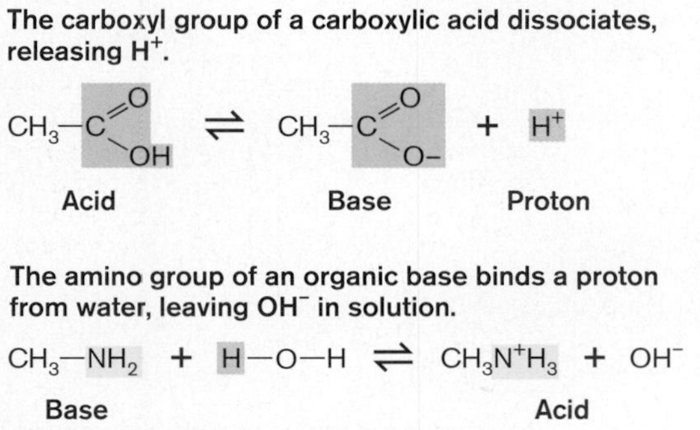

Oxidation-Reduction (Redox) Reactions Transfer Energy

Biomolecules may undergo an important class of chemical reactions termed redox reactions. **Redox reactions** involve the transfer of electrons from one molecule to another or from one atom to another. The molecule that gains electrons becomes reduced, and the molecule that loses electrons becomes oxidized (**Fig. A1.24**). (A useful mnemonic device to remember this is "LEO the lion says GER," where LEO stands for "lose electrons, oxidation" and GER stands for "gain electrons, reduction").

Redox reactions are always coupled. If one atom loses electrons, another atom must gain electrons. Oxygen, with its large electronegativity, usually gains electrons

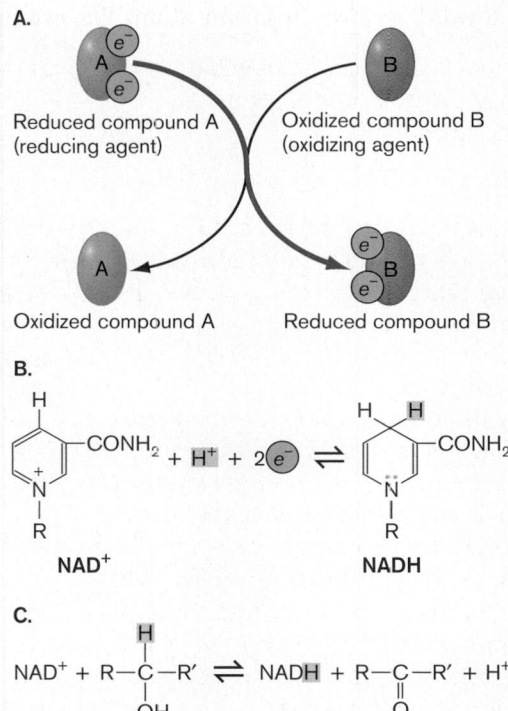

Figure A1.24   Oxidation-reduction reactions.   A. A reducing agent A donates electrons to reduce compound B. Because A loses electrons, it becomes oxidized itself in the process. B is the oxidizing agent. **B.** NADH is a common biological reducing agent. **C.** NAD$^+$ is reduced to NADH as an alcohol is oxidized to an aldehyde.

The carboxyl group of a carboxylic acid dissociates, releasing H^+.

$$CH_3-C\!\!\!\overset{O}{\underset{OH}{\big<}} \rightleftharpoons CH_3-C\!\!\!\overset{O}{\underset{O-}{\big<}} + H^+$$

Acid   Base   Proton

The amino group of an organic base binds a proton from water, leaving OH^- in solution.

$$CH_3-NH_2 + H-O-H \rightleftharpoons CH_3N^+H_3 + OH^-$$

Base   Acid

Figure A1.23   Organic acids and bases.

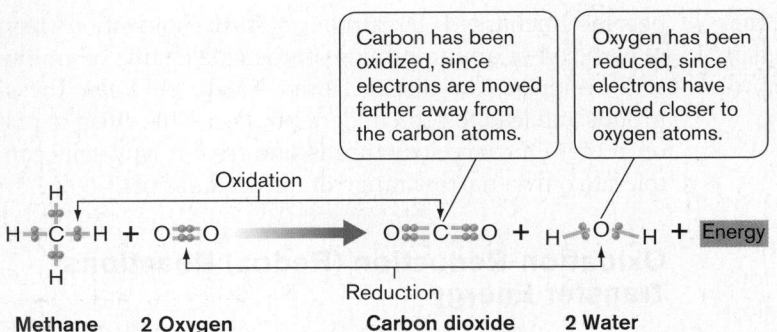

Carbon has been oxidized, since electrons are moved farther away from the carbon atoms.

Oxygen has been reduced, since electrons have moved closer to oxygen atoms.

Figure A1.25   The oxidation of methane.   Bonding electrons were shared equally by carbon and hydrogen in methane, but they are closer to oxygen in carbon dioxide and water. Carbon and hydrogen have been oxidized; oxygen has been reduced.

and becomes reduced in redox reactions. Molecules that become reduced are called oxidizing agents because they cause something else to become oxidized. In contrast, reducing agents can donate electrons, reduce other molecules, and become oxidized themselves in the process. Molecules with reduced carbons contain more energy than their oxidized counterparts. Common reducing agents (electron donors) in the cell are NADH, NADPH, and $FADH_2$. These are all high-energy molecules that can donate electrons.

In addition to achieving a complete transfer of electrons, redox reactions can also occur if electrons are shifted toward or away from an atom. For example, in the burning (oxidation) of methane, electrons move away from the carbon and toward the oxygen (**Fig. A1.25**). Oxygen usually acts as an oxidizing agent (electron acceptor); thus, we expect it to get reduced (gain electrons). By default, then, methane is oxidized to carbon dioxide. Indeed, while the bonding electrons were shared fairly equally between C and H in methane, they are now close to the O in both carbon dioxide and water and farther away from the C and H. Thus, methane has been oxidized and oxygen has been reduced. This reaction releases energy because CO_2 and H_2O are the most stable, lowest-energy forms available when carbon, hydrogen, and oxygen are combined.

Appendix 2

Introductory Cell Biology: Eukaryotic Cells

This appendix presents a review of cell biology principles generally covered in an introductory-level biology class. Cell biology encompasses fundamental principles of cell structure and function. All cells are enclosed by a cell membrane and need to regulate the transport of materials across the membrane. Additional membrane structures are unique to eukaryotic cells.

All eukaryotic cells contain a nucleus, an organelle that houses the DNA. We describe the structure of the nucleus and the processes of chromosome segregation through mitosis and meiosis. We then explore the endomembrane system of organelles that partition the cell into functional compartments, enabling eukaryotic cells to maintain larger sizes than most prokaryotic cells. Eukaryotic subcellular form includes the cytoskeleton, a collection of proteins that provide cell architecture and mediate cell movement. Finally, we discuss mitochondria and chloroplasts, the energy-producing organelles of eukaryotic cells.

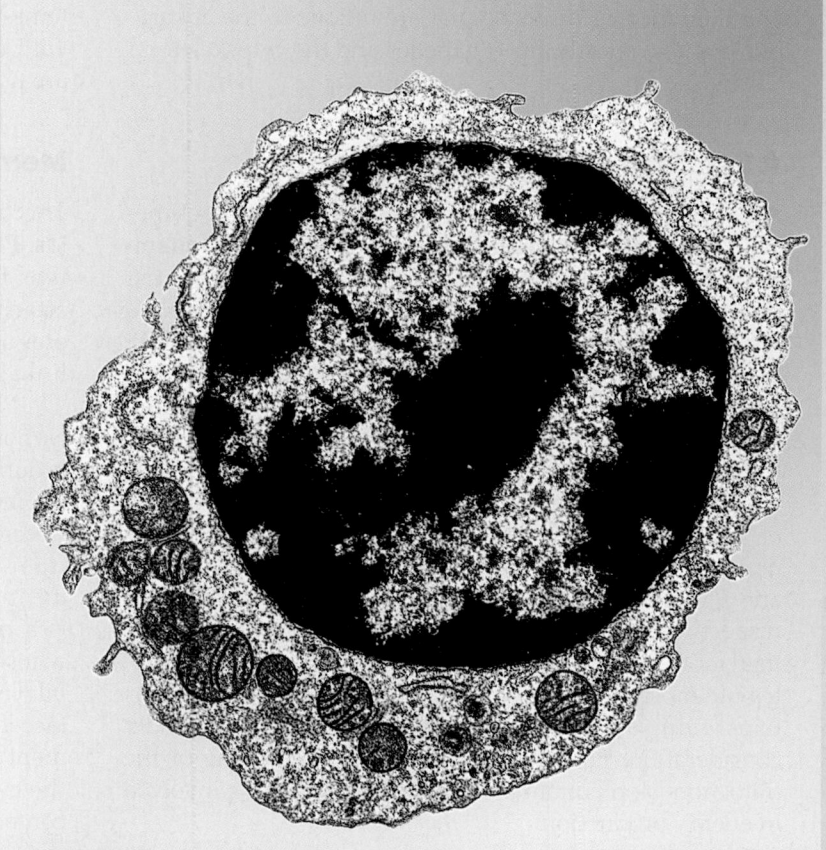

Current Research Highlight

This eukaryotic cell, a lymphocyte white blood cell, shows numerous major organelles, such as a large nucleus (orange) and multiple mitochondria (blue). (TEM, 20,550×) *Source:* ©Gopal Murti/Visuals Unlimited

The cell is the basic unit of life. All organisms consist of either a single cell or a collection of cells. All cells can potentially perform a common set of tasks, including replication, catalysis, and regulation. The structures that enable these functions are found in all cells and include the cell membrane, DNA, and ribosomes (**Fig. A2.1**). Although the detailed organization of these structures differs in cells from the three domains (**Bacteria**, **Archaea**, and **Eukarya**; see **Table A2.1**), the fundamental structure and function of the cell membrane, DNA, and ribosomes are the same in all cells. This appendix describes the cell structures that are found across all three domains of life, and then focuses on those that are unique to the eukaryotic domain, specifically organelles and the cytoskeleton.

A2.1 The Cell Membrane

All cells are enclosed by a **cell membrane** (sometimes called the **plasma membrane** or **cytoplasmic membrane**) that maintains an internal environment distinct from the external environment. The aqueous fluid inside the membrane is called **cytoplasm** (or **cytosol**). Major functions of the cell membrane include the regulated transport of substances into and out of the cell and the reception of signals from the external environment. Membranes are also critical for energy production.

As shown in **Figure A2.2**, membranes consist mainly of lipids and proteins, but also contain some carbohydrates found in hybrid structures such as glycolipids and glycoproteins (sugars joined to lipids or to proteins, respectively). Although the membrane contains more lipid molecules than protein molecules, proteins may contribute most of the mass. The number and nature of proteins within a membrane depend on the membrane under consideration. For example, the inner membrane of the mitochondrion contains a rich array of proteins involved in energy production.

Membrane proteins can be classified based on how they interact with membranes (**Fig. A2.2A**). **Transmembrane proteins** (integral proteins) span the bilayer. Transmembrane proteins are **amphipathic**, meaning they have both hydrophobic and hydrophilic portions; the hydrophilic portions face the cytoplasm and the extracellular environment, and the hydrophobic components span the membrane. The transmembrane domains are often alpha helices containing 15–20 amino acid residues (**Fig. A2.2B**). The portions of the protein that are intracellular, in the membrane, and extracellular depend on the protein and can be quite different for different proteins. **Peripheral membrane proteins** are associated with the cell membrane through noncovalent bonds but are not directly inserted into the bilayer.

Membranes Are Composed of Lipids

The predominant lipids in membranes are phospholipids. **Phospholipids** consist of a core of glycerol, to which two fatty acids and a modified phosphate group are attached via ester linkages (**Fig. A2.3A**). Unlike eukaryotes and bacteria, archaea have phospholipids with ether linkages.

Phospholipids are amphipathic: the fatty acid hydrocarbon tails are hydrophobic, and the phosphate head group is hydrophilic. Amphipathic lipids are most stable in water when the hydrophilic portions interact with water and the hydrophobic portions cluster together away from water. One way phospholipids can achieve stability is by forming a bilayer. Indeed, the cell membrane is a **phospholipid bilayer**, two layers of phospholipids whose hydrocarbon fatty acid tails face the interior of the bilayer and whose charged phospholipid head groups face the aqueous cytoplasm and extracellular environment (**Fig. A2.3B**). The phospholipid layer in contact with the cytoplasm is called the **inner leaflet**, and the layer in contact with the environment is called the **outer leaflet**.

Table A2.1 Comparison of cell structures in the three domains.

Feature	Bacteria	Archaea	Eukarya
Genome	Usually circular DNA	Usually circular DNA	Linear DNA
	Usually one chromosome	Usually one chromosome	Multiple chromosomes, in pairs
	Usually lacks introns		May have introns
Location of DNA	Nucleoid region in cytoplasm	Nucleoid region in cytoplasm	Contained within membrane-enclosed nucleus
Cell membrane	Straight-chain fatty acids ester-linked to glycerol	Branched fatty acids ether-linked to glycerol	Straight-chain fatty acids ester-linked to glycerol
Cell wall	Usually present, composed of peptidoglycan	If present, composed of proteins or pseudopeptidoglycan	If present, composed of cellulose (algae) or chitin (fungi)
Internal membranes	May have energy-transducing lamellae	Uncommon	Extensive membranous organelles

A. Prokaryotic cell (bacteria or archaea)

Flagellum

Ribosomes

Nucleoid

Cytoplasm

Chromosome

Cell membrane

Cell wall

On average, bacteria and archaea are about 10 times smaller than eukaryotic cells in diameter and about 1,000 times smaller than eukaryotic cells in volume.

B. Generalized plant cell

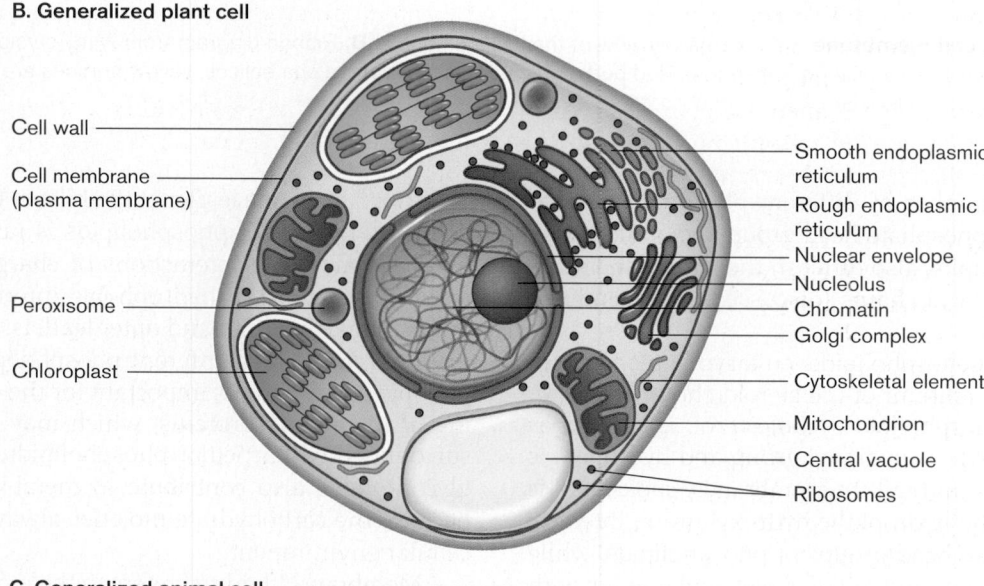

Cell wall

Cell membrane (plasma membrane)

Peroxisome

Chloroplast

Smooth endoplasmic reticulum

Rough endoplasmic reticulum

Nuclear envelope

Nucleolus

Chromatin

Golgi complex

Cytoskeletal element

Mitochondrion

Central vacuole

Ribosomes

C. Generalized animal cell

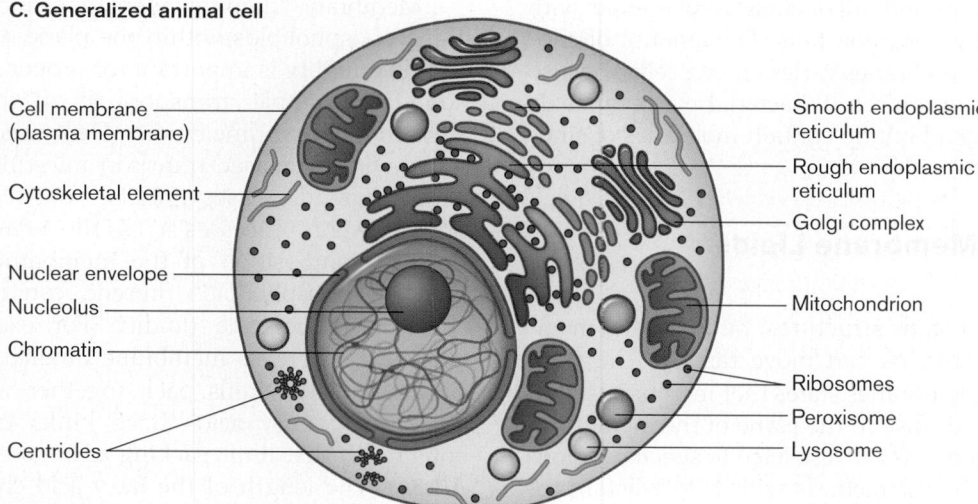

Cell membrane (plasma membrane)

Cytoskeletal element

Nuclear envelope

Nucleolus

Chromatin

Centrioles

Smooth endoplasmic reticulum

Rough endoplasmic reticulum

Golgi complex

Mitochondrion

Ribosomes

Peroxisome

Lysosome

Figure A2.1 The prokaryotic cell and the eukaryotic cell. A. The prokaryotic cell typically contains a single compartment, and its DNA is organized in the nucleoid region. **B, C.** Eukaryotic cells are typically much larger than prokaryotic cells and contain organelles.

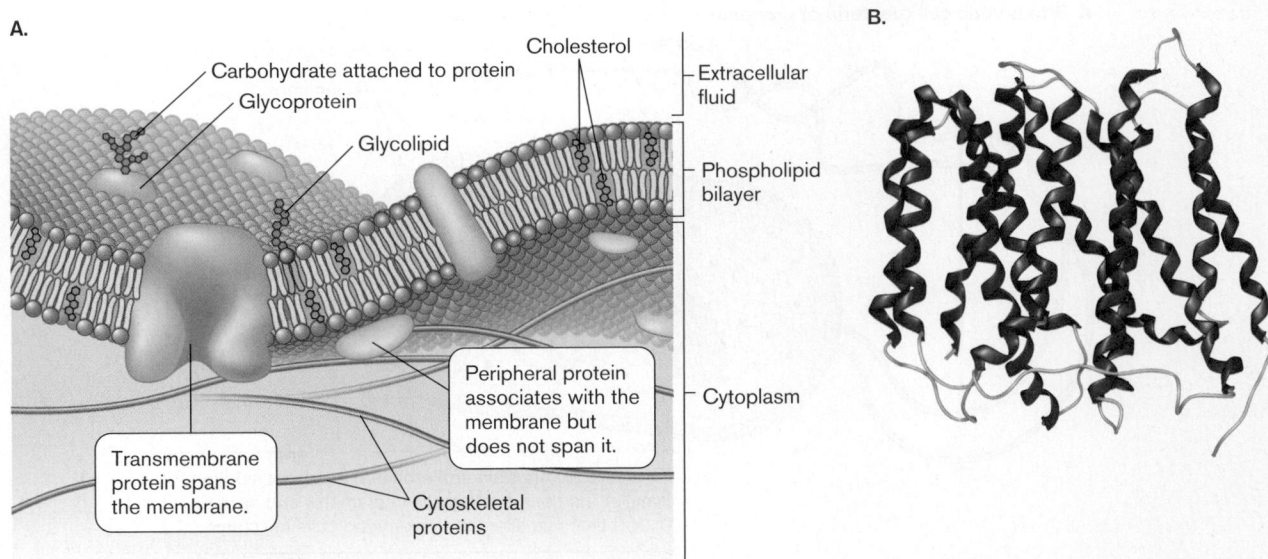

Figure A2.2 The cell membrane. A. A cutaway view of the cell membrane. **B.** Ribbon diagram from X-ray crystallographic data of the *E. coli* EmrD protein, a multidrug transporter. Red portions are transmembrane alpha helices; yellow portions are intracellular and extracellular loops. (PDB code: 2GFP) *Source*: Yin et al. 2006. *Science* **312**:741.

Cells contain a number of different phospholipids that differ in how the phosphate head group is modified (**Fig. A2.3C**). Phospholipids also differ in the length and saturation of the fatty acid chains (discussed in Appendix 1, Section A1.6).

In addition to phospholipids, eukaryotic membranes contain a variable amount of the steroid cholesterol (**Fig. A2.3D**). Like phospholipids, cholesterol is amphipathic, with a hydrophilic hydroxyl group and hydrophobic hydrocarbon rings and tail. In membranes, cholesterol is oriented so that the hydrophilic hydroxyl group interacts with the phosphate head groups of phospholipids, while the hydrophobic rings and tail of cholesterol interact with the phospholipid hydrocarbon tails. The amount of cholesterol present in membranes varies among cells and also among organelles within a cell. Bacterial membranes do not contain cholesterol but do contain molecules of similar form called hopanoids.

Movement of Membrane Lipids and Proteins

Membranes are not static structures; rather, many membrane lipids and proteins can move rapidly. The **fluid mosaic model** of membranes states that membrane components are free to diffuse in the plane of the membrane. Some membrane proteins are restricted to specific regions of the membrane by interactions with cytoskeletal proteins (discussed in Section A2.5).

Although many phospholipids and membrane proteins can move laterally within a leaflet, they do not "flip-flop" from one leaflet of the bilayer to the other (**Fig. A2.3E**). Flip-flop of phospholipids is rare, owing to the highly unfavorable interactions of charged head groups moving through the hydrophobic interior of the membrane. Thus, the inner and outer leaflets of the membrane may be made up of different phospholipids. Such phospholipid asymmetry is important for the correct functioning of membrane proteins, which may work best when surrounded by particular phospholipids. Glycolipids and glycoproteins also contribute to membrane asymmetry, because the carbohydrate moieties always face the extracellular environment.

Membrane "fluidity" refers to the movement of membrane phospholipids within the plane of the membrane, and this fluidity is important for proper membrane function. For example, transport across the membrane is affected by membrane fluidity. Decreased fluidity is associated with decreased transport rates. Because a drop in temperature decreases fluidity, cold temperatures may slow transport processes across the membrane.

The composition of the membrane, especially the types of phospholipids present, can have a dramatic effect on membrane fluidity. For example, saturated fatty acids decrease membrane fluidity because the linear hydrocarbon tails pack together well. In contrast, unsaturated fatty acids have kinks in the hydrocarbon chains that limit packing and increase fluidity (**Fig. A2.3F**). The length of the fatty acid chains also affects fluidity. Phospholipids with longer hydrocarbon chains have increased hydrophobic interactions with neighboring lipids and thus decreased membrane fluidity.

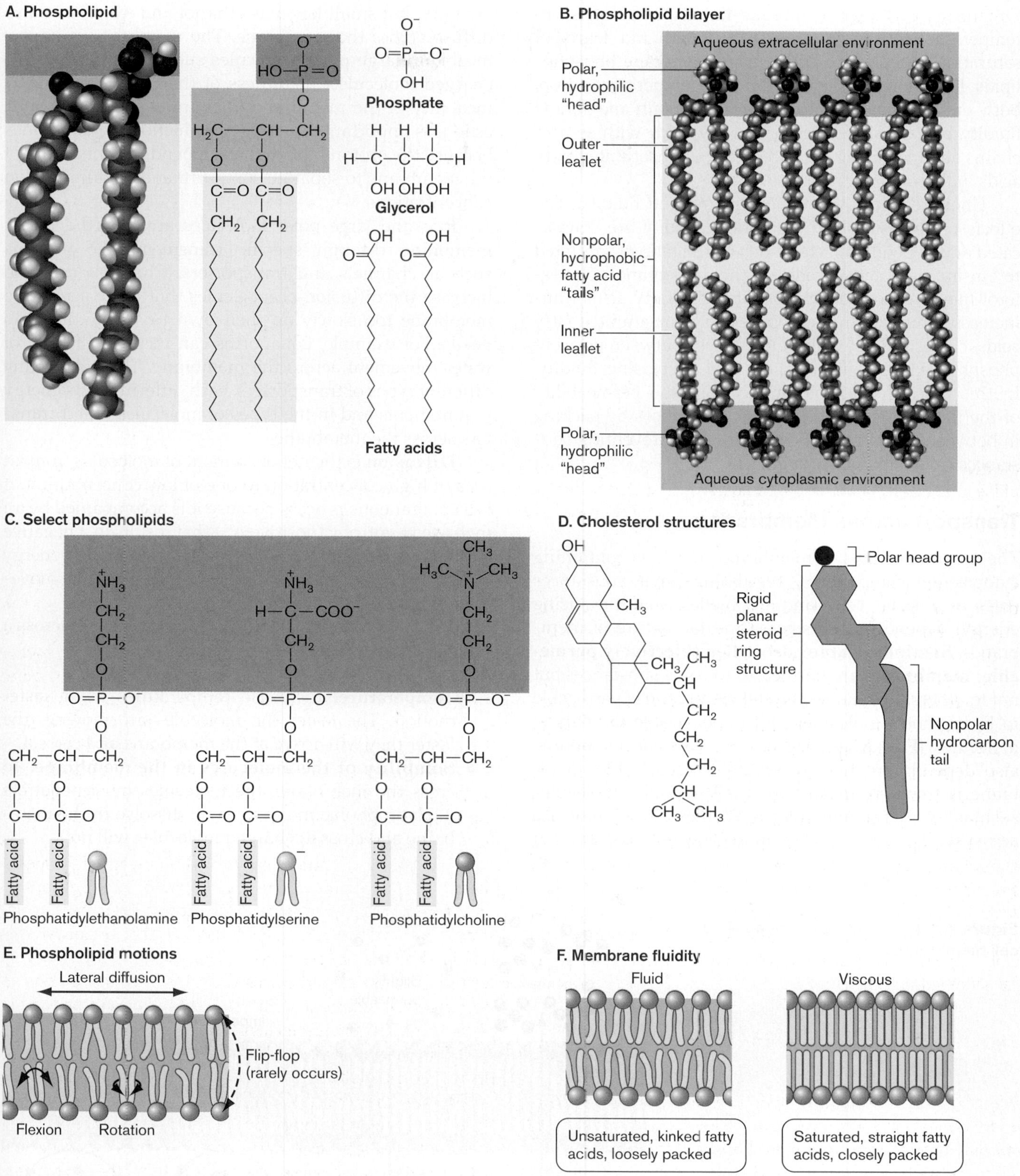

A. Phospholipid

Phosphate

Glycerol

Fatty acids

B. Phospholipid bilayer

Aqueous extracellular environment

Polar, hydrophilic "head"

Outer leaflet

Nonpolar, hydrophobic fatty acid "tails"

Inner leaflet

Polar, hydrophilic "head"

Aqueous cytoplasmic environment

C. Select phospholipids

Phosphatidylethanolamine   Phosphatidylserine   Phosphatidylcholine

D. Cholesterol structures

Polar head group

Rigid planar steroid ring structure

Nonpolar hydrocarbon tail

E. Phospholipid motions

Lateral diffusion

Flip-flop (rarely occurs)

Flexion   Rotation

F. Membrane fluidity

Fluid   Viscous

Unsaturated, kinked fatty acids, loosely packed

Saturated, straight fatty acids, closely packed

Figure A2.3 **Phospholipids, cholesterol, the lipid bilayer, and membrane fluidity.** **A.** Saturated phospholipid. **B.** Orientation of phospholipids in the bilayer. **C.** Some phospholipids present in cell membranes. **D.** Structural formula and schematic drawing of cholesterol. **E.** Motions of phospholipids in membrane bilayers. **F.** The ratio of saturated and unsaturated fatty acids in the phospholipids affects membrane fluidity.

Organisms can alter membrane fluidity in response to temperature stress by changing the length and degree of saturation of fatty acids present in membrane phospholipids. For example, as environmental temperatures drop, both eukaryotes and prokaryotes maintain membrane fluidity by replacing long-chain fatty acids with shorter chains and increasing the percentage of unsaturated fatty acids in their membranes.

Cholesterol also influences membrane fluidity. The effects of cholesterol on membrane fluidity are complicated and depend on factors such as the ratio of saturated to unsaturated fatty acids in the membrane. Cholesterol may prevent packing of saturated fatty acids, thus increasing fluidity. In membranes with unsaturated fatty acids, cholesterol may fill in the spaces between adjacent phospholipids, stabilizing them and decreasing fluidity. In this case, cholesterol can decrease the permeability of the membrane to hydrophobic substances by packing in between the hydrocarbon chains and preventing substances from slipping through.

Transport across Membranes

The major functions of membranes (such as containing cytoplasmic components, regulating what substances enter and leave cells and organelles, and producing energy) depend on the semipermeable nature of membranes. **Semipermeable** (also called **selectively permeable**) **membranes** are permeable to some substances but not to others. In general, the cell membrane is permeable to hydrophobic molecules and impermeable to charged molecules (**Fig. A2.4**). Diffusion across the membrane also depends on the size of the molecule. The membrane is freely permeable to small nonpolar molecules such as O_2. Larger nonpolar molecules can also diffuse across the membrane, albeit more slowly. Molecules that are polar but small (such as ethanol and water) can also diffuse across the membrane. The membrane is impermeable to large polar molecules such as glucose and to charged molecules, regardless of their size. The impermeability of the membrane to charged substances such as ions is important for energy production at membranes because the proton motive force depends on the ability of the membrane to separate compartments of different ion concentrations.

Ions and large polar molecules are moved across a membrane through specific transmembrane proteins such as channels and transporters. Channels can also increase the diffusion of molecules that move across the membrane too slowly on their own to supply the cell's needs. For example, aquaporins can increase the rate of water movement across the membrane. There are many different types of transporters, each differing in its energy requirements and in the types of molecules that it transfers across the membrane.

Diffusion is the net movement of molecules from an area of high concentration to one of low concentration. It is a spontaneous process because it is accompanied by an increase in entropy (positive ΔS) that results in a negative free energy change (ΔG). The process requires no energy input and is brought about by the random, thermal movement of molecules.

Factors that influence diffusion of molecules across a membrane include:

- **Temperature.** Increased temperatures mean faster motion. The faster the molecules are moving, the faster they will arrive at the membrane and cross it.
- **Solubility of the molecules in the membrane.** To cross the membrane, the molecules must penetrate it. Hydrophobic molecules will dissolve in the membrane and cross it; charged molecules will not.

Figure A2.4 Selective permeability of cell membranes.

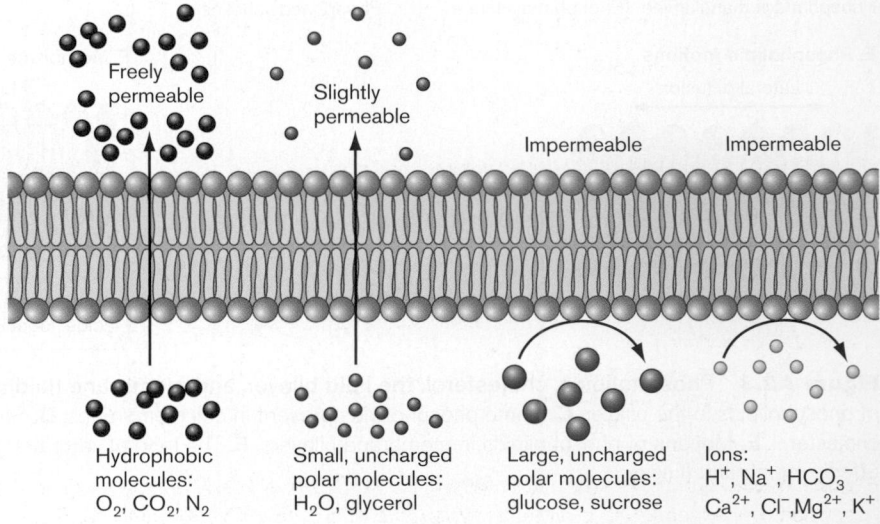

Freely permeable

Slightly permeable

Impermeable

Impermeable

Hydrophobic molecules: O_2, CO_2, N_2

Small, uncharged polar molecules: H_2O, glycerol

Large, uncharged polar molecules: glucose, sucrose

Ions: H^+, Na^+, HCO_3^- Ca^{2+}, Cl^-, Mg^{2+}, K^+

- **Surface area of the membrane.** To cross the membrane, molecules must first encounter it. The chances of this happening are increased with an increase in membrane surface area.
- **Concentration gradient of the dissolved molecules.** A larger concentration gradient speeds up diffusion because the more molecules there are, the more will encounter the membrane and cross.
- **Thickness of the membrane.** Diffusion rates are inversely proportional to the square of the distance the solute must travel across the membrane. The thinner the membrane, the faster the molecules can get across.
- **Mass of the molecule.** Friction between a molecule and its medium is a source of resistance that slows down motion. Larger molecules with more mass experience more resistance and cross the membrane more slowly.

These factors can be expressed as follows:

$$\text{Diffusion rate} \propto \frac{\text{temperature} \times \text{surface area} \times \text{concentration gradient}}{\text{mass} \times \text{distance}^2}$$

Conditions for the diffusion of gases and other substances across the cell membrane will be most favorable when the surface area of the membrane is large, the concentration gradient across the membrane is high, and the membrane is thin.

Transport of Water across the Cell Membrane

Osmosis is the diffusion of water across a selectively permeable membrane from regions of high water concentration (low solute) to regions of low water concentration (high solute). Diffusion rates can be determined with an artificial membrane system as depicted in **Figure A2.5A**. Molecules are added to the left side of the beaker, and samples from the right side are analyzed at various time points for the presence of the test molecule.

Cells must maintain osmotic balance with the surrounding environment. The direction of water movement depends on the concentration of dissolved solutes in the cell relative to the cell's environment. When a cell is in an **isotonic** environment (equal concentrations of dissolved solutes inside the cell and out), there is osmotic balance, and water will enter and exit the cell at equal rates (that is, there is no net movement of water). In a **hypertonic** environment (higher concentration of solutes outside the cell), there is a net loss of water from the cell. The cell shrinks, and the concentration of cell contents increases. In a **hypotonic** environment (lower concentration of solutes outside the cell), there is a net uptake of water by the cell; the cell swells, and the cell components are diluted (**Fig. A2.5B**). If enough water enters, the cell is destroyed by **lysis**, a rupturing of the cell membrane and dispersal of cell contents. Both hypertonic and hypotonic environments can cause other problems for cells. Proteins have

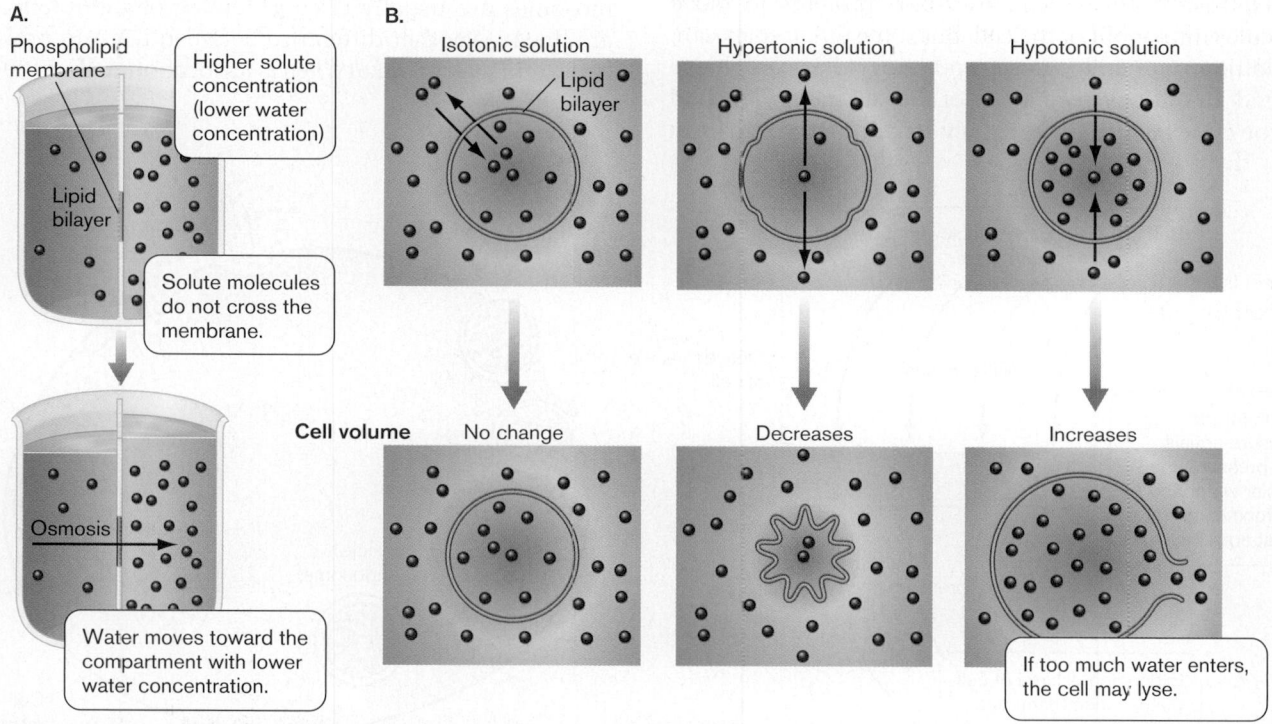

Figure A2.5 Osmosis and water balance. A. Osmosis. **B.** Movement of water across the cell membrane, and shrinkage or expansion of the membrane in isotonic, hypertonic, and hypotonic environments. Black arrows indicate net water movement.

specific salt requirements, and intracellular environments with salt concentrations that are either higher or lower than normal for a cell can cause denaturation of proteins, with potentially fatal results for the cell. Thus, transport of water across the cell membrane must be tightly controlled, and cells need mechanisms that allow them to live in environments that are not isotonic.

Most cells live in environments that are hypotonic. To deal with this challenge, most bacteria and many eukaryotes have a cell wall external to the cell membrane. As water enters by osmosis and pushes against the cell wall (turgor pressure), the wall resists the tension and pushes back with an equal but opposite force known as wall pressure. The wall pressure is an inward pressure exerted by the cell wall against the cell membrane (**Fig. A2.6**). When turgor pressure and wall pressure are equal in magnitude, the cell is at equilibrium with respect to water movement. Organisms that lack a cell wall employ other strategies to deal with hypotonic environments. For example, some freshwater protists that lack a cell wall expel excess water through a contractile vacuole.

In contrast to freshwater microbes, ocean-dwelling organisms face the problem of water loss. Many of these organisms accumulate solutes known as osmolytes to ensure that they are isotonic to the external environment, thus preventing water loss.

Eukaryotes Transport Molecules by Endocytosis and Exocytosis

All cells use diffusion and transport proteins to move molecules into or out of the cell, but some eukaryotes can, in addition, use endocytosis and exocytosis to achieve this end. In **endocytosis**, parts of the cell membrane bud into the cytoplasm and eventually separate from it to form endosomes (**Fig. A2.7**). Endosomes are a type of **vesicle**, a small membranous sphere found within a cell. The interior of these endosomes contains extracellular material. **Phagocytosis** (cell eating) is a form of endocytosis in which large extracellular particles are brought into the cell. **Pinocytosis** (cell drinking) is endocytosis of relatively small volumes of the extracellular fluid. Endocytosis is a controlled, energy-requiring process that relies on many proteins, including cytoskeletal proteins. **Exocytosis** is the reverse of endocytosis. In exocytosis, intracellular vesicles fuse with the cell membrane, and the contents of the vesicles are released to the extracellular environment. Cells can use exocytosis to release wastes.

A2.2 The Nucleus and Mitosis

All cells need to synthesize proteins, and all cells contain DNA, the genetic material that encodes the information needed to specify protein primary structure. The central model of molecular biology holds that DNA is transcribed into messenger RNA (mRNA), and mRNA is translated into protein on ribosomes. Although this model holds for all cells, details in the structure of DNA and ribosomes vary among the three domains of life (see **Table A2.1**). For example, ribosomes always function in protein synthesis, but ribosomes from organisms in different domains differ in their size and their sensitivity to various antibiotics. DNA always encodes the information needed for protein synthesis, but whereas bacterial chromosomes are usually circular, eukaryotic chromosomes are linear. Another difference between DNA in prokaryotes and DNA in eukaryotes is its location within the cell.

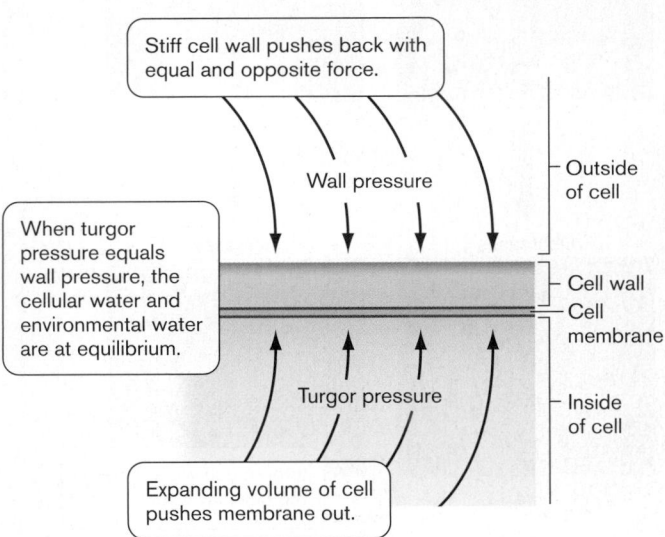

Figure A2.6 Turgor pressure and wall pressure.

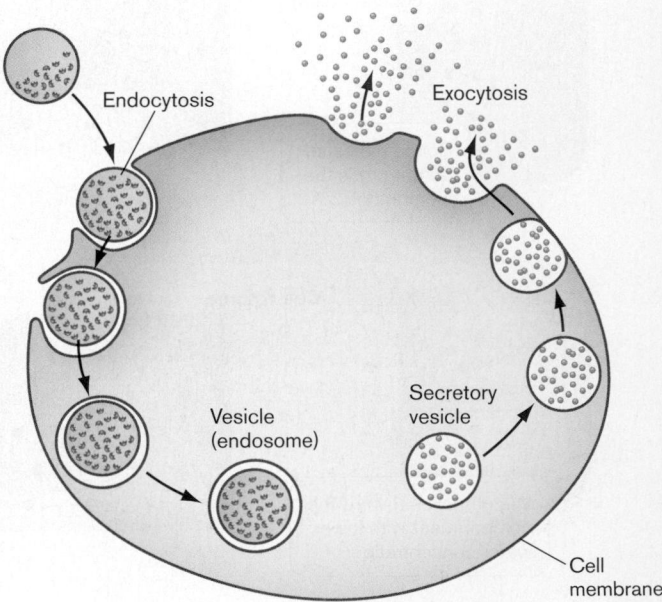

Figure A2.7 Endocytosis and exocytosis.

In bacteria and archaea, DNA is found in an area of the cytoplasm known as the nucleoid, while eukaryotic DNA is contained within a nuclear membrane.

Eukaryotic DNA Is Housed in the Membrane-Enclosed Nucleus

Eukaryotes derive their name ("eukaryote" means "true kernel") from the fact that they possess a **nucleus**, and indeed, the nucleus is often the most prominent feature of eukaryotic cells viewed under a microscope (**Fig. A2.8A**). The nucleus is an **organelle**, an intracellular membrane-enclosed compartment with a specific function. The nucleus contains **chromatin**, a complex of DNA and proteins. The nuclear membrane (envelope) consists of two concentric phospholipid membranes. The outer nuclear membrane is continuous with the membrane of the endoplasmic reticulum (ER), and the space between the two nuclear membranes is continuous with the lumen (inside) of the ER (**Fig. A2.8B**). Nuclei contain a region called the **nucleolus**, where ribosome assembly begins. At the nucleolus, multiple rRNA (ribosomal RNA) genes are transcribed, and the resulting rRNA combines with ribosomal proteins imported into the nucleus from the cytoplasm to form the ribosomal subunits. The ribosomal subunits then need to exit the nucleus.

The nuclear membrane contains nuclear pore complexes (NPCs) that allow for transport of material into and out of the nucleus. Metabolites and small proteins can diffuse through the NPCs, but larger proteins and organelles cannot enter by diffusion. Large proteins that need to enter the nucleus are actively transported in through the NPCs. These selectively imported proteins contain a nuclear localization signal, a sequence of amino acids that acts like a zip code to direct them into the nucleus. In addition to their role in protein import, NPCs also function in exporting mRNAs out of the nucleus.

Eukaryotic Cells Replicate by Mitosis and Meiosis

Cells need to ensure an accurate replication and division of their DNA. Bacteria and archaea replicate by fission, as described in Chapter 3. Eukaryotic cells replicate their nuclear DNA and divide by a completely different process, called mitosis. For sexual reproduction, sex cells with a halved chromosome number are generated by the related process of meiosis. Mitosis and meiosis occur only in eukaryotes, never in bacteria and archaea.

Mitosis is a series of steps that segregates duplicated chromosomes and ensures that each daughter cell receives a copy of the genetic material. When the cell is not undergoing mitosis, it is in interphase (**Fig. A2.9A**). During interphase, the individual chromosomes are long and thin and not visible by a light microscope. Interphase can be divided into three phases: G_1, S, and G_2. Cells that are not committed to dividing are in G_1, the first gap phase. If cell division is to occur, then the chromosomes are replicated during S phase (S for "synthesis"). The duplicated chromosomes, called sister chromatids, remain attached to each other at the centromere (**Fig. A2.9B**). After chromosome replication, a second gap phase, G_2, occurs, after which mitosis can proceed.

Mitosis is divided into four steps (**Fig. A2.9C**):

■ **Prophase.** During prophase, the chromosomes condense and become visible by light microscopy. The

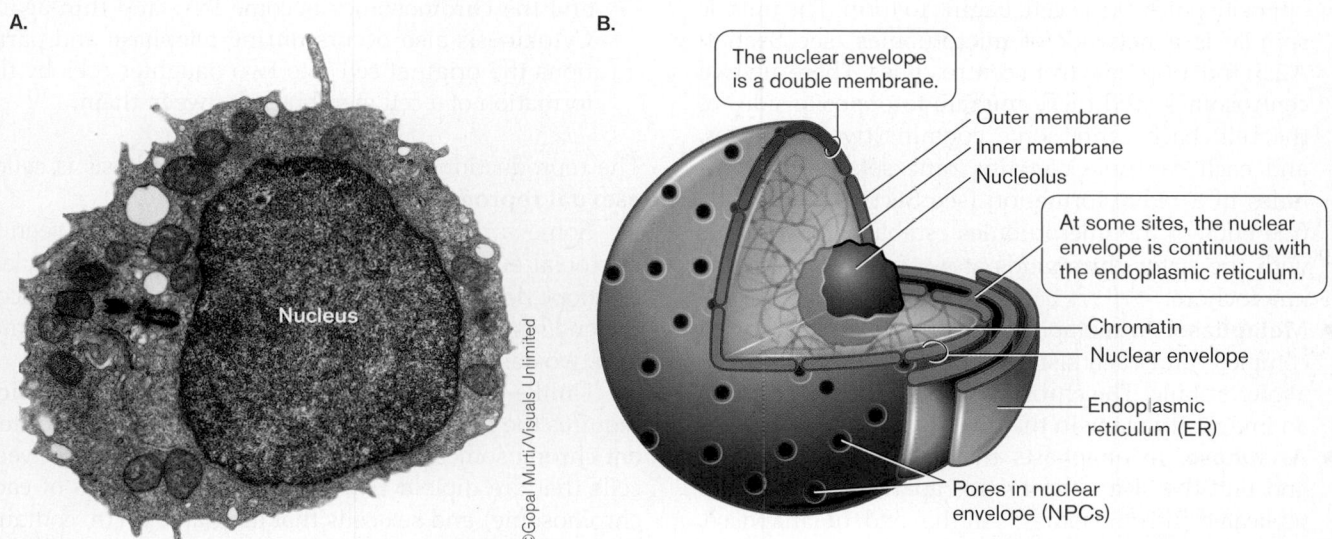

A.

B.

The nuclear envelope is a double membrane.

Outer membrane
Inner membrane
Nucleolus

At some sites, the nuclear envelope is continuous with the endoplasmic reticulum.

Chromatin
Nuclear envelope

Endoplasmic reticulum (ER)

Pores in nuclear envelope (NPCs)

Nucleus

©Gopal Murti/Visuals Unlimited

Figure A2.8   The nucleus.   A. Electron micrograph of a eukaryotic yeast cell showing the prominent nucleus. **B.** Diagram of a nucleus.

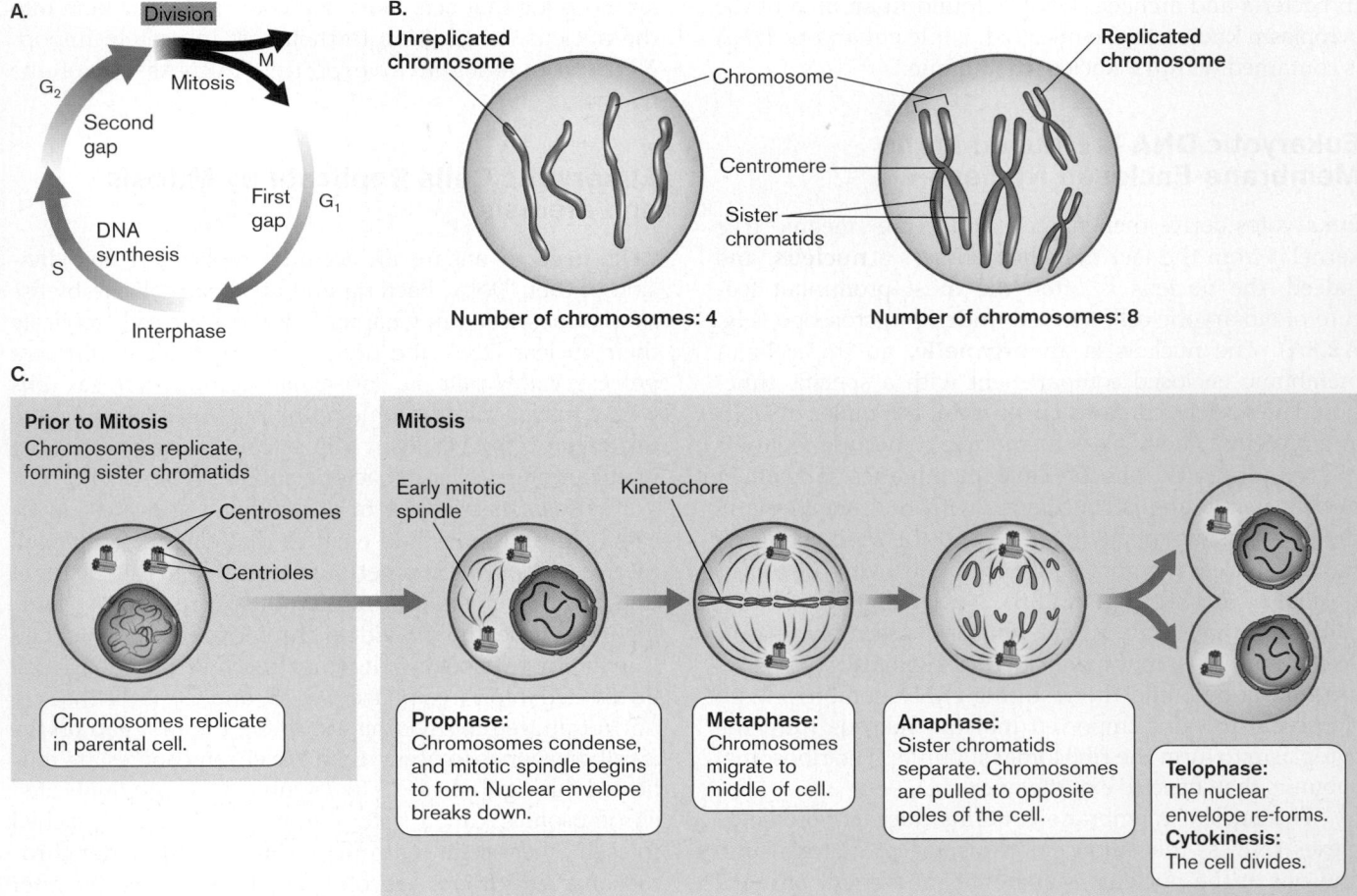

Figure A2.9   The cell cycle and mitosis.   A. Stages of the eukaryotic cell cycle. G₁, S, and G₂ make up interphase (in gray). **B.** Duplication of the chromosomes during S phase. **C.** The phases of mitosis. See text for details.

nuclear membrane may break down. The mitotic spindle, which separates the sister chromatids to opposite poles of the cell, begins to form. The mitotic spindle is a network of microtubules (see Section A2.5) that originate from centrosomes. There are two centrosomes, and these migrate to opposite sides of the cell. Each centrosome contains two centrioles, and each centriole contains nine sets of microtubules in a radial formation (see Section A2.5). The free ends of the microtubules establish connections with the sister chromatids at a structure called the kinetochore.

■ **Metaphase.** At metaphase, the spindle apparatus is complete and each sister chromatid is connected to a microtubule. The chromosomes are arranged along an imaginary plane in the middle of the cell.

■ **Anaphase.** In anaphase, the microtubules shorten and pull the sister chromatids apart, separating the replicated chromosomes. At the end of anaphase, each set of chromosomes is located on opposite sides of the cell.

■ **Telophase.** In this final step of mitosis, the nuclear membrane re-forms around each set of chromosomes and the chromosomes become long and thin again. Cytokinesis also occurs during telophase and partitions the original cell into two daughter cells by the formation of a cell membrane between them.

The reproduction of eukaryotic cells by mitosis is called **asexual reproduction**.

Some variations in the phases of mitosis are seen in microbial eukaryotes. In yeast, for example, the nuclear envelope does not break down. Despite these differences, the end result, the separation of duplicated chromosomes into two daughter cells, is the same.

Unlike asexual reproduction, **sexual reproduction** requires the reassortment of genetic material from different chromosomes. A sexual life cycle alternates between cells that are diploid (2*n*, containing two copies of each chromosome) and sex cells that are haploid (1*n*, containing a single copy of each chromosome). Two of the haploid 1*n* sex cells (called gametes) can join each other by

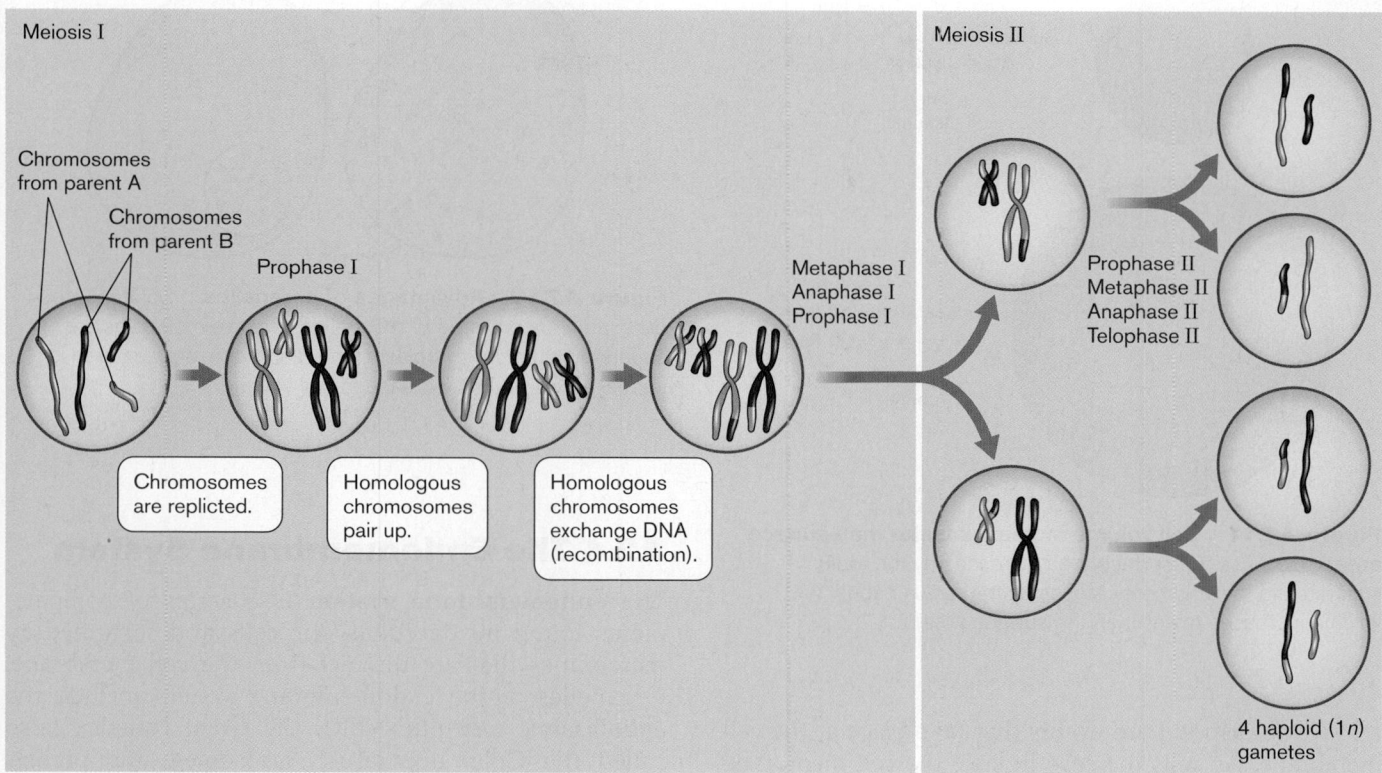

Figure A2.10   Meiosis.   The chromosomes of a diploid (2*n*) cell undergo replication. In meiosis I, homologous chromosomes exchange DNA, and all the pairs are separated between two daughter cells. In meiosis II, homologous pairs are separated to produce haploid (1*n*) gametes.

fertilization to regenerate a diploid cell (called a zygote). The diploid thus possesses two **homologs** of each chromosome—that is, two versions of the same chromosome from two different parents.

The process of gamete formation requires a special modification of mitotic cell division, called **meiosis** (**Fig. A2.10**). Meiosis includes two cell division series, called meiosis I and meiosis II. Like mitosis, meiosis I must be preceded by replication of all chromosomes during S phase. Thus, a diploid (2*n*) cell becomes temporarily 4*n*. Unlike the case in mitosis, in prophase I of meiosis I the replicated chromosomes do not separate; instead, each replicated pair lines up with its homolog, the same chromosome inherited from the other parent. Now the aligned homologs exchange portions of their DNA. This genetic exchange reassorts the traits and increases genetic diversity of the offspring.

Meiosis II is completed by separation of the paired homologs through metaphase I and anaphase II. A short telophase occurs, in which each daughter cell is now diploid (2*n*). Unlike what happens in the telophase of mitosis, in meiosis I no nuclear membrane forms, and

no interphase occurs. Instead, the chromosome pairs immediately separate during meiosis II, which includes prophase II, metaphase II, anaphase II, and telophase II. The result is four haploid (1*n*) gametes, as seen in **Figure A2.10**. Depending on the species, these gametes may develop into specialized forms such as sperm and egg that reunite through fertilization, restoring the diploid form.

A2.3 Problems Faced by Large Cells

Most eukaryotic cells range in size from 10 to 100 micrometers (μm) in diameter, about ten times as large as the typical prokaryotic cell (1–10 μm diameter). Cells face two major challenges as a result of increased cell size. The first problem is that as cells increase in size, their volume (the cytoplasm) increases faster than their surface area (the cell membrane) (**Fig. A2.11**). Cells are filled with metabolically active cytoplasm that requires nutrients and energy and produces wastes. Energy production, nutrient import,

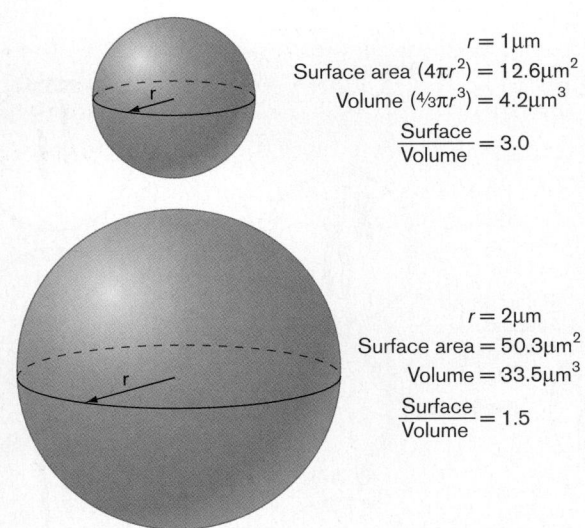

$r = 1\mu m$
Surface area $(4\pi r^2) = 12.6\mu m^2$
Volume $(\frac{4}{3}\pi r^3) = 4.2\mu m^3$
$$\frac{\text{Surface}}{\text{Volume}} = 3.0$$

$r = 2\mu m$
Surface area $= 50.3\mu m^2$
Volume $= 33.5\mu m^3$
$$\frac{\text{Surface}}{\text{Volume}} = 1.5$$

Figure A2.11 Cell volume increases faster than surface area. Because the surface area increases by the radius squared and the volume increases by the radius cubed, the volume increases faster than the surface area.

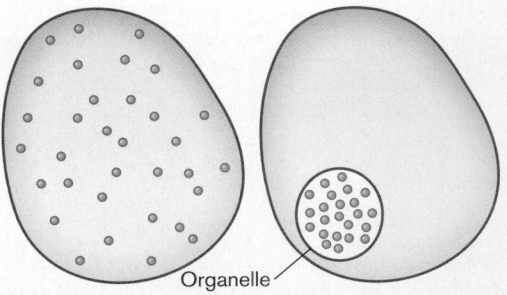

Figure A2.12 Advantages of organelles. Solutes (green dots) are concentrated in organelles, reactive molecules are separated from the cytoplasm, and membrane surface area is increased without an increase in cell volume.

and waste disposal are events that take place at the cell membrane. As cells increase in size, the cell membrane area may not be able to keep up with the demands placed on it by a proportionally larger cytoplasm. The eukaryotic cell's answer to this problem is the endomembrane system, an extensive network of internal membranes that effectively increases the membrane surface area without increasing cell volume.

The second problem associated with an increase in size is related to diffusion. The amount of time it takes a molecule to diffuse a given distance is proportional to the distance squared. For example, if it takes a particular molecule 1 second to diffuse 1 µm, then it takes that same molecule 100 seconds to diffuse 10 µm. Many biochemical reactions depend on the partners in the reaction finding each other by diffusion (that is, their rate is diffusion controlled), so longer diffusion times result in slower reactions. Furthermore, signals received at the cell membrane need to be communicated throughout the cell. In a large cell, the amount of time it takes for a molecule to traverse the cell by diffusion may be too slow for the cell to rapidly adjust to signals it receives from the environment or from other sites within the cell. To deal with this problem, eukaryotic cells possess a cytoskeleton, a group of proteins that, among other things, maintain cell shape and move molecules around the cell, relieving the cell from the need to rely on diffusion for transport.

As we shall see in the next two sections, the endomembrane system and the cytoskeleton serve additional functions for the eukaryotic cell.

A2.4 The Endomembrane System

The **endomembrane system** is a series of compartments found inside eukaryotic cells and separated by membranes that are distinct from the cell membrane. Organelles of the endomembrane system include the endoplasmic reticulum (ER), the Golgi complex (also called the Golgi apparatus), lysosomes, and peroxisomes. Different organelles contain unique subsets of proteins that contribute to their function. To a large extent, the function of the ER and the Golgi complex is to direct to their proper cellular location proteins destined for lysosomes, for the cell membrane, or for secretion from the cell.

Membranous organelles serve many functions for the eukaryotic cell. In general:

■ Endomembranes increase the membrane surface area without increasing cell volume.

■ Separating cellular contents in small, enclosed compartments increases the concentrations of enzymes and their substrates, allowing reactions to proceed faster (**Fig. A2.12**).

■ Organelles can provide different environments that allow disparate reactions to occur simultaneously. For example, proteins are being synthesized by ribosomes in the neutral pH cytoplasm, while at the same time being hydrolyzed within the acidic organelles known as lysosomes.

■ Compartmentalization protects cytoplasmic components from harmful substances. For example, hydrogen peroxide (H_2O_2), a product of cellular oxidation reactions, is produced and converted to water within **peroxisomes**. Localizing the reaction within the peroxisome keeps the toxic peroxide away from other cell components, such as proteins and DNA, that are sensitive to oxidative stress.

Lysosomes Digest Organic Matter

Lysosomes are membrane-enclosed organelles that help eukaryotic cells obtain nourishment from macromolecular nutrients. Lysosomes contain many hydrolytic enzymes (for example, proteases, nucleases, and lipases) and have an acidic pH of about 5. Lysosomes are formed when vesicles containing hydrolytic enzymes and proton pumps bud off from the Golgi complex (**Fig. A2.13A**). Lysosomes then fuse with vesicles containing the material to be digested. Often this material comes from outside the cell via phagocytosis (**Fig. A2.13A** and **B**). Phagocytosis and lysosomal digestion help the eukaryotic cell because they effectively increase the membrane surface area over which nutrients can be absorbed.

Bacteria lack phagocytosis; to obtain nutrition from large molecules in their environment, bacteria must secrete digestive enzymes. The extracellularly digested materials are subsequently transported across the bacterial cell membrane through specific transporters (see Section 4.2). In eukaryotes, by contrast, lysosomes allow for intracellular digestion, and digested material crosses the lysosomal membrane into the cytoplasm. Any waste products left in the lysosome can leave the cell via exocytosis.

The Volume within the ER Is Separate from the Cytoplasm

The endoplasmic reticulum is continuous with the outer nuclear membrane, and the lumen (interior) of the ER is continuous with the space between the two nuclear membranes (**Fig. A2.14A**). In addition, the lumen of the ER is spatially equivalent to the interior spaces of other endomembrane components and to the outside of the cell. This means that material in the ER does not need to cross a membrane to enter these other spaces, and mixing can occur via vesicle fusion. For example, material contained within the lumen of the ER can mix with the contents of the Golgi complex or with the extracellular milieu by fusion of vesicles from the ER with Golgi membrane or the plasma membrane. These topologically equivalent areas, indicated by a common color in **Figure A2.14A**, are completely separated from the cytoplasm by endomembranes, so that the ER can be used to sequester substances that must be held at low concentrations in the cytoplasm—for example, calcium ions.

Smooth ER and Rough ER

There are two morphologically and functionally distinct types of endoplasmic reticulum: smooth ER and rough ER, as shown in the micrograph of **Figure A2.14C**. The

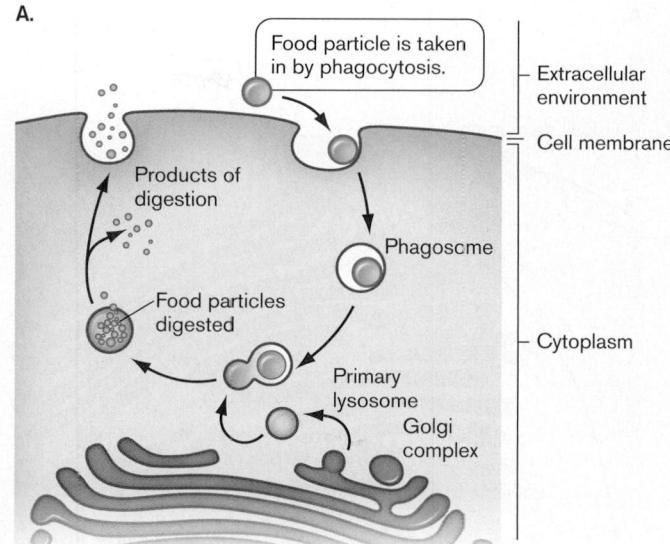

A.

Food particle is taken in by phagocytosis.

Extracellular environment

Cell membrane

Products of digestion

Phagosome

Cytoplasm

Food particles digested

Primary lysosome

Golgi complex

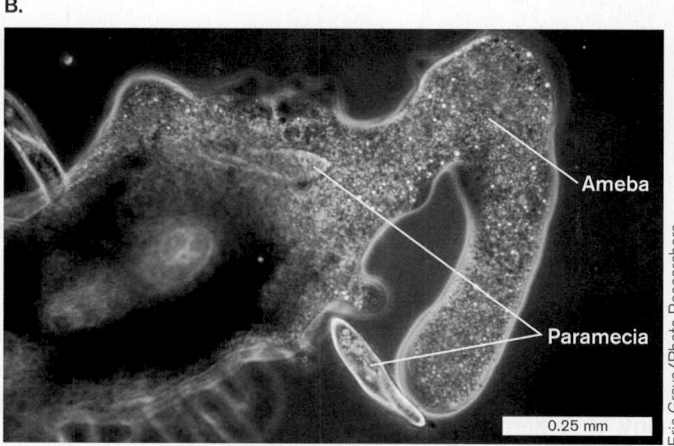

B.

Ameba

Paramecia

0.25 mm

Eric Grave/Photo Researchers

Figure A2.13   Lysosomes.   A. Lysosomes contain hydrolytic enzymes to digest material brought into the cell by phagocytosis. **B.** Amebas engulfing paramecia (light microscopy). The paramecia will be digested with the aid of lysosomes.

smooth ER is the site of lipid synthesis and some detoxification of noxious compounds. The rough ER is the site where transmembrane proteins, secreted proteins, and resident proteins of the ER, Golgi, or lysosomes are translated. The rough ER appears rough because its cytoplasmic surface is studded with ribosomes. These ribosomes are located on the rough ER because the protein being synthesized by the ribosome has a hydrophobic signal sequence on its amino terminus (the first part of the protein translated from the mRNA). A **signal sequence** is a specific sequence of amino acids that directs proteins to a specific cellular location, such as a membrane. The signal sequence that directs proteins to the ER recognizes a

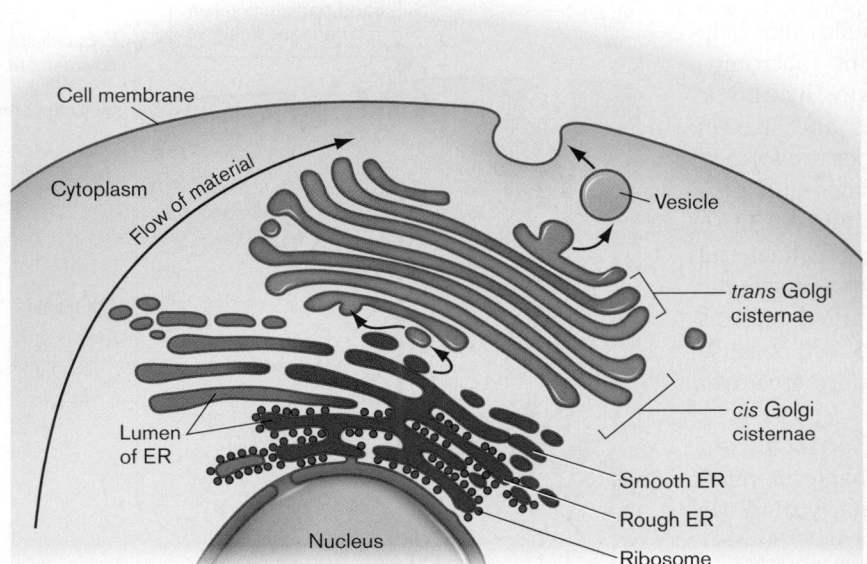

B. Golgi

SPL/Photo Researchers

C. Endoplasmic reticulum

Robert Bolender & Donald Fawcett/
Visuals Unlimited

Rough ER   Smooth ER

Figure A2.14 **The endoplasmic reticulum and Golgi complex.** **A.** The relationship of the ER to other cellular membranes and the flow of material through vesicles from the rough ER to the cell membrane. **B.** Electron micrograph of Golgi cisternae. **C.** Electron micrograph showing rough and smooth ER.

receptor (the **signal recognition particle**) on the rough ER membranes (**Fig. A2.15A**), so the protein-ribosome complex is directed to the surface of the ER membrane. Note that the ribosomes attached to the rough ER are identical to cytoplasmic ribosomes and attach to the ER only transiently because of the types of proteins they are translating—proteins that contain the correct signal sequence.

After the ribosome docks with the signal recognition particle, the protein is threaded through the ER membrane as it is synthesized. Secreted proteins and proteins destined for the lumen of an organelle are threaded completely through the ER membrane and end up in the lumen of the ER (see **Fig. A2.15A**). In contrast, transmembrane proteins are not threaded completely through, and part of the protein spans the membrane (**Fig. A2.15B**). The membrane-spanning regions of transmembrane proteins usually contain a continuous stretch of about 20 hydrophobic amino acids that form an alpha helix with the hydrophobic side chains facing out toward the hydrophobic hydrocarbons of the membrane.

As the nascent polypeptide chains are threaded across the ER membrane, the unfolded proteins are bound by chaperonins. Chaperonins (also known as heat-shock proteins) prevent partially folded proteins from clumping together and help proteins attain their correct tertiary structure. In the ER, proteins may be modified. The

signal sequence that directed the proteins to the ER is usually cleaved off. The environment inside the ER allows disulfide bonds to form between cysteine residues of some proteins. Other ER proteins have oligosaccharide groups covalently attached—a posttranslational modification known as glycosylation. The enzymes that attach the sugars are found only in the ER lumen. In transmembrane proteins of the cell membrane, sugars always face the extracellular environment because the lumen of the ER forms vesicles that open out to the external medium (see **Fig. A2.14A**). Resident ER proteins, those that will stay and function in the ER, are retained inside the ER because they contain a sequence of amino acids that acts as an ER retention signal.

The Golgi Complex Directs Protein Transport

Proteins not retained in the ER move on to the Golgi complex by way of vesicles. The **Golgi complex** consists of separate membrane stacks (cisternae) that each contain unique enzymes. The *cis* face of the Golgi complex is nearest to the ER, and the *trans* face is farthest from the ER, closest to the cell membrane (see **Fig. A2.14A** and **B**). As proteins pass through the cisternae, the carbohydrates on them may be trimmed and modified. These modified carbohydrates can serve as address tags for targeting

A.

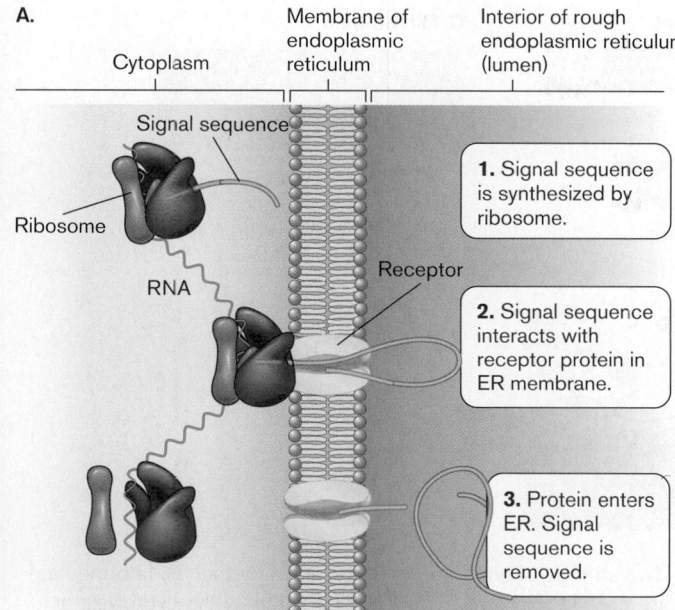

Cytoplasm

Membrane of
endoplasmic
reticulum

Interior of rough
endoplasmic reticulum
(lumen)

Signal sequence

Ribosome

RNA

Receptor

1. Signal sequence
is synthesized by
ribosome.

2. Signal sequence
interacts with
receptor protein in
ER membrane.

3. Protein enters
ER. Signal
sequence is
removed.

B.

Ribosome

Cleavage site

H₂N—

Signal sequence

1. Translocation

H₂N—

Stop-transfer
peptide

2. Translocation
stops.

H₂N—

Mature
transmembrane
protein

COOH

NH₂

3. Cleavage of
start-transfer
peptide.

H₂N—

COOH

Figure A2.15 **Insertion of proteins into the ER.**
A. Proteins are targeted to the ER by a signal sequence on
the growing polypeptide chain. Secreted proteins and proteins
destined for the lumen of an organelle end up in the lumen of
the ER. **B.** Transmembrane proteins are not completely inserted
through the ER membrane, and a portion of the protein remains
within the membrane.

proteins to particular organelles. For example, proteins
tagged with mannose 6-phosphate (mannose phosphorylated on its number 6 carbon) are selectively sent
to lysosomes; that is, vesicles enriched in proteins containing mannose 6-phosphate bud off from the Golgi and
are directed to lysosomes. Proteins not targeted to lysosomes or marked for retention in the Golgi complex may
be sent to the cell membrane. Vesicles leaving the Golgi
complex may fuse with the cell membrane, releasing
their contents to the extracellular environment (see **Fig.
A2.14A**). Transmembrane proteins in these vesicles can
then become part of the cell membrane. Regions of transmembrane proteins that were inside vesicles will face the
extracellular environment, and cytoplasmic portions will
remain cytoplasmic.

A2.5 The Cytoskeleton

Eukaryotic cells contain proteins called intermediate filaments, microfilaments, and microtubules that are collectively termed the **cytoskeleton**. As the name implies,
these proteins serve as a cell skeleton and impart specific
shapes to eukaryotic cells. However, cytoskeletal proteins
are multifunctional and are also involved in whole-cell
movements and movements of substances within the cell.

Microfilaments and Intermediate Filaments

Microfilaments, also known as actin filaments, have a
diameter of 7 nm (**Fig. A2.16A**). They are formed when
individual actin monomers (globular actin, or G-actin)
polymerize, in a process fueled by ATP hydrolysis, to form
chains of filamentous actin (F-actin). Two F-actin chains
twist around each other to form microfilaments that have
a plus end and a minus end. Microfilaments are dynamic
structures, growing and shrinking in a controlled manner. New monomer units add to the plus end and dissociate from the minus end. Whether actin will polymerize
or depolymerize depends on a number of factors, including the concentration of G-actin. The critical concentration is a measure of the ability of actin to polymerize. At
G-actin concentrations below the critical concentration,
F-actin will depolymerize; and at concentrations greater
than the critical concentration, G-actin will polymerize.
The plus end of microfilaments has a critical concentration less than that of the minus end, so actin is preferentially added onto the plus end and removed from the
minus end.

Some microfilaments play a structural role in the
cell to maintain cell shape. These structural microfilaments have protein caps at both ends to prevent changes
in microfilament length. Other microfilaments have

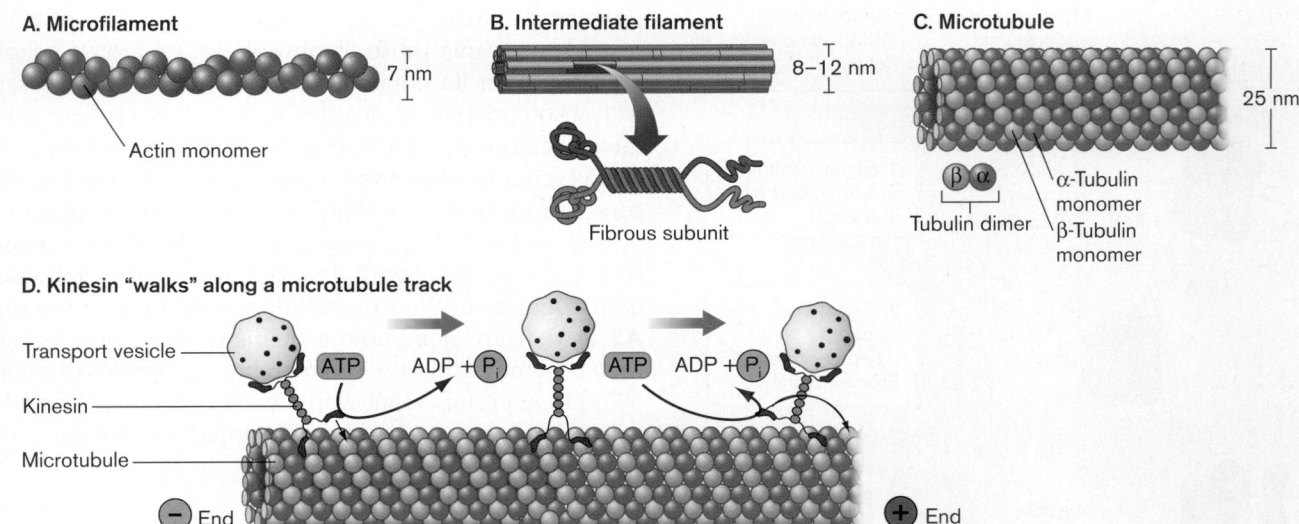

Figure A2.16   Cytoskeletal proteins.   A. Microfilaments consist of two strands of actin polymers twisted together. **B.** Intermediate filaments are ropelike assemblages of various proteins. **C.** Microtubules are polymers of tubulin dimers. **D.** Fueled by the hydrolysis of ATP, the motor protein kinesin can move vesicles or organelles toward the plus end of microtubules.

functions that require dynamic changes in length. For example, the pseudopod movement of an ameba depends on the polymerization of actin at the leading edge of growth. The plus end of the microfilament is located underneath the cell membrane of the extending pseudopod, and polymerization is enhanced by the actin-binding protein profilin. Microfilaments mediate cytoplasmic streaming, a mixing of the cytoplasm that aids diffusion. The protein myosin works with microfilaments to generate the forces needed for cell streaming, pseudopod formation, and cytokinesis, the separation of daughter cells after nuclear division.

Intermediate filaments (Fig. A2.16B) consist of various fibrous proteins that have a diameter of about 10 nm. Intermediate filaments often form a meshwork under the cell membrane and, in cells that lack a cell wall, help impart and maintain cell shape. Intermediate filaments also strengthen the cell by resisting tension placed on the cell membrane. The proteins that make up intermediate filaments vary with cell type. Intermediate filaments are fairly stable and are not thought to undergo acute changes in length the way microfilaments and microtubules do.

Microtubules

Microtubules have a larger diameter (25 nm) than microfilaments and intermediate filaments (**Fig. A2.16C**). The hollow microtubule structure consists of 13 tubulin dimers; one alpha-tubulin protein plus one beta-tubulin protein form one tubulin dimer. Like microfilaments, microtubules have plus (faster-growing) and minus (slower-growing) ends and are dynamic structures that

can polymerize and depolymerize. Polymerization is an energy-requiring process, and the necessary energy is obtained by coupling polymerization to GTP hydrolysis. Microtubules aid movement of substances within the cell and are also involved in powering whole-cell movement by cilia and flagella.

Traffic of proteins through the endomembrane system (see Section A2.4) relies on the controlled movement of vesicles from one cellular compartment to the next. Microtubules provide tracks that can move vesicles from one organelle to the next in an efficient, directed fashion. Working with microtubules to accomplish this are motor proteins. Motor proteins such as kinesin and dynein can capture cargo (for example, vesicles or organelles) and walk them along microtubule tracks in an ATP-dependent process (**Fig. A2.16D**). Kinesin moves cargo toward the plus end of microtubules, while dynein moves cargo toward the minus end. In addition to moving vesicles, microtubules segregate (pull apart) the duplicated chromosomes during mitosis.

Eukaryotic Cilia and Flagella

Cilia and flagella are thin extensions of the cell membrane that can move in a whiplike fashion, driven by interactions between microtubules and the motor protein dynein. **Flagella** (singular, flagellum) are relatively long, and cells usually have only one or two of them; **cilia** (singular, cilium) are shorter and more numerous. Both flagella and cilia can move cells through space. Cilia may also aid in food capture—for example, by sweeping extracellular fluid into the gullet of a paramecium. Eukaryotic

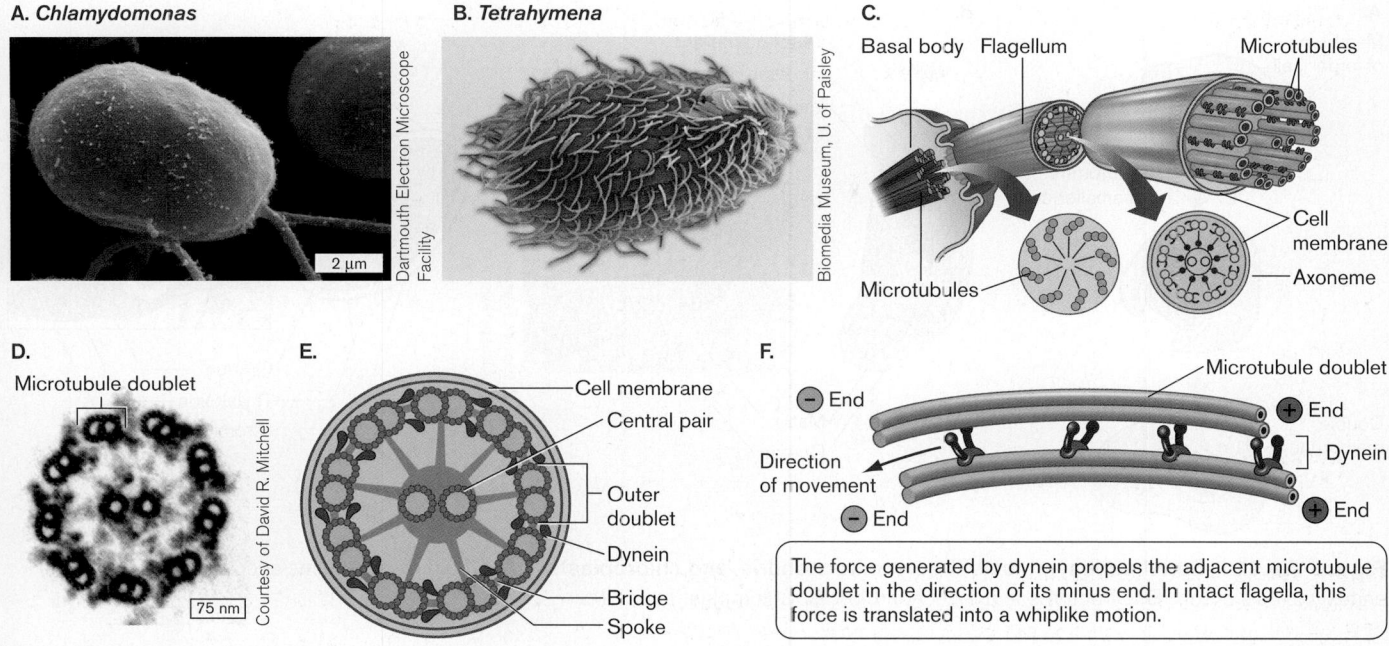

A. *Chlamydomonas*

B. *Tetrahymena*

C. Basal body Flagellum Microtubules Cell membrane Axoneme Microtubules

D. Microtubule doublet

E. Cell membrane Central pair Outer doublet Dynein Bridge Spoke

F. Microtubule doublet ⊖ End ⊕ End Dynein Direction of movement ⊖ End ⊕ End

The force generated by dynein propels the adjacent microtubule doublet in the direction of its minus end. In intact flagella, this force is translated into a whiplike motion.

Figure A2.17 Flagella and cilia. A. The protist *Chlamydomonas* has two long flagella (fluorescence micrograph). **B.** The protist *Tetrahymena* (size 50–70 μm) has numerous cilia (SEM). **C.** Structure of flagella and cilia. **D.** Cross section of axoneme (TEM). **E.** Cross section of axoneme indicating bridges and spokes formed by cross-linking proteins. **F.** Force generated by dynein on a microtubule doublet.

cilia and flagella differ from bacterial cilia and flagella; bacterial flagella depend on the proton motive force to rotate a motor that causes flagellar movement, whereas eukaryotic flagella rely on ATP hydrolysis by dynein to move the flagella in a whiplike fashion.

The study of protists has played a key role in revealing the structure and function of eukaryotic flagella and cilia. Much information about flagellar structure has come from studies of a *Chlamydomonas* species, a unicellular alga that has two long flagella (**Fig. A2.17A**). Dynein was first discovered in the cilia of the unicellular eukaryote *Tetrahymena* (**Fig. A2.17B**). Flagella and cilia have the same structure and mechanism of action, so for convenience we will restrict the following discussion to flagella.

A cross section through a flagellum reveals a central bundle of microtubules called the axoneme (**Fig. A2.17C**). The axoneme originates at a microtubule-organizing center called the basal body, which is similar to the centriole (see Section A2.2). The minus ends of the microtubules are at the basal body; the plus ends at the tip of the flagellum are capped to prevent changes in length.

As shown in the electron micrograph in **Figure A2.17D**, the axoneme has a characteristic arrangement of two central microtubules and nine microtubule doublets around the periphery. The microtubule doublets are connected to each other and to the central microtubules through protein cross-links called bridges and spokes (**Fig. A2.17E**). The motor protein dynein connects adjacent microtubule doublets. If the cell membrane is removed from the surface of a flagellum and the protein cross-links (but not dynein) are dissolved, the microtubule doublets are observed to lengthen in the presence of ATP. This is due to the action of dynein, which slides the microtubule doublets past each other (**Fig. A2.17F**). In intact flagella, the tension imparted as dynein tries to slide microtubules past each other is translated into a bending motion. Controlled cycles of dynein activation and inactivation on opposite sides of the axoneme lead to the whiplike motion of eukaryotic flagella.

A2.6 Mitochondria and Chloroplasts

Mitochondria and chloroplasts are organelles involved in cellular energy production. **Mitochondria** perform oxidative respiration and are found in nearly all eukaryotes. **Chloroplasts** perform photosynthesis and are found only in photosynthetic eukaryotes, such as green algae. Both organelles are thought to have become part of eukaryotic cells through a process of endosymbiosis. The **serial endosymbiosis theory** states that mitochondria and chloroplasts were once free-living bacteria that became ingested, but not digested, by a larger (possibly eukaryotic) cell (**Fig. A2.18A**). A symbiotic relationship developed, with the larger eukaryotic cell providing protection to the intracellular bacterium and the bacterium providing energy to the eukaryote. As we shall see, the structure

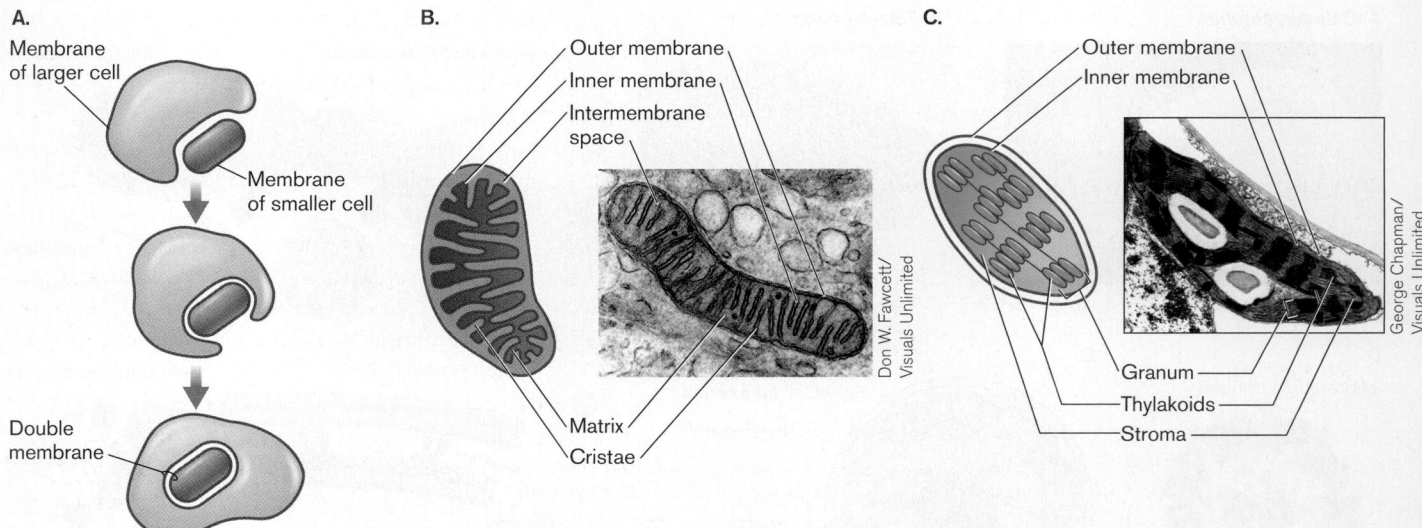

Figure A2.18 Serial endosymbiosis theory, mitochondria, and chloroplasts. A. Origin of organelles according to serial endosymbiosis theory. **B.** Structures in a mitochondrion. **C.** Chloroplast structure.

Table A2.2 Location of oxidative respiration components in bacteria and mitochondria.

Feature	Bacteria	Mitochondria
Electron transport system (ETS)	Cell membrane	Inner membrane
ATP synthase	Cell membrane	Inner membrane
Citric acid cycle	Cytoplasm	Matrix
Proton motive force	Protons diffuse from lower-pH extracellular environment into the cytoplasm across the cell membrane.	Protons diffuse from lower-pH intermembrane space into the matrix across the inner membrane.

of mitochondria and chloroplasts reflects their origin as internalized bacteria.

Mitochondria Produce ATP by Oxidative Respiration

Mitochondria are the powerhouses of the eukaryotic cell. A cell may contain tens or hundreds of mitochondria, depending on its energy needs. Mitochondria have two membranes: an outer membrane and an inner membrane (**Fig. A2.18B**). The inner membrane has numerous infoldings called cristae that increase its surface area. It is thought that as a result of the endocytosis of the bacterium, the inner mitochondrial membrane is derived from the bacterium and the outer membrane is derived from the larger eukaryote. Supporting this idea is the fact that the inner membrane has structural characteristics of a prokaryotic cell membrane, while the outer membrane is similar to the host eukaryotic membrane. For example, 20% of the phospholipids in the inner membrane are cardiolipin, a bacterial phospholipid largely absent from membranes of eukaryotic origin.

Mitochondria contain two distinct compartments: the intermembrane space between the two membranes, and the matrix inside the inner membrane. Different stages of oxidative respiration occur in specific compartments. As predicted by the endosymbiosis theory, the topology of these processes is similar in bacteria and mitochondria (**Table A2.2**). For example, in bacteria and archaea, the tricarboxylic acid cycle occurs in the cytoplasm; in mitochondria, the tricarboxylic acid cycle takes place inside the matrix, the metabolic equivalent of the prokaryotic cytoplasm.

The fact that mitochondria contain their own DNA and ribosomes lends further support to endosymbiosis, since these cell features would be a necessary part of any free-living organism. Both the DNA and ribosomes of mitochondria show similarities with the DNA and ribosomes of bacteria. For example, like prokaryotic DNA, mitochondrial DNA is circular, and mitochondrial ribosomes are sensitive to antibiotics that disrupt prokaryotic ribosomes. Although nuclear DNA encodes some mitochondrial proteins, phylogenetic analysis has shown that these nuclear genes originated from bacteria, probably the engulfed ancestors of mitochondria. Furthermore, the generation of new mitochondria is not tied to replication of the cell, and the division of mitochondria within the cell is similar to the fission seen in prokaryotic cells.

Chloroplasts Perform Photosynthesis

Chloroplasts are organelles found only in photosynthetic eukaryotes. In the light reactions of photosynthesis, chloroplasts convert light energy from the Sun to ATP and reduced NADPH. In the subsequent light-independent reactions, the ATP and NADPH are used to reduce CO_2 to sugar.

Like mitochondria, chloroplasts are probably the result of endosymbiosis. Bacteria related to cyanobacteria are thought to be the prokaryotic partners that gave rise to chloroplasts. Like eukaryotic chloroplasts, cyanobacteria contain chlorophylls *a* and *b* and perform aerobic

photosynthesis. In addition, cyanobacteria contain extensive internal membranes called thylakoids that contain chlorophyll and participate in the light reactions of photosynthesis.

Chloroplasts have three membranes whose topology can be understood in light of endosymbiosis (**Fig. A2.18C**). The outer membrane appears to be derived from the host eukaryotic cell, the inner membrane is equivalent to the bacterial cell membrane, and the thylakoid membrane is derived from the bacterial thylakoid membranes. The region inside the inner membrane is called the stroma and is equivalent to the bacterial cytoplasm. The thylakoid membrane is packed with the chlorophyll pigments that give chloroplasts their green color. ATP and NADPH are produced in the stroma and used there in the light-independent reactions of CO_2 fixation. As would be expected as a result of endosymbiosis, chloroplasts, like mitochondria, contain their own circular DNA and their own ribosomes.

Answers to Thought Questions

Chapter 1

1.1 The minimum size of known microbial cells is about 0.2 μm. Could even smaller cells be discovered? What factors may determine the minimum size of a cell?

ANSWER: The smallest cells known, about 0.2 μm in length, are cell wall–less bacteria called mycoplasmas—for example, *Mycoplasma pneumoniae*, a causative agent of pneumonia. Bacteria might be discovered that are smaller than 0.2 μm, but it is hard to see how their cell components such as ribosomes (about a tenth this size) could fit inside such a small cell. The volume required for DNA and the apparatus of transcription and translation probably sets the lower limit on cell size.

1.2 If viruses are not functional cells, are they truly "alive"?

ANSWER: A traditional definition of a life-form includes the capability for metabolism and homeostasis (maintaining internal conditions of its cytoplasm), as well as reproduction and response to its environment. Viruses reproduce themselves and respond to the environment of the host cell, but they lack metabolism or homeostasis outside their host cell. Nevertheless, viruses such as herpes viruses contain numerous metabolic enzymes that participate in the metabolism of their host. Certain large viruses such as the mimivirus appear to have evolved from cells. Some microbiologists argue that viruses should be considered "alive" if reproduction is the main criterion, and if the viral "environment" is considered the inside of the host cell.

1.3 Why do you think it took so long for humans to connect microbes with infectious disease?

ANSWER: For most of human history we were unaware that microbes existed. Even after microscopy had revealed their existence, the incredible diversity of the microbial world and the difficulties in isolating and characterizing microbial organisms made it difficult to discern the specific effects of microbes. All healthy people contain microbes; and most disease-causing microbes are indistinguishable from normal microbiota by light microscopy. Not all microbial diseases are directly transmittable from human to human; they may require complex cycles with interme-diate hosts, such as the fleas and rats that carry bubonic plague.

1.4 How could you use Koch's postulates to demonstrate the causative agent of influenza? What problems not encountered with anthrax would you need to overcome?

ANSWER: Using Koch's postulates to demonstrate the causative agent of influenza would require an animal model host. Secretions from diseased patients could be applied to different animal species, such as monkeys and mice, in order to find an animal showing signs of the disease. To determine the causative agent of disease, the patient's secretions could be filtered in order to separate bacteria and viruses. Only the filtrate would cause disease, as it contains viruses (relevant to Koch's postulates 1 and 3). Viruses, however, are more difficult to isolate in pure culture than bacteria (postulate 2)—a problem Koch did not address. Furthermore, some viruses, such as HIV (human immunodeficiency virus) have no animal model; they grow only in human cells. Today, viruses are usually isolated in a tissue culture. Once isolated, the virus could be used to inoculate a new host animal (if an animal model exists), or a tissue culture, and determine whether infection results (postulates 3 and 4). Another problem Koch did not address was the detection of infectious agents too small to be observed under a microscope. Today, antibody reactions are used to determine whether an individual has been exposed to a putative pathogen. An antibody test could be used to determine whether healthy and diseased individuals have been exposed to the isolated virus.

1.5 Why do you think some pathogens generate immunity readily, whereas others evade the immune system?

ANSWER: Some pathogens (microbes that cause disease) have external coat proteins that strongly stimulate the immune system and induce production of antibodies. Other pathogens have evolved to avoid the immune system by changing the identity of their external proteins. Immunity also varies greatly with the host's status. The very young and very old generally have weaker immune systems than do people in the prime of life. Some pathogens, such as HIV, will directly attack the host's immune system, limiting the immune response to the pathogen.

1.6   How do you think microbes protect themselves from the antibiotics they produce?

ANSWER: Microbes protect themselves from their antibiotics by producing their own resistance factors. As discussed in later chapters, microbes may synthesize pumps to pump the antibiotics out; or they make altered versions of the target macromolecule, such as the ribosome subunit; or they make enzymes to cleave the antimicrobial substance.

1.7   Why don't all living organisms fix their own nitrogen?

ANSWER: Nitrogen fixation requires a tremendous amount of energy, about 30 molecules of ATP per dinitrogen molecule converted to ammonia (discussed in Chapter 15). In a community containing adequate nitrogen sources, organisms that lose the nitrogen fixation pathway make more efficient use of their energy reserves than do those that spend energy to fix nitrogen from the atmosphere. Another consideration is that nitrogenase is an oxygen-sensitive enzyme, whereas plants, animals, and fungi are aerobes. In order to fix nitrogen, aerobic organisms need to develop complex mechanisms to exclude oxygen from nitrogenase.

1.8   What arguments support the classification of Archaea as a third domain of life? What arguments support the classification of archaea and bacteria together, as prokaryotes, distinct from eukaryotes?

ANSWER: The sequence of 16S rRNA (small-subunit rRNA) and other fundamental genes differs as much between archaea and bacteria as it does between archaea and eukaryotes. The composition of archaeal cell walls and phospholipids is completely distinct from that of bacteria and eukaryotes. Some aspects of gene expression, such as the RNA polymerase complex, are more similar between archaea and eukaryotes than between archaea and bacteria. On the other hand, archaeal and bacterial cells are prokaryotic; they both lack nuclei and complex membranous organelles. Archaeal metabolism and lifestyles are more similar to those of bacteria than to those of eukaryotes. Some archaea and bacteria sharing the same environment, such as high-temperature springs, have undergone horizontal transfer of genes encoding traits such as heat-stable membrane lipids.

1.9   Do you think engineered strains of bacteria should be patentable? What about sequenced genes or genomes?

ANSWER: Microbes—as well as multicellular organisms, such as transgenic mice—have been patented, and the patents have held up in court. DNA sequence per se is not patentable, but specific plans for use of DNA sequence can include the sequence as part of the patent. The reason for granting these patents is to encourage medical research by companies that need to earn a profit. The disadvantage of patents is that they restrict information flow and under-

cut competition. Furthermore, religious and philosophical arguments have been made that patenting live organisms cheapens life. Current laws attempt to reach a balance among these concerns.

Chapter 2

2.1   You have discovered a new kind of microbe, never observed before. What kinds of questions about this microbe might be answered by light microscopy? What questions would be better addressed by electron microscopy?

ANSWER: Light microscopy could answer questions such as: What is the overall shape of this cell? Does it form individual cells or chains? Is the organism motile? Only light microscopy can visualize an organism alive. Electron microscopy can answer questions about internal and external subcellular structures. For example, does a bacterial cell possess external filamentous structures, such as flagella or pili? If the dimensions of the unknown microbe are smaller than the lower limits of a light microscope's resolution, EM may be the only way to observe the organism. Viruses are often characterized by shape, and this shape is observed by electron microscopy.

2.2   Explain what happens to the refracted light wave as it emerges from a piece of glass of even thickness. How do its new speed and direction compare with its original (incident) speed and direction?

ANSWER: The part of the wave front that emerges first travels faster than the portion still in the glass, causing the wave front to bend toward the surface of the glass. Ultimately, the wave travels in the same direction and with the same speed as it did before entering the glass. The path of the emerging light ray is parallel to the path of the light ray entering the glass and is shifted over by an amount dependent on the thickness of the glass. This refraction will alter the path of the beam of light and decrease the amount of light reaching the lens of the microscope. Immersion oil has the same refractive index as glass and will limit the amount of light lost in this way.

2.3   Parabolic lenses are generally "biconvex"—that is, curving outward on both sides. What will happen to parallel light rays that pass through a lens that is concave on both sides? Or a lens that is convex on one side and concave with equal curvature on the other?

ANSWER: When light passes through a lens that is concave (curving inward) on both sides, the light rays diverge within the lens material and then diverge again more steeply on their way out. If the lens is concave on one side and convex with equal curvature on the opposite side, then the light rays will diverge within the lens but will emerge parallel, although slightly farther apart than when they entered.

2.4 In theory, what angle θ would produce the highest resolution? What practical problem would you have in designing a lens to generate this light cone?

ANSWER: In theory, an angle theta (θ) of 90° would produce the highest resolution. However, a 90° angle of theta generates a cone of 180°, which would require the object to sit in the same position as the objective lens—in other words, to have a focal distance of zero. In practice, the cone of light needs to be somewhat less than 180°, in order to allow room for the object and to avoid substantial aberrations (light-distorting properties) in the lens material.

2.5 Under starvation conditions, bacteria such as *Bacillus thuringiensis*, the biological insecticide, package their cytoplasm into spores, leaving behind an empty cell wall. Suppose, under a microscope, you observe what appears to be a hollow cell. How can you tell if the cell is indeed hollow or if it is simply out of focus?

ANSWER: You can tell whether the cell is out of focus or actually hollow by rotating the fine-focus knob to move the objective up and down while observing the specimen carefully. If the hollow shape appears to be the sharpest image possible, it is probably a hollow cell. If the hollow shape turns momentarily into a sharp, dark cell, it was probably out of focus before. You could also use a confocal microscope to visualize the center of the hollow cell.

2.6 Some early observers claimed that the rotary motions observed in bacterial flagella could not be distinguished from whiplike patterns, comparable to the motion of eukaryotic flagella. Can you imagine an experiment to distinguish the two and prove that the flagella rotate? *Hint:* Bacterial flagella can get "stuck" to the microscope slide or coverslip.

ANSWER: To prove that flagella rotate, you can "tether" a bacterium to the microscope slide by getting one of its flagella stuck to the slide. A simple way to tether bacteria is by using a slide coated with anti-flagellin antibody. When the flagellum is stuck to the slide, its motor continues to rotate; thus, the entire cell now rotates. The rotation of the cell body can easily be seen by video microscopy. If the flagella moved in a whiplike fashion, the tethered cell would move back and forth, not rotate.

2.7 What experiment could you devise to determine the actual order of events in DNA movement toward the pole during formation of an endospore?

ANSWER: One way to track the movement of DNA during sporulation would be to stain the DNA with a dye such as DAPI at various stages of sporulation. Alternatively, green fluorescent protein (GFP) fused to a DNA-binding protein could be used to label a specific sequence of DNA and track its position. Another way to determine the order of events in sporulation could be to observe mutant strains of bacteria that contain defects at different points in the sporulation process. (Sporulation is discussed further in Chapter 4.)

2.8 Compare and contrast fluorescence microscopy with dark-field microscopy. What similar advantage do they provide, and how do they differ?

ANSWER: Both dark-field and fluorescence microscopy enable detection (but not resolution) of objects whose dimensions are below the wavelength of light. Dark-field technique is based on light scattering, which detects all small objects without discrimination. Fluorescence, however, provides a means to label specific parts of cells, such as cell membrane or DNA, or particular species of microbes, using fluorescent antibody tags.

2.9 An electron microscope can be focused at successive powers of magnification, as in a light microscope. At each level, the image rotates at an angle of several degrees. Given the geometry of the electron beam, as shown in **Figure 2.37**, why do you think this rotation occurs?

ANSWER: The image rotates because the electron beam is not straight, as for photons, but travels in a spiral through the magnetic field lines. As magnification increases, the spiral expands, and it reaches the image plane at a slightly different angle than before.

2.10 What kinds of research questions could you investigate using SEM? What questions could you answer using TEM?

ANSWER: SEM could be used to examine the surface of cells: Do the cells possess a smooth surface, or does their surface contain protein complexes or bulges that serve special functions? How do pathogens attach to the surface of cells? TEM can be used to determine the intracellular structure of attachment sites, as well as of internal organelles. TEM can also visualize the shape of macromolecular complexes such as flagellar motors or ribosomes.

2.11 What kinds of experiments could prove or disprove the interpretations of the images of "nanobacteria" in blood plasma?

ANSWER: Attempt to observe the proposed "nanobacteria" in the presence of a general antibacterial agent such as sodium azide. If the particles still appear, despite the sodium azide, they cannot be alive and are probably inorganic in nature. Another approach is to use PCR amplification of 16S ribosomal RNA sequences to establish the existence of a novel isolate. So far, the PCR sequences obtained from "nanobacteria" in blood plasma have been identified as those of an environmental microbe that commonly contaminates PCR samples.

Chapter 3

3.1 Which molecules occur in the greatest number in a bacterial cell? The smallest number? Why does a cell contain 100 times as many lipid molecules as strands of RNA?

ANSWER: Inorganic ions occur in the greatest number in a prokaryotic cell (250 million/cell). They are also the smallest in size. DNA molecules are found in the lowest number (one large molecule, branched during replication). A prokaryotic cell contains a hundred times as many lipid molecules as strands of RNA because lipids are small structural molecules, highly packed. RNA molecules are long macromolecules that are either packed into complexes (such as ribosomal RNA), or are temporary information carriers (messenger RNA), present only as needed to make proteins.

3.2 Why does the Svedberg coefficient of the intact ribosome (70S) differ from the sum of the individual subunits of the ribosome (30S and 50S)?

ANSWER: The Svedberg coefficient is a nonlinear function. A particle's mass, density, and shape will determine its S-value. It depends on the frictional forces retarding the particle's movement, which in turn are based on the average cross-sectional area of the particle. The sum of the cross-sectional areas of the two separate subunits is greater than the cross-sectional area of the complex; thus, the sum of the S units (30S + 50S) is greater than that of the intact complex (70S).

3.3 What are the advantages and limitations of biochemical and genetic approaches to deciphering cell structure and function?

ANSWER: Biochemistry is useful to separate particles and identify their functions in isolation. In some cases, particles can also be combined to study the function of a complex such as the cell-free translation system. However, biochemistry may miss aspects of function that are observed only in the intact cell. Genetics reveals the functions of specific genes in the phenotype of a living cell. On its own, however, genetics reveals little about the spatial relationships of gene products working together. Together, genetics and biochemistry reveal a more complete view of how cellular structures function.

3.4 Amino acids have acidic and basic groups that can dissociate. Why are they *not* membrane-permeant weak acids or weak bases? Why do they fail to cross the phospholipid bilayer?

ANSWER: At neutral pH, amino acids each have both a positively charged amine and a negatively charged carboxylate; that is, they can act as either a weak acid or a weak base. Charged ions, no matter what their size, will not freely pass through a plasma membrane. If either charged group becomes neutralized by acid or base, the other group remains charged, so the molecule as a whole will never cross the membrane.

3.5 If the sacculus (cell wall) consists of a single molecule, how do you think it expands as the cell grows?

ANSWER: In principle, the cell wall molecule could be expanded in different ways: (1) adding new material at the ends of the cell; (2) adding new material in the middle, around the equator where the cell divides; or (3) inserting new material throughout the length of the expanding cell. Current evidence from radiolabel incorporation is consistent with the second model (expansion in the middle) for species of cocci such as *Staphylococcus*. Rod-shaped cells such as *Escherichia coli* and *Bacillus subtilis* appear to incorporate new cell wall material throughout the expanding cell surface. Before new material can be inserted, enzymes must first cleave between glycan strands to provide connection points for cross-bridges. For this reason, growing cells are highly susceptible to antibiotics targeting peptidoglycan synthesis; if the wall is cleaved without new material and cross-bridges being inserted, it will eventually fall apart.

3.6 The actual thickness of the cell wall is difficult to determine based solely on electron microscopic observation of the envelope layers. Devise a biochemical experiment that can show the number of layers of peptidoglycan in cells of a given species.

ANSWER: To demonstrate the number of layers of peptidoglycan in a cell wall, determine the percentage by weight of peptidoglycan in the cell sample. Estimate the surface area of cells using scanning EM. Given the known thickness of a layer of peptidoglycan, calculate how many layers would be needed to yield the amount of peptidoglycan present in the cell.

3.7 Why would laboratory culture conditions select for evolution of cells lacking an S-layer?

ANSWER: Degeneration of protective traits is a common problem when conducting research on microbes that can produce 30 generations overnight. Their rapid reproductive rate gives ample opportunity for spontaneous mutations to accumulate over an experimental timescale. In the case of the S-layer, in a laboratory test tube free of predators or viruses, mutant bacteria that fail to produce the thick protein layer would save energy compared to S-layer synthesizers, and would therefore grow faster. Such mutants would quickly take over a rapidly growing population.

3.8 Why would the proteins listed in **Table 3.2** be confined to specific cell fractions? Why could a protein not function everywhere in the cell?

ANSWER: Each protein has evolved a specific function optimized for a specific part of the cell. For example, water-conducting porins are found solely in the inner membrane (cell membrane), which is otherwise impermeable to water. The outer membrane, which is water permeable, is the sole location for specific porins transporting small peptides and sugars. The sugars then need to be taken across the inner membrane by transport proteins that have evolved to function best in this location. Similarly, different chaperones (proteins that aid peptide folding) have evolved to function best in the cytoplasmic environment or in the periplasmic environment, membrane-enclosed regions that differ substantially in pH and ion concentrations. In a different chemical environment of the cell, the protein denatures and loses its functional structure.

3.9   What do you think are the advantages and disadvantages of a contractile vacuole, compared with a cell wall?

ANSWER: The disadvantage of a contractile vacuole is that it requires continual input of energy to bail out the water. On the other hand, the contractile vacuole permits the existence of a cell that is flexible enough to engulf other cells as prey. A cell wall does not require energy (other than the initial energy to synthesize it). A cell wall is inflexible and does not allow another cell to be engulfed as prey.

3.10   How would a magnetotactic species have to behave if it were in the Southern Hemisphere instead of the Northern Hemisphere?

ANSWER: In the Northern Hemisphere, the field lines for magnetic north point downward; in the Southern Hemisphere, the opposite is true. Thus, if downward direction is the aim of magnetotaxis, bacteria existing in the two hemispheres would have to respond oppositely to the magnetic field; in the Southern Hemisphere, anaerobic magnetobacteria swim toward magnetic south. Near the equator, the proportions of north-seeking and south-seeking bacteria are roughly equal.

3.11   Most laboratory strains of *E. coli* and *Salmonella* commonly used for genetic research lack flagella. Why do you think this is the case? How can a researcher maintain a motile strain?

ANSWER: The motility apparatus requires 50 different genes generating different protein components. Cells that acquire mutations eliminating expression of the motility apparatus gain an energy advantage over cells that continue to invest energy in motors. In a natural environment, the nonmotile cells lose out in competition for nutrients, despite their energetic advantage; but in the laboratory, cells are cultured in isotropic environments such as a shaking test tube, where motility confers no advantage. These culture conditions lead to evolutionary degeneration of motility, as they do for the S-layer (see Thought Question 3.7). In order to maintain a motile strain, bacteria are cultured on a soft agar medium containing an attractant nutrient. As cells consume the attractant, they generate a gradient and chemotaxis leads them to swim outward. By subculturing only bacteria from the leading edge of swimming cells, one can maintain a motile strain.

Chapter 4

4.1   In a mixed ecosystem of autotrophs and heterotrophs, what happens when a heterotroph allows the autotroph to grow and begin to make excess organic carbon?

ANSWER: At first the growth of the heterotroph might outpace the growth of the autotroph, using the carbon sources faster than the autotroph can make them. As the organic carbon sources diminish through consumption, growth of the heterotroph decreases, but the CO_2 formed by the heterotroph will allow the autotroph to grow and make more organic carbon. Ultimately, the ecosystem comes into balance.

4.2   In what situation would antiport and symport be passive rather than active transport?

ANSWER: Antiport and symport are passive when both molecules are moving down their concentration gradients. A symport or an antiport system can do work to move a molecule from a low concentration to a high concentration (against a concentration gradient) as long as the cotransported molecule is moving from high to low concentration. When this happens, it is a form of active transport. If a symport or antiport moves both molecules with the concentration gradient (from high concentration to low), it is passive transport, assuming the molecules are moving no faster than the rate of diffusion.

4.3   What kind of transporter, other than an antiporter, could produce electroneutral coupled transport?

ANSWER: Electroneutral coupled transport can occur by symport if molecules of opposite charge are symported— for example, Na^+ flux together with Cl^-.

4.4   What would be the phenotype (growth characteristic) of a cell that lacks the *trp* genes (genes required for the synthesis of tryptophan)? What would be the phenotype of a cell missing the *lac* genes (genes whose products catabolize the carbohydrate lactose)?

ANSWER: The difference lies in the function of the two pathways. The *trp* operon is a biosynthetic operon. Errors in the biosynthetic pathway will lead to a failure to produce tryptophan. Therefore, a *trp* auxotrophic mutant will grow on defined medium *only* if tryptophan is added. The lactose operon involves the catabolism of a carbon source, lactose. If any of these genes are damaged, the cells are no longer able to use lactose as a carbon source. A *lac* mutant will *not* grow on defined medium with lactose as the sole carbon source.

4.5   The addition of sheep blood to agar produces a very rich medium called blood agar. Do you think blood agar can be considered a selective medium? A differential medium? *Hint*: Some bacteria can lyse red blood cells.

ANSWER: Blood agar can be considered differential, because different species growing on blood have different abilities to lyse the red blood cells in the agar. Some do not lyse, others completely lyse red blood cells (secreted hemolysin produces complete clearing around a colony), while still others only partially lyse the blood (the secreted hemolysin produces a greening around the colony). It will therefore differentiate between hemolytic and nonhemolytic bacteria. The medium is very rich and does not prevent the growth of any organism, so it is not considered selective.

4.6   A virus such as influenza virus might produce 800 progeny virus particles from one infected host cell. How would you mathematically represent the exponential growth of the virus? What practical factors might limit such growth?

ANSWER: In theory, the growth rate of the virus would be proportional to 800^n. In practice, however, it is unlikely that the 800 virus particles released from one host cell will find 800 different host cells to infect. Furthermore, it turns out that only a small proportion of the influenza virus progeny are viable (see Chapter 11).

4.7   Suppose you ingest 20 cells of *Salmonella enterica* in a peanut butter cookie. They all survive the stomach and enter the intestine. Suppose further that the sum total of subsequent bacterial replication and death (caused by the host) produces an average generation time of 2 hours and you will feel sick when there are 1,000,000 bacteria. How much time will elapse before you feel sick?

ANSWER: $N = \log_{10}(N_t/N_0)/0.301$;
$\log_{10}(1{,}000{,}000/20)/0.301$; 15.61×2 hours = 31 hours.

4.8   Suppose one cell of the nitrogen fixer *Sinorhizobium meliloti* colonizes a plant root. After 5 days (120 hours), there are 10,000 bacteria fixing N_2 within the plant cells. What is the bacterial doubling time?

ANSWER: 9 hours. Note that this generation time is much longer than if these same organisms were grown in a test tube containing suitable medium. In a suitable laboratory medium the generation time is about 1.5 hours.

4.9   **Figure 4.21C** shows growth curves for different population densities of *Thiobacillus thiooxidans* when the concentration of sulfur in the medium is constant. Draw the growth curves you would expect to see if the initial population density were constant but the concentration of sulfur varied.

ANSWER:

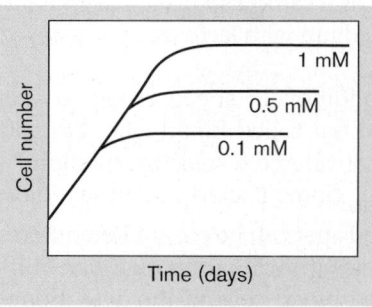

4.10   It takes 40 minutes for a typical *E. coli* cell to completely replicate its chromosome and about 20 minutes to prepare for another round of replication. Yet the organism enjoys a 20-minute generation time growing at 37°C in complex medium. How is this possible? *Hint*: How might the cell overlap the two processes?

ANSWER: After the DNA is replicated about halfway around the chromosome, each daughter half-chromosome initiates a second round of replication, so the time needed to divide from one cell to two is effectively halved. Most cells in a log-phase culture in rich medium actually have four copies of the DNA origin of replication, each with a separate attachment site on the cell envelope, the future midpoint of a cell two generations ahead (see Chapter 3).

4.11   What can happen to the growth curve when a culture medium contains two carbon sources, one a preferred carbon source of growth-limiting concentration and the second, a nonpreferred source?

ANSWER: There are two possibilities. If the enzyme systems needed to utilize both carbon sources are always made, the growth curve will look normal because both will be used simultaneously. Usually the enzyme system for the nonpreferred carbon source is not produced until the preferred source is used up. In this case, a second lag phase will interrupt the exponential phase. This is called a diauxic growth curve and is commonly seen when cells are grown on both glucose and lactose. Lactose is the nonpreferred carbon source and is used second. The second lag phase marks the exhaustion of one nutrient and the gearing up by the cell to use the other (see Chapter 10).

4.12   How would you modify the equations describing microbial growth rate to describe the rate of death?

ANSWER: The death rate applies to a period of declining cell numbers. Therefore, the logarithm of the cell number ratio of N_1 to N_0 will be a negative number, and this factor will need to be preceded by a negative sign to convert it to a positive "halving time," or half-life of the culture.

4.13   Why are cells in log phase larger than cells in stationary phase?

ANSWER: Cells strive to maintain a certain DNA/mass ratio. In so doing, they balance the amount of biochemical processes needed to sustain viability. If the cell mass becomes large relative to the number of copies of a given, critical gene, the amount of enzyme produced may not be sufficient to keep the cell alive and growing. In addition, the DNA/mass ratio serves as a signal to trigger cell division. Thus, when a cell divides faster than it replicates its chromosome, it must start a second round of replication before it finishes the first. This type of replication ensures that at least one chromosome duplication will be complete at the time of division. Because fast-growing cells contain more than one chromosome, they will increase in size to maintain the desired DNA/mass ratio. If the ratio were not maintained, cell division would not occur when needed.

4.14 How might *Streptomyces* and *Actinomyces* species avoid committing suicide when they make their antibiotics?

ANSWER: Bacteria that produce antibiotics need to make defenses against the antibiotic within their own cytoplasm. For example, their genes can express an altered form of the target molecule, such as a ribosomal subunit; or they can make pumps to pump the antibiotic out of the cell.

Chapter 5

5.1 Why haven't cells evolved so that all their enzymes have the same temperature optimum? If they did, wouldn't they grow even more rapidly?

ANSWER: An enzyme's function is not determined by temperature alone. There are other physicochemical constraints based on the variety and complexity of functions different enzymes must carry out. The thousands of different enzyme molecules must work in a coordinated fashion to support the basic functions of life. Having some enzymes work below or above their optimum temperatures will alter the rate of the reactions they catalyze. A population's evolution is based on an entire organism's ability to reproduce, not the speed at which each individual chemical reaction is carried out. The primary goal of a microbe is not just to grow fast but also to survive. Growing too fast could deplete food sources and produce toxic by-products too quickly.

5.2 If microbes lack a nervous system, how can they sense a temperature change?

ANSWER: Most bacteria respond to outside stimuli, such as heat, by altering their gene expression. They sense heat by monitoring the level of misfolded proteins, a consequence of too high a temperature. The mechanism does not perceive heat per se, but recognizes the deleterious effects of moving outside the optimum growth temperature range so that the cell can launch an emergency response. The same mechanisms can sense other environmental stresses that misfold proteins, such as acid stress. See Chapter 10.

5.3 What could be a relatively simple way to grow barophiles in the laboratory?

ANSWER: High pressures can be maintained using a syringe. Scientists would place medium in a stainless steel syringe and maintain pressure with a calibrated vise.

5.4 How might the concept of water availability be used by the food industry to control spoilage?

ANSWER: Food preservation traditionally includes water exclusion by salt, as seen in hams, back bacon, salted fish, and so on; or by high concentration of sugar, as in canned fruit or jellies. The lower a_w prevents microbial growth. Dehydrating foods will also prevent microbial growth.

5.5 In Chapter 4 we stated that an antiporter couples movement of one ion down its concentration gradient with movement of another molecule uphill, against its gradient. If this is true, how could a Na^+/H^+ antiporter work to bring protons into a haloalkaliphile growing in high salt at pH 10? Since the Na^+ concentration is lower inside the cell than outside and H^+ concentration is higher inside than outside (see **Fig. 5.16**), both ions are moving against their gradients.

ANSWER: In this situation the cell has to expend some energy to make the antiporter work. The energy involved is rooted in the charge difference between the inside of the cell (negative charge) and the outside of the cell (positive charge, called delta psi, $\Delta\psi$; see Chapter 14). The antiporter in this case must also exchange a different number of Na^+ and H^+ ions, maintaining an electrical charge difference across the membrane; for example, export $2Na^+$ and import $1H^+$.

5.6 If anaerobes cannot live in oxygen, how do they incorporate oxygen into their cellular components?

ANSWER: They incorporate oxygen from their carbon sources (for example, CO_2 and carbohydrates such as glucose), all of which contain oxygen. This form of oxygen will not damage the cells.

5.7 How can anaerobes grow in the human mouth when there is so much oxygen there?

ANSWER: A synergistic relationship exists between facultatives and anaerobes within a tooth biofilm. The facultatives consume oxygen within the biofilm microenvironment, which allows the anaerobes to grow underneath them.

5.8 What evidence led people to think about looking for anaerobes? *Hint*: Look up Spallanzani, Pasteur, and spontaneous generation on the Internet.

ANSWER: The priest Lazzaro Spallanzani (1729–1799), during his quest to disprove spontaneous generation, said, "Every beast on Earth needs air to live, and I am going to show just how animal these little animals are by putting them in a vacuum and watching them die." He dipped a glass tube into a culture, sealed one end, and attached the other to a vacuum. He was astonished to find that the microbes lived for weeks. He then wrote, "How wonderful this is. For we have always believed there is no living being that can live without the advantages air offers it." Fifty years later, Louis Pasteur observed that air could kill some organisms. After looking at a drop of liquid from a fermentation culture, he wrote, "There is something new here—in the middle of the drop they are lively, going every which way, but on the edge they were stiff as pokers."

5.9 Why would a bacteriostatic antibiotic that only inhibits growth of a pathogenic bacterium be useful for treating an infection? *Hint*: Is the human body a quiet bystander during an infection?

ANSWER: The bacteriostatic agent stops growth of the bacteria and allows the host immune response to kill them.

5.10 How would you test the killing efficacy of an autoclave?

ANSWER: Construct a death curve by measuring survival of a known quantity of spores (for example, *Bacillus stearothermophilus*) after autoclaving for various lengths of time. Spores should be used because they are more resistant to heat than is any vegetative cell. Typically, autoclaves are regularly checked with spore strips that change color once the endosopores are no longer viable.

Chapter 6

6.1 Which viruses do you know that have a narrow host range, and which have a broad host range?

ANSWER: Examples of viruses with a narrow host range include poliovirus (poliomyelitis), which infects only humans and chimpanzees; smallpox virus, which infects only humans; and feline T-cell leukemia virus, which infects only cats. Examples of viruses with a broad host range include rabies virus, which infects numerous species of mammals; and influenza strains, which show preference for particular species but can jump between various mammals and birds.

6.2 What would happen if a virus particle remained intact within a host cell instead of releasing its genome?

ANSWER: The virus would be unable to reproduce, because DNA polymerases could not reach its genome for reproduction and RNA polymerase could not transcribe its genes to make gene products.

6.3 Why do viral capsids take the form of an icosahedron instead of some other polyhedron?

ANSWER: The icosahedron is a polyhedron of 20 triangular faces, the largest number possible. Thus, the icosahedron turns out to be the largest and most economical form to enclose space with a small repeating unit. Natural selection probably favored viruses that could build the largest capsid from the smallest amount of genetic information.

6.4 How can viruses with different kinds of genomes (RNA versus DNA) combine and exchange genetic information?

ANSWER: DNA viruses require messenger RNA intermediates to express their proteins. A DNA virus could acquire the ability to package its RNA transcript in a capsid, rather than its DNA. Alternatively, RNA retroviruses form DNA intermediates within their host cell; these DNA intermediates might recombine with the DNA genome of another virus.

6.5 What are the relative advantages and disadvantages (to a virus) of the slow-release strategy, compared with the strategy of a temperate phage, which alternates between lysis and lysogeny?

ANSWER: A disadvantage of slow release is that the phages can never reproduce progeny phage as rapidly as in a lytic burst. The drain on resources of the host cell infected by a slow-release virus causes it to grow more slowly compared with uninfected cells; in contrast, a lysogenized cell suffers little or no reproductive deficit compared with uninfected cells. An advantage of reproduction by slow release is the continual release of phage, while avoiding the possibility of releasing all particles into an environment where no other host cell exists.

6.6 How could humans evolve to resist rhinovirus infection? Is such evolution likely? Why or why not?

ANSWER: Resistance to all rhinovirus infections might evolve through a mutation in the host gene encoding ICAM-1. The mutation would have to prevent rhinovirus binding without impairing the protein's ability to bind integrin. Such evolution is unlikely because of the importance of integrin binding and because rhinovirus infection is rarely fatal; thus, there is little selection pressure to evolve inherited resistance. Note, however, that the immune system rapidly generates immunity to particular strains of rhinovirus. Over a lifetime, most individuals acquire immunity to many rhinovirus strains but remain susceptible to others.

6.7 What are the advantages and disadvantages to the virus of replication by the host polymerase, compared with using a polymerase encoded by its own genome?

ANSWER: An advantage of using the host polymerase is energetic: The virus avoids the energetic cost of manufacturing a polymerase to package with each virion. This is an advantage to the virus, since its reproductive potential is limited by the energy resources of its host cell. Furthermore, because DNA and RNA polymerases are so central to cell function, the host species is unlikely to evolve a mutant form of the polymerase that resists the virus. On the other hand, the advantage of the virus making its own polymerase is that the viral polymerase can evolve to better meet the needs of its own replication, such as high speed and low accuracy to generate frequent variants.

One disadvantage of a DNA virus using the host cell DNA polymerase is that the virus must gain access to the host cell nucleus where the polymerase is. In addition, if the host cell is fully differentiated, it has exited the cell cycle and is not replicating. If it is not replicating, a DNA polymerase may not be available unless the virus can force the host cell to start going through the cell cycle. These are two problems that the virus will not have to overcome if it brings in its own DNA polymerase.

6.8 Why does bacteriophage reproduction give a step curve, whereas cellular reproduction generates an exponential growth curve? Could you design an experiment in which viruses generate an exponential curve? Under what conditions does the growth of cellular microbes give rise to a step curve?

ANSWER: Lytic viruses appear to make a step curve because the number of progeny per infected cell is 100 or more, released simultaneously. After two or three generations the cell cycles would fall out of synchrony, and the curve would smooth out, but the later cell cycles are rarely observed in practice because by then the supply of host cells is exhausted. If, however, an extremely low ratio of viruses to host cells is provided, the growth of virus particles will eventually generate an exponential curve. By contrast, the growth of cellular microbes is rarely observed during the first few doublings. By the time we measure the population, the cells are all undergoing different stages of division, and the population growth overall generates a smooth exponential curve. But if we observe the growth of a synchronized population of cells, we see a step curve of cell division too.

6.9 Suppose a certain virus depletes the population of an algal bloom. If some of the algae are selected for resistance, will they grow up again and dominate the producer community?

ANSWER: If some algae are resistant to the virus, they will reproduce and avoid infection. But their population growth will still face competition from other algal species that were never hosts of the virus. Furthermore, if the resistant algae form another bloom, eventually some other species of virus will infect them and cut their population again.

Chapter 7

7.1 What do you think happens to two single-stranded DNA molecules isolated from *different* genes when they are mixed together at very high concentrations of salt? *Hint*: High salt concentrations favor bonding between hydrophobic groups.

ANSWER: In high salt conditions, the stacking of hydrophobic bases is so strongly favored that two single strands of DNA will form a duplex no matter what the sequence of base pairs is.

7.2 How do you think the kinetics of denaturation and renaturation depend on DNA concentration?

ANSWER: The speed of denaturation does not depend on DNA concentration, but the speed of renaturation does. The higher the concentration of ssDNA, the more likely it is that complementary sequences will find each other and the faster the duplex can re-form.

7.3 DNA gyrase is essential to cell viability. Why, then, are nalidixic acid–resistant cells that contain mutations in *gyrA* still viable?

ANSWER: The *gyrA* mutations alter only the nalidixic acid binding site on GyrA, not its gyrase activity. In other words, active DNA gyrase is still made, but the drug cannot bind to it.

7.4 Bacterial cells contain many enzymes that can degrade linear DNA. How, then, do linear chromosomes in organisms like *Borrelia burgdorferi* (the causative agent in Lyme disease) avoid degradation?

ANSWER: DNA-digesting exonucleases act on free 5′ or 3′ ends. The *Borrelia* linear chromosomes possess covalently closed hairpin ends called telomeres and do not possess free 5′ or 3′ groups.

7.5 Suppose you have the following capabilities: You can label DNA in a bacterium by growing cells in medium containing either nitrogen ^{14}N or the heavier isotope ^{15}N; you can isolate pure DNA from the organism; and you can subject DNA to centrifugation in a cesium chloride solution, a solution that forms a density gradient when subjected to centrifugal force, thereby separating the light (^{14}N) and heavy (^{15}N) forms of DNA to different locations in the test tube. Given these capabilities, how might you prove that DNA replication is semiconservative?

ANSWER: Grow bacteria in medium containing heavy ^{15}N, so that all the DNA made by the cells will be in the heavy form. Transfer the cells to medium containing only ^{14}N and allow the cells to divide for one generation. Extract the DNA and centrifuge the preparation in cesium chloride. If replication is semiconservative, a hybrid DNA band will be seen at a position located between and equidistant from the point where heavy DNA and light DNA would be located. The hybrid band will be composed of one heavy strand and one light strand, which makes it of intermediate weight (density). This, in fact, was how Meselson and Stahl proved semiconservative replication in 1958, 5 years after the discovery that DNA is double-stranded.

7.6 How fast does *E. coli* DNA polymerase synthesize DNA (in nucleotides per second), given that the genome is 4.6 million base pairs, replication is bidirectional, and the chromosome completes a round of replication in 40 minutes?

ANSWER: It takes *E. coli* 40 minutes (2,400 seconds) to complete the replication of its 4,639,221-bp chromosome. Since Pol III works as a dimer to synthesize both strands simultaneously, we will consider only one strand in the calculation. The rate is 4,639,221 nt per 2,400 seconds = 1,933 nt per second. But there are two replication forks, so each polymerase dimer synthesizes only one-half of the chromosome, which means that each polymerase still synthesizes a remarkable 800–1,000 nt per second.

7.7   Individual cells in a population of *E. coli* typically initiate replication at different times (asynchronous replication). However, depriving the population of a required amino acid can synchronize reproduction of the population. What happens is that ongoing rounds of DNA synthesis finish, but new rounds do not begin. Replication stops until the amino acid is once again added to the medium—an action that triggers simultaneous initiation in all cells; that is, reproduction of the population becomes synchronized. Why?

ANSWER: Initiation requires synthesis of the initiator protein DnaA. Depriving the population of an amino acid prevents protein synthesis, which precludes synthesis of DnaA. Because DnaA is not required to complete already-initiated rounds of replication, all rounds already started are completed, but reinitiation cannot occur. Adding the amino acid once again will allow all cells to simultaneously make DnaA so that initiation is triggered in all cells at the same time.

7.8   The antibiotic rifampin inhibits transcription by RNA polymerase, but not by primase (DnaG). What happens to DNA synthesis if rifampin is added to a synchronous culture?

ANSWER: Initiation of DNA synthesis requires primer transcription at the origin by RNA polymerase, an enzyme sensitive to rifampin. Primase (DnaG), which synthesizes RNA primers in the lagging strand throughout DNA synthesis, is resistant to rifampin. So adding rifampin to a synchronized culture will prevent new rounds of DNA replication but will not affect already-initiated rounds.

7.9   Knowing that *E. coli* possesses restriction endonucleases that cleave DNA from other species, how is it possible to clone a gene from one organism to another without the cloned gene sequence being degraded?

ANSWER: The best way to overcome the restriction barrier is to use a mutant strain of *E. coli* in which its restriction system has been inactivated. DNA entering the cell will not be degraded and, if the modification system is still in place, will be modified. Once modified, that DNA can be safely transferred to other *E. coli* strains that still have their restriction endonucleases.

7.10   PCR is a powerful technique, but one can easily contaminate a sample and amplify the wrong DNA—perhaps sending an innocent person to jail. What might you do to minimize this possibility?

ANSWER: Use sterile, filtered micropipette tips. The filters prevent contamination of the micropipette and subsequent reactions. In addition, perform a control PCR reaction without adding a DNA sample. If the water, buffers, or tubes are contaminated with a given DNA, a PCR product will appear even when DNA template has not been added deliberately to the reaction.

Chapter 8

8.1   If each sigma factor recognizes a different promoter, how does the cell manage to transcribe genes that respond to multiple stresses, each involving a different sigma factor?

ANSWER: In these situations, a given gene will have multiple promoters. Each promoter will be recognized by a different sigma factor and will begin transcription at different distances from the ORF start codon.

8.2   With respect to two different sigma factors with different promoter recognition sequences, predict what would happen to the overall gene expression profile in the cell if one sigma factor were artificially overexpressed. Could there be a detrimental effect on growth?

ANSWER: Since sigma factors compete for the same site on core polymerase, overexpressing one sigma factor could displace the other sigma factor from the RNA polymerase population and compromise expression of those target genes. If those genes were important to survival, the cell could die.

8.3   Why might some genes contain multiple promoters, each one specific for a different sigma factor?

ANSWER: The gene might need to be expressed under multiple conditions at different levels. If a given condition increases expression of an alternate sigma factor, the target gene will need a promoter that the new sigma factor can recognize. As the need disappears and the sigma factor diminishes in concentration, a promoter that uses the housekeeping sigma factor will be needed. For example, the gene for DnaK heat-shock protein has promoters for RpoH (sigma-32) and sigma-70, the housekeeping sigma factor. The level of protein needed during normal growth is supplied by sigma-70. Upon encountering heat stress, the RpoH sigma factor level increases and mediates an increase in DnaK production.

8.4   How might the redundancy of the genetic code be used to establish evolutionary relationships? *Hint*: Different species have different GC content.

ANSWER: The codon preferences of different microorganisms are based in part on their GC content. Thus, an organism with an AT-rich genome will preferentially use codons for a given amino acid that have A's and T's over those with G's and C's. Evolutionarily, finding a long AT-rich region encoding mRNA with AT bias within a chromosome that is otherwise GC-rich suggests that the

AT-rich region was horizontally inherited from illicit DNA exchange with another species. (See Chapter 9.)

8.5 Synthesis of ribosomal RNAs and ribosomal proteins causes a major energy drain on the cell. How might the cell regulate synthesis of these molecules when growth rate slows as a result of amino acid limitation? *Hint*: Predict what happens to any translating ribosome when an amino acid is limiting. Look up RelA protein on the Internet.

ANSWER: As energy levels fall, fewer amino acids are made, which means that fewer charged tRNA molecules are assembled, causing ribosomes to pause during translation until finding the right charged tRNA. This pause causes synthesis of a signal molecule called guanosine tetraphosphate (ppGpp), which interacts with RNA polymerase and selectively stops transcription of the rRNA genes. The decrease in rRNA also stops new ribosome synthesis. As cells divide, the ribosomes will dilute to a lower steady-state level. Since less rRNA is made, fewer partners for ribosomal proteins are available and unassembled ribosomal proteins accumulate. These proteins bind sequences in their own mRNA molecules that are similar to the target sequences on rRNA. By binding to their mRNA, these ribosomal proteins can prevent their own translation. (See Chapter 10.)

8.6 While working as a member of a pharmaceutical company's drug discovery team, you find that a soil microbe snatched from the jungles of South America produces an antibiotic that will kill even the most deadly, drug-resistant form of *Enterococcus faecalis*, which is an important cause of heart valve vegetations in bacterial endocarditis. Your experiments indicate that the compound stops protein synthesis. How could you more precisely determine the antibiotic's mode of action? *Hint*: Can you use mutants resistant to the antibiotic?

ANSWER: One way is to take a culture of sensitive bacteria and isolate resistant mutants (bacteria that are not killed by the antibiotic), purify their ribosomes, and separate the 30S and 50S ribosomal subunits. Cross-mix subunits from sensitive and resistant cells (for example, mix 30S subunits from sensitive cells with 50S subunits from resistant cells). Then, measure protein synthesis by the hybrid ribosomes with and without the drug. If resistance is due to an altered ribosomal protein or RNA, the subunit mix containing the altered component will make protein regardless of whether the drug is present. Once identified, the responsible ribosomal subunits from resistant and sensitive cells can be broken down further into their component parts, reconstituted in hybrid form, and again tested for an ability to make protein in the presence of drug. This reductive approach will likely, but not always, uncover the target ribosomal protein or rRNA.

8.7 How might one gene code for two proteins with different amino acid sequences?

ANSWER: By having two different translation start sites in different reading frames. While this is not a common occurrence, it happens.

8.8 Why involve RNA in protein synthesis? Why not translate directly from DNA?

ANSWER: Because transcription enables the cell to amplify the gene sequence information into multiple copies of RNA. Amplification means that more ribosomes can be engaged in translating the same protein, causing the concentration of the protein to rise more quickly than if only a single gene were used. The transcriptional process also presents an opportunity to regulate production of a protein.

8.9 Codon 45 of a 90-codon gene was changed into a translation stop codon, producing a shortened, truncated protein. What kind of mutant cell could produce a full-length protein from the gene *without* removing the stop codon? *Hint*: What molecule recognizes a codon?

ANSWER: A mutant in which a tRNA gene has been altered by mutation so that the anticodon of the tRNA "sees" the stop codon as an amino acid codon. This mutated tRNA molecule will transfer its amino acid to the peptide chain. The attached amino acid can be used to bridge the gap caused by the stop codon, and a full-length protein is made. These modified tRNAs are called suppressor tRNAs because they suppress the mutant phenotype.

8.10 An ORF 1,200 bp in length could encode a protein of what size and molecular weight? *Hint*: Find the molecular weight of an average amino acid.

ANSWER: There are three bases per codon, so the ORF can encode 400 amino acids. The average molecular weight of an amino acid is 110 Da. Therefore, the hypothetical protein will be approximately 44,000 Da (44 kDa).

Chapter 9

9.1 Transfer of an F factor from an F⁺ cell to an F⁻ cell converts the recipient to F⁺. Why does transfer of an Hfr not do the same?

ANSWER: The last piece of an Hfr to transfer is the F factor and *oriT*. Only rarely will an entire chromosome transfer from one cell to another, so most Hfr transfers do not result in transfer of *oriT* and thus cannot initiate conjugation.

9.2 Would it be easier to demonstrate generalized transduction using a temperate phage (which generates a lysogen) or a lytic (virulent) phage?

ANSWER: Both can be used, but more care must be taken with a virulent phage. The reason is that a lytic phage can kill a potential recipient through superinfection. Superinfection occurs when a normal phage particle and a transducing particle infect the same cell. The normal lytic phage will lyse the cell before a recombinant can be formed. Generalized transduction usually occurs more easily with a temperate phage (like P22 or P1) because a superinfecting temperate phage will usually just lysogenize the cell and not kill it. Therefore, the host DNA delivered by the transducing particle has time to recombine into the recipient genome. Lytic phage, like the variant of the lysogenic P1 phage called P1 vir, commonly used for *E. coli* transduction, can be used in the lab if care is taken to avoid superinfection. (The "vir" designation stands for "virulent" because the mutant P1 phage lost the ability to lysogenize.) If citrate is added moments after mixing P1 vir transducing lysate with a recipient, the ions needed for phage absorption are chelated, thereby limiting superinfection. In nature, where dilution of phage may be very great, there is less chance of superinfection and death. In this situation, transduction can easily be performed by lytic phage.

9.3   How do you think phage DNA containing restriction sites evades the restriction-modification screening systems of its host?

ANSWER: Phage DNA will survive because sometimes the modification enzyme gets to the foreign DNA before the restriction enzyme. Once methylated, the phage DNA will be shielded from restriction and the methylated molecule will replicate unchallenged. If, on the other hand, a foreign DNA fragment (not necessarily a phage) has been modified and the conditions are right, it might recombine into the host chromosome and convey a new character to the strain.

9.4   Each type I restriction-modification system includes separate proteins for restriction (HsdR), modification (HsdM), and sequence recognition (HsdS). Can a plasmid grown in a wild-type strain be used to transform a strain defective in *hsdR*? An *hsdS* mutant? Can a plasmid grown in an *hsdM* mutant strain be transformed into an *hsdR* defective strain?

ANSWER: A plasmid grown in a wild-type strain will be methylated; but methylated or not, it can survive in a restrictionless *hsdR* mutant or an *hsdS* mutant that lacks the recognition protein needed by both the restriction and modification subunits. Since an *hsdM* mutant lacks the modification protein, it cannot protect its genes from restriction endonucleases and would commit suicide, so this last experiment cannot be done.

9.5   In a transductional cross between an $A^+B^+C^+$ genotype donor and an $A^-B^-C^-$ genotype recipient, 100 A^+ recombinants were selected. Of those 100, 15% were also B^+, while 75% were C^+. Is gene B or C closer to gene A?

ANSWER: Gene C, because it was cotransduced with A at the highest frequency.

9.6   You want to calculate the mutation rate for a gene in *Salmonella enterica*, so you dilute an 18-hour broth culture to 1×10^4 cells per milliliter and let it grow to 1×10^9 cells/ml.

At that point you dilute and plate cells on the appropriate agar medium and estimate there were 10,000,000 mutants in the culture. What you don't know is that the original dilution of 1×10^4 cells/ml already had 10 mutants defective in that gene. Does that affect your estimate of mutation rate? If so, how?

ANSWER: Mutation rate is an estimate of how many *new* mutations arise per cell division as a culture grows. The division of preexisting mutants during growth could lead you to estimate a false high mutation rate. By knowing how many mutants existed in the beginning, you can compensate for how many mutants you would have as a result of mutant replication, and subtract that number from the total number of mutants you observe. This adjustment assumes that the growth rates of the mutant and wild-type bacteria are the same. If there were no mutations in the original dilute culture, the mutation rate would be estimated as follows: 1×10^7 mutants / approximately 10^9 cell divisions = 10^{-2} mutation rate. This is a very high rate.

However, it takes about 17 generations (doublings) to go from 1×10^4 cells to 1×10^9 cells. If the diluted culture had 10 mutants to begin with, and if those mutants had the same generation time as the wild-type cells, then 10 cells after 17 generations would amount to about 10^6 cells. Subtracting this calculated number from the actual numbers of mutants found gives the following: 1×10^7 (total mutants) $- 0.1 \times 10^7$ (sibling mutants) = 0.9×10^7 newly derived mutants. So, 0.9×10^7 mutants / 10^9 cell divisions = 9×10^{-3} mutation rate, a rate not much different from the original calculation. But if there were 100 mutants in the original 10,000 cell dilution, the situation would change. Now there would be 1×10^7 sibling mutants, which, subtracted from 1×10^7 total mutants, leaves no new mutants. The mutation rate would seem to be $<10^{-9}$—a huge difference from the previous calculation.

This exercise illustrates the importance of starting with as few cells as possible when calculating mutation rate, of doing controls to determine the number of preexisting mutants, and of performing replicates.

9.7   During transformation in *Streptococcus pneumoniae*, a single strand of donor DNA is transferred into the recipient. This single strand of DNA can be recombined into the recipient's genome at the homologous area. However, if the segment of recipient DNA had a mutation in it relative to the donor DNA strand, the result would be a mismatch after recombination. Explain how, in the face of mismatch repair, the recipient cell ever retains the wild-type sequence from the donor.

ANSWER: Because the donor DNA was probably methylated when transferred, the methyl mismatch repair system should not distinguish between donor and recipient strands. There is equal chance that the system will repair the original recipient sequence (with a mutation) or the donor sequence. However, if the sequence was methylated differently (because there were different numbers of GATC sequences), then the mismatch repair system will preferentially remove the donor strand. The mismatch repair system in *S. pneumoniae* is called Hex and is orthologous to the *E. coli* Mut system.

9.8 It has been reported that hypermutable bacterial strains are overrepresented in clinical isolates. Out of 500 isolates of *Haemophilus influenzae*, for example, 2%–3% were mutator strains having mutation rates 100–1,000 times higher than lab reference strains. Why might mutator strains be beneficial to pathogens?

ANSWER: The mutator strains may speed microbial evolution, which could help the microbe outwit the immune system or escape the effects of administered antibiotics.

9.9 Diagram how a composite transposon like Tn*10* can generate inversions or deletions in target DNA during transposition. *Hint*: This happens when transposition within the chromosome occurs using the inverted repeats

closest to the tetracycline resistance gene (see **Fig. 9.34A**). If you draw it right, you will see that in the end, the tetracycline resistance gene is lost.

ANSWER: The diagram below shows that inversions or deletions occur when the inner rather than outer ends of the inverted repeats are the targets of the transposase. The tetracycline resistance gene is lost, and the chromosome will contain a deletion or inversion, depending on the orientation of the chromosome (looped or unlooped) at the time of transposase action.

9.10 Gene homologs of *dnaK* encoding the heat-shock chaperone HSP70 exist in all three domains of life. All bacteria contain HSP70, but only some species of archaea encode a *dnaK* homolog. The archaeal homologs are closely related to those of bacteria. Knowing this information, how do you suppose *dnaK* genes arose in archaea?

ANSWER: The gene is thought to have been moved by some type of horizontal gene transfer mechanism from domain Bacteria to some members of domain Archaea.

DIAGRAM FOR ANSWER TO THOUGHT QUESTION 9.9

Chapter 10

10.1 If the gene *lacZ* has a nonsense mutation in its open reading frame, will *lacY* be translated?

ANSWER: *LacY* mRNA will still be translated, because each gene in this polycistronic mRNA (*lacZYA*) has its own ribosome-binding site and the *lacY* ribosome-binding site is functional. A transitional stop codon in an upstream gene does not usually affect transcription of the downstream RNA.

10.2 Null mutations completely eliminate the function of a mutated gene. Predict the effect of the following null mutations on the induction of beta-galactosidase, by lactose, and predict whether the *lacZ* gene is expressed at high or low levels in each case: *lacI, lacO, lacP, crp, cya* (the gene encoding adenylate cyclase). What effect will those mutations have on catabolite repression?

ANSWER: Loss of LacI repressor will lead to constitutive expression of *lacZ* but will not affect catabolite repression. A *lacO* mutant will not bind LacI repressor, so the phenotype will mimic that of a *lacI* mutation. A *lacP* mutation will prevent expression of *lacZYA* because RNA polymerase will not bind. Mutations in *crp* or *cya* will prevent catabolite repression, but *lacZYA* induction by lactose will be normal. Without the cAMP-CRP complex, however, expression can never achieve maximal levels.

10.3 Predict what will happen to the expression of *lacZ* under the following partial diploid conditions (see Section 9.2). The genotypes of these strains are presented as follows: chromosomal genes/plasmid gene. (a) *lacI lacO⁺P⁺Z⁺Y⁺A⁺*/plasmid *lacI⁺*; (b) *lacO lacI⁺P⁺Z⁺Y⁺A⁺*/plasmid *lacO⁺*; (c) *crp lacI⁺O⁺P⁺Z⁺Y⁺A⁺*/plasmid *crp⁺*.

ANSWER: (a) The *lacI⁺* gene on the plasmid will produce LacI repressor protein that can diffuse through the cytoplasm, bind chromosomal *lacO*, and repress the *lacZYA* operon. Because the complementing gene and the mutant gene are on different DNA molecules, the gene is said to work in *trans*. (b) Because the *lacO* gene does not produce a diffusible product (for example, protein or RNA), the plasmid *lacO⁺* cannot complement a *lacO* mutation in *trans*, and the strain will not make beta-galactosidase. Thus, the *lacO* gene functions only in *cis*—that is, when it resides next to the gene it regulates. (c) The *crp* gene produces a diffusible protein product, so it can function in *trans* and complement a *crp* mutation. The strain will make beta-galactosidase to the highest level, in the presence of inducer lactose.

10.4 Researchers often use isopropyl-β-D-thiogalactopyranoside (IPTG) rather than lactose to induce the *lacZYA* operon. IPTG structure resembles lactose, which is why it can interact with the LacI repressor, but it is not degraded by beta-galactosidase. Why do you think the use of IPTG is preferred in these studies?

ANSWER: There are at least two reasons. First, the level of IPTG inducer will not change, but the level of lactose inducer will continually decrease as it is consumed. This can affect the kinetics of induction. Second, the act of degrading lactose produces glucose and galactose. Glucose, as the preferred carbon source, will catabolite repress the *lacZYA* operon, once again affecting the kinetics of induction.

10.5 Predict the phenotype of a *spoIIAA* mutant that completely lacks this anti-anti-sigma factor (see **Fig. 10.19**).

ANSWER: The *spoIIAA* mutant will have no way to efficiently remove anti-sigma F factor from sigma F. As a result, sigma F will almost never be active, even in the forespore. The cell can get as far as asymmetrical cell division to make what would be the forespore, but since sigma F and anti-sigma F equally distribute in both compartments, sigma F will not be activated.

10.6 What is the phenotype of a *fljA* mutant? A *fliC* mutant?

ANSWER: A *fljA* mutant lacks the repressor needed to turn off FliC. Thus, a cell will switch back and forth from making H2 flagellin to making both H1 and H2 flagellin. A *fliC* mutant, however, will switch from being motile to nonmotile. In one orientation, the invertible element will allow H2 flagellin to be made; but in the opposite orientation no flagellin will be made, at which point the cell will not be motile.

10.7 Using antibody to flagella, we can tether a cell of *E. coli* to a glass slide via a single flagellum. Looking through a microscope, you will then see the *bacterium* rotate in opposite directions as the flagellar rotor switches from clockwise to counterclockwise rotation and back again. (See the "tethered *E. coli*" video noted at microbiology2.com/links.) Which way will the bacillus rotate when an attractant is added? Now describe what you would expect to observe regarding the interval between clockwise and counterclockwise switching if you tethered the mutants *cheY*, *cheA*, *cheZ*, and *cheR* to the slide and then added attractant.

ANSWER: When an attractant is added to the slide, the rotor will turn counterclockwise for smooth swimming (CheA becomes less active, so there is less CheY-P and there are less frequent tumbles; thus, motor rotation is biased to counterclockwise). But because the flagellum is fixed to the slide, the bacillus will rotate in the *opposite* direction: clockwise. You would expect the following rotating phenotypes from mutants fixed to slides: For mutant *cheY*, the rotor will turn mostly counterclockwise because there is no CheY-P, so there will be longer runs but fixed cells will turn mostly clockwise. Mutant *cheA* will have the same phenotype as *cheY* (no CheY-P, longer runs). For mutant *cheZ*, the rotor will turn mostly clockwise because there is more CheY-P, so there will be frequent tumbles and shorter runs; but the fixed cells will turn counterclockwise. For mutant *cheR*, no methylation of MCPs will reactivate CheA kinase after attractant is added, so there will be less CheY-P; therefore, the rotor will more frequently switch to clockwise, but fixed cells will turn counterclockwise.

10.8 Predict the phenotype of a *glnB* mutant. Will it be a glutamine auxotroph? What about a *glnD* or a *glnE* mutant?

ANSWER: A *glnB* mutant will not dephosphorylate NtrB or remove AMP from GlnA. The cell will continue to make GlnA and fail to inactivate it. Thus, the cell will overproduce glutamine. A *glnD* mutant, however, will not add UMP to GlnB, so GlnB will not direct GlnE to remove AMP from GlnA (remember, GlnA without AMP is active). As a result, GlnA will remain inactive and the cell will likely require glutamine. The same will occur with a *glnE* mutant.

10.9 Genes encoding luciferase can be used as "reporters." What do you think would happen if the promoter for an SOS response gene were fused to the luciferase open reading frame?

ANSWER: It could serve as a real-time biosensor for an environmental stress response. You could fuse the luciferase gene to the *recA* promoter and examine a real-time increase in fluorescence after ultraviolet irradiation by inserting the whole culture into a spectrofluorometer.

10.10   What do you think would happen if a culture were coinoculated with a *Vibrio fischeri luxI* mutant and a *luxA* mutant, neither of which produces light?

ANSWER: The *luxI* mutant would glow. The *luxA* mutant would still make autoinducer as it grew. This autoinducer would accumulate in the culture medium, and then diffuse and enter the *luxI* mutant cells, where it would trigger induction of the *lux* operon and production of luciferase.

10.11   Why might the mass of a protein excised from a gel fail to match any of the ORF molecular weights predicted by the genome sequence?

ANSWER: The protein might be altered by phosphorylation, acetylation, or removal of a leader sequence; or it might be otherwise proteolytically processed. Identification might still be made if certain unmodified fragments of the protein match up with predicted fragments.

Chapter 11

11.1   Why would phage T4 production require a tenfold higher rate of DNA replication than does the host cell? Why would the phage substitute all of its cytosine with an unusual base that requires greater energy to synthesize?

ANSWER: The phage T4 genome needs to replicate itself and generate progeny as fast as possible, in order to use as much of the host cell's resources as possible before the cell deteriorates. By replacing cytosine with a modified base, the phage can avoid degradation of its chromosome by its own endonucleases (which cleave host DNA) or by those of the host (which are made to cleave phage DNA).

11.2   The phage T4 chromosome is a linear piece of DNA, yet the genomic map of T4 (**Fig. 11.3**) is circular. Why?

ANSWER: Each phage head contains just enough space to package slightly more than the length of a genome. The head-filling mechanism of DNA packaging means that successive genomes on a concatemer each get cut at different positions within the genome sequence. Thus, any map of gene order based on recombination linkages between genes, or based on the sequencing of a population of genome molecules, will generate a circular map.

11.3   Which numbered genes from **Figure 11.8** might be mutated in each of the defective phage populations presented in **Figure 11.9D**?

ANSWER: Mutant 1 may have a defect in a gene whose product is needed for tail fiber attachment, such as gene 63. Mutant 2 may be defective for tail fiber assembly (genes 34 or 57). Mutant 3 could be defective for sheath assembly around the internal tube (gene 18).

11.4   Why does influenza virus have to provide its polymerase ready-made, whereas poliovirus does not?

ANSWER: The influenza viral chromosome is (−) strand RNA. When it enters the host cytoplasm, it cannot be translated by ribosomes, and no host polymerase makes RNA from RNA. Therefore, the virion must provide an RNA-dependent RNA polymerase to generate a (+) strand RNA for translation to proteins. Poliovirus, however, injects (+) strand RNA, which can immediately be translated by host ribosomes to generate RNA-dependent RNA polymerase.

11.5   Explain why the expression $(8!/8^8)$ approximates the proportion of infective particles of influenza virus. What assumption might be changed to make the proportion greater or less?

ANSWER: An infective particle can start by packaging any one of the eight chromosomes. Once the first is chosen, seven possibilities remain for the second, six for the third, and so forth. Thus there are 8! ways to obtain a perfect set out of 8^8 total ways to pick eight segments at random. This expression assumes, however, that exactly eight segments are packaged in every virion. If the average number of segments packaged is less than eight, the proportion of infectives falls off. If the number of segments is usually more than eight, the proportion of virions containing at least one of each segment increases.

11.6   How do attachment and entry of HIV resemble attachment and entry of influenza virus? How do attachment and entry differ between these two viruses?

ANSWER: Attachment of HIV requires the envelope spike proteins to bind receptors in the host plasma membrane, just as the influenza envelope protein hemagglutinin binds the sialic acid protein in the host membrane. However, the entry processes differ between the two viruses in that influenza virus induces formation of an endocytic vesicle, whose acidification triggers membrane fusion and release of the core contents into the host cytoplasm. HIV virions, however, do not induce endocytosis and do not require acidification to induce membrane fusion and release of the core into the cytoplasm. In both cases, receptor binding signals major rearrangement of a viral envelope protein so as to insert a fusion peptide into the host membrane: for HIV, the plasma membrane; for influenza virus, the endocytic membrane.

11.7   Compare and contrast the priming of chromosome replication in HIV with the priming mechanisms of poliovirus and influenza virus.

ANSWER: Poliovirus uses a viral protein (VPg) to provide the 3′ OH needed to prime synthesis of RNA. For influenza virus, replication of the RNA genome segments is primed by a viral nucleocapsid protein (NP). Viral mRNA is primed from a capped host mRNA segment, which was obtained by endonuclease clipping of the host mRNA, a process called "cap snatching." Instead of host mRNA, HIV uses host tRNA for primers.

11.8   Compare and contrast the fate of the HSV chromosome with that of the HIV chromosome.

ANSWER: HSV-1 contains a DNA chromosome, which is transported to the nuclear membrane within an intact capsid. HIV contains twin RNA chromosomes, which are released in the cytoplasm upon dissolution of the capsid. The RNA chromosomes of HIV are copied to double-stranded DNA for transport into the nucleus. In HSV-1, the DNA chromosome circularizes and generates concatemeric duplicates by rolling-circle replication. In HIV, the replicated DNA circularizes but immediately integrates into the host chromosome. In both cases, the viral DNA can persist for decades in latent infection.

Chapter 12

12.1   In this postgenomic era, how might you generally design a strategy to construct a strain of *E. coli* that cannot synthesize histidine?

ANSWER: The genomic sequence of *E. coli* is known. We also know which genes encode the steps of histidine biosynthesis. The strategy today would be to engineer a plasmid containing DNA sequences (about 200 bp) that flank the histidine biosynthesis gene you want to delete, but in place of the target gene, you would instead insert an antibiotic resistance gene (for example, kanamycin resistance). The plasmid used would be a "suicide" plasmid that cannot replicate in the strain you wish to mutate but can replicate in other *permissive* strains (strains that produce a specific protein the plasmid needs to replicate). The permissive strain allows you to make a lot of the plasmid. The recombinant plasmid is then moved into the nonpermissive target strain by transformation, and the transformation mixture is plated on nutrient agar medium containing the antibiotic. Since the plasmid cannot replicate in this strain, the only way the target strain can become resistant to the antibiotic is if the chromosomal sequences flanking the drug marker in the plasmid undergo homologous recombination with the same DNA sequences on the actual chromosome. This double recombination event "surgically" removes the target gene from the chromosome and replaces it with the antibiotic resistance marker. In our example, the newly constructed mutant can grow in the presence of antibiotic, but only if histidine is present.

12.2   You have monitored expression of a gene fusion where *lacZ* is fused to the glutamate decarboxylase gene,

gadA. You find a 50-fold increase in expression of *gadA* (based on beta-galactosidase activity) when the cells carrying this fusion are grown in media at pH 5.5 as compared to pH 8. What additional experiment must you do to tell if the control is transcriptional or translational?

ANSWER: You must compare these results with those obtained from an operon fusion to the same gene. If both fusions show an equal 50-fold increase, then the gene is controlled at the transcriptional level. If the gene fusion shows a 50-fold increase in induction but the transcriptional fusion shows no induction, then regulation occurs posttranscriptionally, most likely at the translational level.

12.3   What would you conclude if the Northern blot of the *gad* genes showed no difference in mRNA levels in cells grown at pH 7.7 and pH 5.5 but the Western blot showed more protein at pH 5.5 than at pH 7.7?

ANSWER: Regulatory control would likely be at the posttranscriptional level, through increasing either mRNA stability, translational efficiency, or protein stability.

12.4   Another regulator of glutamate decarboxylase (Gad) production does not affect the production of *gad* mRNA but is required to accumulate Gad protein. What two regulatory mechanisms might account for this phenotype?

ANSWER: Control of translation or control of protein degradation.

12.5   How could you use two-hybrid analysis to determine whether *three* proteins can interact? You suspect that proteins A and C can simultaneously bind to protein B and that this interaction then allows A to interact with C in the complex.

ANSWER: Place three plasmids in the cell. One makes protein A fused to the *GAL4* DNA-binding domain, the second plasmid makes protein C fused to the *GAL4* activator domain, and the third plasmid simply makes protein B. If the yeast cell contains only the first two plasmids, no interaction will occur and *lacZ* will not be activated. If all three plasmids are in the same yeast cell, protein B will be the center of a sandwich, linking the A and C fusion proteins. The fusion proteins can then interact and activate *lacZ*.

12.6   How would you have to modify real-time PCR from standard PCR to quantify the level of a specific mRNA?

ANSWER: You would have to first convert the mRNA into cDNA using the enzyme reverse transcriptase. An oligonucleotide primer hybridizing to the 3′ end of the mRNA is acted on by reverse transcriptase to synthesize DNA (called complementary DNA, or cDNA). Normal real-time PCR techniques can then quantify the cDNA.

12.7   Would insect resistance to an insecticidal protein be a concern when developing a transgenic plant? Why or why not? How would you design a transgenic plant to limit the possibility of insects developing resistance?

ANSWER: Insects have, in fact, developed resistance to single insecticidal proteins. The most common resistance mechanism involves a change in the membrane receptors in the midgut to which activated *Bt* toxins bind. Resistance can be due to a reduced number of *Bt* toxin receptors or to a reduced affinity of the receptor for the toxin. Some insects, such as the spruce budworm, can inactivate specific toxins by precipitating them with a protein complex present in the midgut.

While developing a transgenic plant, several steps could be taken to limit the development of resistance in an insect population. The insecticidal gene could be fused to a promoter that is expressed only when the plant is most susceptible to attack. Alternatively, the gene could be fused to a promoter that is expressed only in a tissue of the plant that is most vulnerable to attack. This would limit the time during which the insects can develop resistance. For instance, cotton plants attacked by bollworms could produce toxin only in young boll tissues, the most important part of the plant. In addition to specifically protecting the critical plant tissue, this strategy would affect only one generation of bollworms, avoiding the constant selection pressure that hastens evolution of resistance. Another technique would be to engineer two different insecticidal proteins into the plant genome that will not exhibit cross-resistance. In other words, even if an insect develops resistance to one toxin, it will still remain susceptible to the second.

12.8 You have just employed phage display technology to select for a protein that tightly binds and blocks the eukaryotic cell receptor targeted by anthrax toxin. When this receptor is blocked, the toxin cannot get into the target cell. You suspect that the protein may be a useful treatment for anthrax. How would you recover the gene following phage display and then express and purify the protein product?

ANSWER: The gene can be cut out of the phage genome using restriction endonucleases or amplified by PCR using known phage DNA sequences that flank the gene. The fragment can then be cloned into a His_6 tag expression vector. This requires that the sequence of the gene be known so that the DNA encoding the His_6 tag can be placed in-frame with the open reading frame of the receptor-blocking protein. The plasmid containing the His_6-tagged protein gene is then induced to overexpress the protein in *E. coli*. The vector will contain an inducible promoter (for example, *lacP*) to drive expression of the gene. The His_6-tagged protein can then be purified by pouring cell extracts over a nickel column, as described in the text.

Chapter 13

13.1 Why can't photosynthesis be driven by microwaves or by X-rays?

ANSWER: The energy of microwaves is a hundred times lower than that of visible light and near-infrared. Microwaves have too little energy to form or break chemical bonds. On the other hand, photons of higher energy, such as X-rays, break molecular bonds indiscriminately, in a manner that enzymes cannot control.

13.2 Methanogens are a small proportion of the microbial community in soil, and the ΔG of methanogenesis is small. Yet methanogens produce large volumes of methane, large enough to contribute significantly as a greenhouse gas to global warming. Why would methanogens produce a relatively large quantity of waste product?

ANSWER: For methanogenesis, the ΔG per reaction cycle is small. Thus, the organism must run numerous cycles in order to store sufficient energy to build biomass.

13.3 The soil bacterium *Geobacter* can metabolize acetic acid by the following reaction:

$$CH_3COOH + 2H_2O \rightarrow 2CO_2 + 4H_2$$

Yet the $\Delta G°'$ is +95 kJ/mol. What could happen in the soil enabling *Geobacter* to grow?

ANSWER: *Geobacter* must grow in the presence of different species of bacteria that oxidize the H_2 produced by *Geobacter*. Removal of H_2 drives forward the acetate catabolism. The sum of the acetate reaction plus the hydrogen oxidation yields a negative $\Delta G°'$. This joint metabolism of two species is called syntrophy. The rate of reaction in this syntrophy is further enhanced at high temperature, which magnifies the term $-\Delta RT \ln k$.

13.4 Linking an amino acid to its cognate tRNA is driven by ATP hydrolysis to AMP (adenosine monophosphate) plus pyrophosphate. Why release PP_i instead of P_i?

ANSWER: The formation of aminoacyl-tRNA must be irreversible until the ribosome is ready to release the tRNA. The pyrophosphate from ATP is immediately cleaved to $2P_i$, preventing the reversal of aminoacyl-tRNA formation.

13.5 In the microbial community of the bovine rumen, the actual ΔG value has been calculated for glucose fermentation to acetate:

$$C_6H_{12}O_6 + 2H_2O \rightarrow 2C_2H_3O_2^- + 2H^+ + 4H_2 + 2CO_2$$
$$\Delta G = -318 \text{ kJ/mol}$$

If the actual ΔG for ATP formation is 44 kJ/mol and each glucose fermentation yields four molecules of ATP, what is the thermodynamic efficiency of energy gain? Where does the lost energy go?

ANSWER: The energy efficiency is $(4 \times 44 \text{ kJ/mol})/(318 \text{ kJ/mol}) \times 100 = 55\%$. The remaining energy is dissipated as heat.

13.6   What would happen to the cell if pyruvate kinase catalyzed PEP conversion to pyruvate but failed to couple this reaction to ATP production?

ANSWER: If PEP conversion to pyruvate occurred without ATP production, then the energy liberated by this reaction would be lost as heat. No energy would be channeled into cell function and growth.

13.7   In the EMP pathway, how are the two water molecules generated?

ANSWER: Each water molecule is removed from 2-phosphoglycerate, converting $R-CH_2OH$ to $R=CH_2$. The water molecule accounts for the water equivalent from incorporation of P_i during oxidation by NAD^+. The P_i ultimately is used to convert one ADP to ATP—a reaction that includes water formation.

13.8   Some bacteria make an enzyme, dihydroxyacetone kinase, that phosphorylates dihydroxyacetone to dihydroxyacetone phosphate. Why would this enzyme be useful?

ANSWER: Bacteria can obtain dihydroxyacetone from their environment using a transporter protein, and then phosphorylate the substrate and direct it into glycolysis. Some bacteria can grow on dihydroxyacetone as a sole carbon source.

13.9   Explain why the ED pathway generates only one ATP, whereas the EMP pathway generates two.

ANSWER: The EMP pathway primes the six-carbon sugar with two phosphoryl groups. The sugar then splits into two three-carbon units (glyceraldehyde 3-phosphate), each of which generates two ATPs for one of the original ATPs. By contrast, the ED pathway phosphorylates the sugar only once before it splits in two. The phosphorylated end yields glyceraldehyde 3-phosphate, which enters the EMP pathway to generate ATP, ending up as pyruvate. The unphosphorylated three-carbon unit yields pyruvate directly, with no ATP.

13.10   Compare the reactions catalyzed by pyruvate dehydrogenase and pyruvate formate lyase (see Section 13.5). What conditions favor each reaction, and why?

ANSWER: The pyruvate dehydrogenase complex (PDC) is favored in the presence of oxygen because the electrons transferred to NADH can enter the electron transport chain, eventually combining with oxygen to release energy. In the absence of oxygen, pyruvate formate lyase is favored to yield fermentation products that can be excreted from the cell without reducing more energy carriers. At high pH, formate and acetate production is especially favorable because the extra acid counteracts alkalinity.

13.11   Suppose a cell is pulse-labeled with ^{14}C-acetate (fed the ^{14}C label briefly, then "chased" with unlabeled acetate). Can you predict what will happen to the level of

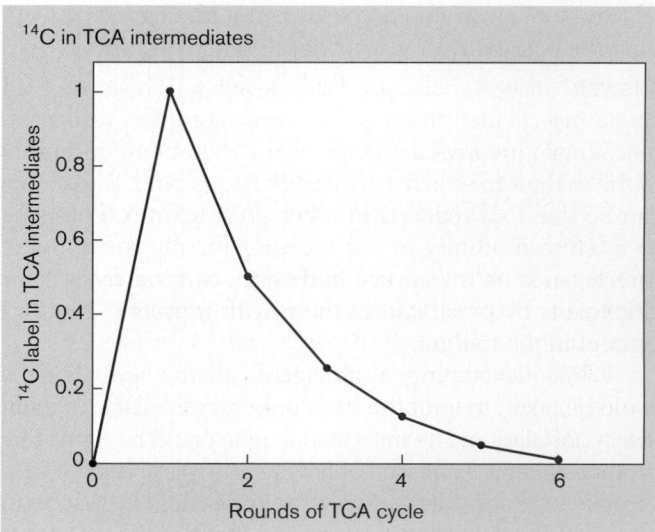

^{14}C in TCA intermediates

radioactivity observed in isolated TCA intermediates? Plot a curve showing your predicted level of radioactivity as a function of the number of rounds of the cycle.

ANSWER: The amount of radioactivity measured in TCA intermediates will rise steeply as labeled acetate is incorporated, and then will decrease by half with each succeeding cycle, as the order of the carbons is randomized by succinate.

13.12   How might the enzymes that catalyze acenaphthene catabolism be discovered?

ANSWER: A genomic library of *Pseudomonas* strain A2279 might be used to transform another species that lacks the ability to degrade acenaphthene. If a transformant capable of degradation is found, the genes can be mapped and the deduced amino acid sequences compared with known enzymes to provide clues as to their mode of action. Other genes in the same operons might be found to catalyze other steps in the catabolic pathways.

Chapter 14

14.1   *Pseudomonas aeruginosa*, a cause of pneumonia in cystic fibrosis patients, oxidizes NADH with nitrate (NO_3^-) at neutral pH. What is the value of $E°'$?

ANSWER: To calculate the reduction potential $E°'$:

$$NADH + H^+ + NO_3^- \rightarrow NAD^+ + NO_2^- + H_2O$$

$$E°' = 320 \text{ mV} + 420 \text{ mV} = 740 \text{ mV}$$

14.2   Could a bacterium obtain energy from succinate as an electron donor with nitrate (NO_3^-) as an electron acceptor?

ANSWER: For nitrate reduction:

$$\text{Succinate} + NO_3^- \rightarrow \text{fumarate} + NO_2^- + H_2O$$

$$E°' = -33 \text{ mV} + 420 \text{ mV} = 387 \text{ mV}$$

Although succinate is a relatively poor electron donor, nitrate is a strong electron acceptor. This reaction should provide energy for bacterial metabolism.

14.3 What do you think happens to $\Delta\psi$ as the cell's external pH increases or decreases? What could happen to Δp of bacteria that are swallowed and enter the extremely acidic stomach?

ANSWER: As external pH changes, ΔpH increases or decreases, affecting the magnitude of Δp. As enteric bacteria enter the stomach, they encounter pH values (pH 1.5–3.0) below their growth range (pH 5.0–9.0). At first, the cell's transmembrane ΔpH may be very large, as the cell tries to maintain its cytoplasmic pH above 5.0, the limit for viability. But since the cell no longer grows, it loses its energy supply and can no longer spend energy to maintain ΔpH. One way to maintain cytoplasmic pH homeostasis is to reverse the electrical potential $\Delta\psi$ (inside positive) so as to drive out some H^+ and maintain a small ΔpH. Oppositely directed ΔpH and $\Delta\psi$ enable the cell to keep cytoplasmic pH high enough to survive when external pH is extremely low. On the other hand, when the external pH is raised by pancreatic secretions, the cell needs to compensate by inverting its ΔpH, and maintaining a relatively large $\Delta\psi$.

14.4 Suppose that de-energized cells ($\Delta p = 0$) with an internal pH 7.6 are placed in a solution at pH 6. What do you predict will happen to the cell's flagella? What does this demonstrate about the function of Δp?

ANSWER: The flagella will rotate, driven solely by the ΔpH component of Δp. This result is consistent with the hypothesis that the transmembrane proton potential Δp drives flagellar rotation.

14.5 What is the advantage of the oxidoreductase transferring electrons to a pool of mobile quinones, which then reduce the terminal reductase (cytochrome complex)? Why does each oxidoreductase not interact directly with a cytochrome complex?

ANSWER: The mobile quinone pool connects diverse electron donors with diverse electron acceptors. If each oxidoreductase had to interact specifically with a different terminal oxidase, the pathways of electron transport would be limited; for example, NADH might only donate electrons to O_2, whereas succinate might only donate electrons to nitrate. Instead, all potential electron donors can be coupled with all potential acceptors.

14.6 Why are most electron transport proteins fixed within the cell membrane? What would happen if they "got loose" in aqueous solution?

ANSWER: If the electron transport proteins came away from the membrane into water solution, they could carry their energized electrons back into the cytoplasm or lose them

outside the cell. In either case, they could no longer convert the electron energy to a proton gradient.

14.7 Why would bacteria evolve membrane-embedded dehydrogenases in their ETS that fail to pump protons even when sufficient donor potential exists?

ANSWER: Dehydrogenases that fail to transport protons despite sufficient reduction potential are needed for cases in which reoxidation of the donor is required but the proton gradient is already high enough to drive generation of ATP. Too high a transmembrane voltage can disrupt the membrane.

14.8 Experiments suggest that NADH dehydrogenase 2 is more often coupled with cytochrome *bo*, and dehydrogenase 1 more often with cytochrome *bd*. Why might this be the case?

ANSWER: The non-proton-pumping oxidoreductase NDH-2 may be used to moderate the amount of proton gradient generated with abundant oxygen. The proton-pumping NDH-1 may be useful under low-oxygen conditions, where the cell is forced to use the high-affinity oxidase (cytochrome *bd*), which fails to pump the extra $2H^+$.

14.9 Would *E. coli* be able to grow in the presence of an uncoupler that eliminates the proton potential supporting ATP synthesis?

ANSWER: Yes, *E. coli* can grow with the proton gradient eliminated, but only with a rich supply of nutrients for substrate phosphorylation to generate ATP (for example, from glycolysis). In addition, the external pH and salt levels must be maintained close to those of the cytoplasm, to minimize the need for ion transport.

14.10 The proposed scheme for uranium removal requires injection of acetate under highly anoxic conditions, with less than 1 part per million (ppm) dissolved oxygen. Why must the acetate be anoxic?

ANSWER: Oxygen is the strongest terminal electron acceptor. If O_2 is present, bacteria will use it preferentially (instead of U^{6+}) to oxidize the acetate to CO_2.

14.11 Hydrogen gas is so light that it rapidly escapes from Earth. Where does all the hydrogen come from to be used for hydrogenotrophy and methanogenesis?

ANSWER: Hydrogen is produced in substantial quantities as a by-product of fermentation. It may seem surprising that organisms would readily excrete quantities of energy-rich H_2, but in the absence of a good electron acceptor (or the enzymes to utilize electron acceptors), hydrogen may be just another waste product. Hydrogen gas trapped underground supports large communities of methanogens and hydrogenotrophs. The human colonic bacteria generate so much hydrogen that all parts of the body show traces of hydrogen gas.

14.12   Suppose you discover bacteria that require a high concentration of Fe^{2+} for photosynthesis. Can you hypothesize what the role of Fe^{2+} may be? How would you test your hypothesis?

ANSWER: The organism uses reduced iron as an electron donor for its photosystem ($Fe^{2+} \rightarrow Fe^{3+}$). To test this hypothesis, grow the organism on a defined concentration of Fe^{2+}. Measure the amount of iron oxidized and the amount of carbon fixed into biomass; if the Fe^{2+} is an electron donor for the photosystem, the two numbers should show a linear correlation.

Chapter 15

15.1   Propose a simple experiment to reveal the key intermediate to receive CO_2.

ANSWER: Add a very short pulse of $[^{14}C]$ CO_2 to the cells. The pulse must contain a level of CO_2 sufficient to begin fixation but insufficient to continue beyond one "turn" of the cycle. In this case, the radiolabel will accumulate in the intermediate that needs to assimilate CO_2 for the next round. This kind of experiment ultimately identified the key intermediate ribulose 1,5-bisphosphate.

15.2   Speculate on why Rubisco catalyzes a competing reaction with oxygen. Why might researchers be unsuccessful in attempting to engineer Rubisco without this reaction?

ANSWER: The oxygenation reaction might have an essential function in regulation of metabolism. For example, it might help prevent excessive reduction of cell components or fixation of too much carbon to be used in biosynthesis. Given the universal existence of the oxygenation reaction in bacterial and chloroplast Rubiscos, it seems unlikely that oxygenation serves no purpose. For this reason, the attempt to engineer Rubisco without oxygenation may not succeed.

15.3   Why does ribulose 1,5-bisphosphate have to contain two phosphoryl groups, whereas the other intermediates of the Calvin cycle contain only one?

ANSWER: Only ribulose 1,5-bisphosphate needs to split into two molecules (3-phosphoglycerate). Each of the two products needs to have its own phosphate as a tag for the enzymes to recognize it within the cycle.

15.4   Which catabolic pathway (see Chapter 13) includes some of the same sugar-phosphate intermediates as those of the Calvin cycle? What might these intermediates in common suggest about the evolution of the two pathways?

ANSWER: The pentose phosphate pathway includes ribulose 5-phosphate, erythrose 4-phosphate, and sedoheptulose 7-phosphate in a similar series of carbon exchanges. Perhaps the pentose phosphate pathway and the Calvin cycle pathway evolved from a common amphibolic pathway of sugar consumption and biosynthesis. Alternatively, the one pathway evolved earlier, and then the sugar intermediates were available for evolution of the second pathway.

15.5   For a given species, uniform thickness of a cell membrane requires uniform chain length of its fatty acids. How do you think chain length may be regulated?

ANSWER: In *E. coli,* the chain length of a growing fatty acid appears to be limited by beta-ketoacyl-ACP synthase, which binds only precursor acyl-ACPs shorter than 18 carbons. Thus, only carbon chains of up to 18 carbons are synthesized.

15.6   Suggest two reasons why transamination is advantageous to cells.

ANSWER: Ammonia is toxic to cells. Transamination enables cells to store amine groups in nontoxic form, readily available for biosynthesis. The availability of multiple enzymes of transamination from different amino acids enables cells to quickly recycle existing resources into the amino acids most needed by the cell in a given environment. For example, if a sudden supply of glutamine appears, cells can immediately distribute its amines into all 20 amino acids.

15.7   Which energy carriers (and how many) are needed to make arginine from 2-oxoglutarate? (See **Fig. 15.25**.)

ANSWER: Arginine biosynthesis requires three ATP molecules and three NADPH molecules (including two for converting two molecules of 2-oxoglutarate to glutamate). An additional ATP is spent converting acetate to acetyl-CoA.

15.8   Why are purines synthesized onto a sugar phosphate base?

ANSWER: Purines are highly hydrophobic, insoluble in the cytoplasm. The ribose phosphate component solubilizes the molecule, enabling synthesis to occur in the cytoplasm, where needed to make RNA and DNA.

15.9   Why are the ribosyl nucleotides synthesized first, and then converted to deoxyribonucleotides as necessary? What does this order suggest about the evolution of nucleic acids?

ANSWER: Ribonucleic acid is believed to be the original chromosomal material of cells. Cells evolved to synthesize RNA first; then later, as DNA was used, pathways evolved to synthesize it by modification of RNA, which the cell already had the ability to make.

Chapter 16

16.1   Why do the lipid components of food experience relatively little breakdown during fermentation?

ANSWER: Lipids are highly reduced molecules, largely hydrocarbon with relatively low oxidizing potential. Thus, lipids cannot undergo as many intramolecular redox reactions as do sugars, which readily generate energy through anaerobic fermentation.

16.2 Why does oxygen allow excessive breakdown of food, compared with anaerobic processes?

ANSWER: Oxygen functions as the terminal electron acceptor for the complete breakdown of all kinds of organic molecules to water and CO_2.

16.3 In an outbreak of listeriosis from unpasteurized cheese, only the refrigerated cheeses were found to cause disease. Why would this be the case?

ANSWER: In the cheeses kept at room temperature, other naturally occurring bacteria outgrew the pathogenic *Listeria*, whereas in the refrigerator only the *Listeria* could grow. (Note, however, that many other potential pathogens, such as *Salmonella*, are inhibited by refrigeration.)

16.4 Cow's milk contains 4% lipid (butterfat). What happens to the lipid during cheese production?

ANSWER: Lipids undergo little catabolism, because the fermentation conditions are anaerobic. During coagulation, lipid droplets become trapped in the network of denatured protein and are largely retained in the bulk of the cheese. "Low-fat" cheeses are made from skim milk, which eliminates the lipids before fermentation.

16.5 In traditional fermented foods, without pure starter cultures, what determines the kind of fermentation that occurs?

ANSWER: The fermentation can be controlled by introducing a crude starter culture obtained from a previous batch of the food product or from a natural source of a particular microbe; for example, rice straw is a source of *Bacillus natto* for natto production. The fermentation type can be manipulated by the addition of factors, such as brine, that retard growth of all but a few strains. In pidan, for example, the high concentration of sodium hydroxide limits bacterial growth to alkali-tolerant strains of *Bacillus*.

16.6 Compare and contrast the role of fermenting organisms in the production of cheese and bread.

ANSWER: In cheese production, fermentation causes major biochemical changes in the food, such as the buildup of acids and the breakdown of proteins to smaller peptides and amino acids. Minor by-products, such as methanethiols and esters, accumulate to levels that confer flavors. In yeast bread, by contrast, the only significant product of fermentation is the carbon dioxide that leavens the dough. The small amount of ethanol produced evaporates during cooking. A form of bread in which extended fermentation does generate flavor is injera, for which the dough ferments for 3 days.

16.7 Compare and contrast the role of low-concentration by-products in the production of cheese and beer.

ANSWER: In both cheese and beer, minor by-products such as esters contribute flavor. Oxidation of esters can lead to off-flavors. In cheese, however, the exclusion of oxygen usually prevents off-flavors. In beer, the yeast requires a low level of oxygen; thus, significant amounts of acetaldehyde and diacetyl are produced and must be eliminated by a secondary fermentation.

16.8 Why would bacteria convert trimethylamine oxide (TMAO) to trimethylamine? Would this kind of spoilage be prevented by exclusion of oxygen?

ANSWER: TMAO acts as a terminal electron acceptor—that is, an alternative to oxygen for anaerobic respiration, as discussed in Chapter 14. Exclusion of oxygen inhibits only aerobic bacteria; TMAO respirers continue to grow and can spoil the fish.

16.9 Is it possible for physical or chemical preservation methods to completely eliminate microbes from food? Explain.

ANSWER: Preservation methods either slow microbial growth or induce microbial death. Microbial death follows a negative exponential curve, as discussed in Chapter 5. In theory, the exponential curve never reaches zero, so total exclusion of microbes is impossible. In practice, there is a high probability of totally eliminating microbes if the treatment time extends several "half-lives" beyond the time at which microbial concentration declines to less than one per total volume.

16.10 Why would different industrial strains or species be used to express different kinds of cloned products?

ANSWER: Different industrial strains have biochemical systems that favor different products. Some fungi naturally possess the highly complex pathways to generate antibiotics, as well as regulatory timing to turn on these pathways after the culture has grown to high population density. On the other hand, bacteria such as *Bacillus subtilis* are the most genetically tractable and predictable in their growth cycles, and the easiest to manipulate to express recombinant products such as human genes.

16.11 Why would an herbicide resistance gene be desirable in an agricultural plant? What long-term problems might be caused by microbial transfer of herbicide resistance genes into plant genomes?

ANSWER: Introduction of an herbicide resistance gene allows application of higher amounts of herbicide to crops in order to control growth of weeds. But the higher concentrations of herbicide may also have greater side effects on animals and on human consumers of the crop. In the long run, the herbicide resistance gene is likely to escape into weed plants through natural genetic transfer mechanisms. Thus, eventually the weeds may require still higher concentrations of herbicide. While the costs versus benefits of new gene modifications remain poorly understood, it must be recognized that all modern crops today are the product of many generations of genetic manipulation.

Chapter 17

17.1   What would have happened to life on Earth if the Sun were a different stellar class, substantially hotter or colder than it is?

ANSWER: If the Sun were hotter, too much ultraviolet and gamma radiation would reach the Earth, breaking chemical bonds of living organisms so rapidly that life could not be sustained. If the Sun were colder, too little radiation with sufficient energy would be available to drive photosynthesis. In either case, life as we know it could not have evolved on Earth.

17.2   Outline the strengths and limitations of each model of the origin of living cells. Which aspects of living cells does each model explain?

ANSWER: The three models are complementary in that each explains aspects of modern cells not addressed by the others. The prebiotic soup model accounts for the major classes of compounds used by cells, such as nucleosides, TCA cycle intermediates, amino acids, and fatty acids. It also suggests the origin of membranes as soap bubble–like micelles. It does not, however, account for the evolution of metabolic pathways and replication of genetic information. The metabolist model accounts for the prevalence of major cellular reactions such as carbon fixation and the TCA cycle. It does not explain the evolution of membranes and genetic material. The RNA world accounts for the central role of RNA in living cells; of all molecular classes, RNA and ribonucleotides probably serve the widest range of functions as information carriers, agents of catalysis, and genetic regulators. The origin of membranes is not addressed.

17.3   Suppose living organisms were found on Mars. How might such a find shed light on the origin and evolution of life on Earth?

ANSWER: If life on Mars showed a completely different basis than that of Earth—for example, it was based on silicon polymers instead of carbon—such a find would support the view that life originated independently on each planet, rather than traveling from one planet to the other; or that both planets were seeded from somewhere else. If life on Mars were based on similar macromolecules, perhaps even showing the same genetic code, this would support the view that life arose on Mars first; or that both planets were seeded from the same source.

17.4   What kind of DNA sequence changes have no effect on gene function?

ANSWER: Base substitutions that do not change the amino acid specified by the codon have no effect on gene function. For example, CUA → CUG still encodes leucine. In addition, a majority of the amino acids in any given protein can be replaced by an amino acid of similar form (for example, leucine → valine) without significantly affecting function of the gene.

17.5   What are the major sources of error in constructing phylogenetic trees?

ANSWER: Phylogenetic trees are affected by variability in the number of substitutions, or rate of mutation in different strains. The tree is distorted by errors in sequence alignment, and by systematic errors due to failure of the fundamental assumptions of the molecular clock. These assumptions include the constant rate of mutation for all branches, constant generation time, and true orthology of the gene chosen (that is, the encoded product has the same function and hence the same degree of selection pressure in all taxa under consideration).

17.6   What are the limits of evidence for horizontal gene transfer in ancestral genomes? What alternative interpretation might be offered?

ANSWER: Horizontal gene transfer is inferred from the appearance of genes in clade A that are absent from other members of the clade but present in clade B. The degree of similarity between genes in the two clades, however, must be high enough to exclude the possibility that the genes in question were retained from a common ancestor of the two clades but lost from other members of clade A. This possibility is difficult to exclude in the case of deep-branching clades, where all genes have had a long time to diverge. For example, the large number of archaeal genes present in deep-branching thermophilic bacteria such as *Thermotoga* may include some inherited from the last common ancestor.

17.7   Based on **Figure 17.26,** what would be the identification of a straight, nonsheathed, Gram-negative bacterium that has sulfur granules, is motile, and is 1.0 μm wide? What would happen if you assigned the bacterium a width of 0.9 μm?

ANSWER: The bacterium would key out as *Beggiatoa* sp. If the cell width were measured as 0.9 μm, however, the identification would proceed down a completely different, wrong track, at the first step of the key. This is one disadvantage of the dichotomous key.

17.8   Besides mitochondria and chloroplasts, what other kinds of entities within cells might have evolved from endosymbionts?

ANSWER: Some of the large "megaplasmids" found in bacteria and protists are as large as genomic chromosomes and contain numerous housekeeping genes. These megaplasmids may have originated as endosymbiotic cells that lost all their membranes through reductive evolution. Similarly, some of the giant viruses, such as mimivirus and smallpox virus, as well as phages such as T4, possess a wide spectrum of housekeeping genes. These viruses may have originated as cellular parasites that underwent reductive evolution.

Chapter 18

18.1 Which taxonomic groups in **Table 18.1** stain Gram-positive and which Gram-negative? Which group contains both Gram-positive and Gram-negative species? For which groups is the Gram stain undefined, and why?

ANSWER: Most Firmicutes and Actinobacteria stain Gram-positive. These bacteria have relatively thick cell walls that retain the stain. The Proteobacteria, Nitrospirae, and Bacteroidetes/Chlorobi groups stain Gram-negative. The Cyanobacteria have an outer membrane and are considered Gram-negative, although their cell walls are very thick. The Deinococcus-Thermus group includes both Gram-positive and Gram-negative staining members. For Chlamydiae and Planctomycetes, the Gram stain is irrelevant because they lack the cell wall that retains the stain. For Spirochetes, many species are too narrow to observe the stain under light microscopy.

18.2 Which groups of species share common structure and physiology within the group? Which groups show extreme structural and physiological diversity?

ANSWER: Cyanobacteria all share oxygenic photosynthesis through thylakoid membranes. Their overall cell structure and organization, however, take diverse forms. Spirochetes all share the common structure of sheathed flexible spiral with internal flagella; most share anaerobic or facultative heterotrophy. Chlamydiae and Planctomycetes groups of species each share general structural features. Other groups, particularly Firmicutes and Actinobacteria, show considerable diversity of form and physiology. The Proteobacteria display more extreme diversity of metabolism than any other division.

18.3 What taxonomic questions are raised by the apparent high rate of gene transfer between archaea and thermophilic bacteria?

ANSWER: If gene transfer results in a species containing a quarter of its genes from organisms outside its domain, such a mosaic genome raises questions of how to define the species and the domain. How can a species be defined if its genome contains large portions from distantly related sources? Other interesting questions relate to the means of gene transfer. How do such distantly related organisms as bacteria and archaea maintain a compatible mechanism of gene transfer?

18.4 What are the relative advantages and disadvantages of propagation by hormogonia, as compared with akinetes?

ANSWER: Hormogonia are motile, and thus capable of active chemotaxis toward a more favorable environment. On the other hand, hormogonia have active metabolism that requires nutrition; if the environment lacks nutrients, the hormogonia will die. Akinete cells can persist until environmental conditions improve, but they cannot actively seek out a new location.

18.5 What are the relative advantages and disadvantages of the different strategies for maintaining separation of nitrogen fixation and photosynthesis?

ANSWER: Temporal separation has the advantage that all cells possess both capacities for nitrogen fixation and photosynthesis. On the other hand, it eliminates the ability of a chain of cells to conduct both processes simultaneously—the benefit of heterocysts. Heterocysts face the problem of operating in close proximity to photosynthetic cells generating toxic oxygen; this problem may be solved by symbiosis with respiring bacteria. Globular clusters of cells can bury their nitrogen fixers within the cluster; this arrangement effectively excludes oxygen, but it may lack flexibility during environmental change. Endosymbiotic nitrogen fixation within a respiring eukaryote is probably the most effective strategy of all, because the host provides oxygen-removing proteins such as leghemoglobin. Endosymbiosis, however, requires the presence of an appropriate host organism.

18.6 Why would *Streptomyces* produce antibiotics targeting other bacteria?

ANSWER: *Streptomyces* species may produce antibiotics to curb the growth of bacterial competitors with smaller genomes and faster rates of reproduction. The lysed cells release nutrients that feed growing mycelia of *Streptomyces*.

18.7 Why might genes for the proteorhodopsin light-powered proton pump be more likely to transfer horizontally than the bacteriochlorophyll-based photosystems PS I and PS II?

ANSWER: Proteorhodopsin requires only the one gene encoding the pump, plus one or two genes to produce retinal. This involves a relatively small amount of sequence to transfer, and the encoded products generate proton potential on their own, without requiring interaction with recipient enzymes. By contrast, PS I and PS II each involves multiple electron carriers that must function together and interact with the recipient electron transport chain.

18.8 Why do you think it took many years of study to realize that *Escherichia coli* and other Proteobacteria can grow as a biofilm?

ANSWER: *E. coli* and its relatives grow exceptionally well in liquid culture. Liquid culture is attractive because it enables quantitative measurement of defined aliquots of microbial population. However, repeated subculturing in liquid medium selects for planktonic (nonbiofilm) cells.

Eventually, the biofilm-forming property may be lost if mutants outgrow the original genotype in liquid medium.

18.9   Compare and contrast the formation of cyano-bacterial akinetes, firmicute endospores, actinomycete arthrospores, and myxococcal myxospores.

ANSWER: An endospore forms as the daughter product (forespore) of a single cell. Within the same cell, endospore development is supported by the motherspore, which disintegrates after release of the endospore. Endospores have tough coatings of calcium picolinate; they are heat resistant. By contrast, arthrospores and myxospores are less durable and are not heat resistant, although they can persist in the environment for an extended period. Arthrospores form through binary fission of actinomycete filaments. Myxospores are formed by a multicellular fruiting body. In all three cases, spore formation can be induced by depletion of nutrients; and the spore-producing entity is left behind to die.

Chapter 19

19.1   If two deeply diverging clades each show a wide range of growth temperature, what does this suggest about the evolution of thermophily or psychrophily?

ANSWER: The two clades diverged before temperature adaptation occurred. Adaptations to high or low temperature must have evolved independently in the two clades.

19.2   What might be the advantages of flagellar motility for a hyperthermophile living in a thermal spring or in a black smoker vent? What would be the advantages of growth in a biofilm?

ANSWER: Flagellar motility enables isolated cells to detect a new nutrient source, or an approximate temperature range, and approach it through chemotaxis. Growth in a biofilm attached to a substrate prevents the microbes from floating away from the nutrient source, or from being carried away in the flow from the vent.

19.3   What problem with cell biochemistry is faced by acidophiles that conduct heterotrophic metabolism?

ANSWER: Heterotrophic metabolism generates fermentation products such as acetate and lactate, which act as permeant acids. Permeant acids become protonated outside the cell, at low pH; the protonated forms then permeate the membrane, returning into the cell. Given the high transmembrane pH difference maintained by *Sulfolobus*, one would expect even small traces of fermentation acids to cross the membrane in the protonated form, and then dissociate and accumulate to toxic levels of organic acids. It is unknown how *Sulfolobus* solves this problem.

19.4   What conclusions might be drawn if viruses of mesophilic archaea are found to have RNA genomes? What if they all have DNA genomes only?

ANSWER: If mesophilic viruses show RNA genomes but thermophiles do not, then it is likely that only double-stranded DNA is sufficiently stable for viruses to persist in the environment of hyperthermophiles. If only double-stranded DNA viruses are found throughout the archaea, this would suggest that all archaeal viruses evolved from viruses infecting a common ancestral cell that was a thermophile. The latter hypothesis would require supporting evidence from archaeal cell physiology and phylogeny.

19.5   What do the multiple metal requirements suggest about the evolution of methanogens?

ANSWER: The requirement for so many different metals may suggest that methanogens evolved in habitats such as geothermal vents where superheated water carries up high concentrations of dissolved metal ions.

19.6   Compare and contrast the metabolic options available for *Pyrococcus* and for the crenarchaeote *Sulfolobus*.

ANSWER: *Sulfolobus* catabolizes sugars and amino acids aerobically, using O_2 as terminal electron acceptor. *Pyrococcus abyssi* catabolizes sugars and amino acids anaerobically, using S^0 as terminal electron acceptor. *Sulfolobus* oxidizes sulfur lithotrophically, from S^{2-} to S^0, SO_3^{2-}, and ultimately SO_4^{2-}. *P. abyssi*, however, reduces sulfur lithotrophically with H_2 to form H_2S.

19.7   Compare and contrast sulfur metabolism in *Pyrococcus* and in *Ferroplasma*.

ANSWER: *Pyrococcus* species reduce S^0 with hydrogens from organic substrates, forming HS^- and H_2S. *Ferroplasma* species oxidize sulfur in the form of FeS_2 (using oxidant Fe^{3+}), forming sulfuric acid. The result is extreme acidification of their environment.

Chapter 20

20.1   Why would yeasts remain unicellular? What are the relative advantages and limitations of hyphae?

ANSWER: Yeast grow in environments with sufficient dissolved nutrients to absorb from the medium. The advantage of forming hyphae is that they enable penetration of other organisms, and hence provide access to nutrients. On the other hand, hyphae formation limits the rate of dispersal of progeny cells. Since yeasts grow in environments where dissolved nutrients can be absorbed from the medium, they do not need to produce hyphae and can proliferate more rapidly than mycelial fungi.

20.2   Why would some fungi conduct asexual reproduction under most conditions? What are the advantages and limitations of sexual reproduction?

ANSWER: The sexual life cycle involves significant genetic and metabolic costs to the organism. Asexual reproduction enables fungi to eliminate an energy drain and produce more offspring using fewer resources. Reductive evolution leading to loss of sexual reproduction might eliminate an energy drain and perhaps enable greater proliferation with fewer resources. On the other hand, the sexual life cycle provides a valuable means of generating diversity through genetic recombination, so that the population may respond to environmental change. Fungi reproducing asexually must rely on mutation and gene transfer by viruses and mobile sequence elements to generate genetic diversity.

20.3 What are the advantages and limitations of motile gametes, as compared to nonmotile spores?

ANSWER: Motile gametes have the advantage of rapid dispersal on their own and the potential for chemotaxis toward a food source or toward a gamete of the opposite mating type. On the other hand, motility uses up energy that could alternatively be invested in production of a greater number of nonmotile gametes. Motile gametes are especially useful in a watery habitat but are of little use in a terrestrial habitat, where air currents or animal hosts must be used for dispersal.

20.4 Compare the life cycle of an ascomycete (**Fig. 20.14C**) with that of a chytridiomycete (**Fig. 20.12C**). How are they similar, and how do they differ?

ANSWER: Both chytridiomycetes and actinomycetes undergo alternation of generations. Each possesses an alternative route of an asexual cycle of mitotic cell proliferation. In the chytridiomycete asexual cycle, the diploid form develops a mycelium, motile zoospores, and cysts. In the actinomycete, however, the haploid form undergoes mitotic divisions. The sexual cycles of the two fungal groups differ structurally. In the chytridiomycete, the zoosporangium forms motile zoospores that develop and give rise to motile gametes, which fertilize each other to form motile zygotes. In the ascomycete, there are no motile forms. Instead of motile gametes, haploid hyphal antheridia undergo cytoplasmic fusion and form fruiting bodies, or asci, within which ascospores develop. The spores will not be motile; they are carried by wind or water.

20.5 How are coralline algae able to grow at greater depths than coral?

ANSWER: Corals generally grow symbiotically with green algae, which contain chlorophyll but lack accessory photopigments. Corallines are red algae, possessing both chlorophyll and phycoerythrins. The latter absorb blue and green light that is not absorbed by green algae, and thus coralline algae penetrate deeper in the water column.

20.6 What might happen when an ameba phagocytoses algae?

ANSWER: If light is available, the algae may be retained as endosymbionts providing energy through photosynthesis. For example, *Chlorarachnion* possesses obligate

chloroplast-bearing endosymbionts descended from green algae. Alternatively, the ameba may digest all but the algal chloroplast, which persists for some time, providing photosynthetic products.

20.7 What kind of habitat would favor a flagellated ameba?

ANSWER: A dilute watery habitat would favor flagella, which allow more rapid propulsion than pseudopods. Pseudopod motility requires a solid substrate.

20.8 Compare and contrast the processes of conjugation in ciliates and bacteria (see Chapter 9).

ANSWER: Conjugation in ciliates is a completely different process from conjugation in bacteria, although the purpose (genetic exchange) is similar. In ciliates, two cells form a bridge allowing cytoplasm to flow directly between them; whereas in bacteria, a donor cell attaches to another (by pili in some cases), and then a protein complex transfers DNA across both cell envelopes, without direct cytoplasmic contact. In bacterial conjugation, DNA is transferred unidirectionally from the donor cell to the recipient, whereas in ciliates there is reciprocal exchange of DNA. A donor bacterium generally transfers only part of its genome, whereas ciliates exchange entire copies of theirs.

20.9 For ciliates, what are the advantages and limitations of conjugation, as compared with gamete production?

ANSWER: The process of conjugation avoids the necessity of dissolving the intricate cell structure of the ciliate in order to form gametes that fuse or fertilize each other. On the other hand, conjugation requires two diploid organisms to find each other and make contact for several hours, during which time feeding is suspended and the pair is vulnerable to predation.

Chapter 21

21.1 From Chapters 13–15, give examples of microbial metabolism that fit patterns of assimilation and dissimilation.

ANSWER: Microbes can assimilate carbon either by reducing carbon dioxide or by oxidizing methane, and they can dissimilate carbon by fermentation and respiration. Nitrogen is assimilated by N_2 fixation, and by incorporation of NH_4^+ into glutamine and glutamate. Nitrogen is dissimilated by deamination of amino acids, and by lithotrophic oxidation.

21.2 How do viruses select for increased diversity of microbial plankton?

ANSWER: Since viruses tend to infect only a narrow host range, their existence favors the evolution of a large number of different species with highly dispersed populations.

Highly dispersed populations minimize the chance of viral transmission from one host to another.

21.3   Design an experiment to test the hypothesis that the presence of mycorrhizae enhances plant growth in nature.

ANSWER: Such an experiment requires a control based on the natural environment, where various unknown factors may be very different from those in the laboratory. One possibility is to compare the growth of seedlings in natural soil versus sterilized natural soil. However, this experiment would not prove that fungi are the cause of enhanced growth in unsterile soil. The sterilization procedure (usually involving heat and pressure) could break down key nutrients in the soil. A follow-up experiment might be to grow the plants in the presence of a fungus inhibitor in sterilized and unsterilized soil.

21.4   High levels of nitrate or ammonium ion corepress the expression of Nod factors. What is the biological advantage of Nod regulation?

ANSWER: Nitrate and ammonium ion are the main forms of nitrogen assimilated by plants. If they are abundant in the soil, the plant does not need rhizobial symbionts; therefore, development of the rhizobia-legume symbiosis is inhibited.

21.5   One unanswered question is, How do symbiotic rhizobia reproduce? Why do bacteroids develop if they cannot proliferate?

ANSWER: Various answers have been proposed. Not all of the invading bacteria become bacteroids; some continue to undergo cell division, particularly within senescing tissues of the plant. These bacteria benefit from plant growth, which is sustained by the bacteroids whose genes they share. Alternatively, the entire plant-bacteroid system may benefit rhizobia that grow just outside the plant, in the rhizosphere.

21.6   Compare and contrast the processes of plant infection by rhizobia and by fungal haustoria.

ANSWER: Both rhizobial bacteria and fungal haustoria penetrate the volume of a plant cell, but they keep the plant cell membrane intact, its invagination always surrounding the invading cell. Rhizobia establish a complex, highly regulated exchange of nutrients with the host, receiving catabolites and oxygen in exchange for ammonium, and cycling the components of amino acids. By contrast, haustoria establish one-way removal of nutrients such as sucrose, while providing no nutrients in return. Fungal pathogens weaken the structure of the host plant and decrease or halt its growth.

21.7   Why does ruminant fermentation leave food value for the animal host? How is the animal able to obtain nourishment from waste products that the microbes could not use?

ANSWER: The rumen interior is anaerobic. In the absence of oxygen as a terminal electron acceptor, microbes are forced to generate waste products in which the electrons are put back onto the electron donors (fermentation; see Chapter 13). When the short-chain fatty acid wastes enter the animal's bloodstream, the blood is full of oxygen, which enables complete digestion to CO_2 and water.

21.8   How do you think cattle feed might be altered or supplemented to decrease methane production?

ANSWER: Several methods have been proposed to limit methanogenesis. One is to feed cattle the antibiotic monensin, an inhibitor of sodium transport, which is required for methanogenesis. Another approach is to feed cattle an electron acceptor for H_2, such as fumarate, which bacteria use to generate short-chain acids instead of methane. These approaches have been used with only partial success—not surprising, given the complexity of the system.

Chapter 22

22.1   Why is oxidation state critical for the acquisition, usability, and potential toxicity of cycled compounds? Cite examples based on your study of microbial metabolism.

ANSWER: Many examples can be cited. In the case of carbon, CO_2 can be fixed by many species with substantial input from photosynthesis or hydrogen donors. The reduced form methane, however, can be assimilated only by methanotrophic bacteria, usually with oxygen as electron acceptor. Nitrogen gas can be assimilated only by nitrogen-fixing bacteria and archaea, whereas NH_4^+ can be assimilated by many plants and microbes. But NH_4^+ can also be oxidized by lithotrophs to NO_3^-, a substance potentially toxic to humans.

22.2   What would happen if wastewater treatment lacked microbial predators? Why would the result be harmful?

ANSWER: Without predators, too many planktonic bacteria would remain in the wastewater after sedimentation of the sludge. The bacteria could be killed by chlorination, but the treated water would have significant BOD (biochemical oxygen demand) because the bacterial remains provide an organic carbon source for respirers.

22.3   How many kinds of biomolecules can you recall that contain nitrogen? What are the usual oxidation states for nitrogen?

ANSWER: Amino acids, nucleotide bases, polyamines for DNA stabilization, peptidoglycan (both amino sugar and peptide chains), and the heme derivatives of cytochromes, chlorophyll, and vitamin B_{12} all include nitrogen (as do many other biochemicals). The oxidation states of nitrogen in living organisms are nearly always reduced, either $R-NH_2$, $R=NH$, or $R-N=R$. An exception is the neurotransmitter NO (nitric oxide).

22.4 In the laboratory, which bacterial genus would likely grow on artificial medium including NH_2OH as the energy source: *Nitrosomonas* or *Nitrobacter*?

ANSWER: *Nitrosomonas* is more likely to utilize NH_2OH, since it performs the intermediate oxidation of NH_2OH during nitrification of ammonia.

22.5 The nitrogen cycle has to be linked with the carbon cycle, since both contribute to biomass. How might the carbon cycle of an ecosystem be affected by increased input of nitrogen?

ANSWER: One hypothesis is that the injection of nitrogen into an ecosystem accelerates growth of producers (phytoplankton in the ocean, or trees in a forest) and therefore facilitates net removal of CO_2 from the atmosphere. Overall, however, the additional fixed carbon ends up dissipated by consumers and decomposers.

22.6 Compare and contrast the cycling of nitrogen and sulfur. How are the cycles similar? How are they different?

ANSWER: Cycling of both nitrogen and sulfur involves interconversion between different oxidation states. Most of these interconversion reactions are performed solely by microbes, many of them solely by bacteria. Examples include nitrification of ammonia and denitrification to N_2, as well as sulfide oxidation and photolysis. In both cases, oxidation produces strong acids (HNO_3, H_2SO_4). The major sources and sinks differ; nitrogen is obtained primarily from the atmosphere as N_2, whereas sulfur is at high levels in the ocean and soil. Sulfur is rarely limiting, whereas nitrogen frequently is. Sulfur participates extensively in phototrophy; nitrogen shows little involvement in phototrophy; phototrophy based on nitrate reduction has been observed.

22.7 Compare and contrast the cycling of nitrogen and phosphorus. How are the cycles similar? How are they different?

ANSWER: Nitrogen and phosphorus are both limiting nutrients in many ecosystems—marine, aquatic, and terrestrial. Addition of either element into an aquatic system may cause algal bloom and eutrophication. On the other hand, the two elements differ in their major sources: the atmosphere for nitrogen, and crustal rock for phosphate. Within biomass, nitrogen exists almost entirely in reduced form, whereas phophorus is entirely oxidized. Phosphorus cycles through the biosphere mainly as inorganic or organic phosphates, whereas nitrogen cycles through a broad range of oxidation states, from NH_3 to NO_3^-.

Chapter 23

23.1 How can an anaerobic microorganism grow on skin or in the mouth, both of which are exposed to air?

ANSWER: Facultative organisms living in proximity to the anaerobes will deplete oxygen in the environment, especially around nooks and crannies (for example, between teeth and gums, gingival pockets) that would ordinarily prevent anaerobes from growing. These small spaces have limited access to oxygen.

23.2 What can a person with mitral valve prolapse do to prevent formation of subacute bacterial endocarditis when visiting the dentist?

ANSWER: Take antibiotics prophylactically. A high dose of antibiotic (usually amoxicillin or a cephalosporin) taken 1 hour before the procedure will produce a high enough blood level to kill bacteria. A patient that is hypersensitive to beta-lactam antibiotics such as amoxicillin will be prescribed an alternate antibiotic, such as azithromycin.

23.3 Why do many Gram-positive microbes that grow on the skin, such as *Staphylococcus epidermidis*, grow poorly or not at all in the gut?

ANSWER: Bile salts present in the intestine (not on the skin) easily gain access to and destroy cytoplasmic membranes of Gram-positive organisms (unless the organism possesses bile hydrolases). Gram-negative microbes have extra protection in the form of an outer membrane and so can survive better in the intestine.

23.4 How might normal biota escape the intestine and cause disease at other body sites?

ANSWER: Normal biota can escape through intestinal perforations resulting from gunshot or knife wounds, surgery, or cancer.

23.5 Why can the colon be considered a fermenter?

ANSWER: The contents are continually flowing through the intestinal tube, with food containing substrates for fermentation ingested at one end and waste containing fermentation products removed from the other.

23.6 Why do defensins have to be so small? Do defensins kill normal microbiota?

ANSWER: They need to be small so they can get through the outer membrane of Gram-negative organisms and the thick peptidoglycan maze of Gram-positive organisms. Defensins do kill normal microbiota and in fact are part of what keeps levels of the normal intestinal microbiota in check. Research suggests that a decrease in intestinal defensin production can lead to an imbalance of gastrointestinal microbes in both number and species. This imbalance appears to contribute to conditions such as inflammatory bowel syndrome (IBS). Pathogens or normal microbiota that penetrate the intestinal mucosa probably encounter higher concentrations of defensins as they do so.

23.7   What happens to all the neutrophils that enter a site of infection once the infection has resolved?

ANSWER: Once bacteria at the site of infection have been killed, the tissue cells in the area stop making the cytokines and chemokines that attracted neutrophils in the first place, so neutrophils stop coming in and some wander out. But the majority of neutrophils undergo a self-programmed cell death (called apoptosis) and are cleared by monocytes in the area by phagocytosis. The average life span of a neutrophil is only 5 days.

23.8   If NK cells can attack infected host cells coated with antibody, why won't neutrophils?

ANSWER: Actually, neutrophils (PMNs) *can* attack infected cells coated with antibody—but the killing mechanism is different from ADCC. Human neutrophils do not make perforin or the other ADCC-related compounds, called granzymes, used by NK cells to kill target cells. In addition to that difference, NK cells possess a type of Fc receptor not found on neutrophils, which means the intracellular signaling pathways are different between NK cells and neutrophils. Neutrophils can, however, be activated when their Fc receptors bind antibody. Activated neutrophils make reactive oxygen products and can release a variety of peptides, including defensins, cathelicidin, and myeloperoxidase, which can all damage target cells.

23.9   If increased fever limits bacterial growth, why do bacteria make pyrogenic toxins?

ANSWER: The pyrogenic toxins have other effects that compromise and damage the host. The toxins can induce cytokines that damage local host cells or confuse the immune system. This can provide the pathogen with nutrients and help it hide from the immune system longer. Pyrogenic toxins include lipopolysaccharide and protein toxins such as toxic shock syndrome toxin (see Chapter 25).

Chapter 24

24.1   Two different stretches of amino acids in a single protein form a three-dimensional antigenic determinant. Will the specific immune response to that three-dimensional antigen also respond to one of the amino acid stretches alone?

ANSWER: Most likely no. It is the three-dimensional shape formed by the two stretches that is recognized as an antigen. A denatured protein that contains both amino acid stretches will not possess the three-dimensional shape of the antigen. However, other specific immune responses involving different subsets of lymphocytes can recognize the separate amino acid stretches that together form the three-dimensional antigenic determinant. As an analogy, take a

computer image of a friend's face and shuffle the facial features. Turn the nose upside down, exchange the eyes with the mouth, and lower the ears. Since you were programmed to respond to the original facial configuration, you likely would not recognize the rearranged face as a whole. But you might find that the nose looks familiar.

24.2   How does a neutralizing antibody that recognizes a viral coat protein prevent infection by the associated virus?

ANSWER: Neutralizing antibodies usually bind attachment proteins on the virus and sterically prevent them from binding to host cell receptors (see **Fig. 24.4**). Some antibodies to enveloped viruses might trigger the complement cascade (see later in the chapter), destroying the membrane.

24.3   Blood group type A individuals do not have B antigen, yet they make anti-B antibodies. If antibody production comes about in response to the presence of a specific antigen, what stimulates the production of anti-A or anti-B antibodies?

ANSWER: The stimulus for making anti-A and anti-B antibodies comes from the lipopolysaccharides and oligopolysaccharides present on commensal bacteria that colonize an infant's gut soon after birth. These molecules are structurally similar to the A and B antigens on RBCs and can stimulate B cells preprogrammed to make anti-A or anti-B antibodies.

24.4   The attachment proteins of different rhinovirus strains bind to ICAM-1. How can all these proteins be immunologically different if they find the same target (ICAM-1)? Why won't antibodies directed against one rhinovirus strain block attachment of other rhinovirus strains?

ANSWER: The ICAM-1 binding sites on different rhinovirus strains have small, but immunologically significant, differences. ICAM-1 is "promiscuous" in its binding specificity toward the rhinovirus attachment proteins, whereas antibodies are more specific. Thus, ICAM-1 protein is like a master key that can fit dozens of different locks (rhinovirus-binding sites). Each lock (binding site) has a very specific key (antibody) that won't unlock any of the other locks. But the master key (ICAM-1) can turn all locks.

24.5   The mother of a newborn was found to be infected with rubella, a viral disease. Infection of the fetus could lead to serious consequences for the newborn. How could you determine whether the newborn was infected in utero?

ANSWER: Since maternal IgM antibodies cannot cross the placenta, finding IgM antibodies to rubella antigens in the newborn's circulation indicates that the fetus was infected and initiated its own immune response. If the newborn has only IgG antibodies to rubella (no IgM antibodies), the child was not infected and maternal IgG crossed the placenta.

24.6 B cells in early stages have both IgM and IgD surface antibodies, but the delta region has no switch region. Why does the delta region have no switch region?

ANSWER: If the delta region had a switch region, the B cell could only make IgD after DNA rearrangement. Then, no one B cell could have IgM and IgD at the same time. Recombination at the DNA level is not involved, because alternative RNA-splicing events after transcription determine whether an IgM or IgD molecule is made. B cells at the early stages have both IgM and IgD surface antibodies.

24.7 Why do individuals with type A blood have anti-B and not anti-A antibodies?

ANSWER: Because the B-cell population that would react to type A antigen was deleted during B-cell maturation.

24.8 Why do immunizations lose their effectiveness over time?

ANSWER: Because memory B cells eventually die. Without some exposure to antigen, those memory cells will not be replaced.

24.9 Transplant rejection is a major consideration when transplanting most tissues because host T_C cells can recognize allotypic MHC on donor cells. So, why are corneas easily transplanted from a donor to just about any other person?

ANSWER: The cornea is not normally vascularized. So even though corneal cells express MHC proteins, circulating host T cells do not have an opportunity to interact with them. The cornea will not be rejected. This is called an immune privileged site.

24.10 Why does attaching a hapten to a carrier protein allow production of antihapten antibodies?

ANSWER: B cells with antihapten surface antibody (as part of the B-cell receptor) can take up hapten but cannot present the hapten to a helper T cell. The same B cell can also take up the hapten bound to a carrier molecule, and because the carrier molecule is larger than the hapten, the B cell will present the carrier epitope to the helper T cell. The helper T cell stimulates the B cell, which was already programmed to make antihapten antibody, to differentiate into plasma and memory B cells.

24.11 Do bone marrow transplants in a patient with severe combined immunodeficiency require immunosuppressive chemotherapy?

ANSWER: No. Since the patient has no T cells to recognize foreign antigens, the transplant is not rejected.

24.12 How can a stem cell be differentiated from a B cell at the level of DNA?

ANSWER: DNA recombination at switch regions will have taken place in the B cell but not the stem cell. Thus, the B cell will have fewer segments for each of the V, D, and J regions, while the stem cell will have all of them. PCR techniques can be used to view these differences.

Chapter 25

25.1 Is a microbe with an LD_{50} of 5×10^4 more or less virulent than a microbe with an LD_{50} of 5×10^7?

ANSWER: Because it takes fewer cells to cause disease, the microbe with the smaller LD_{50} (5×10^4) is the more virulent.

25.2 Antibodies to which subunit of cholera toxin will best protect a person from the toxin's effects?

ANSWER: Antibodies to the B subunit will be more protective. Inactivating the B subunit will prevent binding of the toxin to cell membranes. The A-subunit active site is typically sequestered in these toxins and inaccessible to antibody. Furthermore, once the A subunit has entered a host cell, antibodies cannot enter and neutralize it.

25.3 How might you experimentally determine whether a pathogen secretes an exotoxin?

ANSWER: The microbe can be grown in liquid culture and the cells removed either by centrifugation or filtration. If the organism makes an exotoxin, it may well be present in the cell-free supernatant. The presence of a toxin can be determined by injecting the supernatant into an animal model (for example, mice) and examining the result (death or altered function). Alternatively, the supernatant can be administered to a layer of tissue culture cells and the health of the monolayer noted.

25.4 Would patients with iron overload (excess free iron in the blood) be more susceptible to infection?

ANSWER: With a few exceptions, withholding iron from potential pathogens is a host defense strategy because when iron is plentiful, the microbe does not have to expend energy to get it and so can readily grow. On the other hand, low iron can also be a signal to express various virulence genes, so for some organisms, high iron might hinder infection.

25.5 Use an Internet search engine to determine what other toxins are related to the cholera enterotoxin A subunit.

ANSWER: Go to PubMed (www.ncbi.nlm.nih.gov/pubmed). At the top of the page, select "Protein" from the Search dropdown list. Then type "cholera enterotoxin A subunit" in the text box below that and click **Search**. Click on the appropriate result (any designated "258 aa protein"). Highlight and copy the entire amino acid sequence at the bottom of the page. Then, back at the NCBI home page, click on **BLAST**; select "Microbes"; paste the A-subunit sequence into the box; change the settings in the Query and Database dropdown lists to "Protein"; scroll down the list of organisms and select "Bacteria"; click **BLAST**. On the next page, click **View report**. After some seconds, all the homology results will be displayed. Items with large negative E values represent the most significant homologies. Scroll down the page to find the actual alignments between the query protein and each homolog. Note that the toxin of *Escherichia coli* possesses an A subunit similar to that of cholera.

25.6   Protein and DNA have very different structures. Why would a protein secretion system be derived from a DNA-pumping system?

ANSWER: Conjugation systems actually move DNA that is attached to a pilot protein at the 5′ end. A pilot protein is made by the conjugation system and then binds to the 5′ end of the DNA to be transferred and "pilots" the DNA through the conjugation pore. So a modified conjugation system that moves protein only is not as much of a leap as might initially be thought.

25.7   How can one determine whether a bacterium is an intracellular parasite?

ANSWER: Microscopic examination to see whether bacteria are found within cultured mammalian cells is usually not satisfactory. The difficulty lies in determining whether the organism is inside the host cell or just bound to its surface—or, if it is inside, whether it is a live or dead bacterium. One commonly used approach is to add to infected cell monolayers an antibiotic that can kill the microbe but will not penetrate the mammalian cells. The protein synthesis inhibitor gentamicin is typically used. A bacterium that invades a host cell will gain sanctuary from gentamicin and grow intracellularly. Extracellular bacteria and bacteria attached to the outside of the host cell are killed. Counting viable colony-forming units of bacteria released from the mammalian cells by gentle detergent treatments at various times will reveal if the organism grew intracellularly. Of course, this will work only if the microorganism is not an obligate intracellular parasite.

25.8   Why would killing a host be a bad strategy for a pathogen?

ANSWER: The goal of any microbe is to maintain its species. If a microbe does not have an opportunity to easily spread to a new host, killing its host would be tantamount to suicide.

25.9   Why do rhinovirus infections fail to progress beyond the nasopharynx?

ANSWER: For one thing, rhinoviruses are susceptible to acid pH (pH 3.0), so they are unable to replicate in the gastrointestinal tract. They also grow best at 33°C, which may help explain their predilection for the cooler environs of the nasal mucosa.

Chapter 26

26.1   Why would treatment of some infections require multiple antibiotics?

ANSWER: Although not typically recommended for simple infections, with some organisms multiple antibiotic therapy is required to effect a cure. The organisms that require multidrug therapy are very hardy in vivo and able to grow in the presence of single antibiotics. Active infection with *Mycobacterium tuberculosis*, for example, is typically treated with isoniazid (thought to inhibit synthesis of mycolic acid) and rifampin (an RNA polymerase inhibitor), and sometimes a third drug, pyrazinamide (which inhibits fatty acid synthesis). Using multiple antibiotics will minimize the chance that the organism will develop resistance to any one antibiotic. Multiple antibiotic therapy is also recommended for *Helicobacter pylori* (ulcers, treated with lansoprazole, amoxicillin, and clarithromycin) and methicillin-resistant *Staphylococcus epidermidis* (treated with vancomycin and gentamicin). MRSE does exist and can cause serious infections such as meningitis.

26.2   Explain why patient noncompliance (failure to take drugs as directed) is thought to have led to XDR-TB.

ANSWER: In most cases, an organism causing an infection starts out being susceptible to a given antimicrobial agent. However, the more times an organism divides, the more likely it is that a spontaneous mutation producing drug resistance will arise. The amount of the antibiotic and the duration of its use are calibrated so that all organisms are killed—either by the drug itself or by the immune system. Often the drug is given to stop growth of the bacterium so that the immune system has enough time to actually do the killing. If a patient stops taking an antibiotic before the prescribed end of the treatment, the organism can start to replicate again, renewing the chance that a spontaneous drug-resistant variant will be produced. Even if drug therapy is resumed, the resistant organism will continue to grow and cause disease—and can be transmitted to other individuals.

26.3   Why do you think most urinary tract infections occur in women?

ANSWER: The major reason is anatomy. Most bladder UTIs come from access through the urethra. Since the urethra in men is longer than in women, it usually takes a catheter to introduce bacteria into a male bladder. The trip for the infecting organism in women is much shorter. However, in people older than 50, UTIs become more common in both men and women, with less difference between the sexes. The reason is not clear.

26.4   Urine samples collected from six hospital patients were placed on a table at the nurses' station awaiting pickup from the microbiology lab. Several hours later, a courier retrieved the samples and transported them to the lab. The next day, the lab reported that four of the six patients had UTIs. Would you consider these results reliable? Would you start treatment based on these results?

ANSWER: Because urine is a good growth medium for many bacteria, the delay of several hours in picking up the samples gave the organisms time to replicate and increase their numbers. Consequently, the lab results should be viewed with suspicion.

26.5   Considering that *Neisseria gonorrhoeae* is exquisitely sensitive to ceftriaxone, why do you suppose the patient in this case was also treated with tetracycline? And why can one person be infected repeatedly with *N. gonorrhoeae*?

ANSWER: Because STDs often travel in pairs and because the initial symptomology is similar, the clinician will want to "cover" the patient for possible chlamydia infection. Chlamydias are not susceptible to ceftriaxone. The large, single dose of ceftriaxone (as opposed to smaller multiple injections given over days) was given because patients with gonorrhea are typically poorly compliant and fail to return for subsequent injections. A person who experiences gonorrhea is not protected from reinfection, because the organism's surface antigens undergo phase variations (see Section 10.6) and may down-regulate the immune system (see **Fig 26.13C**).

26.6   Human immunodeficiency virus was discussed in this section and in Chapter 25. Like the plague, it is a blood-borne disease. Why, then, do fleas and mosquitoes fail to transmit HIV?

ANSWER: When insect vectors take a blood meal, they typically defecate or regurgitate simultaneously. So theoretically, they could serve as a vector for HIV. Pathogens such as the West Nile virus that are transmitted by insect vectors actually grow in their insect hosts; however, HIV does not. Because it is unusually fragile, HIV will die quickly. There have been no cases of HIV transmitted by insect vectors.

26.7   Normal cerebrospinal fluid is usually low in protein and high in glucose. The protein and glucose content does not change much during a viral meningitis, but bacterial infection leads to greatly elevated protein and lowered glucose levels. What could account for this?

ANSWER: There are several explanations. Bacteria and infiltrating PMNs will consume glucose, and alterations in the blood-brain barrier can lead to decreased transport of glucose into the spinal fluid. As a result, glucose levels plummet. Growth of the bacteria and infiltration of PMNs account for the increase in CSF protein levels. Viruses are very small and consist mostly of nucleic acids, so even at high numbers they will not significantly add to CSF protein content. Since viruses do not grow in CSF directly, they will not consume glucose. In addition, viral menin-

gitis does not cause a great infiltration of PMNs into the CSF—another reason glucose levels remain high and protein levels remain low.

26.8   Knowing the symptoms of tetanus, what kind of therapy would you use to treat the disease?

ANSWER: At the first sign of muscle spasm, antitoxin should be given. If tetany is severe, muscle relaxants can relieve spasms.

26.9   How do the actions of tetanus toxin and botulinum toxin actually help the bacteria colonize or obtain nutrients?

ANSWER: This is a difficult question to answer. Few scientists have speculated. Recall that the toxins are encoded by genes in resident bacteriophages that became part of the clostridial genome through horizontal transfer from some other source. Since the organisms (vegetative, as well as spores) normally reside in soil, the actual function of these toxins may have something to do with survival in that habitat. The toxin's effect on humans may simply be an unfortunate accident. However, with tetanus toxin, there may be benefit in that muscle spasms could limit oxygen delivery to infected tissues, enabling a more anaerobic environment for growth. Cell death may also release iron or other nutrients useful to *Clostridium tetani*.

26.10   If *Clostridium botulinum* is an anaerobe, how might botulism toxin get into foods?

ANSWER: The most common way botulism toxin gets into food today is via home canning. The canning process involves heating food in jars to very high temperatures. The heat destroys microorganisms and drives out oxygen; both processes help to preserve foods. If the jars are not heated to sterilization conditions, spores of *C. botulinum* will survive. When the jars are cooled for storage at room temperature, the spores germinate; the organism then grows in the anaerobic medium and releases the toxin. When the food is eaten, the toxin is eaten too.

26.11   A patient presenting with high fever and in an extremely weakened state is suspected of having a septicemia. Two sets of blood cultures are taken from different arms. One bottle from each set grows *Staphylococcus aureus*, yet the laboratory report states that the results are inconclusive. New blood cultures are ordered. What might explain this result?

ANSWER: There must be something different about the two strains of staphylococci grown in the separate bottles. For example, they may have different antibiotic susceptibility patterns when tested against a battery of antibiotics. One strain may be susceptible to penicillin while the other strain is resistant. Since the expectation is that a single strain initiated the infection, both isolates should exhibit the same susceptibility pattern. The laboratory suspects contamination of the bottles from separate sources. These organisms are not the source of infection.

26.12   How might you design/construct a more effective vaccine for an antigen (for instance, the *Yersinia pestis* F1 antigen) that would harness the power of a Toll-like receptor (TLR)? *Hint*: TLR5 from Table 23.4.

ANSWER: By genetically fusing DNA encoding a hepatitis antigen to DNA encoding flagellin, a PAMP that is recognized by TLR5 on dendritic cells, you would end up with a chimeric protein that would activate innate immune mechanisms, via TLR5, and improve antigen presentation. The result is a better link between innate and adaptive immune mechanisms.

Chapter 27

27.1   The drug tobramycin is added to a concentration of 1,000 µg/ml in a tube of broth from which serial two-fold dilutions were made. Including the initial tube (tube 1), there are a total of ten tubes. Twenty-four hours after all the tubes are inoculated with *Listeria monocytogenes,* turbidity is observed in tubes 6–10. What is the MIC?

ANSWER: 62.5 µg/ml, the concentration in tube 5, the last tube with no growth. (Relative to tube 1, tube 5 has been diluted 2^4—16-fold dilution: 1,000/500/250/125/62.5.)

27.2   What additional test performed on an MIC series of tubes will tell you whether a drug is bacteriostatic or bactericidal?

ANSWER: Streaking a portion of the broth from the dilution tubes showing no growth. If the drug is bacteriostatic, colonies will form on the agar plate because during streaking, the bacteria are removed from the presence of the drug. If the antibiotic is bactericidal, no colonies will form, because the organisms are dead before plating. This method determines the minimum bactericidal concentration (MBC) of an antibiotic. MBC would be defined as the lowest dilution that did not yield viable cells.

27.3   You are testing whether a new antibiotic will be a good treatment choice for a patient with a staph infection. The Kirby-Bauer test using the organism from the patient shows a zone of inhibition of 15 mm around the disk containing this drug. Clearly, the organism is susceptible. But you conclude from other studies that the drug would not be effective in the patient. What would make you draw this conclusion?

ANSWER: If the average attainable tissue level of the drug is below the MIC, the drug will not be effective.

27.4   When treating a patient for an infection, why would combining a drug such as erythromycin with a penicillin be counterproductive? Erythromycin is described in Section 8.3.

ANSWER: Erythromycin, a bacteriostatic drug, will stop growth, which indirectly stops cell wall synthesis and renders the microbe insensitive to penicillin.

27.5   The enzyme DNA gyrase, a target of the quinolone antibiotics, is an essential protein in DNA replication. The quinolones bind to and inactivate this protein. Research has proved that quinolone-resistant mutants contain mutations in the gene encoding DNA gyrase. If the resistant mutants contain a mutant DNA gyrase and DNA gyrase is essential for growth, why are these mutations not lethal?

ANSWER: The mutations cause changes in DNA gyrase that have little to no effect on function but do prevent the drug from binding. Thus, the mutant organism will continue to twist its DNA and grow well with or without the drug.

27.6   Why might a combination therapy of an aminoglycoside antibiotic and cephalosporin be synergistic?

ANSWER: The two drugs given together could act synergistically because the cephalosporin can weaken the cell wall and allow the aminoglycoside easier access to the cell interior, where it can attack ribosomes. This synergism is especially useful in organisms that have some resistance to both drugs.

27.7   Could genomics ever predict the drug resistance phenotype of a microbe? If so, how?

ANSWER: Yes. If the organism's genome possesses genes whose deduced protein sequences harbor significant similarity to antibiotic resistance proteins from other organisms, one can predict a similar drug resistance. Definitive proof of drug resistance requires actual in vitro testing.

27.8   Fusaric acid is a cation chelator that normally does not penetrate the *E. coli* membrane, which means *E. coli* is typically resistant to this compound. Curiously, cells that develop resistance to tetracycline become *sensitive* to fusaric acid. Resistance to tetracycline is usually the result of an integral membrane efflux pump that pumps tetracycline out of the cell. What might explain the development of fusaric acid sensitivity?

ANSWER: Fusaric acid is imported by the tetracycline efflux pump. This phenomenon can be used to isolate mutants with deletions of transposons encoding tetracycline resistance. Transposons, such as Tn*10*, which carries tetracycline resistance, can spontaneously delete at a frequency of about 10^{-6}, but finding one out of a million tetracycline-susceptible cells is impossible without a positive selection. Fusaric acid provides that positive selection, since the cell with the Tn*10* deletion will be resistant to fusaric acid.

27.9 Mutations in the ribosomal protein S12 (encoded by *rpsL*) confer resistance to streptomycin. Given a cell containing both *rpsL*$^+$ and *rpsL*^{R} genes, would the cell be streptomycin resistant or sensitive? (Recall that genes encoding ribosome proteins for the small subunit are designated *rps*. A + indicates the wild-type allele, while R indicates a gene whose product is resistant to a certain drug.)

ANSWER: This merodiploid cell would contain two sets of ribosomes: One set, containing normal S12, would be sensitive to streptomycin; another set, containing the resistant S12, would be resistant. Because streptomycin causes mistranslation of mRNA on sensitive ribosomes, inappropriate proteins that can kill the cell would still be synthesized. Thus, the cell would remain sensitive to streptomycin. Note, however, that the recessive nature of antibiotic resistance seen in this case is not the norm. Resistance is a dominant trait in a majority of cases.

Chapter 28

28.1 Use **Figure 28.6** to identify the organism from the following case: A sample was taken from a boil located on the arm of a 62-year-old man. Bacteriological examination revealed the presence of Gram-positive cocci that were also catalase-positive, coagulase-positive, and novobiocin resistant.

ANSWER: *Staphylococcus aureus.* The novobiocin test is irrelevant in this situation.

28.2 Why does finding IgM to West Nile virus indicate current infection? Why wouldn't finding IgG do the same?

ANSWER: Upon infection with any organism, IgM antibodies are the first to rise. After a short time the levels of IgM decline as IgG levels rise. IgG, however, can remain in serum for years, making it a poor prognosticator of current infection.

28.3 Why does adding albumin or powdered milk prevent false positives in ELISA?

ANSWER: Antibodies are proteins. They can stick to plastic just as easily as to the antigen being tested. Without blocking all the possible binding sites on plastic with albumin, any antibody from the patient's serum could stick to the plastic instead of to the antigen and react with the secondary enzyme-conjugated antihuman antibody.

28.4 Specific antibodies against an infectious agent can persist for years in the bloodstream, long after the infection resolves. So how is it possible that antibody titers can be used to diagnose recently aquired diseases such as infectious mononucleosis? Couldn't the antibody be from an old infection?

ANSWER: During the course of a disease, the body's immune system increases the amount of antibody made specifically against the infectious agent. Thus, one compares the antibody titer in a blood sample taken from a patient in the active, or acute, phase of disease with the antibody titer several weeks later, when the patient is in the recovery, or convalescent, stage. Seeing a greater-than-fourfold rise in a specific antibody titer (for example, in mononucleosis) indicates that the patient's immune system was responding to the specific agent. Remember, simply finding IgG against an organism or virus in serum indicates only that the patient was exposed to that microbe at some time in the past.

28.5 Two blood cultures, one from each arm, were taken from a patient with high fever. One culture grew *Staphylococcus epidermidis,* but the other blood culture was negative (no organisms grew out). Is the patient suffering from septicemia caused by *S. epidermidis*?

ANSWER: Probably not. *S. epidermidis* is a common inhabitant of the skin and could easily have contaminated the needle when blood was taken from the patient. The fact that only one of the two cultures grew this organism supports this conclusion. If the patient had really been infected with *S. epidermidis,* both blood cultures would have grown this organism.

28.6 A 30-year-old woman with abdominal pain went to her physician. After the examination, the physician asked the patient to collect a midstream urine sample that they would send to the lab across town for analysis. The woman complied and handed the collection cup to the nurse. The nurse placed the cup on a table at the nurses' station. Three hours later, the courier service picked up the specimen and transported it to the laboratory. The next day, the report came back "greater than 200,000 CFUs/ml; multiple colony types; sample unsuitable for analysis." Why was this determination made?

ANSWER: Although the CFU number is high enough to consider relevant, UTIs are typically caused by a single organism. The fact that the lab found many different colony types suggests a problem with specimen collection. In this case, not refrigerating the sample allowed the small number of urethral contaminants to overgrow the specimen.

28.7 Methicillin, a beta-lactam antibiotic, is very useful in treating staphylococcal infections. The development of methicillin-resistant strains of *Staphylococcus aureus* (MRSA) is a very serious development because few antibiotics can kill these strains. Imagine a large metropolitan hospital in which there have been eight serious nosocomial infections with MRSA and you are responsible for determining the source of infection so that it can be eliminated.

How would you accomplish this using common bacteriological and molecular techniques?

ANSWER: Samples from all the affected patients and from the hospital staff would be screened for the presence of *S. aureus* resistant to methicillin. Each strain would then be analyzed for restriction fragment length polymorphisms (RFLPs). Strains from all the patients will likely have identical restriction patterns if they came from the same source. The source is then identified by determining which staff member possesses MRSA with the same pattern. The source may also be inanimate, such as surgical equipment or ventilation apparatus. A connection between patients and specific staff members or instruments would also have to be demonstrated.

28.8   What are some reasons why some diseases spread quickly through a population while others take a long time?

ANSWER: There are several factors. One is mode of transmission; airborne diseases can spread more quickly than food-borne diseases, for instance. Sexually transmitted diseases spread more slowly still. Herd immunity is another factor. Herd immunity is based on the number of individuals within a population that are resistant to a disease. Someone immune to the disease cannot pass it on. The more immune people there are in a population (or herd), the slower is the spread of the epidemic to susceptible people.

28.9   On the ProMED-mail web page (www.promedmail.org), click on **Links** to view outbreaks recorded by WHO. What outbreaks happened throughout the World during the current year?

ANSWER: As of this writing, there were numerous influenza outbreaks, polio in Tajikistan and Angola, yellow fever in the Congo, plague in Peru, and numerous others.

Glossary

2-D gels *or* **2-D PAGE** See **two-dimensional polyacrylamide gel electrophoresis**.

A site See **acceptor site**.

ABC transporter An ATP-powered transport system that contains an ATP-binding cassette.

aberration An imperfection in a lens.

abiotic Produced without living organisms; occurring in the absence of life.

absorption In optics, the capacity of a material to absorb light.

acceptor site (A site) The region of the ribosome that binds an incoming charged tRNA.

accessory protein A protein found in the viral capsid or tegument that is needed early in the viral life cycle.

acid-fast stain A diagnostic stain for mycobacteria, which retain the dye fuchsin because of mycolic acids in the cell wall.

acidophile An organism that grows fastest in acid (generally defined as below pH 5).

ACP See **acyl carrier protein**.

acquired immunodeficiency syndrome (AIDS) A disease caused by HIV that leads to the destruction of T cells and the inability to fight off opportunistic infections.

actin filament See **microfilament**.

Actinobacteria A phylum of high-GC-content Gram-positive bacteria.

actinomycete A member of the group Actinomycetales, an order of Actinobacteria including branched spore formers such as *Streptomyces*, as well as irregularly shaped corynebacteria.

activated sludge Organic material concentrated from wastewater, containing microbes that digest the material to inorganic compounds.

activation energy The energy needed for reactants to reach the transition state between reactants and products.

activator A protein that increases gene transcription.

active transport An energy-requiring process that moves molecules across a membrane against their electrochemical gradient.

acyl carrier protein (ACP) A protein that can carry an acetyl group for anabolic pathways such as fatty acid synthesis.

adaptive immunity Also called *adaptive immune response*. Immune responses activated by a specific antigen and mediated by B cells and T cells.

ADCC See **antibody-dependent cell-mediated cytotoxicity**.

adenosine triphosphate (ATP) A ribonucleotide with three phosphates and the base adenine. It has many functions in the cell, including precursor for RNA synthesis and energy carrier.

adenylate cyclase An enzyme that converts ATP into cyclic adenosine monophosphate (cAMP).

adherence The ability of an organism to attach to a substrate.

adhesin Any cell surface factor that promotes attachment of an organism to a substrate.

ADP-ribosyltransferase A bacterial toxin that enzymatically transfers the ADP-ribose group from NAD^+ to target proteins, altering the target protein's structure and function.

aerial mycelium A hypha that extends above the surface and produces spores at its tip.

aerobic respiration The use of oxygen as the terminal electron acceptor in an electron transport system. A proton gradient is generated and used to drive ATP synthesis.

aerotolerant anaerobe An organism that does not use oxygen for metabolism but can grow in the presence of oxygen.

affinity chromatography A chromatographic technique that utilizes the ability of biological molecules to bind to certain ligands specifically and reversibly.

AFM See **atomic force microscopy**.

agar A polymer of galactose that is used as a gelling agent.

AIDS See **acquired immunodeficiency syndrome**.

Airy disk A bright central point surrounded by rings of light and dark caused by the pattern of interference of spherical wavefronts converging at the focal point.

akinete A specialized spore cell formed by some filamentous cyanobacteria.

alga *pl.* **algae** A microbial eukaryote that contains chloroplasts.

algal bloom An overgrowth of algae on the water surface, caused by an increase in a limiting nutrient.

alkaline fermentation Bacterial fermentation in conjunction with proteolysis and amino acid catabolism that generates ammonia in amounts that raise pH.

alkaliphile An organism with optimal growth in alkali (generally defined as above pH 9).

allergen An antigen that causes an allergic hypersensitivity reaction.

allosteric site A regulatory site on a biological molecule distinct from the ligand/substrate-binding site.

allotype   An amino acid sequence in the antibody constant region that is shared by some, but not all, members of a species.

alternation of generations   A life cycle that alternates between a haploid cell population and a diploid cell population.

alveolar macrophage   A type of macrophage, located in the lung alveoli, that phagocytoses foreign material.

alveolate   A member of the eukaryotic group Alveolata—ciliated or flagellated protists with complex cortical structure.

ameba (*or* amoeba)   A protist that moves via pseudopods.

amensalism   An interaction between species that harms one partner but not the other.

amino acid   The monomer unit of proteins. Each amino acid contains a central carbon covalently bound by a hydrogen, an amino group, a carboxyl group and a side chain. An exception is proline, in which the side chain is cyclized with the central carbon.

aminoacyl-tRNA synthetase   An enzyme that attaches a specific amino acid to the correct tRNA, thereby charging the tRNA.

ammonification   The generation of ammonia from organic nitrogen.

amphibolic   Describing a metabolic pathway that is reversible and can be used for both catabolism and anabolism.

amphipathic   Having both hydrophilic and hydrophobic portions.

amphotericin B   An antifungal drug that binds the fungus-specific sterol ergosterol and destroys membrane integrity.

anabolism   Also called *biosynthesis*. The building up of complex biomolecules from smaller precursors.

anaerobic methane oxidizers (ANME)   A group of archaea that oxidize methane syntrophically with bacteria that reduce sulfate.

anaerobic respiration   The use of a molecule other than oxygen as the final electron acceptor of an electron transport chain.

anammox reaction   The anaerobic oxidation of ammonium to nitrogen gas (using nitrate as electron acceptor); yields energy.

anaphylatoxin   A substance produced as a result of the complement cascade that can trigger degranulation of endothelial cells, mast cells, or phagocytes.

anaphylaxis   A severe hypersensitivity reaction caused by chemically induced contraction of smooth muscles and dilation of capillaries.

anaplerotic reaction   A type of metabolic reaction, occurring in all organisms, that fixes small amounts of CO_2 to regenerate TCA cycle intermediates.

angle of aperture   The width of a light cone (theta, θ) that projects from the midline of a lens. Greater angles of aperture increase resolution.

anion   A negatively charged ion.

ANME   See **anaerobic methane oxidizers**.

annotation   Deciphering genome sequences, including identifying genes and predicting gene function.

antenna complex   A complex of chlorophylls and accessory pigments in the photosynthetic membrane that collects photons and funnels them to a reaction center.

anthropogenic   Caused by humans.

anti-anti-sigma factor   A protein that inhibits an anti-sigma factor, allowing the target sigma factor to participate in initiating transcription.

anti-attenuator stem loop   An mRNA secondary structure whose formation prevents assembly of a downstream transcriptional termination stem loop. The anti-attenuator stem permits transcription of the downstream structural genes.

antibiotic   A molecule that can kill or inhibit the growth of selected microorganisms.

antibody   A host defense protein produced by B cells in response to a specific antigenic determinant. Antibodies bind to their corresponding antigenic determinant.

antibody stain   The attachment of a stain to an antibody to visualize cell components recognized by the antibody with high specificity.

antibody-dependent cell-mediated cytotoxicity (ADCC)   The process by which natural killer cells destroy viral protein expressing antibody-coated host cells.

anticodon   Three nucleotides in the middle loop of a tRNA that base-pair with a codon in mRNA.

antigen   Also called *immunogen*. A compound, recognized as foreign by the cell, that elicits an adaptive immune response.

antigen-presenting cell (APC)   An immune cell that can process antigens into antigenic determinants and display those determinants on the cell surface for recognition by other immune cells.

antigenic determinant   Also called *epitope*. A small segment of an antigen that is capable of eliciting an immune response. An antigen can have many different antigenic determinants.

antigenic shift   A genetic change in a pathogen's surface protein that prevents recognition by antibodies and host immune cells produced in response to the previous version of the protein.

antimicrobial agent   A chemical substance that can kill microbes or slow their growth.

antiparallel   Oriented such that the two strands are in opposite directions. Commonly refers to a nucleic acid double helix with one strand in the 5'-to-3' orientation and the other strand in the 3'-to-5' orientation.

antiporter   A transport protein in which the molecules being transported move in opposite directions across the membrane.

antisense RNA   A noncoding RNA that binds to a complementary sequence of protein-coding RNA and (usually) prevents its translation.

antisepsis   The removal of pathogens from living tissues.

antiseptic   A chemical that kills microbes.

anti-sigma factor   A protein that inhibits a specific sigma factor, preventing transcription initiation.

AP endonuclease   An enzyme that cleaves the DNA backbone at regions missing a nitrogenous base.

AP site   A position in DNA where there is no base attached to the sugar of the backbone.

APC   See **antigen-presenting cell**.

apical complex   A specialized structure that facilitates entry of apicomplexan parasites into host cells.

apicomplexan   A member of the eukaryotic group Apicomplexa—parasitic alveolates that possess an apical complex used for entry into a host cell.

apurinic site   Also called *AP site*. A DNA site missing a purine base because the bond linking the base to the sugar has been hydrolyzed.

Archaea   One of the three domains of life, consisting of organisms with a last common ancestor not shared with members of Bacteria or Eukarya. Organisms are prokaryotic (lacking nuclei, unlike eukaryotes) and possess ether-linked phospholipid membranes (unlike bacteria).

Archaean eon   The second eon (major time period) of Earth's existence, from 3.8 to 2.5 gigayears (Gyr, 10^9 years) before the present. The earliest geological evidence for life dates to this eon.

archaeon *pl.* **archaea**   A prokaryotic organism that is a member of the domain Archaea, distinct from bacteria and eukaryotes.

aromatic   A ring-shaped organic molecule with π orbital electrons delocalized equally around the ring.

arthrospore   A small vegetative cell, produced by mature mycelia, that gets dispersed.

artifact   A structure viewed through a microscope that is incorrectly interpreted.

ascomycete   A member of the eukaryotic group Ascomycota—fungi whose mycelia form paired nuclei. Haploid ascospores are produced in pods called asci.

ascospore   The spore produced by an ascomycete fungus.

ascus *pl.* **asci**   A spore-containing pod produced by ascomycete fungi.

aseptic   Free of microbes.

asexual reproduction   Reproduction of a cell by fission or by mitosis to form identical daughter cells.

assembly   The packaging of a viral genome into the capsid to form a complete virion.

assimilation   An organism's acquisition of an element, such as carbon from CO_2, to build into body parts.

assimilatory nitrate reduction   The uptake of nitrate and reduction to NH_4^+ by plants and bacteria for use in biosynthetic pathways.

atomic force microscopy (AFM)   A technique that maps the three dimensional topography of a object using van der Waals forces between the object and a probe.

atomic mass   The mass (in grams) of one mole of an element.

atomic number   The number of protons in an atom; it is unique for each element.

ATP   See **adenosine triphosphate**.

ATP synthase   A protein complex that synthesizes ATP from ADP and inorganic phosphate using energy derived from the transmembrane proton potential. It is located in the prokaryotic cell membrane and in the mitochondrial inner membrane.

attenuator stem loop   An intramolecular mRNA structure consisting of a base-paired stem connected by a single-stranded loop. The stem loop structure causes transcription to terminate. Its formation requires efficient translation of a leader peptide sequence.

autoclave   A device that uses pressurized steam to sterilize materials by raising the temperature above the boiling point of water at standard pressure.

autoimmune response   A pathology caused by lymphocytes that can react to self antigens.

autoinducer   A secreted molecule that induces quorum-sensing behavior in bacteria.

autophagy   Eukaryotic cell function normally used to degrade damaged organelles. Also used to kill intracellular pathogens.

autoradiography   The visualization of a radioactive probe by exposing the probed material to X-ray film, followed by photographic development of the film.

autotroph   An organism that can reduce carbon dioxide to produce organic carbon for biosynthesis.

axenic growth   The ability of an organism to grow in the absence of any other species, as, for example, in a pure culture.

azidothymidine (AZT)   A nucleotide analog that inhibits reverse transcriptase and was the first drug clinically used to fight HIV infections.

AZT   See **azidothymidine**.

B cell   An adaptive immune cell that can give rise to antibody-producing cells.

B-cell receptor   A B-cell membrane protein complex containing an antibody in association with the Igα and Igβ immunoglobulins.

B-cell tolerance   The exposure of B cells to a high antigen dose, preventing future antibody production against that antigen.

bacillus *pl.* **bacilli**   See **rod**.

bacitracin   A topical antibiotic that affects cell wall synthesis.

bacteremia   A bacterial infection of the blood.

Bacteria   One of the three domains of life, consisting of organisms with a last common ancestor not shared with members of Archaea or Eukarya. Organisms are prokaryotic (lacking nuclei, unlike eukaryotes) and possess ester-linked phospholipid membranes (unlike archaea).

bactericidal   Having the ability to kill bacterial cells.

bacteriochlorophyll   The chlorophyll of anaerobic phototrophs; it absorbs photons most strongly in the far-red end of the light spectrum.

bacteriophage   Also called *phage*. A virus that infects bacteria.

bacteriorhodopsin (BR)   An archaeal membrane-embedded protein that contains retinal and acts as a light-driven proton pump; it is homologous to the bacterial proteorhodopsin.

bacteriostatic   Having the ability to inhibit the growth of bacterial cells.

bacterium *pl.* **bacteria**   A prokaryotic organism that is a member of the domain Bacteria.

bacteroid   Cell wall–less, undividing, differentiated rhizobial cell within a plant cell. The bacteroid provides fixed nitrogen for the plant.

Bacteroidetes   A phylum of Gram-negative bacteria; nearly all members are obligate anaerobes.

banded iron formation (BIF)   A geological formation consisting of layers of oxidized iron (Fe^{3+}), which indicates formation under oxygen-rich conditions.

barophile   Also called *piezophile*. An organism that requires high pressure to grow.

base excision repair (BER) A DNA repair mechanism that cleaves damaged bases off the sugar-phosphate backbone. After endonuclease activity at the AP site, a new, correct DNA strand is synthesized complementary to the undamaged strand.

basidiomycete A member of the eukaryotic group Basidiomycota—fungi that form mushrooms.

basidiospore A haploid spore formed by a basidiomycete through meiosis of a basidium, a reproductive cell of a mushroom.

basophil A white blood cell, stained by basic dyes, that secretes compounds that aid innate immunity.

batch culture The growth of bacteria in a closed system without inputs of nutrients.

benthic organism An organism that lives on the ocean floor or within the sediment.

BER See **base excision repair**.

BIF See **banded iron formation**.

binary fission The process of replication in which one cell divides to form two genetically equivalent daughter cells of equal size.

bioburden The number of microorganisms found in a pharmaceutical product or in a body part.

biochemical oxygen demand (BOD) Also called *biological oxygen demand*. The amount of oxygen removed from an environment by aerobic respiration.

biofilm A community of microbes growing on a solid surface.

biogeochemical cycle The recycling of elements needed for life (such as carbon or nitrogen) through the biotic and abiotic components of the biosphere.

biogeochemistry Also called *geomicrobiology*. The metabolic interactions of microbial communities with the abiotic (mineral) components of their ecosystems.

bioinformatics A discipline at the intersection of biology and computing that analyzes gene and protein sequence data.

biological oxygen demand See **biochemical oxygen demand**.

biological signature See **biosignature**.

biomass The mass found in bodies of living organisms.

bioprospecting The search for organisms with potential commercial applications.

bioremediation The use of microbes to detoxify environmental contaminants.

biosignature Also called *biological signature*. A chemical indicator of life.

biosphere The region containing the sum total of all life on Earth.

biosynthesis See **anabolism**.

biotic Caused by living organisms.

black smoker An oceanic thermal vent containing high concentrations of minerals such as iron sulfide.

blood-brain barrier A selectively permeable membrane made up of tightly packed capillaries that supply blood to the brain and spinal cord. Large molecules and most pathogens cannot permeate the narrow spaces. Fat-soluble (lipophilic) molecules and oxygen can dissolve through the capillary cell membranes and are absorbed into the brain.

BOD See **biochemical oxygen demand**.

borreliosis Also called *Lyme disease*. A tick-borne disease caused by *Borrelia burgdorferi*, which may involve skin lesions and arthritis.

botulism A food-borne disease caused by a *Clostridium botulinum* toxin, involving muscle paralysis.

BR See **bacteriorhodopsin**.

bradykinin A cell signaling molecule that promotes extravasation, activates mast cells, and stimulates pain perception.

bright-field microscopy A type of light microscopy in which the specimen absorbs light and appears dark against a light background.

bubonic plague A disease caused by the bacterium *Yersinia pestis*; characterized by swollen lymph nodes that often turn black.

budding A form of reproduction in which mitosis of the mother cell generates daughter cells of unequal size.

burst size The number of virus particles released from a lysed host cell.

C-reactive protein A peptide that stimulates the complement cascade, induced by cytokines in the liver. Elevated levels in the blood may be associated with heart disease.

calorimetry A technique to measure the amount of heat released or absorbed during a reaction.

Calvin cycle Also called *Calvin-Benson cycle, Calvin-Benson-Bassham cycle*, or *CBB cycle*. The metabolic pathway of carbon fixation in which the CO_2-condensing step is catalyzed by Rubisco. Found in chloroplasts and in many bacteria.

candidate species A newly described microbial isolate that may become accepted as an official species.

cannula *pl.* **cannulae** A narrow tube.

capsid The protein shell that surrounds a virion's nucleic acid.

capsule A slippery outer layer composed of polysaccharides that surrounds the cell envelope of some bacteria.

carbon-concentrating mechanism (CCM) The inducible expression of transporters of CO_2 and bicarbonate (HCO_3^-) into carboxysomes to enhance CO_2 levels near Rubisco.

carbon dioxide fixation The enzymatic covalent incorporation of inorganic carbon dioxide (CO_2) into an organic compound.

carbon monoxide reductase pathway The carbon fixation process in methanogenic archaea, so called because the key enzyme can fix both CO and CO_2.

carbon sink An ecosystem that removes carbon dioxide from the atmosphere, for example by fixation into biomass or by dissolving into marine water.

carboxysome A protein-enclosed compartment containing Rubisco to fix CO_2.

cardiolipin Diphosphatidylglycerol, a double phospholipid linked by glycerol.

carotenoid An accessory photosynthetic pigment that absorbs photons in the green end of the spectrum.

catabolism The cellular breakdown of large molecules into smaller molecules, releasing energy.

catabolite repression The inhibition of transcription of an operon encoding catabolic proteins in the presence of a more favorable catabolite, such as glucose.

catalytic RNA Also called *ribozyme*. An RNA capable of enzymatic reactions.

catenane   Linked rings of DNA found immediately after replication of circular chromosomes.

cation   A positively charged ion.

CBB cycle   See **Calvin cycle**.

CCM   See **carbon-concentrating mechanism**.

CCR   See **chemokine receptor**.

cDNA   See **complementary DNA**.

cell-mediated immunity   A type of adaptive immunity employing mainly T-cell lymphocytes.

cell membrane   Also called *cytoplasmic membrane* or *plasma membrane*. The phospholipid bilayer that encloses the cytoplasm.

cell surface receptor   A transmembrane protein that senses a specific extracellular signal and may be the docking site for a specific virus.

cell wall   A rigid structure external to the cell membrane. The molecular composition depends on the organism; it is composed of peptidoglycan in bacteria.

cellular slime mold   A slime mold in which the individual cells retain their own cell membranes.

CFTR   See **cystic fibrosis transmembrane conductance regulator**.

chain of infection   The serial passage of a pathogenic organism from an infected individual to an uninfected individual, thus transmitting disease.

chancre   A painless, hard lesion due to an inflammatory reaction at the site of infection with *Treponema pallidum*, the causative agent of syphilis.

chaperone   Also called *chaperonin*. A protein that helps other proteins fold into their correct tertiary structure.

cheddared curd   Curd that has been cut and piled in order to remove the liquid whey.

cheese   A solid or semisolid food product prepared by coagulating milk proteins. Its production commonly involves microbial fermentation.

chemiosmotic theory   A theory stating that the products of oxidative metabolism store their energy in an electrochemical gradient that can drive cell processes such as ATP synthesis.

chemoautotroph   Also called *chemolithotroph* or *lithotroph*. An organism that oxidizes inorganic compounds to yield energy and reduce carbon dioxide.

chemoautotrophy   The metabolic oxidation of inorganic compounds to yield energy used to reduce carbon dioxide.

chemoheterotroph   An organism that oxidizes organic compounds to yield energy without the use of light.

chemoheterotrophy   The metabolic oxidation of organic compounds to yield energy without the use of light.

chemokine   An attractant for white blood cells that is produced by damaged tissues.

chemokine receptor (CCR)   A human T-cell membrane protein that binds chemokine hormones but is also used by HIV for attachment and infection.

chemolithotroph   See **chemoautotroph**.

chemolithotrophy   See **lithotrophy**.

chemostat   A continuous culture system in which the introduced medium contains a limiting nutrient.

chemotaxis   The ability of organisms to move toward or away from specific chemicals.

chemotrophy   Metabolism that yields energy from oxidation-reduction reactions without using light energy.

ChIP   See **chromatin immunoprecipitation**.

chiral carbon   A carbon bonded to four different types of functional groups; it can thus take two different forms exhibiting mirror symmetry.

chlamydia   An obligate, intracellular parasitic bacterium of the phylum Chlamydiae.

Chlamydiae   A phylum of intracellular parasitic bacteria that lose most of their cell envelope during intracellular growth and generate multiple spore-like structures that escape to infect the next host.

chloramphenicol   A bacteriostatic antibiotic that acts by inhibiting peptidyltransferase activity of the bacterial ribosome.

Chlorobi   A phylum of Gram-negative bacteria. They are obligate anaerobes, "green sulfur" phototrophs that photolyze sulfides or H_2.

chlorophyll   A magnesium-containing porphyrin pigment that captures light energy at the start of photosynthesis.

chlorophyte   See **green alga**.

chloroplast   An organelle of endosymbiotic origin that conducts oxygenic photosynthesis; found in algae and plant cells.

chlorosome   A membranous photosynthetic organelle found in bacteria of the genus *Chloroflexus*.

cholesterol   A sterol lipid found in eukaryotic cell membranes.

chromatin   Chromosomal DNA complexed with proteins. Usually refers to a eukaryotic chromosome.

chromatin immunoprecipitation (ChIP)   An experimental method used to determine the DNA-binding sites on a chromosome to which a DNA protein binds.

chromophore   A light-absorbing redox cofactor.

chrysophyte   A member of the eukaryotic group Chrysophyceae (also known as "golden algae")—flagellated heterokont protists possessing chloroplasts as secondary endosymbionts.

cilia *sing.* **cilium**   Short, hairlike structures that beat in waves to propel a eukaryotic cell; structurally similar to flagella.

ciliate   An alveolate that has paired cilia.

cistron   A functional unit of RNA, containing the information from a single gene.

citric acid cycle   See **tricarboxylic acid cycle**.

clade   Also called *monophyletic group*. A group of organisms that includes an ancestral species and all of its descendants.

class switching   See **isotype switching**.

classical complement pathway   An antibody-mediated pathway for complement activation.

classification   The recognition of different forms of life and their placement into different categories.

clonal   Giving rise to a population of genetically identical cells, all descendants of a single cell.

clonal selection   The rapid proliferation of a subset of B cells during the primary or secondary antibody responses.

clone library   See **genome library**.

clone pool See **genome library**.

cloning The insertion of DNA into a plasmid where it can be replicated.

cloning vector A small genome that can carry specific genes for cloning.

CoA See **coenzyme A**.

coastal shelf Shallow regions of the ocean, less than 200 meters deep.

coccus *pl.* **cocci** A spherically shaped microbial cell.

codon A set of three nucleotides that encodes a particular amino acid.

coenzyme A (CoA) A nonprotein cellular organic molecule that can carry acetyl groups and participates in metabolism.

coevolution The evolution of two species in response to one another.

cofactor A metallic ion or a coenzyme required by an enzyme to perform normal catalysis.

cointegrate A DNA molecule formed by a single site recombination event joining two participating circular DNA molecules.

cold seep A cold region of the seafloor where seeping hydrogen sulfide and methane feed bacteria and their animal symbionts, such as tube worms.

colony A visible cluster of microbes on a plate, all derived from a single founding microbe.

commensal organism An organism that benefits from, but neither helps nor harms the host. In medical usage, some commensals benefit the host. Commensal organisms are normally found at various nonsterile host body sites.

commensalism An interaction between two different species that benefits only one partner.

competence factor A species-specific secreted bacterial protein that induces competence for transformation.

competent Able to take up DNA from the environment.

complement Innate immunity proteins in the blood that form holes in bacterial membranes, killing the bacteria.

complementary DNA (cDNA) DNA synthesized complementary to an RNA template via reverse transcriptase.

complex medium A nutrient-rich growth solution including undefined chemical components such as beef broth.

complex transposon A transposon containing a gene for the transposase enzyme, which is needed for replicative transposition.

composite transposon A transposon containing genes in addition to those for transposition, such as antibiotic resistance or catabolic functions.

compound microscope A microscope with multiple lenses to compensate for lens aberration and increase magnification.

compromised host An animal with a weakened immune system.

condensation The joining of two molecules to release a water molecule.

condenser A lens that focuses parallel light rays from the light source onto a small area of the specimen to improve the resolution of the objective lens.

conditional lethal mutation A mutation that leads to death under one growth condition but permits growth under a second condition.

confluent A lawn of organisms that have completely covered a surface.

confocal laser scanning microscopy A type of fluorescence microscopy in which the excitation and emitted light are laser beams focused together, producing high-resolution images.

congenital syphilis Syphilis contracted in utero.

conjugation Horizontal gene transmission involving cell-cell contact. In bacteria, pili draw together the donor and recipient cell envelopes, and a protein complex transmits DNA across. In ciliated eukaryotes, a conjugation bridge forms between two cells connecting their cytoplasm, through which micronuclei are exchanged.

conjugative transposon A transposon that can be transferred from one cell to another via conjugation.

consensus sequence A sequence of nucleotides or amino acids with a common function at many nucleic acid or protein positions. Consists of the base pair or amino acid most frequently found at each position in the sequence.

constant region The region of an antibody that defines the class of a heavy chain or a light chain.

consumer An organism that acquires nutrients from producers, either directly or indirectly.

contig A sequence of overlapping fragments of cloned DNA that are contiguous along a chromosome.

continuous culture A culture system in which new medium is continually added to replace old medium.

contractile vacuole An organelle in eukaryotic microbes that pumps water out of the cell.

contrast Differential absorption or reflection of electromagnetic radiation between an object and background that allows the object to be distinguished from the background.

coral bleaching The death or expulsion of coral algal symbionts. One cause is an increase in temperature.

coralline alga An algal species that calcifies its fronds into hardened shapes similar to corals.

core genome A set of genes shared by a group of related bacterial strains, showing stable inheritance.

core particle A viral capsid that encloses its nucleic acid genome and is surrounded by an envelope. In a eukaryotic cell, a core particle consists of a segment of DNA wrapped around a nucleus.

core polysaccharide A sugar chain that attaches to the glucosamine of lipopolysaccharides and extends outside the cell.

coreceptor A cell surface receptor needed for viral entry along with a primary receptor.

corepressor A small molecule that must bind to a repressor to allow the repressor to bind operator DNA.

cortical alveolus One of the vesicles that forms a network in the outer covering of an alveolate.

counterstain A secondary stain used to visualize cells that do not retain the first stain.

coupled transport The movement of a substance against its electrochemical gradient (from lower to higher concentration, or from opposite charge to like charge) using the energy provided by the simultaneous movement of a different chemical down its electrochemical gradient.

covalent bond   A bond between two atoms that share a pair of electrons.

Crenarchaeota   One of the two major divisions of Archaea, containing sulfur thermophiles and marine mesophiles.

CRISPR   Clustered regularly interspaced short palindromic repeats. Consists of short repeated DNA sequences in a bacterial or archaeal genome, derived from previous bacteriophage or viral infection and conferring protection from future infection; considered a prokaryotic "immune system."

cross-bridge   An attachment that links parallel molecules, such as the peptide link between glycan chains in peptidoglycan.

cryo–electron microscopy (cryo-EM)   Also called *electron cryomicroscopy*. Electron microscopy in which the sample is cooled rapidly in a cryoprotectant medium that prevents freezing. The sample does not need to be stained.

cryo–electron tomography   Also called *electron cryotomography*. A method of cryo–electron microscopy in which the electron beam generates multiple views in parallel planes through the specimen.

cryptogamic crust   A low-growing ground cover consisting of an algal-fungal symbiont, similar to lichens.

curd   Coagulated milk proteins produced by the combined action of lactic acid–producing bacteria and stomach enzymes of certain mammals, such as cattle.

Cyanobacteria   A phylum of oxygen-producing photoautotrophic bacteria containing chlorophylls *a* and *b*. They are closely related to chloroplasts.

cyclic photophosphorylation   A photosynthetic process in which chlorophyll serves as both the initial electron donor and the final electron acceptor. ATP is produced via the proton potential from an electron transport system, but no NADPH is generated.

cycloserine   A polypeptide antibiotic that inhibits peptidoglycan synthesis.

cystic fibrosis transmembrane conductance regulator (CFTR)   A chloride channel found in respiratory epithelia. Mutations in the CFTR gene lead to cystic fibrosis.

cytochrome   A membrane protein that donates and receives electrons.

cytokine   A small, secreted host protein that binds to receptors on various endothelial and immune system cells, regulating the cells' responses.

cytoplasm   Also called *cytosol*. The aqueous solution contained by the cell membrane in all cells and outside the nucleus (in eukaryotes).

cytoplasmic membrane   See **cell membrane**.

cytoskeleton   A collection of filamentous proteins that impart structure to and aid movement of cells; in a eukaryote, these include intermediate filaments, microtubules, and microfilaments.

cytosol   See **cytoplasm**.

cytotoxic T cell (T$_C$ cell)   A T cell that expresses CD8 on its cell surface and can secrete toxic proteins such as perforin and granzymes.

D-value   See **decimal reduction time**.

dark-field microscopy   The detection of microbes too small to be resolved by light rays by observing the light they scatter.

dead zone   Also called *zone of hypoxia*. An anoxic region of an ocean, devoid of most fish and invertebrates.

death phase   A period of cell culture, following stationary phase, during which bacteria die faster than they replicate.

death rate   The rate at which cells die; it is exponential during the death phase.

decay-accelerating factor   A host cell membrane protein that stimulates decay of complement factors and prevents their deposition at the cell surface.

decimal reduction time (D-value)   The length of time it takes for a treatment to kill 90% of a microbial population, and hence a measure of the efficacy of the treatment.

decomposer   An organism that consumes dead biomass.

defensin   A type of small, positively charged peptide, produced by animal tissues, that destroys the cell membranes of invading microbes.

defined minimal medium   A solution of known compounds for organismal growth that contains only the minimal components required for growth.

degenerative evolution   See **reductive evolution**.

degranulate   To release antimicrobial granule contents by fusing granule membranes to cytoplasmic or vacuolar membranes.

degron   A degradation signal contained within a protein.

dehalorespiration   The reduction of halogenated organic molecules by H_2.

delayed-type hypersensitivity (DTH)   See **type IV hypersensitivity**.

deletion   The loss of nucleotides from a DNA sequence.

denature   To lose secondary and tertiary structure in a protein or nucleic acid because of high temperature or chemical treatment.

dendritic cell   An antigen-presenting white blood cell that primarily takes up small soluble antigens from its surroundings.

denitrification   Also called *dissimilatory nitrate reduction*. Energy-yielding metabolism in which nitrate (NO_3^-) is reduced to nitrite (NO_2^-), diatomic nitrogen (N_2), and in some cases ammonia (NH_3).

depth of field   A region of the optical column over which a specimen appears in reasonable focus.

derepression   An increase in gene expression caused by the decrease in concentration of a corepressor.

desensitization   A clinical treatment to reduce allergic reactions by exposing patients to small doses of an allergen.

Desulfurococcales   An order of thermophilic archaea (Crenarchaeota) that metabolize sulfur and organic compounds.

detection   The ability to determine the presence of an object.

detritus   Discarded biomass that can be consumed by decomposers.

diaphragm   A device in a microscope to vary the diameter of the light column, changing the amount of light admitted.

diatom   A member of the eukaryotic group Bacillariophyceae—protists known for intricate, silica-containing bipartite shells.

diauxic growth   A biphasic cell growth curve caused by depletion of the favored carbon source and a metabolic switch to the second carbon source.

dichotomous key   A tool for identifying organisms, in which a series of yes/no decisions successively narrows down the possible categories of species.

differential medium   A growth medium that can distinguish between various bacteria based on metabolic differences.

differential stain   A stain that differentiates among objects by staining only particular types of cells or specific subcellular structures.

diffusion   The energy-independent net movement of a substance from a region of high concentration to a region of lower concentration.

dilution streaking   A method of spreading bacteria on a plate in order to obtain colonies arising from an individual bacterium.

dinoflagellate   A member of the eukaryotic group Dinoflagellata—tertiary endosymbiont algae, alveolates with two flagella, one of which is wrapped distinctively around the cell equator.

diphosphatidylglycerol   See **cardiolipin**.

diplococcus   The paired cocci of *Neisseria* species.

direct repeat   Two identical sequences in a DNA molecule, aligned in the same direction.

disinfection   The removal of pathogenic organisms from inanimate surfaces.

dissemination   The movement of virions from the initial site of infection to other regions of the body.

dissimilation   An organism's catabolism or oxidation of nutrients to inorganic minerals that are released into the environment.

dissimilatory denitrification   Metabolic reduction of nitrate or nitrite to yield energy; anaerobic respiration of nitrate or nitrite.

dissimilatory metal reduction   A type of anaerobic respiration that uses metal cations as terminal electron acceptors.

dissimilatory nitrate reduction   See **denitrification**.

DNA cassette   A manipulable fragment of DNA carrying one or more genes that can be transferred from one DNA sequence to another.

DNA control sequence   A region of DNA, such as the promoter region, that controls the expression of structural genes but is not itself transcribed to RNA.

DNA ligase   An enzyme cells use to form a covalent bond at a nick in the phosphodiester backbone. Also used in molecular biology laboratories to join pieces of DNA.

DNA microarray (or microchip)   A microchip containing short DNA sequences corresponding to all the open reading frames in an organism affixed to specific locations. It can be used to measure the amount of specific mRNA molecules transcribed in cells.

DNA reverse-transcribing virus   See **pararetrovirus**.

DNA sequencing   A technique to determine the order of bases in a DNA sample.

dormant   See **viable but nonculturable**.

doubling time   The generation time of bacteria in culture. The amount of time it takes for the population to double.

downstream processing   The recovery and purification of a commercial product produced by industrial microbes.

DTH   See **type IV hypersensitivity**.

E site   See **exit site**.

early gene   A viral gene expressed early in the infectious cycle.

eclipse period   The time after viral genome injection into host cell but before complete virions are formed.

ecosystem   A community of species plus their environment (habitat).

ectomycorrhizae   Mycorrhizae that colonize the surface of plant roots. Their mycelia do not penetrate the root cells.

ectoparasite   A harmful organism that colonizes the surface of a host.

ED pathway   See **Entner-Doudoroff pathway**.

edema   Tissue swelling due to fluid accumulation.

edema factor (EF)   A component of anthrax toxin with adenylate cyclase activity.

EF   See **edema factor**.

electrochemical potential   A type of potential energy formed by the combined concentration gradient of a molecule and the electrical potential across a membrane.

electrogenic   Describing a transport system that results in a net movement of charged molecules across a membrane.

electromagnetic radiation   Energy radiating in the form of alternating electrical and magnetic waves, quantized in photons.

electron acceptor   An oxidized molecule (e.g., NAD^+) that can accept electrons.

electron cryomicroscopy   See **cryo–electron microscopy**.

electron cryotomography   See **cryo–electron tomography**.

electron donor   A reduced molecule (e.g., NADH) that can donate electrons.

electron microscope   A microscope that obtains high resolution and magnification by focusing electron beams on samples using magnetic lenses.

electron microscopy (EM)   A form of microscopy in which a beam of electrons accelerated through a voltage potential is focused by magnetic lenses onto a specimen.

electron transport chain   See **electron transport system**.

electron transport system (ETS)   Also called *electron transport chain*. A collection of membrane proteins that converts the energy of redox reactions into a proton potential.

electronegativity   The affinity of an atom for electrons. The greater the electronegativity, the stronger the attraction for electrons.

electroneutral   Describing a transport system that results in no net change in charge across the membrane.

electrophoresis   A technique to separate charged proteins and nucleic acids based on how rapidly they migrate in an electric field through a gel.

electrophoretic mobility shift assay (EMSA)   A technique to observe DNA-protein interactions based on the ability of a bound protein to slow the voltage-driven migration of DNA through a gel.

electroporation   A laboratory technique that temporarily makes the cell membrane more leaky to allow the uptake of DNA.

elementary body   The endospore-like form of chlamydias transmitted outside host cells.

EM   See **electron microscopy**.

Embden-Meyerhof-Parnas (EMP) pathway   A glycolytic pathway in which glucose 6-phosphate isomerizes to fructose 6-phosphate, ultimately yielding 2 pyruvate, 2 ATP, and 2 NADH.

emission wavelength   The wavelength of light emitted by a fluorescent molecule. It is of a lower energy and longer wavelength than the excitation wavelength.

EMP pathway   See **Embden-Meyerhof-Parnas pathway**.

empty magnification   Magnification without an increase in resolution.

EMSA   See **electrophoretic mobility shift assay**.

endemic   Describing a disease that is always present in a population, although the frequency of infection may be low.

endergonic   Describing a reaction that requires an input of energy to proceed.

endocarditis   An inflammation of the heart's inner lining.

endocytosis   The invagination of the cell membrane to form a vesicle that contains extracellular material.

endogenous retrovirus   A retroelement that contains *gag*, *env*, and *pol* genes.

endolith   A bacterium that grows within the crystals of solid rock.

endomembrane system   A series of membranous organelles that organize uptake, transport, digestion, and expulsion of particles through a eukaryotic cell. Includes endosomes, lysosomes, endoplasmic reticulum, and Golgi apparatus.

endomycorrhizae   Mycorrhizae in which fungal hyphae penetrate plant root cells.

endoparasite   A harmful organism that lives inside a host.

endophyte   An endosymbiont of vascular plants.

endosome   A vesicle formed from the pinching in of the cell membrane.

endospore   A durable, inert, heat-resistant spore that can remain viable for thousands of years.

endosymbiont   An organism that lives as a symbiont inside another organism.

endosymbiosis   An intimate association between different species in which one partner population grows within the body of another organism.

endotoxin   Lipopolysaccharide in the outer membrane of Gram-negative bacteria that becomes toxic to the host after the bacterial cell has lysed.

energy   The ability to do work.

energy carrier   A molecule in the cell, such as ATP or NADH, that serves as energy currency. Energy carriers are produced during catabolic reactions and can be used to drive energy-requiring reactions.

enhancer   A noncoding DNA region in eukaryotes that can lead to activation of transcription when bound by the appropriate transcription factor. Its location on the chromosome can be far removed from the regulated gene.

enriched medium   A growth solution for fastidious bacteria, consisting of complex medium plus additional components.

enrichment culture   The use of selective growth media to allow only certain microbes to grow.

enterotoxin   A protein that damages the intestine of the host and causes diarrhea; produced by some Gram-negative pathogens.

enthalpy   A measure of the heat energy in a system.

Entner-Doudoroff (ED) pathway   A glycolytic pathway in which glucose 6-phosphate is initially oxidized to 6-phosphogluconate, and ultimately yields 1 pyruvate, 1 ATP, 1 NADH and 1 NADPH.

entropy   A measure of the disorder in a system.

envelope   A structure external to the cell membrane, such as a cell wall or outer membrane. For a virus, the envelope is a membrane enclosing the capsid or core particle.

enzyme   A biological catalyst; a protein or RNA that can speed up the progress of a reaction without itself being changed.

eosinophil   A white blood cell that stains with the acidic dye eosin and secretes compounds that facilitate innate immunity.

epidemic   A disease outbreak in which large numbers of individuals in a population become infected over a short time.

epidemiology   The study of factors affecting the health and illness of populations.

epidermis   The outer protective cell layer in most multicellular animals.

episome   A DNA element that can exist as part of the chromosome or independently, as a plasmid.

epitope   See **antigenic determinant**.

EPS   See **exopolysaccharide**.

equilibrium   A dynamic state in which there is no net change in a reaction.

equivalence   The antigen/antibody ratio that leads to immunoprecipitation of large, insoluble complexes.

error-prone repair   Low-accuracy DNA repair mechanisms that allow mutations.

error-proof repair   DNA repair mechanisms that minimize the formation of mutations.

erythema migrans   A bull's-eye rash characteristic of borreliosis (Lyme disease).

essential nutrient   A compound that an organism cannot synthesize and must acquire from the environment in order to survive.

ethanolic fermentation   A fermentation reaction yielding 2 ethanol and 2 CO_2 as products.

ETS   See **electron transport system**.

Eukarya   One of the three domains of life, consisting of organisms with a last common ancestor not shared with members of Archaea or Bacteria. Cells possess nuclei, unlike cells of bacteria and archaea.

eukaryote   An organism whose cells contain a nucleus. All eukaryotes are members of the domain Eukarya.

Eumycota   True fungi, a taxonomic group of opisthokont eukaryotes with chitinous cell walls; the group most closely related to animals.

euphotic zone   Also called *photic zone* or *photic region*. The region of the ocean that receives sunlight capable of supporting photosynthesis.

Euryarchaeota   One of the two major divisions of Archaea, containing methanogens, halophiles, and extreme acidophiles.

eutrophic lake   A lake in which overgrowth of heterotrophic microbes has eliminated oxygen, leading to a decrease in animal life.

eutrophication   A sudden increase of a formerly limiting nutrient in an aquatic environment, leading to overgrowth of algae and grazing bacteria and subsequent oxygen depletion.

excitation wavelength   The wavelength of light that must be absorbed by a molecule in order for the molecule to fluoresce. It is of a higher energy and shorter wavelength than the emission wavelength.

exergonic   Describing a spontaneous reaction that releases free energy.

exit site (E site)   The region of the ribosome that holds the uncharged, exiting tRNA.

exocytosis   Fusion of vesicles with the cell membrane to release vesicle contents extracellularly.

exon   An expressed or protein-coding portion of a eukaryotic gene.

exonuclease   An enzyme that cleaves DNA from the end.

exopolysaccharide (EPS)   A thick extracellular matrix of polysaccharides and entrapped materials that forms around the microbes in a biofilm.

exotoxin   Protein toxin, secreted by bacteria, that kills or damages host cells.

exponential phase   Also called *logarithmic (log) phase*. A period of cell culture during which bacteria grow exponentially at their maximal possible rate given the conditions.

extravasation   The movement of immune cells out of blood vessels and into surrounding infected tissue.

extremophile   An organism that grows only in an extreme environment—that is, an environment including one or more conditions that are "extreme" relative to the conditions for human life.

eyepiece   See **ocular lens**.

F⁻ cell   The DNA recipient cell in conjugation.

F⁺ cell   The DNA donor cell that transmits the fertility factor F⁺ to an F⁻ cell during conjugation.

F factor   See **fertility factor**.

F-prime (F′) plasmid   Also called *F-prime (F′) factor*. A fertility factor plasmid that contains some chromosomal DNA.

F(ab′)₂ region   The amino-terminal variable "arms" of an antibody that bind to a specific antigen.

facilitated diffusion   A process of passive transport across a membrane facilitated by transport proteins.

FACS   See **fluorescence-activated cell sorter**.

factor H   A normal serum protein that prevents the inadvertent activation of complement in the absence of infection.

facultative   Able to grow in the presence or absence of a given environmental factor, such as oxygen.

facultative anaerobe   An organism that can grow in either the presence or absence of oxygen.

facultative intracellular pathogen   A pathogen that can live either inside host cells or outside host cells.

FAD   See **flavin adenine dinucleotide**.

fatty acid synthase complex   A collection of all the enzymes and binding proteins necessary for fatty acid synthesis.

Fc region   The region of an antibody that binds to specific receptors on host cells in an antigen-independent manner. It is found in the carboxy-terminal "tail" region of the antibody.

feline leukemia virus (FeLV)   A retrovirus that is a major cat pathogen.

FeLV   See **feline leukemia virus**.

fermentation   Also called *fermentative metabolism*. The production of ATP via substrate-level phosphorylation, using organic compounds as both electron donors and electron acceptors.

fermented food   Food products that are biochemically modified by microbial growth.

ferredoxin   An iron- and sulfur-containing protein that transfers electrons in electron transport systems.

fertility factor (F factor)   A specific plasmid (transferred by an F⁺ donor cell) that contains the genes needed for pilus formation and DNA export.

filamentous virus   A viral structure type consisting of a helical capsid surrounding a single-stranded nucleic acid.

fimbria *pl.* **fimbriae**   See **pilus**.

Firmicutes   A phylum of low-GC-content Gram-positive bacteria.

fixation   The adherence of cells to a slide by a chemical or heat treatment.

flagellate   A protist that has one or more flagella.

flagellum *pl.* **flagella**   A filamentous structure for motility. In prokaryotes, a helical protein filament attached to a rotary motor; in eukaryotes, an undulating cell membrane–enclosed complex of microtubules and ATP-driven motor proteins.

flavin adenine dinucleotide (FAD)   An energy carrier in the cell that can donate (FADH₂) or accept (FAD⁺) electrons.

flavonoid   A type of signaling molecule released from legumes to attract nitrogen-fixing rhizobia.

fluid mosaic model   A model of the cell membrane in which proteins are free to diffuse laterally within the membrane.

fluorescence   The emission of light from a molecule that absorbed light of a shorter, higher-energy wavelength.

fluorescence resonance energy transfer (FRET)   The detectable transfer of fluorescent energy from one molecule to another. Since the participating molecules must be near each other, FRET can be used to monitor protein-protein interactions in cells and is also used in real-time PCR.

fluorescence-activated cell sorter (FACS)   A device that can count cells and sort them based on differences in fluorescence.

fluorescent-focus assay   An assay to detect viruses that do not kill host cells, based on intracellular detection of viruses using antiviral antibodies or green fluorescent protein–modified viruses.

fluorophore   A fluorescent molecule used to stain specimens for fluorescence microscopy.

focal plane   A plane that contains the focal point for a given lens.

focal point The position at which light rays that pass through a lens intersect.

focus *pl.* **foci** 1. The point at which rays of energy converge; in light microscopy, the convergence of light rays maximizes the clarity of the optical image. 2. A group of cells infected by a virus.

folate Folic acid, a heteroaromatic cofactor that is required by some enzymes.

fomite An inanimate object on which pathogens can be transmitted from one host to another.

food poisoning The presence of human disease–causing microbial pathogens or toxins in food.

food spoilage Microbial changes that render a food unfit or unpalatable for consumption.

food web A network of interactions in which organisms obtain or provide nutrients for each other—for example, by predation or by mutualism.

foraminiferan An ameba with a calcium carbonate shell and a helical arrangement of chambers.

forespore In sporulation of Gram-positive bacteria, the smaller cell compartment formed through asymmetrical cell division; it develops into the endospore.

fossil fuel Ancient organismal remains that have been converted to hydrocarbons through microbial digestion followed by reduction under high pressure underground; extracted and burned by humans for energy.

frameshift mutation A gene mutation involving the insertion or deletion of nucleotides that cause a shift in the codon reading frame.

free energy change See **Gibbs free energy change**.

freeze-drying Also called *lyophilization*. The removal of water from food, by freezing under vacuum, to limit microbial growth.

FRET See **fluorescence resonance energy transfer**.

fruiting body A multicellular fungal or bacterial reproductive structure.

frustule The silica bipartite shell produced by a diatom.

functional genomics A field of molecular biology that utilizes data from genomic projects to describe gene and protein interactions.

functional group A cluster of covalently bound atoms that behaves with specific properties and functions as a unit.

fungus *pl.* **fungi** A heterotrophic opisthokont eukaryote with chitinous cell walls. Includes Eumycota, but traditionally may refer to fungus-like protists such as the oomycetes.

fusion peptide A portion of a viral envelope protein that changes shape to facilitate envelope fusion with the host cell membrane.

gain-of-function mutation A mutation that enhances the activity or allows new activity of a gene product.

GALT See **gut-associated lymphoid tissue**.

gamogony The differentiation of parasitic haploid cells into male and female gametes.

gas vesicle An organelle that traps gases to increase buoyancy of aquatic microbes.

GC content The proportion of an organism's genome consisting of guanine-cytosine base pairs.

GDH See **glutamate dehydrogenase**.

gene fusion Also called *protein fusion* or *translational fusion*. A technique to measure control of gene transcription or translation by inserting a reporter gene into a target gene. The reporter relies on both the promoter and the ribosome-binding site of the target gene.

gene switching Switching between two (out of five) different classes of immunoglobulin genes (e.g., from IgM to IgG) during B cell development.

gene transfer vector A mobile DNA engineered from a virus or plasmid, designed to insert a genetic sequence into the genome of an organism for experimental study or for medical therapy.

generalized recombination Recombination between two DNA molecules that share long regions of DNA homology.

generalized transduction A phage-mediated gene transfer process in which any donor gene can be transferred to a recipient cell.

generation time The species-specific time period for doubling of a population (e.g., by bacterial cell division) in a given environment, assuming no depletion of resources.

genetic analysis Determination of the function of cell RNAs and proteins based on the phenotype of cells in which the gene encoding the RNA or protein is mutated.

genome The complete genetic content of an organism. The sequence of all the nucleotides in a haploid set of chromosomes.

genome library Also called *clone library* or *clone pool*. A population of host bacteria, each of which carries a DNA molecule cloned from an organism's genome. The set of clones contains overlapping DNA fragments representing the entire genome.

genomic island A region of DNA sequence whose properties indicate that it has been transferred from another genome. Usually comprises a set of genes with shared function, such as pathogenicity or symbiosis support.

genus name The Latin name assigned to the taxonomic rank consisting of closely related species.

geochemical cycling The global interconversion of various inorganic and organic forms of elements.

geomicrobiology See **biogeochemistry**.

geosmin A molecule released from decaying *Streptomyces* cells; it causes the characteristic odor of soil and can affect the taste of drinking water.

germ theory of disease The theory that many diseases are caused by microbes.

germicidal Able to kill cells but not spores.

germination The activation of a dormant spore to generate a vegetative cell.

Gibbs free energy change (ΔG) Also called *free energy change* and *Gibbs energy change*. In a chemical reaction, a measure of how much energy available to do work is released or required as the reaction proceeds.

gliding motility The movement of cells individually or as a collective over surfaces using pili.

gluconeogenesis The biosynthesis of glucose from single-carbon compounds.

glucosamine A glucose modified with an amine group.

glutamate dehydrogenase (GDH) An enzyme that condenses NH_4^+ with 2-oxoglutarate to form glutamate. The condensation requires reduction by NADPH.

glutamate synthase (GOGAT) Glutamine:2-oxoglutarate amidotransferase, an enzyme that converts 2-oxoglutarate plus glutamine into two molecules of glutamate.

glutamine synthetase (GS) An enzyme that condenses NH_4^+ with glutamate to form glutamine.

glycan See **polysaccharide**.

glycolysis The catabolic pathway of glucose oxidation to pyruvate, generating ATP and NADH.

glyoxalate bypass An alternative to the tricarboxylic acid cycle, induced under low glucose conditions.

gnotobiotic animal An animal that is germ-free or colonized by a known set of microbes.

GOGAT See **glutamate synthase**.

Golgi complex A series of membrane stacks that modifies proteins and helps sort them to the correct eukaryotic cell compartment.

Gram stain A differential stain that distinguishes between cells that possess a thick cell wall and retain a positively charged stain (Gram-positive) from cells with a thin cell wall and outer membrane that fail to retain the stain (Gram-negative).

Gram-negative Describing cells that do not retain the Gram stain.

Gram-positive Describing cells that do retain the Gram stain and appear dark purple after staining.

gramicidin A peptide antibiotic that disrupts membrane integrity.

granuloma A thick lesion formed around a site of infection.

granzyme A cytotoxic T cell–secreted enzyme that damages target cells.

grazer A first-level consumer, feeding directly on producers.

green alga Also called *chlorophyte*. A member of the eukaryotic group Chlorophyta—microbes that have chloroplasts; primary endosymbionts, closely related to plants (Viridiplantae).

greenhouse effect The trapping of solar radiation heat in the atmosphere by CO_2; a cause of global warming.

griseofulvin An antifungal antibiotic that inhibits cell division.

group translocation A form of active transport in which the transported molecule is modified after it enters the cell, thus keeping a favorable inward concentration gradient for the unmodified extracellular molecule.

growth factor A compound needed for the growth of only certain cells.

growth rate The rate of increase in population number or biomass.

growth rate constant The number of organismal generations per unit time; k.

GS See **glutamine synthetase**.

gut-associated lymphoid tissue (GALT) Lymphatic tissues such as tonsils and adenoids that are found in conjunction with the gastrointestinal tract and contain immune cells.

Haber process Industrial nitrogen fixation, in which dinitrogen is hydrogenated by methane (natural gas) under extreme heat and pressure to form ammonia.

Hadean eon The first eon (major time period) of Earth's existence, from 4.5 to 3.8 gigayears (Gyr, 10^9 years) before the present.

half-life The amount of time it takes for one-half of a radioactive sample to decay.

Haloarchaea A class of extremely halophilic archaea, inhabiting high-salt environments.

Halobacteriales An order of Euryarchaeota that contains the class Haloarchaea.

halophile An organism that requires a high extracellular sodium chloride concentration for optimal growth.

halorhodopsin (HR) An archaeal light-driven chloride pump.

hapten A small compound that must be conjugated to a larger carrier antigen in order to elicit production of an antibody that binds to it.

haustorium *pl.* **haustoria** A bulbous hyphal extension of a fungal plant pathogen into the host cell.

heat-shock protein (HSP) A chaperone protein that is induced by high-temperature stress.

heat-shock response A coordinated response of cells to higher-than-normal temperatures. It includes changes in the membrane and expression of heat-shock genes.

heavy chain The larger of the two protein types comprising an antibody. Each antibody contains two heavy chains and two light chains.

helper T cell (T_H cell) A T cell that expresses CD4 on its cell surface and secretes cytokines that modulate B-cell class switching.

heme An organic molecule containing a ring of conjugated double bonds surrounding an iron atom. It is involved in redox reactions and oxygen binding.

hepatitis An inflammation of the liver, caused by infection or by exposure to a toxic substance.

heteroaromatic Having aromatic rings containing noncarbon atoms, such as the nitrogenous bases found in nucleotides.

heterocyst In filamentous cyanobacteria, a specialized nitrogen-fixing cell that maintains a reduced environment and excludes O_2.

heteroduplex A double-stranded nucleic acid in which the two strands come from different sources.

heterokont A member of the eukaryotic group Heterokonta—protists possessing two flagella of unequal length.

heterolactic fermentation A fermentation reaction in which the products are lactic acid, ethanol, and CO_2.

heterotroph An organism that relies on external sources of organic carbon compounds for biosynthesis.

heterotrophy The use of external sources of organic carbon compounds for biosynthesis.

Hfr A high-frequency recombination bacterial strain, caused by the presence of a chromosomally integrated F factor.

histone A protein that helps compact eukaryotic chromosomes in nucleosomes.

HIV See **human immunodeficiency virus**.

holdfast Adhesion factors secreted by the tip of a stalk to firmly attach an organism to a substrate.

Holliday junction Also called *Holliday structure*. A cross-like configuration of recombining DNA molecules that forms during generalized recombination.

homolog A protein whose amino acid sequence is similar to that of another protein.

homologous Similar, referring to DNA sequences derived from a common ancestral gene.

hopanoid Also called *hopane*. A five-ringed hydrocarbon lipid found in bacterial cell membranes.

horizontal gene transfer Also called *lateral gene transfer* or just *horizontal transfer*. The passage of genes from one cell into another mature cell.

horizontal transmission The transfer of a pathogen from one organism into another, nonprogeny organism.

hormogonium *pl.* **hormogonia** A short motile chain of three to five cells produced by filamentous bacteria to disseminate their cells.

host range The species that can be infected by a given pathogen.

HR See **halorhodopsin**.

HSP See **heat-shock protein**.

human immunodeficiency virus (HIV) A human-specific retrovirus that is the causative agent of AIDS.

humic material Also called *humus*. Phenolic molecules, derived from lignin, that are resistant to degradation and hence very stable in soil.

humoral immunity A type of adaptive immunity mediated by antibodies.

humus See **humic material**.

hybridization The annealing of a nucleic acid strand with another nucleic acid strand containing a complementary sequence of bases. The binding of one nucleic acid strand with a complementary strand.

hydric soil Soil that undergoes periods of anoxic water saturation.

hydrogen bond An electrostatic attraction between a hydrogen bound to an oxygen or nitrogen and a second, nearby oxygen or nitrogen.

hydrogenosome Found in some eukaryotes, an organelle that ferments carbohydrates in a pathway generating H_2; it may have evolved from mitochondria.

hydrogenotrophy The use of molecular hydrogen (H_2) as an electron donor for a variety of electron acceptors.

hydrologic cycle Also called *water cycle*. The cyclic exchange of water between atmospheric water vapor and Earth's bodies of liquid water.

hydrolysis The cleaving of a bond by the addition of a water molecule.

hydrophilic Soluble in water; either ionic or polar.

hydrophobic Insoluble in water; nonpolar.

hydrothermal vent Also called *thermal vent*. An opening in the seafloor through which superheated water arises, carrying high concentrations of reduced minerals such as sulfides.

3-hydroxypropionate cycle A carbon fixation process that condenses hydrated CO_2 and acetyl-CoA to form 3-hydroxypropionate.

hyperthermophile An organism adapted for optimal growth at extremely high temperatures, generally above 80°C.

hypertonic Having more solutes than another environment separated by a semipermeable membrane. Water will tend to flow toward the hypertonic solution.

hypha *pl.* **hyphae** Threadlike filament forming the mycelium of a fungus.

hypotonic Having fewer solutes than another environment separated by a semipermeable membrane. Water will tend to flow away from the hypotonic solution.

hypoxia A state of lower-than-normal oxygen concentration.

icosahedral capsid For a virus, a crystalline protein shell with 20 identical faces, enclosing the nucleic acid.

ID$_{50}$ See **infectious dose**.

identification The recognition of the class of a microbe isolated in pure culture.

idiotype Amino acid differences in hypervariable regions (N terminus of heavy and light chains) within a single antibody class in an individual that allow recognition of different antigens.

IEF See **isoelectric focusing**.

IgA An antibody isotype that contains the alpha heavy chain. It can be secreted and found in tears, saliva, breast milk, and so on.

IgD An antibody isotype that contains the delta heavy chain and is found on B cell membranes.

IgE An antibody isotype that contains the epsilon heavy chain and is involved in degranulation of mast cells.

IgG An antibody isotype that contains the gamma heavy chain. It is found in serum.

IgM The first antibody isotype detected during the early stages of an immune response. It contains the mu heavy chain and is found as a pentamer in serum.

IL-1 See **interleukin-1**.

immediate hypersensitivity See **type I hypersensitivity**.

immersion oil An oil with a refractive index similar to glass that minimizes light ray loss at wide angles, thereby minimizing wave front interference and maximizing resolution.

immune avoidance The changing of cell surface proteins by pathogens to prevent antibody detection and prolong infection.

immune system An organism's cellular defense system against pathogens.

immunity Resistance to a specific disease.

immunization The stimulation of an immune response by deliberate inoculation with a weakened pathogen, in hopes of providing immunity to disease caused by the pathogen.

immunogen See **antigen**.

immunogenicity A measure of the effectiveness of an antigen in eliciting an immune response.

immunoglobulin A member of a family of proteins that contain a 110-amino-acid domain with an internal disulfide bond. Members include antibodies and major histocompatibility proteins.

immunological specificity The ability of antibodies produced in response to a particular epitope to bind that epitope almost exclusively. Antibodies made to one epitope bind only weakly, if at all, to other epitopes.

immunomodulin A protein made by normal microbiota that influences the host immune response by modifying the secretion of host proteins, such as a cytokine.

immunopathogenesis The process by which an immune response or the products of an immune response cause disease.

immunoprecipitation The antibody-mediated cross-linking of antigens to form large, insoluble complexes. Immunoprecipitation is used in research labs and is normally seen only in vitro.

in vivo expression technology (IVET) A laboratory technique to uncover bacterial genes that are expressed only during growth within a host.

index case Also called *patient zero*. The first case of an infectious disease and an important piece of data for helping contain the spread of disease.

indigenous microbiota (microflora) Microbes found naturally in a particular location, often in association with a food substrate.

inducer A molecule that binds to a repressor and prevents repressor binding to the operator sequence.

inducer exclusion The ability of glucose to cause metabolic changes that prevent the cellular uptake of less favorable carbon sources that could cause unnecessary induction.

induction Increased transcription of target genes due to an inducer binding to a repressor and preventing repressor-operator binding.

industrial fermentor The equipment used to grow microbes on an industrial scale.

industrial microbiology The commercial exploitation of microbes.

industrial strain A microbial strain whose characteristics are optimized for industrial use.

infection The growth of a pathogen or parasite in or on a host.

infection cycle The route a pathogen takes as it moves from one host into another.

infection thread A column of rhizobial cells that projects down a tube into a plant root cell during the early stages of the rhizobia-legume symbiosis.

infectious dose (ID_{50}) The number of bacteria or virions required to cause disease symptoms in 50% of an experimental group of hosts.

injera A highly fermented Ethiopian flat bread using the grain teff.

innate immunity See **nonadaptive immunity**.

inner leaflet The layer of the cell membrane phospholipid bilayer that faces the cytoplasm.

inner membrane In Gram-negative bacteria, the membrane in contact with the cytoplasm, equivalent to the cell membrane.

insertion The addition of nucleotides to the middle of a DNA sequence.

insertion sequence (IS) A simple transposable element consisting of a transposase gene flanked by short, inverted-repeat sequences that are the target of transposase.

interference The interaction of two wave fronts. Interference can be additive (amplitudes in phase, constructive) or subtractive (amplitudes out of phase, destructive).

interference microscopy Observation of an object using contrast enhanced by superimposing interference bands upon an image to accentuate small differences in refractive indexes.

interferon A host-secreted immunomodulatory protein that inhibits viral replication.

interleukin 1 (IL-1) A cytokine release by macrophages.

intermediate filament A eukaryotic cytoskeletal protein that is composed of various proteins depending on the cell type.

intimin A pathogenic *E. coli* adhesion protein that binds tightly to an *E. coli*–produced receptor injected into host cells.

intracellular pathogen A pathogen that lives within a host cell.

intron In eukaryotic genes, an intervening sequence that does not code for protein and is spliced out of the mRNA prior to translation.

inversion A flipping of a DNA fragment within a chromosome. It may allow or repress the transcription of a particular gene.

inverted repeat A DNA sequence that is found in an identical but inverted form at two sites on the same double helix (e.g., 5′-ATCGATCGnnnnnnCGATCGAT-3′).

ion gradient A difference in concentration of an ion across a membrane.

ionic bond A bond between ions of positive and negative charge.

IS See **insertion sequence**.

isoelectric focusing (IEF) A technique that separates proteins based on their charge.

isoelectric point The pH at which there is no net charge on an amino acid or a protein.

isolate A microbe that has been obtained from a specific location and grown in pure culture.

isoprenoid A condensed isoprene chain, found in archaeal membrane lipids.

isotonic Being in osmotic balance, having equal concentrations of solutes on both sides of a semipermeable membrane. A cell in an isotonic environment will neither gain nor lose water.

isotope An atom of an element with a specific number of neutrons. For example, carbon-12, carbon-13, and carbon-14 are all isotopes of carbon.

isotope ratio The ratio of amounts of two different isotopes of an element. May serve as a biosignature if the ratio between certain isotopes of a given element is altered by biological activity.

isotype A species-specific antibody class, defined by the structure of the heavy chain. IgG, IgA, and IgE are examples of isotypes.

isotype switching Also called *class switching*. A change in the predominant antibody isotype produced by a cell.

IVET See **in vivo expression technology**.

joule (J) The standard SI unit for energy.

Kaposi's sarcoma A malignancy originating in endothelial or lymphatic cells that is often found in late-stage AIDS patients.

kelp A heterokont protist, also known as "brown algae," that grows as multicellular sheets at the water's surface.

kimchi A popular Korean food based on brine-fermented cabbage.

Kirby-Bauer assay A method for determining antibiotic susceptibility. Antibiotic-impregnated disks are placed on an agar plate whose surface has been confluently inoculated with a test organism. The antibiotic diffuses away from the disk and inhibits growth of susceptible bacteria. The width of the inhibitory zone is proportional to the susceptibility of the organism.

knockout mutation A mutation that completely eliminates the activity of the gene product.

Koch's postulates Four criteria that should be met for a microbe to be designated the causative agent of an infectious disease.

Krebs cycle See **tricarboxylic acid cycle**.

labile toxin (LT) An *E. coli* enterotoxin, destroyed by heat, that increases cellular cAMP concentrations.

lactate fermentation Also called *lactic acid fermentation*. A fermentation reaction that generates lactic acid from reduction of pyruvic acid.

lag phase A period of cell culture, occurring right after bacteria are inoculated into new media, during which there is slow growth or no growth.

lamellar pseudopod A pseudopod with a sheetlike morphology.

laminar flow biological safety cabinet An air filtration appliance that removes pathogenic microbes from within the cabinet.

Langerhans cell A specialized, phagocytic dendritic cell that is the predominant cell type in skin-associated lymphatic tissue.

late gene A viral gene expressed late in the infectious cycle.

late-phase anaphylaxis Anaphylaxis caused by leukotrienes released by eosinophils recruited by mast cells.

latent period The time in the viral life cycle when progeny virions have formed but are still within the host cell.

latent state A period of the infection process during which a pathogenic agent is dormant in the host and cannot be cultured.

lateral gene transfer See **horizontal gene transfer**.

law of mass action The tendency of a reaction at equilibrium to return to equilibrium after the concentrations of products or reactants is altered.

LD$_{50}$ See **lethal dose**.

leaching The process of metal dissolution from ores.

leader sequence A short DNA sequence preceding a structural gene. In amino acid operons it contains multiple codons for the amino acid synthesized by the downstream structural genes. The leader sequence and translating ribosome help determine whether the structural genes are transcribed.

leaflet One of the two lipid layers in a phospholipid bilayer. The inner leaflet of the cell membrane faces the cytoplasm.

leaven For bread dough, to cause to rise by generating air spaces, usually through carbon dioxide production by microbial fermentation.

leghemoglobin An iron-bearing plant protein that sequesters oxygen to maintain an anoxic environment for nitrogenase within cells containing bacteroids.

lentivirus A member of a family of retroviruses that propagate slowly. An example is HIV.

lethal dose (LD$_{50}$) A measure of virulence; the number of bacteria or virions required to kill 50% of an experimental group of hosts.

lethal factor (LF) A component of anthrax toxin that cleaves host protein kinases.

LF See **lethal factor**.

lichen An organism formed by a mutualistic relationship between algae and fungi.

light chain The smaller of the two protein types comprising an antibody. Each antibody contains two heavy chains and two light chains.

light microscopy Observation of a microscopic object based on light absorption and transmission.

lignin A complex aromatic organic compound that forms the key structural support for trees and woody stems.

limiting nutrient A nutrient that is in short supply and limits growth.

lincosamide A class of bacteriostatic antibiotics (e.g., clindamycin).

lipopolysaccharide (LPS) Structurally unique phospholipids found in the outer leaflet of the outer membrane in Gram-negative bacteria. Many are endotoxins.

lithotroph See **chemoautotroph**.

lithotrophy Also called *chemolithotrophy*. Energy-yielding metabolism that uses an inorganic electron donor; usually includes fixation of CO_2 into biomass.

littoral zone The upper water layer of aquatic habitats that light can penetrate.

logarithmic (log) phase See **exponential phase**.

long terminal repeat (LTR) A repeated nucleic acid sequence at the 5′ and 3′ ends of a provirus.

loss-of-function mutation A mutation that eliminates or decreases the function of the gene product.

LPS See **lipopolysaccharide**.

LT See **labile toxin**.

LTR See **long terminal repeat**.

lumen The interior of an intracellular membrane-enclosed compartment.

Lyme disease See **borreliosis**.

lymph node A secondary lymphatic organ, formed by the convergence of lymphatic vessels, that traps foreign particles from local tissue and presents them to resident immune cells.

lymphocyte A mononuclear leukocyte (white blood cell) that is a product of lymphoid tissue and participates in immunity (e.g., B cell and T cell).

lyophilization See **freeze-drying**.

lysate The contents of broken cells; may include virus particles.

lysis The rupture of the cell by a break in the cell membrane.

lysogeny A viral life cycle in which the viral genome integrates into and replicates with the host genome, but retains the ability to initiate host cell lysis.

lysosome An acidic eukaryotic organelle that aids digestion of molecules. Not found in plant cells.

lytic Describing a viral life cycle in which the virus produces new virions and lyses the cell, releasing virions.

M cell A phagocytic innate immune cell (microfold cell) found between intestinal epithelial cells.

MAC See **membrane attack complex**.

MacConkey medium A differential, selective medium that selects for Gram-negative bacteria and can differentiate between lactose fermenters and nonfermenters.

macrolide Any of a group of antibiotics containing a large lactone ring (e.g., erythromycin).

macronucleus A form of nucleus found in ciliates, derived from gene amplification and rearrangement of micronuclear DNA, that contains actively transcribed genes.

macronutrient A nutrient that an organism needs in large quantities.

macrophage A mononuclear, phagocytic, antigen-presenting cell of the immune system.

magnetosome An organelle containing the mineral magnetite that allows microbes to sense a magnetic field.

magnetotaxis The ability to direct motility along magnetic field lines.

magnification An increase in the apparent size of a viewed object as an optical image.

major histocompatibility complex (MHC) Transmembrane cell proteins important for recognizing self and for presenting foreign antigens to the adaptive immune system.

malaria A disease caused by the apicomplexan *Plasmodium falciparum*, transmitted by mosquitoes.

malolactic fermentation Fermentation of κ-malate (a side product of glucose fermentation) by *Oenococcus oeni* bacteria; an important process in winemaking.

marine snow Microbial biofilms on inorganic particles suspended in marine water.

mast cell A white blood cell that secretes proteins that aid nonadaptive immunity. Mast cells reside in connective tissues and mucosa and do not circulate in the bloodstream.

matrix protein A protein found in some viruses that is located between the capsid and the membrane envelope.

MCP See **methyl-accepting chemotaxis protein**.

MDR efflux pump See **multidrug resistance efflux pump**.

mean generation time The reciprocal of the mean growth rate constant; the mean time period for doubling of a population.

mechanical transmission A nonspecific mode of viral entry into damaged tissue.

meiosis A form of cell division by which a diploid eukaryotic cell generates haploid sex cells with recombinant chromosomes.

membrane attack complex (MAC) A cell-destroying pore produced in the membrane of invading bacteria by the host cell complement cascade.

membrane potential Energy stored as an electrical voltage difference across a membrane.

membrane-permeant organic acid See **membrane-permeant weak acid**.

membrane-permeant weak acid An acid that exists in charged and uncharged forms, such as acetic acid. The uncharged form can penetrate the membrane.

membrane-permeant weak base A base that exists in charged and uncharged forms, such as methylamine. The uncharged form can penetrate the membrane.

memory B cell A long-lived type of lymphocyte preprogrammed to produce a specific antibody. After encountering their activating antigen, memory B cells differentiate into antibody-producing plasma cells.

merozoite The form of *Plasmodium falciparum*, the causative agent of malaria, that invades red blood cells.

mesocosm A small, controlled model ecosystem.

mesophile An organism with optimal growth between 20°C and 40°C.

messenger RNA (mRNA) An RNA molecule that encodes a protein.

metabolist model A model of early life in which the central components of intermediary metabolism arose from self-sustaining chemical reactions based on inorganic chemicals.

metagenome The sum of genomes of a community of organisms.

metagenomics The study of community genomes, or metagenomes.

metastatic lesion A lesion of infection, or of cancerous cells, that develops at secondary sites away from the initial site of infection.

methane gas hydrate A crystalline material in which methane molecules are surrounded by a cage of water molecules. This molecular configuration is found in the deep ocean.

methanogen An organism that uses hydrogen to reduce CO_2 and other single-carbon compounds to methane, yielding energy.

methanogenesis An energy-yielding metabolic process that produces methane. It is unique to archaea.

methanotroph An organism that oxidizes methane to yield energy.

methanotrophy The metabolic oxidation of methane to yield energy.

methyl-accepting chemotaxis protein (MCP) A cell-membrane signal transduction protein that becomes methylated during adaptation to a chemotactic signal.

methyl mismatch repair A DNA repair system that fixes misincorporation of a nucleotide after DNA synthesis. The unmethylated daughter strand is corrected to complement the methylated parental strand.

methylotroph An organism that oxidizes reduced single-carbon compounds such as methanol to yield energy. Includes methanotrophs (oxidizing methane).

methylotrophy The metabolic oxidation of single-carbon compounds such as methanol, methylamine, or methane to yield energy.

MHC See **major histocompatibility complex**.

MHC restriction The ability of T cells to recognize only those antigens complexed to self MHC molecules.

MIC See **minimal inhibitory concentration**.

microaerophilic Requiring oxygen at a concentration lower than that of the atmosphere, but unable to grow in high-oxygen environments.

microbe An organism or virus too small to be seen with the unaided human eye.

microbiome The total community of microbes found within a specified environment.

microbiota The normally occurring microbes of a body part or an environmental habitat.

microcolony A small colony of bacteria visible only with the aid of a microscope.

microfilament Also called *actin filament*. A eukaryotic cytoskeletal protein composed of polymerized actin.

microfossil A microscopic fossil in which calcium carbonate deposits have filled in the form of ancient microbial cells.

micrometer (μm) One-millionth (10^{-6}) of a meter.

micronucleus A form of nucleus found in ciliates that contains a diploid set of chromosomes and undergoes meiosis for sexual exchange by conjugation.

micronutrient A nutrient that an organism needs in small quantity, typically a vitamin or a mineral.

microplankton Plankton consisting of cells 20–200 μm in diameter.

microscope A tool that increases the magnification of specimens to enable viewing at higher resolution.

microtubule A eukaryotic cytoskeletal protein composed of polymerized tubulin.

millimeter (mm) One-thousandth (10^{-3}) of a meter.

minimal inhibitory concentration (MIC) The lowest concentration of a drug that will prevent the growth of an organism.

miso A Japanese condiment, made from ground soy and rice, salted and fermented by the mold *Aspergillus oryzae*.

missense mutation A point mutation that alters the sequence of a single codon, leading to a single amino acid substitution in a protein.

mitochondrion *pl.* **mitochondria** An organelle of endosymbiotic origin that produces ATP through the use of an electron transport chain to generate a proton motive force. O_2 is the final electron acceptor to produce H_2O.

mitosis The orderly replication and segregation of eukaryotic chromosomes, usually prior to cell division.

mitosporic fungus A species of fungus that generates spores by mitosis and lacks a known sexual cycle.

mixed-acid fermentation A bacterial fermentation process in which pyruvate is converted to several different organic acids, as well as ethanol, CO_2, and H_2O.

mixotroph An organism capable of both photosynthetic and heterotrophic metabolism.

mixotrophic Having the ability to switch among metabolic strategies, such as heterotrophy and phototrophy, depending on the environmental conditions.

modular enzyme A multifunctional enzyme in which several domains or subunits conduct sequential steps to generate a product.

MOI See **multiplicity of infection**.

molarity A unit of concentration measured as the number of moles of solute per liter of solution.

molecular clock The use of DNA or RNA sequence information to measure the time of divergence among different species.

molecular formula A notation indicating the number and type of atoms in a molecule. For example, H_2O is the molecular formula for water.

molecular mimicry A structural similarity between two different molecules.

monocyte A white blood cell with a single nucleus that can differentiate into macrophages or dendritic cells.

monophyletic Diverging from a common ancestor.

monophyletic group See **clade**.

monosaccharide The monomer unit of sugars. Monosaccharides have a molecular formula of $(CH_2O)_n$.

mordant A chemical binding agent that causes specimens to retain stains better.

mother cell Also called *motherspore*. The larger cell that forms during the asymmetrical cell division leading to spore formation. The mother cell will engulf the forespore.

motility The ability of a microbe to generate self-directed movement.

mRNA See **messenger RNA**.

mucociliary elevator The ciliated mucous lining of the trachea, bronchi, and bronchioles that sweeps foreign particles up and away from the lungs.

Mueller-Hinton agar A specialized, standardized, para-aminobenzoic acid–free media used for the Kirby-Bauer assay.

multidrug resistance (MDR) efflux pump A transmembrane protein pump that can export many different kinds of antibiotics with diverse structure.

multiplex PCR A polymerase chain reaction that uses multiple pairs of oligonucleotide primers to amplify several different DNA sequences simultaneously.

multiplicity of infection (MOI) The ratio of infecting virions to host cells.

murein See **peptidoglycan**.

murein lipoprotein The major lipoprotein that connects the outer membrane of Gram-negative bacteria to the peptidoglycan cell wall.

mutagen A chemical that damages DNA and leads to mutations.

mutation A heritable change in the DNA sequence.

mutation frequency The fraction of mutant cells (defective in a given gene) within the total cell population.

mutation rate The number of mutations introduced into DNA per generation (cell doubling).

mutator strain A strain of cells with a high mutation rate, usually due to a mutation in a DNA repair enzyme.

mutualism A symbiotic relationship in which both partners benefit.

mycelium *pl.* **mycelia** A fungal hypha that projects into the air (aerial mycelium) or into the growth substrate.

mycolic acid One of a diverse class of sugar-linked fatty acids found in the cell envelopes of mycobacteria.

mycology The study of fungi.

mycorrhizae *sing.* **mycorrhiza** Fungi involved in an intimate symbiosis with plant roots, in which nutrients are exchanged.

myxospore A durable spherical cell produced by the fruiting body of myxobacteria.

N-terminal rule The tendency of the N-terminal amino acid of a protein to influence protein stability.

NAD See **nicotinamide adenine dinucleotide**.

Nanoarchaeota A deeply branching division of Archaea; includes thermophilic cells of extremely small size.

patient zero   See **index case**.

PBP   See **penicillin-binding protein**.

PCR   See **polymerase chain reaction**.

PDC   See **pyruvate dehydrogenase complex**.

pelagic zone   The water column of the open ocean, away from the shore and the ocean floor.

penicillin   An antibiotic, produced by the *Penicillium* mold, that blocks cross-bridge formation during peptidoglycan synthesis.

penicillin-binding protein (PBP)   A bacterial protein, involved in cell wall synthesis, that is the target of the penicillin antibiotic.

pentose phosphate shunt (PPS)   An alternate glycolytic pathway in which glucose 6-phosphate is first oxidized and then decarboxylated to ribulose 5-phosphate, ultimately generating 1 ATP and 2 NADPH.

peptide bond   The covalent bond that links two amino acid monomers.

peptidoglycan   Also called *murein*. A polymer of peptide-linked chains of amino sugars; a major component of the bacterial cell wall.

peptidyl-tRNA site (P site)   The ribosomal site that contains the growing protein attached to a tRNA.

peptidyltransferase   The rRNA enzymatic ability to form peptide bonds.

perforin   A cytotoxic protein, secreted by T cells, that forms pores in target cell membranes.

peripheral membrane protein   A protein that is associated with a membrane but does not span the phospholipid bilayer.

permease   A substrate-specific carrier protein in the membrane.

permissive temperature   A temperature at which a temperature-sensitive mutation in a gene is masked, permitting growth of the organism.

peroxisome   A eukaryotic organelle that converts hydrogen peroxide to water.

petechia *pl.* **petechiae**   A pinpoint capillary hemorrhage due to the absence of clotting factors; may indicate the presence of endotoxin.

petri dish   Also called *pour plate*. A round dish with vertical walls covered by an inverted dish of slightly larger diameter. The smaller dish can be filled with a substrate for growing microbes.

PFU   See **plaque-forming unit**.

phage   See **bacteriophage**.

phage display   A technique in which a phage particle contains recombinant coat proteins expressed by genes encoding a coat protein fused to the protein of interest, such as a vaccine antigen.

phagocytosis   A form of endocytosis in which large extracellular particles are brought into the cell.

phagosome   A large intracellular vesicle that forms as a result of phagocytosis.

phase-contrast microscopy   Observation of a microscopic object based on the differences in the refractive index between cell components and the surrounding medium. Contrast is generated as the difference between refracted light and transmitted light shifted out of phase.

phase variation   A gene regulatory mechanism that changes the amino acid sequence of a protein from one antigenic type to another. One mechanism involves site-specific recombination that flips a DNA sequence in a chromosome.

phenol coefficient test   A test of the ability of a disinfectant to kill bacteria; the higher the coefficient, the more effective the disinfectant.

phenol red broth test   A clinical test for particular bacterial fermentation pathways, indicating the presence or absence of particular species, based on fermentative acids changing a pH indicator.

phosphatidate   A negatively charged phosphate head group of a phospholipid.

phosphatidylethanolamine   A type of phospholipid with a positively charged ethanolamine attached to the phosphate group.

phosphatidylglycerol   A type of phospholipid with a glycerol attached to the phosphate group.

phosphodiester bond   The bond that covalently attaches to adjacent nucleotides in a nucleic acid.

phospholipid   The major component of membranes. A typical phospholipid is composed of a core of glycerol to which two fatty acids and a modified phosphate group are attached.

phospholipid bilayer   Two layers of phospholipids; the hydrocarbon fatty acid tails face the interior of the bilayer, and the charged phosphate groups face the cytoplasm and extracellular environment.

phosphorylation   The enzyme-catalyzed addition of a phosphoryl group onto a molecule.

phosphotransferase system (PTS)   A group translocation system that uses phosphoenolpyruvate to transfer phosphoryl groups onto the incoming molecule.

photic zone   See **euphotic zone**.

photoautotroph   An organism that performs photosynthesis, using light energy to reduce carbon dioxide.

photoautotrophy   The metabolic reduction of carbon dioxide using light as an energy source.

photoheterotrophy   The photolysis of organic compounds to yield energy.

photolysis   The first energy-yielding phase of photosynthesis; the light-driven separation of an electron from a molecule coupled to an electron transport system.

photoreactivation   A light-induced, photolyase-catalyzed repair of pyrimidine dimers.

photosynthesis   The metabolic ability to absorb and convert solar energy into chemical energy for biosynthesis; a precise definition includes CO_2 fixation.

photosystem I   A protein complex that harvests light from a chlorophyll, splits an electron from a small molecule such as H_2S or H_2O, and stores energy in the form of NADPH.

photosystem II   A protein complex that splits an electron from bacteriochlorophyll and stores energy in the form of a proton potential.

phototrophy   The use of chemical reactions triggered by the absorption of light to yield energy.

phylogenetic tree   A diagram depicting estimates of the relative amounts of evolutionary divergence among different species.

phylogeny   A measurement of genetic relatedness. The classification of animals based on their genetic relatedness.

phylum *pl.* **phyla**   The taxonomic rank one level below domain; a group of organisms sharing a common ancestor that diverged early from other groups.

phytoplankton   Phototrophic marine bacteria, algae, and protists, the primary producers in pelagic food webs.

picometer (pm)   One-trillionth (10^{-12}) of a meter.

picoplankton   Plankton consisting of cells 0.3–2 μm in diameter.

picornavirus   A member of a family of icosahedral RNA viruses. Examples include poliovirus and rhinovirus.

piezophile   See **barophile**.

pilin   The protein monomer that polymerizes to form a pilus.

pilus *pl.* **pili**   Also called *fimbria*. A straight protein filament composed of a tube of protein monomers that extend from the bacterial cell envelope.

pinocytosis   A form of endocytosis in which only extracellular fluid and small molecules are brought into the cell.

Planctomycetes   A phylum of free-living bacteria that have stalked cells and reproduce by budding. Their nucleoid is surrounded by a membrane.

plankton   Organisms that float in water.

planktonic cell   An isolated cell, growing individually in a liquid without connections to other cells.

plaque   A cell-free zone on a lawn of bacterial cells caused by viral lysis.

plaque assay   An assay to determine the presence of bacteriophages based on their ability to form plaques.

plaque-forming unit (PFU)   A measure of the concentration of phage particles in liquid culture.

plasma cell   A short-lived antibody-producing cell.

plasma membrane   See **cell membrane**.

plasmid   An extrachromosmal genetic element that may be present in some cells.

plasmodesma *pl.* **plasmodesmata**   A membrane channel in plants that connects adjacent plant cells.

plasmodial slime mold   A slime mold in which the mass of aggregating cells becomes a multinucleate single cell.

plasmodium *pl.* **plasmodia**   The giant multinucleate cell formed by a plasmodial slime mold.

pneumonic plague   A highly virulent and contagious *Yersinia pestis* lung infection.

point mutation   A change in a single nucleotide within a nucleic acid sequence.

polar covalent bond   A covalent bond in which the electrons are distributed unequally between two atoms.

poliomyelitis (polio)   The paralytic disease caused by the poliovirus.

poliovirus   A picornavirus that is the causative agent of poliomyelitis.

poliovirus receptor (PVR)   The cell surface receptor unique to humans and simians through which poliovirus gains entry to cells.

polyamine   A molecule containing multiple amine groups, positively charged at neutral pH.

polycistronic   Produced from an operon containing several genes and hence containing several functional sequences (usually encoding different proteins).

polymerase chain reaction (PCR)   A method to amplify DNA in vitro using many cycles of DNA denaturation, primer annealing, and DNA polymerization with a heat-stable polymerase.

polyphyletic   Having multiple evolutionary origins.

polysaccharide   Also called *glycan*. A polymer of sugars.

polysome   A cell structure consisting of multiple ribosomes performing translation on the same mRNA molecule.

porin   A transmembrane protein complex that allows movement of specific molecules across the cell membrane or the outer membrane.

positive selection   Growth of a population under a condition that favors a particular genotype and prevents growth of other genotypes. In immunology: the survival of T cells bearing T-cell receptors (TCRs) that don't recognize self MHC proteins displayed on thymus epithelial cells.

pour plate   See **petri dish**.

PPS   See **pentose phosphate shunt**.

pre-mRNA   See **preliminary mRNA transcript**.

prebiotic soup   A model for the origin of life based on the abiotic formation of fundamental biomolecules and cell structures such as membranes out of a "soup" of nutrients present on early Earth.

predator   A consumer that feeds on grazers.

preliminary mRNA transcript (pre-mRNA)   A eukaryotic messenger RNA prior to intron removal.

primary antibody response   The production of antibodies upon first exposure to a particular antigen. B cells become activated and differentiate into plasma cells and memory B cells.

primary endosymbiont   An organism in a lineage derived from a single endosymbiotic event.

primary pathogen   A disease-causing microbe that can breach the defenses of a healthy host.

primary producer   An organism that produces biomass (reduced carbon) from inorganic carbon sources such as CO_2.

primary recovery   The initial isolation of commercial product from industrial microbes.

primary structure   The first level of organization of polymers, consisting of the linear sequence of monomers—for example, the sequence of amino acids in a protein or nucleotides in a nucleic acid.

primary syphilis   The initial inflammatory reaction (chancre) at the site of infection with *Treponema pallidum*.

primase   An RNA polymerase that synthesizes short RNA primers complementary to a DNA template to launch DNA replication.

primer extension   A technique to determine the 5′ end of an RNA transcript.

prion   An infectious agent that causes propagation of misfolded host proteins; usually consists of a defective version of the host protein.

probabilistic indicator   A means of quickly identifying microbes in the clinical setting, based on a battery of biochemical tests performed simultaneously on an isolated strain.

probiotic A food or nutritional supplement that contains live microorganisms and aims to improve health by promoting beneficial bacteria.

prokaryote An organism whose cell or cells lack a nucleus; includes both bacteria and archaea.

promoter A noncoding DNA regulatory region immediately upstream of a structural gene that is needed for transcription initiation.

proofreading An enzymatic activity of some nucleic acid polymerases that attempts to correct mispaired bases.

prophage A phage genome integrated into a host genome.

propionic acid fermentation The fermentation of lactic acid to propionic acid by *Propionibacterium* species; used in the production of Swiss cheese.

protease inhibitor A molecule that inhibits a protease enzyme; some are used as anti-HIV drugs to block the virally encoded protease needed to complete HIV assembly.

protective antigen (PA) The core subunit of anthrax toxin, so called because immunity to this protein protects against disease.

protein A A *Staphylococcus aureus* cell wall protein that binds to the Fc region of antibodies, hiding the *S. aureus* cells from phagocytes.

protein fusion See **gene fusion**.

Proteobacteria A large, metabolically and morphologically diverse phylum of Gram-negative bacteria.

proteome All the proteins expressed in a cell at a given time. The "complete proteome" includes all the proteins the cell can express under any condition. The "expressed proteome" represents the set of proteins made under a given condition.

proteomics The biological field of proteome analysis.

proteorhodopsin A bacterial membrane-embedded protein that contains retinal and acts as a light-driven proton pump; it is homologous to the archaeal protein bacteriorhodopsin.

protist A single-celled eukaryotic microbe, usually motile; not a fungus.

proton motive force See **proton potential**.

proton potential The potential energy of the concentration gradient of protons (hydrogen ions, H^+) plus the charge difference across a membrane.

protozoan *pl.* **protozoa** A heterotrophic eukaryotic microbe, usually motile, that is not a fungus.

provirus A viral genome that is integrated into the host cell genome.

pseudogene A gene that is no longer functional.

pseudomurein See **pseudopeptidoglycan**.

pseudopeptidoglycan A peptidoglycan-like molecule composed of sugars and peptides that is found in some archaeal cell walls.

pseudopod A locomotory extension of cytoplasm bounded by the cell membrane.

psychrophile An organism with optimal growth at temperatures below 20°C.

PTS See **phosphotransferase system**.

pure culture A culture containing only a single strain or species of microorganism. A large number of microorganisms all descended from a single individual cell.

purine A nitrogenous base with fused rings found in nucleotides; examples are adenine and guanine.

putrefaction Food spoilage due to the decomposition of proteins and amino acids.

PVR See **poliovirus receptor**.

pyrimidine A single-ring nitrogenous base found in nucleotides; examples are cytosine, thymine, and uracil.

pyrogen Any substance that induces fever.

pyrosequencing A method of DNA sequencing that relies on the detection of pyrophosphate released upon nucleotide incorporation.

pyruvate dehydrogenase complex (PDC) The multisubunit enzyme that couples the oxidative decarboxylation of pyruvate to acetyl-CoA and NADH production.

quarantine The separation of infectious individuals from the general population to limit the spread of infection.

quasispecies A collection of isolates (usually viruses) from a common source of infection that have evolved into many different types within one host.

quaternary structure The highest level of organization of proteins, in which multiple polypeptide chains interact and function together.

quinol A reduced electron carrier that can diffuse laterally within membranes.

quinolone A type of antibiotic drug that inhibits DNA synthesis by targeting bacterial topoisomerases such as DNA gyrase.

quinone An oxidized electron carrier that can diffuse laterally within membranes.

quinone pool The oxidized quinones and reduced quinols that diffuse freely within the phospholipid membrane and are able to transfer electrons between many different redox enzymes.

quorum sensing The ability of bacteria to sense the presence of other bacteria via secreted chemical signals called autoinducers.

radiolarian A member of the eukaryotic group Radiolaria—amebas with a silicate shell penetrated by filamentous pseudopods.

rancidity Food spoilage due to the oxidation of fats; may or may not involve microbial activity.

RC See **reaction center**.

reaction center (RC) The chlorophyll molecule that donates its excited electron to an electron transport system.

reading frame The position in a nucleic acid sequence from which triplet codons encode amino acids.

real-time PCR A technique using fluorescence to detect the products of PCR amplification as the reaction progresses, in order to quantify the amount of DNA in a sample.

recombinant DNA DNA that has been combined with other DNA to make novel DNA sequences.

recombination The process by which a donor DNA molecule replaces a segment of a host genome or is inserted into a host genome.

recombination signal sequence (RSS) A DNA region downstream of antibody heavy- and light-chain genes that allows recombination between widely separated gene segments.

recombinational repair A DNA repair mechanism that relies on recombination between an undamaged chromosome and a gap that occurred during replication of damaged DNA.

red alga Also called *rhodophyte*. A unicellular primary alga that contains red pigment.

redox couple The oxidized and reduced states of a compound. For example, NAD$^+$ and NADH form a redox couple.

redox reaction A reaction in which one molecule or functional group becomes reduced and another becomes oxidized.

reductive acetyl-CoA pathway A carbon assimilation pathway in which two CO_2 molecules are condensed and reduced by 2 H_2 molecules to form an acetyl group.

reductive evolution Also called *degenerative evolution*. The loss or mutation of DNA encoding unselected traits.

reductive pentose phosphate cycle See **Calvin cycle**.

reductive (reverse) TCA cycle A CO_2 fixation pathway that generates acetyl-CoA through reversal of TCA cycle reactions. It requires ATP and NADPH.

reflection Deflection of an incident light ray by an object, at an angle equal to the incident angle.

refraction The bending and slowing of light as it passes through a substance.

refractive index The degree to which a substance causes the refraction of light; a ratio of the speed of light in a vacuum to its speed in another medium.

regulatory protein A protein that can bind DNA and modulate transcription in response to a metabolite.

regulatory T cell (Treg) A T cell that regulates the activity of another T cell, usually by suppressing its activity.

regulon A group of genes and operons that is coordinately regulated and shares a common biochemical function.

release factor A molecule that enters a ribosome A site containing an mRNA stop codon and initiates protein cleavage from the tRNA.

replication fork During DNA synthesis, the region of the chromosome that is being unwound.

replisome A complex of DNA polymerase and other accessory molecules that performs DNA replication.

reporter gene A gene, such as *lacZ* (beta-galactosidase) or *gfp* (green fluorescent protein), whose protein product can be easily quantified; commonly used in a gene fusion.

repression The down-regulation of gene transcription.

repressor A regulatory protein that can bind to a specific DNA sequence and inhibit transcription of genes.

reservoir An organism that maintains a virus in an area by serving as a high-titer host.

resolution The smallest distance that two objects can be separated and still be distinguished as separate objects.

resource colimitation A situation in which a population size is limited by the lack of two different resources, such as nitrogen and phosphorus.

respiration The oxidation of reduced organic electron donors through a series of membrane-embedded electron carriers to a final electron acceptor. The energy derived from the redox reactions is stored as an electrochemical gradient across the membrane, which may be harnessed to produce ATP.

response regulator A cytoplasmic protein that is phosphorylated by a sensor kinase and modulates gene transcription depending on its phosphorylation state.

restricted transduction See **specialized transduction**.

restriction endonuclease Also called *restriction enzyme*. A bacterial enzyme that cleaves double-stranded DNA within a specific short sequence, usually a palindrome.

restriction site A DNA sequence recognized and cleaved by a restriction endonuclease.

restrictive temperature A temperature at which a temperature-sensitive mutation in a gene leads to the mutant phenotype, which generally includes failure to grow.

reticulate body The metabolically and reproductively active form of chlamydias.

reticuloendothelial system A collection of cells that can phagocytose and sequester extracellular material.

retinal A vitamin A–related cofactor in opsin proteins; it undergoes a conformational change in response to photon absorption.

retrotransposons A retroelement that contains only partial retroviral sequences but may contain reverse transcriptase to allow further movement into the host genome.

retrovirus Also called *RNA reverse-transcribing virus*. A virus containing a positive single-stranded RNA genome; uses reverse transcriptase to generate a double-stranded DNA.

reverse transcriptase (RT) An enzyme that produces a double-stranded DNA molecule from a single-stranded RNA template.

reversion A mutation that changes a previous mutation back to its original state.

rhinovirus A picornavirus that causes the common cold.

rhizobium *pl.* **rhizobia** A bacterial species of the order Rhizobiales that forms highly specific mutualistic associations with plants in which the bacteria fix nitrogen for the plant.

rhizoplane The surface of plant roots.

rhizosphere The soil environment surrounding plant roots.

Rho-dependent Describing a bacterial transcription termination mechanism that requires Rho protein.

Rho-independent Describing a bacterial transcription termination mechanism that requires a GC-rich region near the transcript terminus.

rhodophyte See **red alga**.

ribosomal RNA (rRNA) RNA molecules that include the scaffolding and catalytic components of ribosomes.

ribosome-binding site Also called *Shine-Dalgarno sequence*. In bacteria, a stretch of nucleotides upstream of the start codon in an mRNA that hybridizes to the 16S rRNA of the ribosome, correctly positioning the mRNA for translation.

ribozyme See **catalytic RNA**.

ribulose 5-phosphate The five-carbon sugar ribulose, phosphorylated at carbon 5.

ripening The aging of cheese.

rise period During viral culture, the time when cells lyse and viral progeny enter the media.

RNA reverse-transcribing virus See **retrovirus**.

RNA world A model of early life in which RNA performed all the informational and catalytic roles of today's DNA and proteins.

RNA-dependent RNA polymerase   An enzyme that produces an RNA complementary to a template RNA strand.

rod   Also called *bacillus*. 1. A bacterium with a linear shape. 2. A photoreceptor cell.

root of a tree   The earliest common ancestor of all members of a phylogenetic tree.

rotavirus   One of a group of nonenveloped dsRNA viruses that cause severe diarrhea in children.

Rous sarcoma virus (RSV)   A retrovirus that was the first virus shown to cause cancer.

rRNA   See **ribosomal RNA**.

RSS   See **recombination signal sequence**.

RSV   See **Rous sarcoma virus**.

RT   See **reverse transcriptase**.

Rubisco   Ribulose 1,5-bisphosphate carbon dioxide reductase/oxidase, the enzyme that catalyzes the carbon fixation step in the Calvin cycle.

rumen   The first chamber of the digestive tract of ruminant animals such as cattle; the main site for microbial digestion of feed.

S-layer   A crystalline protein surface layer replacing or external to the cell wall in many species of archaea and bacteria.

sacculus *pl.* **sacculi**   The bacterial cell wall, consisting of a single covalent molecule.

SALT   See **skin-associated lymphoid tissue**.

sanitation   The safe disposal of wastes hazardous to humans.

saprophyte   A fungal decomposer.

sarcina *pl.* **sarcinae**   A cubical octad cluster of cells formed by septation at right angles to the previous cell division.

sargassum weed   Unrooted kelp forests that float in marine water.

scanning electron microscopy (SEM)   Electron microscopy in which the electron beams scan across the specimen's surface to reveal the three-dimensional topology of the specimen.

scattering   Interaction of light with an object resulting in propagation of spherical light waves at relatively low intensity.

schizogony   Mitotic reproduction of parasitic cells to achieve a large population within a host tissue.

secondary antibody response   A memory B cell–mediated rapid increase in the production of antibodies in response to a repeat exposure to a particular antigen.

secondary endosymbiont   An organism evolved through engulfment of a primary endosymbiont.

secondary metabolite   Also called *secondary product*. A biosynthetic product that is not an essential nutrient but enhances nutrient uptake or inhibits competing species (e.g., an antibiotic).

secondary structure   The second level of organization of polymers, consisting of regular patterns that repeat, such as the double helix in DNA or the beta sheet in proteins.

secondary syphilis   A rash that may appear at some point after the primary latent stage of syphilis.

sedimentation rate   The rate at which particles of a given size and shape travel to the bottom of a tube under centrifugal force. The rate depends on the particle's mass and cross-sectional area.

segmented genome   A viral genome that consists of more than one nucleic acid molecule.

selectin   One of a family of cell adhesion molecules.

selective medium   A medium that allows the growth of certain species or strains of organisms but not others.

selective toxicity   The ability of a drug, at a given dose, to harm the pathogen and not the host.

selectively permeable membrane   See **semipermeable membrane**.

SEM   See **scanning electron microscopy**.

semiconservative   The mode of DNA replication whereby each new double helix contains one old, parental strand and one newly synthesized daughter strand.

semipermeable membrane   Also called *selectively permeable membrane*. A membrane that is permeable to some substances but impermeable to other substances.

sensor kinase   A transmembrane protein that phosphorylates itself in response to an extracellular signal, and transfers the phosphoryl group to a receiver protein.

septation   The formation of a septum, a new section of cell wall and envelope to separate two prokaryotic daughter cells.

septicemia   An infection of the bloodstream.

septicemic plague   Infection of the bloodstream by *Yersinia pestis*.

septum *pl.* **septa**   A plate of cell wall and envelope that forms to separate two daughter cells.

sequela *pl.* **sequelae**   A serious, harmful immunological consequence of bacterial and host antigen cross-reactivity that occurs after the infection itself is over. An example is rheumatic fever.

serial endosymbiosis theory   The theory that mitochondria and chloroplasts were originally free-living prokaryotes that formed an internal symbiosis with early eukaryotes.

serum   The noncell, liquid component of the blood.

sex pilus   A pilus specialized for DNA transfer between bacteria.

sexual reproduction   Reproduction involving the joining of gametes generated by meiosis.

Shine-Dalgarno sequence   See **ribosome-binding site**.

shuttle vector   A plasmid with origins of replication recognized by both bacteria and eukaryotes.

siderophore   A high-affinity iron-binding protein used to scavenge iron from the environment and deliver it to a siderophore-producing organism.

sigma factor   A protein needed to bind RNA polymerase for the initiation of transcription in bacteria.

signal recognition particle (SRP)   A receptor that recognizes the signal sequence of peptides undergoing translation. The complex attaches to the cell membrane of prokaryotes (or the rough endoplasmic reticulum of eukaryotes), where it docks the protein-ribosome complex to the membrane for protein membrane insertion or secretion.

signal sequence   A specific amino acid sequence on the amino terminus of proteins that directs them to the endoplasmic reticulum (of a eukaryote) or the cell membrane (of a prokaryote).

silencer   A eukaryotic DNA element that can lead to decreased transcription when bound by an appropriate transcription factor.

silent mutation   A mutation that does not change the amino acid sequence encoded by an open reading frame. The changed codon encodes the same amino acid as the original codon.

simple stain   A stain that makes an object more opaque, increasing its contrast with the external medium or surrounding tissue.

single-celled protein   An edible microbe of high food value, such as *Spirulina* or some yeasts.

site-specific recombination   Recombination between DNA molecules that do not share long regions of homology but do contain short regions of homology specifically recognized by the recombination enzyme.

skin-associated lymphoid tissue (SALT)   Immune cells, such as dendritic cells, located under the skin that help eliminate bacteria that have breached the skin surface.

sliding clamp   A protein that keeps DNA polymerase affixed to DNA during replication.

slime mold   An organism in which unicellular amebas can aggregate into a fruiting body.

sludge   The solid products of wastewater treatment.

small RNA (sRNA)   Non-protein-coding regulatory RNAs that modulate translation.

small-subunit (SSU) rRNA   In bacteria, *16S rRNA*. A ribosomal RNA found in the small subunit of the ribosome. Its gene is often sequenced for phylogenetic comparisons.

soil   A complex mixture of decaying organic and mineral matter that covers the terrestrial portions of the planet.

solute   Any dissolved molecule.

SOS response   A coordinated cellular response to extensive DNA damage. It includes error-prone repair.

sourdough   An undefined yeast population, derived from a previous batch of dough, that is used in bread production.

Southern blot   A technique to detect specific DNA sequences. Sample DNA segments are separated by gel electrophoresis, transferred to a blot, and probed with a labeled DNA that will hybridize to complementary DNA sequences.

space-filling model   A molecular model that represents the volume of the electron orbitals of the atoms, usually to the limit of the van der Waals radii.

specialized transduction   Also called *restricted transduction*. Transduction in which the phage can transfer only a specific, limited number of donor genes to the recipient cell.

species   A single, specific type of organism, designated by a genus and species name.

species name   The scientific name of a specific type of organism; presented following the genus name, as in *Escherichia* (genus) *coli* (species).

spectrum of activity   The range of pathogens for which an antimicrobial agent is effective.

spike protein   A viral glycoprotein that connects the membrane to the capsid or the matrix and may be involved in viral binding to host cell receptors.

spirochete   A bacterium with a tight, flexible spiral shape; a species of the phylum Spirochetes (Spirochaeta).

Spirochetes   Also called *Spirochaeta*. A phylum of bacteria with a unique morphology: a flexible, extended spiral that twists via intracellular flagella.

spongiform encephalopathy   A brain-wasting disease caused by a prion.

spontaneous generation   The theory, much debated in the nineteenth century, that under current Earth conditions life can arise spontaneously from nonliving matter.

sporangiospore   A haploid spore that can germinate to form a haploid mycelium.

sporangium *pl.* **sporangia**   A fungal organ that disburses non-motile spores via ballistic expulsion.

spore stain   A type of differential stain that is specific for the endospore coat of various bacteria, typically a firmicute species.

sporogony   The development of a parasite's zygote into a spore-like form transmissible to the next host.

spread plate   A method to grow separate bacterial colonies by plating serial dilutions of a liquid culture.

sRNA   See **small RNA**.

SRP   See **signal recognition particle**.

SSU rRNA   See **small-subunit rRNA**.

staining   The process of treating microscopic specimens with a stain to enhance their detection or to visualize specific cell components.

stalk   An extension of the cytoplasm and envelope that attaches a microbe to a substrate.

stalked ciliate   A ciliate that adheres to a substrate and uses its cilia to obtain prey.

staphylococcus *pl.* **staphylococci**   A hexagonal arrangement of cells formed by septation in random orientations.

start codon   A codon (usually AUG) that signals the first amino acid of a protein.

starter culture   A mixture of fermenting microbes added to a food substrate to generate a fermented product.

stationary phase   A period of cell culture, following exponential phase, during which there is no net increase in replication.

sterilization   The destruction of all cells, spores, and viruses on an object.

stick model   A molecular model in which stick lengths represent the distances between bonded pairs of atomic nuclei.

stop codon   One of three codons (UAA, UAG, UGA) that do not encode an amino acid and thus trigger the end of translation.

streptogramin   An antibiotic that binds 23S rRNA and blocks elongation of protein synthesis in bacteria.

strict aerobe   An organism that performs aerobic respiration and can grow only in the presence of oxygen.

strict anaerobe   An organism that cannot grow in the presence of oxygen.

stringent response   A cellular response to idle ribosomes (often indicating low carbon and energy stores) that includes a decrease in rRNA and tRNA production.

stroma   The compartment contained by the inner chloroplast membrane where the light-independent reactions of photosynthesis occur (CO_2 fixation).

stromatolite   A mass of sedimentary layers of limestone produced by a marine microbial community over many years.

structural formula   A representation of molecular structure in which covalent bonds are shown as a line between atoms.

structural gene   A string of nucleotides that encodes a functional RNA molecule.

structural isomer   A molecule with the same molecular formula as a different molecule but a different arrangement of atoms.

subcellular fractionation   A procedure to separate cell components; often includes ultracentrifugation.

substrate-binding protein   An extracytoplasmic protein that binds specific substrates and delivers them to their cognate uptake ABC transporters.

substrate-level phosphorylation   The formation of ATP by the enzymatic transfer of phosphate from a substrate molecule onto ADP.

sulfa drug   An antibiotic that inhibits folic acid synthesis and, thus, nucleotide synthesis.

Sulfolobales   An order of Crenarchaeota; thermophilic sulfur oxidizers.

superantigen   A molecule that directly stimulates T cells without undergoing antigen-presenting cell processing and surface presentation.

supercoil   An extra twist or turn found in DNA, either positive (increases DNA winding) or negative (decreases DNA winding).

supernova   An exploding star that has used up most of the nuclei available for fusion reactions.

Svedberg coefficient   A measure of particle size based on the particle's sedimentation rate in a tube subjected to a high g force.

swarming   A behavior in which some microbial cells differentiate into large swarmer cells and swim together as a unit.

switch region   A repeating DNA sequence in antibody constant-segment genes that serves as a recombination site during isotype switching.

symbiogenesis   An evolutionary process by which two or more species become intimately associated.

symbiont   An organism that lives in a close association with another organism.

symbiosis   The intimate association of two unrelated species.

symbiosome   A bacteroid enclosed by a plant membrane that mediates the exchange of nutrients between the bacteroid and the host cell.

symporter   A transport protein where the molecules being transported move in the same direction across the membrane.

synergism   Cooperation between species in which both species benefit but can grow independently. The cooperation is less intimate than symbiosis.

synthetic medium   A bacterial growth solution that contains defined, known components.

syntrophy   Metabolic cooperation between two different species.

T cell   A small lymphocyte developed in the thymus; participates in the immune response.

tailed phage   A phage such as T4 that contains a genome delivery device called the tail.

tandem repeat   A stretch of directly repeating DNA sequence (direct repeats) without any intervening DNA.

taxonomy   The description of distinct life-forms and their organization into different categories.

T_C cell   See **cytotoxic T cell**.

TCA cycle   See **tricarboxylic acid cycle**.

tegument   The contents of a virion between the capsid and the envelope.

teichoic acid   A chain of phosphodiester-linked glycerol or ribitol that threads through and reinforces the cell wall in Gram-positive bacteria.

telomerase   A eukaryotic enzyme with reverse transcriptase activity that is required to replicate the ends of linear chromosomes to avoid chromosome shortening.

telomere   The end of a linear chromosome, composed of a repeated DNA sequence.

TEM   See **transmission electron microscopy**.

tempeh   A mold-fermented soy product, popular as a food in parts of Asia.

temperate phage   A phage capable of lysogeny.

template strand   A DNA strand (or an RNA strand in some viruses) that is used as a template for the synthesis of mRNA.

terminal electron acceptor   The final electron acceptor at the end of an electron transport system.

termination (*ter*) site   A sequence of DNA that halts replication of DNA by DNA polymerase.

terpenoid   A branched lipid derived from isoprene that is found in hydrocarbon chains of archaeal membranes.

terraforming   The idea of transforming the environment of another planet to make it suitable for life from Earth.

tertiary structure   The third level of organization of polymers; the unique three-dimensional shape of a polymer.

tertiary syphilis   A final stage of syphilis, manifested by cardiovascular and nervous system symptoms.

tetanospasmin   The tetanus-causing potent exotoxin produced by *Clostridium tetani*.

tetraether   A molecule containing four ether links. An example is found in archaeal membranes, when two lipid side chains form ether linkages with a pair of side chains from the other side of the bilayer.

tetrapyrrole   An essential primary product that contains four pyrroles (five-membered rings containing nitrogen), each with one or two double bonds. A precursor for many important cell cofactors, such as chlorophylls and vitamin B_{12}.

T_H cell   See **helper T cell**.

thermal vent   See **hydrothermal vent**.

thermocline   A region of the ocean where temperature decreases steeply with depth, and water density increases.

thermophile   An organism adapted for optimal growth at high temperatures, usually 55°C or higher.

Thermoproteales   An order of crenarchaeotes containing hyperthermophilic organisms.

threshold dose   The concentration of antigen needed to elicit adequate antibody production.

thylakoid   A chlorophyll-containing membrane folded within a phototrophic bacterium or a chloroplast.

Ti plasmid   A plasmid found in tumorigenic strains of *Agrobacterium tumefaciens* that can be used as a vector to introduce DNA into plant cells.

TLR   See **Toll-like receptor**.

tmRNA   A molecule resembling both tRNA and mRNA that rescues ribosomes stalled on damaged mRNAs lacking a stop codon.

TNF-α   See **tumor necrosis factor alpha**.

Toll-like receptor (TLR)   A member of a eukaryotic transmembrane glycoprotein family that recognizes a particular pathogen-associated molecular pattern (PAMP) present on pathogenic microorganisms.

tomography   The acquisition of projected images of a transparent specimen from different angles that are digitally combined to visualize the entire specimen.

topoisomerase   An enzyme that can change the supercoiling of DNA.

total magnification   The magnification of the ocular lens multiplied by the magnification of the objective lens.

transamination   The transfer of an ammonium ion between two metabolites.

transcript   An RNA copy of a DNA template.

transcription   The synthesis of RNA complementary to a DNA template.

transcriptional attenuation   A transcriptional regulatory mechanism in which translation of a leader peptide affects transcription of downstream structural genes.

transcriptional fusion   See **operon fusion**.

transcriptome   The set of transcribed genes in a cell at given time. The "complete transcriptome" includes all the possible RNA transcription products from a given genome. The "expressed transcriptome" is the set of RNAs present during a given condition.

transcytosis   The movement of a cell or substance from one side of a polarized cell to the other side, using an intracellular route.

transduction   The transfer of host genes between bacterial cells via a phage head coat.

transfection   Deliberate transfer of DNA (usually viral) into cells.

transfer RNA (tRNA)   An RNA that carries an amino acid to the ribosome. The anticodon on the tRNA matches the codon on the mRNA.

transform   To convert cultured cells into cancer cells.

transformasome   A bacterial cell membrane protein complex that imports external DNA during transformation.

transformation   The internalization of free DNA from the environment into bacterial cells.

transformed-focus assay   The detection of oncogenic viruses based on their ability to transform cells, generating foci of unrestricted cell growth.

transgene   A gene that has been transferred by genetic engineering techniques from one organism to another.

transglycosylase   An enzyme that links *N*-acetylglucosamine and *N*-acetylmuramic acid into chains during bacterial cell wall synthesis.

transition   A point mutation in which a purine is replaced by a different purine or a pyrimidine is replaced by a different pyrimidine.

translation   The ribosomal synthesis of proteins based on triplet codons present in mRNA.

translational control   A regulatory mechanism that modulates protein production by influencing the translation of mRNA.

translational fusion   See **gene fusion**.

translocation   The energy-dependent movement of the ribosome to the next triplet codon along an mRNA.

transmembrane protein   A protein with a membrane-spanning region.

transmission electron microscopy (TEM)   Electron microscopy in which electron beams are transmitted through a thin specimen to reveal internal structure.

transovarial transmission   Transmission of a pathogen from parent to offspring by infection of the egg cell.

transpeptidase   An enzyme that cross-links the side chains from adjacent peptidoglycan strands during bacterial cell wall synthesis.

transport protein   Also called *transporter*. A membrane protein that moves specific molecules across a membrane.

transposable element   A segment of DNA that can move from one DNA region to another.

transposase   A transposable element–encoded enzyme that catalyzes the transfer of the transposable element from one DNA region to another.

transposition   The process of moving a transposable element from one DNA region to another.

transposon   A transposable DNA element that contains genes in addition to those required for transposition.

transversion   A point mutation in which a purine is replaced by a pyrimidine or vice versa.

Treg   See **regulatory T cell**.

tricarboxylic acid (TCA) cycle   A metabolic cycle that catabolizes the acetyl group from acetyl-CoA to $2CO_2$ with the concomitant production of NADH, $FADH_2$, and ATP.

tRNA   See **transfer RNA**.

trophic level   A level of the food web representing the consumption of biomass of organisms from another level, usually closer to producers.

tropism   The ability of a virus to infect a particular tissue type.

trypanosome   A parasitic excavate protist that has a cortical skeleton of microtubules culminating in a long flagellum.

tumor necrosis factor   A cytokine released by several cell types (e.g., macrophages) in response to cell damage.

tumor necrosis factor alpha (TNF-α)   A cytokine involved in systemic inflammation.

twitching motility   A type of bacterial movement on solid surfaces where a specific pilus extends and retracts.

two-component signal transduction system A message relay system composed of a sensor kinase protein and a response regulator protein that regulates gene expression in response to a signal (usually an extracellular signal).

two-dimensional polyacrylamide gel electrophoresis (2-D PAGE, 2-D gels) A technique to separate proteins based on differences in charge and molecular weight.

type I hypersensitivity Also called *immediate hypersensitivity*. An IgE-mediated allergic reaction that causes degranulation of mast cells within minutes of exposure to the antigen.

type I pilus A pilus that adheres to mannose residues on host cell surfaces.

type II hypersensitivity An immune condition in which antibodies bind to cell surface antigens, triggering cell-mediated cytotoxicity or activation of the complement cascade.

type II secretion A bacterial protein secretion system that uses a type IV pilus-like extraction/retraction mechanism to push proteins out of the cell.

type III hypersensitivity An immune overreaction triggered by IgG antibody binding to soluble antigens.

type III pilus A bacterial pilus that does not bind to mannose but binds to tannic acid.

type III secretion A bacterial protein secretion system that uses a molecular syringe to inject bacterial proteins into the host cytoplasm.

type IV hypersensitivity Also called *delayed-type hypersensitivity (DTH)*. An immune response that develops 24–72 hours after exposure to an antigen that the immune system recognizes as foreign. DTH is triggered by antigen-specific T cells. The response is delayed because the T cells need time to proliferate after being activated by the allergen.

type IV pilus A dynamic pilus that can repeatedly assemble and disassemble; it mediates twitching motility.

ultracentrifuge A machine that exposes samples to high centrifugal forces and can be used to separate subcellular components.

uncoating The release of a viral genome from its capsid, following entry of the virion into a host cell.

uncoupler A molecule that makes a membrane permeable to protons, dissipating the proton motive force and uncoupling electron transport from ATP synthesis.

uncultured Not yet grown in the laboratory.

upstream processing The culturing of industrial microbes to produce large quantities of a desired product.

vaccination Exposure of an individual to a weakened version of a microbe to provoke immunity and prevent development of disease upon reexposure.

VAM See **vesicular-arbuscular mycorrhiza**.

van der Waals force Weak, temporary electrostatic attraction between molecules caused by shifting electron clouds.

vancomycin A glycopeptide antibiotic that inhibits bacterial cell wall synthesis in a mechanism distinct from penicillin inhibition.

variable region The amino-terminal portions of antibody light and heavy chains that confer specificity to antigen binding and define the antibody idiotype.

vasoactive factor A cell signaling molecule that increases capillary permeability.

VBNC See **viable but nonculturable**.

vector 1. An organism (e.g., insect) that can carry infectious agents from one animal to another. 2. In molecular biology, a molecule of DNA into which exogenous DNA can be inserted to be cloned.

vegetative cell A metabolically active, replicating bacterial cell.

vegetative mycelium A branched filament produced by vegetative cells that expands into the substrate.

Verrucomicrobia A phylum of free-living aquatic bacteria with wart-like, protruding structures containing actin.

vertical gene transfer Gene transfer from parent to offspring through reproduction.

vertical transmission In genetics, the passage of genes from parent to offspring. In disease, transmission of a pathogen from parent to offspring.

vesicle A small, membrane-enclosed sphere found within a cell.

vesicular-arbuscular mycorrhizae (VAM) Mutualistic associations between plant roots and certain fungi, involving hyphal penetration of plant root cells.

viable Capable of replicating—for instance, by forming a colony on an agar plate.

viable but nonculturable (VBNC) Also called *dormant*. Metabolically active but unable to replicate to form a colony on a plate by current means of culture.

viral envelope A host-derived membrane that surrounds a virus capsid.

viremia The presence of large numbers of virions in the bloodstream.

virion A virus particle

viroid An infectious naked nucleic acid.

virulence A measure of the severity of a disease caused by a pathogenic agent.

virulence factor A trait of a pathogen that enhances the pathogen's disease-producing capability.

virus A non-cellular particle containing a genome that can replicate only inside a cell.

virus shedding The release of virions from the host organism into the environment.

wastewater treatment A series of wastewater transformations designed to lower biological oxygen demand and eliminate human pathogens before water is returned to local rivers.

water activity A measure of the water that is not bound to solutes and is available for use by organisms.

water cycle See **hydrologic cycle**.

water table The layer of soil that is permanently saturated with water.

Western blot A technique to detect specific proteins. Proteins are subjected to gel electrophoresis, transferred to a blot, and probed with enzyme-linked or fluorescently tagged antibodies that specifically bind the protein of interest.

wet mount   A technique to view living microbes with a microscope by placing the microbes in water on a slide under a coverslip.

wetland   A region of land that undergoes seasonal fluctuations in water level and aeration.

whey   The liquid portion of milk after proteins have precipitated out of solution, usually during cheese production.

Winogradsky column   A tube containing a stratified environment that causes specific microbes to grow at particular levels; a type of enrichment culture for the growth of microbes from wetland environments.

X-ray crystallography   Also called *X-ray diffraction analysis*. A technique to determine the positions of atoms (atomic coordinates) within a molecule or molecular complex, based on the diffraction of X-rays by the molecule.

yeast   A unicellular fungus.

yeast two-hybrid system   An in vivo technique to determine protein-protein interactions in which DNA sequences encoding proteins of interest are fused separately to the DNA-binding and activation domains of a yeast transcription factor. The recombinant yeast is then tested for expression of a reporter gene.

yogurt   A semisolid food produced through acidification of milk by lactic acid–producing bacteria.

zone of hypoxia   See **dead zone**.

zone of inhibition   A region of no bacterial growth on an agar plate due to the diffusion of a test antibiotic. Correlates to the minimal inhibitory concentration.

zoonotic disease   An infection that normally affects animals but can be transmitted to humans.

zoospore   A flagellated reproductive cell produced by chytridiomycete fungi.

zooxanthella *pl.* **zooxanthellae**   A phototrophic dinoflagellate that is a coral endosymbiont.

zygomycete   A member of the eukaryotic group Zygomycota—fungi forming nonmotile haploid gametes that grow toward each other, fusing to form the zygospore.

zygospore   In zygomycetes, the diploid structure formed by the fusion of two gamete-bearing hyphae.

Figure Credits

Figure 1.8B: National Center for Biotechnology Information: Figure: "Genome of Haemophilius Influenzae RD and Segment of Genome Base Pairs" from National Center for Biotechnology Information Genome Database.

Figure 7.18: Fig. 8.21, pp. 209–210 from *Molecular Biology of the Gene*, 5th Ed. by James D. Watson et al. Copyright © 2004 by Pearson Education, Inc. Reprinted by permission.

Figure 7.25: © Campbell and Mullins. 2007. Originally published in *The Journal of Cell Biology*, doi: 10.1083/jcb.200708206.

Special Topic 8.1 Fig. 2: Reprinted by permission from Macmillan Publishers Ltd: Figure from Jin-Der Wen et al., "Following translation by single ribosomes one codon at a time," *Nature*, Vol. **452**, No. 7187, pp. 598–603. Copyright © 2008, Nature Publishing Group.

Figure 8.5C: Katsuhiko S. Murakami et al.: Figure from "Structural Basis of Transcription Initiation: RNA Polymerase Holoenzyme at 4A Resolution," *Science* 296:1285–1290 (17 May 2002). Copyright © 2002, The American Association for the Advancement of Science. Reprinted with permission from AAAS.

Figure 8.10: David L. Nelson & Michael M. Cox: Figure from *Lehninger Principles of Biochemistry*, 3rd Edition by David L. Nelson and Michael M. Cox. Copyright © 1982, 1993, 2000 by Worth Publishers. Used with permission.

Special Topic 10.1 Fig. 2: From Mikkel G. Jorgensen, Kenn Gerdes, "HicA of *Escherichia coli* Defines a Novel Family of Translation-Independent mRNA Interferases in Bacteria and Archaea," *Journal of Bacteriology*, Feb. 2009, Vol. **191**, No. 4: 1194. Copyright © 2009, American Society for Microbiology. Reproduced with permission from American Society for Microbiology.

Figure 11.3: Christopher K. Matthews et al.: Figure from *Bacteriophage T4*, 1983, ASM Press, Washington, DC. Reprinted with permission from ASM Press.

Special Topic 17.2 Fig. 2a: Reprinted by permission from Macmillan Publishers Ltd: Nicole T. Perna et al.: Figure: "Genome of *E. coli* and LEE gene cluster" from "Genome sequence of enterohaemorrhagic *Escherichia coli* O157:H7," *Nature*, Vol. **409**, No. 6819, pp. 529–533. Copyright © 2001, Nature Publishing Group.

Figure 17.23: W. Ford Doolittle: Figure from "Phylogenetic Classification and the Universal Tree," *Science* 284:2124 (25 June 1999). Copyright © 1999, The American Association for the Advancement of Science. Reprinted with permission from AAAS.

Figure 19.16: Karner, DeLong, & Karl: Figure reprinted by permission from Macmillan Publishers Ltd. from "Archael Dominance in the Mesopelagic Zone of the Pacific Ocean," *Nature*, Vol. **409**, No. 507. Copyright © 2001, Nature Publishing Group.

Figure 21.7A: Lynn Margulis: Figure from *Symbiosis in Cell Evolution* by Lynn Margulis. Copyright © 1981, 1993 by W.H. Freeman and Company. Used with permission.

Figure 21.7B: L. R. Cleveland: Figure from "The Protozoa of Termites, Part III" by L.R. Cleveland and B.T. Cleveland. Reprinted by permission.

Figure 21.17A: Raina Maier et al.: Figure from *Environmental Microbiology*, Copyright © 2000 by Academic Press. Reprinted by permission of Elsevier, Ltd.

Figure 22.14: With kind permission from Springer Science+Business Media: *Applied Microbiology and Biotechnology*, "Enrichment of anammox bacteria from marine environment for the construction of a bioremediation reactor," Vol. **77**, No. 5, 2008, pp. 1159–1166, Jun Nakajima et al., Fig. 2. Copyright © 2008, Springer.

Index

tics

ourth edition
uk/james to
uding:

nowledge of

FAQ/support
E-mail us for help
E-mail us feedback

Modern Engineering Mathematics
4th Edition
Glvn James
ISBN: 9780132391443

Learn more about this title.

Instructor resources

Visit the Instructor
Resource Centre

Search catalogue

○ Author
● Title
○ ISBN

Browse catalogue
Advanced search

Bridging chapters

Format: WinZip® .zip (Adobe Acrobat® .pdf once extracted)

Bridging chapters

To download the resources for this title, right-click on the file names above and save
them to your hard disk. For further support, refer to the links in the left-hand menu.

PEARSON
Education

We work with leading authors to develop the strongest
educational materials in engineering, bringing cutting-edge
thinking and best learning practice to a global market.

Under a range of well-known imprints, including
Prentice Hall, we craft high-quality print and
electronic publications which help readers to understand
and apply their content, whether studying or at work.

To find out more about the complete range of our
publishing, please visit us on the World Wide Web at:
www.pearsoned.co.uk

Modern Engineering Mathematics

Fourth Edition

Glyn James   *Coventry University*
and
David Burley   *University of Sheffield*
Dick Clements   *University of Bristol*
Phil Dyke   *University of Plymouth*
John Searl   *University of Edinburgh*
Jerry Wright   *AT&T Shannon Laboratory*

PEARSON
Prentice
Hall

Harlow, England • London • New York • Boston • San Francisco • Toronto • Sydney • Singapore • Hong Kong
Tokyo • Seoul • Taipei • New Delhi • Cape Town • Madrid • Mexico City • Amsterdam • Munich • Paris • Milan

Pearson Education Limited
Edinburgh Gate
Harlow
Essex CM20 2JE
England

and Associated Companies throughout the world

Visit us on the World Wide Web at:
www.pearsoned.co.uk

First published 1992
Second edition 1996
Third edition 2001
Fourth edition published 2007

ISBN: 978-0-13-239144-3

British Library Cataloguing-in-Publication Data
A catalogue record for this book is available from the British Library

10 9 8 7 6 5 4 3 2
11 10 09 08

Typeset by 35 in 10/12pt Times
Printed and bound in Malaysia (CTP-VVP)

The publisher's policy is to use paper manufactured from sustainable forests.

Contents

Chapter 5 Matrix Algebra 296

Chapter 8 Differentiation and Integration 539

Chapter 9   Further Calculus   679

Supporting resources

Visit **www.pearsoned.co.uk/james** to find valuable online resources

Companion Website for students

- 'Bridging chapters', which refresh your knowledge of fundamental topics

For instructors

- Complete, downloadable Solutions Manual
- PowerPoint slides of all figures from the book

For more information please contact your local Pearson Education sales representative or visit **www.pearsoned.co.uk/james**

Preface

As with the previous editions, the range of material covered in this fourth edition is regarded as appropriate for a first level core studies course in mathematics for undergraduate courses in all engineering disciplines. Whilst designed primarily for use by engineering students it is believed that the book is also highly suitable for students of the physical sciences and applied mathematics. Additional material appropriate for second level undergraduate core studies, or possibly elective studies for some engineering disciplines, is contained in the companion text *Advanced Modern Engineering Mathematics*.

The objective of the authoring team remains that of achieving a balance between the development of understanding and the mastering of solution techniques with the emphasis being on the development of the student's ability to use mathematics with understanding to solve engineering problems. Consequently, the book is not a collection of recipes and techniques designed to teach students to solve routine exercises, nor is mathematical rigour introduced for its own sake. To achieve the desired objective the text contains:

- Worked examples
  Approximately 500 worked examples, many of which incorporate mathematical models and are designed both to provide relevance and to reinforce the role of mathematics in various branches of engineering. In response to feedback from users, additional worked examples have been incorporated within this revised edition.
- Applications
  To provide further exposure to the use of mathematical models in engineering practice, each chapter contains sections on engineering applications. These sections form an ideal framework for individual, or group, case study assignments leading to a written report and/or oral presentation; thereby helping to develop the skills of mathematical modelling necessary to prepare for the more open-ended modelling exercises at a later stage of the course.
- Exercises
  There are numerous exercise sections throughout the text and at the end of each chapter there is a comprehensive set of review exercises. While many of the exercise problems are designed to develop skills in mathematical techniques, others are designed to develop understanding and to encourage learning by doing, and some are of an open-ended nature. This book contains over 1200 exercises and answers to all the questions are given. It is hoped that this provision, together with the large number of worked examples and style of presentation,

also makes the book suitable for private or directed study. Again in response to feedback from users, the frequency of exercises sections has been increased and additional questions have been added to many of the sections.

- Numerical methods
  Recognizing the increasing use of numerical methods in engineering practice, which often complement the use of analytical methods in analysis and design and are of ultimate relevance when solving complex engineering problems, there is wide agreement that they should be integrated within the mathematics curriculum. Consequently the treatment of numerical methods is integrated within the analytical work throughout the book.

Much of the feedback from users relates to the role and use of software packages, particularly symbolic algebra packages, in the teaching of mathematics to engineering students. In response use of such packages has been incorporated and is a significant new feature of this new edition. Whilst any appropriate software package can be used, the authors recommend the use of MATLAB or MAPLE and have adopted their use in this text. The basic MATLAB package is primarily a number crunching package; it does not perform symbolic manipulations and cannot undertake algebra containing unknowns. However, such work can be undertaken by its Symbolic Math Toolbox, which uses the symbolic manipulation computational engine from the MAPLE package. Consequently, most of the commands in the Symbolic Math Toolbox are identical to the MAPLE commands, although there are syntax differences in implementation. Throughout this text emphasis will be on the use of MATLAB, interpreted as including the Symbolic Math Toolbox, with reference made to corresponding MAPLE commands and differences in syntax highlighted. MATLAB/MAPLE commands have been introduced and illustrated, as inserts, throughout the text so that their use can be integrated into the teaching and learning processes. Students are strongly encouraged to use one of these packages to check the answers to the examples and exercises. It is stressed that the MATLAB/MAPLE inserts are not intended to be a first introduction of the package to students; it is anticipated that they will receive an introductory course elsewhere and will be made aware of the excellent 'help' facility available. The purpose of incorporating the inserts is not only to improve efficiency in the use of the package but also to provide a facility to help develop a better understanding of the related mathematics. Whilst use of such packages takes the tedium out of arithmetic and algebraic manipulations it is important that they are used to enhance understanding and not to avoid it. It is recognised that not all users of the text will have access to either MATLAB or MAPLE and consequently all the inserts are highlighted and can be 'omitted' without loss of continuity. Throughout the text two icons are used

- An open screen ⌨ indicates that use of a software package would be useful (e.g. for checking solutions) but not essential

- A closed screen ⌨ indicates that the use of a software package is essential or highly desirable.

Feedback, from users of the current edition, on the subject content has been favourable and consequently no new chapters have been introduced. However, in response to the feedback chapters have been reviewed and amended/updated accordingly. Specifically, changes made include:

- redistribution of material both within chapters and across chapters;
- introduction of additional worked examples to achieve a more progressive level of difficulty;
- increase in the frequency of exercise sections and inclusion of additional questions;
- improved format of text to make material more accessible to students.

A comprehensive Solutions Manual is obtainable free of charge to lecturers using this textbook. It will also be available for download via the Web at www.booksites.net/james.

It is also planned to have available online a set of 'Refresher Units' covering topics students should have encountered at school but may not have used for some time.

Acknowledgements

The authoring team is extremely grateful to all the reviewers and users of the text who have provided valuable comments on previous editions of this book. Most of this has been highly constructive and very much appreciated. The team has continued to enjoy the full support of a very enthusiastic production team at Pearson Education and wishes to thank all those concerned. Finally I would like to thank my wife, Dolan, for her full support throughout the preparation of this text and its previous editions.

Glyn James
Coventry
May 2007

About the Authors

Glyn James retired as Dean of the School of Mathematical and Information Sciences at Coventry University in 2001 and is now Emeritus Professor in Mathematics at the University. He graduated from the University College of Wales, Cardiff in the late 1950s, obtaining first class honours degrees in both Mathematics and Chemistry. He obtained a PhD in Engineering Science in 1971 as an external student of the University of Warwick. He has been employed at Coventry since 1964 and held the position of the Head of Mathematics Department prior to his appointment as Dean in 1992. His research interests are in control theory and its applications to industrial problems. He also has a keen interest in mathematical education, particularly in relation to the teaching of engineering mathematics and mathematical modelling. He was co-chairman of the European Mathematics Working Group established by the European Society for Engineering Education (SEFI) in 1982, a past chairman of the Education Committee of the Institute of Mathematics and its Applications (IMA), and a member of the Royal Society Mathematics Education Subcommittee. In 1995 he was chairman of the Working Group that produced the report 'Mathematics Matters in Engineering' on behalf of the professional bodies in engineering and mathematics within the UK. He is also a member of the editorial/advisory board of three international journals. He has published numerous papers and is co-editor of five books on various aspects of mathematical modelling. He is a past Vice-President of the IMA and has also served a period as Honorary Secretary of the Institute. He is a Chartered Mathematician and a Fellow of the IMA.

David Burley retired from the University of Sheffield in 1998. He graduated in mathematics from King's College, University of London in 1955 and obtained his PhD in mathematical physics. After working in the University of Glasgow, he spent most of his academic career in the University of Sheffield, being Head of Department for six years. He has long experience of teaching engineering students and has been particularly interested in encouraging students to construct mathematical models in physical and biological contexts to enhance their learning. His research work has ranged through statistical mechanics, optimization and fluid mechanics. He has particular interest in the flow of molten glass in a variety of situations and the application of results in the glass industry. Currently he is involved in a large project concerning heat transfer problems in the deep burial of nuclear waste.

Dick Clements is Professor in the Department of Engineering Mathematics at Bristol University. He read for the Mathematical Tripos, matriculating at Christ's College,

Cambridge in 1966. He went on to take a PGCE at Leicester University School of Education before returning to Cambridge to research a PhD in Aeronautical Engineering. In 1973 he was appointed Lecturer in Engineering Mathematics at Bristol University and has taught mathematics to engineering students ever since, becoming successively Senior Lecturer, Reader and Professorial Teaching Fellow. He has undertaken research in a wide range of engineering topics but is particularly interested in mathematical modelling and in new approaches to the teaching of mathematics to engineering students. He has published numerous papers and one previous book, *Mathematical Modelling: A Case Study Approach.* He is a Chartered Engineer, a Chartered Mathematician, a member of the Royal Aeronautical Society, a Fellow of the Institute of Mathematics and Its Applications, an Associate Fellow of the Royal Institute of Navigation, and a Fellow of the Higher Education Academy.

Phil Dyke is Professor of Applied Mathematics and Head of School of Mathematics and Statistics at the University of Plymouth. After graduating with first class honours in Mathematics from the University of London, he gained a PhD in coastal sea modelling at Reading in 1972. Since then, Phil Dyke has been a full-time academic initially at Heriot-Watt University teaching engineers followed by a brief spell at Sunderland. He has been at Plymouth since 1984. He still engages in teaching and is actively involved in building mathematical models relevant to environmental issues.

John Searl was Director of the Edinburgh Centre for Mathematical Education at the University of Edinburgh before his recent retirement. As well as lecturing on mathematical education, he taught service courses for engineers and scientists. His most recent research concerned the development of learning environments that make for the effective learning of mathematics for 16–20 year olds. As an applied mathematician who worked collaboratively with (among others) engineers, physicists, biologists and pharmacologists, he is keen to develop the problem solving skills of students and to provide them with opportunities to display their mathematical knowledge within a variety of practical contexts. These contexts develop the extended reasoning needed in all fields of engineering.

Jerry Wright is a Lead Member of Technical Staff at the AT&T Shannon Laboratory, New Jersey, USA. He graduated in Engineering (BSc and PhD at the University of Southampton) and in Mathematics (MSc at the University of London) and worked at the National Physical Laboratory before moving to the University of Bristol in 1978. There he acquired wide experience in the teaching of mathematics to students of engineering, and became Senior Lecturer in Engineering Mathematics. He held a Royal Society Industrial Fellowship for 1994, and is a Fellow of the Institute of Mathematics and its Applications. In 1996 he moved to AT&T Labs (formerly part of Bell labs) to continue his research in spoken language understanding, human/computer dialog systems, and data mining.

1 Numbers, Algebra and Geometry

1.1 Introduction

Mathematics plays an important role in our lives. It is used in everyday activities from buying food to organizing maintenance schedules for aircraft. Through applications developed in various cultural and historical contexts, mathematics has been one of the decisive factors in shaping the modern world. It continues to grow and to find new uses, particularly in engineering and technology.

Mathematics provides a powerful, concise and unambiguous way of organizing and communicating information. It is a means by which aspects of the physical universe can be explained and predicted. It is a problem-solving activity supported by a body of knowledge. Mathematics consists of facts, concepts, skills and thinking processes – aspects that are closely interrelated. It is a hierarchical subject in that new ideas and skills are developed from existing ones. This sometimes makes it a difficult subject for learners who, at every stage of their mathematical development, need to have ready recall of material learned earlier.

In the first two chapters we shall summarize the concepts and techniques that most students will already understand and we shall extend them into further developments in mathematics. There are four key areas of which students will already have considerable knowledge.

- numbers
- algebra
- geometry
- functions

These areas are vital to making progress in engineering mathematics (indeed, they will solve many important problems in engineering). Here we shall aim to consolidate that knowledge, to make it more precise and to develop it. In this first chapter we will deal with the first three topics; functions are considered in Chapter 2.

1.2 Number and arithmetic

1.2.1 Number line

Mathematics has grown from primitive arithmetic and geometry into a vast body of knowledge. The most ancient mathematical skill is counting, using, in the first instance, the natural numbers and later the integers. The term **natural numbers** commonly refers to the set $\mathbb{N} = \{1, 2, 3, \ldots\}$, and the term **integers** to the set $\mathbb{Z} = \{0, 1, -1, 2, -2, 3, -3, \ldots\}$. The integers can be represented as equally spaced points on a line called the **number line** as shown in Figure 1.1. In a computer the integers can be stored exactly. The set of all points (not just those representing integers) on the number line represents the **real numbers** (so named to distinguish them from the complex numbers, which are

Figure 1.1
The number line.

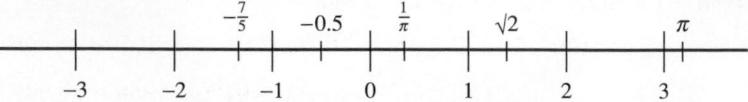

discussed in Chapter 3). The set of real numbers is denoted by $\mathbb{R}$. The general real number is usually denoted by the letter x and we write 'x in $\mathbb{R}$', meaning x is a real number. A real number that can be written as the ratio of two integers, like $\frac{3}{2}$ or $-\frac{7}{5}$, is called a **rational number**. Other numbers, like $\sqrt{2}$ and π, that cannot be expressed in that way are called **irrational numbers**. In a computer the real numbers can be stored only to a limited number of figures. This is a basic difference between the ways in which computers treat integers and real numbers, and is the reason why the computer languages commonly used by engineers distinguish between integer values and variables on the one hand and real number values and variables on the other.

1.2.2 Rules of arithmetic

The basic arithmetical operations of addition, subtraction, multiplication and division are performed subject to the **Fundamental Rules of Arithmetic**. For any three numbers a, b and c:

(a1)   the commutative law of addition

$$a + b = b + a$$

(a2)   the commutative law of multiplication

$$a \times b = b \times a$$

(b1)   the associative law of addition

$$(a + b) + c = a + (b + c)$$

(b2)   the associative law of multiplication

$$(a \times b) \times c = a \times (b \times c)$$

(c1)   the distributive law of multiplication over addition and subtraction

$$(a + b) \times c = (a \times c) + (b \times c)$$
$$(a - b) \times c = (a \times c) - (b \times c)$$

(c2)   the distributive law of division over addition and subtraction

$$(a + b) \div c = (a \div c) + (b \div c)$$
$$(a - b) \div c = (a \div c) - (b \div c)$$

Here the brackets indicate which operation is performed first. These operations are called **binary** operations because they associate with every two members of the set of real numbers a unique third member; for example,

$$2 + 5 = 7 \quad \text{and} \quad 3 \times 6 = 18$$

Example 1.1   Find the value of $(100 + 20 + 3) \times 456$.

Solution   Using the distributive law we have

$$(100 + 20 + 3) \times 456 = 100 \times 456 + 20 \times 456 + 3 \times 456$$
$$= 45\,600 + 9120 + 1368 = 56\,088$$

This, of course, is normally set out in the traditional school arithmetic way:

$$\begin{array}{r} 456 \\ 123 \times \\ \hline 1\,368 \\ 9\,120 \\ 45\,600 \\ \hline 56\,088 \end{array}$$

Example 1.2 Rewrite $(a + b) \times (c + d)$ as the sum of products.

Solution Using the distributive law we have

$$(a + b) \times (c + d) = a \times (c + d) + b \times (c + d)$$
$$= (c + d) \times a + (c + d) \times b$$
$$= c \times a + d \times a + c \times b + d \times b$$
$$= a \times c + a \times d + b \times c + b \times d$$

applying the commutative laws several times.

A further operation used with real numbers is that of **powering**. For example, $a \times a$ is written as a^2, and $a \times a \times a$ is written as a^3. In general the product of n a's where n is a positive integer is written as a^n. (Here the n is called the **index** or **exponent**.) Operations with powering also obey simple rules:

$$a^n \times a^m = a^{n+m} \tag{1.1a}$$

$$a^n \div a^m = a^{n-m} \tag{1.1b}$$

$$(a^n)^m = a^{nm} \tag{1.1c}$$

From rule (1.1b) it follows, by setting $n = m$ and $a \neq 0$, that $a^0 = 1$. It is also convention to take $0^0 = 1$. The process of powering can be extended to include the fractional powers like $a^{1/2}$. Using rule (1.1c),

$$(a^{1/n})^n = a^{n/n} = a^1$$

and we see that

$$a^{1/n} = \sqrt[n]{a}$$

the nth root of a. Also, we can define a^{-m} using rule (1.1b) with $n = 0$, giving

$$1 \div a^m = a^{-m}, \qquad a \neq 0$$

Thus a^{-m} is the reciprocal of a^m. In contrast with the binary operations $+$, $\times$, $-$ and $\div$, which operate on two numbers, the powering operation $(\ )^r$ operates on just one element and is consequently called a **unary** operation. Notice that the fractional power

$$a^{m/n} = (^n\!\sqrt{a})^m = {}^n\!\sqrt{(a^m)}$$

is the nth root of a^m. If n is an even integer, then $a^{m/n}$ is not defined when a is negative. When $^n\!\sqrt{a}$ is an irrational number then such a root is called a **surd**.

Numbers like $\sqrt{2}$ were described by the Greeks as **a-logos**, without a ratio number. An Arabic translator took the alternative meaning 'without a word' and used the arabic word for 'deaf', which subsequently became **surdus**, latin for deaf, when translated from arabic to Latin in the mid-twelfth century.

Example 1.3 Find the values of

(a) $27^{1/3}$ (b) $(-8)^{2/3}$ (c) $16^{-3/2}$

(d) $(-2)^{-2}$ (e) $(-1/8)^{-2/3}$ (f) $(9)^{-1/2}$

Solution (a) $27^{1/3} = {}^3\!\sqrt{27} = 3$

(b) $(-8)^{2/3} = ({}^3\!\sqrt{(-8)})^2 = (-2)^2 = 4$

(c) $16^{-3/2} = (16^{1/2})^{-3} = (4)^{-3} = \frac{1}{4^3} = \frac{1}{64}$

(d) $(-2)^{-2} = \dfrac{1}{(-2)^2} = \frac{1}{4}$

(e) $(-1/8)^{-2/3} = [{}^3\!\sqrt{(-1/8)}]^{-2} = [{}^3\!\sqrt{(-1)}/{}^3\!\sqrt{(8)}]^{-2} = [-1/2]^{-2} = 4$

(f) $(9)^{-1/2} = (3)^{-1} = \frac{1}{3}$

Example 1.4 Express (a) in terms of $\sqrt{2}$ and simplify (b) to (f).

(a) $\sqrt{18} + \sqrt{32} - \sqrt{50}$ (b) $6/\sqrt{2}$ (c) $(1 - \sqrt{3})(1 + \sqrt{3})$

(d) $\dfrac{2}{1 - \sqrt{3}}$ (e) $(1 + \sqrt{6})(1 - \sqrt{6})$ (f) $\dfrac{1 - \sqrt{2}}{1 + \sqrt{6}}$

Solution (a) $\sqrt{18} = \sqrt{(2 \times 9)} = \sqrt{2} \times \sqrt{9} = 3\sqrt{2}$

$\sqrt{32} = \sqrt{(2 \times 16)} = \sqrt{2} \times \sqrt{16} = 4\sqrt{2}$

$\sqrt{50} = \sqrt{(2 \times 25)} = \sqrt{2} \times \sqrt{25} = 5\sqrt{2}$

Thus $\sqrt{18} + \sqrt{32} - \sqrt{50} = 2\sqrt{2}$.

(b) $6/\sqrt{2} = 3 \times 2/\sqrt{2}$

Since $2 = \sqrt{2} \times \sqrt{2}$, we have $6/\sqrt{2} = 3\sqrt{2}$.

(c) $(1 - \sqrt{3})(1 + \sqrt{3}) = 1 + \sqrt{3} - \sqrt{3} - 3 = -2$

(d) Using the result of part (c) $\dfrac{2}{1 - \sqrt{3}}$ can be simplified by multiplying 'top and bottom' by $1 + \sqrt{3}$ (notice the sign change in front of the $\sqrt{\ }$). Thus

$$\frac{2}{1 - \sqrt{3}} = \frac{2(1 + \sqrt{3})}{(1 - \sqrt{3})(1 + \sqrt{3})}$$

$$= \frac{2(1 + \sqrt{3})}{1 - 3}$$

$$= -1 - \sqrt{3}$$

(e) $(1 + \sqrt{6})(1 - \sqrt{6}) = 1 - \sqrt{6} + \sqrt{6} - 6 = -5$

(f) Using the same technique as in part (d) we have

$$\frac{1 - \sqrt{2}}{1 + \sqrt{6}} = \frac{(1 - \sqrt{2})(1 - \sqrt{6})}{(1 + \sqrt{6})(1 - \sqrt{6})}$$

$$= \frac{1 - \sqrt{2} - \sqrt{6} + \sqrt{12}}{1 - 6}$$

$$= -(1 - \sqrt{2} - \sqrt{6} + 2\sqrt{3})/5$$

This process of expressing the irrational number so that all of the surds are in the numerator is called **rationalization**.

When evaluating arithmetical expressions the following rules of precedence are observed:

- the powering operation $(\ )^r$ is performed first
- then multiplication $\times$ and/or division $\div$
- then addition $+$ and/or subtraction $-$

When two operators of equal precedence are adjacent in an expression the left-hand operation is performed first. For example

$$12 - 4 + 13 = 8 + 13 = 21$$

and

$$15 \div 3 \times 2 = 5 \times 2 = 10$$

The precedence rules are overridden by brackets; thus

$$12 - (4 + 13) = 12 - 17 = -5$$

and

$$15 \div (3 \times 2) = 15 \div 6 = 2.5$$

Example 1.5   Evaluate $7 - 5 \times 3 \div 2^2$.

Solution   Following the rules of precedence, we have

$$7 - 5 \times 3 \div 2^2 = 7 - 5 \times 3 \div 4 = 7 - 15 \div 4 = 7 - 3.75 = 3.25$$

1.2.3 Exercises

1 Simplify the following expressions, giving the answers with positive indices and without brackets:

(a) $2^3 \times 2^{-4}$ (b) $2^3 \div 2^{-4}$ (c) $(2^3)^{-4}$

(d) $3^{1/3} \times 3^{5/3}$ (e) $(36)^{-1/2}$ (f) $16^{3/4}$

2 The expression $7 - 2 \times 3^2 + 8$ may be evaluated using the usual implicit rules of precedence. It could be rewritten as $((7 - (2 \times (3^2))) + 8)$ using brackets to make the precedence explicit. Similarly rewrite the following expressions in fully bracketed form:

(a) $21 + 4 \times 3 \div 2$

(b) $17 - 6^{2+3}$

(c) $4 \times 2^3 - 7 \div 6 \times 2$

(d) $2 \times 3 - 6 \div 4 + 3^{2-5}$

3 Express the following in the form $x + y\sqrt{2}$ with x and y rational numbers:

(a) $(7 + 5\sqrt{2})^3$ (b) $(2 + \sqrt{2})^4$

(c) $\sqrt[3]{(7 + 5\sqrt{2})}$ (d) $\sqrt{(\frac{11}{2} - 3\sqrt{2})}$

4 Show that

$$\frac{1}{a + b\sqrt{c}} = \frac{a - b\sqrt{c}}{a^2 - b^2 c}$$

Hence express the following numbers in the form $x + y\sqrt{n}$ where x and y are rational numbers and n is an integer:

(a) $\dfrac{1}{7 + 5\sqrt{2}}$ (b) $\dfrac{2 + 3\sqrt{2}}{9 - 7\sqrt{2}}$

(c) $\dfrac{4 - 2\sqrt{3}}{7 - 3\sqrt{3}}$ (d) $\dfrac{2 + 4\sqrt{5}}{4 - \sqrt{5}}$

5 Find the difference between 2 and the squares of

$$\frac{1}{1}, \frac{3}{2}, \frac{7}{5}, \frac{17}{12}, \frac{41}{29}, \frac{99}{70}$$

(a) Verify that successive terms of the sequence stand in relation to each other as m/n does to $(m + 2n)/(m + n)$. (b) Verify that if m/n is a good approximation to $\sqrt{2}$ then $(m + 2n)/(m + n)$ is a better one, and that the errors in the two cases are in opposite directions. (c) Find the next three terms of the above sequence.

1.2.4 Inequalities

The number line (Figure 1.1) makes explicit a further property of the real numbers – that of **ordering**. This enables us to make statements like 'seven is greater than two' and 'five is less than six'. We represent this using the comparison symbols

> $>$, 'greater than'
> $<$, 'less than'

It also makes obvious two other comparators:

> $=$, 'equals'
> $\neq$, 'does not equal'

These comparators obey simple rules when used in conjunction with the arithmetical operations. For any four numbers a, b, c and d:

($a < b$ and $c < d$)	implies	$a + c < b + d$	**(1.2a)**
($a < b$ and $c > d$)	implies	$a - c < b - d$	**(1.2b)**
($a < b$ and $b < c$)	implies	$a < c$	**(1.2c)**
$a < b$	implies	$a + c < b + c$	**(1.2d)**
($a < b$ and $c > 0$)	implies	$ac < bc$	**(1.2e)**
($a < b$ and $c < 0$)	implies	$ac > bc$	**(1.2f)**
($a < b$ and $ab > 0$)	implies	$\dfrac{1}{a} > \dfrac{1}{b}$	**(1.2g)**

Example 1.6   Show, without using a calculator, that $\sqrt{2} + \sqrt{3} > 2(\sqrt[4]{6})$.

Solution   By squaring we have that

$$(\sqrt{2} + \sqrt{3})^2 = 2 + 2\sqrt{2}\sqrt{3} + 3 = 5 + 2\sqrt{6}$$

Also

$$(2\sqrt{6})^2 = 24 < 25 = 5^2$$

implying that $5 > 2\sqrt{6}$. Thus

$$(\sqrt{2} + \sqrt{3})^2 > 2\sqrt{6} + 2\sqrt{6} = 4\sqrt{6}$$

and, since $\sqrt{2} + \sqrt{3}$ is a positive number, it follows that

$$\sqrt{2} + \sqrt{3} > \sqrt{(4\sqrt{6})} = 2(\sqrt[4]{6})$$

1.2.5 Modulus and intervals

The size of a real number x is called its modulus and is denoted by $|x|$ (or sometimes by mod (x)). Thus

$$|x| = \begin{cases} x & (x \geq 0) \\ -x & (x < 0) \end{cases} \tag{1.3}$$

where the comparator $\geq$ indicates 'greater than or equal to'. (Likewise $\leq$ indicates 'less than or equal to'.)

Geometrically $|x|$ is the distance of the point representing x on the number line from the point representing zero. Similarly $|x - a|$ is the distance of the point representing x on the number line from that representing a.

The set of numbers between two numbers, a and b say, defines an **open interval** on the real line. This is the set $\{x : a < x < b, x \text{ in } \mathbb{R}\}$ and is usually denoted by (a, b). (Set notation will be fully described in Chapter 6; here $\{x : P\}$ denotes the set of all x that have property P.) Here the double-sided inequality means that x is greater than a and less than b; that is, the inequalities $a < x$ and $x < b$ apply simultaneously. An interval that includes the end points is called a **closed interval**, denoted by $[a, b]$, with

$$[a, b] = \{x : a \leq x \leq b, x \text{ in } \mathbb{R}\}$$

Note that the distance between two numbers a and b might either be $a - b$ or $b - a$ depending on which was the larger. An immediate consequence of this is that

$$|a - b| = |b - a|$$

since a is the same distance from b as b is from a.

Example 1.7   Find the values of x so that

$$|x - 4.3| = 5.8$$

Solution $|x - 4.3| = 5.8$ means that the distance between the real numbers x and 4.3 is 5.8 units, but does not tell us whether $x > 4.3$ or whether $x < 4.3$. The situation is illustrated in Figure 1.2, from which it is clear that the two possible values of x are -1.5 and 10.1.

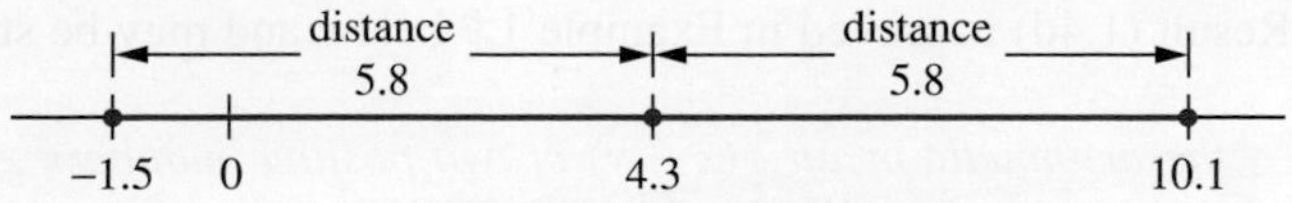

Figure 1.2
Illustration of
$|x - 4.3| = 5.8$.

distance 5.8   distance 5.8

-1.5 0   4.3   10.1

Example 1.8

Express the sets (a) $\{x : |x - 3| < 5, x \text{ in } \mathbb{R}\}$ and (b) $\{x : |x + 2| \leq 3, x \text{ in } \mathbb{R}\}$ as intervals.

Solution

(a) $|x - 3| < 5$ means that the distance of the point representing x on the number line from the point representing 3 is less than 5 units, as shown in Figure 1.3(a). This implies that

$$-5 < x - 3 < 5$$

Adding 3 to each member of this inequality, using rule (1.2d), gives

$$-2 < x < 8$$

and the set of numbers satisfying this inequality is the open interval $(-2, 8)$.

(b) Similarly $|x + 2| \leq 3$, which may be rewritten as $|x - (-2)| \leq 3$, means that the distance of the point x on the number line from the point representing -2 is less than or equal to 3 units, as shown in Figure 1.3(b). This implies

$$-3 \leq x + 2 \leq 3$$

Subtracting 2 from each member of this inequality, using rule (1.2d), gives

$$-5 \leq x \leq 1$$

and the set of numbers satisfying this inequality is the closed interval $[-5, 1]$.

It is easy (and sensible) to check these answers using spot values. For example, putting $x = -4$ in (b) gives $|-4 + 2| < 3$ correctly. Sometimes the sets $|x + 2| \leq 3$ and $|x + 2| < 3$ are described verbally as 'lies in the interval x equals -2 ± 3'.

Figure 1.3
(a) The open interval
$(-2, 8)$. (b) The closed
interval $[-5, 1]$.

(a)

-5 -4 -3 -2 -1 0 1 2 3 4 5 6 7 8 9

(b)

-7 -6 -5 -4 -3 -2 -1 0 1 2 3 4 5 6 7

We note in passing the following results. For any two real numbers x and y:

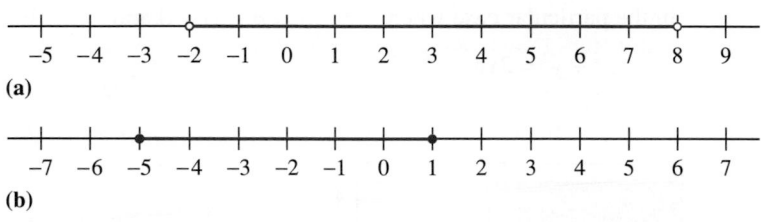

$$|xy| = |x||y| \tag{1.4a}$$

$$|x| < a, a > 0, \quad \text{implies} \quad -a < x < a \tag{1.4b}$$

$$|x + y| \leq |x| + |y|, \quad \text{known as the 'triangle inequality'} \tag{1.4c}$$

$$\tfrac{1}{2}(x + y) \geq \sqrt{(xy)}, \quad \text{when } x \geq 0 \text{ and } y \geq 0 \tag{1.4d}$$

Result (1.4d) is proved in Example 1.9 below and may be stated in words as

> *the arithmetic mean $\frac{1}{2}(x + y)$ of two positive numbers x and y is greater than or equal to the geometric mean $\sqrt{(xy)}$. Equality holds only when $y = x$.*

Results (1.4a) to (1.4c) should be verified by the reader, who may find it helpful to try some particular values first, for example, setting $x = -2$ and $y = 3$ in (1.4c).

Example 1.9

Prove that for any two positive numbers x and y, the arithmetic–geometric inequality

$$\frac{1}{2}(x + y) \geqslant \sqrt{(xy)}$$

holds.

Deduce that $x + \dfrac{1}{x} \geqslant 2$ for any positive number x.

Solution

The quantity xy can be interpreted as the area of a rectangle with sides x and y. The quantity $(x + y)^2$ can be interpreted as the area of a square of side $(x + y)$. Comparing areas in Figure 1.4, where the broken lines cut the square into 4 equal quarters of size A and it is assumed that $x > y$.

From Figure 1.4, we see that

$$(x + y)^2 = x^2 + y^2 + 2xy \tag{1.5}$$

Also, from Figure 1.4, we see that

$$\left.\begin{array}{l} x^2 = A + 2B + C \\ y^2 = A - 2D - C \end{array}\right\} x^2 + y^2 = 2A + 2B - 2D$$

$$xy = A - B + D$$

Since B > D, (B = D + C), it follows that

$$x^2 + y^2 > 2xy$$

In the particular case when $x = y$ then B = D = 0 and

$$x^2 + y^2 = 2xy$$

Figure 1.4
Illustration of
$x^2 + y^2 \geqslant 2xy$.

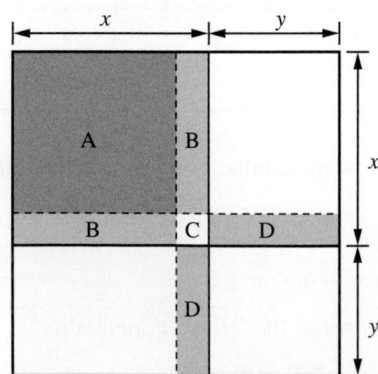

so in general

$$x^2 + y^2 \geqslant 2xy \qquad (1.6)$$

Combining (1.5) and (1.6) we deduce

$$(x + y)^2 \geqslant 4xy$$

and since x and y are both positive we have

$$x + y \geqslant 2\sqrt{(xy)}$$

which is equivalent to

$$\tfrac{1}{2}(x + y) \geqslant \sqrt{(xy)}$$

In the special case when $y = \dfrac{1}{x}$ we have

$$x + \frac{1}{x} \geqslant 2\sqrt{\left(x\frac{1}{x} \right)}$$

that is,

$$x + \frac{1}{x} \geqslant 2$$

1.2.6 Exercises

6 Show that $(\sqrt{5} + \sqrt{13})^2 > 34$ and determine without using a calculator the larger of $\sqrt{5} + \sqrt{13}$ and $\sqrt{3} + \sqrt{19}$.

7 Show the following sets on number lines and express them as intervals:

(a) $\{x : |x - 4| \leqslant 6\}$ (b) $\{x : |x + 3| < 2\}$

(c) $\{x : |2x - 1| \leqslant 7\}$ (d) $\{x : |\tfrac{1}{4}x + 3| < 3\}$

8 Show the following intervals on number lines and express them as sets in the form $\{x : |ax + b| < c\}$ or $\{x : |ax + b| \leqslant c\}$:

(a) $(1, 7)$ (b) $[-4, -2]$

(c) $(17, 26)$ (d) $[-\tfrac{1}{2}, \tfrac{3}{4}]$

9 Given that $a < b$ and $c < d$, which of the following statements are always true?

(a) $a - c < b - d$ (b) $a - d < b - c$

(c) $ac < bd$ (d) $\dfrac{1}{b} < \dfrac{1}{a}$

In each case either prove that the statement is true or give a numerical example to show it can be false.

If, additionally, a, b, c and d are all greater than zero, how does that modify your answer?

10 The average speed for a journey is the distance covered divided by the time taken.

(a) A journey is completed by travelling for the first half of the *time* at speed v_1 and the second half at speed v_2. Find the average speed v_a for the journey in terms of v_1 and v_2.

(b) A journey is completed by travelling at speed v_1 for half the *distance* and at speed v_2 for the second half. Find the average speed v_b for the journey in terms of v_1 and v_2.

Deduce that a journey completed by travelling at two different speeds for equal distances will take longer than the same journey completed at the same two speeds for equal times.

1.3 Algebra

The origins of algebra are to be found in Arabic mathematics as the name suggests, coming from the word *aljabara* meaning 'combination' or 're-uniting'. Algorithms are rules for solving problems in mathematics by standard step-by-step methods. Such methods were first described by the ninth century mathematician Abu Ja'far Mohammed ben Musa from Khwarizm, modern Khiva on the southern border of Uzbekistan. The arabic al-Khwarizm ('from Khwarizm') was latinized to algorithm in the late Middle Ages. Often the letter x is used to denote an unassigned (or free) variable. It is thought that this is a corruption of the script letter *r* abbreviating the latin word *res*, thing. The use of unassigned variables enables us to form mathematical models of practical situations as illustrated in the following example. First we deal with a specific case and then with the general case using unassigned variables.

The idea, first introduced in the seventeenth century, of using letters to represent unspecified quantities led to the development of algebraic manipulation based on the elementary laws of arithmetic. This development greatly enhanced the problem-solving power of mathematics – so much so that it is difficult now to imagine doing mathematics without this resource.

Example 1.10　A pipe has the form of a hollow cylinder as shown in Figure 1.5. Find its mass when

(a) its length is 1.5 m, its external diameter is 205 mm, its internal diameter is 160 mm and its density is 5500 kg m^{-3},

(b) its length is l m, its external diameter is D mm, its internal diameter is d mm and its density is ρ kg m^{-3}.

Solution　(a) Standardizing the units of length, the internal and external diameters are 0.16 m and 0.205 m respectively. The area of cross-section of the pipe is

$$0.25\pi(0.205^2 - 0.160^2)\ \text{m}^2$$

(Reminder: The area of a circle of diameter D is $\pi D^2/4$)
　Hence the volume of the material of the pipe is

$$0.25\pi(0.205^2 - 0.160^2) \times 1.5\ \text{m}^3$$

and the mass (volume × density) of the pipe is

$$0.25 \times 5500 \times \pi(0.205^2 - 0.160^2) \times 1.5\ \text{kg}$$

Evaluating this last expression by calculator gives the mass of the pipe as 106 kg to the nearest kilogram.

(b) The internal and external diameters of the pipe are $d/1000$ and $D/1000$ metres, respectively, so that the area of cross-section is

$$0.25\pi(D^2 - d^2)/1\,000\,000\ \text{m}^2$$

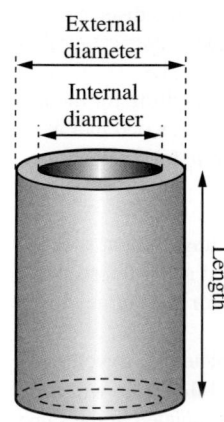

External diameter

Internal diameter

Length

Figure 1.5
Cylindrical pipe of Example 1.10.

The volume of the pipe is

$$0.25\pi l(D^2 - d^2)/10^6 \, \text{m}^3$$

Hence the mass M kg of the pipe of density ρ is given by the formulae

$$M = 0.25\pi\rho l(D^2 - d^2)/10^6 = 2.5\pi\rho l(D + d)(D - d) \times 10^{-5}$$

1.3.1 Algebraic manipulation

Algebraic manipulation made possible concise statements of well-known results, such as

$$(a + b)^2 = a^2 + 2ab + b^2 \tag{1.7a}$$

Previously these results had been obtained by a combination of verbal reasoning and elementary geometry as illustrated in Figure 1.6.

Example 1.11

Prove that

$$ab = \tfrac{1}{4}[(a + b)^2 - (a - b)^2]$$

Given $70^2 = 4900$ and $36^2 = 1296$, calculate 53×17.

Solution

Since

$$(a + b)^2 = a^2 + 2ab + b^2$$

we deduce

$$(a - b)^2 = a^2 - 2ab + b^2$$

and

$$(a + b)^2 - (a - b)^2 = 4ab$$

and

$$ab = \tfrac{1}{4}[(a + b)^2 - (a - b)^2]$$

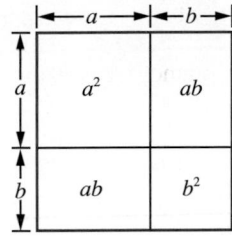

Figure 1.6
Illustration of
$(a + b)^2 = a^2 + 2ab + b^2$.

The result is illustrated geometrically in Figure 1.7. Setting $a = 53$ and $b = 17$, we have

$$53 \times 17 = \tfrac{1}{4}[70^2 - 36^2] = 901$$

This method of calculating products was used by the Babylonians and is sometimes called the 'quarter-squares' algorithm. It has been used in some analogue devices and simulators.

Figure 1.7
Illustration of $ab = \tfrac{1}{4}[(a + b)^2 - (a - b)^2]$.

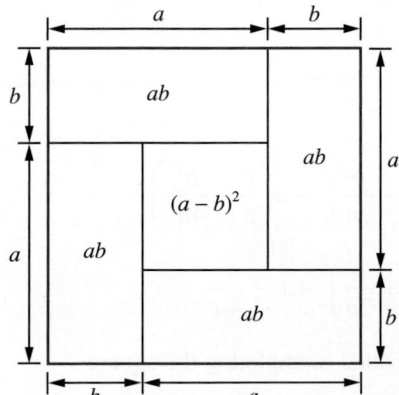

Example 1.12

Show that
$$(a + b + c)^2 = a^2 + b^2 + c^2 + 2ab + 2bc + 2ca$$

Solution

Rewriting $a + b + c$ as $(a + b) + c$ we have
$$((a + b) + c)^2 = (a + b)^2 + 2(a + b)c + c^2 \quad \text{using (1.7a)}$$
$$= a^2 + 2ab + b^2 + 2ac + 2bc + c^2$$
$$= a^2 + b^2 + c^2 + 2ab + 2bc + 2ac$$

Example 1.13

Verify that
$$(x + p)^2 + q - p^2 = x^2 + 2px + q$$

and deduce that
$$ax^2 + bx + c = a\left(x + \frac{b}{2a}\right)^2 + c - \frac{b^2}{4a}$$

Solution

$$(x + p)^2 = x^2 + 2px + p^2$$

so that
$$(x + p)^2 + q - p^2 = x^2 + 2px + q$$

Working in the reverse direction is more difficult
$$ax^2 + bx + c = a\left(x^2 + \frac{b}{a}x + \frac{c}{a}\right)$$

Comparing $x^2 + \dfrac{b}{a}x + \dfrac{c}{a}$ with $x^2 + 2px + q$, we can identify

$$\frac{b}{a} = 2p \quad \text{and} \quad \frac{c}{a} = q$$

Thus we can write
$$ax^2 + bx + c = a[(x + p)^2 + q - p^2]$$

where $p = \dfrac{b}{2a}$ and $q = \dfrac{c}{a}$

giving

$$ax^2 + bx + c = a\left(x + \frac{b}{2a}\right)^2 + a\left(\frac{c}{a} - \frac{b^2}{4a^2}\right)$$

$$= a\left(x + \frac{b}{2a}\right)^2 + c - \frac{b^2}{4a}$$

This algebraic process is called 'completing the square'.

We may summarize the results so far

$$(a + b)^2 = a^2 + 2ab + b^2 \tag{1.7a}$$

$$(a - b)^2 = a^2 - 2ab + b^2 \tag{1.7b}$$

$$a^2 - b^2 = (a + b)(a - b) \tag{1.7c}$$

$$a^2 + bx + c = a\left(x + \frac{b}{2a}\right)^2 + c - \frac{b^2}{4a} \tag{1.7d}$$

As shown in the previous examples the ordinary rules of arithmetic carry over to the generalized arithmetic of algebra. This is illustrated again in the following example.

Example 1.14 Express as a single fraction

(a) $\dfrac{1}{12} - \dfrac{2}{3} + \dfrac{3}{4}$

(b) $\dfrac{1}{(x + 1)(x + 2)} - \dfrac{2}{x + 1} + \dfrac{3}{x + 2}$

Solution (a) The lowest common denominator of these fractions is 12, so we may write

$$\frac{1}{12} - \frac{2}{3} + \frac{3}{4} = \frac{1 - 8 + 9}{12}$$

$$= \frac{2}{12} = \frac{1}{6}$$

(b) The lowest common multiple of the denominators of these fractions is $(x + 1)(x + 2)$ so we may write

$$\frac{1}{(x + 1)(x + 2)} - \frac{2}{x + 1} + \frac{3}{x + 2}$$

$$= \frac{1}{(x + 1)(x + 2)} - \frac{2(x + 2)}{(x + 1)(x + 2)} + \frac{3(x + 1)}{(x + 1)(x + 2)}$$

$$= \frac{1 - 2(x + 2) + 3(x + 1)}{(x + 1)(x + 2)}$$

$$= \frac{1 - 2x - 4 + 3x + 3}{(x + 1)(x + 2)}$$

$$= \frac{x}{(x + 1)(x + 2)}$$

Example 1.15   Use the method of completing the square to manipulate the following quadratic expressions into the form of a number + (or −) the square of a term involving x.

(a) $x^2 + 3x - 7$ (b) $5 - 4x - x^2$

(c) $3x^2 - 5x + 4$ (d) $1 + 2x - 2x^2$

Solution   Remember $(a + b)^2 = a^2 + 2ab + b^2$.

(a) To convert $x^2 + 3x$ into a perfect square we need to add $(\frac{3}{2})^2$. Thus we have

$$x^2 + 3x - 7 = [(x + \tfrac{3}{2})^2 - (\tfrac{3}{2})^2] - 7$$
$$= (x + \tfrac{3}{2})^2 - \tfrac{37}{4}$$

(b) $5 - 4x - x^2 = 5 - (4x + x^2)$

To convert $x^2 + 4x$ into a perfect square we need to add 2^2. Thus we have

$$x^2 + 4x = (x + 2)^2 - 2^2$$

and

$$5 - 4x - x^2 = 5 - [(x + 2)^2 - 2^2] = 9 - (x + 2)^2$$

(c) First we 'take outside' the coefficient of x^2:

$$3x^2 - 5x + 4 = 3(x^2 - \tfrac{5}{3}x + \tfrac{4}{3})$$

Then we rearrange

$$x^2 - \tfrac{5}{3}x = (x - \tfrac{5}{6})^2 - \tfrac{25}{36}$$

so that $3x^2 - 5x + 4 = 3[(x - \tfrac{5}{6})^2 - \tfrac{25}{36} + \tfrac{4}{3}] = 3[(x - \tfrac{5}{6})^2 + \tfrac{23}{36}]$.

(d) Similarly

$$1 + 2x - 2x^2 = 1 - 2(x^2 - x)$$

and

$$x^2 - x = (x - \tfrac{1}{2})^2 - \tfrac{1}{4}$$

so that

$$1 + 2x - 2x^2 = 1 - 2[(x - \tfrac{1}{2})^2 - \tfrac{1}{4}] = \tfrac{3}{2} - 2(x - \tfrac{1}{2})^2$$

The reader should confirm that these results agree with identity (1.7d)

The number 45 can be factorized as $3 \times 3 \times 5$. Any product from 3, 3 and 5 is also a factor of 45. Algebraic expressions can be factorized in a similar fashion. An algebraic expression with more than one term can be factorized if each term contains common factors (either numerical or algebraic). These factors are removed by division from each term and the non-common factors remaining are grouped into brackets.

Example 1.16　　Factorize $xz + 2yz - 2y - x$.

Solution　　There is no common factor to all four terms so we take them in pairs:

$$xz + 2yz - 2y - x = (x + 2y)z - (2y + x)$$
$$= (x + 2y)z - (x + 2y)$$
$$= (x + 2y)(z - 1)$$

Alternatively, we could have written:

$$xz + 2yz - 2y - x = (xz - x) + (2yz - 2y)$$
$$= x(z - 1) + 2y(z - 1)$$
$$= (x + 2y)(z - 1)$$

to obtain the same result.

In many problems we are able to facilitate the solution by factorizing a quadratic expression $ax^2 + bx + c$ 'by-hand', using knowledge of the factors of the numerical coefficients a, b and c.

Example 1.17　　Factorize the expressions

(a) $x^2 + 12x + 35$　　　(b) $2x^2 + 9x - 5$

Solution　　(a) Since

$$(x + \alpha)(x + \beta) = x^2 + (\alpha + \beta)x + \alpha\beta$$

we examine the factors of the constant term of the expression:

$$35 = 5 \times 7 = 35 \times 1$$

and notice that $5 + 7 = 12$ while $35 + 1 = 36$. So we can chose $\alpha = 5$ and $\beta = 7$ and write

$$x^2 + 12x + 35 = (x + 5)(x + 7)$$

(b) Since

$$(mx + \alpha)(nx + \beta) = mnx^2 + (n\alpha + m\beta)x + \alpha\beta$$

we examine the factors of the coefficient of x^2 and of the constant to give the coefficient of x. Here

$$2 = 2 \times 1 \text{ and } -5 = (-5) \times 1 = 5 \times (-1)$$

and we see that

$$2 \times 5 + 1 \times (-1) = 9$$

Thus we can write

$$(2x - 1)(x + 5) = 2x^2 + 9x - 5$$

It is sensible to do a 'spot-check' on the factorization by inserting a sample value of x, for example $x = 1$

$$(1)(6) = 2 + 9 - 5$$

Comment   Some quadratic expressions, for example $x^2 + y^2$, do not have real factors.

The expansion of $(a + b)^2$ in (1.7a) is a special case of a general result for $(a + b)^n$ known as the binomial expansion. This is discussed again in Sections 1.3.4 and 7.7.2. Here we shall look at the cases for $n = 0, 1, \ldots, 6$.

Writing these out, we have

$$(a + b)^0 = 1$$
$$(a + b)^1 = a + b$$
$$(a + b)^2 = a^2 + 2ab + b^2$$
$$(a + b)^3 = a^3 + 3a^2b + 3ab^2 + b^3$$
$$(a + b)^4 = a^4 + 4a^3b + 6a^2b^2 + 4ab^3 + b^4$$
$$(a + b)^5 = a^5 + 5a^4b + 10a^3b^2 + 10a^2b^3 + 5ab^4 + b^5$$
$$(a + b)^6 = a^6 + 6a^5b + 15a^4b^2 + 20a^3b^3 + 15a^2b^4 + 6ab^5 + b^6$$

This table can be extended indefinitely. Each line can easily be obtained from the previous one. Thus, for example,

$$(a + b)^4 = (a + b)(a + b)^3$$
$$= a(a^3 + 3a^2b + 3ab^2 + b^3) + b(a^3 + 3a^2b + 3ab^2 + b^3)$$
$$= a^4 + 3a^3b + 3a^2b^2 + ab^3 + a^3b + 3a^2b^2 + 3ab^3 + b^4$$
$$= a^4 + 4a^3b + 6a^2b^2 + 4ab^3 + b^4$$

The coefficients involved form a pattern of numbers called Pascal's triangle, shown in Figure 1.8. Each number in the interior of the triangle is obtained by summing the numbers to its right and left in the row above, as indicated by the arrows in Figure 1.8. This number pattern had been discovered prior to Pascal by the Chinese mathematician Chu Shih-chieh.

Figure 1.8
Pascal's triangle.

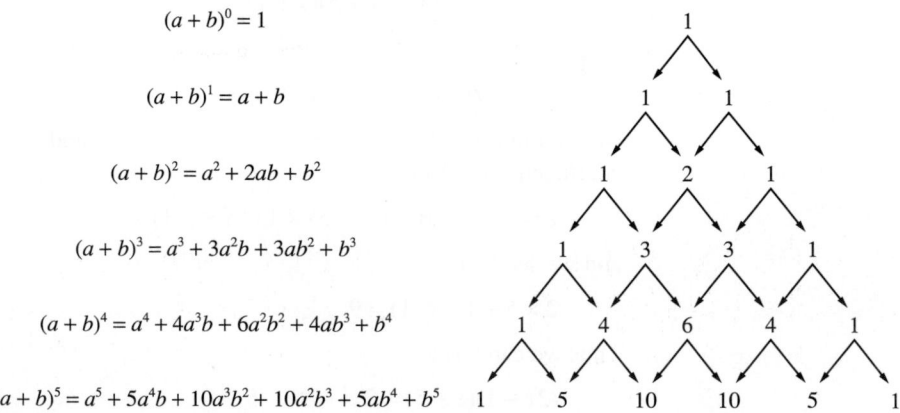

$$(a + b)^0 = 1$$
$$(a + b)^1 = a + b$$
$$(a + b)^2 = a^2 + 2ab + b^2$$
$$(a + b)^3 = a^3 + 3a^2b + 3ab^2 + b^3$$
$$(a + b)^4 = a^4 + 4a^3b + 6a^2b^2 + 4ab^3 + b^4$$
$$(a + b)^5 = a^5 + 5a^4b + 10a^3b^2 + 10a^2b^3 + 5ab^4 + b^5$$

Example 1.18   Expand

(a) $(2x + 3y)^2$ (b) $(2x - 3)^3$ (c) $\left(2x - \dfrac{1}{x}\right)^4$

Solution   (a) Here we use the expansion

$$(a + b)^2 = a^2 + 2ab + b^2$$

with $a = 2x$ and $b = 3y$ to obtain

$$(2x + 3y)^2 = (2x)^2 + 2(2x)(3y) + (3y)^2$$
$$= 4x^2 + 12xy + 9y^2$$

(b) Here we use the expansion

$$(a + b)^3 = a^3 + 3a^2b + 3ab^2 + b^3$$

with $a = 2x$ and $b = -3$ to obtain

$$(2x - 3)^3 = 8x^3 - 36x^2 + 54x - 27$$

(c) Here we use the expansion

$$(a + b)^4 = a^4 + 4a^3b + 6a^2b^2 + 4ab^3 + b^4$$

with $a = 2x$ and $b = -1/x$ to obtain

$$\left(2x - \frac{1}{x}\right)^4 = (2x)^4 + 4(2x)^3(-1/x) + 6(2x)^2(-1/x)^2 + 4(2x)(-1/x)^3 + (-1/x)^4$$
$$= 16x^4 - 32x^2 + 24 - 8/x^2 + 1/x^4$$

1.3.2 Exercises

11   Simplify the following expressions

(a) $x^3 \times x^{-4}$ (b) $x^3 \div x^{-4}$ (c) $(x^3)^{-4}$

(d) $x^{1/3} \times x^{5/3}$ (e) $(4x^8)^{-1/2}$ (f) $\left(\dfrac{3}{2\sqrt{x}}\right)^{-2}$

(g) $\sqrt{x}\left(x^2 - \dfrac{2}{x}\right)$ (h) $\left(5x^{1/3} - \dfrac{1}{2x^{1/3}}\right)^2$

(i) $\dfrac{2x^{1/2} - x^{-1/2}}{x^{1/2}}$ (j) $\dfrac{(a^2b)^{1/2}}{(ab^{-2})^2}$

(k) $(4ab^2)^{-3/2}$

12   Factorize

(a) $x^2y - xy^2$

(b) $x^2yz - xy^2z + 2xyz^2$

(c) $ax - 2by - 2ay + bx$

(d) $x^2 + 3x - 10$

(e) $x^2 - \frac{1}{4}y^2$ (f) $81x^4 - y^4$

13   Simplify

(a) $\dfrac{x^2 - x - 12}{x^2 - 16}$ (b) $\dfrac{x - 1}{x^2 - 2x - 3} - \dfrac{2}{x + 1}$

(c) $\dfrac{1}{x^2 + 3x - 10} + \dfrac{1}{x^2 + 17x + 60}$

(d) $(3x + 2y)(x - 2y) + 4xy$

14   An isosceles trapezium has non-parallel sides of length 20 cm and the shorter parallel side is 30 cm, as illustrated in Figure 1.9. The perpendicular distance between the parallel sides is h cm.

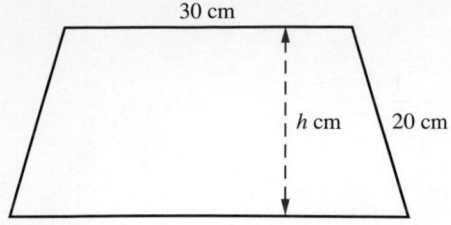

Figure 1.9

Show that the area of the trapezium is $h(30 + \sqrt{(400 - h^2)})\,\text{cm}^2$.

15 An open container is made from a sheet of cardboard of size $200\,\text{mm} \times 300\,\text{mm}$ using a simple fold, as shown in Figure 1.10. Show that the capacity C ml of the box is given by

$$C = x(150 - x)(100 - x)/250$$

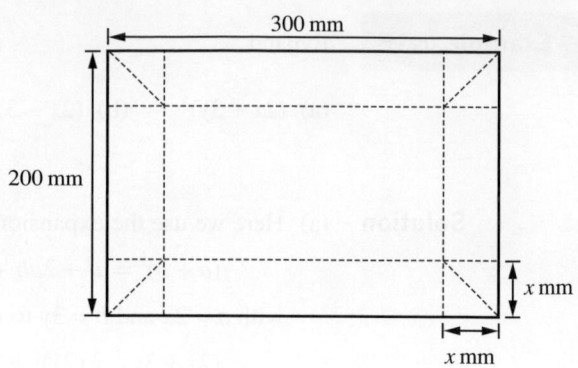

Figure 1.10 Sheet of cardboard of Question 15.

16 Rearrange the following quadratic expressions by completing the square.

(a) $x^2 + x - 12$ (b) $3 - 2x + x^2$

(c) $(x - 1)^2 - (2x - 3)^2$ (d) $1 + 4x - x^2$

1.3.3 Equations, inequalities and identities

It commonly occurs in the application of mathematics to practical problem-solving that the numerical value of an expression involving unassigned variables is specified and we have to find the values of the unassigned variables which yield that value. We illustrate the idea with the elementary examples that follow.

Example 1.19 A hollow cone of base diameter 100 mm and height 150 mm is held upside down and filled with a liquid. The liquid is then transferred to a hollow circular cylinder of base diameter 80 mm. To what height is the cylinder filled?

Solution The situation is illustrated in Figure 1.11. The capacity of the cone is

$\frac{1}{3}$(base area) $\times$ (perpendicular height)

Thus the volume of liquid contained in the cone is

$\frac{1}{3}\pi(50^2)(150) = 125\,000\pi\,\text{mm}^3$

The volume of the liquid in the circular cylinder is

(base area) $\times$ (height) $= \pi(40^2)h\,\text{mm}^3$

where h mm is the height of the liquid in the cylinder. Equating these quantities (assuming no liquid is lost in the transfer) we have

$1600\pi h = 125\,000\pi$

This **equation** enables us to find the value of the unassigned variable h:

$h = 1250/16 = 78.125$

Thus the height of the liquid in the cylinder is 78 to the nearest millimetre.

Figure 1.11
The cone and cylinder
of Example 1.19.

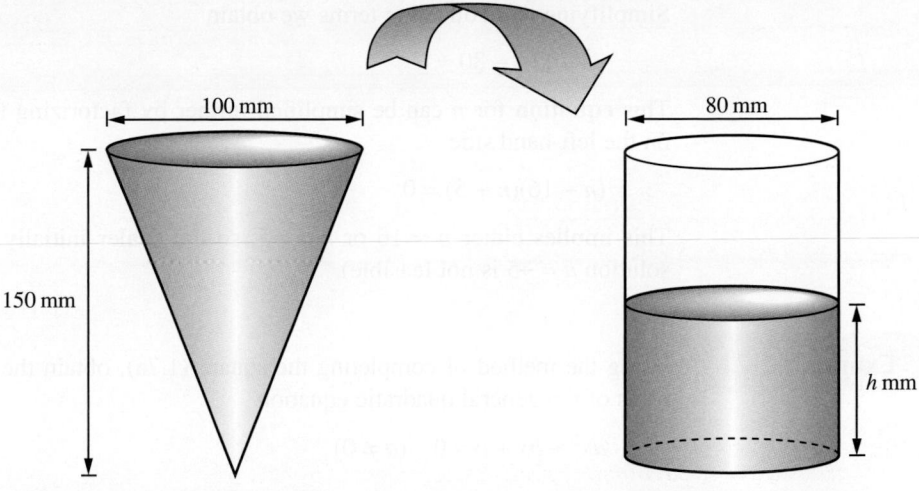

In the previous example we made use of the formula for the volume V of a cone of base diameter D and height H. We normally shorthand this as

$$V = \tfrac{1}{12} \pi D^2 H$$

understanding that the units of measurement are compatible. This formula also tells us the height of such a cone in terms of its volume and base diameter

$$H = \frac{12V}{\pi D^2}$$

This type of rearrangement is common and is generally described as 'changing the subject of the formula'.

Example 1.20

A dealer bought a number of equally priced articles for a total cost of £120. He sold all but one of them, making a profit of £1.50 on each article with a total revenue of £135. How many articles did he buy?

Solution

Let n be the number of articles bought. Then the cost of each article was £$(120/n)$. Since $(n - 1)$ articles were sold the selling price of each article was £$(135/(n - 1))$. Thus the profit per item was

$$£\left\{ \frac{135}{n - 1} - \frac{120}{n} \right\}$$

which we are told is equal to £1.50. Thus

$$\frac{135}{n - 1} - \frac{120}{n} = 1.50$$

This implies

$$135n - 120(n - 1) = 1.50(n - 1)n$$

Dividing both sides by 1.5 gives

$$90n - 80(n - 1) = n^2 - n$$

Simplifying and collecting terms we obtain

$$n^2 - 11n - 80 = 0$$

This **equation** for n can be simplified further by factorizing the quadratic expression on the left-hand side

$$(n - 16)(n + 5) = 0$$

This implies either $n = 16$ or $n = -5$, so the dealer initially bought 16 articles (the solution $n = -5$ is not feasible).

Example 1.21   Using the method of completing the square (1.7a), obtain the formula for finding the roots of the general quadratic equation

$$ax^2 + bx + c = 0 \quad (a \neq 0)$$

Solution   Dividing throughout by a gives

$$x^2 + \frac{b}{a}x + \frac{c}{a} = 0$$

Completing the square leads to

$$\left(x + \frac{b}{2a}\right)^2 + \frac{c}{a} = \left(\frac{b}{2a}\right)^2$$

giving

$$\left(x + \frac{b}{2a}\right)^2 = \frac{b^2}{4a^2} - \frac{c}{a} = \frac{b^2 - 4ac}{4a^2}$$

which on taking the square root gives

$$x + \frac{b}{2a} = +\frac{\sqrt{(b^2 - 4ac)}}{2a} \quad \text{or} \quad -\frac{\sqrt{(b^2 - 4ac)}}{2a}$$

or

$$x = \frac{-b \pm \sqrt{(b^2 - 4ac)}}{2a} \tag{1.8}$$

Here the $\pm$ symbol provides a neat shorthand for the two solutions.

Comments   (a) The formula given in (1.8) makes clear the three cases: where for $b^2 > 4ac$ we have two real roots to the equation, for $b^2 < 4ac$ we have no real roots, and for $b^2 = 4ac$ we have one real root (which is repeated).

(b) The condition for equality of the roots of a quadratic equation occurs in practical applications, and we shall illustrate this in Chapter 2, Example 2.48 after considering the trigonometric functions.

(c) The quadratic equation has many important applications. One, which is of historical significance, concerned the electrical engineer Oliver Heaviside. In 1871 the telephone cable between England and Denmark developed a fault caused by a short circuit under the sea. His task was to locate that fault. The cable had a uniform resistance per unit length. His method of solution was brilliantly simple. The situation can be represented schematically as shown in Figure 1.12.

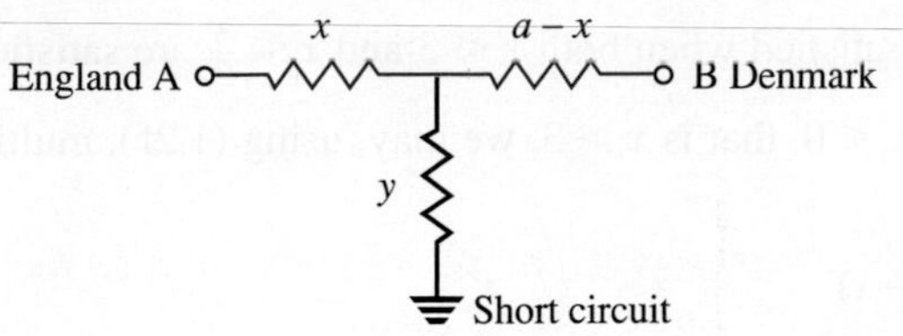

Figure 1.12
The circuit for the telephone line fault.

In the figure the total resistance of the line between A and B is a ohms and is known; x and y are unknown. If we can find x, we can locate the distance along the cable where the fault has occurred. Heaviside solved the problem by applying two tests. First he applied a battery, having voltage E, at A with the circuit open at B, and measured the resulting current I_1. Then he applied the same battery at A but with the cable earthed at B, and again measured the resulting current I_2. Using Ohm's law and the rules for combining resistances in parallel and in series, this yields the pair of equations

$$E = I_1(x + y)$$

$$E = I_2\left[x + \left(\frac{1}{y} + \frac{1}{a-x}\right)^{-1}\right]$$

Writing $b = E/I_1$ and $c = E/I_2$, we can eliminate y from these equations to obtain an equation for x:

$$x^2 - 2cx + c(a + b) - ab = 0$$

which, using (1.8), has solutions

$$x = c \pm \sqrt{[(a - c)(b - c)]}$$

From his experimental data Heaviside was able to predict accurately the location of the fault.

In some problems we have to find the values of unassigned variables such that the value of an expression involving those variables satisfies an inequality condition (that is, it is either greater than, or alternatively less than, a specified value). Solving such inequalities requires careful observance of the rules for inequalities (1.2a–1.2g) set out in Section 1.2.4.

Example 1.22   Find the values of x for which

$$\frac{1}{3 - x} < 2 \tag{1.9}$$

Solution   (a) When $3 - x > 0$, that is $x < 3$, we may, using (1.2e), multiply (1.9) throughout by $3 - x$ to give

$$1 < 2(3 - x)$$

which, using (1.2d, e), reduces to

$$x < \tfrac{5}{2}$$

so that (1.9) is satisfied when both $x < 3$ and $x < \tfrac{5}{2}$ are satisfied; that is, $x < \tfrac{5}{2}$.

(b) When $3 - x < 0$, that is $x > 3$, we may, using (1.2f), multiply (1.9) throughout by $3 - x$ to give

$$1 > 2(3 - x)$$

which reduces to $x > \tfrac{5}{2}$ so that (1.9) is also satisfied when both $x > 3$ and $x > \tfrac{5}{2}$; that is, $x > 3$.

Thus inequality (1.9) is satisfied by values of x in the ranges $x > 3$ and $x < \tfrac{5}{2}$.

Comment   A common mistake made is simply to multiply (1.9) throughout by $3 - x$ to give the answer $x < \tfrac{5}{2}$, forgetting to consider both cases of $3 - x > 0$ and $3 - x < 0$. We shall return to consider this example from the graphical point of view in Example 2.36.

Example 1.23   Find the values of x such that

$$x^2 + 2x + 2 > 50$$

Solution   Completing the square on the left-hand side of the inequality we obtain

$$(x + 1)^2 + 1 > 50$$

which gives

$$(x + 1)^2 > 49$$

Taking the square root of both sides of this inequality we deduce that

either $(x + 1) < -7$ or $(x + 1) > 7$

Note particularly the first of these inequalities. From these we deduce that

$$x^2 + 2x + 2 > 50 \text{ for } x < -8 \text{ or } x > 6$$

The reader should check these results using spot values of x, say $x = -10$ and $x = 10$.

Example 1.24   A food manufacturer found that the sales figure for a certain item depended on its selling price. The company's market research department advised that the maximum number of items that could be sold weekly was 20 000 and that the number sold decreased by 100 for every 1p increase in its price. The total production cost consisted of a set-up cost of £200 **plus** 50p for every item manufactured. What price should the manufacturer adopt?

Solution   The data supplied by the market research department suggests that if the price of the item is p pence, then the number sold would be $20\,000 - 100p$. (So the company would sell none with $p = 200$, when the price is £2.) The production cost in pounds would be $200 + 0.5 \times$ (number sold), so that in terms of p we have the production cost £C given by

$$C = 200 + 0.5(20\,000 - 100p)$$

The revenue £R accrued by the manufacturer for the sales is (number sold) $\times$ (price), which gives

$$R = (20\,000 - 100p)p/100$$

(remember to express the amount in pounds). Thus, the profit £P is given by

$$P = R - C$$

$$= (20\,000 - 100p)p/100 - 200 - 0.5(20\,000 - 100p)$$

$$= -p^2 + 250p - 10\,200$$

Completing the square we have

$$P = 125^2 - (p - 125)^2 - 10\,200$$

$$= 5425 - (p - 125)^2$$

Since $(p - 125)^2 \geqslant 0$, we deduce that $P \leqslant 5425$ and that the maximum value of P is 5425. To achieve this weekly profit, the manufacturer should adopt the price £1.25.

It is important to distinguish between those equalities that are valid for a restricted set of values of the unassigned variable x and those that are true for all values of x. For example

$$(x - 5)(x + 7) = 0$$

is true only if $x = 5$ or $x = -7$. In contrast

$$(x - 5)(x + 7) = x^2 + 2x - 35 \qquad\qquad \textbf{(1.10)}$$

is true for all values of x. The word 'equals' here is being used in subtly different ways. In the first case '=' means 'is numerically equal to'; in the second case '=' means 'is algebraically equal to'. Sometimes we emphasize the different meaning by means of the special symbol $\equiv$, meaning 'algebraically equal to'. (However, it is fairly common practice in engineering to use '=' in both cases.) Such equations are often called **identities**. Identities that involve an unassigned variable x as in (1.10) are valid for all values of x, and we can sometimes make use of this fact to simplify algebraic manipulations.

Example 1.25   Find the numbers A, B and C such that

$$x^2 + 2x - 35 \equiv A(x - 1)^2 + B(x - 1) + C$$

Solution   Method (a): Since $x^2 + 2x - 35 \equiv A(x - 1)^2 + B(x - 1) + C$ it will be true for any value we give to x. So we choose values that make finding A, B and C easy.

Choosing $x = 0$ gives $-35 = A - B + C$
Choosing $x = 1$ gives $-32 = C$
Choosing $x = 2$ gives $-27 = A + B + C$

So we obtain $C = -32$, with $A - B = -3$ and $A + B = 5$. Hence $A = 1$ and $B = 4$ to give the identity

$$x^2 + 2x - 35 \equiv (x - 1)^2 + 4(x - 1) - 32$$

Method (b): Expanding the terms on the right-hand side, we have

$$x^2 + 2x - 35 \equiv Ax^2 + (B - 2A)x + A - B + C$$

The expressions on either side of the equals sign are algebraically equal, which means that the coefficient of x^2 on the left-hand side must equal the coefficient of x^2 on the right-hand side and so on. Thus

$$1 = A$$

$$2 = B - 2A$$

$$-35 = A - B + C$$

Hence we find $A = 1$, $B = 4$ and $C = -32$, as before.
  Note: Method (a) assumes that a valid A, B and C exist.

Example 1.26　Find numbers A, B and C such that

$$\frac{x^2}{x - 1} \equiv Ax + B + \frac{C}{x - 1}, \qquad x \neq 1$$

Solution　Expressing the right-hand side as a single term, we have

$$\frac{x^2}{x - 1} \equiv \frac{(Ax + B)(x - 1) + C}{x - 1}$$

which, with $x \neq 1$, is equivalent to

$$x^2 \equiv (Ax + B)(x - 1) + C$$

Choosing $x = 0$ gives $0 = -B + C$
Choosing $x = 1$ gives $1 = C$
Choosing $x = 2$ gives $4 = 2A + B + C$

Thus we obtain

$$C = 1, B = 1 \text{ and } A = 1, \text{ yielding}$$

$$\frac{x^2}{x - 1} \equiv x + 1 + \frac{1}{x - 1}$$

1.3.4 Exercises

17 Rearrange the following formula to make s the subject

$$m = p\sqrt{\frac{s+t}{s-t}}$$

18 Given $u = \dfrac{x^2+t}{x^2-t}$, find t in terms of u and x.

19 Solve for t

$$\frac{1}{1-t} - \frac{1}{1+t} = 1$$

20 If

$$\frac{3c^2 + 3xc + x^2}{3c^2 + 3yc + y^2} = \frac{yV_1}{xV_2}$$

find the positive value of c when

$$x = 4, \; y = 6, \; V_1 = 120, \; V_2 = 315$$

21 Solve for p the equation

$$\frac{2p+1}{p+5} + \frac{p-1}{p+1} = 2$$

22 A rectangle has a perimeter of 30 m. If its length is twice its breadth, find the length.

23 (a) A4 paper is such that a half sheet has the same shape as the whole sheet. Find the ratio of the lengths of the sides of the paper.

(b) Foolscap paper is such that cutting off a square whose sides equal the shorter side of the paper leaves a rectangle which has the same shape as the original sheet. Find the ratio of the sides of the original page.

24 Find the values of x for which

(a) $\dfrac{5}{x} < 2$ (b) $\dfrac{1}{2-x} < 1$

(c) $\dfrac{3x-2}{x-1} > 2$ (d) $\dfrac{3}{3x-2} > \dfrac{1}{x+4}$

25 Find the values of x for which

$$x^2 < 2 + |x|$$

26 Prove that

(a) $x^2 + 3x - 10 \geqslant -(\tfrac{7}{2})^2$

(b) $18 + 4x - x^2 \leqslant 22$

(c) $x + \dfrac{4}{x} \geqslant 4$ where $x > 0$

(Hint: first complete the square of the left-hand members)

27 Find the values of A and B such that

(a) $\dfrac{1}{(x+1)(x-2)} \equiv \dfrac{A}{x+1} + \dfrac{B}{x-2}$

(b) $3x + 2 \equiv A(x-1) + B(x-2)$

(c) $\dfrac{5x+1}{\sqrt{(x^2+x+1)}} \equiv \dfrac{A(2x+1) + B}{\sqrt{(x^2+x+1)}}$

28 Find the values of A, B and C such that

$$2x^2 - 5x + 12 \equiv A(x-1)^2 + B(x-1) + C$$

1.3.5 Suffix, sigma and pi notation

We have seen in previous sections how letters are used to denote general or unspecified values or numbers. This process has been extended in a variety of ways. In particular, the introduction of suffixes enables us to deal with problems that involve a high degree of generality or whose solutions have the flexibility to apply in a large number of situations. Consider for the moment an experiment involving measuring the temperature of an object (for example, a piece of machinery or a cooling fin in a heat exchanger) at intervals over a period of time. In giving a theoretical description of the experiment we would talk about the total period of time in general terms, say T minutes, and the time interval between measurements as h minutes, so that the total number n of time intervals would be given by T/h. Assuming that the initial and final temperatures are recorded there are $(n + 1)$ measurements. In practice we would obtain a set of experimental results, as illustrated partially in Figure 1.13.

Figure 1.13
Experimental results:
temperature against
lapsed time.

Lapsed time (minutes)	0	5	10	15	...	170	175	180
Temperature (°C)	97.51	96.57	93.18	91.53	...	26.43	24.91	23.57

Here we could talk about the twenty-first reading and look it up in the table. In the theoretical description we would need to talk about any one of the $(n + 1)$ temperature measurements. To facilitate this we introduce a suffix notation. We label the times at which the temperatures are recorded $t_0, t_1, t_2, \ldots, t_n$ where t_0 corresponds to the time when the initial measurement is taken, t_n to the time when the final measurement is taken, and

$$t_1 = t_0 + h, \; t_2 = t_0 + 2h, \; \ldots, \; t_n = t_0 + nh$$

so that $t_n = t_0 + T$. We label the corresponding temperatures by $\theta_0, \theta_1, \theta_2, \ldots, \theta_n$. We can then talk about the general result θ_k as measuring the temperature at time t_k.

In the analysis of the experimental results we may also wish to manipulate the data we have obtained. For example, we might wish to work out the average value of the temperature over the time period. With the 37 specific experimental results given in Figure 1.13 it is possible to compute the average directly as

$$(97.51 + 96.57 + 93.18 + 91.53 + \ldots + 23.57)/37$$

In general, however, we have

$$(\theta_0 + \theta_1 + \theta_2 + \ldots + \theta_n)/(n + 1)$$

A compact way of writing this is to use the **sigma notation** for the extended summation $\theta_0 + \theta_1 + \ldots + \theta_n$. We write

$$\sum_{k=0}^{n} \theta_k \qquad (\Sigma \text{ is the upper-case Greek letter sigma.})$$

to denote

$$\theta_0 + \theta_1 + \theta_2 + \ldots + \theta_n$$

Thus

$$\sum_{k=0}^{3} \theta_k = \theta_0 + \theta_1 + \theta_2 + \theta_3$$

and

$$\sum_{k=5}^{10} \theta_k = \theta_5 + \theta_6 + \theta_7 + \theta_8 + \theta_9 + \theta_{10}$$

The suffix k appearing in the quantity to be summed and underneath the sigma symbol is the 'counting variable' or 'counter'. We may use any letter we please as a counter, provided that it is not being used at the same time for some other purpose. Thus

$$\sum_{i=0}^{3} \theta_i = \theta_0 + \theta_1 + \theta_2 + \theta_3 = \sum_{n=0}^{3} \theta_n = \sum_{j=0}^{3} \theta_j$$

Thus, in general, if $a_0, a_1, a_2, \ldots, a_n$ is a sequence of numbers or expressions, we write

$$\sum_{k=0}^{n} a_k = a_0 + a_1 + a_2 + \ldots + a_n$$

Another shorthand that is sometimes useful is for the extended product $a_0 a_1 a_2 \ldots a_n$, which we write as

$$\prod_{k=0}^{n} a_k = a_0 a_1 a_2 \ldots a_n \qquad (\Pi \text{ is the upper-case Greek letter pi.})$$

Thus

$$\prod_{k=0}^{3} a_k = a_0 a_1 a_2 a_3$$

and

$$\prod_{k=5}^{8} a_k = a_5 a_6 a_7 a_8$$

Example 1.27 Given $a_0 = 1$, $a_1 = 5$, $a_2 = 2$, $a_3 = 7$, $a_4 = -1$ and $b_0 = 0$, $b_1 = 2$, $b_2 = -2$, $b_3 = 11$, $b_4 = 3$, calculate

(a) $\displaystyle\sum_{k=0}^{4} a_k$ (b) $\displaystyle\sum_{i=2}^{3} a_i$ (c) $\displaystyle\sum_{k=1}^{3} a_k b_k$

(d) $\displaystyle\sum_{k=0}^{4} b_k^2$ (e) $\displaystyle\prod_{j=1}^{3} a_j$ (f) $\displaystyle\prod_{k=2}^{4} b_k$

Solution (a) $\displaystyle\sum_{k=0}^{4} a_k = a_0 + a_1 + a_2 + a_3 + a_4$

Substituting the given values for a_k ($k = 0, \ldots, 4$) gives

$$\sum_{k=0}^{4} a_k = 1 + 5 + 2 + 7 + (-1) = 14$$

(b) $\displaystyle\sum_{i=2}^{3} a_i = a_2 + a_3 = 2 + 7 = 9$

(c) $\displaystyle\sum_{k=1}^{3} a_k b_k = a_1 b_1 + a_2 b_2 + a_3 b_3 = (5 \times 2) + (2 \times (-2)) + (7 \times 11) = 83$

(d) $\displaystyle\sum_{k=0}^{4} b_k^2 = b_0^2 + b_1^2 + b_2^2 + b_3^2 + b_4^2 = 0 + 4 + 4 + 121 + 9 = 138$

(e) $\displaystyle\prod_{j=1}^{3} a_j = a_1 a_2 a_3 = 5 \times 2 \times 7 = 70$

(f) $\displaystyle\prod_{k=2}^{4} b_k = b_2 b_3 b_4 = -2 \times 11 \times 3 = -66$

1.3.6 Factorial notation and the binomial expansion

The special extended product of integers

$$1 \times 2 \times 3 \times \ldots \times n = n \times (n-1) \times (n-2) \times \ldots \times 1$$

has a special notation and name. It is called **factorial n** and is denoted by $n!$. Thus with

$$n! = n(n-1)(n-2) \ldots (1)$$

as examples

$$5! = 5 \times 4 \times 3 \times 2 \times 1 \quad \text{and} \quad 8! = 8 \times 7 \times 6 \times 5 \times 4 \times 3 \times 2 \times 1$$

Notice that $5! = 5(4!)$ so that we can write in general

$$n! = (n-1)! \times n$$

This relationship enables us to define $0!$, since $1! = 1 \times 0!$ and $1!$ also equals 1. Thus $0!$ is defined by

$$0! = 1$$

Example 1.28   Evaluate

(a) $4!$ (b) $3! \times 2!$ (c) $6!$ (d) $7!/(2! \times 5!)$

Solution   (a) $4! = 4 \times 3 \times 2 \times 1 = 24$

(b) $3! \times 2! = (3 \times 2 \times 1) \times (2 \times 1) = 12$

(c) $6! = 6 \times 5 \times 4 \times 3 \times 2 \times 1 = 720$

Notice that $2! \times 3! \neq (2 \times 3)!$.

(d) $\dfrac{7!}{2! \times 5!} = \dfrac{7 \times 6 \times 5 \times 4 \times 3 \times 2 \times 1}{2 \times 1 \times 5 \times 4 \times 3 \times 2 \times 1} = \dfrac{7 \times 6}{2} = 21$

Notice that we could have simplified the last item by writing

$$7! = 7 \times 6 \times (5!)$$

then

$$\frac{7!}{2! \times 5!} = \frac{7 \times 6 \times (5!)}{2! \times 5!} = \frac{7 \times 6}{2 \times 1} = 21$$

An interpretation of $n!$ is the total number of different ways it is possible to arrange n different objects in a single line. For example, the word SEAT comprises four different letters, and we can arrange the letters in $4! = 24$ different ways.

SEAT	EATS	ATSE	TSEA
SETA	EAST	ATES	TSAE
SAET	ESAT	AETS	TESA
SATE	ESTA	AEST	TEAS
STAE	ETSA	ASET	TAES
STEA	ETAS	ASTE	TASE

This is because we can choose the first letter in four different ways (S, E, A or T). Once that choice is made, we can choose the second letter in three different ways, then we can choose the third letter in two different ways. Having chosen the first three letters, the last letter is automatically fixed. For each of the four possible first choices, we have three possible choices for the second letter, giving us twelve (4×3) possible choices of the first two letters. To each of these twelve possible choices we have two possible choices of the third letter, giving us twenty-four ($4 \times 3 \times 2$) possible choices of the first three letters. Having chosen the first three letters, there is only one possible choice of last letter. So in all we have $4!$ possible choices.

Example 1.29

In how many ways can the letters of the word REGAL be arranged in a line, and in how many of those do the two letters A and E appear in adjacent positions?

Solution

The word REGAL has five distinct letters so they can be arranged in a line in $5! = 120$ different ways. To find out in how many of those arrangements the A and E appear together, we consider how many arrangements can be made of RGL(AE) and RGL(EA), regarding the bracketed terms as a single symbol. There are $4!$ possible arrangements of both of these, so of the 120 different ways in which the letters of the word REGAL can be arranged, 48 contain the letters A and E in adjacent positions.

The introduction of the factorial notation facilitates the writing down of many complicated expressions. In particular it enables us to write down the general form of the binomial expansion discussed earlier in Section 1.3.1. There we wrote out longhand the expansion of $(a + b)^n$ for $n = 0, 1, 2, \ldots, 6$ and noted the relationship between the coefficients of $(a + b)^n$ and those of $(a + b)^{n-1}$, shown clearly in Pascal's triangle of Figure 1.8.

If

$$(a + b)^{n-1} = c_0 a^{n-1} + c_1 a^{n-2}b + c_2 a^{n-3}b^2 + c_3 a^{n-4}b^3 + \ldots + c_{n-1}b^{n-1}$$

and

$$(a + b)^n = d_0 a^n + d_1 a^{n-1} b + d_2 a^{n-2} b^2 + \ldots + d_{n-1} a b^{n-1} + d_n b^n$$

then, as described on p. 18 when developing Pascal's triangle,

$$c_0 = d_0 = 1, \quad d_1 = c_1 + c_0, \quad d_2 = c_2 + c_1, \quad d_3 = c_3 + c_2, \ldots$$

and in general

$$d_r = c_r + c_{r-1}$$

It is easy to verify that this relationship is satisfied by

$$d_r = \frac{n!}{r!(n-r)!}, \quad c_r = \frac{(n-1)!}{r!(n-1-r)!}, \quad c_{r-1} = \frac{(n-1)!}{(r-1)!(n-1-r+1)!}$$

and it can be shown that the coefficient of $a^{n-r} b^r$ in the expansion of $(a + b)^n$ is

$$\frac{n!}{r!(n-r)!} = \frac{n(n-1)(n-2)\ldots(n-r+1)}{r(r-1)(r-2)\ldots(1)} \tag{1.11}$$

This is a very important result, with many applications. Using it we can write down the general binomial expansion

$$(a + b)^n = \sum_{r=0}^{n} \frac{n!}{r!(n-r)!} a^{n-r} b^r \tag{1.12}$$

The coefficient $\dfrac{n!}{r!(n-r)!}$ is called the **binomial coefficient** and has the special notation

$$\binom{n}{r} = \frac{n!}{r!(n-r)!}$$

Thus we may write

$$(a + b)^n = \sum_{r=0}^{n} \binom{n}{r} a^{n-r} b^r \tag{1.13}$$

which is referred to as the general **binomial expansion**.

Example 1.30   Expand the expression $(2 + x)^5$.

Solution   Setting $a = 2$ and $b = x$ in the general binomial expansion we have

$$(2 + x)^5 = \sum_{r=0}^{5} \binom{5}{r} 2^{5-r} x^r$$

$$= \binom{5}{0} 2^5 + \binom{5}{1} 2^4 x + \binom{5}{2} 2^3 x^2 + \binom{5}{3} 2^2 x^3 + \binom{5}{4} 2 x^4 + \binom{5}{5} x^5$$

$$= (1)(2^5) + (5)(2^4)x + (10)(2^3)x^2 + (10)(2^2)x^3 + (5)(2)x^4 + 1x^5$$

since $\binom{5}{0} = \frac{5!}{0!5!} = 1$, $\binom{5}{1} = \frac{5!}{1!4!} = 5$, $\binom{5}{2} = \frac{5!}{2!3!} = 10$ and so on. Thus

$$(2 + x)^5 = 32 + 80x + 80x^2 + 40x^3 + 10x^4 + x^5$$

1.3.7 Exercises

29 Given $a_0 = 2$, $a_1 = -1$, $a_2 = -4$, $a_3 = 5$, $a_4 = 3$ and $b_0 = 1$, $b_1 = 1$, $b_2 = 2$, $b_3 = -1$, $b_4 = 2$, calculate

(a) $\sum_{k=0}^{4} a_k$ (b) $\sum_{i=1}^{3} a_i$ (c) $\sum_{k=1}^{2} a_k b_k$

(d) $\sum_{j=0}^{4} b_j^2$ (e) $\prod_{k=1}^{3} a_k$ (f) $\prod_{k=1}^{4} b_k$

30 Evaluate

(a) 5!   (b) 3!/4!   (c) 7!/(3! × 4!)

(d) $\binom{5}{2}$ (e) $\binom{9}{3}$ (f) $\binom{8}{4}$

31 Using the general binomial expansion expand the following expressions

(a) $(x - 3)^4$ (b) $(x + \frac{1}{2})^3$

(c) $(2x + 3)^5$ (d) $(3x + 2y)^4$

1.4 Geometry

1.4.1 Coordinates

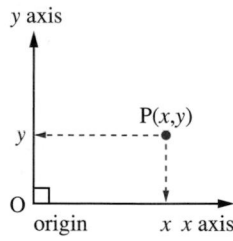

Figure 1.14

In addition to the introduction of algebraic manipulation another innovation made in the seventeenth century was the use of coordinates to represent the position of a point P on a plane as shown in Figure 1.14. Conventionally the point P is represented by an ordered pair of numbers contained in brackets thus: (x, y). This innovation was largely due to Descartes and consequently we often refer to (x, y) as the **cartesian coordinates** of P. This notation is the same as that for an open interval on the number line introduced in Section 1.2.5, but has an entirely separate meaning and the two should not be confused. Whether (x, y) denotes an open interval or a coordinate pair is usually clear from the context.

1.4.2 Straight lines

The introduction of coordinates made possible the algebraic description of the plane curves of classical geometry and the proof of standard results by algebraic methods.

Consider, for example, the point P lying on the line AB as shown in Figure 1.15. Let P divide AB in the ratio $\lambda : 1 - \lambda$. Then AP/AB = λ and, by similar triangles,

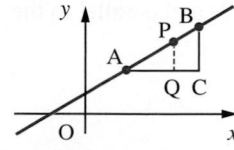

Figure 1.15

$$\frac{AP}{AB} = \frac{PQ}{BC} = \frac{AQ}{AC}$$

Let A, B and P have coordinates (x_0, y_0), (x_1, y_1) and (x, y) respectively, then from the diagram

$$AQ = x - x_0, \ AC = x_1 - x_0, \ PQ = y - y_0, \ BC = y_1 - y_0$$

Thus

$$\frac{PQ}{BC} = \frac{AQ}{AC} \quad \text{implies} \quad \frac{y - y_0}{y_1 - y_0} = \frac{x - x_0}{x_1 - x_0}$$

from which we deduce, after some rearrangement,

$$y = \frac{y_1 - y_0}{x_1 - x_0}(x - x_0) + y_0 \tag{1.14}$$

which represents the equation of a straight line passing through two points (x_0, y_0) and (x_1, y_1).

More simply, the equation of a straight line passing through the two points having coordinates (x_0, y_0) and (x_1, y_1) may be written as

$$y = mx + c \tag{1.15}$$

where $m = \dfrac{y_1 - y_0}{x_1 - x_0}$ is the gradient (slope) of the line and $c = \dfrac{y_0 x_1 - y_1 x_0}{x_1 - x_0}$ is the intercept on the y axis.

Thus equations of the form

$$y = mx + c$$

represent straight lines on the plane and, consequently, are called **linear equations**.

Example 1.31   Find the equation of the straight line that passes through the points $(1, 2)$ and $(3, 3)$.

Solution   Taking $(x_0, y_0) = (1, 2)$ and $(x_1, y_1) = (3, 3)$

$$\text{slope of line} = \frac{y_1 - y_0}{x_1 - x_0} = \frac{3 - 2}{3 - 1} = \frac{1}{2}$$

so from formula (1.14) the equation of the straight line is

$$y = \tfrac{1}{2}(x - 1) + 2$$

which simplifies to

$$y = \tfrac{1}{2}x + \tfrac{3}{2}$$

Example 1.32   Find the equation of the straight line passing through the point $(3, 2)$ and parallel to the line $2y = 3x + 4$. Determine its x and y intercepts.

Solution   Writing $2y = 3x + 4$ as

$$y = \tfrac{3}{2}x + 2$$

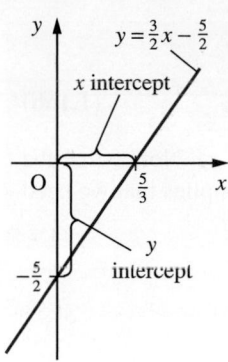

Figure 1.16
The straight line
$2y = 3x - 5$.

we have from (1.15) that the slope of this line is $\frac{3}{2}$. Since the required line is parallel to this line, it will also have a slope of $\frac{3}{2}$. (The slope of the line perpendicular to it is $-\frac{2}{3}$.) Thus from (1.15) it has equation

$$y = \tfrac{3}{2}x + c$$

To determine the constant c, we use the fact that the line passes through the point $(3, 2)$, so that

$$2 - \tfrac{9}{2} + c \qquad \text{giving} \qquad c = -\tfrac{5}{2}$$

Thus the equation of the required line is

$$y = \tfrac{3}{2}x - \tfrac{5}{2} \qquad \text{or} \qquad 2y = 3x - 5$$

The y intercept is $c = -\frac{5}{2}$.

To obtain the x intercept we substitute $y = 0$, giving $x = \frac{5}{3}$, so that the x intercept is $\frac{5}{3}$.

The graph of the line is shown in Figure 1.16.

1.4.3 Circles

A circle is the planar curve whose points are all equidistant from a fixed point called the centre of the circle. The simplest case is a circle centred at the origin with radius r, as shown in Figure 1.17(a). Applying Pythagoras' theorem to triangle OPQ we obtain

$$x^2 + y^2 = r^2$$

(Note that r is a constant.) When the centre of the circle is at the point (a, b), rather than the origin, the equation is

$$(x - a)^2 + (y - b)^2 = r^2 \tag{1.16a}$$

obtained by applying Pythagoras' theorem in triangle O′PN of Figure 1.17(b). This expands to

$$x^2 + y^2 - 2ax - 2by + (a^2 + b^2 - r^2) = 0$$

Figure 1.17
(a) A circle of centre origin, radius r. (b) A circle of centre (a, b), radius r.

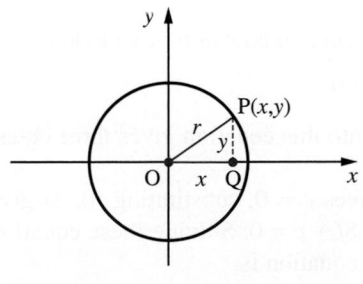

(a)

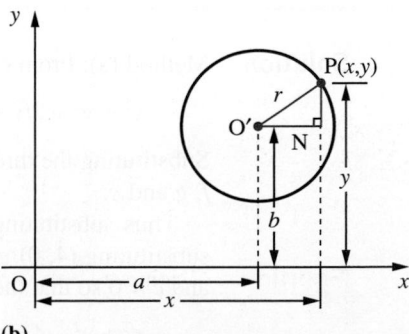

(b)

Thus the general equation

$$x^2 + y^2 + 2fx + 2gy + c = 0 \tag{1.16b}$$

represents a circle having centre $(-f, -g)$ and radius $\sqrt{(f^2 + g^2 - c)}$. Notice that the general circle has three constants f, g and c in its equation. This implies that we need three points to specify a circle completely.

Example 1.33 Find the equation of the circle with centre $(1, 2)$ and radius 3.

Solution Using Pythagoras' theorem, if the point $P(x, y)$ lies on the circle then from (1.16a)

$$(x - 1)^2 + (y - 2)^2 = 3^2$$

Thus

$$x^2 - 2x + 1 + y^2 - 4y + 4 = 9$$

giving the equation as

$$x^2 + y^2 - 2x - 4y - 4 = 0$$

Example 1.34 Find the radius and the coordinates of the centre of the circle whose equation is

$$2x^2 + 2y^2 - 3x + 5y + 2 = 0$$

Solution Dividing through by the coefficient of x^2 we obtain

$$x^2 + y^2 - \tfrac{3}{2}x + \tfrac{5}{2}y + 1 = 0$$

Now completing the square on the x terms and the y terms separately gives

$$(x - \tfrac{3}{4})^2 + (y + \tfrac{5}{4})^2 = \tfrac{9}{16} + \tfrac{25}{16} - 1 = \tfrac{18}{16}$$

Hence, from (1.16a), the circle has radius $(3\sqrt{2})/4$ and centre $(3/4, -5/4)$.

Example 1.35 Find the equation of the circle which passes through the points $(0, 0)$, $(0, 2)$, $(4, 0)$.

Solution Method (a): From (1.16b) the general equation of a circle is

$$x^2 + y^2 + 2fx + 2gy + c = 0$$

Substituting the three points into this equation gives three equations for the unknowns f, g and c.

Thus substituting $(0, 0)$ gives $c = 0$, substituting $(0, 2)$ gives $4 + 4g + c = 0$ and substituting $(4, 0)$ gives $16 + 8f + c = 0$. Solving these equations gives $g = -1$, $f = -2$ and $c = 0$ so that the required equation is

$$x^2 + y^2 - 4x - 2y = 0$$

Figure 1.18
The circle of
Example 1.35.

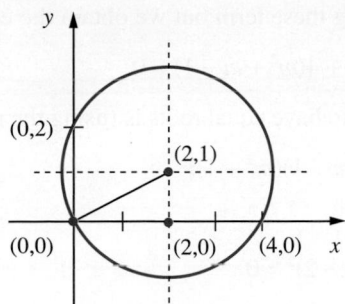

Method (b): From Figure 1.18 using the geometrical properties of the circle, we see that its centre lies at (2, 1) and since it passes through the origin its radius is $\sqrt{5}$. Hence, from (1.16a), its equation is

$$(x - 2)^2 + (y - 1)^2 = (\sqrt{5})^2$$

which simplifies to

$$x^2 + y^2 - 4x - 2y = 0$$

as before.

Example 1.36   Find the point of intersection of the line $y = x - 1$ with the circle $x^2 + y^2 - 4y - 1 = 0$.

Solution   Substituting $y = x - 1$ into the formula for the circle gives

$$x^2 + (x - 1)^2 - 4(x - 1) - 1 = 0$$

which simplifies to

$$x^2 - 3x + 2 = 0$$

This equation may be factored to give

$$(x - 2)(x - 1) = 0$$

so that $x = 1$ and $x = 2$ are the roots. Thus the points of intersection are $(1, 0)$ and $(2, 1)$.

Example 1.37   Find the equation of the tangent at the point (2, 1) of the circle $x^2 + y^2 - 4y - 1 = 0$.

Solution   A tangent is a line, which is the critical case between a line intersecting the circle in two distinct points and it not intersecting at all. We can describe this as the case when the line cuts the circle in two coincident points. Thus the line, which passes through (2, 1) with slope m

$$y = m(x - 2) + 1$$

is a tangent to the circle when the equation

$$x^2 + [m(x - 2) + 1]^2 - 4[m(x - 2) + 1] - 1 = 0$$

has two equal roots. Multiplying these term out we obtain the equation

$$(m^2 + 1)x^2 - 2m(2m + 1)x + 4(m^2 + m - 1) = 0$$

The condition for this equation to have equal roots is (using the result of Example 1.16)

$$4m^2(2m + 1)^2 = 4[4(m^2 + m - 1)(m^2 + 1)]$$

This simplifies to

$$m^2 - 4m + 4 = 0 \quad \text{or} \quad (m - 2)^2 = 0$$

giving the result $m = 2$ and the equation of the tangent $y = 2x - 3$.

1.4.4 Exercises

32 Find the equation of the straight line

(a) with gradient $\frac{3}{2}$ passing through the point (2, 1),

(b) with gradient -2 passing through the point (−2, 3),

(c) passing through the points (1, 2) and (3, 7),

(d) passing through the points (5, 0) and (0, 3),

(e) parallel to the line $3y - x = 5$, passing through (1, 1),

(f) perpendicular to the line $3y - x = 5$, passing through (1, 1).

33 Write down the equation of the circle with centre (1, 2) and radius 5.

34 Find the radius and the coordinates of the centre of the circle with equation

$$x^2 + y^2 + 4x - 6y = 3$$

35 Find the equation of the circle with centre (−2, 3) that passes through (1, −1).

36 Find the equation of the circle that passes through the points (1, 0), (3, 4) and (5, 0).

37 Find the equation of the tangent to the circle

$$x^2 + y^2 - 4x - 1 = 0$$

at the point (1, 2).

38 A rod, 50 cm long, moves in a plane with its ends on two perpendicular wires. Find the equation of the curve followed by its midpoint.

1.4.5 Conics

The circle is one of the conic sections (Figure 1.19) introduced around 200 BC by Apollonius, who published an extensive study of their properties in a textbook that he called *Conics*. He used this title because he visualized them as cuts made by a 'flat' or plane surface when it intersects the surface of a cone in different directions, as illustrated in Figures 1.20(a–d). Note that the conic sections degenerate into a point and straight lines at the extremities, as illustrated in Figures 1.20(e–g). Although at the time of Apollonius his work on conics appeared to be of little value in terms of applications, it has since turned out to have considerable importance. This is primarily due to the fact that the conic sections are the paths followed by projectiles, artificial satellites, moons and the Earth under the influence of gravity around planets or stars. The early Greek astronomers thought that the planets moved in circular orbits, and it was not until 1609 that the German astronomer Johannes Kepler described their paths correctly as being elliptic, with the Sun at one focus. It is quite possible for an orbit to be a curve other

Figure 1.19
Standard equations
of the four conics.

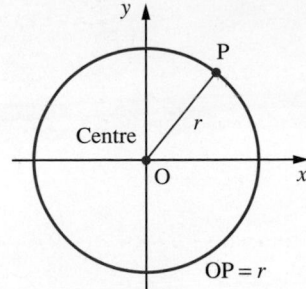

(a) Circle: $x^2 + y^2 = r^2$

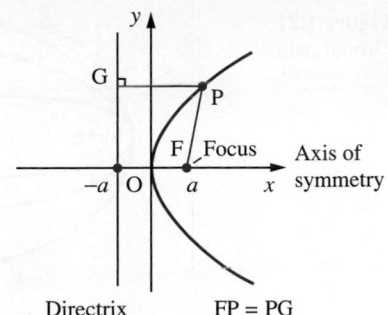

(b) Parabola: $y^2 = 4ax$
$e = 1$

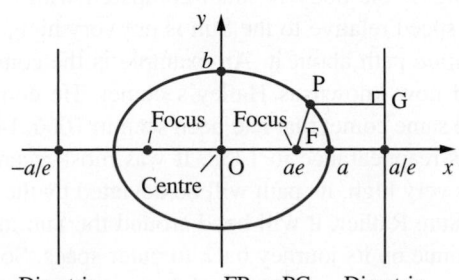

(c) Ellipse: $\dfrac{x^2}{a^2} + \dfrac{y^2}{b^2} = 1$
$b^2 = a^2(1 - e^2)$
and eccentricity $e < 1$

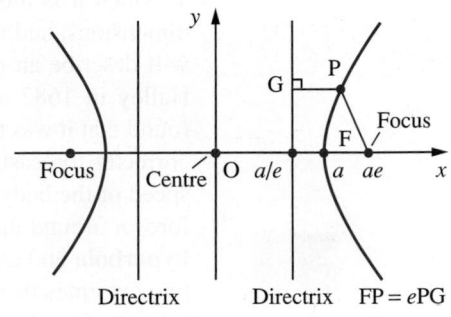

(d) Hyperbola: $\dfrac{x^2}{a^2} - \dfrac{y^2}{b^2} = 1$
$b^2 = a^2(e^2 - 1)$
and eccentricity $e > 1$

Figure 1.20

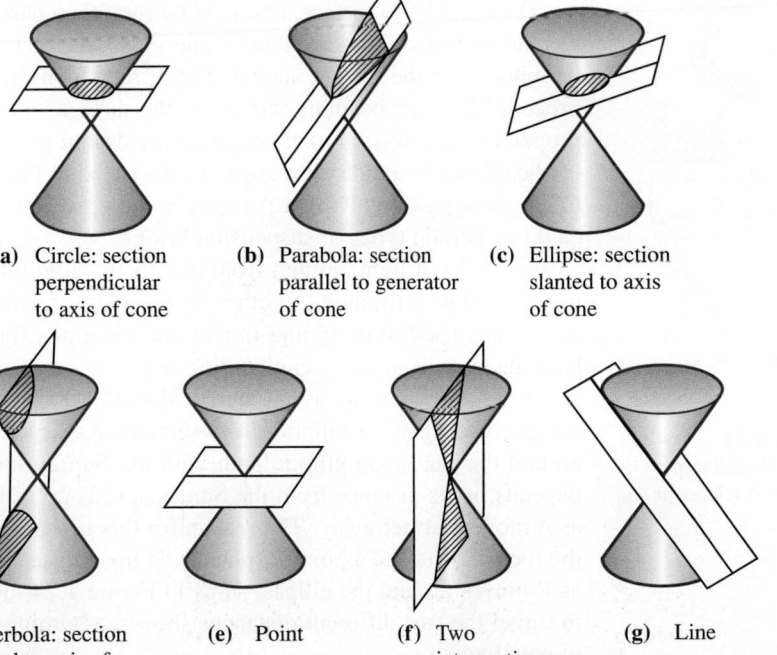

(a) Circle: section
perpendicular
to axis of cone

(b) Parabola: section
parallel to generator
of cone

(c) Ellipse: section
slanted to axis
of cone

(d) Hyperbola: section
slanted to axis of
cone

(e) Point

(f) Two
intersecting
lines

(g) Line

Figure 1.21
Orbital path.

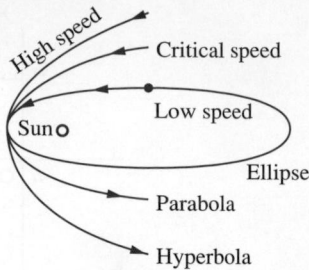

than an ellipse. Imagine a meteorite or comet approaching the Sun from some distant region in space. The path that the body will follow depends very much on the speed at which it is moving. If the body is small compared with the Sun, say of planetary dimensions, and its speed relative to the Sun is not very high, it will never escape and will describe an *elliptic* path about it. An example is the comet observed by Edward Halley in 1682 and now known as Halley's comet. He computed its elliptic orbit, found that it was the same comet that had been seen in 1066, 1456, 1531 and 1607, and correctly forecast its reappearance in 1758. It was most recently seen in 1997. If the speed of the body is very high, its path will be deviated by the Sun but it will not orbit forever around the Sun. Rather, it will bend around the Sun in a path in the form of a **hyperbola** and continue on its journey back to outer space. Somewhere between these two extremes there is a certain critical speed that is just too great to allow the body to orbit the Sun, but not great enough for the path to be a hyperbola. In this case the path is a **parabola**, and once again the body will bend around the Sun and continue on its journey into outer space. These possibilities are illustrated in Figure 1.21.

Other examples of where conic sections appear in engineering practice include the following.

(a) A parabolic surface, obtained by rotating a parabola about its axis of symmetry, has the important property that an energy source placed at the focus will cause rays to be reflected at the surface such that after reflection they will be parallel. Reversing the process, a parallel beam impinging on the surface will be reflected on to the focus. This property is involved in many engineering design projects: for example the design of a car headlamp or a radio telescope, as illustrated in Figures 1.22(a) and (b) respectively. Other examples involving a parabola are the path of a projectile and the shape of the cable on certain types of suspension bridge.

(b) A ray of light emitted from one focus of an elliptic mirror and reflected by the mirror will pass through the other focus as illustrated in Figure 1.23. This property is sometimes used in designing mirror combinations for a reflecting telescope. Ellipses have been used in other engineering designs, such as aircraft wings and stereo styli. Formerly, in order to avoid bursts due to freezing, water pipes were sometimes designed to have an elliptic cross-section. As described earlier, every planet orbits around the Sun in an elliptic path with the Sun at one of its foci. The planet's speed depends on its distance from the Sun; it speeds up as it nears the Sun and slows down as it moves further away. The reason for this is that for an ellipse the line drawn from the focus S (Sun) to a point P (planet) on the ellipse sweeps out areas at a constant rate as P moves around the ellipse. Thus in Figure 1.24 the planet will take the same time to travel the two different distances shown, assuming that the two shaded regions are of equal area.

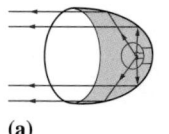

(a)

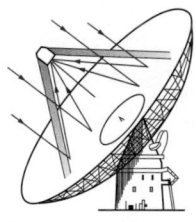

(b)

Figure 1.22
(a) Car headlamp.
(b) Radio telescope.

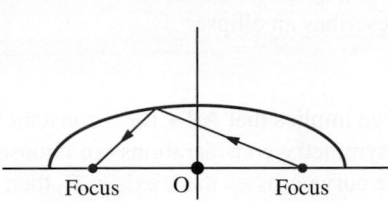

Figure 1.23 Reflection of a ray by an elliptic mirror.

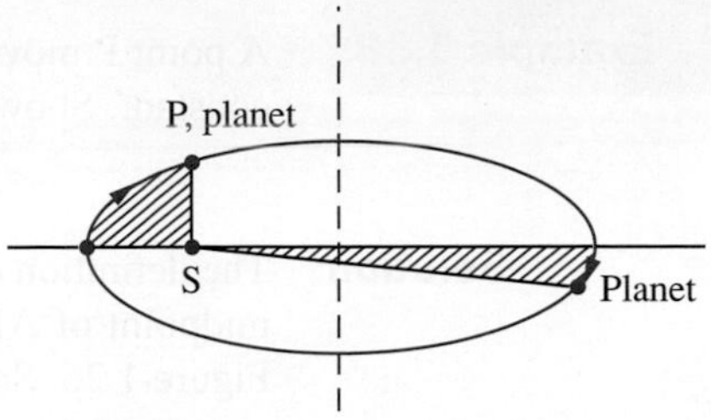

Figure 1.24 Regions of equal area.

(c) Consider a supersonic aircraft flying over land. As it breaks the sound barrier (that is, it travels faster than the speed of sound, which is about 750 mph ($331.4\,\mathrm{m\,s^{-1}}$)), it will create a shock wave, which we hear on the ground as a *sonic boom* – this being one of the major disadvantages of supersonic aircraft. This shock wave will trail behind the aircraft in the form of a cone with the aircraft as vertex. This cone will intersect the ground in a *hyperbolic curve* as illustrated in Figure 1.25. The sonic boom will hit every point on this curve at the same instant of time, so that people living on the curve will hear it simultaneously. No boom will be heard by people living outside this curve, but eventually it will be heard at every point inside it.

Figure 1.25 Sonic boom.

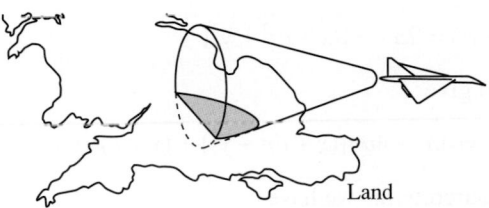

Land

Figure 1.19 illustrates the conics in their standard positions, and the corresponding equations may be interpreted as the standard equations for the four curves. More generally the conic sections may be represented by the general second-order equation

$$ax^2 + by^2 + 2fx + 2gy + 2hxy + c = 0 \tag{1.17}$$

Provided its graph does not degenerate into a point or straight lines, equation (1.17) is representative of

- a circle if $a = b \neq 0$ and $h = 0$
- a parabola if $h^2 = ab$
- an ellipse if $h^2 < ab$
- a hyperbola if $h^2 > ab$

The conics can be defined mathematically in a number of (equivalent) ways as we will illustrate in the next examples.

Example 1.38

A point P moves in such a way that its total distance from two fixed points A and B is constant. Show that it describes an ellipse.

Solution

The definition of the curve implies that $AP + BP = $ constant with the origin O being the midpoint of AB. From symmetry considerations we choose x and y axes as shown in Figure 1.26. Suppose the curve crosses the x axis at P_0 then

$$AP_0 + BP_0 = AB + 2AP_0 = 2OP_0$$

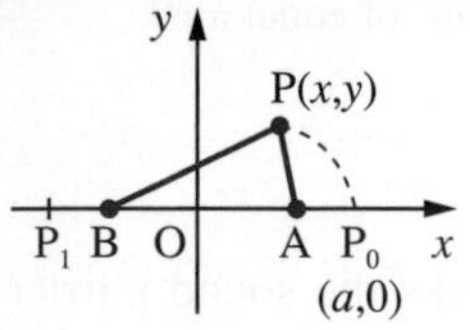

so the constant in the definition is $2OP_0$ and for any point P on the curve

$$AP + BP = 2OP_0$$

Figure 1.26
Path of Example 1.38.

Let $P = (x, y)$, $P_0 = (a, 0)$, $P_1 = (-a, 0)$, $A = (c, 0)$ and $B = (-c, 0)$. Then using Pythagoras' theorem we have

$$AP = \sqrt{[(x - c)^2 + y^2]}$$

$$BP = \sqrt{[(x + c)^2 + y^2]}$$

so that the defining equation of the curve becomes

$$\sqrt{[(x - c)^2 + y^2]} + \sqrt{[(x + c)^2 + y^2]} = 2a$$

To obtain the required equation we need to 'remove' the square root terms. This can only be done by squaring both sides of the equation. First we rewrite the equation as

$$\sqrt{[(x - c)^2 + y^2]} = 2a - \sqrt{[(x + c)^2 + y^2]}$$

and then square to give

$$(x - c)^2 + y^2 = 4a^2 - 4a\sqrt{[(x + c)^2 + y^2]} + (x + c)^2 + y^2$$

Expanding the squared terms we have

$$x^2 - 2cx + c^2 + y^2 = 4a^2 - 4a\sqrt{[(x + c)^2 + y^2]} + x^2 + 2cx + c^2 + y^2$$

Collecting together terms, we obtain

$$a\sqrt{[(x + c)^2 + y^2]} = a^2 + cx$$

Squaring both sides again gives

$$a^2[x^2 + 2cx + c^2 + y^2] = a^4 + 2a^2cx + c^2x^2$$

which simplifies to

$$(a^2 - c^2)x^2 + a^2y^2 = a^2(a^2 - c^2)$$

Noting that $a > c$ we write $a^2 - c^2 = b^2$, to obtain

$$b^2x^2 + a^2y^2 = a^2b^2$$

which yields the standard equation of the ellipse

$$\frac{x^2}{a^2} + \frac{y^2}{b^2} = 1$$

The points A and B are the foci of the ellipse, and the property that the sum of the focal distances is a constant is known as the **string property** of the ellipse since it enables us to draw an ellipse using a piece of string.

For a hyperbola, the *difference* of the focal distances is constant.

Example 1.39

A point moves in such a way that its distance from a fixed point F is equal to its perpendicular distance from a fixed line. Show that it describes a parabola.

Solution

Suppose the fixed line is LL' shown in Figure 1.27, choosing the coordinate axes shown. Since PF = PN for points on the curve we deduce that the curve bisects FM, so that if F is $(a, 0)$, then M is $(-a, 0)$. Let the general point P on the curve have coordinates (x, y). Then by Pythagoras' theorem

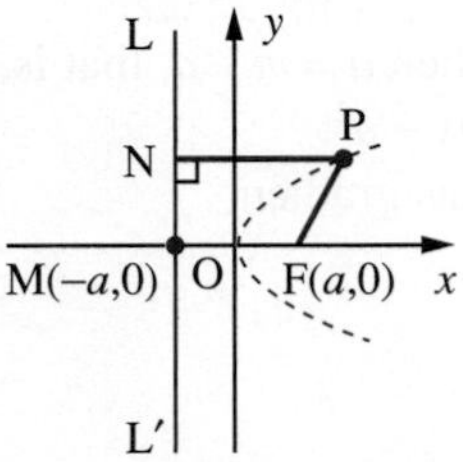

$$PF = \sqrt{[(x - a)^2 + y^2]}$$

Also PN $= x + a$, so that PN = PF implies that

$$x + a = \sqrt{[(x - a)^2 + y^2]}$$

Squaring both sides gives

$$(x + a)^2 = (x - a)^2 + y^2$$

which simplifies to

$$y^2 = 4ax$$

the standard equation of a parabola. The line LL' is called the **directrix** of the parabola.

Figure 1.27
Path of point in
Example 1.31.

Example 1.40

(a) Find the equation of the tangent at the point $(1, 1)$ to the parabola $y = x^2$. Show that it is parallel to the line through the points $(\frac{1}{2}, \frac{1}{4})$, $(\frac{3}{2}, \frac{9}{4})$, which also lie on the parabola.

(b) Find the equation of the tangent at the point (a, a^2) to the parabola $y = x^2$. Show that it is parallel to the line through the points $(a - h, (a - h)^2)$, $(a + h, (a + h)^2)$.

Solution

(a) Consider the general line through $(1, 1)$. It has equation $y = m(x - 1) + 1$. This cuts the parabola when

$$m(x - 1) + 1 = x^2$$

that is, when

$$x^2 - mx + m - 1 = 0$$

Factorizing this quadratic, we have

$$(x - 1)(x - m + 1) = 0$$

giving the roots $x = 1$ and $x = m - 1$.

These two roots are equal when $m - 1 = 1$, that is, when $m = 2$. Hence the equation of the tangent is $y = 2x - 1$.

The line through the points $(\frac{1}{2}, \frac{1}{4})$, $(\frac{3}{2}, \frac{9}{4})$ has gradient

$$\frac{\frac{9}{4} - \frac{1}{4}}{\frac{3}{2} - \frac{1}{2}} = 2$$

so that it is parallel to the tangent at $(1, 1)$.

(b) Consider the general line through (a, a^2). It has equation $y = m(x - a) + a^2$. This cuts the parabola $y = x^2$ when

$$m(x - a) + a^2 = x^2$$

that is, where

$$x^2 - mx + ma - a^2 = 0$$

This factorizes into

$$(x - a)(x - m + a) = 0$$

giving the roots $x = a$ and $x = m - a$. These two roots are equal when $a = m - a$, that is, when $m = 2a$. Thus the equation of the tangent at (a, a^2) is $y = 2ax - a^2$.

The line through the points $(a - h, (a - h)^2)$, $(a + h, (a + h)^2)$ has gradient

$$\frac{(a + h)^2 - (a - h)^2}{(a + h) - (a - h)} = \frac{a^2 + 2ah + h^2 - (a^2 - 2ah + h^2)}{2h}$$

$$= \frac{4ah}{2h} = 2a$$

So the symmetrically disposed chord through $(a - h, (a - h)^2)$, $(a + h, (a + h)^2)$ is parallel to the tangent at $x = a$. This result is true for all parabolas.

1.4.6 Exercises

39 Find the coordinates of the focus and the equation of the directrix of the parabola whose equation is

$$3y^2 = 8x$$

The chord which passes through the focus parallel to the directrix is called the **latus rectum** of the parabola. Show that the latus rectum of the above parabola has length 8/3.

40 For the ellipse $25x^2 + 16y^2 = 400$ find the coordinates of the foci, the eccentricity, the equations of the directrices and the lengths of the semi-major and semi-minor axes.

41 For the hyperbola $9x^2 - 16y^2 = 144$ find the coordinates of the foci and the vertices and the equations of its asymptotes.

1.5 Numbers and accuracy

Arithmetic that only involves integers can be performed to obtain an exact answer (that is, one without rounding errors). In general this is not possible with real numbers and when solving practical problems such numbers are rounded to an appropriate number of digits. In this section we shall review the methods of recording numbers, obtain estimates for the effect of rounding errors in elementary calculations and discuss the implementation of arithmetic on computers.

1.5.1 Representation of numbers

For ordinary everyday purposes we use a system of representation based on ten **numerals**: 0, 1, 2, 3, 4, 5, 6, 7, 8, 9. These ten symbols are sufficient to represent all numbers if a **position notation** is adopted. For whole numbers this means that, starting from the right-hand end of the number, the least significant end, the figures represent the number of units, tens, hundreds, thousands, and so on. Thus one thousand, three hundred and sixty-five is represented by 1365, and two hundred and nine is represented by 209. Notice the role of the 0 in the latter example, acting as a position keeper. The use of a decimal point makes it possible to represent fractions as well as whole numbers. This system uses ten symbols. The number system is said to be 'to base ten' and is called the **decimal** system. Other bases are possible: for example, the Babylonians used a number system to base sixty, a fact that still influences our measurement of time. In some societies a number system evolved with more than one base, a survival of which can be seen in imperial measures (inches, feet, yards, …). For some applications it is more convenient to use a base other than ten. Early electronic computers used **binary** numbers (to base two); modern computers use **hexadecimal** numbers (to base sixteen). For elementary (pen-and-paper) arithmetic a representation to base twelve would be more convenient than the usual decimal notation because twelve has more integer divisors (2, 3, 4, 6) than ten (2, 5).

In a decimal number the positions to the left of the decimal point represent units (10^0), tens (10^1), hundreds (10^2) and so on, while those to the right of the decimal point represent tenths (10^{-1}), hundredths (10^{-2}) and so on. Thus, for example

$$
\begin{array}{ccccccc}
2 & 1 & 4 & \cdot & 3 & & 6 \\
\downarrow & \downarrow & \downarrow & & \downarrow & & \downarrow \\
10^2 & 10^1 & 10^0 & & 10^{-1} & & 10^{-2}
\end{array}
$$

so

$$
214.36 = 2(10^2) + 1(10^1) + 4(10^0) + 3(\tfrac{1}{10}) + 6(\tfrac{1}{100})
$$
$$
= 200 + 10 + 4 + \tfrac{3}{10} + \tfrac{6}{100}
$$
$$
= \tfrac{21436}{100} = \tfrac{5359}{25}
$$

In other number bases the pattern is the same: in base b the position values are b^0, b^1, b^2, … and b^{-1}, b^{-2}, … . Thus in binary (base two) the position values are units, twos, fours, eights, sixteens and so on, and halves, quarters, eighths and so on. In hexadecimal (base sixteen) the position values are units, sixteens, two hundred and fifty-sixes, and so on, and sixteenths, two hundred and fifty-sixths, and so on.

Example 1.41 Write (a) the binary number 1011101_2 as a decimal number and (b) the decimal number 115_{10} as a binary number.

Solution (a) $1011101_2 = 1(2^6) + 0(2^5) + 1(2^4) + 1(2^3) + 1(2^2) + 0(2^1) + 1(2^0)$

$$
= 64_{10} + 0 + 16_{10} + 8_{10} + 4_{10} + 0 + 1_{10}
$$
$$
= 93_{10}
$$

(b) We achieve the conversion to binary by repeated division by 2. Thus

$$115 \div 2 = 57 \quad \text{remainder 1} \quad (2^0)$$
$$57 \div 2 = 28 \quad \text{remainder 1} \quad (2^1)$$
$$28 \div 2 = 14 \quad \text{remainder 0} \quad (2^2)$$
$$14 \div 2 = 7 \quad \text{remainder 0} \quad (2^3)$$
$$7 \div 2 = 3 \quad \text{remainder 1} \quad (2^4)$$
$$3 \div 2 = 1 \quad \text{remainder 1} \quad (2^5)$$
$$1 \div 2 = 0 \quad \text{remainder 1} \quad (2^6)$$

so that

$$115_{10} = 1110011_2$$

Example 1.42 Represent the numbers (a) two hundred and one, (b) two hundred and seventy-five, (c) five and three-quarters and (d) one-third in

(i) decimal form using the figures 0, 1, 2, 3, 4, 5, 6, 7, 8, 9;

(ii) binary form using the figures 0, 1;

(iii) duodecimal (base 12) form using the figures 0, 1, 2, 3, 4, 5, 6, 7, 8, 9, Δ, ε.

Solution (a) two hundred and one

(i) $= 2$ (hundreds) $+ 0$ (tens) and 1 (units) $= 201_{10}$

(ii) $= 1$ (one hundred and twenty-eight) $+ 1$ (sixty-four) $+ 1$ (eight) $+ 1$ (unit)
$= 11001001_2$

(iii) $= 1$ (gross) $+ 4$ (dozens) $+ 9$ (units) $= 149_{12}$

Here the subscripts 10, 2, 12 indicate the number base.

(b) two hundred and seventy-five

(i) $= 2$ (hundreds) $+ 7$ (tens) $+ 5$ (units) $= 275_{10}$

(ii) $= 1$ (two hundred and fifty-six) $+ 1$ (sixteen) $+ 1$ (two) $+ 1$ (unit) $= 100010011_2$

(iii) $= 1$ (gross) $+ 10$ (dozens) $+$ eleven (units) $= 1\Delta\varepsilon_{12}$
(Δ represents ten and ε represents eleven)

(c) five and three-quarters

(i) $= 5$ (units) $+ 7$ (tenths) $+ 5$ (hundredths) $= 5.75_{10}$

(ii) $= 1$ (four) $+ 1$ (unit) $+ 1$ (half) $+ 1$ (quarter) $= 101.11_2$

(iii) $= 5$ (units) $+ 9$ (twelfths) $= 5.9_{12}$

(d) one-third

(i) $= 3$ (tenths) $+ 3$ (hundredths) $+ 3$ (thousandths) $+ \ldots = 0.333 \ldots {}_{10}$

(ii) $= 1$ (quarter) $+ 1$ (sixteenth) $+ 1$ (sixty-fourth) $+ \ldots = 0.010101 \ldots {}_{2}$

(iii) $= 4$ (twelfths) $= 0.4_{12}$

1.5.2 Rounding, decimal places and significant figures

The Fundamental Laws of Arithmetic are, of course, independent of the choice of representation of the numbers. Similarly, the representation of irrational numbers will always be incomplete. Because of these numbers and because some rational numbers have recurring representations (whether the representation of a particular rational number is recurring or not will of course depend on the number base used – see Example 1.42d), any arithmetical calculation will contain errors caused by truncation. In practical problems it is usually known how many figures are meaningful, and the numbers are 'rounded' accordingly. In the decimal representation, for example, the numbers are approximated by the closest decimal number with some prescribed number of figures after the decimal point. Thus, to two decimal places (dp),

$$\pi = 3.14 \quad \text{and} \quad \tfrac{5}{12} = 0.42$$

and to five decimal places

$$\pi = 3.141\,59 \quad \text{and} \quad \tfrac{5}{12} = 0.416\,67$$

Normally this is abbreviated to

$$\pi = 3.141\,59 \text{ (5dp)} \quad \text{and} \quad \tfrac{5}{12} = 0.416\,67 \text{ (5dp)}$$

Similarly

$$\sqrt{2} = 1.4142 \text{ (4dp)} \quad \text{and} \quad \tfrac{2}{3} = 0.667 \text{ (3dp)}$$

In hand computation, by convention, when shortening a number ending with a five we 'round to the even'. For example,

$$1.2345 \quad \text{and} \quad 1.2335$$

are both represented by 1.234 to three decimal places. In contrast, most calculators and computers would 'round up' in the ambiguous case, giving 1.2345 and 1.2335 as 1.235 and 1.234 respectively.

Any number occurring in practical computation will either be given an error bound or be correct to within half a unit in the least significant figure (sf). For example

$$\pi = 3.14 \pm 0.005 \quad \text{or} \quad \pi = 3.14$$

Any number given in scientific or mathematical tables observes this convention. Thus

$$g_0 = 9.806\,65$$

implies

$$g_0 = 9.806\,65 \pm 0.000\,005$$

that is,

$$9.806\,645 < g_0 < 9.806\,655$$

as illustrated in Figure 1.28.

Figure 1.28

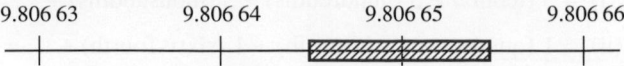

Sometimes the decimal notation may create a false impression of accuracy. When we write that the distance of the Earth from the Sun is ninety-three million miles, we mean that the distance is nearer to 93 000 000 than 94 000 000 or 92 000 000, not that it is nearer to 93 000 000 than 93 000 001 or 92 999 999. This possible misinterpretation of numerical data is avoided by either stating the number of significant figures, giving an error estimate or using scientific notation. In this example the distance d miles is given in the forms

$$d = 93\,000\,000 \ (2\text{sf})$$

or

$$d = 93\,000\,000 \pm 500\,000$$

or

$$d = 9.3 \times 10^7$$

Notice how information about accuracy is discarded by the rounding-off process. The value ninety-three million miles is actually correct to within fifty thousand miles, while the convention about rounded numbers would imply an error bound of five hundred thousand.

The number of significant figures tells us about the relative accuracy of a number when it is related to a measurement. Thus a number given to 3sf is relatively ten times more accurate than one given to 2sf. The number of decimal places, dp, merely tells us the number of digits including leading zeros after the decimal point. Thus

$$2.321 \quad \text{and} \quad 0.000\,059\,71$$

both have 4sf, while the former has 3dp and the latter 8dp.

It is not clear how many significant figures a number like 3200 has. It might be 2, 3 or 4. To avoid this ambiguity it must be written in the form 3.2×10^3 (when it is correct to 2sf) or 3.20×10^3 (3sf) or 3.200×10^3 (4sf). This is usually called **scientific notation**. It is widely used to represent numbers that are very large or very small. Essentially, a number x is written in the form

$$x = a \times 10^n$$

where $1 \leqslant |a| < 10$ and n is an integer. Thus the mass of an electron at rest is 9.11×10^{-28} g, while the velocity of light in a vacuum is 2.9978×10^{10} cm s^{-1}.

Example 1.43   Express the number 150.4152

(a) correct to 1, 2 and 3 dp;   (b) correct to 1, 2 and 3 sf.

Solution   (a) $150.4152 = 150.4$ (1dp)

$\qquad\qquad\qquad\quad = 150.42$ (2dp)

$\qquad\qquad\qquad\quad = 150.415$ (3dp)

(b) $150.4152 = 1.504\,152 \times 10^2$

$\qquad\qquad = 2 \times 10^2 \qquad \text{(1sf)}$

$\qquad\qquad = 1.5 \times 10^2 \qquad \text{(2sf)}$

$\qquad\qquad = 1.50 \times 10^3 \qquad \text{(3sf)}$

1.5.3 Estimating the effect of rounding errors

Numerical data obtained experimentally will often contain rounding errors due to the limited accuracy of measuring instruments. Also, because irrational numbers and some rational numbers do not have a terminating decimal representation, arithmetical operations inevitably contain errors arising from rounding off. The effect of such errors can accumulate in an arithmetical procedure and good engineering computations will include an estimate for it. This process has become more important with the widespread use of computers. When users are isolated from the computational chore, they often fail to develop a sense of the limits of accuracy of an answer. In this section we shall develop the basic ideas for such sensitivity analyses of calculations.

Example 1.44   Compute

(a) $3.142 + 4.126$ (b) $5.164 - 2.341$ (c) 235.12×0.531

Calculate estimates for the effects of rounding errors in each answer and give the answer as a correctly rounded number.

Solution   (a) $3.142 + 4.126 = 7.268$

Because of the convention about rounded numbers, 3.142 represents all the numbers a between 3.1415 and 3.1425, and 4.126 represents all the numbers b between 4.1255 and 4.1265. Thus if a and b are correctly rounded numbers, their sum $a + b$ lies between $c_1 = 7.2670$ and $c_2 = 7.2690$. Rounding c_1 and c_2 to 3dp gives $c_1 = 7.267$ and $c_2 = 7.269$. Since these disagree, we cannot give an answer to 3dp. Rounding c_1 and c_2 to 2dp gives $c_1 = 7.27$ and $c_2 = 7.27$. Since these agree, we can give the answer to 2dp; thus $a + b = 7.27$, as shown in Figure 1.29.

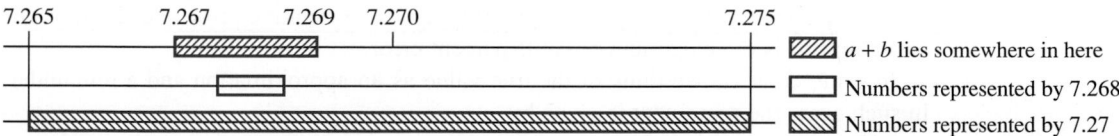

Figure 1.29

(b) $5.164 - 2.341 = 2.823$

Applying the same 'worst case' analysis to this implies that the difference lies between $5.1635 - 2.3415$ and $5.1645 - 2.3405$, that is between 2.8220 and 2.8240. Thus the answer should be written 2.823 ± 0.001 or, as a correctly rounded number, 2.82.

(c) $235.12 \times 0.531 = 124.84872$

Clearly, writing an answer with so many decimal places is unjustified if we are using rounded numbers, but how many decimal places are sensible? Using the 'worst case' analysis again, we deduce that the product lies between 235.115×0.5305 and 235.125×0.5315, that is between $c_1 = 124.7285075$ and $c_2 = 124.9689375$. Thus the answer should be written 124.85 ± 0.13. In this example, because of the place where the number occurs on the number line, c_1 and c_2 only agree when we round them to 3sf (0dp). Thus the product as a correctly rounded number is 125.

A competent computation will contain within it estimates of the effect of rounding errors. Analysing the effect of such errors for complicated expressions has to be approached systematically.

Definitions

(a) The **error** in a value is defined by

error = approximate value − true value

This is sometimes termed the dead error. Notice that the true value equals the approximate value minus the error.

(b) Similarly the **correction** is defined by

true value = approximate value + correction

so that

correction = −error

(c) The **error modulus** is the size of the error, |error|, and the **error bound** (or **absolute error bound**) is the maximum possible error modulus.

(d) The **relative error** is the ratio of the size of the error to the size of the true value:

$$\text{relative error} = \left| \frac{\text{error}}{\text{value}} \right|$$

The **relative error bound** is the maximum possible relative error.

(e) The **percent error** (or percentage error) is $100 \times$ relative error and the **percent error bound** is the maximum possible percent error.

In some contexts we think of the true value as an approximation and a remainder. In such cases the remainder is given by

remainder = −error

= correction

Example 1.45 Give the absolute and relative error bounds of the following correctly rounded numbers

(a) 29.92   (b) −0.01523   (c) 3.9×10^{10}

Solution   (a) The number 29.92 is given to 2dp, which implies that it represents a number within the domain 29.92 ± 0.005. Thus its absolute error bound is 0.005, half a unit of the least significant figure, and its relative error bound is 0.005/29.92 or 0.000 17.

(b) The absolute error bound of $-0.015\,23$ is half a unit of the least significant figure, that is, 0.000 005. Notice that it is a positive quantity. Its relative error bound is 0.000 005/0.015 23 or 0.000 33.

(c) The absolute error bound of 3.9×10^{10} is $0.05 \times 10^{10} = 5 \times 10^{8}$ and its relative error bound is 0.05/3.9 or 0.013.

Usually, because we do not know the true values, we estimate the effects of error in a calculation in terms of the error bounds, the 'worst case' analysis illustrated in Example 1.42. The error bound of a value v is denoted by ε_v.

Consider, first, the sum $c = a + b$. When we add together the two rounded numbers a and b their sum will inherit a rounding error from both a and b. The true value of a lies between $a - \varepsilon_a$ and $a + \varepsilon_a$ and the true value of b lies between $b - \varepsilon_b$ and $b + \varepsilon_b$. Thus the smallest value that the true value of c can have is $a - \varepsilon_a + b - \varepsilon_b$, and its largest possible value is $a + \varepsilon_a + b + \varepsilon_b$. (Remember that ε_a and ε_b are positive.) Thus $c = a + b$ has an error bound

$$\varepsilon_c = \varepsilon_a + \varepsilon_b$$

as illustrated in Figure 1.30. A similar 'worst case' analysis shows that the difference $d = a - b$ has an error bound that is the sum of the error bounds of a and b:

$$d = a - b, \qquad \varepsilon_d = \varepsilon_a + \varepsilon_b$$

Thus for both addition and subtraction the error bound of the result is the sum of the individual error bounds.

Next consider the product $p = a \times b$, where a and b are positive numbers. The smallest possible value of p will be equal to the product of the least possible values of a and b; that is,

$$p > (a - \varepsilon_a) \times (b - \varepsilon_b)$$

Figure 1.30

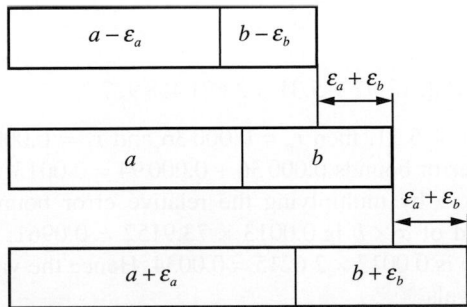

Similarly

$$p < (a + \varepsilon_a) \times (b + \varepsilon_b)$$

Thus, on multiplying out the brackets, we obtain

$$ab - a\varepsilon_b - b\varepsilon_a + \varepsilon_a\varepsilon_b < p < ab + a\varepsilon_b + b\varepsilon_a + \varepsilon_a\varepsilon_b$$

Ignoring the very small term $\varepsilon_a\varepsilon_b$, we obtain an estimate for the error bound of the product:

$$\varepsilon_p = a\varepsilon_b + b\varepsilon_a, \qquad p = a \times b$$

Dividing both sides of the equation by p, we obtain

$$\frac{\varepsilon_p}{p} = \frac{\varepsilon_a}{a} + \frac{\varepsilon_b}{b}$$

Now the relative error of a is defined as the ratio of the error in a to the size of a. The above equation connects the relative error bounds for a, b and p:

$$r_p = r_a + r_b$$

Here $r_a = \varepsilon_a/|a|$ allowing for a to be negative, and so on.

A similar worst case analysis for the quotient $q = a/b$ leads to the estimate

$$r_q = r_a + r_b$$

Thus for both multiplication and division, the relative error bound of the result is the sum of the individual relative error bounds.

These elementary rules for estimating error bounds can be combined to obtain more general results. For example, consider $z = x^2$; then $r_z = 2r_x$. In general, if $z = x^y$, where x is a rounded number and y is exact, then

$$r_z = yr_x$$

Example 1.46   Evaluate 13.92×5.31 and $13.92 \div 5.31$.

Assuming that these values are correctly rounded numbers, calculate error bounds for each answer and write them as correctly rounded numbers which have the greatest possible number of significant digits.

Solution $13.92 \times 5.31 = 73.9152$; $13.92 \div 5.31 = 2.621\,468\,927$

Let $a = 13.92$ and $b = 5.31$, then $r_a = 0.000\,36$ and $r_b = 0.000\,94$, so that $a \times b$ and $a \div b$ have relative error bounds $0.000\,36 + 0.000\,94 = 0.0013$. We obtain the absolute error bound of $a \times b$ by multiplying the relative error bound by $a \times b$. Thus the absolute error bound of $a \times b$ is $0.0013 \times 73.9152 = 0.0961$. Similarly, the absolute error bound of $a \div b$ is $0.0013 \times 2.6215 = 0.0034$. Hence the values of $a \times b$ and $a \div b$ lie in the error intervals

$$73.9152 - 0.0961 < a \times b < 73.9152 + 0.0961$$

and

$$2.6215 - 0.0034 < a \div b < 2.6215 + 0.0034$$

Thus $73.8191 < a \times b < 74.0113$ and $2.6181 < a \div b < 2.6249$.

From these inequalities we can deduce the correctly rounded values of $a \times b$ and $a \div b$:

$$a \times b = 74 \quad \text{and} \quad a \div b = 2.62$$

and we see how the rounding convention discards information. In a practical context, it would probably be more helpful to write:

$$73.81 < a \times b < 74.02$$

and

$$2.618 < a \div b < 2.625$$

Example 1.47

Evaluate

$$6.721 - \frac{4.931 \times 71.28}{89.45}$$

Assuming that all the values given are correctly rounded numbers, calculate an error bound for your answer and write it as a correctly rounded number.

Solution

Using a calculator, the answer obtained is

$$6.721 - \frac{4.931 \times 71.28}{89.45} = 2.791\,635\,216$$

To estimate the effect of the rounding error of the data, we first draw up a tree diagram representing the order in which the calculation is performed. Remember that $+$, $-$, $\times$ and $\div$ are binary operations, so only one operation can be performed at each step. Here we are evaluating

$$a - \frac{b \times c}{d} = e$$

We calculate this as $b \times c = p$, then $p \div d = q$ and then $a - q = e$, as shown in Figure 1.31(a). We set this calculation out in a table as shown in Figure 1.31(b), where the arrows show the flow of the error analysis calculation. Thus the value of e lies between $2.790\,235\ldots$ and $2.793\,035\ldots$, and the answer may be written as 2.7916 ± 0.0015 or as the correctly rounded number 2.79.

The following formulae indicate the way in which errors may accumulate in simple arithmetical calculations. The error bounds given are not always extreme and their behaviour is 'random'. This is discussed later in Example 13.3.1 in the work on Statistics.

Figure 1.31

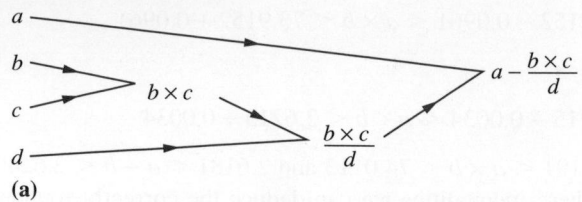

(a)

Label	Value	Absolute error bound	Relative error bound
b	4.931	⟶ 0.0005	⟶ 0.0005/4.931 = 0.0001 ⎫
c	71.28	⟶ 0.005	⟶ 0.005/71.28 = 0.000 07 ⎬
p	351.481 68		0.000 17
d	89.45	⟶ 0.005	⟶ 0.005/89.45 = 0.000 06 ⎬
q	3.929 364 784	⎰ 0.0009 = 0.000 23 × 3.9	⟵ 0.000 23
a	6.721	⎱ 0.0005	
e	2.791 635 216	+ 0.0014	

(b)

1.5.4 Exercises

42 Find the decimal equivalent of 110110.101_2.

43 Find the binary and octal (base eight) equivalents of the decimal number 16 321. Obtain a simple rule that relates these two representations of the number, and hence write down the octal equivalent of 1011100101101_2.

44 Find the binary and octal equivalents of the decimal number 30.6. Does the rule obtained in Question 43 still apply?

45 Use binary arithmetic to evaluate

(a) $100011.011_2 + 1011.001_2$

(b) $111.10011_2 \times 10.111_2$

46 State the numbers of decimal places and significant figures of the following correctly rounded numbers:

(a) 980.665　　　(b) 9.11×10^{-28}

(c) 2.9978×10^{10}　　(d) 2.00×10^{33}

(e) 1.759×10^7　　(f) 6.67×10^{-8}

47 In a right-angled triangle the height is measured as 1 m and the base as 2 m, both measurements being accurate to the nearest centimetre. Using Pythagoras' theorem, the hypotenuse is calculated as 2.236 07 m. Is this a sensible deduction? What other source of error will occur?

48 Determine the error bound and relative error bound for x, where

(a) $x = 35 \min \pm 5 s$

(b) $x = 35 \min \pm 4\%$

(c) $x = 0.58$ and x is correctly rounded to 2dp.

49 A value is calculated to be 12.9576, with a relative error bound of 0.0003. Calculate its absolute error bound and give the value as a correctly rounded number with as many significant digits as possible.

50 Using exact arithmetic, compute the values of the expressions below. Assuming that all the numbers given are correctly rounded, find absolute and relative error bounds for each term in the expressions and for your answers. Give the answers as correctly rounded numbers.

(a) $1.316 - 5.713 + 8.010$

(b) 2.51×1.01

(c) $19.61 + 21.53 - 18.67$

51 Evaluate $12.42 \times 5.675/15.63$, giving your answer as a correctly rounded number with the greatest number of significant figures.

52 Evaluate

$$a + b, \quad a - b, \quad a \times b, \quad a/b$$

for $a = 4.99$ and $b = 5.01$. Give absolute and relative error bounds for each answer.

53 Complete the table below for the computation

$$9.21 + (3.251 - 3.115)/0.112$$

and give the result as the correctly rounded answer with the greatest number of significant figures.

Label	Value	Absolute error bound	Relative error bound
a	3.251		
b	3.115		
$a - b$			
c	0.112		
$(a - b)/c$			
d	9.21		
$d + (a - b)/c$			

54 Evaluate $uv/(u + v)$ for $u = 1.135$ and $v = 2.332$, expressing your answer as a correctly rounded number.

55 Working to 4dp, evaluate

$$E = 1 - 1.65 + \tfrac{1}{2}(1.65)^2 - \tfrac{1}{6}(1.65)^3 + \tfrac{1}{24}(1.65)^4$$

(a) by evaluating each term and then summing,

(b) by 'nested multiplication'

$$E = 1 + 1.65(-1 + 1.65(\tfrac{1}{2} + 1.65(-\tfrac{1}{6} + \tfrac{1}{24}(1.65))))$$

Assuming that the number 1.65 is correctly rounded and that all other numbers are exact, obtain error bounds for both answers.

1.5.5 Computer arithmetic

The error estimate outlined in Example 1.44 is a 'worst case' analysis. The actual error will usually be considerably less than the error bound. For example, the maximum error in the sum of 100 numbers, each rounded to three decimal places, is 0.05. This would only occur in the unlikely event that each value has the greater possible rounding error. In contrast, the chance of the error being as large as one-tenth of this is only about 1 in 20.

When calculations are performed on a computer the situation is modified a little by the limited space available for number storage. Arithmetic is usually performed using floating-point notation. Each number x is stored in the **normal form**

$$x = (\text{sign})b^n(a)$$

where b is the number base, usually 2 or 16, n is an integer, and the **mantissa** a is a fraction with a fixed number of digits such that $1/b \leq a < 1$. As there are a limited number of digits available to represent the mantissa, calculations will involve intermediate rounding. As a consequence, the order in which a calculation is performed may affect the outcome – in other words the Fundamental Laws of Arithmetic may no longer hold! We shall illustrate this by means of an exaggerated example for a small computer using a decimal representation whose capacity for recording numbers is limited to four figures only. In large-scale calculations in engineering such considerations are sometimes important.

Consider a computer with storage capacity for real numbers limited to four figures; each number is recorded in the form $(\pm)10^n(a)$ where the exponent n is an integer, $0.1 \leq a < 1$ and a has four digits. For example,

$$\pi = +10^1(0.3142)$$

$$-\tfrac{1}{3} = -10^0(0.3333)$$

$$5764 = +10^4(0.5764)$$

$$-0.000\,971\,3 = -10^{-3}(0.9713)$$

$$5\,764\,213 = +10^7(0.5764)$$

Addition is performed by first adjusting the exponent of the smaller number to that of the larger, then adding the numbers, which now have the same multiplying power of 10, and lastly truncating the number to four digits. Thus $7.182 + 0.05381$ becomes

$$+10^1(0.7182) + 10^{-1}(0.5381) = 10^1(0.7182) + 10^1(0.005381)$$
$$= 10^1(0.723581)$$
$$= 10^1(0.7236)$$

With $a = 31.68$, $b = -31.54$ and $c = 83.21$, the two calculations $(a + b) + c$ and $(a + c) + b$ yield different results on this computer:

$$(a + b) + c = 83.35, \qquad (a + c) + b = 83.34$$

Notice how the symbol '=' is being used in the examples above. Sometimes it means 'equals to 4sf'. This computerized arithmetic is usually called **floating-point arithmetic**, and the number of digits used is normally specified.

1.5.6 Exercises

56 Two possible methods of adding five numbers are

$$(((a + b) + c) + d) + e$$

and

$$(((e + d) + c) + b) + a$$

Using 4dp floating-point arithmetic, evaluate the sum

$$10^1(0.1000) + 10^1(0.1000) - 10^0(0.5000)$$
$$+ 10^0(0.1667) + 10^{-1}(0.4167)$$

by both methods. Explain any discrepancy in the results.

57 Find $(10^{-2}(0.3251) \times 10^{-5}(0.2011))$ and $(10^{-1}(0.2168) \div 10^2(0.3211))$ using 4-digit floating-point arithmetic.

58 Find the relative error resulting when 4-digit floating-point arithmetic is used to evaluate

$$10^4(0.1000) + 10^2(0.1234) - 10^4(0.1013)$$

1.6 Engineering applications

In this section we illustrate through two examples how some of the results developed in this chapter may be used in an engineering application.

Example 1.48

A continuous belt of length L m passes over two wheels of radii r and R m with their centres a distance l m apart as illustrated in Figure 1.32. The belt is sufficiently tight for any sag to be negligible. Show that L is given approximately by

Figure 1.32
Continuous belt of Example 1.48.

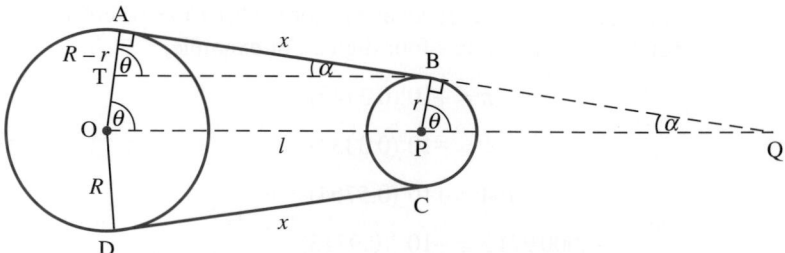

$$L \approx 2[l^2 - (R - r)^2]^{1/2} + \pi(R + r)$$

Find the error inherent in this approximation and obtain error bounds for L given the rounded data $R = 1.5$, $r = 0.5$ and $l = 3.5$.

Solution The length of the belt consists of the straight sections AB and CD and the wraps round the wheels $\widehat{BC}$ and $\widehat{DA}$. From Figure 1.32 it is clear that BT = OP = l and ∠OAB is a right-angle. Also, AT = AO − OT and OT − PB so that AT = $R - r$. Applying Pythagoras' theorem to the triangle TAB gives

$$AB^2 = l^2 - (R - r)^2$$

Since the length of an arc of a circle is the product of its radius and the angle (measured in radians) subtended at the centre (see equation 2.17), the length of wrap $\widehat{DA}$ is given by

$$(2\pi - 2\theta)R$$

where the angle is measured in radians. By geometry, $\theta = \dfrac{\pi}{2} - \alpha$, so that

$$\widehat{DA} = \pi R + 2R\alpha$$

Similarly, the arc $\widehat{BC} = \pi r - 2r\alpha$. Thus the total length of the belt is

$$L = 2[l^2 - (R - r)^2]^{1/2} + \pi(R + r) + 2(R - r)\alpha$$

Taking the length to be given approximately by

$$L \approx 2[l^2 - (R - r)^2]^{1/2} + \pi(R + r)$$

the error of the approximation is given by $-2(R - r)\alpha$, where the angle α is expressed in radians (remember that error = approximation − true value). The angle α is found by elementary trigonometry, since $\sin \alpha = (R - r)/l$. (Trigonometric functions will be reviewed in Section 2.6.)

For the (rounded) data given we deduce, following the procedures of Section 1.5.3, that for $R = 1.5$, $r = 0.5$ and $l = 3.5$ we have an error interval for α of

$$\left[\sin^{-1}\left(\frac{1.45 - 0.55}{3.55}\right), \sin^{-1}\left(\frac{1.55 - 0.45}{3.45}\right) \right] = [0.256, 0.325]$$

Thus $\alpha = 0.29 \pm 0.035$, and similarly $2(R - r)\alpha = 0.572 \pm 0.111$.

Evaluating the approximation for L gives

$$2[l^2 - (R - r)^2]^{1/2} + \pi(R + r) = 12.991 \pm 0.478$$

and the corresponding value for L is

$$L = 13.563 \pm 0.589$$

Thus, allowing for both the truncation error of the approximation and for the rounding errors in the data, the value 12.991 given by the approximation has an error interval [12.974, 14.152]. Its error bound is the larger of $|12.991 - 14.152|$ and $|12.991 - 12.974|$, that is, 1.16. Its relative error is 0.089 and its percent error is 8.9%, where the terminology follows the definitions given in Section 1.5.3.

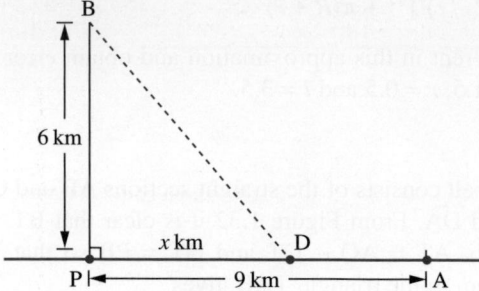

Example 1.49

A cable company is to run an optical cable from a relay station, A, on the shore to an installation, B, on an island, as shown in Figure 1.33. The island is 6 km from the shore at its nearest point, P, and A is 9 km from P measured along the shore. It is proposed to run the cable from A along the shoreline and then underwater to the island. It costs 25% more to run the cable underwater than along the shoreline. At what point should the cable leave the shore in order to minimize the total cost?

Solution

Optimization problems frequently occur in engineering and technology and often their solution is found algebraically.

If the cable leaves the shore at D, a distance x km from P, then the underwater distance is $\sqrt{(x^2 + 36)}$ km and the overland distance is $(9 - x)$ km, assuming $0 < x < 9$. If the overland cost of laying the cable is £c per kilometre, then the total cost £C is given by

$$C(x) = [(9 - x) + 1.25\sqrt{(x^2 + 36)}]c$$

We wish to find the value of x, $0 \leqslant x \leqslant 9$, which minimizes C. To do this we first change the variable x by substituting

$$x = 3\left(t - \frac{1}{t}\right)$$

such that $x^2 + 36$ becomes a perfect square:

$$x^2 + 36 = 36 + 9(t^2 - 2 + 1/t^2)$$
$$= 9(t + 1/t)^2$$

Hence C(x) becomes

$$C(t) = [9 - 3(t - 1/t) + 3.75(t + 1/t)]c$$
$$= [9 + 0.75(t + 9/t)]c$$

Using the arithmetic–geometric inequality $x + y \geqslant 2\sqrt{(xy)}$, see (1.4d), we know that

$$t + \frac{9}{t} \geqslant 6$$

and that the equality occurs where $t = 9/t$, that is where $t = 3$.

Thus the minimum cost is achieved where $t = 3$ and $x = 3(3 - 1/3) = 8$.

Hence the cable should leave the shore after laying the cable 1 km from its starting point at A.

1.7 Review exercises (1–25)

1 (a) A formula in the theory of ventilation is

$$Q = \frac{\sqrt{H}}{K}\sqrt{\frac{A^2D^2}{A^2 + D^2}}$$

Express A in terms of the other symbols.

(b) Solve the equation

$$\frac{1}{x + 2} - \frac{2}{x} = \frac{3}{x - 1}$$

2 Factorize the following

(a) $ax - 2x - a + 2$ (b) $a^2 - b^2 + 2bc - c^2$

(c) $4k^2 + 4kl + l^2 - 9m^2$ (d) $p^2 - 3pq + 2q^2$

(e) $l^2 + lm + ln + mn$

3 (a) Two small pegs are 8 cm apart on the same horizontal line. An inextensible string of length 16 cm has equal masses fastened at either end and is placed symmetrically over the pegs. The middle point of the string is pulled down vertically until it is in line with the masses. How far does each mass rise?

(b) Find an 'acceptable' value of x to three decimal places if the shaded area in Figure 1.34 is 10 square units.

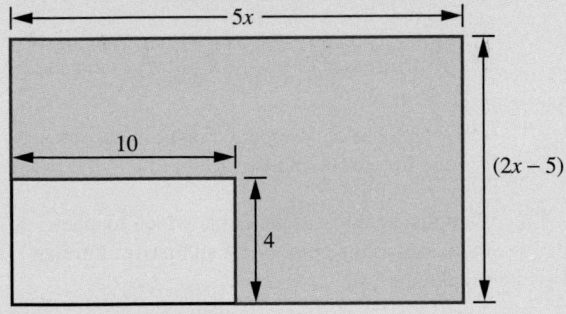

Figure 1.34 Shaded area of Question 3(b).

4 The impedance Z ohms of a circuit containing a resistance R ohms, inductance L henries and capacity C farads, when the frequency of the oscillation is n per second, is given by

$$Z = \sqrt{\left(R^2 + \left(2\pi nL - \frac{1}{2\pi nC}\right)^2\right)}$$

(a) Make L the subject of this formula.

(b) If $n = 50$, $R = 15$ and $C = 10^{-4}$ show that there are two values of L which make $Z = 20$ but only one value of L which will make $Z = 100$. Find the values of Z in each case to two decimal places.

5 Expand out (a) and (b) and rationalize (c) to (e).

(a) $(3\sqrt{2} - 2\sqrt{3})^2$

(b) $(\sqrt{5} + 7\sqrt{3})(2\sqrt{5} - 3\sqrt{3})$

(c) $\dfrac{4 + 3\sqrt{2}}{5 + \sqrt{2}}$

(d) $\dfrac{\sqrt{3} + \sqrt{2}}{2 - \sqrt{3}}$

(e) $\dfrac{1}{1 + \sqrt{2} - \sqrt{3}}$

6 Find integers m and n such that

$$\sqrt{(11 + 2\sqrt{30})} = \sqrt{m} + \sqrt{n}$$

7 Show that

$$\sqrt{(n + 1)} - \sqrt{n} = \frac{1}{\sqrt{(n + 1)} + \sqrt{n}}$$

and deduce that

$$\sqrt{(n + 1)} - \sqrt{n} < \frac{1}{2\sqrt{n}} < \sqrt{n} - \sqrt{(n - 1)}$$

for any integer $n \geqslant 1$. Deduce that the sum

$$\frac{1}{\sqrt{1}} + \frac{1}{\sqrt{2}} + \frac{1}{\sqrt{3}} + \ldots + \frac{1}{\sqrt{(9999)}} + \frac{1}{\sqrt{(10\,000)}}$$

lies between 198 and 200.

8 Express each of the following subsets of $\mathbb{R}$ in terms of intervals:

(a) $\{x : 4x^2 - 3 < 4x, x \text{ in } \mathbb{R}\}$

(b) $\{x : 1/(x + 2) > 2/(x - 1), x \text{ in } \mathbb{R}\}$

(c) $\{x : |x + 1| < 2, x \text{ in } \mathbb{R}\}$

(d) $\{x : |x + 1| < 1 + \frac{1}{2}x, x \text{ in } \mathbb{R}\}$

9 It is known that of all plane curves that enclose a given area, the circle has the least perimeter. Show that if a plane curve of perimeter L encloses an area A then $4\pi A \leqslant L^2$. Verify this inequality for a square and a semicircle.

10 The arithmetic–geometric inequality

$$\frac{x+y}{2} \geqslant \sqrt{xy}$$

implies

$$\left(\frac{x+y}{2}\right)^2 \geqslant xy$$

Use the substitution $x = \frac{1}{2}(a+b)$, $y = \frac{1}{2}(c+d)$, where a, b, c and $d > 0$, to show that

$$\left(\frac{a+b}{2}\right)\left(\frac{c+d}{2}\right) \leqslant \left(\frac{a+b+c+d}{4}\right)^2$$

and hence that

$$\left(\frac{a+b}{2}\right)^2\left(\frac{c+d}{2}\right)^2 \leqslant \left(\frac{a+b+c+d}{4}\right)^4$$

By applying the arithmetic–geometric inequality to the first two terms of this inequality, deduce that

$$abcd \leqslant \left(\frac{a+b+c+d}{4}\right)^4$$

and hence

$$\frac{a+b+c+d}{4} \geqslant \sqrt[4]{abcd}$$

11 Show that if $a < b$, $b > 0$ and $c > 0$ then

$$\frac{a}{b} < \frac{a+c}{b+c} < 1$$

Obtain a similar inequality for the case $a > b$.

12 (a) If $n = n_1 + n_2 + n_3$ show that

$$\binom{n}{n_1}\binom{n_2+n_3}{n_2} = \frac{n!}{n_1!\,n_2!\,n_3!}$$

(This represents the number of ways in which n objects may be divided into three groups containing respectively n_1, n_2 and n_3 objects.)

(b) Expand the following expressions

(i) $\left(1 - \dfrac{x}{2}\right)^5$ (ii) $(3 - 2x)^6$

13 (a) Evaluate $\displaystyle\sum_{n=-2}^{3} [n^{n+1} + 3(-1)^n]$

(b) A square grid of dots may be divided up into a set of L-shaped groups as illustrated in Figure 1.35.

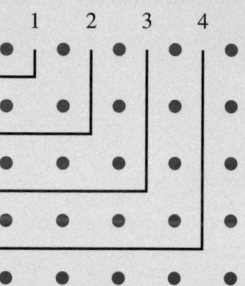

Figure 1.35

How many dots are inside the third L shape? How many extra dots are needed to extend the 3 by 3 square to one of side 4 by 4? How many dots are needed to extend an $(r-1)$ by $(r-1)$ square to one of size r by r? Denoting this number by P_r, use a geometric argument to obtain an expression for $\sum_{r=1}^{n} P_r$ and verify your conclusion by direct calculation in the case $n = 10$.

14 Find the equations of the straight line:

(a) which passes through the points $(-6, -11)$ and $(2, 5)$;

(b) which passes through the point $(4, -1)$ and has gradient $\frac{1}{3}$;

(c) which has the same intercept on the y axis as the line in (b) and is parallel to the line in (a).

15 Find the equation of the circle which touches the y axis at the point $(0, 3)$ and passes through the point $(1, 0)$.

16 Find the centres and radii of the following circles.

(a) $x^2 + y^2 + 2x - 4y + 1 = 0$

(b) $4x^2 - 4x + 4y^2 + 12y + 9 = 0$

(c) $9x^2 + 6x + 9y^2 - 6y = 25$

17 For each of the two parabolas

(i) $y^2 = 8x + 4y - 12$, and

(ii) $x^2 + 12y + 4x = 8$

determine

(a) the coordinates of the vertex;

(b) the coordinates of the focus;

(c) the equation of the directrix;

(d) the equation of the axis of symmetry.

Sketch each parabola.

18 Find the coordinates of the centre and foci of the ellipse with equation

$$25x^2 + 16y^2 - 100x - 256y + 724 = 0$$

What are the coordinates of its vertices and the equations of its directrices? Sketch the ellipse.

19 Find the duodecimal equivalent of the decimal number 10.386 23.

20 Show that if $y = x^{1/2}$ then the relative error bound of y is one-half that of x. Hence complete the table in Figure 1.36.

	Value	Absolute error bound	Relative error bound
a	7.01	0.005	$\longrightarrow$ 0.0007
$\sqrt{a}$	2.6476	0.0009	$\longleftarrow$ 0.000 35
b	52.13		
$\sqrt{b}$			
c	0.010 11		
$\sqrt{c}$			
d	5.631×10^{11}		
$\sqrt{d}$			
Correctly rounded values	$\sqrt{a}$ $\sqrt{b}$ $\sqrt{c}$ $\sqrt{d}$ 2.65		

Figure 1.36

21 Assuming that all the numbers given are correctly rounded, calculate the positive root together with its error bound of the quadratic equation

$$1.4x^2 + 5.7x - 2.3 = 0$$

Give your answer also as a correctly rounded number.

22 The quantities f, u and v are connected by

$$\frac{1}{f} = \frac{1}{u} + \frac{1}{v}$$

Find f when $u = 3.00$ and $v = 4.00$ are correctly rounded numbers. Compare the error bounds obtained for f when

(a) it is evaluated by taking the reciprocal of the sum of the reciprocals of u and v,

(b) it is evaluated using the formula

$$f = \frac{uv}{u + v}$$

23 If the number whose decimal representation is 14 732 has the representation $152\,112_b$ to base b, what is b?

24 A milk carton has capacity 2 pints (1136 ml). It is made from a rectangular waxed card using the net shown in Figure 1.37. Show that the total area A (mm^2) of card used is given by

$$A(h, w) = (2w + 145)(h + 80)$$

with $hw = 113\,600/7$. Show that

$$A(h, w) = C(h, w) + \frac{308\,400}{7}$$

where $C(h, w) = 145h + 160w$.

Use the arithmetic–geometric inequality to show that

$$C(h, w) \geqslant 2\sqrt{(160w \times 145h)}$$

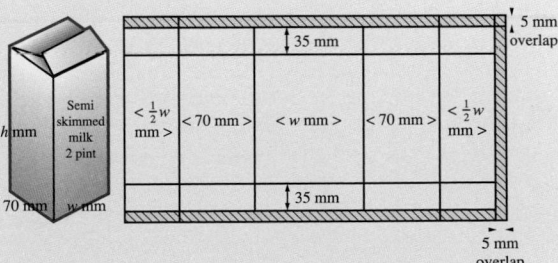

Figure 1.37 Milk carton of Question 24.

with equality when $160w = 145h$. Hence show that the minimum values of $C(h, w)$ and $A(h, w)$ are achieved when $h = 133.8$ and $w = 121.3$. Give these answers to more sensible accuracy.

25 A family of straight lines in the (x, y)-plane is such that each line joins the point $(-p, p)$ on the line $y = -x$ to the point $(10 - p, 10 - p)$ on the line $y = x$, as shown in Figure 1.38, for different values of p. On a piece of graph paper draw the lines corresponding to $p = 1, 2, 3, \dots , 9$. The resulting family is seen to envelop a curve. Show that the line which joins $(-p, p)$ to $(10 - p, 10 - p)$ has equation

$$5y = 5x - px + 10p - p^2$$

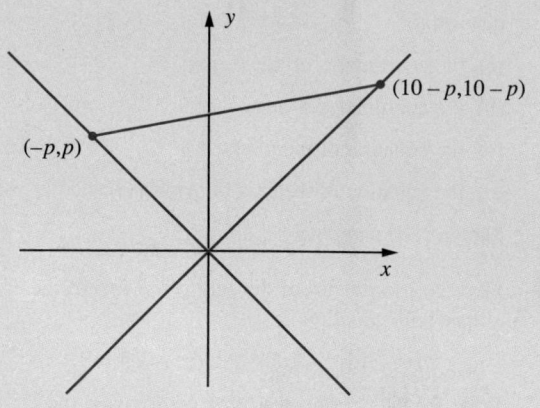

Figure 1.38

Show that two lines of the family pass through the point (x_0, y_0) if $x_0^2 > 20(y_0 - 5)$, but no lines pass through (x_0, y_0) if $x_0^2 < 20(y_0 - 5)$. Deduce that the enveloping curve of the family of straight lines is

$$y = \tfrac{1}{20} x^2 + 5$$

2 Functions

Chapter 2 Contents

2.1 Introduction

As we have remarked in the introductory section of Chapter 1, mathematics provides a means of solving the practical problems that occur in engineering. To do this, it uses concepts and techniques that operate on and within the concepts. In this chapter we shall describe the concept of a function – a concept that is both fundamental to mathematics and intuitive. We shall make the intuitive idea mathematically precise by formal definitions and describe why such formalism is needed for practical problem-solving.

The function concept has taken many centuries to evolve. The intuitive basis for the concept is found in the analysis of cause and effect, which underpins developments in science, technology and commerce. As with many mathematical ideas, many people use the concept in their everyday activities without being aware that they are using mathematics, and many would be surprised if they were told that they were. The abstract manner in which the developed form of the concept is expressed by mathematicians often intimidates learners but the essential idea is very simple. A consequence of the long period of development is that the way in which the concept is described often makes an idiomatic use of words. Ordinary words which in common parlance have many different shades of meaning are used in mathematics with very specific meanings.

The key idea is that of the values of two variable quantities being related. For example, the amount of tax paid depends on the selling price of an item; the deflection of a beam depends on the applied load; the cost of an article varies with the number produced, and so on. Historically, this idea has been expressed in a number of ways. The oldest gave a verbal recipe for calculating the required value. Thus, in the early Middle Ages, a very elaborate verbal recipe was given for calculating the monthly interest payments on a loan which would now be expressed very compactly by a single formula. John Napier, when he developed the logarithm function at the beginning of the seventeenth century, expressed the functional relationship in terms of two particles moving along a straight line. One particle moved with constant velocity and the other with a velocity that depended on its distance from a fixed point on the line. The relationship between the distances travelled by the particles was used to define the logarithms of numbers. This would now be described by the solution of a differential equation. The introduction of algebraic notation led to the representation of functions by algebraic rather than verbal formulae. That produced many theoretical problems. For example, a considerable controversy was caused by Fourier when he used functions that did not have the same algebraic formula for all values of the independent variable. Similarly, the existence of functions that do not have a simple algebraic representation caused considerable difficulties for mathematicians in the early nineteenth century.

2.2 Basic definitions

2.2.1 Concept of a function

The essential idea that flows through all of the developments is that of two quantities whose values are related. One of these variables, the **independent** or **free variable**,

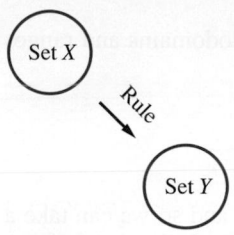

Figure 2.1
Schematic
representation
of a function.

may take any value in a set of values. The value it actually takes fixes uniquely the value of the second quantity, the **dependent** or **slave variable**. Thus for each value of the independent variable there is one and only one value of the dependent variable. The way in which that value is calculated will vary between functions. Sometimes it will be by means of a formula, sometimes by means of a graph and sometimes by means of a table of values. Here the words 'value' and 'quantity' cover many very different contexts, but in each case what we have is two sets of values X and Y and a rule that assigns to each value x in the set X precisely one value y from the set Y. The elements of X and Y need not be numbers, but the essential idea is that to every x in the set X there corresponds exactly one y in the set Y. Whenever this situation arises we say that there is a **function** f that maps the set X to the set Y. Such a function may be illustrated schematically as in Figure 2.1.

We represent a functional relationship symbolically in two ways: either

$$f{:}x \to y \quad (x \text{ in } X)$$

or

$$y = f(x) \quad (x \text{ in } X)$$

The first emphasizes the fact that a function f associates each element (value) x of X with exactly one element (value) y of Y: it 'maps x to y'. The second method of notation emphasizes the dependence of the elements of Y on the elements of X under the function f. In this case the value or variable appearing within the brackets is known as the **argument** of the function; we might say 'the argument x of a function $f(x)$'. In engineering it is more common to use the second notation $y = f(x)$ and to refer to this as the function $f(x)$, while modern mathematics textbooks prefer the mapping notation, on the grounds that it is less ambiguous. The set X is called the **domain** of the function and the set Y is called its **codomain**. Knowing the domain and codomain is important in computing. We need to know the type of variables, whether they are integers or reals, and their size. When $y = f(x)$, y is said to be the **image** of x under f. The set of all images $y = f(x)$, x in X, is called the **image set** or **range** of f. It is not necessary for all elements y of the codomain set Y to be images under f. In the terminology of Chapter 6 the range is a subset of the codomain. We may regard x as being a variable that can be replaced by any element of the set X. The rule giving f is then completely determined if we know $f(x)$, and consequently in engineering it is common to refer to the function as being $f(x)$ rather than f. Likewise we can regard $y = f(x)$ as being a variable. However, while x can freely take any value from the set X, the variable $y = f(x)$ depends on the particular element chosen for x. We therefore refer to x as the **free** or **independent** variable and to y as the **slave** or **dependent** variable. The function $f(x)$ is therefore specified completely by the set of ordered pairs (x, y) for all x in X. For real variables a graphical representation of the function may then be obtained by plotting a graph determined by this set of ordered pairs (x, y), with the independent variable x measured along the horizontal axis and the dependent variable y measured along the vertical axis. Obtaining a good graph by hand is not always easy but there are now available excellent graphics facilities on computers and calculators which assist in the task. Even so some practice is required to ensure that a good choice of 'drawing window' is selected to obtain a meaningful graph.

Example 2.1

For the functions with formulae below, identify their domains, codomains and ranges and calculate values of $f(2), f(-3)$ and $f(-x)$.

(a) $f(x) = 3x^2 + 1$ (b) $f{:}x \rightarrow \sqrt{[(x+4)(3-x)]}$

Solution

(a) The formula for $f(x)$ can be evaluated for all real values of x and so we can take a domain which includes all the real numbers, $\mathbb{R}$. The values obtained are also real numbers so we may take $\mathbb{R}$ as the codomain. The range of $f(x)$ is actually less than $\mathbb{R}$ in this example because the minimum value of $y = 3x^2 + 1$ occurs at $y = 1$ where $x = 0$. Thus the range of f is the set

$$\{x{:}1 \leqslant x, x \text{ in } \mathbb{R}\} = [1, \infty)$$

Notice the convention here that the set is specified using the *dummy* variable x. We could also write $\{y{:}1 \leqslant y, y \text{ in } \mathbb{R}\}$, any letter could be used but conventionally x is used. Using the formula we find that $f(2) = 13$, $f(-3) = 28$ and $f(-x) = 3(-x)^2 + 1 = 3x^2 + 1$.

(b) The formula $f{:}x \rightarrow \sqrt{[(x+4)(3-x)]}$ only gives real values for $-4 \leqslant x \leqslant 3$, since we cannot take square roots of negative numbers. Thus the domain of f is $[-4,3]$. Within its domain the function has real values so that its codomain is $\mathbb{R}$ but its range is less than $\mathbb{R}$. The least value of f occurs at $x = -4$ and $x = 3$ when $f(-4) = f(3) = 0$. The largest value of f occurs at $x = -\frac{1}{2}$ when $f(-\frac{1}{2}) = \sqrt{(35)}/2$.

So the range of f in this example is $[0, \sqrt{(35)}/2]$. Using the formula we have $f(2) = \sqrt{6}$, $f(-3) = \sqrt{6}$, $f(-x) = \sqrt{[(4-x)(x+3)]}$.

Example 2.2

The function $y = f(x)$ is given by the minimum diameter y of a circular pipe that can contain x circular pipes of unit diameter where $x = 1, 2, 3, 4, 5, 6, 7$. Find the domain, codomain and range of $f(x)$.

Solution

This function is illustrated in Figure 2.2

Figure 2.2
Enclosing x circular pipes in a circular pipe.

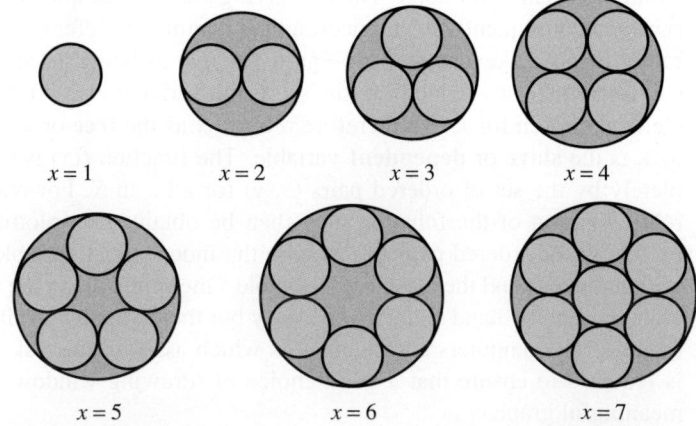

$x = 1$ $x = 2$ $x = 3$ $x = 4$

$x = 5$ $x = 6$ $x = 7$

Here the domain is the set $\{1, 2, 3, 4, 5, 6, 7\}$ and the codomain is $\mathbb{R}$. Calculating the range is more difficult as there is not a simple algebraic formula relating x and y. From geometry we have

$$f(1) = 1, f(2) = 2, f(3) = 1 + 2/\sqrt{3}, f(4) = 1 + \sqrt{2}, f(5) = \tfrac{1}{4}\sqrt{[2(5 - \sqrt{5})]},$$
$$f(6) = 3, f(7) = 3$$

The range of $f(x)$ is the set of these values $\{ f(x) : x = 1, 2, 3, 4, 5, 6, 7\}$.

Example 2.3

The relationship between the temperature T_1 measured in degrees Celsius (°C) and the corresponding temperature T_2 measured in degrees Fahrenheit (°F) is

$$T_2 = \tfrac{9}{5}T_1 + 32$$

Interpreting this as a function with T_1 as the independent variable and T_2 as the dependent variable:

(a) What are the domain and codomain of the function?

(b) What is the function rule?

(c) Plot a graph of the function.

(d) What is the image set or range of the function?

(e) Use the function to convert the following into °F:

    (i) 60°C,   (ii) 0°C,   (iii) −50°C

Solution

(a) Since temperature can vary continuously, the domain is the set $T_1 \geqslant T_0 - -273.16$ (absolute zero). The codomain can be chosen as the set of real numbers $\mathbb{R}$.

(b) The function rule in words is

    multiply by $\tfrac{9}{5}$ and then add 32

or algebraically

$$f(T_1) = \tfrac{9}{5}T_1 + 32$$

(c) Since the domain is the set $T_1 \geqslant T_0$, there must be an image for every value of T_1 on the horizontal axis which is greater than −273.16. The graph of the function is that part of the line $T_2 = \tfrac{9}{5}T_1 + 32$ for which $T_1 > -273.16$ as illustrated in Figure 2.3.

(d) Since each value of T_2 is an image of some value T_1 in its domain, it follows that the range of $f(T_1)$ is the set of real numbers greater than −459.69.

(e) The conversion may be done graphically by reading values of the graph, as illustrated by the broken lines in Figure 2.3, or algebraically using the rule

$$T_2 = \tfrac{9}{5}T_1 + 32$$

giving the values

    (i) 140°F,   (ii) 32°F,   (iii) −58°F

Figure 2.3
Graph of
$T_2 = f(T_1) = \frac{9}{5}T_1 + 32$.

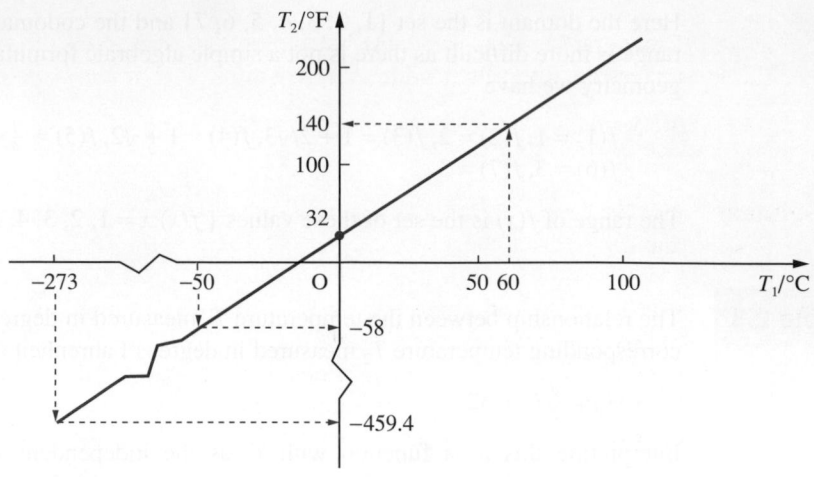

Figure 2.4
Graph of
$y = (x - 1)(x + 2)$.

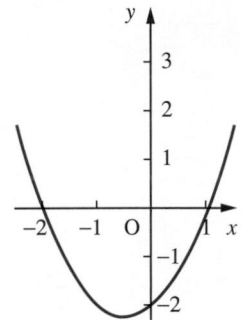

A value of the independent variable for which the value of a function is zero, is called a **zero** of that function. Thus the function $f(x) = (x - 1)(x + 2)$ has two zeros, $x = 1$ and $x = -2$. These correspond to where the graph of the function crosses the x axis, as shown in Figure 2.4. We can see from the diagram that, for this function, its values decrease as the values of x increase from (say) -5 up to $-\frac{1}{2}$, and then its values increase with x. We can demonstrate this algebraically by rearranging the formula for $f(x)$:

$$f(x) = (x - 1)(x + 2)$$
$$= x^2 + x - 2$$
$$= (x + \tfrac{1}{2})^2 - \tfrac{9}{4},$$

From this we can see that $f(x)$ achieves its smallest value ($-\frac{9}{4}$) where $x = -\frac{1}{2}$ and that the value of the function is greater than $-\frac{9}{4}$ both sides of $x = -\frac{1}{2}$ because $(x + \frac{1}{2})^2 \geqslant 0$. The function is said to be a **decreasing function** for $x < -\frac{1}{2}$ and an **increasing function** for $x > -\frac{1}{2}$. More formally, a function is said to be increasing on an interval (a, b) if $f(x_2) > f(x_1)$ when $x_2 > x_1$ for all x_1 and x_2 lying in (a, b). Similarly for decreasing functions, we have $f(x_2) < f(x_1)$ when $x_2 > x_1$.

The value of a function at the point where its behaviour changes from decreasing to increasing is a **minimum** (*plural* **minima**) of the function. Often this is denoted by an asterisk superscript f^* and the corresponding value of the independent variable by x^* so that $f(x^*) = f^*$. Similarly a **maximum** (*plural* **maxima**) occurs when a function changes from being increasing to being decreasing. In many cases the terms maximum and minimum refer to the local values of the function, as illustrated in Example 2.4(a). Sometimes, in practical problems, it is necessary to distinguish between the largest value the function achieves on its domain and the *local maxima* it achieves elsewhere. Similarly for *local minima*. Maxima and minima are jointly referred to as **optimal values** and as **extremal values** of the function.

The point (x^*, f^*) of the graph of $f(x)$ is often called a turning point of the graph, whether it is a maximum or a minimum. These properties will be discussed in more detail in Chapter 8, Sections 8.2.7 and 8.5. For smooth functions as in Figure 2.5, the tangent to the graph of the function is horizontal at a turning point. This property can be used to locate maxima and minima.

Example 2.4

Draw graphs of the functions below, locating their zeros, intervals in which they are increasing, intervals in which they are decreasing and their optimal values.

(a) $y = 2x^3 + 3x^2 - 12x + 32$ (b) $y = (x - 1)^{2/3} - 1$

Solution (a) The graph of the function is shown in Figure 2.5. From the graph we can see that the function has one zero at $x = -4$. It is an increasing function on the intervals $-\infty < x < -2$ and $1 < x < \infty$ and a decreasing function on the interval $-2 < x < 1$. It achieves a maximum value of 52 at $x = -2$ and a minimum value of 25 at $x = 1$. In this example the extremal values at $x = -2$ and $x = 1$ are *local maximum* and *local minimum* values. The function is defined on the set of real numbers $\mathbb{R}$. Thus it does not have finite upper and lower values. If the domain were restricted to $[-4, 4]$, say, then the *global minimum* would be $f(-4) = 0$ and the *global maximum* would be $f(4) = 160$.

Figure 2.5
Graph of $y = 2x^3 + 3x^2 - 12x + 32$.

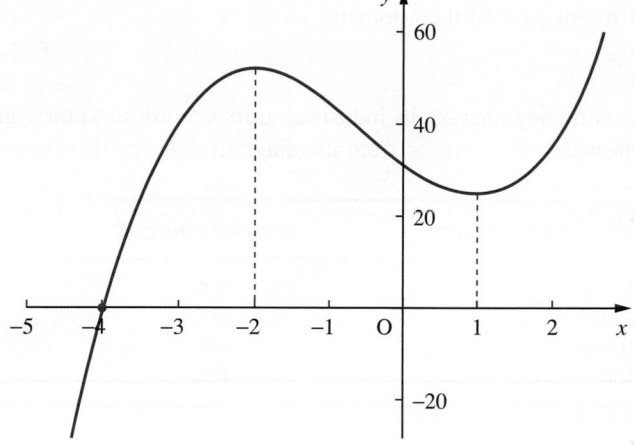

(b) The graph of the function is shown in Figure 2.6. (Note that to evaluate $(x - 1)^{2/3}$ on some calculators/computer packages it has to be expressed as $((x - 1)^2)^{1/3}$ for $x < 1$.)

Figure 2.6
Graph of $y = (x - 1)^{2/3} - 1$.

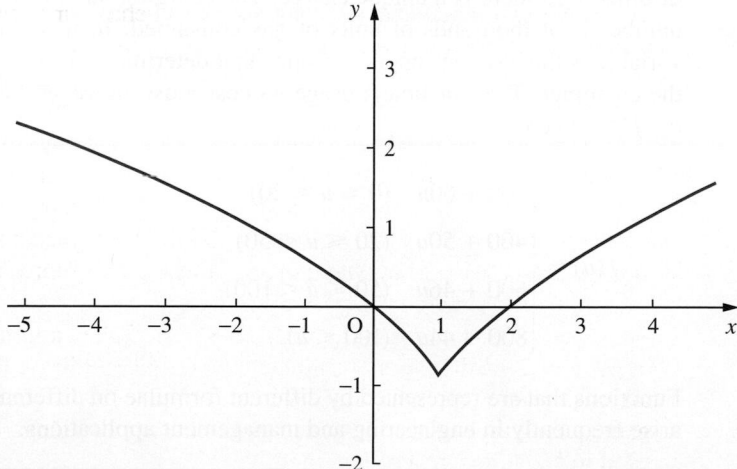

From the graph, we see that the function has two zeros, one at $x = 0$ and the other at $x = 2$. It is a decreasing function for $x < 1$ and an increasing function for $x > 1$. This is obvious algebraically since $(x - 1)^{2/3}$ is greater than or equal to zero. This example also provides an illustration of the behaviour of some algebraic functions at a maximum or minimum value. In contrast to (a) where the function changes from decreasing to increasing at $x = 1$ quite smoothly, in this case the function changes from decreasing to increasing abruptly at $x = 1$. Such a minimum value is called a **cusp**. In this example, the value at $x = 1$ is both a local minimum and a global minimum.

It is important to appreciate the difference between a function and a formula. A function is a mapping that associates one and only one member of the codomain with every member of its domain. It may be possible to express this association, as in Example 2.3, by a formula. Some functions may be represented by different formulae on different parts of their domain.

Example 2.5

A gas company charges its industrial users according to their gas usage. Their tariff is as follows:

Quarterly usage/10^3 units	Standing charge/£	Charge per 10^3 units/£
0–19.999	200	60
20–49.999	400	50
50–99.999	600	46
$\geqslant 100$	800	44

What is the quarterly charge paid by a user?

Solution

The charge £c paid by a user for a quarter's gas is a function, since for any number of units used there is a unique charge. The charging tariff is expressed in terms of the number n of thousands of units of gas consumed. In this situation the independent variable is the gas consumption n since that determines the charge £c which accrues to the customer. The function f: usage $\rightarrow$ cost must, however, be expressed in the form $c = f(u)$, where

$$f(u) = \begin{cases} 200 + 60u & (0 \leqslant u < 20) \\ 400 + 50u & (20 \leqslant u < 50) \\ 600 + 46u & (50 \leqslant u < 100) \\ 800 + 44u & (100 \leqslant u) \end{cases}$$

Functions that are represented by different formulae on different parts of their domains arise frequently in engineering and management applications.

The basic MATLAB package is primarily a number crunching package. It does not perform symbolic manipulations and cannot undertake algebra containing unknowns. However, such work can be undertaken by the Symbolic Math Toolbox, which incorporates many MAPLE commands to implement the algebraic work. Consequently, most of the commands in Symbolic Math Toolbox are identical to the MAPLE commands. In order to use any symbolic variables, such as x and y, in MATLAB these must be declared by entering a command, such as `syms x y;`. MAPLE does not need to construct symbols since these are assumed in the package. Another important difference is that in MAPLE assignment is performed by := rather than = and each statement must end with a semicolon ';'. In MATLAB inserting a semicolon at the end of a statement suppresses display on screen of the output to the command. In MAPLE output to the screen is suppressed using ':'. In this chapter only MATLAB versions of any process are given since a MAPLE user can easily adopt the codes. If there are significant syntax differences these will be noted.

The MATLAB operators for the basic arithmetic operations are + for addition, – for subtraction, * for multiplication, / for division and ^ for power. The colon command `x = a:dx:b` generates an array of numbers which are the values of x between a and b in steps of dx. For example, the command

```
x = 0:0.1:1
```

generates the array

```
x = 0 0.1 0.2 0.3 0.4 0.5 0.6 0.7 0.8 0.9 1.0
```

When using the operations of multiplication, division and power on such arrays *, / and ^ are respectively replaced by `.*`, `./` and `.^` in which the 'dot' implies element by element operations. For example, if $x = [1\ 2\ 3]$ and $y = [4\ -3\ 5]$ are two arrays then `x.*y` denotes the array $[4\ -6\ 15]$ and `x.^2` denotes the array $[1\ 4\ 9]$. Note that to enter an array it must be enclosed within square brackets `[ ]`.

To plot the graph of $y = f(x)$, $a \leqslant x \leqslant b$, an array of x values is first produced and then a corresponding array of y values is produced. Then the command `plot(x,y)` plots a graph of y against x. Check that the sequence of commands

```
x = -5:0.1:3;
y = 2*x.^3 + 3*x.^2 - 12*x + 32;
plot(x,y)
```

plots the graph of Figure 2.5. Entering a further command

```
grid
```

draws gridlines on the existing plot. The following commands may be used for labelling the graph:

```
title('text')    prints 'text' at the top of the plot
xlabel('text')   labels the x-axis with 'text'
ylabel('text')   labels the y-axis with 'text'
```

Example 2.

Soluti

Draw a caref…
design the gr…
used to asses…

5   The initial c…
years, its val…
increase, as s…

t	
Value after	
t years	
Running cos…	
in year t	

2.2.…

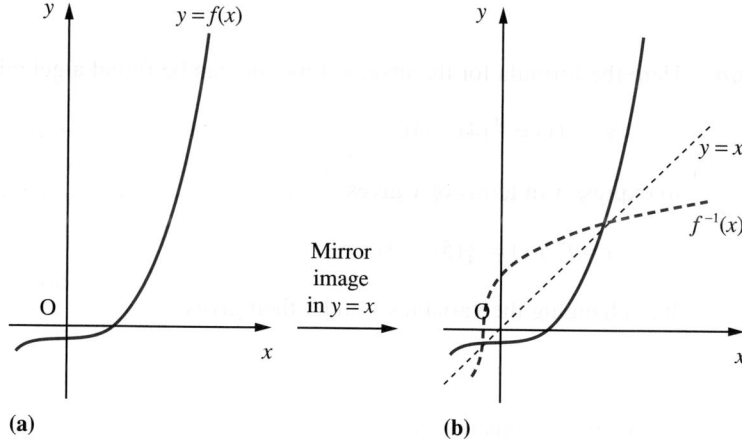

x → $\boxed{f(\bullet)}$ → $y = f$

(a)

$x = f^{-1}(y)$ ← $\boxed{f^{-1}(\bullet)}$ ←

(b)

Figure 2.10
Block diagram of
(a) function and
(b) inverse function

Example 2.7   Obtain the inverse function of $y = f(x) = \dfrac{x+2}{x+1}$, $x \neq -1$.

Solution   We rearrange $y = \dfrac{x+2}{x+1}$ to obtain x in terms of y. (Notice that y is not defined where $x = -1$.) Thus

$$y(x+1) = x+2 \quad \text{so that} \quad x(y-1) = 2-y$$

giving $x = \dfrac{2-y}{y-1}$, $y \neq 1$ (Notice that x is not defined where $y = 1$. Putting $y = 1$ into the formula for y results in the equation $x+1 = x+2$ which is not possible.)

Thus $f^{-1}(x) = \dfrac{2-x}{x-1}$, $x \neq 1$

If we are given the graph of $y = f(x)$ and wish to obtain the graph of the inverse function $y = f^{-1}(x)$ then what we really need to do is interchange the roles of x and y. Thus we need to manipulate the graph of $y = f(x)$ so that the x and y axes are interchanged. This can be achieved by taking the mirror image in the line $y = x$ and relabelling the axes as illustrated in Figures 2.11(a) and (b). It is important to recognize that the graphs of $y = f(x)$ and $y = f^{-1}(x)$ are symmetrical about the line $y = x$, since this property is frequently used in mathematical arguments. Notice that the x and y axes have the same scale.

Figure 2.11
The graph of
$y = f^{-1}(x)$.

(a)   **(b)**

Example 2.8   Obtain the graph of $f^{-1}(x)$ when (a) $f(x) = \frac{9}{5}x + 32$, (b) $f(x) = \dfrac{x+2}{x+1}$, $x \neq -1$, (c) $f(x) = x^2$.

Solution   (a) This is the formula for converting the temperature measured in °C to the temperature in °F and its graph is shown by the blue line in Figure 2.12(a). Reflecting the graph in the line $y = x$ yields the graph of the inverse function $y = g(x) = \frac{5}{9}(x - 32)$ as illustrated by the black line in Figure 2.12(a).

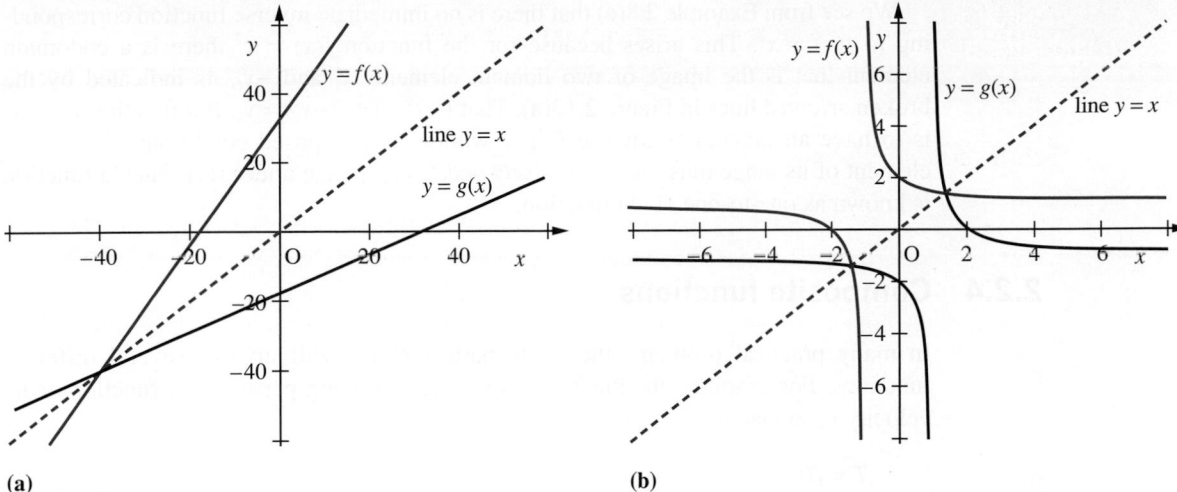

(a)　　　　　　　　　　　　　　　**(b)**

Figure 2.12 (a) Graph of $f(x) = \frac{9}{5}x + 32$ and its inverse $g(x)$, (b) Graph of $f(x) = \dfrac{x+2}{x+1}$ and its inverse $g(x)$.

(b) The graph of $y = f(x) = \dfrac{x+2}{x+1}$, $x \neq -1$ is shown in blue in Figure 2.12(b). The graph of its inverse function $y = g(x) = \dfrac{2-x}{x-1}$, $x \neq 1$ can be seen as the mirror image illustrated in black in Figure 2.12(b).

(c) The graph of $y = x^2$ is shown in Figure 2.13(a). Its mirror image in the line $y = x$ gives the graph of Figure 2.13(b). We note that this graph is not representative of a function according to our definition, since for all values of $x > 0$ there are two images – one positive and one negative – as indicated by the broken line. This follows because $y = x^2$ corresponds to $x = +\sqrt{y}$ or $x = -\sqrt{y}$. In order to avoid this ambiguity, we define the inverse function of $f(x) = x^2$ to be $f^{-1}(x) = +\sqrt{x}$, which corresponds to the upper half of the graph as illustrated in Figure 2.13(c). $\sqrt{x}$ therefore denotes a positive number (cf. calculators), so the range of $\sqrt{x}$ is $x \geqslant 0$. Thus the inverse function of

$$y = f(x) = x^2 \quad (x \geqslant 0)$$

is

$$y = f^{-1}(x) = \sqrt{x}$$

Note that the domain of $f(x)$ had to be restricted to $x \geqslant 0$ in order that an inverse could be defined. In modern usage, the symbol $\sqrt{x}$ denotes a positive number.

Figure 2.13
Graphs of $f(x) = x^2$
and its inverse.

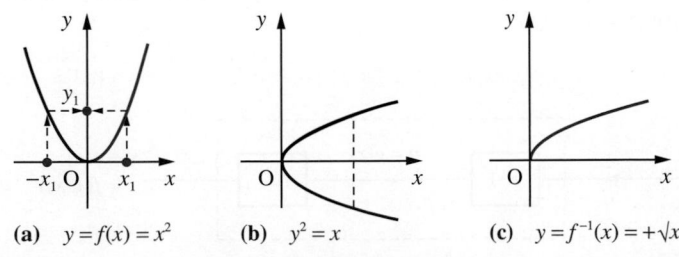

(a) $\quad y = f(x) = x^2$　　　**(b)** $\quad y^2 = x$　　　**(c)** $\quad y = f^{-1}(x) = +\sqrt{x}$

We see from Example 2.8(c) that there is no immediate inverse function corresponding to $f(x) = x^2$. This arises because for the function $f(x) = x^2$ there is a codomain element that is the image of two domain elements x_1 and $-x_1$, as indicated by the broken arrowed lines in Figure 2.13(a). That is, $f(x_1) = f(-x_1) = y_1$. If a function $y = f(x)$ is to have an immediate inverse $f^{-1}(x)$, without any imposed conditions, then *every* element of its range must occur *precisely once* as an image under $f(x)$. Such a function is known as one-to-one (1–1) function.

2.2.4 Composite functions

In many practical problems the mathematical model will involve several different functions. For example, the kinetic energy T of a moving particle is a function of its velocity v, so that

$$T = f(v)$$

Also, the velocity v itself is a function of time t, so that

$$v = g(t)$$

Clearly, by eliminating v, it is possible to express the kinetic energy as a function of time according to

$$T = f(g(t))$$

A function of the form $y = f(g(x))$ is called a **function of a function** or a **composite** of the functions $f(x)$ and $g(x)$. In modern mathematical texts it is common to denote the composite function by $f \circ g$ so that

$$y = f \circ g(x) = f(g(x)) \tag{2.5}$$

We can represent the composite function (2.5) schematically by the block diagram of Figure 2.14, where $u = g(x)$ is called the intermediate variable.

It is important to recognize that the composition of functions is not in general commutative. That is, for two general functions $f(x)$ and $g(x)$

$$f(g(x)) \neq g(f(x))$$

Algebraically, given two functions $y = f(x)$ and $y = g(x)$, the composite function $y = f(g(x))$ may be obtained by replacing x in the expression for $f(x)$ by $g(x)$. Likewise, the composite function $y = g(f(x))$ may be obtained by replacing x in the expression for $g(x)$ by $f(x)$.

Figure 2.14
The composite function $f(g(x))$.

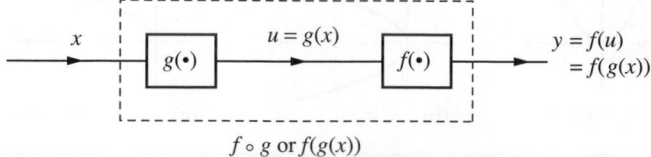

$f \circ g$ or $f(g(x))$

Example 2.9   If $y = f(x) = x^2 + 2x$ and $y = g(x) = x - 1$, obtain the composite functions $f(g(x))$ and $g(f(x))$.

Solution   To obtain $f(g(x))$ replace x in the expression for $f(x)$ by $g(x)$, giving

$$y = f(g(x)) = (g(x))^2 + 2(g(x))$$

But $g(x) = x - 1$, so that

$$y = f(g(x)) = (x - 1)^2 + 2(x - 1)$$
$$= x^2 - 2x + 1 + 2x - 2$$

That is,

$$f(g(x)) = x^2 - 1$$

Similarly,

$$y = g(f(x)) = (f(x)) - 1$$
$$= (x^2 + 2x) - 1$$

That is,

$$g(f(x)) = x^2 + 2x - 1$$

Note that this example confirms the result that in general $f(g(x)) \neq g(f(x))$.

Given a function $y = f(x)$, two composite functions that occur frequently in engineering are

$$y = f(x + k) \quad \text{and} \quad y = f(x - k)$$

where k is a positive constant. As illustrated in Figures 2.15(b) and (c), the graphs of these two composite functions are readily obtained given the graph of $y = f(x)$ as in Figure 2.15(a). The graph of $y = f(x - k)$ is obtained by displacing the graph of $y = f(x)$ by k units to the right, while the graph of $y = f(x + k)$ is obtained by displacing the graph of $y = f(x)$ by k units to the left.

Viewing complicated functions as composites of simpler functions often enables us to 'get to the heart' of a practical problem, and to obtain and understand the solution. For example, recognizing that $y = x^2 + 2x - 3$ is the composite function $y = (x + 1)^2 - 4$, tells us that the function is essentially the squaring function. Its graph is a parabola with minimum point at $x = -1$, $y = -4$ (rather than at $x = 0$, $y = 0$). A similar process of

Figure 2.15
Graphs of $f(x)$,
$f(x - k)$ and $f(x + k)$,
with $k > 0$.

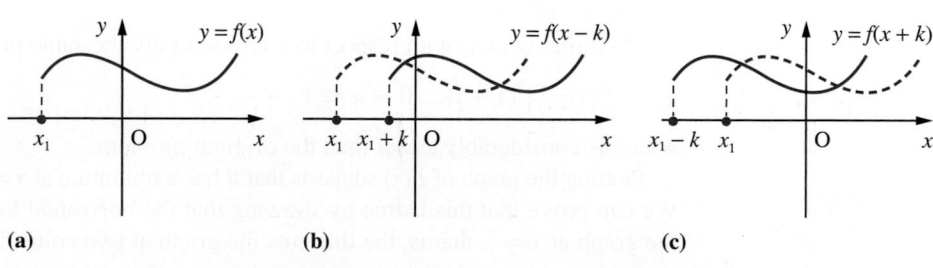

(a)   (b)   (c)

reducing a complicated problem to a simpler one occurred in the solution of the practical problem discussed in Example 1.42 at the end of Chapter 1.

Example 2.10

An open conical container is made from a sector of a circle of radius 10 cm as illustrated in Figure 2.16, with sectional angle θ (radians). The capacity C cm³ of the cone depends on θ. Find the algebraic formula for C in terms of θ and the simplest associated function that could be studied if we wish to maximize C with respect to θ.

Figure 2.16
Conical container
of Example 2.10.

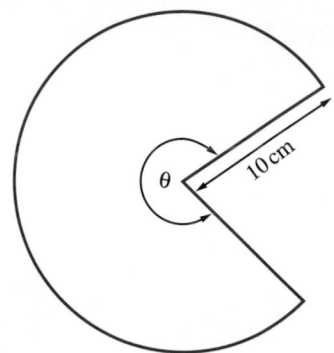 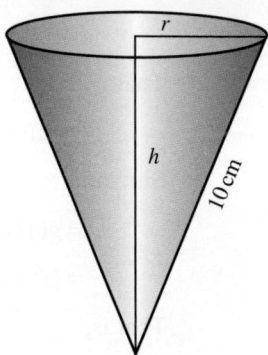

Solution   Let the cone have base radius r cm and height h cm. Then its capacity is given by $C = \frac{1}{3}\pi r^2 h$ with r and h dependent upon the sectorial angle θ (since the perimeter of the sector has to equal the circumference of the base of the cone). Thus, by Pythagoras' theorem,

$$10\theta = 2\pi r \quad \text{and} \quad h^2 = 10^2 - r^2$$

so that

$$C(\theta) = \frac{1}{3}\pi\left(\frac{10\theta}{2\pi}\right)^2\left[10^2 - \left(\frac{10\theta}{2\pi}\right)^2\right]^{1/2}$$

$$= \frac{1000}{3}\pi\left(\frac{\theta}{2\pi}\right)^2\left[1 - \left(\frac{\theta}{2\pi}\right)^2\right]^{1/2}, \quad 0 \leq \theta \leq 2\pi$$

Maximizing $C(\theta)$ with respect to θ is essentially the same problem as maximizing

$$D(x) = x(1-x)^{1/2}, \quad 0 \leq x \leq 1$$

(where $x = (\theta/2\pi)^2$).

Maximizing $D(x)$ with respect to x is essentially the same problem as maximizing

$$E(x) = x^2(1-x), \quad 0 \leq x \leq 1$$

which is considerably easier than the original problem.

Plotting the graph of $E(x)$ suggests that it has a minimum at $x = \frac{2}{3}$ where its value is $\frac{4}{27}$. We can prove that this is true by showing that the horizontal line $y = \frac{4}{27}$ is a tangent to the graph at $x = \frac{2}{3}$; that is, the line cuts the graph at two coincident points at $x = \frac{2}{3}$.

Setting $x^2(1-x) = \frac{4}{27}$ gives $27x^3 - 27x^2 + 4 = 0$ which factorizes into

$$(3x-2)^2(3x+1) = 0$$

Thus the equation has a double root at $x = \frac{2}{3}$ and a single root at $x = -\frac{1}{3}$. Thus $E(x)$ has a maximum at $x = \frac{2}{3}$ and the corresponding optimal value of θ is $2\pi\sqrt{(\frac{2}{3})}$. (In Section 8.5 of Chapter 8 (see also Q5 in Review Exercises 8.13) we shall consider theoretical methods of confirming such results.)

When we compose a function with its inverse function, we usually obtain the identity function $y = x$. Thus from Example 2.6, we have

$$f(x) = \tfrac{1}{5}(4x - 3) \quad \text{and} \quad f^{-1}(x) = \tfrac{1}{4}(5x + 3)$$

and

$$f(f^{-1}(x)) = \tfrac{1}{5}\{4[\tfrac{1}{4}(5x + 3)] - 3\} = x$$

and

$$f^{-1}(f(x)) = \tfrac{1}{4}\{5[\tfrac{1}{5}(4x - 3)] + 3\} = x$$

We need to take care with the exceptional cases that occur, like the square root function, where the inverse function is defined only after restricting the domain of the original function. Thus for $f(x) = x^2$ ($x \geq 0$) and $f^{-1}(x) = \sqrt{x}$ ($x \geq 0$), we obtain

$$f(f^{-1}(x)) = x, \quad \text{for } x \geq 0 \text{ only}$$

and

$$f^{-1}(f(x)) = \begin{cases} x, & \text{for } x \geq 0 \\ -x, & \text{for } x \leq 0 \end{cases}$$

2.2.5 Exercises

7 A function $f(x)$ is defined by $f(x) = \frac{1}{2}(10^x + 10^{-x})$, for x in $\mathbb{R}$. Show that

(a) $2(f(x))^2 = f(2x) + 1$

(b) $2f(x)f(y) = f(x + y) + f(x - y)$

8  Draw separate graphs of the functions f and g where

$$f(x) = (x + 1)^2 \text{ and } g(x) = x - 2$$

The functions F and G are defined by

$$F(x) = f(g(x)) \text{ and } G(x) = g(f(x))$$

Find formulae for $F(x)$ and $G(x)$ and sketch their graphs. What relationships do the graphs of F and G bear to those of f and g?

9 A function f is defined by

$$f(x) = \begin{cases} 0 & (x < -1) \\ x + 1 & (-1 \leq x < 0) \\ 1 - x & (0 \leq x \leq 1) \\ 0 & (x > 1) \end{cases}$$

Sketch on separate diagrams the graphs of $f(x)$, $f(x + \frac{1}{2})$, $f(x + 1)$, $f(x + 2)$, $f(x - \frac{1}{2})$, $f(x - 1)$ and $f(x - 2)$.

10 Find the inverse function (if it is defined) of the following functions:

(a) $f(x) = 2x - 3$ (x in $\mathbb{R}$)

(b) $f(x) = \dfrac{2x - 3}{x + 4}$ (x in $\mathbb{R}$, $x \neq -4$)

(c) $f(x) = x^2 + 1$ (x in $\mathbb{R}$)

If $f(x)$ does not have an inverse function, suggest a suitable restriction of the domain of $f(x)$ that will allow the definition of an inverse function.

11 Show that

$$f(x) = \frac{2x - 3}{x + 4}$$

may be expressed in the form

$$f(x) = g(h(l(x)))$$

where

$$l(x) = x + 4$$

$$h(x) = 1/x$$

$$g(x) = 2 - 11x$$

Interpret this result graphically.

12 The stiffness of a rectangular beam varies directly with the cube of its height and directly with its breadth. A beam of rectangular section is to be cut from a circular log of diameter d. Show that the optimal choice of height and breadth of the beam in terms of its stiffness is related to the value of x which maximizes the function

$$E(x) = x^3(d^2 - x), \quad 0 \leqslant x \leqslant d^2$$

13 A beam is used to support a building as shown in Figure 2.17. The beam has to pass over a 3 m brick wall which is 2 m from the building. Show that the minimum length of the beam is associated with the value of x which minimizes

$$E(x) = (x + 2)^2\left(1 + \frac{9}{x^2}\right)$$

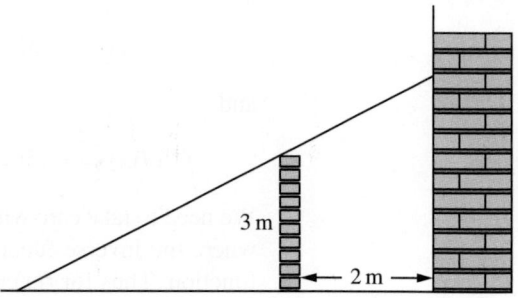

Figure 2.17 Beam of Question 13.

2.2.6 Odd, even and periodic functions

Some commonly occurring functions in engineering contexts have the special properties of oddness or evenness or periodicity. These properties are best understood from the graphs of the functions.

An **even function** is one that satisfies the functional equation

$$f(-x) = f(x)$$

Thus the value of $f(-2)$ is the same as $f(2)$, and so on. The graph of such a function is symmetrical about the y axis, as shown in Figure 2.18.

In contrast, an **odd function** has a graph which is antisymmetrical about the origin, as shown in Figure 2.19, and satisfies the equation

$$f(-x) = -f(x)$$

We notice that $f(0) = 0$ or is undefined.

Polynomial functions like $y = x^4 - x^2 - 1$, involving only even powers of x, are examples of even functions, while those like $y = x - x^5$, involving only odd powers of x, provide examples of odd functions. Of course, not all functions have the property of oddness or evenness.

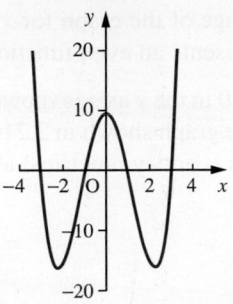

Figure 2.18 Graph of an even function.

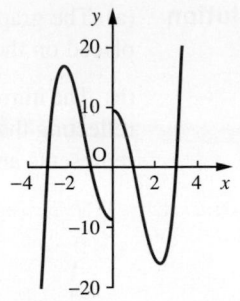

Figure 2.19 Graph of an odd function.

Example 2.11

Which of the functions $y = f(x)$ whose graphs are shown in Figure 2.20 are odd, even or neither odd nor even?

Figure 2.20
Graphs of
Example 2.11.

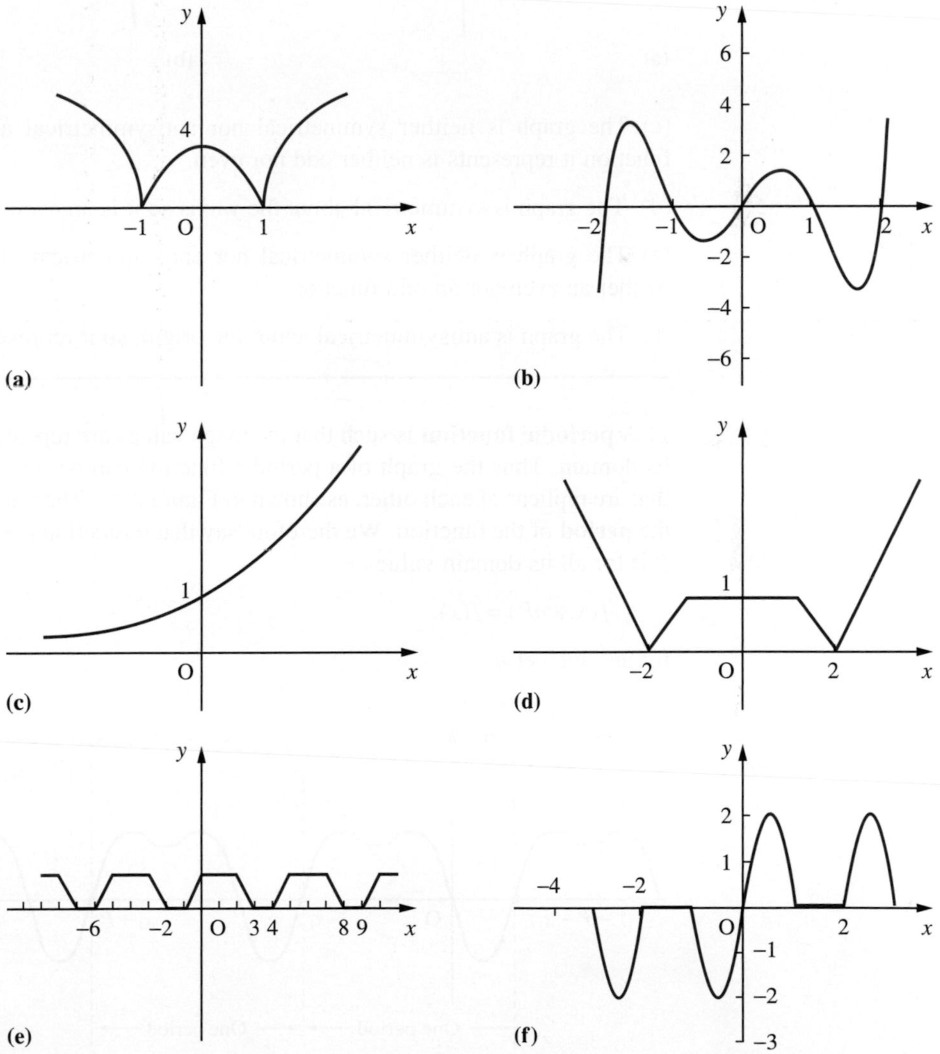

(a)

(b)

(c)

(d)

(e)

(f)

Solution

(a) The graph for $x < 0$ is the mirror image of the graph for $x > 0$ when the mirror is placed on the y axis. Thus the graph represents an even function.

(b) The mirror image of the graph for $x > 0$ in the y axis is shown in Figure 2.21(a). Now reflecting that image in the x axis gives the graph shown in 2.21(b). Thus Figure 2.20(b) represents an odd function since its graph is antisymmetrical about the origin.

Figure 2.21

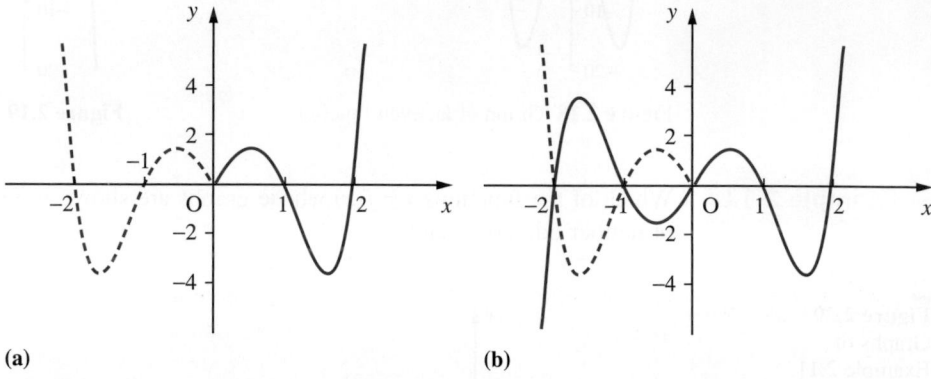

(a)　　　　　　　　　　　　　　**(b)**

(c) The graph is neither symmetrical nor antisymmetrical about the origin, so the function it represents is neither odd nor even.

(d) The graph is symmetrical about the y axis so it is an even function.

(e) The graph is neither symmetrical nor antisymmetrical about the origin, so it is neither an even nor an odd function.

(f) The graph is antisymmetrical about the origin, so it represents an odd function.

A **periodic function** is such that its image values are repeated at regular intervals in its domain. Thus the graph of a periodic function can be divided into 'vertical strips' that are replicas of each other, as shown in Figure 2.22. The width of each strip is called the **period** of the function. We therefore say that a function $f(x)$ is periodic with period P if for all its domain values x

$$f(x + nP) = f(x)$$

for any integer n.

Figure 2.22
A periodic function
of period P.

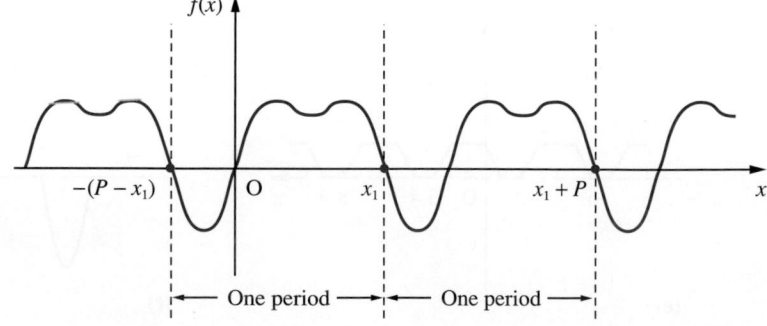

To provide a measure of the number of repetitions per unit of x, we define the **frequency** of a periodic function to be the reciprocal of its period, so that

$$frequency = \frac{1}{period}$$

The Greek letter v ('nu') is usually used to denote the frequency so that $v = 1/P$. The term **circular frequency** is also used in some engineering contexts. This is denoted by the Greek letter ω ('omega') and is defined by

$$\omega = 2\pi v = \frac{2\pi}{P}$$

It is measured in radians per unit of x. When the meaning is clear from the context the adjective 'circular' is commonly omitted.

Example 2.12   A function $f(x)$ has the graph on [0, 1] shown in Figure 2.23. Sketch its graph on [−3, 3] given that

(a) $f(x)$ is periodic with period 1

(b) $f(x)$ is periodic with period 2 and is even

(c) $f(x)$ is periodic with period 2 and is odd

Figure 2.23
$f(x)$ of Example 2.12
defined on [0, 1].

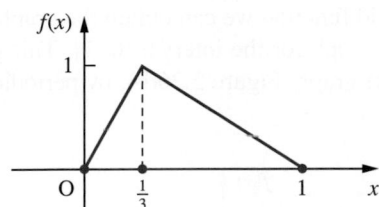

Solution   (a) Since $f(x)$ has period 1, strips of width 1 unit are simply replicas of the graph between 0 and 1. Hence we obtain the graph shown in Figure 2.24.

Figure 2.24
$f(x)$ having period 1.

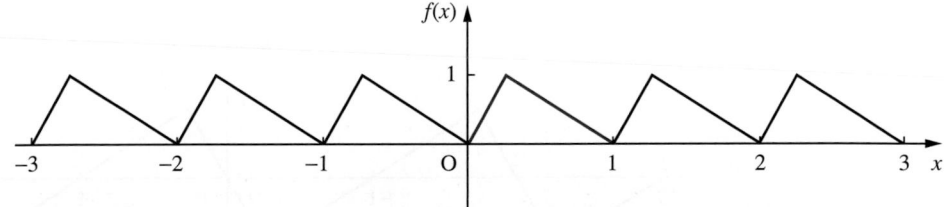

(b) Since $f(x)$ has period 2 we need to establish the graph over a complete period before we can replicate it along the domain of $f(x)$. Since it is an even function and we

Figure 2.25
$f(x)$ periodic with
period 2 and is even.

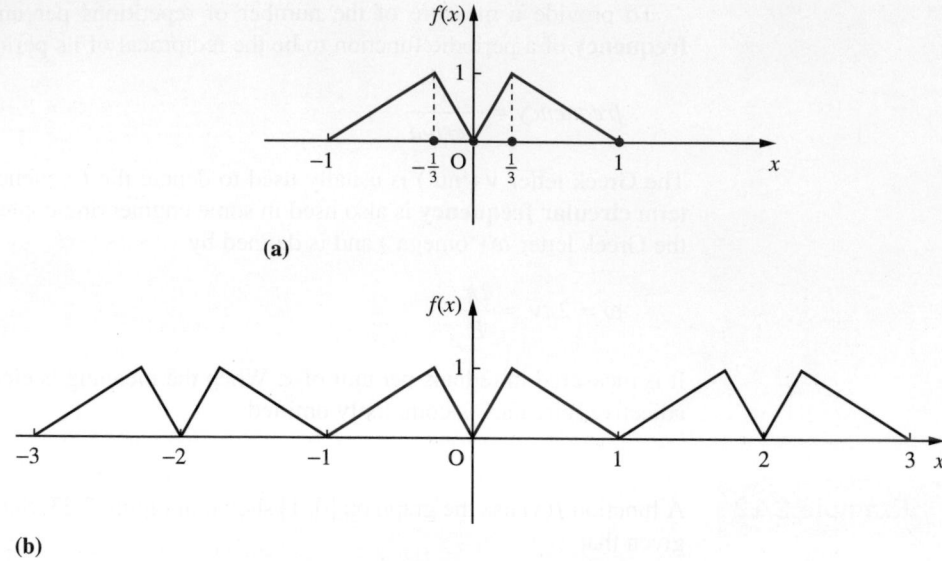

(a)

(b)

know its values between 0 and 1, we also know its values between −1 and 0. We can obtain the graph of $f(x)$ between −1 and 0 by reflecting in the y axis, as shown in Figure 2.25(a). Thus we have the graph over a complete period, from −1 to +1, and so we can replicate along the x axis, as shown in Figure 2.25(b).

(c) Similarly, if $f(x)$ is an odd function we can obtain the graph for the interval [−1, 0] using antisymmetry and the graph for the interval [0, 1]. This gives us Figure 2.26(a) and we then obtain the whole graph, Figure 2.26(b), by periodic extension.

Figure 2.26
$f(x)$ periodic with
period 2 and is odd.

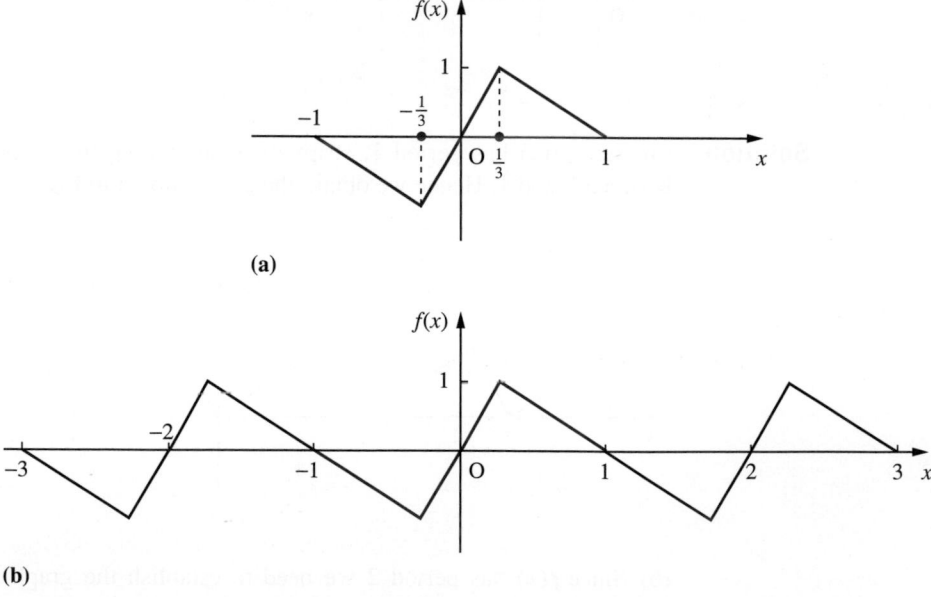

(a)

(b)

2.2.7 Exercises

14 Which of the functions $y = f(x)$ whose graphs are shown in Figure 2.27 are odd, even or neither odd nor even?

15 Three different functions, $f(x)$, $g(x)$ and $h(x)$, have the same graph on [0, 2] as shown in Figure 2.28. On separate diagrams, sketch their graphs for [−4, 4] given that

(a) $f(x)$ is periodic with period 2

(b) $g(x)$ is periodic with period 4 and is even

(c) $h(x)$ is periodic with period 4 and is odd.

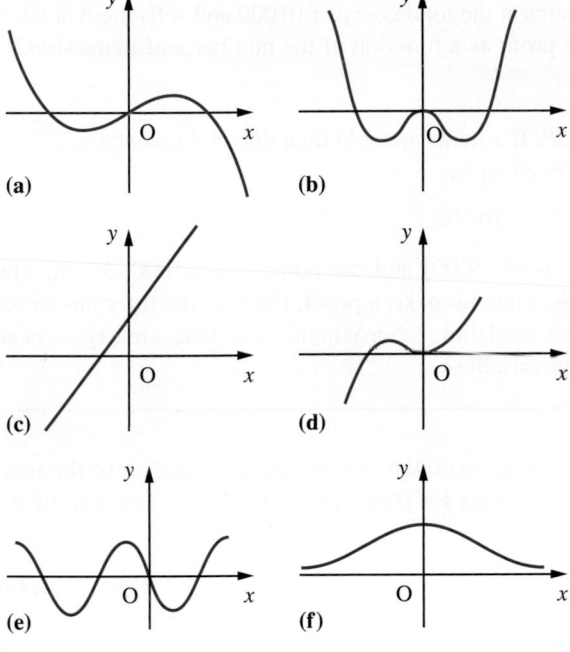

(a)

(b)

(c)

(d)

(e)

(f)

Figure 2.27 Graphs of Question 14.

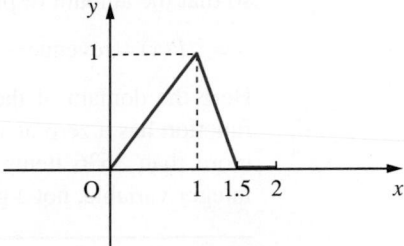

Figure 2.28 Graph of Question 15.

2.3 # Linear and quadratic functions

Among the more commonly used functions in engineering contexts are the linear and quadratic functions. This is because the mathematical models of practical problems often involve linear functions and also because more complicated functions are often well approximated locally by linear or quadratic functions. We will review the properties of these functions and in the process describe some of the contexts in which they occur.

2.3.1 Linear functions

The **linear function** is the simplest function that occurs in practical problems. It has the formula $f(x) = mx + c$ where m and c are constant numbers and x is the unassigned or independent variable as usual. The graph of $f(x)$ is the set of points (x, y) where $y = mx + c$, which is the equation of a straight line on a cartesian coordinate plot (see Section 1.4.2). Hence, the function is called the linear function. An example of a linear function is the conversion of a temperature $T_1\,°C$ to the temperature $T_2\,°F$. Here

$$T_2 = \tfrac{9}{5}T_1 + 32$$

and $m = \tfrac{9}{5}$ with $c = 32$.

To determine the formula for a particular linear function the two constants m and c have to be found. This implies that we need two pieces of information to determine $f(x)$.

Example 2.13

A manufacturer produces 5000 items at the total cost of £10 000 and sells them at £2.75 each. What is the manufacturer's profit as a function of the number x of items sold?

Solution

Let the manufacturer's profit be £P. If x items are sold then the total revenue is £2.75x, so that the amount of profit $P(x)$ is given by

$$P(x) = \text{revenue} - \text{cost} = 2.75x - 10\,000$$

Here the domain of the function is $[0, 5000]$ and the range is $[-10\,000, 3750]$. This function has a zero at $x = 3636\tfrac{4}{11}$. Thus to make a profit, the manufacturer has to sell more than 3636 items. (Note the modelling approximation in that, strictly, x is an integer variable, not a general real variable.)

If we know the values that the function $f(x)$ takes at two values, x_0 and x_1, of the independent variable x we can find the formula for $f(x)$. Let $f(x_0) = f_0$ and $f(x_1) = f_1$, then

$$f(x) = \frac{x - x_1}{x_0 - x_1} f_0 + \frac{x - x_0}{x_1 - x_0} f_1 \qquad \qquad \textbf{(2.6)}$$

This formula is known as **Lagrange's formula**. It is obvious that the function is linear since we can arrange it as

$$f(x) = x\left[\frac{f_1 - f_0}{x_1 - x_0}\right] + \left[\frac{x_1 f_0 - x_0 f_1}{x_1 - x_0}\right]$$

The reader should verify from (2.6) that $f(x_0) = f_0$ and $f(x_1) = f_1$.

Example 2.14

Use Lagrange's formula to find the linear function $f(x)$ where $f(10) = 1241$ and $f(15) = 1556$.

Solution

Taking $x_0 = 10$ and $x_1 = 15$ so that $f_0 = 1241$ and $f_1 = 1556$ we obtain

$$f(x) = \frac{x - 15}{10 - 15}(1241) + \frac{x - 10}{15 - 10}(1556)$$

$$= \frac{x}{5}(1556 - 1241) + 3(1241) - 2(1556)$$

$$= \frac{x}{5}(315) + (3723 - 3112) = 63x + 611$$

The **rate of change** of a function, between two values $x = x_0$ and $x = x_1$ in its domain, is defined by the ratio of the change in the values of the function to the change in the values of x. Thus

$$\text{rate of change} = \frac{\text{change in values of } f(x)}{\text{change in values of } x} = \frac{f(x_1) - f(x_0)}{x_1 - x_0}$$

For a linear function with formula $f(x) = mx + c$ we have

$$\text{rate of change} = \frac{(mx_1 + c) - (mx_0 + c)}{x_1 - x_0}$$

$$= \frac{m(x_1 - x_0)}{x_1 - x_0} = m$$

which is a constant. If we know the rate of change m of a linear function $f(x)$ and the value f_0 at a point $x = x_0$, then we can write the formula for $f(x)$ as

$$f(x) = mx + f_0 - mx_0$$

For a linear function, the slope of the graph is the rate of change of the function.

Example 2.15

The labour cost of producing a certain item is £21 per 10 000 items and the raw materials cost is £4 for 1000 items. Each time a new production run is begun, there is a set-up cost of £8. What is the cost, £$C(x)$, of a production run of x items?

Solution

Here the cost function has a rate of change comprising the labour cost per item (21/10 000) and the materials cost per item (4/1000). Thus the rate of change is 0.0061. We also know that if there is a production run with zero items, there is still a set up cost of £8 so $f(0) = 8$. Thus the required function is

$$C(x) = 0.0061x + 8$$

2.3.2 Least squares fit of a linear function to experimental data

Because the linear function occurs in many mathematical models of practical problems, we often have to 'fit' linear functions to experimental data. That is, we have to find the values of m and c which yield the best overall description of the data. There are two distinct mathematical models that occur. These are given by the functions with formulae

(a) $y = ax$ and   (b) $y = mx + c$

For example, the extension of an ideal spring under load may be represented by a function of type (a), while the velocity of a projectile launched vertically may be represented by a function of type (b).

From experiments we obtain a set of data points (x_k, y_k), $k = 1, 2, \dots, n$. We wish to find the value of the constant(s) of the linear function that best describes the phenomenon the data represents.

Case (a): the theoretical model has the form $y = ax$

The difference between theoretical value ax_k and the experimental value y_k at x_k is $(ax_k - y_k)$. This is the 'error' of the model at $x = x_k$. We define the value of a for which $y = ax$ best represents the data to be that value which minimizes the sum S of the squared errors:

$$S = \sum_{k=1}^{n} (ax_k - y_k)^2$$

(Hence the name 'least squares fit': the squares of the errors are chosen to avoid simple cancellation of two large errors of opposite sign.)

It is easy to find the minimizing value of a since S is essentially a quadratic expression in a. (All the x_k's and y_k's are numbers.) Rewriting, we have

$$S = \sum_{k=1}^{n} (a^2 x_k^2 - 2ax_k y_k + y_k^2)$$

$$= \sum_{k=1}^{n} (a^2 x_k^2) + \sum_{k=1}^{n} (-2ax_k y_k) + \sum_{k=1}^{n} y_k^2$$

$$= a^2 \sum_{k=1}^{n} x_k^2 - 2a \sum_{k=1}^{n} x_k y_k + \sum_{k=1}^{n} y_k^2$$

(Notice the 'taking out' of the common factors a^2 and $-2a$ in these sums.) Writing

$$P = \sum_{k=1}^{n} x_k^2, \quad Q = \sum_{k=1}^{n} x_k y_k \quad \text{and} \quad R = \sum_{k=1}^{n} y_k^2$$

we have

$$S = Pa^2 - 2aQ + R$$

On 'completing the square'

$$S = P\left(a - \frac{Q}{P}\right)^2 + \frac{RP - Q^2}{P}$$

and we see that the minimizing value of a is given by Q/P, when the first term is zero. Thus S is minimized when

$$a = \frac{\displaystyle\sum_{k=1}^{n} x_k y_k}{\displaystyle\sum_{k=1}^{n} x_k^2} \tag{2.7}$$

Example 2.16

Find the value of a which provides the least squares fit to the model $y = ax$ for the data given in Figure 2.29.

Figure 2.29
Data of Example 2.16.

k	1	2	3	4	5	6
x_k	50	100	150	200	250	300
y_k	5	8	9	11	12	15

Solution   From (2.7) the least squares fit is provided by

$$a = \left(\sum_{k=1}^{6} x_k y_k \right) \bigg/ \left(\sum_{k=1}^{6} x_k^2 \right)$$

Here

$$\sum_{k=1}^{6} x_k y_k = 250 + 800 + 1350 + 2200 + 3000 + 4500 = 12\,100$$

and

$$\sum_{k=1}^{6} x_k^2 = 50^2 + 100^2 + 150^2 + 200^2 + 250^2 + 300^2 = 227\,500$$

so that $a = 121/2275 = 0.053$.

Case (b): the theoretical model has the form $y = mx + c$

Analagous to case (a), this can be seen as minimizing the sum

$$S = \sum_{k=1}^{n} (mx_k + c - y_k)^2$$

The algebraic approach to this minimization uses completion of squares in two variables. The details are complicated but are given below. Working through the details provides useful practice and consolidation of the use of the sigma notation.

Multiplying out the terms gives

$$S = m^2 \sum_{k=1}^{n} x_k^2 - 2m \sum_{k=1}^{n} x_k y_k + 2mc \sum_{k=1}^{n} x_k - 2c \sum_{k=1}^{n} y_k + nc^2 + \sum_{k=1}^{n} y_k^2$$

Now $\sum_{k=1}^{n} x_k = n\bar{x}$ and $\sum_{k=1}^{n} y_k = n\bar{y}$, where $\bar{x}$ and $\bar{y}$ are the **mean values** of the x_k's and y_k's respectively, so S can be written

$$S = m^2 \sum_{k=1}^{n} x_k^2 - 2m \sum_{k=1}^{n} x_k y_k + 2mcn\bar{x} - 2cn\bar{y} + nc^2 + \sum_{k=1}^{n} y_k^2$$

Completing the square with terms involving n gives

$$S = n(c - \bar{y} + m\bar{x})^2 + m^2 \left\{ \sum_{k=1}^{n} x_k^2 - n\bar{x}^2 \right\} - 2m \left\{ \sum_{k=1}^{n} x_k y_k - n\bar{x}\bar{y} \right\} + \sum_{k=1}^{n} y_k^2 - n\bar{y}^2$$

Now completing the square with the remaining terms involving m we have

$$S = n(c - \bar{y} + m\bar{x})^2 + p(m - q/p)^2 + r - q^2/p$$

where

$$p = \sum_{k=1}^{n} x_k^2 - n\bar{x}^2 \quad \text{and} \quad q = \sum_{k=1}^{n} x_k y_k - n\bar{x}\bar{y} \quad \text{and} \quad r = \sum_{k=1}^{n} y_k^2 - n\bar{y}^2$$

Thus S is minimized where

$$m = \frac{\displaystyle\sum_{k=1}^{n} x_k y_k - n\bar{x}\bar{y}}{\displaystyle\sum_{k=1}^{n} x_k^2 - n\bar{x}^2} \quad \text{and} \quad c = \bar{y} - m\bar{x} \tag{2.8}$$

To avoid loss of significance, the formula for m is usually expressed in the form

$$m = \frac{\displaystyle\sum_{k=1}^{n} (x_k - \bar{x})(y_k - \bar{y})}{\displaystyle\sum_{k=1}^{n} (x_k - \bar{x})^2} \tag{2.9}$$

We can observe that in this case the best straight line passes through the average data point $(\bar{x}, \bar{y})$, and the best straight line has the formula

$$y = mx + c$$

with $c = \bar{y} - m\bar{x}$.

Example 2.17 Find the values of m and c which provide the least squares fit to the linear model $y = mx + c$ for the data given in Figure 2.30.

Figure 2.30
Data of Example 2.17.

k	1	2	3	4	5
x_k	0	1	2	3	4
y_k	1	1	2	2	3

Solution From (2.9) the least squares fit is provided by

$$m = \frac{\displaystyle\sum_{k=1}^{n} (x_k - \bar{x})(y_k - \bar{y})}{\displaystyle\sum_{k=1}^{n} (x_k - \bar{x})^2}$$

Here $\bar{x} = \frac{1}{5}(10) = 2.0$, $\bar{y} = \frac{1}{5}(9) = 1.8$, $\sum_{k=1}^{n}(x_k - \bar{x})(y_k - \bar{y}) = 5.0$ and $\sum_{k=1}^{n}(x_k - \bar{x})^2 = 10$, so that

$$m = 0.5$$

and hence $c = 1.8 - 0.5(2) = 0.8$.

Thus the best straight line fit to the data is provided by $y = 0.5x + 0.8$.

  See page 109 for MATLAB commands to reproduce the answer.

The formula for case (b) is the one most commonly given on calculators and in computer packages (where it is called **linear regression**). It is important to have a theoretical justification to fitting data to a function, otherwise it is easy to produce nonsense. For example, the data in Example 2.16 actually related to the extension of a soft spring under a load so that it would be inappropriate to fit that data to $y = mx + c$. A non-zero value for c would imply an extension with zero load! A little care is needed when using computer packages. Some use the form $y = ax + b$ and others the form $y = a + bx$ as the basic formula.

2.3.3 Exercises

16 Obtain the formula for the linear functions $f(x)$ such that

(a) $f(0) = 3$ and $f(2) = -1$

(b) $f(-1) = 2$ and $f(3) = 4$

(c) $f(1.231) = 2.791$ and $f(2.492) = 3.112$

17 Calculate the rate of change of the linear functions given by

(a) $f(x) = 3x - 2$

(b) $f(x) = 2 - 3x$

(c) $f(-1) = 2$ and $f(3) = 4$

18 The total labour cost of producing a certain item is £43 per 100 items produced. The raw materials cost £25 per 1000 items. There is a set-up cost of £50 for each production run. Obtain the formula for the cost of a production run of x items.

The manufacturer decides to have a production run of 2000 items. What is its cost? If the items are sold at £1.20 each, write down a formula for the manufacturer's profit if x items are sold. What is the breakeven number of items sold?

19   Find the least squares fit to the linear function $y = ax$ of the data given in Figure 2.31.

k	1	2	3	4	5
x_k	10.1	10.2	10.3	10.4	10.5
y_k	3.10	3.12	3.21	3.25	3.32

Figure 2.31 Table of Question 19.

20 Find the least squares fit to the linear function $y = mx + c$ for the experimental data given in Figure 2.32.

k	1	2	3	4	5
x_k	55	60	65	70	75
y_k	107	109	114	118	123

Figure 2.32 Table of Question 20.

21 On the graph of the line $y = x$, draw the lines $y = 0$, $x = a$ and $x = b$. Show that the area enclosed by these four lines is $\frac{1}{2}(b^2 - a^2)$ (assume $b > a$).

Deduce that this area is the average value of $y = x$ on the interval $[a, b]$ multiplied by the size of that interval.

22 The velocity of an object falling under gravity is $v(t) = gt$ where t is the lapsed time from its release from rest and g is the acceleration due to gravity. Draw a graph of $v(t)$ to show that its average velocity over that time period is $\frac{1}{2}gt$ and deduce that the distance travelled is $\frac{1}{2}gt^2$.

2.3.4 The quadratic function

The general quadratic function has the form

$$f(x) = ax^2 + bx + c$$

where a, b and c are constants and $a \neq 0$. By 'completing the square' we can show that

$$f(x) = a\left[\left(x + \frac{b}{2a}\right)^2 + \frac{4ac - b^2}{4a^2}\right] \qquad \text{(2.10)}$$

which implies that the graph of $f(x)$ is either a 'cup' ($a > 0$) or a 'cap' ($a < 0$), as shown in Figure 2.33, and is a parabola.

We can see that, because the quadratic function has three constants, to determine a specific quadratic function requires three data points. The formula for the quadratic function $f(x)$ taking the values f_0, f_1, f_2 at the values x_0, x_1, x_2, of the independent variable x, may be written in Lagrange's form:

$$f(x) = \frac{(x - x_1)(x - x_2)}{(x_0 - x_1)(x_0 - x_2)}f_0 + \frac{(x - x_0)(x - x_2)}{(x_1 - x_0)(x_1 - x_2)}f_1 + \frac{(x - x_0)(x - x_1)}{(x_2 - x_0)(x_2 - x_1)}f_2$$

$$\text{(2.11)}$$

The right-hand side of this formula is clearly a quadratic function. The reader should spend a few minutes verifying that inserting the values $x = x_0$, x_1 and x_2 yields $f(x_0) = f_0$, $f(x_1) = f_1$ and $f(x_2) = f_2$.

Figure 2.33
(a) $a > 0$; (b) $a < 0$.

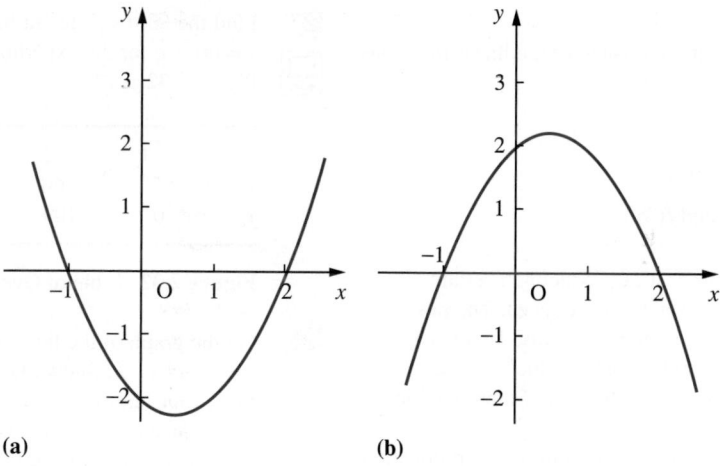

(a)　　　　　(b)

Example 2.18　Find the formula of the quadratic function which satisfies the data points $(1, 2)$, $(2, 4)$ and $(3, 8)$.

Solution　Choose $x_0 = 1$, $x_1 = 2$ and $x_2 = 3$ so that $f_0 = 2$, $f_1 = 4$ and $f_2 = 8$. Then using Lagrange's formula (2.10) we have

$$f(x) = \frac{(x-2)(x-3)}{(1-2)(1-3)}(2) + \frac{(x-1)(x-3)}{(2-1)(2-3)}(4) + \frac{(x-1)(x-2)}{(3-1)(3-2)}(8)$$

$$= (x-2)(x-3) - 4(x-1)(x-3) + 4(x-1)(x-2) = x^2 - x + 2$$

Lagrange's formula is not always the best way to obtain the formula of a quadratic function. Sometimes we wish to obtain the formula as an expansion about a specific point, as illustrated in Example 2.19.

Example 2.19 Find the quadratic function in the form

$$f(x) = A(x-2)^2 + B(x-2) + C$$

which satisfies $f(1) = 2, f(2) = 4, f(3) = 8$.

Solution Setting $x = 1$, 2 and 3 into the formula for $f(x)$ we obtain

$$f(1): A - B + C = 2$$

$$f(2): \qquad\quad C = 4$$

$$f(3): A + B + C = 8$$

from which we quickly find $A = 1, B = 3$ and $C = 4$. Thus

$$f(x) = (x-2)^2 + 3(x-2) + 4$$

The way we express the quadratic function depends on the problem context. The form $f(x) = ax^2 + bx + c$ is convenient for values of x near $x = 0$, while the form $f(x) = A(x - x_0)^2 + B(x - x_0) + C$ is convenient for values of x near $x = x_0$. (The second form here is sometimes called the **Taylor expansion** of $f(x)$ about $x = x_0$.) This is discussed for the general function in Section 9.4, where we make use of the differential calculus to obtain the expansion.

Since we can write $f(x)$ in the form (2.10), we see that when $b^2 > 4ac$ we can factorize $f(x)$ into the product of two linear factors and $f(x)$ has two zeros given as in (1.5) by

$$x = \frac{-b \pm \sqrt{(b^2 - 4ac)}}{2a}$$

When $b^2 < 4ac$, $f(x)$ cannot be factorized and does not have a zero. In this case it is called an **irreducible quadratic function**.

Example 2.20 Complete the squares of the following quadratics and specify which are irreducible.

(a) $y = x^2 + x + 1$ (b) $y = 3x^2 - 2x - 1$

(c) $y = 4 + 3x - x^2$ (d) $y = 2x - 1 - 2x^2$

Solution (a) In this case, $a = b = c = 1$ so that $b^2 - 4ac = -3 < 0$ and we deduce that the quadratic is irreducible. Alternatively, using the method of completing the square we have

$$y = x^2 + x + 1 = (x + \tfrac{1}{2})^2 + \tfrac{3}{4} = (x + \tfrac{1}{2})^2 + (\tfrac{\sqrt{3}}{2})^2$$

Since this is a sum of squares, like $A^2 + B^2$, it cannot, unlike a difference of squares, $A^2 - B^2 = (A - B)(A + B)$, be factorized. Thus this is an irreducible quadratic function.

(b) Here $a = 3$, $b = -2$ and $c = -1$ so that $b^2 - 4ac = 16 > 0$ and we deduce that this is not an irreducible quadratic. Alternatively, completing the square we have:

$$y = 3x^2 - 2x - 1 = 3(x^2 - \tfrac{2}{3}x - \tfrac{1}{3})$$
$$= 3[(x - \tfrac{1}{3})^2 - \tfrac{4}{9}] = 3[(x - \tfrac{1}{3}) - \tfrac{2}{3}][(x - \tfrac{1}{3}) + \tfrac{2}{3}]$$
$$= 3[x - 1][x + \tfrac{1}{3}] = (x - 1)(3x + 1)$$

Thus this is not an irreducible quadratic function.

(c) Here $a = -1$, $b = 3$ and $c = 4$ so that $b^2 - 4ac = 25 > 0$ and we deduce that the quadratic is irreducible. Alternatively, completing the square we have:

$$y = 4 + 3x - x^2 = 4 + \tfrac{9}{4} - (x - \tfrac{3}{2})^2$$
$$= \tfrac{25}{4} - (x - \tfrac{3}{2})^2 = [\tfrac{5}{2} - (x - \tfrac{3}{2})][\tfrac{5}{2} + (x - \tfrac{3}{2})]$$
$$= (4 - x)(1 + x)$$

Thus y is a product of two linear factors and $4 + 3x - x^2$ is not an irreducible quadratic function.

(d) Here $a = -2$, $b = 2$ and $c = -1$ so that $b^2 - 4ac = -4 < 0$ and we deduce that the quadratic is irreducible. Alternatively we may complete the square

$$y = 2x - 1 - 2x^2 = -1 - 2(x^2 - x)$$
$$= -1 + \tfrac{1}{2} - 2(x - \tfrac{1}{2})^2 = -\tfrac{1}{2} - 2(x - \tfrac{1}{2})^2$$
$$= -2[\tfrac{1}{4} + (x - \tfrac{1}{2})^2]$$

Since the term inside the square brackets is the sum of squares, we have an irreducible quadratic function.

The quadratic function

$$f(x) = ax^2 + bx + c$$

has a maximum when $a < 0$ and a minimum when $a > 0$, as illustrated earlier in Figure 2.33. The position and value of that extremal point (that is, of the maximum or the minimum) can be obtained from the completed square form (2.10) of $f(x)$. These occur where

$$x + \frac{b}{2a} = 0$$

Thus, when $a > 0$, $f(x)$ has a minimum value $(4ac - b^2)/(4a)$ where $x = -b/(2a)$. When $a < 0$, $f(x)$ has a maximum value $(4ac - b^2)/(4a)$ at $x = -b/(2a)$.

This result is important in engineering contexts when we are trying to optimize costs or profits or to produce an optimal design. (See Section 2.10.)

Example 2.21   Find the extremal values of the functions

(a) $y = x^2 + x + 1$ (b) $y = 3x^2 - 2x - 1$

(c) $y = 4 + 3x - x^2$ (d) $y = 2x - 1 - 2x^2$

Solution   This uses the completed squares of Example 2.20.

(a) $y = x^2 + x + 1 = (x + \frac{1}{2})^2 + \frac{3}{4}$

Clearly the smallest value y can take is $\frac{3}{4}$ and this occurs when $x + \frac{1}{2} = 0$; that is, when $x = -\frac{1}{2}$.

(b) $y = 3x^2 - 2x - 1 = 3(x - \frac{1}{3})^2 - \frac{4}{3}$

Clearly the smallest value of y occurs when $x = \frac{1}{3}$ and is equal to $-\frac{4}{3}$.

(c) $y = 4 + 3x - x^2 = \frac{25}{4} - (x - \frac{3}{2})^2$

Clearly the largest value y can take is $\frac{25}{4}$ and this occurs when $x = \frac{3}{2}$.

(d) $y = 2x - 1 - 2x^2 = -\frac{1}{2} - 2(x - \frac{1}{2})^2$

Thus the maximum value of y equals $-\frac{1}{2}$ and occurs where $x = \frac{1}{2}$.

Confirm that these results conform with the theory above.

2.3.5 Exercises

23 Find the formulae of the quadratic functions $f(x)$ such that

(a) $f(1) = 3$, $f(2) = 7$ and $f(4) = 19$

(b) $f(-1) = 1$, $f(1) = -1$ and $f(4) = 2$

24 Find the numbers A, B and C such that

$$f(x) = x^2 - 8x + 10$$

$$= A(x - 2)^2 + B(x - 2) + C$$

25 Determine which of the following quadratic functions are irreducible.

(a) $f(x) = x^2 + 2x + 3$ (b) $f(x) = 4x^2 - 12x + 9$

(c) $f(x) = 6 - 4x - 3x^2$ (d) $f(x) = 3x - 1 - 5x^2$

26 Find the maximum or minimum values of the quadratic functions given in Question 25.

27 For what values of x are the values of the quadratic functions below greater than zero?

(a) $f(x) = x^2 - 6x + 8$ (b) $f(x) = 15 + x - 2x^2$

28 A car travelling at u mph has to make an emergency stop. There is an initial reaction time T_1 before the driver applies a constant braking deceleration of a mph². After a further time T_2 the car comes to rest. Show that $T_2 = u/a$ and that the average speed during the braking period is $u/2$. Hence show that the total stopping distance D may be expressed in the form

$$D = Au + Bu^2$$

where A and B depend on T_1 and a.

The stopping distances for a car travelling at 20 mph and 40 mph are 40 feet and 120 feet respectively. Estimate the stopping distance for a car travelling at 70 mph.

A driver sees a hazard 150 feet ahead. What is the maximum possible speed of the car at that moment if a collision is to be avoided?

| 2.4 | Polynomial functions |

A **polynomial function** has the general form

$$f(x) = a_n x^n + a_{n-1} x^{n-1} + \ldots + a_1 x + a_0, \quad x \text{ in } \mathbb{R} \tag{2.12}$$

where n is a positive integer and a_r is a real number called the coefficient of x^r, $r = 0, 1, \ldots, n$. The index n of the highest power of x occurring is called the **degree of the polynomial**. For $n = 1$ we obtain the linear function

$$f(x) = a_1 x + a_0$$

and for $n = 2$ the quadratic function

$$f(x) = a_2 x^2 + a_1 x + a_0$$

and so on.

We obtained in Sections 2.3.1 and 2.3.4 Lagrange's formulae for linear and for quadratic functions. The basic idea of the formulae can be used to obtain a formula for a polynomial of degree n which is such that $f(x_0) = f_0$, $f(x_1) = f_1$, $f(x_2) = f_2$, $\ldots$, $f(x_n) = f_n$. Notice we need $(n + 1)$ values to determine a polynomial of degree n. We can write Lagrange's formula in the form.

$$f(x) = L_0(x)f_0 + L_1(x)f_1 + L_2(x)f_2 + \ldots + L_n(x)f_n$$

where $L_0(x), L_1(x), \ldots, L_n(x)$ are polynomials of degree n such that

$$L_k(x_j) = 0, \quad x_j \neq x_k \text{ (or } j \neq k)$$

$$L_k(x_k) = 1$$

This implies that L_k has the form

$$L_k(x) = \frac{(x - x_0)(x - x_1)(x - x_2) \ldots (x - x_{k-1})(x - x_{k+1}) \ldots (x - x_n)}{(x_k - x_0)(x_k - x_1)(x_k - x_2) \ldots (x_k - x_{k-1})(x_k - x_{k+1}) \ldots (x_k - x_n)}$$

(It is easy to verify that L_k has degree n and that $L_k(x_j) = 0$, $j \neq k$ and $L_k(x_k) = 1$.)

Example 2.22   Find the cubic function such that $f(-3) = 528$, $f(0) = 1017$, $f(2) = 1433$ and $f(5) = 2312$.

Solution   Notice that we need four data points to determine a cubic function. We can write

$$f(x) = L_0(x)f_0 + L_1(x)f_1 + L_2(x)f_2 + L_3(x)f_3$$

where $x_0 = -3$, $f_0 = 528$, $x_1 = 0$, $f_1 = 1017$, $x_2 = 2$, $f_2 = 1433$, $x_3 = 5$ and $f_3 = 2312$. Thus

$$L_0(x) = \frac{(x - 0)(x - 2)(x - 5)}{(-3 - 0)(-3 - 2)(-3 - 5)} = -\tfrac{1}{120}(x^3 - 7x^2 + 10x)$$

$$L_1(x) = \frac{(x + 3)(x - 2)(x - 5)}{(0 + 3)(0 - 2)(0 - 5)} = \tfrac{1}{30}(x^3 - 4x^2 - 11x + 30)$$

$$L_2(x) = \frac{(x+3)(x-0)(x-5)}{(2+3)(2-0)(2-5)} = -\tfrac{1}{30}(x^3 - 2x^2 - 15x)$$

$$L_3(x) = \frac{(x+3)(x-0)(x-2)}{(5+3)(5-0)(5-2)} = \tfrac{1}{120}(x^3 + x^2 - 6x)$$

Notice that each of the L_k's is a cubic function, so that their sum will be a cubic function

$$f(x) = -\tfrac{1}{120}(x^3 - 7x^2 + 10x)(528) + \tfrac{1}{30}(x^3 - 4x^2 - 11x + 30)(1017)$$
$$- \tfrac{1}{30}(x^3 - 2x^2 - 15x)(1433) + \tfrac{1}{120}(x^3 + x^2 - 6x)(2312)$$
$$= x^3 + 10x^2 + 184x + 1017$$

2.4.1   Basic properties

Polynomials have two important mathematical properties.

Property (i)

If two polynomials are equal for all values of the independent variable then corresponding coefficients of the powers of the variable are equal. Thus if

$$f(x) = a_n x^n + a_{n-1} x^{n-1} + \ldots + a_1 x + a_0$$

$$g(x) = b_n x^n + b_{n-1} x^{n-1} + \ldots + b_1 x + b_0$$

and

$$f(x) = g(x) \quad \text{for all } x$$

then

$$a_i = b_i \quad \text{for } i = 0, 1, 2, \ldots, n$$

This property forms the basis of a technique called **equating coefficients**, which will be used in determining partial fractions in Section 2.5.

Property (ii)

Any polynomial with real coefficients can be expressed as a product of linear and irreducible quadratic factors.

Example 2.23

Find the values of A, B and C that ensure that

$$x^2 + 1 = A(x - 1) + B(x + 2) + C(x^2 + 2)$$

for all values of x.

Solution   Multiplying out the right-hand side, we have

$$x^2 + 0x + 1 = Cx^2 + (A + B)x + (-A + 2B + 2C)$$

Using Property (i), we compare, or equate, the coefficients of x^2, x and x^0 in turn to give

$$C = 1$$
$$A + B = 0$$
$$-A + 2B + 2C = 1$$

which we then solve to give

$$A = \tfrac{1}{3}, \quad B = -\tfrac{1}{3}, \quad C = 1$$

Checking, we have

$$Cx^2 + (A + B)x + (-A + 2B + 2C) = x^2 + (\tfrac{1}{3} - \tfrac{1}{3})x + (-\tfrac{1}{3} - \tfrac{2}{3} + 2) = x^2 + 1$$

2.4.2 Factorization

Although Property (ii) was known earlier, the first rigorous proof was published by Gauss in 1799. The result is an 'existence theorem'. It tells us that polynomials can be factored but does not indicate how to find the factors!

Example 2.24

Factorize the polynomials

(a) $x^3 - 3x^2 + 6x - 4$ (b) $x^4 - 16$ (c) $x^4 + 16$

Solution

(a) The function $f(x) = x^3 - 3x^2 + 6x - 4$ clearly has the value zero at $x = 1$. Thus $x - 1$ must be a factor of $f(x)$. We can now divide $x^3 - 3x^2 + 6x - 4$ by $x - 1$ using algebraic division, a process akin to long division of numbers. The process may be set out as follows.

Step 1

$$x - 1)x^3 - 3x^2 + 6x - 4($$

In order to produce the term x^3, $x - 1$ must be multiplied by x^2. Do this and subtract the result from $x^3 - 3x^2 + 6x - 4$.

$$
\begin{array}{r}
x - 1)x^3 - 3x^2 + 6x - 4(x^2 \\
\underline{x^3 - x^2} \\
-2x^2 + 6x - 4
\end{array}
$$

Step 2

Now repeat the process on the polynomial $-2x^2 + 6x - 4$. In this case, in order to eliminate the term $-2x^2$, we must multiply $x - 1$ by $-2x$.

$$
\begin{array}{r}
x - 1)x^3 - 3x^2 + 6x - 4(x^2 - 2x \\
\underline{x^3 - x^2} \\
-2x^2 + 6x - 4 \\
\underline{-2x^2 + 2x} \\
4x - 4
\end{array}
$$

Step 3

Finally we must multiply $x - 1$ by 4 to eliminate $4x - 4$ as follows.

$$x - 1)\overline{x^3 - 3x^2 + 6x - 4}(x^2 - 2x + 4$$

$$\underline{x^3 - x^2}$$

$$-2x^2 + 6x - 4$$

$$\underline{-2x^2 + 2x}$$

$$4x - 4$$

$$\underline{4x - 4}$$

Thus

$$f(x) = (x - 1)(x^2 - 2x + 4)$$

The quadratic factor $x^2 - 2x + 4$ is an **irreducible factor**, as is shown by 'completing the square':

$$x^2 - 2x + 4 = (x - 1)^2 + 3$$

(b) The functions $f_1(x) = x^4$ and $f_2(x) = x^4 - 16$ have similar graphs, as shown in Figures 2.34(a) and (b). It is clear from these graphs that $f_2(x)$ has zeros at two values of x, where $x^4 = 16$: that is, at $x^2 = 4$ ($x^2 = -4$ is not allowed for real x). Thus the zeros of f_2 are at $x = 2$ and $x = -2$, and we can write

$$f_2(x) = x^4 - 16 = (x^2 - 4)(x^2 + 4)$$

$$= (x - 2)(x + 2)(x^2 + 4)$$

(c) The functions $f_1(x) = x^4$ and $f_3(x) = x^4 + 16$ have similar graphs, as shown in Figures 2.34(a) and (c). It is clear from these graphs that $f_3(x)$ does not have any real zeros, so we expect it to be factored into two quadratic terms. We can write

$$x^4 + 16 = (x^2 + 4)^2 - 8x^2$$

which is a difference of squares and may be factored.

$$(x^2 + 4)^2 - 8x^2 = (x^2 + 4)^2 - (x\sqrt{8})^2 = [(x^2 + 4) - x\sqrt{8}][(x^2 + 4) + x\sqrt{8}]$$

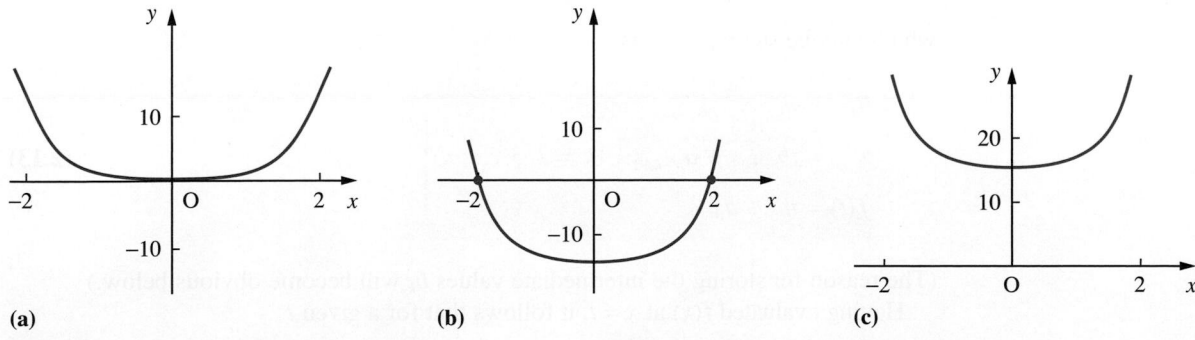

Figure 2.34 Graphs of (a) $y = f_1(x) = x^4$, (b) $y = f_2(x) = x^4 - 16$ and (c) $y = f_3(x) = x^4 + 16$.

Thus we obtain

$$f_3(x) = x^4 + 16 = (x^2 - 2x\sqrt{2} + 4)(x^2 + 2x\sqrt{2} + 4)$$

Since $x^2 \pm 2x\sqrt{2} + 4 = (x \pm \sqrt{2})^2 + 2$, we deduce that these are irreducible quadratics.

2.4.3 Nested multiplication and synthetic division

In Example 2.24(a) we found the image value of the polynomial at $x = 1$ by direct substitution. In general, however, the most efficient way to evaluate the image values of a polynomial function is to use **nested multiplication**. Consider the cubic function

$$f(x) = 4x^3 - 5x^2 + 2x + 3$$

This may be written as

$$f(x) = [(4x - 5)x + 2]x + 3$$

We evaluate this by evaluating each bracketed expression in turn, working from the innermost. Thus to find $f(6)$, the following steps are taken:

(1) Multiply 4 by x and subtract 5; in this case $4 \times 6 - 5 = 19$.
(2) Multiply the result of step 1 by x and add 2; in this case $19 \times 6 + 2 = 116$.
(3) Multiply the result of step 2 by x and add 3; in this case $116 \times 6 + 3 = 699$.

Thus $f(6) = 699$.

On a computer this is performed by means of a simple recurrence relation. To evaluate

$$f(x) = a_n x^n + a_{n-1} x^{n-1} + \ldots + a_0$$

at $x = t$, we use the formulae

$$b_{n-1} = a_n$$
$$b_{n-2} = t b_{n-1} + a_{n-1}$$
$$b_{n-3} = t b_{n-2} + a_{n-2}$$
$$\vdots$$
$$b_1 = t b_2 + a_2$$
$$b_0 = t b_1 + a_1$$
$$f(t) = t b_0 + a_0$$

which may be summarized as

$$\left. \begin{aligned} b_{n-1} &= a_n \\ b_{n-k} &= t b_{n-k+1} + a_{n-k+1} \quad (k = 2, 3, \ldots, n) \\ f(t) &= t b_0 + a_0 \end{aligned} \right\} \tag{2.13}$$

(The reason for storing the intermediate values b_k will become obvious below.)

Having evaluated $f(x)$ at $x = t$, it follows that for a given t

$$f(x) - f(t) = 0$$

at $x = t$; that is, $f(x) - f(t)$ has a factor $x - t$. Thus we can write

$$f(x) - f(t) = (x - t)(c_{n-1}x^{n-1} + c_{n-2}x^{n-2} + \ldots + c_1 x + c_0)$$

Multiplying out the right-hand side, we have

$$f(x) - f(t) = c_{n-1}x^n + (c_{n-2} - tc_{n-1})x^{n-1} + (c_{n-3} - tc_{n-2})x^{n-2} + \ldots + (c_0 - tc_1)x + (-tc_0)$$

so that we may write

$$f(x) = c_{n-1}x^n + (c_{n-2} - tc_{n-1})x^{n-1} + (c_{n-3} - tc_{n-2})x^{n-2} + \ldots + (c_0 - tc_1)x + f(t) - tc_0$$

But

$$f(x) = a_n x^n + a_{n-1}x^{n-1} + a_{n-2}x^{n-2} + \ldots + a_1 x + a_0$$

So, using Property (i) of Section 2.4.1 and comparing coefficients of like powers of x, we have

$$c_{n-1} = a_n$$

$$c_{n-2} - tc_{n-1} = a_{n-1} \quad \text{implying} \quad c_{n-2} = tc_{n-1} + a_{n-1}$$

$$c_{n-3} - tc_{n-2} = a_{n-2} \quad \text{implying} \quad c_{n-3} = tc_{n-2} + a_{n-2}$$

$$\vdots \qquad\qquad \vdots \qquad\qquad \vdots$$

$$c_0 - tc_1 = a_1 \quad \text{implying} \quad c_0 = tc_1 + a_1$$

$$f(t) - tc_0 = a_0 \quad \text{implying} \quad f(t) = tc_0 + a_0$$

Thus c_k satisfies exactly the same formula as b_k, so that the intermediate numbers generated by the method are the coefficients of the quotient polynomial. We can then write

$$f(x) = (b_{n-1}x^{n-1} + b_{n-2}x^{n-2} + \ldots + b_1 x + b_0)(x - t) + f(t) \qquad (2.14)$$

or

$$\frac{f(x)}{x - t} = b_{n-1}x^{n-1} + b_{n-2}x^{n-2} + \ldots + b_1 x + b_0 + \frac{f(t)}{x - t}$$

Result (2.14) tells us that if the polynomial $f(x)$ given in (2.12) is divided by $x - t$ then this results in a quotient polynomial $q(x)$ given by

$$q(x) = b_{n-1}x^{n-1} + \ldots + b_0$$

and a remainder $r = f(t)$ that is independent of x. Because of this property, the method of nested multiplication is sometimes called **synthetic division**.

The coefficients b_i, $i = 0, \ldots, n - 1$, of the quotient polynomial and remainder term $f(t)$ may be determined using the formulae (2.13). The process may be carried out in the following tabular form:

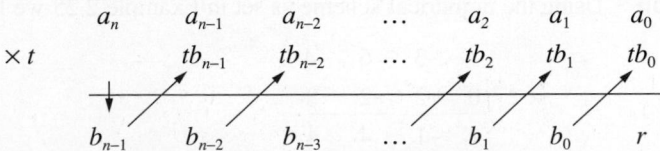

	a_n	a_{n-1}	a_{n-2}	$\ldots$	a_2	a_1	a_0
$\times t$		tb_{n-1}	tb_{n-2}	$\ldots$	tb_2	tb_1	tb_0
	b_{n-1}	b_{n-2}	b_{n-3}	$\ldots$	b_1	b_0	r

After the number below the line is calculated as the sum of the two numbers immediately above it, it is multiplied by t and placed in the next space above the line as indicated by the arrows. This procedure is repeated until all the terms are calculated.

The method of synthetic division could have been used as an alternative to algebraic division in Example 2.24.

Example 2.25   Show that $f(x) = x^3 - 3x^2 + 6x - 4$ is zero at $x = 1$, and hence factorize $f(x)$.

Solution   Using the nested multiplication procedure to divide $x^3 - 3x^2 + 6x - 4$ by $x - 1$ gives the tabular form

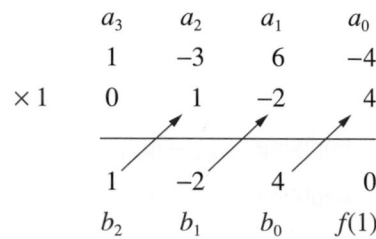

Since the remainder $f(1)$ is zero, it follows that $f(x)$ is zero at $x = 1$. Thus

$$f(x) = (x^2 - 2x + 4)(x - 1)$$

and we have extracted the factor $x - 1$. We may then examine the quadratic factor $x^2 - 2x + 4$ as we did in Example 2.24(a) and show that it is an irreducible quadratic factor.

Sometimes in problem-solving we need to rearrange the formula for the polynomial function as an expansion about a point, $x = a$, other than $x = 0$. That is, we need to find the numbers $A_0, A_1, \ldots, A_n$ such that

$$f(x) = a_n x^n + a_{n-1} x^{n-1} + \ldots + a_1 x + a_0$$
$$= A_n(x - a)^n + A_{n-1}(x - a)^{n-1} + \ldots + A_1(x - a) + A_0$$

This transformation can be achieved using the technique illustrated for the quadratic function in Example 2.19 which depends on the identity property of polynomials. It can be achieved more easily using **repeated synthetic division**, as is shown in Example 2.26.

Example 2.26   Obtain the expansion about $x = 2$ of the function $y = x^3 - 3x^2 + 6x - 4$.

Solution   Using the numerical scheme as set in Example 2.25 we have

$$
\begin{array}{rrrrr}
 & 1 & -3 & 6 & -4 \\
\times 2 & 0 & 2 & -2 & 8 \\
\hline
 & 1 & -1 & 4 & 4 \\
\end{array}
$$

so that

$$x^3 - 3x^2 + 6x - 4 = (x - 2)(x^2 - x + 4) + 4$$

Now repeating the process with $y = x^2 - x + 4$, we have

$$
\begin{array}{r|rrr}
 & 1 & -1 & 4 \\
\times 2 & 0 & 2 & 2 \\
\hline
 & 1 & 1 & 6
\end{array}
$$

so that

$$x^2 - x + 4 = (x - 2)(x + 1) + 6$$

and

$$x^3 - 3x^2 + 6x - 4 = (x - 2)[(x - 2)(x + 1) + 6] + 4$$

Lastly,

$$x + 1 = (x - 2) + 3$$

so that

$$y = (x - 2)[(x - 2)^2 + 3(x - 2) + 6] + 4$$
$$= (x - 2)^3 + 3(x - 2)^2 + 6(x - 2) + 4$$

For hand computation the whole process can be set out as a single table:

$$
\begin{array}{r|rrrr}
 & 1 & -3 & 6 & -4 \\
\times 2 & 0 & 2 & -2 & 8 \\
\hline
 & 1 & -1 & 4 & \vdots 4 \\
\times 2 & 0 & 2 & 2 & \\
\hline
 & 1 & 1 & \vdots 6 & \\
\times 2 & 0 & 2 & & \\
\hline
 & 1 & \vdots 3 & &
\end{array}
$$

Here, then, 1, 3, 6 and 4 provide the coefficients of $(x - 2)^3$, $(x - 2)^2$, $(x - 2)^1$ and $(x - 2)^0$ in the Taylor expansion.

2.4.4 Roots of polynomial equations

Polynomial equations occur frequently in engineering applications, from the identification of resonant frequencies when concerned with rotating machinery to the stability analysis of circuits. It is often useful to see the connections between the roots of a polynomial equation and its coefficients.

Example 2.27

Show that any real roots of the equation

$$x^3 - 3x^2 + 6x - 4 = 0$$

lie between $x = 0$ and $x = 2$.

Solution   From Example 2.26 we know that

$$x^3 - 3x^2 + 6x - 4 \equiv (x - 2)^3 + 3(x - 2)^2 + 6(x - 2) + 4$$

Now if $x > 2$, $(x - 2)^3$, $(x - 2)^2$ and $(x - 2)$ are all positive numbers so that for $x > 2$

$$(x - 2)^3 + 3(x - 2)^2 + 6(x - 2) + 4 > 0$$

Thus $x^3 - 3x^2 + 6x - 4 = 0$ does not have a root that is greater than $x = 2$.

Similarly for $x < 0$, x^3 and x are both negative and $x^3 - 3x^2 + 6x - 4 < 0$ for $x < 0$. Thus $x^3 - 3x^2 + 6x - 4 = 0$ does not have a root that is less than $x = 0$. Hence all the real roots of

$$x^3 - 3x^2 + 6x - 4 = 0$$

lie between $x = 0$ and $x = 2$.

We can generalize the results of Example 2.27. Defining

$$f(x) = \sum_{k=0}^{n} A_n (x - a)^n$$

then the polynomial equation $f(x) = 0$ has no roots greater than $x = a$ if all of the A_k's have the same sign and has no roots less than $x = a$ if the A_k's alternate in sign.

The roots of a polynomial equation are related to its coefficients in more direct ways. Consider, for the moment, the quadratic equation with roots α and β. Then we can write the equation as

$$(x - \alpha)(x - \beta) = 0$$

which is equivalent to

$$x^2 - (\alpha + \beta)x + \alpha\beta = 0$$

Comparing this to the standard quadratic equation we have

$$a(x^2 - (\alpha + \beta)x + \alpha\beta) \equiv ax^2 + bx + c$$

Thus $-a(\alpha + \beta) = b$ and $a\alpha\beta = c$ so that

$$\alpha + \beta = -b/a \quad \text{and} \quad \alpha\beta = c/a$$

This gives us direct links between the sum of the roots of a quadratic equation and its coefficients and between the product of the roots and the coefficients. Similarly, we can show that if α, β and γ are the roots of the cubic equation

$$ax^3 + bx^2 + cx + d = 0$$

then

$$\alpha + \beta + \gamma = -b/a, \quad \alpha\beta + \beta\gamma + \gamma\alpha = c/a, \quad \alpha\beta\gamma = -d/a$$

In general, for the polynomial equation

$$a_n x^n + a_{n-1} x^{n-1} + a_{n-2} x^{n-2} + \ldots + a_1 x + a_0 = 0$$

the sum of the products of the roots, k at a time, is $(-1)^k a_{n-k}/a_n$.

Example 2.28 Show that the roots, α, β of the quadratic equation

$$ax^2 + bx + c = 0$$

may be written in the form

$$\frac{-b - \sqrt{(b^2 - 4ac)}}{2a} \quad \text{and} \quad \frac{2c}{-b - \sqrt{(b^2 - 4ac)}}$$

Obtain the roots of the equation

$$1.0x^2 + 17.8x + 1.5 = 0$$

Assuming the numbers given are correctly rounded, calculate error bounds for the roots.

Solution Using the formula for the roots of a quadratic equation we can select one root, α say, so that

$$\alpha = \frac{-b - \sqrt{(b^2 - 4ac)}}{2a}$$

Then, since $\alpha\beta = c/a$, we have

$$\beta = \frac{c}{a\alpha} = \frac{2c}{-b - \sqrt{(b^2 - 4ac)}}$$

Now consider the equation

$$1.0x^2 + 17.8x + 1.5 = 0$$

whose coefficients are correctly rounded numbers. Using the quadratic formula we obtain the roots

$$\alpha \approx -17.715\,327\,56$$

and

$$\beta \approx -0.084\,672\,44$$

Using the results of Section 1.5.3 we can estimate error bounds for these answers as shown in Figure 2.35. From that table we can see that using the form

$$\frac{-b - \sqrt{(b^2 - 4ac)}}{2a}$$

to estimate α we have an error bound of 0.943, while using

$$\frac{-b + \sqrt{(b^2 - 4ac)}}{2a}$$

Figure 2.35
Estimating error
bounds for roots.

Label	Value	Absolute error bound	Relative error bound
a	1.0	0.05	0.05
b	17.8	0.05	0.0028
c	1.5	0.05	0.0333
b^2	316.84	1.77	0.0056
$4ac$	6.00	0.50	0.0833
$b^2 - 4ac$	310.84	2.27	0.0073
$d = \sqrt{(b^2 - 4ac)}$	17.630 66	0.065	0.0037
$-b - d$	$-35.430\ 66$	0.115	0.0032
$(-b - d)/(2a)$	$-17.715\ 33$	0.943	0.0532
$-b + d$	$-0.169\ 34$	0.115	0.6791
$(-b + d)/(2a)$	$-0.084\ 67$	0.062	0.7291
$2c/(-b - d)$	$-0.084\ 67$	0.003	0.0365

to estimate β we have an error bound of 0.062. As this latter estimate of error is almost as big as the root itself we might be inclined to regard the answer as valueless. But calculating the error bound using the form

$$\beta = \frac{2c}{-b - \sqrt{(b^2 - 4ac)}}$$

gives an estimate of 0.003. Thus we can write

$$\alpha = -17.7 \pm 5\% \text{ and } \beta = -0.085 \pm 4\%$$

The reason for the discrepancy between the two error estimates for β lies in the fact that in the traditional form of the formula we are subtracting two nearly equal numbers, and consequently the error bounds dominate.

Example 2.29

The equation $3x^3 - x^2 - 3x + 1 = 0$ has a root at $x = 1$. Obtain the other two roots.

Solution

If α, β and γ are the roots of the equation then

$$\alpha + \beta + \gamma = \tfrac{1}{3}$$
$$\alpha\beta + \beta\gamma + \gamma\alpha = -\tfrac{3}{3}$$
$$\alpha\beta\gamma = -\tfrac{1}{3}$$

Setting $\alpha = 1$ simplifies these to

$$\beta + \gamma = -\tfrac{2}{3}$$
$$\beta + \gamma + \beta\gamma = -1$$
$$\beta\gamma = -\tfrac{1}{3}$$

Hence $\gamma = -1/(3\beta)$ and $3\beta^2 + 2\beta - 1 = 0$. Factorizing this equation gives

$$(3\beta - 1)(\beta + 1) = 0$$

from which we obtain the solution $x = -1$ and $x = \tfrac{1}{3}$.

The numerical method most often used for evaluating the roots of a polynomial is the Newton–Raphson procedure. This will be described in Chapter 9, Section 9.5.8.

In MATLAB a polynomial is represented by an array of its coefficients, with the highest coefficient listed first. For example, the polynomial function

$$f(x) = x^3 - 5x^2 - 17x + 21$$

is represented by

```
f = [1 -5 -17 21]
```

The roots of the corresponding polynomial equation $f(x) = 0$ are obtained using the command $roots(f)$ so for the above example the command

```
r = roots(f)
```

returns the roots as

```
r = 7.0000
   -3.0000
    1.0000
```

which also indicate that the factors of $f(x)$ are $(x - 7)$, $(x + 3)$ and $(x - 1)$. It is noted that the output gives the roots r as a column array of numbers (and not a row array). If the roots are known and we wish to determine the corresponding polynomial $f(x)$, having unity as the coefficient of its highest power, then use is made of the command $poly(r)$. To use this command the roots r must be specified as a row array; so the commands

```
r = [7 -3 1]
f = poly(r)
```

return the answer

```
f = 1.0000 -5.0000 -17.0000 21.0000
```

indicating that the polynomial is

$$f(x) = x^3 - 5x^2 - 17x + 21$$

To determine the polynomial of degree n that passes through $n + 1$ points we use the command $polyfit(x,y,n)$; which outputs the array of coefficients of a polynomial of order n that fits the pairs (x, y). If the number of points (x, y) is greater than n then the command will give the best fit in the least squares sense. Check that the commands

```
x = [-3 0 2 5]; y = [528 1017 1433 2312];
f = polyfit(x,y,3)
```

reproduce the answer of Example 2.22 and that the commands

```
x = [0 1 2 3 4]; y = [1 1 2 2 3];
polyfit(x,y,1)
```

reproduce the answer to Example 2.17.

Graphs of polynomial functions may be plotted using the commands given earlier on page (71). The result of multiplying two polynomials $f(x)$ and $g(x)$ is obtained using the command $conv(f,g)$, where f and g are the array specification of $f(x)$ and $g(x)$ respectively. With reference to Example 2.25 confirm that the product $f(x) = (x^2 - 2x + 4)(x - 1)$ is obtained using the commands

```
f1 = [1 -2 4]; f2 = [1 -1];
f = conv(f1, f2)
```

The division of two polynomials $f(x)$ and $g(x)$ is obtained, by the process of deconvolution, using the command

```
[Q,R] = deconv(f,g)
```

which produces two outputs Q and R, with Q being the coefficients of the quotient polynomial and R the coefficients of the remainder polynomial. Again with reference to Example 2.25 check that $x^3 - 3x^2 + 6x - 4$ divided by $x - 1$ gives a quotient $x^2 - 2x + 4$ and a remainder of zero.

Using the Symbolic Math Toolbox operations on polynomials may be undertaken in symbolic form. Some useful commands, for carrying out algebraic manipulations, are:

(a) *factor command*

If $f(x)$ is a polynomial function, expressed in symbolic form, with rational coefficients (see Section 1.2.1) then the commands

```
syms x
f = factor(f(x))
```

factorize $f(x)$ as the product of polynomials of lower degree with rational coefficients. For example, to factorize the cubic $f(x) = x^3 - 5x^2 - 17x + 21$ the commands

```
syms x
f = factor(x^3 - 5*x^2 - 17*x + 21)
```

return

```
f = (x - 1)*(x - 7)*(x + 3)
```

Using the *pretty* command

```
pretty(f)
```

returns the more readable display

```
f = (x - 1)(x - 7)(x + 3)
```

Using the *factor* command, confirm the factorization of polynomials (a) and (b) in Example 2.24.

(b) *horner command*

This command transforms a polynomial $f(x)$ expressed in symbolic form into its nested (or Horner) representation. For example the commands

```
syms x
f = horner(4*x^3 - 5*x^2 + 2*x + 3)
```

return

```
f = 3 + (2 + (-5 + 4*x)*x)*x
```

which confirms the nested representation at the outset of Section 2.4.3.

(c) collect command

This collects all the coefficients with the same power of x. For example, if

$$f(x) = 4x(x^2 + 2x + 1) - 5(x(x + 2) - x^3) + (x + 3)^3$$

then the commands

```
syms x
f = collect(4*x*(x^2 + 2*x + 1) - 5*(x*(x + 2) - x^3)
     + (x + 3)^3);
pretty(f)
```

return

```
f = 27 + 10x³ + 12x² + 21x
```

The collect command may also be used to multiply two polynomials. With reference to Example 2.25 the product of the two polynomials $x^2 - 2x + 4$ and $x - 1$ is returned by the commands

```
syms x
f = collect((x - 1)*(x^2 - 2*x + 4));
pretty(f)
```

as

```
f = x³ - 3x² + 6x - 4
```

(d) simplify command

This is a powerful general purpose command that can be used with a wide range of functions. For example, if $f(x) = (9 - x^2)/(3 + x)$ then the commands

```
syms x
f = simplify((9 - x^2)/(3 + x))
```

return

```
f = -x + 3
```

(e) simple command

This command seeks to find a simplification of a symbolic expression so that it has the fewest number of characters; that is it seeks to obtain the shortest form of the expression. The command sometimes improves on the result returned by the simplify command. There is no corresponding command in MAPLE.

(f) expand command

This is another general purpose command which can be used with a wide range of functions. It distributes products over sums and differences. For example, if $f(x) = a(x + y)$ then the commands

```
syms x a y
f = expand(a*(x + y));
pretty(f)
```

return

```
f = ax + ay
```

(g) solve command

If $f(x)$ is a symbolic expression in the variable x (the expression may also include parameters) then the command

```
s = solve (f)
```

seeks to solve the equation $f(x) = 0$, returning the solution in a column array. To solve an equation expressed in the form $f(x) = g(x)$ use is made of the command

```
s = solve('f(x) = g(x)')
```

For example, considering the general quadratic equation $ax^2 + bx + c = 0$ the commands

```
syms x a b c
s = solve(a*x^2 + b*x + c);
pretty(s)
```

return the well known answers (see Example 1.21)

$$\left[1/2 \ \frac{-b + (b^2 - 4ac)^{1/2}}{a} \right]$$

$$\left[1/2 \ \frac{-b - (b^2 - 4ac)^{1/2}}{a} \right]$$

2.4.5 Exercises

 Check your answers using MATLAB or MAPLE whenever possible.

 29 Factorize the following polynomial functions and sketch their graphs:

(a) $x^3 - 2x^2 - 11x + 12$

(b) $x^3 + 2x^2 - 5x - 6$

(c) $x^4 + x^2 - 2$

(d) $2x^4 + 5x^3 - x^2 - 6x$

(e) $2x^4 - 9x^3 + 14x^2 - 9x + 2$

(f) $x^4 + 5x^2 - 36$

30 Find the coefficients A, B, C, D and E such that

$$y = 2x^4 - 9x^3 + 145x^2 - 9x + 2$$
$$= A(x-2)^4 + B(x-2)^3 + C(x-2)^2$$
$$+ D(x-2) + E$$

31 Show that the zeros of

$$y = x^4 - 5x^3 + 5x^2 - 10x + 6$$

lie between $x = 0$ and $x = 5$.

32 Show that the roots α, β of the equation

$$x^2 + 4x + 1 = 0$$

satisfy the equations

$$\alpha^2 + \beta^2 = 14$$
$$\alpha^3 + \beta^3 = -52$$

Hence find the quadratic equations whose roots are

(a) α^2 and β^2 (b) α^3 and β^3

33 Use Lagrange's formula to find the formula for the cubic function that passes through the points (5.2, 6.408), (5.5, 16.125), (5.6, 19.816) and (5.8, 27.912).

34 Find a formula for the quadratic function whose graph passes through the points (1, 403), (3, 471) and (7, 679).

35 (a) Show that if the equation $ax^3 + bx + c = 0$ has a repeated root α then $3a\alpha^2 + b = 0$.

(b) A can is to be made in the form of a circular cylinder of radius r (in cm) and height h (in cm) as shown in Figure 2.36. Its capacity is to be

0.5 l. Show that the surface area A (in cm^2) of the can is

$$A = 2\pi r^2 + \frac{1000}{r}$$

Using the result of (a), deduce that A has a minimum value A^* when $6\pi r^2 - A^* = 0$. Hence find the corresponding values of r and h.

36 A box is made from a sheet of plywood, $2\,\text{m} \times 1\,\text{m}$, with the waste shown in Figure 2.37(a). Find the

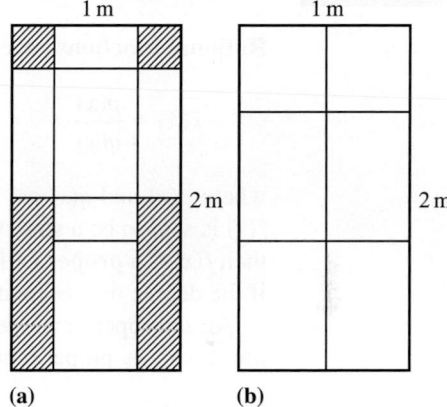

(a)   **(b)**

Figure 2.37

maximum capacity of such a box and compare it with the capacity of the box constructed without the wastage, as shown in Figure 2.37(b).

37 Two ladders, of lengths $12\,\text{m}$ and $8\,\text{m}$, lean against buildings on opposite sides of an alley, as shown in Figure 2.38. Show that the heights x and y

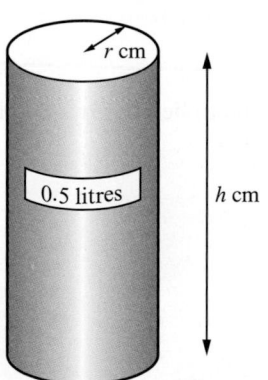

Figure 2.36

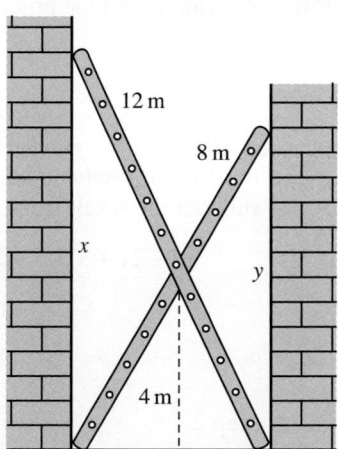

Figure 2.38

(in metres) reached by the tops of the ladders in the positions shown satisfy the equations

$$\frac{1}{x} + \frac{1}{y} = \frac{1}{4} \quad \text{and} \quad x^2 - y^2 = 80$$

Show that x satisfies the equation

$$x^4 - 8x^3 - 80x^2 + 640x - 1280 = 0$$

and that the width of the alley is given by $\sqrt{(12^2 - x_0^2)}$, where x_0 is the positive root of this equation. By first tabulating the polynomial over a suitable domain and then drawing its graph, estimate the value of x_0 and the width of the alley. Check your solution of the quartic (to 2dp) using a suitable software package.

2.5 Rational functions

Rational functions have the general form

$$f(x) = \frac{p(x)}{q(x)}$$

where $p(x)$ and $q(x)$ are polynomials. If the degree of p is less than the degree of q, $f(x)$ is said to be a **strictly proper rational function**. If p and q have the same degree then $f(x)$ is a **proper rational function**. It is said to be an **improper rational function** if the degree of p is greater than the degree of q.

An improper or proper rational function can always be expressed as a polynomial plus a strictly proper rational function, for example, by algebraic division.

Example 2.30

Express the improper rational function

$$f(x) = \frac{3x^4 + 2x^3 - 5x^2 + 6x - 7}{x^2 - 2x + 3}$$

as the sum of a polynomial function and a strictly proper rational function.

Solution

We can record the process of division in a manner similar to that of Example 2.22.

Step 1

$$x^2 - 2x + 3)3x^4 + 2x^3 - 5x^2 + 6x - 7($$

In order to produce the term $3x^4$, $x^2 - 2x + 3$ must be multiplied by $3x^2$. Do this and subtract the result from $3x^4 + 2x^3 - 5x^2 + 6x - 7$.

$$x^2 - 2x + 3)3x^4 + 2x^3 - 5x^2 + 6x - 7(3x^2$$

$$\underline{3x^4 - 6x^3 + 9x^2}$$

$$8x^3 - 14x^2 + 6x - 7$$

Step 2

Now repeat the process on the polynomial $8x^3 - 14x^2 + 6x - 7$. In this case, in order to eliminate the term $8x^3$ we must multiply $x^2 - 2x + 3$ by $8x$.

$$x^2 - 2x + 3 \overline{)3x^4 + 2x^3 - 5x^2 + 6x - 7}(3x^2 + 8x$$

$$\underline{3x^4 - 6x^3 + 9x^2}$$

$$8x^3 - 14x^2 + 6x - 7$$

$$\underline{8x^3 - 16x^2 + 24x}$$

$$2x^2 - 18x - 7$$

Step 3

Finally, to eliminate the $2x^2$ term, we must multiply $x^2 - 2x + 3$ by 2.

$$x^2 - 2x + 3 \overline{)3x^4 + 2x^3 - 5x^2 + 6x - 7}(3x^2 + 8x + 2$$

$$\underline{3x^4 - 6x^3 + 9x^2}$$

$$8x^3 - 14x^2 + 6x - 7$$

$$\underline{8x^3 - 16x^2 + 24x}$$

$$2x^2 - 18x - 7$$

$$\underline{2x^2 - 4x + 6}$$

$$-14x - 13$$

We cannot eliminate the $-14x - 13$ terms, so we have

$$f(x) = 3x^2 + 8x + 2 - \frac{14x + 13}{x^2 - 2x + 3}$$

Any strictly proper rational function can be expressed as a sum of simpler functions whose denominators are linear or irreducible quadratic functions. For example:

$$\frac{x^2 + 1}{(1 + x)(1 - x)(2 + 2x + x^2)} = \frac{1}{1 + x} + \frac{1}{5(1 - x)} - \frac{4x + 7}{5(2 + 2x + x^2)}$$

These simpler functions are called the **partial fractions** of the rational function, and are often useful in the mathematical analysis and design of engineering systems. Notice that strictly the equality above is an identity since it is true for all values of x in the domain of the expressions. Here we are following the common practice of writing = instead of ≡ (as we did in Section 1.3.2.).

The construction of the partial fraction form of a rational function is the inverse process to that of collecting together separate rational expressions into a single rational function. For example:

$$\frac{1}{1 + x} + \frac{1}{5(1 - x)} - \frac{4x + 7}{5(2 + 2x + x^2)}$$

$$= \frac{1(5)(1 - x)(2 + 2x + x^2) + (1 + x)(2 + 2x + x^2) - (1 + x)(1 - x)(4x + 7)}{5(1 + x)(1 - x)(2 + 2x + x^2)}$$

$$= \frac{5(2 - x^2 - x^3) + (2 + 4x + 3x^2 + x^3) - (1 - x^2)(4x + 7)}{5(1 - x^2)(2 + 2x + x^2)}$$

$$= \frac{5(2 - x^2 - x^3) + (2 + 4x + 3x^2 + x^3) - (7 + 4x - 7x^2 - 4x^3)}{5(2 + 2x - x^2 - 2x^3 - x^4)}$$

$$= \frac{5 + 5x^2}{5(2 + 2x - x^2 - 2x^3 - x^4)}$$

$$= \frac{1 + x^2}{2 + 2x - x^2 - 2x^3 - x^4}$$

But it is clear from this example that reversing the process (working backwards from the final expression) is not easy, and we require a different method in order to find the partial fractions of a given function. To describe the method in its full generality is easy but difficult to understand, so we will apply the method to a number of commonly occurring types of function in the next section before stating the general algorithm.

2.5.1 Partial fractions

In this section we will illustrate how proper rational functions of the form $p(x)/q(x)$ may be expressed in partial fractions.

(a) Distinct linear factors

Each distinct linear factor, of the form $(x + \alpha)$, in the denominator $q(x)$ will give rise to a partial fraction of the form $\dfrac{A}{x + \alpha}$, where A is a real constant.

Example 2.31 Express in partial fractions the rational function

$$\frac{3x}{(x - 1)(x + 2)}$$

Solution In this case we have two distinct linear factors $(x - 1)$ and $(x + 2)$ in the denominator so the corresponding partial fractions are of the form

$$\frac{3x}{(x - 1)(x + 2)} = \frac{A}{x - 1} + \frac{B}{x + 2} = \frac{A(x + 2) + B(x - 1)}{(x - 1)(x + 2)}$$

where A and B are constants to be determined. Since both expressions are equal and their denominators are identical we must therefore make their numerators equal, yielding

$$3x = A(x + 2) + B(x - 1)$$

This identity is true for all values of x, so we can find A and B by setting first $x = 1$ and then $x = -2$. So

$$x = 1 \quad \text{gives} \quad 3 = A(3) + B(0) \quad \text{that is} \quad A = 1$$

and

$$x = -2 \quad \text{gives} \quad -6 = A(0) + B(-3) \quad \text{that is} \quad B = 2$$

Thus

$$\frac{3x}{(x-1)(x+2)} = \frac{1}{x-1} + \frac{2}{x+2}$$

When the denominator $q(x)$ of a strictly proper rational function $\dfrac{p(x)}{q(x)}$ is a product of linear factors, as in Example 2.31, there is a quick way of expressing $\dfrac{p(x)}{q(x)}$ in partial fractions.

Considering again Example 2.31, if

$$\frac{3x}{(x-1)(x+2)} = \frac{A}{(x-1)} + \frac{B}{(x-2)}$$

then to obtain A simply **cover up** the factor $(x-1)$ in

$$\frac{3x}{(x-1)(x+2)}$$

and evaluate what is left at $x = 1$, giving

$$A = \frac{3(1)}{(x-1)(1+2)} = 1$$

Likewise, to obtain B **cover up** the factor $(x+2)$ in the left-hand side and evaluate what is left at $x = -2$, giving

$$B = \frac{3(-2)}{(-2-1)(x+2)} = 2$$

Thus, as before,

$$\frac{3x}{(x-1)(x+2)} = \frac{1}{x-1} + \frac{2}{x+2}$$

This method of obtaining partial fractions is called the **cover up rule**.

Example 2.32 Using the cover up rule, express in partial fractions the rational function

$$\frac{2x+1}{(x-2)(x+1)(x-3)}$$

Solution The corresponding partial fractions are of the form

$$\frac{2x+1}{(x-2)(x+1)(x-3)} = \frac{A}{(x-2)} + \frac{B}{(x+1)} + \frac{C}{(x-3)}$$

Using the cover up rule

$$A = \frac{2(2) + 1}{(x - 2)(2 + 1)(2 - 3)} = -\frac{5}{3}$$

$$B = \frac{2(-1) + 1}{(-1 - 2)(x + 1)(-1 - 3)} = -\frac{1}{12}$$

$$C = \frac{2(3) + 1}{(3 - 2)(3 + 1)(x - 3)} = \frac{7}{4}$$

so that

$$\frac{2x + 1}{(x - 2)(x + 1)(x - 3)} = -\frac{\frac{5}{3}}{x - 2} - \frac{\frac{1}{12}}{x + 1} + \frac{\frac{7}{4}}{x - 3}$$

Because it is easy to make an error with this process, it is sensible to check the answers obtained. This can be done by using a 'spot' value to check that the left- and right-hand sides yield the same value. When doing this avoid using $x = 0$ or any of the special values of x that were used in finding the coefficients.

For example, taking $x = 1$ in the partial fraction expansion of Example 2.32, we have

$$\text{left-hand side} \quad = \frac{2(1) + 1}{(1 - 2)(1 + 1)(1 - 3)} = \frac{3}{4}$$

$$\text{right-hand side} \quad = -\frac{\frac{5}{3}}{1 - 2} - \frac{\frac{1}{12}}{1 + 1} + \frac{\frac{7}{4}}{1 - 3} = \frac{3}{4}$$

giving a positive check.

(b) Repeated linear factors

Each k times repeated linear factor, of the form $(x - \alpha)^k$, in the denominator $q(x)$ will give rise to a partial fraction of the form

$$\frac{A_1}{(x - \alpha)} + \frac{A_2}{(x - \alpha)^2} + \ldots + \frac{A_k}{(x - \alpha)^k}$$

where $A_1, A_2, \ldots, A_k$ are real constants.

Example 2.33   Express as partial fractions the rational function

$$\frac{3x + 1}{(x - 1)^2(x + 2)}$$

Solution   In this case the denominator consists of the distinct linear factor $(x + 2)$ and the twice repeated linear factor $(x - 1)$. Thus, the corresponding partial fractions are of the form

$$\frac{3x+1}{(x-1)^2(x+2)} = \frac{A}{(x-1)} + \frac{B}{(x-1)^2} + \frac{C}{(x+2)}$$

$$= \frac{A(x-1)(x+2) + B(x+2) + C(x-1)^2}{(x-1)^2(x+2)}$$

which gives

$$3x+1 = A(x-1)(x+2) + B(x+2) + C(x-1)^2$$

Setting $x = 1$ gives $4 = B(3)$ and $B = \frac{4}{3}$. Setting $x = -2$ gives $-5 = C(-3)^2$ and $C = -\frac{5}{9}$. To obtain A we can give x any other value, so taking $x = 0$ gives

$$1 = (-2)A + 2B + C$$

and substituting the values of B and C gives $A = \frac{5}{9}$. Hence

$$\frac{3x+1}{(x-1)^2(x+2)} = \frac{\frac{5}{9}}{x-1} + \frac{\frac{4}{3}}{(x-1)^2} - \frac{\frac{5}{9}}{(x+2)}$$

(c) Irreducible quadratic factors

Each distinct irreducible quadratic factor, of the form $(ax^2 + bx + c)$, in the denominator $q(x)$ will give rise to a partial fraction of the form

$$\frac{Ax+B}{ax^2+bx+c}$$

where A and B are real constants.

Example 2.34 Express as partial fractions the rational function

$$\frac{5x}{(x^2+x+1)(x-2)}$$

Solution In this case the denominator consists of the distinct linear factor $(x-2)$ and the distinct irreducible quadratic factor (x^2+x+1). Thus, the corresponding partial fractions are of the form

$$\frac{5x}{(x^2+x+1)(x-2)} = \frac{Ax+B}{x^2+x+1} + \frac{C}{x-2} = \frac{(Ax+B)(x-2) + C(x^2+x+1)}{(x^2+x+1)(x-2)}$$

giving

$$5x = (Ax+B)(x-2) + C(x^2+x+1)$$

Setting $x = 2$ enables us to calculate C:

$$10 = (2A+B)(0) + C(7) \quad \text{and} \quad C = \frac{10}{7}$$

Here, however, we cannot select special values of x that give A and B immediately, because $x^2 + x + 1$ is an irreducible quadratic and cannot be factorized. Instead we make use of Property (i) of polynomials, described in Section 2.4.1, which stated that if two polynomials are equal in value for all values of x then the corresponding coefficients are equal. Applying this to

$$5x = (Ax + B)(x - 2) + C(x^2 + x + 1)$$

we see that the coefficient of x^2 on the right-hand side is $A + C$ while that on the left-hand side is zero. Thus

$$A + C = 0 \quad \text{and} \quad A = -C = -\tfrac{10}{7}$$

Similarly the coefficient of x^0 on the right-hand side is $-2B + C$ and that on the left-hand side is zero, and we obtain $-2B + C = 0$, which implies $B = \tfrac{1}{2}C = \tfrac{5}{7}$. Hence

$$\frac{5x}{(x^2 + x + 1)(x - 2)} = \frac{\tfrac{5}{7} - \tfrac{10}{7}x}{x^2 + x + 1} + \frac{\tfrac{10}{7}}{x - 2}$$

Example 2.35   Express as partial fractions the rational function

$$\frac{3x^2}{(x - 1)(x + 2)}$$

Solution   In this example the numerator has the same degree as the denominator.

The first step in such examples is to divide the bottom into the top to obtain a polynomial and a strictly proper rational function. Thus

$$\frac{3x^2}{(x - 1)(x + 2)} = 3 + \frac{6 - 3x}{(x - 1)(x + 2)}$$

We then apply the partial-fraction process to the remainder, setting

$$\frac{6 - 3x}{(x - 1)(x + 2)}$$

$$= \frac{A}{x - 1} + \frac{B}{x + 2}$$

$$= \frac{A(x + 2) + B(x - 1)}{(x - 1)(x + 2)}$$

giving

$$6 - 3x = A(x + 2) + B(x - 1)$$

Setting first $x = 1$ and then $x = -2$ gives $A = 1$ and $B = -4$ respectively. Thus

$$\frac{3x^2}{(x - 1)(x + 2)} = 3 + \frac{1}{x - 1} - \frac{4}{x + 2}$$

Summary of method

In general, the method for finding the partial fractions of a given function $f(x) = p(x)/q(x)$ consists in the following steps.

Step 1: If the degree of p is greater than or equal to the degree of q, divide q into p to obtain

$$f(x) = r(x) + \frac{s(x)}{q(x)}$$

where the degree of s is less than the degree of q.

Step 2: Factorize $q(x)$ fully into real linear and irreducible quadratic factors, collecting together all like factors.

Step 3: Each **linear factor** $ax + b$ in $q(x)$ will give rise to a fraction of the type

$$\frac{A}{ax + b}$$

(Here a and b are known and A is to be found.)

Each **repeated linear factor** $(ax + b)^n$ will give rise to n fractions of the type

$$\frac{A_1}{ax + b} + \frac{A_2}{(ax + b)^2} + \frac{A_3}{(ax + b)^3} + \ldots + \frac{A_n}{(ax + b)^n}$$

Each **irreducible quadratic factor** $ax^2 + bx + c$ in $q(x)$ will give rise to a fraction of the type

$$\frac{Ax + B}{ax^2 + bx + c}$$

Each **repeated irreducible quadratic factor** $(ax^2 + bx + c)^n$ will give rise to n fractions of the type

$$\frac{A_1x + B_1}{ax^2 + bx + c} + \frac{A_2x + B_2}{(ax^2 + bx + c)^2} + \ldots + \frac{A_nx + B_n}{(ax^2 + bx + c)^n}$$

Put $p(x)/q(x)$ (or $s(x)/q(x)$, if that case occurs) equal to the sum of all the fractions involved.

Step 4: Multiply both sides of the equation by $q(x)$ to obtain an identity involving polynomials, from which the multiplying constants of the linear combination may be found (because of Property (i) in Section 2.4.1).

Step 5: To find these coefficients, two strategies are used.

- *Strategy 1*: Choose special values of x that make finding the values of the unknown coefficients easy: for example choose x equal to the roots of $q(x) = 0$ in turn and use the 'cover up' rule.
- *Strategy 2*: Compare the coefficients of like powers of x on both sides of the identity. Starting with the highest and lowest powers usually makes it easier.

Strategy 1 may leave some coefficients undetermined. In that case we complete the process using Strategy 2.

Step 6: Lastly, check the answer either by choosing a test value for x or by putting the partial fractions over a common denominator.

There is no command in MATLAB that will symbolically express rational functions in partial fractions. However use of the *maple* command in MATLAB enables us to access MAPLE commands directly. Thus, adopting the *convert* command in MAPLE a rational function $f(x)$ may be expressed in partial fraction '*pf*' form using, in MATLAB, the commands

```
syms x
pf = maple('convert',f(x),' parfrac',x);
pretty(pf)
```

For example, considering Example 2.32, the commands

```
syms x
pf = maple('convert',(2*x + 1)/((x - 2)*(x + 1)*(x - 3)),
'parfrac',x);
pretty(pf)
```

return

$$-1/12 \ \frac{1}{x+1} \ + \ 7/4 \ \frac{1}{x-3} \ - \ 5/3 \ \frac{1}{x-2}$$

confirming the answer in the example.
For practice check the answers to Examples 2.33–2.35.

2.5.2 Exercises

 Where appropriate check your answers using MATLAB or MAPLE.

38 Express the following improper rational functions as the sum of a polynomial function and a strictly proper rational function

(a) $f(x) = (x^2 + x + 1)/[(x+1)(x-1)]$

(b) $f(x) = (x^5 - x^4 - x + 1)/(x^2 + x + 1)$

39 Express as a single fraction

(a) $\dfrac{1}{x} - \dfrac{2}{x-2} + \dfrac{x-1}{x^2+1}$

(b) $\dfrac{1}{x^3 - 3x^2 + 3x - 1} - \dfrac{1}{x^3 - x^2 - x + 1}$

(c) $\dfrac{x+1}{x^2+1} + \dfrac{1}{x-1} - \dfrac{1}{(x-1)^2} + \dfrac{2}{x-2}$

40 Express as partial fractions

(a) $\dfrac{1}{(x+1)(x-2)}$

(b) $\dfrac{2x-1}{(x+1)(x-2)}$

(c) $\dfrac{x^2-2}{(x+1)(x-2)}$

(d) $\dfrac{x-1}{(x+1)(x-2)^2}$

(e) $\dfrac{1}{(x+1)(x^2+2x+2)}$

(f) $\dfrac{1}{(x+1)(x^2-4)}$

41 Express as partial fractions

(a) $\dfrac{1}{x^2 - 5x + 4}$

(b) $\dfrac{1}{x^3-1}$

(c) $\dfrac{3x-1}{x^3 - 3x - 2}$

(d) $\dfrac{x^2-1}{x^2 - 5x + 6}$

(e) $\dfrac{x^2+x-1}{(x^2+1)^2}$

(f) $\dfrac{18x^2 - 5x + 47}{(x^2+4)(x-1)(x+5)}$

2.5.3 Asymptotes

Sketching the graphs of rational functions gives rise to the concept of an asymptote. To illustrate, let us consider the graph of the function

$$y = f(x) = \frac{x}{1 + x} \quad (x > 0)$$

and that of its inverse

$$y = f^{-1}(x) = \frac{x}{1 - x} \quad (0 \leqslant x < 1)$$

Expressing $x/(x + 1)$ as $(x + 1 - 1)/(x + 1) = 1 - 1/(x + 1)$, we see that as x gets larger and larger $1/(x + 1)$ gets smaller and smaller, so that $x/(x + 1)$ approaches closer and closer to the value 1. This is illustrated in the graph of $y = f(x)$ shown in Figure 2.39(a). The line $y = 1$ is called a **horizontal asymptote** to the curve, and we note that the graph of $f(x)$ approaches this asymptote as $|x|$ becomes large.

Figure 2.39
Horizontal and
vertical asymptotes.

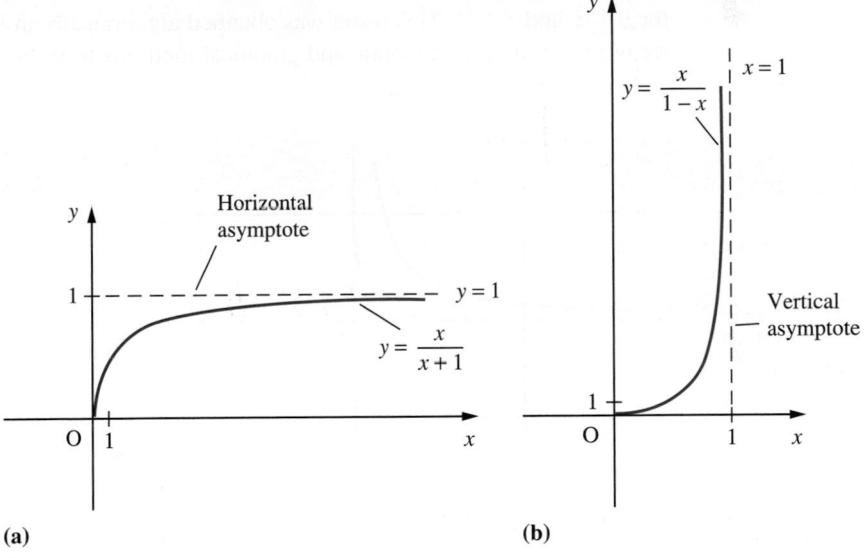

(a)

(b)

The graph of the inverse function $y = f^{-1}(x)$ is shown in Figure 2.39(b), and the line $x = 1$ is called a **vertical asymptote** to the curve.

The existence of asymptotes is a common feature of the graphs of rational functions. They feature in various engineering applications, such as in the plotting of root locus plots in control engineering. In more advanced applications of mathematics to engineering the concept of an asymptote is widely used for the purposes of making approximations. Asymptotes need not necessarily be horizontal or vertical lines; they may be sloping lines or indeed non-linear graphs, as we shall see in Example 2.37.

Example 2.36

Sketch the graph of the function

$$y = \frac{1}{3 - x} \quad (x \neq 3)$$

and find the values of x for which

$$\frac{1}{3 - x} < 2$$

Solution

We can see from the formula for y that the line $x = 3$ is a vertical asymptote of the function. As x gets closer and closer to the value $x = 3$ from the left-hand side (that is, $x < 3$), y gets larger and larger and is positive. As x gets closer and closer to $x = 3$ from the right-hand side (that is, $x > 3$), y is negative and large. As x gets larger and larger, y gets smaller and smaller for both $x > 0$ and $x < 0$, so $y = 0$ is a horizontal asymptote. Thus we obtain the sketch shown in Figure 2.40. By drawing the line $y = 2$ on the sketch, we see at once that

$$\frac{1}{3 - x} < 2$$

for $x < \frac{5}{2}$ and $x > 3$. This result was obtained algebraically in Example 1.22. Generally we use a mixture of algebraic and graphical methods to solve such problems.

Figure 2.40

Graph of $y = \dfrac{1}{3 - x}$.

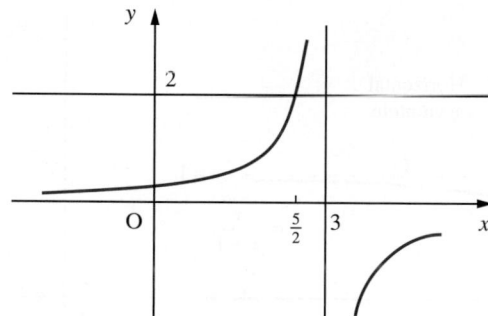

Example 2.37

Sketch the graph of the function

$$y = f(x) = \frac{x^2 - x - 6}{x + 1} \quad (x \neq -1)$$

Solution

We begin the task by locating points at which the function is zero. Now $f(x) = 0$ implies that $x^2 - x - 6 = (x - 3)(x + 2) = 0$, from which we deduce that $x = 3$ and $x = -2$ are zeros of the function. Thus the graph $y = f(x)$ crosses the x axis at $x = -2$ and $x = 3$.

Next we locate the points at which the denominator of the rational function is zero, which in this case is $x = -1$. As x approaches such a point, the value of $f(x)$ becomes infinitely large in magnitude, and the value of the rational function is undefined at such a point. Thus the graph of $y = f(x)$ has a vertical asymptote at $x = -1$. (There is usually a vertical asymptote to the graph of the rational function $y = p(x)/q(x)$ at points where the denominator $q(x) = 0$.)

Next we consider the behaviour of the function as x gets larger and larger, that is as $x \to \infty$ or $x \to -\infty$. To do this, we first simplify the rational function by algebraic division, giving

$$y = f(x) = x - 2 - \frac{4}{x + 1}$$

As $x \to \pm\infty$, $4/(x + 1) \to 0$. Thus, for large values of x, both positive and negative, $4/(x + 1)$ becomes negligible compared with x, so that $f(x)$ tends to behave like $x - 2$. Thus the line $y = x - 2$ is also an asymptote to the graph of $y = f(x)$.

Having located the asymptotes, we then need to find how the graph approaches them. When x is large and positive the term $4/(x + 1)$ will be small but positive, so that $f(x)$ is slightly less than $x - 2$. Hence the graph approaches the asymptote from below. When x is large and negative the term $4/(x + 1)$ is small but negative, so the graph approaches the asymptote from above. To consider the behaviour of the function near $x = -1$, we examine the factorized form

$$y = f(x) = \frac{(x - 3)(x + 2)}{x + 1}$$

When x is slightly less than -1, $f(x)$ is positive. When x is slightly greater than -1, $f(x)$ is negative.

We are now in a position to sketch the graph of $y = f(x)$ as shown in Figure 2.41.

Figure 2.41

Graph of $y = \dfrac{x^2 - x - 6}{x + 1}$.

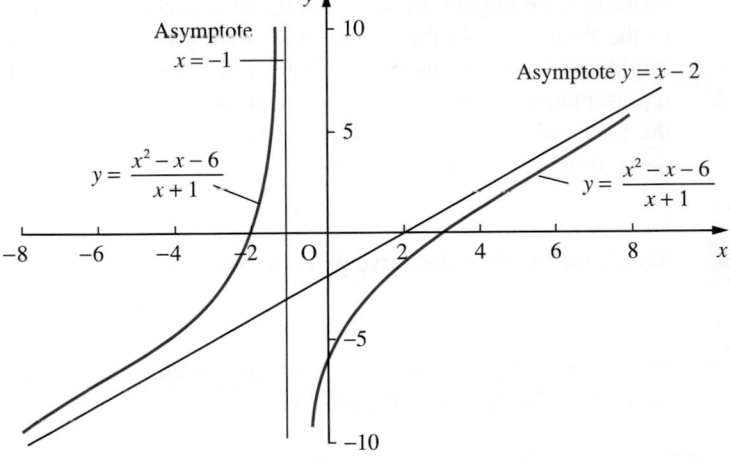

Modern computational aids have made graphing functions much easier but to obtain graphs of a reasonably good quality some preliminary analysis is always necessary. This helps to select the correct range of values for the independent variable and for the function. For example, asking a computer package to plot the function

$$y = \frac{13x^2 - 34x + 25}{x^2 - 3x + 2}$$

Figure 2.42

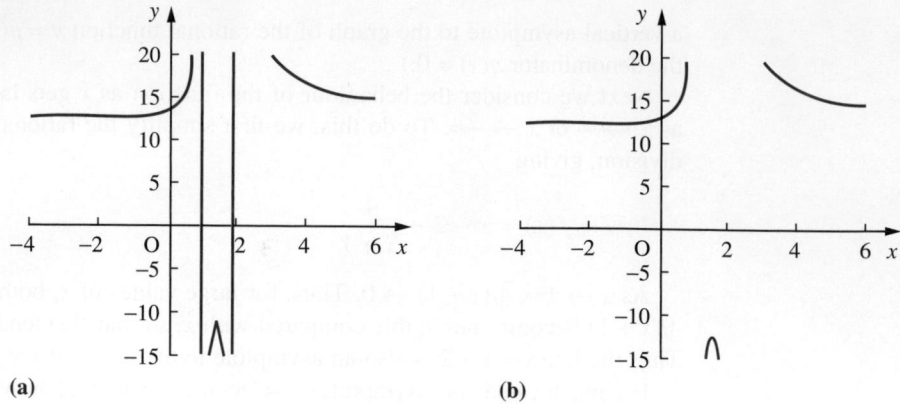

(a)   (b)

without prior analysis might result in the graph shown in Figure 2.42(a). A little analysis shows that the function is undefined at $x = 1$ and 2. Excluding these points from the range of values for x produces the more acceptable plot shown in Figure 2.42(b), although it is not clear from either plot that the graph has a horizontal asymptote $y = 13$. Clearly, much more preliminary work is needed to obtain a good quality graph of the function.

2.5.4 Parametric representation

In some practical situations the equation describing a curve in cartesian coordinates is very complicated and it is easier to specify the points in terms of a parameter. Sometimes this occurs in a very natural way. For example, in considering the trajectory of a projectile, we might specify its height and horizontal displacement separately in terms of the flight time. In the design of a safety guard for a moving part in a machine we might specify the position of the part in terms of an angle it has turned through. Such representation of curves is called **parametric representation** and we will illustrate the idea with an example. Later, in Section 2.6.6, we shall consider the polar form of specifying the equation of a curve.

Example 2.38   Sketch the graph of the curve given by $x = t^3$, $y = t^2$ ($t \in \mathbb{R}$).

Solution   The simplest approach to this type of curve sketching using pencil and paper is to draw up a table of values as in Figure 2.43.

Figure 2.43
Table of values for
Example 2.38.

t	-4	-3	-2	-1	0	1	2	3	4
x	-64	-27	-8	-1	0	1	8	27	64
y	16	9	4	1	0	1	4	9	16

Clearly in this example we need to evaluate x and y at intermediate values of t to obtain a good drawing. A sketch is shown in Figure 2.44.

Figure 2.44
Graph of the
semi-cubical parabola
$x = t^3$, $y = t^2$ ($t \in \mathbb{R}$).

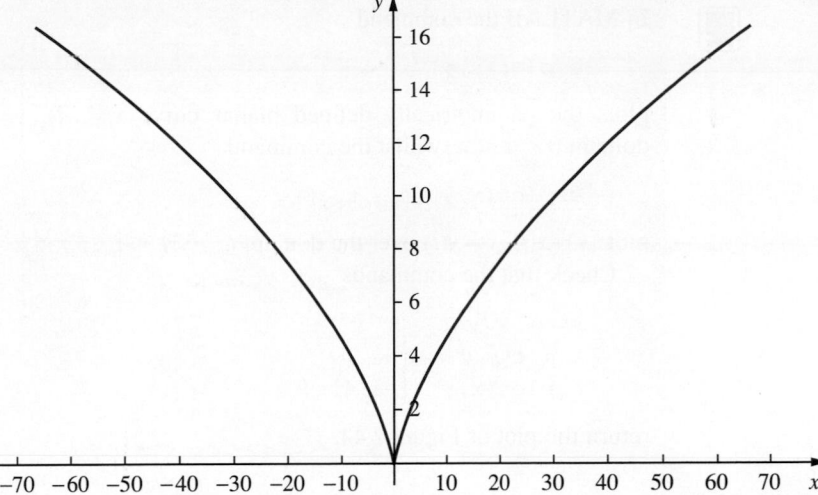

Example 2.39

The horizontal and vertical displacements of a projectile at time t are x and y, respectively, as illustrated in Figure 2.45 where $x = ut$ and $y = vt - \frac{1}{2}gt^2$ where u and v are the initial horizontal and vertical velocities and g is the acceleration due to gravity. Show that its trajectory is a parabola.

Figure 2.45
Path of a projectile.

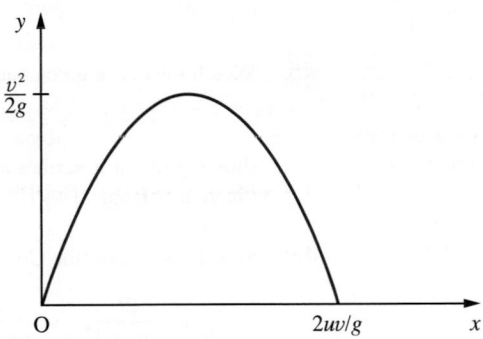

Solution   Since $x = ut$ we may write $t = x/u$. Substituting this into the expression for y gives

$$y = \frac{vx}{u} - \frac{gx^2}{2u^2}$$

which is the equation of a parabola.
   Completing the square we obtain

$$y = \frac{v^2}{2g} - \frac{g}{2u^2}\left(\frac{uv}{g} - x\right)^2$$

from which we can see that the projectile attains its maximum height, $\dfrac{v^2}{2g}$, at $x = uv/g$.

In MATLAB the command

```
ezplot(x,y)
```

plots the parametrically defined planar curve $x = x(t)$, $y = y(t)$ over the default domain $0 < t < \pi$, whilst the command

```
ezplot(x,y,[t_min, t_max])
```

plots $x = x(t)$, $y = y(t)$ over the domain $t_{\min} < t < t_{\max}$.
Check that the commands

```
syms x y t
x = t³; y = t²;
ezplot(x,y, [-4,4] )
```

return the plot of Figure 2.44.

2.5.5 Exercises

 Check the graphs obtained using MATLAB or MAPLE.

42 Plot the graphs of the functions

(a) $y = \dfrac{2 + x}{1 + x}$ (b) $y = \dfrac{1}{2}\left(x + \dfrac{2}{x}\right)$

(c) $y = \dfrac{3x^4 + 12x^2 - 4}{8x^3}$ (d) $y = \dfrac{(x - 1)(x - 2)}{(x + 1)(x - 3)}$

for the domain $-3 \leqslant x \leqslant 3$. Find the points on each graph at which they intersect with the line $y = x$.

43 Sketch the graphs of the functions given below locating their turning points and asymptotes.

(a) $y = \dfrac{x^2 - 8x + 15}{x}$ (b) $y = \dfrac{x + 1}{x - 1}$

(c) $y = \dfrac{x^2 + 5x - 14}{x + 5}$

(*Hint*: writing (a) as

$$y = (\sqrt{x} - \sqrt{(15/x)})^2 + 2\sqrt{15} - 8$$

shows that there is a turning point at $x = \sqrt{15}$.)

44 Plot the curve whose parametric equations are $x = t(t + 4)$, $y = t + 1$. Show that it is a parabola.

45 Sketch the curve given parametrically by

$$x = t^2 - 1, \quad y = t^3 - t$$

showing that it describes a closed curve as t increases from -1 to 1.

46 Sketch the curve (the Cissoid of Diocles) given by

$$x = \dfrac{2t^2}{t^2 + 1}, \quad y = \dfrac{2t^3}{t^2 + 1}$$

Show that the cartesian form of the curve is

$$y^2 = x^3/(2 - x)$$

2.6 Circular functions

The study of circular functions has a long history. The earliest known table of a circular function dates from 425 BCE and was calculated using complicated geometrical methods by the Greek astronomer–mathematician Hipparchus. He calculated the lengths of chords subtended by angles at the centre of a circle from 0° to 60° at intervals of $\frac{1}{2}$° (see Figure 2.46(a)). His work was developed by succeeding generations of Greek

Figure 2.46
(a) Hipparchus: chords as a function of angle, expressed as parts of a radius. (b) Aryabhata: half-chords as a function of angle, expressed as parts of the arc subtended by the angle with $\pi \approx 31\,416/10\,000$.

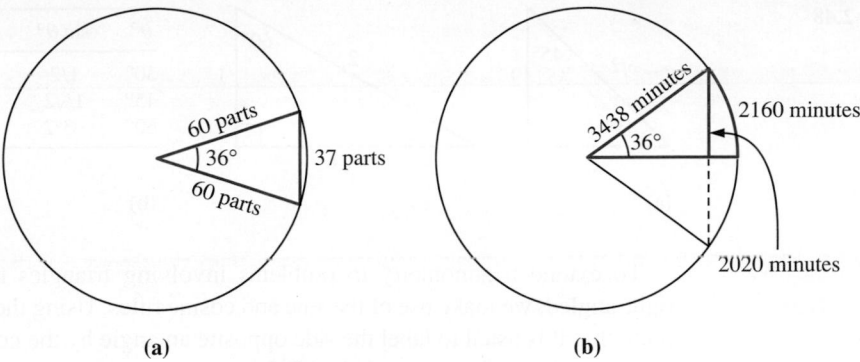

(a)   (b)

mathematicians culminating in the publication in the second century CE of a book by Ptolemy. His book *Syntaxis*, commonly called '*The Great Collection*', was translated first into Arabic, where it became *Al-majisti* and then into Latin, *Almagestus*.

Another contribution came from the Hindu mathematician Aryabhata (about 500 CE) who developed a radial measure related to angle measures and the function we now call the sine function (see Figure 2.46(b)). His work was first translated from Hindu into Arabic and then from Arabic into Latin. The various terms we use in studying these functions reflect this rich history of applied mathematics (360° from the Babylonians through the Greeks, degrees from the Latin *degradus*, minutes from *pars minuta*, sine from the Latin *sinus*, a mistranslation of the Hindu–Arabic *jiva*).

There are two approaches to the definition of the **circular** or **trigonometric functions** and this is reflected in their double name. One approach is static in nature and the other dynamic.

2.6.1 Trigonometric ratios

The static approach began with practical problems of surveying and gave rise to the mathematical problems of triangles and their measurement that we call trigonometry. We consider a right-angled triangle ABC, where $\angle CAB$ is the right-angle, and define the sine, cosine and tangent functions in relation to that triangle. Thus in Figure 2.47 we have

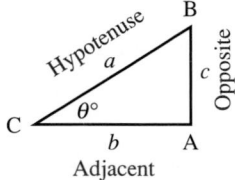

Figure 2.47

$$\text{sine } \theta° = \sin \theta° = \frac{c}{a} = \frac{\text{opposite}}{\text{hypotenuse}}$$

$$\text{cosine } \theta° = \cos \theta° = \frac{b}{a} = \frac{\text{adjacent}}{\text{hypotenuse}}$$

$$\text{tangent } \theta° = \tan \theta° = \frac{c}{b} = \frac{\text{opposite}}{\text{adjacent}}$$

The way in which these functions were defined led to their being called the 'trigonometrical ratios'. The context of the applications implied that the angles were measured in the sexagesimal system (degees, minutes, seconds): for example 35°21′41″ which today is written in the decimal form 35.36°. In modern textbooks this is shown explicitly, writing, for example, sin 30°, or cos 35.36°, or tan $\theta°$, so that the independent variable θ is a pure number. For example, by considering the triangles shown in Figure 2.48(a), we can readily write down the trigonometric ratios for 30°, 45° and 60° as indicated in the table of Figure 2.48(b).

Figure 2.48

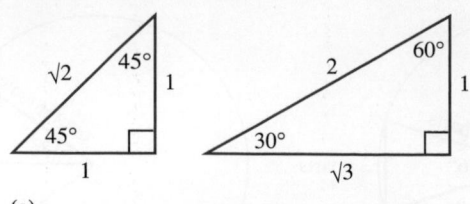

$\theta°$	$\sin \theta°$	$\cos \theta°$	$\tan \theta°$
30°	1/2	$\sqrt{3}/2$	$1/\sqrt{3}$
45°	$1/\sqrt{2}$	$1/\sqrt{2}$	1
60°	$\sqrt{3}/2$	1/2	$\sqrt{3}$

(a)   (b)

To extend trigonometry to problems involving triangles that are not necessarily right-angled, we make use of the sine and cosine rules. Using the notation of Figure 2.49 (note that it is usual to label the side opposite an angle by the corresponding lower-case letter), we have, for any triangle ABC:

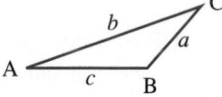

Figure 2.49

The sine rule

$$\frac{a}{\sin A} = \frac{b}{\sin B} = \frac{c}{\sin C} \tag{2.15}$$

The cosine rule

$$a^2 = b^2 + c^2 - 2bc \cos A \tag{2.16}$$

or

$$b^2 = a^2 + c^2 - 2ac \cos B$$

or

$$c^2 = a^2 + b^2 - 2ab \cos C$$

Example 2.40

Consider the surveying problem illustrated in Figure 2.50. The height of the tower is to be determined using the data measured at two points A and B, which are 20 m apart. The angles of elevation at A and B are 28°53′ and 48°51′ respectively.

Figure 2.50
Tower of
Example 2.40.

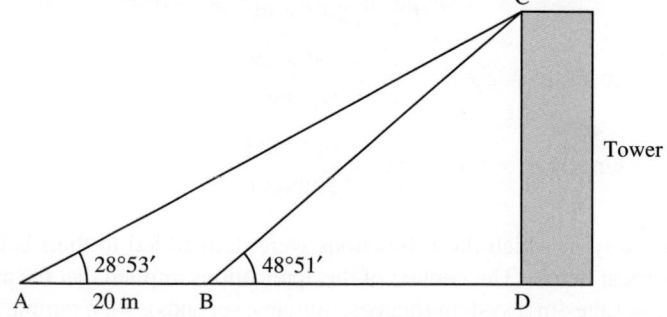

Solution   By elementary geometry

$$\angle ACB = 40°51' - 28°53' = 11°58'$$

Using the sine rule, we have

$$\frac{CB}{\sin(28°53')} = \frac{AB}{\sin(11°58')}$$

so that

$$CB = 20\sin(28°53')/\sin(11°58')$$

The height required CD is given by

$$CD = CB\sin(48°51')$$

$$= 20\sin(28°53') \times \sin(48°51')/\sin(11°58')$$

$$= 30.475$$

Hence the height of the tower is 30.5 m.

2.6.2 Exercises

47 In the triangles shown in Figure 2.51, calculate $\sin\theta°$, $\cos\theta°$ and $\tan\theta°$. Use a calculator to determine the value of θ in each case.

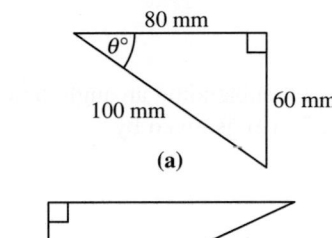

(a)

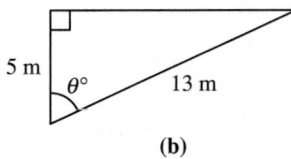

(b)

Figure 2.51

48 In the triangle ABC shown in Figure 2.52, calculate the lengths of the sides AB and BC.

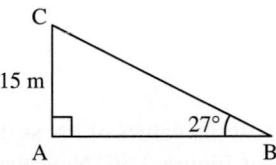

Figure 2.52

49 Calculate the value of θ where

$$\sin\theta° = \sin10°\cos20° + \cos10°\sin20°$$

50 Calculate the value of θ where

$$\cos\theta° = 2\cos^2 30° - 1$$

51 In triangle ABC, angle A is 40°, angle B is 60° and side BC is 20 mm. Calculate the lengths of the remaining two sides.

52 In triangle ABC, the angle C is 35° and the sides AC and BC have lengths 42 mm and 73 mm respectively. Calculate the length of the third side AB.

53 The lower edge of a mural, which is 4 m high, is 2 m above an observer's eye level, as shown in Figure 2.53. Show that the optical angle $\theta°$ is given by

$$\cos\theta° = \frac{12 + d^2}{\sqrt{[(4 + d^2)(36 + d^2)]}}$$

where d m is the distance of the observer from the mural. See Review exercises Question 23.

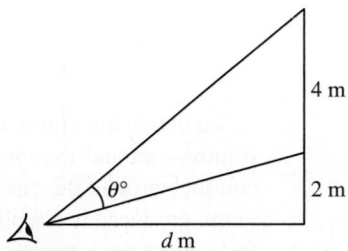

Figure 2.53 Optical angle of mural of Question 53.

2.6.3 Circular functions

The dynamic definition of the functions arises from considering the motion of a point P around a circle as shown in Figure 2.54. Many practical mechanisms involve this mathematical model.

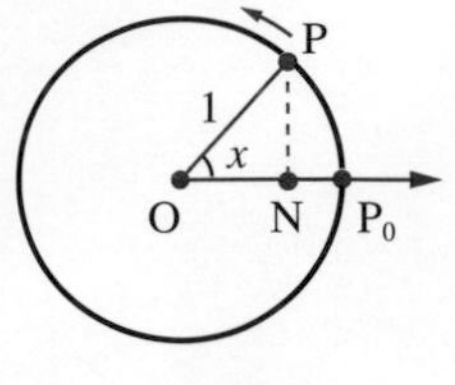

The distance OP is one unit, and the perpendicular distance NP of P from the initial position OP_0 of the rotating radius is the **sine** of the angle $\angle P_0OP$. Note that we are measuring NP positive when P is above OP_0 and negative when P is below OP_0. Similarly, the distance ON defines the **cosine** of $\angle P_0OP$ as being positive when N is to the right of O and negative when it is to the left of O.

Because we are concerned with circles and rotations in these definitions, it is natural to use circular measure so that $\angle P_0OP$, which we denote by x, is measured in radians. In this case we write simply $\sin x$ or $\cos x$, where, as before, x is a pure number. One radian is the angle that, in the notation of Figure 2.54, is subtended at the centre when the arclength P_0P is equal to the radius OP_0. Obviously therefore

$$180° = \pi \text{ radians}$$

a result we can use to convert degrees to radians and vice versa. It also follows from the definition of a radian that

(a) the length of the arc AB shown in Figure 2.55(a), of a circle of radius r, subtending an angle θ radians at the centre of the circle, is given by

$$\text{length of arc} = r\theta \qquad\qquad (2.17)$$

(b) the area of the sector OAB of a circle of radius r, subtending an angle θ radians at the centre of the circle (shown shaded in Figure 2.55(b)), is given by

$$\text{area of sector} = \tfrac{1}{2}r^2\theta \qquad\qquad (2.18)$$

Figure 2.54

Figure 2.55
(a) Arc of a circle.
(b) Sector of a circle.

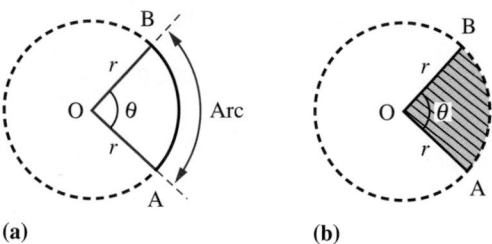

(a)   (b)

To obtain the graph of $\sin x$, we simply need to read off the values of PN as the point P moves around the circle, thus generating the graph of Figure 2.56. Note that as we continue around the circle for a second revolution (that is, as x goes from 2π to 4π) the graph produced is a replica of that produced as x goes from 0 to 2π, the same being true for subsequent intervals of 2π. By allowing P to rotate clockwise around the circle, we see that $\sin(-x) = -\sin x$, so that the graph of $\sin x$ can be extended to negative values of x, as shown in Figure 2.57.

Figure 2.56
Generating the
graph of sin x.

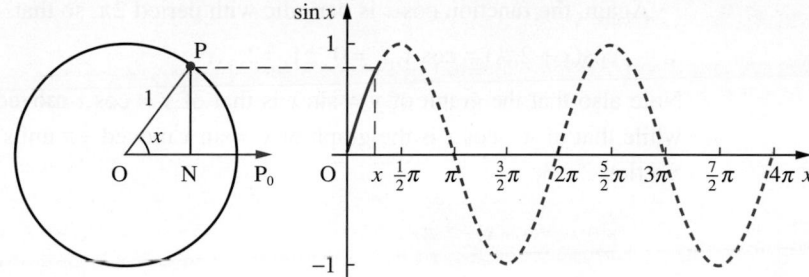

Figure 2.57
Graph of $y = \sin x$.

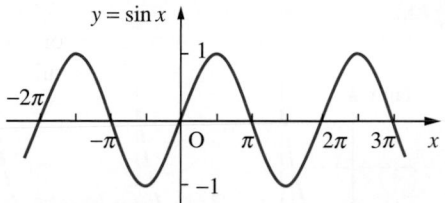

Since the graph replicates itself for every interval of 2π,

$$\sin(x + 2\pi k) = \sin x, \; k = 0, \pm 1, \pm 2, \ldots \tag{2.19}$$

and the function $\sin x$ is said to be **periodic with period** 2π.

To obtain the graph of $y = \cos x$, we need to read off the value of ON as the point P moves around the circle. To make the plotting of the graph easier, we first rotate the circle through $90°$ anticlockwise and then proceed as for $y = \sin x$ to produce the graph of Figure 2.58. By allowing P to rotate clockwise around the circle, we see that $\cos(-x) = \cos x$, so that the graph can be extended to negative values of x, as shown in Figure 2.59.

Figure 2.58
Generating the
graph of cos x.

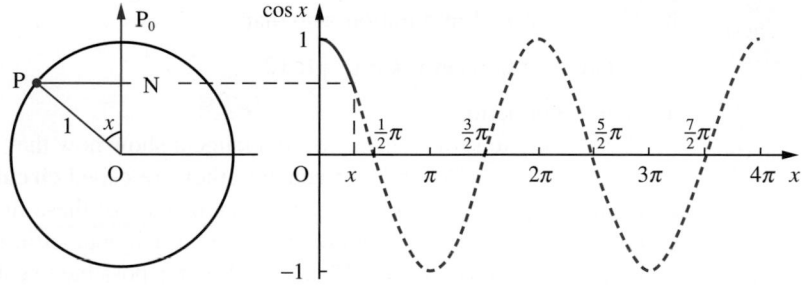

Figure 2.59
Graph of $y = \cos x$.

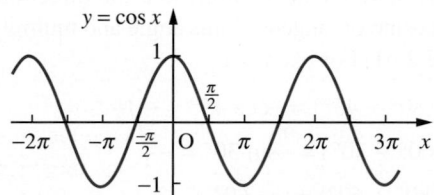

Again, the function $\cos x$ is periodic with period 2π, so that

$$\cos(x + 2\pi k) = \cos x, \; k = 0, \pm 1, \pm 2, \dots \tag{2.20}$$

Note also that the graph of $y = \sin x$ is that of $y = \cos x$ moved $\frac{1}{2}\pi$ units to the right, while that of $y = \cos x$ is the graph of $y = \sin x$ moved $\frac{1}{2}\pi$ units to the left. Thus, from Section 2.2.3,

$$\sin x = \cos(x - \tfrac{1}{2}\pi) \tag{2.21}$$

or

$$\cos x = \sin(x + \tfrac{1}{2}\pi)$$

Figure 2.60

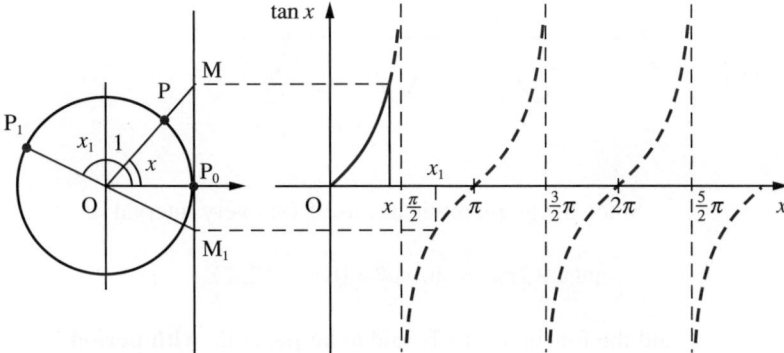

The definition of $\tan x$ is similar, and makes obvious the origin of the name 'tangent' for this function. In Figure 2.60 the rotating radius OP is extended until it cuts the tangent P_0M to the circle at the initial position P_0. The length P_0M is the **tangent** of $\angle P_0OP$. Allowing P to move around the circle, we generate the graph shown in Figure 2.60. Again, by allowing P to move in a clockwise direction, we have $\tan(-x) = -\tan x$, and the graph can readily be extended to negative values of x. In this case the graph replicates itself every interval of duration π so that

$$\tan(x + \pi k) = \tan x, \; k = 0, \pm 1, \pm 2, \dots \tag{2.22}$$

and $\tan x$ is of period π.

These definitions of sine, cosine and tangent show how they are associated with the properties of the circle, and consequently they are called **circular functions**. Often in an engineering context, the static and dynamic uses of these functions occur simultaneously. Consequently, we often refer to them as trigonometric functions.

Using the results (2.19), (2.20) and (2.22), it is possible to calculate the values of the trigonometric functions for angles greater than $\frac{1}{2}\pi$ using their values for angles between zero and $\frac{1}{2}\pi$. The rule is: take the acute angle that the direction makes with the initial direction, find the sine, cosine or tangent of this angle and multiply by $+1$ or -1 according to the scheme of Figure 2.61. For example

$$\cos(135°) = \cos(180° - 45°) = -\cos 45° = -\sqrt{\tfrac{1}{2}}$$

$$\sin(330°) = \sin(360° - 30°) = -\sin 30° = -\tfrac{1}{2}$$

$$\tan(240°) = \tan(180° + 60°) = \tan 60° = \sqrt{3}$$

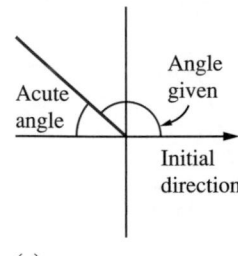

(a)

sine +	all +
cosine −	
tangent −	
tangent +	cosine +
sine −	sine −
cosine −	tangent −

(b)

Figure 2.61

As we frequently move between measuring angles in degrees and in radians, it is important to check that your calculator is in the correct mode.

If the radius OP is rotating with constant angular velocity ω (in rad s^{-1}) about O then $x = \omega t$, where t is the time (in s). The time T taken for one complete revolution is given by $\omega T = 2\pi$; that is, $T = 2\pi/\omega$. This is the **period** of the motion. In one second the radius makes $\omega/2\pi$ such revolutions. This is the **frequency**, v. Its value is given by

$$v = \text{frequency} = \frac{1}{\text{period}} = \frac{\omega}{2\pi}$$

Thus, the function $y = A \sin \omega t$, which is associated with oscillatory motion in engineering, has period $2\pi/\omega$ and **amplitude** A. The term amplitude is used to indicate the maximum distance of the graph of $y = A \sin \omega t$ from the horizontal axis.

Example 2.41

Sketch using the same set of axes the graphs of the functions

(a) $y = 2 \sin t$ (b) $y = \sin t$ (c) $y = \frac{1}{2} \sin t$

and discuss.

Solution

The graphs of the three functions are shown in Figure 2.62. The functions (a), (b) and (c) have amplitudes 2, 1 and $\frac{1}{2}$ respectively. We note that the effect of changing the amplitude is to alter the size of the 'humps' in the sine wave. Note that changing only the amplitude does not alter the points at which the graph crosses the x axis. All three functions have period 2π.

Figure 2.62

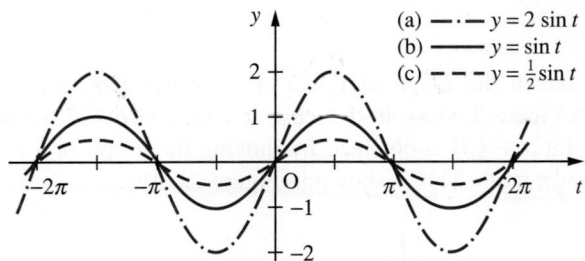

(a) $—\cdot— \;\; y = 2 \sin t$
(b) $———\;\; y = \sin t$
(c) $----\;\; y = \frac{1}{2}\sin t$

Example 2.42

Sketch using the same axes the graphs of the functions

(a) $y = \sin t$ (b) $y = \sin 2t$ (c) $y = \sin \frac{1}{2} t$

and discuss.

Solution

The graphs of the three functions (a), (b) and (c) are shown in Figure 2.63. All three have amplitude 1 and periods 2π, π and 4π respectively. We note that the effect of changing the parameter ω in $\sin \omega t$ is to 'squash' or 'stretch' the basic sine wave $\sin t$. All that happens is that the basic pattern repeats itself less or more frequently; that is, the period changes.

Figure 2.63

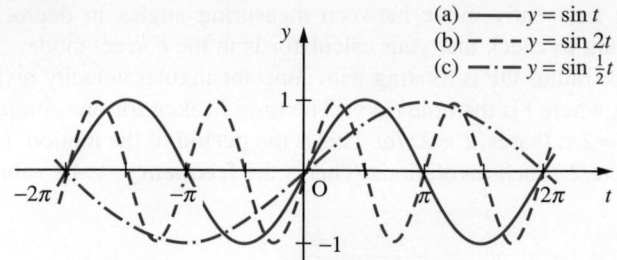

In engineering we frequently encounter the sinusoidal function

$$y = A \sin(\omega t + \alpha), \quad \omega > 0 \tag{2.23}$$

Following the discussion in Section 2.2.4, we have that the graph of this function is obtained by moving the graph of $y = A \sin \omega t$ horizontally:

$$\frac{\alpha}{\omega} \text{ units to the left if } \alpha \text{ is positive}$$

or

$$\frac{|\alpha|}{\omega} \text{ units to the right if } \alpha \text{ is negative}$$

The sine wave of (2.23) is said to 'lead' the sine wave $A \sin \omega t$ when α is positive and to 'lag' it when α is negative.

Example 2.43 Sketch the graph of $y = 3 \sin(2t + \frac{1}{3}\pi)$.

Solution First we sketch the graph of $y = 3 \sin 2t$, which has amplitude 3 and period π, as shown in Figure 2.64(a). In this case $\alpha = \frac{1}{3}\pi$ and $\omega = 2$, so it follows that the graph of $y = 3 \sin(2t + \frac{1}{3}\pi)$ is obtained by moving the graph of $y = 3 \sin 2t$ horizontally to the left by $\frac{1}{6}\pi$ units. This is shown in Figure 2.64(b).

Figure 2.64

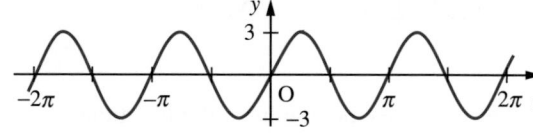

(a) $y = 3 \sin 2t$

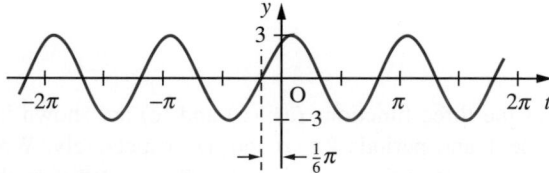

(b) $y = 3 \sin(2t + \frac{1}{3}\pi)$

Example 2.44

Consider the crank and connecting rod mechanism illustrated in Figure 2.65. Determine a functional relationship between the displacement of Q and the angle through which the crank OP has turned.

Figure 2.65
Crank and connecting rod mechanism.

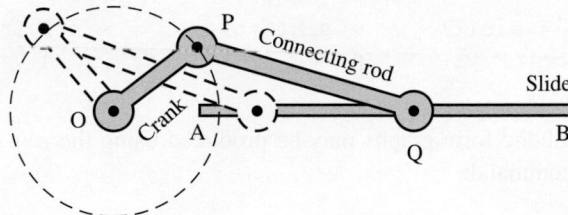

Solution

As the crank OP rotates about O, the other end of the connecting rod moves backwards and forwards along the slide AB. The displacement of Q from its initial position depends on the angle through which the crank OP has turned. A mathematical model for the mechanism replaces the crank and connecting rod, which have thickness as well as length, by straight lines, which have length only, and we consider the motion of the point Q as the line OP rotates about O, with PQ fixed in length and Q constrained to move on the line AB, as shown in Figure 2.66. We can specify the dependence of Q on the angle of rotation of OP by using some elementary trigonometry. Labelling the length of OP as r units, the length of PQ as l units, the length of OQ as y units and the angle $\angle AOP$ as x radians, and applying the cosine formula gives

Figure 2.66
Model of crank and connecting rod.

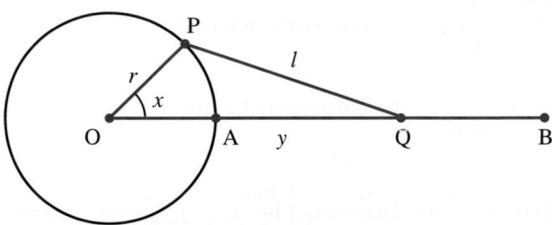

$$l^2 = r^2 + y^2 - 2yr \cos x$$

which implies

$$(y - r \cos x)^2 = l^2 - r^2 + r^2 \cos^2 x$$
$$= l^2 - r^2 \sin^2 x$$

and

$$y = r \cos x + \sqrt{(l^2 - r^2 \sin^2 x)}$$

Thus for any angle x we can calculate the corresponding value of y. We can represent this relationship by means of a graph, as shown in Figure 2.67.

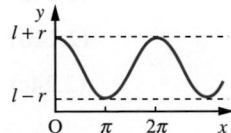

Figure 2.67

In MATLAB the circular functions are represented by $sin(x)$, $cos(x)$ and $tan(x)$ respectively. (Note that MATLAB uses radians in function evaluation.) Also in MATLAB pi (Pi in MAPLE) is a predefined variable representing the quantity π. As an example check that the commands

```
t = -2*pi : pi/90 : 2*pi;
y1 = sin(t); y2 = sin(2*t); y3 = sin(0.5*t);
plot(t, y1, '-',t ,y2, '- -', t, y3, '-.')
```

output the basic plots of Figure 2.63.

In symbolic form graphs may be produced using the $ezplot$ command. Check that the commands

```
syms t
y = sym(3*sin(2*t + pi/3));
ezplot(y,[-2*pi,2*pi] )
grid
```

produce the plot of Figure 2.64(b).

2.6.4 Trigonometric identities

Other circular functions are defined in terms of the three basic functions sine, cosine and tangent. In particular, we have

$$\sec x = \frac{1}{\cos x}, \quad \text{the } \textbf{secant} \text{ function}$$

$$\operatorname{cosec} x = \frac{1}{\sin x}, \quad \text{the } \textbf{cosecant} \text{ function}$$

$$\cot x = \frac{1}{\tan x}, \quad \text{the } \textbf{cotangent} \text{ function}$$

 In MATLAB these are determined by $sec(x)$, $csc(x)$ and $cot(x)$ respectively.

From the basic definitions it is possible to deduce the following trigonometric identities relating the functions.

Triangle identities

$$\cos^2 x + \sin^2 x = 1 \tag{2.24a}$$

$$1 + \tan^2 x = \sec^2 x \tag{2.24b}$$

$$1 + \cot^2 x = \operatorname{cosec}^2 x \tag{2.24c}$$

The first of these follows immediately from the use of Pythagoras' theorem in a right-angled triangle with a unit hypotenuse. Dividing (2.24a) through by $\cos^2 x$ yields identity (2.24b), and dividing through by $\sin^2 x$ yields identity (2.24c).

Compound-angle identities

$$\sin(x + y) = \sin x \cos y + \cos x \sin y \qquad \text{(2.25a)}$$

$$\sin(x - y) = \sin x \cos y - \cos x \sin y \qquad \text{(2.25b)}$$

$$\cos(x + y) = \cos x \cos y - \sin x \sin y \qquad \text{(2.25c)}$$

$$\cos(x - y) = \cos x \cos y + \sin x \sin y \qquad \text{(2.25d)}$$

$$\tan(x + y) = \frac{\tan x + \tan y}{1 - \tan x \tan y} \qquad \text{(2.25e)}$$

$$\tan(x - y) = \frac{\tan x - \tan y}{1 + \tan x \tan y} \qquad \text{(2.25f)}$$

Sum and product identities

$$\sin x + \sin y = 2 \sin \tfrac{1}{2}(x + y) \cos \tfrac{1}{2}(x - y) \qquad \text{(2.26a)}$$

$$\sin x - \sin y = 2 \sin \tfrac{1}{2}(x - y) \cos \tfrac{1}{2}(x + y) \qquad \text{(2.26b)}$$

$$\cos x + \cos y = 2 \cos \tfrac{1}{2}(x + y) \cos \tfrac{1}{2}(x - y) \qquad \text{(2.26c)}$$

$$\cos x - \cos y = -2 \sin \tfrac{1}{2}(x + y) \sin \tfrac{1}{2}(x - y) \qquad \text{(2.26d)}$$

From identities (2.25a), (2.25c) and (2.25e) we can obtain the double-angle formulae.

$$\sin 2x = 2 \sin x \cos x \qquad \text{(2.27a)}$$

$$\cos 2x = \cos^2 x - \sin^2 x \qquad \text{(2.27b)}$$

$$= 2 \cos^2 x - 1 \qquad \text{(2.27c)}$$

$$= 1 - 2 \sin^2 x \qquad \text{(2.27d)}$$

$$\tan 2x = \frac{2 \tan x}{1 - \tan^2 x} \qquad \text{(2.27e)}$$

(Writing $x = \theta/2$ we can obtain similar identities called half-angle formulae.)

Example 2.45

Express $\cos(\pi/2 + 2x)$ in terms of $\sin x$ and $\cos x$.

Solution

Using identity (2.25c) we obtain

$$\cos(\pi/2 + 2x) = \cos \pi/2 \cos 2x - \sin \pi/2 \sin 2x$$

Since $\cos \pi/2 = 0$ and $\sin \pi/2 = 1$, we can simplify to obtain

$$\cos(\pi/2 + 2x) = -\sin 2x$$

Now using the double-angle formula (2.27a), we obtain

$$\cos(\pi/2 + 2x) = -2 \sin x \cos x$$

Example 2.46

Show that

$$\sin(A + B) + \sin(A - B) = 2 \sin A \cos B$$

and deduce that

$$\sin x + \sin y = 2 \sin \tfrac{1}{2}(x + y) \cos \tfrac{1}{2}(x - y)$$

Hence sketch the graph of $y = \sin 4x + \sin 2x$.

Solution

Using identities (2.25a) and (2.25b) we have

$$\sin(A + B) = \sin A \cos B + \cos A \sin B$$

$$\sin(A - B) = \sin A \cos B - \cos A \sin B$$

Adding these two identities gives

$$\sin(A + B) + \sin(A - B) = 2 \sin A \cos B$$

Now setting $A + B = x$ and $A - B = y$, we see that $A = \tfrac{1}{2}(x + y)$ and $B = \tfrac{1}{2}(x - y)$ so that

$$\sin x + \sin y = 2 \sin \tfrac{1}{2}(x + y) \cos \tfrac{1}{2}(x - y)$$

which is identity (2.26a). (The identities (2.26b–d) can be proved in the same manner.)
Applying the formula to

$$y = \sin 4x + \sin 2x$$

we obtain

$$y = 2 \sin 3x \cos x$$

The graphs of $y = \sin 3x$ and $y = \cos x$ are shown in Figures 2.68(a) and (b). The combination of these two graphs yields Figure 2.68(c). This type of combination of oscillations in practical situations leads to the phenomena of 'beats'.

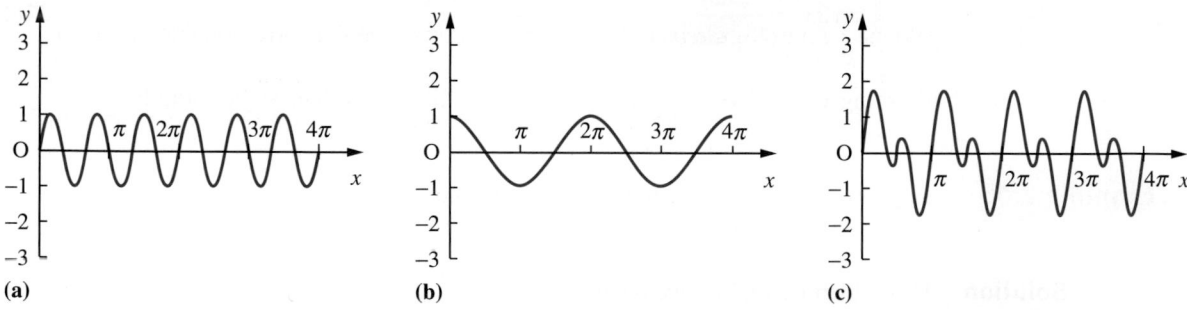

Figure 2.68 (a) graph $y = \sin 3x$ (b) $y = \cos x$ (c) $y = 2 \sin 3x \cos x$

The identities 2.26(a–d) are useful for turning the sum or difference of sines and cosines into a product of sines and/or cosines in many problems. But the reverse process is also useful in others! So we summarize here the expressing of products as sums or differences.

$$\sin x \cos y = \tfrac{1}{2}[\sin(x+y) + \sin(x-y)] \tag{2.28a}$$

$$\cos x \sin y = \tfrac{1}{2}[\sin(x+y) - \sin(x-y)] \tag{2.28b}$$

$$\cos x \cos y = \tfrac{1}{2}[\cos(x+y) + \cos(x-y)] \tag{2.28c}$$

$$\sin x \sin y = -\tfrac{1}{2}[\cos(x+y) - \cos(x-y)] \tag{2.28d}$$

Note the minus sign before the bracket in (2.28d). Before the invention of calculating machines these identities were used to perform multiplications. Commonly the mathematical tables used only tabulated the functions up to 45° to save space so that all four identities were used.

Example 2.47   Solve the equation $2\cos^2 x + 3\sin x = 3$ for $0 \leqslant x \leqslant 2\pi$.

Solution   First we express the equation in terms of $\sin x$ only. This can be done by eliminating $\cos^2 x$ using the identity (2.24a), giving

$$2(1 - \sin^2 x) + 3\sin x = 3$$

which reduces to

$$2\sin^2 x - 3\sin x + 1 = 0$$

This is now a quadratic equation in $\sin x$, and it is convenient to write $\lambda = \sin x$, giving

$$2\lambda^2 - 3\lambda + 1 = 0$$

Factorizing then gives $(2\lambda - 1)(\lambda - 1) = 0$

leading to the two solutions $\lambda = \tfrac{1}{2}$ and $\lambda = 1$

We now return to the fact that $\lambda = \sin x$ to determine the corresponding values of x.

(i) If $\lambda = \tfrac{1}{2}$ then $\sin x = \tfrac{1}{2}$. Remembering that $\sin x$ is positive for x lying in the first and second quadrants and that $\sin \tfrac{1}{6}\pi = \tfrac{1}{2}$, we have two solutions corresponding to $\lambda = \tfrac{1}{2}$, namely $x = \tfrac{1}{6}\pi$ and $x = \tfrac{5}{6}\pi$.

(ii) If $\lambda = 1$ then $\sin x = 1$, giving the single solution $\lambda = \tfrac{1}{2}\pi$.

Thus there are three solutions to the given equation, namely

$$x = \tfrac{1}{6}\pi, \quad \tfrac{1}{2}\pi \quad \text{and} \quad \tfrac{5}{6}\pi$$

Example 2.48   The path of a projectile fired with speed V at an angle α to the horizontal is given by

$$y = x\tan\alpha - \frac{1}{2}\frac{gx^2}{V^2\cos^2\alpha}$$

For fixed V a family of trajectories, for various angles of projection α, is obtained, as shown in Figure 2.69. Find the condition for a point P with coordinates (X, Y) to lie beyond the reach of the projectile.

Solution Given the coordinates (X, Y), the possible angles α of launch are given by the roots of the equation

$$Y = X \tan \alpha - \frac{1}{2} \frac{gX^2}{V^2 \cos^2 \alpha}$$

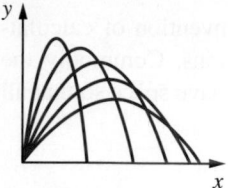

Figure 2.69
Trajectories for
different launch
angles.

Using the trigonometric identity

$$1 + \tan^2 \alpha = \frac{1}{\cos^2 \alpha}$$

gives

$$Y = X \tan \alpha - \frac{1}{2} \frac{gX^2}{V^2} (1 + \tan^2 \alpha)$$

Writing $T = \tan \alpha$, this may be rewritten as

$$(gX^2)T^2 - (2XV^2)T + (gX^2 + 2V^2Y) = 0$$

which is a quadratic equation in T. From (1.8), this equation will have two different real roots if

$$(2XV^2)^2 > 4(gX^2)(gX^2 + 2V^2Y)$$

but no real roots if

$$(2XV^2)^2 < 4(gX^2)(gX^2 + 2V^2Y)$$

Thus the point P(X, Y) is 'safe' if

$$V^4 < g^2X^2 + 2gV^2Y$$

The critical case where the point (X, Y) lies on the curve

$$V^4 = g^2x^2 + 2gV^2y$$

gives us the so-called 'parabola of safety', with the safety region being that above this parabola

$$y = \frac{V^2}{2g} - \frac{gx^2}{2V^2}$$

2.6.5 Amplitude and phase

Often in engineering contexts we are concerned with vibrations of parts of a structure or machine. These vibrations are a response to a periodic external force and will usually have the same frequency as that force. Usually, also, the response will lag behind the exciting force. Mathematically this is often represented by an external force of the form $F \sin \omega t$ with a response of the form $a \sin \omega t + b \cos \omega t$ where a and b are constants dependent on F, ω and the physical characteristics of the system. To find the size of the response we need to write it in the form $A \sin(\omega t + \alpha)$ where

$$A \sin(\omega t + \alpha) = a \sin \omega t + b \cos \omega t$$

This we can always do, as is illustrated in Example 2.49.

Example 2.49 Express $y = 4 \sin 3t - 3 \cos 3t$ in the form $y = A \sin(3t + \alpha)$.

Solution To determine the appropriate values of A and α, we proceed as follows.
Using the identity (2.25a), we have

$$A \sin(3t + \alpha) = A(\sin 3t \cos \alpha + \cos 3t \sin \alpha)$$

$$= (A \cos \alpha) \sin 3t + (A \sin \alpha) \cos 3t$$

Since this must equal the expression

$$4 \sin 3t - 3 \cos 3t$$

for all values of t, the respective coefficients of $\sin 3t$ and $\cos 3t$ must be the same in both expressions, so that

$$4 = A \cos \alpha \tag{2.29}$$

and

$$-3 = A \sin \alpha \tag{2.30}$$

The angle α is shown in Figure 2.70. By Pythagoras' theorem,

$$A = \sqrt{(16 + 9)} = 5$$

and clearly

$$\tan \alpha = -\tfrac{3}{4}$$

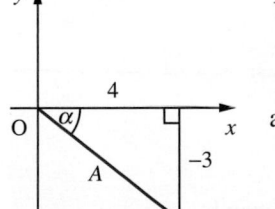

Figure 2.70
The angle α.

The value of α may now be determined using a calculator. However, care must be taken to ensure that the correct quadrant is chosen for α. Since A is taken to be positive, it follows from Figure 2.70 that α lies in the fourth quadrant. Thus, using a calculator, we have $\alpha = -0.64$ rad and

$$y = 4 \sin 3t - 3 \cos 3t = 5 \sin(3t - 0.64)$$

Using the Symbolic Math Toolbox in MATLAB commands such as *expand*, *simplify* and *simple* may be used to manipulate trigonometric functions, and the command *solve* may be used to solve trigonometric equations (these commands have been introduced earlier). Some illustrations are:

(a) The commands

```
syms x y
expand(cos(x + y))
```

return

```
cos(x)*cos(y) - sin(x)*sin(y)
```

(b) The commands

```
syms x
simplify(cos(x)^2 + sin(x)^2)
```

return

```
1
```

(c) The commands

```
syms (x)
simplify(cos(x)^2 - sin(x)^2)
```

return

```
2*cos(x)^2-1
```

whilst the command

```
simple(cos(x)^2 - sin(x)^2)
```

returns

```
cos(2*x)
```

(d) The commands

```
syms x
s = solve('2*cos(x)^2 + 3*sin(x) = 3')
```

return

```
s = 1/2*pi
    1/6*pi
    5/6*pi
```

confirming the answer obtained in Example 2.47.
  If numeric answers are required then use the command

```
double(s)
```

to obtain

```
s = 1.5708
    0.5236
    2.6180
```

2.6.6 Exercises

Check your answers using MATLAB or MAPLE whenever possible.

54 Copy and complete the table in Figure 2.71.

degrees	0	30		60			150	
radians			$\pi/4$		$\pi/2$	$2\pi/3$		π

degrees	210	225	240	270	300	315	330	
radians								2π

Figure 2.71 Conversion table: degrees to radians.

55 Sketch for $-3\pi \leqslant x \leqslant 3\pi$ the graphs of

(a) $y = \sin 2x$ (b) $y = \sin \frac{1}{2}x$

(c) $y = \sin^2 x$ (d) $y = \sin x^2$

(e) $y = \dfrac{1}{\sin x}$ $(x \neq n\pi, \ n = 0, \pm 1, \pm 2, \dots)$

(f) $y = \sin\left(\dfrac{1}{x}\right)$ $(x \neq 0)$

56 Solve the following equations for $0 \leqslant x \leqslant 2\pi$:

(a) $3 \sin^2 x + 2 \sin x - 1 = 0$

(b) $4 \cos^2 x + 5 \cos x + 1 = 0$

(c) $2 \tan^2 x - \tan x - 1 = 0$

(d) $\sin 2x = \cos x$

57 By referring to an equilateral triangle, show that $\cos \frac{1}{3}\pi = \frac{1}{2}\sqrt{3}$ and $\tan \frac{1}{6}\pi = \frac{1}{3}\sqrt{3}$, and find values for $\sin \frac{1}{3}\pi$, $\tan \frac{1}{3}\pi$, $\cos \frac{1}{6}\pi$ and $\sin \frac{1}{6}\pi$. Hence, using the double-angle formulae, find $\sin \frac{1}{12}\pi$, $\cos \frac{1}{12}\pi$ and $\tan \frac{1}{12}\pi$. Using appropriate properties from Section 2.6, calculate

(a) $\sin \frac{2}{3}\pi$ (b) $\tan \frac{7}{6}\pi$ (c) $\cos \frac{11}{6}\pi$

(d) $\sin \frac{5}{12}\pi$ (e) $\cos \frac{7}{12}\pi$ (f) $\tan \frac{11}{12}\pi$

58 Given $s = \sin \theta$, where $\frac{1}{2}\pi < \theta < \pi$, find, in terms of s,

(a) $\cos \theta$ (b) $\sin 2\theta$

(c) $\sin 3\theta$ (d) $\sin \frac{1}{2}\theta$

59 Show that

$$\frac{1 + \sin 2\theta + \cos 2\theta}{1 + \sin 2\theta - \cos 2\theta} = \cot \theta$$

60 Given $t = \tan \frac{1}{2}x$, prove that

(a) $\sin x = \dfrac{2t}{1 + t^2}$

(b) $\cos x = \dfrac{1 - t^2}{1 + t^2}$

(c) $\tan x = \dfrac{2t}{1 - t^2}$

Hence solve the equation

$$2 \sin x - \cos x = 1$$

61 In each of the following, the value of one of the six circular functions is given. Without using a calculator, find the values of the remaining five.

(a) $\sin x = \frac{1}{2}$ (b) $\cos x = -\frac{1}{2}\sqrt{3}$

(c) $\tan x = -1$ (d) $\sec x = \sqrt{2}$

(e) $\operatorname{cosec} x = -2$ (f) $\cot x = \sqrt{3}$

62 Express as a product of sines and/or cosines

(a) $\sin 3\theta + \sin \theta$ (b) $\cos \theta - \cos 2\theta$

(c) $\cos 5\theta + \cos 2\theta$ (d) $\sin \theta - \sin 2\theta$

63 Express as a sum or difference of sines or cosines

(a) $\sin 3\theta \sin \theta$ (b) $\sin 3\theta \cos \theta$

(c) $\cos 3\theta \sin \theta$ (d) $\cos 3\theta \cos \theta$

64 Express in the forms $r \cos(\theta - \alpha)$ and $r \sin(\theta - \beta)$

(a) $\sqrt{3} \sin \theta - \cos \theta$ (b) $\sin \theta - \cos \theta$

(c) $\sin \theta + \cos \theta$ (d) $2 \cos \theta + 3 \sin \theta$

65 Show that $-\frac{3}{2} \leqslant 2 \cos x + \cos 2x \leqslant 3$ for all x, and determine those values of x for which equality holds. Plot the graph of $y = 2 \cos x + \cos 2x$ for $0 \leqslant x \leqslant 2\pi$.

2.6.7 Inverse circular (trigonometric) functions

Considering the inverse of the trigonometric functions, it follows from the definition given in (2.4) that the inverse sine function $\sin^{-1}x$ (also sometimes denoted by arcsin x) is such that

$$\text{if} \quad y = \sin^{-1}x \quad \text{then} \quad x = \sin y$$

Here x should not be interpreted as an angle – rather $\sin^{-1}x$ represents the angle whose sine is x. Applying the procedures for obtaining the graph of the inverse function given in Section 2.2.3 to the graph of $y = \sin x$ (Figure 2.55) leads to the graph shown in Figure 2.72(a). As we explained in Example 2.8, when considering the inverse of $y = x^2$, the graph of Figure 2.72(a) is not representative of a function, since for each value of x in the domain $-1 \leqslant x \leqslant 1$ there are an infinite number of image values (as indicated by the points of intersection of the broken vertical line with the graph). To overcome this problem, we restrict the range of the inverse function $\sin^{-1}x$ to $-\frac{1}{2}\pi \leqslant \sin^{-1}x \leqslant \frac{1}{2}\pi$ and define the inverse sine function by

$$\text{if } y = \sin^{-1}x \text{ then } x = \sin y, \quad \text{where } -\tfrac{1}{2}\pi \leqslant y \leqslant \tfrac{1}{2}\pi \text{ and } -1 \leqslant x \leqslant 1 \qquad (2.31)$$

The corresponding graph is shown in Figure 2.72(b).

Similarly, in order to define the inverse cosine and inverse tangent functions $\cos^{-1}x$ and $\tan^{-1}x$ (also sometimes denoted by arccos x and arctan x), we have to restrict the ranges. This is done according to the following definitions.

$$\text{if } y = \cos^{-1}x \text{ then } x = \cos y, \quad \text{where } 0 \leqslant y \leqslant \pi \text{ and } -1 \leqslant x \leqslant 1 \qquad (2.32)$$

$$\text{if } y = \tan^{-1}x \text{ then } x = \tan y, \quad \text{where } -\tfrac{1}{2}\pi < y < \tfrac{1}{2}\pi \text{ and } x \text{ is any real number} \qquad (2.33)$$

Figure 2.72
Graph of $\sin^{-1}x$.

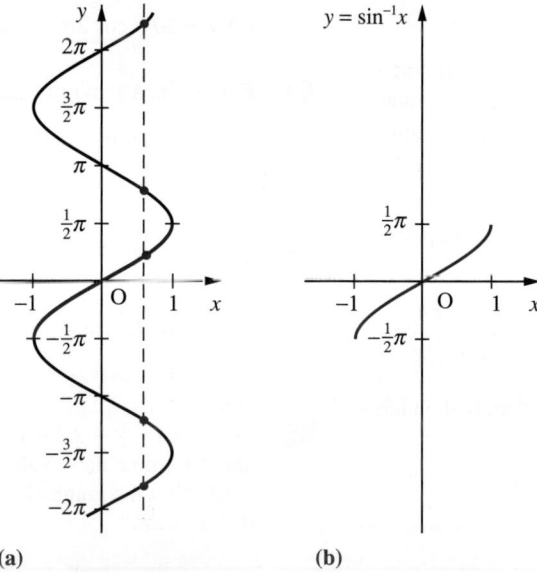

(a)　　　　(b)

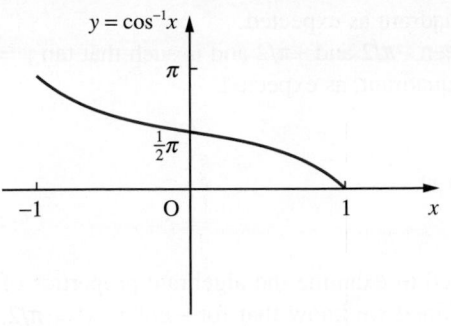

Figure 2.73 Graph of $\cos^{-1}x$.

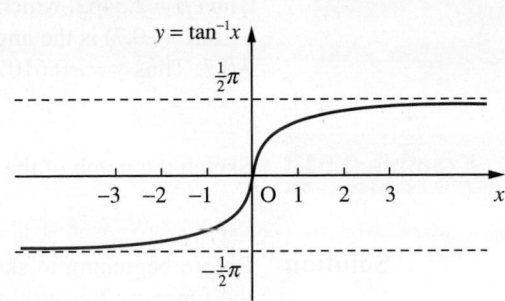

Figure 2.74 Graph of $\tan^{-1}x$.

The corresponding graphs of $y = \cos^{-1}x$ and $y = \tan^{-1}x$ are shown in Figures 2.73 and 2.74, respectively.

In some books (2.31)–(2.33) are called the *principal values* of the inverse functions. A calculator will automatically give these values.

Example 2.50 Evaluate $\sin^{-1}x$, $\cos^{-1}x$, $\tan^{-1}x$ where (a) $x = 0.35$ and (b) $x = -0.7$, expressing the answers correct to 4dp.

Solution (a) $\sin^{-1}(0.35)$ is the angle α which lies between $-\pi/2$ and $+\pi/2$ and is such that $\sin \alpha = 0.35$. Using a calculator we have

$$\sin^{-1}(0.35) = 0.3576 \text{ (4dp)} = 0.1138\pi$$

which clearly lies between $-\pi/2$ and $+\pi/2$.

$\cos^{-1}(0.35)$ is the angle β which lies between 0 and π and is such that $\cos \beta = 0.35$. Using a calculator we obtain

$$\cos^{-1}(0.35) = 1.2132 \text{ (4dp)} = 0.3862\pi$$

which lies between 0 and π.

$\tan^{-1}(0.35)$ is the angle γ which lies between $-\pi/2$ and $+\pi/2$ and is such that $\tan \gamma = 0.35$. Using a calculator we have

$$\tan^{-1}(0.35) = 0.3367 \text{ (4dp)} = 0.1072\pi$$

which lies in the correct range of values.

Notice

$$\frac{\sin^{-1}(0.35)}{\cos^{-1}(0.35)} \neq \tan^{-1}(0.35)$$

(b) $\sin^{-1}(-0.7)$ is the angle α which lies between $-\pi/2$ and $+\pi/2$ and is such that $\sin \alpha = -0.7$. Again using a calculator we obtain

$$\sin^{-1}(-0.7) = -0.7754 \text{ (4dp)}$$

which lies in the correct range of values.

$\cos^{-1}(-0.7)$ is the angle β which lies between 0 and π and is such that $\cos \beta = -0.7$.

Thus $\beta = 2.3462$, which lies in the second quadrant as expected.

$\tan^{-1}(-0.7)$ is the angle γ which lies between $-\pi/2$ and $+\pi/2$ and is such that $\tan \gamma = -0.7$. Thus $\gamma = -0.6107$, lying in the fourth quadrant, as expected.

Example 2.51   Sketch the graph of the function $y = \sin^{-1}(\sin x)$.

Solution   Before beginning to sketch the graph we need to examine the algebraic properties of the function. Because of the way $\sin^{-1}$ is defined we know that for $-\pi/2 \leqslant x \leqslant \pi/2$, $\sin^{-1}(\sin x) = x$. (The function $\sin^{-1}x$ *strictly* is the inverse function of $\sin x$ with the restricted domain $-\pi/2 \leqslant x \leqslant \pi/2$.) We also know that $\sin x$ is an odd function, so that $\sin(-x) = -\sin x$. This implies that $\sin^{-1}x$ is an odd function. In fact, this is obvious from its graph (Figure 2.72(b)). Thus, $\sin^{-1}(\sin x)$ is an odd function. Lastly, since $\sin x$ is a periodic function with period 2π we conclude that $\sin^{-1}(\sin x)$ is also a periodic function of period 2π. Thus, if we can sketch the graph between 0 and π, we can obtain the graph between $-\pi$ and 0 by antisymmetry about $x = 0$ and the whole graph by periodicity elsewhere. Using Figures 2.72(a) and 2.72(b) we can obtain the graph of the function for $0 \leqslant x \leqslant \pi$ as shown in Figure 2.75 (blue). The graph between $-\pi$ and 0 is obtained by antisymmetry about the origin, as shown with the broken line in Figure 2.75, and the whole graph is obtained making use of the piece between $-\pi$ and $+\pi$ and periodicity.

Figure 2.75
Graph of
$y = \sin^{-1}(\sin x)$.

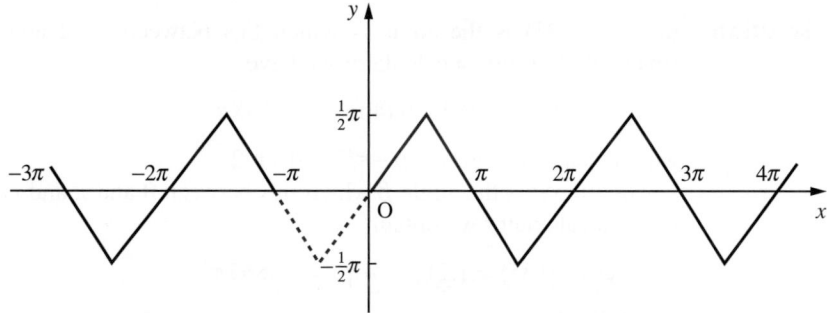

2.6.8   Polar coordinates

In some applications the position of a point P in a plane is represented by its distance r from a fixed point O and the angle θ that the line joining P to O makes with some fixed direction. The pair (r, θ) determine the point uniquely and are called the **polar coordinates** of P. If polar coordinates are chosen sharing the same origin O as rectangular cartesian coordinates and with the angle θ measured from the direction of the Ox axis then, as can be seen from Figure 2.76, the polar coordinates (r, θ) and the cartesian coordinates (x, y) of a point are related by

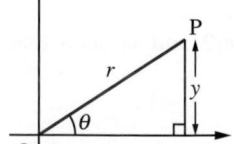

Figure 2.76

$$x = r \cos \theta, \quad y = r \sin \theta \qquad (2.34)$$

and also

$$r = \sqrt{(x^2 + y^2)}, \quad \tan\theta = \frac{y}{x}$$

Note that the origin does not have a well-defined θ. Some care must be taken when evaluating θ using the above formula to ensure that it is located in the correct quadrant. The angle $\tan^{-1}(y/x)$ obtained from tables or a calculator will usually lie between $\pm\frac{1}{2}\pi$ and will give the correct value of θ if P lies in the first or fourth quadrant. If P lies in the second or third quadrant then $\theta = \tan^{-1}(y/x) + \pi$. It is sensible to use the values of $\sin\theta$ and $\cos\theta$ to check that θ lies in the correct quadrant.

Note that the angle θ is positive when measured in an anticlockwise direction and negative when measured in a clockwise direction. Many calculators have rectangular (cartesian) to polar conversion and vice versa.

Example 2.52 (a) Find the polar coordinates of the points whose cartesian coordinates are $(1, 2)$, $(-1, 3)$, $(-1, -1)$, $(1, -2)$, $(1, 0)$, $(0, 2)$, $(0, -2)$.

(b) Find the cartesian coordinates of the points whose polar coordinates are $(3, \pi/4)$, $(2, -\pi/6)$, $(2, -\pi/2)$, $(5, 3\pi/4)$.

Solution (a) Using the formula (2.34) we see that:

$$(x = 1, y = 2) \equiv (r = \sqrt{5}, \theta = \tan^{-1}(2/1) = 1.107)$$

$$(x = -1, y = 3) \equiv (r = \sqrt{10}, \theta = 1.893)$$

$$(x = -1, y = -1) \equiv (r = \sqrt{2}, \theta = 5\pi/4)$$

$$(x = 1, y = -2) \equiv (r = \sqrt{5}, \theta = -1.107)$$

$$(x = 1, y = 0) \equiv (r = 1, \theta = 0)$$

$$(x = 0, y = 2) \equiv (r = 2, \theta = \pi/2)$$

$$(x = 0, y = -2) \equiv (r = 2, \theta = -\pi/2)$$

(Here answers, where appropriate, are given to 3dp.)

(b) Using the formula (2.34) we see that

$$(r = 3, \theta = \pi/4) \equiv (x = 3/\sqrt{2}, y = 3/\sqrt{2})$$

$$(r = 2, \theta = -\pi/6) \equiv (x = \sqrt{3}, y = -1)$$

$$(r = 2, \theta = -\pi/2) \equiv (x = 0, y = -2)$$

$$(r = 5, \theta = 3\pi/4) \equiv (x = -5/\sqrt{2}, y = 5\sqrt{2})$$

To plot a curve specified using polar coordinates we first look for any features, for example symmetry, which would reduce the amount of calculation, and then we draw up a table of values of r against values of θ. This is a tedious process and we usually use a graphics calculator or a computer package to perform the task. There are, however, different conventions in use about polar plotting. Some packages are designed to

Figure 2.77
(a) $r = 2a \cos \theta$, $0 \leqslant \theta \leqslant \pi$, $r \geqslant 0$. (b) $r = 2a \cos \theta$, $0 \leqslant \theta \leqslant \pi$, r unrestricted.

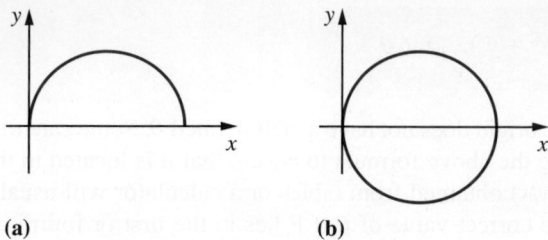

(a)　　　　(b)

plot only points where r is positive, so that plotting $r = 2a \cos \theta$ for $0 \leqslant \theta \leqslant \pi$ would yield Figure 2.77(a) while other packages plot negative values of r treating r as a number line, so that $r = 2a \cos \theta$ for $0 \leqslant \theta \leqslant \pi$ yields Figure 2.77(b).

Example 2.53

Express the equation of the circle

$$(x - a)^2 + y^2 = a^2$$

in polar form.

Solution

Expanding the squared term, the equation of the given circle becomes

$$x^2 + y^2 - 2ax = 0$$

Using the relationships (2.34), we have

$$r^2(\cos^2\theta + \sin^2\theta) - 2ar \cos \theta = 0$$

Using the trigonometric identity (2.24a),

$$r(r - 2a \cos \theta) = 0, \quad -\pi/2 < \theta \leqslant \pi/2$$

Since $r = 0$ gives the point $(0, 0)$, we can ignore this, and the equation of the circle becomes

$$r = 2a \cos \theta, \quad -\pi/2 < \theta \leqslant \pi/2$$

Example 2.54

Sketch the curve whose polar equation is $r = 1 + \cos \theta$.

Solution

The simplest approach when sketching a curve given in polar coordinate form is to draw up a table of values as in Figure 2.78.

Figure 2.78
Table of values for $r = 1 + \cos \theta$.

θ	0	15	30	45	60	75	90	105	120	135	150	165	180
r	2	1.97	1.87	1.71	1.50	1.26	1	0.74	0.50	0.29	0.13	0.03	0

Because it is difficult to measure angles accurately it is easier to convert these values into the cartesian coordinate values using (2.34) when polar coordinate graph paper is not available. The sketch of the curve, a cardioid, is shown in Figure 2.79. Here we have made use of the symmetry of the curve about the line $\theta = 0$, that is the line $y = 0$.

Figure 2.79
The cardinal
$r = 1 + \cos \theta$.

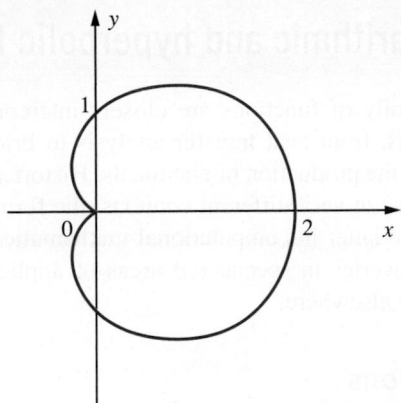

 In MATLAB the inverse circular functions $\sin^{-1}(x)$, $\cos^{-1}(x)$ and $\tan^{-1}(x)$ are denoted by $asin(x)$, $acos(x)$ and $atan(x)$ respectively. (In MAPLE these are denoted by $arcsin$, $arccos$ and $arctan$ respectively.) Using the graphical commands given in page (71) check the graphs of Figures 2.70–2.73.

Symbolically a plot of the polar curve $r = f(\theta)$ is obtained using the command $ezpolar(f)$, over the default domain $0 < \theta < 2\pi$; whilst the command $ezpolar(f,[a,b])$ plots the curve over the domain $a < \theta < b$. Check that the commands

```
syms theta
r = 1 + cos(theta);
ezpolar(r)
```

plot the graph of the cardioid in Example 2.54.

2.6.9 Exercises

66 Evaluate

(a) $\sin^{-1}(0.5)$ (b) $\sin^{-1}(-0.5)$

(c) $\cos^{-1}(0.5)$ (d) $\cos^{-1}(-0.5)$

(e) $\tan^{-1}(\sqrt{3})$ (f) $\tan^{-1}(-\sqrt{3})$

67 Sketch the graph of the functions

(a) $y = \sin^{-1}(\cos x)$

(b) $y = \cos^{-1}(\sin x)$

(c) $y = \cos^{-1}(\cos x)$

(d) $y = \cos^{-1}(\cos x) - \sin^{-1}(\sin x)$

68 If $\tan^{-1}x = \alpha$ and $\tan^{-1}y = \beta$, show that

$$\tan(\alpha + \beta) = \frac{x + y}{1 - xy}$$

Deduce that

$$\tan^{-1}x + \tan^{-1}y = \tan^{-1}\left(\frac{x + y}{1 - xy}\right) + k\pi$$

where $k = -1$, 0, 1 depending on the values of x and y.

69 Sketch the curve with polar form

$$r = 1 + 2 \cos \theta$$

70 Sketch the curve whose polar form is

$$r = 1/(1 + 2 \cos \theta)$$

Show that its cartesian form is

$$3x^2 - 4x - y^2 + 1 = 0$$

2.7 Exponential, logarithmic and hyperbolic functions

The members of this family of functions are closely interconnected. They occur in widely varied applications, from heat transfer analysis to bridge design, from transmission line modelling to the production of chemicals. Historically the exponential and logarithmic functions arose in very different contexts, the former in the calculation of compound interest and the latter in computational mathematics, but, as often happens in mathematics, the discoveries in specialized areas of applicable mathematics have found applications widely elsewhere.

2.7.1 Exponential functions

Functions of the type $f(x) = a^x$ where a is a positive constant (and x is the independent variable as usual) are called **exponential functions**.

The graphs of the exponential functions, shown in Figure 2.80, are similar. By a simple scaling of the x axis, we can obtain the same graphs for $y = 2^x$, $y = 3^x$ and $y = 4^x$, as shown in Figure 2.81. The reason for this is that we can write $3^x = 2^{kx}$ where $k \approx 1.585$ and $4^x = 2^{2x}$. Thus all exponential functions can be expressed in terms of one exponential function. The standard exponential function that is used is $y = e^x$, where e is a special number approximately equal to

$$2.718\,281\,828\,459\,045\,2\ldots$$

Figure 2.80
Graphs of exponential functions.

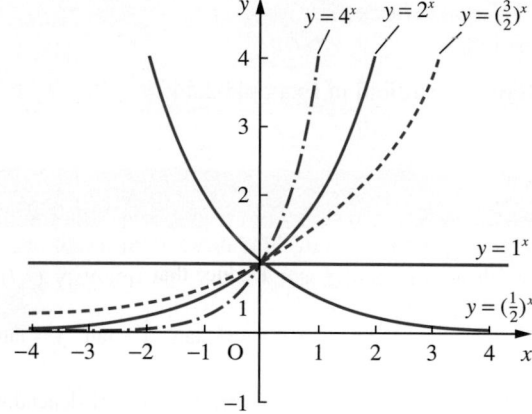

Figure 2.81
Scaled graphs of exponential functions.

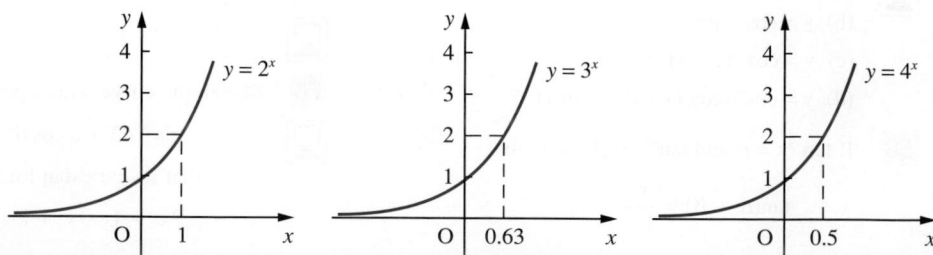

Figure 2.82
The standard
exponential
function $y = e^x$.

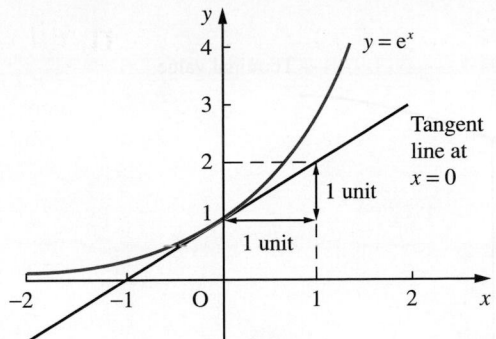

This number e is chosen because the graph of $y = e^x$ (Figure 2.82) has the property that the slope of the tangent at any point on the curve is equal to the value of the function at that point. We shall discuss this property again in Chapter 8 section 8.3.12.

We note that the following properties are satisfied by the exponential function:

$$e^{x_1}e^{x_2} = e^{x_1+x_2} \qquad\qquad\qquad (2.35a)$$

$$e^{x+c} = e^x e^c = Ae^x, \quad \text{where } A = e^c \qquad\qquad (2.35b)$$

$$\frac{e^{x_1}}{e^{x_2}} = e^{x_1-x_2} \qquad\qquad\qquad (2.35c)$$

$$e^{kx} = (e^k)^x - a^x, \quad \text{where } a = e^k \qquad\qquad (2.35d)$$

Often e^x is written as exp x for clarity when 'x' is a complicated expression. For example,

$$e^{(x+1)/(x+2)} = \exp\left(\frac{x+1}{x+2}\right)$$

Example 2.55 A tank is initially filled with 1000 litres of brine containing 0.25 kg of salt/litre. Fresh brine containing 0.5 kg of salt/litre flows in at a rate of 3 litres per second and a uniform mixture flows out at the same rate. The quantity $Q(t)$ kg of salt in the tank t seconds later is given by

$$Q(t) = A + Be^{-3t/1000}$$

Find the values of A and B and sketch a graph of $Q(t)$. Use the graph to estimate the time taken for $Q(t)$ to achieve the value 375.

Solution Initially there is 1000×0.25 kg of salt in the tank so that $Q(0) = 250$. Ultimately the brine in the tank will contain 0.5 kg of salt/litre so that the terminal value of Q will be 500. The terminal value of $A + Be^{-3t/1000}$ is A so we deduce $A = 500$. From initial data we have

$$250 = 500 + Be^0$$

and since $e^0 = 1$, $B = -250$ and

$$Q(t) = 500 - 250e^{-3t/1000}$$

Figure 2.83
The timeline of $Q(t)$.

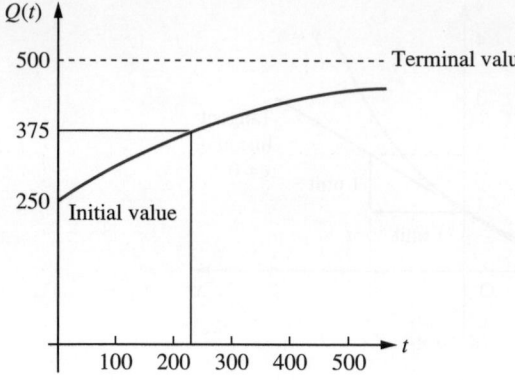

The graph of $Q(t)$ is shown in Figure 2.83. From the graph, an estimate for the time taken for $Q(t)$ to achieve the value 375 is 234 seconds. From the formula this gives $Q(234) = 376.1$. Investigating values near $t = 234$ using a calculator gives the more accurate time 231 seconds.

Example 2.56   The temperature T of a body cooling in an environment, whose unknown ambient temperature is α, is given by

$$T(t) = \alpha + (T_0 - \alpha)e^{-kt}$$

where T_0 is the initial temperature of the body and k is a physical constant. To determine the value of α, the temperature of the body is recorded at two times, t_1 and t_2, where $t_2 = 2t_1$ and $T(t_1) = T_1$, $T(t_2) = T_2$. Show that

$$\alpha = \frac{T_0 T_2 - T_1^2}{T_2 - 2T_1 + T_0}$$

Solution   From the formula for $T(t)$ we have

$$T_1 - \alpha = (T_0 - \alpha)e^{-kt_1}$$

and

$$T_2 - \alpha = (T_0 - \alpha)e^{-2kt_1}$$

Squaring the first of these two equations and then dividing by the second gives

$$\frac{(T_1 - \alpha)^2}{T_2 - \alpha} = \frac{(T_0 - \alpha)^2 e^{-2kt_1}}{(T_0 - \alpha)e^{-2kt_1}}$$

This simplifies to

$$(T_1 - \alpha)^2 = (T_2 - \alpha)(T_0 - \alpha)$$

Multiplying out both sides, we obtain

$$T_1^2 - 2\alpha T_1 + \alpha^2 = T_0 T_2 - (T_0 + T_2)\alpha + \alpha^2$$

which gives

$$(T_0 - 2T_1 + T_2)\alpha = T_0 T_2 - T_1^2$$

Hence the result.

2.7.2 Logarithmic functions

From the graph of $y = e^x$, given in Figure 2.82, it is clear that it is a one-to-one function, so that its inverse function is defined. This inverse is called the **natural logarithm** function and is written as

$$y = \ln x$$

(In some textbooks it is written as $\log_e x$, while in many pure mathematics books it is written simply as $\log x$.) Using the procedures given in Section 2.2.1, its graph can be drawn as in Figure 2.84. From the definition we have

$$\text{if } y = e^x \quad \text{then} \quad x = \ln y \tag{2.36}$$

which implies that

$$\ln e^x = x, \quad e^{\ln y} = y$$

In the same way as there are many exponential functions (2^x, 3^x, 4^x, ...), there are also many logarithmic functions. In general,

$$y = a^x \quad \text{gives} \quad x = \log_a y \tag{2.37}$$

which can be expressed verbally as 'x equals log to base a of y'. (Note that $\log_{10} x$ is often written, except in advanced mathematics books, simply as $\log x$.) Recalling that $a^x = e^{kx}$ for some constant k, we see now that $a^x = (e^k)^x$, so that $a = e^k$ and $k = \ln a$.

From the definition of $\log_a x$ it follows that

Figure 2.84
Graph of $y = \ln x$.

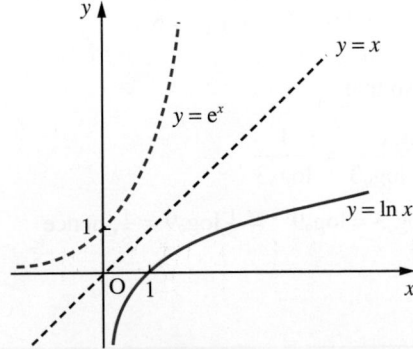

$$\log_a(x_1 x_2) = \log_a x_1 + \log_a x_2 \qquad\qquad \textbf{(2.38a)}$$

$$\log_a\left(\frac{x_1}{x_2}\right) = \log_a x_1 - \log_a x_2 \qquad\qquad \textbf{(2.38b)}$$

$$\log_a x^n = n \log_a x \qquad\qquad \textbf{(2.38c)}$$

$$x = a^{\log_a x} \qquad\qquad \textbf{(2.38d)}$$

$$y^x = a^{x \log_a y} \qquad\qquad \textbf{(2.38e)}$$

$$\log_a x = \frac{\log_b x}{\log_b a} \qquad\qquad \textbf{(2.38f)}$$

Example 2.57   (a) Evaluate $\log_2 32$.

(b) Simplify $\frac{1}{3}\log_2 8 - \log_2\frac{2}{7}$.

(c) Expand $\ln\left(\dfrac{\sqrt{(10x)}}{y^2}\right)$.

(d) Use the change of base formula (2.36f) to evaluate $\dfrac{\log_{10} 32}{\log_{10} 2}$.

(e) Evaluate $\dfrac{\log_3 x}{\log_9 x}$.

Solution   (a) Since $32 = 2^5$, $\log_2 32 = \log_2 2^5 = 5\log_2 2 = 5$, since $\log_2 2 = 1$.

(b) $\frac{1}{3}\log_2 8 - \log_2\frac{2}{7} = \log_2 8^{1/3} - \log_2\frac{2}{7}$

$$= \log_2 2 - [\log_2 2 - \log_2 7] = \log_2 7$$

(c) $\ln\left(\dfrac{\sqrt{(10x)}}{y^2}\right) = \ln(\sqrt{(10x)}) - \ln(y^2) = \frac{1}{2}\ln(10x) - 2\ln y$

$$= \frac{1}{2}\ln(10) + \frac{1}{2}\ln x - 2\ln y$$

(d) $\log_{10} 32 = \log_2 32 \, \log_{10} 2$, hence

$$\frac{\log_{10} 32}{\log_{10} 2} = \log_2 32 = \log_2 2^5 = 5\log_2 2 = 5$$

(e) $\log_9 x = \log_3 x \, \log_9 3$, so that

$$\frac{\log_3 x}{\log_9 x} = \frac{\log_3 x}{\log_3 x \log_9 3} = \frac{1}{\log_9 3}$$

But $3 = 9^{1/2}$ so that $\log_9 3 = \log_9 9^{1/2} = \frac{1}{2}\log_9 9 = \frac{1}{2}$, hence

$$\frac{\log_3 x}{\log_9 x} = 2$$

Despite the fact that these functions occur widely in engineering analysis, they first occurred in computational mathematics. Property (2.38a) transforms the problem of multiplying two numbers to that of adding their logarithms. The widespread use of scientific calculators has now made the computational application of logarithms largely irrelevant. They are, however, still used in the analysis of experimental data.

In MATLAB the exponential and logarithmic functions are represented by

> exponential: $exp(x)$
> natural logarithm ln: $log(x)$
> logarithm to base 10: $log10(x)$

(MAPLE uses $ln(x)$ and $log10(x)$ for the last two respectively and uses $log(x)$ for work with a general base.)

2.7.3 Exercises

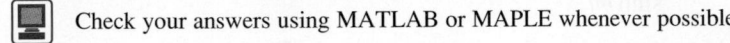

Check your answers using MATLAB or MAPLE whenever possible.

71 Simplify

(a) $(e^2)^3 + e^2 \times e^3 + (e^3)^2$ (b) e^{7x}/e^{3x}

(c) $(e^3)^2$ (d) $exp(3^2)$ (e) $\sqrt{(e^x)}$

72 Sketch the graphs of $y = e^{-2x}$ and $y = e^{-x^2}$ on the same axes. Note that $(e^{-x})^2 \neq e^{-x^2}$.

73 Find the following logarithms *without* using a calculator:

(a) $\log_2 8$ (b) $\log_2 \frac{1}{4}$

(c) $\log_2 \frac{1}{\sqrt{2}}$ (d) $\log_3 81$

(e) $\log_9 3$ (f) $\log_4 0.5$

74 Express in terms of $\ln x$ and $\ln y$

(a) $\ln(x^2 y)$ (b) $\ln \sqrt{(xy)}$ (c) $\ln(x^5/y^2)$

75 Express as a single logarithm

(a) $\ln 14 - \ln 21 + \ln 6$

(b) $4 \ln 2 - \frac{1}{2} \ln 25$

(c) $1.5 \ln 9 - 2 \ln 6$

(d) $2 \ln(2/3) - \ln(8/9)$

76 Simplify (a) $\exp\left\{\frac{1}{2} \ln\left[\frac{1-x}{1+x}\right]\right\}$ (b) $e^{2\ln x}$

77 Sketch carefully the graphs of the functions

(a) $y = 2^x$, $y = \log_2 x$ (on the same axes)

(b) $y = e^x$, $y = \ln x$ (on the same axes)

(c) $y = 10^x$, $y = \log x$ (on the same axes)

78 Sketch the graph of $y = e^{-x} - e^{-2x}$. Prove that the maximum of y is $\frac{1}{4}$ and find the corresponding value of x. Find the two values of x corresponding to $y = \frac{1}{40}$.

79 Express $\ln y$ as simply as possible when

$$y = \frac{(x^2 + 1)^{3/2}}{(x^4 + 1)^{1/3}(x^4 + 4)^{1/5}}$$

2.7.4 Hyperbolic functions

In applications certain combinations of exponential functions recur many times and these combinations are given special names. For example, the mathematical model for the steady state heat transfer in a straight bar leads to an expression for the temperature $T(x)$ at a point distance x from one end, given by

$$T(x) = \frac{T_0(e^{m(l-x)} - e^{-m(l-x)}) + T_1(e^{mx} - e^{-mx})}{e^{ml} - e^{-ml}}$$

where l is the total length of the bar, T_0 and T_1 are the temperatures at the ends and m is a physical constant. To simplify such expressions a family of functions, called the **hyperbolic** functions, is defined as follows:

$$\cosh x = \tfrac{1}{2}(e^x + e^{-x}), \quad \text{the \textbf{hyperbolic cosine}}$$

$$\sinh x = \tfrac{1}{2}(e^x - e^{-x}), \quad \text{the \textbf{hyperbolic sine}}$$

$$\tanh x = \frac{\sinh x}{\cosh x}, \quad \text{the \textbf{hyperbolic tangent}}$$

The abbreviation cosh comes from the original latin name cosinus hyperbolicus; similarly sinh and tanh.

Thus, the expression for $T(x)$ becomes

$$T(x) = \frac{T_0 \sinh m(l - x) + T_1 \sinh mx}{\sinh ml}$$

The reason for the names of these functions is geometric. They bear the same relationship to the hyperbola as the circular functions do to the circle, as shown in Figure 2.85.

Following the pattern of the circular or trigonometric functions, other hyperbolic functions are defined as follows:

$$\operatorname{sech} x = \frac{1}{\cosh x}, \quad \text{the \textbf{hyperbolic secant}}$$

$$\operatorname{cosech} x = \frac{1}{\sinh x} \quad (x \neq 0), \quad \text{the \textbf{hyperbolic cosecant}}$$

$$\coth x = \frac{1}{\tanh x} \quad (x \neq 0), \quad \text{the \textbf{hyperbolic cotangent}}$$

The graphs of $\sinh x$, $\cosh x$ and $\tanh x$ are shown in Figure 2.86, where the black broken lines indicate asymptotes.

Figure 2.85
The analogy between circular and hyperbolic functions. The circle has parametric equations $x = \cos\theta$, $y = \sin\theta$. The hyperbola has parametric equations $x = \cosh t$, $y = \sinh t$.

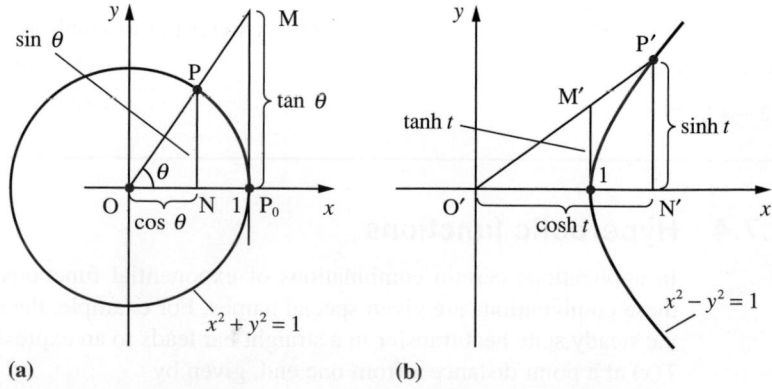

Figure 2.86
Graphs of the
hyperbolic functions.

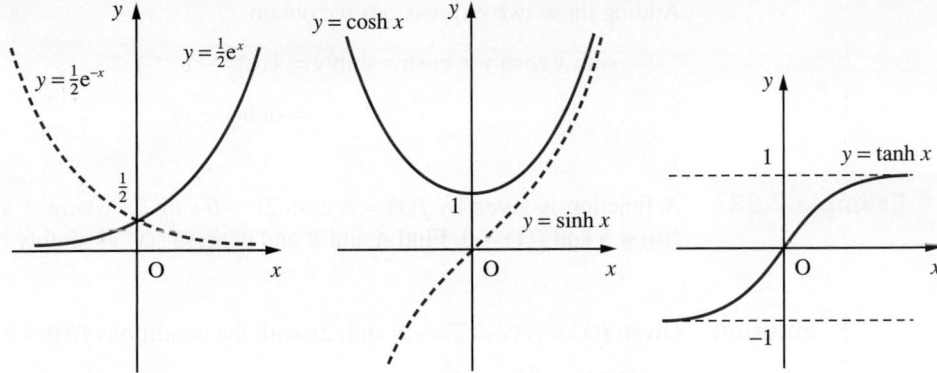

The hyperbolic functions satisfy identities analogous to those satisfied by the circular functions. From their definitions we have

$$
\left.\begin{array}{l}
\cosh x = \frac{1}{2}(e^x + e^{-x}) \\
\sinh x = \frac{1}{2}(e^x - e^{-x})
\end{array}\right\} \tag{2.39}
$$

from which we deduce

$$\cosh x + \sinh x = e^x$$
$$\cosh x - \sinh x = e^{-x}$$

and

$$(\cosh x + \sinh x)(\cosh x - \sinh x) = e^x e^{-x}$$

that is,

$$\cosh^2 x - \sinh^2 x = 1 \tag{2.40}$$

Similarly, we can show that

$$\sinh(x \pm y) = \sinh x \cosh y \pm \cosh x \sinh y \tag{2.41a}$$

$$\cosh(x \pm y) = \cosh x \cosh y \pm \sinh x \sinh y \tag{2.41b}$$

$$\tanh(x \pm y) = \frac{\tanh x \pm \tanh y}{1 \pm \tanh x \tanh y} \tag{2.41c}$$

To prove the first two of these results, it is easier to begin with the expressions on the right-hand sides and replace each hyperbolic function by its exponential form. The third result follows immediately from the previous two by dividing them. Thus

$$\sinh x \cosh y = \frac{1}{4}(e^x - e^{-x})(e^y + e^{-y})$$

$$= \frac{1}{4}(e^{x+y} + e^{x-y} - e^{-x+y} - e^{-x-y})$$

and interchanging x and y we have

$$\cosh x \sinh y = \frac{1}{4}(e^{x+y} + e^{y-x} - e^{-y+x} - e^{-x-y})$$

Adding these two expressions we obtain

$$\sinh x \cosh y + \cosh x \sinh y = \tfrac{1}{2}(e^{x+y} - e^{-x-y})$$
$$= \sinh(x + y)$$

Example 2.58   A function is given by $f(x) = A \cosh 2x + B \sinh 2x$ where A and B are constants and $f(0) = 5$ and $f(1) = 0$. Find A and B and express $f(x)$ as simply as possible.

Solution   Given $f(x) = A \cosh 2x + B \sinh 2x$ with the conditions $f(0) = 5, f(1) = 0$, we see that

$$A(1) + B(0) = 5$$

and

$$A \cosh 2 + B \sinh 2 = 0$$

Hence we have $A = 5$ and $B = -5 \cosh 2/\sinh 2$. Substituting into the formula for $f(x)$ we obtain

$$f(x) = 5 \cosh 2x - 5 \cosh 2 \sinh 2x/\sinh 2$$

$$= \frac{5 \sinh 2 \cosh 2x - 5 \cosh 2 \sinh 2x}{\sinh 2}$$

$$= \frac{5 \sinh(2 - 2x)}{\sinh 2}, \quad \text{using (2.41a)}$$

$$= \frac{5 \sinh 2(1 - x)}{\sinh 2}$$

Osborn's rule

In general, to obtain the formula for hyperbolic functions from the analogous identity for the circular functions, we replace each circular function by the corresponding hyperbolic function and change the sign of every product or implied product of two sines. This result is called **Osborn's rule**. Its justification will be discussed in Section 3.2.7.

Example 2.59   Verify the identity

$$\tanh 2x = \frac{2 \tanh x}{1 + \tanh^2 x}$$

using the definition of $\tanh x$. Confirm that it obeys Osborn's rule.

Solution   From the definition

$$\tanh 2x = \frac{e^{2x} - e^{-2x}}{e^{2x} + e^{-2x}}$$

and

$$1 + \tanh^2 x = 1 + \frac{(e^x - e^{-x})^2}{(e^x + e^{-x})^2} = \frac{(e^x + e^{-x})^2 + (e^x - e^{-x})^2}{(e^x + e^{-x})^2}$$

$$= \frac{2(e^{2x} + e^{-2x})}{(e^x + e^{-x})^2}$$

Thus

$$\frac{2 \tanh x}{1 + \tanh^2 x} = \frac{2(e^x - e^{-x})/(e^x + e^{-x})}{2(e^{2x} + e^{-2x})/(e^x + e^{-x})^2} = \frac{(e^x - e^{-x})(e^x + e^{-x})}{e^{2x} + e^{-2x}}$$

$$= \frac{e^{2x} - e^{-2x}}{e^{2x} + e^{-2x}} = \tanh 2x \text{ as required}$$

The formula for $\tan 2\theta$ from (2.27e) is

$$\tan 2\theta = \frac{2 \tan \theta}{1 - \tan^2 \theta}$$

We see that this has an implied product of two sines ($\tan^2\theta$) so that in terms of hyperbolic functions we have, using Osborn's rule,

$$\tanh 2x = \frac{2 \tanh x}{1 + \tanh^2 x}$$

which confirms the proof above.

Example 2.60   Solve the equation

$$5 \cosh x + 3 \sinh x = 4$$

Solution   The first step in solving problems of this type is to express the hyperbolic functions in terms of exponential functions. Thus we obtain

$$\tfrac{5}{2}(e^x + e^{-x}) + \tfrac{3}{2}(e^x - e^{-x}) = 4$$

On rearranging, this gives

$$4e^x - 4 + e^{-x} = 0$$

or

$$4e^{2x} - 4e^x + 1 = 0$$

which may be written as

$$(2e^x - 1)^2 = 0$$

from which we deduce

$$e^x = \tfrac{1}{2} \text{ (twice)}$$

and hence

$$x = -\ln 2$$

is a repeated root of the equation.

2.7.5 Inverse hyperbolic functions

The inverse hyperbolic functions, illustrated in Figure 2.87, are defined in a completely natural way:

$$y = \sinh^{-1}x \quad (x \text{ in } \mathbb{R})$$

$$y = \cosh^{-1}x \quad (x \geqslant 1, y \geqslant 0)$$

$$y = \tanh^{-1}x \quad (-1 < x < 1)$$

Figure 2.87
Graphs of the inverse hyperbolic functions.

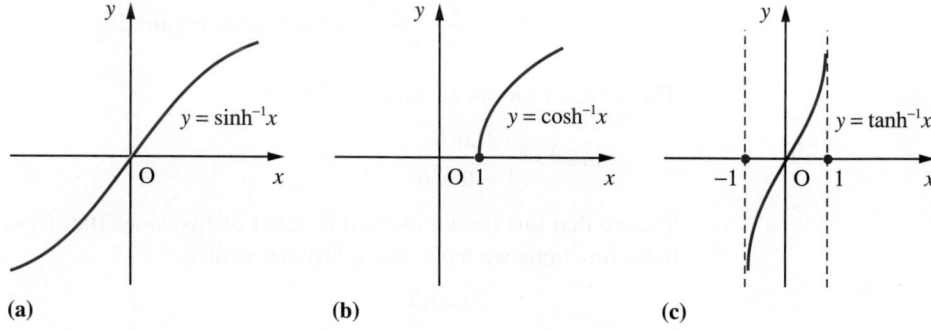

(a)   (b)   (c)

(These are also sometimes denoted as arsinh x, arcosh x and artanh x – *not* arcsinh x, etc.) Note the restriction on the range of the inverse hyperbolic cosine to meet the condition that exactly one value of y be obtained. These functions, not surprisingly, can be expressed in terms of logarithms.

For example,

$$y = \sinh^{-1}x \quad \text{implies} \quad x = \sinh y = \tfrac{1}{2}(e^y - e^{-y})$$

Thus

$$(e^y)^2 - 2x(e^y) - 1 = 0$$

and

$$e^y = x \pm \sqrt{(x^2 + 1)}$$

Since $e^y > 0$, we can discount the negative root, and we have, on taking logarithms,

$$y = \sinh^{-1}x = \ln[x + \sqrt{(x^2 + 1)}] \tag{2.42}$$

Similarly,

$$\cosh^{-1}x = \ln[x + \sqrt{(x^2 - 1)}] \quad (x \geqslant 1) \tag{2.43}$$

and

$$\tanh^{-1}x = \tfrac{1}{2}\ln\left(\frac{1 + x}{1 - x}\right) \quad (-1 < x < 1) \tag{2.44}$$

Example 2.61

Evaluate (to 4sf)

(a) $\sinh^{-1}(0.5)$ (b) $\cosh^{-1}(3)$ (c) $\tanh^{-1}(-2/5)$

using the logarithmic forms of these functions. Check your answers directly using a calculator.

Solution

(a) Using formula (2.42), we have

$$\sinh^{-1}(0.5) = \ln[0.5 + \sqrt{(0.25 + 1)}]$$

$$= \ln(0.5 + 1.118\,034)$$

$$= \ln(1.618\,034)$$

$$= 0.4812$$

(b) Using formula (2.43), we have

$$\cosh^{-1}(3) = \ln(3 + \sqrt{8}) = 1.7627$$

(c) Using formula (2.44), we have

$$\tanh^{-1}(-2/5) = \frac{1}{2}\ln\left(\frac{1 - \frac{2}{5}}{1 + \frac{2}{5}}\right)$$

$$= \frac{1}{2}\ln\left(\frac{5 - 2}{5 + 2}\right)$$

$$= \frac{1}{2}\ln\frac{3}{7} = -0.4236$$

In MATLAB notation associated with the hyperbolic functions is

> hyperbolic cosine: $cosh(x)$
> hyperbolic sine: $sinh(x)$
> hyperbolic tangent: $tanh(x)$
> inverse hyperbolic cosine: $acosh(x)$
> inverse hyperbolic sine: $asinh(x)$
> inverse hyperbolic tangent: $atanh(x)$

with the last three denoted by $arccosh(x)$, $arcsinh(x)$ and $arctanh(x)$ respectively in MAPLE.

As an example the commands

```
syms x
s = solve('5*cosh(x) + 3*sinh(x) = 4')
```

return

```
s = -log(2)
    -log(2)
```

confirming the answer in Example 2.60. (Note that it produces $-log(2)$ twice because it is a repeated root. MAPLE only produces it once.)

2.7.6 Exercises

80 In each of the following exercises a value of one of the six hyperbolic functions of x is given. Find the remaining five.

(a) $\cosh x = \frac{5}{4}$ (b) $\sinh x = \frac{8}{15}$

(c) $\tanh x = -\frac{7}{25}$ (d) $\operatorname{sech} x = \frac{5}{13}$

(e) $\operatorname{cosech} x = -\frac{3}{4}$ (f) $\coth x = \frac{13}{12}$

81 Use Osborn's rule to write down formulae corresponding to

(a) $\tan 3x = \dfrac{(3 - \tan^2 x)\tan x}{1 - 3\tan^2 x}$

(b) $\cos(x + y) = \cos x \cos y - \sin x \sin y$

(c) $\cosh 2x = 1 + 2\sinh^2 x$

(d) $\sin x - \sin y = 2 \sin \frac{1}{2}(x - y)\cos \frac{1}{2}(x + y)$

82 Prove that

(a) $\cosh^{-1} x = \ln[x + \sqrt{(x^2 - 1)}]$ $(x \geqslant 1)$

(b) $\tanh^{-1} x = \frac{1}{2}\ln\left(\dfrac{1 + x}{1 - x}\right)$ $(|x| < 1)$

83 Find to 4dp

(a) $\sinh^{-1} 0.8$

(b) $\cosh^{-1} 2$

(c) $\tanh^{-1}(-0.5)$

84 The speed V of waves in shallow water is given by

$$V^2 = 1.8L \tanh \frac{6.3d}{L}$$

where d is the depth and L the wavelength. If $d = 30$ and $L = 270$, calculate the value of V.

85 The formula

$$\lambda = \frac{\alpha t}{2}\frac{\sinh \alpha t + \sin \alpha t}{\cosh \alpha t - \cos \alpha t}$$

gives the increase in resistance of strip conductors due to eddy currents at power frequencies. Calculate λ when $\alpha = 1.075$ and $t = 1$.

86 The functions

$$f_1(x) = \frac{1}{1 + e^{-x}}, \quad f_2(x) = \frac{1}{2}\tanh \frac{1}{2}x$$

are two different forms of activating functions representing the output of a neuron in a typical neural network. Sketch the graphs of $f_1(x)$ and $f_2(x)$ and show that $f_1(x) - f_2(x) = \frac{1}{2}$.

87 The potential difference E (in V) between a telegraph line and earth is given by

$$E = A \cosh\left(x\sqrt{\frac{r}{R}}\right) + B \sinh\left(x\sqrt{\frac{r}{R}}\right)$$

where A and B are constants, x is the distance in km from the transmitting end, r is the resistance per km of the conductor and R is the insulation resistance per km. Find the values of A and B when the length of the line is 400 km, $r = 8\,\Omega$, $R = 3.2 \times 10^7\,\Omega$ and the voltages at the transmitting and receiving ends are 250 and 200 V respectively.

2.8 Irrational functions

The circular and exponential functions are examples of **transcendental functions**. They cannot be expressed as rational functions, that is, as the quotient of two polynomials. Other irrational functions occur in engineering, and they may be classified either as algebraic or as transcendental functions. For example

$$y = \frac{\sqrt{(x + 1)} - 1}{\sqrt{(x + 1)} + 1} \quad (x \geqslant -1)$$

is an algebraic irrational function. Here y is a root of the algebraic equation

$$xy^2 - 2(2 + x)y + x = 0$$

which has polynomial coefficients in x.

On the other hand, $y = |x|$, although it satisfies $y^2 = x^2$, is not a root of that equation (whose roots are $y = x$ and $y = -x$). The modulus function $|x|$ is an example of a non-algebraic irrational function.

2.8.1 Algebraic functions

In general we have an algebraic function $y = f(x)$ defined when y is the root of a polynomial equation of the form

$$a_n(x)y^n + a_{n-1}(x)y^{n-1} + \ldots + a_1(x)y + a_0(x) = 0$$

Note that here all the coefficients $a_0 \ldots a_n$ may be polynomial functions of the independent variable x. For example, consider

$$y^2 - 2xy - 8x = 0$$

This defines, for $x \geq 0$, two algebraic functions with formulae

$$y = x + \sqrt{(x^2 + 8x)} \quad \text{and} \quad y = x - \sqrt{(x^2 + 8x)}$$

One of these corresponds to $y^2 - 2xy - 8x = 0$ with $y \geq 0$ and the other to $y^2 - 2xy - 8x = 0$ with $y \leq 0$. So, when we specify a function implicitly by means of an equation we often need some extra information to define it uniquely. Often, too, we cannot obtain an explicit algebraic formula for y in terms of x and we have to evaluate the function at each point of its domain by solving the polynomial equation for y numerically.

Care has to be exercised when using algebraic functions in a larger computation in case special values of parameters produce sudden changes in value, as illustrated in Example 2.62.

Example 2.62   Sketch the graphs of the function

$$y = \sqrt{(a + bx^2 + cx^3)/(d - x)}$$

for the domain $-3 < x < 3$ where

(a) $a = 18$, $b = 1$, $c = -1$ and $d = 6$

(b) $a = 0$, $b = 1$, $c = -1$ and $d = 0$

Solution   (a) $y = \sqrt{(18 + x^2 - x^3)/(6 - x)}$

We can see that the term inside the square root is positive only when $18 + x^2 - x^3 > 0$. Since we can factorize this $(18 + x^2 - x^3) = (3 - x)(x^2 + 2x + 6)$ we deduce that y is not defined for $x > 3$. Also, for large negative values of x it behaves like $\sqrt{(-x)}$. A sketch of the graph is shown in Figure 2.88.

(b) $y = -\sqrt{(x^2 - x^3)/x}$

Here we can see that the function is defined for $x \leq 1$, $x \neq 0$. Near $x = 0$, since we can write $x = \sqrt{x^2}$ for $x > 0$ and $x = -\sqrt{x^2}$ for $x < 0$, we see that

$$y = -\sqrt{(1 - x)} \quad \text{for } x > 0$$

Figure 2.88
Graph of $y = \sqrt{(18 + x^2 - x^3)}/(6 - x)$.

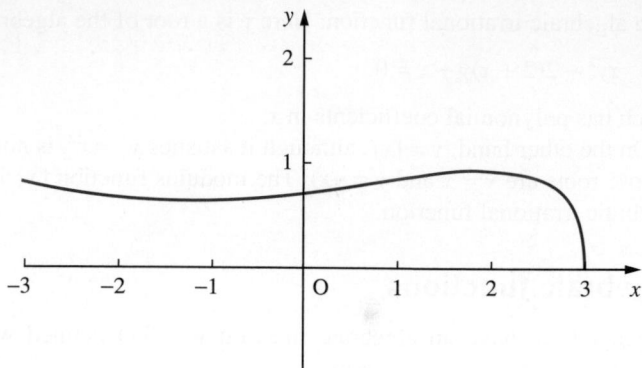

Figure 2.89
Graph of
$y = -\sqrt{(x^2 - x^3)}/x$.

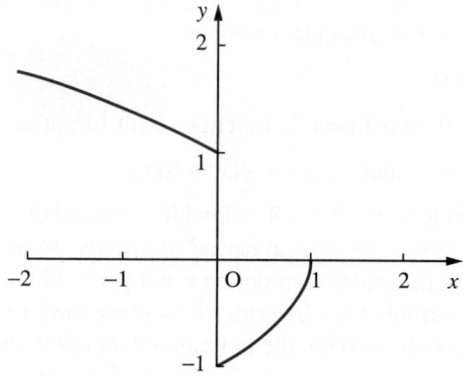

and

$$y = \sqrt{(1 - x)} \quad \text{for } x < 0$$

At $x = 0$ the function is not defined. The graph of the function is shown in Figure 2.89.

2.8.2 Implicit functions

We have seen in Section 2.8.1 that some algebraic functions are defined implicitly because we cannot obtain an algebraic formula for them. This applies to a wider class of functions where we have an equation relating the dependent and independent variables, but where finding the value of y corresponding to a given value of x requires a numerical solution of the equation. Generally we have an equation connecting x and y such as

$$f(x, y) = 0$$

Sometimes we are able to draw a curve which represents the relationship (using algebraic methods), but more commonly we have to calculate for each value of x the corresponding value of y. Most computer graphics packages have an implicit function option which will perform the task efficiently.

Example 2.63   The velocity v and the displacement x of a mass attached to a non-linear spring satisfy the equation

$$v^2 = -4x^2 + x^4 + A$$

where A depends on the initial velocity v_0 and displacement x_0 of the mass. Sketch the graph of v against x where

(a) $x_0 = 1, v_0 = 0$

(b) $x_0 = 3, v_0 = 0$

and interpret your graph.

Solution   (a) With $x_0 = 1, v_0 = 0$ we have $A = 3$ and

$$v^2 = x^4 - 4x^2 + 3 = (x^2 - 3)(x^2 - 1)$$

To sketch the graph by hand it is easiest first to sketch the graph of v^2 against x as shown in Figure 2.90(a). Taking the 'square root' of the graph is only possible for $v^2 \geqslant 0$, but we also know we want that part of the graph which has the initial point (x_0, v_0) on it. So we obtain the closed loop shown in Figure 2.90(b). The arrows on the closed curve indicate the variation of v with x as time increases where the velocity v is positive, the displacement x increases. Where the velocity is negative, the displacement decreases. The closed curve indicates that this motion repeats after completing one circuit of the curve, that is there is a periodic motion.

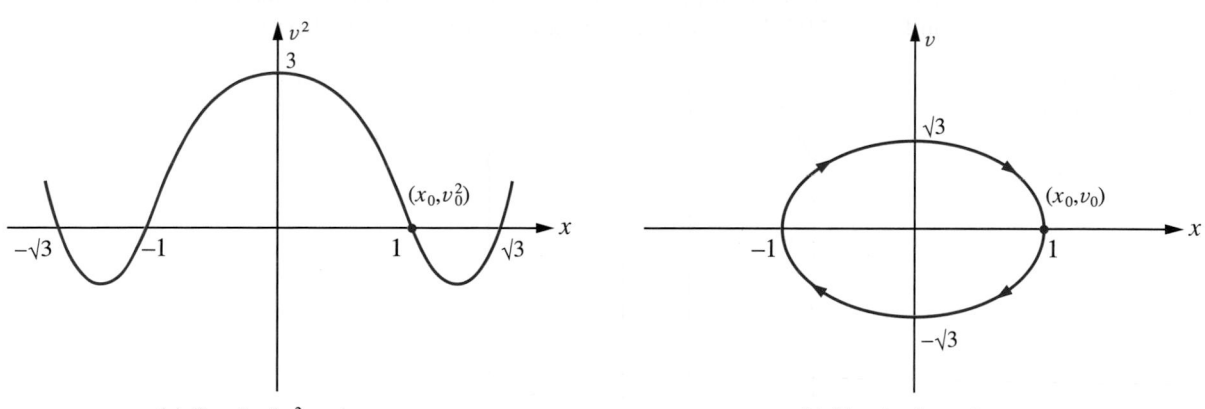

(a) Sketch of v^2 against x.   **(b)** Sketch of v against x.

Figure 2.90 Graphs for Example 2.63(a).

(b) With $x_0 = 3, v_0 = 0$ we have $A = -45$ and

$$v^2 = x^4 - 4x^2 - 45 = (x^2 - 9)(x^2 + 5)$$

Using the same technique as in part (a), we see that when the mass is released from rest at $x = 3$, its displacement increases without a bound and the motion is not periodic. The corresponding graphs are shown in Figures 2.91(a) and (b).

$$f(6) = 1403.2, f(7) = 1385.0, f(8) = 1423.7$$

This shows clearly that the minimum value occurs between $h = 6$ and $h = 8$.

We approximate to $f(h)$ using a local quadratic approximation of the form

$$f(h) \simeq A(h - 7)^2 + B(h - 7) + C$$

Setting $h = 7$ gives $\qquad C = 1385.0$
Setting $h = 6$ gives $\quad A - B + C = 1403.2$
Setting $h = 8$ gives $\quad A + B + C = 1423.7$

Hence $C = 1385.0$, $A = 28.45$ and $B = 10.25$. The minimum of the approximating quadratic function occurs where $h - 7 = -B/(2A)$, that is, at $h = 7 - 0.18 = 6.82$. Thus the optimal choice for h is approximately 6.82 giving a value for $f(h)$ at that point of 1383.5.

The corresponding values for b and l are $b = 27.3$ and $l = 53.7$. Thus we have obtained an optimal design of the container in this special case.

General case

Here we seek the optimal design without restricting the ratio of b to h. For a container of capacity K, we have to minimize $A(l, b, h, t)$ subject to the constraint $C(l, b, h) = K$. Here

$$A(l, b, h, t) = (lb + 6bh + 2hl)t + (2l + 6b + 12h)t^2 + 12t^3$$

and

$$C(l, b, h) = lbh$$

These functions have certain algebraic symmetries that enable us to solve the problem algebraically. Consider the formula for A and set $x = 2h$ and $y = l/3$, then

$$A(l, b, h, t) = 3(by + bx + xy)t + 6(y + b + x)t^2 + 12t^3$$
$$= A^*(y, b, x, t)$$

and

$$C(l, b, h) = 3bxy/2$$

From this we can conclude that if $A^*(y, b, x, t)$ has a minimum value at (y_0, b_0, x_0) for a given value of t, then it has the same value at (x_0, b_0, y_0), (x_0, y_0, b_0), (y_0, x_0, b_0), (b_0, y_0, x_0) and (b_0, x_0, y_0). Assuming that the function has a unique minimum point, we conclude that these six points are the same, that is $b_0 = y_0 = x_0$. Thus we deduce that the minimum occurs where $l = 6h$ and $b = 2h$. Since the capacity is fixed, we have $lbh = K$, which implies that $12h^3 = K$.

Thus the optimal choice for h in the general case is $(\frac{1}{12}K)^{1/3}$.

Returning to the special case where $K = 10\,000$ and $t = 0.4$, we obtain an optimal design when

$$h = 9.41, \quad b = 18.82, \quad l = 56.46$$

using 1330.1 cm^3 of material. Note that the amount of material used is close to that used in the special case where $b = 4h$. This indicates that the design is not sensitive to small errors made during its construction.

2.11 Review exercises (1–23)

 Check your answers using MATLAB or MAPLE whenever possible.

1 The functions f and g are defined by

$$f(x) = x^2 - 4 \quad (x \text{ in } [-20, 20])$$

$$g(x) = x^{1/2} \quad (x \text{ in } [0, 200])$$

Let $h(x)$ and $k(x)$ be the compositions $f \circ g(x)$ and $g \circ f(x)$ respectively. Determine $h(x)$ and $k(x)$. Is the composite function $k(x)$ defined for all x in the domain of $f(x)$? If not, then for what part of the domain of $f(x)$ is $k(x)$ defined?

2 The perimeter of an ellipse depends on the lengths of its major and minor axes, and is given by

$$\text{perimeter} = 2 \times (\text{major axis}) \times E(m)$$

where

$$m = \frac{(\text{major axis})^2 - (\text{minor axis})^2}{(\text{major axis})^2}$$

and E is the function whose graph is given in Figure 2.107.

(a) Calculate the perimeter of the ellipse whose axes are of length 10 cm and 6 cm.

(b) A fairing is to be made from sheet metal bent into the shape of an ellipse of major axis 55 cm and minor axis 13 cm, and is to be of length 2 m. Estimate the area of sheet metal required.

3 The sales volume of a product depends on its price as follows:

Price/£	1.00	1.05	1.10	1.15	1.20	1.25	1.30
Sales/000	8	7	6	5	4	3	2

The cost of production is £1 per unit. Draw up a table showing the sales revenue, the cost and the profits for each selling price, and deduce the selling price to be adopted.

4 A function f is defined by

$$f = \begin{cases} x + 1 & (x < -1) \\ 0 & (-1 \leqslant x \leqslant 1) \\ x - 1 & (x > 1) \end{cases}$$

Draw the graphs of $f(x)$, $f(x - 2)$ and $f(2x)$. The function $g(x)$ is defined as $f(x + 2) - f(2x - 1)$. Draw a graph of $g(x)$.

5 The function $f(x)$ has formula $y = x^2$ for $0 \leqslant x < 1$. Sketch the graphs of $f(x)$ for $-4 < x < 4$ when

(a) $f(x)$ is periodic with period 1

(b) $f(x)$ is even and periodic with period 2

(c) $f(x)$ is odd and periodic with period 2

6 Assuming that all the numbers given are correctly rounded, calculate the positive root together with its error bound of the quadratic equation

$$1.4x^2 + 5.7x - 2.3 = 0$$

Give your answer also as a correctly rounded number.

7 Sketch the functions

(a) $x^2 - 4x + 7$ (b) $x^3 - 2x^2 + 4x - 3$

(c) $\dfrac{x + 4}{x^2 - 1}$ (d) $\dfrac{x^2 - 2x + 3}{x^2 + 2x - 3}$

8 Find the Taylor expansion of

$$x^4 + 3x^3 - x^2 + 2x - 1 \text{ about } x = 1$$

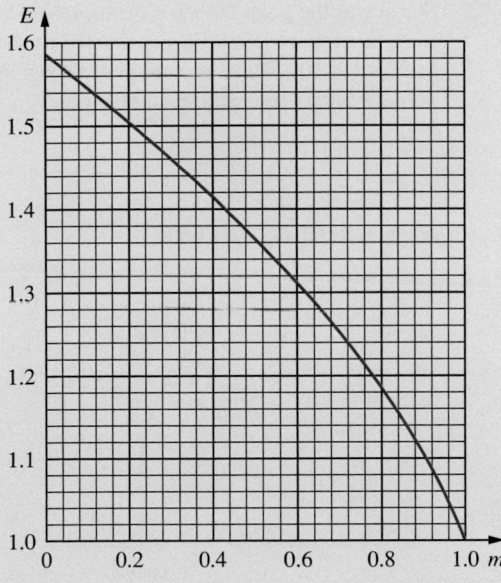

Figure 2.107

9 Find the partial fractions of

(a) $\dfrac{x+2}{(x-1)(x-4)}$ (b) $\dfrac{x^2+4}{(x+1)(x-3)}$

(c) $\dfrac{x^2-2x+3}{(x+2)^2(x-1)}$ (d) $\dfrac{x(2x-1)}{(x^2-x+1)(x+3)}$

10 Express as products of sines and/or cosines

(a) $\sin 2\theta - \sin \theta$ (b) $\cos 2\theta + \cos 3\theta$

(c) $\sin 4\theta - \sin 7\theta$

11 Express in the form $r \sin(\theta - \alpha)$

(a) $4 \sin \theta - 2 \cos \theta$ (b) $\sin \theta + 8 \cos \theta$

(c) $\sqrt{3} \sin \theta + \cos \theta$

12 (a) From the definition of the hyperbolic sine function prove

$$\sinh 3x = 3 \sinh x + 4 \sinh^3 x$$

(b) Sketch the graph of $y = x^3 + x$ carefully, and show that for each value of y there is exactly one value of x. Setting $z = \frac{1}{2}x\sqrt{3}$, show that

$$4z^3 + 3z = \frac{3\sqrt{3}}{2}y$$

and using (a), deduce that

$$x = \frac{2}{\sqrt{3}} \sinh\left[\frac{1}{3} \sinh^{-1}\left(\frac{3\sqrt{3}}{2}y\right)\right]$$

13 The parts produced by three machines along a factory aisle (shown in Figure 2.108 as the x axis) are to be taken to a nearby bench for assembly before they undergo further processing. Each assembly takes one part from each machine. There is a fixed cost per metre for moving any of the parts. Show that if x represents the position of the assembly bench the cost $C(x)$ of moving the parts for each assembled item is given by

$$C(x) \propto d(x)$$

where $d(x) = |x+3| + |x-2| + |x-4|$

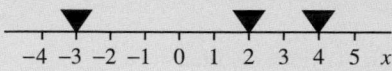

Figure 2.108

Draw the graph of $d(x)$ and find the optimal position of the bench.

14 Sketch the graphs of the functions

(a) $\lfloor \frac{1}{2}x \rfloor - \frac{3}{2}\lfloor \frac{1}{3}x \rfloor$

(b) $xH(x) - (x-1)H(x-1) + (x-2)H(x-2)$

15 Draw up a table of values of the function $f(x) = x^2 e^{-x}$ for $x = -0.1(0.1)1.1$. Determine the maximum error incurred in linearly interpolating for the function $f(x)$ in this table, and hence estimate the value of $f(0.83)$, giving your estimate to an appropriate number of decimal places.

16 By setting $t = \tan \frac{1}{2}x$, find the maximum value of $(\sin x)/(2 - \cos x)$.

17 (a) Show that a root x_0 of the equation

$$x^4 - px^3 + q = 0$$

is a repeated root if and only if

$$4x_0 - 3p = 0$$

(b) The stiffness of a rectangular beam varies with the cube of its height h and directly with its breadth b. Find the section of the beam that can be cut from a circular log of diameter D that has the maximum stiffness.

18 Starting at the point $(x_0, y_0) = (1, 0)$, a sequence of right-angled triangles is constructed as shown in Figure 2.109. Show that the coordinates of the vertices satisfy the recurrence relations

$$x_i = x_{i-1} - w_i y_{i-1}$$

$$y_i = w_i x_{i-1} + y_{i-1}$$

where $w_i = \tan \alpha_i^\circ$, $x_0 = 1$ and $y_0 = 0$.

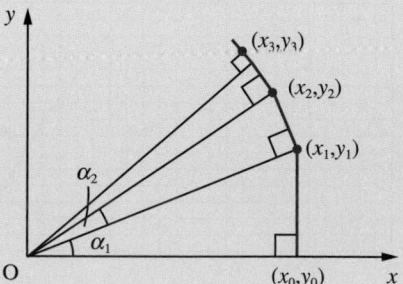

Figure 2.109

Any angle $0° < \theta° < 360°$ can be expressed in the form

$$\theta = \sum_{i=0}^{\infty} n_i \phi_i$$

where $\tan \phi_i° = 10^{-i}$ and n_i is a non-negative integer. Express $\theta = 56.5$ in this form and, using the recurrence relations above, calculate $\sin \theta°$ and $\cos \theta°$ to 5dp. (This method of calculating the trigonometric functions is used in some calculators.)

19 A mechanism consists of the linkage of three rods AB, BC and CD as shown in Figure 2.110, where AB = CD (= a, say), BC = AD = $a\sqrt{2}$, and M is the midpoint. The rods are freely jointed at B and C, and are free to rotate about A and D. Using polar coordinates with their pole O at the midpoint of AD and initial line OD, show that the curve described by M as CD rotates about D

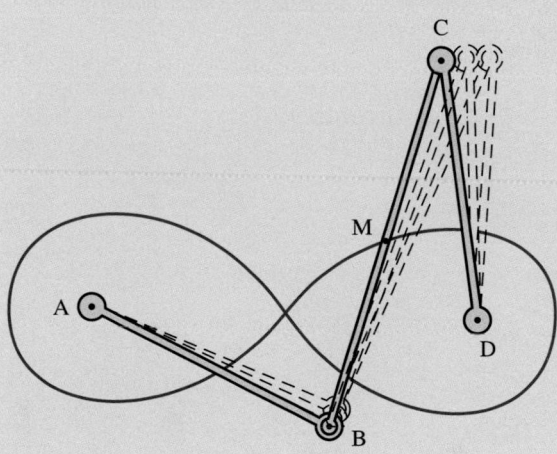

Figure 2.110

is $r^2 = a^2 \cos 2\theta$. Draw a careful graph of this curve, the 'lemniscate' of Bernoulli.

Show that

(a) The cartesian coordinates of M satisfy

$$(x^2 + y^2)^2 = a^2(x^2 - y^2)$$

(b) $AM \times DM = \frac{1}{2}a^2$.

20 Show that the equation

$$r = p/\sin(\theta - \alpha)$$

represents a straight line which cuts the x axis at the angle α and whose perpendicular distance from the origin is p.

21 Use the result of Question 20 to find the polar coordinate representation of the line which passes through the points $(1, 2)$ and $(3, 3)$.

22 Show that the equation

$$r = ep/(1 + e \cos \theta)$$

where e and p are constants represents an ellipse where $0 < e < 1$, a parabola where $e = 1$ and a hyperbola where $e > 1$, the origin of the coordinate system being at a focus of the conic concerned.

23 Continuing question 53 of Exercises 2.6.2, show that

$$\cot \theta = \frac{12 + d^2}{4d}$$

and by applying the arithmetic-geometric inequality to

$$\frac{3}{d} + \frac{d}{4}$$

deduce that $\theta°$ achieves its maximum value where $d = 2\sqrt{3}$.

3 Complex Numbers

Chapter 3 Contents

3.1 Introduction

Complex numbers first arose in the solution of cubic equations in the sixteenth century using a method known as Cardano's solution. This gives the solution of the equation

$$x^3 + qx + r = 0$$

as

$$x = \sqrt[3]{[-\tfrac{1}{2}r + \sqrt{(\tfrac{1}{4}r^2 + \tfrac{1}{27}q^3)}]} + \sqrt[3]{[-\tfrac{1}{2}r - \sqrt{(\tfrac{1}{4}r^2 + \tfrac{1}{27}q^3)}]}$$

which may be verified by direct substitution. This solution gave difficulties when it unexpectedly involved square roots of negative numbers. For example, the equation

$$x^3 - 15x - 4 = 0$$

was known to have three roots. An obvious one is $x = 4$, but the corresponding root obtained using the formula was

$$x = \sqrt[3]{[2 + \sqrt{(-121)}]} + \sqrt[3]{[2 - \sqrt{(-121)}]}$$

Writing in 1572, Bombelli showed that

$$2 + \sqrt{(-121)} = [2 + \sqrt{(-1)}]^3$$

and

$$2 - \sqrt{(-121)} = [2 - \sqrt{(-1)}]^3$$

and so

$$x = [2 + \sqrt{(-1)}] + [2 - \sqrt{(-1)}] = 4$$

as expected. Since

$$\sqrt{(-x)} = \sqrt{(-1)}\sqrt{x}$$

where x is a positive number, the square roots of negative numbers can be represented as a number multiplied by $\sqrt{(-1)}$. Thus $\sqrt{(-121)} = 11\sqrt{(-1)}$, $\sqrt{(-4900)} = 70\sqrt{(-1)}$ and so on. Because the introduction of the special number $\sqrt{(-1)}$ simplified calculations, it quickly gained acceptance by mathematicians. Denoting $\sqrt{(-1)}$ by the letter j, we obtain the general number z where

$$z = x + jy$$

Here x and y are ordinary **real numbers** and obey the Fundamental Rules of Arithmetic. (Most mathematics and physics texts use the letter i instead of j. However, we shall follow the standard engineering practice and use j.) The number z is called a **complex number**. The ordinary processes of arithmetic still apply, but become a little more complicated. As well as simplifying the process of obtaining roots as above, the introduction of $j = \sqrt{(-1)}$ simplified the theory of equations, so that, for example, the quadratic equation

$$ax^2 + bx + c = 0$$

always has two roots

$$x = \frac{-b \pm \sqrt{(b^2 - 4ac)}}{2a}$$

These roots are real numbers when $b^2 \geqslant 4ac$ and complex numbers when $b^2 < 4ac$. Thus, any irreducible quadratic may be factorized into two complex factors. It then follows from property (ii) of the polynomial functions, given in Section 2.4.1, that any polynomial equation of degree n having real coefficients has exactly n roots which may be real or complex. This is a result known as the **Fundamental Theorem of Algebra**, which is also valid for polynomial equations having complex coefficients. Thus

$$x^7 - 7x^5 - 6x^4 + 4x^3 - 28x - 24 = 0$$

is an equation of degree seven and has the roots

$$x = -1, -2, -3, -1 - j, -1 + j, 1 - j, 1 + j$$

As has often been the case, what began as a mathematical curiosity has turned out to be of considerable practical importance, and complex numbers are invaluable in many aspects of engineering analysis. An elementary, but important, application is discussed later in this chapter.

3.2 Properties

To specify a complex number z, we use two real numbers, x and y, and write

$$z = x + jy$$

where $j = \sqrt{(-1)}$, and x is called the **real part** of z and y its **imaginary part**. This is often abbreviated to

$$z = x + jy, \quad \text{where } x = \text{Re}(z) \text{ and } y = \text{Im}(z)$$

Note that the imaginary part of z does *not* include the j. For example, if $z = 3 - j2$ then $\text{Re}(z) = 3$ and $\text{Im}(z) = -2$. If $x = 0$, the complex number is said to be **purely imaginary** and if $y = 0$ it is said to be **purely real**.

3.2.1 The Argand diagram

Geometrically, complex numbers can be represented as points on a plane similar to the way in which real numbers are represented by points on a straight line. The number $z = x + jy$ is represented by the point P with coordinates (x, y), as shown in Figure 3.1. Such a diagram is called an **Argand diagram**, after one of its inventors. The x axis is called the **real axis** and the y axis is called the **imaginary axis**.

Figure 3.1
The Argand diagram:
$z = x + jy$.

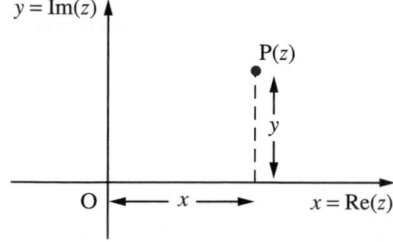

Example 3.1

Represent on an Argand diagram the complex numbers

(a) $3 + j2$ (b) $-5 + j3$ (c) $8 - j5$ (d) $-2 - j3$

Solution

(a) The number $3 + j2$ is represented by the point A(3, 2)

(b) The number $-5 + j3$ is represented by the point B(-5, 3)

(c) The number $8 - j5$ is represented by the point C(8, -5)

(d) The number $-2 - j3$ is represented by the point D(-2, -3)

as shown in Figure 3.2.

Figure 3.2

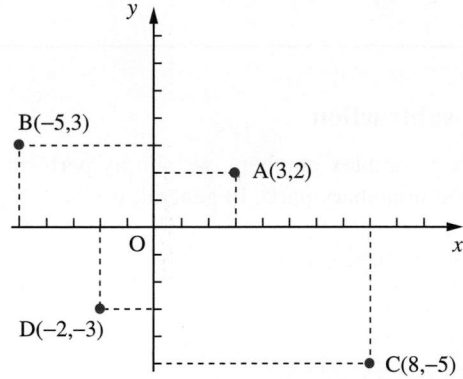

3.2.2 The arithmetic of complex numbers

(i) Equality

If two complex numbers $z_1 = x_1 + jy_1$ and $z_2 = x_2 + jy_2$ are equal then they are represented by the same point on the Argand diagram and it clearly follows that

$$x_1 = x_2 \quad \text{and} \quad y_1 = y_2$$

That is, when two complex numbers are equal we can equate their respective real and imaginary parts.

Example 3.2

If the two complex numbers

$$z_1 = (3a + 2) + j(3b - 1) \quad \text{and} \quad z_2 = (b + 1) - j(a + 2 - b)$$

are equal

(a) find the values of the real numbers a and b, and

(b) write down the real and imaginary parts of z_1 and z_2.

Solution (a) Since $z_1 = z_2$ we can equate their respective real and imaginary parts, giving

$$(3a + 2) = (b + 1) \qquad \text{or} \quad 3a - b = -1$$

and

$$(3b - 1) = -(a + 2 - b) \quad \text{or} \quad a + 2b = -1$$

Solving for a and b then gives

$$a = -\tfrac{3}{7}, \quad b = -\tfrac{2}{7}$$

(b) $\left.\begin{aligned} \text{Re}(z_1) &= 3a + 2 = \tfrac{5}{7} \\ \text{Re}(z_2) &= b + 1 = \tfrac{5}{7} \end{aligned}\right\}$ thus $\text{Re}(z_1) = \text{Re}(z_2) = \tfrac{5}{7}$

$\left.\begin{aligned} \text{Im}(z_1) &= 3b - 1 = -\tfrac{13}{7} \\ \text{Im}(z_2) &= -(a + 2 - b) = -\tfrac{13}{7} \end{aligned}\right\}$ thus $\text{Im}(z_1) = \text{Im}(z_2) = -\tfrac{13}{7}$

(ii) Addition and subtraction

To add or subtract two complex numbers, we simply perform the operations on their corresponding real and imaginary parts. In general, if $z_1 = x_1 + jy_1$ and $z_2 = x_2 + jy_2$ then

$$z_1 + z_2 = (x_1 + x_2) + j(y_1 + y_2)$$

and

$$z_1 - z_2 = (x_1 - x_2) + j(y_1 - y_2)$$

In Chapter 4, Section 4.2.5, we shall interpret complex numbers geometrically as two-dimensional vectors and illustrate how the rules for the addition of vectors can be used to represent addition of complex numbers in the Argand diagram.

Example 3.3 If $z_1 = 3 + j2$ and $z_2 = 5 - j3$ determine

(a) $z_1 + z_2$ (b) $z_1 - z_2$

Solution (a) Adding the corresponding real and imaginary parts gives

$$z_1 + z_2 = (3 + 5) + j(2 - 3) = 8 - j1$$

(b) Subtracting the corresponding real and imaginary parts gives

$$z_1 - z_2 = (3 - 5) + j(2 - (-3)) = -2 + j5$$

(iii) Multiplication

When multiplying two complex numbers the normal rules for multiplying out brackets hold. Thus, in general, if $z_1 = x_1 + jy_1$ and $z_2 = x_2 + jy_2$ then

$$z_1 z_2 = (x_1 + jy_1)(x_2 + jy_2)$$
$$= x_1 x_2 + jy_1 x_2 + jx_1 y_2 + j^2 y_1 y_2$$

Making use of the fact that $j^2 = -1$ then gives

$$z_1 z_2 = x_1 x_2 - y_1 y_2 + j(x_1 y_2 + x_2 y_1)$$

Example 3.4   If $z_1 = 3 + j2$ and $z_2 = 5 + j3$ determine $z_1 z_2$.

Solution $z_1 z_2 = (3 + j2)(5 + j3) = 15 + j10 + j9 + j^2 6$
$$= 15 - 6 + j(10 + 9), \text{ using the fact that } j^2 = -1$$
$$= 9 + j19$$

(iv) Division

The division of two complex numbers is less straightforward. If $z_1 = x_1 + jy_1$ and $z_2 = x_2 + jy_2$, then we use the following technique to obtain the quotient. We multiply 'top and bottom' by $x_2 - jy_2$, giving

$$\frac{z_1}{z_2} = \frac{x_1 + jy_1}{x_2 + jy_2} = \frac{(x_1 + jy_1)(x_2 - jy_2)}{(x_2 + jy_2)(x_2 - jy_2)}$$

Multiplying out 'top and bottom', we obtain

$$\frac{z_1}{z_2} = \frac{(x_1 x_2 + y_1 y_2) + j(x_2 y_1 - x_1 y_2)}{x_2^2 + y_2^2}$$

giving

$$\frac{z_1}{z_2} = \frac{(x_1 x_2 + y_1 y_2)}{x_2^2 + y_2^2} + j\frac{(x_2 y_1 - x_1 y_2)}{x_2^2 + y_2^2}$$

The number $x - jy$ is called the **complex conjugate** of $z = x + jy$ and is denoted by z^*. (Sometimes the complex conjugate is denoted with an overbar as $\bar{z}$.) Note that the complex conjugate z^* is obtained by changing the sign of the imaginary part of z.

Example 3.5   If $z_1 = 3 + j2$ and $z_2 = 5 + j3$ determine $\dfrac{z_1}{z_2}$.

Solution $\dfrac{z_1}{z_2} = \dfrac{3 + j2}{5 + j3}$

Multiplying 'top and bottom' by the conjugate $5 - j3$ of the denominator gives

$$\frac{z_1}{z_2} = \frac{(3 + j2)(5 - j3)}{(5 + j3)(5 - j3)}$$

Multiplying out 'top and bottom' we obtain

$$\frac{3 + j2}{5 + j3} = \frac{(15 + 6) + j(10 - 9)}{(25 + 9) + j(15 - 15)} = \frac{21 + j}{34} = \tfrac{21}{34} + j\tfrac{1}{34}$$

Example 3.6 Find the real and imaginary parts of the complex number $z + 1/z$ for $z = (2 + j)/(1 - j)$.

Solution

$$z = \frac{2 + j}{1 - j} = \frac{(2 + j)(1 + j)}{(1 - j)(1 + j)} = \frac{1 + j3}{2} = \tfrac{1}{2} + j\tfrac{3}{2}$$

then

$$z^{-1} = \frac{2}{1 + j3} = \frac{2(1 - j3)}{(1 + j3)(1 - j3)} = \frac{2 - j6}{10} = \tfrac{1}{5} - j\tfrac{3}{5}$$

so that

$$z + \frac{1}{z} = (\tfrac{1}{2} + j\tfrac{3}{2}) + (\tfrac{1}{5} - j\tfrac{3}{5}) = (\tfrac{1}{2} + \tfrac{1}{5}) + j(\tfrac{3}{2} - \tfrac{3}{5}) = \tfrac{7}{10} + j\tfrac{9}{10}$$

giving

$$\operatorname{Re}\left(z + \frac{1}{z}\right) = \tfrac{7}{10} \quad \text{and} \quad \operatorname{Im}\left(z + \frac{1}{z}\right) = \tfrac{9}{10}$$

3.2.3 Complex conjugate

As we have seen above, the complex conjugate of $z = x + jy$ is $z^* = x - jy$. In the Argand diagram z^* is the mirror image of z in the real or x axis. The following important results are readily deduced.

$$z + z^* = 2x = 2\operatorname{Re}(z)$$
$$z - z^* = 2jy = 2j\operatorname{Im}(z)$$
$$zz^* = (x + jy)(x - jy) = x^2 + y^2$$
$$(z_1 z_2)^* = z_1^* z_2^*$$

(3.1)

with the next to last result indicating that the product of a complex number and its complex conjugate is a real number.

The zeros of an irreducible quadratic function, which has real coefficients, are complex conjugates of each other.

Example 3.7 Express the zeros of $f(x) = x^2 - 6x + 13$ as complex numbers.

Solution The zeros of $f(x)$ are the roots of the equation

$$x^2 - 6x + 13 = 0$$

Using the quadratic formula (1.7) we obtain

$$x = \frac{6 \pm \sqrt{(36 - 52)}}{2} = \frac{6 \pm \sqrt{(-16)}}{2}$$

$$= \frac{6 \pm 4\sqrt{(-1)}}{2} = 3 \pm j4$$

So the two zeros form a conjugate pair.

Example 3.8   Find all the roots of the quartic equation

$$x^4 + 4x^2 + 16 = 0$$

Solution   Rewriting the equation we can achieve a difference of squares which makes possible a first factorization

$$x^4 + 8x^2 + 16 - 4x^2 = (x^2 + 4)^2 - 4x^2$$

$$= [(x^2 + 4) - 2x][(x^2 + 4) + 2x]$$

Now $x^2 - 2x + 4 = (x - 1)^2 + 3$ and $x^2 + 2x + 4 = (x + 1)^2 + 3$, so we obtain the equations

$$x - 1 = \pm j3 \quad \text{and} \quad x + 1 = \pm j3$$

and the four roots of the quartic equation are

$$x = 1 + j3, \ 1 - j3, \ -1 + j3, \ -1 - j3$$

These roots form two conjugate pairs.

Example 3.9   For the complex numbers $z_1 = 5 + j3$ and $z_2 = 3 - j2$ verify the identity

$$(z_1 z_2)^* = z_1^* z_2^*$$

Solution
$$z_1 z_2 = (5 + j3)(3 - j2) = 15 + 6 + j(9 - 10) = 21 - j$$

$$(z_1 z_2)^* = 21 + j$$

$$z_1^* z_2^* = (5 - j3)(3 + j2) = 15 + 6 + j(10 - 9) = 21 + j$$

Thus $(z_1 z_2)^* = z_1^* z_2^*$.

3.2.4 Modulus and argument

As indicated in the Argand diagram of Figure 3.3, the point P is specified uniquely if we know the length of the line OP and the angle it makes with the positive x direction. The length of OP is a measure of the size of z and is called the **modulus** of z, which is usually denoted by mod z or $|z|$. The angle between the positive real axis and OP is called the **argument** of z and is denoted by arg z. Since the polar coordinates (r, θ) and $(r, \theta + 2\pi)$ represent the same point, a convention is used to determine the argument

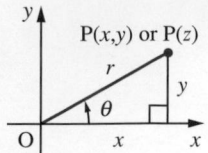

Figure 3.3
Modulus (r) and argument (θ) of the complex number $z = x + jy$.

of z uniquely, restricting its range so that $-\pi < \arg z \leqslant \pi$. (In some textbooks this is referred to as the 'principal value' of the argument.) The argument of the complex number $0 + j0$ is not defined.

Thus from Figure 3.3, $|z|$ and $\arg z$ are given by

$$\left.\begin{array}{l} |z| = r = \sqrt{(x^2 + y^2)} \\[2mm] \arg z = \theta \quad \text{where } \tan \theta = y/x, \, z \neq 0 \end{array}\right\} \tag{3.2}$$

Note that from equations (3.1)

$$zz^* = x^2 + y^2 = |z|^2$$

There are two common mistakes to avoid when calculating $|z|$ and $\arg z$ using (3.2). First note that the modulus of z is the square root of the sum of squares of x and y, *not* of x and jy. The j part of the number has been accounted for in the representation of the Argand diagram. The second common mistake is to place θ in the wrong quadrant. To avoid this, it is advisable when evaluating $\arg z$ to draw a sketch of the Argand diagram showing the location of the number.

Example 3.7 Determine the modulus and argument of

(a) $3 + j2$ (b) $1 - j$ (c) $-1 + j$ (d) $-\sqrt{6} - j\sqrt{2}$

Solution Note that the sketches of the Argand diagrams locating the positions of the complex numbers are given in Figure 3.4(a–d).

(a) $|3 + j2| = \sqrt{(3^2 + 2^2)} = \sqrt{(9 + 4)} = \sqrt{13} = 3.606$

$$\arg(3 + j2) = \tan^{-1}\left(\frac{2}{3}\right) = 0.588$$

(b) $|1 - j| = \sqrt{[1^2 + (-1)^2]} = \sqrt{2} = 1.414$

$$\arg(1 - j) = -\tan^{-1}\left(\frac{1}{1}\right) = -\tfrac{1}{4}\pi$$

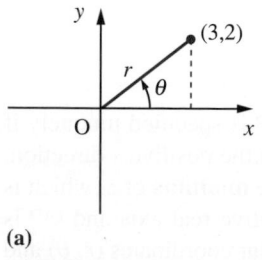

(a)

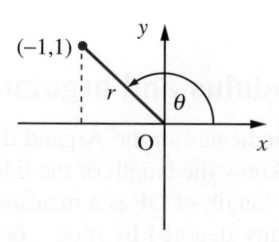

(b)

(c)

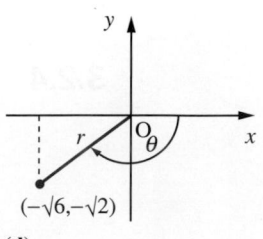

(d)

Figure 3.4

(c) $|-1 + j| = \sqrt{[(-1)^2 + 1^2]} = \sqrt{2} = 1.414$

$$\arg(-1 + j) = \pi - \tan^{-1}\left(\frac{1}{1}\right) = \pi - \tfrac{1}{4}\pi = \tfrac{3}{4}\pi$$

(d) $|-\sqrt{6} - j\sqrt{2}| = \sqrt{(6 + 2)} = \sqrt{8} = 2.828$

$$\arg(-\sqrt{6} - j\sqrt{2}) = -(\pi - \tan^{-1}\tfrac{\sqrt{2}}{\sqrt{6}}) = -(\pi - \tan^{-1}\sqrt{\tfrac{1}{3}}) = -(\pi - \tfrac{1}{6}\pi) = -\tfrac{5}{6}\pi$$

MATLAB handles complex numbers automatically. Either i or j can be used to denote the imaginary part, but in any output to a command MATLAB will always use i. Consequently, to avoid confusion i will be used throughout when using MATLAB so, for example, the complex number $z = 4 + j3$ will be entered as:

```
z = 4 + 3i
```

Note that the i is located after the number 3 and there is no need to insert the multiplication sign * between the 3 and the i (if it is located before then * must be included). However, in some cases it is necessary to insert *; for example, the complex number $z = -\tfrac{1}{2} + j\tfrac{1}{2}$ must be entered as

```
z = -1/2 + (1/2)*i
```

MAPLE is similar to MATLAB in dealing with complex numbers, except it uses I and * is always required.

The complex conjugate z^* of a complex number z is obtained using the command $conj$; for example, to obtain the conjugate of $z = 4 + j3$ enter the commands

MATLAB
```
z = 4 + 3i;
zbar = conj(z)
```
MAPLE
```
z:= 4 + 3*I;
zbar:= conjugate(z);
```

which return

```
zbar = 4 - 3i
```
```
zbar = 4 - 3I
```

The arithmetical operations of addition, subtraction, multiplication and division are carried out by the standard operators +, −, *, and / respectively. For example, if $z_1 = 4 + j3$ and $z_2 = -3 + j2$ then $z_3 = z_1 + z_2$ and $z_4 = z_1/z_2$ are determined as follows:

MATLAB
```
z1 = 4 + 3i;
z2 = -3 + 2i;
z3 = z1 + z2
```
MAPLE
```
z1:= 4 + 3*I;
z2:= -3 + 2*I;
z3:= z1 + z2;
```

return

```
z3 = 1.0000 + 5.0000i
```
```
z3 = 1 + 5I
```

and the further command

```
z4 = z1/z2
```
```
z4:= z1/z2;
```

returns

```
z4 = -0.4615 - 1.3077i
```
```
z4:= -6/13 - 17/13 I
evalf(%);
```

returns

```
z4 = -0.4615 - 1.3077I
```

Note that MAPLE produces exact arithmetic; the command $evalf$ is used to produce the numerical answer. Exact arithmetic may be undertaken in MATLAB using the Symbolic Math Toolbox with the command *double* used to obtain numerical results. For example the commands

```
syms z1 z2 z4
z1 = sym(4 + 3i); z2 = sym(3 + 2i); z4 = z1/z2
```

return

$$z4 = -\frac{6}{13} - \frac{17}{13}*i$$

and

```
double(z4)
```

returns

```
z4 = -0.4615 - 1.3077i
```

The real and imaginary parts of a complex number are determined using the commands *real* and *imag* respectively. Considering Example 3.6 the MATLAB commands

```
z = (2 + i)/(1 - i); z1 = z + 1/z;
real(z1)
```

return the answer *0.7000* and the further command

```
imag(z1)
```

returns the answer *0.9000*; thus confirming the answers obtained in the given solution. MAPLE uses *Re* and *Im*.

To represent complex numbers as points on an Argand diagram check that the following commands reaffirm the solution given in Example 3.1:

```
z1 = 3 + 2i; x = real(z1); y = imag(z1);
plot(x,y,'*')
xlabel('x = Re(z)')
ylabel('y = Im(z)')
hold on
plot([-6,9],[0,0], 'k')
plot([0,0],[-6,4], 'k')
z2 = -5 + 3i; x = real(z2); y = imag(z2);
plot(x,y, '*')
z3 = 8 - 5i; x = real(z3); y = imag(z3);
plot(x,y, '*')
z4 = -2 - 3i; x = real(z4); y = imag(z4);
plot(x,y, '*')
```

To label the points add the additional commands:

```
text(3.2,2 'A(3,2)')
text(-5,3.3, 'B(-5,3)')
text(8.2,-5, 'C(8,-5)')
text(-2,-3.3, 'D(-2,3)')
plot(x,y,'*')
hold off
```

[Note: (1) The `'*'` in the plot commands means that the point will be printed as an asterisk; alternatives include `'.'`, `'x'` and `'+'`.

(2) The *hold on* command holds the current axes for subsequent plots.

(3) The two plot commands following the *hold on* command draw the x- and y-axes with the entry k indicating that the lines are drawn in black (alternatives include b for blue, r for red and g for green).]

Symbolically the MATLAB commands

```
syms x y real
z = x + i*y
```

create symbolic variables x and y that have the additional property that they are real. Then z is a complex variable and can be manipulated as such. For example

conj(z) returns $x - i*y$ and *expand(z*conj(z))* returns $x^2 + y^2$

The modulus and argument (measured in radians) of a complex number z can be calculated directly using the commands *abs* and *angle* respectively (*abs* and *argument* in MAPLE). For example, considering Example 3.10(a) the commands

```
z = 3 + 2i;
modz = abs(z)
```

return

```
modz = 3.6056
```

and the additional command

```
argz = angle(z)
```

returns

```
argz = 0.5880
```

confirming the answers obtained in the given solution. Using these commands check the answers to Examples 3.10(b)–(d).

3.2.5 Exercises

Check your answers using MATLAB or MAPLE whenever possible.

1 Show in an Argand diagram the points representing the following complex numbers:

(a) $1 + j$ (b) $\sqrt{3} - j$

(c) $-3 + j4$ (d) $1 - j\sqrt{3}$

(e) $-1 + j\sqrt{3}$ (f) $-1 - j\sqrt{3}$

(c) $\dfrac{5 - j8}{3 - j4}$ (d) $\dfrac{1 - j}{1 + j}$

(e) $\frac{1}{2}(1 + j)^2$ (f) $(3 - j2)^2$

(g) $\dfrac{1}{5 - j3} - \dfrac{1}{5 + j3}$ (h) $\dfrac{1}{2} - \dfrac{3 - j4}{5 - j8}$

2 Express in the form $x + jy$ where x and y are real numbers

(a) $(5 + j3)(2 - j) - (3 + j)$ (b) $(1 - j2)^2$

3 What is the complex conjugate of:

(a) $2 + j7$ (b) $-3 - j$ (c) $-j6$ (d) $\frac{2}{3} - j\frac{2}{3}$

4 Find the roots of the equations

(a) $x^2 + 2x + 2 = 0$ (b) $x^3 + 8 = 0$

5 Find z such that

$$zz^* + 3(z - z^*) = 13 + j12$$

6 With $z = 2 - j3$, find

(a) jz (b) z^* (c) $1/z$ (d) $(z^*)^*$

7 Find the modulus and argument of each of the complex numbers given in Question 1.

8 Find the complex numbers w, z which satisfy the simultaneous equations

$$4z + 3w = 23$$
$$z + jw = 6 + j8$$

9 For $z = x + jy$ (x and y real) satisfying

$$\frac{2z}{1+j} - \frac{2z}{j} = \frac{5}{2+j}$$

find x and y.

10 Given $z = 2 - j2$ is a root of

$$2z^3 - 9z^2 + 20z - 8 = 0$$

find the remaining roots of the equation.

11 Find the real and imaginary parts of z when

$$\frac{1}{z} = \frac{2}{2+j3} + \frac{1}{3-j2}$$

12 Find $z = z_1 + z_2z_3/(z_2 + z_3)$ when $z_1 = 2 + j3$, $z_2 = 3 + j4$ and $z_3 = -5 + j12$.

13 Find the values of the real numbers x and y which satisfy the equation

$$\frac{2 + x - jy}{3x + jy} = 1 + j2$$

14 Find z_3 in the form $x + jy$, where x and y are real numbers, given that

$$\frac{1}{z_3} = \frac{1}{z_1} + \frac{1}{z_1z_2}$$

where $z_1 = 3 - j4$ and $z_2 = 5 + j2$.

3.2.6 Polar form of a complex number

Figure 3.3 shows that the relationships between (x, y) and (r, θ) are

$$x = r \cos\theta \quad \text{and} \quad y = r \sin\theta$$

Hence the complex number $z = x + jy$ can be expressed in the form

$$z = r\cos\theta + jr\sin\theta = r(\cos\theta + j\sin\theta) \tag{3.3}$$

This is called the **polar form** of the complex number. In engineering it is frequently written as $r \angle \theta$, so that

$$z = r \angle \theta = r(\cos\theta + j\sin\theta)$$

Example 3.11   Express the following complex numbers in polar form.

(a) $12 + j5$ (b) $-3 + j4$ (c) $-4 - j3$

Solution   (a) A sketch of the Argand diagram locating the position of $12 + j5$ is given in Figure 3.5(a). Thus

$$|12 + j5| = \sqrt{(144 + 25)} = 13$$
$$\arg(12 + j5) = \tan^{-1}\tfrac{5}{12} = 0.395$$

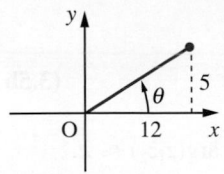

(a)

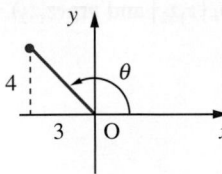

(b)

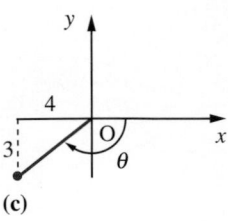

(c)

Figure 3.5

Thus in polar form

$$12 + j5 = 13[\cos(0.395) + j\sin(0.395)]$$

(b) A sketch of the Argand diagram locating the position of $-3 + j4$ is given in Figure 3.5(b). Thus

$$|-3 + j4| = \sqrt{(9 + 16)} = 5$$

$$\arg(-3 + j4) = \pi - \tan^{-1}\tfrac{4}{3} - \pi - 0.9273$$

$$= 2.214$$

Thus in polar form

$$-3 + j4 = 5[\cos(2.214) + j\sin(2.214)]$$

(c) A sketch of the Argand diagram locating the position of $-4 - j3$ is given in Figure 3.5(c). Thus

$$|-4 - j3| = \sqrt{(16 + 9)} = 5$$

$$\arg(-4 - j3) = -(\pi - \tan^{-1}\tfrac{3}{4}) = -(\pi - 0.643)$$

$$= -2.498$$

Thus in polar form

$$-4 - j3 = 5[\cos(-2.498) + j\sin(-2.498)]$$

$$= 5[\cos(2.498) - j\sin(2.498)]$$

using the results $\cos(-t) = \cos t$ and $\sin(-t) = -\sin t$.

Note: Rectangular to polar conversion can be done using a calculator and students are encouraged to check the answers in this way.

Multiplication in polar form

Let

$$z_1 = r_1(\cos\theta_1 + j\sin\theta_1) \quad \text{and} \quad z_2 = r_2(\cos\theta_2 + j\sin\theta_2)$$

then

$$z_1 z_2 = r_1 r_2(\cos\theta_1 + j\sin\theta_1)(\cos\theta_2 + j\sin\theta_2)$$

$$= r_1 r_2[(\cos\theta_1\cos\theta_2 - \sin\theta_1\sin\theta_2) + j(\sin\theta_1\cos\theta_2 + \cos\theta_1\sin\theta_2)]$$

which, on using the trigonometric identities (2.24a, c), gives

$$z_1 z_2 = r_1 r_2[\cos(\theta_1 + \theta_2) + j\sin(\theta_1 + \theta_2)] \tag{3.4}$$

Hence

$$|z_1 z_2| = r_1 r_2 = |z_1||z_2| \tag{3.5a}$$

and

$$\arg(z_1 z_2) = \theta_1 + \theta_2 = \arg z_1 + \arg z_2 \tag{3.5b}$$

When using these results care must be taken to ensure that $-\pi < \arg(z_1 z_2) \leqslant \pi$.

Example 3.12   If $z_1 = -12 + j5$ and $z_2 = -4 + j3$, determine, using (3.5a) and (3.5b), $|z_1 z_2|$ and $\arg(z_1 z_2)$.

Solution

$$|z_1| = \sqrt{(144 + 25)} = \sqrt{(169)} = 13$$

$$\arg(z_1) = \pi - \tan^{-1}\tfrac{5}{12} = \pi - 0.395 = 2.747$$

$$|z_2| = \sqrt{(16 + 9)} = 5$$

$$\arg(z_2) = \pi - \tan^{-1}\tfrac{3}{4} = 2.498$$

Thus from (3.4) and (3.5)

$$|z_1 z_2| = |z_1||z_2| = (13)(5) = 65$$

$$\arg(z_1 z_2) = \arg z_1 + \arg z_2 = 2.747 + 2.498$$

$$= 5.245 \text{ (or } 300.51°)$$

However, this does not express $\arg(z_1 z_2)$ within the defined range $-\pi < \arg \leqslant \pi$. Thus

$$\arg(z_1 z_2) = -2\pi + 5.245 = -1.038$$

Geometrical representation of multiplication by j

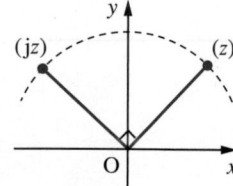

Figure 3.6
Relationship between
z and jz.

Since

$$z = r(\cos \theta + j \sin \theta) \quad \text{and} \quad j = 1(\cos \tfrac{1}{2}\pi + j \sin \tfrac{1}{2}\pi)$$

it follows from (3.4) that

$$jz = r[\cos(\theta + \tfrac{1}{2}\pi) + j \sin(\theta + \tfrac{1}{2}\pi)]$$

Thus the effect of multiplying a complex number by j is to leave the modulus unaltered but to increase the argument by $\tfrac{1}{2}\pi$ as indicated in Figure 3.6. This property is of importance in the application of complex numbers to the theory of alternating current.

Division in polar form

Now

$$\frac{1}{\cos \theta + j \sin \theta} = \frac{1}{\cos \theta + j \sin \theta} \frac{\cos \theta - j \sin \theta}{\cos \theta - j \sin \theta}$$

$$= \frac{\cos \theta - j \sin \theta}{\cos^2 \theta + \sin^2 \theta}$$

$$= \cos \theta - j \sin \theta, \quad \text{since} \quad \cos^2 \theta + \sin^2 \theta = 1$$

Thus if

$$z_1 = r_1(\cos\theta_1 + j\sin\theta_1) \quad \text{and} \quad z_2 = r_2(\cos\theta_2 + j\sin\theta_2)$$

then

$$\frac{z_1}{z_2} = \frac{r_1(\cos\theta_1 + j\sin\theta_1)}{r_2(\cos\theta_2 + j\sin\theta_2)}$$

$$= \frac{r_1}{r_2}(\cos\theta_1 + j\sin\theta_1)(\cos\theta_2 - j\sin\theta_2) \quad \text{(from above)}$$

$$= \frac{r_1}{r_2}[(\cos\theta_1\cos\theta_2 + \sin\theta_1\sin\theta_2) + j(\sin\theta_1\cos\theta_2 - \cos\theta_1\sin\theta_2)]$$

or

$$\frac{z_1}{z_2} = \frac{r_1}{r_2}[\cos(\theta_1 - \theta_2) + j\sin(\theta_1 - \theta_2)] \tag{3.6}$$

using the trigonometric identities (2.25b, d). Hence

$$\left|\frac{z_1}{z_2}\right| = \frac{r_1}{r_2} = \frac{|z_1|}{|z_2|} \tag{3.7}$$

and

$$\arg\left(\frac{z_1}{z_2}\right) = \theta_1 - \theta_2 = \arg z_1 - \arg z_2 \tag{3.8}$$

Again some adjustment may be necessary to ensure that $-\pi < \arg(z_1/z_2) \leq \pi$.

Example 3.13

For the following pairs of complex numbers obtain z_1/z_2 and z_2/z_1.

(a) $z_1 = 4(\cos\pi/2 + j\sin\pi/2)$, $z_2 = 9(\cos\pi/3 + j\sin\pi/3)$

(b) $z_1 = \cos 3\pi/4 + j\sin 3\pi/4$, $z_2 = 2(\cos\pi/8 + j\sin\pi/8)$

Solution

(a) $|z_1| = 4$, $\arg z_1 = \pi/2$; $|z_2| = 9$, $\arg z_2 = \pi/3$

From (3.7)

$$\left|\frac{z_1}{z_2}\right| = \frac{4}{9} \quad \text{and} \quad \left|\frac{z_2}{z_1}\right| = \frac{9}{4}$$

From (3.8)

$$\arg\left(\frac{z_1}{z_2}\right) = \frac{\pi}{2} - \frac{\pi}{3} = \frac{\pi}{6} \quad \text{and} \quad \arg\left(\frac{z_2}{z_1}\right) = \frac{\pi}{3} - \frac{\pi}{2} = -\frac{\pi}{6}$$

Thus $\dfrac{z_1}{z_2} = \dfrac{4}{9}\left(\cos\dfrac{\pi}{6} + j\sin\dfrac{\pi}{6}\right)$

and $\dfrac{z_2}{z_1} = \dfrac{9}{4}\left(\cos\dfrac{\pi}{6} - j\sin\dfrac{\pi}{6}\right)$

(b) $|z_1| = 1$, $\arg z_1 = 3\pi/4$; $|z_2| = 2$, $\arg z_2 = \pi/8$

From (3.7)

$$\left|\dfrac{z_1}{z_2}\right| = \dfrac{1}{2} \quad \text{and} \quad \left|\dfrac{z_2}{z_1}\right| = 2$$

From (3.8)

$$\arg\left(\dfrac{z_1}{z_2}\right) = \dfrac{3\pi}{4} - \dfrac{\pi}{8} = \dfrac{5\pi}{8} \quad \text{and} \quad \arg\left(\dfrac{z_2}{z_1}\right) = \dfrac{\pi}{8} - \dfrac{3\pi}{4} = -\dfrac{5\pi}{8}$$

Thus $\dfrac{z_1}{z_2} = \dfrac{1}{2}\left(\cos\dfrac{5\pi}{8} + j\sin\dfrac{5\pi}{8}\right)$

and $\dfrac{z_2}{z_1} = 2\left(\cos\dfrac{5\pi}{8} - j\sin\dfrac{5\pi}{8}\right)$

Example 3.14 Find the modulus and argument of

$$z = \dfrac{(1 + j2)^2(4 - j3)^3}{(3 + j4)^4(2 - j)^3}$$

Solution

$$|z| = \dfrac{|1 + j2|^2|4 - j3|^3}{|3 + j4|^4|2 - j|^3}$$

$$= \dfrac{[\sqrt{(1 + 4)}]^2[\sqrt{(16 + 9)}]^3}{[\sqrt{(9 + 16)}]^4[\sqrt{(4 + 1)}]^3} = \dfrac{1}{25}\sqrt{5}$$

$\arg z = 2\arg(1 + j2) + 3\arg(4 - j3) - 4\arg(3 + j4) - 3\arg(2 - j)$

$\qquad = 2(1.107) + 3(-0.643) - 4(0.927) - 3(-0.461) = -2.035$

3.2.7 Euler's formula

In Chapter 2, Section 2.7.3, we obtained the result

$$e^x = \cosh x + \sinh x$$

which links the exponential and hyperbolic functions. A similar, but more important, formula links the exponential and circular functions. It is

$$e^{j\theta} = \cos\theta + j\sin\theta \qquad \qquad \textbf{(3.9)}$$

This formula is known as **Euler's formula**. The justification for this definition depends on the following facts.

We know from the properties of the exponential function that

$$e^{j\theta_1}e^{j\theta_2} = e^{j(\theta_1+\theta_2)}$$

When expressed in terms of Euler's formula this becomes

$$(\cos\theta_1 + j\sin\theta_1)(\cos\theta_2 + j\sin\theta_2) = \cos(\theta_1 + \theta_2) + j\sin(\theta_1 + \theta_2)$$

which is just (3.4) with $r_1 = r_2 - 1$.

Similarly

$$\frac{e^{j\theta_1}}{e^{j\theta_2}} = e^{j(\theta_1-\theta_2)}$$

becomes

$$\frac{\cos\theta_1 + j\sin\theta_1}{\cos\theta_2 + j\sin\theta_2} = \cos(\theta_1 - \theta_2) + j\sin(\theta_1 - \theta_2)$$

which is just (3.6) with $r_1 = r_2 = 1$.

Euler's formula enables us to write down the polar form of the complex number z very concisely:

$$z = r(\cos\theta + j\sin\theta) = re^{j\theta} = r \angle \theta \tag{3.10}$$

This is known as the **exponential form** of the complex number z.

Example 3.15

Express the following complex numbers in exponential form

(a) $2 + j3$ (b) $-2 + j$

Solution (a) A sketch of the Argand diagram showing the position of $2 + j3$ is given in Figure 3.7(a).

$$|2 + j3| = \sqrt{(2^2 + 3^2)} = \sqrt{13}$$

$$\arg(2 + j3) = \tan^{-1}(3/2) = 0.9828$$

Thus $2 + j3 = \sqrt{13}e^{j0.9828}$

Figure 3.7
Argand diagrams for
Example 3.15.

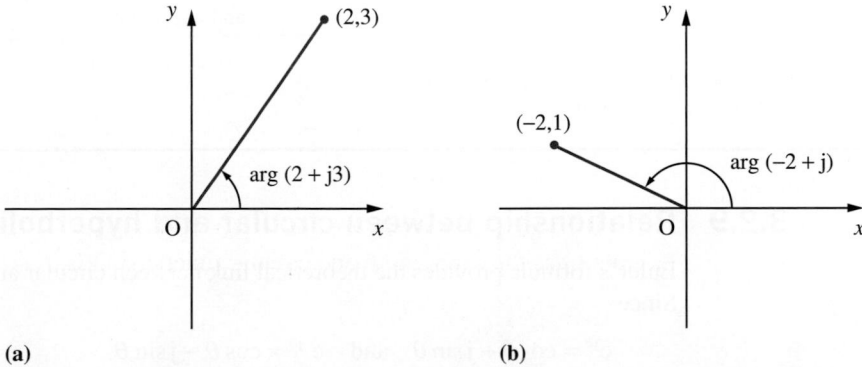

(a) (b)

(b) A sketch of the Argand diagram showing the position of $-2 + j$ is given in Figure 3.7(b).

$$|-2 + j| = \sqrt{5}$$

$$\arg(-2 + j) = \pi - \tan^{-1}(1/2) = 2.6779$$

Thus $-2 + j = \sqrt{5}e^{j2.6779}$

Example 3.16 Express in cartesian form the complex number $e^{2+j\pi/3}$.

Solution $e^{2+j\pi/3} = e^2 e^{j\pi/3} = e^2(\cos \pi/3 + j \sin \pi/3)$

Now $e^2 = 7.3891$, $\cos \pi/3 = 0.5$ and $\sin \pi/3 = 0.8660$, so that

$$e^{2+j\pi/3} = 3.6945 + j6.3991$$

Having determined the modulus r and argument $theta$ of a complex number, its polar form is given in MATLAB by

```
r*(cos(theta) + i*sin(theta))
```

and its exponential form by

```
r*exp(i*theta)
```

3.2.8 Exercises

Check your answers using MATLAB or MAPLE whenever possible.

15 Express in polar form the complex numbers

(a) j (b) 1

(c) -1 (d) $1 - j$

(e) $\sqrt{3} - j\sqrt{3}$ (f) $-2 + j$

(g) $-3 - j2$ (h) $7 - j5$

(i) $(2 - j)(2 + j)$ (j) $(-2 + j7)^2$

16 Express $z = (2 - j)(3 + j2)/(3 - j4)$ in the form $x + jy$ and also in polar form.

17 Given $z_1 = e^{j\pi/4}$ and $z_2 = e^{-j\pi/3}$, find

(a) the arguments of $z_1 z_2^2$ and z_1^3/z_2

(b) the real and imaginary parts of $z_1^2 + jz_2$

18 Given $z_1 = 2e^{j\pi/3}$ and $z_2 = 4e^{-2j\pi/3}$, find the modulus and argument of

(a) $z_1^3 z_2^2$ (b) $z_1^2 z_2^4$ (c) z_1^2/z_2^3

3.2.9 Relationship between circular and hyperbolic functions

Euler's formula provides the theoretical link between circular and hyperbolic functions. Since

$$e^{j\theta} = \cos \theta + j \sin \theta \quad \text{and} \quad e^{-j\theta} = \cos \theta - j \sin \theta$$

we deduce that

$$\cos\theta = \frac{e^{j\theta} + e^{-j\theta}}{2}$$ (3.11a)

and

$$\sin\theta = \frac{e^{j\theta} - e^{-j\theta}}{2j}$$ (3.11b)

In Section 2.7 we defined the hyperbolic functions by

$$\cosh x = \frac{e^x + e^{-x}}{2}$$ (3.12a)

and

$$\sinh x = \frac{e^x - e^{-x}}{2}$$ (3.12b)

Comparing (3.12a, b) with (3.11a, b), we have

$$\cosh jx = \frac{e^{jx} + e^{-jx}}{2} = \cos x$$ (3.13a)

$$\sinh jx = \frac{e^{jx} - e^{-jx}}{2} = j\sin x$$ (3.13b)

so that

$$\tanh jx = j\tan x$$ (3.13c)

Also,

$$\cos jx = \frac{e^{j^2x} + e^{-j^2x}}{2} = \frac{e^{-x} + e^x}{2} = \cosh x$$ (3.14a)

$$\sin jx = \frac{e^{j^2x} - e^{-j^2x}}{2j} = \frac{e^{-x} - e^x}{2j} = j\sinh x$$ (3.14b)

so that

$$\tan jx = j\tanh x$$ (3.14c)

These relationships provide the justification for Osborn's rule used in Section 2.7 for obtaining hyperbolic function identities from those satisfied by circular functions, since whenever a product of two sines occurs, j^2 will also occur.

Using these results we can evaluate functions such as $\sin z$, $\cos z$, $\tan z$, $\sinh z$, $\cosh z$ and $\tanh z$. For example, to evaluate

$$\cos z = \cos(x + jy)$$

we use the identity

$$\cos(A + B) = \cos A \cos B - \sin A \sin B$$

and obtain

$$\cos z = \cos x \cos jy - \sin x \sin jy$$

Using results (3.14a, b), this gives

$$\cos z = \cos x \cosh y - j \sin x \sinh y$$

Example 3.17   Find the values of

(a) $\sin[\frac{1}{4}\pi(1 + j)]$ (b) $\sinh(3 + j4)$

(c) $\tan(\frac{\pi}{4} - j3)$ (d) z such that $\cos z = 2$

Solution   (a) We may use the identity

$$\sin(A + B) = \sin A \cos B + \cos A \sin B$$

and obtain

$$\sin(\tfrac{1}{4}\pi + j\tfrac{1}{4}\pi) = \sin \tfrac{1}{4}\pi \cos j\tfrac{1}{4}\pi + \cos \tfrac{1}{4}\pi \sin j\tfrac{1}{4}\pi$$

Here $\sin \frac{1}{4}\pi$ and $\cos \frac{1}{4}\pi$ are evaluated as usual ($= \sqrt{\frac{1}{2}}$), while we make use of results (3.14a, b) to obtain

$$\cos j\tfrac{1}{4}\pi = \cosh \tfrac{1}{4}\pi \quad \text{and} \quad \sin j\tfrac{1}{4}\pi = j \sinh \tfrac{1}{4}\pi$$

giving

$$\sin[\tfrac{1}{4}\pi(1 + j)] = \sin \tfrac{1}{4}\pi \cosh \tfrac{1}{4}\pi + j \cos \tfrac{1}{4}\pi \sinh \tfrac{1}{4}\pi$$

$$= (0.7071)(1.3246) + j(0.7071)(0.8687)$$

$$= 0.9366 + j0.6142$$

(b) Using the identity

$$\sinh(A + B) = \sinh A \cosh B + \cosh A \sinh B$$

we obtain

$$\sinh(3 + j4) = \sinh 3 \cosh j4 + \cosh 3 \sinh j4$$

which, on using results (3.13a, b), gives

$$\sinh(3 + j4) = \sinh 3 \cos 4 + j \cosh 3 \sin 4$$

$$= (10.0179)(-0.6536) + j(10.0677)(-0.7568)$$

$$= -6.548 - j7.619$$

(c) Using the identity

$$\tan(A - B) = \frac{\tan A - \tan B}{1 + \tan A \tan B}$$

we obtain

$$\tan(\tfrac{1}{4}\pi - j3) = \frac{\tan\tfrac{1}{4}\pi - \tan j3}{1 + \tan\tfrac{1}{4}\pi \tan j3}$$

which, on using result (3.14c) and $\tan\tfrac{1}{4}\pi = 1$, gives

$$\tan(\tfrac{1}{4}\pi - j3) = \frac{1 - j\tanh 3}{1 + j\tanh 3} = \frac{(1 - j\tanh 3)^2}{1 + \tanh^2 3}$$

$$= \frac{1 - \tanh^2 3}{1 + \tanh^2 3} - j\frac{2\tanh 3}{1 + \tanh^2 3}$$

$$= \frac{1}{\cosh^2 3 + \sinh^2 3} - j\frac{2\sinh 3\cosh 3}{\cosh^2 3 + \sinh^2 3}$$

$$= \frac{1}{\cosh 6} + j\frac{\sinh 6}{\cosh 6} = 0.005 - j1.000$$

(d) Writing $z = x + jy$, we have

$$2 = \cos(x + jy)$$

Expanding the right-hand side gives

$$2 = \cos x \cos jy - \sin x \sin jy$$

$$= \cos x \cosh y - \sin x \, (j \sinh y)$$

$$2 = \cos x \cosh y - j \sin x \sinh y$$

Equating real and imaginary parts of each side of this equation gives

$$2 = \cos x \cosh y$$

and

$$0 = \sin x \sinh y$$

The latter equation implies either $\sin x = 0$ or $y = 0$. If $y = 0$ then the first equation implies $2 = \cos x$, so clearly that is not a solution since x is a real number . The alternative, $\sin x = 0$, implies $x = 0, \pm\pi, \pm 2\pi, \pm 3\pi, \dots$, and hence

$$2 = \cos(\pm n\pi) \cosh y, \quad n = 0, 1, 2, \dots$$

This gives

$$2 = \cos n\pi \cosh y$$

$$= (-1)^n \cosh y$$

But $\cosh y \geqslant 1$, so n must be an even number. Thus the values of z such that $\cos z = 2$ are

$$z = \pm 2n\pi \pm j \cosh^{-1} 2, \, n = 0, 1, 2, \dots$$

$$= \pm 2n\pi \pm j(1.3170)$$

3.2.10 Logarithm of a complex number

Consider the equation

$$z = e^w$$

Writing $z = x + jy$ and $w = u + jv$, we have

$$x + jy = e^{u+jv} = e^u e^{jv}$$

$$= e^u(\cos v + j \sin v), \quad \text{by Euler's formula}$$

Equating real and imaginary parts,

$$x = e^u \cos v \quad \text{and} \quad y = e^u \sin v$$

Squaring both these equations and adding gives

$$x^2 + y^2 = e^{2u}(\cos^2 v + \sin^2 v) = e^{2u}$$

so that

$$u = \tfrac{1}{2} \ln(x^2 + y^2) = \ln |z|$$

Dividing the two equations,

$$\tan v = \frac{y}{x}$$

From this and $x = e^u \cos v$

$$v = \arg z + 2n\pi, \quad n = 0, \pm 1, \pm 2, \dots$$

Hence

$$v = \ln |z| + j \arg z + j2n\pi, \quad n = 0, \pm 1, \pm 2, \dots$$

We select just one of these solutions to define for us the logarithm of the complex number z, writing

$$\ln z = \ln |z| + j \arg z \tag{3.15}$$

This is sometimes called its **principal value**.

Example 3.18   Evaluate $\ln(-3 + j4)$ in the form $x + jy$.

Solution

$$|-3 + j4| = \sqrt{(9 + 16)} = 5$$

$$\arg(-3 + j4) = \pi - \tan^{-1}\tfrac{4}{3} = 2.214$$

Thus from (3.15)

$$\ln(-3 + j4) = \ln 5 + j2.214 = 1.609 + j2.214$$

In MATLAB functions of a complex variable can be evaluated as easily as functions of a real variable. For example, in relation to Example 3.17 (a) and (b), entering

$sin((pi/4)*(1 + i))$ returns the answer $0.9366 + 0.6142i$

whilst entering

$sinh(3 + 4i)$ returns the answer $-6.5481 - 7.6192i$

confirming the answers obtained in the given solution. Similarly, considering Example 3.18, entering

$log(-3 + 4i)$ returns the answer $1.6094 + 2.2143i$

confirming the answer obtained in the solution. In MAPLE functions of a complex variable must be evaluated using $evalc$. The result is exact and the numerical values require $evalf$; for example

$evalc(sin((Pi/4)*(1 + I)));$

returns

$\frac{1}{2}\sqrt{2}\cosh(\frac{1}{4}\pi) + \frac{1}{2}\sqrt{2}I\sinh(\frac{1}{4}\pi)$

and $evalf(\%)$; returns $0.9366 + 0.6142I$.

3.2.11 Exercises

Check your answers using MATLAB or MAPLE whenever possible.

19 Using the exponential forms of $\cos\theta$ and $\sin\theta$ given in (3.11a, b), prove the following trigonometric identities:

(a) $\sin(\alpha + \beta) = \sin\alpha\cos\beta + \cos\alpha\sin\beta$

(b) $\sin^3\theta = \frac{3}{4}\sin\theta - \frac{1}{4}\sin 3\theta$

20 Express in the form $x + jy$

(a) $\sin(\frac{5}{6}\pi + j)$ (b) $\cos(j\frac{3}{4})$

(c) $\sinh[\frac{\pi}{3}(1 + j)]$ (d) $\cosh(j\frac{\pi}{4})$

21 Solve $z = x + jy$ when

(a) $\sin z = 2$ (b) $\cos z = j\frac{3}{4}$

(c) $\sin z = 3$ (d) $\cosh z = -2$

22 Show that

(a) $\ln(5 + j12) = \ln 13 + j1.176$

(b) $\ln(-\frac{1}{2} - j\frac{1}{2}\sqrt{3}) = -j\frac{2\pi}{3}$

23 Writing $\tanh(u + jv) = x + jy$, with x, y, u and v real, determine x and y in terms of u and v. Hence evaluate $\tanh(2 + j\frac{1}{4}\pi)$ in the form $x + jy$.

24 In a certain cable of length l the current I_0 at the sending end when it is raised to a potential V_0 and the other end is earthed is given by

$$I_0 = \frac{V_0}{Z_0}\tanh Pl$$

Calculate the value of I_0 when $V_0 = 100$, $Z_0 = 500 + j400$, $l = 10$ and $P = 0.1 + j0.15$.

3.3 Powers of complex numbers

In earlier sections we have discussed the extensions of ordinary arithmetic, including $+, -, \times, \div$, to complex numbers. We now extend the arithmetical operations to include the operation of powers.

3.3.1 De Moivre's theorem

From (3.10) a complex number z may be expressed in terms of its modulus r and argument θ in the exponential form

$$z = re^{j\theta}$$

Using the rules of indices and the property (2.33a) of the exponential function, we have, for any n,

$$z^n = r^n(e^{j\theta})^n = r^n e^{j(n\theta)}$$

so that

$$z^n = r^n(\cos n\theta + j \sin n\theta) \tag{3.16}$$

This result is known as **de Moivre's theorem**.

Example 3.19

Express $1 - j$ in the form $r(\cos \theta + j \sin \theta)$ and hence evaluate $(1 - j)^{12}$.

Solution

From Example 3.7(b)

$$|1 - j| = \sqrt{2} \quad \text{and} \quad \arg(1 - j) = -\tfrac{1}{4}\pi$$

so that

$$1 - j = \sqrt{2}[\cos(-\tfrac{1}{4}\pi) + j \sin(-\tfrac{1}{4}\pi)]$$
$$= \sqrt{2}(\cos \tfrac{1}{4}\pi - j \sin \tfrac{1}{4}\pi)$$

Then

$$(1 - j)^{12} = (\sqrt{2})^{12}(\cos \tfrac{1}{4}\pi - j \sin \tfrac{1}{4}\pi)^{12}$$

which, on using de Moivre's theorem (3.16), gives

$$(1 - j)^{12} = 2^6[\cos(12 \times \tfrac{1}{4}\pi) - j \sin(12 \times \tfrac{1}{4}\pi)]$$
$$= 2^6(\cos 3\pi - j \sin 3\pi)$$
$$= 2^6(-1 - j0)$$
$$= -64$$

Most commonly, we use de Moivre's theorem to find the roots of complex numbers like $\sqrt{z}$ and $\sqrt[3]{z}$. More generally, we want to find $z^{1/n}$, the nth root, where n is a natural number. Setting $w = z^{1/n}$, we see that $z = w^n$, and by (3.16),

$$w^n = R^n(\cos n\phi + j \sin n\phi), \quad \text{where } |w| = R \text{ and arg } w = \phi$$

$$z = r(\cos \theta + j \sin \theta), \qquad \text{where } |z| = r \text{ and arg } z = \theta$$

Comparing real and imaginary parts in the equality $z = w^n$, we deduce that

$$r \cos \theta = R^n \cos n\phi$$

and

$$r \sin \theta = R^n \sin n\phi$$

Squaring and adding these two equations gives $r^2 = R^{2n}$; that is, $R = r^{1/n}$. Substituting this value into the equations gives

$$\cos \theta = \cos n\phi$$

and

$$\sin \theta = \sin n\phi$$

This pair of simultaneous equations has an infinite number of solutions because of the 2π-periodicity of the sine and cosine functions. Thus

$$n\phi = \theta + 2\pi k, \quad \text{where } k \text{ is an integer}$$

and

$$\phi = \frac{\theta}{n} + \frac{2\pi k}{n}, \quad \text{where } k = 0, 1, -1, 2, -2, 3, -3, \ldots$$

Substituting these values for R and ϕ into the formula for w gives

$$z^{1/n} = r^{1/n}\left[\cos\left(\frac{\theta}{n} + \frac{2\pi k}{n}\right) + j\sin\left(\frac{\theta}{n} + \frac{2\pi k}{n}\right)\right] \tag{3.17}$$

where k is an integer. This expression yields exactly n different roots, corresponding to $k = 0, 1, 2, \ldots, n - 1$. The value for $k = n$ is the same as that for $k = 0$, the value for $k = n + 1$ is the same as that for $k = 1$, and so on. The n values of $z^{1/n}$ are equally spaced around a circle of radius $r^{1/n}$ whose centre is the origin of the Argand diagram. Also, the arguments increase in arithmetic progression, so that joining the roots on the circle creates a regular polygon inscribed in the latter.

Equation (3.17) may be written alternatively in the exponential form

$$z^{1/n} = r^{1/n}e^{j(\theta/n + 2\pi k/n)}, \quad k = 0, 1, 2, \ldots, n - 1 \tag{3.18}$$

Example 3.20 Given $z = -\frac{1}{2} + j\frac{1}{2}$, evaluate

(a) $z^{1/2}$ (b) $z^{1/3}$

and display the roots on an Argand diagram.

Solution We first express z in polar form.

Since $r = |z| = \sqrt{(\frac{1}{4} + \frac{1}{4})} = 2^{-1/2}$, and $\theta = \arg(z) = \pi - \tan^{-1}1 = \frac{3}{4}\pi$, we have

$$z = 2^{-1/2}(\cos \tfrac{3}{4}\pi + j \sin \tfrac{3}{4}\pi)$$

Figure 3.8
Roots on an Argand
diagram for
Example 3.20.

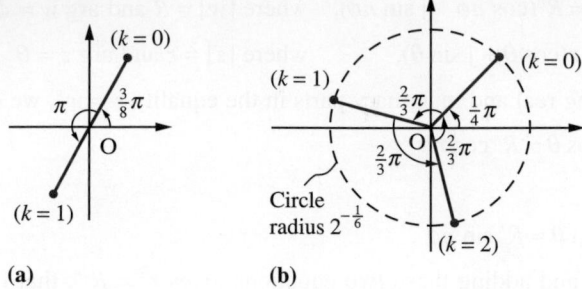

(a)   (b)

(a) From (3.17)

$$z^{1/2} = r^{1/2}\left[\cos\left(\frac{\theta}{2} + \frac{2\pi k}{2}\right) + j\sin\left(\frac{\theta}{2} + \frac{2\pi k}{2}\right)\right], \quad k = 0, 1$$

$$= 2^{-1/4}[\cos(\tfrac{3}{8}\pi + \pi k) + j\sin(\tfrac{3}{8}\pi + \pi k)], \qquad k = 0, 1$$

Thus we have two square roots:

$$z^{1/2} = 2^{-1/4}(\cos \tfrac{3}{8}\pi + j\sin \tfrac{3}{8}\pi) \quad \text{(for } k = 0)$$

and

$$z^{1/2} = 2^{-1/4}(\cos \tfrac{11}{8}\pi + j\sin \tfrac{11}{8}\pi) \quad \text{(for } k = 1)$$

as shown in Figure 3.8(a). These can be evaluated numerically, giving respectively (to 4dp) $z = 0.3218 + j0.7769$ and $z = -0.3218 - j0.7769$.

(b) From (3.17)

$$z^{1/3} = r^{1/3}\left[\cos\left(\frac{\theta}{3} + \frac{2\pi k}{3}\right) + j\sin\left(\frac{\theta}{3} + \frac{2\pi k}{3}\right)\right], \quad k = 0, 1, 2$$

$$= 2^{-1/6}[\cos(\tfrac{1}{4}\pi + \tfrac{2}{3}\pi k) + j\sin(\tfrac{1}{4}\pi + \tfrac{2}{3}\pi k)], \quad k = 0, 1, 2$$

Thus we obtain three cube roots:

$$z^{1/3} = 2^{-1/6}(\cos \tfrac{1}{4}\pi + j\sin \tfrac{1}{4}\pi) \quad \text{(for } k = 0)$$

$$z^{1/3} = 2^{-1/6}(\cos \tfrac{11}{12}\pi + j\sin \tfrac{11}{12}\pi) \quad \text{(for } k = 1)$$

and

$$z^{1/3} = 2^{-1/6}(\cos \tfrac{19}{12}\pi + j\sin \tfrac{19}{12}\pi) \quad \text{(for } k = 2)$$

as shown in Figure 3.8(b). Note that the three roots are equally spaced around a circle of radius $2^{-1/6}$ with centre at the origin.

Formula (3.17) can easily be extended to deal with the general rational power z^p of z.
Let $p = \dfrac{m}{n}$, where n is a natural number and m is an integer, then

$$z^p = (z^{1/n})^m$$

$$= \left\{ r^{1/n} \left[\cos\left(\frac{\theta}{n} + \frac{2\pi k}{n}\right) + j\sin\left(\frac{\theta}{n} + \frac{2\pi k}{n}\right) \right] \right\}^m, \quad k = 0, 1, 2, \dots, (n-1)$$

$$= r^{m/n} \left[\cos\left(\frac{m\theta}{n} + \frac{2\pi km}{n}\right) + j\sin\left(\frac{m\theta}{n} + \frac{2\pi km}{n}\right) \right]$$

$$= r^p [\cos(p\theta + 2\pi kp) + j\sin(p\theta + 2\pi kp)], \quad k = 0, 1, 2, \dots, (n-1)$$

Example 3.21   Evaluate $(-\frac{1}{2} + j\frac{1}{2})^{-2/3}$ and display the roots on an Argand diagram.

Solution   From Example 3.17, we can write

$$-\tfrac{1}{2} + j\tfrac{1}{2} = 2^{-1/2}(\cos\tfrac{3}{4}\pi + j\sin\tfrac{3}{4}\pi)$$

giving

$$z^{-2/3} = r^{-2/3} \left[\cos\left(-\frac{2\theta}{3} - \frac{4\pi k}{3}\right) + j\sin\left(-\frac{2\theta}{3} - \frac{4\pi k}{3}\right) \right], \quad k = 0, 1, 2$$

$$= 2^{1/3} [\cos(-\tfrac{1}{2}\pi - \tfrac{4}{3}\pi k) + j\sin(-\tfrac{1}{2}\pi - \tfrac{4}{3}\pi k)], \quad k = 0, 1, 2$$

Thus we obtain three values:

$$z^{-2/3} = 2^{1/3} [\cos(-\tfrac{1}{2}\pi) + j\sin(-\tfrac{1}{2}\pi)] \quad \text{(for } k = 0)$$

$$z^{-2/3} = 2^{1/3} (\cos\tfrac{1}{6}\pi + j\sin\tfrac{1}{6}\pi) \quad \text{(for } k = 1)$$

and

$$z^{-2/3} = 2^{1/3} (\cos\tfrac{5}{6}\pi + j\sin\tfrac{5}{6}\pi) \quad \text{(for } k = 2)$$

as shown in Figure 3.9.

Figure 3.9
Roots on an Argand
diagram for
Example 3.2.1.

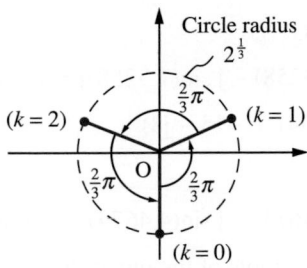

Example 3.22   Solve the quadratic equation

$$z^2 + (2j - 3)z + (5 - j) = 0$$

Solution   Using formula (1.5)

$$z = \frac{-(2j - 3) \pm \sqrt{[(2j - 3)^2 - 4(5 - j)]}}{2}$$

Figure 3.10
The complex
number $-15 - j8$.

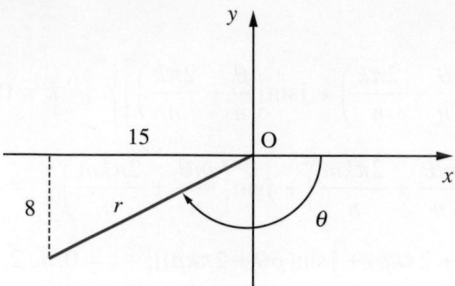

that is,

$$z = \frac{-(2j - 3) \pm \sqrt{(-15 - j8)}}{2} \tag{3.19}$$

Now we need to determine $(-15 - j8)^{1/2}$ so first we express it in polar form. Since

$$|-15 - j8| = \sqrt{[(15)^2 + (8)^2]} = 17$$

and from Figure 3.10

$$\arg(-15 - j8) = -(\pi - \tan^{-1}\tfrac{8}{15})$$

$$= -2.6516$$

we have

$$-15 - j8 = 17[\cos(2.6516) - j\sin(2.6516)]$$

From (3.17)

$$(-15 - j8)^{1/2} = (17)^{1/2}\left[\cos\left(\frac{2.6516}{2} + \frac{2\pi k}{2}\right) - j\sin\left(\frac{2.6516}{2} + \frac{2\pi k}{2}\right)\right]$$

$$= (17)^{1/2}[\cos(1.3258 + \pi k) - j\sin(1.3258 + \pi k)], \quad k = 0, 1$$

Thus we have the two square roots

$$(-15 - j8)^{1/2} = (17)^{1/2}[\cos(1.3258) - j\sin(1.3258)] = 1 - j4 \quad (\text{for } k = 0)$$

(the reader should verify that $(1 - j4)^2 = -15 - j8$)

and

$$(-15 - j8)^{1/2} = (17)^{1/2}[\cos(4.4674) - j\sin(4.4674)] = -1 + j4 \quad (\text{for } k = 1)$$

Substituting back in (3.19) gives the roots of the quadratic as

$$z = 2 - j3 \quad \text{and} \quad 1 + j$$

3.3.2 Powers of trigonometric functions and multiple angles

Euler's formula may be used to express $\sin^n\theta$ and $\cos^n\theta$ in terms of sines and cosines of multiple angles. If $z = \cos\theta + j\sin\theta$ then

$$z^n = \cos n\theta + j\sin n\theta$$

and

$$z^{-n} = \cos n\theta - j \sin n\theta$$

so that

$$z^n + z^{-n} = 2 \cos n\theta \qquad \text{(3.20a)}$$

$$z^n - z^{-n} = 2j \sin n\theta \qquad \text{(3.20b)}$$

Using these results, $\cos^n\theta$ and $\sin^n\theta$ can be expressed in terms of sines and cosines of multiple angles as illustrated in Example 3.23.

Example 3.23

Expand in terms of sines and cosines of multiple angles

(a) $\cos^5\theta$ (b) $\sin^6\theta$

Solution

(a) Using (3.20a) with $n = 1$,

$$(2\cos\theta)^5 = \left(z + \frac{1}{z}\right)^5 = z^5 + 5z^3 + 10z + \frac{10}{z} + \frac{5}{z^3} + \frac{1}{z^5}$$

so that

$$32\cos^5\theta = \left(z^5 + \frac{1}{z^5}\right) + 5\left(z^3 + \frac{1}{z^3}\right) + 10\left(z + \frac{1}{z}\right)$$

which, on using (3.20a) with $n = 5$, 3 and 1, gives

$$\cos^5\theta = \tfrac{1}{32}(2\cos 5\theta + 10 \cos 3\theta + 20 \cos \theta) = \tfrac{1}{16}(\cos 5\theta + 5 \cos 3\theta + 10 \cos \theta)$$

(b) Using (3.20b) with $n = 1$,

$$(2j\sin\theta)^6 = \left(z - \frac{1}{z}\right)^6 = z^6 - 6z^4 + 15z^2 - 20 + \frac{15}{z^2} - \frac{6}{z^4} + \frac{1}{z^6}$$

which, on noting that $j^6 = -1$, gives

$$-64\sin^6\theta = \left(z^6 + \frac{1}{z^6}\right) - 6\left(z^4 + \frac{1}{z^4}\right) + 15\left(z^2 + \frac{1}{z^2}\right) - 20$$

Using (3.20a) with $n = 6$, 4 and 2 then gives

$$\sin^6\theta = -\tfrac{1}{64}(2\cos 6\theta - 12 \cos 4\theta + 30 \cos 2\theta - 20)$$

$$= \tfrac{1}{32}(10 - 15 \cos 2\theta + 6 \cos 4\theta - \cos 6\theta)$$

Conversely, de Moivre's theorem may be used to expand $\cos n\theta$ and $\sin n\theta$, where n is a positive integer, as polynomials in $\cos \theta$ and $\sin \theta$. From the theorem

$$\cos n\theta + j \sin n\theta = (\cos \theta + j \sin \theta)^n$$

we obtain, writing $s = \sin \theta$ and $c = \cos \theta$ for convenience,

$$\cos n\theta + j \sin n\theta = (c + js)^n = c^n + jnc^{n-1}s + j^2\frac{n(n-1)}{2!}c^{n-2}s^2 + \ldots + j^n s^n$$

Equating real and imaginary parts yields

$$\cos n\theta = c^n - \frac{n(n-1)}{2!}c^{n-2}s^2 + \frac{n(n-1)(n-2)(n-3)}{4!}c^{n-4}s^4 + \dots$$

and

$$\sin n\theta = nc^{n-1}s - \frac{n(n-1)(n-2)}{3!}c^{n-3}s^3 + \dots$$

Using the trigonometric identity $\cos^2\theta = 1 - \sin^2\theta$ (so that $c^2 = 1 - s^2$), we see that

(a) $\cos n\theta$ can be expanded in terms of $(\cos \theta)^n$ for any n or in terms of $(\sin \theta)^n$ if n is even;

(b) $\sin n\theta$ can be expanded in terms of $(\sin \theta)^n$ if n is odd.

Example 3.24 Expand $\cos 4\theta$ as a polynomial in $\cos \theta$.

Solution By de Moivre's theorem,

$$(\cos 4\theta + j \sin 4\theta) = (\cos \theta + j \sin \theta)^4 = (c + js)^4$$
$$= c^4 + j4c^3s + j^2 6c^2 s^2 + j^3 4cs^3 + j^4 s^4$$
$$= c^4 + j4c^3s - 6c^2 s^2 - j4cs^3 + s^4$$

Equating real parts,

$$\cos 4\theta = c^4 - 6c^2 s^2 + s^4$$

which on using $s^2 = 1 - c^2$ gives

$$\cos 4\theta = c^4 - 6c^2(1 - c^2) + (1 - c^2)^2 = 8c^4 - 8c^2 + 1$$

Thus

$$\cos 4\theta = 8 \cos^4\theta - 8 \cos^2\theta + 1$$

Note that by equating imaginary parts we could have obtained a polynomial expansion for $\sin 4\theta$.

In MATLAB raising to a power is obtained using the standard operator ^. For example, considering Example 3.19 entering

$(1 - i)\,\hat{}\,12$ returns the answer -64

as determined in the given solution. Considering Example 3.20(a) entering the commands

```
z = -1/2 + (1/2)*i; z1 = z^1/2
```

return

```
z1 = 0.3218 + 0.7769i
```

which is the root corresponding to $k = 0$. From knowledge that the two roots are equally spaced around a circle the second root may be easily written down.

In Example 3.22 the solution may be obtained symbolically using the `solve` command. Entering

```
syms z
solve(z^2 + (2*i - 3)*z + (5 - i))
```

returns the answer

```
2 - 3*i
1 + i
```

which checks with the answer given in the solution.

Expanding in terms of sines and cosines of multiple angles may be undertaken symbolically using the `expand` command. For example, considering Example 3.24 the commands

```
syms theta
expand(cos(4*theta))
```

return the answer

```
8*cos(theta)^4 - 8*cos(theta)^2 + 1
```

which checks with the answer obtained in the given solution.

With the usual small modifications, MAPLE uses the same instructions.

3.3.3 Exercises

 Check your answer using MATLAB or MAPLE whenever possible.

25 Use de Moivre's theorem to calculate the third and fourth powers of the complex numbers

    (a) $1 + j$ (b) $\sqrt{3} - j$ (c) $-3 + j4$

    (d) $1 - j\sqrt{3}$ (e) $-1 + j\sqrt{3}$ (f) $-1 - j\sqrt{3}$

(The moduli and arguments of these numbers were found in Exercises 3.2.5, Question 7.)

26 Expand in terms of multiple angles

    (a) $\cos^4\theta$ (b) $\sin^3\theta$

27 Use the method of Section 3.3.2 to prove the following results:

    (a) $\sin 3\theta = 3 \cos^2\theta \sin\theta - \sin^3\theta$

    (b) $\cos 8\theta = 128 \cos^8\theta - 256 \cos^6\theta + 160 \cos^4\theta - 32 \cos^2\theta + 1$

    (c) $\tan 5\theta = \dfrac{5 \tan\theta - 10 \tan^3\theta + \tan^5\theta}{1 - 10 \tan^2\theta + 5 \tan^4\theta}$

28 Find the three values of $(8 + j8)^{1/3}$ and show them on an Argand diagram.

29 Find the following complex numbers in their polar forms:

    (a) $(\sqrt{3} - j)^{1/4}$ (b) $(j8)^{1/3}$

    (c) $(3 - j3)^{-2/3}$ (d) $(-1)^{1/4}$

    (e) $(2 + j2)^{4/3}$ (f) $(5 - j3)^{-1/2}$

30 Obtain the four solutions of the equation

$$z^4 = 3 - j4$$

giving your answers to three decimal places.

31 Solve the quadratic equation

$$z^2 - (3 + j5)z + j8 - 5 = 0$$

32 Find the values of $z^{1/3}$, where $z = \cos 2\pi + j \sin 2\pi$. Generalize this to an expression for $1^{1/n}$. Hence solve the equations

    (a) $\left[\dfrac{z - 2}{z + 2}\right]^5 = 1$ (*Hint*: first show that there are only 4 roots)

    (b) $(z - 3)^6 - z^6 = 0$

3.4 Loci in the complex plane

A **locus** (plural **loci**) is the set of points that have a specified property. For example, a circle is the locus of the points in a plane that are a fixed distance, its radius, from a fixed point, its centre. The property may be specified in words or algebraically. Loci occur frequently in engineering contexts, from the design of safety guards around moving machinery to the design of aircraft wing sections. The Argand diagram representation of complex numbers as points on a plane often makes it possible to represent complicated loci very concisely in terms of a complex variable and this simplifies the engineering analysis. This occurs in a wide range of engineering problems, from the water percolation through dams to the design of microelectronic devices.

3.4.1 Straight lines

There are many ways in which straight lines may be represented using complex numbers. We will illustrate these with a number of examples.

Example 3.25 Describe the locus of z given by

(a) $\text{Re}(z) = 4$ (b) $\arg(z - 1 - j) = \pi/4$

(c) $\left| \dfrac{z - j2}{z - 1} \right| = 1$ (d) $\text{Im}((1 - j2)z) = 3$

Solution (a) Here $z = 4 + jy$ for any real y, so that the locus is the vertical straight line with equation $x = 4$ illustrated in Figure 3.11(a).

(b) Here $z = 1 + j + r(\cos \pi/4 + j \sin \pi/4)$ for any positive (> 0) real number r, so that the locus is a half-line making an angle $\pi/4$ with the positive x direction with the end point $(1, 1)$ *excluded* (since arg 0 is not defined). Algebraically we can write it as $y = x$, $x > 1$, and it is illustrated in Figure 3.11(b).

(c) The equation, in this case, may be written

$$|z - j2| = |z - 1|$$

Recalling the definition of modulus, we can rewrite this as

$$\sqrt{[x^2 + (y - 2)^2]} = \sqrt{[(x - 1)^2 + y^2]}$$

Squaring both sides and multiplying out, we obtain

$$x^2 + y^2 - 4y + 4 = x^2 - 2x + 1 + y^2$$

which simplifies to

$$y = \tfrac{1}{2}x + \tfrac{3}{4}$$

the equation of a straight line.

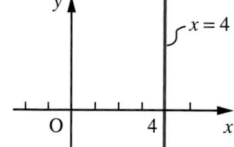

(a) Line $x = 4$

(b) Half-line $y = x$, $x > 1$

Figure 3.11

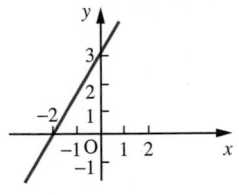

(c) Line $y = \frac{1}{2}x + \frac{3}{4}$

Alternatively, we can interpret $|z - j2|$ as the distance on the Argand diagram from the point $0 + j2$ to the point z, and $|z - 1|$ as the distance from the point $1 + j0$ to the point z, so that

$$|z - j2| = |z - 1|$$

is the locus of points that are equidistant from the two fixed points $(0, 2)$ and $(1, 0)$, as shown in Figure 3.11(c).

(d) Writing $z = x + jy$,

$$(1 - j2)z = (1 - j2)(x + jy) = x + 2y + j(y - 2x)$$

so that $\mathrm{Im}((1 - j2)z) = 3$, implies $y - 2x = 3$.
  Thus $\mathrm{Im}((1 - j2)z) = 3$ describes the straight line

$$y = 2x + 3$$

illustrated in Figure 3.11(d).

(d) Line $y = 2x + 3$

Figure 3.11
continued

3.4.2 Circles

The simplest representation of a circle on the Argand diagram makes use of the fact that $|z - z_1|$ is the distance between the point $z = x + jy$ and the point $z_1 = a + jb$ on the diagram. Thus a circle of radius R and centre (a, b), illustrated in Figure 3.12, may be written

$$|z - z_1| = R$$

We can also write this as $z - z_1 = R\mathrm{e}^{jt}$, where t is a parameter such that

$$-\pi < t \leqslant \pi$$

Figure 3.12
The circle $|z - z_1| = R$.

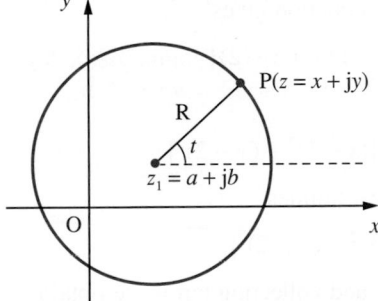

Example 3.26 Find the cartesian equation of the circle

$$|z - (2 + j3)| = 2$$

Solution Now,

$$z - (2 + j3) = (x - 2) + j(y - 3)$$

so that

$$|z - (2 + j3)| = \sqrt{[(x - 2)^2 + (y - 3)^2]}$$

and hence on the circle

$$|z - (2 + j3)| = 2$$

we have

$$\sqrt{[(x - 2)^2 + (y - 3)^2]} = 2$$

which implies

$$(x - 2)^2 + (y - 3)^2 = 4$$

indicating that the circle has centre (2, 3) and radius 2.

This may be written in the standard form

$$x^2 + y^2 - 4x - 6y + 9 = 0$$

This is not the only method of representing a circle, as is shown in the following two examples.

Example 3.27   Find the cartesian equation of the curve whose equation on the Argand diagram is

$$\left| \frac{z - j}{z - 1 - j2} \right| = \sqrt{2}$$

Solution   By expressing it in the form $|z - j| = \sqrt{2}|z - (1 + j2)|$ we can interpret this equation as 'the distance between z and j is $\sqrt{2}$ times the distance between z and $(1 + j2)$', so this is different from Example 3.25(d).

Putting $z = x + jy$ into the equation gives

$$|x + j(y - 1)| = \sqrt{2}|(x - 1) + j(y - 2)|$$

Thus

$$\sqrt{[x^2 + (y - 1)^2]} = \sqrt{2}\sqrt{[(x - 1)^2 + (y - 2)^2]}$$

which, on squaring both sides, implies

$$x^2 + (y - 1)^2 = 2[(x - 1)^2 + (y - 2)^2]$$

Multiplying out the brackets and collecting terms we obtain

$$x^2 + y^2 - 4x - 6y + 9 = 0 \quad \text{or} \quad (x - 2)^2 + (y - 3)^2 = 4$$

which, from (1.14), is the equation of the circle of centre (2, 3), and radius 2.

This is a special case of a general result. If z_1 and z_2 are fixed complex numbers and k is a positive real number, then the locus of z which satisfies $\left| \dfrac{z - z_1}{z - z_2} \right| = k$ is a circle, known as the circle of Apollonius, *unless* $k = 1$. When $k = 1$, the locus is a straight line, as we saw in Example 3.25(d).

Example 3.28   Find the locus of z in the Argand diagram such that

$$\mathrm{Re}[(z - \mathrm{j})/(z + 1)] = 0$$

Solution   Setting $z = x + \mathrm{j}y$, as usual, we obtain

$$\frac{z - \mathrm{j}}{z + 1} = \frac{x + \mathrm{j}(y - 1)}{(x + 1) + \mathrm{j}y} = \frac{[x + \mathrm{j}(y - 1)][(x + 1) - \mathrm{j}y]}{(x + 1)^2 + y^2}$$

Hence $\mathrm{Re}[(z - \mathrm{j})/(z + 1)] = 0$ implies $x(x + 1) + y(y - 1) = 0$.
Rearranging this, we have

$$x^2 + y^2 + x - y = 0$$

and

$$(x + \tfrac{1}{2})^2 + (y - \tfrac{1}{2})^2 = \tfrac{1}{2}$$

Hence the locus of z on the Argand diagram is a circle of centre $(-\tfrac{1}{2}, \tfrac{1}{2})$ and radius $\sqrt{2}/2$.

3.4.3 More general loci

In general we approach the problem of finding the locus of z on the Argand diagram using a mixture of elementary pure geometry and algebraic manipulation of expressions involving $z = x + \mathrm{j}y$. We illustrate this in Example 3.29.

Example 3.29   Find the cartesian equation of the locus of z given by

$$|z + 1| + |z - 1| = 4$$

Solution   The defining equation here may be interpreted as the sum of the distances of the point z from the points 1 and -1 is a constant $(= 4)$. By elementary considerations (Figure 3.13) we can see that the locus passes through $(2, 0)$, $(0, \sqrt{3})$, $(-2, 0)$ and $(0, -\sqrt{3})$. Results from classical geometry would identify the locus as an ellipse with foci at $(1, 0)$ and $(-1, 0)$, using the 'string property' (see Example 1.37). Using algebraic methods, however, we set $z = x + \mathrm{j}y$ into the equation, giving

$$\sqrt{[(x + 1)^2 + y^2]} + \sqrt{[(x - 1)^2 + y^2]} = 4$$

Figure 3.13
The ellipse of
Example 3.29.

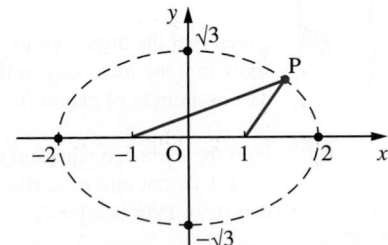

Rewriting this equation as

$$\sqrt{[(x+1)^2 + y^2]} = 4 - \sqrt{[(x-1)^2 + y^2]}$$

and squaring both sides gives

$$(x+1)^2 + y^2 = 16 - 8\sqrt{[(x-1)^2 + y^2]} + (x-1)^2 + y^2$$

This simplifies to give

$$4 - x = 2\sqrt{[(x-1)^2 + y^2]}$$

so that squaring both sides again gives

$$16 - 8x + x^2 = 4[x^2 - 2x + 1 + y^2]$$

which reduces to

$$\frac{x^2}{4} + \frac{y^2}{3} = 1$$

in the standard form of an ellipse.

3.4.4 Exercises

33 Let $z = 8 + j$ and $w = 4 + j4$. Calculate the distance on the Argand diagram from z to w and from z to $-w$.

34 Describe the locus of z when

(a) Re $z = 5$ (b) $|z - 1| = 3$

(c) $\left| \dfrac{z-1}{z+1} \right| = 3$ (d) $\arg(z - 2) = \pi/4$

35 The circle $x^2 + y^2 + 4x = 0$ and the straight line $y = 3x + 2$ are taken to lie on the Argand diagram. Describe the circle and the straight line in terms of z.

36 Identify and sketch the loci on the complex plane given by

(a) $\text{Re}\left(\dfrac{z+j}{z-j} \right) = 1$ (b) $\text{Re}\left(\dfrac{z+j}{z-j} \right) = 2$

(c) $\left| \dfrac{z+j}{z-j} \right| = 3$ (d) $\tan \arg\left(\dfrac{z+j}{z-j} \right) = \sqrt{3}$

(e) $\text{Im}(z^2) = 2$ (f) $|z+j| + |z-1| = 2$

(g) $|z+j| - |z-1| = \frac{1}{2}$ (h) $\arg(z + j2) = \frac{1}{4}\pi$

(i) $\arg(2z - 3) = -\frac{2}{3}\pi$ (j) $|z - j2| = 1$

37 Express as simply as possible the following loci in terms of a complex variable:

(a) $y = 3x - 2$ (b) $x^2 + y^2 + 4x = 0$

(c) $x^2 + y^2 + 2x - 4y - 4 = 0$ (d) $x^2 - y^2 = 1$

38 Find the locus of the point z in the Argand diagram which satisfies the equation

(a) $|z - 1| = 2$ (b) $|2z - 1| = 3$

(c) $|z - 2 - j3| = 4$ (d) $\arg(z) = 0$

(e) $|z - 4| = 3|z + 1|$ (f) $\arg\left(\dfrac{z-1}{z-j} \right) = \frac{1}{2}\pi$

39 Find the cartesian equation of the circle given by

$$\left| \frac{z+j}{z-1} \right| = \sqrt{2}$$

and give two other representations of the circle in terms of z.

40 Given that the argument of $(z - 1)/(z + 1)$ is $\frac{1}{4}\pi$, show that the locus of z in the Argand diagram is part of a circle of centre $(0, 1)$ and radius $\sqrt{2}$.

41 Find the cartesian equation of the locus of the point $z = x + jy$ that moves in the Argand diagram such that $|(z + 1)/(z - 2)| = 2$.

3.5 Functions of a complex variable

In Chapter 2, Section 2.2.1, the basic idea of a function was described. Essentially it involves two sets X and Y and a rule that assigns to every element x in the set X precisely one element y in the set Y. In Chapter 2 we were concerned with real functions so that x and y were real numbers. When the independent variable is a complex number $z = x + jy$ then, in general, a function $f(z)$ of z will have values which are complex numbers. Conventionally $w = u + jv$ is used to denote the dependent variable of a function of a complex variable, thus

$$w = u + jv = f(z) \quad \text{where} \quad z + x + jy$$

Example 3.30 Express u and v in terms of x and y where $w = u + jv$, $z = x + jy$, $w = f(z)$ and

(a) $f(z) = z^2$ (b) $f(z) = \dfrac{z - j}{z + 1}$, $z \neq -1$

Solution (a) When $w = z^2$, we have $u + jv = (x + jy)^2$. This may be rewritten as

$$u + jv = x^2 - y^2 + j2xy$$

so that comparing real and imaginary parts on either side of this equation we have

$$u = x^2 - y^2 \quad \text{and} \quad v = 2xy$$

(b) When $w = \dfrac{z - j}{z + 1}$, we have

$$u + jv = \frac{x + j(y - 1)}{(x + 1) + jy} = \frac{[x + j(y - 1)][(x + 1) - jy]}{(x + 1)^2 + y^2}$$

Hence comparing real and imaginary parts we have

$$u = \frac{x(x + 1) + y(y - 1)}{(x + 1)^2 + y^2} \quad \text{amd} \quad v = \frac{(x + 1)(y - 1) - xy}{(x + 1)^2 + y^2}$$

These may be written as

$$u = \frac{x^2 + y^2 + x - y}{x^2 + y^2 + 2x + 1} \quad \text{amd} \quad v = \frac{y - x - 1}{x^2 + y^2 + 2x + 1}$$

The graphical representation of functions of a complex variable requires two planes, one for the independent variable $z = x + jy$ and another for the dependent variable $w = u + jv$. Thus the function $w = f(z)$ can be regarded as a **mapping** of points on the z plane to points on the w plane. Under such a mapping a region A on the z plane is **transformed** into the region A' on the w plane.

Example 3.31 Find the image on the w plane of the strip between $x = 1$ and $x = 2$ on the z plane under the mapping defined by

$$w = \frac{z + 2}{z}$$

Solution The easiest approach to this problem is firstly to find x in terms of u and v. So solving $w = \dfrac{z+2}{z}$ for z we have

$$z = \frac{2}{w-1}$$

and

$$x + jy = \frac{2}{(u-1) + jv} = \frac{2[(u-1) - jv]}{(u-1)^2 + v^2}$$

Equating real parts then gives

$$x = \frac{2(u-1)}{(u-1)^2 + v^2}$$

The line $x = 1$ maps into

$$1 = \frac{2u - 2}{u^2 - 2u + 1 + v^2}$$

which simplifies to give the circle on the w plane

$$(u-2)^2 + v^2 = 1$$

The line $x = 2$ maps into

$$2 = \frac{2u - 2}{u^2 - 2u + 1 + v^2}$$

which simplifies to give the circle on the w plane

$$(u - \tfrac{3}{2})^2 + v^2 = \tfrac{1}{4}$$

Thus the strip between $x = 1$ and $x = 2$ maps into that portion of the w plane between these two circles as illustrated in Figure 3.14. The point $z = \frac{3}{2}$ maps to $w = \frac{7}{3}$ confirming that the shaded areas correspond.

As will be shown in the companion text *Advanced Modern Engineering Mathematics*, these properties are used to solve steady state potential problems in two dimensions.

Figure 3.14
Transformation of the strip $1 < \operatorname{Re} z < 2$ onto the w plane.

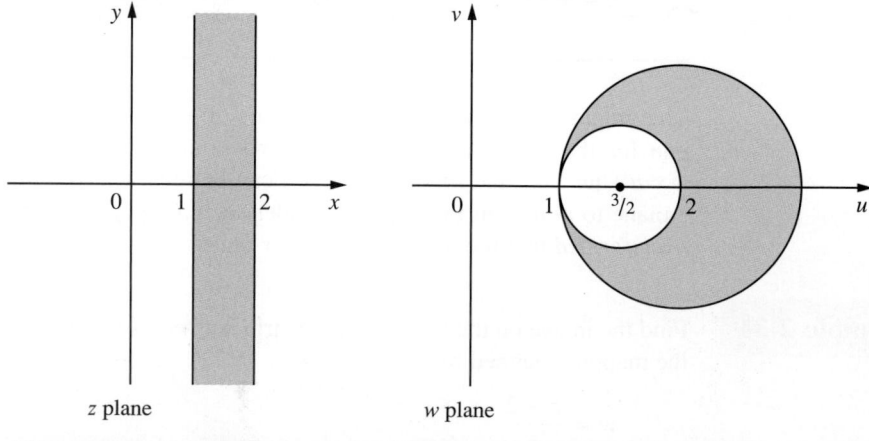

z plane   w plane

3.5.1 Exercises

42 Find u and v in terms of x and y where $w = f(z)$, $z = x + jy$, $w = u + jv$ and

   (a) $f(z) = (1 - j)z$ (b) $f(z) = (z - 1)^2$

   (c) $f(z) = z + \frac{1}{z}$

43 Find the values of the complex numbers a and b such that the function $w = az + b$ maps the point $z = 4j$ to $w = j$ and the point $z = -1$ to the point $w = 1 + j$.

44 Show that the line $y = 1$ on the z plane is transformed into the line $u = 1$ on the w plane by the function $w = (z + j)/(z - j)$.

45 Show that the function $w = (jz - 1)/(z - 1)$ maps the line $y = x$ on the z plane onto the circle

   $$(u - 1)^2 + (v - 1)^2 = 1$$

   on the w plane.

3.6 Engineering application: alternating currents in electrical networks

When an alternating current $i = I \sin \omega t$ (ω is a constant and t is the time) flows in a circuit the corresponding voltage depends on ω and on the resistance, capacitance and inductance of the circuit. (Note that the frequency of the current is $\omega/2\pi$.) For simplicity we shall separate these three elements and consider their effects individually.

For a resistor of resistance R the corresponding voltage is $v = IR \sin \omega t$. This voltage is 'in phase' with the current. It is zero at the same times as i and achieves its maxima at the same times as i, as shown in Figure 3.15. For a capacitor of capacitance C the corresponding voltage is $v = (I/\omega C) \sin(\omega t - \frac{1}{2}\pi)$, as shown in Figure 3.16. Here the voltage 'lags' behind the current by a phase of $\frac{1}{2}\pi$. For an inductor of inductance L the corresponding voltage is $v = \omega L I \sin(\omega t + \frac{1}{2}\pi)$, as shown in Figure 3.17. Here the voltage 'leads' the current by a phase of $\frac{1}{2}\pi$.

Figure 3.15
A resistor of
resistance R.

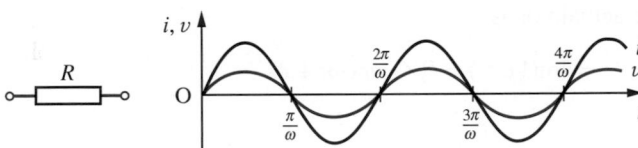

Figure 3.16
A capacitor of
capacitance C.

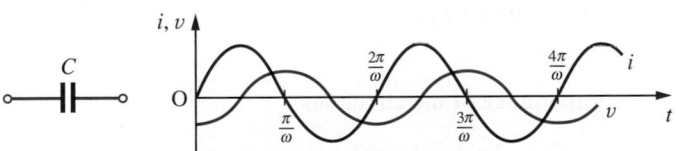

Figure 3.17
An inductor of
inductance L.

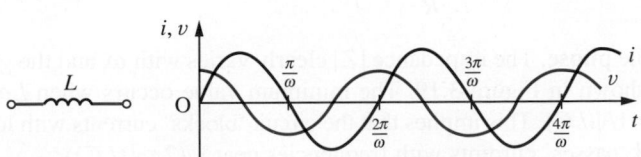

Combining these results to find v in the case of a general network is easily done using the properties of complex numbers. Remembering that $\sin\theta = \text{Im}(e^{j\theta})$, we can summarize the results as

$$v = \begin{cases} \text{Im}(IRe^{j\omega t}) & \text{for a resistor} \\[2mm] \text{Im}\left(\dfrac{I}{\omega C}e^{j(\omega t - \pi/2)}\right) & \text{for a capacitor} \\[2mm] \text{Im}(\omega LI e^{j(\omega t + \pi/2)}) & \text{for an inductor} \end{cases}$$

Now $e^{j\pi/2} = \cos\frac{1}{2}\pi + j\sin\frac{1}{2}\pi = j$ and $e^{-j\pi/2} = -j$, so we may rewrite these as

$$v = \text{Im}(IZe^{j\omega t})$$

where

$$Z = \begin{cases} R & \text{for a resistor} \\[2mm] -\dfrac{j}{\omega C} & \text{for a capacitor} \\[2mm] j\omega L & \text{for an inductor} \end{cases}$$

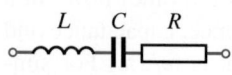

Figure 3.18
A linear *LCR* circuit.

Z is called the **complex impedance** of the element, and $V = IZ$ is the **complex voltage**.

For the general *LCR* circuit shown in Figure 3.18 the complex voltage V is the algebraic sum of the complex voltages of the individual elements; that is,

$$V = IR + j\omega LI - \frac{jI}{\omega C} = IZ$$

where

$$Z = R + j\omega L - \frac{j}{\omega C}$$

The actual voltage

$$v = \text{Im}(Ve^{j\omega t}) = I|Z|\sin(\omega t + \phi)$$

where

$$|Z| = \left[R^2 + \left(L\omega - \frac{1}{C\omega} \right)^2 \right]^{1/2}$$

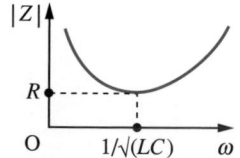

Figure 3.19
The impedance of
an *LCR* circuit.

is the **impedance** of the circuit and

$$\phi = \tan^{-1}\left(\frac{L\omega - 1/C\omega}{R} \right)$$

is the **phase**. The impedance $|Z|$ clearly varies with ω, and the graph of this dependence is shown in Figure 3.19. The minimum value occurs when $L\omega = 1/C\omega$; that is, when $\omega = 1/\sqrt{(LC)}$. This implies that the circuit 'blocks' currents with low and high frequencies, and 'passes' currents with frequencies near $1/(2\pi\sqrt{(LC)})$.

Example 3.32   Calculate the complex impedance of the element shown in Figure 3.20 when an alternating current of frequency 100 Hz flows.

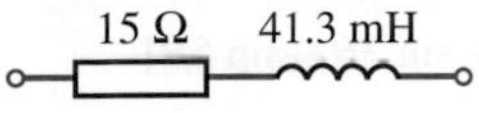

Solution   The complex impedance is the sum of the individual impedances. Thus

$$Z = R + j\omega L$$

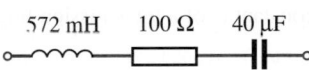

15 Ω   41.3 mH

Figure 3.20
The element of
Example 3.32.

Here $R = 15\,\Omega$, $\omega = 2\pi \times 100\,\text{rad s}^{-1}$ and $L = 41.3 \times 10^{-3}\,\text{H}$, so that

$$Z = 15 + j25.9$$

and $|Z| = 30\,\Omega$ and $\phi = \tfrac{1}{3}\pi$.

3.6.1 Exercises

46   Calculate the complex impedance for the circuit shown in Figure 3.21 when an alternating current of frequency 50 Hz flows.

572 mH   100 Ω   40 µF

Figure 3.21

47   The complex impedance of two circuit elements in series as shown in Figure 3.22(a) is the sum of the complex impedances of the individual elements, and the reciprocal of the impedance of two elements in parallel is the sum of the reciprocals of the individual impendances, as shown in Figure 3.22(b). Use these results to calculate the complex impedance of the network shown in Figure 3.23, where $Z_1 = 1 + j\,\Omega$, $Z_2 = 5 - j5\,\Omega$ and $Z_3 = 1 + j2\,\Omega$.

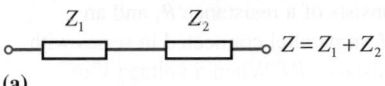

(a) $Z = Z_1 + Z_2$

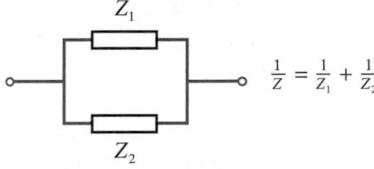

(b)

Figure 3.22

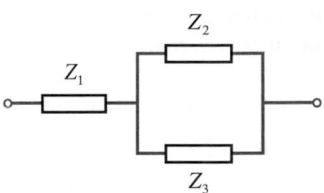

Figure 3.23

3.7 Review exercises (1–34)

Check your answers using MATLAB or MAPLE whenever possible.

1   Let $z = 4 + j3$ and $w = 2 - j$. Calculate

(a) $3z$ (b) w^* (c) zw

(d) z^2 (e) $|z|$ (f) w/z

(g) $z - \tfrac{1}{w}$ (h) $\arg z$ (i) $z^{\frac{3}{2}}$

2   For x and y real solve the equation

$$\frac{jy}{jx+1} - \frac{3y+j4}{3x+y} = 0$$

3   Given $z = (2 + j)/(1 - j)$, find the real and imaginary parts of $z + z^{-1}$.

$-(u + v)$

where $u^3 = \frac{1}{2}r + \sqrt{[\frac{1}{4}r^2 + \frac{1}{27}q^3]}$

and $v^3 = \frac{1}{2}r - \sqrt{[\frac{1}{4}r^2 + \frac{1}{27}q^3]}$

Express the remaining two roots in terms of u, v and ω and find the condition that all three roots are real.

28 ABCD is a square, lettered anticlockwise, on an Argand diagram. If the points A, B represent $3 + j2$, $-1 + j4$ respectively, show that C lies on the real axis, and find the number represented by D and the length of AB.

29 If $z_1 = 3 + j2$ and $z_2 = 1 + j$, and O, P, Q, R represent the numbers 0, z_1, z_1z_2, z_1/z_2 on the Argand diagram, show that RP is parallel to OQ and is half its length.

30 Show that as z describes the circle $z = be^{j\theta}$, $u + jv = z + a^2/z$ describes an ellipse ($a \neq b$). What is the image locus when $a = b$?

31 Show that the function

$$w = \frac{4}{z}$$

where $z = x + jy$ and $w = u + jv$, maps the line $3x + 4y = 1$ in the z plane onto a circle in the w plane and determine its radius and centre.

32 Show that the function

$$w = (1 + j)z + 1$$

where $z = x + jy$ and $w = u + jv$, maps the line $y = 2x - 1$ in the z plane onto a line in the w plane and determine its equation.

33 Show that the function

$$w = \frac{z - 1}{z + 1}$$

where $z = x + jy$ and $w = u + jv$, maps the circle $|z| = 3$ on the z plane onto a circle in the w plane.

Find the centre and radius of this circle in the w plane and indicate, by means of shading on a sketch, the region in the w plane that corresponds to the interior of the circle $|z| = 3$ in the z plane.

34 Show that as θ varies the point $z = a(h + \cos\theta) + ja(k + \sin\theta)$ describes a circle. The Joukowski transformation $u + jv = z + l^2/z$ is applied to this circle to produce an aerofoil shape in the u–v plane. Show that the coordinates of the aerofoil can be written in the form

$$\frac{u}{a} = (h + \cos\theta)$$

$$\times \left\{1 + \frac{l^2}{a^2(1 + h^2 + k^2 + 2h\cos\theta + 2k\sin\theta)}\right\}$$

$$\frac{v}{a} = (k + \sin\theta)$$

$$\times \left\{1 - \frac{l^2}{a^2(1 + h^2 + k^2 + 2h\cos\theta + 2k\sin\theta)}\right\}$$

Taking the case $a = 1$ and $l^2 = 8$, trace the aerofoil where

(a) $h = k = 0$, and show that it is an ellipse;

(b) $h = 0.04$, $k = 0$ and show that it is a symmetrical aerofoil with a blunt leading and trailing edge;

(c) $h = 0$, $k = 0.1$ and show that it is a symmetrical aerofoil (about v axis) with camber;

(d) $h = 0.04$, $k = 0.1$ and show that it is a non-symmetrical aerofoil with camber and rounded leading and trailing edges.

4 Vector Algebra

Chapter 4 Contents

4.1 Introduction

Much of the work of engineers and scientists involves forces. Ensuring the structural integrity of a building or a bridge involves knowing the forces acting on the system and designing the structural members to withstand them. Many have seen the dramatic pictures of the Tacoma bridge disaster (see also Section 10.10.3), when the forces acting on the bridge were not predicted accurately. To analyse such a system requires the use of Newton's laws in a situation where vector notation is essential. Similarly, in a reciprocating engine periodic forces act, and Newton's laws are used to design a crankshaft that will reduce the side forces to zero, thereby minimizing wear on the moving parts. Forces are three-dimensional quantities and provide one of the commonest examples of vectors. Associated with these forces are accelerations and velocities, which can also be represented by vectors. The use of formal mathematical notation and rules becomes progressively more important as problems become complicated and, in particular, in three-dimensional situations. Forces, velocities and accelerations all satisfy rules of addition that identify them as vectors. In this chapter we shall construct an algebraic theory for the manipulation of vectors and see how it can be applied to some simple practical problems.

The ideas behind vectors as formal quantities developed mainly during the nineteenth century, and they became a well-established tool in the twentieth century. Vectors provide a convenient and compact way of dealing with multi-dimensional situations without the problem of writing down every bit of information. They allow the principles of the subject to be developed without being obscured by complicated notation.

It is inconceivable that modern scientists and engineers could work successfully without computers. Since such machines cannot think like an engineer or scientist, they have to be told in a totally precise and formal way what to do. For instance, a robot arm needs to be given instructions on how to position itself to perform a spot weld. Three-dimensional vectors prove to be the perfect way to tell the computer how to specify the position of the workpiece of the robot arm and a set of rules then tells the robot how to move to its working position.

Computers have put a great power at the disposal of the engineer; problems that proved to be impossible 50 years ago are now routine. With the aid of numerical algorithms, equations can often be solved very quickly. The stressing of a large structure or an aircraft wing, the lubrication of shafts and bearings, the flow of sewage in pipes and the flow past the fuselage of an aircraft are all examples of systems that were well understood in principle but could not be analysed until the necessary computer power became available. Algorithms are usually written in terms of vectors and matrices (see Chapter 5), since these form a natural setting for the numerical solution of engineering problems and are also ideal for the computer. It is vital that the manipulation of vectors be understood before embarking on more complex mathematical structures used in engineering computations.

Perhaps the most powerful influence of computers is in their graphical capabilities, which have proved invaluable in displaying the static and dynamic behaviour of systems. We accept this tool without thinking how it works. A simple example shows the complexity. How do we display a box with an open top with 'hidden' lines when we look at it from a given angle? The problem is a complicated three-dimensional one that must be analysed instantly by a computer. Vectors allow us to define lines that can be projected on to the screen, and intersections can then be computed so that the 'hidden' portion can be eliminated. Extending the analysis to a less regular shape is a formidable vector problem. Work of this type is the basis of CAD/CAM systems, which now assist engineers in all stages of the manufacturing process from design to production of a finished product. Such systems typically allow engineers to manipulate the product geometry during initial design, to produce working drawings, to generate toolpaths in the production process and generally to automate a host of previously tedious and time-consuming tasks.

The general development of the theory of vectors is closely associated with coordinate geometry, so we shall introduce a few ideas in the next section that will be used later in the chapter. The comments largely concern the two- and three-dimensional cases, but we shall mention higher-dimensional extensions where they are relevant to later work such as on the theory of matrices. While in two and three dimensions we can appeal to geometrical intuition, it is necessary to work in a much more formal way in higher dimensions, as with many other areas of mathematics.

4.2 Basic definitions and results

4.2.1 Cartesian coordinates

Setting up rectangular cartesian axes $Oxyz$ or $Ox_1x_2x_3$, we define the position of a point by **coordinates** or **components** (x, y, z) or (x_1, x_2, x_3) as indicated in Figure 4.1(a). The indicial notation is particularly important when we consider vectors in many dimensions $(x_1, x_2, \ldots, x_n)$. The axes Ox, Oy, Oz, in that order, are assumed to be right-handed in the sense of Figure 4.1(b), so that a rotation of a right-handed screw from Ox to Oy advances it along Oz, a rotation from Oy to Oz advances it along Ox and a rotation from Oz to Ox advances it along Oy. This is an accepted convention, and it will be seen to be particularly important in Section 4.2.9 when we deal with the vector product.

The length of OP in Figure 4.1(a) is obtained from Pythagoras' theorem as

$$r = (x^2 + y^2 + z^2)^{1/2}$$

The angle $\alpha = \angle POA$ in the right angled triangle OAP is the angle that OP makes with the positive x direction as in Figure 4.2. We can see that

Figure 4.1
(a) Right-handed coordinate axes.
(b) Right-hand rule.

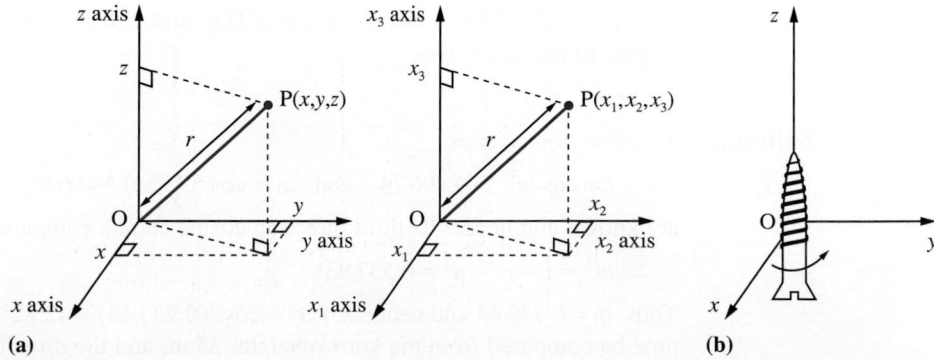

(a)   (b)

Figure 4.2
Direction cosines
of OP, $l = \cos\alpha$,
$m = \cos\beta$, $n = \cos\gamma$.

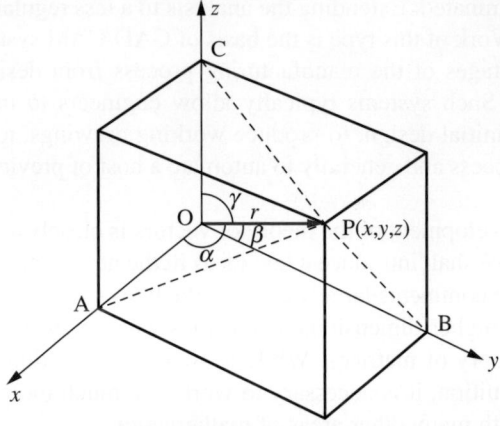

$$l = \cos\alpha = \frac{x}{r}$$

Likewise, β and γ are the angles that OP makes with y and z directions respectively, so

$$m = \cos\beta = \frac{y}{r}, \quad n = \cos\gamma = \frac{z}{r}$$

The triad (l, m, n) are called the **direction cosines** of the line OP. Note that

$$l^2 + m^2 + n^2 = \frac{x^2}{r^2} + \frac{y^2}{r^2} + \frac{z^2}{r^2} = \frac{x^2 + y^2 + z^2}{r^2} = 1$$

Example 4.1   If P has coordinates $(2, -1, 3)$, find the length OP and the direction cosines of OP.

Solution $OP^2 = (2)^2 + (-1)^2 + (3)^2 = 4 + 1 + 9$,   so that $OP = \sqrt{14}$

The direction cosines are

$$l = 2\sqrt{\tfrac{1}{14}}, \quad m = -\sqrt{\tfrac{1}{14}}, \quad n = 3\sqrt{\tfrac{1}{14}}$$

Example 4.2   A surveyor sets up his theodolite on horizontal ground, at a point O, and observes the top of a church spire, as illustrated in Figure 4.3. Relative to axes Oxyz, with Oz vertical, the surveyor measures the angles $\angle TOx = 66°$ and $\angle TOz = 57°$. The church is known to have height 35 m. Find the angle $\angle TOy$ and calculate the coordinates of T with respect to the given axes.

Solution   The direction cosines

$$l = \cos 66° = 0.406\,74 \quad \text{and} \quad n = \cos 57° = 0.544\,64$$

are known and hence the third direction cosine can be computed as

$$m^2 = 1 - l^2 - n^2 = 0.537\,93$$

Thus, $m = 0.733\,44$ and hence $\angle TOy = \cos^{-1}(0.733\,44) = 42.82°$. The length $OT = r$ can now be computed from the known height, 35 m, and the direction cosine n, as

Figure 4.3
Representation of the
axes and church spire
in Example 4.2.

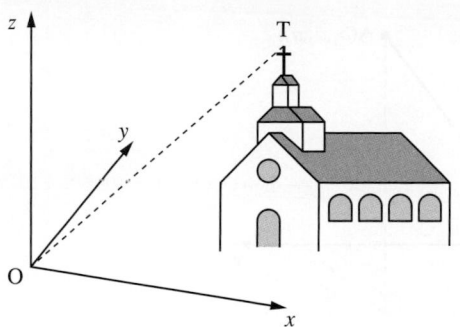

$$\cos 57° = 35/r, \quad \text{so} \quad r = 64.26\,\text{m}$$

The remaining coordinates are obtained from

$$x/r = \cos 66° \quad \text{and} \quad y/r = \cos 42.82°$$

giving $x = r\cos 66° = 26.14$ and $y = r\cos 42.82° = 47.13$
Hence the coordinates of T are (26.14, 47.13, 35).

4.2.2 Scalars and vectors

Quantities like distance or temperature are represented by real numbers in appropriate units, for instance 5 m or 10°C. Such quantities are called **scalars** – they obey the usual rules of real numbers and they have no direction associated with them. However, **vectors** have both a magnitude and a direction associated with them; these include force, velocity and magnetic field. To qualify as vectors, the quantities must have more than just magnitude and direction – they must also satisfy some particular rules of combination. Angular displacement in three dimensions gives an example of a quantity which has a direction and magnitude but which does not add by the addition rules of vectors, so angular displacements are *not* vectors.

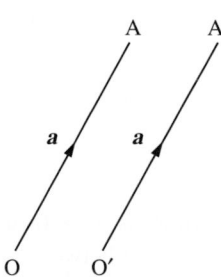

Figure 4.4
Line segments
representing a
vector *a*.

We represent a vector geometrically by a line segment whose length represents the vector's magnitude in some appropriate units and whose direction represents the vector's direction, with the arrowhead indicating the sense of the vector, as shown in Figure 4.4. According to this definition, the starting point of the vector is irrelevant. In Figure 4.4, the two line segments OA and O'A' represent the same vector because their lengths are the same, their directions are the same and the sense of the arrows is the same. Thus each of these vectors is equivalent to the vector through the origin, with A given by its coordinates (a_1, a_2, a_3) as in Figure 4.5. We can therefore represent a vector in a three-dimensional space by an ordered set of three numbers or a 3-tuple. We shall see how this representation is used in Section 4.2.4.

We shall now introduce some of the basic notation and definitions for vectors. The vector of Figure 4.5 is handwritten or typewritten as $\underline{a}$, $\underline{a}$, $\overrightarrow{OA}$. On the printed page bold-face type *a* is used. Using the coordinate definition, the vector could equally be written as (a_1, a_2, a_3). (Note: There are several possible coordinate notations; the traditional one is (a_1, a_2, a_3), but in Section 5.2.1 of Chapter 5 on matrices we shall use an alternative standard notation.)

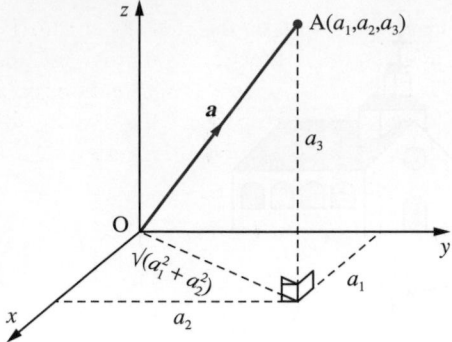

Some basic properties of vectors are:

(a) Equality

As we considered earlier, two vectors *a* and *b* are equal if and only if they have the same modulus and the same direction and sense. We write this in the usual way

$$a = b$$

We shall see in Section 4.2.4 that in component form, two vectors $a = (a_1, a_2, a_3)$ and $b = (b_1, b_2, b_3)$ are **equal** if and only if the components are equal, that is

$$a_1 = b_1, \quad a_2 = b_2, \quad a_3 = b_3$$

(b) Multiplication by a scalar

If λ is a scalar and the vectors are related by $a = \lambda b$ then

if $\lambda > 0$, *a* is a vector in the same direction as *b* with magnitude λ times the magnitude of *b*;

if $\lambda < 0$, *a* is a vector in the opposite direction to *b* with magnitude $|\lambda|$ times the magnitude of *b*.

(c) Parallel vectors

The vectors *a* and *b* are said to be **parallel** or **antiparallel** according as $\lambda > 0$ or $\lambda < 0$ respectively. (Note that we do not insert any multiplication symbol between λ and *b* since the common symbols · and × are reserved for special uses that we shall discuss later.)

(d) Modulus

The **modulus** or **length** or **magnitude** of a vector *a* is written as $|a|$ or $|\overrightarrow{OA}|$ or *a* if there is no ambiguity. A vector with modulus one is called a **unit vector** and is written $\hat{a}$, with the hat (ˆ) indicating a unit vector. Clearly

$$a = |a|\hat{a} \quad \text{or} \quad \hat{a} = \frac{a}{|a|}$$

(e) Zero vector

The **zero** or **null vector** has zero modulus; it is written as **0** or often just as 0 when there is no ambiguity whether it is a vector or not.

Example 4.3

A cyclist travels at a steady 16 km/h on the four legs of his journey. From his origin, O, he travels for one hour in a NE direction to the point A; he then travels due E for half an hour to point B. He then cycles in a NW direction until he reaches the point C, which is due N of his starting point. He returns due S to the starting point. Indicate the path of the cyclist using vectors and calculate the modulus of the vectors along BC and CO.

Solution

The four vectors are shown in Figure 4.6. If $\hat{i}$ and $\hat{j}$ are the unit vectors along the two axes then by property (b)

$$\overrightarrow{AB} = 8\hat{i} \quad \text{and} \quad \overrightarrow{CO} = -L\hat{j}$$

where L is still to be determined. By trigonometry

$$DB = 8 + 16 \sin 45° = 8 + 8\sqrt{2}$$

and hence the modulus of the vector $\overrightarrow{BC}$ is

$$|BC| = \frac{DB}{\cos 45°} = 8\sqrt{2} + 16$$

The modulus L of the vector $\overrightarrow{CO}$ is

$$L - |\overrightarrow{CO}| = CD + DO = (8 + 8\sqrt{2}) + 16 \cos 45° = 8 + 16\sqrt{2}$$

Figure 4.6 Cyclist's path in Example 4.3.

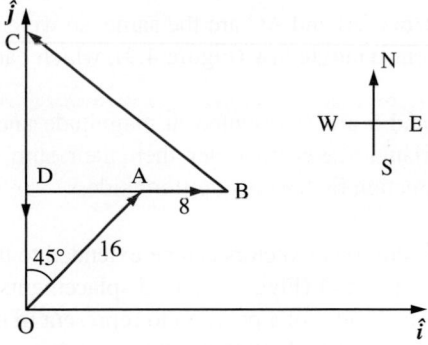

4.2.3 Addition of vectors

Having introduced vectors and their basic properties, it is natural to ask if vectors can be combined. The simplest form of vector combination is addition and it is the definition of addition that finally identifies a vector. Consider the following situation. The helmsman of a small motor boat steers his vessel due east (E) at 4 knots for one hour. The path taken by the boat could be represented by the line OA, or **a**, in Figure 4.7. Unfortunately there is also a tidal stream, **b**, running north-north-east (NNE) at $2\frac{1}{2}$ knots. Where will the boat actually be at the end of one hour?

If we imagine the vessel to be steaming E for one hour through still water, and then lying still in the water and drifting with the tidal stream for one hour, we can see that it will travel from O to A in the first hour and from A to C in the second hour. If, on the

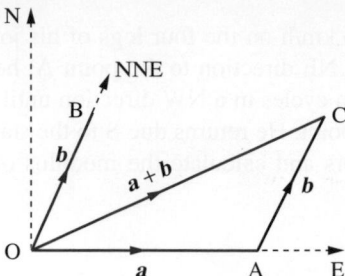

Figure 4.7
Addition of
two vectors.

other hand, the vessel steams due E through water that is simultaneously moving NNE with the tidal stream then the result will be to arrive at C after one hour. The net velocity of the boat is represented by the line OC. Putting this another way, the result of subjecting the boat to a velocity $\overrightarrow{OA}$ and a velocity $\overrightarrow{AC}$ simultaneously is the same as the result of subjecting it to a velocity $\overrightarrow{OC}$. Thus the velocity $\overrightarrow{OC} = \mathbf{a} + \mathbf{b}$ is the sum of the velocity $\overrightarrow{OA} = \mathbf{a}$ and the velocity $\overrightarrow{AC} = \mathbf{b}$.

This leads us to the **parallelogram rule** for vector addition illustrated in Figure 4.8 and stated as follows:

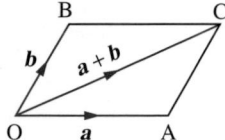

Figure 4.8
Parallelogram rule for
addition of vectors.

> The sum, or resultant, of two vectors $\mathbf{a}$ and $\mathbf{b}$ is found by forming a parallelogram with $\mathbf{a}$ and $\mathbf{b}$ as two adjacent sides. The sum $\mathbf{a} + \mathbf{b}$ is the vector represented by the diagonal of the parallelogram.

In Figure 4.8 the vectors $\overrightarrow{OB}$ and $\overrightarrow{AC}$ are the same, so we can rewrite the parallelogram rule as an equivalent **triangle law** (Figure 4.9), which can be stated as follows:

> If two vectors $\mathbf{a}$ and $\mathbf{b}$ are represented in magnitude and direction by the two sides of a triangle taken in order then their sum is represented in magnitude and direction by the closing third side.

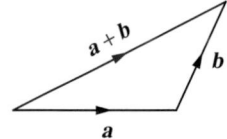

Figure 4.9
Triangle law for
addition of vectors.

The triangle law for the addition of vectors can be extended to the addition of any number of vectors. If from a point O (Figure 4.10), displacements $\overrightarrow{OA}, \overrightarrow{AB}, \overrightarrow{BC}, \dots, \overrightarrow{LK}$ are drawn along the adjacent sides of a polygon to represent in magnitude and direction the vectors $\mathbf{a}, \mathbf{b}, \mathbf{c}, \dots, \mathbf{k}$ respectively then the sum

$$r = a + b + c + \dots + k$$

of these vectors is represented in magnitude and direction by the closing side OK of the polygon, the sense of the sum vector being represented by the arrow in Figure 4.10. This is referred to as the **polygon law** for the addition of vectors.

We now need to look at the usual rules of algebra for scalar quantities to check whether or not they are satisfied for vectors.

(a) Commutative law

$$a + b = b + a$$

This result is obvious from the geometrical definition, and says that order does not matter.

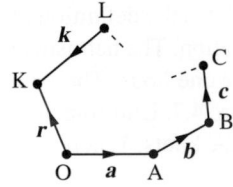

Figure 4.10
Polygon law for
addition of vectors.

(b) Associative law

$$(a + b) + c = a + (b + c)$$

Geometrically, the result can be deduced using the triangle and polygon laws, as shown in Figure 4.11. We see that brackets do not matter and can be omitted.

Figure 4.11
Deduction of
associative law.

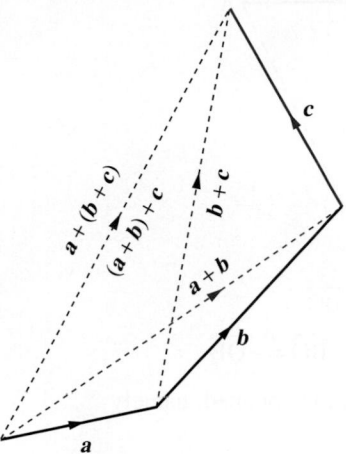

(c) Distributive law

$$\lambda(a + b) = \lambda a + \lambda b$$

The result follows from similar triangles. In Figure 4.12 the side $O'B'$ is just λ times OB in length and in the same direction so $\overrightarrow{O'B'} = \lambda(a + b)$. The triangle law therefore gives the required result since $\overrightarrow{O'B'} = \overrightarrow{O'A'} + \overrightarrow{A'B'} = \lambda a + \lambda b$. This result just says that we can multiply brackets out by the usual laws of algebra.

Figure 4.12
Similar triangles for
the proof of the
distributive law.

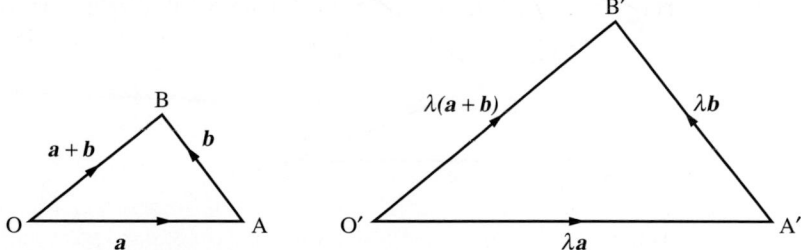

(d) Subtraction
We define subtraction in the obvious way:

$$a - b = a + (-b)$$

This is illustrated geometrically in Figure 4.13. Applying the triangle rule to triangle OAB gives

Figure 4.13
Subtraction of vectors.

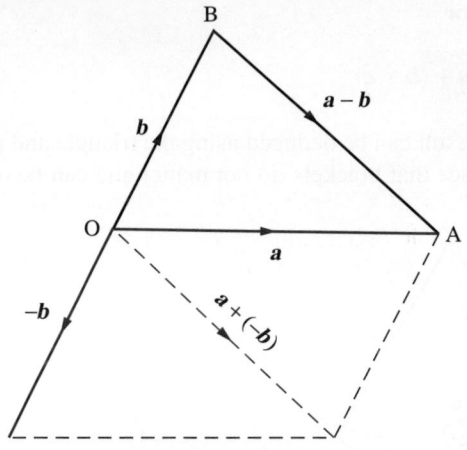

$$\overrightarrow{BA} = \overrightarrow{BO} + \overrightarrow{OA} = \overrightarrow{OA} + \overrightarrow{BO}$$
$$= \overrightarrow{OA} - \overrightarrow{OB} \quad \text{since} \quad \overrightarrow{BO} = -\overrightarrow{OB}$$

from which the important result is obtained, namely

$$\overrightarrow{BA} = \overrightarrow{OA} - \overrightarrow{OB}$$

Example 4.4

From Figure 4.14, evaluate

g in terms of a and b, f in terms of b and c

e in terms of c and d, e in terms of f, g and h

Figure 4.14
Figure of Example 4.4.

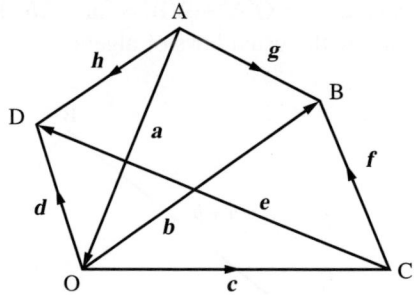

Solution   From the triangle OAB: $\overrightarrow{AB} = \overrightarrow{AO} + \overrightarrow{OB}$ and hence $g = a + b$

From the triangle OBC: $\overrightarrow{CB} = \overrightarrow{OB} - \overrightarrow{OC}$ and hence $f = b - c$

From the triangle OCD: $\overrightarrow{CD} = \overrightarrow{OD} - \overrightarrow{OC}$ and hence $e = d - c$

From the quadrilateral CBAD the polygon rule gives

$\overrightarrow{CD} + \overrightarrow{DA} + \overrightarrow{AB} + \overrightarrow{BC} = 0$ and hence $e + (-h) + g + (-f) = 0$ so $e = f - g + h$

Example 4.5   A quadrilateral OACB is defined in terms of the vectors $\overrightarrow{OA} = \boldsymbol{a}$, $\overrightarrow{OB} = \boldsymbol{b}$ and $\overrightarrow{OC} = \boldsymbol{b} + \frac{1}{2}\boldsymbol{a}$. Calculate the vector representing the other two sides $\overrightarrow{BC}$ and $\overrightarrow{CA}$.

Solution   Now as in rule (d)

$$\overrightarrow{BC} = \overrightarrow{BO} + \overrightarrow{OC} = -\overrightarrow{OB} + \overrightarrow{OC}$$

so

$$\overrightarrow{BC} = \overrightarrow{OC} - \overrightarrow{OB} = (\boldsymbol{b} + \tfrac{1}{2}\boldsymbol{a}) - \boldsymbol{b} = \tfrac{1}{2}\boldsymbol{a}$$

and similarly $\overrightarrow{CA} = \overrightarrow{OA} - \overrightarrow{OC} = \boldsymbol{a} - (\boldsymbol{b} + \tfrac{1}{2}\boldsymbol{a}) = \tfrac{1}{2}\boldsymbol{a} - \boldsymbol{b}$

Example 4.6   A force $\boldsymbol{F}$ has magnitude 2 N and a second force $\boldsymbol{F}'$ has magnitude 1 N and is inclined at an angle of 60° to $\boldsymbol{F}$ as illustrated in Figure 4.15. Find the magnitude of the resultant force $\boldsymbol{R}$ and the angle it makes to the force $\boldsymbol{F}$.

Solution   (i) Now, from Figure 4.15 we have $\boldsymbol{R} = \boldsymbol{F} + \boldsymbol{F}'$, so we require the length OC and the angle CON.

Figure 4.15
Figure of Example 4.6.

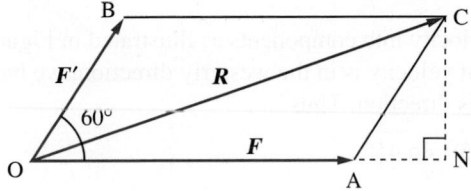

(ii) We first need to calculate CN and AN using trigonometry. Noting that $|\boldsymbol{F}'| = $ OB = AC = 1 we see that

$$CN = AC \sin 60° = \tfrac{\sqrt{3}}{2} \quad \text{and} \quad AN = AC \cos 60° = \tfrac{1}{2}$$

(iii) Noting that $|\boldsymbol{F}| = $ OA = 2 then ON = OA + AN = $\tfrac{5}{2}$. Thus using Pythagoras' theorem

$$OC^2 = ON^2 + CN^2 = \left(\tfrac{\sqrt{3}}{2}\right)^2 + \left(\tfrac{5}{2}\right)^2 = 7$$

and hence the resultant has magnitude $\sqrt{7}$.

(iv) The angle CON is determined from $\tan CON = \dfrac{CN}{ON} = \tfrac{\sqrt{3}}{5}$ giving angle CON = 19.1°.

Example 4.7   An aeroplane is flying at 400 knots in a strong NW wind of 50 knots. The plane wishes to fly due west. In which direction should the pilot fly the plane to achieve this end, and what will be his actual speed over the ground?

Solution The resultant velocity of the plane is the vector sum of 50 knots from the NW direction and 400 knots in a direction $\alpha°$ north of west. In appropriate units the situation is shown in Figure 4.16(a). The vector $\overrightarrow{OA}$ represents the wind velocity and $\overrightarrow{OB}$ represents the aeroplane velocity. The resultant velocity is $\overrightarrow{OP}$, which is required to be due W. We wish to determine the angle α (giving the direction of flight) and magnitude of the resultant velocity (giving the ground speed).

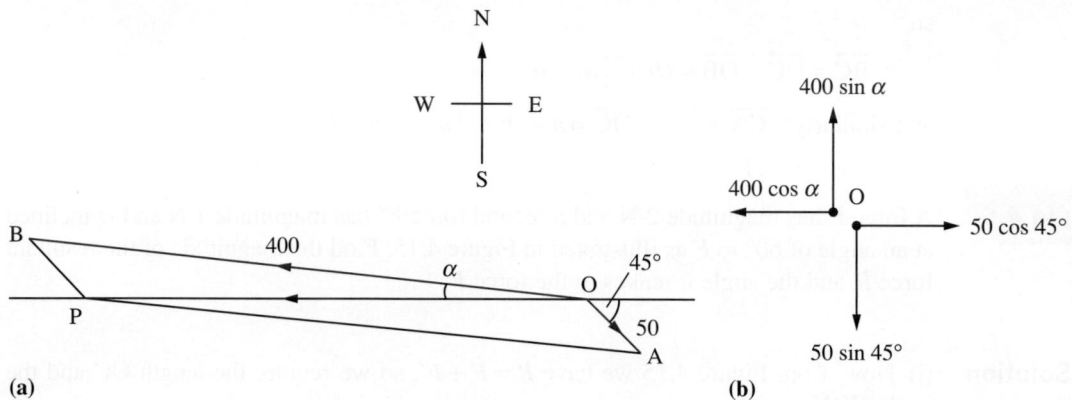

(a) (b)

Figure 4.16 (a) The track of the aeroplane in Example 4.7. (b) Resolving the velocity into components.

Resolving the velocity into components as illustrated in Figure 4.16(b) and recognizing that the resultant velocity is in the westerly direction, we have no resultant velocity perpendicular to this direction. Thus

$$400 \sin \alpha° = 50 \sin 45°$$

so that

$$\alpha = 5.07°$$

The resultant speed due west is

$$400 \cos \alpha° - 50 \cos 45° = 363 \text{ knots}$$

Example 4.8 If ABCD is any quadrilateral, show that $\overrightarrow{AD} + \overrightarrow{BC} = 2\overrightarrow{EF}$, where E and F are the midpoints of AB and DC respectively, and that

$$\overrightarrow{AB} + \overrightarrow{AD} + \overrightarrow{CB} + \overrightarrow{CD} = 4\overrightarrow{XY}$$

where X and Y are the midpoints of the diagonals AC and BD respectively.

Solution Applying the polygon law for the addition of vectors to Figure 4.17,

$$\overrightarrow{EF} = \overrightarrow{EA} + \overrightarrow{AD} + \overrightarrow{DF}$$

and

$$\overrightarrow{EF} = \overrightarrow{EB} + \overrightarrow{BC} + \overrightarrow{CF}$$

Adding these two then gives

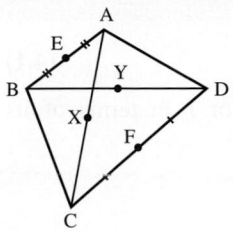

Figure 4.17
Quadrilateral of
Example 4.8.

$$2\overrightarrow{EF} = \overrightarrow{EA} + \overrightarrow{AD} + \overrightarrow{DF} + \overrightarrow{EB} + \overrightarrow{BC} + \overrightarrow{CF}$$

$$= \overrightarrow{AD} + \overrightarrow{BC} + (\tfrac{1}{2}\overrightarrow{BA} + \tfrac{1}{2}\overrightarrow{CD} - \tfrac{1}{2}\overrightarrow{BA} - \tfrac{1}{2}\overrightarrow{CD})$$

since E and F are the midpoints of AB and CD respectively. Thus

$$2\overrightarrow{EF} = \overrightarrow{AD} + \overrightarrow{BC}$$

Also, by the polygon law for addition of vectors,

$$\overrightarrow{XY} = \overrightarrow{XA} + \overrightarrow{AB} + \overrightarrow{BY}$$

and

$$\overrightarrow{XY} = \overrightarrow{XC} + \overrightarrow{CB} + \overrightarrow{BY}$$

Adding and multiplying by two gives

$$4\overrightarrow{XY} = 2\overrightarrow{XA} + 2\overrightarrow{AB} + 2\overrightarrow{BY} + 2\overrightarrow{XC} + 2\overrightarrow{CB} + 2\overrightarrow{BY}$$

$$= 2\overrightarrow{AB} + 2\overrightarrow{CB} + 4\overrightarrow{BY} \qquad (\text{since } \overrightarrow{XA} = -\overrightarrow{XC})$$

$$= 2\overrightarrow{AB} + 2\overrightarrow{CB} + 2\overrightarrow{BD} \qquad (\text{since } \overrightarrow{BD} = 2\overrightarrow{BY})$$

$$= \overrightarrow{AB} + \overrightarrow{CB} + (\overrightarrow{AB} + \overrightarrow{BD}) + (\overrightarrow{CB} + \overrightarrow{BD})$$

so that

$$4\overrightarrow{XY} = \overrightarrow{AB} + \overrightarrow{CB} + \overrightarrow{AD} + \overrightarrow{CD}$$

4.2.4 Cartesian components and basic properties

In Section 4.2.2 we saw that vectors could be written as an ordered set of three numbers or 3-tuple. We shall now explore the properties of these ordered triples and how they relate to the geometrical definitions used in previous sections.

In Figure 4.18, we denote mutually perpendicular unit vectors in the three coordinate directions by i, j and k. (Sometimes the alternative notation $\hat{e}_1$, $\hat{e}_2$ and $\hat{e}_3$ is used.) The notation i, j, k is so standard that the 'hats' indicating unit vectors are usually omitted.

Applying the triangle law to the triangle OXM, we have

$$\overrightarrow{OM} = \overrightarrow{OX} + \overrightarrow{OY} = x\mathbf{i} + y\mathbf{j}$$

Figure 4.18
The component form
of a vector.

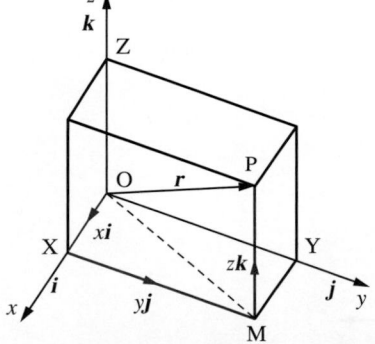

Applying the triangle law to the triangle OMP then yields

$$\overrightarrow{OP} = \overrightarrow{OM} + \overrightarrow{MP} = x\boldsymbol{i} + y\boldsymbol{j} + z\boldsymbol{k} \tag{4.1}$$

The analysis applies to any point, so we can write any vector $\boldsymbol{r}$ in terms of its **components** x, y, z with respect to the unit vectors $\boldsymbol{i}, \boldsymbol{j}, \boldsymbol{k}$ as

$$\boldsymbol{r} = x\boldsymbol{i} + y\boldsymbol{j} + z\boldsymbol{k}$$

Indeed, the vector notation $\boldsymbol{r} = (x, y, z)$ should be intepreted as the vector given in (4.1). In some contexts it is more convenient to use a suffix notation for the coordinates, and

$$(x_1, x_2, x_3) = x_1\hat{\boldsymbol{e}}_1 + x_2\hat{\boldsymbol{e}}_2 + x_3\hat{\boldsymbol{e}}_3$$

is interpreted in exactly the same way. It is assumed that the three basic unit vectors are known, and all vectors in coordinate form are referred to them.

The **modulus** of a vector is just the length OP so from Figure 4.18 we have, using Pythagoras' theorem,

$$|\boldsymbol{r}| = (x^2 + y^2 + z^2)^{1/2}$$

The basic properties of vectors follow easily from the component definition in (4.1).

(a) Equality
Two vectors $\boldsymbol{a} = (a_1, a_2, a_3)$ and $\boldsymbol{b} = (b_1, b_2, b_3)$ are **equal** if and only if the three components are equal, that is

$$a_1 = b_1, \quad a_2 = b_2, \quad a_3 = b_3$$

(b) Zero vector
The zero vector has zero components, so

$$\boldsymbol{0} = (0, 0, 0)$$

(c) Addition
The addition rule is expressed very simply in terms of vector components:

$$\boldsymbol{a} + \boldsymbol{b} = (a_1 + b_1, a_2 + b_2, a_3 + b_3)$$

The equivalence of this definition with the geometrical definition for addition using the parallelogram rule can be deduced from Figure 4.19. We know that $\overrightarrow{OB} = \overrightarrow{AC}$, since

Figure 4.19
Parallelogram rule, x component.

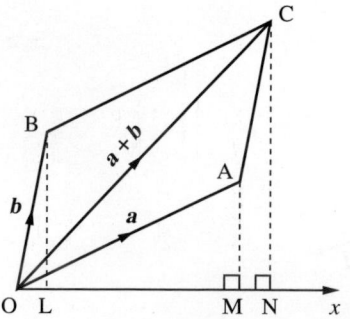

they are equivalent displacements, and hence their x components are the same so that we have OL = MN. Thus if we take the x component of $\boldsymbol{a} + \boldsymbol{b}$

$$(\boldsymbol{a} + \boldsymbol{b})_1 = \text{ON} = \text{OM} + \text{MN} = \text{OM} + \text{OL} = a_1 + b_1$$

the y and z components can be considered in a similar manner, giving $(\boldsymbol{a} + \boldsymbol{b})_2 = a_2 + b_2$ and $(\boldsymbol{a} + \boldsymbol{b})_3 = a_3 + b_3$.

(d) Multiplication by a scalar
If λ is a scalar and the vectors are related by $\boldsymbol{a} = \lambda\boldsymbol{b}$ then the components satisfy

$$a_1 = \lambda b_1, \quad a_2 = \lambda b_2, \quad a_3 = \lambda b_3$$

which follows from the similar triangles of Figure 4.12.

(e) Distributive law
The distributive law in components is simply a restatement of the distributive law for the addition of numbers:

$$\begin{aligned}
\lambda(\boldsymbol{a} + \boldsymbol{b}) &= \lambda(a_1 + b_1, a_2 + b_2, a_3 + b_3) \\
&= (\lambda(a_1 + b_1), \lambda(a_2 + b_2), \lambda(a_3 + b_3)) \\
&= (\lambda a_1 + \lambda b_1, \lambda a_2 + \lambda b_2, \lambda a_3 + \lambda b_3) \\
&= (\lambda a_1, \lambda a_2, \lambda a_3) + (\lambda b_1, \lambda b_2, \lambda b_3) \\
&= \lambda\boldsymbol{a} + \lambda\boldsymbol{b}
\end{aligned}$$

(f) Subtraction
Subtraction is again straightforward and the components are just subtracted from each other:

$$\boldsymbol{a} - \boldsymbol{b} = (a_1 - b_1, a_2 - b_2, a_3 - b_3)$$

The component form of vectors allows problems to be solved algebraically and results can be interpreted either as algebraic ideas or in a geometrical manner. Both these interpretations can be very useful in applications of vectors to engineering.

In MATLAB a vector is inserted as an array within square brackets so, for example, a vector $\boldsymbol{a} = (1, 2, 3)$ is inserted as a = [1 2 3] or a = [1,2,3], where in the latter commas have been used instead of spaces. It is inserted as a := *array*([1,2,3]); in MAPLE, where it is usually necessary to invoke the *linalg* package first. The operations of addition, subtraction and multiplication by a scalar are represented by +, − and * respectively. The magnitude or length of a vector $\boldsymbol{a}$ appears in MATLAB as *norm*(a) and in MAPLE as *norm*(a,2).

Example 4.9

Determine whether constants α and β can be found to satisfy the vector equations

(a) $(2, 1, 0) = \alpha(-2, 0, 2) + \beta(1, 1, 1)$

(b) $(-3, 1, 2) = \alpha(-2, 0, 2) + \beta(1, 1, 1)$

and interpret the results.

Solution (a) For the two vectors to be the same each of the components must be equal, and hence

$$2 = -2\alpha + \beta$$
$$1 = \beta$$
$$0 = 2\alpha + \beta$$

Thus the second equation gives $\beta = 1$ and both of the other two equations give the same value of α, namely $\alpha = -\frac{1}{2}$, so the equations can be satisfied.

(b) A similar argument gives

$$-3 = -2\alpha + \beta$$
$$1 = \beta$$
$$2 = 2\alpha + \beta$$

Again, the second equation gives $\beta = 1$ but the first equation leads to $\alpha = 2$ and the third to $\alpha = \frac{1}{2}$. The equations are now not consistent and no appropriate α and β can be found.

In case (a) the three vectors lie in a plane, and any vector in a plane, including the one given, can be written as the vector sum of the two vectors $(-2, 0, 2)$ and $(1, 1, 1)$ with appropriate multipliers. In case (b), however, the vector $(-3, 1, 2)$ does not lie in the plane of the two vectors $(-2, 0, 2)$ and $(1, 1, 1)$ and can, therefore, never be written as the vector sum of the two vectors $(-2, 0, 2)$ and $(1, 1, 1)$ with appropriate multipliers.

Example 4.10 Given the vectors $\boldsymbol{a} = (1, 1, 1)$, $\boldsymbol{b} = (-1, 2, 3)$ and $\boldsymbol{c} = (0, 3, 4)$, find

(a) $\boldsymbol{a} + \boldsymbol{b}$ (b) $2\boldsymbol{a} - \boldsymbol{b}$ (c) $\boldsymbol{a} + \boldsymbol{b} - \boldsymbol{c}$

(d) the unit vector in the direction of $\boldsymbol{c}$

Solution (a) $\boldsymbol{a} + \boldsymbol{b} = (1 - 1, 1 + 2, 1 + 3) = (0, 3, 4)$

(b) $2\boldsymbol{a} - \boldsymbol{b} = (2 \times 1 - (-1), 2 \times 1 - 2, 2 \times 1 - 3) = (3, 0, -1)$

(c) $\boldsymbol{a} + \boldsymbol{b} - \boldsymbol{c} = (1 - 1 + 0, 1 + 2 - 3, 1 + 3 - 4) = (0, 0, 0) = \boldsymbol{0}$

(d) $|\boldsymbol{c}| = (3^2 + 4^2)^{1/2} = 5$, so

$$\hat{\boldsymbol{c}} = \frac{\boldsymbol{c}}{5} = (0, \tfrac{3}{5}, \tfrac{4}{5})$$

Example 4.11 Given $\boldsymbol{a} = (2, -3, 1) = 2\boldsymbol{i} - 3\boldsymbol{j} + \boldsymbol{k}$, $\boldsymbol{b} = (1, 5, -2) = \boldsymbol{i} + 5\boldsymbol{j} - 2\boldsymbol{k}$ and $\boldsymbol{c} = (3, -4, 3) = 3\boldsymbol{i} - 4\boldsymbol{j} + 3\boldsymbol{k}$

(a) find the vector $\boldsymbol{d} = \boldsymbol{a} - 2\boldsymbol{b} + 3\boldsymbol{c}$;

(b) find the magnitude of $\boldsymbol{d}$ and write down a unit vector in the direction of $\boldsymbol{d}$;

(c) what are the direction cosines of $\boldsymbol{d}$?

Solution   (a) $d = a - 2b + 3c$

$$= (2i - 3j + k) - 2(i + 5j - 2k) + 3(3i - 4j + 3k)$$

$$= (2i - 3j + k) - (2i + 10j - 4k) + (9i - 12j + 9k)$$

$$= (2 - 2 + 9)i + (-3 - 10 - 12)j + (1 + 4 + 9)k$$

that is, $d = 9i - 25j + 14k$.

(b) The magnitude of d is $d = \sqrt{[9^2 + (-25)^2 + 14^2]} = \sqrt{902}$
A unit vector in the direction of d is $\hat{d}$, where

$$\hat{d} = \frac{d}{d} = \frac{9}{\sqrt{902}}i - \frac{25}{\sqrt{902}}j + \frac{14}{\sqrt{902}}k$$

(c) The direction cosines of d are $9/\sqrt{902}$, $-25/\sqrt{902}$ and $14/\sqrt{902}$.

Check that in MATLAB the commands

```
a = [2 -3 1]; b = [1 5 -2]; c = [3 -4 3];
d = a - 2*b + 3*c
```

return the answer given in (a) and that the further command

```
norm(d)
```

gives the magnitude of d as 30.0333. Here MATLAB gives the numeric answer; to obtain the answer in the exact form then the calculation in MATLAB must be done symbolically using the Symbolic Math Toolbox. To do this the vector d must first be expressed in symbolic form using the `sym` command. Since the command `norm` does not appear to be available directly in the Toolbox, use can be made of the `maple` command to access the command in MAPLE. Check that the commands

```
d = sym(d);
maple('norm',d,2)
```

return the answer `902^(1/2)` given in (b).

Example 4.12   A molecule XY_3 has a tetrahedral form; the position vector of the X atom is $(2\sqrt{3} + \sqrt{2}, 0, -2 + \sqrt{6})$ and those of the three Y atoms are

$$\overrightarrow{OY} = (\sqrt{3}, -2, -1), \quad \overrightarrow{OY'} = (\sqrt{3}, 2, -1), \quad \overrightarrow{OY''} = (\sqrt{2}, 0, \sqrt{6})$$

(a) Show that all of the bond lengths are equal.

(b) Show that $\overrightarrow{XY} + \overrightarrow{YY'} + \overrightarrow{Y'Y''} + \overrightarrow{Y''X} = \mathbf{0}$

Solution   (a) $\overrightarrow{XY} = \overrightarrow{OY} - \overrightarrow{OX} = (-\sqrt{3} - \sqrt{2}, -2, 1 - \sqrt{6})$ and the bond length is

$$|\overrightarrow{XY}| = [(-\sqrt{3} - \sqrt{2})^2 + (-2)^2 + (1 - \sqrt{6})^2]^{1/2} = 4$$

$\overrightarrow{YY'} = \overrightarrow{OY'} - \overrightarrow{OY} = (0, 4, 0)$ and clearly the bond length is again 4.

The other four bonds $\overrightarrow{XY'}$, $\overrightarrow{XY''}$, $\overrightarrow{Y'Y''}$, $\overrightarrow{Y''Y}$ are treated in exactly the same way, and each gives a bond length of 4.

(b) Now $\overrightarrow{Y'Y''} = \overrightarrow{OY''} - \overrightarrow{OY'} = (\sqrt{2} - \sqrt{3}, -2, \sqrt{6} + 1)$ and $\overrightarrow{Y''X} = \overrightarrow{OX} - \overrightarrow{OY''}$
$= (2\sqrt{3}, 0, -2)$ so adding the four vectors gives

$$\overrightarrow{XY} + \overrightarrow{YY'} + \overrightarrow{Y'Y''} + \overrightarrow{Y''X}$$

$$= (-\sqrt{3} - \sqrt{2}, -2, 1 - \sqrt{6}) + (0, 4, 0) + (\sqrt{2} - \sqrt{3}, -2, \sqrt{6} + 1) + (2\sqrt{3}, 0, -2)$$

$$= \mathbf{0}$$

and is just a verification of the polygon law.

Example 4.13

Three forces, with units of newtons,

$$\mathbf{F}_1 = (1, 1, 1)$$

$\mathbf{F}_2$ has magnitude 6 and acts in the direction $(1, 2, -2)$

$\mathbf{F}_3$ has magnitude 10 and acts in the direction $(3, -4, 0)$

act on a particle. Find the resultant force that acts on the particle. What additional force must be imposed on the particle to reduce the resultant force to zero?

Solution

The first force is given in the usual vector form. The second two are given in an equally acceptable way but it is necessary to convert the information to the normal vector form so that the resultant can be found by vector addition. First the unit vector in the given direction of $\mathbf{F}_2$ is required

$$|(1, 2, -2)| = (1 + 2^2 + (-2)^2)^{1/2} = 3$$

and hence the unit vector in this direction is $\frac{1}{3}(1, 2, -2)$. Since $\mathbf{F}_2$ is in the direction of this unit vector and has magnitude 6 it can be written $\mathbf{F}_2 = 6(\frac{1}{3}, \frac{2}{3}, -\frac{2}{3}) = (2, 4, -4)$

Similarly for $\mathbf{F}_3$, the unit vector is $\frac{1}{5}(3, -4, 0)$ and hence $\mathbf{F}_3 = (6, -8, 0)$. The resultant force is obtained by vector addition.

$$\mathbf{F} = \mathbf{F}_1 + \mathbf{F}_2 + \mathbf{F}_3 = (1, 1, 1) + (2, 4, -4) + (6, -8, 0) = (9, -3, -3)$$

Clearly to make the resultant force zero, the additional force $(-9, 3, 3)$ must be imposed on the particle.

Example 4.14

Two geostationary satellites have known positions $(0, 0, h)$ and $(0, A, H)$ relative to a fixed set of axes on the earth's surface (which is assumed flat, with the x and y axes lying on the surface and the z axis vertical). Radar signals measure the distance of a ship from the satellites. Find the position of the ship relative to the given axes.

Solution

Figure 4.20 illustrates the situation described with R $(a, b, 0)$ describing the position of the ship and P and Q the positions of the satellites.

The radar signals measure PR and QR which are denoted by p and q respectively. The vectors

Figure 4.20

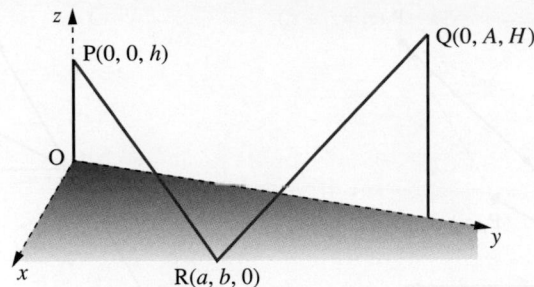

$$\overrightarrow{PR} = \overrightarrow{OR} - \overrightarrow{OP} = (a, b, 0) - (0, 0, h) = (a, b, -h)$$
$$\overrightarrow{QR} = \overrightarrow{OR} - \overrightarrow{OQ} = (a, b, 0) - (0, A, H) = (a, b - A, -H)$$

are easily calculated by the triangle law. The lengths of the two vectors are

$$p^2 = |\overrightarrow{PR}|^2 = a^2 + b^2 + h^2 \quad \text{and} \quad q^2 = |\overrightarrow{QR}|^2 = a^2 + (b - A)^2 + H^2$$

Subtracting gives

$$p^2 - q^2 = A(2b - A) + h^2 - H^2$$

and hence

$$b = (p^2 - q^2 - h^2 + H^2 + A^2)/2A$$

Having calculated b then a can be calculated from

$$a = \pm\sqrt{(p^2 - b^2 - h^2)}$$

Note the ambiguity in sign; clearly it will need to be known which side of the y axis the ship is lying.

Comment In practice the axes will need to be transformed to standard latitude and longitude and the curvature of the earth will need to be taken into consideration. Can the same calculation be used for aircraft? The speed of the ship has been neglected in the calculation above but is the speed of the aircraft important?

4.2.5 Complex numbers as vectors

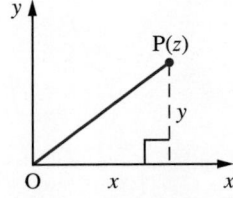

Figure 4.21 Argand diagram representation of $z = x + jy$.

We saw in Chapter 3, Section 3.2, that a complex number $z = x + jy$ can be represented geometrically by the point P in the Argand diagram as illustrated in Figure 4.21. We could equally well represent the point P by the vector $\overrightarrow{OP}$. Hence we can express the complex number z as a two-dimensional vector

$$z = \overrightarrow{OP}$$

With this interpretation of a complex number we can use the parallelogram rule to represent the addition and subtraction of complex numbers geometrically as illustrated in Figures 4.22(a, b).

Figure 4.22
(a) Addition of
complex numbers.
(b) Subtraction of
complex numbers.

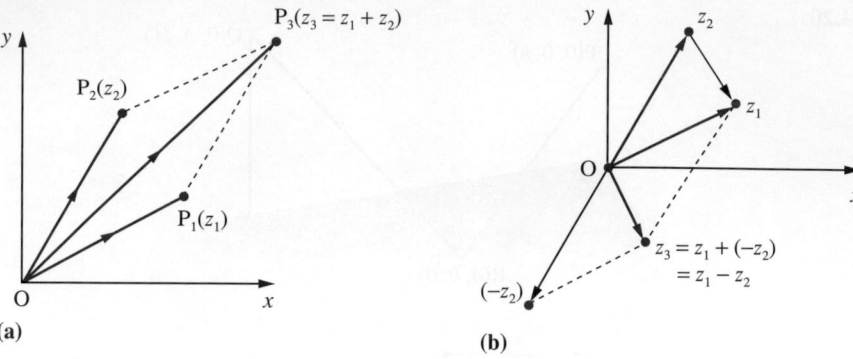

(a)

(b)

Example 4.15

A square is formed in the first and second quadrant with OP as one side of the square and $\overrightarrow{OP} = (1, 2)$. Find the coordinates of the other two vertices of the square.

Solution

The situation is illustrated in Figure 4.23. Using the complex form $\overrightarrow{OP} = 1 + 2j$ the side OQ is obtained by rotating OP through $\pi/2$ radians, then

$$\overrightarrow{OQ} = j(1 + 2j) = -2 + j$$

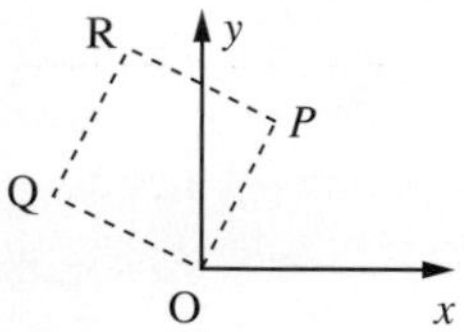

The fourth point R is found by observing that $\overrightarrow{OR}$ is the vector sum of $\overrightarrow{OP}$ and $\overrightarrow{OQ}$, and hence

$$\overrightarrow{OR} = \overrightarrow{OP} + \overrightarrow{OQ} = -1 + j3$$

The four coordinates are therefore

$$(0, 0), (1, 2), (-2, 1) \text{ and } (-1, 3)$$

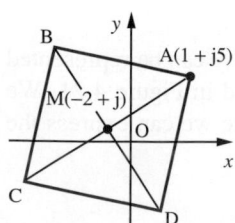

Figure 4.23
Square of
Example 4.15.

Example 4.16

M is the centre of a square with vertices A, B, C and D taken anticlockwise in that order. If, in the Argand diagram, M and A are represented by the complex numbers $-2 + j$ and $1 + j5$ respectively, find the complex numbers represented by the vertices B, C and D.

Solution

Applying the triangle law for addition of vectors of Figure 4.24 gives

$$\overrightarrow{MA} = \overrightarrow{MO} + \overrightarrow{OA}$$

$$= \overrightarrow{OA} - \overrightarrow{OM}$$

$$\equiv (1 + j5) - (-2 + j)$$

$$= 3 + j4$$

Since ABCD is a square,

$$MA = MB = MC = MD$$

$$\angle AMB = \angle BMC = \angle CMD = \angle DMA = \tfrac{1}{2}\pi$$

Figure 4.24
Square of
Example 4.16.

Remembering that multiplying a complex number by j rotates it through $\frac{1}{2}\pi$ radians in an anticlockwise direction, we have

$$\overrightarrow{MB} = j\overrightarrow{MA} \equiv j(3 + j4) = -4 + j3$$

giving

$$\overrightarrow{OB} - \overrightarrow{OM} + \overrightarrow{MB} \equiv (-2 + j) + (-4 + j3) = -6 + j4$$

Likewise

$$\overrightarrow{MC} = j\overrightarrow{MB} \equiv j(-4 + j3) = -3 - j4$$

giving

$$\overrightarrow{OC} = \overrightarrow{OM} + \overrightarrow{MC} \equiv -5 - j3$$

and

$$\overrightarrow{MD} = j\overrightarrow{MC} \equiv j(-3 - j4) = 4 - j3$$

giving

$$\overrightarrow{OD} = \overrightarrow{OM} + \overrightarrow{MD} \equiv 2 - j2$$

Thus the vertices B, C and D are represented by the complex numbers $-6 + j4$, $-5 - j3$ and $2 - j2$ respectively.

4.2.6 Exercises

Check your answers using MATLAB or MAPLE whenever possible.

1 Given $a = (1, 1, 0)$, $b = (2, 2, 1)$ and $c = (0, 1, 1)$, evaluate

(a) $a + b$ (b) $a + \frac{1}{2}b + 2c$ (c) $b - 2a$

(d) $|a|$ (e) $|b|$ (f) $|a - b|$

(g) $\hat{a}$ (h) $\hat{b}$

2 If the position vectors of the points P and Q are $i + 3j - 7k$ and $5i - 2j + 4k$ respectively, find $\overrightarrow{PQ}$ and determine its length and direction cosines.

3 A particle P is acted upon by forces (measured in newtons) $F_1 = 3i - 2j + 5k$, $F_2 = -i + 7j - 3k$, $F_3 = 5i - j + 4k$ and $F_4 = -2j + 3k$. Determine the magnitude and direction of the resultant force acting on P.

4 If $a = 3i - 2j + k$, $b = -2i + 5j + 4k$, $c = -4i + j - 2k$ and $d = 2i - j + 4k$, determine α, β and γ such that

$$d = \alpha a + \beta b + \gamma c$$

5 Prove that the vectors $2i - 4j - k$, $3i + 2j - 2k$ and $5i - 2j - 3k$ can form the sides of a triangle.

Find the lengths of each side of the triangle and show that it is right-angled.

6 The vector $\overrightarrow{OP}$ makes an angle of 60° with the positive x axis and 45° with the positive y axis. Find the possible angles that the vector can make with the z axis.

7 Given the points P(1, −3, 4), Q(2, 2, 1) and R(3, 7, −2), find the vectors $\overrightarrow{PQ}$ and $\overrightarrow{QR}$. Show that P, Q and R lie on a straight line and find the ratio PQ : QR.

8 A cyclist travelling east at 8 kilometres per hour finds that the wind appears to blow directly from the north. On doubling his speed it appears to blow from the north-east. Find the actual velocity of the wind.

9 Relative to a landing stage, the position vectors in kilometres of two boats A and B at noon are

$$3i + j \quad \text{and} \quad i - 2j$$

respectively. The velocities of A and B, which are constant and in kilometres per hour, are

$$10i + 24j \quad \text{and} \quad 24i + 32j$$

Find the distance between the boats t hours after noon and find the time at which this distance is a minimum.

10 If the complex numbers z_1, z_2 and z_3 are represented on the Argand diagram by the points P_1, P_2 and P_3 respectively and

$$\overrightarrow{OP_2} = 2j\overrightarrow{OP_1} \quad \text{and} \quad \overrightarrow{OP_3} = \tfrac{2}{5}j\overrightarrow{P_2P_1}$$

prove that P_3 is the foot of the perpendicular from O on to the line P_1P_2.

11 ABCD is a square, lettered anticlockwise, on an Argand diagram, with A representing $3 + j2$ and B representing $-1 + j4$. Show that C lies on the real axis and find the complex number represented by D and the length of AB.

12 A triangle has vertices A, B, C represented by $1 + j$, $2 - j$ and -1 respectively. Find the point that is equidistant from A, B and C.

13 Given the triangle OAB, where O is the origin, and denoting the midpoints of the opposite sides as O′, A′ and B′, show vectorially that the lines OO′, AA′ and BB′ meet at a point. (Note that this is the result that the medians of a triangle meet at the centroid.)

14 Three weights W_1, W_2 and W_3 hang in equilibrium on the pulley system shown in Figure 4.25. The pulleys are considered to be smooth and the forces add by the rules of vector addition. Calculate θ and ϕ, the angles the ropes make with the horizontal.

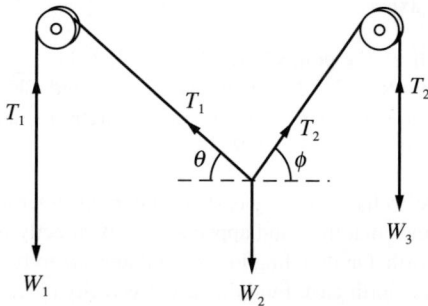

Figure 4.25 Pulley system in Question 14.

15 A telegraph pole OP has three wires connected to it at P. The other ends of the wires are connected to houses at A, B and C. Axes are set up as shown

in Figure 4.26. The points relative to these axes, with distances in metres, are $\overrightarrow{OP} = 8k$, $\overrightarrow{OA} = 20j + 6k$, $\overrightarrow{OB} = -i - 18j + 10k$ and $\overrightarrow{OC} = -22i + 3j + 7k$. The tension in each wire is 900 N. Find the total force acting at P. A tie cable at an angle of 45° is connected to P and fixed in the ground. Where should the ground fixing be placed, and what is the tension required to ensure a zero horizontal resultant force at P?

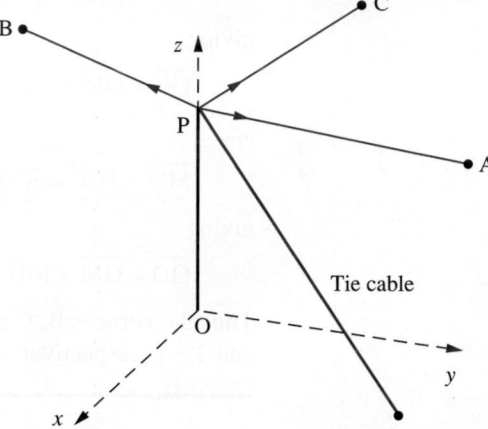

Figure 4.26 The telegraph pole of Question 15.

16 A boom OB carries a load F of magnitude 500 N and is supported by cables BC and BD as shown in Figure 4.27 where the dimensions of the system are given. Determine the tensions in the cables so that equilibrium is maintained and the resultant force at the point B is along OB.

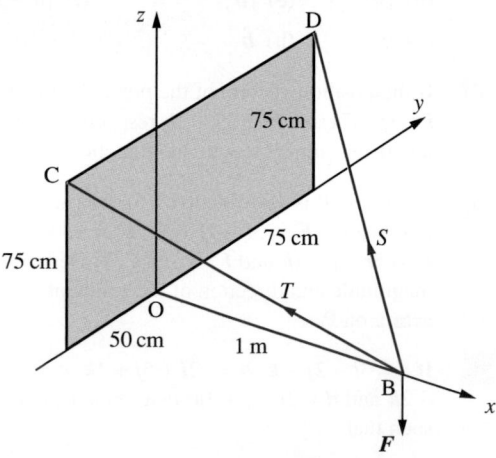

Figure 4.27 Boom supported by cables in Question 16.

4.2.7   The scalar product

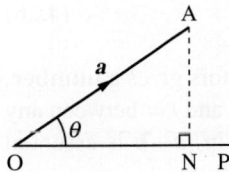

Figure 4.28
The component of **a**
in the direction OP is
ON = |**a**| cos θ.

A natural idea in mathematics, explored in Chapter 1, is not only to add quantities but also to multiply them together. The concept of multiplication of vectors translates into a useful tool for many engineering applications, with two different products of vectors – the 'scalar' and 'vector' products – turning out to be particularly important.

The determination of a component of a vector is a basic procedure in analysing many physical problems. For the vector **a** shown in Figure 4.28 the component of **a** in the direction of OP is just ON = |**a**| cos θ. The component is relevant in the physical context of work done by a force. Suppose the point of application, O, of a constant force **F** is moved along the vector **a** from O to the point A, as in Figure 4.29. The component of **F** in the **a** direction is |**F**| cos θ, and O is moved a distance |**a**|. The work done is defined as the product of the distance moved by the point of application and the component of the force in this direction. It is thus given by

$$\text{work done} = |\mathbf{F}|\,|\mathbf{a}| \cos \theta$$

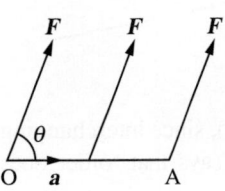

Figure 4.29
The work done by a
constant force **F** with
point of application
moved from O to A
is |**F**||**a**| cos θ.

The definition of the scalar product in geometrical terms takes the form of this expression for the work done by a force. Again there is an equivalent component definition, and both are now presented.

Definition

The **scalar** (or **dot** or **inner**) **product** of two vectors $\mathbf{a} = (a_1, a_2, a_3)$ and $\mathbf{b} = (b_1, b_2, b_3)$ is defined as follows:

In components

$$\mathbf{a} \cdot \mathbf{b} = a_1 b_1 + a_2 b_2 + a_3 b_3 \qquad\qquad\qquad (4.2a)$$

Geometrically

$$\mathbf{a} \cdot \mathbf{b} = |\mathbf{a}|\,|\mathbf{b}| \cos \theta, \quad \text{where } \theta \ (0 \leqslant \theta \leqslant \pi) \text{ is the angle between the two vectors}$$

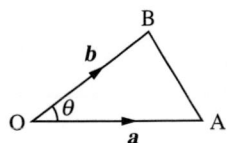

Figure 4.30
Cosine rule for a
triangle; equivalence
of the geometrical and
component definitions
of the scalar product.

Both definitions prove to be useful in different contexts, but to establish the basic rules the component definition is the simpler. The equivalence of the two definitions can easily be established from the cosine rule for a triangle. Using Figure 4.30 the cosine rule (2.16) states

$$AB^2 = OA^2 + OB^2 - 2(OA)(OB) \cos \theta$$

which in appropriate vector or component notation gives

$$(a_1 - b_1)^2 + (a_2 - b_2)^2 + (a_3 - b_3)^2 = (a_1^2 + a_2^2 + a_3^2) + (b_1^2 + b_2^2 + b_3^2)$$
$$- 2|\mathbf{a}|\,|\mathbf{b}| \cos \theta$$

Thus expanding the left-hand side gives

$$a_1^2 - 2a_1 b_1 + b_1^2 + a_2^2 - 2a_2 b_2 + b_2^2 + a_3^2 - 2a_3 b_3 + b_3^2$$
$$= a_1^2 \qquad\quad + b_1^2 + a_2^2 \qquad\quad + b_2^2 + a_3^2 \qquad\quad + b_3^2 - 2|\mathbf{a}|\,|\mathbf{b}| \cos \theta$$

and hence

$$\boldsymbol{a} \cdot \boldsymbol{b} = a_1 b_1 + a_2 b_2 + a_3 b_3 = |\boldsymbol{a}| \, |\boldsymbol{b}| \cos \theta \qquad \text{(4.2b)}$$

Two important points to note are: (i) the scalar product of two vectors gives a **number**. (ii) the scalar product is only defined as the product of two vectors and *not* between any other two quantities. For this reason, the presence of the dot ($\cdot$) in $\boldsymbol{a} \cdot \boldsymbol{b}$ is essential between the two vectors.

Basic rules and properties

The basic rules are now very straightforward to establish.

(a) Commutative law

$$\boldsymbol{a} \cdot \boldsymbol{b} = \boldsymbol{b} \cdot \boldsymbol{a}$$

This rule follows immediately from the component definition (4.2a), since interchanging a_i and b_i does not make any difference to the products. The rule says that 'order does not matter'.

(b) Associative law

The idea of associativity involves the product of three vectors. Since $\boldsymbol{a} \cdot \boldsymbol{b}$ is a scalar, it cannot be dotted with a third vector, so the idea of associativity is not applicable here and $\boldsymbol{a} \cdot \boldsymbol{b} \cdot \boldsymbol{c}$ is not defined.

(c) Distributive law for products with a scalar λ

$$\boldsymbol{a} \cdot (\lambda \boldsymbol{b}) = (\lambda \boldsymbol{a}) \cdot \boldsymbol{b} = \lambda (\boldsymbol{a} \cdot \boldsymbol{b})$$

These results follow directly from the component definition (4.2a). The implication is that scalars can be multiplied out in the normal manner.

(d) Distributive law over addition

$$\boldsymbol{a} \cdot (\boldsymbol{b} + \boldsymbol{c}) = \boldsymbol{a} \cdot \boldsymbol{b} + \boldsymbol{a} \cdot \boldsymbol{c}$$

The proof is straightforward, since

$$\boldsymbol{a} \cdot (\boldsymbol{b} + \boldsymbol{c}) = a_1(b_1 + c_1) + a_2(b_2 + c_2) + a_3(b_3 + c_3)$$
$$= (a_1 b_1 + a_2 b_2 + a_3 b_3) + (a_1 c_1 + a_2 c_2 + a_3 c_3)$$
$$= \boldsymbol{a} \cdot \boldsymbol{b} + \boldsymbol{a} \cdot \boldsymbol{c}$$

Thus the normal rules of algebra apply, and brackets can be multiplied out in the usual way.

(e) Powers of a

One simple point to note is that

$$\boldsymbol{a} \cdot \boldsymbol{a} = a_1^2 + a_2^2 + a_3^2 = |\boldsymbol{a}| \, |\boldsymbol{a}| \cos 0 = |\boldsymbol{a}|^2$$

in agreement with Section 4.2.4. This expression is written $a^2 = a \cdot a$ and, where there is no ambiguity, $a^2 = a^2$ is also used. No other powers of vectors can be constructed, since, as in (b) above, scalar products of more than two vectors do not exist. For the standard unit vectors, i, j and k,

$$i^2 = i \cdot i = 1, \quad j^2 = j \cdot j = 1, \quad k^2 = k \cdot k = 1 \tag{4.3}$$

(f) Perpendicular vectors

It is clear from (4.2b) that if a and b are perpendicular (orthogonal) then $\cos \theta = \cos \frac{1}{2}\pi = 0$, and hence $a \cdot b = 0$, or in component notation

$$a \cdot b = a_1 b_1 + a_2 b_2 + a_3 b_3 = 0$$

However, the other way round, $a \cdot b = 0$, *does not* imply that a and b are perpendicular. There are three possibilities:

$$\text{either } a = 0 \quad \text{or} \quad b = 0 \quad \text{or} \quad \theta = \tfrac{1}{2}\pi$$

It is only when the first two possibilities have been dismissed that perpendicularity can be deduced.

The commonest mistake is to deduce from

$$a \cdot b = a \cdot c$$

that $b = c$. This is only one of three possible solutions – the other two being $a = 0$ and a perpendicular to $b - c$. The rule to follow is that *you can't cancel vectors in the same way as scalars.*

Since the unit vectors i, j and k are mutually perpendicular,

$$i \cdot j = j \cdot k = k \cdot i = 0 \tag{4.4}$$

Using the distributive law over addition, we obtain using (4.3) and (4.4)

$$(a_1, a_2, a_3) \cdot (b_1, b_2, b_3) = (a_1 i + a_2 j + a_3 k) \cdot (b_1 i + b_2 j + b_3 k)$$

$$= a_1 b_1 i \cdot i + a_1 b_2 i \cdot j + a_1 b_3 i \cdot k + a_2 b_1 j \cdot i + a_2 b_2 j \cdot j$$

$$+ a_2 b_3 j \cdot k + a_3 b_1 k \cdot i + a_3 b_2 k \cdot j + a_3 b_3 k \cdot k$$

$$= a_1 b_1 + a_2 b_2 + a_3 b_3$$

which is consistent with the component definition of a scalar product.

Perpendicularity is a very important idea, which is used a great deal in both mathematics and engineering. Pressure acts on a surface in a direction perpendicular to the surface, so that the force per unit area is given by $p\hat{n}$, where p is the pressure and $\hat{n}$ is the unit normal. To perform many calculations, we must be able to find a vector that is perpendicular to another vector. We shall also see that many matrix methods rely on being able to construct a set of mutually orthogonal vectors. Such constructions are not only of theoretical interest, but form the basis of many practical numerical methods used in engineering (see Chapter 6 of the companion text *Advanced Modern Engineering Mathematics*). The whole of the study of Fourier series (considered in Chapter 12), which is central to much of signal processing and is heavily used by electrical engineers, is based on constructing functions that are orthogonal.

In MATLAB the scalar product of two vectors a and b is given by the command $dot(a,b)$. In MAPLE it is given by $innerprod(a,b)$.

Example 4.17

Given the vectors $a = (1, -1, 2)$, $b = (-2, 0, 2)$ and $c = (3, 2, 1)$, evaluate

(a) $a \cdot c$ (b) $b \cdot c$ (c) $(a + b) \cdot c$

(d) $a \cdot (2b + 3c)$ (e) $(a \cdot b)c$

Solution

(a) $a \cdot c = (1 \times 3) + (-1 \times 2) + (2 \times 1) = 3$

(b) $b \cdot c = (-2 \times 3) + (0 \times 2) + (2 \times 1) = -4$

(c) $(a + b) = (1, -1, 2) + (-2, 0, 2) = (-1, -1, 4)$ so that

$(a + b) \cdot c = (-1, -1, 4) \cdot (3, 2, 1) = -3 - 2 + 4 = -1$

(note that $(a + b) \cdot c = a \cdot c + b \cdot c$)

(d) $a \cdot (2b + 3c) = (1, -1, 2) \cdot [(-4, 0, 4) + (9, 6, 3)]$

$= (1, -1, 2) \cdot (5, 6, 7) = (5 - 6 + 14) = 13$

(note that $2(a \cdot b) + 3(a \cdot c) = 4 + 9 = 13$)

(e) $(a \cdot b)c = [(1, -1, 2) \cdot (-2, 0, 2)](3, 2, 1) = [-2 + 0 + 4](3, 2, 1)$

$= 2(3, 2, 1) = (6, 4, 2)$

(note that $a \cdot b$ is a scalar, so $(a \cdot b)c$ is a vector parallel or antiparallel to c)

Check that in MATLAB the commands

```
a = [1 -1 2]; b = [-2 0 2]; c = [3 2 1];
dot(a,c), dot(b,c), dot(a + b,c), dot(a,2*b + 3*c),
dot(a,b)*c
```

return the answers given in this example.

Example 4.18

Find the angle between the vectors $a = (1, 2, 3)$ and $b = (2, 0, 4)$.

Solution

By definition

$$a \cdot b = |a| \, |b| \cos \theta = a_1 b_1 + a_2 b_2 + a_3 b_3$$

We have in the right-hand side

$$(1, 2, 3) \cdot (2, 0, 4) = 2 + 0 + 12 = 14$$

Also

$$|(1, 2, 3)| = \sqrt{(1^2 + 2^2 + 3^2)} = \sqrt{14}$$

and

$$|(2, 0, 4)| = \sqrt{(2^2 + 0^2 + 4^2)} = \sqrt{20}$$

Thus, from the definition of the scalar product,

$$14 = \sqrt{(14)}\sqrt{(20)} \cos \theta$$

giving

$$\theta = \cos^{-1}\sqrt{\tfrac{7}{10}}$$

Example 4.19   Given $a = (1, 0, 1)$ and $b = (0, 1, 0)$, show that $a \cdot b = 0$, and interpret this result.

Solution $a \cdot b = (1, 0, 1) \cdot (0, 1, 0) = 0$

Since $|a| \neq 0$ and $|b| \neq 0$, the two vectors are perpendicular. We can see this result geometrically, since a lies in the x–z plane and b is parallel to the y axis.

Example 4.20   The three vectors

$$a = (1, 1, 1), \quad b = (3, 2, -3) \quad \text{and} \quad c = (-1, 4, -1)$$

are given. Show that $a \cdot b = a \cdot c$ and interpret the result.

Solution   Now $a \cdot b = 1 \times 3 + 1 \times 2 - 1 \times 3 = 2$

and $a \cdot c = 1 \times (-1) + 1 \times 4 + 1 \times (-1) = 2$

so the two scalar products are clearly equal. Certainly $b \neq c$ since they are given to be unequal and a is non-zero so the conclusion from

$$a \cdot (b - c) = 0$$

is that the vectors a and $(b - c) = (4, -2, -2)$ are perpendicular.

Example 4.21   In a triangle ABC show that the perpendiculars from the vertices to the opposite sides intersect in a point.

Solution   Let the perpendiculars AD and BE meet in O as indicated in Figure 4.31, and choose O to be the origin. Define $\overrightarrow{OA} = a$, $\overrightarrow{OB} = b$ and $\overrightarrow{OC} = c$. Then

AD perpendicular to BC implies $a \cdot (b - c) = 0$

BE perpendicular to AC implies $b \cdot (c - a) = 0$

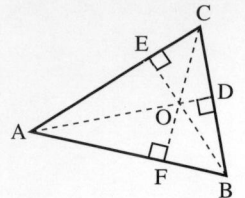

Figure 4.31
The altitudes of a triangle meet in a point (Example 4.21).

Hence, adding,

$$a \cdot b - a \cdot c + b \cdot c - b \cdot a = 0$$

so

$$b \cdot c - a \cdot c = c \cdot (b - a) = 0$$

This statement implies that $b - a$ is perpendicular to c or AB is perpendicular to CF, as required. The case $b - a = 0$ is dismissed, since then the triangle would collapse. The case $c = 0$ implies that C is at O; the triangle is then right-angled and the result is trivial.

Example 4.22

Find the work done by the force $F = (3, -2, 5)$ in moving a particle from a point P to a point Q having position vectors $(1, 4, -1)$ and $(-2, 3, 1)$ respectively.

Solution

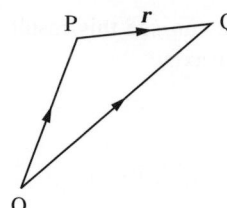

Figure 4.32
Triangle law for Example 4.22.

Applying the triangle law to Figure 4.32, we have the displacement of the particle given by

$$r = \overrightarrow{PQ} = \overrightarrow{PO} + \overrightarrow{OQ} = \overrightarrow{OQ} - \overrightarrow{OP}$$

$$= (-2, 3, 1) - (1, 4, -1) = (-3, -1, 2)$$

Then the work done by the force F is

$$F \cdot r = (3, -2, 5) \cdot (-3, -1, 2) = -9 + 2 + 10$$

$$= 3 \text{ units}$$

The **component** of a vector in a given direction was discussed at the start of this section, and, as indicated in Figure 4.28, the component of F in the a direction is $|F| \cos \theta$. Taking $\hat{a}$ to be the unit vector in the a direction,

$$F \cdot \hat{a} = |F| |\hat{a}| \cos \theta = |F| \cos \theta$$

$$= \text{the component of } F \text{ in the } a \text{ direction}$$

Example 4.23

Find the component of the vector $F = (2, -1, 3)$ in

(a) the i direction

(b) the direction $(\frac{1}{3}, \frac{2}{3}, \frac{2}{3})$

(c) the direction $(4, 2, -1)$

Solution

(a) The direction i is represented by the vector $(1, 0, 0)$, so the component of F in the i direction is

$$F \cdot (1, 0, 0) = (2, -1, 3) \cdot (1, 0, 0) = 2$$

(note how this result just picks out the x component and agrees with the usual idea of a component).

(b) Since $\sqrt{(\frac{1}{9} + \frac{4}{9} + \frac{4}{9})} = 1$, the vector $(\frac{1}{3}, \frac{2}{3}, \frac{2}{3})$ is a unit vector. Thus the component of $\boldsymbol{F}$ in the direction $(\frac{1}{3}, \frac{2}{3}, \frac{2}{3})$ is

$$\boldsymbol{F} \cdot (\tfrac{1}{3}, \tfrac{2}{3}, \tfrac{2}{3}) = \tfrac{2}{3} - \tfrac{2}{3} + 2 = 2$$

(c) Since $\sqrt{(16 + 4 + 1)} \neq 1$, the vector $(4, 2, -1)$ is not a unit vector. Therefore we must first compute its magnitude as

$$\sqrt{(4^2 + 2^2 + 1^2)} = \sqrt{21}$$

indicating that a unit vector in the direction of $(4, 2, -1)$ is $(4, 2, -1)/\sqrt{21}$. Thus the component of $\boldsymbol{F}$ in the direction of $(4, 2, -1)$ is

$$\boldsymbol{F} \cdot (4, 2, -1)/\sqrt{21} = 3/\sqrt{21}$$

4.2.8 Exercises

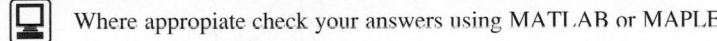

Where appropiate check your answers using MATLAB or MAPLE.

17 Given that $\boldsymbol{u} = (4, 0, -2)$, $\boldsymbol{v} = (3, 1, -1)$, $\boldsymbol{w} = (2, 1, 6)$ and $\boldsymbol{s} = (1, 4, 1)$, evaluate

(a) $\boldsymbol{u} \cdot \boldsymbol{v}$ (b) $\boldsymbol{v} \cdot \boldsymbol{s}$

(c) $\hat{\boldsymbol{w}}$ (d) $(\boldsymbol{v} \cdot \boldsymbol{s})\hat{\boldsymbol{u}}$

(e) $(\boldsymbol{u} \cdot \boldsymbol{w})(\boldsymbol{v} \cdot \boldsymbol{s})$ (f) $(\boldsymbol{u} \cdot \boldsymbol{i})\boldsymbol{v} + (\boldsymbol{w} \cdot \boldsymbol{s})\boldsymbol{k}$

18 Given $\boldsymbol{u}$, $\boldsymbol{v}$, $\boldsymbol{w}$ and $\boldsymbol{s}$ as for Question 17, find

(a) the angle between $\boldsymbol{u}$ and $\boldsymbol{w}$

(b) the angle between $\boldsymbol{v}$ and $\boldsymbol{s}$

(c) the value of λ for which the vectors $\boldsymbol{u} + \lambda \boldsymbol{k}$ and $\boldsymbol{v} - \lambda \boldsymbol{i}$ are perpendicular

(d) the value of μ for which the vectors $\boldsymbol{w} + \mu \boldsymbol{i}$ and $\boldsymbol{s} - \mu \boldsymbol{i}$ are perpendicular.

19 Find the work done by the force $\boldsymbol{F} = (-2, -1, 3)$ in moving a particle from the point P to the point Q having position vectors $(-1, 2, 3)$ and $(1, -3, 4)$ respectively.

20 Find the resolved part in the direction of the vector $(3, 2, 1)$ of a force of 5 units acting in the direction of the vector $(2, -3, 1)$.

21 Find the value of t that makes the angle between the two vectors $\boldsymbol{a} = (3, 1, 0)$ and $\boldsymbol{b} = (t, 0, 1)$ equal to $45°$.

22 For any four points A, B, C and D in space, prove that

$$(\overrightarrow{DA} \cdot \overrightarrow{BC}) + (\overrightarrow{DB} \cdot \overrightarrow{CA}) + (\overrightarrow{DC} \cdot \overrightarrow{AB}) = 0$$

23 If $(\boldsymbol{c} - \frac{1}{2}\boldsymbol{a}) \cdot \boldsymbol{a} = (\boldsymbol{c} - \frac{1}{2}\boldsymbol{b}) \cdot \boldsymbol{b} = 0$, prove that the vector $\boldsymbol{c} - \frac{1}{2}(\boldsymbol{a} + \boldsymbol{b})$ is perpendicular to $\boldsymbol{a} - \boldsymbol{b}$.

24 Prove that the line joining the points $(2, 3, 4)$ and $(1, 2, 3)$ is perpendicular to the line joining the points $(1, 0, 2)$ and $(2, 3, -2)$.

25 Show that the diagonals of a rhombus intersect at right-angles.

26 (a) A batch of bricks weighing $10\,\text{N}$ is lifted from storage, taken to be the point $(0, 0, 0)$, against gravity to a point on the first floor of the building with coordinates $(0, 4, 5)\,\text{m}$. Gravity acts in the $(-z)$ direction and the x, y directions are at ground level. Calculate the work done in raising the bricks.

(b) A straight wall is to be built with p bricks in each layer. The weight of each brick is W newtons and it has thickness $h\,\text{m}$. Neglecting the thickness of the mortar, estimate the work done in raising the bricks from ground level to build a wall of height $nh\,\text{m}$. Show that the work done increases linearly with p but as the square of n.

27 Find the equation of a circular cylinder with the origin on the axis of the cylinder, the unit vector $\boldsymbol{a}$ along the axis and radius R.

28 A cube has corners with coordinates $(0, 0, 0)$, $(1, 0, 0)$, $(0, 1, 0)$, $(1, 1, 0)$, $(0, 0, 1)$, $(1, 0, 1)$, $(0, 1, 1)$ and $(1, 1, 1)$. Find the vectors representing the diagonals of the cube and hence find the length of the diagonals and the angle between the diagonals.

29 A lifeboat hangs from a davit as shown in Figure 4.33 with the x direction, the vertical part

of the davit and the arm of the davit being mutually perpendicular. The rope is fastened to the deck at a distance X from the davit. It is known that the maximum force in the x direction that the davit can withstand is 200 N. If the weight supported is 500 N and the pulley system is a single loop so that the tension is 250 N, then determine the maximum value that X can take.

30 A simple derrick is constructed as in Figure 4.34 with axes set up as indicated. The wires AP and BP are in tension, and the arm of the derrick, PC, is loaded with a weight W at C. The x and y components of the forces at P are always in equilibrium. Determine the range of the angle θ that will ensure that the tensions T_1 and T_2 are always positive and hence the wires will not slacken.

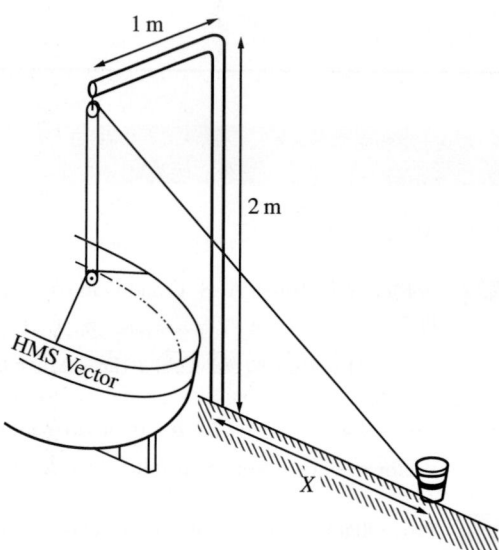

Figure 4.33 Davit in Question 29.

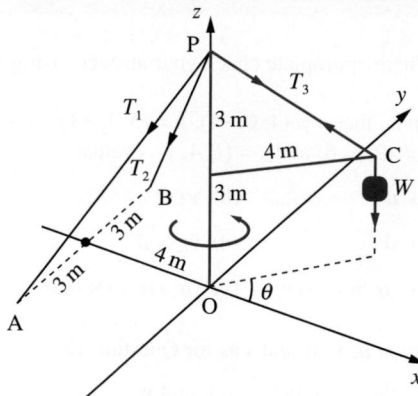

Figure 4.34 Simple derrick in Question 30.

4.2.9 The vector product

The **vector** or **cross product** was developed during the nineteenth century, its main practical use being to define the moment of a force in three dimensions. It is generally only in three dimensions that the vector product is used. The adaptation for two-dimensional vectors is of restricted scope, since for two-dimensional problems, where all vectors are confined to a plane, the direction of the vector product is always perpendicular to that plane.

Definition

Given two vectors $\boldsymbol{a}$ and $\boldsymbol{b}$, we define the vector product geometrically as

$$\boldsymbol{a} \times \boldsymbol{b} = |\boldsymbol{a}| |\boldsymbol{b}| \sin \theta \, \hat{\boldsymbol{n}}$$

(4.5)

Figure 4.35
Vector product $a \times b$, right-hand rule.

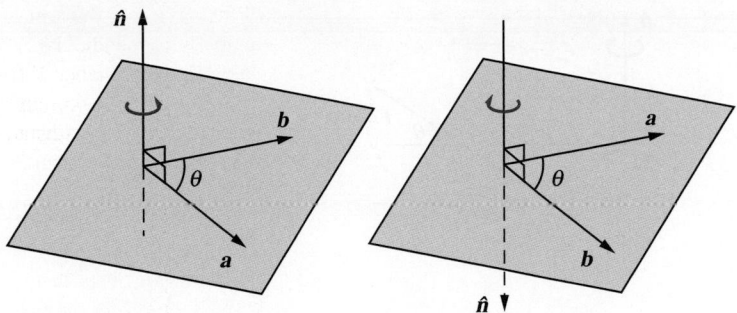

where θ is the angle between a and b ($0 \leqslant \theta \leqslant \pi$), and $\hat{n}$ is the unit vector perpendicular to both a and b such that a, b, $\hat{n}$ form a right-handed set – see Figure 4.35 and the definition at the beginning of Section 4.2.1.

It is important to recognize that the vector product of two vectors is itself a vector. The alternative notation $a \wedge b$ is also sometimes used to denote the vector product, but this is less common since the similar wedge symbol $\wedge$ is also used for other purposes (see e.g. Chapter 6, Section 6.4.2).

There are wide ranging applications of the vector product.

Motion of a charged particle in a magnetic field

- If a charged particle has velocity v and moves in a magnetic field H then the particle experiences a force perpendicular to both v and H, which is proportional to $v \times H$. It is this force that is used to direct the beam in a television tube.
- Similarly a wire moving with velocity v in a magnetic field H produces a current proportional to $v \times H$, see Figure 4.36, thus converting mechanical energy into electric current, and provides the principle of the **dynamo**.
- For an **electric motor** the idea depends on the observation that an electric current C in a wire that lies in a magnetic field H produces a mechanical force proportional to $C \times H$, again see Figure 4.36. Thus electrical energy is converted to a mechanical force.

Figure 4.36
In a magnetic field H, (i) motion of the wire in the v direction creates a current in the $H \times v$ (dynamo), (ii) a current C causes motion v in the $C \times H$ direction (electric motor).

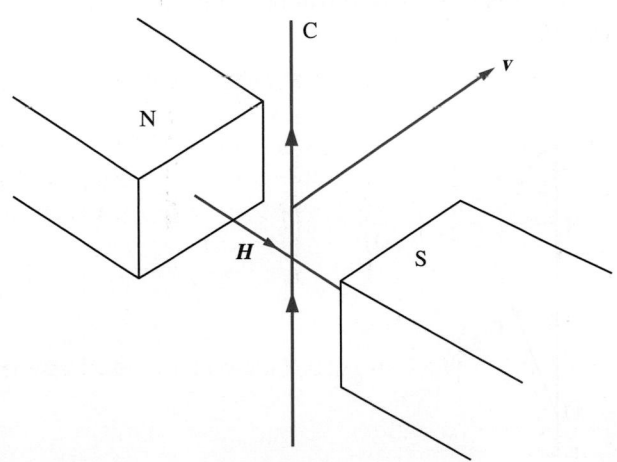

Figure 4.37
Moment of a force.

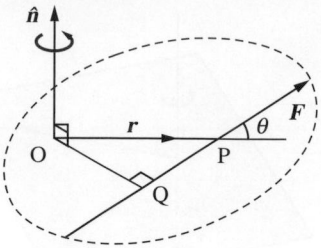

Moment of a force

The moment or torque of a force $\boldsymbol{F}$ provides the classical application of the vector product in a mechanical context. Although moments are easy to define in two dimensions, the extension to three dimensions is not so easy. In vector notation, however, if the force passes through the point P and $\overrightarrow{\text{OP}} = \boldsymbol{r}$, as illustrated in Figure 4.37, then the moment $\boldsymbol{M}$ of the force about O is simply defined as

$$\boldsymbol{M} = \boldsymbol{r} \times \boldsymbol{F} = |\boldsymbol{r}||\boldsymbol{F}|\sin\theta\,\hat{\boldsymbol{n}} = \text{OQ}|\boldsymbol{F}|\hat{\boldsymbol{n}} \tag{4.6}$$

This is a vector in the direction of the normal $\hat{\boldsymbol{n}}$, and moments add by the usual parallelogram law.

Angular velocity of a rigid body

A further application of the vector product relates to rotating bodies. Consider a rigid body rotating with angular speed ω (in rad s^{-1}) about a fixed axis LM that passes through a fixed point O as illustrated in Figure 4.38. A point P of the rigid body having position vector $\boldsymbol{r}$ relative to O will move in a circular path whose plane is perpendicular to OM and whose centre N is on OM. If NQ is a fixed direction and the angle QNP is equal to χ then

$$\text{the magnitude of angular velocity} = \frac{d\chi}{dt} = \omega$$

(Note that we have used here the idea of a derivative, which will be introduced in Chapter 8.) The velocity $\boldsymbol{v}$ of P will be in the direction of the tangent shown and will have magnitude

Figure 4.38
Angular velocity of a rigid body.

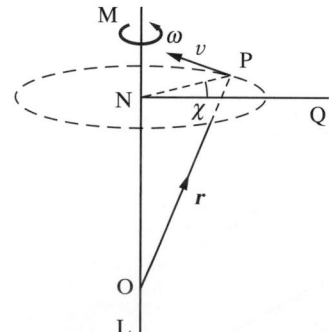

$$v = \text{NP} \frac{d\chi}{dt} = \text{NP}\omega$$

If we define $\boldsymbol{\omega}$ to be a vector of magnitude ω and having direction along the axis of rotation, in the sense in which the rotation would drive a right-handed screw, then

$$v = \boldsymbol{\omega} \times \boldsymbol{r} \tag{4.7}$$

correctly defines the velocity of P in both magnitude and direction. This vector $\boldsymbol{\omega}$ is called the **angular velocity** of the rigid body.

Area of parallelogram and a triangle

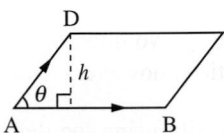

Figure 4.39
Representation of a parallelogram.

Geometrically we have from Figure 4.39 that the area of a parallelogram ABCD is given by

$$\text{area} = h|\overrightarrow{AB}| = |\overrightarrow{AD}|\sin\theta|\overrightarrow{AB}| = |\overrightarrow{AD} \times \overrightarrow{AB}|$$

Note also that the area of the triangle ABD is $\frac{1}{2}|\overrightarrow{AD} \times \overrightarrow{AB}|$, which corresponds to the result

$$\text{area of triangle ABD} = \tfrac{1}{2}(AD)(AB)\sin\theta$$

We now examine the properties of vector products in order to determine whether or not the usual laws of algebra apply.

Basic properties

(a) Anti-commutative law

$$\boldsymbol{a} \times \boldsymbol{b} = -(\boldsymbol{b} \times \boldsymbol{a})$$

This follows directly from the right-handedness of the set in the geometrical definition (4.5), since $\hat{\boldsymbol{n}}$ changes direction when the order of multiplication is reversed. Thus the vector product does not commute, but rather anti-commutes, unlike the multiplication of scalars or the scalar product of two vectors. Therefore the order of multiplication matters when using the vector product. For example, it is important that the moment of a force is calculated as $\boldsymbol{M} = \boldsymbol{r} \times \boldsymbol{F}$ and *not* $\boldsymbol{F} \times \boldsymbol{r}$.

(b) Non-associative multiplication

Since the vector product of two vectors is a vector, we can take the vector product with a third vector, and associativity can be tested. It turns out to *fail in general*, and

$$\boldsymbol{a} \times (\boldsymbol{b} \times \boldsymbol{c}) \neq (\boldsymbol{a} \times \boldsymbol{b}) \times \boldsymbol{c}$$

except in special cases, such as when $\boldsymbol{a} = 0$. This can be seen to be the case from geometrical considerations using the definition (4.5). The vector $\boldsymbol{b} \times \boldsymbol{c}$ is perpendicular to both $\boldsymbol{b}$ and $\boldsymbol{c}$, and is thus perpendicular to the plane containing $\boldsymbol{b}$ and $\boldsymbol{c}$. Also, by definition, $\boldsymbol{a} \times (\boldsymbol{b} \times \boldsymbol{c})$ is perpendicular to $\boldsymbol{b} \times \boldsymbol{c}$, and is therefore in the plane of $\boldsymbol{b}$ and $\boldsymbol{c}$. Similarly, $(\boldsymbol{a} \times \boldsymbol{b}) \times \boldsymbol{c}$ is in the plane of $\boldsymbol{a}$ and $\boldsymbol{b}$. Hence, in general, $\boldsymbol{a} \times (\boldsymbol{b} \times \boldsymbol{c})$ and $(\boldsymbol{a} \times \boldsymbol{b}) \times \boldsymbol{c}$ are different vectors.

Since the associative law does not hold in general, we never write $a \times b \times c$, since it is ambiguous. Care must be taken to maintain the correct order and thus brackets must be inserted when more than two vectors are involved in a vector product.

(c) Distributive law over multiplication by a scalar

The definition (4.5) shows trivially that

$$a \times (\lambda b) = \lambda(a \times b) = (\lambda a) \times b$$

and the usual algebraic rule applies.

(d) Distributive law over addition

$$a \times (b + c) = (a \times b) + (a \times c)$$

This law holds for the vector product. It can be proved geometrically using the definition (4.5). The proof, however, is rather protracted and is omitted here.

(e) Parallel vectors

It is obvious from the definition (4.5) that if a and b are parallel or antiparallel then $\theta = 0$ or π, so that $a \times b = 0$, and this includes the case $a \times a = 0$. We note, however, that if $a \times b = 0$ then we have three possible cases: either $a = 0$ or $b = 0$ or a and b are parallel. As with the scalar product, if we have $a \times b = a \times c$ then we cannot deduce that $b = c$. We first have to show that $a \neq 0$ and that a is not parallel to $b - c$.

(f) Cartesian form

From the definition (4.5), it clearly follows that the three unit vectors, i, j and k parallel to the coordinate axes satisfy

$$i \times i = j \times j = k \times k = 0$$

$$i \times j = k, \quad j \times k = i, \quad k \times i = j \tag{4.8}$$

Note the cyclic order of these latter equations. Using these results, we can obtain the cartesian or component form of the vector product. Taking

$$a = (a_1, a_2, a_3) = a_1 i + a_2 j + a_3 k$$

and

$$b = (b_1, b_2, b_3) = b_1 i + b_2 j + b_3 k$$

then, using rules (c), (d) and (a),

$$a \times b = (a_1 i + a_2 j + a_3 k) \times (b_1 i + b_2 j + b_3 k)$$

$$= a_1 b_1 (i \times i) + a_1 b_2 (i \times j) + a_1 b_3 (i \times k) + a_2 b_1 (j \times i) + a_2 b_2 (j \times j)$$

$$+ a_2 b_3 (j \times k) + a_3 b_1 (k \times i) + a_3 b_2 (k \times j) + a_3 b_3 (k \times k)$$

$$= a_1 b_2 k + a_1 b_3 (-j) + a_2 b_1 (-k) + a_2 b_3 i + a_3 b_1 j + a_3 b_2 (-i)$$

so that

$$a \times b = (a_2 b_3 - a_3 b_2)i + (a_3 b_1 - a_1 b_3)j + (a_1 b_2 - a_2 b_1)k \qquad (4.9)$$

The cartesian form (4.9) can be more easily remembered in its determinant form (actually an accepted misuse of the determinant form)

$$a \times b = \begin{vmatrix} i & j & k \\ a_1 & a_2 & a_3 \\ b_1 & b_2 & b_3 \end{vmatrix} = i \begin{vmatrix} a_2 & a_3 \\ b_2 & b_3 \end{vmatrix} - j \begin{vmatrix} a_1 & a_3 \\ b_1 & b_3 \end{vmatrix} + k \begin{vmatrix} a_1 & a_2 \\ b_1 & b_2 \end{vmatrix}$$

$$= (a_2 b_3 - b_2 a_3)i - (a_1 b_3 - b_1 a_3)j + (a_1 b_2 - b_1 a_2)k \qquad (4.10)$$

This notation is so convenient that we use it here before formally introducing determinants in the next chapter.

An alternative way to work out the cross product, which is easy to memorize, is to write the vectors (a, b, c) and (A, B, C) twice and read off the components by taking the products as indicated in Figure 4.40.

Figure 4.40
Gives the three components as
$bC - cB, cA - aC,$
$aB - bA.$

In MATLAB the vector product of two vectors a and b is given by the command $cross(a,b)$. In MAPLE it is given by $crossprod(a,b)$.

Example 4.24   Given the vectors $a = (2, 1, 0)$, $b = (2, -1, 1)$ and $c = (0, 1, 1)$, evaluate

(a) $a \times b$ (b) $(a \times b) \times c$ (c) $(a \cdot c)b - (b \cdot c)a$

(d) $b \times c$ (e) $a \times (b \times c)$ (f) $(a \cdot c)b - (a \cdot b)c$

Solution

(a) $a \times b = \begin{vmatrix} i & j & k \\ 2 & 1 & 0 \\ 2 & -1 & 1 \end{vmatrix} = i \begin{vmatrix} 1 & 0 \\ -1 & 1 \end{vmatrix} - j \begin{vmatrix} 2 & 0 \\ 2 & 1 \end{vmatrix} + k \begin{vmatrix} 2 & 1 \\ 2 & -1 \end{vmatrix} = (1, -2, -4)$

(b) $(a \times b) \times c = \begin{vmatrix} i & j & k \\ 1 & -2 & -4 \\ 0 & 1 & 1 \end{vmatrix} = i \begin{vmatrix} -2 & -4 \\ 1 & 1 \end{vmatrix} - j \begin{vmatrix} 1 & -4 \\ 0 & 1 \end{vmatrix} + k \begin{vmatrix} 1 & -2 \\ 0 & 1 \end{vmatrix} = (2, -1, 1)$

(c) $(a \cdot c)b - (b \cdot c)a = 1b - 0a = (2, -1, 1)$
    (Note that (b) and (c) give the same result.)

(d) $\boldsymbol{b} \times \boldsymbol{c} = (-2, -2, 2)$

(e) $\boldsymbol{a} \times (\boldsymbol{b} \times \boldsymbol{c}) = (2, -4, -2)$
(Note that (b) and (e) do not give the same result and the cross product is *not* associative.)

(f) $(\boldsymbol{a} \cdot \boldsymbol{c})\boldsymbol{b} - (\boldsymbol{a} \cdot \boldsymbol{b})\boldsymbol{c} = 1\boldsymbol{b} - 3\boldsymbol{c} = (2, -4, -2)$
(Note that (e) and (f) give the same result.)

Check that in MATLAB the commands

```
a = [2 1 0]; b = [2 -1 1]; c = [0 1 1];
cross(a,b)
cross(cross(a,b),c)
```

return the answers to (a) and (b).

Example 4.25   Find a unit vector perpendicular to the plane of the vectors $\boldsymbol{a} = (2, -3, 1)$ and $\boldsymbol{b} = (1, 2, -4)$.

Solution   A vector perpendicular to the plane of the two vectors is the vector product

$$\boldsymbol{a} \times \boldsymbol{b} = \begin{vmatrix} \boldsymbol{i} & \boldsymbol{j} & \boldsymbol{k} \\ 2 & -3 & 1 \\ 1 & 2 & -4 \end{vmatrix} = (10, 9, 7)$$

whose modulus is

$$|\boldsymbol{a} \times \boldsymbol{b}| = \sqrt{(100 + 81 + 49)} = \sqrt{230}$$

Hence a unit vector perpendicular to the plane of $\boldsymbol{a}$ and $\boldsymbol{b}$ is $(10/\sqrt{230}, 9/\sqrt{230}, 7/\sqrt{230})$.

Example 4.26   Find the area of the triangle having vertices at P(1, 3, 2), Q(−2, 1, 3) and R(3, −2, −1).

Solution   We have seen in Figure 4.39 that the area of the parallelogram formed with sides PQ and PR is $|\overrightarrow{PQ} \times \overrightarrow{PR}|$ so the area of the triangle PQR is $\frac{1}{2}|\overrightarrow{PQ} \times \overrightarrow{PR}|$. Now

$$\overrightarrow{PQ} = (-2 - 1, 1 - 3, 3 - 2) = (-3, -2, 1)$$

and

$$\overrightarrow{PR} = (3 - 1, -2 - 3, -1 - 2) = (2, -5, -3)$$

so that

$$\overrightarrow{PQ} \times \overrightarrow{PR} = \begin{vmatrix} \boldsymbol{i} & \boldsymbol{j} & \boldsymbol{k} \\ -3 & -2 & 1 \\ 2 & -5 & -3 \end{vmatrix} = (11, -7, 19)$$

Hence the area of the triangle PQR is

$$\tfrac{1}{2}|\overrightarrow{PQ} \times \overrightarrow{PR}| = \tfrac{1}{2}\sqrt{(121 + 49 + 361)} = \tfrac{1}{2}\sqrt{531} \approx 11.52 \text{ square units.}$$

Example 4.27

Four vectors are constructed corresponding to the four faces of a tetrahedron. The magnitude of a vector is equal to the area of the corresponding face and its direction is the outward perpendicular to the face, as in Figure 4.41. Show that the sum of the four vectors is zero.

Figure 4.41
(a) Tetrahedron
for Example 4.27;
(b) triangle from (a).

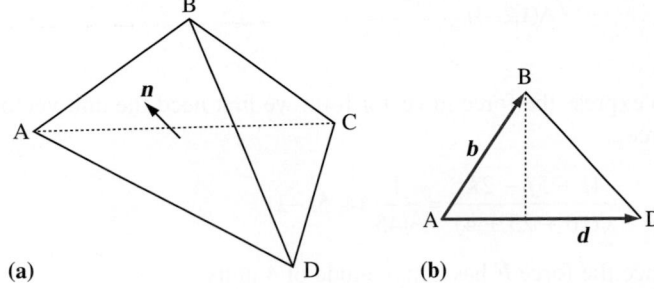

Solution

In Figure 4.41(a) let $\overrightarrow{AB} = \boldsymbol{b}$, $\overrightarrow{AC} = \boldsymbol{c}$ and $\overrightarrow{AD} = \boldsymbol{d}$. The outward perpendicular to triangle ABD is parallel to

$$\boldsymbol{n} = \overrightarrow{AD} \times \overrightarrow{AB} = \boldsymbol{d} \times \boldsymbol{b}$$

and the unit vector in the outward normal direction is

$$\hat{\boldsymbol{n}} = \frac{\boldsymbol{d} \times \boldsymbol{b}}{|\boldsymbol{d} \times \boldsymbol{b}|}$$

From Figure 4.41(b) the area of triangle ABD follows from the definition of cross product as

$$\text{area} = \tfrac{1}{2}\text{AD}(\text{AB} \sin \theta) = \tfrac{1}{2}|\boldsymbol{d} \times \boldsymbol{b}|$$

so the vector we require is

$$\boldsymbol{v}_1 = \text{area} \times \hat{\boldsymbol{n}} = \tfrac{1}{2}\boldsymbol{d} \times \boldsymbol{b}$$

In a similar manner for triangles ACB and ADC the vectors are

$$\boldsymbol{v}_2 = \tfrac{1}{2}\boldsymbol{b} \times \boldsymbol{c} \quad \text{and} \quad \boldsymbol{v}_3 = \tfrac{1}{2}\boldsymbol{c} \times \boldsymbol{d}$$

For the fourth face BCD the appropriate vector is

$$\boldsymbol{v}_4 = \tfrac{1}{2}\overrightarrow{BD} \times \overrightarrow{BC} = \tfrac{1}{2}(\boldsymbol{d} - \boldsymbol{b}) \times (\boldsymbol{c} - \boldsymbol{b}) = \tfrac{1}{2}(\boldsymbol{d} \times \boldsymbol{c} - \boldsymbol{d} \times \boldsymbol{b} - \boldsymbol{b} \times \boldsymbol{c})$$

Adding the four vectors $\boldsymbol{v}_1$, $\boldsymbol{v}_2$, $\boldsymbol{v}_3$ and $\boldsymbol{v}_4$ together gives the zero vector.

Example 4.28

A force of 4 units acts through the point P(2, 3, −5) in the direction of the vector (4, 5, −2). Find its moment about the point A(1, 2, −3). See Figure 4.42.

What are the moments of the force about axes through A parallel to the coordinate axes?

Figure 4.42
Moment of the force F
about the point A in
Example 4.28.

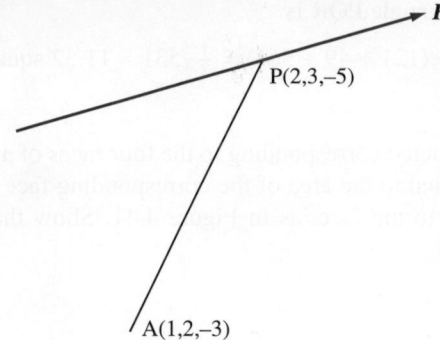

P(2,3,–5)

A(1,2,–3)

Solution To express the force in vector form we first need the unit vector in the direction of the force.

$$\frac{4i + 5j - 2k}{\sqrt{(16 + 25 + 4)}} = \frac{1}{\sqrt{45}} (4, 5, -2)$$

Since the force F has a magnitude of 4 units

$$F = \frac{4}{\sqrt{45}} (4, 5, -2)$$

The position vector of P relative to A is

$$\overrightarrow{AP} = (1, 1, -2)$$

Thus from (4.6) the moment M of the force about A is

$$M = \overrightarrow{AP} \times F = \frac{4}{\sqrt{45}} \begin{vmatrix} i & j & k \\ 1 & 1 & -2 \\ 4 & 5 & -2 \end{vmatrix}$$

$$= (32/\sqrt{45}, -24/\sqrt{45}, 4/\sqrt{45})$$

The moments about axes through A parallel to the coordinate axes are $32/\sqrt{45}$, $-24/\sqrt{45}$ and $4/\sqrt{45}$.

Example 4.29 A rigid body is rotating with an angular velocity of $5\,\text{rad s}^{-1}$ about an axis in the direction of the vector $(1, 3, -2)$ and passing through the point A$(2, 3, -1)$. Find the linear velocity of the point P$(-2, 3, 1)$ of the body.

Solution A unit vector in the direction of the axis of rotation is $\dfrac{1}{\sqrt{14}} (1, 3, -2)$. Thus the angular velocity vector of the rigid body is

$$\omega = (5/\sqrt{14})(1, 3, -2)$$

The position vector of P relative to A is

$$\overrightarrow{AP} = (-2 - 2, 3 - 3, 1 + 1) = (-4, 0, 2)$$

Thus from (4.7) the linear velocity of P is

$$v = \boldsymbol{\omega} \times \overrightarrow{\text{AP}} = \frac{5}{\sqrt{14}} \begin{vmatrix} \boldsymbol{i} & \boldsymbol{j} & \boldsymbol{k} \\ 1 & 3 & -2 \\ -4 & 0 & 2 \end{vmatrix}$$

$$= (30/\sqrt{14},\ 30/\sqrt{14},\ 60/\sqrt{14})$$

Example 4.30   A trapdoor is raised and lowered by a rope attached to one of its corners. The rope is pulled via a pulley fixed to a point A, 50 cm above the hinge as shown in Figure 4.43. If the trapdoor is uniform and of weight 20 N, what is the tension required to lift the door?

Figure 4.43
Trapdoor in
Example 4.30.

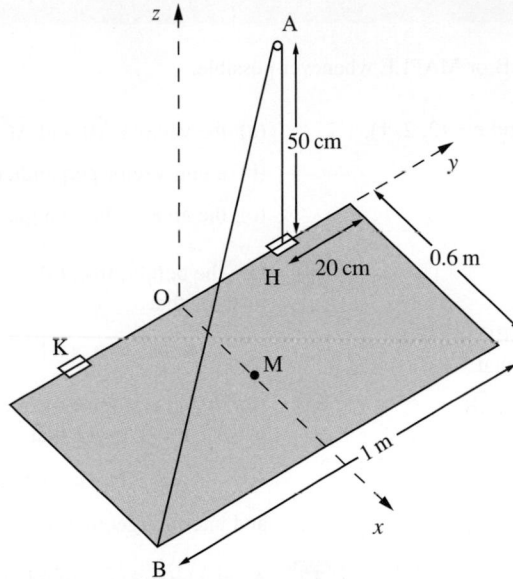

Solution   From the data given we can calculate various vectors immediately.

$$\overrightarrow{\text{OA}} = (0, 30, 50), \quad \overrightarrow{\text{OB}} = (60, -50, 0), \quad \overrightarrow{\text{OH}} = (0, 30, 0)$$

If M is the midpoint of the trapdoor then

$$\overrightarrow{\text{OM}} = (30, 0, 0)$$

The forces acting are the tension T in the rope along BA, the weight W through M in the $-z$ direction and reactions R and S at the hinges. Now

$$\overrightarrow{\text{AB}} = \overrightarrow{\text{OB}} - \overrightarrow{\text{OA}} = (60, -80, -50)$$

so that $|\overrightarrow{\text{AB}}| = 112$, and hence

$$T = -T(60, -80, -50)/112$$

Taking moments about the hinge H, we first note that there is no moment of the reaction at H. For the remaining forces

$$M_H = \overrightarrow{HM} \times W + \overrightarrow{HB} \times T + \overrightarrow{HK} \times R$$

$$= (30, -30, 0) \times (0, 0, -20) + (60, -80, 0) \times (60, -80, -50)(-T/112) + \overrightarrow{HK} \times R$$

$$= (600, 600, 0) + T(-35.8, -26.8, 0) + \overrightarrow{HK} \times R$$

Since we require the moment about the y axis, we take the scalar product of M_H and j. The vector $\overrightarrow{HK}$ is along j, so $j \cdot (\overrightarrow{HK} \times R)$ must be zero. Thus the j component of M_H must be zero as the trapdoor just opens; that is,

$$0 = 600 - 26.8T$$

so

$$T = 22.4 \text{ N}$$

4.2.10 Exercises

Check your answers using MATLAB or MAPLE whenever possible.

31 Given $p = (1, 1, 1)$, $q = (0, -1, 2)$ and $r = (2, 2, 1)$, evaluate

(a) $p \times q$ (b) $p \times r$

(c) $r \times q$ (d) $(p \times r) \cdot q$

(e) $q \cdot (r \times p)$ (f) $(p \times r) \times q$

32 Show that the area of the triangle ABC in Figure 4.44 is $\frac{1}{2} |\overrightarrow{AB} \times \overrightarrow{AC}|$. Show that

$$\overrightarrow{AB} \times \overrightarrow{AC} = \overrightarrow{BC} \times \overrightarrow{BA} = \overrightarrow{CA} \times \overrightarrow{CB}$$

and hence deduce the sine rule

$$\frac{\sin A}{a} = \frac{\sin B}{b} = \frac{\sin C}{c}$$

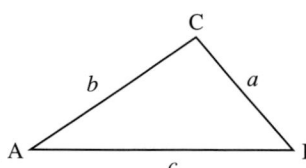

Figure 4.44
Sine rule: Section 2.6.1.

33 Prove that

$$(a - b) \times (a + b) = 2(a \times b)$$

and interpret geometrically.

34 The points A, B and C have coordinates $(1, -1, 2)$, $(9, 0, 8)$ and $(5, 0, 5)$ relative to rectangular cartesian axes. Find

(a) the vectors $\overrightarrow{AB}$ and $\overrightarrow{AC}$

(b) a unit vector perpendicular to the triangle ABC

(c) the area of the triangle ABC.

35 Use the definitions of the scalar and vector products to show that

$$|a \cdot b|^2 + |a \times b|^2 = a^2 b^2$$

36 If a, b and c are three vectors such that $a + b + c = 0$, prove that

$$a \times b = b \times c = c \times a$$

and interpret geometrically.

37 A rigid body is rotating with angular velocity 6 rad s^{-1} about an axis in the direction of the vector $(3, -2, 1)$ and passing through the point A$(3, -2, 5)$. Find the linear velocity of the point P$(3, -2, 1)$ on the body.

38 A force of 4 units acts through the point P$(4, -1, 2)$ in the direction of the vector $(2, -1, 4)$. Find its moment about the point A$(3, -1, 4)$.

39 The moment of a force F acting at a point P about a point O is defined to be a vector M perpendicular to the plane containing F and the point O such that $|M| = p|F|$, where p is the perpendicular distance from O to the line of action of r. Figure 4.45 illustrates such a force F. Show that the perpendicular distance from O to the line of action

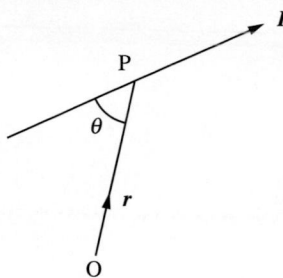

Figure 4.45
Moment of force F about O.

of F is $|r|\sin\theta$, where r is the position vector of P. Hence deduce that $M = r \times F$. Show that the moment of F about O is the same for any point P on the line of action of F.

Forces $(1, 0, 0)$, $(1, 2, 0)$ and $(1, 2, 3)$ act through the points $(1, 1, 1)$, $(0, 1, 1)$ and $(0, 0, 1)$ respectively:

(a) find the moment of each force about the origin;

(b) find the moment of each force about the point $(1, 1, 1)$;

(c) find the total moment of the three forces about the point $(1, 1, 1)$.

40 Find a unit vector perpendicular to the plane of the two vectors $(2, -1, 1)$ and $(3, 4, -1)$. What is the sine of the angle between these two vectors?

41 Prove that the shortest distance of a point P from the line through the points A and B is

$$\frac{|\overrightarrow{AP} \times \overrightarrow{AB}|}{|\overrightarrow{AB}|}$$

A satellite is stationary at P(2, 5, 4) and a warning signal is activated if any object comes within a distance of 3 units. Determine whether a rocket moving in a straight line passing through A(1, 5, 2) and B(3, −1, 5) activates the warning signal.

42 The position vector r, with respect to a given origin O, of a charged particle of mass m and charge e at time t is given by

$$r = \left(\frac{Et}{B} + a\sin(\omega t)\right)i + a\cos(\omega t)j + ctk$$

where E, B, a and ω are constants. The corresponding velocity and acceleration are

$$v = \left(\frac{E}{B} + a\omega\cos(\omega t)\right)i - a\omega\sin(\omega t)j + ck$$

$$f = -a\omega^2\sin(\omega t)i - a\omega^2\cos(\omega t)j$$

For the case when $B = Bk$, show that the equation of motion

$$mf = e(Ej + v \times B)$$

is satisfied provided ω is chosen suitably.

4.2.11 Triple products

In Example 4.24 products of several vectors were computed: the product $(a \times b) \cdot c$ is called the **triple scalar product** and the product $(a \times b) \times c$ is called the **triple vector product**.

Triple scalar product

The triple scalar product is of interest because of its geometrical interpretation. Looking at Figure 4.46, we see that

$$a \times b = |a||b|\sin\theta k$$

$$= (\text{area of the parallelogram OACB})k$$

Thus, by definition,

$$(a \times b) \cdot c = (\text{area of OACB})k \cdot c$$

$$= (\text{area of OACB})|k||c|\cos\phi$$

$$= (\text{area of OACB})h \quad (\text{where } h \text{ is the height of the parallelepiped})$$

$$= \text{volume of the parallelepiped}$$

Figure 4.46
Triple scalar product as the volume of a parallelepiped.

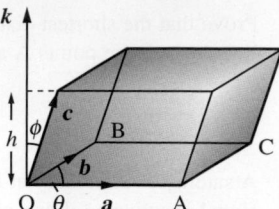

Considering $(\boldsymbol{a} \times \boldsymbol{b}) \cdot \boldsymbol{c}$ to be the volume of the parallelepiped mounted on $\boldsymbol{a}, \boldsymbol{b}, \boldsymbol{c}$ has several useful consequences.

(a) If two of the vectors $\boldsymbol{a}, \boldsymbol{b}$ and $\boldsymbol{c}$ are parallel then $(\boldsymbol{a} \times \boldsymbol{b}) \cdot \boldsymbol{c} = 0$. This follows immediately since the parallelepiped collapses to a plane and has zero volume. In particular,

$$(\boldsymbol{a} \times \boldsymbol{b}) \cdot \boldsymbol{a} = 0 \quad \text{and} \quad (\boldsymbol{a} \times \boldsymbol{b}) \cdot \boldsymbol{b} = 0$$

(b) If the three vectors are coplanar then $(\boldsymbol{a} \times \boldsymbol{b}) \cdot \boldsymbol{c} = 0$. The same reasoning as in (a) gives this result.

(c) If $(\boldsymbol{a} \times \boldsymbol{b}) \cdot \boldsymbol{c} = 0$ then either $\boldsymbol{a} = 0$ or $\boldsymbol{b} = 0$ or $\boldsymbol{c} = 0$ or two of the vectors are parallel or the three vectors are coplanar.

(d) In the triple scalar product the dot $\cdot$ and the cross $\times$ can be interchanged:

$$(\boldsymbol{a} \times \boldsymbol{b}) \cdot \boldsymbol{c} = \boldsymbol{a} \cdot (\boldsymbol{b} \times \boldsymbol{c})$$

since it is easily checked that they measure the same volume mounted on $\boldsymbol{a}, \boldsymbol{b}, \boldsymbol{c}$. If we retain the same cyclic order of the three vectors then we obtain

$$\boldsymbol{a} \cdot (\boldsymbol{b} \times \boldsymbol{c}) = \boldsymbol{b} \cdot (\boldsymbol{c} \times \boldsymbol{a}) = \boldsymbol{c} \cdot (\boldsymbol{a} \times \boldsymbol{b}) \tag{4.11}$$

(e) In cartesian form the scalar triple product can be written as the determinant

$$\boldsymbol{a} \cdot (\boldsymbol{b} \times \boldsymbol{c}) = \begin{vmatrix} a_1 & a_2 & a_3 \\ b_1 & b_2 & b_3 \\ c_1 & c_2 & c_3 \end{vmatrix} \tag{4.12}$$

$$= a_1 b_2 c_3 - a_1 b_3 c_2 - a_2 b_1 c_3 + a_2 b_3 c_1 + a_3 b_1 c_2 - a_3 b_2 c_1$$

Example 4.31   Find λ so that $\boldsymbol{a} = (2, -1, 1), \boldsymbol{b} = (1, 2, -3)$ and $\boldsymbol{c} = (3, \lambda, 5)$ are coplanar.

Solution   None of these vectors are zero or parallel, so by property (b) the three vectors are coplanar if $(\boldsymbol{a} \times \boldsymbol{b}) \cdot \boldsymbol{c} = 0$. Now

$$\boldsymbol{a} \times \boldsymbol{b} = (1, 7, 5)$$

so

$$(\boldsymbol{a} \times \boldsymbol{b}) \cdot \boldsymbol{c} = 3 + 7\lambda + 25$$

This will be zero, and the three vectors coplanar, when $\lambda = -4$.

Example 4.32

In a triangle OAB the sides $\overrightarrow{OA} = \boldsymbol{a}$ and $\overrightarrow{OB} = \boldsymbol{b}$ are given. Find the point P, with $\boldsymbol{c} = \overrightarrow{OP}$, where the perpendicular bisectors of the two sides intersect. Hence prove that the perpendicular bisectors of the sides of a triangle meet at a point.

Solution Let $\hat{\boldsymbol{k}}$ be the unit vector perpendicular to the plane of the triangle; the situation is illustrated in Figure 4.47.

Figure 4.47
Perpendicular bisectors in Example 4.32.

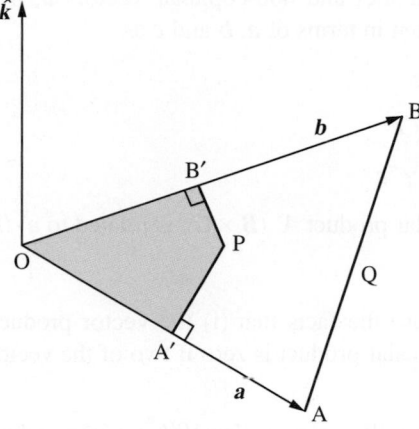

Now

$$\overrightarrow{OP} = \overrightarrow{OA'} + \overrightarrow{A'P} = \tfrac{1}{2}\boldsymbol{a} + \alpha\hat{\boldsymbol{k}} \times \boldsymbol{a}$$

for some α, since the vector $\hat{\boldsymbol{k}} \times \boldsymbol{a}$ is in the direction perpendicular to $\boldsymbol{a}$. Similarly

$$\overrightarrow{OP} = \overrightarrow{OB'} + \overrightarrow{B'P} = \tfrac{1}{2}\boldsymbol{b} + \beta\hat{\boldsymbol{k}} \times \boldsymbol{b}$$

Subtracting these two equations

$$\tfrac{1}{2}\boldsymbol{a} + \alpha\hat{\boldsymbol{k}} \times \boldsymbol{a} = \tfrac{1}{2}\boldsymbol{b} + \beta\hat{\boldsymbol{k}} \times \boldsymbol{b}$$

Take the dot product of this equation with $\boldsymbol{b}$, which eliminates the final term, since $\boldsymbol{b} \cdot (\hat{\boldsymbol{k}} \times \boldsymbol{b}) = 0$, and gives

$$\tfrac{1}{2}\boldsymbol{b} \cdot (\boldsymbol{b} - \boldsymbol{a}) = \alpha\boldsymbol{b}\cdot(\hat{\boldsymbol{k}} \times \boldsymbol{a})$$

Hence α has been computed in terms of the known data, so assuming $\boldsymbol{b} \cdot (\hat{\boldsymbol{k}} \times \boldsymbol{a}) \neq 0$

$$\overrightarrow{OP} = \tfrac{1}{2}\boldsymbol{a} + \frac{\tfrac{1}{2}\boldsymbol{b} \cdot (\boldsymbol{b} - \boldsymbol{a})}{\boldsymbol{b} \cdot (\hat{\boldsymbol{k}} \times \boldsymbol{a})}\hat{\boldsymbol{k}} \times \boldsymbol{a}$$

We now need to check that PQ is perpendicular to AB:

$$\overrightarrow{AB} = \overrightarrow{OB} - \overrightarrow{OA} = \boldsymbol{b} - \boldsymbol{a}$$

and

$$\overrightarrow{PQ} = \overrightarrow{OQ} - \overrightarrow{OP} = \tfrac{1}{2}(\boldsymbol{a} + \boldsymbol{b}) - \tfrac{1}{2}\boldsymbol{a} - \frac{\tfrac{1}{2}\boldsymbol{b} \cdot (\boldsymbol{b} - \boldsymbol{a})}{\boldsymbol{b} \cdot (\hat{\boldsymbol{k}} \times \boldsymbol{a})}\hat{\boldsymbol{k}} \times \boldsymbol{a}$$

Now take the dot product of these two vectors

$$\left[\frac{1}{2}\boldsymbol{b} - \frac{\frac{1}{2}\boldsymbol{b}\cdot(\boldsymbol{b}-\boldsymbol{a})}{\boldsymbol{b}\cdot(\hat{\boldsymbol{k}}\times\boldsymbol{a})}\hat{\boldsymbol{k}}\times\boldsymbol{a}\right]\cdot(\boldsymbol{b}-\boldsymbol{a}) = \frac{1}{2}\boldsymbol{b}\cdot(\boldsymbol{b}-\boldsymbol{a}) - \frac{\frac{1}{2}\boldsymbol{b}\cdot(\boldsymbol{b}-\boldsymbol{a})}{\boldsymbol{b}\cdot(\hat{\boldsymbol{k}}\times\boldsymbol{a})}\boldsymbol{b}\cdot(\hat{\boldsymbol{k}}\times\boldsymbol{a}) = 0$$

Since neither $\overrightarrow{PQ}$ nor $\overrightarrow{AB}$ is zero, the two vectors must therefore be perpendicular. Hence the three perpendicular bisectors of the sides of a triangle meet at a point.

Example 4.33

Three non-zero, non-parallel and non-coplanar vectors $\boldsymbol{a}$, $\boldsymbol{b}$ and $\boldsymbol{c}$ are given. Three further vectors are written in terms of $\boldsymbol{a}$, $\boldsymbol{b}$ and $\boldsymbol{c}$ as

$$A = \alpha\boldsymbol{a} + \beta\boldsymbol{b} + \gamma\boldsymbol{c}$$

$$B = \alpha'\boldsymbol{a} + \beta'\boldsymbol{b} + \gamma'\boldsymbol{c}$$

$$C = \alpha''\boldsymbol{a} + \beta''\boldsymbol{b} + \gamma''\boldsymbol{c}$$

Find how the triple scalar product $A \cdot (B \times C)$ is related to $\boldsymbol{a} \cdot (\boldsymbol{b} \times \boldsymbol{c})$.

Solution

To find the result we use the facts that (i) the vector product of identical vectors is zero and (ii) the triple scalar product is zero if two of the vectors in the product are the same. Now

$$\begin{aligned}
A \cdot (B \times C) &= (\alpha\boldsymbol{a} + \beta\boldsymbol{b} + \gamma\boldsymbol{c}) \cdot [(\alpha'\boldsymbol{a} + \beta'\boldsymbol{b} + \gamma'\boldsymbol{c}) \times (\alpha''\boldsymbol{a} + \beta''\boldsymbol{b} + \gamma''\boldsymbol{c})] \\
&= (\alpha\boldsymbol{a} + \beta\boldsymbol{b} + \gamma\boldsymbol{c}) \cdot [\alpha'\beta''\boldsymbol{a} \times \boldsymbol{b} + \alpha'\gamma''\boldsymbol{a} \times \boldsymbol{c} + \beta'\alpha''\boldsymbol{b} \times \boldsymbol{a} \\
&\qquad\qquad + \beta'\gamma''\boldsymbol{b} \times \boldsymbol{c} + \gamma'\alpha''\boldsymbol{c} \times \boldsymbol{a} + \gamma'\beta''\boldsymbol{c} \times \boldsymbol{b}] \\
&= (\alpha\boldsymbol{a} + \beta\boldsymbol{b} + \gamma\boldsymbol{c}) \cdot [(\alpha'\beta'' - \beta'\alpha'')\boldsymbol{a} \times \boldsymbol{b} + (\beta'\gamma'' - \gamma'\beta'')\boldsymbol{b} \times \boldsymbol{c} \\
&\qquad\qquad + (\gamma'\alpha'' - \alpha'\gamma'')\boldsymbol{c} \times \boldsymbol{a}] \\
&= \gamma(\alpha'\beta'' - \beta'\alpha'')\boldsymbol{c} \cdot \boldsymbol{a} \times \boldsymbol{b} + \alpha(\beta'\gamma'' - \gamma'\beta'')\boldsymbol{a} \cdot \boldsymbol{b} \times \boldsymbol{c} \\
&\quad + \beta(\gamma'\alpha'' - \alpha'\gamma'')\boldsymbol{b} \cdot \boldsymbol{c} \times \boldsymbol{a} \\
&= (\boldsymbol{a} \cdot \boldsymbol{b} \times \boldsymbol{c})[\alpha(\beta'\gamma'' - \gamma'\beta'') + \beta(\gamma'\alpha'' - \alpha'\gamma'') + \gamma(\alpha'\beta'' - \beta'\alpha'')]
\end{aligned}$$

The result can be written most conveniently in determinant form (see Section 5.3 of the next chapter) as

$$A \cdot (B \times C) = \begin{vmatrix} \alpha & \beta & \gamma \\ \alpha' & \beta' & \gamma' \\ \alpha'' & \beta'' & \gamma'' \end{vmatrix} (\boldsymbol{a} \cdot \boldsymbol{b} \times \boldsymbol{c})$$

Triple vector product

For the triple vector product we shall show in general that

$$(\boldsymbol{a} \times \boldsymbol{b}) \times \boldsymbol{c} = (\boldsymbol{a} \cdot \boldsymbol{c})\boldsymbol{b} - (\boldsymbol{b} \cdot \boldsymbol{c})\boldsymbol{a} \tag{4.13}$$

as suggested in Example 4.24. We have from (4.9)

$$\boldsymbol{a} \times \boldsymbol{b} = (a_2b_3 - a_3b_2, \, a_3b_1 - a_1b_3, \, a_1b_2 - a_2b_1)$$

and hence

$$(\boldsymbol{a} \times \boldsymbol{b}) \times \boldsymbol{c} = ((a_3b_1 - a_1b_3)c_3 - (a_1b_2 - a_2b_1)c_2,$$
$$(a_1b_2 - a_2b_1)c_1 - (a_2b_3 - a_3b_2)c_3,$$
$$(a_2b_3 - a_3b_2)c_2 - (a_3b_1 - a_1b_3)c_1)$$

The first component of this vector is

$$a_3c_3b_1 - b_3c_3a_1 - b_2c_2a_1 + a_2c_2b_1 = (a_1c_1 + a_2c_2 + a_3c_3)b_1 - (b_1c_1 + b_2c_2 + b_3c_3)a_1$$
$$= (\boldsymbol{a} \cdot \boldsymbol{c})b_1 - (\boldsymbol{b} \cdot \boldsymbol{c})a_1$$

Treating the second and third components similarly, we find

$$(\boldsymbol{a} \times \boldsymbol{b}) \times \boldsymbol{c} = ((\boldsymbol{a} \cdot \boldsymbol{c})b_1 - (\boldsymbol{b} \cdot \boldsymbol{c})a_1, \, (\boldsymbol{a} \cdot \boldsymbol{c})b_2 - (\boldsymbol{b} \cdot \boldsymbol{c})a_2, \, (\boldsymbol{a} \cdot \boldsymbol{c})b_3 - (\boldsymbol{b} \cdot \boldsymbol{c})a_3)$$
$$= (\boldsymbol{a} \cdot \boldsymbol{c})\boldsymbol{b} - (\boldsymbol{b} \cdot \boldsymbol{c})\boldsymbol{a}$$

In a similar way we can show that

$$\boldsymbol{a} \times (\boldsymbol{b} \times \boldsymbol{c}) = (\boldsymbol{a} \cdot \boldsymbol{c})\boldsymbol{b} - (\boldsymbol{a} \cdot \boldsymbol{b})\boldsymbol{c} \qquad\qquad (4.14)$$

We can now see why the associativity of the vector product does not hold in general. The vector in (4.13) is in the plane of $\boldsymbol{b}$ and $\boldsymbol{a}$, while the vector in (4.14) is in the plane of $\boldsymbol{b}$ and $\boldsymbol{c}$; hence they are not in the same planes in general, as we inferred geometrically in Section 4.2.9. Consequently, in general

$$\boldsymbol{a} \times (\boldsymbol{b} \times \boldsymbol{c}) \neq (\boldsymbol{a} \times \boldsymbol{b}) \times \boldsymbol{c}$$

so use of brackets is essential.

Example 4.34 If $\boldsymbol{a} = (3, -2, 1)$, $\boldsymbol{b} = (-1, 3, 4)$ and $\boldsymbol{c} = (2, 1, -3)$, confirm that

$$\boldsymbol{a} \times (\boldsymbol{b} \times \boldsymbol{c}) = (\boldsymbol{a} \cdot \boldsymbol{c})\boldsymbol{b} - (\boldsymbol{a} \cdot \boldsymbol{b})\boldsymbol{c}$$

Solution

$$\boldsymbol{b} \times \boldsymbol{c} = \begin{vmatrix} \boldsymbol{i} & \boldsymbol{j} & \boldsymbol{k} \\ -1 & 3 & 4 \\ 2 & 1 & -3 \end{vmatrix} = (-13, 5, -7)$$

$$\boldsymbol{a} \times (\boldsymbol{b} \times \boldsymbol{c}) = \begin{vmatrix} \boldsymbol{i} & \boldsymbol{j} & \boldsymbol{k} \\ 3 & -2 & 1 \\ -13 & 5 & -7 \end{vmatrix} = (9, 8, -11)$$

$$(a \cdot c)b - (a \cdot b)c = [(3)(2) + (-2)(1) + (1)(-3)](-1, 3, 4)$$
$$- [(3)(-1) + (-2)(3) + (1)(4)](2, 1, -3)$$
$$= (-1, 3, 4) + 5(2, 1, -3)$$
$$= (9, 8, -11)$$

thus confirming the result

$$a \times (b \times c) = (a \cdot c)b - (a \cdot b)c$$

Example 4.35

Verify that $a \times (b \times c) \neq (a \times b) \times c$ for the three vectors $a = (1, 0, 0)$, $b = (-1, 2, 0)$ and $c = (1, 1, 1)$.

Solution

Evaluate the cross products in turn:

$$b \times c = (-1, 2, 0) \times (1, 1, 1) = (2, 1, -3)$$

and therefore

$$a \times (b \times c) = (1, 0, 0) \times (2, 1, -3) = (0, 3, 1)$$

Similarly for the right-hand side

$$a \times b = (1, 0, 0) \times (-1, 2, 0) = (0, 0, 2)$$

and hence

$$(a \times b) \times c = (0, 0, 2) \times (1, 1, 1) = (-2, 2, 0)$$

Clearly for these three vectors $a \times (b \times c) \neq (a \times b) \times c$.

Example 4.36

The vectors a, b and c and the scalar p satisfy the equations

$$a \cdot b = p \quad \text{and} \quad a \times b = c$$

and a is not parallel to b. Solve for a in terms of the other quantities and give a geometrical interpretation of the result.

Solution

First evaluate the cross product of the second equation with b

$$b \times (a \times b) = b \times c$$

gives

$$(b \cdot b)a - (b \cdot a)b = b \times c$$

and hence, using $a \cdot b = p$, and collecting the terms

$$a = \frac{pb + b \times c}{|b|^2}$$

Since $b \times c$ is in the plane of a and b, any vector in the plane can be written as a linear combination of b and $b \times c$. The expression for a gives the values of the coefficients in the linear combination.

4.2.12   Exercises

Check your answers using MATLAB or MAPLE whenever possible.

43 Find the volume of the parallelepiped whose edges are represented by the vectors $(2, -3, 4)$, $(1, 3, -1)$, $(3, -1, 2)$.

44 Prove that the vectors $(3, 2, -1)$, $(5, -7, 3)$ and $(11, -3, 1)$ are coplanar.

45 Find the constant λ such that the three vectors $(3, 2, -1)$, $(1, -1, 3)$ and $(2, -3, \lambda)$ are coplanar.

46 Prove that the four points having position vectors $(2, 1, 0)$, $(2, -2, -2)$, $(7, -3, -1)$ and $(13, 3, 5)$ are coplanar.

47 Given $p = (1, 4, 1)$, $q = (2, 1, -1)$ and $r = (1, -3, 2)$, find

(a) a unit vector perpendicular to the plane containing p and q,

(b) a unit vector in the plane containing $p \times q$ and $p \times r$ that has zero x component.

48 Show that if a is any vector and $\hat{u}$ any unit vector then

$$a = (a \cdot \hat{u})\hat{u} + \hat{u} \times (a \times \hat{u})$$

and draw a diagram to illustrate this relation geometrically.

The vector $(3, -2, 6)$ is resolved into two vectors along and perpendicular to the line whose direction cosines are proportional to $(1, 1, 1)$. Find these vectors.

49 Three vectors u, v, w are expressed in terms of the three vectors l, m, n in the form

$$u = u_1 l + u_2 m + u_3 n$$

$$v = v_1 l + v_2 m + v_3 n$$

$$w = w_1 l + w_2 m + w_3 n$$

Show that

$$u \cdot (v \times w) = \lambda l \cdot (m \times n)$$

and evaluate λ.

50 Forces $F_1, F_2, \dots, F_n$ act at the points $r_1, r_2, \dots, r_n$ respectively. The total force and the total moment about the origin O are

$$F = \sum F_i \quad \text{and} \quad G = \sum r_i \times F_i$$

Show that for any other origin O′ the moment is given by

$$G' = G + \overrightarrow{O'O} \times F$$

If O′ lies on the line

$$\overrightarrow{OO'} = r = \alpha(F \times G) + tF$$

find the constant α that ensures that G' is parallel to F. This line is called the central axis of the system of forces.

51 Extended exercise on products of four vectors.

(a) Use (4.11) to show

$$(a \times b) \cdot (c \times d) = [(a \times b) \times c] \cdot d$$

and use (4.13) to simplify the expression on the right-hand side.

(b) Use (4.13) to show that

$$(a \times b) \times (a \times c) = [a \cdot (a \times c)]b$$
$$- [b \cdot (a \times c)]a$$

and show that the right-hand side can be simplified to

$$[(a \times b) \cdot c]a$$

(c) Use (4.14) to show that

$$a \times [b \times (a \times c)]$$
$$= a \times [(b \cdot c)a - (b \cdot a)c]$$

and simplify the right-hand side further. Note that the product is different from the result in (b), verifying that the position of the brackets matters in cross products.

(d) Use the result in (a) to show that

$$(l \times m) \cdot (l \times n) = l^2(m \cdot n) - (l \cdot m)(l \cdot n)$$

Take l, m and n to be unit vectors along the sides of a regular tetrahedron. Deduce that the angle between two faces of the tetrahedron is $\cos^{-1}\frac{1}{3}$.

The vector treatment of the geometry of lines and planes

4.3.1 Vector equation of a line

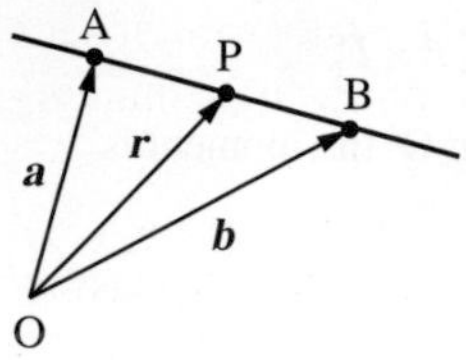

Figure 4.48
Line AB in terms of
$r = \overrightarrow{OP}$.

Take an arbitrary origin O and let $\overrightarrow{OA} = a$, $\overrightarrow{OB} = b$ and $\overrightarrow{OP} = r$, as in Figure 4.48. If P is any point on the line then

$$\overrightarrow{OP} = \overrightarrow{OA} + \overrightarrow{AP}, \quad \text{by the triangle law}$$

giving

$$r = a + t\overrightarrow{AB} \quad (\text{since } \overrightarrow{AP} \text{ is a multiple of } \overrightarrow{AB})$$

$$= a + t(b - a) \quad (\text{since } a + \overrightarrow{AB} = b)$$

Thus the equation of the line is

$$r = (1 - t)a + tb \tag{4.15}$$

As t varies from $-\infty$ to $+\infty$, the point P sweeps along the line, with $t = 0$ corresponding to point A and $t = 1$ to point B.

Since $\overrightarrow{OP} = \overrightarrow{OA} + \overrightarrow{AP} = \overrightarrow{OA} + t\overrightarrow{AB}$, we have $r = a + t(b - a)$. If we write $c = b - a$ then we have an alternative intepretation of a line through A in the direction c:

$$r = a + tc \tag{4.16}$$

The cartesian or component form of this equation is

$$\frac{x - a_1}{c_1} = \frac{y - a_2}{c_2} = \frac{z - a_3}{c_3} (= t) \tag{4.17}$$

where $a = (a_1, a_2, a_3)$ and $c = (c_1, c_2, c_3)$. Alternatively the cartesian equation of (4.15) may be written in the form

$$\frac{x - a_1}{b_1 - a_1} = \frac{y - a_2}{b_2 - a_2} = \frac{z - a_3}{b_3 - a_3} (= t)$$

where $a = (a_1, a_2, a_3)$ and $b = (b_1, b_2, b_3)$ are two points on the line. If any of the denominators is zero, then both forms of the equation of a line are interpreted as the corresponding numerator is zero.

Example 4.37   Find the equation of the lines L_1 through the points $(0, 1, 0)$ and $(1, 3, -1)$ and L_2 through $(1, 1, 1)$ and $(-1, -1, 1)$. Do the two lines intersect and, if so, at what point?

Solution   From (4.15) L_1 has the equation

$$r = (0, 1 - t, 0) + (t, 3t, -t) = (t, 1 + 2t, -t)$$

and L_2 has the equation

$$r = (1 - s, 1 - s, 1 - s) + (-s, -s, s) = (1 - 2s, 1 - 2s, 1)$$

Note that the cartesian equation of L_2 reduces to $x = y$; $z = 1$. The two lines intersect if it is possible to find s and t such that

$$t = 1 - 2s, \quad 1 + 2t = 1 - 2s, \quad -t = 1$$

Solving two of these equations will give the values of s and t. If these values satisfy the remaining equation then the lines intersect; however, if they do not satisfy the remaining equation then the lines do not intersect. In this particular case, the third equation gives $t = -1$ and the first equation $s = 1$. Putting these values into the second equation the left-hand side equals -1 and the right-hand side equals -1 so the equations are all satisfied and therefore the lines intersect. Substituting back into either equation, the point of intersection is $(-1, -1, 1)$.

Example 4.38   The position vectors of the points A and B are

$$(1, 4, 6) \quad \text{and} \quad (3, 5, 7)$$

Find the vector equation of the line AB and find the points where the line intersects the coordinate planes.

Solution   The line has equation

$$r = (1, 4, 6) + t(2, 1, 1)$$

or in components

$$x = 1 + 2t$$

$$y = 4 + t$$

$$z = 6 + t$$

Thus the line meets the y–z plane when $x = 0$ and hence $t = -\frac{1}{2}$ and the point of intersection with the plane is $(0, \frac{7}{2}, \frac{11}{2})$.

The line meets the z–x plane when $y = 0$ and hence $t = -4$ and the point of intersection with the plane is $(-7, 0, 2)$.

The line meets the x–y plane when $z = 0$ and hence $t = -6$ and the point of intersection with the plane is $(-11, -2, 0)$.

Example 4.39   The line L_1 passes through the points with position vectors

$$(5, 1, 7) \quad \text{and} \quad (6, 0, 8)$$

and the line L_2 passes through the points with position vectors

$$(3, 1, 3) \quad \text{and} \quad (-1, 3, \alpha)$$

Find the value of α for which the two lines L_1 and L_2 intersect.

Solution Using the vector form:
From (4.15) the equations of the two lines can be written in vector form as

$$L_1: \quad \boldsymbol{r} = (5, 1, 7) + t(1, -1, 1)$$

$$L_2: \quad \boldsymbol{r} = (3, 1, 3) + s(-4, 2, \alpha - 3)$$

These two lines intersect if t, s and α can be chosen so that the two vectors are equal, that is they have the same components. Thus

$$5 + t = 3 - 4s$$

$$1 - t = 1 + 2s$$

$$7 + t = 3 + s(\alpha - 3)$$

The first two of these equations are simultaneous equations for t and s. Solving gives $t = 2$ and $s = -1$. Putting these values into the third equation

$$9 = 3 - (\alpha - 3) \Rightarrow \alpha = -3$$

and it can be checked that the point of intersection is $(7, -1, 9)$.

Using the cartesian form:
Equation (4.17) gives the equations of the lines as

$$L_1: \quad \frac{x - 5}{6 - 5} = \frac{y - 1}{0 - 1} = \frac{z - 7}{8 - 7}$$

$$L_2: \quad \frac{x - 3}{-1 - 3} = \frac{y - 1}{3 - 1} = \frac{z - 3}{\alpha - 3}$$

The two equations for x and y are

$$x - 5 = 1 - y$$

$$\tfrac{1}{4}(3 - x) = \tfrac{1}{2}(y - 1)$$

and are solved to give $x = 7$ and $y = -1$. Putting in these values, the equations for z and α become

$$z - 7 = 2$$

$$\frac{z - 3}{\alpha - 3} = -1$$

which give $z = 9$ and $\alpha = -3$.

Example 4.40 A tracking station observes an aeroplane at two successive times to be

$$(-500, 0, 1000) \quad \text{and} \quad (400, 400, 1050)$$

relative to axes x in an easterly direction, y in a northerly direction and z vertically upwards, with distances in metres. Find the equation of the path of the aeroplane. Control advises the aeroplane to change course from its present position to level flight at the current height and turn easterly through an angle of 90°; what is the equation of the new path?

Figure 4.49
Path of aeroplane in
Example 4.40.

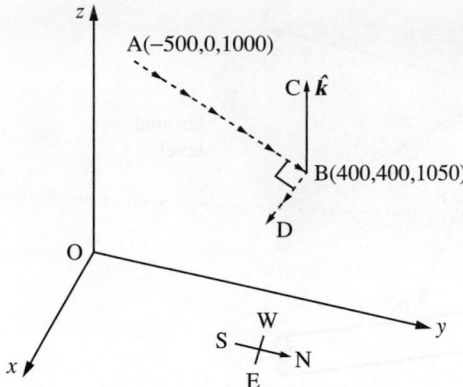

Solution The situation is illustrated in Figure 4.49. The equation of the path of the aeroplane is

$$r = (-500, 0, 1000) + t(900, 400, 50)$$

The new path starts at the point $(400, 400, 1050)$. The vector $\overrightarrow{AB} \times \hat{k}$ is a vector in the direction $\overrightarrow{BD}$ which is perpendicular to $\hat{k}$, and is therefore horizontal, and at $90°$ to AB in the easterly direction. Thus we have $90°$ turn to horizontal flight. Since

$$(900, 400, 0) \times k = (400, -900, 0)$$

the new path is

$$r = (400, 400, 1050) + s(400, -900, 0)$$

Equating the components

$$x = 400 + 400s$$
$$y = 400 - 900s$$
$$z = 1050$$

In cartesian coordinates the equations are

$$9x + 4y = 5200$$
$$z = 1050$$

Example 4.41 It is necessary to drill to an underground pipeline in order to undertake repairs, so it is decided to aim for the nearest point from the measuring point. Relative to axes x, y in the horizontal ground and with z vertically downwards, remote measuring instruments locate two points on the pipeline at

$$(20, 20, 30) \quad \text{and} \quad (0, 15, 32)$$

with distances in metres. Find the nearest point on the pipeline from the origin O.

Solution The situation is illustrated in Figure 4.50. The direction of the pipeline is

$$d = (0, 15, 32) - (20, 20, 30) = (-20, -5, 2).$$

Thus any point on the pipeline will have position vector

$$r = (20, 20, 30) + t(-20, -5, 2)$$

Figure 4.50
Pipeline of
Example 4.41

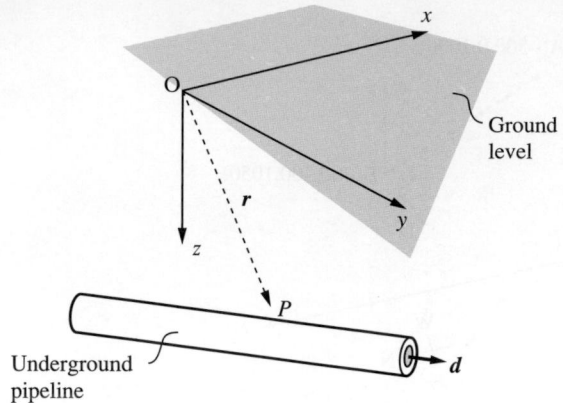

Ground
level

r

z

P

Underground
pipeline

d

for some t. Note that this is just the equation of the line given in (4.15). At the shortest distance from O to the pipeline the vector $r = \overrightarrow{OP}$ is perpendicular to d, so $r \cdot d = 0$ gives the required condition to evaluate t. Thus

$$(-20, -5, 2) \cdot [(20, 20, 30) + t(-20, -5, 2)] = 0$$

and hence $-440 + 429t = 0$. Putting this value back into r gives

$$r = (-0.51, 14.87, 32.05)$$

Note that the value of t is close to 1 so the optimum point is not far from the second of the points located.

Example 4.42 Find the shortest distance between the two skew lines

$$\frac{x}{3} = \frac{y-9}{-1} = \frac{z-2}{1} \quad \text{and} \quad \frac{x+6}{-3} = \frac{y+5}{2} = \frac{z-10}{4}$$

Also determine the equation of the common perpendicular. (Note that two lines are said to be skew if they do not intersect and are not parallel.)

Solution In vector form the equations of the lines are

$$r = (0, 9, 2) + t(3, -1, 1)$$

and

$$r = (-6, -5, 10) + s(-3, 2, 4)$$

The shortest distance between the two lines will be their common perpendicular, see Figure 4.51. Let P_1 and P_2 be the end points of the common perpendicular, having position vectors r_1 and r_2 respectively, where

$$r_1 = (0, 9, 2) + t_1(3, -1, 1)$$

and

$$r_2 = (-6, -5, 10) + t_2(-3, 2, 4)$$

Figure 4.51
Skew lines in
Example 4.42.

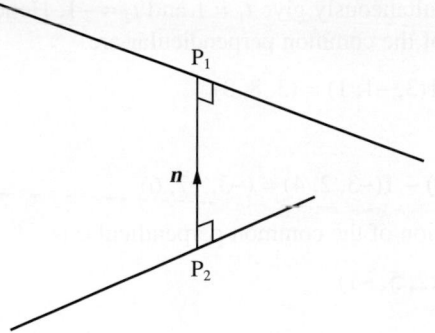

Then the vector $\overrightarrow{P_2P_1}$ is given by

$$\overrightarrow{P_2P_1} = r_1 - r_2 = (6, 14, -8) + t_1(3, -1, 1) - t_2(-3, 2, 4) \qquad (4.18)$$

Since $(3, -1, 1)$ and $(-3, 2, 4)$ are vectors in the direction of each of the lines, it follows that a vector n perpendicular to both lines is

$$n = (-3, 2, 4) \times (3, -1, 1) = (6, 15, -3)$$

So a unit vector perpendicular to both lines is

$$\hat{n} = (6, 15, -3)/\sqrt{270} = (2, 5, -1)/\sqrt{30}$$

Thus we can also express $\overrightarrow{P_2P_1}$ as

$$\overrightarrow{P_2P_1} = d\hat{n}$$

where d is the shortest distance between the two lines.

Equating the two expressions for $\overrightarrow{P_2P_1}$ gives

$$(6, 14, -8) + t_1(3, -1, 1) - t_2(-3, 2, 4) = (2, 5, -1)d/\sqrt{30}$$

Taking the scalar product throughout with the vector $(2, 5, -1)$ gives

$$(6, 14, -8) \cdot (2, 5, -1) + t_1(3, -1, 1) \cdot (2, 5, -1) - t_2(-3, 2, 4) \cdot (2, 5, -1)$$

$$= (2, 5, -1) \cdot (2, 5, -1)d/\sqrt{30}$$

which reduces to

$$90 + 0t_1 + 0t_2 = 30d/\sqrt{30}$$

giving the shortest distance between the two lines as

$$d = 3\sqrt{30}$$

To obtain the equation of the common perpendicular, we need to find the coordinates of either P_1 or P_2 – and to achieve this we need to find the value of either t_1 or t_2. We therefore take the scalar product of (4.18) with $(3, -1, 1)$ and $(-3, 2, 4)$ in turn, giving respectively

$$11t_1 + 7t_2 = 4$$

and

$$-7t_1 - 29t_2 = 22$$

which on solving simultaneously give $t_1 = 1$ and $t_2 = -1$. Hence the coordinates of the end points P_1 and P_2 of the common perpendicular are

$$\mathbf{r}_1 = (0, 9, 2) + 1(3, -1, 1) = (3, 8, 3)$$

and

$$\mathbf{r}_2 = (-6, -5, 10) - 1(-3, 2, 4) = (-3, -7, 6)$$

From (4.16) the equation of the common perpendicular is

$$\mathbf{r} = (3, 8, 3) + s(2, 5, -1)$$

or in cartesian form

$$\frac{x - 3}{2} = \frac{y - 8}{5} = \frac{z - 3}{-1} = s$$

MAPLE contains a geometry package which takes a bit of time to master but which can solve many coordinate geometry problems. For the current problem the code is given: note that printing has been largely suppressed, but replace ':' by ';' at the end of statements for more information.

```
with (geom3d):
point (A, [0, 9, 2]): v:= [3, -1, 1]: line
(L1, [A, v]): detail (L1);
point (B, [-6, -5, 10]): w:= [-3, 2, 4]: line
(L2, [B, w]): detail (L2);
distance (L1, L2);              (gives result 3√30 in the text)
z:= Equation (L1, t): y:= Equation (L2, s):
with (linalg):
m:= innerprod (z - y, v): n:= innerprod (z - y, w):
solve ({m, n}, {s, t});        (gives solution t = 1 and s = −1)
point (P, eval (z, t = 1)): point (Q, eval (y, s = -1)):
line (L3, [P, Q]): detail (L3);  (gives the required equation of
                                  the common perpendicular)
```

Example 4.43   A box with an open top and unit side length is observed from the direction (a, b, c) as in Figure 4.52. Determine the part of OC that is visible.

Solution   The line or ray through $Q(0, 0, \alpha)$ parallel to the line of sight has the equation

$$\mathbf{r} = (0, 0, \alpha) + t(a, b, c)$$

where $0 \leq \alpha \leq 1$ to ensure that Q lies between O and C. The line RS passes through $R(1, 0, 1)$ and is in the direction $(0, 1, 0)$ so from (4.16) it has the equation

$$\mathbf{r} = (1, 0, 1) + s(0, 1, 0)$$

The ray that intersects RS must therefore satisfy

$$ta = 1, \quad tb = s, \quad \alpha = 1 - \frac{c}{a}$$

Figure 4.52
Looking for hidden
lines in Example 4.43.

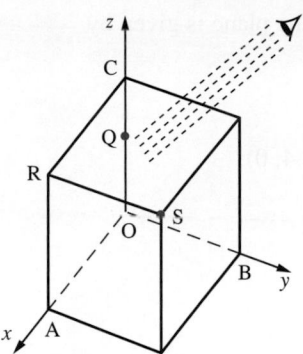

Note that if $c = 0$ then we are looking parallel to the open top and can only see the point C. If $c < 0$ then we are looking up at the box; since $\alpha > 1$, we cannot see any of side OC, so the line is hidden. If, however, $c > a$ then the solution gives α to be negative, so that all of the side OC is visible. For $0 < c < a$ the parameter α lies between 0 and 1, and only part of the line is visible. A similar analysis needs to be performed for the other sides of the open top. Other edges of the box also need to be analysed to check whether or not they are visible to the ray.

4.3.2 Vector equation of a plane

To obtain the equation of a plane, we use the result that the line joining any two points in the plane is perpendicular to the normal to the plane as illustrated in Figure 4.53. The vector $\mathbf{n}$ is perpendicular to the plane, $\mathbf{a}$ is the position vector of a given point A in the plane and $\mathbf{r}$ is the position vector of any point P on the plane. The vector $\overrightarrow{AP} = \mathbf{r} - \mathbf{a}$ is perpendicular to $\mathbf{n}$, and hence

$$(\mathbf{r} - \mathbf{a}) \cdot \mathbf{n} = 0$$

so that

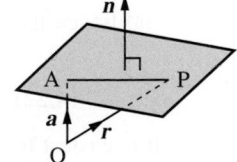

$$\mathbf{r} \cdot \mathbf{n} = \mathbf{a} \cdot \mathbf{n} \quad \text{or} \quad \mathbf{r} \cdot \mathbf{n} = p \qquad (4.19)$$

Figure 4.53
Equation of a plane;
$\mathbf{n}$ is perpendicular to
the plane.

is the general form for the **equation of a plane** with normal $\mathbf{n}$. In the particular case when $\mathbf{n}$ is a unit vector, p in (4.19) represents the perpendicular distance from the origin to the plane. In cartesian form we take $\mathbf{n} = (\alpha, \beta, \gamma)$, and the equation becomes

$$\alpha x + \beta y + \gamma z = p \qquad (4.20)$$

which is just a linear relation between the variables x, y and z.

Example 4.44

Find the equation of the plane through the three points

$$\mathbf{a} = (1, 1, 1), \quad \mathbf{b} = (0, 1, 2) \quad \text{and} \quad \mathbf{c} = (-1, 1, -1)$$

Solution

The vectors $\mathbf{a} - \mathbf{b} = (1, 0, -1)$ and $\mathbf{a} - \mathbf{c} = (2, 0, 2)$ will lie in the plane. The normal $\mathbf{n}$ to the plane can thus be constructed as $(\mathbf{a} - \mathbf{b}) \times (\mathbf{a} - \mathbf{c})$, giving

$$\mathbf{n} = (1, 0, -1) \times (2, 0, 2) = (0, -4, 0)$$

Thus from (4.19) the equation of the plane is given by

$$r \cdot n = a \cdot n$$

or

$$r \cdot (0, -4, 0) = (1, 1, 1) \cdot (0, -4, 0)$$

giving

$$r \cdot (0, -4, 0) = -4$$

In cartesian form

$$(x, y, z) \cdot (0, -4, 0) = -4$$

or simply $y = 1$.

Example 4.45

A metal has a simple cubic lattice structure so that the atoms lie on the lattice points given by

$$r = a(l, m, n)$$

where a is the lattice spacing and l, m, n are integers. The metallurgist requires to identify the points that lie on two lattice planes

LP_1 through $a(0, 0, 0)$, $a(1, 1, 0)$ and $a(0, 1, 2)$

LP_2 through $a(0, 0, 2)$, $a(1, 1, 0)$ and $a(0, 1, 0)$

Solution

The direction perpendicular to LP_1 is $(1, 1, 0) \times (0, 1, 2) = (2, -2, 1)$ and hence the equation of LP_1 is

$$r \cdot (2, -2, 1) = 0 \quad \text{or in cartesian form} \quad 2x - 2y + z = 0 \tag{4.21}$$

The direction perpendicular to LP_2 is $(1, 1, -2) \times (0, 1, -2) = (0, 2, 1)$ and hence the equation of LP_2 is

$$r \cdot (0, 2, 1) = 2 \quad \text{or in cartesian form} \quad 2y + z = 2 \tag{4.22}$$

Points that lie on both lattice planes must satisfy both (4.21) and (4.22). It is easiest to solve these equations in their cartesian form. The coordinates must be integers so take $y = m$, then z can easily be calculated from (4.22) as

$$z = 2 - 2m$$

and then x is computed from (4.21) to be $x = 2m - 1$.

Hence the required points all lie on a line and take the form

$$r = a(2m - 1, m, 2 - 2m)$$

where m is an integer.

Example 4.46

Find the point where the plane

$$r \cdot (1, 1, 2) = 3$$

meets the line

$$r = (2, 1, 1) + \lambda(0, 1, 2)$$

Solution   At the point of intersection r must satisfy both equations, so

$$[(2, 1, 1) + \lambda(0, 1, 2)] \cdot (1, 1, 2) = 3$$

or

$$5 + 5\lambda = 3$$

so

$$\lambda = -\tfrac{2}{5}$$

Substituting back into the equation of the line gives the point of intersection as

$$r = (2, \tfrac{3}{5}, \tfrac{1}{5})$$

Example 4.47   Find the equation of the line of intersection of the two planes $x + y + z = 5$ and $4x + y + 2z = 15$.

Solution   In vector form the equations of the two planes are

$$r \cdot (1, 1, 1) = 5$$

and

$$r \cdot (4, 1, 2) = 15$$

The required line lies in both planes, and is therefore perpendicular to the vectors $(1, 1, 1)$ and $(4, 1, 2)$, which are normal to the individual planes. Hence a vector c in the direction of the line is

$$c = (1, 1, 1) \times (4, 1, 2) = (1, 2, -3)$$

To find the equation of the line, it remains only to find the coordinates of any point on the line. To do this, we are required to find the coordinates of a point satisfying the equation of the two planes. Taking $x = 0$, the corresponding values of y and z are given by

$$y + z = 5 \quad \text{and} \quad y + 2z = 15$$

that is, $y = -5$ and $z = 10$. Hence it can be checked that the point $(0, -5, 10)$ lies in both planes and is therefore a point on the line. From (4.16) the equation of the line is

$$r = (0, -5, 10) + t(1, 2, -3)$$

or in cartesian form

$$\frac{x}{1} = \frac{y + 5}{2} = \frac{z - 10}{-3} = t$$

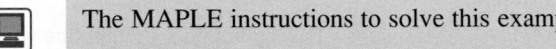

The MAPLE instructions to solve this example are

```
with (geom3d):
plane (P1, x + y + z = 5, [x, y, z]): plane (P2, 4*x
+ y + 2*z = 15, [x, y, z]): intersection (L, P1, P2):
detail (L);
```

Example 4.48   Find the perpendicular distance from the point P(2, −3, 4) to the plane $x + 2y + 2z = 13$.

Solution   In vector form the equation of the plane is

$$r \cdot (1, 2, 2) = 13$$

and a vector perpendicular to the plane is

$$n = (1, 2, 2)$$

Thus from (4.16) the equation of a line perpendicular to the plane and passing through P(2, −3, 4) is

$$r = (2, -3, 4) + t(1, 2, 2)$$

This will meet the plane when

$$r \cdot (1, 2, 2) = (2, -3, 4) \cdot (1, 2, 2) + t(1, 2, 2) \cdot (1, 2, 2) = 13$$

giving

$$4 + 9t = 13$$

so that

$$t = 1$$

Thus the line meets the plane at N having position vector

$$r = (2, -3, 4) + 1(1, 2, 2) = (3, -1, 6)$$

Hence the perpendicular distance is

$$PN = \sqrt{[(3 - 2)^2 + (-1 + 3)^2 + (6 - 4)^2]} = 3$$

4.3.3 Exercises

 Many of the exercises can be checked using the geom3d package in MAPLE.

52   If A and B have position vectors (1, 2, 3) and (4, 5, 6) respectively, find

   (a) the direction vector of the line through A and B,

   (b) the vector equation of the line through A and B,

   (c) the cartesian equation of the line.

53   Find the vector equation of the plane that passes through the points (1, 2, 3), (2, 4, 5) and (4, 5, 6). What is its cartesian equation?

54   Show that the line joining (2, 3, 4) to (1, 2, 3) is perpendicular to the line joining (1, 0, 2) to (2, 3, −2).

55   Prove that the lines $r = (1, 2, -1) + t(2, 2, 1)$ and $r = (-1, -2, 3) + s(4, 6, -3)$ intersect, and find the coordinates of their point of intersection. Also find the acute angle between the lines.

56   Find the vector equation of the plane that contains the line $r = a + \lambda b$ and passes through the point with position vector c.

57   P is a point on a straight line with position vector $r = a + tb$. Show that

$$r^2 = a^2 + 2a \cdot bt + b^2 t^2$$

By completing the square, show that r^2 is a minimum for the point P for which $t = -a \cdot b / b^2$.

Show that at this point $\overrightarrow{OP}$ is perpendicular to the line $r = a + tb$. (This proves the well-known result that the shortest distance from a point to a line is the length of the perpendicular from that point to the line.)

58 The line of intersection of two planes $r \cdot n_1 = p_1$ and $r \cdot n_2 = p_2$ lies in both planes. It is therefore perpendicular to both n_1 and n_2. Give an expression for this direction, and so show that the equation of the line of intersection may be written as $r = r_0 + t(n_1 \times n_2)$, where r_0 is any vector satisfying $r_0 \cdot n_1 = p_1$ and $r_0 \cdot n_2 = p_2$. Hence find the line of intersection of the planes $r \cdot (1, 1, 1) = 5$ and $r \cdot (4, 1, 2) = 15$.

59 Find the equation of the line through the point $(1, 2, 4)$ and in the direction of the vector $(1, 1, 2)$. Find where this line meets the plane $x + 3y - 4z = 5$.

60 Find the acute angle between the planes $2x + y - 2z = 5$ and $3x - 6y - 2z = 7$.

61 Given that $a = (3, 1, 2)$ and $b = (1, -2, -4)$ are the position vectors of the points P and Q respectively, find

(a) the equation of the plane passing through Q and perpendicular to PQ,

(b) the distance from the point $(-1, 1, 1)$ to the plane obtained in (a).

62 Find the equation of the line joining $(1, -1, 3)$ to $(3, 3, -1)$. Show that it is perpendicular to the plane $2x + 4y - 4z = 5$, and find the angle that the line makes with the plane $12x - 15y + 16z = 10$.

63 Find the vector equation of the line through the points with position vectors $a = (2, 0, -1)$ and $b = (1, 2, 3)$. Write down the equivalent cartesian coordinate form. Does this line intersect the line through the points $c = (0, 0, 1)$ and $d = (1, 0, 1)$?

64 Find the shortest distance between the two lines

$$r = (4, -2, 3) + t(2, 1, -1)$$

and

$$r = (-7, -2, 1) + s(3, 2, 1)$$

65 Find the equation of the plane through the line

$$r = (1, -3, 4) + t(2, 1, 1)$$

and parallel to the line

$$r = s(1, 2, 3)$$

66 Find the equation of the line through $P(-1, 0, 1)$ that cuts the line $r = (3, 2, 1) + t(1, 2, 2)$ at right-angles at Q. Also find the length PQ and the equation of the plane containing the two lines.

67 Show that the equation of the plane through the points P_1, P_2 and P_3 with position vectors r_1, r_2 and r_3 respectively takes the form

$$r \cdot [(r_1 \times r_2) + (r_2 \times r_3) + (r_3 \times r_1)] = r_1 \cdot (r_2 \times r_3)$$

4.4 Engineering application: spin-dryer suspension

Vectors are at their most powerful when dealing with complicated three-dimensional situations. Geometrical and physical intuition are often difficult to use, and it becomes necessary to work quite formally to analyse such situations. For example, the front suspension of a motor car has two struts supported by a spring-and-damper system and subject to a variety of forces and torques from both the car and the wheels. To analyse the stresses and the vibrations in the various components of the structure is non-trivial, even in a two-dimensional version; the true three-dimensional problem provides a testing exercise for even the most experienced automobile engineer. In the present text a much simpler situation is analysed to illustrate the use of vectors.

4.4.1 Point-particle model

As with the car suspension, many machines are mounted on springs to isolate vibrations. A typical example is a spin-dryer, which consists of a drum connected to the

Figure 4.54
The particle P is attached by equal springs to the eight corners of the cube.

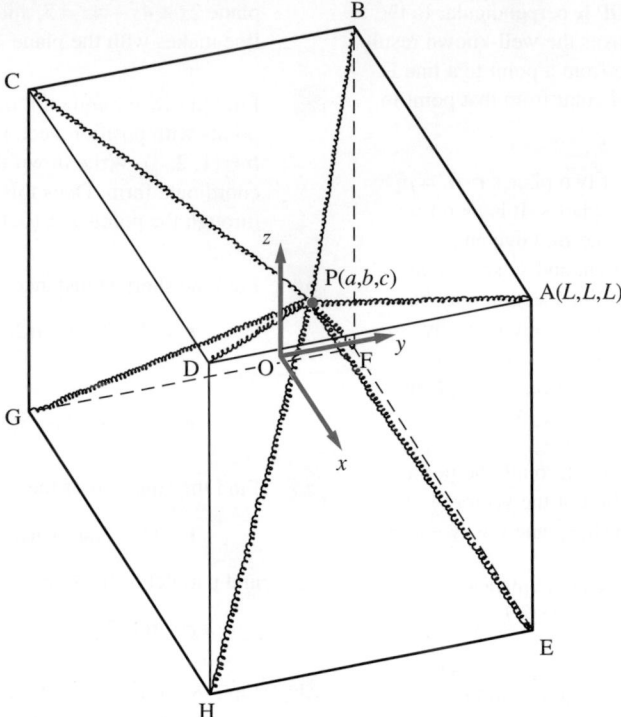

casing by heavy springs. Oscillations can be very severe when spinning at high speed, and it is essential to know what forces are transmitted to the casing and hence to the mounts. Before the dynamical situation can be analysed, it is necessary to compute the restoring forces on the drum when it is displaced from its equilibrium position. This is a static problem that is best studied using vectors.

We model the spin-dryer as a heavy point particle connected to the eight corners of the casing by springs (Figure 4.54). The drum has weight W and the casing is taken to be a cube of side $2L$. The springs are all equal, having spring constant k and natural length $L\sqrt{3}$. Thus when the drum is at the midpoint of the cube the springs are neither compressed nor extended.

The particle is displaced from its central position by a small amount (a, b, c), where the natural coordinates illustrated in Figure 4.54 are used; the origin is at the centre of the cube and the axes are parallel to the sides. What is required is the total force acting on the particle arising from the weight and the springs. Clearly, this information is needed before any dynamical calculations can be performed. It will be assumed that the displacements are sufficiently small that squares $(a/L)^2$, $(b/L)^2$, $(c/L)^2$ and higher powers are neglected.

Consider a typical spring PA. The tension in the spring is assumed to obey Hooke's law that the force is along PA and has magnitude proportional to extension. $\overrightarrow{PA}/|\overrightarrow{PA}|$ is the unit vector in the direction along PA, and $|\overrightarrow{PA}| - L\sqrt{3}$ is the extension of the spring over its natural length $L\sqrt{3}$ so in vector form the tension can be written as

$$T_A = k \frac{\overrightarrow{PA}}{|\overrightarrow{PA}|}(|\overrightarrow{PA}| - L\sqrt{3}) \tag{4.23}$$

where k is the proportionality constant.

Now

$$\overrightarrow{PA} = \overrightarrow{OA} - \overrightarrow{OP} = (L - a, L - b, L - c)$$

so calculating the modulus squared gives

$$|\overrightarrow{PA}|^2 = (L - a)^2 + (L - b)^2 + (L - c)^2$$

$$= 3L^2 - 2L(a + b + c) + \text{quadratic terms}$$

Thus

$$|\overrightarrow{PA}| = \left[1 - \frac{2}{3L}(a + b + c)\right]^{1/2} L\sqrt{3}$$

and, on using the binomial expansion (see equation (7.16)) and neglecting quadratic and higher terms, we obtain

$$|\overrightarrow{PA}| = \left[1 - \frac{1}{3L}(a + b + c)\right] L\sqrt{3}$$

Putting the information acquired back into (4.23) gives

$$T_A = kL \frac{(1 - a/L, 1 - b/L, 1 - c/L)}{[1 - (a + b + c)/3L]L\sqrt{3}} \frac{(-1)(a + b + c)L\sqrt{3}}{3L}$$

and by expanding again, using the binomial expansion to first order in a/L and so on, we obtain

$$T_A = -\tfrac{1}{3}k(a + b + c)(1, 1, 1)$$

Similar calculations give

$$T_B = -\tfrac{1}{3}k(-a + b + c)(-1, 1, 1)$$

$$T_C = -\tfrac{1}{3}k(-a - b + c)(-1, -1, 1)$$

$$T_D = -\tfrac{1}{3}k(a - b + c)(1, -1, 1)$$

$$T_E = -\tfrac{1}{3}k(a + b - c)(1, 1, -1)$$

$$T_F = -\tfrac{1}{3}k(-a + b - c)(-1, 1, -1)$$

$$T_G = -\tfrac{1}{3}k(-a - b - c)(-1, -1, -1)$$

$$T_H = -\tfrac{1}{3}k(a - b - c)(1, -1, -1)$$

The total spring force is therefore obtained by adding these eight tensions together:

$$T = -\tfrac{8}{3}k(a, b, c)$$

The restoring force is therefore towards the centre of the cube, as expected, in the direction PO and with magnitude $\tfrac{8}{3}k$ times the length of PO.

When the weight is included, the total force is

$$F = (-\tfrac{8}{3}ka, -\tfrac{8}{3}kb, -\tfrac{8}{3}kc - W)$$

If the drum just hangs in equilibrium then $F = 0$, and hence

$$a = b = 0 \quad \text{and} \quad c = -\frac{3W}{8k}$$

Typical values are $W = 400\,\text{N}$ and $k = 10\,000\,\text{N m}^{-1}$, and hence

$$c = -3 \times 400/8 \times 10\,000 = -0.015\,\text{m}$$

so that the centre of the drum hangs 1.5 cm below the midpoint of the centre of the casing.

It is clear that the model used in this section is an idealized one, but it is helpful in describing how to calculate spring forces in complicated three-dimensional static situations. It also gives an idea of the size of the forces involved and the deflections. The next major step is to put these forces into the equations of motion of the drum; this, however, requires a good knowledge of calculus – and, in particular, of differential equations – so it is not appropriate at this point. You may wish to consider this problem after studying the relevant chapters later in this book. A more advanced model must include the fact that the drum is of finite size.

4.5 Engineering application: cable stayed bridge

One of the standard methods of supporting bridges is with cables. Readers will no doubt be familiar with suspension bridges such as the Golden Gate in the USA, the Humber bridge in the UK and the Tsing Ma bridge in Hong Kong with their spectacular form. Cable stayed bridges are similar in that they have towers and cables that support a roadway but they are not usually on such a grand scale as suspension bridges. They are often used when the foundations can only support a single tower at one end of the roadway. They are commonly seen on bridges over motorways and footbridges over steep narrow valleys.

In any of the situations described it is essential that information is available on the tension in the wire supports and the forces on the towers. The geometry is fully three-dimensional and quite complicated. Vectors provide a logical and efficient way of dealing with the situation.

4.5.1 A simple stayed bridge

There are many configurations that stayed bridges can take; they can have one or more towers and a variety of arrangements of stays. In Figure 4.55 a simple example of a cable stayed footbridge is illustrated. It is constructed with a central vertical pillar with four ties attached by wires to the sides of the pathway.

Relative to the axes, with the z axis vertical, the various points are given, in metres, as A(5, −2, 0.5), B(10, 2, 1), C(15, −2, 1.5), D(20, 2, 1) and S(0, 0, 10). Assuming the weight is evenly distributed, there is an equivalent weight of 2 tonnes at each of the four points A, B, C and D. An estimate is required of the tensions in the wires and force at the tie point S.

The vectors along the ties can easily be evaluated

$$\overrightarrow{AS} = (-5, 2, 9.5), \quad \overrightarrow{BS} = (-10, -2, 9)$$
$$\overrightarrow{CS} = (-15, 2, 8.5), \quad \overrightarrow{DS} = (-20, -2, 9)$$

Figure 4.55
Model of a stayed
bridge.

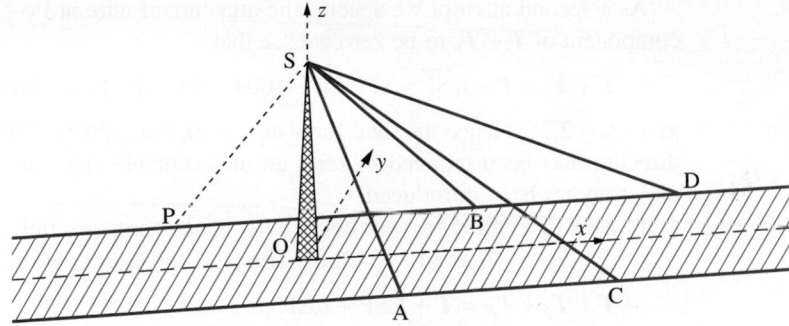

The tension at S in the tie AS can be written $T_A = t_A \overrightarrow{SA}$. Assuming the whole system is in equilibrium, the vertical components at A must be equal

$$T_A \cdot k = 2 \quad \text{and hence} \quad t_A = \frac{2}{9.5}$$

and the four tensions can be computed similarly.

$$T_A = \tfrac{2}{9.5}(5, -2, -9.5) = (1.052, -0.421, -2) \quad \text{and} \quad |T_A| = 2.299 \text{ tonnes}$$

$$T_B = \tfrac{2}{9}(10, 2, -9) = (2.222, 0.444, -2) \quad \text{and} \quad |T_B| = 3.022 \text{ tonnes}$$

$$T_C = \tfrac{2}{8.5}(15, -2, -8.5) = (3.529, -0.471, -2) \quad \text{and} \quad |T_C| = 4.084 \text{ tonnes}$$

$$T_D = \tfrac{2}{9}(20, 2, -9) = (4.444, 0.444, -2) \quad \text{and} \quad |T_D| = 4.894 \text{ tonnes}$$

The total force acting at the tie point S is

$$T = T_A + T_B + T_C + T_D = (11.25, -0.004, -8)$$

Thus with straightforward addition of vectors we have been able to compute the tensions and the total force on the tower.

The question now is how to compensate for the total force on the tower and to try to ensure that it is subject to zero force or a force as small as possible. Suppose that it is decided to have just a single compensating tie wire attached to S and to one side on the pathway at P. It is assumed that on this side of the footbridge the pathway is flat and lies in the x–y plane. Where should we position the attachment of the compensating wire so that it produces zero horizontal force at S?

Let the attachment point P on the side of the footbridge be $(-a, 2, 0)$ so that the tension in the compensating cable is

$$T_P = t_P \overrightarrow{SP} = t_P(-a, 2, -10)$$

We require the y component of $(T + T_P)$ to be zero so that

$$2t_P - 0.004 = 0 \quad \text{and hence} \quad t_P = 0.002$$

which in turn gives for the x component

$$at_P = 11.248 \quad \text{and hence} \quad a = 5624 \text{ metres!}$$

Clearly the answer is ridiculous and either more than one compensating cable must be used or the y component can be neglected completely since the force in this direction is only 4 kg.

As a second attempt we specify the attachment wire at P(–5, 2, 0). Requiring the x component of $T + T_P$ to be zero we see that

$$T + T_P = T + t_P \overrightarrow{SP} = (11.25, -0.004, -8) + t_P(-5, 2, -10)$$

gives $t_P = 2.25$. Hence the total force at S is $(0, 4.5, -30.5)$. Although the force in the x direction has been reduced to zero, an unacceptable side force on the tower in the y direction has been introduced.

In a further effort, we introduce two equal compensating wires connected to the points P(–5, –2, 0) and P′(–5, 2, 0). The total force at S is now

$$T + T_P + T_{P'} = T + t_P \overrightarrow{SP} + t_P \overrightarrow{SP'}$$

$$= (11.25, -0.004, -8) + t_P(-5, 2, -10) + t_P(-5, -2, -10)$$

Now choosing $t_P = 1.125$ gives a total force $(0, -0.004, -30.5)$. We now have a satisfactory resolution of the problem with the only significant force being in the downwards direction.

The different forms of stayed bridge construction will require a similar analysis to obtain an estimate of the forces involved. The example given should be viewed as illustrative.

4.6 Review exercises (1–24)

Check your answers using MATLAB or MAPLE whenever possible.

1. Given that $a = 3i - j - 4k$, $b = -2i + 4j - 3k$ and $c = i + 2j - k$, find

   (a) the magnitude of the vector $a + b + c$,

   (b) a unit vector parallel to $3a - 2b + 4c$,

   (c) the angles between the vectors a and b and between b and c,

   (d) the position vector of the centre of mass of particles of masses 1, 2 and 3 placed at points A, B and C with position vectors a, b and c respectively.

2. If the vertices X, Y and Z of a triangle have position vectors

   $$x = (2, 2, 6), \quad y = (4, 6, 4) \quad \text{and} \quad z = (4, 1, 7)$$

   relative to the origin O, find

   (a) the midpoint of the side XY of the triangle,

   (b) the area of the triangle,

   (c) the volume of the tetrahedron OXYZ.

3. The vertices of a tetrahedron are the points

   $$W(2, 1, 3), \quad X(3, 3, 3), \quad Y(4, 2, 4) \quad \text{and} \quad Z(3, 3, 5)$$

   Determine

   (a) the vectors $\overrightarrow{WX}$ and $\overrightarrow{WY}$,

   (b) the area of the face WXZ,

   (c) the volume of the tetrahedron WXZY,

   (d) the angles between the faces WXY and WYZ.

4. Given $a = (-1, -3, -1)$, $b = (q, 1, 1)$ and $c = (1, 1, q)$ determine the values of q for which

   (a) a is perpendicular to b

   (b) $a \times (b \times c) = 0$

5. Given the vectors $a = (2, 1, 2)$ and $b = (-3, 0, 4)$, evaluate the unit vectors $\hat{a}$ and $\hat{b}$. Use these unit vectors to find a vector that bisects the angle between a and b.

6. A triangle, ABC, is inscribed in a circle, centre O, with AOC as a diameter of the circle. Take $\overrightarrow{OA} = a$ and $OB = b$. By evaluating $\overrightarrow{AB} \cdot \overrightarrow{CB}$ show that angle ABC is a right angle.

7. According to the inverse square law, the force on a particle of mass m_1 at the point P_1 due to a particle of mass m_2 at the point P_2 is given by

   $$\gamma \frac{m_1 m_2}{r^2} \hat{r} \quad \text{where } r = \overrightarrow{P_1 P_2}$$

Particles of mass $3m$, $3m$, m are fixed at the points A(1, 0, 1), B(0, 1, 2) and C(2, 1, 2) respectively. Show that the force on the particle at A due to the presence of B and C is

$$\frac{2\gamma m^2}{\sqrt{3}}(-1, 2, 2)$$

8 Show that the vector a which satisfies the vector equation

$$a \times (i + 2j) = -2i + j + k$$

must take the form $a = (\alpha, 2\alpha - 1, 1)$. If in addition the vector a makes an angle $\cos^{-1}(\frac{1}{3})$ with the vector $(i - j + k)$ show that there are now two such vectors that satisfy both conditions.

9 The electric field at a point having position vector r, due to a charge e at R, is $e(r - R)/|r - R|^3$. Find the electric field E at the point P(2, 1, 1) given that there is a charge e at each of the points (1, 0, 0), (0, 1, 0) and (0, 0, 1).

10 Given that $\overrightarrow{OP} = (3, 1, 2)$ and $\overrightarrow{OQ} = (1, -2, -4)$ are the position vectors of the points P and Q respectively find

(a) the equation of the plane passing through Q and perpendicular to PQ,

(b) the perpendicular distance from the point (−1, 1, 1) to the plane.

11 (a) Determine the equation of the plane that passes through the points (1, 2, −2), (−1, 1, −9) and (2, −2, −12). Find the perpendicular distance from the origin to this plane.

(b) Calculate the area of the triangle whose vertices are at the points (1, 1, 0), (1, 0, 1) and (0, 1, 1).

12 Find the point P on the line L through the points

A(5, 1, 7)   and   B(6, 0, 8)

and the point Q on the line M through the points

C(3, 1, 3)   and   D(−1, 3, 3)

such that the line through P and Q is perpendicular to both lines L and M. Verify that P and Q are at a distance $\sqrt{6}$ apart, and find the point where the line through P and Q intersects the coordinate plane Oxy.

13 The angular momentum vector H of a particle of mass m is defined by

$$H = r \times (mv)$$

where $v = \omega \times r$.
  Using the result

$$a \times (b \times c) = (a \cdot c)b - (a \cdot b)c$$

show that if r is perpendicular to ω then $H = mr^2\omega$.
  Given that $m = 100$, $r = 0.1(i + j + k)$ and $\omega = 5i + 5j - 10k$ calculate

(a) $(r \cdot \omega)$ (b) H

14 A particle of mass m, charge e and moving with velocity v in a magnetic field of strength H is known to have acceleration

$$\frac{e}{mc}(v \times H)$$

where c is the speed of light. Show that the component of acceleration parallel to H is zero.

15 A force F is of magnitude 14 N and acts at the point A(3, 2, 4) in the direction of the vector $-2i + 6j + 3k$. Find the moment of the force about the point B(1, 5, −2). Find also the angle between F and $\overrightarrow{AB}$.

16 Points A, B, C have coordinates (1, 2, 1), (−1, 1, 3) and (−2, −2, −2) respectively.
  Calculate the vector product $\overrightarrow{AB} \times \overrightarrow{AC}$, the angle BAC and a unit vector perpendicular to the plane containing A, B and C. Hence obtain

(a) the equation of the plane ABC,

(b) the equation of a second plane, parallel to ABC, and containing the point D(1, 1, 1),

(c) the shortest distance between the point D and the plane containing A, B and C.

17 A plane Π passes through the three non-collinear points A, B and C having position vectors a, b and c respectively. Show that the parametric vector equation of the plane Π is

$$r = a + \lambda(b - a) + \mu(c - a)$$

The plane Π passes through the points (−3, 0, 1), (5, −8, −7) and (2, 1, −2) and the plane Θ passes through the points (3, −1, 1), (1, −2, 1) and (2, −1, 2). Find the parametric vector equation of Π and the normal vector equation of Θ, and hence show that their line of intersection is

$$r = (1, -4, -3) + t(5, 1, -3)$$

where t is a scalar variable.

18 Two skew lines L_1, L_2 have respective equations

$$\frac{x+3}{4} = \frac{y-3}{-1} = \frac{z-2}{1} \quad \text{and}$$

$$\frac{x-1}{2} = \frac{y-5}{1} = \frac{z+3}{2}$$

Obtain the equation of a plane through L_1 parallel to L_2 and show that the shortest distance between the lines is 6.

19 A particle P of constant mass m moves in a viscous medium under the influence of a uniform gravitational force $-mg\boldsymbol{j}$, where g is a scalar constant and $\boldsymbol{j}$ is a constant unit vector. The medium offers a resistance to the motion of the particle proportional to its momentum (with proportionality constant K). The solution of the equation of motion of P gives the velocity as

$$\boldsymbol{v} = \boldsymbol{V}e^{-Kt} - \frac{g}{K}(1 - e^{-Kt})\boldsymbol{j}$$

where $\boldsymbol{V}$ is the velocity at time $t = 0$, and the position vector as

$$\boldsymbol{r} = \boldsymbol{A} - \frac{\boldsymbol{V}e^{-Kt}}{K} - \frac{gt}{K}\boldsymbol{j} - \frac{ge^{-Kt}}{K^2}\boldsymbol{j}$$

where $\boldsymbol{A}$ is some constant vector.

At time $t = 0$, a projectile is introduced into the atmosphere at a height H above the ground with a speed U parallel to the ground. Assuming that the atmosphere behaves as a viscous medium as described above, show that at time t the path of the projectile will be inclined at an angle α to the horizontal, where

$$\alpha = \tan^{-1}\left[\frac{g(e^{Kt} - 1)}{KU}\right]$$

Show also that the time of flight T to hitting the ground is a solution of the equation

$$T + \frac{1}{K}(e^{-KT} - 1) = \frac{HK}{g}$$

and find the horizontal distance travelled in terms of T.

20 The three vectors $\boldsymbol{a} = (1, 0, 0)$, $\boldsymbol{b} = (1, 1, 0)$ and $\boldsymbol{c} = (1, 1, 1)$ are given. Evaluate

(a) $\boldsymbol{a} \times \boldsymbol{b}, \boldsymbol{b} \times \boldsymbol{c}, \boldsymbol{c} \times \boldsymbol{a}$

(b) $\boldsymbol{a} \cdot (\boldsymbol{b} \times \boldsymbol{c})$

For the vector $\boldsymbol{d} = (2, -1, 2)$ calculate

(c) the parameters α, β, γ in the expression

$$\boldsymbol{d} = \alpha\boldsymbol{a} + \beta\boldsymbol{b} + \gamma\boldsymbol{c}$$

(d) the parameters p, q, r in the expression

$$\boldsymbol{d} = p\boldsymbol{a} \times \boldsymbol{b} + q\boldsymbol{b} \times \boldsymbol{c} + r\boldsymbol{c} \times \boldsymbol{a}$$

and show that

$$p = \frac{\boldsymbol{c} \cdot \boldsymbol{d}}{\boldsymbol{a} \cdot (\boldsymbol{b} \times \boldsymbol{c})}, \quad q = \frac{\boldsymbol{a} \cdot \boldsymbol{d}}{\boldsymbol{a} \cdot (\boldsymbol{b} \times \boldsymbol{c})} \quad \text{and}$$

$$r = \frac{\boldsymbol{b} \cdot \boldsymbol{d}}{\boldsymbol{a} \cdot (\boldsymbol{b} \times \boldsymbol{c})}$$

21 Given the line with parametric equation

$$\boldsymbol{r} = \boldsymbol{a} + \lambda\boldsymbol{d}$$

show that the perpendicular distance p from the origin to this line can take either of the forms

(i) $p = \dfrac{|\boldsymbol{a} \times \boldsymbol{d}|}{|\boldsymbol{d}|}$ (ii) $p = \left|\boldsymbol{a} - \dfrac{\boldsymbol{a} \cdot \boldsymbol{d}}{\boldsymbol{d} \cdot \boldsymbol{d}}\boldsymbol{d}\right|$

Find the parametric equation of the straight line through the points

$$A(1, 0, 2) \quad \text{and} \quad B(2, 3, 0)$$

and determine

(a) the length of the perpendicular from the origin to the line,

(b) the point at which the line intersects the y–z plane,

(c) the coordinates of the foot of the perpendicular to the line from the point $(1, 1, 1)$.

22 Given the three non-coplanar vectors $\boldsymbol{a}, \boldsymbol{b}, \boldsymbol{c}$, and defining $v = \boldsymbol{a} \cdot \boldsymbol{b} \times \boldsymbol{c}$, three further vectors are defined as

$$\boldsymbol{a}' = \boldsymbol{b} \times \boldsymbol{c}/v \quad \boldsymbol{b}' = \boldsymbol{c} \times \boldsymbol{a}/v \quad \boldsymbol{c}' = \boldsymbol{a} \times \boldsymbol{b}/v$$

Show that

$$\boldsymbol{a} = \boldsymbol{b}' \times \boldsymbol{c}'/v' \quad \boldsymbol{b} = \boldsymbol{c}' \times \boldsymbol{a}'/v' \quad \boldsymbol{c} = \boldsymbol{a}' \times \boldsymbol{b}'/v'$$

where

$$v' = \boldsymbol{a}' \cdot \boldsymbol{b}' \times \boldsymbol{c}'$$

Deduce that

$$\boldsymbol{a} \cdot \boldsymbol{a}' = \boldsymbol{b} \cdot \boldsymbol{b}' = \boldsymbol{c} \cdot \boldsymbol{c}' = 1$$

$$\boldsymbol{a} \cdot \boldsymbol{b}' = \boldsymbol{a} \cdot \boldsymbol{c}' = \boldsymbol{b} \cdot \boldsymbol{a}' = \boldsymbol{b} \cdot \boldsymbol{c}' = \boldsymbol{c} \cdot \boldsymbol{a}'$$

$$= \boldsymbol{c} \cdot \boldsymbol{b}' = 0$$

If a vector is written in terms of a, b, c as

$$r = \alpha a + \beta b + \gamma c$$

evaluate α, β, γ in terms of a', b' and c'.

(*Note*: these sets of vectors are called **reciprocal sets** and are widely used in crystallography and materials science.)

23 An unbalanced machine can be approximated by two masses, 2 kg and 1.5 kg, placed at the ends A and B respectively of light rods OA and OB of lengths 0.7 m and 1.1 m. The point O lies on the axis of rotation and OAB forms a plane perpendicular to this axis; OA and OB are at right-angles. The machine rotates about the axis with an angular velocity ω, which gives a centrifugal force $mr\omega^2$ for a mass m and rod length r. Find the unbalanced force at the axis. To balance the machine a mass of 1 kg is placed at the end of a light rod OC so that C is coplanar with OAB. Determine the position of C.

24 In an automated drilling process three holes are drilled simultaneously into the centre of the faces of a block of side 0.2 m, as shown in Figure 4.56. The force exerted by each drill is 25 N, and the couples applied during the drilling are 0.2 N m in the directions indicated. Find the resultant force on the system and the moment of the forces and couples about the corner A. (Take x, y, z axes along the sides of the block, with origin at A.)

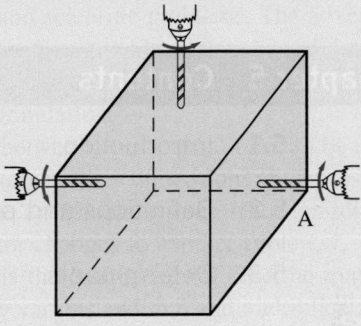

Figure 4.56 Drilling into a block, Question 24.

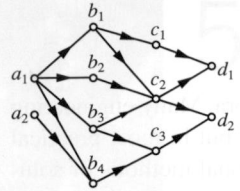

Figure 5.1
Simple electrical
network.

The analysis of circuits can be very complex. In modern VLSI (Very Large Scale Integration) systems there can be many hundreds of connections. Circuit simulation programs use matrix theory, and have proved to be a most important tool for the analysis of the electrical performance of integrated circuits. Such systems have wide use in CAD (Computer Aided Design) work. The detailed structure of the matrices gives information about the characteristics of the system and provides an indication of how to tackle the solution of the matrix equations. To give an example of a simple electrical network (or road system for that matter) we consider the connections in the situation illustrated in Figure 5.1. They can be represented by arrays with entries of one if a link exists and zero otherwise, thus:

$$
\begin{array}{cccc}
 & b_1 & b_2 & b_3 & b_4 \\
a_1 & 1 & 1 & 1 & 1 \\
a_2 & 0 & 0 & 0 & 1
\end{array}
\qquad
\mathbf{P} = \begin{bmatrix} 1 & 1 & 1 & 1 \\ 0 & 0 & 0 & 1 \end{bmatrix}
$$

$$
\begin{array}{ccc}
 & c_1 & c_2 & c_3 \\
b_1 & 1 & 1 & 0 \\
b_2 & 0 & 1 & 0 \\
b_3 & 0 & 1 & 1 \\
b_4 & 0 & 0 & 1
\end{array}
\qquad
\mathbf{Q} = \begin{bmatrix} 1 & 1 & 0 \\ 0 & 1 & 0 \\ 0 & 1 & 1 \\ 0 & 0 & 1 \end{bmatrix}
$$

$$
\begin{array}{cc}
 & d_1 & d_2 \\
c_1 & 1 & 0 \\
c_2 & 1 & 1 \\
c_3 & 0 & 1
\end{array}
\qquad
\mathbf{R} = \begin{bmatrix} 1 & 0 \\ 1 & 1 \\ 0 & 1 \end{bmatrix}
$$

We now have a concise numerical way of representing the diagram in Figure 5.1. The arrays can be written in matrix form as $\mathbf{P}$, $\mathbf{Q}$ and $\mathbf{R}$, and we can perform algebraic operations on them. The product of matrices will be defined in Section 5.2.4; it is found that the entries in the product $\mathbf{PQ}$ give the number of different paths from a_i to c_j, and those in $\mathbf{PQR}$ the numbers of different paths from a_i to d_k.

The matrix product is the most interesting property in the theory, since it enables complicated sets of equations to be written in a convenient and compact way. For instance, three ores are known to contain fractions of Pb, Fe, Cu and Mn as indicated in Figure 5.2. If we mix the ores so that there are x_1 kg of ore 1, x_2 kg of ore 2 and x_3 kg of ore 3 then we can compute the amount of each element as

Figure 5.2
Table of fractions in
each kilogram of ore.

	Ore 1	Ore 2	Ore 3
Pb	0.1	0.2	0.3
Fe	0.2	0.3	0.3
Cu	0.6	0.2	0.2
Mn	0.1	0.3	0.2

$$\left.\begin{array}{l} \text{amount of Pb} = A_{Pb} = 0.1x_1 + 0.2x_2 + 0.3x_3 \\ \text{amount of Fe} = A_{Fe} = 0.2x_1 + 0.3x_2 + 0.3x_3 \\ \text{amount of Cu} = A_{Cu} = 0.6x_1 + 0.2x_2 + 0.2x_3 \\ \text{amount of Mn} = A_{Mn} = 0.1x_1 + 0.3x_2 + 0.2x_3 \end{array}\right\} \qquad \textbf{(5.1)}$$

We can rewrite the array in Figure 5.2 as a matrix

$$\textbf{A} = \begin{bmatrix} 0.1 & 0.2 & 0.3 \\ 0.2 & 0.3 & 0.3 \\ 0.6 & 0.2 & 0.2 \\ 0.1 & 0.3 & 0.2 \end{bmatrix}$$

and if we define the vectors

$$\textbf{M} = \begin{bmatrix} A_{Pb} \\ A_{Fe} \\ A_{Cu} \\ A_{Mn} \end{bmatrix} \quad \text{and} \quad \textbf{X} = \begin{bmatrix} x_1 \\ x_2 \\ x_3 \end{bmatrix}$$

then the equations can be written in matrix form

$$\textbf{M} = \textbf{AX}$$

with the product interpreted as in (5.1). The matrix $\textbf{A}$ has 4 rows and 3 columns, so it is called a 4×3 (read '4 by 3') matrix. $\textbf{M}$ and $\textbf{X}$ are called column vectors; they are 4×1 and 3×1 matrices respectively.

5.2 Definitions and properties

We can look at the 'ore' problem in a different context. In Section 4.3.2 of Chapter 4, we saw that the equation of a plane can be written in the form

$$\alpha x + \beta y + \gamma z = p$$

where α, β, γ and p are constants. The four planes

$$\left.\begin{array}{l} 4x + 2y + z = 7 \\ 2x + y - z = 5 \\ x + 2y + 2z = 3 \\ 3x - 2y - z = 0 \end{array}\right\} \qquad \textbf{(5.2)}$$

meet in a single point. What are the coordinates of that point? Obviously they are those values of x, y and z that satisfy all four of (5.2) simultaneously.

Equations (5.2) provide an example of a mathematical problem that arises in a wide range of engineering problems: the simultaneous solution of a set of linear equations,

$$\begin{bmatrix} x \\ y \\ z \end{bmatrix} \quad \text{and} \quad \begin{bmatrix} x \\ y \end{bmatrix}$$

will be denoted by X. (Vectors and matrices are further distinguished here by the use of a 'serif' bold face for the former (e.g. b) and a 'sans serif' bold face for the latter (e.g. A).) As an example of the notation used consider the matrix A and vector b

$$A = \begin{bmatrix} 0 & -1 & 2 \\ 3 & 0 & 1 \end{bmatrix} \quad \text{and} \quad b = \begin{bmatrix} 0.15 \\ 1.11 \\ -3.01 \end{bmatrix}$$

The matrix A is a 2×3 matrix with elements $a_{11} = 0$, $a_{12} = -1$, $a_{13} = 2$, $a_{21} = 3$ and so on. The vector b is a column vector with elements $b_1 = 0.15$, $b_2 = 1.11$ and $b_3 = -3.01$.

In a square matrix of order n the diagonal containing the elements $a_{11}, a_{22}, \dots, a_{nn}$ is called the **principal**, **main** or **leading** diagonal. The sum of the elements of the leading diagonal is called the **trace** of the square matrix A, that is

$$\text{trace } A = a_{11} + a_{22} + \dots + a_{nn} = \sum_{i=1}^{n} a_{ii}$$

A **diagonal matrix** is a square matrix that has its only non-zero elements along the leading diagonal. (It may have zeros on the leading diagonal also.)

$$\begin{bmatrix} a_{11} & 0 & 0 & \dots & 0 \\ 0 & a_{22} & 0 & \dots & 0 \\ 0 & 0 & a_{33} & \dots & 0 \\ \vdots & \vdots & \vdots & & \vdots \\ 0 & 0 & 0 & \dots & a_{nn} \end{bmatrix}$$

An important special case of a diagonal matrix is the **unit matrix** or **identity matrix** I, for which $a_{11} = a_{22} = \dots = a_{nn} = 1$.

$$I = \begin{bmatrix} 1 & 0 & 0 & \dots & 0 \\ 0 & 1 & 0 & \dots & 0 \\ 0 & 0 & 1 & \dots & 0 \\ \vdots & \vdots & \vdots & & \vdots \\ 0 & 0 & 0 & \dots & 1 \end{bmatrix}$$

The unit matrix can be written conveniently in terms of the Kronecker delta. This is defined as

$$\delta_{ij} = \begin{cases} 1 & \text{if} \quad i = j \\ 0 & \text{if} \quad i \neq j \end{cases}$$

The unit matrix thus has elements δ_{ij}. The notation $\boldsymbol{I}_n$ is sometimes used to denote the $n \times n$ unit matrix where its size is important or not clear.

The **zero** or **null matrix** is the matrix with every element zero, and is written as either 0 or $\boldsymbol{0}$. Sometimes a zero matrix of order $m \times n$ is written $\boldsymbol{O}_{m\times n}$.

The **transposed matrix** $\boldsymbol{A}^T$ of (5.4) is the matrix with elements $b_{ij} = a_{ji}$ and is written in full as the $n \times m$ matrix

$$\boldsymbol{A}^T = \begin{bmatrix} a_{11} & a_{21} & a_{31} & \cdots & a_{m1} \\ a_{12} & a_{22} & a_{32} & \cdots & a_{m2} \\ \vdots & \vdots & \vdots & & \vdots \\ a_{1n} & a_{2n} & a_{3n} & \cdots & a_{mn} \end{bmatrix}$$

This is just the matrix in (5.4) with rows and columns interchanged. We may note from (5.5) that

$$\boldsymbol{b}^T = [b_1 \quad b_2 \quad \cdots \quad b_m] \quad \text{and} \quad \boldsymbol{c}^T = \begin{bmatrix} c_1 \\ c_2 \\ \vdots \\ c_n \end{bmatrix}$$

so that a column vector is transposed to a row vector and vice versa.

If a square matrix is such that $\boldsymbol{A}^T = \boldsymbol{A}$ then $a_{ij} = a_{ji}$, and the elements are therefore symmetric about the diagonal. Such a matrix is called a **symmetric matrix**; symmetric matrices play important roles in many computations. If $\boldsymbol{A}^T = -\boldsymbol{A}$, so that $a_{ij} = -a_{ji}$, the matrix is called **skew-symmetric** or **antisymmetric**. Obviously the diagonal elements of a skew-symmetric matrix satisfy $a_{ii} = -a_{ii}$ and so must all be zero.

A few examples will illustrate these definitions:

$$\boldsymbol{A} = \begin{bmatrix} 2 & 3 \\ 1 & 2 \\ 4 & 5 \end{bmatrix} \text{ is a } 3 \times 2 \text{ matrix}$$

$$\boldsymbol{A}^T = \begin{bmatrix} 2 & 1 & 4 \\ 3 & 2 & 5 \end{bmatrix} \text{ is a } 2 \times 3 \text{ matrix}$$

$$\boldsymbol{B} = \begin{bmatrix} 1 & 2 & 3 \\ 2 & 3 & 4 \\ 3 & 4 & 5 \end{bmatrix} \text{ is a symmetric } 3 \times 3 \text{ matrix}$$

trace $B = 1 + 3 + 5 = 9$

$$C = \begin{bmatrix} 0 & 7 & -1 \\ -7 & 0 & 4 \\ 1 & -4 & 0 \end{bmatrix} \text{ and } C^T = \begin{bmatrix} 0 & -7 & 1 \\ 7 & 0 & -4 \\ -1 & 4 & 0 \end{bmatrix} \text{ are skew-symmetric}$$

3×3 matrices

$$D = \begin{bmatrix} 2 & 0 & 0 \\ 0 & 3 & 0 \\ 0 & 0 & 4 \end{bmatrix} \text{ is a } 3 \times 3 \text{ diagonal matrix}$$

trace $D = 2 + 3 + 4 = 9$

$$I = \begin{bmatrix} 1 & 0 & 0 & 0 \\ 0 & 1 & 0 & 0 \\ 0 & 0 & 1 & 0 \\ 0 & 0 & 0 & 1 \end{bmatrix} \text{ is the } 4 \times 4 \text{ unit matrix (sometimes written } I_4)$$

5.2.2 Basic operations of matrices

(a) Equality

Two matrices A and B are said to be **equal** if and only if all their elements are the same, $a_{ij} = b_{ij}$ for $1 \leqslant i \leqslant m$, $1 \leqslant j \leqslant n$, and this equality is written as

$$A = B$$

Note that this requires the two matrices to be of the same order $m \times n$.

(b) Addition and subtraction

Addition of matrices is straightforward; we can only add an $m \times n$ matrix to another $m \times n$ matrix, and an element of the sum is the sum of the corresponding elements. If A has elements a_{ij} and B has elements b_{ij} then $A + B$ has elements $a_{ij} + b_{ij}$.

$$\begin{bmatrix} a_{11} & a_{12} & a_{13} & \cdots \\ a_{21} & a_{22} & a_{23} & \cdots \\ \vdots & \vdots & \vdots & \end{bmatrix} + \begin{bmatrix} b_{11} & b_{12} & b_{13} & \cdots \\ b_{21} & b_{22} & b_{23} & \cdots \\ \vdots & \vdots & \vdots & \end{bmatrix}$$

$$= \begin{bmatrix} a_{11} + b_{11} & a_{12} + b_{12} & a_{13} + b_{13} & \cdots \\ a_{21} + b_{21} & a_{22} + b_{22} & a_{23} + b_{23} & \cdots \\ \vdots & \vdots & \vdots & \end{bmatrix}$$

Similarly for subtraction, $A - B$ has elements $a_{ij} - b_{ij}$.

(c) Multiplication by a scalar

The matrix $\lambda\mathbf{A}$ has elements λa_{ij}; that is, we just multiply each element by the scalar λ

$$\lambda \begin{bmatrix} a_{11} & a_{12} & a_{13} & \cdots \\ a_{21} & a_{22} & a_{23} & \cdots \\ \vdots & \vdots & \vdots & \end{bmatrix} = \begin{bmatrix} \lambda a_{11} & \lambda a_{12} & \lambda a_{13} & \cdots \\ \lambda a_{21} & \lambda a_{22} & \lambda a_{23} & \cdots \\ \vdots & \vdots & \vdots & \end{bmatrix}$$

(d) Properties of the transpose

From the definition, the transpose of a matrix is such that

$$(\mathbf{A} + \mathbf{B})^{\mathrm{T}} = \mathbf{A}^{\mathrm{T}} + \mathbf{B}^{\mathrm{T}}$$

Similarly, we observe that

$$(\mathbf{A}^{\mathrm{T}})^{\mathrm{T}} = \mathbf{A}$$

so that transposing twice gives back the original matrix.

We may note as a special case of this result that for a square matrix $\mathbf{A}$

$$(\mathbf{A}^{\mathrm{T}} + \mathbf{A})^{\mathrm{T}} = (\mathbf{A}^{\mathrm{T}})^{\mathrm{T}} + \mathbf{A}^{\mathrm{T}} = \mathbf{A} + \mathbf{A}^{\mathrm{T}}$$

and hence $\mathbf{A}^{\mathrm{T}} + \mathbf{A}$ must be a symmetric matrix. This proves to be a very useful result, which we shall see used in several places. Similarly, $\mathbf{A} - \mathbf{A}^{\mathrm{T}}$ is a skew-symmetric matrix, so that any square matrix $\mathbf{A}$ may be expressed as the sum of a symmetric and a skew-symmetric matrix:

$$\mathbf{A} = \tfrac{1}{2}(\mathbf{A} + \mathbf{A}^{\mathrm{T}}) + \tfrac{1}{2}(\mathbf{A} - \mathbf{A}^{\mathrm{T}})$$

(e) Basic rules of addition

Because the usual rules of arithmetic are followed in the definitions of the sum of matrices and of multiplication by scalars, the

commutative law	$\mathbf{A} + \mathbf{B} = \mathbf{B} + \mathbf{A}$
associative law	$(\mathbf{A} + \mathbf{B}) + \mathbf{C} = \mathbf{A} + (\mathbf{B} + \mathbf{C})$

and

distributive law	$\lambda(\mathbf{A} + \mathbf{B}) = \lambda\mathbf{A} + \lambda\mathbf{B}$

all hold for matrices.

Example 5.1   Let

$$A = \begin{bmatrix} 1 & 2 & 1 \\ 1 & 1 & 2 \\ 1 & 1 & 1 \end{bmatrix}, \quad B = \begin{bmatrix} 2 & 1 \\ 1 & 0 \\ 1 & 1 \end{bmatrix}, \quad C = \begin{bmatrix} 0 & 1 & 1 \\ 0 & 0 & 1 \\ 1 & 0 & 0 \end{bmatrix}$$

Find, where possible, (a) $A + B$, (b) $A + C$, (c) $C - A$, (d) $3A$, (e) $4B$, (f) $C + B$, (g) $3A + 2C$, (h) $A^T + A$ and (i) $A + C^T + B^T$.

Solution   (a) $A + B$ is not possible.

(b) $A + C = \begin{bmatrix} 1+0 & 2+1 & 1+1 \\ 1+0 & 1+0 & 2+1 \\ 1+1 & 1+0 & 1+0 \end{bmatrix} = \begin{bmatrix} 1 & 3 & 2 \\ 1 & 1 & 3 \\ 2 & 1 & 1 \end{bmatrix}$

(c) $C - A = \begin{bmatrix} 0-1 & 1-2 & 1-1 \\ 0-1 & 0-1 & 1-2 \\ 1-1 & 0-1 & 0-1 \end{bmatrix} = \begin{bmatrix} -1 & -1 & 0 \\ -1 & -1 & -1 \\ 0 & -1 & -1 \end{bmatrix}$

(d) $3A = \begin{bmatrix} 3 & 6 & 3 \\ 3 & 3 & 6 \\ 3 & 3 & 3 \end{bmatrix}$

(e) $4B = \begin{bmatrix} 8 & 4 \\ 4 & 0 \\ 4 & 4 \end{bmatrix}$

(f) $C + B$ is not possible.

(g) $3A + 2C = \begin{bmatrix} 3 & 6 & 3 \\ 3 & 3 & 6 \\ 3 & 3 & 3 \end{bmatrix} + \begin{bmatrix} 0 & 2 & 2 \\ 0 & 0 & 2 \\ 2 & 0 & 0 \end{bmatrix} = \begin{bmatrix} 3 & 8 & 5 \\ 3 & 3 & 8 \\ 5 & 3 & 3 \end{bmatrix}$

(h) $A^T + A = \begin{bmatrix} 1 & 1 & 1 \\ 2 & 1 & 1 \\ 1 & 2 & 1 \end{bmatrix} + \begin{bmatrix} 1 & 2 & 1 \\ 1 & 1 & 2 \\ 1 & 1 & 1 \end{bmatrix} = \begin{bmatrix} 2 & 3 & 2 \\ 3 & 2 & 3 \\ 2 & 3 & 2 \end{bmatrix}$

(Note that this matrix is symmetric.)

(i) $A + C^T + B^T$ is not possible.

Example 5.2 A local roadside cafe serves beefburgers, eggs, chips and beans in four combination meals:

Slimmers	–	150 g chips	100 g beans	1 burger
Normal	1 egg	250 g chips	150 g beans	1 burger
Jumbo	2 eggs	350 g chips	200 g beans	2 burgers
Veggie	1 egg	200 g chips	150 g beans	–

A party orders 1 slimmer, 4 normal, 2 jumbo and 2 veggie meals. What is the total amount of materials that the kitchen staff need to cook? One of the customers sees the size of a jumbo meal and changes his order to a normal meal. How much less material will the kitchen staff need?

Solution The meals written in matrix form are

$$
s = \begin{bmatrix} 0 \\ 150 \\ 100 \\ 1 \end{bmatrix}, \quad n = \begin{bmatrix} 1 \\ 250 \\ 150 \\ 1 \end{bmatrix},
$$

$$
j = \begin{bmatrix} 2 \\ 350 \\ 200 \\ 2 \end{bmatrix}, \quad v = \begin{bmatrix} 1 \\ 200 \\ 150 \\ 0 \end{bmatrix}
$$

and hence the kitchen requirements are

$$
s + 4n + 2j + 2v = \begin{bmatrix} 10 \\ 2250 \\ 1400 \\ 9 \end{bmatrix}
$$

The change in requirements is

$$
j - n = \begin{bmatrix} 2 \\ 350 \\ 200 \\ 2 \end{bmatrix} - \begin{bmatrix} 1 \\ 250 \\ 150 \\ 1 \end{bmatrix} = \begin{bmatrix} 1 \\ 100 \\ 50 \\ 1 \end{bmatrix}
$$

less materials needed.

Although this may appear to be a rather trivial example, the basic problem is identical to any production process that requires a supply of parts.

All the basic matrix operations may be implemented in MATLAB and MAPLE using simple commands.

MATLAB
A matrix is entered as an array, with row elements separated by a space (or a comma) and each row of elements separated by a semicolon. Thus for example

```
A = [1 2 3; 4 0 5; 7 6 2]
```

gives A as

```
A =
   1  2  3
   4  0  5
   7  6  2
```

The transpose of a matrix is written A', with an apostrophe.

```
A' =
   1  4  7
   2  0  6
   3  5  2
```

and *trace*(A) produces the obvious answer = 3.

Having specified two matrices *A* and *B* the usual operations are written

```
C = A + B,  C = A - B
```

and multiplication with a scalar as

```
C = 2*A + 3*B
```

MAPLE
There are several ways of setting up arrays in MAPLE the simplest is to use the linear algebra package

```
with (linalg):
array([list of elements]);
```

Thus for example

```
A:=array([[1,2,3],[4,0,5],
[7,6,2]]);
```

produces

```
       1  2  3
A =    4  0  5
       7  6  2
```

The transpose and trace are obtained from

```
transpose(A); and trace(A);
```

Having specified two matrices *A* and *B* the usual operations are written

```
C := A + B; and C := A - B;
```

Because MAPLE is a symbolic package, the *evaluation* of the multiplication of a matrix by a scalar requires the command

```
C:= evalm(2*A + 3*B);
```

5.2.3 Exercises

Check your answers using MATLAB or MAPLE whenever possible.

1 Given the matrices

$$a = \begin{bmatrix} 1 \\ 2 \\ 0 \end{bmatrix}, \quad b = [0 \quad 1 \quad 1],$$

$$C = \begin{bmatrix} 3 & 2 & 1 \\ 1 & 2 & -1 \end{bmatrix}, \quad D = \begin{bmatrix} 5 & 6 \\ 7 & 8 \\ 9 & 10 \end{bmatrix}$$

evaluate, where possible, (a) $a + b$, (b) $b^{\mathrm{T}} + a$, (c) $b + C^{\mathrm{T}}$, (d) $C + D$, (e) $D^{\mathrm{T}} + C$.

2 Given the matrices

$$A = \begin{bmatrix} 1 & 0 & 1 \\ 2 & 2 & 0 \\ 0 & 1 & 2 \end{bmatrix} \quad \text{and}$$

$$B = \begin{bmatrix} -1 & 1 & 0 \\ 1 & 0 & -1 \\ 0 & -1 & 1 \end{bmatrix}$$

evaluate C in the three cases.

(a) $C = A + B$ (b) $2A + 3C = 4B$

(c) $A - C = B + C$

3 Solve for the matrix X

$$X - 2\begin{bmatrix} 3 & 2 & -1 \\ 7 & 2 & 6 \end{bmatrix} = \begin{bmatrix} -2 & 3 & 1 \\ 4 & 6 & 2 \end{bmatrix}$$

4 If

$$A = \begin{bmatrix} 1 & 2 & -3 \\ 5 & 0 & 2 \\ 1 & -1 & 1 \end{bmatrix}, \quad B = \begin{bmatrix} 3 & -1 & 2 \\ 4 & 2 & 5 \\ 2 & 0 & 5 \end{bmatrix}$$

and $\quad C = \begin{bmatrix} 4 & 0 & -2 \\ 5 & 3 & 1 \\ 2 & 5 & 4 \end{bmatrix}$

(a) show that

$$\text{trace}(A + B) = \text{trace } A + \text{trace } B$$

(b) find D so that $A + D = C$.

(c) verify the associative law

$$(A + B) + C = A + (B + C)$$

5 Find the values of x, y, z and t from the equation

$$\begin{bmatrix} x & y - x + t \\ t - z & z - 1 \end{bmatrix} = \begin{bmatrix} 1 & 2 \\ 0 & 1 \end{bmatrix}$$

6 Find the values of α, β, γ that satisfy

$$\alpha \begin{bmatrix} 1 \\ 0 \\ 0 \end{bmatrix} + \beta \begin{bmatrix} 1 \\ -1 \\ 0 \end{bmatrix} + \gamma \begin{bmatrix} 0 \\ 1 \\ -1 \end{bmatrix} = \begin{bmatrix} 0 \\ 3 \\ -2 \end{bmatrix}$$

7 Given the matrix

$$A = \lambda \begin{bmatrix} 1 & 0 \\ 0 & 1 \end{bmatrix} + \mu \begin{bmatrix} 1 & 1 \\ 0 & 1 \end{bmatrix} + v \begin{bmatrix} 0 & 0 \\ 0 & 1 \end{bmatrix}$$

(a) find the value of λ, μ, v so that $A = \begin{bmatrix} 0 & -1 \\ 0 & 3 \end{bmatrix}$;

(b) show that no solution is possible if

$$A = \begin{bmatrix} 1 & -1 \\ 1 & 0 \end{bmatrix}.$$

8 Market researchers are testing customers' preferences for five products. There are four researchers who are allocated to different groups: researcher R_1 deals with men under 40, R_2 deals with men over 40, R_3 deals with women under 40 and R_4 deals with women over 40. They return their findings as a vector giving the number of customers with first preference for a particular product.

Product	R_1	R_2	R_3	R_4
a	23	32	28	39
b	34	22	33	21
c	18	21	22	17
d	9	15	10	12
e	16	10	7	11

Find the average over the whole sample. The company decides that their main target is older women so they weight the returns in the ratio $1 : 1 : 2 : 3$; find the weighted average.

9 A newspaper shop has sales organized in matrix form as

	Mon	Tues	Weds	Thur	Fri
Post	272	331	309	284	348
News	356	402	389	312	458
Bulletin	157	200	179	196	110

and wishes to increase the sales by 15%. Find the additional number of papers required, rounded down to the nearest integer and the projected total sales.

10 A builder's yard organizes its stock in the form of a vector

Bricks – type A
Bricks – type B
Bricks – type C
Bags of cement
Tons of sand

The current stock, S, and the minimum stock, M, required to avoid running out of materials, are given as

$$S = \begin{bmatrix} 45\,750 \\ 23\,600 \\ 17\,170 \\ 462 \\ 27 \end{bmatrix} \quad \text{and} \quad M = \begin{bmatrix} 5000 \\ 4000 \\ 3500 \\ 100 \\ 10 \end{bmatrix}$$

The firm has five lorries which take materials from stock for deliveries; Lorry1 makes three deliveries in the day with the same load each time; Lorry2 makes two deliveries in the day with the same load each time; the other lorries make one delivery. The loads are

$$L_1 = \begin{bmatrix} 5500 \\ 0 \\ 3800 \\ 75 \\ 3 \end{bmatrix} \quad L_2 = \begin{bmatrix} 2500 \\ 1500 \\ 0 \\ 40 \\ 2 \end{bmatrix} \quad L_3 = \begin{bmatrix} 7500 \\ 2000 \\ 1500 \\ 0 \\ 3 \end{bmatrix}$$

$$L_4 = \begin{bmatrix} 0 \\ 4000 \\ 2500 \\ 20 \\ 2 \end{bmatrix} \quad L_5 = \begin{bmatrix} 2000 \\ 0 \\ 1500 \\ 15 \\ 0 \end{bmatrix}$$

How much material has gone from stock, what is the current stock position and has any element gone below the minimum?

5.2.4 Matrix multiplication

The most important property of matrices as far as their practical applications are concerned is the multiplication of one matrix by another. We saw informally in Section 5.2 how multiplication arose and how to define multiplication of a matrix and a vector. The idea can be extended further by looking again at change of axes. Take

$$z_1 = a_{11}y_1 + a_{12}y_2, \quad y_1 = b_{11}x_1 + b_{12}x_2$$
$$z_2 = a_{21}y_1 + a_{22}y_2, \quad y_2 = b_{21}x_1 + b_{22}x_2$$

Then we can ask for the transformation from the zs to the xs. This we can do by straight substitution:

$$z_1 = (a_{11}b_{11} + a_{12}b_{21})x_1 + (a_{11}b_{12} + a_{12}b_{22})x_2$$
$$z_2 = (a_{21}b_{11} + a_{22}b_{21})x_1 + (a_{21}b_{12} + a_{22}b_{22})x_2$$

If we write the first two transformations as

$$A = \begin{bmatrix} a_{11} & a_{12} \\ a_{21} & a_{22} \end{bmatrix} \quad \text{and} \quad B = \begin{bmatrix} b_{11} & b_{12} \\ b_{21} & b_{22} \end{bmatrix}$$

then the composite transformation is written

$$AB = \begin{bmatrix} a_{11}b_{11} + a_{12}b_{21} & a_{11}b_{12} + a_{12}b_{22} \\ a_{21}b_{11} + a_{22}b_{21} & a_{21}b_{12} + a_{22}b_{22} \end{bmatrix}$$

and this is precisely how we define the matrix product.

Definition

If A is an $m \times p$ matrix with elements a_{ij} and B a $p \times n$ matrix with elements b_{ij} then we define the **product** $C = AB$ as the $m \times n$ matrix with components

$$c_{ij} = \sum_{k=1}^{p} a_{ik}b_{kj} \quad \text{for } i = 1, \dots, m \quad \text{and} \quad j = 1, \dots, n$$

Illustration

$z_1 = y_1 + 3y_2$
$z_2 = 2y_1 - y_2$

$y_1 = -x_1 + 2x_2$
$y_2 = 2x_1 - x_2$

Substitute to get

$z_1 = 5x_1 - x_2$
$z_2 = -4x_1 + 5x_2$

In matrix form

$$A = \begin{bmatrix} 1 & 3 \\ 2 & -1 \end{bmatrix}$$

$$B = \begin{bmatrix} -1 & 2 \\ 2 & -1 \end{bmatrix}$$

$$AB = \begin{bmatrix} 5 & -1 \\ -4 & 5 \end{bmatrix}$$

In pictorial form, the ith row of A is multiplied term by term with the jth column of B and the products are added to form the ijth component of C. This is commonly referred to as the 'row-by-column' method of multiplication. Clearly, in order for multiplication to be possible, A must have p columns and B must have p rows otherwise the product AB is not defined.

$$
i \rightarrow \begin{bmatrix} & \vdots & \\ \cdots & c_{ij} & \cdots \\ & \vdots & \end{bmatrix} = i \rightarrow \begin{bmatrix} a_{i1} & a_{i2} & \cdots & a_{ip} \end{bmatrix} \begin{bmatrix} b_{1j} \\ b_{2j} \\ \vdots \\ b_{pj} \end{bmatrix}
$$

Example 5.3

Given

$$
A = \begin{bmatrix} 1 & 1 & 0 \\ 2 & 0 & 1 \end{bmatrix}, \quad B = \begin{bmatrix} 2 & 0 \\ 0 & 1 \\ 1 & 3 \end{bmatrix},
$$

$$
b = \begin{bmatrix} -1 \\ 2 \end{bmatrix}, \quad c = \begin{bmatrix} 1 \\ 1 \\ -1 \end{bmatrix} \quad \text{and} \quad C = \begin{bmatrix} 1 & -2 \\ -1 & 2 \\ -2 & 4 \end{bmatrix}
$$

find (a) AB, (b) BA, (c) Bb, (d) $A^{\mathrm{T}}b$, (e) $c^{\mathrm{T}}(A^{\mathrm{T}}b)$ and (f) AC.

Solution

(a) $$AB = \begin{bmatrix} 1 & 1 & 0 \\ 2 & 0 & 1 \end{bmatrix} \begin{bmatrix} 2 & 0 \\ 0 & 1 \\ 1 & 3 \end{bmatrix} = \begin{bmatrix} \text{row}\,1 \times \text{col}\,1 & \text{row}\,1 \times \text{col}\,2 \\ \text{row}\,2 \times \text{col}\,1 & \text{row}\,2 \times \text{col}\,2 \end{bmatrix}$$

$$
= \begin{bmatrix} (1)(2) + (1)(0) + (0)(1) & (1)(0) + (1)(1) + (0)(3) \\ (2)(2) + (0)(0) + (1)(1) & (2)(0) + (0)(1) + (1)(3) \end{bmatrix} = \begin{bmatrix} 2 & 1 \\ 5 & 3 \end{bmatrix}
$$

(b) $$BA = \begin{bmatrix} 2 & 0 \\ 0 & 1 \\ 1 & 3 \end{bmatrix} \begin{bmatrix} 1 & 1 & 0 \\ 2 & 0 & 1 \end{bmatrix} = \begin{bmatrix} 2+0 & 2+0 & 0+0 \\ 0+2 & 0+0 & 0+1 \\ 1+6 & 1+0 & 0+3 \end{bmatrix} = \begin{bmatrix} 2 & 2 & 0 \\ 2 & 0 & 1 \\ 7 & 1 & 3 \end{bmatrix}$$

(Note that BA is not equal to AB.)

(c) $$Bb = \begin{bmatrix} 2 & 0 \\ 0 & 1 \\ 1 & 3 \end{bmatrix} \begin{bmatrix} -1 \\ 2 \end{bmatrix} = \begin{bmatrix} -2 \\ 2 \\ 5 \end{bmatrix}$$

(d) $A^T b = \begin{bmatrix} 1 & 2 \\ 1 & 0 \\ 0 & 1 \end{bmatrix} \begin{bmatrix} -1 \\ 2 \end{bmatrix} = \begin{bmatrix} 3 \\ -1 \\ 2 \end{bmatrix}$

(e) $c^T(A^T b) = \begin{bmatrix} 1 & 1 & -1 \end{bmatrix} \begin{bmatrix} 3 \\ -1 \\ 2 \end{bmatrix} = [0] = 0$

(Note that this matrix is the zero 1×1 matrix, which can just be written 0.)

(f) $AC = \begin{bmatrix} 1 & 1 & 0 \\ 2 & 0 & 1 \end{bmatrix} \begin{bmatrix} 1 & -2 \\ -1 & 2 \\ -2 & 4 \end{bmatrix} = \begin{bmatrix} 0 & 0 \\ 0 & 0 \end{bmatrix} = 0$

(Note that the product AC is zero even though neither A nor C is zero.)

Example 5.4   If

$$A = \begin{bmatrix} 1 & 2 & 0 \\ 1 & 1 & 0 \\ 2 & 1 & 1 \end{bmatrix} \quad \text{and} \quad X = \begin{bmatrix} x \\ y \\ z \end{bmatrix}$$

evaluate (a) $X^T X$, (b) AX, (c) $X^T(AX)$ and (d) $\frac{1}{2} X^T[(A^T + A)X]$.

Solution

(a) $X^T X = \begin{bmatrix} x & y & z \end{bmatrix} \begin{bmatrix} x \\ y \\ z \end{bmatrix} = x^2 + y^2 + z^2$

(b) $AX = \begin{bmatrix} 1 & 2 & 0 \\ 1 & 1 & 0 \\ 2 & 1 & 1 \end{bmatrix} \begin{bmatrix} x \\ y \\ z \end{bmatrix} = \begin{bmatrix} x + 2y \\ x + y \\ 2x + y + z \end{bmatrix}$

(c) $X^T(AX) = \begin{bmatrix} x & y & z \end{bmatrix} \begin{bmatrix} x + 2y \\ x + y \\ 2x + y + z \end{bmatrix} = (x^2 + 2xy) + (yx + y^2) + (2xz + yz + z^2)$

$= x^2 + y^2 + z^2 + 3xy + 2xz + yz$

(d) $\frac{1}{2}(A^T + A) = \begin{bmatrix} 1 & \frac{3}{2} & 1 \\ \frac{3}{2} & 1 & \frac{1}{2} \\ 1 & \frac{1}{2} & 1 \end{bmatrix}$

and

$$\tfrac{1}{2}(\mathbf{A}^{\mathrm{T}} + \mathbf{A})\mathbf{X} = \begin{bmatrix} 1 & \tfrac{3}{2} & 1 \\ \tfrac{3}{2} & 1 & \tfrac{1}{2} \\ 1 & \tfrac{1}{2} & 1 \end{bmatrix} \begin{bmatrix} x \\ y \\ z \end{bmatrix} = \begin{bmatrix} x + \tfrac{3}{2}y + z \\ \tfrac{3}{2}x + y + \tfrac{1}{2}z \\ x + \tfrac{1}{2}y + z \end{bmatrix}$$

Therefore

$$\tfrac{1}{2}\mathbf{X}^{\mathrm{T}}[(\mathbf{A}^{\mathrm{T}} + \mathbf{A})\mathbf{X}] = \begin{bmatrix} x & y & z \end{bmatrix} \begin{bmatrix} x + \tfrac{3}{2}y + z \\ \tfrac{3}{2}x + y + \tfrac{1}{2}z \\ x + \tfrac{1}{2}y + z \end{bmatrix}$$

$$= x^2 + y^2 + z^2 + 3xy + 2xz + yz$$

(Note that this is the same as the result of part (c).)

There are several points to note from the preceding examples. One-by-one matrices are just numbers, so the square brackets become redundant and are usually omitted. The expression $\mathbf{X}^{\mathrm{T}}\mathbf{X}$ just gives the square of the length of the vector $\mathbf{X}$ in the usual sense, namely $\mathbf{X}^{\mathrm{T}}\mathbf{X} = x^2 + y^2 + z^2$. Similarly,

$$\mathbf{X}^{\mathrm{T}}\mathbf{X}' = \begin{bmatrix} x & y & z \end{bmatrix} \begin{bmatrix} x' \\ y' \\ z' \end{bmatrix} = xx' + yy' + zz'$$

which is the usual **scalar** or **inner product**, here written in matrix form. The expression $\mathbf{AX}$ gives a column vector with linear expressions as its elements. Using Example 5.4(b), we can rewrite the linear equations

$$x + 2y \quad = 3$$
$$x + y \quad = 4$$
$$2x + y + z = 5$$

as

$$\begin{bmatrix} 1 & 2 & 0 \\ 1 & 1 & 0 \\ 2 & 1 & 1 \end{bmatrix} \begin{bmatrix} x \\ y \\ z \end{bmatrix} = \begin{bmatrix} 3 \\ 4 \\ 5 \end{bmatrix}$$

which may be written in the standard matrix form for linear equations as

$$\mathbf{AX} = \mathbf{b}$$

It is also important to realize that if $\mathbf{AB} = 0$ it does not follow that either $\mathbf{A}$ or $\mathbf{B}$ is zero. In Example 5.3(f) we saw that the product $\mathbf{AC} = 0$, but neither $\mathbf{A}$ nor $\mathbf{C}$ is the zero matrix.

Solution

Now $\boldsymbol{BC} = \begin{bmatrix} 0 & 0 & 1 \\ 0 & 2 & 3 \\ 1 & 2 & 3 \end{bmatrix} \begin{bmatrix} 2 & 3 \\ -1 & 2 \\ -3 & 1 \end{bmatrix} = \begin{bmatrix} -3 & 1 \\ -11 & 7 \\ -9 & 10 \end{bmatrix}$ and

$\boldsymbol{A(BC)} = \begin{bmatrix} 1 & -1 & 1 \\ -2 & 0 & 3 \\ 0 & 1 & -2 \end{bmatrix} \begin{bmatrix} -3 & 1 \\ -11 & 7 \\ -9 & 10 \end{bmatrix} = \begin{bmatrix} -1 & 4 \\ -21 & 28 \\ 7 & -13 \end{bmatrix}$

Likewise $\boldsymbol{AB} = \begin{bmatrix} 1 & -1 & 1 \\ -2 & 0 & 3 \\ 0 & 1 & -2 \end{bmatrix} \begin{bmatrix} 0 & 0 & 1 \\ 0 & 2 & 3 \\ 1 & 2 & 3 \end{bmatrix} = \begin{bmatrix} 1 & 0 & 1 \\ 3 & 6 & 7 \\ -2 & -2 & -3 \end{bmatrix}$ and

$\boldsymbol{(AB)C} = \begin{bmatrix} 1 & 0 & 1 \\ 3 & 6 & 7 \\ -2 & -2 & -3 \end{bmatrix} \begin{bmatrix} 2 & 3 \\ -1 & 2 \\ -3 & 1 \end{bmatrix} = \begin{bmatrix} -1 & 4 \\ -21 & 28 \\ 7 & -13 \end{bmatrix}$

Thus the associative law is satisfied for these three matrices. For the distributive law we need to evaluate

$\boldsymbol{(A + B)C} = \begin{bmatrix} 1 & -1 & 2 \\ -2 & 2 & 6 \\ 1 & 3 & 1 \end{bmatrix} \begin{bmatrix} 2 & 3 \\ -1 & 2 \\ -3 & 1 \end{bmatrix} = \begin{bmatrix} -3 & 3 \\ -24 & 4 \\ -4 & 10 \end{bmatrix}$

and

$\boldsymbol{AC + BC} = \begin{bmatrix} 1 & -1 & 1 \\ -2 & 0 & 3 \\ 0 & 1 & -2 \end{bmatrix} \begin{bmatrix} 2 & 3 \\ -1 & 2 \\ -3 & 1 \end{bmatrix} + \begin{bmatrix} 0 & 0 & 1 \\ 0 & 2 & 3 \\ 1 & 2 & 3 \end{bmatrix} \begin{bmatrix} 2 & 3 \\ -1 & 2 \\ -3 & 1 \end{bmatrix}$

$= \begin{bmatrix} 0 & 2 \\ -13 & -3 \\ 5 & 0 \end{bmatrix} + \begin{bmatrix} -3 & 1 \\ -11 & 7 \\ -9 & 10 \end{bmatrix} = \begin{bmatrix} -3 & 3 \\ -24 & 4 \\ -4 & 10 \end{bmatrix}$

The two matrices are equal so the distributive law is verified for the three given matrices.

Example 5.7 Show that the transformation

$$\begin{bmatrix} x' \\ y' \end{bmatrix} = \begin{bmatrix} \cos\theta & \sin\theta \\ -\sin\theta & \cos\theta \end{bmatrix} \begin{bmatrix} x \\ y \end{bmatrix}$$

with $\theta = 60°$, maps the square with corners $\begin{bmatrix} 1 \\ 1 \end{bmatrix}$, $\begin{bmatrix} 1 \\ 2 \end{bmatrix}$, $\begin{bmatrix} 2 \\ 2 \end{bmatrix}$ and $\begin{bmatrix} 2 \\ 1 \end{bmatrix}$ on to a square.

Solution   Substituting the given vectors in turn for $\begin{bmatrix} x \\ y \end{bmatrix}$ into the equation

$$\begin{bmatrix} x' \\ y' \end{bmatrix} = \begin{bmatrix} 0.5 & 0.8660 \\ -0.8660 & 0.5 \end{bmatrix} \begin{bmatrix} x \\ y \end{bmatrix}$$

we find the following vectors for $\begin{bmatrix} x' \\ y' \end{bmatrix}$

$$\begin{bmatrix} 1.366 \\ -0.366 \end{bmatrix}, \quad \begin{bmatrix} 2.232 \\ 0.134 \end{bmatrix}, \quad \begin{bmatrix} 2.732 \\ -0.732 \end{bmatrix} \quad \text{and} \quad \begin{bmatrix} 1.866 \\ -1.232 \end{bmatrix}$$

Plotting these points on the plane as in Figure 5.4, we see that the square has been rotated through an angle of 60° about the origin. It is left as an exercise for the reader to verify the result.

This type of analysis forms the basis of manipulation of diagrams on a computer screen, and is used in many CAD/CAM situations.

Figure 5.4
Transformation of a
square in Example 5.7.

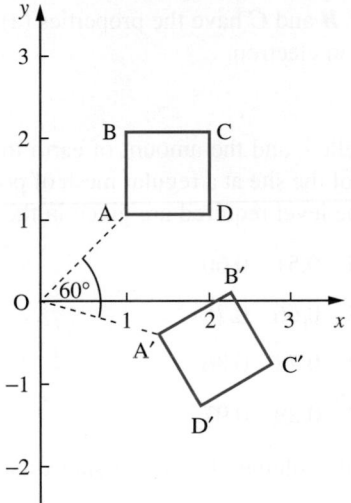

Example 5.8   In quantum mechanics the components of the spin of an electron can be repesented by the Pauli matrices

$$A = \begin{bmatrix} 0 & 1 \\ 1 & 0 \end{bmatrix}, \quad B = \begin{bmatrix} 0 & -j \\ j & 0 \end{bmatrix}, \quad C = \begin{bmatrix} 1 & 0 \\ 0 & -1 \end{bmatrix}$$

Show that

(a) the matrices anticommute

$$AB + BA = 0, \quad BC + CB = 0, \quad CA + AC = 0$$

(b) $AB - BA = 2jC$, $\quad BC - CB = 2jA$, $\quad CA - AC = 2jB$

(c) $AB = jC$, $\quad BC = jA$, $\quad CA = jB$

Solution (a) $AB = \begin{bmatrix} 0 & 1 \\ 1 & 0 \end{bmatrix}\begin{bmatrix} 0 & -j \\ j & 0 \end{bmatrix} = \begin{bmatrix} j & 0 \\ 0 & -j \end{bmatrix}$ and $BA = \begin{bmatrix} 0 & -j \\ j & 0 \end{bmatrix}\begin{bmatrix} 0 & 1 \\ 1 & 0 \end{bmatrix} = \begin{bmatrix} -j & 0 \\ 0 & j \end{bmatrix}$

so

$$AB + BA = 0$$

and the other two results follow similarly.

(b) From part (a)

$$AB - BA = \begin{bmatrix} j & 0 \\ 0 & -j \end{bmatrix} - \begin{bmatrix} -j & 0 \\ 0 & j \end{bmatrix} = \begin{bmatrix} 2j & 0 \\ 0 & -2j \end{bmatrix} = 2jC$$

and again the other two results follow similarly.

(c) These results can be obtained directly from part (a) since AB has already been calculated, similarly for BC and CA.

Note: that this example illustrates the use of matrices that have complex elements. Pauli discovered that the matrices A, B and C have the properties (a), (b) and (c) required of the components of the spin of an electron.

Example 5.9 A rectangular site is to be levelled, and the amount of earth that needs to be removed must be determined. A survey of the site at a regular mesh of points 10 m apart is made. The heights in metres above the level required are given in the following table.

0	0.31	0.40	0.45	0.51	0.60
0.12	0.33	0.51	0.58	0.66	0.75
0.19	0.38	0.60	0.69	0.78	0.86
0.25	0.46	0.68	0.77	0.89	0.97

It is known that the approximate volume of a cell of side x and with corner heights of a, b, c and d is

$$V = \tfrac{1}{4}x^2(a + b + c + d)$$

Write the total approximate volume in matrix form and hence estimate the volume to be removed.

Solution Note that for the first row of cells the volume is

$$25(\ 0 \quad +0.31 \quad +0.31+0.40 \quad +0.40+0.45 \quad +0.45+0.51 \quad +0.51+0.60$$
$$+0.12+0.33 \quad +0.33+0.51 \quad +0.51+0.58 \quad +0.58+0.66 \quad +0.66+0.75)$$

$$= 25[0 + 2(0.31 + 0.40 + 0.45 + 0.51) + 0.60]$$
$$+ 25[0.12 + 2(0.33 + 0.51 + 0.58 + 0.66) + 0.75]$$

The second and third rows of cells are dealt with in a similar manner, so that, when we compute the total volume, we need to multiply the corner values by 1, the other side values by 2 and the centre values by 4. In matrix form this multiplication can be performed as

$$[1 \quad 2 \quad 2 \quad 1] \begin{bmatrix} 0 & 0.31 & 0.40 & 0.45 & 0.51 & 0.60 \\ 0.12 & 0.33 & 0.51 & 0.58 & 0.66 & 0.75 \\ 0.19 & 0.38 & 0.60 & 0.69 & 0.78 & 0.86 \\ 0.25 & 0.46 & 0.68 & 0.77 & 0.86 & 0.97 \end{bmatrix} \begin{bmatrix} 1 \\ 2 \\ 2 \\ 2 \\ 2 \\ 1 \end{bmatrix}$$

 This can be checked by multiplying the matrices out. The checking can be done on one of the symbolic manipulation packages, such as MAPLE or the Symbolic Math Toolbox of MATLAB, by putting in general symbols for the matrix and verifying that, after the matrix multiplications, the elements are multiplied by the stated factors. Performing the calculation and multiplying by the 25 gives the total volume as $816.5\,\mathrm{m}^3$.

A similar analysis can be applied to other situations – all that is needed is measured heights and a matrix multiplication routine on a computer to deal with the large amount of data that would be required. For other mesh shapes, or even irregular meshes, the method is similar, but the multiplying vectors will need careful calculation.

Example 5.10

A contractor makes two products P_1 and P_2. The four components required to make the products are subcontracted out and each of the components is made up from three ingredients A, B and C as follows:

Component	Units of A	Units of B	Units of C	Make-up cost and profit for subcontractor
1 requires	5	4	3	10
2 requires	2	1	1	7
3 requires	0	1	3	5
4 requires	3	4	1	2

The cost per unit of the ingredients A, B and C are a, b and c respectively. The contractor makes the product P_1 with 2 of component 1, 3 of component 2 and 4 of component 4, the make-up cost is 15; product P_2 requires 1 of component 1, 1 of component 2, 1 of component 3 and 2 of component 4, the make-up cost is 12. Find the cost to the contractor for P_1 and P_2. What is the change in costs if a increases to $(a + 1)$? It is found that the 5 units of A required for component 1 can be reduced to 4. What is the effect on the costs?

Solution

The information presented can be written naturally in matrix form. Let C_1, C_2, C_3 and C_4 be the cost the subcontractor charges the contractor for the four components, then the cost C_1 is computed as $C_1 = 5a + 4b + 3c + 10$. This expression is the first row of the matrix equation

$$
\begin{bmatrix} C_1 \\ C_2 \\ C_3 \\ C_4 \end{bmatrix} = \begin{bmatrix} 5 & 4 & 3 \\ 2 & 1 & 1 \\ 0 & 1 & 3 \\ 3 & 4 & 1 \end{bmatrix} \begin{bmatrix} a \\ b \\ c \end{bmatrix} + \begin{bmatrix} 10 \\ 7 \\ 5 \\ 2 \end{bmatrix}
$$

and the other three costs follow in a similar manner. Now let p_1, p_2 be the costs of producing the final products. The costs are constructed in exactly the same way as

$$
\begin{bmatrix} p_1 \\ p_2 \end{bmatrix} = \begin{bmatrix} 2 & 3 & 0 & 4 \\ 1 & 1 & 1 & 2 \end{bmatrix} \begin{bmatrix} C_1 \\ C_2 \\ C_3 \\ C_4 \end{bmatrix} + \begin{bmatrix} 15 \\ 12 \end{bmatrix}
$$

Substituting gives

$$
\begin{bmatrix} p_1 \\ p_2 \end{bmatrix} = \begin{bmatrix} 2 & 3 & 0 & 4 \\ 1 & 1 & 1 & 2 \end{bmatrix} \begin{bmatrix} 5 & 4 & 3 \\ 2 & 1 & 1 \\ 0 & 1 & 3 \\ 3 & 4 & 1 \end{bmatrix} \begin{bmatrix} a \\ b \\ c \end{bmatrix} + \begin{bmatrix} 2 & 3 & 0 & 4 \\ 1 & 1 & 1 & 2 \end{bmatrix} \begin{bmatrix} 10 \\ 7 \\ 5 \\ 2 \end{bmatrix} + \begin{bmatrix} 15 \\ 12 \end{bmatrix}
$$

or

$$
\begin{bmatrix} p_1 \\ p_2 \end{bmatrix} = \begin{bmatrix} 28 & 27 & 13 \\ 13 & 14 & 9 \end{bmatrix} \begin{bmatrix} a \\ b \\ c \end{bmatrix} + \begin{bmatrix} 64 \\ 38 \end{bmatrix}
$$

Thus a simple matrix formulation gives a convenient way of coding the data. If a is increased to $(a + 1)$ then multiplying out shows that p_1 increases by 28 and p_2 by 13. If the 5 in the first matrix is reduced to 4 then the costs will be

$$
\begin{bmatrix} p_1 \\ p_2 \end{bmatrix} = \begin{bmatrix} 26 & 27 & 13 \\ 12 & 14 & 9 \end{bmatrix} \begin{bmatrix} a \\ b \\ c \end{bmatrix} + \begin{bmatrix} 64 \\ 38 \end{bmatrix}
$$

so p_1 is reduced by $2a$ and p_2 by a.

A similar approach can be used in more complicated, realistic situations. Storing and processing the information is convenient, particularly in conjunction with a computer package or spreadsheet.

Example 5.11

The tape of a tape recorder passes the reading head at a constant speed v. On the feed reel there is a length of tape L left and it has radius R. If the rev counter is set to zero and the thickness of the tape is h find how the radius and the length of the reel vary with the number of revolutions n.

Solution

Let t_n, R_n and L_n denote the time, radius and length after n revolutions. It is given that $t_0 = 0$, $R_0 = R$ and $L_0 = L$. At stage n after one further revolution a thickness h is peeled off the radius, the length is reduced by $2\pi R_n$ and the time has advanced by $2\pi R_n/v$. A table summarizes these remarks.

Revs (n)	*Time* (t_n)	*Radius* (R_n)	*Length remaining* (L_n)
0	$t_0 = 0$	$R_0 = R$	$L_0 = L$
1	$t_1 = 2\pi R_0/v + t_0$	$R_1 = R_0 - h$	$L_1 = L_0 - 2\pi R_0$
2	$t_2 = 2\pi R_1/v + t_1$	$R_2 = R_1 - h$	$L_2 = L_1 - 2\pi R_1$
...	...	...	...
...	...	...	...
$n + 1$	$t_{n+1} = 2\pi R_n/v + t_n$	$R_{n+1} = R_n - h$	$L_{n+1} = L_n - 2\pi R_n$

This data can be written conveniently in matrix notation as

$$\begin{bmatrix} t_{n+1} \\ R_{n+1} \\ L_{n+1} \end{bmatrix} = \begin{bmatrix} 1 & 2\pi/v & 0 \\ 0 & 1 & 0 \\ 0 & -2\pi & 1 \end{bmatrix} \begin{bmatrix} t_n \\ R_n \\ L_n \end{bmatrix} - \begin{bmatrix} 0 \\ h \\ 0 \end{bmatrix}$$

or identifying the matrices in an obvious manner as

$$X_{n+1} = A X_n - \Delta$$

Successive substitution gives

$$X_{n+1} = A(A X_{n-1} - \Delta) - \Delta = A^2 X_{n-1} - (A + I)\Delta$$

$$= A^2(A X_{n-2} - \Delta) - (A + I)\Delta = A^3 X_{n-2} - (A^2 + A + I)\Delta$$

$$\ldots$$

$$= A^{n+1} X_0 - (A^n + A^{n-1} + \ldots + A + I)\Delta \qquad (5.6)$$

The required values can now be determined by evaluating the right-hand side of this equation. Repeated products of A gives

$$A^n = \begin{bmatrix} 1 & 2\pi n/v & 0 \\ 0 & 1 & 0 \\ 0 & -2\pi n & 1 \end{bmatrix}$$

and the sum

$$A^n + A^{n-1} + \ldots + A + I = \begin{bmatrix} n+1 & 2\pi S/v & 0 \\ 0 & n+1 & 0 \\ 0 & -2\pi S & n+1 \end{bmatrix}$$

can be computed, where

$$S = 1 + 2 + 3 + \ldots + n = \tfrac{1}{2}n(n+1)$$

Writing out equation (5.6) in full gives

$$\begin{bmatrix} t_{n+1} \\ R_{n+1} \\ L_{n+1} \end{bmatrix} = \begin{bmatrix} 1 & 2\pi(n+1)/v & 0 \\ 0 & 1 & 0 \\ 0 & -2\pi(n+1) & 1 \end{bmatrix} \begin{bmatrix} 0 \\ R \\ L \end{bmatrix} - \begin{bmatrix} n+1 & \pi n(n+1)/v & 0 \\ 0 & n+1 & 0 \\ 0 & -\pi n(n+1) & n+1 \end{bmatrix} \begin{bmatrix} 0 \\ h \\ 0 \end{bmatrix}$$

Thus

$$R_{n+1} = R - (n+1)h$$

$$t_{n+1} = \frac{2\pi}{v}(n+1)(R - \tfrac{1}{2}nh)$$

$$L_{n+1} = L - 2\pi(n+1)(R - \tfrac{1}{2}nh)$$

which gives the required result.

Note that the length depends on time in a linear fashion but that the relation between the radius and the time is quadratic

$$t_{n+1} = \frac{\pi}{vh}(R - R_{n+1})(R + R_{n+1} + h)$$

Example 5.12

Find the values of x that make the matrix Z^5 a diagonal matrix, where

$$Z = \begin{bmatrix} x & 0 & 0 \\ 0 & x & 1 \\ 0 & -1 & 0 \end{bmatrix}$$

Solution   Although this problem can be done by hand it is tedious and a MAPLE solution is given.

```
with (linalg):
Z:= array ([[x, 0, 0], [0, x, 1], [0, -1, 0]]):
Z5:= simplify (multiply (Z, Z, Z, Z, Z));
```

$$Z5 := \begin{bmatrix} x^5 & 0 & 0 \\ 0 & 3x - 4x^3 + x^5 & 1 - 3x^2 + x^4 \\ 0 & -1 + 3x^2 - x^4 & 2x - x^3 \end{bmatrix}$$

```
evalf (solve ({Z5 [2, 3] = 0}, {x}));
{x = 1.618}, {x = -0.618}, {x = 0.618}, {x = -1.618}
```

Using MATLAB's Symbolic Math Toolbox the commands

```
syms x
Z = syms([x 0 0; 0 x 1; 0 -1 0]);
Z5 = Z^5; simplify(Z5);
pretty(ans)
```

produce the same matrix as above. The additional commands

```
solve(1 - 3*x^2 + x^4); double(ans)
```

produce the same values of *x*.

5.2.7 Exercises

Check the answers to the exercises using MATLAB or MAPLE whenever possible.

17 Given the matrices

$$A = \begin{bmatrix} 1 & 0 & 1 \\ 2 & 1 & 2 \end{bmatrix}, \quad B = \begin{bmatrix} 0 & 1 \\ 1 & 0 \\ 0 & 1 \end{bmatrix},$$

$$C = \begin{bmatrix} 2 & 1 \\ -1 & 2 \end{bmatrix}$$

evaluate where possible

AB, BA, BC, CB, CA, AC

18 For the matrices

$$A = \begin{bmatrix} 1 & 1 \\ 0 & 1 \end{bmatrix} \quad \text{and} \quad B = \begin{bmatrix} 0 & 1 \\ 1 & 0 \end{bmatrix}$$

(a) evaluate $(A + B)^2$ and $A^2 + 2AB + B^2$

(b) evaluate $(A + B)(A - B)$ and $A^2 - B^2$

Repeat the calculations with the matrices

$$A = \begin{bmatrix} 1 & 2 \\ 5 & 2 \end{bmatrix} \quad \text{and} \quad B = \begin{bmatrix} 2 & -2 \\ -5 & 1 \end{bmatrix}$$

and explain the differences between the results for the two sets.

19 Show that for a square matrix $(A^2)^T = (A^T)^2$.

20 Show that AA^T is a symmetric matrix.

21 Find all the 2×2 matrices that commute (that is $AB = BA$) with $\begin{bmatrix} 1 & -1 \\ 0 & 2 \end{bmatrix}$.

22 A matrix with *m* rows and *n* columns is said to be of type $m \times n$. Give simple examples of matrices **A** and **B** to illustrate the following situations:

(a) **AB** is defined but **BA** is not,

(b) **AB** and **BA** are both defined but have different type,

(c) **AB** and **BA** are both defined and have the same type but are unequal.

23 Given

$$A = \begin{bmatrix} 1 & 3 & 2 \\ 2 & -1 & 0 \\ 1 & 4 & 1 \end{bmatrix}$$

determine a symmetric matrix **C** and a skew-symmetric matrix **D** such that

$$A = C + D$$

24 Given the matrices

$$a = [3 \quad 2 \quad -1], \quad b = \begin{bmatrix} 11 \\ 0 \\ 2 \end{bmatrix} \quad \text{and}$$

$$C = \begin{bmatrix} 4 & 1 & 1 \\ -1 & 7 & -3 \\ -1 & 3 & 5 \end{bmatrix}$$

determine the elements of **G** where

$$(ab)I + C^2 = C^T + G$$

and I is the unit matrix.

25 A firm allocates staff into four categories: welders, fitters, designers and administrators. It is estimated that for their three main products the time spent, in hours, on each item is given in the following matrix.

	Boiler	Water tank	Holding frame
Welder	2	0.75	1.25
Fitter	1.4	0.5	1.75
Designer	0.3	0.1	0.1
Admin	0.1	0.25	0.3

The wages, pension contributions and overheads, in £ per hour, are known to be

	Welder	Fitter	Designer	Administrator
Wages	12	8	20	10
Pension	1	0.5	2	1
O/heads	0	0	1	3

Write the problem in matrix form and use matrix products to find the total cost of producing 10 boilers, 25 water tanks and 35 frames.

26 Given

$$A = \begin{bmatrix} 1 & 1 & 1 \\ 2 & 1 & 2 \\ -2 & 1 & -1 \end{bmatrix}$$

evaluate A^2 and A^3. Verify that

$$A^3 - A^2 - 3A + I = 0$$

27 Given

$$A = \begin{bmatrix} 5 & -2 & 0 \\ -2 & 6 & 2 \\ 0 & 2 & 7 \end{bmatrix} \quad \text{and} \quad X = \begin{bmatrix} x_1 \\ x_2 \\ x_3 \end{bmatrix}$$

show that

$$X^T A X = 27 \tag{5.7}$$

implies that

$$5x_1^2 + 6x_2^2 + 7x_3^2 - 4x_1x_2 + 4x_2x_3 = 27$$

Under the transformation

$$X = BY$$

show that (5.7) becomes

$$Y^T(B^T AB)Y = 27$$

If

$$B = \begin{bmatrix} 2 & 2 & -1 \\ 2 & -1 & 2 \\ -1 & 2 & 2 \end{bmatrix} \quad \text{and} \quad Y = \begin{bmatrix} y_1 \\ y_2 \\ y_3 \end{bmatrix}$$

evaluate $B^T AB$, and hence show that

$$y_1^2 + 2y_2^2 + 3y_3^2 = 1$$

28 Figure 5.5 shows a Wheatstone bridge circuit. Kirchhoff's law states that the total current entering a junction is equal to the total current leaving it. Ohm's law in a circuit states that the imposed voltage in the circuit is the sum of current times resistance in the sections of the circuit. Write down the equations for the system and put them into the matrix form

$$A \begin{bmatrix} i_1 \\ i_2 \\ \vdots \\ i_6 \end{bmatrix} = b$$

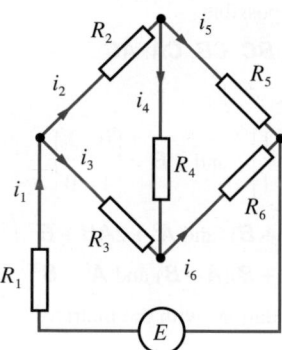

Figure 5.5 Wheatstone bridge circuit

29 A well-known problem concerns a mythical country that has three cities, A, B and C, with a total population of 2400. At the end of each year it is decreed that all people must move to another city, half to one and half to the other. If a, b and c are the populations in the cities A, B and C respectively, show that in the next year the populations are given by

$$\begin{bmatrix} a' \\ b' \\ c' \end{bmatrix} = \begin{bmatrix} 0 & \frac{1}{2} & \frac{1}{2} \\ \frac{1}{2} & 0 & \frac{1}{2} \\ \frac{1}{2} & \frac{1}{2} & 0 \end{bmatrix} \begin{bmatrix} a \\ b \\ c \end{bmatrix}$$

Supposing that the three cities have initial populations of 600, 800 and 1000, what are the populations after 10 years and after a very long time (a package such as MATLAB is ideal for the calculations)? (Note that this example is a version of a **Markov chain** problem. Markov chains have applications in many areas of science and engineering.)

30 Find values of h, k, l and m so that $\boldsymbol{A} \neq 0$, $\boldsymbol{B} \neq 0$, $\boldsymbol{A}^2 = \boldsymbol{A}$, $\boldsymbol{B}^2 = \boldsymbol{B}$ and $\boldsymbol{AB} = 0$, where

$$\boldsymbol{A} = h \begin{bmatrix} 1 & 1 & 1 \\ 1 & 1 & 1 \\ 1 & 1 & 1 \end{bmatrix} \quad \text{and}$$

$$\boldsymbol{B} = \begin{bmatrix} k & -l & -l \\ -l & m & m \\ -l & m & m \end{bmatrix}$$

31 The Königsberg bridge problem concerns trying to follow a path across all the bridges to the islands in a river, as shown in Figure 5.6, without going over any bridge twice. Defining three matrices in an obvious way,

the adjacency matrix $\quad \boldsymbol{A} = \begin{bmatrix} 0 & 2 & 0 & 1 \\ 2 & 0 & 2 & 1 \\ 0 & 2 & 0 & 1 \\ 1 & 1 & 1 & 0 \end{bmatrix}$

the degree matrix $\quad \boldsymbol{D} = \begin{bmatrix} 3 & 0 & 0 & 0 \\ 0 & 5 & 0 & 0 \\ 0 & 0 & 3 & 0 \\ 0 & 0 & 0 & 3 \end{bmatrix}$

and the vertex-arc matrix

$$\boldsymbol{B} = \begin{bmatrix} 1 & 1 & 0 & 0 & 0 & 1 & 0 \\ 1 & 1 & 1 & 1 & 1 & 0 & 0 \\ 0 & 0 & 1 & 1 & 0 & 0 & 1 \\ 0 & 0 & 0 & 0 & 1 & 1 & 1 \end{bmatrix}$$

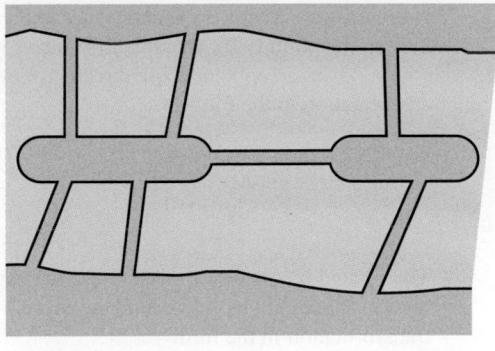

(a)

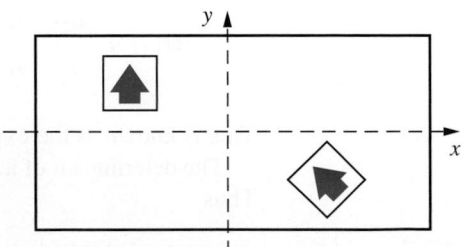

(b)

Figure 5.6 The Königsberg bridge problem.

part of the solution involves showing that

$$\boldsymbol{BB}^{\mathrm{T}} = \boldsymbol{A} + \boldsymbol{D}$$

Verify this result.

32 A computer screen has dimensions 20 cm × 30 cm. Axes are set up at the centre of the screen as illustrated in Figure 5.7. A box containing an arrow

Figure 5.7 Manipulation of a computer screen in Question 32.

has dimensions 2 cm × 2 cm and is situated with its centre at the point (−16, 10). It is first to be rotated through 45° in an anticlockwise direction. Find this transformation in the form

$$\begin{bmatrix} x' + 16 \\ y' - 10 \end{bmatrix} = A \begin{bmatrix} x + 16 \\ y - 10 \end{bmatrix}$$

The rotated box is now moved to a new position with its centre at (16, −10). Find the overall transformation in the form

$$\begin{bmatrix} x'' \\ y'' \end{bmatrix} = \begin{bmatrix} a \\ b \end{bmatrix} + B \begin{bmatrix} x \\ y \end{bmatrix}$$

33 Given the matrix

$$A = \begin{bmatrix} 0 & 1 & 0 & 0 & 0 & 0 & 0 & 0 \\ 0 & 0 & 0 & 1 & 0 & 0 & 0 & 0 \\ 0 & 0 & 1 & 0 & 0 & 0 & 0 & 0 \\ 1 & 0 & 0 & 0 & 0 & 0 & 0 & 0 \\ 0 & 0 & 0 & 0 & 0 & 0 & 0 & 1 \\ 0 & 0 & 0 & 0 & 1 & 0 & 0 & 0 \\ 0 & 0 & 0 & 0 & 0 & 0 & 1 & 0 \\ 0 & 0 & 0 & 0 & 0 & 1 & 0 & 0 \end{bmatrix}$$

it is known that $A^n = I$, the unit matrix, for some n; find this value.

5.3 Determinants

The idea of a determinant is closely related to that of a square matrix and is crucial to the solution of linear equations. We shall deal here mainly with 2 × 2 and 3 × 3 determinants.

Given the square matrices

$$A = \begin{bmatrix} a_{11} & a_{12} \\ a_{21} & a_{22} \end{bmatrix} \quad \text{and} \quad B = \begin{bmatrix} a_{11} & a_{12} & a_{13} \\ a_{21} & a_{22} & a_{23} \\ a_{31} & a_{32} & a_{33} \end{bmatrix}$$

the **determinant** of A, denoted by det A or $|A|$, is given by

$$|A| = a_{11}a_{22} - a_{12}a_{21} \tag{5.8}$$

For the 3 × 3 matrix B

$$|B| = a_{11}\begin{vmatrix} a_{22} & a_{23} \\ a_{32} & a_{33} \end{vmatrix} - a_{12}\begin{vmatrix} a_{21} & a_{23} \\ a_{31} & a_{33} \end{vmatrix} + a_{13}\begin{vmatrix} a_{21} & a_{22} \\ a_{31} & a_{32} \end{vmatrix} \tag{5.9}$$

This is known as the expansion of the determinant along the first row.

The determinant of a 1 × 1 matrix, $A = [a]$, having a single entry a is simply its entry. Thus

$$|A| = a$$

It is important that this be distinguished from mod a which is also written as $|a|$.

Example 5.13   Evaluate the third-order determinant

$$\begin{vmatrix} 1 & 2 & 4 \\ -1 & 0 & 3 \\ 3 & 1 & -2 \end{vmatrix}$$

Solution   Expanding along the first row as in (5.9), we have

$$\begin{vmatrix} 1 & 2 & 4 \\ -1 & 0 & 3 \\ 3 & 1 & -2 \end{vmatrix} = 1\begin{vmatrix} 0 & 3 \\ 1 & -2 \end{vmatrix} - 2\begin{vmatrix} -1 & 3 \\ 3 & -2 \end{vmatrix} + 4\begin{vmatrix} -1 & 0 \\ 3 & 1 \end{vmatrix}$$

$$= 1[(0)(-2) - (1)(3)] - 2[(-1)(-2) - (3)(3)]$$

$$+ 4[(-1)(1) - (3)(0)] \quad \text{(using (5.8))}$$

$$= 1(-3) - 2(-7) + 4(-1)$$

$$= 7$$

If we take a determinant and delete row i and column j then the determinant remaining is called the **minor** M_{ij}. In general we can take *any* row (or column) and evaluate an $n \times n$ determinant $|\mathbf{A}|$ as

$$|\mathbf{A}| = \sum_{j=1}^{n} (-1)^{i+j} a_{ij} M_{ij} \tag{5.10}$$

The fact that the determinant is the same for *any i* requires detailed proof.

The sign associated with a minor is given in the array

$$\begin{vmatrix} + & - & + & - & + & \cdots \\ - & + & - & + & - & \cdots \\ + & - & + & - & + & \cdots \\ \vdots & \vdots & \vdots & \vdots & \vdots \end{vmatrix}$$

A minor multiplied by the appropriate sign is called the **cofactor** A_{ij} of the element, so

$$A_{ij} = (-1)^{i+j} M_{ij}$$

and thus

$$|\mathbf{A}| = \sum_{j} a_{ij} A_{ij}$$

Example 5.14   Evaluate the minors and cofactors of the determinant

$$|A| = \begin{vmatrix} 3 & 4 & 5 \\ 6 & -4 & 2 \\ 2 & -1 & 1 \end{vmatrix}$$

associated with the first row, and hence evaluate the determinant.

Solution

$$\begin{vmatrix} 3 & 4 & 5 \\ 6 & -4 & 2 \\ 2 & -1 & 1 \end{vmatrix} \rightarrow \begin{vmatrix} -4 & 2 \\ -1 & 1 \end{vmatrix} = -4 - (-2) = -2$$

Element a_{11} has minor $M_{11} = -2$ and cofactor $A_{11} = -2$.

$$\begin{vmatrix} 3 & 4 & 5 \\ 6 & -4 & 2 \\ 2 & -1 & 1 \end{vmatrix} \rightarrow \begin{vmatrix} 6 & 2 \\ 2 & 1 \end{vmatrix} = 2$$

Element a_{12} has minor $M_{12} = 2$ and cofactor $A_{12} = -2$.

$$\begin{vmatrix} 3 & 4 & 5 \\ 6 & -4 & 2 \\ 2 & -1 & 1 \end{vmatrix} \rightarrow \begin{vmatrix} 6 & -4 \\ 2 & -1 \end{vmatrix} = 2$$

Element a_{13} has minor $M_{13} = 2$ and cofactor $A_{13} = 2$. Thus the determinant is

$$|A| = 3 \times (-2) + 4 \times (-2) + 5 \times 2 = -4$$

It may be checked that the same result is obtained by expanding along any row (or column), care being taken to incorporate the correct signs.

The properties of determinants are not always obvious, and are often quite difficult to prove in full generality. The commonly useful row operations are as follows.

(a) Two rows (or columns) equal

$$|A| = \begin{vmatrix} a_{11} & a_{12} & a_{13} \\ a_{21} & a_{22} & a_{23} \\ a_{21} & a_{22} & a_{23} \end{vmatrix} = a_{11}\begin{vmatrix} a_{22} & a_{23} \\ a_{22} & a_{23} \end{vmatrix} - a_{12}\begin{vmatrix} a_{21} & a_{23} \\ a_{21} & a_{23} \end{vmatrix} + a_{13}\begin{vmatrix} a_{21} & a_{22} \\ a_{21} & a_{22} \end{vmatrix} = 0$$

Thus if two rows (or columns) are the same, the determinant is zero.

(b) Multiple of a row by a scalar

$$|B| = \begin{vmatrix} \lambda a_{11} & \lambda a_{12} & \lambda a_{13} \\ a_{21} & a_{22} & a_{23} \\ a_{31} & a_{32} & a_{33} \end{vmatrix} = \lambda |A|$$

The proof of this follows immediately from the definition. A consequence of (a) and (b) is that if any row (or column) is a multiple of another row (or column) then the determinant is zero.

(c) Interchange of two rows (or columns)

Consider $|A|$ and $|B|$ in which rows 1 and 2 are interchanged

$$|A| = \begin{vmatrix} a_{11} & a_{12} & a_{13} \\ a_{21} & a_{22} & a_{23} \\ a_{31} & a_{32} & a_{33} \end{vmatrix} \quad \text{and} \quad |B| = \begin{vmatrix} a_{21} & a_{22} & a_{23} \\ a_{11} & a_{12} & a_{13} \\ a_{31} & a_{32} & a_{33} \end{vmatrix}$$

Expanding $|A|$ by the first row,

$$|A| = a_{11} \begin{vmatrix} a_{22} & a_{23} \\ a_{32} & a_{33} \end{vmatrix} - a_{12} \begin{vmatrix} a_{21} & a_{23} \\ a_{31} & a_{33} \end{vmatrix} + a_{13} \begin{vmatrix} a_{21} & a_{22} \\ a_{31} & a_{32} \end{vmatrix}$$

and $|B|$ by the second row

$$|B| = -a_{11} \begin{vmatrix} a_{22} & a_{23} \\ a_{32} & a_{33} \end{vmatrix} + a_{12} \begin{vmatrix} a_{21} & a_{23} \\ a_{31} & a_{33} \end{vmatrix} - a_{13} \begin{vmatrix} a_{21} & a_{22} \\ a_{31} & a_{32} \end{vmatrix}$$

Thus

$$|A| = -|B|$$

so that interchanging two rows changes the sign of the determinant. Entirely similar results apply changing two columns.

(d) Addition rule

Expanding by the first row

$$\begin{vmatrix} a_{11} + b_{11} & a_{12} + b_{12} & a_{13} + b_{13} \\ a_{21} & a_{22} & a_{23} \\ a_{31} & a_{32} & a_{33} \end{vmatrix}$$

$$= (a_{11} + b_{11})A_{11} + (a_{12} + b_{12})A_{12} + (a_{13} + b_{13})A_{13}$$

$$= (a_{11}A_{11} + a_{12}A_{12} + a_{13}A_{13}) + (b_{11}A_{11} + b_{12}A_{12} + b_{13}A_{13})$$

$$= \begin{vmatrix} a_{11} & a_{12} & a_{13} \\ a_{21} & a_{22} & a_{23} \\ a_{31} & a_{32} & a_{33} \end{vmatrix} + \begin{vmatrix} b_{11} & b_{12} & b_{13} \\ a_{21} & a_{22} & a_{23} \\ a_{31} & a_{32} & a_{33} \end{vmatrix}$$

It should be noted that $|A + B|$ is *not* equal to $|A| + |B|$ in *general*.

(e) Adding multiples of rows (or columns)

Consider

$$|A| = \begin{vmatrix} a_{11} & a_{12} & a_{13} \\ a_{21} & a_{22} & a_{23} \\ a_{31} & a_{32} & a_{33} \end{vmatrix}$$

Then

$$|B| = \begin{vmatrix} a_{11} + \lambda a_{21} & a_{12} + \lambda a_{22} & a_{13} + \lambda a_{23} \\ a_{21} & a_{22} & a_{23} \\ a_{31} & a_{32} & a_{33} \end{vmatrix}$$

$$= \begin{vmatrix} a_{11} & a_{12} & a_{13} \\ a_{21} & a_{22} & a_{23} \\ a_{31} & a_{32} & a_{33} \end{vmatrix} + \lambda \begin{vmatrix} a_{21} & a_{22} & a_{23} \\ a_{21} & a_{22} & a_{23} \\ a_{31} & a_{32} & a_{33} \end{vmatrix} \quad \text{(using (d) and then (b))}$$

$$= |A| \quad \text{(since, by (a), the second determinant is zero)}$$

This means that adding multiples of rows (or columns) together makes no difference to the determinant.

(f) Transpose

$$|A^{\mathsf{T}}| = |A|$$

This just states that expanding by the first row or the first column gives the same result.

(g) Product

$$|AB| = |A||B|$$

This result is difficult to prove generally, but it can be verified rather tediously for the 2×2 or 3×3 cases. For the 2×2 case

$$|A||B| = (a_{11}a_{22} - a_{12}a_{21})(b_{11}b_{22} - b_{12}b_{21})$$

$$= a_{11}a_{22}b_{11}b_{22} - a_{11}a_{22}b_{12}b_{21} - a_{12}a_{21}b_{11}b_{22} + a_{12}a_{21}b_{12}b_{21}$$

and

$$|AB| = \begin{vmatrix} a_{11}b_{11} + a_{12}b_{21} & a_{11}b_{12} + a_{12}b_{22} \\ a_{21}b_{11} + a_{22}b_{21} & a_{21}b_{12} + a_{22}b_{22} \end{vmatrix}$$

$$= (a_{11}b_{11} + a_{12}b_{21})(a_{21}b_{12} + a_{22}b_{22}) - (a_{11}b_{12} + a_{12}b_{22})(a_{21}b_{11} + a_{22}b_{21})$$

$$= a_{11}a_{22}b_{11}b_{22} - a_{11}a_{22}b_{12}b_{21} - a_{12}a_{21}b_{11}b_{22} + a_{12}a_{21}b_{12}b_{21}$$

Example 5.15

Evaluate the 3×3 determinants

(a) $\begin{vmatrix} 1 & 0 & 1 \\ 0 & 1 & 2 \\ 1 & 1 & 0 \end{vmatrix}$, (b) $\begin{vmatrix} 1 & 0 & 1 \\ 1 & 1 & 0 \\ 0 & 1 & 2 \end{vmatrix}$, (c) $\begin{vmatrix} 1 & 1 & 0 \\ 0 & 1 & 1 \\ 1 & 0 & 2 \end{vmatrix}$, (d) $\begin{vmatrix} 1 & 0 & 1 \\ 0 & 2 & 4 \\ 3 & 3 & 0 \end{vmatrix}$

Solution

(a) Expand by the first row:

$$\begin{vmatrix} 1 & 0 & 1 \\ 0 & 1 & 2 \\ 1 & 1 & 0 \end{vmatrix} = 1\begin{vmatrix} 1 & 2 \\ 1 & 0 \end{vmatrix} - 0\begin{vmatrix} 0 & 2 \\ 1 & 0 \end{vmatrix} + 1\begin{vmatrix} 0 & 1 \\ 1 & 1 \end{vmatrix} = -2 - 0 - 1 = -3$$

(b) Expand by the first column:

$$\begin{vmatrix} 1 & 0 & 1 \\ 1 & 1 & 0 \\ 0 & 1 & 2 \end{vmatrix} = 1\begin{vmatrix} 1 & 0 \\ 1 & 2 \end{vmatrix} - 1\begin{vmatrix} 0 & 1 \\ 1 & 2 \end{vmatrix} + 0\begin{vmatrix} 0 & 1 \\ 1 & 0 \end{vmatrix} = 2 + 1 + 0 = 3$$

Note that (a) and (b) are the same determinant, but with two rows interchanged. The result confirms property (c) just stated above.

(c) Expand by the third row:

$$\begin{vmatrix} 1 & 1 & 0 \\ 0 & 1 & 1 \\ 1 & 0 & 2 \end{vmatrix} = 1\begin{vmatrix} 1 & 0 \\ 1 & 1 \end{vmatrix} - 0\begin{vmatrix} 1 & 0 \\ 0 & 1 \end{vmatrix} + 2\begin{vmatrix} 1 & 1 \\ 0 & 1 \end{vmatrix} = 1 - 0 + 2 = 3$$

Note that the matrix associated with the determinant in (c) is just the transpose of the matrix associated with the determinant in (b).

(d) $\begin{vmatrix} 1 & 0 & 1 \\ 0 & 2 & 4 \\ 3 & 3 & 0 \end{vmatrix} = 2\begin{vmatrix} 1 & 0 & 1 \\ 0 & 1 & 2 \\ 3 & 3 & 0 \end{vmatrix} = 6\begin{vmatrix} 1 & 0 & 1 \\ 0 & 1 & 2 \\ 1 & 1 & 0 \end{vmatrix} = -18$

Note that we have used the multiple of a row rule on two occasions; the final determinant is the same as (a).

In MATLAB and MAPLE the determinant of a matrix A is given by the command $det(A)$. Considering Example 5.15(d) the MATLAB commands

```
A = [1 0 1; 0 2 4; 3 3 0];
det(A)
```

return the answer -18.

Example 5.16   Given the matrices

$$A = \begin{bmatrix} 1 & 2 & 3 \\ 2 & 3 & 4 \\ 4 & 5 & 6 \end{bmatrix} \quad \text{and} \quad B = \begin{bmatrix} 1 & 0 & 1 \\ 1 & 1 & 1 \\ 1 & 2 & 3 \end{bmatrix}$$

evaluate (a) $|A|$, (b) $|B|$ and (c) $|AB|$.

Solution   (a) $|A| = 1 \begin{vmatrix} 3 & 4 \\ 5 & 6 \end{vmatrix} - 2 \begin{vmatrix} 2 & 4 \\ 4 & 6 \end{vmatrix} + 3 \begin{vmatrix} 2 & 3 \\ 4 & 5 \end{vmatrix}$

$$= 1 \times (-2) - 2 \times (-4) + 3 \times (-2) = 0$$

(b) $|B| = \begin{vmatrix} 1 & 0 & 1 \\ 1 & 1 & 1 \\ 1 & 2 & 3 \end{vmatrix}$

$$= \begin{vmatrix} 1 & 0 & 0 \\ 1 & 1 & 0 \\ 1 & 2 & 2 \end{vmatrix} \quad \text{(subtracting column 1 from column 3)}$$

$$= 1 \begin{vmatrix} 1 & 0 \\ 2 & 2 \end{vmatrix} = 2 \quad \text{(expanding by first row)}$$

(c) $|AB| = \begin{vmatrix} 6 & 8 & 12 \\ 9 & 11 & 17 \\ 15 & 17 & 27 \end{vmatrix}$

$$= 6[(11)(27) - (17)(17)] - 8[(9)(27) - (17)(15)] + 12[(9)(17) - (11)(15)]$$

$$= 48 + 96 - 144 = 0$$

We can use properties (a)–(e) to reduce the amount of computation involved in evaluating a determinant. We introduce as many zeros as possible into a row or column, and then expand along that row or column.

Example 5.17   Evaluate

$$D = \begin{vmatrix} 1 & 1 & 1 & 1 \\ 1 & 1+a & 1 & 1 \\ 1 & 1 & 1+b & 1 \\ 1 & 1 & 1 & 1+c \end{vmatrix}$$

Solution

$$D = \begin{vmatrix} 1 & 0 & 0 & 0 \\ 1 & a & 0 & 0 \\ 1 & 0 & b & 0 \\ 1 & 0 & 0 & c \end{vmatrix} \quad \text{(by subtracting col. 1 from col. 2, col. 3 and col. 4)}$$

$$= 1 \begin{vmatrix} a & 0 & 0 \\ 0 & b & 0 \\ 0 & 0 & c \end{vmatrix} \quad \text{(by expanding by the top row)}$$

$$= a \begin{vmatrix} b & 0 \\ 0 & c \end{vmatrix} = abc$$

A solution using MAPLE is given by the commands

```
with (linalg):
E:= array ([[1, 1, 1, 1], [1, 1+a, 1, 1],
[1, 1, 1+b, 1], [1, 1, 1, 1+c]]):
det(E);
```

Using MATLAB's Symbolic Math Toolbox the commands

```
syms a b c
E = sym([1 1 1 1; 1 1+a 1 1; 1 1 1+b 1; 1 1 1 1+c]);
det(E);
pretty(ans)
```

return the answer abc.

A point that should be carefully noted concerns large determinants; they are extremely difficult and time-consuming to evaluate (using the basic definition (5.10) for an $n \times n$ determinant involves $n!(n-1)$ multiplications). This is a problem even with computers – which in fact use alternative methods. If at all possible, evaluation of large determinants should be avoided. They do, however, play a central role in matrix theory.

The cofactors $A_{11}, A_{12}, \ldots$ defined earlier have the property that

$$|\mathbf{A}| = a_{11}A_{11} + a_{12}A_{12} + a_{13}A_{13}$$

Consider the expression $a_{21}A_{11} + a_{22}A_{12} + a_{23}A_{13}$. In determinant form we have

$$a_{21}A_{11} + a_{22}A_{12} + a_{23}A_{13} = \begin{vmatrix} a_{21} & a_{22} & a_{23} \\ a_{21} & a_{22} & a_{23} \\ a_{31} & a_{32} & a_{33} \end{vmatrix} = 0$$

since two rows are identical. Similarly,

$$a_{31}A_{11} + a_{32}A_{12} + a_{33}A_{13} = 0$$

In general it can be shown

$$\sum_k a_{ik} A_{jk} = \begin{cases} |\boldsymbol{A}| & \text{if } i = j \\ 0 & \text{if } i \neq j \end{cases} \tag{5.11}$$

and, expanding by columns,

$$\sum_k a_{ki} A_{kj} = \begin{cases} |\boldsymbol{A}| & \text{if } i = j \\ 0 & \text{if } i \neq j \end{cases} \tag{5.12}$$

A numerical example illustrates these points.

Example 5.18 Illustrate the use of cofactors in the expansion of determinants on the matrix

$$\boldsymbol{A} = \begin{bmatrix} 1 & 2 & 3 \\ 6 & 5 & 4 \\ 7 & 8 & 1 \end{bmatrix}$$

Solution The cofactors are evaluated as

$$A_{11} = \begin{vmatrix} 5 & 4 \\ 8 & 1 \end{vmatrix} = -27, \quad A_{12} = -\begin{vmatrix} 6 & 4 \\ 7 & 1 \end{vmatrix} = 22, \quad A_{13} = \begin{vmatrix} 6 & 5 \\ 7 & 8 \end{vmatrix} = 13$$

and continuing in the same way

$$A_{21} = 22, A_{22} = -20, A_{23} = 6, A_{31} = -7, A_{32} = 14 \text{ and } A_{33} = -7$$

A selection of the evaluations in equation (5.11), that is expansion by rows, is

$$a_{11}A_{11} + a_{12}A_{12} + a_{13}A_{13} = 1 \times (-27) + 2 \times 22 + 3 \times 13 = 56$$

$$a_{21}A_{11} + a_{22}A_{12} + a_{23}A_{13} = 6 \times (-27) + 5 \times 22 + 4 \times 13 = 0$$

$$a_{31}A_{21} + a_{32}A_{22} + a_{33}A_{23} = 7 \times 22 + 8 \times (-20) + 1 \times 6 = 0$$

and in equation (5.12), that is expansion by columns, is

$$a_{11}A_{12} + a_{21}A_{22} + a_{31}A_{32} = 1 \times 22 + 6 \times (-20) + 7 \times 14 = 0$$

$$a_{12}A_{12} + a_{22}A_{22} + a_{32}A_{32} = 2 \times 22 + 5 \times (-20) + 8 \times 14 = 56$$

$$a_{13}A_{11} + a_{23}A_{21} + a_{33}A_{31} = 3 \times (-27) + 4 \times 22 + 1 \times (-7) = 0$$

The other expansions in equations (5.11) and (5.12) can be verified on this example. It may be noted that the determinant of the matrix is 56.

A matrix with particularly interesting properties is the **adjoint** or **adjugate matrix**, which is defined as the transpose of the matrix of cofactors; that is,

$$\text{adj } \mathbf{A} = \begin{bmatrix} A_{11} & A_{12} & A_{13} \\ A_{21} & A_{22} & A_{23} \\ A_{31} & A_{32} & A_{33} \end{bmatrix}^{\text{T}}$$

(5.13)

If we now calculate $\mathbf{A}$ (adj $\mathbf{A}$), we have

$$[\mathbf{A} (\text{adj } \mathbf{A})]_{ij} = \sum_k a_{ik}(\text{adj } \mathbf{A})_{kj} = \sum_k a_{ik}A_{jk}$$

$$= \begin{cases} |\mathbf{A}| & \text{if } i = j \quad \text{(from (5.11))} \\ 0 & \text{if } i \neq j \end{cases}$$

So

$$\mathbf{A}(\text{adj } \mathbf{A}) = \begin{bmatrix} |\mathbf{A}| & 0 & 0 \\ 0 & |\mathbf{A}| & 0 \\ 0 & 0 & |\mathbf{A}| \end{bmatrix} = |\mathbf{A}|\mathbf{I}$$

(5.14)

and we have thus discovered a matrix that when multiplied by $\mathbf{A}$ gives a scalar times the unit matrix.

If $\mathbf{A}$ is a square matrix of order n then, taking determinants on both sides of (5.14),

$$|\mathbf{A}| |\text{adj } \mathbf{A}| = |\mathbf{A}(\text{adj } \mathbf{A})| = ||\mathbf{A}|\mathbf{I}_n| = |\mathbf{A}|^n$$

If $|\mathbf{A}| \neq 0$, it follows that

$$|\text{adj } \mathbf{A}| = |\mathbf{A}|^{n-1}$$

(5.15)

a result known as **Cauchy's theorem**.

It is also the case that

$$\text{adj}(\mathbf{AB}) = (\text{adj } \mathbf{B})(\text{adj } \mathbf{A})$$

(5.16)

so in taking the adjoint of a product the order is reversed.

An important piece of notation that has significant implications for the solution of sets of linear equations concerns whether or not a matrix has zero determinant. A square matrix $\mathbf{A}$ is called **non-singular** if $|\mathbf{A}| \neq 0$ and **singular** if $|\mathbf{A}| = 0$.

Example 5.19

Derive the adjoint of the 2×2 matrices

$$\mathbf{A} = \begin{bmatrix} 1 & 3 \\ 2 & 8 \end{bmatrix} \quad \text{and} \quad \mathbf{B} = \begin{bmatrix} -1 & 2 \\ -3 & -4 \end{bmatrix}$$

and verify the results in equations (5.14), (5.15) and (5.16).

Solution The cofactors are very easy to evaluate in the 2×2 case: for the matrix A

$$A_{11} = 8, A_{12} = -2, A_{21} = -3 \quad \text{and} \quad A_{22} = 1$$

and for the matrix B

$$B_{11} = -4, B_{12} = 3, B_{21} = -2 \quad \text{and} \quad B_{22} = -1$$

The adjoint or adjugate matrices can be written down immediately as

$$\text{adj } A = \begin{bmatrix} 8 & -3 \\ -2 & 1 \end{bmatrix} \quad \text{and} \quad \text{adj } B = \begin{bmatrix} -4 & -2 \\ 3 & -1 \end{bmatrix}$$

Now (5.14) gives

$$A(\text{adj } A) = \begin{bmatrix} 1 & 3 \\ 2 & 8 \end{bmatrix}\begin{bmatrix} 8 & -3 \\ -2 & 1 \end{bmatrix} = \begin{bmatrix} 2 & 0 \\ 0 & 2 \end{bmatrix} = 2I$$

$$B(\text{adj } B) = \begin{bmatrix} -1 & 2 \\ -3 & -4 \end{bmatrix}\begin{bmatrix} -4 & -2 \\ 3 & -1 \end{bmatrix} = \begin{bmatrix} 10 & 0 \\ 0 & 10 \end{bmatrix} = 10I$$

so the property is satisfied and the determinants are 2 and 10 respectively. For equation (5.15) we have $n = 2$ so

$$|\text{adj } A| = \begin{vmatrix} 8 & -3 \\ -2 & 1 \end{vmatrix} = 2 \quad \text{and} \quad |\text{adj } B| = \begin{vmatrix} -4 & -2 \\ 3 & -1 \end{vmatrix} = 10$$

as required.

Evaluating the matrices in (5.16)

$$\text{adj}(AB) = \text{adj}\begin{bmatrix} -10 & -10 \\ -26 & -28 \end{bmatrix} = \begin{bmatrix} -28 & 10 \\ 26 & -10 \end{bmatrix}$$

and

$$\text{adj } B \text{ adj } A = \begin{bmatrix} -4 & -2 \\ 3 & -1 \end{bmatrix}\begin{bmatrix} 8 & -3 \\ -2 & 1 \end{bmatrix} = \begin{bmatrix} -28 & 10 \\ 26 & -10 \end{bmatrix}$$

and the statement is clearly verified. It is left as an exercise to show that the product of the matrices the other way round, adj A adj B, gives a totally different matrix.

Example 5.20 Given

$$A = \begin{bmatrix} 1 & 1 & 2 \\ 2 & 0 & 1 \\ 3 & 1 & 1 \end{bmatrix}$$

determine adj A and show that $A(\text{adj } A) = (\text{adj } A)A = |A|I$.

Solution The matrix of cofactors is

$$
\begin{bmatrix}
\begin{vmatrix} 0 & 1 \\ 1 & 1 \end{vmatrix} & -\begin{vmatrix} 2 & 1 \\ 3 & 1 \end{vmatrix} & \begin{vmatrix} 2 & 0 \\ 3 & 1 \end{vmatrix} \\
-\begin{vmatrix} 1 & 2 \\ 1 & 1 \end{vmatrix} & \begin{vmatrix} 1 & 2 \\ 3 & 1 \end{vmatrix} & -\begin{vmatrix} 1 & 1 \\ 3 & 1 \end{vmatrix} \\
\begin{vmatrix} 1 & 2 \\ 0 & 1 \end{vmatrix} & -\begin{vmatrix} 1 & 2 \\ 2 & 1 \end{vmatrix} & \begin{vmatrix} 1 & 1 \\ 2 & 0 \end{vmatrix}
\end{bmatrix}
=
\begin{bmatrix}
-1 & 1 & 2 \\
1 & -5 & 2 \\
1 & 3 & -2
\end{bmatrix}
$$

so, from (5.13)

$$
\operatorname{adj} \boldsymbol{A} =
\begin{bmatrix}
-1 & 1 & 2 \\
1 & -5 & 2 \\
1 & 3 & -2
\end{bmatrix}^{\mathrm{T}}
=
\begin{bmatrix}
-1 & 1 & 1 \\
1 & -5 & 3 \\
2 & 2 & -2
\end{bmatrix}
$$

$$
\boldsymbol{A}(\operatorname{adj} \boldsymbol{A}) =
\begin{bmatrix}
1 & 1 & 2 \\
2 & 0 & 1 \\
3 & 1 & 1
\end{bmatrix}
\begin{bmatrix}
-1 & 1 & 1 \\
1 & -5 & 3 \\
2 & 2 & -2
\end{bmatrix}
=
\begin{bmatrix}
4 & 0 & 0 \\
0 & 4 & 0 \\
0 & 0 & 4
\end{bmatrix}
$$

$$
(\operatorname{adj} \boldsymbol{A})\boldsymbol{A} =
\begin{bmatrix}
-1 & 1 & 1 \\
1 & -5 & 3 \\
2 & 2 & -2
\end{bmatrix}
\begin{bmatrix}
1 & 1 & 2 \\
2 & 0 & 1 \\
3 & 1 & 1
\end{bmatrix}
=
\begin{bmatrix}
4 & 0 & 0 \\
0 & 4 & 0 \\
0 & 0 & 4
\end{bmatrix}
$$

Since $|\boldsymbol{A}| = 4$ the last result then follows.

In MAPLE the adjoint of a matrix $\boldsymbol{A}$ is determined by the command adj(A). There appears to be no equivalent command in either MATLAB or its Symbolic Math Toolbox. However, the *maple* command in the Toolbox may be used to access the command in MAPLE, having first expressed the matrix $\boldsymbol{A}$ in symbolic form using the *sym* command. Consequently in MATLAB's Symbolic Math Toolbox the adjoint is determined by the commands

```
A = sym(A);
adjA = maple('adj', A)
```

Check that in MATLAB the commands

```
A = [1 1 2; 2 0 1; 3 1 1];
A = sym(A);
adjA = maple('adj',A)
```

return the first answer in Example 5.20.

5.3.1 Exercises

 Check your answers using MATLAB or MAPLE whenever possible.

34 Find all the minors and cofactors of the determinant

$$\begin{vmatrix} 1 & 2 & 3 \\ 1 & 0 & 1 \\ 1 & 1 & 1 \end{vmatrix}$$

Hence evaluate the determinant.

35 Evaluate the determinants of the following matrices:

(a) $\begin{bmatrix} 1 & 7 \\ 4 & 9 \end{bmatrix}$ (b) $\begin{bmatrix} 1 & 4 & 3 \\ 2 & -4 & 1 \\ 3 & 2 & -6 \end{bmatrix}$

(c) $\begin{bmatrix} 2 & -1 & 3 \\ 4 & 2 & 9 \\ 1 & 3 & -4 \end{bmatrix}$ (d) $\begin{bmatrix} 1 & 0 & 1 \\ 0 & 1 & 0 \\ 1 & 0 & 2 \end{bmatrix}$

(e) $\begin{bmatrix} 1 & -1 & 0 \\ 1 & 1 & 1 \\ 0 & 1 & -1 \end{bmatrix}$

36 Given the matrix

$$A = \begin{bmatrix} 1 & 0 & -1 \\ 1 & 0 & 1 \\ 2 & 2 & 2 \end{bmatrix}$$

determine $|A|$, $|AA^T|$, $|A^2|$ and $|A + A|$.

37 Find a series of row manipulations that takes

$$\begin{vmatrix} 1 & 0 & 1 \\ 2 & 1 & 0 \\ 0 & 1 & 1 \end{vmatrix} \text{ to } -\begin{vmatrix} 2 & 1 & 0 \\ 0 & -\frac{1}{2} & 1 \\ 0 & 0 & 3 \end{vmatrix} \text{ and hence evaluate}$$

the determinant.

38 Determine adj A when

$$A = \begin{bmatrix} a & b \\ c & d \end{bmatrix}$$

39 Determine adj A when

$$A = \begin{bmatrix} 2 & 1 & 1 \\ 3 & 2 & 2 \\ 1 & 1 & 2 \end{bmatrix}$$

Check that $A(\text{adj } A) = (\text{adj } A)A = |A|I$.

40 For the matrix

$$A = \begin{bmatrix} 2 & 0 \\ 3 & 1 \end{bmatrix}$$

evaluate $|A|$, $\text{adj}(A)$, $B = \dfrac{\text{adj}(A)}{|A|}$ and AB.

41 Show that the matrix

$$B = \begin{bmatrix} 1 & 0 & 2 \\ 3 & 4 & 0 \\ 6 & -2 & 1 \end{bmatrix}$$

is non-singular and verify Cauchy's theorem, namely $|\text{adj } B| = |B|^2$.

42 If $|A| = 0$ deduce that $|A^n| = 0$ for any integer n.

43 Given

$$A = \begin{bmatrix} 2 & -1 & 0 \\ -4 & 3 & -1 \\ 1 & -1 & 1 \end{bmatrix} \text{ and}$$

$$B = \begin{bmatrix} 1 & 0 & 2 \\ 3 & 4 & 0 \\ 6 & -2 & 1 \end{bmatrix}$$

verify that $\text{adj}(AB) = (\text{adj } B)(\text{adj } A)$.

44 Find the values of λ that make the following determinants zero:

(a) $\begin{vmatrix} 2 - \lambda & 7 \\ 4 & 6 - \lambda \end{vmatrix}$

(b) $\begin{vmatrix} 1 & 3 - \lambda & 4 \\ 4 - \lambda & 2 & -1 \\ 1 & \lambda - 6 & 2 \end{vmatrix}$

(c) $\begin{vmatrix} 0 & 2 - \lambda & 0 \\ 2 - \lambda & 4 & 1 \\ 2 & -3 & \lambda - 4 \end{vmatrix}$

45 Evaluate the determinants of the square matrices

(a) $\begin{bmatrix} 0.42 & 0.31 & -0.16 \\ 0.17 & -0.22 & 0.63 \\ 0.89 & 0.93 & 0.41 \end{bmatrix}$

(b) $\begin{bmatrix} 5 & 4 & 1 & 1 \\ 4 & 5 & 1 & 1 \\ 1 & 1 & 4 & 2 \\ 1 & 1 & 2 & 4 \end{bmatrix}$

46 Show that the area of a triangle with vertices (x_1, y_1), (x_2, y_2) and (x_3, y_3) is given by the absolute value of

$$\frac{1}{2} \begin{vmatrix} 1 & x_1 & y_1 \\ 1 & x_2 & y_2 \\ 1 & x_3 & y_3 \end{vmatrix}$$

Refer to Question 32 in Exercises (4.2.10) and the definition of the vector product in Section 4.2.9.

47 Show that $x + x^2 - 2x^3$ is a factor of the determinant D where

$$D = \begin{vmatrix} 0 & x & 2 & x^2 \\ -x & 0 & 1 & x^3 \\ -2 & -1 & 0 & 1 \\ -x^2 & -x^3 & -1 & 0 \end{vmatrix}$$

and hence express D as a product of linear factors.

48 Show that

$$\begin{vmatrix} x & a & b \\ x^2 & a^2 & b^2 \\ a + b & x + b & x + a \end{vmatrix}$$

$$= (b - a)(x - a)(x - b)(x + a + b)$$

Such an exercise can be solved in two lines of code of a symbolic manipulation package such as MAPLE or MATLAB's Symbolic Math Toolbox.

49 Verify that if A is a symmetric matrix then so is adj A.

50 If A is a skew-symmetric $n \times n$ matrix, verify that adj A is symmetric or skew-symmetric according to whether n is odd or even.

5.4 The inverse matrix

In Section 5.3 we constructed adj A and saw that it had interesting properties in relation to the unit matrix. We also saw, in Example 5.5, that we had a method of solving linear equations if we could construct B such that $AB = I$. These ideas can be brought together to provide a comprehensive theory of the solution of linear equations, which we will consider in Section 5.5.

Given a square matrix A, if we can construct a matrix B such that

$$BA = AB = I$$

then **we call B the inverse of A and write it as A^{-1}**. From (5.14)

$$A(\text{adj } A) = |A|I$$

so that we have gone a long way to constructing the inverse. We have two cases:

- If A is non-singular then $|A| \neq 0$ and

$$A^{-1} = \frac{\text{adj } A}{|A|}$$

- If A is singular then $|A| = 0$ and it can be shown that the inverse A^{-1} does not exist.

If the inverse exists then it is unique. Suppose for a given A we have two inverses B and C. Then

$$AB = BA = I, \quad AC = CA = I$$

and therefore

$$AB = AC$$

Pre-multiplying by C, we have

$$C(AB) = C(AC)$$

But matrix multiplication is associative, so we can write this as

$$(CA)B = (CA)C$$

Hence

$$IB = IC \quad (\text{since } CA = I)$$

and so

$$B = C$$

The inverse is therefore unique.

It should be noted that if both A and B are square matrices then $AB = I$ if and only if $BA = I$.

Example 5.21 Find A^{-1} and B^{-1} for the matrices

(a) $A = \begin{bmatrix} 1 & 2 \\ 2 & 3 \end{bmatrix}$ and (b) $B = \begin{bmatrix} 5 & 2 & 4 \\ 3 & -1 & 2 \\ 1 & 4 & -3 \end{bmatrix}$

Solution

(a) $\text{adj } A = \begin{bmatrix} 3 & -2 \\ -2 & 1 \end{bmatrix}^{\text{T}} = \begin{bmatrix} 3 & -2 \\ -2 & 1 \end{bmatrix}$ and $|A| = -1$

so that

$$A^{-1} = \frac{\text{adj } A}{|A|} = \begin{bmatrix} -3 & 2 \\ 2 & -1 \end{bmatrix}$$

(b) $\operatorname{adj} \boldsymbol{B} = \begin{bmatrix} -5 & 11 & 13 \\ 22 & -19 & -18 \\ 8 & 2 & -11 \end{bmatrix}^{\mathrm{T}} = \begin{bmatrix} -5 & 22 & 8 \\ 11 & -19 & 2 \\ 13 & -18 & -11 \end{bmatrix}$ and $|\boldsymbol{B}| = 49$

so that

$$\boldsymbol{B}^{-1} = \frac{1}{49} \begin{bmatrix} -5 & 22 & 8 \\ 11 & -19 & 2 \\ 13 & -18 & -11 \end{bmatrix}$$

In both cases it can be checked that $\boldsymbol{A}\boldsymbol{A}^{-1} = \boldsymbol{I}$ and $\boldsymbol{B}\boldsymbol{B}^{-1} = \boldsymbol{I}$.

Finding the inverse of a 2×2 matrix is very easy, since for

$$\boldsymbol{A} = \begin{bmatrix} a & b \\ c & d \end{bmatrix}$$

$$\boldsymbol{A}^{-1} = \frac{1}{ad - bc} \begin{bmatrix} d & -b \\ -c & a \end{bmatrix} \quad \text{(provided that } ad - bc \neq 0\text{)}$$

Unfortunately there is no simple extension of this result to higher-order matrices. On the other hand, in most practical situations the inverse itself is rarely required – it is the solution of the corresponding linear equations that is important. To understand the power and applicability of the various methods of solution of linear equations, the role of the inverse is essential. The consideration of the adjoint matrix provides a theoretical framework for this study, but as a practical method for finding the inverse of a matrix it is virtually useless, since, as we saw earlier, it is so time-consuming to compute determinants.

To find the inverse of a product of two matrices, the order is reversed:

$$(\boldsymbol{AB})^{-1} = \boldsymbol{B}^{-1}\boldsymbol{A}^{-1} \tag{5.17}$$

(provided that $\boldsymbol{A}$ and $\boldsymbol{B}$ are invertible). To prove this, let $\boldsymbol{C} = \boldsymbol{B}^{-1}\boldsymbol{A}^{-1}$. Then

$$\boldsymbol{C}(\boldsymbol{AB}) = (\boldsymbol{B}^{-1}\boldsymbol{A}^{-1})(\boldsymbol{AB}) = \boldsymbol{B}^{-1}(\boldsymbol{A}^{-1}\boldsymbol{A})\boldsymbol{B} = \boldsymbol{B}^{-1}\boldsymbol{I}\boldsymbol{B} = \boldsymbol{B}^{-1}\boldsymbol{B} = \boldsymbol{I}$$

and thus

$$\boldsymbol{C} = \boldsymbol{B}^{-1}\boldsymbol{A}^{-1} = (\boldsymbol{AB})^{-1}$$

Since matrices do not commute in general $\boldsymbol{A}^{-1}\boldsymbol{B}^{-1} \neq \boldsymbol{B}^{-1}\boldsymbol{A}^{-1}$.

In MATLAB the inverse of a matrix $\boldsymbol{A}$ is determined by the command $inv(\text{A})$; first expressing $\boldsymbol{A}$ in symbolic form using the sym command if the Symbolic Math Toolbox is used. In MAPLE the corresponding command is $inverse(\text{A})$.

Example 5.22   Given

$$A = \begin{bmatrix} 1 & 2 \\ 2 & 1 \end{bmatrix} \quad \text{and} \quad B = \begin{bmatrix} 0 & 1 \\ 1 & 1 \end{bmatrix}$$

evaluate $(AB)^{-1}$, $A^{-1}B^{-1}$, $B^{-1}A^{-1}$ and show that $(AB)^{-1} = B^{-1}A^{-1}$.

Solution

$$A^{-1} = \begin{bmatrix} -\frac{1}{3} & \frac{2}{3} \\ \frac{2}{3} & -\frac{1}{3} \end{bmatrix}, \quad B^{-1} = \begin{bmatrix} -1 & 1 \\ 1 & 0 \end{bmatrix}$$

$$AB = \begin{bmatrix} 2 & 3 \\ 1 & 3 \end{bmatrix}, \quad (AB)^{-1} = \begin{bmatrix} 1 & -1 \\ -\frac{1}{3} & \frac{2}{3} \end{bmatrix}$$

$$A^{-1}B^{-1} = \begin{bmatrix} -\frac{1}{3} & \frac{2}{3} \\ \frac{2}{3} & -\frac{1}{3} \end{bmatrix} \begin{bmatrix} -1 & 1 \\ 1 & 0 \end{bmatrix} = \begin{bmatrix} 1 & -\frac{1}{3} \\ -1 & \frac{2}{3} \end{bmatrix}$$

$$B^{-1}A^{-1} = \begin{bmatrix} -1 & 1 \\ 1 & 0 \end{bmatrix} \begin{bmatrix} -\frac{1}{3} & \frac{2}{3} \\ \frac{2}{3} & -\frac{1}{3} \end{bmatrix} = \begin{bmatrix} 1 & -1 \\ -\frac{1}{3} & \frac{2}{3} \end{bmatrix} = (AB)^{-1}$$

Example 5.23   Given the two matrices

$$A = \begin{bmatrix} 0 & -\frac{3}{5} & 0 \\ \frac{5}{3} & 0 & -\frac{5}{3} \\ 0 & 6 & -6 \end{bmatrix} \quad \text{and} \quad T = \begin{bmatrix} 0.6 & 0.3 & 0.1 \\ 1 & 1 & 0.5 \\ 1.2 & 1.5 & 1 \end{bmatrix}$$

show that the matrix $T^{-1}AT$ is diagonal.

Solution   The inverse is best computed using MATLAB or a similar package. It may be verified by direct multiplication that

$$T^{-1} = \frac{1}{6} \begin{bmatrix} 25 & -15 & 5 \\ -40 & 48 & -20 \\ 30 & -54 & 30 \end{bmatrix}$$

The further multiplications give

$$\frac{1}{6} \begin{bmatrix} 25 & -15 & 5 \\ -40 & 48 & -20 \\ 30 & -54 & 30 \end{bmatrix} \begin{bmatrix} 0 & -\frac{3}{5} & 0 \\ \frac{5}{3} & 0 & -\frac{5}{3} \\ 0 & 6 & -6 \end{bmatrix} \begin{bmatrix} 0.6 & 0.3 & 0.1 \\ 1 & 1 & 0.5 \\ 1.2 & 1.5 & 1 \end{bmatrix} = \begin{bmatrix} -1 & 0 & 0 \\ 0 & -2 & 0 \\ 0 & 0 & -3 \end{bmatrix}$$

This technique is an important one mathematically (see the companion text *Advanced Modern Engineering Mathematics*) since it provides a method of uncoupling a system of coupled equations. Practically it is the process used to reduce a physical system to principal axes; in elasticity it provides the principal stresses in a body.

Check that in MATLAB the commands

```
T = [0.6 0.3 0.1; 1 1 0.5; 1.2 1.5 1];
inv(T)
```

return the inverse T^{-1} in the numeric form

$$\begin{bmatrix} 4.1667 & -2.5000 & 0.8333 \\ -6.6667 & 8.0000 & -3.3333 \\ 5.0000 & -9.0000 & 5.0000 \end{bmatrix}$$

Check also that, using the Symbolic Math Toolbox, the exact form given in the solution is obtained using the commands

```
T = [0.6 0.3 0.1; 1 1 0.5; 1.2 1.5 1];
T = sym(T);
inv(T)
```

5.4.1 Exercises

Check your answers to the exercises using MATLAB.

51 Determine whether the following matrices are singular or non-singular *and* find the inverse of the non-singular matrices:

(a) $\begin{bmatrix} 1 & 2 \\ 2 & 1 \end{bmatrix}$ (b) $\begin{bmatrix} 1 & 2 & 3 \\ 2 & 2 & 1 \\ 5 & 6 & 5 \end{bmatrix}$

(c) $\begin{bmatrix} 1 & 0 & 0 & 1 \\ 0 & 1 & 0 & 1 \\ 0 & 0 & 1 & 1 \\ 0 & 0 & 0 & 1 \end{bmatrix}$ (d) $\begin{bmatrix} 1 & 0 & 1 \\ 0 & 1 & 0 \\ 1 & 0 & 1 \end{bmatrix}$

52 Find the inverses of the matrices

(a) $\begin{bmatrix} 1 & 1 & 1 \\ 0 & 1 & 0 \\ 0 & 0 & 1 \end{bmatrix}$ (b) $\begin{bmatrix} 1 & 0 & 0 & 0 \\ 0 & 2 & 0 & 0 \\ 0 & 0 & 3 & 0 \\ 0 & 0 & 0 & 4 \end{bmatrix}$

(c) $\begin{bmatrix} 1 & j \\ -j & 2 \end{bmatrix}$ (d) $\begin{bmatrix} 1 & 2 & 3 \\ 0 & 1 & 2 \\ 2 & 3 & 1 \end{bmatrix}$

53 Verify that

$$A = \begin{bmatrix} 0 & 1 & -3 \\ 2 & 1 & 0 \\ 1 & -2 & 1 \end{bmatrix}$$

has an inverse

$$A^{-1} = \frac{1}{13} \begin{bmatrix} 1 & 5 & 3 \\ -2 & 3 & -6 \\ -5 & 1 & -2 \end{bmatrix}$$

and hence solve the equation

$$ACA = \begin{bmatrix} 2 & 1 & 1 \\ 0 & 2 & 3 \\ 2 & 1 & 0 \end{bmatrix}$$

54 (a) If a square matrix A satisfies $A^2 = A$ and has an inverse show that A is the unit matrix.

(b) Show that $A = \begin{bmatrix} 1 & 0 \\ 0 & 0 \end{bmatrix}$ satisfies $A^2 = A$.

(*Note*: matrices that satisfy $A^2 = A$ are called idempotent.)

55 If

$$A = \begin{bmatrix} 1 & 0 & 0 \\ 1 & 1 & 0 \\ 0 & 2 & 1 \end{bmatrix}, \quad B = \begin{bmatrix} 1 & 4 & -1 \\ 0 & 2 & 1 \\ 0 & 0 & 2 \end{bmatrix}$$

and $C = \begin{bmatrix} 1 & 4 & -1 \\ 1 & 6 & 0 \\ 0 & 4 & 4 \end{bmatrix}$

show that $AB = C$. Find the inverse of A and B and hence of C.

(*Note*: this is an example of a powerful method called **LU decomposition**.)

56 Given the matrix

$$A = \begin{bmatrix} 1 & 1 & 0 \\ 0 & 1 & 2 \\ 1 & 2 & 3 \end{bmatrix}$$

and the elementary matrices

$$E_1 = \begin{bmatrix} 1 & 0 & 0 \\ 0 & 1 & 0 \\ -1 & 0 & 1 \end{bmatrix} \quad E_2 = \begin{bmatrix} 1 & 0 & 0 \\ 0 & 1 & 0 \\ 0 & -1 & 1 \end{bmatrix}$$

$$E_3 = \begin{bmatrix} 1 & 0 & 0 \\ 0 & 1 & -2 \\ 0 & 0 & 1 \end{bmatrix} \quad E_4 = \begin{bmatrix} 1 & -1 & 0 \\ 0 & 1 & 0 \\ 0 & 0 & 1 \end{bmatrix}$$

evaluate E_1A, E_2E_1A, $E_3E_2E_1A$ and $E_4E_3E_2E_1A$ and hence find the inverse of A.

(*Note*: the elementary matrices manipulate the rows of the matrix A.)

57 For the matrix

$$A = \begin{bmatrix} 1 & 2 & 2 \\ 2 & 1 & 2 \\ 2 & 2 & 1 \end{bmatrix}$$

show that $A^2 - 4A - 5I = 0$ and hence that $A^{-1} = \frac{1}{5}(A - 4I)$. Calculate A^{-1} from this result. Further show that the inverse of A^2 is given by $\frac{1}{25}(21I - 4A)$ and evaluate.

58 Given

$$A = \begin{bmatrix} 1 & 0 & 2 \\ 6 & 4 & 0 \\ 6 & -2 & 1 \end{bmatrix} \quad \text{and} \quad B = \begin{bmatrix} 5 & 2 & 4 \\ 3 & -1 & 2 \\ 1 & 4 & -3 \end{bmatrix}$$

find A^{-1} and B^{-1}. Verify that $(AB)^{-1} = B^{-1}A^{-1}$.

59 Given the matrices

$$A = \begin{bmatrix} 1 & 0 & 0 & 0 \\ 0 & 0 & 1 & 0 \\ 0 & 1 & 0 & 0 \\ 0 & 0 & 0 & 1 \end{bmatrix} \quad \text{and} \quad B = \begin{bmatrix} 1 & 0 & 0 & 0 \\ 0 & 0 & 1 & 0 \\ 0 & 0 & 0 & 1 \\ 0 & 1 & 0 & 0 \end{bmatrix}$$

show that $A^2 = I$ and $B^3 = I$, and hence find A^{-1}, B^{-1} and $(AB)^{-1}$.

(*Note*: the matrices A and B in this exercise are examples of **permutation matrices**, since

$$A \begin{bmatrix} a_1 \\ a_2 \\ a_3 \\ a_4 \end{bmatrix} = \begin{bmatrix} a_1 \\ a_3 \\ a_2 \\ a_4 \end{bmatrix}$$

and the suffixes are just permuted; B has similar properties.)

5.5 Linear equations

Although matrices are of great importance in themselves, their practical importance lies in the solution of sets of linear equations. Such sets of equations occur in a wide range of scientific and engineering problems. In the first part of this section we shall consider whether or not a solution exists, and then in Sections 5.5.2 and 5.5.4 we shall look at practical methods of solution.

We now make some definitive statements about the solution of the system of simultaneous linear equations.

$$\left.\begin{array}{r} a_{11}x_1 + a_{12}x_2 + \ldots + a_{1n}x_n = b_1 \\ a_{21}x_1 + a_{22}x_2 + \ldots + a_{2n}x_n = b_2 \\ \vdots \qquad\qquad \vdots \\ a_{n1}x_1 + a_{n2}x_2 + \ldots + a_{nn}x_n = b_n \end{array}\right\} \tag{5.18}$$

or, in matrix notation,

$$\begin{bmatrix} a_{11} & a_{12} & \ldots & a_{1n} \\ a_{21} & a_{22} & \ldots & a_{2n} \\ \vdots & & & \\ a_{n1} & a_{n2} & \ldots & a_{nn} \end{bmatrix} \begin{bmatrix} x_1 \\ x_2 \\ \vdots \\ x_n \end{bmatrix} = \begin{bmatrix} b_1 \\ b_2 \\ \vdots \\ b_n \end{bmatrix}$$

that is,

$$\boldsymbol{AX} = \boldsymbol{b} \tag{5.19}$$

where $\boldsymbol{A}$ is the matrix of coefficients and $\boldsymbol{X}$ the vector of unknowns. If $\boldsymbol{b} = 0$ the equations are called **homogeneous**, while if $\boldsymbol{b} \neq 0$ they are called **nonhomogeneous** (or **inhomogeneous**). There are several cases to consider.

Case (a) $b \neq 0$ and $|A| \neq 0$

We know that $\boldsymbol{A}^{-1}$ exists, and hence

$$\boldsymbol{A}^{-1}\boldsymbol{AX} = \boldsymbol{A}^{-1}\boldsymbol{b}$$

so that

$$\boldsymbol{X} = \boldsymbol{A}^{-1}\boldsymbol{b} \tag{5.20}$$

and we have a unique solution to (5.18) and (5.19).

Case (b) $b = 0$ and $|A| \neq 0$

Again $\boldsymbol{A}^{-1}$ exists, and the homogeneous equations

$$\boldsymbol{AX} = 0$$

give

$$A^{-1}AX = A^{-1}0 \quad \text{or} \quad X = 0$$

We therefore only have the **trivial solution** $X = 0$.

Case (c) $b \neq 0$ and $|A| = 0$

The inverse matrix does not exist, and this is perhaps the most complicated case. We have two possibilities: either we have no solution or we have infinitely many solutions. A simple example will illustrate the situation. The equations

$$\left.\begin{array}{r} 3x + 2y = 2 \\ 3x + 2y = 6 \end{array}\right\}, \quad \text{or} \quad \begin{bmatrix} 3 & 2 \\ 3 & 2 \end{bmatrix}\begin{bmatrix} x \\ y \end{bmatrix} = \begin{bmatrix} 2 \\ 6 \end{bmatrix}$$

are clearly inconsistent, and no solution exists. However, in the case of

$$\left.\begin{array}{r} 3x + 2y = 2 \\ 6x + 4y = 4 \end{array}\right\}, \quad \text{or} \quad \begin{bmatrix} 3 & 2 \\ 6 & 4 \end{bmatrix}\begin{bmatrix} x \\ y \end{bmatrix} = \begin{bmatrix} 2 \\ 4 \end{bmatrix}$$

where one equation is a multiple of the other, we have infinitely many solutions: $x = \lambda$, $y = 1 - \frac{3}{2}\lambda$ is a solution for any value of λ.

The same behaviour is observed for problems involving more than two variables, but the situation is then much more difficult to analyse. The problem of determining whether or not a set of equations has a solution will be discussed in Section 5.6.

Case (d) $b = 0$ and $|A| = 0$

As in case (c), we have infinitely many solutions. For instance, the case of two equations takes the form

$$px + qy = 0$$

$$\alpha px + \alpha qy = 0$$

so that $|A| = 0$ and we find a solution $x = \lambda$, $y = -p\lambda/q$ if $q \neq 0$. If $q = 0$ then $x = 0$, $y = \lambda$ is a solution.

This case is one of the most important, since we can deduce the important general result that *the equation*

$$AX = 0$$

has a non-trivial solution if and only if $|A| = 0$.

Example 5.24

Write the five sets of equations in matrix form and decide whether they have or do not have a solution.

(a) $2x + y = 5$
      $x - 2y = -5$

(b) $2x + y = 0$
      $x - 2y = 0$

(c) $-3x + 6y = 15$
      $x - 2y = -5$

(d) $-3x + 6y = 10$
      $x - 2y = -5$

(e) $-3x + 6y = 0$
      $x - 2y = 0$

Solution (a) In matrix form the equations are $\begin{bmatrix} 2 & 1 \\ 1 & -2 \end{bmatrix} \begin{bmatrix} x \\ y \end{bmatrix} = \begin{bmatrix} 5 \\ -5 \end{bmatrix}$. The determinant of the matrix has the value -5 and the right-hand side is non-zero so the problem is of the type **Case (a)** and hence has a unique solution, namely $x = 1$, $y = 3$.

(b) In matrix form the equations are $\begin{bmatrix} 2 & 1 \\ 1 & -2 \end{bmatrix} \begin{bmatrix} x \\ y \end{bmatrix} = \begin{bmatrix} 0 \\ 0 \end{bmatrix}$. The determinant of the matrix has the value -5 and the right-hand side is now zero so the problem is of the type **Case (b)** and hence only has the trivial solution, namely $x = 0$, $y = 0$.

(c) In matrix form the equations are $\begin{bmatrix} -3 & 6 \\ 1 & -2 \end{bmatrix} \begin{bmatrix} x \\ y \end{bmatrix} = \begin{bmatrix} 15 \\ -5 \end{bmatrix}$. The determinant of the matrix is now zero and the right-hand side is non-zero so the problem is of the type **Case (c)** and hence the solution is not so easy. Essentially the first equation is just (-3) times the second equation so a solution can be computed. A bit of rearrangement soon gives $x = 2t - 5$, $y = t$ for any t, and thus there are infinitely many solutions to this set of equations.

(d) In matrix form the equations are $\begin{bmatrix} -3 & 6 \\ 1 & -2 \end{bmatrix} \begin{bmatrix} x \\ y \end{bmatrix} = \begin{bmatrix} 10 \\ -5 \end{bmatrix}$. The determinant of the matrix is zero again and the right-hand side is non-zero so the problem is once more of the type **Case (c)** and hence the solution is not so easy. The left-hand side of the first equation is (-3) times the second equation but the right-hand side is only (-2) times the second equation so the equations are inconsistent and there is no solution to this set of equations.

(e) In matrix form the equations are $\begin{bmatrix} -3 & 6 \\ 1 & -2 \end{bmatrix} \begin{bmatrix} x \\ y \end{bmatrix} = \begin{bmatrix} 0 \\ 0 \end{bmatrix}$. The determinant of the matrix is zero again and the right-hand side is also zero so the problem is of the type **Case (d)** and hence a non-trivial solution can be found. It can be seen that $x = 2s$ and $y = s$ gives the solution for any s.

Example 5.25 Find a solution of

$$x + y + z = 6$$
$$x + 2y + 3z = 14$$
$$x + 4y + 9z = 36$$

Solution Expressing the equations in matrix form $\mathbf{AX} = \mathbf{b}$

$$\begin{bmatrix} 1 & 1 & 1 \\ 1 & 2 & 3 \\ 1 & 4 & 9 \end{bmatrix} \begin{bmatrix} x \\ y \\ z \end{bmatrix} = \begin{bmatrix} 6 \\ 14 \\ 36 \end{bmatrix}$$

we have

$$|A| = \begin{vmatrix} 1 & 1 & 1 \\ 1 & 2 & 3 \\ 1 & 4 & 9 \end{vmatrix} = \begin{vmatrix} 1 & 0 & 0 \\ 1 & 1 & 2 \\ 1 & 3 & 8 \end{vmatrix} = 2 \neq 0 \quad \text{(subtracting column 1 from columns 2 and 3)}$$

so that a solution does exist and is unique. The inverse of A can be computed as

$$A^{-1} = \begin{bmatrix} 3 & -\frac{5}{2} & \frac{1}{2} \\ -3 & 4 & -1 \\ 1 & -\frac{3}{2} & \frac{1}{2} \end{bmatrix}$$

and hence, from (5.20),

$$X = \begin{bmatrix} x \\ y \\ z \end{bmatrix} = A^{-1} \begin{bmatrix} 6 \\ 14 \\ 36 \end{bmatrix} = \begin{bmatrix} 1 \\ 2 \\ 3 \end{bmatrix}$$

so the solution is $x = 1$, $y = 2$ and $z = 3$.

Example 5.26 Find the values of k for which the equations

$$x + 5y + 3z = 0$$
$$5x + y - kz = 0$$
$$x + 2y + kz = 0$$

have a non-trivial solution.

Solution The matrix of coefficients is

$$A = \begin{bmatrix} 1 & 5 & 3 \\ 5 & 1 & -k \\ 1 & 2 & k \end{bmatrix}$$

For a non-zero solution, $|A| = 0$. Hence

$$0 = |A| = \begin{vmatrix} 1 & 5 & 3 \\ 5 & 1 & -k \\ 1 & 2 & k \end{vmatrix} = 27 - 27k$$

Thus the equations have a non-trivial solution if $k = 1$; if $k \neq 1$, the only solution is $x = y = z = 0$. For $k = 1$ a simple calculation gives $x = \lambda$, $y = -2\lambda$ and $z = 3\lambda$ for any λ.

Example 5.27 Find the values of λ and the corresponding column vector X such that

$$(A - \lambda I)X = 0$$

has a non-trivial solution, given

$$A = \begin{bmatrix} 3 & 1 \\ -2 & 0 \end{bmatrix}$$

Solution We require

$$0 = |A - \lambda I| = \begin{vmatrix} 3 - \lambda & 1 \\ -2 & -\lambda \end{vmatrix}$$

$$= -3\lambda + \lambda^2 + 2 = (\lambda - 2)(\lambda - 1)$$

Non-trivial solutions occur only if $\lambda = 1$ or 2.
  If $\lambda = 1$,

$$\begin{bmatrix} 2 & 1 \\ -2 & -1 \end{bmatrix}\begin{bmatrix} x \\ y \end{bmatrix} = 0, \quad \text{so } X = \begin{bmatrix} x \\ y \end{bmatrix} = \alpha\begin{bmatrix} 1 \\ -2 \end{bmatrix} \quad \text{for any } \alpha$$

  If $\lambda = 2$,

$$\begin{bmatrix} 1 & 1 \\ -2 & -2 \end{bmatrix}\begin{bmatrix} x \\ y \end{bmatrix} = 0, \quad \text{so } X = \begin{bmatrix} x \\ y \end{bmatrix} = \beta\begin{bmatrix} 1 \\ -1 \end{bmatrix} \quad \text{for any } \beta$$

(Note that the problem described here is an important one. The λ and X are called **eigenvalues** and **eigenvectors**, which are introduced in Section 5.7.)

It is possible to write down the solution of a set of equations explicitly in terms of the cofactors of a matrix. However, as a method for computing the solution, this is extremely inefficient; a set of ten equations, for example, will require 4×10^8 multiplications – which takes a long time even on modern computers. The method is of great theoretical interest though. Consider the set of equations

$$\left. \begin{array}{l} a_{11}x_1 + a_{12}x_2 + a_{13}x_3 = b_1 \\ a_{21}x_1 + a_{22}x_2 + a_{23}x_3 = b_2 \\ a_{31}x_1 + a_{32}x_2 + a_{33}x_3 = b_3 \end{array} \right\} \tag{5.21}$$

Denoting the matrix of coefficients by A and recalling the definitions of the cofactors in Section 5.3, we multiply the equations by A_{11}, A_{21} and A_{31} respectively, and add to give

$$(a_{11}A_{11} + a_{21}A_{21} + a_{31}A_{31})x_1 + (a_{12}A_{11} + a_{22}A_{21} + a_{32}A_{31})x_2$$

$$+ (a_{13}A_{11} + a_{23}A_{21} + a_{33}A_{31})x_3$$

$$= b_1A_{11} + b_2A_{21} + b_3A_{31}$$

Using (5.12), we obtain

$$|A|x_1 + 0x_2 + 0x_3 = b_1A_{11} + b_2A_{21} + b_3A_{31}$$

The right-hand side can be written as a determinant, so

$$|A|x_1 = \begin{vmatrix} b_1 & a_{12} & a_{13} \\ b_2 & a_{22} & a_{23} \\ b_3 & a_{32} & a_{33} \end{vmatrix}$$

The other x_i follow similarly, and we derive **Cramer's rule** that a solution of (5.21) is

$$x_1 = |A|^{-1} \begin{vmatrix} b_1 & a_{12} & a_{13} \\ b_2 & a_{22} & a_{23} \\ b_3 & a_{32} & a_{33} \end{vmatrix},$$

$$x_2 = |A|^{-1} \begin{vmatrix} a_{11} & b_1 & a_{13} \\ a_{21} & b_2 & a_{23} \\ a_{31} & b_3 & a_{33} \end{vmatrix},$$

$$x_3 = |A|^{-1} \begin{vmatrix} a_{11} & a_{12} & b_1 \\ a_{21} & a_{22} & b_2 \\ a_{31} & a_{32} & b_3 \end{vmatrix}$$

Again it should be stressed that this rule should not be used as a computational method because of the large effort required to evaluate determinants.

Example 5.28

A function $u(x, y)$ is known to take values u_1, u_2 and u_3 at the points (x_1, y_1), (x_2, y_2) and (x_3, y_3) respectively. Find the linear interpolating function

$$u = a + bx + cy$$

within the triangle having its vertices at these three points.

Solution

To fit the data to the linear interpolating function

$$\begin{aligned} u_1 &= a + bx_1 + cy_1 \\ u_2 &= a + bx_2 + cy_2 \quad \text{or in matrix form} \\ u_3 &= a + bx_3 + cy_3 \end{aligned} \qquad \begin{bmatrix} u_1 \\ u_2 \\ u_3 \end{bmatrix} = \begin{bmatrix} 1 & x_1 & y_1 \\ 1 & x_2 & y_2 \\ 1 & x_3 & y_3 \end{bmatrix} \begin{bmatrix} a \\ b \\ c \end{bmatrix}$$

The values of a, b and c can be obtained from Cramer's rule as

$$a = \begin{vmatrix} u_1 & x_1 & y_1 \\ u_2 & x_2 & y_2 \\ u_3 & x_3 & y_3 \end{vmatrix} / \det(A), \quad b = \begin{vmatrix} 1 & u_1 & y_1 \\ 1 & u_2 & y_2 \\ 1 & u_3 & y_3 \end{vmatrix} / \det(A) \quad \text{and}$$

$$c = \begin{vmatrix} 1 & x_1 & u_1 \\ 1 & x_2 & u_2 \\ 1 & x_3 & u_3 \end{vmatrix} / \det(\boldsymbol{A})$$

where $\boldsymbol{A}$ is the matrix of coefficients. The interpolation formula is now known. In finite-element analysis the evaluation of interpolation functions, such as the one described, is of great importance. Finite elements are central to many large-scale calculations in all branches of engineering; an introduction is given in Chapter 9 of the companion text *Advanced Modern Engineering Mathematics*.

Example 5.29

Solve the matrix equation $\boldsymbol{AX} = \boldsymbol{c}$ where

$$\boldsymbol{A} = \begin{bmatrix} 4 & 1 & 0 & 0 & 0 & 0 & 0 & 0 & 0 & 0 \\ 1 & 4 & 1 & 0 & 0 & 0 & 0 & 0 & 0 & 0 \\ 1 & 0 & 4 & 1 & 0 & 0 & 0 & 0 & 0 & 0 \\ 1 & 0 & 0 & 4 & 1 & 0 & 0 & 0 & 0 & 0 \\ 1 & 0 & 0 & 0 & 4 & 1 & 0 & 0 & 0 & 0 \\ 1 & 0 & 0 & 0 & 0 & 4 & 1 & 0 & 0 & 0 \\ 1 & 0 & 0 & 0 & 0 & 0 & 4 & 1 & 0 & 0 \\ 1 & 0 & 0 & 0 & 0 & 0 & 0 & 4 & 1 & 0 \\ 1 & 0 & 0 & 0 & 0 & 0 & 0 & 0 & 4 & 1 \\ 1 & 0 & 0 & 0 & 0 & 0 & 0 & 0 & 0 & 4 \end{bmatrix} \quad \text{and} \quad \boldsymbol{c} = \begin{bmatrix} 1 \\ 2 \\ 3 \\ 4 \\ 5 \\ 5 \\ 4 \\ 3 \\ 2 \\ 1 \end{bmatrix}$$

Solution   The solution of such a problem is beyond the scope of hand computation; Cramer's rule, evaluation of the adjoint and direct evaluation of the inverse are all impracticable. Even the more practical methods in the next sections struggle with this size of problem if hand computation is tried. A computer package must be used. In MATLAB the relevant instructions are given.

```
b = zeros (10, 10);
for i = 1 : 9, b (i, i) = 4; b (i, i + 1) = 1; b (i + 1, 1)
= 1; end
b (10, 10) = 4;
c = [1; 2; 3; 4; 5; 5; 4; 3; 2; 1];
b\c
```

gives the solution

```
0.1685  0.3258  0.5282  0.7188  0.9563  1.0063  0.8063
0.6064  0.4059  0.2079
```

5.5.1 Exercises

 Check your answers using MATLAB or MAPLE whenever possible.

60 Solve the matrix equation $AX = b$ for the vector X in the following:

(a) $A = \begin{bmatrix} 2 & 3 \\ 5 & -2 \end{bmatrix}$ $b = \begin{bmatrix} 8 \\ 1 \end{bmatrix}$

(b) $A = \begin{bmatrix} 1 & 0 & 0 \\ 2 & -1 & 0 \\ 2 & 2 & 2 \end{bmatrix}$ $b = \begin{bmatrix} 1 \\ 6 \\ -6 \end{bmatrix}$

(c) $A = \begin{bmatrix} 2 & 1 \\ 0 & 1 \end{bmatrix}\begin{bmatrix} 1 & 0 \\ 3 & 1 \end{bmatrix}$ $b = -\begin{bmatrix} 3 \\ 1 \end{bmatrix}$

(d) $A = \begin{bmatrix} 1 & 0 & 0 & 0 \\ 0 & 3 & 1 & 0 \\ 0 & 1 & 2 & 0 \\ 0 & 0 & 0 & 1 \end{bmatrix}$ $b = \begin{bmatrix} 4 \\ 11 \\ 7 \\ 1 \end{bmatrix}$

61 If

$$A = \begin{bmatrix} \cos\alpha & -\sin\alpha \\ \sin\alpha & \cos\alpha \end{bmatrix}$$

show that

$$A^{-1} = \begin{bmatrix} \cos\alpha & \sin\alpha \\ -\sin\alpha & \cos\alpha \end{bmatrix}$$

and hence solve for the vector X in the equation

$$\begin{bmatrix} \cos\frac{\pi}{8} & -\sin\frac{\pi}{8} \\ \sin\frac{\pi}{8} & \cos\frac{\pi}{8} \end{bmatrix} X = \begin{bmatrix} \cos\frac{\pi}{4} \\ \sin\frac{\pi}{4} \end{bmatrix}$$

62 Solve the complex matrix equation

$$\begin{bmatrix} 1 & j & 0 \\ 0 & 1 & 0 \\ j & 0 & j \end{bmatrix} X = \begin{bmatrix} 0 \\ 1 \\ 0 \end{bmatrix}$$

63 Find the inverse of the matrix

$$A = \begin{bmatrix} -1 & 2 & 1 \\ 0 & 1 & -2 \\ 1 & 4 & -1 \end{bmatrix}$$

and hence solve the equations

$$-x + 2y + z = 2$$
$$y - 2z = -3$$
$$x + 4y - z = 4$$

64 Show that there are two values of α for which the equations

$$\alpha x - 3y + (1 + \alpha)z = 0$$
$$2x + y - \alpha z = 0$$
$$(\alpha + 2)x - 2y + \alpha z = 0$$

have non-trivial solutions. Find the solutions corresponding to these two values of α.

65 If

$$A = \begin{bmatrix} -3 & 1 & -1 \\ 1 & -5 & 1 \\ -1 & 1 & -3 \end{bmatrix}$$

find the values of λ for which the equation $AX = \lambda X$ has non-trivial solutions.

66 Given the matrix

$$A = \begin{bmatrix} 1 & a & -1 \\ a & -2 & 2 \\ -1 & 1 & a \end{bmatrix}$$

(a) solve $|A| = 0$ for real a;

(b) if $a = 2$, find A^{-1} and hence solve

$$A \begin{bmatrix} x \\ y \\ z \end{bmatrix} = \begin{bmatrix} 1 \\ 0 \\ 2 \end{bmatrix}$$

(c) if $a = 0$, find the general solution of

$$A \begin{bmatrix} x \\ y \\ z \end{bmatrix} = \begin{bmatrix} 0 \\ 0 \\ 0 \end{bmatrix}$$

(d) if $a = 1$, show that

$$A \begin{bmatrix} x \\ y \\ z \end{bmatrix} = 2 \begin{bmatrix} x \\ y \\ z \end{bmatrix}$$

can be solved for non-zero x, y and z.

 67 Use MATLAB or a similar package to find the inverse of the matrix

$$\begin{bmatrix} 6 & 2 & 1 & 0 & 0 & 0 \\ 2 & 6 & 2 & 1 & 0 & 0 \\ 1 & 2 & 6 & 2 & 1 & 0 \\ 0 & 1 & 2 & 6 & 2 & 1 \\ 0 & 0 & 1 & 2 & 6 & 2 \\ 0 & 0 & 0 & 1 & 2 & 6 \end{bmatrix}$$

and hence solve the matrix equation

$$AX = c$$

where $c^T = [1\ 0\ 0\ 0\ 0\ 1]$.

68 In finite-element calculations the bilinear function

$$u(x, y) = a + bx + cy + dxy$$

is commonly used for interpolation over a quadrilateral and data is always stored in matrix form. If the function fits the data $u(0, 0) = u_1$, $u(p, 0) = u_2$, $u(0, q) = u_3$ and $u(p, q) = u_4$ at the four corners of a rectangle, use matrices to find the coefficients a, b, c and d.

 69 In an industrial process water flows through three tanks in succession as illustrated in Figure 5.8.

The tanks have unit cross-section and have heads of water x, y and z respectively. The rate of inflow into the first tank is u, the flowrate in the tube connecting tanks 1 and 2 is $6(x - y)$, the flowrate in the tube connecting tanks 2 and 3 is $5(y - z)$ and the rate of outflow from tank 3 is $4.5z$.

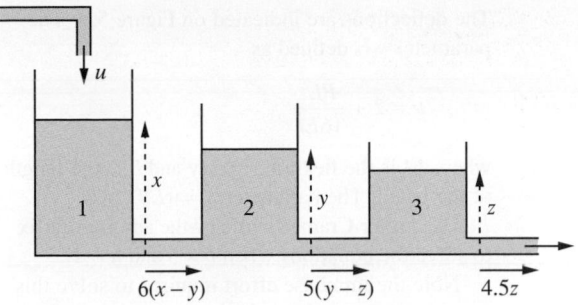

Figure 5.8 Flow through three tanks in Question 69.

Show that the equations of the system in the steady flow situation are

$$u = 6x - 6y$$
$$0 = 6x - 11y + 5z$$
$$0 = 5y - 9.5z$$

and hence find x, y and z.

 70 A function is known to fit closely to the approximate function

$$f(z) = \frac{az + b}{cz + 1}$$

It is fitted to the three points $(z = 0, f = 1)$, $(z = 0.5, f = 1.128)$ and $(z = 1.3, f = 1.971)$. Show that the parameters satisfy

$$\begin{bmatrix} 1 \\ 1.128 \\ 1.971 \end{bmatrix} = \begin{bmatrix} 0 & 1 & 0 \\ 0.5 & 1 & -0.5640 \\ 1.3 & 1 & -2.562 \end{bmatrix} \begin{bmatrix} a \\ b \\ c \end{bmatrix}$$

Find a, b and c and hence the approximating function (use of MATLAB is recommended). Check the value $f(1) = 1.543$. (Note that the values were chosen from tables of cosh z.)

The method described here is a simple example of a powerful approximation method.

 71 A cantilever beam bends under a uniform load w per unit length and is subject to an axial force P at its free end. For small deflections a numerical approximation to the shape of the beam is given by the set of equations

$$-vy_1 + y_2 \qquad\qquad = -u$$
$$y_1 - vy_2 + y_3 \qquad\quad = -4u$$
$$y_2 - vy_3 + y_4 = -9u$$
$$2y_3 - vy_4 = -16u$$

The deflections are indicated on Figure 5.9. The parameter v is defined as

$$v = 2 + \frac{PL^2}{16EI}$$

where EI is the flexural rigidity and L is the length of the beam. The parameter $u = wL^4/32EI$.

Use either Cramer's rule or the adjoint matrix to solve the equations when $v = 3$ and $u = 1$.

Note the immense effort required to solve this very simple problem using these methods. In later sections much more efficient methods will be described. A computer package such as MATLAB should be used to check the results.

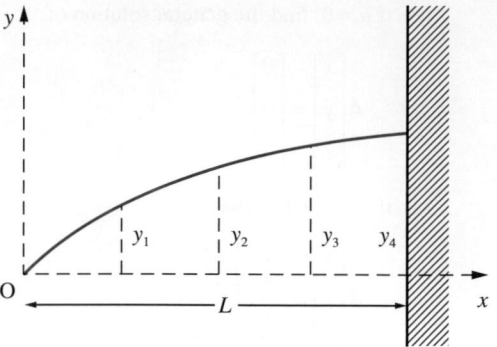

Figure 5.9 Cantilever beam in Question 71.

5.5.2 The solution of linear equations: elimination methods

The idea behind elimination techniques can be seen by considering the solution of two simultaneous equations

$$x + 2y = 4$$

$$2x + y = 5$$

Subtract $2 \times$ (equation 1) from (equation 2) to give

$$x + 2y = 4$$

$$-3y = -3$$

Divide the second equation by -3

$$x + 2y = 4$$

$$y = 1$$

From the second of these equations $y = 1$ and substituting into the first equation gives $x = 2$.

This example illustrates the basic technique for the solution of a set of linear equations by Gaussian elimination, which is very straightforward in principle. However, it needs considerable care to ensure that the calculations are carried out efficiently. Given n linear equations in the variables $x_1, x_2, \ldots, x_n$, we solve in a series of steps:

(1)   We solve the first equation for x_1 in terms of $x_2, \ldots, x_n$, and eliminate x_1 from the remaining equations.

(2)   We then solve the second equation of the remaining set for x_2 in terms of $x_3, \ldots, x_n$ and eliminate x_2 from the remaining equations.

(3)   We repeat the process in turn on $x_3, x_4, \ldots$ until we arrive at a final equation for x_n, which we can then solve.

(4)   We substitute back to get in turn $x_{n-1}, x_{n-2}, \ldots, x_1$.

For a small number of variables, say two, three or four, the method is easy to apply, and efficiency is not of the highest priority. In most science and engineering problems we are normally dealing with a large number of variables – a simple stability analysis of a vibrating system can lead to seven or eight variables, and a plate-bending problem could easily give rise to several hundred variables.

As a further example of the basic technique, we solve

$$x_1 + x_2 \qquad = 3 \tag{5.22}$$

$$2x_1 + x_2 + x_3 = 7 \tag{5.23}$$

$$x_1 + 2x_2 + 3x_3 = 14 \tag{5.24}$$

First, we eliminate x_1:

(5.22) gives $\qquad x_1 = 3 - x_2$ (5.22′)

(5.23) gives $\qquad 2(3 - x_2) + x_2 + x_3 = 7, \quad$ or $\quad -x_2 + x_3 = 1$ (5.23′)

(5.24) gives $\qquad (3 - x_2) + 2x_2 + 3x_3 = 14, \quad$ or $\quad x_2 + 3x_3 = 11$ (5.24′)

Secondly we eliminate x_2:

(5.23′) gives $\qquad x_2 = x_3 - 1$ (5.23″)

(5.24′) gives $\qquad (x_3 - 1) + 3x_3 = 11, \quad$ or $\quad 4x_3 = 12$ (5.24″)

Equation (5.24″) gives $x_3 = 3$; we put this into (5.23″) to obtain $x_2 = 2$; we then put this into (5.22′) to obtain $x_1 = 1$. Thus the values $x_1 = 1$, $x_2 = 2$ and $x_3 = 3$ give a solution to the original problem.

Equations (5.22)–(5.24) in matrix form become

$$\begin{bmatrix} 1 & 1 & 0 \\ 2 & 1 & 1 \\ 1 & 2 & 3 \end{bmatrix} \begin{bmatrix} x_1 \\ x_2 \\ x_3 \end{bmatrix} = \begin{bmatrix} 3 \\ 7 \\ 14 \end{bmatrix}$$

The elimination procedure has reduced the equations to (5.22), (5.23′) and (5.24″), which in matrix form become

$$\begin{bmatrix} 1 & 1 & 0 \\ 0 & -1 & 1 \\ 0 & 0 & 4 \end{bmatrix} \begin{bmatrix} x_1 \\ x_2 \\ x_3 \end{bmatrix} = \begin{bmatrix} 3 \\ 1 \\ 12 \end{bmatrix}$$

Essentially the elimination has brought the equations to **upper-triangular** form (that is, a form in which the matrix of coefficients has zeros in every position below the diagonal), which are then very easy to solve.

Elimination procedures rely on the manipulation of equations or, equivalently, the rows of the matrix equation. There are various **elementary row operations** used which do not alter the solution of the equations:

(a) multiply a row by a constant,
(b) interchange any two rows,
(c) add or subtract one row from another.

To illustrate these, we take the matrix equation

$$\begin{bmatrix} 1 & 1 & 0 \\ 2 & 1 & 1 \\ -1 & 2 & 3 \end{bmatrix} \begin{bmatrix} x_1 \\ x_2 \\ x_3 \end{bmatrix} = \begin{bmatrix} 3 \\ 7 \\ 12 \end{bmatrix}$$

which has the solution $x_1 = 1$, $x_2 = 2$, $x_3 = 3$.

Multiplying the first row by 2 (a row operation of type (a)) yields

$$\begin{bmatrix} 2 & 2 & 0 \\ 2 & 1 & 1 \\ -1 & 2 & 3 \end{bmatrix} \begin{bmatrix} x_1 \\ x_2 \\ x_3 \end{bmatrix} = \begin{bmatrix} 6 \\ 7 \\ 12 \end{bmatrix}$$

Interchanging rows 1 and 3 (a row operation of type (b)) yields

$$\begin{bmatrix} -1 & 2 & 3 \\ 2 & 1 & 1 \\ 1 & 1 & 0 \end{bmatrix} \begin{bmatrix} x_1 \\ x_2 \\ x_3 \end{bmatrix} = \begin{bmatrix} 12 \\ 7 \\ 3 \end{bmatrix}$$

Subtracting row 1 from row 2 (a row operation of type (c)) yields

$$\begin{bmatrix} 1 & 1 & 0 \\ 1 & 0 & 1 \\ -1 & 2 & 3 \end{bmatrix} \begin{bmatrix} x_1 \\ x_2 \\ x_3 \end{bmatrix} = \begin{bmatrix} 3 \\ 4 \\ 12 \end{bmatrix}$$

In each case we see that the solution of the modified equations is still $x_1 = 1$, $x_2 = 2$, $x_3 = 3$.

Elimination procedures use repeated applications of (a), (b) and (c) in some systematic manner until the equations are processed into a required form such as the upper-triangular equations

$$\begin{bmatrix} a_{11} & a_{12} & a_{13} & \dots & a_{1n} \\ 0 & a_{22} & a_{23} & \dots & a_{2n} \\ 0 & 0 & a_{33} & \dots & a_{3n} \\ \vdots & \vdots & \vdots & & \vdots \\ 0 & 0 & 0 & \dots & a_{nn} \end{bmatrix} \begin{bmatrix} x_1 \\ x_2 \\ x_3 \\ \vdots \\ x_n \end{bmatrix} = \begin{bmatrix} b_1 \\ b_2 \\ b_3 \\ \vdots \\ b_n \end{bmatrix} \qquad (5.25)$$

The solution of the equations in upper-triangular form can be written as

$$x_n = b_n/a_{nn}$$

$$x_{n-1} = (b_{n-1} - a_{n-1,n}x_n)/a_{n-1,n-1}$$

$$x_{n-2} = (b_{n-2} - a_{n-2,n}x_n - a_{n-2,n-1}x_{n-1})/a_{n-2,n-2}$$

$$\vdots$$

$$x_1 = (b_1 - a_{1n}x_n - a_{1,n-1}x_{n-1} - \dots - a_{12}x_2)/a_{11}$$

A pseudocode procedure implementing these equations is shown in Figure 5.10. The elementary row operations and the elimination technique are illustrated in Example 5.30.

Figure 5.10
Procedure to solve
the upper-triangular
system (5.25).

```
procedure uppertriangular (A, b, n → x)
{A is an n × n matrix, b and x are vectors with n elements}
    x(n)←b(n)/A(n, n)
    for j is n – 1   to 1 by –1 do
        sum←b(j)
        for i is n    to j + 1 by –1 do
            sum←sum – A(j, i) * x(i)
        endfor
        x(j)←sum/A(j, j)
    endfor
endprocedure
```

Example 5.30

Use elementary row operations and elimination to solve the set of linear equations

$$x + 2y + 3z = 10$$

$$-x + y + z = 0$$

$$y - z = 1$$

Solution   In matrix form the equations are: $\begin{bmatrix} 1 & 2 & 3 \\ -1 & 1 & 1 \\ 0 & 1 & -1 \end{bmatrix} \begin{bmatrix} x \\ y \\ z \end{bmatrix} = \begin{bmatrix} 10 \\ 0 \\ 1 \end{bmatrix}$

Add row 1 to row 2: $\begin{bmatrix} 1 & 2 & 3 \\ 0 & 3 & 4 \\ 0 & 1 & -1 \end{bmatrix} \begin{bmatrix} x \\ y \\ z \end{bmatrix} = \begin{bmatrix} 10 \\ 10 \\ 1 \end{bmatrix}$

Divide row 2 by 3: $\begin{bmatrix} 1 & 2 & 3 \\ 0 & 1 & \frac{4}{3} \\ 0 & 1 & -1 \end{bmatrix} \begin{bmatrix} x \\ y \\ z \end{bmatrix} = \begin{bmatrix} 10 \\ \frac{10}{3} \\ 1 \end{bmatrix}$

Subtract row 2 from row 3: $\begin{bmatrix} 1 & 2 & 3 \\ 0 & 1 & \frac{4}{3} \\ 0 & 0 & -\frac{7}{3} \end{bmatrix} \begin{bmatrix} x \\ y \\ z \end{bmatrix} = \begin{bmatrix} 10 \\ \frac{10}{3} \\ -\frac{7}{3} \end{bmatrix}$

Divide row 3 by $(-\frac{7}{3})$: $\begin{bmatrix} 1 & 2 & 3 \\ 0 & 1 & \frac{4}{3} \\ 0 & 0 & 1 \end{bmatrix} \begin{bmatrix} x \\ y \\ z \end{bmatrix} = \begin{bmatrix} 10 \\ \frac{10}{3} \\ 1 \end{bmatrix}$

The equations are now in a standard upper-triangular form for the application of the
back substitution procedure described formally in Figure 5.10.

From the third row $\qquad z = 1$

From the second row $\qquad y = \frac{10}{3} - \frac{4}{3}z = 2$

From the first row $\qquad x = 10 - 2y - 3z = 3$

It remains to undertake the operations in the example in a routine and logical manner to make the method into one of the most powerful techniques, called **elimination methods,** available for the solution of sets of linear equations. The method is available on all computer packages. Such packages are excellent at undertaking the rather tedious arithmetic and some will even illustrate the computational detail also. They are well worth mastering. However, writing and checking your own procedures, following the pseudocode in Figure 5.10 for instance, is a powerful learning tool and gives great understanding of the method, the difficulties and errors involved in a method.

Tridiagonal or Thomas algorithm

Because of the ease of solution of upper-triangular systems, many methods use the general strategy of reducing the equations to this form. As an example of this strategy, we shall look at a **tridiagonal system,** which takes the form

$$
\begin{bmatrix}
a_1 & b_1 & 0 & 0 & 0 & 0 & \cdots & 0 \\
c_2 & a_2 & b_2 & 0 & 0 & 0 & \cdots & 0 \\
0 & c_3 & a_3 & b_3 & 0 & 0 & \cdots & 0 \\
0 & 0 & c_4 & a_4 & b_4 & 0 & \cdots & 0 \\
 & & & & & & & \vdots \\
 & & & & & & 0 & \\
\vdots & & & & & & & \\
0 & \cdots & & 0 & c_{n-1} & a_{n-1} & b_{n-1} & \\
0 & \cdots & & 0 & 0 & c_n & a_n &
\end{bmatrix}
\begin{bmatrix}
x_1 \\ x_2 \\ \\ \vdots \\ \\ \\ x_{n-1} \\ x_n
\end{bmatrix}
=
\begin{bmatrix}
d_1 \\ d_2 \\ \\ \vdots \\ \\ \\ d_{n-1} \\ d_n
\end{bmatrix}
\qquad (5.26)
$$

or

$$
\begin{aligned}
a_1 x_1 + b_1 x_2 &= d_1 \\
c_2 x_1 + a_2 x_2 + b_2 x_3 &= d_2 \\
c_3 x_2 + a_3 x_3 + b_3 x_4 &= d_3 \\
\ddots \qquad \ddots \qquad &\vdots \\
c_n x_{n-1} + a_n x_n &= d_n
\end{aligned}
$$

First we eliminate x_1:

$$
\begin{aligned}
x_1 + b_1' x_2 &= d_1' \\
a_2' x_2 + b_2 x_3 &= d_2' \\
c_3 x_2 + a_3 x_3 + b_3 x_4 &= d_3
\end{aligned}
$$

and so on

Figure 5.11
Tridiagonal or Thomas algorithm for the solution of (5.26).

```
procedure tridiagonal (a, b, c, d, n→a, b, d, x)
    {a, b, c, d and x are vectors with n elements}
    for i is 1 to n − 1 do
    b(i)←b(i)/a(i)
    d(i)←d(i)/a(i)
    a(i + 1)←a(i + 1) − c(i + 1) * b(i)
    d(i + 1)←d(i + 1) − c(i + 1) * d(i)
    endfor {elimination stage}
        x(n)←d(n)/a(n)
        for i is n − 1 to 1 by −1 do
            x(i)←d(i) − b(i) * x(i + 1)
        endfor {back substitution}
endprocedure
```

where

$$b_1' = \frac{b_1}{a_1}, \quad d_1' = \frac{d_1}{a_1}, \quad a_2' = a_2 - c_2 b_1' \quad \text{and} \quad d_2' = d_2 - c_2 d_1'$$

Next we eliminate x_2:

$$x_1 + b_1' x_2 \qquad\qquad\qquad = d_1'$$
$$x_2 + b_2'' x_3 \qquad\qquad\quad = d_2''$$
$$a_3'' x_3 + b_3 x_4 \qquad\quad = d_3''$$
$$c_4 x_3 + a_4 x_4 + b_4 x_5 \;\; = d_4$$

and so on

where

$$b_2'' = \frac{b_2}{a_2'}, \quad d_2'' = \frac{d_2'}{a_2'}, \quad a_3'' = a_3 - c_3 b_2'' \quad \text{and} \quad d_3'' = d_3 - c_3 d_2''$$

We can proceed to eliminate all the variables down to the nth. We have then converted the problem to an upper-triangular form, which can be solved by the procedure in Figure 5.10. A pseudocode procedure to solve (5.26) called the **tridiagonal or Thomas algorithm** is shown in Figure 5.11. The algorithm is written so that each primed value, when it is computed, replaces the previous value. Similarly the double-primed values replace the primed values. This is called **overwriting**, and reduces the storage required to implement the algorithm on a computer. It should be noted, however, that the algorithm is written for clarity and not minimum storage or maximum efficiency. The algorithm is very widely used; it is exceptionally fast and requires very little storage. Again writing your own procedure from Figure 5.11 can greatly enhance the understanding of the method.

Example 5.31

Use the tridiagonal procedure to solve

$$\begin{bmatrix} 2 & 1 & 0 & 0 \\ 1 & 2 & 1 & 0 \\ 0 & 1 & 2 & 1 \\ 0 & 0 & 1 & 2 \end{bmatrix} \begin{bmatrix} x \\ y \\ z \\ t \end{bmatrix} = \begin{bmatrix} 1 \\ 1 \\ 1 \\ -2 \end{bmatrix}$$

Solution   The sequence of matrices is given by

$$
\begin{bmatrix} 1 & \frac{1}{2} & 0 & 0 \\ 0 & \frac{3}{2} & 1 & 0 \\ 0 & 1 & 2 & 1 \\ 0 & 0 & 1 & 2 \end{bmatrix} \begin{bmatrix} x \\ y \\ z \\ t \end{bmatrix} = \begin{bmatrix} \frac{1}{2} \\ \frac{1}{2} \\ 1 \\ -2 \end{bmatrix}, \qquad
\begin{bmatrix} 1 & \frac{1}{2} & 0 & 0 \\ 0 & 1 & \frac{2}{3} & 0 \\ 0 & 0 & \frac{4}{3} & 1 \\ 0 & 0 & 1 & 2 \end{bmatrix} \begin{bmatrix} x \\ y \\ z \\ t \end{bmatrix} = \begin{bmatrix} \frac{1}{2} \\ \frac{1}{3} \\ \frac{2}{3} \\ -2 \end{bmatrix}
$$

$$
\begin{bmatrix} 1 & \frac{1}{2} & 0 & 0 \\ 0 & 1 & \frac{2}{3} & 0 \\ 0 & 0 & 1 & \frac{3}{4} \\ 0 & 0 & 0 & \frac{5}{4} \end{bmatrix} \begin{bmatrix} x \\ y \\ z \\ t \end{bmatrix} = \begin{bmatrix} \frac{1}{2} \\ \frac{1}{3} \\ \frac{1}{2} \\ -\frac{5}{2} \end{bmatrix}
$$

The elimination stage is now complete, and we substitute back to give

$$t = -2, \ z = \tfrac{1}{2} - \tfrac{3}{4}t = 2, \ y = \tfrac{1}{3} - \tfrac{2}{3}z = -1 \quad \text{and} \quad x = \tfrac{1}{2} - \tfrac{1}{2}y = 1$$

so that the complete solution is $x = 1$, $y = -1$, $z = 2$, $t = -2$.

Although the Thomas algorithm is efficient, the procedure in Figure 5.11 is not fool-proof, as illustrated by the simple example

$$
\begin{bmatrix} -1 & 1 & 0 \\ 1 & -1 & 1 \\ 0 & 1 & -1 \end{bmatrix} \begin{bmatrix} x \\ y \\ z \end{bmatrix} = \begin{bmatrix} -1 \\ 2 \\ 1 \end{bmatrix}
$$

After the first step we have

$$
\begin{bmatrix} 1 & -1 & 0 \\ 0 & 0 & 1 \\ 0 & 1 & -1 \end{bmatrix} \begin{bmatrix} x \\ y \\ z \end{bmatrix} = \begin{bmatrix} 1 \\ 1 \\ 1 \end{bmatrix}
$$

The next step divides by the diagonal element a_{22}. Since this element is zero, the method crashes to a halt. There is a perfectly good solution, however, since simply interchanging the last two rows,

$$
\begin{bmatrix} 1 & -1 & 0 \\ 0 & 1 & -1 \\ 0 & 0 & 1 \end{bmatrix} \begin{bmatrix} x \\ y \\ z \end{bmatrix} = \begin{bmatrix} 1 \\ 1 \\ 1 \end{bmatrix}
$$

gives an upper-triangular matrix with the obvious solution $z = 1$, $y = 2$, $x = 3$. It is clear that checks must be put into the algorithm to prevent such failures.

Gaussian elimination

Since most matrix equations are not tridiagonal, we should like to extend the idea to a general matrix

$$
\begin{bmatrix}
a_{11} & a_{12} & a_{13} & \cdots & a_{1n} \\
a_{21} & a_{22} & a_{23} & \cdots & a_{2n} \\
\vdots & \vdots & \vdots & & \vdots \\
a_{n1} & a_{n2} & a_{n3} & \cdots & a_{nn}
\end{bmatrix}
\begin{bmatrix}
x_1 \\ x_2 \\ \vdots \\ x_n
\end{bmatrix}
=
\begin{bmatrix}
b_1 \\ b_2 \\ \vdots \\ b_n
\end{bmatrix}
\tag{5.27}
$$

The result of doing this is a method known as **Gaussian elimination**. It is a little more involved than the Thomas algorithm. First we eliminate x_1:

$$
\begin{bmatrix}
1 & a'_{12} & a'_{13} & \cdots & a'_{1n} \\
0 & a'_{22} & a'_{23} & \cdots & a'_{2n} \\
0 & a'_{32} & a'_{33} & \cdots & a'_{3n} \\
\vdots & \vdots & \vdots & & \vdots \\
0 & a'_{n2} & a'_{n3} & \cdots & a'_{nn}
\end{bmatrix}
\begin{bmatrix}
x_1 \\ x_2 \\ \vdots \\ x_n
\end{bmatrix}
=
\begin{bmatrix}
b'_1 \\ b'_2 \\ \vdots \\ b'_n
\end{bmatrix}
$$

where

$$
a'_{12} = \frac{a_{12}}{a_{11}}, \quad a'_{13} = \frac{a_{13}}{a_{11}}, \quad \ldots, \quad a'_{1n} = \frac{a_{1n}}{a_{11}}, \quad b'_1 = \frac{b_1}{a_{11}}
$$

$$
a'_{22} = a_{22} - a_{21}a'_{12}, \quad a'_{23} = a_{23} - a_{21}a'_{13}, \quad \ldots, \quad b'_2 = b_2 - a_{21}b'_1
$$

$$
a'_{32} = a_{32} - a_{31}a'_{12}, \quad a'_{33} = a_{33} - a_{31}a'_{13}, \quad \ldots, \quad b'_3 = b_3 - a_{31}b'_1
$$

and so on.

Generally these can be written as

$$
a'_{1j} = \frac{a_{1j}}{a_{11}}, \quad j = 1, \ldots, n \quad b'_1 = \frac{b_1}{a_{11}}
$$

$$
\left.\begin{aligned}
a'_{ij} &= a_{ij} - a_{i1}a'_{ij} \\
b'_i &= b_i - a_{i1}b'_1
\end{aligned}\right\}, \quad i = 2, \ldots, n \quad \text{and} \quad j = 2, \ldots, n
$$

We now operate in an identical manner on the $(n-1) \times (n-1)$ submatrix, formed by ignoring row 1 and column 1, and repeat the process until the equations are of upper-triangular form. At the general step in the algorithm the equations will take the form

$$
\begin{bmatrix}
1 & * & * & * & & & & \cdots & * \\
0 & 1 & * & * & & & & \cdots & * \\
0 & 0 & 1 & * & & & & \cdots & * \\
0 & 0 & & \ddots & & & & & \\
\vdots & \vdots & & \ddots & & & & & \vdots \\
0 & \cdots & & 0 & 1 & * & \cdots & & * \\
0 & \cdots & & & 0 & a_{ii} & \cdots & & a_{in} \\
\vdots & & & & 0 & \vdots & \ddots & & \vdots \\
& & & & & \vdots & & & \\
0 & \cdots & & & 0 & a_{ni} & \cdots & & a_{nn}
\end{bmatrix}
\begin{bmatrix}
x_1 \\ x_2 \\ \vdots \\ \vdots \\ x_n
\end{bmatrix}
=
\begin{bmatrix}
* \\ * \\ \vdots \\ *
\end{bmatrix}
\tag{5.28}
$$

Figure 5.12
Elimination procedure
for (5.27).

```
procedure eliminate (A, b, n→A, b)
    {A is an n × n matrix and b is a vector with n elements}
    for i is 1 to n − 1 do
        {a segment will be inserted here later}
        b(i)←b(i)/A(i, i)
        for j is i to n do
            A(i, j)←A(i, j)/A(i, i)
            for k is i + 1 to n do
                A(k, j)←A(k, j) − A(k, i)*A(i, j)
            endfor
        endfor
        for k is i + 1 to n do
            b(k)←b(k) − A(k, i)*b(i)
        endfor
    endfor
endprocedure
```

Again overwriting avoids the need for introducing primed symbols; the algorithm is shown in Figure 5.12.

The general Gaussian elimination procedure would then put the eliminate and upper-triangular procedures together in a program

```
read (file, A, b, n)
    eliminate (A, b, n→A, b)
    uppertriangular (A, b, n→x)
write (vdu, x)
```

This algorithm, sharing the merits of the Thomas algorithm, is very widely used by engineers to solve linear equations.

In Example 5.30 the basic elimination technique was illustrated but now the two procedures *eliminate* and *uppertriangular* have reduced the method to one of routine. The major problem is to perform the arithmetic accurately.

Example 5.32   Using elimination and back substitution solve the equations

$$\begin{bmatrix} 2 & 3 & 4 \\ 1 & 2 & 3 \\ 1 & 4 & 5 \end{bmatrix} \begin{bmatrix} x \\ y \\ z \end{bmatrix} = \begin{bmatrix} 1 \\ 1 \\ 2 \end{bmatrix}$$

Solution   From the method in Figure 5.12 the steps are

Divide first row by 2: $\begin{bmatrix} 1 & \frac{3}{2} & 2 \\ 1 & 2 & 3 \\ 1 & 4 & 5 \end{bmatrix} \begin{bmatrix} x \\ y \\ z \end{bmatrix} = \begin{bmatrix} \frac{1}{2} \\ 1 \\ 2 \end{bmatrix}$

Subtract row 1 from row 2 and row 3:
$$\begin{bmatrix} 1 & \frac{3}{2} & 2 \\ 0 & \frac{1}{2} & 1 \\ 0 & \frac{5}{2} & 3 \end{bmatrix} \begin{bmatrix} x \\ y \\ z \end{bmatrix} = \begin{bmatrix} \frac{1}{2} \\ \frac{1}{2} \\ \frac{3}{2} \end{bmatrix}$$

Divide second row by $\frac{1}{2}$:
$$\begin{bmatrix} 1 & \frac{3}{2} & 2 \\ 0 & 1 & 2 \\ 0 & \frac{5}{2} & 3 \end{bmatrix} \begin{bmatrix} x \\ y \\ z \end{bmatrix} = \begin{bmatrix} \frac{1}{2} \\ 1 \\ \frac{3}{2} \end{bmatrix}$$

Subtract $\frac{5}{2} \times$ (row 2) from row 3:
$$\begin{bmatrix} 1 & \frac{3}{2} & 2 \\ 0 & 1 & 2 \\ 0 & 0 & -2 \end{bmatrix} \begin{bmatrix} x \\ y \\ z \end{bmatrix} = \begin{bmatrix} \frac{1}{2} \\ 1 \\ -1 \end{bmatrix}$$

Divide row 3 by (−2):
$$\begin{bmatrix} 1 & \frac{3}{2} & 2 \\ 0 & 1 & 2 \\ 0 & 0 & 1 \end{bmatrix} \begin{bmatrix} x \\ y \\ z \end{bmatrix} = \begin{bmatrix} \frac{1}{2} \\ 1 \\ \frac{1}{2} \end{bmatrix}$$

The elimination procedure is now complete and the back substitution (from Figure 5.10) is applied to the upper-triangular matrix.

From the third row $z = \frac{1}{2}$

From the second row $y = 1 - 2z = 0$

From the first row $x = \frac{1}{2} - \frac{3}{2}y - 2z = -\frac{1}{2}$

so the solution is $x = -\frac{1}{2}, y = 0, z = \frac{1}{2}$.

Example 5.33   Solve

$$\begin{bmatrix} 1 & 2 & 3 & 1 \\ 2 & 1 & 1 & 1 \\ 1 & 2 & 1 & 0 \\ 0 & 1 & 1 & 2 \end{bmatrix} \begin{bmatrix} x \\ y \\ z \\ t \end{bmatrix} = \begin{bmatrix} 5 \\ 3 \\ 4 \\ 0 \end{bmatrix}$$

Solution   The elimination sequence is

$$\begin{bmatrix} 1 & 2 & 3 & 1 \\ 0 & -3 & -5 & -1 \\ 0 & 0 & -2 & -1 \\ 0 & 1 & 1 & 2 \end{bmatrix} \begin{bmatrix} x \\ y \\ z \\ t \end{bmatrix} = \begin{bmatrix} 5 \\ -7 \\ -1 \\ 0 \end{bmatrix}, \quad \begin{bmatrix} 1 & 2 & 3 & 1 \\ 0 & 1 & \frac{5}{3} & \frac{1}{3} \\ 0 & 0 & -2 & -1 \\ 0 & 0 & -\frac{2}{3} & \frac{5}{3} \end{bmatrix} \begin{bmatrix} x \\ y \\ z \\ t \end{bmatrix} = \begin{bmatrix} 5 \\ \frac{7}{3} \\ -1 \\ -\frac{7}{3} \end{bmatrix}$$

$$\begin{bmatrix} 1 & 2 & 3 & 1 \\ 0 & 1 & \frac{5}{3} & \frac{1}{3} \\ 0 & 0 & 1 & \frac{1}{2} \\ 0 & 0 & 0 & 2 \end{bmatrix} \begin{bmatrix} x \\ y \\ z \\ t \end{bmatrix} = \begin{bmatrix} 5 \\ \frac{7}{3} \\ \frac{1}{2} \\ -2 \end{bmatrix}$$

and application of the upper-triangular procedure gives $t = -1, z = 1, y = 1$ and $x = 1$.

It is clear again that if, in the algorithm shown in Figure 5.12, A(i, i) is zero at any time, the method will fail. It is, in fact, also found to be beneficial to the stability of the method to have A(i, i) as large as possible. Thus in (5.28) it is usual to perform a 'partial pivoting' so that the largest value in the column, $\max_{i \leqslant p \leqslant n} |A(p, i)|$, is chosen and the equations are swapped around to make this element the pivot. In Figure 5.12 the following segment of program would need to be inserted at the point indicated:

> {find $\max_{i \leqslant p \leqslant n} |A(p, i)|$}
>
> {interchange row i with row p_{max}}

In practical computer implementations of the algorithm the elements of the rows would not be swapped explicitly. Instead, a pointer system would be used to implement a technique known as **indirect addressing**, which allows much faster computations. The interested reader is referred to texts on computer programming techniques for a full explanation of this method.

In a hand-computation version of this elimination procedure there are methods that maintain running checks and minimize the amount of writing. In this book the emphasis is on a computer implementation, and the hand computations are provided to illustrate the principle of the method. It is a powerful learning technique to write your own programs, but the practising professional engineer will normally use procedures from a computer software library, where these are available.

In MATLAB the instruction [L, U] = lu(A) provides in *U* the eliminated matrix. However the method used in MATLAB always uses partial pivoting and the method only works for a square matrix. The instruction A\b will give the solution to the matrix equation in one step. The MAPLE package can deal with any size matrix, so the right-hand side of the matrix equation should be appended to *A* and hence included in the elimination, and the instruction gausselim(A); provides the elimination but without partial pivoting. The instruction gaussjord(A); uses a much more subtle elimination process – see any advanced textbook on numerical linear algebra – and gives the solution in the most convenient form. The instruction backsub(B) is also available for the back substitution.

Example 5.34   Solve the matrix equation

$$\begin{bmatrix} 1 & 2 & 3 & 1 \\ 2 & 1 & 1 & 1 \\ 1 & 3 & 1 & 0 \\ 0 & 1 & 1 & 2 \end{bmatrix} \begin{bmatrix} x \\ y \\ z \\ t \end{bmatrix} = \begin{bmatrix} 4 \\ 3 \\ 2 \\ 1 \end{bmatrix}$$

by Gaussian elimination with partial pivoting.

Solution The sequence is as follows. We first interchange rows 1 and 2 and eliminate:

$$\begin{bmatrix} 2 & 1 & 1 & 1 \\ 1 & 2 & 3 & 1 \\ 1 & 3 & 1 & 0 \\ 0 & 1 & 1 & 2 \end{bmatrix}\begin{bmatrix} x \\ y \\ z \\ t \end{bmatrix} = \begin{bmatrix} 3 \\ 4 \\ 2 \\ 1 \end{bmatrix} \rightarrow \begin{bmatrix} 1 & \frac{1}{2} & \frac{1}{2} & \frac{1}{2} \\ 0 & \frac{3}{2} & \frac{5}{2} & \frac{1}{2} \\ 0 & \frac{5}{2} & \frac{1}{2} & -\frac{1}{2} \\ 0 & 1 & 1 & 2 \end{bmatrix}\begin{bmatrix} x \\ y \\ z \\ t \end{bmatrix} = \begin{bmatrix} \frac{3}{2} \\ \frac{5}{2} \\ \frac{1}{2} \\ 1 \end{bmatrix}$$

We then interchange rows 2 and 3 and eliminate:

$$\begin{bmatrix} 1 & \frac{1}{2} & \frac{1}{2} & \frac{1}{2} \\ 0 & \frac{5}{2} & \frac{1}{2} & -\frac{1}{2} \\ 0 & \frac{3}{2} & \frac{5}{2} & \frac{1}{2} \\ 0 & 1 & 1 & 2 \end{bmatrix}\begin{bmatrix} x \\ y \\ z \\ t \end{bmatrix} = \begin{bmatrix} \frac{3}{2} \\ \frac{1}{2} \\ \frac{5}{2} \\ 1 \end{bmatrix} \rightarrow \begin{bmatrix} 1 & \frac{1}{2} & \frac{1}{2} & \frac{1}{2} \\ 0 & 1 & \frac{1}{5} & -\frac{1}{5} \\ 0 & 0 & \frac{11}{5} & \frac{4}{5} \\ 0 & 0 & \frac{4}{5} & \frac{11}{5} \end{bmatrix}\begin{bmatrix} x \\ y \\ z \\ t \end{bmatrix} = \begin{bmatrix} \frac{3}{2} \\ \frac{1}{5} \\ \frac{11}{5} \\ \frac{4}{5} \end{bmatrix}$$

There is no need to interchange rows at this stage, and the elimination proceeds immediately:

$$\begin{bmatrix} 1 & \frac{1}{2} & \frac{1}{2} & \frac{1}{2} \\ 0 & 1 & \frac{1}{5} & -\frac{1}{5} \\ 0 & 0 & 1 & \frac{4}{11} \\ 0 & 0 & 0 & \frac{21}{11} \end{bmatrix}\begin{bmatrix} x \\ y \\ z \\ t \end{bmatrix} = \begin{bmatrix} \frac{3}{2} \\ \frac{1}{5} \\ 1 \\ 0 \end{bmatrix}$$

Back substitution now gives $t = 0$, $z = 1$, $y = 0$ and $x = 1$.

Ill-conditioning

Elimination methods are not without their difficulties, and the following example will highlight some of them.

Example 5.35 Solve, by elimination, the equations

(a) $\begin{bmatrix} 2 & 1 \\ 1 & 0.5001 \end{bmatrix}\begin{bmatrix} x \\ y \end{bmatrix} = \begin{bmatrix} 0.3 \\ 0.6 \end{bmatrix}$ (b) $\begin{bmatrix} 2 & 1 \\ 1 & 0.4999 \end{bmatrix}\begin{bmatrix} x \\ y \end{bmatrix} = \begin{bmatrix} 0.3 \\ 0.6 \end{bmatrix}$

Solution Keeping the calculations parallel,

(a) $\begin{bmatrix} 1 & 0.5 \\ 1 & 0.5001 \end{bmatrix}\begin{bmatrix} x \\ y \end{bmatrix} = \begin{bmatrix} 0.15 \\ 0.6 \end{bmatrix}$ (b) $\begin{bmatrix} 1 & 0.5 \\ 1 & 0.4999 \end{bmatrix}\begin{bmatrix} x \\ y \end{bmatrix} = \begin{bmatrix} 0.15 \\ 0.6 \end{bmatrix}$

$\begin{bmatrix} 1 & 0.5 \\ 0 & 0.0001 \end{bmatrix}\begin{bmatrix} x \\ y \end{bmatrix} = \begin{bmatrix} 0.15 \\ 0.45 \end{bmatrix}$ $\begin{bmatrix} 1 & 0.5 \\ 0 & -0.0001 \end{bmatrix}\begin{bmatrix} x \\ y \end{bmatrix} = \begin{bmatrix} 0.15 \\ 0.45 \end{bmatrix}$

with solution with solution

$y = 4500, \qquad x = -2249.85$ $y = -4500, \qquad x = 2250.15$

In Example 5.35 simple equations that have only marginally different coefficients have wildly different solutions. This situation is likely to cause problems so it must be analysed carefully. To do so in full detail is not appropriate here, but the problem is clearly connected with taking differences of numbers that are almost equal: $0.5001 - 0.5 = 0.0001$.

Systems of equations that exhibit such awkward behaviour are called **ill-conditioned**. It is not straightforward to identify ill-conditioning in matrices involving many variables, but an example will illustrate the difficulties in the two-variable case. Suppose we solve

$$2x + \quad y = 0.3$$

$$x - \alpha y = 0$$

where $\alpha = 1 \pm 0.05$ has some error in its value. We easily obtain $x = 0.3\alpha/(1 + 2\alpha)$ and $y = 0.3/(1 + 2\alpha)$, and putting in the range of α values we get $0.0983 \leqslant x \leqslant 0.1016$ and $0.0968 \leqslant y \leqslant 0.1034$. Thus an error of $\pm 5\%$ in the value of α produces an error of $\pm 2\%$ in x and an error of $\pm 3\%$ in y.

If we now try to solve

$$2x + \quad y = 0.3$$

$$x + \alpha y = 0.3$$

where $\alpha = 0.4 \pm 0.05$, then we get the solution $x = 0.3(1 - \alpha)/(1 - 2\alpha)$, $y = -0.3/(1 - 2\alpha)$. Putting in the range of α values now gives $0.65 \leqslant x \leqslant 1.65$ and $-3 \leqslant y \leqslant -1$, and an error of $\pm 12\%$ in the value of α produces errors in x and y of up to 100%.

Figure 5.13 illustrates these equations geometrically. We see that a small change in the slope of the line $x - \alpha y = 0$ makes only a small difference in the solution. However, changing the slope of the line $x + \alpha y = 0.3$ makes a large difference, because the lines are nearly parallel. Identifying such behaviour for higher-dimensional problems is not at all easy. Sets of equations of this kind do occur in engineering contexts, so the difficulties outlined here should be appreciated. In each of the ill-conditioned cases we have studied, the determinant of the system is 'small':

$$\begin{vmatrix} 2 & 1 \\ 1 & 0.5001 \end{vmatrix} = 0.0002,$$

$$\begin{vmatrix} 2 & 1 \\ 1 & 0.4999 \end{vmatrix} = -0.0002$$

$$\begin{vmatrix} 2 & 1 \\ 1 & 0.4 \pm 0.05 \end{vmatrix} = -0.2 \pm 0.1$$

Thus the equations are 'nearly singular' – and this is one means of identifying the problem. However, the reader should refer to a more advanced book on numerical analysis to see how to identify and deal with ill-conditioning in the general case.

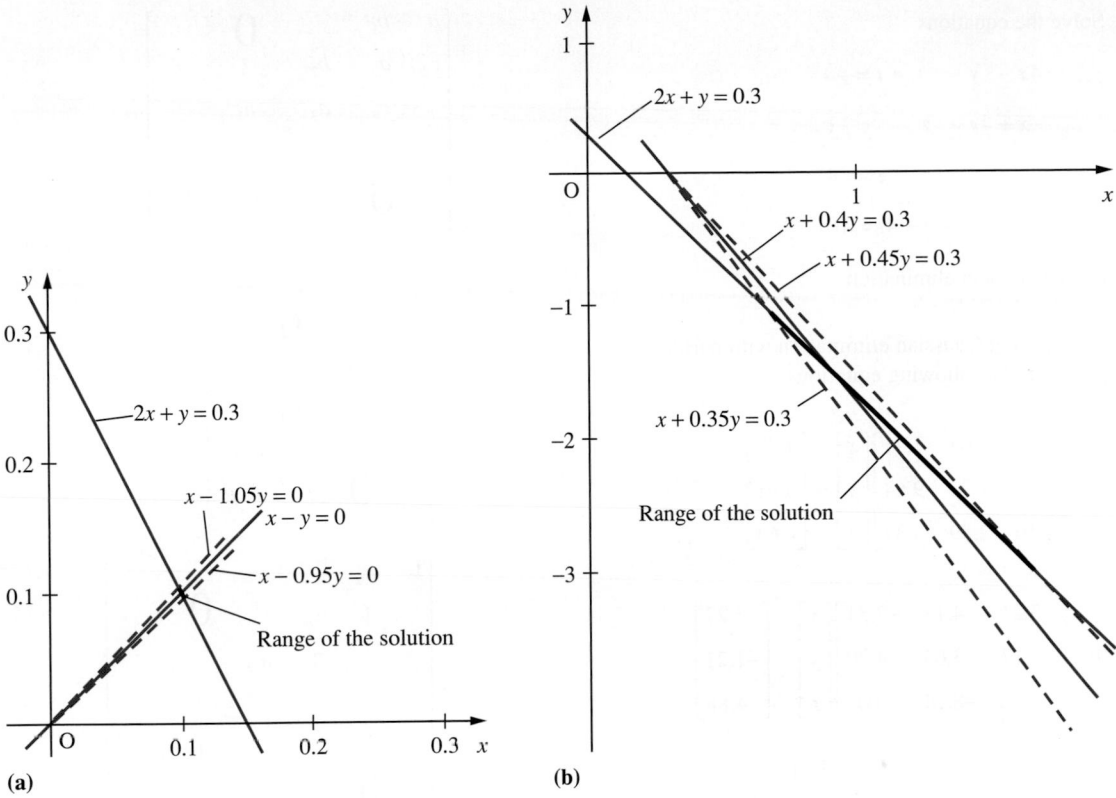

Figure 5.13 Solution of (a) $2x + y = 0.3$, $x - \alpha y = 0$ with $\alpha = 1 \pm 0.05$; and (b) $2x + y = 0.3$, $x + \alpha y = 0.3$ with $\alpha = 0.4 \pm 0.05$. The heavy black lines indicate the ranges of the solutions.

5.5.3 Exercises

Most of these exercises will require MATLAB or MAPLE for their solution. To appreciate the elimination method, hand computation should be tried on the first few exercises.

72 Use elimination with and/or without partial pivoting, to solve the equations

(a)
$$\begin{bmatrix} 1 & 3 & 2 \\ 2 & 1 & 4 \\ 3 & -1 & 5 \end{bmatrix} \begin{bmatrix} x \\ y \\ z \end{bmatrix} = \begin{bmatrix} 1 \\ 2 \\ 1 \end{bmatrix}$$

(b)
$$\begin{bmatrix} 0 & 1 & 1 \\ 3 & -1 & 1 \\ 1 & 1 & -3 \end{bmatrix} \begin{bmatrix} x \\ y \\ z \end{bmatrix} = \begin{bmatrix} 6 \\ -7 \\ -13 \end{bmatrix}$$

(c)
$$\begin{bmatrix} 1 & 2 & 4 \\ -6 & 2 & 10 \\ 2 & 8 & 7 \end{bmatrix} \begin{bmatrix} x \\ y \\ z \end{bmatrix} = \begin{bmatrix} 0 \\ 1 \\ 0 \end{bmatrix}$$

73 Solve the equations
$$\begin{aligned}
4x - y & & & = 2 \\
-x + 4y - z & & & = 5 \\
-y + 4z - t & & & = 3 \\
-z + 4t & & & = 10
\end{aligned}$$

using the tridiagonal algorithm.

74 Solve the equations

$$4x - y \qquad - t = -4$$

$$-x + 4y - z \qquad = 1$$

$$-y + 4z - t = 4$$

$$-x \qquad - z + 4t = 10$$

using Gaussian elimination.

75 Solve, using Gaussian elimination with partial pivoting, the following equations:

(a) $\begin{bmatrix} 1.17 & 2.64 & 7.41 \\ 3.37 & 1.22 & 9.64 \\ 4.10 & 2.89 & 3.37 \end{bmatrix} \begin{bmatrix} x \\ y \\ z \end{bmatrix} = \begin{bmatrix} 1.27 \\ 3.91 \\ 4.63 \end{bmatrix}$

(b) $\begin{bmatrix} 3.21 & 4.18 & -2.31 \\ -4.17 & 3.63 & 4.20 \\ 1.88 & -8.14 & 0.01 \end{bmatrix} \begin{bmatrix} x \\ y \\ z \end{bmatrix} = \begin{bmatrix} 3.27 \\ -1.21 \\ 4.88 \end{bmatrix}$

(c) $\begin{bmatrix} 1 & 7 & 2 & -1 \\ 11 & 4 & -3 & 9 \\ 7 & 6 & 4 & -2 \\ 5 & 8 & -5 & 3 \end{bmatrix} \begin{bmatrix} x \\ y \\ z \\ t \end{bmatrix} = \begin{bmatrix} 12 \\ -12 \\ 7 \\ -7 \end{bmatrix}$

76 The two almost identical matrix equations are given

$$\begin{bmatrix} 0.11 & 0.19 & 0.10 \\ 0.49 & -0.31 & 0.21 \\ 1.55 & -0.70 & 0.70 \end{bmatrix} \begin{bmatrix} x \\ y \\ z \end{bmatrix} = \begin{bmatrix} 1 \\ 1 \\ 1 \end{bmatrix} \quad \text{and}$$

$$\begin{bmatrix} 0.11 & 0.19 & 0.10 \\ 0.49 & -0.31 & 0.21 \\ 1.55 & -0.70 & 0.71 \end{bmatrix} \begin{bmatrix} x \\ y \\ z \end{bmatrix} = \begin{bmatrix} 1 \\ 1 \\ 1 \end{bmatrix}$$

Use MATLAB or MAPLE to show that the solutions are wildly different. Evaluate the determinants of the two 3×3 matrices.

77 Show that a tridiagonal matrix can be written in the form

$$\begin{bmatrix} a_1 & b_1 & & & & \mathbf{0} \\ c_2 & a_2 & b_2 & & & \\ & c_3 & a_3 & b_3 & & \\ & & \ddots & \ddots & \ddots & \\ \mathbf{0} & & & c_{n-1} & a_{n-1} & b_{n-1} \\ & & & & c_n & a_n \end{bmatrix}$$

$$= \begin{bmatrix} l_{11} & & & & \mathbf{0} \\ l_{21} & l_{22} & & & \\ & l_{32} & l_{33} & & \\ & & \ddots & \ddots & \\ \mathbf{0} & & & l_{n,n-1} & l_{nn} \end{bmatrix}$$

$$\times \begin{bmatrix} 1 & u_{12} & & & & \\ & 1 & u_{23} & & \mathbf{0} & \\ & & 1 & u_{34} & & \\ & & & \ddots & \ddots & \\ & & & & 1 & u_{n-1,n} \\ \mathbf{0} & & & & & 1 \end{bmatrix}$$

A matrix that has zeros in every position below the diagonal is called an **upper-triangular matrix** and one with zeros everywhere above the diagonal is called a **lower-triangular matrix**. A matrix that only has non-zero elements in certain diagonal lines is called a **banded matrix**. In this case we have shown that a tridiagonal matrix can be written as the product of a lower-triangular banded matrix and an upper-triangular banded matrix.

78 The cantilever beam in Question 71 (Exercises 5.5.1) and illustrated in Figure 5.9 was solved using very inefficient methods. Solve the same equations using the Thomas algorithm.

If the displacements are not small then the equations are more complicated. They take the form

$$\left. \begin{array}{l} -v_1 y_1 + \quad y_2 \qquad\qquad\qquad = -uw_1 \\ y_1 - v_2 y_2 + \quad y_3 \qquad\qquad = -4uw_2 \\ \qquad y_2 - v_3 y_3 + \quad y_4 = -9uw_3 \\ \qquad\qquad 2y_3 - v_4 y_4 = -16u \end{array} \right\} \quad \textbf{(5.29)}$$

where

$$w_1 = \left[1 + 4\left(\frac{y_2}{L}\right)^2\right]^{3/2}$$

$$w_2 = \left[1 + 4\left(\frac{y_3 - y_1}{L}\right)^2\right]^{3/2}$$

and

$$w_3 = \left[1 + 4\left(\frac{y_4 - y_2}{L}\right)^2\right]^{3/2}$$

and putting $k = PL^2/16EI$

$$v_1 = 2 + kw_1, \quad v_2 = 2 + kw_2$$

$$v_3 = 2 + kw_3, \quad v_4 = 2 + k$$

Solve the equations iteratively, calculating w_i and v_i from the previous iteration and then solving the tridiagonal scheme. Use the same values as before, namely $k = 1$ and $u = 1$, and take $L = 2$. A full solution of this exercise will require the use of a computer package such as MATLAB.

79 A wire is loaded with equal weights W at nine uniformly spaced points as illustrated in Figure 5.14. The wire is sufficiently taut that the tension T may be considered to be constant. The end points are at the same level, so that $u_0 = u_{10} = 0$ and the system is symmetrical about

its midpoint. The equations to determine the displacements u_i are

$$W = (T/d)(2u_1 - u_2 \qquad\qquad)$$
$$W = (T/d)(-u_1 + 2u_2 - u_3 \qquad)$$
$$W = (T/d)(\qquad -u_2 + 2u_3 - u_4 \qquad)$$
$$W = (T/d)(\qquad\qquad -u_3 + 2u_4 - u_5)$$
$$W = (T/d)(\qquad\qquad\qquad - 2u_4 + 2u_5)$$

Taking $Wd/T = l$, calculate u_i/l for $i - 1, \dots, 5$.

80 A ladder network is shown in Figure 5.15. The driver is an a.c. voltage of $E = E_0 e^{j\omega t}$ and the currents are taken to be $I_p = Z_p e^{j\omega t}$. The equations satisfied by the Z_p are

$$jE_0\omega = -\tfrac{1}{2}LZ_0\omega^2 + \frac{1}{C}(Z_0 - Z_1)$$

$$0 = -LZ_p\omega^2 + \frac{1}{C}(-Z_{p-1} + 2Z_p - Z_{p+1})$$

$$\text{for } p = 1, \dots, n - 1$$

$$0 = -\tfrac{1}{2}LZ_n\omega^2 + \frac{1}{C}(Z_n - Z_{n-1})$$

Take $n = 3$ and solve for Z_3. Evaluate the effective resistance in the final circuit as $|E_0/Z_3|$. Plot this resistance against ω and interpret the graph obtained. Note that the matrices contain complex numbers. (See Review exercises 7.12, Question 21.) Although it is possible to solve this exercise by hand, the use of MATLAB's Symbolic Math Toolbox or MAPLE, is recommended.

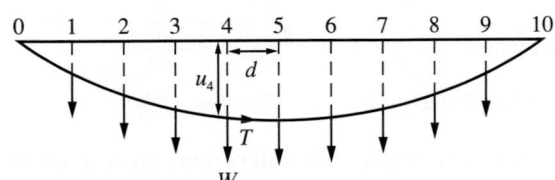

Figure 5.14 Loaded wire.

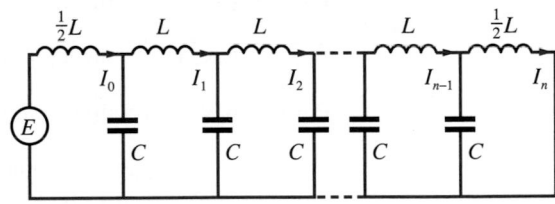

Figure 5.15 Ladder network.

5.5.4 The solution of linear equations: iterative methods

An alternative and very popular way of solving linear equations is by iteration. This has the attraction of being easy to program. In practice, the availability of efficient procedures in computer libraries means that elimination methods are usually preferred for small problems. However, when the number of variables gets large, say several hundred, elimination methods struggle because the matrices can contain 10^6 or more elements. Problems of such size commonly occur in those scientific and engineering computations that require numerical solution on a mesh. Typically, in a turbine flow, we have a three-dimensional fluid flow problem that would need to be solved for three velocities and pressure on a $30 \times 30 \times 30$ mesh. The problem would require the solution of a $27\,000 \times 27\,000$ matrix equation. The saving feature of such problems is that it is very common for almost all the entries in the matrix to be zero. Matrices in which the large majority of elements are zero are called **sparse matrices**. Unless there is special structure to the equations, elimination will quickly destroy the sparseness. On the other hand, iterative methods only have to deal with the non-zero terms, so there is considerable computational saving. As usual, there is a price to pay:

(a) it is not always easy to decide when the method has converged;
(b) if the method takes a very large number of iterations to converge, any savings are quickly consumed.

A simple example will illustrate the way the method proceeds; in this example exact fractions will be used.

To solve the equations

$$4x + y = 2$$

$$x + 4y = -7$$

we first rearrange them as

$$x = \tfrac{1}{4}(2 - y)$$

$$y = \tfrac{1}{4}(-7 - x)$$

and start with $x = 0$, $y = 0$.

Putting these values into the right-hand side gives $\qquad x = \tfrac{1}{2}, \quad y = -\tfrac{7}{4}$

Putting these new values into the right-hand side gives $\quad x = \tfrac{15}{16}, \quad y = -\tfrac{15}{8}$

Putting these new values into the right-hand side gives $\quad x = \tfrac{31}{32}, \quad y = -\tfrac{127}{64}$

Putting these new values into the right-hand side gives $\quad x = \tfrac{255}{256}, y = -\tfrac{255}{128}$

Putting these new values into the right-hand side gives $\quad x = \tfrac{511}{512}, y = -\tfrac{2047}{1024}$

Performing the same procedure repeatedly, normally called **iteration**, gives a set of numbers that appear to be tending to the solution $x = 1$, $y = -2$.

This particular example shows the strength of the method but we are not always so fortunate, as illustrated in the next example.

Consider the tridiagonal equations in Example 5.31:

$$
\begin{aligned}
2x + \ y \qquad\qquad\;\; &= 1 \\
x + 2y + \ z \qquad\; &= 1 \\
y + 2z + \ t &= 1 \\
z + 2t &= -2
\end{aligned}
$$

Iteration	0	1	2	3	4	5	6	7	8	9	10
x	0	0.5	0.25	0.5	0.5	0.6562	0.6719	0.7734	0.7852	0.8516	0.8594
y	0	0.5	0	0	−0.3125	−0.3437	−0.5469	−0.5703	−0.7031	−0.7187	−0.8057
z	0	0.5	0.75	1.1250	1.1875	1.4375	1.4687	1.6328	1.6523	1.7598	1.7725
t	0	−1	−1.25	−1.3750	−1.5625	−1.5937	−1.7187	−1.7344	−1.8164	−1.8262	−1.8799

Figure 5.16 Iterative solution of (5.30) using Jacobi iteration.

We can rearrange these as

$$x = \tfrac{1}{2}(1 - y)$$
$$y = \tfrac{1}{2}(1 - x - z)$$
$$z = \tfrac{1}{2}(1 - y - t)$$
$$t = \tfrac{1}{2}(-2 - z)$$

or

$$\begin{bmatrix} x \\ y \\ z \\ t \end{bmatrix} = \frac{1}{2}\begin{bmatrix} 0 & -1 & 0 & 0 \\ -1 & 0 & -1 & 0 \\ 0 & -1 & 0 & -1 \\ 0 & 0 & -1 & 0 \end{bmatrix}\begin{bmatrix} x \\ y \\ z \\ t \end{bmatrix} + \frac{1}{2}\begin{bmatrix} 1 \\ 1 \\ 1 \\ -2 \end{bmatrix} \qquad (5.30)$$

Suppose we start with $x = y = z = t = 0$. We substitute these into the right-hand side and evaluate the new x, y, z and t; we then substitute the new values back in and repeat the process. Such iteration gives the results shown in Figure 5.16. This shows values going depressingly slowly to the solution 1, −1, 2, −2: even after 20 iterations the values are 0.9381, −0.9767, 1.9727, −1.9856. The method just described is called the **Jacobi method**, and can be written, using superscripts as iteration counters, in the form

$$x^{(r+1)} = \tfrac{1}{2}(1 - y^{(r)})$$
$$y^{(r+1)} = \tfrac{1}{2}(1 - x^{(r)} - z^{(r)})$$
$$z^{(r+1)} = \tfrac{1}{2}(1 - y^{(r)} - t^{(r)})$$
$$t^{(r+1)} = \tfrac{1}{2}(-2 - z^{(r)})$$

An obvious step is to use the new values as soon as they are available. In the two-variable example the same equations

$$x = \tfrac{1}{4}(2 - y)$$
$$y = \tfrac{1}{4}(-7 - x)$$

are used and the same starting point $x = 0$, $y = 0$ is used. The iteration proceeds slightly differently:

Put the values 0, 0 in the first equation $\Rightarrow x = \tfrac{1}{2}$

Put the values $\tfrac{1}{2}$, 0 in the second equation $\Rightarrow y = -\tfrac{15}{8}$

Put the values $\tfrac{1}{2}$, $-\tfrac{15}{8}$ in the first equation $\Rightarrow x = \tfrac{31}{32}$

Put the values $\tfrac{31}{32}$, $-\tfrac{15}{8}$ in the second equation $\Rightarrow y = -\tfrac{255}{128}$

Put the values $\tfrac{31}{32}$, $-\tfrac{255}{128}$ in the first equation $\Rightarrow x = \tfrac{511}{512}$

and continue in the same way. It can be seen that already the convergence is very much faster.

Iteration	0	1	2	3	4	5	6	7	8	9	10
x	0	0.5	0.375	0.4375	0.6172	0.7480	0.8350	0.8920	0.9293	0.9537	0.9697
y	0	0.25	0.125	−0.2344	−0.4961	−0.6699	−0.7839	−0.8586	−0.9074	−0.9394	−0.9603
z	0	0.375	1.0312	1.3750	1.5918	1.7329	1.8252	1.8856	1.9251	1.9510	1.9679
t	0	−1.1875	−1.5156	−1.6875	−1.7959	−1.8665	−1.9126	−1.9426	−1.9626	−1.9755	−1.9840

Figure 5.17 Iterative solution of (5.30) using Gauss–Seidel iteration.

In the second example we use the new values of x, y, z and t as soon as they are calculated: the method is called **Gauss–Seidel iteration**. This can be written as

$$x^{(r+1)} = \tfrac{1}{2}(1 - y^{(r)})$$
$$y^{(r+1)} = \tfrac{1}{2}(1 - x^{(r+1)} - z^{(r)})$$
$$z^{(r+1)} = \tfrac{1}{2}(1 - y^{(r+1)} - t^{(r)})$$
$$t^{(r+1)} = \tfrac{1}{2}(-2 - z^{(r+1)})$$

The calculation now yields the results shown in Figure 5.17. We see that, after the ten iterations quoted, the solution obtained by Gauss–Seidel iteration is within 4% of the actual solution whereas that obtained by Jacobi iteration still has an error of about 20%. The Gauss–Seidel method is both faster and more convenient for computer implementation. Within 20 iterations the Gauss–Seidel solution is accurate to three decimal places. A pseudocode algorithm for this method applied to the present problem is shown in Figure 5.18 and is easily implemented in MATLAB.

Although the two iteration methods have been described in terms of a particular example, the method is quite general. To solve

$$AX = b$$

we rewrite

$$A = D + L + U$$

where D is diagonal, L only has non-zero elements below the diagonal and U only has non-zero elements above the diagonal, so that

Figure 5.18
Algorithm to implement the Gauss–Seidel iteration.

```
read(vdu,eps,kmax)
k←0
x←0;y←0;z←0;t←0
repeat
   k←k + 1
   xold←x;   x←(1 − y)/2
   yold←y;      y←(1 − x − z)/2
   zold←z;      z←(1 − y − t)/2
   told←t;    t←(−2 − z)/2
until ((abs(x − xold) < eps)and(abs(y − yold) < eps)and(abs(z − zold) < eps)
      and(abs(t − told) < eps))or (k > kmax)
```

$$A = \begin{bmatrix} a_{11} & & & & \\ & a_{22} & & \mathbf{0} & \\ & & a_{33} & & \\ & \mathbf{0} & & \ddots & \\ & & & & a_{nn} \end{bmatrix} + \begin{bmatrix} 0 & & & & \\ a_{21} & 0 & & \mathbf{0} & \\ a_{31} & a_{32} & 0 & & \\ \vdots & \vdots & & \ddots & \\ a_{n1} & a_{n2} & \cdots & a_{n,n-1} & 0 \end{bmatrix}$$

$$+ \begin{bmatrix} 0 & a_{12} & a_{13} & \cdots & a_{1n} \\ & 0 & a_{23} & \cdots & a_{2n} \\ & & \ddots & & \vdots \\ & \mathbf{0} & & 0 & a_{n-1,n} \\ & & & & 0 \end{bmatrix}$$

The Jacobi method is written in this notation as

$$DX^{(r+1)} = -(L + U)X^{(r)} + b$$

and the Gauss–Seidel method as

$$DX^{(r+1)} = -LX^{(r+1)} - UX^{(r)} + b$$

(Remember that $X^{(r)}$ denotes the rth iteration of X, not X raised to the power r.)

By changing the method slightly, we have been able to speed up the method, so it is natural to ask if it can be speeded up even further. A popular method for doing this is **successive over-relaxation (SOR)**. This anticipates what the x_i values might be and overshoots the values obtained by Gauss–Seidel iteration. The new value of each component of the vector $X^{(r+1)}$ is taken to be

$$wx_i^{(r+1)} + (1 - w)x_i^{(r)} \tag{5.31}$$

which is the weighted average of the previous value and the new value given by Gauss–Seidel iteration. In the two-variable example the weighted average rearranges the equations as

$$x = w[\tfrac{1}{4}(2 - y)] + (1 - w)x = x + w[\tfrac{1}{4}(2 - y - 4x)]$$

$$y = w[\tfrac{1}{4}(-7 - x)] + (1 - w)y = y + w[\tfrac{1}{4}(-7 - x - 4y)]$$

The convergence for this example is so rapid that the enhanced convergence of SOR is hardly worth the effort; an optimum value of $w = 1.05$ reduces the convergence, to six significant figures, from seven to six iterations. However, for most problems the improved convergence is significant. Note that $w = 1$ gives the Gauss–Seidel method.

If we repeat the calculation for (5.30) including (5.31) with $w = 1.4$, we obtain the results shown in Figure 5.19. It may be noted that the iterations converge even faster than the two previous methods, with a solution accurate to about 0.1% after ten steps. The optimum value of w is of great interest, and specialist books on numerical analysis

Iteration	0	1	2	3	4	5	6	7	8	9	10
x	0	0.7	0.273	0.5643	1.1060	1.0195	1.0226	1.0055	0.9956	0.9995	0.9995
y	0	0.21	0.0378	−0.9025	−1.0885	−1.0434	−1.0208	−0.9969	−0.9967	−0.9990	−0.9997
z	0	0.553	1.7033	1.9646	2.0930	2.0320	2.0019	1.9979	1.9971	1.9996	2.0001
t	0	−1.7871	−1.8775	−2.0242	−2.0554	−2.0002	−2.0013	−1.9981	−1.9988	−2.0002	−2.0000

Figure 5.19 Iterative solution of (5.30) with SOR factor $w = 1.4$ in (5.31).

SOR factor w	0.2	0.4	0.6	0.8	1.0	1.2	1.4	1.6	1.8
Iterations required for convergence	>50	>50	47	34	26	21	17	29	>50

Figure 5.20 Variation of rate of convergence with SOR factor w.

give details of how this can be computed (for example, *Applied Linear Algebra*, Peter Olver and Cheri Shakiban (2005), Pearson). Usually the best approach is a heuristic one – experiment with w to find a value that gives the fastest convergence. For 'one-off' problems this is hardly worth the effort so long as convergence is achieved, but in many scientific and engineering problems the same calculation may be done many hundreds of times, so the optimum value of w can reduce calculation time by half or more. For the current problem the number of iterations required to give four-decimal-place accuracy is shown in Figure 5.20.

It can be shown that outside the region $0 < w < 2$ the method will diverge but that inside it may or may not converge. The case $w < 1$ is called **under-relaxation** and $w > 1$ is called **over-relaxation**. In straightforward problems w in the range 1.2–1.8 usually gives the most rapid convergence, and this is normally the region to explore as a first guess. In the problem studied a value of $w = 1.4$ gives just about the fastest convergence, requiring only about two-thirds of the iterations required for the Gauss–Seidel method. In some physical problems, however, under-relaxation is required in order to avoid too rapid variation from iteration to iteration.

Great care must be taken with iterative methods, and convergence for some equations can be particularly difficult. Considerable experience is needed in looking at sets of equations to decide whether or not convergence can be expected, and often – even for the experienced mathematician – the answer is 'try it and see'. One simple test that will guarantee convergence is to test whether the matrix is **diagonally dominant**. This means that the magnitude of a diagonal element is larger than or equal to the sum of the magnitudes of the off-diagonal elements in that row, or $|a_{ii}| \geq \sum_{\substack{j=1 \\ i \neq j}}^{n} |a_{ij}|$ for each i. If the system is not diagonally dominant, the iteration method may or may not converge.

A detailed analysis of the convergence of iterative methods is not possible without a study of eigenvalues, and can be found in specialist numerical analysis books.

Iterative methods described in this section are fairly easy to program and an implementation in MATLAB, or similar package, is highly suitable.

5.5.5 Exercises

(*Note*: All of these exercises are best solved using a computer matrix package such as MATLAB.)

81 Solve the equations in Question 73 (Exercises 5.5.3) using Jacobi iteration starting from the estimate $X = [1 \quad 1 \quad 1 \quad 1]^T$. How accurate is the solution obtained after five iterations?

82 Solve the equations in Question 74 (Exercises 5.5.3) using Gauss–Seidel iteration, starting from the estimate $X = [1 \quad 0 \quad 0 \quad 0]^T$. How accurate is the solution obtained after three iterations?

83 Write a computer program in MATLAB or similar package to obtain the solution, by SOR, to the equations in Question 75 (Exercises 5.5.3). Determine the optimum SOR factor for each equation.

84 Use an SOR program to solve the equations

$$x - 0.7y = -4$$
$$-0.7x + y - 0.7z = 34$$
$$-0.7y + z = -44$$

so that successive iterations differ by no more than 1 in the fourth decimal place. Find an SOR factor that produces this convergence in less than 50 iterations.

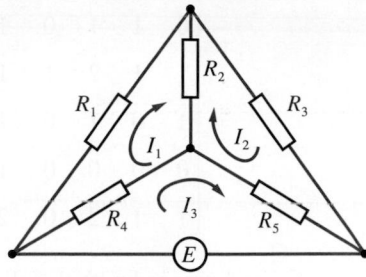

Figure 5.21 Circuit for Question 85.

85 Show that the circuit in Figure 5.21 has equations

$$\begin{bmatrix} R_1 + R_2 + R_4 & -R_2 & -R_4 \\ -R_2 & R_3 + R_5 + R_2 & -R_5 \\ -R_4 & -R_5 & R_4 + R_5 \end{bmatrix} \begin{bmatrix} I_1 \\ I_2 \\ I_3 \end{bmatrix}$$
$$= \begin{bmatrix} 0 \\ 0 \\ E \end{bmatrix}$$

Take $R_1 = 1$, $R_2 = 2$, $R_3 = 2$, $R_4 = 2$ and $R_5 = 3$ (all in Ω) and $E = 1.5$ V. Show that the equations are diagonally dominant, and hence solve the equations by an iterative method.

86 Solve the 10×10 matrix equation in Example 5.29 using an iterative method starting from $X = [1 \quad 1 \quad 1 \quad 1 \quad 1 \quad 1 \quad 1 \quad 1 \quad 1 \quad 1]^T$. Verify that a solution to four-figure accuracy can be obtained in less than ten iterations.

5.6 Rank

The solution of sets of linear equations has been considered in Section 5.5. Provided the determinant of a matrix is non-zero, we can obtain explicit solutions in terms of the inverse matrix. However, when we looked at cases with zero determinant the results were much less clear. The idea of the **rank** of a matrix helps to make these results more precise. Unfortunately, rank is not an easy concept, and it is usually difficult to compute. We shall take an informal approach that is not fully general but is sufficient to deal with the cases (c) and (d) of Section 5.5. The method we shall use is to take the Gaussian elimination procedure described in Figure 5.12 (Section 5.5.2) and examine the consequences for a zero-determinant situation.

If we start with the equations

$$
\begin{bmatrix}
1 & 0 & 1 & 1 & 0 & 0 \\
0 & 1 & 1 & 0 & 1 & 2 \\
1 & 1 & 2 & 1 & 1 & 2 \\
1 & 0 & 1 & 0 & 1 & 3 \\
0 & 0 & 0 & 0 & 1 & 3 \\
1 & 1 & 2 & 0 & 2 & 5
\end{bmatrix}
\begin{bmatrix}
x_1 \\ x_2 \\ . \\ . \\ . \\ x_6
\end{bmatrix}
=
\begin{bmatrix}
1 \\ 1 \\ 2 \\ 0 \\ 0 \\ 1
\end{bmatrix}
\tag{5.32}
$$

and proceed with the elimination, the first and second steps are quite normal:

$$
\begin{bmatrix}
1 & 0 & 1 & 1 & 0 & 0 \\
0 & 1 & 1 & 0 & 1 & 2 \\
0 & 1 & 1 & 0 & 1 & 2 \\
0 & 0 & 0 & -1 & 1 & 3 \\
0 & 0 & 0 & 0 & 1 & 3 \\
0 & 1 & 1 & -1 & 2 & 5
\end{bmatrix}
\begin{bmatrix}
x_1 \\ x_2 \\ . \\ . \\ . \\ x_6
\end{bmatrix}
=
\begin{bmatrix}
1 \\ 1 \\ 1 \\ -1 \\ 0 \\ 0
\end{bmatrix},
\quad
\begin{bmatrix}
1 & 0 & 1 & 1 & 0 & 0 \\
0 & 1 & 1 & 0 & 1 & 2 \\
0 & 0 & 0 & 0 & 0 & 0 \\
0 & 0 & 0 & -1 & 1 & 3 \\
0 & 0 & 0 & 0 & 1 & 3 \\
0 & 0 & 0 & -1 & 1 & 3
\end{bmatrix}
\begin{bmatrix}
x_1 \\ x_2 \\ . \\ . \\ . \\ x_6
\end{bmatrix}
=
\begin{bmatrix}
1 \\ 1 \\ 0 \\ -1 \\ 0 \\ -1
\end{bmatrix}
$$

The next step in the elimination procedure looks for a non-zero entry in the third column on or below the diagonal element. All the entries are zero – so the procedure, as it stands, fails. To overcome the problem, we just proceed to the next column and repeat the normal sequence of operations. We interchange the third and fourth rows and perform the elimination on column 4. Finally we interchange rows 4 and 5 to give

$$
\begin{bmatrix}
1 & 0 & 1 & 1 & 0 & 0 \\
0 & 1 & 1 & 0 & 1 & 2 \\
0 & 0 & 0 & 1 & -1 & -3 \\
0 & 0 & 0 & 0 & 1 & 3 \\
0 & 0 & 0 & 0 & 0 & 0 \\
0 & 0 & 0 & 0 & 0 & 0
\end{bmatrix}
\begin{bmatrix}
x_1 \\ x_2 \\ . \\ . \\ . \\ x_6
\end{bmatrix}
=
\begin{bmatrix}
1 \\ 1 \\ 1 \\ 0 \\ 0 \\ 0
\end{bmatrix}
\tag{5.33}
$$

To perform the back substitution we put $x_6 = \mu$. Then

row 4   gives $x_5 = -3x_6 = -3\mu$

row 3   gives $x_4 = 1 + x_5 + 3x_6 = 1$

put $x_3 = \lambda$

row 2   gives $x_2 = 1 - x_3 - x_5 - 2x_6 = 1 - \lambda + \mu$

row 1   gives $x_1 = 1 - x_3 - x_4 = -\lambda$

Thus our solution is

$$x_1 = -\lambda, \quad x_2 = 1 - \lambda + \mu, \quad x_3 = \lambda, \quad x_4 = 1, \quad x_5 = -3\mu, \quad x_6 = \mu$$

The equations have been reduced to **echelon form**, and it is clear that the same process can be followed for any matrix.

In general we use the elementary row operations, introduced in Section 5.5.2, to manipulate the equation or matrix to **echelon form**:

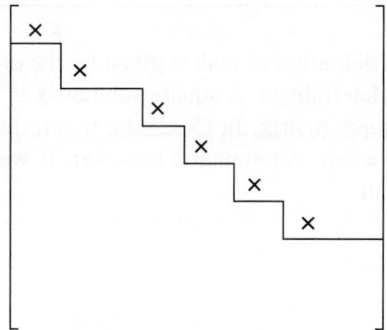

Below the line all the entries are zero, and the leading element, marked ×, in each row above the line is non-zero. The row operations do not change the solution to the set of equations corresponding to the matrix.

When this procedure is applied to a non-singular matrix, the method reduces to that shown in Figure 5.12, the final matrix has non-zero diagonal elements, and back substitution gives a unique solution. When the determinant is zero, as in (5.32), the elimination gives a matrix with some zeros in the diagonal and some zero rows, as in (5.33). The number of non-zero rows in the echelon form is called the **rank** of the matrix, rank **A**; in the case of the matrix in (5.32) and that derived from it by row manipulation (5.33), we have rank **A** = 4.

Example 5.36   Find the rank of the matrices

$$\text{(a)} \begin{bmatrix} 1 & 1 & -1 \\ 2 & -1 & 2 \\ 0 & -3 & 4 \end{bmatrix} \quad \text{(b)} \begin{bmatrix} 1 & -1 & 1 \\ -2 & 2 & -2 \\ -1 & 1 & -1 \end{bmatrix} \quad \text{(c)} \begin{bmatrix} 1 & 0 & 0 \\ 0 & 1 & 1 \\ 2 & 0 & 1 \end{bmatrix}$$

Solution   Using the usual elimination method gives in each case

$$\text{(a)} \begin{bmatrix} 1 & 1 & -1 \\ 2 & -1 & 2 \\ 0 & -3 & 4 \end{bmatrix} \rightarrow \begin{bmatrix} 1 & 1 & -1 \\ 0 & -3 & 4 \\ 0 & -3 & 4 \end{bmatrix} \rightarrow \begin{bmatrix} 1 & 1 & -1 \\ 0 & -3 & 4 \\ 0 & 0 & 0 \end{bmatrix} \Rightarrow \text{rank 2}$$

$$\text{(b)} \begin{bmatrix} 1 & -1 & 1 \\ -2 & 2 & -2 \\ -1 & 1 & -1 \end{bmatrix} \rightarrow \begin{bmatrix} 1 & -1 & 1 \\ 0 & 0 & 0 \\ 0 & 0 & 0 \end{bmatrix} \Rightarrow \text{rank 1}$$

(c) $\begin{bmatrix} 1 & 0 & 0 \\ 0 & 1 & 1 \\ 2 & 0 & 1 \end{bmatrix} \rightarrow \begin{bmatrix} 1 & 0 & 0 \\ 0 & 1 & 1 \\ 0 & 0 & 1 \end{bmatrix} \Rightarrow \text{rank } 3$

The more common definition of rank is given by the order of the largest square sub-matrix with non-zero determinant. A square submatrix is formed by deleting rows and columns to form a square matrix. In (5.32) the 6×6 determinant is zero and all the 5×5 submatrices have zero determinant; however, if we delete columns 3 and 6 and rows 3 and 4, we obtain

$$\begin{bmatrix} 1 & 0 & 1 & 0 \\ 0 & 1 & 0 & 1 \\ 0 & 0 & 0 & 1 \\ 1 & 1 & 0 & 2 \end{bmatrix}$$

which has determinant equal to one, hence confirming that the matrix is of rank 4. To show equivalence of the two definitions is not straightforward and is omitted here. To determine the rank of a matrix, it is very much easier to look at the echelon form.

If we find any of the rows of the echelon matrix to be zero then, for consistency, the corresponding right-hand sides of the matrix equation must also be zero. The elementary row operations reduce the equation to echelon form, so that the equations take the form

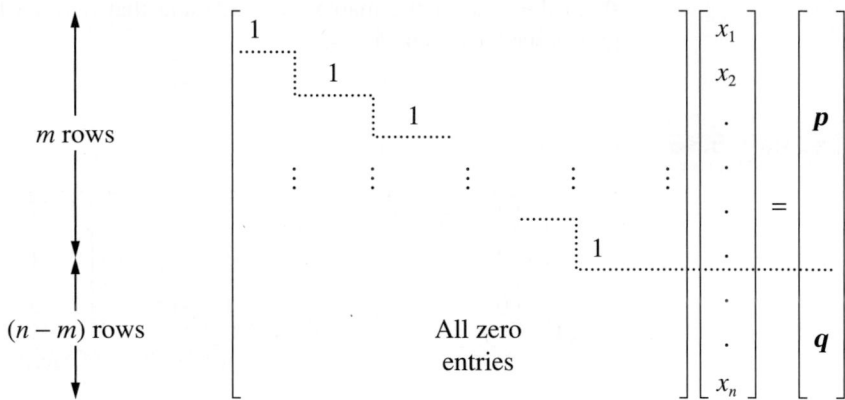

where p is a vector with m elements and q is a vector with $(n - m)$ elements. Note that each of the m non-zero rows will have a leading non-zero entry of 1 but this entry will not necessarily be on a diagonal, as illustrated for example in (5.33). Three statements follow from this reduction.

(i)   The matrix has rank $(A) = m$.
(ii)   If $q \neq 0$ then the equations are inconsistent.
(iii) If $q = 0$ then the equations are consistent and have a solution. In addition it can be shown that the solution has $(n - m)$ free parameters.

Writing the equations as

$$AX = b \tag{5.34}$$

we define the **augmented matrix** $(A:b)$ as the matrix A with the b column added to it. When reduced to echelon form the matrix and the augmented matrix take the form

$$A =$$

$$\text{and } (A:b) =$$

and the solution of the equations can be written in terms of *rank*. It is easy to see from the echelon form that A and $(A:b)$ must have the same rank to ensure consistency. The original equations must have the same property, so we can state the results (c) and (d) of Section 5.5 more clearly in terms of rank.

> If A and the augmented matrix $(A:b)$ have different rank then we have no solution to the equations (5.34). If the two matrices have the same rank then a solution exists, and furthermore it can be shown that the solution will contain a number of free parameters equal to $n -$ rank A.

The calculation of rank is not easy, so, while the result is rigorous, it is not simple to apply. Reducing equations to echelon form tells us immediately the rank of the associated matrix, and gives a constructive method of solution. There is a large amount of arithmetic in the reduction, but if the solution is required then this is inevitable anyway. The numerical calculation of rank does not normally entail reduction to echelon form; rather more advanced methods such as singular value decomposition are used.

The instruction `rank(A)` evaluates the rank of an $m \times n$ matrix A in both MATLAB and MAPLE.

Example 5.37 Reduce the following equations to echelon form, calculate the rank of the matrices and find the solutions of the equations (if they exist):

(a) $\begin{bmatrix} 0 & 1 & 1 & 0 \\ 1 & 0 & 3 & 2 \\ 2 & 1 & 5 & 4 \\ 1 & -2 & 0 & 2 \end{bmatrix} \begin{bmatrix} x_1 \\ x_2 \\ x_3 \\ x_4 \end{bmatrix} = \begin{bmatrix} 1 \\ 3 \\ 7 \\ 2 \end{bmatrix}$

(b) $\begin{bmatrix} 1 & 0 & -1 & 1 & -1 \\ 0 & 1 & 1 & -1 & 1 \\ 1 & 1 & 0 & 0 & 0 \\ 2 & 3 & 1 & -1 & 1 \\ 2 & 2 & 0 & 0 & 0 \end{bmatrix} \begin{bmatrix} x_1 \\ x_2 \\ x_3 \\ x_4 \\ x_5 \end{bmatrix} = \begin{bmatrix} 0 \\ 1 \\ 1 \\ 3 \\ 2 \end{bmatrix}$

Solution (a) Rows 1 and 3 are interchanged, and the elimination then proceeds as follows:

$\begin{bmatrix} 2 & 1 & 5 & 4 \\ 1 & 0 & 3 & 2 \\ 0 & 1 & 1 & 0 \\ 1 & -2 & 0 & 2 \end{bmatrix} \begin{bmatrix} x_1 \\ x_2 \\ x_3 \\ x_4 \end{bmatrix} = \begin{bmatrix} 7 \\ 3 \\ 1 \\ 2 \end{bmatrix} \rightarrow \begin{bmatrix} 1 & \frac{1}{2} & \frac{5}{2} & 2 \\ 0 & -\frac{1}{2} & \frac{1}{2} & 0 \\ 0 & 1 & 1 & 0 \\ 0 & -\frac{5}{2} & -\frac{5}{2} & 0 \end{bmatrix} \begin{bmatrix} x_1 \\ x_2 \\ x_3 \\ x_4 \end{bmatrix} = \begin{bmatrix} \frac{7}{2} \\ -\frac{1}{2} \\ 1 \\ -\frac{3}{2} \end{bmatrix}$

interchange row 2 and row 4

$\rightarrow \begin{bmatrix} 1 & \frac{1}{2} & \frac{5}{2} & 2 \\ 0 & -\frac{5}{2} & -\frac{5}{2} & 0 \\ 0 & 1 & 1 & 0 \\ 0 & -\frac{1}{2} & \frac{1}{2} & 0 \end{bmatrix} \begin{bmatrix} x_1 \\ x_2 \\ x_3 \\ x_4 \end{bmatrix} = \begin{bmatrix} \frac{7}{2} \\ -\frac{3}{2} \\ 1 \\ -\frac{1}{2} \end{bmatrix}$

eliminate elements in column 2

$\rightarrow \begin{bmatrix} 1 & \frac{1}{2} & \frac{5}{2} & 2 \\ 0 & 1 & 1 & 0 \\ 0 & 0 & 0 & 0 \\ 0 & 0 & 1 & 0 \end{bmatrix} \begin{bmatrix} x_1 \\ x_2 \\ x_3 \\ x_4 \end{bmatrix} = \begin{bmatrix} \frac{7}{2} \\ \frac{3}{5} \\ \frac{2}{5} \\ -\frac{1}{5} \end{bmatrix}$

interchange row 3 and row 4

$\rightarrow \begin{bmatrix} 1 & \frac{1}{2} & \frac{5}{2} & 2 \\ 0 & 1 & 1 & 0 \\ 0 & 0 & 1 & 0 \\ 0 & 0 & 0 & 0 \end{bmatrix} \begin{bmatrix} x_1 \\ x_2 \\ x_3 \\ x_4 \end{bmatrix} = \begin{bmatrix} \frac{7}{2} \\ \frac{3}{5} \\ -\frac{1}{5} \\ \frac{2}{5} \end{bmatrix}$

The rank of the matrix is 3 while that of the augmented matrix $(A : b)$ is 4, so the equations represented by the matrix equation (a) are not consistent. Note that the last row cannot be satisfied and hence the equations have no solution.

(b) Interchanging the first and last rows, making the pivot 1 and performing the first elimination, we obtain

$$\begin{bmatrix} 1 & 1 & 0 & 0 & 0 \\ 0 & 1 & 1 & -1 & 1 \\ 0 & 0 & 0 & 0 & 0 \\ 0 & 1 & 1 & -1 & 1 \\ 0 & -1 & -1 & 1 & -1 \end{bmatrix} \begin{bmatrix} x_1 \\ . \\ . \\ . \\ x_5 \end{bmatrix} = \begin{bmatrix} 1 \\ 1 \\ 0 \\ 1 \\ -1 \end{bmatrix} \rightarrow \begin{bmatrix} 1 & 1 & 0 & 0 & 0 \\ 0 & 1 & 1 & -1 & 1 \\ 0 & 0 & 0 & 0 & 0 \\ 0 & 0 & 0 & 0 & 0 \\ 0 & 0 & 0 & 0 & 0 \end{bmatrix} \begin{bmatrix} x_1 \\ . \\ . \\ . \\ x_5 \end{bmatrix} = \begin{bmatrix} 1 \\ 1 \\ 0 \\ 0 \\ 0 \end{bmatrix}$$

The matrix and the augmented matrix both have rank 2, so the equations are consistent and we can compute the solution

$$x_1 = 1 - \lambda, \quad x_2 = \lambda, \quad x_3 = 1 - \lambda + \mu - \nu, \quad x_4 = \mu, \quad x_5 = \nu$$

As expected, the solution contains three free parameters, since the order of the equation is 5 and the rank is 2.

In most practical problems that reduce to the solution of linear equations, it is usual that there are n independent variables to be computed from n equations. This is not always the case and the resulting matrix form is *not* square. A geometrical example of four equations and three unknowns was described in equation (5.2). The idea of a determinant is only sensible if matrices are square so the simple results about the solution of the equations cannot be used. However, the ideas of elementary row operations, reduction to echelon form and rank still hold and the existence or non-existence of solutions can be written in terms of these concepts. Some examples will illustrate the possible situations that can occur.

Underspecified sets of equations

Here there are more variables than equations.

Case (a)

Solve $\begin{bmatrix} 1 & 1 & 1 \\ 1 & 2 & 3 \end{bmatrix} \begin{bmatrix} x \\ y \\ z \end{bmatrix} = \begin{bmatrix} 1 \\ 2 \end{bmatrix}$

Subtract row 1 from row 2: $\begin{bmatrix} 1 & 1 & 1 \\ 0 & 1 & 2 \end{bmatrix} \begin{bmatrix} x \\ y \\ z \end{bmatrix} = \begin{bmatrix} 1 \\ 1 \end{bmatrix}$

The elimination is now complete and the back substitution starts

Put $z = t$

From row 2 $y = 1 - 2t$

From row 1 $x = 1 - y - z = t$

so the full solution is

$$x = t, \quad y = 1 - 2t, \quad z = t$$

for any t. Note that rank $(\mathbf{A}) =$ rank $(\mathbf{A} : \mathbf{b}) = 2$ and $n = 3$ so the solution has one free parameter.

Case (b)

Solve
$$\begin{bmatrix} 1 & 1 & 1 \\ 2 & 2 & 2 \end{bmatrix} \begin{bmatrix} x \\ y \\ z \end{bmatrix} = \begin{bmatrix} 1 \\ 1 \end{bmatrix}$$

Subtract $2 \times$ (row 1) from row 2:
$$\begin{bmatrix} 1 & 1 & 1 \\ 0 & 0 & 0 \end{bmatrix} \begin{bmatrix} x \\ y \\ z \end{bmatrix} = \begin{bmatrix} 1 \\ -1 \end{bmatrix}$$

and it is clear that rank $(\mathbf{A}) = 1$ and rank $(\mathbf{A} : \mathbf{b}) = 2$ so there is no solution. Obviously the last row is inconsistent. Although this example may be seen to be almost trivial since the equations are obviously inconsistent [$x + y + z = 1$ and $2(x + y + z) = 1$], in larger systems the situation is hardly ever obvious.

Overspecified sets of equations

Here there are more equations than variables.

Case (c)

Solve
$$\begin{bmatrix} 1 & 1 \\ 1 & 2 \\ 1 & 3 \end{bmatrix} \begin{bmatrix} x \\ y \end{bmatrix} = \begin{bmatrix} -1 \\ 0 \\ 1 \end{bmatrix}$$

Subtract row 1 from rows 2 and 3:
$$\begin{bmatrix} 1 & 1 \\ 0 & 1 \\ 0 & 2 \end{bmatrix} \begin{bmatrix} x \\ y \end{bmatrix} = \begin{bmatrix} -1 \\ 1 \\ 2 \end{bmatrix}$$

Subtract $2 \times$ (row 2) from row 3:
$$\begin{bmatrix} 1 & 1 \\ 0 & 1 \\ 0 & 0 \end{bmatrix} \begin{bmatrix} x \\ y \end{bmatrix} = \begin{bmatrix} -1 \\ 1 \\ 0 \end{bmatrix}$$

It can be observed that rank $(\mathbf{A}) =$ rank $(\mathbf{A} : \mathbf{b}) = 2$ and that the equations are consistent since the last row contains all zeros. Since $n = 2$ a unique solution is obtained as $x = -2$ and $y = 1$ using back substitution.

However, for overspecified equations the more common situation is that no solution is possible.

Case (d)

Solve $\begin{bmatrix} 1 & 1 \\ 1 & 2 \\ 1 & 3 \end{bmatrix} \begin{bmatrix} x \\ y \end{bmatrix} = \begin{bmatrix} 0 \\ 1 \\ -2 \end{bmatrix}$

Subtract row 1 from rows 2 and 3: $\begin{bmatrix} 1 & 1 \\ 0 & 1 \\ 0 & 2 \end{bmatrix} \begin{bmatrix} x \\ y \end{bmatrix} = \begin{bmatrix} 0 \\ 1 \\ -2 \end{bmatrix}$

Subtract $2 \times$ (row 2) from row 3: $\begin{bmatrix} 1 & 1 \\ 0 & 1 \\ 0 & 0 \end{bmatrix} \begin{bmatrix} x \\ y \end{bmatrix} = \begin{bmatrix} 0 \\ 1 \\ -4 \end{bmatrix}$

The equations are now clearly inconsistent since the last row says $0 = -4$ and rank $(\boldsymbol{A}) = 2$, rank $(\boldsymbol{A} : \boldsymbol{b}) = 3$ confirms this observation.

The existence or non-existence of solutions can be deduced from the echelon form and hence the idea of rank, and we can understand the solution of matrix equations involving non-square matrices. If $\boldsymbol{A}$ is a $p \times q$ matrix and $\boldsymbol{b}$ a $p \times 1$ *column vector, the matrix equation* $\boldsymbol{AX} = \boldsymbol{b}$ represents p linear equations in q variables. The rank of a matrix, being the number of non-zero rows in the echelon form of the matrix, cannot exceed p. On the other hand, the row reduction process will produce an echelon form with at most q non-zero rows. Hence the rank of a $p \times q$ matrix cannot exceed the smaller of p and q. There are two possible cases:

(i) $p < q$: Here there are more variables than equations. The rank of $\boldsymbol{A}$ must be less than the number of variables. If rank $(\boldsymbol{A} : \boldsymbol{b}) >$ rank $\boldsymbol{A}$, the equations are inconsistent and there is no solution, as in case (b). If rank $(\boldsymbol{A} : \boldsymbol{b}) =$ rank $\boldsymbol{A}$, as in case (a), there is a solution, which must contain $q -$ rank $\boldsymbol{A}$ free parameters.

(ii) $p > q$: Here there are more equations than variables. The rank of $\boldsymbol{A}$ cannot exceed the number of variables. If rank $(\boldsymbol{A} : \boldsymbol{b}) >$ rank $\boldsymbol{A}$, as in case (d), the equations are inconsistent and there is no solution. If rank $(\boldsymbol{A} : \boldsymbol{b}) =$ rank $\boldsymbol{A}$, as in case (c), some of the equations are redundant and there is a solution containing $q -$ rank $\boldsymbol{A}$ free parameters.

5.6.1 Exercises

  Check your answers using MATLAB or MAPLE whenever possible.

87 Find the rank of $\boldsymbol{A}$ and of the augmented matrix $(\boldsymbol{A} : \boldsymbol{b})$. Solve $\boldsymbol{AX} = \boldsymbol{b}$ where possible and check that there are $(n - \text{rank}(\boldsymbol{A}))$ free parameters.

(a) $\boldsymbol{A} = \begin{bmatrix} 1 & 2 \\ 2 & 1 \end{bmatrix}$ $\boldsymbol{b} = \begin{bmatrix} 0 \\ 1 \end{bmatrix}$

(b) $\boldsymbol{A} = \begin{bmatrix} 1 & 0 \\ 0 & 0 \end{bmatrix}$ $\boldsymbol{b} = \begin{bmatrix} 0 \\ 1 \end{bmatrix}$

(c) $\boldsymbol{A} = \begin{bmatrix} 1 & 0 & 0 \\ 0 & 1 & 1 \\ 0 & 1 & 1 \end{bmatrix}$ $\boldsymbol{b} = \begin{bmatrix} 1 \\ 0 \\ 0 \end{bmatrix}$

(d) $A = \begin{bmatrix} 1 & 0 & 1 \\ 0 & 1 & 0 \end{bmatrix}$ $b = \begin{bmatrix} 2 \\ 1 \end{bmatrix}$

(e) $A = \begin{bmatrix} 1 & 0 \\ 0 & 1 \\ 1 & 0 \end{bmatrix}$ $b = \begin{bmatrix} 1 \\ 0 \\ 0 \end{bmatrix}$

(f) $A = \begin{bmatrix} 0 & 0 & 0 & 1 \\ 0 & 0 & 2 & 0 \\ 0 & 3 & 0 & 0 \\ 4 & 0 & 0 & 0 \end{bmatrix}$ $b = \begin{bmatrix} 1 \\ 0 \\ 1 \\ 0 \end{bmatrix}$

88 Find the rank of the coefficient matrix and of the augmented matrix in the matrix equation

$$\begin{bmatrix} 1 & 1-\alpha \\ \alpha & -2 \end{bmatrix}\begin{bmatrix} x \\ y \end{bmatrix} = \begin{bmatrix} \alpha^2 \\ \alpha \end{bmatrix}$$

For each value of α, find, where possible, the solution of the equation.

89 Find the rank of the matrices

(a) $\begin{bmatrix} 2 & 1 & 1 & 1 \\ 4 & 2 & 2 & 3 \\ 0 & 0 & 0 & 1 \\ -2 & -1 & -1 & 0 \end{bmatrix}$, (b) $\begin{bmatrix} 1 & 1 & 1 & 1 \\ 2 & 1 & 2 & 1 \\ 0 & 1 & 0 & 1 \\ 1 & 0 & 1 & 1 \end{bmatrix}$

90 Reduce the matrices in the following equations to echelon form, determine their ranks and solve the equations, if a solution exists:

(a) $\begin{bmatrix} 1 & 2 & 3 \\ 3 & 2 & 1 \\ 1 & 1 & 1 \end{bmatrix}\begin{bmatrix} x \\ y \\ z \end{bmatrix} = \begin{bmatrix} 8 \\ 4 \\ 3 \end{bmatrix}$

(b) $\begin{bmatrix} 1 & 2 & -1 & 1 \\ 1 & 1 & 0 & 0 \\ 0 & 1 & -1 & 1 \\ 1 & 0 & 1 & -1 \end{bmatrix}\begin{bmatrix} x \\ y \\ z \\ t \end{bmatrix} = \begin{bmatrix} 0 \\ 1 \\ -1 \\ 1 \end{bmatrix}$

91 By obtaining the order of the largest square submatrix with non-zero determinant determine the rank of the matrix

$A = \begin{bmatrix} 1 & 1 & 0 & 1 \\ 1 & 0 & 0 & 1 \\ 0 & 1 & 0 & 0 \\ 1 & 1 & 1 & 1 \end{bmatrix}$

Reduce the matrix to echelon form and confirm your result. Check the rank of the augmented matrix $(A : b)$, where $b^{\mathrm{T}} = [-1 \ \ 0 \ \ -1 \ \ 0]$. Does the equation $AX = b$ have a solution?

92 Solve, where possible, the following matrix equations:

(a) $\begin{bmatrix} 1 & 3 & 4 \\ -1 & 3 & 4 \end{bmatrix}\begin{bmatrix} x \\ y \\ z \end{bmatrix} = \begin{bmatrix} 1 \\ 3 \end{bmatrix}$

(b) $\begin{bmatrix} 2 & 1 \\ 4 & 6 \\ 3 & 5 \end{bmatrix}\begin{bmatrix} x \\ y \end{bmatrix} = \begin{bmatrix} 1 \\ 4 \\ -2 \end{bmatrix}$

(c) $\begin{bmatrix} 1 & 4 & 7 & -3 \\ -2 & 3 & -6 & 1 \\ 0 & 11 & 8 & -5 \end{bmatrix}\begin{bmatrix} x \\ y \\ z \\ t \end{bmatrix} = \begin{bmatrix} 1 \\ 3 \\ 5 \end{bmatrix}$

(d) $\begin{bmatrix} 2 & 1 & 4 \\ 3 & 2 & 9 \\ 4 & 1 & 3 \\ 3 & 3 & 3 \end{bmatrix}\begin{bmatrix} x \\ y \\ z \end{bmatrix} = \begin{bmatrix} 1 \\ 4 \\ -2 \\ -3 \end{bmatrix}$

93 In a fluid flow problem there are five natural parameters. These have dimensions in terms of length L, mass M and time T as follows:

velocity $= V = LT^{-1}$, density $= \rho = ML^{-3}$

distance $= D = L$, gravity $= g = LT^{-2}$

and

viscosity $= \mu = ML^{-1}T^{-1}$

To determine how many non-dimensional parameters can be constructed, seek values of p, q, r, s and t so that

$$V^p \rho^q D^r g^s \mu^t$$

is dimensionless. Write the equations for p, q, r, s and t in matrix form and show that the resulting 3×5 matrix has rank 3. Thus there are two parameters that can be chosen independently.

By choosing these appropriately, show that they correspond to the Reynolds number $Re = V\rho D/\mu$ and the Froude number $Fr = Dg/V^2$.

Repeat a similar dimensional analysis when heat transfer is included.

 94 Four points in a three-dimensional space have coordinates (x_i, y_i, z_i) for $i = 1, \ldots, 4$. From the rank of the matrix

$$\begin{bmatrix} x_1 & y_1 & z_1 & 1 \\ x_2 & y_2 & z_2 & 1 \\ x_3 & y_3 & z_3 & 1 \\ x_4 & y_4 & z_4 & 1 \end{bmatrix}$$

determine whether the points lie on a plane or a line or whether there are other possibilities.

95 A popular method of numerical integration – see the work in Chapter 8 – involves Gaussian integration; it is used in finite-element calculations which are well used in most of engineering. As a simple example, the numerical integral over the interval $-1 \leqslant x \leqslant 1$ is written

$$\int_{-1}^{1} f(x)\mathrm{d}x = C_1 f(x_1) + C_2 f(x_2)$$

and the formula is made exact for the four functions $f = 1, f = x, f = x^2$ and $f = x^3$, so it must be accurate for all cubics. This leads to the four equations

$$C_1 + C_2 = 2$$
$$C_1 x_1 + C_2 x_2 = 0$$
$$C_1 x_1^2 + C_2 x_2^2 = \tfrac{2}{3}$$
$$C_1 x_1^3 + C_2 x_2^3 = 0$$

Use Gaussian elimination to reduce the equations and hence deduce that the equations are only consistent if x_1 and x_2 are chosen at the 'Gauss' points $\pm\frac{1}{\sqrt{3}}$.

5.7 The eigenvalue problem

A problem that leads to a concept of crucial importance in many branches of mathematics and its applications is that of seeking non-trivial solutions $X \neq 0$ to the matrix equation

$$AX = \lambda X$$

This is referred to as the eigenvalue problem; values of the scalar λ for which non-trivial solutions exist are called **eigenvalues** and the corresponding solutions $X \neq 0$ are called the **eigenvectors**. We saw an example of eigenvalues in Example 5.27. Such problems arise naturally in many branches of engineering. For example, in vibrations the eigenvalues and eigenvectors describe the frequency and mode of vibration respectively, while in mechanics they represent principal stresses and the principal axes of stress in bodies subjected to external forces. Eigenvalues also play an important role in the stability analysis of dynamical systems and are central to the evaluation of energy levels in quantum mechanics.

5.7.1 The characteristic equation

The set of simultaneous equations

$$AX = \lambda X \tag{5.35}$$

where A is an $n \times n$ matrix and $X = [x_1 \quad x_2 \quad \ldots \quad x_n]^{\mathrm{T}}$ is an $n \times 1$ column vector can be written in the form

$$(\lambda I - A)X = 0 \tag{5.36}$$

where I is the identity matrix. The matrix equation (5.36) represents simply a set of homogeneous equations, and we know that a non-trivial solution exists if

$$c(\lambda) = |\lambda I - A| = 0 \tag{5.37}$$

Here $c(\lambda)$ is the expansion of the determinant and is a polynomial of degree n in λ, called the **characteristic polynomial** of A. Thus

$$c(\lambda) = \lambda^n + c_{n-1}\lambda^{n-1} + c_{n-2}\lambda^{n-2} + \ldots + c_1\lambda + c_0$$

and the equation $c(\lambda) = 0$ is called the **characteristic equation** of A. We note that this equation can be obtained just as well by evaluating $|A - \lambda I| = 0$; however, the form (5.37) is preferred for the definition of the characteristic equation, since the coefficient of λ^n is then always +1.

In many areas of engineering, particularly in those involving vibration or the control of processes, the determination of those values of λ for which (5.36) has a non-trivial solution (that is, a solution for which $x \neq 0$) is of vital importance. These values of λ are precisely the values that satisfy the characteristic equation, and are called the eigenvalues of A.

Example 5.38   Find the characteristic equation and the eigenvalues of the matrix

$$A = \begin{bmatrix} -2 & 1 \\ 1 & -2 \end{bmatrix}$$

Solution   Equation (5.37) gives

$$0 = |\lambda I - A| = \begin{vmatrix} \lambda + 2 & -1 \\ -1 & \lambda + 2 \end{vmatrix} = (\lambda + 2)^2 - 1$$

so the characteristic equation is

$$\lambda^2 + 4\lambda + 3 = 0$$

The roots of this equation, namely $\lambda = -1$ and -3, give the eigenvalues.

Example 5.39   Find the characteristic equation for the matrix

$$A = \begin{bmatrix} 1 & 1 & -2 \\ -1 & 2 & 1 \\ 0 & 1 & -1 \end{bmatrix}$$

Solution   By (5.37), the characteristic equation for A is the cubic equation

$$c(\lambda) = \begin{vmatrix} \lambda - 1 & -1 & 2 \\ 1 & \lambda - 2 & -1 \\ 0 & -1 & \lambda + 1 \end{vmatrix} = 0$$

Expanding the determinant along the first column gives

$$c(\lambda) = (\lambda - 1) \begin{vmatrix} \lambda - 2 & -1 \\ -1 & \lambda + 1 \end{vmatrix} - \begin{vmatrix} -1 & 2 \\ -1 & \lambda + 1 \end{vmatrix}$$

$$= (\lambda - 1)[(\lambda - 2)(\lambda + 1) - 1] - [2 - (\lambda + 1)]$$

Thus

$$c(\lambda) = \lambda^3 - 2\lambda^2 - \lambda + 2 = 0$$

is the required characteristic equation.

For matrices of large order, determining the characteristic polynomial by direct expansion of $|\lambda I - A|$ is unsatisfactory in view of the large number of terms involved in the determinant expansion but alternative procedures are available.

5.7.2 Eigenvalues and eigenvectors

The roots of the characteristic equation (5.37) are called the eigenvalues of the matrix A (the terms latent roots, proper roots and characteristic roots are also sometimes used). By the Fundamental Theorem of Algebra, a polynomial equation of degree n has exactly n roots, so that the matrix A has exactly n eigenvalues λ_i, $i = 1, 2, \ldots , n$. These eigenvalues may be real or complex, and not necessarily distinct. Corresponding to each eigenvalue λ_i, there is a non-zero solution $x = e_i$ of (5.36); e_i is called the eigenvector of A corresponding to the eigenvalue λ_i. (Again the terms latent vector, proper vector and characteristic vector are sometimes seen, but are generally obsolete.) We note that if $x = e_i$ satisfies (5.36) then any scalar multiple $\beta_i e_i$ of e_i also satisfies (5.36), so that the eigenvector e_i may only be determined to within a scalar multiple.

Example 5.40 Verify that $\begin{bmatrix} 1 \\ 1 \end{bmatrix}$ and $\begin{bmatrix} 1 \\ -1 \end{bmatrix}$ are eigenvectors of the matrix

$$A = \begin{bmatrix} -2 & 1 \\ 1 & -2 \end{bmatrix}$$

Solution The matrix is the same as the one given in Example 5.38 so we would expect that these eigenvectors correspond to the eigenvalues -1 and -3. To verify the fact we must check that equation (5.35) is satisfied. Now for the first column vector

$$\begin{bmatrix} -2 & 1 \\ 1 & -2 \end{bmatrix} \begin{bmatrix} 1 \\ 1 \end{bmatrix} = \begin{bmatrix} -1 \\ -1 \end{bmatrix} = -1 \begin{bmatrix} 1 \\ 1 \end{bmatrix}$$

so $\begin{bmatrix} 1 \\ 1 \end{bmatrix}$ is an eigenvector corresponding to the eigenvector -1.

For the second column vector

$$\begin{bmatrix} -2 & 1 \\ 1 & -2 \end{bmatrix}\begin{bmatrix} 1 \\ -1 \end{bmatrix} = \begin{bmatrix} -3 \\ 3 \end{bmatrix} = -3\begin{bmatrix} 1 \\ -1 \end{bmatrix}$$

so $\begin{bmatrix} 1 \\ -1 \end{bmatrix}$ is an eigenvector corresponding to the eigenvector -3.

Example 5.41   Find the eigenvalues and eigenvectors of the matrix $\boldsymbol{A} = \begin{bmatrix} 0 & -1 \\ 1 & 0 \end{bmatrix}$.

Solution   To find the eigenvalues use equation (5.37)

$$0 = |\lambda\boldsymbol{I} - \boldsymbol{A}| = \begin{vmatrix} \lambda & 1 \\ -1 & \lambda \end{vmatrix} = \lambda^2 + 1$$

This characteristic equation has two roots $\lambda = +j$ and $-j$ which are the eigenvalues, in this case complex. Note that in general eigenvalues are complex although in most of the remaining examples in this section they have been constructed to be real.

To obtain the eigenvectors use equation (5.36).

For the eigenvalue $\lambda = j$ then (5.36) gives

$$(\lambda\boldsymbol{I} - \boldsymbol{A})\begin{bmatrix} a \\ b \end{bmatrix} = \begin{bmatrix} j & 1 \\ -1 & j \end{bmatrix}\begin{bmatrix} a \\ b \end{bmatrix} = 0$$

or in expanded form

$$\begin{array}{l} ja + b = 0 \\ -a + jb = 0 \end{array}$$ with solution $a = j$ and $b = 1$

and hence the eigenvector corresponding to $\lambda = j$ is $\begin{bmatrix} j \\ 1 \end{bmatrix}$.

For the eigenvalue $\lambda = -j$ then (5.36) gives

$$(\lambda\boldsymbol{I} - \boldsymbol{A})\begin{bmatrix} c \\ d \end{bmatrix} = \begin{bmatrix} -j & 1 \\ -1 & -j \end{bmatrix}\begin{bmatrix} c \\ d \end{bmatrix} = 0$$

or in expanded form

$$\begin{array}{l} -jc + d = 0 \\ -c - jd = 0 \end{array}$$ with solution $c = 1$ and $d = j$

and hence the eigenvector corresponding to $\lambda = -j$ is $\begin{bmatrix} 1 \\ j \end{bmatrix}$.

Example 5.42

Determine the eigenvalues and eigenvectors for the matrix A of Example 5.39.

Solution

$$A = \begin{bmatrix} 1 & 1 & -2 \\ -1 & 2 & 1 \\ 0 & 1 & -1 \end{bmatrix}$$

The eigenvalues λ_i of A satisfy the characteristic equation $c(\lambda) = 0$, and this has been obtained in Example 5.39 as the cubic

$$\lambda^3 - 2\lambda^2 - \lambda + 2 = 0$$

which can be solved to obtain the eigenvalues λ_1, λ_2 and λ_3.

Alternatively, it may be possible, using the determinant form $|A - \lambda I|$, to carry out suitable row and/or column operations to factorize the determinant.

In this case

$$|A - \lambda I| = \begin{vmatrix} 1 - \lambda & 1 & -2 \\ -1 & 2 - \lambda & 1 \\ 0 & 1 & -1 - \lambda \end{vmatrix}$$

and adding column 1 to column 3 gives

$$\begin{vmatrix} 1 - \lambda & 1 & -1 - \lambda \\ -1 & 2 - \lambda & 0 \\ 0 & 1 & -1 - \lambda \end{vmatrix} = -(1 + \lambda) \begin{vmatrix} 1 - \lambda & 1 & 1 \\ -1 & 2 - \lambda & 0 \\ 0 & 1 & 1 \end{vmatrix}$$

Subtracting row 3 from row 1 gives

$$-(1 + \lambda) \begin{vmatrix} 1 - \lambda & 0 & 0 \\ -1 & 2 - \lambda & 0 \\ 0 & 1 & 1 \end{vmatrix} = -(1 + \lambda)(1 - \lambda)(2 - \lambda)$$

Setting $|A - \lambda I| = 0$ gives the eigenvalues as $\lambda_1 = 2$, $\lambda_2 = 1$ and $\lambda_3 = -1$. The order in which they are written is arbitrary, but for consistency we shall adopt the convention of taking λ_1, λ_2, ... , λ_n in decreasing order.

Having obtained the eigenvalues λ_i ($i = 1, 2, 3$), the corresponding eigenvectors e_i are obtained by solving the appropriate homogeneous equations

$$(A - \lambda_i I)e_i = 0 \tag{5.38}$$

When $i = 1$, $\lambda_1 = 2$ and (5.38) is

$$\begin{bmatrix} -1 & 1 & -2 \\ -1 & 0 & 1 \\ 0 & 1 & -3 \end{bmatrix} \begin{bmatrix} e_{11} \\ e_{12} \\ e_{13} \end{bmatrix} \equiv 0$$

that is,

$$-e_{11} + e_{12} - 2e_{13} = 0$$

$$-e_{11} + 0e_{12} + e_{13} = 0$$

$$0e_{11} + e_{12} - 3e_{13} = 0$$

leading to the solution

$$\frac{e_{11}}{-1} = \frac{-e_{12}}{3} = \frac{e_{13}}{-1} = \beta_1$$

where β_1 is an arbitrary non-zero scalar. Thus the eigenvector e_1 corresponding to the eigenvalue $\lambda_1 = 2$ is

$$e_1 = \beta_1[1 \quad 3 \quad 1]^T$$

As a check, we can compute

$$\boldsymbol{Ae}_1 = \beta_1 \begin{bmatrix} 1 & 1 & -2 \\ -1 & 2 & 1 \\ 0 & 1 & -1 \end{bmatrix} \begin{bmatrix} 1 \\ 3 \\ 1 \end{bmatrix} = \beta_1 \begin{bmatrix} 2 \\ 6 \\ 2 \end{bmatrix} = 2\beta_1 \begin{bmatrix} 1 \\ 3 \\ 1 \end{bmatrix} = \lambda_1 e_1$$

and thus conclude that our calculation was correct.

When $i = 2$, $\lambda_2 = 1$ and we have to solve

$$\begin{bmatrix} 0 & 1 & -2 \\ -1 & 1 & 1 \\ 0 & 1 & -2 \end{bmatrix} \begin{bmatrix} e_{21} \\ e_{22} \\ e_{23} \end{bmatrix} = 0$$

that is,

$$0e_{21} + e_{22} - 2e_{23} = 0$$

$$-e_{21} + e_{22} + e_{23} = 0$$

$$0e_{21} + e_{22} - 2e_{23} = 0$$

leading to the solution

$$\frac{e_{21}}{-3} = \frac{-e_{22}}{2} = \frac{e_{23}}{-1} = \beta_2$$

where β_2 is an arbitrary scalar. Thus the eigenvector e_2 corresponding to the eigenvalue $\lambda_2 = 1$ is

$$e_2 = \beta_2[3 \quad 2 \quad 1]^T$$

Again a check could be made by computing $\boldsymbol{Ae}_2$.

Finally, when $i = 3$, $\lambda_3 = -1$ and we obtain from (5.38)

$$\begin{bmatrix} 2 & 1 & -2 \\ -1 & 3 & 1 \\ 0 & 1 & 0 \end{bmatrix} \begin{bmatrix} e_{31} \\ e_{32} \\ e_{33} \end{bmatrix} = 0$$

that is,

$$2e_{31} + e_{32} - 2e_{33} = 0$$

$$-e_{31} + 3e_{32} + e_{33} = 0$$

$$0e_{31} + e_{32} + 0e_{33} = 0$$

and hence

$$\frac{e_{31}}{-1} = \frac{e_{32}}{0} = \frac{e_{33}}{-1} = \beta_3$$

Here again β_3 is an arbitrary scalar, and the eigenvector e_3 corresponding to the eigenvalue λ_3 is

$$e_3 = \beta_3 [1 \quad 0 \quad 1]^T$$

The calculation can be checked as before. Thus we have found that the eigenvalues of the matrix **A** are 2, 1 and −1, with corresponding eigenvectors

$$\beta_1 [1 \quad 3 \quad 1]^T, \quad \beta_2 [3 \quad 2 \quad 1]^T \quad \text{and} \quad \beta_3 [1 \quad 0 \quad 1]^T$$

respectively.

Since in Example 5.42 the β_i, $i = 1, 2, 3$, are arbitrary, it follows that there are an infinite number of eigenvectors, scalar multiples of each other, corresponding to each eigenvalue. Sometimes it is convenient to scale the eigenvectors according to some convention. A convention frequently adopted is to **normalize** the eigenvectors so that they are uniquely determined up to a scale factor of ±1. The normalized form of an eigenvector $e = [e_1 \quad e_2 \quad \dots \quad e_n]^T$ is denoted by $\hat{e}$ and is given by

$$\hat{e} = \frac{e}{|e|}$$

where

$$|e| = \sqrt{(e_1^2 + e_2^2 + \dots + e_n^2)}$$

For example, for the matrix **A** of Example 5.42, the normalized forms of the eigenvectors are

$$\hat{e}_1 = [1/\sqrt{11} \quad 3/\sqrt{11} \quad 1/\sqrt{11}]^T, \quad \hat{e}_2 = [3/\sqrt{14} \quad 2/\sqrt{14} \quad 1/\sqrt{14}]^T$$

and

$$\hat{e}_3 = [1/\sqrt{2} \quad 0 \quad 1/\sqrt{2}]^T$$

However, throughout the text, unless otherwise stated, the eigenvectors will always be presented in their 'simplest' form, so that for the matrix of Example 5.42 we take $\beta_1 = \beta_2 = \beta_3 = 1$ and write

$$e_1 = [1 \quad 3 \quad 1]^T, \quad e_2 = [3 \quad 2 \quad 1]^T \quad \text{and} \quad e_3 = [1 \quad 0 \quad 1]^T$$

Example 5.43   Find the eigenvalues and eigenvectors of

$$A = \begin{bmatrix} \cos\theta & -\sin\theta \\ \sin\theta & \cos\theta \end{bmatrix}$$

Solution   Now

$$|\lambda I - A| = \begin{vmatrix} \lambda - \cos\theta & \sin\theta \\ -\sin\theta & \lambda - \cos\theta \end{vmatrix}$$

$$= \lambda^2 - 2\lambda\cos\theta + \cos^2\theta + \sin^2\theta = \lambda^2 - 2\lambda\cos\theta + 1$$

So the eigenvalues are the roots of

$$\lambda^2 - 2\lambda\cos\theta + 1 = 0$$

that is,

$$\lambda = \cos\theta \pm j\sin\theta$$

Solving for the eigenvectors as in Example 5.42, we obtain

$$e_1 = [1 \quad -j]^T \quad \text{and} \quad e_2 = [1 \quad j]^T$$

In Examples 5.41 and 5.43 we see that eigenvalues can be complex numbers, and that the eigenvectors may have complex components. This situation arises when the characteristic equation has complex (conjugate) roots.

 For a $n \times n$ matrix A the MATLAB command p=poly(A) generates an $n + 1$ element row vector whose elements are the coefficients of the characteristic polynomial of A, the coefficients being ordered in descending powers. The eigenvalues of A are the roots of the polynomial and are generated using the command roots(p). The command

```
[M,S]=eig(A)
```

generates the normalized eigenvectors of A as the columns of the matrix M and its corresponding eigenvalues as the diagonal elements of the diagonal matrix S (M and S are called respectively the modal and spectral matrices of A). In the absence of the left-hand arguments, the command eig(A) by itself simply generates the eigenvalues of A.

For the matrix A of Example 5.42 the commands

```
A=[1 1 -2; -1 2 1; 0 1 -1];
[M,S]=eig(A)
```

generate the output

$$M = \begin{matrix} 0.3015 & -0.8018 & 0.7071 \\ 0.9045 & -0.5345 & 0.0000 \\ 0.3015 & -0.2673 & 0.7071 \end{matrix} \qquad S = \begin{matrix} 2.0000 & 0 & 0 \\ 0 & 1.0000 & 0 \\ 0 & 0 & -1.0000 \end{matrix}$$

These concur with our calculated answers, with $\beta_1 = 0.3015$, $\beta_2 = -0.2673$ and $\beta_3 = 0.7071$.

Using the Symbolic Math Toolbox in MATLAB the matrix **A** may be converted from numeric into symbolic form using the command A=sym(A). Then its symbolic eigenvalues and eigenvectors are generated using the sequence of commands

```
A=[1 1 -2; -1 2 1; 0 1 -1];
A=sym(A);
[M, S]=eig(A)
```

as

```
M=[3, 1, 1]      S=[1, 0, 0]
  [2, 3, 0]        [0, 2, 0]
  [1, 1, 1]        [0, 0, -1]
```

In MAPLE eigenvectors(A); produces the corresponding results using the linalg package.

5.7.3 Exercises

Check your answers using MATLAB or MAPLE.

96 Obtain the characteristic polynomials of the matrices

(a) $\begin{bmatrix} 2 & -1 \\ -1 & 2 \end{bmatrix}$ (b) $\begin{bmatrix} 2 & 1 \\ 1 & 1 \end{bmatrix}$

(c) $\begin{bmatrix} 1 & 2 & 3 \\ 0 & 2 & 3 \\ 0 & 0 & 3 \end{bmatrix}$ (d) $\begin{bmatrix} 1 & 2 & 0 \\ 0 & 2 & 2 \\ 0 & 1 & 3 \end{bmatrix}$

(e) $\begin{bmatrix} 3 & 2 & 1 \\ 4 & 5 & -1 \\ 2 & 3 & 4 \end{bmatrix}$ (f) $\begin{bmatrix} 2 & 1 \\ -1 & a \end{bmatrix}$

and hence evaluate the eigenvalues of the matrices.

97 Find the eigenvalues and corresponding eigenvectors of the matrices

(a) $\begin{bmatrix} 1 & 1 \\ 1 & 1 \end{bmatrix}$ (b) $\begin{bmatrix} 1 & 2 \\ 3 & 2 \end{bmatrix}$

(c) $\begin{bmatrix} 1 & 0 & -4 \\ 0 & 5 & 4 \\ -4 & 4 & 3 \end{bmatrix}$ (d) $\begin{bmatrix} 1 & 1 & 2 \\ 0 & 2 & 2 \\ -1 & 1 & 3 \end{bmatrix}$

(e) $\begin{bmatrix} 5 & 0 & 6 \\ 0 & 11 & 6 \\ 6 & 6 & -2 \end{bmatrix}$ (f) $\begin{bmatrix} 1 & -1 & 0 \\ 1 & 2 & 1 \\ -2 & 1 & -1 \end{bmatrix}$

(g) $\begin{bmatrix} 4 & 1 & 1 \\ 2 & 5 & 4 \\ -1 & -1 & 0 \end{bmatrix}$ (h) $\begin{bmatrix} 1 & -4 & -2 \\ 0 & 3 & 1 \\ 1 & 2 & 4 \end{bmatrix}$

5.7.4 Repeated eigenvalues

In the examples considered so far the eigenvalues λ_i ($i = 1, 2, \dots$) of the matrix A have been distinct, and in such cases the corresponding eigenvectors can be found. The matrix A is then said to have a full set of independent eigenvectors. It is clear that the roots of the characteristic equation $c(\lambda)$ may not all be distinct; and when $c(\lambda)$ has $p \leq n$ distinct roots, $c(\lambda)$ may be factorized as

$$c(\lambda) = (\lambda - \lambda_1)^{m_1}(\lambda - \lambda_2)^{m_2} \dots (\lambda - \lambda_p)^{m_p}$$

indicating that the root $\lambda = \lambda_i$, $i = 1, 2, \dots, p$, is a root of order m_i, where the integer m_i is called the **algebraic multiplicity** of the eigenvalue λ_i. Clearly $m_1 + m_2 + \dots + m_p = n$. When a matrix A has repeated eigenvalues, the question arises as to whether it is possible to obtain a full set of independent eigenvectors for A. We first consider some examples to illustrate the situation.

Example 5.44

Determine the eigenvalues and corresponding eigenvectors of the matrices

(a) $A = \begin{bmatrix} 1 & 0 \\ 0 & 1 \end{bmatrix}$ (b) $B = \begin{bmatrix} 1 & 1 \\ 0 & 1 \end{bmatrix}$

Solution

(a) The eigenvalues of A are obtained from

$$0 = |\lambda I - A| = \begin{vmatrix} \lambda - 1 & 0 \\ 0 & \lambda - 1 \end{vmatrix} = (\lambda - 1)^2$$

giving the value 1 repeated twice.

The eigenvectors we calculate from

$$0 = (I - A)\begin{bmatrix} a \\ b \end{bmatrix} = \begin{bmatrix} 0 & 0 \\ 0 & 0 \end{bmatrix}\begin{bmatrix} a \\ b \end{bmatrix}$$

which is clearly satisfied by any values of a and b. Thus taking

$$\begin{bmatrix} a \\ b \end{bmatrix} = a\begin{bmatrix} 1 \\ 0 \end{bmatrix} + b\begin{bmatrix} 0 \\ 1 \end{bmatrix}$$

it can be seen that there are two independent eigenvectors $\begin{bmatrix} 1 \\ 0 \end{bmatrix}$ and $\begin{bmatrix} 0 \\ 1 \end{bmatrix}$. Any linear combination of the two vectors is also an eigenvector. Geometrically this corresponds to the fact that the unit matrix maps *every* vector on to itself.

(b) The eigenvalues of B are obtained from

$$0 = |\lambda I - B| = \begin{vmatrix} \lambda - 1 & -1 \\ 0 & \lambda - 1 \end{vmatrix} = (\lambda - 1)^2$$

giving the value 1 repeated twice.

The eigenvectors we calculate from

$$0 = (\boldsymbol{I} - \boldsymbol{B})\begin{bmatrix} c \\ d \end{bmatrix} = \begin{bmatrix} 0 & -1 \\ 0 & 0 \end{bmatrix}\begin{bmatrix} c \\ d \end{bmatrix} = \begin{bmatrix} -d \\ 0 \end{bmatrix}$$

Thus $d = 0$ and there is *only one* eigenvector $\begin{bmatrix} 1 \\ 0 \end{bmatrix}$ and, of course, any multiple of this vector.

We note from Example 5.44 that the evaluation of eigenvectors is much more complicated when there are multiple eigenvalues. The idea of rank, introduced in Section 5.6, is required to sort out the complications but the details are left to the companion text *Advanced Modern Engineering Mathematics*.

The following two 3×3 examples illustrate similar points.

Example 5.45 Determine the eigenvalues and corresponding eigenvectors of the matrix

$$\boldsymbol{A} = \begin{bmatrix} 3 & -3 & 2 \\ -1 & 5 & -2 \\ -1 & 3 & 0 \end{bmatrix}$$

Solution We find the eigenvalues from

$$\begin{vmatrix} 3 - \lambda & -3 & 2 \\ -1 & 5 - \lambda & -2 \\ -1 & 3 & -\lambda \end{vmatrix} = 0$$

as $\lambda_1 = 4$, $\lambda_2 = \lambda_3 = 2$.

The eigenvectors are obtained from

$$(\boldsymbol{A} - \lambda \boldsymbol{I})e_i = 0 \tag{5.39}$$

and when $\lambda = \lambda_1 = 4$, we obtain from (5.39)

$$e_1 = [1 \quad -1 \quad -1]^{\mathrm{T}}$$

When $\lambda = \lambda_2 = \lambda_3 = 2$, (5.39) becomes

$$\begin{bmatrix} 1 & -3 & 2 \\ -1 & 3 & -2 \\ -1 & 3 & -2 \end{bmatrix}\begin{bmatrix} e_{21} \\ e_{22} \\ e_{23} \end{bmatrix} = 0$$

so that the corresponding eigenvector is obtained from the single equation

$$e_{21} - 3e_{22} + 2e_{23} = 0 \tag{5.40}$$

Clearly we are free to choose any two of the components e_{21}, e_{22} or e_{23} at will, with the remaining one determined by (5.40). Suppose we set $e_{22} = \alpha$ and $e_{23} = \beta$; then (5.41) means that $e_{21} = 3\alpha - 2\beta$, and thus

$$e_2 = [3\alpha - 2\beta \quad \alpha \quad \beta]^T$$

$$= \alpha \begin{bmatrix} 3 \\ 1 \\ 0 \end{bmatrix} + \beta \begin{bmatrix} -2 \\ 0 \\ 1 \end{bmatrix} \tag{5.41}$$

Now $\lambda = 2$ is an eigenvalue of multiplicity 2, and we seek, if possible, two independent eigenvectors defined by (5.41). Setting $\alpha = 1$ and $\beta = 0$ yields

$$e_2 = [3 \quad 1 \quad 0]^T$$

and setting $\alpha = 0$ and $\beta = 1$ gives a second vector

$$e_3 = [-2 \quad 0 \quad 1]^T$$

These two vectors are independent and of the form defined by (5.41), and it is clear that many other choices are possible. However, any other choices of the form (5.41) will be linear combinations of e_2 and e_3 as chosen above. For example, $e = [1 \quad 1 \quad 1]$ satisfies (5.41), but $e = e_2 + e_3$.

In this example, although there was a repeated eigenvalue of algebraic multiplicity 2, it was possible to construct two independent eigenvectors corresponding to this eigenvalue. Thus the matrix **A** has three and only three independent eigenvectors.

The MATLAB commands for Example 5.45

```
A=[3 -3 2; -1 5 -2; -1 3 0];
[M, S]=eig(A)
```

generate

```
       0.5774   -0.5774   -0.7513        4.0000    0        0
M=-0.5774   -0.5774    0.1735     S=0         2.0000    0
    -0.5774   -0.5774    0.6361        0         0        2.0000
```

Clearly the first column of **M** (corresponding to the eighenvalue $\lambda_1 = 4$) is a scalar multiple of e_1. The second and third columns of **M** (corresponding to the repeated eigenvalue $\lambda_2 = \lambda_3 = 2$) are not scalar multiples of e_2 and e_3. However, both satisfy (5.39) and are equally acceptable as a pair of linearly independent eigenvectors corresponding to the repeated eigenvalue. It is left as an exercise to show that both are linear combinations of e_2 and e_3.

Check that in symbolic form the commands

```
A=sym(A);
[M, S]=eig(A)
```

generate

```
M=[-1,  3,  -2]     S=[4,  0,  0]
   [1,  1,  0]        [0,  2,  0]
   [1,  0,  1]        [0,  0,  2]
```

In the linalg package of MAPLE, `eigenvectors(A);` produces the corresponding results.

Example 5.46 Determine the eigenvalues and corresponding eigenvectors of the matrix

$$A = \begin{bmatrix} 1 & 2 & 2 \\ 0 & 2 & 1 \\ -1 & 2 & 2 \end{bmatrix}$$

Solution Solving $|A - \lambda I| = 0$ gives the eigenvalues as $\lambda_1 = \lambda_2 = 2$, $\lambda_3 = 1$. The eigenvector corresponding to the non-repeated or simple eigenvalue $\lambda_3 = 1$ is easily found as

$$e_3 = [1 \quad 1 \quad -1]^T$$

When $\lambda = \lambda_1 = \lambda_2 = 2$, the corresponding eigenvector is given by

$$(A - 2I)e_1 = 0$$

that is, as the solution of

$$-e_{11} + 2e_{12} + 2e_{13} = 0 \tag{i}$$

$$e_{13} = 0 \tag{ii}$$

$$-e_{11} + 2e_{12} \qquad = 0 \tag{iii}$$

From (ii) we have $e_{13} = 0$, and from both (i) and (iii) it follows that $e_{11} = 2e_{12}$. We deduce that there is only one independent eigenvector corresponding to the repeated eigenvalue $\lambda = 2$, namely

$$e_1 = [2 \quad 1 \quad 0]^T$$

and in this case the matrix A does not possess a full set of independent eigenvectors.

We see from Examples 5.44–5.46 that if an $n \times n$ matrix A has repeated eigenvalues then a full set of n independent eigenvectors may or may not exist.

5.7.5 Exercises

 Check your answers using MATLAB or MAPLE whenever possible.

98 Find the eigenvalues and eigenvectors of the matrices

$$\begin{bmatrix} 3 & 0 \\ 0 & 3 \end{bmatrix}, \quad \begin{bmatrix} 2 & 0 \\ 1 & 2 \end{bmatrix}, \quad \begin{bmatrix} 3 & \frac{1}{4} \\ -1 & 2 \end{bmatrix}, \quad \begin{bmatrix} \frac{1}{2} & \frac{1}{4} \\ -1 & -\frac{1}{2} \end{bmatrix}$$

99 Obtain the eigenvalues and corresponding eigenvectors of the matrices

(a) $\begin{bmatrix} 2 & 2 & 1 \\ 1 & 3 & 1 \\ 1 & 2 & 2 \end{bmatrix}$

(b) $\begin{bmatrix} 0 & -2 & -2 \\ -1 & 1 & 2 \\ -1 & -1 & 2 \end{bmatrix}$

(c) $\begin{bmatrix} 4 & 6 & 6 \\ 1 & 3 & 2 \\ -1 & -5 & -2 \end{bmatrix}$

(d) $\begin{bmatrix} 7 & -2 & -4 \\ 3 & 0 & -2 \\ 6 & -2 & -3 \end{bmatrix}$

100 Given that $\lambda = 1$ is a three-times repeated eigenvalue of the matrix

$$A = \begin{bmatrix} -3 & -7 & -5 \\ 2 & 4 & 3 \\ 1 & 2 & 2 \end{bmatrix}$$

determine how many independent eigenvectors correspond to this value of λ. Determine a corresponding set of independent eigenvectors.

101 Given that $\lambda = 1$ is a twice-repeated eigenvalue of the matrix

$$A = \begin{bmatrix} 2 & 1 & -1 \\ -1 & 0 & 1 \\ -1 & -1 & 2 \end{bmatrix}$$

determine a set of independent eigenvectors.

102 Find all the eigenvalues and eigenvectors of the matrix

$$\begin{bmatrix} 1 & 0 & 0 & 2 \\ 0 & 2 & 0 & 0 \\ 0 & 0 & 2 & 0 \\ 2 & 0 & 0 & 1 \end{bmatrix}$$

5.7.6 Some useful properties of eigenvalues

The following basic properties of the eigenvalues $\lambda_1, \lambda_2, \dots, \lambda_n$ of an $n \times n$ matrix A are sometimes useful. The results are readily proved from either the definition of eigenvalues as the values of λ satisfying (5.35), or by comparison of corresponding characteristic polynomials (5.37). Consequently, the proofs are left to Exercise 103.

Property 1

The sum of the eigenvalues of A is

$$\sum_{i=1}^{n} \lambda_i = \text{trace } A = \sum_{i=1}^{n} a_{ii}$$

Property 2

The product of the eigenvalues of **A** is

$$\prod_{i=1}^{n} \lambda_i = \det \mathbf{A}$$

where det **A** denotes the determinant of the matrix **A**.

Property 3

The eigenvalues of the inverse matrix $\mathbf{A}^{-1}$, provided it exists, are

$$\frac{1}{\lambda_1}, \quad \frac{1}{\lambda_2}, \quad \dots, \quad \frac{1}{\lambda_n}$$

Property 4

The eigenvalues of the transposed matrix $\mathbf{A}^{\mathrm{T}}$ are

$$\lambda_1, \quad \lambda_2, \quad \dots, \quad \lambda_n$$

as for the matrix **A**.

Property 5

If k is a scalar then the eigenvalues of $k\mathbf{A}$ are

$$k\lambda_1, \quad k\lambda_2, \quad \dots, \quad k\lambda_n$$

Property 6

If k is a scalar and **I** the $n \times n$ identity (unit) matrix then the eigenvalues of $\mathbf{A} \pm k\mathbf{I}$ are respectively

$$\lambda_1 \pm k, \quad \lambda_2 \pm k, \quad \dots, \quad \lambda_n \pm k$$

Property 7

If k is a positive integer then the eigenvalues of $\mathbf{A}^k$ are

$$\lambda_1^k, \quad \lambda_2^k, \quad \dots, \quad \lambda_n^k$$

Property 8

As a consequence of Properties 5 and 7, any polynomial in A

$$A^m + \alpha_{m-1}A^{m-1} + \ldots + \alpha_1 A + \alpha_0 I$$

has eigenvalues

$$\lambda_i^m + \alpha_{m-1}\lambda_i^{m-1} + \ldots + \alpha_1\lambda_i + \alpha_0 \qquad \text{for} \qquad i = 1, 2, \ldots, n$$

5.7.7 Symmetric matrices

A square matrix A is said to be **symmetric** if $A^T = A$. Such matrices form an important class and arise in a variety of practical situations. Two important results concerning the eigenvalues and eigenvectors of such matrices can be proved:

(i) The eigenvalues of a real symmetric matrix are real.
(ii) For an $n \times n$ real symmetric matrix it is always possible to find n independent eigenvectors $e_1, e_2, \ldots, e_n$ that are mutually orthogonal so that $e_i^T e_j = 0$ for $i \neq j$.

If the orthogonal eigenvectors of a symmetric matrix are normalized as

$$\hat{e}_1, \hat{e}_2, \ldots, \hat{e}_n$$

then the **inner (scalar) product** is

$$\hat{e}_i^T \hat{e}_j = \delta_{ij} \quad (i, j = 1, 2, \ldots, n)$$

where δ_{ij} is the Kronecker delta defined in Section 5.2.1.

The set of normalized eigenvectors of a symmetric matrix therefore form an orthonormal set (that is, they form a mutually orthogonal normalized set of vectors).

Example 5.47

Obtain the eigenvalues and corresponding orthogonal eigenvectors of the symmetric matrix

$$A = \begin{bmatrix} 2 & 2 & 0 \\ 2 & 5 & 0 \\ 0 & 0 & 3 \end{bmatrix}$$

and show that the normalized eigenvectors form an orthonormal set.

Solution

The eigenvalues of A are $\lambda_1 = 6$, $\lambda_2 = 3$ and $\lambda_3 = 1$, with corresponding eigenvectors

$$e_1 = [1 \quad 2 \quad 0]^T, \quad e_2 = [0 \quad 0 \quad 1]^T, \quad e_3 = [-2 \quad 1 \quad 0]^T$$

which in normalized form are

$$\hat{e}_1 = [1 \quad 2 \quad 0]^T/\sqrt{5}, \quad \hat{e}_2 = [0 \quad 0 \quad 1]^T, \quad \hat{e}_3 = [-2 \quad 1 \quad 0]^T/\sqrt{5}$$

Evaluating the inner products, we see that, for example,

$$\hat{e}_1^T \hat{e}_1 = \tfrac{1}{5} + \tfrac{4}{5} + 0 = 1, \quad \hat{e}_1^T \hat{e}_3 = -\tfrac{2}{5} + \tfrac{2}{5} + 0 = 0$$

and that

$$\hat{e}_i^T \hat{e}_j = \delta_{ij} \quad (i, j = 1, 2, 3)$$

confirming that the eigenvectors form an orthonormal set.

5.7.8 Exercises

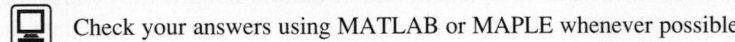 Check your answers using MATLAB or MAPLE whenever possible.

103 Verify Properties 1–8 of Section 5.7.6.

104 Given that the eigenvalues of the matrix

$$A = \begin{bmatrix} 4 & 1 & 1 \\ 2 & 5 & 4 \\ -1 & -1 & 0 \end{bmatrix}$$

are 5, 3 and 1:

(a) confirm Properties 1–4 of Section 5.7.6;

(b) taking $k = 2$, confirm Properties 5–8 of Section 5.7.6.

105 Determine the eigenvalues and corresponding eigenvectors of the symmetric matrix

$$A = \begin{bmatrix} -3 & -3 & -3 \\ -3 & 1 & -1 \\ -3 & -1 & 1 \end{bmatrix}$$

and verify that the eigenvectors are mutually orthogonal.

106 The 3×3 symmetric matrix A has eigenvalues 6, 3 and 2. The eigenvectors corresponding to the eigenvalues 6 and 3 are $[1 \quad 1 \quad 2]^T$ and $[1 \quad 1 \quad -1]^T$ respectively. Find an eigenvector corresponding to the eigenvalue 2.

107 Verify that the matrix

$$A = \begin{bmatrix} -\tfrac{3}{20} & \tfrac{1}{5} \\ \tfrac{1}{5} & \tfrac{3}{20} \end{bmatrix}$$

has eigenvalues $\pm\tfrac{1}{4}$ and corresponding eigenvectors $X = \begin{bmatrix} 2 \\ -1 \end{bmatrix}$ and $Y = \begin{bmatrix} 1 \\ 2 \end{bmatrix}$. What are the eigenvalues of A^n? Show that any vector $Z = \begin{bmatrix} a \\ b \end{bmatrix}$ can be written as $Z = \alpha X + \beta Y$ and hence deduce that $A^n Z \to 0$ as $n \to \infty$.

5.8 Engineering application: spring systems

The vibration of many mechanical systems can be modelled very satisfactorily by spring and damper systems. The shock absorbers and springs of a motor car give one of the simplest practical examples. On a more fundamental level, the vibration of the atoms or molecules of a solid can be modelled by a lattice containing atoms or molecules that interact with each other through spring forces. The model gives a detailed understanding of the structure of the solid and the strength of interactions and has practical applications in such areas as the study of impurities or 'doped' materials in semiconductor physics.

The motion of these systems demands the use of Newton's equations, which in turn require the calculus. We shall look at methods of solution in Chapters 10 and 11. In this case study we shall not consider vibrations but shall restrict our attention to the static situation. This is the first step in the solution of vibrational systems. Even here, we shall see that matrices and vectors allow a systematic approach to the more complicated situation.

5.8.1 A two-particle system

We start with the very simple situation illustrated in Figure 5.22. Two masses are connected by springs of stiffnesses k_1, k_2 and k_3 and of natural lengths l_1, l_2 and l_3 that are fixed to the walls at A and B, with distance AB = L. It is required to calculate the equilibrium values of x_1 and x_2. We use Hooke's law – that force is proportional to extension – to calculate the tension:

$$T_1 = k_1(x_1 - l_1)$$
$$T_2 = k_2(x_2 - x_1 - l_2)$$
$$T_3 = k_3(L - x_2 - l_3)$$

Figure 5.22
Two-particle system.

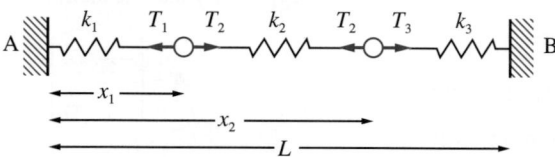

Since the forces are in equilibrium,

$$k_1(x_1 - l_1) = k_2(x_2 - x_1 - l_2)$$
$$k_2(x_2 - x_1 - l_2) = k_3(L - x_2 - l_3)$$

We have two simultaneous equations in the two unknowns, which can be written in matrix form as

$$\begin{bmatrix} k_1 + k_2 & -k_2 \\ -k_2 & k_2 + k_3 \end{bmatrix} \begin{bmatrix} x_1 \\ x_2 \end{bmatrix} = \begin{bmatrix} k_1 l_1 - k_2 l_2 \\ k_2 l_2 - k_3 l_3 + k_3 L \end{bmatrix}$$

It is easy to invert 2×2 matrices, so we can compute the solution as

$$\begin{bmatrix} x_1 \\ x_2 \end{bmatrix} = \frac{1}{(k_1 + k_2)(k_2 + k_3) - k_2^2} \begin{bmatrix} k_2 + k_3 & k_2 \\ k_2 & k_1 + k_2 \end{bmatrix} \begin{bmatrix} k_1 l_1 - k_2 l_2 \\ k_2 l_2 - k_3 l_3 + k_3 L \end{bmatrix}$$

If we take the simplest situation when $k_1 = k_2 = k_3$ and $l_1 = l_2 = l_3$ then we obtain the obvious solution $x_1 = \frac{1}{3}L$, $x_2 = \frac{2}{3}L$.

5.8.2 An *n*-particle system

In the simplest situation, described in Section 5.8.1, matrix notation is convenient but not really necessary. If we try to extend the problem to many particles and many springs

Figure 5.23
n-particle system.

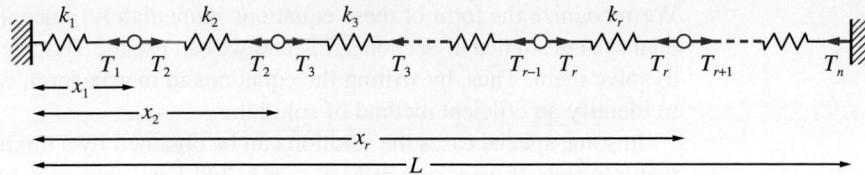

then such notation simplifies the statement of the problem considerably. Consider the problem illustrated in Figure 5.23. From Hooke's law

$$T_1 = k_1(x_1 - l_1)$$

$$T_2 = k_2(x_2 - x_1 - l_2)$$

$$T_3 = k_3(x_3 - x_2 - l_3)$$

$$\vdots$$

$$T_r = k_r(x_r - x_{r-1} - l_r)$$

$$\vdots$$

$$T_n = k_n(L - x_{n-1} - l_n)$$

The equilibrium equations for each 'unit' are

$$k_1(x_1 - l_1) = k_2(x_2 - x_1 - l_2)$$

$$k_2(x_2 - x_1 - l_2) = k_3(x_3 - x_2 - l_3)$$

$$\vdots$$

$$k_r(x_r - x_{r-1} - l_r) = k_{r+1}(x_{r+1} - x_r - l_{r+1})$$

$$\vdots$$

$$k_{n-1}(x_{n-1} - x_{n-2} - l_{n-1}) = k_n(L - x_{n-1} - l_n)$$

In matrix form, these become

$$\begin{bmatrix} k_1 + k_2 & -k_2 & & & & & \\ -k_2 & k_2 + k_3 & -k_3 & & & \Large 0 & \\ & -k_3 & k_3 + k_4 & -k_4 & & & \\ & & \ddots & \ddots & & \ddots & \\ & \Large 0 & & & -k_{n-2} & k_{n-2} + k_{n-1} & -k_{n-1} \\ & & & & & -k_{n-1} & k_{n-1} + k_n \end{bmatrix} \begin{bmatrix} x_1 \\ x_2 \\ \\ \vdots \\ \\ x_{n-1} \end{bmatrix}$$

$$= \begin{bmatrix} k_1 l_1 - k_2 l_2 \\ k_2 l_2 - k_3 l_3 \\ \vdots \\ k_{n-2} l_{n-2} - k_{n-1} l_{n-1} \\ k_{n-1} l_{n-1} - k_n l_n + k_n L \end{bmatrix}$$

We recognize the form of these equations immediately, since they constitute a tridiagonal system studied in Section 5.5.2, and we can use the Thomas algorithm (Figure 5.11) to solve them. Thus, by writing the equations in matrix form, we are immediately able to identify an efficient method of solution.

In some special cases the solution can be obtained by a mixture of insight and physical intuition. If we take $k_1 = k_2 = \ldots = k_n$ and $l_1 = l_2 = \ldots = l_n$ and the couplings are all the same then the equations become

$$
\begin{bmatrix}
2 & -1 & & & & & & \\
-1 & 2 & -1 & & & \text{\huge 0} & & \\
& -1 & 2 & -1 & & & & \\
& & -1 & 2 & & & & \\
& & & & \ddots & \ddots & \ddots & \\
& \text{\huge 0} & & & -1 & 2 & -1 \\
& & & & & -1 & 2
\end{bmatrix}
\begin{bmatrix}
x_1 \\ x_2 \\ \\ \vdots \\ \\ \\ x_{n-1}
\end{bmatrix}
=
\begin{bmatrix}
0 \\ 0 \\ \\ \vdots \\ \\ 0 \\ L
\end{bmatrix}
$$

We should expect all the spacings to be uniform, so we seek a solution $x_1 = \alpha$, $x_2 = 2\alpha$, $x_3 = 3\alpha$, …. The first $n - 2$ equations are satisfied identically, as expected, and the final equation in matrix formulation gives $[-(n-2) + 2(n-1)]\alpha = L$. Thus $\alpha = L/n$, and our intuitive solution is justified.

In a second special case where a simple solution is possible, we assume one of the couplings to be a 'rogue'. We take $k_1 = k_2 = \ldots = k_{r-1} = k_{r+1} = \ldots = k_n = k$, $k_r = k'$ and $l_1 = l_2 = \ldots = l_n = l$. If we divide all the equations in the matrix by k and write $\lambda = k'/k$ then the matrix takes the form

$$
\begin{bmatrix}
2 & -1 & & & & & & & & & \\
-1 & 2 & -1 & & & & \text{\huge 0} & & & & \\
& -1 & 2 & -1 & & & & & & & \\
& & \ddots & \ddots & & \ddots & & & & & \\
& & & -1 & 2 & -1 & & & & & \\
& & & & -1 & 1+\lambda & -\lambda & & & & \\
& & & & & -\lambda & 1+\lambda & -1 & & & \\
& & & & & & -1 & 2 & -1 & & \\
& & & & & & & \ddots & \ddots & \ddots & \\
& & \text{\huge 0} & & & & & & & & -1 \\
& & & & & & & & & -1 & 2
\end{bmatrix}
\begin{bmatrix}
x_1 \\ x_2 \\ \vdots \\ \\ \\ x_{r-1} \\ x_r \\ \vdots \\ \\ \\ x_{n-1}
\end{bmatrix}
=
\begin{bmatrix}
0 \\ 0 \\ \vdots \\ \\ 0 \\ l(1-\lambda) \\ l(\lambda-1) \\ \vdots \\ 0 \\ \vdots \\ L
\end{bmatrix}
$$

A reasonable assumption is that the spacings between 'good' links are all the same. Thus we try a solution of the form

$$x_1 = a, \quad x_2 = 2a, \quad \ldots, \quad x_{r-1} = (r-1)a, \quad x_r = b$$

$$x_{r+1} = b + a, \quad x_{r+2} = b + 2a, \quad \ldots, \quad x_{n-1} = b + (n-1-r)a$$

It can be checked that the matrix equation is satisfied except for the $(r-1)$th, rth and $(n-1)$th rows. These give respectively

$$-\lambda b + a(-\lambda + 1 + \lambda r) = l(1 - \lambda)$$

$$\lambda b + a(\lambda - 1 - \lambda r) = l(\lambda - 1)$$

and

$$b + a(n - r) = L$$

The first two of these are identical, so we have two equations in the two unknowns, a and b, to solve. We obtain

$$a = \frac{L - l(1 - \lambda^{-1})}{n - (1 - \lambda^{-1})}, \qquad b = \frac{rL - (1 - \lambda^{-1})[L - (n - r)l]}{n - (1 - \lambda^{-1})}$$

We note that if $\lambda = 1$ then the solution reduces to the previous one, as expected.

The solution just obtained gives the deformation due to a single rogue coupling. Although this problem is of limited interest, its two- and three-dimensional extensions are of great interest in the theory of crystal lattices. It is possible to determine the deformation due to a single impurity, to compute the effect of two or more impurities and how close they have to be to interact with each other. These are problems with considerable application in materials science.

5.9 Engineering application: steady heat transfer through composite materials

5.9.1 Introduction

In many practical situations heat is transferred through several layers of different materials. Perhaps the simplest example is a double glazing unit, which comprises a layer of glass, a layer of air and another layer of glass. The thermal properties and the thicknesses of the individual layers are known but what is required is the overall thermal properties of the composite unit. How do the overall properties depend on the components? Which parameters are the most important? How sensitive is the overall heat transfer to changes in each of the components?

A second example looks at the thickness of a furnace wall. A furnace wall will comprise three layers: refractory bricks for heat resistance, insulating bricks for heat insulation and steel casing for mechanical protection. Such a furnace is enormously expensive to construct so it is important that the thickness of the wall is minimized subject to acceptable heat losses, working within the serviceable temperatures and known thickness constraints. The basic problem is again to construct a model that will give some idea how heat is transferred through such a composite material.

The basic properties of heat conduction will be discussed, and it will then be seen that matrices give a natural method of solving the theoretical equations of composite layers.

5.9.2 Heat conduction

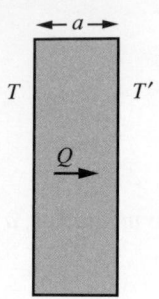

Figure 5.24
Heat transfer
through layers.

In its full generality heat conduction forms a part of partial differential equations (see Chapter 9 of *Advanced Modern Engineering Mathematics* 3rd edition). However, for current purposes a simplified one-dimensional version is sufficient. The theory is based on the well-established **Fourier law**:

> Heat transferred per unit area is proportional to the temperature gradient.

Provided a layer is not too thick and the thermal properties do not vary, then the temperature varies linearly across the solid. If Q is the amount of heat transferred per unit area from left, at temperature T, to right, at temperature T', as shown in Figure 5.24, then this law can be written mathematically

$$Q = -k\frac{T' - T}{a}$$

where k is the proportionality constant, called the thermal conductivity, a is the thickness of the layer and the minus sign is to ensure that heat is transferred from hot to cold.

For the conduction *through* an interface between two solids with good contact, as in the situation of the furnace wall, it is assumed that

(i) The temperatures at each side of the interface are equal.
(ii) The heat transferred out of the left side is equal to the heat transferred into the right side.

With the Fourier law and these interface conditions the multilayer situation can be analysed satisfactorily, provided, of course, the heat flow remains one-dimensional and steady.

5.9.3 The three-layer situation

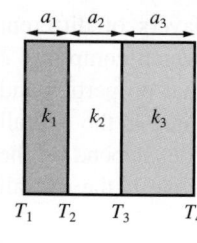

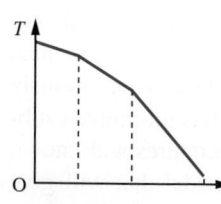

Figure 5.25
Temperature
distribution across
three layers.

Let the three layers have thicknesses a_1, a_2 and a_3 and thermal conductivities k_1, k_2 and k_3, as illustrated in Figure 5.25. At the interfaces the temperatures are taken to be T_1, T_2, T_3 and T_4. The simplest problem to study is to fix the temperatures T_1 and T_4 at the edges and determine how the temperatures T_2 and T_3 depend on the known parameters.

From the specification of the problem the temperatures at the interfaces are specified so it only remains to satisfy the heat transfer condition across the interface.

At the first interface:

$$\frac{k_1}{a_1}(T_2 - T_1) = \frac{k_2}{a_2}(T_3 - T_2)$$

and at the second interface:

$$\frac{k_2}{a_2}(T_3 - T_2) = \frac{k_3}{a_3}(T_4 - T_3)$$

It turns out to be convenient to let $u_1 = \dfrac{a_1}{k_1}$, $u_2 = \dfrac{a_2}{k_2}$ and so on. The equations then become

$$u_2(T_2 - T_1) = u_1(T_3 - T_2)$$

$$u_3(T_3 - T_2) = u_2(T_4 - T_3)$$

or in matrix form

$$\begin{bmatrix} (u_1 + u_2) & -u_1 \\ -u_3 & (u_2 + u_3) \end{bmatrix} \begin{bmatrix} T_2 \\ T_3 \end{bmatrix} = u_2 \begin{bmatrix} T_1 \\ T_4 \end{bmatrix}$$

The determinant of the matrix is easily calculated as $u_2(u_1 + u_2 + u_3)$, which is non-zero so a solution can be computed as

$$\begin{bmatrix} T_2 \\ T_3 \end{bmatrix} = \frac{1}{(u_1 + u_2 + u_3)} \begin{bmatrix} (u_2 + u_3) & u_1 \\ u_3 & (u_1 + u_2) \end{bmatrix} \begin{bmatrix} T_1 \\ T_4 \end{bmatrix}$$

Thus the temperatures T_2 and T_3 are now known, and any required properties can be deduced.

For the furnace problem described in Section 5.9.1 the following data is known:

$$T_1 = 1650 \text{ K} \quad \text{and} \quad T_4 = 300 \text{ K}$$

and

	Maximum working temperature (K)	Thermal conductivity at 100 K (W m⁻¹K⁻¹)	Thermal conductivity at 2000 K (W m⁻¹K⁻¹)
Refractory brick	1700	3.1	6.2
Insulating brick	1400	1.6	3.1
Steel	–	45.2	45.2

It may be noted that the thermal conductivity depends on the temperature but in these calculations it is assumed constant (a more sophisticated analysis is required to take these variations into account). Average values $k_1 = 5$, $k_2 = 2.5$ and $k_3 = 45.2$ are chosen. The required temperatures are evaluated as

$$T_2 = \frac{(0.4a_2 + 0.022a_3)1650 + (0.2a_1)300}{0.2a_1 + 0.4a_2 + 0.022a_3}$$

$$T_3 = \frac{(0.022a_3)1650 + (0.2a_1 + 0.4a_2)300}{0.2a_1 + 0.4a_2 + 0.022a_3}$$

A typical question that would be asked is how to minimize the thickness (or perhaps the cost) subject to appropriate constraints. For instance find

$$\min(a_1 + a_2 + a_3)$$

subject to

$$\frac{k_3}{a_3}(300 - T_3) < 50\,000 \qquad \text{(allowable heat loss at the right-hand boundary)}$$

$$T_2 < 1400 \qquad \text{(below the maximum working temperature)}$$

$$a_1 > 0.1 \qquad \text{(must have a minimum refractory thickness)}$$

The problem is beyond the scope of the present book, but it illustrates the type of question that can be answered.

3 At a point in an elastic continuum the matrix representation of the infinitesimal strain tensor referred to axes $Ox_1x_2x_3$ is

$$E = \begin{bmatrix} 1 & -3 & \sqrt{2} \\ -3 & 1 & -\sqrt{2} \\ \sqrt{2} & -\sqrt{2} & 4 \end{bmatrix}$$

If i, j and k are unit vectors in the direction of the $Ox_1x_2x_3$ coordinate axes, determine the normal strain in the direction of

$$n = \tfrac{1}{2}(i - j + \sqrt{2}k)$$

and the shear strain between the directions n and

$$m = \tfrac{1}{2}(-i + j + \sqrt{2}k)$$

(Note that, using matrix notation, the normal strain is En, and the shear strain between two directions is $m^T En$.)

4 Express the determinant

$$\begin{vmatrix} \alpha & \beta & \gamma \\ \beta\gamma & \gamma\alpha & \alpha\beta \\ -\alpha + \beta + \gamma & \alpha - \beta + \gamma & \alpha + \beta - \gamma \end{vmatrix}$$

as a product of linear factors.

5 Determine the values of θ for which the system of equations

$$x + y + z = 1$$

$$x + 2y + 4z = \theta$$

$$x + 4y + 10z = \theta^2$$

possesses a solution, and for each such value find all solutions.

6 Given

$$A = \begin{bmatrix} 1 & 1 & 1 \\ 2 & 1 & 2 \\ -2 & 1 & -1 \end{bmatrix}$$

evaluate A^2 and A^3. Verify that

$$A^3 - A^2 - 3A + I = 0$$

where I is the unit matrix of order 3. Using this result, or otherwise, find the inverse A^{-1} of A, and hence solve the equations

$$x + y + z = 3$$

$$2x + y + 2z = 7$$

$$-2x + y - z = 6$$

7 (a) If $P = \dfrac{1}{3}\begin{bmatrix} 2 & 1 & 2 \\ -2 & 2 & 1 \\ -1 & -2 & 2 \end{bmatrix}$ write down the

transpose matrix P^T. Calculate PP^T and hence show that $P^T = P^{-1}$. What does this mean about the solution of the matrix equation $Px = b$?

(b) The matrix $F = \begin{bmatrix} I_x & I_{xy} & Q_x \\ I_{xy} & I_y & Q_y \\ Q_x & Q_y & A \end{bmatrix}$ occurs in the

structural analysis of an arch. If

$$B = \begin{bmatrix} 1 & 0 & -Q_x/A \\ 0 & 1 & -Q_y/A \\ 0 & 0 & 1 \end{bmatrix}$$

find $E = BFB^T$ and show that it is a symmetric matrix.

8 (a) If the matrix $A = \begin{bmatrix} 1 & 0 & 0 \\ 1 & -1 & 0 \\ 1 & -2 & 1 \end{bmatrix}$ show that $A^2 = I$

and derive the elements of a square matrix B which satisfies

$$BA = \begin{bmatrix} 1 & 4 & 3 \\ 0 & 2 & 1 \\ -1 & 0 & 0 \end{bmatrix}$$

(b) Find suitable values for k in order that the following system of linear simultaneous equations are consistent:

$$6x + (k - 6)y = 3$$

$$2x + y = 5$$

$$(2k + 1)x + 6y = 1$$

9   Express the system of linear equations

$$3x - y + 4z = 13$$

$$5x + y - 3z = 5$$

$$x - y + z = 3$$

in the form $AX = b$, where A is a 3×3 matrix and X, b are appropriate column matrices.

(a) Find adj A, $|A|$ and A^{-1} and hence solve the system of equations.

(b) Find a matrix Y which satisfies the equation

$$AYA^{-1} = 22A^{-1} + 2A$$

(c) Find a matrix Z which satisfies the equation

$$AZ = 44I_3 - A + AA^T$$

where I_3 is the 3×3 identity matrix.

10 (a) Using the method of Gaussian elimination, find the solution of the equation

$$\begin{bmatrix} 1 & 2 & 4 & 8 \\ 2 & 7 & 13 & 25 \\ -1 & 1 & 5 & 9 \\ 2 & 1 & 11 & 24 \end{bmatrix} \begin{bmatrix} x_1 \\ x_2 \\ x_3 \\ x_4 \end{bmatrix} = \begin{bmatrix} 19 \\ 57 \\ 16 \\ 52 \end{bmatrix}$$

Hence evaluate the determinant of the matrix in the equation.

(b) Solve by the method of Gaussian elimination

$$\begin{bmatrix} 1 & 1 & -1 & 1 \\ 2 & 3 & -3 & 3 \\ -1 & 1 & 0 & 0 \\ 2 & 3 & -1 & 2 \end{bmatrix} \begin{bmatrix} x_1 \\ x_2 \\ x_3 \\ x_4 \end{bmatrix} = \begin{bmatrix} 4 \\ 11 \\ 1 \\ 13 \end{bmatrix}$$

with partial pivoting.

11   Rearrange the equations

$$x_1 - x_2 + 3x_3 = 8$$

$$4x_1 + x_2 - x_3 = 3$$

$$x_1 + 2x_2 + x_3 = 8$$

so that they are diagonally dominant to ensure convergence of the Gauss–Seidel method. Write a MATLAB program to obtain the solution of these equations using this method, starting from

$(0, 0, 0)$. Compare your solution with that from a program when the equations are not rearranged. Use SOR, with $\omega = 1.3$, to solve the equations. Is there any improvement?

12 Find the rank of the matrix

$$\begin{bmatrix} 0 & c & b & a \\ -c & 0 & a & b \\ -b & -a & 0 & c \\ -a & -b & -c & 0 \end{bmatrix}$$

where $b \neq 0$ and $a^2 + c^2 = b^2$.

13 For a given set of discrete data points (x_i, f_i) $(i = 0, 1, 2, \ldots, n)$, show that the coefficients a_k $(k = 0, 1, \ldots, n)$ fitted to the polynomial

$$y(x) = \sum_{k=0}^{n} a_k x^k$$

are given by the solution of the equations written in the matrix form as

$$Aa = f$$

where

$$A = \begin{bmatrix} 1 & x_0 & x_0^2 & \cdots & x_0^n \\ 1 & x_1 & x_1^2 & \cdots & x_1^n \\ \vdots & \vdots & & & \vdots \\ 1 & x_n & x_n^2 & \cdots & x_n^n \end{bmatrix}$$

$$a = [a_0 \quad a_1 \quad \cdots \quad a_n]^T$$

$$f = [f_0 \quad f_1 \quad \cdots \quad f_n]^T$$

(See Question 102 in Exercises 2.9.2 of Chapter 2 for the Lagrange interpolation solution of these equations for the case $n = 3$.)

The following data is taken from the tables of the Airy function $f(x) = Ai(-x)$:

x	1	1.5	2.3	3.0	3.9
$f(x)$	0.535 56	0.464 26	0.026 70	−0.378 81	−0.147 42

Estimate from the polynomial approximation the values of $f(2.0)$ and $f(3.5)$.

14 Data is fitted to a cubic

$$f = ax^3 + bx^2 + cx + d$$

with the slope of the curve given by

$$f' = 3ax^2 + 2bx + c$$

If $f_1 = f(x_1)$, $f_2 = f(x_2)$, $f'_1 = f'(x_1)$ and $f'_2 = f'(x_2)$, show that fitting the data gives the matrix equation for a, b, c and d as

$$\begin{bmatrix} f_1 \\ f_2 \\ f'_1 \\ f'_2 \end{bmatrix} = \begin{bmatrix} x_1^3 & x_1^2 & x_1 & 1 \\ x_2^3 & x_2^2 & x_2 & 1 \\ 3x_1^2 & 2x_1 & 1 & 0 \\ 3x_2^2 & 2x_2 & 1 & 0 \end{bmatrix} \begin{bmatrix} a \\ b \\ c \\ d \end{bmatrix}$$

Use Gaussian elimination to evaluate a, b, c and d. For the case

x	f	f'
0.4	0.327 54	0.511 73
0.8	0.404 90	−0.054 14

evaluate a, b, c and d. Plot the cubic and estimate the maximum value of f in the region $0 < x < 1$. Note that this exercise forms the basis of one of the standard methods for finding the maximum of a function $f(x)$ numerically.

15 The transformation $y = AX$ where

$$A = \frac{1}{9}\begin{bmatrix} 8 & -1 & -4 \\ 4 & 4 & 7 \\ 1 & -8 & 4 \end{bmatrix}$$

$$y = \begin{bmatrix} y_1 \\ y_2 \\ y_3 \end{bmatrix} \quad \text{and} \quad X = \begin{bmatrix} x_1 \\ x_2 \\ x_3 \end{bmatrix}$$

takes a point with coordinates (x_1, x_2, x_3) into a point with coordinates (y_1, y_2, y_3). Show that the coordinates of the points that transform into themselves satisfy the matrix equation $BX = 0$, where $B = A - I$, with I the identity matrix. Find the rank of B and hence deduce that for points which transform into themselves

$$[x_1 \quad x_2 \quad x_3] = \alpha[-3 \quad -1 \quad 1]$$

where α is a parameter.

Find AA^T. What is the inverse of A? If $y_1 = 3$, $y_2 = -1$ and $y_3 = 2$, determine the values of x_1, x_2 and x_3 under this transformation.

16 (a) If

$$A = \begin{bmatrix} 1 & 0 & 1 & 0 \\ 2 & 1 & 2 & 1 \\ 1 & -2 & 2 & -2 \\ 2 & 0 & 3 & 1 \end{bmatrix}$$

verify that

$$A^{-1} = \begin{bmatrix} 6 & -2 & -1 & 0 \\ -5 & 3 & 1 & -1 \\ -5 & 2 & 1 & 0 \\ 3 & -2 & -1 & 1 \end{bmatrix}$$

(b) Use the inverse matrix given in (a) to solve the system of linear equations $AX = b$ in which

$$b^T = [5 \quad -5 \quad -4 \quad 4]$$

17 When a body is deformed in a certain manner, the particle at point x moves to AX, where

$$X = \begin{bmatrix} x \\ y \\ z \end{bmatrix} \quad \text{and} \quad A = \begin{bmatrix} 1 & -2 & 0 \\ -2 & 3 & 0 \\ 0 & 0 & 2 \end{bmatrix}$$

(a) Where would the point $\begin{bmatrix} 2 \\ 1 \\ 1 \end{bmatrix}$ move to?

(b) Find the point from which the particle would move to the point $\begin{bmatrix} 2 \\ 1 \\ 1 \end{bmatrix}$.

18 Find the eigenvalues and the normalized eigenvectors of the matrices

(a) $\begin{bmatrix} 4 & 1 & 1 \\ 2 & 1 & -1 \\ -2 & 2 & 4 \end{bmatrix}$ (b) $\begin{bmatrix} 1 & -1 & 2 \\ -2 & 0 & 5 \\ 6 & -3 & 6 \end{bmatrix}$

(c) $\begin{bmatrix} 5 & -2 & 0 \\ -2 & 6 & 2 \\ 0 & 2 & 7 \end{bmatrix} = \mathbf{C}$

In (c) write the normalized eigenvectors as the columns of the matrix $\mathbf{U}$ and show that $\mathbf{U}^{\mathrm{T}}\mathbf{C}\mathbf{U}$ is a diagonal matrix with the eigenvalues in the diagonal.

19 The vector $[1 \quad 0 \quad 1]^{\mathrm{T}}$ is an eigenvector of the symmetric matrix

$$\begin{bmatrix} 6 & -1 & 3 \\ -1 & 7 & \alpha \\ 3 & \alpha & \beta \end{bmatrix}$$

Find the values of α and β and find the corresponding eigenvalue.

20 Show that the matrix $\begin{bmatrix} -1 & 0 & 2 \\ 0 & 1 & 0 \\ 2 & 0 & -1 \end{bmatrix}$ has eigenvalues 1, 1 and −3. Find the corresponding eigenvectors. Is there a full set of three independent eigenvectors?

21 A colony of insects is observed at regular intervals and comprises four age groups containing n_1, n_2, n_3, n_4 insects in the groups. At the end of an interval, of the n_1 in group 1 some have died and $(1 - \beta_1)n_1$ become the new group 2. Similarly $(1 - \beta_2)n_2$ of group 2 become the new group 3 and $(1 - \beta_3)n_3$ of group 3 become the new group 4. All group 4 die out at the end of the interval. Groups 2, 3 and 4 produce $\alpha_2 n_2$, $\alpha_3 n_3$ and $\alpha_4 n_4$ infant insects that enter group 1. Show that the changes from one interval to the next can be written

$$\begin{bmatrix} n_1 \\ n_2 \\ n_3 \\ n_4 \end{bmatrix}_{\text{new}} = \begin{bmatrix} 0 & \alpha_2 & \alpha_3 & \alpha_4 \\ 1-\beta_1 & 0 & 0 & 0 \\ 0 & 1-\beta_2 & 0 & 0 \\ 0 & 0 & 1-\beta_3 & 0 \end{bmatrix} \begin{bmatrix} n_1 \\ n_2 \\ n_3 \\ n_4 \end{bmatrix}_{\text{old}}$$

Take $\alpha_3 = 0.5$, $\alpha_4 = 0.25$, $\beta_1 = 0.2$, $\beta_2 = 0.25$ and $\beta_3 = 0.5$. Try the values $\alpha_2 = 0.77$, 0.78, 0.79 and check whether the population grows or dies out

over many intervals starting from an initial

$$\text{population} \begin{bmatrix} 100 \\ 90 \\ 50 \\ 30 \end{bmatrix}.$$

Find the eigenvalues in the three cases and check the magnitudes of the eigenvalues. Is there any connection between survival and eigenvalues?

Realistic populations can be modelled using this approach, the matrices are called Leslie matrices.

22 (a) Find the eigenvalues λ_1, λ_2 and the normalized eigenvectors $\mathbf{X}_1$, $\mathbf{X}_2$ of the matrix $\mathbf{A} = \begin{bmatrix} 2 & 1 \\ 1 & 2 \end{bmatrix}$. Check that

$$\mathbf{A} = \lambda_1 \mathbf{X}_1 \mathbf{X}_1^{\mathrm{T}} + \lambda_2 \mathbf{X}_2 \mathbf{X}_2^{\mathrm{T}}$$

(b) Use MATLAB or MAPLE to repeat a similar calculation for the three eigenvalues and normalized eigenvectors of

$$\mathbf{B} = \begin{bmatrix} -1 & 1 & 0 \\ 1 & 0 & 1 \\ 0 & 1 & -2 \end{bmatrix}$$

(*Note*: The process described in this question calculates the spectral decomposition of a symmetric matrix.)

23 In Section 5.7.7 it was stated that a symmetric matrix $\mathbf{A}$ has real eigenvalues λ_1, λ_2, ... , λ_n (written in descending order) and corresponding orthonormal eigenvectors $\mathbf{e}_1$, $\mathbf{e}_2$, ... , $\mathbf{e}_n$, that is $\mathbf{e}_i^{\mathrm{T}}\mathbf{e}_j = \delta_{ij}$. In consequence any vector can be written as

$$\mathbf{X} = c_1 \mathbf{e}_1 + c_2 \mathbf{e}_2 + ... + c_n \mathbf{e}_n$$

Deduce that

$$\frac{\mathbf{X}^{\mathrm{T}}\mathbf{A}\mathbf{X}}{\mathbf{X}^{\mathrm{T}}\mathbf{X}} \leq \lambda_1 \tag{5.42}$$

so that a lower bound of the largest eigenvalue has been found. The left-hand side of (5.42) is called the Rayleigh quotient.

It is known that the matrix $\begin{bmatrix} 0 & 1 & 0 & 0 \\ 1 & 0 & 1 & 0 \\ 0 & 1 & 0 & 1 \\ 0 & 0 & 1 & 0 \end{bmatrix}$ has a largest eigenvalue of $\frac{1}{2}(1 + \sqrt{5})$. Check that the result (5.42) holds for any vector of your choice.

24 A rotation of a set of rectangular cartesian axes $\Phi(Ox_1x_2x_3)$ to a set $\Phi'(Ox'_1x'_2x'_3)$ is described by the matrix $L = (l_{ij})$ $(i, j = 1, 2, 3)$, where l_{ij} is the cosine of the angle between Ox'_i and Ox_j. Show that L is such that

$$LL^T = I$$

and that the coordinates of a point in space referred to the two sets of axes are related by

$$X' = LX$$

where $X' = [x'_1 \ x'_2 \ x'_3]^T$ and $X = [x_1 \ x_2 \ x_3]^T$. Prove that

$$x'^2_1 + x'^2_2 + x'^2_3 = x^2_1 + x^2_2 + x^2_3$$

Describe the relationship between the axes Φ and Φ', given that

$$L = \begin{bmatrix} \frac{1}{2} & 0 & \frac{1}{2}\sqrt{3} \\ 0 & 1 & 0 \\ -\frac{1}{2}\sqrt{3} & 0 & \frac{1}{2} \end{bmatrix}$$

The axes Φ' are now rotated through 45° about Ox'_3 in the sense from Ox'_1 to Ox'_2 to form a new set Φ''. Show that the angle θ between the line OP and the axis Ox''_1, where P is the point with coordinates $(1, 2, -1)$ referred to the original system Φ, is

$$\theta = \cos^{-1}\left(\frac{5\sqrt{3} - 3}{12}\right)$$

25 A car is at rest on horizontal ground as shown in Figure 5.27. The weight W acts through the centre of gravity, and the springs have stiffness constants k_1 and k_2 and natural lengths a_1 and a_2. Show that the height z and the angle θ (assumed too small) satisfy the matrix equation

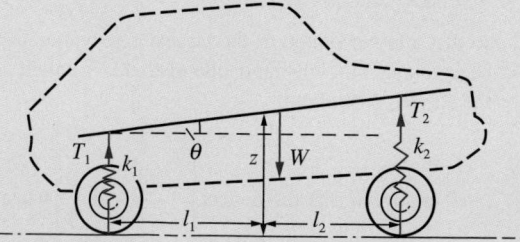

Figure 5.27 Car at rest on horizontal ground.

$$\begin{bmatrix} -W + a_1k_1 + a_2k_2 \\ l_1k_1a_1 - l_2k_2a_2 \end{bmatrix}$$

$$= \begin{bmatrix} k_1 + k_2 & -l_1k_1 + l_2k_2 \\ l_1k_1 - l_2k_2 & -l^2_1k_1 - l^2_2k_2 \end{bmatrix} \begin{bmatrix} z \\ \theta \end{bmatrix}$$

Obtain reasonable values for the various parameters to ensure that $\theta = 0$.

26 In the circuit in Figure 5.28(a) show that the equations can be written

$$\begin{bmatrix} E_1 \\ I_1 \end{bmatrix} = \begin{bmatrix} 1 & Z_1 \\ 0 & 1 \end{bmatrix} \begin{bmatrix} E_2 \\ I_2 \end{bmatrix}$$

and that in Figure 5.28(b) they take the form

$$\begin{bmatrix} E_1 \\ I_1 \end{bmatrix} = \begin{bmatrix} 1 & 0 \\ 1/Z_2 & 1 \end{bmatrix} \begin{bmatrix} E_2 \\ I_2 \end{bmatrix}$$

Dividing the circuit in Figure 5.28(c) into blocks, with the output from one block inputting to the next block, analyse the relation between I_1, E_1 and I_2, E_2.

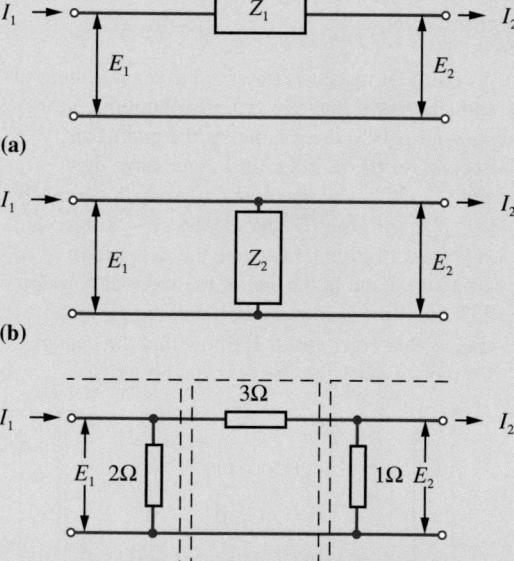

Figure 5.28

6 An Introduction to Discrete Mathematics

6.1 Introduction

The term 'discrete mathematics' is often seen as describing a new and exciting area of mathematics with applications to digital electronics. Virtually everyone these days knows that personal computers operate using digital electronics, and previously analogue systems such as radio and television transmissions are also turning digital. Digital systems are less prone to signal loss through dissipation, attenuation and interference through noise than traditional analogue systems. The ability of digital systems to handle the vast quantity of information required to reproduce high-resolution graphics in a very efficient and cost-effective way is a consequence of this. Another consequence of digitization is greater security due to less penetrable encryption algorithms based on the discrete mathematics of number systems. The present and the future are therefore most definitely digital, and digital systems make use of discrete mathematics. The ironic fact is that discrete mathematics itself is remarkably old. In fact it pre-dates calculus, which might be called 'continuous mathematics'. All counting is discrete mathematics. However, it was only in the nineteenth and twentieth centuries that mathematicians like George Boole (1816–1864) gave a rigorous basis to set theory. The work of Bertrand Russell (1872–1970) and Alfred North Whitehead (1861–1947), and later Kurt Gödel (1906–1978), on logic and the foundation of mathematics, which was to have a great effect on the development of mathematics in the twentieth century, was intimately connected with questions of set theory. This material is now seen to be of great relevance to engineering. Electronic engineers have for a long time required knowledge of Boolean algebra in order to understand the principles of switching circuits. The computer is now very much part of engineering: processes are computer controlled, manufacturing by robots is now commonplace and design is computer aided. Engineers now have a duty to understand how to check the correctness of the algorithms that design, build and repair. In order to do this, branches of discrete mathematics such as propositional logic have to be part of the core curriculum for engineers and not optional extras. This chapter develops the mathematics required in a logical and systematic way, beginning with sets and applications to manufacturing, moving on to switching circuits and applications to electronics, and then to propositional calculus and applications to computing.

6.2 Set theory

The concept of a set is a relatively recent one in that it was born in the past hundred years. In the past few decades it has gained in popularity, and now forms part of school mathematics – this is natural, since the concepts involved, although they may seem unfamiliar initially, are not difficult.

Set theory is concerned with identifying one or more common characteristics among objects. We introduce basic concepts and set operations first, and then examine some applications. The largest areas of application deserve sections to themselves; however, in this section we apply set theory fundamentals to the manufacture and efficient assembly of components.

6.2.1   Definitions and notation

A **set** is a collection of objects, which are called the **elements** or **members** of the set. We shall denote sets by capital letters such as A, S and X, and elements of a set by lower-case letters such as a, s and x. The notation $\in$ is used as follows: if an element a is contained in a set S then we write

$$a \in S$$

which is read 'a belongs to S'. If b does not belong to S then the symbol $\notin$ is used:

$$b \notin S$$

read as 'b does not belong to S'.

A **finite set** is one that contains only a finite number of elements, while an **infinite set** is one consisting of an infinite number of elements. For example,

(i)   the months of the year form a finite set, while
(ii)   the set consisting of all integers is an infinite set.

If we wish to indicate the composition of S then there are two ways of doing this. The first method is suitable only for finite sets, and involves listing the elements of the set between open and closed braces as, for example, in

$$S = \{a, b, c, d, e, f\}$$

which denotes the set S consisting only of the six elements a, b, c, d, e and f.

The second method involves giving a rule by which all elements of the set can be determined. The notation

$$S = \{x : x \text{ has property } P\}$$

will be used to denote the set of all elements x that have the property P. For example,

(i) $S = \{N : N \in Z, N \leqslant 500\}$

is the set of integers that are less than or equal to 500, and

(ii) $S = \{x : x^2 - x - 6 = 0, x \in \mathbb{R}\}$

is the set containing only the two elements 3 and -2.

An example of an infinite set would be

$$S = \{x : 0 \leqslant x \leqslant 1, x \in \mathbb{R}\}$$

which denotes all real numbers that lie in the range 0 to 1, including 0 and 1 themselves.

Very seldom are we satisfied with the type of statement 'S is the set of all fruit' beloved of early school mathematics.

> Two sets A and B are said to be **equal** if every element of each is also an element of the other. For such sets we write $A = B$; otherwise we write $A \neq B$.

For example,

$$A = \{3, 4\} \quad \text{and} \quad B = \{x : x^2 - 7x + 12 = 0\}$$

are two equal sets.

If every element of a set A is also an element of the set B then A is said to be a **subset** of B or, alternatively, B is a **superset** of A. The statement 'A is a subset of B' is written $A \subset B$, while the statement 'B is a superset of A' is written $B \supset A$. The negations of these two statements are written as $A \not\subset B$ and $B \not\supset A$ respectively. Note that if $A \subset B$ and $B \subset A$ then $A = B$, since every element of A is an element of B and vice versa. Thus the definition of a subset does not exclude the possibility of the two sets being equal. If $A \subset B$ and $A \ne B$ then A is said to be a **proper subset** of B. In order to distinguish between a **subset** and a **proper subset**, we shall use the notation $A \subseteq B$ to denote 'A is a subset of B' and $A \subset B$ to denote 'A is a proper subset of B'. For example,

$$A = \{a, b, c\} \quad \text{is a proper subset of} \quad B = \{a, b, c, d, e, f\}$$

A set containing no elements is called the **empty** or **null** set, and is denoted by $\varnothing$. For example,

$$A = \{x : x^2 = 25, \, x \text{ even}\}$$

is an example of a null set, so $A = \varnothing$. It is noted that the empty set may be considered to be a subset of any set.

In most applications it is possible to define sensibly a universal set U that contains all the elements of interest. For example, when dealing with sets of integers, the universal set is the set of all integers, while in two-dimensional geometry the universal set contains all the points in the plane. In such cases we can define the complement of a set A: if all the elements of a set A are removed from the universal set U then the elements that remain in U form the **complement** of A, which is denoted by $\bar{A}$. Thus the sets A and $\bar{A}$ have no elements in common, and we may write

$$\bar{A} = \{x : x \in U, \, x \notin A\}$$

Relations between sets can be illustrated by schematic drawings called **Venn diagrams**, in which each set is represented as the interior of a closed region (normally drawn as a circle) of the plane. It is usual to represent the universal set by a surrounding rectangle. For example, $A \subset B$ and $\bar{A}$ are illustrated by the Venn diagrams of Figures 6.1(a) and (b) respectively.

Figure 6.1

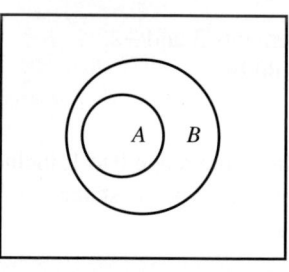

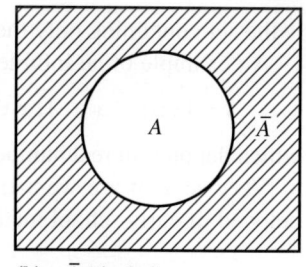

(a) $A \subset B$

(b) $\bar{A}$ (shaded)

6.2.2   Union and intersection

If A and B are two sets, related to the same universal set U, then we can combine A and B to form new sets in the following two different ways.

Figure 6.2

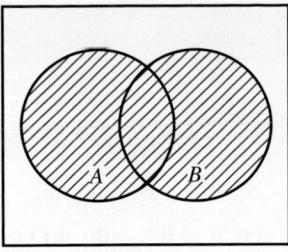

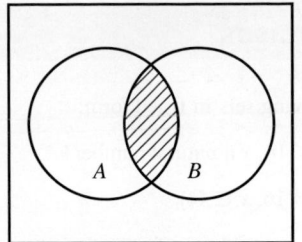

(a) $A \cup B$ (shaded)   **(b)** $A \cap B$ (shaded)

Union

The union of two sets A and B is a third set containing all the elements of A and all the elements of B. It is denoted by $A \cup B$, read as 'A union B'. Thus

$$A \cup B = \{x : x \in A \quad \text{or} \quad x \in B\}$$

where 'or' in this context is used in the inclusive sense: x is an element of A, or B, or both.

Intersection

The intersection of two sets A and B is a third set containing all the elements that belong to both A and B. It is denoted by $A \cap B$, read as 'A intersection B'. Thus

$$A \cap B = \{x : x \in A \quad \text{and} \quad x \in B\}$$

These two definitions are illustrated by the Venn diagrams of Figures 6.2(a) and (b). It is clear from the illustration that union and intersection are commutative, so that

$$A \cup B = B \cup A$$

and

$$A \cap B = B \cap A$$

If the two sets A and B have no elements in common then $A \cap B = \emptyset$: the sets A and B are said to be **disjoint**.

Since union ($\cup$) and intersection ($\cap$) combine two sets from within the same universal set U to form a third set in U, they are called **binary** operations on U. On the other hand, operations on a single set A, such as forming the complement $\bar{A}$, are called **unary** operations on U. It is worthwhile noting at this stage the importance of the words 'or', 'and' and 'not' in the definitions of union, intersection and complementation, and we shall return to this when considering applications in later sections. It is also worth noting that the numerical solutions to the examples and exercises that follow can be checked using MAPLE.

Example 6.1   If $A = \{3, 4, 5, 6\}$ and $B = \{1, 5, 7, 9\}$, determine

(a) $A \cup B$ (b) $A \cap B$

Solution   (a) $A \cup B = \{1, 3, 4, 5, 6, 7, 9\}$

(b) $A \cap B = \{5\}$

6.2.3 Exercises

1 Express the following sets in listed form:

$A = \{x : x < 10, x \text{ a natural number}\}$

$B = \{x : x^2 = 16, x \in \mathbb{R}\}$

$C = \{x : 4 < x < 11, x \text{ an integer}\}$

$D = \{x : 0 < x < 28, x \text{ an integer divisible by } 4\}$

2 For the sets A, B, C and D of Question 1 list the sets $A \cup B, A \cap B, A \cup C, A \cap C, B \cup D, B \cap D$ and $B \cap C$.

3 If $A = \{1, 3, 5, 7, 9\}$, $B = \{2, 4, 6, 8, 10\}$ and $C = \{1, 4, 5, 8, 9\}$, list the sets $A \cup B, A \cap C, A \cap B, B \cup C$ and $B \cap C$.

4 Illustrate the following sets using Venn diagrams:

$\bar{A} \cap \bar{B}, \bar{A} \cup \bar{B}, \overline{A \cap B}, \overline{A \cup B}, A \cap \bar{B}$

5 Given

$A = \{N : N \text{ an integer } 1 \leqslant N \leqslant 10\}$

$B = \{N : N \text{ an even integer}, N \leqslant 20\}$

and

$C = \{N : N = 2^n, n \text{ an integer}, 1 \leqslant n \leqslant 5\}$

determine the following:

(a) $A \cup B$ (b) $A \cap B$

(c) $A \cup C$ (d) $A \cap C$

6 For the sets defined in Question 5 check whether the following statements are true or false:

(a) $A \cap B \supseteq A \cap C$

(b) $A \cup B \supseteq C$

(c) $A \cup B \subseteq C$

7 If the universal set is the set of all integers less than or equal to 32, and A and B are as in Question 5, interpret

(a) $\bar{A}$ (b) $\overline{A \cup B}$ (c) $\bar{A} \cap \bar{B}$

(d) $\overline{A \cap B}$ (e) $\bar{A} \cup \bar{B}$

8 (a) If $A \subset B$ and $A \subset \bar{B}$, show that $A = \varnothing$.

(b) If $A \subset B$ and $C \subset D$, show that $(A \cup C) \subset (B \cup D)$ and illustrate the result using a Venn diagram.

6.2.4 Algebra of sets

In Section 6.2.2 we saw that, given two sets A and B, the operations $\cup$ and $\cap$ could be used to generate two further sets $A \cup B$ and $A \cap B$. These two new sets can then be combined with a third set C, associated with the same universal set U as the sets A and B, to form four further sets

$$C \cup (A \cup B), C \cap (A \cup B), C \cup (A \cap B), C \cap (A \cap B)$$

and the compositions of these sets are clearly indicated by the shaded regions in the Venn diagrams of Figure 6.3.

Clearly, by using various combinations of the binary operations $\cup$ and $\cap$ and the unary operation of complementation ($\bar{}$), many further sets can be generated. In practice, it is useful to have rules that enable us to simplify expressions involving $\cup$, $\cap$ and ($\bar{}$). In this section we develop such rules, which form the basis of the algebra of sets. In the next section we then proceed to show the analogy between this algebra and the algebra of switching circuits, which is widely used by practising engineers.

Given the three sets A, B and C, belonging to the same universal set U, we have already seen that the operations $\cup$ and $\cap$ are commutative, so that we have the following.

Figure 6.3

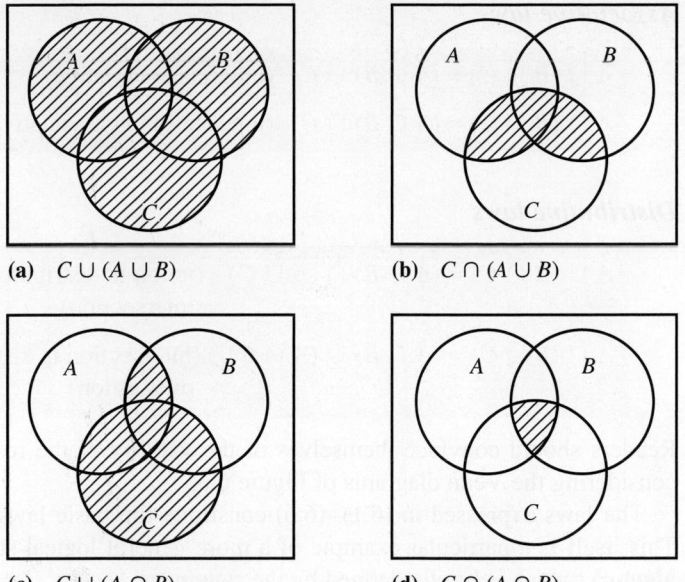

(a) $C \cup (A \cup B)$ **(b)** $C \cap (A \cup B)$

(c) $C \cup (A \cap B)$ **(d)** $C \cap (A \cap B)$

Commutative laws

$A \cup B = B \cup A$ (union is commutative)

$A \cap B = B \cap A$ (intersection is commutative)

(6.1)

It follows directly from the definitions that we have the

Idempotent laws

$A \cup A = A$ (union is idempotent)

$A \cap A = A$ (intersection is idempotent)

(6.2)

Identity laws

$A \cup \varnothing = A$ ($\varnothing$ is an identity relative to union)

$A \cap U = A$ (U is an identity relative to intersection)

(6.3)

Complementary laws

$A \cup \bar{A} = U$

$A \cap \bar{A} = \varnothing$

(6.4)

In addition, it can be shown that the following associative and distributive laws hold:

Associative laws

$$A \cup (B \cup C) = (A \cup B) \cup C \quad \text{(union is associative)}$$
$$A \cap (B \cap C) = (A \cap B) \cap C \quad \text{(intersection is associative)}$$

(6.5)

Distributive laws

$$A \cup (B \cap C) = (A \cup B) \cap (A \cup C) \quad \text{(union is distributive over intersection)}$$
$$A \cap (B \cup C) = (A \cap B) \cup (A \cap C) \quad \text{(intersection is distributive over union)}$$

(6.6)

Readers should convince themselves of the validity of the results (6.5) and (6.6) by considering the Venn diagrams of Figure 6.3.

The laws expressed in (6.1)–(6.6) constitute the basic laws of the algebra of sets. This itself is a particular example of a more general logical structure called **Boolean algebra**, which is briefly defined by the statement

A class of members (equivalent to sets here) together with two binary operations (equivalent to union and intersection) and a unary operation (equivalent to complementation) is a Boolean algebra provided the operations satisfy the equivalent of the commutative laws (6.1), the identity laws (6.3), the complementary laws (4.4) and the distributive laws (6.6).

We note that it is therefore not essential to include the idempotent laws (6.2) and associative laws (6.5) in the basic rules of the algebra of sets, since these are readily deducible from the others. The reader should, at this stage, reflect on and compare the basic rules of the algebra of sets with those associated with conventional numerical algebra in which the binary operations are addition (+) and multiplication (×), and the identity elements are zero (0) and unity (1). It should be noted that in numerical algebra there is no unary operation equivalent to complementation, the idempotency laws do not hold, and that addition is not distributive over multiplication.

While the rules (6.1)–(6.6) are sufficient to enable us to simplify expressions involving $\cup$, $\cap$ and ($\bar{\ }$) the following, known as the De Morgan laws, are also useful in practice.

De Morgan laws

$$\overline{A \cup B} = \bar{A} \cap \bar{B}$$
$$\overline{A \cap B} = \bar{A} \cup \bar{B}$$

(6.7)

The first of these laws states 'the complement of the union of two sets is the intersection of the two complements', while the second states that 'the complement of the intersection of two sets is the union of the two complements'. The validity of the results is illustrated by the Venn diagrams of Figure 6.4, and they are such that they enable us to negate or invert expressions.

If we look at the pairs of laws in each of (6.1)–(6.6) and replace $\cup$ by $\cap$ and interchange $\varnothing$ and U in the first law in each pair then we get the second law in each pair.

Figure 6.4

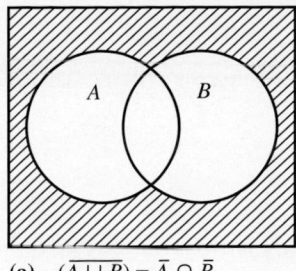

(a) $(\overline{A \cup B}) = \bar{A} \cap \bar{B}$

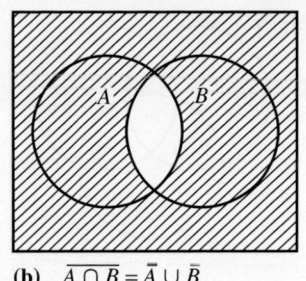

(b) $\overline{A \cap B} = \bar{A} \cup \bar{B}$

Conversely, if we replace $\cap$ and $\cup$ and interchange $\varnothing$ and U in the second law of each pair, we get the first law. This important observation is embedded in the **principle of duality**, which states that if any statement involving $\cup$, $\cap$ and $(\bar{\ })$ is true for all sets then the dual statement (obtained by replacing $\cup$ by $\cap$, $\varnothing$ by U, and U by $\varnothing$) is also true for all sets. This holds for inclusion, with duality existing between $\subset$ and $\supset$.

Example 6.2   Using the laws (6.1)–(6.6), verify the statement

$$(\overline{A \cap B}) \cup (\bar{A} \cap \bar{B} \cap C) \cup A = U$$

stating clearly the law used in each step.

Solution   Starting with the left-hand side, we have

$$
\begin{aligned}
\text{LHS} &= (\overline{A \cap B}) \cup (\overline{\bar{A} \cap \bar{B} \cap C}) \cup A \\
&= (\bar{B} \cup \bar{A}) \cup (\bar{\bar{A}} \cup \bar{\bar{B}} \cup \bar{C}) \cup A && \text{(De Morgan laws)} \\
&= (\bar{B} \cup \bar{A}) \cup (A \cup B \cup \bar{C}) \cup A && (\bar{\bar{A}} = A) \\
&= \bar{A} \cup (A \cup A) \cup (\bar{B} \cup B) \cup \bar{C} && \text{(associative and commutative)} \\
&= (\bar{A} \cup A) \cup (\bar{B} \cup B) \cup \bar{C} && \text{(idempotent)} \\
&= (U \cup U) \cup \bar{C} && \text{(complementary)} \\
&= U \cup \bar{C} && \text{(idempotent)} \\
&= U && \text{(definition of union)} \\
&= \text{RHS}
\end{aligned}
$$

Example 6.3   When carrying out a survey on the popularity of three different brands X, Y and Z of washing powder, 100 users were interviewed, and the results were as follows: 30 used brand X only, 22 used brand Y only, 18 used brand Z only, 8 used brands X and Y, 9 used brands X and Z, 7 used brands Z and Y and 14 used none of the brands.

(a) How many users used brands X, Y and Z?

(b) How many users used brands X and Z but not brand Y?

Figure 6.5

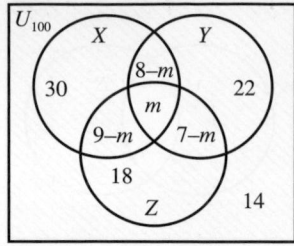

Solution   We can regard the users using brands X, Y and Z as being elements of the sets X, Y and Z respectively. If we denote the number of users using brands X, Y and Z by m then we can illustrate all the given information by the Venn diagram of Figure 6.5. We are then in a position to answer the two given questions.

(a) Since 14 users used none of the three brands, we have that $100 - 14 = 86$ users used one or more of the brands, so

$$\text{number of elements of } X \cup Y \cup Z = 86$$

Thus, from the Venn diagram,

$$30 + (8 - m) + m + (9 - m) + 22 + (7 - m) + 18 = 86$$

$$94 - 2m = 86$$

giving $m = 4$

indicating that 4 users use all three brands X, Y and Z.

(b) The number of users using brands X and Z and not Y is the number of elements in $(X \cap Z) \cap \bar{Y}$, which is the region indicated as having $9 - m$ elements in the Venn diagram. Thus the required answer is $9 - m = 9 - 4 = 5$ users.

Example 6.4   A company manufactures cranes. There are three basic types of crane, labelled A, B and C. Each crane is assembled from a subassembly set $\{a, b, c, d, e, f\}$ as follows:

A is assembled from $\{a, b, c, d\}$
B is assembled from $\{a, c, f\}$
C is assembled from $\{b, d, e\}$

In turn, the subassemblies are manufactured from basic components $\{p, q, r, s, t, u, v, w, x, y\}$ as follows:

a is manufactured from $\{p, q, r, s\}$
b is manufactured from $\{q, r, t, v\}$
c is manufactured from $\{p, r, s, t\}$
d is manufactured from $\{p, w, y\}$
e is manufactured from $\{u, x\}$
f is manufactured from $\{p, r, u, v, x, y\}$

(a) Give the make-up of the following subassemblies:

(i) $a \cup b$,   (ii) $a \cup c \cup f$,   (iii) $d \cup e$

(b) Given that A is made in Newcastle, and B and C are made in Birmingham, what components need to be available on both sites?

Solution   The solution of this problem is a reasonably straightforward application of set theory. From the definitions of *a*, *b*, *c*, *d*, *e* and *f* given, and the fact that the union of two sets contains those items that are either in one or the other or both, the following can be written down:

(i) $a \cup b \quad = \{p, q, r, s, t, v\}$

(ii) $a \cup c \cup f = \{p, q, r, s, t, u, v, x, y\}$

(iii) $d \cup e \quad = \{p, u, w, x, y\}$

This solves (a).

Now, *A* is made from subassemblies $\{a, b, c, d\}$, whereas *B* and *C* require $\{a, b, c, d, e, f\}$ in all of them. Inspection of those components required to make all six subassemblies reveals that subassemblies *a*, *b*, *c* and *d* do not require components *u* and *x*. Therefore only components *u* and *x* need not be made available in both sites. Using the notation of set theory, the solution to (b) is that the components that constitute

$$a \cup b \cup c \cup d$$

have to be available on both sites, or equivalently

$$\overline{a \cup b \cup c \cup d}$$

need only be available at the Birmingham site.

Comment   Of course, Example 6.4, which took much longer to state than to solve, is far too simple to represent a real situation. In a real crane manufacturing company there will be perhaps 20 basic types, and in a car production plant only a few basic types but far more than three hierarchies. However, what this example does is show how set theory can be used for sort purposes. It should also be clear that set theory, being precise, is ideally suited as a framework upon which to build a user-friendly computer program (an expert system) that can answer questions equivalent to part (b) of Example 6.4, when questioned by, for example, a managing director.

6.2.5 Exercises

9 If *A*, *B* and *C* are the sets $\{2, 5, 6, 7, 10\}$, $\{1, 3, 4, 7, 9\}$ and $\{2, 3, 5, 8, 9\}$ respectively, verify that

(a) $A \cap (B \cap C) = (A \cap B) \cap C$

(b) $A \cap (B \cup C) = (A \cap B) \cup (A \cap C)$

10 Using the rules of set algebra, verify the absorption rules

(a) $X \cup (X \cap Y) = X$ (b) $X \cap (X \cup Y) = X$

11 Using the laws of set algebra, simplify the following:

(a) $A \cap (\bar{A} \cup B)$ (b) $(\bar{A} \cup \bar{B}) \cap (A \cap B)$

(c) $(A \cup B) \cap (A \cup \bar{B})$ (d) $(\bar{A} \cap \bar{B}) \cup (A \cup B)$

(e) $(A \cup B \cup C) \cap (A \cup B \cup \bar{C}) \cap (A \cup \bar{B})$

(f) $(A \cup B \cup C) \cap (A \cup (B \cap C))$

(g) $(A \cap B \cap C) \cap (A \cup B \cup \bar{C}) \cup (A \cup \bar{B})$

12 Defining the difference $A - B$ between two sets *A* and *B* belonging to the same universal set *U* to be the set of elements of *A* that are not elements of *B*, that is $A - B = A \cap \bar{B}$, verify the following properties:

(a) $U - A = \bar{A}$ (b) $(A - B) \cup B = A \cup B$

(c) $C \cap (A - B) = (C \cap A) - (C \cap B)$

(d) $(A \cup B) \cup (B - A) = A \cup B$

Illustrate the identities using Venn diagrams.

13 If $n(X)$ denotes the number of elements of a set X, verify the following results, which are used for checking the results of opinion polls:

(a) $n(A \cup B \cup C) = n(A) + n(B) + n(C)$
$$- n(A \cap B) - n((A \cup B) \cap C)$$

(b) $n((A \cup B) \cap C) = n(A \cap C) + n(B \cap C)$
$$- n(A \cap B \cap C)$$

(c) $n(A \cup B \cup C) = n(A) + n(B) + n(C)$
$$- n(A \cap B) - n(B \cap C)$$
$$- n(C \cap A) + n(A \cap B \cap C)$$

Here the sets A, B and C belong to the same universal set U.

14 In carrying out a survey of the efficiency of lights, brakes and steering of motor vehicles, 100 vehicles were found to be defective, and the reports on them were as follows:

no. of vehicles with defective lights	= 35
no. of vehicles with defective brakes	= 40
no. of vehicles with defective steering	= 41
no. of vehicles with defective lights and brakes	= 8
no. of vehicles with defective lights and steering	= 7
no. of vehicles with defective brakes and steering	= 6

Use a Venn diagram to determine

(a) how many vehicles had defective lights, brakes and steering,

(b) how many vehicles had defective lights only.

15 On carrying out a later survey on the efficiency of the lights, brakes and steering on the 100 vehicles of Question 14, the report was as follows:

no. of vehicles with defective lights	= 42
no. of vehicles with defective brakes	= 30
no. of vehicles with defective steering	= 28
no. of vehicles with defective lights and brakes	= 8
no. of vehicles with defective lights and steering	= 10
no. of vehicles with defective brakes and steering	= 5

no. of vehciles with defective lights, brakes and steering	= 3

Use a Venn diagram to determine

(a) how many vehicles were non-defective,

(b) how many vehicles had defective lights only.

16 An analysis of 100 personal injury claims made upon a motor insurance company revealed that loss or injury in respect of an eye, an arm or a leg occurred in 30, 50 and 70 cases respectively. Claims involving the loss or injury to two of these members numbered 42. How many claims involved loss or injury to all three members? (You may assume that one or other of the three members was mentioned in each of the 100 claims.)

17 Bright Homes plc has warehouses in three different locations, L_1, L_2 and L_3, for making replacement windows. There are three different styles, called 'standard', 'executive' and 'superior':

standard units require parts B, C and D;
executive units require parts B, C, D and E;
superior units require parts A, B, C and F.

The parts A, B, C, D, E and F are made from components a, b, c, d, e, f, g, h and i as follows:

A is made from $\{a, b, c\}$
B is made from $\{c, d, e, f\}$
C is made from $\{c, e, f, g, h\}$
D is made from $\{b, e, h\}$
E is made from $\{c, h, i\}$
F is made from $\{b, c, f, i\}$

(a) If the universal set is the set of all components $\{a, b, c, d, e, f, g, h, i\}$, write down the following:

$$\bar{C}, \quad \overline{B \cup C}, \quad \bar{B} \cap \bar{C}, \quad A \cap B \cap D,$$

$$A \cup F, \quad D \cup (E \cap F), \quad (D \cup E) \cap F$$

(b) New parts $B \cup C$, $C \cup E$ and $D \cup E \cup F$ are to be made; what are their components?

(c) Standard units are made at L_1, L_2 and L_3. Executive units are made at L_1 and L_2 only. Superior units are made at L_3 only. What basic components are needed at each location?

6.3 Switching and logic circuits

Throughout engineering, extensive use is made of switches. This is now truer than ever, since personal computers and miniaturized electronic devices have found their way into practically every branch of engineering. A switch is either on or off: denoted by the digits 1 or 0. We shall see that the analysis of circuits containing switches provides a natural vehicle for the use of algebra of sets introduced in the last section.

6.3.1 Switching circuits

Consider a simple 'on–off' switch, which we shall denote by a lower-case letter such as p and illustrate as in Figure 6.6. Such a switch is a two-state device in that it is either **closed** (or 'on') or **open** (or 'off'). We denote a closed contact by 1 and an open contact by 0, so that the variable p can only take one of the two values 1 or 0, with

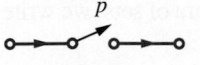

p

Figure 6.6
An 'on–off' switch.

$p = 1$ denoting a closed contact (or 'on' switch), so that a current is able to flow through it

and

$p = 0$ denoting an open contact (or 'off' switch), so that a current cannot flow through it

A **switching circuit** will consist of an energy source or input, for example a battery, and an output, for example a light bulb, together with a number of switches p, q, r and so on. Two switches may be combined together in two basic ways, namely by a series connection or by a parallel connection as illustrated in Figures 6.7 and 6.8 respectively.

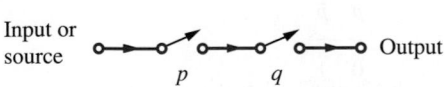

Input or
source ——o Output

p q

Figure 6.7 Two switches in series.

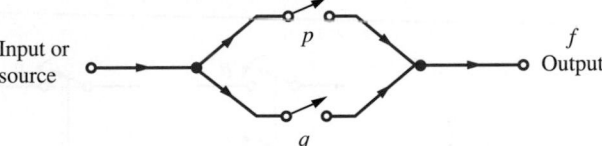

Input or
source

p

q

f
Output

Figure 6.8 Two switches in parallel.

Associated with such a circuit is a **switching function** or **Boolean function** f of the variables contained in the circuit. This is a binary function with

$f = 1$ denoting that the entire circuit is closed

and

$f = 0$ denoting that the entire circuit is open

Clearly the states of f depend upon the states of the individual switches comprising the circuit, so we need to know how to write down an expression for f. For the series circuit of Figure 6.7 there are four possible states:

(a) p open, q open   (b) p open, q closed

(c) p closed, q open   (d) p closed, q closed

	p	q	f
Case (a)	0	0	0
Case (b)	0	1	0
Case (c)	1	0	0
Case (d)	1	1	1

Figure 6.9
Truth table for series connection $f = p \cdot q$.

p	q	f
0	0	0
0	1	1
1	0	1
1	1	1

Figure 6.10
Truth table for parallel connection $f = p + q$.

and it is obvious that current will flow through the circuit from input to output only if both switches p *and* q are closed. In tabular form the state of the circuit may be represented by the **truth table** of Figure 6.9.

Drawing an analogy with use of the word 'and' in the algebra of sets we write

$$f = p \cdot q$$

with $p \cdot q$ being read as 'p and q' (sometimes the dot is omitted and $p \cdot q$ is written simply as pq). Here the 'multiplication' or dot symbol is used in an analogous manner to $\cap$ in the algebra of sets.

When we connect two switches p and q in parallel, as in Figure 6.8, the state of the circuit may be represented by the truth table of Figure 6.10, and it is clear that current will flow through the circuit if either p or q is closed or if they are both closed.

Again, drawing an analogy with the use of the word 'or' in the algebra of sets, we write

$$f = p + q$$

read as 'p or q', with the $+$ symbol used in an analogous manner to $\cup$ in the algebra of sets.

So far we have assumed that the two switches p and q act independently of one another. However, two switches may be connected to one another so that

> they open and close simultaneously

or

> the closing (opening) of one switch will open (close) the other

This is illustrated in Figures 6.11(a) and (b) respectively. We can easily accommodate the situation of Figure 6.11(a) by denoting both switches by the same letter. To accommodate the situation of Figure 6.11(b), we define the **complement switch** $\bar{p}$ (or p') of a switch p to be a switch always in the state opposite to that of p. The action of the complement switch is summarized in the truth table of Figure 6.12.

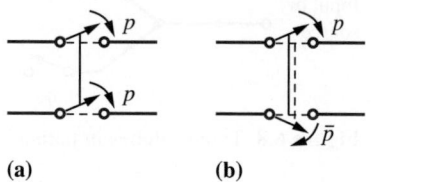

(a)   **(b)**

Figure 6.11 Two switches not acting independently.

p	$\bar{p}$
0	1
1	0

Figure 6.12 Truth table for complementary switch.

6.3.2 Algebra of switching circuits

We can use the operations $\cdot$, $+$ and $(\bar{\ })$ to write down the Boolean function f for complex switching circuits. The states of such circuits may then be determined by constructing truth tables.

Example 6.5

Draw up the truth table that determines the state of the switching circuit given by the Boolean function

$$f = (p \cdot \bar{q}) + (\bar{p} \cdot q)$$

Figure 6.13
Truth table for
$f = (p \cdot \bar{q}) + (\bar{p} \cdot q)$.

p	q	$\bar{p}$	$\bar{q}$	$p \cdot \bar{q}$	$\bar{p} \cdot q$	$(p \cdot \bar{q}) + (\bar{p} \cdot q)$
0	0	1	1	0	0	0
0	1	1	0	0	1	1
1	0	0	1	1	0	1
1	1	0	0	0	0	0

Solution   The required truth table is shown in Figure 6.13. This circuit is interesting in that it is closed (that is, there is a current flow at the output) only if the two switches p and q are in different states. We will see later that it corresponds to the EXCLUSIVE OR function in logic circuits.

By constructing the appropriate truth table, it is readily shown that the operations $\cdot$, $+$ and ($\bar{\ }$) satisfy the following laws, analogous to results (6.1)–(6.6) for the algebra of sets:

Commutative laws

$$p + q = q + p, \quad p \cdot q = q \cdot p$$

Idempotent laws

$$p + p = p, \quad p \cdot p = p$$

Identity laws

$$p + 0 = p \quad (0 \text{ is the identity relative to } +), \quad p + 1 = 1$$
$$p \cdot 1 = p \quad (1 \text{ is the identity relative to } \cdot), \quad p \cdot 0 = 0$$

Complementary laws

$$p + \bar{p} = 1, \quad p \cdot \bar{p} = 0$$

Associative laws

$$p + (q + r) = (p + q) + r, \quad p \cdot (q \cdot r) = (p \cdot q) \cdot r$$

Distributive laws

$$p + (q \cdot r) = (p + q) \cdot (p + r), \quad p \cdot (q + r) = p \cdot q + p \cdot r$$

These rules form the basis of the algebra of switching circuits, and it is clear that it is another example of a Boolean algebra, with $+$ and $\cdot$ being the two binary operations, $\overline{(\ )}$ being the unary operation, and 0 and 1 the identity elements. It follows that the results developed for the algebra of sets carry through to the algebra of switching circuits, with equivalence between $\cup$, $\cap$, $\overline{(\ )}$, $\varnothing$, U and $+$, $\cdot$, $\overline{(\ )}$, 0, 1 respectively. Using these results, complicated switching circuits may be reduced to simpler equivalent circuits.

Example 6.6

Construct truth tables to verify the De Morgan laws for the algebra of switching circuits analogous to (6.7) for the algebra of sets.

Solution

The analogous De Morgan laws for the switching circuits are

$$\overline{p+q} = \bar{p} \cdot \bar{q} \quad \text{and} \quad \overline{p \cdot q} = \bar{p} + \bar{q}$$

the validity of which is verified by the truth tables of Figures 6.14(a) and (b).

Figure 6.14
Truth tables for
De Morgan laws.

p	q	$\bar{p}$	$\bar{q}$	$p+q$	$\overline{p+q}$	$\bar{p} \cdot \bar{q}$
0	0	1	1	0	1	1
0	1	1	0	1	0	0
1	0	0	1	1	0	0
1	1	0	0	1	0	0

(a) $\overline{p+q} = \bar{p} \cdot \bar{q}$

p	q	$\bar{p}$	$\bar{q}$	$p \cdot q$	$\overline{p \cdot q}$	$\bar{p} + \bar{q}$
0	0	1	1	0	1	1
0	1	1	0	0	1	1
1	0	0	1	0	1	1
1	1	0	0	1	0	0

(b) $\overline{p \cdot q} = \bar{p} + \bar{q}$

Example 6.7

Simplify the Boolean function

$$f = p + p \cdot q \cdot r + \bar{p} \cdot \bar{q}$$

stating the law used in each step of the simplification.

Solution $f = p + p \cdot q \cdot r + \bar{p} \cdot \bar{q}$

$\qquad = p \cdot 1 + p \cdot (q \cdot r) + \bar{p} \cdot \bar{q}$ (identity, $p \cdot 1 = p$, and associative)

$\qquad = p \cdot (1 + (q \cdot r)) + \bar{p} \cdot \bar{q}$ (distributive, $p \cdot (1 + (q \cdot r)) = p \cdot 1 + p \cdot (q \cdot r)$)

$\qquad = p \cdot 1 + \bar{p} \cdot \bar{q}$ (identity, $1 + (q \cdot r) = 1$)

$\qquad = p + \bar{p} \cdot \bar{q}$ (identity, $p \cdot 1 = p$)

$\qquad = (p + \bar{p}) \cdot (p + \bar{q})$ (distributive, $p + (\bar{p} \cdot \bar{q}) = (p + \bar{p}) \cdot (p + \bar{q})$)

$\qquad = 1 \cdot (p + \bar{q})$ (complementary, $p + \bar{p} = 1$)

that is,

$\qquad f = p + \bar{q}$ (identity, $1 \cdot (p + \bar{q}) = p + \bar{q}$)

Example 6.8   A machine contains three fuses p, q and r. It is desired to arrange them so that if p blows then the machine stops, but if p does not blow then the machine only stops when both q and r have blown. Derive the required fuse circuit.

Solution   In this case we can regard the fuses as being switches, with '1' representing fuse intact (current flows) and '0' representing the fuse blown (current does not flow). We are then faced with the problem of designing a circuit given a statement of its requirements. To do this, we first convert the specified requirements into logical specification in the form of a truth table. From this, the Boolean function representing the machine is written down. This may then be simplified using the algebraic rules of switching circuits to determine the simplest appropriate circuit.

Denoting the state of the machine by f (that is, $f = 1$ denotes that the machine is operating, and $f = 0$ denotes that it has stopped), the truth table of Figure 6.15 summarizes the state f in relation to the states of the individual fuses. We see from the last two columns that the machine is operating when it is in either of the three states

$\qquad p \cdot q \cdot r$ or $p \cdot q \cdot \bar{r}$ or $p \cdot \bar{q} \cdot r$

Thus it may be represented by the Boolean function

$\qquad f = p \cdot q \cdot r + p \cdot q \cdot \bar{r} + p \cdot \bar{q} \cdot r$

Simplifying this expression gives

$\qquad f = (p \cdot r) \cdot (q + \bar{q}) + p \cdot q \cdot \bar{r}$ (distributive)

$\qquad = p \cdot r + p \cdot q \cdot \bar{r}$ (complementary)

$\qquad = p \cdot (r + q \cdot \bar{r})$ (distributive)

$\qquad = p \cdot ((r + q) \cdot (r + \bar{r}))$ (distributive)

$\qquad = p \cdot (r + q) \cdot 1$ (complementary)

$\qquad = p \cdot (r + q)$ (identity)

Thus a suitable layout of the three fuses is as given in Figure 6.16.

In the case of this simple example we could have readily drawn the required layout from the problem specification. However, it serves to illustrate the procedure that could be adopted for a more complicated problem.

p	q	r	f	State of circuit
1	1	1	1	$p \cdot q \cdot r$
1	1	0	1	$p \cdot q \cdot \bar{r}$
1	0	1	1	$p \cdot \bar{q} \cdot r$
1	0	0	0	$p \cdot \bar{q} \cdot \bar{r}$
0	1	1	0	$\bar{p} \cdot q \cdot r$
0	1	0	0	$\bar{p} \cdot q \cdot \bar{r}$
0	0	1	0	$\bar{p} \cdot \bar{q} \cdot r$
0	0	0	0	$\bar{p} \cdot \bar{q} \cdot \bar{r}$

Figure 6.15

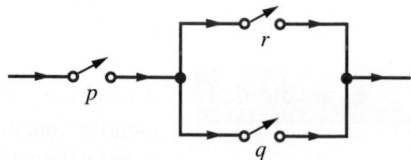

Figure 6.16

Example 6.9

In a large hall there are three electrical switches next to the three doors to operate the central lights. The three switches operate alternatively; that is, each can switch on or switch off the lights. Design a suitable switching circuit.

Solution

The light state f is either '1' (light on) or '0' (light off). Denoting the three switches by p, q and r, the state of f as it relates to the states of the three switches is given in the truth table of Figure 6.17, remembering that operating any switch turns the light off if

Figure 6.17

p	q	r	f	State of circuit
1	1	1	1	$p \cdot q \cdot r$
1	1	0	0	$p \cdot q \cdot \bar{r}$
1	0	1	0	$p \cdot \bar{q} \cdot r$
1	0	0	1	$p \cdot \bar{q} \cdot \bar{r}$
0	1	1	0	$\bar{p} \cdot q \cdot r$
0	1	0	1	$\bar{p} \cdot q \cdot \bar{r}$
0	0	1	1	$\bar{p} \cdot \bar{q} \cdot r$
0	0	0	0	$\bar{p} \cdot \bar{q} \cdot \bar{r}$

it was on and turns the light on if it was off. We arbitrarily set $p = q = r = 1$ and $f = 1$ initially. We see from the last two columns that the light is on ($f = 1$) when the circuit is in either of the four states

$$p \cdot q \cdot r \quad \text{or} \quad p \cdot \bar{q} \cdot \bar{r} \quad \text{or} \quad \bar{p} \cdot q \cdot \bar{r} \quad \text{or} \quad \bar{p} \cdot \bar{q} \cdot r$$

Thus the required circuit is specified by the Boolean function

$$f = p \cdot q \cdot r + p \cdot \bar{q} \cdot \bar{r} + \bar{p} \cdot q \cdot \bar{r} + \bar{p} \cdot \bar{q} \cdot r$$

In this case it is not possible to simplify f any further, and in order to design the corresponding switching circuit we need to use two 1-pole, 2-way switches and one 2-pole, 2-way switch (or intermediate switch), as illustrated in Figure 6.18(a). The four possible combinations leading to 'light on' are shown in Figures 6.18(b), (c), (d) and (e) respectively.

Figure 6.18

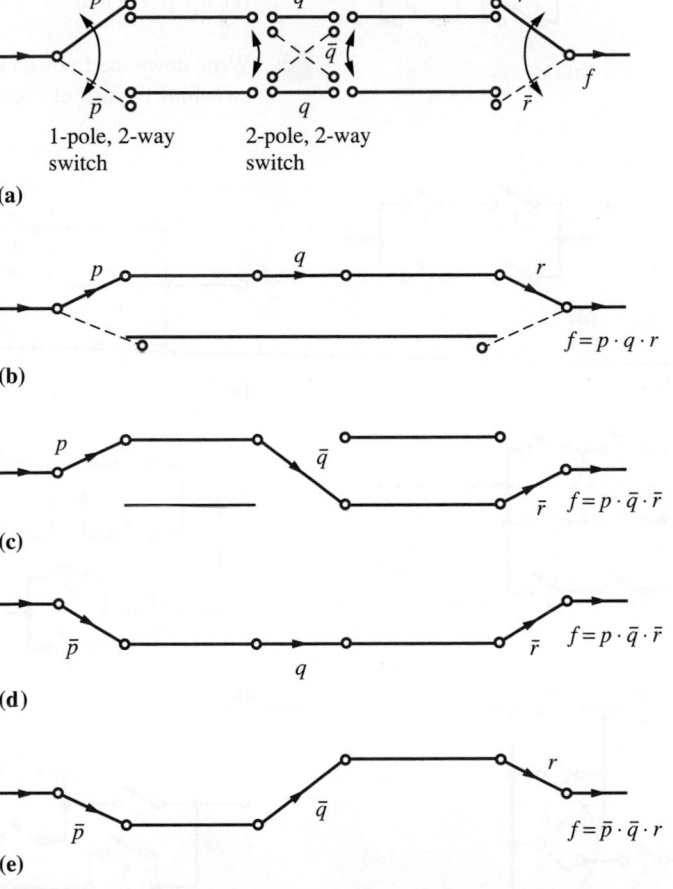

6.3.3 Exercises

18 By setting up truth tables, find the possible values of the following Boolean functions:

(a) $p \cdot (q \cdot p)$ (b) $p + (q + p)$

(c) $(p + q) \cdot (\bar{p} \cdot \bar{q})$

(d) $[(\bar{p} + \bar{q})(\bar{r} + \bar{p})] + (r + p)$

19 Figure 6.19 shows six circuits. Write down a Boolean function that represents each by using truth tables.

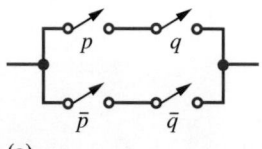

(a)

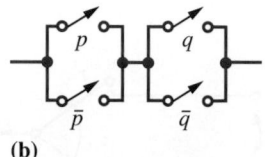

(b)

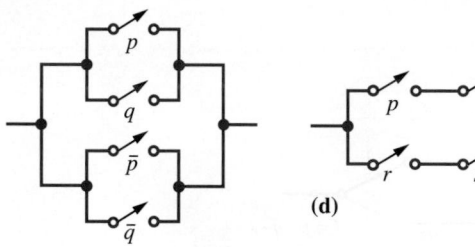

(c)   **(d)**

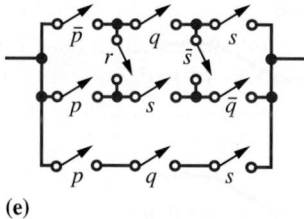

(e)

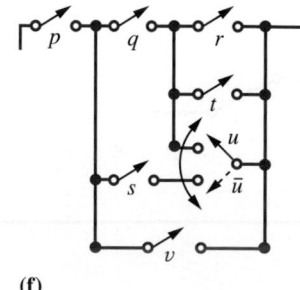

(f)

Figure 6.19

20 Use the De Morgan laws to negate the function

$$f = (p + q) \cdot (\bar{r} \cdot s) \cdot (q + \bar{t})$$

21 Give a truth table for the expression

$$f = \bar{p} \cdot q \cdot \bar{r} + \bar{p} \cdot q \cdot r + p \cdot \bar{q} \cdot \bar{r} + p \cdot q \cdot r$$

22 Simplify the following Boolean functions, stating the law used in each step of the simplification:

(a) $p \cdot (\bar{p} + p \cdot q)$ (b) $r \cdot (\overline{p + \bar{q} \cdot \bar{r}})$

(c) $(\overline{p \cdot \bar{q} + \bar{p} \cdot q})$ (d) $p + q + r + \bar{p} \cdot q$

(e) $(\overline{p \cdot q}) + (\overline{\bar{p} \cdot q \cdot r}) + p$

(f) $q + p \cdot r + p \cdot q + r$

23 Write down the Boolean functions for the switching circuits of Figure 6.20.

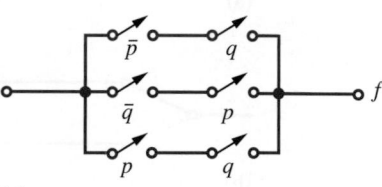

(a)

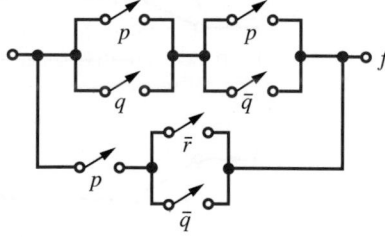

(b)

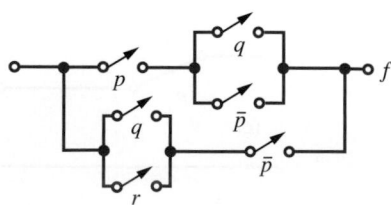

(c)

Figure 6.20

24 Draw the switching circuit corresponding to the following Boolean functions:

(a) $f = (p + q) \cdot r + s \cdot t$ (b) $(p + q) \cdot (r + \bar{p})$

(c) $p \cdot q + \bar{p} \cdot q$ (d) $p \cdot (q + \bar{p}) + (q + r) \cdot \bar{p}$

25 Four engineers J, F, H and D are checking a rocket. Each engineer has a switch that he or she presses in the event of discovering a fault. Show how these must be wired to a warning lamp, in the countdown control room, if the lamp is to light only under the following circumstances:

 (i) D discovers a fault,

 (ii) any two of J, F and H discover a fault.

26 In a public discussion a chairman asks questions of a panel of three. If to a particular question a majority of the panel answer 'yes' then a light will come on, while if to a particular question a majority of the panel answer 'no' then a buzzer will sound. The members of the panel record their answers by means of a two-position switch having position '1' for 'yes' and position '0' for 'no'. Design a suitable circuit for the discussion.

27 Design a switching circuit that can turn a lamp 'on' or 'off' at three different locations independently.

28 Design a switching circuit containing three independent contacts for a machine so that the machine is turned on when any two, but not three, of the contacts are closed.

29 The operation of a machine is monitored on a set of three lamps A, B and C, each of which at any given instant is either 'on' or 'off'. Faulty operation is indicated by each of the following conditions:

(a) when both A and B are off,

(b) when all lamps are on,

(c) when B is on and either A is off or C is on.

Simplify these conditions by describing as concisely as possible the state of the lamps that indicates faulty operation.

6.3.4 Logic circuits

As indicated in Section 6.3.1, a switch is a two-state device, and the algebra of switching circuits developed in Section 6.3.2 is equally applicable to systems involving other such devices. In this section we consider how the algebra may be applied to logic circuit design.

In logic circuit design the two states denoted by '1' and '0' usually denote HIGH and LOW voltage respectively (positive logic), although the opposite convention can be used (negative logic). The basic building blocks of logic circuits are called **logic gates**. These represent various standard Boolean functions. First let us consider the logic gates corresponding to the binary operation of 'and' and 'or' and the unary operation of complementation. We shall illustrate this using two inputs, although in practice more can be used.

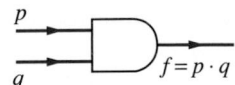

Figure 6.21
AND gate.

AND gate

The AND gate is commonly represented diagrammatically as in Figure 6.21, and corresponds to the Boolean function

$$f = p \cdot q \quad \text{(read '}p \text{ and } q\text{')}$$

$f = 1$ (output HIGH) if and only if the inputs p and q are simultaneously in state 1 (both inputs HIGH). For all other input combinations f will be zero. The corresponding truth table is as in Figure 6.9, with 1 denoting HIGH voltage and 0 denoting LOW voltage.

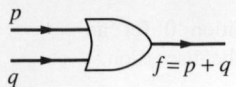

Figure 6.22
OR gate.

OR gate

The OR gate is represented diagrammatically as in Figure 6.22, and corresponds to the Boolean function

$$f = p + q \quad \text{(read 'p or q')}$$

In this case $f = 1$ (HIGH output) if either p or q or both are in state 1 (at least one input HIGH). $f = 0$ (LOW output) if and only if inputs are simultaneously 0. The corresponding truth table is as in Figure 6.10.

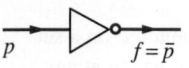

Figure 6.23
NOT gate.

NOT gate

The NOT gate is represented diagrammatically as in Figure 6.23, and corresponds to the Boolean function

$$f = \bar{p} \quad \text{(read 'not p')}$$

When the input is in state 1 (HIGH), the output is in state 0 (LOW) and vice versa. The corresponding truth table is as in Figure 6.12.

With these interpretations of $\cdot$, $+$, $(\bar{})$, 0 and 1, the rules developed in Section 6.3.2 for the algebra of switching circuits are applicable to the analysis and design of logic circuits.

Example 6.10   Build a logic circuit to represent the Boolean function

$$f = \bar{p} \cdot q + p$$

Solution   We first use a NOT gate to obtain $\bar{p}$ then an AND gate to generate $\bar{p} \cdot q$, and finally an OR gate to represent f. The resulting logic circuit is shown in Figure 6.24.

Figure 6.24
Logic circuit
$f = \bar{p} \cdot q + p$.

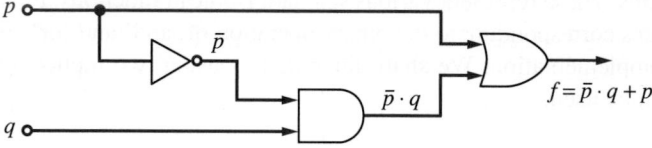

Example 6.11   Build a logic circuit to represent the Boolean function

$$f = (p + \bar{q}) \cdot (r + s \cdot q)$$

Solution   Adopting a similar procedure to the previous example leads to the logic circuit of Figure 6.25.

Figure 6.25
Logic circuit
$f = (p + \bar{q}) \cdot (r + s \cdot q)$.

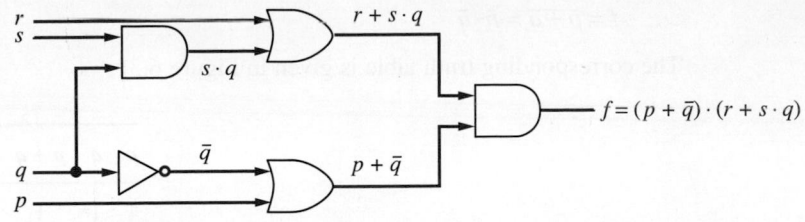

So far we have considered the three logic gates AND, OR and NOT and indicated how these can be used to build a logic circuit representative of a given Boolean function. We now introduce two further gates, which are invaluable in practice and are frequently used.

NAND gate

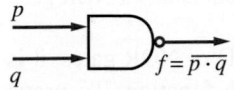

Figure 6.26
NAND gate $f = \overline{p \cdot q}$.

The NAND (or 'NOT AND') gate is represented diagrammatically in Figure 6.26, and corresponds to the function

$$f = \overline{p \cdot q}$$

The small circle on the output line of the gate symbol indicates negation or NOT. Thus the gate negates the AND gate, and is equivalent to the logic circuit of Figure 6.27.

The corresponding truth table is given in Figure 6.28.

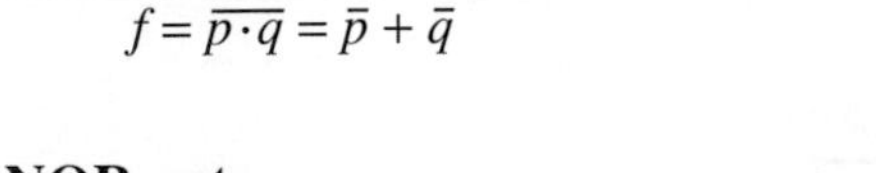

p	q	$p \cdot q$	f
1	1	1	0
1	0	0	1
0	1	0	1
0	0	0	1

Figure 6.27 Equivalent circuit to NAND gate.

Figure 6.28 Truth table for NAND gate.

Note that, using De Morgan laws, the Boolean function for the NAND gate may also be written as

$$f = \overline{p \cdot q} = \bar{p} + \bar{q}$$

NOR gate

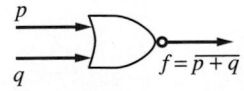

Figure 6.29
NOR gate $f = \overline{p + q}$.

The NOR (or 'NOT OR') gate is represented diagrammatically as in Figure 6.29, and corresponds to the Boolean function

$$f = \overline{p + q}$$

Again we have equivalence with the logic circuit of Figure 6.30, and (using the De Morgan laws) with the Boolean function

$$f = \overline{p+q} = \bar{p} \cdot \bar{q}$$

The corresponding truth table is given in Figure 6.31.

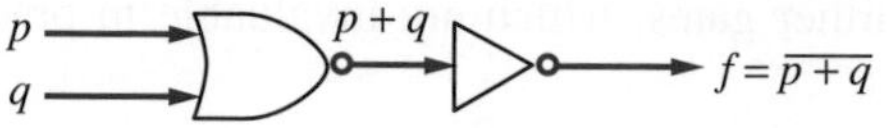

p	q	$p+q$	f
1	1	1	0
1	0	1	0
0	1	1	0
0	0	0	1

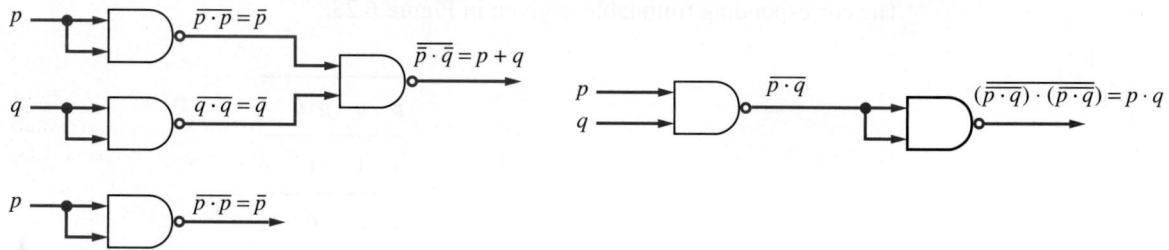

Figure 6.30 Equivalent circuit to NOR gate.

Figure 6.31 Truth table for NOR gate.

It is of interest to recognize that, using either one of the NAND or NOR gates, it is possible to build a logic circuit to represent any given Boolean function. To prove this, we have to show that, using either gate, we can implement the three basic Boolean functions $p + q$, $p \cdot q$ and $\bar{p}$. This is illustrated in Figure 6.32 for the NAND gate; the illustration for the NOR gate is left as an exercise for the reader.

Figure 6.32 Basic Boolean functions using NAND gates.

Example 6.12 Using only NOR gates, build a logic circuit to represent the Boolean function

$$f = \bar{p} \cdot q + p \cdot \bar{q}$$

Solution The required logic circuit is illustrated in Figure 6.33.

Figure 6.33

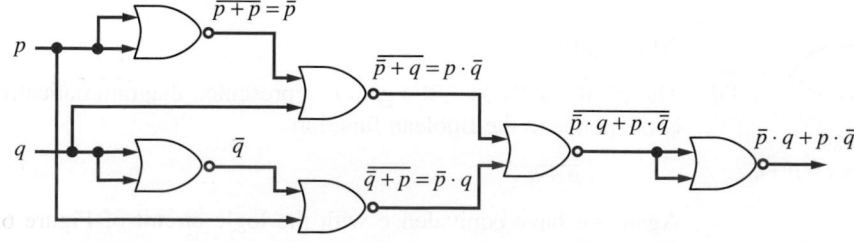

We note that the Boolean function considered in Example 6.12 is the same as that considered in Example 6.5, where its truth table was constructed, indicating that the output is in state 1 only if the two inputs are in different states. This leads us to defining a further logic gate used in practice.

EXCLUSIVE OR gate

The EXCLUSIVE OR gate is represented diagrammatically as in Figure 6.34, and corresponds to the Boolean function

$$f = \bar{p} \cdot q + p \cdot \bar{q}$$

Figure 6.34
EXCLUSIVE OR
gate.

$$p$$
$$q$$
$$f = \bar{p} \cdot q + p \cdot \bar{q}$$

As indicated above, $f = 1$ (output HIGH) only if the inputs p and q are in different states; that is, either p or q is in state 1 but *not both*. It therefore corresponds to the everyday exclusive usage of the word 'OR' where it is taken to mean 'one or the other but not both'. On the other hand, the OR gate introduced earlier is used in the sense 'one or the other or both', and could more precisely be called the INCLUSIVE OR gate.

Although present technology is such that a logic circuit consisting of thousands of logic gates may be incorporated in a single silicon chip, the design of smaller equivalent logic circuits is still an important problem. As for switching circuits, simplification of a Boolean function representation of a logic circuit may be carried out using the algebraic rules given in Section 6.3.2. More systematic methods are available for carrying out such simplification. For Boolean expressions containing not more than six variables the pictorial approach of constructing Karnaugh maps is widely used by engineers. An alternative algebraic approach, which is well suited for computer implementation, is to use the Quine–McCluskey algorithm. For details of such methods the reader is referred to specialist texts on the subject.

6.3.5   Exercises

30   Write down the Boolean function for the logic blocks of Figure 6.35. Simplify the functions as far as possible and draw the equivalent logic block.

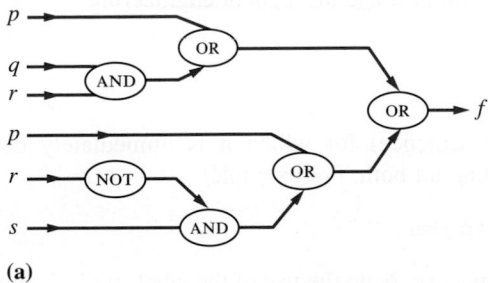

(a)

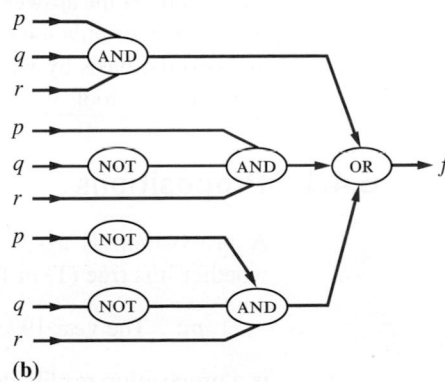

(b)

Figure 6.35

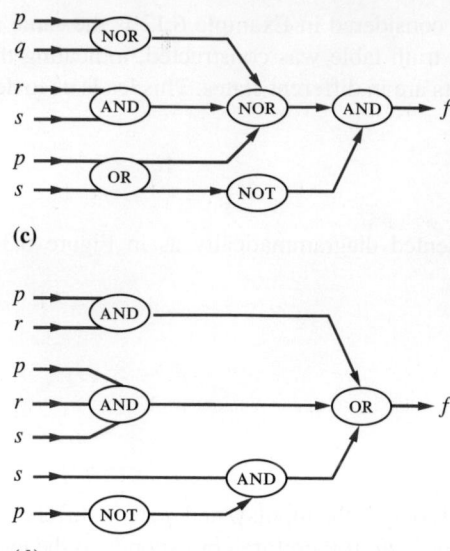

(c)

(d)

Figure 6.35 *continued*

31 Simplify the following Boolean functions and sketch the logic block corresponding to both the given and simplified functions:

(a) $(\bar{p}\cdot q + p\cdot \bar{q})\cdot(\bar{p}+\bar{q})\cdot(p+q)$

(b) $\bar{r}\cdot\bar{p}\cdot\bar{q}+\bar{r}\cdot\bar{p}\cdot q+r\cdot\bar{p}\cdot\bar{q}$

(c) $\bar{p}\cdot\bar{q}+r\cdot\bar{p}\cdot s+\bar{p}\cdot\bar{q}\cdot s$

(d) $(p+q)\cdot(p+r)+r\cdot(p+q\cdot r)$

(e) $(\bar{p}+\bar{q})\cdot(\bar{p}+q)\cdot(p+q)$

6.4 ## Propositional logic and methods of proof

In the last section we dealt with switches that are either off or on. These lend themselves naturally to the application of set algebra. On the other hand, everyday use of English contains many statements that are neither obviously true nor false: for example 'Chilly for the time of year, isn't it?' There are, however, some statements that are immediately either true or false: for example, 'In 2004 the Summer Olympics were held in Athens, Greece' (true) or 'All children watch too much television' (false). Propositional logic can be used to analyse, simplify and establish the equivalence of statements. Applications of propositional logic include the efficient operation of computer-based expert systems, where the user may phrase questions differently or answer in different ways, and yet the answers are logically equivalent. Propositional logic leads naturally to the precise formulation of the proof of statements that, though important in themselves, are also the basis by which computer programs can be made more efficient. Thus we shall develop tools with a vast potential for use throughout engineering.

6.4.1 Propositions

A proposition is a statement (or sentence) for which it is immediately decidable whether it is true (T) or false (F), but not both. For example

p_1:   The year 1973 was a leap year

is a proposition readily decidable as false. Note the use of the label 'p_1: ...', so that the overall statement is read 'p_1 is the statement: "The year 1973 was a leap year"'.

Since when considering propositions we are concerned with statements that are decidable as true or false, we obviously exclude all questions and commands. Also excluded are assertions that involve subjective value judgements or opinions such as

r: The Director of the company is overpaid

Statements such as

m: He was Prime Minister of England

n: The number $x + 3$ is divisible by 3

that involve pronouns (he, she, and so on) or a mathematical variable, are not readily decidable as true or false, and are therefore not propositions. However, as soon as the pronoun or variable is specified (or quantified in some way) then the statements are decidable as true or false and become propositions. Statements such as m and n are examples of **predicates**.

Given any statement p, there is always an associated statement called the **negation** of p. We denote this by $\tilde{p}$, read as 'not p'. (The notations $\neg p$ and $\sim p$ are also sometimes used.) For example, the negation of the proposition p_1 above is the proposition

$\tilde{p}_1$:   The year 1973 was not a leap year

which is decidable as true, the opposite truth value to p_1. In general the negation $\tilde{p}$ of a statement p always has precisely the opposite truth value to that of p itself. The truth values of both p and $\tilde{p}$ are given in the truth table shown in Figure 6.36.

p	$\tilde{p}$
T	F
F	T

Figure 6.36
Truth table for $\tilde{p}$.

Example 6.13

List A is a list of propositions, while list B is a list of sentences that are not propositions.

(a) Determine the truth values of the propositions in list A and state their negation statements.

(b) Explain why the sentences in list B are not propositions.

List A:

(a) Everyone can say where they were when President J. F. Kennedy was assassinated

(b) $2^n = n^2$ for some $n \in \mathbb{N}$, where $\mathbb{N}$ is the set of natural numbers

(c) The number 5 is negative

(d) $2^{89\,301} + 1$ is a prime number

(e) Air temperatures were never above 0°C in February 1935 in Bristol, UK

List B:

(a) Maths is fun

(b) Your place or mine?

(c) $y - x = x - y$

(d) Why am I reading this?

(e) Flowers are more interesting than calculus

(f) n is a prime number

(g) He won an Olympic medal

Solution   First of all, let us examine list A.

(a) This is obviously false. Besides those with poor memories or those from remote parts of the world, not everyone had been born in 1963.

(b) This is true (for $n = 2$).

(c) This is obviously false.

(d) There is no doubt that this is either true or false, but only specialists would know which (it is true).

(e) This is true, but again specialist knowledge is required before this can be verified.

All statements in list A are propositions because they are either true or false, never both. The negation predicates for list A are as follows.

(a) Not everyone can say where they were when President J. F. Kennedy was assassinated.

(b) $2^n \neq n^2$ for all $n \in \mathbb{N}$.

(c) The number 5 is not negative.

(d) $2^{89\,301} + 1$ is not a prime number.

(e) Air temperature was above 0°C at some time in February 1935 in Bristol, UK.

The sentences in list B are not propositions, for the following reasons.

(a) This is a subjective judgement. I think maths is fun (most of the time) – you probably do not!

(b) This is a question, and thus cannot be a proposition.

(c) This can easily be made into a proposition by the addition of the phrase 'for some real numbers x and y'. It is then true (whenever $x = y$).

(d) This is the same category as (b), a question.

(e) This is a subjective statement in the same category as (a).

(f) This is a predicate, since it will become a proposition once n is specified.

(g) Again, this is a predicate, since once we know who 'he' is, the statement will be certainly either true or false and hence be a proposition.

6.4.2   Compound propositions

When we combine simple statements together by such words as 'and', 'or' and so on we obtain compound statements. For example,

m: Today is Sunday and John has gone to church

n: Mary is 35 years old or Mary is 36 years old

constitute compound statements, with the constituent simple statements being respectively

m_1: Today is Sunday, m_2:   John has gone to church

n_1:   Mary is 35 years old, n_2:   Mary is 36 years old

As for switching circuits, we can again draw an analogy between the use of the words 'or' and 'and' in English and their use in the algebra of sets to form the union $A \cup B$ and intersection $A \cap B$ of two sets A and B. Drawing on the analogy the word 'or' is used to mean 'at least one statement' and the word 'and' to mean 'both statements'. The symbolism commonly used in propositional logic is to adopt the symbol $\vee$ (analogous to $\cup$) for 'or' and the symbol $\wedge$ (analogous to $\cap$) for 'and'. Thus in symbolic form the statements m and n may be written in terms of their constituent simple statements as

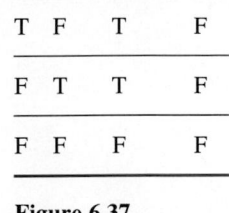

p	q	$p \vee q$	$p \wedge q$
T	T	T	T
T	F	T	F
F	T	T	F
F	F	F	F

Figure 6.37
Truth table for $p \vee q$
and $p \wedge q$.

$$m = m_1 \wedge m_2 \quad (m_1 \text{ and } m_2)$$
$$n = n_1 \vee n_2 \quad (n_1 \text{ or } n_2)$$

In general for two statements p and q the truth values of the compound statements

$p \vee q$ (meaning 'p or q' and called the **disjunction** of p, q)

$p \wedge q$ (meaning 'p and q' and called the **conjunction** of p, q)

are as given in the truth table of Figure 6.37.

Here are two examples that use compound statements and also make use of $\tilde{p}$ meaning 'not p' and

$$p \rightarrow q$$

meaning p implies q. There will be more about this kind of compound statement $p \rightarrow q$ (p implies q) when we deal with proof in Section 6.4.5.

Example 6.14   Let A, B and C be the following propositions:

   A: It is frosty
   B: It is after 11.00 a.m.
   C: Jim drives safely

(a) Translate the following statements into logical statements using the notation of this section.

   (i) It is not frosty.
   (ii) It is frosty and after 11.00 a.m.
   (iii) It is not frosty, it is before 11.00 a.m. and Jim drives safely.

(b) Translate the following into English sentences:

(i) $A \wedge B$,   (ii) $\tilde{A} \rightarrow C$,   (iii) $A \wedge \tilde{B} \rightarrow \tilde{C}$,   (iv) $\tilde{A} \vee B \rightarrow C$

Solution   (a)   (i) is the negation of A, so is written $\tilde{A}$.
      (ii) is A AND B, written $A \wedge B$.
      (iii) is slightly more involved, but is a combination of NOT A, NOT B AND C, and so is written $\tilde{A} \wedge \tilde{B} \wedge C$.

(b)   (i) $A \wedge B$ is A AND B; that is, 'It is frosty and it is after 11.00 a.m.'
      (ii) $\tilde{A} \rightarrow C$ is NOT A implies C; that is, 'It is not frosty, therefore Jim drives safely'.

(iii) $A \wedge \tilde{B} \rightarrow \tilde{C}$ is A AND NOT B implies NOT C; that is, 'It is frosty and before 11.00 a.m.; therefore Jim does not drive safely.'

(iv) $\tilde{A} \vee B \rightarrow C$ is NOT A or B implies C; that is, 'It is not frosty or it is after 11.00 a.m.; therefore Jim drives safely.'

Example 6.15 (Adapted from Exercise 5.15 in K.A. Ross and C.R.B. Wright, *Discrete Mathematics*, Prentice Hall, Englewood Cliffs, NJ, 1988.) In a piece of software, we have the following three propositions:

P: The flag is set
Q: $I = 0$
R: Subroutine S is completed

Translate the following into symbols:

(a) If the flag is set then $I = 0$.

(b) Subroutine S is completed if the flag is set.

(c) The flag is set if subroutine S is not completed.

(d) Whenever $I = 0$, the flag is set.

(e) Subroutine S is completed only if $I = 0$.

(f) Subroutine S is completed only if $I = 0$ or the flag is set.

Solution Most of the answers can be given with minimal explanation. The reader should check each and make sure each is understood before going further.

(a) $P \rightarrow Q$ (that is, P implies Q)

(b) $P \rightarrow R$ (that is, P implies R)

(c) $\tilde{R} \rightarrow P$ (that is, NOT R implies P)

Note that the logical expression is sometimes, as in (b) and (c), the 'other way round' from the English sentence. This reflects the adaptability of the English language, but can be a pitfall for the unalert student.

(d) $Q \rightarrow P$ (that is, Q implies P)

(e) $R \rightarrow Q$ (that is, R implies Q)

(f) This is really two statements owing to the presence of the (English, not logical) 'or'. 'S is completed only if $I = 0$' is written in logical symbols as (e) $R \rightarrow Q$. So including 'the flag is set' as a logical alternative gives

$$(R \rightarrow Q) \vee P$$

as the logical interpretation of (f). Alternatively, we can interpret the phrase '$I = 0$ or the flag is set' logically first as $Q \vee P$, then combine this with 'subroutine S is completed' to give

$$R \rightarrow (Q \vee P)$$

Now these two logical expressions are not the same. The sentence (f) may seem harmless; however, some extra punctuation or rephrasing is required before it is rendered unambiguous. One version could read:

(f) Subroutine S is completed only if either $I = 0$ or the flag is set (or both).

This is $R \rightarrow (Q \vee P)$.

Another could read:

(f) Subroutine S is completed only if $I = 0$, or the flag is set (or both).

This is $(R \rightarrow Q) \vee P$.

Part (f) highlights the fact that there is no room for sloppy thought in this branch of engineering mathematics.

6.4.3 Algebra of statements

In the same way as we used $\cup$, $\cap$ and $(\overline{})$ to generate complex expressions for sets we can use $\vee$, $\wedge$ and $\sim$ to form complex compound statements by constructing truth tables.

Example 6.16   Construct the truth table determining the truth values of the compound proposition

$$p \vee (p \wedge q)$$

Solution   The truth table is shown in Figure 6.38. Note that this verifies the analogous absorption law for set algebra of Question 10 (Exercises 6.2.5).

Figure 6.38
Truth table for
$p \vee (p \wedge q)$.

p	q	$p \wedge q$	$p \vee (p \wedge q)$
T	T	T	T
T	F	F	T
F	T	F	F
F	F	F	F

The statements are said to be **equivalent** (or more precisely **logically equivalent**) if they have the same truth values. Again, to show that two statements are equivalent, we simply need to construct the truth table for each statement and compare truth values. For example, from Example 6.16 we see that the two statements

$$p \vee (p \wedge q) \quad \text{and} \quad p$$

are equivalent. The symbolism ≡ is used to denote equivalent statements, so we can write

$$p \vee (p \wedge q) \equiv p$$

By constructing the appropriate truth tables, the following laws, analogous to the results (6.1), (6.2), (6.5) and (6.6) for set algebra, are readily verified:

Commutative laws

$$p \vee q \equiv q \vee p, \quad p \wedge q \equiv q \wedge p$$

Idempotent laws

$$p \vee p \equiv p, \quad p \wedge p \equiv p$$

Associative laws

$$p \vee (q \vee r) \equiv (p \vee q) \vee r, \quad p \wedge (q \wedge r) \equiv (p \wedge q) \wedge r$$

Distributive laws

$$p \vee (q \wedge r) \equiv (p \vee q) \wedge (p \vee r), \quad p \wedge (q \vee r) \equiv (p \wedge q) \vee (p \wedge r)$$

To develop a complete parallel with the algebra of sets, we need to identify two unit elements analogous to $\varnothing$ and U, relative to $\vee$ and $\wedge$ respectively.

Relative to $\vee$, we need to identify a statement s such that

$$p \vee s \equiv p$$

for any statement p. Clearly s must have a false value under all circumstances, and an example of such a statement is

$$s \equiv q \wedge \tilde{q}$$

where q is any statement, as evidenced by the truth table of Figure 6.39(a). Such a statement that is false under all circumstances is called a **contradiction**, and its role in the algebra of statements is analogous to the role of the empty set $\varnothing$ in the algebra of sets.

Relative to $\wedge$, we need to identify a statement t such that

$$p \wedge t \equiv p$$

for any statement p. Clearly, t must have a true truth value under all circumstances, and an example of such a statement is

$$t \equiv q \vee \tilde{q}$$

for any statement q, as evidenced by the truth table of Figure 6.39(b). Such a statement that is true under all circumstances is called a **tautology** and its role in the algebra of statements is analogous to that of the universal set U in the algebra of sets.

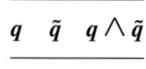

q	$\tilde{q}$	$q \wedge \tilde{q}$
T	F	F
F	T	F

(a) Contradiction

q	$\tilde{q}$	$q \wedge \tilde{q}$
T	F	T
F	T	T

(b) Tautology

Figure 6.39

Introducing the tautology and contradiction statements t and s respectively leads to the identity and complementary laws

Identity laws

$$p \vee s \equiv p \quad (s \text{ is the identity relative to } \vee)$$
$$p \wedge t \equiv p \quad (t \text{ is the identity relative to } \wedge)$$

Complementary laws

$$p \vee \tilde{p} \equiv t, \quad p \wedge \tilde{p} \equiv s$$

analogous to (6.3) and (6.4) for set algebra.

It then follows that the algebra of statements is another example of a Boolean algebra, with $\vee$ and $\wedge$ being the two binary operations, $\sim$ being the unary operation and s and t the identity elements. Consequently all the results developed for the algebra of sets carry through to the algebra of statements with equivalence between $\cup$, $\cap$, $\overline{(\ )}$, $\varnothing$, $\cup$ and $\vee$, $\wedge$, $\sim$, s, t respectively. These rules may then be used to reduce complex statements to simpler compound statements. These rules of the algebra of statements form the basis of propositional logic.

Example 6.17

Construct a truth table to verify the De Morgan laws for the algebra of statements analogous to (6.1) for the algebra of sets.

Solution

The analogous De Morgan laws for statements are the negations

$$\widetilde{(p \vee q)} \equiv \tilde{p} \wedge \tilde{q}, \qquad \widetilde{(p \wedge q)} \equiv \tilde{p} \vee \tilde{q}$$

whose validity is verified by the tables displayed in Figures 6.40 and 6.41.

p	q	$\tilde{p}$	$\tilde{q}$	$p \vee q$	$\widetilde{p \vee q}$	$\tilde{p} \wedge \tilde{q}$
T	T	F	F	T	F	F
T	F	F	T	T	F	F
F	T	T	F	T	F	F
F	F	T	T	F	T	T

Figure 6.40 Truth table for $\widetilde{p \vee q} \equiv \tilde{p} \wedge \tilde{q}$.

p	q	$\tilde{p}$	$\tilde{q}$	$p \wedge q$	$\widetilde{p \wedge q}$	$\tilde{p} \vee \tilde{q}$
T	T	F	F	T	F	F
T	F	F	T	F	T	T
F	T	T	F	F	T	T
F	F	T	T	F	T	T

Figure 6.41 Truth table for $\widetilde{p \wedge q} \equiv \tilde{p} \vee \tilde{q}$.

6.4.4 Exercises

32 Negate the following propositions:

(a) Fred is my brother.

(b) 12 is an even number.

(c) There will be gales next winter.

(d) Bridges collapse when design loads are exceeded.

33 Determine the truth values of the following propositions:

(a) The world is flat.

(b) $2^n + n$ is a prime number for some integer n.

(c) $a^2 = 0$ implies $a = 0$ for all $a \in \mathbb{N}$.

(d) $a + bc = (a + b)(a + c)$ for real numbers a, b and c.

34 Determine which of the following are propositions and which are not. For those that are, determine their truth values.

(a) $x + y = y + x$ for all $x, y \in \mathbb{R}$.

(b) $\mathbf{AB} = \mathbf{BA}$, where $\mathbf{A}$ and $\mathbf{B}$ are square matrices.

(c) Academics are absent-minded.

(d) I think that the world is flat.

(e) Go fetch a policeman.

(f) Every even integer greater than 4 is the sum of two prime numbers. (This is Goldbach's conjecture.)

35 Let A, B and C be the following propositions:

A: It is raining
B: The sun is shining
C: There are clouds in the sky

Translate the following into logical notation:

(a) It is raining and the sun is shining.

(b) If it is raining then there are clouds in the sky.

(c) If it is not raining then the sun is not shining and there are clouds in the sky.

(d) If there are no clouds in the sky then the sun is shining.

36 Let A, B and C be as in Question 35. Translate the following logical expressions into English sentences:

(a) $A \wedge B \to C$ (b) $(A \to C) \to B$

(c) $\tilde{A} \to (B \vee C)$ (d) $(\widetilde{A \vee B}) \wedge C$

37 Consider the ambiguous sentence

$$x^2 = y^2 \text{ implies } x = y \text{ for all } x \text{ and } y$$

(a) Make the sentence into a proposition that is true.

(b) Make the sentence into a proposition that is false.

6.4.5 Implications and proofs

A third type of compound statement of importance in propositional logic is that of **implication**, which lies at the heart of a mathematical argument. We have already met it briefly in Example 6.14, but here we give its formal definition. If p and q are two statements then we write the implication compound statement as

> If p then q

which asserts that the truth of p guarantees the truth of q. Alternatively, we say

> p implies q

and adopt the symbolism $p \to q$ (the notation $p \Rightarrow q$ is also commonly in use).

The truth table corresponding to $p \to q$ is given in Figure 6.42. From the truth table we see that $p \to q$ is false only when p is true and q is false. At first the observation that $p \to q$ is true whenever p is false may appear strange, but a simple example should

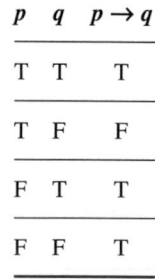

p	q	$p \to q$
T	T	T
T	F	F
F	T	T
F	F	T

Figure 6.42
Truth table for $p \to q$.

convince you. Suppose that prior to interviews for a senior management post within a company, the candidate states

If I am appointed then company profits will rise

This is clearly an implication statement $p \rightarrow q$, with the statements p and q being

p: I am appointed

q: Company profits will rise

If the candidate is not appointed (that is, p is 'false') then the statement made by the candidate is not false – independently of whether or not the company profits will rise. Hence $p \rightarrow q$ must be 'true'.

Example 6.18

Use truth tables to show that the following are tautologies:

(a) $A \rightarrow A$, (b) $A \wedge (A \rightarrow B) \rightarrow B$

A	$A \rightarrow A$
F	T
T	T

Figure 6.43 Truth table for $A \rightarrow A$.

A	B	$A \rightarrow B$	$A \wedge (A \rightarrow B)$	$[A \wedge (A \rightarrow B)] \rightarrow B$
F	F	T	F	T
F	T	T	F	T
T	F	F	F	T
T	T	T	T	T

Figure 6.44 Truth table for $[A \wedge (A \rightarrow B)] \rightarrow B$.

Solution (a) The truth table in Figure 6.43 is easily constructed, and shows that, no matter whether A is true or false, $A \rightarrow A$ is true. It is thus a tautology.

(b) The truth table shown in Figure 6.44 can be drawn, and we see that all the entries in the last column are true and the outcome of $A \wedge (A \rightarrow B) \rightarrow B$ is always true; it is thus a tautology.

The implication statement

$q \rightarrow p$

is called the **converse** of the statement $p \rightarrow q$, and it is perfectly possible for one to be true and the other to be false. For example, if p and q are defined by the statements

p: I go for a walk in the rain

q: I get wet

then the implication statements $p \rightarrow q$ and $q \rightarrow p$ are

If I go for a walk in the rain then I get wet

and

       If I am getting wet then I am going for a walk in the rain

respectively. The first, $p \to q$, is true but the second, $q \to p$, is false (I could be taking a shower).

      An implication statement that asserts both $p \to q$ and $q \to p$ is called **double implication**, and is denoted by

    $p \leftrightarrow q$

which may be expressed verbally as

      p if and only if q

or 'p is a necessary and sufficient condition for q'. Again the notation $p \Leftrightarrow q$ is also frequently used to represent double implication.

      It thus follows that $p \leftrightarrow q$ is defined to be

    $(p \to q) \wedge (q \to p)$

and its truth table is given in Figure 6.45.

Figure 6.45
Truth table for $p \leftrightarrow q$.

p	q	$p \to q$	$q \to p$	$p \leftrightarrow q$
T	T	T	T	T
T	F	F	T	F
F	T	T	F	F
F	F	T	T	T

      From Figure 6.45 we see that $p \leftrightarrow q$ is true if p and q have the same truth values, and is false if p and q have different truth values. It therefore follows that

    $(p \leftrightarrow q) \leftrightarrow (p \equiv q)$

meaning that each of the statements $p \leftrightarrow q$ and $p \equiv q$ implies the other. We must be careful when interpreting implication when negation statements are involved. A commonly made mistake is to assume that if the implication

    $p \to q$

is valid then the implication

    $\tilde{p} \to \tilde{q}$

is also valid. A little thought should convince you that this is not necessarily the case. This can be confirmed by reconsidering the previous example, when the negations $\tilde{p}$ and $\tilde{q}$ would be

    $\tilde{p}$:   I do not go for a walk in the rain

    $\tilde{q}$:   I do not get wet

and $\tilde{p} \to \tilde{q}$ is

      If I do not go for a walk in the rain then I do not get wet

Figure 6.46
The equivalence of
$p \rightarrow q$ and $\tilde{q} \rightarrow \tilde{p}$.

p	q	$\tilde{p}$	$\tilde{q}$	$p \rightarrow q$	$\tilde{q} \rightarrow \tilde{p}$
T	T	F	F	T	T
T	F	F	T	F	F
F	T	T	F	T	T
F	F	T	T	T	T

(This is obviously false, since someone could throw a bucket of water over me.) $p \rightarrow q$ and $\tilde{p} \rightarrow \tilde{q}$ so have different truth values. The construction of the two truth tables will establish this rigorously. On the other hand, the implication statements $p \rightarrow q$ and $\tilde{q} \rightarrow \tilde{p}$ are equivalent, as can be seen from the truth table in Figure 6.46. The implication $\tilde{q} \rightarrow \tilde{p}$ is called the **contrapositive form** of the implication $p \rightarrow q$.

In mathematics we need to establish beyond any doubt the truth of statements. If we denote by p a type of statement called a **hypothesis** and by q a second type of statement called a **conclusion** then the implication $p \rightarrow q$ is called a **theorem**.

In general, p can be formed from several statements; there is, however, usually only one conclusion in a theorem. A sequence of propositions that end with a conclusion, each proposition being regarded as valid, is called a **proof**. In practice, there are three ways of proving a theorem. These are direct proof, indirect proof and proof by induction. **Direct proof** is, as its name suggests, directly establishing the conclusion by a sequence of valid implementations. Here is an example of direct proof.

Example 6.19   If $a, b, c, d \in \mathbb{R}$, prove that the inverse of the 2×2 matrix

$$\boldsymbol{A} = \begin{bmatrix} a & b \\ c & d \end{bmatrix} \quad (ad \neq bc) \quad \text{is} \quad \frac{1}{ad - bc}\begin{bmatrix} d & -b \\ -c & a \end{bmatrix}$$

Solution   This has already been done in Chapter 5, Section 5.4. In the context of propositional logic, we conveniently split the proof as follows.

H_1:   If there exists a 2×2 matrix $\boldsymbol{B}$ such that

$$\boldsymbol{AB} = \boldsymbol{BA} = \boldsymbol{I}_2$$

where $\boldsymbol{I}_2$ is the 2×2 identity matrix, then $\boldsymbol{B}$ is the inverse of $\boldsymbol{A}$

$$H_2: \quad \begin{bmatrix} \alpha & 0 \\ 0 & \alpha \end{bmatrix} = \alpha\begin{bmatrix} 1 & 0 \\ 0 & 1 \end{bmatrix} = \alpha\boldsymbol{I}_2$$

$$H_3: \quad \begin{bmatrix} a & b \\ c & d \end{bmatrix}\begin{bmatrix} d & -b \\ -c & a \end{bmatrix} = \begin{bmatrix} d & -b \\ -c & a \end{bmatrix}\begin{bmatrix} a & b \\ c & d \end{bmatrix} = \begin{bmatrix} ad - bc & 0 \\ 0 & ad - bc \end{bmatrix}$$

Using H_2, we deduce that

$$\begin{bmatrix} ad - bc & 0 \\ 0 & ad - bc \end{bmatrix} = (ad - bc)\begin{bmatrix} 1 & 0 \\ 0 & 1 \end{bmatrix}$$

$$= (ad - bc)\mathbf{I}_2$$

Dividing by $ad - bc$ then gives the result.

In this proof, H_1 is a definition and hence true, H_2 and H_3 are properties of matrices established in Chapter 5. (It is possible to split H_3 into arithmetical hypotheses detailing the process of matrix multiplication.) Hence

$$H_1 \wedge H_2 \wedge H_3 \text{ implies } \mathbf{A}^{-1} = \frac{1}{ad - bc}\begin{bmatrix} d & -b \\ -c & a \end{bmatrix}$$

hence establishing that the right-hand side is the inverse of $\mathbf{A}$.

We have seen that $p \rightarrow q$ and $\tilde{q} \rightarrow \tilde{p}$ are logically equivalent. The use of this in a proof sometimes makes the arguments easier to follow, and we call this an **indirect proof**. Here is an example of this.

Example 6.20 Prove that if $a + b \geq 15$ then either $a \geq 8$ or $b \geq 8$, where a and b are integers.

Solution Let p, q and r be the statements

$$p: \quad a + b \geq 15 \qquad q: \quad a \geq 8 \qquad r: \quad b \geq 8$$

Then the negations of these statements are

$$\tilde{p}: \quad a + b < 14 \qquad \tilde{q}: \quad a < 7 \qquad \tilde{r}: \quad b < 7$$

The statement to be proved can be put into logical notation

$$p \rightarrow (q \vee r)$$

This is equivalent to

$$\widetilde{(q \vee r)} \rightarrow \tilde{p}$$

or, using the De Morgan laws,

$$(\tilde{q} \wedge \tilde{r}) \rightarrow \tilde{p}$$

If we prove the truth of this implication statement then we have also proved that

$$p \rightarrow (q \vee r)$$

We have

$$\tilde{q} \wedge \tilde{r}: \quad a < 7 \text{ and } b < 7$$

$$\tilde{p}: \quad a + b < 14$$

Hence $\tilde{p} \wedge \tilde{r} \to \tilde{p}$ is

$a < 7$ and $b < 7$ implies $a + b < 14$ for integers a and b

which is certainly true.

We have thus proved that $p \to (q \vee r)$, as required.

Another indirect form of proof is **proof by contradiction**. Instead of proving 'p is true' we prove '$\tilde{p}$ is false'. An example of this kind of indirect proof follows.

Example 6.21   Prove that $\sqrt{2}$ is irrational.

Solution   Let p be the statement

p: $\sqrt{2}$ is irrational

then $\tilde{p}$ is the statement

$\tilde{p}$: $\sqrt{2}$ is rational

Here are the arguments establishing that $\tilde{p}$ is 'false'. If $\sqrt{2}$ is rational then there are integers m and n, with no common factor, such that

$$\sqrt{2} = \frac{m}{n}$$

Squaring this gives

$$2 = \frac{m^2}{n^2} \quad \text{or} \quad m^2 = 2n^2$$

This implies that m^2 is an even number, and therefore so is m. Hence

$m = 2k$ with k an integer

So

$$m^2 = 4k^2$$

However, since $n^2 = \frac{1}{2}m^2$, this implies

$$n^2 = 2k^2$$

and therefore n^2 is also even, which means that n is even. But if both m and n are even, they have the factor 2 (at least) in common. We thus have a contradiction, since we have assumed that m and n have no common factors. Thus $\tilde{p}$ must be false. If $\tilde{p}$ is false then p is true, and hence we have proved that $\sqrt{2}$ is irrational.

The final method of proof we shall examine is **proof by induction**. If $p_1, p_2, \ldots,$ $p_n, \ldots$ is a sequence of propositions, n is a natural number and

(a) p_1 is true (the **basis for induction**)

(b)   if p_n is true then p_{n+1} is true (the **induction hypothesis**) then p_n is true for all n **by induction**. Proof by induction is used extensively by mathematicians to establish formulae. Here is such an example.

Example 6.22   Use mathematical induction to show that

$$1 + 2 + \ldots + N = \tfrac{1}{2}N(N + 1) \tag{6.8}$$

for any natural number N.

Solution   Let us follow the routine for proof by induction.
First of all, we set $N = 1$ in the proposition (6.8):

$$1 = \tfrac{1}{2}1(1 + 1)$$

which is certainly true. Now we set $N = n$ in (6.8) and assume the statement is true:

$$1 + 2 + \ldots + n = \tfrac{1}{2}n(n + 1) \tag{6.9}$$

We now have to show that

$$1 + 2 + \ldots + n + (n + 1) = \tfrac{1}{2}(n + 1)(n + 2) \tag{6.10}$$

which is the proposition (6.8) with N replaced by $n + 1$. If we add $n + 1$ to both sides of (6.9) then the right-hand side becomes

$$\tfrac{1}{2}n(n + 1) + (n + 1)$$

which can be rewritten as

$$(\tfrac{1}{2}n + 1)(n + 1) = \tfrac{1}{2}(n + 1)(n + 2)$$

thus establishing the proof of the induction hypothesis. The truth of (6.8) then follows by induction.

6.4.6 Exercises

38   The **counterexample** is a good way of disproving assertions. (Examples can *never* be used as proof.) Find counterexamples for the following assertions:

(a) $2^n - 1$ is a prime for every $n \geqslant 2$

(b) $2^n + 3^n$ is a prime for all $n \in \mathbb{N}$

(c) $2^n + n$ is prime for every positive odd integer n

39   Give the converse and contrapositive for each of the following propositions:

(a) $A \to (B \land C)$

(b) If $x + y = 1$ then $x^2 + y^2 \geqslant 1$

(c) If $2 + 2 = 4$ then $3 + 3 = 9$

40   Construct the truth tables for the following:

(a) $A \land \tilde{A}$ (b) $\tilde{A} \lor \tilde{B}$

(c) $(A \land B) \to C$ (d) $\widetilde{(A \land B) \to C}$

41   Prove or disprove the following:

(a) $(B \to A) \leftrightarrow (A \land B)$

(b) $(A \land B) \to (A \to B)$

(c) $(A \land B) \to (A \lor B)$

Note that to *disprove* a tautology, only one line of a truth table is required.

42   Use contradiction to show that $\sqrt{3}$ is irrational.

43 Prove or disprove the following:

(a) The sum of two even integers is an even integer.

(b) The sum of two odd integers is an odd integer.

(c) The sum of two primes is never a prime.

(d) The sum of three consecutive integers is divisible by 3.

Indicate the methods of proof where appropriate.

44 Prove that the number of primes is infinite by contradiction.

45 Use induction to establish the following results:

(a) $\sum_{k=1}^{n} k^2 = \frac{1}{6}n(n+1)(2n+1)$ (n a natural number)

(b) $4 + 10 + 16 + \ldots$

$+ (6n - 2) = n(3n + 1)$ ($n \in \mathbb{N}$)

(c) $(2n + 1) + (2n + 3) + (2n + 5) + \ldots$

$+ (4n - 1) = 3n^2$ (n a natural number)

(d) $1^3 + 2^3 + \ldots$

$+ n^3 = (1 + 2 + \ldots + n)^2$ (n a natural number)

(*Hint*: use $1 + 2 + \ldots + n = \frac{1}{2}n(n+1)$, established in the text.)

46 Prove that $11^n - 4^n$ is divisible by 7 for all natural numbers n.

47 Consider the following short procedure:

Step 1: Let $S = 1$
Step 2: Print S
Step 3: Replace S by $S + 2\sqrt{S} + 1$ and go back to step 2.

List the first four printed values of S, and prove by induction that $S = n^2$ the nth time the procedure reaches step 2.

6.5 Engineering application: expert systems

In the early 1960s many people believed that machines could be made to think and that computers that could, for instance, automatically translate text from one language to another or make accurate medical diagnoses would soon be available. The problems associated with creating machines that could undertake these tasks are well illustrated by the story (possibly apocryphal but none the less salutary) of the early language-translating machine that was asked to translate the English sentence 'The spirit is willing but the flesh is weak', into Russian. The machine's attempt was found to read, in Russian, 'The vodka is very strong but the meat has gone off'. Problems such as these and the growing appreciation of the sheer magnitude of the computing power needed to undertake these intelligent tasks (an effect often referred to as the 'combinatorial explosion') finally resulted in the realization that thinking machines were further away than some scientists had thought. Interest waned for 20 years until, in the early 1980s, advances both in our understanding of theoretical issues in computer software and in the design of computer hardware again brought the achievement of intelligent tasks by computers nearer reality.

The modern approach to producing intelligent machines (or at least machines that seem intelligent) is through 'expert systems'. The basis of an expert system is a database of facts and rules together with an 'inference engine', that is, a computer program that matches some query with the known facts and rules and determines the answer to the query. The essence of the 'intelligence' of an expert system is the way in which the inference engine is able to combine the known facts, using the given rules together with the general methods of proof that we discussed in Section 6.4, to answer queries that could not be answered by direct interrogation of the database of facts. The theoretical basis of these systems lies in propositional logic and predicate calculus. The facts and rules of the expert system's database loosely correspond to the concepts of proposition and predicate that we discussed in Section 6.4.

Expert systems that are able to answer routine queries in certain restricted areas of knowledge are now in everyday use in industry, commerce and public service. Such systems can, for instance, help tax lawyers advise clients, help geologists assess the results of seismographic tests, or advise disabled people on the benefits to which they are entitled. Nearer home, the same techniques are used in computer programs that can help with the routine drudgery of mathematics, differentiating, integrating and manipulating expressions with a speed and accuracy that humans cannot match. It is easy to envisage that expert systems that can undertake some of the work of the design engineer or design building structures and carry out the routine tasks of architecture (routeing cables and pipework within a building for instance) cannot be far away. Here we shall give more of the flavour of expert systems by an example in the domain of family relationships. Imagine that an expert system has a set of facts about the relationships in a certain family such as those shown in Figure 6.47. It is easy for a human to deduce that the family tree is that shown in Figure 6.48 (assuming, of course, that no one in the family has been married more than once and that all the children were born within wedlock). From the family tree a human could ascertain the truth of some further statements about the family. For instance, it is obvious that the statement 'Peter is the grandfather of David' is true and that the statement 'Alan is the brother of Robert' is false.

(1) Peter is the father of Robert
(2) James is the father of Alan
(3) Anne is the mother of Robert
(4) Anne is the mother of Melanie
(5) Lilian is the mother of David
(6) Robert is the father of Jennifer
(7) James is the brother of Peter
(8) Lilian is the wife of Robert
(9) Alan is the son of Martha

Figure 6.47 A short database of facts about family relationships.

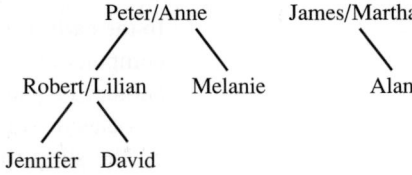

Figure 6.48 The family tree deduced from the facts in the database.

An expert system can equally well be designed to evaluate the truth of such statements. In order to do so it needs, as well as the facts, some rules about how relationships combine. A typical set of rules is shown in Figure 6.49. If we were to ask if the statement 'Peter is a grandparent of David' is true the expert system might reason as follows:

From fact (1) Peter is the father of Robert;
therefore, from rule (1), Peter is a parent of Robert.

From fact (8) Lilian is the wife of Robert;
therefore, from rule (4), Lilian is the spouse of Robert;
therefore, from rule (7), Robert is the spouse of Lilian.

From fact (5) Lilian is the mother of David;
therefore, from rule (2), Lilian is a parent of David.

Figure 6.49
A short database of
rules about family
relationships.

(1) If X is the father of Y
      then X is a parent of Y
(2) If X is the mother of Y
      then X is a parent of Y
(3) If X is a parent of Y
  and Y is a parent of Z
      then X is a grandparent of Z
(4) If X is the wife of Y
      then X is the spouse of Y
(5) If X is the husband of Y
      then X is the spouse of Y
(6) If X is the spouse of Y
  and Y is a parent of Z
      then X is a parent of Z
(7) If X is the spouse of Y
      then Y is the spouse of X

Now it has been proved that Robert is the spouse of Lilian
and that Lilian is a parent of David;
therefore, from rule (6), Robert is a parent of David.

Finally, it has been proved that Peter is a parent of Robert
and Robert is a parent of David;
therefore, from rule (3), Peter is a grandparent of David.

A little more is needed to deduce that Peter is the grandfather of David and this is left as an exercise for the reader.

Of course, the expert system needs a way of determining which rule to try to apply next in seeking to prove the truth of the query. That is the role of the part of the program called the inference engine – the inference engine attempts to prove the truth of the query by using rules in the most effective order and in such a way as to leave no possible path to a proof unexplored. In many expert systems this is achieved by using a search algorithm.

It is interesting to ask how such an expert system can prove that some assertion ('Alan is the brother of Robert' for instance) is false. Most expert systems tackle this by exhaustively trying every possible way of proving that the assertion is true. Then, if this fails, to most expert systems, it actually means merely that, given the facts and rules at the disposal of the expert systems, the assertion cannot be proved to be true. There are obviously dangers in this approach, since an incomplete database may lead an expert system to classify as false an assertion that, given more complete data, can be shown to be true. If, for instance, we were to ask the family expert system if the statement 'Alan is the cousin of Robert' is true, the expert system would allege it was not. On the other hand, if we gave the system some further, more sophisticated, rules about relationships then it would be able to deduce that the statement is actually true.

6.6   Engineering application: control

We consider a simplified model of a container for chemical reactions and design a circuit that involves four variables: upper and lower contacts for each of the temperature and pressure gauges. The control of the reaction within the container is managed using a mixing motor, a cooling-water valve, a heating device and a safety valve. We will analyse the control of the reaction given the following data and notation:

T_L = lower temperature, T_u = upper temperature

p_L = lower pressure, p_u = upper pressure

m = mixing motor, c = cooling-water valve

h = heating device, s = safety valve

$T_L = 0$, $T_u = 0$ temperature is too low

$T_L = 1$, $T_u = 0$ temperature is correct

$T_L = 1$, $T_u = 1$ temperature is too high

$p_L = 0$, $p_u = 0$ pressure is too low

$p_L = 1$, $p_u = 0$ pressure is correct

$p_L = 1$, $p_u = 1$ pressure is too high

$m = 0, 1$ mixing motor is off, on

$c = 0, 1$ cooling-water valve is off, on

$h = 0, 1$ heating is off, on

$s = 0, 1$ safety valve is closed, open

Figure 6.50 shows the container. The table in Figure 6.51 gives nine states – three initial states, three normal states and three danger states – exemplified by the pressure in the vessel. From this table we can write down that

$$s = T_L \cdot T_u \cdot p_L \cdot p_u$$

Figure 6.50

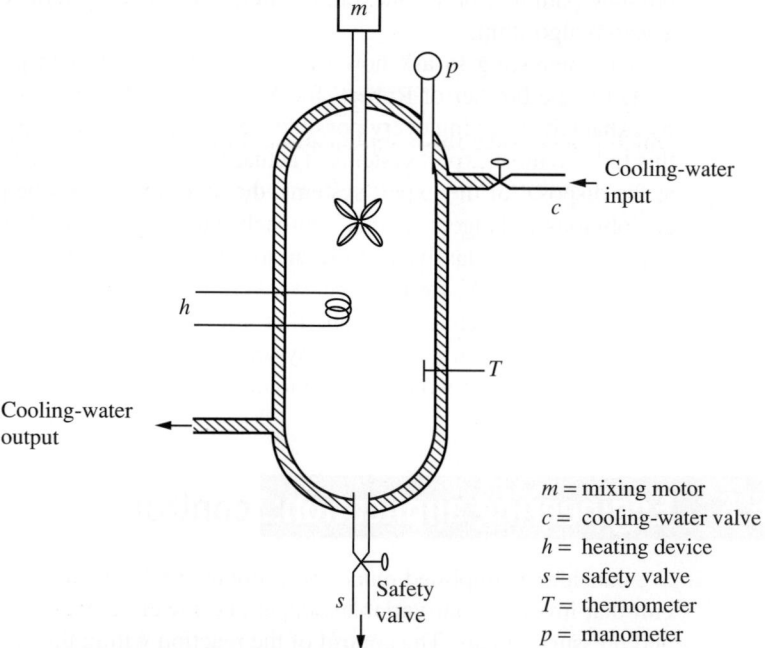

m = mixing motor
c = cooling-water valve
h = heating device
s = safety valve
T = thermometer
p = manometer

Figure 6.51

	T_L	T_u	p_L	p_u	h	c	m	s	*Comments*
Initial state (low pressure)	0	0	0	0	1	0	1	0	Gauges off; switch on motor and heater
	1	0	0	0	1	0	0	0	Correct temperature; switch off motor
	1	1	0	0	0	0	1	0	Temperature too high; heater off, motor on
Normal state (pressure acceptable)	0	0	1	0	1	0	0	0	Cold; heater on
	1	0	1	0	0	0	0	0	Normal; heater off
	1	1	1	0	0	1	1	0	Hot; motor on, cooling water in
Danger state (pressure high)	0	0	1	1	0	0	1	0	Low temperature; motor on
	1	0	1	1	0	1	1	0	Normal temperature; motor on, cooling water in
	1	1	1	1	0	1	1	1	High temperature; $c = m = s = 1$ to try to prevent an explosion!

that is, the safety valve is only open when the temperature and pressure are too high. The Boolean expressions for h, c and m are obtained by taking the union of the rows of T_u, T_L, p_u and p_L that have 1 under the columns headed h, c and m respectively. Hence

$$h = (\overline{T}_L \cdot \overline{T}_u \cdot \bar{p}_L \cdot \bar{p}_u) + (T_L \cdot \overline{T}_u \cdot \bar{p}_L \cdot \bar{p}_u) + (\overline{T}_L \cdot \overline{T}_u \cdot p_L \cdot \bar{p}_u)$$

$$= (\overline{T}_L + T_L) \cdot (\overline{T}_u \cdot \bar{p}_L \cdot \bar{p}_u) + (\overline{T}_L \cdot \overline{T}_u \cdot p_L \cdot \bar{p}_u)$$

using the distributive law, $\overline{T}_u \cdot \bar{p}_L \cdot \bar{p}_u$ being a common factor

$$= 1 \cdot (\overline{T}_u \cdot \bar{p}_u) \cdot (\bar{p}_L + (\overline{T}_L \cdot p_L))$$

$$= (\overline{T}_u \cdot \bar{p}_u) \cdot (\bar{p}_L + \overline{T}_L)$$

which is a considerable simplification. Similarly, c is given by

$$c = (T_L \cdot T_u \cdot p_L \cdot \bar{p}_u) + (T_L \cdot \overline{T}_u \cdot p_L \cdot p_u) + (T_L \cdot T_u \cdot p_L \cdot p_u)$$

Combining the first and last, and using $p_u + \bar{p}_u = 1$, gives

$$c = (T_L \cdot T_u \cdot p_L) + (T_L \cdot \overline{T}_u \cdot p_L \cdot p_u)$$

$$= (T_L \cdot p_L) \cdot (T_u + (\overline{T}_u \cdot p_u))$$

$$= (T_L \cdot p_L) \cdot (T_u + p_u)$$

Finally, for m, which has six entries as 1, we get the more complicated expression

$$m = (\overline{T}_L \cdot \overline{T}_u \cdot \bar{p}_L \cdot \bar{p}_u) + (T_L \cdot T_u \cdot \bar{p}_L \cdot \bar{p}_u) + (T_L \cdot T_u \cdot p_L \cdot \bar{p}_u)$$

$$+ (\overline{T}_L \cdot \overline{T}_u \cdot p_L \cdot p_u) + (T_L \cdot \overline{T}_u \cdot p_L \cdot p_u) + (T_L \cdot T_u \cdot p_L \cdot p_u)$$

Labelling these brackets 1, ..., 6, and leaving 1 and 6 alone, we note that 2 and 3 combine since $T_L \cdot T_u \cdot \bar{p}_u$ is common, and 4 and 5 combine since $\overline{T}_u \cdot p_L \cdot p_u$ is common; hence

Figure 6.52

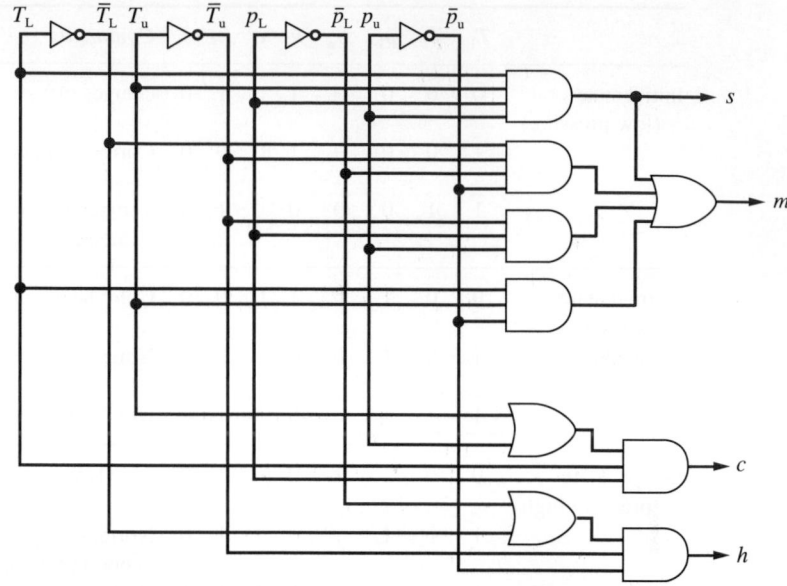

$$m = (\overline{T}_L \cdot \overline{T}_u \cdot \overline{p}_L \cdot \overline{p}_u) + (T_L \cdot T_u \cdot p_L \cdot p_u) + (T_L \cdot T_u \cdot \overline{p}_u) + (\overline{T}_u \cdot p_L \cdot p_u)$$

Hence we can draw the control of the vessel in terms of the switching circuit in Figure 6.52.

6.7 Review exercises (1–23)

1 If $U = \{1, 2, 3, 4, 5, 6, 7, 8, 9\}$, $A = \{2, 4, 6\}$, $B = \{1, 3, 5, 7\}$ and $C = \{2, 3, 4, 7, 8\}$ find the sets

(a) $\overline{A \cup B}$ (b) $C - A$ (c) $\overline{C} \cap \overline{B}$

2 Let $A = \{n \in \mathbb{N}, n \leqslant 11\}$

$B = \{n \in \mathbb{N}, n \text{ is even and } n \leqslant 20\}$

$C = \{10, 11, 12, 13, 14, 15, 17, 20\}$

Write down the sets

(a) $A \cap B$ (b) $A \cap B \cap C$

(c) $A \cup (B \cap C)$

and verify that $(A \cup B) \cap (A \cup C) = A \cup (B \cap C)$.

3 If A, B and C are defined as in Question 2, and the universal set is the set of all integers less than or equal to 20, find the following sets:

(a) $\overline{A}$ (b) $\overline{A} \cup \overline{B}$

(c) $\overline{A \cup B}$ (d) $A \cap (\overline{B \cup C})$

Verify the De Morgan laws for A and B.

4 The sets A and B are defined by

$A = \{x : x^2 + 6 = 5x \quad \text{or} \quad x^2 + 2x = 8\}$

$B = \{2, 3, 4\}$

Which of the following statements is true?

(a) $A \neq B$

(b) $A = B$

Give reasons for your answers.

5 (a) Simplify the Boolean functions

$f = (A \cap \overline{B} \cap C) \cup (A \cap (B \cup \overline{C}))$

$g = (\overline{(\overline{A} \cup \overline{B}) \cap C}) \cap ((\overline{C} \cup A) \cap B)$

(b) Draw Venn diagrams to verify that

$(A \cap \overline{B}) \cup (\overline{A} \cap B) = A \cup B$

if and only if $A \cap B = \varnothing$.

6 In an election there are three candidates and 800 voters. The voters may exercise one, two or three votes each. The following results were obtained:

Votes cast	240	400	500
Candidate	A	B	C

Voters	110	90	200	50
Candidates	B and C	A and C	A and B	A and B and C

Show that these results are inconsistent if all the voters use at least one vote.

7   Draw switching circuits to establish the truth of the following laws:

(a) $p + p \cdot q = p$

(b) $p + \bar{p} \cdot q = p + q$

(c) $p \cdot q + p \cdot r = p \cdot (q + r)$

(d) $(p + q) \cdot (p + r) = p + q \cdot r$

Use these to simplify the expression

$$s = p \cdot \bar{p} + p \cdot q + \bar{p} \cdot r + q \cdot r$$

so that s only contains two pairs of products added.

8   Write down, in set-theoretical notation, expressions corresponding to the outputs in (a) Figure 6.53 and (b) Figure 6.54.

Inputs
A B C D

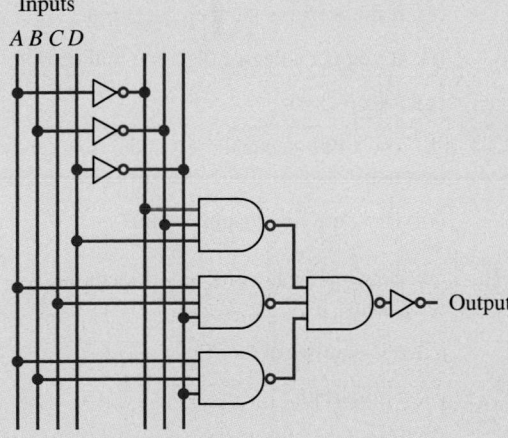

Figure 6.53

Inputs
B   C

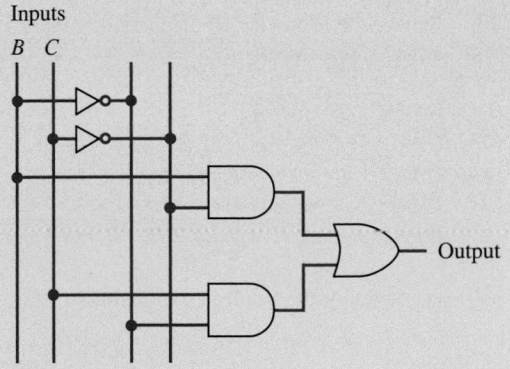

Output

Figure 6.54

9   Draw a switching circuit with inputs x, y, z and u to correspond to the following expressions:

(a) $(x \cdot y \cdot z \cdot u) + (\bar{x} \cdot \bar{y} \cdot z \cdot u) + (x \cdot \bar{u})$

(b) $(\bar{x} \cdot \bar{y}) + (\bar{z} \cdot \bar{u}) + (x \cdot y \cdot z)$

(c) $(x \cdot y \cdot z \cdot u) + (\bar{x} \cdot \bar{y} \cdot z \cdot u) + (x \cdot \bar{y} \cdot \bar{z} \cdot u)$

$\quad + (\bar{x} \cdot y \cdot z \cdot \bar{u}) + (\bar{x} \cdot y \cdot \bar{z} \cdot u)$

$\quad + (x \cdot \bar{y} \cdot z \cdot \bar{u})$

$\quad + (\bar{x} \cdot \bar{y} \cdot \bar{z} \cdot u)$

For (c) establish the output for the input states

(i) $x = y = 1, \quad z = u = 0$

(ii) $x = 1, \quad y = z = u = 0$

10   Write down truth tables for the following expressions:

(a) $p \wedge q$ 　　(b) $p \vee q$ 　　(c) $p \rightarrow q$

The contrapositive of the conditional statement $p \rightarrow q$ is defined as $\tilde{q} \rightarrow \tilde{p}$.

(d) Use truth tables to show that

$$\tilde{q} \rightarrow \tilde{p} \equiv p \rightarrow q$$

(e) Use truth tables to evaluate the status of the expression

$$(p \vee q) \wedge (\overline{\tilde{p} \wedge q}) \rightarrow p$$

(f) By taking the contrapositive of this conditional statement and using (d) together with the De Morgan laws (see Example 6.6) show that

$$\tilde{p} \rightarrow (\tilde{p} \wedge \tilde{q}) \vee (\tilde{p} \wedge q)$$

is a tautology

11 Reduce the following Boolean expressions by taking complements:

(a) $\overline{[\overline{(\overline{p \cdot q}) \cdot p}][\overline{(\overline{p \cdot q}) \cdot q}]}$

(b) $\overline{\overline{(p + q + \bar{r}) \cdot (\overline{p \cdot q} + \bar{r} \cdot s)} + \bar{q} \cdot r \cdot s}$

(c) $\overline{\overline{(p \cdot q \cdot r + q \cdot \bar{r} \cdot s)} + \overline{(\bar{q} \cdot r \cdot s + \bar{q} \cdot \bar{r} \cdot \bar{s} + q \cdot r \cdot \bar{s})}}$

12 (a) Simplify the Boolean expressions

(i) $p \cdot r + p \cdot q \cdot r + q \cdot \bar{r} \cdot s + \bar{q} \cdot r \cdot \bar{s} + p \cdot q \cdot r \cdot s$

(ii) $\overline{[(\bar{p} + q) \cdot (\bar{r} + s)] \cdot (\bar{s} + p) + r}$

(b) Show the Boolean function $p \cdot q + \bar{p} \cdot r$ on a Venn diagram.

13 A lift (elevator) services three floors. On each floor there is a call button to call the lift. It is assumed that at the moment of call the cabin is stationary at one of the three floors. Using these six input variables, determine a control that moves the motor in the right direction for the current situation. (*Hint*: There are 24 combinations to consider.)

14 There are four people on a TV game show. Each has a 'Yes/No' button for recording opinions. The display must register 'Yes' or 'No' according to a majority vote.

(a) Derive a truth table for the above.

(b) Write down the Boolean expression for the output.

(c) Simplify this expression and suggest a suitable circuit.

(d) If there is a tie, the host has a 'casting vote'. Modify the above circuit to indicate this.

15 Consider the following logical statements:

(a) Mike never smokes dope.

(b) Rick smokes if, and only if, Mike and Vivian are present.

(c) Neil smokes under all conditions – even by himself.

(d) Vivian smokes if, and only if, Mike is not present.

The police raid: determine the state of there being no dope smoking in terms of M, R, N and V's presence (Mike, Rick, Neil and Vivian respectively).

16 Find the explicit Boolean function for the logic circuit of Figure 6.55. Show that the function simplifies to $f = q \cdot \bar{r}$ and draw two different simplified circuits which may be used to represent the circuit.

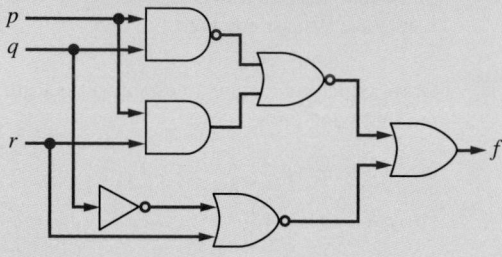

Figure 6.55

17 Which of the following statements are propositions? For those that are not, say why and suggest ways of changing them so that they become propositions. For those that are, comment on their truth value.

(a) Julius Caesar was prime minister of Great Britain.

(b) Stop hitting me.

(c) Turn right at the next roundabout.

(d) The Moon is made of green cheese.

(e) If the world is flat then $3 + 3 = 6$.

(f) If you get a degree then you will be rich.

(g) $x + y + z = 0$.

(h) The 140th decimal digit in the representation of π is 8.

(i) There are five Platonic solids.

18 (a) Draw up truth tables to represent the statements

(i) p is equivalent to q

(ii) p implies q

(b) Using the algebra of statements, represent the truth of the statements below in tabular form and hence determine whether they are true or false:

(i) If p implies q, and r implies q, then either r implies p or p implies r.

(ii) If p is equivalent to q, and q is equivalent to r, then p implies r.

19 A panel light in the control room of a satellite launching site is to go on if the pressure in both the oxidizer and fuel tanks is equal to or above a required minimum value and there are 15 minutes or less to 'lift-off', or if the pressure in the oxidizer tank is equal to or above the required minimum value and the pressure in the fuel tank is below the required minimum value but there are more than 15 minutes to 'lift-off', or if the pressure in the oxidizer tank is below the required minimum value but there are more than 15 minutes to 'lift-off'. By using a truth table, write down a Boolean expression to represent the state of the panel light. Minimize the Boolean function.

20 In the control problem of Section 6.6 show that h may also be expressed as

$$h = \overline{T_u + p_u + p_L \cdot T_L}$$

Compare the resulting control switching circuit with that of Figure 6.52.

21 Write down all subsets of the set $A = \{p, q, r, s\}$ that contain the product of *four* of p, q, r, s or their complement. Represent these on a Venn diagram. [The ideas are pursued through Karnaugh maps which are outside the scope of this text.]

22 State the converse and contrapositive of each of the following statements:

(a) If the train is late, I will not go.

(b) If you have enough money, you will retire.

(c) I cannot do it unless you are there too.

(d) If you go, so will I.

23 An island is inhabited by two tribes of vicious cannibals and, sadly, you are a prisoner of one of them. One tribe always tell the truth, the other tribe always lie. Unfortunately both tribes look identical. They will answer 'yes' or 'no' to a single question they will allow you. The God of one tribe is female, the God of the other tribe is male, and if you correctly state the sex of their God they will set you free. Use truth tables to help you formulate a question that will enable you to survive.

7 Sequences, Series and Limits

Chapter 7	Contents

7.1   Introduction

In the analysis of practical problems certain mathematical ideas and techniques appear in many different contexts. One such idea is the concept of a sequence. Sequences occur in management activities such as the determination of programmes for the maintenance of hardware or production schedules for bulk products. They also arise in investment plans and financial control. They are intrinsic to computing activities, since the most important feature of computers is their ability to perform sequences of instructions quickly and accurately. Sequences are of great importance in the numerical methods that are essential for modern design and the development of new products. As well as illustrating these basic applications, we shall show how these simple ideas lead to the idea of a limit, which is a prerequisite for a proper understanding of the calculus and numerical methods. Without that understanding, it is not possible to form mathematical models of real problems, to solve them or to interpret their solutions adequately. At the same time, we shall illustrate some of the elementary properties of the standard functions described in Chapter 2 and how they link together, and we shall look forward to further applications in more advanced engineering applications, in particular to the work on Z transforms contained in the companion text *Advanced Modern Engineering Mathematics*.

7.2   Sequences and series

7.2.1   Notation

Consider a function f whose domain is the set of whole numbers $\{0, 1, 2, 3, \ldots\}$. The set of values of the function $\{f(0), f(1), f(2), f(3), \ldots\}$ is called a **sequence**. Usually we denote the values using a subscript, so that $f(0) = f_0, f(1) = f_1, f(2) = f_2$, and so on. Often we list the elements of a sequence in order on the assumption that the first in the list is f_0, the second is f_1 and so on. For example, we may write

'Consider the sequence 1, 1, 2, 3, 5, 8, 13, 21, 34, …'

implying $f_0 = 1, f_5 = 8, f_8 = 34$ and so on. In this example the continuation dots … are used to imply that the sequence does not end. Such a sequence is called an **infinite** sequence to distinguish it from **finite** or **terminating** sequences. The finite sequence $\{f_0, f_1, \ldots, f_n\}$ is often denoted by $\{f_k\}_{k=0}^{n}$ and the infinite sequence by $\{f_k\}_{k=0}^{\infty}$. When the context makes the meaning clear, the notation is further abbreviated to $\{f_k\}$. Here the letter k is used as the 'counting' variable. It is a **dummy variable** in the sense that we could replace it by any other letter and not change the result. Often n and r are used as dummy variables.

Example 7.1   A bank pays interest at a fixed rate of 8.5% per year, compounded annually. A customer deposits the fixed sum of £1000 into an account at the beginning of each year. How much is in the account at the beginning of each of the first four years?

Solution   Let £x_n denote the amount in the account at the beginning of the $(n + 1)$th year. Then

Amount at beginning of first year $x_0 = 1000$

Amount at beginning of 2nd year $x_1 = 1000(1 + \frac{8.5}{100}) + 1000 = 2085$

Amount at beginning of 3rd year $x_2 = 2085(1 + 0.085) + 1000 = 3262.22$

Amount at beginning of 4th year $x_3 = 3262.22(1.085) + 1000 = 4539.51$

We can see that in general

$$x_n = 1.085x_{n-1} + 1000$$

This is a **recurrence relation**, which gives the value of each element of the sequence in terms of the value of the previous element.

Example 7.2   Consider the ducting of a number of cables of the same diameter d. The diameter D_n of the smallest duct with circular cross-section depends on the number n of cables to be enclosed, as shown in Figure 7.1:

$$D_0 = 0, \quad D_1 = d, \quad D_2 = 2d, \quad D_3 = (1 + 2/\sqrt{3})d, \quad D_4 = (1 + \sqrt{2})d$$

$$D_5 = \tfrac{1}{4}\sqrt{[2(5 - \sqrt{5})]}d, \quad D_6 = 3d, \quad D_7 = 3d, \ldots$$

Thus the duct diameters form a sequence of values $\{D_1, D_2, D_3, \ldots\} = \{D_n\}_{n=1}^{\infty}$.

Figure 7.1
Enclosing a
number of cables
in a circular duct.

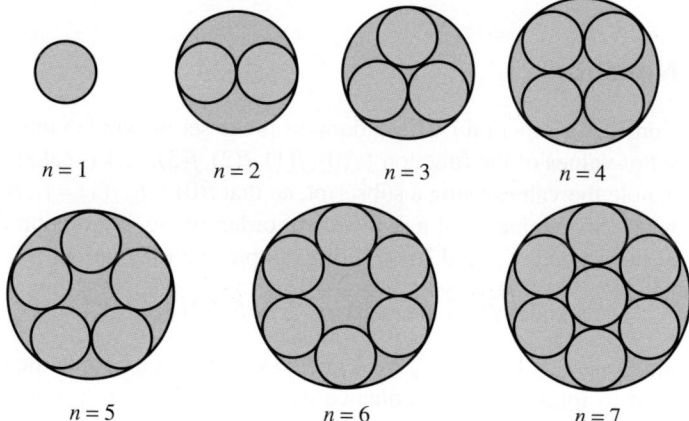

$n = 1$ $n = 2$ $n = 3$ $n = 4$

$n = 5$ $n = 6$ $n = 7$

Example 7.3   A computer simulation of the crank and connecting rod mechanism considered in Example 2.36 evaluates the position of the end Q of the connecting rod at equal intervals of the angle $x°$. Given that the displacement y of Q satisfies

$$y = r\cos x° + \sqrt{(l^2 - r^2 \sin^2 x°)}$$

find the sequence of values of y where $r = 5$, $l = 10$ and the interval between successive values of $x°$ is $1°$.

Solution   In this example the independent variable x is restricted to the sequence of values $\{0, 1, 2, \ldots, 360\}$. The corresponding sequence of values of y can be calculated from the formula

$$y_k = 5 \cos k° + \sqrt{(100 - 25 \sin^2 k°)}$$

$$= 5[\cos k° + \sqrt{(4 - \sin^2 k°)}]$$

$$= 5[\cos k° + \sqrt{(3 + \cos^2 k°)}]$$

Thus

$$\{y_k\}_{k=0}^{360} = \{15, 14.999, 14.995, 14.990, \ldots, 14.999, 15\}$$

Notice how in Example 7.3 we did not list every element of the sequence. Instead, we relied on the formula for y_k to supply the value of a particular element in the sequence. In Example 7.1 we could use the recurrence relation to determine the elements of the sequence. In Example 7.2, however, there is no formula or recurrence relation that enables us to work out the elements of the sequence. These three examples are representative of the general situation.

A **series** is an extended sum of terms. For example, a very simple series is the sum

$$1 + 2 + 3 + 4 + 5 + 6 + 7 + 8 + 9 + 10 + 11$$

When we look for a general formula for summing such series, we effectively turn it into a sequence, writing, for example, the sum to eleven terms as S_{11} and the sum to n terms as S_n, where

$$S_n = 1 + 2 + 3 + \ldots + n = \sum_{k=1}^{n} k$$

Series often occur in the mathematical analysis of practical problems and we give some important examples later in this chapter.

7.2.2 Graphical representation of sequences

Sequences, as remarked earlier, are functions whose domains are the whole numbers. We can display their properties using a conventional graph with the independent variable (now an integer n) represented as points along the positive x axis. This will show the behaviour of the sequence for low values of n but will not display the whole behaviour adequately. An alternative approach displays the terms of the sequence against the values of $1/n$. This enables us to see the whole sequence but in a rather 'telescoped' manner. When the terms of a sequence are generated by a recurrence relation a third method, known as a **cobweb diagram**, is available to us. We will illustrate these three methods in the examples below.

Example 7.4   Calculate the sequence $\left\{1 + \dfrac{(-1)^n}{n}\right\}_{n=1}^{10}$ and illustrate the answer graphically.

Solution By means of a calculator we can obtain the terms of the sequence explicitly (to 2dp) as

$$\{0, 1.50, 0.67, 1.25, 0.80, 1.17, 0.86, 1.12, 0.89, 1.10\}$$

The graph of this function is strictly speaking the set of points

$$\{(1, 0), (2, 1.5), (3, 0.67), \ldots, (10, 1.1)\}$$

These can be displayed on a graph as isolated points but it is more helpful to the reader to join the points by straight line segments, as shown in Figure 7.2. The figure tells us that the values of the sequence oscillate about the value 1, getting closer to it as n increases.

Example 7.5 Calculate the sequence $\{n^{1/n}\}_{n=4}^{10}$ and show the points $\{(1/n, n^{1/n})\}_{n=4}^{10}$ on a graph.

Solution Using a calculator we obtain (to 2dp)

$$\{n^{1/n}\}_{n=4}^{10} = \{1.41, 1.38, 1.35, 1.32, 1.30, 1.28, 1.26\}$$

and the set of points is

$$\{(1/n, n^{1/n})\}_{n=4}^{10} = \{(0.25, 1.41), (0.2, 1.38), \ldots, (0.1, 1.26)\}$$

In Figure 7.3 these points are displayed with a smooth curve drawn through them. The graph suggests that as n increases (i.e. $1/n$ decreases), $n^{1/n}$ approaches the value 1.

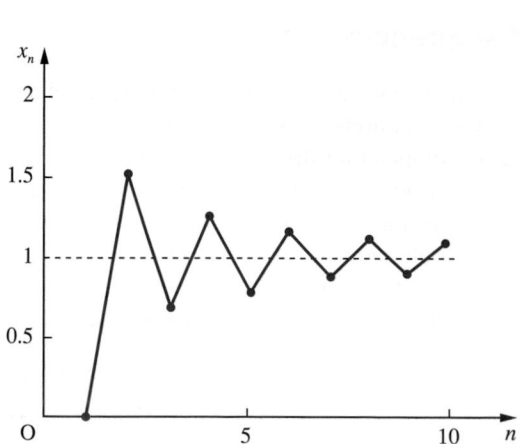

Figure 7.2 Graph of the sequence defined by $x_n = 1 + (-1)^n/n$.

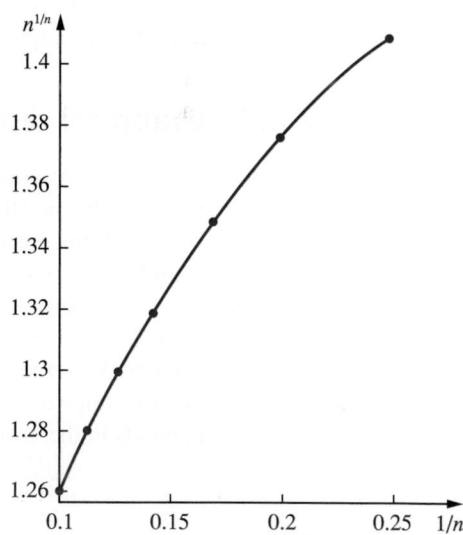

Figure 7.3 Graph of the points $\{(\frac{1}{n}, n^{1/n})\}_{n=4}^{10}$.

Figure 7.4
(a) Graphs of
$y = (x + 10)/$
$(5x + 1)$ and $y = x$.
(b) Construction of
the sequence defined
by $x_{n+1} = (x_n + 10)/$
$(5x_n + 1)$, $x_0 = 1$.

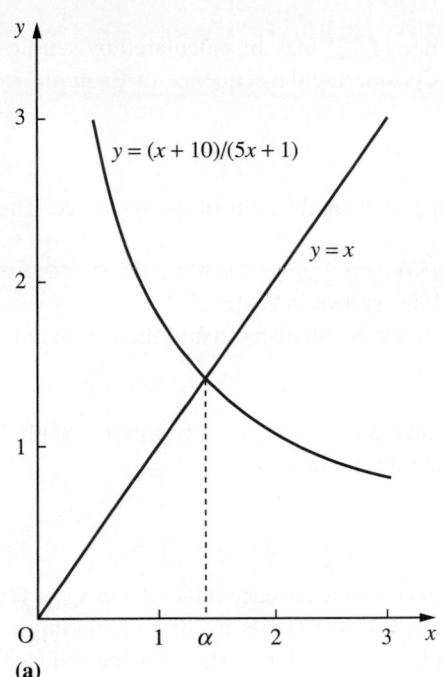

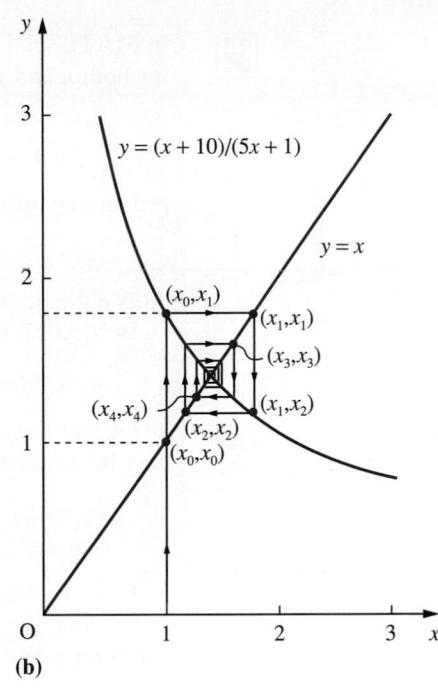

(a)

(b)

Example 7.6　Calculate the sequence $\{x_n\}_{n=0}^6$ where $x_0 = 1$ and $x_{n+1} = \dfrac{x_n + 10}{5x_n + 1}$.

Solution　Using a calculator we obtain (to 2dp) the values of the sequence

$$\{x_n\}_{n=0}^6 = \{1, 1.83, 1.16, 1.64, 1.27, 1.54, 1.33\}$$

We can display this sequence very effectively using a **cobweb diagram**. To construct this we first draw the graphs of $y = (x + 10)/(5x + 1)$ and $y = x$ as shown in Figure 7.4(a). Then we construct the points of the sequence by starting at $x = x_0 = 1$. Drawing a vertical line through $x = 1$, we cut $y = x$ at x_0 and $y = (x + 10)/(5x + 1)$ at $y = x_1$. Now drawing the horizontal line through (x_0, x_1) we find it cuts $y = x$ at x_1. Next we draw the vertical line through (x_1, x_1) to locate x_2 and so on as shown in Figure 7.4(b). We can see from this diagram that as n increases, x_n approaches the point of intersection of the two graphs, that is the value α where

$$\alpha = \frac{\alpha + 10}{5\alpha + 1} \quad (\alpha > 0)$$

This gives $\alpha = \sqrt{2}$. The value α is termed the **fixed point** of the iteration. Setting $x_n = \alpha$ returns the value $x_{n+1} = \alpha$.

As we have seen, three different methods can be used for representing sequences graphically. The choice of method will depend on the problem context.

In MATLAB a sequence $\{f_n\}_{n=a}^{n=b}$ may be calculated by setting up an array of values for both n and $y = f_n$. Considering the sequence of Example 7.4, the commands

```
n = 1:1:10;
y = 1 + ((-1).^n)./n
```

produce an array for the calculated values of the sequence. The additional command

```
plot(n,y '*')
```

plots a graph of the points shown in Figure 7.2.

In MAPLE the sequence is calculated using the command

```
evalf(seq(1 + (-1)^n/n, n = 1..10));
```

Using the `maple` command in MATLAB's Symbolic Math Toolbox the sequence may be calculated using the commands

```
syms n
maple('evalf(seq(1 + (-1)^n/n,n = 1..10))')
```

If the sequence is given as a recurrence relationship $x_{n+1} = f(x_n)$, as in Example 7.6, then it may be calculated in MATLAB by first expressing f as an `inline` object and then using a simple `for-end` loop. Thus for the sequence of Example 7.6 the commands

```
f = inline('(x + 10)/(5*x + 1)')
x = 1; y(1) = x;
for n = 1:6
y(n + 1) = f(x); x = y(n + 1);
end
double(y)
```

return the answer:

```
1.0000 1.8333 1.1639 1.2669 1.5361 1.3290
```

and the additional command

```
plot(y '-')
```

plots a graph of the sequence.

7.2.3 Exercises

1 Write down x_1, x_2 and x_3 for the sequences defined by

(a) $x_n = \dfrac{n^2}{n + 2}$

(b) $x_{n+1} = x_n + 4,\ x_0 = 2$

(c) $x_{n+1} = \dfrac{-x_n}{4},\ x_0 = 256$

2 On the basis of the evidence of the first four terms give a recurrence relation for the sequence

$$\{5,\ 15/8,\ 45/64,\ 135/512,\ \dots\}$$

3 A sequence is defined by $x_n = pn + q$ where p and q are constants. If $x_2 = 7$ and $x_8 = -11$, find p and q and write down

(a) the first four terms of the sequence;

(b) the defining recurrence relation for the sequence.

4 Triangular numbers (T_n) are defined by the number of dots that occur when arranged in equilateral triangles as shown in Figure 7.5. Show that $T_n = \frac{1}{2}n(n + 1)$ for every positive integer n.

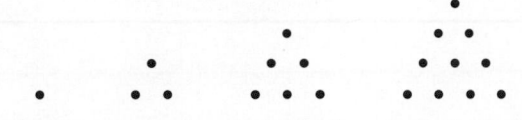

Figure 7.5 Triangular numbers.

5 A detergent manufacturer wishes to forecast their future sales. Their market research department assesses that their 'Number One' brand has 20% of the potential market at present. They also estimate that 15% of those who bought 'Number One' in a given month will buy a different detergent in the following month and that 35% of those who bought a rival brand will buy 'Number One' in the next month. Show that their share P_n% of the market in the nth month satisfies the recurrence relation

$$P_{n+1} = 35 + 0.5P_n, \quad \text{with } P_0 = 20$$

Find the values of P_n for $n = 1, 2, 3$ and 4 and illustrate them on an appropriate diagram.

6 (a) If $x_r = r(r - 1)(2r - 5)$, calculate $\sum_{r=0}^{4} x_r$

(b) If $x_r = r^{r+1} + 3(-1)^r$, calculate $\sum_{r=1}^{5} x_r$

(c) If $x_r = r^2 - 3r + 1$, calculate $\sum_{r=2}^{6} x_r$

7 A precipitate at the bottom of a beaker of capacity V always retains about it a volume v of liquid. What percentage of the original solution remains about it after it has been washed n times by filling the beaker with distilled water and emptying it?

8 A certain process in statistics involves the following steps S_i ($i = 1, 2, \ldots , 6$):

S_1: Selecting a number from the set $T = \{x_1, x_2, \ldots , x_n\}$

S_2: Subtracting 10 from it

S_3: Squaring the result

S_4: Repeating steps S_1–S_3 with the remaining numbers in T

S_5: Adding the results obtained at stage S_3 of each run through

S_6: Dividing the result of S_5 by n

Express the final outcome algebraically using Σ notation.

9 Newton's recurrence formula for determining the root of a certain equation is

$$x_{n+1} = \frac{x_n^2 - 1}{2x_n - 3}$$

Taking $x_0 = 3$ as your initial approximation, obtain the root correct to 4sf.

By setting $x_{n+1} = x_n = \alpha$ show that the fixed points of the iteration are given by the equation $\alpha^2 - 3\alpha + 1 = 0$.

10 Calculate the terms of the sequence

$$\left\{ \frac{n^4}{n^4 + n^3 + 1} \right\}_{n=0}^{5}$$

and show them on graphs similar to Figures 7.2 and 7.3.

11 Calculate the sequence $\{x_n\}_{n=0}^{6}$ where

$$x_{n+1} = \frac{x_n + 2}{x_n + 1}, \quad x_0 = 1$$

Show the sequence using a cobweb diagram similar to Figure 7.4.

12 A steel ball-bearing drops on to a smooth hard surface from a height h. The time to the first impact is $T = \sqrt{(2h/g)}$ where g is the acceleration due to gravity. The times between successive bounces are $2eT, 2e^2T, 2e^3T, \ldots$, where e is the coefficient of restitution between the ball and the surface ($0 < e < 1$). Find the total time taken up to the fifth bounce. If $T = 1$ and $e = 0.1$, show in a diagram the times taken up to the first, second, third, fourth and fifth bounces and estimate how long the total motion lasts.

13 Consider the following puzzle: how many single, loose, smooth 30 cm bricks are necessary to form a single leaning pile with no part of the bottom brick under the top brick? Begin by considering

a pile of 2 bricks. The top brick cannot project further than 15 cm without collapse. Then consider a pile of 3 bricks. Show that the top one cannot project further than 15 cm beyond the second one and that the second one cannot project further than 7.5 cm beyond the bottom brick (so that the maximum total lean is $(\frac{1}{2} + \frac{1}{4})30$ cm). Show that the maximum total lean for a pile of 4 bricks is $(\frac{1}{2} + \frac{1}{4} + \frac{1}{6})30$ cm and deduce that for a pile of n bricks it is $(\frac{1}{2} + \frac{1}{4} + \frac{1}{6} + \ldots + \frac{1}{2n+2})$ 30 cm. Hence solve the puzzle.

7.3 Finite sequences and series

In this section we consider some finite sequences and series that are frequently used in engineering.

7.3.1 Arithmetical sequences and series

An **arithmetical sequence** is one in which the difference between successive terms is a constant number. Thus, for example, $\{2, 5, 8, 11, 14\}$ and $\{2, 0, -2, -4, -6, -8, -10\}$ define arithmetical sequences. In general an arithmetical sequence has the form $\{a + kd\}_{k=0}^{n-1}$ where a is the first term, d is the common difference and n is the number of terms in the sequence. Thus, in the first example above, $a = 2$, $d = 3$ and $n = 5$, and in the second example, $a = 2$, $d = -2$ and $n = 7$. (The old name for such sequences was **arithmetical progressions**.) The sum of the terms of an arithmetical sequence is an **arithmetical series**. The general arithmetical series is

$$S_n = a + (a + d) + (a + 2d) + \ldots + [a + (n-1)d] = \sum_{k=0}^{n-1}(a + kd) \qquad (7.1)$$

To obtain an expression for the sum of the n terms in this series, write the series in the reverse order,

$$S_n = \quad a \quad + \quad (a + d) \quad + \quad (a + 2d) \quad + \ldots + [a + (n-1)d]$$

$$S_n = [a + (n-1)d] + [a + (n-2)d] + [a + (n-3)d] + \ldots + \quad a$$

Summing the two series then gives

$$2S_n = [2a + (n-1)d] + [2a + (n-1)d] + [2a + (n-1)d] + \ldots + [2a + (n-1)d]$$

giving the sum S_n of the first n terms of an arithmetical series as

$$S_n = \tfrac{1}{2}n[2a + (n-1)d] = \tfrac{1}{2}n(\text{first term} + \text{last term}) \qquad (7.2)$$

The result is illustrated geometrically for $n = 6$ in Figure 7.6, where the breadth of each rectangle is unity and the area under each shaded step is equal to a term of the series. In particular, when $a = 1$ and $d = 1$,

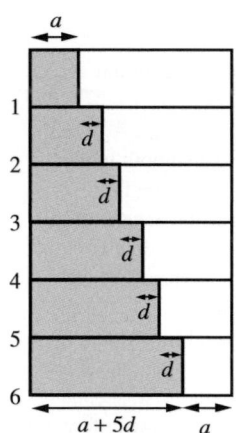

Figure 7.6

$$S_6 = \sum_{k=0}^{5}(a + kd)$$

$$= \tfrac{1}{2} \times 6 \times (2a + 5d).$$

$$S_n = 1 + 2 + \ldots + n = \sum_{k=1}^{n} k = \tfrac{1}{2}n(n + 1) \qquad (7.3)$$

Example 7.7 How many terms of the arithmetical series 11, 15, 19, etc. will give a sum of 341?

Solution In this particular case the first term $a = 11$ and the common difference $d = 4$. We need to find the number of terms n such that the sum S_n is 341. Using the result in (7.2)

$$S_n = 341 = \tfrac{1}{2}n[2(11) + (n-1)(4)]$$

leading to

$$4n^2 + 18n - 682 = 0$$

or

$$(4n + 62)(n - 11) = 0$$

giving

$$n = 11 \quad \text{or} \quad n = -\tfrac{31}{2}$$

Since $n = -\tfrac{31}{2}$ is not a whole number, the number of terms required is $n = 11$.

Example 7.8 A contractor agrees to sink a well 40 metres deep at a cost of £30 for the first metre, £35 for the second metre and increasing by £5 for each subsequent metre.

(a) What is the total cost of sinking the well?

(b) What is the cost of drilling the last metre?

Solution (a) The total cost constitutes an arithmetical series whose terms are the cost per metre. Thus, taking $a = 30$, $d = 5$ and $n = 40$ in (7.2) gives the total cost

$$S_n = \text{£}\tfrac{40}{2}[2(30) + (40 - 1)5] = \text{£}5100$$

(b) The cost of drilling the last metre is given by the 40th term of the series. Since the nth term is $a + (n-1)d$, the cost of drilling the last metre $= 30 + (40-1)5 = \text{£}225$.

7.3.2 Geometric sequences and series

A **geometric sequence** is one in which the ratio of successive terms is a constant number. Thus, for example, $\{2, 4, 8, 16, 32\}$ and $\{2, -1, \tfrac{1}{2}, -\tfrac{1}{4}, \tfrac{1}{8}, -\tfrac{1}{16}, \tfrac{1}{32}\}$ define geometric sequences. In general a geometric sequence has the form $\{ar^k\}_{k=0}^{n-1}$ where a is the first term, r is the common ratio and n is the number of terms in the sequence. Thus, in the first example above, $a = 2$, $r = 2$, $n = 5$ and, in the second example, $a = 2$, $r = -\tfrac{1}{2}$ and $n = 7$. (The old name for such sequences was **geometric progressions**.) The sum of the terms of a geometric sequence is called a **geometric series**. The general geometric series has the form

$$S_n = a + ar + ar^2 + ar^3 + \ldots + ar^{n-1}$$

To obtain the sum S_n of the first n terms of the series we multiply S_n by the common ratio r, to obtain

$$rS_n = ar + ar^2 + \ldots + ar^{n-1} + ar^n$$

Subtracting this from S_n then gives

$$S_n - rS_n = a - ar^n$$

so that

$$(1 - r)S_n = a(1 - r^n)$$

Thus for $r \neq 1$, the sum of the first n terms is

$$S_n = \sum_{k=0}^{n-1} ar^k = \frac{a(1 - r^n)}{1 - r} \qquad\qquad (7.4)$$

Clearly, for the particular case of $r = 1$ the sum is $S_n = an$.

The geometric series is very important. It has many applications in practical problems as well as within mathematics.

Example 7.9   In its publicity material an insurance company guarantees that, for a fixed annual premium payable at the beginning of each year for a period of 25 years, the return will be at least equivalent to the premiums paid, together with 3% per annum compound interest. For an annual premium of £250 what is the guaranteed sum at the end of 25 years?

Solution   The first-year premium earns interest for 25 years and thus guarantees

$$£250(1 + 0.03)^{25}$$

The second-year premium earns interest for 24 years and thus guarantees

$$£250(1 + 0.03)^{24}$$

$$\vdots$$

The final-year premium earns interest for 1 year and thus guarantees

$$£250(1 + 0.03)$$

Thus, the total sum guaranteed is

$$£250[(1.03) + (1.03)^2 + \ldots + (1.03)^{25}]$$

The term inside the square brackets is a geometric series. Thus, taking $a = 1.03, r = 1.03$ and $n = 25$ in (7.4) gives

$$\text{Guaranteed sum} = £250\left[1.03\frac{(1.03^{25} - 1)}{(1.03 - 1)}\right] \approx £9388.$$

7.3.3 Other finite series

In addition to the arithmetical and geometric series, there are other finite series that occur in engineering applications for which an expression can be obtained for the sum of the first n terms. We shall illustrate this in Examples 7.10 and 7.11.

Example 7.10   Consider the sum-of-squares series

$$S_n = 1^2 + 2^2 + 3^2 + \ldots + n^2 = \sum_{k=1}^{n} k^2$$

Obtain an expression for the sum of this series.

Solution   There are various methods for finding the sum. A method that can be generalized makes use of the identity

$$(k + 1)^3 - k^3 = 3k^2 + 3k + 1$$

Thus

$$\sum_{k=1}^{n} [(k + 1)^3 - k^3] = \sum_{k=1}^{n} (3k^2 + 3k + 1)$$

The left-hand side equals

$$2^3 - 1^3 + 3^3 - 2^3 + 4^3 - 3^3 + \ldots + (n + 1)^3 - n^3 = (n + 1)^3 - 1$$

The right-hand side equals

$$3 \sum_{k=1}^{n} k^2 + 3 \sum_{k=1}^{n} k + \sum_{k=1}^{n} 1$$

Now

$$\sum_{k=1}^{n} k = \tfrac{1}{2}n(n + 1) \quad \text{from (7.3)} \quad \text{and} \quad \sum_{k=1}^{n} 1 = n$$

so that

$$(n + 1)^3 - 1 = 3 \sum_{k=1}^{n} k^2 + \frac{3n}{2}(n + 1) + n$$

whence

$$\sum_{k=1}^{n} k^2 = \tfrac{1}{6}n(n + 1)(2n + 1) \tag{7.5}$$

This method can be generalized to obtain the sum of other similar series. For example, to find the sum of cubes series $\sum_{k=1}^{n} k^3$, we would consider $(k + 1)^4 - k^4$ and so on.

Example 7.11 Obtain the sum of the series

$$S_n = \frac{1}{1 \cdot 2} + \frac{1}{2 \cdot 3} + \frac{1}{3 \cdot 4} + \ldots + \frac{1}{n(n+1)} = \sum_{k=1}^{n} \frac{1}{k(k+1)}$$

Solution The technique for summing this series is to express the general term in its partial fractions:

$$\frac{1}{k(k+1)} = \frac{1}{k} - \frac{1}{k+1}$$

Then

$$S_n = \sum_{k=1}^{n} \frac{1}{k} - \sum_{k=1}^{n} \frac{1}{k+1}$$

$$= \left(1 + \frac{1}{2} + \frac{1}{3} + \ldots + \frac{1}{n}\right) - \left(\frac{1}{2} + \frac{1}{3} + \ldots + \frac{1}{n} + \frac{1}{n+1}\right)$$

$$= 1 - \frac{1}{n+1}$$

giving

$$S_n = \frac{n}{n+1}$$

There are many other similar series that can be summed by expressing the general term in its partial fractions. Some examples are given in Exercises 7.3.4.

Example 7.12 Obtain the sum of the series

$$S_n = 1 + 2r + 3r^2 + 4r^3 + \ldots + nr^{n-1} = \sum_{k=1}^{n} kr^{k-1}$$

Solution The technique for summing this *arithmetico-geometric* series is similar to that for summing geometric series. We multiply S_n by r and then subtract the result from S_n. Thus

$$rS_n = r + 2r^2 + 3r^3 + \ldots + nr^n$$

and

$$(1 - r)S_n = 1 + r + r^2 + r^3 + \ldots r^{n-1} - nr^n$$

The first n terms on the right-hand side of this equation form a geometric series and using result (7.4) we can write

$$(1 - r)S_n = \frac{1 - r^n}{1 - r} - nr^n$$

Hence

$$S_n = \frac{1 - r^n - nr^n(1-r)}{(1-r)^2} = \frac{1 - (n+1)r^n + nr^{n+1}}{(1-r)^2} \tag{7.6}$$

This method can be generalized to obtain the sum of other similar series, for example

$$\sum_{k=1}^{n} (2k+1)r^{k-1} \quad \text{and} \quad \sum_{k=1}^{n} k^2 r^{k-1}$$

Example 7.13   Sum the series

$$S_n = 1 + \cos\theta + \cos 2\theta + \ldots + \cos(n-1)\theta$$

Solution   The easiest way of summing this series is to recall the formula

$$e^{j\theta} = \cos\theta + j\sin\theta$$

Then we can write

$$S_n = Re\{1 + e^{j\theta} + e^{j2\theta} + e^{j3\theta} + \ldots + e^{j(n-1)\theta}\}$$

The series inside the brackets is a geometric series with common ratio $e^{j\theta}$ and using result (7.4) we obtain

$$S_n = Re\left\{\frac{1 - e^{jn\theta}}{1 - e^{j\theta}}\right\}$$

rearranging the expression inside the brackets we have

$$S_n = Re\left\{\frac{e^{j(n-\frac{1}{2})\theta} - e^{-j\frac{1}{2}\theta}}{e^{j\frac{1}{2}\theta} - e^{-j\frac{1}{2}\theta}}\right\}$$

$$= Re\left\{\frac{\cos(n-\frac{1}{2})\theta + j\sin(n-\frac{1}{2})\theta - \cos\frac{1}{2}\theta + j\sin\frac{1}{2}\theta}{2j\sin\frac{1}{2}\theta}\right\}$$

$$= \frac{1}{2}\left\{\frac{\sin(n-\frac{1}{2})\theta}{\sin\frac{1}{2}\theta} + 1\right\}$$

The same method can be used to show that $\displaystyle\sum_{k=1}^{n} \sin k\theta = \sin(n+1)\theta\sin(n\theta)/\sin\frac{1}{2}\theta$.

Symbolic summation may be achieved in MATLAB using the $symsum$ command. MAPLE command is similar with minor syntax differences. For example, to sum of the series in Example 7.10 we have

MATLAB	MAPLE
`syms x k n`	
`s = symsum(k^2,1,n);`	`sum('k^2','k' = 1..n):`
`s = factor(s);`	`factor(%);`
`pretty(ans)`	

returns

$$s = \tfrac{1}{6}n(n + 1)(2n + 1)$$

Similarly, considering Example 7.11

```
syms x k n
s = symsum(1/(k*(k + 1)),1,n);        sum('1/(k*(k + 1))',
s = simplify(s)                        'k' = 1..n);
```

returns

$$s = n/(n + 1)$$

$$-\frac{1}{n + 1} + 1$$

(Note: When using MAPLE it is recommended and often necessary (see MAPLE help) that both f and k be enclosed in single quotes to prevent premature evaluation (for example, k may have a previous value). Thus the common format is `sum('f', 'k' = m..n))`.)

7.3.4 Exercises

14 (a) Find the fifth and tenth terms of the arithmetical sequence whose first and second terms are 4 and 7.
(b) The first and sixth terms of a geometric sequence are 5 and 160 respectively. Find the intermediate terms.

15 An individual starts a business and loses £150k in the first year, £120k in the second year and £90k in the third year. If the improvement continues at the same rate, find the individual's total profit or loss at the end of 20 years.
After how many years would the losses be just balanced by the gains?

16 Show that

$$\frac{1}{1 + \sqrt{x}}, \quad \frac{1}{1 - x}, \quad \frac{1}{1 - \sqrt{x}}$$

are in arithmetical progression and find the nth term of the sequence of which these are the first three terms.

17 The area of a circle of radius 1 is a transcendental number (that is, a number that cannot be obtained by the process of solving algebraic equations) denoted by the Greek letter π. To calculate its value, we may use a limiting process in which π is the limit of a sequence of known numbers. The method used by Archimedes was to inscribe in the circle a sequence of regular polygons.

As the number of sides increased, so the polygon 'filled' the circle. Show, by use of the trigonometric identity $\cos 2\theta = 1 - 2\sin^2\theta$, that the area a_n of an inscribed regular polygon of n sides satisfies the equation

$$2\left(\frac{a_{2n}}{n}\right)^2 = 1 - \sqrt{\left[1 - \left(\frac{2a_n}{n}\right)^2\right]} \quad (n \geqslant 4)$$

Show that $a_4 = 2$ and use the recurrence relation to find a_{64}.

18 A **harmonic sequence** is a sequence with the property that every three consecutive terms (a, b and c, say) of the sequence satisfy

$$\frac{a}{c} = \frac{a - b}{b - c}$$

Prove that the reciprocals of the terms of a harmonic sequence form an arithmetical progression. Hence find the intermediate terms of a harmonic sequence of 8 terms whose first and last terms are $\frac{2}{3}$ and $\frac{2}{17}$ respectively.

19 The price of houses increases at 10% per year. Show that the price P_n in the nth year satisfies the recurrence relation

$$P_{n+1} = 1.1P_n$$

A house is currently priced at £80 000. What was its price two years ago? What will be its price in five years' time? After how many years will its price be double what it is now?

20 Evaluate each of the following sums:

(a) $1 + 2 + 3 + \dots + 152 + 153$

(b) $1^2 + 2^2 + 3^2 + \dots + 152^2 + 153^2$

(c) $\frac{1}{2} + \frac{1}{4} + \frac{1}{8} + \dots + (\frac{1}{2})^{152} + (\frac{1}{2})^{153}$

(d) $2 + 6 + 18 + \dots + 2(3)^{152} + 2(3)^{153}$

(e) $1 \cdot 2 + 2 \cdot 3 + 3 \cdot 4 + \dots + 152 \cdot 153 + 153 \cdot 154$

(f) $\dfrac{1}{1 \cdot 2} + \dfrac{1}{2 \cdot 3} + \dfrac{1}{3 \cdot 4} + \dots + \dfrac{1}{152 \cdot 153} + \dfrac{1}{153 \cdot 154}$

21 A certain bacterium propagates itself by subdividing, creating four additional bacteria, each identical to the parent bacterium. If the bacteria subdivide in this manner n times, then, assuming that none of the bacteria die, the number of bacteria present after each subdivision is given by the sequence $\{B_k\}_{k=0}^n$, where

$$B_k = \frac{4^{k+1} - 1}{3}$$

Three such bacteria subdivide n times and none of the bacteria die. The total number of bacteria is then 1 048 575. How many times did the bacteria divide?

22 By considering the sum

$$\sum_{k=1}^{n} [(k + 1)^4 - k^4]$$

show that

$$\sum_{k=1}^{n} k^3 = [\tfrac{1}{2}n(n + 1)]^2$$

23 The repayment instalment of a fixed rate, fixed period loan may be calculated by summing the *present values* of each instalment. This sum must

equal the amount borrowed. The present value of an instalment £x paid after k years where $r\%$ is the rate of interest is

$$£\frac{x}{(1 + r/100)^k}$$

Thus £1000 borrowed over n years at $r\%$ satisfies the equation

$$1000 = \frac{x}{1 + r/100} + \frac{x}{(1 + r/100)^2} + \dots$$
$$+ \frac{x}{(1 + r/100)^n}$$

Find x in terms of r and n, and compute its value when $r = 10$ and $n = 20$.

24 Consider the series

$$S_n = \frac{1}{2} + \frac{2}{4} + \frac{3}{8} + \dots + \frac{n}{2^n}$$

Show that

$$\tfrac{1}{2}S_n = \frac{1}{4} + \frac{2}{8} + \frac{3}{16} + \dots + \frac{n}{2^{n+1}}$$

and hence that

$$S_n - \tfrac{1}{2}S_n = \frac{1}{2} + \frac{1}{4} + \frac{1}{8} + \frac{1}{16} + \dots + \frac{1}{2^n} + \frac{n}{2^{n+1}}$$

Hence sum the series.

25 Consider the general arithmetico-geometric series

$$S_n = a + (a + d)r + (a + 2d)r^2 + \dots$$
$$+ [a + (n - 1)d]r^{n-1}$$

Show that

$$(1 - r)S_n = a + dr + dr^2 + \dots$$
$$+ dr^{n-1} - [a + (n - 1)d]r^n$$

and find a simple expression for S_n.

7.4 Recurrence relations

We saw in Example 7.1 that sometimes the elements of a sequence satisfy a recurrence relation such that the value of an element x_n of a sequence $\{x_k\}$ can be expressed in terms of the values of earlier elements of the sequence. In general we may have a formula of the form

$$x_n = f(x_{n-1}, x_{n-2}, \ldots, x_1, x_0)$$

In this section we are going to consider two commonly occurring types of recurrence relation. These will provide sufficient background to make possible the solution of more difficult problems.

7.4.1 First-order linear recurrence relations with constant coefficients

These relations have the general form

$$x_{n+1} = ax_n + b_n, \quad n = 0, 1, 2, \ldots$$

where a is constant and b_n is a known sequence. The simplest case that occurs is when $b_n = 0$, when the relation reduces to

$$x_{n+1} = ax_n \tag{7.7}$$

This is called a **homogeneous relation** and every solution is a geometric sequence of the form

$$x_n = Aa^n \tag{7.8}$$

This is called the **general solution** of (7.7) since A is a constant which may be given any value. To determine the value of A we require more information about the sequence. For example, if we know the value of x_0 (say C) then $C = Aa^0$, which gives the value of A.

A slightly more difficult example is

$$x_{n+1} = ax_n + b \tag{7.9}$$

where b is a constant as well as a.

If the first term of the sequence is $x_0 = C$, as before, then

$$x_1 = aC + b$$

$$x_2 = ax_1 + b = a(aC + b) + b = Ca^2 + b(1 + a)$$

$$x_3 = ax_2 + b = a[Ca^2 + b(1 + a)] + b = Ca^3 + b(1 + a + a^2)$$

and so on.

In general, we obtain

$$x_n = Ca^n + \left(\frac{1 - a^n}{1 - a}\right)b, \quad a \neq 1$$

Rearranging, we can express this as

$$x_n = Aa^n + \frac{b}{1 - a}, \quad a \neq 1 \tag{7.10}$$

where $A = C - b/(1 - a)$. After the next example we will see that this solution (and that of more general problems) can be obtained more quickly by an alternative method.

Notice that Aa^n is the general solution of the homogeneous relation (7.7) and that $x_n = b/(1 - a)$, for all n, satisfies the full recurrence relation $x_{n+1} = ax_n + b$, so that it is a **particular solution** of the relation.

Example 7.14   Calculate the fixed annual payments £B required to amortize a debt of £D over N years, when the rate of interest is fixed at $100i\%$.

Solution   Let £d_n denote the debt after n years. Then, following the same argument as in Example 7.1, $d_0 = D$ and

$$d_{n+1} = (1 + i)d_n - B$$

This is similar to the recurrence relation (7.9) but with $a = (1 + i)$ an $b = -B$. Hence, using (7.10) we can write the general solution as

$$d_n = A(1 + i)^n - \frac{B}{1 - (1 + i)} = A(1 + i)^n + B/i$$

In addition, we know that $d_0 = D$ so that $D = A + B/i$ and thus the particular solution is given by

$$d_n = (D - B/i)(1 + i)^n + B/i$$

We require the value of B so that the debt is zero after N years, that is $d_N = 0$. Thus

$$0 = (D - B/i)(1 + i)^N + B/i$$

Solving this equation for B gives

$$B = \frac{iD(1 + i)^N}{(1 + i)^N - 1} = iD/[1 - (1 + i)^{-N}]$$

as the required payment.

In summary, we have that the general solution to the first-order recurrence relation

$$x_{n+1} = ax_n + b$$

can be expressed as the sum of the **general solution** of the reduced relation

$$x_{n+1} = ax_n$$

and a **particular solution** of the full relation (7.9).

This is true for linear recurrence relations in general, that is, recurrence relations of the form

$$x_{n+1} = a_n x_n + a_{n-1} x_{n-1} + \ldots + a_1 x_1 + a_0$$

where the coefficients a_k are independent of the x_k but may depend on n. The property is easy to show in full generality but the same proof holds for the simplest case (7.9) above.

Suppose we can identify one particular solution p_n of (7.9) so that

$$p_{n+1} = ap_n + b$$

Now we seek a function q_n which complements p_n in such a way that

$$x_n = p_n + q_n$$

is the general solution of (7.9). Substituting x_n into this relation gives

$$p_{n+1} + q_{n+1} = ap_n + aq_n + b$$

Since $p_{n+1} = ap_n + b$, this implies that

$$q_{n+1} = aq_n$$

From (7.8), the general solution of this relation is

$$q_n = Aa^n$$

where A is a constant. Thus the general solution of (7.9) is

$$x_n = p_n + Aa^n$$

Because q_n complements p_n to form the general solution, it is usually called the **complementary solution**. As we have seen, with first-order recurrence relations, we can always find the complementary solution. Thus we are left with the task of finding the particular solution p_n. The method for finding p_n depends on the term b, as we illustrate in Example 7.15.

Indeed, the property of the general solution being the sum of a particular solution and a complementary solution applies to all linear systems, both continuous and discrete. We will meet it again in Chapter 10 when considering the general solution of linear ordinary differential equations.

Example 7.15

Find the general solutions of the recurrence relations

(a) $x_{n+1} = 3x_n + 4$ (b) $x_{n+1} = x_n + 4$

(c) $x_{n+1} = \alpha x_n + C\beta^n$ (d) $x_{n+1} = \alpha x_n + C\alpha^n$ (α, β, C given constants)

Solution

(a) First we try to find any function of n which will satisfy the relation. Since it contains the constant term 4, it is common sense to see if a constant K can be found which satisfies the relation. (Then all terms will be constants.) Setting $x_n = K$ implies $x_{n+1} = K$ and we have

$$K = 3K + 4$$

which gives $K = -2$. Thus, in this case, we can choose $p_n = -2$. Next we find the complementary solution q_n, which is the general solution of

$$x_{n+1} = 3x_n$$

From (7.8) we can see that $q_n = A3^n$ where A is a constant. Thus the general solution of (a) is

$$x_n = -2 + A3^n$$

(b) The basic steps are the same for this relation. We first find a particular solution p_n of the relation. Then we find the complementary solution q_n, so that $x_n = p_n + q_n$ is the general solution. In this case trying $x_n = K$ leads nowhere, since we obtain the inconsistent equation $K = K + 4$. Trying something a little more complicated than just a constant, we set $x_n = Kn$ and $x_{n+1} = K(n + 1)$ and we have

$$K(n + 1) = Kn + 4$$

which yields $K = 4$ and $p_n = 4n$. The general solution of $x_{n+1} = x_n$ is $q_n = A1^n$, so that the general solution of (b) is

$$x_n = 4n + A$$

(c) Since the recurrence relation has the term $C\beta^n$, it is natural to expect a solution of the form $K\beta^n$, where K is a constant, to satisfy the relation. Setting $x_n = K\beta^n$ gives

$$K\beta^{n+1} = \alpha K\beta^n + C\beta^n$$

Dividing through by β^n gives $K\beta = \alpha K + C$, from which we deduce $K = C/(\beta - \alpha)$ provided that $\beta \neq \alpha$. Thus we deduce the particular solution

$$p_n = C\beta^n/(\beta - \alpha)$$

The complementary solution q_n is the general solution of

$$x_{n+1} = \alpha x_n$$

which, using (7.7), is $q_n = A\alpha^n$. Hence the general solution of (c) is

$$x_n = C\beta^n/(\beta - \alpha) + A\alpha^n$$

(d) This is the special case of (c) where $\beta = \alpha$. If we set $p_n = K\alpha^n$, we obtain the equation $K\alpha^{n+1} = K\alpha^{n+1} + C\alpha^n$, which can only be true if $C = 0$. (We see then that p_n is the solution of $x_{n+1} = \alpha x_n$, that is, it is the complementary solution.) As in case (b), we instead seek a solution of the form $p_n = Kn\alpha^n$, so that $p_{n+1} = K(n + 1)\alpha^{n+1}$ and

$$K(n + 1)\alpha^{n+1} = \alpha Kn\alpha^n + C\alpha^n$$

This last equation gives $K = C/\alpha$. Hence the general solution of (d) is

$$x_n = Cn\alpha^{n-1} + A\alpha^n$$

where A is an arbitrary constant.

7.4.2 Exercises

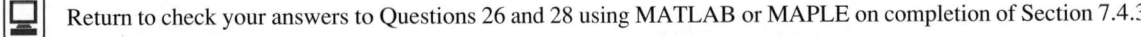

Return to check your answers to Questions 26 and 28 using MATLAB or MAPLE on completion of Section 7.4.3.

26 Find the general solutions of the recurrence relations

(a) $x_{n+1} = 2x_n - 3$ (b) $x_{n+1} = 3x_n + 10n$

(c) $x_{n+1} = -x_n + (\frac{1}{2})^n$ (d) $x_{n+1} = 2x_n + 3 \times 2^n$

27 If a debt is amortized by equal annual payments of amount B, and if interest is charged at rate i per

annum, then the debt after n years, d_n, satisfies $d_{n+1} = (1 + i)d_n - B$, where $d_0 = D$, the initial debt.

Show that $d_n = D(1 + i)^n + B\dfrac{1 - (1 + i)^n}{i}$

and deduce that to clear the debt on the Nth

payment we must take $B = \dfrac{Di}{1 - (1 + i)^{-N}}$.

If £10 000 is borrowed at an interest rate of

0.12 (= 12%) per annum, calculate (to the nearest £) the appropriate annual payment which will amortize the debt at the end of 10 years.

For this annual payment calculate the amount of the debt d_n for $n = 1, 2, \ldots, 10$ (use the recurrence rather than its solution, and record your answers to the nearest £) and calculate the first differences for this sequence. Comment briefly on the behaviour of the first differences.

 28 Find the general solution of the linear recurrence relation

$$(n + 1)^2 x_{n+1} - n^2 x_n = 1, \quad \text{for } n \geqslant 1$$

(*Hint*: the coefficients are not constants. Use the substitution $z_n = n^2 x_n$ to find a constant coefficient equation for z_n. Find the general solution for z_n and hence for x_n.)

7.4.3 Second-order linear recurrence relations with constant coefficients

A second-order linear recurrence with constant coefficients has the form

$$x_{n+2} = ax_{n+1} + bx_n + c_n \tag{7.11}$$

If $c_n = 0$ for all n, then the relation is said to be **homogeneous**. As before, the solution of (7.11) can be expressed in the form

$$x_n = p_n + q_n$$

where p_n is any solution which satisfies (7.11) while q_n is the general solution of the associated homogeneous recurrence relation

$$x_{n+2} = ax_{n+1} + bx_n \tag{7.12}$$

Let α and β be the two roots of the algebraic equation

$$\lambda^2 = a\lambda + b$$

so that $\alpha^{n+2} = a\alpha^{n+1} + b\alpha^n$ and $\beta^{n+2} = a\beta^{n+1} + b\beta^n$, which imply that $y_n = \alpha^n$ and $y_n = \beta^n$ are particular solutions of (7.12). Since $(\lambda - \alpha)(\lambda - \beta) = 0$ implies $\lambda^2 = (\alpha + \beta)\lambda - \alpha\beta$ we may rewrite (7.12) as

$$x_{n+2} = (\alpha + \beta)x_{n+1} - \alpha\beta x_n$$

Rearranging the relation, we have

$$x_{n+2} - \alpha x_{n+1} = \beta(x_{n+1} - \alpha x_n)$$

Substituting $t_n = x_{n+1} - \alpha x_n$, this becomes

$$t_{n+1} = \beta t_n$$

with general solution, from (7.8), $t_n = C\beta^n$ where C is any constant.
Thus

$$x_{n+1} - \alpha x_n = C\beta^n$$

which, using the results of Example 7.15(c) and (d), has the general solution

$$x_n = \begin{cases} C\beta^n/(\beta - \alpha) + A\alpha^n, & \alpha \neq \beta \\ Cn\alpha^{n-1} + A\alpha^n, & \alpha = \beta \end{cases}$$

Since C is any constant, we can rewrite this in the neater form

$$x_n = \begin{cases} A\alpha^n + B\beta^n, & \alpha \neq \beta \\ A\alpha^n + Bn\alpha^n, & \alpha = \beta \end{cases} \tag{7.13}$$

where A and B are arbitrary constants. Thus (7.13) gives the general solution of (7.12) where α and β are the roots of the equation

$$\lambda^2 = a\lambda + b$$

This is called the **characteristic equation** of the recurrence relation; the Greek letter *lambda* λ is used as the unknown instead of x to avoid confusion.

Example 7.16   Find the solution of the Fibonacci recurrence relation

$$x_{n+2} = x_{n+1} + x_n$$

given $x_0 = 1$, $x_1 = 1$.

Solution   The characteristic equation of the recurrence relation is

$$\lambda^2 = \lambda + 1$$

which has roots $\lambda_1 = (1 + \sqrt{5})/2$ and $\lambda_2 = (1 - \sqrt{5})/2$.
   Hence its general solution is

$$x_n = A\left(\frac{1 + \sqrt{5}}{2}\right)^n + B\left(\frac{1 - \sqrt{5}}{2}\right)^n$$

Since $x_0 = 1$, we deduce $1 = A + B$

Since $x_1 = 1$, we deduce $1 = A\left(\frac{1 + \sqrt{5}}{2}\right) + B\left(\frac{1 - \sqrt{5}}{2}\right)$

Solving these simultaneous equations gives

$$A = (1 + \sqrt{5})/(2\sqrt{5}) \text{ and } B = -(1 - \sqrt{5})/(2\sqrt{5})$$

and hence

$$x_n = \frac{1}{\sqrt{5}}\left[\left(\frac{1 + \sqrt{5}}{2}\right)^{n+1} - \left(\frac{1 - \sqrt{5}}{2}\right)^{n+1}\right]$$

defining the Fibonacci sequence explicitly.

We have seen that we can always find the complementary solution q_n of the recurrence relation (7.11)

$$x_{n+2} = ax_{n+1} + bx_n + c_n$$

The general solution of this relation is the sum of a particular solution p_n of the relation and its complementary solution q_n. The problem, then, is how to find one solution p_n. Here we will use methods based on experience and trial and error.

Example 7.17   Find all the solutions of

(a) $x_{n+2} = \frac{7}{2}x_{n+1} - \frac{3}{2}x_n + 12$, where $x_0 = x_1 = 1$ (b) $x_{n+2} = \frac{7}{2}x_{n+1} - \frac{3}{2}x_n + 12n$

(c) $x_{n+2} = \frac{7}{2}x_{n+1} - \frac{3}{2}x_n + 3(2^n)$

Solution   (a) First we find the general solution of the associated homogeneous relation $x_{n+2} = \frac{7}{2}x_{n+1} - \frac{3}{2}x_n$ which has characteristic equation $\lambda^2 = \frac{7}{2}\lambda - \frac{3}{2}$ with roots $\lambda = 3$ and $\lambda = \frac{1}{2}$. Thus, the complementary solution is

$$x_n = A3^n + B(\tfrac{1}{2})^n$$

Next we find a particular solution of

$$x_{n+2} = \tfrac{7}{2}x_{n+1} - \tfrac{3}{2}x_n + 12$$

We try the simplest possible function $x_n = K$ (for all n). Then, if this is a solution, we have

$$K = \tfrac{7}{2}K - \tfrac{3}{2}K + 12$$

giving $K = -12$.

Thus $p_n = -12$ and the general solution is

$$x_n = -12 + A3^n + B(\tfrac{1}{2})^n$$

Applying the initial data $x_0 = 1$, $x_1 = 1$ gives two equations for the arbitrary constants A and B

$$A + B - 12 = 1$$

$$3A + \tfrac{1}{2}B - 12 = 1$$

from which we deduce $A = 13/5$ and $B = 52/5$. Thus the particular solution which fits the initial data is

$$x_n = \tfrac{13}{5}3^n + \tfrac{52}{5}\tfrac{1}{2^n} - 12$$

(b) This has the same complementary solution as (a) so we have only to find a particular solution. We try the function $x_n = Kn + L$ where K and L are constants. Substituting into the recurrence relation gives

$$K(n + 2) + L \equiv \tfrac{7}{2}[K(n + 1) + L] - \tfrac{3}{2}[Kn + L] + 12n$$

Thus

$$Kn + 2K + L \equiv 2Kn + \tfrac{7}{2}K + 2L + 12n$$

Comparing coefficients of n gives

$$K = 2K + 12$$

so that $K = -12$.

Comparing the terms independent of n gives

$$2K + L = \tfrac{7}{2}K + 2L$$

so that $L = -\tfrac{3}{2}K = 18$, and the general solution required is

$$x_n = -12n + 18 + A3^n + B(\tfrac{1}{2})^n$$

(c) This has the same complementary function as (a) so we only need to find a particular solution. To find this we try $x_n = K2^n$, giving

$$K(2^{n+2}) = \tfrac{7}{2}K(2^{n+1}) - \tfrac{3}{2}K(2^n) + 3(2^n)$$

so that

$$(4 - 7 + \tfrac{3}{2})K(2^n) = 3(2^n)$$

Hence $K = -2$ and the general solution required is

$$x_n = -2(2^n) + A3^n + B/2^n$$

Difference equations can be solved directly in MAPLE using the `rsolve` command. For example, considering Example 7.17(a) the general solution is given by the command

```
rsolve({x(n + 2) - 7/2*x(n + 1) + 3/2*x(n) = 12},x(n));
```

as

$$-(\tfrac{1}{5}x(0) - \tfrac{2}{5}x(1))3^n - \tfrac{1}{2}(-\tfrac{12}{5}x(0) + \tfrac{4}{5}x(1))(\tfrac{1}{2})^n + \tfrac{12}{5}(3)^n$$
$$+ (\tfrac{48}{5})(\tfrac{1}{2})^n - 12$$

which is equivalent to the given solution, with $A = (-1/5x(0) - 2/5x(1) + 12/5)$ and $B = (6/5x(0) + 2/5x(1) - 48/5)$. Given initial conditions $x(0) = 1$ *and* $x(1) = 1$, then these are incorporated directly in the command

```
rsolve({x(n + 2) - 7/2*x(n + 1) + 3/2*x(n)
= 12,x(0) = 1,x(1) = 1},x(n));
```

to give the particular solution

$$\tfrac{13}{5}3^n + \tfrac{52}{5}(\tfrac{1}{2})^n - 12$$

In MATLAB's Symbolic Math Toolbox there is no equivalent command so we make use of the `maple` command to access the MAPLE kernel. Check that the command

```
maple('rsolve({x(n + 2) - 7/2*x(n + 1) + 3/2*x(n)
= 12,x(0) = 1,x(1) = 1},x(n))')
```

returns the same answer as above.

As further examples we consider Examples 7.15(a) and 7.16. For 7.15(a) the commands

```
syms x n
maple('rsolve({x(n + 1) - 3*x(n) = 4, x(n))')
```

return

```
ans = x(0)*3^n - 2 + 2*3^n
```

This corresponds to the answer given in the solution with $A = (x(0) + 2)$.
For 7.16 the commands

```
syms x n
maple('rsolve({x(n + 2) - x(n + 1) - x(n)
= 0,x(0) = 1,x(1) = 1, x(n))')
```

return

```
ans = (1/10*5^(1/2) + 1/2)*(1/2 + 1/2*5^(1/2))^n
+ (1/2 - 1/10*5^(1/2))*(1/2 - 1/2*5^(1/2))^n
```

Simple rearrangement gives

$$\frac{1}{\sqrt{5}}\left(\frac{1}{2} + \frac{\sqrt{5}}{2}\right)\left(\frac{1}{2} + \frac{\sqrt{5}}{2}\right)^n + \frac{1}{\sqrt{5}}\left(\frac{\sqrt{5}}{2} - \frac{1}{2}\right)\left(\frac{1}{2} - \frac{\sqrt{5}}{2}\right)^n$$

which reduces to the answer given in the solution.

When the roots of the characteristic equation are complex numbers, the general solution of the homogeneous recurrence relation has a different form, as illustrated in Example 7.18.

Example 7.18   Show that the general solution of the recurrence relation

$$x_{n+2} = 6x_{n+1} - 25x_n$$

may be expressed in the form

$$x_n = 5^n(A\cos n\theta + B\sin n\theta)$$

where θ is such that $\sin\theta = \frac{4}{5}$ and $\cos\theta = \frac{3}{5}$.

Solution   The characteristic equation

$$\lambda^2 = 6\lambda - 25$$

has the (complex) roots $\lambda = 3 + j4$ and $\lambda = 3 - j4$ so that we can write the general solution in the form

$$x_n = A(3 + j4)^n + B(3 - j4)^n$$

Now writing the complex numbers in polar form we have

$$x_n = A(re^{j\theta})^n + B(re^{-j\theta})^n$$

where $r^2 = 3^2 + 4^2$ and $\tan\theta = \frac{4}{3}$ with $0 < \theta < \pi/2$ (or $\cos\theta = \frac{3}{5}$, $\sin\theta = \frac{4}{5}$). This can be simplified to give

$$x_n = A(5^n e^{jn\theta}) + B(5^n e^{-jn\theta}) = A5^n(\cos n\theta + j\sin n\theta) + B5^n(\cos n\theta - j\sin n\theta)$$

$$= (A + B)5^n \cos n\theta + j(A - B)5^n \sin n\theta$$

Here A and B are arbitrary complex constants, so their sum and difference are also arbitrary constants and we can write

$$x_n = P5^n \cos n\theta + Q5^n \sin n\theta$$

giving the form required. (Since P and Q are constants we can replace them by A and B if we wish.)

Example 7.19   Find the solution of the recurrence relation

$$x_{n+2} + 2x_n = 0$$

which satisfies $x_0 = 1$, $x_1 = 2$.

Solution   Here the characteristic equation is

$$\lambda^2 + 2 = 0$$

and has roots $\pm j\sqrt{2}$, so that we can write the general solution in the form

$$x_n = A(j\sqrt{2})^n + B(-j\sqrt{2})^n$$

Since $e^{j\pi/2} = \cos\dfrac{\pi}{2} + j\sin\dfrac{\pi}{2} = j$, we can rewrite the solution as

$$x_n = A(\sqrt{2})^n e^{jn\pi/2} + B(\sqrt{2})^n e^{-jn\pi/2}$$

$$= A2^{n/2}\left(\cos\frac{n\pi}{2} + j\sin\frac{n\pi}{2}\right) + B2^{n/2}\left(\cos\frac{n\pi}{2} - j\sin\frac{n\pi}{2}\right)$$

$$= (A + B)2^{n/2}\cos\frac{n\pi}{2} + j(A - B)2^{n/2}\sin\frac{n\pi}{2}$$

$$= P2^{n/2}\cos\frac{n\pi}{2} + Q2^{n/2}\sin\frac{n\pi}{2}$$

We can find the values of P and Q by applying the initial data $x_0 = 1$, $x_1 = 2$, giving

$$P = 1 \text{ and } 2^{1/2}Q = 2$$

Hence the required solution is

$$x_n = 2^{n/2}\cos\frac{n\pi}{2} + 2^{(n+1)/2}\sin\frac{n\pi}{2}$$

If complex roots are involved then using the command `evalc` alongside `rsolve` attempts to express complex exponentials in terms of trigonometric functions, leading in most cases to simplified answers. Considering Example 7.19 the MATLAB commands

```
syms x n
maple('evalc(rsolve({x(n + 2) + 2*x(n) = 0,x(0)
= 1,x(1) = 2}, x(n)))')
```

return the answer

```
2^(1/2*n)*cos(1/2*n*pi) + 2^(1/2*n)*sin(1/2*n*pi)*2^(1/2)
```

which reduces to

$$2^{n/2}\cos(n\pi/2) + 2^{(n+1)/2}\sin(n\pi/2)$$

Check that for the equation of Example 7.18 the MATLAB commands

```
syms x n
maple('evalc(rsolve({x(n + 2) - 6*x(n + 1) + 25*x(n)
= 0}, x(n)))')
```

subject to noting that $\exp(n*\log 5) = 5^n$, $\operatorname{atan}(4/3) = \theta$ and the collection of terms, produce the answer

$$x(0)5^n\cos(n\theta) + (1/4x(1) - 3/4x(0))5^n\sin(n\theta))$$

which is of the required form.

The general result corresponding to that obtained in Example 7.16 is that if the roots of the characteristic equation can be written in the form

$$\lambda = u \pm \mathrm{j}v$$

where u, v are real numbers, then the general solution of the homogeneous recurrence relation is

$$x_n = r^n(A \cos n\theta + B \sin n\theta)$$

where $r = \sqrt{(u^2 + v^2)}$, $\cos \theta = u/r$, $\sin \theta = v/r$ and A and B are arbitrary constants.

Recurrence relations are sometimes called **difference equations**. This name is used since we can rearrange the relations in terms of the differences of unknown sequence x_n. Thus

$$x_{n+1} = ax_n + b$$

can be rearranged as

$$\Delta x_n = (a - 1)x_n + b$$

where $\Delta x_n = x_{n+1} - x_n$.

Similarly, after some algebraic manipulation, we may write

$$x_{n+2} = ax_{n+1} + bx_n + c$$

as

$$\Delta^2 x_n = (a - 2)\Delta x_n + (a + b - 1)x_n + c$$

where

$$\Delta^2 x_n = \Delta x_{n+1} - \Delta x_n = x_{n+2} - 2x_{n+1} + x_n$$

The method for solving second-order linear recurrence relations with constant coefficients is summarized in Figure 7.7.

Figure 7.7
Summary:
second-order linear
recurrence relation
with constant
coefficients.

Homogeneous case:

$$x_{n+2} = ax_{n+1} + bx_n \tag{1}$$

(i) Solve the characteristic equation.
(ii) Write down the general solution for x_n from the table:

Roots of characteristic equation	General solution (A and B are arbitrary constants)
Real α, β and $\alpha \neq \beta$	$A\alpha^n + B\beta^n$
Real α, β and $\alpha = \beta$	$(A + Bn)\alpha^n$
Non-real α, $\beta = u \pm jv$	$(u^2 + v^2)^{n/2}(A \cos n\theta + B \sin n\theta)$ where $\cos\theta = u/(u^2 + v^2)^{1/2}$, $\sin\theta = v/(u^2 + v^2)^{1/2}$

Nonhomogeneous case:

$$x_{n+2} = ax_{n+1} + bx_n + c_n \text{ where } c_n \text{ is a known sequence.} \tag{2}$$

 (i) Find the general solution of the associated homogeneous problem (1).
 (ii) Find a particular solution of (2).
(iii) The general solution of (2) is the sum of (i) and (ii).

To find a particular solution to (2) substitute a likely form of particular solution into (2). If the correct form has been chosen then comparing coefficients will be enough to determine the values of the constants in the trial solution. Here are some suitable forms of particular solutions:

c_n	7	$3n + 5$	$2n^2 + 3n + 8$	$3\cos(7n) + 5\sin(7n)$	6^n	$n5^n$
p_n	C	$Cn + D$	$Cn^2 + Dn + E$	$C\cos(7n) + D\sin(7n)$	$C6^n$	$5^n(C + Dn)$

In solving problems, note that the top line of the table involves any *known* constants (these will be different from problem to problem), while the bottom line involves *unknown* constants, C, D, E, which must be determined by substituting the trial form into the nonhomogeneous relation.

An exceptional case arises when the suggested form for p_n already is present in the general solution of the associated homogeneous problem. If this happens, just multiply the suggested form by n (and if that does not work, by n repeatedly until it does).

7.4.4 Exercises

 Check your answers using MATLAB or MAPLE whenever possible.

29 Evaluate the expression $2x_{n+2} - 7x_{n+1} + 3x_n$ when x_n is defined for all $n \geq 0$ by

(a) $x_n = 3^n$ (b) $x_n = 2^n$

(c) $x_n = 2^{-n}$ (d) $x_n = 3(-2)^n$

Which of (a) to (d) are solutions of the following recurrence relation?

$$2x_{n+2} - 7x_{n+1} + 3x_n = 0$$

30 Show, by substituting them into the recurrence relation, that $x_n = 2^n$ and $x_n = (-1)^n$ are two solutions of $x_{n+2} - x_{n+1} - 2x_n = 0$. Verify similarly that $x_n = A(2^n) + B(-1)^n$ is also a solution of the recurrence relation for all constants A and B.

31 Obtain the general solutions of

(a) $Y_{n+2} - 7Y_{n+1} + 10Y_n = 0$

(b) $u_{n+2} - u_{n+1} - 6u_n = 0$

(c) $25T_{n+2} = -T_n$

(d) $p_{n+2} - 5p_{n+1} = 5(p_{n+1} - 5p_n)$

(e) $2E_{n+2} = E_{n+1} + E_n$

32 Solve the nonhomogeneous problems (use parts of Question 31):

(a) $Y_{n+2} - 7Y_{n+1} + 10Y_n = 1$, $Y_0 = 5/4$, $Y_1 = 2$

(b) $2E_{n+2} - E_{n+1} - E_n = 1$, $E_0 = 2$, $E_1 = 0$

(c) $u_{n+2} - u_{n+1} - 6u_n = n$ (general solution only)

33 Show that the characteristic equation for the recurrence relation $x_{n+2} - 2ax_{n+1} + a^2x_n = 0$, where a is a non-zero constant, has two equal roots $\lambda = a$.

(a) Verify (by substituting into the relation) that $x_n = (A + Bn)a^n$ is a solution for all constants A and B.

(b) Find the particular solution which satisfies $x_0 = 1$, $x_1 = 0$. (Your answer will involve a, of course.)

(c) Find the particular solution for which $x_0 = 3$, $x_{10} = 20$.

34 Let x be a constant such that $|x| < 1$. Find the solution of

$$T_{n+2} - 2xT_{n+1} + T_n = 0, \quad T_0 = 1, \quad T_1 = x$$

Find T_2, T_3 and T_4 also directly by recursion and deduce that $\cos(2\cos^{-1}x) = 2x^2 - 1$ and express $\cos(3\cos^{-1}x)$ and $\cos(4\cos^{-1}x)$ as polynomials in x.

35 A topic from information theory: imagine an information transmission system that uses an alphabet consisting of just two symbols 'dot' and 'dash', say. Messages are transmitted by first encoding them into a string of these symbols, and no other symbols (e.g. blank spaces) are allowed. Each symbol requires some length of time for its transmission. Therefore, for a fixed total time duration only a finite number of different message strings is possible. Let N_t denote the number of different message strings possible in t time units.

(a) Suppose that dot and dash each require one time unit for transmission. What is the value of N_1? Why is $N_{t+1} = 2N_t$ for all $t \geq 1$? Write down a simple formula for N_t for $t \geq 1$.

(b) Suppose instead that dot requires one unit of time for transmission while dash requires two units. What are the values of N_1 and N_2? Justify the relation $N_{t+2} = N_{t+1} + N_t$ for $t \geq 1$. Hence write down a formula for N_t in terms of t. (*Hint*: the general solution of Fibonacci recurrence is given in Example 7.16.)

7.5 Limit of a sequence

In Section 7.2.1 the idea of a sequence and the associated notation were described. We shall now develop the concept of a limit of a sequence and then discuss the properties of sequences that have limits (termed convergent sequences) and methods for evaluating those limits algebraically and numerically.

7.5.1   Convergent sequences

In Example 7.6, we obtained the following sequence of approximations (working to 2dp) for $\sqrt{2}$:

$$x_0 = 1, \quad x_1 = 1.83, \quad x_2 = 1.16, \quad x_3 = 1.64$$

Continuing with the process, we obtain

$$x_{22} = 1.41, \quad x_{23} = 1.41$$

and

$$x_n = 1.41 \quad \text{for } n \geqslant 22$$

The terms x_{22} and x_{23} of the sequence are indistinguishable to two decimal places; in other words, their difference is less than a rounding error. This situation is shown clearly in Figure 7.4(b). This phenomenon occurs with many sequences, and we say that the sequence **tends to a limit** or **has a limiting value** or **converges** or **is convergent**. While it is clear in the above example what we mean by saying that the sequence converges to $\sqrt{2}$, we need a precise definition for all the cases that may occur.

In general, a sequence $\{a_k\}_{k=0}^{\infty}$ has the limiting value a as n becomes large if, given a small positive number ε (no matter how small), a_n differs from a by less than ε for all sufficiently large n. More concisely,

> $a_n \to a$ as $n \to \infty$ if, given any $\varepsilon > 0$, there is a number N such that $|a_n - a| < \varepsilon$ for all $n > N$

Here the $\to$ stands for 'tends to the value' or 'converges to the limit'. An alternative notation for $a_n \to a$ as $n \to \infty$ is

$$\lim_{n \to \infty} a_n = a$$

Diagrammatically, this means that the terms of the sequence lie between $y = a - \varepsilon$ and $y = a + \varepsilon$ for $n > N$, as shown in Figure 7.8.

Note that the limit of a sequence need not actually be an element of the sequence. For example $\{n^{-1}\}_{n=1}^{\infty}$ has limit 0, but 0 does not occur in the sequence.

Returning to the square-root example discussed above, we have

$$x_n \to \sqrt{2} \quad \text{as} \quad n \to \infty$$

Figure 7.8
Convergence
of $\{a_n\}$ to limit a.

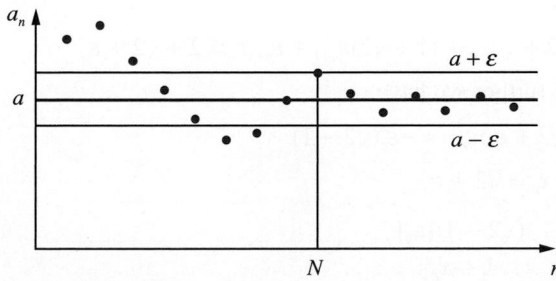

Figure 7.9
Convergence
of $\{x_n\}$ to $\sqrt{2}$.

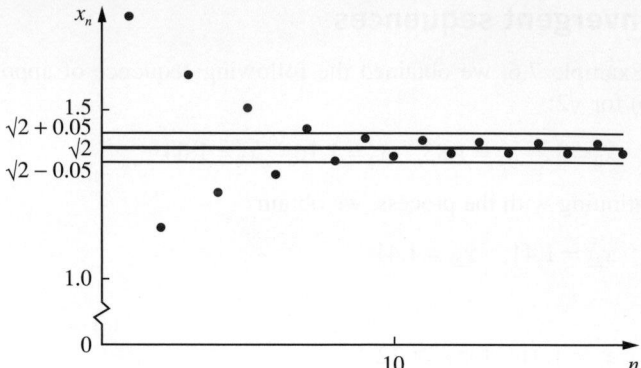

It is clear from the terms of the sequence that for an error bound of 0.05 we need $n > 8$ (see Figure 7.9). Thus $\sqrt{2} = 1.4$ (to 1dp). However, to prove convergence in the formal sense, we have to be able to say how many terms we need to take in order to obtain a specified level of precision. Suppose we need an answer correct to 10dp, or 100dp, or whatever; we must be able to give the corresponding value of N in the definition of convergence. Finding an expression for N is not often easy.

We shall illustrate the type of methods used by finding an expression for N for a classical method for calculating $\sqrt{2}$. This uses the iteration

$$x_{n+1} = \frac{2 + x_n}{1 + x_n} \quad \text{with } x_0 = 1$$

This produces the rational approximations

$$\left\{ 1, \frac{3}{2}, \frac{7}{5}, \frac{17}{12}, \frac{41}{29}, \frac{99}{70}, \dots \right\}$$

The last given approximation has an error of less than 0.0001. Suppose we require an approximation which is correct to p decimal places, then we need to find an N such that

$$|x_n - \sqrt{2}| < 0.5 \times 10^{-p}$$

for $n > N$. Writing $\varepsilon_n = x_n - \sqrt{2}$ so that $x_0 = \sqrt{2} + \varepsilon_0$, $x_1 = \sqrt{2} + \varepsilon_1$, $\dots$, $x_{n+1} = \sqrt{2} + \varepsilon_{n+1}$ (and so on) we have

$$\sqrt{2} + \varepsilon_{n+1} = \frac{2 + \sqrt{2} + \varepsilon_n}{1 + \sqrt{2} + \varepsilon_n}$$

Multiplying across, we have

$$(\sqrt{2} + \varepsilon_{n+1})(1 + \sqrt{2} + \varepsilon_n) = 2 + \sqrt{2} + \varepsilon_n$$

which gives

$$\sqrt{2} + 2 + \sqrt{2}\varepsilon_n + (1 + \sqrt{2})\varepsilon_{n+1} + \varepsilon_{n+1}\varepsilon_n = 2 + \sqrt{2} + \varepsilon_n$$

Simplifying further we have

$$(1 + \sqrt{2} + \varepsilon_n)\varepsilon_{n+1} = -\varepsilon_n(\sqrt{2} - 1)$$

Thus, since $x_n = \sqrt{2} + \varepsilon_n$

$$|\varepsilon_{n+1}| = \frac{(\sqrt{2} - 1)|\varepsilon_n|}{1 + x_n}$$

Since $x_n \geqslant 1$ and $\sqrt{2} < 1.5$, this implies

$$|\varepsilon_{n+1}| < \frac{0.5}{2}|\varepsilon_n| < 0.25|\varepsilon_n|$$

Since $x_0 = 1$ we have $|\varepsilon_0| < \frac{1}{2}$, so that $|\varepsilon_1| < 0.25(\frac{1}{2})$, $|\varepsilon_2| < 0.25^2(\frac{1}{2})$, ... and $|\varepsilon_n| < 0.25^n(\frac{1}{2})$.

Hence if we require $|\varepsilon_n| < 0.5 \times 10^{-p}$, for $n > N$, then we may find m such that

$$0.25^m(\tfrac{1}{2}) < 0.5 \times 10^{-p}$$

or

$$\frac{1}{4^m} < \frac{1}{10^p}$$

which implies $4^m > 10^p$.

Taking logarithms to base 10, this gives

$$m > p/\log 4$$

Then choose N to be the greatest integer not greater than m, that is, $N = \lfloor p/\log 4 \rfloor$. Thus, to guarantee 10dp, we need to evaluate at most $\lfloor 10/\log 4 \rfloor = 16$ iterations, which you may verify on your calculator.

7.5.2 Properties of convergent sequences

As we have seen in the $\sqrt{2}$ example, it is usually difficult and tedious to prove the convergence of a sequence from first principles. Normally we are able to compute the limit of a sequence from simpler sequences by means of very simple rules based on the properties of convergent sequences. These are:

(a) Every convergent sequence is bounded; that is, if $\{a_n\}_{n=0}^{\infty}$ is convergent then there is a positive number M such that $|a_n| < M$ for all n.

(b) If $\{a_n\}$ has limit a, and $\{b_n\}$ has limit b, then
  (i) $\{a_n + b_n\}$ has limit $a + b$
  (ii) $\{a_n - b_n\}$ has limit $a - b$
  (iii) $\{a_n b_n\}$ has limit ab
  (iv) $\{a_n/b_n\}$ has limit a/b, for $b_n \neq 0$, $b \neq 0$.

We illustrate the technique in Example 7.20.

Example 7.20

Find the limits of the sequence $\{x_n\}_{n=0}^{\infty}$ defined by

(a) $x_n = \dfrac{n}{n+1}$ (b) $x_n = \dfrac{2n^2 + 3n + 1}{5n^2 + 6n + 2}$

Solution

(a) With $x_n = n/(n+1)$, we generate the sequence $\{0, \frac{1}{2}, \frac{2}{3}, \frac{3}{4}, \frac{4}{5}, \dots\}$. From these values it seems clear that $x_n \to 1$ as $n \to \infty$. This can be proved by rewriting x_n as

$$x_n = 1 - \frac{1}{n+1}$$

and we make $1/(n + 1)$ as small as we please by taking n sufficiently large.

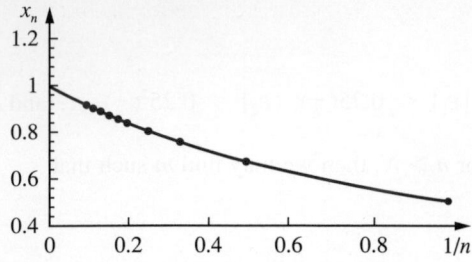

Figure 7.10 Sequence $x_n = n/(n + 1)$ plotted against $1/n$.

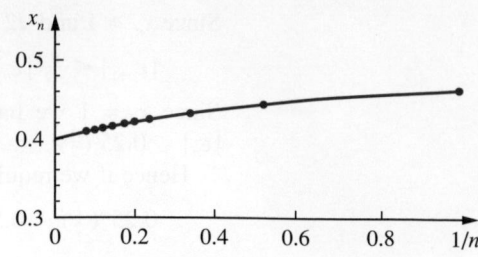

Figure 7.11 Sequence $x_n = \dfrac{2n^2 + 3n + 1}{5n^2 + 6n + 2}$ plotted against $1/n$.

Alternatively, we write

$$x_n = \frac{1}{1 + 1/n}$$

Now $1/n \to 0$ as $n \to \infty$. Hence, by the property (b)(i), $1 + 1/n \to 1$ and so, by the property (b)(iv),

$$\frac{1}{1 + 1/n} \to 1 \quad \text{as } n \to \infty$$

as illustrated in Figure 7.10.

(b) For

$$x_n = \frac{2n^2 + 3n + 1}{5n^2 + 6n + 2}$$

the easiest approach is to divide both numerator and denominator by the highest power of n occurring and use the fact that $1/n \to 0$ as $n \to \infty$. Thus

$$x_n = \frac{2 + 3/n + 1/n^2}{5 + 6/n + 2/n^2}$$

The limits of numerator and denominator are 2 and 5 (using the property (b)(i) repeatedly), and so $x_n \to \frac{2}{5}$ as $n \to \infty$ (using (b)(iv)). This is shown clearly in Figure 7.11.

Example 7.21 Show that the ratio x_n of successive terms of the Fibonacci sequence satisfies the recurrence relation

$$x_{n+1} = 1 + 1/x_n, \quad x_0 = 1$$

Calculate the first few terms of this sequence and find the value of its limit.

Solution The Fibonacci sequence was defined in Example 7.16 as

$$f_{n+2} = f_{n+1} + f_n \quad \text{with} \quad f_0 = f_1 = 1$$

Defining $x_n = f_{n+1}/f_n$ gives $f_{n+2} = x_{n+1} \times f_{n+1}$ and $f_n = f_{n+1}/x_n$, so that the recurrence relation becomes

$$x_{n+1}f_{n+1} = f_{n+1} + f_{n+1}/x_n$$

and dividing through by f_{n+1} we have

$$x_{n+1} = 1 + 1/x_n$$

Also, $x_0 = f_1/f_0 = 1/1 = 1$.

Using the recurrence relation, we obtain the sequence

$$\{1, 2, 1.5, 1.6667, 1.6, 1.625, 1.6154, 1.6190, \dots\}$$

The numerical results suggest a limiting value near 1.62. Indeed, the oscillatory nature of the sequence suggests $1.6154 < x_n < 1.6190$ for $n > 8$, which implies a limit value $x = 1.62$ correct to 2dp.

In this case we can check this conclusion, for if $x_n \to x$ as $n \to \infty$ then $x_{n+1} \to x$ also, and so the recurrence relation yields

$$x = 1 + \frac{1}{x}, \quad \text{with } x > 0$$

Thus $x^2 - x - 1 = 0$, which implies $x = \frac{1}{2}(1 + \sqrt{5})$ or $x = \frac{1}{2}(1 - \sqrt{5})$. Since the sequence has positive values only, it is clear that the appropriate root is $x = \frac{1}{2}(1 + \sqrt{5}) = 1.62$ (to 2dp).

This limiting value is called the **golden number** and is often denoted by the Greek letter tau τ. A rectangle the ratio of whose sides is the golden number is said to be the most pleasing aesthetically, and this has often been adopted by architects as a basis of design.

7.5.3 Computation of limits

The examples considered so far tend to create the impression that all sequences converge, but this is not so. An important sequence that illustrates this is the geometric sequence

$$a_n = r^n, \quad r \text{ constant}$$

For this sequence we have

$$\lim_{n \to \infty} a_n = \begin{cases} 0 & (-1 < r < 1) \\ 1 & (r = 1) \end{cases}$$

If $r > 1$, the sequence increases without bound as $n \to \infty$, and we say it **diverges**. If $r = -1$, the sequence takes the values -1 and 1 alternately, and there is no limiting value. If $r < -1$, the sequence is unbounded and the terms alternate in sign.

Often in computational applications of sequences the limit of the sequence is not known, so that it is not possible to apply the formal definition to determine the number of terms N we need to take in order to obtain a specified level of precision. If we do not know the limit a, to which a sequence $\{a_n\}$ converges, then we cannot measure $|a_n - a|$. In the computational context, when we apply a recurrence relation to find a solution to a problem, we say that the sequence $\{a_n\}$ has converged to its limit when all subsequent terms yield the same value of the approximation required. In other words, we say that the sequence of finite terms is convergent if, for any n and $m > N$,

$$|a_n - a_m| < \varepsilon$$

where the bound ε is specified. Thus a sequence tends to a limit if all the terms of the sequence for $n > N$ are restricted to an interval that can be made arbitrarily small by choosing N sufficiently large. This is called **Cauchy's test for convergence**.

In many practical problems we need to find a numerical estimate for the limit of a sequence. A graphical method for this is to sketch the graph defined by the points $\{(1/n, a_n) : n = 1, 2, 3, \dots\}$ and then extrapolate from it, since $1/n \to 0$ as $n \to \infty$. If greater precision is required than can be obtained in this way, an effective numerical procedure is a form of repeated linear extrapolation due to Aitken. We illustrate the procedure in Example 7.22.

Example 7.22　Examine the convergence of the sequence $\{a_n\}_{n=1}^{\infty}$, $a_n = (1 + 1/n)^n$.

Solution　It can be shown that $\lim_{n\to\infty} a_n = e$, but convergence is rather slow. In fact,

$$a_1 = 2, \quad a_2 = 2.2500, \quad a_3 = 2.3704, \quad a_4 = 2.4414, \quad \dots$$

$$a_8 = 2.5658, \quad \dots, \quad a_{16} = 2.6379, \quad \dots, \quad a_{32} = 2.6770, \quad \dots$$

$$a_{64} = 2.6973, \quad \dots, \quad \text{and} \quad e = 2.7183 \text{ to 4dp}$$

Now consider the two terms corresponding to $n = 16$ and $n = 32$ and set $x_n = 1/n$. Then

$$n = 16 \quad \text{gives} \quad x_{16} = 0.0625 \quad \text{and} \quad a_{16} = 2.6379$$

$$n = 32 \quad \text{gives} \quad x_{32} = 0.031\,25 \quad \text{and} \quad a_{32} = 2.6770$$

We wish to find the value corresponding to $x = 0$. To estimate this, we may use linear extrapolation as shown in Figure 7.12. This gives

$$b_{16,32} = \frac{x_{16}a_{32} - x_{32}a_{16}}{x_{16} - x_{32}} = 2.7161$$

Note that $b_{16,32}$ is a better estimate for e than either a_{16} or a_{32}.

Figure 7.12
Linear extrapolation
for the limit of a
sequence.

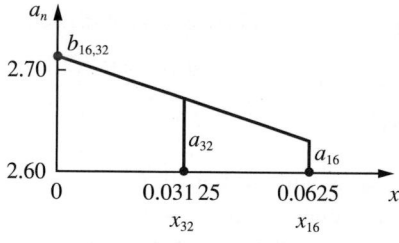

In MATLAB's Symbolic Math Toolbox the limit as $n \to \infty$ of the sequence defined by $x_n = f(n)$ is determined by the commands

```
syms n
limit(f_n, n, inf)
```

the corresponding command in MAPLE being:

```
limit(f_n, n = infinity);
```

As illustrative examples consider Examples 7.20(a) and 7.22:

MATLAB	MAPLE

```
syms n
limit(n/(n + 1), n,inf)
```
```
limit(n/(n + 1),
n = infinity);
```

returns *ans* 1

```
limit((1 + 1/n)^n, n, inf)
```
```
limit((1 + 1/n)^n,
n = infinity);
```

returns

```
exp(1)
```

e

7.5.4 Exercises

Check your answers using MATLAB or MAPLE whenever possible.

36 Calculate the first six terms of each of the following sequences $\{a_n\}$ and draw a graph of a_n versus $1/n$. (Some care is needed in choosing the scale of the *y* axis.) What is the behaviour of a_n as $n \to \infty$?

(a) $a_n = \dfrac{n}{n^2 + 1}$ ($n \geqslant 1$)

(b) $a_n = \dfrac{3n^2 + 2n + 1}{6n^2 + 5n + 2}$ ($n \geqslant 1$)

(c) $a_n = (2n)^{1/n}$ ($n \geqslant 1$)

(d) $a_n = \left(1 + \dfrac{1}{2n}\right)^n$ ($n \geqslant 1$)

(e) $a_n = \sqrt{(1 + a_{n-1})}$, $a_1 = 1$ ($n \geqslant 2$)

(f) $a_n = \dfrac{n}{2} \sin \dfrac{2\pi}{n}$ ($n \geqslant 1$)

(Comment: Part (f) is the area of a regular polygon of *n* sides inscribed in a circle of unit radius.)

37 Calculate the first six terms of each of the following sequences $\{a_n\}$ and draw a graph of a_n against *n*. What is the behaviour of a_n as $n \to \infty$?

(a) $a_n = \dfrac{n^2 + 1}{n + 1}$ ($n \geqslant 0$)

(b) $a_n = (\sin \frac{1}{2} n\pi)^n$ ($n \geqslant 1$)

(c) $a_n = 3/a_{n-1}$, $a_0 = 1$ ($n \geqslant 1$)

38 Find the least value of *N* such that when $n \geqslant N$,

(a) $n^2 + 2n > 100$ (b) $\dfrac{n^2}{2^n} < \dfrac{1}{1000}$

(c) $\dfrac{1}{n} - \dfrac{(-1)^n}{n^2} < 0.000\,001$

(d) $\sqrt{(n + 1)} - \sqrt{n} < \frac{1}{10}$

(e) $\dfrac{n^2 + 2}{n^2 - 1} - 1 < 0.01$

39 What is the long-term share of the detergent market achieved by the brand 'Number One', described in Question 5 (Exercises 7.2.3)?

40 A **linearly convergent** sequence has the property that

$$a_n - a = \lambda(a_{n-1} - a) \quad \text{for all } n$$

where λ is a constant and $a = \lim\limits_{n \to \infty} a_n$. Show that

$$a_{n+1} - a = \lambda(a_n - a)$$

Deduce that

$$\frac{a_{n+1} - a}{a_n - a} = \frac{a_n - a}{a_{n-1} - a}$$

and show that

$$a = a_{n-1} - \frac{(a_n - a_{n-1})^2}{a_{n+1} - 2a_n + a_{n-1}}$$

This is known as **Aitken's estimate** for the limit of a sequence.

Compute the first four terms of the sequence

$$a_0 = 2, \quad a_{n+1} = \tfrac{1}{5}(3 + 4a_n^2 - a_n^3) \quad (n \geqslant 0)$$

and estimate the limit of the sequence.

7.6 Infinite series

Infinite series occur in a large variety of practical problems, from estimating the long-term effects of pollution to the stability analysis of the motions of machinery parts. They also occur in the development of computer algorithms for the numerical solution of practical problems. In this section we will consider the underlying ideas. Care has to be exercised when dealing with infinite series, since it is easy to generate fallacious results. For example, consider the infinite series

$$S = 1 - 2 + 4 - 8 + 16 - 32 + \ldots$$

Then we can write

$$2S = 2 - 4 + 8 - 16 + 32 - 64 + \ldots$$

and adding these two results, we obtain

$$3S = 1 \quad \text{or} \quad S = \tfrac{1}{3}$$

which is clearly wrong. Such blunders, however, are not always so glaringly obvious, so we have to develop simple methods for determining whether an infinite series sums to a finite value and for obtaining or estimating that value.

7.6.1 Convergence of infinite series

As we discussed in Section 7.2.1, series and sequences are closely connected. When the sum S_n of a series of n terms tends to a limit as $n \to \infty$, the series is **convergent**. When we can express S_n in a simple form, it is usually easy to establish whether or not the series converges. To find the sum of an infinite series, the sequence of partial sums $\{S_n\}$ is taken to the limit.

Example 7.23

Examine the following series for convergence:

(a) $1 + 3 + 5 + 7 + 9 + \ldots + (2k + 1) + \ldots$

(b) $1^2 + 2^2 + 3^2 + 4^2 + 5^2 + \ldots + k^2 + \ldots$

(c) $1 + \dfrac{1}{2} + \dfrac{1}{4} + \dfrac{1}{8} + \dfrac{1}{16} + \ldots + \dfrac{1}{2^k} + \ldots$

(d) $\dfrac{1}{1 \cdot 2} + \dfrac{1}{2 \cdot 3} + \dfrac{1}{3 \cdot 4} + \dfrac{1}{4 \cdot 5} + \ldots + \dfrac{1}{(k + 1)(k + 2)} + \ldots$

Solution

(a) This is an arithmetic series so we can write its finite sum as a simple formula

$$S_n = \sum_{k=0}^{n-1}(2k + 1) = 1 + 3 + 5 + \ldots + (2n - 1) = n^2 \quad (n \text{ terms})$$

It is clear from this that $S_n \to \infty$ as $n \to \infty$ and the series does not converge to a limit. It is a **divergent** series.

(b) As we saw in Example 7.10

$$S_n = 1^2 + 2^2 + 3^2 + \ldots + n^2 = \tfrac{1}{6}n(n+1)(2n+1) \quad (n \text{ terms})$$

As n becomes large, so does S_n, and $S_n \to \infty$ as $n \to \infty$. Hence the series is divergent.

(c) $S_n = 1 + \tfrac{1}{2} + \tfrac{1}{4} + \ldots + \dfrac{1}{2^{n-1}} \quad (n \text{ terms})$

This is a geometric series with common ratio $\tfrac{1}{2}$. Using the formula (7.4) with $a = 1$ and $r = \tfrac{1}{2}$ gives

$$S_n = \frac{1 - \dfrac{1}{2^n}}{1 - \dfrac{1}{2}} = 2\left(1 - \frac{1}{2^n}\right)$$

As $n \to \infty$, $\dfrac{1}{2^n} \to 0$, so that $S_n \to 2$. Hence the series converges to the sum 2.

(d) We showed in Example 7.11 that

$$S_n = \frac{1}{1 \cdot 2} + \frac{1}{2 \cdot 3} + \ldots + \frac{1}{n(n+1)} = 1 - \frac{1}{n+1}$$

As $n \to \infty$, $1/(n+1) \to 0$, so that $S_n \to 1$. Hence the series converges to the sum 1.

Among the elementary series, the geometric series is the most important.

$$S_n = a + ar + ar^2 + \ldots + ar^{n-1} \quad (n \text{ terms})$$

$$= \frac{a(1 - r^n)}{1 - r}$$

$$= \frac{a}{1 - r} - \frac{ar^n}{1 - r}$$

Since $r^n \to 0$ as $n \to \infty$ when $|r| < 1$, we conclude that $S_n \to a/(1 - r)$ where $|r| < 1$ and the series is convergent. Where $|r| \geq 1$, the series is divergent. These results are used in many applications and the sum of the infinite series is

$$S = a + ar + ar^2 + ar^3 + \ldots = \frac{a}{1 - r}, \quad |r| < 1 \tag{7.14}$$

Similarly

$$S_n = a + 2ar + 3ar^2 + \ldots + nar^{n-1}$$

$$= \frac{a}{(1 - r)^2} - (n + 1)ar^n + anr^{n-1}$$

$$\to \frac{a}{(1 - r)^2} \quad \text{as } n \to \infty$$

Summation may be carried out in MATLAB using the `symsum` command. For Example 7.23(a) the sum of the first n terms is determined by the commands

```
syms k n
sn = symsum(2*k + 1,0,n)
```

as

```
sn = n^2
```

which tends to infinity as $n \to \infty$ so it is a divergent series.

For Example 7.23(d) the sum to infinity is determined by the commands

```
syms k
sinf = symsum(1/((k + 1)*(k + 2)),0,inf)
```

as `sn = 1`, so it is a convergent series.

7.6.2 Tests for convergence of positive series

The convergence or divergence of the series discussed in Example 7.23 was established by considering the behaviour of the partial sum S_n as $n \to \infty$. In many cases, however, it is not possible to express S_n in a closed form. When this occurs, the convergence or divergence of the series is established by means of a test. Two tests are commonly used.

(a) Comparison test

Suppose we have a series, $\sum_{k=0}^{\infty} c_k$, of positive terms ($c_k \geq 0$, all k) which is known to be convergent. If we have another series, $\sum_{k=0}^{\infty} u_k$, of positive terms such that $u_k \leq c_k$ for all k then $\sum_{k=0}^{\infty} u_k$ is convergent also.

Also, if $\sum_{k=0}^{\infty} c_k$ diverges and $u_k \geq c_k \geq 0$ for all k, then $\sum_{k=0}^{\infty} u_k$ also diverges.

Example 7.24　Examine for convergence the series

(a) $1 + \dfrac{1}{1!} + \dfrac{1}{2!} + \dfrac{1}{3!} + \dfrac{1}{4!} + \dots + \dfrac{1}{n!} + \dots$　　(the **factorial series**)

(b) $1 + \dfrac{1}{2} + \dfrac{1}{3} + \dfrac{1}{4} + \dots + \dfrac{1}{n} + \dots$　　(the **harmonic series**)

Solution　(a) We can establish the convergence of the series (a) by considering its partial sum

$$A_n = 1 + \frac{1}{1!} + \frac{1}{2!} + \frac{1}{3!} + \frac{1}{4!} + \dots + \frac{1}{n!}$$

Each term of this series is less than or equal to the corresponding term of the series

$$C_n = 1 + 1 + \frac{1}{2} + \frac{1}{2^2} + \frac{1}{2^3} + \dots + \frac{1}{2^{n-1}}$$

This geometric series may be summed to give

$$C_n = 3 - \frac{1}{2^{n-1}}$$

Thus

$$A_n < 3 - \frac{1}{2^{n-1}}$$

which implies that, since all the terms of the series are positive numbers, A_n tends to a limit less than 3 as $n \to \infty$. Thus the series is convergent.

(b) The divergence of the series (b) is similarly established.

$$1 + \tfrac{1}{2} + \tfrac{1}{3} + \tfrac{1}{4} + \tfrac{1}{5} + \tfrac{1}{6} + \tfrac{1}{7} + \tfrac{1}{8} + \tfrac{1}{9} + \dots$$

Collecting together successive groups of two, four, eight, ... terms, we have

$$1 + \tfrac{1}{2} + (\tfrac{1}{3} + \tfrac{1}{4}) + (\tfrac{1}{5} + \tfrac{1}{6} + \tfrac{1}{7} + \tfrac{1}{8}) + (\tfrac{1}{9} + \dots + \tfrac{1}{16}) + (\tfrac{1}{17} + \dots$$

which may be compared with the series

(c) $1 + \tfrac{1}{2} + (\tfrac{1}{4} + \tfrac{1}{4}) + (\tfrac{1}{8} + \tfrac{1}{8} + \tfrac{1}{8} + \tfrac{1}{8}) + (\tfrac{1}{16} + \dots + \tfrac{1}{16}) + (\tfrac{1}{32} + \dots$

Each term of the rearranged (b) is greater than or at least equal to the corresponding term of the series (c), and so the 'sum' of the series (b) is greater than the 'sum' of the series (c), which is

$$1 + \tfrac{1}{2} + \tfrac{1}{2} + \tfrac{1}{2} + \tfrac{1}{2} + \dots$$

on summing the terms in brackets and which is clearly divergent.

Note that the harmonic series is divergent despite the fact that its nth term tends to zero as $n \to \infty$.

The harmonic series is the borderline case for divergence/convergence of the series

$$S(r) = \sum_{k=1}^{\infty} \frac{1}{k^r}$$

For $r > 1$, this series converges; for $r \leqslant 1$, it diverges as shown in the table below, where the values have been calculated using the *symsum* command (followed by the *double* command) in MATLAB's Symbolic Math Toolbox

r	1	1.01	1.05	1.10	1.20	1.50	2
$S(r)$	∞	100.58	20.58	0.58	5.59	2.61	1.64

(b) d'Alembert's ratio test

Suppose we have a series of positive terms, $\sum_{k=0}^{\infty} u_k$, and also $\lim\limits_{n \to \infty} \dfrac{u_{n+1}}{u_n} = l$ exists.

Then the series is convergent if $l < 1$ and divergent if $l > 1$. If $l = 1$, we are not able to decide, using this test, whether the series converges or diverges.

The proof of this result is straightforward. Assume that $\lim\limits_{n \to \infty} \dfrac{u_{n+1}}{u_n} = l < 1$ and choose r to be any number between l and 1. Then since the values of u_{n+1}/u_n, when n is sufficiently large, differ from l by as little as we please, we have

$$\frac{u_{n+1}}{u_n} < r$$

for $n \geqslant N$. Thus

$$u_{N+1} < r u_N, \quad u_{N+2} < r^2 u_N, \ldots$$

Thus, from and after the term u_N of the series, the terms do not exceed those of the convergent geometric series

$$u_N(1 + r + r^2 + r^3 + \ldots)$$

Hence $\sum_{k=0}^{\infty} u_k$ converges.

It is left as an exercise for the reader to show that the series diverges when $l > 1$.

Example 7.25   Use d'Alembert's test to determine whether the following series are convergent.

(a) $\displaystyle\sum_{k=0}^{\infty} \frac{2^k}{k!}$ (b) $\displaystyle\sum_{k=0}^{\infty} \frac{2^k}{(k+1)^2}$

Solution   (a) Let $u_k = \dfrac{2^k}{k!}$, then

$$\frac{u_{n+1}}{u_n} = \frac{2^{n+1}}{(n+1)!} \Big/ \frac{2^n}{n!} = \frac{2}{n+1}$$

which tends to zero as $n \to \infty$. Thus $l = 0$ and the series is convergent.

(b) Here

$$l = \lim_{n \to \infty}\left[\frac{2^{n+1}}{(n+2)^2} \Big/ \frac{2^n}{(n+1)^2}\right] = 2$$

so that the series diverges.

A necessary condition for convergence of all series is that the terms of the series must tend to zero as $n \to \infty$. Thus a simple test for divergence is:

If $u_n \to u \neq 0$ as $n \to \infty$, then $\sum_{k=0}^{\infty} u_k$ is divergent

Notice, however, that $u_n \to 0$ as $n \to \infty$ does not guarantee that $\sum_{k=0}^{\infty} u_k$ is convergent. To prove that, we need more information. (Recall, for example, the harmonic series $1 + \frac{1}{2} + \frac{1}{3} + \frac{1}{4} + \ldots$, of Example 7.24, which is divergent.)

Example 7.26   Show that the series $\frac{1}{2} + \frac{2}{3} + \frac{3}{4} + \ldots$ is divergent.

Solution   Here $u_k = \dfrac{k}{k+1}$, so that d'Alembert's ratio test does not give a conclusion (since $l = 1$). However, we note that $u_n = 1 - \dfrac{1}{n+1}$, so that $u_n \to 1$ as $n \to \infty$, from which we conclude that $\sum_{k=1}^{\infty} u_k$ diverges.

7.6.3   The absolute convergence of general series

In practical problems, we are concerned with series which may have both positive and negative terms. **Absolutely convergent** series are a special case of such series. Consider the general series

$$S = \sum_{k=0}^{\infty} u_k$$

which may have both positive and negative terms u_k. If the associated series

$$T = \sum_{k=0}^{\infty} |u_k|$$

is convergent then S is convergent and is said to be **absolutely convergent**. If it is impossible to obtain a value for the limit of the partial sum T_n, we must use some other test to determine the convergence (or divergence) of T. A simple test for absolute convergence of a series $\sum_{k=1}^{\infty} u_k$ is a natural extension of d'Alembert's ratio test.

If $\lim\limits_{n \to \infty} \left| \dfrac{u_{n+1}}{u_n} \right| < 1$ then $\sum\limits_{k=0}^{\infty} u_k$ is absolutely convergent

If $\lim\limits_{n \to \infty} \left| \dfrac{u_{n+1}}{u_n} \right| > 1$ then $\sum\limits_{k=0}^{\infty} u_k$ is divergent

If $\lim\limits_{n \to \infty} \left| \dfrac{u_{n+1}}{u_n} \right| = 1$ then no conclusion is possible

Absolutely convergent series have the following useful properties:

(a) the insertion of brackets into the series does not alter its sum;

(b) the rearrangment of the series does not alter its sum;

(c) the product of two absolutely convergent series $A = \sum a_n$ and $B = \sum b_n$ is an absolutely convergent series C, where

$$C = a_1 b_1 + (a_2 b_1 + a_1 b_2) + (a_3 b_1 + a_2 b_2 + a_1 b_3) + (a_4 b_1 + a_3 b_2 + a_2 b_3 + a_1 b_4) + \ldots$$

There are convergent series that are not absolutely convergent; that is $\sum_{k=1}^{\infty} u_k$ converges but $\sum_{k=0}^{\infty} |u_k|$ diverges. The most common series of this type are alternating series. Here the u_k alternate in sign. If, in addition, the terms decrease in size and tend to zero,

$$|u_n| < |u_{n-1}| \quad \text{for all } n, \quad \text{with } u_n \to 0 \quad \text{as} \quad n \to \infty$$

then the series converges. Thus

$$\sum_{k=1}^{\infty} (-1)^{k+1} \frac{1}{k} = 1 - \tfrac{1}{2} + \tfrac{1}{3} - \tfrac{1}{4} + \tfrac{1}{5} - \tfrac{1}{6} + \dots$$

which we write as

$$\sum_{k=1}^{\infty} (-1)^{k+1} \frac{1}{k} = 1 + (-\tfrac{1}{2}) + (\tfrac{1}{3}) + (-\tfrac{1}{4}) + (\tfrac{1}{5}) + (-\tfrac{1}{6}) + \dots$$

converges. Its sum is $\ln 2$, as we shall show in Section 7.7. The associated series of positive terms, $\sum_{k=1}^{\infty}(1/k)$, diverges of course (see Example 7.24b).

7.6.4 Exercises

41   Decide which of the following geometric series are convergent.

(a) $2 + \dfrac{2}{3} + \dfrac{2}{9} + \dfrac{2}{27} + \dots + \dfrac{2}{3^k} + \dots$

(b) $4 - 2 + 1 - \dfrac{1}{2} + \dots + \dfrac{(-1)^k 4}{2^k} + \dots$

(c) $10 + 11 + \tfrac{121}{10} + \tfrac{1331}{100} + \dots + 10(\tfrac{11}{10})^k + \dots$

(d) $1 - \tfrac{5}{4} + \tfrac{25}{16} - \tfrac{125}{64} + \dots + (\tfrac{-5}{4})^k + \dots$

42   Show that if

$$T_n = a + 2ar + 3ar^2 + 4ar^3 + \dots + nar^{n-1}$$

then $(1 - r)T_n = a + ar + ar^2 + \dots + ar^{n-1} - nar^n$
Deduce that

$$T_n = \frac{a(1 - r^n)}{(1 - r)^2} - \frac{nar^n}{1 - r}$$

Show that if $|r| < 1$, then $T_n \to a/(1 - r)^2$ as $n \to \infty$. Hence sum the infinite series

$$1 + \frac{2}{3} + \frac{1}{3} + \frac{4}{27} + \frac{5}{81} + \dots + \frac{k}{3^{k-1}} + \dots$$

43   For each of the following series find the sum of the first N terms, and, by letting $N \to \infty$, show that the infinite series converges and state its sum.

(a) $\tfrac{2}{1 \cdot 3} + \tfrac{2}{3 \cdot 5} + \tfrac{2}{5 \cdot 7} + \dots$

(b) $\tfrac{1}{1} + \tfrac{2}{2} + \tfrac{3}{2^2} + \tfrac{4}{2^3} + \tfrac{5}{2^4} + \dots$

(c) $\tfrac{1}{1 \cdot 2 \cdot 3} + \tfrac{1}{2 \cdot 3 \cdot 4} + \tfrac{1}{3 \cdot 4 \cdot 5} + \dots$

44   Which of the following series are convergent?

(a) $\displaystyle\sum_{k=1}^{\infty} (-1)^k$ (b) $-\tfrac{2}{3} + \tfrac{3}{4} - \tfrac{4}{5} + \dots$

(c) $\displaystyle\sum_{k=0}^{\infty} \frac{1}{3^k + 1}$

45   By comparison with the series $\sum_{k=2}^{\infty}[1/k(k - 1)]$ and $\sum_{k=2}^{\infty}[1/k(k + 1)]$, show that $S = \sum_{k=2}^{\infty}(1/k^2)$ is convergent and $\tfrac{1}{2} < S < 1$. (In fact, $\sum_{k=1}^{\infty}(1/k^2) = S + 1 = \tfrac{1}{6}\pi^2$.)

46   Show that $0.\dot{5}\dot{7}$ (that is, $0.575\,757\,\dots$) may be expressed as $57 \times 10^{-2} + 57 \times 10^{-4} + \dots$, and so $0.\dot{5}\dot{7} = 57\sum_{r=1}^{\infty}100^{-r}$. Hence express $0.\dot{5}\dot{7}$ as a rational number. Use a similar method to express as rational numbers:

(a) $0.4\dot{1}\dot{3}$ (b) $0.101\,010\dots$

(c) $0.999\,999\dots$ (d) $17.231\,723\,172\,3\dots$

47   Consider the series $\sum_{r=1}^{\infty} k^{-p}$. By means of the inequalities ($p > 0$)

$$\frac{1}{2^p} + \frac{1}{3^p} < \frac{2}{2^p}$$

$$\frac{1}{4^p} + \frac{1}{5^p} + \frac{1}{6^p} + \frac{1}{7^p} < \frac{4}{4^p}$$

$$\frac{1}{8^p} + \frac{1}{9^p} + \frac{1}{10^p} + \frac{1}{11^p} + \frac{1}{12^p} + \frac{1}{13^p}$$

$$+ \frac{1}{14^p} + \frac{1}{15^p} < \frac{8}{8^p}$$

and so on, deduce that the series is convergent for $p > 1$. Show that it is divergent for $p \leqslant 1$.

48 Two attempts to evaluate the sum $\sum_{k=1}^{\infty} k^{-4}$ are made on a computer working to 8 digits. The first evaluates the sum

$$1 + \frac{1}{2^4} + \frac{1}{3^4} + \frac{1}{4^4} + \ldots + \frac{1}{72^4}$$

from the left; the second evaluates it from the right. The first method yields the result 1.082 320 2, the second 1.082 322 1. Which is the better approximation and why?

49 Show that

$$\sum_{k=1}^{2n} \frac{(-1)^{k+1}}{k^4} = \sum_{k=1}^{2n} \frac{1}{k^4} - \frac{1}{8} \sum_{k=1}^{n} \frac{1}{k^4}$$

and deduce that

$$\sum_{k=1}^{\infty} \frac{(-1)^{k+1}}{k^4} = \frac{7}{8} \sum_{k=1}^{\infty} \frac{1}{k^4}$$

Deduce that the modulus of error in the estimate for the sum $\sum_{k=1}^{\infty} k^{-4}$ obtained by computing $\frac{8}{7} \sum_{k=1}^{N} (-1)^k k^{-4}$ is less than $\frac{8}{7} (N+1)^{-4}$.

7.7 Power series

Power series frequently occur in the solution of practical problems, as we shall see in Chapter 9, Sections 9.4.2, 9.8 and elsewhere. Often they are used to determine the sensitivity of systems to small changes in design parameters, to examine whether such systems are stable when small variations occur (as they always will in real life). The basic mathematics involved in power series is a natural extension of the series considered earlier.

> A series of the type
>
> $$a_0 + a_1 x + a_2 x^2 + a_3 x^3 + \ldots + a_n x^n + \ldots$$
>
> where the $a_0, a_1, a_2, \ldots$ are independent of x is called a **power series**.

7.7.1 Convergence of power series

Power series will, in general, converge for certain values of x and diverge elsewhere. Applying d'Alembert's ratio test to the above series, we see that it is absolutely convergent when

$$\lim_{n \to \infty} \left| \frac{a_{n+1} x^{n+1}}{a_n x^n} \right| < 1$$

Thus the series converges if

$$|x| \lim_{n \to \infty} \left| \frac{a_{n+1}}{a_n} \right| < 1$$

that is, if

$$|x| < \lim_{n \to \infty} \left| \frac{a_n}{a_{n+1}} \right|$$

Denoting $\lim\limits_{n \to \infty} |a_n/a_{n+1}|$ by r, we see that the series is absolutely convergent for $-r < x < r$ and divergent for $x < -r$ and $x > r$. The limit r is called the **radius of convergence** of the series. The behaviour at $x = \pm r$ has to be determined by other methods.

The various cases that occur are shown in Example 7.27.

Example 7.27

Find the radius of convergences of the series

(a) $\sum\limits_{n=1}^{\infty} \dfrac{x^n}{n}$ (b) $\sum\limits_{n=1}^{\infty} n^n x^n$

Solution

(a) Here $a_n = 1/n$, so that $|a_n/a_{n+1}| = (n + 1)/n$ and $r = 1$. Thus the domain of absolute convergence of the series is $-1 < x < 1$. The series diverges for $|x| > 1$ and for $x = 1$. At $x = -1$ the series is

$$-1 + \tfrac{1}{2} - \tfrac{1}{3} + \tfrac{1}{4} + \dots$$

which is convergent to $\ln \tfrac{1}{2}$ (see Section 7.6.3 and formula (7.15) below). Thus the series

$$\sum_{n=1}^{\infty} \frac{x^n}{n} = x + \tfrac{1}{2}x^2 + \tfrac{1}{3}x^3 + \tfrac{1}{4}x^4 + \dots$$

is convergent for $-1 \leqslant x < 1$.

(b) Here $a_n = n^n$ and

$$\left| \frac{a_n}{a_{n+1}} \right| = \frac{n^n}{(n + 1)^{n+1}} = \left(\frac{n}{n + 1} \right)^n \frac{1}{n + 1}$$

Now

$$\left(\frac{n}{n + 1} \right)^n = \frac{1}{(1 + 1/n)^n} \to e^{-1} \quad \text{as } n \to \infty \text{ (see Example 7.22)}$$

and

$$\frac{1}{n + 1} \to 0 \quad \text{as } n \to \infty$$

so that $a_n/a_{n+1} \to 0$ as $n \to \infty$. Thus the series converges only at $x = 0$, and diverges elsewhere.

7.7.2 Special power series

Power series may be added, multiplied and divided within their common domains of convergence (provided the denominator is non-zero within this common domain) to give power series that are convergent, and these properties are often exploited to express a given power series in terms of standard series and to obtain power series expansions of complicated functions.

Four elementary power series that are of widespread use are

(a) The geometric series

$$\frac{1}{1+x} = 1 - x + x^2 - x^3 + \ldots + (-1)^n x^n + \ldots \quad (-1 < x < 1) \tag{7.15}$$

(b) The binomial series

$$(1+x)^r = 1 + \binom{r}{1}x + \binom{r}{2}x^2 + \binom{r}{3}x^3 + \ldots + \binom{r}{n}x^n + \ldots \quad (-1 < x < 1) \tag{7.16}$$

where

$$\binom{r}{n} = \frac{r(r-1)\ldots(r-n+1)}{1 \cdot 2 \cdot 3 \cdot \ldots \cdot n}$$

is the binomial coefficient.

In series (7.16) r is any real number. When r is a positive integer, N say, the series terminates at the term x^N and we have the binomial expansion discussed in Chapter 1. When r is not a positive integer, the series does not terminate.

We can see that setting $r = -1$ gives

$$(1+x)^{-1} = \frac{1}{1+x} = 1 + \frac{(-1)}{1}x + \frac{(-1)(-2)}{1 \cdot 2}x^2 + \frac{(-1)(-2)(-3)}{1 \cdot 2 \cdot 3}x^3 + \ldots$$

which simplifies to the geometric series

$$\frac{1}{1+x} = 1 - x + x^2 - x^3 + \ldots$$

Comment (The series is often written as $(1+x)^{-1} = 1 - x + x^2 - x^3 + O(x^4)$, where $O(x^4)$ means terms involving powers of x greater than or equal to 4.)

Similarly,

$$(1+x)^{-2} = 1 + \frac{(-2)}{1}x + \frac{(-2)(-3)}{1 \cdot 2}x^2 + \frac{(-2)(-3)(-4)}{1 \cdot 2 \cdot 3}x^3 + \ldots$$

which simplifies to the arithmetical–geometric series

$$\frac{1}{(1+x)^2} = 1 - 2x + 3x^2 - 4x^3 + \ldots$$

(Compare Exercises 7.6.4, Question 42.) So the geometric series may be thought of as a special case of the binomial series.

Example 7.28   Obtain the power series expansions of

(a) $\dfrac{1}{\sqrt{(1 - x^2)}}$ (b) $\dfrac{1}{(1 - x)(1 + 3x)}$ (c) $\dfrac{\ln(1 + x)}{1 + x}$

Solution   (a) Using the binomial series (7.16) with $n = -\frac{1}{2}$ gives

$$\frac{1}{\sqrt{(1 + x)}} = (1 + x)^{-1/2}$$

$$= 1 + \frac{(-\frac{1}{2})}{1!}x + \frac{(-\frac{1}{2})(-\frac{3}{2})}{2!}x^2 + \frac{(-\frac{1}{2})(-\frac{3}{2})(-\frac{5}{2})}{3!}x^3 + \dots \quad (-1 < x < 1)$$

Now replacing x with $-x^2$ gives the required result

$$\frac{1}{\sqrt{(1 - x^2)}} = 1 + \frac{(\frac{1}{2})}{1!}x^2 + \frac{(\frac{1}{2})(\frac{3}{2})}{2!}x^4 + \frac{(\frac{1}{2})(\frac{3}{2})(\frac{5}{2})}{3!}x^6 + \dots \quad (-1 < x < 1)$$

$$= 1 + \frac{1}{2}x^2 + \frac{1 \cdot 3}{2 \cdot 4}x^4 + \frac{1 \cdot 3 \cdot 5}{2 \cdot 4 \cdot 6}x^6 + \dots \quad (-1 < x < 1)$$

(b) Expressed in partial fractions

$$\frac{1}{(1 - x)(1 + 3x)} = \frac{\frac{1}{4}}{1 - x} + \frac{\frac{3}{4}}{1 + 3x}$$

From the table of Figure 7.13

$$\frac{1}{1 - x} = 1 + x + x^2 + x^3 + \dots + x^n + \dots \quad (-1 < x < 1)$$

and replacing x by $3x$ in (7.15) gives

$$\frac{1}{1 + 3x} = 1 - (3x) + (3x)^2 - (3x)^3 + \dots + (-1)^n(3x)^n + \dots \quad (-\tfrac{1}{3} < x < \tfrac{1}{3})$$

Thus

$$\frac{1}{(1 - x)(1 + 3x)}$$

$$= \tfrac{1}{4}[1 + x + x^2 + x^3 + \dots] + \tfrac{3}{4}[1 - 3x + 9x^2 - 27x^3 + \dots] \quad (-\tfrac{1}{3} < x < \tfrac{1}{3})$$

$$= 1 - 2x + 7x^2 - 20x^3 + \dots + \tfrac{1}{4}(1 + (-1)^n 3^{n+1})x^n + \dots \quad (-\tfrac{1}{3} < x < \tfrac{1}{3})$$

(c) Using the series for $\ln(1 + x)$ and $(1 + x)^{-1}$ from (7.18) and (7.15),

$$\frac{\ln(1 + x)}{1 + x} = (x - \tfrac{1}{2}x^2 + \tfrac{1}{3}x^3 - \tfrac{1}{4}x^4 + \dots)(1 - x + x^2 - x^3 + \dots)$$

$$= x - (1 + \tfrac{1}{2})x^2 + (1 + \tfrac{1}{2} + \tfrac{1}{3})x^3 - (1 + \tfrac{1}{2} + \tfrac{1}{3} + \tfrac{1}{4})x^4 + \dots$$

$$(-1 < x < 1)$$

There is no separate command for power series expansion in either MATLAB or MAPLE. However, in MAPLE the command *series* is more general and can cope with many situations; it does not tell us how the series is constructed but may be used for checking answers. As illustrations, for Example 7.28(a) the command

```
series(1/sqrt(1 - x^2), x = 0);
```

returns the answer

$$1 + \tfrac{1}{2}x^2 + \tfrac{3}{8}x^4 + \tfrac{5}{16}x^6 + O(x^8)$$

and for Example 7.28(c) the command

```
series(ln(1 + x)/(1 + x), x = 0);
```

returns the answer

$$x - \tfrac{3}{2}x^2 + \tfrac{11}{6}x^3 - \tfrac{25}{12}x^4 + \tfrac{137}{60}x^5 + O(x^6)$$

Making use of the *maple* command the *series* command may be used in MATLAB's Symbolic Math Toolbox. Check that the commands

```
syms x
maple('series((1/sqrt(1 - x^2), x = 0))')
```

return the same answer for Example 7.28(a).

Power series are examples of Maclaurin series dealt with later in Section 9.5.2 of Chapter 9. They can be obtained using the *taylor* command *taylor(f,n)*. For example, the first five terms of the series expansion for Example 7.28(b) are determined by the commands

```
syms x
f = 1/((1 - x)*(1 + 3*x));
taylor(f,5);
pretty(ans)
```

as

$$1 - 2x + 7x^2 - 20x^3 + 61x^4$$

In MAPLE the commands

```
series(1/((1 - x)*(1 + 3*x)), x = 0); and
taylor(1/((1 - x)*(1 + 3*x)), x = 0);
```

produce the answer

$$1 - 2x + 7x^2 - 20x^3 + 61x^4 - 182x^5 + O(x^6)$$

The inverse process of expressing the sum of a power series in terms of the elementary functions is often difficult or impossible, but when it can be achieved it usually results in dramatic simplification of a practical problem.

Example 7.29

Sum the series

(a) $1^2 + 2^2x + 3^2x^2 + 4^2x^3 + 5^2x^4 + \dots$ (b) $1 + \dfrac{x}{2!} + \dfrac{x^2}{4!} + \dfrac{x^3}{6!} + \dfrac{x^4}{8!} + \dots$

Solution

(a) Set

$$S = 1^2 + 2^2x + 3^2x^2 + 4^2x^3 + 5^2x^4 + \dots + (n+1)^2x^n + \dots$$

Then

$$xS = 1^2x + 2^2x^2 + 3^2x^3 + 4^2x^4 + \dots + n^2x^n + \dots$$

and subtracting this from S gives

$$(1 - x)S = 1^2 + (2^2 - 1^2)x + (3^2 - 2^2)x^2 + (4^2 - 3^2)x^3 + \dots + [(n+1)^2 - n^2]x^n + \dots$$

$$= 1 + 3x + 5x^2 + 7x^3 + \dots + (2n + 1)x^n + \dots$$

$$= (1 + x + x^2 + x^3 + \dots + x^n \dots) + 2(x + 2x^2 + 3x^3 + \dots + nx^n + \dots)$$

$$= (1 + x + x^2 + x^3 + \dots + x^n + \dots) + 2x(1 + 2x + 3x^2 + 4x^3 + \dots$$
$$+ nx^{n-1} + \dots)$$

The first bracket is a geometric series of ratio x and sums to $\dfrac{1}{1-x}$, $|x| < 1$.

The second bracket is an arithmetico-geometric series of ratio x and sums to $\dfrac{1}{(1-x)^2}$, $|x| < 1$. Thus

$$(1 - x)S = \frac{1}{1-x} + \frac{2x}{(1-x)^2} = \frac{1+x}{(1-x)^2}$$

and

$$S = \frac{1+x}{(1-x)^2} \quad (-1 < x < 1)$$

(b) Summing this series relies on recognizing its similarity to the series for the hyperbolic cosine:

$$\cosh x = 1 + \frac{x^2}{2!} + \frac{x^4}{4!} + \frac{x^6}{6!} + \dots \quad (-\infty < x < \infty)$$

Replacing x by $\sqrt{x}$ gives

$$\cosh \sqrt{x} = 1 + \frac{x}{2!} + \frac{x^2}{4!} + \frac{x^3}{6!} + \dots \quad (-\infty < x < \infty)$$

and thus the series is summed.

Series may be summed in MATLAB using the *symsum* command. For Example (7.29) the series sums are

```
syms x k
Sa = symsum(k*x^(k - 1),k,1,inf)
```

giving

```
Sa = 1/(x - 1)^2
Sb = symsum(x^(k - 1)/'factorial(2*(k - 1))',k,1,inf)
```

giving

```
Sb = cosh(x^(1/2))
```

Commands in MAPLE are the same except for minor syntax differences; use can be made of (2*(k - 1))! So the command

```
sum('x^(k - 1)/(2*(k - 1))!', 'k' = 1..infinity);
```

returns the answer $cosh(\sqrt{x})$.

7.7.3 Exercises

Check your answers using MATLAB or MAPLE whenever possible.

50 For what values of x are the following series convergent?

(a) $\displaystyle\sum_{n=1}^{\infty} (2n - 1)x^n$

(b) $\displaystyle\sum_{n=0}^{\infty} (-1)^n \frac{x^{2n}}{(2n + 1)!}$

(c) $\displaystyle\sum_{n=1}^{\infty} \frac{x^n}{n(n + 1)}$

(d) $\displaystyle\sum_{n=1}^{\infty} \frac{n^2}{1 + n^2} x^n$

51 From known series deduce the following:

(a) $\dfrac{1}{1 + x^2} = 1 - x^2 + x^4 - x^6 + \dots$

(b) $\frac{1}{2} \ln \dfrac{1 + x}{1 - x} = x + \frac{1}{3}x^3 + \frac{1}{5}x^5 + \frac{1}{7}x^7 + \dots$

(c) $\dfrac{1}{(1 + x)^2} = 1 - 2x + 3x^2 - 4x^3 + 5x^4 - \dots$

(d) $\sqrt{(1 - x)} = 1 - \frac{1}{2}x - \frac{1}{8}x^2 + \frac{1}{16}x^3 - \frac{5}{128}x^4 - \dots$

(e) $\dfrac{1}{(1 - 2x)(2 + x)} = \frac{1}{2} + \frac{3}{4}x + \frac{13}{8}x^2 + \frac{51}{16}x^3 + \dots$

(f) $\dfrac{1}{(1 - x)(1 + x^2)} = 1 + x + x^4 + x^5 + \dots$

In each case give the general term and the radius of convergence.

52 Calculate the binomial coefficients

(a) $\dbinom{5}{2}$ (b) $\dbinom{-2}{3}$

(c) $\dbinom{1/2}{3}$ (d) $\dbinom{-1/2}{4}$

53 From known series deduce the following (the general term is not required):

(a) $\tan x = x + \frac{1}{3}x^3 + \frac{2}{15}x^5 + \dots$

(b) $\cos^2 x = 1 - \dfrac{2x^2}{2!} + \dfrac{2^3 x^4}{4!} - \dfrac{2^5 x^6}{6!} + \dots$

(c) $e^x \cos x = 1 + x - \dfrac{2x^3}{3!} - \dfrac{2^2 x^4}{4!} - \dfrac{2^2 x^5}{5!} + \dots$

(d) $\ln(1 + \sin x) = x - \frac{1}{2}x^2 + \frac{1}{6}x^3 + \frac{1}{12}x^4 - \dots$

54 Show that

$$\frac{1}{1-x} = 1 + x + x^2 + \ldots + x^{n-1} + \frac{x^n}{1-x} \quad (x \neq 1)$$

Hence derive a polynomial approximation to $(1-x)^{-1}$ with an error that, in modulus, is less than 0.5×10^{-4} for $0 \leqslant x \leqslant 0.25$.

Using nested multiplication, calculate from your approximation the reciprocal of 0.84 to 4dp, and compare your answer with the value given by your calculator. How many multiplications are needed in this case?

55 Find the sums of the following power series:

(a) $\displaystyle\sum_{k=0}^{\infty} (-1)^k 2^k x^{2k}$

(b) $1 + \frac{1}{2}x + \frac{1 \cdot 3}{2 \cdot 4}x^2 + \frac{1 \cdot 3 \cdot 5}{2 \cdot 4 \cdot 6}x^3 + \frac{1 \cdot 3 \cdot 5 \cdot 7}{2 \cdot 4 \cdot 6 \cdot 8}x^4 + \ldots$

(c) $\displaystyle\sum_{k=1}^{\infty} \frac{x^k}{k(k+1)}$

(d) $\frac{1}{2}x^2 + \frac{2}{3}x^3 + \frac{3}{4}x^4 + \frac{4}{5}x^5 + \ldots$

56 A regular polygon of n sides is inscribed in a circle of unit diameter. Show that its perimeter p_n is given by

$$p_n = n \sin \frac{\pi}{n}$$

Using the series expansion for sine, prove that

$$\pi = p_n + \frac{\pi^3}{3!}\frac{1}{n^2} - \frac{\pi^5}{5!}\frac{1}{n^4} + \ldots$$

and deduce that

$$\pi = \frac{1}{3}(4p_{2n} - p_n) + \frac{1}{4}\frac{\pi^5}{5!}\frac{1}{n^4} + \ldots$$

Given $p_{12} = 3.1058$ and $p_{24} = 3.1326$, use this result to obtain a better estimate of π.

7.8 Functions of a real variable

So far in this chapter we have concentrated on sequences and series. The terms of a sequence may be seen as defining a function whose domain is a subset of integers, such as N. We now turn to the fundamental properties that are essential to mathematical modelling and problem-solving, but we shall also be developing some basic mathematics that is necessary for later chapters.

7.8.1 Limit of a function of a real variable

The notion of limit can be extended in a natural way to include functions of a real variable:

> A function $f(x)$ is said to approach a limit l as x approaches the value a if, given any small positive quantity ε, it is possible to find a positive number δ such that $|f(x) - l| < \varepsilon$ for all x satisfying $0 < |x - a| < \delta$.

Less formally, this means that we can make the value of $f(x)$ as close as we please to l by taking x sufficiently close to a. Note that, using the formal definition, there is no need to evaluate $f(a)$; indeed, $f(a)$ may or may not equal l. The limiting value of f as $x \to a$ depends only on nearby values!

Example 7.30

Using a calculator, examine the values of $f(x)$ near $x = 0$ where

$$f(x) = \frac{x}{1 - \sqrt{(1 + x)}}, \quad x \neq 0$$

What is the value of $\lim_{x \to 0} f(x)$?

Solution

Note that $f(x)$ is not defined where $x = 0$. At nearby values of x we can calculate $f(x)$, and some values are shown in Figure 7.14.

Figure 7.14
Values of $f(x)$ to 6dp.

x	−0.1	−0.01	−0.001	0.001	0.01	0.1
$f(x)$	−1.948 683	−1.994 987	−1.999 500	−2.000 500	−2.004 988	−2.048 809

It seems that as x gets close to the value of 0, $f(x)$ gets close to the value of −2. Indeed, it can be proved that for $0 < |x| < 2\varepsilon - \varepsilon^2$, $|f(x) + 2| < \varepsilon$, so that

$$\lim_{x \to 0} f(x) = -2.$$

Comment

Notice that this is a rather artificial example to illustrate the idea and theory. In this case we can rewrite the formula for $f(x)$ to give

$$f(x) = \frac{x(1 + \sqrt{(1 + x)})}{(1 - \sqrt{(1 + x)})(1 + \sqrt{(1 + x)})}$$

which gives

$$f(x) = \frac{x(1 + \sqrt{(1 + x)})}{1 - (1 + x)} = -(1 + \sqrt{(1 + x)})$$

It is clear from this that $f(x) \to -2$ as $x \to 0$.

The elementary rules for limits (listed in Section 7.5.2) carry over from those of sequences, and these enable us to evaluate many limits by reduction to standard cases. Some common standard limits are

(i) $\quad \lim_{x \to a} \dfrac{x^r - a^r}{x - a} = r a^{r-1}, \quad$ where r is a real number

(ii) $\quad \lim_{x \to 0} \dfrac{\sin x}{x} = 1, \quad$ where x is in radians

(iii) $\quad \lim_{h \to 0} (1 + xh)^{1/h} = e^x$

These results can be deduced from the results of Section 7.7.2. For instance, consider $x^r - a^r$. Since $x \to a$, set $x = a + h$. Then as $x \to a$, $h \to 0$. We have

$$x^r - a^r = a^r\left(1 + \frac{h}{a}\right)^r - a^r \quad (a \neq 0)$$

Expanding $(1 + h/a)^r$ by the binomial series (7.16), we have

$$x^r - a^r = \frac{r}{1!}ha^{r-1} + \frac{r(r-1)}{2!}h^2a^{r-2} + \frac{r(r-1)(r-2)}{3!}h^3a^{r-3} + \ldots$$

But $x - a = h$, so

$$\frac{x^r - a^r}{x - a} = ra^{r-1} + \frac{r(r-1)}{2!}ha^{r-2} + \ldots$$

and letting $h \to 0$ yields the result (i)

$$\lim_{x \to a} \frac{x^r - a^r}{x - a} = ra^{r-1}$$

(When $a = 0$, the result is obtained trivially.)

The result (ii) is obtained even more simply. The series expansion

$$\sin x = x - \frac{x^3}{3!} + \frac{x^5}{5!} - \ldots$$

gives

$$\lim_{x \to 0} \frac{\sin x}{x} = 1$$

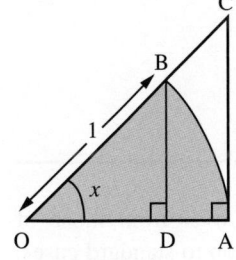

A geometric interpretation of (ii) is given in Figure 7.15. OAB is a sector of a circle of unit radius with angle x (measured in radians). Then

the area of $\triangle OBD <$ area of sector OBA $<$ area of $\triangle OCA$

Algebraically, we have

$$\tfrac{1}{2}\sin x \cos x < \tfrac{1}{2}x < \tfrac{1}{2}\tan x$$

Considering $x > 0$, we may write this as

$$1 < \frac{\sin x}{x} < \frac{1}{\cos x}$$

Figure 7.15
Geometric
interpretation of
$\lim_{x \to 0} \dfrac{\sin x}{x} = 1$.

As $x \to 0$, $\cos x \to 1$, so that $\dfrac{\sin x}{x} \to 1$ also.

The result (iii) is obtained from a binomial series:

$$(1 + xh)^{1/h} = 1 + \frac{1}{1!}\frac{1}{h}xh + \frac{1}{2!}\frac{1}{h}\left(\frac{1}{h} - 1\right)(xh)^2 + \frac{1}{3!}\frac{1}{h}\left(\frac{1}{h} - 1\right)\left(\frac{1}{h} - 2\right)(xh)^3 + \ldots$$

$$= 1 + \frac{x}{1!} + \frac{x^2(1 - h)}{2!} + \frac{x^3(1 - h)(1 - 2h)}{3!} + \ldots$$

and, as $h \to 0$,

$$(1 + xh)^{1/h} \to 1 + \frac{x}{1!} + \frac{x^2}{2!} + \frac{x^3}{3!} + \ldots = e^x$$

Example 7.31   Evaluate the following limits:

(a) $\displaystyle\lim_{x \to 0} \frac{\sqrt{(1 + x^2)} - 1}{x^2}$ (b) $\displaystyle\lim_{x \to 0} \frac{1 - \cos x}{x^2}$

Solution   (a) Method 1: Expand $\sqrt{(1 + x^2)}$ by the binomial series (7.16), giving

$$\sqrt{(1 + x^2)} = (1 + x^2)^{1/2} = 1 + \tfrac{1}{2}x^2 - \tfrac{1}{8}x^4 + \ldots$$

so that

$$\frac{\sqrt{(1 + x^2)} - 1}{x^2} = \frac{\tfrac{1}{2}x^2 - \tfrac{1}{8}x^4 + \ldots}{x^2} = \tfrac{1}{2} - \tfrac{1}{8}x^2 + \ldots$$

Thus

$$\lim_{x \to 0} \frac{\sqrt{(1 + x^2)} - 1}{x^2} = \tfrac{1}{2}$$

Method 2: Multiply numerator and denominator by $\sqrt{(1 + x^2)} + 1$, giving

$$\frac{[\sqrt{(1 + x^2)} - 1][\sqrt{(1 + x^2)} + 1]}{x^2[\sqrt{(1 + x^2)} + 1]} = \frac{(1 + x^2) - 1}{x^2[\sqrt{(1 + x^2)} + 1]} = \frac{1}{\sqrt{(1 + x^2)} + 1}$$

Now let $x \to 0$, to obtain

$$\lim_{x \to 0} \frac{\sqrt{(1 + x^2)} - 1}{x^2} = \tfrac{1}{2}$$

(b) Method 1: Replace $\cos x$ by its power series expansion,

$$\cos x = 1 - \frac{x^2}{2!} + \frac{x^4}{4!} - \frac{x^6}{6!} - \ldots$$

giving

$$\frac{1 - \cos x}{x^2} = \frac{x^2/2! - x^4/4! + x^6/6! - \ldots}{x^2} = \frac{1}{2!} - \frac{x^2}{4!} + \frac{x^4}{6!} - \ldots$$

Thus

$$\lim_{x \to 0} \frac{1 - \cos x}{x^2} = \tfrac{1}{2}$$

Method 2: Using the half-angle formula for $\cos x$ (see 2.7d), we have

$$1 - \cos x = 2 \sin^2 \tfrac{1}{2}x$$

so

$$\frac{1 - \cos x}{x^2} = \frac{2 \sin^2 \frac{1}{2}x}{x^2} = 2\left(\frac{\sin \frac{1}{2}x}{x}\right)^2 = \frac{1}{2}\left(\frac{\sin \varphi}{\varphi}\right)^2 \quad \text{where } \varphi = \frac{1}{2}x$$

On letting $x \to 0$, we have $\varphi \to 0$ and $(\sin \varphi)/\varphi \to 1$, so that

$$\frac{1 - \cos x}{x^2} \to \frac{1}{2}$$

Example 7.32

The volume of a sphere of radius a is $4\pi a^3/3$. Show that the volume of material used in constructing a hollow sphere of interior radius a and exterior radius $a + t$ is

$$V = 4\pi(3a^2t + 3at^2 + t^3)/3$$

Deduce that the surface area of the sphere of radius a is $S = 4\pi a^2$ and show that it is equal to the area of the curved surface of the enclosing cylinder.

Solution

$$V = \frac{4\pi}{3}[(a + t)^3 - a^3] = \frac{4\pi}{3}[a^3 + 3a^2t + 3at^2 + t^3 - a^3]$$

$$= \frac{4\pi}{3}(3a^2t + 3at^2 + t^3)$$

The volume V is approximately the surface area S of the interior sphere times the thickness t. That is

$$V = St + O(t^2)$$

Hence

$$S = V/t + O(t) = \frac{4\pi}{3}(3a^2 + 3at + t^2) + O(t)$$

Now proceeding to the limit as $t \to 0$, gives

$$S = 4\pi a^2$$

The radius of the enclosing cylinder is a and its height is $2a$ so that its curved surface area is

$$(2\pi a) \times (2a) = 4\pi a^2$$

as shown in Figure 7.16.

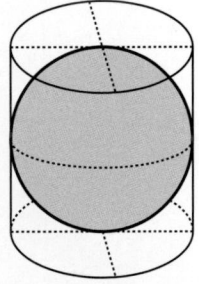

Figure 7.16
Enclosing cylinder in
Example 7.32.

7.8.2 One-sided limits

In some applications we have to use one-sided limits, for example

$$\lim_{x \to 0+} \sqrt{x} = 0 \quad \text{(as } x \text{ tends to zero 'from above')}$$

In this example, $\lim\limits_{x \to 0-} \sqrt{x}$ (as x tends to zero 'from below') does not exist, since no negative numbers are in the domain of $\sqrt{x}$. When we write

$$\lim_{x \to a} f(x) = l$$

we mean that

$$\lim_{x \to a-} f(x) = \lim_{x \to a+} f(x) = l$$

Example 7.33   Sketch the graph of the function $f(x)$ where

$$f(x) = \frac{\sqrt{(x^2 - x^3)}}{x}, \quad x \neq 0 \text{ and } x < 1$$

and show that $\lim\limits_{x \to 0} f(x)$ does not exist.

Solution   Notice that the function is not defined for $x = 0$. A sketch of the function is given for $-1 \leqslant x \leqslant 1$, $x \neq 0$ in Figure 7.17. From that diagram we see that $f(x) \to -1$ as $x \to 0$ from below and $f(x) \to +1$ as $x \to 0$ from above. Since the existence of a limit requires the same value whether we approach from above or below we deduce that $\lim\limits_{x \to 0} f(x)$ does not exist.

Figure 7.17
Graph of
$$y = \frac{\sqrt{(x^2 - x^3)}}{x}.$$

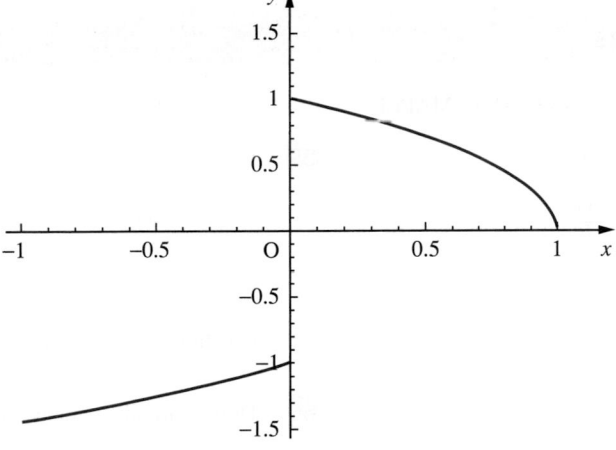

Symbolically in MATLAB limits are determined using the following commands:

$\lim\limits_{x \to a} f(x)$ by `limit(f,x,a)` or `limit(f,a)`

$\lim\limits_{x \to a^-} f(a)$ by `limit(f,x,a, 'left')`

and

$\lim\limits_{x \to a^+} f(a)$ by `limit(f,x,a, 'right')`

Figure 7.19
The oscillation
of a function.

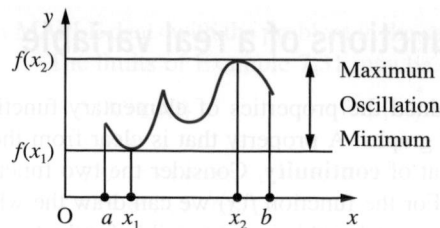

(d) If $a \leqslant x_1 \leqslant x_2 \leqslant x_3 \leqslant \ldots \leqslant x_n < b$, there is an $X \in [a, b]$ such that

$$f(X) = \frac{f(x_1) + f(x_2) + \ldots + f(x_n)}{n}$$

This property is known as the **average value theorem**.

(e) Given $\varepsilon > 0$, the interval $[a, b]$ can be divided into a number of intervals in each of which the oscillation of the function is less than ε.

(f) Given $\varepsilon > 0$, there is a subdivision of $[a, b]$, $a = x_0 < x_1 < x_2 < \ldots < x_n = b$, such that in each subinterval (x_i, x_{i+1})

$$\left| f(x) - \left[f_i + (x - x_i)\frac{f_{i+1} - f_i}{x_{i+1} - x_i} \right] \right| < \varepsilon, \, f_i = f(x_i)$$

That is, by making a subtabulation that is sufficiently fine, we can represent $f(x)$ locally by linear interpolation to within any prescribed error bound.

(g) Given $\varepsilon > 0$, $f(x)$ can be approximated on the interval $[a, b]$ by a polynomial of suitable degree such that

$$|f(x) - p_n(x)| < \varepsilon \quad \text{for } x \in [a, b]$$

This is known as **Weierstrass' theorem**. Note, however, that the theorem does not tell us how to obtain $p_n(x)$.

The properties of limits listed in Section 7.5.2 enable us to determine the continuity of functions formed by combining continuous functions. Thus if $f(x)$ and $g(x)$ are continuous functions then so are the functions

(a) $af(x)$, where a is a constant
(b) $f(x) + g(x)$
(c) $f(x)g(x)$
(d) $f(x)/g(x)$, except where $g(x) = 0$

Also the composite function $f(g(x))$ is continuous at x_0 if $g(x)$ is continuous at x_0 and $f(x)$ is continuous at $x = g(x_0)$.

Some of the properties of continuous functions are illustrated in Example 7.34 and in Exercises 7.9.4.

Example 7.34   Show that $f(x) = 2x/(1 + x^2)$ for $x \in \mathbb{R}$ is continuous on its whole domain. Find its maximum and minimum values and show that it attains every value between these extrema.

Figure 7.20
Graph of $2x/(1 + x^2)$.

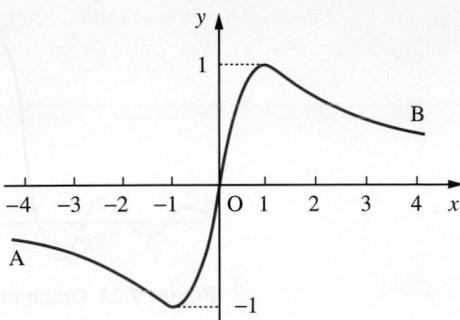

Solution The graph of the function is shown in Figure 7.20, from which we can see that the part
shown is a continuous curve. That is to say, we can put a pencil at point A at the left-hand
end of the graph and trace along the whole length of the curve to reach the point B at the
right-hand end without lifting the pencil from the page. We can prove this more formally
as follows. Select any point x_0 of the domain of the function. Then we have to show that
$|f(x) - f(x_0)|$ can be made as small as we please by taking x sufficiently close to x_0. Now

$$\frac{2x}{1 + x^2} - \frac{2x_0}{1 + x_0^2} = \frac{2x(1 + x_0^2) - 2x_0(1 + x^2)}{(1 + x^2)(1 + x_0^2)} = \frac{2(1 - xx_0)(x - x_0)}{(1 + x^2)(1 + x_0^2)}$$

$$\rightarrow 0 \quad \text{as } x \rightarrow x_0$$

This implies that $f(x)$ is continuous at x_0, and since x_0 is any point of the domain, it
follows that $f(x)$ is continuous for all x.

   Then to show that the function takes a given value y we have to solve the equation
$y = f(x)$ for x in terms of y. So that in this example

$$y = \frac{2x}{1 + x^2} \quad \text{gives} \quad yx^2 - 2x + y = 0$$

where we are now solving the equation for x in terms of y. Hence we obtain

$$x = \frac{1 \pm \sqrt{(1 - y^2)}}{y}, \quad y \neq 0 \text{ and } -1 \leq y \leq 1$$

This gives two values of x for each $y \in (-1, 1)$, $y \neq 0$. Clearly $y = 0$ is also attained
for $x = 0$. The maximum and minimum values for y are 1 and -1 respectively, and the
corresponding values of x are 1 and -1. Thus $f(x)$ is a continuous function on its
domain, and it attains its maximum and minimum values and every value in between.

7.9.2 Continuous and discontinuous functions

The technique used to show that $f(x)$ is a continuous function in Example 7.34 can
be used to show that polynomials, rational functions (except where the denominator is
zero) and many transcendental functions are continuous on their domains. We frequently
make use of the properties of continuous functions unconsciously in problem-solving!
For example, in solving equations we trap the root between two points x_1 and x_2 where
$f(x_1) < 0$ and $f(x_2) > 0$ and conclude that the root we seek lies between x_1 and x_2. The

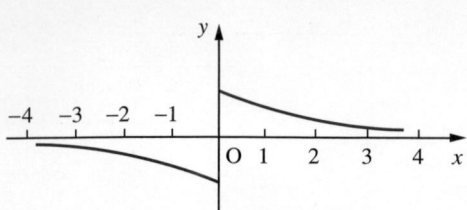

Figure 7.21 Graph of $\tan^{-1}(1/x)$, $x \neq 0$.

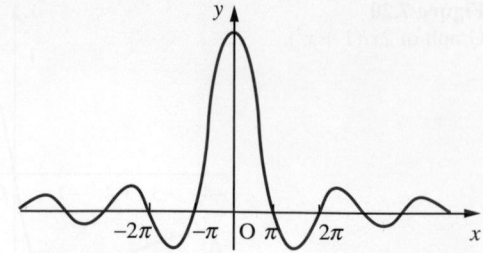

Figure 7.22 Graph of sinc $x = (\sin x)/x$.

need for continuity here is shown by the graph of $y = \tan^{-1}(1/x)$ (Figure 7.21). There is no value of x corresponding to $y = 0$, despite the facts that $\tan^{-1}(1/0.01)$ is positive and $\tan^{-1}[1/(-0.01)]$ is negative.

Similarly, when locating the maximum or minimum value of a function $y = f(x)$, in many practical situations we would be content with a solution that yields a value close to the true optimum value, and property (e) above tells us we can make that value as close as we please. Sometimes we use the continuity idea to fill in 'gaps' in function definitions. A simple example of this is $f(x) = (\sin x)/x$ for $x \neq 0$. This function is defined everywhere except at $x = 0$. We can extend it to include $x = 0$ by insisting that it be continuous at $x = 0$. Since $(\sin x)/x \rightarrow 1$ as $x \rightarrow 0$, defining $f(x)$ as

$$f(x) = \begin{cases} \dfrac{\sin x}{x} & (x \neq 0) \\ 1 & (x = 0) \end{cases} \tag{7.19}$$

yields a function with no 'gaps' in its domain. The function $f(x)$ in (7.19) is known as the **sinc function**; that is,

$$\text{sinc } x = \begin{cases} \dfrac{\sin x}{x} & (x \neq 0) \\ 1 & (x = 0) \end{cases}$$

and its graph is drawn in Figure 7.22. This function has important applications in engineering, particularly in digital signal analysis. See the chapter on Fourier transforms in the companion text *Advanced Modern Engineering Mathematics*.

Of course it is not always possible to fill in 'gaps' in function definitions. The function

$$g(x) = \frac{x}{\sin x} \quad (x \neq n\pi,\ n = 0, \pm1, \pm2, \dots)$$

can have its domain extended to include the points $x = n\pi$, but it will always have a discontinuity at those points (except perhaps $x = 0$). Thus

$$f(x) = \begin{cases} \dfrac{x}{\sin x} & (x \neq n\pi,\ n = 0, \pm1, \pm2, \dots) \\ 1 & (x = 0) \\ 0 & (x = n\pi,\ n = \pm1, \pm2, \dots) \end{cases}$$

yields a function that is defined everywhere but is discontinuous at an infinite set of points.

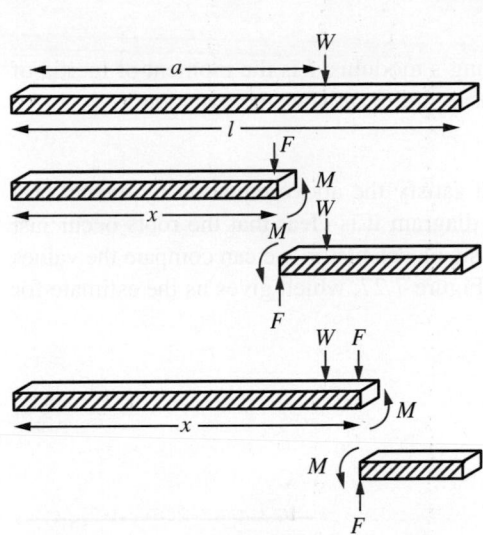

Figure 7.23 A beam, hinged at both ends, carrying a point load.

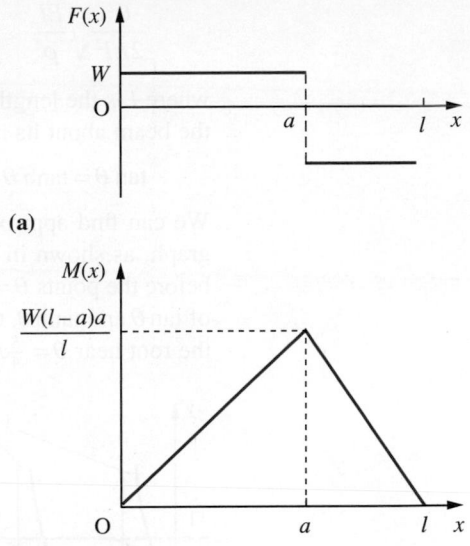

(a)

(b)

Figure 7.24 (a) The shear force and (b) the bending moment for a freely hinged beam.

In the analysis of practical problems we frequently use functions that have different formulae on different parts of their domain. For example, consider a beam of length l that is freely hinged at both ends and carries a concentrated load W at $x = a$, as shown in Figure 7.23. Then the shear force F is given by

$$F(x) = \begin{cases} W - Wa/l & (0 < x < a) \\ -Wa/l & (a \leq x < l) \end{cases}$$

and is sketched in Figure 7.24(a).

The bending moment M is

$$M(x) = \begin{cases} W(l - a)x/l & (0 < x \leq a) \\ W(l - x)a/l & (a \leq x < l) \end{cases}$$

and is sketched in Figure 7.24(b). (The terms 'shear force' and 'bending moment' are discussed in the next chapter in Example 8.7.)

Notice here that F has a finite discontinuity at $x = a$ while M is continuous there.

7.9.3 Numerical location of zeros

Many practical engineering problems may involve the determination of the points at which a function takes a specific value (often zero) or the points at which it takes its maximum or minimum values. There are many different numerical procedures for solving such problems and we shall illustrate the technique by considering its application to the analysis of structural vibration.

This is a very common problem in engineering. To avoid resonance effects, it is necessary to calculate the natural frequencies of vibration of a structure. For a beam built in at one end and simply supported at the other as shown in Figure 7.25 the natural frequencies are given by

Figure 7.25
A beam built in at one end and simply supported at the other.

$$\frac{\theta^2}{2\pi l^2}\sqrt{\frac{EI}{\rho}}$$

where l is the length of the beam, E is Young's modulus, I is the moment of inertia of the beam about its neutral axis, ρ is its density and θ satisfies the equation

$$\tan\theta = \tanh\theta$$

We can find approximate values for θ that satisfy the above equation by means of a graph, as shown in Figure 7.26. From the diagram it is clear that the roots occur just before the points $\theta = 0,\ \frac{5}{4}\pi,\ \frac{9}{4}\pi,\ \frac{13}{4}\pi,\ \dots$. Using a calculator, we can compare the values of $\tan\theta$ and $\tanh\theta$, to produce the table of Figure 7.27, which gives us the estimate for the root near $\theta = \frac{5}{4}\pi$ as 3.925 ± 0.005.

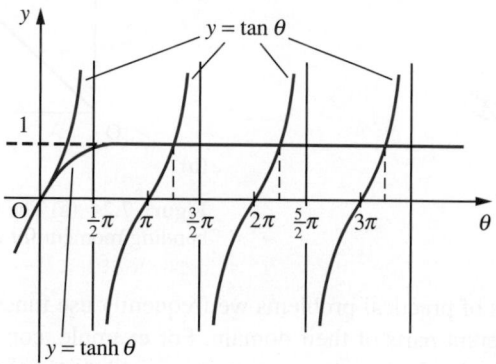

Figure 7.26 The roots of the equation $\tan\theta = \tanh\theta$.

Figure 7.27 Table of values.

θ	$\tan\theta$	$\tanh\theta$
3.90	0.9474	0.9992
3.91	0.9666	0.9992
3.92	0.9861	0.9992
3.93	1.0060	0.9992

If we require a more precise answer than this provides, we can resort to a finer subtabulation. In some problems this can be very tedious and time-consuming. A better strategy is to use an **interval-halving** or **bisection method**. We know that the root lies between $\theta_1 = 3.92$ and $\theta_2 = 3.93$. We work out the value of the functions at the midpoint of this interval, $\theta_3 = 3.925$, and determine whether the root lies between θ_1 and θ_3 or between θ_3 and θ_2. The process is then repeated on the subinterval that contains the root, and so on until sufficient precision is obtained.

The process is set out in tabular form in Figure 7.28. Note the renaming of the end points of the root-bracketing interval at each step, so that the interval under scrutiny is always denoted by $[\theta_1, \theta_2]$. After five applications we have $\theta = 3.92672 \pm (0.005/2^5)$.

Figure 7.28
Solution of
$\tan\theta - \tanh\theta = 0$
by the bisection
method.

θ_1	$f(\theta_1)$	θ_2	$f(\theta_2)$	θ_m	$f(\theta_m)$
3.92	−0.013098	3.93	0.006808	3.925	−0.003195
3.925	−0.003195	3.93	0.006808	3.9275	0.001794
3.925	−0.003195	3.9275	0.001794	3.92625	−0.000703
3.92625	−0.000703	3.9275	0.001794	3.926875	0.000545
3.92625	−0.000703	3.926875	0.000545	3.9265625	−0.000079

A refinement of the bisection method is the **method of false position** (also known as *regula falsa*). To solve the equation $f(x) = 0$, given x_1 and x_2 such that $f(x_1) > 0$ and $f(x_2) < 0$ and $f(x)$ is continuous in (x_1, x_2), the bisection method takes the point

Figure 7.29
Solution of
$\tan \theta - \tanh \theta = 0$
by *regula falsa*.

θ_1	$f(\theta_1)$	θ_2	$f(\theta_2)$	$\dfrac{\theta_1 f(\theta_2) - \theta_2 f(\theta_1)}{f(\theta_2) - f(\theta_1)}$	$f\left(\dfrac{\theta_1 f(\theta_2) - \theta_2 f(\theta_1)}{f(\theta_2) - f(\theta_1)}\right)$
3.92	−0.013 098	3.93	0.006 808	3.926 580	−0.000 045
3.926 580	−0.000 045	3.93	0.006 808	3.926 602	−0.000 000

$\frac{1}{2}(x_1 + x_2)$ as the next estimate of the root. The method of false position uses linear interpolation to derive the next estimate of the root. The straight line joining the points $(x_1, f(x_1))$ and $(x_2, f(x_2))$ is given by

$$\frac{y - f(x_1)}{f(x_2) - f(x_1)} = \frac{x - x_1}{x_2 - x_1}$$

This line cuts the x axis where

$$x = \frac{x_1 f(x_2) - x_2 f(x_1)}{f(x_2) - f(x_1)}$$

so this is the new estimate of the root. This method usually converges more rapidly than the bisection method. The computation of the root of $\tan \theta - \tanh \theta = 0$ in the interval (3.92, 3.93) is shown in Figure 7.29. Notice how, as a result of the first step, the estimate of the root is $\theta = 3.926\,580$ and $f(3.926\,580) = 0.000\,045$. The root is now bracketed in the interval (3.926 580, 3.93), and the method is repeated. In two steps we have an estimate of the root giving a value of $f(\theta) < 10^{-6}$. This obviously converges much faster than the bisection method.

Both the bisection method and the method of false position are **bracketing methods** – the root is known to lie in an interval of steadily decreasing size. As such, they are guaranteed to converge to a solution. An alternative method of solution for an equation $f(x) = 0$ is to devise a scheme producing a convergent sequence whose limit is the root of the equation. Such **fixed point iteration methods** are based on a relation of the form $x_{n+1} = g(x_n)$. If $\lim_{n \to \infty} x_n = \alpha$, say, then evidently $\alpha = g(\alpha)$. The simplest way to devise an iterative scheme for the solution of an equation $f(x) = 0$ is to find some rearrangement of the equation in the form $x = g(x)$. Then, if the scheme $x_{n+1} = g(x_n)$ converges, the limit will be a root of $f(x) = 0$.

We can arrange the equation $\tan \theta = \tanh \theta$ in the form

$$\theta = \tan^{-1}(\tanh \theta) + k\pi \quad (k = 0, \pm 1, \pm 2, \dots)$$

If we take $k = 1$ and $\theta_0 = \frac{5}{4}\pi$ we obtain, using the iteration scheme,

$$\theta_n = \tan^{-1}(\tanh \theta_{n-1}) + \pi$$

the sequence

$$\theta_0 = 3.926\,991, \quad \theta_1 = 3.926\,603, \quad \theta_2 = 3.926\,602, \quad \theta_3 = 3.926\,602$$

and the root is $\theta = 3.926\,602$ to 6dp. (Taking other values of k will, of course, give schemes that converge to other roots of $\tan \theta = \tanh \theta$.)

The disadvantage of such iterative schemes is that not all of them converge. We shall return to this topic in Section 9.3.2.

7.9.4 Exercises

 Check your answers using MATLAB or MAPLE whenever possible.

62 Draw sketches and discuss the continuity of

(a) $\dfrac{|x|}{x}$ (b) $\dfrac{x-1}{2-x}$

(c) $\tanh\dfrac{1}{x}$ (d) $\lfloor 1-x^2\rfloor$

63 Find upper and lower bounds obtained by

(a) $2x^2 - 4x + 7$ $(0 \leqslant x \leqslant 2)$

(b) $-x^2 + 4x - 1$ $(0 \leqslant x \leqslant 3)$

in the appropriate domains. Draw sketches to illustrate your answers.

64 Use the intermediate value theorem to show that the equation

$$x^3 + 10x^2 + 8x - 50 = 0$$

has roots between 1 and 2, between −4 and −3 and between −9 and −8. Find the root between 1 and 2 to 2dp using the bisection method.

65 Show that the equation $3^x = 3x$ has a root in the interval (0.7, 0.9). Use the intermediate value theorem and the method of *regula falsa* to find this root to 3dp.

66 Show that the equation

$$x^3 - 3x + 1 = 0$$

has three roots α, β and γ, where $\alpha < -1$, $0 < \beta < 1$ and $\gamma > 1$. For which of these is the iterative scheme

$$x_{n+1} = \tfrac{1}{3}(x_n^3 + 1)$$

convergent? Calculate the roots to 3dp.

67 The cubic equation $x^3 + 2x - 2 = 0$ can be written as

(a) $x = 1 - \tfrac{1}{2}x^3$ (b) $x = \dfrac{2}{2 + x^2}$

(c) $x = (2 - 2x)^{1/3}$

Determine which of the corresponding iteration processes converges most rapidly to find the real root of the equation. Hence calculate the root to 3dp.

68 Show that the iteration

$$x_{n+1} = \frac{1}{3}\left(2x_n + \frac{a}{x_n^2}\right)$$

converges to the limit $a^{1/3}$. Use the formula with $a = 157$ and $x_0 = 5$ to compare x_1 and x_2.

Show that the error ε_n in the nth iterate is given by $\varepsilon_{n+1} \approx \varepsilon_n^2 / x_{n-1}$, where $x_n = a^{1/3} + \varepsilon_n$. Hence estimate the error in x_1 obtained above.

69 The periods of natural vibrations of a cantilever are given by

$$\frac{2\pi l^2}{\theta^2}\sqrt{\frac{\rho}{EI}}$$

where l, E, I and ρ are physical constants dependent on the shape and material of the cantilever and θ is a root of the equation

$$\cosh\theta\cos\theta = -1$$

Examine this equation graphically. Estimate its lowest root α_0 and obtain an approximation for the kth root α_k. Compare the two iterations:

$$\theta_{n+1} = \cosh^{-1}(-\sec\theta_n)$$

and

$$\theta_{n+1} = \cos^{-1}(-\operatorname{sech}\theta_n)$$

Which should be used to find an improved approximation to x_0?

7.10 Engineering application: insulator chain

The voltage V_k at the kth pin of the insulator chain shown in Figure 7.30 satisfies the recurrence relation

$$V_{k+2} - \left(2 + \frac{C_2}{C_1}\right)V_{k+1} + V_k = 0$$

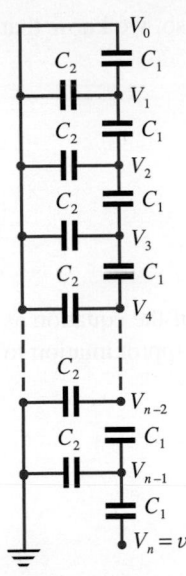

Figure 7.30

with $V_0 = 0$ and $V_n = v$, the amplitude of the voltage applied at the head of the chain. The characteristic equation for this recurrence relation is

$$\lambda^2 - \left(2 + \frac{C_2}{C_1}\right)\lambda + 1 = 0$$

which has real roots

$$\lambda_{1,2} = 1 + \frac{C_2}{2C_1} \pm \sqrt{\left[\frac{C_2}{C_1}\left(1 + \frac{C_2}{4C_1}\right)\right]}$$

Thus, the general solution is

$$V_k = A\lambda_1^k + B\lambda_2^k$$

Applying the condition $V_0 = 0$ gives

$$A + B = 0$$

Applying the condition $V_n = v$ gives

$$A\lambda_1^n + B\lambda_2^n = v$$

Hence $B = -A$ and $A = v/(\lambda_1^n - \lambda_2^n)$ and

$$V_k = \frac{v(\lambda_1^k - \lambda_2^k)}{\lambda_1^n - \lambda_2^n}$$

In a typical insulator chain $C_2/C_1 = 0.1$ and $n = 10$. It is left to the reader to calculate V_k/v for $k = 1, 2, \ldots, 9$.

7.11 Engineering application: approximating functions and Padé approximants

In Section 2.9.1 we introduced linear and quadratic interpolation as a means of obtaining estimates of the values of functions in between known values. Often in engineering applications it is of considerable importance to obtain good approximations to functions. In this section we shall show how what we have learned about power series representation can be used to produce a type of approximate representation of a function widely used by engineers, for example, when approximating exponentials by rational functions in modelling time delays in control systems. The approach is attributed to Padé and is based on the matching of series expansion.

Example 7.35   Obtain an approximation to the function e^{-x} in the form

$$e^{-x} \approx \frac{a + bx + cx^2}{A + Bx + Cx^2}$$

and find an estimate for the error.

Solution   Assuming an exact match at $x = 0$, we deduce at once that $a = A$. Also, we know that $1/e^x = e^{-x}$, and assuming a similar relation for the approximation

$$\frac{A - Bx + Cx^2}{a - bx + cx^2} \equiv \frac{a + bx + cx^2}{A + Bx + Cx^2}$$

This holds if we choose $A = a$ (as above), $B = -b$ and $C = c$, giving

$$e^{-x} \approx \frac{A - Bx + Cx^2}{A + Bx + Cx^2}$$

We can see from this that it would be possible to express both sides of the equation as power series in x (at least in a restricted domain). We can rewrite the approximation to make it exact:

$$(A + Bx + Cx^2)e^{-x} = (A - Bx + Cx^2) + px^3 + qx^4 + rx^5 + \ldots$$

where p, q, ... are to be found.

Replacing e^{-x} by its power series representation, we have

$$(A + Bx + Cx^2)(1 - x + \tfrac{1}{2}x^2 - \tfrac{1}{6}x^3 + \tfrac{1}{24}x^4 - \tfrac{1}{120}x^5 + \ldots)$$
$$= A - Bx + Cx^2 + px^3 + qx^4 + rx^5 + \ldots$$

Multiplying out the left-hand side and collecting terms, we obtain

$$A + (B - A)x + (\tfrac{1}{2}A - B + C)x^2 + (-\tfrac{1}{6}A + \tfrac{1}{2}B - C)x^3$$
$$+ (\tfrac{1}{24}A - \tfrac{1}{6}B + \tfrac{1}{2}C)x^4 + (-\tfrac{1}{120}A + \tfrac{1}{24}B - \tfrac{1}{6}C)x^5 + \ldots$$
$$= A - Bx + Cx^2 + px^3 + qx^4 + rx^5 + \ldots$$

Comparing the coefficients of like powers of x on either side of this equation gives

$$A = A$$
$$B - A = -B$$
$$\tfrac{1}{2}A - B + C = C$$
$$-\tfrac{1}{6}A + \tfrac{1}{2}B - C = p$$
$$\tfrac{1}{24}A - \tfrac{1}{6}B + \tfrac{1}{2}C = q$$
$$-\tfrac{1}{120}A + \tfrac{1}{24}B - \tfrac{1}{6}C = r$$

and so on.

We see from this that there is not a unique solution for A, B and C, but that we may choose them (or some of them) arbitrarily. Taking $A = 1$ gives $B = \tfrac{1}{2}$ and $\tfrac{1}{12} - C = p$. Setting $p = 0$ will make the error term smaller near $x = 0$, so we adopt that choice, giving $C = \tfrac{1}{12}$. This gives $q = 0$ and $r = -\tfrac{1}{720}$. Thus

$$(1 + \tfrac{1}{2}x + \tfrac{1}{12}x^2)e^{-x} = (1 - \tfrac{1}{2}x + \tfrac{1}{12}x^2) - \tfrac{1}{720}x^5 + \ldots$$

so that

$$e^{-x} = \frac{1 - \tfrac{1}{2}x + \tfrac{1}{12}x^2}{1 + \tfrac{1}{2}x + \tfrac{1}{12}x^2} - \frac{\tfrac{1}{720}x^5 + \ldots}{1 + \tfrac{1}{2}x + \tfrac{1}{12}x^2}$$

$$= \frac{12 - 6x + x^2}{12 + 6x + x^2} - \tfrac{1}{720}x^5 + \ldots$$

The principal term of the error, $-\frac{1}{720}x^5$, enables us to decide the domain of usefulness of the approximation. For example, if we require an approximation correct to 4dp, we need $\frac{1}{720}x^5$ to be less then $\frac{1}{2} \times 10^{-4}$. Thus the approximation

$$e^{-x} \approx \frac{12 - 6x + x^2}{12 + 6x - x^2}$$

yields answers correct to 4dp for $|x| < 0.51$.

This particular approximation is used by control engineers in order to enable them to apply linear systems techniques to the analysis and design of systems characterizing a time delay in their dynamics. Since the degree of both the numerator and denominator is 2, this is referred to as the (2, 2) Padé approximant.

As an extended exercise, the reader should obtain the following (1, 1) and (3, 3) Padé approximants:

$$e^{-x} \approx \frac{2 - x}{2 + x} \quad \text{and} \quad e^{-x} \approx \frac{120 - 60x + 12x^2 - x^3}{120 + 60x + 12x^2 + x^3}$$

7.12 Review exercises (1–25)

Check your answers using MATLAB or MAPLE whenever possible.

1. There are two methods of assessing the value of a wasting asset. The first assumes that it decreases each year by a fixed amount; the second assumes that it depreciates by a fixed percentage.

    A piece of equipment costs £1000 and has a 'lifespan' of six years after which its scrap value is £100. Estimate the value of the equipment by both methods for the intervening years.

2. A machine that costs £1000 has a working life of three years, after which it is valueless and has to be replaced. It saves the owner £500 per year while it is in use. Show that the true total saving £S to the owner over the three years is

$$S = 500\left[\frac{1}{1 + r/100} + \frac{1}{(1 + r/100)^2} + \frac{1}{(1 + r/100)^3}\right]$$
$$- 1000$$

    where $r\%$ is the current rate of interest. Estimate S for $r = 5$, 10, 15 and 20. When does the machine truly save the owner money?

3. An economic model for the supply $S(P)$ and demand $D(P)$ of a product at a market price of P is given by

$$D(P) = 2 - P$$
$$S(P) = \tfrac{1}{2} + \tfrac{1}{2}P$$

and

$$D(P_{t+1}) = S(P_t)$$

(so that supply lags behind demand by one time unit). Show that

$$P_{t+1} - 1 = -\tfrac{1}{2}(P_t - 1)$$

and deduce that

$$P_t = 1 + (-\tfrac{1}{2})^t(P_0 - 1)$$

Find the particular solution of the recurrence relation corresponding to $P_0 = 0.8$ and sketch it in a cobweb diagram. What is the steady-state price of the product?

4. Show that

$$\sum_{k=1}^{n} \frac{1}{T_k} = \frac{T_{n-1} + T_n}{T_n}$$

where T_k is the kth triangular number. (See Question 4 in Exercises 7.2.3.)

5. Find the general solutions of the following linear recurrence relations.

    (a) $f_{n+2} - 5f_{n+1} + 6f_n = 0$ (b) $f_{n+2} - 4f_{n+1} + 4f_n = 0$

    (c) $f_{n+2} - 5f_{n+1} + 6f_n = 4^n$ (d) $f_{n+2} - 5f_{n+1} + 6f_n = 3^n$

6 Suppose that consumer spending in period t, C_t, is related to personal income two periods earlier, I_{t-2}, by

$$C_t = 0.875I_{t-2} - 0.2C_{t-1} \quad (t \geqslant 2)$$

Deduce that if personal income increases by a factor 1.05 each period, that is

$$I_{t+1} = 1.05I_t$$

then $I_t = 1.05^t I_0$ and hence

$$C_t = (C_1 - 0.7I_0)(-0.2)^{t-1} + 0.7I_0(1.05)^{t-1}$$

Describe the behaviour of C_t in the long run.

7 An economist believes that the price P_t of a seasonal commodity in period t, satisfies the recurrence relation

$$P_{t+2} = 2(P_{t+1} - P_t) + C \quad (t \geqslant 0)$$

where C is a positive constant.
   Show that

$$P_t = A(1 + j)^t + B(1 - j)^t + C$$

where A and B are complex conjugate constants. Noting that $1 \pm j = \sqrt{2}(\cos\frac{\pi}{4} \pm j\sin\frac{\pi}{4})$, explain why the economist is mistaken.

8 The cobweb model applied to agricultural commodities assumes that current supply depends on prices in the previous season. If P_t denotes market price in any period and Q_{St}, Q_{Dt} supply and demand in that period, then

$$Q_{Dt} = 180 - 0.75P_t$$

$$Q_{St} = -30 + 0.3P_{t-1} \quad \text{where } P_0 = 220$$

   Find the market price and comment on its form.

9 Solve for National Income, Y_t, the set of recurrence relations

$$Y_t = 1 + C_t + I_t$$

$$C_t = \tfrac{1}{2}Y_{t-1}$$

$$I_t = 2(C_t - C_{t-1})$$

Comment on your solution.

10 A sequence is defined by

$$a_k = 1 + \frac{1}{2} + \frac{1}{3} + \ldots + \frac{1}{k} - \ln k \quad (k = 1, 2, \ldots)$$

Given $a_{10} = 0.626\,383$, $a_{16} = 0.608\,140$ and $a_{20} = 0.602\,009$ estimate $\gamma = \lim_{n \to \infty} a_n$, using repeated linear extrapolation. (γ is known as **Euler's constant**.)

11 Discuss the convergence of

(a) $\dfrac{2}{1^2} + \dfrac{3}{2^2} + \dfrac{4}{3^2} + \dfrac{5}{4^2} + \ldots$

(b) $\displaystyle\sum_{k=1}^{\infty} \frac{k^p}{k!}$ (all p)

(c) $\frac{1}{11} - \frac{2}{13} + \frac{3}{15} - \frac{4}{17} + \ldots$

(d) $1 - \frac{1}{3} + \frac{1}{5} - \frac{1}{7} + \ldots$

12 Express the following recurring decimal numbers in the form p/q where p and q are integers:

(a) $1.231\,231\,23\ldots$ (b) $0.429\,429\,429\ldots$

(c) $0.101\,101\,101\ldots$ (d) $0.517\,251\,72\ldots$

13 Determine which of the following series are convergent:

(a) $\displaystyle\sum_{n=0}^{\infty} \frac{1}{n^2 + 1}$ (b) $\displaystyle\sum_{n=1}^{\infty} \frac{n + 2}{n^2}$

(c) $\displaystyle\sum_{n=1}^{\infty} \frac{n - 1}{2n^5 - 1}$ (d) $\displaystyle\sum_{n=1}^{\infty} \frac{n - 1}{n^2 + n - 3}$

14 A rational function $f(x)$ has the following power series representation for $-1 < x < 1$:

$$f(x) = 1^2 x + 2^2 x^2 + 3^2 x^3 + 4^2 x^4 + \ldots$$

Find a closed-form expression for $f(x)$.

15 Find the values of a and b such that

$$\tan x = \frac{ax}{1 + bx^2} + cx^5 + O(x^7)$$

giving the value of c. (The series for $\tan x$ is given in Question 53(a) in Exercises 7.7.3.)
   For what values of x will the approximation

$$\tan x = \frac{ax}{1 + bx^2}$$

be valid to 4dp? Use the approximation to calculate $\tan 0.29$ and $\tan 0.295$, and compare your answers with the values given by your calculator. Comment on your results.

[Here $O(x^7)$ mean terms involving powers of x greater than or equal to 7.]

16 The function $f(x) = \sinh^{-1}x$ has the power series expansion

$$\sinh^{-1}x = x - \frac{1}{2}\frac{x^3}{3} + \frac{1 \cdot 3}{2 \cdot 4}\frac{x^5}{5} - \frac{1 \cdot 3 \cdot 5}{2 \cdot 4 \cdot 6}\frac{x^7}{7} + \ldots$$

Obtain polynomial approximations for $\sinh^{-1} x$ for $-0.5 < x < 0.5$ such that the truncation error is less than (a) 0.005 and (b) 0.000 05.

17 A chord of a circle is half a mile long and supports an arc whose length is 1 foot longer (1 mile = 5280 feet). Show that the angle θ subtended by the arc at the centre of the circle satisfies

$$\sin \tfrac{1}{2}\theta = \frac{1320}{2641}\theta$$

Use the series expansion for sine to obtain an approximate solution of this equation, and estimate the maximum height of the arc above its chord.

18 A machine is purchased for £3600. The annual running cost of the machine is initially £1800, but rises annually by 10%. After x years its secondhand value is £$3600\mathrm{e}^{-0.35x}$. Show that the average annual cost £C (including depreciation) after x years is given by

$$C = \frac{3600(1 - \mathrm{e}^{-0.35x})}{x} + 90(19 + x)$$

Show graphically that the machine should be replaced after about 4 years, and use an iterative method to refine this estimate.

19 Consider the sequence ϕ_n defined by

$$\phi_n = \frac{1}{2}\left[\left(1 + \frac{1}{n}\right)^n + \left(1 - \frac{1}{n}\right)^{-n}\right]$$

Show that $\phi_n \to \mathrm{e}$ as $n \to \infty$. Using the power series expansions of $\ln(1 + x)$ and e^x, show that

$$\left(1 + \frac{1}{n}\right)^n = \exp\left[n \ln\left(1 + \frac{1}{n}\right)\right]$$

$$= \mathrm{e}\left(1 - \frac{1}{2n} + \frac{11}{24n^2} - \frac{7}{16n^3} + \dots\right)$$

and deduce that

$$\phi_n = \mathrm{e}\left(1 + \frac{11}{24n^2} + \dots\right)$$

Evaluate ϕ_{64} and ϕ_{128} (without using the y^x key of your calculator), and use extrapolation to estimate the value of e.

20 A beam of weight W per unit length is simply supported at the same level at $(N + 1)$ equidistant points, the extreme supports being at the ends of

the beam. The bending moment M_k at the kth support satisfies the recurrence relation

$$M_{k+2} + 4M_{k+1} + M_k = \tfrac{1}{2} Wa^2$$

where a is the distance between the supports and $M_0 = 0$ and $M_N = 0$. (This is a consequence of Clapeyron's theorem of three moments.) Show that if the sequences $\{A_k\}_{k=0}^N$ and $\{B_k\}_{k=0}^N$ are calculated by the recurrences

$$A_0 = A_1 = 0$$
$$A_{k+2} + 4A_{k+1} + A_k = 1 \quad (k = 0, 1, \dots, N-2)$$

and

$$B_0 = 0, \quad B_1 = 1$$
$$B_{k+2} + 4B_{k+1} + B_k = 0 \quad (k = 0, 1, \dots, N-2)$$

then the solution of the bending-moment problem is given by

$$M_k = \tfrac{1}{2} Wa^2 A_k + M_1 B_k \quad (k = 0, \dots, N)$$

with $\tfrac{1}{2}Wa^2 A_N + M_1 B_N = 0$ determining the value of M_1.

Perform the calculation for the case where $N = 8$, $a = 1$ and $W = 25$.

21 A complex voltage E is applied to the ladder network of Figure 7.31. Show that the (complex) mesh currents I_k satisfy the equations

$$\tfrac{1}{2}L\omega \mathrm{j} I_0 - \frac{\mathrm{j}}{C\omega}(I_0 - I_1) = E$$

$$L\omega \mathrm{j} I_k - \frac{\mathrm{j}}{C\omega}(I_k - I_{k+1}) + \frac{\mathrm{j}}{C\omega}(I_{k-1} - I_k) = 0$$

$$(k = 1, \dots, N-1) \quad \textbf{(7.20)}$$

$$\tfrac{1}{2}L\omega \mathrm{j} I_n + \frac{\mathrm{j}}{C\omega}(I_{n-1} - I_n) = 0$$

(See Section 3.6 for the application of complex numbers to alternating circuits.)

Show that $I_k = A(\mathrm{e}^\theta)^k = A\mathrm{e}^{k\theta}$ satisfies (7.20) provided that $\cosh \theta = 1 - \tfrac{1}{2}LC\omega^2$. Note that this equation yields two values for θ, so that in general I_k may be written as

$$I_k = A\mathrm{e}^{k\theta} + B\mathrm{e}^{-k\theta}$$

where A and B are independent of k. Using the special equations for I_0 and I_n, obtain the values of A and B and prove that

$$I_k = \mathrm{j}EC\omega \frac{\cosh(n - k)\theta}{\sinh \theta \sinh n\theta}$$

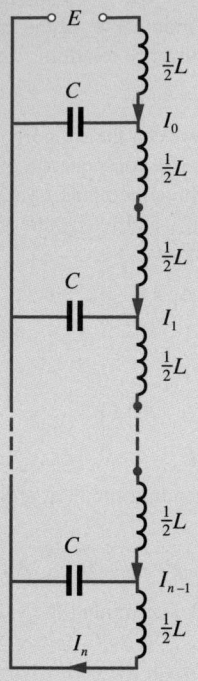

Figure 7.31

22 A lightweight beam of length l is clamped horizontally at both ends. It carries a concentrated load W at a distance a from one end ($x = 0$). The shear force F and bending moment M at the point x on the beam are given by

$$F = \begin{cases} \dfrac{W(l-a)^2(l+2a)}{l^3} & (0 < x < a) \\[3mm] \dfrac{-Wa^2(3l-2a)}{l^3} & (a < x < l) \end{cases}$$

and

$$M = \begin{cases} \dfrac{W(l-a)^2[al - x(l+2a)]}{l^3} & (0 < x < a) \\[3mm] \dfrac{Wa^2[al - 2l^2 + x(3l-2a)]}{l^3} & (a < x < l) \end{cases}$$

Draw the graphs of these functions. Use Heaviside functions to obtain single formulae for M and F.

23 (a) Show that

$$\sin 4\theta = 4\sin\theta\,(1 - 2\sin^2\theta)\sqrt{(1 - \sin^2\theta)},$$
$$-\pi/2 < \theta < \pi/2$$

and explain why there is a restriction on the domain of θ.

(b) Use the binomial expansion to show that

$$\sqrt{(1 - \sin^2\theta)} = 1 - \tfrac{1}{2}\sin^2\theta - \tfrac{1}{8}\sin^4\theta$$
$$- \tfrac{1}{16}\sin^6\theta + A\sin^8\theta + \ldots\ldots$$

giving the value of A.

(c) Show that $\sin 4\theta$ can be expressed in the form

$$\sin 4\theta = 4\sin\theta - 10\sin^3\theta + \tfrac{7}{2}\sin^5\theta$$
$$+ \tfrac{3}{4}\sin^7\theta + \ldots$$

24 The series

$$\sum_{k=0}^{\infty} \frac{1}{(2k+1)^2} = 1 + \frac{1}{3^2} + \frac{1}{5^2} + \frac{1}{7^2} + \ldots$$

sums to the value $\pi^2/8$. Lagrange's formula for linear interpolation is

$$f(x) \simeq \frac{(x-x_0)f_1 - (x-x_1)f_0}{x_1 - x_0}$$

By setting $x = 1/n$ and $f(x) = S_n$ where

$$S_n = 1 + \frac{1}{3^2} + \frac{1}{5^2} + \ldots + \frac{1}{(2n-1)^2},$$

show that

$$S_\infty \simeq \frac{qS_q - pS_p}{q - p}$$

where p and q are integers. Choosing $p = 5$ and $q = 10$, estimate the value of $\pi^2/8$.

25 The expression $\dfrac{x}{1 + ax^2}$ is to be used as an approximation to $\tfrac{1}{2}\ln[(1+x)/(1-x)]$ on $-1 < x < 1$, by choosing a suitable value for the constant a. Show that

$$\frac{x}{1 + ax^2} - \tfrac{1}{2}\ln\left(\frac{1+x}{1-x}\right)$$
$$= -(\tfrac{1}{3} + a)x^3 - (\tfrac{1}{5} - a^2)x^5 - (\tfrac{1}{7} + a^3)x^7 + \ldots$$

for $|x| < R$ giving the value of R. The error in this approximation is dominated by the first term in this expansion. Obtain the value of a which makes this term equal zero and compute the corresponding value of the coefficient of x^5.

Draw on the same diagram, the graphs of

$$y = \tfrac{1}{2}\ln\left(\frac{1+x}{1-x}\right), \quad y = x + \tfrac{1}{3}x^3$$

and $\quad y = \dfrac{3x}{3 - x^2} \quad$ for $0 \leqslant x \leqslant 1$

8 Differentiation and Integration

8.1 Introduction

Many of the practical situations that engineers have to analyse involve quantities that are varying. Whether it is the temperature of a coolant, the voltage on a transmission line or the torque on a turbine blade, the mathematical tools for performing such analyses are the same. One of the most successful of these is **calculus**, which involves two fundamental operations: differentiation and integration. Historically, integration was discovered first, and indeed some of the ideas and results date back over 2000 years to when the Greeks developed the **method of exhaustion** to evaluate the area of a region bounded on one side by a curve – a method used by Archimedes (287–212 BC) to obtain the exact formula for the area of a circle. Differentiation was discovered very much later, during the seventeenth century, in relation to the problem of determining the tangent at an arbitrary point on a curve. Its characteristic features were probably first used by Fermat in 1638 to find the maximum and minimum points of some special functions. He noticed that tangents must be horizontal at some points, and developed a method for finding them by slightly changing the variable in a single algebraic equation and then letting the change 'disappear'. The connection between the two processes of determining the area under a curve and obtaining a tangent at a point on a curve was first realized in 1663 by Barrow, who was Newton's professor at Cambridge University. However, it was Newton (1642–1727) and Leibniz (1646–1716), working independently, who fully recognized the implications of this relationship. This led them to develop the calculus as a way of dealing with change and motion. Exploitation of their work resulted in an era of tremendous mathematical activity, much of which was motivated by the desire to solve applied problems, particularly by Newton, whose accomplishments were immense and included the formulation of the laws of gravitation. The calculus was put on a firmer mathematical basis in the nineteenth century by Cauchy and Riemann. It remains today one of the most powerful mathematical tools used by engineers, and in this chapter and the next we shall review its basic ideas and techniques and show their application both in the formulation of mathematical models of practical problems and in their solution.

In recent years we have seen significant developments in symbolic algebra packages, such as MAPLE and the Symbolic Math Toolbox in MATLAB, which are capable of performing algebraic manipulation, including the calculation of derivatives and integrals. To the inexperienced, this development may appear to eliminate the need for engineers to be able to carry out even basic operations in calculus by hand. This, however, is far from the truth. If engineers are to apply the powerful techniques associated with the calculus to the design and analysis of industrial problems then it is essential that they have a sound grounding of differentiation and integration. First, this allows effective formulation, comprehension and analysis of mathematical models. Secondly, it provides the basis for understanding symbolic algebra packages, particularly when specific forms of results are desired. In order to acquire this understanding it is necessary to have a certain degree of fluency in the manipulation of associated basic techniques. It is the objective of this chapter and the next to provide the minimum requirements for this. At the same time, students should be given the opportunity to develop their skills in the use of a symbolic algebra package and, whenever appropriate, be encouraged to check their answers to the exercises using such a package.

8.2 Differentiation

Here we shall introduce the concept of differentiation and illustrate its role in some problem-solving and modelling situations.

8.2.1 Rates of change

Consider an object moving along a straight line with constant velocity u (in m s^{-1}). The distance s (in metres) travelled by the object in time t (in seconds) is given by the formula $s = ut$. The distance–time graph of this motion is the straight line shown in Figure 8.1. Note that the velocity u is the rate of change of distance with respect to time, and that on the distance–time graph it is the gradient (slope) of the straight line representing the relationship between the distance travelled and the time elapsed. This, of course, is a special case where the velocity is constant and the distance travelled is a linear function of time. Even when the velocity varies with time, however, it is still given by the gradient of the distance–time graph, although it then varies from point to point along the curve.

Figure 8.1
Distance–time graph for constant velocity u.

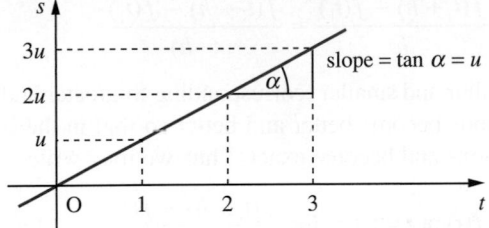

Consider the distance–time graph shown in Figure 8.2(a). Suppose we wish to find the velocity at the time $t = t_1$. The velocity at $t = t_1$ is given by the gradient of the graph at $t = t_1$. To find that we can enlarge that piece of the graph near $t = t_1$, as shown in Figure 8.2(b) and (c). We recall that continuous functions have the property that locally they may be approximated by linear functions (see Section 7.9.1, property f). We see that as we increase the magnification, that is, zooming closer, the graph takes on the

Figure 8.2
(a) Distance–time graph, (b) enlargement of outer rectangle surrounding (t_1, s_1) and (c) enlargement of inner rectangle surrounding (t_1, s_1).

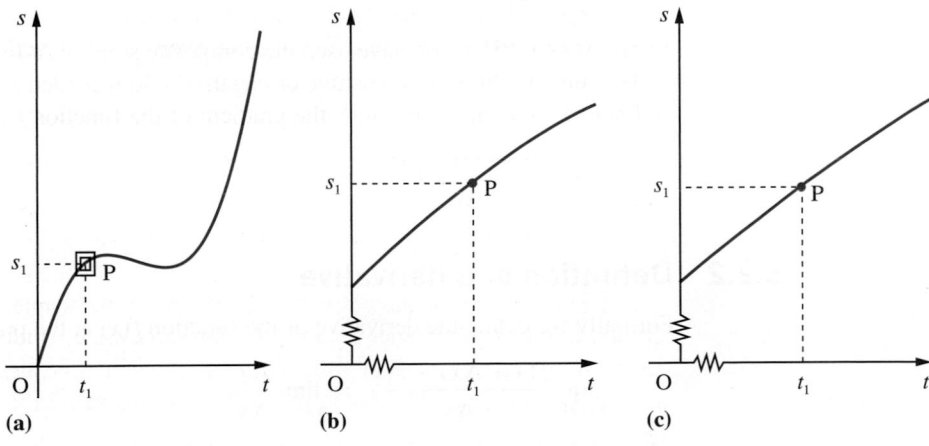

Figure 8.3
Section of the
distance–time graph.

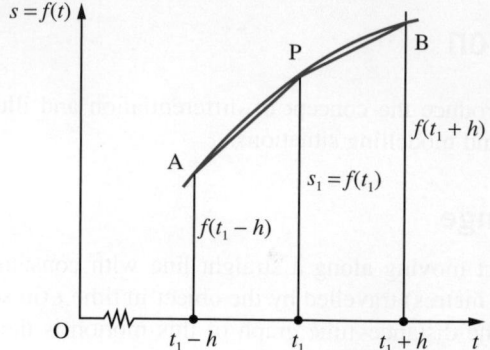

appearance of a straight line through the point $P(t_1, s_1)$. The gradient of that straight line (in the limit) gives us the gradient of the graph at P. Consider that section of the graph contained in the rectangle whose sides parallel to the s axis are $t = t_1 + h$ and $t = t_1 - h$ where h is a positive (small) number, as shown in Figure 8.3. If we denote the function relating distance and time by $f(t)$, then $s_1 = f(t_1)$ and we can approximate the gradient of the function $f(t)$ at the point P by the gradients of either of the chords AP or BP. Thus

$$\text{gradient} \simeq \frac{f(t_1 + h) - f(t_1)}{h} \simeq \frac{f(t_1 - h) - f(t_1)}{(-h)}$$

As h becomes smaller and smaller (corresponding to greater and greater magnifications) these approximations become better and better so that in the limit ($h \to 0$) they cease being approximations and become exact. Thus we may write

$$(\text{gradient of } f(t) \text{ at } t = t_1) = \lim_{h \to 0} \frac{f(t_1 + h) - f(t_1)}{h}$$

$$= \lim_{h \to 0} \frac{f(t_1 - h) - f(t_1)}{(-h)}$$

Here we specified $h > 0$, which means that the former limit is the limit from above and the latter is the limit from below of the expression

$$\frac{f(t_1 + \Delta t) - f(t)}{\Delta t}$$

where $\Delta t \to 0$. (Here we have used the composite symbol Δt to indicate a small change in the value of t. It may be positive or negative.) So provided that the limits from above and below have the same value, the gradient of the function $f(t)$ is defined at $t = t_1$ by

$$\lim_{\Delta t \to 0} \frac{f(t_1 + \Delta t) - f(t)}{\Delta t}$$

8.2.2   Definition of a derivative

Formally we define the derivative of the function $f(x)$ at the point x to be

$$\lim_{\Delta x \to 0} \frac{f(x + \Delta x) - f(x)}{\Delta x} = \lim_{\Delta x \to 0} \frac{\Delta f}{\Delta x}$$

where $\Delta f = f(x + \Delta x) - f(x)$ is the change in $f(x)$ corresponding to the change Δx in x.

Two notations are used for the derivative. One uses a composite symbol, $\dfrac{df}{dx}$, and the other uses a prime, $f'(x)$, so that

$$\frac{df}{dx} = f'(x) = \lim_{\Delta x \to 0} \frac{\Delta f}{\Delta x} = \lim_{\Delta x \to 0} \frac{f(x + \Delta x) - f(x)}{\Delta x} \tag{8.1}$$

In terms of the function $y = f(x)$, we write $\Delta y = \Delta f$ and $y + \Delta y = y(x + \Delta x)$ and

$$\frac{dy}{dx} = \lim_{\Delta x \to 0} \frac{\Delta y}{\Delta x} \tag{8.2}$$

Example 8.1 Using the definition of a derivative given in (8.1), find $f'(x)$ when $f(x)$ is

(a) x^2 (b) $\dfrac{1}{x}$ (c) $mx + c$ (m, c constants)

Solution (a) With $f(x) = x^2$, $f(x + \Delta x) = (x + \Delta x)^2 = x^2 + 2x\,\Delta x + (\Delta x)^2$

so that $\dfrac{\Delta f}{\Delta x} = \dfrac{f(x + \Delta x) - f(x)}{\Delta x} = \dfrac{2x\,\Delta x + (\Delta x)^2}{\Delta x} = 2x + \Delta x$

Thus, from (8.1), the derivative of $f(x)$ is

$$\frac{df}{dx} = f'(x) = \lim_{\Delta x \to 0} \frac{\Delta f}{\Delta x} = \lim_{\Delta x \to 0} (2x + \Delta x) = 2x$$

so that $\dfrac{d}{dx}(x^2) = 2x$

(b) With $f(x) = \dfrac{1}{x}$, $f(x + \Delta x) = \dfrac{1}{x + \Delta x}$

so that $\dfrac{\Delta f}{\Delta x} = \dfrac{f(x + \Delta x) - f(x)}{\Delta x} = \dfrac{\left[\dfrac{1}{x + \Delta x} - \dfrac{1}{x} \right]}{\Delta x} = \left[\dfrac{x - x - \Delta x}{\Delta x(x + \Delta x)x} \right]$

$$= \left[\frac{-1}{x^2 + x\,\Delta x} \right]$$

Thus, from (8.1), the derivative of $f(x)$ is

$$\frac{df}{dx} = \lim_{\Delta x \to 0} \frac{\Delta f}{\Delta x} = \lim_{\Delta x \to 0} \left[\frac{-1}{x^2 + x\,\Delta x} \right] = -\frac{1}{x^2}$$

so that $\dfrac{d}{dx}(x^{-1}) = -1x^{-2}$

(c) with $f(x) = mx + c$, $f(x + \Delta x) = m(x + \Delta x) + c$

so that $\dfrac{\Delta f}{\Delta x} = \dfrac{f(x + \Delta x) - f(x)}{\Delta x} = \dfrac{m\Delta x}{\Delta x} = m$

Thus, from (8.1), the derivative of $f(x)$ is

$$\frac{\mathrm{d}f}{\mathrm{d}x} = \lim_{\Delta x \to 0} \frac{\Delta f}{\Delta x} = m$$

so the gradient of the function $f(x) = mx + c$ is the same as that of the straight line $y = mx + c$, as we would expect.

8.2.3 Interpretation as the slope of a tangent

The definition is illustrated graphically in Figure 8.4, where Δx denotes a small incremental change in the independent variable x and Δf is the corresponding incremental change in $f(x)$. P and Q are the points on the graph with coordinates $(x, f(x))$, $(x + \Delta x, f(x + \Delta x))$ respectively. The slope of the line segment PQ is

$$\frac{\Delta f}{\Delta x} = \frac{f(x + \Delta x) - f(x)}{\Delta x}$$

In the limit as Δx tends to zero the point Q approaches P, and the line segment becomes the tangent to the curve at P, whose slope is given by the derivative

$$\frac{\mathrm{d}f}{\mathrm{d}x} = \lim_{\Delta x \to 0} \frac{\Delta f}{\Delta x}$$

Figure 8.4
Illustration of derivative as slope of a tangent.

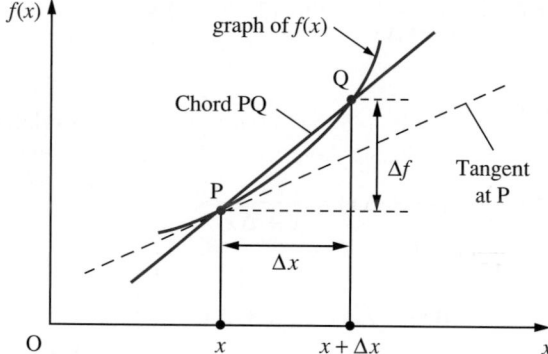

Summary

If $y = f(x)$ then the derivative of $f(x)$ is defined by

$$\frac{\mathrm{d}y}{\mathrm{d}x} = \frac{\mathrm{d}f}{\mathrm{d}x} = f'(x) = \lim_{\Delta x \to 0} \frac{\Delta y}{\Delta x} = \lim_{\Delta x \to 0} \frac{\Delta f}{\Delta x} = \lim_{\Delta x \to 0} \frac{f(x + \Delta x) - f(x)}{\Delta x}$$

The derivative may be interpreted as

(a) the rate of change of the function $y = f(x)$ with respect to x, or

(b) the slope of the tangent at the point (x, y) on the graph of $y = f(x)$.

Example 8.2   Consider the function $f(x) = 25x - 5x^2$. Find

(a) the derivative of $f(x)$ from first principles,

(b) the rate of change of $f(x)$ at $x = 1$,

(c) the equation of the tangent to the graph of $f(x)$ at the point $(1, 20)$.

(d) the equation of the normal to the graph of $f(x)$ at the point $(1, 20)$.

Solution   (a) $f(x) = 25x - 5x^2$

$$f(x + \Delta x) = 25(x + \Delta x) - 5(x + \Delta x)^2 = 25x + 25\Delta x - 5x^2 - 10x\,\Delta x - 5(\Delta x)^2$$

so that

$$\frac{\Delta f}{\Delta x} = \frac{f(x + \Delta x) - f(x)}{\Delta x} = \frac{25\Delta x - 10x\,\Delta x - 5(\Delta x)^2}{\Delta x} = 25 - 10x - 5\Delta x$$

Thus the derivative of $f(x)$ is

$$\frac{df}{dx} = f'(x) = \lim_{\Delta x \to 0} \frac{\Delta f}{\Delta x} = \lim_{\Delta x \to 0} (25 - 10x - 5\Delta x) = 25 - 10x$$

(b) The rate of change of $f(x)$ at $x = 1$ is $f'(1) = 15$.

(c) The slope of the tangent to the graph of $f(x)$ at $(1, 20)$ is $f'(1) = 15$. Remembering from equation (1.14) that the equation of a line passing through a point (x_1, y_1) and having slope m is

$$y - y_1 = m(x - x_1)$$

we have the equation of the tangent to the graph of $y = f(x)$ at $(1, 20)$ is

$$y - 20 = 15(x - 1)$$

or

$$y = 15x + 5$$

(d) The slope n of the normal to the graph is given by the relation $mn = -1$, where m is the slope of the tangent. This is illustrated in Figure 8.5. Thus in this example the slope of the normal at $(1, 20)$ is $-1/15$ and hence the equation of the normal at $(1, 20)$ is

$$y - 20 = -\frac{1}{15}(x - 1)$$

or

$$y = \frac{1}{15}(301 - x)$$

Figure 8.5
Relationship between
slopes of the tangent
and normal to a plane
curve.

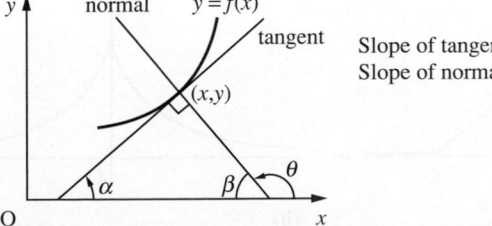

Slope of tangent $= \tan \alpha$
Slope of normal $= \tan \theta = -\tan \beta$
$\qquad\qquad\quad = -\tan(90 - \alpha)$
$\qquad\qquad\quad = -\dfrac{1}{\text{slope of tangent}}$

8.2.4 Differentiable functions

The formal definition of the derivative of $f(x)$ implies that the limits from below and above are equal. In some cases this does not happen. For example, the function $f(x) = \sqrt{(1 + \sin x)}$ is such that its two limits are

$$\lim_{\Delta x \to 0-} \frac{f(3\pi/2 + \Delta x) - f(3\pi/2)}{\Delta x} = \frac{-1}{\sqrt{2}}$$

$$\lim_{\Delta x \to 0+} \frac{f(3\pi/2 + \Delta x) - f(3\pi/2)}{\Delta x} = \frac{1}{\sqrt{2}}$$

Clearly the derivative of the function is not defined at $x = 3\pi/2$ (the two limits above are sometimes referred to as 'left-hand' and 'right-hand' derivatives, respectively).

The graph of $y = \sqrt{(1 + \sin x)}$ is shown in Figure 8.6, and it is clear that at $x = 3\pi/2$ a unique tangent cannot be drawn to the graph of the function. This is not surprising since from the interpretation of the derivative as the slope of the tangent, it follows that for a function $f(x)$ to be **differentiable** at $x = a$, the graph of $f(x)$ must have a unique, non-vertical well-defined tangent at $x = a$. Otherwise the limit

$$\lim_{\Delta x \to 0} \frac{f(a + \Delta x) - f(a)}{\Delta x}$$

does not exist. We say that a function $f(x)$ is differentiable if it is differentiable at all points in its domain. For practical purposes it is sufficient to interpret a differentiable function as one having a smooth continuous graph with no sharp corners. Engineers frequently refer to such functions as being 'well behaved'. Clearly the function having the graph shown in Figure 8.7(a) is differentiable at all points except $x = x_1$ and $x = x_2$, since a unique tangent cannot be drawn at these points. Similarly, the function having the graph shown in Figure 8.7(b) is differentiable at all points except at $x = 0$.

Figure 8.6
The graph of
$y = \sqrt{(1 + \sin x)}$.

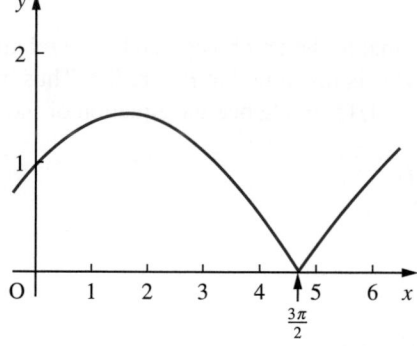

Figure 8.7

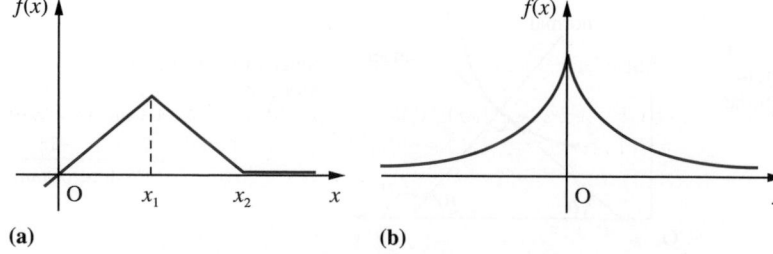

(a)   (b)

8.2.5 Speed, velocity and acceleration

Considering the motion of the object in Section 8.2.1 enables us to distinguish between the terms **speed** and **velocity**. In everyday usage we talk of speed rather than velocity, and always regard it as being positive or zero. As we saw in Chapter 4, velocity is a vector quantity and has a direction associated with it, while speed is a scalar quantity, being the magnitude or modulus of the velocity. When s and v are measured horizontally, the object will have a positive velocity when travelling to the right and a negative velocity when travelling to the left. Throughout its motion, the speed of the particle will be positive or zero. Likewise, **acceleration** a, being the rate of change of velocity with respect to time, is a vector quantity and is determined by

$$a(t) = \frac{\mathrm{d}v}{\mathrm{d}t}$$

Example 8.3 A particle is thrown vertically upwards into the air. Its height s (in m) above the ground after time t (in seconds) is given by

$$s = 25t - 5t^2$$

(a) What height does the particle reach?

(b) What is its velocity when it returns to hit the ground?

(c) What is its acceleration?

Solution Since velocity v is rate of change of distance s with time t we have

$$v = \frac{\mathrm{d}s}{\mathrm{d}t}$$

In this particular example

$$s(t) = 25t - 5t^2 \tag{8.3}$$

so, from Example 8.2,

$$v(t) = \frac{\mathrm{d}s}{\mathrm{d}t} = 25 - 10t \tag{8.4}$$

(a) When the particle reaches its maximum height, it will be momentarily at rest, so that its velocity will be momentarily zero. From (8.4) this will occur when $t = 25/10 = 2.5$. Then, from (8.3), the height reached at this instant is

$$s(2.5) = 25 \times \tfrac{5}{2} - 5 \times (\tfrac{5}{2})^2 = \tfrac{125}{4}$$

That is, the maximum height reached by the particle is 31.25 m.

(b) First we need to find the time at which the particle will return to hit the ground. This will occur when the height s is again zero, which from (8.3) is when $t = 5$. Then, from (8.4), the velocity of the particle when it hits the ground is

$$v(5) = -25$$

That is, when it returns to hit the ground, the particle will be travelling at $25\,\mathrm{m\,s^{-1}}$, with the negative sign indicating that it is travelling downwards, since s and v are measured upwards.

(c) The acceleration, a, is the rate of change of velocity with respect to time. Thus, from (8.4)

$$a(t) = \frac{\mathrm{d}v}{\mathrm{d}t} = -10$$

that is, $a \approx -g = -9.806\,65$ $(\mathrm{m\,s^{-2}})$, the acceleration due to gravity.

The MATLAB Symbolic Math Toolbox and MAPLE provide commands to do the basic operations of calculus and many of these will be introduced in this chapter. If $y = f(x)$ then we denote $\mathrm{d}y/\mathrm{d}x$ by dy or df (we could use any name such as, for example, dydx or dybydx). Denoting Δx by h the derivative of $f(x)$, as given in (8.1), is determined by the commands

MATLAB

```
syms h x
df = limit((f(x + h) -
f(x))/h,h,0)
```

MAPLE

```
f:= x -> f(x);
df:= limit((f(x + h) -
f(x))/h,h = 0);
```

For example, if $f(x) = 25x - 5x^2$ then its derivative df is determined by the MATLAB commands

```
syms h x
df = limit(((25*(x + h) - 5*(x + h)^2) -
(25*x - 5*x^2))/h,h,0)
```

as

```
df = 25 - 10*x
```

which checks with the answer obtained in the solution to Example 8.2(a).

8.2.6 Exercises

Check your answers using MATLAB or MAPLE whenever possible.

1 Using the definition of a derivative given in (8.1), find $f'(x)$ when $f(x)$ is

(a) a constant K (b) x (c) $x^2 - 2$

(d) x^3 (e) $\sqrt{x}$ (f) $1/(1 + x)$

(d) the value of m such that the points of intersection found in (c) are coincident,

(e) the equation of the tangent to the graph of $f(x)$ at the point $(1, -15)$.

2 Consider the function $f(x) = 2x^2 - 5x - 12$. Find

(a) the derivative of $f(x)$ from first principles,

(b) the rate of change of $f(x)$ at $x = 1$,

(c) the points at which the line through $(1, -15)$ with slope m cuts the graph of $f(x)$,

3 Consider the function $f(x) = 2x^3 - 3x^2 + x + 3$. Find

(a) the derivative of $f(x)$ from first principles,

(b) the rate of change of $f(x)$ at $x = 1$,

(c) the points at which the line through $(1, 3)$ with slope m cuts the graph of $f(x)$,

(d) the values of m such that two of the points of intersection found in (c) are coincident,

(e) the equations of the tangents to the graph of $f(x)$ at $x = 1$ and $x = \frac{1}{4}$.

4 Show from first principles that the derivative of

$$f(x) = ax^2 + bx + c$$

is

$$f'(x) = 2ax + b$$

Hence confirm the result of Section 2.3.4 (page 94) and using the calculus method verify the results of Example 2.21 (page 97).

5 Show that if $f(x) = ax^3 + bx^2 + cx + d$, then

$$f(x + \Delta x) = ax^3 + bx^2 + cx + d + (3ax^2 + 2bx + c)\Delta x$$
$$+ (3ax + b)(\Delta x)^2 + a(\Delta x)^3$$

Deduce that

$$f'(x) = 3ax^2 + 2bx + c.$$

6 The displacement–time graph for a vehicle is given by

$$s(t) = \begin{cases} t, & 0 \leqslant t \leqslant 1 \\ t^2 - t + 1, & 1 \leqslant t \leqslant 2 \\ 3t - 3, & 2 \leqslant t \leqslant 3 \\ 9 - t, & 3 \leqslant t \leqslant 9 \end{cases}$$

Obtain the formula for the velocity–time graph.

7 Consider the function $f(x) = \sqrt{(1 + \sin x)}$. Show that $f(3\pi/2 \pm h) = \sqrt{2} \sin \frac{1}{2}h$ $(h > 0)$ and deduce that $f'(x)$ does not exist at $x = 3\pi/2$.

8.2.7 Mathematical modelling using derivatives

We have seen that the gradient of a tangent to the graph $y = f(x)$ can be expressed as a derivative, but derivatives have much wider application than just this. Any quantity that can be expressed as a limit of the form (8.1) can be represented by a derivative, and such quantities arise in many practical situations. Because gradients of tangents to graphs can be expressed as derivatives, it follows that we can always interpret a derivative geometrically as the slope of a tangent to a graph. In Example 8.3 we saw that the particle reached its maximum height 31.25 m when $t = 2.5$ s. This maximum height occurred when $v = ds/dt = 0$. This implies that the tangent to the graph of **distance** against **time** was horizontal. In general at a maximum or minimum of a function, its derivative is zero and its tangent horizontal (as discussed in Section 2.2.1). This is discussed fully later, in Section 8.5.

Example 8.4   Suppose that a tank initially contains 80 litres of pure water. At a given instant (taken to be $t = 0$) a salt solution containing 0.25 kg of salt per litre flows into the tank at the rate of 8 litres min^{-1}. The liquid in the tank is kept homogeneous by constant stirring. Also, at time $t = 0$ liquid is allowed to flow out from the tank at the rate of 12 litres min^{-1}. Show that the amount of salt $x(t)$ (in kg) in the tank at time t (min) $\geqslant 0$ is determined by the mathematical model

$$\frac{dx(t)}{dt} + \frac{3x(t)}{20 - t} = 2 \quad (t < 20)$$

Solution   The situation is illustrated in Figure 8.8. Since $x(t)$ denotes the amount of salt in the tank at time $t \geqslant 0$, the rate of increase of the amount of salt in the tank is dx/dt, and is given by

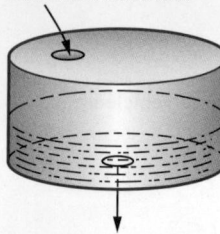

Flow in: 8 litres min⁻¹

Flow out: 12 litres min⁻¹

Figure 8.8
Water tank of
Example 8.4.

$$\frac{dx}{dt} = \text{rate of inflow of salt} - \text{rate of outflow of salt} \tag{8.5}$$

The rate of inflow of salt is $(0.25\,\text{kg litre}^{-1})\,(8\,\text{litres min}^{-1})\ = 2\,\text{kg min}^{-1}$.

The rate of outflow of salt is $c \times (\text{rate of outflow of liquid}) = c \times 12\,\text{litres min}^{-1}$

$$= 12c \quad (\text{in kg min}^{-1})$$

where $c(t)$ is the concentration of salt in the tank (in kg litre⁻¹). The concentration at time t is given by

$$c(t) = \frac{\text{amount of salt in the tank at time } t}{\text{volume of liquid in the tank at time } t}$$

After time t (in min) $8t$ litres have entered the tank and $12t$ litres have left. Also, at $t = 0$ there were 80 litres in the tank. Therefore the volume V of liquid in the tank at time t is given by

$$V(t) = 80 - (12t - 8t) = (80 - 4t)$$

(Note that $V(t) \geq 0$ only if $t \leq 20\,\text{min}$; after this time the liquid will flow out as quickly as it flows in and none will accumulate in the tank.) Thus the concentration $c(t)$ is given by

$$c(t) = \frac{x(t)}{V(t)} = \frac{x(t)}{80 - 4t}$$

so that

$$\text{rate of outflow of salt} = 12 \times \frac{x(t)}{80 - 4t} = \frac{3x(t)}{20 - t}$$

Substituting back into (8.5) gives the rate of increase as

$$\frac{dx}{dt} = 2 - \frac{3x}{20 - t}$$

or

$$\frac{dx(t)}{dt} + \frac{3x(t)}{20 - t} = 2$$

This equation involving the derivative of $x(t)$ is called a *differential equation*, and in Chapter 10 (Review exercises 10.13, Question 9) we shall show how it can be solved to give the quantity $x(t)$ of salt in the tank at time t.

Example 8.5

In a suspension bridge a roadway, of length $2l$, is suspended by vertical hangers from cables carried by towers at the ends of the span, as illustrated in Figure 8.9(a). The lowest points of the cables are a distance h below the top of the supporting towers. Find an equation which represents the line shape of the cables.

Solution

To solve this problem we have to make some simplifying assumptions. We assume that the roadway is massive compared to the cables, so that the weight W of the roadway is the dominant factor in determining the shape of the cables. Secondly, we assume

Figure 8.9
(a) Schematic diagram for a suspension bridge. (b) Forces acting on the cable between A and B.

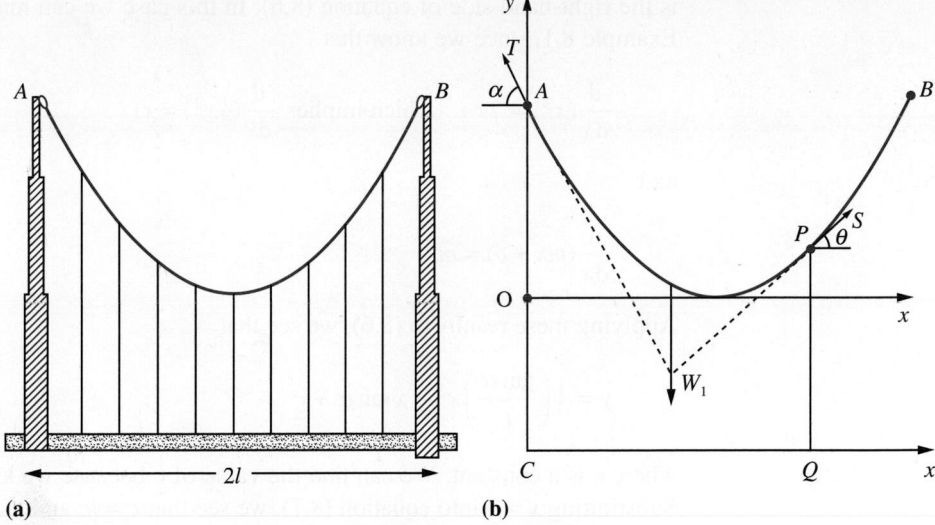

(a) (b)

that the weight of the roadway is uniformly distributed along its length, and if the hangers are equally spaced they can be adjusted in length so that they carry equal vertical loads.

We solve this problem using elementary statics because at each point P on the cable the forces are in equilibrium. Figure 8.9(b) shows the forces acting on the part of the cable between A and P. These are the weight W_1 of the roadway between C and Q ($W_1 = Wx/2l$, where $x = CQ$), the tension T in the cable acting at the angle α at A and the tension S in the cable acting at the angle θ at P.

Resolving forces horizontally, we have $T \cos \alpha = S \cos \theta$.

Resolving forces vertically, we have $T \sin \alpha + S \sin \theta = \dfrac{Wx}{2l}$.

Eliminating S between these equations gives $T \sin \alpha + T \cos \alpha \tan \theta = \dfrac{Wx}{2l}$.

Also, we know that the total weight W of the roadway is supported by the tensions at A and B, so that $2T \sin \alpha = W$. Hence, substituting, we obtain

$$\frac{W}{2 \sin \alpha} \sin \alpha + \frac{W}{2 \sin \alpha} \cos \alpha \tan \theta = \frac{Wx}{2l}$$

giving

$$\tan \theta = x \frac{\tan \alpha}{l} - \tan \alpha$$

Now $\tan \theta$ is the slope of the curve at P (the tensions act along the direction of the tangent at each point of the curve), so that

$$\frac{dy}{dx} = x \frac{\tan \alpha}{l} - \tan \alpha \qquad (8.6)$$

using the coordinate system shown in Figure 8.9(b). This is another example of a differential equation. To solve this equation we have to find the function whose derivative is the right-hand side of equation (8.6). In this case we can make use of the results of Example 8.1, since we know that

$$\frac{d}{dx}(x^2) = 2x \quad \text{(which implies } \frac{d}{dx}(\tfrac{1}{2}x^2) = x)$$

and

$$\frac{d}{dx}(mx + c) = m$$

Applying these results to (8.6), we see that

$$y = \tfrac{1}{2}\left(\frac{\tan \alpha}{l}\right)x^2 - x \tan \alpha + c \tag{8.7}$$

where c is a constant. We can find the value of c because we know that $y = h$ at $x = 0$. Substituting $x = 0$ into equation (8.7), we see that $c = h$, and the solution becomes

$$y = \tfrac{1}{2}\left(\frac{\tan \alpha}{l}\right)x^2 - x \tan \alpha + h \tag{8.8}$$

But we also know that $y = 0$ where $x = l$. This enables us to find the value of $\tan \alpha$. Substituting $x = l$ into equation (8.8), we have

$$0 = \tfrac{1}{2}l \tan \alpha - l \tan \alpha + h$$

which implies $\tan \alpha = 2h/l$. Thus the shape of the supporting cable is given by

$$y = \frac{hx^2}{l^2} - \frac{2hx}{l} + h$$
$$= h(x - l)^2/l^2$$

indicating that the points of attachment of the hangers to the cable lie on a parabolic curve.

Example 8.6

A radio telescope has the shape of a paraboloid of revolution. Show that all the radio waves arriving in a direction parallel to its axis of symmetry are reflected to pass through the same point on that axis of symmetry.

Solution

The diagram in Figure 8.10 shows a section of the paraboloid through its axis of symmetry. We choose the coordinate system such that the equation of the parabola shown is $y = x^2$. Let AP represent the path of a radio signal travelling parallel to the y axis. At P it is reflected to pass through the point B on the y axis. The laws of reflection state that $\angle APN = \angle BPN$, where PN is the normal to the curve at P. Now given the coordinates (a, a^2) of the point P we have to find the coordinates $(0, b)$ of the point B. From the diagram we can see that if $\angle PTQ = \theta$, then $\angle PQT = \pi/2 - \theta$, which implies that $\angle ONP = \theta$. Since AP is parallel to NB, we see that $\angle APN = \theta$ and hence $\angle BPN = \theta$. This implies that $\angle PBN = \pi - 2\theta$. With all of these angles known we can calculate the coordinates of B. From the diagram

Figure 8.10
Section of
paraboloid through
axis of symmetry.

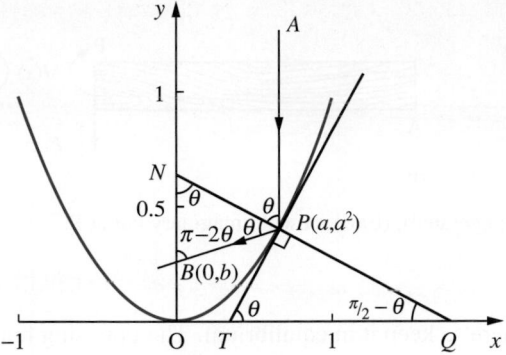

$$\tan \angle NBP = \frac{a}{a^2 - b}$$

Since $\angle NBP = \pi - 2\theta$, this implies $\tan 2\theta = \frac{a}{b - a^2}$. Also

$$\tan \theta = \left(\frac{dy}{dx}\right)_{x=a} = 2a$$

and since $\tan 2\theta = \frac{2 \tan \theta}{1 - \tan^2 \theta}$, identity 2.27e, we obtain

$$\frac{4a}{1 - 4a^2} = \frac{a}{b - a^2}$$

This gives $b = \frac{1}{4}$. Notice that the value of b is independent of a. Thus all the reflected rays pass through $(0, \frac{1}{4})$. As was indicated in Section 1.4.5, this property is important in many engineering design projects.

Example 8.7   Show that the shear force F acting in a beam is related to the bending moment M by

$$F = \frac{dM}{dx}$$

Solution   A beam is a horizontal structural member which carries loads. These induce forces and stresses inside the beam in transmitting the loads to the supports. For design safety two internal quantities are used, the shear force F and the bending moment M. At each point along the beam the forces are in equilibrium. These forces can be thought of as acting along the beam and (vertically) perpendicular to it. When a beam bends its upper surface is compressed and its lower surface is stretched, so that forces on the upper and lower surfaces at any point along the beam are acting in opposite directions. There will, of course, be a line within the beam, which is neither stretched nor compressed. This is the neutral axis. The situation is illustrated in Figure 8.11(a). To analyse the situation we imagine cutting the beam at a point P along its length and examine the forces which

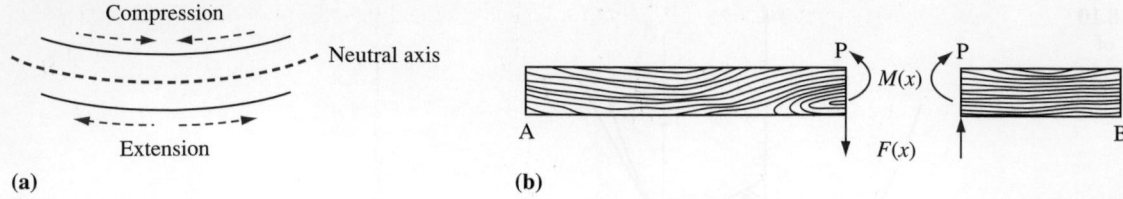

Figure 8.11 (a) The bending of a beam (exaggerated); (b) beam with imaginary cut at P.

are necessary there to keep it in equilibrium. The opposing horizontal forces at P, give a moment M about the neutral axis, as shown in Figure 8.11(b). This is called the **bending moment**. As the beam is in equilibrium there will be an equal and opposite bending moment on the other side of our imaginary cut. In the same way the vertical forces F balance at the cut. This vertical force is called the **shear force**. The shear force and bending moment are important in consideration of design safety.

We find the shear force F at a point distance x from the left-hand end of the beam by considering the vertical equilibrium of forces for the left-hand portion of the beam, and we find the bending moment M by looking at the balance of moments of force for that left-hand portion (see Figure 8.12(a)). The force F is the sum of the forces acting vertically on AP, and M is the sum of moments.

Consider the small element of the beam of length Δx between P and Q shown in Figure 8.12(b). Then examining the balance of moments about Q we see that

$$M(x + \Delta x) = M(x) + \Delta x F(x)$$

so that

$$M(x + \Delta x) - M(x) = \Delta x F(x)$$

giving

$$F(x) = \frac{M(x + \Delta x) - M(x)}{\Delta x}$$

Now letting $\Delta x \to 0$, we obtain

$$F(x) = \frac{dM}{dx}$$

as required.

Figure 8.12
(a) Horizontal beam.
(b) Element of the beam.

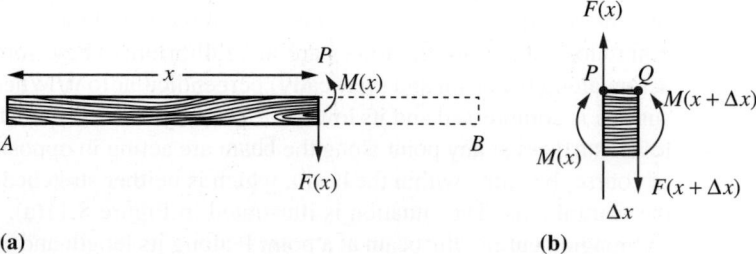

We saw in Section 7.9.2 that for a freely hinged beam with a point load W at $x = a$

$$M(x) = \begin{cases} W(l-a)x/l & 0 < x \leqslant a \\ W(l-x)a/l & a \leqslant x < l \end{cases}$$

$$F(x) = \begin{cases} W - Wa/l & 0 < x < a \\ -Wa/l & a \leqslant x < l \end{cases}$$

It is left to the reader to verify that these satisfy the equation relating $M(x)$ and $F(x)$.

Example 8.8

An open box, illustrated in Figure 8.13(a), is made from an A4 sheet of card using the folds shown in Figure 8.13(b). Find the dimensions of the tray which maximize its capacity.

Figure 8.13
(a) The open box.
(b) The net of an open box used commercially.

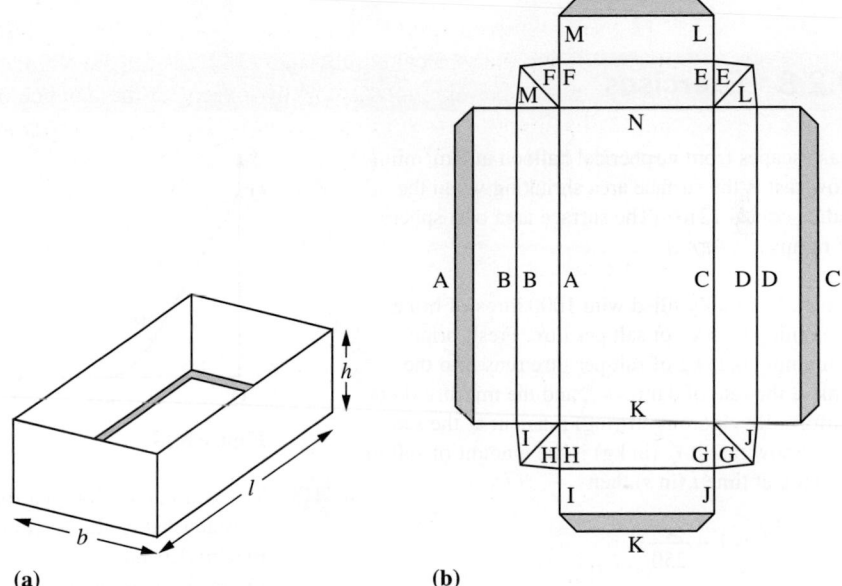

(a)   **(b)**

Solution

Boxes like these are used commercially for food sales. Packaging is expensive so that manufacturers often try to design a container that has the biggest capacity for a standard size of cardboard. An A4 sheet has size 210×297 mm.

Allowing 10 mm flaps as stiffeners, shaded in the diagram, and denoting the length, breadth and height by l, b and h (in mm) respectively, we have

$$l + 4h + 20 = 297$$

$$b + 4h + 20 = 210$$

and the capacity C is $l \times b \times h$ mm^3. Thus

$$C(h) = (277 - 4h)(190 - 4h)h = 52\,630x - 1868x^2 + 16x^3$$

The maximum capacity C^* occurs where $C'(h) = 0$.

It can be shown from first principles, see Question 5 of Exercises 8.2.6, that the general cubic function

$$f(x) = ax^3 + bx^2 + cx + d$$

has derivative

$$f'(x) = 3ax^2 + 2bx + c$$

In this example $a = 16$, $b = -1868$, $c = 52\,630$, $d = 0$ and $x = h$. Thus

$$C'(h) = 52\,630 - 3736h + 48h^2$$

so that the value h^* of h which yields the maximum capacity is $h^* = 18.47$ where $C'(h^*) = 0$. We can verify that it is a maximum by showing that

$$C'(18.4) > 0 \quad \text{and} \quad C'(18.5) < 0$$

8.2.8 Exercises

8   Gas escapes from a spherical balloon at $2\,\text{m}^3\,\text{min}^{-1}$. How fast is the surface area shrinking when the radius equals $12\,\text{m}$? (The surface area of a sphere of radius r is $4\pi r^2$.)

9   A tank is initially filled with 1000 litres of brine, containing $0.15\,\text{kg}$ of salt per litre. Fresh brine containing $0.25\,\text{kg}$ of salt per litre runs into the tank at the rate of 4 litres s^{-1}, and the mixture (kept uniform by vigorous stirring) runs out at the same rate. Show that if Q (in kg) is the amount of salt in the tank at time t (in s) then

$$\frac{\text{d}Q}{\text{d}t} = 1 - \frac{Q}{250}$$

10   The bending moment $M(x)$ for a beam of length l is given by $M(x) = W(2x - l)^3/8l^2$, $0 \leqslant x \leqslant l$. Find the formula for the shear force F. (See Example 8.7.)

11   A small weight is dragged across a horizontal plane by a string PQ of length a, the end P being attached to the weight while the end Q is made to move steadily along a fixed line perpendicular to the original position of PQ. Choosing the coordinate axes so that Oy is that fixed line and Ox passes through the initial position of P as shown in Figure 8.14, show that the curve $y = y(x)$ described by P is such that

$$\frac{\text{d}y}{\text{d}x} = -\frac{\sqrt{(a^2 - x^2)}}{x}$$

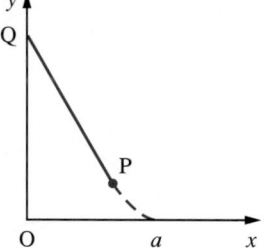

Figure 8.14

12   The limiting tension in a rope wound round a capstan (that is, the tension when the rope is about to slip) depends on the angle of wrap θ as shown in Figure 8.15. Show that an increase $\Delta\theta$ in the angle of wrap produces a corresponding increase ΔT in the value of the limiting tension such that

$$\Delta T \approx \mu T \Delta\theta$$

where μ is the coefficient of friction. Deduce $\text{d}T/\text{d}\theta$.

Figure 8.15

13 A chemical dissolves in water at a rate jointly proportional to the amount undissolved and to the difference between the concentration in the solution and that in the saturated solution. Initially none of the chemical is dissolved in the water. Show that the amount $x(t)$ of undissolved chemical satisfies the differential equation

$$\frac{dx}{dt} = kx(M - x_0 + x)$$

where k is a constant, M is the amount of the chemical in the saturated solution and $x_0 = x(0)$.

14 The rate at which a solute diffuses through a membrane is proportional to the area and to the concentration difference across the membrane. A solution of concentration C flows down a tube with constant velocity v. The solute diffuses through the wall of the tube into an ambient solution of the same solute of a lower fixed concentration C_0. If the tube has constant circular cross-section of radius r, show that at distance x along the tube the concentration $C(x)$ satisfies the differential equation

$$\frac{dC}{dx} = -\frac{2k}{rv}(C - C_0)$$

where k is a constant.

15 A lecture theatre having volume $1000\,m^3$ is designed to seat 200 people. The air is conditioned continuously by an inflow of fresh air at the constant rate V (in $m^3\,min^{-1}$). An average person generates $980\,cm^3$ of CO_2 per minute, while fresh air contains 0.04% of CO_2 by volume. Show that the percentage concentration x of CO_2 by volume in the lecture theatre at time t (in min) after the audience enters satisfies the differential equation

$$1000\frac{dx}{dt} = 19.6 + 0.04V - Vx(t)$$

If initially $x(0) = 0.04$, show that $x(t)$ is an increasing function of t for $t > 0$. Deduce that the maximum $x*$ of $x(t)$ is given by

$$x* = (19.6 + 0.04V)/V$$

If the specification is that x does not exceed 0.06 (that is, 50% increase above fresh air), deduce that V must be chosen so that $V > 980$. Comment on this result.

16 Consider the chemical reaction

$$A + B \rightarrow X$$

Let x be the amount of product X, and a and b the initial amounts of A and B (with x, a and b in mol). The rate of reaction is proportional to the product of the uncombined amounts of A and B remaining. Express this relationship in terms of $\frac{dx}{dt}$, x, a and b.

17 A wire of length l metres is bent so as to form the boundary of a sector of a circle of radius r metres and angle θ radians. Show that

$$\theta = \frac{l - 2r}{r}$$

and prove that the area of the sector is greatest when the radius is $l/4$.

18 A manufacturer found that the sales figure for a certain item depended on the selling price. The market research department found that the maximum number of items that could be sold was 20 000 and that the number actually sold decreased by 100 for every 1p increase in price. The total cost of production of the items consisted of a set-up cost of £200 plus 50p per item manufactured. Show that the profit y pence as a function of the selling price x pence is

$$y = 25\,000x - 100x^2 - 1\,020\,000$$

What price should be adopted to maximize profits, and how many items are produced?

8.3 Techniques of differentiation

In this section we will obtain the derivatives of some basic functions from 'first principles', that is, using the definition of a derivative given in (8.1), and will show how we obtain the derivatives of other functions using the basic results and some elementary rules. The rules themselves may be derived from the basic definition of the differentiation process. In practice, we make use of a very few basic facts, which, together with the rules, enable us to differentiate a wide variety of functions.

8.3.1 Basic rules of differentiation

To enable us to exploit the basic derivatives as we obtain them, we will first obtain the rules which make that exploitation possible. These rules you should know 'by heart'.

Rule 1 (constant multiplication rule)

If $y = f(x)$ and k is a constant then

$$\frac{d}{dx}(ky) = k\frac{dy}{dx} = kf'(x)$$

Rule 2 (sum rule)

If $u = f(x)$ and $v = g(x)$ then

$$\frac{d}{dx}(u + v) = \frac{du}{dx} + \frac{dv}{dx} = f'(x) + g'(x)$$

Rule 3 (product rule)

If $u = f(x)$ and $v = g(x)$ then

$$\frac{d}{dx}(uv) = u\frac{dv}{dx} + v\frac{du}{dx} = f(x)g'(x) + g(x)f'(x)$$

Rule 4 (quotient rule)

If $u = f(x)$ and $v = g(x)$ then

$$\frac{d}{dx}\left(\frac{u}{v}\right) = \frac{v(du/dx) - u(dv/dx)}{v^2} = \frac{g(x)f'(x) - f(x)g'(x)}{[g(x)]^2}$$

Verification of rules

Rule 1 follows directly from the definition given in (8.1), for if

$$g(x) = kf(x), \quad k \text{ constant}$$

$$g(x + \Delta x) = kf(x + \Delta x)$$

and

$$\Delta g = g(x + \Delta x) - g(x) = k[f(x + \Delta x) - f(x)]$$

so that

$$\frac{dg}{dx} = \lim_{\Delta x \to 0} \frac{\Delta g}{\Delta x} = \lim_{\Delta x \to 0} k\left[\frac{f(x + \Delta x) - f(x)}{\Delta x}\right] = kf'(x)$$

using the properties of limits given in Section 7.5.2.

Likewise, for *Rule 2*, if

$$h(x) = f(x) + g(x)$$

$$h(x + \Delta x) = f(x + \Delta x) + g(x + \Delta x)$$

and

$$\Delta h = h(x + \Delta x) - h(x) = [f(x + \Delta x) - f(x)] + [g(x + \Delta x) - g(x)]$$

so that

$$\frac{dh}{dx} = \lim_{\Delta x \to 0} \frac{\Delta h}{\Delta x} = \lim_{\Delta x \to 0} \left[\frac{f(x + \Delta x) - f(x)}{\Delta x} \right] + \lim_{\Delta x \to 0} \left[\frac{g(x + \Delta x) - g(x)}{\Delta x} \right]$$

using the properties of limits given in Section 7.5.2.

Thus

$$\frac{dh}{dx} = f'(x) + g'(x)$$

It also readily follows that if

$$y = f(x) - g(x)$$

then

$$\frac{dy}{dx} = f'(x) - g'(x)$$

To verify *Rule 3* consider Figure 8.16. For any value of x, the area y of the rectangle ABCD is

$$y = u(x)v(x)$$

Increasing x by the increment Δx changes u and v by amounts Δu and Δv, respectively, giving

$$u(x + \Delta x) = u(x) + \Delta u \quad \text{and} \quad v(x + \Delta x) = v(x) + \Delta v$$

From the diagram we see that the corresponding increment in y is given by

$$\Delta y = \text{area EFCB} + \text{area CHJD} + \text{area CFGH}$$

$$= u\,\Delta v + v\,\Delta u + \Delta u\,\Delta v$$

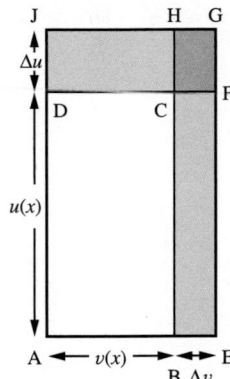

Figure 8.16

so that

$$\frac{dy}{dx} = \lim_{\Delta x \to 0} \frac{\Delta y}{\Delta x} = \lim_{\Delta x \to 0} \left[u\frac{\Delta v}{\Delta x} + v\frac{\Delta u}{\Delta x} + \frac{\Delta u \Delta v}{\Delta x} \right]$$

leading to the result

$$\frac{dy}{dx} = u(x)v'(x) + u'(x)v(x)$$

since Δu and $\Delta v \to 0$ as $\Delta x \to 0$.

Rule 4 may then be deduced from Rule 3, for if

$$y = \frac{u}{v}$$

where $u = f(x)$ and $v = g(x)$, then $u = yv$ and Rule 3 gives

$$\frac{du}{dx} = y\frac{dv}{dx} + v\frac{dy}{dx}$$

$$= \frac{u}{v}\frac{dv}{dx} + v\frac{dy}{dx} \quad \text{on substituting for } y$$

Rearranging then gives the required result

$$\frac{dy}{dx} = \frac{v\dfrac{du}{dx} - u\dfrac{dv}{dx}}{v^2}$$

8.3.2 Derivative of x^r

Using the definition of a derivative given in (8.1) and following the procedure of Example 8.1, we can proceed to obtain the derivative of the power function $f(x) = x^r$ when r is a real number.

Since $f(x) = x^r$ we have

$$f(x + \Delta x) = (x + \Delta x)^r$$

Using the binomial series (7.16) from Section 7.7.2 we have

$$(x + \Delta x)^r = x^r\left(1 + \frac{\Delta x}{x}\right)^r, \quad x \neq 0$$

$$= x^r\left[1 + r\frac{\Delta x}{x} + \frac{1}{2}r(r - 1)\left(\frac{\Delta x}{x}\right)^2 + \ldots\right]$$

$$= x^r + rx^{r-1}\Delta x + \frac{1}{2}r(r - 1)x^{r-2}(\Delta x)^2 + \ldots$$

so that

$$\frac{\Delta f}{\Delta x} = \frac{f(x + \Delta x) - f(x)}{\Delta x} = rx^{r-1} + \frac{1}{2}r(r - 1)x^{r-2}(\Delta x) + \ldots$$

Now letting $\Delta x \to 0$ we have that

$$\frac{df}{dx} = \frac{d(x^r)}{dx} = \lim_{\Delta x \to 0}\frac{\Delta f}{\Delta x} = rx^{r-1}$$

leading to the general result

$$\frac{d}{dx}(x^r) = rx^{r-1}, \, r \in \mathbb{R} \tag{8.9}$$

Note that the solutions of Example 8.1 satisfy this general result.

Note also that (8.9) implies that if k is a constant then $\dfrac{\mathrm{d}k}{\mathrm{d}x} = 0$, which is as expected since the derivative measures the rate of change of the function.

Check that result (8.9) is determined by the following MATLAB commands:

```
syms h r x
df = limit(((x + h)^r   x^r)/h,h,0);
pretty(df)
```

Example 8.9 Using result (8.9) find $f'(x)$ when $f(x)$ is

(a) $\sqrt{x}$ (b) $\dfrac{1}{x^5}$ (c) $\dfrac{1}{\sqrt[3]{x}}$

Solution (a) Taking $r = \frac{1}{2}$ in (8.9) gives

$$\frac{\mathrm{d}}{\mathrm{d}x}(\sqrt{x}) = \frac{\mathrm{d}}{\mathrm{d}x}(x^{1/2}) = \tfrac{1}{2}x^{-1/2} = \frac{1}{2\sqrt{x}}$$

(b) Taking $r = -5$ in (8.9) gives

$$\frac{\mathrm{d}}{\mathrm{d}x}\left(\frac{1}{x^5}\right) = \frac{\mathrm{d}}{\mathrm{d}x}(x^{-5}) = -5x^{-6} = -\frac{5}{x^6}$$

(c) Taking $r = -\frac{1}{3}$ in (8.9) gives

$$\frac{\mathrm{d}}{\mathrm{d}x}\left(\frac{1}{\sqrt[3]{x}}\right) = \frac{\mathrm{d}}{\mathrm{d}x}(x^{-1/3}) = -\tfrac{1}{3}x^{-4/3} = -\frac{1}{3}\left(\frac{1}{\sqrt[3]{x^4}}\right)$$

Example 8.10 Using the result (8.9) and the rules of Section 8.3.1, find $f'(x)$ where $f(x)$ is

(a) $8x^4 - 4x^2$ (b) $(2x^2 + 5)(x^2 + 3x + 1)$ (c) $4x^7(x^2 - 3x)$

(d) $(x + 1)\sqrt{x}$ (e) $\dfrac{\sqrt{x}}{x + 1}$ (f) $\dfrac{x^3 + 2x + 1}{x^2 + 1}$

Solution (a) Using Rule 1 and result (8.9) with $r = 4$, we have

$$\frac{\mathrm{d}}{\mathrm{d}x}(8x^4) = 32x^3$$

Similarly

$$\frac{\mathrm{d}}{\mathrm{d}x}(4x^2) = 8x$$

Using Rule 2, we have

$$\frac{d}{dx}(8x^4 - 4x^2) = 32x^3 - 8x$$

(b) Using the result (8.9) with Rules 1 and 2, we obtain

$$\frac{d}{dx}(2x^2 + 5) = 4x \quad \text{and} \quad \frac{d}{dx}(x^2 + 3x + 1) = 2x + 3$$

Taking $u = 2x^2 + 5$ and $v = x^2 + 3x + 1$ in Rule 3, we obtain

$$\frac{d}{dx}[(2x^2 + 5)(x^2 + 3x + 1)] = 4x(x^2 + 3x + 1) + (2x^2 + 5)(2x + 3)$$

Multiplying out these terms we obtain

$$f'(x) = 8x^3 + 18x^2 + 14x + 15$$

(c) Using Rule 3 with $u(x) = 4x^7$ and $v(x) = (x^2 - 3x)$ we obtain

$$f'(x) = (28x^6)(x^2 - 3x) + (4x^7)(2x - 3)$$

$$= 36x^8 - 96x^7$$

$$= 12x^7(3x - 8)$$

Alternatively, we could write $f(x) = 4x^9 - 12x^8$ giving $f'(x) = 36x^8 - 96x^7$ directly. It is important to look at expressions carefully before applying the rules. Sometimes there are quicker routes to the solution.

(d) Multiplying out we have

$$f(x) = x\sqrt{x} + \sqrt{x} = x^{3/2} + x^{1/2}$$

so that

$$f'(x) = \tfrac{3}{2}x^{1/2} + \tfrac{1}{2}x^{-1/2} = \frac{(3x + 1)}{2\sqrt{x}}$$

(e) Using Rule 4 with $u = \sqrt{x}$ and $v = x + 1$ so that $u'(x) = \tfrac{1}{2}x^{-1/2}$ and $v'(x) = 1$, we obtain

$$f'(x) = \frac{\tfrac{1}{2}x^{-1/2}(x + 1) - x^{1/2}(1)}{(x + 1)^2} = \frac{1 - x}{2(1 + x)^2\sqrt{x}}$$

(f) Using Rule 4 with $u = x^3 + 2x + 1$ and $v = x^2 + 1$, we obtain

$$f'(x) = \frac{(3x^2 + 2)(x^2 + 1) - (x^3 + 2x + 1)(2x)}{(x^2 + 1)^2}$$

$$= \frac{x^4 + x - 2x + 2}{(x^2 + 1)^2}$$

Symbolically in MATLAB, if $y = f(x)$ then its derivative, with respect to x, is determined using the $diff(y)$ or $diff(y,x)$ commands (either can be used as y is a function of only one variable). Thus the derivative is determined by the commands

```
syms x y
y = f(x);  dy = diff(y)
```

The corresponding commands in MAPLE are

```
y:= f(x);  dy:= diff(y,x);
```

In both cases the $simplify$ command may be used to simplify the answer returned by the $diff$ command. To illustrate we consider Example 8.10(b), for which the commands

MATLAB   MAPLE
```
syms x y
y = (2*x^2 + 5)*(x^2 + 3*x + 1);        y:= (2*x^2 + 5)*
                                             (x^2 + 3*x + 1);
dy = diff(y)                              dy:= diff(y,x);
```

return

```
dy = 4*x*(x^2 + 3*x + 1) +              dy:= 4x(x^2 + 3x + 1) +
(2*x^2 + 5)*(2*x + 3)                     (2x^2 + 5)(2x + 3)
```

Using the $simplify$ command, coupled with the $pretty$ command for MATLAB, we have that the commands

```
dy = simplify(dy);                       dy:= simplify(dy);
pretty(dy)
```

return the derivative as

$$dy = 8x^3 + 18x^2 + 14x + 15$$

In practice we often anticipate the need to simplify and use the single command $simplify(diff(y))$ in MATLAB and $simplify(diff(y,x));$ in MAPLE. Considering Example 8.10(e), the MATLAB commands

```
syms x y
y = sqrt(x)/(x + 1);  dy = simplify(diff(y));
pretty(dy)
```

return the derivative as

$$dy = -1/2 \; \frac{x - 1}{x^{1/2}(x + 1)^2}$$

For practice check the answers to sections (a), (c), (d) and (f) of Example 8.10 using MATLAB or MAPLE.

8.3.3 Exercises

19 Differentiate the function f where $f(x)$ is

(a) x^9 (b) $\sqrt{(x^3)}$ (c) $-4x^2$

(d) $4x^4 + 2x^5$ (e) $4x^3 + x - 8$ (f) $1/(2x^2)$

(g) $x + \sqrt{x}$ (h) $2x^{7/2}$ (i) $1/(3x^3)$

20 Differentiate the function f where $f(x)$ is

(a) $(3x^4 - 1)(x^2 + 5x)$ (b) $(5x + 1)(x^3 + 3x - 6)$

(c) $(7x + 3)(\sqrt{x} + 1/\sqrt{x})$ (d) $(3 - 2x)(2x - 9/x)$

(e) $(\sqrt{x} - 1/\sqrt{x})(x - 1/x)$

(f) $(x^2 + x + 1)(2x^2 + x - 1)$

21 Differentiate the function f where $f(x)$ is

(a) $(3x^2 + x + 1)/(x^3 + 1)$ (b) $\sqrt{(2x)}/(x^2 + 4)$

(c) $(x + 1)/(x^2 + 1)$ (d) $x^{2/3}/(x^{1/3} + 1)$

(e) $(x^2 + 1)/(x + 1)$

(f) $(2x^2 - x + 1)/(x^2 - 2x + 2)$

22 Differentiate the function f where $f(x)$ is

(a) $(ax + b)(cx + d)$ (b) $(ax + b)/(cx + d)$

(c) $3ax^2 + 5bx + c$ (d) $ax^2/(bx + c)$

23 A fruit juice manufacturer wishes to design a carton that has a square face as shown in Figure 8.17(a). The carton is to contain 1 litre of juice and is made from a rectangular sheet of waxed cardboard by folding it into a rectangular tube and sealing down the edge and then folding and sealing the top and bottom. To make the carton airtight and robust for handling, an overlap of at least 0.5 cm is needed. The net for the carton is shown in Figure 8.17(b).

(a)

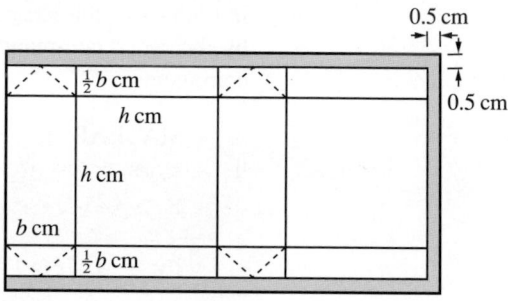

(b)

Figure 8.17 Carton of Question 23.

Show that the amount $A(h)$ cm^2 of card used is given by

$$A(h) = \left[h + \frac{1000}{h^2} + 1\right]\left[2h + \frac{2000}{h^2} + 0.5\right]$$

Verify that

$$A(h) = 2\left[h + \frac{1000}{h^2} + \frac{5}{8}\right]^2 - \frac{9}{32}$$

By finding the value $x*$ of x which minimizes $y = x + \frac{1000}{x^2}$, find the value $h*$ of h which minimizes $A(h)$.

8.3.4 Differentiation of polynomial functions

Using the result

$$\frac{d}{dx}(x^r) = rx^{r-1}$$

given in equation (8.9), together with the rules developed in Section 8.3.1, we proceed in this and following sections to find the derivatives of a range of algebraic functions.

It is a simple matter to find the derivative of the polynomial function

$$f(x) = a_0 + a_1 x + a_2 x^2 + a_3 x^3 + \ldots + a_{n-1} x^{n-1} + a_n x^n = \sum_{r=0}^{n} a_r x^r \tag{8.10}$$

where n is a non-negative integer and the coefficients a_r, $r = 0, 1, \ldots, n$, are real numbers.

Using the constant multiplication rule together with the sum rule we may differentiate term by term to give

$$f'(x) = a_1 + 2a_2 x + 3a_3 x^2 + \ldots + (n-1)a_{n-1} x^{n-2} + na_n x^{n-1} = \sum_{r=1}^{n} ra_r x^{r-1}$$

Example 8.11   If $y = 2x^4 - 2x^3 - x^2 + 3x - 2$, find $\dfrac{dy}{dx}$.

Solution   Differentiating term by term, using the sum rule, gives

$$\frac{dy}{dx} = \frac{d}{dx}(2x^4) - \frac{d}{dx}(2x^3) - \frac{d}{dx}(x^2) + \frac{d}{dx}(3x) - \frac{d}{dx}(2)$$

which on using the constant multiplication rule gives

$$\frac{dy}{dx} = 2\frac{d}{dx}(x^4) - 2\frac{d}{dx}(x^3) - \frac{d}{dx}(x^2) + 3\frac{d}{dx}(x) - \frac{d}{dx}(2)$$

$$= 2(4x^3) - 2(3x^2) - (2x) + 3(1) - 0$$

so that

$$\frac{dy}{dx} = 8x^3 - 6x^2 - 2x + 3$$

Example 8.12   The distance s metres moved by a body in t seconds is given by

$$s = 2t^3 - 1.5t^2 - 6t + 12$$

Determine the velocity and acceleration after 2 seconds.

Solution $s = 2t^3 - 1.5t^2 - 6t + 12$

The velocity v (m s^{-1}) is given by $v = \dfrac{ds}{dt}$ so that

$$v = \frac{ds}{dt} = 2(3t^2) - 1.5(2t) - 6(1) = 6t^2 - 3t - 6$$

When $t = 2$ seconds

$$v = 6(4) - 3(2) - 6 = 12$$

so that the velocity after 2 seconds is $12\,\text{m s}^{-1}$.

The acceleration a (m s^{-2}) is given by $a = \dfrac{dv}{dt}$ so that

$$a = \frac{dv}{dt} = \frac{d}{dt}(6t^2 - 3t - 6) = 12t - 3$$

When $t = 2$ seconds

$$a = 12(2) - 3 = 21$$

so that the acceleration after 2 seconds is $21 \, \text{m s}^{-2}$.

Sometimes polynomial functions are not expressed in the standard form of (8.10), $f(x) = (2x + 5)^3$ and $f(x) = (3x - 1)^2(x + 2)^3$ being such examples. Such cases will be considered in Section 8.3.6 where the differentiation of composite functions will be discussed.

The derivatives of polynomial functions can be evaluated numerically by a simple extension of the method of synthetic division (or nested multiplication) which is used for evaluating the function itself. We saw in Section 2.4.3 that

$$f(x) = a_n x^n + a_{n-1} x^{n-1} + \dots + a_1 x + a_0$$

could be written as $f(x) = g(x)(x - c) + f(c)$ where

$$g(x) = b_{n-1} x^{n-1} + b_{n-2} x^{n-2} + \dots + b_1 x + b_0$$

and where the coefficients $b_{n-1}, \dots, b_0$ were generated in the process of nested multiplication. Differentiating $f(x)$ with respect to x using the product rule gives

$$f'(x) = g'(x)(x - c) + g(x)(1)$$

so that

$$f'(c) = g'(c)(0) + g(c)(1) = g(c)$$

Thus we can evaluate $f'(c)$ by applying the nested multiplication method again but this time to $g(x)$.

Example 8.13

Evaluate $f(2)$ and $f'(2)$ for the polynomial function

$$f(x) = 2x^4 - 2x^3 - x^2 + 3x - 2$$

Solution

$$
\begin{array}{r|rrrrl}
 & 2 & -2 & -1 & 3 & -2 \\
\times 2 & 0 & 4 & 4 & 6 & 18 \\
\hline
 & 2 & 2 & 3 & 9 & 16 = f(2) \\
\times 2 & 0 & 4 & 12 & 30 \\
\hline
 & 2 & 6 & 15 & 39 = f'(2)
\end{array}
$$

This method of evaluating the function and its derivative is very efficient and is often used in computer packages which require finding the roots of polynomial equations.

8.3.5 Differentiation of rational functions

As we saw in Section 2.5, rational functions have the general form

$$f(x) = \frac{p(x)}{q(x)}$$

where $p(x)$ and $q(x)$ are polynomials. To obtain the derivatives of such functions we make use of the constant multiplication, sum and quotient rules as illustrated in Example 8.14.

Example 8.14 Find the derivative of the following functions of x:

(a) $\dfrac{3x + 2}{2x^2 + 1}$ (b) $\dfrac{2x + 3}{x^2 + x + 1}$

(c) $x^3 + 2x^2 - \dfrac{1}{x} + \dfrac{1}{x^2} + 3, \quad x \neq 0$

Solution (a) Taking $u = 3x + 2$ and $v = 2x^2 + 1$ gives

$$\frac{du}{dx} = 3 \quad \text{and} \quad \frac{dv}{dx} = 4x$$

so, from the quotient rule,

$$\frac{d}{dx}\left[\frac{3x + 2}{2x^2 + 1}\right] = \frac{v\left(\dfrac{du}{dx}\right) - u\left(\dfrac{dv}{dx}\right)}{v^2}$$

$$= \frac{(2x^2 + 1)3 - (3x + 2)4x}{(2x^2 + 1)^2}$$

$$= \frac{-(6x^2 + 8x - 3)}{(2x^2 + 1)^2}$$

(b) Taking $u = 2x + 3$ and $v = x^2 + x + 1$

$$\frac{du}{dv} = 2 \quad \text{and} \quad \frac{dv}{dx} = 2x + 1$$

so, from the quotient rule,

$$\frac{d}{dx}\left[\frac{2x + 3}{x^2 + x + 1}\right] = \frac{(x^2 + x + 1)(2) - (2x + 3)(2x + 1)}{(x^2 + x + 1)^2}$$

$$= -\frac{(2x^2 + 6x + 1)}{(x^2 + x + 1)^2}$$

(c) In this case we can express the function as

$$y = x^3 + 2x^2 - x^{-1} + x^{-2} + 3, \quad x \neq 0$$

and differentiate term by term to give

$$\frac{dy}{dx} = 3x^2 + 2(2x) - (-x^{-2}) + (-2x^{-3}) + 0 = 3x^2 + 4x + x^{-2} - 2x^{-3}$$

$$= 3x^2 + 4x + \frac{1}{x^2} - \frac{2}{x^3}, \quad x \neq 0$$

8.3.6 Differentiation of composite functions

As mentioned earlier, to differentiate many functions we need a further rule to deal with composite functions.

Rule 5 (composite-function or chain rule)

If $z = g(x)$ and $y = f(z)$ then

$$\frac{dy}{dx} = \frac{dy}{dz}\frac{dz}{dx} = f'(z)g'(x)$$

Verifying *Rule 5* is a little more difficult than Rules 1–4. For an increment Δx in x, let Δz and Δy be the corresponding increments in z and y, respectively. It then follows from definition (8.1) that

$$\Delta z = g'(x)\Delta x + \varepsilon_1 \Delta x \tag{8.11}$$

where $\varepsilon_1 \to 0$ as $\Delta x \to 0$. Likewise

$$\Delta y = f'(z)\Delta z + \varepsilon_2 \Delta z \tag{8.12}$$

where $\varepsilon_2 \to 0$ as $\Delta z \to 0$. Combining (8.11) and (8.12) then gives

$$\Delta y = [f'(z) + \varepsilon_2][g'(x) + \varepsilon_1]\Delta x$$

so that

$$\frac{\Delta y}{\Delta x} = f'(z)g'(x) + \varepsilon_2 g'(x) + \varepsilon_1 f'(z) + \varepsilon_1\varepsilon_2$$

As $\Delta x \to 0$ so does $\Delta z \to 0$, $\varepsilon_1 \to 0$ and $\varepsilon_2 \to 0$ and

$$\frac{dy}{dx} = \lim_{\Delta x \to 0} \frac{\Delta y}{\Delta x} = f'(z)g'(x)$$

as required.

Adapting Figure 2.12 (Section 2.2.4), the chain rule

$$\frac{dy}{dx} = \frac{dy}{dz} \cdot \frac{dz}{dx}$$

may be represented as in Figure 8.18.

Figure 8.18
The chain rule of
differentiation.

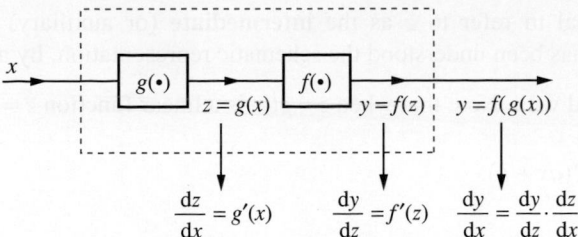

$$\frac{dz}{dx} = g'(x) \qquad \frac{dy}{dz} = f'(z) \qquad \frac{dy}{dx} = \frac{dy}{dz}\cdot\frac{dz}{dx}$$

Example 8.15   Find $\dfrac{dy}{dx}$ when y is

(a) $(5x^2 + 11)^9$ (b) $\sqrt{(3x^2 + 1)}$

Solution   (a) In this case we could expand out $(5x^2 + 11)^9$ and treat it as a polynomial of degree 18. However, it is advantageous to view it as a composite function, as represented in Figure 8.19(a). Thus, taking

$$y = z^9 \qquad \text{and} \quad z = 5x^2 + 11$$

$$\frac{dy}{dz} = 9z^8 \quad \text{and} \quad \frac{dz}{dx} = 10x$$

so, by the chain rule,

$$\frac{dy}{dx} = \frac{dy}{dz}\frac{dz}{dx} = 9(5x^2 + 11)^8(10x) = 90x(5x^2 + 11)^8$$

(b) The composite function $y = \sqrt{(3x^2 + 1)}$ may be represented as in Figure 8.19(b). Thus, taking

$$y = \sqrt{z} = z^{1/2} \qquad\qquad \text{and} \quad z = 3x^2 + 1$$

$$\frac{dy}{dz} = \tfrac{1}{2}z^{-1/2} = \frac{1}{2\sqrt{z}} \qquad \text{and} \quad \frac{dz}{dx} = 6x$$

so, by the chain rule,

$$\frac{dy}{dx} = \frac{dy}{dz}\frac{dz}{dx} = \frac{1}{2\sqrt{(3x^2 + 1)}}6x = \frac{3x}{\sqrt{(3x^2 + 1)}}$$

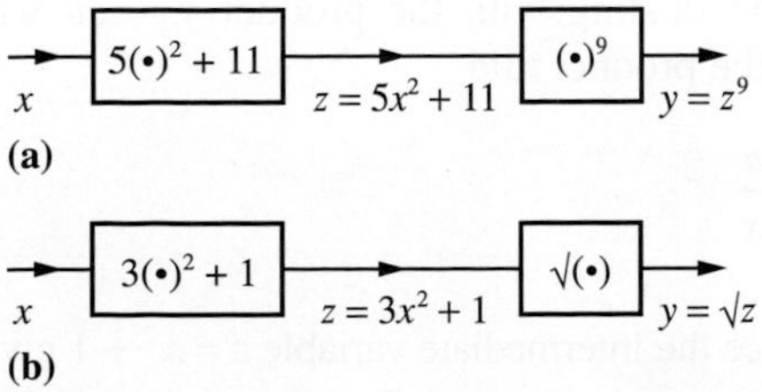

Figure 8.19
(a) Representation
of $y = (5x^2 + 11)^9$.
(b) Representation
of $y = \sqrt{(3x^2 + 1)}$.

It is usual to refer to z as the **intermediate** (or auxiliary) variable, and once the process has been understood the schematic representation, by a block diagram, stage is dispensed with. Note that when $z = g(z)$ is a linear function $z = ax + b$, then $\dfrac{dz}{dx} = a$ and

$$\frac{dy}{dx} = af'(ax + b).$$

Example 8.16 Find $\dfrac{dy}{dx}$ when y is

(a) $(3x^3 - 2x^2 + 1)^5$ (b) $\dfrac{1}{(5x^2 - 2)^7}$

(c) $(x^2 + 1)^3 \sqrt{(x - 1)}$ (d) $\dfrac{\sqrt{(2x + 1)}}{(x^2 + 1)^3}$

Solution (a) Introducing the intermediate variable $z = 3x^3 - 2x^2 + 1$ we have

$$y = z^5 \quad \text{and} \quad z = 3x^3 - 2x^2 + 1$$

$$\frac{dy}{dz} = 5z^4 \quad \text{and} \quad \frac{dz}{dx} = 9x^2 - 4x$$

so, by the chain rule,

$$\frac{dy}{dx} = \frac{dy}{dz}\frac{dz}{dx} = 5(3x^3 - 2x^2 + 1)^4(9x^2 - 4x)$$

$$= 5x(9x - 4)(3x^3 - 2x^2 + 1)^4$$

(b) Introducing the intermediate variable $z = 5x^2 - 2$ we have

$$y = \frac{1}{z^7} = z^{-7} \quad \text{and} \quad z = 5x^2 - 2$$

$$\frac{dy}{dz} = -7z^{-8} = -\frac{7}{z^8} \quad \text{and} \quad \frac{dz}{dx} = 10x$$

so, by the chain rule,

$$\frac{dy}{dx} = \frac{dy}{dz}\frac{dz}{dx} = -\frac{7}{(5x^2 - 2)^8}10x = -\frac{70x}{(5x^2 - 2)^8}$$

(c) In this case we are dealing with the product $y = uv$ where $u = (x^2 + 1)^3$ and $v = \sqrt{(x - 1)}$. Then by the product rule

$$\frac{dy}{dx} = u\frac{dv}{dx} + v\frac{du}{dx}$$

To find $\dfrac{du}{dx}$ we introduce the intermediate variable $z = x^2 + 1$ giving $u = z^3$ and so, by the chain rule,

$$\frac{du}{dx} = \frac{du}{dz}\frac{dz}{dx} = 3(x^2 + 1)^2 2x = 6x(x^2 + 1)^2$$

Likewise, to find $\dfrac{dv}{dx}$ we introduce the intermediate variable $w = x - 1$ giving $v = \sqrt{w}$ $= w^{1/2}$ and $w = x - 1$, so by the chain rule

$$\frac{dv}{dx} = \frac{1}{2\sqrt{(x-1)}}$$

It then follows from the product rule that

$$u \qquad \frac{dv}{dx} \qquad v \qquad \frac{du}{dx}$$

$$\downarrow \qquad \downarrow \qquad \downarrow \qquad \downarrow$$

$$\frac{dy}{dx} = (x^2 + 1)^3\frac{1}{2\sqrt{(x-1)}} + \sqrt{(x-1)}\,6x(x^2 + 1)^2$$

$$= \frac{(x^2 + 1)^2}{2\sqrt{(x-1)}}[(x^2 + 1) + 12x(x - 1)]$$

$$= \frac{(x^2 + 1)^2(13x^2 - 12x + 1)}{2\sqrt{(x-1)}}$$

(d) In this case we are dealing with the quotient

$$y = \frac{u}{v}$$

where $u = \sqrt{(2x + 1)}$ and $v = (x^2 + 1)^3$. Then by the quotient rule

$$\frac{dy}{dx} = \frac{v\left(\dfrac{du}{dx}\right) - u\left(\dfrac{dv}{dx}\right)}{v^2}$$

To find $\dfrac{du}{dx}$ we introduce the intermediate variable $z = 2x + 1$, giving $u = z^{1/2}$ and $z = 2x + 1$, so by the chain rule

$$\frac{du}{dx} = \frac{du}{dz}\frac{dz}{dx} = \frac{1}{\sqrt{(2x + 1)}}$$

To find $\dfrac{dv}{dx}$ we introduce the intermediate variable $w = x^2 + 1$, giving $v = w^3$ and $w = x^2 + 1$, so by the chain rule

$$\frac{dv}{dx} = \frac{dv}{dw}\frac{dw}{dx} = 3(x^2 + 1)^2 2x = 6x(x^2 + 1)^2$$

Then, by the quotient rule,

$$
\begin{array}{cccc}
v & \dfrac{du}{dx} & u & \dfrac{dv}{dx} & v^2 \\
\downarrow & \downarrow & \downarrow & \downarrow & \\
\end{array}
$$

$$
\frac{dy}{dx} = \frac{(x^2+1)^3\dfrac{1}{\sqrt{(2x+1)}} - \sqrt{(2x+1)}\,6x(x^2+1)^2}{(x^2+1)^6} \;\longleftarrow\; = \frac{1-6x-11x^2}{(x^2+1)^4\sqrt{(2x+1)}}
$$

Both MATLAB and MAPLE can handle the functions covered in previous sections. To illustrate consider Examples 8.14(a), 8.15(b) and 8.16(c). For 8.14(a) the commands

MATLAB

```
syms x y
y = (3*x + 2)/(2*x^2 + 1);
dy = simplify(diff(y))
pretty(dy)
```

MAPLE

```
y:= (3*x + 2)/(2*x^2 + 1);
dy:= simplify(diff(y,x));
```

return the derivative as

$$
dy = -\frac{6x^2 - 3 + 8x}{(2x^2 + 1)^2}
$$

For Example 8.15(b) the commands

```
syms x y
y = sqrt(3*x^2 + 1);
dy = diff(y);
pretty(dy)
```

```
y:= sqrt(3*x^2 + 1);
dy:= diff(y,x);
```

return the derivative as

$$
dy = 3\frac{x}{(3x^2 + 1)^{1/2}}
$$

$$
dy = \frac{3x}{\sqrt{(3x^2 + 1)}}
$$

For Example 8.16(c) the commands

```
syms x y
y = (x^2 + 1)^3*sqrt(x - 1);
dy = simplify(diff(y));
pretty(dy)
```

```
y:= (x^2 + 1)^3*sqrt(x - 1);
dy:= simplify(diff(y,x));
```

return the derivative as

$$
dy = \frac{1}{2}\frac{(x^2 + 1)^2(13x^2 - 12x + 1)}{(x - 1)^{1/2}}
$$

$$
dy = \frac{(x^2 + 1)^2(13x^2 - 12x + 1)}{2\sqrt{(x - 1)}}
$$

For practice check the answers to the remaining sections of Examples 8.14–8.16.

8.3.7 Differentiation of inverse functions

The algebraic and graphical properties of inverse functions were described in Section 2.2.3 earlier. It is often useful to be able to express the derivative of an inverse function in terms of the derivatives of the original function from which it came. To do this we use the following rule

Rule 6 (inverse-function rule)

If $y = f^{-1}(x)$ then $x = f(y)$ and

$$\frac{dy}{dx} = \frac{1}{dx/dy} = \frac{1}{f'(y)}$$

Rule 6 may be readily deduced from graphical considerations. Since the graph of $y = f^{-1}(x)$ is the mirror image of the graph of $y = f(x)$ in the line $y = x$ (see Figure 8.20) it follows that since $\tan\theta = \cot(\frac{1}{2}\pi - \theta)$, the gradient with respect to the x direction is

$$\frac{dy}{dx} = \frac{1}{\text{gradient with respect to the } y \text{ direction}} = \frac{1}{dx/dy}$$

We will be making use of this rule in following sections, but a simple example is to consider the function defined by $y = x^{1/3}$. Then $x = y^3$ and $\dfrac{dx}{dy} = 3y^2$. Using the inverse-function rule we obtain

$$\frac{dy}{dx} = 1 \bigg/ \frac{dx}{dy} = \frac{1}{3y^2}, \qquad y \neq 0$$

Since $y = x^{1/3}$ we deduce that $\dfrac{dy}{dx} = \frac{1}{3}x^{-2/3}$, $x \neq 0$ (agreeing with the general result

$\dfrac{d}{dx}(x^r) = rx^{r-1}$)

Figure 8.20
Derivative of
inverse function.

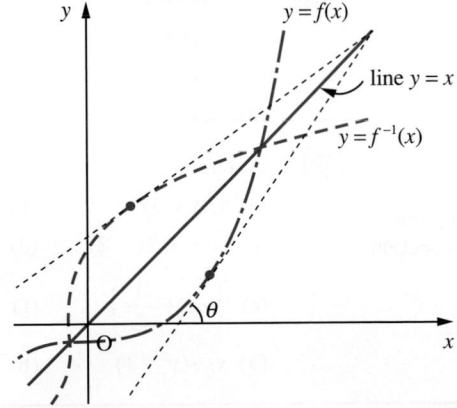

8.3.8 Exercises

 Check your answers using MATLAB or MAPLE whenever possible.

24 Differentiate the function f where $f(x)$ is

(a) $5x^2 - 2x + 1$ (b) $4x^3 + x - 8$

(c) $x^{24} + 3$ (d) $(5x + 3)^9$

(e) $(4x - 2)^7$ (f) $(1 - 3x)^6$

(g) $(3x^2 - x + 1)^3$

(h) $(4x^3 - 2x + 1)^6$

(i) $(1 + x - x^4)^5$

25 Differentiate the function f where $f(x)$ is

(a) $(2x + 4)^7(3x - 2)^5$ (b) $(5x + 1)^3(3 - 2x)^4$

(c) $(\frac{1}{2}x + 2)^2(x + 3)^4$

(d) $(x^2 + x - 2)(3x^2 - 5x + 1)$

(e) $(x^4 - 3x + 1)(6x^2 + 5)$

(f) $(x^2 + x + 1)^2(x^3 + 2x^2 + 1)^4$

(g) $(x^5 + 2x + 1)^3(2x^2 + 3x - 1)^4$

(h) $(2x + 1)^3(7 - x)^5$ (i) $(x^2 + 4x + 1)(3x + 1)^5$

26 The algebraic function

$$y = \frac{\sqrt{(1 + x)} - 1}{\sqrt{(1 + x)} + 1}, \quad x > -1$$

is a root of the equation

$$xy^2 - 2(2 + x)y + x = 0$$

Show that

$$\frac{dy}{dx} = -\frac{(y - 1)^3}{4(y + 1)}, \quad |y| < 1$$

27 An open water conduit is to be cut in the shape of an isosceles trapezium and lined with material which is available in a standard width of 1 metre as shown in Figure 8.21.

To achieve maximum potential capacity, the designer has to maximize the area of cross-section $A(b)$. Show that

$$A(b) = [(1 + b)^3(1 - b)]^{1/2}$$

and that this is maximized when $b = 0.5$.

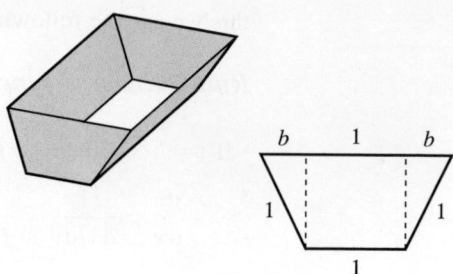

Figure 8.21 Water conduit of Question 19.

28 A carton is made from a sheet of A4 card (210 mm × 297 mm) using the net shown in Figure 8.22. Find the dimensions that yield the largest capacity.

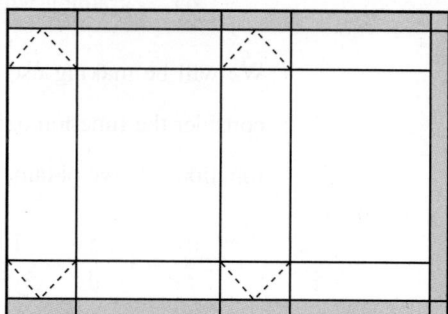

□ 5 mm overlap for seal

Figure 8.22 Net used in Question 28.

29 Differentiate:

(a) $\sqrt{(1 + 2x)}$ (b) $(x + 2)\sqrt{x}$ (c) $x\sqrt{(x + 2)}$

30 Differentiate:

(a) $x\sqrt{(4 + x^2)}$ (b) $x\sqrt{(9 - x^2)}$

(c) $(x + 1)\sqrt{(x^2 + 2x + 3)}$ (d) $x^{2/3} - x^{1/4}$

(e) $\sqrt[3]{(x^2 + 1)}$ (f) $x(2x - 1)^{1/3}$

31 Differentiate:

(a) $1/(x + 3)^2$ (b) $x/(x + 1)$

(c) $(x - 3)/(x - 2)$ (d) $1/(x^2 - 4x + 1)$

(e) $\left(\sqrt{x} + \frac{1}{\sqrt{x}}\right)^2$ (f) $x/(x^2 + 5x + 6)$

(g) $x/\sqrt{(x^2 - 1)}$ (h) $(2x + 1)^2/(3x^2 + 1)^3$

8.3.9 Differentiation of circular functions

Taking $f(x) = \sin x$ and using the sum identity (2.26b) gives from the formal definition (8.1)

$$f'(x) = \lim_{\Delta x \to 0} \frac{\sin(x + \Delta x) - \sin x}{\Delta x} = \lim_{\Delta x \to 0} \frac{\cos(x + \frac{1}{2}\Delta x)\sin(\frac{1}{2}\Delta x)}{\frac{1}{2}\Delta x}$$

Since, from Section 7.8.1 remembering that x is measured in radians here,

$$\lim_{\Delta x \to 0} \frac{\sin(\frac{1}{2}\Delta x)}{\frac{1}{2}\Delta x} = 1 \quad \text{and} \quad \lim_{\Delta x \to 0} \cos(x + \frac{1}{2}\Delta x) = \cos x$$

we have

$$f'(x) = \frac{\mathrm{d}}{\mathrm{d}x}(\sin x) = \cos x \tag{8.13}$$

Likewise, using the sum identity (2.25d), we have

$$\frac{\cos(x + \Delta x) - \cos x}{\Delta x} = \frac{-\sin(x + \frac{1}{2}\Delta x)\sin(\frac{1}{2}\Delta x)}{\frac{1}{2}\Delta x}$$

from which we deduce using (8.1) that

$$\frac{\mathrm{d}}{\mathrm{d}x}(\cos x) = -\sin x \tag{8.14}$$

 As an exercise check results (8.13) and (8.14) using MATLAB or MAPLE.

Since $\tan x = \sin x / \cos x$, we take $u = \sin x$ and $v = \cos x$, giving

$$\frac{\mathrm{d}u}{\mathrm{d}x} = \cos x \quad \text{and} \quad \frac{\mathrm{d}v}{\mathrm{d}x} = -\sin x$$

Then, from the quotient rule,

$$\frac{\mathrm{d}}{\mathrm{d}x}(\tan x) = \frac{(\cos x)(\cos x) - (\sin x)(-\sin x)}{\cos^2 x}$$

$$= \frac{\cos^2 x + \sin^2 x}{\cos^2 x} = \frac{1}{\cos^2 x}$$

That is,

$$\frac{\mathrm{d}}{\mathrm{d}x}(\tan x) = \sec^2 x \tag{8.15}$$

Since $\sec x = 1/\cos x$, we take $u = 1$ and $v = \cos x$ in the quotient rule to give

$$\frac{\mathrm{d}}{\mathrm{d}x}(\sec x) = \frac{(\cos x)(0) - (1)(-\sin x)}{\cos^2 x} = \frac{1}{\cos x}\frac{\sin x}{\cos x}$$

That is,

$$\frac{d}{dx}(\sec x) = \sec x \tan x \tag{8.16}$$

Since $\operatorname{cosec} x = 1/\sin x$, following the same procedure as above we obtain

$$\frac{d}{dx}(\operatorname{cosec} x) = -\operatorname{cosec} x \cot x \tag{8.17}$$

Since $\cot x = \cos x/\sin x$, taking $u = \cos x$ and $v = \sin x$ and using the quotient rule gives

$$\frac{d}{dx}(\cot x) = -\operatorname{cosec}^2 x \tag{8.18}$$

Taking $y = \sin^{-1} x$, we have $x = \sin y$, so that

$$\frac{dx}{dy} = \cos y$$

Then, from the inverse-function rule,

$$\frac{dy}{dx} = \frac{1}{\cos y}$$

Using the identity $\cos^2 y = 1 - \sin^2 y$, this simplifies to

$$\frac{d}{dx}(\sin^{-1} x) = \frac{1}{\sqrt{(1 - x^2)}}, \quad |x| < 1 \tag{8.19}$$

(Note that we have taken the positive square root, since from Figure 2.70(b) the derivative must be positive.)

Taking $y = \cos^{-1} x$, we have $x = \cos y$, so that

$$\frac{dy}{dx} = \frac{1}{dx/dy} = -\frac{1}{\sin y}$$

which, using the identity $\sin^2 y = 1 - \cos^2 y$, reduces to

$$\frac{d}{dx}(\cos^{-1} x) = -\frac{1}{\sqrt{(1 - x^2)}}, \quad |x| < 1 \tag{8.20}$$

(Note from Figure 2.71 that the derivative is negative.)

Taking $y = \tan^{-1} x$, we have $x = \tan y$, so that

$$\frac{dy}{dx} = \frac{1}{dx/dy} = \frac{1}{\sec^2 y}$$

Using the identity $1 + \tan^2 y = \sec^2 y$, this reduces to

$$\frac{d}{dx}(\tan^{-1} x) = \frac{1}{1 + x^2} \tag{8.21}$$

Summary

$$\frac{d}{dx}(\sin x) = \cos x, \quad \frac{d}{dx}(\cos x) = -\sin x$$

$$\frac{d}{dx}(\tan x) = \sec^2 x, \quad \frac{d}{dx}(\sec x) = \sec x \tan x$$

$$\frac{d}{dx}(\operatorname{cosec} x) = -\operatorname{cosec} x \cot x, \quad \frac{d}{dx}(\cot x) = -\operatorname{cosec}^2 x$$

$$\frac{d}{dx}(\sin^{-1}x) = \frac{1}{\sqrt{(1 - x^2)}}, \quad |x| < 1$$

$$\frac{d}{dx}(\cos^{-1}x) = \frac{-1}{\sqrt{(1 - x^2)}}, \quad |x| < 1$$

$$\frac{d}{dx}(\tan^{-1}x) = \frac{1}{1 + x^2}$$

Example 8.17   Find $\dfrac{dy}{dx}$ when y is given by:

(a) $\sin(2x + 3)$ (b) $x^2 \cos x$ (c) $\dfrac{\sin 2x}{x^2 + 2}$

(d) $\sec 6x$ (e) $x \tan 2x$ (f) $\sin^{-1}6x$

(g) $x^2 \cos^{-1}x$ (h) $\tan^{-1}\dfrac{2x}{1 + x^2}$

Solution   (a) Introducing the intermediate variable $z = 2x + 3$, we have $y = \sin z$ and $z = 2x + 3$, so by the chain rule

$$\frac{dy}{dx} = \frac{dy}{dz}\frac{dz}{dx} = \cos(2x + 3)2 = 2\cos(2x + 3)$$

Note that this result could have been written using the particular linear case of the composite-function rule given in Section 8.3.6 after Example 8.15.

(b) Taking $u = x^2$ and $v = \cos x$ gives

$$\frac{du}{dx} = 2x \quad \text{and} \quad \frac{dv}{dx} = -\sin x$$

so by the product rule

$$\frac{dy}{dx} = -x^2 \sin x + 2x \cos x$$

(c) Taking $u = \sin 2x$ and $v = x^2 + 2$ gives

$$\frac{du}{dx} = 2\cos 2x \quad \text{and} \quad \frac{dv}{dx} = 2x$$

so by the quotient rule

$$\frac{dy}{dx} = \frac{2(x^2 + 2)\cos 2x - 2x\sin 2x}{(x^2 + 2)^2}$$

(d) Introducing the intermediate variable $z = 6x$ we have

$y = \sec z$ and $z = 6x$ so by the chain rule

$$\frac{dy}{dx} = \frac{dy}{dz}\frac{dz}{dx} = 6\sec 6x\tan 6x$$

(e) Taking $u = x$ and $v = \tan 2x$ gives

$$\frac{du}{dx} = 1 \quad \text{and} \quad \frac{dv}{dx} = 2\sec^2 2x$$

where the chain rule, with intermediate variable $z = 2x$, has been used to find $\dfrac{dv}{dx}$. Then by the product rule

$$\frac{dy}{dx} = 2x\sec^2 2x + \tan 2x$$

(f) Introducing the intermediate variable $z = 6x$ we have $y = \sin^{-1}z$ and $z = 6x$

$$\frac{dy}{dz} = \frac{1}{\sqrt{(1 - z^2)}} \quad \text{and} \quad \frac{dz}{dx} = 6, \quad |z| < 1$$

so by the chain rule

$$\frac{dy}{dx} = \frac{dy}{dz}\frac{dz}{dx} = \frac{6}{\sqrt{(1 - 36x^2)}}, \quad |x| < \tfrac{1}{6}$$

(g) Taking $u = x^2$ and $v = \cos^{-1}x$ gives

$$\frac{du}{dx} = 2x \quad \text{and} \quad \frac{dv}{dx} = -\frac{1}{\sqrt{(1 - x^2)}}$$

so by the product rule

$$\frac{dy}{dx} = -\frac{x^2}{\sqrt{(1 - x^2)}} + 2x\cos^{-1}x$$

(h) Introducing the intermediate variable $z = \dfrac{2x}{1 + x^2}$ we have

$$y = \tan^{-1}z \quad \text{and} \quad z = \frac{2x}{1 + x^2}$$

$$\frac{dy}{dz} = \frac{1}{1 + z^2} \quad \text{and} \quad \frac{dz}{dx} = \frac{(1 + x^2)2 - 2x(2x)}{(1 + x^2)^2} = \frac{2(1 - x^2)}{(1 + x^2)^2}$$

so from the chain rule

$$\frac{\mathrm{d}y}{\mathrm{d}x} = \frac{\mathrm{d}y}{\mathrm{d}z}\frac{\mathrm{d}z}{\mathrm{d}x} = \frac{1}{1 + \left(\dfrac{2x}{1 + x^2}\right)^2} \cdot \frac{2(1 - x^2)}{(1 + x^2)^2} = \frac{2(1 - x^2)}{(1 + x^2)^2 + (2x)^2}$$

$$= \frac{2(1 - x^2)}{x^4 + 6x^2 + 1}$$

8.3.10 Extended form of the chain rule

Sometimes there are more than two component functions involved in a composite function. For example, consider the composite function

$$y = f(w), \quad w = g(z), \quad z = h(x)$$

which may be represented schematically by the block diagram of Figure 8.23. To obtain the derivative $\dfrac{\mathrm{d}y}{\mathrm{d}x}$ we first consider y as a composite function of h and the 'dotted box', giving, on applying the chain rule,

$$\frac{\mathrm{d}y}{\mathrm{d}x} = \frac{\mathrm{d}y}{\mathrm{d}z}\frac{\mathrm{d}z}{\mathrm{d}x}$$

Figure 8.23
Composite function
containing three
component functions.

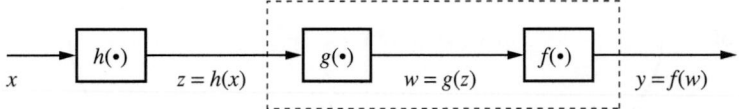

Reapplying the chain rule, this time with z as the domain variable, gives

$$\frac{\mathrm{d}y}{\mathrm{d}z} = \frac{\mathrm{d}y}{\mathrm{d}w}\frac{\mathrm{d}w}{\mathrm{d}z}$$

which on back substitution gives

$$\frac{\mathrm{d}y}{\mathrm{d}x} = \frac{\mathrm{d}y}{\mathrm{d}w}\frac{\mathrm{d}w}{\mathrm{d}z}\frac{\mathrm{d}z}{\mathrm{d}x}$$

as the extended form of the chain rule.

Example 8.18　Find $\dfrac{\mathrm{d}y}{\mathrm{d}x}$ when y is given by

(a) $\sin^2(x^2 + 1)$　　(b) $\cos^{-1}\sqrt{(1 - x^2)}$

Solution (a) Introducing the intermediate variables $z = x^2 + 1$ and $w = \sin z$, then

$$y = w^2, \qquad w = \sin z, \qquad z = x^2 + 1$$

$$\frac{dy}{dw} = 2w, \qquad \frac{dw}{dz} = \cos z, \qquad \frac{dz}{dx} = 2x$$

so by the extended chain rule

$$\frac{dy}{dx} = \frac{dy}{dw} \frac{dw}{dz} \frac{dz}{dx} = (2w)(\cos z)(2x)$$

Since $z = x^2 + 1$ and $w = \sin z = \sin(x^2 + 1)$,

$$\frac{dy}{dx} = 4x \sin(x^2 + 1) \cos(x^2 + 1)$$

(b) Introducing the intermediate variables $z = 1 - x^2$ and $w = \sqrt{z}$, then

$$y = \cos^{-1} w, \qquad w = z^{1/2}, \qquad z = 1 - x^2$$

$$\frac{dy}{dw} = -\frac{1}{\sqrt{(1 - w^2)}}, \qquad \frac{dw}{dz} = \tfrac{1}{2} z^{-1/2}, \qquad \frac{dz}{dx} = -2x$$

Since $z = 1 - x^2$ and $w = \sqrt{(1 - x^2)}$ we have, by the extended chain rule,

$$\frac{dy}{dx} = \frac{dy}{dw} \frac{dw}{dz} \frac{dz}{dx} = \left(-\frac{1}{\sqrt{[1 - (1 - x^2)]}} \right) \left(\frac{1}{2\sqrt{(1 - x^2)}} \right) (-2x) = \frac{1}{\sqrt{(1 - x^2)}}$$

Here we have assumed $0 < x < 1$. If $-1 < x < 0$ the derivative is $-1/\sqrt{(1 - x^2)}$. The function has no derivative at $x = 0$. This is illustrated in Figure 8.24.

Figure 8.24 Graph of $y = \cos^{-1}(1 - x^2)$.

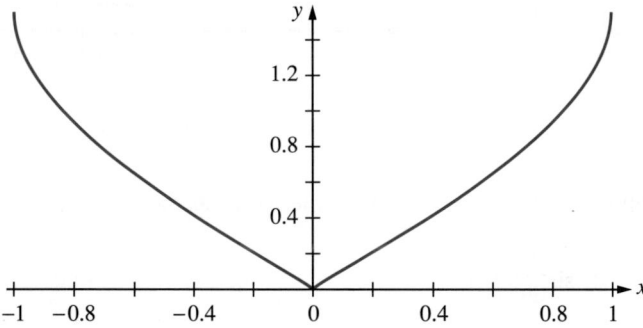

For practice use MATLAB or MAPLE to check the answers to Examples 8.17 and 8.18. As illustrative examples we consider Examples 8.17(c and h) and Example 8.18(a). For Example 8.17(c) the MATLAB commands

```
syms x y
y = sin(2*x)/(x^2 + 2); dy = simplify(diff(y)); pretty(dy)
```

return the derivative as

$$dy = 2 \frac{2\cos(2x) + \cos(2x)x^2 - \sin(2x)x}{(2 + x^2)^2}$$

For Example 8.17(h) the commands

MATLAB	MAPLE

```
syms x y
y = atan(2*x/(1 + x^2));        y:= arctan(2*x/(1 + x^2)):
dy = simplify(diff(y));         dy:= simplify(diff(y,x));
pretty(dy)
```

return the derivative as

$$dy = -2\frac{x^2 - 1}{x^4 + 6x^2 + 1}$$

For Example 8.18(a) the MATLAB commands

```
syms x y
y = (sin(x^2 + 1))^2; dy = diff(y); pretty(dy)
```

return the answer

$$4\sin(x^2 + 1)\cos(x^2 + 1)x$$

8.3.11 Exercises

Check your answers using MATLAB or MAPLE whenever possible.

32 Differentiate with respect to x:

(a) $\sin(3x - 2)$ (b) $\cos^4 x$

(c) $\cos^2 3x$ (d) $\sin 2x \cos 3x$

(e) $x \sin x$ (f) $\sqrt{(2 + \cos 2x)}$

(g) $a \cos(x + \theta)$ (h) $\tan 4x$

33 Differentiate with respect to x:

(a) $\sin^{-1}(x/2)$ (b) $\cos^{-1}(5x)$

(c) $\sqrt{(1 + x^2)}\tan^{-1}x$ (d) $\sin^{-1}((x - 1)/2)$

(e) $\tan^{-1}3x$

(f) $\sqrt{(1 - x^2)}\sin^{-1}x$

34 A cone of semi-vertical angle θ is inscribed in a sphere of radius a. Show that the volume of the cone is

$$V = \tfrac{8}{3}\pi a^3 \sin^2\theta \cos^4\theta$$

Hence prove that the cone of maximum volume that can be inscribed in a sphere of given radius is $\tfrac{8}{27}$th of the volume of the sphere.

8.3.12 Differentiation of exponential and related functions

The formal definition (8.1) gives the derivative of e^x as

$$\frac{d}{dx}(e^x) = \lim_{\Delta x \to 0}\frac{e^{x+\Delta x} - e^x}{\Delta x} = \lim_{\Delta x \to 0}\frac{e^x(e^{\Delta x} - 1)}{\Delta x}$$

$$= e^x \lim_{\Delta x \to 0}\frac{1 + \Delta x + (\Delta x)^2/2! + (\Delta x)^3/3! + \dots - 1}{\Delta x}$$

(using (7.16))

$$= e^x \lim_{\Delta x \to 0}[1 + \tfrac{1}{2}\Delta x + \tfrac{1}{6}(\Delta x)^2 + \dots]$$

so that

$$\frac{d}{dx}(e^x) = e^x \tag{8.22}$$

Thus the exponential function (to base e) has the special property that it is its own derivative. This was described in Section 2.7.1 earlier.

Taking $y = \ln x$, we have $x = e^y$ so that

$$\frac{dx}{dy} = e^y$$

Then, from the inverse-function rule,

$$\frac{dy}{dx} = \frac{1}{e^y} = \frac{1}{x}$$

That is

$$\frac{d}{dx}(\ln x) = \frac{1}{x}, \quad x > 0 \tag{8.23}$$

Example 8.19   Find $\dfrac{dy}{dx}$ when y is given by:

(a) $x^2 e^x$ (b) $3e^{-2x}$ (c) $\dfrac{\ln x}{x^2}$

(d) $\ln(x^2 + 1)$ (e) $e^{-x}(\sin x + \cos x)$

Solution   (a) Taking $u = x^2$ and $v = e^x$

$$\frac{du}{dx} = 2x \quad \text{and} \quad \frac{dv}{dx} = e^x$$

Then by the product rule

$$\frac{dy}{dx} = x^2 e^x + 2x e^x = x(x + 2)e^x$$

(b) Introducing the intermediate variable $z = -2x$ then

$$y = 3e^z \qquad \text{and} \quad z = -2x$$

$$\frac{dy}{dz} = 3e^z \quad \text{and} \quad \frac{dz}{dx} = -2$$

so by the chain rule

$$\frac{dy}{dx} = \frac{dy}{dz}\frac{dz}{dx} = (3e^{-2x})(-2) = -6e^{-2x}$$

(c) Taking $u = \ln x$ and $v = x^2$ gives

$$\frac{du}{dx} = \frac{1}{x} \quad \text{and} \quad \frac{dv}{dx} = 2x \quad (x \neq 0)$$

so by the quotient rule

$$\frac{d}{dx}\left(\frac{\ln x}{x^2}\right) = \frac{(1/x)x^2 - (\ln x)(2x)}{x^4}$$

$$= \frac{1 - 2\ln x}{x^3}$$

(d) Introducing the intermediate variable $z = x^2 + 1$ then

$$y = \ln z \quad \text{and} \quad z = x^2 + 1$$

$$\frac{dy}{dz} = \frac{1}{z} \quad \text{and} \quad \frac{dz}{dx} = 2x$$

so by the chain rule

$$\frac{dy}{dx} = \frac{dy}{dz}\frac{dz}{dx} = \frac{1}{x^2 + 1}(2x) = \frac{2x}{x^2 + 1}$$

(e) Taking $u = e^{-x}$ and $v = \sin x + \cos x$

$$\frac{du}{dx} = -e^{-x} \quad \text{and} \quad \frac{dv}{dx} = \cos x - \sin x$$

Then by the product rule

$$\frac{dy}{dx} = e^{-x}(\cos x - \sin x) + (\sin x + \cos x)(-e^{-x})$$

$$= -2e^{-x}\sin x$$

The hyperbolic functions, introduced in Section 2.7.4, are closely related to the exponential function and their derivatives are readily deduced. From their definitions

$$\frac{d}{dx}(\sinh x) = \frac{d}{dx}\left[\frac{e^x - e^{-x}}{2}\right] = \tfrac{1}{2}(e^x + e^{-x}) = \cosh x \qquad \text{(8.24a)}$$

$$\frac{d}{dx}(\cosh x) = \frac{d}{dx}\left[\frac{e^x + e^{-x}}{2}\right] = \tfrac{1}{2}(e^x - e^{-x}) = \sinh x \qquad \text{(8.24b)}$$

$$\frac{d}{dx}(\tanh x) = \frac{d}{dx}\left[\frac{\sinh x}{\cosh x}\right] = \frac{(\cosh x)(\cosh x) - (\sinh x)(\sinh x)}{\cosh^2 x}$$

$$= \frac{1}{\cosh^2 x} = \text{sech}^2 x \qquad \text{(8.24c)}$$

$$\frac{d}{dx}(\text{sech } x) = \frac{d}{dx}\left[\frac{1}{\cosh x}\right] = \frac{-\sinh x}{\cosh^2 x} = -\text{sech } x \tanh x \qquad \text{(8.24d)}$$

$$\frac{d}{dx}(\text{cosech } x) = \frac{d}{dx}\left[\frac{1}{\sinh x}\right] = -\text{cosech } x \coth x \qquad \text{(8.24e)}$$

$$\frac{d}{dx}(\coth x) = \frac{d}{dx}\left[\frac{\cosh x}{\sinh x}\right] = -\text{cosech}^2 x \qquad \text{(8.24f)}$$

Following the same procedure as for the inverse circular functions in Section 8.3.9 the following derivatives of the inverse hyperbolic functions are readily obtained.

$$\frac{d}{dx}(\sinh^{-1} x) = \frac{1}{\sqrt{(1 + x^2)}} \qquad \text{(8.25a)}$$

$$\frac{d}{dx}(\cosh^{-1} x) = \frac{1}{\sqrt{(x^2 - 1)}}, \quad x > 1 \qquad \text{(8.25b)}$$

$$\frac{d}{dx}(\tanh^{-1} x) = \frac{1}{1 - x^2}, \quad |x| < 1 \qquad \text{(8.25c)}$$

Example 8.20   Find $\dfrac{dy}{dx}$ when y is given by

(a) $\tanh 2x$ (b) $\cosh^2 x$ (c) $e^{-3x} \sinh 3x$ (d) $\sinh^{-1}\left[\dfrac{3x}{4}\right]$

Solution   (a) Introducing the intermediate variable $z = 2x$ gives

$$y = \tanh z \qquad \text{and} \quad z = 2x$$

$$\frac{dy}{dz} = \text{sech}^2 z \quad \text{and} \quad \frac{dz}{dx} = 2$$

so by the chain rule

$$\frac{dy}{dx} = 2\,\text{sech}^2(2x)$$

(b) Introducing the intermediate variable $z = \cosh x$ gives

$$y = z^2 \qquad \text{and} \quad z = \cosh x$$

$$\frac{dy}{dz} = 2z \quad \text{and} \quad \frac{dz}{dx} = \sinh x$$

so by the chain rule

$$\frac{dy}{dx} = 2 \cosh x \sinh x = \sinh 2x$$

(c) Taking $u = e^{-3x}$ and $v = \sinh 3x$

gives using the chain rule

$$\frac{du}{dx} = -3e^{-3x} \quad \text{and} \quad \frac{dv}{dx} = 3\cosh 3x$$

so by the product rule

$$\frac{dy}{dx} = (e^{-3x})(3\cosh 3x) + (\sinh 3x)(-3e^{-3x}) = 3e^{-3x}(\cosh 3x - \sinh 3x)$$

$$= 3e^{-3x}(e^{-3x}) = 3e^{-6x}$$

(d) Introducing the intermediate variable $z = \frac{3}{4}x$ gives

$$y = \sinh^{-1} z \quad \text{and} \quad z = \frac{3}{4}x$$

$$\frac{dy}{dz} = \frac{1}{\sqrt{(1 + z^2)}} \quad \text{and} \quad \frac{dz}{dx} = \frac{3}{4}$$

so by the chain rule

$$\frac{dy}{dx} = \frac{3}{4} \cdot \frac{1}{\sqrt{(1 + \frac{9}{16}x^2)}} = \frac{3}{\sqrt{(16 + 9x^2)}}$$

For practice, use MATLAB or MAPLE to check the answers to Examples 8.19 and 8.20. As illustrative examples we consider Examples 8.19(a) and 8.20(c). For Example 8.19(a) the MATLAB commands

```
syms x y
y = (x^2)*exp(x); dy = simplify(diff(y)); pretty(dy)
```

return the derivative as

```
dy = xexp(x)(2 + x)
```

For Example 8.20(c) the commands

MATLAB

```
syms x y
y = exp(-3*x)*sinh(3*x);
dy = simplify(diff(y));
pretty(df)
```

MAPLE

```
y:= exp(-3*x)*sinh(3*x):
dy:= simplify(diff(y,x));
```

return the derivative as

```
dy = -3exp(-3x)sinh3x +
3 exp(3x)cosh3x
```

```
dy:= -3e^(-3x)(sinh(3x) -
cosh(3x))
```

In MAPLE the answer may be converted to exponential form and then simplified using the command $simplify(convert(\%,exp))$; to give the answer $3e^{(-6x)}$.

8.3.13 Exercises

Check your answers using MATLAB or MAPLE whenever possible.

35 Differentiate with respect to x:

(a) e^{2x} (b) $e^{-x/2}$

(c) $\exp(x^2 + x)$ (d) $x^2 e^{5x}$

(e) $(3x + 2)e^{-x}$ (f) $e^x/(1 + e^x)$

(g) $\sqrt{(1 + e^x)}$ (h) e^{ax+b}

36 Differentiate with respect to x:

(a) $\ln(2x + 3)$ (b) $\ln(x^2 + 2x + 3)$

(c) $\ln[(x - 2)/(x - 3)]$ (d) $\dfrac{1}{x}\ln x$

(e) $\ln[(2x + 1)/(1 - 3x)]$ (f) $\ln[(x + 1)x]$

37 Differentiate with respect to x:

(a) $\sinh 3x$ (b) $\tanh 4x$ (c) $x^3 \cosh 2x$

(d) $\ln[\cosh \frac{1}{2}x]$ (e) $\cos x \cosh x$ (f) $1/\cosh x$

38 Differentiate with respect to x:

(a) $\sinh^{-1} 2x$ (b) $\cosh^{-1}(2x^2 - 1)$

(c) $\tanh^{-1}(1/x)$ (d) $\sqrt{(1 + x^2)} \sinh^{-1} x$

(e) $\sqrt{(4 - x^2)} - 2\cosh^{-1}(2/x)$

(f) $\tanh^{-1} x/(1 + x^2)$

39 Draw a careful sketch of $y = e^{-ax} \sin \omega x$ where a and ω are positive constants. What is the ratio of the heights of successive maxima of the function?

40 The line AB joins the points $A(a, 0)$, $B(0, b)$ on the x and y axes respectively and passes through the point $(8, 27)$. Find the positions of A and B which minimize the length of AB.

41 Sketch the curve $y = e^{-x^2}$. Find the rectangle inscribed under the curve having one edge on the x axis, which has maximum area.

42 Show that $y = 9e^{-9t}/(10 - e^{-9t})$ satisfies the differential equation

$$\frac{dy}{dt} = -y(9 + y)$$

43 A sky diver's downward velocity $v(t)$ is given by

$$v(t) = u(1 - e^{-\alpha t})/(1 + e^{-\alpha t})$$

Where u and α are constants. What is the terminal velocity achieved? When does the sky diver achieve half that velocity and what is the acceleration then?

8.3.14 Parametric and implicit differentiation

The chain rule is used with the inverse-function rule to evaluate derivatives when a function is specified **parametrically**.

In general, if a function is defined by $y = f(x)$, where $x = g(t)$ and $y = h(t)$ and t is a parameter, then

$$\frac{dy}{dx} = \frac{dy}{dt} \bigg/ \frac{dx}{dt} \quad \text{or} \quad \frac{dy}{dx} = \frac{dy}{dt}\frac{dt}{dx} \tag{8.26}$$

Example 8.21   The function $y = f(x)$ is defined by $x = t^3$, $y = t^2$ $(t \in \mathbb{R})$. Find dy/dx.

Solution   The graph of $f(x)$ is shown in Figure 8.25. There are many ways in which dy/dx may be evaluated. The simplest uses the result (8.26). In this case

Figure 8.25
The graph of
$\{(x, y): y = t^2, x = t^3,$
$-\infty < t < \infty\}.$

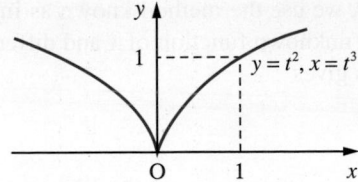

$$\frac{dy}{dt} = 2t \quad \text{and} \quad \frac{dx}{dt} = 3t^2$$

so that

$$\frac{dy}{dx} = \frac{dy}{dt} \bigg/ \frac{dx}{dt} = \frac{2}{3t} \quad (t \neq 0)$$

This gives the result in terms of t. In terms of x, it may be written as

$$\frac{dy}{dx} = \tfrac{2}{3}x^{-1/3} \quad (x \neq 0)$$

In terms of x and y, we have

$$\frac{dy}{dx} = \frac{2y}{3x} \quad (x \neq 0)$$

Note from Figure 8.25 that the graph does not have a well-defined tangent at $x = 0$, so the derivative does not exist at this point; that is, the function is not differentiable at $x = 0$.

We can also obtain these results directly. Eliminating t between the defining equations for x and y, we have

$$y = x^{2/3}$$

Differentiating with respect to x gives

$$\frac{dy}{dx} = \tfrac{2}{3}x^{-1/3}$$

Parametric differentiation is achieved using the MATLAB commands (or comparable MAPLE commands)

```
syms x y t
x = x(t); y = y(t); dx = diff(x,t); dy = diff(y,t);
dydx = dy/dx
```

The chain rule may also be used to differentiate functions expressed in an implicit form. For example, the function of Example 8.21 may be expressed implicitly, by eliminating t, as

$$y^3 = x^2$$

To obtain the derivative dy/dx, we use the method known as **implicit differentiation**. In this method we treat y as an unknown function of x and differentiate both sides term by term with respect to x. This gives

$$\frac{d}{dx}(y^3) = \frac{d}{dx}(x^2)$$

Now y^3 is a composite function of x, with y being the intermediate variable, so the chain rule gives

$$\frac{d}{dx}(y^3) = \frac{d}{dy}(y^3)\frac{dy}{dx} = 3y^2\frac{dy}{dx}$$

Then, substituting back, we have

$$3y^2\frac{dy}{dx} = 2x$$

giving

$$\frac{dy}{dx} = \frac{2x}{3y^2} = \frac{2y}{3x} \quad \text{(on substituting for } y^3\text{)}$$

Example 8.22 Find $\dfrac{dy}{dx}$ when $x^2 + y^2 + xy = 1$.

Solution Differentiating both sides, term by term, gives

$$\frac{d}{dx}(x^2) + \frac{d}{dx}(y^2) + \frac{d}{dx}(xy) = \frac{d}{dx}(1)$$

Recognizing that y is a function of x and taking care over the product term xy, the chain rule gives

$$2x + \frac{d}{dy}(y^2)\frac{dy}{dx} + x\frac{dy}{dx} + y = 0$$

$$2x + 2y\frac{dy}{dx} + x\frac{dy}{dx} + y = 0$$

leading to

$$\frac{dy}{dx} = -\frac{(2x + y)}{(x + 2y)}$$

Implicit differentiation is useful in calculating the slopes of tangents and normals to curves specified implicitly, such as in Example 8.22. Having obtained the slope of the tangent at the point (x, y) as $\dfrac{dy}{dx}$, the slope of the normal to the curve at the corresponding point is $-1/$(slope of tangent) as inferred from Figure 8.5.

Example 8.23 Find the equations of the tangent and normal to the curve having equation $x^2 + y^2 - 3xy + 4 = 0$ at the point (2, 4).

Solution Differentiating the equation implicitly with respect to x

$$2x + 2y\frac{dy}{dx} - 3x\frac{dy}{dx} - 3y = 0$$

gives

$$\frac{dy}{dx} = \frac{3y - 2x}{2y - 3x}$$

This represents the slope of the tangent at the point (x, y) on the curve. Thus the slope of the tangent at the point (2, 4) is

$$\left[\frac{dy}{dx}\right]_{(2,4)} = \frac{12 - 4}{8 - 6} = 4$$

Remembering from equation (1.14) that the equation of a line passing through a point (x, y) and having slope m is $y - y_1 = m(x - x_1)$ we have that the equation of the tangent to the graph at (2, 4) is

$$(y - 4) = 4(x - 2) \quad \text{or} \quad y = 4x - 4$$

The slope of the normal at (2, 4) is $-\frac{1}{4}$ so it has equation

$$y - 4 = -\frac{1}{4}(x - 2) \quad \text{or} \quad 4y = 18 - x$$

Example 8.24 Find the slope of the tangents to the circle

$$x^2 + y^2 - 2x + 4y - 20 = 0$$

at the points A(1, 3), B(4, 2) and C(−2, −6).

Solution The circle defined by the equation is shown in Figure 8.26, together with the three points A, B and C. Clearly this equation does not define a function in general, but near specific points we can restrict it so that it behaves locally like a function. To compute the slopes of the tangents, we differentiate the equation defining the curve with respect

Figure 8.26
Graph of the circle $x^2 + y^2 - 2x + 4y - 20 = 0$.

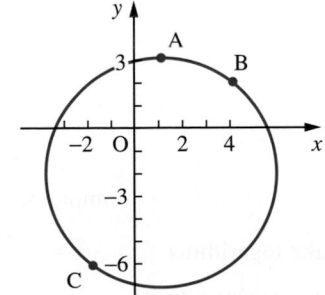

to x implicitly, and then we insert the x and y coordinates of the points. Thus in this example we have

$$2x + 2y\frac{dy}{dx} - 2 + 4\frac{dy}{dx} = 0$$

giving

$$\frac{dy}{dx} = \frac{1 - x}{2 + y} \quad (y \neq -2)$$

Then at A the slope is zero, at B the slope is $-\frac{3}{4}$, and at C the slope is $-\frac{3}{4}$. Note that dy/dx is not defined at $y = -2$. There are two corresponding points: $(-4, -2)$ and $(6, -2)$. At these points the curve has a vertical tangent.

The implicit differentiation rule can be used in a double way to obtain derivatives of functions of the form $f(x)^{g(x)}$ as illustrated in Example 8.25.

Example 8.25 Find the derivative of the function

$$f(x) = (\sin x)^x \quad (x \in (0, \pi))$$

Solution The simplest way of dealing with this is first to take logarithms. Thus $y = (\sin x)^x$ gives

$$\ln y = x \ln \sin x$$

Then differentiating implicitly with respect to x, remembering that y is a function of x, gives

$$\frac{1}{y}\frac{dy}{dx} = \ln \sin x + x\frac{\cos x}{\sin x} = \ln \sin x + x \cot x$$

and so

$$\frac{dy}{dx} = (\ln \sin x + x \cot x)(\sin x)^x$$

Sometimes the technique used in Example 8.25 is described as **logarithmic differentiation**. It is useful for differentiating complicated functions.

Example 8.26 Differentiate with respect to x

$$y = \frac{(x - 2)^3(x + 3)^9}{\sqrt{(x^2 + 1)}}$$

Solution To simplify the process, we first take logarithms

$$\ln y = 3 \ln(x - 2) + 9 \ln(x + 3) - \tfrac{1}{2}\ln(x^2 + 1)$$

Then differentiating with respect to x gives

$$\frac{1}{y}\frac{dy}{dx} = \frac{3}{x-2} + \frac{9}{x+3} - \frac{1}{2}\frac{2x}{x^2+1}$$

$$= \frac{3(x+3)(x^2+1) + 9(x-2)(x^2+1) - x(x-2)(x+3)}{(x-2)(x+3)(x^2+1)}$$

$$= \frac{11x^3 - 10x^2 + 18x - 9}{(x-2)(x+3)(x^2+1)}$$

Hence

$$\frac{dy}{dx} = (11x^3 - 10x^2 + 18x - 9)(x-2)^2(x+3)^8/(x^2+1)^{3/2}$$

8.3.15 Exercises

Check your answers using MATLAB or MAPLE whenever possible.

44 The equations $x = t \sin t$, $y = t \cos t$ are the parametric equations for a spiral. Find $\dfrac{dy}{dx}$ in terms of t.

45 A curve is defined parametrically by the equations

$$x = 2\cos\theta + \cos 2\theta$$

$$y = 2\sin\theta - \sin 2\theta$$

Draw a sketch of the curve for $0 \leqslant \theta \leqslant 2\pi$. Find the equation of the tangent to the curve at the point where $\theta = \pi/4$.

46 Find the equation of the tangent, at the point $(0, 4)$, to the curve defined by

$$y^3 x + y + 7x^4 = 4$$

47 Find the value of $\dfrac{dy}{dx}$ at the point $(1, -1)$ on the curve given by the equation

$$x^3 - y^3 - xy - x = 0$$

48 Differentiate with respect to x:

(a) 10^x (b) 2^{-x} (c) $\dfrac{(x-1)^{7/2}(x+1)^{1/2}}{x^2+2}$

49 Use logarithmic differentiation to prove that

$$\frac{d}{dx}(y_1 y_2 \ldots y_n)$$

$$= \sum_{k=1}^{n} (y_1 y_2 \ldots y_{k-1} y_{k+1} \ldots y_n) y_k'$$

Hence differentiate $x^3 e^{-2x} \sin \pi x$.

50 The equation of a curve is

$$xy^3 - 2x^2 y^2 + x^4 - 1 = 0$$

Show that the tangent to the curve at the point $(1, 2)$ has a slope of unity. Hence write down the equation of the tangent to the curve at this point. What are the coordinates of the points at which this tangent crosses the coordinate axes?

51 A **cycloid** is a curve traced out by a point p on the rim of a wheel as it rolls along the ground. Using the coordinate system shown in Figure 8.27, show that the curve has the parametric representation

$$x = a(\theta - \sin\theta), \quad y = a(1 - \cos\theta)$$

where θ is the angle through which the wheel has turned.

Draw a sketch of the curve.

Find the gradient of the curve at a general point (x, y).

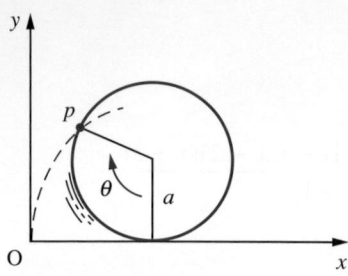

Figure 8.27

If the wheel rotates at a constant speed, with $\theta = \omega t$, where ω is constant and t is the time, show that the speed V of the point on the rim is given by

$$V(t) = 2a\omega |\sin \tfrac{1}{2}\omega t|$$

52 Find the slope of the tangent to the lemniscate

$$(x^2 + y^2)^2 = a^2(x^2 - y^2)$$

at the point (x, y). (See Review exercises 2.11, Question 19.)

53 Use logarithmic differentiation to differentiate

(a) $(\ln x)^x$ (b) $x^{\ln x}$

(c) $(1 - x^2)^{1/2}(2x^2 + 3)^{-4/3}$

54 Using logarithmic differentiation, find the derivatives of

(a) $x^3 e^{-2x} \ln x$ (b) $\dfrac{1}{x}e^x \sin 2x$

8.4   Higher derivatives

The derivative df/dx of function $f(x)$ is itself a function and may be differentiable. The derivative of a derivative is called the **second derivative**, and is written as

$$\frac{d^2 f}{dx^2} \quad \text{or} \quad f''(x) \quad \text{or} \quad f^{(2)}(x)$$

This may in turn be differentiated, yielding **third derivatives** and so on. In general, the **nth derivative** is written as

$$\frac{d^n f}{dx^n} \quad \text{or} \quad f^{(n)}(x)$$

8.4.1   The second derivative

In mechanics the second derivative of the displacement of an object with respect to time is its acceleration and this is used in the mathematical modelling of problems in mechanics using the law:

$$\text{mass} \times \text{acceleration} = \text{applied force}$$

Example 8.27   Find the second derivative of the functions given by

(a) $y = 3x^4 - 2x^2 + x - 1$ (b) $y = x/(x^2 + 1)$

(c) $y = e^{-x} \sin 2x$ (d) $y = \dfrac{\ln x}{x}$

Solution   (a) Differentiating once gives

$$\frac{dy}{dx} = 12x^3 - 4x + 1$$

and differentiating a second time gives

$$\frac{d^2y}{dx^2} = 36x^2 - 4$$

(b) This simply requires two differentiations, as above,

$$\frac{dy}{dx} = \frac{1(x^2 + 1) - 2x(x)}{(x^2 + 1)^2} = \frac{1 - x^2}{(x^2 + 1)^2}$$

Then $\dfrac{d^2y}{d^2x} = \dfrac{(-2x)(x^2 + 1)^2 - 2(2x)(x^2 + 1)(1 - x^2)}{(x^2 + 1)^4}$

$$= \frac{(-2x)(x^2 + 1) - 4x(1 - x^2)}{(x^2 + 1)^3} = \frac{2x(x^2 - 3)}{(x^2 + 1)^3}$$

(c) This simply requires two differentiations. Applying the product rule, we have

$$\frac{dy}{dx} = (e^{-x})(2\cos 2x) + (\sin 2x)(-e^{-x}) = e^{-x}(2\cos 2x - \sin 2x)$$

Applying the rule again we have

$$\frac{d^2y}{dx^2} = (e^{-x})(-4\sin 2x - 2\cos 2x) + (2\cos 2x - \sin 2x)(-e^{-x})$$

$$= -e^{-x}(3\sin 2x + 4\cos 2x)$$

(d) Again this simply requires two differentiations. Applying the quotient rule, we have

$$\frac{dy}{dx} = \frac{(1/x)x - \ln x}{x^2} = \frac{1 - \ln x}{x^2}$$

Applying the rule again, we obtain

$$\frac{d^2y}{dx^2} = \frac{-(1/x)x^2 - (1 - \ln x)(2x)}{x^4} = \frac{2\ln x - 3}{x^3} \qquad (x \neq 0)$$

The second derivative is obtained using the commands

MATLAB	MAPLE
`d2y = diff(y,2)`	`d2y:= diff(y,x,x);`

and the third derivative by the commands

`d3y = diff(y,3)`	`d3y:= diff(y,x,x,x);` or
	`d3y:= diff(y,x$3);`

and so on for higher derivatives.

Considering Example 8.27(c) the commands

MATLAB

```
syms x y
y = exp(-x)*sin(2*x);
d2y = simplify(diff(y,2));
pretty(d2y)
```

MAPLE

```
y:= exp(-x)*sin(2*x):
d2y:= diff(y,x,x);
```

return the second derivative as

```
d2y = -exp(-x)(3sin(2x) +
4cos(2x))
```

```
d2y:= -3e^(-x) sin(2x) -
4 e^(-x) cos(2x)
```

Example 8.28

Show that

$$y = e^{-t}(A \cos t + B \sin t) + 2 \sin 2t - \cos 2t$$

satisfies the equation

$$\frac{d^2y}{d^2t} + 2\frac{dy}{dt} + 2y = 10 \cos 2t$$

Solution

Differentiating y twice with respect to t gives

$$\frac{dy}{dt} = e^{-t}[(A-B)\cos t + (A+B)\sin t] + 4\cos 2t + 2\sin 2t$$

$$\frac{d^2y}{d^2t} = e^{-t}[-2B\cos t + 2A\sin t] - 8\sin 2t + 4\cos 2t$$

Thus

$$\frac{d^2y}{d^2t} + 2\frac{dy}{dt} + 2y = 10\cos 2t$$

When determining the second derivative using parametric or implicit differentiation care must be taken to ensure correct use of the chain rule. The approach is illustrated in Example 8.29.

Example 8.29

Find $\frac{d^2y}{dx^2}$ when y is given by

(a) $y = t^2, x = t^3$ (b) $x^2 + y^2 - 2x + 4y - 20 = 0$

Solution

(a) Here $y = t^2$ and $x = t^3$ gives, as in Example 8.21,

$$\frac{dy}{dx} = \frac{2}{3}\frac{1}{t} \quad (t \neq 0)$$

Differentiating again, using the chain rule, gives

$$\frac{d^2y}{dx^2} = \frac{d}{dx}\left(\frac{dy}{dx}\right) = \frac{d}{dt}\left(\frac{dy}{dx}\right)\frac{dt}{dx} \quad \text{(this is an important step)}$$

$$= \frac{d}{dt}\left(\frac{dy}{dx}\right)\Big/\frac{dx}{dt}$$

$$= \frac{\frac{2}{3}(-1/t^2)}{3t^2} = -\frac{2}{9}\frac{1}{t^4}$$

(b) Here x and y are related by the equation

$$x^2 + y^2 - 2x + 4y - 20 = 0$$

so that as in Example 8.24

$$2x + 2y\frac{dy}{dx} - 2 + 4\frac{dy}{dx} = 0$$

and

$$(y + 2)\frac{dy}{dx} + x - 1 = 0$$

Differentiating a second time gives

$$\left(\frac{dy}{dx}\right)\frac{dy}{dx} + (y + 2)\frac{d^2y}{dx^2} + 1 = 0$$

using the product rule and remembering that

$$\frac{d}{dx}\left(\frac{dy}{dx}\right) = \frac{d^2y}{dx^2}$$

After rearrangement, we have

$$\frac{d^2y}{dx^2} = -\frac{1 + (dy/dx)^2}{y + 2} \quad (y \neq -2)$$

and substituting

$$\frac{dy}{dx} = \frac{1 - x}{2 + y}$$

into the right-hand side gives eventually

$$\frac{d^2y}{dx^2} = -\frac{x^2 + y^2 - 2x + 4y + 5}{(2 + y)^3}$$

This may be further simplified, using the original equation, to give

$$\frac{d^2y}{dx^2} = -\frac{25}{(2 + y)^3} \quad (y \neq -2)$$

Further results for higher derivatives are developed in Exercises 8.4. These are on the whole straightforward extensions of previous work. One result that sometimes causes blunders is the extension of the inverse-function rule to higher derivatives.

We know that

$$\frac{dx}{dy} = 1 \bigg/ \frac{dy}{dx}$$

To find the second derivative of x with respect to y needs a little care:

$$\frac{d^2x}{dy^2} = \frac{d}{dy}\left(\frac{dx}{dy}\right) = \frac{d}{dy}\left[\left(\frac{dy}{dx}\right)^{-1}\right]$$

$$= \frac{d}{dx}\left[\left(\frac{dy}{dx}\right)^{-1}\right]\frac{dx}{dy} \quad \text{(using the chain rule)}$$

$$= \left[-\frac{d}{dx}\left(\frac{dy}{dx}\right)\bigg/\left(\frac{dy}{dx}\right)^2\right]\left(1\bigg/\frac{dy}{dx}\right)$$

Thus

$$\frac{d^2x}{dy^2} = -\frac{d^2y}{dx^2}\bigg/\left(\frac{dy}{dx}\right)^3$$

8.4.2   Exercises

Check your answers using MATLAB or MAPLE.

55   Find $\dfrac{d^2y}{dx^2}$ when y is given by:

(a) $x^3\sqrt{(1+x^2)}$

(b) $\ln(x^2 + x + 1)$

(c) $y^3x + y + 7x^4 = 4$

(d) $x^3 - y^3 - xy - x = 0$

56   Find $\dfrac{d^2y}{dx^2}$ when x and y are given by:

(a) $x = t\sin t$ and $y = t\cos t$

(b) $x = 2\cos t + \cos 2t$ and $y = 2\sin t - \sin 2t$

57   If $y = 3e^{2x}\cos(2x - 3)$, verify that

$$\frac{d^2y}{dx^2} - 4\frac{dy}{dx} + 8y = 0$$

58   If $y = (\sin^{-1}x)^2$, prove that

$$(1 - x^2)\left(\frac{dy}{dx}\right)^2 = 4y$$

and deduce that

$$(1 - x^2)\frac{d^2y}{dx^2} - x\frac{dy}{dx} - 2 = 0$$

59   (a) If $y = x^2 + 1/x^2$, find dy/dx and d^2y/dx^2. Hence show that

$$x^2\frac{d^2y}{dx^2} + 4x\frac{dy}{dx} + 2y = 12x^2$$

(b) If $x = \tan t$ and $y = \cot t$, show that

$$\frac{d^2y}{dx^2} + 2y\frac{dy}{dx} = 0$$

60   If $x = a(\theta - \sin\theta)$ and $y = a(1 - \cos\theta)$, find dy/dx and d^2y/dx^2.

61 Find dy/dx in terms of t for the curve with parametric representation

$$x = \frac{1-t}{1+2t} \quad y = \frac{1-2t}{1+t}$$

Show that

$$\frac{d^2y}{dx^2} = -\frac{2}{3}\left(\frac{1+2t}{1+t}\right)^3$$

and find a similar expression for d^2x/dy^2.

62 Confirm that the point $(1, 1)$ lies on the curve with equation $x^3 - y^2 + xy - x^2 = 0$ and find the values of dy/dx and d^2y/dx^2 at that point.

63 Find $f^{(4)}(x)$ and $f^{(n)}(x)$ for the following functions $f(x)$:

(a) e^{3x} (b) $\ln(x + 2)$

(c) $\dfrac{1}{1 - x^2}$

64 Find the fourth derivative of $f(x) = \sin(ax + b)$ and verify that $f^{(n)}(x) = a^n \sin(ax + b + \frac{1}{2}n\pi)$.

65 Prove that

$$\frac{d^n}{dx^n}(e^{ax}\sin bx) = (a^2 + b^2)^{n/2}\,e^{ax}\sin(bx + n\theta)$$

where $\cos\theta = a/\sqrt{(a^2 + b^2)}$, $\sin\theta = b/\sqrt{(a^2 + b^2)}$.

66 If $y = u(x)v(x)$, prove that

(a) $y^{(2)}(x) = u^{(2)}(x)v(x) + 2u^{(1)}(x)v^{(1)}(x) + u(x)v^{(2)}(x)$

(b) $y^{(3)}(x) = u^{(3)}(x)v(x) + 3u^{(2)}(x)v^{(1)}(x)$

$$+ 3u^{(1)}(x)v^{(2)}(x) + u(x)v^{(3)}(x)$$

Hence prove **Leibniz' theorem** for the nth derivative of a product:

$$y^{(n)}(x) = u^{(n)}(x)v(x) + \binom{n}{1}u^{(n-1)}(x)v^{(1)}(x)$$

$$+ \binom{n}{2}u^{(n-2)}(x)v^{(2)}(x) + \dots + u(x)v^{(n)}(x)$$

67 Use Leibniz' theorem (Question 66) to find the following:

(a) $\dfrac{d^5}{dx^5}(x^2\sin x)$ (put $u = \sin x$, $v = x^2$)

(b) $\dfrac{d^4}{dx^4}(xe^{-x})$ (c) $\dfrac{d^3}{dx^3}[x^2(3x + 1)^{12}]$

8.4.3 Curvature of plane curves

The second derivative d^2f/dx^2 represents the rate of change of df/dx as x increases; geometrically, this gives us information as to how the slope of the tangent to the graph of $y = f(x)$ is changing with increasing x.

- If $d^2f/dx^2 > 0$ then df/dx is increasing as x increases, and the tangent rotates in an anticlockwise direction as we move along the horizontal axis, as illustrated in Figure 8.28(a).
- If $d^2f/dx^2 < 0$ then df/dx is decreasing as x increases, and the tangent rotates in a clockwise direction as we move along the horizontal axis, as illustrated in Figure 8.28(b).

Also note that when $d^2f/dx^2 > 0$, the graph of $y = f(x)$ is 'concave up', and when $d^2f/dx^2 < 0$ the graph is 'concave down'. Thus the sign of d^2f/dx^2 relates to the concavity of the graph; we shall use this information in section 8.5.1 to define a point of inflection.

The **curvature** κ of a plane curve, having equation $y = f(x)$, at any point is the rate at which the curve is bending or curving away from the tangent at that point. In other words, the curvature measures the rate at which the tangent to the curve changes as it moves along the curve. This implies that it will depend on d^2f/dx^2 in some way.

Take two points P and Q on the curve $y = f(x)$ and a distance Δs apart *measured along the curve*. Then, with the notation of Figure 8.29(a), the average curvature of the

Figure 8.28
Rates of change of
$\dfrac{dy}{dx}$ as x increases.

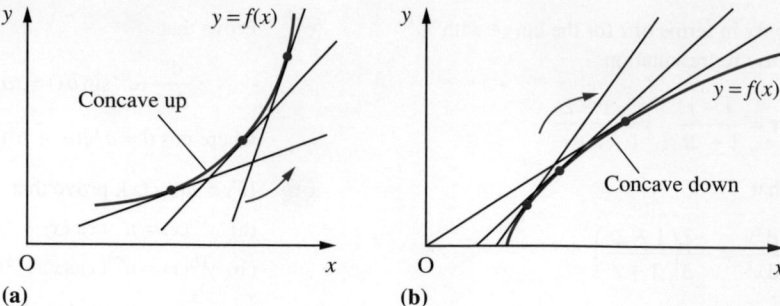

Figure 8.29
Curvature and
radius of curvature.

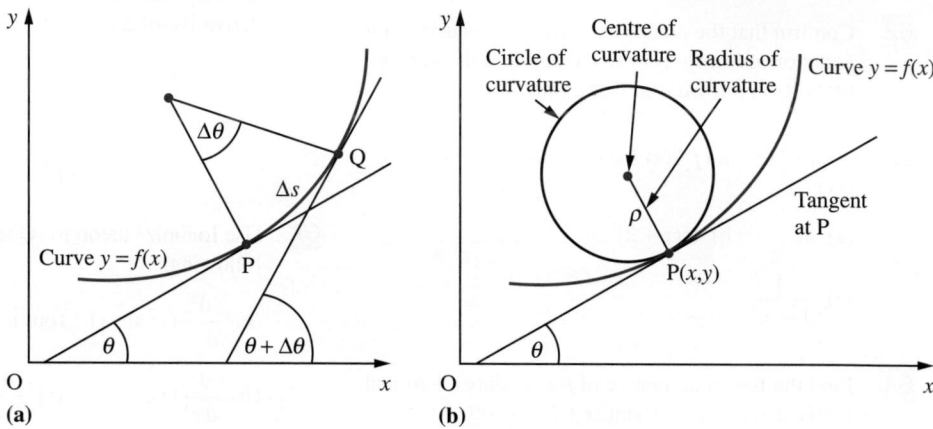

curve PQ is $\Delta\theta/\Delta s$. We then define the curvature κ of the curve at the point P to be the absolute value of the average curvature as Q approaches P. That is,

$$\kappa = \left| \lim_{\Delta s \to 0} \frac{\Delta\theta}{\Delta s} \right| = \left| \frac{d\theta}{ds} \right| \tag{8.27}$$

If we now construct a circle, as shown in Figure 8.29(b), so that it

- has the same tangent at P as $y = f(x)$,
- lies on the same side of the tangent as $y = f(x)$ and
- has the same curvature κ as $y = f(x)$ at P

then this is called the **circle of curvature** at P. Its radius ρ is called the **radius of curvature** at P and is given by

$$\rho = \text{radius of curvature} = \frac{1}{\kappa}$$

The centre of the circle is called the **centre of curvature** at P. Clearly the curvature is zero when the radius of curvature is infinite.

In order to obtain the curvature of a curve given by an equation of the form $y = f(x)$, we must obtain a more usable formula than (8.27). Since

$$\tan\theta = \text{slope of the tangent at P} = \frac{dy}{dx}$$

differentiating with respect to s, using the chain rule, gives

$$\sec^2\theta \frac{d\theta}{ds} = \frac{d^2y}{dx^2}\frac{dx}{ds}$$

so that

$$\frac{d\theta}{ds} = \frac{d^2y/dx^2}{[1+(dy/dx)^2]}\frac{dx}{ds} \tag{8.28}$$

using the trigonometric identity $1 + \tan^2\theta = \sec^2\theta$.

We shall see in Section 8.9.6 that

$$\frac{ds}{dx} = \sqrt{\left[1+\left(\frac{dy}{dx}\right)^2\right]}$$

so, from the inverse-function rule,

$$\frac{dx}{ds} = \frac{1}{ds/dx} = \frac{1}{\sqrt{[1+(dy/dx)^2]}}$$

which on substituting into (8.28) gives the formula

$$\kappa = \left|\frac{d\theta}{ds}\right| = \frac{|d^2y/dx^2|}{[1+(dy/dx)^2]^{3/2}} \tag{8.29}$$

If we denote the coordinates of the centre of curvature by (X, Y) then it follows from Figure 8.29(b) that

$$X = x - \rho\sin\theta, \quad Y = y + \rho\cos\theta$$

Since $\tan\theta = dy/dx$, it follows that

$$\sin\theta = \frac{dy/dx}{\sqrt{[1+(dy/dx)^2]}}, \quad \cos\theta = \frac{1}{\sqrt{[1+(dy/dx)^2]}}$$

Using these results together with $\rho = 1/\kappa$, with κ from (8.29), gives the coordinates of the centre of curvature as

$$X = x - \frac{dy}{dx}\left[1+\left(\frac{dy}{dx}\right)^2\right]\bigg/\frac{d^2y}{dx^2}, \quad Y = y + \left[1+\left(\frac{dy}{dx}\right)^2\right]\bigg/\frac{d^2y}{dx^2} \tag{8.30}$$

Although these results have been deduced for the curve of Figure 8.29(b), which at the point P(x, y) has positive slope $(dy/dx > 0)$ and is concave upwards $(d^2y/dx^2 > 0)$, it can be shown that these are valid in all cases.

It is left as an exercise for the reader to show that if the curve $y = f(x)$ is given in parametric form

$$x = g(t), \quad y = h(t)$$

then the curvature κ is given by

$$\kappa = \left|\frac{dg}{dt}\frac{d^2h}{dt^2} - \frac{d^2g}{dt^2}\frac{dh}{dt}\right|\bigg/\left[\left(\frac{dg}{dt}\right)^2 + \left(\frac{dh}{dt}\right)^2\right]^{3/2} \tag{8.31}$$

8.4.4 Exercises

68 Find the radius of curvature at the point $(2, 8)$ on the curve $y = x^3$.

69 Show that the radius of curvature at the origin to the curve

$$x^3 + y^3 + 2x^2 - 4y + 3x = 0$$

is $\frac{125}{64}$.

70 Find the radius of curvature and the coordinates of the centre of curvature of the curve

$$y = (11 - 4x)/(3 - x)$$

at the point $(2, 3)$.

71 Find the radius of curvature at the point where $\theta = \frac{1}{3}\pi$ on the curve defined parametrically by

$$x = 2\cos\theta, \quad y = \sin\theta$$

8.5 Applications to optimization problems

In many industrial situations the role of management is to make decisions that will lead to the most effective use of the resources available. These decisions seldom affect the whole operation in one sweeping decision, but are usually a chain of small decisions: organizing stock control, designing a product, pricing it, servicing equipment and so on. Effective management seeks to optimize the constituent parts of the whole operation. A wide variety of mathematical techniques is used to solve such optimization problems. Here, and later in Section 9.4.9, we consider methods based on the methods and concepts of calculus.

8.5.1 Optimal values

The basic idea is that the **optimal value** of a differentiable function $f(x)$ (that is, its **maximum** or **minimum value**) generally occurs where its derivative is zero; that is, where

$$f'(x) = 0$$

As can be seen from Figure 8.30, this is a necessary condition, since at a maximum or minimum value of the function its graph has a horizontal tangent. Figure 8.30 does, however, show that these extremal values are generally only local maximum or minimum values, corresponding to turning points on the graph, so some care must be

Figure 8.30
Maximum and minimum values.

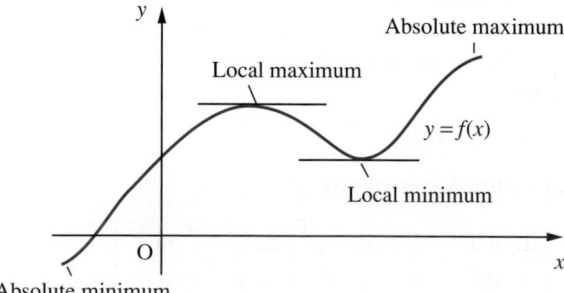

Figure 8.31
Graph with horizontal
tangents.

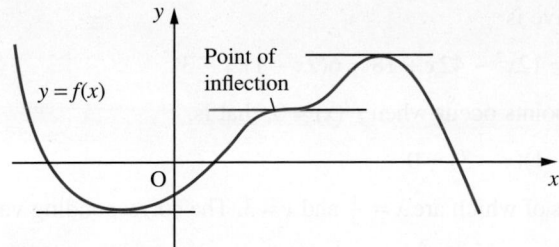

Figure 8.32
Graph of $f(x) = x^{2/3}$,
with minimum at
$x = 0$.

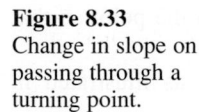

exercised in using the horizontal tangent as a test for an optimal value. In seeking the extremal values of a function it is also necessary to check the end points (if any) of the domain of the function.

Figure 8.31 gives another illustration of why care must be exercised: at some **points of inflection** – that is, points where the graph crosses its own tangent – the tangent may be horizontal.

A third reason for caution is that a function may have an optimal value at a point where its derivative does not exist. A simple example of this is given by $f(x) = x^{2/3}$, whose graph is shown in Figure 8.32.

Having determined the **critical or stationary points** where $f'(x) = 0$, we need to be able to determine their character or nature; that is, whether they correspond to a local maximum, a local minimum or a point of inflection of the function $f(x)$. We can do this by examining values of $f'(x)$ close to and on either side of the critical point. From Figure 8.33 we see that

- if the value of $f'(x)$, the slope of the tangent, changes from positive to negative as we pass from left to right through a stationary point then the latter corresponds to a **local maximum**;
- if the value of $f'(x)$ changes from negative to positive as we pass from left to right through a stationary point then the latter corresponds to a **local minimum**;
- if $f'(x)$ does not change sign as we pass through a stationary point then the latter corresponds to a **point of inflection**.

Figure 8.33
Change in slope on
passing through a
turning point.

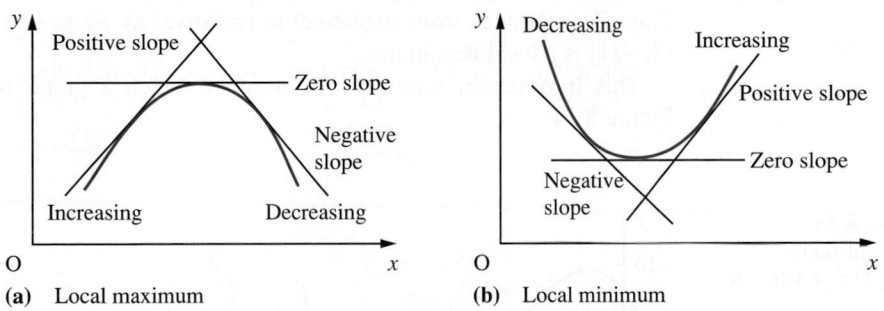

(a) Local maximum   **(b)** Local minimum

Example 8.30   Determine the stationary points of the function

$$f(x) = 4x^3 - 21x^2 + 18x + 6$$

and examine their nature.

Solution   The derivative is

$$f'(x) = 12x^2 - 42x + 18 = 6(2x - 1)(x - 3)$$

Stationary points occur when $f'(x) = 0$; that is,

$$6(2x - 1)(x - 3) = 0$$

the solutions of which are $x = \frac{1}{2}$ and $x = 3$. The corresponding values of the function are

$$f(\tfrac{1}{2}) = 4(\tfrac{1}{8}) - 21(\tfrac{1}{4}) + 18(\tfrac{1}{2}) + 6 = \tfrac{41}{4}$$

and

$$f(3) = 4(27) - 21(9) + 18(3) + 6 = -21$$

so that the stationary points of $f(x)$ are

$$(\tfrac{1}{2}, \tfrac{41}{4}) \quad \text{and} \quad (3, -21)$$

In order to investigate their nature, we use the procedure outlined above.

(a) Considering the point $(\tfrac{1}{2}, \tfrac{41}{4})$: if x is a little less than $\frac{1}{2}$ then $2x - 1 < 0$ and $x - 3 < 0$, so that

$$f'(x) = 6(2x - 1)(x - 3) = \text{(negative)(negative)} = \text{(positive)}$$

while if x is a little greater than $\frac{1}{2}$ then $2x - 1 > 0$ and $x - 3 < 0$, so that

$$f'(x) = \text{(positive)(negative)} = \text{(negative)}$$

Thus $f'(x)$ changes from (positive) to (negative) as we pass through the point so that $(\tfrac{1}{2}, \tfrac{41}{4})$ is a local maximum.

(b) Considering the point $(3, -21)$: if x is a little less than 3 then $2x - 1 > 0$ and $x - 3 < 0$, so that

$$f'(x) = \text{(positive)(negative)} = \text{(negative)}$$

while if x is a little greater than 3 then $2x - 1 > 0$ and $x - 3 > 0$, so that

$$f'(x) = \text{(positive)(positive)} = \text{(positive)}$$

Thus $f'(x)$ changes from (negative) to (positive) as we pass through the point so that $(3, -21)$ is a local minimum.

This information may now be used to sketch a graph of $f(x)$ as illustrated in Figure 8.34.

Figure 8.34
Graph of $f(x) = 4x^3 - 21x^2 + 18x + 6$.

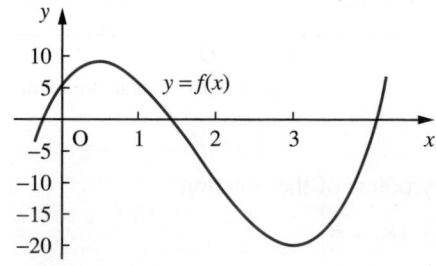

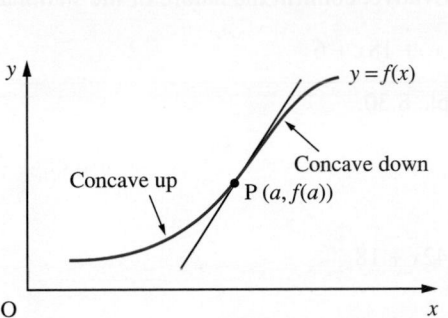

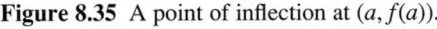

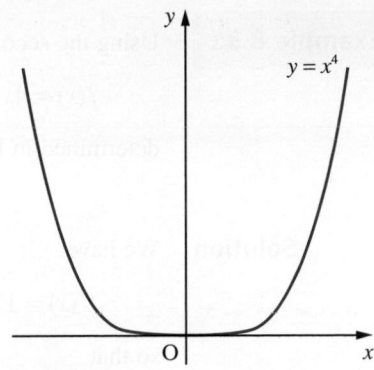

Figure 8.35 A point of inflection at $(a, f(a))$.

Figure 8.36 Graph of $f(x) = x^4$, illustrating the local minimum at $x = 0$.

An alternative approach to determining the nature of a stationary point is to calculate the value of the second derivative $f''(x)$ at the point. Recall from Section 8.4.1 that $f''(x)$ determines the rate of change of $f'(x)$. Suppose that $f(x)$ has a stationary point at $x = a$, so that $f'(a) = 0$. Then, provided $f''(a)$ is defined, either $f''(a) < 0, f''(a) = 0$ or $f''(a) > 0$.

If $f''(a) < 0$ then $f'(x)$ is decreasing at $x = a$; and since $f'(a) = 0$, it follows that $f'(x) > 0$ for values of x just less than a and $f'(x) < 0$ for values of x just greater than a. We therefore conclude that $x = a$ corresponds to a local maximum. Note that this concurs with our observation in Section 8.4.1 that the sign of $f''(x)$ determines the concavity of the graph of $f(x)$. Since the graph is concave down at a local maximum, $f''(a) \leqslant 0$. The equality case is discussed further in Section 9.4.9.

Similarly, we can argue that if $f''(a) > 0$ then the stationary point $x = a$ corresponds to a local minimum. Again this concurs with our observation that the graph is concave up at a local minimum.

Summarizing, we have

> - the function $f(x)$ has a local maximum at $x = a$ provided $f'(a) = 0$ and $f''(a) < 0$;
> - the function $f(x)$ has a local minimum at $x = a$ provided $f'(a) = 0$ and $f''(a) > 0$.

If $f''(a) = 0$, we cannot assume that $x = a$ corresponds to a point of inflection, and we must revert to considering the sign of $f'(x)$ on either side of the stationary point. As mentioned earlier, at a point of inflection the graph crosses its own tangent, or, in other words, the concavity of the graph changes. Since the concavity is determined by the sign of $f''(x)$, it follows that $f''(x) = 0$ at a point of inflection and that $f''(x)$ changes sign as we pass through the point. Note, as illustrated by the graph of Figure 8.35, that it is not necessary for $f'(x) = 0$ at a point of inflection. If, as illustrated in Figure 8.31, $f'(x) = 0$ at a point of inflection then it is a **stationary point of inflection**. It does not follow, however, that if $f'(a) = 0$ and $f''(a) = 0$ then $x = a$ is a point of inflection. An example of when this is not the case is $y = x^4$, which, as illustrated in Figure 8.36, has a local minimum at $x = 0$ even though both dy/dx and d^2y/dx^2 are zero at $x = 0$. It is for this reason that we must take care and revert to considering the sign of $f'(x)$ on either side. We shall return to reconsider these conditions in Section 9.4.9 following consideration of Taylor series.

Example 8.31

Using the second derivative, confirm the nature of the stationary points of the function

$$f(x) = 4x^3 - 21x^2 + 18x + 6$$

determined in Example 8.30.

Solution

We have

$$f'(x) = 12x^2 - 42x + 18$$

so that

$$f''(x) = 24x - 42$$

At the stationary point $(\frac{1}{2}, \frac{41}{4})$

$$f''(\tfrac{1}{2}) = 12 - 42 = -30 < 0$$

confirming that it corresponds to a local maximum.
  At the stationary point $(3, -21)$

$$f''(3) = 72 - 42 = 30 > 0$$

confirming that it corresponds to a local minimum.
  Note also that $f''(x) = 0$ at $x = \frac{7}{4}$ and that $f''(x) < 0$ for $x < \frac{7}{4}$ and $f''(x) > 0$ for $x > \frac{7}{4}$. Thus $(\frac{7}{4}, -\frac{43}{8})$ is a point of inflection (but not a stationary point of inflection), which is clearly identifiable in the graph of Figure 8.34.

Considering the cubic of Examples 8.30 and 8.31 the stationary points may be investigated using the following MATLAB commands

```
syms x y
y = 4*x^3 - 21*x^2 + 18*x + 6; dy = diff(y); solve(dy)
```

The last command solves $dy = 0$ to obtain the x-coordinates 3 and $1/2$ of the stationary points, and the commands

```
y1 = subs(y,x,3)
y2 = subs(y,x,1/2)
```

determine the corresponding y coordinates -21 and 10.25; so that the stationary points are $(3,-21)$ and $(1/2,10.25)$. The commands

```
d2y = diff(y,2);
subs(d2y,x,3)
subs(d2y,x,1/2)
```

return the value of the second derivative at each stationary point as 30 and -30 respectively; thus confirming that $(3,-21)$ is a local minimum and that

$(1/2, 10.25)$ is a local maximum. Finally, to illustrate the results the plot of the cubic is given by the command

```
ezplot(y,[-1,5]).
```

The corresponding commands in MAPLE are

```
y:= 4*x^3 - 21*x^2 + 18*x + 6; dy:= diff(y,x);
solve(dy = 0,x);
y1:= subs(x = 3,y); y2:= subs(x = 1/2,y);
d2y:= diff(y,x,x); subs(x - 3,d2y); subs(x = 1/2,d2y);
plot(y,x = -1..5);
```

Example 8.32   Determine the stationary values of the function

$$f(x) = x^2 - 6x + \frac{82}{x} + \frac{45}{x^2}, \quad x \neq 0$$

Solution   The derivative is

$$f'(x) = 2x - 6 - \frac{82}{x^2} - \frac{90}{x^3}$$

and $f'(x) = 0$ when

$$2x^4 - 6x^3 - 82x - 90 = 0$$

Factorizing we have

$$2(x^2 - 4x - 5)(x^2 + x + 9) = 0$$

or

$$2(x + 1)(x - 5)(x^2 + x + 9) = 0$$

So the real roots are $x = 5$ and $x = -1$.

At $x = 5, f(5) = 66/5$. To decide whether this is a maximum or minimum we examine the value of $f''(x)$ at $x = 5$.

$$f''(x) = 2 + \frac{164}{x^3} + \frac{270}{x^4} \quad \text{and} \quad f''(5) > 0$$

Thus $f(5) = 66/5$ is a minimum value of the function.

At $x = -1, f(-1) = -30$ and $f''(-1) > 0$ so that $f(-1) = -30$ is also a minimum of the function.

Note that $f(x)$ has an asymptote $x = 0$ and behaves like $(x - 3)^2$ where $|x|$ is very large.

In many applications we know for practical reasons that a particular problem has a minimum (or maximum) solution. If the equation $f'(x) = 0$ is satisfied by only one sensible value of x then that value must determine the unique minimum (or maximum) we are seeking. We will illustrate using three simple examples.

Example 8.33

A manufacturer has to supply N items per month at a uniform daily rate. Each time a production run is started it costs $£c_1$, the 'set-up' cost. In addition, each item costs $£c_2$ to manufacture. To avoid unnecessarily high production costs, the manufacturer decides to produce a large quantity q in one run and store it until the contract calls for delivery. The cost of storing each item is $£c_3$ per month. What is the optimal size of a production run?

Solution

As the contract calls for a monthly supply of N items, we need to look for a production run size that will minimize the total monthly cost to the manufacturer.

The costs the manufacturer incurs are the production costs and the storage costs. The production cost for a production run of q items is

$$£(c_1 + c_2 q)$$

This production run will satisfy the contract for q/N months, so the monthly production cost will be

$$£\frac{c_1 + c_2 q}{q/N} = £\left(\frac{c_1}{q} + c_2\right)N$$

To this must be added the monthly storage cost, which will be $£\frac{1}{2}q c_3$, since the stock is depleted at a uniform rate and the average stock size is $\frac{1}{2}q$. Thus the total monthly cost $£C$ is given by

$$C = \left(\frac{c_1}{q} + c_2\right)N + \frac{1}{2}q c_3$$

which has a graph similar to that shown in Figure 8.37.

To find the value q^* of q that minimizes C, we differentiate the expression for C with respect to q and set the derivative equal to zero:

$$\frac{dC}{dq} = \frac{-c_1 N}{q^2} + \frac{1}{2}c_3$$

and

$$\frac{dC}{dq} = 0 \quad \text{implies} \quad \frac{-c_1 N}{(q^*)^2} + \frac{1}{2}c_3 = 0$$

and hence

$$q^* = \sqrt{\left(\frac{2c_1 N}{c_3}\right)}$$

This quantity is called the **economic lot size**.

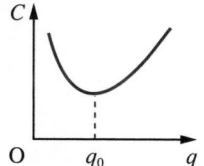

Figure 8.37
Monthly cost versus run size.

Optimization plays an important role in design, and in Example 8.34 we illustrate this by applying it to the relatively easy problem of designing a milk carton.

Example 8.34   A milk retailer wishes to design a milk carton that has a square cross-section, as illustrated in Figure 8.38(a), and is to contain two pints of milk (2 pints ≡ 1.136 litres). The carton is to be made from a rectangular sheet of waxed cardboard, by folding into a square tube and sealing down the edge, and then folding and sealing the top and bottom. To make the resulting carton airtight and robust for handling, an overlap of at least 5 mm is needed. The procedure is illustrated in Figure 8.38(b). As the milk retailer will be using a large number of such cartons, there is a requirement to use the design that is least expensive to produce. In particular the retailer desires the design that minimizes the amount of waxed cardboard used.

Solution   If, as illustrated in Figure 8.38(a), the final dimensions of the container are $h \times b \times b$ (all in mm) then the area of waxed cardboard required is

$$A = (4b + 5)(h + b + 10) \tag{8.32}$$

Since the capacity of the carton is fixed at two pints (1.136 litres), the values of h and b must be such that

$$\text{volume} = hb^2 = 1\,136\,000 \, \text{mm}^3 \tag{8.33}$$

Substituting (8.33) back into (8.32) gives

$$A = (4b + 5)\left(\frac{1\,136\,000}{b^2} + b + 10\right)$$

To find the value of b that minimizes A, we differentiate A with respect to b to obtain $A'(b)$ and then set $A'(b) = 0$. Differentiating gives

$$A'(b) = 8b + 45 - \frac{4\,544\,000}{b^2} - \frac{11\,360\,000}{b^3}$$

so the required value of b is given by the root of the equation

$$8b^4 + 45b^3 - 4\,544\,000b - 11\,360\,000 = 0$$

A straightforward tabulation of this polynomial, or use of a suitable software package, yields a root at $b = 81.8$. From (8.33) the corresponding value of h is $h = 169.8$. Thus the optimal design of the milk carton will have dimensions $81.8\,\text{mm} \times 81.8\,\text{mm} \times 169.8\,\text{mm}$.

Figure 8.38
The construction of a milk carton.

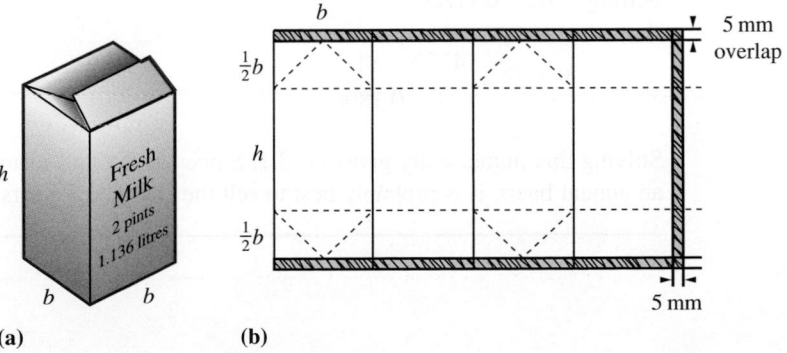

(a)   (b)

Optimization problems also occur in programmes for replacing equipment and machinery in industry. We will illustrate this by a more commonplace decision: the best policy for replacing a car.

Example 8.35

For a particular model of car, bought in 1987 for £4750, the second-hand value after t years is given fairly accurately by the formula

$$\text{price} = £e^{8.41-0.189t}$$

The running costs of the car increase as the car gets older, so after t years the annual running cost is $£(917 + 163t)$. When should it be replaced?

Solution

The accumulated running cost for the car over t years is

$$£\sum_{r=0}^{t-1} (917 + 163r) = £917t + £163[1 + 2 + 3 + \ldots + (t-1)]$$

$$= £917t + £\tfrac{163}{2}(t-1)t \quad \left(\text{using } \sum_{r=1}^{n} r = \tfrac{1}{2}n(n+1) \right)$$

$$= £(835.5 + 81.5t)t$$

The total cost of the car clearly includes depreciation as well as running costs, so the average annual cost $£C$ of the car is given by

$$C = \frac{4750 - e^{8.41-0.189t} + (835.5 + 81.5t)t}{t}$$

To find the optimal time for replacing the car, we find the value of t that minimizes C. Differentiating C with respect to t gives

$$C'(t) = -\frac{1}{t^2}(4750 - e^{8.41-0.189t}) + \frac{1}{t}(0.189e^{8.41-0.189t}) + 81.5$$

Setting $C'(t) = 0$ gives

$$e^{8.41-0.189t} = \frac{4750 - 81.5t^2}{1 + 0.189t}$$

Solving this numerically gives $t = 3.1$. Since car tax and insurance are usually paid on an annual basis, it is probably best to sell the car after 3 years!

8.5.2 Exercises

Check your answers using MATLAB or MAPLE whenever possible.

72 Find the stationary values of the following functions and determine their nature. In each case also find the point of inflection and sketch a graph of the function.

(a) $f(x) = 2x^3 - 5x^2 + 4x - 1$

(b) $f(x) = x^3 + 6x^2 - 15x + 51$

73 Find the stationary values of the following functions, distinguishing carefully between them. In each case sketch a graph of the function.

(a) $f(x) = \dfrac{3x}{(x-1)(x-4)}$

(b) $f(x) = 2e^{-x}(x-1)^3$

(c) $f(x) = x^2 e^{-x}$

74 Consider the can shown in Figure 8.39, which has capacity 500 ml. The cost of manufacture is proportional to the amount of metal used, which in turn is proportional to the surface area of the can. Ignoring the overlaps necessary for the manufacture of the can, find the diameter and height of the can which minimizes its cost.

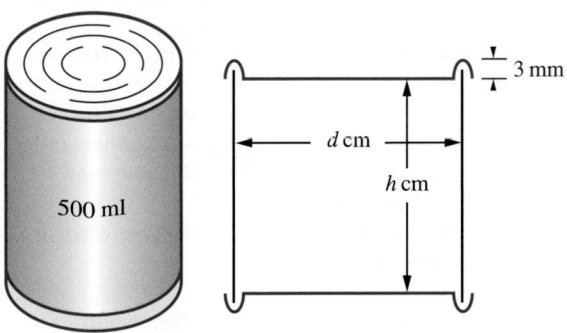

Figure 8.39 Can of Questions 74 and 75.

75 Consider again the can shown in Figure 8.39. Allowing for an overlap of 6 mm top and bottom surfaces to give a rim of 3 mm on the can, show that the area $A\,\text{mm}^2$ of metal used is given by

$$A(d) = \pi(d^2 + 3.6d + 1.44)/2 + 2000/d$$

where d cm is the diameter of the can.

Show that the value of d^* which minimizes the area of the can satisfies the equation

$$\pi d^2(d + 1.8) = 2000$$

Calculate d^* and the corresponding value of the height of the can.

76 In an underwater telephone cable the ratio of the radius of the core to the thickness of the protective sheath is denoted by x. The speed v at which a signal is transmitted is proportional to $x^2 \ln(1/x)$. Show that

$$\frac{dv}{dx} = Kx\left[2\ln\left(\frac{1}{x}\right) - 1\right]$$

where K is some constant, and hence deduce the stationary values of v. Distinguish between these stationary values and show that the speed is greatest when $x = 1/\sqrt{e}$.

77 A closed hollow vessel is in the form of a right-circular cone, together with its base, and is made of sheet metal of negligible thickness. Express the total surface area S in terms of the volume V and the semi-vertical angle θ of the cone. Show that for a given volume the total area of the surface is a minimum if $\sin\theta = \frac{1}{3}$. Find the value of S if $V = \frac{8}{3}\pi a^3$.

78 A numerical method which is more efficient than repeated subtabulation for obtaining the optimal solution is the following **bracketing method**. The initial tabulation locates an interval in which the solution occurs. The optimal solution is then estimated by optimizing a suitable quadratic approximation.

Consider again the milk carton problem, Example 8.34. Calculate $A(70)$, $A(80)$ and $A(90)$ and deduce that a minimum occurs in $[70, 90]$. Next find numbers p, q and r such that

$$C(b) = p(b - 80)^2 + q(b - 80) + r$$

satisfies $C(70) = A(70)$, $C(80) = A(80)$ and $C(90) = A(90)$. The minimum of C occurs at $80 - q/(2p)$. Show that this yields the estimate $b = 82.2$. Evaluate $A(82.2)$ and deduce that the solution lies in the interval $[80, 90]$. Next repeat the process using the values $A(80)$, $A(82.2)$ and $A(90)$ and show that the solution lies in the interval $[80, 83.1]$. Apply the method once more to obtain an improved estimate of the solution.

79 A pipeline is to be laid from a point A on one bank of a river of width 1 unit to a point B 2 units downstream on the opposite bank, as shown in Figure 8.40. Because it costs more to lay the pipe under water than on dry land, it is proposed to take it in a straight line across the river to a point C and then along the river bank to B. If it costs $\alpha\%$ more to lay a given length of pipe under the river than along the bank, write down a formula for the cost of the pipeline, specifying the domain of the function carefully. What recommendation would you make about the position of C when (a) $\alpha = 25$, (b) $\alpha = 10$?

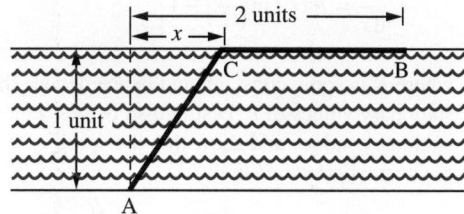

Figure 8.40

80 Cross-current extraction methods are used in many chemical processes. Solute is extracted from a stream of solvent by repeated washings with water. The solvent stream is passed consecutively through a sequence of extractors, in each of which a cross-current of wash water, flowing at a determined rate, carries out some of the solute. The aim is to choose the individual wash flowrates in such a way as to extract as much solute as possible by the end, the total flow of wash water being fixed.

Consider the three-state extractor process shown in Figure 8.41, where c, x, y and z are the solute concentrations in the main stream, and αx, αy and αz are the solute concentrations in the effluent wash-water streams, with α a constant. The solute balance equations for the extractors are

$$Q(c - x) = u\alpha x$$

$$Q(x - y) = v\alpha y$$

$$Q(y - z) = w\alpha z$$

The total wash-water flowrate is W, so that

$$u + v + w = W$$

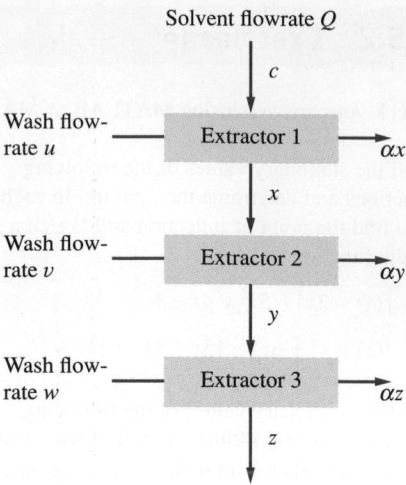

Figure 8.41

We wish to find u, v and w such that the outflow concentration z is minimized.

This is an example of **dynamic programming**. The key to its solution is the **Principle of Optimality**, which states that an optimal programme has the property that, whatever the initial state and decisions, the remaining decisions must constitute an optimal policy with respect to the state resulting from the initial decision. This means we solve the problem first for a one-extractor process, then a two-extractor process, then a three-extractor process, and so on.

For a one-stage process x is minimized when $u = W$, giving $x^* = Qc/(Q + \alpha W)$.

For a two-stage process $y = Qx/(Q + \alpha v)$, where $x = \frac{1}{2}W$ with $v = \frac{1}{2}W$, giving $y^* = Q^2c/(Q + \frac{1}{2}\alpha W)^2$.

For a three-stage process, $z = Q^2x/[Q + \frac{1}{2}\alpha(W - u)]^2$, where $x = Qc/(Q + \alpha u)$. Show that z is minimized when $u = \frac{1}{3}W$, with $v = w = \frac{1}{3}W$, giving $z^* = Q^3c/(Q + \frac{1}{3}\alpha W)^3$.

Generalize your answers to the case when n extractors are used.

81 The management of resources often requires a chain of decisions similar to that described in Question 80. Consider the harvesting policy for a large forest. The profit produced from the sale of felled timber is proportional to the square root of the volume sold, while the volume of standing timber increases in proportion to itself year on year. Use the technique outlined in Question 80 to produce a 10-year harvesting programme for a forest.

8.6 Numerical differentiation

Although the formula

$$f'(x) = \lim_{\Delta x \to 0} \frac{f(x + \Delta x) - f(x)}{\Delta x} = \lim_{\Delta x \to 0} \frac{\Delta f}{\Delta x}$$

provides the definition of the derivative of $f(x)$, it does not provide a good basis for evaluating $f'(x)$ numerically. This is because it provides a one-sided approximation of the gradient at x as shown in Figure 8.42. When we set $\Delta x = h$ (> 0), we obtain the slope of the chord PR. When we set $\Delta x = -h$ (< 0), we obtain the slope of the chord QP. Clearly the chord QR offers a better approximation to the tangent at P. A second reason why the formal definition of a derivative yields a poor approximation is that the evaluation of derivatives involves the division of a small quantity Δf by a second small quantity Δx. This process magnifies the rounding errors involved in calculating Δf from the values of $f(x)$, a process that worsens as $\Delta x \to 0$. This phenomenon is called **ill-conditioning**. Generally speaking, numerical differentiation is a process in which accuracy is lost and the 'noise' caused by experimental error is magnified.

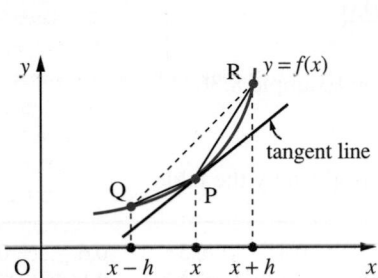

Figure 8.42 Approximations to the tangent at P.

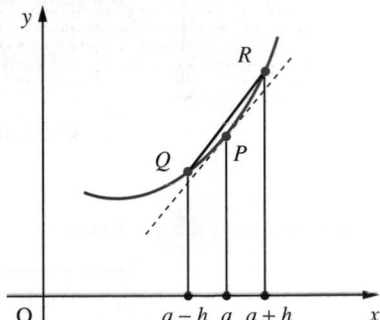

Figure 8.43 Chord approximation.

8.6.1 The chord approximation

This method uses the slope of a chord QR symmetrically disposed about x to approximate the slope of the tangent at x, as shown in Figure 8.43. Thus

$$f'(x) \approx \frac{f(x + h) - f(x - h)}{2h} = \phi(h)$$

Thus when the function is specified graphically, a value of h is chosen, and at a series of points along the curve the quotient $\phi(h)$ is calculated. When the function is given as a table of values, we do not have control of the value of h, but the same approximation is used using the tabular interval as h. Consequently, to estimate the value of the derivative $f'(a)$ at $x = a$ we use the approximation

$$\frac{f(a + h) - f(a - h)}{2h} = \phi(h)$$

as the basis for an extrapolation. For almost all functions commonly occurring in engineering applications

$$f'(a) = \phi(h) + \text{terms involving powers of } h \text{ greater than or equal to } h^2$$

For example, considering $f(x) = x^3$

$$\phi(h) = \frac{(a+h)^3 - (a-h)^3}{2h} = \frac{6a^2h + 2h^3}{2h} = 3a^2 + h^2$$

Similarly, for $f(x) = x^4$

$$\phi(h) = 4a^3 + 4ah^2$$

In general, we may write

$$f'(a) = \phi(h) + Ah^2 + \text{terms involving higher powers of } h$$

where A is independent of h. Interval-halving gives

$$f'(a) = \phi(\tfrac{1}{2}h) + \tfrac{1}{4}A'h^2 + \text{terms involving higher powers of } h$$

where $A' \approx A$. Hence we obtain a better estimate for $f'(a)$ by extrapolation, eliminating the terms involving h^2:

$$f'(a) \approx \tfrac{1}{3}[4\phi(\tfrac{1}{2}h) - \phi(h)] \tag{8.34}$$

We illustrate this technique in Example 8.36.

Example 8.36   Estimate $f'(0.5)$, where $f(x)$ is given by the table

x	0.1	0.2	0.3	0.4	0.5	0.6	0.7	0.8	0.9
$f(x)$	0.0998	0.1987	0.2955	0.3894	0.4794	0.5646	0.6442	0.7174	0.7833

Solution   Using the data provided, taking $h = 0.4$ and 0.2, we obtain

$$\phi(0.4) = \frac{0.7833 - 0.0998}{0.8} = 0.8544$$

and

$$\phi(0.2) = \frac{0.6442 - 0.2955}{0.4} = 0.8718$$

Hence, by extrapolation, we have, using (8.34),

$$f'(0.5) \approx (4 \times 0.8718 - 0.8544)/3 = 0.8776$$

The tabulated function is actually $\sin x$, so that in this illustrative example we can compare the estimate with the true value $\cos 0.5$, and we find that the answer is correct to 4dp.

In general, any numerical procedure is subject to two types of error. One is due to the accumulation of rounding errors within a calculation, while the other is due to the nature of the approximation formula (the truncation error). In this example the truncation error is of order h^2 for $\phi(h)$, but we do not have an estimate for the truncation error for the extrapolated estimate for $f'(a)$. This will be discussed in the next chapter following the introduction of the Taylor series (see Exercises 9.4.6, Question 23). The effect of the rounding errors on the answer can be assessed, however, and, using the methods of Chapter 1, we see that the maximum effect of the rounding errors on the answer in Example 8.36 is $\pm 2.5 \times 10^{-4}$.

8.6.2 Exercises

82 Use the chord approximation to obtain two estimates for $f'(1.2)$ using $h = 0.2$ and $h = 0.1$ where $f(x)$ is given in the table below.

x	1.0	1.1	1.2	1.3	1.4
$f(x)$	1.000	1.008	1.061	1.192	1.414

Use extrapolation to obtain an improved approximation.

83 Use your calculator (in radian mode) to calculate the quotient $\{f(x + h) - f(x - h)\}/(2h)$ for $f(x) = \sin x$, where $x = 0(0.1)1.0$ and $h = 0.001$. Compare your answers with $\cos x$.

84   Consider the function $f(x) = x e^x$, tabulated below:

x	0.96	0.97	0.98	0.99	1.00
$f(x)$	2.5072	2.5588	2.6112	2.6643	2.7183

x	1.01	1.02	1.03	1.04
$f(x)$	2.7731	2.8287	2.8851	2.9424

(a) Find, *exactly*, $f'(1)$ and $f''(1)$.

(b) Use the tabulated values and the formula

$$f'(a) \simeq (f(a + h) - f(a - h))/2h$$

to estimate $f'(1)$, for various h. Compute the errors involved and comment on the results.

(c) Repeat (b) for $f''(1)$ using

$$f''(a) \simeq (f(a + h) - 2f(a) + f(a - h))/h^2$$

85 Use the following table of $f(x) = (e^x - e^{-x})/2$ to estimate $f'(1.0)$ by means of an extrapolation method.

x	0.2	0.6	0.8	1.2	1.4	1.8
$f(x)$	0.2013	0.6367	0.8881	1.5095	1.9043	2.9422

Compare your answer with $(e + e^{-1})/2 = 1.5431$ correct to 4dp.

86 Investigate the effect of using a smaller value for h in Example 8.36. Show that $\phi(0.1)$ gives a poorer estimate for $f'(0.5)$ and the error bound for the consequent extrapolation $[4\phi(0.1) - \phi(0.2)]/3$ is 7×10^{-4}.

8.7 Integration

In this section we shall introduce the concept of integration and illustrate its role in problem-solving and modelling situations.

8.7.1 Basic ideas and definitions

Consider an object moving along a line with constant velocity u (in m s^{-1}). The distance s (in m) travelled by the object between times t_1 and t_2 (in s) is given by

$$s = u(t_2 - t_1)$$

Figure 8.44
Velocity–time graph
for an object moving
with constant velocity
u. The shaded area
shows the distance
travelled by the object
between times t_1 and t_2.

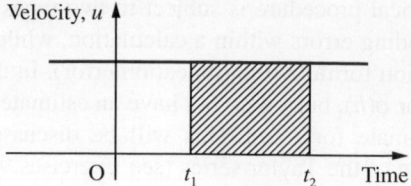

Figure 8.45
A velocity–time
graph and two
piecewise-constant
approximations to it.

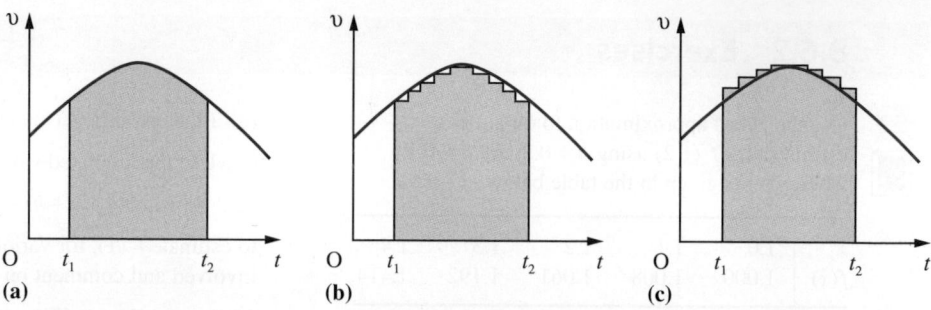

This is the area 'under' the graph of the velocity function between $t = t_1$ and $t = t_2$, as
shown in Figure 8.44. This, of course, deals with the special case where the velocity is
a constant function. However, even when the velocity varies with time, the area under
the velocity graph still gives the distance travelled. Consider the velocity graph shown
in Figure 8.45(a). We can approximate the velocity–time graph by a series of small
horizontal lines that lie either entirely below the curve (as in Figure 8.45(b)) or entirely
above it (Figure 8.45(c)). An object moving such that its velocity–time graph is (b)
would always be slower at a particular time than an object with velocity–time graph (a),
so that the distance it covers is less than that of the object with graph (a). Similarly, an
object with velocity–time graph (c) will cover a greater distance than an object with
graph (a). Thus

distance with graph (b) < distance with graph (a) < distance with graph (c)

In cases (b) and (c), because the velocities are piecewise-constant, the distances covered
are represented by the areas under the graphs between $t = t_1$ and $t = t_2$. So we have

area under graph (b) < area under graph (a) < area under graph (c)

If the horizontal steps of graphs (b) and (c) are made very small, the difference between
the areas for the approximating graphs (b) and (c) becomes very small. In other words,
the distance for graph (a) is just the area under the graph between $t = t_1$ and $t = t_2$.

This is one of many practical problems that involve this process of area evaluation
at some stage in their solution. This process is called **integration**: the summing together
of all the parts that make up a given area. The area under the graph is called the
integral of the function. For some functions it is possible to obtain formulae for their
integrals; for others we have to be content with numerical approximations.

Formally, we define the integral of the function $f(x)$ between $x = a$ and $x = b$ to be

$$\lim_{\substack{n \to \infty \\ \Delta x \to 0}} \sum_{r=1}^{n} f(x_r^*) \Delta x_{r-1}$$

where $a = x_0 < x_1 < x_2 < \ldots < x_{n-1} < x_n = b$ are the points of subdivision of the interval $[a, b]$,

$$\Delta x_{r-1} = x_r - x_{r-1}, \Delta x = \max(\Delta x_0, \Delta x_1, \ldots, \Delta x_{n-1}) \text{ and } x_{r-1} \leq x_r^* \leq x_r$$

Here we have used the special notation

$$\lim_{\substack{n \to \infty \\ \Delta x \to 0}}$$

to emphasize that $n \to \infty$ and $\Delta x \to 0$ simultaneously. The value of the integral is independent of both the method of subdivision of $[a, b]$ and the choices of x_r^*.

The usual notation for the integral is

$$\int_a^b f(x)\,dx$$

where the integration symbol $\int$ is an elongated S, standing for 'summation'. The dx is called the **differential** of x, and a and b are called the **limits of integration**. The function $f(x)$ being integrated is the **integrand**.

Figure 8.46
Strip about typical value $x = x_r^*$.

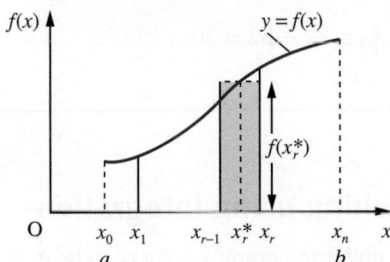

The process is illustrated in Figure 8.46, where the area under the graph of $f(x)$ for $x \in [a, b]$ has been subdivided into n vertical strips (by which we strictly mean that the area has been approximated by the n vertical strips). The area of a typical strip is given by

$$f(x_r^*)(x_r - x_{r-1}) = f(x_r^*)\Delta x_{r-1}$$

where $x_{r-1} \leq x_r^* \leq x_r$ and $\Delta x_{r-1} = x_r - x_{r-1}$. Thus the area under the graph can be approximated by

$$\sum_{r=1}^{n} f(x_r^*)\Delta x_{r-1}$$

This approximation becomes closer to the exact area as the number of strips is increased and their widths decreased. In the limiting case as $n \to \infty$ and $\Delta x \to 0$ this leads to the exact area being given by

$$A = \int_a^b f(x)\,dx$$

so that

$$\int_a^b f(x)\mathrm{d}x = \lim_{\substack{n\to\infty \\ \Delta x\to 0}} \sum_{r=1}^n f(x_r^*)\Delta x_{r-1}$$

(8.35)

In line with the definition of an integral, we note that if the graph of $f(x)$ is below the x axis then the summation involves products of negative ordinates with positive widths, so that areas below the x axis must be interpreted as being negative.

Example 8.37

By considering the area under the graph of $y = x + 3$, evaluate the integral $\int_{-5}^5 (x + 3)\mathrm{d}x$.

Solution

The area under the graph is shown hatched in Figure 8.47, with the area A_1 being negative and the area A_2 positive. In each case the areas are triangular, so that

$$A_1 = -\tfrac{1}{2} \times 2 \times 2 = -2$$

and

$$A_2 = \tfrac{1}{2} \times 8 \times 8 = 32$$

Thus

$$\int_{-5}^5 (x + 3)\mathrm{d}x = A_1 + A_2 = -2 + 32 = 30$$

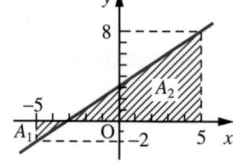

Figure 8.47

8.7.2 Mathematical modelling using integration

We have seen that the area under the graph $y = f(x)$ can be expressed as an integral, but integrals have a much wider application. Any quantity that can be expressed in the form of the limit of a sum as in (8.35) can be represented by an integral, and this occurs in many practical situations. Because areas can be expressed as integrals, it follows that we can always interpret an integral geometrically as an area under a graph.

Example 8.38

What is the volume of a pyramid with square base, of side 4 metres, and height 6 metres?

Solution

Imagine the pyramid of Figure 8.48(a) is cut into horizontal slices of thickness Δh, as shown in Figure 8.48(b), and then sum their volumes to give the volume of the pyramid.
From Figure 8.48(b) the volume of the slice is

$$\Delta V_k = \text{area square flat face} \times \text{thickness} = 4d_k^2\Delta h$$

where $2d_k$ is the length of one side of the square slice. The length of the side is related to the height h_k of the slice above the base and using similar triangle relation (see Figure 8.48(c)) we have

Figure 8.48
Pyramid of
Example 8.38.

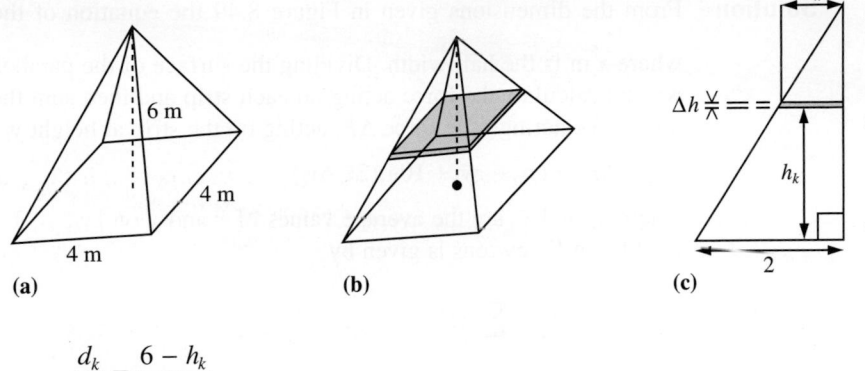

(a)   **(b)**   **(c)**

$$\frac{d_k}{2} = \frac{6 - h_k}{6}$$

Thus the slice has volume

$$\Delta V_k = \frac{4(6 - h_k)^2}{9} \Delta h$$

The volume of all slices is

$$\sum_{k=1}^{n} \Delta V_k = \sum_{k=1}^{n} \frac{4(6 - h_k)^2}{9} \Delta h$$

Proceeding to the limit ($n \to \infty$, $\Delta h \to 0$) as in (8.35) gives the volume of the pyramid as the integral

$$V = \int_0^6 \tfrac{4}{9}(6 - h)^2 dh$$

We will see later, in Example 8.42, that $V = 32$ and the volume is 32 m³.

Example 8.39

A reservoir is created by constructing a dam across a glacial valley. Its wet face is vertical and has approximately the shape of a parabola as shown in Figure 8.49. The water pressure p (Pascals) varies with depth according to

$$p = p_0 - gy + 10g$$

where p_0 is the pressure at the surface, g is the acceleration due to gravity and y metres is the height from the bottom of the parabola as shown in the figure. Calculate the total force acting on the wet face of the dam.

Figure 8.49
Schematic
representation of dam
showing strip of width
Δy_k at height y_k.

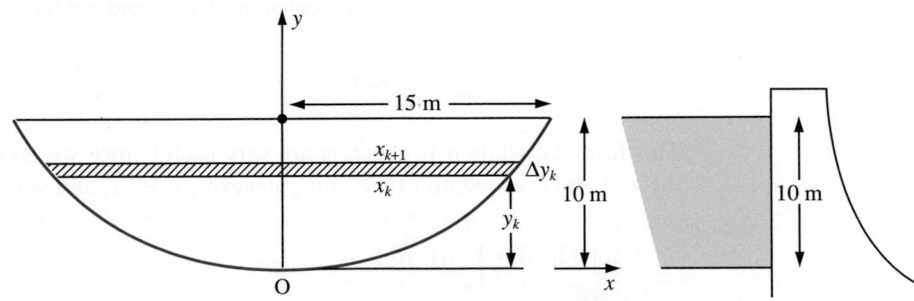

Solution From the dimensions given in Figure 8.49 the equation of the parabola is $y = \dfrac{2x^2}{45}$, where x m is the half width. Dividing the surface of the parabola into horizontal strips we can calculate the force acting on each strip and then sum these forces to obtain the total force acting. The force ΔF_k acting on the strip at height y_k is

$$\Delta F_k = (p_0 - g\bar{y}_k + 10g)(2\bar{x}_k \Delta y_k)$$

where $\bar{x}_k$ and $\bar{y}_k$ are the average values of x and y on $[y_k, y_{k+1}]$. Thus, using (8.35), the total force F newtons is given by

$$F = \lim_{\substack{n \to \infty \\ \Delta y \to 0}} \sum_{k=1}^{n} \Delta F_k$$

$$= \lim_{\substack{n \to \infty \\ \Delta y \to 0}} \sum_{k=1}^{n} (p_0 - g\bar{y}_k + 10g)(2\bar{x}_k \Delta y_k)$$

$$= \int_0^{10} 2x(p_0 - gy + 10g)\,dy$$

Now $y = \dfrac{2x^2}{45}$ so that $2x = (90y)^{1/2}$ and we may rewrite the expression for F as

$$F = \int_0^{10} 3\sqrt{10}(p_0 + 10g - gy)\sqrt{(y)}\,dy$$

Later in Example 8.43 we will show that $F = 200p_0 + 800g$.

Example 8.40 A beam of length l is freely hinged at both ends and carries a distributed load $w(x)$ where

$$w(x) = \begin{cases} 4Wx/l^2 & 0 \leqslant x \leqslant l/2 \\ 4W(l - x)/l^2 & l/2 \leqslant x \leqslant l \end{cases}$$

Show that the total load is W and find the shear force at a point on the beam.

Solution To find the total load on the beam we divide the interval $(0, l)$ into n subintervals of length Δx, so that $x_k = k\Delta x$ and $\Delta x = l/n$. Then the load on the subinterval (x_k, x_{k+1}) is $w(x_k^*)\Delta x$ where $w(x_k^*)$ is the average value of $w(x)$ in that subinterval. The total load on the beam is the sum of all such elementary loads, and we have

$$\text{total load} = \sum_{k=0}^{n-1} w(x_k^*)\Delta x$$

This formula, while it is exact, is not very useful since we do not know the values of the x_k^*'s. By proceeding to the limit, however, $x_k^* \to x_k$ and we obtain the formula

$$\text{total load} = \int_0^l w(x)\,dx$$

Figure 8.50
Non-uniform
load on a beam.

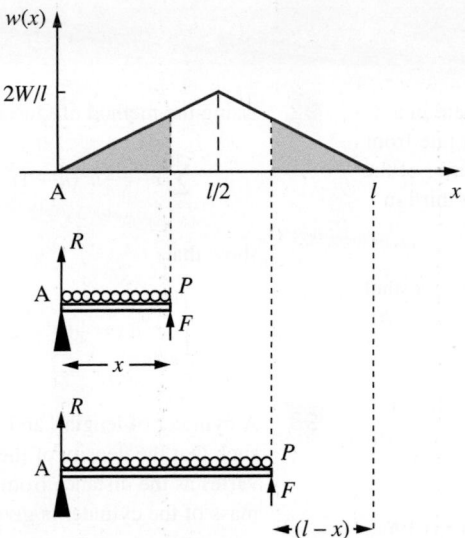

Now the integral $\int_0^l w(x)\mathrm{d}x$ is the area under the curve $y = w(x)$ between $x = 0$ and $x = l$, and by considering the graph of $w(x)$ shown in Figure 8.50 we see that this is W.

From the symmetry of the loading and the end conditions we see that the reactions at the supports at both ends are equal (to R, say). Then the vertical forces must balance for equilibrium giving

$$2R = W$$

for equilibrium. To find the shear force F we have to consider the vertical equilibrium of the portion of the beam to the left of P. Thus

$R + F =$ load between A and P which is represented by the area under the graph between A and P

Consideration of the areas under the graph of $y = w(x)$ for x shows that

$$R + F = \begin{cases} \frac{1}{2}x(4Wx/l^2) & 0 \leqslant x \leqslant l/2 \\ W - \frac{1}{2}(l - x)[4W(l - x)/l^2] & l/2 \leqslant x \leqslant l \end{cases}$$

This simplifies as

$$R + F = \begin{cases} 2Wx^2/l^2 & 0 \leqslant x \leqslant l/2 \\ W - 2W(l - x)^2/l^2 & l/2 \leqslant x \leqslant l \end{cases}$$

Thus

$$F = \begin{cases} 2Wx^2/l^2 - W/2 & 0 \leqslant x \leqslant l/2 \\ W/2 - 2W(l - x)^2/l^2 & l/2 \leqslant x \leqslant l \end{cases}$$

8.7.3 Exercises

87 Two hot-rodders, Alan and Brian, compete in a drag race. Each accelerates at a constant rate from a standing start. Alan covers the last quarter of the course in 3 s, while Brian covers the last third in 4 s. Who wins and by what time margin?

88 Show that the area under the graph of the constant function $f(x) = 1$ between $x = a$ and $x = b$ $(a < b)$ is given by $b - a$.

89 Show that the area under the graph of the linear function $f(x) = x$ between $x = a$ and $x = b$ $(a < b)$ is given by $\frac{1}{2}(b^2 - a^2)$.

90 Draw the graph of the function $f(x) = 2x - 1$ for $-3 < x < 3$. By considering the area under the graph, evaluate the integral $\int_{-3}^{3} (2x - 1)\mathrm{d}x$.

91 Using n strips of equal width, show that the area under the graph $y = x^2$ between $x = 0$ and $x = c$ satisfies the inequality

$$h^3 \sum_{r=1}^{n-1} r^2 < \text{area} < h^3 \sum_{r=1}^{n} r^2$$

and deduce

(a) $\displaystyle\int_{0}^{c} x^2\,\mathrm{d}x = \frac{1}{3}c^3$ (b) $\displaystyle\int_{a}^{b} x^2\,\mathrm{d}x = \frac{1}{3}(b^3 - a^3)$

(c) $\displaystyle\int_{a}^{b} x^{1/2}\,\mathrm{d}x = \frac{2}{3}(b^{3/2} - a^{3/2})$

(Recall that $\sum_{r=1}^{n} r^2 = \frac{1}{6}n(n + 1)(2n + 1)$.)

92 Using the method of Question 91 and the fact that

$$\sum_{r=1}^{n} r^3 = \frac{1}{4}n^2(n + 1)^2$$

show that

$$\int_{a}^{b} x^3\,\mathrm{d}x = \frac{1}{4}(b^4 - a^4)$$

93 A cylinder of length l and diameter D is constructed such that the density of the material comprising it varies as the distance from the base. Show that the mass of the cylinder is given by

$$\int_{0}^{l} \frac{1}{4}KD^2\pi x\,\mathrm{d}x$$

where K is a proportionality constant.

94 A beam of length l is freely hinged at both ends and carries a distributed load $w(x)$ where

$$w = \begin{cases} 4W/l & 0 \leqslant x \leqslant l/4 \\ 0 & l/4 < x \leqslant l \end{cases}$$

Find the shear force at a point on the beam.

95 A hemi-spherical vessel has internal radius 0.5 m. It is initially empty. Water flows in at a constant rate of 1 litre per second. Find an expression for the depth of the water after t seconds.

8.7.4 Definite and indefinite integrals

We have seen that the area under the graph $y = f(x)$ between $x = a$ and $x = b$ is given by the integral

$$\int_{a}^{b} f(x)\mathrm{d}x$$

Clearly, this area depends on the values of a and b as well as on the function $f(x)$. Thus the integral of a function $f(x)$ may be regarded as a function of a and b. If we replace the number b by the variable x, we obtain a function, F say, that is the area under the graph between a and x, as shown in Figure 8.51. This type of integral is called an

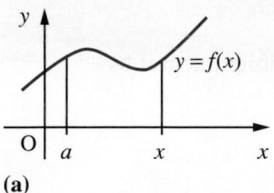 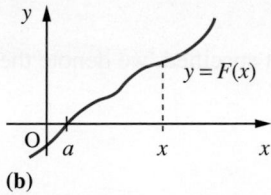

Figure 8.51 (a) Graph of $y = f(x)$. (b) Graph of $\int_a^x f(t)dt$.

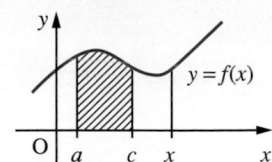

Figure 8.52

indefinite integral to distinguish it from integrals with fixed a and b, which are called **definite integrals**. We have defined F by the relation

$$F(x) = \int_a^x f(t)dt$$

Notice here that the dummy variable t, used as the integrator, is chosen to be different from the variable x on which the function F depends.

If a different lower limit is chosen, a different function is obtained, say G:

$$G(x) = \int_c^x f(t)dt$$

By interpreting an integral as the area under a curve, we see from Figure 8.52 that this new function differs from F only by a constant. This follows since

$$F(x) - G(x) = \int_a^x f(t)dt - \int_c^x f(t)dt = \int_a^c f(t)dt$$

which is a definite integral having a constant value representing the area under the graph between a and c, shown shaded in Figure 8.52.

For example, using the definition of an integral, we can show that

$$\int_a^b (t^2 - 1)dt = \tfrac{1}{3}(b^3 - a^3) - (b - a)$$

so that

$$\int_a^x (t^2 - 1)dt = \tfrac{1}{3}(x^3 - a^3) - (x - a) = \tfrac{1}{3}x^3 - x + (a - \tfrac{1}{3}a^3)$$

Giving a the values 1 and 2 leads to the two functions

$$F(x) = \int_1^x (t^2 - 1)dt = \tfrac{1}{3}x^3 - x + \tfrac{2}{3}$$

and

$$G(x) = \int_2^x (t^2 - 1)dt = \tfrac{1}{3}x^3 - x - \tfrac{2}{3}$$

In fact, all indefinite integrals of $f(x) = x^2 - 1$ are of the general form

$\frac{1}{3}x^3 - x + \text{constant}$

When the lower limit is not specified, we denote the indefinite integral by

$$\int f(x)\mathrm{d}x \quad \text{or} \quad \int^x f(t)\mathrm{d}t$$

and include the constant as an arbitrary **constant of integration**. Thus

$$\int (x^2 - 1)\mathrm{d}x = \tfrac{1}{3}x^3 - x + c$$

where c is the arbitrary constant of integration.

It is important to recognize that an indefinite integral is itself a function, while a definite integral is a number.

Figure 8.53
(a) Graph of $f(x)$.
(b) Graph of

$$F(x) = \int_0^x f(t)\mathrm{d}(t).$$

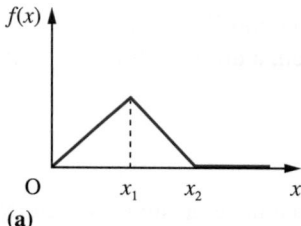

(a)

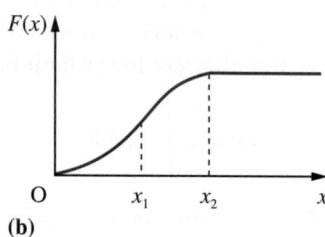

(b)

We noted in Section 8.2.4 that a function could only be differentiated at points where its graph had a unique tangent, and that, for example, the function represented by the graph of Figure 8.7(a), reproduced as Figure 8.53(a), is not differentiable at domain values $x = x_1$ and $x = x_2$. However, such functions are integrable, with the corresponding indefinite integrals being functions having 'smooth' graphs. For example, the graph of the indefinite integral $F(x)$ of the function $f(x)$ shown in Figure 8.53(a) has the form shown in Figure 8.53(b). For this reason, engineers often refer to integration as being a 'smoothing' process and an integrator is frequently incorporated within a system design in order to ensure 'smoother' operation.

We can express definite integrals in terms of indefinite integrals. Thus

$$\int_a^b f(x)\mathrm{d}x = g(b) - g(a) \quad \text{where} \quad g(x) = \int f(x)\mathrm{d}x$$

This is often denoted by

$$\int_a^b f(x)\mathrm{d}x = [\,g(x)\,]_a^b$$

a notation introduced by Fourier. Thus, for example,

$$\int_1^5 (x^2 - 1)\mathrm{d}x = [\tfrac{1}{3}x^3 - x + c]_1^5 = [\tfrac{125}{3} - 5 + c] - [\tfrac{1}{3} - 1 + c] = 37\tfrac{1}{3}$$

When evaluating definite integrals, the constant of integration can be omitted, since it cancels out in the arithmetic.

8.7.5 The Fundamental Theorem of Calculus

From Questions 88, 89 and 91 (Exercises 8.7.3) we have

$$\int_a^b 1\,dx = b - a, \qquad \text{giving} \quad \int 1\,dx = x + \text{constant}$$

$$\int_a^b x\,dx = \tfrac{1}{2}(b^2 - a^2), \quad \text{giving} \quad \int x\,dx = \tfrac{1}{2}x^2 + \text{constant}$$

$$\int_a^b x^2\,dx = \tfrac{1}{3}(b^3 - a^3), \quad \text{giving} \quad \int x^2\,dx = \tfrac{1}{3}x^3 + \text{constant}$$

The comparable results for differentiation are

$$\frac{d}{dx}(k) = 0, \quad k \text{ constant}$$

$$\frac{d}{dx}(x) = 1$$

$$\frac{d}{dx}(x^2) = 2x$$

Using the sum and constant multiplication rules for differentiation from Section 8.3.1,

$$\frac{d}{dx}[f(x) + k] = \frac{d}{dx}f(x) + \frac{d}{dx}(k) = \frac{d}{dx}f(x), \quad k \text{ constant}$$

$$\frac{d}{dx}[kf(x)] = k\frac{d}{dx}f(x)$$

the above results may be combined to give

$$\frac{d}{dx}\left(\int 1\,dx\right) = \frac{d}{dx}(x + \text{constant}) = 1$$

$$\frac{d}{dx}\left(\int x\,dx\right) = x$$

$$\frac{d}{dx}\left(\int x^2\,dx\right) = x^2$$

These results suggest a more general result:

> The process of differentiation is the inverse of that of integration.

This conjecture is also supported by elementary applications of the processes. We obtained the distance travelled by an object by integrating its velocity function. We obtained the velocity of an object by differentiating its distance function. The general result is called the **Fundamental Theorem of Integral and Differential Calculus**, and may be stated in the form of the following theorem.

Theorem 8.1	The indefinite integral $F(x)$ of a continuous function $f(x)$ always possesses a derivative $F'(x)$, and, moreover, $F'(x) = f(x)$.

Proof The formula for $F(x)$ may be written as

$$F(x) = \int_a^x f(t)\,\mathrm{d}t, \quad \text{where } a \text{ is a constant}$$

The quotient

$$\frac{F(x+h) - F(x)}{h}$$

may be written in terms of $f(x)$ as

$$\frac{F(x+h) - F(x)}{h} = \frac{\displaystyle\int_a^{x+h} f(t)\,\mathrm{d}t - \int_a^x f(t)\,\mathrm{d}t}{h} = \frac{1}{h}\int_x^{x+h} f(t)\,\mathrm{d}t$$

Consider the case when h is positive. The function $f(x)$ is continuous, and so it is bounded on $[x, x+h]$. Suppose it attains its upper bound at x_1, as shown in Figure 8.54, and its lower bound at x_2. Then by considering the area under the graph, we see that

Figure 8.54

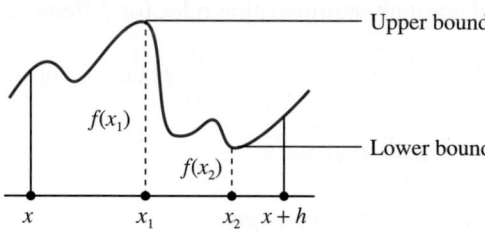

$$hf(x_2) \leqslant \int_x^{x+h} f(t)\,\mathrm{d}t \leqslant hf(x_1)$$

which implies that

$$f(x_2) \leqslant \frac{1}{h}\int_x^{x+h} f(t)\,\mathrm{d}t \leqslant f(x_1)$$

or equivalently

$$f(x_2) \leqslant \frac{F(x+h) - F(x)}{h} \leqslant f(x_1)$$

As $h \to 0$, $x_2 \to x$ and $x_1 \to x$, and we obtain the result

$$F'(x) = f(x)$$

(The proof when h is negative is similar.)

end of theorem

This theorem is of fundamental importance, and is used repeatedly in practical problem-solving using calculus.

8.7.6 Exercise

96 Using the Fundamental Theorem of Integral and Differential Calculus, evaluate the following integrals:

(a) $\int x^6 \, dx$, noting that $\dfrac{d}{dx} x^7 = 7x^6$

(b) $\int e^{3x} \, dx$, noting that $\dfrac{d}{dx} e^{3x} = 3e^{3x}$

(c) $\int \sin 5x \, dx$, noting that $\dfrac{d}{dx} \cos 5x$
$$= -5 \sin 5x$$

(d) $\int (2x+1)^3 \, dx$, noting that $\dfrac{d}{dx}(2x+1)^4$
$$= 8(2x+1)^3$$

(e) $\int \sec^2 3x \, dx$, noting that $\dfrac{d}{dx}(\tan 3x)$
$$= 3 \sec^2 3x$$

(f) $\int \dfrac{2}{x} \, dx$, noting that $\dfrac{d}{dx} \ln x = \dfrac{1}{x}$

(g) $\int \dfrac{3}{x^2} \, dx$, noting that $\dfrac{d}{dx}\left(\dfrac{1}{x}\right) = -\dfrac{1}{x^2}$

(h) $\int \cos 2x \, dx$, noting that $\dfrac{d}{dx} \sin 2x$
$$= 2 \cos 2x$$

(i) $\int \sec 4x \tan 4x \, dx$, noting that $\dfrac{d}{dx} \sec 4x$
$$= 4 \sec 4x \tan 4x$$

(j) $\int \sqrt{(4x-1)} \, dx$, noting that $\dfrac{d}{dx}(4x-1)^{3/2}$
$$= 6(4x-1)^{1/2}$$

8.8 Techniques of integration

In this section we consider some of the methods available for determining the integrals of functions. Again we shall concentrate on developing techniques, leaving problem-solving applications for later in both this chapter and the rest of the book.

8.8.1 Integration as antiderivative

Applying the Fundamental Theorem of Calculus to some of the standard derivatives deduced in Section 8.3, we deduce the integrals given in Figure 8.55. Note that we have used the notation

$$\ln|x| = \begin{cases} \ln x, & x > 0 \\ \ln(-x), & x < 0 \end{cases}$$

To help extend the number of functions that can be integrated analytically, using the results of Figure 8.55, the following rules may be used.

Figure 8.55
Some standard integrals.

$f(x)$	$\int f(x)\mathrm{d}x$
	Here c is a constant of integration
$x^n \quad (n \neq -1)$	$\dfrac{x^{n+1}}{n+1} + c$
$\dfrac{1}{x}$	$\left. \begin{array}{l} \ln x \quad + c \ (x > 0) \\ \ln(-x) + c \ (x < 0) \end{array} \right\} = \ln|x| + c$
$\sin x$	$-\cos x + c$
$\cos x$	$\sin x + c$
e^x	$\mathrm{e}^x + c$
$\sec^2 x$	$\tan x + c$
$\dfrac{1}{\sqrt{(1-x^2)}}, |x| < 1$	$\sin^{-1} x + c$
$\dfrac{1}{1+x^2}$	$\tan^{-1} x + c$

Rule 1 (scalar-multiplication rule)
If k is a constant then

$$\int k f(x)\mathrm{d}x = k \int f(x)\mathrm{d}x$$

Rule 2 (sum rule)

$$\int [f(x) \pm g(x)]\mathrm{d}x = \int f(x)\mathrm{d}x \pm \int g(x)\mathrm{d}x$$

Rule 3 (linear composite rule)
If a and b are constants and $F'(x) = f(x)$ then

$$\int f(ax+b)\mathrm{d}x = \frac{1}{a}F(ax+b) + \text{constant}, \quad a \neq 0$$

Rule 4 (inverse-function rule)
If $y = f^{-1}(x)$, so that $x = f(y)$, then

$$\int f^{-1}(x)\mathrm{d}x = xy - \int f(y)\mathrm{d}y$$

Rules 1–3 follow directly from the definition of an integral, while rule 4 may be demonstrated graphically as illustrated in Figure 8.56.

Figure 8.56
Illustration of
$\int f^{-1}(x)\,dx =$
$xy - \int f(y)\,dy.$

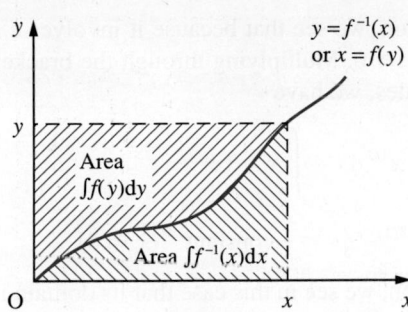

Example 8.41   Using the inverse-function rule, obtain the integrals of

(a) $\sin^{-1}x$ (b) $\ln x$

Solution   (a) If $y = \sin^{-1}x$ then $x = \sin y$ and

$$\int \sin^{-1}x\,dx = xy - \int \sin y\,dy$$

$$= xy + \cos y + \text{constant}$$

which, on using the identity $\sin^2 y + \cos^2 y = 1$, gives

$$\int \sin^{-1}x\,dx = x\sin^{-1}x + \sqrt{(1 - x^2)} + \text{constant}$$

since $\cos y \geqslant 0$ on the domain of $\sin^{-1}x$.

(b) If $y = \ln x$ then $x = e^y$, and

$$\int \ln x\,dx = xy - \int e^y\,dy = xy - e^y + \text{constant}$$

$$= x\ln x - x + \text{constant}$$

since $e^{\ln x} = x$.

Example 8.42   Find the indefinite integrals of

(a) $6x^4 + 4x - \dfrac{3}{x}$ (b) $(2 - x)\sqrt{x}$ (c) $\sqrt{(5x + 2)}$ (d) $\dfrac{x + 1}{x}$

Solution   (a) Using the scalar-multiplication and sum rules,

$$\int \left(6x^4 + 4x - \frac{3}{x}\right)dx = 6\int x^4\,dx + 4\int x\,dx - 3\int \frac{1}{x}\,dx$$

$$= \tfrac{6}{5}x^5 + 2x^2 - 3\ln|x| + \text{constant}$$

using the standard integrals of Figure 8.53.

(b) Looking at the function, we see that because it involves a square root, its domain is restricted to values of $x \geq 0$. Multiplying through the brackets and using the scalar-multiplication and sum rules, we have

$$\int (2-x)\sqrt{x}\,dx = 2\int x^{1/2}\,dx - \int x^{3/2}\,dx$$

$$= \tfrac{4}{3}x^{3/2} - \tfrac{2}{5}x^{5/2} + \text{constant} \quad (x \geq 0)$$

(c) Examining the function, we see in this case that its domain is restricted to values of x greater than or equal to $-\tfrac{2}{5}$. We note that the formula is the square root of a linear function, and so we use the linear composite rule to obtain its integral. Thus, since

$$\int \sqrt{x}\,dx = \tfrac{2}{3}x^{3/2} + \text{constant}$$

we obtain

$$\int \sqrt{(5x+2)}\,dx = \tfrac{1}{5}[\tfrac{2}{3}(5x+2)^{3/2}] + \text{constant}$$

$$= \tfrac{2}{15}(5x+2)^{3/2} + \text{constant} \quad (x \geq -\tfrac{2}{5})$$

(d) In this case we see that the function is defined except at $x = 0$. Expressing $(x+1)/x$ as $1 + 1/x$ and using the sum rule, we obtain

$$\int \frac{x+1}{x}\,dx = \int \left(1 + \frac{1}{x}\right)dx = \int 1\,dx + \int \frac{1}{x}\,dx$$

$$= \begin{cases} x + \ln x + \text{constant} & (x > 0) \\ x + \ln(-x) + \text{constant} & (x < 0) \end{cases}$$

$$= x + \ln|x| + \text{constant}$$

MATLAB's Symbolic Math Toolbox and MAPLE can evaluate both indefinite and definite integrals. If $y = f(x)$ then the MATLAB command $int(y)$ returns the indefinite integral of $f(x)$, provided it exists in closed form. (Symbolic integration is more difficult than symbolic differentiation and difficulties can arise in computing the integral.) Thus in MATLAB the indefinite integral of $y = f(x)$ is returned using the commands

```
syms x y
y = f(x);  int(y)
```

To determine the definite integral of $y = f(x)$ from $x = a$ to $x = b$ the last command is replaced by

```
int(y,a,b)
```

The corresponding commands in MAPLE are

```
y:= f(x);  int(y,x);  int(y,x = a..b);
```

Note that MATLAB and MAPLE do not supply a constant of integration when evaluating indefinite integrals. To illustrate we consider Examples 8.42(a) and (c). For Example 8.42(a) the commands

MATLAB

```
syms x y
y = 6*x^4 + 4*x - 3/x;
int(y);
pretty(ans)
```

MAPLE

```
y:= 6*x^4 + 4*x - 3/x;
int(y,x);
```

return the integral as

$$6/5x^5 + 2x^2 - 3\log(x)$$

For Example 8.42(c) the commands

```
syms x y
y = sqrt(5*x + 2);
int(y);
pretty(ans)
```

```
y:= sqrt(5*x + 2);
int(y,x);
```

return the integral as

$$2/15(5x + 2)^{3/2} \qquad \frac{2}{15}(5x + 2)^{(3/2)}$$

For practice check the answers to Examples 8.42(b) and (d) using MATLAB or MAPLE.

Example 8.43 Evaluate the integrals

(a) $\displaystyle\int_0^6 \tfrac{4}{9}(6 - h)^2 \, dh$ (b) $\displaystyle\int_0^{10} 3\sqrt{10}(p_0 + 10g - gy)\sqrt{(y)} \, dy$

Solution (a) Notice first of all that the label used for the integrating variable in a **definite** integral does not affect the value of the integral. It is a dummy variable. Thus

$$\int_0^6 \tfrac{4}{9}(6 - h)^2 \, dh = \int_0^6 \tfrac{4}{9}(6 - x)^2 \, dx$$

Expanding the integrand, we have

$$\int_0^6 \tfrac{4}{9}(6 - x)^2 \, dx = \int_0^6 \tfrac{4}{9}(36 - 12x + x^2) \, dx = \tfrac{4}{9}[36x - 6x^2 + \tfrac{1}{3}x^3]_0^6$$

$$= \tfrac{4}{9}[36 \times 6 - 6 \times 36 + \tfrac{1}{3} \times 216] - \tfrac{4}{9}[0]$$

$$= \tfrac{4}{9}[72] = 32$$

as predicted in Example 8.37

(b) $\displaystyle\int_0^{10} 3\sqrt{10}(p_0 + 10g - gy)\sqrt{(y)}dy = 3\sqrt{10}\int_0^{10} [(p_0 + 10g)y^{1/2} - gy^{3/2}]dy$

$$= 3\sqrt{10}[\tfrac{2}{3}(p_0 + 10g)y^{3/2} - \tfrac{2}{5}gy^{5/2}]_0^{10}$$

$$= 3\sqrt{10}[\tfrac{2}{3}(p_0 + 10g)10^{3/2} - \tfrac{2}{5}g10^{5/2}] - 0$$

$$= 200p_0 + 800g$$

as predicted in Example 8.38.

We have seen in Example 8.42 how a carefully chosen rearrangement of the integrand makes it possible to evaluate non-standard integrals. This technique is widely used in finding integrals of products of sines and cosines and of quadratic functions and their square roots. The first case uses the trigonometric sum identities (Section 2.6.4) and the second uses 'completion of the square'.

Example 8.44   Find the indefinite integrals of

(a) $\sin(5x + 1)\cos(x + 2)$ (b) $\dfrac{1}{x^2 + 10x + 50}$

Solution   (a) First we express the product as the sum of two sine terms:

$$\sin(5x + 1)\cos(x + 2) = \tfrac{1}{2}[\sin(6x + 3) + \sin(4x - 1)]$$

Then we evaluate the integral, using the scalar-multiplication, sum and linear composite rules:

$$\int \sin(5x + 1)\cos(x + 2)dx = \int \frac{1}{2}[\sin(6x + 3) + \sin(4x - 1)]dx$$

$$= \frac{1}{2}\int \sin(6x + 3)dx + \frac{1}{2}\int \sin(4x - 1)dx$$

$$= -\frac{1}{12}\cos(6x + 3) - \frac{1}{8}\cos(4x - 1) + \text{constant}$$

recalling that

$$\int \sin x\,dx = -\cos x + \text{constant}$$

(b) Rewriting the quadratic term by completing the square (1.7d) gives

$$\int \frac{1}{x^2 + 10x + 50}dx = \int \frac{1}{(x + 5)^2 + 5^2}dx = \int \frac{1}{25[(\tfrac{1}{5}x + 1)^2 + 1]}dx$$

Recalling that

$$\int \frac{1}{x^2 + 1} dx = \tan^{-1}x + \text{constant}$$

we have, using the linear composite and scalar-multiplication rules,

$$\int \frac{1}{x^2 + 10x + 50} dx = \frac{1}{25} \frac{\tan^{-1}(\frac{1}{5}x + 1)}{1/5} + \text{constant} = \frac{1}{5} \tan^{-1}\frac{x + 5}{5} + \text{constant}$$

Finally, we consider the use of partial fractions in evaluating integrals of rational functions. Partial fractions are so frequently used to evaluate such integrals that one talks of the **partial fraction method of integration**.

Example 8.45 Using partial fractions, evaluate the integrals

(a) $\int \frac{6}{x^2 - 2x - 8} dx$ (b) $\int \frac{9}{(x - 1)(x + 2)^2} dx$ (c) $\int_0^6 \frac{1}{x^2 + 5x + 6} dx$

Solution (a) Factorizing the denominator as $x^2 - 2x - 8 = (x + 2)(x - 4)$, we can express the integrand in terms of its partial fractions:

$$\frac{6}{x^2 - 2x - 8} = \frac{6}{(x + 2)(x - 4)} = \frac{-1}{x + 2} + \frac{1}{x - 4}$$

Thus

$$\int \frac{6}{x^2 - 2x - 8} dx = \int \frac{-1}{x + 2} dx + \int \frac{1}{x - 4} dx$$

$$= -\ln|x + 2| + \ln|x - 4| + \text{constant}$$

$$= \ln\left|\frac{x - 4}{x + 2}\right| + \text{constant}$$

(b) In partial fractions we have

$$\frac{9}{(x - 1)(x + 2)^2} = \frac{1}{x - 1} + \frac{-1}{x + 2} + \frac{-3}{(x + 2)^2}$$

Then

$$\int \frac{9}{(x - 1)(x + 2)^2} dx = \int \frac{1}{x - 1} dx - \int \frac{1}{x + 2} dx - \int \frac{3}{(x + 2)^2} dx$$

$$= \ln|x - 1| - \ln|x + 2| + 3(x + 2)^{-1} + \text{constant}$$

$$= \ln\left|\frac{x - 1}{x + 2}\right| + \frac{3}{x + 2} + \text{constant}$$

(c) In partial fractions we have

$$\frac{1}{x^2 + 5x + 6} = \frac{1}{x + 2} - \frac{1}{x + 3}$$

so that

$$\int_0^6 \frac{1}{x^2 + 5x + 6} dx = \int_0^6 \left[\frac{1}{x + 2} - \frac{1}{x + 3} \right] dx$$

$$= [\ln(x + 2) - \ln(x + 3)]_0^6$$

$$= \ln(\tfrac{8}{2}) - \ln(\tfrac{9}{3}) = \ln 4 - \ln 3 = \ln(\tfrac{4}{3})$$

Confirm the answers to Examples 8.44 and 8.45 using MATLAB or MAPLE. As examples consider Examples 8.44(a) and 8.45(c). For Example 8.44(a) the commands

MATLAB
```
syms x y
y = sin(5*x + 1)*cos(x + 2);
int(y);
pretty(ans)
```

MAPLE
```
y:= sin(5*x + 1)*cos(x + 2):
int(y,x);
```

return the integral as

$$-1/12cos(6x + 3) - 1/8cos(4x - 1)$$

For Example 8.45(c) the commands

```
syms x y
y = 1/(x^2 + 5*x + 6);
int(y,0,6)
```

```
y:= 1/(x^2 + 5*x + 6):
int(y,x = 0..6);
```

return the answer

$$-log(3) + 2*log(2) \qquad\qquad -ln(3) + 2ln(2)$$

(Note: This answer may be further simplified to $\log 2^2 + \log 3 = \log(4/3)$, as given in the solution)
To obtain a numerical answer enter the command

```
double(ans)                    evalf(%);
```

to give the answer 0.2877 to 4dp

Note in the preceding examples how the problems are systematically reduced to simpler ones by rearranging the integrand and carefully applying the basic rules together with standard integrals. The ability to do this comes only with practice of the type offered by the exercises at the end of this section.

In addition to the rules given earlier, two further results follow immediately from the basic definition of an integral. These are

$$\int_a^b f(x)\,dx = -\int_b^a f(x)\,dx$$

and

$$\int_a^b f(x)\,dx = \int_a^c f(x)\,dx + \int_c^b f(x)\,dx$$

(Thus we may break the interval $[a, b]$ into convenient subintervals if the function is defined piecewise as illustrated in Example 8.46.)

Example 8.46 Evaluate

(a) $\displaystyle\int_{-1}^2 |x|\,dx$ (b) $\displaystyle\int_0^{10} H(x-5)\,dx$

where H is the Heaviside step function given by (2.45).

Solution The areas involved are illustrated in Figure 8.57.

(a) Since

$$|x| = \begin{cases} -x & (x \leqslant 0) \\ x & (x \geqslant 0) \end{cases}$$

we split the integral at $x = 0$ and write

$$\int_{-1}^2 |x|\,dx = \int_{-1}^0 -x\,dx + \int_0^2 x\,dx = [-\tfrac{1}{2}x^2]_{-1}^0 + [\tfrac{1}{2}x^2]_0^2 = \tfrac{5}{2}$$

(b) Since $H(x-5)$ has a discontinuity at $x = 5$, we write

$$\int_0^{10} H(x-5)\,dx = \int_0^5 H(x-5)\,dx + \int_5^{10} H(x-5)\,dx$$

$$= \int_0^5 0\,dx + \int_5^{10} 1\,dx = 5$$

These results can be readily confirmed by inspection of the relevant areas.

Figure 8.57

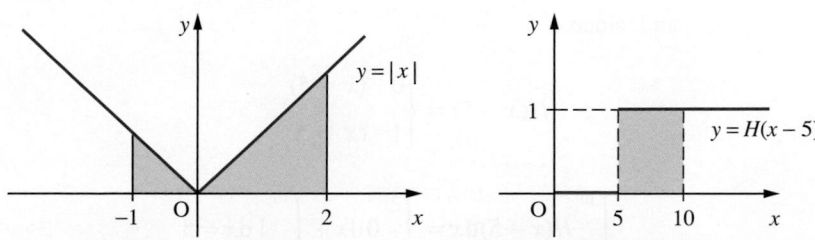

Figure 8.58
Piecewise-continuous
function.

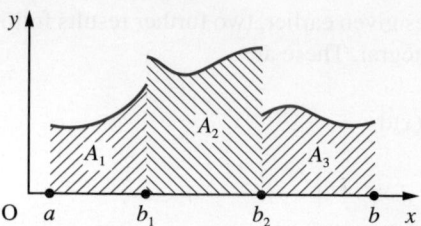

We see from this last example that it is sometimes possible to integrate functions even if they have discontinuities. This is possible provided that there are only a finite number of finite discontinuities within the domain of integration and that elsewhere the function is continuous and bounded. To illustrate this, consider the function $f(x)$ illustrated in Figure 8.58 where

$$y = f(x) = \begin{cases} f_1(x) & (a \leqslant x < b_1) \\ f_2(x) & (b_1 < x < b_2) \\ f_3(x) & (b_2 < x \leqslant b) \end{cases}$$

Such a function is called a **piecewise-continuous function**. Interpreting the integral as the area under the curve, we have

$$\int_a^b f(x)\mathrm{d}x = A_1 + A_2 + A_3$$

but in this case we interpret the individual areas as

$$\int_a^b f(x)\mathrm{d}x = \int_a^{b_1^-} f_1(x)\mathrm{d}x + \int_{b_1^+}^{b_2^-} f_2(x)\mathrm{d}x + \int_{b_2^+}^b f_3(x)\mathrm{d}x$$

where, as before, b_1^- signifies approaching b_1 from the left and b_1^+ signifies approaching b_1 from the right (see Section 7.8.1). It is in this sense that we evaluated $\int_0^{10} H(x-5)\mathrm{d}x$ in Example 8.45, and – strictly speaking – we should have written

$$\int_0^{10} H(x-5)\mathrm{d}x = \int_0^{5^-} H(x-5)\mathrm{d}x + \int_{5^+}^{10} H(x-5)\mathrm{d}x$$

and, since

$$H(x-5) = \begin{cases} 0 & (x < 5) \\ 1 & (x \geqslant 5) \end{cases}$$

$$\int_0^{10} H(x-5)\mathrm{d}x = \int_0^{5^-} 0\,\mathrm{d}x + \int_{5^+}^{10} 1\,\mathrm{d}x = 5$$

Considering Example 8.46(a) the commands

MATLAB	MAPLE
`syms x y`	
`y = abs(x);`	`y:= abs(x);`
`int(y,-1,2)`	`int(y,x = -1..2);`

return the answer $5/2$

and, for Example 8.46(b), the commands

`syms x y`	
`y = sym('Heaviside(x - 5)')`	
`int(y,0,10)`	`int(Heaviside(x - 5), x = 0..10);`

return the answer 5

Example 8.47

As shown in Example 8.7, the bending moment M and shear force F acting in a beam satisfy the differential equation

$$F = \frac{dM}{dx}$$

In Example 8.40, we showed that for a continuously non-uniformly loaded beam which is freely hinged at both ends the shear force F is given by

$$F(x) = \begin{cases} 2Wx^2/l^2 - W/2 & 0 \leqslant x \leqslant l/2 \\ W/2 - 2W(l - x)^2/l^2 & l/2 \leqslant x \leqslant l \end{cases}$$

Given that $M = 0$ at $x = 0$, find an expression for $M(x)$ at a general point.

Solution   Since $\dfrac{dM}{dx} = F(x)$ with $M(0) = 0$ we deduce by the Fundamental Theorem

$$M(x) = \int_0^x F(t)\,dt$$

In evaluating this integral we have to remember that $F(x)$ is defined separately on $(0, l/2)$ and $(l/2, l)$.

For $x < l/2$, we have

$$M(x) = \int_0^x (2Wt^2/l^2 - W/2)\,dt$$

For $x > l/2$, we have

$$M(x) = \int_0^{l/2} (2Wt^2/l^2 - W/2)\,dt + \int_{l/2}^x (W/2 - 2W(l - t)^2/l^2)\,dt$$

Thus

$$M(x) = \begin{cases} \dfrac{Wx}{6l^2}(4x^2 - 3l^2) & 0 \leqslant x \leqslant l/2 \\[3mm] \dfrac{W(l-x)}{6l^2}(4(l-x)^2 - 3l^2) & l/2 \leqslant x \leqslant l \end{cases}$$

8.8.2 Exercises

Check your answers using MATLAB or MAPLE whenever possible.

97 Find the indefinite integrals of

(a) $3x^{2/3}$

(b) $\sqrt{(2x)}$

(c) $2x^3 - 2x^2 + \dfrac{1}{x} - 2$

(d) $2e^x + 3\cos 2x$

(e) $x^2 + 3e^x - \dfrac{1}{x^2}$

(f) $(2x+1)^3$

(g) $(1-2x)^{1/3}$

(h) $(2x^2+1)^3$

(i) $\cos(2x+1)$

(j) 2^x (*Hint*: $2 = e^{\ln 2}$)

(k) $\sin 3x \cos 5x$

(l) $\cos 7x \cos 5x$

98 Using partial fractions, integrate

(a) $\dfrac{x}{x^2 - 3x - 4}$

(b) $\dfrac{x}{(x-2)^2}$

(c) $\dfrac{1}{x(x+1)}$

(d) $\dfrac{x}{x^2 + 2x + 1}$

(e) $\dfrac{1}{x^2 - 1}$

(f) $\dfrac{1}{x^2(x-1)}$

(g) $\dfrac{1}{x(x-1)(x-2)}$

(h) $\dfrac{1}{1 + x - 2x^2}$

(i) $\dfrac{2x^3}{x^3 - 1}$

(j) $\dfrac{3x^3 - 3x^2 + 4x - 2}{x(x-1)(x^2+1)}$

(k) $\dfrac{9}{(x-1)(x+2)^2}$

(l) $\dfrac{x^2 - 2x + 3}{(x-1)(x^2 - x - 1)}$

99 Evaluate the definite integrals

(a) $\displaystyle\int_2^3 \dfrac{x\,dx}{\sqrt{(x+1)}}$

(b) $\displaystyle\int_0^1 x(x-1)^{11}dx$

(c) $\displaystyle\int_0^\pi \sin 5x \sin 6x\,dx$

(d) $\displaystyle\int_0^\pi \sin^2 5x\,dx$

(e) $\displaystyle\int_1^2 \left(x^{3/2} - \dfrac{1}{x^2}\right)dx$

(f) $\displaystyle\int_0^{\pi/2} \sin x\,dx$

(g) $\displaystyle\int_0^2 \dfrac{dx}{\sqrt{(3 + 2x - x^2)}}$

(h) $\displaystyle\int_0^1 \dfrac{2\,dx}{(x+1)^2(x^2+1)}$

(*Hint*: Replace the x in (a) by $(x+1) - 1$ and in (b) by $(x-1) + 1$.)

100 Express $12/(x-3)(x+1)$ in partial fractions and hence show that

$$\int_4^6 \dfrac{12}{(x-3)(x+1)}\,dx = 3\ln\tfrac{15}{7}$$

101 Find the indefinite integrals of

(a) x^{-2}

(b) $(x+1)^{-1/3}$

(c) $\dfrac{4x^3 - 7x^2 + 1}{x^2}$

(d) $\sin x + \cos x$

(e) $\sin^2 x$

(f) $\cos^2 x$

(g) $\cosh^2 x$

(h) $\sinh(5x+1)$

(i) $\dfrac{1}{9 - 16x^2}$

(j) $\dfrac{1}{\sqrt{(2x - x^2)}}$

(k) $\dfrac{1}{\sqrt{(1 - 9x^2)}}$

(l) $\dfrac{1}{\sqrt{(4 - x^2)}}$

(m) $\dfrac{1}{\sqrt{(1 - x - x^2)}}$

(n) $\dfrac{1}{\sqrt{[x(1-x)]}}$

(o) $\dfrac{1}{\sqrt{(5 + 4x - x^3)}}$

(p) $\dfrac{1}{x^2 + 6x + 13}$

102 Evaluate

(a) $\displaystyle\int_0^3 |x-2|\,\mathrm{d}x$ (b) $\displaystyle\int_0^5 (x-2)H(x-2)\,\mathrm{d}x$

(c) $\displaystyle\int_0^3 \lfloor x\rfloor\,\mathrm{d}x$ (d) $\displaystyle\int_0^3 \mathrm{FRACPT}(x)\,\mathrm{d}x$

(e) $\displaystyle\int_0^3 x\lfloor x\rfloor\,\mathrm{d}x$

103 The function $f(x)$ is periodic with period 1 and is defined on $[0, 1]$ by

$$f(x) = 1 \qquad 0 \leqslant x < \tfrac{1}{2}$$
$$f(x) = -1 \qquad \tfrac{1}{2} \leqslant x < 1$$

Sketch its graph and obtain the graph of

$$g(x) = \int_0^x f(t)\,\mathrm{d}t$$

for $-4 \leqslant x \leqslant 4$. Show that $g(x)$ is a periodic function of period 1.

104 Draw the graph of the function $f(x)$ defined by

$$f(x) = \int_0^x \sin^{-1}(\sin t)\,\mathrm{d}t$$

for $-2\pi < x < 2\pi$ (see Example 2.51).

8.8.3 Integration by parts

The product rule for differentiation

$$\frac{\mathrm{d}}{\mathrm{d}x}(uv) = \frac{\mathrm{d}u}{\mathrm{d}x}v + u\frac{\mathrm{d}v}{\mathrm{d}x}$$

may also be used for integration after a little rearrangement. From the above we have

$$u\frac{\mathrm{d}v}{\mathrm{d}x} = \frac{\mathrm{d}}{\mathrm{d}x}(uv) - v\frac{\mathrm{d}u}{\mathrm{d}x}$$

and on integrating we have

$$\int u\frac{\mathrm{d}v}{\mathrm{d}x}\,\mathrm{d}x = uv - \int v\frac{\mathrm{d}u}{\mathrm{d}x}\,\mathrm{d}x$$

We may use this result to determine an integral when the integrand is the product of the two functions. The method is called **integration by parts**. The procedure is to choose one term of the product to be u and the other to be $\mathrm{d}v/\mathrm{d}x$. We then calculate $\mathrm{d}u/\mathrm{d}x$ and v, and the hope is that the resulting integral on the right-hand side is easier than the one we started with. We shall illustrate the method with a few examples.

Example 8.48   Find the indefinite integrals of

(a) $x\ln x$ (b) $x^2\cos x$ (c) $\mathrm{e}^x\sin 2x$

Solution   (a) With this integral, we set

$$u = \ln x \quad\text{and}\quad \frac{\mathrm{d}v}{\mathrm{d}x} = x$$

giving

$$\frac{du}{dx} = \frac{1}{x} \quad \text{and} \quad v = \tfrac{1}{2}x^2$$

Note: There is no need to introduce a constant of integration when determining v. Substituting in the formula for integration by parts gives

$$dv/dx \; u \qquad\qquad v \quad u \qquad\qquad v \; du/dx$$
$$\downarrow\;\downarrow \qquad\qquad \downarrow\;\downarrow \qquad\qquad \downarrow\;\downarrow$$

$$\int x \ln x \, dx = (\tfrac{1}{2}x^2) \ln x - \int (\tfrac{1}{2}x^2)\left(\frac{1}{x}\right) dx = \tfrac{1}{2}x^2 \ln x - \int \tfrac{1}{2}x \, dx$$

$$= \tfrac{1}{2}x^2 \ln x - \tfrac{1}{4}x^2 + \text{constant}$$

(b) Since differentiation reduces the squared term to a linear one, leading to some simplification, we choose

$$u = x^2 \quad \text{and} \quad \frac{dv}{dx} = \cos x$$

so that

$$\frac{du}{dx} = 2x \quad \text{and} \quad v = \sin x$$

Integration by parts then gives

$$u \quad dv/dx \qquad u \quad v \qquad\qquad v \; du/dx$$
$$\downarrow \qquad \downarrow \qquad\quad \downarrow\;\downarrow \qquad\qquad \downarrow\;\downarrow$$

$$\int x^2 \cos x \, dx = x^2(\sin x) - \int (\sin x)(2x) dx = x^2 \sin x - 2\int x \sin x \, dx$$

We now apply the same technique to the last integral, taking

$$u = x \quad \text{and} \quad \frac{dv}{dx} = \sin x$$

to give

$$\int x \sin x \, dx = (x)(-\cos x) - \int (-\cos x)(1) dx = -x \cos x + \sin x + \text{constant}$$

Substituting back gives

$$\int x^2 \cos x \, dx = x^2 \sin x - 2(-x \cos x + \sin x) + \text{constant}$$

$$= x^2 \sin x - 2 \sin x + 2x \cos x + \text{constant}$$

(c) In this case it is not obvious that any choice of u and v will result in a simpler integral. Setting

$$u = \sin 2x \quad \text{and} \quad \frac{dv}{dx} = e^x$$

(only because integrating $\sin 2x$ will mean dividing by 2 and getting clumsy fractions!) gives

$$\frac{du}{dx} = 2 \cos 2x \quad \text{and} \quad v = e^x$$

Integration by parts then gives

$$\int e^x \sin 2x \, dx = e^x \sin 2x - \int e^x (2 \cos 2x) dx$$

$$= e^x \sin 2x - 2 \int e^x \cos 2x \, dx$$

which has produced no simplification at all. We repeat the process, however, on the last integral, taking care to integrate the part we integrated the first time and to differentiate the part we differentiated the first time. Thus we take

$$u = \cos 2x \quad \text{and} \quad \frac{dv}{dx} = e^x$$

giving

$$\int e^x \cos 2x \, dx = e^x \cos 2x - \int e^x (-2 \sin 2x) dx$$

$$= e^x \cos 2x + 2 \int e^x \sin 2x \, dx$$

Substituting in the previous expression, we obtain

$$\int e^x \sin 2x \, dx = e^x \sin 2x - 2 \left(e^x \cos 2x + 2 \int e^x \sin 2x \, dx \right)$$

Hence

$$5 \int e^x \sin 2x \, dx = e^x (\sin 2x - 2 \cos 2x)$$

so

$$\int e^x \sin 2x \, dx = \tfrac{1}{5} e^x (\sin 2x - 2 \cos 2x) + \text{constant}$$

For Example 8.48(b) the MATLAB commands

```
syms x y
y = (x^2)*cos(x); int(y); pretty(ans)
```

return the integral as $x^2\sin(x) - 2\sin(x) + 2x\cos(x)$, which checks with the given solution.

For practice check the answers to Examples 8.48(a) and (c) using MATLAB or MAPLE.

8.8.4 Exercises

Check your answers using MATLAB or MAPLE whenever possible

105 Use integration by parts to find the indefinite integrals of

(a) $x \sin x$ (b) xe^{3x} (c) $x^3 \ln x$

(d) $e^{-2x}\sin 3x$ (e) $x \tan^{-1}x$ (f) $x \cos 2x$

106 Show that

$$\int f(x)\,dx = xf(x) - \int xf'(x)\,dx$$

Use this result to integrate

(a) $\sin^{-1}x$ (b) $\ln x$ (c) $\cosh^{-1}x$ (d) $\tan^{-1}x$

107 Using integration by parts, evaluate the definite integrals

(a) $\displaystyle\int_0^{\pi/2} x^2 \sin x\,dx$

(b) $\displaystyle\int_1^3 x^2 \ln x\,dx$

(c) $\displaystyle\int_0^1 xe^{3x}\,dx$

8.8.5 Integration by substitution

The composite-function rule for differentiation

$$\frac{d}{dx}[f(g(x))] = f'(g(x))g'(x)$$

can be used to evaluate some integrals. Reversing the differentiation process, we may write

$$\int f'(g(x))g'(x)\,dx = f(g(x)) + \text{constant}$$

The key step here is identifying the function $g(x)$. This will not be unique: different choices of $g(x)$ may differ by a constant. To make the process of manipulation easier to follow, it is usual to set $t = g(x)$, so that the integral becomes

$$\int f'(g(x))g'(x)\,dx = \int f'(t)\frac{dt}{dx}\,dx = \int f'(t)\,dt = f(t) + \text{constant}$$

$$= f(g(x)) + \text{constant} \quad \text{(on back substitution)}$$

This technique for evaluating integrals is called the **substitution method**; we shall illustrate its use with a number of examples.

Example 8.49   Find the indefinite integrals

(a) $\displaystyle\int 2x\sqrt{(x^2+3)}\,dx$ (b) $\displaystyle\int \frac{x+1}{x^2+2x+2}\,dx$

Solution   (a) Comparison with the general form above suggests that we take

$$g(x) = x^2 + 3, \quad \text{with } g'(x) = 2x$$

Setting $t = x^2 + 3$, the integral becomes

$$\int 2x\sqrt{(x^2+3)}\,dx = \int \frac{dt}{dx}\sqrt{t}\,dx = \int t^{1/2}dt$$

$$= \tfrac{2}{3}t^{3/2} + \text{constant} = \tfrac{2}{3}(x^2+3)^{3/2} + \text{constant}$$

(b) Comparison with the general form suggests that we choose

$$g(x) = x^2 + 2x + 2, \quad \text{with } g'(x) = 2x + 2$$

This necessitates a slight modification of the integral giving

$$\int \frac{x+1}{x^2+2x+2}\,dx = \tfrac{1}{2}\int \frac{2x+2}{x^2+2x+2}\,dx = \tfrac{1}{2}\int \frac{1}{t}\,dt$$

where $t = x^2 + 2x + 2$ and $dt = (2x+2)dx$. Thus

$$\int \frac{x+1}{x^2+2x+2}\,dx = \tfrac{1}{2}\ln t + \text{constant} = \tfrac{1}{2}\ln(x^2+2x+2) + \text{constant}$$

This example is a special case of a commonly occurring form when the integrand can be written as

$$\frac{\text{derivative of denominator}}{\text{denominator}}$$

so that the integral is the logarithm of the denominator.

The choice of substitution $t = g(x)$ that results in a convenient simplification of the integral is not always dictated by the form of the composite-function rule given above. Sometimes the integrand contains terms that suggest an initial substitution, at least, in the hope of simplification.

Example 8.50   Find the indefinite integral

$$\int \frac{1}{2 + \sqrt{(1 - x)}}\, dx$$

Solution   The source of the difficulty with this integral is the square-root term in the denominator. We try to simplify the integral by the substitution $t = \sqrt{(1 - x)}$. Thus $x = 1 - t^2$ and $dx/dt = -2t$, giving

$$\int \frac{1}{2 + \sqrt{(1 - x)}}\, dx = \int \frac{1}{2 + t}\frac{dx}{dt}\, dt = \int \frac{1}{2 + t}(-2t)\, dt$$

$$= \int \frac{-2t}{2 + t}\, dt = 2 \int \left(\frac{2}{2 + t} - 1 \right) dt$$

$$= 4\ln(2 + t) - 2t + \text{constant}$$

$$= 4\ln[2 + \sqrt{(1 - x)}] - 2\sqrt{(1 - x)} + \text{constant}$$

The choice of such substitutions is not always immediately obvious. We shall consider a further example and then give a list of substitutions commonly used to simplify integrals.

Example 8.51   Find the indefinite integral $\int \sqrt{(1 - x^2)}\, dx$, $0 \leqslant x \leqslant 1$.

Solution   Based on our experience with Example 8.49, we are tempted to try to remove the square-root term using the substitution

$$u = \sqrt{(1 - x^2)}$$

Then $u^2 = 1 - x^2$ and $2u = -2x\, dx/du$, so that

$$\frac{dx}{du} = -\frac{u}{x} = -\frac{u}{\sqrt{(1 - u^2)}}$$

giving

$$\int \sqrt{(1 - x^2)}\, dx = -\int \frac{u^2\, du}{\sqrt{(1 - u^2)}}$$

which leaves us with an integral more complicated than the one with which we started.

Thus in this case the simple substitution does not work, and we need to look for a more sophisticated substitution, bearing in mind that what we wish to do is to remove the awkward square-root term $\sqrt{(1 - x^2)}$. Noting that $\cos^2\theta = 1 - \sin^2\theta$ we try the substitution $x = \sin\theta$ so that $\dfrac{dx}{d\theta} = \cos\theta$ giving

$$\int \sqrt{(1 - x^2)}\,dx = \int \sqrt{(1 - \sin^2\theta)}\,\frac{dx}{d\theta}\,d\theta$$

$$= \int \cos\theta\cos\theta\,d\theta$$

$$= \int \cos^2\theta\,d\theta$$

which looks simpler than the original integral but is not immediately integrable.
    Using the double-angle trigonometric identity (see 2.27c)

$$\cos 2\theta = 2\cos^2\theta - 1$$

we obtain

$$\int \sqrt{(1 - x^2)}\,dx = \int \tfrac{1}{2}(1 + \cos 2\theta)\,d\theta$$

$$= \tfrac{1}{2}\theta + \tfrac{1}{4}\sin 2\theta + \text{constant}$$

This gives the answer in terms of θ rather than the original variable x. Since $\theta = \sin^{-1}x$, back substitution gives

$$\int \sqrt{(1 - x^2)}\,dx = \tfrac{1}{2}\sin^{-1}x + \tfrac{1}{4}\sin(2\sin^{-1}x) + \text{constant}$$

or, since $\sin 2\theta = 2\sin\theta\cos\theta = 2\sin\theta\sqrt{(1 - \sin^2\theta)}$, we may write this in the alternative form

$$\int \sqrt{(1 - x^2)}\,dx = \tfrac{1}{2}\sin^{-1}x + \tfrac{1}{2}x\sqrt{(1 - x^2)} + \text{constant}$$

Figure 8.59 shows a number of substitutions that are often used in the evaluation of $\int f(x)\,dx$. This list is not exhaustive. There are many special cases, some of which are given in Exercises 8.8.6.

When using substitution methods with definite integrals, it is usually best to change the limits of the integral when the integrating variable is changed. This saves returning to the original variable, which can sometimes be very tedious. In general, setting $x = g(t)$ gives

$$\int_a^b f(x)\,dx = \int_{g^{-1}(a)}^{g^{-1}(b)} f(g(t))g'(t)\,dt$$

$$= \int_{t_a}^{t_b} h(t)\,dt, \quad \text{where } a = g(t_a),\ b = g(t_b) \text{ and } h(t) = f(g(t))g'(t)$$

Example 8.52

Using the substitution $u = \sqrt{(x + 2)}$, evaluate the definite integral

$$\int_{-2}^{2} \frac{\sqrt{(x + 2)}}{x + 6} \, dx$$

Solution

Setting $u = \sqrt{(x + 2)}$, or $u^2 = x + 2$, gives $2u \, du = dx$. Regarding limits, when $x = -2$, $u = 0$ and when $x = 2$, $u = \sqrt{4} = 2$.

Making the substitution gives

$$\int_{-2}^{2} \frac{\sqrt{(x + 2)}}{x + 6} \, dx = \int_{0}^{2} \frac{u}{u^2 + 4} 2u \, du = \int_{0}^{2} \frac{2u^2}{u^2 + 4} \, du = \int_{0}^{2} 2 - \frac{8}{u^2 + 4} \, du$$

$$= \left[2u - 4\tan^{-1}\frac{u}{2} \right]_{0}^{2} = 4 - \pi$$

For Example 8.50 the MATLAB commands

```
syms x y
y = 1/(2 + sqrt(1 - x)); int(y); pretty(ans)
```

return the integral as

```
2log(-x - 3) - 2(1 - x)^1/2 - 2log(-2 + (1 - x)^1/2)
+ 2log(2 + (1 - x)^1/2)
```

Some algebraic manipulation is necessary to obtain the answer in the form given in the solution. Collecting the `log` terms gives

$$-2(1 - x)^{1/2} + 2\log \frac{(-x - 3)[2 + (1 - x)^{1/2}]}{[-2 + (1 - x)^{1/2}]}$$

Multiplying 'top and bottom' of the log term by $(2 + (1 - x)^{1/2})$ and subsequent cancelling of the $(-x - 3)$ term gives the answer in the form given in the solution.

The corresponding MAPLE commands

```
y:= 1/(2 + sqrt(1 - x)); int(y,x);
```

return the integral as

```
2ln(-x - 3) - 2√(1 - x) + 4arctanh (½√(1 - x))
```

Using the command

```
convert(%, ln);
```

the answer is expressed in the logarithmic form

```
2ln(-x - 3) - 2√(1 - x) + 2ln(2 + √(1 - x))
- 2ln(2 - √(1 - x))
```

This example clearly emphasizes the fact that integrals can be the same even if they look totally different. For practice check the answers to Examples 8.48, 8.50 and 8.51 using MATLAB or MAPLE.

Figure 8.59
Substitutions for
evaluation of $\int f(x)dx$.

If $f(x)$ contains	try	
$\sqrt{(a^2 - x^2)}$	$x = a \sin\theta,$	$\dfrac{dx}{d\theta} = a\cos\theta$
	or $x = a \tanh u,$	$\dfrac{dx}{du} = a\,\text{sech}^2 u$
$\sqrt{(a^2 + x^2)}$	$x = a \sinh u,$	$\dfrac{dx}{du} = a\cosh u$
	or $x = a \tan\theta,$	$\dfrac{dx}{d\theta} = a\sec^2\theta$
$\sqrt{(x^2 - a^2)}$	$x = a \cosh u$	$\dfrac{dx}{du} = a\sinh u$
	or $x = a \sec\theta$	$\dfrac{dx}{d\theta} = a\sec\theta\tan\theta$
Circular functions	$s = \sin x,$	$\dfrac{ds}{dx} = \cos x$
	or $c = \cos x,$	$\dfrac{dc}{dx} = -\sin x$
	or $t = \tan\frac{1}{2}x,$	
	$\left(\sin x = \dfrac{2t}{1 + t^2},\quad \cos x = \dfrac{1 - t^2}{1 + t^2},\quad \dfrac{dx}{dt} = \dfrac{2}{1 + t^2}\right)$	
Hyperbolic functions	$u = e^x,$	$\dfrac{du}{dx} = e^x$
	or $s = \sinh x,$	$\dfrac{ds}{dx} = \cosh x$
	or $c = \cosh x,$	$\dfrac{dc}{dx} = \sinh x$
	or $t = \tanh\frac{1}{2}x,$	$\dfrac{dt}{dx} = \frac{1}{2}\,\text{sech}^2\frac{1}{2}x$

8.8.6 Exercises

 Check your answers using MATLAB or MAPLE whenever possible.

108 Using the substitution method, integrate the following functions:

(a) $x\sqrt{(1 + x^2)}$ (b) $\cos x \sin^3 x$ (c) $\dfrac{x}{(1 + x^2)^2}$

(d) $\dfrac{x}{\sqrt{(x^2 - 1)}}$ (e) $\dfrac{2x + 3}{x^2 + 3x + 2}$ (f) $\dfrac{1}{x \ln x}$

109 Find the values of the constants a and b such that

$$\frac{3x + 2}{x^2 + 2x + 5} = \frac{a(2x + 2)}{x^2 + 2x + 5} + \frac{b}{x^2 + 2x + 5}$$

and hence find its integral. (Note that $(d/dx)(x^2 + 2x + 5) = 2x + 2$.)

110 Use the technique of Question 109 to integrate

(a) $\dfrac{x+1}{x^2+4x+5}$ (b) $\dfrac{2x+3}{\sqrt{(5+4x-x^2)}}$

(c) $\dfrac{\sin x}{\sin x + \cos x}$

111 Use the given substitutions to integrate the following functions:

(a) $x^3\sqrt{(1+x^2)}$,   with $t=\sqrt{(1+x^2)}$

(b) $\sin^3 x \cos^5 x$,   with $t=\sin x$

(c) $\dfrac{x}{(1+x^2)^2}$,   with $x=\tan t$

(d) $\dfrac{3}{x\sqrt{(x^2+9)}}$,   with $t=\dfrac{1}{x}$

(e) $\dfrac{x}{\sqrt{(4-x^2)}}$,   with $x=2\sin t$

(f) $\dfrac{1}{3+\sqrt{x}}$,   with $t=\sqrt{x}$

112 Use an appropriate substitution to integrate the following functions:

(a) $\dfrac{1}{1+\sqrt{(1+x)}}$ (b) $\sin^2 x \cos^3 x$ (c) $\sin\sqrt{x}$

113 Show that $t=\tan\frac{1}{2}x$ implies

$$\sin x = \frac{2t}{1+t^2},$$

$$\cos x = \frac{1-t^2}{1+t^2}$$

and

$$dx = \frac{2}{1+t^2}\,dt$$

Hence integrate

(a) $\operatorname{cosec} x$ (b) $\sec x$

(c) $\dfrac{1}{3+4\sin x}$ (d) $\dfrac{1}{5\sin x + 12\cos x}$

114 Using the substitution $u=\sqrt{x}-1$, evaluate the definite integral

$$\int_4^9 \frac{dx}{(\sqrt{x}-1)\sqrt{x}}$$

115 Evaluate the following definite integrals using the given substitutions:

(a) $\displaystyle\int_1^4 \frac{e^{\sqrt{x}}}{\sqrt{x}}\,dx$,   with $u=\sqrt{x}$

(b) $\displaystyle\int_{-2}^2 \frac{x+6}{\sqrt{(x+2)}}\,dx$,   with $u=\sqrt{(x+2)}$

(c) $\displaystyle\int_0^{\sqrt{3}} \frac{\tan^{-1}x}{1+x^2}\,dx$,   with $u=\tan^{-1}x$

(d) $\displaystyle\int_{1/6}^{1/2} \frac{dx}{(5+6x)^3}$,   with $u=5+6x$

116 In Question 11 (Exercises 8.2.8) the equation of the path of P was found to be such that

$$\frac{dy}{dx}=\frac{\sqrt{(a^2-x^2)}}{x}, \quad \text{with } y=0 \text{ at } x=a$$

Use the substitution $x=a\operatorname{sech}u$ to integrate this differential equation and show that

$$y = \ln\left[\frac{a+\sqrt{(a^2-x^2)}}{x}\right]-\sqrt{(a^2-x^2)}$$

This curve is called a **tractrix**.

8.9   Applications of integration

In this section we consider some other situations in which integration is widely used.

8.9.1   Volume of a solid of revolution

Imagine rotating the plane area A under the graph of the function $f(x)$, $x \in [a, b]$, of Figure 8.60 through a complete revolution about the x axis. The result would be to

Figure 8.60
Plane area rotated.

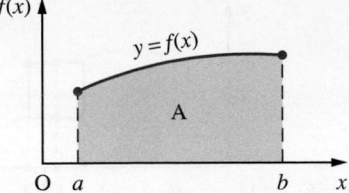

Figure 8.61
Solid of revolution.

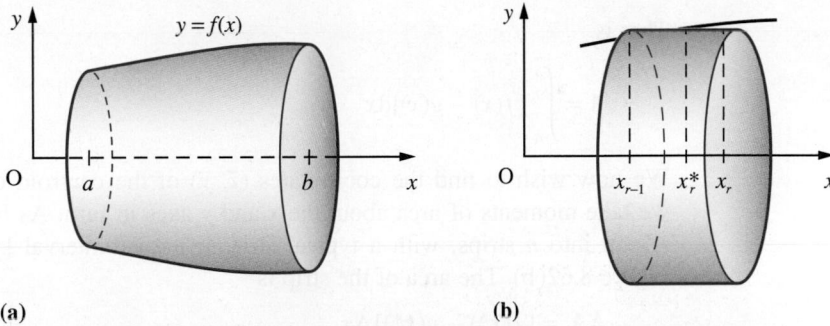

(a)   (b)

generate a solid having the x axis as axis of symmetry as shown in Figure 8.61(a): this is called a **solid of revolution**. If we wish to determine the volume of this solid, we proceed as in Section 8.7.1 and subdivide the rotating area into n vertical strips. When a typical strip within the subinterval $[x_{r-1}, x_r]$ is rotated through a revolution about the x axis, it will generate a thin disc of radius $f(x_r^*)$ (with $x_{r-1} < x_r^* < x_r$) and thickness Δx_{r-1} as shown in Figure 8.61(b). The volume of the disc is given by

$$\Delta V_r = \pi [f(x_r^*)]^2 \Delta x_{r-1}$$

Thus the volume of the solid can be approximated by

$$V \approx \sum_{r=1}^{n} \Delta V_r = \pi \sum_{r=1}^{n} [f(x_r^*)]^2 \Delta x_{r-1}$$

Again this approximation is closer to the exact volume as the number of strips is increased. Thus in the limiting case as $n \to \infty$ and $\Delta x \to 0$, $\Delta x = \max_r \Delta x_r$, it leads to the volume being given by

$$V = \lim_{\substack{n \to \infty \\ \Delta x \to 0}} \pi \sum_{r=1}^{n} [f(x_r^*)]^2 \Delta x_{r-1} = \pi \int_a^b [f(x)]^2 \mathrm{d}x \tag{8.36}$$

8.9.2 Centroid of a plane area

Consider the plane region of Figure 8.62(a) bounded between the graphs of the two continuous functions $f(x)$ and $g(x)$ on the interval $x \in [a, b]$, with $g(x) \leqslant f(x)$ on the interval. The area A of this region is clearly given by

$$A = \text{area under the graph of } f(x) - \text{area under the graph of } g(x)$$

$$= \int_a^b f(x)\mathrm{d}x - \int_a^b g(x)\mathrm{d}x$$

Figure 8.62

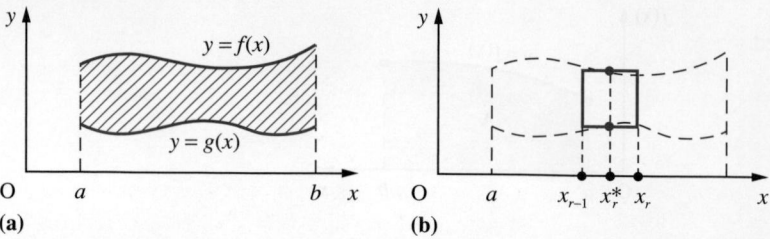

(a)   (b)

That is

$$A = \int_a^b [f(x) - g(x)]\mathrm{d}x \tag{8.37}$$

We now wish to find the coordinates $(\bar{x}, \bar{y})$ of the centroid of this area. To do this, we take moments of area about the x and y axes in turn. As before, we subdivide the region into n strips, with a typical strip in the subinterval $[x_{r-1}, x_r]$ being shown in Figure 8.62(b). The area of the strip is

$$\Delta A_r = [f(x_r^*) - g(x_r^*)]\Delta x_{r-1}$$

and the moment of this area about the y axis is

$$\Delta M_{y_r} = x_r^*\Delta A_r = x_r^*[f(x_r^*) - g(x_r^*)]\Delta x_{r-1}$$

Thus the sum of the moments of the n strips about the y axis is

$$\sum_{r=1}^{n} \Delta M_{y_r} = \sum_{r=1}^{n} x_r^*\Delta A_r = \sum_{r=1}^{n} x_r^*[f(x_r^*) - g(x_r^*)]\Delta x_{r-1}$$

Proceeding to the limit $n \to \infty$, $\Delta x \to 0$, $\Delta x = \max_r \Delta x_r$, we have the moment of the plane area about the y axis being given by

$$M_y = \lim_{\substack{n\to\infty \\ \Delta x \to 0}} \sum_{r=1}^{n} x_r^*[f(x_r^*) - g(x_r^*)]\Delta x_{r-1} = \int_a^b x[f(x) - g(x)]\mathrm{d}x$$

Since the x coordinate of the centroid of the plane area is $\bar{x}$, it follows that the moment of the area about the y axis is also given by

$$M_y = A\bar{x}$$

Equating, we have

$$\bar{x} = \frac{1}{A} \int_a^b x[f(x) - g(x)]\mathrm{d}x \tag{8.38}$$

where the area A is given by (8.37).

Likewise, taking moments about the x axis,

$$M_x = A\bar{y} = \lim_{\substack{n\to\infty \\ \Delta x \to 0}} \left[\sum_{r=1}^{n} \tfrac{1}{2}f(x_r^*)f(x_r^*)\Delta x_{r-1} - \sum_{r=1}^{n} \tfrac{1}{2}g(x_r^*)g(x_r^*)\Delta x_{r-1} \right]$$

$$= \lim_{\substack{n\to\infty \\ \Delta x \to 0}} \tfrac{1}{2} \sum_{r=1}^{n} \{[f(x_r^*)]^2 - [g(x_r^*)]^2\}\Delta x_{r-1} = \tfrac{1}{2} \int_a^b \{[f(x)]^2 - [g(x)]^2\}\mathrm{d}x$$

giving

$$\bar{y} = \frac{1}{2A} \int_a^b \{[f(x)]^2 - [g(x)]^2\} dx \tag{8.39}$$

where A again is given by (8.37).

In the particular case when $g(x)$ is the x axis, we find that the centroid of the plane area bounded by $f(x)$ ($x \in [a, b]$) and the x axis has coordinates

$$\bar{x} = \frac{1}{A} \int_a^b xf(x)dx, \quad \bar{y} = \frac{1}{2A} \int_a^b [f(x)]^2 dx \tag{8.40}$$

8.9.3 Centre of gravity of a solid of revolution

Proceeding as in Section 8.9.2, we can obtain the coordinates $(\bar{X}, \bar{Y})$ of the centre of gravity of the solid of revolution generated by $f(x)$ ($x \in [a, b]$) and shown in Figure 8.58. By symmetry, it lies on the x axis, so that

$$\bar{Y} = 0$$

Taking moments about the y axis gives

$$V\bar{X} = \lim_{\substack{n \to \infty \\ \Delta x \to 0}} \pi \sum_{r=1}^n x_r^*[f(x_r^*)]^2 \Delta x_{r-1} = \pi \int_a^b x[f(x)]^2 dx \tag{8.41}$$

giving

$$\bar{X} = \frac{\pi}{V} \int_a^b x[f(x)]^2 dx \tag{8.42}$$

where the volume V is given by (8.36).

8.9.4 Mean values

In many engineering applications we need to know the mean value of a continuously varying quantity. When dealing with a sequence of values we can compute the mean value simply by adding the values together and then dividing by the number of values taken. When dealing with a continuously varying quantity, we cannot do that directly. Using integration, however, we are able to calculate the mean value.

Consider the function $f(x)$ on the interval $[a, b]$ and divide the interval into n equal strips of width h so that $nh = b - a$. Now evaluate the function at the midpoint of each strip. Formally, let $x_k = a + kh$ be the points of subdivision, so that the points of evaluation are $f(x_k^*)$ where $x_k^* = x_k + h/2$. Then the mean value (m.v.) of $f(x)$ on $[a, b]$ is approximately

$$\text{m.v.}(f(x)) \approx \frac{1}{n} \sum_{k=0}^{n-1} f(x_k^*) = \frac{1}{b-a} \sum_{k=0}^{n-1} f(x_k^*)h$$

Now allowing $n \to \infty$ (with $h \to 0$), the summation becomes an integral and the approximation becomes exactly true. Thus

Figure 8.63
Mean value
of a function
$y = f(x)$, $x \in [a, b]$.

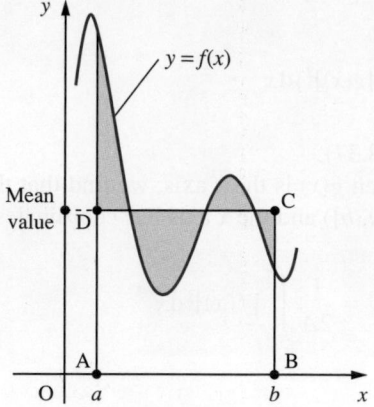

$$\text{m.v.}(f(x)) = \frac{1}{b-a} \int_a^b f(x)\,dx \tag{8.43}$$

The graphical representation of this makes the situation quite clear. In Figure 8.63, the sum of the shaded areas above the line $y = $ (mean value) is equal to the sum of the shaded areas below it, so that the area of the rectangle ABCD is the same as the area between the curve and the x axis.

8.9.5 Root mean square values

In some contexts the computation of the mean value of a function is not useful, for example the mean of an alternating current is zero but that does not imply it is not dangerous! To deal with such situations we use the root mean square (r.m.s.) of the function $f(x)$. Literally this is the square root of the mean value of $[f(x)]^2$. Thus we can write

$$[\text{r.m.s.}(f(x))]^2 = \frac{1}{b-a} \int_a^b [f(x)]^2\,dx \tag{8.44}$$

Although the obvious applications of root mean square values are in electrical engineering, they also occur in the application of statistics to engineering contexts (as standard deviations of continuously distributed random variables). They also occur in the design of gyroscopes and in mechanics, where the 'radius of gyration' is in effect the root mean of moments about an axis.

8.9.6 Arclength and surface area

In many practical problems we are required to work out the length of a curve or the surface area generated by rotating a curve. The formula for the length s of a curve with formula $y = f(x)$ between two points corresponding to $x = a$ and $x = b$ is obtained using the basic idea of integration. Let Δs_k be the element of arclength between $x = x_k$ and $x = x_{k+1}$. Then for a curve that is concave upwards as in Figure 8.64 we deduce that

$$\Delta x_k \sec \theta_k \leqslant \Delta s_k \leqslant \Delta x_k \sec \theta_{k+1}$$

Figure 8.64
(a) Curve $y = f(x)$.
(b) Element of arclength.

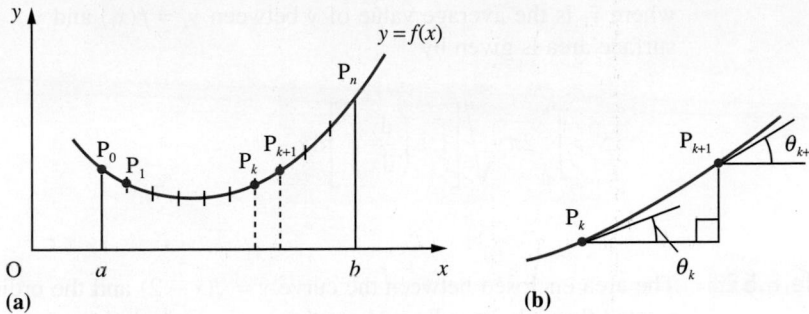

(a) (b)

where θ_k and θ_{k+1} are the angles of slope made by the tangents to the curve at P_k and P_{k+1}. Thus the length s of the curve between $x = a$ and $x = b$ satisfies the inequality

$$\sum_{k=0}^{n-1} \Delta x_k \sec\theta_k \leqslant s = \sum_{k=0}^{n-1} \Delta s_k \leqslant \sum_{k=0}^{n-1} \Delta x_k \sec\theta_{k+1}$$

Letting $n \to \infty$ and max $\Delta x_k \to 0$ yields the inequality

$$\int_a^b \sec\theta\, dx \leqslant s \leqslant \int_a^b \sec\theta\, dx$$

from which we deduce that

$$s = \int_a^b \sec\theta\, dx$$

A similar analysis for curves that are concave downwards yields the same result.
   We can express $\sec\theta$ in terms of dy/dx by means of the identity

$$\sec^2\theta = 1 + \tan^2\theta$$

Here $\tan\theta = dy/dx$, so, using the convention that s increases with x, we obtain

$$\sec\theta = \sqrt{\left[1 + \left(\frac{dy}{dx}\right)^2\right]}$$

so that the length of the curve is

$$s = \int_a^b \sqrt{\left[1 + \left(\frac{dy}{dx}\right)^2\right]}\, dx \qquad (8.45)$$

The surface area S generated by s when it is rotated through 2π radians about the x axis is calculated in a similar way. The element of arc Δs_k generates an element of surface area ΔS_k, where

$$\Delta S_k = 2\pi \bar{y}_k \Delta s_k$$

where $\bar{y}_k$ is the average value of y between $y_k = f(x_k)$ and $y_{k+1} = f(x_{k+1})$. Thus the total surface area is given by

$$S = \int_a^b 2\pi y \sqrt{\left[1 + \left(\frac{dy}{dx}\right)^2\right]} dx \qquad (8.46)$$

Example 8.53

The area enclosed between the curve $y = \sqrt{(x-2)}$ and the ordinates $x = 2$ and $x = 5$ is rotated through 2π radians about the x axis. Calculate

(a) the rotating area and the coordinates of its centroid:

(b) the volume of the solid of revolution generated and the coordinates of its centre of gravity.

Solution

The rotating area is the shaded region shown in Figure 8.65.

(a) The rotating area is given by

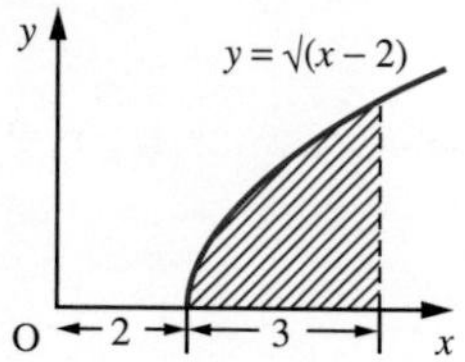

Figure 8.65
Rotating area.

$$A = \int_2^5 y \, dx = \int_2^5 (x-2)^{1/2} dx$$

$$= [\tfrac{2}{3}(x-2)^{3/2}]_2^5 = 2\sqrt{3} \text{ square units}$$

If we denote the coordinates of the centroid of the area by $(\bar{x}, \bar{y})$ then, from (8.37),

$$\bar{x} = \frac{1}{A} \int_2^5 xy \, dx = \frac{1}{A} \int_2^5 x(x-2)^{1/2} dx = \frac{1}{A} \int_2^5 [(x-2)^{3/2} + 2(x-2)^{1/2}] dx$$

$$= \frac{1}{A}\left[\frac{2}{5}(x-2)^{5/2} + \frac{4}{3}(x-2)^{3/2}\right]_2^5 = \frac{1}{A}\left[\frac{2}{5}(3)^{5/2} + \frac{4}{3}(3)^{3/2}\right] = \frac{1}{A}\frac{38}{5}\sqrt{3}$$

Inserting the value $A = 2\sqrt{3}$ obtained earlier gives $\bar{x} = \frac{19}{5}$.
Likewise, from (8.40),

$$\bar{y} = \frac{1}{A} \int_2^5 \tfrac{1}{2} y^2 dx = \frac{1}{A} \int_2^5 \tfrac{1}{2}(x-2) dx$$

$$= \frac{1}{A}[\tfrac{1}{4}(x-2)^2]_2^5 = \frac{9}{4A}$$

Inserting $A = 2\sqrt{3}$ then gives $\bar{y} = \frac{3}{8}\sqrt{3}$ so that the coordinates of the centroid are $(\frac{19}{5}, \frac{3}{8}\sqrt{3})$.

(b) From (8.36) the volume V of the solid of revolution formed is

$$V = \pi \int_2^5 y^2 dx = \pi \int_2^5 (x-2) dx$$

$$= \pi[\tfrac{1}{2}x^2 - 2x]_2^5 = \tfrac{9}{2}\pi \text{ cubic units}$$

If we denote the coordinates of the centre of gravity of the solid of revolution by $(\bar{X}, \bar{Y})$ then, from (8.41) and (8.42),

$$\bar{Y} = 0$$

and

$$\bar{X} = \frac{\pi}{V} \int_2^5 xy^2 \, dx = \frac{\pi}{V} \int_2^5 x(x-2) \, dx$$

$$= \frac{\pi}{V} [\tfrac{1}{3}x^3 - x^2]_2^5 = \frac{\pi}{V} [(\tfrac{125}{3} - 25) - (\tfrac{8}{3} - 4)]$$

$$= \frac{18\pi}{V}$$

Inserting the value $V = \tfrac{9}{2}\pi$ obtained earlier gives $\bar{X} = 4$ so that the coordinates of the centre of gravity are $(4, 0)$.

Example 8.54   An electric current i is given by the expression

$$i = I \sin \theta$$

where I is a constant. Find the root mean square value of the current over the interval $0 \le \theta \le 2\pi$.

Solution   Using (8.44) the r.m.s. value of the given current is given by

$$(\text{r.m.s.}\,i)^2 = \frac{1}{2\pi - 0} \int_0^{2\pi} I^2 \sin^2 \theta \, d\theta$$

$$= \frac{I^2}{2\pi} \int_0^{2\pi} \tfrac{1}{2}(1 - \cos 2\theta) \, d\theta$$

$$= \frac{I^2}{4\pi} [\theta - \tfrac{1}{2}\sin 2\theta]_0^{2\pi} = \frac{I^2}{4\pi} 2\pi = \tfrac{1}{2}I^2$$

so that

$$\text{r.m.s. current} = \sqrt{(\tfrac{1}{2}I^2)} = I/\sqrt{2}$$

Example 8.55   A parabolic reflector is formed by rotating the part of the curve $y = \sqrt{x}$ between $x = 0$ and $x = 1$ about the x axis. What is the surface area of the reflector?

Solution   The parabolic reflector is shown in Figure 8.66. Since $y = x^{1/2}$,

$$\frac{dy}{dx} = \frac{1}{2}x^{-1/2} = \frac{1}{2\sqrt{x}}$$

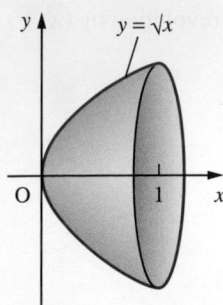

Figure 8.66
Parabolic reflector.

so that, using (8.46), the surface area S of the reflector is given by

$$S = 2\pi \int_0^1 y \sqrt{\left[1 + \left(\frac{dy}{dx}\right)^2\right]} dx$$

$$= 2\pi \int_0^1 \sqrt{x} \sqrt{\left(1 + \frac{1}{4x}\right)} dx$$

$$= 2\pi \int_0^1 \sqrt{x} \frac{\sqrt{(4x+1)}}{2\sqrt{x}} dx = \pi \int_0^1 \sqrt{(4x+1)}\, dx$$

$$= \pi \left[\tfrac{1}{4}\tfrac{2}{3}(4x+1)^{3/2}\right]_0^1 = \tfrac{1}{6}\pi\,(5^{3/2} - 1) \text{ square units}$$

Example 8.56

The curve described by the cable of the suspension bridge shown in Figure 8.67 is given by

$$y = \frac{hx^2}{l^2} - \frac{2h}{l}x + h$$

where x is the distance measured from one end of the bridge. What is the length of the cable?

Figure 8.67
Suspension bridge.

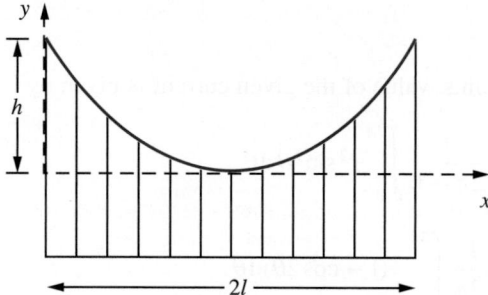

Solution   Here the equation of the curve is

$$y = h\left(\frac{x}{l} - 1\right)^2 \quad \text{so that} \quad \frac{dy}{dx} = \frac{2h}{l}\left(\frac{x}{l} - 1\right)$$

Using (8.45), the length s of the cable is

$$s = \int_0^{2l} \sqrt{\left[1 + \frac{4h^2}{l^2}\left(\frac{x}{l} - 1\right)^2\right]} dx$$

This integral can be simplified by putting

$$t = \frac{2h}{l}\left(\frac{x}{l} - 1\right)$$

Thus

$$s = \frac{l^2}{2h} \int_{-2h/l}^{2h/l} \sqrt{(1 + t^2)}\,dt = \frac{l^2}{h} \int_{0}^{2h/l} \sqrt{(1 + t^2)}\,dt \quad \text{(from symmetry)}$$

This can be further simplified by putting $t = \sinh u$, giving

$$s = \frac{l^2}{h} \int_{0}^{\sinh^{-1}(2h/l)} \cosh^2 u\,du = \frac{l^2}{2h} \int_{0}^{\sinh^{-1}(2h/l)} (\cosh 2u + 1)\,du$$

$$= \frac{l^2}{2h} \left[\tfrac{1}{2}\sinh 2u + u \right]_{0}^{\sinh^{-1}(2h/l)} = \frac{l^2}{2h} \left[\sinh u \cosh u + u \right]_{0}^{\sinh^{-1}(2h/l)}$$

$$= \frac{l^2}{2h} \left[\frac{2h}{l} \sqrt{\left(1 + \frac{4h^2}{l^2}\right)} + \sinh^{-1}\left(\frac{2h}{l}\right) \right]$$

That is,

$$s = \sqrt{(l^2 + 4h^2)} + \frac{l^2}{2h} \sinh^{-1}\left(\frac{2h}{l}\right)$$

Example 8.57   Find the equation of the curve described by a heavy cable hanging, without load, under gravity, from two equally high points.

Solution   Consider the cable illustrated in Figure 8.68. Let T be the tension acting at a point P that is a horizontal distance x from the axis of symmetry, as shown, and let the tangent to the curve at P make an angle θ to the horizontal. If s is the length of the curve between A and P, and T_0 is the tension at A, then resolving the forces acting on the length of cable between A and P horizontally and vertically gives

$$T_0 = T \cos\theta \quad \text{and} \quad s\rho g = T \sin\theta$$

where ρ is the line density of the cable and g is the acceleration due to gravity. Dividing these equations, we obtain

$$\tan\theta = \frac{s}{c}$$

Figure 8.68
Heavy hanging cable.

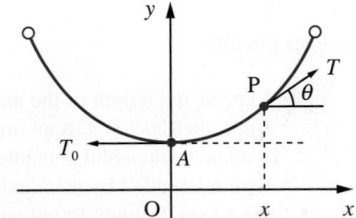

where $c = T_0/\rho g$. In terms of x and y, this equation, using (8.45), implies that the co-ordinates of P satisfy

$$y'(x) = \frac{1}{c} \int_0^x \sqrt{[1 + (y'(t))^2]} \, dt$$

To solve this equation to obtain the equation of the curve, we first differentiate it with respect to x, giving

$$y''(x) = \frac{1}{c} \sqrt{[1 + (y'(x))^2]}$$

with $dy/dx = 0$ at $x = 0$. This may be rewritten as

$$\frac{d^2y/dx^2}{\sqrt{[1 + (dy/dx)^2]}} = \frac{1}{c}$$

and integrating with respect to x, using the substitution $dy/dx = \sinh u$, and remembering that $(d/dx)(dy/dx) = d^2y/dx^2$, gives

$$\sinh^{-1}\left(\frac{dy}{dx}\right) = \frac{x}{c} + A$$

Since $dy/dx = 0$ at $x = 0$, we deduce that $A = 0$ and

$$\frac{dy}{dx} = \sinh \frac{x}{c}$$

This is easy to integrate, giving

$$y = c \cosh \frac{x}{c} + B$$

The value of B is fixed by the value of y at $x = 0$. This may be chosen quite arbitrarily without changing the shape of the curve. Choosing $y(0) = c$ gives a neat answer (with $B = 0$):

$$y = c \cosh \frac{x}{c}$$

Note that this curve, called the **catenary**, is different from the shape of the cable of a suspension bridge, which is a parabola. The catenary has many applications, including the design of roofs and arches.

8.9.7 Exercises

 Check your answers using MATLAB or MAPLE whenever possible.

117 Find the volume generated when the plane figure bounded by the curve $xy = x^3 + 3$, the x axis and the ordinates at $x = 1$ and $x = 2$ is rotated about the x axis through one complete revolution.

118 Express the length of the arc of the curve $y = \sin x$ from $x = 0$ to $x = \pi$ as an integral. Also find the volume of the solid generated by revolving the region bounded by the x axis and this arc about the x axis through 2π radians.

119 (a) Sketch the curve whose equation is

$$y = (x - 2)(x - 1)$$

Show that the volume generated when the finite area between the curve and the x axis is rotated through 2π radians about the x axis is $\pi/30$.

(b) Show that the curved surface generated by the revolution about the x axis of the portion of the curve $y^2 = 4ax$ included between the origin and the ordinate $x = 3a$ is $\frac{56}{3}\pi a^2$.

120 A curve is represented parametrically by

$$x(t) = 3t - t^3, \quad y(t) = 3t^2 \quad (0 \leqslant t \leqslant 1)$$

Find the volume and surface area of the solid of revolution generated when the curve is rotated about the x axis through 2π radians.

121 The electrical resistance R (in Ω) of a rheostat at a temperature θ (in °C) is given by $R = 38(1 + 0.004\theta)$. Find the average resistance of the rheostat as the temperature varies uniformly from 10°C to 40°C.

122 The area enclosed between the x axis, the curve $y = x(2 - x)$ and the ordinates $x = 1$ and $x = 2$ is rotated through 2π radians about the x axis. Calculate

(a) the rotating area and the coordinates of its centroid;

(b) the volume of the solid of revolution formed and the coordinates of its centre of gravity.

123 Show that the area enclosed between the x axis, the curve $4y = x^2 - 2\ln x$ and the coordinates $x = 1$ and $x = 3$ is $\frac{1}{6}(19 - 9\ln 3)$.

124 The speed V of a rocket at a time t after launch is given by

$$V = at^2 + b$$

where a and b are constants. The average speed over the first second was $10\,\mathrm{m\,s^{-1}}$, and that over the next second was $50\,\mathrm{m\,s^{-1}}$. Determine the values of a and b. What was the average speed over the third second?

125 Find the centroid of the area bounded by $y^2 = 4x$ and $y = 2x$ and also the centroid of the volume obtained by revolving this area about the x axis.

8.10 Numerical evaluation of integrals

In many practical problems the functions that have to be integrated are often specified by a graph or by a table of values. Even when the function is given analytically, it often cannot be integrated to give an answer in terms of simple functions. Also, in many engineering and scientific problems it is often known in advance that the value of an integral is only required to a certain precision and the use of an approximate method can avoid considerable unwanted labour. In all these cases we have to evaluate the integrals numerically. There are many ways of doing this, varying from the simplest square-counting for working out the area under a graph to sophisticated computer procedures. In this section we shall develop a simple numerical method known as the trapezium rule, which is the basis of many computer algorithms, and a hand computation method known as Simpson's rule.

8.10.1 The trapezium rule

The simplest methods return to the initial ideas about integration introduced in Section 8.7.1. As indicated in Figure 8.69, they involve slicing up the area to be found into a number of strips of equal width, approximating the area of each strip in some way; the sum of these approximations then gives the final numerical result.

The points of subdivision of the domain of integration $[a, b]$ are labelled $x_0, x_1, \ldots, x_n$, where $x_0 = a$, $x_n = b$, $x_r = x_0 + rh$ ($r = 0, 1, 2, \ldots, n$), and the width of each strip

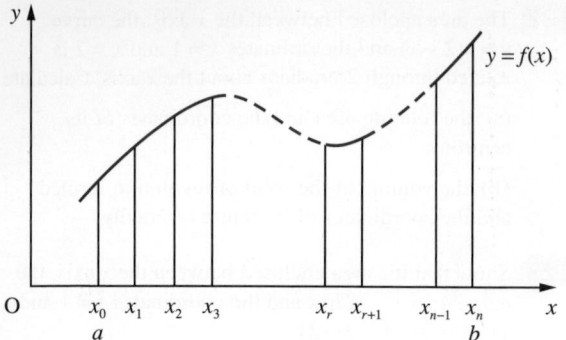

Figure 8.69 Slicing up an area into vertical strips of equal width.

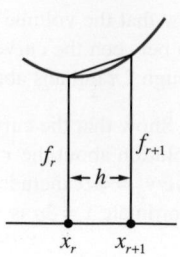

Figure 8.70 Trapezium approximation to area of strip.

is $h = (b - a)/n$. The value of the integrand $f(x)$ at these points is, as usual, denoted by $f_r = f(x_r)$. A basic method for numerical integration approximates the area of each strip by the area of the trapezium formed when the upper end is replaced by the chord of the graph, as shown in Figure 8.70.

By the sum rule of integration

$$\int_a^b f(x)\mathrm{d}x = \int_{x_0}^{x_1} f(x)\mathrm{d}x + \int_{x_1}^{x_2} f(x)\mathrm{d}x + \ldots + \int_{x_{n-1}}^{x_n} f(x)\mathrm{d}x = \sum_{r=0}^{n-1} \int_{x_r}^{x_{r+1}} f(x)\mathrm{d}x$$

From Figure 8.70 we can see that the approximate area of the rth strip is

$$\tfrac{1}{2}(f_r + f_{r+1})h$$

so that

$$\int_a^b f(x)\mathrm{d}x \approx \sum_{r=0}^{n-1} \tfrac{1}{2}(f_r + f_{r+1})h = \tfrac{1}{2}h \sum_{r=0}^{n-1} (f_r + f_{r+1})$$

$$= \tfrac{1}{2}h[(f_0 + f_1) + (f_1 + f_2) + \ldots + (f_{n-1} + f_n)]$$

That is,

$$\int_a^b f(x)\mathrm{d}x \approx h(\tfrac{1}{2}f_0 + f_1 + f_2 + \ldots + f_{n-1} + \tfrac{1}{2}f_n) \tag{8.47}$$

This approximation method is called the **trapezium rule**. As we shall see below, the best method for using it is given in formula (8.48).

Example 8.58

Evaluate the integral $\int_1^2 (1/x)\mathrm{d}x$ to 5dp, using the trapezium rule.

Solution

This integral is one of the standard integrals given in Figure 8.55, and so can be evaluated analytically. Its value is $\ln 2 = 0.693\,147$ to 6dp. This enables us, in this illustrative example, to check our methods. Usually, of course, the value of the integral is not known beforehand, and assessing the accuracy of the estimate obtained using the trapezium rule is an important aspect of the evaluation.

The first decision to be made in the numerical procedure is that of how many strips should be used; that is, what value n should have. A large number of strips may yield a good approximation to each strip, but will involve a lot of calculation, with the possibility of consequent rounding error accumulation. A small number of strips will obviously involve a large error in the approximation to the area of each strip. We shall investigate the situation.

First of all, we shall introduce the notation $T(h)$ to denote the approximation to the value of the integral given by the trapezium rule using strips of width h. Obviously, ignoring the possible effects of rounding errors, we expect

$$\lim_{h \to 0} T(h) = \int_1^2 \frac{1}{x} \, dx$$

Taking $n = 1$ gives $h = (2 - 1)/n = 1$, $x_0 = 1$ and $x_1 = 2$. This gives the estimate

$$\int_1^2 \frac{1}{x} \, dx = \tfrac{1}{2}(1)(f_0 + f_1) = T(1)$$

Here $f_0 = 1$ and $f_1 = 0.5$, so that $T(1) = 0.75$. This estimate for the value of the integral has an error of $0.75 - 0.693 = +0.057$.

Taking $n = 2$ gives $h = 0.5$, $x_0 = 1$, $x_2 = 2$ and $x_1 = 1.5$. Note that x_0 and x_2 are the two points used before, but now relabelled. This gives the estimate

$$T(0.5) = (0.5)[f_1 + \tfrac{1}{2}(f_0 + f_2)]$$

where $f_0 = 1$, $f_1 = 0.666\,667$ and $f_2 = 0.5$, so that $T(0.5) = 0.708\,333$. This estimate has an error of $+0.015$, so by doubling the number of strips, we have reduced the error by a factor of nearly four.

Taking $n = 4$ gives $h = 0.25$, $x_0 = 1$, $x_4 = 2$, $x_1 = 1.25$, $x_2 = 1.5$ and $x_3 = 1.75$. Note that three of these points were used in the previous calculation. This value of n gives the estimate

$$T(0.25) = (0.25)[f_1 + f_2 + f_3 + \tfrac{1}{2}(f_0 + f_4)]$$

where $f_0 = 1$, $f_1 = 0.8$, $f_2 = 0.666\,667$, $f_3 = 0.571\,429$ and $f_4 = 0.5$, so that $T(0.25) = 0.697\,024$. This estimate has an error of $+0.004$, so by doubling the number of strips, we have again reduced the error by a factor of four.

Continuing this process, with $n = 8$, we obtain the estimate $T(0.125) = 0.694\,122$, with an error of $+0.001$.

Based on these four calculations, we can estimate the values of n and h that will give an answer correct to 5dp; that is, with an absolute error less than $0.000\,005$. If we continue the process of doubling the number of strips, reducing the error by a factor of four each time, we shall obtain an answer with the required accuracy when $n = 128$. With this large number of strips, we clearly need to organize the calculation to do it as economically as possible. Looking back at the previous calculations, we see that at each new value of n we almost double the number of points at which the integrand has to be evaluated, but as can be seen from Figure 8.71, at half of these points it has been evaluated in previous calculations.

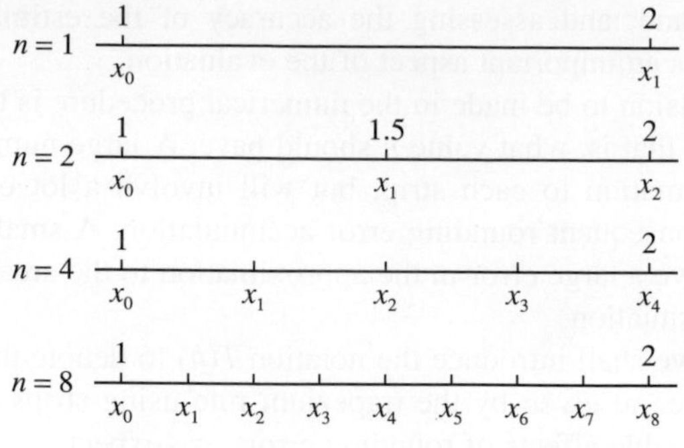

Figure 8.71
Points at which
integrand is evaluated.

Taking into account the effect of interval-halving on h, we can reduce the amount of calculation to evaluate $T(h)$ by making use of the result obtained for $T(2h)$:

$$T(h) = h[f_1 + f_2 + f_3 + \ldots + f_{n-1} + \tfrac{1}{2}(f_0 + f_n)], \quad h = \frac{b-a}{n}$$

$$T(2h) = (2h)[f_2 + f_4 + \ldots + f_{n-2} + \tfrac{1}{2}(f_0 + f_n)]$$

Here f_0, f_2, f_4, $\ldots$, f_n were all calculated previously, in the evaluation of $T(2h)$. Rearranging, we have

$$T(h) = h(f_1 + f_3 + f_5 + \ldots + f_{n-1}) + \tfrac{1}{2}(2h)[f_2 + f_4 + \ldots + f_{n-2} + \tfrac{1}{2}(f_0 + f_n)]$$

Thus

$$T(h) = h(f_1 + f_3 + f_5 + \ldots + f_{n-1}) + \tfrac{1}{2}T(2h) \tag{8.48}$$

(remembering that if h is the strip width for n intervals then $2h$ is the strip width for $\tfrac{1}{2}n$ intervals).

This formula enables us to perform the calculations economically, but we can exploit it in a more subtle way.

We have seen that halving the strip width reduces the error by a factor of approximately four. This means that the error is proportional to h^2. In fact, this behaviour is typical of the application of the trapezium rule to the evaluation of many kinds of integrals and we can use it to obtain a more accurate estimate of the value of the integral. Since the error is proportional to h^2, we can write

$$T(h) - \int_1^2 \frac{1}{x} \, dx = Ah^2$$

where $h = 1/n$ and A is some number that, in general, will depend upon n but will remain bounded as n becomes large. A similar formula holds for $T(2h)$:

$$T(2h) - \int_1^2 \frac{1}{x} \, dx = 4A'h^2$$

where h has the same value as before and $A' \approx A$. These two formulae enable us to estimate the error in the approximation for the integral. Subtracting them gives

$$3Ah^2 \approx T(2h) - T(h)$$

so that the approximation $T(h)$ to the integral has an error estimate of $\frac{1}{3}[T(2h) - T(h)]$. Thus in the calculation above the estimated error for $T(0.125)$ is

$$\tfrac{1}{3}(0.697\,024 - 0.694\,122) = +0.000\,967$$

as we found before. This means that we can estimate the error in the usual situation of not knowing (unlike in this example) the true value of the integral. It also enables us to obtain a better approximation. Subtracting the estimated error from $T(h)$ gives the improved approximation (Richardson's extrapolation)

$$\int_1^2 \frac{1}{x}\,\mathrm{d}x \approx T(h) - \tfrac{1}{3}[T(2h) - T(h)]$$

Alternatively we may write

$$\int_1^2 \frac{1}{x}\,\mathrm{d}x \approx \tfrac{1}{3}[4T(h) - T(2h)]$$

Using the values for $T(0.25)$ and $T(0.125)$ obtained above, we have

$$\int_1^2 \frac{1}{x}\,\mathrm{d}x \approx 0.694\,122 - 0.000\,967 = 0.693\,15$$

which is correct to 5dp. In general, of course, we could not know how good an approximation this extrapolated value is, and the usual practice is to continue interval-halving until two successive extrapolated values agree to the accuracy required. Not all integrals will converge as quickly as in this example. For example $\int_0^1 \sqrt{x}\,\mathrm{d}x$ requires a large number of evaluations to achieve reasonable accuracy. The reason for the slow convergence of the approximation to $\int_0^1 \sqrt{x}\,\mathrm{d}x$ compared with that of $\int_1^2 (1/x)\mathrm{d}x$ is readily seen from Figure 8.72.

Figure 8.72
(a) Graph of $y = \sqrt{x}$.
(b) Graph of $y = 1/x$.

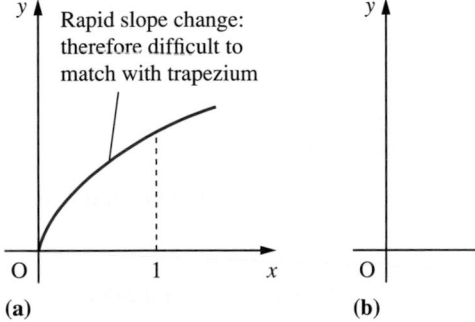

The trapezium rule as given in (8.47) is implemented in MATLAB using the commands

```
a = lower limit;  b = upper limit;  n = number of strips;
h = (b - a)/n;  x = (a:h:b)'
```

(which outputs the x_i values as a column array)

```
y = f(x)
```

(which outputs the corresponding values of y as a column array)

```
h*trapz(y)
```

Considering the integral in Example 8.58 and taking 8 strips then the commands

```
a = 1; b = 2; n = 8; h = (b - a)/n; x = (a:h:b)';
y = 1./x;
```

(Note use of ./ as we are dealing with arrays.)

```
h*trapz(y)
```

return the answer 0.6941 to 4dp, which checks with the value of $T(0.125)$ in the given solution.

The corresponding commands in MAPLE are:

```
with(student):
y:= f(x): trapezoid(y, x = a..b, n): evalf (%);
```

Example 8.59

Evaluate the integral $\int_0^1 \sqrt{(1 + x^2)}\,dx$ to 5dp, using the trapezium rule and extrapolation.

Solution

As before, we begin with just one strip, so that $h = 1$ and $T(1) = \frac{1}{2}(f_0 + f_1)$, where $f_0 = f(x_0) = f(0) = 1.000\,000$ and $f_1 = f(x_1) = f(1) = 1.414\,214$. Thus $T(1) = 1.207\,107$. Next we set $h = \frac{1}{2}$, and we calculate one new value of the integrand at $x = \frac{1}{2}$, giving a new $f_1 = \frac{1}{2}\sqrt{5} = 1.118\,034$ and

$$T(0.5) = hf_1 + \tfrac{1}{2}T(1)$$

$$= 0.5 \times 1.118\,034 + 0.603\,554$$

$$= 1.162\,570$$

An estimate for the error in $T(0.5)$ is

$$\tfrac{1}{3}[T(1) - T(0.5)] = 0.014\,846$$

and a better approximation for the value of the integral is given by

$$1.162\,570 - 0.014\,846 = 1.147\,724$$

Next we interval-halve again, giving $h = 0.25$, and calculate new values of the integrand (at $x = 0.25$ and $x = 0.75$):

$$f_1 = f(0.25) = 1.030\,776 \quad \text{and} \quad f_3 = f(0.75) = 1.25$$

Thus

$$T(0.25) = h(f_1 + f_3) + \tfrac{1}{2}T(0.5) = 1.151\,479$$

with an error estimate of $\tfrac{1}{3}[T(0.5) - T(0.25)] = 0.003\,697$ and an extrapolated value

$$1.151\,479 - 0.003\,697 = 1.147\,782$$

At this stage we can see that the value of the integral is 1.148 to 3dp. We continue interval-halving, giving: for $h = 0.125$, $T(0.125) = 1.148714$, with an error estimate of 0.000922 and an extrapolated value 1.147793; and for $h = 0.0625$, $T(0.0625) = 1.148714$, with an error estimate of 0.000230 and an extrapolated value 1.147793. Thus the extrapolated values agree to 6dp, so that we can write

$$\int_0^1 \sqrt{(1+x^2)}\,dx = 1.14779$$

with confidence that the value is correct to the number of decimal places given.

8.10.2 Simpson's rule

The interval-halving algorithm developed in Section 8.10.1 is the appropriate algorithm to use for automatic computation. It is easy to program and is computationally efficient when used with extrapolation. It is, however, cumbersome for hand computation. For pencil and paper calculations a method that has been commonly used is equivalent to the extrapolated result obtained in Section 8.10.1 but does not give any estimate of error or permit easy interval-halving to check the accuracy of the result.

The trapezium rule approximation to $\int_a^b f(x)dx$ using one strip is

$$T_1 = \tfrac{1}{2}(b-a)[f(a) + f(b)]$$

and that using two strips is

$$T_2 = \frac{b-a}{4}[f(a) + 2f\left(\frac{a+b}{2}\right) + f(b)]$$

The extrapolation based on these two estimates is

$$S = [4T_2 - T_1]/3$$

$$= \frac{(b-a)}{6}[f(a) + 4f\left(\frac{a+b}{2}\right) + f(b)]$$

The formula provides the basic approximation for the area under the curve between $x = a$ and $x = b$. It can be shown to be the area under the parabola which passes through the three points $(a, f(a))$, $((a+b)/2, f((a+b)/2))$ and $(b, f(b))$.

Now consider the interval $[a, b]$ divided into n equal strips of width h where n is an even number. Then we may write

$$\int_a^b f(x)dx = \int_{x_0}^{x_2} f(x)\,dx + \int_{x_2}^{x_4} f(x)dx + \int_{x_4}^{x_6} f(x)dx + \ldots + \int_{x_{n-2}}^{x_n} f(x)dx$$

where $x_k = a + kh$.

Applying the basic formula to each of the integrals on the right-hand side yields the approximation

$$\int_a^b f(x)dx \approx \frac{h}{3}[f_0 + 4f_1 + f_2] + \frac{h}{3}[f_2 + 4f_3 + f_4] + \frac{h}{3}[f_4 + 4f_5 + f_6]$$

$$+ \ldots + \frac{h}{3}[f_{n-2} + 4f_{n-1} + f_n]$$

$$\int_a^b f(x)\mathrm{d}x \approx \frac{h}{3}[f_0 + 4f_1 + 2f_2 + 4f_3 + 2f_4 + \ldots + 2f_{n-2} + 4f_{n-1} + f_n] \tag{8.49}$$

or in words

> The integral is approximately one-third the step size times the sum of four times the odd ordinates plus twice the even ordinates plus first and last ordinates.

This is referred to as **Simpson's rule** and a pencil and paper calculation would be set out as shown in Example 8.60.

There is no command in MATLAB for implementing Simpson's rule (8.49), in which the number of strips is specified. Instead the package incorporates the command $quad(f,a,b)$, which tries to approximate the integral of the scalar-valued function $f = f(x)$ from a to b to within an error of $1.e^{-6}$ using recursive adaptive Simpson quadrature. There is no need to specify the number of strips and the method is somewhat hidden from the user. It is an efficient approach to evaluate an integral numerically but is of limited value as a learning tool. When using the $quad$ command the function $f(x)$ must be expressed as an $inline$ function with the array operations $.*, ./$ and $.^$ used in its specification, so that it can be evaluated with a vector argument. As an illustration we consider the integral of Example 8.58, for which the commands

```
f = inline('(1 + x.^2).^(1/2)'); quad(f, 0,1)
```

return the answer 1.1478.

In MAPLE Simpson's rule (8.49) is evaluated by the commands

```
with(student):
f:= x->f(x); simpson(f(x),x = a..b,n); evalf(%);
```

For the integral of Example (8.58) the commands

```
with(student):
f:= 1/x; simpson(f,x = 1..2,8); evalf(%);
```

return the answer $.6931545307$.

MAPLE also has the facility to produce a sequence of answers corresponding to an array of values for the number of strips; for example the commands

```
nn = [4,8,12];
seq(evalf(simpson(1/x,x = 1..2,n)),n = nn);
```

return the sequence of answers

```
.6932539681, .6931545307, .6931486622
```

This facility provides a valuable learning tool; it is also available for the trapezium rule using the command

```
seq(evalf(trapezoidal(1/x,x = 1..2,n)), n = nn);
```

Example 8.60

Figure 8.73 shows a longitudinal section PQ of rough ground through which a straight horizontal road is to be cut. The width of the road is to be 10 m, and the sides of the cutting and embankment slope at 2 horizontal to 1 vertical. Estimate the net volume of earth removed in making the road.

Figure 8.73
Cross-section with distances above or below datum at 200 m intervals (not to scale).

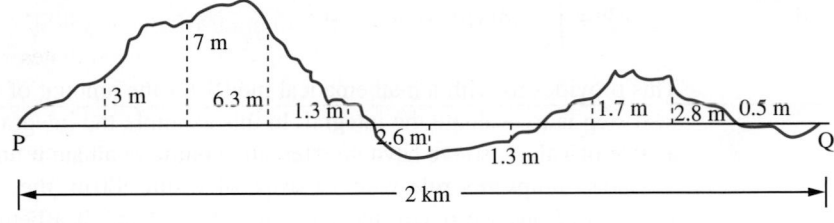

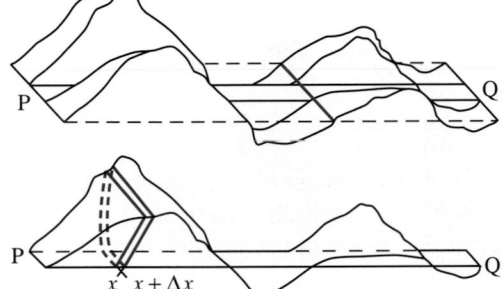

Figure 8.74 Volume of soil to be removed in road construction.

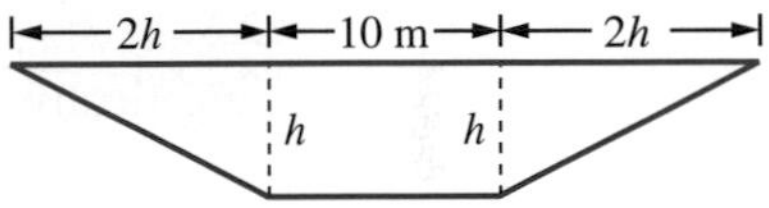

Figure 8.75 Cross-section of cutting with sides sloping at 1 in 2.

Solution

In this case we are not dealing with a solid of revolution, and so cannot use (8.36) to find the volume. Instead, we slice the volume up, estimate the volume of each slice and then add all the individual volumes together as illustrated in Section 8.7.1. The volume above the datum PQ is counted as positive and that below the datum as negative, so that infill on site is accounted for automatically.

Consider the 'slice' between the points at distances x and $x + \Delta x$ from P as shown in Figure 8.74. The volume of this slice is $\bar{A}\Delta x$ where $\bar{A}$ is the average cross-sectional area between x and $x + \Delta x$. The cross-sectional area A depends on the height h of the soil above the datum line PQ. This relationship is given by

$$A = (2h + 10)h$$

as shown in Figure 8.75.

The height h depends on the distance x along the road, so that we can construct a table of values for A as a function of x, as shown in Figure 8.76.

Figure 8.76
Cross-sectional area versus distance.

x	0	200	400	600	800	1000	1200	1400	1600	1800	2000
h	0.0	3.0	7.0	6.3	1.3	−2.6	−1.3	1.7	2.8	−0.5	0.0
A	0.0	48.0	168.0	142.4	16.4	−39.5	−16.4	22.8	43.7	−5.5	0.0

The total volume V of soil removed from the site is the sum of the volumes of the individual slices:

$$V = \sum A(\bar{x})\Delta x$$

where $A(\bar{x})$ is given by

$$A(\bar{x}) = \bar{A} \quad (x \leqslant \bar{x} \leqslant x + \Delta x)$$

Letting the number of slices tend to infinity while making their thicknesses all tend to zero gives V in the form of an integral:

$$V = \int_0^{2000} A(x)\mathrm{d}x$$

This provides us with a mathematical model for the amount of soil to be removed: the next step is to evaluate the integral. In this example the integrand is known only from a table of values, so we have no alternative but to evaluate it numerically.

Using Simpson's rule with 10 strips of width 200 m, the calculation is shown in Figure 8.77 and we obtain the estimate $7.3 \times 10^4 \, \mathrm{m}^3$. If a better estimate is required, more data will have to be collected.

Figure 8.77
Simpson's rule
'paper and pencil'
calculation.

Odds	Evens	First and Last
48.0	168.0	0.0
142.4	16.4	0.0
−39.5	−16.4	0.0
22.8	43.7	423.4
−5.5	211.7 × 2	672.8
168.2 × 4		1096.2 × $\frac{200}{3}$
		73 080.0

8.10.3 Exercises

 Check your answers using MATLAB or MAPLE.

126 Use the trapezium rule to evaluate $\int_0^{0.8} e^{-x^2}\mathrm{d}x$. Take the step size h equal to 0.8, 0.4, 0.2, 0.1 in turn and use extrapolation to improve the accuracy of your answer.

127 Use the trapezium rule, with interval-halving and extrapolation, to evaluate

$$\int_0^1 \log(\cosh x)\mathrm{d}x \quad \text{to 4dp}$$

128 An ellipse has parametric equations $x = \cos t$, $y = \frac{1}{2}\sqrt{3}\sin t$. Show that the length of its circumference is given by

$$2\int_0^{\pi/2} \sqrt{(3 + \sin^2 t)}\mathrm{d}t$$

This integral cannot be evaluated in terms of elementary functions. Use the trapezium rule with interval-halving to evaluate it to 6dp.

129 The capacity of a battery is measured by $\int i \, \mathrm{d}t$, where i is the current. Estimate, using Simpson's rule, the capacity of a battery whose current was measured over an 8 h period with the results shown below:

Time/h	0	1	2	3	4	5	6	7	8
Current/A	25.2	29.0	31.8	36.5	33.7	31.2	29.6	27.3	28.6

130 The speed $V(t)\,\mathrm{m\,s}^{-1}$ of a vehicle at time t s is given by the table below. Use Simpson's rule to estimate the distance travelled over the eight seconds.

t	0	1	2	3	4	5	6	7	8
$V(t)$	0	0.63	2.52	5.41	9.02	13.11	16.72	18.75	20.15

131 Use Simpson's rule with $h = 0.1$ to estimate

$$\int_0^1 \sqrt{(1 + x^3)}\mathrm{d}x$$

(Notice that by this method you have no way of knowing how accurate your estimate is.)

8.11 Engineering application: design of prismatic channels

The mean velocity V of flow in straight prismatic channels is proportional to $(A/p)^r$, where A is the cross-sectional area of the flow, p is the wetted perimeter and r is approximately a constant $(\frac{7}{12})$. Given the channel section for minimum flows (that is, A_0 and p_0), the objective is to design a channel such that V has the same value for all larger discharges.

Assume a symmetric channel cross-section as shown in Figure 8.78, where A_0 and p_0 are the minimum flow values of A and p. Let the shape of the channel be given by $x = f(y)$. (Note that in this application y, the height of the surface above the datum line, is the independent variable.) Then we want to find the function $f(y)$ such that the mean flow velocity is independent of y. This implies because it is proportional to $(A/p)^r$ that

Figure 8.78
Channel cross-section.

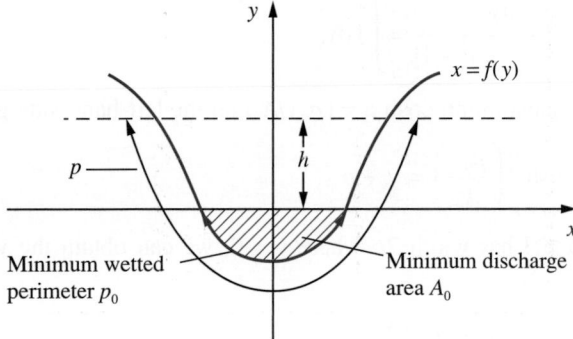

$$\frac{A}{p} = \frac{A_0}{p_0}$$

The area A is given by the integral of $f(y)$. Thus

$$A = A_0 + 2\int_0^h x\,\mathrm{d}y$$

where $x = f(y)$ and $h > 0$.

Using (8.45), the wetted perimeter p is given by

$$p = p_0 + 2\int_0^h \sqrt{\left[1 + \left(\frac{\mathrm{d}x}{\mathrm{d}y}\right)^2\right]}\,\mathrm{d}y \quad (h > 0)$$

Since $A/A_0 = p/p_0$, we deduce that

$$1 + \frac{2}{A_0}\int_0^h x\,\mathrm{d}y = 1 + \frac{2}{p_0}\int_0^h \sqrt{\left[1 + \left(\frac{\mathrm{d}x}{\mathrm{d}y}\right)^2\right]}\,\mathrm{d}y$$

Rearranging the integrals under a common integral sign gives

$$\int_0^h \left\{\frac{x}{A_0} - \frac{1}{p_0}\sqrt{\left[1 + \left(\frac{\mathrm{d}x}{\mathrm{d}y}\right)^2\right]}\right\}\mathrm{d}y = 0 \quad (h > 0)$$

Since this is true for all $h > 0$, it implies that the integrand must be identically zero. Thus $x = f(y)$ satisfies the differential equation

$$\frac{x}{A_0} = \frac{1}{p_0}\sqrt{\left[1 + \left(\frac{dx}{dy}\right)^2\right]}$$

which, assuming $\dfrac{dx}{dy} \geqslant 0$, implies

$$\frac{dx}{dy} = \sqrt{\left[\left(\frac{p_0 x}{A_0}\right)^2 - 1\right]}$$
(8.50)

Integrating with respect to y then gives

$$\int \frac{dx}{\sqrt{[(p_0 x/A_0)^2 - 1]}} = \int 1\, dy$$

Using the substitution $\cosh u = (p_0 x/A_0)$ on the left-hand side gives

$$\frac{A_0}{p_0}\cosh^{-1}\left(\frac{p_0 x}{A_0}\right) = y + c$$

If the channel has width $2b$ where $y = 0$, we can obtain the value of the constant of integration c as

$$c = \frac{A_0}{p_0}\cosh^{-1}\left(\frac{p_0 b}{A_0}\right)$$

and deduce the formula for a suitable channel shape as

$$y = \frac{A_0}{p_0}\left[\cosh^{-1}\left(\frac{p_0 x}{A_0}\right) - \cosh^{-1}\left(\frac{p_0 b}{A_0}\right)\right]$$

This solution, however, is not unique and we note that the differential equation (8.50) is also satisfied by

$$x = \frac{A_0}{p_0}$$

As an exercise, use this information to show that the general solution may take the form of either of the cross-sections shown in Figures 8.79(a) and (b). Notice that the line shape in Figure 8.79(b) does not have $\dfrac{dx}{dy} \geqslant 0$.

Figure 8.79

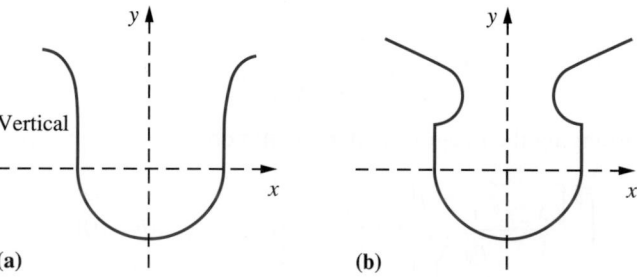

(a)   (b)

Vertical

8.12 Engineering application: harmonic analysis of periodic functions

Periodic functions occur frequently in practical problems and in natural phenomena like tidal systems. Rotating parts of machinery produce vibrations, which may become dangerous when resonance occurs. Indeed such a resonance led to the failure of the Tacoma Road bridge (see Section 10.10.3). Periodic motions usually involve several frequencies of vibrations at the same time and the method of finding the amplitude of each frequency is called **harmonic analysis** (see also Chapter 12).

Consider, for example, the crank and connecting rod mechanism discussed in Chapter 2, Example 2.44. The displacement function of the slider is

$$y = r \cos x + \sqrt{(l^2 - r^2 \sin^2 x)}$$

where r is the radius of the crank, l the length of the connecting rod and x (radians) is the angle turned through. The motion is periodic but is not a simple sinusoid. It involves many harmonics and we may write

$$y = a_0 + a_1 \cos x + a_2 \cos 2x + a_3 \cos 3x + \dots$$

where the a's are constants and we choose a cosine series since y is an even function $y(-x) = y(x)$. To simplify the problem, we take a special case with $r = 1$ and $l = 3$. Then

$$y = \cos x + \sqrt{(8 + \cos^2 x)}$$

A graph of y is shown in Figure 8.80.

The displacement y has period 2π and its mean value $\bar{y}$ is given by

$$\bar{y} = \frac{1}{2\pi} \int_0^{2\pi} [\cos x + \sqrt{(8 + \cos^2 x)}] \, dx$$

The contribution of $\cos x$ to the value of the integral over a complete period is zero, so this simplifies to

$$\bar{y} = \frac{1}{2\pi} \int_0^{2\pi} \sqrt{(8 + \cos^2 x)} \, dx$$

Figure 8.80 Graph of $y = \cos x + \sqrt{(8 + \cos^2 x)}$

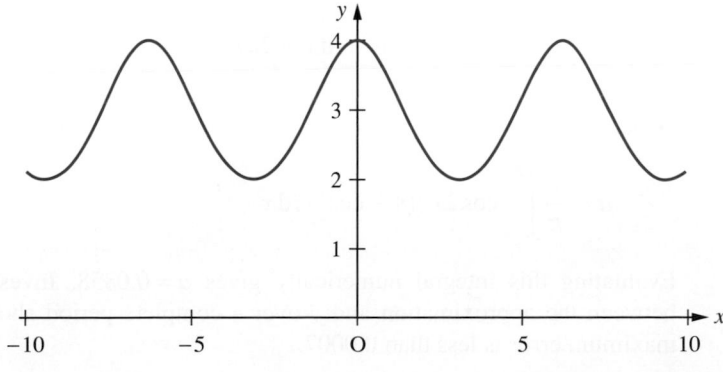

This integral cannot be evaluated analytically. Using the trapezium rule with step sizes $\pi/2$, $\pi/4$ and $\pi/8$ together with Richardson's extrapolation we obtain

$$\bar{y} = 2.9148$$

We now seek an approximation to $y(x)$ having the form

$$y(x) \simeq \bar{y} + \cos x + a \cos 2x$$

such that the integral of the squared error over a complete period is as small as possible. That is, a is chosen so that

$$\frac{d}{da}\left\{ \int_{-\pi}^{\pi} [\bar{y} + \cos x + a \cos 2x - y(x)]^2 dx \right\} = 0$$

Expanding the integrand this gives

$$\frac{d}{da}\left\{ \int_{-\pi}^{\pi} [\bar{y}^2 + a^2 \cos^2 2x + 8 + \cos^2 x + 2\bar{y}a \cos 2x \right.$$

$$\left. - 2\bar{y}\sqrt{(8 + \cos^2 x)} - 2a \cos 2x \sqrt{(8 + \cos^2 x)}]dx \right\} = 0$$

This tidies up to

$$\frac{d}{da}\left\{ \int_{-\pi}^{\pi} [\bar{y}^2 + 8 + \cos^2 x - 2\bar{y}\sqrt{(8 + \cos^2 x)}]dx + 2a\bar{y}\int_{-\pi}^{\pi} \cos 2x \, dx \right.$$

$$\left. - 2a \int_{-\pi}^{\pi} \cos 2x \sqrt{(8 + \cos^2 x)} \, dx + a^2 \int_{-\pi}^{\pi} \cos^2 2x \, dx \right\} = 0$$

The first integral inside the curly brackets is independent of a, and so differentiates to zero. The second integral is zero in value and the last integral has value π. Thus differentiating with respect to a gives

$$-2\int_{-\pi}^{\pi} \cos 2x \sqrt{(8 + \cos^2 x)} \, dx + 2a\pi = 0$$

Hence

$$a = \frac{1}{\pi} \int_{-\pi}^{\pi} \cos 2x \sqrt{(8 + \cos^2 x)} \, dx$$

Evaluating this integral numerically gives $a = 0.0858$. Investigating the difference between the approximation and y over a complete period shows that the size of the maximum error is less than 0.0007.

8.13 Review exercises (1–39)

 Check your answers using MATLAB or MAPLE whenever possible.

1 Differentiate the following expressions, giving your answers as simply as possible.

(a) e^{x^2+x}

(b) $\dfrac{x^3}{(3-x)^2}$

(c) $\sin(5x-1)$

(d) $(\tan x)^x$

(e) $\cos^{-1}\sqrt{(1-x^2)}$

(f) $\dfrac{1}{\sqrt{(x+1)}}$

(g) $\sin^{-1}\dfrac{1}{\sqrt{(1+x^2)}}$

(h) $\dfrac{1}{(x-1)(x+2)}$

(i) $\sin(3x+1)$

(j) $x^3 \ln x$

(k) $\dfrac{x^3}{(3-x^2)}$

(l) $\tan^{-1}(e^{-2x})$

(m) $\sqrt{(1+\cosh x)}$

(n) $(x^2+1)\sin 2x$

(o) $\dfrac{x-1}{(x+2)^2}$

(p) $e^{\sqrt{x}}$

(q) $\ln \tan x$

(r) $\dfrac{(2x-1)^{3/2}}{(x+1)^5}$

(s) $x \sin x$

(t) e^{x^2}

(u) 2^x

(v) $\dfrac{\cos x}{1+\sin x}$

(w) $\sin^{-1}\left(\dfrac{1}{x}\right)$

(x) $x^3 \cos 2x$

(y) $\sqrt{(x^3+x+3)}$

(z) $\dfrac{e^{-x}}{1+x}$

2 Evaluate

(a) $\displaystyle\int x^{1/2}\ln x\,dx$

(b) $\displaystyle\int \dfrac{(2x+3)\,dx}{x^2+2x+2}$

(c) $\displaystyle\int_0^3 \dfrac{1}{x}\lfloor x\rfloor\,dx$

(d) $\displaystyle\int_{1/2}^1 \dfrac{x\sin^{-1}x\,dx}{\sqrt{(1-x^2)}}$

(set $x=\sin t$)

(e) $\displaystyle\int \dfrac{x\,dx}{(x-1)(x-2)}$

(f) $\displaystyle\int \tan^4 x\,dx$

(g) $\displaystyle\int_0^1 \sqrt{(4-3x^2)}\,dx$

(h) $\displaystyle\int \dfrac{dx}{\sqrt{(4-9x^2)}}$

(i) $\displaystyle\int \dfrac{x^2\,dx}{\sqrt{(x^3-1)}}$

(j) $\displaystyle\int \dfrac{(x^2+1)\,dx}{x+1}$

(k) $\displaystyle\int \dfrac{dx}{x^2+6x+13}$

(l) $\displaystyle\int \sqrt{x}\,\sin\sqrt{x}\,dx$

(m) $\displaystyle\int_0^2 \text{FRACPT}(x)\,dx$

(n) $\displaystyle\int_0^1 \sinh^2 x\,dx$

(o) $\displaystyle\int (1-3x)^9\,dx$

(p) $\displaystyle\int \sin 3x \sin 2x\,dx$

(q) $\displaystyle\int \ln 2x\,dx$

(r) $\displaystyle\int xe^{-x^2/2}\,dx$

(s) $\displaystyle\int \dfrac{dx}{\sqrt{(4x^2-9)}}$

(t) $\displaystyle\int_{-1}^4 \dfrac{(3x-1)\,dx}{\sqrt{(4+3x-x^2)}}$

(u) $\displaystyle\int_0^1 \dfrac{x\,dx}{(x+1)(x^2+1)}$

(v) $\displaystyle\int (4-3x)^4\,dx$

(w) $\displaystyle\int \cos 2x \cos 3x\,dx$

(x) $\displaystyle\int \sin^{-1}x\,dx$

(y) $\displaystyle\int x^2 e^{-x}\,dx$

(z) $\displaystyle\int \dfrac{dx}{1+x+x^2}$

3 Find the equation of the tangent and normal at the point (1, 4) to the curve whose equation is

$$y = 2x^4 - 3x^3 + 5x^2 + 3x - 3$$

4 Find the equation of the tangent to the curve $x^2 - 3xy + 2y^2 = 3$ at the point (1, 2) and the equation of the normal to the curve $y = x^3 - x^2$ at the point (1, 0). Find the distance of the point of intersection of these lines from the point $(-1, 2)$.

5 With reference to Example 2.10, Chapter 2, confirm that the function

$$E(x) = x^2(1-x), \qquad 0 \le x \le 1$$

has maximum value when $x = 2/3$.

6 Find the turning points on the curve

$$y = 2x^3 - 5x^2 + 4x - 1$$

and determine their nature. Find the point of inflection and sketch the graph of the curve.

39   (a) A curve (an oval) is defined by the formulae

$$x(\theta) = \cos^4\theta, \; y(\theta) = \cos^3\theta \, \sin\theta$$

Complete the table below for values of x and y to 2 decimal places.

θ	0	0.23				0.57
x	1.0	0.9	0.8	0.7	0.6	0.5
y	0.00					0.32

θ					$\pi/2$
x	0.4	0.3	0.2	0.1	0.0
y					0.00

Use these data to draw the oval on graph paper.

(b) Show that the volume of the body whose surface is generated by rotating the curve in part (a) about the x-axis (an ovaloid) is $\pi/15$.

(c) Assuming that the ovaloid generated in part (b) has uniform density, show that its centre of mass is at the point $(15/28, 0)$.

(d) Show that the tangent to the curve in part (a) at the point $(\cos^4\theta, \cos^3\theta \, \sin\theta)$ has the equation

$$y - \cos^3\theta \, \sin\theta$$
$$= (4\sin^2\theta - 1)(x - \cos^4\theta)/(4\sin\theta \, \cos\theta)$$

Deduce the turning points of the curve and show that the breadth (that is, distance between maximum and minimum values of y) of the oval is $3\sqrt{3}/4$.

(e) Show that the normal to the curve in part (a) at the point $(\cos^4\theta, \cos^3\theta \, \sin\theta)$ has the equation

$$y - \cos^3\theta \, \sin\theta$$
$$= 4\sin\theta \, \cos\theta \, (x - \cos^4\theta)/(4\cos^2\theta - 3)$$

(f) For what values of θ does the normal to the curve found in (e) pass through the centre of mass?

(g) Show that the distance from a point on the surface of the ovaloid to its centre of mass has stationary values where $\theta = 0$, $\cos^{-1}(\sqrt{(5/7)})$, $\pi/2$, $\cos^{-1}(-\sqrt{(5/7)})$, and π, and classify their nature.

(h) If the ovaloid is to rest in *stable* equilibrium on a horizontal plane, which points on the generating oval correspond to possible points of contact with the plane?

9 Further Calculus

Chapter 9	Contents	

9.1 Introduction

In Chapter 8 we discussed the fundamental ideas and concepts of integral and differential calculus and applied them to various practical problems. We also developed techniques for solving problems using calculus. In this chapter we shall extend the techniques developed in Chapter 8 to deal with a wide range of problems and develop the theory to enable us to understand the numerical techniques widely used in practical problem-solving. We shall introduce multivariable calculus and use it to solve problems in optimization.

9.2 Improper integrals

When we considered the definite integral $\int_a^b f(x)\,\mathrm{d}x$ in Chapter 8 and showed its equivalence with an area under a curve, it was assumed that the integrand $f(x)$ was continuous, or at least piecewise-continuous, over the closed domain of integration $[a, b]$. To illustrate a possible consequence of this not being the case, consider the apparent definite integral $\int_{-1}^1 (1/x^2)\,\mathrm{d}x$. If we proceed in a mechanistic way and follow the usual procedure, we should write

$$\int_{-1}^1 \frac{1}{x^2}\,\mathrm{d}x = \left[\frac{-1}{x}\right]_{-1}^1 = -2$$

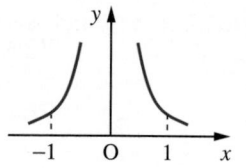

Figure 9.1
Graph of $f(x) = 1/x^2$.

However, if we plot the graph of $f(x) = 1/x^2$, as in Figure 9.1, it is clear that this is not correct, since it implies that the area under a curve that lies entirely above the x axis is negative. So where have we gone wrong? The answer lies in the fact that $f(x) = 1/x^2$ has an **infinite discontinuity** or **singularity** (that is, it is unbounded) at $x = 0$. As a consequence, the region under the curve over the domain of integration $[-1, 1]$ is unbounded, and our integration process was invalid.

In this section we consider the conditions under which the integral $\int_a^b f(x)\,\mathrm{d}x$ exists when either

(a) the integrand $f(x)$ becomes unbounded (that is, $f(x)$ has an infinite discontinuity) at some point within the domain of integration, or

(b) the domain of integration is infinite (that is, either a or b or both are infinite).

Such integrals are called **improper integrals**, and are encountered in many contexts in engineering. For example, the period of a simple pendulum of length l released from rest with angle α is given by

$$2\sqrt{\left(\frac{l}{g}\right)}\int_0^\alpha \frac{1}{\sqrt{(\sin^2\frac{\alpha}{2} - \sin^2\frac{\theta}{2})}}\,\mathrm{d}\theta$$

where g is the acceleration due to gravity. The integrand is infinite at $x = \alpha$; yet we know that the answer is meaningful from elementary physics. Further examples are met when Laplace transforms are introduced in Chapter 11.

9.2.1 Integrand with an infinite discontinuity

Suppose that the lower limit $x = a$ is the only point of infinite discontinuity of $f(x)$ in $[a, b]$. Then we define

$$\int_a^b f(x)\,dx = \lim_{X \to a^+} \int_X^b f(x)\,dx \qquad\qquad (9.1)$$

provided that the limit exists. Otherwise $\int_a^b f(x)\,dx$ has no meaning.

Similarly, if the upper limit $x = b$ is the only point of infinite discontinuity in $[a, b]$, we define

$$\int_a^b f(x)\,dx = \lim_{X \to b^-} \int_a^X f(x)\,dx \qquad\qquad (9.2)$$

provided that the limit exists. Otherwise $\int_a^b f(x)\,dx$ has no meaning.

Example 9.1   Evaluate the following, if they are defined:

(a) $\displaystyle\int_0^1 x^{-2/3}\,dx$ (b) $\displaystyle\int_0^1 \frac{dx}{\sqrt{(1-x^2)}}$ (c) $\displaystyle\int_0^1 \ln x\,dx$ (d) $\displaystyle\int_0^1 \frac{dx}{x^2}$

Solution   (a) Here the integral has an infinite discontinuity at the lower limit $x = 0$, and we consider

$$\lim_{X \to 0^+} \int_X^1 x^{-2/3}\,dx = \lim_{X \to 0^+} [3x^{1/3}]_X^1 = \lim_{X \to 0^+} (3 - 3X^{1/3}) = 3$$

Since the limit exists, it follows from (9.1) that

$$\int_0^1 x^{-2/3}\,dx = 3$$

(b) Here the discontinuity in the integrand occurs at the upper limit $x = 1$, and so we consider

$$\lim_{X \to 1^-} \int_0^X \frac{dx}{\sqrt{(1-x^2)}} = \lim_{X \to 1^-} [\sin^{-1}x]_0^X = \lim_{X \to 1^-} (\sin^{-1}X) = \tfrac{1}{2}\pi$$

Since the limit exists, it follows from (9.2) that

$$\int_0^1 \frac{dx}{\sqrt{(1-x^2)}} = \tfrac{1}{2}\pi$$

(c) Again the integrand has an infinite discontinuity at the lower limit $x = 0$, and so we consider

$$\lim_{X \to 0^+} \int_X^1 \ln x \, dx = \lim_{X \to 0^+} [x \ln x - x]_X^1 \quad \text{(integrating by parts)}$$

$$= \lim_{X \to 0^+} (X - X \ln X - 1)$$

$$= -1 \quad \text{(since } X \ln X \to 0 \text{ as } X \to 0^+, \text{ question 61, Section 7.8.3)}$$

Since the limit exists, it follows from (9.1) that

$$\int_0^1 \ln x \, dx = -1$$

(d) In this case the integrand has an infinite discontinuity at the lower limit $x = 0$, and we consider the limit

$$\lim_{X \to 0^+} \int_X^1 \frac{dx}{x^2} = \lim_{X \to 0^+} \left[\frac{-1}{x} \right]_X^1 = \lim_{X \to 0^+} \left(\frac{1}{X} - 1 \right)$$

This becomes infinite as $X \to 0$ and so the integral has no meaning.

If the integrand $f(x)$ has an infinite discontinuity at $x = c$, where $a < c < b$, then we define

$$\int_a^b f(x) dx = \lim_{X \to 0^+} \int_a^{c-X} f(x) dx + \lim_{X \to 0^+} \int_{c+X}^b f(x) dx \qquad \textbf{(9.3)}$$

provided that both limits on the right-hand side exist. Otherwise $\int_a^b f(x) dx$ is not defined.

Example 9.2   Confirm that $\int_{-1}^1 (1/x^2) dx$ is not defined.

Solution   This is the apparent integral considered in the introductory discussion, where we saw that following the usual integration techniques in a mechanistic sense led to a ridiculous answer. In this case the integrand has an infinite discontinuity at $x = 0$, so, following (9.3), we consider the two limits

$$\lim_{X \to 0^+} \int_{-1}^{-X} \frac{dx}{x^2} \quad \text{and} \quad \lim_{X \to 0^+} \int_X^1 \frac{dx}{x^2}$$

From the solution to Example 9.1(d) it is clear that both these tend to infinity, so that neither limit exists and the integral $\int_{-1}^1 (1/x^2) dx$ is not defined.

Both MATLAB and MAPLE can evaluate such integrals. Considering Example 9.1

(a) The MATLAB commands

```
syms x y
y = x^(-2/3);  int(y,0,1)
```

return the answer 3.

(b) The MAPLE commands

```
y:= 1/sqrt(1 - x^2);  int(y,x = 0..1);
```

return the answer $\frac{1}{2}\pi$.

(c) The MATLAB commands

```
syms x
int(log(x),0,1)
```

return the answer -1.

(d) The MAPLE command

```
int(1/x^2,x = 0..1);
```

returns infinity.

As an exercise consider how MATLAB or MAPLE deals with Example 9.2.

The numerical evaluation of integrals whose integrands have infinite discontinuities will clearly cause numerical problems. Often such integrals can be evaluated by first changing the integrand by means of a substitution, as illustrated in the following example.

Example 9.3   Obtain the value of the integral

$$T(\alpha) = 2\sqrt{\left(\frac{l}{g}\right)} \int_0^\pi \frac{d\theta}{\sqrt{(\sin^2\frac{\alpha}{2} - \sin^2\frac{\theta}{2})}}$$

where $\alpha = \pi/3$.

Solution   This integral has an integrand which is unbounded at $\theta = \alpha$. In this case we can 'remove' the difficulty by the substitution $\sin\dfrac{\theta}{2} = \sin\dfrac{\alpha}{2}\sin\phi$. Then

$$\tfrac{1}{2}\cos\frac{\theta}{2}d\theta = \sin\frac{\alpha}{2}\cos\phi\,d\phi$$

with $\theta = 0$ corresponding to $\phi = 0$ and $\theta = \alpha$ corresponding to $\phi = \pi/2$ and

$$T(\alpha) = 2\sqrt{\left(\frac{l}{g}\right)} \int_0^{\pi/2} \frac{1}{\sin\frac{1}{2}\alpha\sqrt{(1 - \sin^2\phi)}} \cdot \frac{2\sin\frac{1}{2}\alpha\cos\phi}{\sqrt{(1 - \sin^2\frac{1}{2}\alpha\sin^2\phi)}}\,d\phi$$

$$= 4\sqrt{\left(\frac{l}{g}\right)} \int_0^{\pi/2} \frac{1}{\sqrt{(1 - \sin^2\frac{1}{2}\alpha\sin^2\phi)}}\,d\phi$$

When $\alpha = \pi/3$, we have

$$T(\tfrac{\pi}{3}) = 8\sqrt{\left(\frac{l}{g}\right)}\int_0^{\pi/2}\frac{1}{\sqrt{(4-\sin^2\phi)}}\,d\phi = 8\sqrt{\left(\frac{l}{g}\right)}\int_0^{\pi/2}\frac{1}{\sqrt{(3+\cos^2\phi)}}\,d\phi$$

Applying the trapezium rule (with 4 intervals) gives $T(\tfrac{\pi}{3}) = 13.4864\sqrt{\left(\frac{l}{g}\right)}$.

9.2.2 Infinite integrals

The second case, where the domain of integration is infinite, is dealt with in a similar manner. We define

$$\int_a^\infty f(x)\,dx = \lim_{X\to\infty}\int_a^X f(x)\,dx \tag{9.4}$$

if that limit exists. Otherwise $\int_a^\infty f(x)\,dx$ has no meaning.

Example 9.4 Evaluate the following:

(a) $\displaystyle\int_1^\infty x^{-3/2}\,dx$ (b) $\displaystyle\int_0^\infty\frac{dx}{1+x^2}$ (c) $\displaystyle\int_0^\infty e^{-x}\sin x\,dx$ (d) $\displaystyle\int_{-\infty}^\infty e^{3x}\exp(-e^x)\,dx$

Solution (a) $\displaystyle\int_1^\infty x^{-3/2}\,dx = \lim_{X\to\infty}\int_1^X x^{-3/2}\,dx = \lim_{X\to\infty}[-2x^{-1/2}]_1^X = \lim_{X\to\infty}(2-2X^{-1/2}) = 2$

(b) $\displaystyle\int_0^\infty\frac{dx}{1+x^2} = \lim_{X\to\infty}\int_0^X\frac{dx}{1+x^2} = \lim_{X\to\infty}[\tan^{-1}x]_0^X = \lim_{X\to\infty}(\tan^{-1}X) = \tfrac{1}{2}\pi$

(c) $\displaystyle\int_0^\infty e^{-x}\sin x\,dx = \lim_{X\to\infty}\int_0^X e^{-x}\sin x\,dx$

$$= \lim_{X\to\infty}[-\tfrac{1}{2}e^{-x}(\cos x+\sin x)]_0^X \quad\text{(integration by parts)}$$

$$= \lim_{X\to\infty}[\tfrac{1}{2}-\tfrac{1}{2}e^{-X}(\cos X+\sin X)]$$

$$= \tfrac{1}{2}$$

The indefinite integral is obtained using integration by parts, as in Example 8.47(c). It can be verified by direct differentiation.

(d) Here we simplify the integral by the substitution $t = e^x$, so that $x \to -\infty$ gives $t = 0$, $x \to \infty$ gives $t \to \infty$ and $dt = e^x\,dx$. The integral becomes

$$\int_{-\infty}^\infty e^{3x}\exp(-e^x)\,dx = \int_0^\infty t^2 e^{-t}\,dt = [-t^2 e^{-t} - 2t e^{-t} - 2e^{-t}]_0^{T\to\infty}$$

using integration by parts twice.
Hence

$$\int_{-\infty}^{\infty} e^{3x} \exp(-e^x)\,dx = 2$$

Again such integrals may be evaluated directly by MATLAB and MAPLE. To illustrate we consider Examples 9.4(b) and (d). For 9.4(b) the commands

MATLAB	MAPLE
`syms x y`	
`y = 1/(1 + x^2);`	`y:= 1/(1 + x^2);`
`int(y,0,inf)`	`int(y,x = 0..infinity);`

return the answer

`1/2*pi`	$\frac{1}{2}\pi$

and for 9.4(d) the commands

`syms x`	
`int(exp(3*x)*`	`int(exp(3*x)*`
`exp(-exp(x)),-inf,inf)`	`exp(-exp(x)),-infinity..infinity);`

return the answer 2

For practice check the answers to Examples 9.4(a) and (c).

9.2.3 Exercise

Check your answer using MATLAB or MAPLE whenever possible.

1 Evaluate the following improper integrals.

(a) $\int_0^1 (-x \ln x)\,dx$

(b) $\int_0^{\infty} x \exp(-x^2)\,dx$

(c) $\int_0^{\infty} x^2 e^{-2x}\,dx$

(d) $\int_{-\infty}^{\infty} e^x \exp(-e^x)\,dx$

(e) $\int_0^1 x^2(1 - x^3)^{-1/2}\,dx$

(f) $\int_0^1 (x - 1)/\sqrt{x}\,dx$

(g) $\int_0^{\frac{\pi}{2}} \frac{\sin x}{\sqrt{\cos x}}\,dx$

(h) $\int_0^{\frac{\pi}{2}} \cos x \sin^{-1/3} x\,dx$

(i) $\int_0^{\infty} \frac{x}{1 + x^4}\,dx$

Some theorems with applications to numerical methods

There are a number of theorems involving integration and differentiation that are useful in understanding why certain numerical methods are better than others and in devising new methods. They are also useful in the more mundane tasks of assessing the effect of data error when evaluating functions and probing the accuracy of analytical approximations to functions. We shall now briefly consider such theorems and indicate their potential uses. Deriving the results is not easy and the reader may prefer to omit the proofs. The results, however, have many practical implications and should be studied carefully.

9.3.1 Rolle's theorem and the first mean value theorems

The simplest result is the following

Theorem 9.1 **Rolle's theorem**

If the function $f(x)$ is continuous on the domain $[a, b]$ and differentiable on (a, b) with $f(a) = f(b)$ then there is at least one point $x = c$ in (a, b) such that $f'(c) = 0$.

end of theorem

The validity of this theorem can be easily illustrated geometrically as shown in Figure 9.2, since what the theorem tells us is that it is possible to find at least one point on the curve $y = f(x)$ between the values $x = a$ and $x = b$ where the tangent is parallel to the x axis; that is, there must exist at least one maximum or minimum between $x = a$ and $x = b$.

In Chapter 7, Section 7.9.1, we discussed the properties of continuous functions. All continuous functions are integrable, and this fact enables us to calculate the mean value of a continuous function over a given domain, say $[a, b]$. The mean value is given by

$$\frac{1}{b-a}\int_a^b f(x)\,\mathrm{d}x$$

Figure 9.2
Four examples of
Rolle's theorem.

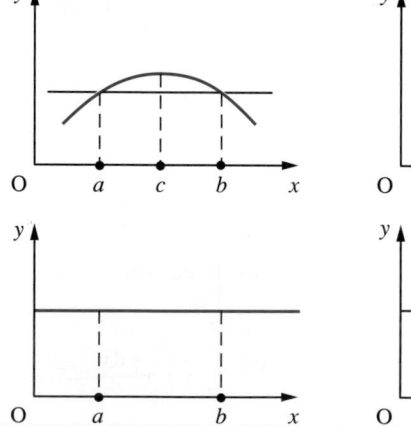
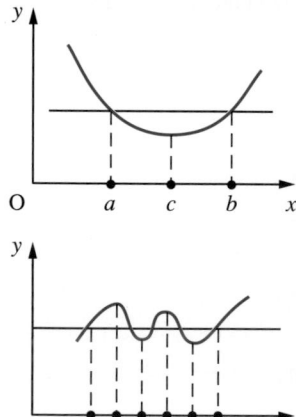

Clearly the mean value of $f(x)$ lies between its maximum and minimum values on the domain $[a, b]$ and, from the intermediate value theorem (Property (c), Section 7.9.1), we deduce that there is a point $x = c$ in the interval $[a, b]$ such that (see 8.43)

$$f(c) = \text{mean value of } f(x) = \frac{1}{b - a} \int_a^b f(x)\mathrm{d}x$$

This result is referred to as the first mean value theorem of integral calculus and may be stated as follows.

Theorem 9.2

The first mean value theorem of integral calculus

If the function $f(x)$ is continuous over the domain $[a, b]$ then there exists at least one point $x = c$, with $a < c < b$, such that

$$f(c) = \frac{1}{b - a} \int_a^b f(x)\mathrm{d}x$$

end of theorem

This theorem is illustrated geometrically in Figure 9.3(a).

If $f(x)$ is a differentiable function then

$$\int_a^b f'(x)\mathrm{d}x = f(b) - f(a)$$

Applying Theorem 9.2 to $f'(x)$ gives

$$\int_a^b f'(x)\mathrm{d}x = (b - a)f'(c), \quad \text{with } a < c < b$$

and hence, by equating the two values of $\int_a^b f'(x)\mathrm{d}x$,

$$\frac{f(b) - f(a)}{b - a} = f'(c)$$

This result is referred to as the first mean value theorem of differential calculus, and may be stated as follows.

Figure 9.3
The first mean value theorems: (a) $f(c_i)$ is the mean value of $f(x)$ ($a \leqslant x \leqslant b$); (b) the chord PQ is parallel to the tangents at $x = c_i$.

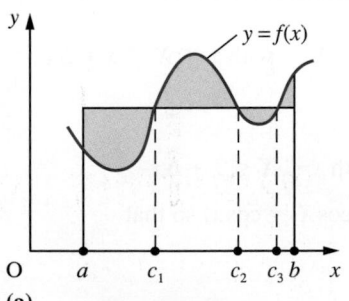
(a)

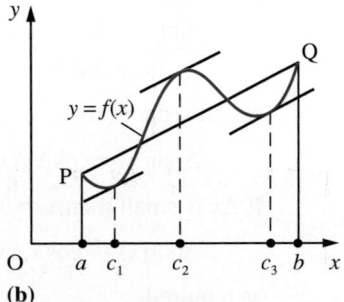
(b)

Theorem 9.3	**First mean value theorem of differential calculus**

If the function $f(x)$ is continuous on the domain $[a, b]$ and differentiable on (a, b) then there exists at least one point $x = c$, with $a < c < b$, such that

$$\frac{f(b) - f(a)}{b - a} = f'(c)$$

end of theorem

It is this theorem that is normally referred to as the first mean value theorem. Geometrically, it implies that at some point on the interval $[a, b]$ the slope of the tangent to the graph of $f(x)$ is parallel to the chord between the end points $x = a$ and $x = b$ of the graph, as shown in Figure 9.2(b).

An immediate application of Theorem 9.3 is in the estimation of the effect of rounding errors in the independent variable x on the calculated value of the dependent variable $y = f(x)$. If ε_x is the error bound for x then the error bound for y is ε_y, where

$$\varepsilon_y = \max_{x - \varepsilon_x < x^* < x + \varepsilon_x} |f(x^*) - f(x)|$$

Applying the first mean value theorem with $a = x$, $b = x^*$ gives

$$|f(x^*) - f(x)| = |x^* - x| \, |f'(c)|$$

with c lying between x and x^*. Since $f'(c) \approx f'(x)$, we have

$$\varepsilon_y \approx \max_{x - \varepsilon_x < x^* < x + \varepsilon_x} |f'(x)(x^* - x)| = |f'(x)|\varepsilon_x \qquad (9.5)$$

We illustrate this by Example 9.5.

Example 9.5	Show that

$$\Delta(\sin x) \approx \cos x \, \Delta x$$

and hence estimate an error bound for $\sin a$, where $a = 1.935$ (3dp). Compare the error interval obtained with $[\sin 1.9355, \sin 1.9345]$. Express $\sin a$ as a correctly rounded number with the maximum number of decimal places.

Solution	The difference $\Delta(\sin x)$ is given by

$$\Delta(\sin x) = \sin(x + \Delta x) - \sin x$$

Since $(d/dx) \sin x = \cos x$, application of Theorem 9.3 gives

$$\frac{\sin(x + \Delta x) - \sin x}{(x + \Delta x) - x} = \cos X, \quad \text{with } x < X < x + \Delta x$$

which reduces to

$$\Delta(\sin x) = \cos X \, \Delta x, \quad \text{with } x < X < x + \Delta x$$

If Δx is small then $x \approx X$ and $\cos X \approx \cos x$, so that

$$\Delta(\sin x) \approx \cos x \, \Delta x$$

as required.

Setting $x = a$ gives $\Delta(\sin a) \approx \cos a\, \Delta a$, and hence, using (9.5), an error bound estimate for $\sin a$ is

$$\varepsilon_{\sin a} = |\cos a|\, \varepsilon_a$$

In this example $a = 1.935$ and $\varepsilon_a = 0.0005$, so that

$$\varepsilon_{\sin a} = |\cos 1.935|(0.0005) = |-0.3562|(0.0005) = 0.000\,18$$

Thus

$$\sin a = \sin 1.935 \pm 0.000\,18 = 0.934\,41 \pm 0.000\,18$$

which spans the interval $[0.934\,23, 0.934\,59]$.

Now $\sin 1.9355 = 0.934\,23$ and $\sin 1.9345 = 0.934\,59$, so that in this example the estimate of the error interval and the error interval are the same to 5dp.

Thus

$$\sin a = 0.9344 \pm 0.0002$$

or

$$\sin a = 0.93$$

9.3.2 Convergence of iterative schemes

In Chapter 7, Section 7.9.3, the solution of equations by iteration was discussed. We now consider the convergence of such iterative schemes. As before, suppose that an iteration for the root $x = \alpha$ of the equation $f(x) = 0$ is given by

$$x_{n+1} = g(x_n) \quad (n = 0, 1, 2, \dots)$$

where $\alpha = g(\alpha)$. When we use such an iteration we need a rule which tells us when to stop the process. The usual practice is to stop the iteration when the difference $|x_{n+1} - x_n|$ between two successive iterates is sufficiently small; that is, when it is less than half-a-unit of the least significant figure required in the answer.

There are two separate issues here: one concerns the convergence of the iteration formula to the root, and the other concerns the 'stopping' mechanism. In practical computation, the rule of stopping an iteration is important because it vitally affects the accuracy of the estimate of the root of the equation.

Convergence process

To examine the convergence of the iteration to the root α, we estimate $|x_{n+1} - \alpha|$ as $n \to \infty$. Now

$$x_{n+1} = g(x_n) \quad \text{and} \quad \alpha = g(\alpha)$$

so that

$$x_{n+1} - \alpha = g(x_n) - g(\alpha)$$

Using the mean value theorem 9.3, this may be written as

$$x_{n+1} - \alpha = (x_n - \alpha)g'(X_n)$$

where X_n lies in the interval (x_n, α), assuming $x_n < \alpha$. Writing $\varepsilon_n = x_n - \alpha$, we obtain

$$|\varepsilon_{n+1}| \leq r|\varepsilon_n|$$

where $r = |g'(x)|_{max}$ in the neighbourhood of $x = \alpha$. By comparison with the geometric sequence, we deduce that $\varepsilon_n \to 0$, as $n \to \infty$, if $0 < r < 1$ and that, provided we start near $x = \alpha$, the iteration converges if $|g'(x)| < 1$ near $x = \alpha$. Note that the more horizontal the graph of $g(x)$ near the root, the smaller r is and hence the more rapid is the convergence. We will discuss this further in Section 9.4.7.

Stopping process

The 'stopping' rule can be investigated similarly. The rule says that the iteration is stopped when $|x_{n+1} - x_n| < \varepsilon$, where ε is the maximum acceptable error. We therefore seek a relationship between $|x_{n+1} - \alpha|$ and $|x_{n+1} - x_n|$.

Now we can rewrite $x_{n+1} - \alpha$ as

$$x_{n+1} - \alpha = (x_{n+1} - x_{n+2}) + (x_{n+2} - x_{n+3}) + (x_{n+3} - x_{n+4}) + \ldots + (x_{n+k} - \alpha)$$

Since $x_n \to \alpha$ as $n \to \infty$, it follows that $x_{n+k} \to \alpha$ as $k \to \infty$, since all the previous terms on the right-hand side tend to zero. The terms on the right-hand side can be thought of as the corrections made to successive iterates in the process. We may therefore write

$$x_{n+1} - \alpha = \sum_{k=1}^{\infty} (x_{n+k} - x_{n+k+1}) \tag{9.6}$$

Using the first mean value theorem 9.3, we have

$$x_{n+1} - x_{n+2} = g(x_n) - g(x_{n+1})$$
$$= (x_n - x_{n+1})g'(X_n)$$

where X_n lies in the interval (x_n, x_{n+1}), assuming $x_n < x_{n+1}$. By repeated application of this result, we have

$$x_{n+2} - x_{n+3} = (x_{n+1} - x_{n+2})g'(X_{n+1}) = (x_n - x_{n+1})g'(X_n)g'(X_{n+1})$$
$$\vdots$$

leading to

$$x_{n+k} - x_{n+k+1} = (x_n - x_{n+1})g'(X_n)g'(X_{n+1}) \ldots g'(X_{n+k-1}) \tag{9.7}$$

If, as before, $|g'(x)| < r < 1$ in the neighbourhood of $x = \alpha$ then we obtain from (9.6)

$$|x_{n+1} - \alpha| \leq \sum_{k=1}^{\infty} |x_{n+k} - x_{n+k+1}|$$

$$\leq \sum_{k=1}^{\infty} |x_n - x_{n+1}|r^k \qquad \text{(using (9.7) with } |g'(x)| < r)$$

$$= |x_n - x_{n+1}| \sum_{k=1}^{\infty} r^k$$

$$= \frac{r}{1 - r}|x_n - x_{n+1}|$$

using the expression for the sum of a geometric progression given in (7.14). Hence

$$|x_{n+1} - \alpha| < \frac{r\varepsilon}{1 - r}$$

Thus $|x_{n+1} - \alpha| < \varepsilon$ provided that $r < \frac{1}{2}$, and the 'stopping' rule is valid provided that $|g'(x)| < \frac{1}{2}$ near the root $x = \alpha$. In many practical problems it is necessary to estimate r by

$$|x_{n+1} - x_n|/|x_n - x_{n-1}| = |g(x_n) - g(x_{n-1})|/|x_n - x_{n-1}|$$

Clearly, the smaller the value of r, the more rapid is the convergence. Note, however, that this discussion has ignored the effects of rounding errors on the computation so that the result above has been shown only for exact arithmetic.

Example 9.6 Show that the iteration

$$\theta_{n+1} = \tan^{-1}(\tanh \theta_n), \quad \text{with } \theta_0 = \tfrac{5}{4}\pi \approx 3.9$$

considered in Section 7.9.3 is convergent to the root near $\theta = 3.9$ of the equation $\tan \theta = \tanh \theta$ (see Figure 9.4).

Figure 9.4
Roots of the equation
$\tan \theta = \tanh \theta$.

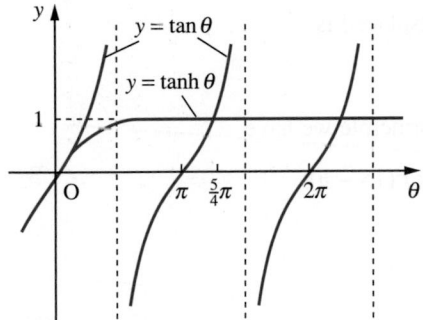

Solution Here the iteration function has formula

$$g(\theta) = \tan^{-1}(\tanh \theta)$$

with derivative

$$g'(\theta) = \frac{1}{1 + \tanh^2\theta}\,\text{sech}^2\theta$$

$$= \frac{1}{\cosh^2\theta + \sinh^2\theta} = \frac{1}{\cosh 2\theta}$$

Near $\theta = 3.9$, $\cosh 2\theta \approx 1220$, so $|g'(\theta)|$ is small (in fact $r < 0.004$) and the method converges.

Example 9.7

A spherical wooden ball floats in water as illustrated in Figure 9.5. Its diameter is 10 cm and its density is 0.8 g cm^{-3}. Find the depth h cm to which it sinks.

Figure 9.5
Floating ball of
Example 9.7.

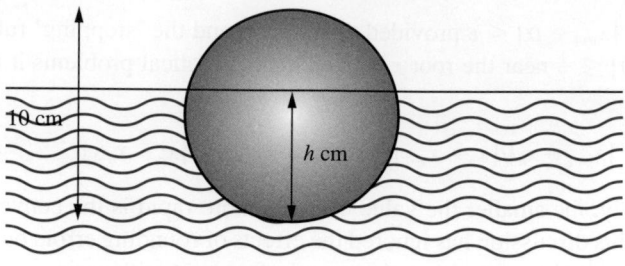

Solution

Archimedes shouted '$\varepsilon \upsilon \rho \eta \kappa \alpha$!' when he realized that the weight of a floating body must balance the weight of water it displaces. In this case we have the weight of the ball is

$$\tfrac{4}{3}\pi(5)^3 \times 0.8 \text{ g}$$

The volume of a zone depth h of a sphere of radius r is

$$\tfrac{1}{3}\pi h^2(3r - h)$$

so the weight of water displaced is

$$\tfrac{1}{3}\pi h^2(15 - h)\,\text{g}$$

Hence by Archimedes' principle we have

$$\tfrac{4}{3}\pi \times 125 \times \tfrac{4}{5} = \tfrac{1}{3}\pi h^2(15 - h)$$

that is

$$400 = h^2(15 - h)$$

Graphing $y = (x - 15)x^2 + 400$ shows that there is a root near $x = 7$. To find the root more accurately we can construct an iteration. For example

$$h_{n+1} = [(h_n^3 + 400)/15]^{1/2}$$

Starting with $h_0 = 7.00$, we obtain the iterates given in the table below.

n	0	1	2	3	4	5	6	7	8	9	10
h_n	7.00	7.04	7.06	7.08	7.10	7.11	7.11	7.12	7.12	7.12	7.12

With this set of iterates we would be tempted to conclude that the root is 7.11 or 7.12. In fact the correct answer is 7.13. This example shows the importance of the size of the derivative of the iteration function. In this case it is 0.7 near the root and there is danger of premature termination of the process. Clearly it is not of vital importance here but the example illustrates the danger of using an iteration without due care.

9.3.3 Exercises

2 By means of sketches of the graphs $y = 1/x$ and $y = \tan x$, show that the equation $x \tan x = 1$ has a root between $x = 0$ and $x = \frac{1}{2}\pi$ and an infinity of roots near $x = k\pi$, where $k = 1, 2, 3, \dots$. Deduce which of the two iterations

(a) $x_{n+1} = \cot x_n$ (b) $x_{n+1} = \tan^{-1}(1/x_n) + k\pi$

is convergent to the roots, and use it to locate the smallest positive root to 6dp.

3 If $\alpha = f(\alpha)$ but the iteration $x_{n+1} = f(x_n)$ fails to converge to the root α, under what condition on $f(x)$ will the iteration $x_{n+1} = f^{-1}(x_n)$ converge?

4 Show the cubic equation $x^3 - 2x - 1 = 0$ has a root near $x = 2$. Prove that the iteration

$$x_{n+1} = \frac{1}{2}(x_n^3 - 1)$$

fails to converge to that root. Devise a simple iteration formula for the root of the equation, and use it to find the root to 6dp.

5 The equation $f(x) = 0$ has a root at $x = \alpha$. Show that rewriting the equation as $x = x + \lambda f(x)$, where λ is a constant, yields a convergent iteration for α if $\lambda = -1/f'(x_0)$ and x_0 is sufficiently close to α.

Use this method to devise an iteration for the root near $x = 2$ of the equation $x^3 - 2x - 1 = 0$.

6 Consider the iteration defined by

$$x_{n+1} = \frac{1}{3}(x_n^3 + 2)$$

Show that

(a) if $0 < x_0 < 1$ then the iteration tends to a limit as $n \to \infty$;

(b) if $x_0 > 1$ then the iteration is divergent. Explain this behaviour.

7 Consider the iteration

$$x_{n+1} = \frac{2 + 30x_n - x_n^2}{30}, \quad x_0 = 1.5$$

Working to 2dp, obtain the first three iterates. Then continue to obtain the following six iterates. From the numerical evidence what do you estimate as the limit of the sequence?

Assuming that the sequence has a limit near 1.5, obtain its value algebraically and then explain the phenomena observed above.

9.4 ## Taylor's theorem and related results

A question that frequently arises in both engineering and mathematical problem-solving is the behaviour of a solution when one (or more) of the parameters in the problem statement is changed. This occurs in sensitivity analysis when we examine solutions for their dependence on errors in the original data. It is also relevant to analysing the equilibrium of structures. One of the mathematical tools for such analyses is Taylor's theorem. In this section we shall develop the theorem and then use it to solve problems in design and numerical methods.

9.4.1 Taylor polynomials and Taylor's theorem

In Section 2.9.1 we discussed the use of interpolating functions to approximate functions specified by a table of values. The simplest case was linear interpolation. With this, we require a different formula between successive tabular points. Another approach to the problem of function approximation is to construct a polynomial that, together with its derivatives, takes the same values as those of the function and its derivatives at a particular point in the domain. That is, we seek a polynomial $p(x)$ such that

$$p(a) = f(a), \quad p'(a) = f'(a), \quad p''(a) = f''(a), \ldots$$

The idea is illustrated by Example 9.8.

Example 9.8

Find a polynomial approximation to the function $f(x)$ such that

$$f(0) = 3, \quad f'(0) = 4, \quad f''(0) = -10 \quad \text{and} \quad f'''(0) = 12$$

Solution

In this example we have information about the value of the function and its first three derivatives at $x = 0$. This means that we can form an approximating polynomial of degree 3

$$p(x) = a + bx + cx^2 + dx^3$$

and determine the values of a, b, c and d from the information given.
  Setting $p(0) = f(0)$ gives $a = 3$.
  Differentiating gives

$$p'(x) = b + 2cx + 3dx^2$$

and on setting $p'(0) = f'(0) = 4$, we have $b = 4$.
  Differentiating again gives

$$p''(x) = 2c + 6dx$$

and on setting $p''(0) = f''(0) = -10$, we have $c = -5$.
  Differentiating again gives

$$p'''(x) = 6d$$

and on setting $p'''(0) = f'''(0) = 12$, we have $d = 2$.
  Thus the approximating polynomial is

$$p(x) = 3 + 4x - 5x^2 + 2x^3$$

The technique used in Example 9.8 can be applied at points other than $x = 0$, as shown in Example 9.9.

Example 9.9

Find a polynomial approximation to $f(x)$ such that

$$f(1) = 4, \quad f'(1) = 0, \quad f''(1) = 2 \quad \text{and} \quad f'''(1) = 12$$

Solution

Because the information concerns the value of the function and its derivatives at the point $x = 1$, we look for a polynomial in powers of $x - 1$. So in this case we are seeking an approximation in the form

$$p(x) = a + b(x - 1) + c(x - 1)^2 + d(x - 1)^3$$

Setting $x = 1$ in $p(x)$ and its derivatives gives, in turn,

$$p(1) = a = 4 \qquad p'(1) = b = 0$$
$$p''(1) = 2c = 2 \qquad p'''(1) = 6d = 12$$

Thus the required approximation is

$$p(x) = 4 + 0(x - 1) + 1(x - 1)^2 + 2(x - 1)^3 = 4 + (x - 1)^2 + 2(x - 1)^3$$

Such polynomial approximations to functions are called **Taylor polynomials**. In general, we can write the nth-degree Taylor polynomial approximation to the function $f(x)$, given the value of the function and its derivatives at $x = a$, in the form

$$f(x) \approx p_n(x)$$

where

$$p_n(x) = f(a) + \frac{x - a}{1!}f'(a) + \frac{(x - a)^2}{2!}f''(a) + \frac{(x - a)^3}{3!}f'''(a) + \dots$$
$$+ \frac{(x - a)^n}{n!}f^{(n)}(a) \tag{9.8}$$

Clearly, $p_n(a) = f(a)$, and also the first n derivatives of $p_n(x)$ match the first n derivatives of $f(x)$ at $x = a$.

The approximation of $f(x)$ given in (9.8) can be made exact by writing

$$f(x) = p_n(x) + R_n(x) \tag{9.9}$$

where $R_n(x)$ is the **remainder**. The remainder term can be expressed in many different forms, with the simplest, known as **Lagrange's form**, being

$$R_n(x) = \frac{(x - a)^{n+1}}{(n + 1)!}f^{(n+1)}(a + \theta h)$$

where $h = x - a$ and $0 < \theta < 1$.

The result (9.9) constitutes **Taylor's theorem**, which may be stated as follows.

Theorem 9.4 **Taylor's theorem**

If $f(x), f'(x), \dots, f^{(n)}(x)$ exist and are continuous on the closed domain $[a, x]$ and $f^{(n+1)}(x)$ exists on the open domain (a, x) then there exists a number θ, with $0 < \theta < 1$, such that

$$f(x) = f(a) + \frac{x - a}{1!}f'(a) + \frac{(x - a)^2}{2!}f''(a) + \dots$$
$$+ \frac{(x - a)^n}{n!}f^{(n)}(a) + \frac{(x - a)^{n+1}}{(n + 1)!}f^{(n+1)}(a + \theta h) \tag{9.10}$$

where $h = x - a$.

end of theorem

Taylor's theorem is in fact a natural extension of the first mean value theorem (Theorem 9.3), and it is sometimes referred to as the ***n*th mean value theorem**. It may be proved by repeated use of Rolle's theorem (Theorem 9.1), but, since the proof does not add to our understanding of how to apply the result to the solution of engineering problems, it is not developed here.

9.4.2 Taylor and Maclaurin series

An alternative form of the Taylor polynomial (9.10) is obtained when we replace x in the expansion by $a + x$. Then we obtain a polynomial in x, rather than $x - a$, namely

$$f(x + a) = f(a) + \frac{x}{1!}f'(a) + \frac{x^2}{2!}f''(a) + \frac{x^3}{3!}f'''(a) + \ldots$$

$$+ \frac{x^n}{n!}f^{(n)}(a) + R_n(x) \tag{9.11}$$

where

$$R_n(x) = \frac{x^{n+1}}{(n+1)!}f^{(n+1)}(a + \theta x), \quad \text{with } 0 < \theta < 1$$

Equation (9.11) is called the **Taylor polynomial expansion of** $f(x)$ **about** $x = a$.

The remainder $R_n(x)$ represents the error involved in approximating $f(x)$ by the polynomial

$$f(a) + \frac{x}{1!}f'(a) + \frac{x^2}{2!}f''(a) + \ldots + \frac{x^n}{n!}f^{(n)}(a)$$

If $R_n(x) \to 0$ as $n \to \infty$ then we may represent $f(x)$ by the power series

$$f(x + a) = f(a) + \frac{x}{1!}f'(a) + \frac{x^2}{2!}f''(a) + \ldots = \sum_{n=0}^{\infty} \frac{x^n}{n!}f^{(n)}(a) \tag{9.12}$$

The power series (9.12) is called the **Taylor series expansion of** $f(x)$ **about** $x = a$. We saw in Section 7.7.1 that a power series may have a restricted domain of convergence. Similarly, $R_n(x)$ may tend to zero as $n \to \infty$ only for a restricted interval of values of x or not at all. In that case the power series given by (9.12) will only represent the function $f(x)$ in that interval of convergence.

Setting $a = 0$ in (9.13) leads to the special case

$$f(x) = f(0) + \frac{x}{1!}f'(0) + \frac{x^2}{2!}f''(0) + \ldots = \sum_{n=0}^{\infty} \frac{x^n}{n!}f^{(n)}(0) \tag{9.13}$$

which is known as the **Maclaurin series expansion of** $f(x)$.

Example 9.10   Find the Maclaurin series expansion of $e^x \sin x$.

Solution   Since $f(x) = e^x \sin x$,

$$f'(x) = e^x(\sin x + \cos x)$$

This may be rewritten (see Section 2.6.5) as

$$f'(x) = \sqrt{2}e^x \sin(x + \tfrac{1}{4}\pi)$$

so the process of differentiation is equivalent to multiplying by $\sqrt{2}$ and adding $\tfrac{1}{4}\pi$ to the argument of the sine function. Thus we can write the second derivative directly as

$$f''(x) = (\sqrt{2})^2 e^x \sin(x + 2 \times \tfrac{1}{4}\pi) = 2e^x \cos x$$

and so on for higher derivatives, giving in general

$$f^{(k)}(x) = (\sqrt{2})^k e^x \sin(x + \tfrac{1}{4}k\pi)$$

Putting $x = 0$ gives $f(0) = 0$, $f^{(1)}(0) = 1$, $f^{(2)}(0) = 2$, $f^{(3)}(0) = 2$, $f^{(4)}(0) = 0$, $f^{(5)}(0) = -4$, $f^{(6)}(0) = -8$, ..., which, on substituting into (9.13), gives

$$e^x \sin x = 0 + x(1) + \frac{1}{2!}x^2(2) + \frac{1}{3!}x^3(2) + \frac{1}{4!}x^4(0) + \frac{1}{5!}x^5(-4) + \ldots$$

$$= x + x^2 + \tfrac{1}{3}x^3 - \tfrac{1}{30}x^5 + \ldots$$

It remains to show that $R_n(x) \to 0$ as $n \to \infty$. Since

$$R_n(x) = \frac{x^{n+1}}{(n+1)!} f^{(n+1)}(\theta x), \quad \text{with } 0 < \theta < 1$$

we have in this particular example

$$R_n(x) = \frac{x^{n+1}}{(n+1)!} (\sqrt{2})^{n+1} e^{\theta x} \sin[\theta x + \tfrac{1}{4}(n+1)\pi]$$

$$= \frac{(x\sqrt{2})^{n+1}}{(n+1)!} e^{\theta x} \sin[\theta x + \tfrac{1}{4}(n+1)\pi], \quad \text{with } 0 < \theta < 1$$

Now $(x\sqrt{2})^{n+1}/(n+1)! \to 0$ as $n \to \infty$, and $|\sin[\theta x + \tfrac{1}{4}(n+1)\pi]| \leqslant 1$, and so

$$R_n(x) \to 0 \quad \text{as } n \to \infty \quad \text{for all } x$$

Thus the Maclaurin expansion of $e^x \sin x$ is

$$e^x \sin x = x + x^2 + \tfrac{1}{3}x^3 - \tfrac{1}{30}x^5 + \ldots$$

In practice it is rarely the case that we obtain the Maclaurin series expansion of a function by direct calculation of the derivatives as in Example 9.10. More commonly, we obtain such series by the manipulation of known standard Maclaurin series as we did in Section 7.7.2. Most of the standard series were given in Figure 7.13. For convenience, we reproduce some of them in Figure 9.6.

(a) $\quad (1 + x)^r = 1 + rx + \dfrac{r(r-1)x^2}{2!} + \dfrac{r(r-1)(r-2)x^3}{3!} + \ldots + \dfrac{r(r-1)\ldots(r-n+1)}{n!}x^n + \ldots \quad (-1 < x < 1, r \in \mathbb{R})$

(b) $\quad e^x = 1 + \dfrac{x}{1!} + \dfrac{x^2}{2!} + \dfrac{x^3}{3!} + \ldots + \dfrac{x^n}{n!} + \ldots \quad$ (all x)

(c) $\quad \sin x = x - \dfrac{x^3}{3!} + \dfrac{x^5}{5!} - \ldots + \dfrac{(-1)^n x^{2n+1}}{(2n+1)!} + \ldots \quad$ (all x)

(d) $\quad \cos x = 1 - \dfrac{x^2}{2!} + \dfrac{x^4}{4!} - \ldots + \dfrac{(-1)^n x^{2n}}{(2n)!} + \ldots \quad$ (all x)

(e) $\ln(1 + x) = x - \dfrac{x^2}{2} + \dfrac{x^3}{3} - \dfrac{x^4}{4} + \ldots + \dfrac{(-1)^n x^{n+1}}{n+1} + \ldots \quad (-1 < x \leqslant 1)$

(f) $\quad \tan x = x + \dfrac{x^3}{3} + \dfrac{2x^5}{15} + \dfrac{17x^7}{315} + \ldots \quad (-\tfrac{1}{2}\pi < x < \tfrac{1}{2}\pi)$

(g) $\quad \sinh x = x + \dfrac{x^3}{3!} + \dfrac{x^5}{5!} + \ldots + \dfrac{x^{2n+1}}{(2n+1)!} + \ldots \quad$ (all x)

(h) $\quad \cosh x = 1 + \dfrac{x^2}{2!} + \dfrac{x^4}{4!} + \ldots + \dfrac{x^{2n}}{(2n)!} + \ldots \quad$ (all x)

Figure 9.6 Some standard Maclaurin series expansions.

In MATLAB the command $taylor(f,n,x,a)$ returns $(n-1)$th order Taylor series expansion of $f(x)$ about $x = a$, while the command $taylor(f,n,x)$ returns the Maclaurin series expansion of $f(x)$. The corresponding command in MAPLE is $taylor(f,x = a,n)$; (Note that in this case we need to specify $x = 0$ or $x = a$ or similar and if n is not specified then 6 is the default value). Considering Example 9.10 the commands

MATLAB	MAPLE
`syms x`	
`f = taylor(exp(x)*sin(x),`	`taylor(exp(x)*sin(x),`
`6,x);`	`x = 0);`
`pretty(ans)`	

return the answers

$$x + x^2 + 1/3x^3 - 1/30x^5 \qquad x + x^2 + \tfrac{1}{3}x^3 - \tfrac{1}{30}x^5 + O(x^6)$$

which check with the answer given in the solution.

To obtain the first three terms of the corresponding series about $x = a$ the commands

```
syms x a
f = taylor(exp(x)*sin(x),        taylor(exp(x)*sin(x),
3,x,a);                          x = a,3);
```

return the answer

$$e^a \sin(a) + (e^a \cos(a) + e^a \sin(a))(x - a)$$
$$+ e^a \cos(a)(x - a)^2 + O((x - a)^3)$$

with e^a expressed as $exp(a)$ in the MATLAB response.

Example 9.11 Using the Maclaurin series expansions of e^x and $\sin x$, confirm the Maclaurin series expansion of $e^x \sin x$ obtained in Example 9.10.

Solution From entries (b) and (c) of Figure 9.6

$$e^x = 1 + \frac{x}{1!} + \frac{x^2}{2!} + \frac{x^3}{3!} + \dots \quad \text{(all } x)$$

$$\sin x = x - \frac{x^3}{3!} + \frac{x^5}{5!} - \dots \quad \text{(all } x)$$

As indicated in Section 7.7.2, we can multiply two power series within their common domain of convergence, giving in this case

$$e^x \sin x = \left(1 + \frac{x}{1!} + \frac{x^2}{2!} + \frac{x^3}{3!} + \frac{x^4}{4!} + \dots\right)\left(x - \frac{x^3}{3!} + \frac{x^5}{5!} - \dots\right)$$

$$= x + x^2 + x^3(\tfrac{1}{2} - \tfrac{1}{6}) + x^4(\tfrac{1}{6} - \tfrac{1}{6}) + x^5(\tfrac{1}{120} + \tfrac{1}{24} - \tfrac{1}{12}) + \dots$$

$$= x + x^2 + \tfrac{1}{3}x^3 - \tfrac{1}{30}x^5 + \dots \quad \text{(all } x)$$

which is the series obtained in Example 9.10.

Example 9.12 Obtain the binomial expansion of $(1 - x^2)^{-1/2}$ and deduce a power series expansion for $\sin^{-1}x$.

Solution From entry (a) of Figure 9.5.

$$(1 + x)^r = 1 + rx + \frac{r(r - 1)x^2}{2!} + \frac{r(r - 1)(r - 2)x^3}{3!} + \dots \quad (|x| < 1)$$

To obtain the expansion of $(1 - x^2)^{-1/2}$, we need to set $r = -\tfrac{1}{2}$ and replace x by $-x^2$. We shall do this in two steps. First setting $r = -\tfrac{1}{2}$ gives

$$(1 + x)^{-1/2} = 1 + \frac{-\tfrac{1}{2}}{1}x + \frac{(-\tfrac{1}{2})(-\tfrac{3}{2})}{1 \cdot 2}x^2 + \frac{(-\tfrac{1}{2})(-\tfrac{3}{2})(-\tfrac{5}{2})}{1 \cdot 2 \cdot 3}x^3 + \frac{(-\tfrac{1}{2})(-\tfrac{3}{2})(-\tfrac{5}{2})(-\tfrac{7}{2})}{1 \cdot 2 \cdot 3 \cdot 4}x^4 + \dots$$

$$= 1 - \tfrac{1}{2}x + \frac{1 \cdot 3}{2 \cdot 4}x^2 - \frac{1 \cdot 3 \cdot 5}{2 \cdot 4 \cdot 6}x^3 + \frac{1 \cdot 3 \cdot 5 \cdot 7}{2 \cdot 4 \cdot 6 \cdot 8}x^4 + \dots \quad (|x| < 1)$$

Then, replacing x by $-x^2$, we have

$$(1 - x^2)^{-1/2} = 1 - \tfrac{1}{2}(-x^2) + \frac{1 \cdot 3}{2 \cdot 4}(-x^2)^2 - \frac{1 \cdot 3 \cdot 5}{2 \cdot 4 \cdot 6}(-x^2)^3 + \frac{1 \cdot 3 \cdot 5 \cdot 7}{2 \cdot 4 \cdot 6 \cdot 8}(-x^2)^4 + \dots$$

giving the required binomial expansion

$$(1 - x^2)^{-1/2} = 1 + \tfrac{1}{2}x^2 + \frac{1 \cdot 3}{2 \cdot 4}x^4 + \frac{1 \cdot 3 \cdot 5}{2 \cdot 4 \cdot 6}x^6 + \frac{1 \cdot 3 \cdot 5 \cdot 7}{2 \cdot 4 \cdot 6 \cdot 8}x^8 + \dots$$

$$= 1 + \tfrac{1}{2}x^2 + \tfrac{3}{8}x^4 + \tfrac{5}{16}x^6 + \tfrac{35}{128}x^8 + \dots \quad (|x| < 1) \tag{9.14}$$

Now

$$\int_0^x \frac{dt}{\sqrt{(1-t^2)}} = \sin^{-1}x$$

and so, integrating the series (9.14) term by term, we obtain

$$\sin^{-1}x = x + \tfrac{1}{6}x^3 + \tfrac{3}{40}x^5 + \tfrac{5}{112}x^7 + \dots \quad (|x| < 1)$$

Notice that in Example 9.12 we have integrated a power series to obtain the expansion of another function. In general, we may integrate and differentiate power series within their domains of absolute convergence.

For Example 9.12 check that the commands

MATLAB	MAPLE
`syms x`	
`taylor((1 - x^2)^(-1/2),`	`taylor((1 - x^2)^(-1/2),`
`9,x);`	`x = 0,9);`
`pretty(ans)`	

both return the answer given in (9.12) and that the additional commands

`int(ans);`	`int(%,x);`
`pretty(ans)`	

return the integrated series for sin⁻¹x.
(Note: In both cases the square root term could be entered as `1/sqrt(1 - x^2)` and this is often preferred.)

Example 9.13

The continuous belt of Example 1.40 has length L given by

$$L = 2[l^2 - (R - r)^2]^{1/2} + \pi(R + r) + 2(R - r)\sin^{-1}\left(\frac{R - r}{l}\right)$$

Show that when $R - r \ll l$, a good approximation to L is given by

$$L \simeq 2l + \pi(R + r) + (R - r)^2/l$$

Solution

Taking the first and last term of the formula for L separately we obtain

$$2[l^2 - (R - r)^2]^{1/2} = 2l\left[1 - \left(\frac{R - r}{l}\right)^2\right]^{1/2}$$

$$= 2l\left[1 - \tfrac{1}{2}\left(\frac{R - r}{l}\right)^2 + \frac{\tfrac{1}{2}(-\tfrac{1}{2})}{1 \cdot 2}\left(\frac{R - r}{l}\right)^4 - \dots\right]$$

and

$$2(R - r)\sin^{-1}\left(\frac{R - r}{l}\right) = 2(R - r)\left[\left(\frac{R - r}{l}\right) + \frac{1}{6}\left(\frac{R - r}{l}\right)^3 + \dots\right]$$

Hence

$$L = 2l + \pi(R + r) + \frac{(R - r)^2}{l} + \frac{1}{12}\frac{(R - r)^4}{l^3} + \dots$$

Thus when $l \gg R - r$, we have

$$L \simeq 2l + \pi(R + r) + (R - r)^2/l$$

See Question 18 for an examination of the error.

9.4.3 L'Hôpital's rule

Sometimes we need to find limits of the form

$$\lim_{x \to a} \frac{f(x)}{g(x)}$$

where $f(a) = g(a) = 0$. Even though such a limit may be defined, it cannot be found by substituting $x = a$, since this produces the indeterminate form 0/0. Using Taylor's theorem 9.4, we can formulate a rule for obtaining such limits if they exist.

Using Taylor's series, we may write

$$\frac{f(x)}{g(x)} = \frac{f(a) + (x - a)f'(a) + \frac{1}{2}(x - a)^2 f''(a) + \dots}{g(a) + (x - a)g'(a) + \frac{1}{2}(x - a)^2 g''(a) + \dots}$$

$$= \frac{f'(a) + \frac{1}{2}(x - a)f''(a) + \dots}{g'(a) + \frac{1}{2}(x - a)g''(a) + \dots} \qquad \text{since } f(a) = g(a) = 0, \, x \neq a$$

Hence

$$\lim_{x \to a} \frac{f(x)}{g(x)} = \frac{f'(a)}{g'(a)}$$

provided $g'(a) \neq 0$. This is known as **L'Hôpital's rule**.

It may be that $f'(a)/g'(a)$ is also indeterminate. Consequently, when applying L'Hôpital's rule to obtain the limit

$$\lim_{x \to a} \frac{f(x)}{g(x)}$$

we must repeat the process of differentiating $f(x)$ and $g(x)$ each time we have the indeterminate form 0/0 at $x = c$. If, however, at any stage in the process one or other of the derivatives is non-zero at $x = a$ then we must stop the process, since the rule will no longer apply. In such cases the limit is either zero or infinite or does not exist; for example, $\lim_{x \to 0} \dfrac{1}{x}$ does not exist.

Example 9.14   Using L'Hôpital's rule, obtain the limits

(a) $\lim\limits_{x\to 0}\dfrac{\sin x - x}{x^3}$ (b) $\lim\limits_{x\to 0}\dfrac{1 - \cos x}{x + x^2}$

Solution   (a) Since $(\sin x - x)/x^3$ takes the indeterminate form $0/0$ at $x = 0$, we apply L'Hôpital's rule to give

$$\lim_{x\to 0}\frac{\sin x - x}{x^3} = \lim_{x\to 0}\frac{\cos x - 1}{3x^2} \qquad \text{(again } 0/0 \text{ at } x = 0)$$

$$= \lim_{x\to 0}\frac{-\sin x}{6x} \qquad \text{(again } 0/0 \text{ at } x = 0)$$

$$= \lim_{x\to 0}\frac{-\cos x}{6} = -\tfrac{1}{6}$$

so that

$$\lim_{x\to 0}\frac{\sin x - x}{x^3} = -\tfrac{1}{6}$$

(b) Since $(1 - \cos x)/(x + x^2)$ takes the form $0/0$ at $x = 0$, we apply L'Hôpital's rule to give

$$\lim_{x\to 0}\frac{1 - \cos x}{x + x^2} = \lim_{x\to 0}\frac{\sin x}{1 + 2x} = 0$$

Note that in this case the limit is zero since $(\sin x)/(1 + 2x)$ takes the form $0/1$ at $x = 0$. If we mistakenly proceeded to apply the rule once again, we should obtain

$$\lim_{x\to 0}\frac{1 - \cos x}{x + x^2} = \lim_{x\to 0}\frac{\sin x}{1 + 2x} = \lim_{x\to 0}\frac{\cos x}{2} = \frac{1}{2}$$

an incorrect answer, since the rule was not applicable. The reader may have noticed that both of these limits can be readily evaluated using Maclaurin series.

9.4.4   Exercises

8   Show that if $f(x) = e^{\cos x}$ then

$$f'(x) = -f(x)\sin x$$

and find $f(0)$ and $f'(0)$. Differentiating the expression for $f'(x)$, obtain $f''(x)$ in terms of $f(x)$ and $f'(x)$, and find $f''(0)$. Repeating the process, obtain $f^{(n)}(0)$ for $n = 3, 4, 5$ and 6, and hence obtain the Maclaurin polynomial of degree six for $f(x)$. Confirm your answer by obtaining the series using the Maclaurin expansions of e^x and $\cos x$.

9   A function $y = y(x)$ satisfies the equation

$$\frac{dy}{dx} = y - x + 1$$

with $y = 1$ when $x = 0$. By repeated differentiation, show that $y^{(n)}(0) = 1$ ($n \geqslant 2$), and find the Maclaurin series for y.

10   An alternative approach to Question 9 uses the method of successive approximation, rewriting the equation as

$$y_{n+1}(x) = 1 + \int_0^x [y_n(t) - t + 1]\,dt,$$

$$\text{with } y_0(x) = y(0) = 1$$

Putting $y_0(x) = 1$ into the integral, show that

$$y_1(x) = 1 + 2x - \tfrac{1}{2}x^2$$

$$y_2(x) = 1 + 2x + \tfrac{1}{2}x^2 - \tfrac{1}{6}x^3$$

and find y_3 and y_4.

11 Show that the binomial expansion of $(1 + x)^{-1}$ is

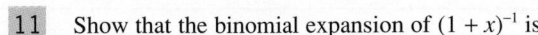

$$(1 + x)^{-1} = 1 - x + x^2 - x^3 + \dots \quad (-1 < x < 1)$$

Hence find the Maclaurin series expansion of $\tan^{-1}x$.

12 Use the series for $\sin x$ and $\cos x$ to obtain the Maclaurin series for $\tan x$ as far as the term in x^7. Deduce the series for $\ln \cos x$.

13 Show that

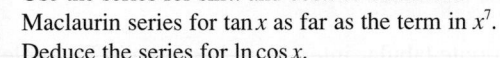

$$\coth x = \frac{1}{x}\left(1 + \tfrac{1}{3}x^2 - \tfrac{1}{45}x^4 + \tfrac{2}{945}x^6 - \dots\right)$$

14 The field strength H of a magnet at a point on the axis at a distance x from its centre is given by

$$H = \frac{M}{2l}\left[\frac{1}{(x - l)^2} - \frac{1}{(x + l)^2}\right]$$

where $2l$ is the length of the magnet and M is its moment. Show that if l is very small compared with x then

$$H \approx \frac{2M}{x^3}$$

15 Using the Maclaurin series expansions of e^x and $\cos x$, show that

$$\lim_{x \to 0}\left(\frac{e^x + e^{-x} - 2}{2\cos 2x - 2}\right) = -\tfrac{1}{4}$$

16 Show that

$$\ln\left(\frac{\sin x}{x}\right) \approx -\tfrac{1}{6}x^2 - \tfrac{1}{180}x^4$$

if powers of x greater than x^5 are neglected.

17 By expanding e^{-x^2} as a Maclaurin series, show that

$$\int_0^{1/2} e^{-x^2}\,dx \approx 0.461$$

18 Considering the problem of Example 9.13, for what values of l does the approximation

$$L \approx 2l + \frac{(R - r)^2}{l} + 3.14(R + r)$$

have a percentage error of less than 0.05% when $R = 5$ and $r = 4$?

19 Using L'Hôpital's rule, find the following limits:

(a) $\displaystyle\lim_{x \to 2}\frac{x^3 - 3x - 2}{x^3 - 8}$ (b) $\displaystyle\lim_{x \to 0}\frac{1 - (1 - x)^{1/4}}{x}$

(c) $\displaystyle\lim_{x \to \pi}\frac{\sin 3x}{\sin 2x}$ (d) $\displaystyle\lim_{x \to 1}\left(\frac{3}{x^3 - 1} - \frac{1}{x - 1}\right)$

(e) $\displaystyle\lim_{x \to 0}\frac{x\cos x - \sin x}{x^3}$ (f) $\displaystyle\lim_{x \to \pi/2}\frac{1 - \sin x}{\ln \sin x}$

20 Consider again the design of the milk carton discussed in Example 8.34. Show that if the overlap used in its construction is x mm instead of 5 mm, the objective function that must be minimized is

$$f(b) = (4b + x)\left(\frac{1\,136\,000}{b^2} + b + 2x\right)$$

Show that when $x = 0$, the optimal value for b is $b_0^* = 10(568)^{1/3}$. The optimal value b^* depends on x. Obtain the Maclaurin series expansion for b^* as far as the term in x^2 and discuss the effect of the overlap size on the design of the carton. (*Hint*: let $b^* \approx b_0 + b_1 x + b_2 x^2$.)

9.4.5 Interpolation revisited

In Chapter 2, Section 2.9.1, we developed the idea of linear interpolation and showed that the approximation

$$f(x) \simeq f_i + \frac{x - x_i}{x_{i+1} - x_i}(f_{i+1} - f_i)$$

gave a value for $f(x)$ which was as accurate as the original data when $|\Delta^2 f_i|$ is less than 4 units of the least significant figure. In many applications, it is easier to express this condition in terms of the second derivative rather than the second difference.

Now

$$\Delta^2 f_i = f(x_i + h) - 2f(x_i) + f(x_i - h)$$

Replacing $f(x_i + h)$ and $f(x_i - h)$ by their Taylor expansions about $x = x_i$, we have (after some cancelling of terms)

$$\Delta^2 f_i = h^2 f''(x_i) + \frac{h^4}{12} f''''(x_i) + \dots$$

The leading term provides a good estimate for $\Delta^2 f_i$ so that the condition for accurate linear interpolation becomes

$$h^2 |f''(x)| < 4 \text{ units of the least significant figure}$$

This enables us to choose an appropriate tabular interval, as is shown in Example 9.15.

Example 9.15

The function $f(x) = e^{-x}$ is to be tabulated to 4dp on the interval [0, 0.5]. Find the maximum tabular interval such that the resulting table is suitable for linear interpolation to 4dp; that is, to yield an interpolated value which is as accurate as the tabulated value.

Solution

Here we require that

$$h^2 |f''(x)| < 4 \times 0.0001$$

Since $f(x) = e^{-x}$ we deduce that $f''(x) = e^{-x}$. On the interval [0, 0.5], the maximum value of e^{-x} occurs at $x = 0$, where $e^0 = 1$. Thus we need the largest value of h such that

$$h^2 < 4 \times 0.0001$$

Hence $h < 0.02$, so that the largest tabular interval is 0.02.

9.4.6 Exercises

21 A table for e^x is required for use with linear interpolation to 6dp. It is tabulated for values of x from $x = 0$ to $x = X$ at intervals of 0.001. What is the largest possible value of X?

22 A table for $\tan x$ is required for use with linear interpolation to 6dp. It is tabulated for values of x from $x = 0$ to $x = 1$ at intervals of h rad. What is the largest possible value of h?

23 In Section 8.6 we discussed the process of numerical differentiation using the approximation

$$\phi(h) = \frac{f(a + h) - f(a - h)}{2h}$$

Using the Taylor series for $f(a + h)$ and $f(a - h)$ about $x = a$, show that

$$f'(a) = \phi(h) - \frac{h^2}{3!} f^{(3)}(a) - \frac{h^4}{5!} f^{(5)}(a) - \dots$$

and deduce that

$$f'(a) = \tfrac{1}{3}[4\phi(\tfrac{1}{2}h) - \phi(h)] + \frac{1}{4}\frac{h^4}{5!} f^{(5)}(a) + \dots$$

Writing $\psi(h) = \tfrac{1}{3}[4\phi(\tfrac{1}{2}h) - \phi(h)]$, show that $\frac{1}{15}[16\psi(\tfrac{1}{2}h) - \psi(h)]$ yields an approximation to $f'(a)$ with truncation error $O(h^6)$. Apply this extrapolation procedure to find $f'(1)$ when $f(x) = \cosh x$, taking $h = 0.4$, 0.2 and 0.1, working to as many decimal places as your calculator will permit.

9.4.7   The convergence of iterations revisited

In Section 9.4.2 we analysed the convergence of an iteration $x_{n+1} = g(x_n)$ for the root α of an equation $f(x) = 0$. We can use the Taylor expansion to analyse the **rate of convergence** of such schemes. Setting $x_n = \alpha + \varepsilon_n$, so that ε_n is the error after n iterations, we have

$$\alpha + \varepsilon_{n+1} = g(\alpha + \varepsilon_n)$$

Expanding $g(\alpha + \varepsilon_n)$ about $x = \alpha$, using the Taylor series (9.12), gives

$$g(\alpha + \varepsilon_n) = \alpha + \varepsilon_{n+1} = g(\alpha) + \frac{\varepsilon_n}{1!}g'(\alpha) + \frac{\varepsilon_n^2}{2!}g''(\alpha) + \frac{\varepsilon_n^3}{3!}g'''(\alpha) + \dots \qquad \textbf{(9.15)}$$

Since α is a root of the equation $f(x) = 0$, we have $\alpha = g(\alpha)$ and (9.15) simplifies to

$$\varepsilon_{n+1} = \frac{\varepsilon_n}{1!}g'(\alpha) + \frac{\varepsilon_n^2}{2!}g''(\alpha) + \frac{\varepsilon_n^3}{3!}g'''(\alpha) + \dots \qquad \textbf{(9.16)}$$

If $g'(\alpha) \neq 0$ then ε_{n+1} is proportional to ε_n, and we have a first-order process. If $g'(\alpha) = 0$ and $g''(\alpha) \neq 0$ then ε_{n+1} is proportional to ε_n^2, and we have a second-order process, and so on.

Example 9.16   The equation $x \tan x = 4$ has an infinite number of roots. To find the root near $x = 1$, we may use the iteration

$$x_{n+1} = \tan^{-1}\left(\frac{4}{x_n}\right)$$

Show that this is a first order process. Starting with $x_0 = 1$, find x_3 and assess its accuracy.

Solution   Here $g(x) = \tan^{-1}(4/x)$, so that

$$g'(x) = \frac{-4}{x^2 + 16}$$

Which is non-zero for all x, i.e. $g'(\alpha) \neq 0$. Thus the iteration is a first-order process. Starting with $x_0 = 1$, we obtain, working to 4dp, the following table.

n	x_n	$4/x_n$	$\tan^{-1}(4/x_n)$
0	1.0000	4.0000	1.3258
1	1.3258	3.0170	1.2507
2	1.2507	3.1982	1.2678
3	1.2678		

From (9.16) we can assess the accuracy of x_n using

$$\varepsilon_{n+1} = \varepsilon_n g'(\alpha) + \dots$$

and approximating ε_n by $x_n - x_{n+1}$ and α by x_3. Thus in this case we have

$$\varepsilon_3 \approx g'(x_3)(x_2 - x_3) = \frac{-4}{16 + (1.2678)^2}(-0.0171) = 0.0039$$

so that the root is 1.26 to 3sf.

9.4.8 Newton–Raphson procedure

One of the most popular techniques used by engineers for solving non-linear equations is the **Newton–Raphson procedure**. The basic idea is that if x_0 is an approximation to the root $x = \alpha$ of the equation $f(x) = 0$ then a closer approximation will be given by the point $x = x_1$ where the tangent to the graph at $x = x_0$ cuts the x axis, as shown in Figure 9.7.

Figure 9.7
The Newton–Raphson
root-finding method.

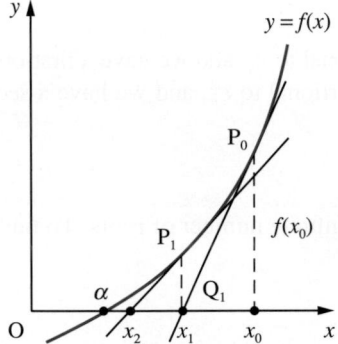

From the definition of the derivative

$$f'(x_0) = \text{slope of } P_0Q_1 = \frac{f(x_0)}{x_0 - x_1}$$

which can be rearranged to give

$$x_1 = x_0 - \frac{f(x_0)}{f'(x_0)}$$

Taking x_1 as the new approximation to the root $x = \alpha$ and repeating the procedure, as illustrated in Figure 9.7, we obtain the closer aproximation

$$x_2 = x_1 - \frac{f(x_1)}{f'(x_1)}$$

and so on. In general, we may write

$$x_{n+1} = x_n - \frac{f(x_n)}{f'(x_n)} \quad (n = 0, 1, 2, \dots) \tag{9.17}$$

Equation (9.7) is known as the Newton–Raphson iteration procedure for obtaining an approximation to the root of $f(x) = 0$. Note that if $f'(x_n) = 0$ then (9.7) cannot be used to obtain x_{n+1}. This is because the tangent to the graph of $y = f(x)$ at $x = x_n$ will be parallel to the horizontal x axis.

Comparing with the general iteration $x_{n+1} = g(x_n)$, we see that in the case of the Newton–Raphson procedure (9.19) the iteration function is

$$g(x) = x - \frac{f(x)}{f'(x)}$$

which, using the quotient rule, has derivative

$$g'(x) = 1 - \frac{[f'(x)]^2 - f(x)f''(x)}{[f'(x)]^2} = \frac{f(x)f''(x)}{[f'(x)]^2}$$

Since α is a root of $f(x) = 0$, we have $f(\alpha) = 0$, giving

$$g'(\alpha) = 0$$

so the procedure is not a first-order process. Differentiating again and substituting $x = \alpha$, we obtain

$$g''(\alpha) = \frac{f''(\alpha)}{f'(\alpha)}$$

and we have a second-order process provided that $f'(\alpha) \neq 0$. If $f''(\alpha) = 0, f'(\alpha) \neq 0$ then we have a third- or higher-order process. When $f(x) = 0$ has a repeated root at $x = \alpha$, $g'(\alpha)$ has the indeterminate form 0/0, and the analysis fails. Repeated roots cause numerical as well as theoretical problems.

Example 9.17

The equation $x \tan x = 4$ was considered earlier in Example 9.16. Apply the Newton–Raphson method to find the root near $x = 1$.

Solution

First, we rewrite the equation in the more convenient (for differentiation) form

$$x \sin x - 4 \cos x = 0$$

Then taking $f(x) = x \sin x - 4 \cos x$ we have $f'(x) = x \cos x + 5 \sin x$. Using the iteration

$$x_{n+1} = x_n - f(x_n)/f'(x_n), \quad x_0 = 1$$

gives the values (to 9dp)

1.000 000 000
1.277 976 731
1.264 600 951
1.264 591 571
1.264 591 571

so that after four iterations we obtain an answer correct to 9dp.

Example 9.18

Find the root of

$$8.0000x^4 + 0.4500x^3 - 4.5440x - 0.1136 = 0$$

near $x = 0.8$ to 4sf.

Figure 9.8
Iteration for the root
of the equation
$8.0000x^4 + 0.4500x^3 - 4.5440x - 0.1136 = 0$.

n	x_n	$f(x_n)$	$f'(x_n)$	$-f_n/f'_n$
0	0.8000	−0.241 600	12.7040	0.019 018
1	0.8190	0.011 436	13.9408	−0.000 820
2	0.8182	0.000 340	13.8876	−0.000 022
3	0.8182			

Solution In this particular example

$$f(x) = 8.0000x^4 + 0.4500x^3 - 4.5440x - 0.1136$$

giving

$$f'(x) = 32.0000x^3 + 1.3500x^2 - 4.5440$$

When iterating for the root using the Newton–Raphson procedure (9.17), it is usual to present the calculations in tabular form as shown in Figure 9.8 for this particular example. To 4sf the root is given by $x = 0.8182$. When using the Newton–Raphson method, it is recommended that the iteration formula is *not* tidied up into a single expression but is left in the 'approximation minus error' format. Tidying up may lead to ill-conditioning of the numerical procedure.

There are no built in programs for Newton–Raphson in either MATLAB or MAPLE. The method is basically a numerical procedure so MATLAB seems to be the obvious package to use. You will need to develop a little program, as illustrated below for Example 9.18.

```
% Set up initial data and put initial results into R
e = 0.0001; acc = 1; x = .8; f = 8*x^4 + .45*x^3 -
4.544*x - 0.1136; fd = 32*x^3 + 1.35*x^2 - 4.544;
R = [x;f;fd];
% Now iterate until acc is less than e and add results
to R
while acc>e xold = x;
x = x - f/fd; f = 8*x^4 + .45*x^3 - 4.544*x - 0.1136;
fd = 32*x^3 + 1.35*x^2 - 4.5440;
R = [R [x;f;fd]];
acc = abs(x - xold);
end
R
```

which returns

```
R =
      0.8000    0.8190    0.8182    0.8182
     -0.2416    0.0117    0.0000    0.0000
     12.7040   13.9420   13.8863   13.8861
```

(Note: Small discrepancies with answers given in Figure 9.8 are due to the number of decimal places being retained during working.)

9.4.9 Optimization revisited

In Section 8.5 we indicated that we would return to reconsider the conditions for determining the nature of stationary points following the introduction of the Taylor series.

If a minimum value of a differentiable function $f(x)$ occurs at $x = a$ then the difference $f(a + h) - f(a)$ will be positive for all small h. However, from the Taylor series (9.12)

$$f(a + h) - f(a) = hf'(a) + \frac{1}{2!}h^2 f''(a) + \frac{1}{3!}h^3 f'''(a) + \dots$$

and the sign of the expression on the right-hand side depends on the sign of h. It will change sign as h changes sign unless $f'(a) = 0$, in which case the sign depends on the sign of $f''(a)$. Thus a necessary condition for the minimum to occur at $x = a$ is that $f'(a) = 0$, and a necessary and sufficient condition for a minimum of $f(x)$ at $x = a$ is $f'(a) = 0$ and $f''(a) > 0$. Similarly, the maxima of differentiable functions occur when $f'(a) = 0$ and $f''(a) < 0$. If $f'(a) = 0$ and $f''(a) = 0$, we may have a maximum or minimum value or a point of inflection. If $f'(a) = f''(a) = 0$, a necessary condition for a minimum or maximum at $x = a$ is $f'''(a) = 0$, and so on. However, it is important to remember that a function may have an optimal value at a point where its derivative does not exist, as illustrated in Figure 9.9. A numerical scheme for locating the optimal point of a function using the Newton–Raphson procedure can be established. The resulting iteration

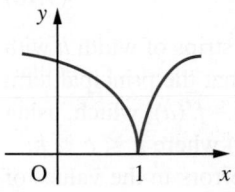

Figure 9.9
$y = (x - 1)^{2/3}$ has a minimum at $x = 1$ but it is not differentiable here.

$$x_{n+1} = x_n - \frac{f'(x_n)}{f''(x_n)}$$

is, however, rarely used in practice. Generally, bracketing methods are used similar to that described in Question 78 (Exercises 8.5.2).

9.4.10 Exercises

 24 Given below are three methods for calculating $\sqrt{2}$ by iteration. Find the order of each process and discuss their numerical properties.

(a) $x_{n+1} = 1 + 1/(1 + x_n)$ (b) $x_{n+1} = \frac{1}{2}(x_n + 2/x_n)$

(c) $x_{n+1} = (3x_n^4 + 12x_n^2 - 4)/(8x_n^3)$

 25 Use the Newton–Raphson iteration procedure to find the real root of $x^3 - 6x^2 + 9x + 1 = 0$ to 4dp.

26 Use the Newton–Raphson method to find the two positive roots of $x^4 - 4x^3 - 12x^2 + 32x + 28 = 0$.

27 The iteration $x_{n+1} = x_n(3 - 3ax_n + a^2 x_n^2)$ may be used to calculate the reciprocal of a, that is, to solve $ax = 1$. Show that this is a third-order process with $\varepsilon_{n+1} = a^2 \varepsilon_n^3$. Apply the iteration with $a = 1.735$, starting with $x_0 = 0.5$, and prove that x_2 is correct to 8dp.

9.4.11 Numerical integration

A remarkable mathematical result that follows from the Taylor series is known as the **Euler–Maclaurin formula**:

$$\int_a^b f(x)\,dx = \frac{b - a}{2}[f(b) + f(a)] - \frac{(b - a)^2}{12}[f'(b) - f'(a)]$$

$$+ \frac{(b - a)^4}{720}[f^{(3)}(b) - f^{(3)}(a)] - \frac{(b - a)^6}{30\,240}[f^{(5)}(b) - f^{(5)}(a)] \dots$$

Subdividing the interval $[a, b]$ into n equal strips of width h, we have

$$\int_a^b f(x)dx = \sum_{r=0}^{n-1} \int_{x_r}^{x_{r+1}} f(x)dx, \quad x_r = a + rh$$

Applying the formula to each term in the summation, we obtain the trapezium rule together with a power series expansion of the truncation error in terms of h:

$$\int_a^b f(x)dx = \tfrac{1}{2}h(f_0 + 2f_1 + 2f_2 + \ldots + 2f_{n-1} + f_n) - \tfrac{1}{12}h^2(f_n' - f_0')$$

$$+ \tfrac{1}{720}h^4(f_n^{(3)} - f_0^{(3)}) - \tfrac{1}{30\,240}h^6(f_n^{(5)} - f_0^{(5)}) + \ldots$$

$$= T(h) + \alpha_1 h^2 + \alpha_2 h^4 + \alpha_3 h^6 + \ldots \tag{9.18}$$

where $T(h)$ is the trapezium approximation to the integral using n strips of width h with $nh = b - a$, and the α's are independent of h. From this we see that the principal term of the **global truncation** error for the approximation is $\tfrac{1}{12}h^2[f'(b) - f'(a)]$ which, using the first mean value theorem 9.3, may be written $\tfrac{1}{12}h^2(b - a)f''(c)$ where $a < c < b$.

This analysis makes no allowance for the effect of rounding errors in the values of f_i ($i = 0, 1, \ldots, n$). A simple estimate of these is

$$h(\tfrac{1}{2} + \underbrace{1 + 1 + \ldots + 1}_{n-1 \text{ terms}} + \tfrac{1}{2}) \times (\tfrac{1}{2} \text{ unit of the least significant figure})$$

$$= nh(\tfrac{1}{2} \text{ unit of the least significant figure})$$

$$= (b - a)(\tfrac{1}{2} \text{ unit of the least significant figure})$$

This result assumes a fixed number of decimal places in the values of the integrand, and is suitable for calculator work. For computers, when h is small and n large, there is the problem of loss of significant digits when adding a large number of almost-equal numbers.

Example 9.19

In Example 8.57 the integral $\int_1^2 (1/x)dx$ was estimated using the trapezium rule with $h = \tfrac{1}{4}$ and tabulating the integrand to 6dp. Estimate an error bound for the answer obtained.

Solution

Here $f(x) = 1/x$, $a = 1$ and $b = 2$. The global error is given by

$$\tfrac{1}{12}(b - a)h^2 f''(X), \quad \text{with } a \leqslant X \leqslant b$$

so that in this example it is

$$\tfrac{1}{12}(1)(0.25)^2 \frac{2}{X^3}, \quad \text{with } 1 \leqslant X \leqslant 2$$

The largest possible value this can take is when $X = 1$, so we obtain an estimate for the truncation error of 0.010. The rounding-error effect, 0.000 000 5, is negligible compared with this. The error bound we have now calculated safely overestimates the actual error 0.004 obtained in the calculation.

Returning to the full Euler–Maclaurin expansion (9.18), using $2n$ strips of width $\frac{1}{2}h$, we obtain

$$\int_a^b f(x)\,dx = T(\tfrac{1}{2}h) + \tfrac{1}{4}\alpha_1 h^2 + \tfrac{1}{16}\alpha_2 h^4 + \tfrac{1}{64}\alpha_3 h^6 + \dots \tag{9.19}$$

Eliminating the α_1 terms from (9.18) and (9.19) (by subtracting the former from $4 \times$ the latter, and dividing the result by 3) gives

$$\int_a^b f(x)\,dx = \tfrac{1}{3}[4T(\tfrac{1}{2}h) - T(h)] - \tfrac{1}{4}\alpha_2 h^4 - \tfrac{5}{16}\alpha_3 h^6 - \dots$$

Thus the estimate $\frac{1}{3}[4T(\frac{1}{2}h) - T(h)]$ is more accurate than either $T(\frac{1}{2}h)$ or $T(h)$ taken separately. This implies that the truncation error for Simpson's rule is proportional to h^4, which explains why it is a good method for hand computation (as opposed to automatic computation).

9.4.12 Exercises

28 Simpson's rule for the numerical evaluation of an integral is

$$\int_a^b f(x)\,dx \approx \frac{b-a}{n}(f_0 + 4f_1 + 2f_2 + \dots$$
$$+ 2f_{n-2} + 4f_{n-1} + f_n)$$

where n is an even number. The global truncation error is

$$\frac{(b-a)^5}{180n^4}f^{(4)}(c), \quad \text{with } a < c < b$$

If $f(x) = \ln \cosh x$ and $a = 0$, $b = 0.5$, show that $|f^{(4)}(x)| < 2$ for $0 \le x \le 0.5$ and deduce that the global truncation error will be less than $1/(2880n^4)$.

If $f(x)$ is tabulated to 4dp, show that the accumulated rounding error using the formula is less than $1/40\,000$, and find n such that, using the formula, the integral $\int_0^{0.5} \ln \cosh x\,dx$ would be evaluated correctly to 4dp.

29 (a) Use the trapezium rule with $h = 0.25$ to evaluate $\int_0^1 \sqrt{x}\,dx$. Compare your answer with the exact value, $\frac{2}{3}$.

(b) Put $x = t^2$ in the integral and again evaluate it using the trapezium rule with four strips. Compare your answer with the exact value and with the answer found in (a).

(c) Examine the global truncation errors in both cases and draw some general conclusions.

30 The trapezium rule estimate for $\int_0^1 e^{x^2}dx$ with $h = 0.25$ is $1.490\,68$ to 5dp. Estimate the size of the global truncation error in this approximation and show that

$$1.40 \le \int_0^1 e^{x^2}dx < 1.48$$

What value of h will give an answer correct to 4dp?

31 Show that the composite trapezium rule with step length h yields the approximation

$$\int_0^1 e^x\,dx \approx \tfrac{1}{2}h(e-1)\coth\left(\frac{h}{2}\right)$$

Using the series expansion for $\coth x$

$$\coth x = \frac{1}{x}\left(1 + \tfrac{1}{3}x^2 - \tfrac{1}{45}x^4 + \tfrac{2}{945}x^6 - \dots\right)$$

obtain the approximation

$$\int_0^1 e^x\,dx \approx (e-1)(1 + \tfrac{1}{12}h^2 - \tfrac{1}{720}h^4$$
$$+ \tfrac{1}{30\,240}h^6 - \dots)$$

Compare this answer with the Euler–Maclaurin theorem.

9.5 Calculus of vectors

In mechanics the vectors describing a dynamic system are time-dependent. Such vectors may be integrated and differentiated in a natural extension of the same processes for scalar quantities. In this section we briefly introduce the relevant definitions.

9.5.1 Differentiation and integration of vectors

The formal definition gives the derivative of a vector $\boldsymbol{v}(t)$ as

$$\frac{\mathrm{d}\boldsymbol{v}}{\mathrm{d}t} = \lim_{\Delta t \to 0} \frac{\boldsymbol{v}(t + \Delta t) - \boldsymbol{v}(t)}{\Delta t}$$

so if $\boldsymbol{v} = (v_1(t), v_2(t), v_3(t))$ then

$$\frac{\mathrm{d}\boldsymbol{v}}{\mathrm{d}t} = \left(\frac{\mathrm{d}v_1}{\mathrm{d}t}, \frac{\mathrm{d}v_2}{\mathrm{d}t}, \frac{\mathrm{d}v_3}{\mathrm{d}t} \right)$$

For example, the position vector $\boldsymbol{r}(t) = (x(t), y(t), z(t))$ of a particle may be differentiated with respect to time t to give its velocity $\boldsymbol{v}(t)$ as

$$\boldsymbol{v}(t) = \frac{\mathrm{d}\boldsymbol{r}}{\mathrm{d}t} = \left(\frac{\mathrm{d}x}{\mathrm{d}t}, \frac{\mathrm{d}y}{\mathrm{d}t}, \frac{\mathrm{d}z}{\mathrm{d}t} \right)$$

Differentiating again gives the acceleration of the particle as

$$\boldsymbol{f}(t) = \frac{\mathrm{d}\boldsymbol{v}}{\mathrm{d}t} = \frac{\mathrm{d}^2\boldsymbol{r}}{\mathrm{d}t^2} = \left(\frac{\mathrm{d}^2x}{\mathrm{d}t^2}, \frac{\mathrm{d}^2y}{\mathrm{d}t^2}, \frac{\mathrm{d}^2z}{\mathrm{d}t^2} \right)$$

When differentiating a vector with respect to time, it is conventional to use a 'dot' notation and write

$$\frac{\mathrm{d}\boldsymbol{r}}{\mathrm{d}t} = \dot{\boldsymbol{r}} \quad \text{and} \quad \frac{\mathrm{d}^2\boldsymbol{r}}{\mathrm{d}t^2} = \ddot{\boldsymbol{r}}$$

The usual rules of differentiation may be deduced from this definition.

(a) $\dfrac{\mathrm{d}}{\mathrm{d}t}[\boldsymbol{u}(t) + \boldsymbol{v}(t)] = \dfrac{\mathrm{d}\boldsymbol{u}}{\mathrm{d}t} + \dfrac{\mathrm{d}\boldsymbol{v}}{\mathrm{d}t}$

(b) $\dfrac{\mathrm{d}}{\mathrm{d}t}[\lambda(t)\boldsymbol{v}(t)] = \dfrac{\mathrm{d}\lambda}{\mathrm{d}t}\boldsymbol{v}(t) + \lambda(t)\dfrac{\mathrm{d}\boldsymbol{v}}{\mathrm{d}t}$, where $\lambda(t)$ is a scalar function

(c) $\dfrac{\mathrm{d}}{\mathrm{d}t}[\boldsymbol{u}(t) \cdot \boldsymbol{v}(t)] = \dfrac{\mathrm{d}\boldsymbol{u}}{\mathrm{d}t} \cdot \boldsymbol{v}(t) + \boldsymbol{u}(t) \cdot \dfrac{\mathrm{d}\boldsymbol{v}}{\mathrm{d}t}$

(d) $\dfrac{\mathrm{d}}{\mathrm{d}t}[\boldsymbol{u}(t) \times \boldsymbol{v}(t)] = \dfrac{\mathrm{d}\boldsymbol{u}}{\mathrm{d}t} \times \boldsymbol{v}(t) + \boldsymbol{u}(t) \times \dfrac{\mathrm{d}\boldsymbol{v}}{\mathrm{d}t}$, note importance of order

Example 9.20

Sketch the curve

$$\boldsymbol{r} = \sin t\, \boldsymbol{i} + \cos t\, \boldsymbol{j}$$

Calculate

(a) $\dfrac{\mathrm{d}\boldsymbol{r}}{\mathrm{d}t}$ (b) $\dfrac{\mathrm{d}^2\boldsymbol{r}}{\mathrm{d}t^2}$ (c) $\left|\dfrac{\mathrm{d}\boldsymbol{r}}{\mathrm{d}t}\right|$ (d) $\dfrac{\mathrm{d}}{\mathrm{d}t}(|\boldsymbol{r}|)$

Solution   A sketch of the curve is shown in Figure 9.10. It is a circle with centre at the origin and of unit radius.

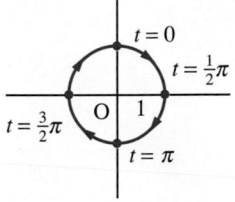

Figure 9.10

(a) $\dfrac{\mathrm{d}\boldsymbol{r}}{\mathrm{d}t} = \dfrac{\mathrm{d}}{\mathrm{d}t}(\sin t)\boldsymbol{i} + \dfrac{\mathrm{d}}{\mathrm{d}t}(\cos t)\boldsymbol{j} = \cos t\,\boldsymbol{i} - \sin t\,\boldsymbol{j}$

(b) $\dfrac{\mathrm{d}^2\boldsymbol{r}}{\mathrm{d}t^2} = \dfrac{\mathrm{d}}{\mathrm{d}t}(\cos t)\boldsymbol{i} - \dfrac{\mathrm{d}}{\mathrm{d}t}(\sin t)\boldsymbol{j} = -\sin t\,\boldsymbol{i} - \cos t\,\boldsymbol{j}$

(c) $\left|\dfrac{\mathrm{d}\boldsymbol{r}}{\mathrm{d}t}\right| = (\cos^2 t + \sin^2 t)^{1/2} = 1$

(d) $|\boldsymbol{r}| = (\sin^2 t + \cos^2 t)^{1/2} = 1$

so that

$$\frac{\mathrm{d}}{\mathrm{d}t}(|\boldsymbol{r}|) = \frac{\mathrm{d}}{\mathrm{d}t}(1) = 0$$

Note that

$$\frac{\mathrm{d}}{\mathrm{d}t}(|\boldsymbol{r}|) \neq \left|\frac{\mathrm{d}\boldsymbol{r}}{\mathrm{d}t}\right|$$

In the same way, the integration of a vector $\boldsymbol{v}(t)$ with respect to the variable t is usually performed in terms of its components:

$$\int \boldsymbol{v}(t)\mathrm{d}t = \int (v_1(t),\, v_2(t),\, v_3(t))\,\mathrm{d}t$$

$$= \left(\int v_1(t)\,\mathrm{d}t, \int v_2(t)\,\mathrm{d}t, \int v_3(t)\,\mathrm{d}t \right)$$

Of course, the arbitrary constant of integration is now a vector constant $\boldsymbol{c} = (c_1, c_2, c_3)$.

Example 9.21   Given

$$\frac{\mathrm{d}^2\boldsymbol{r}}{\mathrm{d}t^2} = -g\boldsymbol{k} \quad \text{with} \quad \boldsymbol{r}(0) = 0 \quad \text{and} \quad \dot{\boldsymbol{r}}(0) = \boldsymbol{V}$$

find $\boldsymbol{r}(t)$. Obtain the locus of the point P, such that $\overrightarrow{\mathrm{OP}} = \boldsymbol{r}$, in terms of x and z when $\boldsymbol{V} = (u, 0, v)$.

Solution This is the equation of motion of a projectile under gravity. Integrating the equation once gives

$$\frac{d\boldsymbol{r}}{dt} = -gt\boldsymbol{k} + \boldsymbol{c}$$

Since $\dot{\boldsymbol{r}}(0) = \boldsymbol{V}$, we have

$$\boldsymbol{c} = \boldsymbol{V} \quad \text{and} \quad \frac{d\boldsymbol{r}}{dt} = -gt\boldsymbol{k} + \boldsymbol{V}$$

Integrating a second time gives

$$\boldsymbol{r}(t) = \boldsymbol{V}t - \tfrac{1}{2}gt^2\boldsymbol{k} + \boldsymbol{a}$$

Since $\boldsymbol{r}(0) = 0$, we have $\boldsymbol{a} = 0$, giving

$$\boldsymbol{r} = \boldsymbol{V}t - \tfrac{1}{2}gt^2\boldsymbol{k}$$

Now $\boldsymbol{r} = (x, y, z)$, so that when $\boldsymbol{V} = (u, 0, v)$, we have

$$(x, y, z) = (u, 0, v)t + (0, 0, -\tfrac{1}{2}gt^2)$$

$$= (ut, 0, vt) + (0, 0, -\tfrac{1}{2}gt^2)$$

$$= (ut, 0, vt - \tfrac{1}{2}gt^2)$$

Thus

$$x = ut, \quad y = 0 \quad \text{and} \quad z = vt - \tfrac{1}{2}gt^2$$

Substituting $t = x/u$ into the equation for z gives, after some rearrangement,

$$z = \tfrac{1}{2}\frac{v^2}{g} - \frac{g}{2u^2}\left(x - \frac{uv}{g}\right)^2$$

This is a parabola with vertex at $(uv/g, 0, v^2/2g)$.

9.5.2 Exercises

32 If $\boldsymbol{r} = (t, t^2, t^3)$, find $\dot{\boldsymbol{r}}(t)$ and $\ddot{\boldsymbol{r}}(t)$.

33 Given the vector

$$\boldsymbol{r} = (1 + t)\boldsymbol{i} + t^2\boldsymbol{j} + \tfrac{2}{3}t^3\boldsymbol{k}$$

evaluate $d\boldsymbol{r}/dt$ and write it in the form

$$\frac{d\boldsymbol{r}}{dt} = f(t)\,\hat{\boldsymbol{T}}(t)$$

where $\hat{\boldsymbol{T}}$ is the unit tangent direction. Calculate $d\hat{\boldsymbol{T}}/dt$ in its simplest form and show that it is perpendicular to $\hat{\boldsymbol{T}}$.

34 In polar coordinates (r, θ), the unit vectors $\hat{\boldsymbol{r}}$ and $\hat{\boldsymbol{\theta}}$ are defined as in Figure 9.11. Show that

$$\hat{\boldsymbol{r}} = \cos\theta\,\boldsymbol{i} + \sin\theta\,\boldsymbol{j}$$

$$\hat{\boldsymbol{\theta}} = -\sin\theta\,\boldsymbol{i} + \cos\theta\,\boldsymbol{j}$$

Hence from the definition $\boldsymbol{r} = r\hat{\boldsymbol{r}}$ show that

$$\frac{d\boldsymbol{r}}{dt} = \frac{dr}{dt}\hat{\boldsymbol{r}} + r\omega\hat{\boldsymbol{\theta}} \quad \text{where} \quad \omega = \frac{d\theta}{dt}$$

Deduce that

$$\frac{d\hat{\boldsymbol{r}}}{dt} = \omega\hat{\boldsymbol{\theta}} \quad \text{and} \quad \frac{d\hat{\boldsymbol{\theta}}}{dt} = -\omega\hat{\boldsymbol{r}}$$

and

$$\frac{d^2\boldsymbol{r}}{dt^2} = \left(\frac{d^2r}{dt^2} - r\omega^2\right)\hat{\boldsymbol{r}} + \left(2\omega\frac{dr}{dt} + r\frac{d\omega}{dt}\right)\hat{\boldsymbol{\theta}}$$

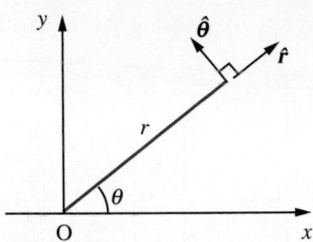

Figure 9.11

measured from a fixed point P_0 on the curve so that s increases as t increases, then

$$\left|\frac{\mathrm{d}\boldsymbol{r}}{\mathrm{d}t}\right| = \frac{\mathrm{d}s}{\mathrm{d}t}$$

Deduce that $\dfrac{\mathrm{d}\boldsymbol{r}}{\mathrm{d}s}$ is a unit tangent vector to the curve at $\boldsymbol{r}(t)$ and that (using the result of Question 48), $\dfrac{\mathrm{d}\boldsymbol{r}}{\mathrm{d}s}$ and $\dfrac{\mathrm{d}^2\boldsymbol{r}}{\mathrm{d}s^2}$ are perpendicular. Show that

$$\left|\frac{\mathrm{d}^2\boldsymbol{r}}{\mathrm{d}s^2}\right| = |\kappa|$$

where κ is the curvature of the curve at that point.

35 Show that if the vector $\boldsymbol{a}(t) = f(t)\boldsymbol{i} + g(t)\boldsymbol{j}$ has constant magnitude, then $\boldsymbol{a}$ and $\dfrac{\mathrm{d}\boldsymbol{a}}{\mathrm{d}t}$ are perpendicular.

36 A curve is given parametrically by $\boldsymbol{r}(t) = f(t)\boldsymbol{i} + g(t)\boldsymbol{j}$. Show that, if s is the length of an arc

---

9.6 **Functions of several variables**

In many applications we use functions of several independent variables, for example, the velocity of a fluid at a point depends on its space coordinates, the temperature in a heat furnace depends upon its position and so on. The basic ideas of calculus apply to functions of several variables as well as to functions of one variable. Of course, because more variables are involved, the notation and technical detail are more complicated but the essential ideas are the same. In the remainder of this chapter we will explore the extension of the process and ideas of differentiation to functions of several independent variables. As we shall see below, the rate of change of the function with respect to its variables can be expressed in terms of the rates of change of the function with respect to each of the independent variables separately.

9.6.1 Representation of functions of two variables

For functions of two independent variables, we are able to extend the ideas of a function of one variable. We use three coordinate axes, conventionally setting x and y as the independent variables and $z = f(x, y)$ as the dependent variable. Instead of a function being represented by a curve in two dimensions, now a function is represented by a surface in three dimensions as illustrated in Figure 9.12(a) for the function $f(x, y) = 3x - x^3 - y^2$. Often it is easier to understand the behaviour of a function by sketching its **contours** (or **level curves**), that is, the curves defined by $f(x, y) = c$ for various values of the constant c as shown in Figure 9.12(b) for the same function. Such plots are readily produced using MATLAB or MAPLE.

Figure 9.12
(a) Surface
$f(x, y) = 3x - x^3 - y^2$.
(b) Contours
$3x - x^3 - y^2 = c$.

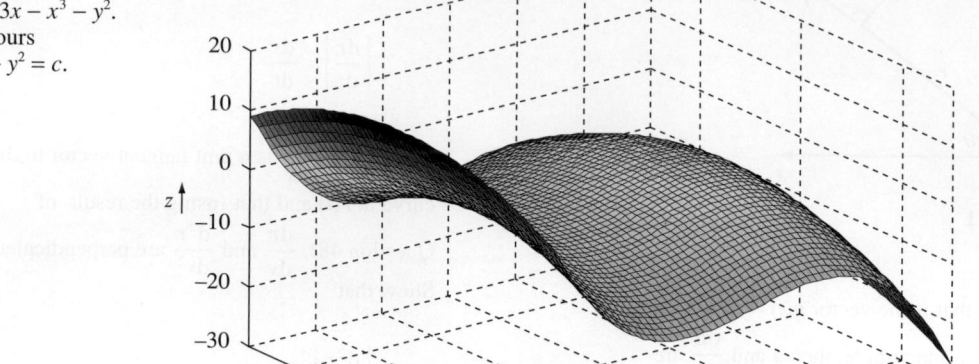

(a)

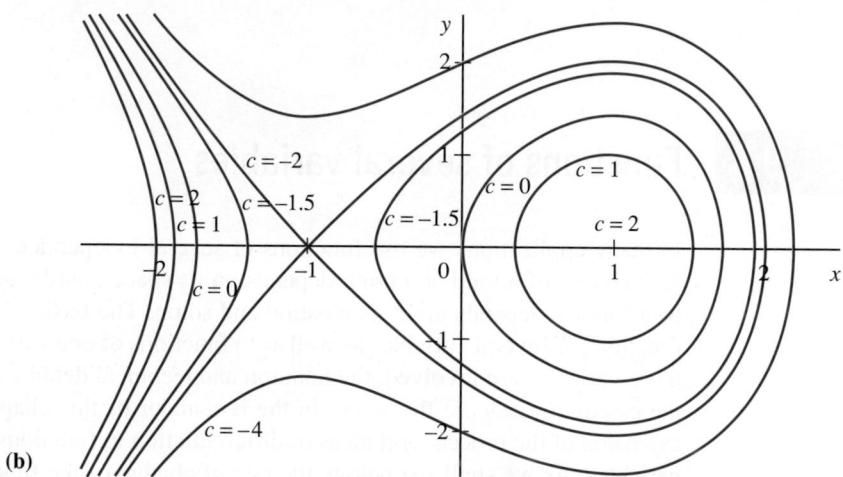

(b)

Using the Symbolic Math Toolbox in MATLAB the commands

```
syms x y
ezsurf(f(x,y))
```

where $f(x, y)$ is a symbolic expression expressed in terms of x and y, plot the surface $z = f(x, y)$ over the default domain $-2\pi < x < 2\pi$, $-2\pi < y < 2\pi$, with the computational grid being chosen according to the amount of variation that occurs. If we wish to specify the domain then we use the command

```
ezsurf (f(x,y),domain)
```

where the domain is specified as either the 4-array $[a, b, c, d]$, with $a \leqslant x \leqslant b, c \leqslant y \leqslant b$, or the 2-array $[a, b]$ with $a \leqslant x \leqslant b, a \leqslant y \leqslant b$.

In MAPLE the surface $f(x, y)$, $a \leqslant x \leqslant b$, $c \leqslant y \leqslant b$ is obtained using the command

```
plot3d(f(x,y), x = a..b, y = c..d);
```

where $f(x, y)$ is expressed in terms of x and y.

Likewise, in MATLAB the commands

```
syms x y
ezcontour(f(x,y))
```

plot the contour of $f(x, y)$ over the default domain $-2\pi < x < 2\pi$, $-2\pi < y < 2\pi$. The domain may be specified using the command

```
ezcontour(f(x,y), domain)
```

where the domain may be the 4-array or 2-array specified above for `ezsurf`. The corresponding commands in MAPLE are

```
with(plots):
contourplot(f(x,y), x = a..b, y = c..d);
```

9.6.2 Partial derivatives

Given a function of one variable, $f(x)$, we recall from Section 8.2.2 that the derivative was defined by

$$\frac{\mathrm{d}f}{\mathrm{d}x} = \lim_{\Delta x \to 0} \frac{\Delta f}{\Delta x} = \lim_{\Delta x \to 0} \left[\frac{f(x + \Delta x) - f(x)}{\Delta x} \right]$$

and that this was a measure of the rate of change of the value of the function $f(x)$ with respect to its variable (or argument) x. For a function of several variables it is also useful to know how the function changes when one, some or all of the variables change. To achieve this we define the **partial derivatives** of a function.

First, we consider a function $f(x, y)$ of the two variables x and y. The partial derivative $\dfrac{\partial f}{\partial x}$, of $f(x, y)$ with respect to x is its derivative with respect to x treating the value of y as being constant. Thus

$$\frac{\partial f}{\partial x} = \left[\frac{\mathrm{d}f}{\mathrm{d}x} \right]_{y = \mathrm{const}} = \lim_{\Delta x \to 0} \left[\frac{f(x + \Delta x, y) - f(x, y)}{\Delta x} \right]$$

Likewise, the partial derivative, $\dfrac{\partial f}{\partial y}$, of $f(x, y)$ with respect to y is its derivative with respect to y treating the value of x as being constant, so that

$$\frac{\partial f}{\partial y} = \left[\frac{\mathrm{d}f}{\mathrm{d}y} \right]_{x = \mathrm{const}} = \lim_{\Delta y \to 0} \left[\frac{f(x, y + \Delta y) - f(x, y)}{\Delta y} \right]$$

The process of obtaining the partial derivatives is called **partial differentiation**. Note the use of 'curly dees', which is to distinguish between partial differentiation and ordinary differentiation. In writing, care must be taken to distinguish between $\dfrac{df}{dx}$, $\dfrac{\Delta f}{\Delta x}$ and $\dfrac{\partial f}{\partial x}$, all of which have different meanings.

A concise notation is sometimes used for partial derivatives; as an alternative to the 'curly dee', we write

$$f_x = \frac{\partial f}{\partial x} \quad \text{and} \quad f_y = \frac{\partial f}{\partial y}$$

It should be noted, however, that subscripts often have other connotations, so care should be taken in using them in this way.

If we write $z = f(x, y)$ then the partial derivatives may also be written as

$$\frac{\partial z}{\partial x}, \frac{\partial z}{\partial y} \quad \text{or} \quad z_x, z_y$$

Summary

The **partial derivatives** of the function $z = f(x, y)$ with respect to the variables x and y respectively are given by

$$\frac{\partial f}{\partial x} = f_x = \frac{\partial z}{\partial x} = z_x = \lim_{\Delta x \to 0} \left[\frac{f(x + \Delta x, y) - f(x, y)}{\Delta x} \right] \tag{9.20}$$

$$\frac{\partial f}{\partial y} = f_y = \frac{\partial z}{\partial y} = z_y = \lim_{\Delta y \to 0} \left[\frac{f(x, y + \Delta y) - f(x, y)}{\Delta y} \right] \tag{9.21}$$

Finding partial derivatives is no more difficult than finding derivatives of functions of one variable, with the constant multiplication, sum, product and quotient rules having counterparts for partial derivatives. Note, however, that, despite the notation, partial derivatives do not behave like fractions. For example, $\dfrac{\partial x}{\partial z} \neq 1 \Big/ \left(\dfrac{\partial z}{\partial x} \right)$ (see Exercises 9.6.7, Question 62).

Example 9.22   Find $\dfrac{\partial f}{\partial x}$ and $\dfrac{\partial f}{\partial y}$ where $f(x, y)$ is given by

(a) $3x^2 + 2xy + y^3$ (b) $(y^2 + x)e^{-xy}$

Solution   (a) $f(x, y) = 3x^2 + 2xy + y^3$

To find $\dfrac{\partial f}{\partial x}$, we differentiate $f(x, y)$ with respect to x regarding y as a constant. Thus we obtain

$$\frac{\partial f}{\partial x} = \frac{\partial}{\partial x}(3x^2) + \frac{\partial}{\partial x}(2xy) + \frac{\partial}{\partial x}(y^3)$$

$$= 3\frac{d}{dx}(x^2) + 2y\frac{d}{dx}(x) + 0 \quad \text{(Note: term in brackets involves } x \text{ only)}$$

$$= 6x + 2y$$

Similarly,

$$\frac{\partial f}{\partial y} = \frac{\partial}{\partial y}(3x^2) + \frac{\partial}{\partial y}(2xy) + \frac{\partial}{\partial y}(y^3)$$

$$= 0 + 2x\frac{d}{dy}(y) + \frac{d}{dy}(y^3) = 2x + 3y^2$$

(b) $f(x, y) = (y^2 + x)e^{-xy}$

Using the product rule, differentiating with respect to x, regarding y as a constant, gives

$$\frac{\partial f}{\partial x} = (e^{-xy})\frac{\partial}{\partial x}(y^2 + x) + (y^2 + x)\frac{\partial}{\partial x}(e^{-xy})$$

$$= (e^{-xy})(1) + (y^2 + x)(-ye^{-xy})$$

$$= (1 - y^3 - xy)e^{-xy}$$

Similarly,

$$\frac{\partial f}{\partial y} = (e^{-xy})\frac{\partial}{\partial y}(y^2 + x) + (y^2 + x)\frac{\partial}{\partial y}(e^{-xy})$$

$$= (e^{-xy})(2y) + (y^2 + x)(-xe^{-xy})$$

$$= (2y - xy^2 - x^2)e^{-xy}$$

Example 9.23   Find $\partial f/\partial x$ and $\partial f/\partial y$ when $f(x, y)$ is

(a) $xy^2 + 3xy - x + 2$ (b) $\sin(x^2 - 3y)$

Solution   (a) Taking $f(x, y) = xy^2 + 3xy - x + 2$ and differentiating with respect to x, keeping y fixed, gives

$$\frac{\partial f}{\partial x} = f_x = y^2 + 3y - 1$$

Differentiating with respect to y, keeping x fixed, gives

$$\frac{\partial f}{\partial y} = f_y = 2xy + 3x$$

(b) Taking $f(x, y) = \sin(x^2 - 3y)$ and applying the composite-function rule, we obtain

$$\frac{\partial f}{\partial x} = \cos(x^2 - 3y)\frac{\partial}{\partial x}(x^2 - 3y) = \cos(x^2 - 3y)2x$$

$$= 2x\cos(x^2 - 3y)$$

and

$$\frac{\partial f}{\partial y} = \cos(x^2 - 3y)\frac{\partial}{\partial y}(x^2 - 3y) = -3\cos(x^2 - 3y)$$

Although we have introduced partial derivatives in the context of functions of two variables, the concept may be readily extended to obtain the partial derivatives of a function of as many variables as we please. Thus for a function $f(x_1, x_2, \ldots, x_n)$ of n variables the partial derivative with respect to x_i is given by

$$f_{x_i} = \frac{\partial f}{\partial x_i} = \lim_{\Delta x_i \to 0} \frac{f(x_1, x_2, \ldots, x_{i-1}, x_i + \Delta x_i, x_{i+1}, \ldots, x_n) - f(x_1, x_2, \ldots x_i, x_n)}{\Delta x_i}$$

and is obtained by differentiating the function with respect to x_i with all the other $n - 1$ variables kept constant.

Example 9.24

Find the partial derivatives of

$$f(x, y, z) = xyz^2 + 3xy - z$$

with respect to x, y and z.

Solution

Differentiating $f(x, y, z)$ with respect to x, keeping y and z fixed, gives

$$f_x = \frac{\partial f}{\partial x} = yz^2 + 3y$$

Differentiating $f(x, y, z)$ with respect to y, keeping x and z fixed, gives

$$f_y = \frac{\partial f}{\partial y} = xz^2 + 3x$$

Differentiating $f(x, y, z)$ with respect to z, keeping x and y fixed, gives

$$f_z = \frac{\partial f}{\partial z} = xy(2z) + 0 - 1 = 2xyz - 1$$

The partial derivatives f_x and f_y of the function $f(x, y)$, with respect to x and y respectively, are given by the commands

MATLAB	MAPLE
`syms x y`	
`f = f(x,y)`	`f:= f(x,y);`
`fx = diff(f,x)`	`fx:= diff(f,x);`
`fy = diff(f,y)`	`fy:= diff(f,y);`

Considering Example 9.22(b). The commands

MATLAB	MAPLE
`syms x y`	
`f = (y^2 + x)*exp(-x*y);`	`f:= (y^2 + x)*exp(-x*y);`
`fx = diff(f,x);`	`fx:= diff(f,x);`
`fx = simplify(fx);`	
`pretty (fx)`	

return the answer

$$-exp(-xy)(-1 + y^3 + xy) \qquad -e^{(-xy)}(-1 + y^3 + xy)$$

with the additional commands

```
fy = diff(f,y);              fy:= diff(f,y);
fy = simplify(fy);
pretty(fy)
```

returning the answer

$$-exp(-xy)(-2y + xy^2 + x^2) \qquad -e^{(-xy)}(-2y + xy^2 + x^2)$$

The commands for partial derivatives can readily be extended to functions of more than two variables. For example, considering Example 9.24 the MATLAB commands

```
syms x y z
f = x*y*z^2 + 3*x*y - z;
fx = diff(f,x); pretty(fx)  return the answer  yz² + 3y
fy = diff(f,y); pretty(fy)  return the answer  xz² + 3x,  and
fz = diff(fz); pretty(fz)  return the answer  2xyz - 1
```

For practice check the answers to Examples 9.22(a) and 9.23(a) and (b).

9.6.3 Directional derivatives

Consider a function of two variables $z = f(x, y)$. This may be represented as a surface in three dimensions as shown in Figure 9.13.

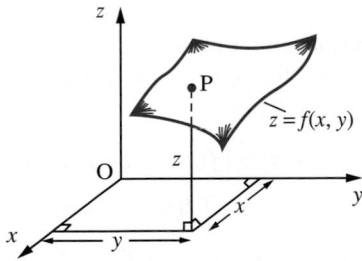

Figure 9.13 Surface $z = f(x, y)$.

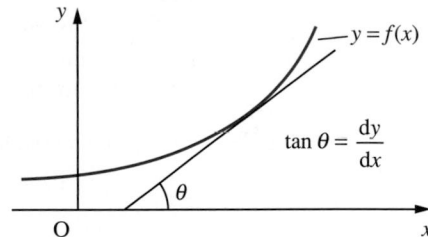

Figure 9.14 Tangent to the graph of $y = f(x)$.

We recall from Chapter 8 that the derivative of a function $f(x)$ of one variable measures the slope of the tangent to the graph of the function, as illustrated in Figure 9.14. In the case of a function of two variables, because $z = f(x, y)$ defines a surface in three dimensions, there is no unique meaning of 'slope' unless we specify the direction in which it is to be measured. In general, the slope will be different for different directions. Now consider two points P and Q on the surface $z = f(x, y)$, as shown in Figure 9.15 and let P′ and Q′ be their projections on the x–y plane. To simplify, set P′Q′ = l; then the coordinates of P′ and Q′ are given by

$$(x, y, 0) \quad \text{and} \quad (x + l\cos\alpha, y + l\sin\alpha, 0)$$

Figure 9.15
Directional derivative.

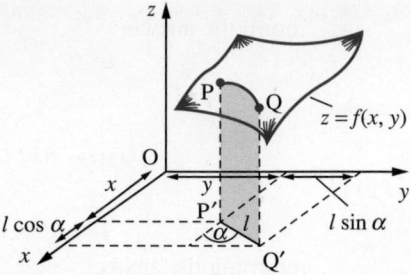

respectively, where α is the angle that P'Q' makes with the positive x direction. The slope of the line PQ is then

$$\frac{f(x + l \cos \alpha, y + l \sin \alpha) - f(x, y)}{l}$$

and the slope of the surface at P in the direction of $\overrightarrow{PQ}$ is the limit of this quotient as $l \to 0$. Denoting this slope by $m_\alpha(x, y)$, we have

$$m_\alpha(x, y) = \lim_{l \to 0} \frac{f(x + l \cos \alpha, y + l \sin \alpha) - f(x, y)}{l}$$

Here the subscript α indicates the direction with respect to which the slope is measured, and the (x, y) shows the point at which it is evaluated. Essentially, we have reduced the problem of a function of two variables to a function of one variable by fixing the direction along which we allow x and y to vary. It would be very clumsy to have to perform the calculation this way every time we wish to work out the rate of change or slope of the function. To simplify the process, we shall show how to represent the slope m_α in terms of two standard slopes: one in the x direction and the other in the y direction.

To do this, we rearrange the numerator of the quotient as a sum of terms, one showing the change in $f(x, y)$ due to the change $l \cos \alpha$ in x, the other showing the change in $f(x, y)$ due to the change $l \sin \alpha$ in y. Thus

$$f(x + l \cos \alpha, y + l \sin \alpha) - f(x, y) = [f(x + l \cos \alpha, y + l \sin \alpha) - f(x, y + l \sin \alpha)]$$
$$+ [f(x, y + l \sin \alpha) - f(x,y)]$$

and
$$m_\alpha(x, y) = \lim_{l \to 0} \frac{f(x + l \cos \alpha, y + l \sin \alpha) - f(x, y + l \sin \alpha)}{l \cos \alpha} \cos \alpha$$

$$+ \lim_{l \to 0} \frac{f(x, y + l \sin \alpha) - f(x, y)}{l \sin \alpha} \sin \alpha$$

$$= p(x, y) \cos \alpha + q(x, y) \sin \alpha$$

where $p(x, y)$ and $q(x, y)$ are the values of the respective limits

$$p(x, y) = \lim_{l \to 0} \frac{f(x + l \cos \alpha, y + l \sin \alpha) - f(x, y + l \sin \alpha)}{l \cos \alpha}$$

$$q(x, y) = \lim_{l \to 0} \frac{f(x, y + l \sin \alpha) - f(x, y)}{l \sin \alpha}$$

Examining the numerator of $p(x, y)$, we see that the 'y value' in both terms is the same, $y + l \sin \alpha$, and also that $l \sin \alpha \to 0$ as $l \to 0$. In contrast, the 'x value' in the terms differs by $l \cos \alpha$. Denoting this by Δx we may write

$$p(x, y) = \lim_{\Delta x \to 0} = \frac{f(x + \Delta x, y + \Delta x \tan \alpha) - f(x, y + \Delta x \tan \alpha)}{\Delta x}$$

which simpifies to

$$p(x, y) = \lim_{\Delta x \to 0} \frac{f(x + \Delta x, y) - f(x, y)}{\Delta x} = \frac{\partial f}{\partial x} \qquad (9.22)$$

In the same way,

$$q(x, y) = \lim_{\Delta y \to 0} \frac{f(x, y + \Delta y) - f(x, y)}{\Delta y} = \frac{\partial f}{\partial y} \qquad (9.23)$$

and we may then write the slope in the direction at an angle α to the x axis as

$$m_\alpha(x, y) = \frac{\partial f}{\partial x} \cos \alpha + \frac{\partial f}{\partial y} \sin \alpha \qquad (9.24)$$

Example 9.25   Find the partial derivatives of $f(x, y) = x^2 y^3 + 3y + x$ with respect to x and y, and the slope of the function in the direction at an angle α to the x axis.

Solution   To find the partial derivative of $f(x, y)$ with respect to x, we differentiate $f(x, y)$ with respect to x, keeping y constant. Thus

$$\frac{\partial f}{\partial x} = 2xy^3 + 1$$

Similarly, we obtain the partial derivative with respect to y by differentiating $f(x, y)$ with respect to y, keeping x constant. Thus

$$\frac{\partial f}{\partial y} = 3x^2 y^2 + 3$$

The general expression for the slope of the surface $z = f(x, y)$ in the direction at an angle α to the x axis is

$$m_\alpha(x, y) = \frac{\partial f}{\partial x} \cos \alpha + \frac{\partial f}{\partial y} \sin \alpha$$

So for this function we have

$$m_\alpha(x, y) = (2xy^3 + 1)\cos \alpha + (3x^2 y^2 + 3)\sin \alpha$$

Since in evaluating $\partial f / \partial x$ we consider only the variation of $f(x, y)$ in the x direction, $\partial f / \partial x$ gives the slope of the surface $z = f(x, y)$ at the point (x, y) in the x direction ($\alpha = 0$ in (9.24)). Similarly, $\partial f / \partial y$ gives the slope in the y direction ($\alpha = \frac{1}{2}\pi$ in (9.24)). This is illustrated in Figure 9.16.

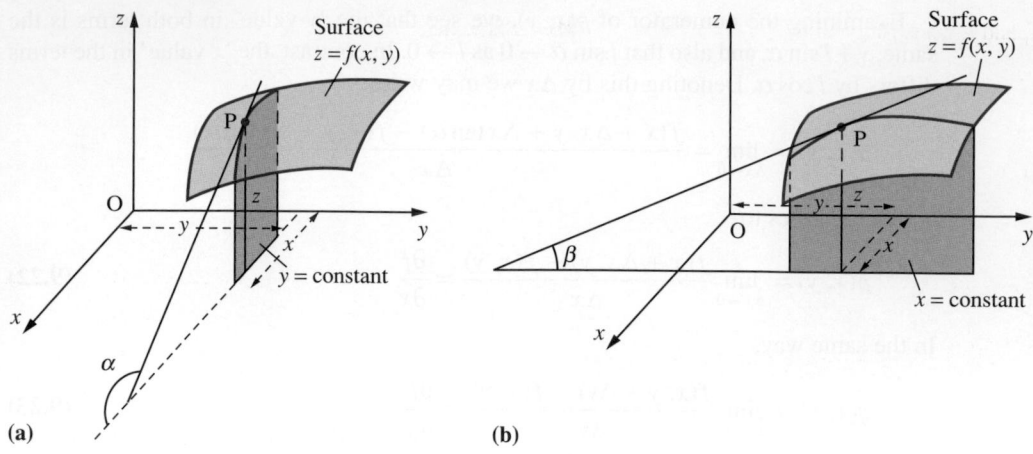

(a)   **(b)**

Figure 9.16 Geometrical illustration of partial derivatives (a) $\dfrac{\partial f}{\partial x} = \tan \alpha$ and (b) $\dfrac{\partial f}{\partial y} = \tan \beta$.

Thus if we know $\partial f/\partial x$ and $\partial f/\partial y$, we can calculate the slope $m_\alpha(x, y)$ of the function in any given direction using (9.24). This is called the **directional derivative** of $f(x, y)$, and may be regarded as the projection of the vector $(\partial f/\partial x, \partial f/\partial y)$ onto the direction represented by the unit vector $(\cos \alpha, \sin \alpha)$, so that $(\cos \alpha, \sin \alpha)$ is a unit vector in the direction of the required derivative. Thus we may express $m_\alpha(x, y)$ as the scalar product

$$m_\alpha(x, y) = \left(\frac{\partial f}{\partial x}, \frac{\partial f}{\partial y} \right) \cdot (\cos \alpha, \sin \alpha)$$

9.6.4 Exercises

Check your answers using MATLAB or MAPLE whenever possible.

37 Obtain from first principles the partial derivatives $\partial f/\partial x$ and $\partial f/\partial y$ of the function $f(x, y)$ at the point $(1, 2)$, where

$$f(x, y) = 2x^2 - xy + y^2$$

38 Obtain from first principles the partial derivatives $\partial f/\partial x$ and $\partial f/\partial y$ of the function $f(x, y)$ at the general point (x, y) where

$$f(x, y) = x \cos y$$

39 Find $\partial f/\partial x$ and $\partial f/\partial y$ when $f(x, y)$ is

(a) $x^3 y + 2x^2 + 9y^2 + xy + 10$

(b) $(x + y^2)^3$ (c) $(3x^2 + y^2 + 2xy)^{1/2}$

40 Find $\partial f/\partial x$ and $\partial f/\partial y$ when $f(x, y)$ is

(a) $e^{xy} \cos x$ (b) $\dfrac{x}{x^2 + y^2}$ (c) $\dfrac{x + y}{x^2 + 2y^2 + 6}$

41 Find the gradient of $f(x, y) = x^2 + 2y^2 - 3x + 2y$ at the point (x, y) in the direction making an angle α with the positive x direction. What is the value of the gradient at $(2, -1)$ when $\alpha = \frac{1}{6}\pi$? What values of α give the largest gradient at $(2, -1)$?

The level curve of $f(x, y)$ through $(2, -1)$ is given by $f(x, y) = f(2, -1)$. This defines the relationship between x and y on the curve. Show that the tangent to the level curve at $(2, -1)$ is perpendicular to the direction of maximum gradient at that point and parallel to the direction of zero gradient.

42 Find $\partial z/\partial x$ and $\partial z/\partial y$ when $z(x, y)$ satisfies

(a) $x^2 + y^2 + z^2 = 10$ (b) $xyz = x - y + z$

43 Show that $z = x^2 y^2 / (x^2 + y^2)$ satisfies the differential equation

$$x \frac{\partial z}{\partial x} + y \frac{\partial z}{\partial y} = 2z$$

44 Find f_x, f_y and f_z when $f(x, y, z)$ is

(a) $x^2y + 3yxz - 2z^3x^2y$ (b) $e^{2z}\cos xy$

45 Show that

$$f(x, y, z) = (x^2 + y^2 + z^2)^{-1/2}$$

satisfies

$$xf_x + yf_y + zf_z = -f(x, y, z)$$

46 Show that

$$f(x, y, z) = x + \frac{x - y}{y - z}$$

satisfies

$$f_x + f_y + f_z = 1$$

9.6.5 The chain rule

As can be seen from Examples 9.22–9.24, the rules and results of ordinary differentiation carry over to partial differentiation. In particular, the composite-function rule still holds, but in a modified form. Consider the two-variable case where $z = f(x, y)$ and x and y are themselves functions of two independent variables, s and t. Then z itself is also a function of s and t, say $F(s, t)$, and we can find its derivatives using a composite-function rule that gives the rates of change of z with respect to s and t in terms of the rates of change of z with respect to x and y and the rates of change of x and y with respect to s and t. Thus

$$\frac{\partial z}{\partial s} = \frac{\partial z}{\partial x}\frac{\partial x}{\partial s} + \frac{\partial z}{\partial y}\frac{\partial y}{\partial s} \quad \text{and} \quad \frac{\partial z}{\partial t} = \frac{\partial z}{\partial x}\frac{\partial x}{\partial t} + \frac{\partial z}{\partial y}\frac{\partial y}{\partial t} \tag{9.25}$$

or, in vector–matrix form,

$$\begin{bmatrix} \dfrac{\partial z}{\partial s} & \dfrac{\partial z}{\partial t} \end{bmatrix} = \begin{bmatrix} \dfrac{\partial z}{\partial x} & \dfrac{\partial z}{\partial y} \end{bmatrix} \begin{bmatrix} \dfrac{\partial x}{\partial s} & \dfrac{\partial x}{\partial t} \\ \dfrac{\partial y}{\partial s} & \dfrac{\partial y}{\partial t} \end{bmatrix}$$

This result is often called the **chain rule**. The proof is straightforward. Consider $\partial z/\partial s$, given by

$$\frac{\partial z}{\partial s} = \lim_{\Delta s \to 0} \frac{F(s + \Delta s, t) - F(s, t)}{\Delta s}$$

The point $(s + \Delta s, t)$ in the s–t plane will correspond to the point $(x + \Delta x, y + \Delta y)$ in the x–y plane, while (s, t) corresponds to (x, y). Thus

$$\frac{\partial z}{\partial s} = \lim_{\Delta s \to 0} \frac{f(x + \Delta x, y + \Delta y) - f(x, y)}{\Delta s}$$

$$= \lim_{\Delta s \to 0} \frac{f(x + \Delta x, y + \Delta y) - f(x, y + \Delta y)}{\Delta x}\frac{\Delta x}{\Delta s}$$

$$+ \lim_{\Delta s \to 0} \frac{f(x, y + \Delta y) - f(x, y)}{\Delta y}\frac{\Delta y}{\Delta s}$$

$$= \frac{\partial f}{\partial x}\frac{\partial x}{\partial s} + \frac{\partial f}{\partial y}\frac{\partial y}{\partial s}$$

We can similarly prove the result for $\partial z/\partial t$.

It may happen, of course, that x and y are functions of one variable only or of three variables or more. In all these cases the chain rule still applies when the functions involved are differentiable.

Example 9.26　Find $\partial T/\partial r$ and $\partial T/\partial \theta$ when

$$T(x, y) = x^3 - xy + y^3$$

and

$$x = r\cos\theta \quad \text{and} \quad y = r\sin\theta$$

Solution　By the chain rule (9.25),

$$\frac{\partial T}{\partial r} = \frac{\partial T}{\partial x}\frac{\partial x}{\partial r} + \frac{\partial T}{\partial y}\frac{\partial y}{\partial r}$$

In this example

$$\frac{\partial T}{\partial x} = 3x^2 - y \quad \text{and} \quad \frac{\partial T}{\partial y} = -x + 3y^2$$

and

$$\frac{\partial x}{\partial r} = \cos\theta \quad \text{and} \quad \frac{\partial y}{\partial r} = \sin\theta$$

so that

$$\frac{\partial T}{\partial r} = (3x^2 - y)\cos\theta + (-x + 3y^2)\sin\theta$$

Substituting for x and y in terms of r and θ gives

$$\frac{\partial T}{\partial r} = 3r^2(\cos^3\theta + \sin^3\theta) - 2r\cos\theta\sin\theta$$

Similarly,

$$\frac{\partial T}{\partial \theta} = (3x^2 - y)(-r\sin\theta) + (-x + 3y^2)r\cos\theta$$

$$= 3r^3(\sin\theta - \cos\theta)\cos\theta\sin\theta + r^2(\sin^2\theta - \cos^2\theta)$$

Example 9.27　Find dH/dt when

$$H(t) = \sin(3x - y)$$

and

$$x = 2t^2 - 3 \quad \text{and} \quad y = \tfrac{1}{2}t^2 - 5t + 1$$

Solution   We note that x and y are functions of t only, so that the chain rule (9.5) becomes

$$\frac{dH}{dt} = \frac{\partial H}{\partial x}\frac{dx}{dt} + \frac{\partial H}{\partial y}\frac{dy}{dt}$$

Note the mixture of partial and ordinary derivatives. H is a function of the one variable t, but its dependence is expressed through the two variables x and y.

Substituting for the derivatives involved, we have

$$\frac{dH}{dt} = 3[\cos(3x - y)]4t - [\cos(3x - y)](t - 5)$$

$$= (11t + 5)\cos(3x - y)$$

$$= (11t + 5)\cos(\tfrac{11}{2}t^2 + 5t - 10)$$

Example 9.28   The base radius r cm of a right-circular cone increases at $2\,\text{cm s}^{-1}$ and its height h cm at $3\,\text{cm s}^{-1}$. Find the rate of increase in its volume when $r = 5$ and $h = 15$.

Solution   The volume V of a cone having base radius r and height h is

$$V = \tfrac{1}{3}\pi r^2 h$$

We wish to determine dV/dt given dr/dt and dh/dt. Applying the chain rule (9.25) gives

$$\frac{dV}{dt} = \frac{\partial V}{\partial r}\frac{dr}{dt} + \frac{\partial V}{\partial h}\frac{dh}{dt}$$

Now

$$\frac{\partial V}{\partial r} = \tfrac{2}{3}\pi rh, \quad \frac{\partial V}{\partial h} = \tfrac{1}{3}\pi r^2, \quad \frac{dr}{dt} = 2, \quad \frac{dh}{dt} = 3$$

so that

$$\frac{dV}{dt} = \tfrac{4}{3}\pi rh + \pi r^2$$

When $r = 5\,\text{cm}$ and $h = 15\,\text{cm}$, the rate of increase in volume is

$$\frac{dV}{dt} = (\tfrac{4}{3}\pi \times 5 \times 15 + \pi \times 5^2)\,\text{cm}^3\,\text{s}^{-1}$$

$$= 125\pi\,\text{cm}^3\,\text{s}^{-1}$$

The chain rule can be readily handled in both MATLAB and MAPLE. Considering Example 9.26, in MATLAB the solution may be developed as follows:

The commands

```
syms x y r theta
T = x^3 - x*y + y^3; Tx = diff(T,x); Ty = diff(T,y);
x = r*cos(theta); y = r*sin(theta);
xr = diff(x,r), xtheta = diff(x,theta); yr = diff(y,r);
ytheta = diff(y,theta);
Tr = Tx*xr + Ty*yr
```

return

```
Tr = (3*x^2 - y)*cos(theta) + (-x + 3*y^2)*sin(theta)
```

To substitute for x and y in terms of r and *theta* we make use of the `eval` command, with

```
eval(Tr); pretty(ans)
```

returning the answer

```
(3r²cos(theta)² - rsin(theta))cos(theta) + (-rcos(theta)
+ 3r²sin(theta)²)sin(theta)
```

which readily reduces to the answer given in the solution.

Similarly the commands

```
Ttheta = Tx*xtheta + Ty*ytheta;
eval(Ttheta); pretty(ans)
```

return the answer

```
(-3r²cos(theta)² + rsin(theta))rsin(theta) + (-rcos(theta)
+ 3r²sin(theta)²)rcos(theta)
```

which also readily reduces to the answer given in the solution.

MAPLE solves this problem much more efficiently using the commands

```
T:= (x,y) -> x^3 - x*y + y^3;
diff(T(r*cos(theta), r*sin(theta)), r);
diff(T(r*cos(theta), r*sin(theta)), theta);
collect(%,r);
```

returning the final answer

```
(-3cos(θ)²sin(θ) + 3sin(θ)²cos(θ))r³ + (sin(θ)² - cos(θ)²)r²
```

9.6.6 Exercises

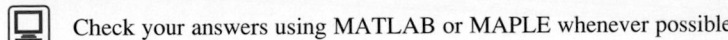

Check your answers using MATLAB or MAPLE whenever possible.

47 Find dz/dt when

(a) $z^2 = x^2 + y^2$, $x = t^2 + 1$ and $y = t - 1$

(b) $z = x^2t^2$ and $x^2 + 3xt + 2t^2 = 1$

48 Find $\partial f/\partial s$ and $\partial f/\partial t$ when $f(x, y) = e^x \cos y$, $x = s^2 - t^2$ and $y = 2st$.

49 Show that if $u = xy$, $v = xy$ and $z = f(u, v)$ then

(a) $x\dfrac{\partial z}{\partial x} - y\dfrac{\partial z}{\partial y} = (x - y)\dfrac{\partial z}{\partial u}$

(b) $\dfrac{\partial z}{\partial x} - \dfrac{\partial z}{\partial y} = (y - x)\dfrac{\partial z}{\partial v}$

50 Show that if $z = x^n f(u)$, where $u = y/x$, then

$$x\frac{\partial z}{\partial x} + y\frac{\partial z}{\partial y} = nz$$

Verify this result for $z = x^4 + 2y^4 + 3xy^3$.

51 Show that, if f is a function of the independent variables x and y, and the latter are changed to independent variables u and v where $u = e^{y/x}$ and $v = x^2 + y^2$, then

(a) $x\dfrac{\partial f}{\partial x} + y\dfrac{\partial f}{\partial y} = 2v\dfrac{\partial f}{\partial v}$

(b) $x^3\dfrac{\partial f}{\partial y} - x^2y\dfrac{\partial f}{\partial x} = uv\dfrac{\partial f}{\partial u}$

52 In a right-angled triangle a cm and b cm are the sides containing the right-angle. a is increasing at $2\,\mathrm{cm\,s^{-1}}$ and b is increasing at $3\,\mathrm{cm\,s^{-1}}$. Calculate the rate of change of (a) the area and (b) the hypotenuse when $a = 5$ and $b = 3$.

53 Show that the total surface area S of a closed cone of base radius r cm and perpendicular height h cm is given by

$$S = \pi r^2 + \pi r\sqrt{(r^2 + h^2)}$$

If r and h are each increasing at the rate of $0.25\,\mathrm{cm\,s^{-1}}$, find the rate at which S is increasing at the instant when $r = 3$ and $h = 4$.

54 A particle moves such that its position at time t is given by $\mathbf{r} = (t, t^2, t^3)$. Find the rate of change of the distance of the particle from the origin.

55 Find $\partial f/\partial s$ and $\partial f/\partial t$ where

$$f(x, y) = x^2 + 2y^2$$

and $x = e^{-s} + e^{\,t}$ and $y = e^{-s} - e^{-t}$.

9.6.7 Successive differentiation

Consider the function $f(x, y)$ with partial derivatives $\partial f/\partial x$ and $\partial f/\partial y$. In general, these partial derivatives will themselves be functions of x and y, and thus may themselves be differentiated to yield second derivatives. We write

$$\frac{\partial}{\partial x}\left(\frac{\partial f}{\partial x}\right) = \frac{\partial^2 f}{\partial x^2} = f_{xx}$$

$$\frac{\partial}{\partial y}\left(\frac{\partial f}{\partial x}\right) = \frac{\partial^2 f}{\partial y \partial x} = \frac{\partial}{\partial y}(f_x) = f_{xy}$$

$$\frac{\partial}{\partial x}\left(\frac{\partial f}{\partial y}\right) = \frac{\partial^2 f}{\partial x \partial y} = \frac{\partial}{\partial x}(f_y) = f_{yx}$$

and

$$\frac{\partial}{\partial y}\left(\frac{\partial f}{\partial y}\right) = \frac{\partial^2 f}{\partial y^2} = f_{yy}$$

There are some functions for which the mixed second derivatives are not equal, that is

$$\frac{\partial^2 f}{\partial x \partial y} \neq \frac{\partial^2 f}{\partial y \partial x}$$

and the order of differentiation is therefore important, but for most of the functions that occur in engineering problems, when the second derivatives are usually continuous functions, these mixed derivatives are the same in value. In a similar manner we can define higher-order partial derivatives

$$\frac{\partial^{m+n} f}{\partial x^m \partial y^n}$$

Example 9.29   Find the second partial derivatives of $f(x, y) = x^2y^3 + 3y + x$.

Solution   We found in Example 9.25 that

$$\frac{\partial f}{\partial x} = 2xy^3 + 1 \quad \text{and} \quad \frac{\partial f}{\partial y} = 3x^2y^2 + 3$$

Differentiating again, we obtain

$$\frac{\partial}{\partial x}\left(\frac{\partial f}{\partial x}\right) = \frac{\partial^2 f}{\partial x^2} = 2y^3, \quad \frac{\partial}{\partial y}\left(\frac{\partial f}{\partial x}\right) = \frac{\partial^2 f}{\partial y \partial x} = 6xy^2$$

$$\frac{\partial}{\partial y}\left(\frac{\partial f}{\partial y}\right) = \frac{\partial^2 f}{\partial y^2} = 6x^2y, \quad \frac{\partial}{\partial x}\left(\frac{\partial f}{\partial y}\right) = \frac{\partial^2 f}{\partial x \partial y} = 6xy^2$$

Note that in this example

$$\frac{\partial^2 f}{\partial x \partial y} = \frac{\partial^2 f}{\partial y \partial x}$$

In MATLAB and MAPLE second-order partial derivatives can be obtained by suitably differentiating the first-order partial derivatives already found. Thus in MATLAB the second-order partial derivatives of $f(x, y)$ are given by

```
fxx = diff(fx,x), fxy = diff(fx,y), fyy = diff(fy,y),
fyx = diff(fy,x)
```

Alternatively, the non-mixed derivatives can be obtained directly using the commands

```
fxx = diff(f,x,2), fyy = diff(f,y,2)
```

which can be extended to higher-order partial derivatives.
The corresponding commands in MAPLE are

```
fxx:= diff(f,x,x); fxy:= diff(f,x,y);
fyy:= diff(f,y,y);
```

Considering Example 9.29 the MATLAB commands

```
syms x y
f = x^2*y^3 + 3*y + x;
fx = diff(f,x); fy = diff(f,y); fxx = diff(fx,x)  return
fxx = 2*y^3
fxy = diff(fx,y)  returns  fxy = 6*x*y^2
fyy = diff(fy,y)  returns  fyy = 6*x^2*y
fyx = diff(fy,x)  returns  fyx = 6*x*y^2
```

Example 9.30

Find the second partial derivatives of

$$f(x, y, z) = xyz^2 + 3xy - z$$

Solution

In Example 9.24 we obtained the first partial derivatives as

$$f_x = \frac{\partial f}{\partial x} = yz^2 + 3y, \quad f_y = \frac{\partial f}{\partial y} = xz^2 + 3x, \quad f_z = \frac{\partial f}{\partial z} = 2xyz - 1$$

Differentiating again, we obtain

$$\frac{\partial}{\partial x}\left(\frac{\partial f}{\partial x}\right) = \frac{\partial^2 f}{\partial x^2} = f_{xx} = 0, \quad \frac{\partial}{\partial y}\left(\frac{\partial f}{\partial x}\right) = \frac{\partial^2 f}{\partial y \partial x} = f_{xy} = z^2 + 3$$

$$\frac{\partial}{\partial z}\left(\frac{\partial f}{\partial x}\right) = \frac{\partial^2 f}{\partial z \partial x} = f_{xz} = 2yz, \quad \frac{\partial}{\partial x}\left(\frac{\partial f}{\partial y}\right) = \frac{\partial^2 f}{\partial x \partial y} = f_{yx} = z^2 + 3$$

$$\frac{\partial}{\partial y}\left(\frac{\partial f}{\partial y}\right) = \frac{\partial^2 f}{\partial y^2} = f_{yy} = 0, \quad \frac{\partial}{\partial z}\left(\frac{\partial f}{\partial y}\right) = \frac{\partial^2 f}{\partial z \partial y} = f_{yz} = 2xz$$

$$\frac{\partial}{\partial x}\left(\frac{\partial f}{\partial z}\right) = \frac{\partial^2 f}{\partial x \partial z} = f_{zx} = 2yz, \quad \frac{\partial}{\partial y}\left(\frac{\partial f}{\partial z}\right) = \frac{\partial^2 f}{\partial y \partial z} = f_{zy} = 2xz$$

$$\frac{\partial}{\partial z}\left(\frac{\partial f}{\partial z}\right) = \frac{\partial^2 f}{\partial z^2} = f_{zz} = 2xy$$

Note that, as expected,

$$f_{xy} = f_{yx}, \quad f_{xz} = f_{zx} \quad \text{and} \quad f_{yz} = f_{zy}$$

Example 9.31

$f(x, y)$ is a function of two variables x and y that we wish to change to variables s and t, where

$$s = x^2 - y^2, \quad t = xy$$

Determine f_{xx} and f_{yy} in terms of $s, t, f_s, f_t, f_{ss}, f_{tt}$ and f_{st}. Show that

$$f_{xx} + f_{yy} = \sqrt{(s^2 + 4t^2)}(4f_{ss} + f_{tt})$$

Solution   Using the chain rule,

$$f_x = \frac{\partial f}{\partial x} = \frac{\partial f}{\partial s}\frac{\partial s}{\partial x} + \frac{\partial f}{\partial t}\frac{\partial t}{\partial x} = 2x\frac{\partial f}{\partial s} + y\frac{\partial f}{\partial t}$$

$$f_y = \frac{\partial f}{\partial y} = \frac{\partial f}{\partial s}\frac{\partial s}{\partial y} + \frac{\partial f}{\partial t}\frac{\partial t}{\partial y} = -2y\frac{\partial f}{\partial s} + x\frac{\partial f}{\partial t}$$

Differentiating f_x with respect to x gives

$$f_{xx} = \frac{\partial}{\partial x}\left(2x\frac{\partial f}{\partial s} + y\frac{\partial f}{\partial t} \right)$$

$$= 2\frac{\partial f}{\partial s} + 2x\frac{\partial}{\partial x}\left(\frac{\partial f}{\partial s} \right) + y\frac{\partial}{\partial x}\left(\frac{\partial f}{\partial t} \right) \quad \text{(using the product rule)}$$

Repeated use of the chain rule as indicated above leads to

$$f_{xx} = 2\frac{\partial f}{\partial s} + 2x\left[\frac{\partial}{\partial s}\left(\frac{\partial f}{\partial s} \right)\frac{\partial s}{\partial x} + \frac{\partial}{\partial t}\left(\frac{\partial f}{\partial s} \right)\frac{\partial t}{\partial x} \right] + y\left[\frac{\partial}{\partial s}\left(\frac{\partial f}{\partial t} \right)\frac{\partial s}{\partial x} + \frac{\partial}{\partial t}\left(\frac{\partial f}{\partial t} \right)\frac{\partial t}{\partial x} \right]$$

$$= 2f_s + 2x(2xf_{ss} + yf_{st}) + y(2xf_{ts} + yf_{tt})$$

which, on assuming $f_{st} = f_{ts}$, gives

$$f_{xx} = 2f_s + 4x^2 f_{ss} + y^2 f_{tt} + 4xy f_{st} \tag{9.26}$$

Following a similar procedure, we can determine f_{yy}. Differentiating f_y with respect to y gives

$$f_{yy} = \frac{\partial}{\partial y}(-2yf_s + xf_t)$$

$$= -2f_s - 2y\frac{\partial}{\partial y}(f_s) + x\frac{\partial}{\partial y}(f_t)$$

$$= -2f_s - 2y\left[\frac{\partial}{\partial s}(f_s)\frac{\partial s}{\partial y} + \frac{\partial}{\partial t}(f_s)\frac{\partial t}{\partial y} \right] + x\left[\frac{\partial}{\partial s}(f_t)\frac{\partial s}{\partial y} + \frac{\partial}{\partial t}(f_t)\frac{\partial t}{\partial y} \right]$$

$$= -2f_s - 2y(-2yf_{ss} + xf_{st}) + x(-2yf_{ts} + xf_{tt})$$

giving

$$f_{yy} = -2f_s + 4y^2 f_{ss} + x^2 f_{tt} - 4xy f_{st} \tag{9.27}$$

Adding (9.26) and (9.27), we obtain

$$f_{xx} + f_{yy} = 4(x^2 + y^2)f_{ss} + (x^2 + y^2)f_{tt}$$

$$= (x^2 + y^2)(4f_{ss} + f_{tt})$$

$$= \sqrt{[(x^2 - y^2)^2 + 4x^2 y^2]}(4f_{ss} + f_{tt})$$

which leads to the required result

$$f_{xx} + f_{yy} = \sqrt{(s^2 + 4t^2)}(4f_{ss} + f_{tt})$$

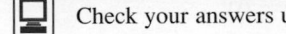

9.6.8 Exercises

Check your answers using MATLAB or MAPLE whenever possible.

56 Verify that

$$f(x, y) = \frac{x}{x^2 + y^2}$$

satisfies the equation

$$\frac{\partial^2 f}{\partial x^2} + \frac{\partial^2 f}{\partial y^2} = 0$$

57 Find the value of the constant a if $V(x, y) = x^3 + axy^2$ satisfies

$$\frac{\partial^2 V}{\partial x^2} + \frac{\partial^2 V}{\partial y^2} = 0$$

58 Verify that

$$\frac{\partial^2 f}{\partial x \partial y} = \frac{\partial^2 f}{\partial y \partial x}$$

in the cases
(a) $f(x, y) = x^2 \cos y$ (b) $f(x, y) = \sinh x \cos y$

59 Show that

$$V(x, y, z) = \frac{1}{z} \exp\left(-\frac{x^2 + y^2}{4z}\right)$$

satisfies the differential equation

$$\frac{\partial^2 V}{\partial x^2} + \frac{\partial^2 V}{\partial y^2} = \frac{\partial V}{\partial z}$$

60 Prove that $z = xf(x + y) + yF(x + y)$, where f and F are arbitrary functions, satisfies the equation

$$z_{xx} + z_{yy} = 2z_{xy}$$

61 Show that, if $z = xe^{Kxy}$, where K is a constant, then

$$xz_x - yz_y = z \quad \text{and} \quad xz_{xx} - yz_{xy} = 0$$

62 If $u = ax + by$ and $v = bx - ay$, where a and b are constants, obtain $\partial u/\partial x$ and $\partial v/\partial y$. By expressing x and y in terms of u and v, obtain $\partial x/\partial u$ and $\partial y/\partial v$ and deduce that

$$\frac{\partial u}{\partial x} \frac{\partial x}{\partial u} = \frac{a^2}{a^2 + b^2}$$

$$\frac{\partial v}{\partial y} \frac{\partial y}{\partial v} = \frac{a^2 + b^2}{a^2}$$

Show also that

$$\frac{\partial^2 f}{\partial x \partial y} = ab\left(\frac{\partial^2 f}{\partial u^2} - \frac{\partial^2 f}{\partial v^2}\right) + (b^2 - a^2)\frac{\partial^2 f}{\partial u \partial v}$$

63 Find the values of the constants a and b such that $u = x + ay$, $v = x + by$ transforms

$$9\frac{\partial^2 f}{\partial x^2} - 9\frac{\partial^2 f}{\partial x \partial y} + 2\frac{\partial^2 f}{\partial y^2} = 0$$

into

$$\frac{\partial^2 f}{\partial u \partial v} = 0$$

64 Regarding u and v as functions of x and y and defined by the equations

$$x = e^u \cos v, \qquad y = e^u \sin v$$

show that

(a) $\dfrac{\partial u}{\partial x} \dfrac{\partial x}{\partial u} = \cos^2 v = \dfrac{\partial v}{\partial y} \dfrac{\partial y}{\partial v}$

(b) $\dfrac{\partial^2 z}{\partial x^2} + \dfrac{\partial^2 z}{\partial y^2} = e^{-2u}\left(\dfrac{\partial^2 z}{\partial u^2} - \dfrac{\partial^2 z}{\partial v^2}\right)$

where z is a twice-differentiable function of u and v.

9.6.9 The total differential and small errors

Consider a function $u = f(x, y)$ of two variables x and y. Let Δx and Δy be increments in the values of x and y. Then the corresponding increment in u is given by

$$\Delta u = f(x + \Delta x, y + \Delta y) - f(x, y)$$

We rewrite this as two terms: one showing the change in u due to the change in x, and the other showing the change in u due to the change in y. Thus

$$\Delta u = [f(x + \Delta x, y + \Delta y) - f(x, y + \Delta y)] + [f(x, y + \Delta y) - f(x, y)]$$

Dividing the first bracketed term by Δx and the second by Δy gives

$$\Delta u = \frac{f(x + \Delta x, y + \Delta y) - f(x, y + \Delta x)}{\Delta x}\Delta x + \frac{f(x, y + \Delta y) - f(x, y)}{\Delta y}\Delta y$$

From the definition of the partial derivative, we may approximate this expression by

$$\Delta u \approx \frac{\partial f}{\partial x}\Delta x + \frac{\partial f}{\partial y}\Delta y$$

We define the **differential** du by the equation

$$du = \frac{\partial f}{\partial x}\Delta x + \frac{\partial f}{\partial y}\Delta y \tag{9.28}$$

By setting $f(x, y) = f_1(x, y) = x$ and $f(x, y) = f_2(x, y) = y$ in turn in (9.28), we see that

$$dx = \frac{\partial f_1}{\partial x}\Delta x + \frac{\partial f_1}{\partial y}\Delta y = \Delta x \quad \text{and} \quad dy = \Delta y$$

so that for the independent variables, increments and differentials are equal. For the dependent variable we have

$$du = \frac{\partial f}{\partial x}dx + \frac{\partial f}{\partial y}dy \tag{9.29}$$

We see that the differential du is an approximation to the change Δu in $u = f(x, y)$ resulting from small changes Δx and Δy in the independent variables x and y; that is,

$$\Delta u \approx du = \frac{\partial f}{\partial x}dx + \frac{\partial f}{\partial y}dy = \frac{\partial f}{\partial x}\Delta x + \frac{\partial f}{\partial y}\Delta y \tag{9.30}$$

a result illustrated in Figure 9.17.

This extends to functions of as many variables as we please, provided that the partial derivatives exist. For example, for a function of three variables (x, y, z) defined by $u = f(x, y, z)$ we have

$$\Delta u \approx du = \frac{\partial f}{\partial x}dx + \frac{\partial f}{\partial y}dy + \frac{\partial f}{\partial z}dz$$

$$= \frac{\partial f}{\partial x}\Delta x + \frac{\partial f}{\partial y}\Delta y + \frac{\partial f}{\partial z}\Delta z$$

Figure 9.17
Total differential.

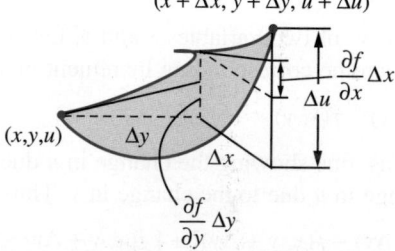

The differential of a function of several variables is often called a **total differential**, emphasizing that it shows the variation of the function with respect to small changes in *all* the independent variables.

Example 9.32   Find the total differential of $u(x, y) = x^2y^3$.

Solution   Taking partial derivatives we have

$$\frac{\partial u}{\partial x} = 2xy^3 \quad \text{and} \quad \frac{\partial u}{\partial y} = 3x^2y^2$$

Hence, using (9.29)

$$du = 2xy^3\,dx + 3x^2y^2dy$$

All physical measurements are subject to error, and a calculated quantity usually depends on several measurements. It is very important to know the degree of accuracy that can be relied upon in a quantity that has been calculated. The total differential can be used to estimate error bounds for quantities calculated from experimental results or from data that is subject to errors. This is illustrated in Example 9.33.

Example 9.33   The volume $V\,\text{cm}^3$ of a circular cylinder of radius $r\,\text{cm}$ and height $h\,\text{cm}$ is given by $V = \pi r^2 h$. If $r = 3 \pm 0.01$ and $h = 5 \pm 0.005$ find the greatest possible error in the calculation of V and compare it with the estimate obtained using the total differential.

Solution   The total differential is

$$dV = \frac{\partial V}{\partial r}\,dr + \frac{\partial V}{\partial h}\,dh = 2\pi rh\,dr + \pi r^2\,dh$$

Then from (9.30)

$$\Delta V \approx dV = 2\pi rh\,dr + \pi r^2\,dh = \pi r(2h\,\Delta r + r\,\Delta h)$$

When $r = 3$ and $h = 5$, we are given that $\Delta r = \pm 0.01$ and $\Delta h = \pm 0.005$, so that

$$\Delta V \approx \pm 3\pi(10 \times 0.01 + 3 \times 0.005)$$

giving

$$\Delta V \approx \pm 0.345\pi$$

(It should be noted that Δr, Δh, ΔV represent maximum errors.) The calculated volume V is subject to a maximum positive error of

$$\{(3.01)^2(5.005) - 45\}\pi = 0.3458\pi$$

and a maximum negative error of

$$\{(2.99)^2(4.995) - 45\}\pi = -0.3442\pi$$

Thus the approximation gives a good guide to the accuracy of the result.

Example 9.34

Two variables, x and y, are related by $y = ae^{-bx}$, where a and b are constants. The values of a and b are determined from experimental data and have relative error bounds p and q respectively. What is the relative error bound for a value of y calculated using the formula with these values of a and b?

Solution

Note that in this example it is assumed that the value of x is known exactly. We are given $y = ae^{-bx}$, where a and b are approximations with errors Δa and Δb, which are unknown but are such that

$$\left|\frac{\Delta a}{a}\right| \leq p \quad \text{and} \quad \left|\frac{\Delta b}{b}\right| \leq q$$

The formula for the total differential gives

$$dy = \frac{\partial y}{\partial a}da + \frac{\partial y}{\partial b}db$$

For the independent variables a and b the increments and the differentials are the same quantity, so that $da = \Delta a$ and $db = \Delta b$. Also, from the given formula for y, we have

$$\frac{\partial y}{\partial a} = e^{-bx} \quad \text{and} \quad \frac{\partial y}{\partial b} = -xae^{-bx}$$

Thus, from (9.28),

$$dy = e^{-bx}\Delta a - xae^{-bx}\Delta b$$

and division by y gives

$$\frac{dy}{y} = \frac{\Delta a}{a} - bx\frac{\Delta b}{b}$$

Hence

$$\left|\frac{dy}{y}\right| \leq \left|\frac{\Delta a}{a}\right| + |bx|\left|\frac{\Delta b}{b}\right| \leq p + |bx|q$$

Since $\Delta y \approx dy$, we obtain an estimate for the relative error bound for y as $p + |bx|q$.

9.6.10 Exercises

65 The function z is defined by

$$z(x, y) = x^2y - 3y$$

Find Δz and dz when $x = 4$, $y = 3$, $\Delta x = -0.01$ and $\Delta y = 0.02$.

66 An open box has internal dimensions $2\,\text{m} \times 1.25\,\text{m} \times 0.75\,\text{m}$. It is made of sheet metal 4 mm thick.

Find the actual volume of metal used and compare it with the approximate volume found using the differential of the capacity of the box.

67 The angle of elevation of the top of a tower is found to be $30° \pm 0.5°$ from a point $300 \pm 0.1\,\text{m}$ on a horizontal line through the base of the tower. Estimate the height of the tower.

68 The equations

$$x + 2y + 3z + 4u = -3$$

$$x^2 + y^2 + z^2 + u^2 = 10$$

$$x^3 + y^3 + z^3 + u^3 = 0$$

define u as a function of y if x and z are eliminated. Find $\mathrm{d}u/\mathrm{d}y$ when $x = 1$, $y = -1$, $z = 2$, $u = -2$.

69 The acceleration f of a piston is given by

$$f = r\omega^2\left(\cos\theta + \frac{r}{L}\cos 2\theta\right)$$

When $\theta = \frac{1}{6}\pi$ radians and when $r/L = \frac{1}{2}$, calculate the approximate percentage error in the calculated value of f if the values of both r and ω are 1% too small.

70 The area of a triangle ABC is calculated using the formula

$$S = \tfrac{1}{2}bc\sin A$$

and it is known that b, c and A are measured correctly to within 1%. If the angle A is measured as 45°, prove that the percentage error in the calculated value of S is not more than about 2.8%.

71 The angular deflection θ of a beam of electrons in a cathode-ray tube due to a magnetic field is given by

$$\theta = K\frac{HL}{V^{1/2}}$$

where H is the intensity of the magnetic field, L is the length of the electron path, V is the accelerating voltage and K is a constant. If errors of up to ±0.2% are present in each of the measured H, L and V, what is the greatest possible percentage error in the calculated value of θ (assume that K is known accurately)?

72 In a coal processing plant the flow V of slurry along a pipe is given by

$$V = \frac{\pi p r^4}{8\eta l}$$

If r and l both increase by 5%, and p and η decrease by 10% and 30% respectively, find the approximate percentage change in V.

9.6.11 Exact differentials

Differentials sometimes arise naturally when modelling practical problems. An example in fluid dynamics is given in Section 9.9. When this occurs, it is often possible to analyse the problem further by testing to see if the expression in which the differentials occur is a total differential. Consider the equation

$$P(x, y)\mathrm{d}x + Q(x, y)\mathrm{d}y = 0$$

connecting x, y and their differentials. The left-hand side of this equation is said to be an **exact differential** if there is a function $f(x, y)$ such that

$$\mathrm{d}f = P(x, y)\mathrm{d}x + Q(x, y)\mathrm{d}y$$

Now we know that

$$\mathrm{d}f = \frac{\partial f}{\partial x}\mathrm{d}x + \frac{\partial f}{\partial y}\mathrm{d}y$$

so if $f(x, y)$ exists then

$$P(x, y) = \frac{\partial f}{\partial x} \quad \text{and} \quad Q(x, y) = \frac{\partial f}{\partial y}$$

For functions with continuous second derivatives we have

$$\frac{\partial^2 f}{\partial x \partial y} = \frac{\partial^2 f}{\partial y \partial x}$$

Thus if $f(x, y)$ exists then

$$\frac{\partial P}{\partial y} = \frac{\partial Q}{\partial x}$$

(9.31)

This gives us a test for the existence of $f(x, y)$, but does not tell us how to find it! The technique for finding $f(x, y)$ is shown in Example 9.39.

Example 9.35   Show that

$$(6x + 9y + 11)\mathrm{d}x + (9x - 4y + 3)\mathrm{d}y$$

is an exact differential and find the relationship between y and x given

$$\frac{\mathrm{d}y}{\mathrm{d}x} = -\frac{6x + 9y + 11}{9x - 4y + 3}$$

and the condition $y = 1$ when $x = 0$.

Solution   In this example

$$P(x, y) = 6x + 9y + 11 \quad \text{and} \quad Q(x, y) = 9x - 4y + 3$$

First we test whether the expression is an exact differential. In this example

$$\frac{\partial P}{\partial y} = 9 \quad \text{and} \quad \frac{\partial Q}{\partial x} = 9$$

so from (9.31) we have an exact differential. Thus we know that there is a function $f(x, y)$ such that

$$\frac{\partial f}{\partial x} = 6x + 9y + 11, \quad \frac{\partial f}{\partial y} = 9x - 4y + 3$$

(9.32), (9.33)

Integrating (9.32) with respect to x, keeping y constant (that is, reversing the partial differentiation process), we have

$$f(x, y) = 3x^2 + 9xy + 11x + g(y)$$

(9.34)

Note that the 'constant' of integration is a function of y. You can check that this expression for $f(x, y)$ is correct by differentiating it partially with respect to x. But we also know from (9.33) the partial derivative of $f(x, y)$ with respect to y, and this enables us to find $g'(y)$. Differentiating (9.34) partially with respect to y and equating it to (9.33), we have

$$\frac{\partial f}{\partial y} = 9x + \frac{\mathrm{d}g}{\mathrm{d}y} = 9x - 4y + 3$$

(Note that since g is a function of y only we use $\mathrm{d}g/\mathrm{d}y$ rather than $\partial g/\partial y$.) Thus

$$\frac{\mathrm{d}g}{\mathrm{d}y} = -4y + 3$$

so, on integrating,

$$g(y) = -2y^2 + 3y + C$$

Substituting back into (9.34) gives

$$f(x, y) = 3x^2 + 9xy + 11x - 2y^2 + 3y + C$$

Now we are given that

$$\frac{dy}{dx} = -\frac{6x + 9y + 11}{9x - 4y + 3}$$

which implies that

$$(6x + 9y + 11)dx + (9x - 4y + 3)dy = 0$$

which in turn implies that

$$3x^2 + 9xy + 11x - 2y^2 + 3y + A = 0$$

The arbitrary constant A is fixed by applying the given condition $y = 1$ when $x = 0$, giving $A = -1$. Thus x and y satisfy the equation

$$3x^2 + 9xy + 11x - 2y^2 + 3y = 1$$

9.6.12 Exercises

73 Determine which of the following are exact differentials of a function, and find, where appropriate, the corresponding function.

(a) $(y^2 + 2xy + 1)dx + (2xy + x^2)dy$

(b) $(2xy^2 + 3y\cos 3x)dx + (2x^2y + \sin 3x)dy$

(c) $(6xy - y^2)dx + (2xe^y - x^2)dy$

(d) $(z^3 - 3y)dx + (12y^2 - 3x)dy + 3xz^2 dz$

74 Find the value of the constant λ such that

$$(y\cos x + \lambda \cos y)dx + (x\sin y + \sin x + y)dy$$

is the exact differential of a function $f(x, y)$. Find the corresponding function $f(x, y)$ that also satisfies the condition $f(0, 1) = 0$.

75 Show that the differential

$$g(x, y) = (10x^2 + 6xy + 6y^2)dx$$
$$+ (9x^2 + 4xy + 15y^2)dy$$

is not exact, but that a constant m can be chosen so that

$$(2x + 3y)^m g(x, y)$$

is equal to dz, the exact differential of a function $z = f(x, y)$. Find $f(x, y)$.

9.7 Taylor's theorem for functions of two variables

In this section we extend Taylor's theorem for one variable (Theorem 9.4) to a function of two variables and apply it to unconstrained and constrained optimization problems.

9.7.1 Taylor's theorem

First we consider a function of two variables. Suppose $f(x, y)$ is a function all of whose nth-order partial derivatives exist and are continuous on some circular domain D with centre (a, b). Then, if $(a + h, b + k)$ lies in D, we have

$$f(a + h, b + k) = f(a, b) + \frac{1}{1!}\left(h\frac{\partial}{\partial x} + k\frac{\partial}{\partial y}\right)f(a, b) + \frac{1}{2!}\left(h\frac{\partial}{\partial x} + k\frac{\partial}{\partial y}\right)^2 f(a, b)$$

$$+ \ldots + \frac{1}{(n-1)!}\left(h\frac{\partial}{\partial x} + k\frac{\partial}{\partial y}\right)^{n-1} f(a, b)$$

$$+ \frac{1}{n!}\left(h\frac{\partial}{\partial x} + k\frac{\partial}{\partial y}\right)^n f(a + \theta h, b + \theta k) \qquad (9.35)$$

where $0 < \theta < 1$. Here we have introduced the notation

$$\left(h\frac{\partial}{\partial x} + k\frac{\partial}{\partial y}\right)^r f(a, b)$$

to represent the value of the expression

$$h^r \frac{\partial^r f}{\partial x^r} + \binom{r}{1}h^{r-1}k\frac{\partial^r f}{\partial x^{r-1}\partial y} + \binom{r}{2}h^{r-2}k^2\frac{\partial^r f}{\partial x^{r-2}\partial y^2} + \ldots$$

$$+ \binom{r}{r-1}hk^{r-1}\frac{\partial^r f}{\partial x \partial y^{r-1}} + k^r\frac{\partial^r f}{\partial y^r}$$

at the point (a, b).

This result is obtained by repeated use of the chain rule. Setting $x = a + ht$ and $y = b + kt$, where $0 \leq t \leq 1$, we obtain

$$g(t) = f(a + ht, b + kt)$$

which is a function of one variable, so that, from Theorem 9.4, it has a Taylor expansion

$$g(t) = g(0) + \frac{t}{1!}g'(0) + \frac{t^2}{2!}g''(0) + \ldots + \frac{t^{n-1}}{(n-1)!}g^{(n-1)}(0) + \frac{t^n}{n!}g^{(n)}(\theta t)$$

where $0 \leq \theta \leq 1$. The derivatives of g are found using the chain rule:

$$g' = \frac{dg}{dt} = \frac{dx}{dt}\frac{\partial f}{\partial x} + \frac{dy}{dt}\frac{\partial f}{\partial y} = h\frac{\partial f}{\partial x} + k\frac{\partial f}{\partial y} = \left(h\frac{\partial}{\partial x} + k\frac{\partial}{\partial y}\right)f$$

$$g'' = \frac{d^2 g}{dt^2} = \frac{d}{dt}\left(h\frac{\partial}{\partial x} + k\frac{\partial}{\partial y}\right)f = \left(h\frac{\partial}{\partial x} + k\frac{\partial}{\partial y}\right)\left(h\frac{\partial}{\partial x} + k\frac{\partial}{\partial y}\right)f$$

$$= \left(h\frac{\partial}{\partial x} + k\frac{\partial}{\partial y}\right)^2 f$$

and, in general,

$$g^{(r)} = \frac{d^r g}{dt^r} = \left(h \frac{\partial}{\partial x} + k \frac{\partial}{\partial y} \right)^r f \quad (r = 0, 1, 2, \dots, n)$$

Putting $t = 1$ into the Taylor expansion of g gives the required result.

The same method can be used to extend the result to as many variables as we please. For the function $f(\boldsymbol{x})$, where $\boldsymbol{x} = (x_1, x_2, \dots, x_n)$, we have

$$f(\boldsymbol{a} + \boldsymbol{h}) = f(\boldsymbol{a}) + \sum_{i=1}^{n} h_i \frac{\partial f}{\partial x_i}(\boldsymbol{a}) + \frac{1}{2!} \left(\sum_{i=1}^{n} h_i \frac{\partial}{\partial x_i} \right)^2 f(\boldsymbol{a}) + \dots$$

$$+ \frac{1}{(m-1)!} \left(\sum_{i=1}^{n} h_i \frac{\partial}{\partial x_i} \right)^{m-1} f(\boldsymbol{a}) + \frac{1}{m!} \left(\sum_{i=1}^{n} h_i \frac{\partial}{\partial x_i} \right)^{m} f(\boldsymbol{a} + \theta \boldsymbol{h}) \quad \textbf{(9.36)}$$

where $0 \leqslant \theta \leqslant 1$, provided that all the partial derivatives exist and are continuous.

By setting $h = x - a$ and $k = y - b$ in (9.35), we have the following alternative form of the Taylor expansion:

$$f(x, y) = f(a, b) + \frac{1}{1!} \left[(x-a) \frac{\partial}{\partial x} + (y-b) \frac{\partial}{\partial y} \right] f(a, b)$$

$$+ \frac{1}{2!} \left[(x-a) \frac{\partial}{\partial x} + (y-b) \frac{\partial}{\partial y} \right]^2 f(a, b)$$

$$+ \dots$$

$$+ \frac{1}{n!} \left[(x-a) \frac{\partial}{\partial x} + (y-b) \frac{\partial}{\partial y} \right]^n f(a + \theta(x-a), b + \theta(y-b))$$

$$\textbf{(9.37)}$$

which is referred to as the **Taylor expansion** of $f(x, y)$ about the point (a, b).

Example 9.36   Obtain the Taylor series of the function $f(x, y) = \sin xy$ about the point $(1, \frac{1}{3}\pi)$, neglecting terms of degree three and higher.

Solution   From (9.37) the required series is

$$f(x, y) = f(1, \tfrac{1}{3}\pi) + \frac{1}{1!} \left[(x-1) \frac{\partial}{\partial x} + (y - \tfrac{1}{3}\pi) \frac{\partial}{\partial y} \right] f(1, \tfrac{1}{3}\pi)$$

$$+ \frac{1}{2!} \left[(x-1) \frac{\partial}{\partial x} + (y - \tfrac{1}{3}\pi) \frac{\partial}{\partial y} \right]^2 f(1, \tfrac{1}{3}\pi) \dots$$

Since $f(x, y) = \sin xy$, $f(1, \tfrac{1}{3}\pi) = \dfrac{\sqrt{3}}{2}$. Also,

$$\frac{\partial f}{\partial x} = y \cos xy \qquad \text{giving} \qquad \left(\frac{\partial f}{\partial x}\right)_{(1,\pi/3)} = \tfrac{1}{6}\pi$$

$$\frac{\partial f}{\partial y} = x \cos xy \qquad \text{giving} \qquad \left(\frac{\partial f}{\partial y}\right)_{(1,\pi/3)} = \tfrac{1}{2}$$

$$\frac{\partial^2 f}{\partial x^2} = -y^2 \sin xy \qquad \text{giving} \qquad \left(\frac{\partial^2 f}{\partial x^2}\right)_{(1,\pi/3)} = -\tfrac{1}{18}\pi^2\sqrt{3}$$

$$\frac{\partial^2 f}{\partial x \, \partial y} = \cos xy - xy \sin xy \quad \text{giving} \quad \left(\frac{\partial^2 f}{\partial x \partial y}\right)_{(1,\pi/3)} = \tfrac{1}{2} - \tfrac{1}{6}\pi\sqrt{3}$$

$$\frac{\partial^2 f}{\partial y^2} = -x^2 \sin xy \qquad \text{giving} \qquad \left(\frac{\partial^2 f}{\partial y^2}\right)_{(1,\pi/3)} = -\tfrac{1}{2}\sqrt{3}$$

Hence, neglecting terms of degree three and higher,

$$\sin xy \approx \frac{\sqrt{3}}{2} + \tfrac{1}{6}\pi(x-1) + \tfrac{1}{2}(y - \tfrac{1}{3}\pi) - \tfrac{1}{36}\pi^2\sqrt{3}(x-1)^2$$

$$+ (\tfrac{1}{2} - \tfrac{1}{6}\pi\sqrt{3})(x-1)(y - \tfrac{1}{3}\pi) - \tfrac{1}{4}\sqrt{3}(y - \tfrac{1}{3}\pi)^2$$

There appears to be no command in MATLAB for determining directly the Taylor series expansion of $f(x, y)$ about a point (a, b). In MAPLE the first n terms in such an expansion may be obtained using the multivariable Taylor command

```
readlib(mtaylor):
mtaylor(f(x,y), [x = a,y = b], n);
```

For example, considering Example 9.36, the MAPLE commands

```
readlib(mtaylor):
mtaylor(sin(x*y), [x = 1, y = Pi/3],3);
```

return the first three terms of the series as

$$\tfrac{1}{2}\sqrt{3} + \tfrac{1}{2}y - \tfrac{1}{6}\pi + \tfrac{1}{6}(x - 1)\pi - \tfrac{1}{36}\sqrt{(3)}\pi^2(x - 1)^2 +$$
$$(\tfrac{1}{2} - \tfrac{1}{6}\sqrt{3}\pi)(y - \tfrac{1}{3}\pi)(x - 1) - \tfrac{1}{4}\sqrt{3}(y - \tfrac{1}{3}\pi)^2$$

which checks with the answer given in the solution.

Using the `maple` command such an expansion may be obtained in MATLAB using the command

```
maple('mtaylor(f(x,y),[x = a,y = b],n)')
```

Considering Example 9.36 check that the MATLAB commands

```
syms x y
f = sin(x*y);
s = maple('mtaylor(sin(x*y),[x = 1,y = pi/3],3)')
```

return the same answer as above.

9.7.2 Optimization of unconstrained functions

In Section 8.5 we considered the problem of determining the maximum and minimum values of a function $f(x)$ of one variable. We now turn our attention to obtaining the maximum and minimum values of a function $f(x, y)$ of two variables. Geometrically $z = f(x, y)$ represents a surface in three-dimensional space, with z being the height of the surface above the x–y plane. Suppose that $f(x, y)$ has a local maximum value at the point (a, b) as illustrated in Figure 9.18(a). Then for all possible (small) values of h and k

$$f(a, b) > f(a + h, b + k)$$

so that the difference (increment)

$$\Delta f = f(a + h, b + k) - f(a, b)$$

is negative. Then, provided that the partial derivatives exist and are continuous, using Taylor's theorem we can express Δf in terms of the partial derivatives of $f(x, y)$ evaluated at (a, b):

$$\Delta f = \left(h\frac{\partial f}{\partial x} + k\frac{\partial f}{\partial y} \right)_{(a,b)} + \frac{1}{2!}\left(h^2\frac{\partial^2 f}{\partial x^2} + 2hk\frac{\partial^2 f}{\partial x \partial y} + k^2\frac{\partial^2 f}{\partial y^2} \right)_{(a,b)} + \dots$$

where h and k may be negative or positive numbers. Since h and k are small, the sign of Δf depends on the sign of

$$\left(h\frac{\partial f}{\partial x} + k\frac{\partial f}{\partial y} \right)_{(a,b)}$$

That is, the sign of Δf depends on the values of h and k. But for a maximum value of $f(x, y)$ at (a, b) the sign of Δf must be negative whatever the values of h and k. This implies that for a maximum to occur at (a, b), $\partial f/\partial x$ and $\partial f/\partial y$ must be zero there.

If $f(x, y)$ has a local minimum at (a, b), as illustrated in Figure 9.18(b), then

$$f(a, b) < f(a + h, b + k)$$

and, using the above argument, we find that for a local minimum to occur at (a, b), $\partial f/\partial x$ and $\partial f/\partial y$ must again be zero.

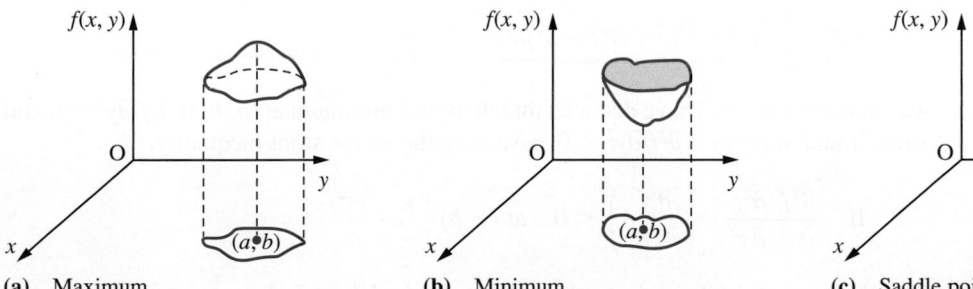

(a) Maximum   **(b)** Minimum   **(c)** Saddle point (maximum for curve C_2, minimum for curve C_1)

Figure 9.18

Thus a first necessary condition for a maximum or a minimum is

$$\frac{\partial f}{\partial x} = \frac{\partial f}{\partial y} = 0 \quad \text{at } (a, b)$$

In terms of differentials, this means that

$$df = 0 \quad \text{at } (a, b)$$

Points at which this occurs are called **stationary points** of the function and the values of the function at those points are called its **stationary values**. When this condition is satisfied, we have

$$\Delta f = \frac{1}{2!}\left(h^2 \frac{\partial^2 f}{\partial x^2} + 2hk \frac{\partial^2 f}{\partial x \partial y} + k^2 \frac{\partial^2 f}{\partial y^2} \right)_{(a,b)} + \dots$$

Putting

$$R = \frac{\partial^2 f}{\partial x^2}, \quad S = \frac{\partial^2 f}{\partial x \partial y} \quad \text{and} \quad T = \frac{\partial^2 f}{\partial y^2}$$

we deduce that the sign of Δf depends on the sign of the second differential

$$d^2 f = Rh^2 + 2Shk + Tk^2$$

Rearranging, we have, provided that $R \neq 0$,

$$d^2 f = \frac{1}{R}(R^2 h^2 + 2RShk + RTk^2) = \frac{1}{R}[(Rh + Sk)^2 + (RT - S^2)k^2]$$

If $RT - S^2 > 0$, the sign of $d^2 f$ is independent of the values of h and k; while if $RT - S^2 < 0$, its sign depends on those values. Thus a second necessary condition for a maximum or minimum value to occur at (a, b) is that

$$\frac{\partial^2 f}{\partial x^2} \frac{\partial^2 f}{\partial y^2} - \left(\frac{\partial^2 f}{\partial x \partial y} \right)^2 = f_{xx} f_{yy} - f_{xy}^2 \geq 0 \quad \text{at } (a, b)$$

Note that $f_{xx} f_{yy} - f_{xy}^2 = \begin{vmatrix} f_{xx} & f_{xy} \\ f_{yx} & f_{yy} \end{vmatrix}$. If strict inequality is satisfied, the sign of Δf depends on $R = \partial^2 f/\partial x^2$. If $\partial^2 f/\partial x^2 > 0$, there is a minimum at (a, b). If $\partial^2 f/\partial x^2 < 0$, there is a maximum at (a, b).

By expressing $d^2 f$ as

$$d^2 f = \frac{1}{T}[(TK + Sh)^2 + (RT - S^2)h^2], \quad T \neq 0$$

we could equally well have deduced that there is a minimum at (a, b) if $\partial^2 f/\partial y^2 > 0$ and a maximum at (a, b) if $\partial^2 f/\partial y^2 < 0$, assuming the above strict inequality.

$$\text{If} \quad \frac{\partial^2 f}{\partial x^2} \frac{\partial^2 f}{\partial y^2} - \left(\frac{\partial^2 f}{\partial x \partial y} \right)^2 < 0 \quad \text{at } (a, b)$$

then the sign of Δf depends on the values of h and k, and along some paths through (a, b) the function has a maximum value while along other paths it has a minimum value. Such a point is called a **saddle point**, as illustrated in Figure 9.18(c).

Figure 9.19
Nature of stationary
points: (a) saddle and
(b) maximum or
minimum.

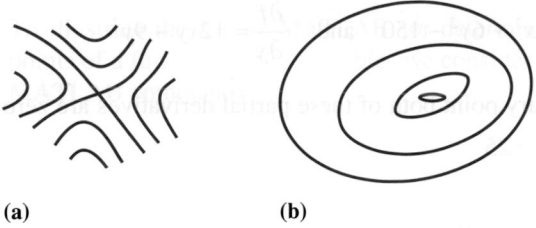

(a)   (b)

The contours of a function often show clearly where maximum or minimum or saddle points occur, as illustrated in Figure 9.19.

Summary

(1) A necessary condition for the function $f(x, y)$ to have a stationary value at (a, b) is that

$$\frac{\partial f}{\partial x} = 0 \quad \text{and} \quad \frac{\partial f}{\partial y} = 0 \quad \text{at } (a, b)$$

(2) If $\quad \dfrac{\partial^2 f}{\partial x^2} \dfrac{\partial^2 f}{\partial y^2} - \left(\dfrac{\partial^2 f}{\partial x \partial y}\right)^2 > 0 \quad$ and $\quad \dfrac{\partial^2 f}{\partial x^2}$ or $\dfrac{\partial^2 f}{\partial y^2} < 0 \quad$ at (a, b)

then the stationary point is a local maximum.

(3) If $\quad \dfrac{\partial^2 f}{\partial x^2} \dfrac{\partial^2 f}{\partial y^2} - \left(\dfrac{\partial^2 f}{\partial x \partial y}\right)^2 > 0 \quad$ and $\quad \dfrac{\partial^2 f}{\partial x^2}$ or $\dfrac{\partial^2 f}{\partial y^2} > 0 \quad$ at (a, b)

then the stationary point is a local minimum.

(4) If $\quad \dfrac{\partial^2 f}{\partial x^2} \dfrac{\partial^2 f}{\partial y^2} - \left(\dfrac{\partial^2 f}{\partial x \partial y}\right)^2 < 0 \quad$ at (a, b)

then the stationary point is a saddle point.

(5) If $\quad \dfrac{\partial^2 f}{\partial x^2} \dfrac{\partial^2 f}{\partial y^2} - \left(\dfrac{\partial^2 f}{\partial x \partial y}\right)^2 = 0 \quad$ at (a, b)

we cannot draw a conclusion, and the point may be a maximum, minimum or saddle point. Further investigation is required, and it may be necessary to consider the third-order terms in the Taylor series.

Example 9.37   Find the stationary points of the function

$$f(x, y) = 2x^3 + 6xy^2 - 3y^3 - 150x$$

and determine their nature.

89 Find the maximum and minimum values of

$$f(x, y) = 4x + y + y^2$$

where (x, y) lies on the circle $x^2 + y^2 + 2x + y = 1$.

90 Obtain the stationary value of $2x + y + 2z + x^2 - 3z^2$ subject to the two constraints $x + y + z = 1$ and $2x - y + z = 2$.

9.8   Engineering application: deflection of a built-in column

In this section we consider an example in which the techniques developed in Section 9.4 may be used to solve an engineering problem.

The deflection $y(x)$ of a column buckling under its own weight satisfies the differential equation

$$EI\frac{d^3y}{dx^3} + wx\frac{dy}{dx} = 0 \tag{9.53}$$

where E is Young's modulus, I the second moment of area of the cross-section and w is the weight per unit run of the column. The deflection of the built-in column shown in Figure 9.23 also satisfies the conditions

$$\frac{d^2y}{dx^2} = 0 \quad \text{at } x = 0$$

and

$$y = 0 \quad \text{and} \quad \frac{dy}{dx} = 0 \quad \text{at } x = l$$

where l is the length of the column. We need to find the greatest height attainable for the column without collapse.

To make the algebraic manipulations easier, we first simplify the differential equation. Putting $x = ct$ gives

$$\frac{dy}{dx} = \frac{dy}{dt}\frac{dt}{dx} = \frac{1}{c}\frac{dy}{dt}$$

$$\frac{d^2y}{dx^2} = \frac{d}{dt}\left(\frac{1}{c}\frac{dy}{dt}\right)\frac{dt}{dx} = \frac{1}{c^2}\frac{d^2y}{dt^2}$$

and

$$\frac{d^3y}{dx^3} = \frac{d}{dt}\left(\frac{1}{c^2}\frac{d^2y}{dt^2}\right)\frac{dt}{dx} = \frac{1}{c^3}\frac{d^3y}{dt^3}$$

which on substituting into (9.53) transforms it to

$$\frac{EI}{c^3}\frac{d^3y}{dt^3} + wt\frac{dy}{dt} = 0$$

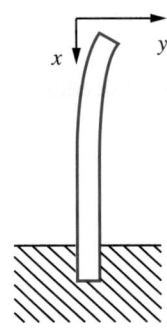

Figure 9.23
Deflection of a column.

so that choosing $c^3 = EI/w$ and setting $f(t) = dy/dt$ simplifies the equation further to

$$\frac{d^2 f}{dt^2} + tf = 0 \tag{9.54}$$

with the conditions

$$\frac{df}{dt} = 0 \quad \text{at } t = 0 \quad \text{and} \quad f(t) = 0 \quad \text{at } t = l(EI/w)^{-1/3} = T$$

Assuming that $f(t)$ has a Maclaurin series expansion, we may write it as

$$f(t) = a_0 + a_1 t + a_2 t^2 + a_3 t^3 + \ldots + a_n t^n + \ldots$$

Differentiating this, we have

$$f'(t) = a_1 + 2a_2 t + 3a_3 t^2 + \ldots + na_n t^{n-1} + (n + 1)a_{n+1} t^n + \ldots$$

and

$$f''(t) = 2a_2 + 6a_3 t + 12a_4 t^2 + \ldots + n(n - 1)a_n t^{n-2} + \ldots$$

Since $f'(0) = 0$, we deduce at once that $a_1 = 0$. Since $f(t)$ satisfies the differential equation (9.54), we deduce on substitution that

$$2a_2 + 6a_3 t + 12a_4 t^2 + \ldots + n(n - 1)a_n t^{n-2} + \ldots$$
$$= -a_0 t - a_1 t^2 - a_2 t^3 - \ldots - a_n t^{n+1} - \ldots$$

This expression is true for all values of t, with $0 < t < T$, so we deduce from Property (i) of polynomials given in Section 2.4.1 that the coefficients of each power of t on each side of the equation are equal. That is,

$$2a_2 = 0 \qquad \text{(coefficient of } t^0\text{)}$$
$$6a_3 = -a_0 \qquad \text{(coefficient of } t^1\text{)}$$
$$12a_4 = -a_1 \qquad \text{(coefficient of } t^2\text{)}$$

and so on. In general, the coefficient of t^r (obtained by setting $n - 2 = r$ on the left-hand side and $n + 1 = r$ on the right-hand side) yields

$$(r + 2)(r + 1)a_{r+2} = -a_{r-1}$$

This recurrence relation enables us to calculate a_{r+3} in terms of a_r as

$$a_{r+3} = \frac{-a_r}{(r + 3)(r + 2)} \quad (r = 0, 1, 2, \ldots) \tag{9.55}$$

Thus

$$a_3 = \frac{-a_0}{3 \cdot 2} \quad (r = 0), \quad a_4 = \frac{-a_1}{4 \cdot 3} \quad (r = 1)$$

$$a_5 = \frac{-a_2}{5 \cdot 4} \quad (r = 2), \quad a_6 = \frac{-a_3}{6 \cdot 5} \quad (r = 3)$$

$$a_7 = \frac{-a_4}{7 \cdot 6} \quad (r = 4), \quad a_8 = \frac{-a_5}{8 \cdot 7} \quad (r = 5)$$

and so on.

Since we deduced earlier, using the condition $f'(0) = 0$, that $a_1 = 0$, some terms can be eliminated immediately, and we have

$$a_4 = 0, \quad a_7 = 0, \quad a_{10} = 0, \quad a_{13} = 0, \quad \ldots$$

Since $a_2 = 0$ (from the coefficient of t^0), we have

$$a_5 = 0, \quad a_8 = 0, \quad \ldots$$

We are therefore left with

$$f(t) = a_0 + a_3 t^3 + a_6 t^6 + a_9 t^9 + \ldots$$

Substituting for $a_3, a_6, a_9, \ldots$ in terms of a_0, using (9.55) gives

$$f(t) = a_0 \left(1 - \frac{1}{3!} t^3 + \frac{1 \cdot 4}{6!} t^6 - \frac{1 \cdot 4 \cdot 7}{9!} t^9 + \frac{1 \cdot 4 \cdot 7 \cdot 10}{12!} t^{12} - \ldots \right)$$

So far we have only applied the condition at $t = 0$. Now we apply the condition at $t = T$, namely $f(T) = 0$. This gives

$$a_0 \left(1 - \frac{1}{3!} T^3 + \frac{4}{6!} T^6 - \frac{4 \cdot 7}{9!} T^9 \ldots \right) = 0$$

so that either

$$a_0 = 0 \quad \text{or} \quad 1 - \frac{1}{3!} T^3 + \frac{4}{6!} T^6 - \frac{4 \cdot 7}{9!} T^9 + \ldots = 0$$

This means that there is no deflection ($a_0 = 0$) unless

$$1 - \frac{1}{3!} T^3 + \frac{4}{6!} T^6 - \frac{4 \cdot 7}{9!} T^9 + \ldots = 0$$

The smallest value of T that satisfies this equation gives the critical height of the column. At that height the value of a_0 becomes arbitrary (and non-zero), and the column buckles. A first approximation to the critical value of T can be found by solving the quadratic equation (in T^3)

$$1 - \frac{1}{3!} T^3 + \frac{4}{6!} T^6 = 1 - \tfrac{1}{6} T^3 + \tfrac{1}{180} (T^3)^2 = 0$$

giving $T^3 = 8.292$. This may be refined using the Newton–Raphson procedure (9.19), eventually giving the critical length L in terms of E, I and w:

$$L = 1.99 (EI/w)^{1/3}$$

The detailed calculation is left as an exercise for the reader.

9.9　Engineering application: streamlines in fluid dynamics

As we mentioned in Section 9.6.10, differentials often occur in mathematical modelling of practical problems. An example occurs in fluid dynamics. Consider the case of steady-state incompressible fluid flow in two dimensions. Using rectangular cartesian coordinates (x, y) to describe a point in the fluid, let u and v be the velocities of the fluid

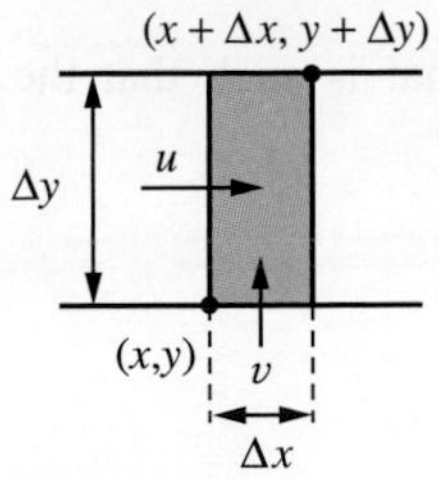

$(x + \Delta x, y + \Delta y)$

Figure 9.24
Flow through
rectangular element.

in the x and y directions respectively. Then by considering the flow in and flow out of a small rectangle, as shown in Figure 9.24, per unit time, we obtain a differential relationship between $u(x, y)$ and $v(x, y)$ that models the fact that no fluid is lost or gained in the rectangle; that is, the fluid is conserved.

The velocity of the fluid $\boldsymbol{q}$ is a vector point function. The values of its components u and v depend on the spatial coordinates x and y. The flow into the small rectangle in unit time is

$$u(x, \bar{y})\Delta y + v(\bar{x}, y)\Delta x$$

where $\bar{x}$ lies between x and $x + \Delta x$, and $\bar{y}$ lies between y and $y + \Delta y$. Similarly, the flow out of the rectangle is

$$u(x + \Delta x, \bar{y})\Delta y + v(\tilde{x}, y + \Delta y)\Delta x$$

where $\tilde{x}$ lies between x and $x + \Delta x$ and $\tilde{y}$ lies between y and $y + \Delta y$. Because no fluid is created or destroyed within the rectangle, we may equate these two expressions, giving

$$u(x, \bar{y})\Delta y + v(\bar{x}, y)\Delta x = u(x + \Delta x, \bar{y})\Delta y + v(\tilde{x}, y + \Delta y)\Delta x$$

Rearranging, we have

$$\frac{u(x + \Delta x, \bar{y}) - u(x, \bar{y})}{\Delta x} + \frac{v(\tilde{x}, y + \Delta y) - v(\bar{x}, y)}{\Delta y} = 0$$

Letting $\Delta x \to 0$ and $\Delta y \to 0$ gives the **continuity equation**

$$\frac{\partial u}{\partial x} + \frac{\partial v}{\partial y} = 0$$

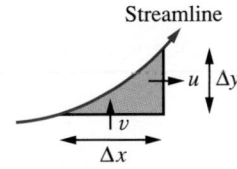

Streamline

Figure 9.25
Streamline.

The fluid actually flows along paths called **streamlines** so that there is no flow across a streamline. Thus from Figure 9.25 we deduce that

$$v\,\Delta x = u\,\Delta y$$

and hence

$$v\,\mathrm{d}x - u\,\mathrm{d}y = 0$$

The condition for this expression to be an exact differential is

$$\frac{\partial}{\partial y}(v) = \frac{\partial}{\partial x}(-u)$$

or

$$\frac{\partial u}{\partial x} + \frac{\partial v}{\partial y} = 0$$

This is satisfied for incompressible flow since it is just the continuity equation, so that we deduce that there is a function $\psi(x, y)$, called the **stream function**, such that

$$v = \frac{\partial \psi}{\partial x} \quad \text{and} \quad u = -\frac{\partial \psi}{\partial y}$$

It follows that if we are given u and v, as functions of x and y, that satisfy the continuity equation then we can find the equations of the streamlines given by $\psi(x, y) = $ constant.

Example 9.41 Find the stream function $\psi(x, y)$ for the incompressible flow that is such that the velocity q at the point (x, y) is

$$(-y/(x^2 + y^2), \; x/(x^2 + y^2))$$

Solution From the definition of the stream function, we have

$$u(x, y) = -\frac{\partial \psi}{\partial y} \quad \text{and} \quad v(x, y) = \frac{\partial \psi}{\partial x}$$

provided that

$$\frac{\partial u}{\partial x} + \frac{\partial v}{\partial y} = 0$$

Here we have

$$u = \frac{-y}{x^2 + y^2} \quad \text{and} \quad v = \frac{x}{x^2 + y^2}$$

so that

$$\frac{\partial u}{\partial x} = \frac{2xy}{(x^2 + y^2)^2} \quad \text{and} \quad \frac{\partial v}{\partial y} = -\frac{2yx}{(x^2 + y^2)^2}$$

confirming that

$$\frac{\partial u}{\partial x} + \frac{\partial v}{\partial y} = 0$$

Integrating

$$\frac{\partial \psi}{\partial y} = -u(x, y) = \frac{y}{x^2 + y^2}$$

with respect to y, keeping x constant, gives

$$\psi(x, y) = \tfrac{1}{2} \ln(x^2 + y^2) + g(x)$$

Differentiating partially with respect to x gives

$$\frac{\partial \psi}{\partial x} = \frac{x}{x^2 + y^2} + \frac{dg}{dx}$$

Since it is known that

$$\frac{\partial \psi}{\partial x} = v(x, y) = \frac{x}{x^2 + y^2}$$

we have

$$\frac{dg}{dx} = 0$$

which on integrating gives

$$g(x) = C$$

Figure 9.26
A vortex.

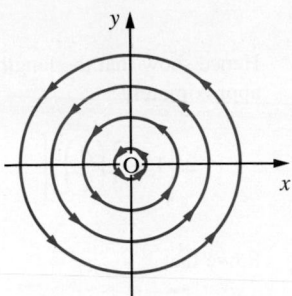

where C is a constant. Substituting back into the expression obtained for $\psi(x, y)$, we have

$$\psi(x, y) = \tfrac{1}{2}\ln(x^2 + y^2) + C$$

A streamline of the flow is given by the equation $\psi(x, y) = k$, where k is a constant. After a little manipulation this gives

$$x^2 + y^2 = a^2 \quad \text{and} \quad \ln a = k - C$$

and the corresponding streamlines are shown in Figure 9.26. This is an example of a **vortex**.

9.10 Review exercises (1–35)

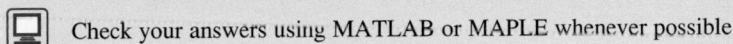 Check your answers using MATLAB or MAPLE whenever possible.

1 Use the Newton–Raphson method to find the root of

$$e^x - x^2 + 3x - 2 = 0$$

in the interval $0 \leqslant x \leqslant 1$. Start with $x = 0.5$ and give the root correct to 4dp.

2 The deflection at the midpoint of a uniform beam of length l, flexural rigidity EI and weight per unit length w, subject to an axial force P, is

$$d = \frac{w}{m^2 P}(\sec \tfrac{1}{2}ml - 1) - \frac{wl^2}{8P}$$

where $m^2 = P/EI$. On making the substitution $\theta = \tfrac{1}{2}ml$, show that

$$d = \frac{wl^4}{32EI} \frac{2\sec\theta - 2 - \theta^2}{\theta^4}$$

As the force P is relaxed, the deflection should reduce to that of a beam sagging under its own weight. By first representing $\sec\theta$ by its Maclaurin series expansion, show that

$$\lim_{\theta \to 0} d = \frac{5wl^4}{384EI}$$

3 Using the Maclaurin series expansion of e^x, determine the Maclaurin series expansion of $x/(e^x - 1)$ as far as the term in x^4, and hence obtain the approximation

$$\int_0^1 \frac{x}{e^x - 1}\,dx \approx \frac{311}{400}$$

4 Use L'Hôpital's rule to find

$$\lim_{x \to 1} \frac{\ln x}{x^2 - 1}$$

5   Determine

$$\lim_{x \to 2} \frac{2 \sin kx - x \sin 2k}{2(4 - x^2)}$$

where k is a constant.

6   Show that the equation

$$x^3 - 2x - 5 = 0$$

has a root in the neighbourhood of $x = 2$ and find it to three significant figures using the Newton–Raphson method.

7   (a) Obtain the Maclaurin series expansions of $\sinh x$ and $\cosh x$.

(b) A telegraph wire is stretched between two poles at the same height and a distance $2l$ apart. The sag at the midpoint is h. If the axes are taken as shown in Figure 9.27, it can be shown that the equation of the curve followed by the wire is

$$y = c \cosh \frac{x}{c}$$

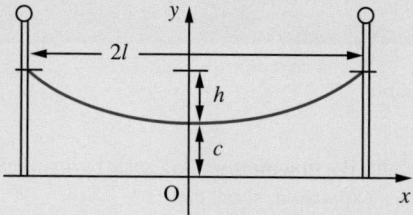

Figure 9.27 Telegraph wire of Question 7.

where c is an undetermined constant (see Example 8.56).

(i) Show that the length $2s$ of the wire is given by

$$2s = 2c \sinh \frac{l}{c}$$

(ii) If the wire is taut, so that h/c is small, it can be shown that l/c is also small. Ignoring powers of l/c higher than the second, show that

$$\frac{h^2}{l^2} \approx \frac{1}{4}\left(\frac{l}{c}\right)^2$$

Hence show that the length of the wire is approximately

$$2l\left[1 + \frac{2}{3}\left(\frac{h}{l}\right)^2\right]$$

8   Prove that

$$\int_0^\infty \operatorname{sech} x \, dx = \pi$$

and deduce $\int_0^1 \operatorname{sech}^{-1} x \, dx$.

9   Evaluate

(a) $\displaystyle\int_1^\infty \frac{1}{x^3} \, dx$ (b) $\displaystyle\int_0^\infty \frac{1}{x^2 + 2x + 2} \, dx$

(c) $\displaystyle\int_0^\infty x e^{-4x} \, dx$ (d) $\displaystyle\int_1^\infty \frac{\ln x}{x^3} \, dx$

(e) $\displaystyle\int_0^\infty e^{-2x} \cos x \, dx$ (f) $\displaystyle\int_0^\infty e^{-2x} \cosh x \, dx$

10   Evaluate

(a) $\displaystyle\int_0^8 x^{-1/3} \, dx$ (b) $\displaystyle\int_{3/2}^6 \frac{1}{\sqrt{(2x - 3)}} \, dx$

(c) $\displaystyle\int_0^1 \ln x \, dx$

stating in each case the value of x for which the integrand becomes unbounded.

11   Use the Taylor series to show that the principal term of the truncation error of the approximation

$$f''(a) \approx [f(a + h) - 2f(a) + f(a - h)]/h^2$$

is $\frac{1}{12} h^2 f^{(4)}(a)$.

Consider the function $f(x) = x e^x$. Estimate $f''(1)$ using the approximation above with $h = 0.01$, and $h = 0.02$. Compare your answer with the true value.

12   A particle moves in three-dimensional space such that its position at time t (seconds) is given by the vector $(4 \cos t, 4 \sin t, 3)$ where distance is measured in metres. Find the magnitude of its velocity and acceleration.

13 The acceleration a (m s^{-2}) of a particle at time t (s) is given by $a = (1 + t)i + t^2 j + 2k$. At $t = 0$ its displacement r is zero and its velocity v (m s^{-1}) is $i - j$. Find its displacement at time t.

14 The temperature gradient u at a point in a solid is

$$u(x, t) = t^{-1/2} e^{-x^2/4kt}$$

where k is a constant. Verify that

$$\frac{\partial^2 u}{\partial x^2} = \frac{1}{k} \frac{\partial u}{\partial t}$$

15 Show that the surfaces defined by

$$z^2 = \tfrac{1}{2}(x^2 + y^2) - 1$$

and

$$z = 1/xy$$

intersect, and that they do so orthogonally.

16 The height h of the top of a pylon is calculated by measuring its angle of elevation α at a point a distance s horizontally from the base of the pylon. Find the error in h due to small errors in s and α. If s and α are taken as 20 m and 30° respectively when the correct values are 19.8 m and 30.2°, find the error and the relative error in the calculated height.

17 The resistance of a length of wire is given by

$$R = \frac{k\rho L}{D^2}$$

where k is a constant. L is increasing at a rate of 0.4% min^{-1}, ρ is increasing at a rate of 0.01% min^{-1} and D is decreasing at a rate of 0.1% min^{-1}. At what percentage rate is the resistance R increasing?

18 The deflection H of a metal structure can be calculated using the formula

$$H = \sqrt{\left(\frac{I\rho^4 D^2 L^{3/2}}{20g} \right)}$$

where I, ρ, D and L are the moment of inertia, density, diameter and length respectively, and g is the acceleration due to gravity. If the value of H is to remain unaltered when I increases by 0.1%, ρ by 0.2% and D decreases by 0.3%, what percentage change in L is required?

19 In the calculation of the power in an a.c. circuit using the formula $W = EI \cos \phi$, errors of +1% in I, −0.7% in E and +2% in ϕ occur. Find the percentage error in the calculated value of W when $\phi = \frac{1}{3} \pi$ rad.

20 (a) Prove that $u = x^3 - 3xy^2$ satisfies

$$\frac{\partial^2 u}{\partial x^2} + \frac{\partial^2 u}{\partial y^2} = 0$$

(b) Given

$$u = x^2 \tan^{-1}\left(\frac{y}{x}\right) - y^2 \tan^{-1}\left(\frac{x}{y}\right)$$

evaluate

$$x \frac{\partial u}{\partial x} + y \frac{\partial u}{\partial y}$$

in terms of u.

21 Verify that $z = \ln \sqrt{(x^2 - y^2)}$ satisfies the equation

$$\left(\frac{\partial z}{\partial x}\right)^2 + \frac{\partial^2 z}{\partial y \partial x} + \left(\frac{\partial z}{\partial y}\right)^2 = \frac{1}{(x - y)^2}$$

22 (a) Find the value of the positive constant c for which the function

$$y = \frac{k}{2\pi} \sin\left(\frac{\pi x}{k}\right) \sin\left(\frac{2\pi t}{k}\right)$$

satisfies the equation

$$c^2 \frac{\partial^2 y}{\partial x^2} = \frac{\partial^2 y}{\partial t^2}$$

(b) V is a function of the independent variables x and y. Given that $x = r \cos \theta$ and $y = r \sin \theta$, find $\partial V/\partial \theta$ and $\partial V/\partial r$ in terms of $\partial V/\partial x$ and $\partial V/\partial y$, and hence show that

$$\frac{\partial V}{\partial y} = \frac{1}{r}\left(r \sin \theta \frac{\partial V}{\partial r} + \cos \theta \frac{\partial V}{\partial \theta} \right)$$

and

$$\frac{\partial V}{\partial x} = \frac{1}{r}\left(r \cos \theta \frac{\partial V}{\partial r} - \sin \theta \frac{\partial V}{\partial \theta} \right)$$

23 A curve C in three dimensions is given parametrically by $(x(t), y(t), z(t))$, where t is a real parameter, with $a \leqslant t \leqslant b$. Show that the equation

of the tangent line at a point P on this curve where $t = t_0$ is given by

$$\frac{x - x_0}{x_0'} = \frac{y - y_0}{y_0'} = \frac{z - z_0}{z_0'}$$

where $x_0 = x(t_0)$, $x_0' = x'(t_0)$, and so on.

Hence find the equation of the tangent line to the circular helix

$$x = a \cos t, \quad y = a \sin t, \quad z = at$$

at $t = \frac{1}{4}\pi$ and show that the length of the helix between $t = 0$ and $t = \frac{1}{2}\pi$ is $\pi a/\sqrt{2}$.

24 Show that $u = f(x + y) + g(x - y)$ satisfies the differential equation

$$\frac{\partial^2 u}{\partial x^2} - \frac{\partial^2 u}{\partial y^2} = 0$$

25 Show that if

$$\phi(x, t) = \frac{f(z)}{\sqrt{t}} \quad \text{and} \quad z = \frac{x}{2\sqrt{t}}$$

then

$$\frac{\partial \phi}{\partial t} = -\frac{zf'(z) + f(z)}{2t\sqrt{t}}$$

and find a similar expression for $\partial^2 \phi/\partial x^2$.

Deduce that if

$$\frac{\partial^2 \phi}{\partial x^2} = \frac{1}{k}\frac{\partial \phi}{\partial t}$$

then

$$kf''(z) + 2zf'(z) + 2f(z) = 0$$

26 Water waves move in the direction of the x axis with speed c. Their height h at time t is given by

$$h(t) = a \sin(x - ct)$$

where a is a constant. A small cork floats on the water and is blown by the wind in the direction of the x axis with constant velocity U. Show that the vertical acceleration of the cork at time t is given by

$$\frac{d^2 h}{dt^2} = -(U - c)^2 h$$

27 The components of velocity of an inviscid incompressible fluid in the x and y directions are u and v respectively, where

$$u = \frac{x^2 - y^2}{(x^2 + y^2)^2} \quad \text{and} \quad v = \frac{2xy}{(x^2 + y^2)^2}$$

Find the stream function $\psi(x, y)$ such that

$$d\psi = v\,dx - u\,dy$$

and verify that it satisfies Laplace's equation

$$\frac{\partial^2 \psi}{\partial x^2} + \frac{\partial^2 \psi}{\partial y^2} = 0$$

28 Show that the function

$$f(x, y) = x^2 y^2 - 5x^2 - 8xy - 5y^2$$

has one maximum and four saddle points. Sketch the part of the surface $z = f(x, y)$ that lies in the first quadrant.

29 Determine the position and nature of the stationary points on the surface

$$z = e^{-(x+y)}(3x^2 + y^2)$$

30 A trough of capacity $1\,m^3$ is to be made from sheet metal in the shape shown in Figure 9.28. Calculate the dimensions that use the least amount of metal. *Hint*: Set $y = xY$ and $z = xZ$ and show that the area of sheet metal needed is

$$\frac{2(1 + Y\cos\theta)Y\sin\theta + (2Y + 1)Z}{[(1 + Y\cos\theta)YZ\sin\theta]^{2/3}}$$

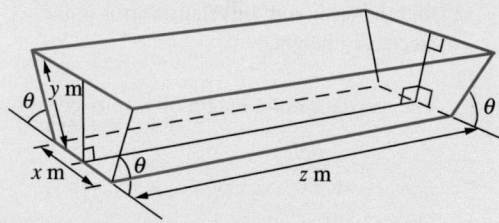

Figure 9.28 Trough of Question 30.

31 Find the critical points of the function

$$z = 12xy - 3xy^2 - x^3$$

and identify the character of each point.

32 Find the local maxima and minima of the function

$$f(x, y) = y^2 - 8x + 17$$

subject to the constraint

$$x^2 + y^2 = 9$$

33 A non-linear spring has a restoring force which is proportional to the cube of the displacement x.

The period T of oscillation from an initial displacement a is given by

$$T = 4\sqrt{2} \int_0^a \frac{1}{\sqrt{(a^4 - x^4)}}\, dx$$

Use the substitution $x^2 = a^2 \sin\theta$ to transform this integral to give

$$T = \frac{2\sqrt{2}}{a} \int_0^{\pi/2} \sin^{-1/2}\theta\, d\theta$$

Use the recurrence relation

$$(r + 1)F(r) = (r + 2)F(r + 2)$$

where

$$F(r) = \int_0^{\pi/2} \sin^r x\, dx,\ r > -1$$

to show that

$$T = \frac{42\sqrt{2}}{5a} \int_0^{\pi/2} \sin^{7/2}\theta\, d\theta$$

and use the trapezium rule to evaluate this integral.

34 The period T of oscillation of a simple pendulum of length l is given by

$$T = 4\sqrt{\left(\frac{l}{g}\right)} \int_0^{\pi/2} \frac{1}{\sqrt{(1 - \sin^2\frac{1}{2}\alpha \sin^2\phi)}}\, d\phi$$

by expanding the integrand as a power series in $\sin^2\frac{1}{2}\alpha$ show that

$$T = 2\pi\sqrt{\left(\frac{l}{g}\right)} [1 + 4\sin^2\tfrac{1}{2}\alpha + \tfrac{9}{64}\sin^4\tfrac{1}{2}\alpha + \ldots$$

35 (a) An oil tanker runs aground on a reef and its tanks rupture. Assuming that the oil forms a layer of uniform thickness on the sea and that the rate of spill is constant, show that the rate at which the radius r of the outer boundary of the oil spill increases in still water is proportional to $1/r$.

(b) The spillage takes place in a current flowing north with constant speed V. Assuming that the velocity of the oil with the current is the vector sum of the velocity of the oil in still water and the velocity of the current, show that the velocity (u, v) of the oil at the point (x, y) relative to the stricken tanker is given by

$$u = \frac{kx}{x^2 + y^2}, \quad v = V + \frac{ky}{x^2 + y^2}$$

where k is a constant of proportionality and the x and y axes are drawn in the easterly and northerly directions (see Figure 9.29).

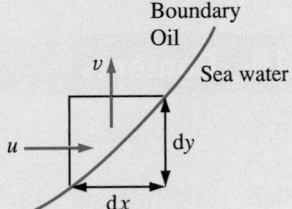

Figure 9.29

(c) Deduce that the most southerly point $(0, -c)$ reached by the oil slick is given by $c = k/V$.

(d) Show that, after a large interval of time, the oil slick occupies a region whose boundary $y = f(x)$ is the solution of the differential equation

$$\frac{dy}{dx} = \frac{x^2 + y^2 + cy}{cx}$$

that also satisfies the condition $y = -c$ at $x = 0$.

(e) Use the substitution $y = xz$ to transform the differential equation and initial conditions of part (d) to the differential equation

$$c\frac{dz}{dx} = 1 + z^2 \tag{9.56}$$

where $z \to -\infty$ as $x \to 0^+$.

(f) Show that the solution of (9.56) together with the boundary condition is $z = -\cot\dfrac{x}{c}$. Hence find y and sketch its graph.

10 Introduction to Ordinary Differential Equations

Chapter 10   Contents

<table>
<tr><td>**10.1**</td><td></td></tr>
</table>

10.1 Introduction

The essential role played by mathematical models in both engineering analysis and engineering design has been noted earlier in this book. It often happens that, in creating a mathematical model of a physical system, we need to express such relationships as 'the acceleration of A is directly proportional to B' or 'changes in D produce proportionate changes in E with constant of proportionality F'. Such statements naturally give rise to equations involving derivatives and integrals of the variables in the model as well as the variables themselves. Equations which introduce derivatives are called **differential equations**, those which introduce integrals are called **integral equations** and those which introduce both are called **integro-differential equations**. Generally speaking, integral and integro-differential equations are rather more difficult to solve than purely differential ones. This chapter starts with a discussion of the general characteristics of differential equations and then deals with ways of solving first-order differential equations. It is concluded by an examination of the solution of differential equations of second and higher orders.

Firstly, though, we will give some example of engineering problems which give rise to differential equations. As we will see, two of these (Sections 10.2.2 and 10.2.3) give rise to first-order equations and the other two (Sections 10.2.1 and 10.2.4) give rise to second-order equations. There are many other systems of engineering importance which give rise to second- or even higher-order differential equations. Section 10.10 develops an important engineering application of second-order constant-coefficient differential equations – the analysis of vibrations or oscillations. This analysis has very wide applications in engineering practice. A complete appreciation of this material requires that readers are already familiar with the material in Sections 10.8 and 10.9 but those who wish to have some understanding of the engineering importance of second-order constant-coefficient differential equations before beginning their study of the mathematical material may read quickly through that section before beginning their detailed study of the chapter.

10.2 Engineering examples

10.2.1 The take-off run of an aircraft

Aeronautical engineers need to be able to predict the length of runway that an aircraft will require to take off safely. To do this, a mathematical model of the forces acting on the aircraft during the take-off run is constructed, and the relationships holding between the forces are identified. Figure 10.1 shows an aircraft and the forces acting on it. If the mass of the aircraft is m, gravity causes a downward force mg. There is a ground reaction force through the wheels, denoted by G, and an aerodynamic lift force L. The engines provide a thrust T, which is opposed by an aerodynamic drag D and a rolling resistance from contact with the ground R. Since the aircraft is rolling along the runway, it is not accelerating vertically, so the vertical forces are in balance and the vertical equation of motion yields

$$L + G = mg \tag{10.1}$$

Figure 10.1
Forces on an
aircraft during
the take-off run.

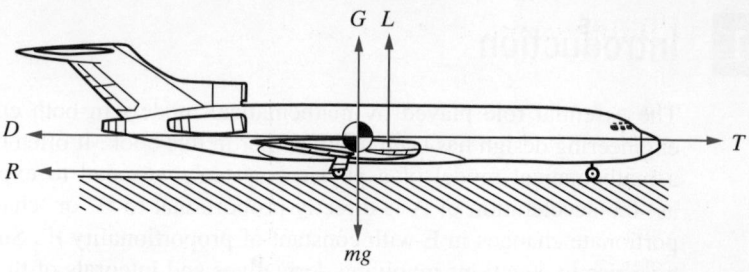

On the other hand, the aircraft is accelerating along the runway, so the horizontal equation of motion is

$$T - D - R = m\frac{\mathrm{d}^2 s}{\mathrm{d}t^2} \qquad (10.2)$$

where s is the distance the aircraft has travelled along the runway.

We know from experimental evidence that both aerodynamic lift and aerodynamic drag forces on a body vary roughly as the square of the velocity of the airflow relative to the body. We shall therefore choose to model the lift and drag forces as proportional to velocity squared. The rolling resistance is also known to be roughly proportional to the reaction force between ground and aircraft. Thus we make the modelling assumptions

$$L = \alpha v^2, \quad D = \beta v^2 \quad \text{and} \quad R = \mu G$$

Substituting for L, D and R in (10.1) and (10.2) and eliminating G results in the equation

$$m\frac{\mathrm{d}^2 s}{\mathrm{d}t^2} - (\mu\alpha - \beta)v^2 + \mu mg = T$$

or, replacing v by $\mathrm{d}s/\mathrm{d}t$,

$$m\frac{\mathrm{d}^2 s}{\mathrm{d}t^2} - (\mu\alpha - \beta)\left(\frac{\mathrm{d}s}{\mathrm{d}t}\right)^2 = T - \mu mg \qquad (10.3)$$

Thus our model of the aircraft travelling along the runway provides an equation relating the first and second time derivatives of the distance travelled by the aircraft, the thrust provided by the engines and various constants – the model is expressed as a differential equation for the distance s travelled along the runway. The model is not yet really complete, since we have not specified how the thrust varies. The thrust could, of course, vary with time (the pilot could open or close the throttles during the take-off run), and may also vary with the forward speed of the aircraft. On the other hand, we could just assume that thrust is constant. Also, the constants m, μ, α and β need to be determined. This information might be provided by measurements on the aircraft or on scale models of it, by other calculations or by engineers' estimates.

Once the model is complete it could be used, for instance, to predict the length of runway needed by the aircraft to attain flying speed. Flying speed is, of course, the speed at which the lift (αv^2) is equal to the weight (mg) of the aircraft. For a real aircraft our model would probably need to be made more elaborate, including, for instance, the angle of attack of the wing, which would change during the take-off run as the balance between aerodynamic and ground forces changed and as the pilot (or autopilot) changed the control surface settings.

10.2.2   Domestic hot-water supply

The second example involves modelling the heating of water in a hot-water storage tank. Figure 10.2 shows schematically an 'indirect' domestic hot-water tank. In this design of a hot-water system the central heating boiler, or other primary source of heat, supplies hot water to a calorifier (which takes the form of a coiled pipe) inside the hot-water storage tank. The main mass of water in the tank is then heated by the hot water passing through the calorifier coil. We wish to calculate how quickly the hot water in the tank will heat up.

Figure 10.2
An 'indirect'
hot-water tank.

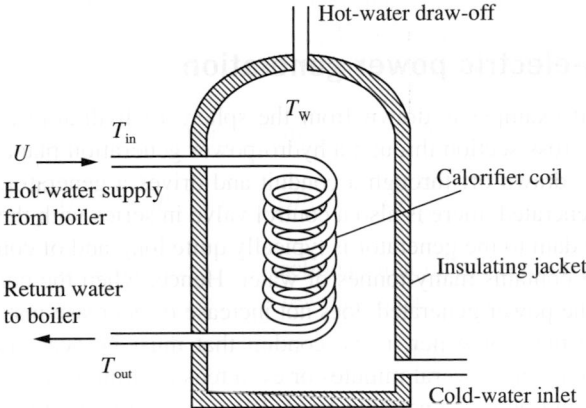

We shall assume that, to a good approximation, during heating convection ensures that the main mass of water in the tank is well mixed and at a uniform temperature T_w. The heating water flows into the calorifier at a speed U at a temperature T_{in}. The outflow from the calorifier is at temperature T_{out}. The cross-sectional area of the calorifier tube is A. The mass flowrate of heating water through the calorifier is therefore $\rho A U$, where ρ is the density of water, and the rate of heat loss from the heating water is $\rho A U (T_{in} - T_{out})c$, where c is the specific heat of water. The heat capacity of the main mass of water in the tank is $\rho V c$, where V is the volume of the tank, and so the rate of gain of heat in the main mass of water is given by

$$\rho V c \frac{dT_w}{dt}$$

The tank is well insulated, so, to a first approximation, we shall assume that the heat loss from the external shell of the tank is negligible. The rate of heat gain of the main mass of water is therefore equal to the rate of heat loss from the heating water; that is,

$$AU(T_{in} - T_{out}) = V\frac{dT_w}{dt} \tag{10.4}$$

where it is assumed that no hot water is being drawn off.

We should also expect that the difference in temperature of the heating water flowing in and that flowing out of the calorifier will be greater the cooler the mass of water in the tank. If we assume direct proportionality of these two quantities, we may express this modelling assumption as

$$T_{in} - T_{out} = \alpha(T_{in} - T_w) \tag{10.5}$$

the independent variables are x and y and the dependent variable is f. In the ordinary differential equation

$$\frac{d^2 f}{dx^2} - 4x\frac{df}{dx} = \cos 2x$$

the independent variable is x and the dependent variable is f. In the pair of coupled ordinary differential equations

$$4\frac{dx}{dt} + 3\frac{dy}{dt} - x + 2y = \cos t$$

$$6\frac{dx}{dt} - 2\frac{dy}{dt} - 2x + y = 2\sin t$$

the independent variable is t and the dependent variables are x and y.

10.3.3   The order of a differential equation

Another classification of differential equations is in terms of their order. The **order of a differential equation** is the degree of the highest derivative that occurs in the equation. In the case of partial differential equations the degree of a mixed derivative is the total of the degrees of differentiation with respect to each of the independent variables. The order of an equation is not affected by any power to which the derivatives may be raised.

Example 10.3

$$\frac{\partial f}{\partial x} + \frac{\partial f}{\partial y} = 4x^2 + 2y$$

is a first-order partial differential equation.

$$\frac{\partial^3 f}{\partial x \partial y^2} + \frac{\partial^2 f}{\partial x^2} - x\frac{\partial f}{\partial y} = 4x^2 + 2y$$

is a third-order partial differential equation.

$$\frac{d^2 f}{dx^2} - 4x\frac{df}{dx} = \cos 2x$$

is a second-order ordinary differential equation. The coupled ordinary differential equations

$$4\frac{dx}{dt} + 3\frac{dy}{dt} - x + 2y = \cos t$$

$$6\frac{dx}{dt} - 2\frac{dy}{dt} - 2x + y = 2\sin t$$

are both first-order equations as is the equation

$$\left(\frac{dx}{dt}\right)^2 + 4\frac{dx}{dt} = 0$$

despite the term in $(dx/dt)^2$.

10.3.4 Linear and nonlinear differential equations

Differential equations are also classified as linear or nonlinear. We may informally define **linear equations** as those in which the dependent variable or variables and their derivatives do not occur as products, raised to powers or in nonlinear functions. We shall meet a more formal definition of a linear differential equation in Section 10.8. **Nonlinear equations** are those that are not linear. Linear equations are an important category, since they have useful simplifying properties. Many of the nonlinear equations that occur in engineering science cannot be solved easily as they stand, but can be solved, for practical engineering purposes, by the process of replacing them with linear equations that are a close approximation – at least in some region of interest – and then studying the solution of the linear approximation. We shall see more of this later.

Example 10.4

$$\frac{\partial f}{\partial x} + \frac{\partial f}{\partial y} = 4x^2 + 2y$$

and

$$\frac{\partial^3 f}{\partial x \partial y^2} = 4x^2 + 2y$$

are both linear partial differential equations.

$$\frac{d^2 f}{dx^2} - 4x \frac{df}{dx} = \cos 2x$$

and the coupled differential equations

$$4\frac{dx}{dt} + 3\frac{dy}{dt} - x + 2y = \cos t$$

$$6\frac{dx}{dt} - 2\frac{dy}{dt} - 2x + y = 2\sin t$$

are linear ordinary differential equations.

$$\left(\frac{dx}{dt}\right)^2 + 4\frac{dx}{dt} = 0$$

$$\frac{d^2 x}{dt^2} + x\frac{dx}{dt} = 4\sin t$$

$$4\frac{dx}{dt} + \sin x = 0$$

are all nonlinear differential equations, the first because the derivative dx/dt is squared, the second because of the product between the dependent variable x and its derivative, and the third because of the nonlinear function, $\sin x$, of the dependent variable.

10.3.5 Homogeneous and nonhomogeneous equations

There is a further classification that can be applied to linear equations: the distinction between homogeneous and nonhomogeneous equations. In all the examples we have presented so far the differential equations have been arranged so that all terms containing the dependent variable occur on the left-hand side of the equality sign, and those terms that involve only the independent variable and constant terms occur on the right-hand side. This is a standard way of arranging terms, and aids in the identification of equations. Specifically, when linear equations are arranged in this way, those in which the right-hand side is zero are called **homogeneous equations** and those in which it is non-zero are **nonhomogeneous equations**. Expressed another way, each term in a homogeneous equation involves the dependent variable or one of its derivatives. In a nonhomogeneous equation there is at least one term that does not contain the independent variable or any of its derivatives.

Example 10.5

$$\frac{\partial f}{\partial x} + \frac{\partial f}{\partial y} = 4x^2 + 2y$$

is a nonhomogeneous partial differential equation, whereas

$$\frac{\partial^2 f}{\partial x \partial y} = 0$$

is a homogeneous one. The equations

$$\frac{\mathrm{d}x}{\mathrm{d}t} + 4x = 0$$

and

$$4\frac{\mathrm{d}x}{\mathrm{d}t} + (\sin t)x = 0$$

are both homogeneous ordinary differential equations, while

$$\frac{\mathrm{d}^2 x}{\mathrm{d}t^2} + t\frac{\mathrm{d}x}{\mathrm{d}t} = 4\sin t$$

and

$$\frac{\mathrm{d}^2 f}{\mathrm{d}x^2} - 4x\frac{\mathrm{d}f}{\mathrm{d}x} = \cos 2x$$

are both nonhomogeneous ordinary differential equations.

Example 10.6

Classify the equations (10.3), (10.6), (10.7) and (10.9) derived in the engineering examples of Section 10.2.

Solution

(a) Equation (10.3) is a second-order nonlinear ordinary differential equation whose dependent variable is s and whose independent variable is t.

(b) Equation (10.6) is a first-order linear nonhomogeneous ordinary differential equation whose dependent variable is T_w and whose independent variable is t.

(c) Equation (10.7) is a first-order linear nonhomogeneous ordinary differential equation whose dependent variable is Q and whose independent variable is t.

(d) Equation (10.9) is a second-order linear homogeneous ordinary differential equation whose dependent variable is i and whose independent variable is t.

10.3.6 Exercises

1 State the order of each of the following differential equations and name the dependent and independent variables. Classify each equation as partial or ordinary and as linear homogeneous, linear nonhomogeneous or nonlinear differential equations.

(a) $\dfrac{dx}{dt} + 2x = 0$

(b) $\dfrac{\partial y}{\partial t} + \dfrac{\partial x}{\partial t} = 0$

(c) $\dfrac{d^2 x}{dt^2} + 2\dfrac{dx}{dt} + 3x = 0$

(d) $\dfrac{\partial x}{\partial t} + c\dfrac{\partial^2 y}{\partial t^2} = 0$

(e) $\left(\dfrac{dx}{dt}\right)^2 + x = 0$

(f) $\dfrac{\partial y}{\partial t}\dfrac{\partial x}{\partial t} - x = 0$

(g) $\dfrac{dx}{dt} + 2x = t^2$

(h) $\dfrac{d^2 x}{dt^2} + \dfrac{dx}{dt} - 4x = \cos t + e^t$

2 Classify the following differential equations as partial or ordinary and as linear homogeneous, linear nonhomogeneous or nonlinear differential equations, state their order and name the dependent and independent variables.

(a) $\dfrac{\partial^2 f}{\partial x^2} + y\dfrac{\partial f}{\partial x}\dfrac{\partial f}{\partial y} + \sin y = 0$

(b) $\dfrac{d^2 p}{dz^2}\dfrac{dp}{dz} + (\sin z)p = \ln z$

(c) $\dfrac{\partial^2 h}{\partial x \partial y} + x^2 h = 0$

(d) $\dfrac{d^2 s}{dt^2} + (\sin t)\dfrac{ds}{dt} + (t + \cos t)s = e^t$

(e) $\dfrac{\partial x}{\partial u} + v\dfrac{\partial x}{\partial v} = uv$

(f) $\left(\dfrac{d^3 p}{dy^3}\right)^{1/2} + 4\dfrac{d^2 p}{dy^2} - 6\dfrac{dp}{dy} + 8p = 0$

(g) $\dfrac{dr}{dz} + z^2 = 0$

(h) $\dfrac{\partial f}{\partial x}\dfrac{\partial f}{\partial y}\dfrac{\partial f}{\partial z} + xyzf = 0$

(i) $\dfrac{dx}{dt} = f(t)x$

(j) $\dfrac{dx}{dt} = f(t)x + g(t)$

(k) $a(x)\dfrac{\partial^2 f}{\partial x^2} + b(x, y)\dfrac{\partial^2 f}{\partial x \partial y} + c(y)\dfrac{\partial^2 f}{\partial y^2} = p(x, y)$

(l) $\dfrac{d^3 p}{dq^3} + \dfrac{d^2 p}{dq^2}p + 4q^2 = 0$

(m) $\dfrac{d^2 x}{dy^2} = \dfrac{y}{x^2 - 1}$

(n) $\dfrac{\partial^2 y}{\partial p \partial t} + (\sin t)\dfrac{\partial y}{\partial p} + (\cos p)\dfrac{\partial y}{\partial t} = 0$

(o) $(\sin z)\dfrac{dy}{dz} + \dfrac{\cos z}{z}y = 0$

10.4   Solving differential equations

So far we have said that differential equations are equations which express relationships between a dependent variable and the derivatives of that variable with respect to one or more independent variables. We are now going to study some methods of solving differential equations. First, though, we should give some thought to exactly what form we expect that solution to take.

When we solve an algebraic equation we expect the solution to be a number (e.g. the solution of the equation $4x + 9 = 7$ is $x = -\frac{1}{2}$) or, perhaps, a set of numbers (e.g. the solution of a cubic polynomial equation like $x^3 - 5x^2 + 8x - 12 = 0$ is that x is one of a set of three real or complex numbers). Again, equations involving vectors and matrices have solutions that are constant vectors or one of a set of constant vectors. Differential equations, on the other hand, are equations involving not a simple scalar or vector variable but a function and its derivatives. The solution of a differential equation is, therefore, not a single value (or one from a set of values) but a function (or a family of functions). With this in mind let us proceed.

10.4.1   Solution by inspection

The solution to some differential equations can be obtained by recalling some results about differentiation.

Example 10.7

Faced with the differential equation

$$\frac{\mathrm{d}x}{\mathrm{d}t} = -4x \tag{10.10}$$

we might recall that if $x(t) = e^{-4t}$ then

$$\frac{\mathrm{d}x}{\mathrm{d}t} = -4e^{-4t} = -4x$$

In other words, the function $x(t) = e^{-4t}$ is a solution of the differential equation.

Example 10.8

The differential equation

$$\frac{\mathrm{d}^2x}{\mathrm{d}t^2} + \lambda^2 x = 0 \tag{10.11}$$

may be solved by recollecting that

$$\frac{\mathrm{d}^2}{\mathrm{d}t^2}(\sin \alpha t) = -\alpha^2 \sin \alpha t$$

Therefore, the function $x(t) = \sin \lambda t$ satisfies the differential equation.

Many differential equations can be solved by inspection in a similar manner to Examples 10.7 and 10.8. Solution by inspection requires the recognition of the equation and its connection to a familiar result in differentiation. It is therefore dependent upon experience and inspiration, and for this reason is only practical for solving the simplest differential equations.

MATLAB and MAPLE can both readily solve differential equations like these. In MATLAB, using the Symbolic Math Toolbox, analytic solutions of differential equations are computed using the *dsolve* command. The letter D denotes differentiation. The dependent variable is that preceded by D, whilst the default independent variable is *t*. Thus the general solution of the first-order differential equation

$$\frac{\mathrm{d}x}{\mathrm{d}t} = f(t, x)$$

is given by the commands

syms x
dsolve('equation')

In MAPLE the routine used is also called *dsolve* thus

dsolve(equation)

So, to solve Example 10.7, we would use

dsolve('Dx = -4*x')

in MATLAB and

dsolve(diff(x(t),t) = -4*x(t));

in MAPLE. The answer returned by MATLAB is C1*exp(-4*t) and by MAPLE is _C1*exp(-4*t). In each case C1 (or _C1) indicates an arbitrary constant. The reason for this will become apparent in the next section. To solve Example 10.8 we could use

dsolve('D2x + lambda^2*x','t')

in MATLAB and

dsolve(diff(x(t),t,t) + lambda^2*x(t));

in MAPLE. Notice here that MATLAB requires a second argument to specify what is the independent variable (otherwise how does the package know that lambda is not the independent variable?) whereas MAPLE does not require this because it has

effectively been specified in the expression diff(x(t),t,t) for $\dfrac{\mathrm{d}^2 x}{\mathrm{d}t^2}$.

10.4.2 General and particular solutions

Examples 10.7 and 10.8 also illustrate a pitfall of solving equations in this way. The function $x(t) = \mathrm{e}^{-4t}$ is certainly a solution of the equation in Example 10.7, but so is the function $x(t) = A\mathrm{e}^{-4t}$, where A is an arbitrary constant. The function $x(t) = \sin \lambda t$ is certainly a solution of the equation in Example 10.8, but so is the function $x(t) = A \sin \lambda t + B \cos \lambda t$, where A and B are arbitrary constants. Differential equations in general have this property – the most general function that will satisfy the differential equation contains one or more arbitrary constants. Such a function is known as the **general solution** of the differential equation. Giving particular numerical values to the constants in the general solution results in a **particular solution** of the equation. The general solution normally contains a number of arbitrary constants equal to the order of the differential equation.

Example 10.9

Find the general solution of the differential equation

$$\frac{d^2x}{dt^2} = t - 3e^{3t}$$

Solution

This differential equation can be solved by twice integrating both sides, remembering that each time we integrate the righthand side an unknown constant of integration is introduced. Thus integrating

$$\frac{d^2x}{dt^2} = t - 3e^{3t}$$

twice we have

$$\frac{dx}{dt} = \tfrac{1}{2}t^2 - e^{3t} + A$$

and

$$x(t) = \tfrac{1}{6}t^3 - \tfrac{1}{3}e^{3t} + At + B$$

This solution contains two arbitrary constants, A and B. The equation $\dfrac{d^2x}{dt^2} = t - 3e^{3t}$ is a second-order differential equation so, as a general rule, we would expect two constants.

When solving differential equations, we should, as a rule, seek the most general solution that is compatible with the constraints imposed by the problem. If we do not do this, we run the risk of neglecting some feature of the problem which may have serious implications for the performance, efficiency or even safety of the engineering equipment or system being analysed.

10.4.3 Boundary and initial conditions

The arbitrary constants in the general solution of a differential equation can often be determined by the application of other conditions.

Example 10.10

Find the function $x(t)$ that satisfies the differential equation

$$\frac{dx}{dt} = -4x$$

and that has the value 2.5 when $t = 0$.

Solution   We noted in Section 10.4.2 that $x(t) = Ae^{-4t}$ is a solution of the differential equation

$$\frac{\mathrm{d}x}{\mathrm{d}t} = -4x$$

(This can be checked by differentiating $x(t)$ to find $\dfrac{\mathrm{d}x}{\mathrm{d}t}$ and substituting into the differential equation.) But the solution $x(t) = Ae^{-4t}$ does not have the value 2.5 at $t = 0$ as required by the example. We can impose the boundary condition by $x(0) = 2.5$ to give

$$Ae^{-4 \times 0} = Ae^{-0} = A = 2.5$$

so the solution that satisfies the boundary condition is $x(t) = 2.5e^{-4t}$.

Additional conditions on the solution of a differential equation such as that in Example 10.10 are called **boundary conditions**. In the special case in which all the boundary conditions are given at the same value of the independent variable the boundary conditions are called **initial conditions**. In many circumstances it is convenient to consider a differential equation as incomplete until the boundary conditions have been specified. A differential equation together with its boundary conditions is referred to as a **boundary-value problem**, unless the boundary conditions satisfy the requirements for being initial conditions, in which case the differential equation together with its boundary conditions is referred to as an **initial-value problem**.

Example 10.11   Find the function $x(t)$ that satisfies the initial-value problem

$$\frac{\mathrm{d}^2 x}{\mathrm{d}t^2} + \lambda^2 x = 0 \quad x(0) = 4, \quad \frac{\mathrm{d}x}{\mathrm{d}t}(0) = 3, \quad \lambda \neq 0$$

Solution   We know from Section 10.4.2 that the general solution of this differential equation is

$$x(t) = A \sin \lambda t + B \cos \lambda t$$

We can confirm this by differentiating $x(t)$ twice and substituting into the differential equation to demonstrate that this $x(t)$ does satisfy the differential equation. With this $x(t)$ we have

$$\frac{\mathrm{d}x}{\mathrm{d}t} = \lambda A \cos \lambda t - \lambda B \sin \lambda t$$

Applying the initial conditions gives rise to the equations

$$0A + 1B = 4$$

$$\lambda A + 0B = 3$$

and hence to the solution

$$x(t) = \frac{3}{\lambda} \sin \lambda t + 4 \cos \lambda t$$

which is the particular solution of the initial value problem.

Example 10.12

Find the function $x(t)$ that satisfies the boundary-value problem

$$\frac{d^2x}{dt^2} + \lambda^2 x = 0 \quad x(0) = 4, \quad \frac{dx}{dt}\left(\frac{\pi}{\lambda}\right) = 3, \quad \lambda \neq 0$$

Solution

As in the previous example, the general solution of the differential equation is

$$x(t) = A \sin \lambda t + B \cos \lambda t$$

and so

$$\frac{dx}{dt} = \lambda A \cos \lambda t - \lambda B \sin \lambda t$$

Applying the boundary conditions gives rise to the equations

$$0A + 1B = 4$$
$$-\lambda A + 0B = 3$$

and hence to the particular solution

$$x(t) = -\frac{3}{\lambda} \sin \lambda t + 4 \cos \lambda t$$

Obviously, since a first-order differential equation has only one arbitrary constant in its solution, only one boundary condition is needed to determine the constant, and so the boundary condition of a first-order equation can always be treated as an initial condition. For higher-order equations (and for sets of coupled first-order equations) the distinction between initial-value and boundary-value problems is an important one, not least because, generally speaking, initial-value problems are easier to solve than boundary-value problems.

MATLAB and MAPLE can both solve initial and boundary value problems. In MAPLE the differential equation and its boundary conditions must be presented as a set (indicated by placing curly brackets round the list of equation and boundary conditions), thus for Example 10.11 the solution is given by the commands

```
ode:= diff(x(t),t,t) + lambda^2*x(t)
dsolve({ode,x(0) = 4,D(x)(Pi/lambda) = 3});
```

Notice that the derivative boundary condition uses D(x) to denote $\frac{dx}{dt}$. In MATLAB the boundary conditions are defined by a separate list of boundary conditions, thus for Example 10.11 we have

```
dsolve('D2x + lambda^2*x','x(0) = 4,Dx(Pi/lambda) = 3','t')
```

In MATLAB we define the derivative boundary condition using Dx to denote $\frac{dx}{dt}$.

In both MAPLE and MATLAB an initial value problem would be solved in exactly the same way; the difference would be that both boundary conditions would be defined at the same value of the independent variable.

10.4.4   Analytical and numerical solution

We have seen that some differential equations are so simple that they can be solved by inspection, given a reasonable knowledge of differentiation. There are many differential equations that are not amenable to solution in this way. For some of these we may be able, by the use of more complex mathematical techniques, to find a solution that expresses a functional relationship between the dependent and independent variables. We say that such equations have an **analytical solution**. In the case of other equations we may not be able to find a solution in such a form – either because no suitable mathematical technique for finding the solution exists or because there is no analytical solution. In these cases the only way of solving the equation is by the use of numerical techniques, leading to a **numerical solution**.

An analytical solution is almost always preferable to a numerical one. This is chiefly because an analytical solution is a mathematical function, and so the numerical value of the dependent variable can be computed for any value of the independent variable. In contrast with this, a numerical solution takes the form of a table giving the values of the dependent variable at a discrete set of values of the independent variable. The value of the dependent variable corresponding to any value of the independent variable not included in that discrete set can only be computed by interpolation from the table (or by repeating the whole numerical solution process, making sure the desired value of the independent variable is included in the solution set).

If the differential equation being solved contains parameters (such as the constant λ in Example 10.8) then an analytical solution of the equation will contain that parameter. The behaviour of the solution of the equation as the parameter value changes can be readily understood. For a numerical solution the parameter must be given a specific numerical value before the solution is computed. The numerical solution will then be valid only for that value of the parameter. If the behaviour of the solution as the parameter value is changed is of interest then the equation must be solved repeatedly using different parameter values.

When we obtain an analytical solution of a differential equation without its associated boundary conditions, the arbitrary constants in the solution are effectively parameters of the solution. A numerical solution to a differential equation cannot be obtained unless the boundary conditions are specified. This is one reason why it is sometimes convenient to refer to the whole problem (differential equation and boundary conditions) as a unit rather than consider the differential equation separately from its boundary conditions.

Another reason for preferring an analytical solution to a numerical one when such a solution is available is that the work required to obtain a numerical solution is generally much greater than that required to obtain an analytical one. On the other hand, most of this greater quantity of work can be delegated to a computer (and this may sometimes be considered to be an argument for numerical solutions being preferable to analytical ones).

Finally, it should be pointed out that this somewhat simplified overview of the contrast between analytical and numerical solutions of differential equations is becoming increasingly blurred by the availability of computerized symbolic manipulation systems (often known as computer algebra systems). We shall, in the remainder of this chapter, be studying methods for both the numerical and analytical solution of ordinary differential equations.

10.4.5 Exercises

3 Give the general solution of the following differential equations. In each case state how many arbitrary constants you expect to find in the general solution. Are your expectations confirmed in practice?

(a) $\dfrac{dx}{dt} = 4t^2$

(b) $\dfrac{d^2x}{dt^2} = t^3 - 2t$

(c) $\dfrac{d^2x}{dt^2} = e^{4t}$

(d) $\dfrac{dx}{dt} = -6x$

(e) $\dfrac{d^3x}{dt^3} = \dfrac{2}{t^3} + \sin 5t$

(f) $\dfrac{d^2x}{dt^2} = 8x$

4 For each of the following differential equation problems state how many arbitrary constants you would expect to find in the most general solution satisfying the problem. Find the solution and check whether your expectation is confirmed.

(a) $\dfrac{d^2x}{dt^2} = 4t$, $x(0) = 2$

(b) $\dfrac{d^2x}{dt^2} = \sin 2t$, $x(\tfrac{1}{4}\pi) = 2$, $x(\tfrac{3}{4}\pi) = 2$

(c) $\dfrac{dx}{dt} = 4$

(d) $\dfrac{dx}{dt} + 2t = 0$, $x(1) = 1$

(e) $\dfrac{d^2x}{dt^2} = 2e^{-2t}$, $x(0) = a$

(f) $\dfrac{dx}{dt} - 2\sin 2t = 0$

(g) $\dfrac{dx}{dt} = 2x$, $x(0) = 1$

(h) $\dfrac{d^2x}{dt^2} - x = 0$, $x(0) = 0$, $x(1) = 1$

5 State which of the following problems are **under-determined** (that is, have insufficient boundary conditions to determine all the arbitrary constants in the general solution) and which are **fully determined**. In the case of fully determined problems state which are boundary-value problems

and which are initial-value problems. (Do not attempt to solve the differential equations.)

(a) $4x\dfrac{d^2x}{dt^2} + \left(2t^2 - \dfrac{1}{x}\right)\dfrac{dx}{dt} - 4x^2 t = 0$, $x(0) = 4$

(b) $\left(\dfrac{d^3x}{dt^3}\right)^2 + t\dfrac{d^2x}{dt^2} - x\left(\dfrac{dx}{dt}\right)^2 = 0$

$x(0) = 0$, $\dfrac{dx}{dt}(0) = 1$, $x(2) = 0$

(c) $\left(\dfrac{dx}{dt}\right)^2 - x^2 = \sin t$, $x(0) = a$

(d) $\dfrac{d^4x}{dt^4} + 4\dfrac{d^3x}{dt^3} - 2\dfrac{d^2x}{dt^2} + \dfrac{dx}{dt} - 4x = e^t$

$x(0) = 1$, $x(2) = 0$

(e) $\dfrac{d^2x}{dt^2} - 2t\dfrac{dx}{dt} = t^2 - 4$, $x(0) = 1$, $x(2) = 0$

(f) $\dfrac{d^2x}{dt^2} + 2x\left(\dfrac{dx}{dt}\right)^2 - \dfrac{x}{t} = 0$

$x(1) = 0$, $\dfrac{dx}{dt}(1) = 4$

(g) $\left(\dfrac{d^2x}{dt^2}\right)^2 + 2t = 0$, $\dfrac{dx}{dt}(2) = 1$

(h) $\dfrac{d^3x}{dt^3}\dfrac{dx}{dt} + x\dfrac{d^2x}{dt^2} = 2t^2$

$x(0) = 0$, $\dfrac{dx}{dt}(0) = 0$

(i) $\left(\dfrac{d^3x}{dt^3}\right)^{1/2} + t\dfrac{d^2x}{dt^2} + x\dfrac{dx}{dt} - \dfrac{x}{t} = 0$

$x(1) = 1$, $\dfrac{dx}{dt}(1) = 0$, $\dfrac{d^2x}{dt^2}(3) = 0$

(j) $\dfrac{dx}{dt} = (x - t)^2$, $x(4) = 2$

(k) $\dfrac{d^2x}{dt^2} - 4\dfrac{dx}{dt} + 4x = \cos t$, $x(1) = 0$, $x(3) = 0$

(1) $\dfrac{1}{t}\dfrac{d^3x}{dt^3} - t^2\left(\dfrac{dx}{dt}\right)^2 + x\left(\dfrac{dx}{dt}\right)^{1/2} - (t^2+4)x = 0$

$$x(0) = 0, \quad \frac{dx}{dt}(0) = U, \quad \frac{d^2x}{dt^2}(0) = 0$$

6 A uniform horizontal beam OA, of length a and weight w per unit length, is clamped horizontally at O and freely supported at A. The transverse displacement y of the beam is governed by the differential equation

$$EI\frac{d^2y}{dx^2} = \tfrac{1}{2}w(a-x)^2 - R(a-x)$$

where x is the distance along the beam measured from O, R is the reaction at A, and E and I are physical constants. At O the boundary conditions are $y(0) = 0$ and $\dfrac{dy}{dx}(0) = 0$. Solve the differential equation. What is the boundary condition at A? Use this boundary condition to determine the reaction R. Hence find the maximum transverse displacement of the beam.

All of the differential equations in Questions 3, 4 and 6 of Exercises 10.4.5 could be solved using MATLAB or MAPLE. For example, using MAPLE, Questions 3(c) would be solved by

```
dsolve(diff(x(t),t,t) = exp(4*t));
```

Question 4(g) could be solved in MATLAB as

```
dsolve('Dx = 2*x','x(0) = 1','t')
```

and the first part of Question 6 would be solved by

```
ode:= E*I*diff(y(x),x,x) = 1/2*w*(a - x)^2 - R*(a - x);
dsolve({ode,y(0) = 0,D(y)(0) = 0});
```

or by

```
ode = 'E*I*D2y = 1/2*w*(a - x)^2 - R*(a - x)'
dsolve(ode,'y(0) = 0,Dy(0) = 0','x')
```

in MAPLE and MATLAB respectively. Now, for practice, use MAPLE or MATLAB to check your solutions to Questions 3, 4 and 6.

10.5   First-order ordinary differential equations

For the next three sections of this chapter we are going to concentrate our attention on the solution of first-order differential equations. This is not as restrictive as it might at first sight seem, since higher-order differential equations can, using a technique that we shall meet later in this chapter, be expressed as sets of coupled first-order differential equations. Some of the methods used for the solution of first-order equations, particularly the numerical techniques, are also applicable to such sets of coupled first-order equations, and thus may be used to solve higher-order differential equations.

10.5.1   A geometrical perspective

Most first-order differential equations can be expressed in the form

$$\frac{dx}{dt} = f(t, x) \tag{10.12}$$

Figure 10.5
The direction field
for the equation
$dx/dt = x(1 - x)t$.

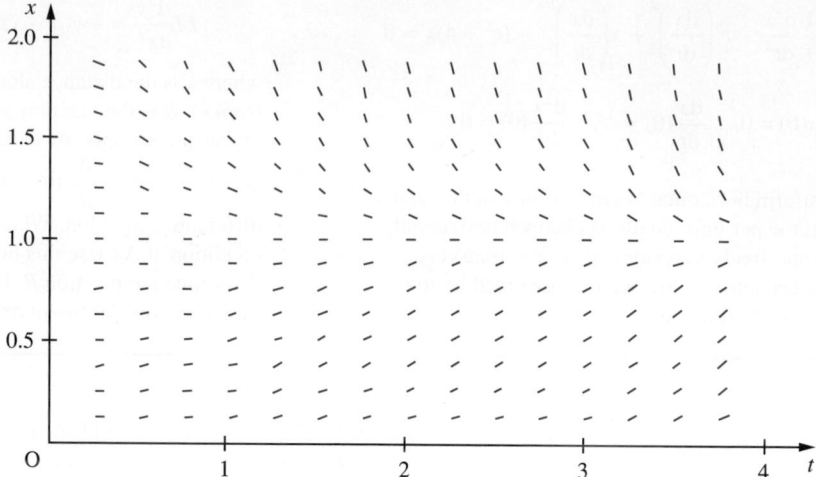

It is true that there are some nonlinear equations which cannot be reduced to this form, but they are relatively uncommon in engineering applications, and the treatment of such oddities is beyond the scope of this text. Expressing the equation in this form means that, for any point in the t–x plane for which $f(t, x)$ is defined, we can compute the value of dx/dt at that point. If we then do this for a grid of points in the t–x plane, we can draw a picture such as Figure 10.5. At each point a short line segment with gradient dx/dt is drawn. Such a diagram is called the **direction field** of the differential equation. Obviously, there is a gradient direction at every point of the t–x plane, but it is equally obviously only practical to draw in a finite number of them as we have done in Figure 10.5. The equation whose direction field is drawn in Figure 10.5 is in fact

$$\frac{dx}{dt} = x(1 - x)t$$

but the same process could be carried out for any equation expressible in the form (10.12).

A solution of the differential equation is a function relating x and t (that is, a curve in the t–x plane) which satisfies the differential equation. Since the solution function satisfies the differential equation, the solution curve has the property that its gradient is the same as the direction of the direction field of the equation at every point on the curve; in other words, the direction field consists of line segments that are tangential to the solution curves. With this insight, it is then fairly easy to infer what the solution curves of the equation whose direction field is shown in Figure 10.5 must look like. Some typical solution curves are shown in Figure 10.6.

By continuing this process, we could cover the whole t–x plane with an infinite number of different solution curves. Each solution curve is a particular solution of the differential equation. Since we are considering first-order equations, we expect the general solution to contain one unknown constant. Giving a specific value to that constant derives, from the general solution, one or other of the particular solution curves. In other words, the general solution, with its unknown constant, represents a **family of solution curves**. The curves drawn in Figure 10.6 are particular members of that family.

Figure 10.6
Solutions of
$dx/dt = x(1 - x)t$
superimposed on
its direction field.

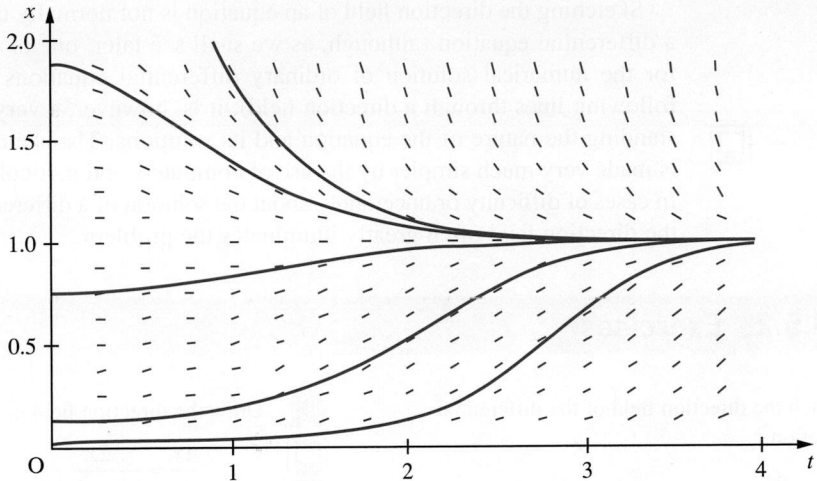

Example 10.13 Sketch the direction field of the differential equation

$$\frac{dx}{dt} = -\frac{1}{2}x$$

Verify that $x(t) = Ce^{-t/2}$ is the general solution of the differential equation. Find the particular solution that satisfies $x(0) = 2$ and sketch it on the direction field. Do the same with the solution for which $x(3) = -1$.

Solution The direction field is shown in Figure 10.7. Substituting the function $x(t) = Ce^{-t/2}$ into the equation immediately verifies that it is a solution. The initial condition $x(0) = 2$ implies $C = 2$. The condition $x(3) = -1$ implies $C = -e^{3/2}$. Both of these curves are shown on Figure 10.7, and are readily seen to be in the direction of the direction field at every point.

Figure 10.7
The direction field and
some solution curves
of $dx/dt = -x/2$.

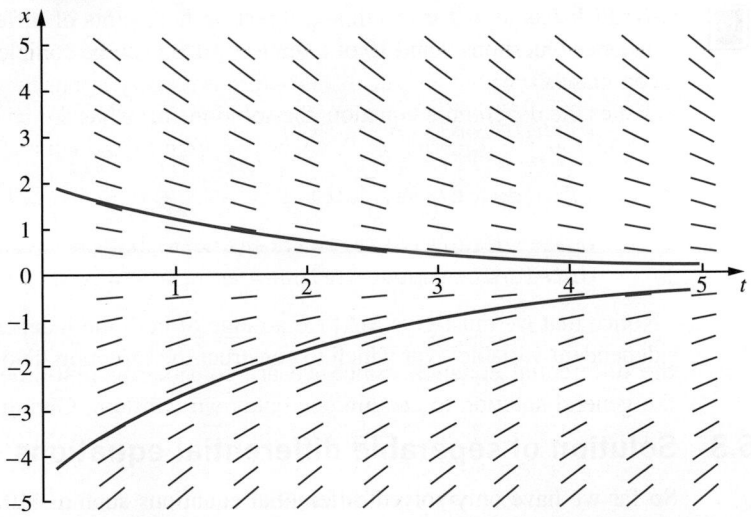

Sketching the direction field of an equation is not normally used as a way of solving a differential equation (although, as we shall see later, one of the simplest techniques for the numerical solution of ordinary differential equations may be interpreted as following lines through a direction field). It is, however, a very valuable aid to understanding the nature of the equation and its solutions. The sketching of direction fields is made very much simpler by the use of computers and particularly computer graphics. In cases of difficulty or uncertainty about the solution of a differential equation, sketching the direction field often greatly illuminates the problem.

10.5.2 Exercises

7 Sketch the direction field of the differential equation

$$\frac{\mathrm{d}x}{\mathrm{d}t} = -2t$$

Find the solution of the equation. Sketch the particular solutions for which $x(0) = 2$, and for which $x(2) = -3$, and check that these are consistent with your direction field.

8 Sketch the direction field of the differential equation

$$\frac{\mathrm{d}x}{\mathrm{d}t} = t - x$$

Verify that $x = t - 1 + Ce^{-t}$ is the solution of the equation. Sketch the solution curve for which $x(0) = 2$, and that for which $x(4) = 0$, and check that these are consistent with your direction field.

9 Draw the direction field of the equation

$$\frac{\mathrm{d}x}{\mathrm{d}t} = -\frac{2x}{t - 3}$$

Sketch some of the solution curves suggested by the direction field. Verify that the general solution of the equation is $x = C/(t - 3)^2$ and check that the members of this family resemble the solution curves you have sketched on the direction field.

10 Draw the direction field of the equation

$$\frac{\mathrm{d}x}{\mathrm{d}t} = \frac{1 - t}{t}x$$

Sketch some of the solution curves suggested by the direction field. Verify that the general solution of the equation is $x = Cte^{-t}$ and check that the members of this family resemble the solution curves you have sketched on the direction field.

MAPLE has tools for examining direction field plots of differential equations. For instance Questions 9 and 10 of Exercises 10.5.2 can be completed with the following commands

```
with(DETools):
ode:= diff(x(t),t) = -2*x(t)/(t - 3);
dfieldplot(ode, x(t),t = -2..2,x = -3..3);

ode:= diff(x(t),t) = (1-t)*x(t)/t
dfieldplot(ode, x(t),t = -2..4,x = -3..3)
```

Notice that we must give MAPLE a range of both the independent variable and the dependent variable over which to construct the direction field.

10.5.3 Solution of separable differential equations

So far we have only solved differential equations such as (10.10) and (10.11) whose solution is immediately obvious. We are now going to introduce some techniques that allow us to solve somewhat more difficult equations. These techniques are basically ways of manipulating differential equations into forms in which their solutions become obvious. The first method applies to equations that take what is known as a separable form. If the function $f(t, x)$ in the first-order differential equation

$$\frac{dx}{dt} = f(t, x)$$

is such that the equation can be manipulated (by algebraic operations) into the form

$$g(x)\frac{dx}{dt} = h(t) \tag{10.13}$$

then the equation is called a **separable equation**. We may find an expression for the solution of such equations by the following argument.

Integrating both sides of 10.13 with respect to t we have

$$\int g(x)\frac{dx}{dt}\,dt = \int h(t)\,dt \tag{10.14}$$

Now let

$$G(x) = \int g(x)\,dx$$

Then

$$\frac{dG(x)}{dx} = g(x)$$

and

$$\frac{d}{dt}G(x) = \frac{dG(x)}{dx}\frac{dx}{dt} = g(x)\frac{dx}{dt}$$

Integrating both sides of this equation with respect to t we have

$$G(x) = \int g(x)\frac{dx}{dt}\,dt$$

Hence we have

$$\int g(x)\frac{dx}{dt}\,dt = G(x) = \int g(x)\,dx$$

Finally substituting $\int g(x)\,dx$ for $\int g(x)\dfrac{dx}{dt}\,dt$ in 10.14 we have

$$\int g(x)\,dx = \int h(t)\,dt \tag{10.15}$$

so we have demonstrated that if a differential equation can be manipulated into the form of 10.13 then 10.15 holds. If the functions $g(x)$ and $h(t)$ are integrable then 10.15 leads to a solution of the differential equation.

Example 10.14   Solve the equation

$$\frac{dx}{dt} = 4xt, \quad x > 0$$

Solution   This equation can be written as

$$\frac{1}{x}\frac{dx}{dt} = 4t$$

and so is a separable equation. The solution is given by

$$\int \frac{dx}{x} = \int 4t\, dt$$

That is,

$$\ln x = 2t^2 + C$$

or

$$x = e^{2t^2 + C} = e^{2t^2} e^C$$

$$= C'e^{2t^2}, \quad \text{where } C' = e^C$$

Note: The cases $x < 0$ and $x = 0$ can be solved by allowing C' to be negative and zero respectively.

Note that a constant of integration has been introduced. We might expect such constants as a result of the integration of both left- and right-hand sides. However, if two constants had been introduced, they could then have been combined into one constant either on the left- or the right-hand side of the equation, so only one constant is actually necessary.

10.5.4 Exercises

11 Find the general solutions of the following differential equations:

(a) $\dfrac{dx}{dt} = kx$ (b) $\dfrac{dx}{dt} = 6xt^2$

(c) $\dfrac{dx}{dt} = \dfrac{bx}{t}$ (d) $\dfrac{dx}{dt} = \dfrac{a}{xt}$

12 Find the solutions of the following initial-value problems:

(a) $\dfrac{dx}{dt} = \dfrac{\sin t}{x^2}$, $x(0) = 4$

(b) $t^2 \dfrac{dx}{dt} = \dfrac{1}{x}$, $x(4) = 9$

13 Find the general solutions of the following differential equations:

(a) $\sqrt{t}\dfrac{dx}{dt} = \sqrt{x}$ (b) $\dfrac{dx}{dt} = (1 + \sin t)\cot x$

(c) $\dfrac{dx}{dt} = xte^{t^2}$ (d) $x^2 \dfrac{dx}{dt} = e^t$

(e) $\dfrac{dx}{dt} = ax(x - 1)$ (f) $x\dfrac{dx}{dt} = \sin t$

14 Find the solutions of the following initial-value problems:

(a) $\dfrac{dx}{dt} = \dfrac{t^2 + 1}{x + 2}$, $x(0) = -2$

(b) $t(t - 1)\dfrac{dx}{dt} = x(x + 1)$, $\quad x(2) = 2$

(c) $\dfrac{dx}{dt} = (x^2 - 1)\cos t$, $\quad x(0) = 2$

(d) $\dfrac{dx}{dt} = e^{x+t}$, $\quad x(0) = a$

(e) $\dfrac{dx}{dt} = \dfrac{4\ln t}{x^2}$, $\quad x(1) = 0$

15 A chemical reaction is governed by the differential equation

$$\frac{dx}{dt} = K(5 - x)^2$$

where $x(t)$ is the concentration of the chemical at time t. The initial concentration is zero and the concentration at time 5 s is found to be 2. Determine the reaction rate constant K and find the concentration at time 10 s and 50 s. What is the ultimate value of the concentration?

16 A skydiver's vertical velocity is governed by the differential equation

$$m\frac{dv}{dt} = mg - Kv^2$$

where K is the skydiver's coefficient of drag. If the skydiver leaves her aeroplane at time $t = 0$ with zero vertical velocity find at what time she reaches half her final velocity.

17 A chemical A is formed by an irreversible reaction from chemicals B and C. Assuming that the amounts of B and C are adequate to sustain the reaction, the amount of A formed at time t is governed by the differential equation

$$\frac{dA}{dt} = K(1 - \alpha A)^7$$

If no A is present at time $t = 0$ find an expression for the amount of A present at time t.

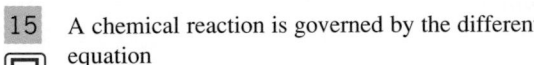

Either MAPLE or MATLAB can be used to solve any of the equations above. Sometimes the answers given may differ in exact form from those given in the 'Answers to Exercises' section at the end of this book. For instance both MAPLE and MATLAB give three answers to Question 14(e). This is because the differential equation can be solved to show $x(t)^3 = 12[t(\ln(t) - 1) + 1]$. There are then, of course, three cube roots of a real quantity, one real and two complex conjugates. Sometimes, the physical origins of a problem will indicate that the real root is the one of interest. In the answers in this chapter, where multiple roots exist, only the principal root is usually given.

10.5.5 Solution of differential equations of $\dfrac{dx}{dt} = f\left(\dfrac{x}{t}\right)$ form

Some differential equations, while not being in separable form, can be transformed, by means of a substitution, into separable equations. The best known example of this is a differential equation of the form

$$\frac{dx}{dt} = f\left(\frac{x}{t}\right) \tag{10.16}$$

(Note: equations of the form (10.16) are sometimes called 'homogeneous equations', but this use of the term homogeneous is different from the definition of homogeneous equations which we gave in Section 10.3.5.)

If the substitution $y = x/t$ is made then, since $x = yt$ and therefore, by the rule for differentiation of a product,

$$\frac{dx}{dt} = t\frac{dy}{dt} + y$$

we obtain

$$t\frac{dy}{dt} + y = f(y)$$

That is,

$$\frac{1}{f(y) - y}\frac{dy}{dt} = \frac{1}{t}$$

which is an equation of separable form.

Example 10.15   Solve the equation

$$t^2\frac{dx}{dt} = x^2 + xt, \quad t > 0, x \neq 0$$

Solution   Dividing both sides of the equation by t^2 results in

$$\frac{dx}{dt} = \frac{x^2}{t^2} + \frac{x}{t}$$

which is of the form (10.16). Making the substitution $y = x/t$ results in

$$t\frac{dy}{dt} + y = y^2 + y$$

that is,

$$\frac{1}{y^2}\frac{dy}{dt} = \frac{1}{t}$$

which is of separable form. The solution of this equation is given by

$$\int\frac{dy}{y^2} = \int\frac{dt}{t}$$

that is,

$$-\frac{1}{y} = \ln t + C \quad \text{or} \quad y = \frac{-1}{\ln t + C} = \frac{x}{t}$$

so

$$x = \frac{-t}{\ln t + C}$$

Note: The requirement $t > 0$ and $x \neq 0$ means that it is valid to divide throughout by t, and later by y, in the solution process. Solutions can be obtained without these restrictions and this is left as an exercise for the reader.

10.5.6 Exercises

 Again either MATLAB or MAPLE may be used to check your answers to all the following questions.

18 Find the general solutions of the following differential equations:

(a) $xt\dfrac{dx}{dt} = x^2 + t^2$ (b) $x^2\dfrac{dx}{dt} = \dfrac{t^3 + x^3}{t}$

(c) $t\dfrac{dx}{dt} = \dfrac{x^2 + xt}{t}$

(b) $xt\dfrac{dx}{dt} = 2(x^2 + t^2)$, $x(2) = -1$

(c) $t\dfrac{dx}{dt} = te^{-x/t} + x$, $x(2) = 4$

(d) $xt\dfrac{dx}{dt} = t^2 e^{-x^2/t^2} + x^2$, $x(1) = 2$

(e) $t^2\dfrac{dx}{dt} = x^2 + 2xt$, $x(1) = 4$

19 Find the solution of the following initial-value problem:

$$x^3 t\frac{dx}{dt} = t^4 + x^4, \quad x(1) = 4$$

20 Find the general solutions of the following differential equations:

(a) $2xt\dfrac{dx}{dt} = -x^2 - t^2$ (b) $t\dfrac{dx}{dt} = x + t\sin^2\left(\dfrac{x}{t}\right)$

(c) $t\dfrac{dx}{dt} = \dfrac{3t^2 - x^2}{t - 2x}$ (d) $t\dfrac{dx}{dt} = x + t\tan\left(\dfrac{x}{t}\right)$

(e) $\dfrac{dx}{dt} = \dfrac{x + t}{x - t}$ (f) $t\dfrac{dx}{dt} = x + te^{x/t}$

21 Find the solutions of the following initial-value problems:

(a) $\dfrac{dx}{dt} = \dfrac{x^3 - xt^2}{t^3}$, $x(1) = 2$

22 Show that, by making the substitution $y = at + bx + c$, equations of the form

$$\frac{dx}{dt} = f(at + bx + c)$$

can be reduced to separable form. Hence find the general solutions of the following differential equations:

(a) $\dfrac{dx}{dt} = \dfrac{t - x + 2}{t - x + 3}$ (b) $2\dfrac{dx}{dt} = -\dfrac{(t + 2x)}{t + 2x + 1}$

(c) $\dfrac{dx}{dt} = \dfrac{1 - 2x - t}{4x + 2t}$ (d) $\dfrac{dx}{dt} = \dfrac{x - t + 2}{x - t + 1}$

(e) $\dfrac{dx}{dt} = 2t + x + 2$ (f) $2\dfrac{dx}{dt} = 2x - t + 5$

(g) $\dfrac{dx}{dt} = 4t^2 + 4xt + x^2 - 2$

10.5.7 Solution of exact differential equations

Some first-order differential equations are of a form (or can be manipulated into a form) that is called **exact**. Since such equations can be solved readily, it would be useful to be able to recognize them or, better still, to have a test for them. In this section we shall see how exact equations are solved, and develop a test that allows us to recognize them.

The solution of exact equations depends on the following observation: if $h(t, x)$ is a function of the variables x and t, and the variable x is itself a function of t, then, by the chain rule of differentiation,

$$\frac{dh}{dt} = \frac{\partial h}{\partial x}\frac{dx}{dt} + \frac{\partial h}{\partial t}$$

Now if a first-order differential equation is of the form

$$p(t, x)\frac{dx}{dt} + q(t, x) = 0 \tag{10.17}$$

and a function $h(t, x)$ can be found such that

$$\frac{\partial h}{\partial x} = p(t, x) \quad \text{and} \quad \frac{\partial h}{\partial t} = q(t, x) \tag{10.18}$$

then (10.17) is equivalent to the equation

$$\frac{dh}{dt} = 0$$

and the solution must be

$$h(t, x) = C$$

Example 10.16   Solve the differential equation

$$2xt \frac{dx}{dt} + x^2 - 2t = 0$$

Solution   If $h(t, x) = x^2 t - t^2$ then

$$\frac{\partial h}{\partial x} = 2xt \quad \text{and} \quad \frac{\partial h}{\partial t} = x^2 - 2t$$

so the differential equation takes the form

$$\frac{d}{dt}(x^2 t - t^2) = 0$$

and the solution is

$$x^2 t - t^2 = C$$

Assuming $t > 0$ and $C > 0$ the solution can be written as

$$x = \pm \sqrt{\left(t + \frac{C}{t}\right)}$$

Thus we can solve equations of the form (10.17) provided that we can guess a function $h(t, x)$ that satisfies the conditions (10.18). If such a function is not immediately obvious, there are two possibilities: first there is no such function and, secondly, there is such a function but we don't see what it is. We shall now develop a test that enables us to answer the question of whether an appropriate function $h(t, x)$ exists and a procedure that enables us to find such a function if it does exist. If

$$\frac{\partial h}{\partial x} = p(t, x) \quad \text{and} \quad \frac{\partial h}{\partial t} = q(t, x)$$

then

$$\frac{\partial p}{\partial t} = \frac{\partial^2 h}{\partial x \partial t} = \frac{\partial q}{\partial x}$$

so, for a function $h(t, x)$ satisfying (10.18) to exist, the functions $p(t, x)$ and $q(t, x)$ must satisfy

$$\frac{\partial p}{\partial t} = \frac{\partial q}{\partial x} \qquad (10.19)$$

If $p(t, x)$ and $q(t, x)$ do not satisfy this condition then there is no point in seeking a function $h(t, x)$ satisfying (10.18).

If $p(t, x)$ and $q(t, x)$ do satisfy (10.19), how do we find the function $h(t, x)$ that satisfies (10.18) and thus solve the equation (10.17)? It may be that, as in Example 10.15, the function is obvious. If not, it can be obtained by solving the two equations (10.18) independently and then comparing the answers, as in Example 10.16.

Example 10.17   Solve the differential equation

$$(\ln \sin t - 3x^2)\frac{\mathrm{d}x}{\mathrm{d}t} + x \cot t + 4t = 0$$

Solution   First, since

$$\frac{\partial}{\partial t}(\ln \sin t - 3x^2) = \cot t = \frac{\partial}{\partial x}(x \cot t + 4t)$$

an appropriate function $h(t, x)$ may exist. Now

$$\frac{\partial h}{\partial x} = \ln \sin t - 3x^2 \quad \text{gives} \quad h = x \ln \sin t - x^3 + C_1(t)$$

and

$$\frac{\partial h}{\partial t} = x \cot t + 4t \quad \text{gives} \quad h = x \ln \sin t + 2t^2 + C_2(x)$$

where $C_1(t)$ and $C_2(x)$ are arbitrary functions of t and x respectively. These functions play the same role in the integration of partial differential relations as the arbitrary constants do in the integration of ordinary differential relations. Comparing the two results, we see that

$$h(t, x) = x \ln \sin t - x^3 + 2t^2$$

satisfies (10.18) and so the solution of the differential equation is

$$x \ln \sin t - x^3 + 2t^2 = C$$

Notice that the solution, in this case, is not an explicit expression for $x(t)$ in terms of t, but an implicit equation relating $x(t)$ and t, to be precise, a cubic polynomial in x with coefficients which are functions of t.

If an initial condition had been given, say $x(\frac{1}{2}\pi) = 3$, we would impose that initial condition on the implicit equation resulting in a value for the constant of integration C, thus

$$x(\tfrac{1}{2}\pi) = 3 \text{ giving}$$

$$3 \ln \sin \tfrac{1}{2}\pi - 3^3 + 2 \times 0^2 = C \text{ or } 0 - 27 + 0 = C$$

so $\quad x^3 - x \ln \sin t - 2t^2 - 27 = 0$

Both MAPLE and MATLAB can solve differential equations of the exact differential type. One drawback of systems such as MAPLE and MATLAB is that they may seek an explicit solution and, in doing so, give an answer which is of a more complex form and correspondingly less easily comprehended than the solution which might be obtained by a human. For instance, we can use MAPLE or MATLAB to solve Example 10.17. In MAPLE we would use

```
ode:= (ln(sin(t)) - 3*x(t)^2)*diff(x(t),t) +
x(t)*cot(t) + 4*t;
dsolve(ode);
```

and an equivalent form in MATLAB. The solution given in Example 10.17 is an implicit one which takes the form of a cubic function of $x(t)$, $x(t)^3 - \ln[\sin(t)]x(t) - 2t^2 + C = 0$. There is a general method for solving cubic algebraic equations which the computer algebra packages use to derive an explicit form for $x(t)$. There are, of course, three roots of the cubic equation, all of which are much less immediately understandable than the implicit solution given above.

All of the questions in Exercises 10.5.8 may be tackled with MAPLE or MATLAB. Some of the solutions derived in that way will appear different from those given in the Answers section but, with some persistence, all can be shown equivalent. Use of the `simplify` command can often be helpful.

10.5.8 Exercises

Check your answers using MATLAB or MAPLE whenever possible.

23 For each of the following differential equations determine whether they are exact equations and, if so, find the general solutions:

(a) $x\dfrac{dx}{dt} + t = 0$

(b) $x\dfrac{dx}{dt} - t = 0$

(c) $(x+t)\dfrac{dx}{dt} + x - t = 0$

(d) $(x - t^2)\dfrac{dx}{dt} - 2xt = 0$

(e) $(x - t)\dfrac{dx}{dt} - x + t - 1 = 0$

(f) $(2x + t)\dfrac{dx}{dt} + x + 2t = 0$

24 Find the solution of the following initial-value problems:

(a) $(x - 1)\dfrac{dx}{dt} + t + 1 = 0, \quad x(0) = 2$

(b) $(2x + t)\dfrac{dx}{dt} + x - t = 0, \quad x(0) = -1$

(c) $(2 - xt^2)\dfrac{dx}{dt} - x^2t = 0, \quad x(1) = 2$

(d) $\cos t\dfrac{dx}{dt} - x\sin t + 1 = 0, \quad x(0) = 2$

25 For each of the following differential equations determine whether they are exact, and, if so, find the general solution:

(a) $(x + t)\dfrac{dx}{dt} - x + t = 0$

(b) $\sqrt{t}\dfrac{\mathrm{d}x}{\mathrm{d}t} - xt = 0$

(c) $[\sin(x + t) + x\cos(x + t)]\dfrac{\mathrm{d}x}{\mathrm{d}t} + x\cos(x + t) = 0$

(d) $\sin(xt)\dfrac{\mathrm{d}x}{\mathrm{d}t} + \cos xt = 0$

(e) $(1 + te^{xt})\dfrac{\mathrm{d}x}{\mathrm{d}t} + xe^{xt} = 0$

(f) $2(x + \sqrt{t})\dfrac{\mathrm{d}x}{\mathrm{d}t} + \dfrac{x}{\sqrt{t}} + 1 = 0$

(g) $te^{-xt}\dfrac{\mathrm{d}x}{\mathrm{d}t} - xe^{xt} = 0$

(h) $\dfrac{t}{x + t}\dfrac{\mathrm{d}x}{\mathrm{d}t} + \dfrac{t}{x + t} + \ln(x + t) = 0$

26 Find the solutions of the following initial-value problems:

(a) $\cos(x + t)\left(\dfrac{\mathrm{d}x}{\mathrm{d}t} + 1\right) + 1 = 0, \quad x(0) = \tfrac{1}{2}\pi$

(b) $3(x + 2t)^{1/2}\dfrac{\mathrm{d}x}{\mathrm{d}t} + 6(x + 2t)^{1/2} + 1 = 0,$

$x(-1) = 6$

(c) $x(x^2 - t^2)\dfrac{\mathrm{d}x}{\mathrm{d}t} - t(x^2 - t^2) + 1 = 0, \quad x(0) = -1$

(d) $\dfrac{1}{x + t}\dfrac{\mathrm{d}x}{\mathrm{d}t} + \dfrac{1}{x + t} - \dfrac{1}{t^2} = 0, \quad x(2) = 2$

27 What conditions on the constants a, b, e and f must be satisfied for the differential equation

$$(ax + bt)\dfrac{\mathrm{d}x}{\mathrm{d}t} + ex + ft = 0$$

to be exact, and what is the solution of the equation when they are satisfied?

28 What conditions on the functions $g(t)$ and $h(t)$ must be satisfied for the differential equation

$$g(t)\dfrac{\mathrm{d}x}{\mathrm{d}t} + h(t)x = 0$$

to be exact, and what is the solution of the equation when they are satisfied?

29 For what value of k is the function $(x + t)^k$ an integrating factor for the differential equation

$$[(x + t)\ln(x + t) + x]\dfrac{\mathrm{d}x}{\mathrm{d}t} + x = 0?$$

30 For what value of k is the function t^k an integrating factor for the differential equation

$$(t^2\cos xt)\dfrac{\mathrm{d}x}{\mathrm{d}t} + 3\sin xt + xt\cos xt = 0?$$

10.5.9 Solution of linear differential equations

In Section 10.3.4 we defined linear differential equations. The most general first-order linear differential equation must have the form

$$\frac{\mathrm{d}x}{\mathrm{d}t} + p(t)x = r(t) \tag{10.20}$$

where $p(t)$ and $r(t)$ are arbitrary functions of the independent variable t. We shall first see how to solve the slightly simpler equation

$$\frac{\mathrm{d}x}{\mathrm{d}t} + p(t)x = 0 \tag{10.21}$$

If we multiply this equation throughout by a function $g(t)$, the resulting equation

$$g(t)\frac{\mathrm{d}x}{\mathrm{d}t} + g(t)p(t)x = 0$$

will be exact if

$$\frac{\partial g}{\partial t} = \frac{\partial}{\partial x}(gpx)$$

Since g and p are functions of t only, this reduces to

$$\frac{dg}{dt} = gp$$

which is a separable equation with solution

$$\int \frac{dg}{g} = \int p(t)dt$$

That is,

$$\ln g = \int p(t)dt$$

or

$$g(t) = e^{k(t)}, \quad \text{where } k(t) = \int p(t)dt$$

Hence, multiplying (10.21) throughout by $g(t)$, we obtain

$$e^{k(t)}\frac{dx}{dt} + p(t)e^{k(t)}x = 0$$

or

$$\frac{d}{dt}(e^{k(t)}x) = 0, \quad \text{since} \quad \frac{d}{dt}(e^{k(t)}) = e^{k(t)}\frac{d}{dt}(k(t)) = p(t)e^{k(t)}$$

Hence, integrating with respect to t, we have

$$e^{k(t)}x = C$$

so the solution can be written as

$$x = Ce^{-k(t)}$$

The function $g(t)$ is called the **integrating factor** for the differential equation. This name expresses the property that, whilst

$$\frac{dx}{dt} + p(t)x$$

is not an exact integral, the expression

$$g(t)\frac{dx}{dt} + g(t)p(t)x$$

is an exact integral. In other words $g(t)$ is a factor which makes the expression integrable.

This technique can, in fact, be used on the full equation (10.20). In that case, multiplying by the integrating factor $g(t)$, we obtain

$$e^{k(t)} \frac{dx}{dt} + p(t) e^{k(t)} x = e^{k(t)} r(t)$$

or

$$\frac{d}{dt} (e^{k(t)} x) = e^{k(t)} r(t)$$

Then, integrating with respect to t, we have

$$e^{k(t)} x = \int e^{k(t)} r(t) \, dt + C$$

and the solution

$$x = e^{-k(t)} \left[\int e^{k(t)} r(t) \, dt + C \right] \tag{10.22}$$

Thus (10.22) is an analytical solution of (10.20). The form of the solution can be simplified considerably if $\int p(t) \, dt$ has a simple analytical form, as in Examples 10.18 and 10.19.

Example 10.18   Solve the first-order linear differential equation

$$\frac{dx}{dt} + tx = t$$

Solution   We have shown that the integrating factor for a linear differential equation is

$$g(t) = e^{k(t)} \quad \text{where} \quad k(t) = \int p(t) \, dt$$

In this case

$$p(t) = t \quad \text{so} \quad k(t) = \int t \, dt = \tfrac{1}{2} t^2 \quad \text{and} \quad g(t) = e^{\frac{1}{2} t^2}$$

Multiplying both sides of the differential equation by this integrating factor we have

$$e^{\frac{1}{2} t^2} \frac{dx}{dt} + t e^{\frac{1}{2} t^2} x = t e^{\frac{1}{2} t^2}$$

Now the left-hand side is a perfect differential (the form of the integrating factor is chosen to make this so), so the differential equation can be written

$$\frac{d}{dt}(e^{\frac{1}{2}t^2}x) = te^{\frac{1}{2}t^2}$$

and integrating both sides of the equation with respect to t we have

$$e^{\frac{1}{2}t^2}x = \int te^{\frac{1}{2}t^2}dt = e^{\frac{1}{2}t^2} + C$$

Finally, dividing both sides by $e^{\frac{1}{2}t^2}$ we find

$$x(t) = 1 + Ce^{-\frac{1}{2}t^2}$$

Note: In evaluating $\int t\,dt$ for the integrating factor we have taken the constant of integration to be zero. Any other value of the constant of integration would also produce a valid (but more complicated!) integrating factor.

Example 10.19   Solve the first-order linear initial-value problem

$$\frac{dx}{dt} + \frac{1}{t}x = t, \quad x(2) = \tfrac{1}{3}$$

Solution   We have shown that the integrating factor for a linear differential equation is

$$g(t) = e^{k(t)} \quad \text{where} \quad k(t) = \int p(t)\,dt$$

In this case

$$p(t) = \frac{1}{t} \quad \text{so} \quad k(t) = \int \frac{1}{t}d(t) = \ln t \quad \text{and} \quad g(t) = e^{\ln t} = t$$

Multiplying both sides of the differential equation by this integrating factor we have

$$t\frac{dx}{dt} + x = t^2$$

Now the left-hand side is a perfect differential, so the differential equation can be written

$$\frac{d}{dt}(tx) = t^2$$

and integrating both sides of the equation with respect to t we have

$$tx = \int t^2\,dt = \tfrac{1}{3}t^3 + C$$

Now, dividing both sides by t, we find

$$x(t) = \tfrac{1}{3}t^2 + \frac{C}{t}$$

The initial value $x(2) = \tfrac{1}{3}$ so we must have

$$\tfrac{1}{3} = \tfrac{1}{3}4 + \frac{C}{2} \quad \text{or} \quad C = -2$$

So, finally,

$$x(t) = \tfrac{1}{3}t^2 - \frac{2}{t}$$

Again both MAPLE and MATLAB can be used to solve first-order linear differential equations. The preceding examples and the exercises in the next section can all be tackled in this way. It is worth observing that this is not at all unexpected. Computer algebra packages derive their results by following standard mathematical methods, which have been programmed by the package designers. All of the analytical methods for solving differential equations described in this chapter are well known and certainly included in the spectrum of methods incorporated into the dsolve and related routines used by MAPLE and by MATLAB. MAPLE provides a facility to see 'inside' the workings of the dsolve routine. The commands

```
infolevel[dsolve]:= 3:
ode:= t*diff(x(t),t) + x(t) = t^2;
dsolve({ode,x(2) = 1/3});
```

cause MAPLE to give a commentary on the different methods it is trying out in order to solve the differential equation. In this case it almost immediately identifies the equation as '1st order linear' and solves it by that method.

10.5.10 Solution of the Bernoulli differential equations

Differential equations of the form

$$\frac{\mathrm{d}x}{\mathrm{d}t} + p(t)x = q(t)x^\alpha$$

are called Bernoulli differential equations. If the index α is 0 or 1 then the equation reduces to

$$\alpha = 0, \quad \frac{\mathrm{d}x}{\mathrm{d}t} + p(t)x = q(t)$$

$$\alpha = 1, \quad \frac{\mathrm{d}x}{\mathrm{d}t} + [p(t) - q(t)]x = 0$$

Both these forms are linear, first-order, differential equations which we can solve by the method of Section 10.5.9. But if α does not take either of these values then the equation is non-linear. However, these equations can be reduced to a linear form by a substitution. Let

$$y(t) = x(t)^{1-\alpha}$$

then

$$\frac{dy}{dt} = (1 - \alpha)x^{-\alpha}\frac{dx}{dt}$$

giving

$$\frac{dx}{dt} = \frac{x^\alpha}{1 - \alpha}\frac{dy}{dt}$$

Substituting for $\dfrac{dx}{dt}$ in the original differential equation $\dfrac{dx}{dt} + p(t)x = q(t)x^\alpha$ we have

$$\frac{x^\alpha}{1 - \alpha}\frac{dy}{dt} + p(t)x = q(t)x^\alpha$$

Now dividing throughout by x^α we have

$$\frac{1}{1 - \alpha}\frac{dy}{dt} + p(t)x^{1-\alpha} = q(t)$$

But $y(t) = x(t)^{1-\alpha}$ so substituting for $x(t)^{1-\alpha}$ and multiplying throughout by $(1 - \alpha)$ we obtain

$$\frac{dy}{dt} + (1 - \alpha)p(t)y = (1 - \alpha)q(t)$$

which is a linear differential equation for $y(t)$. Hence we can solve the equation for $y(t)$ using the method of Section 10.5.9.

Example 10.20   Solve the differential equation

$$t^2 x - t^3\frac{dx}{dt} = x^4\cos t$$

Solution   Firstly we rearrange the equation into canonical form

$$\frac{dx}{dt} - \frac{1}{t}x = \frac{\cos t}{t^3}x^4$$

We recognise this as a Bernoulli differential equation with index $\alpha = 4$, so we make the substitution

$$y(t) = x(t)^{1-4} = x(t)^{-3}$$

giving

$$\frac{dy}{dt} = -3x^{-4}\frac{dx}{dt} \quad \text{and} \quad \frac{dx}{dt} = -\frac{x^4}{3}\frac{dy}{dt}$$

Substituting into the equation we have

$$-\frac{x^4}{3}\frac{dy}{dt} - \frac{1}{t}x = -\frac{\cos t}{t^3}x^4 \quad \text{so that} \quad \frac{dy}{dt} + \frac{3}{t}x^{-3} = \frac{3\cos t}{t^3}$$

and substituting y for x^{-3} we have

$$\frac{dy}{dt} + \frac{3}{t}y = \frac{3\cos t}{t^3}$$

This is now seen to be a linear equation so the integrating factor $g(t)$ is obtained by the standard method

$$p(t) = \frac{3}{t} \quad \text{giving} \quad k(t) = \int \frac{3}{t}dt = 3\ln t \quad \text{and} \quad g(t) = e^{3\ln t} = \left(e^{\ln t}\right)^3 = t^3$$

Multiplying both sides of the differential equation by this integrating factor we have

$$t^3\frac{dy}{dt} + 3t^2y = 3\cos t$$

Now the left-hand side is a perfect differential, so the differential equation can be written

$$\frac{d}{dt}(t^3y) = 3\cos t$$

and integrating both sides of the equation with respect to t we have

$$t^3y = \int 3\cos t\, dt = 3\sin t + C$$

Finally, substituting for $y(t)$ to obtain a solution for $x(t)$ we have

$$\frac{t^3}{x^3} = 3\sin t + C \quad \text{giving} \quad x(t)^3 = \frac{t^3}{3\sin t + C}$$

so that

$$x(t) = \sqrt[3]{\left(\frac{t^3}{3\sin t + C}\right)}$$

10.5.11 Exercises

 Check your answers using MATLAB or MAPLE whenever possible.

31 Find the solution of the following differential equations:

(a) $\dfrac{dx}{dt} + 3x = 2$ (b) $\dfrac{dx}{dt} - 4x = t$

(c) $\dfrac{dx}{dt} + 2x = e^{-4t}$ (d) $\dfrac{dx}{dt} + tx = -2t$

32 Find the solution of the following initial-value problems:

(a) $\dfrac{dx}{dt} - 2x = 3$, $x(0) = 2$

(b) $\dfrac{dx}{dt} + 3x = t$, $x(0) = 1$

(c) $\dfrac{dx}{dt} - \dfrac{x}{t} = t^2 - 3, \quad x(1) = -1$

33 Find the solutions of the following differential equations:

(a) $\dfrac{dx}{dt} - x = t + 2t^2$

(b) $\dfrac{dx}{dt} - 4tx = t^3$

(c) $\dfrac{dx}{dt} + \dfrac{2x}{t} = \cos t$

(d) $t\dfrac{dx}{dt} + 4x = e^t$

(e) $\dfrac{dx}{dt} - (2 \cot 2t)x = \cos t$

(f) $\dfrac{dx}{dt} + 6t^2 x = t^2 + 2t^5$

(g) $\dfrac{dx}{dt} - \dfrac{x}{t^2} = \dfrac{4}{t^2}$

34 Find the solutions of the following initial-value problems:

(a) $\dfrac{dx}{dt} - 2t(2x - 1) = 0, \quad x(0) = 0$

(b) $\dfrac{dx}{dt} = -x \ln t, \quad x(1) = 2$

(c) $\dfrac{dx}{dt} + 5x - t = e^{-2t}, \quad x(-1) = 0$

(d) $t^2 \dfrac{dx}{dt} - 1 + x = 0, \quad x(2) = 2$

(e) $\dfrac{dx}{dt} - \dfrac{1 - 2x}{t} = 4t + e^t, \quad x(1) = 0$

(f) $\dfrac{dx}{dt} + (x - U) \sin t = 0, \quad x(\pi) = 2U$

35 Solve (10.6), which arose from the model of the heating of the water in a domestic hot-water storage tank developed in Section 10.2.2. If the water in the tank is initially at 10°C and T_{in}

is 80°C, what is the ratio of the times taken for the water in the tank to reach 60°C, 70°C and 75°C?

36 Solve (10.7), which arose from the model of a hydro-electric power station developed in Section 10.2.3. The setting of the control valve is represented in the model by the value of the parameter γ. Derive an expression for the discharge $Q(t)$ following a sudden increase in the valve opening such that the parameter γ changes from γ_0 to $\tfrac{1}{2}\gamma_0$.

37 Find the solutions of the following differential equations:

(a) $\dfrac{dx}{dt} + \dfrac{1}{t}x = \dfrac{1}{x^2}$

(b) $\dfrac{dx}{dt} + 2x = tx^2$

(c) $\dfrac{dx}{dt} - x = \dfrac{e^t}{x}$

(d) $\dfrac{dx}{dt} + \dfrac{2}{t}x = x^2$

38 Find the solutions of the following initial-value problems:

(a) $\dfrac{dx}{dt} + \dfrac{1}{t}x = t^2 x^3, \quad x(1) = 1$

(b) $\dfrac{dx}{dt} + 3x = x^3, \quad x(0) = 6$

(c) $\dfrac{dx}{dt} + x = \sin t\, x^4, \quad x(0) = -1$

(d) $\dfrac{dx}{dt} - \dfrac{3}{t}x = \dfrac{1}{x^2}, \quad x(-1) = 1$

10.6 Numerical solution of first-order ordinary differential equations

Having met, in the last few sections, some techniques that may yield analytical solutions for first-order ordinary differential equations, we are now going to see how first-order ordinary differential equations can be solved numerically. In this chapter we will only study the simplest such method, Euler's method. Many more sophisticated (but also more complex) methods exist which yield solutions more efficiently, but space precludes their inclusion in this introductory treatment.

10.6.1   A simple solution method: Euler's method

In Section 10.5.1 we met the concept of the direction field of a differential equation

$$\frac{dx}{dt} = f(t, x)$$

We noted that solutions of the differential equation are curves in the t–x plane to which the direction field lines are tangential at every point. This immediately suggests that a curve representing a solution can be obtained by sketching on the direction field a curve that is always tangential to the lines of the direction field. In Figure 10.8 a way of systematically constructing an approximation to such a curve is shown.

Starting at some point (t_0, x_0), a straight line with gradient equal to the value of the direction field at that point, $f(t_0, x_0)$, is drawn. This line is followed to a point with abscissa $t_0 + h$. The ordinate at this point is $x_0 + hf(t_0, x_0)$, which we shall call X_1. The value of the direction field at this new point is calculated, and another straight line from this point with the new gradient is drawn. This line is followed as far as the point with abscissa $t_0 + 2h$. The process can be repeated any number of times, and a curve in the t–x plane consisting of a number of short straight line segments is constructed. The curve is completely defined by the points at which the line segments join, and these can obviously be described by the equations

$$t_1 = t_0 + h, \qquad X_1 = x_0 + hf(t_0, x_0)$$
$$t_2 = t_1 + h, \qquad X_2 = X_1 + hf(t_1, X_1)$$
$$t_3 = t_2 + h, \qquad X_3 = X_2 + hf(t_2, X_2)$$
$$\vdots \qquad\qquad \vdots$$
$$t_{n+1} = t_n + h, \qquad X_{n+1} = X_n + hf(t_n, X_n)$$

These define, mathematically, the simplest method for integrating first-order differential equations. It is called **Euler's method**. Solutions are constructed step by step, starting from some given starting point (t_0, x_0). For a given t_0 each different x_0 will give rise to

Figure 10.8
The construction of a numerical solution of the equation $dx/dt = f(t, x)$.

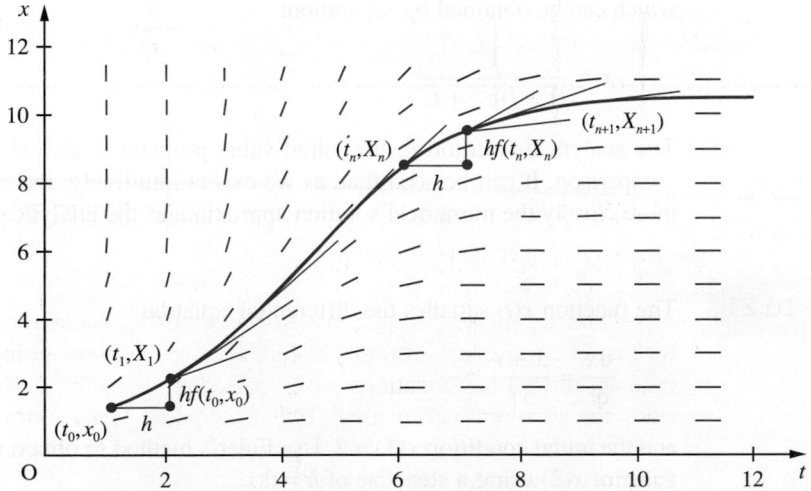

804 INTRODUCTION TO ORDINARY DIFFERENTIAL EQUATIONS

Figure 10.9
The Euler-method
solutions of
$dx/dt = x^2 t e^{-t}$ for
$h = 0.05, 0.025$
and 0.0125.

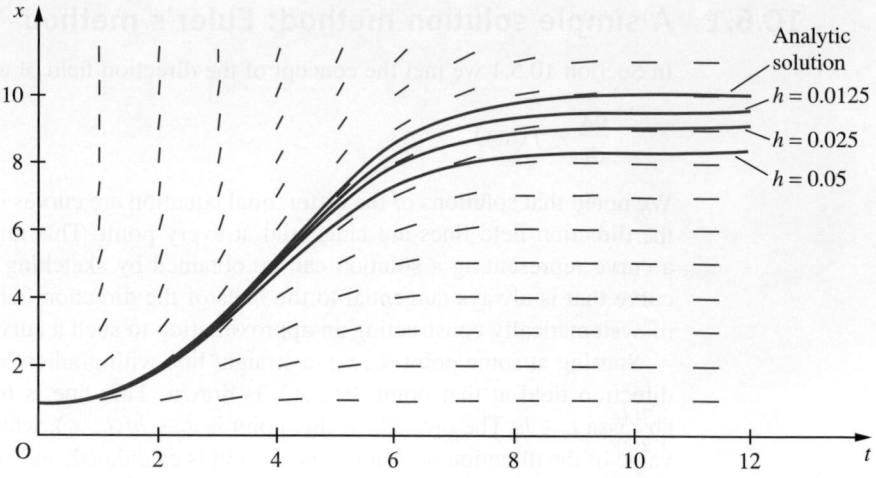

a different solution curve. These curves are all solutions of the differential equation, but each corresponds to a different initial condition.

The solution curves constructed using this method are obviously not exact solutions but only approximations to solutions, because they are only tangential to the direction field at certain points. Between these points, the curves are only approximately tangential to the direction field. Intuitively, we expect that, as the distance for which we follow each straight line segment is reduced, the curve we are constructing will become a better and better approximation to the exact solution. The increment h in the independent variable t along each straight-line segment is called the **step size** used in the solution. In Figure 10.9 three approximate solutions of the initial-value problem

$$\frac{dx}{dt} = x^2 t e^{-t}, \quad x(0) = 0.91 \tag{10.23}$$

for step sizes $h = 0.05, 0.025$ and 0.0125 are shown. These steps are sufficiently small that the curves, despite being composed of a series of short straight lines, give the illusion of being smooth curves. The equation (10.23) actually has an analytical solution, which can be obtained by separation:

$$x = \frac{1}{(1 + t)e^{-t} + C}$$

The analytical solution to the initial-value problem is also shown in Figure 10.9 for comparison. It can be seen that, as we expect intuitively, the smaller the step size the more closely the numerical solution approximates the analytical solution.

Example 10.21   The function $x(t)$ satisfies the differential equation

$$\frac{dx}{dt} = \frac{x + t}{xt}$$

and the initial condition $x(1) = 2$. Use Euler's method to obtain an approximation to the value of $x(2)$ using a step size of $h = 0.1$.

Solution The solution is obtained step by step as set out in Figure 10.10. The approximation $X(2) = 3.1162$ results.

Figure 10.10
Computational results
for Example 10.21.

t	X	$X + t$	Xt	$h\dfrac{X+t}{Xt}$
1.0000	2.0000	3.0000	2.0000	0.1500
1.1000	2.1500	3.2500	2.3650	0.1374
1.2000	2.2874	3.4874	2.7449	0.1271
1.3000	2.4145	3.7145	3.1388	0.1183
1.4000	2.5328	3.9328	3.5459	0.1109
1.5000	2.6437	4.1437	3.9656	0.1045
1.6000	2.7482	4.3482	4.3971	0.0989
1.7000	2.8471	4.5471	4.8400	0.0939
1.8000	2.9410	4.7410	5.2939	0.0896
1.9000	3.0306	4.9306	5.7581	0.0856
2.0000	3.1162			

10.6.2 Analysing Euler's method

We have introduced Euler's method via an intuitive argument from a geometrical understanding of the problem. Euler's method can be seen in another light – as an application of Taylor series. The Taylor series given in Section 9.5.2 applied to a function $x(t)$ gives

$$x(t + h) = x(t) + h\frac{\mathrm{d}x}{\mathrm{d}t}(t) + \frac{h^2}{2!}\frac{\mathrm{d}^2 x}{\mathrm{d}t^2}(t) + \frac{h^3}{3!}\frac{\mathrm{d}^3 x}{\mathrm{d}t^3}(t) + \ldots \quad \textbf{(10.24)}$$

Using this formula, we could, in theory, given the value of $x(t)$ and all the derivatives of x at t, compute the value of $x(t + h)$ for any given h. If we choose a small value for h then the Taylor series truncated after a finite number of terms will provide a good approximation to the value of $x(t + h)$. Euler's method can be interpreted as using the Taylor series truncated after the second term as an approximation to the value of $x(t + h)$.

In order to distinguish between the exact solution of a differential equation and a numerical approximation to the exact solution (and it should be appreciated that all numerical solutions, however accurate, are only approximations to the exact solution), we shall now make explicit the convention that we used in the last section. The exact solution of a differential equation will be denoted by a lower-case letter and a numerical approximation to the exact solution by the corresponding capital letter. Thus, truncating the Taylor series, we write

$$X(t + h) = x(t) + h\frac{\mathrm{d}x}{\mathrm{d}t}(t) = x(t) + hf(t, x) \quad \textbf{(10.25)}$$

Applying this truncated Taylor series, starting at the point (t_0, x_0) and denoting $t_0 + nh$ by t_n, we obtain

$$X(t_1) = X(t_0 + h) = x(t_0) + hf(t_0, x_0)$$
$$X(t_2) = X(t_1 + h) = X(t_1) + hf(t_1, X_1)$$
$$X(t_3) = X(t_2 + h) = X(t_2) + hf(t_2, X_2)$$

and so on

which is just the Euler-method formula obtained in Section 10.6.1. As an additional abbreviated notation, we shall adopt the convention that $x(t_0 + nh)$ is denoted by x_n, $X(t_0 + nh)$ by X_n, $f(t_n, x_n)$ by f_n, and $f(t_n, X_n)$ by F_n. Hence we may express Euler's method, in general terms, as the recursive rule

$$X_0 = x_0$$

$$X_{n+1} = X_n + hF_n \quad (n \geq 0)$$

The advantage of viewing Euler's method as an application of Taylor series in this way is that it gives us a clue to obtaining more accurate methods for the numerical solution of differential equations. It also enables us to analyse in more detail how accurate Euler's method may be expected to be. We can abbreviate (10.24) to

$$x(t + h) = x(t) + hf(t, x) + O(h^2)$$

where $O(h^2)$ covers all the terms involving powers of h greater than or equal to h^2. Combining this with (10.25), we see that

$$X(t + h) = x(t + h) + O(h^2) \tag{10.26}$$

(Note that in obtaining this result we have used the fact that signs are irrelevant in determining the order of terms; that is, $-O(h^p) = O(h^p)$.) Equation (10.26) expresses the fact that at each step of the Euler process the value of $X(t + h)$ obtained has an error of order h^2, or, to put it another way, the formula used is accurate as far as terms of order h. For this reason Euler's method is known as a **first-order method**. The exact size of the error is, as we intuitively expected, dependent on the size of h, and decreases as h decreases. Since the error is of order h^2, we expect that halving h, for instance, will reduce the error at each step by a factor of four.

This does not, unfortunately, mean that the error in the solution of the initial-value problem is reduced by a factor of four. To understand why this is so, we argue as follows. Starting from the point (t_0, x_0) and using Euler's method with a step size h to obtain a value of $X(t_0 + 4)$, say, requires $4/h$ steps. At each step an error of order h^2 is incurred. The total error in the value of $X(t_0 + 4)$ will be the sum of the errors incurred at each step, and so will be $4/h$ times the value of a typical step error. Hence the total error is of the order of $(4/h)O(h^2)$; that is, the total error is $O(h)$. From this argument we should expect that if we compare solutions of a differential equation obtained using Euler's method with different step sizes, halving the step size will halve the error in the solution. Examination of Figure 10.9 confirms that this expectation is roughly correct in the case of the solutions presented there.

Example 10.22   Let X_a denote the approximation to the solution of the initial-value problem

$$\frac{dx}{dt} = \frac{x^2}{t + 1}, \quad x(0) = 1$$

obtained using Euler's method with a step size $h = 0.1$, and X_b that obtained using a step size of $h = 0.05$. Compute the values of $X_a(t)$ and $X_b(t)$ for $t = 0.1, 0.2, \ldots, 1.0$. Compare these values with the values of $x(t)$, the exact solution of the problem. Compute the ratio of the errors in X_a and X_b.

Solution The exact solution, which may be obtained by separation, is

$$x = \frac{1}{1 - \ln(t + 1)}$$

The numerical solutions X_a and X_b and their errors are shown in Figure 10.11. Of course, in this figure the values of X_a are recorded at every step whereas those of X_b are only recorded at alternate steps.

Again, the final column of Figure 10.11 shows that our expectations about the effects of halving the step size when using Euler's method to solve a differential equation are confirmed. The ratio of the errors is not, of course, exactly one-half, because there are some higher-order terms in the errors, which we have ignored.

Figure 10.11
Computational results for Example 10.22.

t	X_a	X_b	$x(t)$	$\lvert x - X_a \rvert$	$\lvert x - X_b \rvert$	$\dfrac{\lvert x - X_b \rvert}{\lvert x - X_a \rvert}$
0.000 00	1.000 00	1.000 00	1.000 00			
0.100 00	1.100 00	1.102 50	1.105 35	0.005 35	0.002 85	0.53
0.200 00	1.210 00	1.216 03	1.222 97	0.012 97	0.006 95	0.54
0.300 00	1.332 01	1.342 94	1.355 68	0.023 67	0.012 75	0.54
0.400 00	1.468 49	1.486 17	1.507 10	0.038 61	0.020 92	0.54
0.500 00	1.622 52	1.649 52	1.681 99	0.059 47	0.032 47	0.55
0.600 00	1.798 03	1.837 91	1.886 81	0.088 78	0.048 90	0.55
0.700 00	2.000 08	2.057 92	2.130 51	0.130 42	0.072 59	0.56
0.800 00	2.235 40	2.318 57	2.425 93	0.190 53	0.107 36	0.56
0.900 00	2.513 01	2.632 51	2.792 16	0.279 15	0.159 65	0.57
1.000 00	2.845 39	3.018 05	3.258 89	0.413 50	0.240 84	0.58

Both MAPLE and MATLAB can be used to obtain numerical solutions of differential equations. Both encapsulate highly sophisticated numerical methods, which enable the production of very accurate numerical solutions. The Euler method described above is the simplest numerical method available and it might be considered somewhat perverse to use MAPLE or MATLAB to obtain an Euler method solution when much more accurate methods are available within the packages. Nonetheless, the numerical results in column X_a of Figure 10.11 could be obtained in MAPLE as follows.

```
odeprob:= {diff(x(t),t) = x(t)^2/(t + 1),x(0) = 1};
oseq:= array([seq(0.1*i,i = 0..10)]):
oput:= dsolve(odeprob,numeric,
        method = classical[foreuler],
        output = oseq,stepsize = 0.1);
evalm(oput[2,1]);
```

The results in column X_b of Figure 10.11 could be obtained by changing the stepsize argument in the dsolve routine to stepsize=0.05. A more extensive programming effort would be required to obtain the same numerical results through MATLAB.

10.6.3 Using numerical methods to solve engineering problems

In Example 10.22 the errors in the values of X_a and X_b are quite large (up to about 14% in the worst case). While carrying out computations with large errors such as these is quite useful for illustrating the mathematical properties of computational methods, in engineering computations we usually need to keep errors very much smaller. Exactly how small they must be is largely a matter of engineering judgement. The engineer must decide how accurately a result is needed for a given engineering purpose. It is then up to that engineer to use the mathematical techniques and knowledge available to carry out the computations to the desired accuracy. The engineering decision about the required accuracy will usually be based on the use that is to be made of the result. If, for instance, a preliminary design study is being carried out then a relatively approximate answer will often suffice, whereas for final design work much more accurate answers will normally be required. It must be appreciated that demanding greater accuracy than is actually needed for the engineering purpose in hand will usually carry a penalty in time, effort or cost.

Let us imagine that, for the problem posed in Example 10.22, we had decided we needed the value of $x(1)$ accurate to 1%. In the cases in which we should normally resort to numerical solution we should not have the analytical solution available, so we must ignore that solution. We shall suppose then that we had obtained the values of $X_a(1)$ and $X_b(1)$ and wanted to predict the step size we should need to use to obtain a better approximation to $x(1)$ accurate to 1%. Knowing that the error in $X_b(1)$ should be approximately one-half the error in $X_a(1)$ suggests that the error in $X_b(1)$ will be roughly the same as the difference between the errors in $X_a(1)$ and $X_b(1)$, which is the same as the difference between $X_a(1)$ and $X_b(1)$; that is, 0.172 66. One percent of $X_b(1)$ is roughly 0.03, that is, roughly one-sixth of the error in $X_b(1)$. Hence we expect that a step size roughly one-sixth of that used to obtain X_b will suffice; that is, a step size $h = 0.008\,33$. In practice, of course, we shall round to a more convenient non-recurring decimal quantity such as $h = 0.008$. This procedure is closely related to the Aitken extrapolation procedure introduced in Section 7.5.3 for estimating limits of convergent sequences and series.

Example 10.23   Compute an approximation $X(1)$ to the value of $x(1)$ satisfying the initial-value problem

$$\frac{dx}{dt} = \frac{x^2}{t+1}, \quad x(0) = 1$$

by using Euler's method with a step size $h = 0.008$.

Solution   It is worth commenting here that the calculations performed in Example 10.22 could reasonably be carried out on any hand-held calculator, but this new calculation requires 125 steps. To do this is on the boundaries of what might reasonably be done on a hand-held calculator, and is more suited to a computer. Repeating the calculation with a step size $h = 0.008$ produces the result $X(1) = 3.213\,91$.

We had estimated from the evidence available (that is, values of $X(1)$ obtained using step sizes $h = 0.1$ and 0.05) that the step size $h = 0.008$ should provide a value of $X(1)$ accurate to approximately 1%. Comparison of the value we have just computed with the exact solution shows that it is actually in error by approximately 1.4%. This does not quite meet the target of 1% that we set ourselves. This example therefore serves, first, to illustrate how, given two approximations to $x(1)$ derived using Euler's method with different step sizes, we can estimate the step size needed to compute an approximation within a desired accuracy, and, secondly, to emphasize that the estimate of the appropriate step size is only an *estimate*, and will not *guarantee* an approximate solution to the problem meeting the desired accuracy criterion. If we had been more conservative and rounded the estimated step size down to, say, 0.005, we should have obtained $X(1) = 3.23043$, which is in error by only 0.9% and would have met the required accuracy criterion.

Since we have mentioned in Example 10.23 the use of computers to undertake the repetitious calculations involved in the numerical solution of differential equations, it is also worth commenting briefly on the writing of computer programs to implement those numerical solution methods. While it is perfectly possible to write informal, unstructured programs to implement algorithms such as Euler's method, a little attention to planning and structuring a program well will usually be amply rewarded – particularly in terms of the reduced probability of introducing 'bugs'. Another reason for careful structuring is that, in this way, parts of programs can often be written in fairly general terms and can be re-used later for other problems. The two pseudocode algorithms in Figures 10.12 and 10.13 will both produce the table of results in Example 10.22. The pseudocode program of Figure 10.12 is very specific to the problem posed, whereas that of Figure 10.13 is more general, better structured, and more expressive of the structure of mathematical problems. It is generally better to aim at the style of Figure 10.13.

Both the MAPLE and MATLAB packages include a procedural programming language with all the basic structures of such languages. The pseudocode algorithms in Figures 10.12 and 10.13 can be implemented as programs in both MAPLE and MATLAB. But, again, it would be perverse to do so when very much more sophisticated numerical algorithms are packaged within the standard procedures of both languages. Nevertheless, either package could be used as a programming environment for implementing simple programs to complete Questions 43–45 in Exercises 10.6.4.

Figure 10.12
A poorly structured algorithm for Example 10.22.

```
x1←1
x2←1
write(printer,0,1,1,1)
for i is 1 to 10 do
    x1←x1 + 0.1*x1*x1/((i − 1)*0.1 + 1)
    x2←x2 + 0.05*x2*x2/((i − 1)*0.1 + 1)
    x2←x2 + 0.05*x2*x2/((i − 1)*0.1 + 1.05)
    x←1/(1 − ln(i*0.1 + 1))
    write(printer,0.1*i,x1,x2,x,x − x1,x − x2,(x − x2)/(x − x1))
endfor
```

Figure 10.13
A better structured algorithm for Example 10.22.

```
initial_time←0
final_time←1
initial_x←1
step←0.1
t←initial_time
x1←initial_x
x2←initial_x
h1←step
h2←step/2
write(printer,initial_time,x1,x2,initial_x)
repeat
    euler(t,x1,h1,1→x1)
    euler(t,x2,h2,2→x2)
    t←t + h
    x←exact_solution(t,initial_time,initial_x)
    write(printer,t,x1,x2,x,abs(x − x1),abs(x − x2),abs((x − x2)/(x− x1)))
until t ⩾ final_time

procedure euler(t_old,x_old,step,number→x_new)
    temp_x←x_old
    for i is 0 to number − 1 do
        temp_x←temp_x + step*derivative(t_old + step*i,temp_x)
    endfor
    x_new←temp_x
endprocedure

procedure derivative(t,x → derivative)
    derivative←x*x/(t+1)
endprocedure

procedure exact_solution(t,t0,x0→exact_solution)
    c←ln(t0 + 1) + 1/x0
    exact_solution←1/(c − ln(t + 1))
endprocedure
```

10.6.4 Exercises

39 Find the value of $X(0.3)$ for the initial-value problem

$$\frac{dx}{dt} = x - 2t, \quad x(0) = 1$$

using Euler's method with steps of $h = 0.1$.

40 Find the value of $X(0.25)$ for the initial-value problem

$$\frac{dx}{dt} = xt, \quad x(0) = 2$$

using Euler's method with steps of $h = 0.05$.

41 Find the value of $X(1)$ for the initial-value problem

$$\frac{dx}{dt} = \frac{x}{2\sqrt{(t + x)}}, \quad x(0.5) = 1$$

using Euler's method with step size $h = 0.1$.

42   Find the value of $X(0.5)$ for the initial-value problem

$$\frac{dx}{dt} = \frac{4 - t}{t + x}, \quad x(0) = 1$$

using Euler's method with step size $h = 0.05$.

43 Denote the Euler-method solution of the initial-value problem

$$\frac{dx}{dt} = \frac{xt}{t^2 + 2}, \quad x(1) = 2$$

using step size $h = 0.1$ by $X_a(t)$, and that using $h = 0.05$ by $X_b(t)$. Find the values of $X_a(2)$ and $X_b(2)$. Estimate the error in the value of $X_b(2)$, and suggest a value of step size that would provide a value of $X(2)$ accurate to 0.1%. Find the value of $X(2)$ using this step size. Find the exact solution of the initial-value problem, and determine the actual magnitude of the errors in $X_a(2)$, $X_b(2)$ and your final value of $X(2)$.

44 Denote the Euler-method solution of the initial-value problem

$$\frac{dx}{dt} = \frac{1}{xt}, \quad x(1) = 1$$

using step size $h = 0.1$ by $X_a(t)$, and that using $h = 0.05$ by $X_b(t)$. Find the values of $X_a(2)$ and $X_b(2)$. Estimate the error in the value of $X_b(2)$, and suggest a value of step size that would provide a value of $X(2)$ accurate to 0.2%. Find the value of $X(2)$ using this step size. Find the exact solution of the initial-value problem, and determine the actual magnitude of the errors in $X_a(2)$, $X_b(2)$ and your final value of $X(2)$.

45 Denote the Euler-method solution of the initial-value problem

$$\frac{dx}{dt} = \frac{1}{\ln x}, \quad x(1) = 1.2$$

using step size $h = 0.05$ by $X_a(t)$, and that using $h = 0.025$ by $X_b(t)$. Find the values of $X_a(1.5)$ and $X_b(1.5)$. Estimate the error in the value of $X_b(1.5)$, and suggest a value of step size that would provide a value of $X(1.5)$ accurate to 0.25%. Find the value of $X(1.5)$ using this step size. Find the exact solution of the initial-value problem, and determine the actual magnitude of the errors in $X_a(1.5)$, $X_b(1.5)$ and your final value of $X(1.5)$.

10.7 Engineering application: analysis of damper performance

In this section we shall carry out a modest engineering design exercise that will illustrate the modelling of an engineering problem using first-order differential equations and the solution of that problem using the techniques we have met so far in this chapter.

A small engineering company produces, among other artefacts, hydraulic dampers for specialized applications. One of the test rigs used by the company to check the quality and consistency of the operational characteristics of its output is illustrated in Figure 10.14. A carriage carrying a mass, which can be altered to suit the damper under test, is projected along a track of very low friction at a carefully controlled speed. At the end of the track the carriage impacts into a buffer which is connected to the damper under test. Immediately prior to impact the carriage passes through a pair of photocells whose output is used to measure the carriage speed accurately. The mass of

Figure 10.14
The damper test apparatus.

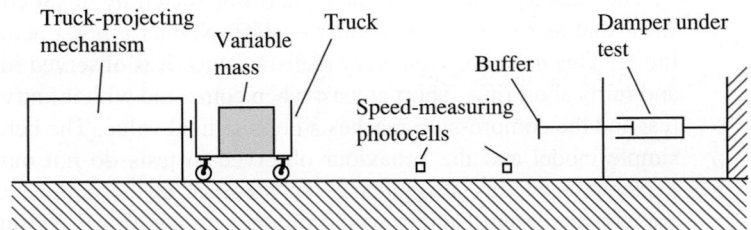

the buffer is very small compared with the mass of the carriage and test weight. The time/displacement history of the damper as it is compressed by the impact of the carriage is recorded digitally. The apparatus can produce time/displacement graphs and time/compression speed graphs for dampers on test.

In order to interpret the test results, the company needs to know how a damper should, in theory, behave under such a test. The simplest classical model of a damper assumes that the resistance of the damper is proportional to the velocity of compression. Since the mass of the buffer and damper components is small compared with the mass of the test apparatus carriage, it is reasonable to assume that, on impact, the moving components of the buffer and damper accelerate instantaneously to the velocity of the carriage, with negligible loss of speed on the part of the carriage. Since the track is of very low friction, it will be assumed that the only force decelerating the carriage is that provided by the damper (this also means assuming the carriage is not moving sufficiently fast for air resistance to have a significant effect). With these assumptions, the equation of motion of the carriage is

$$m\frac{\mathrm{d}v}{\mathrm{d}t} = -kv, \quad v(0) = U \tag{10.27}$$

where m is the mass of the carriage, $v(t)$ is its speed and k is the damper constant. Time is measured from the moment of impact and U is the impact speed of the carriage. The damper constant describes the force produced by the damper per unit speed of compression (and, for double-acting dampers, extension). The design engineer can adjust this constant by altering the internal design and dimensions of the damper. Equation (10.27) can be solved on sight, or by separation. The solution is

$$v = Ce^{-\lambda t}, \quad \text{with } \lambda = k/m$$

which, upon substituting in the initial conditions, becomes

$$v = Ue^{-\lambda t} \tag{10.28}$$

Writing $v = \mathrm{d}x/\mathrm{d}t$, where x is the compression of the damper and is taken as zero initially, this equation can be expressed as

$$\frac{\mathrm{d}x}{\mathrm{d}t} = Ue^{-\lambda t}, \quad x(0) = 0$$

This can be integrated directly, giving the solution, after substitution of the initial condition,

$$x = \frac{U}{\lambda}(1 - e^{-\lambda t}) \tag{10.29}$$

The velocity and displacement curves predicted by this model, (10.28) and (10.29), show that as $t \to \infty$, $v \to 0$ and $x \to U/\lambda$. Neither v nor x actually ever achieve these limits! This does not seem very realistic, since it is observed in tests that, after a finite and fairly short time (short at least when compared with infinity), the carriage comes to rest and the compression reaches a definite final value. The behaviour predicted by the simple model and the behaviour observed in tests do not quite agree. One possible explanation of this mismatch is the presence in the damper of friction between the components. Such friction would produce an additional resistance in the damper that does not vary with the speed of compression. The force resisting compression might therefore be better modelled as $kv + b$, where b is some constant force, rather than just kv. The compression of such a damper would be described by the equation

$$m\frac{dv}{dt} = -kv - b, \quad v(0) = U \tag{10.30}$$

Equation (10.30) is a linear first-order equation whose solution is

$$v = Ce^{-\lambda t} - \frac{b}{\lambda m}$$

or, substituting in the initial conditions,

$$v = Ue^{-\lambda t} - \frac{b}{\lambda m}(1 - e^{-\lambda t}) \tag{10.31}$$

This can be integrated again to provide displacement as a function of time:

$$x = \frac{1}{\lambda}\left(U + \frac{b}{\lambda m}\right)(1 - e^{-\lambda t}) - \frac{bt}{\lambda m} \tag{10.32}$$

Equation (10.31) predicts that the compression velocity of the damper will be zero when

$$t = \frac{1}{\lambda}\ln\left(\frac{b + \lambda Um}{b}\right) \tag{10.33}$$

at which time the compression of the damper will be

$$x = \frac{U}{\lambda} - \frac{b}{\lambda^2 m}\ln\left(\frac{b + \lambda Um}{b}\right) \tag{10.34}$$

This model therefore seems more realistic.

Figures 10.15 and 10.16 show the velocity and displacement curves represented by (10.28) and (10.29) and (10.31) and (10.32) for a test in which the carriage carries a mass of $2\,\text{kg}$ and travels at $1.5\,\text{m s}^{-1}$ at impact, the damper has a damping constant $25\,\text{N s m}^{-1}$ and the constant frictional force in the damper amounts to $1.5\,\text{N}$.

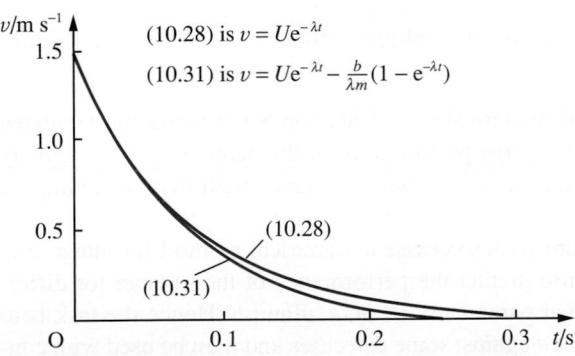

Figure 10.15 The predicted velocity–time curves for the damper test, both with and without the constant friction term.

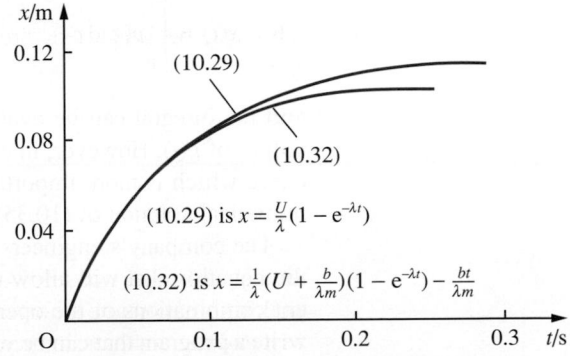

Figure 10.16 The predicted displacement–time curves for the damper test, both with and without the constant friction term.

The company perceives that one of the disadvantages of the classical hydraulic damper is, as may be inferred from Figure 10.15, that the largest force, and hence the largest deceleration of the damped object, is produced early in the history of the impact, when the velocity is largest. This means that the object, whatever it is, must be able to withstand this high deceleration. If a damper could be designed that produced a more even force over the deceleration process, the maximum deceleration experienced by an object being stopped from a given speed in a given distance would be reduced. The company's designers think they may have a solution to this problem – they have devised a new pattern of damper with a patent internal mechanism such that the damping constant increases as the damper operates. The effect of this mechanism is that, during any given operating cycle, the damping constant may be expressed as $k(1 + at)$, where t is the time elapsed in the operating cycle. The internal mechanism is such that in a short time after an operating cycle the effective damper constant returns to its initial state and the damper is ready for another operating cycle.

A model of this new design of damper is provided by the equation

$$m\frac{\mathrm{d}v}{\mathrm{d}t} = -k(1 + at)v - b, \quad v(0) = U \tag{10.35}$$

This is a linear first-order differential equation. Applying the appropriate solution method gives the solution as

$$v = -\frac{b}{m}\mathrm{e}^{-\lambda g(t)} \int \mathrm{e}^{\lambda g(t)}\,\mathrm{d}t, \quad \text{where } g(t) = t + \tfrac{1}{2}at^2$$

The integral in this solution does not result in a simple expression for $v(t)$, although it can be expressed in terms of a standard tabulated function called the error function. However, (10.35) can be solved numerically to produce $v(t)$ in tabulated form. Although we will not obtain $x(t)$ immediately using this method, we could readily derive $x(t)$ from the tabulated values of $v(t)$. Since

$$v = \frac{\mathrm{d}x}{\mathrm{d}t}$$

we can integrate both sides of this equation to obtain

$$x(t) = \int_0^t v(\tau)\,\mathrm{d}\tau$$

and the integral can be evaluated numerically (see Section 8.10) using the tabulated values of $v(t)$. However, in evaluating the performance of the damper it is the velocity curve which is more important and we will content ourselves with demonstrating the numerical solution of (10.35).

The company's engineers would wish to devise a numerical method for integrating the equation that will allow them to predict the performance of the damper for different combinations of the operational parameters U, m, k, a and b. Hence the task is to write a program that can be validated against some test cases and then be used with considerable confidence in other circumstances. If the value of a is taken to be 0 then the program to solve (10.35) should produce the same results as the analytical solution (10.31) of (10.30). This provides an appropriate test for the adequacy of the method and

Figure 10.17
Computational results for the damper design problem.

t	V_a	V_b	$V_a - V_b$	(10.31)
0.000	1.50000	1.50000		1.50000
0.020	1.15476	1.15477	0.00001	1.15478
0.040	0.88592	0.88594	0.00001	0.88595
0.060	0.67658	0.67660	0.00002	0.67662
0.080	0.51357	0.51359	0.00002	0.51361
0.100	0.38663	0.38665	0.00002	0.38667
0.120	0.28779	0.28781	0.00002	0.28782
0.140	0.21082	0.21084	0.00001	0.21085
0.160	0.15089	0.15090	0.00001	0.15091
0.180	0.10421	0.10423	0.00001	0.10424
0.200	0.06787	0.06788	0.00001	0.06789
0.220	0.03957	0.03958	0.00001	0.03959
0.240	0.01754	0.01754	0.00001	0.01755
0.260	0.00038	0.00038	0.00001	0.00039
0.280	−0.01298	−0.01298	0.00001	−0.01297

step size chosen. A program written to integrate equation (10.35) by Euler's method produced the results in the table in Figure 10.17. Several test runs of the program were undertaken using different step sizes, and results using $h = 0.00001$ (V_a) and $h = 0.000005$ (V_b) together with analytical solution (10.31) are shown in the figure.

It can be seen that the results using a step size of $h = 0.000005$ are in agreement with the analytical solution to at least 4 decimal places, and the agreement between the two numerical solutions V_a and V_b is good to 4dp. This agreement suggests that the accuracy of the numerical solution is adequate.

It therefore seems that a step size of $h = 0.000005$ will produce results that are accurate to at least 4dp and probably more. Using this step size, the (v, t) traces shown in Figure 10.18 were produced. First, for comparison, the predicted result of a test on a standard damper described by (10.30) is shown. Secondly, the predicted result of a test with a new model of damper with a parameter $a = 4$ is shown. It can be seen that the modified damper stops the carriage in a shorter time than the original model. The

Figure 10.18
Comparison of velocity–time curves for the damper test.

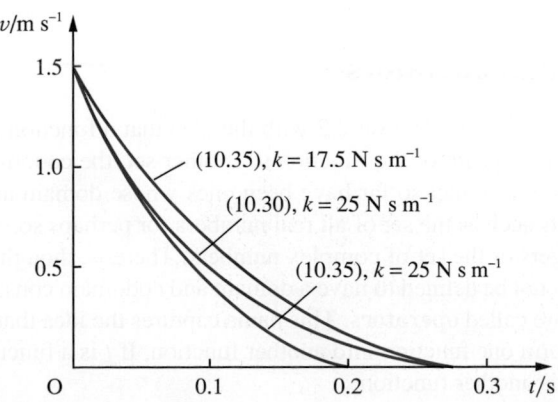

velocity–time trace is also slightly straighter, indicating that the design objective of making the deceleration more nearly uniform has been, at least in part, achieved. The third trace shown is for a new model damper with the basic damper constant k reduced to 17.5 and the parameter a kept at 4. This damper is able to halt the carriage in the same time as the original unmodified damper, but, in so doing, the maximum deceleration is somewhat smaller. This is the advantage of the new design that the company hope to exploit in the market.

In this section we have seen how differential equations and numerical solution methods can be used to provide an analytical tool that the company can now use as a routine design tool for predicting the performance of a new model damper with any given combination of parameters. Such a tool is an invaluable aid to the designer, whose task will usually be to specify appropriate parameters to meet an operational requirement specified by a client, for instance something like 'to be capable of halting a mass M travelling at velocity U within a time T while subjecting it to a deceleration of no more than D'.

It should also be commented here that we have completed the numerical work in this example using Euler's method. In practice it would be far better to use a more sophisticated method, which would yield a solution of equivalent accuracy while using a much larger time step and therefore much less computing effort. Although the difference for a single computation would be very small (and, therefore, considerably outweighed by the additional programming effort of implementing a more complex method), if we were undertaking a large number of comparative runs or creating a design tool which would be used by many engineers over a long period of time then such issues would be important.

10.8 Linear differential equations

Having dealt, in the last three sections, with first-order differential equations we will now turn our attention to differential equations of higher orders. To begin with we will restrict our attention to linear differential equations.

In Section 10.3.4 we defined the concept of linearity and mentioned that the solutions of linear equations have important simplifying properties. In this section we are going to study these simplifying properties in more detail. Before we do so, however, it is helpful to define some new notation.

10.8.1 Differential operators

We are familiar from Section 2.2 with the idea that a function is a mapping from a set known as the domain of the function to another set, the codomain of the function. The functions we have met so far have been ones whose domain and codomain have been familiar sets such as the set of all real numbers (or perhaps some subset of that set), the set of integers or the set of complex numbers. There is, though, no reason why a function should not be defined to have a domain and codomain consisting of functions. Such functions are called **operators**. This name captures the idea that operators are functions that transform one function into another function. If f is a function and ϕ is an operator then $\phi[f]$ is another function.

Example 10.24

Let the set A be the set of functions on the real numbers; that is, functions whose domain and codomain are both the real numbers. The operator ϕ has domain A and is defined by

$$\phi[f(t)] = f(t)^2$$

In other words, the effect of an operator ϕ on a function f is defined by specifying the function $\phi[f(t)]$. Thus for the ϕ defined here

$$\phi[3t^2 - 2t + 4] = 9t^4 - 12t^3 + 28t^2 - 16t + 16$$

and

$$\phi[\sin t - t] = \sin^2 t - 2t \sin t + t^2$$

Example 10.25

The operator ϕ is defined by

$$\phi[f(t)] = tf(t)^2 - 4f(t) + t^2$$

Then $tg(t)^4 - 4g(t)^2 + t^2$ may be expressed as $\phi[g(t)^2]$ and $te^{2t} - 4e^t + t^2$ may be expressed as $\phi[e^t]$.

Where no ambiguity is likely to result, it is permissible and conventionally acceptable to write $\phi[f(t)]$ as $\phi f(t)$; that is, to omit the square brackets.

We may view the operation of differentiation as transforming a differentiable function to another function, its derivative. When we are going to take this view, we often write the differentiation symbol separately from the function on which it will operate; for instance, we write

$$\frac{dx}{dt} \quad \text{as} \quad \frac{d}{dt}[x] \quad \text{or} \quad \frac{d^2x}{dt^2} \quad \text{as} \quad \frac{d^2}{dt^2}[x]$$

This notation is already familiar in those contexts in which we habitually write such expressions as

$$\frac{d}{dt}[f(t)g(t)] = \frac{df}{dt}g + f\frac{dg}{dt}$$

In such contexts we refer to the symbol d/dt as a differential operator.

Example 10.26

Let the operator ϕ be defined by

$$\phi[f(t)] = \frac{d}{dt}f(t)$$

Then we have

$$\phi[t^2] = 2t, \quad \phi[\sin t] = \cos t, \quad \phi[4t^3 - \tan t] = 12t^2 - \sec^2 t, \quad \text{and so on}$$

Using this notation, a differential equation may be expressed as an operator equation.

Example 10.27 Let the operator L be defined by

$$L[f(t)] = \frac{d^2 f}{dt^2} - (\sin t)\frac{df}{dt} + e^t f$$

The differential equation

$$\frac{d^2 f}{dt^2} - (\sin t)\frac{df}{dt} + e^t f = t^4$$

may, using the operator notation, be written as

$$L[f(t)] = t^4$$

In Section 10.3.5 we introduced the concept of homogeneous and nonhomogeneous linear differential equations and mentioned the convention whereby differential equations are usually written with the terms involving the dependent variable on the left-hand side and those not involving it on the right-hand side. When written in this way, a homogeneous equation can be characterized as an equation of the form

$$L[x(t)] = 0$$

and a nonhomogeneous one as an equation of the form

$$L[x(t)] = f(t)$$

where L is the differential operator of the equation.

10.8.2 Linear differential equations

Returning now to linear and nonlinear equations, we see that linear ones can be more precisely and compactly defined as those for which the operator satisfies

$$L[ax_1 + bx_2] = aL[x_1] + bL[x_2] \tag{10.36}$$

for all functions x_1 and x_2 and all constants a and b.

Example 10.28 The equation

$$\frac{d^2 x}{dt^2} + 4t\frac{dx}{dt} - (\sin t)x = \cos t$$

is a linear differential equation. Identify the operator of the equation and show that (10.36) holds for this operator.

Solution The operator is

$$L \equiv \frac{d^2}{dt^2} + 4t\frac{d}{dt} - \sin t$$

Hence we have

$$L[ax_1 + bx_2] = \frac{d^2}{dt^2}[ax_1 + bx_2] + 4t\frac{d}{dt}[ax_1 + bx_2] - (\sin t)(ax_1 + bx_2)$$

$$= a\frac{d^2x_1}{dt^2} + b\frac{d^2x_2}{dt^2} + 4t\left(a\frac{dx_1}{dt} + b\frac{dx_2}{dt}\right) - (a\sin t)x_1 - (b\sin t)x_2$$

$$= a\left[\frac{d^2x_1}{dt^2} + 4t\frac{dx_1}{dt} - (\sin t)x_1\right] + b\left[\frac{d^2x_2}{dt^2} + 4t\frac{dx_2}{dt} - (\sin t)x_2\right]$$

$$= aL[x_1] + bL[x_2]$$

Equation (10.36) is the strict mathematical definition of linearity for any type of operator, and the definition we gave earlier in Section 10.3.4 is considerably less satisfactory mathematically. The formal definition of a linear differential equation is therefore any differential equation whose differential operator is linear in the sense of (10.36).

We said before that linear differential equations are an important subcategory of differential equations because they have particularly useful simplifying properties. The most important simplifying property can be summed up in the following principle:

Linearity principle: if x_1 and x_2 are both solutions of the homogeneous linear differential equation $L[x] = 0$ then so is $ax_1 + bx_2$, where a and b are arbitrary constants.

This result follows directly from the definition of a linear operator. Since x_1 and x_2 are solutions of the differential equation, we have

$$L[x_1] = 0 \quad \text{and} \quad L[x_2] = 0$$

Since the equation is linear, we have

$$L[ax_1 + bx_2] = aL[x_1] + bL[x_2] = 0$$

Therefore $ax_1 + bx_2$ is a solution of the equation $L[x] = 0$.

Example 10.29

We noted in 10.4.2 that the general solution of the equation

$$\frac{d^2x}{dt^2} + \lambda^2x = 0$$

is

$$x = A \sin \lambda t + B \cos \lambda t$$

This solution can be interpreted in the light of the linearity principle. Let $x_1 = \sin \lambda t$ and $x_2 = \cos \lambda t$. Then x_1 and x_2 are solutions of the differential equation. The equation is linear, so we know that $Ax_1 + Bx_2$ is also a solution.

A formal proof of this result is not straightforward, and is not given here. We may, however, argue for its plausibility in the following way. Since the equation is linear, repeated application of the linearity principle shows that $A_1x_1 + A_2x_2 + \ldots + A_px_p$ is a solution of $L[x] = 0$. The expression $A_1x_1 + A_2x_2 + \ldots + A_px_p$ has p arbitrary constants and, since $x_1, x_2, \ldots, x_p$ are linearly independent, there is no way of rewriting the expression to reduce the number of arbitrary constants. Hence $A_1x_1 + A_2x_2 + \ldots + A_px_p$ has the characteristics of the general solution of the differential equation.

The relatively simple structure of the general solution of a homogeneous linear differential equation has now been exposed. The general solution of a nonhomogeneous equation is only slightly more complex. It is given by the following result:

> **General solution of a linear nonhomogeneous equation:** let
>
> $$L[x] = f(t)$$
>
> be a nonhomogeneous linear differential equation. If x^* is *any* solution of this equation and x_c is a solution of the equivalent homogeneous equation
>
> $$L[x] = 0$$
>
> then $x^* + x_c$ is also a solution of the nonhomogeneous equation.

This result is relatively straightforward to prove. By definition of x^* and x_c, we have

$$L[x^*] = f(t) \quad \text{and} \quad L[x_c] = 0$$

Since L is a linear operator, we have

$$L[x^* + x_c] = L[x^*] + L[x_c] = f(t) + 0 = f(t)$$

Hence $x^* + x_c$ is a solution of $L[x] = f(t)$.

It follows from this that finding the general solution of a nonhomogeneous linear differential equation can be reduced to the problem of finding any solution of the nonhomogeneous equation and adding to it the general solution of the equivalent homogeneous equation. The resulting expression is a solution of the nonhomogeneous equation containing the appropriate number of arbitrary constants, and so is the general solution. The first part of the solution (the 'any solution' of the nonhomogeneous equation, x^*) is known as a **particular integral** and the second part of the solution (the general solution of the equivalent homogeneous equation, x_c) is called the **complementary function**. The reader should note the similarity in structure with the general solution of linear recurrence relations developed in Section 7.4.

Example 10.32   Find the general solution of the differential equation

$$\frac{d^2x}{dt^2} + \lambda^2 x = 4t^3, \quad \lambda > 0$$

Solution   A particular integral of the equation is

$$x = \frac{4}{\lambda^2}t^3 - \frac{24}{\lambda^4}t$$

which you can check by direct substitution.

The complementary function is the general solution of the equation

$$\frac{d^2x}{dt^2} + \lambda^2 x = 0$$

that is,

$$x = A \sin \lambda t + B \cos \lambda t$$

Hence the general solution of

$$\frac{d^2x}{dt^2} + \lambda^2 x = 4t^3$$

is

$$x = \frac{4}{\lambda^2}t^3 - \frac{24}{\lambda^4}t + A \sin \lambda t + B \cos \lambda t$$

Example 10.33   Find the general solution of the boundary-value problem

$$\frac{d^2x}{dt^2} - k^2 x = \sin 2t, \quad k > 0, \quad x(0) = 0, \; x\left(\frac{\pi}{4}\right) = 0$$

Solution   A particular integral of the equation is

$$x = -\frac{\sin 2t}{4 + k^2}$$

which again can be checked by direct substitution. The complementary function is the general solution of the equation

$$\frac{d^2x}{dt^2} - k^2 x = 0$$

that is

$$x = Ae^{kt} + Be^{-kt}$$

Hence the general solution of

$$\frac{d^2x}{dt^2} - k^2 x = \sin 2t$$

is

$$x = -\frac{\sin 2t}{4 + k^2} + Ae^{kt} + Be^{-kt}$$

Now, imposing the boundary conditions gives two equations from which we obtain values for the two arbitrary constants in the general solution.

$$x(0) = 0 \quad \text{implies} \quad -\frac{\sin 0}{4 + k^2} + Ae^0 + Be^0 = 0$$

$$x\left(\frac{\pi}{4}\right) = 0 \quad \text{implies} \quad -\frac{\sin\left(\frac{1}{2}\pi\right)}{4 + k^2} + Ae^{k\pi/4} + Be^{-k\pi/4} = 0$$

Again MAPLE implements the method which we have just developed for non-homogeneous, linear, constant coefficient, differential equations of arbitrary order (and, of course, this method is therefore also available in the MATLAB Symbolic Math Toolbox). Any of the Examples 10.40–10.45 can be solved using either package. For instance, Example 10.45 would be solved by

```
ode:= diff(x(t),t$4) - 2*diff(x(t),t$3) +
5*diff(x(t),t$2) - 8*diff(x(t),t) + 4*x(t) = exp(t);
sol:= dsolve(ode);
```

or, using MATLAB, by

```
ode = 'D4x - 2*D3x + 5*D2x - 8*Dx + 4*x = exp(t)'
dsolve(ode,'t')
```

Solutions of any of the questions in Exercises 10.9.4 may readily be checked using either package.

10.9.4 Exercises

Check your answers using MATLAB or MAPLE whenever possible.

62 Find the general solution of the following differential equations:

(a) $\dfrac{d^2x}{dt^2} - 2\dfrac{dx}{dt} - 3x = t$

(b) $\dfrac{d^2x}{dt^2} - 2\dfrac{dx}{dt} - 5x = t^2 - 2t$

(c) $\dfrac{d^2x}{dt^2} - \dfrac{dx}{dt} - x = 5e^t$

(h) $3\dfrac{d^2x}{dt^2} + 3\dfrac{dx}{dt} - x = t^2 + e^{-2t}$

(i) $\dfrac{d^2x}{dt^2} + 2\dfrac{dx}{dt} - 3x = 5e^{-3t} + \sin 2t$

(j) $\dfrac{d^2x}{dt^2} + 16x = 1 + 2\sin 4t$

(k) $\dfrac{d^2x}{dt^2} - 4\dfrac{dx}{dt} = 7 - 3e^{4t}$

63 Find the general solutions of the following differential equations:

(a) $\dfrac{d^2x}{dt^2} - 3\dfrac{dx}{dt} + 4x = \cos 4t - 2\sin 4t$

(b) $9\dfrac{d^2x}{dt^2} - 12\dfrac{dx}{dt} + 4x = e^{-3t}$

(c) $2\dfrac{d^2x}{dt^2} + 4\dfrac{dx}{dt} - 7x = 7\cos 2t$

(d) $\dfrac{d^2x}{dt^2} + \dfrac{dx}{dt} + 4x = 5t - 7$

(e) $16\dfrac{d^2x}{dt^2} + 8\dfrac{dx}{dt} + x = t + 6$

(f) $\dfrac{d^2x}{dt^2} - 8\dfrac{dx}{dt} + 16x = -3\sin 3t$

(g) $\dfrac{d^2x}{dt^2} - 4\dfrac{dx}{dt} + 7x = e^{-5t}$

64 Show that the characteristic equation of the differential equation

$$\frac{d^4x}{dt^4} - 3\frac{d^3x}{dt^3} - 5\frac{d^2x}{dt^2} + 9\frac{dx}{dt} - 2x = 0$$

is

$$(m^2 + m - 2)(m^2 - 4m + 1) = 0$$

and hence find the general solutions of the equations

(a) $\dfrac{d^4x}{dt^4} - 3\dfrac{d^3x}{dt^3} - 5\dfrac{d^2x}{dt^2} + 9\dfrac{dx}{dt} - 2x = \cos 2t$

(b) $\dfrac{d^4x}{dt^4} - 3\dfrac{d^3x}{dt^3} - 5\dfrac{d^2x}{dt^2} + 9\dfrac{dx}{dt} - 2x = e^{2t} + e^{-2t}$

(c) $\dfrac{d^4x}{dt^4} - 3\dfrac{d^3x}{dt^3} - 5\dfrac{d^2x}{dt^2} + 9\dfrac{dx}{dt} - 2x = t^2 - 1 + e^{-t}$

65 Show that the characteristic equation of the differential equation

$$\frac{d^3x}{dt^3} - 9\frac{d^2x}{dt^2} + 27\frac{dx}{dt} - 27x = 0$$

is

$$(m - 3)^3 = 0$$

and hence find the general solutions of the equations

(a) $\dfrac{d^3x}{dt^3} - 9\dfrac{d^2x}{dt^2} + 27\dfrac{dx}{dt} - 27x = \cos t - \sin t + t$

(b) $\dfrac{d^3x}{dt^3} - 9\dfrac{d^2x}{dt^2} + 27\dfrac{dx}{dt} - 27x = e^t$

(c) $\dfrac{d^3x}{dt^3} - 9\dfrac{d^2x}{dt^2} + 27\dfrac{dx}{dt} - 27x = e^{3t} + t$

10.10 Engineering application: second-order linear constant-coefficient differential equations

In this section we are going to show how simple mathematical models of a variety of engineering systems give rise to second-order linear constant-coefficient differential equations. We shall also investigate the major features of the solutions of such models.

10.10.1 Free oscillations of elastic systems

If a wooden plank or a metal beam is attached firmly to a rigid foundation at one end with its other end projecting and unsupported as shown in Figure 10.20 then the imposition of a force on the free end, or equivalently the placing of a heavy object on it, will cause the plank or beam to bend under the load. The greater the force or load, the greater will be the deflection. If the load is moderate then the plank or beam will spring back to its original position when the load is removed. If the load is great enough, the plank will eventually break. The metal beam, on the other hand, may either deform permanently (so that it does not return to its original position when the load is removed) or fracture, depending on the type of metal. Experiments on planks or beams such as described here have revealed that for beams made of a wide variety of materials there is commonly a

Figure 10.20
The deflection of a cantilever by a load.

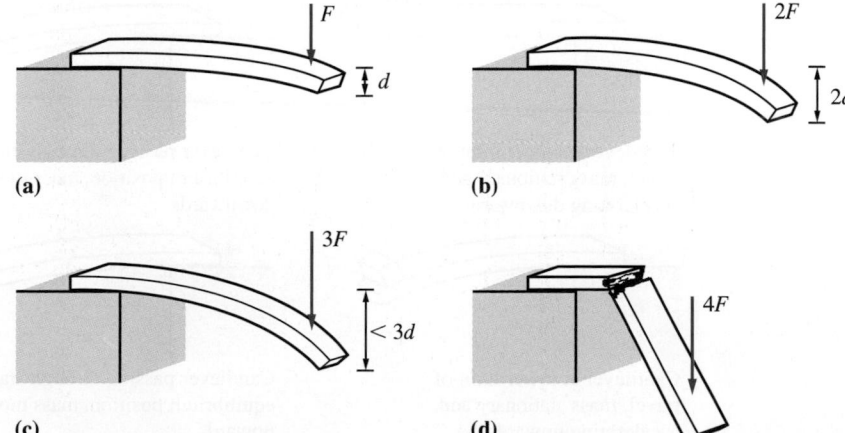

range of loads for which the deflection of the beam is roughly proportional to the load applied (Figures 10.20 (a), (b)). When the load becomes large enough, however, there is usually a region in which the deflection increases either less rapidly or more rapidly than the load (Figure 10.20 (c)), and finally a load beyond which the beam either breaks or is permanently deformed (Figure 10.20 (d)).

A beam that is fixed rigidly at one end and designed to support a load of some sort on the other end is called a cantilever. There are many common everyday and engin-eering applications of cantilevers. One with which most readers will be familiar is a diving springboard. Engineering applications include such things as warehouse hoists, the wings of aircraft and some types of bridges. For most of these applications the cantilever is designed to operate with small deflections; that is, the size and material of construction of the cantilever will be chosen by the designer so that, under the greatest anticipated load, the deflection of the cantilever will be small. Within this regime, the deflection of the tip of the cantilever will be proportional to the load applied. In the notation of Figure 10.20, we can write

$$d = \frac{1}{k}F \tag{10.40}$$

where d is the deflection of the cantilever, F is the load applied and k is a constant. Equation (10.40) essentially expresses a mathematical model of the cantilever, albeit a very simple one. The model is valid for applied loads such that the deflection of the cantilever remains within the linear range (where the deflection is proportional to load), and would not be valid for larger loads leading to nonlinear deflections, permanent distortions and breakages.

Equation (10.40) can also be used to investigate the dynamic behaviour of can-tilevers. So far, we have assumed that the cantilever is in equilibrium under the applied load. Such situations, in which the cantilever is not moving, are called **static**. The term **dynamic** is conventionally used to describe situations and analyses in which the deflection of the cantilever is not constant in time. When the deflection of a cantilever is either greater than or less than the static deflection under the same load, the cantilever exerts a net force accelerating the mass back towards its equilibrium position. As a result, the deflection of the cantilever oscillates about the static equilibrium position. The situation is illustrated in Figure 10.21.

Figure 10.21
The dynamic behaviour of a loaded cantilever.

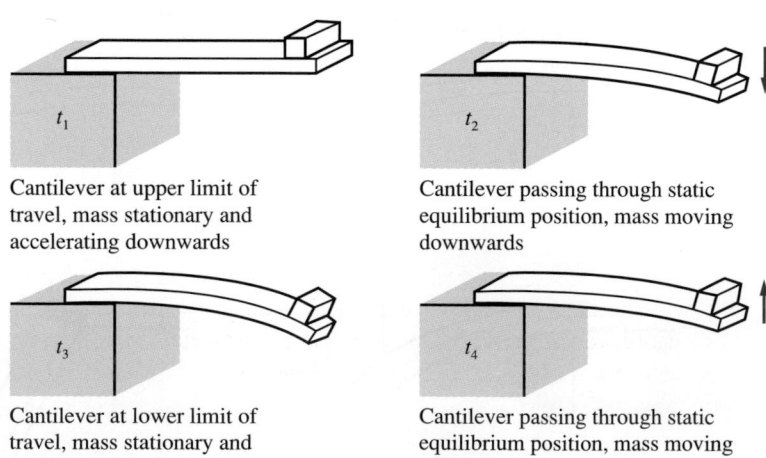

Cantilever at upper limit of travel, mass stationary and accelerating downwards

Cantilever passing through static equilibrium position, mass moving downwards

Cantilever at lower limit of travel, mass stationary and accelerating upwards

Cantilever passing through static equilibrium position, mass moving upwards

Such oscillations can be analysed fairly readily. If the mass supported on the end of the cantilever is large compared with the mass of the cantilever itself, the effect of the cantilever is merely to apply a force to the mass. The vertical equation of motion of the mass is then

$$m\frac{d^2x}{dt^2} = mg - F$$

where x is the instantaneous deflection of the tip of the cantilever below the horizontal, m is the mass and F is the upward force exerted on the mass by the cantilever due to its bending. But the restoring force, provided the deflection of the cantilever remains small enough at all times during the motion, is given by (10.40). Thus the motion is governed by the equation

$$m\frac{d^2x}{dt^2} = mg - kx$$

This equation, rearranged in the form

$$\frac{d^2x}{dt^2} + \frac{kx}{m} = g \tag{10.41}$$

is recognizable as a second-order linear nonhomogeneous constant-coefficient equation. In the static case, when the load is not moving, the solution of this equation is $x = mg/k$. This is, of course, also a particular integral for (10.41). The complementary function for (10.41) is

$$x = A\cos \omega t + B\sin \omega t, \quad \text{where } \omega^2 = k/m$$

The complete solution of (10.41) is therefore

$$x = \frac{mg}{k} + A\cos \omega t + B\sin \omega t \tag{10.42}$$

The constants A and B could of course be determined if suitable initial conditions were provided. What is at least as important − if not more so for the engineer − is to understand the physical meaning of the solution (10.42). This is more easily done if (10.42) is slightly rearranged. Taking $C = (A^2 + B^2)^{1/2}$ and $\tan \delta = B/A$, so that

$$A = C\cos \delta \quad \text{and} \quad B = C\sin \delta$$

(10.42) becomes

$$x = \frac{mg}{k} + C\cos(\omega t - \delta) \tag{10.43}$$

In physical terms this equation implies that the deflection of the cantilever takes the form of periodic oscillations of angular frequency ω and constant amplitude C about the position of static equilibrium of the cantilever (the position at which $kx = mg$).

The interested reader can check the accuracy of this description by constructing a cantilever from a flexible wooden or plastic ruler (the flexible plastic type is the most effective). The ruler should be held firmly by one end so that it projects over the edge of a desk or table, and the free end loaded with a sufficient mass of plasticine or other suitable material. The static equilibrium position is easily found. If the end of the ruler is displaced from this position and released, the plasticine-loaded end will be found to

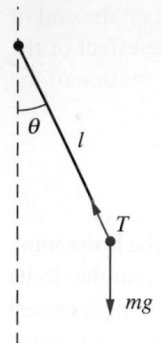

Figure 10.22
A pendulum.

vibrate up and down around the equilibrium position. If the mass of plasticine is increased, the frequency of the vibration will be found to decrease as predicted by the relation $\omega^2 = k/m$.

A cantilever is not the only engineering system that gives rise to linear constant-coefficient equations. The pendulum shown in Figure 10.22 can be analysed thus. It is of length l and carries a mass m at its free end. If the mass of the pendulum arm is very small compared with m then, resolving forces at right-angles to the pendulum arm, the equation of motion of the mass is

$$ml\frac{d^2\theta}{dt^2} = -mg\sin\theta$$

This is a second-order nonlinear differential equation, but if the displacement from the equilibrium position (in which the pendulum hangs stationary and vertically below the pivot) is small then $\sin\theta \approx \theta$ and the equation becomes

$$\frac{d^2\theta}{dt^2} + \frac{g}{l}\theta = 0 \tag{10.44}$$

The solution of this equation is

$$\theta = A\cos\omega t + B\sin\omega t, \quad \text{with } \omega^2 = g/l \tag{10.45}$$

In other words, the pendulum's displacement from its equilibrium position oscillates sinusoidally with a frequency that decreases as the pendulum increases in length but is independent of the mass of the pendulum bob.

The buoy (or floating oil drum or similar) shown in Figure 10.23 also gives rise to a second-order linear constant-coefficient equation. Suppose the immersed depth of the buoy is z. Its mass (which is concentrated near the bottom of the buoy in order that it should float upright and not tip over) is m. We know, by Archimedes' principle, that the water in which the buoy floats exerts an upthrust on the buoy equal to the weight of the water displaced by the latter. If the cross-sectional area of the buoy is A and the density of the water is ρ, the upthrust will be ρAzg. Hence the equation of motion is

$$m\frac{d^2z}{dt^2} = mg - \rho Azg$$

that is,

$$\frac{d^2z}{dt^2} + \frac{\rho Ag}{m}z = g \tag{10.46}$$

Equation (10.46) has particular integral $z = m/\rho A$ and complementary function

$$z = A\cos\omega t + B\sin\omega t, \quad \text{with } \omega^2 = \rho Ag/m$$

Figure 10.23
A floating buoy.

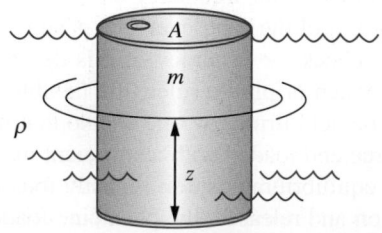

so the complete solution is

$$z = \frac{m}{\rho A} + A\cos \omega t + B\sin \omega t, \quad \text{with } \omega^2 = \frac{\rho A g}{m} \tag{10.47}$$

As in the case of the cantilever, the particular integral of the equation corresponds to the static equilibrium solution (when the buoy is floating just sufficiently immersed that the upthrust exerted by the water equals the weight of the buoy), and the complementary function describes oscillations of the buoy about this position. In this case the buoy oscillates with constant amplitude and a frequency that decreases as the mass of the buoy increases and increases as the density of the water and/or the cross-sectional area of the buoy increases.

10.10.2 Free oscillations of damped elastic systems

Equations (10.42), (10.45) and (10.47) all describe oscillations of constant amplitude. In reality, in all the situations described, a vibrating cantilever, an oscillating pendulum and a bobbing buoy, experience leads us to expect that the oscillations or vibrations are of decreasing amplitude, so that the motion eventually decays away and the system finally comes to rest in its static equilibrium position. This suggests that the mathematical models of the situation that we constructed in Section 10.10.1, and which are represented by (10.41), (10.44) and (10.46), are inadequate in some way.

What has been ignored in each case is the effect of dissipation of energy. Suppose, in the case of the pendulum, the motion of the pendulum were opposed by air resistance. The work which the pendulum does against the air resistance represents a continuous loss of energy, as a result of which the amplitude of oscillation of the pendulum decreases until it finally comes to rest. The situation is illustrated in Figure 10.24. The forces acting on the pendulum mass are gravity, air resistance (which opposes motion) and the tension in the pendulum arm. Resolving these forces perpendicular to the pendulum arm results in the equation of motion.

$$ml\frac{\mathrm{d}^2\theta}{\mathrm{d}t^2} = -R - mg\sin\theta$$

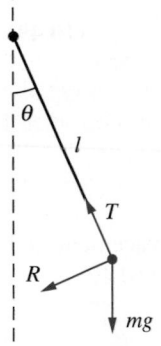

Figure 10.24
A pendulum with
air resistance.

If the air resistance is assumed to be proportional to the speed of the pendulum mass then, since the speed of the mass is $l(\mathrm{d}\theta/\mathrm{d}t)$, we have

$$R = kl\frac{\mathrm{d}\theta}{\mathrm{d}t}$$

Hence

$$ml\frac{\mathrm{d}^2\theta}{\mathrm{d}t^2} = -kl\frac{\mathrm{d}\theta}{\mathrm{d}t} - mg\sin\theta$$

or, assuming θ is small so that $\sin\theta \approx \theta$ and rearranging the terms,

$$\frac{\mathrm{d}^2\theta}{\mathrm{d}t^2} + \frac{k}{m}\frac{\mathrm{d}\theta}{\mathrm{d}t} + \frac{g}{l}\theta = 0 \tag{10.48}$$

This is a second-order linear constant-coefficient differential equation. It should be noted that the assumption that air resistance is proportional to speed is not the only possible assumption. For very slow-moving objects air resistance may well be more nearly constant, while for very fast-moving objects air resistance is usually taken to be proportional to the square of speed, which is a much better description of reality for fast-moving objects. For objects moving at modest speeds, however, the assumption that air resistance is proportional to speed is commonly adopted.

In the case of the cantilever and the buoy, also, we might assume that there is a resistance to motion that is proportional to the speed of motion. Again these are not the only possible assumptions, but they are ones that, under appropriate circumstances, are reasonable. The guiding principle when modelling physical systems such as these is to identify the physical source of the resistance and try to describe its behaviour. This is not a problem of mathematics but rather one of mathematical modelling, in which engineers must use their knowledge of physics and engineering as well as of mathematics in order to arrive at an appropriate mathematical description of reality.

Constructing models of a whole host of other engineering situations also leads to equations similar to (10.48). Basically, any situation in which the motion of some mass is caused by the sum of a force opposing displacement that is proportional to the displacement from some fixed position and a force that resists motion and is proportional to the speed of motion gives rise to an equation of the form

$$m\frac{d^2x}{dt^2} = -\mu\frac{dx}{dt} - \lambda x$$

that is,

$$\frac{d^2x}{dt^2} + p\frac{dx}{dt} + qx = 0 \tag{10.49}$$

where

$$p = \frac{\mu}{m} \quad \text{and} \quad q = \frac{\lambda}{m}$$

We must have $p > 0$ and $q > 0$, because the two forces oppose displacement and motion respectively. We know from Section 10.10.1 that the solution of (10.49) is

$$x(t) = Ae^{m_1 t} + Be^{m_2 t}$$

where m_1 and m_2 are the roots of the characteristic equation

$$m^2 + pm + q = 0$$

For reasons that will become apparent, it is convenient to put (10.49) into the standard form

$$\frac{d^2x}{dt^2} + 2\zeta\omega\frac{dx}{dt} + \omega^2 x = 0 \tag{10.50}$$

where, because $p, q > 0$, so are ζ and ω. The characteristic equation is then $m^2 + 2\zeta\omega m + \omega^2 = 0$, whose roots are

$$m = \begin{cases} -\zeta\omega \pm (\zeta^2 - 1)^{1/2}\omega & (\zeta > 1) \\ -\omega \text{ (twice)} & (\zeta = 1) \\ -\zeta\omega \pm j(1 - \zeta^2)^{1/2}\omega & (0 < \zeta < 1) \end{cases}$$

and the solution of (10.50) is therefore

$$x = A \exp\{-[\zeta - (\zeta^2 - 1)^{1/2}]\omega t\} + B \exp\{-[\zeta + (\zeta^2 - 1)^{1/2}]\omega t\} \quad (\zeta > 1) \qquad \textbf{(10.51a)}$$

$$x = e^{-\omega t}(At + B) \qquad (\zeta = 1) \qquad \textbf{(10.51b)}$$

$$x = e^{-\zeta \omega t}\{A \cos[(1 - \zeta^2)^{1/2}\omega t] + B \sin[(1 - \zeta^2)^{1/2}\omega t]\} \qquad (0 < \zeta < 1) \quad \textbf{(10.51c)}$$

The first point to note about these solutions is that, since $\zeta > 0$ and $\omega > 0$, we have $x \to 0$ as $t \to \infty$ in all cases. Figure 10.25 shows the typical form of the solution (10.51) for various values of ζ. Variation of ω will only change the scale along the horizontal axis. For $0 < \zeta < 1$ the solution takes an oscillatory form with decaying amplitude. For $\zeta > 1$ the solution has the form of an exponential decay. The larger ζ, the slower is the final decay, since the exponential coefficient $\zeta - (\zeta^2 - 1)^{1/2} \to 0$ as $\zeta \to \infty$. In Figure 10.25 the envelopes of the oscillatory solutions are shown as broken lines. If the envelope of the oscillatory decay is compared with the solutions for $\zeta > 1$ it is quickly apparent that the most rapid decay is when $\zeta = 1$. It is now apparent why we chose to take (10.50) as the standard form for the description of second-order damped systems. The parameter ω is the **natural frequency** of the system, that is, the frequency with which it would oscillate in the absence of damping, and the parameter ζ is the **damping parameter** of the system. When $\zeta = 1$, the decay of the motion of the system to its equilibrium state is as fast as is possible. For this reason, $\zeta = 1$ is referred to as **critical damping**. When $\zeta < 1$, the motion described by the equation decays to its equilibrium

Figure 10.25
The motion of damped second-order systems.

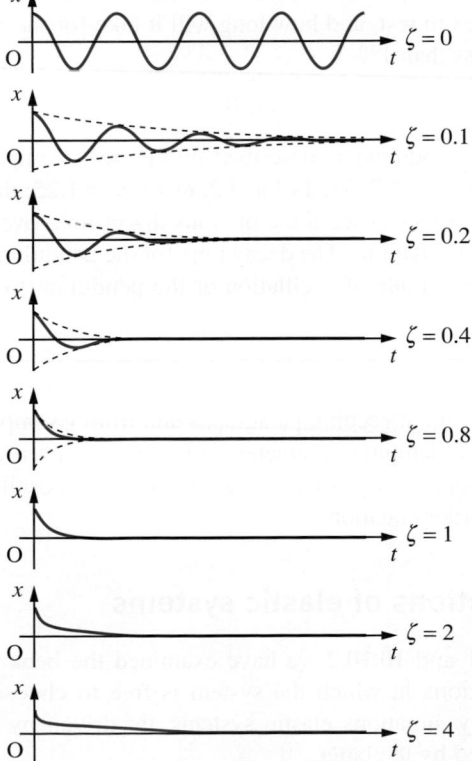

state in an oscillatory manner, passing through the equilibrium position on a number of occasions before coming to rest. For this reason, the motion is described as **under-damped**. When $\zeta > 1$, the motion described by the equation decays to the equilibrium position in a direct manner, but less rapidly than for a critically damped system. In this case the motion is described as **over-damped**.

For the under-damped case an engineering rule of thumb that is commonly used is that when $\zeta = 0.3$ the system shows three *discernible* overshoots before settling down. That is not to say that there are only three overshoots – on the contrary, there are an infinite number – but by the fourth and subsequent overshoots the amplitude of the oscillations has decayed to less than 2% of the initial amplitude. When $\zeta = 0.5$, there are two discernible overshoots (the third and subsequent ones have amplitude less than $\frac{1}{2}$% of initial); and when $\zeta = 0.7$, there is only one significant overshoot.

Another rule of thumb relates to the envelope containing the response. The response is contained within an envelope defined by the function $e^{-\zeta\omega t}$. Now $e^{-3} = 0.0498$ and $e^{-4.5} = 0.0111$; so when $\zeta\omega t = 3$, the amplitude of the response will have fallen to approximately 5% of its original amplitude; and when $\zeta\omega t = 4.5$, it will have fallen to roughly 1% of its original amplitude. For this reason, $t = 1/\zeta\omega$ is called the **decay time** of the system, and engineers use the rule of thumb that response falls to 5% in three decay times and 1% in four and a half decay times.

Example 10.46

A pendulum of mass 4 kg, length 2 m and an air resistance coefficient of 5 N s m^{-1} is released from an initial position in which it makes an angle of 20° with the vertical. Assuming that this angle is small enough for the small-angle approximation to be made in the equation of motion, how many oscillations will be obviously observable before the pendulum comes to rest, and how long will it take for the amplitude of the motion to have fallen to less than 1°?

Solution

The motion of the pendulum is described by (10.48). Comparing this with (10.50), we see that $\omega = (g/l)^{1/2} = 2.215$ rad s^{-1} and $2\zeta\omega = k/m = 1.25$; that is, $\zeta = 0.282$. Hence, since $\zeta \approx 0.3$, we expect to see three obvious discernible overshoots (one and a half complete cycles of oscillation). The decay time for the pendulum is $1/\zeta\omega = 2m/k = 1.6$ s, so we expect the amplitude of oscillation of the pendulum to fall to 5% of its initial amplitude in 4.8 s.

It is evident from the preceding paragraphs and from Example 10.46 that the natural frequency ω and the damping parameter ζ of a system are a very convenient way of summarizing the properties of any physical system whose oscillations are described by a damped second-order equation.

10.10.3 Forced oscillations of elastic systems

In Sections 10.10.1 and 10.10.2 we have examined the behaviour of elastic systems undergoing oscillations in which the system is free to choose its own frequency of oscillation. In many situations elastic systems are driven by some external force at a frequency imposed by the latter.

A familiar example of such a situation is the vibration of lamp posts in strong winds. The lightweight tubular metal lamp posts that have frequently been installed by highway authorities since the 1960s are a form of cantilever. The vertical post is rigidly mounted in the ground, and carries at its top a lamp apparatus. The lamp is effectively a concentrated mass, though it would probably not be sufficiently massive to allow the assumption, which we made for the cantilevers treated in Sections 10.10.1 and 10.10.2, that the mass of the post is small compared with the mass of the lamp. Nonetheless, if the top of the lamp post were to be pulled to one side and released, the post would certainly vibrate. The frequency of that vibration would be a function of the stiffness (restoring force per unit lateral tip displacement) of the lamp post and the mass of both the post and the lamp apparatus carried at the top. The stiffer the lamp post, the higher would be the frequency of vibration. The more massive the post and the lamp apparatus, the lower the frequency.

When the wind blows past the lamp post, aerodynamic effects (known as vortex shedding) result in an oscillating side-force on the lamp post. The frequency of this side-force is a function of the wind speed and the diameter of the lamp post. There is no reason why the frequency of oscillation of the wind-induced side-force should coincide with the frequency of the oscillations that result if the top of the lamp post is displaced sideways and released to vibrate freely. Under the influence of the oscillating side-force, such lamp posts commonly vibrate from side to side in time with the oscillating side-force. As the wind speed changes, so the frequency of the side-force and therefore of the lamp post's vibrations changes. Other types of lamp post, notably the reinforced concrete type and the older cast-iron lamp posts, do not seem to exhibit this behaviour. This can be explained in terms of their greater stiffness, as we shall see later.

Oscillations of elastic systems in which the system is free to adopt its own natural frequency of vibration are called **free vibrations**, while those caused by oscillating external forces (and in which the system must vibrate at the frequency of the external forcing) are called **forced vibrations**.

Other large structures can also be forced to oscillate by the wind blowing past them, just like lamp posts. Large modern factory chimneys made of steel or aluminium sections bolted together and stayed by wires exhibit this type of vibration, as do the suspension cables and hangers of suspension bridges and the overhead power transmission lines of electricity grid systems. The legs of offshore oil rigs can be forced to vibrate by ocean currents and waves. The wings of an aircraft (which, being mounted rigidly in the fuselage of the aircraft, are also a form of cantilever) may vibrate under aerodynamic loads, particularly from atmospheric turbulence. Large pieces of static industrial machinery are usually bolted down to the ground. If such fastening is subjected to a large load, it will usually give a little, so the attachment of the machinery to the floor must be considered as elastic. If the machinery, when in operation, produces an internal side-load (such as an out-of-balance rotor would produce) then the machinery is seen to rock from side to side on its mountings at the frequency of the internally generated side-loading. This effect can often be observed in the rocking vibrations of a car engine when it is idling in a stationary car. It is well known that bodies of men or women marching are ordered to break step when passing over bridges. If they did not, the regular footfalls of the whole group would create a periodic force on the bridge. The dangers of such regular forces will become apparent in our analysis. All these situations are similar in nature to the forced vibrations of the lamp post under the influence of the wind. In most of them the oscillations induced by the side-force are potentially

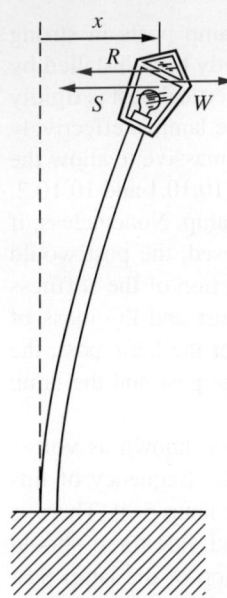

Figure 10.26
The forces acting on a
vibrating lamp post.

disastrous, and must be understood by the engineer so that engineering artefacts may be designed to avoid the destructive effects of forced vibrations.

A simple model of the vibrations of a lamp post can be constructed as shown in Figure 10.26. The lamp apparatus, of mass m, is displaced from its equilibrium position by a distance x. The structure of the cantilever results in a restoring force S and air resistance in a restoring force R. The wind load (which, remember, is not a force in the direction of the wind but rather an oscillatory side-force) is W. If the displacement x is small and the displacement velocity is not too great then we may reasonably assume

$$S = kx \quad \text{and} \quad R = \lambda \frac{dx}{dt}$$

Making the somewhat unrealistic assumption that the mass of the lamp post itself is small compared with the mass of the lamp apparatus, the equation of motion of the lamp is seen to be

$$m\frac{d^2x}{dt^2} = -\lambda\frac{dx}{dt} - kx + W$$

The wind-induced force W is oscillatory, so we shall assume that it is of the form

$$W = W_0 \cos \Omega t$$

Hence the equation of motion becomes

$$m\frac{d^2x}{dt^2} + \lambda\frac{dx}{dt} + kx = W_0 \cos \Omega t \tag{10.52}$$

which is a second-order linear nonhomogeneous constant-coefficient differential equation. In order to facilitate the interpretation of the result, we shall replace (10.52) with the equivalent equation

$$\frac{d^2x}{dt^2} + 2\zeta\omega\frac{dx}{dt} + \omega^2 x = F \cos \Omega t \tag{10.53}$$

The particular integral for (10.53) is obtained by assuming the form $A \cos \Omega t + B \sin \Omega t$, and is found to be

$$\frac{(\omega^2 - \Omega^2)F \cos \Omega t + 2\zeta\omega\Omega F \sin \Omega t}{(\omega^2 - \Omega^2)^2 + 4\zeta^2\omega^2\Omega^2} \tag{10.54a}$$

or equivalently

$$\frac{F}{[(\omega^2 - \Omega^2)^2 + 4\zeta^2\omega^2\Omega^2]^{1/2}} \cos(\Omega t - \delta) \tag{10.54b}$$

where

$$\delta = \tan^{-1}\left(\frac{2\zeta\omega\Omega}{\omega^2 - \Omega^2}\right)$$

The complementary function is of course the solution of the homogeneous equivalent of (10.52), which is just (10.50). The complementary function is therefore given by (10.51). The motion of a damped second-order system in response to forcing by a force

Figure 10.27
The response of damped second-order systems to sinusoidal forcing.

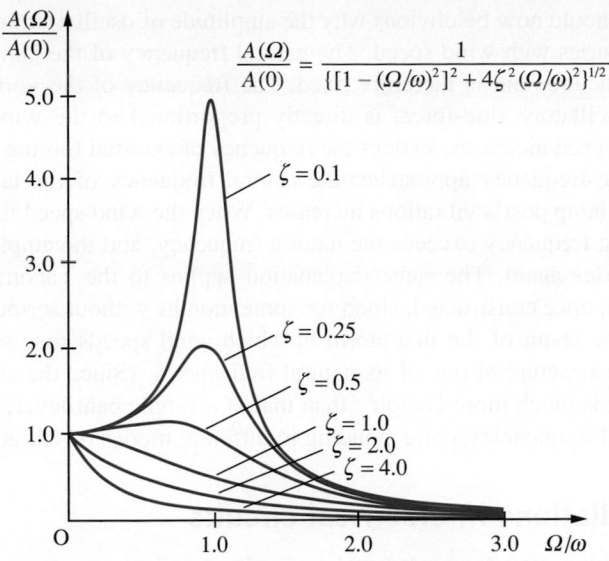

$$\frac{A(\Omega)}{A(0)} = \frac{1}{\{[1 - (\Omega/\omega)^2]^2 + 4\zeta^2(\Omega/\omega)^2\}^{1/2}}$$

$F \cos \Omega t$ is therefore the sum of (10.51) and (10.54a) or (10.54b). In Section 10.10.2 we saw that (10.51) is, for positive ζ, always a decaying function of time. The complementary function for (10.53) therefore represents a motion that decays to nothing with time, and is therefore called a **transient solution**. The particular integral, on the other hand, does not decay, but continues at a steady amplitude for as long as the forcing remains. The long-term response of a damped second-order system to forcing by a force $F \cos \Omega t$ is therefore to oscillate at the forcing frequency Ω with amplitude

$$A(\Omega) = \frac{1}{[(\omega^2 - \Omega^2)^2 + 4\zeta^2\omega^2\Omega^2]^{1/2}} \tag{10.55}$$

times the amplitude of the forcing term. This is called the **steady-state response** of the system. Evidently, the amplitude of the steady-state response changes as the frequency Ω of the forcing changes. In Figure 10.27 the form of the response amplitude $A(\Omega)$ as a function of Ω is shown for a range of values of ζ. Obviously, the characteristics of the response of a damped second-order system to forcing depend crucially on the damping. For lightly damped systems (ζ near to 0) the response has a definite maximum near to ω, the natural frequency of the system. For more heavily damped systems the peak response is smaller, and for large enough ζ the peak disappears altogether.

The significance of this is that systems subjected to an oscillatory external force at a frequency near to the natural frequency of the system will, unless they are sufficiently heavily damped, respond with large-amplitude motion. This phenomenon is known as **resonance**. Resonance can cause catastrophic failure of the structure of a system. The history of engineering endeavour contains many examples of structures that have failed because they have been subjected to some external exciting force with a frequency near to one of the natural frequencies of vibration of the structure. Perhaps the most famous example of such a failure is the collapse in 1941 of the suspension bridge at Tacoma Narrows in the USA. This failure, due to wind-induced oscillations, was recorded on film and provides a salutary lesson for all engineers. Similar forces have destroyed factory chimneys, power transmission lines and aircraft.

It should now be obvious why the amplitude of oscillation of the tubular metal lamp post varies with wind speed. The natural frequency of the lamp post is determined by its structure, and is therefore fixed. The frequency of the vortex shedding, and so of the oscillatory side-force, is directly proportional to the wind speed. Hence, as the wind speed increases, so does the frequency of external forcing of the lamp post. As the forcing frequency approaches the natural frequency of the lamp post, the amplitude of the lamp post's vibrations increases. When the wind speed increases sufficiently, the forcing frequency exceeds the natural frequency, and the amplitude of the oscillations decreases again. The same explanation applies to the Tacoma Narrows bridge. The bridge, once constructed, stood for some months without serious difficulty. The failure was the result of the first storm in which wind speeds rose sufficiently to excite the bridge structure at one of its natural frequencies. (Since the structure of a suspension bridge is much more complex than that of a simple cantilever, such a bridge has many natural frequencies, corresponding to different modes of vibration.)

10.10.4 Oscillations in electrical circuits

In Section 10.2.4 we analysed a simple electrical circuit composed of a resistor, a capacitor and an inductor. In that case we considered what happened when a switch was thrown in a circuit containing a d.c. voltage source. If an alternating voltage signal is applied to a similar circuit the equation governing the resulting oscillations also turns out to be a second-order linear differential equation.

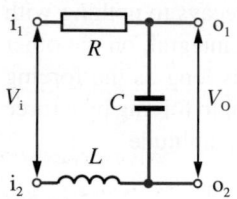

Figure 10.28
An LCR electrical circuit.

Consider the circuit shown in Figure 10.28. Suppose a voltage V_i is applied across the input terminals i_1 and i_2. The voltage drop across the inductor is $L(di/dt)$, that across the capacitor is $\int (i/C)dt$ and that across the resistor is Ri. Kirchhoff's laws (or the principle of conservation of charge) tell us that the current in each component must be the same. The voltage across the output terminals o_1 and o_2 is $\int (i/C)dt = V_o$. Hence we have

$$L\frac{di}{dt} + Ri + \frac{1}{C}\int i\,dt = V_i$$

with

$$V_o = \frac{1}{C}\int i\,dt$$

That is,

$$LC\frac{d^2 V_o}{dt^2} + RC\frac{dV_o}{dt} + V_o = V_i$$

or

$$\frac{d^2 V_o}{dt^2} + \frac{R}{L}\frac{dV_o}{dt} + \frac{V_o}{LC} = \frac{V_i}{LC} \qquad\qquad \textbf{(10.56)}$$

This is a second-order linear nonhomogeneous constant-coefficient differential equation. If the signal V_i is of the form $V\cos\Omega t$ then we essentially have forced oscillations of a second-order system again. If we write

$$\omega^2 = \frac{1}{LC}, \quad 2\zeta\omega = \frac{R}{L} \quad \text{and} \quad F = \frac{V}{LC}$$

then (10.56) takes the standard form of (10.53), and we can infer that the voltage V_o will be sinusoidal with amplitude

$$\frac{A(\Omega)}{LC}V$$

where

$$A(\Omega) = \frac{1}{[(\omega^2 - \Omega^2)^2 + 4\zeta^2\omega^2\Omega^2]^{1/2}}$$

Thus when a sinusoidal voltage waveform is applied to the input terminals of the circuit, the voltage appearing at the output terminals is also a sinusoidal waveform, but one whose amplitude, relative to the input waveform amplitude, depends on the frequency of the input. A circuit that has this property is of course called a **filter**.

The form of $A(\Omega)$ will depend on ω and ζ, which in turn are determined by the values of L, R and C. The latter could be chosen so that ζ is small. In that case the circuit provides a large output when the input frequency Ω is near some frequency ω (which is determined by the choice of L and C) and a smaller output otherwise. This is a **tuned circuit** or a **bandpass filter**. If L, R and C are chosen so that ζ is larger (say near unity) then the circuit provides a larger output for small Ω and a smaller output for larger Ω. Such a circuit is a **low-pass filter**.

In this section, we have seen how problems in two very different areas of engineering – one mechanical and the other electrical – both give rise to very similar equations. Our knowledge of the form of the solutions of the equation is applicable to either area. This is a good example of the unifying properties of mathematics in engineering science. There are many other applications of the theory of the solution of second-order linear constant-coefficient differential equations in engineering.

It is also worth commenting here that filters of the type that we have described in this section are called **passive filters** since they use only inductors, resistors and capacitors – components that are referred to as **passive components**. Modern practice in electrical engineering involves the use of **active components** such as operational amplifiers in filter design, such filters being known as **active filters**. The analysis of the operation of active filters is more complex than that of passive filters. While, for many applications, active filters have displaced passive filters in modern practice, there are also many applications in which passive filters remain the norm.

10.10.5 Exercises

66 Find the damping parameters and natural frequencies of the systems governed by the following second-order linear constant-coefficient differential equations:

(a) $\dfrac{d^2x}{dt^2} + 6\dfrac{dx}{dt} + 9x = 0$

(b) $\dfrac{d^2x}{dt^2} + 4\dfrac{dx}{dt} + 7x = 0$

67 Determine the values of the appropriate parameters needed to give the systems governed by the following second-order linear constant-coefficient differential equations the damping parameters and natural frequencies stated:

(a) $\dfrac{d^2x}{dt^2} + 2a\dfrac{dx}{dt} + bx = 0$, $\zeta = 0.5$, $\omega = 2$

(b) $\dfrac{d^2x}{dt^2} + p\dfrac{dx}{dt} + qx = 0$, $\zeta = 1.4$, $\omega = 0.5$

(c) $\dfrac{d^2x}{dt^2} + \beta\dfrac{dx}{dt} + \gamma x = 0$, $\zeta = 1$, $\omega = 1.1$

68 Find the damping parameters and natural frequencies of the systems governed by the following second-order linear constant-coefficient differential equations:

(a) $\dfrac{d^2x}{dt^2} + 2a\dfrac{dx}{dt} + 16p^2x = 0$

(b) $2\dfrac{d^2x}{dt^2} + 14\dfrac{dx}{dt} + \dfrac{1}{\alpha}x = 0$

(c) $2.41\dfrac{d^2x}{dt^2} + 1.02\dfrac{dx}{dt} + 7.63x = 0$

(d) $\dfrac{1}{\eta}\dfrac{d^2x}{dt^2} + 40\dfrac{dx}{dt} + 25\eta x = 0$

(e) $1.88\dfrac{d^2x}{dt^2} + 4.71\dfrac{dx}{dt} + 0.48x = 0$

69 Determine the values of the appropriate parameters needed to give the systems governed by the following second-order linear constant-coefficient differential equations the damping parameters and natural frequencies stated:

(a) $\dfrac{d^2x}{dt^2} + \alpha\dfrac{dx}{dt} + \beta x = 0, \quad \zeta = 0.5, \quad \omega = \pi$

(b) $\dfrac{d^2x}{dt^2} + a\dfrac{dx}{dt} + bx = 0, \quad \zeta = 0.1, \quad \omega = 2\pi$

(c) $4\dfrac{d^2x}{dt^2} + q\dfrac{dx}{dt} + rx = 0, \quad \zeta = 1, \quad \omega = 1$

(d) $a\dfrac{d^2x}{dt^2} + b\dfrac{dx}{dt} + 14x = 0, \quad \zeta = 2, \quad \omega = 2\pi$

70 The function $A(\Omega)$ is as given by (10.55) and shown in Figure 10.27. Show that $A(\Omega)$ has a simple maximum point when $\zeta < \sqrt{\frac{1}{2}}$. Let the value of Ω for which this maximum occurs be Ω_{max}. Find Ω_{max} as a function of ζ and ω, and also find $A(\Omega_{max})$.

For $\zeta > \sqrt{\frac{1}{2}}$, $A(\Omega)$ has no maximum, but does have a single point of inflection. Show, by consideration of Figure 10.27, that $|dA/d\Omega|$ is a maximum at the point of inflection. Let Ω_c be the value of Ω for which the point of inflection occurs. Show that Ω_c satisfies the equation

$$3\Omega^6 + 5\beta\omega^2\Omega^4 + (4\beta^2 - 3)\omega^4\Omega^2 - \beta\omega^6 = 0$$

where $\beta = 2\zeta^2 - 1$. Hence show that for $\zeta = \sqrt{\frac{1}{2}}$ the greatest value of $|dA/d\Omega|$ occurs when $\Omega = \omega$ and is $1/(\sqrt{2}\omega^3)$. Also find the greatest values of $|dA/d\Omega|$ when $\zeta = \sqrt{(\frac{1}{2} + \frac{1}{6}\sqrt{3})}$ and when $\zeta = 1$.

Show that $|d^2A(0)/d\Omega^2|$ is minimized when $\zeta = \sqrt{\frac{1}{2}}$. The two values of ζ that minimize the maxima of $|dA/d\Omega|$ and $|d^2A(0)/d\Omega^2|$ respectively are important, particularly in control theory, since, in different senses, they maximize the flatness of the response function $A(\Omega)$.

71 An underwater sensor is mounted below the keel of the fast patrol boat shown in Figure 10.29. The supporting bracket is of cylindrical cross-section (diameter 0.04 m), and so is subject to an oscillating side-force due to vortex shedding. The bracket is of negligible mass compared with the sensor itself, which has a mass of 4 kg. The bracket has a tip displacement stiffness of 25 000 N m^{-1}. The frequency of the oscillating side-force is SU/d, where U is the speed of the vessel through the water, d is the diameter of the supporting bracket and S is the Strouhal number for vortex shedding from a circular cylinder. S has the value 0.20 approximately. At what speed will the frequency of the side-force coincide with the natural frequency of the sensor and mounting?

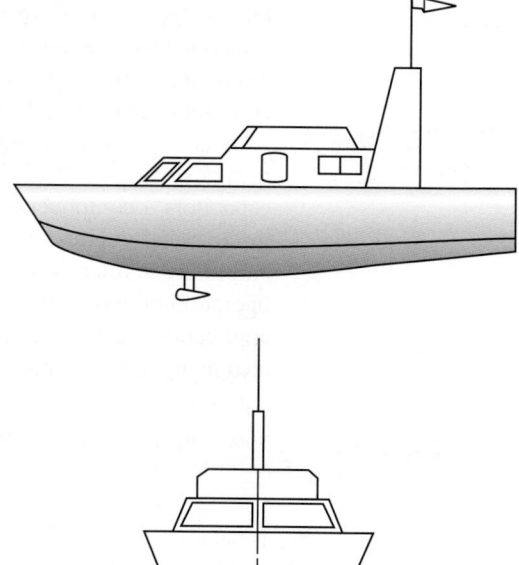

Figure 10.29 An underwater sensor mounting.

72 The piece of machinery shown in Figure 10.30 is mounted on a solid foundation in such a way that the mounting may be characterized as a rigid pivot and two stiff springs as shown. A damper is

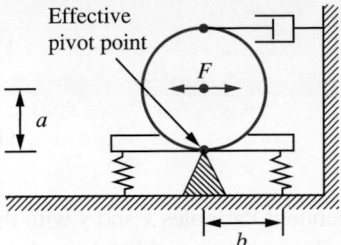

Figure 10.30 A compliantly mounted piece of machinery.

 73 Figure 10.31 shows a radio tuner circuit. Show that the natural frequency and damping parameters of the circuit are $1/\sqrt{(LC)}$ and

$$\frac{1}{2}\left(\frac{L}{C}\right)^{1/2}\left(\frac{1}{R_1}+\frac{1}{R_2}\right)$$

respectively. If $R_1 = 300\,\Omega$ and $R_2 = 50\,\Omega$ what value should L have, and over what range should C be adjustable in order that the circuit have a damping factor of $\zeta = 0.1$ and can be tuned to the medium waveband (505–1605 kHz)?

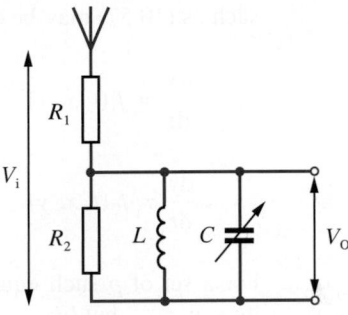

Figure 10.31 A radio tuner circuit.

connected between the machine and an adjacent strong point. The mass of the machine is 500 kg, the length $a = 1$ m, the length $b = 1.2$ m and the spring stiffness is 8000 N m^{-1}. The moment of inertia of the machine about the pivot point is $2ma^2$. The machine generates internally a side-force F that may be approximated as $F_0 \cos 2\pi f t$. As the machine runs up to speed, the frequency f increases from 0 to 6 Hz. What is the minimum damper coefficient that will prevent the machine from vibrating with any amplitude greater than twice its zero-frequency amplitude $A(0)$ during a run-up?

10.11 Numerical solution of second- and higher-order differential equations

Obviously, the classes of second- and higher-order differential equations that can be solved analytically, while representing an important subset of the totality of such equations, are relatively restricted. Just as for first-order equations, those for which no analytical solution exists can still be solved by numerical means. The numerical solution of second- and higher-order equations does not, in fact, need any significant new mathematical theory or technique.

10.11.1 Numerical solution of coupled first-order equations

In Section 10.6 we met Euler's method for the numerical solution of equations of the form

$$\frac{\mathrm{d}x}{\mathrm{d}t} = f(t, x)$$

that is, first-order differential equations involving a single dependent variable and a single independent variable. In Section 10.3 we noted that it was possible to have sets of coupled first-order equations, each involving the same independent variable but with more than one dependent variable. An example of this type of equation set is

$$\frac{\mathrm{d}x}{\mathrm{d}t} = x - y^2 + xt \tag{10.57a}$$

$$\frac{\mathrm{d}y}{\mathrm{d}t} = 2x^2 + xy - t \tag{10.57b}$$

This is a pair of differential equations in the dependent variables x and y with the independent variable t. The derivative of each of the dependent variables depends not only on itself and on the independent variable t, but also on the other dependent variable. Neither of the equations can be solved in isolation or independently of the other – both must be solved simultaneously, or side by side. A pair of coupled differential equations such as (10.57) may be characterized as

$$\frac{\mathrm{d}x}{\mathrm{d}t} = f_1(t, x, y) \tag{10.58a}$$

$$\frac{\mathrm{d}y}{\mathrm{d}t} = f_2(t, x, y) \tag{10.58b}$$

For a set of p such equations it is convenient to denote the dependent variables not by x, y, z, ... but by $x_1, x_2, x_3, \ldots, x_p$ and to denote the set of equations by

$$\frac{\mathrm{d}x_i}{\mathrm{d}t} = f_i(t, x_1, x_2, \ldots, x_p) \quad (i = 1, 2, \ldots, p)$$

or equivalently, using vector notation,

$$\frac{\mathrm{d}}{\mathrm{d}t}[\boldsymbol{x}] = \boldsymbol{f}(t, \boldsymbol{x})$$

where $\boldsymbol{x}(t)$ is a vector function of t given by

$$\boldsymbol{x}(t) = [x_1(t) \quad x_2(t) \quad \ldots \quad x_p(t)]^{\mathrm{T}}$$

$\boldsymbol{f}(t, \boldsymbol{x})$ is a vector-valued function of the scalar variable t and the vector variable $\boldsymbol{x}$.

Euler's method for the solution of a single differential equation takes the form

$$X_{n+1} = X_n + hf(t_n, X_n)$$

If we were to try to apply this method to (10.58a), we should obtain

$$X_{n+1} = X_n + hf_1(t_n, X_n, Y_n)$$

In other words, the value of X_{n+1} depends not only on t_n and X_n but also on Y_n. In the same way, we would obtain

$$Y_{n+1} = Y_n + hf_2(t_n, X_n, Y_n)$$

for Y_{n+1}. In practice, this means that to solve two coupled differential equations, we must advance the solution of both equations simultaneously in the manner shown in Example 10.47.

Example 10.47

Find the value of $X(1.4)$ satisfying the following initial-value problem:

$$\frac{\mathrm{d}x}{\mathrm{d}t} = x - y^2 + xt, \quad x(1) = 0.5$$

$$\frac{\mathrm{d}y}{\mathrm{d}t} = 2x^2 + xy - t, \quad y(1) = 1.2$$

using Euler's method with time step $h = 0.1$.

Solution

The right-hand sides of the two equations will be denoted by $f_1(t, x, y)$ and $f_2(t, x, y)$ respectively, so

$$f_1(t, x, y) = x - y^2 + xt \quad \text{and} \quad f_2(t, x, y) = 2x^2 + xy - t$$

The initial condition is imposed at $t = 1$, so t_n will denote $1 + nh$, X_n will denote $X(1 + nh)$, and Y_n will denote $Y(1 + nh)$. Then we have

$$X_1 = x_0 + hf_1(t_0, x_0, y_0) \qquad\qquad Y_1 = y_0 + hf_2(t_0, x_0, y_0)$$

$$= 0.5 + 0.1f_1(1, 0.5, 1.2) \qquad\quad = 1.2 + 0.1f_2(1, 0.5, 1.2)$$

$$= 0.4560 \qquad\qquad\qquad\qquad\quad = 1.2100$$

for the first step. The next step is therefore

$$X_2 = X_1 + hf_1(t_1, X_1, Y_1) \qquad\qquad Y_2 = Y_1 + hf_2(t_1, X_1, Y_1)$$

$$= 0.4560 \qquad\qquad\qquad\qquad\qquad = 1.2100$$

$$\quad + 0.1f_1(1.1, 0.4560, 1.2100) \qquad\quad + 0.1f_2(1.1, 0.4560, 1.2100)$$

$$= 0.4054 \qquad\qquad\qquad\qquad\qquad = 1.1968$$

and the third step is

$$X_3 = 0.4054 \qquad\qquad\qquad\qquad\quad Y_3 = 1.1968$$

$$\quad + 0.1f_1(1.2, 0.4054, 1.1968) \qquad\quad + 0.1f_2(1.2, 0.4054, 1.1968)$$

$$= 0.3513 \qquad\qquad\qquad\qquad\qquad = 1.1581$$

Finally, we obtain

$$X_4 = 0.3513 + 0.1f_1(1.3, 0.3513, 1.1581)$$

$$= 0.2980$$

Hence we have $X(1.4) = 0.2980$.

It should be obvious from Example 10.47 that the main drawback of extending Euler's method to sets of differential equations is the additional labour and tedium of the computations. Intrinsically, the computations are no more difficult, merely much more laborious – a prime example of a problem ripe for computerization.

10.11.2 State-space representation of higher-order systems

The solution of differential equation initial-value problems of order greater than one can be reduced to the solution of a set of first-order differential equations. This is achieved by a simple transformation, illustrated by Example 10.48.

Example 10.48

The initial-value problem

$$\frac{d^2x}{dt^2} + x^2t\frac{dx}{dt} - xt^2 = \frac{1}{2}t^2, \quad x(0) = 1.2, \quad \frac{dx}{dt}(0) = 0.8$$

can be transformed into two coupled first-order differential equations by introducing an additional variable

$$y = \frac{dx}{dt}$$

With this definition, we have

$$\frac{d^2x}{dt^2} = \frac{dy}{dt}$$

and so the differential equation becomes

$$\frac{dy}{dt} + x^2ty - xt^2 = \frac{1}{2}t^2$$

Thus the original differential equation can be replaced by a pair of coupled first-order differential equations, together with initial conditions:

$$\frac{dx}{dt} = y, \quad x(0) = 1.2$$

$$\frac{dy}{dt} = -x^2ty + xt^2 + \frac{1}{2}t^2, \quad y(0) = 0.8$$

This process can be extended to transform a pth-order initial-value problem into a set of p first-order equations, each with an initial condition. Once the original equation has been transformed in this way, its solution by numerical methods is just the same as if it had been a set of coupled equations in the first place.

Example 10.49

Find the value of $X(0.2)$ satisfying the initial-value problem

$$\frac{d^3x}{dt^3} + xt\frac{d^2x}{dt^2} + t\frac{dx}{dt} - t^2x = 0, \quad x(0) = 1, \quad \frac{dx}{dt}(0) = 0.5, \quad \frac{d^2x}{dt^2}(0) = -0.2$$

using Euler's method with step size $h = 0.05$.

Solution Since this is a third-order equation, we need to introduce two new variables:

$$y = \frac{dx}{dt} \quad \text{and} \quad z = \frac{dy}{dt} = \frac{d^2x}{dt^2}$$

Then the equation is transformed into a set of three first-order differential equations

$$\frac{dx}{dt} = y \qquad\qquad x(0) = 1$$

$$\frac{dy}{dt} = z \qquad\qquad y(0) = 0.5$$

$$\frac{dz}{dt} = -xtz - ty + t^2x \quad z(0) = -0.2$$

Applied to the set of differential equations

$$\frac{dx}{dt} = f_1(t, x, y, z)$$

$$\frac{dy}{dt} = f_2(t, x, y, z)$$

$$\frac{dz}{dt} = f_3(t, x, y, z)$$

the Euler scheme is of the form

$$X_{n+1} = X_n + hf_1(t_n, X_n, Y_n, Z_n)$$

$$Y_{n+1} = Y_n + hf_2(t_n, X_n, Y_n, Z_n)$$

$$Z_{n+1} = Z_n + hf_3(t_n, X_n, Y_n, Z_n)$$

In this case, therefore, we have

$$X_0 = x_0 = 1$$

$$Y_0 = y_0 = 0.5$$

$$Z_0 = z_0 = -0.2$$

$$f_1(t_0, X_0, Y_0, Z_0) = Y_0 = 0.5000$$

$$f_2(t_0, X_0, Y_0, Z_0) = Z_0 = -0.2000$$

$$f_3(t_0, X_0, Y_0, Z_0) = -X_0t_0Z_0 - t_0Y_0 + t_0^2X_0$$

$$= -1.0000 \times 0 \times (-0.2000) - 0 \times 0.5000 + 0^2 \times 1.0000$$

$$= 0.0000$$

$$X_1 = 1.0000 + 0.05 \times 0.5000 = 1.0250$$

$$Y_1 = 0.5000 + 0.05 \times (-0.2000) = 0.4900$$

$$Z_1 = -0.2000 + 0.05 \times 0.0000 = -0.2000$$

$$f_1(t_1, X_1, Y_1, Z_1) = Y_1 = 0.4900$$

$$f_2(t_1, X_1, Y_1, Z_1) = Z_1 = -0.2000$$

$$f_3(t_1, X_1, Y_1, Z_1) = -X_1 t_1 Z_1 - t_1 Y_1 + t_1^2 X_1$$

$$= -1.0250 \times 0.05 \times (-0.2000) - 0.05 \times 0.4900$$

$$+ 0.05^2 \times 1.0250 = -0.0117$$

$$X_2 = 1.0250 + 0.05 \times 0.4900 = 1.0495$$

$$Y_2 = 0.4900 + 0.05 \times (-0.2000) = 0.4800$$

$$Z_2 = -0.2000 + 0.05 \times (-0.0117) = -0.2005$$

Proceeding similarly we have

$$X_3 = 1.0495 + 0.05 \times 0.4800 = 1.0735$$

$$Y_3 = 0.4800 + 0.05 \times (-0.2005) = 0.4700$$

$$Z_3 = -0.2005 + 0.05 \times (-0.0165) = -0.2013$$

$$X_4 = 1.0735 + 0.05 \times 0.4700 = 1.0970$$

$$Y_4 = 0.4700 + 0.05 \times (-0.2013) = 0.4599$$

$$Z_4 = -0.2013 + 0.05 \times (-0.0139) = -0.2018$$

Hence $X(0.2) = X_4 = 1.0970$. It should be obvious by now that computations like these are sufficiently tedious to justify the effort of writing a computer program to carry out the actual arithmetic. The essential point for the reader to grasp is not the mechanics but the principle whereby methods for the solution of first-order differential equations (and this includes the more sophisticated methods as well as Euler's method) can be extended to the solution of sets of equations and hence to higher-order equations.

We noted earlier that both MAPLE and MATLAB can be used to obtain numerical solutions of differential equations and commented that they both implement very accurate methods of solution which are much more sophisticated than the Euler method illustrated here. But MAPLE could be used to obtain an Euler method solution of the third-order differential equation in Example 10.49, as follows:

```
ode:= diff(x(t),t$3) + x(t)*t*diff(x(t),t$2) +
        t*diff(x(t),t) - x(t)*t^2 = 0:
odeprob:= {ode,x(0) = 1,D(x)(0) = 0.5,D(D(x))(0) =
        -0.2};
oseq:= array([seq(0.05*i,i = 0..4)]);
oput:= dsolve(odeprob,numeric,
        method = classical[foreuler],
        output = oseq,stepsize = 0.05):
evalm(oput[2,1]);
```

MAPLE is equally able to solve the same equation presented in state space form, thus

```
ode1:= diff(x(t),t) = y(t):
ode2:= diff(y(t),t) = z(t):
ode3:= diff(z(t),t) = -x(t)*t*z(t) - t*y(t) + x(t)*t^2:
odeprob:= {ode1,ode2,ode3,x(0) = 1,y(0) = 0.5,z(0) =
       -0.2};
oseq:= array([seq(0.05*i,i - 0..4)]);
oput:= dsolve(odeprob,numeric,
       method = classical[foreuler],
       output = oseq,stepsize = 0.05):
evalm(oput[2,1]);
```

In fact, if we set the `infolevel` system variable thus

```
infolevel[dsolve]:= 3:
```

before calling `dsolve` to integrate the differential equation in the third-order form above, we discover that MAPLE first translates it into state space form just as we have done in Example 10.49!

10.11.3 Exercises

74 Transform the following initial-value problems into sets of first-order differential equations with appropriate initial conditions:

(a) $\dfrac{d^2x}{dt^2} + 6(x^2 - t)\dfrac{dx}{dt} - 4xt = 0,$

$x(0) = 1, \quad \dfrac{dx}{dt}(0) = 2$

(b) $\dfrac{d^2x}{dt^2} - \sin\left(\dfrac{dx}{dt}\right) + 4x = 0,$

$x(0) = 0, \quad \dfrac{dx}{dt}(0) = 0$

75 Find the value of $X(0.3)$ for the initial-value problem

$$\dfrac{d^2x}{dt^2} + x^2\dfrac{dx}{dt} + x = \sin t,$$

$x(0) = 0, \quad \dfrac{dx}{dt}(0) = 1$

using Euler's method with step size $h = 0.1$.

76 Transform the following initial-value problems into sets of first-order differential equations with appropriate initial conditions:

(a) $\dfrac{d^2x}{dt^2} + 4(x^2 - t^2)^{1/2} = 0,$

$x(1) = 2, \quad \dfrac{dx}{dt}(1) = 0.5$

(b) $\dfrac{d^3x}{dt^3} + t\dfrac{d^2x}{dt^2} + 6e^t\dfrac{dx}{dt} - x^2t = e^{2t},$

$x(0) = 1, \quad \dfrac{dx}{dt}(0) = 2, \quad \dfrac{d^2x}{dt^2}(0) = 0$

(c) $\dfrac{d^3x}{dt^3} + t\dfrac{d^2x}{dt^2} + x^2 = \sin t,$

$x(1) = 1, \quad \dfrac{dx}{dt}(1) = 0, \quad \dfrac{d^2x}{dt^2}(1) = -2$

(d) $\left(\dfrac{d^3x}{dt^3}\right)^{1/2} + t\dfrac{d^2x}{dt^2} + x^2t^2 = 0,$

$x(2) = 0, \quad \dfrac{dx}{dt}(2) = 0, \quad \dfrac{d^2x}{dt^2}(2) = 2$

(e) $\dfrac{d^4x}{dt^4} + x\dfrac{d^2x}{dt^2} + x^2 = \ln t, \quad x(0) = 0,$

$\dfrac{dx}{dt}(0) = 0, \quad \dfrac{d^2x}{dt^2}(0) = 4, \quad \dfrac{d^3x}{dt^3}(0) = -3$

(f) $\dfrac{d^4x}{dt^4} + \left(\dfrac{dx}{dt} - 1\right)\dfrac{d^3x}{dt^3} + \dfrac{dx}{dt} - (xt)^{1/2}$

$= t^2 + 4t - 5,$

$x(0) = a, \dfrac{dx}{dt}(0) = 0, \dfrac{d^2x}{dt^2}(0) = b, \dfrac{d^3x}{dt^3}(0) = 0$

77 Use Euler's method to compute an approximation $X(0.65)$ to the solution $x(0.65)$ of the initial-value problem

$$\dfrac{d^3x}{dt^3} + \dfrac{d^2x}{dt^2}(x - t) + \left(\dfrac{dx}{dt}\right)^2 - x^2 = 0,$$

$$x(0.5) = -1, \quad \dfrac{dx}{dt}(0.5) = 1, \quad \dfrac{d^2x}{dt^2}(0.5) = 2$$

using a step size of $h = 0.05$.

78 Write a computer program to solve the initial-value problem

$$\dfrac{d^2x}{dt^2} + x^2\dfrac{dx}{dt} + x = \sin t,$$

$$x(0) = 0, \quad \dfrac{dx}{dt}(0) = 1$$

using Euler's method. Use your program to find the value of $X(0.4)$ using steps of $h = 0.01$ and

$h = 0.005$. Hence estimate the accuracy of your value of $X(0.4)$ and estimate the step size that would be necessary to obtain a value of $X(0.4)$ accurate to 4dp.

79 A water treatment plant deals with a constant influx Q of polluted water with pollutant concentration s_0. The treatment tank contains bacteria which consume the pollutant and protozoa which feed on the bacteria, thus keeping the bacteria from increasing too rapidly and overwhelming the system. If the concentration of the bacteria and the protozoa are denoted by b and p the system is governed by the differential equations

$$\dfrac{ds}{dt} = r(s_0 - s) - \alpha m\dfrac{bs}{1 + s}$$

$$\dfrac{db}{dt} = -rb + m\dfrac{bs}{1 + s} - \beta n\dfrac{bp}{1 + p}$$

$$\dfrac{dp}{dt} = -rp + n\dfrac{bp}{1 + p}$$

Write a program to solve these equations numerically.

Measurements have determined that the (biological) parameters α, m, β and n have the values 0.5, 1.0, 0.8 and 0.1 respectively. The parameter r is a measure of the inflow rate of polluted water and s_0 is the level of pollutant. Using the initial conditions $s(0) = 0$, $b(0) = 0.2$ and $p(0) = 0.05$ determine the final steady level of pollutant if $r = 0.05$ and $s_0 = 0.4$. What effect does doubling the inflow rate (r) have?

The solution of Questions 77, 78 and 79 could be accomplished using MAPLE. Taking Question 78 as an example

```
ode:= diff(x(t),t$2) + x(t)^2*diff(x(t),t) +
        x(t) = sin(t):
odeprob:= {ode,x(0) = 0,D(x)(0) = 1};
oseq:= array([seq(0.1*i,i = 0..4)]);
oput1:= dsolve(odeprob,numeric,
        method = classical[foreuler],
        output = oseq,stepsize = 0.01);
oput2:= dsolve(odeprob,numeric,
        method = classical[foreuler],
        output = oseq,stepsize = 0.005);
```

The values of X(0.4) using step sizes of 0.01 and 0.005 are found to be 0.398022 and 0.397919 to 6sf respectively. This enables us to predict, using Richardson extrapolation, that a step size of approximately $h = 0.0024$ or smaller would be required to obtain the specified accuracy. In fact, as we have already noted, the MAPLE `dsolve/numeric` procedure can integrate differential equations numerically using much more sophisticated methods and providing answers to a specified accuracy. The `dsolve/numeric` procedure uses methods similar to the Richardson extrapolation method to achieve this.

10.12 Qualitative analysis of second-order differential equations

Sometimes it is easier or more convenient to discover the qualitative properties of the solutions of a differential equation than to solve it completely. In some cases this qualitative knowledge is just as useful as a complete solution. In other cases the qualitative knowledge is more illuminating than a quantitative solution, particularly if the only quantitative solutions that can be derived are numerical ones. One technique that is very useful in this context is the **phase-plane plot**.

10.12.1 Phase-plane plots

The second-order nonlinear differential equation

$$\frac{d^2 x}{dt^2} + \mu(x^2 - 1)\frac{dx}{dt} + \lambda x = 0$$

is known as the Van der Pol oscillator. It has properties that are typical of many non-linear oscillators. The equation has no simple analytical solution, so, if we wish to investigate its properties, we must resort to a numerical computation. The equation can readily be recast in state-space form as described in Section 10.11.2 and solved by Euler's method described in Section 10.6.

Figure 10.32 shows displacement and velocity plots for a Van der Pol oscillator with $\lambda = 40$ and $\mu = 3$. The initial conditions used were $x(0) = 0.05$ and $(dx/dt)(0) = 0$. It can be seen that initially the amplitude of the displacement oscillations grows quite rapidly, but after about three cycles this rapid growth stops and the displacement curve appears to settle into a periodically repeating pattern. Similar comments could be made about the velocity curve. Is the Van der Pol oscillator tending towards some fixed cyclical pattern?

This question can be answered much more easily if the displacement and velocity curves are plotted in a different way. Instead of plotting each individually against time, we plot velocity against displacement as in Figure 10.33. Such a plot is called a phase-plane plot. Figure 10.33(a) shows the same data as plotted in Figure 10.32. Time increases in the direction shown by the arrows, the plot starting at the point (0.05, 0) and spiralling outwards. From this plot it is easy to see that the fourth and fifth cycles of the oscillations are nearly indistinguishable. Continuing the computations for a larger number of cycles would confirm that, after an initial period, the oscillations settle down into a cyclical pattern. The pattern is called a **limit cycle**. The Van der Pol oscillator has the property that the limit cycle is independent of the initial conditions chosen (but

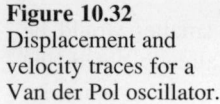

Figure 10.32
Displacement and
velocity traces for a
Van der Pol oscillator.

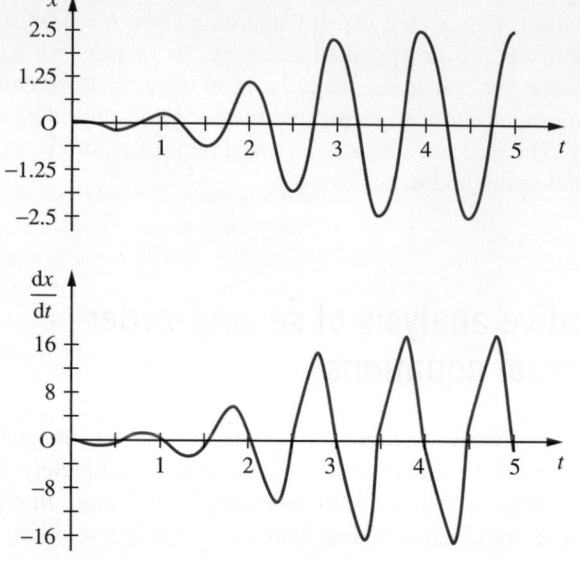

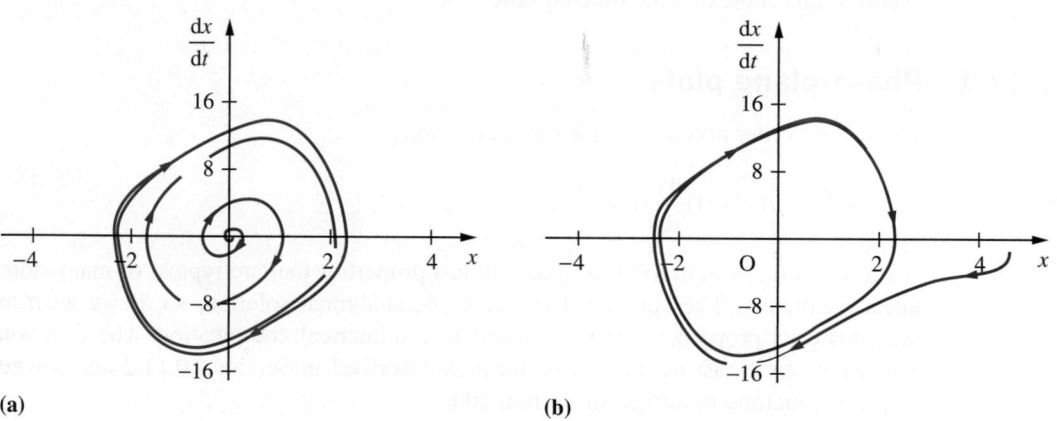

(a)

(b)

Figure 10.33 Phase-plane plots for Van der Pol oscillators – two different initial conditions.

depends on the parameters μ and λ. Figure 10.33(b) shows a phase-plane plot of the oscillations of the Van der Pol oscillator, starting from the initial condition $(4.5, 0)$. The interested reader may wish to explore the Van der Pol oscillator further – perhaps by writing a computer program to solve the equation and plotting solution paths in the phase plane for a number of other initial conditions. Exploration of this type will confirm that the limit cycle is independent of initial conditions, and exploration of other values of μ and λ will show how the limit cycle varies as these parameters change.

Other equations will of course produce different solution paths in the phase plane. The second-order linear constant-coefficient equation

$$\frac{d^2x}{dt^2} + \mu\frac{dx}{dt} + \lambda x = 0$$

yields a phase-plane plot like that shown in Figure 10.34(a). In that particular case the parameters have the values $\mu = 1.5$ and $\lambda = 40$. Other values of μ and λ that result

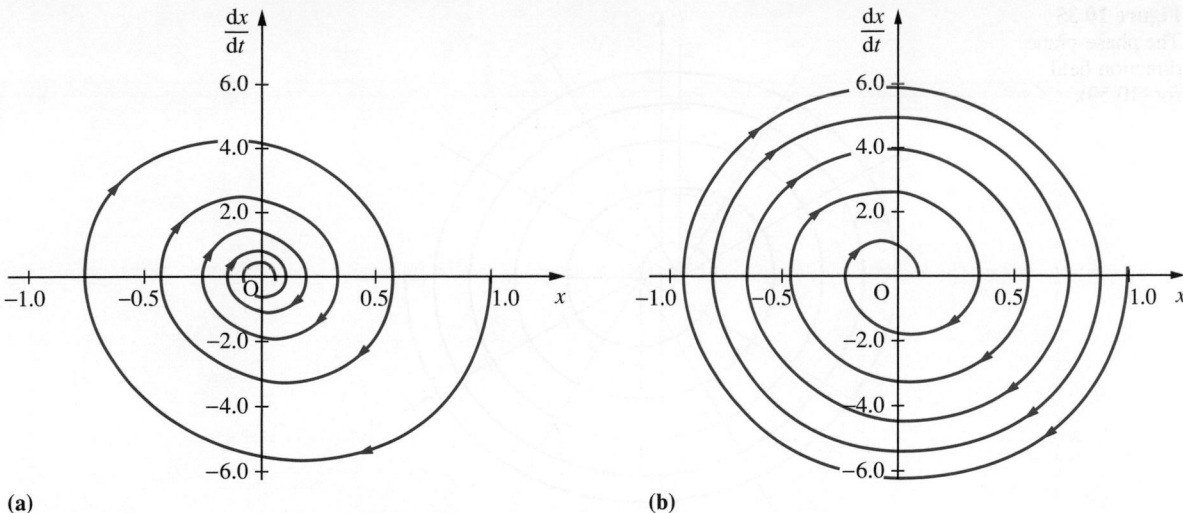

(a)

(b)

Figure 10.34 Phase-plane plots for some second-order oscillators.

in decaying oscillatory solutions of the equation yield similar spiral phase-plane plots tending towards the origin as $t \to \infty$. Such a plot is typical of any system whose behaviour is oscillatory and decaying. For instance, Figure 10.34(b) shows the phase-plane plot of the nonlinear second-order equation

$$\frac{d^2x}{dt^2} + \mu \operatorname{sgn}\left(\frac{dx}{dt}\right) + \lambda x = 0 \tag{10.59}$$

with $\mu = 3$ and $\lambda = 40$ (recall that the function $\operatorname{sgn}(x)$ takes the value 1 if $x \geqslant 0$ and -1 if $x < 0$). The general characters of Figures 10.34(a), (b) are similar. The difference between the two equations is manifest in the difference between the pattern of changing spacing of successive turns of the spirals.

The utility of phase-plane plotting is not restricted to enhancing the understanding of numerical solutions of differential equations. Second-order differential equations which can be expressed in the form

$$\frac{d^2x}{dt^2} = f\left(x, \frac{dx}{dt}\right)$$

arise in mathematical models of many engineering systems. An equation of this form can be expressed as

$$\frac{dv}{dx} = \frac{f(x, v)}{v}, \quad \text{where } v = \frac{dx}{dt}$$

The derivative dv/dx is of course just the gradient of the solution path in the phase plane. Hence we can sketch the path in the phase plane of the solutions of a second-order differential equation of this type without actually obtaining the solution. This provides a useful qualitative insight into the form of solution that might be expected.

As an example, consider (10.59). This may be expressed as

$$\frac{dv}{dx} = -\frac{\mu \operatorname{sgn}(v) + \lambda x}{v}$$

Figure 10.35
The phase-plane
direction field
for (10.59).

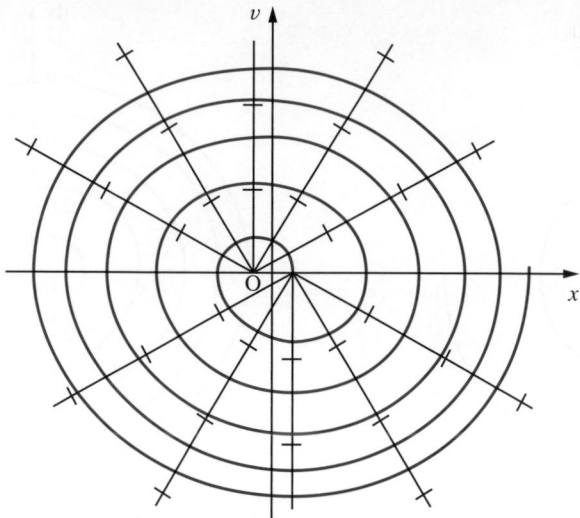

Thus the gradient of the solution path in the phase plane is equal to k for all points on the curve

$$v = -\frac{\lambda}{k}x - \frac{\mu \operatorname{sgn}(v)}{k}, \quad k \neq 0$$

These curves are of course a family of straight lines. Hence we can construct a diagram similar to the direction-field diagrams described in Section 10.5.1. The phase-plane direction-field diagram is shown, with the solution path from Figure 10.34(b) superimposed upon it, in Figure 10.35.

This technique can also be used for equations for which the lines of constant gradient in the phase plane are not straight. Example 10.50 illustrates this.

MAPLE provides tools to assist in the construction of phase-plane plots. One such tool is the `phaseportrait` procedure. The following MAPLE commands produce a diagram similar to Figure 10.33(a) but with a direction field shown in addition to the solution curve.

```
with(DEtools):
vdp2:= {diff(x(t),t) = v(t),
       diff(v(t),t) = -3*(x(t)^2 - 1)*v(t) - 40*x(t)};
phaseportrait(vdp2,{x(t),v(t)},t = 0..10,
       [[x(0) = 0.05,v(0) = 0]],x = -2..2,stepsize = 0.01);
```

Notice that it is necessary to load the `DEtools` package in order to access the `phaseportrait` procedure. Also the second-order differential equation is converted into a state space form for use in `phaseportrait`.

The `phaseportrait` procedure can be used to produce a diagram similar to Figure 10.36 for Example 10.50, and also to check your solutions to all parts of Question 78.

Example 10.50   Draw a phase-plane direction field for the equation

$$\frac{d^2x}{dt^2} + 1.5\left(\frac{dx}{dt}\right)^3 + 40x = 0 \qquad\qquad (10.60)$$

Hence sketch the solution path of the equation that starts from the initial conditions $x = 1$, $dx/dt = 0$.

Solution   Equation (10.60) can be expressed as

$$v\frac{dv}{dx} = -1.5v^3 - 40x$$

so the curve on which the solution-path gradient is equal to k is given by

$$x = -\tfrac{1}{40}(kv + 1.5v^3)$$

Thus, as shown in Figure 10.36, the curves of constant solution-path gradient are in this case cubic functions of v. The solution path of the equation starting from the point $(1,0)$ is sketched.

Figure 10.36
The phase-plane
direction field
for (10.60).

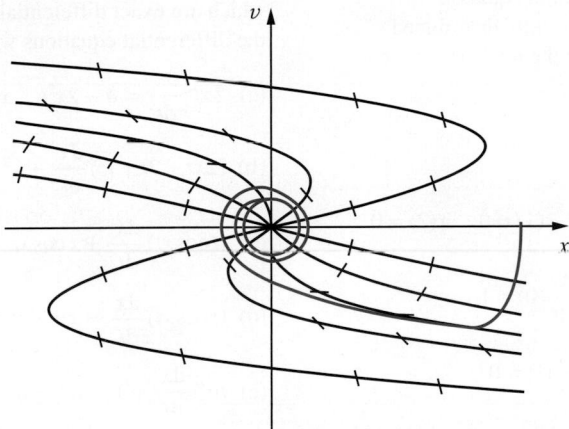

10.12.2 Exercises

80 Draw phase-plane direction fields for the following equations and sketch the form you would expect the solution paths to take, starting from the points $(x, v) = (1, 0)$, $(0, 1)$, $(-1, 0)$ and $(0, -1)$ in each case:

(a) $\dfrac{d^2x}{dt^2} + \dfrac{dx}{dt} + x^3 = 0$

(b) $\dfrac{d^2x}{dt^2} + \dfrac{dx}{dt} + \operatorname{sgn}(x) = 0$

(c) $\dfrac{d^2x}{dt^2} + \dfrac{dx}{dt} + x^2\operatorname{sgn}(x) = 0$

(d) $\dfrac{d^2x}{dt^2} + \operatorname{sgn}\left(\dfrac{dx}{dt}\right) + 2\operatorname{sgn}(x) = 0$

81 For each of the problems in Question 78 solve the differential equation numerically and check that the solutions you obtain are similar to your sketch solutions.

10.13 Review exercises (1–35)

 Whenever possible check your answers using MATLAB or MAPLE.

1 Classify each of the following as partial or ordinary and as linear homogeneous, linear nonhomogeneous or nonlinear differential equations, state the order of the equations and name the dependent and independent variables:

(a) $\dfrac{d^2x}{dt^2} + x\dfrac{dx}{dt} + x^2 = 0$

(b) $\dfrac{\partial^3 h}{\partial t \partial v^2} + \dfrac{\partial^2 h}{\partial t^2} + a\dfrac{\partial h}{\partial v} = 0$

(c) $\dfrac{dz}{dx} + 4z^2 = \sin x$

(d) $\dfrac{d^3p}{ds^3} + 4s\dfrac{d^2p}{ds^2} + s^2 = \cos as$

2 Classify the following differential equation problems as under-determined, fully determined or over-determined, and solve them where possible:

(a) $\dfrac{d^2x}{dt^2} = t, \quad x(0) = 1$

(b) $\dfrac{d^3x}{dt^3} - t = 0, \quad x(0) = 0, \quad x(1) = 0, \quad x(2) = 0$

(c) $\dfrac{dx}{dt} = \sin t, \quad x(0) = 0, \quad \dfrac{dx}{dt}(0) = 1$

(d) $\dfrac{d^2x}{dt^2} = e^{4t}, \quad x(0) = 0, \quad \dfrac{dx}{dt}(1) = 0$

3 Sketch the direction field of the differential equation

$$\dfrac{dx}{dt} = ax(1 - x^2)$$

and sketch the form of solution suggested by the direction field. Solve the equation and confirm that the solution supports the inferences you made from the direction field.

4 Solve the following differential equation problems.

(a) $\dfrac{dx}{dt} + \dfrac{\cos t}{\sin x} = 0, \quad x(0) = -\pi$

(b) $t\dfrac{dx}{dt} - e^{-x} = 0, \quad x(1) = 2$

(c) $\dfrac{dx}{dt} = xt^2, \quad x(2) = 1$

(d) $t\dfrac{dx}{dt} = \dfrac{t}{\sin(x/t)} + x, \quad x(1) = 1$

(e) $\dfrac{dx}{dt} = \dfrac{8t - x}{2x + t}, \quad x(0) = 2$

(f) $t\dfrac{dx}{dt} + x\ln t = x(\ln x + 1), \quad x(1) = 2$

(g) $t\dfrac{dx}{dt} = x - t, \quad x(1) = 3$

(h) $\dfrac{dx}{dt} = \dfrac{x - 7t}{x - t}, \quad x(0) = 2$

5 For each of the following problems, determine which are exact differentials, and hence solve the differential equations where possible:

(a) $2xt^2\dfrac{dx}{dt} = a - 2x^2t, \quad x(1) = 2$

(b) $(2xt + 2t + t^2)\dfrac{dx}{dt} + x^2 + 2tx = 0, \quad x(2) = 2$

(c) $(t\cos xt)\dfrac{dx}{dt} + x\cos xt + 1 = 0, \quad x(\pi) = 0$

(d) $(t\cos xt)\dfrac{dx}{dt} - x\cos xt = 0, \quad x(\pi) = 0$

(e) $te^{xt}\dfrac{dx}{dt} + 1 + xe^{xt} = 0, \quad x(2) = 4$

6 Solve the following differential equation problems:

(a) $\dfrac{dx}{dt} - 2x = t, \quad x(0) = 2$

(b) $\dfrac{dx}{dt} + 2tx = (t - \tfrac{1}{2})e^{-t}, \quad x(0) = 1$

(c) $\dfrac{dx}{dt} + 3x = e^{2t}, \quad x(0) = 2$

(d) $\dfrac{dx}{dt} + x\sin t = \sin t, \quad x(\pi) = e$

7 Solve the differential equation

$$\dfrac{dx}{dt} = \left(\dfrac{xt}{x^2 + t^2}\right)^{1/2}, \quad x(0) = 1$$

to find the value of $X(0.4)$ using Euler's method with step size 0.1 and 0.05. By comparing these two estimates of $x(0.4)$, estimate the accuracy of the better of the two values that you have obtained and also the step size you would need to use in order to calculate an estimate of $x(0.4)$ accurate to 2dp.

8 Solve the differential equation

$$\frac{dx}{dt} = \sin t^2, \quad x(0) = 2$$

to find the value of $X(0.25)$ using Euler's method with steps of size 0.05 and 0.025. By comparing these two estimates of $x(0.25)$, estimate the accuracy of the better of the two values that you have obtained and also the step size you would need to use in order to calculate an estimate of $x(0.25)$ accurate to 3dp.

9 Solve the differential equation

$$\frac{dx}{dt} + \frac{3x}{20 - t} = 2$$

obtained in Example 8.4 to determine the amount $x(t)$ of salt in the tank at time t minutes. Initially the tank contains pure water.

10 An open vessel is in the shape of a right circular cone of semi-vertical angle 45° with axis vertical and apex downwards. At time $t = 0$ the vessel is empty. Water is pumped in at a constant rate $p\,\text{m}^3\,\text{s}^{-1}$ and escapes through a small hole at the vertex at a rate $ky\,\text{m}^3\,\text{s}^{-1}$ where k is a positive constant and y is the depth of water in the cone.
   Given that the volume of a circular cone is $\pi r^2 h/3$, where r is the radius of the base and h its vertical height, show that

$$\pi y^2 \frac{dy}{dt} = p - ky$$

Deduce that the water level reaches the value $y = p/(2k)$ at time

$$t = \frac{\pi p^2}{k^3}\left(\ln 2 - \frac{5}{8}\right)$$

11 Stefan's law states that the rate of change of temperature of a body due to radiation of heat is

$$\frac{dT}{dt} = -k(T^4 - T_0^4)$$

where T is the temperature of the body, T_0 is the temperature of the surrounding medium (both measured in K) and k is a constant. Show that the solution of this differential equation is

$$2\tan\left(\frac{T}{T_0}\right) + \ln\left(\frac{T + T_0}{T - T_0}\right) = 4T_0^3(kt + C)$$

Show that, when the temperature difference between the body and its surroundings is small, Stefan's law can be approximated by Newton's law of cooling

$$\frac{dT}{dt} = -\alpha(T - T_0)$$

and find α in terms of k and T_0.

12 A motor under load generates heat internally at a constant rate H and radiates heat, in accordance with Newton's law of cooling, at a rate $k\theta$, where k is a constant and θ is the temperature difference of the motor over its surroundings. With suitable non-dimensionalization of time the temperature of the motor is given by the differential equation

$$\frac{d\theta}{dt} = H - k\theta$$

Given that $\theta = 0$ and $d\theta/dt = 10$ when $t = 0$ and $\theta = 60$ when $t = 10$ show that

(a) the ultimate rise in temperature is $\theta = 10/k$

(b) k is a solution of the equation $e^{-10k} = 1 - 6k$

(c) $t = 10 + \dfrac{1}{k}\ln\left(\dfrac{10 - 60k}{10 - k\theta}\right)$

13 A linear cam is to be made whose rate of rise (as it moves in the negative x direction) at the point (x, y) on the profile is equal to one half of the gradient of the line joining (x, y) to a fixed point on the cam (x_0, y_0). Show that the cam profile is a solution of the differential equation

$$\frac{dy}{dx} = \frac{y - y_0}{2(x - x_0)}$$

and hence find its equation. Sketch the cam profile.

14 Radioactive elements decay at a constant rate per unit mass of the element. Show that such decays obey equations of the form

$$\frac{dm}{dt} = -km$$

where k is the decay rate of the element and m is the mass of the element present. The half life of an element is the time taken for one half of any given mass of the element to decay. Find the relationship between the decay constant k and the half life of an element.

15 In Section 10.2.4 we showed that the equation governing the current flowing in a series LRC electrical circuit is (equation 10.9)

$$L\frac{d^2i}{dt^2} + R\frac{di}{dt} + \frac{1}{C}i = 0$$

Show, by a similar method, that the equation governing the current flowing in a series LR circuit containing a voltage source E is

$$L\frac{di}{dt} + Ri = E$$

At time $t = 0$ a switch is closed applying a d.c. potential of V to an initially quiescent series LR circuit consisting of an inductor L and a resistor R. Show that the current flowing in the circuit is

$$i(t) = \frac{V}{R}(1 - e^{-Rt/L})$$

and hence find the time needed for the current to reach 95% of its final value.

16 The tread of a car tyre wears more rapidly as it becomes thinner. The tread-wear rate, measured in mm per 10 000 miles, may be modelled as

$$a + b(d - t)^2$$

where d is the initial tread depth, t is the current tread depth and a and b are constants. A tyre company takes measurements on a new design of tyre whose initial tread depth is 8 mm. When the tyre is new its wear rate is found to be 1.03 mm per 10 000 miles run and when the tread depth is reduced to 4 mm the wear rate is 3.43 mm per 10 000 miles. Assuming that a tyre is discarded when the tread depth has been reduced to 2 mm what is its estimated life?

17 Express each of the following differential equations in the form

$$L[x(t)] = f(t)$$

(a) $\dfrac{d^2x}{dt^2} + (\sin t)\dfrac{dx}{dt} - 9x + \cos t = 0$

(b) $\dfrac{d^3x}{dt^3} + t\dfrac{d^2x}{dt^2} + t^2\dfrac{dx}{dt} - 4t\dfrac{dx}{dt} + e^t + x = 0$

(c) $\dfrac{dx}{dt} = e^t + e^{-t}x$

(d) $\dfrac{d^2x}{dt^2} = \cos \Omega t - 4x$

(e) $t^2\dfrac{d^3x}{dt^3} + \ln(t^2 + 4) = \dfrac{1}{t^2 + 2t + 4}\dfrac{dx}{dt}$

18 For each of the following pairs of operators calculate the operator LM − ML; hence state which of the pairs are commutative (that is, satisfy $LMx(t) = MLx(t)$):

(a) $L = \dfrac{d}{dt} + \sin t, \quad M = \dfrac{d}{dt} - \cos t$

(b) $L = \dfrac{d}{dt} + 4, \quad M = \dfrac{d}{dt} + 9$

(c) $L = \dfrac{d}{dt} + \sin t + 2, \quad M = \dfrac{d}{dt} + \sin t - 2$

(d) $L = \dfrac{d^2}{dt^2} + 2t^2 - 9, \quad M = \dfrac{d^2}{dt^2} + 2t^2 + t$

19 What conditions must the functions $f(t)$ and $g(t)$ satisfy in order for the following operator pairs to be commutative?

(a) $L = \dfrac{d}{dt} + f(t), \quad M = \dfrac{d}{dt} + g(t)$

(b) $L = \dfrac{d^2}{dt^2} + f(t), \quad M = \dfrac{d^2}{dt^2} + g(t)$

20 Find the general solution of the following differential equations:

(a) $\dfrac{d^2x}{dt^2} - 3\dfrac{dx}{dt} + 2x = \sin t$

(b) $\dfrac{d^3x}{dt^3} - 7\dfrac{dx}{dt} + 6x = t$

(c) $\dfrac{d^3x}{dt^3} - 7\dfrac{dx}{dt} + 6x = e^{2t}$

(d) $\dfrac{dx}{dt} - 4x = e^{4t}$

(e) $\dfrac{d^2x}{dt^2} + 3\dfrac{dx}{dt} + \tfrac{13}{4}x = t^2$

(f) $\dfrac{d^2x}{dt^2} + 3\dfrac{dx}{dt} + \tfrac{13}{4}x = \sin t$

(g) $\dfrac{d^3x}{dt^3} - 5\dfrac{d^2x}{dt^2} + 2\dfrac{dx}{dt} + 8x = t^2 - t$

(h) $\dfrac{d^2x}{dt^2} - 2\dfrac{dx}{dt} + 5x = e^{-t}$

(i) $\dfrac{d^3x}{dt^3} - 5\dfrac{d^2x}{dt^2} + 2\dfrac{dx}{dt} + 8x = e^{2t} + e^t$

(j) $\dfrac{d^2x}{dt^2} - 2\dfrac{dx}{dt} + 5x = t + e^t \cos 2t$

21 Solve the following initial-value problems:

(a) $\dfrac{d^2x}{dt^2} + 2\dfrac{dx}{dt} + 5x = 1, \quad x(0) = 0, \quad \dfrac{dx}{dt}(0) = 0$

(b) $3\dfrac{d^2x}{dt^2} - 2\dfrac{dx}{dt} - x = 2t - 1,$

$x(0) = 7, \quad \dfrac{dx}{dt}(0) = 2$

(c) $\dfrac{d^2x}{dt^2} + 2\dfrac{dx}{dt} + x = 4\cos 2t,$

$x(0) = 0, \quad \dfrac{dx}{dt}(0) = 2$

(d) $\dfrac{d^2x}{dt^2} - \dfrac{dx}{dt} = -2e^{2t}, \quad x(0) = 0, \quad \dfrac{dx}{dt}(0) = 1$

(e) $\dfrac{d^2x}{dt^2} - 3\dfrac{dx}{dt} + 2x = 2e^{-4t},$

$x(0) = 0, \quad \dfrac{dx}{dt}(0) = 1$

(f) $\dfrac{d^3x}{dt^3} + 5\dfrac{d^2x}{dt^2} + 17\dfrac{dx}{dt} + 13x = 1,$

$x(0) = 1, \quad \dfrac{dx}{dt}(0) = 1, \quad \dfrac{d^2x}{dt^2}(0) = 0$

22 Find the damping parameters and natural frequencies of the systems governed by the following second-order linear constant-coefficient differential equations:

(a) $\dfrac{d^2x}{dt^2} + 7\dfrac{dx}{dt} + 2x = 0$

(b) $\dfrac{d^2x}{dt^2} + p\dfrac{dx}{dt} + p^{1/2}x = 0$

(c) $\dfrac{d^2x}{dt^2} + 2aq\dfrac{dx}{dt} + \tfrac{1}{2}qx = 0$

(d) $\dfrac{d^2x}{dt^2} + 14\dfrac{dx}{dt} + 2\alpha x = 0$

23 Determine the values of the appropriate parameters needed to give the systems governed by the following second-order linear constant-coefficient differential equations the damping parameters and natural frequencies stated:

(a) $\dfrac{d^2x}{dt^2} + \dfrac{a}{2}\dfrac{dx}{dt} + bx = 0, \quad \zeta = 0.25, \quad \omega = 2$

(b) $\dfrac{d^2x}{dt^2} + a\dfrac{dx}{dt} + bx = 0, \quad \zeta = 2, \quad \omega = \pi$

(c) $a\dfrac{d^2x}{dt^2} + 4\dfrac{dx}{dt} + cx = 0, \quad \zeta = 0.5, \quad \omega = 2$

(d) $p\dfrac{d^2x}{dt^2} + q^2\dfrac{dx}{dt} + 6x = 0, \quad \zeta = 1.2, \quad \omega = 0.2$

24 Show that by making the substitution

$$v = \dfrac{dx}{dt}$$

the equation

$$\dfrac{d^2x}{dt^2} + \dfrac{dx}{dt} = 1$$

may be expressed as

$$\dfrac{dv}{dt} + v = 1$$

Show that the solution of this equation is $v = 1 + Ce^{-t}$ and hence find $x(t)$.

This technique is a standard method for solving second-order differential equations in which the dependent variable itself does not appear explicitly. Apply the same method to obtain the solutions of the differential equations:

(a) $\dfrac{d^2x}{dt^2} = 4\dfrac{dx}{dt} + e^{-2t}$

(b) $\dfrac{d^2x}{dt^2} - \left(\dfrac{dx}{dt}\right)^2 = 1$

(c) $t\dfrac{d^2x}{dt^2} = 2\dfrac{dx}{dt}$

25 Using the method introduced in Question 24, find the solutions of the following initial-value problems:

(a) $\dfrac{d^2x}{dt^2} + k\dfrac{dx}{dt} = t^2$, $x(0) = 0$, $\dfrac{dx}{dt}(0) = 1$

(b) $\dfrac{d^2x}{dt^2} = \left(\dfrac{dx}{dt}\right)^2 e^{-kt}$, $x(0) = 0$, $\dfrac{dx}{dt}(0) = U$

(c) $(t^2 + 4)\dfrac{d^2x}{dt^2} = 2t\dfrac{dx}{dt}$, $x(1) = 0$, $\dfrac{dx}{dt}(1) = 2$

(d) $\dfrac{d^2x}{dt^2} + 4\dfrac{dx}{dt} = \sin t$, $x(\pi) = 0$, $\dfrac{dx}{dt}(\pi) = 1$

26 Show that by making the substitution

$$v = \frac{dx}{dt}$$

and noting that

$$\frac{d^2x}{dt^2} = \frac{dv}{dt} = \frac{dv}{dx}\frac{dx}{dt} = v\frac{dv}{dx}$$

the equation

$$\frac{d^2x}{dt^2} = x\frac{dx}{dt}$$

may be expressed as

$$v\frac{dv}{dx} = xv$$

Show that the solution of this equation is $v = \frac{1}{2}x^2 + C$ and hence find $x(t)$.

This technique is a standard method for solving second-order differential equations in which the independent variable does not appear explicitly. Apply the same method to obtain the solutions of the differential equations:

(a) $\dfrac{d^2x}{dt^2} = p\dfrac{dx}{dt}$

(b) $\dfrac{d^2x}{dt^2} = \left(\dfrac{dx}{dt}\right)^2$

(c) $\dfrac{d^2x}{dt^2} = \left(\dfrac{dx}{dt}\right)^2\left(2x - \dfrac{1}{x}\right)$

27 Using the method introduced in Question 26, find the solutions of the following initial-value problems:

(a) $x\dfrac{d^2x}{dt^2} = p\left(\dfrac{dx}{dt}\right)^2$, $x(0) = 4$, $\dfrac{dx}{dt}(0) = 1$

(b) $\dfrac{d^2x}{dt^2} = \dfrac{dx}{dt}e^x$, $x(1) = 1$, $\dfrac{dx}{dt}(1) = 0$

(c) $\dfrac{d^2x}{dt^2} = x^2\dfrac{dx}{dt}$, $x(0) = 2$, $\dfrac{dx}{dt}(0) = \dfrac{8}{3}$

(d) $\dfrac{d^2x}{dt^2} + \dfrac{1}{2}\left(\dfrac{dx}{dt}\right)^2 = x$, $x(0) = 1$, $\dfrac{dx}{dt}(0) = 0$

28 Equation (10.3), arising from the model of the take-off run of an aircraft developed in Section 10.2.1, can be solved by the techniques introduced in Exercises 24 and 26. Assuming that the thrust is constant find the speed of the aircraft both as a function of time and of distance run along the ground. The take-off speed of the aircraft is denoted by V_2. Find expressions for the length of runway required and the time taken by the aircraft to become airborne in terms of take-off speed.

29 Find the values of $X(t)$ for t up to 2, where $X(t)$ is the solution of the differential equation problem

$$\frac{d^3x}{dt^3} + \left(\frac{d^2x}{dt^2}\right)^2 + 4\left(\frac{dx}{dt}\right)^2 - xt = \sin t,$$

$$x(1) = 0.2, \quad \frac{dx}{dt}(1) = 1, \quad \frac{d^2x}{dt^2}(1) = 0$$

using Euler's method with step size $h = 0.025$. Repeat the computation with $h = 0.0125$. Hence estimate the accuracy of the value of $X(2)$ given by your solution.

30 The end of a chain, coiled near the edge of a horizontal surface, falls over the edge. If the friction between the chain and the horizontal surface is negligible and the chain is inextensible then, when a length x of chain has fallen, the equation of motion is

$$\frac{d}{dt}(mxv) = mgx$$

where m is the mass per unit length of the chain, g is gravitational acceleration and v is the velocity of the falling length of the chain. If the mass per unit length of the chain is constant show that this equation can be expressed as

$$xv\frac{dv}{dx} + v^2 = gx$$

and, by putting $y = v^2$, show that $v = \sqrt{(2gx/3)}$.

31 A simple mass spring system, subject to light damping, is vibrating under the action of a periodic force $F \cos pt$. The equation of motion is

$$\frac{d^2x}{dt^2} + 2\frac{dx}{dt} + 4x = F \cos pt$$

where F and p are constants.

Solve the differential equation for the displacement $x(t)$. Show that one part of the solution tends to zero as $t \to \infty$ and show that the amplitude of the steady-state solution is

$$F[(4 - p^2)^2 + 4p^2]^{-1/2}.$$

Hence show that resonance occurs when $p = \sqrt{2}$.

32 An alternating emf of $E \sin \omega t$ volt is supplied to a circuit containing an inductor of L henry, a resistor of R ohm and a capacitor of C farad in series. The differential equation satisfied by the current i amp and the charge q coulomb on the capacitor is

$$L\frac{di}{dt} + Ri + \frac{q}{C} = E \sin \omega t$$

Using $i = dq/dt$ obtain a second-order differential equation satisfied by i. Find the resistance if it is just large enough to prevent natural oscillations. For this value of R and $\omega = (LC)^{-1/2}$ prove that

$$i = \frac{E}{2K}(\sin \omega t - \omega t\, e^{-\omega t})$$

where $K^2 = L/C$, when the current and charge on the capacitor are both zero at time $t = 0$.

The following three questions are intended to be open-ended – there is no single 'correct' answer. They should be approached in an enquiring frame of mind, with the objective of disovering, by use of mathematical knowledge and technique, something more about how the physical world functions. The questions are designed to use primarily mathematical knowledge introduced in this chapter.

33 A truck of mass m moves along a horizontal test track subject only to a force resisting motion that is proportional to its speed. At time $t = 0$ the truck passes a reference point moving with speed U. Find the velocity of the truck both as a function of time and as a function of displacement from the reference point. Find the displacement of the truck from the reference point as a function of time.

Repeat these calculations for similar trucks subject to resistance forces proportional to

(a) square root of speed,

(b) square of speed,

(c) cube of speed.

How long does the truck take to come to rest in each case? Draw plots of velocity against displacement in each case. Explain, in qualitative terms, the behaviour of the truck under each type of resistance.

How would you model mathematically a truck that is subject to a small constant resistance plus a resistance proportional to its speed? How far would such a truck travel before coming to rest, and how long would it take to do so? Can you repeat these calculations for trucks subject to a small constant resistance plus a resistance proportional to speed squared or speed cubed?

What general conclusions can you draw about the type of terms that it is sensible to use in mathematical models of engineering systems to describe resistance to motion?

34 Figure 10.37 shows a system that serves as a simplified model of the phenomenon of 'tool chatter'. The mass A rests on a moving belt and is connected to a rigid support by a spring. The coefficient of sliding friction between the belt and the mass is less than the coefficient of static friction. When the spring is uncompressed, the mass moves to the right with the belt. As it does so, the spring is compressed until the force exerted by the spring exceeds the maximum static frictional force available. The mass then starts to slide. The spring force slows the mass, brings it

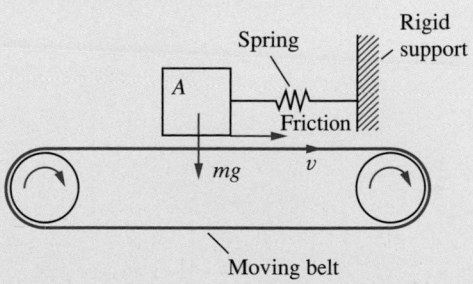

Figure 10.37 Diagram of a model of the 'tool chatter' phenomenon.

to rest, and then accelerates it back along the belt so that it moves leftwards. As it does so, the compression in the spring is reduced, the force of sliding friction slows the mass to rest, and then accelerates it so that its velocity is directed to the right. When its velocity matches that of the belt, sliding ceases and static friction takes over again.

Thus the mass undergoes a cyclic process of being pushed forwards by static friction until the spring is sufficiently compressed and then being flung backwards by the stored energy in the spring until the energy is dissipated. Analyse the model, determining such quantities as how the amplitude and frequency of motion of the mass depend on the coefficients and static friction and the other physical parameters.

35 The second-order linear nonhomogeneous constant-coefficient differential equation

$$\frac{d^2x}{dt^2} + 2\zeta\omega\frac{dx}{dt} + \omega^2 x = F\cos\Omega t$$

(often referred to as a **forced harmonic oscillator**) has a response $A(\Omega)F\cos(\Omega t - \delta)$, where $A(\Omega)$ is often called the **frequency response** (strictly it is the *amplitude response* or *gain spectrum*) and is given by (10.55) and shown in Figure 10.27. How does the frequency response of the second-order nonlinear nonhomogeneous constant-coefficient differential equation

$$\frac{d^2x}{dt^2} + 2\zeta\omega\left|\frac{dx}{dt}\right|\frac{dx}{dt} + \omega^2 x = F\cos\Omega t$$

differ from that of the linear one?

11 Introduction to Laplace Transforms

Introduction

Laplace transform methods have a key role to play in the modern approach to the analysis and design of engineering systems. The stimulus for developing these methods was the pioneering work of the English electrical engineer Oliver Heaviside (1850–1925) in developing a method for the systematic solution of ordinary differential equations with constant coefficients. Heaviside was concerned with solving practical problems, and his method was based mainly on intuition, lacking mathematical rigour: consequently it was frowned upon by theoreticians at the time. However, Heaviside himself was not concerned with rigorous proofs, and was satisfied that his method gave the correct results. Using his ideas, he was able to solve important practical problems that could not be dealt with using classical methods. This led to many new results in fields such as the propagation of currents and voltages along transmission lines.

Because it worked in practice, Heaviside's method was widely accepted by engineers. As its power for problem-solving became more and more apparent, the method attracted the attention of mathematicians, who set out to justify it. This provided the stimulus for rapid developments in many branches of mathematics, including improper integrals, asymptotic series and transform theory. Research on the problem continued for many years before it was eventually recognized that an integral transform developed by the French mathematician Pierre Simon de Laplace (1749–1827) almost a century before provided a theoretical foundation for Heaviside's work. It was also recognized that the use of this integral transform provided a more systematic alternative for investigating differential equations than the method proposed by Heaviside. It is this alternative approach that is the basis of the **Laplace transform method**.

We have already come across instances where a mathematical transformation has been used to simplify the solution of a problem. For example, the logarithm is used to simplify multiplication and division problems. To multiply or divide two numbers, we transform them into their logarithms, add or subtract these, and then perform the inverse transformation (that is, the antilogarithm) to obtain the product or quotient of the original numbers. The purpose of using a transformation is to create a new domain in which it is easier to handle the problem being investigated. Once results have been obtained in the new domain, they can be inverse-transformed to give the desired results in the original domain.

The Laplace transform is an example of a class called **integral transforms**, and it takes a function $f(t)$ of one variable t (which we shall refer to as **time**) into a function $F(s)$ of another variable s (the **complex frequency**). Another integral transform widely used by engineers is the **Fourier transform**, which is dealt with in the companion text *Advanced Modern Engineering Mathematics*. The attraction of the Laplace transform is that it transforms *differential* equations in the t (time) domain into *algebraic* equations in the s (frequency) domain. Solving differential equations in the t domain therefore reduces to solving algebraic equations in the s domain. Having done the latter for the desired unknowns, their values as functions of time may be found by taking inverse transforms. Another advantage of using the Laplace transform for solving differential equations is that initial conditions play an essential role in the transformation process, so they are automatically incorporated into the solution. This constrasts with the classical approach considered in Chapter 10, where the initial conditions are only introduced when the unknown constants of integration are determined. The Laplace transform is

Figure 11.1
Schematic
representation of a
system.

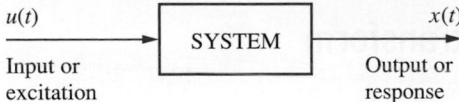

therefore an ideal tool for solving initial-value problems such as those occurring in the investigation of electrical circuits and mechanical vibrations.

The Laplace transform finds particular application in the field of signals and linear systems analysis. A distinguishing feature of a system is that when it is subjected to an excitation (input), it produces a response (output). When the input $u(t)$ and output $x(t)$ are functions of a single variable t, representing time, it is normal to refer to them as **signals**. Schematically, a system may be represented as in Figure 11.1. The problem facing the engineer is that of determining the system output $x(t)$ when it is subjected to an input $u(t)$ applied at some instant of time, which we can take to be $t = 0$. The relationship between output and input is determined by the laws governing the behaviour of the system. If the system is linear and time-invariant then the output is related to the input by a linear differential equation with constant coefficients, and we have a standard initial-value problem, which is amenable to solution using the Laplace transform.

While many of the problems considered in this chapter can be solved by the classical approach of Chapter 10, the Laplace transform leads to a more unified approach and provides the engineer with greater insight into system behaviour. In practice, the input signal $u(t)$ may be a discontinuous or periodic function, or even a pulse, and in such cases the use of the Laplace transform has distinct advantages over the classical approach. Also, more often than not, an engineer is interested not only in system analysis but also in system synthesis or design. Consequently, an engineer's objective in studying a system's response to specific inputs is frequently to learn more about the system with a view to improving or controlling it so that it satisfies certain specifications. It is in this area that the use of the Laplace transform is attractive, since by considering the system response to particular inputs, such as a sinusoid, it provides the engineer with powerful graphical methods for system design that are relatively easy to apply and widely used in practice.

In modelling the system by a differential equation, it has been assumed that both the input and output signals can vary at any instant of time; that is, they are functions of a continuous time variable (note that this does not mean that the signals themselves have to be continuous functions of time). Such systems are called **continuous-time systems**, and it is for investigating these that the Laplace transform is best suited. With the introduction of computer control into system design, signals associated with a system may only change at discrete instants of time. In such cases the system is said to be a **discrete-time system**, and is modelled by a difference equation rather than a differential equation. Such systems are dealt with using the z transform considered in the companion text, *Advanced Modern Engineering Mathematics*.

In this chapter we restrict our consideration to simply introducing the Laplace transform and to illustrating its use in solving differential equations. Its more extensive role in engineering applications is dealt with in the companion text.

There is some overlap in the material covered in this chapter and in Chapter 10, particularly in relation to the modelling aspects of applications to electrical circuits and mechanical vibrations. This overlap has been included so that the two approaches to solving differential equations can be studied independently of each other.

11.2 The Laplace transform

11.2.1 Definition and notation

We define the Laplace transform of a function $f(t)$ by the expression

$$\mathcal{L}\{f(t)\} = \int_0^\infty e^{-st}f(t)\,dt \qquad (11.1)$$

where s is a complex variable and e^{-st} is called the **kernel** of the transformation.

It is usual to represent the Laplace transform of a function by the corresponding capital letter, so that we write

$$\mathcal{L}\{f(t)\} = F(s) = \int_0^\infty e^{-st}f(t)\,dt \qquad (11.2)$$

An alternative notation in common use is to denote $\mathcal{L}\{f(t)\}$ by $\bar{f}(s)$ or simply $\bar{f}$.

Before proceeding, there are a few observations relating to the definition (11.2) worthy of comment.

(a) The symbol $\mathcal{L}$ denotes the **Laplace transform operator**; when it operates on a function $f(t)$, it transforms it into a function $F(s)$ of the complex variable s. We say the operator transforms the function $f(t)$ in the t domain (usually called the **time domain**) into the function $F(s)$ in the s domain (usually called the **complex frequency domain**, or simply the **frequency domain**). This relationship is depicted graphically in Figure 11.2, and it is usual to refer to $f(t)$ and $F(s)$ as a **Laplace transform pair**, written as $\{f(t), F(s)\}$.

Figure 11.2
The Laplace transform operator.

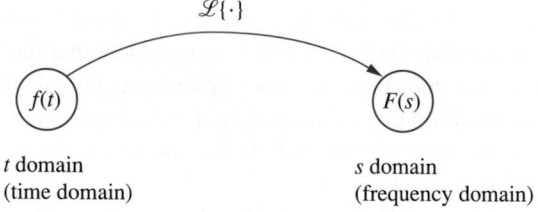

$\mathcal{L}\{\cdot\}$

$f(t)$ $\qquad\qquad\qquad$ $F(s)$

t domain $\qquad\qquad$ s domain
(time domain) $\qquad\quad$ (frequency domain)

(b) Because the upper limit in the integral is infinite, the domain of integration is infinite. Thus the integral is an example of an **improper integral**, as introduced in Chapter 9, Section 9.2; that is,

$$\int_0^\infty e^{-st}f(t)\,dt = \lim_{T\to\infty}\int_0^T e^{-st}f(t)\,dt$$

This immediately raises the question of whether or not the integral converges, an issue we shall consider in Section 11.2.3.

(c) Because the lower limit in the integral is zero, it follows that when taking the Laplace transform, the behaviour of $f(t)$ for negative values of t is ignored or suppressed. This means that $F(s)$ contains information on the behaviour of $f(t)$ only for $t \geq 0$, so that the Laplace transform is not a suitable tool for investigating problems in which values of $f(t)$ for $t < 0$ are relevant. In most engineering applications this does not cause any problems, since we are then concerned with physical systems for which the functions we are dealing with vary with time t. An attribute of physical realizable systems is that they are **non-anticipatory** in the sense that there is no output (or response) until an input (or excitation) is applied. Because of this causal relationship between the input and output, we define a function $f(t)$ to be **causal** if $f(t) = 0$ ($t < 0$). In general, however, unless the domain is clearly specified, a function $f(t)$ is normally intepreted as being defined for all real values, both positive and negative, of t. Making use of the Heaviside unit step function $H(t)$ (see also Chapter 2, Section 2.8.3), where

$$H(t) = \begin{cases} 0 & (t < 0) \\ 1 & (t \geq 0) \end{cases}$$

we have

$$f(t)H(t) = \begin{cases} 0 & (t < 0) \\ f(t) & (t \geq 0) \end{cases}$$

Thus the effect of multiplying $f(t)$ by $H(t)$ is to convert it into a causal function. Graphically, the relationship between $f(t)$ and $f(t)H(t)$ is as shown in Figure 11.3.

Figure 11.3
Graph of $f(t)$ and its causal equivalent function.

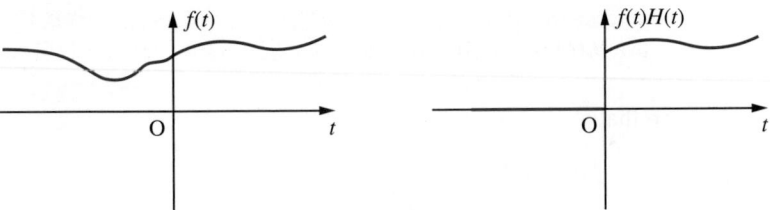

It follows that the corresponding Laplace transform $F(s)$ contains full information on the behaviour of $f(t)H(t)$. Consequently, strictly speaking one should refer to $\{ f(t)H(t), F(s)\}$ rather than $\{ f(t), F(s)\}$ as being a Laplace transform pair. However, it is common practice to drop the $H(t)$ and assume that we are dealing with causal functions.

(d) If the behaviour of $f(t)$ for $t < 0$ is of interest then we need to use the alternative **two-sided** or **bilateral Laplace transform** of the function $f(t)$, defined by

$$\mathscr{L}_\text{B}\{ f(t)\} = \int_{-\infty}^{\infty} \text{e}^{-st}f(t)\text{d}t \tag{11.3}$$

The Laplace transform defined by (11.2), with lower limit zero, is sometimes referred to as the **one-sided** or **unilateral Laplace transform** of the function $f(t)$. In this chapter we shall concern ourselves only with the latter transform, and refer to it simply as the Laplace transform of the function $f(t)$. Note that when $f(t)$ is a causal function,

$$\mathscr{L}_\text{B}\{ f(t)\} = \mathscr{L}\{ f(t)\}$$

11.2.2 Transforms of simple functions

In this section we obtain the Laplace transformations of some simple functions.

Example 11.1 Determine the Laplace transform of the function

$$f(t) = c$$

where c is a constant.

Solution Using the definition (11.2),

$$\mathcal{L}(c) = \int_0^\infty e^{-st} c \, dt = \lim_{T \to \infty} \int_0^T e^{-st} c \, dt$$

$$= \lim_{T \to \infty} \left[-\frac{c}{s} e^{-st} \right]_0^T = \frac{c}{s} \left(1 - \lim_{T \to \infty} e^{-sT} \right)$$

Taking $s = \sigma + j\omega$, where σ and ω are real,

$$\lim_{T \to \infty} e^{-sT} = \lim_{T \to \infty} (e^{-(\sigma + j\omega)T}) = \lim_{T \to \infty} e^{-\sigma T}(\cos \omega T - j\sin \omega T)$$

A finite limit exists provided that $\sigma = \mathrm{Re}(s) > 0$, when the limit is zero. Thus, provided that $\mathrm{Re}(s) > 0$, the Laplace transform is

$$\mathcal{L}(c) = \frac{c}{s}, \quad \mathrm{Re}(s) > 0$$

so that

$$\left. \begin{aligned} f(t) &= c \\ F(s) &= \frac{c}{s} \end{aligned} \right\} \quad \mathrm{Re}(s) > 0 \tag{11.4}$$

constitute an example of a Laplace transform pair.

Example 11.2 Determine the Laplace transform of the ramp function

$$f(t) = t$$

Solution From the definition (11.2),

$$\mathcal{L}\{t\} = \int_0^\infty e^{-st} t \, dt = \lim_{T \to \infty} \int_0^T e^{-st} t \, dt$$

$$= \lim_{T \to \infty} \left[-\frac{t}{s} e^{-st} - \frac{e^{-st}}{s^2} \right]_0^T = \frac{1}{s^2} - \lim_{T \to \infty} \frac{T e^{-sT}}{s} - \lim_{T \to \infty} \frac{e^{-sT}}{s^2}$$

Following the same procedure as in Example 11.1, limits exist provided that $\mathrm{Re}(s) > 0$, when

$$\lim_{T \to \infty} \frac{Te^{-sT}}{s} = \lim_{T \to \infty} \frac{e^{-sT}}{s^2} = 0$$

Thus, provided that $\mathrm{Re}(s) > 0$,

$$\mathscr{L}\{t\} = \frac{1}{s^2}$$

giving us the Laplace transform pair

$$\left. \begin{array}{l} f(t) = t \\[2mm] F(s) = \dfrac{1}{s^2} \end{array} \right\} \quad \mathrm{Re}(s) > 0 \tag{11.5}$$

Example 11.3   Determine the Laplace transform of the one-sided exponential function

$$f(t) = e^{kt}$$

Solution   The definition (11.2) gives

$$\mathscr{L}\{e^{kt}\} = \int_0^\infty e^{-st}e^{kt}\,dt = \lim_{T \to \infty} \int_0^T e^{-(s-k)t}\,dt$$

$$= \lim_{T \to \infty} \frac{-1}{s-k}[e^{-(s-k)t}]_0^T = \frac{1}{s-k}\left(1 - \lim_{T \to \infty} e^{-(s-k)T}\right)$$

Writing $s = \sigma + j\omega$, where σ and ω are real, we have

$$\lim_{T \to \infty} e^{-(s-k)T} = \lim_{T \to \infty} e^{-(\sigma-k)T}e^{-j\omega T}$$

If k is real, then, provided that $\sigma = \mathrm{Re}(s) > k$, the limit exists, and is zero. If k is complex, say $k = a + jb$, then the limit will also exist, and be zero, provided that $\sigma > a$ (that is, $\mathrm{Re}(s) > \mathrm{Re}(k)$). Under these conditions, we then have

$$\mathscr{L}\{e^{kt}\} = \frac{1}{s-k}$$

giving us the Laplace transform pair

$$\left. \begin{array}{l} f(t) = e^{kt} \\[2mm] F(s) = \dfrac{1}{s-k} \end{array} \right\} \quad \mathrm{Re}(s) > \mathrm{Re}(k) \tag{11.6}$$

Example 11.4   Determine the Laplace transforms of the sine and cosine functions

$$f(t) = \sin at, \quad g(t) = \cos at$$

where a is a real constant.

Entering the further commands

```
simple(ans);
pretty(ans)
```

returns

```
2/(s² + 6s + 13)
```

as an alternative form of the answer. Note that the last two commands could be replaced by the single command `pretty(simple(ans))`.

In MAPLE the commands

```
with(inttrans):
laplace(exp(-3*t)*sin(2*t),t,s);
```

return the answer

$$2\frac{1}{(s+3)^2 + 4}$$

There is no 'simple' command in MAPLE.

The function $\mathrm{e}^{-3t}\sin 2t$ in Example 11.10 is a member of a general class of functions called **damped sinusoids**. These play an important role in the study of engineering systems, particularly in the analysis of vibrations. For this reason, we add the following two general members of the class to our standard library of Laplace transform pairs:

$$\mathcal{L}\{\mathrm{e}^{-kt}\sin at\} = \frac{a}{(s+k)^2 + a^2}, \quad \mathrm{Re}(s) > -k \tag{11.9}$$

$$\mathcal{L}\{\mathrm{e}^{-kt}\cos at\} = \frac{s+k}{(s+k)^2 + a^2}, \quad \mathrm{Re}(s) > -k \tag{11.10}$$

where in both cases k and a are real constants.

Property 11.3: Derivative-of-transform property

This property relates operations in the time domain to those in the transformed s domain, but initially we shall simply look upon it as a method of increasing our repertoire of Laplace transform pairs. The property is also sometimes referred to as the **multiplication-by-t** property. A statement of the property is contained in the following theorem.

Theorem 11.3 Derivative of transform

If $f(t)$ is a function having Laplace transform

$$F(s) = \mathcal{L}\{f(t)\}, \quad \mathrm{Re}(s) > \sigma_\mathrm{c}$$

then the functions $t^n f(t)$ ($n = 1, 2, \dots$) also have Laplace transforms, given by

$$\mathcal{L}\{t^n f(t)\} = (-1)^n \frac{\mathrm{d}^n F(s)}{\mathrm{d}s^n}, \quad \mathrm{Re}(s) > \sigma_\mathrm{c}$$

Proof   By definition,

$$\mathcal{L}\{f(t)\} = F(s) = \int_0^\infty e^{-st}f(t)\mathrm{d}t$$

so that

$$\frac{\mathrm{d}^n F(s)}{\mathrm{d}s^n} = \frac{\mathrm{d}^n}{\mathrm{d}s^n}\int_0^\infty e^{-st}f(t)\mathrm{d}t$$

Owing to the convergence properties of the improper integral involved, we can interchange the operations of differentiation and integration and differentiate with respect to s under the integral sign. Thus

$$\frac{\mathrm{d}^n F(s)}{\mathrm{d}s^n} = \int_0^\infty \frac{\partial^n}{\partial s^n}[e^{-st}f(t)]\mathrm{d}t$$

which, on carrying out the repeated differentiation, gives

$$\frac{\mathrm{d}^n F(s)}{\mathrm{d}s^n} = (-1)^n\int_0^\infty e^{-st}t^n f(t)\mathrm{d}t = (-1)^n\mathcal{L}\{t^n f(t)\}, \quad \mathrm{Re}(s) > \sigma_c$$

the region of convergence remaining unchanged.

end of theorem

In other words, Theorem 11.3 says that differentiating the transform of a function with respect to s is equivalent to multiplying the function itself by $-t$. As with the previous properties, we can now use this result to add to our list of Laplace transform pairs.

Example 11.11   Determine $\mathcal{L}\{t\sin 3t\}$.

Solution   Using the result (11.7),

$$\mathcal{L}\{\sin 3t\} = F(s) = \frac{3}{s^2 + 9}, \quad \mathrm{Re}(s) > 0$$

so, by the derivative theorem,

$$\mathcal{L}\{t\sin 3t\} = -\frac{\mathrm{d}F(s)}{\mathrm{d}s} = \frac{6s}{(s^2 + 9)^2}, \quad \mathrm{Re}(s) > 0$$

In MATLAB the commands

```
syms s t
laplace(t*sin(3*t))
```

return

```
ans = 1/(s^2 + 9)*sin(2*atan(3/5))
```

Example 11.16   Find

$$\mathscr{L}^{-1}\left\{\frac{1}{(s+3)(s-2)}\right\}$$

Solution   First $1/(s+3)(s-2)$ is resolved into partial fractions, giving

$$\frac{1}{(s+3)(s-2)} = \frac{-\frac{1}{5}}{s+3} + \frac{\frac{1}{5}}{s-2}$$

Then, using the result $\mathscr{L}^{-1}\{1/(s+a)\} = e^{-at}$ together with the linearity property, we have

$$\mathscr{L}^{-1}\left\{\frac{1}{(s+3)(s-2)}\right\} = -\tfrac{1}{5}\mathscr{L}^{-1}\left\{\frac{1}{s+3}\right\} + \tfrac{1}{5}\mathscr{L}^{-1}\left\{\frac{1}{s-2}\right\} = -\tfrac{1}{5}e^{-3t} + \tfrac{1}{5}e^{2t}$$

Using MATLAB or MAPLE the commands

MATLAB	MAPLE
`syms s t`	`with(inttrans):`
`ilaplace(1/((s + 3) *`	`ilaplace(1/((s + 3) *`
`(s - 2)));`	`(s - 2)),s,t);`
`pretty(ans)`	

return the anwers

$$-1/5exp(-3t) + 1/5exp(2t) \qquad -\tfrac{1}{5}e^{(-3t)} + \tfrac{1}{5}e^{(2t)}$$

Example 11.17   Find

$$\mathscr{L}^{-1}\left\{\frac{s+1}{s^2(s^2+9)}\right\}$$

Solution   Resolving $(s+1)/s^2(s^2+9)$ into partial fractions gives

$$\frac{s+1}{s^2(s^2+9)} = \frac{\frac{1}{9}}{s} + \frac{\frac{1}{9}}{s^2} - \tfrac{1}{9}\frac{s+1}{s^2+9}$$

$$= \frac{\frac{1}{9}}{s} + \frac{\frac{1}{9}}{s^2} - \tfrac{1}{9}\frac{s}{s^2+3^2} - \tfrac{1}{27}\frac{3}{s^2+3^2}$$

Using the results in Figure 11.5, together with the linearity property, we have

$$\mathscr{L}^{-1}\left\{\frac{s+1}{s^2(s^2+9)}\right\} = \tfrac{1}{9} + \tfrac{1}{9}t - \tfrac{1}{9}\cos 3t - \tfrac{1}{27}\sin 3t$$

Using MATLAB or MAPLE check that the answer can be verified using the following commands:

MATLAB	MAPLE
| ```
syms s t
ilaplace((s + 1)/
(s^2*(s^2 + 9)));
pretty(ans)
``` | ```
with(inttrans):
invlaplace((s + 1)/
(s^2*(s^2 + 9)),s,t);
``` |

11.2.9   Inversion using the first shift theorem

In Theorem 11.2 we saw that if $F(s)$ is the Laplace transform of $f(t)$ then, for a scalar a, $F(s - a)$ is the Laplace transform of $e^{at}f(t)$. This theorem normally causes little difficulty when used to obtain the Laplace transforms of functions, but it does frequently lead to problems when used to obtain inverse transforms. Expressed in the inverse form, the theorem becomes

$$\mathcal{L}^{-1}\{F(s - a)\} = e^{at}f(t)$$

The notation

$$\mathcal{L}^{-1}\{[F(s)]_{s \to s-a}\} = e^{at}[f(t)]$$

where $F(s) = \mathcal{L}\{f(t)\}$ and $[F(s)]_{s \to s-a}$ denotes that s in $F(s)$ is replaced by $s - a$, may make the relation clearer.

Example 11.18   Find

$$\mathcal{L}^{-1}\left\{\frac{1}{(s + 2)^2}\right\}$$

Solution

$$\frac{1}{(s + 2)^2} = \left[\frac{1}{s^2}\right]_{s \to s+2}$$

and, since $1/s^2 = \mathcal{L}\{t\}$, the shift theorem gives

$$\mathcal{L}^{-1}\left\{\frac{1}{(s + 2)^2}\right\} = te^{-2t}$$

Check the answer using MATLAB or MAPLE.

Example 11.19   Find

$$\mathcal{L}^{-1}\left\{\frac{2}{s^2 + 6s + 13}\right\}$$

Solution

$$\frac{2}{s^2 + 6s + 13} = \frac{2}{(s+3)^2 + 4} = \left[\frac{2}{s^2 + 2^2}\right]_{s \to s+3}$$

and, since $2/(s^2 + 2^2) = \mathcal{L}\{\sin 2t\}$, the shift theorem gives

$$\mathcal{L}^{-1}\left\{\frac{2}{s^2 + 6s + 13}\right\} = e^{-3t}\sin 2t$$

The MATLAB commands

```
syms s t
ilaplace(2/(s^2 + 6*s + 13);
pretty(simple(ans))
```

return

```
ans = -1/2i(exp((-3 + 2i)t) - exp((-3 - 2i)t))
```

The MAPLE commands

```
with(inttrans):
invlaplace(2/(s^2 + 6*s + 13),s,t);
simplify(%);
```

return the same answer.

To obtain the same format as provided in the solution further manipulation is required as follows:

$$1/2i[-e^{-3t}e^{2it} + e^{-3t}e^{-2it}] = e^{-3t}((e^{2it} - e^{-2it})/(2i)) = e^{-3t}\sin 2t$$

Example 11.20

Find

$$\mathcal{L}^{-1}\left\{\frac{s + 7}{s^2 + 2s + 5}\right\}$$

Solution

$$\frac{s + 7}{s^2 + 2s + 5} = \frac{s + 7}{(s+1)^2 + 4} = \frac{(s+1)}{(s+1)^2 + 4} + 3\frac{2}{(s+1)^2 + 4}$$

$$= \left[\frac{s}{s^2 + 2^2}\right]_{s \to s+1} + 3\left[\frac{2}{s^2 + 2^2}\right]_{s \to s+1}$$

Since $s/(s^2 + 2^2) = \mathcal{L}\{\cos 2t\}$ and $2/(s^2 + 2^2) = \mathcal{L}\{\sin 2t\}$, the shift theorem gives

$$\mathcal{L}^{-1}\left\{\frac{s + 7}{s^2 + 2s + 5}\right\} = e^{-t}\cos 2t + 3e^{-t}\sin 2t$$

Example 11.21

Find

$$\mathcal{L}^{-1}\left\{\frac{1}{(s+1)^2(s^2 + 4)}\right\}$$

Solution Resolving $1/(s + 1)^2(s^2 + 4)$ into partial fractions gives

$$\frac{1}{(s + 1)^2(s^2 + 4)} = \frac{\frac{2}{25}}{s + 1} + \frac{\frac{1}{5}}{(s + 1)^2} - \frac{1}{25}\frac{2s + 3}{s^2 + 4}$$

$$= \frac{\frac{2}{25}}{s + 1} + \frac{1}{5}\left[\frac{1}{s^2}\right]_{s \to s+1} - \frac{2}{25}\frac{s}{s^2 + 2^2} - \frac{3}{50}\frac{2}{s^2 + 2^2}$$

Since $1/s^2 = \mathscr{L}\{t\}$, the shift theorem, together with the results in Figure 11.5, gives

$$\mathscr{L}^{-1}\left\{\frac{1}{(s + 1)^2(s^2 + 4)}\right\} = \tfrac{2}{25}e^{-t} + \tfrac{1}{5}e^{-t}t - \tfrac{2}{25}\cos 2t - \tfrac{3}{50}\sin 2t$$

 Check your answers to Examples 11.20 and 11.21 using MATLAB or MAPLE.

11.2.10 Exercise

 Check your answers using MATLAB or MAPLE.

4 Find $\mathscr{L}^{-1}\{F(s)\}$ when $F(s)$ is given by

(a) $\dfrac{1}{(s + 3)(s + 7)}$

(b) $\dfrac{s + 5}{(s + 1)(s - 3)}$

(c) $\dfrac{s - 1}{s^2(s + 3)}$

(d) $\dfrac{2s + 6}{s^2 + 4}$

(e) $\dfrac{1}{s^2(s^2 + 16)}$

(f) $\dfrac{s + 8}{s^2 + 4s + 5}$

(g) $\dfrac{s + 1}{s^2(s^2 + 4s + 8)}$

(h) $\dfrac{4s}{(s - 1)(s + 1)^2}$

(i) $\dfrac{s + 7}{s^2 + 2s + 5}$

(j) $\dfrac{3s^2 - 7s + 5}{(s - 1)(s - 2)(s - 3)}$

(k) $\dfrac{5s - 7}{(s + 3)(s^2 + 2)}$

(l) $\dfrac{s}{(s - 1)(s^2 + 2s + 2)}$

(m) $\dfrac{s - 1}{s^2 + 2s + 5}$

(n) $\dfrac{s - 1}{(s - 2)(s - 3)(s - 4)}$

(o) $\dfrac{3s}{(s - 1)(s^2 - 4)}$

(p) $\dfrac{36}{s(s^2 + 1)(s^2 + 9)}$

(q) $\dfrac{2s^2 + 4s + 9}{(s + 2)(s^2 + 3s + 3)}$

(r) $\dfrac{1}{(s + 1)(s + 2)(s^2 + 2s + 10)}$

11.3 Solution of differential equations

We first consider the Laplace transforms of derivatives and integrals, and then apply these to the solution of differential equations.

11.3.1 Transforms of derivatives

If we are to use Laplace transform methods to solve differential equations, we need to find convenient expressions for the Laplace transforms of derivatives such as df/dt, d^2f/dt^2 or, in general, d^nf/dt^n. By definition,

$$\mathcal{L}\left\{\frac{df}{dt}\right\} = \int_0^\infty e^{-st}\frac{df}{dt}\,dt$$

Integrating by parts, we have

$$\mathcal{L}\left\{\frac{df}{dt}\right\} = [e^{-st}f(t)]_0^\infty + s\int_0^\infty e^{-st}f(t)\,dt$$

$$= -f(0) + sF(s)$$

that is,

$$\mathcal{L}\left\{\frac{df}{dt}\right\} = sF(s) - f(0) \tag{11.12}$$

In taking the Laplace transform of a derivative we have assumed that $f(t)$ is continuous at $t = 0$, so that $f(0^-) = f(0) = f(0^+)$. In the companion text *Advanced Modern Engineering Mathematics* there are occasions when $f(0^-) \neq f(0^+)$ and we have to revert to a more generalized calculus to resolve the problem.

The advantage of using the Laplace transform when dealing with differential equations can readily be seen, since it enables us to replace the operation of differentiation in the time domain by a simple algebraic operation in the s domain.

Note that to deduce the result (11.12), we have assumed that $f(t)$ is continuous, with a piecewise-continuous derivative df/dt, for $t \geqslant 0$ and that it is also of exponential order as $t \to \infty$.

Likewise, if both $f(t)$ and df/dt are continuous on $t \geqslant 0$ and are of exponential order as $t \to \infty$, and d^2f/dt^2 is piecewise-continuous for $t \geqslant 0$, then

$$\mathcal{L}\left\{\frac{d^2f}{dt^2}\right\} = \int_0^\infty e^{-st}\frac{d^2f}{dt^2}\,dt = \left[e^{-st}\frac{df}{dt}\right]_0^\infty + s\int_0^\infty e^{-st}\frac{df}{dt}\,dt$$

$$= -\left[\frac{df}{dt}\right]_{t=0} + s\mathcal{L}\left\{\frac{df}{dt}\right\}$$

which, on using (11.11), gives

$$\mathcal{L}\left\{\frac{d^2f}{dt^2}\right\} = -\left[\frac{df}{dt}\right]_{t=0} + s[sF(s) - f(0)]$$

leading to the result

$$\mathcal{L}\left\{\frac{d^2f}{dt^2}\right\} = s^2F(s) - sf(0) - \left[\frac{df}{dt}\right]_{t=0} = s^2F(s) - sf(0) - f^{(1)}(0) \tag{11.13}$$

Clearly, provided that $f(t)$ and its derivatives satisfy the required conditions, this procedure may be extended to obtain the Laplace transform of $f^{(n)}(t) = d^nf/dt^n$ in the form

$$\mathcal{L}\{f^{(n)}(t)\} = s^n F(s) - s^{n-1}f(0) - s^{n-2}f^{(1)}(0) - \ldots - f^{(n-1)}(0)$$

$$= s^n F(s) - \sum_{i=1}^{n} s^{n-i} f^{(i-1)}(0) \qquad (11.14)$$

a result that may be readily proved by induction.

Again it is noted that in determining the Laplace transform of $f^{(n)}(t)$ we have assumed that $f^{(n-1)}(t)$ is continuous.

11.3.2 Transforms of integrals

In some applications the behaviour of a system may be represented by an **integro-differential equation**, which is an equation containing both derivatives and integrals of the unknown variable. For example, the current i in a series electrical circuit consisting of a resistance R, an inductance L and capacitance C, and subject to an applied voltage E, is given by

$$L\frac{di}{dt} + iR + \frac{1}{C}\int_0^t i(\tau)\mathrm{d}\tau = E$$

To solve such equations directly, it is convenient to be able to obtain the Laplace transform of integrals such as $\int_0^t f(\tau)\mathrm{d}\tau$.

Writing

$$g(t) = \int_0^t f(\tau)\mathrm{d}\tau$$

we have

$$\frac{\mathrm{d}g}{\mathrm{d}t} = f(t), \quad g(0) = 0$$

Taking Laplace transforms,

$$\mathcal{L}\left\{\frac{\mathrm{d}g}{\mathrm{d}t}\right\} = \mathcal{L}\{f(t)\}$$

which, on using (11.12), gives

$$sG(s) = F(s)$$

or

$$\mathcal{L}\{g(t)\} = G(s) = \frac{1}{s}F(s) = \frac{1}{s}\mathcal{L}\{f(t)\}$$

leading to the result

$$\mathcal{L}\left\{\int_0^t f(\tau)\mathrm{d}\tau\right\} = \frac{1}{s}\mathcal{L}\{f(t)\} = \frac{1}{s}F(s) \qquad (11.15)$$

Example 11.22 Obtain

$$\mathcal{L}\left\{\int_0^t (\tau^3 + \sin 2\tau)\mathrm{d}\tau\right\}$$

Solution In this case $f(t) = t^3 + \sin 2t$, giving

$$F(s) = \mathcal{L}\{f(t)\} = \mathcal{L}\{t^3\} + \mathcal{L}\{\sin 2t\}$$

$$= \frac{6}{s^4} + \frac{2}{s^2 + 4}$$

so, by (11.15),

$$\mathcal{L}\left\{\int_0^t (\tau^3 + \sin 2\tau)\mathrm{d}\tau\right\} = \frac{1}{s}F(s) = \frac{6}{s^5} + \frac{2}{s(s^2 + 4)}$$

11.3.3 Ordinary differential equations

Having obtained expressions for the Laplace transforms of derivatives, we are now in a position to use Laplace transform methods to solve ordinary linear differential equations with constant coefficients, which were introduced in Chapter 10. To illustrate this, consider the general second-order linear differential equation

$$a\frac{\mathrm{d}^2 x}{\mathrm{d}t^2} + b\frac{\mathrm{d}x}{\mathrm{d}t} + cx = u(t) \quad (t \geqslant 0) \tag{11.16}$$

subject to the initial conditions $x(0) = x_0$, $\dot{x}(0) = v_0$ where as usual a dot denotes differentiation with respect to time, t. Such a differential equation may model the dynamics of some system for which the variable $x(t)$ determines the **response** of the system to the **forcing** or **excitation** term $u(t)$. The terms **system input** and **system output** are also frequently used for $u(t)$ and $x(t)$ respectively. Since the differential equation is linear and has constant coefficients, a system characterized by such a model is said to be a **linear time-invariant system**.

Taking Laplace transforms of each term in (11.16) gives

$$a\mathcal{L}\left\{\frac{\mathrm{d}^2 x}{\mathrm{d}t^2}\right\} + b\mathcal{L}\left\{\frac{\mathrm{d}x}{\mathrm{d}t}\right\} + c\mathcal{L}\{x\} = \mathcal{L}\{u(t)\}$$

which on using (11.12) and (11.13) leads to

$$a[s^2 X(s) - sx(0) - \dot{x}(0)] + b[sX(s) - x(0)] + cX(s) = U(s)$$

Rearranging, and incorporating the given initial conditions, gives

$$(as^2 + bs + c)X(s) = U(s) + (as + b)x_0 + av_0$$

so that

$$X(s) = \frac{U(s) + (as + b)x_0 + av_0}{as^2 + bs + c} \tag{11.17}$$

Equation (11.17) determines the Laplace transform $X(s)$ of the response, from which, by taking the inverse transform, the desired time response $x(t)$ may be obtained.

Before considering specific examples, there are a few observations worth noting at this stage.

(a) As we have already noted in Section 11.3.1, a distinct advantage of using the Laplace transform is that it enables us to replace the operation of differentiation by an algebraic operation. Consequently, by taking the Laplace transform of each term in a differential equation, it is converted into an algebraic equation in the variable s. This may then be rearranged using algebraic rules to obtain an expression for the Laplace transform of the response; the desired time response is then obtained by taking the inverse transform.

(b) The Laplace transform method yields the complete solution to the linear differential equation, with the initial conditions automatically included. This contrasts with the classical approach adopted in Chapter 10, in which the general solution consists of two components, the **complementary function** and the **particular integral**, with the initial conditions determining the undetermined constants associated with the complementary function. When the solution is expressed in the general form (11.17), upon inversion the term involving $U(s)$ leads to a particular integral while that involving x_0 and v_0 gives a complementary function. A useful side issue is that an explicit solution for the transient is obtained that reflects the initial conditions.

(c) The Laplace transform method is ideally suited for solving initial-value problems; that is, linear differential equations in which all the initial conditions $x(0)$, $\dot{x}(0)$, and so on, at time $t = 0$ are specified. The method is less attractive for boundary-value problems, when the conditions on $x(t)$ and its derivatives are not all specified at $t = 0$, but some are specified at other values of the independent variable. It is still possible, however, to use the Laplace transform method by assigning arbitrary constants to one or more of the initial conditions and then determining their values using the given boundary conditions.

(d) It should be noted that the denominator of the right-hand side of (11.17) is the left-hand side of (11.16) with the operator d/dt replaced by s. The denominator equated to zero also corresponds to the auxiliary equation or characteristic equation used in the classical approach. Given a specific initial-value problem, the process of obtaining a solution using Laplace transform methods is fairly straightforward, and is illustrated by Example 11.23.

Example 11.23   Solve the differential equation

$$\frac{d^2 x}{dt^2} + 5\frac{dx}{dt} + 6x = 2e^{-t} \quad (t \geqslant 0)$$

subject to the initial conditions $x = 1$ and $dx/dt = 0$ at $t = 0$.

Solution   Taking Laplace transforms

$$\mathcal{L}\left\{\frac{d^2 x}{dt^2}\right\} + 5\mathcal{L}\left\{\frac{dx}{dt}\right\} + 6\mathcal{L}\{x\} = 2\mathcal{L}\{e^{-t}\}$$

leads to the transformed equation

$$[s^2X(s) - sx(0) - \dot{x}(0)] + 5[sX(s) - x(0)] + 6X(s) = \frac{2}{s+1}$$

which on rearrangement gives

$$(s^2 + 5s + 6)X(s) = \frac{2}{s+1} + (s+5)x(0) + \dot{x}(0)$$

Incorporating the given initial conditions $x(0) = 1$ and $\dot{x}(0) = 0$ leads to

$$(s^2 + 5s + 6)X(s) = \frac{2}{s+1} + s + 5$$

That is,

$$X(s) = \frac{2}{(s+1)(s+2)(s+3)} + \frac{s+5}{(s+3)(s+2)}$$

Resolving the rational terms into partial fractions gives

$$X(s) = \frac{1}{s+1} - \frac{2}{s+2} + \frac{1}{s+3} + \frac{3}{s+2} - \frac{2}{s+3}$$

$$= \frac{1}{s+1} + \frac{1}{s+2} - \frac{1}{s+3}$$

Taking inverse transforms gives the desired solution

$$x(t) = e^{-t} + e^{-2t} - e^{-3t} \quad (t \geqslant 0)$$

In principle the procedure adopted in Example 11.23 for solving a second-order linear differential equation with constant coefficients is readily carried over to higher-order differential equations. A general nth-order linear differential equation may be written as

$$a_n\frac{d^n x}{dt^n} + a_{n-1}\frac{d^{n-1}x}{dt^{n-1}} + \ldots + a_0 x = u(t) \quad (t \geqslant 0) \tag{11.18}$$

where a_n, a_{n-1}, $\ldots$, a_0 are constants, with $a_n \neq 0$. This may be written in the more concise form

$$q(D)x(t) = u(t) \tag{11.19}$$

where D denotes the operator d/dt and $q(D)$ is the polynomial

$$q(D) = \sum_{r=0}^{n} a_r D^r$$

The objective is then to determine the response $x(t)$ for a given forcing function $u(t)$ subject to the given set of initial conditions

$$D^r x(0) = \left[\frac{d^r x}{dt^r} \right]_{t=0} = c_r \quad (r = 0, 1, \ldots, n-1)$$

Taking Laplace transforms in (11.19) and proceeding as before leads to

$$X(s) = \frac{p(s)}{q(s)}$$

where

$$p(s) = U(s) + \sum_{r=0}^{n-1} c_r \sum_{i=r+1}^{n} a_i s^{i-r-1}$$

Then, in principle, by taking the inverse transform, the desired response $x(t)$ may be obtained as

$$x(t) = \mathcal{L}^{-1}\left\{ \frac{p(s)}{q(s)} \right\}$$

For high-order differential equations the process of performing this inversion may prove to be rather tedious, and matrix methods may be used as indicated in Chapter 6 of the companion text, *Advanced Modern Engineering Mathematics*.

To conclude this section, further worked examples are developed in order to help consolidate understanding of this method for solving linear differential equations.

Example 11.24 Solve the differential equation

$$\frac{d^2 x}{dt^2} + 6\frac{dx}{dt} + 9x = \sin t \quad (t \geq 0)$$

subject to the initial conditions $x = 0$ and $dx/dt = 0$ at $t = 0$.

Solution Taking the Laplace transforms

$$\mathcal{L}\left\{ \frac{d^2 x}{dt^2} \right\} + 6\mathcal{L}\left\{ \frac{dx}{dt} \right\} + 9\mathcal{L}\{x\} = \mathcal{L}\{\sin t\}$$

leads to the equation

$$[s^2 X(s) - sx(0) - \dot{x}(0)] + 6[sX(s) - x(0)] + 9X(s) = \frac{1}{s^2 + 1}$$

which on rearrangement gives

$$(s^2 + 6s + 9)X(s) = \frac{1}{s^2 + 1} + (s + 6)x(0) + \dot{x}(0)$$

Incorporating the given initial conditions $x(0) = \dot{x}(0) = 0$ leads to

$$X(s) = \frac{1}{(s^2 + 1)(s + 3)^2}$$

Resolving into partial fractions gives

$$X(s) = \tfrac{3}{50}\frac{1}{s+3} + \tfrac{1}{10}\frac{1}{(s+3)^2} + \tfrac{2}{25}\frac{1}{s^2+1} - \tfrac{3}{50}\frac{s}{s^2+1}$$

that is,

$$X(s) = \tfrac{3}{50}\frac{1}{s+3} + \tfrac{1}{10}\left[\frac{1}{s^2}\right]_{s\to s+3} + \tfrac{2}{25}\frac{1}{s^2+1} - \tfrac{3}{50}\frac{s}{s^2+1}$$

Taking inverse transforms, using the shift theorem, leads to the desired solution

$$x(t) = \tfrac{3}{50}e^{-3t} + \tfrac{1}{10}te^{-3t} + \tfrac{2}{25}\sin t - \tfrac{3}{50}\cos t \quad (t \geqslant 0)$$

In MATLAB, using the Symbolic Math Toolbox, the command $dsolve$ computes symbolic solutions to differential equations. The letter D denotes differentiation whilst the symbols $D2, D3, \ldots, DN$ denote the 2nd, 3rd, $\ldots$, Nth derivatives respectively. The dependent variable is that preceded by D whilst the default independent variable is t. The independent variable can be changed from t to another symbolic variable by including that variable as the last input variable. The initial conditions are specified by additional equations, such as $Dx(0) = 6$. If the initial conditions are not specified the solution will contain constants of integration such as $C1$ and $C2$.

For the differential equation of Example 11.24 the MATLAB commands

```
syms x t
x = dsolve(D2x + 6*Dx + 9*x = sin(t)', 'x(0) = 0,Dx(0) = 0');
pretty(simple(x))
```

return the solution

```
x = -3/50cos(t) + 2/25sin(t) + 3/50(1/exp(t)³) +
1/10(t/exp(t)³)
```

It is left as an exercise to express $1/exp(t)^3$ as e^{-3t}.

In MAPLE the command $dsolve$ is also used and the commands

```
ode2:= diff(x(t),t,t) + 6*diff(x(t),t) + 9*x(t) = sin(t);
dsolve({ode2, x(0) = 0, D(x)(0) = 0}, x(t));
```

return the solution

$$x(t) = \tfrac{3}{50}e^{(-3t)} + \tfrac{1}{10}e^{(-3t)}t - \tfrac{3}{50}\cos(t) + \tfrac{2}{25}\sin(t)$$

If the initial conditions were not specified then the command

```
dsolve({ode2}, x(t));
```

returns the solution

$$x(t) = e^{(-3t)}\_C1 + e^{(-3t)}t\_ C2 - \tfrac{3}{50}\cos(t) + \tfrac{2}{25}\sin(t)$$

In MAPLE it is also possible to specify solution by Laplace method and the command

```
dsolve({ode2, x(0) = 0, D(x)(0) = 0}, x(t), method =
laplace);
```

also returns the solution

$$x(t) = -\tfrac{3}{50}\cos(t) + \tfrac{2}{25}\sin(t) + \tfrac{1}{50}e^{(-3t)}(5t + 3)$$

and, when initial conditions are not specified, the command

```
dsolve({ode2},x(t), method = laplace);
```

returns the solution

$$x(t) = -\tfrac{3}{50}\cos(t) + \tfrac{2}{25}\sin(t) + \tfrac{1}{50}e^{(-3t)}(50 \ tD(x)(0) + $$
$$150 \ t \ x(0) + 5t + 50x(0) + 3)$$

Example 11.25 Solve the differential equation

$$\frac{d^3x}{dt^3} + 5\frac{d^2x}{dt^2} + 17\frac{dx}{dt} + 13x = 1 \quad (t \geq 0)$$

subject to the initial conditions $x = dx/dt = 1$ and $d^2x/dt^2 = 0$ at $t = 0$.

Solution Taking Laplace transforms

$$\mathscr{L}\left\{\frac{d^3x}{dt^3}\right\} + 5\mathscr{L}\left\{\frac{d^2x}{dt^2}\right\} + 17\mathscr{L}\left\{\frac{dx}{dt}\right\} + 13\mathscr{L}\{x\} = \mathscr{L}\{1\}$$

leads to the equation

$$s^3X(s) - s^2x(0) - s\dot{x}(0) - \ddot{x}(0) + 5[s^2X(s) - sx(0) - \dot{x}(0)]$$

$$+ 17[sX(s) - x(0)] + 13X(s) = \frac{1}{s}$$

which on rearrangement gives

$$(s^3 + 5s^2 + 17s + 13)X(s) = \frac{1}{s} + (s^2 + 5s + 17)x(0) + (s + 5)\dot{x}(0) + \ddot{x}(0)$$

Incorporating the given initial conditions $x(0) = \dot{x}(0) = 1$ and $\ddot{x}(0) = 0$ leads to

$$X(s) = \frac{s^3 + 6s^2 + 22s + 1}{s(s^3 + 5s^2 + 17s + 13)}$$

Clearly $s + 1$ is a factor of $s^3 + 5s^2 + 17s + 13$, and by algebraic division we have

$$X(s) = \frac{s^3 + 6s^2 + 22s + 1}{s(s + 1)(s^2 + 4s + 13)}$$

Resolving into partial fractions,

$$X(s) = \frac{\tfrac{1}{13}}{s} + \frac{\tfrac{8}{5}}{s + 1} - \frac{1}{65}\frac{44s + 7}{s^2 + 4s + 13} = \frac{\tfrac{1}{13}}{s} + \frac{\tfrac{8}{5}}{s + 1} - \frac{1}{65}\frac{44(s + 2) - 27(3)}{(s + 2)^2 + 3^2}$$

Taking inverse transforms, using the shift theorem, leads to the solution

$$x(t) = \tfrac{1}{13} + \tfrac{8}{5}e^{-t} - \tfrac{1}{65}e^{-2t}(44\cos 3t - 27\sin 3t) \quad (t \geq 0)$$

Confirm that the answer may be checked using the commands

```
syms x t
x = dsolve('D3x + 5*D2x + 17*Dx + 13*x = 1','x(0) =
1,D2x(0) = 0');
pretty(simple(x))
```

in MATLAB, or the commands

```
ode3:= diff(x(t), t$3) + 5*diff(x(t), t$2) +
17*diff(x(t),t) + 13*x(t) = 1;
dsolve({ode3,x(0) = 1,D(x)(0) = 1,(D@@2)(x)(0) =
0},x(t),method = laplace);
```

in MAPLE.

11.3.4 Exercise

Check your answers using MATLAB or MAPLE.

5 Using Laplace transform methods, solve for $t \geq 0$ the following differential equations, subject to the specified initial conditions.

(a) $\dfrac{dx}{dt} + 3x = e^{-2t}$ subject to $x = 2$ at $t = 0$

(b) $3\dfrac{dx}{dt} - 4x = \sin 2t$ subject to $x = \frac{1}{3}$ at $t = 0$

(c) $\dfrac{d^2x}{dt^2} + 2\dfrac{dx}{dt} + 5x = 1$

   subject to $x = 0$ and $\dfrac{dx}{dt} = 0$ at $t = 0$

(d) $\dfrac{d^2y}{dt^2} + 2\dfrac{dy}{dt} + y = 4\cos 2t$

   subject to $y = 0$ and $\dfrac{dy}{dt} = 2$ at $t = 0$

(e) $\dfrac{d^2x}{dt^2} - 3\dfrac{dx}{dt} + 2x = 2e^{-4t}$

   subject to $x = 0$ and $\dfrac{dx}{dt} = 1$ at $t = 0$

(f) $\dfrac{d^2x}{dt^2} + 4\dfrac{dx}{dt} + 5x = 3e^{-2t}$

   subject to $x = 4$ and $\dfrac{dx}{dt} = -7$ at $t = 0$

(g) $\dfrac{d^2x}{dt^2} + \dfrac{dx}{dt} - 2x = 5e^{-t}\sin t$

   subject to $x = 1$ and $\dfrac{dx}{dt} = 0$ at $t = 0$

(h) $\dfrac{d^2y}{dt^2} + 2\dfrac{dy}{dt} + 3y = 3t$

   subject to $y = 0$ and $\dfrac{dy}{dt} = 1$ at $t = 0$

(i) $\dfrac{d^2x}{dt^2} + 4\dfrac{dx}{dt} + 4x = t^2 + e^{-2t}$

   subject to $x = \frac{1}{2}$ and $\dfrac{dx}{dt} = 0$ at $t = 0$

(j) $9\dfrac{d^2x}{dt^2} + 12\dfrac{dx}{dt} + 5x = 1$

   subject to $x = 0$ and $\dfrac{dx}{dt} = 0$ at $t = 0$

(k) $\dfrac{d^2x}{dt^2} + 8\dfrac{dx}{dt} + 16x = 16\sin 4t$

   subject to $x = -\frac{1}{2}$ and $\dfrac{dx}{dt} = 1$ at $t = 0$

(l) $9\dfrac{d^2y}{dt^2} + 12\dfrac{dy}{dt} + 4y = e^{-t}$

   subject to $y = 1$ and $\dfrac{dy}{dt} = 1$ at $t = 0$

(m) $\dfrac{d^3x}{dt^3} - 2\dfrac{d^2x}{dt^2} - \dfrac{dx}{dt} + 2x = 2 + t$

subject to $x = 0$, $\dfrac{dx}{dt} = 1$ and $\dfrac{d^2x}{dt^2} = 0$ at $t = 0$

(n) $\dfrac{d^3x}{dt^3} + \dfrac{d^2x}{dt^2} + \dfrac{dx}{dt} + x = \cos 3t$

subject to $x = 0$, $\dfrac{dx}{dt} = 1$ and $\dfrac{d^2x}{dt^2} = 1$ at $t = 0$

11.3.5 Simultaneous differential equations

In engineering we frequently encounter systems whose characteristics are modelled by a set of simultaneous linear differential equations with constant coefficients. The method of solution is essentially the same as that adopted in Section 11.3.3 for solving a single differential equation in one unknown. Taking Laplace transforms throughout, the system of simultaneous differential equations is transformed into a system of simultaneous algebraic equations, which are then solved for the transformed variables; inverse transforms then give the desired solutions.

Example 11.26   Solve for $t \geqslant 0$ the simultaneous first-order differential equations

$$\frac{dx}{dt} + \frac{dy}{dt} + 5x + 3y = e^{-t} \tag{11.20}$$

$$2\frac{dx}{dt} + \frac{dy}{dt} + x + y = 3 \tag{11.21}$$

subject to the initial conditions $x = 2$ and $y = 1$ at $t = 0$.

Solution   Taking Laplace transforms in (11.20) and (11.21) gives

$$sX(s) - x(0) + sY(s) - y(0) + 5X(s) + 3Y(s) = \frac{1}{s+1}$$

$$2[sX(s) - x(0)] + sY(s) - y(0) + X(s) + Y(s) = \frac{3}{s}$$

Rearranging and incorporating the given initial conditions $x(0) = 2$ and $y(0) = 1$ leads to

$$(s+5)X(s) + (s+3)Y(s) = 3 + \frac{1}{s+1} = \frac{3s+4}{s+1} \tag{11.22}$$

$$(2s+1)X(s) + (s+1)Y(s) = 5 + \frac{3}{s} = \frac{5s+3}{s} \tag{11.23}$$

Hence, by taking Laplace transforms, the pair of simultaneous differential equations (11.20) and (11.21) in $x(t)$ and $y(t)$ has been transformed into a pair of simultaneous algebraic equations (11.22) and (11.23) in the transformed variables $X(s)$ and $Y(s)$. These algebraic equations may now be solved simultaneously for $X(s)$ and $Y(s)$ using standard algebraic techniques.

Solving first for $X(s)$ gives

$$X(s) = \frac{2s^2 + 14s + 9}{s(s+2)(s-1)}$$

Resolving into partial fractions,

$$X(s) = -\frac{\frac{9}{2}}{s} - \frac{\frac{11}{6}}{s+2} + \frac{\frac{25}{3}}{s-1}$$

which on inversion gives

$$x(t) = -\frac{9}{2} - \frac{11}{6}e^{-2t} + \frac{25}{3}e^{t} \quad (t \geqslant 0) \tag{11.24}$$

Likewise, solving for $Y(s)$ gives

$$Y(s) = \frac{s^3 - 22s^2 - 39s - 15}{s(s+1)(s+2)(s-1)}$$

Resolving into partial fractions,

$$Y(s) = \frac{\frac{15}{2}}{s} + \frac{\frac{1}{2}}{s+1} + \frac{\frac{11}{2}}{s+2} - \frac{\frac{25}{2}}{s-1}$$

which on inversion gives

$$y(t) = \frac{15}{2} + \frac{1}{2}e^{-t} + \frac{11}{2}e^{-2t} - \frac{25}{2}e^{t} \quad (t \geqslant 0)$$

Thus the solution to the given pair of simultaneous differential equations is

$$\left.\begin{array}{l} x(t) = -\frac{9}{2} - \frac{11}{6}e^{-2t} + \frac{25}{3}e^{t} \\[2mm] y(t) = \frac{15}{2} + \frac{1}{2}e^{-t} + \frac{11}{2}e^{-2t} - \frac{25}{2}e^{t} \end{array}\right\} \quad (t \geqslant 0)$$

Note: When solving a pair of first-order simultaneous differential equations such as (11.20) and (11.21), an alternative approach to obtaining the value of $y(t)$ having obtained $x(t)$ is to use (11.20) and (11.21) directly.

Eliminating dy/dt from (11.20) and (11.21) gives

$$2y = \frac{dx}{dt} - 4x - 3 + e^{-t}$$

Substituting the solution obtained in (11.24) for $x(t)$ gives

$$2y = \left(\frac{11}{3}e^{-2t} + \frac{25}{3}e^{t}\right) - 4\left(-\frac{9}{2} - \frac{11}{6}e^{-2t} + \frac{25}{3}\right)e^{t} - 3 + e^{-t}$$

leading as before to the solution

$$y = \frac{15}{2} + \frac{1}{2}e^{-t} + \frac{11}{2}e^{-2t} - \frac{25}{2}e^{t}$$

A further alternative is to express (11.22) and (11.23) in matrix form and solve for $X(s)$ and $Y(s)$ using Gaussian elimination.

In MATLAB the solution to the pair of simultaneous differential equations of Example 11.26 may be obtained using the commands

```
syms x y t
[x,y] = dsolve('Dx + Dy + 5*x + 3*y = exp(-t)',
'2*Dx + Dy + x + y = 3',
'x(0) = 2,y(0) = 1')
```

which return

```
x = -11/6*exp(-2*t) + 25/3*exp(t)-9/2
y = -25/2*exp(t) + 11/2*exp(-2*t) + 15/2 + 1/2*exp(-t)
```

These can then be expressed in typeset form using the commands `pretty(x)` and `pretty(y)`. In MAPLE the commands

```
ode1:= D(x)(t) + D(y)(t) + 5*x(t) + 3*y(t) = exp(-t);
ode2:= 2*D(x)(t) + D(y)(t) + x(t) + y(t) = 3;
dsolve({ode1,ode2, x(0) = 2, y(0) = 1},{x(t),y(t)});
```

return

$$\{x(t) = -\tfrac{11}{6}e^{(-2t)} + \tfrac{25}{3}e^t - \tfrac{9}{2}, \ y(t) = -\tfrac{25}{2}e^t + \tfrac{11}{2}e^{(-2t)} +$$
$$\tfrac{15}{2} + \tfrac{1}{2}e^{(-t)}\}$$

In principle, the same procedure as used in Example 11.26 can be employed to solve a pair of higher-order simultaneous differential equations or a larger system of differential equations involving more unknowns. However, the algebra involved can become quite complicated, and matrix methods are usually preferred.

11.3.6 Exercise

Check your answers using MATLAB or MAPLE.

6 Using Laplace transform methods, solve for $t \geqslant 0$ the following simultaneous differential equations subject to the given initial conditions.

(a) $2\dfrac{\mathrm{d}x}{\mathrm{d}t} - 2\dfrac{\mathrm{d}y}{\mathrm{d}t} - 9y = \mathrm{e}^{-2t}$

$2\dfrac{\mathrm{d}x}{\mathrm{d}t} + 4\dfrac{\mathrm{d}y}{\mathrm{d}t} + 4x - 37y = 0$

subject to $x = 0$ and $y = \frac{1}{4}$ at $t = 0$

(b) $\dfrac{\mathrm{d}x}{\mathrm{d}t} + 2\dfrac{\mathrm{d}y}{\mathrm{d}t} + x - y = 5\sin t$

$2\dfrac{\mathrm{d}x}{\mathrm{d}t} + 3\dfrac{\mathrm{d}y}{\mathrm{d}t} + x - y = \mathrm{e}^t$

subject to $x = 0$ and $y = 0$ at $t = 0$

(c) $\dfrac{\mathrm{d}x}{\mathrm{d}t} + \dfrac{\mathrm{d}y}{\mathrm{d}t} + 2x + y = \mathrm{e}^{-3t}$

$\dfrac{\mathrm{d}y}{\mathrm{d}t} + 5x + 3y = 5\mathrm{e}^{-2t}$

subject to $x = -1$ and $y = 4$ at $t = 0$

(d) $3\dfrac{\mathrm{d}x}{\mathrm{d}t} + 3\dfrac{\mathrm{d}y}{\mathrm{d}t} - 2x = \mathrm{e}^t$

$\dfrac{\mathrm{d}x}{\mathrm{d}t} + 2\dfrac{\mathrm{d}y}{\mathrm{d}t} - y = 1$

subject to $x = 1$ and $y = 1$ at $t = 0$

(e) $3\dfrac{\mathrm{d}x}{\mathrm{d}t} + \dfrac{\mathrm{d}y}{\mathrm{d}t} - 2x = 3\sin t + 5\cos t$

$2\dfrac{\mathrm{d}x}{\mathrm{d}t} + \dfrac{\mathrm{d}y}{\mathrm{d}t} + y = \sin t + \cos t$

subject to $x = 0$ and $y = -1$ at $t = 0$

(f) $\dfrac{\mathrm{d}x}{\mathrm{d}t} + \dfrac{\mathrm{d}y}{\mathrm{d}t} + y = t$

$\dfrac{\mathrm{d}x}{\mathrm{d}t} + 4\dfrac{\mathrm{d}y}{\mathrm{d}t} + x = 1$

subject to $x = 1$ and $y = 0$ at $t = 0$

(g) $2\dfrac{\mathrm{d}x}{\mathrm{d}t} + 3\dfrac{\mathrm{d}y}{\mathrm{d}t} + 7x = 14t + 7$

$5\dfrac{\mathrm{d}x}{\mathrm{d}t} - 3\dfrac{\mathrm{d}y}{\mathrm{d}t} + 4x + 6y = 14t - 14$

subject to $x = y = 0$ at $t = 0$

(h) $\dfrac{d^2x}{dt^2} = y - 2x$

subject to $x = \frac{7}{4}$, $y = 1$, $dx/dt = 0$ and $dy/dt = 0$ at $t = 0$

$\dfrac{d^2y}{dt^2} = x - 2y$

(j) $2\dfrac{d^2x}{dt^2} - \dfrac{d^2y}{dt^2} - \dfrac{dx}{dt} - \dfrac{dy}{dt} = 3y - 9x$

subject to $x = 4$, $y = 2$, $dx/dt = 0$ and $dy/dt = 0$ at $t = 0$

$2\dfrac{d^2x}{dt^2} - \dfrac{d^2y}{dt^2} + \dfrac{dx}{dt} + \dfrac{dy}{dt} = 5y - 7x$

(i) $5\dfrac{d^2x}{dt^2} + 12\dfrac{d^2y}{dt^2} + 6x = 0$

subject to $x = dx/dt = 1$ and $y = dy/dt = 0$ at $t = 0$

$5\dfrac{d^2x}{dt^2} + 16\dfrac{d^2y}{dt^2} + 6y = 0$

11.4 Engineering applications: electrical circuits and mechanical vibrations

To illustrate the use of Laplace transforms, we consider here their application to the analysis of electrical circuits and vibrating mechanical systems. Since initial conditions are automatically taken into account in the transformation process, the Laplace transform is particularly attractive for examining the transient behaviour of such systems. Although electrical circuits and mechanical vibrations were considered in Chapter 10, we shall review here the modelling aspects in each case. This is to enable the two chapters to be studied independently of each other.

  Using the commands adapted in the previous sections MATLAB or MAPLE can be used throughout this section to confirm answers obtained.

11.4.1 Electrical circuits

Passive electrical circuits are constructed of three basic elements: **resistors** (having resistance R, measured in ohms Ω), **capacitors** (having capacitance C, measured in farads F) and **inductors** (having inductance L, measured in henries H), with the associated variables being **current** $i(t)$ (measured in amperes A) and **voltage** $v(t)$ (measured in volts V). The current flow in the circuit is related to the charge $q(t)$ (measured in coulombs C) by the relationship

$$i = \frac{dq}{dt}$$

Conventionally, the basic elements are represented symbolically as in Figure 11.7.

Figure 11.7
Constituent elements of an electrical circuit.

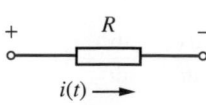

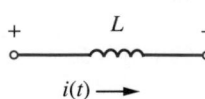

(a) Resistor   **(b)** Capacitor   **(c)** Inductor

The relationship between the flow of current $i(t)$ and the voltage drops $v(t)$ across these elements at time t are

voltage drop across resistor $= Ri$ (Ohm's law)

voltage drop across capacitor $= \dfrac{1}{C}\displaystyle\int i\,\mathrm{d}t = \dfrac{q}{C}$

The interaction between the individual elements making up an electrical circuit is determined by **Kirchhoff's laws**:

Law 1
The algebraic sum of all the currents entering any junction (or node) of a circuit is zero.

Law 2
The algebraic sum of the voltage drops around any closed loop (or path) in a circuit is zero.

Use of these laws leads to circuit equations, which may then be analysed using Laplace transform techniques.

Example 11.27

The *LCR* circuit of Figure 11.8 consists of a resistor R, a capacitor C and an inductor L connected in series together with a voltage source $e(t)$. Prior to closing the switch at time $t = 0$, both the charge on the capacitor and the resulting current in the circuit are zero. Determine the charge $q(t)$ on the capacitor and the resulting current $i(t)$ in the circuit at time t given that $R = 160\,\Omega$, $L = 1\,\mathrm{H}$, $C = 10^{-4}\,\mathrm{F}$ and $e(t) = 20\,\mathrm{V}$.

Figure 11.8
LCR circuit of
Example 11.27.

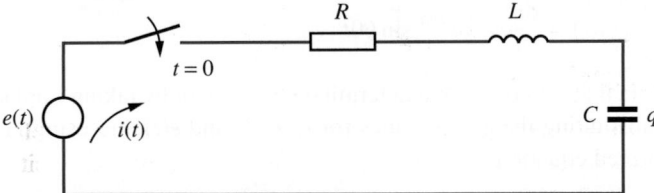

Solution Applying Kirchhoff's second law to the circuit of Figure 11.8 gives

$$Ri + L\frac{\mathrm{d}i}{\mathrm{d}t} + \frac{1}{C}\int i\,\mathrm{d}t = e(t) \qquad\qquad \textbf{(11.25)}$$

or, using $i = \mathrm{d}q/\mathrm{d}t$,

$$L\frac{\mathrm{d}^2q}{\mathrm{d}t^2} + R\frac{\mathrm{d}q}{\mathrm{d}t} + \frac{1}{C}q = e(t)$$

Substituting the given values for L, R, C and $e(t)$ gives

$$\frac{\mathrm{d}^2q}{\mathrm{d}t^2} + 160\frac{\mathrm{d}q}{\mathrm{d}t} + 10^4q = 20$$

Taking Laplace transforms throughout leads to the equation

$$(s^2 + 160s + 10^4)Q(s) = [sq(0) + \dot{q}(0)] + 160q(0) + \frac{20}{s}$$

where $Q(s)$ is the transform of $q(t)$. We are given that $q(0) = 0$ and $\dot{q}(0) = i(0) = 0$, so that this reduces to

$$(s^2 + 160s + 10^4)Q(s) = \frac{20}{s}$$

that is,

$$Q(s) = \frac{20}{s(s^2 + 160s + 10^4)}$$

Resolving into partial fractions gives

$$Q(s) = \frac{\frac{1}{500}}{s} - \frac{1}{500}\frac{s + 160}{s^2 + 160s + 10^4}$$

$$= \frac{1}{500}\left[\frac{1}{s} - \frac{(s + 80) + \frac{4}{3}(60)}{(s + 80)^2 + (60)^2}\right]$$

$$= \frac{1}{500}\left[\frac{1}{s} - \left[\frac{s + \frac{4}{3} \times 60}{s^2 + 60^2}\right]_{s \to s + 80}\right]$$

Taking inverse transforms, making use of the shift theorem (Theorem 11.2), gives

$$q(t) = \frac{1}{500}(1 - e^{-80t}\cos 60t - \frac{4}{3}e^{-80t}\sin 60t)$$

The resulting current $i(t)$ in the circuit is then given by

$$i(t) = \frac{dq}{dt} = \frac{1}{3}e^{-80t}\sin 60t$$

Note that we could have determined the current by taking Laplace transforms in (11.25). Substituting the given values for L, R, C and $e(t)$ and using (11.15) leads to the transformed equation

$$160I(s) + sI(s) + \frac{10^4}{s}I(s) = \frac{20}{s}$$

that is,

$$I(s) = \frac{20}{(s^2 + 80)^2 + 60^2} \quad (= sQ(s) \quad \text{since} \quad q(0) = 0)$$

which, on taking inverse transforms, gives as before

$$i(t) = \frac{1}{3}e^{-80t}\sin 60t$$

Example 11.28

In the parallel network of Figure 11.9 there is no current flowing in either loop prior to closing the switch at time $t = 0$. Deduce the currents $i_1(t)$ and $i_2(t)$ flowing in the loops at time t.

Figure 11.9
Parallel circuit of
Example 11.28.

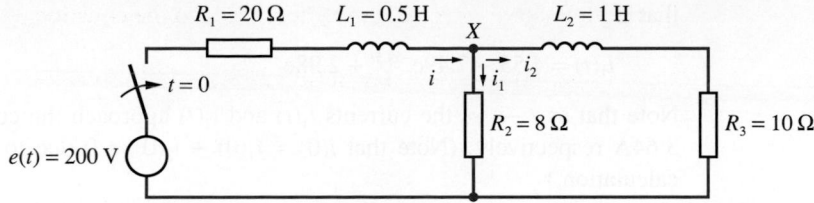

Solution   Applying Kirchhoff's first law to node X gives

$$i = i_1 + i_2$$

Applying Kirchhoff's second law to each of the two loops in turn gives

$$R_1(i_1 + i_2) + L_1\frac{d}{dt}(i_1 + i_2) + R_2i_1 = 200$$

$$L_2\frac{di_2}{dt} + R_3i_2 - R_2i_1 = 0$$

Substituting the given values for the resistances and inductances gives

$$\left.\begin{aligned}
\frac{di_1}{dt} + \frac{di_2}{dt} + 56i_1 + 40i_2 &= 400 \\
\\
\frac{di_2}{dt} - 8i_1 + 10i_2 &= 0
\end{aligned}\right\} \qquad \textbf{(11.26)}$$

Taking Laplace transforms and incorporating the initial conditions $i_1(0) = i_2(0) = 0$ leads to the transformed equations

$$(s + 56)I_1(s) + (s + 40)I_2(s) = \frac{400}{s} \qquad \textbf{(11.27)}$$

$$-8I_1(s) + (s + 10)I_2(s) = 0 \qquad \textbf{(11.28)}$$

Hence

$$I_2(s) = \frac{3200}{s(s^2 + 74s + 880)} = \frac{3200}{s(s + 59.1)(s + 14.9)}$$

Resolving into partial fractions gives

$$I_2(s) = \frac{3.64}{s} + \frac{1.22}{s + 59.1} - \frac{4.86}{s + 14.9}$$

which, on taking inverse transforms, leads to

$$i_2(t) = 3.64 + 1.22e^{-59.1t} - 4.86e^{-14.9t}$$

From (11.26),

$$i_1(t) = \tfrac{1}{8}\left(10i_2 + \frac{di_2}{dt}\right)$$

that is,

$$i_1(t) = 4.55 - 7.49e^{-59.1t} + 2.98e^{-14.9t}$$

Note that as $t \to \infty$, the currents $i_1(t)$ and $i_2(t)$ approach the constant values 4.55 and 3.64A respectively. (Note that $i(0) = i_1(0) + i_2(0) \neq 0$ due to rounding errors in the calculation.)

Example 11.29   A voltage $e(t)$ is applied to the primary circuit at time $t = 0$, and mutual induction M drives the current $i_2(t)$ in the secondary circuit of Figure 11.10. If, prior to closing the switch, the currents in both circuits are zero, determine the induced current $i_2(t)$ in the secondary circuit at time t when $R_1 = 4\,\Omega$, $R_2 = 10\,\Omega$, $L_1 = 2\,H$, $L_2 = 8\,H$, $M = 2\,H$ and $e(t) = 28 \sin 2t\,V$.

Figure 11.10
Circuit of
Example 11.29.

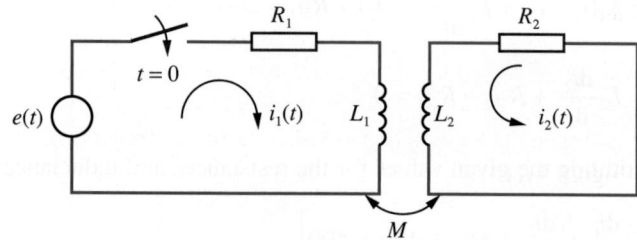

Solution   Applying Kirchhoff's second law to the primary and secondary circuits respectively gives

$$R_1 i_1 + L_1 \frac{di_1}{dt} + M \frac{di_2}{dt} = e(t)$$

$$R_2 i_2 + L_2 \frac{di_2}{dt} + M \frac{di_1}{dt} = 0$$

Substituting the given values for the resistances, inductances and applied voltage leads to

$$2\frac{di_1}{dt} + 4i_1 + 2\frac{di_2}{dt} = 28 \sin 2t$$

$$2\frac{di_1}{dt} + 8\frac{di_2}{dt} + 10i_2 = 0$$

Taking Laplace transforms and noting that $i_1(0) = i_2(0) = 0$ leads to the equations

$$(s + 2)I_1(s) + sI_2(s) = \frac{28}{s^2 + 4} \tag{11.29}$$

$$sI_1(s) + (4s + 5)I_2(s) = 0 \tag{11.30}$$

Solving for $I_2(s)$ yields

$$I_2(s) = -\frac{28s}{(3s + 10)(s + 1)(s^2 + 4)}$$

Resolving into partial fractions gives

$$I_2(s) = -\frac{\frac{45}{17}}{3s + 10} + \frac{\frac{4}{5}}{s + 1} + \frac{7}{85}\frac{s - 26}{s^2 + 4}$$

Taking inverse Laplace transforms gives the current in the secondary circuit as

$$i_2(t) = \tfrac{4}{5}e^{-t} - \tfrac{15}{17}e^{-10t/3} + \tfrac{7}{85}\cos 2t - \tfrac{91}{85}\sin 2t$$

As $t \to \infty$, the current will approach the sinusoidal response

$$i_2(t) = \tfrac{7}{85}\cos 2t - \tfrac{91}{85}\sin 2t$$

11.4.2 Mechanical vibrations

Mechanical translational systems may be used to model many situations, and involve three basic elements: **masses** (having mass M, measured in kg), **springs** (having spring stiffness K, measured in N m^{-1}) and **dampers** (having damping coefficient B, measured in N s m^{-1}). The associated variables are **displacement** $x(t)$ (measured in m) and **force** $F(t)$ (measured in N). Conventionally, the basic elements are represented symbolically as in Figure 11.11.

Figure 11.11
Constituent elements
of a translational
mechanical system.

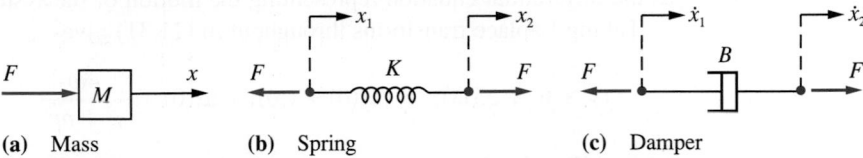

(a) Mass   **(b)** Spring   **(c)** Damper

Assuming we are dealing with ideal springs and dampers (that is, assuming that they behave linearly), the relationships between the forces and displacements at time t are

mass: $F = M\dfrac{d^2x}{dt^2} = M\ddot{x}$ (Newton's law)

spring: $F = K(x_2 - x_1)$ (Hooke's law)

damper: $F = B\left(\dfrac{dx_2}{dt} - \dfrac{dx_1}{dt}\right) = B(\dot{x}_2 - \dot{x}_1)$

Using these relationships leads to the system equations, which may then be analysed using Laplace transform techniques.

Example 11.30

The mass of the mass–spring–damper system of Figure 11.12(a) is subjected to an externally applied periodic force $F(t) = 4 \sin \omega t$ at time $t = 0$. Determine the resulting displacement $x(t)$ of the mass at time t, given that $x(0) = \dot{x}(0) = 0$, for the two cases

(a) $\omega = 2$ (b) $\omega = 5$

In the case $\omega = 5$, what would happen to the response if the damper were missing?

Figure 11.12
Mass–spring–damper
system of
Example 11.30.

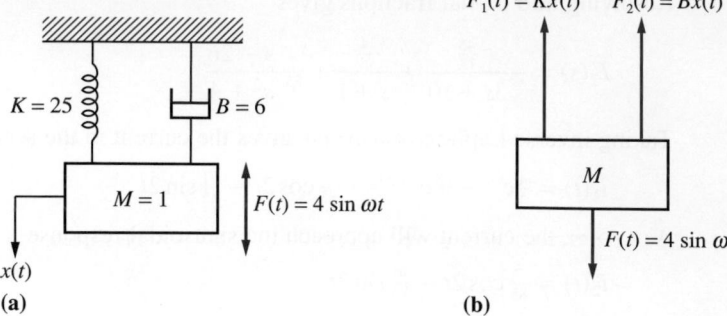

(a)

(b)

Solution As indicated in Figure 11.12(b), the forces acting on the mass M are the applied force $F(t)$ and the restoring forces F_1 and F_2 due to the spring and damper respectively. Thus, by Newton's law,

$$M\ddot{x}(t) = F(t) - F_1(t) - F_2(t)$$

Since $M = 1$, $F(t) = 4\sin\omega t$, $F_1(t) = Kx(t) = 25x(t)$ and $F_2(t) = B\dot{x}(t) = 6\dot{x}(t)$, this gives

$$\ddot{x}(t) + 6\dot{x}(t) + 25x(t) = 4\sin\omega t \qquad (11.31)$$

as the differential equation representing the motion of the system.

Taking Laplace transforms throughout in (11.31) gives

$$(s^2 + 6s + 25)X(s) = [sx(0) + \dot{x}(0)] + 6x(0) + \frac{4\omega}{s^2 + \omega^2}$$

where $X(s)$ is the transform of $x(t)$. Incorporating the given initial conditions $x(0) = \dot{x}(0) = 0$ leads to

$$X(s) = \frac{4\omega}{(s^2 + \omega^2)(s^2 + 6s + 25)} \qquad (11.32)$$

In case (a), with $\omega = 2$, (11.32) gives

$$X(s) = \frac{8}{(s^2 + 4)(s^2 + 6s + 25)}$$

which, on resolving into partial fractions, leads to

$$X(s) = \tfrac{4}{195}\frac{-4s + 14}{s^2 + 4} + \tfrac{2}{195}\frac{8s + 20}{s^2 + 6s + 25}$$

$$= \tfrac{4}{195}\frac{-4s + 14}{s^2 + 4} + \tfrac{2}{195}\frac{8(s + 3) - 4}{(s + 3)^2 + 16}$$

Taking inverse Laplace transforms gives the required response

$$x(t) = \tfrac{4}{195}(7\sin 2t - 4\cos 2t) + \tfrac{2}{195}e^{-3t}(8\cos 4t - \sin 4t) \qquad (11.33)$$

In case (b), with $\omega = 5$, (11.32) gives

$$X(s) = \frac{20}{(s^2 + 25)(s^2 + 6s + 25)} \qquad (11.34)$$

that is,

$$X(s) = \frac{-\frac{2}{15}s}{s^2 + 25} + \frac{1}{15}\frac{2(s+3) + 6}{(s+3)^2 + 16}$$

which, on taking inverse Laplace transforms, gives the required response

$$x(t) = -\tfrac{2}{15}\cos 5t + \tfrac{1}{15}e^{-3t}(2\cos 4t + \tfrac{3}{2}\sin 4t) \qquad\qquad \textbf{(11.35)}$$

If the damping term were missing then (11.34) would become

$$X(s) = \frac{20}{(s^2 + 25)^2} \qquad\qquad \textbf{(11.36)}$$

By Theorem 11.3,

$$\mathscr{L}\{t\cos 5t\} = -\frac{\mathrm{d}}{\mathrm{d}s}\mathscr{L}\{\cos 5t\} = -\frac{\mathrm{d}}{\mathrm{d}s}\left(\frac{s}{s^2 + 25}\right)$$

that is,

$$\mathscr{L}\{t\cos 5t\} = -\frac{1}{s^2 + 25} + \frac{2s^2}{(s^2 + 25)^2} = \frac{1}{s^2 + 25} - \frac{50}{(s^2 + 25)^2}$$

$$= \tfrac{1}{5}\mathscr{L}\{\sin 5t\} - \frac{50}{(s^2 + 25)^2}$$

Thus, by the linearity property (11.10),

$$\mathscr{L}\{\tfrac{1}{5}\sin 5t - t\cos 5t\} = \frac{50}{(s^2 + 25)^2}$$

so that taking inverse Laplace transforms in (11.36) gives the response as

$$x(t) = \tfrac{2}{25}(\sin 5t - 5t\cos 5t)$$

Because of the term $t\cos 5t$, the response $x(t)$ is unbounded as $t \to \infty$. This arises because in this case the applied force $F(t) = 4\sin 5t$ is in **resonance** with the system (that is, the vibrating mass), whose natural oscillating frequency is $5/2\pi$ Hz, equal to that of the applied force. Even in the presence of damping, the amplitude of the system response is maximized when the applied force is approaching resonance with the system. (This is left as an exercise for the reader.) In the absence of damping we have the limiting case of **pure resonance**, leading to an unbounded response. As noted in Chapter 10, Section 10.10.3, resonance is of practical importance, since, for example, it can lead to large and strong structures collapsing under what appears to be a relatively small force.

Example 11.31   Consider the mechanical system of Figure 11.13(a), which consists of two masses $M_1 = 1$ and $M_2 = 2$, each attached to a fixed base by a spring, having constants $K_1 = 1$ and $K_3 = 2$ respectively, and attached to each other by a third spring having constant $K_2 = 2$. The system is released from rest at time $t = 0$ in a position in which M_1 is displaced 1 unit to the left of its equilibrium position and M_2 is displaced 2 units to the right of its equilibrium position. Neglecting all frictional effects, determine the positions of the masses at time t.

Figure 11.13
Two-mass system of
Example 11.31.

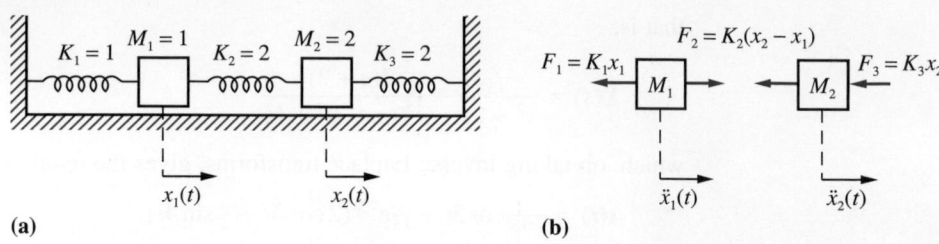

(a)

(b)

Solution Let $x_1(t)$ and $x_2(t)$ denote the displacements of the masses M_1 and M_2 respectively from their equilibrium positions. Since frictional effects are neglected, the only forces acting on the masses are the restoring forces due to the springs, as shown in Figure 11.13(b). Applying Newton's law to the motions of M_1 and M_2 respectively gives

$$M_1\ddot{x}_1 = F_2 - F_1 = K_2(x_2 - x_1) - K_1x_1$$

$$M_2\ddot{x}_2 = -F_3 - F_2 = -K_3x_2 - K_2(x_2 - x_1)$$

which, on substituting the given values for M_1, M_2, K_1, K_2 and K_3, gives

$$\ddot{x}_1 + 3x_1 - 2x_2 = 0 \tag{11.37}$$

$$2\ddot{x}_2 + 4x_2 - 2x_1 = 0 \tag{11.38}$$

Taking Laplace transforms leads to the equations

$$(s^2 + 3)X_1(s) - 2X_2(s) = sx_1(0) + \dot{x}_1(0)$$

$$-X_1(s) + (s^2 + 2)X_2(s) = sx_2(0) + \dot{x}_2(0)$$

Since $x_1(t)$ and $x_2(t)$ denote displacements to the right of the equilibrium positions, we have $x_1(0) = -1$ and $x_2(0) = 2$. Also, the system is released from rest, so that $\dot{x}_1(0) = \dot{x}_2(0) = 0$. Incorporating these initial conditions, the transformed equations become

$$(s^2 + 3)X_1(s) - 2X_2(s) = -s \tag{11.39}$$

$$-X_1(s) + (s^2 + 2)X_2(s) = 2s \tag{11.40}$$

Hence

$$X_2(s) = \frac{2s^3 + 5s}{(s^2 + 4)(s^2 + 1)}$$

Resolving into partial fractions gives

$$X_2(s) = \frac{s}{s^2 + 1} + \frac{s}{s^2 + 4}$$

which, on taking inverse Laplace transforms, leads to the response

$$x_2(t) = \cos t + \cos 2t$$

Substituting for $x_2(t)$ in (11.38) gives

$$x_1(t) = 2x_2(t) + \ddot{x}_2(t)$$

$$= 2\cos t + 2\cos 2t - \cos t - 4\cos 2t$$

that is,

$$x_1(t) = \cos t - 2\cos 2t$$

Thus the positions of the masses at time t are

$$x_1(t) = \cos t - 2\cos 2t$$

$$x_2(t) = \cos t + \cos 2t$$

11.4.3 Exercises

Check your answers using MATLAB or MAPLE whenever possible.

7   Use the Laplace transform technique to find the transforms $I_1(s)$ and $I_2(s)$ of the respective currents flowing in the circuit of Figure 11.14, where $i_1(t)$ is that through the capacitor and $i_2(t)$ that through the resistance. Hence, determine $i_2(t)$. (Initially, $i_1(0) = i_2(0) = q_1(0) = 0$.) Sketch $i_2(t)$ for large values of t.

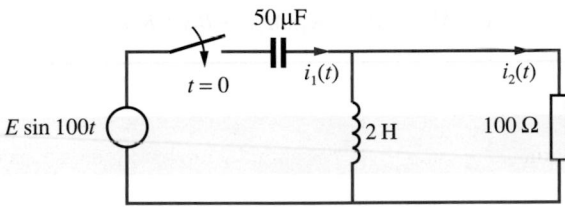

Figure 11.14 Circuit of Question 7.

8   At time $t = 0$, with no currents flowing, a voltage $v(t) = 10\sin t$ is applied to the primary circuit of a transformer that has a mutual inductance of 1 H, as shown in Figure 11.15. Denoting the current flowing at time t in the secondary circuit by $i_2(t)$, show that

$$\mathcal{L}\{i_2(t)\} = \frac{10s}{(s^2 + 7s + 6)(s^2 + 1)}$$

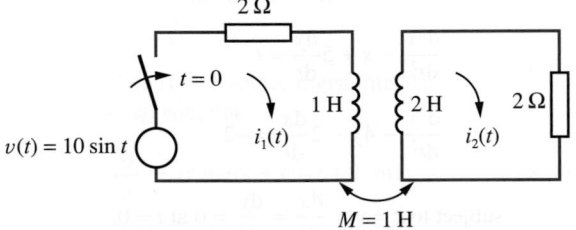

Figure 11.15 Circuit of Question 8.

and deduce that

$$i_2(t) = -e^{-t} + \tfrac{12}{37}e^{-6t} + \tfrac{25}{37}\cos t + \tfrac{35}{37}\sin t$$

9   In the circuit of Figure 11.16 there is no energy stored (that is, there is no charge on the capacitors and no current flowing in the inductances) prior to the closure of the switch at time $t = 0$. Determine $i_1(t)$ for $t > 0$ for a constant applied voltage $E_0 = 10$ V.

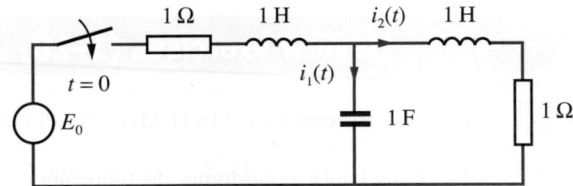

Figure 11.16 Circuit of Question 9.

10   Determine the displacements of the masses M_1 and M_2 in Figure 11.13 at time $t > 0$ when

$$M_1 = M_2 = 1$$

$$K_1 = 1,\ K_2 = 3 \quad \text{and} \quad K_3 = 9$$

What are the natural frequencies of the system?

11   When testing the landing-gear unit of a space vehicle, drop tests are carried out. Figure 11.17 is a schematic model of the unit at the instant when it first touches the ground. At this instant the spring is fully extended and the velocity of the mass is $\sqrt{(2gh)}$, where h is the height from which the unit has been dropped. Obtain the equation representing the displacement of the mass at time $t > 0$ when $M = 50$ kg, $B = 180$ N s m^{-1} and

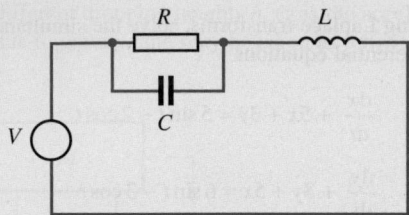

Figure 11.20 Circuit of Question 12.

with the resistance R as shown in Figure 11.20. Show that the current i flowing through the resistance R is given by

$$LCR\frac{d^2i}{dt^2} + L\frac{di}{dt} + Ri = V$$

Suppose that

(i) $V = 0$ for $t < 0$ and $V = E$ (constant) for $t \geq 0$

(ii) $L = 2R^2C$

(iii) $CR = 1/2n$

and show that the equation reduces to

$$\frac{d^2i}{dt^2} + 2n\frac{di}{dt} + 2n^2i = 2n^2\frac{E}{R}$$

Hence, assuming that $i = 0$ and $di/dt = 0$ when $t = 0$, use Laplace transforms to obtain an expression for i in terms of t.

13 Show that the currents in the coupled circuits of Figure 11.21 are determined by the simultaneous differential equations

$$L\frac{di_1}{dt} + R(i_1 - i_2) + Ri_1 = E$$

$$L\frac{di_2}{dt} + Ri_2 - R(i_1 - i_2) = 0$$

Find i_1 in terms of t, L, E and R, given that $i_1 = 0$ and $di_1/dt = E/L$ at $t = 0$, and show that $i_1 \simeq \frac{2}{3}$ E/R for large t. What does i_2 tend to for large t?

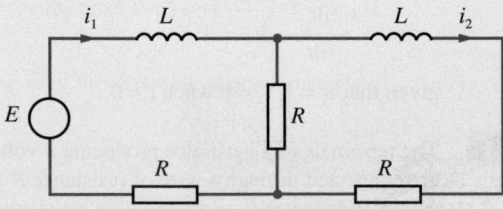

Figure 11.21 Circuit of Question 13.

14 A system consists of two unit masses lying in a straight line on a smooth surface and connected together to two fixed points by three springs. When a sinusoidal force is applied to the system, the displacements $x_1(t)$ and $x_2(t)$ of the respective masses from their equilibrium positions satisfy the equations

$$\frac{d^2x_1}{dt^2} = x_2 - 2x_1 + \sin 2t$$

$$\frac{d^2x_2}{dt^2} = -2x_2 + x_1$$

Given that the system is initially at rest in the equilibrium position ($x_1 = x_2 = 0$), use the Laplace transform method to solve the equations for $x_1(t)$ and $x_2(t)$.

15 (a) Obtain the inverse Laplace transforms of

(i) $\dfrac{s + 4}{s^2 + 2s + 10}$ (ii) $\dfrac{s - 3}{(s - 1)^2(s - 2)}$

(b) Use Laplace transforms to solve the differential equation

$$\frac{d^2y}{dt^2} + 2\frac{dy}{dt} + y = 3te^{-t}$$

given that $y = 4$ and $dy/dt = 2$, when $t = 0$.

16 (a) Determine the inverse Laplace transform of

$$\frac{5}{s^2 - 14s + 53}$$

(b) The equation of motion of the moving coil of a galvanometer when a current i is passed through it is of the form

$$\frac{d^2\theta}{dt^2} + 2K\frac{d\theta}{dt} + n^2\theta = \frac{n^2i}{K}$$

where θ is the angle of deflection from the 'no-current' position and n and K are positive constants. Given that i is a constant and $\theta = 0 = d\theta/dt$ when $t = 0$, obtain an expression for the Laplace transform of θ.

In constructing the galvanometer, it is desirable to have it critically damped (that is, $n = K$). Use the Laplace transform method to solve the differential equation in this case, and sketch the graph of θ against t for positive values of t.

17 Two cylindrical water tanks are connected as shown in Figure 11.22. Initially there are 250 litres in the top tank and 50 litres in the bottom tank. At time $t = 0$ the valve between the two tanks and the valve at the bottom of the lower tank are opened. The flowrate through each of these valves is proportional to the volume of water in the tank immediately above the valve, the constant of proportionality being 0.1 for both valves. Denoting the volume in the top tank by v_1 and the volume in the bottom tank by v_2 show that the following differential equations are satisfied.

$$\frac{dv_1}{dt} = -0.1v_1$$

$$\frac{dv_2}{dt} + 0.1v_2 = 0.1v_1$$

(a) Use Laplace transforms to determine v_1 and v_2.

(b) Find the time taken for the volume of water in the top tank to reach 10% of its starting value.

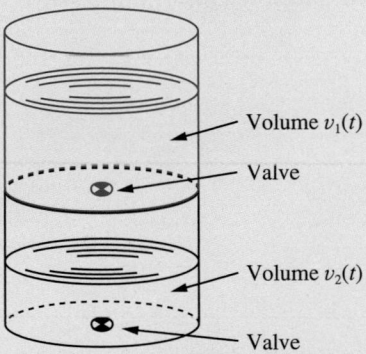

Figure 11.22 Cylindrical tanks of Question 17.

18 In order to transport sensitive equipment a crate is installed inside a truck on damped springs, as shown in Figure 11.23. The suspension system of the truck, including the tyres, may be modelled as a damped spring. The various spring and damper constants are indicated in the figure. The masses of the crate and truck are M_1 and M_2 respectively and their displacements from equilibrium are respectively $x_1(t)$ and $x_2(t)$. The vertical displacement of the truck as it traverses a bumpy road may be modelled by applying a force $u(t)$ to the truck.

Show that the motion of the crate and truck may be modelled by the differential equations

$$M_1\ddot{x}_1 = K_1(x_2 - x_1) + B_1(\dot{x}_2 - \dot{x}_1)$$

$$M_2\ddot{x}_2 = u - (K_1 + K_2)x_2 + K_1x_1$$
$$\qquad - (B_1 + B_2)\dot{x}_2 + B_1\dot{x}_1$$

For the particular case where $M_1 = 1$, $M_2 = 3$, $K_1 = 2$, $K_2 = 1$, $B_1 = 3$, $B_2 = 2$ and $u(t) = \sin t$, and the initial conditions at time $t = 0$ are $x_1 = x_2 = \dot{x}_1 = 0$, $\dot{x}_2 = 2$, show that the Laplace transform of $x_1(t)$ is

$$X_1(s) = \frac{18s^3 + 12s^2 + 21s + 14}{(3s^4 + 14s^3 + 15s^2 + 7s + 2)(s^2 + 1)}$$

(*Note*: Using an appropriate software package, such as MATLAB/SIMULINK, the model developed may be used as the basis for simulation studies of various scenarios.)

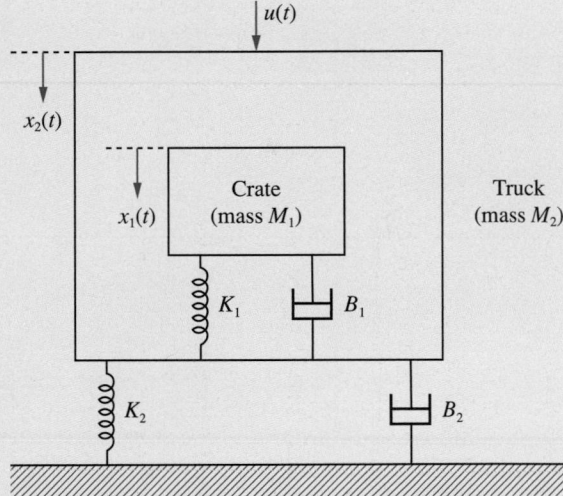

Figure 11.23 Transport crate of Question 18.

12 Introduction to Fourier Series

Chapter 12   Contents

12.1 Introduction

The representation of a function in the form of a series is fairly common practice in mathematics. Probably the most familiar expansions are power series of the form

$$f(x) = \sum_{n=0}^{\infty} a_n x^n$$

in which the resolved components or **base set** comprise the power functions

$$1, x, x^2, x^3, \ldots, x^n, \ldots$$

For example, we recall that the exponential function may be represented by the infinite series

$$\mathrm{e}^x = 1 + x + \frac{x^2}{2!} + \frac{x^3}{3!} + \ldots + \frac{x^n}{n!} + \ldots = \sum_{n=0}^{\infty} \frac{x^n}{n!}$$

There are frequently advantages in expanding a function in such a series, since the first few terms of a good approximation are easy to deal with. For example, term-by-term integration or differentiation may be applied or suitable function approximations can be made.

Power functions comprise only one example of a base set for the expansions of functions: a number of other base sets may be used. In particular, a **Fourier series** is an expansion of a periodic function $f(t)$ of period $T = 2\pi/\omega$ in which that base set is the set of sine functions, giving an expanded representation of the form

$$f(t) = A_0 + \sum_{n=1}^{\infty} A_n \sin(n\omega t + \phi_n)$$

Although the idea of expanding a function in the form of such a series had been used by Bernoulli, D'Alembert and Euler (*c.* 1750) to solve problems associated with the vibration of strings, it was Joseph Fourier (1768–1830) who developed the approach to a stage where it was generally useful. Fourier, a French physicist, was interested in heat-flow problems: given an initial temperature at all points of a region, he was concerned with determining the change in the temperature distribution over time. When Fourier postulated in 1807 that an arbitrary function $f(x)$ could be represented by a trigonometric series of the form

$$\sum_{n=0}^{\infty} (A_n \cos nkx + B_n \sin nkx)$$

the result was considered so startling that it met considerable opposition from the leading mathematicians of the time, notably Laplace, Poisson and, more significantly, Lagrange, who is regarded as one of the greatest mathematicians of all time. They questioned his work because of its lack of rigour, and it was probably this opposition that delayed the publication of Fourier's work, his classic text *Théorie Analytique de la Chaleur* (The Analytical Theory of Heat) not appearing until 1822. This text has since become the source for the modern methods of solving practical problems associated with partial differential equations subject to prescribed boundary conditions. In addition to heat flow, this class of problems includes structural vibrations, wave propagation and diffusion, which are discussed in the companion text *Advanced Modern Engineering Mathematics*. The task of giving Fourier's work a more rigorous mathematical underpinning was undertaken later by Dirichlet (*c.* 1830) and subsequently Riemann, his successor at the University of Göttingen.

In addition to its use in solving boundary-value problems associated with partial differential equations, Fourier series analysis is central to many other applications in engineering, such as the analysis and design of oscillating and nonlinear systems. This chapter is intended to provide only an introduction to Fourier series, with a more detailed treatment, including consideration of frequency spectra, oscillating and nonlinear systems, and generalized Fourier series, being given in *Advanced Modern Engineering Mathematics*.

12.2 Fourier series expansion

In this section we develop the Fourier series expansion of periodic functions and discuss how closely they approximate the functions. We also indicate how symmetrical properties of the function may be taken advantage of in order to reduce the amount of mathematical manipulation involved in determining the Fourier series. First, for continuity, we review the properties of periodic functions considered in Section 2.2.6.

12.2.1 Periodic functions

A function $f(t)$ is said to be **periodic** if its image values are repeated at regular intervals in its domain. Thus the graph of a periodic function can be divided into 'vertical strips' that are replicas of each other, as illustrated in Figure 12.1. The interval between two successive replicas is called the **period** of the function. We therefore say that a function $f(t)$ is periodic with period T if, for all its domain values t,

$$f(t + mT) = f(t)$$

for any integer m.

To provide a measure of the number of repetitions per unit of t, we define the **frequency** of a periodic function to be the reciprocal of its period, so that

$$\text{frequency} = \frac{1}{\text{period}} = \frac{1}{T}$$

Figure 12.1
A periodic function
with period T.

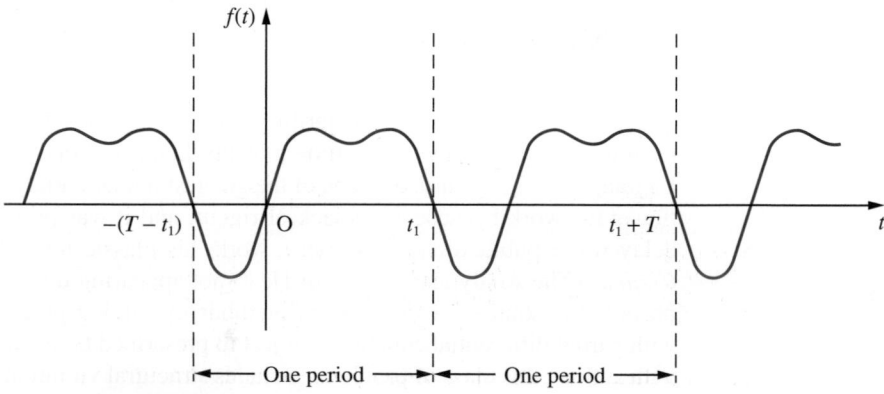

The term **circular frequency** is also used in engineering, and is defined by

$$\text{circular frequency} = 2\pi \times \text{frequency} = \frac{2\pi}{T}$$

and is measured in radians per second. It is common to drop the term 'circular' and refer to this simply as the frequency when the context is clear.

12.2.2 Fourier's theorem

This theorem states that a periodic function that satisfies certain conditions can be expressed as the sum of a number of sine functions of different amplitudes, phases and periods. That is, if $f(t)$ is a periodic function with period T then

$$f(t) = A_0 + A_1 \sin(\omega t + \phi_1) + A_2 \sin(2\omega t + \phi_2) + \ldots$$
$$+ A_n \sin(n\omega t + \phi_n) + \ldots \tag{12.1}$$

where the As and ϕs are constants and $\omega = 2\pi/T$ is the frequency of $f(t)$. The term $A_1 \sin(\omega t + \phi_1)$ is called the **first harmonic** or the **fundamental mode**, and it has the same frequency ω as the parent function $f(t)$. The term $A_n \sin(n\omega t + \phi_n)$ is called the **nth harmonic**, and it has frequency $n\omega$, which is n times that of the fundamental. A_n denotes the **amplitude** of the nth harmonic and ϕ_n is its **phase angle**, measuring the lag or lead of the nth harmonic with reference to a pure sine wave of the same frequency.

Since

$$A_n \sin(n\omega t + \phi_n) \equiv (A_n \cos \phi_n) \sin n\omega t + (A_n \sin \phi_n) \cos n\omega t$$

$$\equiv b_n \sin n\omega t + a_n \cos n\omega t$$

where

$$b_n = A_n \cos \phi_n, \quad a_n = A_n \sin \phi_n \tag{12.2}$$

the expansion (12.1) may be written as

$$f(t) = \tfrac{1}{2}a_0 + \sum_{n=1}^{\infty} a_n \cos n\omega t + \sum_{n=1}^{\infty} b_n \sin n\omega t \tag{12.3}$$

where $a_0 = 2A_0$ (we shall see later that taking the first term as $\tfrac{1}{2}a_0$ rather than a_0 is a convenience that enables us to make a_0 fit a general result). The expansion (12.3) is called the **Fourier series expansion** of the function $f(t)$, and the as and bs are called the **Fourier coefficients**. In electrical engineering it is common practice to refer to a_n and b_n respectively as the **in-phase** and **phase quadrature components** of the nth harmonic, this terminology arising from the use of the phasor notation $e^{jn\omega t} = \cos n\omega t + j \sin n\omega t$. Clearly, (12.1) is an alternative representation of the Fourier series with the amplitude and phase of the nth harmonic being determined from (12.2) as

$$A_n = \sqrt{(a_n^2 + b_n^2)}, \quad \phi_n = \tan^{-1}\left(\frac{a_n}{b_n}\right)$$

with care being taken over choice of quadrant.

12.2.3 The Fourier coefficients

Before proceeding to evaluate the Fourier coefficients, we first state the following integrals, in which $T = 2\pi/\omega$:

$$\int_d^{d+T} \cos n\omega t \, dt = \begin{cases} 0 & (n \neq 0) \\ T & (n = 0) \end{cases} \tag{12.4}$$

$$\int_d^{d+T} \sin n\omega t \, dt = 0 \quad \text{(all } n) \tag{12.5}$$

$$\int_d^{d+T} \sin m\omega t \sin n\omega t \, dt = \begin{cases} 0 & (m \neq n) \\ \tfrac{1}{2}T & (m = n \neq 0) \end{cases} \tag{12.6}$$

$$\int_d^{d+T} \cos m\omega t \cos n\omega t \, dt = \begin{cases} 0 & (m \neq n) \\ \tfrac{1}{2}T & (m = n \neq 0) \end{cases} \tag{12.7}$$

$$\int_d^{d+T} \cos m\omega t \sin n\omega t \, dt = 0 \quad \text{(all } m \text{ and } n) \tag{12.8}$$

The results (12.4)–(12.8) constitute the **orthogonality relations** for sine and cosine functions, and show that the set of functions

$$\{1, \cos \omega t, \cos 2\omega t, \ldots, \cos n\omega t, \sin \omega t, \sin 2\omega t, \ldots, \sin n\omega t\}$$

is an orthogonal set of functions on the interval $d \leq t \leq d + T$. The choice of d is arbitrary in these results, it only being necessary to integrate over a period of duration T.

Integrating the series (12.3) with respect to t over the period $t = d$ to $t = d + T$, and using (12.4) and (12.5), we find that each term on the right-hand side is zero except for the term involving a_0; that is, we have

$$\int_d^{d+T} f(t) \, dt = \tfrac{1}{2}a_0 \int_d^{d+T} dt + \sum_{n=1}^{\infty} \left(a_n \int_d^{d+T} \cos n\omega t \, dt + b_n \int_d^{d+T} \sin n\omega t \, dt \right)$$

$$= \tfrac{1}{2}a_0(T) + \sum_{n=1}^{\infty} [a_n(0) + b_n(0)]$$

$$= \tfrac{1}{2}Ta_0$$

Thus

$$\frac{1}{2}a_0 = \frac{1}{T} \int_d^{d+T} f(t) \, dt$$

and we can see that the constant term $\tfrac{1}{2}a_0$ in the Fourier series expansion represents the mean value of the function $f(t)$ over one period. For an electrical signal it represents the bias level or DC (direct current) component. Hence

$$a_0 = \frac{2}{T} \int_d^{d+T} f(t) \, dt \tag{12.9}$$

To obtain this result, we have assumed that term-by-term integration of the series (12.3) is permissible. This is indeed so because of the convergence properties of the series – its validity is discussed in detail in more advanced texts.

To obtain the Fourier coefficient a_n ($n \neq 0$), we multiply (12.3) throughout by $\cos m\omega t$ and integrate with respect to t over the period $t = d$ to $t = d + T$, giving

$$\int_d^{d+T} f(t) \cos m\omega t \, dt = \tfrac{1}{2} a_0 \int_d^{d+T} \cos m\omega t \, dt + \sum_{n=1}^{\infty} a_n \int_d^{d+T} \cos n\omega t \cos m\omega t \, dt$$

$$+ \sum_{n=1}^{\infty} b_n \int_d^{d+T} \cos m\omega t \sin n\omega t \, dt$$

Assuming term-by-term integration to be possible, and using (12.4), (12.7) and (12.8), we find that, when $m \neq 0$, the only non-zero integral on the right-hand side is the one that occurs in the first summation when $n = m$. That is, we have

$$\int_d^{d+T} f(t) \cos m\omega t \, dt = a_m \int_d^{d+T} \cos m\omega t \cos m\omega t \, dt = \tfrac{1}{2} a_m T$$

giving

$$a_m = \frac{2}{T} \int_d^{d+T} f(t) \cos m\omega t \, dt$$

which, on replacing m by n, gives

$$a_n = \frac{2}{T} \int_d^{d+T} f(t) \cos n\omega t \, dt \tag{12.10}$$

The value of a_0 given in (12.9) may be obtained by taking $n = 0$ in (12.10), so that we may write

$$a_n = \frac{2}{T} \int_d^{d+T} f(t) \cos m\omega t \, dt \quad (n = 0, 1, 2, \dots) \tag{12.11}$$

This explains why the constant term in the Fourier series expansion was taken as $\tfrac{1}{2} a_0$ and not a_0, since this ensures compatibility of the results (12.9) and (12.10). Although a_0 and a_n satisfy the same formula, it is usually safer to work them out separately.

Finally, to obtain the Fourier coefficients b_n, we multiply (12.3) throughout by $\sin m\omega t$ and integrate with respect to t over the period $t = d$ to $t = d + T$, giving

$$\int_d^{d+T} f(t) \sin m\omega t \, dt = \tfrac{1}{2} a_0 \int_d^{d+T} \sin m\omega t \, dt$$

$$+ \sum_{n=1}^{\infty} \left(a_n \int_d^{d+T} \sin m\omega t \cos n\omega t \, dt + b_n \int_t^{d+T} \sin m\omega t \sin n\omega t \, dt \right)$$

Assuming term-by-term integration to be possible, and using (12.5), (12.6) and (12.8), we find that the only non-zero integral on the right-hand side is the one that occurs in the second summation when $m = n$. That is, we have

$$\int_d^{d+T} f(t) \sin m\omega t \, dt = b_m \int_d^{d+T} \sin m\omega t \sin m\omega t \, dt = \tfrac{1}{2} b_m T$$

giving, on replacing m by n,

$$b_n = \frac{2}{T} \int_d^{d+T} f(t) \sin n\omega t \, dt \quad (n = 1, 2, 3, \dots) \tag{12.12}$$

The equations (12.11) and (12.12) giving the Fourier coefficients are known as **Euler's formulae**.

Summary

In summary, we have shown that if a periodic function $f(t)$ of period $T = 2\pi/\omega$ can be expanded as a Fourier series then that series is given by

$$f(t) = \tfrac{1}{2} a_0 + \sum_{n=1}^{\infty} a_n \cos n\omega t + \sum_{n=1}^{\infty} b_n \sin n\omega t \tag{12.3}$$

where the coefficients are given by Euler's formulae

$$a_n = \frac{2}{T} \int_d^{d+T} f(t) \cos n\omega t \, dt \quad (n = 0, 1, 2, \dots) \tag{12.11}$$

$$b_n = \frac{2}{T} \int_d^{d+T} f(t) \sin n\omega t \, dt \quad (n = 1, 2, 3, \dots) \tag{12.12}$$

The limits of integration in Euler's formulae may be specified over any period, so that the choice of d is arbitrary, and may be made in such a way as to help in the calculation of a_n and b_n. In practice, it is common to specify $f(t)$ over either the period $-\tfrac{1}{2}T < t < \tfrac{1}{2}T$ or the period $0 < t < T$, leading respectively to the limits of integration being $-\tfrac{1}{2}T$ and $\tfrac{1}{2}T$ (that is, $d = -\tfrac{1}{2}T$) or 0 and T (that is, $d = 0$).

It is also worth noting that an alternative approach may simplify the calculation of a_n and b_n. Using the formula

$$e^{jn\omega t} = \cos n\omega t + j \sin n\omega t$$

we have

$$a_n + jb_n = \frac{2}{T} \int_d^{d+T} f(t) e^{jn\omega t} \, dt \tag{12.13}$$

Evaluating this integral and equating real and imaginary parts on each side gives the values of a_n and b_n. This approach is particularly useful when only the amplitude $|a_n + jb_n|$ of the nth harmonic is required.

12.2.4 Functions of period 2π

If the period T of the periodic function $f(t)$ is taken to be 2π then $\omega = 1$, and the series (12.3) becomes

$$f(t) = \tfrac{1}{2}a_0 + \sum_{n=1}^{\infty} a_n \cos nt + \sum_{n=1}^{\infty} b_n \sin nt \qquad \text{(12.14)}$$

with the coefficients given by

$$a_n = \frac{1}{\pi} \int_d^{d+2\pi} f(t) \cos nt \, \mathrm{d}t \quad (n = 0, 1, 2, \dots) \qquad \text{(12.15)}$$

$$b_n = \frac{1}{\pi} \int_d^{d+2\pi} f(t) \sin nt \, \mathrm{d}t \quad (n = 1, 2, \dots) \qquad \text{(12.16)}$$

While a unit frequency may rarely be encountered in practice, consideration of this particular case reduces the amount of mathematical manipulation involved in determining the coefficients a_n and b_n. Also, there is no loss of generality in considering this case, since if we have a function $f(t)$ of period T, we may write $t_1 = 2\pi t/T$, so that

$$f(t) \equiv f\!\left(\frac{Tt_1}{2\pi}\right) \equiv F(t_1)$$

where $F(t_1)$ is a function of period 2π. That is, by a simple change of variable, a periodic function $f(t)$ of period T may be transformed into a periodic function $F(t_1)$ of period 2π. Thus, in order to develop an initial understanding and to discuss some of the properties of Fourier series, we shall first consider functions of period 2π, returning to functions of period other than 2π in Section 12.2.10.

Example 12.1 Obtain the Fourier series expansion of the periodic function $f(t)$ of period 2π defined by

$$f(t) = t \quad (0 < t < 2\pi), \qquad f(t) = f(t + 2\pi)$$

Solution A sketch of the function $f(t)$ over the interval $-4\pi < t < 4\pi$ is shown in Figure 12.2.

Figure 12.2
Sawtooth wave
of Example 12.1.

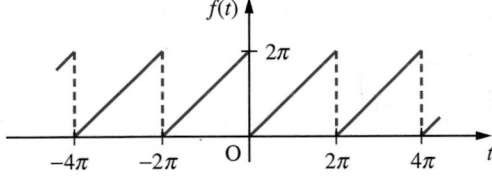

Since the function is periodic we only need to sketch it over one period, the pattern being repeated for other periods. Using (12.15) to evaluate the Fourier coefficients a_0 and a_n gives

$$a_0 = \frac{1}{\pi} \int_0^{2\pi} f(t)\,\mathrm{d}t = \frac{1}{\pi} \int_0^{2\pi} t\,\mathrm{d}t = \frac{1}{\pi}\left[\frac{t^2}{2}\right]_0^{2\pi} = 2\pi$$

and

$$a_n = \frac{1}{\pi} \int_0^{2\pi} f(t) \cos nt \, dt \quad (n = 1, 2, \dots)$$

$$= \frac{1}{\pi} \int_0^{2\pi} t \cos nt \, dt$$

which, on integration by parts, gives

$$a_n = \frac{1}{\pi} \left[t\frac{\sin nt}{n} + \frac{\cos nt}{n^2} \right]_0^{2\pi} = \frac{1}{\pi} \left(\frac{2\pi}{n} \sin 2n\pi + \frac{1}{n^2} \cos 2n\pi - \frac{\cos 0}{n^2} \right) = 0$$

since $\sin 2n\pi = 0$ and $\cos 2n\pi = \cos 0 = 1$. Note the need to work out a_0 separately from a_n in this case. The formula (12.16) for b_n gives

$$b_n = \frac{1}{\pi} \int_0^{2\pi} f(t) \sin nt \, dt \quad (n = 1, 2, \dots)$$

$$= \frac{1}{\pi} \int_0^{2\pi} t \sin nt \, dt$$

which, on integration by parts, gives

$$b_n = \frac{1}{\pi} \left[-\frac{t}{n} \cos nt + \frac{\sin nt}{n^2} \right]_0^{2\pi}$$

$$= \frac{1}{\pi} \left(-\frac{2\pi}{n} \cos 2n\pi \right) \quad (\text{since } \sin 2n\pi = \sin 0 = 0)$$

$$= -\frac{2}{n} \quad (\text{since } \cos 2n\pi = 1)$$

Hence from (12.14) the Fourier series expansion of $f(t)$ is

$$f(t) = \pi - \sum_{n=1}^{\infty} \frac{2}{n} \sin nt$$

or, in expanded form,

$$f(t) = \pi - 2 \left(\sin t + \frac{\sin 2t}{2} + \frac{\sin 3t}{3} + \dots + \frac{\sin nt}{n} + \dots \right)$$

Example 12.2   A periodic function $f(t)$ with period 2π is defined by

$$f(t) = t^2 + t \quad (-\pi < t < \pi), \qquad f(t) = f(t + 2\pi)$$

Sketch a graph of the function $f(t)$ for values of t from $t = -3\pi$ to $t = 3\pi$ and obtain a Fourier series expansion of the function.

Solution A graph of the function $f(t)$ for $-3\pi < t < 3\pi$ is shown in Figure 12.3.

Figure 12.3
Graph of the function
$f(t)$ of Example 12.2.

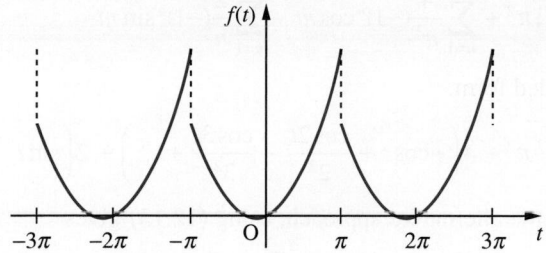

From (12.15) we have

$$a_0 = \frac{1}{\pi} \int_{-\pi}^{\pi} f(t)\,dt = \frac{1}{\pi} \int_{-\pi}^{\pi} (t^2 + t)\,dt = \tfrac{2}{3}\pi^2$$

and

$$a_n = \frac{1}{\pi} \int_{-\pi}^{\pi} f(t) \cos nt\,dt \quad (n = 1, 2, 3, \dots)$$

$$= \frac{1}{\pi} \int_{-\pi}^{\pi} (t^2 + t) \cos nt\,dt$$

which, on integration by parts, gives

$$a_n = \frac{1}{\pi} \left[\frac{t^2}{n} \sin nt + \frac{2t}{n^2} \cos nt - \frac{2}{n^3} \sin nt + \frac{t}{n} \sin nt + \frac{1}{n^2} \cos nt \right]_{-\pi}^{\pi}$$

$$= \frac{1}{\pi} \frac{4\pi}{n^2} \cos n\pi \quad \left(\text{since } \sin n\pi = 0 \text{ and } \left[\frac{1}{n^2} \cos nt \right]_{-\pi}^{\pi} = 0 \right)$$

$$= \frac{4}{n^2}(-1)^n \qquad (\text{since } \cos n\pi = (-1)^n)$$

From (12.16)

$$b_n = \frac{1}{\pi} \int_{-\pi}^{\pi} f(t) \sin nt\,dt \quad (n = 1, 2, 3, \dots)$$

$$= \frac{1}{\pi} \int_{-\pi}^{\pi} (t^2 + t) \sin nt\,dt$$

which, on integration by parts, gives

$$b_n = \frac{1}{\pi} \left[-\frac{t^2}{n} \cos nt + \frac{2t}{n^2} \sin nt + \frac{2}{n^3} \cos nt - \frac{t}{n} \cos nt + \frac{1}{n^2} \sin nt \right]_{-\pi}^{\pi}$$

$$= -\frac{2}{n} \cos n\pi = -\frac{2}{n}(-1)^n \quad (\text{since } \cos n\pi = (-1)^n)$$

Hence from (12.14) the Fourier series expansion of $f(t)$ is

$$f(t) = \tfrac{1}{3}\pi^2 + \sum_{n=1}^{\infty} \frac{4}{n^2}(-1)^n \cos nt - \sum_{n=1}^{\infty} \frac{2}{n}(-1)^n \sin nt$$

or, in expanded form,

$$f(t) = \tfrac{1}{3}\pi^2 + 4\left(-\cos t + \frac{\cos 2t}{2^2} - \frac{\cos 3t}{3^2} + \dots\right) + 2\left(\sin t - \frac{\sin 2t}{2} + \frac{\sin 3t}{3} \dots\right)$$

To illustrate the alternative approach, using (12.13) gives

$$a_n + jb_n = \frac{1}{\pi}\int_{-\pi}^{\pi} f(t)e^{jnt}\,dt = \frac{1}{\pi}\int_{-\pi}^{\pi}(t^2 + t)e^{jnt}\,dt$$

$$= \frac{1}{\pi}\left(\left[\frac{t^2 + t}{jn}e^{jnt}\right]_{-\pi}^{\pi} - \int_{-\pi}^{\pi}\frac{2t + 1}{jn}e^{jnt}\,dt\right)$$

$$= \frac{1}{\pi}\left[\frac{t^2 + t}{jn}e^{jnt} - \frac{2t + 1}{(jn)^2}e^{jnt} + \frac{2e^{jnt}}{(jn)^3}\right]_{-\pi}^{\pi}$$

Since

$$e^{jn\pi} = \cos n\pi + j\sin n\pi = (-1)^n$$
$$e^{-jn\pi} = \cos n\pi - j\sin n\pi = (-1)^n$$

and

$$1/j = -j$$

$$a_n + jb_n = \frac{(-1)^n}{\pi}\left(-j\frac{\pi^2 + \pi}{n} + \frac{2\pi + 1}{n^2} + \frac{j2}{n^3} + j\frac{\pi^2 - \pi}{n} - \frac{1 - 2\pi}{n^2} - \frac{j2}{n^3}\right)$$

$$= (-1)^n\left(\frac{4}{n^2} - j\frac{2}{n}\right)$$

Equating real and imaginary parts gives, as before,

$$a_n = \frac{4}{n^2}(-1)^n, \quad b_n = -\frac{2}{n}(-1)^n$$

A periodic function $f(t)$ may be specified in a piecewise fashion over a period, or, indeed, it may only be piecewise-continuous over a period, as illustrated in Figure 12.4. In order to calculate the Fourier coefficients in such cases, it is necessary to break up the range of integration in Euler's formulae to correspond to the various components of the function. For example, for the function shown in Figure 12.4, $f(t)$ is defined in the interval $-\pi < t < \pi$ by

$$f(t) = \begin{cases} f_1(t) & (-\pi < t < -p) \\ f_2(t) & (-p < t < q) \\ f_3(t) & (q < t < \pi) \end{cases}$$

Figure 12.4
Piecewise-continuous
function over a period.

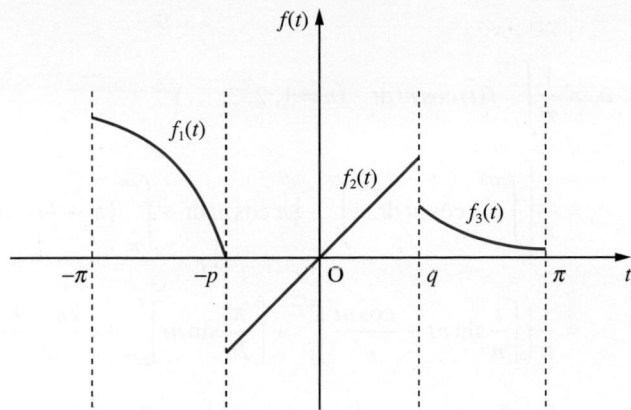

and is periodic with period 2π. Euler's formulae (12.15) and (12.16) for the Fourier coefficients become

$$a_n = \frac{1}{\pi}\left[\int_{-\pi}^{-p} f_1(t)\cos nt\,dt + \int_{-p}^{q} f_2(t)\cos nt\,dt + \int_{q}^{\pi} f_3(t)\cos nt\,dt\right]$$

$$b_n = \frac{1}{\pi}\left[\int_{-\pi}^{-p} f_1(t)\sin nt\,dt + \int_{-p}^{q} f_2(t)\sin nt\,dt + \int_{q}^{\pi} f_3(t)\sin nt\,dt\right]$$

Example 12.3

A periodic function $f(t)$ of period 2π is defined within the period $0 \leq t \leq 2\pi$ by

$$f(t) = \begin{cases} t & (0 \leq t \leq \tfrac{1}{2}\pi) \\ \tfrac{1}{2}\pi & (\tfrac{1}{2}\pi \leq t \leq \pi) \\ \pi - \tfrac{1}{2}t & (\pi \leq t \leq 2\pi) \end{cases}$$

Sketch a graph of $f(t)$ for $-2\pi \leq t \leq 3\pi$ and find a Fourier series expansion of it.

Solution

A graph of the function $f(t)$ for $-2\pi \leq t \leq 3\pi$ is shown in Figure 12.5.

Figure 12.5
Graph of the function
$f(t)$ of Example 12.3.

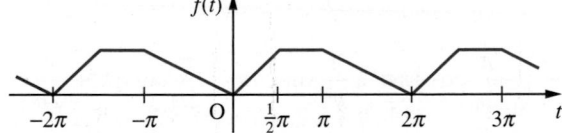

From (12.15),

$$a_0 = \frac{1}{\pi}\int_0^{2\pi} f(t)dt = \frac{1}{\pi}\left[\int_0^{\pi/2} t\,dt + \int_{\pi/2}^{\pi} \tfrac{1}{2}\pi\,dt + \int_{\pi}^{2\pi}\left(\pi - \tfrac{1}{2}t\right)dt\right] = \frac{5}{8}\pi$$

and

$$a_n = \frac{1}{\pi} \int_0^{2\pi} f(t) \cos nt \, dt \quad (n = 1, 2, 3, \dots)$$

$$= \frac{1}{\pi} \left[\int_0^{\pi/2} t \cos nt \, dt + \int_{\pi/2}^{\pi} \tfrac{1}{2}\pi \cos nt \, dt + \int_{\pi}^{2\pi} \left(\pi - \tfrac{1}{2}t \right) \cos nt \, dt \right]$$

$$= \frac{1}{\pi} \left(\left[\frac{t}{n} \sin nt + \frac{\cos nt}{n^2} \right]_0^{\pi/2} + \left[\frac{\pi}{2n} \sin nt \right]_{\pi/2}^{\pi} + \left[\frac{2\pi - t}{2} \frac{\sin nt}{n} - \frac{\cos nt}{2n^2} \right]_{\pi}^{2\pi} \right)$$

$$= \frac{1}{\pi} \left(\frac{\pi}{2n} \sin \tfrac{1}{2}n\pi + \frac{1}{n^2} \cos \tfrac{1}{2}n\pi - \frac{1}{n^2} - \frac{\pi}{2n} \sin \tfrac{1}{2}n\pi - \frac{1}{2n^2} + \frac{1}{2n^2} \cos n\pi \right)$$

$$= \frac{1}{2\pi n^2} (2 \cos \tfrac{1}{2}n\pi - 3 + \cos n\pi)$$

that is,

$$a_n = \begin{cases} \dfrac{1}{\pi n^2}[(-1)^{n/2} - 1] & \text{(even } n) \\[2mm] -\dfrac{2}{\pi n^2} & \text{(odd } n) \end{cases}$$

From (12.16),

$$b_n = \frac{1}{\pi} \int_0^{2\pi} f(t) \sin nt \, dt \quad (n = 1, 2, 3, \dots)$$

$$= \frac{1}{\pi} \left[\int_0^{\pi/2} t \sin nt \, dt + \int_{\pi/2}^{\pi} \tfrac{1}{2}\pi \sin nt \, dt + \int_{\pi}^{2\pi} \left(\pi - \tfrac{1}{2}t \right) \sin nt \, dt \right]$$

$$= \frac{1}{\pi} \left(\left[-\frac{t}{n} \cos nt + \frac{1}{n^2} \sin nt \right]_0^{\pi/2} + \left[-\frac{\pi}{2n} \cos nt \right]_{\pi/2}^{\pi} \right.$$

$$\left. + \left[\frac{t - 2\pi}{2n} \cos nt - \frac{1}{2n^2} \sin nt \right]_{\pi}^{2\pi} \right)$$

$$= \frac{1}{\pi} \left(-\frac{\pi}{2n} \cos \tfrac{1}{2}n\pi + \frac{1}{n^2} \sin \tfrac{1}{2}n\pi - \frac{\pi}{2n} \cos n\pi + \frac{\pi}{2n} \cos \tfrac{1}{2}n\pi + \frac{\pi}{2n} \cos n\pi \right)$$

$$= \frac{1}{\pi n^2} \sin \tfrac{1}{2}n\pi$$

$$= \begin{cases} 0 & \text{(even } n) \\[2mm] \dfrac{(-1)^{(n-1)/2}}{\pi n^2} & \text{(odd } n) \end{cases}$$

Hence from (12.14) the Fourier series expansion of $f(t)$ is

$$f(t) = \tfrac{5}{16}\pi - \frac{2}{\pi}\left(\cos t + \frac{\cos 3t}{3^2} + \frac{\cos 5t}{5^2} + \ldots\right)$$

$$- \frac{2}{\pi}\left(\frac{\cos 2t}{2^2} + \frac{\cos 6t}{6^2} + \frac{\cos 10t}{10^2} + \ldots\right)$$

$$+ \frac{1}{\pi}\left(\sin t - \frac{\sin 3t}{3^2} + \frac{\sin 5t}{5^2} - \frac{\sin 7t}{7^2} + \ldots\right)$$

A major use of the MATLAB Symbolic Math Toolbox and MAPLE, when dealing with Fourier series, is to avoid the tedious and frequently error prone integration involved in determining the coefficients a_n and b_n. It is therefore advisable to use them to check the accuracy of integration. To illustrate we shall consider Examples 12.2 and 12.3.

In MAPLE n may be declared to be an integer using the command

```
assume(n,integer);
```

which helps with simplification of answers. There is no comparable command in MATLAB so, when using the Symbolic Math Toolbox, we shall use the command

```
maple('assume (n,integer)')
```

Considering Example 12.2 the MATLAB commands

```
syms t n
maple('assume (n,integer)');
int((t^2 + t)*cos(n*t),-pi,pi)/pi
```

return the value of a_n as

```
4*(-1)^n/n^2
```

Entering the command $pretty(ans)$ gives a_n in the form $4\dfrac{(-1)^{n\sim}}{n\sim^2}$, where n~ indicates that n is an integer. Likewise the commands

```
int((t^2 + t)*sin(n*t),-pi,pi)/pi;
pretty(ans)
```

returns b_n as

$$-2\frac{(-1)^{n\sim}}{n\sim}$$

thus checking with the values given in the solution.

The corresponding commands in MAPLE are

```
assume(n,integer);
int((t^2 + t)*cos(n*t), t = -Pi..Pi)/Pi;
```

returning the value of a_n as

$$4\frac{(-1)^{n\sim}}{n\sim^2}$$

with the further command

```
int((t^2 + t)*sin(n*t), t = -Pi..Pi)/Pi;
```

returning the value of b_n as

$$-2\frac{(-1)^{n\sim}}{n\sim}$$

again checking with the values given in the solution.

In Example 12.3 we are dealing with a piecewise function, which can be specified using the $piecewise$ command. In MATLAB the commands

```
syms t n
maple('assume (n,integer)');
f = ('PIECEWISE([t,t<= 1/2*pi], [1/2*pi,1/2*pi-t<= 0 and
t-pi< = 0],[pi-1/2t*t,pi<=t])');
int(f*cos(n*t),0,2*pi)/pi;
pretty(ans)
```

return the value of a_n as

$$\frac{1}{2}\frac{-3 + 2\ \cos(1/2\ pi\ n\ \sim) + (-1)^{n\sim}}{n\sim^2\ pi}$$

with the further commands

```
int(f*sin(n*t),0,2*pi)/pi;
pretty(ans)
```

returning the value of b_n as $\dfrac{\sin(1/2\ pi\ n\ \sim)}{n\sim^2\ pi}$.

In MAPLE the commands

```
f:= simplify(piecewise(t<= Pi/2,t,(t>= Pi/2 and
t<= Pi),Pi/2,t>= Pi,Pi-t/2));
ff:= unapply(f,t);
assume(n,integer);
an:= int(ff(t)*cos(n*t), t = 0..Pi)/Pi;
bn:= int(ff(t)*sin(n*t), t = 0..Pi)/Pi;
```

return the same values as MATLAB above for a_n and b_n.

An alternative approach to using the $piecewise$ command is to express the function in terms of Heavyside functions.

12.2.5 Even and odd functions

Noting that a particular function possesses certain symmetrical properties enables us both to tell which terms are absent from a Fourier series expansion of the function and to simplify the expressions determining the remaining coefficients. In this section we consider even and odd function symmetries, while in Section 12.2.6 we shall consider symmetry due to even and odd harmonics.

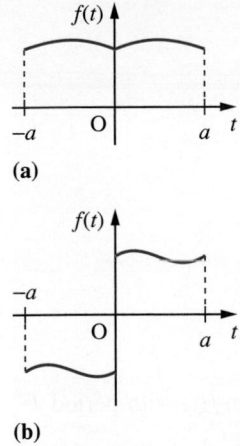

Figure 12.6
Graphs of (a) an
even function and
(b) an odd function.

First we review the properties of even and odd functions, considered in Section 2.2.6, that are useful for determining the Fourier coefficients. If $f(t)$ is an even function then $f(t) = f(-t)$ for all t, and the graph of the function is symmetrical about the vertical axis as illustrated in Figure 12.6(a). From the definition of integration, it follows that if $f(t)$ is an even function then

$$\int_{-a}^{a} f(t)\,dt = 2\int_{0}^{a} f(t)\,dt$$

If $f(t)$ is an odd function then $f(t) = -f(-t)$ for all t, and the graph of the function is symmetrical about the origin; that is, there is opposite-quadrant symmetry, as illustrated in Figure 12.6(b). It follows that if $f(t)$ is an odd function then

$$\int_{-a}^{a} f(t)\,dt = 0$$

The following properties of even and odd functions are also useful for our purposes:

(a)   the *sum* of two (or more) *odd* functions is an *odd* function;
(b)   the *product* of two *even* functions is an *even* function;
(c)   the *product* of two *odd* functions is an *even* function;
(d)   the *product* of an *odd* and an *even* function is an *odd* function;
(e)   the *derivative* of an *even* function is an *odd* function;
(f)   the *derivative* of an *odd* function is an *even* function.

(Noting that t^{even} is even and t^{odd} is odd helps one to remember (a)–(f).)

Using these properties, and taking $d = -\tfrac{1}{2}T$ in (12.11) and (12.12), we have the following:

(i) If $f(t)$ is an *even* periodic function of period T then

$$a_n = \frac{2}{T}\int_{-T/2}^{T/2} f(t)\cos n\omega t\,dt = \frac{4}{T}\int_{0}^{T/2} f(t)\cos n\omega t\,dt$$

using property (b), and

$$b_n = \frac{2}{T}\int_{-T/2}^{T/2} f(t)\sin n\omega t\,dt = 0$$

using property (d).

Thus the Fourier series expansion of an even periodic function $f(t)$ with period T consists of cosine terms only and, from (12.3), is given by

$$f(t) = \tfrac{1}{2}a_0 + \sum_{n=1}^{\infty} a_n \cos n\omega t \tag{12.17}$$

with

$$a_n = \frac{4}{T}\int_{0}^{T/2} f(t)\cos n\omega t \quad (n = 0, 1, 2, \dots) \tag{12.18}$$

(ii) If $f(t)$ is an *odd* periodic function of period T then

$$a_n = \frac{2}{T} \int_{-T/2}^{T/2} f(t) \cos n\omega t \, dt = 0$$

using property (d), and

$$b_n = \frac{2}{T} \int_{-T/2}^{T/2} f(t) \sin n\omega t \, dt = \frac{4}{T} \int_{0}^{T/2} f(t) \sin n\omega t \, dt$$

using property (c).

Thus the Fourier series expansion of an odd periodic function $f(t)$ with period T consists of sine terms only and, from (12.3), is given by

$$f(t) = \sum_{n=1}^{\infty} b_n \sin n\omega t \qquad \text{(12.19)}$$

with

$$b_n = \frac{4}{T} \int_{0}^{T/2} f(t) \sin n\omega t \, dt \quad (n = 1, 2, 3, \dots) \qquad \text{(12.20)}$$

Example 12.4   A periodic function $f(t)$ with period 2π is defined within the period $-\pi < t < \pi$ by

$$f(t) = \begin{cases} -1 & (-\pi < t < 0) \\ 1 & (0 < t < \pi) \end{cases}$$

Find its Fourier series expansion.

Solution   A sketch of the function $f(t)$ over the interval $-4\pi < t < 4\pi$ is shown in Figure 12.7.

Figure 12.7
Square wave of
Example 12.4.

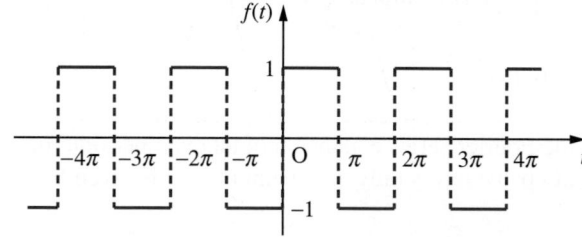

Clearly $f(t)$ is an odd function of t, so that its Fourier series expansion consists of sine terms only. Taking $T = 2\pi$, that is $\omega = 1$, in (12.19) and (12.20), the Fourier series expansion is given by

$$f(t) = \sum_{n=1}^{\infty} b_n \sin nt$$

with

$$b_n = \frac{2}{\pi} \int_0^\pi f(t) \sin nt \, dt \quad (n = 1, 2, 3, \dots)$$

$$= \frac{2}{\pi} \int_0^\pi 1 \sin nt \, dt = \frac{2}{\pi} \left[-\frac{1}{n} \cos nt \right]_0^\pi$$

$$= \frac{2}{n\pi}(1 - \cos n\pi) = \frac{2}{n\pi}[1 - (-1)^n]$$

$$= \begin{cases} 4/n\pi & (\text{odd } n) \\ 0 & (\text{even } n) \end{cases}$$

Thus the Fourier series expansion of $f(t)$ is

$$f(t) = \frac{4}{\pi} \left(\sin t + \frac{1}{3} \sin 3t + \frac{1}{5} \sin 5t + \dots \right) = \frac{4}{\pi} \sum_{n=1}^\infty \frac{\sin(2n - 1)t}{2n - 1} \qquad (12.21)$$

Example 12.5   A periodic function $f(t)$ with period 2π is defined as

$$f(t) = t^2 \quad (-\pi < t < \pi), \qquad f(t) = f(t + 2\pi)$$

Obtain a Fourier series expansion for it.

Solution   A sketch of the function $f(t)$ over the interval $-3\pi < t < 3\pi$ is shown is Figure 12.8.

Figure 12.8
The function $f(t)$
of Example 12.5.

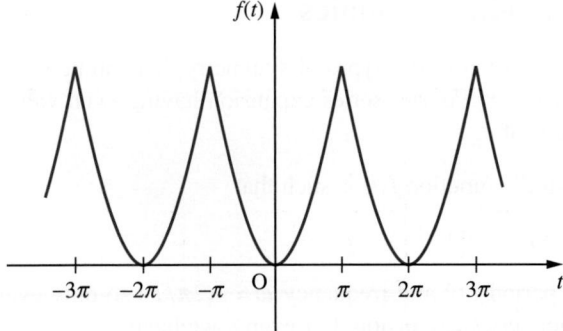

Clearly, $f(t)$ is an even function of t, so that its Fourier series expansion consists of cosine terms only. Taking $T = 2\pi$, that is $\omega = 1$, in (12.17) and (12.18) the Fourier series expansion is given by

$$f(t) = \tfrac{1}{2}a_0 + \sum_{n=1}^\infty a_n \cos nt$$

with

$$a_0 = \frac{2}{\pi} \int_0^{\pi} f(t)\,dt = \frac{2}{\pi} \int_0^{\pi} t^2\,dt = \tfrac{2}{3}\pi^2$$

and

$$a_n = \frac{2}{\pi} \int_0^{\pi} f(t)\cos nt\,dt \quad (n = 1, 2, 3, \dots)$$

$$= \frac{2}{\pi} \int_0^{\pi} t^2 \cos nt\,dt$$

$$= \frac{2}{\pi} \left[\frac{t^2}{n}\sin nt + \frac{2t}{n^2}\cos nt - \frac{2}{n^3}\sin nt \right]_0^{\pi}$$

$$= \frac{2}{\pi}\left(\frac{2\pi}{n^2}\cos n\pi \right) = \frac{4}{n^2}(-1)^n$$

since $\sin n\pi = 0$ and $\cos n\pi = (-1)^n$. Thus the Fourier series expansion of $f(t) = t^2$ is

$$f(t) = \tfrac{1}{3}\pi^2 + 4 \sum_{n=1}^{\infty} \frac{(-1)^n}{n^2}\cos nt \tag{12.22}$$

or, writing out the first few terms,

$$f(t) = \tfrac{1}{3}\pi^2 - 4\cos t + \cos 2t - \tfrac{4}{9}\cos 3t + \dots$$

12.2.6 Even and odd harmonics

In this section we consider types of symmetry that can be identified in order to eliminate terms from the Fourier series expansion having even values of n (including $n = 0$) or odd values of n.

(a) If a periodic function $f(t)$ is such that

$$f(t + \tfrac{1}{2}T) = f(t)$$

then it has period $T/2$ and frequency $\omega = 2(2\pi/T)$ so only even harmonics are present in its Fourier series expansion. For even n we have

$$a_n = \frac{4}{T} \int_0^{T/2} f(t)\cos n\omega t\,dt \tag{12.23}$$

$$b_n = \frac{4}{T} \int_0^{T/2} f(t)\sin n\omega t\,dt \tag{12.24}$$

An example of such a function is given in Figure 12.9(a).

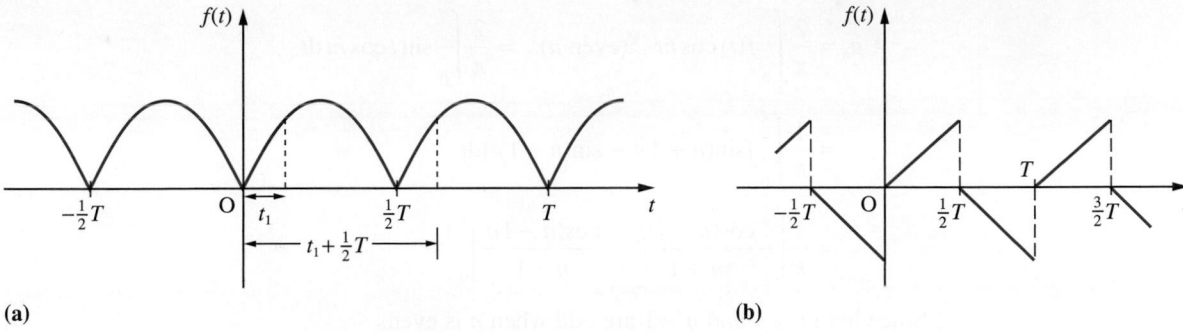

Figure 12.9 Functions having Fourier series with (a) only even harmonics and (b) only odd harmonics.

(b) If a periodic function $f(t)$ with period T is such that

$$f(t + \tfrac{1}{2}T) = -f(t)$$

then only odd harmonics are present in its Fourier series expansion. For odd n

$$a_n = \frac{4}{T} \int_0^{T/2} f(t) \cos n\omega t \, dt \tag{12.25}$$

$$b_n = \frac{4}{T} \int_0^{T/2} f(t) \sin n\omega t \, dt \tag{12.26}$$

An example of such a function is shown in Figure 12.9(b).

The square wave of Example 12.4 is such that $f(t + \pi) = -f(t)$, so that, from (b), its Fourier series expansion consists of only odd harmonics. Since it is also an odd function, it follows that its Fourier series expansion consists only of odd-harmonic sine terms, which is confirmed by the result (12.21).

Example 12.6   Obtain the Fourier series expansion of the rectified sine wave

$$f(t) = |\sin t|$$

Solution   A sketch of the wave over the interval $-\pi < t < 2\pi$ is shown in Figure 12.10. Clearly, $f(t + \pi) = f(t)$ so that only even harmonics are present in the Fourier series expansion. Since the function is also an even function of t, it follows that the Fourier series expansion will consist only of even-harmonic cosine terms. Taking $T = 2\pi$, that is $\omega = 1$, in (12.23), the coefficients of the even harmonics are given by

Figure 12.10
Rectified wave
$f(t) = |\sin t|$.

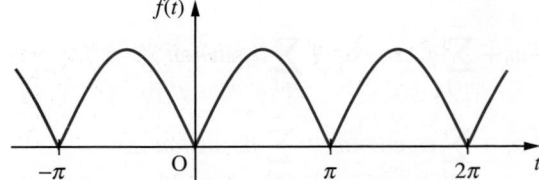

$$a_n = \frac{2}{\pi} \int_0^\pi f(t) \cos nt \quad (\text{even } n) \quad = \frac{2}{\pi} \int_0^\pi \sin t \cos nt \, dt$$

$$= \frac{1}{\pi} \int_0^\pi [\sin(n+1)t - \sin(n-1)t] dt$$

$$= \frac{1}{\pi} \left[-\frac{\cos(n+1)t}{n+1} + \frac{\cos(n-1)t}{n-1} \right]_0^\pi$$

Since both $n + 1$ and $n - 1$ are odd when n is even,

$$\cos(n+1)\pi = \cos(n-1)\pi = -1$$

so that

$$a_n = \frac{1}{\pi} \left[\left(\frac{1}{n+1} - \frac{1}{n-1} \right) - \left(-\frac{1}{n+1} + \frac{1}{n-1} \right) \right] = -\frac{4}{\pi} \frac{1}{n^2 - 1}$$

Thus the Fourier series expansion of $f(t)$ is

$$f(t) = \tfrac{1}{2}a_0 + \sum_{\substack{n=2 \\ (n \text{ even})}}^\infty a_n \cos nt = \frac{2}{\pi} - \frac{4}{\pi} \sum_{\substack{n=2 \\ (n \text{ even})}}^\infty \frac{1}{n^2 - 1} \cos nt$$

$$= \frac{2}{\pi} - \frac{4}{\pi} \sum_{n=1}^\infty \frac{1}{4n^2 - 1} \cos 2nt$$

or, writing out the first few terms,

$$f(t) = \frac{2}{\pi} - \frac{4}{\pi} (\tfrac{1}{3} \cos 2t + \tfrac{1}{15} \cos 4t + \tfrac{1}{35} \cos 6t + \dots)$$

12.2.7 Linearity property

The linearity property as applied to Fourier series may be stated in the form of the following theorem.

Theorem 12.1 If $f(t) = lg(t) + mh(t)$, where $g(t)$ and $h(t)$ are periodic functions of period T and l and m are arbitrary constants, then $f(t)$ has a Fourier series expansion in which the coefficients are the sums of the coefficients in the Fourier series expansions of $g(t)$ and $h(t)$ multiplied by l and m respectively.

Proof Clearly $f(t)$ is periodic with period T. If the Fourier series expansions of $g(t)$ and $h(t)$ are

$$g(t) = \tfrac{1}{2}a_0 + \sum_{n=1}^\infty a_n \cos n\omega t + \sum_{n=1}^\infty b_n \sin n\omega t$$

$$h(t) = \tfrac{1}{2}\alpha_0 + \sum_{n=1}^\infty \alpha_n \cos n\omega t + \sum_{n=1}^\infty \beta_n \sin n\omega t$$

then, using (12.11) and (12.12), the Fourier coefficients in the expansion of $f(t)$ are

$$A_n = \frac{2}{T} \int_d^{d+T} f(t) \cos n\omega t \, dt = \frac{2}{T} \int_d^{d+T} [lg(t) + mh(t)] \cos n\omega t \, dt$$

$$= \frac{2l}{T} \int_d^{d+T} g(t) \cos n\omega t \, dt + \frac{2m}{T} \int_d^{d+T} h(t) \cos n\omega t \, dt$$

$$= la_n + m\alpha_n$$

and

$$B_n = \frac{2}{T} \int_d^{d+T} f(t) \sin n\omega t \, dt = \frac{2l}{T} \int_d^{d+T} g(t) \sin n\omega t \, dt + \frac{2m}{T} \int_d^{d+T} h(t) \sin n\omega t \, dt$$

$$= lb_n + m\beta_n$$

confirming that the Fourier series expansion of $f(t)$ is

$$f(t) = \tfrac{1}{2}(la_0 + m\alpha_0) + \sum_{n=1}^{\infty} (la_n + m\alpha_n) \cos n\omega t + \sum_{n=1}^{\infty} (lb_n + m\beta_n) \sin n\omega t$$

end of theorem

Example 12.7   Suppose that $g(t)$ and $h(t)$ are periodic functions of period 2π and are defined within the period $-\pi < t < \pi$ by

$$g(t) = t^2, \quad h(t) = t$$

Determine the Fourier series expansions of both $g(t)$ and $h(t)$ and use the linearity property to confirm the expansion obtained in Example 12.2 for the periodic function $f(t)$ defined within the period $-\pi < t < \pi$ by $f(t) = t^2 + t$.

Solution   The Fourier series of $g(t)$ is given by (12.22) as

$$g(t) = \tfrac{1}{3}\pi^2 + 4\sum_{n=1}^{\infty} \frac{(-1)^n}{n^2} \cos nt$$

Recognizing that $h(t) = t$ is an odd function of t, we find, taking $T = 2\pi$ and $\omega = 1$ in (12.19) and (12.20), that its Fourier series expansion is

$$h(t) = \sum_{n=1}^{\infty} b_n \sin nt$$

where

$$b_n = \frac{2}{\pi} \int_0^{\pi} h(t) \sin nt \, dt \quad (n = 1, 2, 3, \ldots)$$

$$= \frac{2}{\pi} \int_0^{\pi} t \sin nt \, dt = \frac{2}{\pi} \left[-\frac{t}{n} \cos nt + \frac{\sin nt}{n^2} \right]_0^{\pi}$$

$$= -\frac{2}{n}(-1)^n$$

recognizing again that $\cos n\pi = (-1)^n$ and $\sin n\pi = 0$. Thus the Fourier series expansion of $h(t) = t$ is

$$h(t) = -2\sum_{n=1}^{\infty} \frac{(-1)^n}{n}\sin nt \qquad (12.27)$$

Using the linearity property, we find, by combining (12.12) and (12.27), that the Fourier series expansion of $f(t) = g(t) + h(t) = t^2 + t$ is

$$f(t) = \tfrac{1}{3}\pi^2 + 4\sum_{n=1}^{\infty} \frac{(-1)^n}{n^2}\cos nt - 2\sum_{n=1}^{\infty} \frac{(-1)^n}{n}\sin nt$$

which conforms to the series obtained in Example 12.2.

12.2.8 Convergence of the Fourier series

So far we have concentrated our attention on determining the Fourier series expansion corresponding to a given periodic function $f(t)$. In reality, this is an exercise in integration, since we merely have to compute the coefficients a_n and b_n using Euler's formulae (12.11) and (12.12) and then substitute these values into (12.3). We have not yet considered the question of whether or not the Fourier series thus obtained is a valid representation of the periodic function $f(t)$. It should not be assumed that the existence of the coefficients a_n and b_n in itself implies that the associated series converges to the function $f(t)$.

A full discussion of the convergence of a Fourier series is beyond the scope of this book and we shall confine ourselves to simply stating a set of conditions which ensures that $f(t)$ has a convergent Fourier series expansion. These conditions, known as **Dirichlet's conditions**, may be stated in the form of Theorem 12.2.

Theorem 12.2 **Dirichlet's conditions**

If $f(t)$ is a bounded periodic function that in any period has

(a) a finite number of isolated maxima and minima, and

(b) a finite number of points of finite discontinuity

then the Fourier series expansion of $f(t)$ converges to $f(t)$ at all points where $f(t)$ is continuous and to the average of the right- and left-hand limits of $f(t)$ at points where $f(t)$ is discontinuous (that is, to the mean of the discontinuity).

end of theorem

Example 12.8 Give reasons why the functions

(a) $\dfrac{1}{3-t}$ (b) $\sin\left(\dfrac{1}{t-2}\right)$

do not satisfy Dirichlet's conditions in the interval $0 < t < 2\pi$.

Solution
(a) The function $f(t) = 1/(3 - t)$ has an infinite discontinuity at $t = 3$, which is within the interval, and therefore does not satisfy the condition that $f(t)$ must only have *finite* discontinuities within a period (i.e. it is bounded).

(b) The function $f(t) = \sin[1/(t - 2)]$ has an infinite number of maxima and minima in the neighbourbood of $t = 2$, which is within the interval, and therefore does not satisfy the requirement that $f(t)$ must have only a finite number of isolated maxima and minima within one period.

The conditions of Theorem 12.2 are sufficient to ensure that a representative Fourier series expansion of $f(t)$ exists. However, they are not necessary conditions for convergence, and it does not follow that a representative Fourier series does not exist if they are not satisfied. Indeed, necessary conditions on $f(t)$ for the existence of a convergent Fourier series are not yet known. In practice, this does not cause any problems, since for almost all conceivable practical applications the functions that are encountered satisfy the conditions of Theorem 12.2 and therefore have representative Fourier series.

Another issue of importance in practical applications is the rate of convergence of a Fourier series, since this is an indication of how many terms must be taken in the expansion in order to obtain a realistic approximation to the function $f(t)$ it represents. Obviously, this is determined by the coefficients a_n and b_n of the Fourier series and the manner in which these decrease as n increases.

In an example, such as Example 12.1, in which the function $f(t)$ is only piecewise-continuous, exhibiting jump discontinuities, the Fourier coefficients decrease as $1/n$, and it may be necessary to include a large number of terms to obtain an adequate approximation to $f(t)$. In an example, such as Example 12.3, in which the function is a continuous function but has discontinuous first derivatives (owing to the sharp corners), the Fourier coefficients decrease as $1/n^2$, and so one would expect the series to converge more rapidly. Indeed, this argument applies in general, and we may summarize as follows:

(a) if $f(t)$ is only piecewise-continuous then the coefficients in its Fourier series representation decrease as $1/n$;

(b) if $f(t)$ is continuous everywhere but has discontinuous first derivatives then the coefficients in its Fourier series representation decrease as $1/n^2$;

(c) if $f(t)$ and all its derivatives up to that of the rth order are continuous but the $(r + 1)$th derivative is discontinuous then the coefficients in its Fourier series representation decrease as $1/n^{r+2}$.

These observations are not surprising, since they simply tell us that the smoother the function $f(t)$, the more rapidly will its Fourier series representation converge.

To illustrate some of these issues related to convergence we return to Example 12.4, in which the Fourier series (12.21) was obtained as a representation of the square wave of Figure 12.7.

Since (12.21) is an infinite series, it is clearly not possible to plot a graph of the result. However, by considering finite partial sums, it is possible to plot graphs of approximations to the series. Denoting the sum of the first N terms in the infinite series by $f_N(t)$, that is

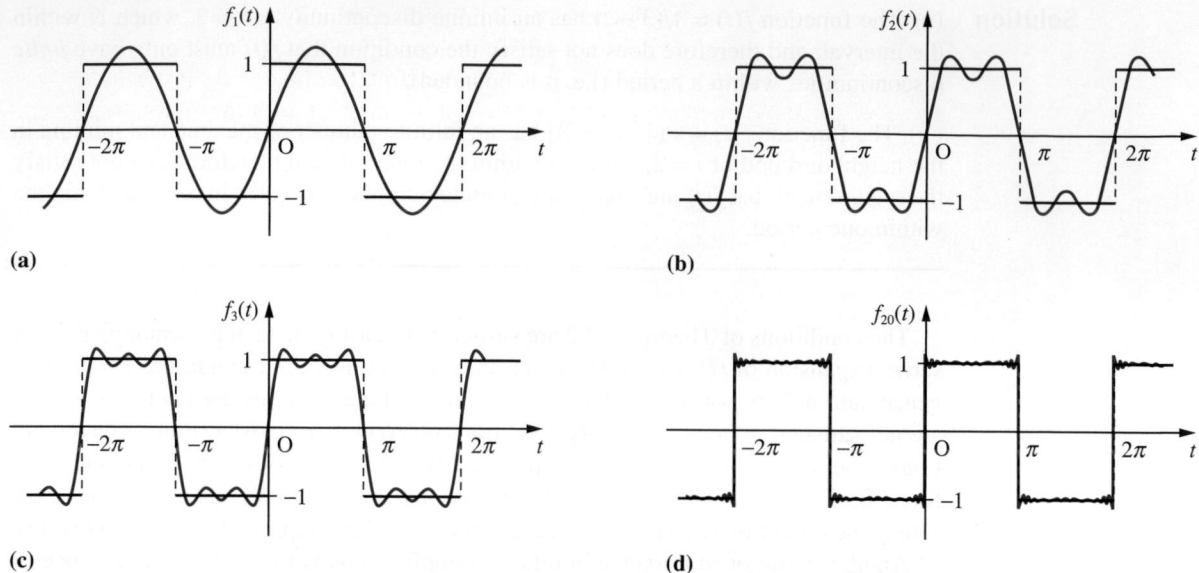

Figure 12.11 Plots of $f_N(t)$ for a square wave; (a) $N = 1$; (b) 2; (c) 3; (d) 20.

$$f_N(t) = \frac{4}{\pi} \sum_{n=1}^{N} \frac{\sin(2n - 1)t}{2n - 1} \tag{12.28}$$

the graphs of $f_N(t)$ for $N = 1$, 2, 3 and 20 are as shown in Figure 12.11. It can be seen that at points where $f(t)$ is continuous the approximation of $f(t)$ by $f_N(t)$ improves as N increases, confirming that the series converges to $f(t)$ at all such points. It can also be seen that at points of discontinuity of $f(t)$, which occur at $t = \pm n\pi$ ($n = 0$, 1, 2, ...), the series converges to the mean value of the discontinuity, which in this particular example is $\frac{1}{2}(-1 + 1) = 0$. As a consequence, the equality sign in (12.21) needs to be interpreted carefully. Although such use may be acceptable, in the sense that the series converges to $f(t)$ for values of t where $f(t)$ is continuous, this is not so at points of discontinuity. To overcome this problem, the symbol ~ (read as 'behaves as' or 'represented by') rather than = is frequently used in the Fourier series representation of a function $f(t)$, so that (12.21) is often written as

$$f(t) \sim \sum_{n=1}^{\infty} \frac{\sin(2n - 1)t}{2n - 1}$$

In the companion text *Advanced Modern Engineering Mathematics* it is shown that the Fourier series converges to $f(t)$ in the sense that the integral of the square of the difference between $f(t)$ and $f_N(t)$ is minimized and tends to zero as $N \to \infty$.

We note that convergence of the Fourier series is slowest near a point of discontinuity, such as the one that occurs at $t = 0$. Although the series does converge to the mean value of the discontinuity (namely zero) at $t = 0$, there is, as indicated in Figure 12.11(d), an undershoot at $t = 0^-$ (that is, just to the left of $t = 0$) and an overshoot at $t = 0^+$ (that is, just to the right of $t = 0$). This non-smooth convergence of the Fourier series leading to the occurrence of an undershoot and an overshoot at points of discontinuity of $f(t)$ is a characteristic of all Fourier series representing discontinuous functions, not only that of the square wave of Example 12.4, and is known as **Gibbs'**

phenomenon after the American physicist J. W. Gibbs (1839–1903). The magnitude of the undershoot/overshoot does not diminish as $N \to \infty$ in (12.28), but simply gets 'sharper' and 'sharper', tending to a spike. In general, the magnitude of the undershoot and overshoot together amount to about 18% of the magnitude of the discontinuity (that is, the difference in the values of the function $f(t)$ to the left and right of the discontinuity). It is important that the existence of this phenomenon be recognized, since in certain practical applications these spikes at discontinuities have to be suppressed by using appropriate smoothing factors.

To reproduce the plots of Figure 12.11 and see how the series converges as N increases use the following MATLAB commands:

```
t=pi/100*[300:300];
f=0;
T=[-3*pi -2*pi -2*pi -pi -pi 0 0 pi pi 2*pi 2*pi 3*pi];
y=[-1 -1 1 1 -1 -1 1 1 -1 -1 1 1];
for n=1:20
f=f+4/pi*sin((2*n-1)*t)/(2*n-1);
plot(T,y,t,f,[-3*pi 3*pi],[0,0],'k-',[0,0],
[-1.3 1.3],'k-')
axis([-3*pi,3*pi,-inf,inf]),pause
end
```

The `pause` command has been included to give you an opportunity to view the plots at the end of each step. Press any key to proceed.

Theoretically, we can use the series (12.21) to obtain an approximation to π. This is achieved by taking $t = \frac{1}{2}\pi$, when $f(t) = 1$; (12.21) then gives

$$1 = \frac{4}{\pi} \sum_{n=1}^{\infty} \frac{\sin \frac{1}{2}(2n-1)\pi}{2n-1}$$

leading to

$$\pi = 4(1 - \tfrac{1}{3} + \tfrac{1}{5} - \tfrac{1}{7} + \ldots) = 4 \sum_{n=1}^{\infty} \frac{(-1)^{n+1}}{2n-1}$$

For practical purposes, however, this is not a good way of obtaining an approximation to π, because of the slow rate of convergence of the series.

12.2.9 Exercises

Check evaluation of the integrals using MATLAB or MAPLE whenever possible.

1 In each of the following a periodic function $f(t)$ of period 2π is specified over one period. In each case sketch a graph of the function for $-4\pi \leqslant t \leqslant 4\pi$ and obtain a Fourier series representation of the function.

(a) $f(t) = \begin{cases} -\pi & (-\pi < t < 0) \\ t & (0 < t < \pi) \end{cases}$

(b) $f(t) = \begin{cases} t + \pi & (-\pi < t < 0) \\ 0 & (0 < t < \pi) \end{cases}$

(c) $f(t) = 1 - \dfrac{t}{\pi}$ $(0 \leqslant t \leqslant 2\pi)$

(d) $f(t) = \begin{cases} 0 & (-\pi \leqslant t \leqslant -\frac{1}{2}\pi) \\ 2\cos t & (-\frac{1}{2}\pi \leqslant t \leqslant \frac{1}{2}\pi) \\ 0 & (\frac{1}{2}\pi \leqslant t \leqslant \pi) \end{cases}$

(e) $f(t) = \cos\frac{1}{2}t$ $(-\pi < t < \pi)$

(f) $f(t) = |t|$ $(-\pi < t < \pi)$

(g) $f(t) = \begin{cases} 0 & (-\pi \leqslant t \leqslant 0) \\ 2t - \pi & (0 < t \leqslant \pi) \end{cases}$

(h) $f(t) = \begin{cases} -t + e^t & (-\pi \leqslant t < 0) \\ t + e^t & (0 \leqslant t < \pi) \end{cases}$

2 Obtain the Fourier series expansion of the periodic function $f(t)$ of period 2π defined over the period $0 \leqslant t \leqslant 2\pi$ by

$$f(t) = (\pi - t)^2 \quad (0 \leqslant t \leqslant 2\pi)$$

Use the Fourier series to show that

$$\tfrac{1}{12}\pi^2 = \sum_{n=1}^{\infty} \frac{(-1)^{n+1}}{n^2}$$

3 The charge $q(t)$ on the plates of a capacitor at time t is as shown in Figure 12.12. Express $q(t)$ as a Fourier series expansion.

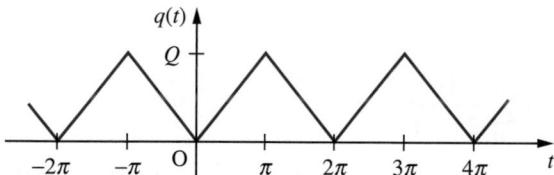

Figure 12.12 Plot of the charge $q(t)$ in Question 3.

4 The clipped response of a half-wave rectifier is the periodic function $f(t)$ of period 2π defined over the period $0 \leqslant t \leqslant 2\pi$ by

$$f(t) = \begin{cases} 5\sin t & (0 \leqslant t \leqslant \pi) \\ 0 & (\pi \leqslant t \leqslant 2\pi) \end{cases}$$

Express $f(t)$ as a Fourier series expansion.

5 Show that the Fourier series representing the periodic function $f(t)$, where

$$f(t) = \begin{cases} \pi^2 & (-\pi < t < 0) \\ (t - \pi)^2 & (0 < t < \pi) \end{cases}$$

$$f(t + 2\pi) = f(t)$$

is

$$f(t) = \tfrac{2}{3}\pi^2 + \sum_{n=1}^{\infty} \left[\frac{2}{n^2}\cos nt + \frac{(-1)^n}{n}\pi \sin nt \right]$$

$$- \frac{4}{\pi} \sum_{n=1}^{\infty} \frac{\sin(2n-1)t}{(2n-1)^3}$$

Use this result to show that

(a) $\displaystyle\sum_{n=1}^{\infty} \frac{1}{n^2} = \tfrac{1}{6}\pi^2$ (b) $\displaystyle\sum_{n=1}^{\infty} \frac{(-1)^{n+1}}{n^2} = \tfrac{1}{12}\pi^2$

6 A periodic function $f(t)$ of period 2π is defined within the domain $0 \leqslant t \leqslant \pi$ by

$$f(t) = \begin{cases} t & (0 \leqslant t \leqslant \frac{1}{2}\pi) \\ \pi - t & (\frac{1}{2}\pi \leqslant t \leqslant \pi) \end{cases}$$

Sketch a graph of $f(t)$ for $-2\pi < t < 4\pi$ for the two cases where

(a) $f(t)$ is an even function

(b) $f(t)$ is an odd function

Find the Fourier series expansion that represents the even function for all values of t, and use it to show that

$$\tfrac{1}{8}\pi^2 = \sum_{n=1}^{\infty} \frac{1}{(2n-1)^2}$$

7 A periodic function $f(t)$ of period 2π is defined within the period $0 \leqslant t \leqslant 2\pi$ by

$$f(t) = \begin{cases} 2 - t/\pi & (0 \leqslant t \leqslant \pi) \\ t/\pi & (\pi \leqslant t \leqslant 2\pi) \end{cases}$$

Draw a graph of the function for $-4\pi \leqslant t \leqslant 4\pi$ and obtain its Fourier series expansion.

By replacing t by $t - \frac{1}{2}\pi$ in your answer, show that the periodic function $f(t - \frac{1}{2}\pi) - \frac{3}{2}$ is represented by a sine series of odd harmonics.

12.2.10 Functions of period T

Although all the results have been related to periodic functions having period T, all the examples we have considered so far have involved periodic functions of period 2π. This was done primarily for ease of manipulation in determining the Fourier coefficients while becoming acquainted with Fourier series. As mentioned in Section 12.2.4, functions having unit frequency (that is, of period 2π) are rarely encountered in practice, and in this section we consider examples of periodic functions having periods other than 2π.

Example 12.9   A periodic function $f(t)$ of period 4 (that is, $f(t + 4) = f(t)$) is defined in the range $-2 < t < 2$ by

$$f(t) = \begin{cases} 0 & (-2 < t < 0) \\ 1 & (0 < t < 2) \end{cases}$$

Sketch a graph of $f(t)$ for $-6 \leqslant t \leqslant 6$ and obtain a Fourier series expansion for the function.

Solution   A graph of $f(t)$ for $-6 \leqslant t \leqslant 6$ is shown is Figure 12.13.

Figure 12.13
The function $f(t)$
of Example 12.9.

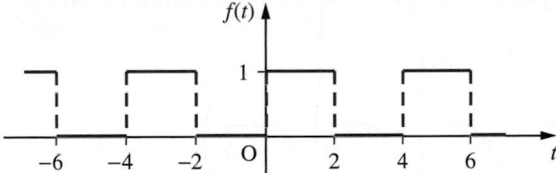

Taking $T = 4$ in (12.11) and (12.12), we have

$$a_0 = \tfrac{1}{2} \int_{-2}^{2} f(t)\,dt = \tfrac{1}{2}\left(\int_{-2}^{0} 0\,dt + \int_{0}^{2} 1\,dt \right) = 1$$

$$a_n = \tfrac{1}{2} \int_{-2}^{2} f(t) \cos \tfrac{1}{2}n\pi t \,dt \quad (n = 1, 2, 3, \dots)$$

$$= \tfrac{1}{2}\left(\int_{-2}^{0} 0\,dt + \int_{0}^{2} \cos \tfrac{1}{2}n\pi t \,dt \right) = 0$$

and

$$b_n = \tfrac{1}{2} \int_{-2}^{2} f(t) \sin \tfrac{1}{2}n\pi t \,dt \quad (n = 1, 2, 3, \dots)$$

$$= \tfrac{1}{2}\left(\int_{-2}^{0} 0\,dt + \int_{0}^{2} \sin \tfrac{1}{2}n\pi t \,dt \right) = \frac{1}{n\pi}(1 - \cos n\pi) = \frac{1}{n\pi}[1 - (-1)^n]$$

$$= \begin{cases} 0 & (\text{even } n) \\ 2/n\pi & (\text{odd } n) \end{cases}$$

Thus, from (12.10), the Fourier series expansion of $f(t)$ is

$$f(t) = \tfrac{1}{2} + \frac{2}{\pi}(\sin\tfrac{1}{2}\pi t + \tfrac{1}{3}\sin\tfrac{3}{2}\pi t + \tfrac{1}{5}\sin\tfrac{5}{2}\pi t + \dots)$$

$$= \tfrac{1}{2} + \frac{2}{\pi}\sum_{n=1}^{\infty}\frac{1}{2n-1}\sin\tfrac{1}{2}(2n-1)\pi t$$

Example 12.10 A periodic function $f(t)$ of period 2 is defined by

$$f(t) = \begin{cases} 3t & (0 < t < 1) \\ 3 & (1 < t < 2) \end{cases}$$

$$f(t+2) = f(t)$$

Sketch a graph of $f(t)$ for $-4 \leqslant t \leqslant 4$ and determine a Fourier series expansion for the function.

Solution A graph of $f(t)$ for $-4 \leqslant t \leqslant 4$ is shown in Figure 12.14.

Figure 12.14
The function $f(t)$
of Example 12.10.

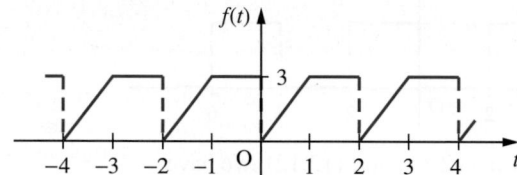

Taking $T = 2$ in (12.11) and (12.12), we have

$$a_0 = \tfrac{2}{2}\int_0^2 f(t)\,dt = \int_0^1 3t\,dt + \int_1^2 3\,dt = \tfrac{9}{2}$$

$$a_n = \tfrac{2}{2}\int_0^2 f(t)\cos\frac{n\pi t}{1}\,dt \quad (n = 1, 2, 3, \dots)$$

$$= \int_0^1 3t\cos n\pi t\,dt + \int_1^2 3\cos n\pi t\,dt$$

$$= \left[\frac{3t\sin n\pi t}{n\pi} + \frac{3\cos n\pi t}{(n\pi)^2}\right]_0^1 + \left[\frac{3\sin n\pi t}{n\pi}\right]_1^2$$

$$= \frac{3}{(n\pi)^2}(\cos n\pi - 1)$$

$$= \begin{cases} 0 & (\text{even } n) \\ -6/(n\pi)^2 & (\text{odd } n) \end{cases}$$

and

$$b_n = \frac{2}{2} \int_0^2 f(t) \sin \frac{n\pi t}{1} \, dt \quad (n = 1, 2, 3, \ldots)$$

$$= \int_0^1 3t \sin n\pi t \, dt + \int_1^2 3 \sin n\pi t \, dt$$

$$= \left[-\frac{3\cos n\pi t}{n\pi} + \frac{3\sin n\pi t}{(n\pi)^2} \right]_0^1 + \left[-\frac{3\cos n\pi t}{n\pi} \right]_1^2 = -\frac{3}{n\pi}\cos 2n\pi = -\frac{3}{n\pi}$$

Thus, from (12.10), the Fourier series expansion of $f(t)$ is

$$f(t) = \frac{9}{4} - \frac{6}{\pi^2} (\cos \pi t + \tfrac{1}{9} \cos 3\pi t + \tfrac{1}{25} \cos 5\pi t + \ldots)$$

$$- \frac{3}{\pi} (\sin \pi t + \tfrac{1}{2} \sin 2\pi t + \tfrac{1}{3} \sin 3\pi t + \ldots)$$

$$= \frac{9}{4} - \frac{6}{\pi^2} \sum_{n=1}^{\infty} \frac{\cos(2n-1)\pi t}{(2n-1)^2} - \frac{3}{\pi} \sum_{n=1}^{\infty} \frac{\sin n\pi t}{n}$$

12.2.11 Exercises

8 Find a Fourier series expansion of the periodic function

$$f(t) = t \quad (-l < t < l)$$

$$f(t + 2l) = f(t)$$

9 A periodic function $f(t)$ of period $2l$ is defined over one period by

$$f(t) = \begin{cases} -\dfrac{K}{l}(l + t) & (-l < t < 0) \\[2mm] \dfrac{K}{l}(l - t) & (0 < t < l) \end{cases}$$

Determine its Fourier series expansion and illustrate graphically for $-3l < t < 3l$.

10 A periodic function of period 10 is defined within the period $-5 < t < 5$ by

$$f(t) = \begin{cases} 0 & (-5 < t < 0) \\ 3 & (0 < t < 5) \end{cases}$$

Determine its Fourier series expansion and illustrate graphically for $-12 < t < 12$.

11 Passing a sinusoidal voltage $A \sin \omega t$ through a half-wave rectifier produces the clipped sine wave shown in Figure 12.15. Determine a Fourier series expansion of the rectified wave.

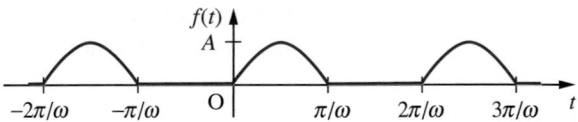

Figure 12.15 Rectified sine wave of Question 11.

12 Obtain a Fourier series expansion of the periodic function

$$f(t) = t^2 \quad (-T < t < T)$$

$$f(t + 2T) = f(t)$$

and illustrate graphically for $-3T < t < 3T$.

13 Determine a Fourier series representation of the periodic voltage $e(t)$ shown in Figure 12.16.

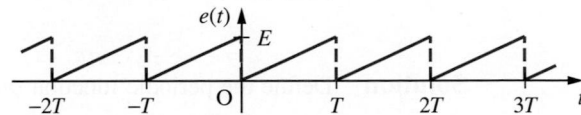

Figure 12.16 Voltage $e(t)$ of Question 13.

12.3 Functions defined over a finite interval

One of the requirements of Fourier's theorem is that the function to be expanded be periodic. Therefore a function $f(t)$ that is not periodic cannot have a Fourier series representation that converges to it *for all values* of t. However, we can obtain a Fourier series expansion that represents a *non-periodic* function $f(t)$ that is defined only over a finite time interval $0 \leqslant t \leqslant \tau$. This is a facility that is frequently used to solve problems in practice, particularly boundary-value problems involving partial differential equations, such as the consideration of heat flow along a bar or the vibrations of a string. Various forms of Fourier series representations of $f(t)$, valid only in the interval $0 \leqslant t \leqslant \tau$, are possible, including series consisting of cosine terms only or series consisting of sine terms only. To obtain these, various periodic extensions of $f(t)$ are formulated.

12.3.1 Full-range series

Suppose the given function $f(t)$ is defined only over the finite time interval $0 \leqslant t \leqslant \tau$. Then, to obtain a full-range Fourier series representation of $f(t)$ (that is, a series consisting of both cosine and sine terms), we define the **periodic extension** $\phi(t)$ of $f(t)$ by

$$\phi(t) = f(t) \quad (0 < t < \tau)$$

$$\phi(t + \tau) = \phi(t)$$

The graphs of a possible $f(t)$ and its periodic extension $\phi(t)$ are shown in Figures 12.17(a) and (b) respectively.

Provided that $f(t)$ satisfies Dirichlet's conditions in the interval $0 \leqslant t \leqslant \tau$, the new function $\phi(t)$, of period τ, will have a convergent Fourier series expansion. Since, within the particular period $0 < t < \tau$, $\phi(t)$ is identical with $f(t)$, it follows that this Fourier series expansion of $\phi(t)$ will be representative of $f(t)$ within this interval.

Figure 12.17
Graphs of a function defined only over (a) a finite interval $0 \leqslant t \leqslant \tau$ and (b) its periodic extension.

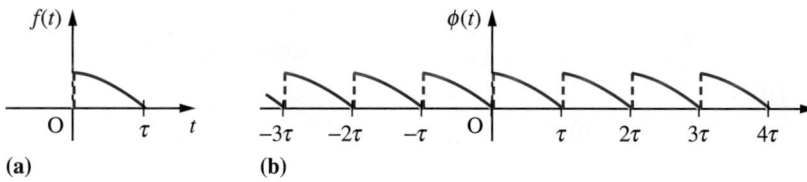

(a)   (b)

Example 12.11

Find a full-range Fourier series expansion of $f(t) = t$ valid in the finite interval $0 < t < 4$. Draw graphs of both $f(t)$ and the periodic function represented by the Fourier series obtained.

Solution

Define the periodic function $\phi(t)$ by

$$\phi(t) = f(t) = t \quad (0 < t < 4)$$

$$\phi(t + 4) = \phi(t)$$

Figure 12.18
The functions $f(t)$ and $\phi(t)$ of Example 12.11.

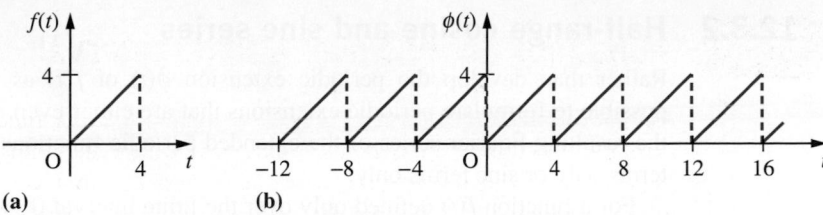

(a)   **(b)**

Then the graphs of $f(t)$ and its periodic extension $\phi(t)$ are as shown in Figures 12.18(a) and (b) respectively. Since $\phi(t)$ is a periodic function with period 4, it has a convergent Fourier series expansion. Taking $T = 4$ in (12.11) and (12.12), the Fourier coefficients are determined as

$$a_0 = \tfrac{1}{2}\int_0^4 f(t)\,dt = \tfrac{1}{2}\int_0^4 t\,dt = 4$$

$$a_n = \tfrac{1}{2}\int_0^4 f(t)\cos\tfrac{1}{2}n\pi t\,dt \quad (n = 1, 2, 3, \dots)$$

$$= \tfrac{1}{2}\int_0^4 t\cos\tfrac{1}{2}n\pi t\,dt = \tfrac{1}{2}\left[\frac{2t}{n\pi}\sin\tfrac{1}{2}n\pi t + \frac{4}{(n\pi)^2}\cos\tfrac{1}{2}n\pi t\right]_0^4 = 0$$

and

$$b_n = \tfrac{1}{2}\int_0^4 f(t)\sin\tfrac{1}{2}n\pi t\,dt \quad (n = 1, 2, 3, \dots)$$

$$= \tfrac{1}{2}\int_0^4 t\sin\tfrac{1}{2}n\pi t\,dt = \tfrac{1}{2}\left[-\frac{2t}{n\pi}\cos\tfrac{1}{2}n\pi t + \frac{4}{(n\pi)^2}\sin\tfrac{1}{2}n\pi t\right]_0^4 = -\frac{4}{n\pi}$$

Thus, by (12.10), the Fourier series expansion of $\phi(t)$ is

$$\phi(t) = 2 - \frac{4}{\pi}\left(\sin\tfrac{1}{2}\pi t + \tfrac{1}{2}\sin\pi t + \tfrac{1}{3}\sin\tfrac{3}{2}\pi t + \tfrac{1}{4}\sin 2t + \tfrac{1}{5}\sin\tfrac{5}{2}\pi t + \dots\right)$$

$$= 2 - \frac{4}{\pi}\sum_{n=1}^{\infty}\frac{1}{n}\sin\tfrac{1}{2}n\pi t$$

Since $\phi(t) = f(t)$ for $0 < t < 4$, it follows that this Fourier series is representative of $f(t)$ within this interval, so that

$$f(t) = t = 2 - \frac{4}{\pi}\sum_{n=1}^{\infty}\frac{1}{n}\sin\tfrac{1}{2}n\pi t \quad (0 < t < 4) \tag{12.29}$$

It is important to appreciate that this series converges to t only within the interval $0 < t < 4$. For values of t outside this interval it converges to the periodic extended function $\phi(t)$. Again convergence is to be interpreted in the sense of Theorem 12.2, so that at the end points $t = 0$ and $t = 4$ the series does not converge to t but to the mean of the discontinuity in $\phi(t)$, namely the value 2.

12.3.2 Half-range cosine and sine series

Rather than develop the periodic extension $\phi(t)$ of $f(t)$ as in Section 12.3.1, it is possible to formulate periodic extensions that are either even or odd functions, so that the resulting Fourier series of the extended periodic functions consist either of cosine terms only or sine terms only.

For a function $f(t)$ defined only over the finite interval $0 \leqslant t \leqslant \tau$ its **even periodic extension** $F(t)$ is the even periodic function defined by

$$F(t) = \begin{cases} f(t) & (0 < t < \tau) \\ f(-t) & (-\tau < t < 0) \end{cases}$$

$$F(t + 2\tau) = f(t)$$

As an illustration, the even periodic extension $F(t)$ of the function $f(t)$ shown in Figure 12.17(a) (redrawn in Figure 12.19(a)) is shown in Figure 12.19(b).

Figure 12.19
(a) A function $f(t)$.
(b) Its even periodic extension $F(t)$.

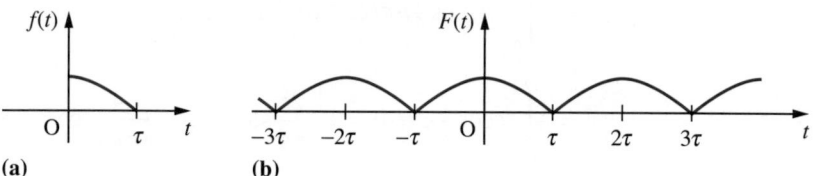

(a)   **(b)**

Provided that $f(t)$ satisfies Dirichlet's conditions in the interval $0 < t < \tau$, since it is an even function of period 2τ, it follows from Section 12.2.5 that the even periodic extension $F(t)$ will have a convergent Fourier series representation consisting of cosine terms only and given by

$$F(t) = \tfrac{1}{2}a_0 + \sum_{n=1}^{\infty} a_n \cos \frac{n\pi t}{\tau} \tag{12.30}$$

where

$$a_n = \frac{2}{\tau} \int_0^\tau f(t) \cos \frac{n\pi t}{\tau}\, dt \quad (n = 0, 1, 2, \ldots) \tag{12.31}$$

Since, within the particular interval $0 < t < \tau$, $F(t)$ is identical with $f(t)$, it follows that the series (12.30) also converges to $f(t)$ within this interval.

For a function $f(t)$ defined only over the finite interval $0 \leqslant t \leqslant \tau$, its **odd periodic extension** $G(t)$ is the odd periodic function defined by

$$G(t) = \begin{cases} f(t) & (0 < t < \tau) \\ -f(-t) & (-\tau < t < 0) \end{cases}$$

$$G(t + 2\tau) = G(t)$$

Again, as an illustration, the odd periodic extension $G(t)$ of the function $f(t)$ shown in Figure 12.17(a) (redrawn in Figure 12.20(a)) is shown in Figure 12.20(b).

Figure 12.20
(a) A function $f(t)$.
(b) Its odd periodic
extension $G(t)$.

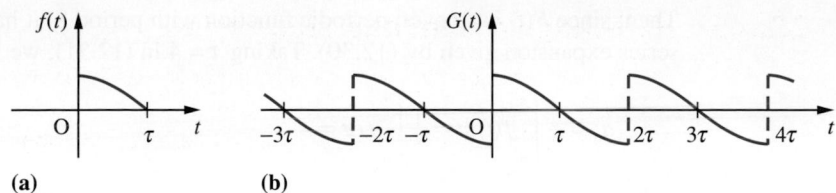

(a)   (b)

Provided that $f(t)$ satisfies Dirichlet's conditions in the interval $0 < t < \tau$, since it is an odd function of period 2τ, it follows from Section 12.2.5 that the odd periodic extension $G(t)$ will have a convergent Fourier series representation consisting of sine terms only and given by

$$G(t) = \sum_{n=1}^{\infty} b_n \sin \frac{n\pi t}{\tau} \tag{12.32}$$

where

$$b_n = \frac{2}{\tau} \int_0^\tau f(t) \sin \frac{n\pi t}{\tau} dt \quad (n = 1, 2, 3, ...) \tag{12.33}$$

Again, since, within the particular interval $0 < t < \tau$, $G(t)$ is identical with $f(t)$, it follows that the series (12.32) also converges to $f(t)$ within this interval.

We note that both the even and odd periodic extensions $F(t)$ and $G(t)$ are of period 2τ, which is twice the length of the interval over which $f(t)$ is defined. However, the resulting Fourier series (12.30) and (12.32) are based only on the function $f(t)$, and for this reason are called the **half-range Fourier series expansions** of $f(t)$. In particular, the even half-range expansion $F(t)$, (12.30), is called the **half-range cosine series expansion** of $f(t)$, while the odd half-range expansion $G(t)$, (12.32), is called the **half-range sine series expansion** of $f(t)$.

Example 12.12

For the function $f(t) = t$ defined only in the interval $0 < t < 4$, and considered in Example 12.11, obtain

(a) a half-range cosine series expansion,

(b) a half-range sine series expansion.

Draw graphs of $f(t)$ and of the periodic functions represented by the two series obtained for $-20 < t < 20$.

Solution   (a) Half-range cosine series. Define the periodic function $F(t)$ by

$$F(t) = \begin{cases} f(t) = t & (0 < t < 4) \\ f(-t) = -t & (-4 < t < 0) \end{cases}$$

$$F(t + 8) = F(t)$$

Then, since $F(t)$ is an even periodic function with period 8, it has a convergent Fourier series expansion given by (12.30). Taking $\tau = 4$ in (12.31), we have

$$a_0 = \frac{2}{4} \int_0^4 f(t)dt = \frac{1}{2} \int_0^4 t\, dt = 4$$

$$a_n = \frac{2}{4} \int_0^4 f(t) \cos \tfrac{1}{4} n\pi t\, dt \quad (n = 1, 2, 3, \dots)$$

$$= \frac{1}{2} \int_0^4 t \cos \tfrac{1}{4} n\pi t\, dt = \frac{1}{2} \left[\frac{4t}{n\pi} \sin \tfrac{1}{4} n\pi t + \frac{16}{(n\pi)^2} \cos \tfrac{1}{4} n\pi t \right]_0^4$$

$$= \frac{8}{(n\pi)^2} (\cos n\pi - 1) = \begin{cases} 0 & (\text{even } n) \\ -16/(n\pi)^2 & (\text{odd } n) \end{cases}$$

Then, by (12.30), the Fourier series expansion of $F(t)$ is

$$F(t) = 2 - \frac{16}{\pi^2} \left(\cos \tfrac{1}{4}\pi t + \tfrac{1}{3^2} \cos \tfrac{3}{4}\pi t + \tfrac{1}{5^2} \cos \tfrac{5}{4}\pi t + \dots \right)$$

or

$$F(t) = 2 - \frac{16}{\pi^2} \sum_{n=1}^{\infty} \frac{1}{(2n-1)^2} \cos \tfrac{1}{4}(2n-1)\pi t$$

Since $F(t) = f(t)$ for $0 < t < 4$, it follows that this Fourier series is representative of $f(t)$ within this interval. Thus the half-range cosine series expansion of $f(t)$ is

$$f(t) = t = 2 - \frac{16}{\pi^2} \sum_{n=1}^{\infty} \frac{1}{(2n-1)^2} \cos \tfrac{1}{4}(2n-1)\pi t \quad (0 < t < 4) \tag{12.34}$$

(b) Half-range sine series. Define the periodic function $G(t)$ by

$$G(t) = \begin{cases} f(t) = t & (0 < t < 4) \\ -f(-t) = t & (-4 < t < 0) \end{cases}$$

$$G(t + 8) = G(t)$$

Then, since $G(t)$ is an odd periodic function with period 8, it has a convergent Fourier series expansion given by (12.32). Taking $\tau = 4$ in (12.33), we have

$$b_n = \frac{2}{4} \int_0^4 f(t) \sin \tfrac{1}{4} n\pi t\, dt \quad (n = 1, 2, 3, \dots)$$

$$= \frac{1}{2} \int_0^4 t \sin \tfrac{1}{4} n\pi t\, dt = \frac{1}{2} \left[-\frac{4t}{n\pi} \cos \tfrac{1}{4} n\pi t + \frac{16}{(n\pi)^2} \sin \tfrac{1}{4} n\pi t \right]_0^4$$

$$= -\frac{8}{n\pi} \cos n\pi = -\frac{8}{n\pi}(-1)^n$$

Thus, by (12.32), the Fourier series expansion of $G(t)$ is

$$G(t) = \frac{8}{\pi}(\sin\tfrac{1}{4}\pi t - \tfrac{1}{2}\sin\tfrac{1}{2}\pi t + \tfrac{1}{3}\sin\tfrac{3}{4}\pi t - \dots)$$

or

$$G(t) = \frac{8}{\pi}\sum_{n=1}^{\infty}\frac{(-1)^{n+1}}{n}\sin\tfrac{1}{4}n\pi t$$

Since $G(t) = f(t)$ for $0 < t < 4$, it follows that this Fourier series is representative of $f(t)$ within this interval. Thus the half-range sine series expansion of $f(t)$ is

$$f(t) = t = \frac{8}{\pi}\sum_{n=1}^{\infty}\frac{(-1)^{n+1}}{n}\sin\tfrac{1}{4}n\pi t \quad (0 < t < 4) \tag{12.35}$$

Graphs of the given function $f(t)$ and of the even and odd periodic expansions $F(t)$ and $G(t)$ are given in Figures 12.21(a), (b) and (c) respectively.

Figure 12.21
The functions $f(t)$, $F(t)$ and $G(t)$ of Example 12.12.

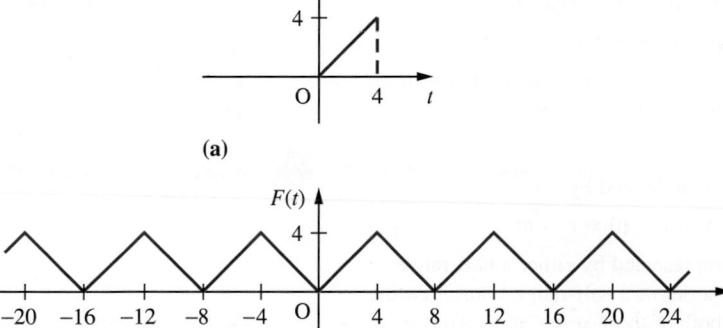

(a)

(b)

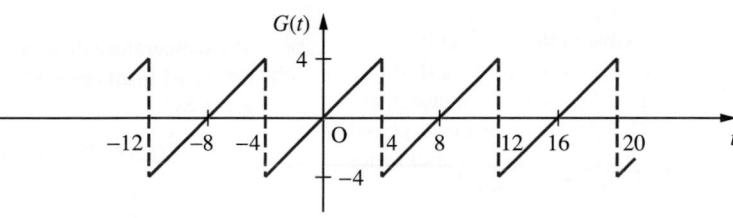

(c)

It is important to realize that the three different Fourier series representations (12.29), (12.34) and (12.35) are representative of the function $f(t) = t$ only within the defined interval $0 < t < 4$. Outside this interval the three Fourier series converge to the three different functions $\phi(t)$, $F(t)$ and $G(t)$, illustrated in Figures 12.18(b), 12.21(b) and 12.21(c) respectively.

12.3.3 Exercises

14 Show that the half-range Fourier sine series expansion of the function $f(t) = 1$, valid for $0 < t < \pi$, is

$$f(t) = \frac{4}{\pi} \sum_{n=1}^{\infty} \frac{\sin(2n-1)t}{2n-1} \quad (0 < t < \pi)$$

Sketch the graphs of both $f(t)$ and the periodic function represented by the series expansion for $-3\pi < t < 3\pi$.

15 Determine the half-range cosine series expansion of the function $f(t) = 2t - 1$, valid for $0 < t < 1$. Sketch the graphs of both $f(t)$ and the periodic function represented by the series expansion for $-2 < t < 2$.

16 The function $f(t) = 1 - t^2$ is to be represented by a Fourier series expansion over the finite interval $0 < t < 1$. Obtain a suitable

(a) full-range series expansion,

(b) half-range sine series expansion,

(c) half-range cosine series expansion.

Draw graphs of $f(t)$ and of the periodic functions represented by each of the three series for $-4 < t < 4$.

17 A function $f(t)$ is defined by

$$f(t) = \pi t - t^2 \quad (0 \le t \le \pi)$$

and is to be represented by either a half-range Fourier sine series or a half-range Fourier cosine series. Find both of these series and sketch the graphs of the functions represented by them for $-2\pi < t < 2\pi$.

18 A tightly stretched flexible uniform string has its ends fixed at the points $x = 0$ and $x = l$. The midpoint of the string is displaced a distance a, as shown in Figure 12.22. If $f(x)$ denotes the displaced profile of the string, express $f(x)$ as a Fourier series expansion consisting only of sine terms.

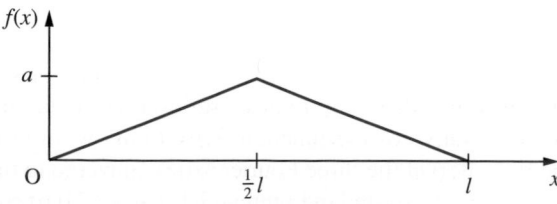

Figure 12.22 Displaced string of Question 18.

19 Repeat Question 18 for the case where the displaced profile of the string is as shown in Figure 12.23.

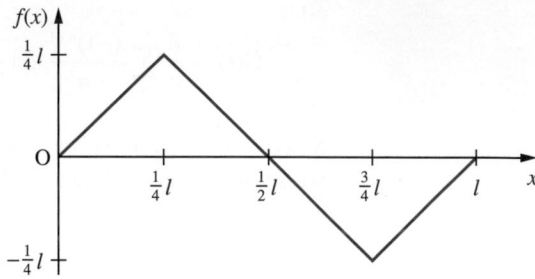

Figure 12.23 Displaced string of Question 19.

20 A function $f(t)$ is defined on $0 \le t \le \pi$ by

$$f(t) = \begin{cases} \sin t & (0 \le t < \tfrac{1}{2}\pi) \\ 0 & (\tfrac{1}{2}\pi \le t \le \pi) \end{cases}$$

Find a half-range Fourier series expansion of $f(t)$ on this interval. Sketch a graph of the function represented by the series for $-2\pi \le t \le 2\pi$.

21 A function $f(t)$ is defined on the interval $-l \le x \le l$ by

$$f(x) = \frac{A}{l}(|x| - l)$$

Obtain a Fourier series expansion of $f(x)$ and sketch a graph of the function represented by the series for $-3l \le x \le 3l$.

22 The temperature distribution $T(x)$ at a distance x, measured from one end, along a bar of length L is given by

$$T(x) = Kx(L - x) \quad (0 \le x \le L), \quad K = \text{constant}$$

Express $T(x)$ as a Fourier series expansion consisting of sine terms only.

23 Find the Fourier series expansion of the function $f(t)$ valid for $-1 < t < 1$, where

$$f(t) = \begin{cases} 1 & (-1 < t < 0) \\ \cos \pi t & (0 < t < 1) \end{cases}$$

To what value does this series converge when $t = 1$?

12.4 Differentiation and integration of Fourier series

It is inevitable that the desire to obtain the derivative or the integral of a Fourier series will arise in some applications. Since the smoothing effects of the integration process tend to eliminate discontinuities, whereas the process of differentiation has the opposite effect, it is not surprising that the integration of a Fourier series is more likely to be possible than its differentiation. We shall not pursue the theory in depth here; rather we shall state, without proof, two theorems concerned with the term-by-term integration and differentiation of Fourier series, and make some observations on their use.

12.4.1 Integration of a Fourier series

Theorem 12.3

A Fourier series expansion of a periodic function $f(t)$ that satisfies Dirichlet's conditions may be integrated term by term, and the integrated series converges to the integral of the function $f(t)$.

end of theorem

According to this theorem, if $f(t)$ satisfies Dirichlet's conditions in the interval $-\pi \leqslant t \leqslant \pi$ and has a Fourier series expansion

$$f(t) = \tfrac{1}{2}a_0 + \sum_{n=1}^{\infty} (a_n \cos nt + b_n \sin nt)$$

then for $-\pi \leqslant t_1 < t \leqslant \pi$

$$\int_{t_1}^{t} f(t)\mathrm{d}t = \int_{t_1}^{t} \tfrac{1}{2}a_0\,\mathrm{d}t + \sum_{n=1}^{\infty} \int_{t_1}^{t} (a_n \cos nt + b_n \sin nt)\mathrm{d}t$$

$$= \tfrac{1}{2}a_0(t - t_1) + \sum_{n=1}^{\infty} \left[\frac{b_n}{n}(\cos nt_1 - \cos nt) + \frac{a_n}{n}(\sin nt - \sin nt_1) \right]$$

Because of the presence of the term $\tfrac{1}{2}a_0 t$ on the right-hand side, this is clearly not a Fourier series expansion of the integral on the left-hand side. However, the result can be rearranged to be a Fourier series expansion of the function

$$g(t) = \int_{t_1}^{t} f(t)\mathrm{d}t - \tfrac{1}{2}a_0 t$$

Example 12.13 serves to illustrate this process. Note also that the Fourier coefficients in the new Fourier series are $-b_n/n$ and a_n/n, so, from the observations made in Section 12.2.8, the integrated series converges faster than the original series for $f(t)$. If the given function $f(t)$ is piecewise-continuous, rather than continuous, over the interval $-\pi \leqslant t \leqslant \pi$ then care must be taken to ensure that the integration process is carried out properly over the various subintervals. Again, Example 12.14 serves to illustrate this point.

Example 12.13   From Example 12.5, the Fourier series expansion of the function

$$f(t) = t^2 \quad (-\pi \le t \le \pi), \qquad f(t + 2\pi) = f(\pi)$$

is

$$t^2 = \tfrac{1}{3}\pi^2 + 4\sum_{n=1}^{\infty} \frac{(-1)^n \cos nt}{n^2} \quad (-\pi \le t \le \pi)$$

Integrating this result between the limits $-\pi$ and t gives

$$\int_{-\pi}^{t} t^2 \, dt = \int_{-\pi}^{t} \tfrac{1}{3}\pi^2 \, dt + 4\sum_{n=1}^{\infty} \int_{-\pi}^{t} \frac{(-1)^n \cos nt}{n^2} \, dt$$

that is,

$$\tfrac{1}{3}t^3 = \tfrac{1}{3}\pi^2 t + 4\sum_{n=1}^{\infty} \frac{(-1)^n \sin nt}{n^3} \quad (-\pi \le t \le \pi)$$

Because of the term $\tfrac{1}{3}\pi^2 t$ on the right-hand side, this is clearly not a Fourier series expansion. However, rearranging, we have

$$t^3 - \pi^2 t = 12\sum_{n=1}^{\infty} \frac{(-1)^n \sin nt}{n^2}$$

and now the right-hand side may be taken to be the Fourier series expansion of the function

$$g(t) = t^3 - \pi^2 t \quad (-\pi \le t \le \pi)$$

$$g(t + 2\pi) = g(t)$$

Example 12.14   Integrate term by term the Fourier series expansion obtained in Example 12.4 for the square wave

$$f(t) = \begin{cases} -1 & (-\pi < t < 0) \\ 1 & (0 < t < \pi) \end{cases}$$

$$f(t + 2\pi) = f(t)$$

illustrated in Figure 12.7.

Solution   From (12.21), the Fourier series expansion for $f(t)$ is

$$f(t) = \frac{4}{\pi} \frac{\sin(2n - 1)t}{2n - 1}$$

We now need to integrate between the limits $-\pi$ and t and, owing to the discontinuity in $f(t)$ at $t = 0$, we must consider separately values of t in the intervals $-\pi < t < 0$ and $0 < t < \pi$.

Case (i), interval $-\pi < t < 0$. Integrating (12.21) term by term, we have

$$\int_{-\pi}^{t} (-1)dt = \frac{4}{\pi} \sum_{n=1}^{\infty} \int_{-\pi}^{t} \frac{\sin(2n-1)t}{(2n-1)} dt$$

that is,

$$-(t + \pi) = -\frac{4}{\pi} \sum_{n=1}^{\infty} \left[\frac{\cos(2n-1)t}{(2n-1)^2} \right]_{-\pi}^{t}$$

$$= -\frac{4}{\pi} \left[\sum_{n=1}^{\infty} \frac{\cos(2n-1)t}{(2n-1)^2} + \sum_{n=1}^{\infty} \frac{1}{(2n-1)^2} \right]$$

It can be shown that

$$\sum_{n=1}^{\infty} \frac{1}{(2n-1)^2} = \frac{1}{8}\pi^2$$

(see Exercises 12.2.9, Question 6), so that the above simplifies to

$$-t = \tfrac{1}{2}\pi - \frac{4}{\pi} \sum_{n=1}^{\infty} \frac{\cos(2n-1)t}{(2n-1)^2} \quad (-\pi < t < 0) \tag{12.36}$$

Case (ii), interval $0 < t < \pi$. Integrating (12.21) term by term, we have

$$\int_{-\pi}^{0} (-1)dt + \int_{0}^{t} 1\,dt = \frac{4}{\pi} \sum_{n=1}^{\infty} \int_{-\pi}^{t} \frac{\sin(2n-1)t}{(2n-1)} dt$$

giving

$$t = \tfrac{1}{2}\pi - \frac{4}{\pi} \sum_{n=1}^{\infty} \frac{\cos(2n-1)t}{(2n-1)^2} \quad (0 < t < \pi) \tag{12.37}$$

Taking (12.36) and (12.37) together, we find that the function

$$g(t) = |t| = \begin{cases} -t & (-\pi < t < 0) \\ t & (0 < t < \pi) \end{cases}$$

$$g(t + 2\pi) = g(t)$$

has a Fourier series expansion

$$g(t) = |t| = \tfrac{1}{2}\pi - \frac{4}{\pi} \sum_{n=1}^{\infty} \frac{\cos(2n-1)t}{(2n-1)^2}$$

12.4.2 Differentiation of a Fourier series

Theorem 12.4

If $f(t)$ is a periodic function that satisfies Dirichlet's conditions then its derivative $f'(t)$, wherever it exists, may be found by term-by-term differentiation of the Fourier series of $f(t)$ if and only if the function $f(t)$ is continuous everywhere and the function $f'(t)$ has a Fourier series expansion (that is, $f'(t)$ satisfies Dirichlet's conditions).

end of theorem

It follows from Theorem 12.4 that if the Fourier series expansion of $f(t)$ is differentiable term by term then $f(t)$ must be periodic at the end points of a period (owing to the condition that $f(t)$ must be continuous everywhere). Thus, for example, if we are dealing with a function $f(t)$ of period 2π and defined in the range $-\pi < t < \pi$ then we must have $f(-\pi) = f(\pi)$. To illustrate this point, consider the Fourier series expansion of the function

$$f(t) = t \quad (-\pi < t < \pi)$$

$$f(t + 2\pi) = f(t)$$

which, from Example 12.7, is given by

$$f(t) = 2(\sin t - \tfrac{1}{2}\sin 2t + \tfrac{1}{3}\sin 3t - \tfrac{1}{4}\sin 4t + \dots)$$

Differentiating term by term, we have

$$f'(t) = 2(\cos t - \cos 2t + \cos 3t - \cos 4t + \dots)$$

If this differentiation process is valid then $f'(t)$ must be equal to unity for $-\pi < t < \pi$. Clearly this is not the case, since the series on the right-hand side does not converge for any value of t. This follows since the nth term of the series is $2(-1)^{n+1}\cos nt$ and does not tend to zero as $n \to \infty$.

If $f(t)$ is continuous everywhere and has a Fourier series expansion

$$f(t) = \tfrac{1}{2}a_0 + \sum_{n=1}^{\infty}(a_n\cos nt + b_n\sin nt)$$

then, from Theorem 12.4, provided that $f'(t)$ satisfies the required conditions, its Fourier series expansion is

$$f'(t) = \sum_{n=1}^{\infty}(nb_n\cos nt - na_n\sin nt)$$

In this case the Fourier coefficients of the derived expansion are nb_n and na_n, so, in contrast to the integrated series, the derived series will converge more slowly than the original series expansion for $f(t)$.

Example 12.15

Consider the process of differentiating term by term the Fourier series expansion of the function

$$f(t) = t^2 \quad (-\pi \le t \le \pi), \qquad f(t + 2\pi) = f(t)$$

Solution   From Example 12.5, the Fourier series expansion of $f(t)$ is

$$t^2 = \tfrac{1}{3}\pi^2 + 4 \sum_{n=1}^{\infty} \frac{(-1)^n \cos nt}{n^2} \quad (-\pi \leqslant t \leqslant \pi)$$

Since $f(t)$ is continuous within and at the end points of the interval $-\pi \leqslant t \leqslant \pi$, we may apply Theorem 12.4 to obtain

$$t = 2 \sum_{n=1}^{\infty} \frac{(-1)^{n+1} \sin nt}{n} \quad (-\pi \leqslant t \leqslant \pi)$$

which conforms with the Fourier series expansion obtained for the function

$$f(t) = t \quad (-\pi < t < \pi), \qquad f(t + 2\pi) = f(t)$$

in Example 12.7.

12.4.3 Exercises

24 Show that the periodic function

$$f(t) = t \quad (-T < t < T)$$

$$f(t + 2T) = f(t)$$

has a Fourier series expansion

$$f(t) = \frac{2T}{\pi} \left(\sin \frac{\pi t}{T} - \tfrac{1}{2} \sin \frac{2\pi t}{T} + \tfrac{1}{3} \sin \frac{3\pi t}{T} \right.$$

$$\left. - \tfrac{1}{4} \sin \frac{4\pi t}{T} + \ldots \right)$$

By term-by-term integration of this series, show that the periodic function

$$g(t) = t^2 \quad (-T < t < T)$$

$$g(t + 2T) = g(t)$$

has a Fourier series expansion

$$g(t) = \tfrac{1}{3} T^2 - \frac{4T^2}{\pi^2} \left(\cos \frac{\pi t}{T} - \tfrac{1}{2^2} \cos \frac{2\pi t}{T} \right.$$

$$\left. + \tfrac{1}{3^2} \cos \frac{3\pi t}{T} - \tfrac{1}{4^2} \cos \frac{4\pi t}{T} + \ldots \right)$$

(*Hint*: A constant of integration must be introduced; it may be evaluated as the mean value over a period.)

25 The periodic function

$$h(t) = \pi^2 - t^2 \quad (-\pi < t < \pi)$$

$$h(t + 2\pi) = h(t)$$

has a Fourier series expansion

$$h(t) = \tfrac{2}{3}\pi^2 + 4(\cos t - \tfrac{1}{2^2} \cos 2t + \tfrac{1}{3^2} \cos 3t \ldots)$$

By term-by-term differentiation of this series, confirm the series obtained for $f(t)$ in Question 24 for the case when $T = \pi$.

26 (a) Suppose that the derivative $f'(t)$ of a periodic function $f(t)$ of period 2π has a Fourier series expansion

$$f'(t) = \tfrac{1}{2}A_0 + \sum_{n=1}^{\infty} A_n \cos nt + \sum_{n=1}^{\infty} B_n \sin nt$$

Show that

$$A_0 = \frac{1}{n}[f(\pi^-) - f(-\pi^+)]$$

$$A_n = (-1)^n A_0 + nb_n$$

$$B_n = -na_n$$

where a_0, a_n and b_n are the Fourier coefficients of the function $f(t)$.

(b) In Example 12.7 we saw that the periodic function

$$f(t) = t^2 + t \quad (-\pi < t < \pi)$$

$$f(t + 2\pi) = f(t)$$

has a Fourier series expansion

$$f(t) = \tfrac{1}{3}\pi^2 + \sum_{n=1}^{\infty} \frac{4}{n^2}(-1)^n \cos nt$$

$$- \sum_{n=1}^{\infty} \frac{2}{n}(-1)^n \sin nt$$

Differentiate this series term by term, and explain why it is not a Fourier expansion of the periodic function

$$g(t) = 2t + 1 \quad (-\pi < t < \pi)$$

$$g(t + 2\pi) = g(t)$$

(c) Use the results of (a) to obtain the Fourier series expansion of $g(t)$ and confirm your solution by direct evaluation of the coefficients using Euler's formulae.

12.5 Engineering application: analysis of a slider–crank mechanism

Figure 12.24(a) represents a slider–crank mechanism. The crank OP rotates about O, and P is connected to Q, which is constrained so that it slides along OB, A special case when OP = 1 m and PQ = 3 m is shown in Figure 12.24(b). The distance OQ is x when the angle QOP is θ and is given by (see Example 2.44)

$$x(\theta) = \cos \theta + \sqrt{(9 - \sin^2\theta)}$$

Here we recap the work of Section 8.12 and extend it to illustrate a general result concerning Fourier series.

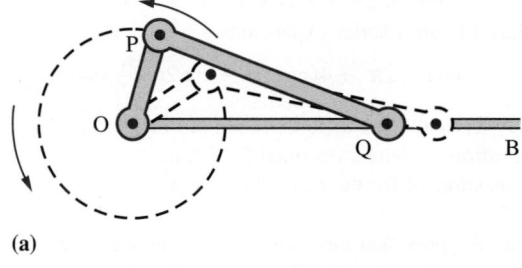

(a)

(b)

Figure 12.24 Slider–crank mechanism.

It is clear both from the basic geometry of the mechanism and from the formula for x that $x(\theta)$ is an even periodic function. This implies it can be represented by a Fourier series of the form

$$x(\theta) = a_0 + a_1 \cos \theta + a_2 \cos 2\theta + a_3 \cos 3\theta + \dots$$

Such mechanisms usually form parts of larger pieces of equipment, so it is important to know the sizes of the coefficients a_k during the design process to avoid dangerous motions due to resonance.

The process of obtaining the values of the coefficients a_k is called harmonic analysis. Truncating the Fourier series, we can obtain an approximation to $x(\theta)$ in the form

$$x(\theta) \approx a_0 + a_1 \cos \theta + a_2 \cos 2\theta + a_3 \cos 3\theta + a_4 \cos 4\theta$$

We wish to determine the values of $a_0, \dots, a_4$ so that we obtain the best approximation possible. We achieve this by choosing $a_0, \dots, a_4$ in such a way that the total squared error over a complete period is a minimum. Because $x(\theta)$ is an even function, this simplifies to finding the values of $a_0, \dots, a_4$ that minimize the integral

$$I(a_0, a_1, a_2, a_3, a_4) = \int_0^\pi [a_0 + a_1 \cos\theta + a_2 \cos 2\theta + a_3 \cos 3\theta + a_4 \cos 4\theta - x(\theta)]^2 \, d\theta$$

Thus we want $a_0, \ldots, a_4$ such that

$$\frac{\partial I}{\partial a_k} = 0 \quad (k = 0, 1, \ldots, 4)$$

Taking the case $k = 0$, this yields

$$\int_0^\pi 2[a_0 + a_1 \cos\theta + a_2 \cos 2\theta + a_3 \cos 3\theta + a_4 \cos 4\theta - x(\theta)] d\theta = 0$$

which reduces to

$$\int_0^\pi [a_0 - x(\theta)] d\theta = 0$$

on using the integration properties of $\cos k\theta$ on $(0, \pi)$ for $k = 1, \ldots, 4$. Thus

$$\int_0^\pi a_0 \, d\theta = \int_0^\pi x(\theta) d\theta$$

giving

$$a_0 = \frac{1}{\pi} \int_0^\pi [\cos\theta + \sqrt{(9 - \sin^2\theta)}] d\theta = \frac{2}{\pi} \int_0^{\pi/2} \sqrt{(9 - \sin^2\theta)} \, d\theta$$

on using the symmetry properties of the integrand about $x = \frac{1}{2}\pi$. This integral has to be evaluated numerically, and, using the trapezium rule, we obtain the value $a_0 = 2.9148$.

Similarly, $\partial I/\partial a_1 = 0$ gives

$$\int_0^\pi 2[a_0 + a_1 \cos\theta + a_2 \cos 2\theta + a_3 \cos 3\theta + a_4 \cos 4\theta - x(\theta)]\cos\theta \, d\theta = 0$$

which reduces to

$$\int_0^\pi [a_1 \cos^2\theta - x(\theta) \cos\theta] d\theta = 0$$

Thus

$$\int_0^\pi a_1 \cos^2\theta \, d\theta = \int_0^\pi x(\theta) \cos\theta \, d\theta$$

which gives

$$\tfrac{1}{2}\pi a_1 = \int_0^\pi [\cos^2\theta + \cos\theta \sqrt{(9 - \sin^2\theta)}] d\theta = \tfrac{1}{2}\pi$$

on using the symmetry properties of the integrand. Thus $a_1 = 1$.

Continuing in the same fashion, we obtain

$$\int_0^\pi a_2 \cos^2 2\theta \, d\theta = \int_0^\pi x(\theta) \cos 2\theta \, d\theta$$

$$\int_0^\pi a_3 \cos^2 3\theta \, d\theta = \int_0^\pi x(\theta) \cos 3\theta \, d\theta$$

and

$$\int_0^\pi a_4 \cos^2 4\theta \, d\theta = \int_0^\pi x(\theta) \cos 4\theta \, d\theta$$

from which we deduce

$$a_2 = \frac{2}{\pi} \int_0^\pi \cos 2\theta \sqrt{(9 - \sin^2 \theta)} \, d\theta$$

$$a_3 = \frac{2}{\pi} \int_0^\pi \cos 3\theta \sqrt{(9 - \sin^2 \theta)} \, d\theta = 0$$

$$a_4 = \frac{2}{\pi} \int_0^\pi \cos 4\theta \sqrt{(9 - \sin^2 \theta)} \, d\theta$$

Calculating the integrals for a_2 and a_4 numerically, we obtain the 'least-squares approximation' for $x(\theta)$ in the form

$$x(\theta) \approx 2.9148 + \cos\theta + 0.0858 \cos 2\theta - 0.0006 \cos 4\theta$$

We could continue the process to find a_5 and a_6. Notice that the coefficients are just those found by the standard formulae for the coefficients of a Fourier series so that we do not have to go through the above minimizing process every time. Indeed the example can be generalized to show that the truncated Fourier series provides the 'best' approximation to a periodic function.

Using the trapezium rule the MATLAB commands

```
a = 0; b = pi/2; n = 8;h = (b - a)/n; x = (a:h:b);
y = 2.*((9 - (sin(x)).^2).^(1/2))./pi;
h*trapz(y)
```

return the answer 2.9148.

Likewise the commands

```
a = 0; b = pi; n = 8; h = (b-a)/n; x = (a:h:b);
y = 2.*cos(2.*x).*((9 - (sin(x)).^2).^(1/2))./pi;
h*trapz(y)
```

return the answer 0.0858.

and the commands

```
a = 0;b = pi; n = 8; h = (b - a)/n; x = (a:h:b);
y = 2.*cos(4.*x).*((9 - (sin(x)).^2).^(1/2))./pi;
h*trapz(y)
```

return the answer -6.3119e - 004.

An interesting alternative approach is to attempt to evaluate the integrals in symbolic form. Using the Symbolic Math Toolbox the integrals can be evaluated symbolically, with some of the answers expressed in terms of elliptic functions. Using the `double` command all answers can be expressed in numeric form. For example considering the integral for a_0 the commands

```
syms x y
y = 2*cos(2*x)*sqrt(9 - (sin(x))^2)/pi;
int(y,0,pi);
double(ans)
```

return the answer `0.0858`.

12.6 Review exercises (1–21)

 Check your answers using MATLAB or MAPLE whenever possible.

1 A periodic function $f(t)$ is defined by

$$f(t) = \begin{cases} t^2 & (0 \leqslant t < \pi) \\ 0 & (\pi < t \leqslant 2\pi) \end{cases}$$

$$f(t + 2\pi) = f(t)$$

Obtain a Fourier series expansion of $f(t)$ and deduce that

$$\tfrac{1}{6}\pi^2 = \sum_{r=1}^{\infty} \frac{1}{r^2}$$

2 Determine the full-range Fourier series expansion of the even function $f(t)$ of period 2π defined by

$$f(t) = \begin{cases} \tfrac{2}{3}t & (0 \leqslant t \leqslant \tfrac{1}{3}\pi) \\ \tfrac{1}{3}(\pi - t) & (\tfrac{1}{3}\pi \leqslant t \leqslant \pi) \end{cases}$$

To what value does the series converge at $t = \tfrac{1}{3}\pi$?

3 A function $f(t)$ is defined for $0 \leqslant t \leqslant \tfrac{1}{2}T$ by

$$f(t) = \begin{cases} t & (0 \leqslant t \leqslant \tfrac{1}{4}T) \\ \tfrac{1}{2}T - t & (\tfrac{1}{4}T \leqslant t \leqslant \tfrac{1}{2}T) \end{cases}$$

Sketch odd and even functions that have a period T and are equal to $f(t)$ for $0 \leqslant t \leqslant \tfrac{1}{2}T$.

(a) Find the half-range Fourier sine series of $f(t)$.

(b) To what value will the series converge for $t = -\tfrac{1}{4}T$?

(c) What is the sum of the following series?

$$S = \sum_{r=1}^{\infty} \frac{1}{(2r-1)^2}$$

4 The magnetomotive force, y, in the air gap of an alternator can be represented approximately by a graph of the form shown in Figure 12.25. Find a Fourier series for y, explaining beforehand, with reasons, any special characteristics you would expect to find.

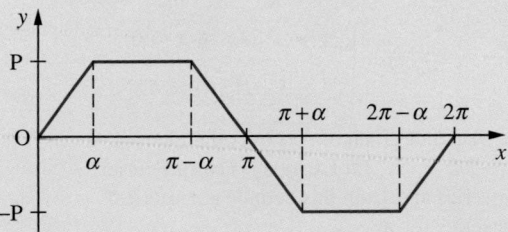

Figure 12.25 Data for $f(t)$ in Question 4.

5 Prove that if $g(x)$ is an odd function and $f(x)$ an even function of x, the product $g(x)[c + f(x)]$ is an odd function if c is a constant.

A periodic function with period 2π is defined by

$$F(\theta) = \tfrac{1}{12}\theta(\pi^2 - \theta^2)$$

in the interval $-\pi \leqslant \theta \leqslant \pi$. Show that the Fourier series representation of the function is

$$F(\theta) = \sum_{n=1}^{\infty} \frac{(-1)^{n+1}}{n^3} \sin n\theta$$

6 A repeating waveform of period 2π is described by

$$f(t) = \begin{cases} \pi + t & (-\pi \leqslant t \leqslant -\tfrac{1}{2}\pi) \\ -t & (-\tfrac{1}{2}\pi \leqslant t \leqslant \tfrac{1}{2}\pi) \\ t - \pi & (\tfrac{1}{2}\pi \leqslant t \leqslant \pi) \end{cases}$$

Sketch the waveform over the range $t = -2\pi$ to $t = 2\pi$ and find the Fourier series representation of $f(t)$, making use of any properties of the waveform that you can identify before any integration is performed.

7   A function $f(x)$ is periodic of period 2π and is defined by

$$f(x) = \begin{cases} -2x & (-\pi < x \le 0) \\ 2x & (0 < x \le \pi) \end{cases}$$

Sketch a graph of $f(x)$ from -2π to 3π and prove that

$$f(x) = \pi - \frac{8}{\pi} \sum_{n=0}^{\infty} \frac{1}{(2n+1)^2} \cos(2n+1)x$$

Hence show that

$$\tfrac{1}{8}\pi^2 = 1 + \sum_{n=1}^{\infty} \frac{1}{(2n+1)^2}$$

8   A function $f(x)$ of period 2π is defined in the interval $-\pi \le x \le \pi$ by

$$f(x) = \begin{cases} \tfrac{1}{2}\pi + x & (-\pi \le x \le 0) \\ \tfrac{1}{2}\pi - x & (0 \le x \le \pi) \end{cases}$$

Sketch a graph of $f(x)$ over the interval $-3\pi \le x \le 3\pi$. Express $f(x)$ as a Fourier series and from this deduce a numerical series for π.

9   A periodic function of period 2π is defined for $0 \le x \le 2\pi$ by

$$f(x) = \begin{cases} x & (0 \le x \le \tfrac{1}{2}\pi) \\ \tfrac{1}{2}\pi & (\tfrac{1}{2}\pi < x \le \pi) \\ -\tfrac{1}{2}\pi & (\pi < x \le \tfrac{3}{2}\pi) \\ x - 2\pi & (\tfrac{3}{2}\pi \le x \le 2\pi) \end{cases}$$

Sketch $f(x)$ for $-2\pi \le x \le 4\pi$ and show that its Fourier series representation is

$$f(x) = \left(1 + \frac{2}{\pi}\right)\sin x - \tfrac{1}{2}\sin 2x$$

$$+ \tfrac{1}{3}\left(1 - \frac{2}{3\pi}\right)\sin 3x - \tfrac{1}{4}\sin 4x + \ldots$$

Express this series in a general form.

10 A waveform is defined by $V(t) = 10e^{-3t}$ for $0 \le t < 0.4$ and $V(t) = V(t - 0.4)$ for all t. Sketch the graphs of V, dV/dt and $\int_0^t V\,dt$.

Express V as a Fourier series and show that the amplitude of the nth harmonic is about $2.22/n$.

11 A function $f(x)$ is defined in the interval $-1 \le x \le 1$ by

$$f(x) = \begin{cases} 1/2\varepsilon & (-\varepsilon < x < \varepsilon) \\ 0 & (-1 \le x < -\varepsilon;\ \varepsilon < x \le 1) \end{cases}$$

Sketch a graph of $f(x)$ and show that a Fourier series expansion of $f(x)$ valid in the interval $-1 \le x \le 1$ is given by

$$f(x) = \tfrac{1}{2} + \sum_{n=1}^{\infty} \frac{\sin n\pi\varepsilon}{n\pi\varepsilon} \cos n\pi x$$

12 Show that the half-range Fourier sine series for the function

$$f(t) = \left(1 - \frac{t}{\pi}\right)^2 \quad (0 \le t \le \pi)$$

is

$$f(t) = \sum_{n=1}^{\infty} \frac{2}{n\pi}\left\{1 - \frac{2}{n^2\pi^2}[1 - (-1)^n]\right\}\sin nt$$

13 Find a half-range Fourier sine and Fourier cosine series for $f(x)$ valid in the interval $0 < x < \pi$ when $f(x)$ is defined by

$$f(x) = \begin{cases} x & (0 \le x \le \tfrac{1}{2}\pi) \\ \pi - x & (\tfrac{1}{2}\pi \le x \le \pi) \end{cases}$$

Sketch the graph of the Fourier series obtained for $-2\pi < x \le 2\pi$.

14 A function $f(x)$ is periodic of period 2π and is defined by $f(x) = e^x$ $(-\pi < x < \pi)$. Sketch the graph of $f(x)$ from $x = -2\pi$ to $x = 2\pi$ and prove that

$$f(x) = \frac{2\sinh\pi}{\pi}\left[\tfrac{1}{2} + \sum_{n=1}^{\infty} \frac{(-1)^n}{1+n^2}(\cos nx - n\sin nx)\right]$$

15 A function $f(t)$ is defined on $0 < t < \pi$ by

$$f(t) = \pi - t$$

Find

(a) a half-range Fourier sine series, and

(b) a half-range Fourier cosine series

for $f(t)$ valid for $0 < t < \pi$.

Sketch the graphs of the functions represented by each series for $-2\pi < t < 2\pi$.

16 A periodic function $f(t)$ of period 2 is defined in the interval $-1 < t < 1$ by

$$f(t) = 1 - t^2$$

Sketch a graph of $f(t)$ for $-3 < t < 3$ and obtain a Fourier series expansion for it.

17 (a) Without actually finding the series state what terms you would expect to find in the Fourier series for the following periodic functions of period 2π.

(i) $f(t) = \sin^2 t$, $\quad -\pi \leqslant t \leqslant \pi$

(ii) $f(t) = 3e^{-t}$, $\quad -\pi \leqslant t \leqslant \pi$

(iii) $f(t) = \begin{cases} 0, & -\pi < t < 0 \\ 1, & 0 < t < \pi \end{cases}$

(b) Find, up to and including the term in $\cos 4t$, the Fourier half-range cosine series for the function defined by

$$f(t) = \begin{cases} t^2, & 0 < t < \pi/2 \\ 0, & \pi/2 < t < \pi \end{cases}$$

18 (a) A periodic function $f(t)$, of period 2π, is defined in $-\pi \leqslant t \leqslant \pi$ by

$$f(t) = \begin{cases} -t & (-\pi \leqslant t \leqslant 0) \\ t & (0 \leqslant t \leqslant \pi) \end{cases}$$

Obtain a Fourier series expansion for $f(t)$.

(b) By formally differentiating the series obtained in (a), obtain the Fourier series expansion of the periodic square wave

$$g(t) = \begin{cases} -1 & (-\pi < t < 0) \\ 0 & (t = 0) \\ 1 & (0 < t < \pi) \end{cases}$$

$$g(t + 2\pi) = g(t)$$

Check the validity of your result by determining directly the Fourier series expansion of $g(t)$.

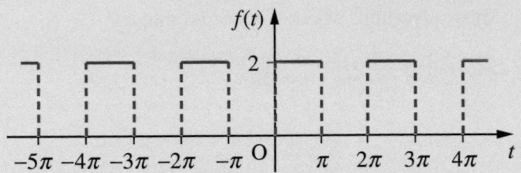

Figure 12.26 Waveform $f(t)$ of Question 19.

19 The periodic waveform $f(t)$ shown in Figure 12.26 may be written as

$$f(t) = 1 + g(t)$$

where $g(t)$ represents an odd function.

(a) Sketch the graph of $g(t)$.

(b) Obtain the Fourier series expansion for $g(t)$, and hence write down the Fourier series expansion for $f(t)$.

20 Show that the Fourier series

$$\frac{1}{2}\pi - \frac{4}{\pi} \sum_{n=1}^{\infty} \frac{\cos(2n-1)t}{(2n-1)^2}$$

represents the function $f(t)$, of period 2π, given by

$$f(t) = \begin{cases} t & (0 \leqslant t \leqslant \pi) \\ -t & (-\pi \leqslant t \leqslant 0) \end{cases}$$

Deduce that, apart from a transient component (that is, a complementary function that dies away as $t \to \infty$), the differential equation

$$\frac{dx}{dt} + x = f(t)$$

has the solution

$$x = \tfrac{1}{2}\pi - \frac{4}{\pi} \sum_{n=1}^{\infty} \frac{\cos(2n-1)t + (2n-1)\sin(2n-1)t}{(2n-1)^2[1 + (2n-1)^2]}$$

21 Show that if $f(t)$ is a periodic function of period 2π and

$$f(t) = \begin{cases} t/\pi & (0 < t < \pi) \\ (2\pi - t)/\pi & (\pi < t < 2\pi) \end{cases}$$

then

$$f(t) = \tfrac{1}{2} - \frac{4}{\pi^2} \sum_{n=0}^{\infty} \frac{\cos(2n+1)t}{(2n+1)^2}$$

Show also that, when ω is not an integer,

$$y = \frac{1}{2\omega^2}(1 - \cos \omega t)$$

$$-\frac{4}{\pi^2}\sum_{n=1}^{\infty}\frac{\cos(2n+1)t - \cos \omega t}{(2n+1)^2[\omega^2 - (2n+1)^2]}$$

satisfies the differential equation

$$\frac{d^2y}{dt^2} + \omega^2 y = f(t)$$

subject to the initial conditions $y = dy/dt = 0$ at $t = 0$.

13 Data Handling and Probability Theory

Chapter 13 Contents

13.1 Introduction

Many events in our lives are subject to chance – by which we mean that they are not entirely predictable. To some extent, we can choose where we live and what sort of work we do, but even so we cannot be sure what sort of neighbours or workmates we shall have: noisy, generous, friendly and so on. In a similar way, experiments in all branches of science and engineering involve unpredictable outcomes that may be expressed either as a quality such as 'turned green' or 'exploded', or numerically in terms of mass, resistance or any standard unit. In contrast with everyday life, an 'experiment' is repeated many times, so that the limited predictability of the various outcomes can emerge as a pattern within the disorder. The subject of statistics is about extracting that pattern and drawing useful conclusions from it, and the theoretical foundation for this is contained in the theory of probability.

Engineers, in particular, are immersed in data throughout their working lives. The term 'data' is used rather loosely to refer to numerical information of all kinds, including for example the specification of a machine or part. For our present purposes, however, we shall use **data** to refer to the set of measured outcomes of an experiment. Engineering is a discipline founded upon experiment, and engineers need to know how to process their experimental data and how to assess the results of others' experiments.

The aim of statistics is to extract useful information from the data. This information can take many forms. If the aim of an experiment is to assist with the making of a decision then the people conducting the experiment will have in mind a question to which they would like the answer, and ideally the question (and its possible answers) will be expressed as simply and clearly as possible in ordinary language. On the other hand, the aim of an experiment may be to calibrate an instrument or to measure some unknown quantity, in which case the conclusions of the experiment will be numerical.

Sometimes all that is needed is to plot the data in a suitable way that makes the message clear. The information is then conveyed in graphical form to the reader. Unfortunately, it very often turns out that the data is rather ambiguous, the conclusions are not obvious and the data must be analysed in a more mathematical way. In this case the conclusion (which relates directly to the purpose of the experiment) cannot be stated with 100% confidence. This issue is taken up in the companion text *Advanced Modern Engineering Mathematics*, where the mathematical methods of statistics are introduced. In the present chapter we shall see how the data may be plotted to good effect, and then go on to cover the essential probability theory without which the proper statistical practices (in engineering and elsewhere) would be impossible.

Whilst MATLAB and MAPLE have some statistical functionality there are specialist statistical packages, such as Minitab, that are more appropriate for use in the teaching of statistical work. Consequently, MATLAB and MAPLE commands are not developed in this chapter. Almost all the statistical calculations in this chapter could also be carried out using spreadsheet software such as Microsoft Excel. In addition, the statistical package R, which is available on the web (http://www.r-project.org), is increasingly popular and runs on a variety of platforms.

13.2 The raw material of statistics

13.2.1 Experiments and sampling

A statistician requires data to work on, and data is usually obtained by experiment – but not any old experiment will do. The most common type of statistical experiment involves taking a **sample** from a **population** and drawing some conclusions about the whole population from the results for the sample. In general, in statistical work the population that is the object of study is assumed to be very large and rather uniform with respect to certain characteristics of interest. The sample that is drawn for investigation is much smaller. The size of the sample governs the confidence with which statements about the characteristics of the population can be made.

Ideally, the entire population would be studied, but this may be impractical for reasons of expense, ethics or destructiveness of tests:

(a) *Expense*: the population may be too large or the cost per individual may be high.

(b) *Ethics*: in medical experiments involving animals or people the aim is to use the smallest sample size that is compatible with obtaining a dependable result.

(c) *Destructiveness*: destructive testing of components, for example breaking stress or lifetime, obviously precludes using the whole population.

The quality of the sample is also important. Imagine an opinion poll in which all the people interviewed were professional engineers. The results would be of interest to someone investigating the voting intentions of this particular group, but such a poll might be a poor indicator of the result of the next general election. Now imagine an opinion poll conducted in a large hall, with a microphone passed from person to person. The intimidating nature of this situation would prevent many respondents from giving a truthful answer, particularly if the poll involved politically, socially or morally sensitive issues. These two examples demonstrate the fundamental requirements of any sampling experiment, including an opinion poll: the sample must be **representative** and successive observations must be **independent**.

13.2.2 Histograms of data

After gathering the data together, the first step is often to display it graphically using a histogram or pie chart. Computer packages are often very useful for this. For example, Figure 13.1 contains some data describing the performance of two prototype car engines: a series of running times (in minutes at constant speed on 1 litre of standard fuel) and ambient temperatures at the times of the tests, for each engine. Two questions that are easy to state, and which might be answerable from this data, are

(a) Is the fuel consumption of one engine different from that of the other?

(b) Does fuel consumption depend upon ambient temperature?

These questions are actually related, as can be seen in *Advanced Modern Engineering Mathematics*, where this example is discussed at some length. For the moment, we shall see what can be learned just by plotting the data.

Figure 13.1
Car engine test data.

	Engine A				Engine B		
Time	Temp.	Time	Temp.	Time	Temp.	Time	Temp.
27.7	24	24.1	7	24.9	13	24.3	17
24.3	25	23.1	14	21.4	19	24.5	16
23.7	18	23.4	16	24.1	18	26.1	18
22.1	15	23.1	9	27.5	19	27.7	14
21.8	19	24.1	14	27.5	21	24.3	19
24.7	16	28.6	23	25.7	17	26.1	5
23.4	17	20.2	14	24.9	17	24.0	17
21.6	14	25.7	18	23.3	19	24.9	18
24.5	18	24.6	18	22.5	21	26.7	23
26.1	20	24.0	12	28.5	12	27.3	28
24.8	15	24.9	18	25.9	17	23.9	18
23.7	15	21.9	20	26.9	13	23.1	10
25.0	22	25.1	16	27.7	17	25.5	25
26.9	18	25.7	16	25.4	23	24.9	22
23.7	19	23.5	11	25.3	30	25.9	16

The first thing to observe is that the measured running times are rather erratic, even taking temperature into account. The six tests of engine A at 18°C produced results ranging from 23.7 min to 26.9 min. This situation is typical, and is not necessarily the result of sloppy experimental practice or inaccurate equipment (though such failings should not be condoned where they are easily avoided). There are practical limitations on the design and conduct of experiments that preclude making measurements to ultimate precision, and mean that certain causal factors that might influence the results are not measured at all. In this series of engine tests the actual quantity of fuel would have varied a little around 1 litre, the condition of the engine oil would have been different from one test to another, and so on.

Figures 13.2(a) and (b) are **histograms** of the running times for engines A and B respectively. The data has been grouped into classes, and the height of each bar indicates the number in the class. Values falling on a boundary are counted in the upper class. The width of each bar is the same, and is chosen to reveal the overall shape of the data. A histogram with too many small classes is very erratic, whereas one with too few large classes has no structure. It is typical for a histogram to span the data with about eight to ten classes.

Figure 13.2(c) shows the two histograms superimposed. It is fairly clear that there is a difference here, and that the running times for engine B tend to be longer than those for engine A. However, just from the histograms, it is difficult to be precise about the amount of the difference, or to assess the confidence with which one could state that a difference exists.

Figure 13.3 contains corresponding histograms for the temperatures. This time the results are much more similar. It is easy to imagine that if a relatively small subset of the sample had given different results from those obtained then there would have been no difference at all between the histograms. This difference could therefore be attributed to chance.

Figure 13.2
Histograms of running
times: (a) engine A;
(b) engine B;
(c) superimposed.

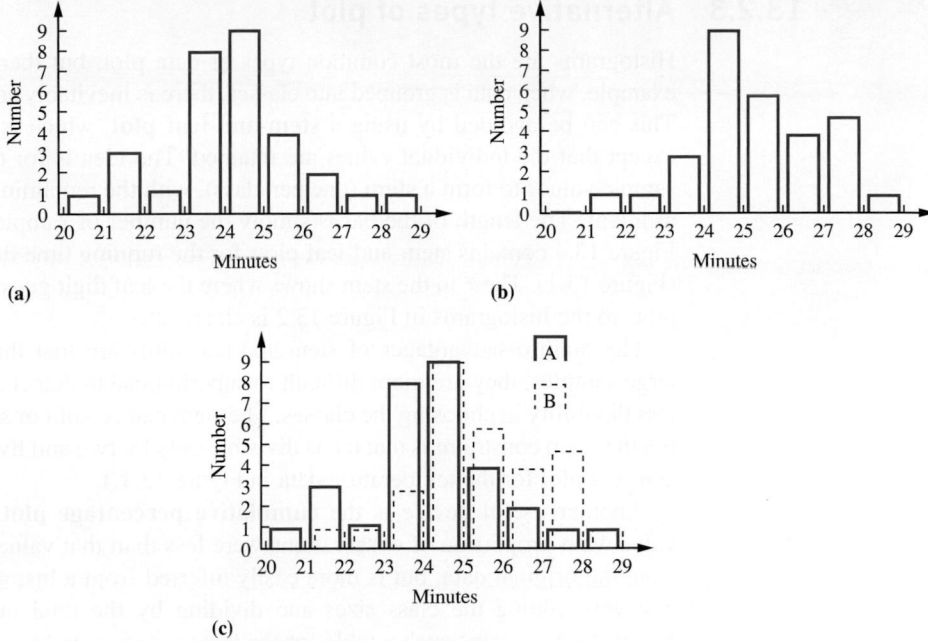

Figure 13.3
Histograms of
temperatures: (a)
engine A; (b) engine
B; (c) superimposed.

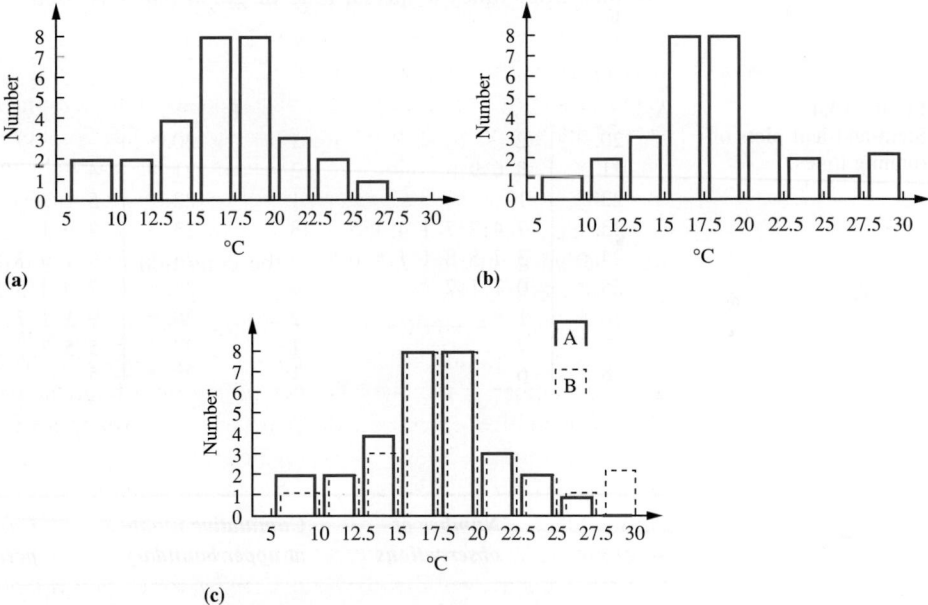

Somewhere in between these two situations is one for which a difference just exists but is not obvious. It is in dealing with this type of situation (which is quite common) that the powerful mathematical discipline of statistics is important. No analysis of the data will definitely settle the question of whether or not the populations differ, but the extent of the evidence for a difference can be assessed, and this may be invaluable if a decision has to be made.

13.2.3 Alternative types of plot

Histograms are the most common types of data plot, but there are many others. For example, when data is grouped into classes, there is inevitably some loss of information. This can be avoided by using a **stem-and-leaf plot**, which is similar to a histogram except that the individual values are retained. The idea is for the leading digits in the sample values to form a **stem** (one per class), with the remaining digits entering the bar as a **leaf**. The length of the bar is simply the number of sample values with that stem. Figure 13.4 contains stem-and-leaf plots for the running time data for engines A and B (Figure 13.1). The * in the stem shows where the leaf digit goes. The similarity of these plots to the histograms in Figure 13.2 is clear.

The main disadvantages of stem-and-leaf plots are that they are less suitable for large samples, they are more difficult to superimpose to detect differences, and there is less flexibility in choosing the classes. The stem can be split or several stems conjoined, but the main constraint is that ten is divisible only by two and five. (Try drawing a stem-and-leaf plot for the temperature data in Figure 13.1.)

Another useful device is the **cumulative percentage plot**, which shows for any value what proportion of observations were less than that value. This can be drawn up from the original data, but is more easily inferred from a histogram of classes by successively adding the class sizes and dividing by the total number of observations. Figure 13.5 contains such a table for the temperature data for engine A in Figure 13.1. The cumulative percentage is plotted in Figure 13.6(a). The S shape is typical for plots like this. Sometimes a special kind of graph paper is used for which the probability

Figure 13.4
Stem-and-leaf plots of running times.

A:				B:		
20. *	2	1		20. *		0
21. *	8 6 9	3		21. *	4	1
22. *	1	1		22. *	5	1
23. *	7 4 7 7 1 4 1 5	8		23. *	3 9 1	3
24. *	3 7 5 8 1 1 6 0 9	9		24. *	9 1 9 3 5 3 0 9 9	9
25. *	0 7 1 7	4		25. *	7 9 4 3 5 9	6
26. *	1 9	2		26. *	9 1 1 7	4
27. *	7	1		27. *	5 5 7 7 3	5
28. *	6	1		28. *	5	1

Figure 13.5
Cumulative percentages for temperature data.

Class range	Number of observations	Cumulative number at upper boundary	Cumulative percentage
0–9.9	2	2	6.7
10.0–12.4	2	4	13.3
12.5–14.9	4	8	26.7
15.0–17.4	8	16	53.3
17.5–19.9	8	24	80.0
20.0–22.4	3	27	90.0
22.5–24.9	2	29	96.7
25.0–27.4	1	30	100.0

Figure 13.6
Cumulative percentage
plots: (a) linear scale;
(b) normal scale.

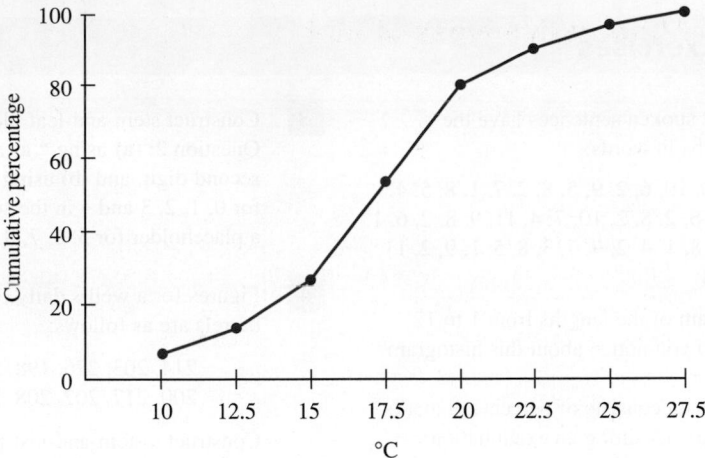

(a)

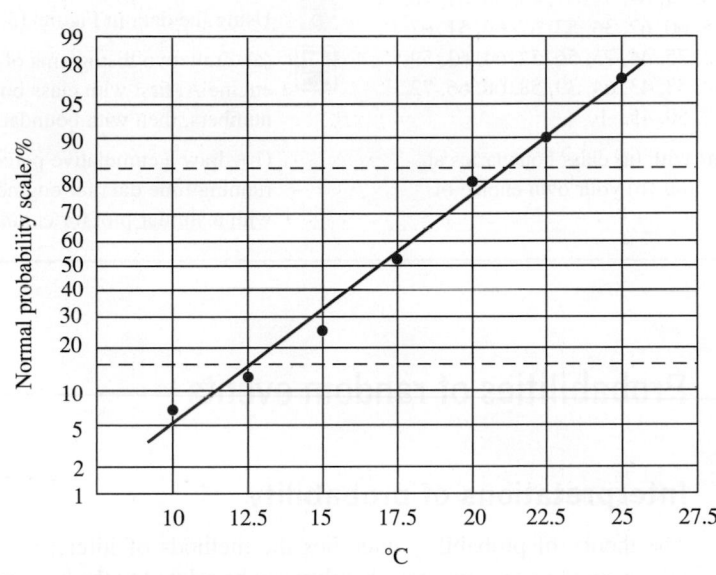

(b)

scale is nonlinear, as shown in Figure 13.6(b). This is known as **normal probability paper**, and is useful for testing whether the data has a particularly important kind of profile called the normal distribution, which will be introduced in Section 13.5.3. If the data is normal, the plot should fit a straight line.

The presentation of data (often using sophisticated graphics) is very important in communicating results, and is often referred to as **descriptive statistics**. In *Advanced Modern Engineering Mathematics* we introduce **inferential statistics**. This means using mathematical methods to analyse the data with a view to answering certain important questions. Inferential statistics is more powerful than descriptive statistics because of the capacity to extract conclusions and quote the confidence with which they are asserted. In the rest of this chapter the necessary theory of probability will be covered so that the statistical methods can be built upon it.

13.2.4 Exercises

1 A sample of 52 spoken sentences have the following lengths in words:

> 7, 3, 8, 6, 10, 6, 2, 9, 5, 8, 2, 7, 1, 8, 5, 4,
> 12, 9, 3, 6, 2, 8, 2, 10, 7, 4, 11, 9, 8, 2, 6, 1,
> 3, 11, 7, 8, 1, 4, 2, 9, 7, 3, 8, 5, 1, 9, 2, 11,
> 6, 7, 3, 8

Draw a histogram of the lengths from 1 to 12 words. What do you notice about this histogram?

2 The following data consists of percentage marks achieved by students sitting an examination:

> 47, 51, 75, 58, 70, 73, 63, 60, 60, 54, 60,
> 67, 50, 60, 74, 69, 51, 67, 49, 66, 61, 46,
> 66, 57, 55, 60, 62, 36, 52, 67, 62, 51, 62,
> 62, 59, 52, 75, 44, 75, 56, 52, 64, 63, 59,
> 54, 57, 68, 53, 43, 64, 39, 58, 68, 66, 72,
> 46, 58, 52, 50, 45

Draw histograms with (a) class boundaries at intervals of five, and (b) your own choice of class boundaries.

3 Construct stem-and-leaf plots for the data in Question 2: (a) using * as a placeholder for the second digit, and (b) using * as a placeholder for 0, 1, 2, 3 and 4 in the second digit and + as a placeholder for 5, 6, 7, 8 and 9.

4 Figures for a well's daily production of oil in barrels are as follows:

> 214, 203, 226, 198, 243, 225, 207, 203, 208,
> 200, 217, 202, 208, 212, 205, 220

Construct a stem-and-leaf plot with stem labels $19*, 20*, \ldots, 24*$.

5 Using the data in Figure 13.1,

(a) draw two histograms of temperatures for engine A, first with class boundaries at even numbers, then with boundaries at multiples of five;

(b) draw a cumulative percentage plot for the running time data for engine A and compare it with a similar plot for engine B.

13.3 Probabilities of random events

13.3.1 Interpretations of probability

The theory of probability underlies the methods of inference used in statistical situations, and the concept of probability can be related to the histogram of data. The height of each bar determines the proportion of the sample that fell into the corresponding class. One way to think of probability is to assume that as a larger and larger sample is taken (ignoring the practical objections raised in Section 13.2.1), the histogram will stabilize and the class proportions will converge to the 'true' probability figures. This concept of probability is of an objective quantity that applies to each observation and measures (in a relative way) how likely it is to fall into the corresponding class. Like the speed of sound and the density of gold, it is known only imperfectly because of our limited capacity to do experiments.

An alternative concept of probability that is important in decision-making and expert systems involves **degree of belief**. This is highly subjective, because it will depend upon the individual (or group) concerned and will vary with past experience. This seems unscientific at first sight, and there is much resistance to this notion, but there are many situations where experiments are unrepeatable in principle and no 'large-sample proportion' approach is applicable. The outcome of an election is uncertain, and it is not unreasonable to say that some outcomes are 'more probable' than others, but the actual election can take place only once. It seems that one is forced into a subjective view of the uncertainties, but the probability figures that emerge must obey certain rules in order to be consistent. Advocates of subjective probability have shown that these rules are the same as those obeyed by the sample proportions.

The formal theory of probability admits a number of 'interpretations', of which these objective and subjective interpretations are by far the most important. For engineering students it is most appropriate to keep the first interpretation – that of probability as an idealized proportion – in mind when studying the theory.

13.3.2 Sample space and events

The first step is to introduce some terminology that allows us to be clear in describing what is observed in an experiment. The language used is that of set theory, introduced in Chapter 6, because this provides a natural way of describing how observed events combine and separate.

Let the set of all possible outcomes of an experiment be called the **sample space** and denote it by S. An **event** is any subset of S. One (initially unspecified) outcome is considered to be the **actual outcome** and an event is said to **occur** if it contains the actual outcome.

Example 13.1

If the experiment is to roll an ordinary six-faced die and observe the numerical value of the outcome then the sample space will be the set $\{1, 2, 3, 4, 5, 6\}$. The event 'the outcome is an even number' will be represented by the set $\{2, 4, 6\}$. If the actual outcome is a 5 then the event 'the outcome is an even number' has not occurred, because $5 \notin \{2, 4, 6\}$.

Example 13.2

If the experiment is to toss a fair coin twice then the sample space may be represented as the set $\{HH, HT, TH, TT\}$. The event 'both tosses yield the same result' is represented by the set $\{HH, TT\}$. If the actual outcome is TT then the event 'both tosses yield the same result' has occurred, because $TT \in \{HH, TT\}$.

The sample space S therefore contains everything that can occur, and may be discrete or continuous. It is **discrete** if the possible outcomes can be written as a list: for instance, for somebody's birthday $S = \{1 \text{ January}, 2 \text{ January}, \ldots, 31 \text{ December}\}$. The list does not need to be finite. **Continuous** sample spaces arise when experiments involve measurements of some continuous variable such as a person's height or the voltage in a circuit. Then parentheses (rather than curly brackets) are used to denote an open interval such as $S = (0, 10)$, and square brackets to denote a closed interval (see Section 1.2.5). Of course we can only measure to limited precision in practice, so the possible values could be listed, but this is a rather arbitrary technical limitation.

Events are what we observe, in general. It is not always necessary – and for a continuous observation it is impossible – to know the actual outcome. If you want to send someone a birthday card then it is sufficient to know that her birthday occurs 'about the end of June'. Events range from S itself (the certain event) to the empty set $\varnothing$ (the impossible event). Most interesting events are in-between: neither certain nor impossible, but with a reasonable chance of containing the actual outcome.

The usual operations of set theory (Section 6.2) apply to events. If A and B are events then so are the following:

(a) *union*: $A \cup B$ corresponding to 'A or B occurs';
(b) *intersection*: $A \cap B$ corresponding to 'A and B occur';
(c) *complement*: $S - A$ corresponding to 'not-A occurs'.

The complement of A is also written $\bar{A}$.

Using this language, it is possible to describe situations in which at least one of two events occurs (union), or where both events occur (intersection), or where an event fails to occur (complement). Since so much of our everyday experience is structured in this way, this should be a fairly natural starting-point for the theory.

13.3.3 Axioms of probability

The next step is to associate a real number $P(A)$ with each event $A \subseteq S$, called the **probability** of that event. (Strictly speaking, assigning probabilities to arbitrary subsets of a continuous sample space is not possible, but there are ways around this and we shall ignore this technical limitation here.) These numbers must satisfy the rules prompted by the interpretations discussed in Section 13.3.1. The following three rules are referred to as the **axioms of probability**, and lay the foundation for the whole theory:

(1) The certain event S has probability one: $P(S) = 1$.
(2) All probabilities are non-negative: $P(A) \geqslant 0$.
(3) Addition rule: if A and B are disjoint events (so that $A \cap B = \varnothing$) then

$$P(A \cup B) = P(A) + P(B)$$

If probability is regarded as an idealized proportion then clearly its maximum value must be one, it must be non-negative, and the addition rule describes how proportions behave in exclusive situations: for instance, if 5% of units of a brand of power supply produce a voltage that is too low and 8% produce a voltage that is too high then the proportion that produces a voltage that is either too low or too high must be 13%. The three axioms are therefore exactly what we should intuitively expect. What is remarkable is that they are also sufficient. Further rules of probability follow from the axioms:

(4) Complement rule: $P(S - A) = 1 - P(A)$.
(5) $P(\varnothing) = 0$.
(6) If $A \subseteq B$ then $P(A) \leqslant P(B)$.
(7) General addition rule:

$$P(A \cup B) = P(A) + P(B) - P(A \cap B)$$

The complement rule (4) follows immediately from axioms (1) and (3) using the fact that a set does not intersect with its complement:

$$P(A \cup \bar{A}) = P(S) = 1 = P(A) + P(\bar{A})$$

Figure 13.7
Venn diagram
illustrating general
addition rule.

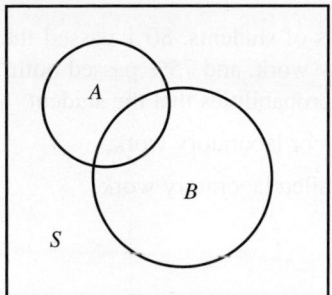

The general addition rule (7) can be illustrated using a Venn diagram as in Figure 13.7. Imagine the probability as a unit mass, spread out (unevenly) over S. Because of the overlap between A and B, adding their probabilities makes the probability of the intersection contribute twice to the total, so this has to be subtracted to compensate.

Example 13.3

A fair six-sided die is tossed. Find the probability of the event 'even number or number less than four'.

Solution

The sample space is $S = \{1, 2, 3, 4, 5, 6\}$, with all values equally likely. Since $P(1) + P(2) + P(3) + P(4) + P(5) + P(6)$ must sum to one, we must have

$$P(1) = P(2) = \ldots = P(6) = \tfrac{1}{6}$$

Figure 13.8
Sample space for
Example 13.3.

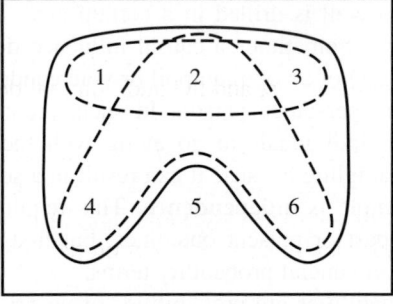

The various events are shown in Figure 13.8. Using the general addition rule,

$$P(\text{even or less than four}) = P(\text{even}) + P(\text{less than four})$$
$$- P(\text{even number less than four})$$
$$= P(\{2, 4, 6\}) + P(\{1, 2, 3\}) - P(\{2\})$$
$$= \tfrac{1}{2} + \tfrac{1}{2} - \tfrac{1}{6} = \tfrac{5}{6}$$

This result also follows from the complement rule:

$$P(\{1, 2, 3, 4, 6\}) = P(\text{not } \{5\}) = 1 - P(\{5\}) = \tfrac{5}{6}$$

Example 13.4

During the assessment of a class of students, 80% passed the examination in mathematics, 85% passed in laboratory work, and 75% passed both. For a student chosen at random from the class, find the probabilities that the student

(a) passed in either mathematics or laboratory work,

(b) passed in mathematics but failed laboratory work,

(c) failed in both.

Solution

Let M and L denote passes in mathematics and laboratory work respectively.

(a) By the general addition rule,

$$P(M \cup L) = P(M) + P(L) - P(M \cap L) = 0.8 + 0.85 - 0.75 = 0.9$$

(b) The group of students who passed in mathematics consists of those who passed in both together with those who passed in mathematics but failed in laboratory work, so that

$$P(M \cap \bar{L}) = P(M) - P(M \cap L) = 0.8 - 0.75 = 0.05$$

(c) De Morgan's law (Section 6.2.4, equations (6.7)), together with the result of (a), gives

$$P(\bar{M} \cap \bar{L}) = 1 - P(M \cup L) = 1 - 0.9 = 0.1$$

13.3.4 Conditional probability

Information sometimes arrives in stages, and this happens whenever the outcome of one experiment (or part of an experiment) is relevant to another outcome subsequent to it. For example, the outcome of a seismic survey tells an oil company something about the chances of finding oil if a well is drilled in a certain area, but has no direct causal influence on that discovery. Sometimes a causal influence does exist, as in the case (mentioned in Section 13.2.1) of an opinion poll or vote conducted in a large hall with a microphone passed from person to person. In such circumstances there is strong psychological pressure on individuals to go along with the majority. This is very undesirable in statistical sampling because it can result in a serious bias, so one of the essential features of a sample is **independence**. This requires that future outcomes should *not* depend upon past or present outcomes, but first we need to express the possibility of dependence in general probability terms.

From the start, all probabilities are probabilities of *events* (see Section 13.3.3), so suppose in general that an event A is known to have occurred (representing the existing information). The probabilities of possible future events are now measured relative to the fact that A has occurred. This event must therefore encompass all possibilities compatible with the known information, and can effectively be regarded as a new, revised, sample space in the light of that information. This is the key to understanding the definition and examples that follow.

The **conditional probability of B given A** is defined as

$$P(B|A) = \frac{P(A \cap B)}{P(A)}, \quad \text{where } P(A) > 0$$

Figure 13.9
Venn diagram for
conditional
probability.

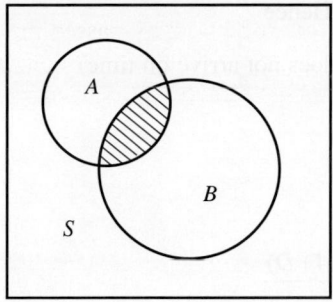

This represents the new probability of B given that A has occurred, and depends upon the probability of the intersection as shown in Figure 13.9.

Example 13.5   Someone tosses a die, covers it up and tells you that the number shown is less than four. How does this change the probability that the number is even?

Solution   From the definition,

$$P(\text{even number} \mid \text{less than four}) = \frac{P(\text{even number less than four})}{P(\text{number less than four})}$$

$$= \frac{P(\{2\})}{P(\{1, 2, 3\})} = \frac{\frac{1}{6}}{\frac{1}{2}}$$

$$= \frac{1}{3}$$

The information that the number is less than four causes the sample space to shrink to $\{1, 2, 3\}$, and only one entry in this set is even. The outcomes are equally likely, so the probability that the number is even drops from one-half to one-third.

Example 13.6   The probability that a regularly scheduled flight departs on time is $P(D) = 0.83$, the probability that it arrives on time is $P(A) = 0.92$, and the probability that it both departs and arrives on time is $P(A \cap D) = 0.78$. Find the probability that a plane

(a) arrives on time given that it departed on time,

(b) did not depart on time given that it fails to arrive on time.

Solution   (a) This is straightforward from the definition:

$$P(\text{arrives on time} \mid \text{departed on time}) = P(A \mid D)$$

$$= \frac{P(A \cap D)}{P(D)} = 0.94$$

(b) First, using De Morgan's law (Section 6.2.4, equations (6.7)), we have

$$P(\bar{A} \cap \bar{D}) = 1 - P(A \cup D)$$

$$= 1 - P(A) - P(D) + P(A \cap D)$$

by the general addition rule (7). Hence

$P(\text{did not depart on time} \mid \text{does not arrive on time})$

$$= P(\bar{D} \mid \bar{A})$$

$$= \frac{P(\bar{A} \cap \bar{D})}{P(\bar{A})}$$

$$= \frac{1 - P(A) - P(D) + P(A \cap D)}{1 - P(A)}$$

$$= 0.375$$

In a sense it is cheating to refer to a conditional 'probability' until it is clear that this quantity actually satisfies the axioms of probability. It is a useful exercise to show that this is the case. Consider the three axioms in turn:

(1)   The event A can be considered as the new sample space, and

$$P(A \mid A) = \frac{P(A)}{P(A)} = 1$$

(2) $P(B \mid A) \geqslant 0$ because $P(A \cap B) \geqslant 0$ and $P(A) > 0$

(3)   If $B \cap C = \varnothing$ then $(B \cap A) \cap (C \cap A) = \varnothing$,   and so

$$P[(B \cap A) \cup (C \cap A)] = P(B \cap A) + P(C \cap A)$$

Since $(B \cap A) \cup (C \cap A) = (B \cup C) \cap A$ we have that

$$P(B \cup C \mid A) = \frac{P[(B \cup C) \cap A]}{P(A)} = \frac{P(B \cap A) + P(C \cap A)}{P(A)}$$

$$= P(B \mid A) + P(C \mid A)$$

So conditional probability satisfies the axioms and therefore the various further rules such as the complement rule (4) and the general addition rule (7). Also, because a conditional probability is itself a probability, it is possible to conditionalize again. Thus if the probabilities in the definition are all conditioned upon another event C, we have

$$P(B \mid A \cap C) = \frac{P(A \cap B \mid C)}{P(A \mid C)}$$

Example 13.7   Suppose that on a small tropical island there are only two kinds of day: sunny days and rainy days. The probability that a sunny day is followed by a rainy day is 0.6, and the probability that a rainy day is followed by another rainy day is 0.8. The weather on any day depends upon the previous day's weather but not upon any earlier days. Find the probability that if Thursday is rainy then it will be sunny on Saturday.

Figure 13.10
Sequences of events
for Example 13.7.

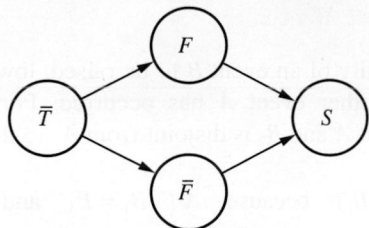

Solution Let T, F and S denote the events that Thursday, Friday and Saturday are sunny respectively (see Figure 13.10). All probabilities must be conditioned upon the assumption that Thursday is rainy. The first step is

$$P(S \mid \overline{T}) = P(S \cap F \mid \overline{T}) + P(S \cap \overline{F} \mid \overline{T})$$

because conditional probabilities obey the addition rule. Also,

$$P(S \cap F \mid \overline{T}) = P(S \mid F \cap \overline{T})P(F \mid \overline{T})$$

from the definition. Now

$$P(F \mid \overline{T}) = P(\text{sunny} \mid \text{rainy})$$

$$= 1 - 0.8 = 0.2$$

because conditional probabilities obey the complement rule. The assumption that the influence of the weather does not extend beyond the previous day implies that

$$P(S \mid F \cap \overline{T}) = P(S \mid F) = P(\text{sunny} \mid \text{sunny})$$

$$= 1 - 0.6 = 0.4$$

(This is known as **independence**, and is described in Section 13.3.5.) Similarly,

$$P(S \cap \overline{F} \mid \overline{T}) = P(S \mid \overline{F} \cap \overline{T})P(\overline{F} \mid \overline{T})$$

$$= P(S \mid \overline{F})P(\overline{F} \mid \overline{T})$$

$$= P(\text{sunny} \mid \text{rainy})P(\text{rainy} \mid \text{rainy})$$

$$= (1 - 0.8)(0.8)$$

Thus

$$P(S \mid \overline{T}) = (0.2)(0.4) + (0.2)(0.8) = 0.24$$

which is the answer required.

We shall not use conditional probabilities very much in this chapter, but the idea of a probability that is conditional upon another event is pervasive. For the moment, however, we must use conditional probability (rather paradoxically) to express the absence of interaction between the events.

13.3.5 Independence

It is possible for the probability of an event B to be raised, lowered or left unchanged by the information that another event A has occurred. For the events shown in Figure 13.11, B_1 is a subset of A and B_2 is disjoint from A, so that

$$P(B_1 \mid A) = \frac{P(B_1)}{P(A)} \geq P(B_1) \quad \text{because} \quad A \cap B_1 = B_1 \quad \text{and} \quad P(A) \leq 1$$

and

$$P(B_2 \mid A) = 0 < P(B_2) \quad \text{because} \quad A \cap B_2 = \varnothing$$

Figure 13.11
Venn diagram illustrating conditional probabilities.

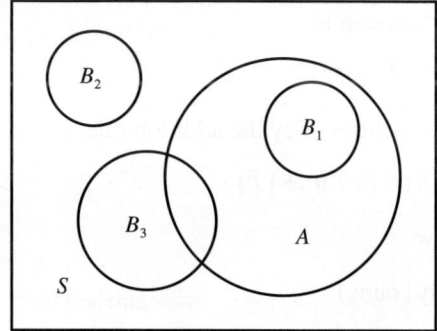

The probability of B_3 could go either way, or remain unchanged, depending on the probability of the intersection. The situation where the probability is unchanged assumes a special importance.

Events A and B are called **independent** when

$$P(B \mid A) = P(B)$$

In this situation A conveys effectively no information about B. From the definition of conditional probability in Section 13.3.4 it follows that

$$P(A \cap B) = P(A)P(B)$$

The joint probability is the product of the separate probabilities. This shows that independence is symmetric between the two events, so we also have

$$P(A \mid B) = P(A)$$

Example 13.8

Items from a production line can have defects A or B. Some items have both, some just one, but most have neither. Tables (a) and (b) show two alternative sets of joint probabilities:

(a)	B	$\bar{B}$	$Total$
A	0.02	0.08	0.10
$\bar{A}$	0.18	0.72	0.90
Total	0.20	0.80	1.00

(b)	B	$\bar{B}$	$Total$
A	0.06	0.04	0.10
$\bar{A}$	0.14	0.76	0.90
Total	0.20	0.80	1.00

Test for independence in each case.

Solution The row and column totals shown in the tables are the respective probabilities for the two defects individually, for example

$$P(A \cap B) + P(A \cap \bar{B}) = P(A)$$

and these figures are the same for both tables. It is easy to see that independence holds for (a) but not for (b); for example, the probability of both defects together is

$$P(A \cap B) = 0.02 = P(A)P(B) \quad \text{for (a)}$$

but

$$P(A \cap B) = 0.06 > P(A)P(B) \quad \text{for (b)}$$

The probability of the combination is greater in (b) than would be expected from the product of the separate probabilities, which suggests that the two defects are causally related in some way.

In general, for any number of independent events the probabilities multiply:

$$P(A_1 \cap A_2 \cap \dots \cap A_n) = P(A_1)P(A_2) \dots P(A_n)$$

This is called the **product rule** and must be distinguished from the addition rule, which applies (in its basic form, axiom (3)) to exclusive events. Independent events cannot be exclusive unless the probability of at least one of them is zero.

Example 13.9 A card is selected at random from an ordinary pack of 52 playing cards. Find the probabilities that the card drawn is

(a) an ace and a club,   (b) an ace or a club,

(c) an ace and a king,   (d) an ace or a king.

Solution Let the events that the card is an ace, king and club be denoted by A, K and C respectively.

(a) The events that the card is an ace and that it is a club are independent because there are the same numbers of cards for each suit. These events are not exclusive unless the ace of clubs happens to be missing. Thus

$$P(A \cap C) = P(A)P(C) = (\tfrac{1}{13})(\tfrac{1}{4}) = \tfrac{1}{52}$$

(b) The general addition rule for events (Section 13.3.3) gives

$$P(A \cup C) = P(A) + P(C) - P(A \cap C)$$
$$= \tfrac{1}{13} + \tfrac{1}{4} - \tfrac{1}{52} = \tfrac{4}{13}$$

(c) The events that the card is an ace and that it is a king are mutually exclusive, so

$$P(A \cap K) = 0$$

(d) By the third axiom of probability (Section 13.3.3),

$$P(A \cup K) = \tfrac{1}{13} + \tfrac{1}{13} = \tfrac{2}{13}$$

Example 13.10   If two fair dice are tossed, find the probability of at least one six occurring.

Solution   We shall assume that the throws are causally independent in that the outcome for one die does not relate in any way to the outcome for the other. They will then be statistically independent, and, by the complement and product rules,

$$P(\text{at least one six}) = 1 - P(\text{no six})$$

$$= 1 - P(\text{first die not six})P(\text{second die not six})$$

$$= 1 - \left(\tfrac{5}{6}\right)^2 = \tfrac{11}{36}$$

Example 13.11   If n people are independently selected, how large does n have to be before there is a better than even chance that at least two of them have the same birthday (not necessarily in the same year and ruling out February 29th)? Assume that all possibilities are equally likely.

Solution   The method of solution to this problem is similar to that for Example 13.10.

$$P(\text{at least two with the same birthday})$$

$$= 1 - P(\text{all different birthdays})$$

$$= 1 - P(\text{2nd different from 1st})P(\text{3rd different from 1st and 2nd})$$

$$\ldots P(n\text{th different from 1st}, \ldots, (n-1)\text{th})$$

$$= 1 - \frac{364}{365}\frac{363}{365}\ldots\frac{366-n}{365}$$

$$= 0.507 \quad \text{when } n = 23$$

Many people are surprised to find that the answer to Example 13.11 is so small, but this shows that our subjective expectations sometimes have to give way when the rules of probability are properly applied.

In connection with this example, a number of fallacies that 'appear to be most prevalent and injurious to the susceptible gambler' have been identified (R. A. Epstein, *The Theory of Gambling and Statistical Logic*, Academic Press, New York, 1977, p. 393) among which is

> *A tendency to interpret the probability of successive independent events as additive rather than multiplicative. Thus the chance of throwing a given number on a die is considered twice as large with two throws as it is with a single throw.*

The addition and product rules apply in different circumstances and must not be confused.

13.3.6 Exercises

6 If S is the set {bolt, nut, washer, screw, bracket, flange}, and A and B are sets {bracket, nut, flange} and {bolt, bracket} respectively, then what combinations of A and B produce the following sets as outcomes?

(a) {bracket}

(b) {flange, bracket, bolt, nut}

(c) {washer, bolt, screw}

(d) {screw, flange, nut, bolt, washer}

7 Let the sample space S and three events be defined as S = {car, bus, train, bicycle, motorcycle, boat, aeroplane}, A = {bus, train, aeroplane}, B = {train, car, boat}, C = {bicycle}. List the elements of the sets corresponding to the following events:

(a) $\bar{A}$ (b) $A \cap B \cap \bar{C}$ (c) $(\bar{A} \cup B) \cap (\bar{A} \cap C)$

8 If A and B are mutually exclusive events and $P(A) = 0.2$ and $P(B) = 0.5$, find

(a) $P(A \cup B)$ (b) $P(\bar{A})$ (c) $P(\bar{A} \cap B)$

9 From a pack of 52 cards a card is withdrawn at random and not replaced. A second card is then drawn. What is the probability that the first card is an ace and the second card a king?

10 Two ordinary six-faced dice are tossed. Write down the sample space of all possible combinations of values. What is the probability that the two values are the same? What is the probability that they differ by at most 1?

11 The personnel manager of a manufacturing plant claims that among the 400 employees, 312 got a pay rise last year, 248 got increased pension benefits, 173 got both and 43 got neither. Explain why this claim should be questioned.

12 If a card is drawn from a well-shuffled pack of 52 playing cards, what is the probability of drawing

(a) a red king

(b) a 3, 4, 5 or 6

(c) a black card

(d) a red ace or a black queen?

13 In a single throw of two dice, what is the probability of getting

(a) a total of 5,

(b) a total of at most 5,

(c) a total of at least 5?

14 Suppose that you roll a pair of ordinary dice repeatedly until you get either a total of seven or a total of 10. What is the probability that the total then is seven?

15 The 'odds' in favour of an event A are quoted as 'a to b' if and only if $P(A) = a/(a + b)$. The 'odds against' are then 'b to a' (which is the usual way to quote odds in betting situations).

(a) If an insurance company quotes odds of 3 to 1 in favour of an individual 70 years of age surviving another 10 years, what is the corresponding probability?

(b) If the probability of a successful transplant operation is $\frac{1}{8}$, what are the odds against success?

16 Two fair coins are tossed once. Find the conditional probability that both coins show heads given that

(a) the first coin shows a head,

(b) at least one coin shows a head.

17 During the repair of a large number of car engines it was found that part number 100 was changed in 36% and part number 101 in 42% of the cases, and that both parts were changed in 30% of cases. Is the replacement of part 100 connected with that of part 101? Find the probability that in repairing an engine for which part 100 has been changed it will also be necessary to replace part 101.

18 If $P(A) = 0.3$, $P(B) = 0.4$ and $P(B|A) = 0.5$, find

(a) $P(A \cap B)$ (b) $P(A \cup B)$ (c) $P(B|\bar{A})$

19 Three people work independently at deciphering a message in code. The probabilities that they will decipher it are $\frac{1}{5}$, $\frac{1}{4}$ and $\frac{1}{3}$. What is the probability that the message will be deciphered?

20 Part of an electric circuit consists of three elements K, L and M in series. Probabilities of failure for elements K and M during operating time t are 0.1 and 0.2 respectively. Element L itself consists of three sub-elements L_1, L_2 and L_3 in parallel, with failure probabilities 0.4, 0.7 and 0.5 respectively, during the same operating time t. Find the probability of failure of the circuit during time t, assuming that all failures of elements are independent.

21   A system can fail (event C) because of two possible causes (events A and B). The probabilities of A, B and $A \cap B$ are known, together with the probabilities of failure given A, given B and given $A \cap B$. Express the following in terms of these known quantities:

(a) $P(A \cup B)$ (b) $P(C \mid A \cap \bar{B})$

(c) $P(C \mid A \cup B)$

22   An advertising agency notes that approximately one in 50 potential buyers of a product sees a given magazine advertisement and one in five sees the corresponding advertisement on television. One in 100 sees both. One in three of those who have seen the advertisement purchase the product, and one in 10 of those who haven't seen it also purchase the product. What is the probability that a randomly selected potential customer will purchase the product?

23   On an infinite chess-board with each side of a square equal to d, a coin of diameter $2r < d$ is thrown at random. Find the probabilities that

(a) the coin falls entirely in the interior of one of the squares,

(b) the coin intersects no more than one side of a square.

13.4   Random variables

13.4.1   Introduction and definition

Now that the foundation of probability theory has been laid, we can begin to consider the data that originates in typical experiments – in particular, numerical data from observations of random variables. It is quite possible for non-numerical outcomes to be of interest, for instance in an experiment where a machine's possible faults might consist of the set {overheated, jammed, misaligned}. Even then, the experiment is likely to be repeated a number of times, and the count for each outcome gives rise to numerical data that can be treated statistically. For the moment, however, let us assume that the outcomes themselves take numerical values.

A **random variable** consists of a sample space of possible numerical values together with a probability over those values.

Random variables vary in their degree of advance predictability. As the following four examples show, the probabilities of the possible values are very dispersed for some random variables, but highly concentrated for others:

(a)   *The toss of a die.* No die is perfect, but for this random variable the probabilities of the six values are almost equal.

(b)   *Next month's rainfall.* Unless you live in a part of the world that has a very constant climate, the amount of rain that falls in March, say, varies from year to year quite considerably. The probabilities are not quite so dispersed as for the die toss, but there is a high degree of uncertainty.

(c)   *A flight delay.* Here there is a high probability of at most a short delay, but a small probability of a very long delay. The probabilities are relatively concentrated.

(d) *The time of tomorrow's sunrise.* Knowing your latitude, longitude, altitude, the date, the direction of sunrise and the height above sea level of the horizon in that direction, you could predict the time very precisely. There would be some small uncertainty because of atmospheric refraction.

The behaviour of a random variable is determined by the profile of its probability distribution. We shall now enlarge upon this for the two common types. The notation convention is to denote a random variable by a capital letter, say X, and an observed value by the corresponding lower-case letter, then x.

13.4.2 Discrete random variables

The distinction between discrete and continuous random variables is inherited from that for sample spaces (Section 13.3.2). First we shall consider the discrete case.

The random variable X, say, has a list of possible values $v_1, v_2, \ldots, v_m$ with probabilities $P(X = v_1), \ldots, P(X = v_m)$ of equalling these values. In other words, each actual value x of X is equal to v_i for some $i = 1, \ldots, m$, and we allow m to be infinite if required. This can be regarded as an idealization of the histogram of data in Section 13.2.2, where m is the number of classes. Typical examples are die tosses, birthdays, and the numbers of defective components in a batch from a production line.

In general, the behaviour of a discrete random variable can be represented graphically by means of a **probability function**

$$P_X(x) = P(X = x) \quad (-\infty < x < +\infty)$$

and illustrated in Figure 13.12(a) for Example 13.12. Also useful is the **distribution function** $F_x(x)$ defined as

$$F_X(x) = P(X \leq x) \quad (-\infty < x < +\infty)$$

Figure 13.12
Example 13.12: (a) probability function; (b) distribution function.

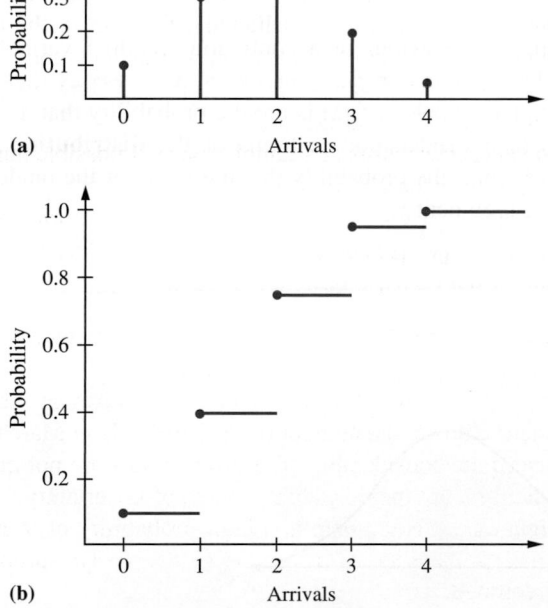

and illustrated in Figure 13.12(b). This definition is based on the fact that the set of points in the sample space for which $X \leqslant x$ constitutes an event, and the probability of this event (as a function of x) forms the distribution function. Sometimes this is referred to as the **cumulative distribution function**, because it measures the cumulative probability up to (and including) the value of its argument.

Example 13.12

The number of ships arriving at a container terminal during any one day can be any integer from zero to four, with respective probabilities 0.1, 0.3, 0.35, 0.2, 0.05. Plot the probability and distribution functions.

Solution

The probability function is shown in Figure 13.12(a). The function has zero value except at the five integer points. The value of the distribution function at any point x is the sum of the probabilities to the left of and including x. This is shown in Figure 13.12(b). The function is discontinuous, with steps occurring at the integer points, and the value at each integer includes the probability of that integer. This is indicated by the blob at each step.

The distribution function will be discussed further after the other class of random variables has been introduced.

13.4.3 Continuous random variables

A continuous random variable X can take any value within some interval (v_1, v_2). If this interval is not already infinite, we define the random variable to have zero probability for any value outside it, and hence extend the domain of definition to $(-\infty, +\infty)$. Typical examples are a person's height and weight, component lifetimes, and all measured quantities expressed in units of mass, length, time, temperature, resistance and so on.

In general, the behaviour of a continuous random variable X is described by a probability density function $f_X(x)$ for $-\infty < x < +\infty$ as illustrated in Figure 13.13. As will be explained below, $f_X(x)$ is not the probability that $X = x$: instead, the density function has to be understood in terms of the **distribution function** $F_X(x)$, which measures (as before) the probability that the value of the random variable is less than or equal to the argument x:

$$F_X(x) = P(X \leqslant x) \quad (-\infty < x < +\infty)$$

Figure 13.13
Typical probability
density function.

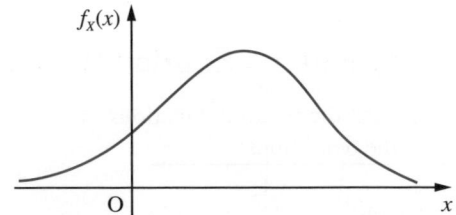

In this case, because there are no discrete steps in probability, $F_X(x)$ is continuous and differentiable, and its derivative is called the **probability density function** $f_X(x)$:

$$f_X(x) = \frac{d}{dx}[F_X(x)]$$

The significance of the density function is that it indicates for a continuous random variable the concentration of possible observed values along the real axis. This interpretation will be clarified in Section 13.4.4.

Example 13.13

The lifetime of an electronic component (in thousands of hours) is a continuous random variable with density function

$$f_X(x) = \begin{cases} \frac{1}{2}e^{-x/2} & (x \geqslant 0) \\ 0 & (x < 0) \end{cases}$$

(This is an example of an **exponential distribution** with parameter $\frac{1}{2}$.) Plot the distribution and density functions.

Solution   Integrating the density function gives the distribution function (Figure 13.14):

Figure 13.14
An exponential distribution (Example 13.13).

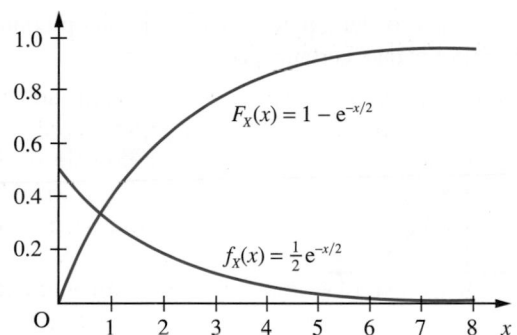

$$F_X(x) = \int_0^x \frac{1}{2}e^{-z/2}\, dz = [-e^{-z/2}]_0^x = 1 - e^{-x/2}$$

for $x \geqslant 0$, and zero for $x < 0$. The variable z is a dummy variable used for integration. The distribution and density functions show that most components have short lifetimes, but a small proportion can survive for much longer.

13.4.4   Properties of density and distribution functions

In order to use the density and distribution functions, we need the following results, which are immediate from the definitions:

(a) $\lim\limits_{x \to -\infty} F_X(x) = 0$ and $\lim\limits_{x \to +\infty} F_X(x) = 1$

Clearly it is impossible for a random variable to have a value less than $-\infty$, and it is certain to have a value less than $+\infty$.

(b) If $x_1 < x_2$ then $F_X(x_1) \leq F_X(x_2)$

Here the event that $X \leq x_1$ is a subset of the event that $X \leq x_2$, so the probability of the latter must be at least as great as that of the former. From results (a) and (b) it follows that at any point the distribution function is either constant or else increasing, ultimately from its lower limit of zero (at $-\infty$) to its upper limit of one (at $+\infty$).

(c) $P(x_1 < X \leq x_2) = F_X(x_2) - F_X(x_1)$

For any random variable the difference between the values of the distribution function at two points is the probability that a value of the random variable will lie between those two points (or is equal to the upper one). For a continuous random variable this is also the area under the density function between those points, by virtue of the relationship between the functions:

$$P(x_1 < X \leq x_2) = \int_{x_1}^{x_2} f_X(z)\mathrm{d}z$$

as illustrated in Figure 13.15. This crucial result expresses the significance of the density function and leads to another feature of continuous random variables that should be clearly understood. Setting $x_1 = x_2 = x$, we see that the probability $P(X = x)$ that the random variable has a value exactly equal to x is zero for any x, because the integral is over a domain of length zero. This is in sharp contrast to discrete random variables, which can *only* take certain specific values.

Figure 13.15
Probability of interval from density function.

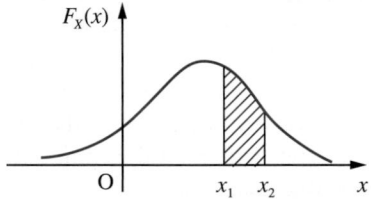

(d) $\displaystyle\int_{-\infty}^{+\infty} f_X(x)\mathrm{d}x = 1$

The total area under the density function must be unity because the random variable must have a value somewhere.

Example 13.14

For the distribution of component lifetimes in Example 13.13 find the proportion of components that last longer than 6000 hours.

Solution Using the distribution function,

$$P(X > 6) = 1 - P(X \leq 6) = 1 - F_X(6)$$
$$= 1 - [1 - e^{-6/2}] = e^{-3} \approx 0.05$$

In other words, approximately one in 20 components lasts longer than 6000 hours.

Example 13.15 Two people have agreed to meet in a definite place between six and seven o'clock. Their actual times of arrival are independent and entirely random (no arrival time more likely than any other) within the hour. Find

(a) the density function of the time that the first person arriving has to wait, and

(b) the probability that the meeting will occur if the first person to arrive does not wait for longer than 15 minutes.

Solution (a) The sample space can be regarded as a unit square as depicted in Figure 13.16(a).

Figure 13.16
(a) Sample space for Example 13.15.
(b) Density and distribution functions for waiting time.

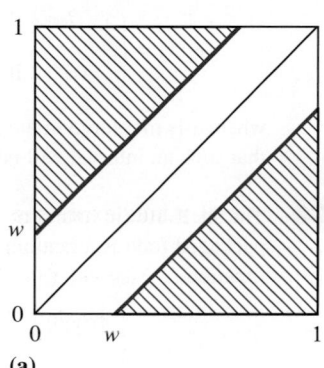

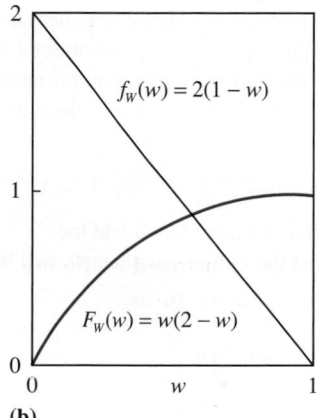

$f_W(w) = 2(1 - w)$

$F_W(w) = w(2 - w)$

(a) **(b)**

Each point represents a pair of arrival times, each measured as part of one hour from six o'clock. Because all arrival times are equally likely for each person and because they arrive independently, all points in the unit square are equally likely. Because the total probability must be one, this implies that the probability of any subset of points is simply equal to the area of that subset. Points along the diagonal lines offset along either axis by a distance w correspond to a waiting time for the first person arriving (W, say) equal to w, because the difference between the arrival times is constant along such lines. The shaded area therefore represents a waiting time greater than w. Putting the two triangles together, we obtain a square of side $1 - w$, so the probability that the waiting time exceeds w is given by

$$P(W > w) = (1 - w)^2$$

The complement of this gives the distribution function of waiting time:

$$F_W(w) = P(W \leqslant w) = 1 - (1 - w)^2 = 2w - w^2$$

and, by differentiation, the density function is

$$f_W(w) = 2(1 - w)$$

both functions being for w between zero and one and illustrated in Figure 13.16(b).

(b) The probability that the meeting will occur is the probability that the waiting time does not exceed 15 minutes:

$$P(W \leqslant \tfrac{1}{4}) = F_W(\tfrac{1}{4}) = \tfrac{7}{16}$$

13.4.5 Exercises

24 Find the distribution of the sum of the numbers when a pair of dice is tossed.

25 At the 18th hole of a golf course the probability that a golfer will score a par four is 0.55, the probability of one under is 0.17, of two under is 0.03, of one over is 0.2 and of two over is 0.05. Plot the (cumulative) distribution function.

26 A difficult assembly process must be undertaken, and the probability of success at each attempt is 0.2. The distribution of the number of independent attempts needed to achieve success is given by the product rule as

$$P(X = k) = (0.2)(0.8)^{k-1} \quad (k = 1, 2, 3, \dots)$$

Plot the distribution function and find the probabilities that the number of attempts will be

(a) less than four,

(b) between three and five.

27 Suppose that a coin is tossed three times and that the random variable W represents the number of heads minus the number of tails.

(a) List the elements of the sample space S for the three tosses of the coin, and to each sample point assign a value w of W.

(b) Find the probability distribution of W, assuming that the coin is fair.

(c) Find the probability distribution of W, assuming that the coin is biased so that a head is twice as likely to occur as a tail.

28 If the probability density function of a random variable X is given by

$$f_X(x) = \begin{cases} c/\sqrt{x} & (0 < x < 4) \\ 0 & (\text{elsewhere}) \end{cases}$$

where c is a constant, find

(a) the value of c,

(b) the distribution function,

(c) $P(X > 1)$.

29 The time interval (X) between successive earthquakes of a certain magnitude has an exponential distribution with density function given by

$$f_X(x) = \begin{cases} \frac{1}{90} e^{-x/90} & \text{if } x \geqslant 0 \\ 0 & \text{if } x < 0 \end{cases}$$

where x is measured in days. Find the probability that such an interval will not exceed 30 days.

30 The shelf life (in hours) of a certain perishable packaged food is a random variable with density function

$$f_X(x) = \begin{cases} 20\,000(x + 100)^{-3} & (x > 0) \\ 0 & (\text{otherwise}) \end{cases}$$

Find the probabilities that one of these packages will have a shelf life of

(a) at least 200 hours,

(b) at most 100 hours,

(c) between 80 and 120 hours.

31 The wave amplitude X on the sea surface often has the following (Rayleigh) distribution:

$$f_X(x) = \begin{cases} \dfrac{x}{a} \exp\left(\dfrac{-x^2}{2a} \right) & (x > 0) \\ 0 & (\text{otherwise}) \end{cases}$$

where a is a positive constant. Find the distribution function and hence the probability that a wave amplitude will exceed 5.5 m when $a = 6$.

13.4.6 Measures of location and dispersion

The observable properties of a random variable are determined by its distribution of probabilities (if discrete) or density function (if continuous), but this amount of information is difficult to extract from data. One common approach that is rather simpler is to assume that the random variable is one of a class whose distribution is specified by a formula and which often arises in practice, such as the binomial, Poisson or normal. These distributions will be covered in Section 13.5. Another common approach is to characterize the random variable in terms of two numbers: a measure of **location** ('typical' value) and a measure of **dispersion** ('spread' about that value). In practice, both approaches are used together, with the measures of location and dispersion often providing the parameters for the formula of the distribution.

Mean, median and mode

There are three common measures of location, the most important of which is the **mean**. For a random variable X this is usually given the symbol μ_X and is defined as

$$\mu_X = \begin{cases} \displaystyle\sum_{k=1}^{m} v_k P(X = v_k) & \text{if } X \text{ is discrete} \\[2em] \displaystyle\int_{-\infty}^{+\infty} x f_X(x)\,\mathrm{d}x & \text{if } X \text{ is continuous} \end{cases}$$

This represents a weighted sum of the possible values of X, with weights reflecting their relative likelihood of occurrence, and is effectively the 'centre of gravity' of the distribution.

Another measure of location that is often used is the **median**. For a continuous random variable X this is the point m_X for which

$$P(X \leqslant m_X) = F_X(m_X) = \tfrac{1}{2}$$

In other words, there are equal chances of X being greater than the median or less than the median. For a discrete random variable the median may not be unique, and is any point for which

$$P(X \leqslant m_X) \geqslant \tfrac{1}{2} \quad \text{and} \quad P(X \geqslant m_X) \geqslant \tfrac{1}{2}$$

The median of a distribution does not coincide with the mean unless the distribution has an axis of symmetry, in which case both measures lie on it.

The third measure of location is the **mode**, which is any point for which the probability function $P_X(\text{mode})$ (if discrete) or the density function $f_X(\text{mode})$ (if continuous) is an overall maximum. The mode can therefore be regarded as the most likely value of X to be observed. The mean, median and mode can all differ (see for example Figure 13.17), and can occur in any order.

Figure 13.17
Mode, median and mean for a particular distribution.

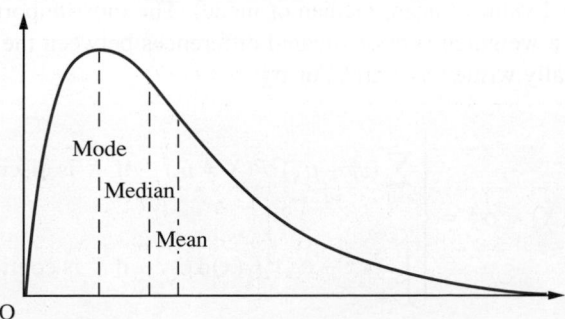

Example 13.16

Find the mean, median and mode for

(a) a simple die toss,

(b) the number of ship arrivals (Example 13.12),

(c) the lifetime distribution (Example 13.13).

Solution

(a) For the toss of a fair die

$$\mu_X = \sum_{k=1}^{6} \frac{1}{6}k = 3.5$$

The median is any point in the interval [3, 4], and each possible value is a mode.

(b) For the number of ship arrivals

$$\mu_X = (0)(0.1) + (1)(0.3) + (2)(0.35) + (3)(0.2) + (4)(0.05) = 1.8$$

The median is two because $P(X \leqslant 2) = 0.75$ and $P(X \geqslant 2) = 0.6$, both of which exceed one-half. The mode is also equal to two because this is the most likely value.

(c) For the lifetime distribution

$$\mu_X = \int_0^\infty \frac{1}{2}x e^{-x/2} \, dx$$

which we integrate by parts to obtain

$$\mu_X = [-x e^{-x/2}]_0^\infty + \int_0^\infty e^{-x/2} \, dx = [-2e^{-x/2}]_0^\infty = 2$$

The median is given by

$$F_X(m_X) = 1 - e^{-m_X/2} = \frac{1}{2}$$

from which $m_X = 1.386$. The mode, however, is zero because this is the peak of the density function.

Variance, standard deviation and quartiles

There are two approaches to measuring the variation of random variables around their central values (mean, median or mode). The most important such measure is the **variance**, a weighted sum of squared differences between the possible values and the mean, usually written as $\mathrm{Var}(X)$ or σ_X^2:

$$\mathrm{Var}(X) = \sigma_X^2 = \begin{cases} \displaystyle\sum_{k=1}^{m} (v_k - \mu_X)^2 P(X = v_k) & \text{if } X \text{ is discrete} \\[2em] \displaystyle\int_{-\infty}^{+\infty} (x - \mu_X)^2 f_X(x) \, dx & \text{if } X \text{ is continuous} \end{cases}$$

This is analogous to 'moment of inertia': it measures how tightly concentrated the possible values are about the mean (centre of gravity). One undesirable feature is that the squaring operation changes the units, so that a random variable measured, say, in volts will have a variance in volts-squared. The remedy is to use the **standard deviation** σ_X, which is defined as the square root of the variance.

The alternative approach to measuring dispersion is to exploit the distribution function $F_X(x)$. Suppose for simplicity that X is a continuous random variable. We have already defined the median by

$$F_X(m_X) = \tfrac{1}{2}$$

The points q_1 and q_3 where

$$F_X(q_1) = \tfrac{1}{4} \quad \text{and} \quad F_X(q_3) = \tfrac{3}{4}$$

are called **quartiles**, and the median can also be described as a quartile q_2. These quartiles divide the range of possible values of X into four successive intervals, for each of which the probability of X falling into the interval is one-quarter. In fact, a finer subdivision into 100 equally likely intervals is also used, the dividing points being called **percentiles**. The first quartile q_1 is then the 25th percentile, and so on. The 10th and 90th percentiles are also known as the first and ninth **deciles**, d_1 and d_9.

The most common measure of dispersion apart from variance (or standard deviation) is the **interquartile range** $q_3 - q_1$. Sometimes the **semi-interquartile range** or **quartile deviation** $\tfrac{1}{2}(q_3 - q_1)$ is quoted instead. The **interdecile range** $d_9 - d_1$ is also used.

Example 13.17　Find the variance and standard deviation for each of the random variables in Example 13.16, and the interquartile range for the lifetime distribution.

Solution　(a) For the toss of a fair die, using $\mu_X = 3.5$,

$$\sigma_X^2 = \left[\sum_{k=1}^{6} \tfrac{1}{6}(k - 3.5)^2 \right] = 2.917$$

from which $\sigma_X = 1.708$.

(b) For the number of ship arrivals, using $\mu_X = 1.8$,

$$\sigma_X^2 = (0 - 1.8)^2(0.1) + (1 - 1.8)^2(0.3) + (2 - 1.8)^2(0.35)$$
$$+ (3 - 1.8)^2(0.2) + (4 - 1.8)^2(0.05) = 1.060$$

from which $\sigma_X = 1.030$.

(c) For the lifetime distribution, using $\mu_X = 2$,

$$\sigma_X^2 = \int_0^\infty \tfrac{1}{2}(x - 2)^2 e^{-x/2} \, dx$$

which we integrate by parts to obtain

$$\sigma_X^2 = -[(x-2)^2 e^{-x/2}]_0^\infty + \int_0^\infty 2(x-2)e^{-x/2}\,dx$$

$$= 4 - [4(x-2)e^{-x/2}]_0^\infty + \int_0^\infty 4e^{-x/2}\,dx = 4$$

from which $\sigma_X = 2$. The quartiles q_1 and q_3 are the solutions of

$$1 - e^{-x/2} = \tfrac{1}{4} \quad \text{and} \quad 1 - e^{-x/2} = \tfrac{3}{4}$$

respectively, from which $q_1 = 0.575$ and $q_3 = 2.773$, and the interquartile range is therefore $2.773 - 0.575 = 2.198$.

13.4.7 Expected values

The mean and variance are special cases of expected values for a random variable. In general, the **expected value** of a function $h(X)$ of a random variable X is

$$E[h(X)] = \begin{cases} \displaystyle\sum_{k=1}^{m} h(v_k)P(X = v_k) & \text{if } X \text{ is discrete} \\[2ex] \displaystyle\int_{-\infty}^{+\infty} h(x)f_X(x)\,dx & \text{if } X \text{ is continuous} \end{cases}$$

As before, this is a weighted combination of the possible values. The mean and variance are retrieved by taking $h(X) = X$ and $h(X) = (X - \mu_X)^2$ respectively.

Expected values have many applications. One immediate application is in a useful alternative expression for the variance, obtained by expanding the square. If X is continuous then

$$\sigma_X^2 = \int_{-\infty}^{+\infty} (x - \mu_X)^2 f_X(x)\,dx$$

$$= \int_{-\infty}^{+\infty} x^2 f_X(x)\,dx - 2\mu_X \int_{-\infty}^{+\infty} x f_X(x)\,dx + \mu_X^2 \int_{-\infty}^{+\infty} f_X(x)\,dx$$

$$= E(X^2) - \mu_X^2$$

In other words, the variance is the expected value (or mean) of the square minus the square of the mean. The same is true when X is discrete, by a similar proof.

Example 13.18   Find the mean and standard deviation of the waiting time in Example 13.15.

Solution   The mean waiting time is

$$\mu_W = \int_0^1 2(w - w^2)\,dw = \tfrac{1}{3}$$

The mean square is

$$E(W^2) = \int_0^1 2(w^2 - w^3)\mathrm{d}w = \tfrac{1}{6}$$

so the standard deviation is

$$\sigma_W = \sqrt{[\tfrac{1}{6} - (\tfrac{1}{3})^2]} = 0.236$$

These translate to 20 minutes for the mean and about 14 minutes for the standard deviation.

13.4.8 Independence of random variables

It is possible for two different random variables to be measured for the same object: for example, a person's height and weight. Individually, these random variables have distributions, mean values and variances, which apply to a particular population, but this is not the whole story. It is clear that taller people tend to be heavier than shorter people (although obviously there are exceptions). In this case we say that these variables are **dependent** upon each other (to some degree). The notion of dependence is basically the same as that applying to events, discussed in Section 13.3.4. Furthermore, just as events can be independent (Section 13.3.5), so can random variables. For example, it is plausible that a person's birthday and telephone number are not related in any way, and are therefore independent random variables. Nothing is likely to be learnt about the one from an observation of the other.

For independent events we have the rule that the joint probability is the product of the separate probabilities:

$$P(A \cap B) = P(A)P(B)$$

For independent discrete random variables a similar rule applies:

$$P(X = u_i \cap Y = v_j) = P(X = u_i)P(Y = v_j)$$

where $u_1, u_2, \ldots, u_k$ are the possible values of X and $v_1, v_2, \ldots, v_m$ are the possible values of Y. This effectively specifies a **joint distribution** for the two random variables. If we sum over the possible values of Y then we find

$$\sum_{j=1}^m P(X = u_i \cap Y = v_j) = \sum_{j=1}^m P(X = u_i)P(Y = v_j)$$

$$= P(X = u_i) \sum_{j=1}^m P(Y = v_j)$$

$$= P(X = u_i)$$

Thus the individual probability of one value u_i of X can be obtained from the joint distribution by summing over all values v_j of Y, with X fixed at u_i. In fact, this is true even when the random variables are dependent.

Example 13.19

A new plant at a manufacturing site has to be first installed and then commissioned. The times required for these two stages depend upon different random factors, and can therefore be regarded as independent. Based on past experience, the respective distributions for X (installation time) and Y (commissioning time), both in days, are as follows:

u_i	3	4	5	6
$P(X = u_i)$	0.1	0.4	0.3	0.2

v_j	2	3	4
$P(Y = v_j)$	0.50	0.35	0.15

Find the joint distribution for X and Y, and the probability that the total time will not exceed seven days.

Solution

Because the random variables are independent, the joint distribution is given by the product of the separate distributions:

$$P(X = u_i \cap Y = v_j) = P(X = u_i)P(Y = v_j)$$

with the following result:

Joint probability		u_i			
		3	4	5	6	Total
	2	0.050	0.200	0.150	0.100	0.50
v_j	3	0.035	0.140	0.105	0.070	0.35
	4	0.015	0.060	0.045	0.030	0.15
Total		0.10	0.40	0.30	0.20	1.00

Note that the row and column totals give the individual distributions for X and Y. The probability that the total time will not exceed seven days is given by the sum of those joint probabilities above the stepped broken line:

$$P(X + Y \leqslant 7) = P(X = 3 \cap Y = 2) + P(X = 3 \cap Y = 3) + P(X = 3 \cap Y = 4)$$
$$+ P(X = 4 \cap Y = 2) + P(X = 4 \cap Y = 3) + P(X = 5 \cap Y = 2)$$
$$= 0.050 + 0.035 + 0.015 + 0.200 + 0.140 + 0.150 = 0.59$$

13.4.9 Scaling and adding random variables

Example 13.19 has introduced the idea of a sum of random variables, itself a random quantity. The distribution of this quantity can be deduced from the joint distribution; thus in Example 13.19 the probability that the total time (installation plus commissioning) will take exactly seven days is

$$P(X + Y = 7) = P(X = 3 \cap Y = 4) + P(X = 4 \cap Y = 3) + P(X = 5 \cap Y = 2)$$
$$= 0.015 + 0.140 + 0.150 = 0.305$$

A similar calculation can be done for every possible value from the minimum (five days) to the maximum (ten days), and the distribution is then complete.

Example 13.20   Find the distribution of total time for the situation described in Example 13.19, and the expected value of this time.

Solution   Proceeding as described above, we obtain the following distribution:

w_k	5	6	7	8	9	10
$P(X + Y = w_k)$	0.050	0.235	0.305	0.265	0.115	0.030

The expected value is then

$$E(X + Y) = \sum_{w_k} w_k P(X + Y = w_k) = 7.25$$

There is, however, an easier way to arrive at the mean of a sum of random variables. The separate means (or expected values) of X and Y are easily found from the values given in Example 13.19:

$$E(X) = \sum_{u_i} u_i P(X = u_i) = 4.60$$

$$E(Y) = \sum_{v_j} v_j P(Y = v_j) = 2.65$$

It has turned out that

$$E(X + Y) = E(X) + E(Y)$$

The mean of the sum of random variables is the sum of the means. That this is a general result is shown as follows:

$$E(X + Y) = \sum_{i=1}^{k} \sum_{j=1}^{m} (u_i + v_j) P(X = u_i \cap Y = v_j)$$

$$= \sum_{i=1}^{k} u_i \left[\sum_{j=1}^{m} P(X = u_i \cap Y = v_j) \right] + \sum_{j=1}^{m} v_j \left[\sum_{i=1}^{k} P(X = u_i \cap Y = v_j) \right]$$

$$= \sum_{i=1}^{k} u_i P(X = u_i) + \sum_{j=1}^{m} v_j P(Y = v_j)$$

$$= E(X) + E(Y)$$

The double summation is over all the possible values of $X + Y$ (which are $u_i + v_j$) times the probability of each combination, and the result in Section 13.4.8 that summing a joint probability over the values of one variable gives the probability of the other variable has been used. Furthermore, because this also holds for dependent variables (as can be seen in *Advanced Modern Engineering Mathematics*), the mean of a sum of random variables is always equal to the sum of the means, whether they are dependent or not.

For the variance of a sum it is not quite so simple. If the mean of X is μ_X and the mean of Y is μ_Y then

$$\mathrm{Var}(X + Y) = E\{[(X + Y) - (\mu_X + \mu_Y)]^2\} = E\{[(X - \mu_X) + (Y - \mu_Y)]^2\}$$

$$= E\{(X - \mu_X)^2 + (Y - \mu_Y)^2 + 2(X - \mu_X)(Y - \mu_Y)\}$$

$$= E\{(X - \mu_X)^2\} + E\{(Y - \mu_Y)^2\} + E\{2(X - \mu_X)(Y - \mu_Y)\}$$

The first two terms on the right-hand side are $\mathrm{Var}(X)$ and $\mathrm{Var}(Y)$ respectively. The third term (which is actually called the **covariance**) is a measure of dependence, and it is shown in *Advanced Modern Engineering Mathematics* that this is always zero for independent variables. Hence if X and Y are independent, the variance of a sum is equal to the sum of the variances:

$$\mathrm{Var}(X + Y) = \mathrm{Var}(X) + \mathrm{Var}(Y)$$

These results for the mean and variance of sums of random variables extend naturally to any number of variables, and apply whether the variables are discrete or continuous.

If we add a constant (c, say) to a random variable X, it follows immediately from the definitions in Section 13.4.6 that the same constant is added to the mean, but the variance does not change. If X is a continuous random variable, with density function $f_X(x)$, say, then

$$E(X + c) = \int_{-\infty}^{+\infty} (x + c)f_X(x)\mathrm{d}x = \int_{-\infty}^{+\infty} xf_X(x)\mathrm{d}x + c\int_{-\infty}^{+\infty} f_X(x)\mathrm{d}x$$

$$= \mu_X + c$$

$$\mathrm{Var}(X + c) = \int_{-\infty}^{+\infty} [(x + c) - (\mu_X + c)]^2 f_X(x)\mathrm{d}x$$

$$= \int_{-\infty}^{+\infty} (x - \mu_X)^2 f_X(x)\mathrm{d}x = \sigma_X^2$$

If we multiply a random variable X by a constant c, the mean is multiplied by c and the variance by c^2:

$$E(cX) = \int_{-\infty}^{+\infty} cxf_X(x)\mathrm{d}x = c\mu_X$$

$$\mathrm{Var}(cX) = \int_{-\infty}^{+\infty} (cx - c\mu_X)^2 f_X(x)\mathrm{d}x = c^2\int_{-\infty}^{+\infty} (x - \mu_X)^2 f_X(x)\mathrm{d}x = c^2\sigma_X^2$$

All of these results hold whether X is continuous or discrete.

Example 13.21   If a mean temperature is 58°F, what is the mean temperature in degrees Celsius?

Solution   If T_F and T_C denote temperatures in Fahrenheit and Celsius respectively then

$$T_C = \tfrac{5}{9}(T_F - 32)$$

so

$$E(T_C) = \tfrac{5}{9}[E(T_F) - 32] = \tfrac{5}{9}(58 - 32) = 13.4°C$$

13.4.10   Measures from sample data

We can now return to the consideration of data, which is the object of the whole exercise. Given that the exact distribution of a random quantity under investigation is usually not known in an experimental context but that the mean and variance at least would be useful characteristics of it, it is reasonable to try to estimate these from the data. Experience shows that quite good estimates of mean and variance can be obtained even from rather small samples, whereas a much larger sample is needed before the histogram gives a good approximation to the whole shape of the true distribution.

Sample average and variance

For a sample $\{X_1, \ldots, X_n\}$ of data, the **sample average** and **sample variance** are defined as

$$\bar{X} = \frac{1}{n}\sum_{i=1}^{n} X_i \quad \text{and} \quad S_X^2 = \frac{1}{n}\sum_{i=1}^{n}(X_i - \bar{X})^2$$

respectively. The **sample standard deviation** is the square root of the sample variance.

The average of the sample, and the average squared deviation from the sample average, are easy to work out from the data, and characterize the data in location and dispersion. It turns out that these approximate the true figures of mean and variance, and the approximations improve as $n \to \infty$ in a sense to be made precise below.

By expanding the square in the formula for the sample variance (as in Section 13.4.7 for the true variance) it is easy to show that an alternative expression (which is useful for hand calculation) is

$$S_X^2 = \overline{X^2} - (\bar{X})^2$$

that is, the average of the square minus the square of the average. When small samples are used in statistics (this is considered in *Advanced Modern Engineering Mathematics*), a different definition of sample variance must be adopted:

$$S_{X,n-1}^2 = \frac{1}{n-1}\sum_{i=1}^{n}(X_i - \bar{X})^2$$

The difference between the two definitions is relatively small. Many scientific calculators provide functions to work out sample average and both forms of sample variance or standard deviation.

Example 13.22   A die was tossed 24 times, producing the following results:

$$4, 6, 2, 4, 2, 1, 5, 1, 3, 1, 3, 4, 5, 4, 3, 1, 6, 5, 6, 3, 1, 2, 4, 6$$

Find the sample average and standard deviation.

Solution   The average score over the 24 tosses is

$$\overline{X} = 3.42$$

The average of the squares is 13.667, so the standard deviation is

$$S_X = \sqrt{[13.667 - (3.42)^2]} = 1.73$$

These figures are close to the theoretical values worked out in Examples 13.16 and 13.17.

An issue first raised in Section 1.5 is important here: to how many places of decimals should these results be quoted? The actual average of the data in this example is 3.4166 ..., but the results should be stated with no more significant digits than can be justified statistically. The average might be quoted as 3, 3.4, 3.42, 3.417 and so on, but the appropriate precision depends upon the sample size n.

The sample average itself is a random variable; it has a mean and variance, and it follows from the results in Section 13.4.9 that

$$\mathrm{Var}(\overline{X}) = \mathrm{Var}\left(\frac{X_1 + \ldots + X_n}{n}\right) = \frac{1}{n^2}\,\mathrm{Var}(X_1 + \ldots + X_n) = \frac{n\sigma_X^2}{n^2} = \frac{\sigma_X^2}{n}$$

since the random variables, X_i, can be reasonably assumed independent here.

The larger the sample size, the smaller the variance of $\overline{X}$ and the greater the precision, but to quantify the precision we also need a value for σ_X. Usually all we have is the estimate S_X, but this is also a random variable and subject to error. It can be shown that in many situations a (rather rough) indication of the accuracy of S_X as an estimate of σ_X is that its relative error (see Section 1.5.3) varies inversely with $\sqrt{(2n)}$:

$$\frac{|S_X - \sigma_X|}{\sigma_X} \approx \frac{1}{\sqrt{(2n)}}$$

Returning to Example 13.22, with $n = 24$, the percentage error in S_X is estimated at 14%, so the error in S_X is likely to be of order 0.2, and the second decimal place has no meaning. The error in $S_X/\sqrt{n}$ is correspondingly of order 0.05 in its value of 0.35. The results of Example 13.22 can therefore be stated more properly as

$$\overline{X} = 3.4 \quad \text{(with likely error of order 0.4)}$$

$$S_X = 1.7 \quad \text{(with likely error of order 0.2)}$$

In practice, these high standards of honesty are not always maintained, and it is very important not to be misled by the spurious precision with which results are often quoted.

Example 13.23

Measured values of resistance (in Ω) for 12 nominally $100\,\Omega$ resistors were as follows:

$$106, 98, 95, 109, 99, 102, 101, 108, 94, 99, 96, 102$$

Find the sample average and both forms of sample variance and standard deviation.

Solution

The average of the 12 figures is

$$\bar{X} = 100.75$$

which is slightly high but close to the nominal figure, and much closer to that figure than a 'typical' value from the data. The two results for sample variance and standard deviation are

$$S_X^2 = 22.9, \quad S_X = 4.8$$

and

$$S_{X,n-1}^2 = 25, \quad S_{X,n-1} = 5$$

Despite the small sample size, the difference between the two versions of sample standard deviation is not very large. Furthermore, following the above discussion, an error of order 1.0 is likely in the standard deviation ($\sqrt{\frac{1}{24}}$ is about 20%), so there is no point in distinguishing them, even to the first decimal place. The value of $S_X/\sqrt{n}$ is 1.4, with a likely error of order 0.3, so the average should properly be stated as

$$\bar{X} = 101 \quad \text{(with likely error of order 1.5)}$$

It is worth noting that in Example 13.23 the distribution of the random variable (resistance) is not known, but the sample provides useful information about the mean value and the variability about that value.

As mentioned above, many scientific calculators will work these results out automatically. There are also many statistical packages that run on computers of all sizes, and they will do the same. Alternatively, Figure 13.18 contains a pseudocode listing of an efficient program to compute the sample average and standard deviation of a set of data $X_1, \ldots, X_n$. The algorithm used works as follows.

Let M_k and Q_k respectively represent the average of the first k observations and the sum of squares of deviations of the first k observations about their average:

$$M_k = \frac{1}{k}\sum_{i=1}^{k} X_i \quad \text{and} \quad Q_k = \sum_{i=1}^{k} (X_i - M_k)^2$$

The program exploits the following recursion relations, which are proved in D. Cooke, A. H. Craven and G. M. Clarke, *Statistical Computing in Pascal* (Edward Arnold, London, 1985), pp. 54–5 (© 1985 Edward Arnold Ltd. Reproduced by permission of Edward Arnold (Publishers) Ltd).

Figure 13.18
Pseudocode listing for
sample average and
variance.

```
{Program to compute the sample average and standard deviation,
x(k) is the array of data,
n is the sample size,
xbar is the sample average,
sx and sxn_1 are the two versions of standard deviation,
Mk and Qk hold running totals,
notation as in Section 13.4.10.}


Mk ← 0
Qk ← 0
for k is 1 to n do
    diff ← x(k) − Mk
    Mk ← ((k − 1)*Mk + x(k))/k
    Qk ← Qk + (1 − 1/k) *diff*diff
endfor
xbar ← Mk
sx ← square_root(Qk/n)
sxn_1 ← square_root(Qk/(n − 1))
```

$$M_k = \frac{1}{k}[(k-1)M_{k-1} + X_k]$$

and

$$Q_k = Q_{k-1} + \left(1 - \frac{1}{k}\right)(X_k - M_{k-1})^2$$

Finally,

$$\overline{X} = M_n, \quad S_X^2 = \frac{Q_n}{n} \quad \text{and} \quad S_{X,n-1}^2 = \frac{Q_n}{n-1}$$

The use of this recurrence method avoids having to make two passes through the data
(as required for the original definition of the sample variance), and also avoids the loss
of precision involved in subtracting two quantities that often turn out in practice to be
large in magnitude and similar in value (as required by the alternative expression).

Sample median and range

The sample average and standard deviation are not the only measures of location and
dispersion derived from data. Suppose that the data $\{X_1, \ldots, X_n\}$ are ordered so that

$$X_{(1)} \leq X_{(2)} \leq \ldots \leq X_{(n)}$$

Then a **sample median** that provides an estimate of the true median (Section 13.4.6)
can be defined as

$$\text{sample median} = \begin{cases} X_{(k)} & (\text{odd } n = 2k - 1) \\ \frac{1}{2}[X_{(k)} + X_{(k+1)}] & (\text{even } n = 2k) \end{cases}$$

A common measure of dispersion (especially for small samples) is the **sample range** $X_{(n)} - X_{(1)}$, the difference between the largest and smallest elements of the data set, often used in quality control. The ideas of quartiles and percentiles (Section 13.4.6) can also be applied to data, based on the cumulative percentages (Section 13.2.3).

Example 13.24   Find the sample median and range for the data in Examples 13.22 and 13.23.

Solution   For the die toss the sorted data is

$$1, 1, 1, 1, 1, 2, 2, 2, 3, 3, 3, 3, 4, 4, 4, 4, 4, 5, 5, 5, 6, 6, 6, 6$$

so the sample median is $\frac{1}{2}(3 + 4) = 3.5$, and the sample range is 5.
    For the resistors the sorted data is

$$94, 95, 96, 97, 98, 99, 101, 101, 102, 106, 108, 109$$

so the sample median is $\frac{1}{2}(99 + 101) = 100$, and the sample range is 15.

13.4.11 Exercises

32 Suppose that the probability distribution for the number of days required to ship a package from London to New York is as follows:

Number of days	2	3	4	5	6	7
Probability	0.05	0.20	0.35	0.25	0.1	0.05

Find the mean of this distribution, and the probability that a particular package arrives in less than five days.

33 The distribution of the daily number of malfunctions of a certain computer is given by the following table:

Number of malfunctions	0	1	2	3	4	5	6
Probability	0.17	0.29	0.27	0.16	0.07	0.03	0.01

Find the mean, the median and the standard deviation of this distribution.

34 Find the average sentence length for the sentences with lengths given in Question 1 in Exercises 13.2.4.

35 The distribution of the number X of independent attempts needed to achieve the first success when the probability of success is 0.2 at each attempt is given by

$$P(X = k) = (0.2)(0.8)^{k-1} \quad (k = 1, 2, 3, \dots)$$

(see Question 26 in Exercises 13.4.5). Find the mean, the median and the standard deviation for this distribution.

36 You arrive at a railway station knowing only that trains leave for your destination at intervals of one hour. Find the mean and standard deviation of your waiting time.

37 A random variable X has the linear distribution given by

$$f_X(x) = \begin{cases} a - bx & (0 \leqslant x \leqslant \frac{a}{b}) \\ 0 & (\text{otherwise}) \end{cases}$$

where a and b are constants. Show that

(a) $a = \sqrt{2b}$ (b) the median of X is $(\sqrt{2} - 1)/\sqrt{b}$

38 Suppose that the running distance (in thousands of kilometres) that car owners get from a tyre is a random variable with density function

$$f_X(x) = \begin{cases} \frac{1}{30}e^{-x/30} & (x > 0) \\ 0 & (x \leqslant 0) \end{cases}$$

Find

(a) the probability that one of these tyres will last at most $19\,000\,\text{km}$,

(b) the mean and standard deviation of X, and

(c) the median and interquartile range of X.

39 If the probability density of the random variable X is

$$f_X(x) = \begin{cases} 30x^2(1-x)^2 & (0 < x < 1) \\ 0 & (\text{otherwise}) \end{cases}$$

find the probability that X will take a value within two standard deviations of its mean.

40 The distribution of downtime T for breakdowns of a computer system is given by

$$f_T(t) = \begin{cases} a^2 t e^{-at} & (t > 0) \\ 0 & (\text{otherwise}) \end{cases}$$

where a is a positive constant. The cost of downtime derived from the disruption resulting from breakdowns rises exponentially with T:

cost factor $= h(T) = e^{bT}$

Show that the expected cost factor for downtime is $[a/(a-b)]^2$, provided that $a > b$.

41 The mean times for completion of tasks A and B are four and six hours respectively. A particular project involves three tasks of type A and two of type B, all to be performed in succession. What is the expected time for completion of the project? Also, if the standard deviations for A and B are one and two hours respectively, and if all project times are independent, what is the standard deviation of the completion time?

42 An inspection of 12 specimens of material from inside a reactor vessel revealed the following percentages of impurities:

2.3, 1.9, 2.1, 2.8, 2.3, 3.6, 1.4, 1.8, 2.1, 3.2, 2.0, 1.9

Find (a) the sample average and both versions of the sample standard deviation, (b) the sample median and range.

43 Find the sample average, standard deviation, median and range for the following sample of component lifetimes (in thousands of hours):

5.6, 4.1, 6.0, 5.8, 5.2, 4.3, 6.4, 5.5, 6.0, 5.1, 4.9, 4.2, 4.8, 6.8, 5.6, 5.2, 7.3, 5.4, 4.7, 5.9, 5.0, 6.3, 4.4, 6.0

44 Find the sample averages and standard deviations for the engine performance data in Figure 13.1.

45 In a problem similar to that in Question 35 the probability of success at the first attempt is 0.2 but the probability of failure at each subsequent attempt (if needed) is half of that for the previous attempt. Find the mean number of attempts needed to achieve the first success.

46 Find the median and the mode for the Rayleigh distribution

$$f_X(x) = \begin{cases} \dfrac{x}{a} \exp\left(-\dfrac{x^2}{2a}\right) & (x > 0) \\ 0 & (\text{otherwise}) \end{cases}$$

(see Question 31 in Exercises 13.4.5). Also show that the mean is given by

$$\mu_X = \int_0^\infty \exp\left(-\frac{x^2}{2a}\right) dx$$

which can be shown to be $\sqrt{(\tfrac{1}{2}\pi a)}$. Compare these quantities when $a = 6$, and find the interquartile range.

47 Two people are separately attempting to succeed at a particular task, and each will continue attempting until success is achieved. The probability of success of each attempt for person A is p, and that for person B is q, all attempts being independent. What is the probability that person B will achieve success with no more attempts than person A does?

$$\left(Hint: \sum_{i=0}^{n-1} x^i = \frac{1 - x^n}{1 - x} \right)$$

48 Sample values that are several standard deviations away from the sample average are called **outliers**. They are often just measurement or transcription errors, but they can bias a statistical calculation. Which of the following data are more than three sample standard deviations away from the average?

19.4, 18.1, 25.6, 18.2, 20.6, 25.0, 21.8, 15.5, 26.3, 15.8, 18.7, 19.3, 22.3, 20.9, 24.2, 21.4, 23.2, 21.4, 47.1, 23.6, 46.3, 21.2, 27.5, 20.8, 24.7, 25.9, 25.8, 33.4, 30.9, 24.5

13.5 Important practical distributions

A lot of information is required to specify the exact distribution of a random variable, and even more to specify the joint distribution of two or more variables. The mean, variance and covariance are useful measures of the most important properties of random variables, namely location, dispersion and dependence, which can realistically be estimated from data. These measures are of great value in statistics, as can be seen in the companion text *Advanced Modern Engineering Mathematics*. Another short cut is provided by the various classes of distributions that are often used in statistical practice. The user has to supply the values of certain essential parameters, perhaps using estimates of mean and variance to do so, and then the probability distribution is determined by a formula. Experience shows that these classes of distributions (which are idealized in mathematical form) do approximate very well to the actual distributions in many practical situations.

The most important of these classes of distributions are the binomial, Poisson and normal. In this section we cover these, with a particular view towards the statistical applications to follow.

13.5.1 The binomial distribution

Consider first a simple coin-tossing experiment, or any other random situation where only two outcomes are possible. We shall refer to these outcomes as 'success' and 'failure', but any other pair of terms (appropriate to the context) will do. Imagine tossing the coin (or performing the general experiment) n times and counting the number of successes. Clearly the sample space for this random variable Y, say, is $S = \{0, 1, \ldots, n\}$ with values near the middle of the range being more probable than values near the ends. It is this distribution that is sought.

> A **Bernoulli trial** is a single observation of a random variable X, say, that can take the values zero or one:
>
> $$P(X = 1) = p \quad \text{and} \quad P(X = 0) = 1 - p$$
>
> for some success probability p.

The mean and variance of X are easily derived:

$$E(X) = 1(p) + 0(1 - p) = p$$

and

$$E(X^2) = 1^2(p) + 0^2(1 - p) = p$$

Hence

$$\sigma_X^2 = p - p^2 = p(1 - p)$$

Now let $\{X_1, \ldots, X_n\}$ denote n independent Bernoulli trials, each with success probability p. The number of successes is

$$Y = X_1 + \ldots + X_n$$

Suppose in general that $Y = k$, where $0 \leqslant k \leqslant n$. Then k of the X_i values are equal to one and $n - k$ are equal to zero. The probability of this occurring is

$$p^k(1 - p)^{n-k}$$

by the product rule (because the separate outcomes are independent).

There are many ways in which the k successes can be distributed among the n trials. For instance, if $n = 5$ and $k = 3$, the result might be $\{1, 1, 0, 1, 0\}$ or $\{0, 1, 1, 1, 0\}$ or $\{1, 0, 0, 1, 1\}$, and so on. As far as we are concerned, these are all equivalent, since we are interested only in the total number of successes and not their particular arrangement among the trials. The number of possible arrangements of the k successes among the n trials is given by the binomial coefficient (see Section 7.7.2)

$$\binom{n}{k} = \frac{n!}{(n - k)!\,k!}$$

Each arrangement of successes is exclusive of every other, so the addition rule of probabilities gives us the distribution

$$P(Y = k) = \binom{n}{k} p^k(1 - p)^{n-k} \quad (k = 0, \ldots, n)$$

This is the general form of the **binomial distribution**, with parameters n and p. The mean and variance of the binomial distribution are

$$E(Y) = np \quad \text{and} \quad \text{Var}(Y) = np(1 - p)$$

(this follows from the mean and variance of the Bernoulli random variable and the results on the mean and variance of sums of random variables in Section 13.4.9). Two typical binomial distributions can be seen in Figure 13.19.

Figure 13.19
Binomial distributions:
(a) $n = 12$, $p = 0.2$;
(b) $n = 12$, $p = 0.5$.

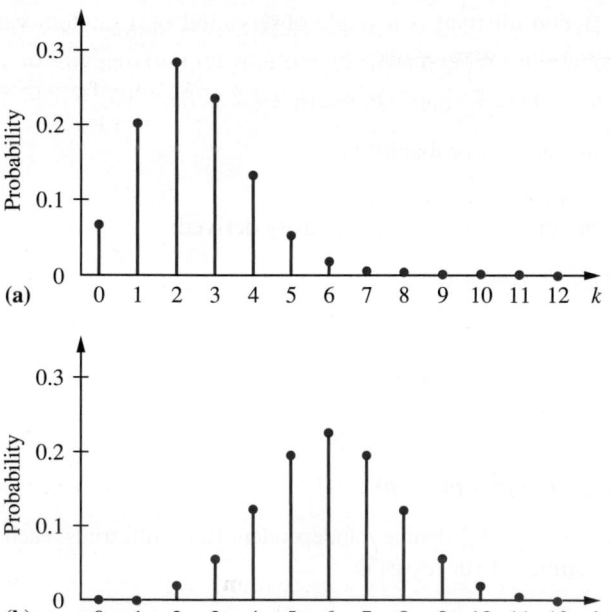

Example 13.25   A component supplier claims that 95% of its catalogue items are in stock at any time. A particular order for 20 different components is returned with three items missing as being out of stock. Is this likely, given the supplier's claim?

Solution   Each item can be either in stock or out of stock at any time, and the probability of each item being out of stock is 5%. The binomial distribution therefore applies and

$$P(k \text{ out of stock}) = \binom{20}{k} 0.05^k 0.95^{20-k}$$

so

$$P(3 \text{ or more out of stock}) = P(3) + P(4) + \ldots + P(20)$$
$$= 1 - P(0) - P(1) - P(2)$$
$$= 1 - 0.3585 - 0.3774 - 0.1887$$
$$= 0.0755$$

This is unlikely, given the supplier's claim.

There are several points to note about this simple example. The assumed figure of 5% probability of being out of stock is prompted by the supplier's claim, but in reality this will be an average figure, both between components (some may be out of stock more often than others because of supply difficulties) and over time (for the same reason). The independence assumption may not be true – if for instance a consignment of several similar types of components is awaited from a manufacturer and several of these are included in the order.

Most importantly, the probability worked out in the solution is that of three *or more* being out of stock, a result *at least as extreme* as that observed. Any result may have a low probability. What matters here is how far into the 'tail' of the distribution the actual result lies, and this is assessed by the total probability from there to the maximum value of k, which is 20. Note the use of the complement rule to simplify the calculation.

13.5.2   The Poisson distribution

The binomial distribution becomes unwieldy for large values of its parameter n, as illustrated in Examples 13.26 and 13.27. Another discrete distribution that often serves as a useful approximation to the binomial is the following:

$$P(X = k) = \frac{\lambda^k e^{-\lambda}}{k!} \quad (k = 0, 1, 2, \ldots)$$

This is the general form of the **Poisson distribution**, with parameter λ. It is shown in *Advanced Modern Engineering Mathematics* that the mean and variance of the Poisson distribution are both equal to λ. These can be derived directly from the definition, but are more easily obtained by using the **moment generating function**, which will also be considered there. Also using this technique, it can be shown that for large n and small

Figure 13.20
Binomial and Poisson
distributions.

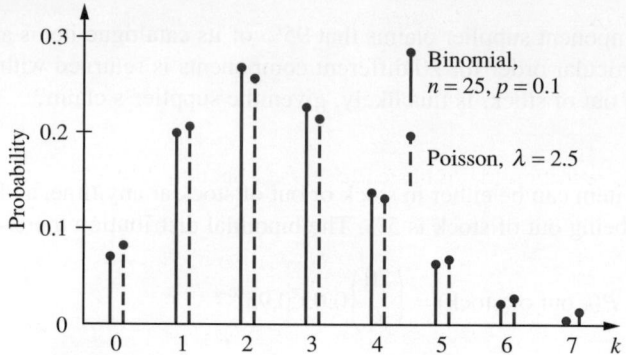

p the Poisson distribution approximates the binomial with $\lambda = np$. As a guide, the Poisson approximation can be used if $n \geqslant 25$ and $p \leqslant 0.1$. This is illustrated numerically in Figure 13.20, where binomial and Poisson distributions are compared for $n = 25$, $p = 0.1$ and $\lambda = 2.5$.

Example 13.26

If 0.04% of cars break down while driving through a certain tunnel, find the probability that at most two break down out of 2000 cars entering the tunnel on a given day.

Solution The true distribution of breakdowns is binomial:

$$P(k \text{ breakdowns}) = \binom{2000}{k}(0.0004)^k(0.9996)^{2000-k}$$

for which $P(0) = 0.449\,26$, $P(1) = 0.359\,55$ and $P(2) = 0.143\,81$, and

$$P(\text{at most two breakdowns}) = P(0) + P(1) + P(2) = 0.952\,61$$

Because n is large and p small, the Poisson aproximation can also be used, with $\lambda = np = 0.8$, so that

$$P(\text{at most two breakdowns}) \approx e^{-\lambda}(1 + \lambda + \tfrac{1}{2}\lambda^2) = 0.952\,58$$

The Poisson calculation is easier, and the agreement is very good. It would not normally be appropriate to quote such an answer to five significant digits but it is only with such precision that the difference between the two distributions shows up.

 Despite its ease of use compared with the binomial distribution some calculations with the Poisson distribution are difficult, especially those involving long summations. The following recurrence formulae are useful both for hand calculation and in computer programs. If X has a Poisson distribution with parameter λ, the successive Poisson probabilities are given by

$$P(X = k) = \frac{\lambda P(X = k - 1)}{k} \qquad (k = 1, 2, \dots)$$

with $P(X = 0) = e^{-\lambda}$. Furthermore, a cumulative property such as

$$P(X \leqslant 3) = e^{-\lambda}\left(1 + \lambda + \frac{\lambda^2}{2!} + \frac{\lambda^3}{3!}\right)$$

can be rewritten in the nested form (see Section 2.4)

$$P(X \leqslant 3) = e^{-\lambda}\left(\left(\left(\frac{\lambda}{3} + 1\right)\frac{\lambda}{2} + 1\right)\lambda + 1\right)$$

This approach can be generalized as follows: let

$$G_n = \frac{\lambda}{n} + 1 \quad \text{and} \quad G_{k-1} = \frac{\lambda}{k-1}G_k + 1 \quad (k = n, n-1, \ldots, 2)$$

then

$$P(X \leqslant n) = e^{-\lambda}G_1$$

Example 13.27

A machine produces components that have defect A with probability 0.015 and defect B with probability 0.020, the two defects being independent. If 54 components are packed into a batch, what is the (approximate) probability that the batch contains at least 50 components without defects?

Solution

By the complement and product rules, the probability that a component will have neither defect is

$$P(\bar{A} \cap \bar{B}) = [1 - P(A)][1 - P(B)] = 0.9653$$

so the probability that a component will have at least one defect is 0.0347. If a batch contains at least 50 good components then it contains at most four defective ones, and, from the binomial distribution,

$$P(\text{at most four defective}) = \sum_{k=0}^{4}\binom{54}{k}(0.0347)^k(0.9653)^{54-k}$$

This is rather unwieldy, so we use the Poisson approximation with $\lambda = (54)(0.0347) = 1.874$. The successive values $G_4, \ldots, G_1$ in the recurrence formula above are 1.469, 1.917, 2.797 and 6.241, and hence

$$P(\text{at most four defective}) \approx 6.241e^{-1.874} = 0.958$$

In other words, about one batch in 24 will contain less than 50 good components.

The binomial and Poisson are discrete distributions, which have the widest application among all discrete random variables. The Poisson distribution is especially useful to engineers because of its importance in statistical quality control. This will be introduced in Section 13.6, but we now turn to the most important of the continuous distributions.

13.5.3 The normal distribution

One class of distributions is awarded the name 'normal' because of the regularity with which random continuous data is found to obey it. This is no coincidence. The central limit theorem (Section 13.5.4) provides an explanation in terms of cumulative independent random parts adding up to a normal whole, a situation that is of great value in statistical inference (considered in *Advanced Modern Engineering Mathematics*). The normal distribution also serves as an approximation to the binomial distribution that complements the Poisson approximation.

The normal distribution has two parameters, which can be shown (see Question 60 in Exercises 13.5.7) to be the mean and standard deviation, so the appropriate symbols μ_X and σ_X are used.

> A continuous random variable X has a **normal distribution** with mean μ_X and variance σ_X^2 if
>
> $$f_X(x) = \frac{1}{\sigma_X \sqrt{(2\pi)}} \exp\left[-\frac{1}{2}\left(\frac{x - \mu_X}{\sigma_X}\right)^2\right] \quad (-\infty < x < +\infty, \; \sigma_X > 0)$$

The density function is symmetrical about μ_X and has the bell-shaped form shown in Figure 13.21. This distribution is also sometimes referred to by its more traditional name: the **Gaussian distribution**.

The need to declare that a random variable has a normal distribution (with a specified mean and variance) is so common that a special notation exists for the purpose:

$$X \sim N(\mu_X, \sigma_X^2)$$

Calculations involving the normal distribution are complicated by the fact that there is no simple expression for the integral of the density function on an arbitrary interval; in other words, the distribution function $F_X(x)$ does not have a simple explicit form. Instead, tables of this function are used. In fact, only a single table is needed: that for the special case of a normal distribution with a mean of zero and a variance of one.

> The **standard normal** cumulative distribution function is
>
> $$\Phi(z) = \frac{1}{\sqrt{(2\pi)}} \int_{-\infty}^{z} e^{-x^2/2} \, dx$$

Figure 13.21
The normal density and distribution functions (for $\mu_X = 0$ and $\sigma_X = 1$).

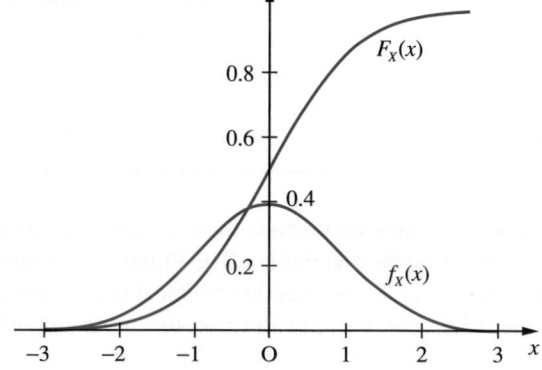

This function is usually tabulated only for $z \geq 0$; for $z < 0$ the symmetry implies that

$$\Phi(-z) = 1 - \Phi(z)$$

A typical table of the standard normal function $\Phi(z)$ is provided in Figure 13.22.

Figure 13.22
Table of the standard normal cumulative distribution function $\Phi(z)$.

z	.00	.01	.02	.03	.04	.05	.06	.07	.08	.09
.0	.5000	.5040	.5080	.5120	.5160	.5199	.5239	.5279	.5319	.5359
.1	.5398	.5438	.5478	.5517	.5557	.5596	.5636	.5675	.5714	.5753
.2	.5793	.5832	.5871	.5910	.5948	.5987	.6026	.6064	.6103	.6141
.3	.6179	.6217	.6255	.6293	.6331	.6368	.6406	.6443	.6480	.6517
.4	.6554	.6591	.6628	.6664	.6700	.6736	.6772	.6808	.6844	.6879
.5	.6915	.6950	.6985	.7019	.7054	.7088	.7123	.7157	.7190	.7224
.6	.7257	.7291	.7324	.7357	.7389	.7422	.7454	.7486	.7517	.7549
.7	.7580	.7611	.7642	.7673	.7704	.7734	.7764	.7794	.7823	.7852
.8	.7881	.7910	.7939	.7967	.7995	.8023	.8051	.8078	.8106	.8133
.9	.8159	.8186	.8212	.8238	.8264	.8289	.8315	.8340	.8365	.8389
1.0	.8413	.8438	.8461	.8485	.8508	.8531	.8554	.8577	.8599	.8621
1.1	.8643	.8665	.8686	.8708	.8729	.8749	.8770	.8790	.8810	.8830
1.2	.8849	.8869	.8888	.8907	.8925	.8944	.8962	.8980	.8997	.9015
1.3	.9032	.9049	.9066	.9082	.9099	.9115	.9131	.9147	.9162	.9177
1.4	.9192	.9207	.9222	.9236	.9251	.9265	.9279	.9292	.9306	.9319
1.5	.9332	.9345	.9357	.9370	.9382	.9394	.9406	.9418	.9429	.9441
1.6	.9452	.9463	.9474	.9484	.9495	.9505	.9515	.9525	.9535	.9545
1.7	.9554	.9564	.9573	.9582	.9591	.9599	.9608	.9616	.9625	.9633
1.8	.9641	.9649	.9656	.9664	.9671	.9678	.9686	.9693	.9699	.9706
1.9	.9713	.9719	.9726	.9732	.9738	.9744	.9750	.9756	.9761	.9767
2.0	.9772	.9778	.9783	.9788	.9793	.9798	.9803	.9808	.9812	.9817
2.1	.9821	.9826	.9830	.9834	.9838	.9842	.9846	.9850	.9854	.9857
2.2	.9861	.9864	.9868	.9871	.9875	.9878	.9881	.9884	.9887	.9890
2.3	.9893	.9896	.9898	.9901	.9904	.9906	.9909	.9911	.9913	.9916
2.4	.9918	.9920	.9922	.9925	.9927	.9929	.9931	.9932	.9934	.9936
2.5	.9938	.9940	.9941	.9943	.9945	.9946	.9948	.9949	.9951	.9952
2.6	.9953	.9955	.9956	.9957	.9959	.9960	.9961	.9962	.9963	.9964
2.7	.9965	.9966	.9967	.9968	.9969	.9970	.9971	.9972	.9973	.9974
2.8	.9974	.9975	.9976	.9977	.9977	.9978	.9979	.9979	.9980	.9981
2.9	.9981	.9982	.9982	.9983	.9984	.9984	.9985	.9985	.9986	.9986
3.0	.9987	.9987	.9987	.9988	.9988	.9989	.9989	.9989	.9990	.9990
3.1	.9990	.9991	.9991	.9991	.9992	.9992	.9992	.9992	.9993	.9993
3.2	.9993	.9993	.9994	.9994	.9994	.9994	.9994	.9995	.9995	.9995
3.3	.9995	.9995	.9995	.9996	.9996	.9996	.9996	.9996	.9996	.9997
3.4	.9997	.9997	.9997	.9997	.9997	.9997	.9997	.9997	.9997	.9998

z	1.282	1.645	1.960	2.326	2.576	3.090	3.291	3.891	4.417
$\Phi(z)$	.90	.95	.975	.99	.995	.999	.9995	.99995	.999995
$2[1 - \Phi(z)]$	.20	.10	.05	.02	.01	.002	.001	.0001	.00001

For any random variable X, whether normal or not, subtracting the mean gives a random variable whose mean is zero:

$$E\{X - \mu_X\} = 0$$

The variance is not changed by this subtraction, but then dividing by the standard deviation gives a variable with a variance of one:

$$\text{Var}\left(\frac{X - \mu_X}{\sigma_X}\right) = 1$$

(this follows from the results in Section 13.4.9). It is a property of the normal distribution, not shared by most distributions, that the result of this operation is still normal. It is usual to denote the new random variable by the letter Z:

$$Z = \frac{X - \mu_X}{\sigma_X}$$

This is then a **standard normal** random variable, to which the table applies. Conversely, any normal random variable can be considered to have been obtained from a standard normal random variable by multiplying by the required standard deviation and adding the mean:

$$X = \sigma_X Z + \mu_X$$

It follows that we can use the one table for the standard normal for all calculations involving normal variates.

Example 13.28   If $X \sim N(4, 4)$, find

(a) $P(X \leqslant 6.7)$

(b) the constant c such that $P(X > c) = 0.1$.

Solution   (a) $P(X \leqslant 6.7) = P\left(\dfrac{X - 4}{2} \leqslant \dfrac{6.7 - 4}{2}\right)$

$$= P(Z \leqslant 1.35) = 0.9115 \quad \text{(from Figure 13.22)}$$

(b) If $P(X > c) = 0.1$ then $P(X \leqslant c) = 0.9$, so that

$$P\left(\frac{X - 4}{2} \leqslant \frac{c - 4}{2}\right) = P\left(Z \leqslant \frac{c - 4}{2}\right) = 0.9$$

from which $\frac{1}{2}(c - 4) = 1.282$ (using Figure 13.22); hence $c = 6.564$.

Example 13.28 shows that the standard normal table can be used in either direction: either to find the probability of an interval or to find the interval that gives a particular probability.

Example 13.29

The burning time X of an experimental rocket is a random variable having (approximately) a normal distribution with mean 600 s and standard deviation 25 s. Find the probability that such a rocket will burn for

(a) less than 550 s,   (b) more than 640 s.

Solution

Using the normal table as appropriate,

(a) $P(X < 550) = P\left(\dfrac{X - 600}{25} < \dfrac{550 - 600}{25}\right) = P(Z < -2)$

$$= \Phi(-2) = 1 - \Phi(2) = 0.0228$$

(b) $P(X > 640) = P\left(\dfrac{X - 600}{25} > \dfrac{640 - 600}{25}\right) = P(Z > 1.6)$

$$= 1 - \Phi(1.6) = 0.0548$$

13.5.4 The central limit theorem

The practical methods of statistical inference have foundations in probability theory, and the fundamental assumption underlying many of these methods is that the data has a distribution that is normal. Some statistical methods are **robust** in the sense that they work reliably even under moderate violations of their assumptions, but it is unsatisfactory to rely heavily upon this. If normality of the data were exceptional then this would severely limit the scope of those methods that assume it. Fortunately (and as the name implies), the normal distribution arises very frequently in practice; the reason for this will be explained in this section.

Continuous measurements of random phenomena such as noise in electronic circuits or wave elevation on the sea surface give rise to graphs of the form shown in Figure 13.23. If the signal is sampled at regular intervals and a histogram of values built up, it is often found that the histogram closely approximates to a normal density curve. Physically, there are many separate independent random components adding up to produce the measured signal, and it is the total that is normal. There are many sources of noise in an electronic circuit and there are many separate waves on the sea. That the cumulative effect of these, which are often not individually normal, is to produce a total that has that special character is the substance of the following result, which is proved in *Advanced Modern Engineering Mathematics*.

Figure 13.23
Continuous signal with normal distribution.

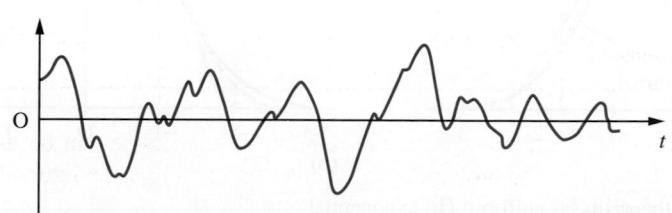
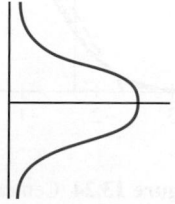

Theorem 13.1	**Central limit theorem**

If $\{X_1, \ldots, X_n\}$ are independent and identically distributed random variables (the distribution being arbitrary), each with mean μ_X and variance σ_X^2, and if

$$W_n = \frac{X_1 + \ldots + X_n}{n} \quad \text{and} \quad Z_n = \frac{X_1 + \ldots + X_n - n\mu_X}{\sigma_X \sqrt{n}}$$

then, as $n \to \infty$, the distributions of W_n and Z_n tend to $W_n \sim N(\mu_X, \sigma_X^2/n)$ and $Z_n \sim N(0, 1)$ respectively.

end of theorem

Loosely speaking and with certain exceptions, the sum of independent identically distributed random variables tends to a normal distribution. The following points should be noted.

(a) The standard normal is obtained by subtracting the mean of the total and dividing by the standard deviation.

(b) The distributions converge to the normal in the sense that the cumulative distribution functions converge. This ensures that all observational properties of Z_n will be standard normal for sufficiently large n.

(c) How large n has to be before the normal approximation is good depends upon the underlying population. If the distribution of the variables X_i is symmetric about the mean then convergence to the normal is rapid. Figure 13.24(a) shows the distributions of the uniform random variable X with density function

$$f_X(x) = \tfrac{1}{2}\sqrt{\tfrac{1}{3}} \quad (-\sqrt{3} \leqslant x \leqslant \sqrt{3})$$

(which has mean zero and variance one), together with those for Z_2 and Z_4. The normal distribution is also shown. Figure 13.24(b) shows similar results for the exponential random variable X with density function

$$f_X(x) = e^{-(x+1)} \quad (x \geqslant -1)$$

(which has mean zero and variance one), together with Z_5 and Z_{25}. Convergence is clearly more rapid for the symmetric distribution.

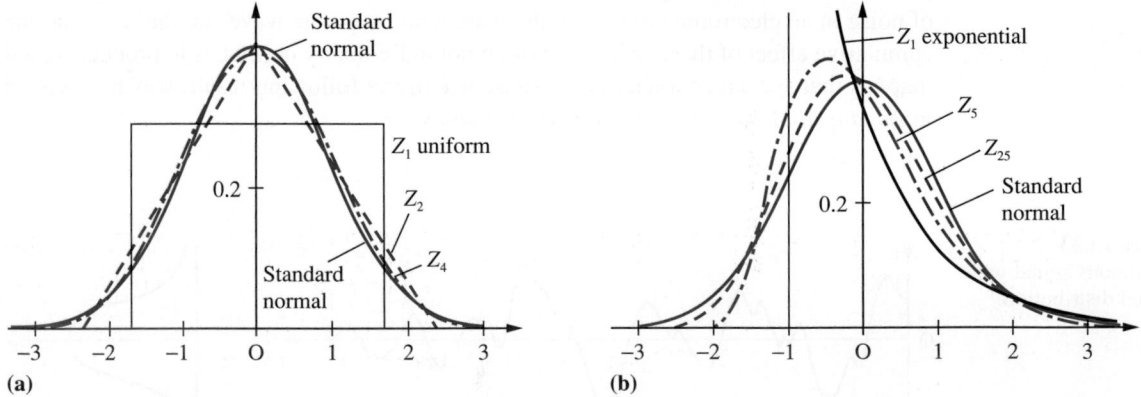

Figure 13.24 Central limit theorem: (a) uniform; (b) exponential.

(d) The theorem can be generalized so that the random variables X_i do not need to be identically distributed, which is usually not the case in physical situations.

(e) Even where the data of an experiment is not normally distributed, the central limit theorem implies that the sample average has a normal distribution for large samples. Much valuable statistics exploits this fact.

Example 13.30

In a quality control scheme at a factory, batches of components are accepted or rejected depending on the number of defective items counted in a sample. Rejected batches are inspected and all defective items are replaced with good ones. From the machine reliability statistics it has been calculated that the probabilities of three, four, five, six, and seven defective items in a rejected batch are 0.3, 0.4, 0.2, 0.08 and 0.02 respectively. Fifty rejected batches produced a total of 221 defective items. Does this suggest that the machines are producing more defective items than they should?

Solution

If X represents the number of defective items in a rejected batch, the mean and standard deviation are given by

$$\mu_X = 3(0.3) + 4(0.4) + 5(0.2) + 6(0.08) + 7(0.02) = 4.12$$

$$\sigma_X = \sqrt{[9(0.3) + 16(0.4) + 25(0.2) + 36(0.08) + 49(0.02) - (4.12)^2]} = 0.9928$$

By the central limit theorem, the aggregate count of defectives Y in 50 rejected batches will be approximately normal, and

$$P(Y \geqslant 221) = P\left(\frac{Y - 50\mu_X}{\sigma_X\sqrt{50}} \geqslant \frac{221 - 50\mu_X}{\sigma_X\sqrt{50}}\right) = 1 - \Phi(2.137) = 0.0163$$

This probability is rather small, so the performance of the machines must come under suspicion. In fact, a rather more accurate answer to this problem is obtained by making a continuity correction, as explained in Section 13.5.5, but the conclusion is the same.

Example 13.30 is typical of many applications of the central limit theorem. The underlying distribution is certainly not normal, but it is reasonable to assume that the aggregate is approximately normal.

Example 13.31

In Section 1.5.5 it was noted that the maximum error that would occur in the sum of 100 numbers, each of which was rounded to three decimal places, is 0.05. Find the probability of the error in the sum exceeding 0.005 in magnitude, and the expected magnitude of the error.

Solution

We assume that the error in a number rounded to 3dp may be anything between -0.0005 and $+0.0005$, with all values in the range equally likely. In other words, the error in each value is a uniform random variable, X, say, with

$$f_X(x) = \begin{cases} 1000 & (-0.0005 < x < +0.0005) \\ 0 & \text{(otherwise)} \end{cases}$$

from which the mean and variance are given by

$$\mu_X = \int_{-0.0005}^{+0.0005} 1000x \, dx = 0$$

$$E(X^2) = \int_{-0.0005}^{+0.0005} 1000x^2 \, dx = 8.333 \times 10^{-8}$$

$$\sigma_X^2 = E(X^2) - \mu_X^2 = 8.333 \times 10^{-8}$$

The error in the sum is a random variable $Y = X_1 + \ldots + X_{100}$. By the central limit theorem, approximately

$$Y \sim N(100\mu_X, \, 100\sigma_X^2) = N(0, \, 8.333 \times 10^{-6})$$

so

$$P(Y > 0.005) \approx P\left(Z > \frac{0.005}{0.002\,89}\right)$$

$$= P(Z > 1.732) = 1 - \Phi(1.732)$$

The error in the sum will exceed 0.005 in magnitude if $Y > 0.005$ or $Y < -0.005$, so, by symmetry,

$$P(|Y| > 0.005) \approx 2[1 - \Phi(1.732)]$$

$$= 2(1 - 0.9584)$$

$$= 0.0832$$

Thus, because the errors tend to cancel each other out, there is only one chance in 12 of the error reaching even $\frac{1}{10}$ of its maximum possible value. Furthermore, the expected value of the error magnitude is

$$E(|Y|) \approx \int_{-\infty}^{+\infty} \frac{|y|}{\sigma_Y \sqrt{(2\pi)}} e^{-y^2/2\sigma_Y^2} dy = 2 \int_0^{\infty} \frac{y}{\sigma_Y \sqrt{(2\pi)}} e^{-y^2/2\sigma_Y^2} dy$$

$$= \frac{1}{\sigma_Y \sqrt{(2\pi)}} \int_0^{\infty} e^{-w/2\sigma_Y^2} dw \quad \text{(by substitution of } w = y^2)$$

$$= \frac{2\sigma_Y^2}{\sigma_Y \sqrt{(2\pi)}} = \sigma_Y \sqrt{\left(\frac{2}{\pi}\right)}$$

With $\sigma_Y^2 = 8.333 \times 10^{-6}$, this gives $E(|Y|) \approx 0.0023$, which is less than $\frac{1}{20}$ of the maximum possible value.

13.5.5   Normal approximation to the binomial

One immediate corollary of the central limit theorem is that the normal distribution can be used to approximate the binomial distribution when n is sufficiently large. This follows from the definition of a binomial random variable as a sum of Bernoulli random

Figure 13.25
Continuous
approximation to a
discrete distribution.

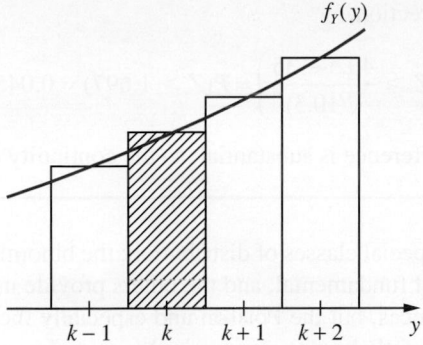

variables (see Section 13.5.1). All that has to be done is to choose the parameters of the normal distribution to match the mean np and variance $np(1 - p)$. As a rule, the normal approximation can be used when $n \geq 25$ and $0.1 \leq p \leq 0.9$. For values of p outside this range the Poisson approximation can be used.

It may seem surprising that the normal distribution, which is continuous, can be used to approximate a discrete distribution, given the very different character of these two types of random variable. The approximation of a discrete distribution X, say, by a continuous one Y works in the manner indicated in Figure 13.25. The probability that X takes the integer value k is approximated by the area under the density function $f_Y(y)$ between $k - 0.5$ and $k + 0.5$. Similarly, the following integral approximates to the probability that X exceeds k:

$$P(X > k) \approx \int_{k+0.5}^{\infty} f_Y(y)\mathrm{d}y$$

By the same token we would have

$$P(X \geq k) \approx \int_{k-0.5}^{\infty} f_Y(y)\mathrm{d}y$$

This use of a half-integer shift in the limit of integration is called the **continuity correction** and gives a more accurate result.

Example 13.32   If 70% of airline passengers using a particular route are members of a frequent-flyer club, find the probability that out of a sample of 50 chosen independently, more than 40 will be members of a frequent-flyer club.

Solution   Let X represent the number who are members of a frequent-flyer club. The conditions for a binomial distribution are met, and the mean and variance are $50(0.7) = 35$ and $50(0.7)(0.3) = 10.5$ respectively. With no continuity correction, we have

$$P(X > 40) = P\left(\frac{X - 35}{\sqrt{(10.5)}} > \frac{40 - 35}{\sqrt{(10.5)}}\right)$$

$$\approx P(Z > 1.543) \quad \text{(where } Z \text{ is standard normal)}$$

$$= 1 - \Phi(1.543) = 0.061$$

With the continuity correction,

$$P(X > 40) \approx P\left(Z > \frac{40.5 - 35}{\sqrt{(10.5)}}\right) = P(Z > 1.697) = 0.045$$

As a percentage, this difference is substantial, so the continuity correction is important.

We now have three special classes of distribution: the binomial, Poisson and normal. The binomial is the most fundamental, and the others provide useful approximations to it in different circumstances, but the Poisson and especially the normal also have very important applications of their own. Increasingly in engineering and in all parts of industry, there are problems arising that involve these and other distributions but which it is not practical to solve without the aid of a computer.

13.5.6 Random variables for simulation

Computer simulations are very widely used in research, design and training. Perhaps the best known is the flight simulator upon which pilots receive much of their training. Simulations are used in research and design wherever a system is too complex for a complete solution to a problem to be obtained theoretically, or where a solution can be obtained but its completeness or accuracy is open to question.

Simulations are **deterministic** if what occurs at any time is completely determined by the state of the system. In contrast, they are **stochastic** if what occurs at any time can be influenced by a chance element that is inherently unpredictable. Stochastic simulations therefore require that random variables (or outcomes) be generated within the program. This may seem a hopeless requirement, considering that computer programs are sequences of deterministic instructions running on deterministic hardware. However, it is possible to generate sequences of numbers that are deterministic and repeatable but that have the appearance of being random. These **pseudo-random numbers** are very useful for simulations, and for other purposes such as the so-called Monte Carlo numerical methods.

Most modern computers contain a software facility for generating pseudo-random numbers with a uniform distribution on the interval (0, 1):

$$f_U(u) = \begin{cases} 1 & (0 < u < 1) \\ 0 & (\text{otherwise}) \end{cases}$$

The successive variables $\{U_1, U_2, \dots\}$ appear to be uncorrelated, and, although there is some structure in the sequence (and indeed the sequence will eventually repeat itself), it is rare for these deficiencies to cause problems in practice.

Random variables with non-uniform distributions are obtained from the sequence $\{U_1, U_2, \dots\}$ by applying various transformations. Figure 13.26 contains pseudocode listings for generating the most common random variables. In each case it is assumed that the system function 'rnd' returns a uniform (0, 1) value, which is stored in the variable U. The variable X contains the required value of the random variable. The binomial is based on the Bernoulli, the Poisson on the exponential, and the normal on the central limit theorem. For a full explanation of how these work see S. J. Yakowitz, *Computational Probability and Simulation* (Addison-Wesley, 1977). Computer packages such as Minitab also often contain facilities for generating random data.

Figure 13.26
Pseudocode listings
for non-uniform
random variables.

```
{Bernoulli random variable X, parameter p.}
U ← rnd
if U < p then X ← 1 else X ← 0 endif

{Binomial random variable X, parameters n,p.}
X ← 0
for i is 1 to n do
    U ← rnd
    if U < p then X ← X + 1 endif
endfor

{Exponential random variable X, parameter L, uses log function to base e.}
U ← rnd
X ← − (log(U))/L

{Poisson random variable X, parameter L.}
X ← −1
W ← 1
P0 ← exp(−L)
repeat
    X ← X + 1
    U ← rnd
    W ← W*U
until W < P0

{Normal random variable X, parameters mean, sd.}
T ← 0
for i is 1 to 12 do
    U ← rnd
    T ← T + U
endfor
X ← sd*(T − 6) + mean
```

13.5.7 Exercises

49 Eight babies are born in a hospital on a particular day. Find the probability that exactly half of them are boys. (The probability that a baby is a boy is actually slightly greater than one-half, but you can take it as exactly one-half for this exercise.)

50 A town has five fire engines operating independently, each of which spends 94% of the time in its station awaiting a call. Find the probability that at least three fire engines are available when needed.

51 The probability of issuing a drill of high brittleness (a reject) is 0.02. Drills are packed in boxes of 100 each. What is the probability that the number of defective drills is no greater than two?

52 If Z is a random variable having the standard normal distribution, find the probabilities that Z will have a value

(a) greater than 1.14,

(b) less than −0.36,

(c) between −0.46 and −0.09,

(d) between −0.58 and 1.12.

53 Assume that

(a) an aircraft can land safely if at least half of its engines are working,

(b) the probability of an engine failing is 0.1, and

(c) engine failures are independent.

Which is safer, a four-engine plane or a two-engine plane?

54 If on average one in 20 of a certain type of column will fail under a given axial load, what are the probabilities that among 16 such columns, (a) at most two, (b) at least four will fail?

55 A machine makes components, and the probability that a component is defective is p. If components are packed in cartons of 20, what value of p will ensure that 90% of cartons contain at most one defective component?

56 If on average 7% of airline passengers order special meals, find the approximate probability that on a particular flight carrying 85 passengers, eight or more will order special meals.

57 A Geiger counter and a source of radioactive particles are so situated that the probability that a particle emanating from the radioactive source will be registered by the counter is 1/10 000. Assume that during the time of observation, 30 000 particles emanated from the source. What is the probability that the number of particles registered was (a) zero, (b) three, (c) more than five?

58 Assume that in the composition of a book there exists a constant probability 0.0001 that an arbitrary letter will be set incorrectly. After composition, the proofs are read by a proofreader, who discovers 90% of the errors. After the proofreader, the author discovers half of the remaining errors. Find the probability that in a book with 500 000 printing symbols there remain after this no more than six unnoticed errors.

59 Suppose that the actual amount of cement that a filling machine puts into 'six-kilogram' bags is a normal random variable with $\sigma = 0.05\,\text{kg}$. If only 3% of bags are to contain less than 6 kg, what must be the mean fill of the bags?

60 Prove by making the substitution $u = (x - \mu_X)/\sigma_X$ in the integrals concerned that the mean and variance of the normal distribution are μ_X and σ_X^2 respectively. (*Hint*: for the variance, integrate $u[u\exp(-u^2/2)]$ by parts.)

61 If 23% of all patients with high blood pressure have bad side-effects from a certain kind of medicine, use the normal approximation to the binomial to find the probability that among 120 patients with high blood pressure treated with this medicine, more than 32 will have bad side-effects.

62 In firing at a target, a marksman scores at each shot either 10, 9, 8, 7, or 6, with respective probabilities 0.5, 0.3, 0.1, 0.05, 0.05. If he fires 100 shots, what is the approximate probability that his aggregate score exceeds 940?

63 A fleet car operator has n cars, each of which has probability 8% of being broken down on any particular day. Find the smallest value of n that gives probability 90% that at least 40 cars will be available for use on any one day.

64 The diameter of ball bearings produced by a machine is a random variable having a normal distribution with mean 6.00 mm and standard deviation 0.02 mm. If the diameter tolerance is ±1%, find the proportion of ball bearings produced that are out of tolerance. After several years' use, machine wear has the effect of increasing the standard deviation, although the mean diameter remains constant. The manufacturer decides to replace the machine when 2% of its output is out of tolerance. What is the standard deviation when this happens?

65 A major airline operates 350 flights a day throughout the world. The probability that a flight will be delayed for more than one hour, for any reason, is 0.7%. If more than four flights suffer such delays in any one day, the implications for route organization and crewing become serious. Call such a day a 'flap-day'. Using approximations as appropriate, find the probabilities that

(a) any particular day is a flap-day,

(b) two flap-days (not more) occur in one week,

(c) more than 50 flap-days occur in a year of 365 days.

13.6 Engineering application: quality control

This is a topic of particular relevance to engineers, because the statistical methods of quality control are widely and increasingly used in industry in order to promote the reliability of products. Orders have been won and lost because one manufacturer has implemented quality control in the workplace more than another and the purchaser has used this as a criterion when deciding where to place the order.

Quality control statistics is not particularly difficult, but (as usual) it rests on fundamental results such as the Poisson approximation to the binomial distribution (Section 13.5.2). The methods apply mainly to mass production systems where quality can be measured numerically.

This section will introduce the use of control charts for continuous monitoring of quality, rather than the more traditional batch inspection plans, and control charts are further discussed in *Advanced Modern Engineering Mathematics*.

13.6.1 Attribute control charts

Manufactured items that are elaborate and therefore expensive can each be tested thoroughly before dispatch to consumers, but such item-by-item testing must be ruled out for low-level components on grounds of cost. Some defective items are bound to slip through, and the objective of quality control is to keep the proportion of these within acceptable and agreed limits.

Variations occur in the quality of a product, caused either by variations in the raw material or input or by variations in processing. Quality is monitored by regular testing of samples of output. Assume for now that the test consists in counting the number within the sample which pass or fail according to some performance criterion. A small proportion of defective items in the output is permitted while the process is said to be **in control**. If the actual proportion of defectives rises to an unacceptable level, the process is said to be **out of control**, and the counts of defectives in the samples would be expected to rise. We should like to detect this as soon as possible when it occurs, but without incurring the expense of a large number of false alarms while the process is actually in control.

An essential aid to the quality controller is the **Shewhart control chart**, which is a plot of the successive counts of defective items against sample number. Figure 13.27 is

Figure 13.27
Attribute control chart for Example 13.33.

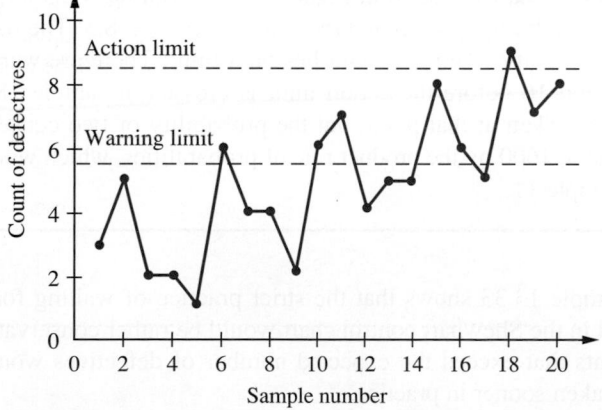

an example of such a chart. Also shown on the chart are two limits on the counts of defectives, corresponding to probabilities of one in 40 and one in 1000 of a sample count falling outside the limit if the process is in control. These are called **warning** and **action limits** respectively, and are denoted by c_W and c_A.

Any sample point falling outside the action limit would normally result in the process being suspended and the problem corrected. Roughly one in 40 sample points will fall outside the warning limit purely by chance, but if this occurs repeatedly or if there is a clear trend upwards in the counts of defectives then action may well be taken before the action limit itself is crossed.

To obtain the warning and action limits, we use the Poisson approximation to the binomial. If the acceptable proportion of defective items is p, usually small, and the sample size is n then for a process in control the defective count C, say, will be a binomial random variable with parameters n and p. Provided that n is not too small, the Poisson approximation can be used (Section 13.5.2):

$$P(C \geqslant c) \approx \sum_{k=c}^{n} \frac{(np)^k e^{-np}}{k!}$$

Equating this to $\frac{1}{40}$ and then to $\frac{1}{1000}$ gives equations that can be solved for the warning limit c_W and the action limit c_A respectively, in terms of the product np. This is the basis of the table shown in Figure 13.28, which enables c_W and c_A to be read directly from the value of np.

Example 13.33

Regular samples of 50 are taken from a process making electronic components, for which an acceptable proportion of defectives is 5%. Successive counts of defectives in each sample are as follows:

Sample	1	2	3	4	5	6	7	8	9	10	11	12	13	14	15	16	17	18	19	20
Count	3	5	2	2	1	6	4	4	2	6	7	4	5	5	8	6	5	9	7	8

At what point would the decision be taken to stop and correct the process?

Solution

The control chart is shown in Figure 13.27. From $np = 2.5$ and Figure 13.28 we have the warning limit $c_W = 5.5$ and the action limit $c_A = 8.5$. The half-integer values are to avoid ambiguity when the count lies on a limit. There are warnings at samples 6, 10, 11, 15 and 16 before the action limit is crossed at sample 18. Strictly, the decision should be taken at that point, but the probability of two consecutive warnings is less than one in 1600 by the product rule of probabilities, which would justify taking action after sample 11.

Example 13.33 shows that the strict practice of waiting for the action limit to be crossed in the Shewhart control chart would be rather conservative. The long sequence of counts that exceed the expected number of defectives would lead to the decision being taken sooner in practice.

Figure 13.28
Shewhart attribute control limits: n is sample size, p is probability of defect, c_W is warning limit and c_A is action limit.

c_W or c_A	np for c_W	np for c_A
1.5	<0.44	<0.13
2.5	0.44–0.87	0.13–0.32
3.5	0.87–1.38	0.32–0.60
4.5	1.38–1.94	0.60–0.94
5.5	1.94–2.53	0.94–1.33
6.5	2.53–3.16	1.33–1.77
7.5	3.16–3.81	1.77–2.23
8.5	3.81–4.48	2.23–2.73
9.5	4.48–5.17	2.73–3.25
10.5	5.17–5.87	3.25–3.79
11.5	5.87–6.59	3.79–4.35
12.5	6.59–7.31	4.35–4.93
13.5	7.31–8.05	4.93–5.52
14.5	8.05–8.80	5.52–6.12
15.5	8.80–9.55	6.12–6.74
16.5	9.55–10.31	6.74–7.37
17.5	10.31–11.08	7.37–8.01
18.5	11.08–11.85	8.01–8.66
19.5	11.85–12.63	8.66–9.31
20.5	12.63–13.42	9.31–9.98
21.5	13.42–14.21	9.98–10.65
22.5	14.21–15.00	10.65–11.33
23.5	15.00–15.80	11.33–12.02
24.5	15.80–16.61	12.02–12.71
25.5	16.61–17.41	12.71–13.41
26.5	17.41–18.23	13.41–14.11
27.5	18.23–19.04	14.11–14.82
28.5	19.04–19.86	14.82–15.53
29.5	19.86–20.68	15.53–16.25
30.5		16.25–16.98
31.5		16.98–17.70
32.5		17.70–18.44
33.5		18.44–19.17
34.5		19.17–19.91
35.5		19.91–20.66

13.6.2 United States standard attribute charts

The control chart decribed above, with action and warning limits set by probability of exceedance, is the standard practice in the United Kingdom. In the United States the practice is rather different in that there is usually no warning limit and that action limit (called the **upper control limit**, **UCL**) is set at three standard deviations above the mean. Because the count of defectives is binomial with mean np and variance $np(1 - p)$, this means that

$$UCL = np + 3[np(1 - p)]^{1/2}$$

Example 13.34   Find the UCL and apply it to the data in Example 13.33.

Solution   From $n = 50$ and $p = 0.05$ we infer that UCL = 7.1, which is between the warning limit c_W and the action limit c_A in Example 13.33. The decision to correct the process would be taken after the 15th sample, the first to exceed the UCL.

Sometimes a **lower limit control**, LCL, is defined at three standard deviations below the mean:

$$LCL = np - 3[np(1 - p)]^{1/2}$$

If this is positive, it can be used to test whether the proportion defective in the output is falling significantly below the expected value.

Control charts are also useful for monitoring the output of a manufacturing process where quality depends upon a numerical measure such as dimension, weight or resistance. Charts that are more powerful than the Shewhart charts at detecting variations in the output are also used. These topics are covered in *Advanced Modern Engineering Mathematics*.

13.6.3 Exercises

66 It is intended that 90% of electronic devices emerging from a machine should pass a simple on-the-spot quality test. The numbers of defectives among samples of 50 taken by successive shifts are as follows:

5, 8, 11, 5, 6, 4, 9, 7, 12, 9, 10, 14

Find the action and warning limits, and the sample number at which an out-of-control decision is taken. Also find the UCL (United States practice) and the sample number for action.

67 Thirty-two successive samples of 100 castings each, taken from a production line, contained numbers of defectives as follows:

3, 3, 5, 3, 5, 0, 3, 1, 3, 5, 4, 2, 4, 3, 5, 4, 3, 4, 5, 6, 5, 6, 4, 4, 7, 5, 4, 8, 5, 6, 6, 7

If the proportion defective is to be maintained at 0.02, use the Shewhart method (both UK and US standard) to indicate whether this proportion is being maintained, and if not then after how many samples action should be taken.

13.7 Engineering application: clustering of rare events

13.7.1 Introduction

To conclude this chapter, we shall apply some of the probability theory covered so far to an investigation of a serious problem, or rather a family of problems. Failures of engineering systems or structures are rare events, but they have serious consequences. If a number of similar failures occur and a link between them can be found then it may be possible to anticipate and prevent future failures. One aspect of this is the detection of regional variations in the number of failures, which may provide clues as to possible causes.

Problems like this, and their associated difficulties, arise in many fields, and the lessons learned from analysing one can often be applied to others. Some typical examples are as follows:

(a)   near-misses between two aircraft in flight,
(b)   collisions or capsizing of ships at sea,
(c)   accidents involving road vehicles,
(d)   occurrences of environmentally induced diseases such as leukaemia.

These problems can be looked at in various ways, but there is an approach that applies to all of them because of the following common elements:

(a)   a very large number of potential cases,
(b)   a very small proportion of these become actual cases,
(c)   a possible common cause,
(d)   regional variations in the common cause if it exists.

Common causes for (a) and (b) that vary regionally could be dangerous weather conditions or inadequate control over routes taken; for (c) they could be inadequate lighting or hazard warnings, and for (d) they could be proximity to a nuclear installation. For each problem it is important to identify the common cause if it exists – and one clue to its existence is the regional variation. Also, for each problem the main difficulty is the rarity of the cases, and it is this that makes an analysis using probability theory useful.

This case study is expressed in terms of a survey of near-misses in aircraft operations, but the analysis could be applied to any of the above examples. The figures are hypothetical, but the method of analysis is realistic.

13.7.2   Survey of near-misses between aircraft

Suppose that the major airlines cooperate in a survey of near-misses during a period of one year. The region being studied is divided up into 1000 areas in such a way that there are on average 200 flights per year through each area. Suppose that the total number of flights is 200 000 and that the total number of near-misses logged by the pilots is 120. Although a near-miss involves two aircraft, it is recorded as a single incident. At the end of the year the data is examined and two areas in particular stand out. In one area A four incidents occurred in a total of 400 flights, and in another area B two incidents occurred in a total of 150 flights.

The question that it is natural to ask is whether there are any areas, in particular these two, in which the number of near-misses is greater than can be accounted for by chance. If any such area exists, it can be examined to see what makes it special, and this may lead to the discovery of a common cause and appropriate action being taken. To approach this, we shall first assume that the probability that a near-miss will occur is the same for every flight and in every area. This probability is taken to be the total number of near-misses divided by the total number of flights, which gives $p = 6 \times 10^{-4}$.

To assess how unlikely the figures for areas A and B are we need to calculate the probability of the given number *or more* of incidents, as explained in the discussion following Example 13.25. If we assume that the probability p applies independently for every flight then for area A, using the binomial distribution, we have

$$P(4 \text{ or more incidents}) = 1 - \sum_{k=0}^{3} \binom{400}{k} p^k (1 - p)^{400-k}$$

which gives 1.13×10^{-4}. Alternatively, using the Poisson approximation (Section 13.5.2),

$$P(4 \text{ or more incidents}) \approx 1 - \sum_{k=0}^{3} \frac{\lambda^k e^{-\lambda}}{k!} \quad (\text{with } \lambda = 400p = 0.24)$$

$$= 1 - e^{-\lambda} \left(\left(\left(\frac{\lambda}{3} + 1 \right) \frac{\lambda}{2} + 1 \right) \lambda + 1 \right)$$

$$= 1.14 \times 10^{-4}$$

which is close to the exact figure. Similarly, for area B

$$P(2 \text{ or more incidents}) = 1 - (1 - p)^{150} - 150p(1 - p)^{149} = 3.79 \times 10^{-3}$$

$$\approx 1 - e^{-\lambda}(1 + \lambda) \quad (\text{with } \lambda = 150p = 0.09)$$

$$= 3.82 \times 10^{-3}$$

The effectiveness of the Poisson approximation to the binomial is clear from these results.

The four incidents in area A are seen to be much less likely to be due to chance than the two incidents in area B. This is interesting, because common sense would suggest comparing the proportions in the respective areas, which are 1% for A and 1.33% for B. Despite having the lower proportion of incidents, area A provides stronger evidence for a regional anomaly, by more than an order of magnitude.

The extent of anomaly can be judged from the probability that at least one of the 1000 areas in the region will give a result at least as extreme as those observed. After all, with 1000 opportunities for a rare event to occur, the probability that it will occur in at least one of them is significantly enhanced. Using the complement and product rules we have

$P(\text{at least one event with probability } 1.13 \times 10^{-4} \text{ in 1000 areas})$

$= 1 - P \text{ (no such events)}$

$= 1 - (1 - 1.13 \times 10^{-4})^{1000} = 0.107$

Similarly,

$P(\text{at least one event with probability } 3.79 \times 10^{-3} \text{ in 1000 areas})$

$= 1 - (1 - 3.79 \times 10^{-3})^{1000} = 0.978$

The area B result has a high probability of occurring by chance, somewhere within the 1000 areas. However, there is only one chance in ten that a result as improbable as that in area A would occur anywhere, assuming a constant value of p. It seems that the true probability of a near-miss is higher in that area. Although the number of incidents is small, quite a firm conclusion has been reached.

The most interesting point about this analysis is that the comparison of proportions, which is the most obvious way of judging the results, is so misleading. The reason why it doesn't work is that (as can be seen in *Advanced Modern Engineering Mathematics*) the variance of a sample proportion depends upon the size of the sample, the denominator in that proportion. Dividing the number of incidents by the number of flights in an attempt to normalize the data fails to eliminate the number of flights as a variable, because of its lingering influence on the statistics.

This is as far as the mathematical analysis can proceed. It cannot point to any particular cause without further data. Tracking down the reason for the anomaly can be very difficult in situations like this, but at least the search can be focused on an area. The local weather, the operating procedures and technical support of the flight controllers, and natural sources of interference in the navigational equipment would all be under suspicion.

13.7.3 Exercises

68 A third area in the near-miss survey recorded five incidents in 800 flights. Should this area also be regarded as unusually risky?

69 Two adjacent areas recorded two incidents in 250 flights and one incident in 85 flights respectively. Test the combination of the two areas.

13.8 Review exercises (1–13)

1 A continuous random variable X has probability density function given by

$$f_X(x) = \begin{cases} \dfrac{c}{x^4} & \text{for } x \geq 1 \\ 0 & \text{for } x < 1 \end{cases}$$

where c is constant. Find

(a) the value of the constant c,

(b) the cumulative distribution function of X,

(c) $P(X > 2)$,

(d) the mean of X,

(e) the standard deviation of X.

2 If there are 720 personal computers in an office building and they each break down independently with probability 0.002 per working day, use the Poisson approximation to the binomial distribution to find the probability that more than four of these computers will break down in any one working day.

3 The City Engineer's department installs 10 000 fluorescent lamp bulbs in street lamp standards. The bulbs have an average life of 7000 operating hours with a standard deviation of 400 hours. Assuming that the life of the bulbs, L, is a normal random variable, what number of bulbs might be expected to have failed after 6000 operating hours? If the engineer wishes to adopt a routine replacement policy which ensures that no more than 5% of the bulbs fail before their routine replacement, after how long should the bulbs be replaced?

4 The binomial is a special case of the more general **multinomial distribution**:

$$P(n_1, \ldots, n_k) = \frac{n!}{n_1! \ldots n_k!} (p_1)^{n_1} \ldots (p_k)^{n_k}$$

where $p_1 + \ldots + p_k = 1$ and $n_1 + \ldots + n_k = n$. Each observation of a random variable has k possible outcomes, with probabilities $p_1, \ldots, p_k$, and the observed total numbers of each possible outcome after n independent observations are made are respectively $n_1, \ldots, n_k$. Suppose that 60% of calls to a telephone banking enquiry service are for account balance requests, 20% are for payment confirmations, 10% are for transfer requests, and 10% are to open new accounts. Find the probability that out of 20 calls to this service there will be 10 balance requests, five payment confirmations, three transfers and two new accounts.

5   A manufacturer has agreed to dispatch small servomechanisms in cartons of 100 to a distributor. The distributor requires that 90% of cartons contain at most one defective servomechanism. Assuming the Poisson approximation to the binomial distribution, write down an equation for the Poisson parameter λ such that the distributor's requirements are just satisfied. Solve by trial and error (approximate solution 0.5), and hence find the required proportion of manufactured servomechanisms that must be satisfactory.

6   Ten thousand numbers are to be added, each rounded to the sixth decimal place. Assuming that the errors arising from rounding the numbers are mutually independent and uniformly distributed on $(-0.5 \times 10^{-6}, +0.5 \times 10^{-6})$, find the limits in which the total error will lie with probability 95%.

7   Suppose that X is a continuous random variable with mean μ_X and variance σ_X^2. By separating the integral in the definition of σ_X^2 into three parts and substituting the respective bounds for $(x - \mu_X)^2$ as follows

$$(x - \mu_X)^2 \geqslant \begin{cases} (k\sigma_X)^2 & \text{on } (-\infty, \mu_X - k\sigma_X) \\ 0 & \text{on } (\mu_X - k\sigma_X, \mu_X + k\sigma_X) \\ (k\sigma_X)^2 & \text{on } (\mu_X + k\sigma_X, +\infty) \end{cases}$$

where k is a constant, prove **Chebyshev's theorem**

$$P(|x - \mu_X| > k\sigma_X) \leqslant k^{-2}$$

Deduce that for every continuous random variable X the probability is at least $\frac{8}{9}$ that X will take a value within three standard deviations of the mean.

8   The function

$$\Gamma(\alpha) = \int_0^\infty y^{\alpha-1} e^{-y} \, dy \quad (\alpha > 0)$$

is known as the **gamma function**, and the probability density function

$$f_X(x) = \begin{cases} [\Gamma(\alpha)]^{-1} \lambda^\alpha x^{\alpha-1} e^{-\lambda x} & (x > 0) \\ 0 & (\text{otherwise}) \end{cases}$$

defines the **gamma distribution**. Prove that

(a) $\mu_X = \alpha/\lambda$,

(b) $\sigma_X^2 = \alpha/\lambda^2$.

9   If $X_1, \ldots, X_n$ are independent exponentially distributed random variables, each with parameter λ, prove that the random variable whose value is given by the minimum of $\{X_1, \ldots, X_n\}$ also has an exponential distribution, with parameter $n\lambda$. In particular, if a complex piece of machinery consists of six parts, each of which has an exponential distribution of time to failure with mean 2000 hours, and if the machine fails as soon as any of its parts fail, find the probability that the time to failure exceeds 300 hours.

10   Find the expected value of the maximum of four independent exponential random variables, each with parameter λ. In particular, if the time taken for a routine test and service of a jet aircraft engine has an exponential distribution with a mean of three hours, find the mean time to complete a four-engine aircraft if the service times are independent.

11   In the game of craps, two dice are tossed. A total of 7 or 11 wins immediately, a total of 2, 3 or 12 loses. For remaining outcomes, both dice are tossed repeatedly until either a total of 7 appears, which loses, or the original number, which wins. Show that the overall probability of winning is approximately 0.493.

12   A large number N of people are subjected to a blood investigation to test for the presence of an illegal drug. This investigation is carried out by mixing the blood of k persons at a time and testing the mixture. If the result of the analysis is negative then this is sufficient for all k persons. If the result is positive then the blood of each person must be analysed separately, making $k + 1$ analyses in all. Assume that the probability p of a positive result is the same for each person and that the results of the analyses are independent. Find the expected number of analyses, and minimize with respect to k. In particular, find the optimum value of k when

$p = 0.01$, and the expected saving compared with a separate analysis for all N people.

13 Error-correcting codes are widely used for data transmission. A message consisting of N binary bits is partitioned into blocks of k bits, and each block is transmitted with some additional parity bits giving a total of n bits per block. The parity bits are used at the receiving end to correct any errors that occur in transmission (bits that get inverted, including the parity bits themselves). Some error-correcting codes can correct only a single error per block; others can correct up to two errors. The number $n - k$ of parity bits is chosen as small as possible to satisfy the relationship:

$$2^{n-k} \geqslant n + 1 \quad \text{(single-error-correcting code)}$$

or

$$2^{n-k} \geqslant n^2 + 1 \quad \text{(double-error-correcting code)}$$

(a) Suppose that transmission errors occur independently at an average rate of 1% of bits transmitted. For data blocks k of 4, 8, 16, 32 and 64 bits, find the value of n and the probability of more errors occurring than the code can correct. Do this for single- and double-error-correcting codes.

(b) Find for each type of code the largest block size k that allows a total of $N = 64$ data bits to be transmitted with at least 95% probability of correct overall interpretation at the receiving end. Compare the total numbers of bits transmitted in each case.

Appendix I Tables

AI.1 Some useful results

Algebraic processes

$$(a + b)^2 = a^2 + 2ab + b^2$$

$$(a - b)^2 = a^2 - 2ab + b^2$$

$$a^2 - b^2 = (a + b)(a - b) \qquad \text{'difference of two squares'}$$

$$ax^2 + bx + c = a\left(x + \frac{b}{2a}\right)^2 + c - \frac{b^2}{4a} \qquad \text{'completing the square'}$$

Quadratic equation

The general form of the quadratic equation is

$$ax^2 + bx + c = 0, \; a \neq 0$$

and its solution is given by

$$x = \frac{-b \pm \sqrt{(b^2 - 4ac)}}{2a}$$

- If $b^2 > 4ac$ the equation has two real roots.
- If $b^2 = 4ac$ the equation has a real root which is repeated.
- If $b^2 < 4ac$ the equation has two complex roots, which are complex conjugates.

Rules of indices

(1) $a^m a^n = a^{m+n}$ (2) $\dfrac{a^m}{a^n} = a^{m-n}$ (3) $(a^n)^m = a^{nm}$ (4) $a^0 = 1$

(5) $a^{1/n} = \sqrt[n]{a}$ (6) $a^{-n} = \dfrac{1}{a^n}$ (7) $a^{m/n} = \sqrt[n]{a^m}$

Logarithmic formulae

Definition
If $y = a^x$ then $x = \log_a y$ expressed verbally as 'x is equal to log to base a of y'.

Rules

(1) $\log_a(xy) = \log_a x + \log_a y$ *'log of product equal to sum of logs'*

(2) $\log_a(\frac{x}{y}) = \log_a x - \log_a y$ *'log of quotient equal difference of logs'*

(3) $\log_a x^n = n \log_a x$

Useful results

(1) $x = a^{\log_a x}$

(2) $\log_a x = \dfrac{\log_b x}{\log_b a}$ *'change of base'*

When the logarithm of x is to base e then it is denoted by $\ln x$ and called the natural logarithm.

Hyperbolic functions

Definitions

$$\cosh x = \tfrac{1}{2}(e^x + e^{-x}) \qquad\qquad \sinh x = \tfrac{1}{2}(e^x - e^{-x})$$

$$\tanh x = \frac{\sinh x}{\cosh x} \qquad\qquad \operatorname{sech}x = \frac{1}{\cosh x}$$

$$\operatorname{cosech}x = \frac{1}{\sinh x}, \ (x \neq 0) \qquad \coth x = \frac{1}{\tanh x}, \ (x \neq 0)$$

Logarithmic form of inverses

$$\sinh^{-1} x = \ln[x + \sqrt{(x^2 + 1)}]$$

$$\cosh^{-1} x = \ln[x + \sqrt{(x^2 - 1)}], \quad (x \geq 1)$$

$$\tanh^{-1} x = \tfrac{1}{2}\ln\left(\frac{1+x}{1-x}\right), \quad (-1 < x < 1)$$

Arithmetic sequence (progression)

- An arithmetic sequence is a sequence of terms of the form

    $$a, a+d, a+2d, a+3d, \dots$$

    where a is called the first term and d the common difference.
- The *nth* term of the sequence is given by $a + (n-1)d$.
- The sum of the terms of an arithmetic sequence is called an arithmetic series and the sum of the first n terms is

    $$S_n = n[2a + (n-1)d]/2$$

Geometric sequence (progression)

- A geometric sequence is a sequence of terms of the form

$$a, ar, ar^2, ar^3, \ldots$$

where a is called the first term and r the common ratio.
- The *nth* term of the sequence is given by ar^{n-1}.
- The sum of the terms of a geometric sequence is called a geometric series and the sum of the first n terms is

$$S_n = a(1 - r^n)/(1 - r) \text{ if } r \neq 1 \quad \text{and} \quad S_n = an \text{ if } r = 1$$

The binomial series

- The binomial expansion of the function $(1 + x)^r$, where r is any real number is given by

$$(1 + x)^r = 1 + rx + \frac{1}{2!}r(r - 1)x^2 + \frac{1}{3!}r(r - 1)(r - 2)x^3 + \ldots$$

- The $(n + 1)^{\text{th}}$ term of the series is $\dfrac{1}{n!}r(r - 1)(r - 2) \ldots (r - n + 1)x^r$.
- If r is a positive integer then the series terminates after $r + 1$ terms.
- If r is not a positive integer then the expansion is valid only if $|x| < 1$.

- To expand $(a + x)^r$ then first express it in the form $a^r\left(1 + \dfrac{x}{a}\right)^r$ and then expand as above.

Taylor and Maclaurin series

The Taylor series expansion of $f(x)$ about $x = a$ is

$$f(x + a) = f(a) + \frac{x}{1!}f'(a) + \frac{x^2}{2!}f''(a) + \ldots = \sum_{n=0}^{\infty} \frac{x^n}{n!}f^{(n)}(a)$$

The expansion is only valid within the region of convergence of the infinite series. In the special case when $a = 0$ we have the Maclaurin series expansion of $f(x)$

$$f(x) = f(0) + \frac{x}{1!}f'(0) + \frac{x^2}{2!}f''(0) + \ldots = \sum_{n=0}^{\infty} \frac{x^n}{n!}f^{(n)}(0)$$

Some standard Maclaurin series expansions

$$e^x = 1 + \frac{x}{1!} + \frac{x^2}{2!} + \frac{x^3}{3!} + \ldots + \frac{x^n}{n!} + \ldots \quad \text{(all } x\text{)}$$

$$\sin x = x - \frac{x^3}{3!} + \frac{x^5}{5!} - \ldots + \frac{(-1)^n x^{2n+1}}{(2n + 1)!} + \ldots \quad \text{(all } x\text{)}$$

$$\cos x = 1 - \frac{x^2}{2!} + \frac{x^4}{4!} - \ldots + \frac{(-1)^n x^{2n}}{(2n)!} + \ldots \quad \text{(all } x\text{)}$$

$$\ln(1 + x) = x - \frac{x^2}{2} + \frac{x^3}{3} - \frac{x^4}{4} + \ldots + \frac{(-1)^n x^{n+1}}{n + 1} + \ldots \quad (-1 < x \leq 1)$$

$$\tan x = x + \frac{x^3}{3} + \frac{2x^5}{15} + \frac{17x^7}{315} + \dots \quad (-\tfrac{1}{2}\pi < x < \tfrac{1}{2}\pi)$$

$$\sinh x = x + \frac{x^3}{3!} + \frac{x^5}{5!} + \dots + \frac{x^{2n+1}}{(2n+1)!} + \dots \quad \text{(all } x)$$

$$\cosh x = 1 + \frac{x^2}{2!} + \frac{x^4}{4!} + \dots + \frac{x^{2n}}{(2n)!} + \dots \quad \text{(all } x)$$

AI.2 Trigonometric identities

$$\cos^2 x + \sin^2 x = 1$$

$$1 + \tan^2 x = \sec^2 x$$

$$1 + \cot^2 x = \operatorname{cosec}^2 x$$

$$\sin(x + y) = \sin x \cos y + \cos x \sin y$$

$$\sin(x - y) = \sin x \cos y - \cos x \sin y$$

$$\cos(x + y) = \cos x \cos y - \sin x \sin y$$

$$\cos(x - y) = \cos x \cos y + \sin x \sin y$$

$$\tan(x + y) = \frac{\tan x + \tan y}{1 - \tan x \tan y}$$

$$\tan(x - y) = \frac{\tan x - \tan y}{1 + \tan x \tan y}$$

$$\sin 2x = 2 \sin x \cos x$$

$$\cos 2x = \cos^2 x - \sin^2 x$$

$$= 1 - 2 \sin^2 x$$

$$= 2 \cos^2 x - 1$$

$$\sin x + \sin y = 2 \sin \tfrac{1}{2}(x + y) \cos \tfrac{1}{2}(x - y)$$

$$\sin x - \sin y = 2 \cos \tfrac{1}{2}(x + y) \sin \tfrac{1}{2}(x - y)$$

$$\cos x + \cos y = 2 \cos \tfrac{1}{2}(x + y) \cos \tfrac{1}{2}(x - y)$$

$$\cos x - \cos y = -2 \sin \tfrac{1}{2}(x + y) \sin \tfrac{1}{2}(x - y)$$

$$\sin x \cos y = \tfrac{1}{2}[\sin(x + y) + \sin(x - y)]$$

$$\cos x \sin y = \tfrac{1}{2}[\sin(x + y) - \sin(x - y)]$$

$$\cos x \cos y = \tfrac{1}{2}[\cos(x + y) + \cos(x - y)]$$

$$\sin x \sin y = \tfrac{1}{2}[\cos(x - y) - \cos(x + y)]$$

$$\sin 3x = 3 \sin x - 4 \sin^3 x$$

$$\cos 3x = 4 \cos^3 x - 3 \cos x$$

AI.3 Derivatives and integrals

y	dy/dx	$\int y\,dx$
x^n	nx^{n-1}	$x^{n+1}/(n+1) \quad (n \neq -1)$
$1/x$	$-1/x^2$	$\ln|x|$
$\sin x$	$\cos x$	$-\cos x$
$\cos x$	$-\sin x$	$\sin x$
$\tan x$	$\sec^2 x$	$-\ln|\cos x|$
$\sec x$	$\sec x \tan x$	$\ln|\sec x + \tan x|$
$\cot x$	$-\text{cosec}^2 x$	$\ln|\sin x|$
$\text{cosec}\, x$	$-\text{cosec}\, x \cot x$	$-\ln|\text{cosec}\, x + \cot x|$
$\sin^{-1}x$	$\dfrac{1}{\sqrt{(1-x^2)}}$	$x\sin^{-1}x + \sqrt{(1-x^2)}$
$\cos^{-1}x$	$\dfrac{-1}{\sqrt{(1-x^2)}}$	$x\cos^{-1}x - \sqrt{(1-x^2)}$
$\tan^{-1}x$	$\dfrac{1}{1+x^2}$	$x\tan^{-1}x - \frac{1}{2}\ln(1+x^2)$
$\sec^{-1}x$	$\dfrac{1}{x\sqrt{(x^2-1)}}$	$x\sec^{-1}x - \ln|x + \sqrt{(x^2-1)}|$ $(0 < \sec^{-1}x < \frac{1}{2}\pi)$
$\text{cosec}^{-1}x$	$\dfrac{-1}{x\sqrt{(x^2-1)}}$	$(x\,\text{cosec}^{-1}x + \ln|x + \sqrt{(x^2-1)}|)$ $(0 < \text{cosec}^{-1}x < \frac{1}{2}\pi)$
$\cot^{-1}x$	$\dfrac{-1}{1+x^2}$	$x\cot^{-1}x + \frac{1}{2}\ln(1+x^2)$
e^{ax}	ae^{ax}	e^{ax}/a
a^x	$a^x \ln a$	$a^x/\ln a$
$\sinh x$	$\cosh x$	$\cosh x$
$\cosh x$	$\sinh x$	$\sinh x$
$\tanh x$	$\text{sech}^2 x$	$\ln\cosh x$
$\text{sech}\, x$	$-\text{sech}\, x \tanh x$	$\tan^{-1}(\sinh x)$
$\text{cosech}\, x$	$-\text{cosech}\, x \coth x$	$\ln|\tanh\frac{1}{2}x|$
$\coth x$	$-\text{cosech}^2 x$	$\ln|\sinh x|$
$\sinh^{-1}x$	$\dfrac{1}{\sqrt{(1+x^2)}}$	$x\sinh^{-1}x - \sqrt{(1+x^2)}$
$\cosh^{-1}x$	$\dfrac{1}{\sqrt{(x^2-1)}}$	$x\cosh^{-1}x - \sqrt{(x^2-1)}$
$\tanh^{-1}x$	$\dfrac{1}{1-x^2}$	$x\tanh^{-1}x + \frac{1}{2}\ln(1-x^2)$ $(|x| < 1)$
$\text{sech}^{-1}x$	$\dfrac{-1}{x\sqrt{(1-x^2)}}$	$x\,\text{sech}^{-1}x + \sin^{-1}x$
$\text{cosech}^{-1}x$	$\dfrac{-1}{x\sqrt{(1+x^2)}}$	$x\,\text{cosech}^{-1}x + \sinh^{-1}x$
$\coth^{-1}x$	$\dfrac{1}{1-x^2}$	$x\coth^{-1}x + \frac{1}{2}\ln(x^2-1)$ $(|x| > 1)$
$\ln x$	$1/x$	$x\ln x - x$

| AI.4 | **Some useful standard integrals** |

$f(x)$, $(a > 0)$	$\int f(x)\,dx$
$\dfrac{1}{a^2 + x^2}$	$\dfrac{1}{a}\tan^{-1}\left(\dfrac{x}{a}\right)$
$\dfrac{1}{\sqrt{(a^2 - x^2)}}$	$\sin^{-1}\left(\dfrac{x}{a}\right)$
$\dfrac{1}{\sqrt{(a^2 + x^2)}}$	$\sinh^{-1}\left(\dfrac{x}{a}\right)$ or $\ln\left[x + \sqrt{(x^2 + a^2)}\right]$
$\dfrac{1}{\sqrt{(x^2 - a^2)}}$	$\cosh^{-1}\left(\dfrac{x}{a}\right)$ or $\ln\left[x + \sqrt{(x^2 - a^2)}\right]$
$\dfrac{1}{a^2 - x^2}$ for $|x| < a$	$\dfrac{1}{2a}\ln\left(\dfrac{a + x}{a - x}\right)$
$\dfrac{1}{x^2 - a^2}$ for $|x| > a$	$\dfrac{1}{2a}\ln\left(\dfrac{x - a}{x + a}\right)$

Answers to Exercises

CHAPTER 1

Exercises

1 (a) $1/2$ (b) 2^7 (c) $1/2^{12}$
   (d) 3^2 (e) $1/6$ (f) 2^3

2 (a) $21 + ((4 \times 3) \div 2)$ (b) $(17 - (6^{(2+3)}))$
   (c) $(4 \times (2^3)) - ((7 \div 6) \times 2)$
   (d) $((2 \times 3) - (6 \div 4)) + 3^{(2-5)}$

3 (a) $1393 + 985\sqrt{2}$ (b) $68 + 48\sqrt{2}$
   (c) $1 + \sqrt{2}$ (d) $-1 + \frac{3}{2}\sqrt{2}$

4 (a) $-7 + 5\sqrt{2}$ (b) $-\frac{60}{17} - \frac{41}{17}\sqrt{2}$
   (c) $\frac{5}{11} - \frac{1}{11}\sqrt{3}$ (d) $\frac{28}{11} + \frac{18}{11}\sqrt{5}$

5   $\frac{239}{169}, \frac{577}{408}, \frac{1393}{985}$

6   $\sqrt{3} + \sqrt{19} > \sqrt{5} + \sqrt{13}$

7 (a) $-2 \leqslant x \leqslant 10, [-2, 10]$
   (b) $-5 < x < -1, (-5, -1)$
   (c) $-3 \leqslant x \leqslant 4, [-3, 4]$
   (d) $-24 < x < 0, (-24, 0)$

8 (a) $\{x: |x - 4| < 3\}$ (b) $\{x: |x + 3| \leqslant 1\}$
   (c) $\{x: |2x - 43| < 9\}$ (d) $\{x: |8x - 1| \leqslant 5\}$

9 (b) only
   (b), (c) and (d) true

10 (a) $v_a = \frac{1}{2}(v_1 + v_2)$ (b) $v_b = \dfrac{2v_1v_2}{v_1 + v_2}$

11 (a) x^{-1} (b) x^7 (c) x^{-12} (d) x^2
   (e) $1/(2x^4)$ (f) $4x/9$ (g) $x^{5/2} - 2x^{-1/2}$
   (h) $25x^{2/3} + 1/(4x^{2/3}) - 5$ (i) $2 - 1/x$
   (j) $a^{-1}b^{9/2}$ (k) $1/(8b^3a^{3/2})$

12 (a) $xy(x - y)$ (b) $xyz(x - y + 2z)$
   (c) $(a + b)(x - 2y)$ (d) $(x + 5)(x - 2)$
   (e) $(x + \frac{1}{2}y)(x - \frac{1}{2}y)$ (f) $(9x^2 + y^2)(3x - y)(3x + y)$

13 (a) $(x + 3)/(x + 4)$ (b) $(5 - x)/[(x - 3)(x + 1)]$
   (c) $2/[(x - 2)(x + 12)]$ (d) $3x^2 - 4y^2$

16 (a) $(x + \frac{1}{2})^2 - \frac{49}{4}$ (b) $(x - 1)^2 + 2$
   (c) $\frac{1}{3} - 3(x - \frac{5}{3})^2$ (d) $5 - (x - 2)^2$

17   $s = (m^2 + p^2)t/(m^2 - p^2), m^2 \neq p^2$

18   $t = (u - 1)x^2/(u + 1), u \neq -1$

19   $-1 \pm \sqrt{2}$

20   $\frac{4}{3}$

21   $7, -2$

22   $10\,\text{m}$

23 (a) $\sqrt{2}$ (b) $(1 + \sqrt{5})/2$

24 (a) $x < 0$ and $x > 5/2$
   (b) $x < 1$ and $x > 2$
   (c) $x < 0$ and $x > 1$
   (d) $x < -4$ and $x > \frac{2}{3}$

25   $-2 < x < 2$

27 (a) $A = -\frac{1}{3}, B = \frac{1}{3}$ (b) $A = 8, B = -5$
   (c) $A = \frac{5}{2}, B = -\frac{3}{2}$

28   $A = 2, B = -1, C = 9$

29 (a) 5 (b) 0 (c) -9 (d) 11 (e) 20 (f) -4

30 (a) 120 (b) $\frac{1}{4}$ (c) 35 (d) 10 (e) 84 (f) 70

31 (a) $x^4 - 12x^3 + 54x^2 - 108x + 81$
   (b) $x^3 + \frac{3}{2}x^2 + \frac{3}{4}x + \frac{1}{8}$
   (c) $32x^5 + 240x^4 + 720x^3 + 1080x^2 + 810x + 243$
   (d) $81x^4 + 216x^3y + 216x^2y^2 + 96xy^3 + 16y^4$

32 (a) $y = \frac{3}{2}x - 2$ (b) $y = -2x - 1$
   (c) $y = \frac{5}{2}x - \frac{1}{2}$ (d) $y = -\frac{3}{5}x + 3$
   (e) $y = \frac{1}{3}x + \frac{2}{3}$ (f) $y = -3x + 4$

33   $(x - 1)^2 + (y - 2)^2 = 25$

34   $4, (-2, 3)$

35   $x^2 + y^2 + 4x - 6y = 12$

36   $x^2 + y^2 - 6x - 3y + 5 = 0$

37 $2y = x + 3$

38 $x^2 + y^2 = 25^2$

39 $x = \frac{2}{3}, x = -\frac{2}{3}$

40 $(0, 3), (0, -3), \frac{3}{5}, y = \frac{25}{3}, y = -\frac{25}{3}, 10, 8$

41 $(-5, 0), (5, 0), (-4, 0), (4, 0), y = \frac{3x}{4}, y = -\frac{3x}{4}$

42 54.625_{10}

43 $11111111000001_2, 37\,701_8$
$13\,455_8$

44 $11110.100110011001\ldots_2$
$36.463\,146\,31\ldots_8$
Yes

45 (a) $101\,110.100_2$ (b) $10\,101.110\,101\,01_2$

46 (a) 3dp, 6sf   (b) 30dp, 3sf   (c) 0dp, 5sf
(d) 0dp, 3sf   (e) 0dp, 4sf   (f) 10dp, 3sf

47 The answer claims unjustified accuracy: hypotenuse = 2.236 ± 0.007 m. The angles are also subject to error.

48 (a) Absolute error bound is $\frac{1}{12}$ min, relative error bound is $\frac{1}{420}$
(b) Absolute error bound is 1.4 min, relative error bound is 0.04
(c) Absolute error bound is 0.005, relative error bound is $\frac{1}{116}$

49 0.0039, 12.9

50 (a) 3.613 ± 0.0015, relative error bound 0.0004, 3.61
(b) 2.5351 ± 0.0176, relative error bound 0.007, 2.5
(c) 22.47 ± 0.015, relative error bound 0.0007, 22.5

51 4.51

52 $10.00 \pm 0.01, \frac{1}{1000}$
$-0.02 \pm 0.01, \frac{1}{2}$
$24.9999 \pm 0.05, \frac{1}{500}$
$0.996\,008 \pm 0.002, \frac{1}{500}$

53

Label	Value	Absolute error bound	Relative error bound
a	3.251	0.0005	
b	3.115	0.0005	
$a - b$	0.136	0.001	0.0074
c	0.112	0.0005	0.0045
$(a - b)/c$	1.2143	0.0145	0.0119
d	9.21	0.005	
$d + (a - b)/c$	10.4243	0.0195	

Result: 10.4

54 $0.7634 \pm 0.000\,72, 0.76$

55 (a) 0.2713 ± 0.0237 (b) 0.2715 ± 0.0072

56 $10^1(0.2709), 10^1(0.2708)$
The second result is more accurate since by adding the small numbers together first their combination is given its proper weight.

57 $10^{-8}(0.6538), 10^{-3}(0.6752)$

58 0.5

1.7 Review exercises

1 (a) $A = \pm QKD/\sqrt{(HD^2 - Q^2K^2)}$
(b) $-(9 \pm \sqrt{145})/8$

2 (a) $(x - 1)(a - 2)$ (b) $(a - b + c)(a + b - c)$
(c) $(2k + l - 3m)(2k + l + 3m)$
(d) $(p - q)(p - 2q)$ (e) $(l + n)(l + m)$

3 (a) 1 cm   (b) 3.812

4 (a) $L = \left(\frac{1}{2\pi nC} \pm \sqrt{(Z^2 - R^2)}\right)\Big/(2\pi n)$

(b) 0.1434; 0.0592; 0.4160
(L must be positive from practical considerations)

5 (a) $30 - 12\sqrt{6}$ (b) $-53 + 11\sqrt{15}$
(c) $\frac{1}{23}(14 + 11\sqrt{2})$ (d) $3 + 2\sqrt{2} + 2\sqrt{3} + \sqrt{6}$
(e) $\frac{1}{2} + \frac{1}{4}\sqrt{2} + \frac{1}{4}\sqrt{6}$

6 5, 6

8 (a) $(-\frac{1}{2}, \frac{3}{2})$ (b) $(-\infty, -5) \cup (-2, 1)$
(c) $(-3, 1)$ (d) $(-\frac{4}{3}, 0)$

11 $\dfrac{a}{b} > \dfrac{a + c}{b + c} > 1$

12 (b) (i) $1 - \frac{5}{2}x + \frac{5}{2}x^2 - \frac{5}{4}x^3 + \frac{5}{16}x^4 - \frac{1}{32}x^5$
(ii) $729 - 2916x + 4860x^2 - 4320x^3 + 2160x^4 - 576x^5 + 64x^6$

13 (a) 90.5
(b) $P_1 = 1, P_2 = 3, P_3 = 5, P_r = (2r - 1)$,
$$\sum_{r=1}^{n} P_r = n^2$$

14 (a) $y = 2x + 1$ (b) $y = (x - 7)/3$ (c) $y = 2x - \frac{7}{3}$

15 $(y - 3)^2 + (x - 5)^2 = 25$

16 (a) $(-1, 2), 2$ (b) $(\frac{1}{2}, -\frac{3}{2}), \frac{1}{2}$ (c) $(-\frac{1}{3}, \frac{1}{3}), \sqrt{3}$

17 (i) (a) $(1, 2)$ (b) $(3, 2)$ (c) $x = -1$ (d) $y = 2$
(ii) (a) $(-2, 1)$ (b) $(-2, -2)$ (c) $y = 4$ (d) $x = -2$

18 $(2, 8), (2, 5), (2, 11), (2, 3), (2, 13), y = -\frac{1}{3}, y = \frac{49}{3}$

19 $\Delta.477\,4\Delta\,774\Delta\,\ldots\,_{12}$, where $\Delta_{12} = 10_{10}$

20

	Value	Absolute error bound	Relative error bound	
a	7.01	0.005	→ 0.0007	
$\sqrt{a}$	2.6476	0.0009	← 0.00035	
b	52.13	0.005	→ 0.000096	
$\sqrt{b}$	7.220111	0.000347	← 0.000048	
c	0.01011	0.000005	→ 0.000495	
$\sqrt{c}$	0.100548	0.000025	← 0.00025	
d	5.631×10^{11}	0.5×10^{8}	→ 0.0000888	
$\sqrt{d}$	7.504×10^{5}	0.33×10^{2}	← 0.0000444	

	$\sqrt{a}$	$\sqrt{b}$	$\sqrt{c}$	$\sqrt{d}$
Correctly rounded values	2.65	7.22	0.101	7.504×10^{5}

21 0.37 ± 0.07

22 1.714   (a) 0.0026, (b) 0.0075

23 6

CHAPTER 2

Exercises

1 (a) $[-5, 5], \mathbb{R}, [0, 5], 0, 3, \sqrt{(25 - x^2)}$
  (b) $\mathbb{R}, \mathbb{R}, \mathbb{R}, 2, -1, \sqrt[3]{(3 - x)}$

2 $A = 2x(5 + |x|)$

x/m	0	1	2	3	4	5
Area/m^2	0	12	28	48	72	100

$A(-2) = -28$, area of cutting

3

r/m	0.10	0.15	0.20	0.25	0.30	0.35	0.40
A/m^2	3.05	2.12	1.71	1.53	1.47	1.50	1.59

$r* = 0.32$ according to worked answer: estimated from a graph (not drawn).

5 5 years

6 (a) 0, 2; increasing for $x > 1$, decreasing for $x < 1$, minimum at $x = 1$
  (b) $-\frac{5}{2}$, 2; increasing for $x < -1$ and $x > 2$, decreasing for $-1 < x < 2$, maximum at $x = -1$, minimum at $x = 2$

  (c) Increasing on $-1 < x < 0$ and $x > +1$
    Decreasing on $x < -1$ and $0 < x < +1$
    Maximum at $(0, 0)$, minimum at $(-1, -1)$ and $(1, -1)$
  (d) Increasing on $x < 1$, decreasing on $x > 1$, maximum at $(1, -1)$

8 $F(x) = (x - 1)^2$: $f(x)$ shifted by 2 units in positive x direction
  $G(x) = (x + 1)^2 - 2$: $f(x)$ shifted by 2 units in negative y direction

10 (a) $\frac{1}{2}(x + 3)$ (b) $\dfrac{4x + 3}{2 - x}$
  (c) Restriction of domain to $[0, \infty)$
    $\sqrt{(x - 1)}, x \geq 1$

14 (a) odd   (b) even   (c) neither
  (d) neither   (e) odd   (f) even

16 (a) $3 - 2x$
  (b) $\frac{1}{2}x + \frac{5}{2}$
  (c) $0.255x + 2.478$ (3dp)

17 (a) 3
  (b) -3
  (c) $\frac{1}{2}$

18 £$(50 + 0.455x)$, £960, £$(1.20x - 960)$, 800

19 $a = 0.311$

20 $m = 0.82, c = 60.9$

23 (a) $\frac{2}{3}x^2 + 2x + \frac{1}{3}$
  (b) $\frac{2}{5}x^2 - x - \frac{2}{5}$

24 $(x - 2)^2 - 4(x - 2) - 2$

25 (a) irreducible   (b) not irreducible
  (c) not irreducible   (d) irreducible

26 (a) minimum at $x = -1$ of 2
  (b) minimum at $x = \frac{3}{2}$ of 0
  (c) maximum at $x = -\frac{2}{3}$ of $\frac{22}{3}$
  (d) maximum at $x = \frac{3}{10}$ of $-\frac{11}{20}$

27 (a) $x < 2$ and $x > 4$
  (b) $-\frac{5}{2} < x < 3$

28 315 feet, 46 mph

29 (a) $(x - 1)(x + 3)(x - 4)$
  (b) $(x + 1)(x - 2)(x + 3)$
  (c) $(x - 1)(x + 1)(x^2 + 2)$
  (d) $x(x - 1)(2x + 3)(x + 2)$
  (e) $(x - 1)^2(2x - 1)(x - 2)$
  (f) $(x^2 + 9)(x - 2)(x + 2)$

30 2, 7, 139, 527, 524

31 $y = (x-5)^4 + 15(x-5)^3 + 80(x-5)^2 + 165(x-5) + 81$. Since coefficients are all positive, zeros of y must all lie to the left of $x = 5$, i.e. $x < 5$. Hence the zeros of y lie between $x = 0$ and $x = 5$.

32 (a) $x^2 - 14x + 1 = 0$
(b) $x^2 + 52x + 1 = 0$

33 $x^3 - 5x^2 + 1$

34 $3x^2 + 22x + 378$

35 (b) $r = 10/(4\pi)^{1/3}$, $h = 20/(4\pi)^{1/3}$

36 $0.096 \, \text{m}^3$, $0.1875 \, \text{m}^3$

37 $x_0 = 10.94$, width of alley $= 4.92 \, \text{m}$

38 (a) $1 + (x+2)/[(x+1)(x-1)]$
(b) $x^3 - 2x^2 + x + 1 - 3x/(x^2 + x + 1)$

39 (a) $(-5x^2 + x - 2)/[x(x-2)(x^2+1)]$
(b) $2/[(x-1)^3(x+1)]$
(c) $(4x^4 - 11x^3 + 10x^2 - 5x + 4)/[(x^2+1)(x-1)^2(x-2)]$

40 (a) $\dfrac{\frac{1}{3}}{x-2} - \dfrac{\frac{1}{3}}{x+1}$

(b) $\dfrac{1}{x-2} + \dfrac{1}{x+1}$

(c) $1 + \dfrac{\frac{2}{3}}{x-2} + \dfrac{\frac{1}{3}}{x+1}$

(d) $\dfrac{\frac{1}{3}}{(x-2)^2} + \dfrac{\frac{2}{9}}{x-2} - \dfrac{\frac{2}{9}}{x+1}$

(e) $\dfrac{1}{x+1} - \dfrac{x+1}{x^2+2x+2}$

(f) $\dfrac{-\frac{1}{3}}{x+1} + \dfrac{\frac{1}{12}}{x-2} + \dfrac{\frac{1}{4}}{x+2}$

41 (a) $\dfrac{\frac{1}{3}}{x-4} - \dfrac{\frac{1}{3}}{x-1}$

(b) $\dfrac{\frac{1}{3}}{x-1} - \dfrac{\frac{1}{3}x + \frac{2}{3}}{x^2 + x + 1}$

(c) $\dfrac{\frac{5}{9}}{x-2} + \dfrac{\frac{4}{3}}{(x+1)^2} - \dfrac{\frac{5}{9}}{x+1}$

(d) $1 - \dfrac{3}{x-2} + \dfrac{8}{x-3}$

(e) $\dfrac{1}{x^2+1} + \dfrac{x-2}{(x^2+1)^2}$

(f) $\dfrac{x+1}{x^2+4} + \dfrac{2}{x-1} - \dfrac{3}{x+5}$

42 (a) $(\sqrt{2}, \sqrt{2}), (-\sqrt{2}, -\sqrt{2})$
(b) $(\sqrt{2}, \sqrt{2}), (-\sqrt{2}, -\sqrt{2})$
(c) $(-\sqrt{\frac{2}{5}}, -\sqrt{\frac{2}{5}}), (\sqrt{\frac{2}{5}}, \sqrt{\frac{2}{5}}), (\sqrt{2}, \sqrt{2}), (-\sqrt{2}, -\sqrt{2})$
(d) does not intersect on domain

43 (a) asymptotes: $y = x - 8$, $x = 0$,
maximum $(\sqrt{15}, -8 + 2\sqrt{15})$,
minimum $(-\sqrt{15}, -8 - 2\sqrt{15})$
(b) asymptotes: $y = 1$, $x = 1$ (c) $y = x$, $x = -5$

44 $y = -1 \pm \sqrt{(x+4)}$, $(y+1)^2 = x + 4$

47 0.6, 0.8, 0.75, $36.87° = 36°52'12''$;
$\frac{12}{13}, \frac{5}{13}, 2.4, 67.38° = 67°22'48''$

48 $AB = 29.44$, $BC = 33.04 \, \text{m}$

49 30

50 60

51 $AB = 30.6 \, \text{mm}$, $AC = 26.9 \, \text{mn}$

52 45.5 mm

54

degrees	0°	30°	45°	60°	90°	120°	150°	180°
radians	0	$\frac{1}{6}\pi$	$\frac{1}{4}\pi$	$\frac{1}{3}\pi$	$\frac{1}{2}\pi$	$\frac{2}{3}\pi$	$\frac{5}{6}\pi$	π

degrees	210°	225°	240°	270°	300°	315°	330°	360°
radians	$\frac{7}{6}\pi$	$\frac{5}{4}\pi$	$\frac{4}{3}\pi$	$\frac{3}{2}\pi$	$\frac{5}{3}\pi$	$\frac{7}{4}\pi$	$\frac{11}{6}\pi$	2π

56 (a) 0.3398, 2.8018, $\frac{3}{2}\pi$
(b) 1.8235, 4.4597, π
(c) 2.6779, 5.8195, $\frac{1}{4}\pi$, $\frac{5}{4}\pi$
(d) $\frac{1}{2}\pi$, $\frac{3}{2}\pi$, $\frac{1}{6}\pi$, $\frac{5}{6}\pi$

57 $\frac{1}{2}\sqrt{3}, \sqrt{3}, \frac{1}{2}\sqrt{3}, \frac{1}{2}$
$\frac{1}{2}\sqrt{(2-\sqrt{3})}, \frac{1}{2}\sqrt{(2+\sqrt{3})}, 2 - \sqrt{3}$
(a) $\frac{1}{2}\sqrt{3}$ (b) $1/\sqrt{3}$
(c) $\frac{1}{2}\sqrt{3}$ (d) $\frac{1}{2}\sqrt{(2+\sqrt{3})}$
(e) $-\frac{1}{2}\sqrt{(2-\sqrt{3})}$ (f) $-(2-\sqrt{3})$

58 (a) $-\sqrt{(1-s^2)}$ (b) $-2s\sqrt{(1-s^2)}$
(c) $s(3-4s^2)$ (d) $\sqrt{\{\frac{1}{2}[1+\sqrt{(1-s^2)}]\}}$

60 $x = n\pi$ ($n = \pm1, \pm3, \dots$)
and $x = 0.9273 + 2n\pi$ ($n = 0, \pm1, \pm2, \dots$)

61

	a	b	c	d	e	f
$\sin x$	$\frac{1}{2}$	$\pm\frac{1}{2}$	$\pm\sqrt{\frac{1}{2}}$	$\pm\sqrt{\frac{1}{2}}$	$-\frac{1}{2}$	$\pm\frac{1}{2}$
$\cos x$	$\pm\frac{1}{2}\sqrt{3}$	$-\frac{1}{2}\sqrt{3}$	$\pm\sqrt{\frac{1}{2}}$	$\sqrt{\frac{1}{2}}$	$\pm\frac{2}{\sqrt{3}}$	$\pm\frac{1}{2}\sqrt{3}$
$\tan x$	$\pm\sqrt{\frac{1}{3}}$	$\pm\sqrt{\frac{1}{3}}$	-1	±1	$\pm\sqrt{\frac{1}{3}}$	$\sqrt{\frac{1}{3}}$
$\operatorname{cosec} x$	2	±2	$\pm\sqrt{2}$	$\pm\sqrt{2}$	-2	±2
$\sec x$	$\pm2\sqrt{\frac{1}{3}}$	$-2\sqrt{\frac{1}{3}}$	$\pm\sqrt{2}$	$\sqrt{2}$	$\pm\frac{2}{3}$	$\pm2\sqrt{\frac{1}{3}}$
$\cot x$	$\pm\sqrt{3}$	$\pm\sqrt{3}$	-1	±1	$\pm\sqrt{3}$	$\sqrt{3}$

62 (a) $2\sin 2\theta \cos\theta$ (b) $2\sin\frac{3}{2}\theta \sin\frac{1}{2}\theta$
(c) $2\cos\frac{7}{2}\theta \cos\frac{3}{2}\theta$ (d) $-2\cos\frac{3}{2}\theta \sin\frac{1}{2}\theta$

63 (a) $\frac{1}{2}(\cos 2\theta - \cos 4\theta)$ (b) $\frac{1}{2}(\sin 4\theta + \sin 2\theta)$
(c) $\frac{1}{2}(\sin 4\theta - \sin 2\theta)$ (d) $\frac{1}{2}(\cos 4\theta + \cos 2\theta)$

64 (a) $2\cos(\theta - \frac{2}{3}\pi)$, $2\sin(\theta - \frac{1}{6}\pi)$
(b) $\sqrt{2}\cos(\theta - \frac{3}{4}\pi)$, $\sqrt{2}\sin(\theta - \frac{1}{4}\pi)$
(c) $\sqrt{2}\cos(\theta - \frac{1}{4}\pi)$, $\sqrt{2}\sin(\theta - \frac{7}{4}\pi)$
(d) $\sqrt{13}\cos(\theta - 0.9828)$, $\sqrt{13}\sin(\theta - 5.6952)$

65 $x = 2n\pi,\ 2n\pi \pm \frac{2}{3}\pi$ $(n = 0, \pm 1, \pm 2, \dots)$

66 (a) $\pi/6$ (b) $-\pi/6$ (c) $\pi/3$
(d) $2\pi/3$ (e) $\pi/3$ (f) $-\pi/3$

69

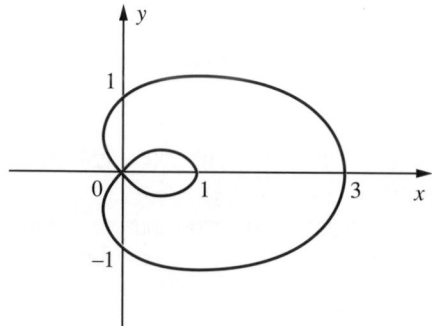

70

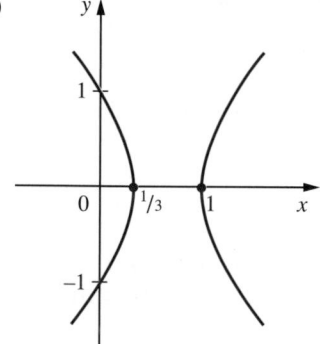

71 (a) $(2e + 1)e^5$ (b) e^{4x} (c) e^6
(d) e^9 (e) $e^{x/2}$

73 (a) 3 (b) -2 (c) $-\frac{1}{2}$
(d) 4 (e) $\frac{1}{2}$ (f) $-\frac{1}{2}$

74 (a) $2\ln x + \ln y$ (b) $\frac{1}{2}\ln x + \frac{1}{2}\ln y$
(c) $5\ln x - 2\ln y$

75 (a) $\ln 4$ (b) $\ln 3.2$ (c) $\ln 0.75$ (d) $\ln 0.5$

76 (a) $\sqrt{[(1-x)/(1+x)]}$ (b) x^2

78 $\ln(20 \pm 6\sqrt{10}) = 3.6629,\ 0.025\,99$

79 $\frac{3}{2}\ln(x^2 + 1) - \frac{1}{3}\ln(x^4 + 1) - \frac{1}{5}\ln(x^4 + 4)$

80

	a	*b*	*c*	*d*	*e*	*f*
$\sinh x$	$\pm\frac{3}{4}$	$\frac{8}{15}$	$\frac{7}{24}$	$\pm\frac{12}{5}$	$-\frac{4}{3}$	$\frac{12}{5}$
$\cosh x$	$\frac{5}{4}$	$\frac{17}{15}$	$\frac{25}{24}$	$\frac{13}{5}$	$\frac{5}{3}$	$\frac{13}{5}$
$\tanh x$	$\pm\frac{3}{5}$	$\frac{8}{17}$	$-\frac{7}{25}$	$\pm\frac{12}{13}$	$-\frac{4}{5}$	$\frac{12}{13}$
$\operatorname{cosech} x$	$\pm\frac{4}{3}$	$\frac{15}{8}$	$\pm\frac{24}{7}$	$\pm\frac{5}{12}$	$-\frac{3}{4}$	$\frac{5}{12}$
$\operatorname{sech} x$	$\frac{4}{5}$	$\frac{15}{17}$	$\frac{24}{25}$	$\frac{5}{13}$	$\frac{3}{5}$	$\frac{5}{13}$
$\coth x$	$\pm\frac{5}{3}$	$\pm\frac{17}{8}$	$-\frac{25}{7}$	$\pm\frac{13}{12}$	$-\frac{3}{4}$	$\frac{13}{12}$

81 (a) $\tanh 3x = \dfrac{(3 + \tanh^2 x)\tanh x}{1 + 3\tanh^2 x}$
(b) $\cosh(x + y) = \cosh x \cosh y + \sinh x \sinh y$
(c) $\cos 2x = 1 - 2\sin^2 x$
(d) $\sinh x - \sinh y = 2\sinh \frac{1}{2}(x - y)\cosh \frac{1}{2}(x + y)$

83 (a) 0.7327 (b) 1.3170 (c) 0.5493

84 17.1383 (4dp)

85 1.0074 (4dp)

87 $A = 250$, $B = -273.26$

90 (a) Cusp at $x = 0$, maximum at $x = 4$, asymptote $y = -x$
(b) Minimum at $x = 2$, asymptotes $y = \pm\sqrt{x} - 1$, $x = 1$

93 $\dfrac{ax}{l}H(x) + \dfrac{2a}{l}(l - x)H(x - l) - \dfrac{a}{l}(2l - x)H(x - 2l)$

94 $x[1 - H(x)] - (x - 1)H(x - 1)$

95 $\operatorname{INTPT}(x + \frac{1}{2})$

98 0.9401, 0.005, 0.9425

99 0.04, 0.16, 0.01, 0.006 25

100 0.3081, 0.2829, 16.79

101 0.2954, 0.2688, 17.10

102 $f(x) = -\frac{1}{84}x^3 + \frac{85}{84}x$

103

x	3045	3051	3058	3064	3070	3077	3083
y		14.50	14.51	14.52	14.53	14.54	14.55

2.11 Review exercises

1 $h(x) = x - 4$ $x \in [0, 200]$
$k(x) = (x^2 - 4)^{1/2}$ $x \in [-20, -2] \cup [2, 20]$

2 (a) 25.6 cm (b) 2.35 m²

3

Price/£	1.00	1.05	1.10	1.15	1.20	1.25	1.30
Sales/000	8	7	6	5	4	3	2
Revenue/£000	8	7.35	6.60	5.75	4.80	3.75	2.60
Profit/£000	0	0.35	0.60	0.75	0.80	0.75	0.60

4 $g(x) = \begin{cases} 3x + 3 & x \leqslant -3 \\ 2x & -3 < x \leqslant -1 \\ 3x + 1 & -1 < x \leqslant 0 \\ x + 1 & 0 < x \leqslant 1 \\ 3x - 1 & x > 1 \end{cases}$

6 0.37 ± 0.005

8 $(x - 1)^4 + 7(x - 1)^3 + 14(x - 1)^2 + 13(x - 1) + 4$

9 (a) $\dfrac{2}{x - 4} - \dfrac{1}{x - 1}$

(b) $1 - \dfrac{\frac{5}{4}}{x + 1} + \dfrac{\frac{13}{4}}{x - 3}$

(c) $\dfrac{2}{9(x - 1)} + \dfrac{7}{9(x + 2)} - \dfrac{11}{3(x + 2)^2}$

(d) $\dfrac{\frac{1}{13}(5x - 7)}{x^2 - x + 1} + \dfrac{\frac{21}{13}}{x + 3}$

10 (a) $2 \cos \frac{3}{2}\theta \sin \frac{1}{2}\theta$

(b) $2 \cos \frac{5}{2}\theta \cos \frac{1}{2}\theta$

(c) $-2 \sin \dfrac{3\theta}{2} \cos \dfrac{11\theta}{2}$

11 (a) $2\sqrt{5} \sin(\theta - \alpha), \alpha = \tan^{-1}\frac{1}{2}$

(b) $\sqrt{65} \sin(\theta - \alpha), \alpha = -\tan^{-1}8$

(c) $2 \sin(\theta + \frac{1}{6}\pi)$

13 2

15 0.0025, 0.300

16 $\sqrt{\frac{1}{3}}$

17 (b) $\frac{1}{2}D\sqrt{3}$

18 1, 2, 0, 1, 3, 7, 5, 5, …
0.833 89, 0.551 94

21 $r = 3/(2 \sin \theta - \cos \theta)$

CHAPTER 3

Exercises

2 (a) 10   (b) $-3 - j4$

(c) $\frac{1}{25}(47 - j4)$ (d) $-j$

(e) j   (f) $5 - j12$

(g) $j\frac{3}{17}$ (h) $-\frac{1}{178}(5 + j8)$

3 (a) $2 - j7$ (b) $-3 + j$ (c) $j6$ (d) $\frac{2}{3} + j\frac{2}{3}$

4 (a) $-1 + j, -1 - j$

(b) $-2, 1 + j\sqrt{3}, 1 - j\sqrt{3}$

5 $\pm 3 + j2$

6 (a) $3 + j2$ (b) $2 + j3$

(c) $\frac{1}{13}(2 + j3)$ (d) $2 - j3$

7 (a) $\sqrt{2}, \frac{1}{4}\pi$ (b) $2, -\frac{1}{6}\pi$

(c) $5, \pi - \tan^{-1}\frac{4}{3}$ (d) $2, -\frac{1}{3}\pi$

(e) $2, \frac{2}{3}\pi$ (f) $2, -\frac{2}{3}\pi$

8 $w = 5 - j4, z = 2 + j3$

9 $x = \frac{1}{2}, y = -\frac{3}{2}$

10 $2 + j2, \frac{1}{2}$

11 $\frac{1}{5}(7 + j4)$

12 $\frac{1}{130}(451 + j878)$

13 $x = \frac{1}{4}, y = -\frac{3}{4}$

14 $\frac{11}{4} - j\frac{13}{4}$

15 (a) $1 \angle \pi/2$

(b) $1 \angle 0$

(c) $1 \angle \pi$

(d) $\sqrt{2} \angle -\pi/4$

(e) $\sqrt{6} \angle -\pi/4$

(f) $\sqrt{5} \angle (\pi - \tan^{-1}\frac{1}{2})$

(g) $\sqrt{13} \angle (\tan^{-1}\frac{2}{3} - \pi)$

(h) $\sqrt{74} \angle (-\tan^{-1}\frac{5}{7})$

(i) $5 \angle 0$

(j) $53 \angle (\tan^{-1}\frac{28}{45} - \pi)$

16 $\frac{4}{5} + j\frac{7}{5}, \frac{1}{5}\sqrt{65} \angle \tan^{-1}\frac{7}{4}$

17 (a) $-\frac{5}{12}\pi, -\frac{11}{12}\pi$ (b) $\frac{1}{2}\sqrt{3} + j\frac{3}{2}$

18 (a) $128, -\frac{1}{3}\pi$ (b) $1024, 0$ (c) $\frac{1}{16}, \frac{2\pi}{3}$

20 (a) $\frac{1}{2} \cosh 1 - j\frac{1}{2}\sqrt{3} \sinh 1$

(b) $\cosh \frac{3}{4}$

(c) $\frac{1}{2} \sinh \dfrac{\pi}{3} + j\dfrac{\sqrt{3}}{2} \cosh \dfrac{\pi}{3}$

(d) $\dfrac{1}{\sqrt{2}}$

21 (a) $\frac{1}{2}(4n + 1)\pi + j \cosh^{-1}2$

(b) $\frac{1}{2}(2n + 1)\pi + j(-1)^{n+1} \sinh^{-1}\frac{3}{4}$

(c) $\frac{1}{2}(4n + 1)\pi + j \cosh^{-1}3$

(d) $\cosh^{-1}2 + j(2n + 1)\pi$

23 $x = \dfrac{\tanh u \sec^2 v}{1 + \tanh^2 u \tan^2 v}$

$y = \dfrac{\tan v \, \text{sech}^2 u}{1 + \tanh^2 u \tan^2 v}$

$$\frac{2\tanh 2}{1+\tanh^2 2} + j\frac{\text{sech}^2 2}{1+\tanh^2 2}$$
$$= 0.9994 + j0.0366$$

24 $0.1645 - j0.1214$

25 (a) $-2 + j2, -4$ (b) $-j8, -8 - j8\sqrt{3}$
(c) $117 + j44, -527 + j336$ (d) $-8, -8 + j8\sqrt{3}$
(e) $8, -8 + j8\sqrt{3}$ (f) $8, -8 - j8\sqrt{3}$

26 (a) $\frac{1}{8}\cos 4\theta + \frac{1}{2}\cos 2\theta + \frac{3}{8}$
(b) $\frac{3}{4}\sin\theta - \frac{1}{4}\sin 3\theta$

28 $2^{7/6}\angle(\frac{\pi}{12} + \frac{2}{3}k\pi), k = 0, 1, 2$

29 (a) $2^{1/4}\angle(-\frac{1}{24}\pi + \frac{1}{2}k\pi), k = 0, 1, 2, 3$
(b) $2\angle(\frac{1}{6}\pi + \frac{2}{3}k\pi), k = 0, 1, 2$
(c) $18^{-1/3}\angle(\frac{1}{6}\pi - \frac{4}{3}k\pi), k = 0, 1, 2$
(d) $1\angle(\frac{1}{4}\pi + \frac{1}{2}k\pi), k = 0, 1, 2, 3$
(e) $4\angle(\frac{1}{3}\pi + \frac{8}{3}k\pi), k = 0, 1, 2$
(f) $34^{-1/4}\angle(\frac{1}{2}\tan^{-1}\frac{3}{5} - k\pi), k = 0, 1$

30 $1.455 - j0.344, 0.344 + j1.455, -1.455 + j0.344,$
$-0.344 - j1.455$

31 $2.529 + j2.743, 0.471 + j2.257$

32 $1, -\frac{1}{2} \pm j\frac{1}{2}\sqrt{3}$
$$\cos\frac{2k\pi}{n} + j\sin\frac{2k\pi}{n}, k = 1, 2, \ldots, n$$
(a) $j2\cot\frac{1}{5}k\pi, k = 1, \ldots, 4$
(b) $\frac{3}{2}(1 + j\cot\frac{1}{6}k\pi), k = 1, \ldots, 5$

33 $5, 13$

34 (a) $x = 5$, a straight line
(b) circle centre $(1, 0)$, radius 3
(c) circle centre $(-\frac{5}{4}, 0)$, radius $\frac{3}{4}$
(d) half-line, $y = x - 2, x > 2$

35 circle is $|z + 2| = 2$
line is Re $((3 + j) + z) = -2$

36 (a) Straight line, $y = 1$
(b) Circle, centre $(0, 2)$, radius 1
(c) Circle, centre $(0, \frac{5}{4})$, radius $\frac{3}{4}$
(d) Circle, centre $(\sqrt{\frac{1}{3}}, 0)$, radius $2\sqrt{\frac{1}{3}}$
(e) Rectangular hyperbola, $xy = 1$
(f) Ellipse, foci at $(1, 0), (0, -1)$, through $(0, 0)$
(g) Hyperbola, foci at $(1, 0), (0, -1)$
(h) Half-line, $y = x - 2, x > 2$
(i) Half-line, $y = \sqrt{3}x - \frac{3}{2}\sqrt{3}, x < \frac{3}{2}$
(j) Circle, centre $(0, 2)$, radius 1

37 (a) Re$[(3 + j)z] = 2$ (b) $|z + 2| = 2$
(c) $|z + 1 - j2| = 3$ (d) Re$(z^2) = 1$

38 (a) Circle, centre $(1, 0)$, radius 2
(b) Circle, centre $(\frac{1}{2}, 0)$, radius $\frac{3}{2}$
(c) Circle, centre $(2, 3)$, radius 4
(d) Half-line, $y = 0, x > 0$
(e) Circle, centre $(-\frac{13}{8}, 0)$, radius $\frac{15}{8}$
(f) Semicircle, centre $(\frac{1}{2}, -\frac{1}{2})$, radius $\frac{1}{2}\sqrt{2}$, through $(0, 0)$

39 $x^2 + y^2 - 4x - 2y + 1 = 0, |z - 2 - j| = 2,$
$$\arg\left(\frac{z - j}{z - 4 - j}\right) = \pm\frac{\pi}{2}$$

40 Part of $x^2 + (y - 1)^2 = 2$

41 $(x - 3)^2 + y^2 = 4$

42 (a) $u = x + y, v = y - x$
(b) $u = (x - 1)^2 - y^2, v = 2(x - 1)y$
(c) $u = x(x^2 + y^2 + 1), v = y(x^2 + y^2 - 1)$

43 $a = (j - 2)/5, b = 3(1 + 2j)/5$

46 $100 + j100.12$

47 $\frac{8}{3} + j\frac{8}{3}$

3.6 Review exercises

1 (a) $12 + j9$ (b) $2 + j$ (c) $11 + j2$
(d) $7 + j24$ (e) 5 (f) $(1 - j2)/5$
(g) $(18 + j14)/5$ (h) $\tan^{-1}(3/4) = 0.6435$
(i) $5\sqrt{5}[\cos(0.9653 + 3k\pi) + j\sin(0.9653 + 3k\pi)],$
   $k = 0, 1$

2 $x = \pm\frac{3}{2}, y = \pm 2$

3 $\frac{1}{10}(7 + j9)$

4 (a) Circle centre $(-\frac{1}{3}, \frac{4}{3})$, radius $\frac{2}{3}\sqrt{2}$
(b) Re$\left(\dfrac{1}{z - 2}\right) = -\frac{1}{2}$

6 Centre $(R_2, \frac{1}{2}\omega L)$, radius $\frac{1}{2}\omega L$

7 (a) $32\cos^6\theta - 48\cos^4\theta + 18\cos^2\theta - 1$

13 $419.8 - j238.8, 0.5928 \times 10^{-3} + j1.0518 \times 10^{-3}$

14 $1 + j3, \frac{1}{5}(3 + j11), \frac{1}{5}(7 + j11)$

15 Mod $\frac{25}{13}$, arg $= -154°17' = -2.6927$ rad

16 (a) $0.22 \pm j0.49$ (b) $1.44 + j1.57$ (c) $10.48 + j19.74$
(d) $0.80 + j0.46$ (e) $1.09 + j0.83$

19 $\theta = \tan^{-1}\left[\dfrac{2R_0 X - 2RX_0}{R^2 + X^2 - R_0^2 - X_0^2}\right]$

20 $4.46 - j2.06$

21 (a) $0.7974 + j0.3685$ (b) $r = 0.8784$
$\theta = 24°49' = 0.4329$ rad, 1.098

23 (a) $-0.04 + j0.28$　　(b) $\pm(0.35 + j0.40)$
　　(c) $0.92 + j0.27$　　(d) $-1.26 + j1.71$
　　(e) $-0.04 + j0.28$

24 $1\angle18°26'$, $1\angle108°26'$, $1\angle198°26'$, $1\angle288°26'$

25 $2^{1/6}e^{j(1/9+k/3)\pi}$, $k = 0, \ldots, 5$

27 $-(\omega u + \omega^2 v)$, $-(\omega^2 u + \omega v)$; $\frac{1}{4}r^2 \leqslant -\frac{1}{27}q^3$

28 $1 - j2$, $2\sqrt{5}$

31 Circle $u^2 + v^2 - 12u + 16v = 0$
　　Centre $(6, -8)$, radius 10

32 $v + 3u = 5$

33 Circle $u^2 + v^2 - \frac{5}{2}u + 1 = 0$; Centre $(\frac{5}{4}, 0)$, radius $\frac{3}{4}$; Maps to region outside circle

CHAPTER 4

Exercises

1 (a) $(3, 3, 1)$　　(b) $(2, 4, \frac{5}{2})$　　(c) $(0, 0, 1)$
　　(d) $\sqrt{2}$　　(e) 3　　(f) $\sqrt{3}$
　　(g) $(\sqrt{\frac{1}{2}}, \sqrt{\frac{1}{2}}, 0)$　　(h) $(\frac{2}{3}, \frac{2}{3}, \frac{1}{3})$

2 $\overrightarrow{PQ} = (4, -5, 11)$, $|\overrightarrow{PQ}| = 9\sqrt{2}$
　　direction cosines $4/(9\sqrt{2})$, $-5/(9\sqrt{2})$, $11/(9\sqrt{2})$

3 $\sqrt{134}$ N, $(7, 2, 9)/\sqrt{134}$

4 $\alpha = 4$　$\beta = 1$　$\gamma = 2$

5 $\sqrt{21}$　$\sqrt{17}$　$\sqrt{38}$

6 $60°$ or $-60°$ to the positive z axis

7 $\overrightarrow{PQ} = \overrightarrow{QR} = (1, 5, -3)$ and $PQ : QR = 1 : 1$

8 $8\sqrt{2}$ kilometres per hour from the NW

9 distance $= 13/5$, $t = 1/5$

11 $1 - j2$　length $= \sqrt{20}$

12 $\frac{1}{2} - j\frac{1}{2}$

14 $\theta = \sin^{-1}\left(\dfrac{W_1^2 + W_2^2 - W_3^2}{2W_1W_2}\right)$

　　$\phi = \sin^{-1}\left(\dfrac{W_2^2 + W_3^2 - W_1^2}{2W_2W_3}\right)$

15 $F = (-940, 124, -31)$ N
　　$(7.93\,\text{m}, -1.04\,\text{m}, 0\,\text{m})$, $T = 1342$ N

16 $T = 539$ N　$S = 389$ N

17 (a) 14　　(b) 6　　(c) $(2, 1, 6)/\sqrt{41}$
　　(d) $(12, 0, -6)/\sqrt{5}$　(e) -24　　(f) $(12, 4, 8)$

18 (a) $98.0° = 1.711$ rad　　(b) $64.8° = 1.130$ rad
　　(c) $\frac{14}{5}$　　(d) 3 or -4

19 4 units

20 $\frac{5}{14}$

21 $\sqrt{5}/2$

26 (a) 50 N m　　(b) $\frac{1}{2}Whpn(n + 1)$

27 $r^2 - (r \cdot \hat{a})^2 = R^2$

28 $\sqrt{3}$, $70.5°$ or 1.23 rad

29 $|X| \leqslant 2.98$ m

30 $|\theta| \leqslant 36.9°$ or 0.644 rad

31 (a) $(3, -2, -1)$　　(b) $(-1, 1, 0)$　　(c) $(5, -4, -2)$
　　(d) -1　　(e) 1　　(f) $(2, 2, 1)$

34 (a) $(8, 1, 6)$, $(4, 1, 3)$　(b) $(-\frac{3}{5}, 0, \frac{4}{5})$　(c) $\frac{5}{2}$

37 $(48, 72, 0)/\sqrt{14}$

38 $(-8, -32, -4)/\sqrt{21}$

39 (a) $(0, 1, -1)$　$(-2, 1, -1)$　$(-2, 1, 0)$
　　(b) $(0, 0, 0)$　$(0, 0, -2)$　$(-3, 3, -1)$
　　(c) $(-3, 3, -3)$

40 $\pm(-3, 5, 11)/\sqrt{155}$　0.9968

41 Distance $= 1.92$

42 $m\omega = eB$

43 15

45 8

47 (a) $(-5, 3, -7)/\sqrt{83}$　　(b) $(0, 1, -4)/\sqrt{17}$

48 $\frac{7}{3}(1, 1, 1)$　$\frac{1}{3}(2, -13, 11)$

49 $\begin{vmatrix} u_1 & u_2 & u_3 \\ v_1 & v_2 & v_3 \\ w_1 & w_2 & w_3 \end{vmatrix}$

50 $\alpha = -1/F^2$

51 (a) $(c \cdot a)(b \cdot d) - (c \cdot b)(d \cdot a)$
　　(c) $-(a \cdot b)(a \times c)$

52 (a) $(3, 3, 3)$　　(b) $(1 + s, 2 + s, 3 + s)$
　　(c) $x - 1 = y - 2 = z - 3$

53 $r \cdot (0, -1, 1) = 1$　$-y + z = 1$

55 $(3, 4, 0)$, $43.5° = 0.759$ rad

56 $r \cdot [b \times (c - a)] = a \cdot (b \times c)$

58 $r = (0, -5, 10) + \lambda(1, 2, -3)$

59 $r = (1, 2, 4) + t(1, 1, 2)$　$(-\frac{5}{2}, -\frac{3}{2}, -3)$

60 $79.0° = 1.38\,\text{rad}$

61 (a) $\boldsymbol{r}\cdot(2, 3, 6) = -28$ (b) 5

62 $\boldsymbol{r} = (1 + 2t, -1 + 4t, 3 - 4t)$, $41.8° = 0.729\,\text{rad}$

63 $\boldsymbol{r} = (2 - t, 2t, -1 + 4t)$
$2 - x = \frac{1}{2}y = \frac{1}{4}(1 + z)$, no intersection

64 $\sqrt{35}$

65 $\boldsymbol{r}\cdot(1, -5, 3) = 28$

66 $\boldsymbol{r} = (-1 + 14t, t, 1 - 8t)$
$\frac{2}{3}\sqrt{29}$, $\boldsymbol{r}\cdot(-18, 36, -27) = -9$

4.5 Review exercises

1 (a) $\sqrt{93}$ (b) $(17, -3, -10)/\sqrt{398}$
    (c) $85.8° = 1.50\,\text{rad}$, $47.0° = 0.820\,\text{rad}$
    (d) $(2, 13, -13)/6$

2 (a) $(3, 4, 5)$ (b) $\sqrt{35}$ (c) $34/3$

3 (a) $(1, 2, 0)$, $(2, 1, 1)$ (b) $\sqrt{5}$
    (c) 1   (d) $112.2° = 1.96\,\text{rad}$

4 (a) -4 (b) 1 or -4

5 $(\frac{2}{3}, \frac{1}{3}, \frac{2}{3})$, $(-\frac{3}{5}, 0, \frac{4}{5})$; $(1, 5, 22)$

8 $(1, 1, 1)$, $(-5, -11, 1)$

9 $E = e(0.550, 0.282, 0.282)$

10 (a) $2x + 3y + 6z + 28 = 0$ (b) 5

11 (a) $2x + 3y - z = 10$; $10/\sqrt{14}$
    (b) $\sqrt{3}/2$

12 $P(2, 4, 4)$, $Q(1, 2, 3)$
    $(-2, -4, 0)$

13 (a) 0   (b) $15(1, 1, -2)$

15 $(-90, -36, 12)$, $85.3°$ or $1.49\,\text{rad}$

16 $(11, -12, 5)$; $76.8°$; $(-11, 12, -5)/\sqrt{290}$
    (a) $-11x + 12y - 5z = 8$
    (b) $-11x + 12y - 5z = -4$ (c) $12/\sqrt{290}$

17 $\boldsymbol{r} = (-3, 0, 1) + \lambda(8, -8, -8) + \mu(5, 1, -3)$
    $\boldsymbol{r}\cdot(-1, 2, -1) = -6$

18 $x + 2y - 2z = -1$

19 $x = \dfrac{U}{K}(1 - e^{-KT})$

20 (a) $(0, 0, 1)$, $(1, -1, 0)$, $(0, 1, -1)$
    (b) 1   (c) $3, -3, 2$ (d) $3, 2, 1$

21 $\boldsymbol{r} = (2 - t, 3 - 3t, 2t)$
    (a) $\sqrt{(61/14)}$ (b) $(0, -3, 4)$ (c) $(19, 15, 18)/14$

22 $\alpha = \boldsymbol{r}\cdot\boldsymbol{a}'$ $\beta = \boldsymbol{r}\cdot\boldsymbol{b}'$ $\gamma = \boldsymbol{r}\cdot\boldsymbol{c}'$

23 Taking $\boldsymbol{i}$ along OA and $\boldsymbol{j}$ along OB then
    $F = \omega^2(1.4, 1.65)$ and $OC = (-1.4, -1.65)\,\text{m}$.

24 $(0, 0, -25)\,\text{N}$ $(-2.5, 2.5, -0.2)\,\text{N}\,\text{m}$

CHAPTER 5

Exercises

1 (a) not possible   (b) $\begin{bmatrix} 1 \\ 3 \\ 1 \end{bmatrix}$ (c) not possible

    (d) not possible   (e) $\begin{bmatrix} 8 & 9 & 10 \\ 7 & 10 & 9 \end{bmatrix}$

2 (a) $\begin{bmatrix} 0 & 1 & 1 \\ 3 & 2 & -1 \\ 0 & 0 & 3 \end{bmatrix}$ (b) $\begin{bmatrix} -2 & \frac{4}{3} & -\frac{2}{3} \\ 0 & -\frac{4}{3} & -\frac{4}{3} \\ 0 & -2 & 0 \end{bmatrix}$

    (c) $\begin{bmatrix} 1 & -\frac{1}{2} & \frac{1}{2} \\ \frac{1}{2} & 1 & \frac{1}{2} \\ 0 & 1 & \frac{1}{2} \end{bmatrix}$

3 $\begin{bmatrix} 4 & 7 & -1 \\ 18 & 10 & 14 \end{bmatrix}$

4 (b) $\begin{bmatrix} 3 & -2 & 1 \\ 0 & 3 & -1 \\ 1 & 6 & 3 \end{bmatrix}$

5 $1, 1, 2, 2,$

6 $\alpha = 1$ $\beta = -1$ $\gamma = 2$

7 (a) $\lambda = 1$ $\mu = -1$ $\nu = 3$

8 Average $= \begin{bmatrix} 30.5 \\ 27.5 \\ 19.5 \\ 11.5 \\ 11.0 \end{bmatrix}$ Weighted average $= \begin{bmatrix} 32.57 \\ 26.43 \\ 19.14 \\ 11.43 \\ 10.43 \end{bmatrix}$

9 $\begin{bmatrix} 40 & 49 & 46 & 42 & 52 \\ 53 & 60 & 58 & 46 & 68 \\ 23 & 30 & 26 & 29 & 16 \end{bmatrix}$

    $\begin{matrix} 312 & 380 & 355 & 326 & 400 \\ 409 & 462 & 447 & 358 & 526 \\ 180 & 230 & 205 & 225 & 126 \end{matrix}$

10 $\begin{bmatrix} 31\,000 \\ 9\,000 \\ 16\,900 \\ 340 \\ 18 \end{bmatrix}$ $\begin{bmatrix} 14\,750 \\ 14\,600 \\ 270 \\ 122 \\ 9 \end{bmatrix}$ Bricks – type C and sand

11 $\begin{bmatrix} 2 & 2 & -1 \\ 2 & 2 & -1 \end{bmatrix}\begin{bmatrix} 0 & 2 \\ 0 & 2 \end{bmatrix}$

$\begin{bmatrix} 1 & 3 \\ 1 & 3 \\ 1 & 1 \end{bmatrix}\begin{bmatrix} 2 & 2 & 2 \\ 2 & 2 & 2 \\ -2 & -2 & -2 \end{bmatrix}$

$\begin{bmatrix} 2 & 2 \\ 2 & 2 \\ -1 & -1 \end{bmatrix}$

12 (a) No, yes, yes, yes, no, no

(b) $\begin{bmatrix} 9 & 7 \\ 7 & 12 \end{bmatrix}\begin{bmatrix} 13 & 18 \\ 8 & 5 \end{bmatrix}\begin{bmatrix} 9 & 5 \\ 18 & 2 \end{bmatrix}$

(c) $\begin{bmatrix} 37 & 33 \\ 26 & 36 \\ 29 & 28 \end{bmatrix}$

13 $AB = BA = \begin{bmatrix} 2 & 0 \\ 0 & 2 \end{bmatrix}$, $X = \begin{bmatrix} 2 \\ -\frac{5}{2} \end{bmatrix}$

16 $x^2 + y^2 + z^2$
$x^2 + 4y^2 + 7z^2 + 5xy + 8xz + 11yz$
$\left.\begin{array}{l} x + 2y + 3z = 2 \\ 3x + 4y + 5z = 3 \\ 5x + 6y + 7z = 4 \end{array}\right\}$

17 $AB = \begin{bmatrix} 0 & 2 \\ 1 & 4 \end{bmatrix}$, $BA = \begin{bmatrix} 2 & 1 & 2 \\ 1 & 0 & 1 \\ 2 & 1 & 2 \end{bmatrix}$, $BC = \begin{bmatrix} -1 & 2 \\ 2 & 1 \\ -1 & 2 \end{bmatrix}$

CB not defined, $CA = \begin{bmatrix} 4 & 1 & 4 \\ 3 & 2 & 3 \end{bmatrix}$, AC not defined

18 (a) $\begin{bmatrix} 3 & 4 \\ 2 & 3 \end{bmatrix}$, $\begin{bmatrix} 4 & 4 \\ 2 & 2 \end{bmatrix}$ (b) $\begin{bmatrix} -1 & 2 \\ 0 & 1 \end{bmatrix}$, $\begin{bmatrix} 0 & 2 \\ 0 & 0 \end{bmatrix}$

both $\begin{bmatrix} 9 & 0 \\ 0 & 9 \end{bmatrix}$ both $\begin{bmatrix} -3 & 12 \\ 30 & 3 \end{bmatrix}$

First set does not commute, second set commutes

21 $\begin{bmatrix} b+d & b \\ o & d \end{bmatrix}$

22 (a) $A = \begin{bmatrix} 1 & 1 & 1 \\ 1 & 1 & 1 \end{bmatrix}$ $B = \begin{bmatrix} 1 \\ 1 \\ 1 \end{bmatrix}$

(b) $A = \begin{bmatrix} 1 & 1 & 1 \end{bmatrix}$ $B = \begin{bmatrix} 1 \\ 1 \\ 1 \end{bmatrix}$

(c) $A = \begin{bmatrix} 1 & 0 \\ 0 & 0 \end{bmatrix}$ $B = \begin{bmatrix} 0 & 0 \\ 1 & 1 \end{bmatrix}$

23 $A = \begin{bmatrix} 1 & \frac{5}{2} & \frac{3}{2} \\ \frac{5}{2} & -1 & 2 \\ \frac{3}{2} & 2 & 1 \end{bmatrix} + \begin{bmatrix} 0 & \frac{1}{2} & \frac{1}{2} \\ -\frac{1}{2} & 0 & -2 \\ -\frac{1}{2} & 2 & 0 \end{bmatrix}$

24 $\begin{bmatrix} 41 & 15 & 7 \\ -9 & 63 & -40 \\ -13 & 38 & 41 \end{bmatrix}$

25 £2273.88

26 $\begin{bmatrix} 1 & 3 & 2 \\ 0 & 5 & 2 \\ 2 & -2 & 1 \end{bmatrix}\begin{bmatrix} 3 & 6 & 5 \\ 6 & 7 & 8 \\ -4 & 1 & -3 \end{bmatrix}$

28 $\begin{bmatrix} R_1 & R_2 & 0 & 0 & R_5 & 0 \\ 0 & R_2 & -R_3 & R_4 & 0 & 0 \\ 0 & 0 & 0 & R_4 & -R_5 & R_6 \\ 1 & -1 & -1 & 0 & 0 & 0 \\ 0 & 1 & 0 & -1 & -1 & 0 \\ 0 & 0 & -1 & -1 & 0 & 1 \end{bmatrix}\begin{bmatrix} i_1 \\ i_2 \\ i_3 \\ i_4 \\ i_5 \\ i_6 \end{bmatrix} = \begin{bmatrix} E \\ 0 \\ 0 \\ 0 \\ 0 \\ 0 \end{bmatrix}$

29 $\begin{bmatrix} 799.8 \\ 800 \\ 800.2 \end{bmatrix}\begin{bmatrix} 800 \\ 800 \\ 800 \end{bmatrix}$

30 $h = \frac{1}{3}, k = \frac{2}{3}, l = \frac{1}{3}, m = \frac{1}{6}$

32 $A = \begin{bmatrix} \sqrt{\frac{1}{2}} & -\sqrt{\frac{1}{2}} \\ \sqrt{\frac{1}{2}} & \sqrt{\frac{1}{2}} \end{bmatrix}$

$\begin{bmatrix} a \\ b \end{bmatrix} = \begin{bmatrix} 16 \\ -10 \end{bmatrix} + A\begin{bmatrix} 16 \\ -10 \end{bmatrix}$ and $B = A$

33 $n = 3$

34 Minors = $\begin{matrix} -1 & 0 & 1 \\ -1 & -2 & -1 \\ 2 & -2 & -2 \end{matrix}$

Cofactors = $\begin{matrix} -1 & 0 & 1 \\ 1 & -2 & 1 \\ 2 & 2 & -2 \end{matrix}$

$|A| = 2$

35 (a) −19　(b) 130　(c) −65
(d) 1　(e) −3

36 −4, 16, 16, −32

37 3

38 $\begin{bmatrix} d & -b \\ -c & a \end{bmatrix}$

39 $\begin{bmatrix} 2 & -1 & 0 \\ -4 & 3 & -1 \\ 1 & -1 & 1 \end{bmatrix}$

40 $2, \begin{bmatrix} 1 & 0 \\ -3 & 2 \end{bmatrix}, \frac{1}{2}\begin{bmatrix} 1 & 0 \\ -3 & 2 \end{bmatrix}, \begin{bmatrix} 1 & 0 \\ 0 & 1 \end{bmatrix}$

44 (a) −1.6569, 9.6569
(b) 4.6667 ± j0.623 61
(c) 2, 3 ± j

45 (a) −0.1884　(b) 100

47 $x^2(2x + 1)^2(x - 1)^2$

51 Non-singular, singular, non-singular, singular

$\begin{bmatrix} -\frac{1}{3} & \frac{2}{3} \\ \frac{2}{3} & -\frac{1}{3} \end{bmatrix} \begin{bmatrix} 1 & 0 & 0 & -1 \\ 0 & 1 & 0 & -1 \\ 0 & 0 & 1 & -1 \\ 0 & 0 & 0 & 1 \end{bmatrix}$

52 $\begin{bmatrix} 1 & -1 & -1 \\ 0 & 1 & 0 \\ 0 & 0 & 1 \end{bmatrix}, \begin{bmatrix} 1 & 0 & 0 & 0 \\ 0 & \frac{1}{2} & 0 & 0 \\ 0 & 0 & \frac{1}{3} & 0 \\ 0 & 0 & 0 & \frac{1}{4} \end{bmatrix},$

$\begin{bmatrix} 2 & -j \\ j & 1 \end{bmatrix}, \frac{1}{3}\begin{bmatrix} 5 & -7 & -1 \\ -4 & 5 & 2 \\ 2 & -1 & -1 \end{bmatrix}$

53 $\frac{1}{169}\begin{bmatrix} -100 & 98 & -92 \\ -47 & -79 & -50 \\ 6 & -87 & -8 \end{bmatrix}$

55 $\begin{bmatrix} 1 & 0 & 0 \\ -1 & 1 & 0 \\ 2 & -2 & 1 \end{bmatrix}, \frac{1}{4}\begin{bmatrix} 4 & -8 & 6 \\ 0 & 2 & -1 \\ 0 & 0 & 2 \end{bmatrix}, \frac{1}{4}\begin{bmatrix} 24 & -20 & 3 \\ -4 & 4 & -1 \\ 4 & -4 & 2 \end{bmatrix}$

56 $\begin{bmatrix} 1 & 1 & 0 \\ 0 & 1 & 2 \\ 0 & 1 & 3 \end{bmatrix}, \begin{bmatrix} 1 & 1 & 0 \\ 0 & 1 & 2 \\ 0 & 0 & 1 \end{bmatrix}, \begin{bmatrix} 1 & 1 & 0 \\ 0 & 1 & 0 \\ 0 & 0 & 1 \end{bmatrix},$

$\begin{bmatrix} 1 & 0 & 0 \\ 0 & 1 & 0 \\ 0 & 0 & 1 \end{bmatrix}, \begin{bmatrix} -1 & -3 & 2 \\ 2 & 3 & -2 \\ -1 & -1 & 1 \end{bmatrix}$

57 $\frac{1}{5}\begin{bmatrix} -3 & 2 & 2 \\ 2 & -3 & 2 \\ 2 & 2 & -3 \end{bmatrix}, \begin{bmatrix} 0.68 & -0.32 & -0.32 \\ -0.32 & 0.68 & -0.32 \\ -0.32 & -0.32 & 0.68 \end{bmatrix}$

58 $\frac{1}{68}\begin{bmatrix} -4 & 4 & 8 \\ 6 & 11 & -12 \\ 36 & -2 & -4 \end{bmatrix}, \frac{1}{49}\begin{bmatrix} -5 & 22 & 8 \\ 11 & -19 & 2 \\ 13 & -18 & -11 \end{bmatrix}$

59 $A = A^{-1}, B^{-1} = \begin{bmatrix} 1 & 0 & 0 & 0 \\ 0 & 0 & 0 & 1 \\ 0 & 1 & 0 & 0 \\ 0 & 0 & 1 & 0 \end{bmatrix}$

$(AB)^{-1} = \begin{bmatrix} 1 & 0 & 0 & 0 \\ 0 & 0 & 0 & 1 \\ 0 & 0 & 1 & 0 \\ 0 & 1 & 0 & 0 \end{bmatrix}$

60 (a) $\begin{bmatrix} 1 \\ 2 \end{bmatrix}$　(b) $\begin{bmatrix} 1 \\ -4 \\ 0 \end{bmatrix}$　(c) $\begin{bmatrix} -1 \\ 2 \end{bmatrix}$　(d) $\begin{bmatrix} 4 \\ 3 \\ 2 \\ 1 \end{bmatrix}$

61 $\begin{bmatrix} \cos(-\frac{\pi}{8}) \\ -\sin(-\frac{\pi}{8}) \end{bmatrix}$

62 $\begin{bmatrix} -j \\ 1 \\ j \end{bmatrix}$

63 $\frac{1}{12}\begin{bmatrix} -7 & -6 & 5 \\ 2 & 0 & 2 \\ 1 & -6 & 1 \end{bmatrix}$

$x = 2, y = 1, z = 2$

64 $\alpha = 1 \quad x = \lambda \quad y = 5\lambda \quad z = 7\lambda$
$\alpha = -6 \quad x = \mu \quad y = -2\mu \quad z = 0$

65 $-6, -3, -2$

66 (a) $a = 0$ (b) $\dfrac{1}{9}\begin{bmatrix} 1 \\ 7 \\ 6 \end{bmatrix}$ (c) $\begin{bmatrix} \lambda \\ \lambda \\ \lambda \end{bmatrix}$ (d) $\begin{bmatrix} -2 \\ 1 \\ 3 \end{bmatrix}$

67
$\begin{bmatrix}
0.1896 & -0.0604 & -0.0167 & 0.0167 & -0.0021 & -0.0021 \\
-0.0604 & 0.2088 & -0.0551 & -0.0218 & 0.0171 & -0.0021 \\
-0.0167 & -0.0551 & 0.2103 & -0.0564 & -0.0218 & 0.0167 \\
0.0167 & -0.0218 & -0.0564 & 0.2103 & -0.0551 & -0.0167 \\
-0.0021 & 0.0171 & -0.0218 & -0.0551 & 0.2088 & -0.0604 \\
-0.0021 & -0.0021 & 0.0167 & -0.0167 & -0.0604 & 0.1896
\end{bmatrix}$

$\begin{bmatrix}
0.1875 \\
-0.0625 \\
0 \\
0 \\
-0.0625 \\
0.1875
\end{bmatrix}$

68 $a = u_1 \quad b = (-u_1 + u_2)/p \quad c = (-u_1 + u_3)/q$
$d = (u_1 - u_2 - u_3 + u_4)/pq$

69 $x = 0.5889u \quad y = 0.4222u \quad z = 0.2222u$

70 $a = -0.4011 \quad b = 1 \quad c = -0.5825$
$f(1) = 1.4345$

71 $y_1 = 1.8936 \quad y_2 = 4.6809 \quad y_3 = 8.1489 \quad y_4 = 10.7660$

72 (a) $\begin{bmatrix} -3 \\ 0 \\ 2 \end{bmatrix}$ (b) $\begin{bmatrix} -3 \\ 2 \\ 4 \end{bmatrix}$ (c) $\begin{bmatrix} -\frac{3}{25} \\ \frac{1}{150} \\ \frac{2}{75} \end{bmatrix}$

73 $x = 1 \quad y = 2 \quad z = 2 \quad t = 3$

74 $x = -0.0833 \quad y = 0.7083 \quad z = 1.9167 \quad t = 2.9583$

75 (a) $\begin{bmatrix} 1.1602 \\ -0.0515 \\ 0.0065 \end{bmatrix}$ (b) $\begin{bmatrix} 2.6844 \\ 0.0234 \\ 2.3569 \end{bmatrix}$ (c) $\begin{bmatrix} -1.3424 \\ 1.2860 \\ 2.4458 \\ 0.5511 \end{bmatrix}$

76 $\begin{bmatrix} -74.17 \\ -25.54 \\ 140.11 \end{bmatrix}$, det $= 0.002\,725$ $\begin{bmatrix} -142.53 \\ -50.52 \\ 262.77 \end{bmatrix}$, det $= 0.001\,453$

78 $y_1 = 1.028 \quad y_2 = 4.004 \quad y_3 = 8.993 \quad y_4 = 11.329$

79 $4.5, 8, 10.5, 12, 12.5$

80 $\left|\dfrac{E_0}{Z_3}\right| = \sqrt{\left(\dfrac{L}{C}\right)\dfrac{1}{\alpha}[(2 - \alpha^2)^2(1 - \frac{1}{2}\alpha^2)^2}$
$\qquad\qquad - 2(2 - \alpha^2)(1 - \frac{1}{2}\alpha^2) - (1 - \frac{1}{2}\alpha^2)^2 + 1]$
where $\alpha^2 = LC\omega^2$

81 Solution: 1, 2, 2, 3
After 5 iterations: 0.989, 1.99, 1.98, 3.00

82 Solution: $-0.083, 0.708, 1.917, 2.958$
After 3 iterations: $-0.189, 0.634, 1.868, 2.920$

83 (a) 0.8   (b) 1.1   (c) no convergence

84 $\begin{bmatrix} 10 \\ 20 \\ -30 \end{bmatrix}$

There is no convergence in 50 iterations, even from a

starting value of $\begin{bmatrix} 10.1 \\ 19.9 \\ -29.9 \end{bmatrix}$

85 $I_1 = 0.5172, I_2 = 0.4914, I_3 = 0.8017$

86 0.1685, 0.3258, 0.5282, 0.7188, 0.9563, 1.0063,
0.8063, 0.6064, 0.4059, 0.2079

87 (a) 2, 2, [2/3, −1/3]   (b) 1, 2, inconsistent
(c) 2, 2, [1, −t, t]   (d) 2, 2, [2 − t, 1, t]
(e) 2, 3, inconsistent   (f) 4, 4, [0, 1/3, 0, 1]

88 $\alpha = 2$ gives 1, 2 and inconsistent equations; $\alpha = -1$
gives 1, 1 solution $[1 - 2t, t]$; otherwise solution is
$\left[\dfrac{-\alpha}{\alpha - 2}, \dfrac{(\alpha - 1)\alpha}{\alpha - 2}\right]$

89 (a) 2   (b) 3

90 (a) Rank = 2, $(-2 + t, 5 - 2t, t)$
(b) Rank = 2, no solution

91 Rank = 3, rank = 3; $(\mu, -1, 1, -\mu)$

92 (a) $x = -1, y = \frac{1}{3}(2 - 4\mu), z = \mu$
(b) No solution
(c) $x = \frac{1}{11}(-9 - 45\lambda + 13\mu), y = \frac{1}{11}(5 - 8\lambda + 5\mu)$
$z = \lambda, t = \mu$
(d) Unique solution $x = -1, y = -1, z = 1$

94 Rank = 4 implies points not coplanar; rank = 3
implies the points lie on a plane; rank = 2 implies the
points lie on a line; rank = 1 implies the four points
are identical

96 (a) $\lambda^2 - 4\lambda + 3$, eigenvalues 3, 1
(b) $\lambda^2 - 3\lambda + 1$, eigenvalues 2.618, 0.382

(c) $\lambda^3 - 6\lambda^2 + 11\lambda - 6$, eigenvalues 3, 2, 1

(d) $\lambda^3 - 6\lambda^2 + 9\lambda - 4$, eigenvalues 4, 1, 1

(e) $\lambda^3 - 12\lambda^2 + 40\lambda - 35$, eigenvalues 7, 3.618, 1.382

(f) $\lambda^2 - (2 + a)\lambda + 1 + 2a$, eigenvalues $1 + \frac{1}{2}a \pm \frac{1}{2}\sqrt{(a^2 - 4a)}$

97 (a) 2, 0; $\begin{bmatrix} 1 \\ 1 \end{bmatrix}$, $\begin{bmatrix} 1 \\ -1 \end{bmatrix}$ (b) 4, −1; $\begin{bmatrix} 2 \\ 3 \end{bmatrix}$, $\begin{bmatrix} -1 \\ 1 \end{bmatrix}$

(c) 9, 3, −3; $\begin{bmatrix} -1 \\ 2 \\ 2 \end{bmatrix}$, $\begin{bmatrix} 2 \\ 2 \\ -1 \end{bmatrix}$, $\begin{bmatrix} 2 \\ -1 \\ 2 \end{bmatrix}$

(d) 3, 2, 1; $\begin{bmatrix} 2 \\ 2 \\ 1 \end{bmatrix}$, $\begin{bmatrix} 1 \\ 1 \\ 0 \end{bmatrix}$, $\begin{bmatrix} 0 \\ 2 \\ -1 \end{bmatrix}$

(e) 14, 7, −7; $\begin{bmatrix} 2 \\ 6 \\ 3 \end{bmatrix}$, $\begin{bmatrix} 6 \\ -3 \\ 2 \end{bmatrix}$, $\begin{bmatrix} 3 \\ 2 \\ -6 \end{bmatrix}$

(f) 2, 1, −1; $\begin{bmatrix} 1 \\ -1 \\ -1 \end{bmatrix}$, $\begin{bmatrix} 1 \\ 0 \\ -1 \end{bmatrix}$, $\begin{bmatrix} 1 \\ 2 \\ -7 \end{bmatrix}$

(g) 5, 3, 1; $\begin{bmatrix} -2 \\ -3 \\ 1 \end{bmatrix}$, $\begin{bmatrix} 1 \\ -1 \\ 0 \end{bmatrix}$, $\begin{bmatrix} 0 \\ -1 \\ 1 \end{bmatrix}$

(h) 4, 3, 1; $\begin{bmatrix} 2 \\ -1 \\ -1 \end{bmatrix}$, $\begin{bmatrix} 2 \\ -1 \\ 0 \end{bmatrix}$, $\begin{bmatrix} 4 \\ 1 \\ -2 \end{bmatrix}$

98 Eigenvalues 3, 3   eigenvectors [1, 0], [0, 1]
Eigenvalues 3, 3   eigenvector [0, 1]
Eigenvalues 5/2, 5/2 eigenvector [1, −2]
Eigenvalues 0, 0   eigenvector [1, −2]

99 (a) 5, 1, 1; $\begin{bmatrix} 1 \\ 1 \\ 1 \end{bmatrix}$, $\begin{bmatrix} -2 \\ 1 \\ 0 \end{bmatrix}$, $\begin{bmatrix} -1 \\ 0 \\ 1 \end{bmatrix}$

(b) 2, 2, −1; $\begin{bmatrix} -1 \\ 1 \\ 0 \end{bmatrix}$, $\begin{bmatrix} 8 \\ 1 \\ 3 \end{bmatrix}$

(c) 2, 2, 1; $\begin{bmatrix} 3 \\ 1 \\ -2 \end{bmatrix}$, $\begin{bmatrix} 4 \\ 1 \\ -3 \end{bmatrix}$

(d) 2, 1, 1; $\begin{bmatrix} 2 \\ 1 \\ 2 \end{bmatrix}$, $\begin{bmatrix} 0 \\ -2 \\ 1 \end{bmatrix}$, $\begin{bmatrix} 1 \\ 3 \\ 0 \end{bmatrix}$

100 One eigenvector $\begin{bmatrix} -3 \\ 1 \\ 1 \end{bmatrix}$

101 2, 1, 1; $\begin{bmatrix} -1 \\ 1 \\ 1 \end{bmatrix}$, $\begin{bmatrix} -1 \\ 1 \\ 0 \end{bmatrix}$, $\begin{bmatrix} 1 \\ 0 \\ 1 \end{bmatrix}$

102 3, [1, 0, 0, 1]; 2, [0, 1, 0, 0] and [0, 0, 1, 0];
−1, [−1, 0, 0, 1]

105 3, 2, −6; $\begin{bmatrix} -1 \\ 1 \\ 1 \end{bmatrix}$, $\begin{bmatrix} 0 \\ 1 \\ -1 \end{bmatrix}$, $\begin{bmatrix} 2 \\ 1 \\ 1 \end{bmatrix}$

106 $\begin{bmatrix} 1 \\ -1 \\ 0 \end{bmatrix}$

5.9 Review exercises

1 (a) $\begin{bmatrix} 12 & 18 & -40 \\ 0 & 0 & 8 \\ 0 & 0 & 4 \end{bmatrix}\begin{bmatrix} 12 & 0 & 0 \\ 18 & 0 & 0 \\ -40 & 8 & 4 \end{bmatrix}$

(b) $\begin{bmatrix} 8 & 9 & -7 \\ 0 & 2 & 0 \\ 0 & 0 & 5 \end{bmatrix}\begin{bmatrix} 12 & 12 & -6 \\ 0 & 4 & 0 \\ 0 & 0 & 16 \end{bmatrix}$

$\begin{bmatrix} 4 & 14 & -3 \\ 0 & 0 & 5 \\ 0 & 0 & 4 \end{bmatrix}$

2 $\lambda = -1, \mu = 2$
$\lambda = 2, \mu = -1$

3 Normal strain = $\begin{bmatrix} 3 \\ -3 \\ 3\sqrt{2} \end{bmatrix}$

Shear strain = 0

4 $(\alpha - \beta)(\beta - \gamma)(\gamma - \alpha)(\alpha + \beta + \gamma)$

5 $\theta = 1: (1 + 2\alpha, -3\alpha, \alpha)$
$\theta = 2: (2\alpha, 1 - 3\alpha, \alpha)$

6 $A^2 = \begin{bmatrix} 1 & 3 & 2 \\ 0 & 5 & 2 \\ 2 & -2 & 1 \end{bmatrix}$ $A^3 = \begin{bmatrix} 3 & 6 & 5 \\ 6 & 7 & 8 \\ -4 & 1 & -3 \end{bmatrix}$

$A^{-1} = \begin{bmatrix} 3 & -2 & -1 \\ 2 & -1 & 0 \\ -4 & 3 & 1 \end{bmatrix}$ $X = \begin{bmatrix} -11 \\ -1 \\ 15 \end{bmatrix}$

7 (a) $P^T = \dfrac{1}{3}\begin{bmatrix} 2 & -2 & -1 \\ 1 & 2 & -2 \\ 2 & 1 & 2 \end{bmatrix}$, the solution $x = P^{-1}b$

exists

(b) $E = \begin{bmatrix} I_x - \dfrac{Q_x^2}{A} & I_{xy} - \dfrac{Q_xQ_y}{A} & 0 \\ I_{xy} - \dfrac{Q_xQ_y}{A} & I_y - \dfrac{Q_y^2}{A} & 0 \\ 0 & 0 & A \end{bmatrix}$

8 (a) $B = \begin{bmatrix} 8 & -10 & 3 \\ 3 & -4 & 1 \\ -1 & 0 & 0 \end{bmatrix}$

(b) $k = 8.2316, k = -1.9316$

9 (a) $\begin{bmatrix} -2 & -3 & -1 \\ -8 & -1 & 29 \\ -6 & 2 & 8 \end{bmatrix}, -22, \dfrac{1}{22}\begin{bmatrix} 2 & 3 & 1 \\ 8 & 1 & -29 \\ 6 & -2 & -8 \end{bmatrix}$,

$z = 2, y = 1, x = 2$

(b) $Y = \begin{bmatrix} 8 & 1 & 9 \\ 18 & 3 & -35 \\ 8 & -4 & -6 \end{bmatrix}$

(c) $Z = \begin{bmatrix} 6 & 11 & 3 \\ 15 & 2 & -59 \\ 16 & -7 & -16 \end{bmatrix}$

10 (a) 3, 0, 2, 1; det = 12

(b) 1, 2, 3, 4

11 1, 2, 3

12 If $c \neq 0$ then rank = 2
if $c = 0$ then rank = 1

13 $a = \begin{bmatrix} 0.0051 \\ 0.9712 \\ -0.3931 \\ -0.0760 \\ 0.0283 \end{bmatrix}$ $f(2) = 0.2200$ $f(3.5) = -0.4228$

14 $a = 0.4424, b = -1.5037, c = 1.5023, d = -0.0611$
max at $x = 0.74, f = 0.4065$

15 rank $B = 2$, $AA^T = I$, $A^{-1} = A^T$
$x_1 = 2.444, x_2 = -2.556, x_3 = -1.222$

16 (b) $x_1 = 44, x_2 = -48, x_3 = -39, x_4 = 33$

17 (a) $\begin{bmatrix} 0 \\ -1 \\ 2 \end{bmatrix}$ (b) $\begin{bmatrix} -8 \\ -5 \\ \frac{1}{2} \end{bmatrix}$

18 (a) 4, 3, 2; $\begin{bmatrix} 0.5774 \\ 0.5774 \\ -0.5774 \end{bmatrix}, \begin{bmatrix} 0.1961 \\ 0.5883 \\ -0.7845 \end{bmatrix}, \begin{bmatrix} 0 \\ 0.7071 \\ -0.7071 \end{bmatrix}$

(b) 5, 3, −1; $\begin{bmatrix} 0.2033 \\ 0.6505 \\ 0.7318 \end{bmatrix}, \begin{bmatrix} 0.1374 \\ 0.8242 \\ 0.5494 \end{bmatrix}, \begin{bmatrix} -0.4472 \\ -0.8944 \\ 0 \end{bmatrix}$

(c) 9, 6, 3; $\dfrac{1}{3}\begin{bmatrix} -1 \\ 2 \\ 2 \end{bmatrix}, \dfrac{1}{3}\begin{bmatrix} 2 \\ -1 \\ 2 \end{bmatrix}, \dfrac{1}{3}\begin{bmatrix} 2 \\ 2 \\ -1 \end{bmatrix}$

19 $\lambda = 9$ $\alpha = 1$ $\beta = 6$

20 $\begin{bmatrix} 1 \\ 0 \\ 1 \end{bmatrix}, \begin{bmatrix} 0 \\ 1 \\ 0 \end{bmatrix}, \begin{bmatrix} -1 \\ 0 \\ 1 \end{bmatrix}$

21 After 100 iterations rounded down to the nearest integer

70	98	136	
56	78	109	and the largest eigenvalues are
42	59	81	0.9963, 0.9996, 1.0029
21	29	40	

22 (a) 3, 1; $\begin{bmatrix} 0.7071 \\ 0.7071 \end{bmatrix}, \begin{bmatrix} 0.7071 \\ -0.7071 \end{bmatrix}$

(b) 0.8794, −1.3473, −2.5321;

$\begin{bmatrix} 0.4491 \\ 0.8440 \\ 0.2931 \end{bmatrix}, \begin{bmatrix} 0.8440 \\ -0.2931 \\ -0.4491 \end{bmatrix}, \begin{bmatrix} 0.2931 \\ -0.4491 \\ 0.8440 \end{bmatrix}$

26 $E_1 = 4E_2 + 3I_2$; $I_1 = 3E_2 + \frac{5}{2}I_2$

CHAPTER 6

Exercises

1 $A = \{1, 2, 3, 4, 5, 6, 7, 8, 9\}$
$B = \{-4, 4\}$
$C = \{5, 6, 7, 8, 9, 10\}$
$D = \{4, 8, 12, 16, 20, 24\}$

2 $A \cup B = \{-4, 1, 2, 3, 4, 5, 6, 7, 8, 9\}$
$A \cap B = \{4\}$
$A \cup C = \{n \in \mathbb{N}: 1 \leqslant n \leqslant 10\}$
$A \cap C = \{n \in \mathbb{N}: 5 \leqslant n \leqslant 9\}$
$B \cup D = \{-4, 4, 8, 12, 16, 20, 24\}$
$B \cap D = \{4\}$
$B \cap C = \varnothing$

3 $A \cup B = \{n \in \mathbb{N}: 1 \leqslant n \leqslant 10\}$
$A \cap C = \{1, 5, 9\}$
$A \cap B = \varnothing$
$B \cup C = \{1, 2, 4, 5, 6, 8, 9, 10\}$
$B \cap C = \{4, 8\}$

4

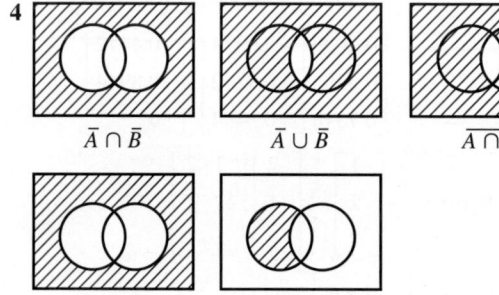

$\overline{A} \cap \overline{B}$ $\overline{A} \cup \overline{B}$ $\overline{A \cap B}$

$\overline{A \cup B}$ $A \cap \overline{B}$

5 (a) $A \cup B = \{1, 2, 3, 4, 5, 6, 7, 8, 9, 10, 12, 14, 16, 18, 20\}$
(b) $A \cap B = \{2, 4, 6, 8, 10\}$
(c) $A \cup C = \{1, 2, 3, 4, 5, 6, 7, 8, 9, 10, 16, 32\}$
(d) $A \cap C = \{2, 4, 8\}$

6 (a) True   (b) False   (c) False

7 (a) $\{n \in \mathbb{N}: 11 \leqslant n \leqslant 32\}$
(b) $\{11, 13, 15, 17, 19, 21, 22, 23, 24, 25, 26, 27, 28, 29, 30, 31, 32\}$
(c) $\overline{A} = \{n \in \mathbb{N}: 11 \leqslant n \leqslant 32\}$
$\overline{B} = \{1, 3, 5, 7, 9, 11, 13, 15, 17, 19, 20, 21, 22, 23, 24, 25, 26, 27, 28, 29, 30, 31, 32\}$
(d) $\overline{A \cap B} = \{1, 3, 5, 7, 9, 11, 12, 13, 14, 15, 16, 17, 18, 19, 20, 21, 22, 23, 24, 25, 26, 27, 28, 29, 30, 31, 32\}$
(e) $\overline{A} \cup \overline{B} = \overline{A \cap B}$ (see (d))

11 (a) $A \cap B$ (b) $\varnothing$ (c) A
(d) U (e) A (f) $A \cup (B \cap C)$
(g) $A \cap (B \cap C)$

14 (a) 5   (b) 25

15 (a) 20   (b) 27

16 4

17 (a) $\overline{C} = \{a, b, d, i\}, \overline{B \cup C} = \{a, b, i\}$
$\overline{B} \cap \overline{C} = \{a, b, i\}, A \cap B \cap D = \varnothing$
$A \cup F = \{a, b, c, f, i\}, D \cup (E \cap F) = \{b, c, e, h, i\},$
$(D \cup E) \cap F = \{b, c, i\}$
(b) $B \cup C = \{c, d, e, f, g, h\}, C \cup E = \{c, e, f, g, h, i\}$
$D \cup E \cup F = \{b, c, e, f, h, i\}$
(c) $L_1: \{b, c, d, e, f, g, h, i\}$
$L_2: \{b, c, d, e, f, g, h, i\}$
$L_3:$ all elements

18 (a) 1 if $p = 1, q = 1$; 0 otherwise
(b) 0 if $p = 0, q = 0$; 1 otherwise
(c) 0   (d) 1

19 (a) $p \cdot q + \overline{p} \cdot \overline{q}$ (b) $(p + \overline{p}) \cdot (q + \overline{q})$
(c) $p + q + \overline{p} + \overline{q}$ (d) $p \cdot q + r \cdot s$
(e) $\overline{p} \cdot q \cdot s + \overline{p} \cdot \overline{q} \cdot r \cdot s + p \cdot q \cdot r \cdot s + p \cdot \overline{q} \cdot s + p \cdot q \cdot s$
(f) $p \cdot q \cdot r + p \cdot q \cdot t + p \cdot q \cdot u + p \cdot s \cdot \overline{u} + p \cdot v$

20 $\overline{p} \cdot \overline{q} + r + \overline{s} + \overline{q} \cdot t$

21

p	q	r	$\overline{p}$	$\overline{q}$	$\overline{r}$	$\overline{p} \cdot q \cdot \overline{r}$	$\overline{p} \cdot q \cdot r$	$p \cdot \overline{q} \cdot \overline{r}$	$p \cdot q \cdot r$	f
0	0	0	1	1	1	0	0	0	0	0
0	0	1	1	1	0	0	0	0	0	0
0	1	0	1	0	1	1	0	0	0	1
0	1	1	1	0	0	0	1	0	0	1
1	0	0	0	1	1	0	0	1	0	1
1	0	1	0	1	0	0	0	0	0	0
1	1	0	0	0	1	0	0	0	0	0
1	1	1	0	0	0	0	0	0	1	1

$f = \overline{p} \cdot q \cdot \overline{r} + \overline{p} \cdot q \cdot r + p \cdot \overline{q} \cdot \overline{r} + p \cdot q \cdot r$

22 (a) $p \cdot q$ (b) $\overline{p} \cdot r$
(c) $p \cdot q + \overline{p} \cdot \overline{q}$ (d) $p + q + r$
(e) 1   (f) $q + r$

23 (a) $\overline{p} \cdot q + p \cdot \overline{q} + p \cdot q$
(b) $(p + q) \cdot (p + \overline{q}) + p \cdot (\overline{r} + \overline{q})$
(c) $p \cdot (q + \overline{p}) + (q + r) \cdot \overline{p}$

24

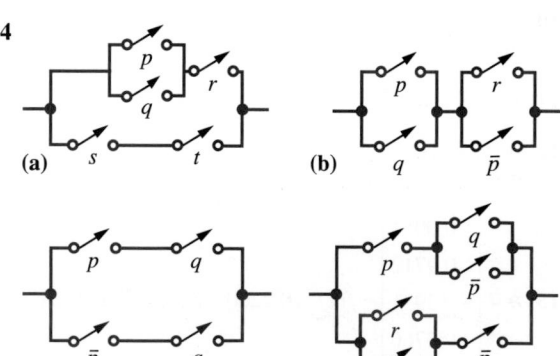

(a)   (b)

(c)   (d)

25

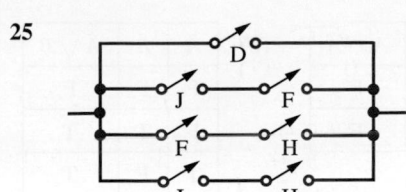

26

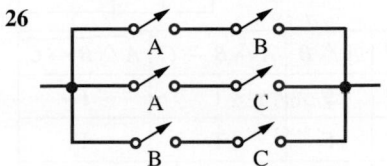

(A, B and C are the panel of three)

27

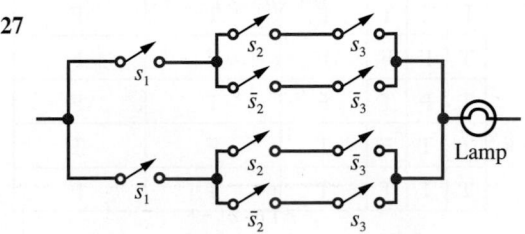

(s_1, s_2 and s_3 are the three switches)

28

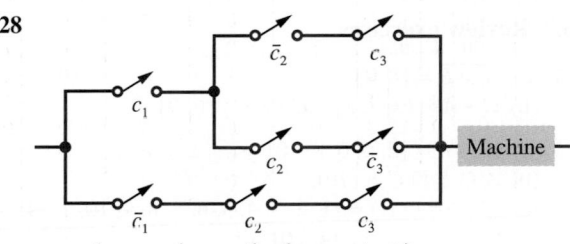

(c_1, c_2 and c_3 are the three contacts)

29

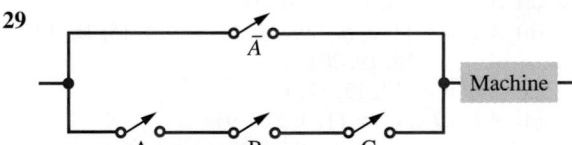

30 (a) $p + q$

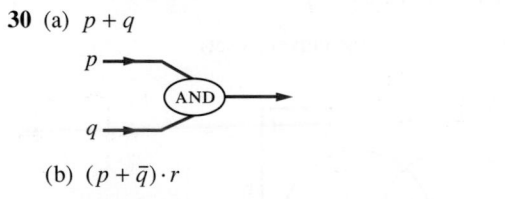

(b) $(p + \bar{q}) \cdot r$

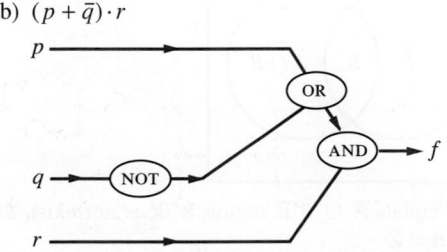

(c) 0

(d) $p \cdot r + s \cdot \bar{p}$

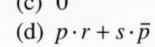

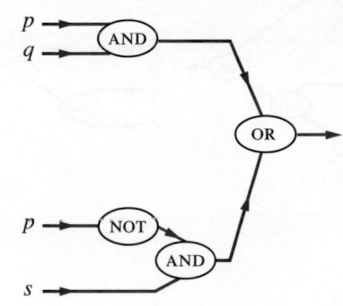

31

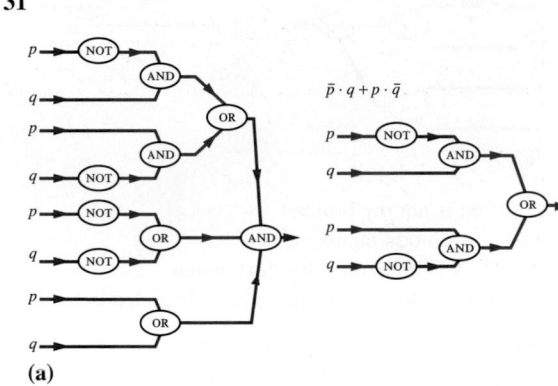

(a)

$\bar{p} \cdot q + p \cdot \bar{q}$

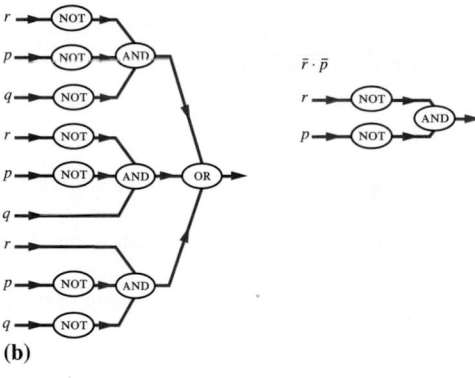

(b)

$\bar{r} \cdot \bar{p}$

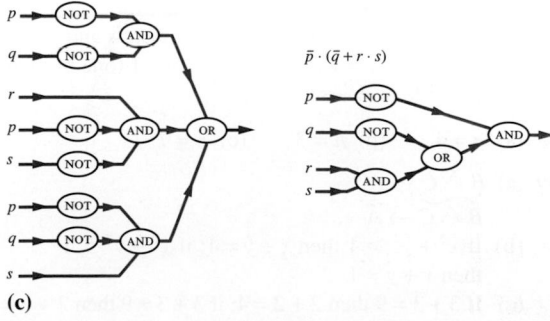

(c)

$\bar{p} \cdot (\bar{q} + r \cdot s)$

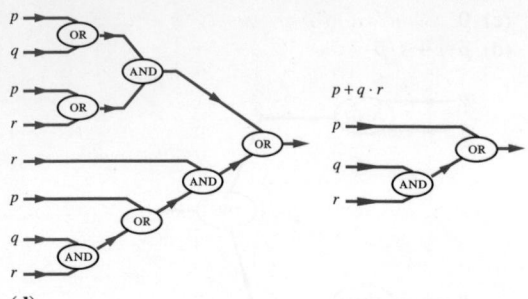

(d)

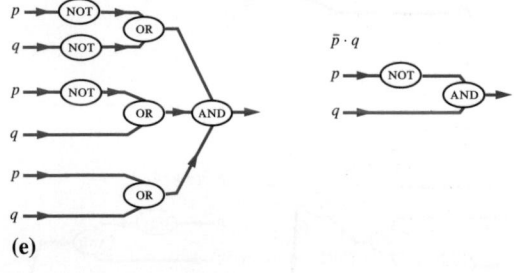

(e)

32 (a) Fred is not my brother
(b) 12 is an odd number
(c) There will be no gales next winter
(d) Bridges do not collapse when design loads are exceeded

33 (a) F   (b) T   (c) T   (d) F

34 (a) T   (b) F
(c)–(e) are not propositions
(f) Truth value is not known

35 (a) $A \wedge B$ (b) $A \rightarrow C$
(c) $\tilde{A} \rightarrow (\tilde{B} \wedge C)$ (d) $\tilde{C} \rightarrow B$

36 (a) It is raining and the sun is shining therefore there are clouds in the sky
(b) It is raining therefore there are clouds in the sky and hence the sun is shining
(c) If it is not raining then the sun is shining or there are clouds in the sky
(d) It is not the case that it rains and the sun shines, and there are clouds in the sky

37 (a) $x^2 = y^2 \rightarrow x = y$ for positive numbers x and y
(b) $x^2 = y^2 \rightarrow x = y$ for $x = 1$ and $y = -1$ (one of many possible answers)

38 (a) $n = 4$ (b) $n = 3$ (c) $n = 7$

39 (a) $B \wedge C \rightarrow A$
$\widetilde{B \wedge C} \rightarrow \tilde{A}$
(b) If $x^2 + y^2 \geqslant 1$ then $x + y = 1$; if $x^2 + y^2 < 1$ then $x + y \neq 1$
(c) If $3 + 3 = 9$ then $2 + 2 = 4$; if $3 + 3 \neq 9$ then $2 + 2 \neq 4$

40 (a)

A	B	$A \wedge \tilde{A}$
T	F	F
F	T	F

(b)

$\tilde{A}$	$\tilde{B}$	$\tilde{A} \vee \tilde{B}$
F	F	F
F	T	T
T	F	T
T	T	T

(c) and (d)

A	B	C	$A \wedge B$	$A \wedge B \rightarrow C$	$\overline{A \wedge B \rightarrow C}$
F	F	F	F	T	F
F	F	T	F	T	F
F	T	F	F	T	F
F	T	T	F	T	F
T	F	F	F	T	F
T	F	T	F	T	F
T	T	F	T	F	T
T	T	T	T	T	F

47 1, 4, 9, 16

6.7 Review exercises

1 (a) $\overline{A \cup B} = \{8, 9\}$
(b) $C - A = \{3, 7, 8\}$ $\overline{C} \cap \overline{B} = \{6, 9\}$

2 (a) $A \cap B = \{2, 4, 6, 8, 10\}$
(b) $A \cap B \cap C = \{10\}$
(c) $A \cup (B \cap C) = \{1, 2, 3, 4, 5, 6, 7, 8, 9, 10, 11, 12, 14, 20\}$

3 (a) $\overline{A} = \{n \in \mathbb{N}: 11 < n \leqslant 20\}$
(b) $\overline{A} \cup B = \{1, 3, 5, 7, 9, 11, 12, 13, 14, 15, 16, 17, 18, 19, 20\}$
(c) $\overline{A \cup B} = \{13, 15, 17, 19\}$
(d) $A \cap (\overline{B \cup C}) = \{1, 3, 5, 7, 9\}$

4 Statement (a) is true

5 (a) $f = A$ $g = U$ (the universal set)
(b)

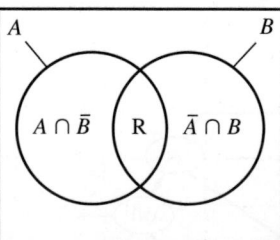

Only equals $A \cup B$ if region R does not exist, that is $A \cap B = \varnothing$

7
(a)
(b)
(c)
(d)

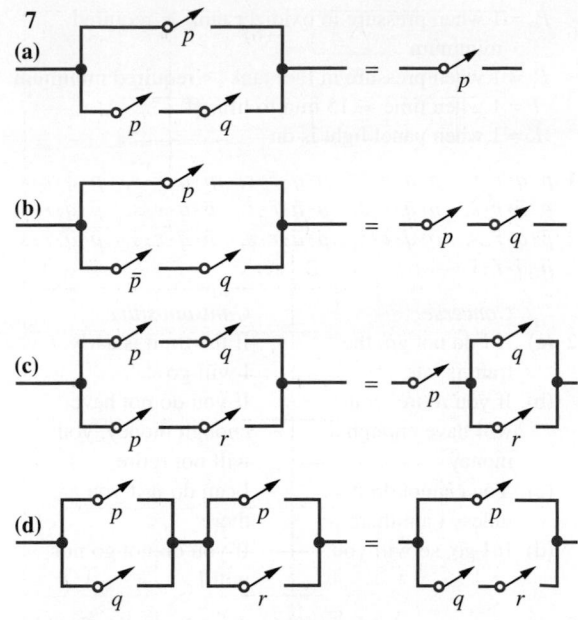

8 (a) $(\overline{A \cap \bar{B} \cap D}) \cap (A \cap C \cap \bar{D}) \cap (\overline{A \cap B \cap \bar{D}})$
(b) $(B \cap \bar{C}) \cup (C \cap \bar{B})$

9 (a) $x \cdot y \cdot z \cdot u + \bar{x} \cdot \bar{y} \cdot z \cdot u + x \cdot \bar{u}$

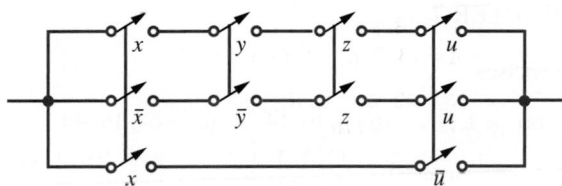

(b) $\bar{x} \cdot \bar{y} + \bar{z} \cdot \bar{u} + x \cdot y \cdot z$

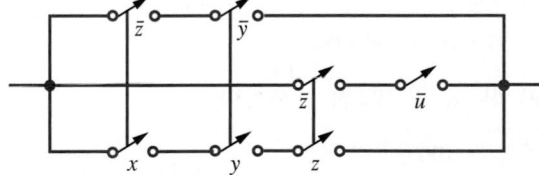

(c) (i) 0   (ii) 0

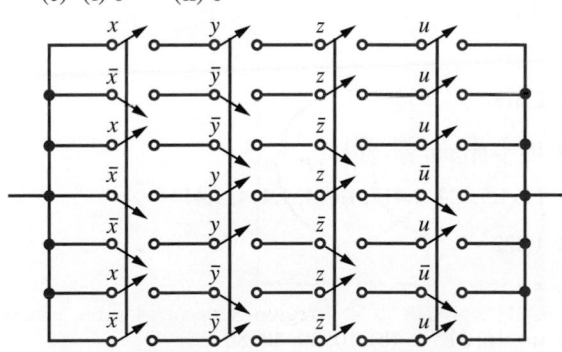

10 (a)

p	q	$p \wedge q$
T	T	T
T	F	F
F	T	F
F	F	F

(b)

p	q	$p \vee q$
T	T	T
T	F	T
F	T	T
F	F	F

(c)

p	q	$p \rightarrow q$
T	T	T
T	F	F
F	T	T
F	F	T

$\boxed{A}$ $\boxed{B}$ $\boxed{C}$

(e)

p	q	$\bar{p}$	$\bar{p} \wedge q$	$\overline{\bar{p} \wedge q}$	$p \vee q$	$\boxed{A} \wedge \boxed{B}$	$\boxed{C} \rightarrow p$
T	T	F	F	T	T	T	T
T	F	F	F	T	T	T	T
F	T	T	T	F	T	F	T
F	F	T	F	T	F	F	T

Hence $(p \vee q) \wedge (\overline{\bar{p} \wedge q}) \rightarrow p$ is a tautology.

11 (a) $\bar{q} \cdot p + \bar{p} \cdot q$ (b) $p + q + \bar{r}$
(c) $p \cdot q \cdot r + q \cdot \bar{r} \cdot s$

12 (a) (i) $p \cdot r + q \cdot \bar{r} \cdot s + \bar{q} \cdot r \cdot \bar{s}$
(ii) $p \cdot \bar{q} \cdot \bar{r} + \bar{p} \cdot \bar{r} \cdot s$

13 $C_1 \cdot \bar{C}_2 \cdot \bar{F}_1 \cdot F_2 \cdot \bar{F}_3 + C_1 \cdot \bar{C}_3 \cdot \bar{F}_1 \cdot \bar{F}_2 \cdot F_3$
$+ C_1 \cdot C_2 \cdot \bar{C}_3 \cdot \bar{F}_1 \cdot \bar{F}_2 \cdot F_3$
where C_i = call button on floor i
and $F_i = 1$ if lift is on floor, 0 otherwise

14 (a) Let the four people be labelled A, B, C, D.
The truth table is then as given below:

A	B	C	D	Yes	No	Tie
0	0	0	0	0	1	0
0	0	0	1	0	1	0
0	0	1	0	0	1	0
0	0	1	1	0	0	1
0	1	0	0	0	1	0
0	1	0	1	0	0	1
0	1	1	0	0	0	1
0	1	1	1	1	0	0
1	0	0	0	0	1	0
1	0	0	1	0	0	1
1	0	1	0	0	0	1
1	0	1	1	1	0	0
1	1	0	0	0	0	1
1	1	0	1	1	0	0
1	1	1	0	1	0	0
1	1	1	1	1	0	0

Extracting from this table those inputs that cause a Yes, No or Tie (Y, N or T) we have
(b) $Y = \bar{A} \cdot B \cdot C \cdot D + A \cdot \bar{B} \cdot C \cdot D + A \cdot B \cdot \bar{C} \cdot D$
$+ A \cdot B \cdot C \cdot \bar{D} + A \cdot B \cdot C \cdot D$
$N = \bar{A} \cdot \bar{B} \cdot \bar{C} \cdot \bar{D} + \bar{A} \cdot \bar{B} \cdot \bar{C} \cdot D$
$+ \bar{A} \cdot \bar{B} \cdot C \cdot \bar{D} + \bar{A} \cdot B \cdot \bar{C} \cdot \bar{D} + A \cdot \bar{B} \cdot \bar{C} \cdot \bar{D}$
$T = \bar{A} \cdot \bar{B} \cdot C \cdot D + \bar{A} \cdot B \cdot \bar{C} \cdot D + \bar{A} \cdot B \cdot C \cdot \bar{D}$
$+ A \cdot \bar{B} \cdot \bar{C} \cdot D + A \cdot \bar{B} \cdot C \cdot \bar{D} + A \cdot B \cdot \bar{C} \cdot \bar{D}$
(c) $Y = A \cdot B \cdot D + A \cdot B \cdot C + A \cdot C \cdot D + B \cdot C \cdot D$
$N = \bar{A} \cdot \bar{B} \cdot \bar{C} + \bar{A} \cdot \bar{B} \cdot \bar{D} + \bar{A} \cdot \bar{C} \cdot \bar{D} + \bar{B} \cdot \bar{C} \cdot \bar{D}$
T does not simplify

(d) To modify the circuit we introduce the chairman's vote E. If N denotes No and Y denotes Yes, the new circuit must have the output

$$N_{new} = (N_{old} + T) \cdot \bar{E}$$

$$Y_{new} = (Y_{old} + T) \cdot E$$

where $T =$ Tie. Hence the modified circuit will be

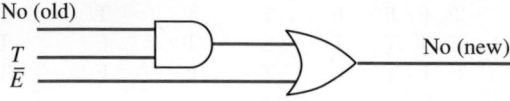

No (old)

T

$\bar{E}$

No (new)

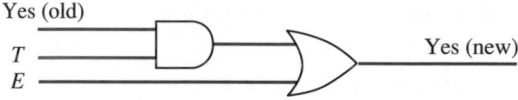

Yes (old)

T

E

Yes (new)

A tie is now impossible.

15 $\bar{N} \cdot (\bar{V} + \bar{R} \cdot M)$

i.e. no dope smoking occurs if Neil is absent and *either* Vivian is absent *or* Mike is present *and* Rick is absent.

16 $p \cdot q \cdot \overline{\bar{p} \cdot r} + q \cdot \bar{r} = q \cdot \bar{r}$

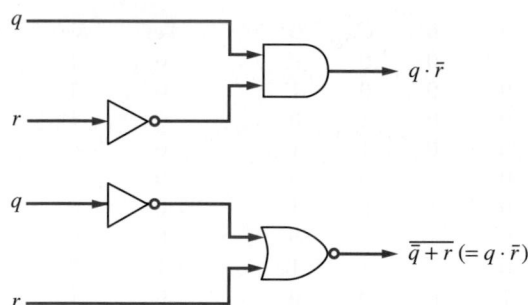

q

r

$q \cdot \bar{r}$

q

r

$\overline{\bar{q} + r} (= q \cdot \bar{r})$

17 (a) F (b) No (c) No

(d) F (e) F (f) F

(g) No (h) F (i) T

18 (a) (i)

p	q	$p \to q$	$q \to p$	$p \leftrightarrow q$
T	T	T	T	T
T	F	F	T	F
F	T	T	F	F
F	F	T	T	T

(ii)

q	p	$p \to q$
T	T	T
T	F	F
T	F	F
F	F	T

(b) (i) False (ii) False

19 $\bar{P}_O \cdot \bar{T} + P_O \cdot \bar{P}_F \cdot \bar{T} + P_O \cdot P_F \cdot T$
$= \bar{P}_O \cdot \bar{T} + \bar{P}_F \cdot \bar{T} + P_O \cdot P_F \cdot T$ is minimal

$P_O = 1$ when pressure in oxidizer tank $\geq$ required minimum

$P_F = 1$ when pressure in fuel tank $\geq$ required minimum

$T = 1$ when time ≤ 15 min to lift-off

$L = 1$ when panel light is on

21 $p \cdot q \cdot r \cdot s, \quad p \cdot q \cdot r \cdot \bar{s}, \quad p \cdot q \cdot \bar{r} \cdot s, \quad p \cdot q \cdot \bar{r} \cdot \bar{s}, \quad p \cdot \bar{q} \cdot r \cdot s,$
$p \cdot \bar{q} \cdot r \cdot \bar{s}, \quad p \cdot \bar{q} \cdot \bar{r} \cdot \bar{s}, \quad p \cdot \bar{q} \cdot \bar{r} \cdot \bar{s}, \quad \bar{p} \cdot q \cdot r \cdot s, \quad \bar{p} \cdot q \cdot r \cdot \bar{s},$
$\bar{p} \cdot q \cdot \bar{r} \cdot s, \quad \bar{p} \cdot q \cdot \bar{r} \cdot \bar{s}, \quad \bar{p} \cdot \bar{q} \cdot r \cdot s, \quad \bar{p} \cdot \bar{q} \cdot r \cdot \bar{s}, \quad \bar{p} \cdot \bar{q} \cdot \bar{r} \cdot s,$
$\bar{p} \cdot \bar{q} \cdot \bar{r} \cdot \bar{s}$

	Converse	*Contrapositive*
22 (a)	If I do not go, the train is late	If the train is early, I will go
(b)	If you retire, you will have enough money	If you do not have enough money, you will not retire
(c)	You cannot do it unless I am there	I can do it if you are there
(d)	If I go, so will you	If you do not go nor will I

23 'If you were a member of the other tribe, what would you answer if I asked you if your God was male?' The answer is then definitely false!

CHAPTER 7

Exercises

1 (a) $\frac{1}{3}, 1, \frac{9}{5}$ (b) 6, 10, 14 (c) $-64, 16, -4$

2 $x_{n+1} = \frac{3}{8} x_n, x_0 = 5$

3 $p = -3, q = 13, x_0 = 13, x_1 = 10, x_2 = 7, x_3 = 4$
$x_{n+1} = x_n - 3$

5 45, 57.5, 63.75, 66.875

6 (a) 40 (b) 16736 (c) 35

7 $\left(\dfrac{v}{V}\right)^{n+1} \times 100$

8 $\dfrac{1}{n} \displaystyle\sum_{l=1}^{n} (x_l - 10)^2$

9 2.618

10 $\{0, \frac{1}{3}, \frac{16}{25}, \frac{81}{109}, \frac{256}{321}, \frac{625}{751}\}$

11 $\{1, 1.5, 1.4, 1.417, 1.414, 1.414, 1.414\}$

12 1.222

13 5

14 (a) 16, 31 (b) 10, 20, 40, 80

15 £2700k, 11

16 $\dfrac{1 + (n - 2)\sqrt{x}}{1 - x}$

17 3.1365 (4dp)

18 $\frac{2}{5}, \frac{2}{7}, \frac{2}{9}, \frac{2}{11}, \frac{2}{13}, \frac{2}{15}$

19 £66 116, £128 841, after 7.3 years

20 (a) 11 781 (b) 1 205 589 (c) $1 - (\frac{1}{2})^{153}$
(d) $3^{154} - 1$ (e) 1 217 370 (f) $\frac{153}{154}$

21 9

23 $x = \dfrac{10r}{1 - (1 + \frac{1}{100}r)^{-n}}$, 117.46

24 $2 - \dfrac{2 + n}{2^n}$

25 $\dfrac{a(1 - r^n)}{1 - r} + \dfrac{dr}{(1 - r)^2}[1 + nr^{n-1} - (n - 1)r^n]$

26 (a) $A2^n + 3$ (b) $A3^n - 5(n + \frac{1}{2})$
(c) $A(-1)^n + \frac{2}{3}(\frac{1}{2})^n$ (d) $A2^n + \frac{3}{2}n2^n$

27 £1770, {10 000, 9430, 8792, 8077, 7276, 6379, 5375, 4250, 2989, 1578, −3}

28 $(A + n)/n^2$

29 (a) 0 (b) -3×2^n (c) 0 (d) $75 \times (-2)^n$
(a) and (c) satisfy recurrence relation

31 (a) $A5^n + B2^n$ (b) $A3^n + B(-2)^n$
(c) $(\frac{1}{5})^n\left(A\cos\dfrac{n\pi}{2} + B\sin\dfrac{n\pi}{2}\right)$
(d) $A5^n + Bn5^n$
(e) $A(-\frac{1}{2})^n + B$

32 (a) $\frac{1}{4} + \frac{13}{12}(5^n) - \frac{1}{12}(2^n)$ (b) $\frac{4}{9} + \frac{14}{9}(-\frac{1}{2})^n + \frac{1}{3}n$
(c) $-\frac{1}{6}n - \frac{1}{36} + A3^n + B(-2)^n$

33 (b) $(1 - n)a^n$ (c) $(3 + (2a^{-10} - 0.3)n)a^n$

34 $T_2 = 2x^2 - 1$, $T_3 = 4x^3 - 3x$, $T_4 = 8x^4 - 8x^2 + 1$

35 (a) $N_t = 2^t$
(b) $N_t = \left[\left(\dfrac{1 + \sqrt{5}}{2}\right)^{t+1} - \left(\dfrac{1 - \sqrt{5}}{2}\right)^{t+1}\right]\Big/\sqrt{5}$

36 (a) 0.5, 0.4, 0.3, 0.2353, 0.1923, 0.1632; $\to 0$
(b) 0.4615, 0.4722, 0.4789, 0.4831, 0.4859, 0.4879; $\to 0.5$
(c) 2, 2, 1.817, 1.682, 1.585, 1.513; $\to 1$
(d) 1.5, 1.5625, 1.5880, 1.6018, 1.6105, 1.6165; $\to e^{1/2} = 1.6487$

(e) 1.4142, 1.5538, 1.5981, 1.6119, 1.6161, 1.6174; $\to \frac{1}{2}(1 + \sqrt{5})$
(f) 0, 0, 1.2990, 2, 2.3776, 2.5981; $\to \pi$

37 (a) 1, 1, $\frac{5}{3}$, 2.5, 3.4, 4.3333; diverges to infinity
(b) 1, 0, −1, 0, 1, 0; oscillates between 1, 0, −1
(c) 1, 3, 1, 3, 1, 3; oscillates between 1 and 3

38 (a) 10 (b) 19 (c) 1 000 002
(d) 25 (e) 18

39 70%

40 2.2, 2.324, 2.418 996, 2.450 262
Estimate = 2.465 5011
Limit = 2.465 571

41 (a) convergent (b) convergent
(c) divergent (d) divergent

42 $\frac{9}{4}$

43 (a) $1 - \dfrac{1}{2N + 1}$, 1 (b) $4 - \dfrac{2 + N}{2^{N-1}}$, 4
(c) $\dfrac{1}{4} - \dfrac{1}{2(N + 1)(N + 2)}$, $\dfrac{1}{4}$

44 (a) divergent (b) divergent (c) convergent

46 $\frac{19}{33}$; (a) $\frac{413}{999}$ (b) $\frac{10}{99}$ (c) 1 (d) $\frac{172\,300}{9999}$

48 1.082 322 1; summation from right allows full account to be taken of the accumulative effect of small terms

50 (a) $|x| < 1$ (b) $x \in \mathbb{R}$ (c) $|x| \leq 1$ (d) $|x| < 1$

51 (a) $(-x^2)^r$ $(|x| < 1)$ (b) $\dfrac{x^{2r+1}}{2r + 1}$ $(|x| < 1)$
(c) $(-1)^r(r + 1)x^r$ $(|x| < 1)$
(d) $-\dfrac{(2r - 3)!}{2^{2r-2}r!(r - 2)!}x^r$ $r > 1$ $(|x| < 1)$
(e) $\dfrac{1}{10}\left[2^{r+2} + \dfrac{(-1)^r}{2^r}\right]$ $(|x| < \frac{1}{2})$
(f) $(1 + x)x^{4r}$ $(|x| < 1)$

52 (a) $\dfrac{5 \cdot 4}{1 \cdot 2}$ (b) $\dfrac{(-2)(-3)(-4)}{1 \cdot 2 \cdot 3}$
(c) $\dfrac{(\frac{1}{2})(-\frac{1}{2})(-\frac{3}{2})}{1 \cdot 2 \cdot 3}$ (d) $\dfrac{(-\frac{1}{2})(-\frac{3}{2})(-\frac{5}{2})(-\frac{7}{2})}{1 \cdot 2 \cdot 3 \cdot 4}$

54 $n = 8$, 1.1905, six multiplications

55 (a) $(1 + 2x^2)^{-1}$ (b) $(1 - x)^{-1/2}$
(c) $1 + \dfrac{1 - x}{x}\ln(1 - x)$ (d) $\dfrac{x^2}{1 - x^2}\ln(1 - x^2)$

56 3.1415

57 (a) $\frac{1}{3}$ (b) $-\frac{2}{3}$ (c) 2   (d) 0

58 (a) 3   (b) $\frac{1}{2}$

59 (a) -1 (b) 1   (c) $n-1$ (d) n

62 (a) Undefined at $x = 0$, continuous for $x \neq 0$
   (b) Infinite discontinuity at $x = 2$, continuous for $x \neq 2$
   (c) Finite discontinuity at $x = 0$, continuous for $x \neq 0$
   (d) Finite discontinuities at $x = \pm\sqrt{n}$, $n = 0, 1, 2, \ldots$

63 (a) Upper bound is 7, lower bound 5
   (b) Upper bound is 3, lower bound -1

64 1.75

65 0.830

66 $\alpha = -1.879$, $\beta = 0.347$, $\gamma = 1.532$

67 (a) is convergent, (b) convergent and (c) divergent
   (b) 0.771

68 5.4267, 5.3949, $\varepsilon_1 \simeq 0.05$

69 $\alpha_0 \approx 1.9$, $\alpha_k \approx \frac{1}{2}(2k+1)\pi$
   $\theta_{n+1} = \cos^{-1}(-\mathrm{sech}\,\theta_n)$, $\alpha_0 = 1.8751$

7.12 Review exercises

1 1000, 850, 700, 550, 400, 250, 100
  1000, 681, 464, 316, 215, 147, 100

2 £361, £243, £141, £53 for $r < 23.375$

3 $1 - 0.2(-\frac{1}{2})^r$,   1

5 (a) $A2^n + B3^n$ (b) $(A + Bn)2^n$
   (c) $A2^n + B3^n + \frac{1}{2}4^n$ (d) $A2^n + B3^n + 3^{n-1}n$

8 $200 + 20(-\frac{2}{5})^r$

9 $2 + A\cos n\theta + B\sin n\theta$, $\tan\theta = \dfrac{\sqrt{7}}{3}$

10 $\gamma \approx 0.577\,235$; compare the true value 0.577 216

11 (a) divergent   (b) convergent
   (c) divergent   (d) convergent

12 (a) $\frac{410}{333}$ (b) $\frac{143}{333}$ (c) $\frac{101}{999}$ (d) $\frac{1724}{3333}$

13 (a) convergent   (b) divergent
   (c) convergent   (d) divergent

14 $\dfrac{x(1 + x)}{(1 - x)^3}$

15 $a = 1$, $b = -\frac{1}{3}$, $c = \frac{1}{45}$
   $|x| < 0.2954$
   $\tan 0.29$ is given to 4dp; $\tan 0.295$ has an error of $\frac{1}{2}$ unit
   in 1dp, but when rounded to 4dp gives an error of 1 unit.

16 $x - \frac{1}{6}x^3$
   $x - \frac{1}{6}x^3 + \frac{3}{40}x^5 - \frac{5}{112}x^7 + \frac{35}{1152}x^9$

17 0.095, 31

19 2.718 586 07
   2.718 357 88
   2.718 281 81

20

r	0	1	2	3	4	5	6	7	8
M_r	0	2.64	1.93	2.13	2.06	2.13	1.93	2.64	0

22 $F = \dfrac{W(l-a)^2(l+2a)}{l^3} - WH(x-a)$

   $M = \dfrac{W(l-a)^2 a}{l^2} - \dfrac{W(l-a)^3(l+2a)x}{l^3}$
      $+ W(x-a)H(x-a)$

23 (a) positive values of $\sqrt{(1 - \sin^2\theta)}$
   (b) $A = -5/128$

24 1.233577

25 $a = -\frac{1}{3}$, $-4/45$

CHAPTER 8

Exercises

1 (a) 0   (b) 1   (c) $2x$
   (d) $3x^2$ (e) $\dfrac{1}{2\sqrt{x}}$ (f) $\dfrac{-1}{(1+x)^2}$

2 (a) $4x - 5$ (b) -1
   (c) $(1, -15)$, $(\frac{1}{2}m + \frac{3}{2}, \frac{1}{2}m^2 + \frac{1}{2}m - 15)$
   (d) -1 (e) $y = -x - 14$

3 (a) $6x^2 - 6x + 1$ (b) 1
   (c) $(1, 3)$, $(\frac{1}{4}[1 \pm \sqrt{(1 + 8m)}], \frac{1}{4}[12 - 3m \pm m\sqrt{(1 + 8m)}])$
   (d) $m = 1, -1/8$ (e) $y = x + 2$, $y = -0.125x + 3.125$

4 (a) minimum at $x = -1/2$ (b) minimum at $x = 1/3$
   (c) maximum at $x = 3/2$ (d) maximum at $x = 1/2$

5 $3ax^2 + 2bx + c$

6 $v(t) = \begin{cases} 1, & 0 \leq t < 1 \\ 2t - 1, & 1 \leq t \leq 2 \\ 3, & 2 \leq t < 3 \\ -1, & 3 \leq t \leq 9 \end{cases}$

8 $\frac{1}{3}m^2\,\mathrm{min}^{-1}$

10 $3W(2x - l)^2/4l^2$

12 μT

16 $\dfrac{\mathrm{d}x}{\mathrm{d}t} \propto (a - x)(b - x)$

17 $r = l/4$

18 £1.25, 7500

19 (a) $9x^8$,   (b) $\frac{3}{2}\sqrt{x}$,   (c) $-8x$,
  (d) $16x^3 + 10x^4$,   (e) $12x^2 + 1$,   (f) $-1/x^3$,
  (g) $1 + 1/(2\sqrt{x})$,   (h) $7x^{5/2}$,   (i) $-1/x^4$

20 (a) $18x^5 + 75x^4 - 2x - 5$,
  (b) $20x^3 + 3x^2 + 30x - 27$,
  (c) $(21x^2 + 10x - 3)/(x^{3/2})$,   (d) $6 - 8x + 27/x^2$,
  (e) $(3x^3 - x^2 + x - 3)/(2x^{5/2})$,   (f) $8x^3 + 9x^2 + 4x$

21 (a) $(-3x^4 - 2x^3 - 3x^2 + 6x + 1)/(x^3 + 1)^2$,
  (b) $(4 - 3x^2)/[(x^2 + 4)^2\sqrt{(2x)}]$,
  (c) $(1 - 2x - x^2)/(x^2 + 1)^2$,
  (d) $(x^{1/3} + 2)/[3x^{1/3}(1 + x^{1/3})^2]$,
  (e) $(x^2 + 2x - 1)/(1 + x^2)^2$,
  (f) $3x(2 - x)/(x^2 - 2x + 2)^2$

22 $2ac + ad + bc$, $(ad - bc)/(cx + d)^2$, $6ax + 5b$,
  $(bax^2 + 2acx)/(bx + c)^2$

23 $10^3\sqrt{2}$

24 (a) $10x - 2$ (b) $12x^2 + 1$ (c) $24x^{23}$
  (d) $45(5x + 3)^8$ (e) $28(4x - 2)^6$
  (f) $-18(1 - 3x)^5$ (g) $3(6x - 1)(3x^2 - x + 1)^2$
  (h) $6(12x^2 - 2)(4x^3 - 2x + 1)^5$
  (i) $-5(4x^3 - 1)(1 + x - x^4)^4$

25 (a) $512(x + 2)^6(3x - 2)^4(9x + 4)$
  (b) $(5x + 1)^2(3 - 2x)^3(37 - 70x)$
  (c) $(\frac{1}{2}x + 2)(x + 3)^3(3x + 11)$
  (d) $12x^3 - 6x^2 - 20x + 11$
  (e) $36x^5 + 20x^3 - 54x^2 + 12x - 15$
  (f) $2(x^2 + x + 1)(x^3 + 2x^2 + 1)^3 \times$
    $(8x^4 + 19x^3 + 16x^2 + 10x + 1)$
  (g) $(x^5 + 2x + 1)^2(2x^2 + 3x - 1)^3 \times$
    $(46x^6 + 57x^5 - 15x^4 + 44x^2 + 58x + 6)$
  (h) $(2x + 1)^2(7 - x)^4(37 - 16x)$
  (i) $(3x + 1)^4(21x^2 + 74x + 19)$

28 $b = 56$, $h = 144$, $w = 90$

29 (a) $1/\sqrt{(1 + 2x)}$ (b) $(3x + 2)/(2\sqrt{x})$
  (c) $(3x + 4)/[2\sqrt{(x + 2)}]$

30 (a) $(2x^2 + 4)/\sqrt{(4 + x^2)}$
  (b) $(9 - 2x^2)/\sqrt{(9 - x^2)}$
  (c) $(2x^2 + 4x + 4)/\sqrt{(x^2 + 2x + 3)}$
  (d) $\frac{2}{3}x^{-1/3} - \frac{1}{4}x^{-3/4}$ (e) $\frac{2x}{3}(x^2 + 1)^{-2/3}$
  (f) $(8x - 3)/[3(2x - 1)^{2/3}]$

31 (a) $-2/(x + 3)^3$ (b) $1/(x + 1)^2$ (c) $1/(x - 2)^2$
  (d) $2(2 - x)/(x^2 - 4x + 1)^2$ (e) $1 - 1/x^2$
  (f) $(6 - x^2)/(x^2 + 5x + 6)^2$ (g) $-1/(x^2 - 1)^{3/2}$
  (h) $2(2x + 1)(2 - 9x - 12x^2)/(3x^2 + 1)^4$

32 (a) $3\cos(3x - 2)$ (b) $-4\cos^3x\sin x$
  (c) $-6\cos 3x\sin 3x = -3\sin 6x$
  (d) $\frac{5}{2}\cos 5x - \frac{1}{2}\cos x$ (e) $\sin x + x\cos x$
  (f) $-\sin 2x/\sqrt{(2 + \cos 2x)}$ (g) $-a\sin(x + \theta)$
  (h) $4\sec^2 4x$

33 (a) $1/\sqrt{(4 - x^2)}$ (b) $-5/\sqrt{(1 - 25x^2)}$
  (c) $(x\tan^{-1}x + 1)/\sqrt{(1 + x^2)}$ (d) $1/\sqrt{(3 + 2x - x^2)}$
  (e) $3/(1 + 9x^2)$ (f) $1 - \dfrac{x\sin^{-1}x}{\sqrt{(1 - x^2)}}$

34 $\frac{8}{27}$ (volume of sphere)

35 (a) $2e^{2x}$ (b) $-\frac{1}{2}e^{-x/2}$
  (c) $(2x + 1)\exp(x^2 + x)$ (d) $xe^{5x}(5x + 2)$
  (e) $e^{-x}(1 - 3x)$ (f) $-e^x/(1 + e^x)^2$
  (g) $\frac{1}{2}e^x/\sqrt{(1 + e^x)}$ (h) ae^{ax+b}

36 (a) $2/(2x + 3)$ (b) $2(x + 1)/(x^2 + 2x + 3)$
  (c) $1/(x - 2) - 1/(x - 3)$ (d) $(1 - \ln x)/x^2$
  (e) $5/[(2x + 1)(1 - 3x)]$ (f) $(2x + 1)/[x(x + 1)]$

37 (a) $3\cosh 3x$ (b) $4\operatorname{sech}^2 4x$
  (c) $3x^2\cosh 2x + 2x^3\sinh 2x$ (d) $\frac{1}{2}\tanh\frac{1}{2}x$
  (e) $\cos x\sinh x - \sin x\cosh x$ (f) $-\sinh x/\cosh^2 x$

38 (a) $2/\sqrt{(4 + x^2)}$ (b) $2/\sqrt{(x^2 - 1)}$ (c) $1/(1 - x^2)$
  (d) $1 + \dfrac{x\sinh^{-1}x}{\sqrt{(1 + x^2)}}$ (e) $\sqrt{(4 - x^2)}/x$
  (f) $[(1 + x^2) - 2x\tanh^{-1}x(1 - x^2)]/[(1 - x^2)(1 + x^2)^2]$

39 $e^{-2\pi a/\omega}$

40 $a = 26$, $b = 39$

41 horizontal side $1/\sqrt{2}$

43 u, $\frac{1}{\alpha}\ln 3$, $3\alpha u/2$

44 $(1 - t\tan t)/(\tan t + t)$

45 $y = 1 - (\sqrt{2} - 1)x$

46 $y = 4 - 64x$

47 $\frac{3}{4}$

48 (a) $10^x\ln 10$ (b) $-2^{-x}\ln 2$
  (c) $(2x^3 + 3x^2 + 6x + 6)(x - 1)^{5/2}[(x + 1)^{1/2}/(x^2 + 2)^2]$

49 $x^2e^{-2x}((3 - 2x)\sin \pi x + \pi x\cos \pi x)$

50 $y = x + 1$; $(0, 1)$, $(-1, 0)$

51 $\cot\frac{1}{2}\theta$

52 $-\dfrac{x(2x^2 + 2y^2 - a^2)}{y(2x^2 + 2y^2 + a^2)}$

53 (a) $(\ln x)^{x-1}[1 + (\ln x)\ln\ln x]$ (b) $(2\ln x)x^{\ln x-1}$
  (c) $-\frac{5}{3}x(5 - 2x^2)(1 - x^2)^{-1/2}(2x^2 + 3)^{-7/3}$

54 (a) $[(3 - 2x)\ln x + 1]x^2 e^{-2x}$

(b) $[(x - 1)\sin 2x + 2x\cos 2x]\dfrac{e^x}{x^2}$

55 (a) $(6 + 19x^2 + 12x^4)x/(1 + x^2)^{3/2}$

(b) $(1 - 2x - 2x^2)/(1 + x + x^2)^2$

(c) $-(84x^2 + 6y^2y' + 6xyy'^2)/(1 + 3xy^2)$ where $y' = -(28x^3 + y^3)/(1 + 3xy^2)$

(d) $(6x - 2y' - 6yy'^2)/(3y^2 + x)$ where $y' = (1 + y - 3x^2)/(3y^2 - x)$

56 (a) $-(2 + t^2)/(\sin t + t\cos t)^3$

(b) $\frac{1}{8}\operatorname{cosec}\dfrac{3t}{2}\sec^3\dfrac{t}{2}$

59 (a) $2x - \dfrac{2}{x^3}, \quad 2 + \dfrac{6}{x^4}$

60 $\cot\dfrac{\theta}{2}, -\dfrac{a}{y^2}$

61 $\left(\dfrac{1 + 2t}{1 + t}\right)^2, \dfrac{2(1 + t)^3}{3(1 + 2t)^3}$

62 $2, 0$

63 (a) $3^4 e^{3x}, 3^n e^{3x}$

(b) $-\dfrac{6}{(2 + x)^4}, (-1)^{n-1}(n - 1)!/(2 + x)^n$

(c) $\dfrac{12}{(1 + x)^5} + \dfrac{12}{(1 - x)^5},$

$\frac{1}{2}n!\left[\dfrac{1}{(1 - x)^{n+1}} + \dfrac{(-1)^n}{(1 + x)^{n+1}}\right]$

64 $a^4 \sin(ax + b)$

67 (a) $(x^2 - 20)\cos x + 10x\sin x$

(b) $(x - 4)e^{-x}$

(c) $216(273x^2 + 39x + 1)(3x + 1)^9$

68 $\frac{1}{12}(145)^{3/2}$

70 $\sqrt{2}, (1, 2)$

71 $13^{3/2}/16$

72 (a) Minimum $(1, 0)$, maximum $(\frac{2}{3}, \frac{1}{27})$, inflection $(\frac{5}{6}, \frac{1}{54})$

(b) Minimum $(1, 43)$, maximum $(-5, 151)$, inflection $(-2, 97)$

73 (a) Minimum $(-2, -\frac{1}{3})$, maximum $(2, -3)$, inflection $(-(4 - 3\sqrt[3]{4})/(\sqrt[3]{4} - 1), 3(\sqrt[3]{4} + 1)/(\sqrt[3]{4} - 1))$

(b) Maximum $(4, 54, e^{-4})$

(c) Minimum $(0, 0)$, maximum $(2, 4e^{-2})$, inflections $(2 \pm \sqrt{2}, (4 \pm 2\sqrt{2})e^{-2\mp\sqrt{2}})$

74 $d = 10^3\sqrt{(2/\pi)}, \quad h = 10^3\sqrt{(2/\pi)}$

75 $d = 8.0$ (1dp), $\quad h = 9.9$

76 $(0, 0)$ minimum, $(1/\sqrt{e}, K/(2e))$ maximum.

(Note: $\dfrac{dv}{dx}$ not defined at $x = 0$ but $\dfrac{dv}{dx} \to 0$ as $x \to 0+$)

77 $S = 8\pi a^2$

78 $b \in [80, 82.2]$

79 (a) $x = \frac{4}{3}$ (b) $x = 2$

80 Distribute wash water equally

81 In year k a volume $(1 - \alpha)/(1 - \alpha^{11-k})$ of standing timber should be felled, α growth factor

82 $1.035, 0.92, 0.88$

84 (a) $5.436 = 2e, 8.155 = 3e$

(b) 5.440 ($h = 0.01$), error depends on h^2

(c) 8.00 ($h = 0.01$)

85 1.5432

87 Brian by $(6\sqrt{3} - 4\sqrt{6})\,$s

90 -6

94 $F = \begin{cases} 7W/8 - 4Wx/l & 0 < x < l/4 \\ -W/8 & l/4 < x < l \end{cases}$

95 Depth h satisfies $1000\pi h^2(3 - 2h) = 6t$

96 (a) $\frac{1}{7}x^7 + c$ (b) $\frac{1}{3}e^{3x} + c$

(c) $-\frac{1}{5}\cos 5x + c$ (d) $\frac{1}{8}(2x + 1)^4 + c$

(e) $\frac{1}{3}\tan 3x + c$ (f) $2\ln|x| + c$

(g) $-\dfrac{3}{x} + c$ (h) $\frac{1}{2}\sin 2x + c$

(i) $\frac{1}{4}\sec 4x + c$ (j) $\frac{1}{6}(4x - 1)^{3/2} + c$

97 (a) $\frac{9}{5}x^{5/3} + c$ (b) $\frac{2}{3}\sqrt{2}x^{3/2} + c$

(c) $\frac{1}{2}x^4 - \frac{2}{3}x^3 + \ln|x| - 2x + c$

(d) $2e^x + \frac{3}{2}\sin 2x + c$ (e) $\frac{1}{3}x^3 + 3e^x + \dfrac{1}{x} + c$

(f) $\frac{1}{8}(2x + 1)^4 + c$ (g) $-\frac{3}{8}(1 - 2x)^{4/3} + c$

(h) $\frac{8}{7}x^7 + \frac{12}{5}x^5 + 2x^3 + x + c$

(i) $\frac{1}{2}\sin(2x + 1) + c$ (j) $\dfrac{2^x}{\ln 2} + c$

(k) $\frac{1}{16}(4\cos 2x - \cos 8x) + c$

(l) $\frac{1}{24}(\sin 12x + 6\sin 2x) + c$

98 (a) $\frac{1}{5}[\ln|x + 1| + 4\ln|x - 4|] + c$

(b) $\ln|x - 2| - \dfrac{2}{(x - 2)} + c$

(c) $\ln\left|\dfrac{x}{x + 1}\right| + c$ (d) $\ln|x + 1| + \dfrac{1}{x + 1} + c$

(e) $\frac{1}{2}\ln\left|\frac{x-1}{x+1}\right| + c$ (f) $\ln\left|\frac{x-1}{x}\right| + \frac{1}{x} + c$

(g) $\frac{1}{2}\ln\left|\frac{x(x-2)}{(x-1)^2}\right| + c$ (h) $\frac{1}{3}\ln\left|\frac{1+2x}{1-x}\right| + c$

(i) $2x + \frac{2}{3}\ln|x-1| - \frac{1}{3}\ln|x^2+2+1| +$

$\frac{10}{3\sqrt{3}}\tan^{-1}\left(\frac{2x+1}{\sqrt{3}}\right) + c$

(j) $2\ln|x| + \ln(x-1) - \tan^{-1}x + c$

(k) $\ln\left|\frac{x-1}{x+2}\right| + \frac{3}{x+2} + c$

(l) $-2\ln|x-1| + (\frac{1}{2} + \frac{3}{2}\sqrt5)\ln|x - \frac{1}{2} - \frac{1}{2}\sqrt5|$

$- (\frac{3}{2}\sqrt5 - \frac{1}{2})\ln|x - \frac{1}{2} + \frac{1}{2}\sqrt5| + c$

99 (a) $\frac{4}{3}$ (b) $-\frac{1}{156}$ (c) 0 (d) $\frac{1}{2}\pi$

(e) $2^{7/2}/5 - \frac{9}{10}$ (f) 1 (g) $\frac{1}{3}\pi$ (h) $\frac{1}{2}\ln 2 + \frac{1}{2}$

101 (a) $-\frac{1}{x} + c$ (b) $\frac{3}{2}(x+1)^{2/3} + c$

(c) $2x^2 - 7x - \frac{1}{x} + c$ (d) $\sin x - \cos x + c$

(e) $\frac{1}{2}x - \frac{1}{4}\sin 2x + c$ (f) $\frac{1}{2}x + \sin 2x + c$

(g) $\frac{1}{2}x + \frac{1}{4}\sinh 2x + c$ (h) $\frac{1}{5}\cosh(5x+1) + c$

(i) $\frac{1}{24}\ln\left|\frac{3+4x}{3-4x}\right| + c$ (j) $\sin^{-1}(x-1) + c$

(k) $\frac{1}{3}\sin^{-1}3x + c$ (l) $\sin^{-1}\frac{1}{2}x + c$

(m) $\sin^{-1}\frac{(2x+1)}{\sqrt5} + c$ (n) $\sin^{-1}(2x-1) + c$

(o) $\sin^{-1}(x-2)/3 + c$ (p) $\frac{1}{2}\tan^{-1}\frac{1}{2}(x+3) + c$

102 (a) $\frac{5}{2}$ (b) $\frac{9}{2}$ (c) 3 (d) $\frac{3}{2}$ (e) $\frac{13}{2}$

103

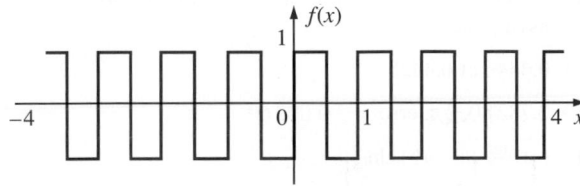

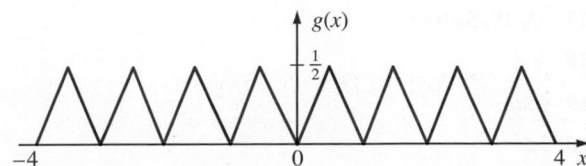

104

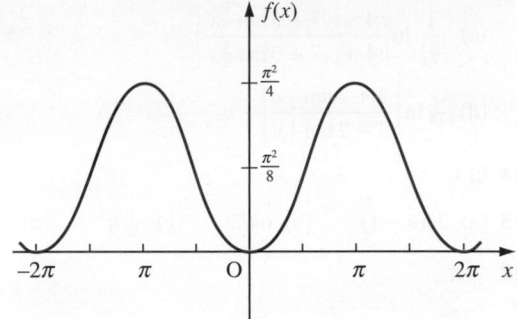

105 (a) $-x\cos x + \sin x + c$

(b) $\frac{1}{9}(3x-1)e^{3x} + c$

(c) $\frac{1}{16}x^4(4\ln|x| - 1) + c$

(d) $-\frac{1}{13}e^{-2x}(3\cos 3x + 2\sin 3x) + c$

(e) $\frac{1}{2}[(x^2+1)\tan^{-1}x] - \frac{1}{2}x + c$

(f) $\frac{1}{4}[(2x\sin 2x + \cos 2x) + c$

106 (a) $x\sin^{-1}x + \sqrt{(1-x^2)} + c$ (b) $x\ln|x| - x + c$

(c) $x\cosh^{-1}x - \sqrt{(x^2-1)} + c$

(d) $x\tan^{-1}x - \frac{1}{2}\ln(x^2+1) + c$

107 (a) $\pi - 2$ (b) $9\ln 3 - \frac{26}{9}$ (c) $\frac{1}{9}(2e^3+1)$

108 (a) $\frac{1}{3}(1+x^2)^{3/2} + c$ (b) $\frac{1}{4}\sin^4 x + c$

(c) $-\frac{1}{2(1+x^2)} + c$ (d) $\sqrt{(x^2-1)} + c$

(e) $\ln|x^2 + 3x + 2| + c$ (f) $\ln|\ln|x|| + c$

109 $a = \frac{3}{2}, b = -1$

$\frac{3}{2}\ln(x^2 + 2x + 5) - \frac{1}{2}\tan^{-1}\frac{1}{2}(x+1) + c$

110 (a) $\frac{1}{2}\ln(x^2 + 4x + 5) - \tan^{-1}(x+2) + c$

(b) $-2\sqrt{(5+4x-x^2)} + 7\sin^{-1}[(x-2)/3] + c$

(c) $\frac{1}{2}x - \frac{1}{2}\ln|\sin x + \cos x| + c$

111 (a) $\frac{1}{15}(3x^2-2)(1+x^2)^{3/2} + c$

(b) $\frac{1}{24}(6 - 8\sin^2 x + 3\sin^4 x)\sin^4 x + c$

(c) $\frac{1}{2}\left(\frac{x}{1+x^2} + \tan^{-1}x\right) + c$

(d) $-\sinh^{-1}\left(\frac{3}{x}\right) + c$ (e) $-\sqrt{(4-x^2)} + c$

(f) $2\sqrt x - 6\ln(3 + \sqrt x) + c$

112 (a) $2\sqrt{(1+x)} - 2\ln[1 + \sqrt{(1+x)}] + c$

(b) $\frac{1}{15}\sin^3 x(5 - 3\sin^2 x) + c$

(c) $2\sin\sqrt x - 2\sqrt x\cos\sqrt x + c$

113 (a) $\ln|\tan\frac{1}{2}x|$

(b) $\ln\left|\frac{1 + \tan\frac{1}{2}x}{1 - \tan\frac{1}{2}x}\right| + c$

(c) $\dfrac{1}{\sqrt{7}} \ln \left| \dfrac{4 - \sqrt{7} + 3\tan\frac{1}{2}x}{4 + \sqrt{7} + 3\tan\frac{1}{2}x} \right| + c$

(d) $\frac{1}{13} \ln \left| \dfrac{2 + 3\tan\frac{1}{2}x}{3 - 2\tan\frac{1}{2}x} \right| + c$

114 $\ln 4$

115 (a) $2e(e-1)$ (b) $64/3$ (c) $\frac{1}{18}\pi^2$ (d) $\frac{7}{6912}$

117 $\frac{197}{10}\pi$

118 $2\displaystyle\int_0^{\pi/2} \sqrt{(1 + \cos^2 x)}\,dx,\ \frac{1}{2}\pi^2$

120 $\frac{54}{35}\pi,\ \frac{48}{5}\pi$

121 $41.8\,\Omega$

122 (a) $\frac{2}{3},\ (\frac{11}{8}, \frac{2}{5})$ (b) $\frac{8}{15}\pi,\ (\frac{21}{16}, 0)$

124 $20,\ \frac{10}{3},\ 130$

125 $(\frac{2}{5}, 1),\ (\frac{1}{2}, 0)$

126 $0.6109,\ 0.6463,\ 0.6549,\ 0.6569;\ 0.6577$

127 0.1526

128 $5.869\,849$

129 $246\,\mathrm{A\,h}$

130 76.09

131 1.1114 (4dp)

8.10 Review exercises

1 (a) $(2x + 1)e^{x^2+x}$ (b) $\dfrac{x^2(9 - x)}{(3 - x)^3}$

(c) $5\cos(5x - 1)$

(d) $(\ln\tan x + 2x\,\mathrm{cosec}\,2x)(\tan x)^x$

(e) $\dfrac{1}{\sqrt{(1 - x^2)}}$ (f) $-\frac{1}{2}(1 + x)^{-3/2}$

(g) $-\dfrac{1}{1 + x^2}$ (h) $-\dfrac{(2x + 1)}{(x - 1)^2(x + 2)^2}$

(i) $3\cos(3x + 1)$ (j) $x^2(1 + 3\ln x)$

(k) $\dfrac{x^2(9 - x^2)}{(3 - x^2)^2}$ (l) $-\mathrm{sech}\,2x$

(m) $\sqrt{\frac{1}{2}}\sinh\frac{1}{2}x$ (n) $2x\sin 2x + 2(x^2 + 1)\cos 2x$

(o) $\dfrac{4 - x}{(x + 2)^3}$ (p) $\dfrac{e^{\sqrt{x}}}{2\sqrt{x}}$

(q) $2\,\mathrm{cosec}\,2x$ (r) $\dfrac{(8 - 7x)(2x - 1)^{1/2}}{(x + 1)^6}$

(s) $\sin x + x\cos x$ (t) $2xe^{x^2}$ (u) $2^x\ln 2$

(v) $-\dfrac{1}{1 + \sin x}$ (w) $-\dfrac{1}{x\sqrt{(x^2 - 1)}}$

(x) $x^2(3\cos 2x - 2x\sin 2x)$

(y) $\dfrac{3x^2 + 1}{2\sqrt{(x^3 + x + 3)}}$ (z) $-\dfrac{e^{-x}(2 + x)}{(1 + x)^2}$

2 (a) $\frac{2}{9}x^{3/2}(3\ln x - 2) + c$

(b) $\ln(x^2 + 2x + 2) + \tan^{-1}(x + 1) + c$

(c) $\ln\frac{9}{2}$ (d) $\frac{1}{2} + \frac{1}{4}\sqrt{\frac{1}{3}}\pi$

(e) $\ln\left| \dfrac{(x - 2)^2}{x - 1} \right| + c$ (f) $\frac{1}{3}\tan^3 x - \tan x + x + c$

(g) $\frac{1}{2} + \frac{2}{3}\sqrt{\frac{1}{3}}\pi$ (h) $\frac{1}{3}\sin^{-1}\frac{3}{2}x + c$

(i) $\frac{2}{3}\sqrt{(x^3 - 1)} + c$

(j) $\frac{1}{2}x^2 - x + 2\ln|x + 1| + c$

(k) $\frac{1}{2}\tan^{-1}\frac{1}{2}(x + 3) + c$

(l) $2(2 - x)\cos\sqrt{x} + 4\sqrt{x}\sin\sqrt{x} + c$

(m) 1 (n) $\frac{1}{4}(\sinh 2 - 2)$

(o) $-\frac{1}{30}(1 - 3x)^{10} + c$

(p) $\frac{1}{2}\sin x - \frac{1}{10}\sin 5x + c$

(q) $x(\ln 2x - 1) + c$ (r) $-e^{-x^2/2} + c$

(s) $\frac{1}{2}\cosh^{-1}\frac{2}{3}x + c$

(t) $7\pi/2$

(u) $\frac{\pi}{8} - \frac{1}{4}\ln 2$

(v) $-\frac{1}{15}(4 - 3x)^5 + c$

(w) $\frac{1}{10}\sin 5x + \frac{1}{2}\sin x + c$

(x) $x\sin^{-1}x + \sqrt{(1 - x^2)} + c$

(y) $-e^{-x}(x^2 + 2x + 2) + c$

(z) $2\sqrt{\frac{1}{3}}\tan^{-1}[\sqrt{\frac{1}{3}}(2x + 1)] + c$

3 $y = 12x - 8,\ y = \frac{1}{12}(49 - x)$

4 $y = \frac{1}{5}(4x + 6),\ y = 1 - x,\ \frac{8}{9}\sqrt{2}$

6 Maximum $(\frac{2}{3}, \frac{1}{27})$, minimum $(1, 0)$, inflection at $x = \frac{5}{6}$

7 Maximum 10.55 when $\theta = 4.42$ and 1.28, minimum 1.45 when $\theta = 2.85$ and 5.99

8 (a) $\dfrac{wL^4}{16EI}$ (b) $\dfrac{L}{2}(1 \pm \sqrt{\frac{1}{3}})$

9 $L = 100\,\mathrm{m},\ W = (200/\pi)\,\mathrm{m}$

10 Local minimum $(0, 0)$, local maximum $(3, -3)$ asymptotes $x = 2,\ x = 6,\ y = 1$

11 0.4446 cf 0.4425

13 $0.782\,80,\ \frac{1}{4}\pi,$ error $= -0.002\,60$

15 (a) $\frac{1076}{15}$ (b) $\ln\frac{7}{6}$

16 $-\frac{1}{4}\sec^4 t,\ -\frac{1}{4}\sec t\,\mathrm{cosec}^3 t$

17 $-3,\ 18,\ 5\sqrt{10}/9$

18 $\frac{8}{15}$

19 $\frac{3}{4}\pi ab$

20 $\frac{3}{2}\pi$

22 $\frac{1}{2}(\sinh^{-1}2 + 2\sqrt{5})$

23 $\frac{8}{15}$

24 (a) $3\pi a^2$ (b) $4a$
   (c) cycloid has cusps at these values   (d) $8a$

25 (a) $\frac{13}{15} - \frac{1}{4}\pi$ (b) $\frac{5}{12} - \frac{1}{2}\ln 2$

29 0.785, 0.626, 0.624; 2.62

30 $\pi/2, \frac{1}{6}\pi(5^{3/2} - 1)$

33 (a) $\frac{54}{35}\pi, \frac{48}{5}\pi$ (b) $(\frac{\sqrt{3}}{2} - \frac{\pi}{6}, \frac{5}{6}\pi - \frac{1}{2}\sqrt{3})$

35 (a) 5.21×10^6 (b) 7.76×10^6

38 (d) 1.910
   (i) 0.000, 0.191, 0.375, 0.541, 0.682, 0.798, 0.888,
   0.953, 0.995, 1.008

39 (a)

θ	0	0.23	0.33	0.42	0.49	0.57	0.65
x	1.0	0.9	0.8	0.7	0.6	0.5	0.4
y	0.00	0.21	0.27	0.31	0.32	0.32	0.30

θ	0.74	0.84	0.97	$\frac{1}{2}\pi$
x	0.3	0.2	0.1	0.0
y	0.27	0.22	0.15	0.00

   (f) $0, \pi, \frac{1}{2}\pi, \cos^{-1}(\sqrt{(5/7)}), \cos^{-1}(-\sqrt{(5/7)})$
   (h) $(\frac{25}{49}, \pm 5\sqrt{10}/49)$

CHAPTER 9

Exercises

1 (a) 1/4   (b) 1/2   (c) 1/4   (d) 1
   (e) 2/3   (f) −4/3   (g) 2   (h) 3/2
   (i) $\pi/4$

2 (b) is convergent, 0.860 334

3 $|f'(x)| > 1$ near $x = \alpha$

4 1.618 034

5 $x_{n+1} = x_n - \frac{1}{10}(x_n^3 - 2x - 1)$

6 (a) $f'(x) < 1 \ (x < 1)$ (b) $f'(x) > 1 \ (x > 1)$

7 1.5, 1.49, 1.48, 1.48, 1.47, 1.47, 1.46, 1.46, 1.45, 1.45;
   $\sqrt{2} = 1.41$

8 $e(1 - \frac{1}{2}x^2 + \frac{1}{6}x^4 - \frac{31}{720}x^6 + \dots)$

9 $1 + 2x + \frac{1}{2}x^2 + \frac{1}{6}x^3 + \dots = x + e^x$

10 $y_3 = 1 + 2x + \frac{1}{2}x^2 + \frac{1}{6}x^3 - \frac{1}{24}x^4$
   $y_4 = 1 + 2x + \frac{1}{2}x^2 + \frac{1}{6}x^3 + \frac{1}{24}x^4 - \frac{1}{120}x^5$

11 $x - \frac{1}{3}x^3 + \frac{1}{5}x^5 - \frac{1}{7}x^7 + \dots$ $(-1 \leqslant x \leqslant 1)$

12 $x + \frac{1}{3}x^3 + \frac{2}{15}x^5 + \frac{17}{315}x^7 + \dots$
   $\ln \cos x = -\left\{\frac{1}{6}x^2 + \frac{1}{12}x^4 + \frac{1}{45}x^6 + \frac{17}{2520}x^8 + \dots\right\}$

18 $l > 18$

19 (a) $\frac{3}{4}$ (b) $\frac{1}{4}$ (c) $-\frac{3}{2}$ (d) −1   (e) $-\frac{1}{3}$ (f) −1

20 $b_0 = 82.82, b_1 = -\frac{5}{24}, b_2 = -0.0018$

21 $X = \ln 4 \approx 1.386$

22 0.0006

23 1.175 201 21

24 (a) 1st order   (b) 2nd order   (c) 3rd order

25 −0.1038

26 2.732 051, 4.872 977

27 0.576 368 88

28 $n = 2$, 0.0203

29 (a) 0.643 283   (b) 0.6875

30 $0.4627, h = \frac{1}{256}$

32 $(1, 2t, 3t^2), (0, 2, 6t)$

33 $\dfrac{d\mathbf{r}}{dt} = (1 + 2t^2)\hat{\mathbf{T}}(t)$, where

$$\hat{\mathbf{T}}(t) = \frac{1}{1 + 2t^2}\mathbf{i} + \frac{2t}{1 + 2t^2}\mathbf{j} + \frac{2t^2}{1 + 2t^2}\mathbf{k}$$

37 2, 3

38 $\cos y, -x \sin y$

39 (a) $3x^2y + 4x + y, \ x^3 + 18y + x$
   (b) $3(x + y^2)^2, \ 6y(x + y^2)^2$
   (c) $\dfrac{3x + y}{(3x^2 + y^2 + 2xy)^{1/2}}, \dfrac{y + x}{(3x^2 + y^2 + 2xy)^{1/2}}$

40 (a) $e^{xy}(y\cos x - \sin x), \ e^{xy}x\cos x$
   (b) $\dfrac{y^2 - x^2}{(x^2 + y^2)^2}, \ -\dfrac{2xy}{(x^2 + y^2)^2}$
   (c) $\dfrac{-x^2 - 2xy + 2y^2 + 6}{(x^2 + 2y^2 + 6)^2}, \ \dfrac{x^2 - 4xy - 2y^2 + 6}{(x^2 + 2y^2 + 6)^2}$

41 $-1 + \frac{1}{2}\sqrt{3}, -\tan^{-1}2$

42 (a) $-x/z, -y/z$ (b) $\dfrac{1 - yz}{xy - 1}, \dfrac{1 + xz}{1 - xy}$

44 (a) $2xy + 3yz - 4z^3xy, \ x^2 + 3xz - 2x^2z^3,$
   $3yx - 6z^2x^2y$
   (b) $-ye^{2z}\sin xy, \ -xe^{2z}\sin xy, \ 2e^{2z}\cos xy$

47 (a) $\dfrac{2t^3 + 3t - 1}{\sqrt{(t^4 + 3t^2 - 2t + 1)}}$
   (b) $4xt(x^2 - 2t^2)/(2x + 3t)$

48 $2se^x\cos y - 2te^x\sin y$
   $-2te^x\cos y - 2se^x\sin y$

52 (a) 10.5   (b) $19\sqrt{\frac{1}{34}}$

53 $\frac{19}{5}\pi\,\mathrm{cm\,s^{-1}}$

54 $\sqrt{(1 + 4t^2 + 9t^4)}$

55 $-6\mathrm{e}^{-2s} + 2\mathrm{e}^{-s-t},\ -6\mathrm{e}^{-2t} + 2\mathrm{e}^{-s-t}$

57 -3

63 $a = 3,\ b = \frac{3}{2}$

65 0.018 702, 0.02

66 $0.029\,65\,\mathrm{m^3},\ 0.0295\,\mathrm{m^3}$

67 $173 \pm 4\,\mathrm{m}$

68 $-\frac{2}{3}$

69 3%

71 0.5%

72 35% increase

73 (a) $xy^2 + x^2y + x + c$
(b) $x^2y^2 + y\sin 3x + c$
(c) Not exact
(d) $z^3x - 3xy + 4y^3$

74 $-1,\ y\sin x - x\cos y + \frac{1}{2}(y^2 - 1)$

75 $m = 2$
$8x^5 + 36x^4y + 62x^3y^2 + 63x^2y^3 + 54xy^4 + 27y^5 + c$

76 (a) $(0, 0)$, maximum; $(10, 0)$, saddle
(b) $(0, 0)$, maximum   (c) $(-1, 3)$, saddle
(d) $(-1, \frac{3}{2})$, saddle; $(1, \frac{3}{2})$ minimum
(e) $(0, -1)$, saddle; $(0, 3)$, saddle; $(-1, 1)$, maximum
(f) Minimum at $(\frac{1}{2}, \frac{1}{3})$; degenerate and stationary sets $x = 0$ and $y = 0$
(g) $(1, 1)$, minimum

78 Maximum at $(0, 0)$; saddle at $(\frac{1}{3}, \frac{1}{3})$

79 $N = 2000,\ n = 2000,\ P = 250$

80 Minimum at $(2/3, 4/3)$

81 $a = \dfrac{20(\pi^2 - 16)}{\pi^5},\ b = \dfrac{12(20 - \pi^2)}{\pi^4}$

82 $x = 2,\ y = 2$

83 Minimum $T = -\frac{1}{4}$ at $(\frac{1}{2}, 0)$
maximum $T = \frac{9}{4}$ at $(-\frac{1}{2}, \pm\frac{1}{2}\sqrt{3})$

84 $x = \frac{200}{3},\ \theta = \frac{1}{3}\pi$

85 $(\frac{2}{3}, \frac{4}{3})$

86 $(\frac{1}{3}, \frac{1}{2}, \frac{1}{6})$

87 1, 2

88 $(-1/3, -2/3, -2/3)$

89 $-41/4, 7/4$

90 $\frac{285}{92}$

9.11 Review exercises

1 0.2575

4 $\frac{1}{2}$

5 $\frac{1}{8}(\sin 2k - 2k\cos 2k)$

6 2.09

7 (a) For these series see Section 6.3.5.

8 $\pi/2$

9 (a) $\frac{1}{2}$ (b) $\pi/4$ (c) $\frac{1}{16}$ (d) $\frac{1}{4}$ (e) $\frac{2}{5}$ (f) $\frac{2}{3}$

10 (a) $6, x = 0$ (b) $3, x = \frac{3}{2}$ (c) $-1, x = 0$

11 8.155 299, 8.154 959, 8.154 845

12 $4\,\mathrm{m\,s^{-1}},\ 4\,\mathrm{m\,s^{-2}}$

13 $(\frac{1}{2}t^2 + \frac{1}{6}t^3 + t)\boldsymbol{i} + (\frac{1}{12}t^4 - t)\boldsymbol{j} + t^2\boldsymbol{k}$

16 $-0.21, 0.01$

17 0.61%

18 -0.2%

19 -3.33%

20 (b) $2u$

22 (a) 2

25 $f''(z)/(4t\sqrt{t})$

27 $-y/(x^2 + y^2) + \text{const}$

28 Maximum at $(0, 0)$, saddle points at $(3, 3)$, $(-3, -3)$, $(1, -1)$, $(-1, 1)$

29 Minimum at $(0, 0)$, saddle at $(\frac{1}{2}, \frac{3}{2})$

30 $x = (\frac{2}{3})^{2/3},\ y = (\frac{2}{3})^{2/3},\ z = 2^{2/3} \cdot 3^{-1/6}$

31 Saddle at $(0, 0)$ and $(0, 4)$, maximum at $(2, 2)$, minimum at $(-2, 2)$

32 $x = 0,\ y = \pm 3$ (max); $y = 0,\ x = \pm 3$ (min)

33 $7.4163/a$

35 $y = -x\cot\frac{x}{c}$

CHAPTER 10

Exercises

1 (a) First-order, dependent variable x, independent variable t, linear, homogeneous, ordinary differential equation

(b) First-order, dependent variables x and y, independent variable t, linear, homogeneous, partial differential equation

(c) Second-order, dependent variable x, independent variable t, linear, homogeneous, ordinary differential equation

(d) Second-order, dependent variables x and y, independent variable t, linear, homogeneous, partial differential equation

(e) First-order, dependent variable x, independent variable t, nonlinear, ordinary differential equation

(f) First-order, dependent variables x and y, independent variable t, nonlinear, partial differential equation

(g) First-order, dependent variable x, independent variable t, linear, nonhomogeneous, ordinary differential equation

(h) Second-order, dependent variable x, independent variable t, linear, nonhomogeneous, ordinary differential equation

2 (a) Second-order nonlinear partial differential equation, dependent variable f, independent variables x and y

(b) Second-order nonlinear ordinary differential equation, dependent variable p, independent variable z

(c) Second-order linear homogeneous partial differential equation, dependent variable h, independent variables x and y

(d) Second-order linear nonhomogeneous ordinary differential equation, dependent variable s, independent variable t

(e) First-order linear nonhomogeneous partial differential equation, dependent variable x, independent variables u and v

(f) Third-order nonlinear ordinary differential equation, dependent variable p, independent variable y

(g) First-order linear nonhomogeneous ordinary differential equation, dependent variable r, independent variable z

(h) First-order nonlinear partial differential equation, dependent variable f, independent variables x, y and z

(i) First-order linear homogeneous ordinary differential equation, dependent variable x, independent variable t

(j) First-order linear nonhomogeneous ordinary differential equation, dependent variable x, independent variable t

(k) Second-order linear nonhomogeneous partial differential equation, dependent variable f, independent variables x and y

(l) Third-order nonlinear ordinary differential equation, dependent variable p, independent variable q

(m) Second-order nonlinear ordinary differential equation, dependent variable x, independent variable y

(n) Second-order linear homogeneous partial differential equation, dependent variable y, independent variables p and t

(o) First-order linear homogeneous ordinary differential equation, dependent variable y, independent variable z

3 (a) $x(t) = \frac{4}{3}t^3 + C$ (b) $x(t) = \frac{1}{20}t^5 - \frac{1}{3}t^3 + Ct + D$

(c) $x(t) = \frac{1}{16}e^{4t} + Ct + D$ (d) $x(t) = Ae^{-6t}$

(e) $x(t) = \ln t + \frac{1}{125}\cos 5t + Ct^2 + Dt + E$

(f) $x(t) = Ae^{2\sqrt{2}t} + Be^{-2\sqrt{2}t}$

4 (a) $x(t) = \frac{2}{3}t^3 + Ct + 2$ (b) $x(t) = -\frac{1}{4}\sin 2t - \dfrac{t}{\pi} + \frac{5}{2}$

(c) $x(t) = 4t + D$ (d) $x(t) = 2 - t^2$

(e) $x(t) = \frac{1}{2}e^{-2t} + Ct + a - \frac{1}{2}$ (f) $x(t) = C - \cos 2t$

(g) $x(t) = e^{2t}$ (h) $x(t) = \dfrac{e}{e^2 - 1}(e^t - e^{-t})$

5 (a) Under-determined

(b) Fully determined, boundary-value problem

(c) Fully determined, initial-value problem

(d) Under-determined

(e) Fully determined, boundary-value problem

(f) Fully determined, initial-value problem

(g) Under-determined

(h) Under-determined

(i) Fully determined, boundary-value problem

(j) Fully determined, initial-value problem

(k) Fully determined, boundary-value problem

(l) Fully determined, initial-value problem

6 $y(x) = \dfrac{1}{24EI}[w(a - x)^4 - 4R(a - x)^3$
$$+ 4a^2(aw - 3R)x - a^3(aw - 4R)]$$
At A the boundary condition is $y(a) = 0$ so $R = 3aw/8$
Maximum displacement is $y = 0.005\,42\,wa^4/EI$

11 (a) $x(t) = Ce^{kt}$ (b) $x(t) = Ce^{2t^3}$

(c) $x(t) = Ct^b$ (d) $x(t) = (2a\ln t + C)^{1/2}$

12 (a) $x(t) = (67 - 3\cos t)^{1/3}$ (b) $x(t) = \left(\dfrac{163}{2} - \dfrac{2}{t}\right)^{1/2}$

13 (a) $x(t) = (t^{1/2} + C)^2$ (b) $x(t) = \cos^{-1}(Ce^{\cos t - t})$

(c) $x(t) = C\exp(\frac{1}{2}e^{t^2})$ (d) $x(t) = (3e^t + C)^{1/3}$

(e) $x(t) = (1 - Ce^{at})^{-1}$ (f) $x(t) = (C - 2\cos t)^{1/2}$

14 (a) $x(t) = -2 \pm (\frac{2}{3}t^3 + 2t)^{1/2}$

(b) $x(t) = \dfrac{4(t - 1)}{4 - t}$ (c) $x(t) = \dfrac{3 + e^{2\sin t}}{3 - e^{2\sin t}}$

(d) $x(t) = -\ln(1 + e^{-a} - e^t)$

(e) $x(t) = [12(t\ln t - t + 1)]^{1/3}$

15 $K = 2/75$, $x(10) = 20/7$, $x(50) = 100/23$, $x \to 5$ as $t \to \infty$

16 $t = \sqrt{(m/Kg)}\tanh^{-1}\frac{1}{2}$

17 $A(t) = \dfrac{1}{\alpha}[1 - (1 + 6\alpha Kt)^{-1/6}]$

18 (a) $x(t) = \pm t\sqrt{[2\ln(Ct)]}$ (b) $x(t) = t(\ln Ct^3)^{1/3}$

(c) $x(t) = \dfrac{-t}{\ln Ct}$

19 $x(t) = \pm t(4\ln t + 256)^{1/4}$

20 (a) $x(t) = \dfrac{t}{\sqrt{3}}\left(\dfrac{C}{t^3} - 1\right)^{1/2}$ (b) $x(t) = t\cot^{-1}(\ln(1/Ct))$

(c) $x(t) = \dfrac{t}{2}\left[1 \pm \left(\dfrac{C}{t} - 11\right)^{1/2}\right]$ (d) $x(t) = t\sin^{-1}Ct$

(e) $x(t) = t \pm (2t^2 + D)^{1/2}$ (f) $x(t) = -t\ln(-\ln Ct)$

21 (a) $x(t) = \dfrac{2t}{\sqrt{(2 - t^4)}}$ (b) $x(t) = \pm\frac{1}{4}(9t^2 - 32)^{1/2}$

(c) $x(t) = t\ln(\ln\frac{1}{2}t + e^2)$ (d) $x(t) = \pm t[\ln(\ln t^2 + e^4)]^{1/2}$

(e) $x(t) = \dfrac{4t^2}{5 - 4t}$

22 (a) $x(t) = t + 3 \pm (2t + C)^{1/2}$

(b) $x(t) = \frac{1}{2}[\pm(2t + C)^{1/2} - t - 1]$

(c) $x(t) = \frac{1}{2}[\pm(2t + C)^{1/2} - t]$

(d) $x(t) = t - 1 \pm (2t + C)^{1/2}$

(e) $x(t) = Ae^t - 2t - 4$

(f) $x(t) = \frac{1}{2}t - 2 + Ae^t$ (g) $x(t) = -\dfrac{1}{t + C} - 2t$

23 (a) $x(t) = \pm\sqrt{(C - t^2)}$ (b) $x(t) = \pm\sqrt{(C + t^2)}$

(c) $x(t) = -t \pm \sqrt{(C + 2t^2)}$ (d) $x(t) = t^2 \pm \sqrt{(C + t^4)}$

(e) $\frac{1}{2}x^2 - xt + \frac{1}{2}t^2 - t = C$ (f) $x^2 + xt + t^2 = C$

24 (a) $x(t) = 1 \pm \sqrt{(1 - 2t - t^2)}$

(b) $x(t) = \frac{1}{2}[-t \pm \sqrt{(3t^2 + 4)}]$

(c) $x(t) = \dfrac{2}{t^2}(1 \pm (1 - t^2)^{1/2})$ (d) $x(t) = \dfrac{2 - t}{\cos t}$

25 (a) Not exact (b) Not exact

(c) $x\sin(x + t) = C$ (d) Not exact

(e) $x + e^{xt} = C$ (f) $(x + \sqrt{t})^2 = C$

(g) Not exact (h) $t\ln(x + t) = C$

26 (a) $x(t) = \sin^{-1}(1 - t) - t$

(b) $x(t) = [\frac{1}{2}(15 - t)]^{2/3} - 2t$

(c) $x(t) = \pm[t^2 \pm (1 - 4t)^{1/2}]^{1/2}$

(d) $x(t) = 4\exp\left(\dfrac{1}{2} - \dfrac{1}{t}\right) - t$

27 Must have $b = e$; then $ax^2 + 2bxt + ft^2 + C = 0$

28 Must have $h(t) = dg/dt$; then $x = -C/g(t)$

29 Must have $k = -1$; then $x\ln(x + t) + C = 0$

30 Must have $k = 2$; then $x = [\sin^{-1}(C/t^3)]/t$

31 (a) $x(t) = \frac{2}{3} + Ce^{-3t}$

(b) $x(t) = -\frac{1}{4}t - \frac{1}{16} + Ce^{4t}$

(c) $x(t) = -\frac{1}{2}e^{-4t} + Ce^{-2t}$

(d) $x(t) = Ce^{-t^2/2} - 2$

32 (a) $x(t) = -\frac{3}{2} + \frac{7}{2}e^{2t}$

(b) $x(t) = \frac{1}{3}t - \frac{1}{9} + \frac{10}{9}e^{-3t}$

(c) $x(t) = \frac{1}{2}t^3 - 3t\ln t - \frac{3}{2}t$

33 (a) $x(t) = Ce^t - 2t^2 - 5t - 5$

(b) $x(t) = -\frac{1}{4}t^2 - \frac{1}{8} + Ce^{2t^2}$

(c) $x(t) = \left(1 - \dfrac{2}{t^2}\right)\sin t + \dfrac{2}{t}\cos t + \dfrac{C}{t^2}$

(d) $x(t) = \left(\dfrac{1}{t} - \dfrac{3}{t^2} + \dfrac{6}{t^3} - \dfrac{6}{t^4}\right)e^t + \dfrac{C}{t^4}$

(e) $x(t) = \frac{1}{2}\sin 2t\ln(\tan\frac{1}{2}t) + C\sin 2t$

(f) $x(t) = \frac{1}{3}t^2 + Ce^{-2t^3}$ (g) $x(t) = Ce^{-1/t} - 4$

34 (a) $x(t) = \frac{1}{2}(1 - e^{2t^2})$ (b) $x(t) = 2e^{t-1}t^{-t}$

(c) $x(t) = \frac{1}{5}t - \frac{1}{25} + \frac{1}{3}e^{-2t} + \dfrac{18 - 25e^2}{75e^5}e^{-5t}$

(d) $x(t) = 1 + e^{1/t - 1/2}$

(e) $x(t) = \frac{1}{2} + t^2 + \left(1 - \dfrac{2}{t} + \dfrac{2}{t^2}\right)e^t - \dfrac{1}{t^2}(\frac{3}{2} + e)$

(f) $x(t) = U(1 + e^{1+\cos t})$

35 $T(t) = T_{in} + Ce^{-AU\alpha t/V}$

36 $Q(t) = \dfrac{2\alpha\rho gh(1 - e^{-pt})}{(2 + 2\alpha\beta + \alpha\gamma_0)} + \dfrac{\alpha\rho ghe^{-pt}}{(1 + \alpha\beta + \alpha\gamma_0)}$

where $p = \dfrac{(2 + 2\alpha\beta + \alpha\gamma_0)A}{2\alpha\rho d}$

37 (a) $x(t) = \dfrac{1}{2t}\sqrt[3]{(6t^4 + C)}$

(b) $x(t) = \dfrac{4}{1 + 2t + Ce^{2t}}$

(c) $x(t) = \pm\sqrt{(Ce^{2t} - 2e^t)}$

(d) $x(t) = \dfrac{1}{t(1 + Ct)}$

38 (a) $x(t) = \dfrac{1}{t\sqrt{(3 - 2t)}}$

(b) $x(t) = \dfrac{6}{\sqrt{(12 - 11e^{6t})}}$

(c) $x(t) = \sqrt[3]{\left(\dfrac{10}{3\cos(t) + 9\sin(t) - 13e^{3t}}\right)}$

(d) $x(t) = -\frac{1}{2}\sqrt{(5t^9 + 3t)}$

39 $X(0.3) = 1.269\,000$

40 $X(0.25) = 2.050\,439$

41 $X(1) = 1.2029$

42 $X(0.5) = 2.1250$

43 $X_a(2) = 2.811\,489$, $X_b(2) = 2.819\,944$

44 $X_a(2) = 1.573\,065$, $X_b(2) = 1.558\,541$

45 $X_a(1.5) = 2.241\,257$, $X_b(1.5) = 2.206\,232$

46 (a) $L = \dfrac{d}{dt} + t^2$ (b) $L = \dfrac{d}{dt} - 6t^2$

(c) $L = \dfrac{d}{dt} - k$

47 (a) independent (b) dependent

48 (a) $k_1 = 2$, $k_2 = -2$, $k_3 = -1$
(b) $k_1 = 1$, $k_2 = -1$, $k_3 = -1$, $k_4 = 1$

49 (a) $L = \dfrac{d}{dt} - f(t)$ (b) $L = \dfrac{d^3}{dt^3} + \sin t\dfrac{d^2}{dt^2} + 4t^2$

(c) $L = \dfrac{d^2}{dt^2} + \sin t\dfrac{d}{dt} - t - \cos t$

(d) $L = \sin t\dfrac{d}{dt} - \dfrac{\cos t}{t}$ (e) $L = \dfrac{d}{dt} - \dfrac{b}{t}$

(f) $L = \dfrac{d}{dt} - te^{t^2}$ (g) $L = t^2\dfrac{d^2}{dt^2} + (2t - t^2)\dfrac{d}{dt} - t$

(h) $L = t\dfrac{d^2}{dt^2} + 3\dfrac{d}{dt} - t$

50 (a) dependent (b) independent
(c) independent (d) independent
(e) dependent (f) dependent
(g) independent (h) dependent
(i) dependent (j) independent
(k) independent

51 (a) $2, -1, 1, 1$ (b) $-3, 3, 2, 1$
(c) $0, 2, -1$ (d) $1, -1, 0, 1$
(e) $0, -1, 0, 1, 6$

52 (a) $x(t) = A + Bt + Ct^2 + Dt^3$
(b) $x(t) = Ae^{pt} + Be^{-pt}$
(c) $x(t) = A\cos pt + B\sin pt + C\cosh pt + D\sinh pt$
(d) $x(t) = A + Be^{-2t}$
(e) $x(t) = A + B\cos 2t + C\sin 2t$
(f) $x(t) = Ae^{-t} + Bte^{-t}$
(g) $x(t) = Ae^t + Bte^t + Ce^{-t}$

53 $LM = \dfrac{1}{t}\dfrac{d^3}{dt^3} - \left(\dfrac{2}{t^2} + 4 + e^t\right)\dfrac{d^2}{dt^2}$
$\qquad + \left(\dfrac{2}{t^3} + \dfrac{4}{t} + 6t + (4t - 2)e^t\right)\dfrac{d}{dt}$
$\qquad + (4t - 6t^2 - 1)e^t$

$ML = \dfrac{1}{t}\dfrac{d^3}{dt^3} - (e^t + 4)\dfrac{d^2}{dt^2} + \left(6t - \dfrac{4}{t} + 4te^t\right)\dfrac{d}{dt}$
$\qquad - 6t^2e^t + 12$

54 $LM = f_1f_2\dfrac{d^2}{dt^2} + \left(f_1\dfrac{df_2}{dt} + f_1g_2 + f_2g_1\right)\dfrac{d}{dt}$
$\qquad + f_1\dfrac{dg_2}{dt} + g_1g_2$

$ML = f_1f_2\dfrac{d^2}{dt^2} + \left(f_2\dfrac{df_1}{dt} + f_1g_2 + f_2g_1\right)\dfrac{d}{dt}$
$\qquad + f_2\dfrac{dg_1}{dt} + g_1g_2$

55 (a) $x(t) = Ae^t + Be^{3t/2}$
(b) $x(t) = e^{-t}(A\cos 2t + B\sin 2t)$
(c) $x(t) = Ae^t + Be^{-4t}$
(d) $x(t) = e^{2t}(A\cos 3t + B\sin 3t)$

56 (a) $x(t) = \frac{1}{7}(3e^t - 10e^{-2t/5})$
(b) $x(t) = e^{3t}(2\cos t - 6\sin t)$
(c) $x(t) = \frac{1}{2}(e^{3t} - e^t)$

57 (a) $x(t) = e^{t/4}[A\cos(\frac{1}{4}\sqrt{27}t) + B\sin(\frac{1}{4}\sqrt{27}t)]$
(b) $x(t) = Ae^{(\sqrt{13}-3)t} + Be^{-(\sqrt{13}+3)t}$
(c) $x(t) = e^{-t/2}[A\cos(\frac{1}{2}\sqrt{3}t) + B\sin(\frac{1}{2}\sqrt{3}t)]$
(d) $x(t) = Ae^{4t} + Bte^{4t}$ (e) $Ae^t + Be^{2t/3} + Ce^{-2t/3}$
(f) $x(t) = Ae^{-t} + e^t[B\cos(2\sqrt{2}t) + C\sin(2\sqrt{2}t)]$
(g) $x(t) = A + e^t[B\cos(\sqrt{2}t) + C\sin(\sqrt{2}t)]$

58 $x(t) = e^t(A\cos t + B\sin t + C\cos 2t + D\sin 2t)$

59 (a) $x(t) = e^{t/2}[\cos(\frac{1}{2}\sqrt{5}t) - \sqrt{\frac{1}{5}}\sin(\frac{1}{2}\sqrt{5}t)]$
(b) $x(t) = 2(t - 1)e^{2(t-1)}$
(c) $x(t) = e^{-5t/2}[\cos(\frac{1}{2}\sqrt{7}t) + \sqrt{\frac{1}{7}}\sin(\frac{1}{2}\sqrt{7}t)]$
(d) $x(t) = \frac{1}{6}(7t + 33)e^{-(t+3)/3}$
(e) $x(t) = \frac{7}{2}e^t - 4e^{2t} + \frac{3}{2}e^{3t}$
(f) $x(t) = (2 - 5t + 4t^2)e^{-2(t-1)}$

60 $x(t) = e^{t/2}[A\cos(\frac{1}{2}\sqrt{3}t) + B\sin(\frac{1}{2}\sqrt{3}t) + Ct\cos(\frac{1}{2}\sqrt{3}t)$
$\qquad + Dt\sin(\frac{1}{2}\sqrt{3}t)]$

61 $x(t) = Ae^{4t} + Be^{-t} + Cte^{-t} + Dt^2e^{-t}$

62 (a) $x(t) = \frac{2}{9} - \frac{1}{3}t + Ae^{-t} + Be^{3t}$
(b) $x(t) = -\frac{1}{5}t^2 + \frac{14}{25}t - \frac{38}{125} + Ae^{(1+\sqrt{6})t} + Be^{(1-\sqrt{6})t}$
(c) $x(t) = -5e^t + Ae^{(\sqrt{5}+1)t/2} + Be^{-(\sqrt{5}-1)t/2}$

63 (a) $x(t) = -\frac{1}{8}\cos 4t + \frac{1}{24}\sin 4t$
$\qquad + e^{3t/2}[A\cos(\frac{1}{2}\sqrt{7}t) + B\sin(\frac{1}{2}\sqrt{7}t)]$

(b) $x(t) = \frac{1}{121}e^{-3t} + Ae^{2t/3} + Bte^{2t/3}$

(c) $x(t) = -\frac{105}{289}\cos 2t + \frac{56}{289}\sin 2t + Ae^{-(1-\frac{3}{\sqrt{2}})t}$
$+ Be^{-(1+\frac{3}{\sqrt{2}})t}$

(d) $x(t) = \frac{5}{4}t - \frac{33}{16} + e^{-t/2}[A\cos(\frac{1}{2}\sqrt{15}t) + B\sin(\frac{1}{2}\sqrt{15}t)]$

(e) $x(t) = t - 2 + Ae^{-t/4} + Bte^{-t/4}$

(f) $x(t) = -\frac{72}{625}\cos 3t - \frac{21}{625}\sin 3t + Ae^{4t} + Bte^{4t}$

(g) $x(t) = \frac{1}{52}e^{-5t} + e^{2t}[A\cos(\sqrt{3}t) + B\sin(\sqrt{3}t)]$

(h) $x(t) = -t^2 - 6t - 24 + \frac{1}{5}e^{-2t} + Ae^{(\sqrt{7}/\sqrt{3}-1)t/2}$
$+ Be^{-(\sqrt{7}/\sqrt{3}+1)t/2}$

(i) $x(t) = -\frac{5}{4}te^{-3t} - \frac{4}{65}\cos 2t - \frac{7}{65}\sin 2t + Ae^{t} + Be^{-3t}$

(j) $x(t) = \frac{1}{16} - \frac{1}{4}t\cos 4t + A\cos 4t + B\sin 4t$

(k) $x(t) = -\frac{7}{4}t - \frac{3}{4}te^{4t} + A + Be^{4t}$

64 (a) $x(t) = \frac{17}{1460}\cos 2t + \frac{21}{1460}\sin 2t + Ae^{t} + Be^{-2t}$
$+ Ce^{(2+\sqrt{3})t} + De^{(2-\sqrt{3})t}$

(b) $x(t) = -\frac{1}{12}e^{2t} - \frac{4}{39}te^{-2t} + Ae^{t} + Be^{-2t} + Ce^{(2+\sqrt{3})t}$
$+ De^{(2-\sqrt{3})t}$

(c) $x(t) = -\frac{1}{2}t^2 - \frac{9}{2}t - \frac{69}{4} - \frac{1}{12}e^{-t} + Ae^{t} + Be^{-2t}$
$+ Ce^{(2+\sqrt{3})t} + De^{(2-\sqrt{3})t}$

65 (a) $x(t) = \frac{1}{125}\cos t + \frac{11}{250}\sin t - \frac{1}{27}(t+1)$
$+ (A + Bt + Ct^2)e^{3t}$

(b) $x(t) = -\frac{1}{8}e^{t} + (A + Bt + Ct^2)e^{3t}$

(c) $x(t) = \frac{1}{6}t^3e^{3t} - \frac{1}{27}(t+1) + (A + Bt + Ct^2)e^{3t}$

66 (a) $\omega = 3,\ \zeta = 1$ (b) $\omega = \sqrt{7},\ \zeta = 2\sqrt{\frac{1}{7}}$

67 (a) $a = 1,\ b = 4$ (b) $p = 1.4,\ q = 0.25$
(c) $\beta = 2.2,\ \gamma = 1.21$

68 (a) $\omega = 4p,\ \zeta = \dfrac{a}{4p}$ (b) $\omega = \dfrac{1}{\sqrt{(2\alpha)}},\ \zeta = 7\sqrt{(\frac{1}{2}\alpha)}$

(c) $\omega = 1.78,\ \zeta = 0.12$ (d) $\omega = 5\eta,\ \zeta = 4$
(e) $\omega = 0.51,\ \zeta = 2.48$

69 (a) $\alpha = \pi,\ \beta = \pi^2;$ (b) $a = 0.4\pi,\ b = 4\pi^2$

(c) $q = 8,\ r = 4$ (d) $a = \dfrac{7}{2\pi^2},\ b = \dfrac{28}{\pi}$

70 $\Omega_{max} = \omega\sqrt{(1 - 2\zeta^2)}$, only exists if $\zeta^2 < \frac{1}{2}$

$$A(\Omega_{max}) = \frac{1}{2\zeta\omega^2\sqrt{(1 - \zeta^2)}}$$

71 $2.52\,\mathrm{m\,s^{-1}}$ (approximately 5 knots)

72 $\mu > 621\,\mathrm{N\,m^{-1}\,s}$

73 $73\,\mathrm{pF} > C > 7\,\mathrm{pF}$

74 (a) $\dfrac{dx}{dt} = v,$ $x(0) = 1$

$\dfrac{dv}{dt} = 4xt - 6(x^2 - t)v,$ $v(0) = 2$

(b) $\dfrac{dx}{dt} = v,$ $x(0) = 0$

$\dfrac{dv}{dt} = \sin v - 4x,$ $v(0) = 0$

75 $X(0.3) = 0.299\,90$

76 (a) $\dfrac{dx}{dt} = v,$ $x(1) = 2$

$\dfrac{dv}{dt} = -4\sqrt{(x^2 - t^2)},$ $v(1) = 0.5$

(b) $\dfrac{dx}{dt} = v,$ $x(0) = 1$

$\dfrac{dv}{dt} = w,$ $v(0) = 2$

$\dfrac{dw}{dt} = e^{2t} + x^2t - 6e^{t}v - tw,$ $w(0) = 0$

(c) $\dfrac{dx}{dt} = v,$ $x(1) = 1$

$\dfrac{dv}{dt} = w,$ $v(1) = 0$

$\dfrac{dw}{dt} = \sin t - x^2 - tw,$ $w(1) = -2$

(d) $\dfrac{dx}{dt} = v,$ $x(2) = 0$

$\dfrac{dv}{dt} = w,$ $v(2) = 0$

$\dfrac{dw}{dt} = (x^2t^2 + tw)^2,$ $w(2) = 2$

(e) $\dfrac{dx}{dt} = v,$ $x(0) = 0$

$\dfrac{dv}{dt} = w,$ $v(0) = 0$

$\dfrac{dw}{dt} = u,$ $w(0) = 4$

$\dfrac{du}{dt} = \ln t - x^2 - xw,$ $u(0) = -3$

(f) $\dfrac{dx}{dt} = v,$ $x(0) = a$

$\dfrac{dv}{dt} = w,$ $v(0) = 0$

$\dfrac{dw}{dt} = u,$ $w(0) = b$

$\dfrac{du}{dt} = t^2 + 4t - 5 + \sqrt{(xt)} - v - (v - 1)u,$ $u(0) = 0$

77 $X(0.65) = -0.834\,63$

78 $X_{0.01}(0.4) = 0.398\,022$
$X_{0.005}(0.4) = 0.397\,919$
step size required is < 0.0024
$X_{0.002}(0.4) = 0.397\,856$

79 s tends to around 6.3%. With double the inflow s tends to about 11.1%.

10.13 Review exercises

1 (a) Second-order nonlinear ordinary differential equation, dependent variable x, independent variable t

(b) Third-order linear homogeneous partial differential equation, dependent variable h, independent variables t and v

(c) First-order nonlinear ordinary differential equation, dependent variable z, independent variable x

(d) Third-order linear nonhomogeneous ordinary differential equation, dependent variable p, independent variable s

2 (a) Under-determined, $x = \frac{1}{6}t^3 + At + 1$

(b) Fully determined, $x(t) = \frac{1}{24}t^4 - \frac{7}{24}t^2 + \frac{1}{4}t$

(c) Over-determined, no solution exists

(d) Fully determined, $x = \frac{1}{16}e^{4t} - \frac{1}{4}te^4 - \frac{1}{16}$

3 $x(t) = \dfrac{C}{\sqrt{(C + e^{-2at})}}$

4 (a) $x(t) = \cos^{-1}(\sin t - 1)$ (b) $x(t) = \ln(\ln t + e^2)$

(c) $x(t) = e^{(t^3-8)/3}$ (d) $x(t) = t\cos^{-1}(\cos 1 - \ln t)$

(e) $x(t) = -\frac{1}{2}[t \pm \sqrt{(17t^2 + 16)}]$ (f) $x(t) = t2^t$

(g) $x(t) = t(3 - \ln t)$ (h) $x(t) = t \pm \sqrt{(4 - 6t^2)}$

5 (a) $x(t) = \dfrac{\sqrt{[4 + a(t - 1)]}}{t}$ (b) not exact

(c) $x(t) = \dfrac{\sin^{-1}(\pi - t)}{t}$ (d) not exact

(e) $x(t) = \dfrac{\ln(2 + e^8 - t)}{t}$

6 (a) $x(t) = \frac{9}{4}e^{2t} - \frac{1}{2}t - \frac{1}{4}$ (b) $x(t) = \frac{1}{2}(e^{-t} + e^{-t^2})$

(c) $x(t) = \frac{1}{5}(e^{2t} + 9e^{-3t})$ (d) $x(t) = 1 + (e - 1)e^{\cos t + 1}$

7 $X_{0.1}(0.4) = 1.125\,583$, $X_{0.05}(0.4) = 1.142\,763$ Richardson extrapolation estimates the error as 0.017 180, so, to obtain an error less than 5×10^{-3}, a step less than 0.0146 should be used

8 $X_{0.05}(0.25) = 2.003\,749$, $X_{0.025}(0.25) = 2.004\,452$ Richardson extrapolation estimates the error as 0.000 703, so, to obtain an error less than 5×10^{-4}, a step less than 0.0179 should be used

9 $x(t) = (20 - t) - \dfrac{(20 - t)^3}{400}$

11 $\alpha = 4kT_0^3$

13 $y(t) = y_0 + C\sqrt{(x - x_0)}$

14 Half life is $\ln 2/k$

15 Time to 95% of final value is $\ln(20)L/R$

16 Tyre life is approximately 29 500 miles

17 (a) $L = \dfrac{d^2}{dt^2} + \sin t \dfrac{d}{dt} - 9$, $f(t) = -\cos t$

(b) $L = \dfrac{d^3}{dt^3} + t\dfrac{d^2}{dt^2} + t(t - 4)\dfrac{d}{dt} + 1$, $f(t) = -e^t$

(c) $L = \dfrac{d}{dt} - e^{-t}$, $f(t) = e^t$

(d) $L = \dfrac{d^2}{dt^2} + 4$, $f(t) = \cos \Omega t$

(e) $L = t^2\dfrac{d^3}{dt^3} - \dfrac{1}{t^2 + 2t + 4}\dfrac{d}{dt}$, $f(t) = -\ln(t^2 + 4)$

18 (a) $\sin t - \cos t$ (b) 0 (commutative)

(c) 0 (commutative)   (d) $2\dfrac{d}{dt}$

19 (a) $\dfrac{df}{dt} = \dfrac{dg}{dt}$ (b) $\dfrac{df}{dt} = \dfrac{dg}{dt}$ and $\dfrac{d^2f}{dt^2} = \dfrac{d^2g}{dt^2}$

20 (a) $x(t) = Ae^t + Be^{2t} + \frac{1}{10}\sin t + \frac{3}{10}\cos t$

(b) $x(t) = Ae^t + Be^{2t} + Ce^{-3t} + \frac{1}{6}t + \frac{7}{36}$

(c) $x(t) = Ae^t + Be^{2t} + Ce^{-3t} + \frac{1}{5}te^{2t}$

(d) $x(t) = Ae^{4t} + te^{4t}$

(e) $x(t) = e^{-3t/2}(A\sin t + B\cos t) + \frac{4}{13}t^2 - \frac{96}{169}t + \frac{736}{2197}$

(f) $x(t) = e^{-3t/2}(A\sin t + B\cos t) - \frac{16}{75}\cos t + \frac{4}{25}\sin t$

(g) $x(t) = Ae^{2t} + Be^{4t} + Ce^{-t} + \frac{1}{8}t^2 - \frac{3}{16}t + \frac{13}{64}$

(h) $x(t) = e^t(A\cos 2t + B\sin 2t) + \frac{1}{8}e^{-t}$

(i) $x(t) = Ae^{2t} + Be^{4t} + Ce^{-t} - \frac{1}{6}te^{2t} + \frac{1}{6}e^t$

(j) $x(t) = e^t(A\cos 2t + B\sin 2t) + \frac{1}{5}t + \frac{2}{25} + \frac{1}{4}te^t\sin 2t$

21 (a) $x(t) = \frac{1}{5}(1 - e^{-t}\cos 2t - \frac{1}{2}e^{-t}\sin 2t)$

(b) $x(t) = -2t + 5 + \frac{7}{2}e^t - \frac{3}{2}e^{-t/3}$

(c) $x(t) = (12e^{-t} + 30te^{-t} - 12\cos 2t + 16\sin 2t)/25$

(d) $x(t) = 3e^t - 2 - e^{2t}$

(e) $x(t) = -\frac{7}{5}e^t + \frac{4}{3}e^{2t} + \frac{1}{15}e^{-4t}$

(f) $x(t) = \frac{1}{13} + \frac{8}{5}e^{-t} - \dfrac{e^{-2t}}{65}(44\cos 3t - 27\sin 3t)$

22 (a) $\omega = \sqrt{2}$, $\zeta = \dfrac{7}{2\sqrt{2}}$ (b) $\omega = p^{1/4}$, $\zeta = \frac{1}{2}p^{3/4}$

(c) $\omega = \dfrac{\sqrt{q}}{\sqrt{2}}$, $\zeta = a\sqrt{(2q)}$

(d) $\omega = \sqrt{(2\alpha)}$, $\zeta = \dfrac{7}{\sqrt{(2\alpha)}}$

23 (a) $a = 2$, $b = 4$ (b) $a = 4\pi$, $b = \pi^2$

(c) $a = 2$, $c = 8$ (d) $p = 150$, $q = 6\sqrt{2}$

24 $x(t) = t - Ce^{-t} + D$
(a) $x(t) = \frac{1}{12}e^{-2t} + Ce^{4t} + D$
(b) $x(t) = -\ln(\cos(t + C)) + D$
(c) $x(t) = Ct^3 + D$

25 (a) $x(t) = \frac{t^3}{3k} - \frac{t^2}{k^2} + \frac{2t}{k^3} + \frac{(k^3 - 2)}{k^4}(1 - e^{-kt})$

(b) $x(t) = \frac{U}{k - U}\ln\left(\frac{U}{k} + \frac{k - U}{k}e^{kt}\right)$

(c) $x(t) = \frac{2}{15}t^3 + \frac{8}{5}t - \frac{26}{15}$

(d) $x(t) = -\frac{4}{17}\cos t - \frac{1}{17}\sin t - \frac{4}{17}e^{-4(t-\pi)}$

26 $x(t) = C\tan(\frac{1}{2}Ct + D)$
(a) $x(t) = Ae^{pt} + B$
(b) $x(t) = -\ln(t + C) + D$
(c) $x(t) = \pm\sqrt{(C - \ln(D - t))}$

27 (a) $x(t) = \left[\frac{(1 - p)t + 4}{4^p}\right]^{\frac{1}{1-p}}$

(b) $x(t) = 1$
(c) $x(t) = (\frac{1}{4} - \frac{2}{3}t)^{-1/2}$
(d) $x(t) = \frac{1}{2}t^2 + 1$

28 Length of runway is $\frac{m}{2(\mu\alpha - \beta)}\ln\left[\frac{\mu\alpha - \beta}{T - \mu mg}V_2^2 + 1\right]$

Time to take off is $\frac{m}{\sqrt{((\mu\alpha - \beta)(T - \mu mg))}}$

$$\arctan\left[\sqrt{\left(\frac{\mu\alpha - \beta}{T - \mu mg}\right)}V_2\right]$$

29 $X_{0.025}(2) = 0.847\,035$, $X_{0.0125}(2) = 0.844\,066$
Richardson extrapolation estimates the error as
0.002 969, so we have $X(2) = 0.84$

32 $R = 2\sqrt{\dfrac{L}{C}}$

CHAPTER 11

Exercises

1 (a) $\dfrac{s}{s^2 - 4}$, $\mathrm{Re}(s) > 2$ (b) $\dfrac{2}{s^3}$, $\mathrm{Re}(s) > 0$

(c) $\dfrac{3s + 1}{s^2}$, $\mathrm{Re}(s) > 0$ (d) $\dfrac{1}{(s + 1)^2}$, $\mathrm{Re}(s) > -1$

2 (a) 5   (b) −3   (c) 0   (d) 3   (e) 2
(f) 0   (g) 0   (h) 0   (i) 2   (j) 3

3 (a) $\dfrac{5s - 3}{s^2}$, $\mathrm{Re}(s) > 0$

(b) $\dfrac{42}{s^4} - \dfrac{6}{s^2 + 9}$, $\mathrm{Re}(s) > 0$

(c) $\dfrac{3s - 2}{s^2} + \dfrac{4s}{s^2 + 4}$, $\mathrm{Re}(s) > 0$

(d) $\dfrac{s}{s^2 - 9}$, $\mathrm{Re}(s) > 3$

(e) $\dfrac{2}{s^2 - 4}$, $\mathrm{Re}(s) > 2$

(f) $\dfrac{5}{s + 2} + \dfrac{3}{s} - \dfrac{2s}{s^2 + 4}$, $\mathrm{Re}(s) > 0$

(g) $\dfrac{4}{(s + 2)^2}$, $\mathrm{Re}(s) > -2$

(h) $\dfrac{4}{s^2 + 6s + 13}$, $\mathrm{Re}(s) > -3$

(i) $\dfrac{2}{(s + 4)^3}$, $\mathrm{Re}(s) > -4$

(j) $\dfrac{36 - 6s + 4s^2 - 2s^3}{s^4}$, $\mathrm{Re}(s) > 0$

(k) $\dfrac{2s + 15}{s^2 + 9}$, $\mathrm{Re}(s) > 0$

(l) $\dfrac{s^2 - 4}{(s^2 + 4)^2}$, $\mathrm{Re}(s) > 0$

(m) $\dfrac{18s^2 - 54}{(s^2 + 9)^3}$, $\mathrm{Re}(s) > 0$

(n) $\dfrac{2}{s^3} - \dfrac{3s}{s^2 + 16}$, $\mathrm{Re}(s) > 0$

(o) $\dfrac{2}{(s + 2)^3} + \dfrac{s + 1}{s^2 + 2s + 5} + \dfrac{3}{s}$, $\mathrm{Re}(s) > 0$

4 (a) $\frac{1}{4}(e^{-3t} - e^{-7t})$ (b) $-e^{-t} + 2e^{3t}$
(c) $\frac{4}{9} - \frac{1}{3}t - \frac{4}{9}e^{-3t}$ (d) $2\cos 2t + 3\sin 2t$
(e) $\frac{1}{64}(4t - \sin 4t)$ (f) $e^{-2t}(\cos t + 6\sin t)$
(g) $\frac{1}{8}(1 - e^{-2t}\cos 2t + 3e^{-2t}\sin 2t)$ (h) $e^t - e^{-t} - 2te^{-t}$
(i) $e^{-t}(\cos 2t + 3\sin 2t)$ (j) $\frac{1}{2}e^t - 3e^{2t} + \frac{11}{2}e^{3t}$
(k) $-2e^{-3t} + 2\cos(\sqrt{2}t) - \sqrt{\frac{1}{2}}\sin(\sqrt{2}t)$
(l) $\frac{1}{5}e^t - \frac{1}{5}e^{-t}(\cos t - 3\sin t)$
(m) $e^{-t}(\cos 2t - \sin 2t)$ (n) $\frac{1}{2}e^{2t} - 2e^{3t} + \frac{3}{2}e^{-4t}$
(o) $-e^t + \frac{3}{2}e^{2t} - \frac{1}{2}e^{-2t}$ (p) $4 - \frac{9}{2}\cos t + \frac{1}{2}\cos 3t$
(q) $9e^{-2t} - e^{-3t/2}[7\cos(\frac{1}{2}\sqrt{3}t) - \sqrt{3}\sin(\frac{1}{2}\sqrt{3}t)]$
(r) $\frac{1}{9}e^{-t} - \frac{1}{10}e^{-2t} - \frac{1}{90}e^{-t}(\cos 3t + 3\sin 3t)$

5 (a) $x(t) = e^{-2t} + e^{-3t}$
(b) $x(t) = \frac{35}{78}e^{4t/3} - \frac{3}{26}(\cos 2t + \frac{2}{3}\sin 2t)$
(c) $x(t) = \frac{1}{5}(1 - e^{-t}\cos 2t - \frac{1}{2}e^{-t}\sin 2t)$
(d) $y(t) = \frac{1}{25}(12e^{-t} + 30te^{-t} - 12\cos 2t + 16\sin 2t)$
(e) $x(t) = -\frac{7}{5}e^t + \frac{4}{3}e^{2t} + \frac{1}{15}e^{-4t}$
(f) $x(t) = e^{-2t}(\cos t + \sin t + 3)$
(g) $x(t) = \frac{13}{12}e^t - \frac{1}{3}e^{-2t} + \frac{1}{4}e^{-t}(\cos 2t - 3\sin 2t)$

(h) $y(t) = -\frac{2}{3} + t + \frac{2}{3}e^{-t}[\cos(\sqrt{2}t) + \sqrt{\frac{1}{2}}\sin(\sqrt{2}t)]$

(i) $x(t) = (\frac{1}{8} + \frac{3}{4}t)e^{-2t} + \frac{1}{2}t^2e^{-2t} + \frac{3}{8} - \frac{1}{2}t + \frac{1}{4}t^2$

(j) $x(t) = \frac{1}{5} - \frac{1}{5}e^{-2t/3}(\cos\frac{1}{3}t + 2\sin\frac{1}{3}t)$

(k) $x(t) = te^{-4t} - \frac{1}{2}\cos 4t$

(l) $y(t) = e^{-t} + 2te^{-2t/3}$

(m) $x(t) = \frac{5}{4} + \frac{1}{2}t - e^t + \frac{5}{12}e^{2t} - \frac{2}{3}e^{-t}$

(n) $x(t) = \frac{9}{20}e^{-t} - \frac{7}{16}\cos t + \frac{25}{16}\sin t - \frac{1}{80}\cos 3t$
$\quad - \frac{3}{80}\sin 3t$

6 (a) $x(t) = \frac{1}{4}(\frac{15}{4}e^{3t} - \frac{11}{4}e^t - e^{-2t})$, $y(t) = \frac{1}{8}(3e^{3t} - e^t)$

(b) $x(t) = 5\sin t + 5\cos t - e^t - e^{2t} - 3$
$\quad y(t) = 2e^t - 5\sin t + e^{2t} - 3$

(c) $x(t) = 3\sin t - 2\cos t + e^{-2t}$
$\quad y(t) = -\frac{7}{2}\sin t + \frac{9}{2}\cos t - \frac{1}{2}e^{-3t}$

(d) $x(t) = \frac{3}{2}e^{t/3} - \frac{1}{2}e^t$, $y(t) = -1 + \frac{1}{2}e^t + \frac{3}{2}e^{t/3}$

(e) $x(t) = 2e^t + \sin t - 2\cos t$
$\quad y(t) = \cos t - 2\sin t - 2e^t$

(f) $x(t) = -3 + e^t + 3e^{-t/3}$
$\quad y(t) = t - 1 - \frac{1}{2}e^t + \frac{3}{2}e^{-t/3}$

(g) $x(t) = 2t - e^t + e^{-2t}$, $y(t) = t - \frac{7}{2} + 3e^t + \frac{1}{2}e^{-2t}$

(h) $x(t) = 3\cos t + \cos(\sqrt{3}t)$
$\quad y(t) = 3\cos t - \cos(\sqrt{3}t)$

(i) $x(t) = \cos(\sqrt{\frac{3}{10}}t) + \frac{3}{4}\cos(\sqrt{6}t)$
$\quad y(t) = \frac{5}{4}\cos(\sqrt{\frac{3}{10}}t) - \frac{1}{4}\cos(\sqrt{6}t)$

(j) $x(t) = \frac{1}{3}e^t + \frac{2}{3}\cos 2t + \frac{1}{3}\sin 2t$
$\quad y(t) = \frac{2}{3}e^t - \frac{2}{3}\cos 2t - \frac{1}{3}\sin 2t$

7 $I_1(s) = \dfrac{E_1(50 + s)s}{(s^2 + 10^4)(s + 100)^2}$

$I_2(s) = \dfrac{Es^2}{(s^2 + 10^4)(s + 100)^2}$

$i_2(t) = E(-\frac{1}{200}e^{-100t} + \frac{1}{2}te^{-100t} + \frac{1}{200}\cos 100t)$

9 $i_1(t) = 20\sqrt{\frac{1}{7}}e^{-t/2}\sin(\frac{1}{2}\sqrt{7}t)$

10 $x_1(t) = -\frac{3}{10}\cos(\sqrt{3}t) - \frac{7}{10}\cos(\sqrt{13}t)$
$\quad x_2(t) = -\frac{1}{10}\cos(\sqrt{3}t) + \frac{21}{10}\cos(\sqrt{13}t)$, $\sqrt{3}$, $\sqrt{13}$

11.5 Review exercises

1 (a) $x(t) = \cos t + \sin t - e^{-2t}(\cos t + 3\sin t)$

(b) $x(t) = -3 + \frac{13}{7}e^t + \frac{15}{7}e^{-2t/5}$

2 (a) $e^{-t} - \frac{1}{2}e^{-2t} - \frac{1}{2}e^{-t}(\cos t + \sin t)$

(b) $i(t) = 4e^{-t} - 3e^{-2t}$
$\quad + V[e^{-t} - \frac{1}{2}e^{-2t} - \frac{1}{2}e^{-t}(\cos t + \sin t)]$

3 $x(t) = -t + 5\sin t - 2\sin 2t$,
$\quad y(t) = 1 - 2\cos t + \cos 2t$

4 $\frac{1}{5}(\cos t + 2\sin t)$
$e^{-t}[(x_0 - \frac{1}{5})\cos t + (x_1 + x_0 - \frac{3}{5})\sin t]$
$\sqrt{\frac{1}{5}}$, 63.4° lag

6 (a) (i) $\dfrac{s\cos\phi - \omega\sin\phi}{s^2 + \omega^2}$

(ii) $\dfrac{s\sin\phi + \omega(\cos\phi + \sin\phi)}{s^2 + 2\omega s + \omega^2}$

(b) $\frac{1}{20}(\cos 2t + 2\sin 2t) + \frac{1}{20}e^{-2t}(39\cos 2t + 47\sin 2t)$

7 (a) $e^{-2t}(\cos 3t - 2\sin 3t)$

(b) $y(t) = 2 + 2\sin t - 5e^{-2t}$

8 $x(t) = e^{-8t} + \sin t$, $y(t) = e^{-8t} - \cos t$

9 $q(t) = \frac{1}{500}(5e^{-100t} - 2e^{-200t}) - \frac{1}{500}(3\cos 100t - \sin 100t)$,
current leads by approximately 18.5°

10 $x(t) = \frac{29}{20}e^{-t} + \frac{445}{1212}e^{-t/5} + \frac{1}{3}e^{-2t}$
$\quad - \frac{1}{505}(76\cos 2t - 48\sin 2t)$

11 (a) $\theta(t) = \frac{1}{100}(4e^{-4t} + 10te^{-4t} - 4\cos 2t + 3\sin 2t)$

(b) $i_1(t) = \frac{1}{7}(e^{4t} + 6e^{-3t})$, $i_2 = \frac{1}{7}(e^{-3t} - e^{4t})$

12 $i(t) = \dfrac{E}{R}[1 - e^{-nt}(\cos nt - \sin nt)]$

13 $i_1(t) = \dfrac{E(4 - 3e^{-Rt/L} - e^{-3Rt/L})}{6R}$, $i_2(t) \to E/3R$

14 $x_1(t) = \frac{1}{3}[\sin t - 2\sin 2t + \sqrt{3}\sin(\sqrt{3}t)]$
$\quad x_2(t) = \frac{1}{3}[\sin t + \sin 2t - \sqrt{3}\sin(\sqrt{3}t)]$

15 (a) (i) $e^{-t}(\cos 3t + \sin 3t)$ (ii) $e^t - e^{2t} + 2te^t$

(b) $y(t) = \frac{1}{2}e^{-t}(8 + 12t + t^3)$

16 (a) $\frac{5}{2}e^{7t}\sin 2t$

(b) $\dfrac{n^2 i}{Ks(s^2 + 2Ks + n^2)}$, $\theta(t) = \frac{i}{K}(1 - e^{-Kt}) - ite^{-Kt}$

17 (a) $v_1 = 250e^{-0.1t}$, $v_2 = (50 + 25t)e^{-0.1t}$

(b) $t = 23.026$

CHAPTER 12

Exercises

1 (a) $f(t) = -\dfrac{1}{4}\pi - \dfrac{2}{\pi}\sum_{n=1}^{\infty}\dfrac{\cos(2n-1)t}{(2n-1)^2}$

$\quad + \sum_{n=1}^{\infty}\left[\dfrac{3\sin(2n-1)t}{2n-1} - \dfrac{\sin 2nt}{2n}\right]$

(b) $f(t) = \dfrac{1}{4}\pi + \dfrac{2}{\pi}\sum_{n=1}^{\infty}\dfrac{\cos(2n-1)t}{(2n-1)^2} - \sum_{n=1}^{\infty}\dfrac{\sin nt}{n}$

(c) $f(t) = \dfrac{2}{\pi} \displaystyle\sum_{n=1}^{\infty} \dfrac{\sin nt}{n}$

(d) $f(t) = \dfrac{2}{\pi} + \dfrac{4}{\pi} \displaystyle\sum_{n=1}^{\infty} \dfrac{(-1)^{n+1} \cos 2nt}{4n^2 - 1}$

(e) $f(t) = \dfrac{2}{\pi} + \dfrac{4}{\pi} \displaystyle\sum_{n=1}^{\infty} \dfrac{(-1)^{n+1} \cos nt}{4n^2 - 1}$

(f) $f(t) = \dfrac{1}{2}\pi - \dfrac{4}{\pi} \displaystyle\sum_{n=1}^{\infty} \dfrac{\cos(2n-1)t}{(2n-1)^2}$

(g) $f(t) = -\dfrac{4}{\pi} \displaystyle\sum_{n=1}^{\infty} \dfrac{\cos(2n-1)t}{(2n-1)^2} - \displaystyle\sum_{n=1}^{\infty} \dfrac{\sin 2nt}{n}$

(h) $f(t) = \left(\dfrac{1}{2}\pi + \dfrac{1}{\pi} \sinh \pi \right)$

$\qquad + \dfrac{2}{\pi} \displaystyle\sum_{n=1}^{\infty} \left[\dfrac{(-1)^n - 1}{n^2} + \dfrac{(-1)^n \sinh \pi}{n^2 + 1} \right] \cos nt$

$\qquad - \dfrac{2}{\pi} \displaystyle\sum_{n=1}^{\infty} \dfrac{n(-1)^n}{n^2 + 1} \sinh \pi \sin nt$

2 $f(t) = \dfrac{1}{3}\pi^2 + 4 \displaystyle\sum_{n=1}^{\infty} \dfrac{\cos nt}{n^2}$

Taking $t = \pi$ gives the required result.

3 $q(t) = Q \left[\dfrac{1}{2} - \dfrac{4}{\pi^2} \displaystyle\sum_{n=1}^{\infty} \dfrac{\cos(2n-1)t}{(2n-1)^2} \right]$

4 $f(t) = \dfrac{5}{\pi} + \dfrac{5}{2} \sin t - \dfrac{10}{\pi} \displaystyle\sum_{n=1}^{\infty} \dfrac{\cos 2nt}{4n^2 - 1}$

5 Taking $t = 0$ and $t = \pi$ gives the required answers.

6 $f(t) = \dfrac{1}{4}\pi - \dfrac{2}{\pi} \displaystyle\sum_{n=1}^{\infty} \dfrac{\cos(4n-2)t}{(2n-1)^2}$

Taking $t = 0$ gives the required series.

7 $f(t) = \dfrac{3}{2} + \dfrac{4}{\pi^2} \displaystyle\sum_{n=1}^{\infty} \dfrac{\cos(2n-1)t}{(2n-1)^2}$

Replacing t by $t - \tfrac{1}{2}\pi$ gives the following sine series of odd harmonics:

$f\left(t - \dfrac{1}{2}\pi \right) - \dfrac{3}{2} = -\dfrac{4}{\pi^2} \displaystyle\sum_{n=1}^{\infty} \dfrac{(-1)^n \sin(2n-1)t}{(2n-1)^2}$

8 $f(t) = \dfrac{2l}{\pi} \displaystyle\sum_{n=1}^{\infty} \dfrac{(-1)^{n+1}}{n} \sin \dfrac{n\pi t}{l}$

9 $f(t) = \dfrac{2K}{\pi} \displaystyle\sum_{n=1}^{\infty} \dfrac{1}{n} \sin \dfrac{n\pi t}{l}$

10 $f(t) = \dfrac{3}{2} + \dfrac{6}{\pi} \displaystyle\sum_{n=1}^{\infty} \dfrac{1}{(2n-1)} \sin \dfrac{(2n-1)\pi t}{5}$

11 $v(t) = \dfrac{A}{\pi} \left(1 + \dfrac{1}{2}\pi \sin \omega t - 2 \displaystyle\sum_{n=1}^{\infty} \dfrac{\cos 2n\omega t}{4n^2 - 1} \right)$

12 $f(t) = \dfrac{1}{3}T^2 + \dfrac{4T^2}{\pi^2} \displaystyle\sum_{n=1}^{\infty} \dfrac{(-1)^n}{n^2} \cos \dfrac{n\pi t}{T}$

13 $e(t) = \dfrac{E}{2} \left(1 - \dfrac{2}{\pi} \displaystyle\sum_{n=1}^{\infty} \dfrac{1}{n} \sin \dfrac{2\pi nt}{T} \right)$

15 $f(t) = -\dfrac{8}{\pi^2} \displaystyle\sum_{n=1}^{\infty} \dfrac{1}{(2n-1)^2} \cos(2n-1)\pi t$

16 (a) $f(t) = \dfrac{2}{3} - \dfrac{1}{\pi^2} \displaystyle\sum_{n=1}^{\infty} \dfrac{1}{n^2} \cos 2n\pi t + \dfrac{1}{\pi} \displaystyle\sum_{n=1}^{\infty} \dfrac{1}{n} \sin 2n\pi t$

(b) $f(t) = \dfrac{1}{\pi} \displaystyle\sum_{n=1}^{\infty} \dfrac{1}{n} \sin 2n\pi t$

$\qquad + \dfrac{2}{\pi} \displaystyle\sum_{n=1}^{\infty} \left[\dfrac{1}{2n-1} + \dfrac{4}{\pi^2(2n-1)^3} \right]$

$\qquad \times \sin(2n-1)\pi t$

(c) $f(t) = \dfrac{2}{3} + \dfrac{4}{\pi^2} \displaystyle\sum_{n=1}^{\infty} \dfrac{(-1)^{n+1}}{n^2} \cos n\pi t$

17 $f(t) = \dfrac{1}{6}\pi^2 - \displaystyle\sum_{n=1}^{\infty} \dfrac{1}{n^2} \cos 2nt$

$f(t) = \dfrac{8}{\pi} \displaystyle\sum_{n=1}^{\infty} \dfrac{1}{(2n-1)^3} \sin(2n-1)t$

18 $f(x) = \dfrac{8a}{\pi^2} \displaystyle\sum_{n=1}^{\infty} \dfrac{(-1)^{n+1}}{(2n-1)^2} \sin \dfrac{(2n-1)\pi x}{l}$

19 $f(x) = \dfrac{2l}{\pi^2} \displaystyle\sum_{n=1}^{\infty} \dfrac{(-1)^{n+1}}{(2n-1)^2} \sin \dfrac{2(2n-1)\pi x}{l}$

20 $f(t) = \dfrac{1}{2} \sin t + \dfrac{4}{\pi} \displaystyle\sum_{n=1}^{\infty} \dfrac{n(-1)^{n+1}}{4n^2 - 1} \sin 2nt$

21 $f(x) = -\dfrac{1}{2}A - \dfrac{4A}{\pi^2} \displaystyle\sum_{n=1}^{\infty} \dfrac{1}{(2n-1)^2} \cos \dfrac{(2n-1)\pi x}{l}$

22 $T(x) = \dfrac{8KL^2}{\pi^3} \displaystyle\sum_{n=1}^{\infty} \dfrac{1}{(2n-1)^3} \sin \dfrac{(2n-1)\pi x}{L}$

23 $f(t) = \dfrac{1}{2} + \dfrac{1}{2} \cos \pi t + \dfrac{4}{\pi} \displaystyle\sum_{n=1}^{\infty} \dfrac{1}{4n^2 - 1} \sin 2n\pi t$

$\qquad - \dfrac{2}{\pi} \displaystyle\sum_{n=1}^{\infty} \dfrac{1}{2n-1} \sin(2n-1)\pi t$

26 (c) $1 + 4 \displaystyle\sum_{n=1}^{\infty} \dfrac{(-1)^{n+1}}{n} \sin nt$

12.6 Review exercises

1 $f(t) = \frac{1}{6}\pi^2 + \sum_{n=1}^{\infty} \frac{2}{n^2}(-1)^n \cos nt$

$\quad + \sum_{n=1}^{\infty}\left[\frac{\pi}{2n-1} - \frac{4}{\pi(2n-1)^3}\right]\sin(2n-1)t$

$\quad - \sum_{n=1}^{\infty}\frac{\pi}{2n}\sin 2nt$

Taking $T = \pi$ gives the required sum.

2 $f(t) = \frac{1}{9}\pi$

$\quad + \frac{2}{\pi}\sum_{n=1}^{\infty}\frac{1}{n^2}\left\{\cos\frac{1}{3}n\pi - \frac{1}{3}[2 + (-1)^n]\right\}\cos nt; \frac{2}{9}\pi$

3 (a) $f(t) = \frac{2T}{\pi^2}\sum_{n=1}^{\infty}\frac{(-1)^{n+1}}{(2n-1)^2}\sin\frac{2(2n-1)\pi t}{T}$

$\quad$ (b) $-\frac{1}{4}T$

$\quad$ (c) Taking $t = \frac{1}{4}T$ gives $S = \frac{1}{8}\pi^2$

4 $y = \frac{4P}{\pi\alpha}\sum_{n=1}^{\infty}\frac{1}{(2n-1)^2}\sin(2n-1)\alpha\sin(2n-1)x$

6 $f(t) = \frac{4}{\pi}\sum_{n=1}^{\infty}\frac{(-1)^n\sin(2n-1)t}{(2n-1)^2}$

8 $f(x) = \frac{4}{\pi}\sum_{n=1}^{\infty}\frac{\cos(2n-1)x}{(2n-1)^2}$

Taking $x = 0$ gives

$\pi^2 = 8\sum_{n=1}^{\infty}\frac{1}{(2n-1)^2}$

9 $f(x) = \sum_{n=1}^{\infty}\frac{1}{(2n-1)}\left[1 + \frac{2(-1)^{n+1}}{\pi(2n-1)}\right]\sin(2n-1)x$

$\quad - \sum_{n=1}^{\infty}\frac{1}{2n}\sin 2nx$

10 $V = \frac{25}{3}(1 - e^{-1.2}) + \sum_{n=1}^{\infty}\frac{50(1 - e^{-1.2})}{9 + 25n^2\pi^2}$

$\quad \times (3\cos 5n\pi t + 5n\pi\sin 5n\pi t)$

Amplitude of the nth harmonic is

$\frac{50(1 - e^{-1.2})}{\sqrt{(9 + 25n^2\pi^2)}} \approx \frac{50(1 - e^{-1.2})}{5n\pi} \approx \frac{2\cdot 22}{n}$

13 $f(x) = \frac{4}{\pi}\sum_{n=1}^{\infty}\frac{(-1)^{n+1}}{(2n-1)^2}\sin(2n-1)x$

$\quad f(x) = \frac{1}{4}\pi - \frac{2}{\pi}\sum_{n=1}^{\infty}\frac{\cos 2(2n-1)x}{(2n-1)^2}$

15 (a) $f(t) = \sum_{n=1}^{\infty}\frac{2}{n}\sin nt$

$\quad$ (b) $f(t) = \frac{1}{2}\pi + \frac{4}{\pi}\sum_{n=1}^{\infty}\frac{1}{(2n-1)^2}\cos(2n-1)t$

16 $f(t) = \frac{2}{3} + \frac{4}{\pi^2}\sum_{n=1}^{\infty}\frac{(-1)^{n+1}}{n^2}\cos n\pi t$

17 (a) (i) a constant term and cosine terms with even harmonics

$\quad\quad$ (ii) constant, cosine and sine terms present

$\quad\quad$ (iii) a constant term and sine terms with odd harmonics

$\quad$ (b) $f(t) = \frac{\pi^2}{24} + \frac{2}{\pi}\left(\frac{\pi^2}{4} - 2\right)\cos t - \frac{1}{2}\cos 2t$

$\quad\quad - \frac{2}{\pi}\left(\frac{\pi^2}{12} - \frac{2}{27}\right)\cos 3t + \frac{1}{8}\cos 4t$

18 (a) $f(t) = \frac{1}{2}\pi - \frac{4}{\pi}\sum_{n=1}^{\infty}\frac{1}{(2n-1)^2}\cos(2n-1)t$

$\quad$ (b) $g(t) = \frac{4}{\pi}\sum_{n=1}^{\infty}\frac{1}{2n-1}\sin(2n-1)t$

19 $g(t) = \frac{4}{\pi}\sum_{n=1}^{\infty}\frac{1}{2n-1}\sin(2n-1)t$

$\quad f(t) = 1 + g(t)$

CHAPTER 13

Exercises

6 (a) $A \cap B$ $\quad$ (b) $A \cup B$
$\quad$ (c) $S - A$ $\quad$ (d) $S - (A \cap B)$

7 (a) {car, bicycle, motorcycle, boat}
$\quad$ (b) {train} $\quad$ (c) {car, motorcycle, boat}

8 (a) 0.7 $\quad$ (b) 0.8 $\quad$ (c) 0.5

9 $\frac{16}{2652}$

10 $P(\text{same values}) = \frac{1}{6}$, $P(\text{differ by at most }1) = \frac{4}{9}$

12 (a) $\frac{1}{26}$ $\quad$ (b) $\frac{4}{13}$ $\quad$ (c) $\frac{1}{2}$ $\quad$ (d) $\frac{1}{13}$

13 (a) $\frac{1}{9}$ $\quad$ (b) $\frac{5}{18}$ $\quad$ (c) $\frac{5}{6}$

14 $P(\text{total} = 7 | 7 \text{ or } 10) = \frac{2}{3}$

15 (a) $\frac{3}{4}$ $\quad$ (b) 7 to 1

16 (a) $\frac{1}{2}$ $\quad$ (b) $\frac{1}{3}$

17 $\frac{5}{6}$

18 (a) 0.15 $\quad$ (b) 0.55 $\quad$ (c) 0.357

19 0.6

20 0.381

21 (a) $P(A) + P(B) - P(A \cap B)$

(b) $\dfrac{P(C|A)P(A) - P(C|A \cap B)P(A \cap B)}{P(A) - P(A \cap B)}$

(c) $\dfrac{P(C|A)P(A) + P(C|B)P(B) - P(C|A \cap B)P(A \cap B)}{P(A) + P(B) - P(A \cap B)}$

22 0.149

23 (a) $\left(1 - \dfrac{2r}{d}\right)^2$ (b) $1 - \left(\dfrac{2r}{d}\right)^2$

24 $P(2) = \frac{1}{36}$ $P(3) = \frac{2}{36}, \dots, P(7) = \frac{6}{36}$
$P(8) = \frac{5}{36}, \dots, P(12) = \frac{1}{36}$

26 (a) 0.488   (b) 0.3123

27 (b) $P(-3) = \frac{1}{8}$, $P(-1) = \frac{3}{8}$
$P(1) = \frac{3}{8}$, $P(3) = \frac{1}{8}$
(c) $P(-3) = \frac{1}{27}$, $P(-1) = \frac{6}{27}$
$P(1) = \frac{12}{27}$, $P(3) = \frac{8}{27}$

28 (a) $\frac{1}{4}$

(b) $F_X(x) = \begin{cases} 0 & (x < 0) \\ \frac{1}{2}\sqrt{x} & (0 \leqslant x \leqslant 4) \\ 1 & (x > 4) \end{cases}$

(c) $\frac{1}{2}$

29 $P(X \leqslant 30) = 0.28$

30 (a) $\frac{1}{9}$ (b) $\frac{3}{4}$ (c) 0.102

31 $1 - \exp(-x^2/2a)$, 0.0804

32 mean = 4.5, P(less than 5 days) = 0.6

33 mean = 1.8, median = 2,
standard deviation = 1.34

34 Average length = 5.88

35 mean = 5, median = 3
standard deviation = 4.47

36 mean = 30 minutes, standard deviation = 17.3 min

38 (a) 0.47   (b) $\mu_X = 30$, $\sigma_X = 30$
(c) median = 20.8, $q_3 - q_1 = 33.0$

39 0.969

41 24 hours, 3.32 hours

42 (a) $\bar{X} = 2.28$, $S_X = 0.60$, $S_{X,n-1} = 0.63$
(b) sample median = 2.1, range = 2.2

43 $\bar{X} = 5.44$, $S_X = 0.81$, median = 5.45, range = 3.2

44 $\bar{A} = 24.2$, $S_A = 1.76$, $\bar{T} = 16.7$, $S_T = 4.00$
$\bar{B} = 25.4$, $S_B = 1.66$, $\bar{U} = 18.1$, $S_U = 4.93$
where A, T are time, temperature for A, and B, U are
time, temperature for B

45 2.19

46 median = $\sqrt{[2a \ln 2]}$, mode = $\sqrt{a}$
$a = 6$: mean = 3.07, median = 2.88, mode = 2.45,
$q_3 - q_1 = 2.22$

47 $q/(p + q - pq)$

48 47.1 and 46.3

49 P(4 boys) = 0.273

50 0.998

51 0.677

52 (a) 0.1271   (b) 0.3594
(c) 0.1413   (d) 0.5876

53 4 engines

54 (a) 0.957   (b) 0.0071

55 0.027

56 P(8 or more) = 0.249

57 (a) 0.050   (b) 0.224   (c) 0.084

58 0.986

59 6.09

61 0.144

62 0.011

63 46

64 0.3%, 0.0258

65 (a) 0.102   (b) 0.128   (c) 0.011

66 Warning 9.5, action 13.5, sample 12
UCL = 11.4, sample 9

67 UK sample 28, US sample 25

68 P(at least one such area) = 0.133

69 P(at least one such area) = 0.688

13.8 Review exercises

1 (a) 3   (b) $F_X(x) = \begin{cases} 0 & \text{for } x \leqslant 1 \\ 1 - x^{-3} & \text{for } x > 1 \end{cases}$

(c) $\frac{1}{8}$ (d) $\frac{3}{2}$ (e) $\sqrt{3}/2$

2 0.0159

3 60, 6342 hours

4 $P(10, 5, 3, 2) = 0.009$

5 $e^{-\lambda}(1 + \lambda) > 0.9$, proportion = 0.0053

6 $\pm 5.66 \times 10^{-5}$

9 0.407

10 $E(\text{minimum}) = \dfrac{25}{12\lambda}$

$\qquad\qquad\quad = \dfrac{25}{4}$ hours when $\lambda = \dfrac{1}{3}$

12 $E(\text{number of analyses}) = N\,[1 - (1 - p)^k] + \dfrac{N}{k}$

$\qquad\qquad\qquad\qquad\quad = 0.196\,N$ when $k = 11$

13 (a) *single*

$\quad k = 4$: $n = 7$, $P(\text{error}) = 0.0020$
$\quad k = 8$: $n = 12$, $P(\text{error}) = 0.0062$

$k = 16$: $n = 21$, $P(\text{error}) = 0.0185$
$k = 32$: $n = 38$, $P(\text{error}) = 0.0555$
$k = 64$: $n = 71$, $P(\text{error}) = 0.1588$

double
$k = 4$: $n = 11$, $P(\text{error}) = 0.0002$
$k = 8$: $n = 17$, $P(\text{error}) = 0.0006$
$k = 16$: $n = 26$, $P(\text{error}) = 0.0022$
$k = 32$: $n = 43$, $P(\text{error}) = 0.0092$
$k = 64$: $n = 77$, $P(\text{error}) = 0.0424$

(b) *single*: $k = 8$, so total 96 bits
$\quad$ *double*: $k = 64$, so total 77 bits

Index

D

J

K

N

T